CHILTON'S
SUV SERVICE
MANUAL
2003 Edition

CHILTON *AUTOMOTIVE* *INFORMATION*

PUBLISHED BY **W. G. NICHOLS, INC.**

Manufactured in USA, © 2002 W. G. Nichols, 1025 Andrew Drive, West Chester, PA 19380
This book is developed by W. G. Nichols, Inc. and Chilton editors under the terms of an exclusive
sales distribution agreement with Delmar Thomson Learning.
ISBN 0-8019-9359-8
1234567890 1098765432

Table of Contents

Unit Repair Sections

Table of Contents

Model Index

EDITORIAL POLICY

Manufacturer and Model Coverage

This Manual does not seek to cover every make and model that is currently available on the market. Rather, the Chilton Editorial Staff makes judicious decisions as to which makes and models warrant coverage. Those that are included herein represent Chilton's judgement as to the makes and models that make up 90% of vehicles that will be presented to the average technician for diagnosis and repair. In general, this Manual does not cover:

- Exotics (e.g. Rolls-Royce, Dodge Viper; Alfa Romeo, etc.)
- OEM's with no U.S. presence (e.g. Fiat)
- OEM's that have not sold enough units to be a factor in the repair market.

Model Year Information

Every effort is made to gather current data for use in this Manual. Data is acquired from manufacturers at the time when each OEM chooses to release it. Different manufacturers choose to release their new model information at different times of the year. Indeed, the same manufacturer can be early one season with information, and then late the next season. As a result, not all models are equally current when each edition of this Manual goes to press. You will note that the Editorial Staff has taken care to indicate the currency of coverage for each model.

Safety Notice

Proper service and repair procedures are vital to the safe, reliable operation of all motor vehicles, as well as the personal safety of those performing the repairs. This manual outlines procedures for servicing and repairing vehicles using safe effective methods. The procedures contain many NOTES, WARNINGS and CAUTIONS which should be followed along with standard safety procedures to eliminate the possibility of personal injury or improper service which could damage the vehicle or compromise its safety.

It is important to note that repair procedures and techniques, tools and parts for servicing vehicles, as well as the skill and experience of the individual performing the work vary widely. It is not possible to anticipate all of the conceivable ways or conditions under which vehicles may be serviced, or to provide cautions as to all of the possible hazards that may result. Standard and accepted safety precautions and equipment should be used when handling toxic or flammable fluids, and safety goggles or other protection should be used during cutting, grinding, chiseling, prying, or any other process that can cause material removal or projectiles.

Some procedures require the use of tools specially designed for a specific purpose. Before substituting another tool or procedure, you must be completely satisfied that neither your personal safety, nor the performance of the vehicle will be endangered.

Although information in this manual is based on industry sources and is as complete as possible at the time of publication, the possibility exists that some vehicle manufacturers made later changes which could not be included here. Information on very late models may not be available in some circumstances. While striving for total accuracy, Nichols Publishing cannot assume responsibility for any errors, changes, or omissions that may occur in the compilation of this data.

LOCATING AND USING INFORMATION

Organization

The Table of Contents, located at the front of the book, lists each Unit Repair Section and Model Specific section in this manual.

To find where a particular model specific section is located in the book, you need only look in the Table of Contents. Once you have found the proper section, you may wish to find where specific procedures are located in that section. Turn to the Index at the front of the model specific section. At the upper left-hand side is a listing of the main topics within that section and the page number on which they may be found. Following the main topics is an alphabetical listing of all of the procedures within the section and their page numbers.

The Model Index, located just after the Table of Contents in the beginning of this manual, may also be used to locate the specific section for any vehicle model covered in this manual.

Specifications

Specifications charts for all models covered in this book are located in Chapter 1. They include: Vehicle and Engine Identification, General Engine Specifications, Engine Tune-Up Specifications, Capacities, Valve Specifications, Crankshaft & Connecting Rod Specifications, Piston & Ring Specifications, Engine Fastener Torque Specifications, Brake Specifications, Maintenance Interval Specifications, Ball Joint Specifications, Wheel & Tire Specifications, and Wheel Alignment Specifications.

Unit Repair Sections

The three Unit Repair Sections are written to cover all applicable models for the specific system or component, unless specifically noted otherwise. The procedures covered in the URS are not repeated in the model specific sections, therefore, refer to the URS for the service procedures for the applicable systems or components. Refer to the Table of Contents for URS coverage.

Model Specific Sections

The model specific sections are grouped by manufacturer and arranged in alphabetical order. The text and illustrations that comprise the service procedures in each model specific section are arranged in the following order of systems and components: Engine Repair (Gasoline, then Diesel if applicable), Fuel System (Gasoline, then Diesel if applicable), Drive Train, Steering and Suspension.

All illustrations are located as close as possible to the applicable procedure. Procedures are for all models in the particular section unless specifically noted otherwise.

Part Numbers

Part numbers listed in this book are not recommendations by Nichols Publishing for any product by brand name. They are references that can be used with interchanges manuals and aftermarket supplier catalogs to locate each brand supplier's discrete part number.

Special Tools

Special tools are recommended by the vehicle manufacturer to perform their specific job. Use has been kept to a minimum, but where absolutely necessary, they are referred to in the text by the part number of the tool manufacturer. These tools may be purchased, under the appropriate part number, from your local dealer or regional distributor, or an equivalent tool can be purchased locally from a tool supplier or parts outlet. Before substituting any tool for the one recommended, read the previous Safety Notice.

ACKNOWLEDGMENTS

This publication contains material that is reproduced and distributed under a license from Ford Motor Company. No further reproduction or distribution of the Ford Motor Company material is allowed without the expressed written permission from Ford Motor Company.

Portions of the material contained herein have been reprinted with permission of General Motors Corporation, Service Technology Group.

Nichols Publishing would like to express thanks to all of the fine companies who participate in the production of our books. Hand tools supplied by Craftsman are used during all phases of our vehicle teardown and photography. Many of the fine specialty tools used in our procedures were provided courtesy of Lisle Corporation. Lincoln Automotive Products (1 Lincoln Way, St. Louis, MO 63120) has provided their industrial shop equipment, including jacks (engine, transmission and floor), engine stands, fluid and lubrication tools, as well as shop presses. Rotary Lifts, the largest automobile lift manufacturer in the world, offering the biggest variety of surface and in-ground lifts available (1-800-640-5438 or www.Rotary-Lift.com), have fulfilled our shop's lift needs. Much of our shop's electronic testing equipment was supplied by Universal Enterprises Inc. (UEI).

SPECIFICATIONS

1

ACURA
MDX

ENGINE AND VEHICLE IDENTIFICATION CHART

		Engine Code						Model Year	
Code	Liters (cc)	Cu. In.	Cyl.	Fuel Sys.	Engine Type	Eng. Mfg.		Code ①	Year
J35A3	3.5 (3471)	212	6	SMFI	SOHC	Honda		1	2001
								2	2002
								3	2003

SOHC: Single Overhead Cam

SMFI: Sequential Multi-port Fuel Injection

① 10th position of VIN

93591C01

GENERAL ENGINE SPECIFICATIONS

Year	Model	Engine Displacement Liters (cc)	Engine ID/VIN	Fuel System Type	Net Horsepower @ rpm	Net Torque @ rpm (ft. lbs.)	Bore x Stroke (in.)	Compression Ratio	Oil Pressure @ rpm
2001	MDX	3.5 (3471)	J35A3	SMFI	210@5200	229@4300	3.50x3.66	9.4:1	71@3000

SMFI: Sequential Multi-port Fuel Injection

93591C02

ENGINE TUNE-UP SPECIFICATIONS

Year	Engine Displacement Liters (cc)	Engine ID/VIN	Spark Plug Gap (in.)	Ignition Timing (deg.)		Fuel Pump (psi)	Idle Speed (rpm)		Valve Clearance (in.)	
				MT	AT		MT	AT	In.	Ex.
2001	3.5 (3471)	J35A1	0.039-0.043	—	8-12B	32-40	—	680-780	0.008-0.009	0.011-0.013

NOTE: The Vehicle Emission Control Information label often reflects changes made during production and must be used if they differ from this chart.

NOTE: The fuel pressure readings are given with the vacuum hose connected to the regulator and the engine running

B: Before top dead center

93591C03

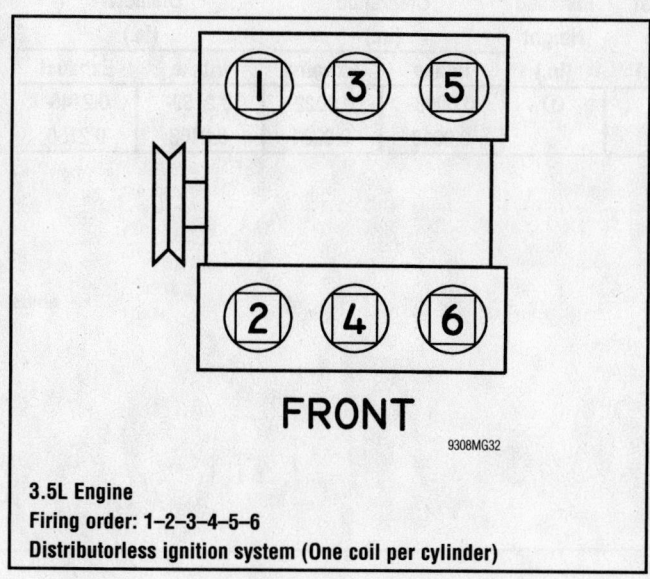

FRONT

9308MG32

3.5L Engine
Firing order: 1–2–3–4–5–6
Distributorless ignition system (One coil per cylinder)

CAPACITIES

Year	Model	Engine Displacement Liters (cc)	Engine ID/VIN	Engine Oil with Filter (qts.)	Transmission (pts.)		Transfer Case (pts.)	Drive Axle		Fuel Tank (gal.)	Cooling System (qts.)
					5-Spd	Auto.		Front (pts.)	Rear (pts.)		
2001	MDX	3.5 (3471)	J35A3	5.0	—	9.0	—	—	—	19.2	8.0

NOTE: All capacities are approximate. Add fluid gradually and check to be sure a proper fluid level is obtained.

93591C04

VALVE SPECIFICATIONS

Year	Engine Displacement Liters (cc)	Engine ID/VIN	Seat Angle (deg.)	Face Angle (deg.)	Spring Test Pressure (lbs. @ in.)	Spring Installed Height (in.)	Stem-to-Guide Clearance (in.)		Stem Diameter (in.)	
							Intake	Exhaust	Intake	Exhaust
2001	3.5 (3471)	J35A3	45	45	NA	①	0.0008-0.0018	0.0022-0.0031	0.2159-0.2163	0.2146-0.2150

NA: Not Available

① Valve spring free length:

Intake: 1.9713 in.

Exhaust: 2.1060 in.

93591C05

CRANKSHAFT AND CONNECTING ROD SPECIFICATIONS

All measurements are given in inches

| Year | Engine Displacement Liters (cc) | Engine ID/VIN | Crankshaft | | | | Connecting Rod | | |
			Main Brg. Journal Dia.	Main Brg. Oil Clearance	Shaft End-play	Thrust on No.	Journal Diameter	Oil Clearance	Side Clearance
2001	3.5 (3471)	J35A3	2.8337-2.8346	0.0008-0.0017	0.0040-0.0140	3	2.1644-2.1654	0.0008-0.0017	0.0060-0.0140

93591C06

PISTON AND RING SPECIFICATIONS

All measurements are given in inches

| Year | Engine Displacement Liters (cc) | Engine ID/VIN | Piston Clearance | Ring Gap | | | Ring Side Clearance | | |
				Top Compression	Bottom Compression	Oil Control	Top Compression	Bottom Compression	Oil Control
2001	3.5 (3471)	J35A1	0.0006-0.0016	0.0080-0.0140	0.0160-0.0220	0.0080-0.0280	0.0014-0.0024	0.0012-0.0022	NA

NA: Not Available

93591C07

TORQUE SPECIFICATIONS
All readings in ft. lbs.

Year	Engine Displacement Liters (cc)	Engine ID/VIN	Cylinder Head Bolts	Main Bearing Bolts	Rod Bearing Bolts	Crankshaft Damper Bolts	Flywheel Bolts	Manifold Intake	Manifold Exhaust	Spark Plugs	Lug Nut
2001	3.5 (3471)	J35A1	①	②	③	181	54	16	23	13	80

NOTE: Dip main bearing bolts and crankshaft damper bolt in clean engine oil prior to tightening.

① Step 1: 29 ft. lbs.
   Step 2: 51 ft. lbs.
   Step 3: 72 ft. lbs.

② 11mm bolt 56 ft. lbs.
   10mm bolt 36 ft. lbs.

③ Step 1: 14 ft. lbs.
   Step 2: 90 degrees

93591C08

WHEEL ALIGNMENT

Year	Model		Caster Range (+/-Deg.)	Caster Preferred Setting (Deg.)	Camber Range (+/-Deg.)	Camber Preferred Setting (Deg.)	Toe-in (in.)	Steering Axis Inclination (Deg.)
2001	MDX	F	1.00	1.00	1.00	30	0+/-1/16	—
		R	—	—	0	30	0+/-1/16	—

93591C09

TIRE, WHEEL AND BALL JOINT SPECIFICATIONS

| Year | Model | OEM Tires | | Tire Pressures (psi) | | Wheel Size | Ball Joint Inspection |
		Standard	Optional	Front	Rear		
2001	MDX	P235/65R17	None	32	32	R17	NS

OEM: Original Equipment Manufacturer

PSI: Pounds Per Square Inch

NS: Not specified by manufacturer

93591C10

BRAKE SPECIFICATIONS
Acura MDX
All measurements in inches unless noted

| Year | Model | | Brake Disc | | | Brake Drum Diameter | | | Minimum Lining Thickness | | Brake Caliper | |
			Original Thickness	Minimum Thickness	Maximum Runout	Original Inside Diameter	Max. Wear Limit	Maximum Machine Diameter	Front	Rear	Bracket Bolts (ft. lbs.)	Mounting Bolts (ft. lbs.)
2001	MDX	F	1.100	1.020	0.004	—	—	—	0.060	—	80	27
		R	0.430	0.350	0.004	—	—	—	—	0.41	41	27

F: Front

R: Rear

93591C11

Timing belt service is covered in Section 3 of this manual

SCHEDULED MAINTENANCE INTERVALS
ACURA—MDX

TO BE SERVICED	TYPE OF SERVICE	VEHICLE MILEAGE INTERVAL (x1000)															
		7.5	15	22.5	30	37.5	45	52.5	60	67.5	75	82.5	90	97.5	105	112.5	120
Accessory drive belts	I & A				✓				✓				✓				✓
Air cleaner element	R				✓				✓				✓				✓
Brake fluid	R	Every 3 years															
Brake hoses & lines (including ABS)	I		✓		✓		✓		✓		✓		✓		✓		✓
Cooling system hoses & connections	I		✓		✓		✓		✓		✓		✓		✓		✓
Engine coolant ①	R						✓						✓				
Engine oil	R	✓	✓	✓	✓	✓	✓	✓	✓	✓	✓	✓	✓	✓	✓	✓	✓
Engine oil and coolant levels	I	Inspect at each fuel stop															
Engine oil filter	R		✓		✓		✓		✓		✓		✓		✓		✓
Exhaust system	I		✓		✓		✓		✓		✓		✓		✓		✓
Fluid levels and condition	I		✓		✓		✓		✓		✓		✓		✓		✓
Front and rear brakes	I		✓		✓		✓		✓		✓		✓		✓		✓
Fuel lines & connection	I		✓		✓		✓		✓		✓		✓		✓		✓
Halfshaft boots	I		✓		✓		✓		✓		✓		✓		✓		✓
Idle speed	I & A														✓		
Parking brake system	I & A		✓		✓		✓		✓		✓		✓		✓		✓
Rear differential fluid	R	✓			✓				✓				✓				✓
Rotate and inspect tires	I	✓	✓	✓	✓	✓	✓	✓	✓	✓	✓	✓	✓	✓	✓	✓	✓
Spark plugs	R														✓		
Supplemental Restrain system (SRS)	I	Inspect the SRS 10 years after production															
Suspension components	I		✓		✓		✓		✓		✓		✓		✓		✓
Tie rod ends, steering gear box & boots	I		✓		✓		✓		✓		✓		✓		✓		✓
Timing belt	R														✓		
Transmission fluid	R						✓				✓				✓		
Valve clearance	I	Adjust if valves are noisy															
Water pump	S/I														✓		

R: Replace   I: Inspect   A: Adjust

① Every 12,000 miles or 10 years, then every 60,000 miles or 5 years

FREQUENT OPERATION MAINTENANCE (SEVERE SERVICE)

If a vehicle is operated under any of the following conditions it is considered severe service:

- Towing a trailer or using a camper or car-top carrier.
- Repeated short trips of less than 5 miles in temperatures below freezing, or trips of less than 10 miles in any temperature.
- Extensive idling or low-speed driving for long distances as in heavy commercial use, such as delivery, taxi or police cars.
- Operating on rough, muddy or salt-covered roads.
- Operating on unpaved or dusty roads.
- Driving in extremely hot (over 90°) conditions.

Air cleaner element: replace every 15,000 miles

Engine oil and filter: replace every 3750 miles or 6 months, whichever occurs first.

Timing belt: replace every 60,000 miles if the vehicle is regularly driven in temperatures above 110°F or below -20°F, or if frequently towing a trailer.

Transmission fluid: replace every 30,000 miles.

Rear differential fluid: replace every 60,000 miles.

Front and rear brakes: inspect every 7500 miles or 6 months, whichever occurs first.

Locks and hinges: lubricate every 15,000 miles.

Tie rods, steering gear box, boots: inspect every 7500 miles or 6 months, whichever occurs first.

Suspension components: inspect every 7500 miles or 6 months, whichever occurs first.

Halfshaft boots: inspect every 7500 miles or 6 months, whichever occurs first.

93591C12

SCHEDULED MAINTENANCE INTERVALS
ACURA
MDX

The following should be used as a guide when determining the amount of work required for a particular service. In estimating how long a particular Scheduled Maintenance Service should take, please observe the following:

- Labor Time is time based on field research and data supplied by the vehicle manufacturer.
- Labor time operations are given in hours and tenths of an hour.
- All labor operations are to be used as a guide.

Mechanic Skill Level Codes:
(A) PRECISION: Highly skilled with multiple certification.
(B) GENERAL: Normally skilled with certification.
(C) MAINTENANCE: Semi-skilled working on certification.

	LABOR TIME
7500 Mile Service (C)	
All Models	1.1
15000 Mile Service (C)	
All Models	1.6
22500 Mile Service (C)	
All Models	1.1
30000 Mile Service (B)	
All Models	2.5
37500 Mile Service (C)	
All Models	1.1
45000 Mile Service (B)	
All Models	3.3

	LABOR TIME
52500 Mile Service (C)	
All Models	1.1
60000 Mile Service (B)	
All Models	2.5
67500 Mile Service (C)	
All Models	1.1
75000 Mile Service (B)	
All Models	2.5
82500 Mile Service (C)	
All Models	1.1
90000 Mile Service (B)	
All Models	2.5

	LABOR TIME
97500 Mile Service (C)	
All Models	1.1
105000 Mile Service (B)	
All Models	3.8
Replace spark plugs add	
Replace timing belt add	
112500 Mile Service (C)	
All Models	1.1
120000 Mile Service (B)	
All Models	2.5

93551C13

Heater Core replacement is covered in Section 2 of this manual

ACURA
SLX

ENGINE AND VEHICLE IDENTIFICATION

			Engine					Model Year	
Code	Liters (cc)	Cu. In.	Cyl.	Fuel Sys.	Engine Type	Eng. Mfg.		Code ①	Year
X	3.5 (3494)	213	6	SMFI	DOHC	Isuzu		X	1999

SMFI: Sequential Multi-port Fuel Injection

DOHC :Double Overhead Camshaft

① 10th position of VIN

93591CAA

GENERAL ENGINE SPECIFICATIONS

Year	Model	Engine Displacement Liters (cc)	Engine Series (ID/VIN)	Fuel System	Net Horsepower @ rpm	Net Torque @ rpm (ft. lbs.)	Bore x Stroke (in.)	Com-pression Ratio	Oil Pressure @ rpm
1999	SLX	3.5 (3494)	6VE1/X	SMFI	215@5400	230@3000	3.68x3.35	9.1:1	60-80@3000

SMFI: Sequential Multi-port Fuel Injection

93591CAB

ENGINE TUNE-UP SPECIFICATIONS

Year	Engine Displacement Liters (cc)	Engine ID/VIN	Spark Plug Gap (in.)	Ignition Timing (deg.) ①		Fuel Pump (psi)	Idle Speed (rpm) ①		Valve Clearance ②	
				MT	AT		MT	AT	In.	Ex.
1999	3.5 (3494)	6VE1/X	0.040-0.043	—	20B	48-55	—	750	0.009-0.013	0.010-0.014

NOTE: The Vehicle Emission Control Information label reflects production specification changes and must be used if different from this chart.

B: Before top dead center

① Controlled by the ECM and is not adjustable.

② Measure with engine cold

93591CAC

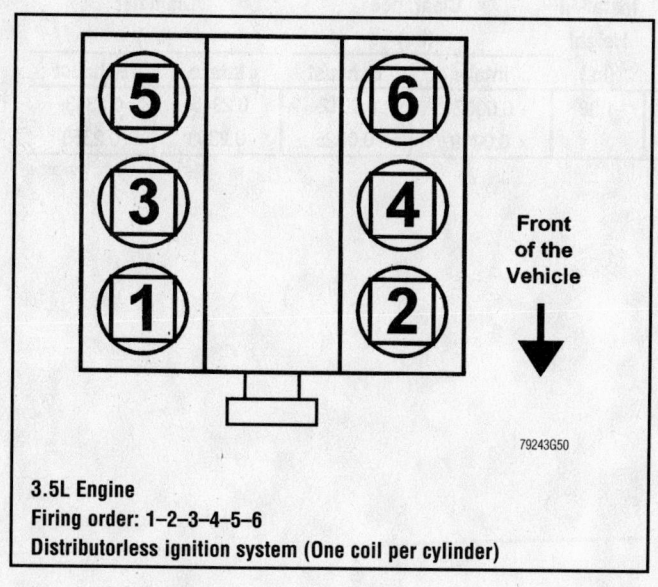

3.5L Engine
Firing order: 1–2–3–4–5–6
Distributorless ignition system (One coil per cylinder)

79243G50

Front of the Vehicle

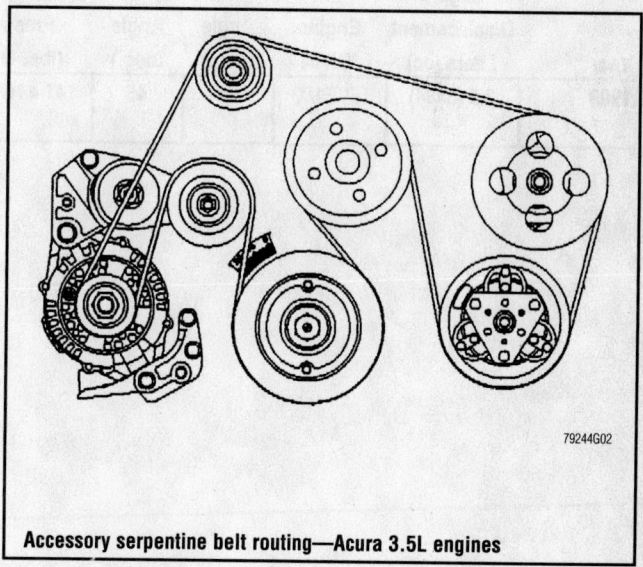

Accessory serpentine belt routing—Acura 3.5L engines

79244G02

Brake service is covered in Section 4 of this manual

CAPACITIES

Year	Model	Engine Displacement Liters (cc)	Engine ID/VIN	Engine Oil with Filter (qts.)	Trans- mission (pts.)	Transfer Case (pts.)	Drive Axle		Fuel Tank (gal.)	Cooling System (qts.)
							Front (pts.)	Rear (pts.)		
1999	SLX	3.5 (3494)	6VE1/X	5.0	18.2	3.0 ①	3.0	6.4	22.5	7.4

NOTE: All capacities are approximate. Add fluid gradually and check to ensure a proper level has been reached.

① 4.0 pts. if equipped with Torque On Demand (TOD)

93591CAD

VALVE SPECIFICATIONS

Year	Engine Displacement Liters (cc)	Engine ID/VIN	Seat Angle (deg.)	Face Angle (deg.)	Spring Test Pressure (lbs. @ in.)	Spring Installed Height (in.)	Stem-to-Guide Clearance (in.)		Stem Diameter (in.)	
							Intake	Exhaust	Intake	Exhaust
1999	3.5 (3494)	6VE1/X	45	45	41-44 @ 1.38	1.38	0.0002- 0.0009	0.0012- 0.0025	0.2346- 0.2353	0.2343- 0.2350

93591CAE

CRANKSHAFT AND CONNECTING ROD SPECIFICATIONS
All measurements are given in inches.

Year	Engine Displacement Liters (cc)	Engine ID/VIN	Crankshaft				Connecting Rod		
			Main Brg. Journal Dia.	Main Brg. Oil Clearance	Shaft End-play	Thrust on No.	Journal Diameter	Oil Clearance	Side Clearance
1999	3.5 (3494)	6VE1/X	2.5165-2.5170	0.0007-0.0017	0.0024-0.0094	3	2.1229-2.1235	0.0010-0.0023	0.0050-0.0150

93591CAF

PISTON AND RING SPECIFICATIONS
All measurements are given in inches.

Year	Engine Displacement Liters (cc)	Engine ID/VIN	Piston Clearance	Ring Gap			Ring Side Clearance		
				Top Compression	Bottom Compression	Oil Control	Top Compression	Bottom Compression	Oil Control
1999	3.5 (3494)	6VE1/X	0.0012-0.0020	0.0118-0.0157	0.0177-0.0236	0.0059-0.0177	0.0006-0.0015	0.0006-0.0015	NA

NA: Not Available

93591CAG

For complete Engine Mechanical specifications, see Section 1 of this manual

TORQUE SPECIFICATIONS
All readings in ft. lbs.

Year	Engine Displacement Liters (cc)	Engine ID/VIN	Cylinder Head Bolts	Main Bearing Bolts	Rod Bearing Bolts	Crankshaft Damper Bolts	Flywheel Bolts	Manifold Intake	Manifold Exhaust	Spark Plugs	Lug Nuts
1999	3.5 (3494)	6VE1/X	①	②	40	123	40	18	38	13	87

① Step 1: 21 ft. lbs.
   Step 2: 47 ft. lbs.

② Main bearing cap bolts: 29 ft. lbs.
   Oil gallery bolts: 21 ft. lbs. plus 55-65 degree turn
   Buttress bolts: 29 ft. lbs.

93591CAH

WHEEL ALIGNMENT

Year	Model	Caster Range (+/-Deg.)	Caster Preferred Setting (Deg.)	Camber Range (+/-Deg.)	Camber Preferred Setting (Deg.)	Toe-in (in.)	Steering Axis Inclination (Deg.)
1999	SLX	0.75	+2.16	0.50	0	0+/-0.08	—

93591CAI

TIRE, WHEEL AND BALL JOINT SPECIFICATIONS

| Year | Model | OEM Tires | | Tire Pressures (psi) | | Wheel Size | Ball Joint Inspection |
		Standard	Optional	Front	Rear		
1999	SLX	245/70R16	None	29	29	7-JJ	U: 4-28 ①
							L: 4-55

OEM: Original Equipment Manufacturer

PSI: Pounds Per Square Inch

L: Lower

U: Upper

① Torque required in inch lbs. to rotate ball joint when removed from the knuckle

93591CAJ

BRAKE SPECIFICATIONS
All measurements in inches unless noted

| Year | Model | | Brake Disc | | | | Brake Drum Diameter | | | Minimum Lining Thickness | | Brake Caliper | |
			Original Thickness	Machine Thickness	Discard Thickness	Maximum Runout	Original Inside Diameter	Max. Wear Limit	Maximum Machine Diameter	Front	Rear	Bracket Bolts (ft. lbs.)	Mounting Bolts (ft. lbs.)
1999	SLX	F	1.020	0.983	0.969	0.005	—	—	—	0.039	—	115	54
		R	0.710	0.668	0.654	0.005	8.27	8.32	8.32	—	0.039	76	32

NA: Not Available

F: Front

R: Rear

93591CAK

For Accessory Drive Belt illustrations, see Section 1 of this manual

SCHEDULED MAINTENANCE INTERVALS

ACURA—SLX

TO BE SERVICED	TYPE OF SERVICE	VEHICLE MILEAGE INTERVAL (x1000)															
		7.5	15	22.5	30	37.5	45	52.5	60	67.5	75	82.5	90	97.5	105	112.5	120
Accelerator linkage ①	L	✓	✓	✓	✓	✓	✓	✓	✓	✓	✓	✓	✓	✓	✓	✓	✓
Accessory drive belts ②	S/I				✓				✓				✓				✓
Air cleaner filter	R				✓				✓				✓				✓
Auto cruise control linkage & hose ③	S/I		✓		✓		✓		✓		✓		✓		✓		✓
Automatic transmission fluid level ③	S/I	✓		✓		✓		✓		✓		✓		✓		✓	
Battery fluid level ③	S/I	✓	✓	✓	✓	✓	✓	✓	✓	✓	✓	✓	✓	✓	✓	✓	✓
Body and chassis ①	L	✓	✓	✓	✓	✓	✓	✓	✓	✓	✓	✓	✓	✓	✓	✓	✓
Brake fluid level ③	S/I	✓	✓	✓	✓	✓	✓	✓	✓	✓	✓	✓	✓	✓	✓	✓	✓
Brake lines & hoses ③	S/I	✓	✓	✓	✓	✓	✓	✓	✓	✓	✓	✓	✓	✓	✓	✓	✓
Brake pedal play ③	S/I		✓		✓		✓		✓		✓		✓		✓		✓
Clutch fluid level ③	S/I	✓	✓	✓	✓	✓	✓	✓	✓	✓	✓	✓	✓	✓	✓	✓	✓
Clutch lines & hose ③	S/I				✓				✓				✓				✓
Clutch pedal free-play ③	S/I		✓		✓		✓		✓		✓		✓		✓		✓
Clutch pedal spring, bushing and clevis pin ①	S/I		✓		✓		✓		✓		✓		✓		✓		✓
Cooling and heating system hoses ③	S/I		✓		✓		✓		✓		✓		✓		✓		✓
Driveshaft flange torque ③	S/I	✓		✓		✓		✓		✓		✓		✓		✓	
Drum and disc brakes ③	S/I		✓		✓		✓		✓		✓		✓		✓		✓
Engine coolant	R				✓				✓				✓				✓
Engine coolant level ③	S/I	✓	✓	✓	✓	✓	✓	✓	✓	✓	✓	✓	✓	✓	✓	✓	✓
Engine oil & filter ③	R	✓	✓	✓	✓	✓	✓	✓	✓	✓	✓	✓	✓	✓	✓	✓	✓
Exhaust system ③	S/I	✓	✓	✓	✓	✓	✓	✓	✓	✓	✓	✓	✓	✓	✓	✓	✓
Front and rear axle lubricant	R		✓		✓				✓				✓				✓
Front and rear driveshafts ①	S/I	✓	✓	✓	✓	✓	✓	✓	✓	✓	✓	✓	✓	✓	✓	✓	✓
Front wheel bearings	S/I & L				✓				✓				✓				✓
Fuel lines & tank cap ③	S/I								✓								✓
Inspect for fluid leaks ③	S/I	✓	✓	✓	✓	✓	✓	✓	✓	✓	✓	✓	✓	✓	✓	✓	✓
Key lock cylinder ③	L		✓		✓		✓		✓		✓		✓		✓		✓
Manual transmission and transfer case fluid	R		✓		✓				✓				✓				✓
Parking brake system ③	S/I		✓		✓		✓		✓		✓		✓		✓		✓
Power steering fluid	R				✓				✓				✓				✓
Radiator core and A/C condenser	S/I & C								✓								✓
Rotate tires	S/I	✓	✓	✓	✓	✓	✓	✓	✓	✓	✓	✓	✓	✓	✓	✓	✓
Shift-on-the-fly system gear fluid ③	S/I		✓		✓		✓		✓		✓		✓		✓		✓
Spark plugs	R	Every 100,000 miles.															

93591CAL

SCHEDULED MAINTENANCE INTERVALS
ACURA—SLX

TO BE SERVICED	TYPE OF SERVICE	VEHICLE MILEAGE INTERVAL (x1000)															
		7.5	15	22.5	30	37.5	45	52.5	60	67.5	75	82.5	90	97.5	105	112.5	120
Starter safety switch ③	S/I	✓	✓	✓	✓	✓	✓	✓	✓	✓	✓	✓	✓	✓	✓	✓	✓
Steering operation ③	S/I	✓	✓	✓	✓	✓	✓	✓	✓	✓	✓	✓	✓	✓	✓	✓	✓
Suspension & steering ③	S/I	✓	✓	✓	✓	✓	✓	✓	✓	✓	✓	✓	✓	✓	✓	✓	✓
Throttle linkage ③	S/I		✓		✓		✓		✓		✓		✓		✓		✓
Timing belt	R										✓						
Tires and wheels ③	S/I	✓	✓	✓	✓	✓	✓	✓	✓	✓	✓	✓	✓	✓	✓	✓	✓
Valve clearance	A									✓							✓

R: Replace   S/I: Service or Inspect   L: Lubricate   A: Adjust   C: Clean

① Perform this at the mileage indicated or every 6 months, whichever occurs first.

② Perform this at the mileage indicated or every 24 months, whichever occurs first.

③ Perform this at the mileage indicated or every 12 months, whichever occurs first.

FREQUENT OPERATION MAINTENANCE (SEVERE SERVICE)

If a vehicle is operated under any of the following conditions it is considered severe service:

- Towing a trailer or using a camper or car-top carrier.

- Repeated short trips of less than 5 miles in temperatures below freezing, or trips of less than 10 miles in any temperature.

- Extensive idling or low-speed driving for long distances as in heavy commercial use, such as delivery, taxi or police cars.

- Operating on rough, muddy or salt-covered roads.

- Operating on unpaved or dusty roads.

- Frequent operation in temperatures above 90°F.

Air cleaner element: replace every 15,000 miles

Engine oil and filter: replace every 3000 miles or 3 months, whichever occurs first.

Automatic transmission fluid: replace every 20,000 miles.

Rear axle lubricant: replace every 15,000 miles.

93591CAM

For Tire, Wheel and Ball Joint specifications, see Section 1 of this manual.

SCHEDULED MAINTENANCE INTERVALS
ACURA
SLX

The following should be used as a guide when determining the amount of work required for a particular service.
In estimating how long a particular Scheduled Maintenance Service should take, please observe the following:

- Labor Time is time based on field research and data supplied by the vehicle manufacturer.
- Labor time operations are given in hours and tenths of an hour.
- All labor operations are to be used as a guide.

Mechanic Skill Level Codes:
(A) PRECISION: Highly skilled with multiple certification.
(B) GENERAL: Normally skilled with certification.
(C) MAINTENANCE: Semi-skilled working on certification.

	LABOR TIME		LABOR TIME		LABOR TIME
7500 Mile Service (C)		**45000 Mile Service (B)**		**90000 Mile Service (B)**	
1998-02	2.7	1998-02	3.8	1998-02	6.7
15000 Mile Service (B)		**52500 Mile Service (C)**		**97500 Mile Service (C)**	
1998-02	3.5	1998-02	2.7	1998-02	2.7
22500 Mile Service (C)		**60000 Mile Service (B)**		**105000 Mile Service (C)**	
1998-02	2.7	1998-02	8.8	1998-02	3.8
30000 Mile Service (B)		**67500 Mile Service (C)**		**112500 Mile Service (C)**	
1998-02	6.7	1998-02	2.7	1998-02	2.7
37500 Mile Service (C)		**75000 Mile Service (B)**		**120000 Mile Service (B)**	
1998-02	2.7	1998-02	5.7	1998-02	8.8
		82500 Mile Service (C)			
		1998-02	2.7		

93551CRR

BMW
X5

ENGINE AND VEHICLE IDENTIFICATION

		Engine						Model Year	
Code ①	Liters (cc)	Cu. In.	Cyl.	Fuel Sys.	Engine Type	Eng. Mfg.		Code ②	Year
M54	3.0 (2979)	182	6	SMPI	DOHC	BMW		Y	2000
M62	4.4 (4398)	268	8	SMPI	DOHC	BMW			

DOHC: Double Overhead Camshaft

① 8th position of VIN

② 10th position of VIN

93591CFA

GENERAL ENGINE SPECIFICATIONS

Year	Model	Engine Displacement Liters (cc)	Engine Series (ID/VIN)	Fuel System	Net Horsepower @ rpm	Net Torque @ rpm (ft. lbs.)	Bore x Stroke (in.)	Com-pression Ratio	Oil Pressure @ rpm
2000	X5	3.0 (2979)	M54	SMPI	225@5900	214@3500	3.31x3.53	10.2:1	7.4@700
	X5	4.4 (4398)	M62	SMPI	282@6400	324@3600	3.62x3.26	10.0:1	7.4@580

SMPI: Sequential Multi-port Fuel Injection

93591CFB

For Wheel Alignment specifications, see Section 1 of this manual

ENGINE TUNE-UP SPECIFICATIONS

Year	Engine Displacement Liters (cc)	Engine ID/VIN	Spark Plug Gap (in.)	Ignition Timing (deg.)	Fuel Pump (psi)	Idle Speed (rpm)	Valve Clearance	
							In.	Ex.
2000	3.0 (2979)	M54	0.024-0.028	①	48-54	②	HYD	HYD
	4.4 (4398)	M62	0.024-0.028	①	48-54	②	HYD	HYD

NOTE: The Vehicle Emission Control Information label often reflects specification changes made during production. The label figures must be used if they differ from those in this chart.

HYD: Hydraulic

① Ignition timing is regulated by the Electronic Control Module (ECM), and cannot be adjusted.

② Idle speed is controled by the Electronic Control Module (ECM), and cannot be adjusted.

93591CFC

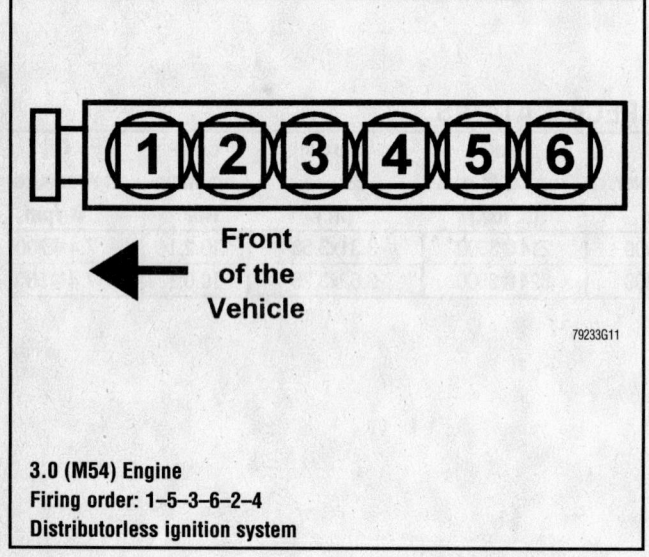

3.0 (M54) Engine
Firing order: 1–5–3–6–2–4
Distributorless ignition system

79233G11

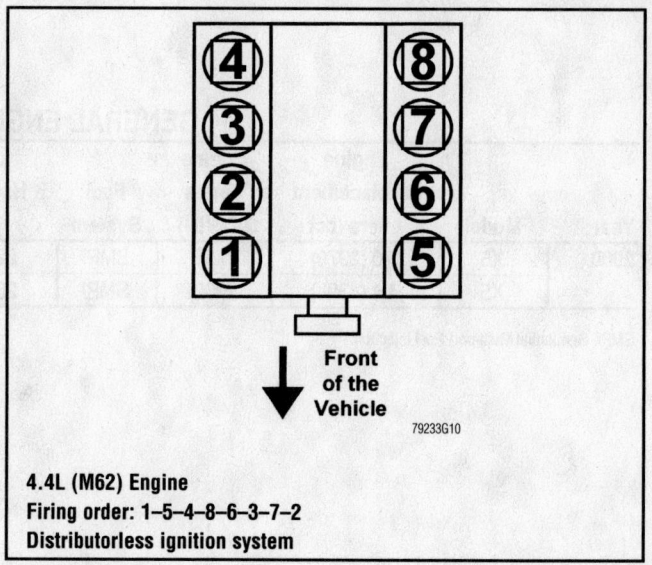

4.4L (M62) Engine
Firing order: 1–5–4–8–6–3–7–2
Distributorless ignition system

79233G10

CAPACITIES

Year	Model	Engine Displacement Liters (cc)	Engine ID/VIN	Engine Oil with Filter (qts.)	Automatic Transaxle (qts.)	Manual Transaxle (qts.)	Rear Drive Axle (pts.)	Fuel Tank (gal.)	Cooling System (qts.)
2000	X5	3.0 (2979)	M54	7.0	7.0	3.2	3.4	24.6	11.1
		4.4 (4398)	M62	7.9	11.7	—	3.4	24.6	13.3

NOTE: All capacities are approximate. Add fluid gradually and check to be sure a proper fluid level is obtained.

93591CFD

VALVE SPECIFICATIONS

Year	Engine Displacement Liters (cc)	Engine ID/VIN	Seat Angle (deg.)	Face Angle (deg.)	Spring Test Pressure (lbs. @ in.)	Spring Installed Height (in.)	Stem-to-Guide Clearance (in.)		Stem Diameter (in.)	
							Intake	Exhaust	Intake	Exhaust
2000	3.0 (2979)	M54	45	45	NA	NA	0.0197	0.0197	0.2372-0.2340	0.2378-0.2384
	4.4 (4398)	M62	45	45	NA	NA	0.0197	0.0197	0.2156-0.2159	0.2146-0.2150

NA: Not Available

93591CFE

For Maintenance Interval recommendations, see Section 1 of this manual

CRANKSHAFT AND CONNECTING ROD SPECIFICATIONS
All measurements are given in inches.

Year	Engine Displacement Liters (cc)	Engine ID/VIN	Crankshaft				Connecting Rod		
			Main Brg. Journal Dia.	Main Brg. Oil Clearance	Shaft End-play	Thrust on No.	Journal Diameter	Oil Clearance	Side Clearance
2000	3.0 (2979)	M54	①	0.0007-0.0029	0.0031-7 0.0064	5	1.7720-1.7706	0.0007-0.0022	0.0060-0.0160
	4.4 (4398)	M62	①	0.0007-0.0018	0.0033-0.0101	3	1.8901-1.8887	0.0007-0.0022	0.0060-0.0196

① Standard yellow: 2.3615-2.3618 inches
Standard green: 2.3613-2.3615 inches
Standard white: 2.3611-2.3613 inches

93591CFF

PISTON AND RING SPECIFICATIONS
All measurements are given in inches.

Year	Engine Displacement Liters (cc)	Engine ID/VIN	Piston Clearance	Ring Gap			Ring Side Clearance		
				Top Compression	Bottom Compression	Oil Control	Top Compression	Bottom Compression	Oil Control
2000	3.0 (2979)	M54	0.0004-0.0016	0.0039-0.0118	0.0078-0.0157	0.0098-0.0197	0.0008-0.0024	0.0012-0.0026	0.0007-0.0024
	4.4 (4398)	M62	0.0002-0.0015	0.0039-0.0118	0.0078-0.0157	0.0078-0.0354	0.0008-0.0024	0.0008-0.0024	NA

93591CFG

TORQUE SPECIFICATIONS
All readings in ft. lbs.

Year	Engine Displacement Liters (cc)	Engine ID/VIN	Cylinder Head Bolts	Main Bearing Bolts	Rod Bearing Bolts	Crankshaft Damper Bolts	Flywheel Bolts	Manifold Intake	Manifold Exhaust	Spark Plugs	Lug Nuts
2000	3.0 (2979)	M54	①	NA	②	100	88	17	15	15	95
	4.4 (4398)	M62	①	NA	③	100	NA	17	15	15	95

NA: Not Available

① Step 1: 22 ft. Lbs.
  Step 2: Plus 80 degrees
  Step 3: Plus 80 degrees

② Step 1: 15 ft. Lbs.
  Step 2: Plus 45 degrees

③ Step 1: 15 ft. Lbs.
  Step 2: Plus 80 degrees

93591CFH

WHEEL ALIGNMENT

Year	Model		Caster Range (+/-Deg.)	Caster Preferred Setting (Deg.)	Camber Range (+/-Deg.)	Camber Preferred Setting (Deg.)	Toe-in (in.)	Steering Axis Inclination (Deg.)
2000	X5	F	0.50	+0.83	0.42	-0.20	0.30+/-0.13	—
		R	—	—	0.33	+0.83	0.30+/-0.13	—

93591CFI

For Tune-up, Capacities and Firing orders, see Section 1 of this manual

TIRE, WHEEL AND BALL JOINT SPECIFICATIONS

Year	Model	OEM Tires		Tire Pressures (psi)		Wheel Size	Ball Joint Inspection
		Standard	Optional	Front	Rear		
2000	X5	P235/65R17	P255/55R18	32	32	7.5J/8.5J	NA
			F P255/50R19			F 9J	
			R P285/45R19			R 10J	

OEM: Original Equipment Manufacturer

PSI: Pounds Per Square Inch

STD: Standard

OPT: Optional

NA: Not Available

F: Front

R: Rear

93591CFJ

BRAKE SPECIFICATIONS
All measurements in inches unless noted

Year	Model		Brake Disc			Brake Drum Diameter			Min. Lining Thickness	Caliper Guide Pin Bolts (ft. lbs.)
			Original Thickness	Minimum Thickness	Maximum Run-out	Original Inside Diameter	Max. Wear Limit	Maximum Machine Diameter		
2000	M54	F	0.803	①	0.005	—	—	—	0.118	25
		R	②	①	0.005	8.63-8.65	NA	NA	①	25
	M62	F	1.118	①	0.007	—	—	—	118.000	25
		R	0.409	①	0.007	8.63-8.65	NA	NA	①	25

NA: Not Available

F: Front

R: Rear

① Minimum thickness is stamped in the brake disc shell

② Solid brake rotor: 0.409 inches

Ventilated brake rotor: 0.724 inches

93591CFK

SCHEDULED MAINTENANCE INTERVALS
BMW—X5

TO BE SERVICED	TYPE OF SERVICE	SERVICE INTERVALS			
		INITIAL 1200 MILES	OIL SERVICE	INSPECTION I	INSPECTION II
Oil level	S/I	✓			
Engine oil	R	✓			
Engine oil & filter	R①		✓	✓	✓
Engine air cleaner element	R②				✓
Spark plugs	R				✓
Fuel filter	R③				✓
Fuel, vapor lines & fuel cap	S/I	✓		✓	✓
Cooling system	S/I	✓		✓	✓
Exhaust pipe & muffler	S/I	✓		✓	✓
Catalytic converter & shielding	S/I	✓		✓	✓
Throttle linkage	S/I			✓	✓
Engine (check for leakage)	S/I	✓			
Engine drive belts	S/I				✓
Maintenance Indicators	RE		④	✓	✓
Engine coolant	R			⑤	⑤
Oxygen sensor	R⑥				
Brake & clutch fluids ⑥	S/I			✓	✓
Brake pads & discs	S/I			✓	✓
Parking brake system	S/I			✓	✓
Power steering system	S/I			✓	✓
Rear axle fluid	S/I			✓	✓
Steering play, suspension track rods, front axle joints, steering linkage & joint disc	S/I			✓	✓
Transmission fluid/oil	S/I			✓	✓
Wheel centering hubs	S/I			✓	✓
Rear axle fluid	R		✓		✓
OBD system for codes	S/I	✓		✓	✓

R: Replace   S/I: Service or Inspect   RE: Reset

Note: BMW does not rely solely on vehicle mileage to determine service intervals. An on-board diagnostic center, monitors engine operating conditions, along with mileage, to determine the most effective maintenance intervals. The information is then conveyed to the driver through the service indicator lights, located in the center of the instrument panel.

Note: Maintenance and most wear items are covered by the manufacturer. Refer to the operator's manual for additional information.

① On vehicles operated less than 6200 miles per year, more frequent service may be required.

② Replace more frequently if vehicle is operated in dusty conditions.

③ Recommended service for California models, required for all other models.

④ Reset the oil service indicator lights only.

⑤ Replace every 2 years with inspection service.

⑥ Replace every 100,000 miles on all models.

FREQUENT OPERATION MAINTENANCE (SEVERE SERVICE)

If a vehicle is operated under any of the following conditions it is considered severe service

- Extremely dusty areas.

- 50% or more of the vehicle operation is in 32°C (90°F) or higher temperatures, or constant operation in temperatures below 0°C (32°F).

- Prolonged idling (vehicle operation in stop and go traffic).

- Frequent short running periods (engine does not warm to normal operating temperatures).

- Police, taxi, delivery usage or trailer towing usage.

93591CFL

For complete service labor times, order Nichols' Chilton Labor Guide

CHRYSLER CORP.
Dodge Durango

ENGINE AND VEHICLE IDENTIFICATION

Engine							Model Year	
Code ①	Liters (cc)	Cu. In.	Cyl.	Fuel Sys.	Engine Type	Eng. Mfg.	Code ②	Year
N	4.7 (4701)	287	6	SMFI	SOHC	Chrysler	X	1999
X	3.9 (3916)	238	6	SMFI	OHV	Chrysler	Y	2000
Y	5.2 (5208)	318	8	SMFI	OHV	Chrysler	1	2001
Z	5.9 (5899)	360	8	SMFI	OHV	Chrysler	2	2002
							3	2003

SOHC: Single overhead camshaft

OHV: Overhead Valve

SMFI: Sequential Multi-port Fuel Injection

① 8th position of VIN

② 10th position of VIN

93591C36

GENERAL ENGINE SPECIFICATIONS

Year	Model	Engine Displacement Liters (cc)	Engine Series (ID/VIN)	Fuel System	Net Horsepower @ rpm	Net Torque @ rpm (ft. lbs.)	Bore x Stroke (in.)	Compression Ratio	Oil Pressure @ rpm
1999	Durango	3.9 (3906)	X	SMFI	175@4800	220@3200	3.91x3.31	9.1:1	30-80@3000
		5.2 (5208)	Y	SMFI	220@4400	300@3200	3.91x3.31	9.1:1	30-80@3000
		5.9 (5899)	Z	SMFI	230@4000	330@3250	4.00x3.58	9.1:1	30-80@3000
2000	Durango	4.7 (4701)	N	SMFI	235@4800	295@3200	3.66x3.40	9.3:1	25@3000
		5.2 (5208)	Y	SMFI	220@4400	300@3200	3.91x3.31	9.1:1	30-80@3000
		5.9 (5899)	Z	SMFI	230@4000	330@3250	4.00x3.58	9.1:1	30-80@3000
2001	Durango	4.7 (4701)	N	SMFI	235@4800	295@3200	3.66x3.40	9.3:1	25@3000
		5.9 (5899)	Z	SMFI	230@4000	330@3250	4.00x3.58	9.1:1	30-80@3000
2002	Durango	4.7 (4701)	N	SMFI	235@4800	295@3200	3.66x3.40	9.3:1	25@3000
		5.9 (5899)	Z	SMFI	230@4000	330@3250	4.00x3.58	9.1:1	30-80@3000

SMFI: Sequential Multi-port Fuel Injection

93591C37

GASOLINE ENGINE TUNE-UP SPECIFICATIONS

Year	Engine Displacement Liters (cc)	Engine ID/VIN	Spark Plug Gap (in.)	Ignition Timing (deg.)	Fuel Pump (psi)	Idle Speed (rpm)	Valve Clearance	
							Intake	Exhaust
1999	3.9 (3916)	X	0.040	①	44.2-54.2	②	HYD	HYD
	5.2 (5208)	Y	0.040	①	44.2-54.2	②	HYD	HYD
	5.9 (5899)	Z	0.040	①	44.2-54.2	②	HYD	HYD
2000	4.7 (4701)	N	0.040	①	47-51	②	HYD	HYD
	5.2 (5208)	Y	0.040	①	44.2-54.2	②	HYD	HYD
	5.9 (5899)	Z	0.040	①	44.2-54.2	②	HYD	HYD
2001	4.7 (4701)	N	0.040	①	47-51	②	HYD	HYD
	5.9 (5899)	Z	0.040	①	44.2-54.2	②	HYD	HYD
2002	4.7 (4701)	N	0.040	①	47-51	②	HYD	HYD
	5.9 (5899)	Z	0.040	①	44.2-54.2	②	HYD	HYD

NOTE: The Vehicle Emission Control Information (VECI) label often reflects specification changes made during production.

The label figures must be used if they differ from those in this chart.

HYD: Hydraulic

① Ignition timing is controlled by the PCM and is not adjustable.

② Idle speed is controlled by the PCM and is not adjustable

93591C38

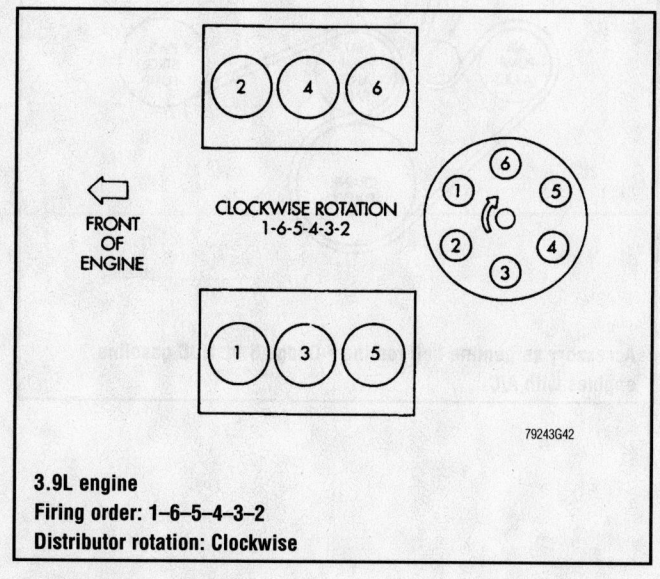

3.9L engine
Firing order: 1–6–5–4–3–2
Distributor rotation: Clockwise

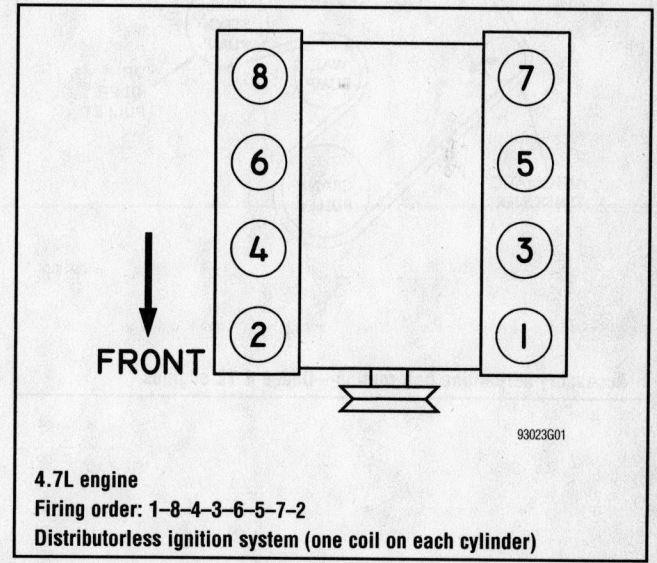

4.7L engine
Firing order: 1–8–4–3–6–5–7–2
Distributorless ignition system (one coil on each cylinder)

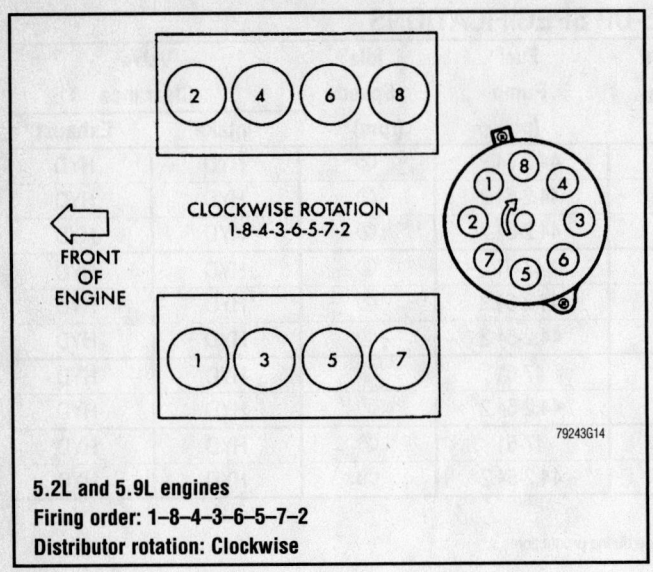

5.2L and 5.9L engines
Firing order: 1–8–4–3–6–5–7–2
Distributor rotation: Clockwise

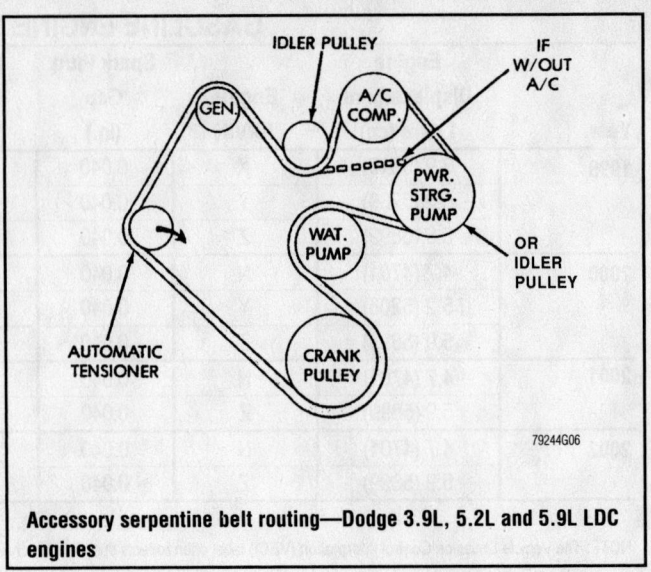

Accessory serpentine belt routing—Dodge 3.9L, 5.2L and 5.9L LDC engines

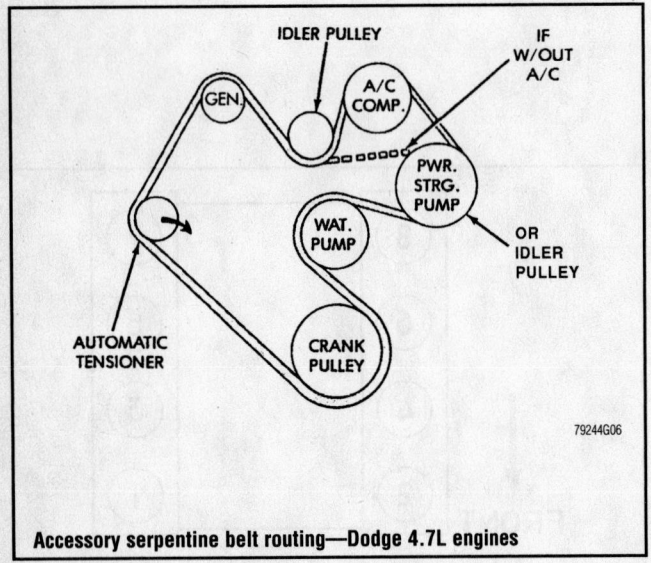

Accessory serpentine belt routing—Dodge 4.7L engines

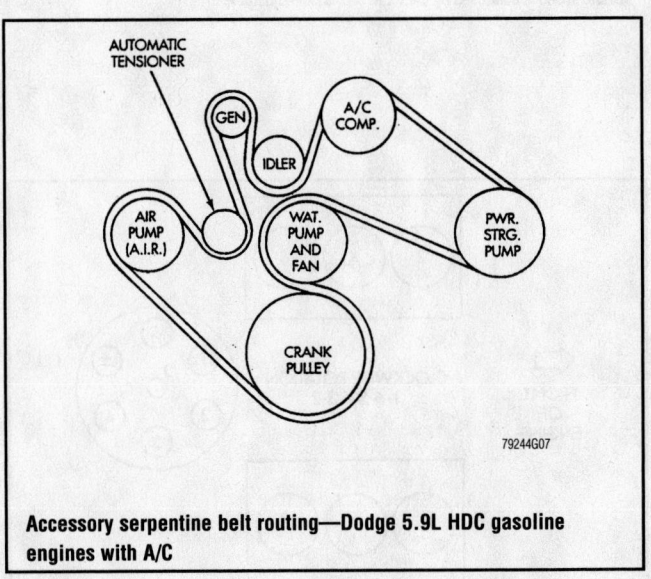

Accessory serpentine belt routing—Dodge 5.9L HDC gasoline engines with A/C

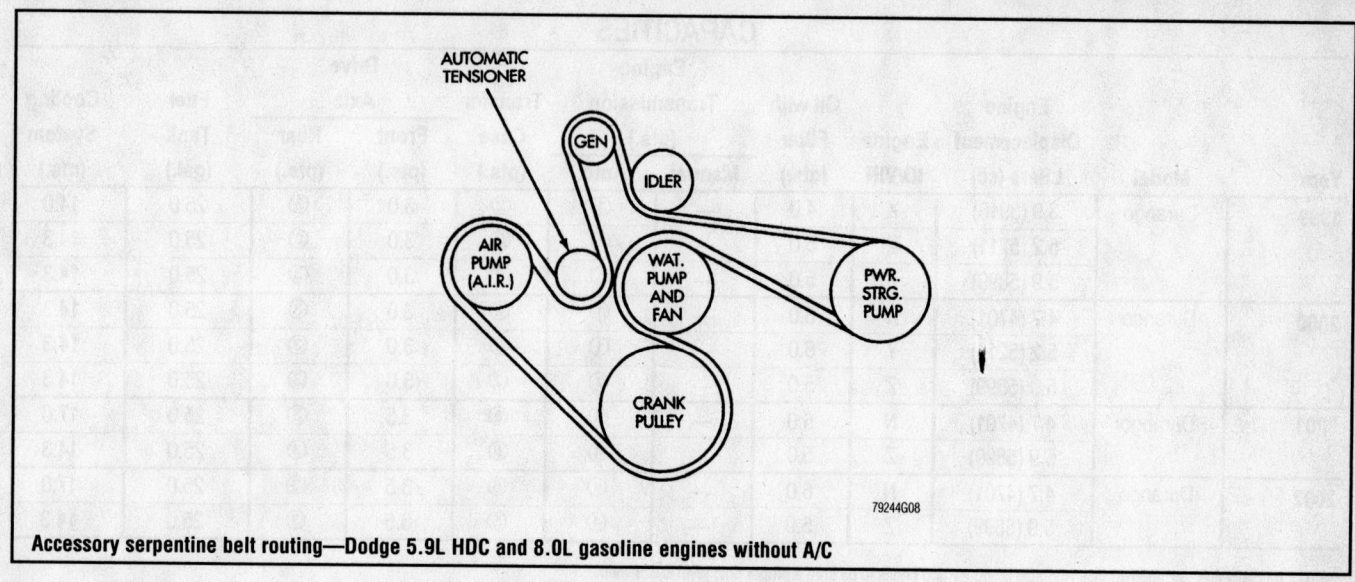

Accessory serpentine belt routing—Dodge 5.9L HDC and 8.0L gasoline engines without A/C

DRIVE BELTS - 4.7L (Continued)

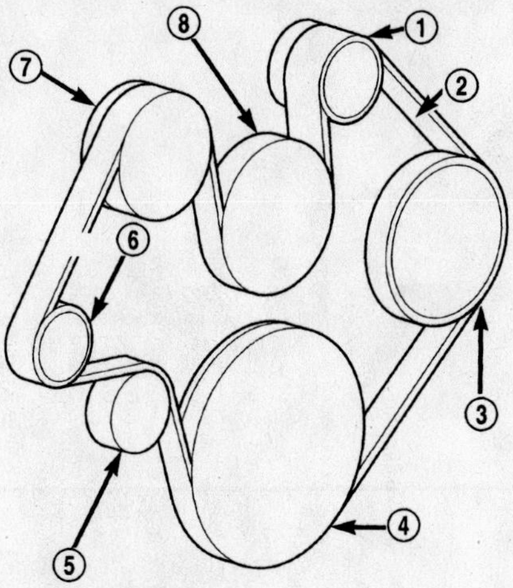

1 - GENERATOR PULLEY
2 - ACCESSORY DRIVE BELT
3 - POWER STEERING PUMP PULLEY
4 - CRANKSHAFT PULLEY
5 - IDLER PULLEY
6 - TENSIONER
7 - A/C COMPRESSOR PULLEY
8 - WATER PUMP PULLEY

93551CHH

Accessory serpentine belt routing—Dodge 4.7L engines

Timing belt service is covered in Section 3 of this manual

CAPACITIES

Year	Model	Engine Displacement Liters (cc)	Engine ID/VIN	Oil with Filter (qts.)	Engine Transmission (pts.) Manual	Engine Transmission (pts.) Auto.	Transfer Case (pts.)	Drive Axle Front (pts.)	Drive Axle Rear (pts.)	Fuel Tank (gal.)	Cooling System (qts.)
1999	Durango	3.9 (3916)	X	4.0	—	①	②	3.0	③	25.0	14.0
		5.2 (5211)	Y	5.0	—	①	②	3.0	③	25.0	14.3
		5.9 (5899)	Z	5.0	—	①	②	3.0	③	25.0	14.3
2000	Durango	4.7 (4701)	N	6.0	—	①	②	3.0	③	25.0	14.3
		5.2 (5211)	Y	5.0	—	①	②	3.0	③	25.0	14.3
		5.9 (5899)	Z	5.0	—	①	②	3.0	③	25.0	14.3
2001	Durango	4.7 (4701)	N	6.0	—	④	⑤	3.5	③	25.0	17.0
		5.9 (5899)	Z	5.0	—	④	⑤	3.5	③	25.0	14.3
2002	Durango	4.7 (4701)	N	6.0	—	④	⑤	3.5	③	25.0	17.0
		5.9 (5899)	Z	5.0	—	④	⑤	3.5	③	25.0	14.3

NOTE: All capacities are approximate. Add fluid gradually and check to be sure a proper fluid level is obtained.

① Fluid drain/filter service: 8.0 pts.

   Overhaul dry fill: 20-23 pts.

② NV231: 2.5 pts.

   NV231-HD: 2.5 pts.

   NV242: 3.0 pts.

③ The following values include 0.25 pt. of friction

   modifier for LSD axles.

   8.25 axle: 4.4 pts.

   9.25 axle: 4.9 pts.

④ Fluid drain/filter service: 8.0 pts.

   Overhaul, 46RE: 20 pts.

   Overhaul, 45RFE: 28 pts.

⑤ NV233: 2.5 pts.

   NV244: 2.85 pts.

93591C39

VALVE SPECIFICATIONS

Year	Engine Displacement Liters (cc)	Engine ID/VIN	Seat Angle (deg.)	Face Angle (deg.)	Spring Test Pressure (lbs. @ in.)	Spring Installed Height (in.)	Stem-to-Guide Clearance (in.)		Stem Diameter (in.)	
							Intake	Exhaust	Intake	Exhaust
1999	3.9 (3916)	X	44.25-44.75	43.25-43.75	200@1.212	1.640	0.0010-0.0030	0.0010-0.0030	0.3110-0.3120	0.3110-0.3120
	5.2 (5208)	Y	44.25-44.75	43.25-43.75	200@1.212	1.640	0.0010-0.0030	0.0010-0.0030	0.3110-0.3120	0.3110-0.3120
	5.9 (5899)	Z	44.25-44.75	43.25-43.75	200@1.212	1.640	0.0010-0.0030	0.0020-0.0040	0.3720-0.3730	0.3710-0.3720
2000	4.7 (4701)	N	44.5-45	45-45.5	176.2-192.4 @1.1532	1.602	0.0011-0.0017	0.0019-0.0039	0.2728-0.2739	0.2717-0.2728
	5.2 (5208)	Y	44.25-44.75	43.25-43.75	200@1.212	1.640	0.001-0.003	0.0010-0.0030	0.3110-0.3120	0.3110-0.3120
	5.9 (5899)	Z	44.25-44.75	43.25-43.75	200@1.212	1.640	0.0010-0.0030	0.0020-0.0040	0.3720-0.3730	0.3710-0.3720
2001	4.7 (4701)	N	44.5-45	45-45.5	176.7-193.3 @1.1670	1.601	0.0008-0.0028	0.0019-0.0039	0.2729-0.2739	0.2717-0.2728
	5.9 (5899)	Z	44.25-44.75	43.25-43.75	200@1.212	1.640	0.0010-0.0030	0.0020-0.0040	0.3720-0.3730	0.3710-0.3720
2002	4.7 (4701)	N	44.5-45	45-45.5	176.7-193.3 @1.1670	1.601	0.0008-0.0028	0.0019-0.0039	0.2729-0.2739	0.2717-0.2728
	5.9 (5899)	Z	44.25-44.75	43.25-43.75	200@1.212	1.640	0.0010-0.0030	0.0020-0.0040	0.3720-0.3730	0.3710-0.3720

93591C40

Heater Core replacement is covered in Section 2 of this manual

CRANKSHAFT AND CONNECTING ROD SPECIFICATIONS

All measurements are given in inches.

Year	Engine Displacement Liters (cc)	Engine ID/VIN	Crankshaft			Thrust on No.	Connecting Rod		
			Main Brg. Journal Dia.	Main Brg. Oil Clearance	Shaft End-play		Journal Diameter	Oil Clearance	Side Clearance
1999	3.9 (3916)	X	2.4995-2.5005	0.0005-0.0015	0.0020-0.0070	2	2.1240-2.1250	0.0005-0.0022	0.0060-0.0140
	5.2 (5211)	Y	2.4995-2.5005	0.0005-0.0015	0.0020-0.0070	2	2.1240-2.1250	0.0005-0.0022	0.0060-0.0140
	5.9 (5899)	Z	2.8095-2.8105	0.0005-0.0015	0.0020-0.0070	2	2.1240-2.1250	0.0005-0.0022	0.0060-0.0140
2000	4.7 (4701)	N	2.4996-2.5005	0.0002-0.0013	0.0021-0.0112	2	2.0076-2.0082	0.0004-0.0019	0.0040-0.0138
	5.2 (5211)	Y	2.4995-2.5005	0.0005-0.0015	0.0020-0.0070	2	2.1240-2.1250	0.0005-0.0022	0.0060-0.0140
	5.9 (5899)	Z	2.8095-2.8105	0.0005-0.0015	0.0020-0.0070	2	2.1240-2.1250	0.0005-0.0022	0.0060-0.0140
2001	4.7 (4701)	N	2.4996-2.5005	0.0008-0.0021	0.0021-0.0112	2	2.0076-2.0082	0.0006-0.0022	0.0040-0.0138
	5.9 (5899)	Z	2.8095-2.8105	①	0.0020-0.0070	2	2.1240-2.1250	0.0005-0.0022	0.0060-0.0140
2002	4.7 (4701)	N	2.4996-2.5005	0.0008-0.0021	0.0021-0.0112	2	2.0076-2.0082	0.0006-0.0022	0.0040-0.0138
	5.9 (5899)	Z	2.8095-2.8105	①	0.0020-0.0070	2	2.1240-2.1250	0.0005-0.0022	0.0060-0.0140

① No. 1: 0.0005-0.0015
Nos. 2-5: 0.0005-0.0020

93591C41

PISTON AND RING SPECIFICATIONS
All measurements are given in inches.

Year	Engine Displacement Liters (cc)	Engine VIN	Piston Clearance	Ring Gap			Ring Side Clearance		
				Top Compression	Bottom Compression	Oil Control	Top Compression	Bottom Compression	Oil Control
1999	3.9 (3916)	X	0.0005-0.0015	0.0100-0.0200	0.0100-0.0200	0.0020-0.0080	0.0015-0.0030	0.0015-0.0030	0.1515-0.1565
	5.2 (5211)	Y	0.0005-0.0015	0.0100-0.0200	0.0100-0.0200	0.0100-0.0500	0.0015-0.0030	0.0015-0.0030	0.0020-0.0080
	5.9 (5899)	Z	0.0005-0.0015	0.0120-0.0220	0.0220-0.0310	0.0150-0.0550	0.0016-0.0033	0.0016-0.0033	0.0020-0.0080
2000	4.7 (4701)	N	0.0008-0.0020	0.0146-0.0249	0.0146-0.0249	0.0100-0.0500	0.0020-0.0041	0.0016-0.0032	0.0007-0.0091
	5.2 (5211)	Y	0.0005-0.0015	0.0100-0.0200	0.0100-0.0200	0.0100-0.0500	0.0015-0.0030	0.0015-0.0030	0.0020-0.0080
	5.9 (5899)	Z	0.0005-0.0015	0.0120-0.0220	0.0220-0.0310	0.0150-0.0550	0.0016-0.0033	0.0016-0.0033	0.0020-0.0080
2001	4.7 (4701)	N	0.0014	0.0146-0.0249	0.0146-0.0249	0.0099-0.0300	0.0020-0.0037	0.0016-0.0031	0.0175-0.0185
	5.9 (5899)	Z	0.0005-0.0015	0.0120-0.0220	0.0220-0.0310	0.0150-0.0550	0.0016-0.0033	0.0016-0.0033	0.0020-0.0080
2002	4.7 (4701)	N	0.0014	0.0146-0.0249	0.0146-0.0249	0.0099-0.0300	0.0020-0.0037	0.0016-0.0031	0.0175-0.0185
	5.9 (5899)	Z	0.0005-0.0015	0.0120-0.0220	0.0220-0.0310	0.0150-0.0550	0.0016-0.0033	0.0016-0.0033	0.0020-0.0080

93591C42

Brake service is covered in Section 4 of this manual

TORQUE SPECIFICATIONS
All readings in ft. lbs.

Year	Engine Displacement Liters (cc)	Engine ID/VIN	Cylinder Head Bolts	Main Bearing Bolts	Rod Bearing Bolts	Crankshaft Damper Bolts	Flexplate Bolts	Manifold		Spark Plugs	Lug Nuts
								Intake	Exhaust		
1999	3.9 (3916)	X	①	85	45	18	23	②	25	30	100
	5.2 (5211)	Y	①	85	45	18	55	②	25	30	100
	5.9 (5899)	Z	①	85	45	18	55	②	25	30	100
2000	5.2 (5211)	Y	①	85	45	18	55	②	25	30	100
	4.7 (4701)	N	③	④	15⑤	130	45	9	18	27	100
	5.9 (5899)	Z	①	85	45	18	55	②	25	30	100
2001	4.7 (4701)	N	⑥	⑦	⑧	130	45	⑨	18	20	100
	5.9 (5899)	Z	⑩	85	45	18	55	⑪	25	30	100
2002	4.7 (4701)	N	⑥	⑦	⑧	130	45	⑨	18	20	100
	5.9 (5899)	Z	⑩	85	45	18	55	⑪	25	30	100

① 1-6 and 8-10: 110 ft. lbs.

  Bolt 7: 100 ft. lbs.

② Refer to illustration in text section

  Step 1: 24 inch lbs.

  Step 2: 48 inch lbs.

  Step 3: 84 inch lbs.

  Step 4: Bolts A1-A5 to 20 ft. lbs.

③ M11 bolts: 60 ft. lbs.

  M8 bolts: 250 inch lbs.

④ Bed plate bolt sequence. Refer to illustration in text section

  Step 1: Bolts 1-10 to 25 inch lbs.

  Step 2: Bolts 1-10 plus 90 degrees

  Step 3: Bolts A-K to 40 ft. lbs.

⑤ Plus 110 degrees

⑥ See text

⑦ Bed plate bolt sequence. Refer to illustration in text section

  Step 1: Bolts A-L to 40 ft. lbs.

  Step 2: Bolts 1-10 25 inch lbs.

  Step 3: Bolts 1-10 plus 90 degrees

  Step 4: Bolts A1-A6 20 ft. lbs.

⑧ 20 ft. lbs. plus 90 degrees

⑨ 105 inch lbs.

⑩ Step 1: 50 ft. lbs.

  Step 2: 105 ft. lbs.

⑪ See illustration in text section
  Step 1: 1-4 to 72 inch lbs. in 12 inch lb. Increments
  Step 2: bolts 5-12: 72 inch lbs.
  Step 3: Check that all bolts are at 72 inch lbs.
  Step 4: All bolts, in sequence, to 12 ft. lbs.
  Step 5: Check that all bolts are at 12 ft. lbs.

93591C43

WHEEL ALIGNMENT

Year	Model	Caster Range (+/-Deg.)	Caster Preferred Setting (Deg.)	Camber Range (+/-Deg.)	Camber Preferred Setting (Deg.)	Toe-in (in.)	Steering Axis Inclination (Deg.)
1999	2WD	0.50	+3.10	0.50	-0.25	0.10+/-0.06	—
	4WD	0.50	+3.30	0.50	-0.25	0.10+/-0.06	—
2000	2WD	0.50	+3.10	0.50	-0.25	0.10+/-0.06	—
	4WD	0.50	+3.30	0.50	-0.25	0.10+/-0.06	—
2001	2WD	0.50	+3.10	0.50	-0.25	0.10+/-0.06	—
	4WD	0.50	+3.30	0.50	-0.25	0.10+/-0.06	—
2002	2WD	0.50	+3.10	0.50	-0.25	0.10+/-0.06	—
	4WD	0.50	+3.30	0.50	-0.25	0.10+/-0.06	—

93591C44

TIRE, WHEEL AND BALL JOINT SPECIFICATIONS

Year	Model	OEM Tires Standard	OEM Tires Optional	Tire Pressures (psi) Front	Tire Pressures (psi) Rear	Wheel Size	Ball Joint Inspection
1999	Durango	P235/75R15XL	31x10.5R15LT	35	35	6.5-JJ	0.060 in. ①
2000	Durango	P235/75R15XL	31x10.5R15LT	35	35	6.5-JJ	0.060 in. ①
2001	Durango	P235/75R15XL	P255/65R16 P265/80R16 P275/60R17	NA	NA	NA	0.060 in. ①
2002	Durango	P235/75R15XL	P255/65R16 P265/80R16 P275/60R17	NA	NA	NA	0.060 in. ①

NA: Information not available

OEM: Original Equipment Manufacturer

PSI: Pounds Per Square Inch

STD: Standard

OPT: Optional

① Both upper and lower

93591C45

For complete Engine Mechanical specifications, see Section 1 of this manual

BRAKE SPECIFICATIONS
All measurements in inches unless noted

| Year | Model | Brake Disc | | | Brake Drum | | | Minimum Lining Thickness | | Brake Caliper | |
		Original Thickness	Minimum Thickness	Maximum Run-out	Original Inside Diameter	Max. Wear Limit	Maximum Machine Diameter	Front	Rear	Bracket Bolts (ft. lbs.)	Mounting Bolts (ft. lbs.)
1999	Durango	0.900	0.890	0.004	11.00	①	①	②	③	47	22
2000	Durango	0.900	0.890	0.004	11.00	①	①	②	③	47	22
2001	Durango	0.900	0.890	0.004	11.00	①	①	②	③	47	22
2002	Durango	0.900	0.890	0.004	11.00	①	①	②	③	47	22

NA: Not Available

① Maximum allowable drum diameter, either from wear or machining, is stamped on the drum.

② Riveted brake pads: 0.0625 in.
Bonded brake pads: 0.1875 in.

③ Riveted brake shoes: 0.031 in.
Bonded brake shoes: 0.0625 in.

93591C46

SCHEDULED MAINTENANCE INTERVALS
1999 DODGE DURANGO

TO BE SERVICED	TYPE OF SERVICE	VEHICLE MILEAGE INTERVAL (x1000)													
		7.5	15	22.5	30	37.5	45	52.5	60	67.5	75	82.5	90	97.5	105
Engine oil & filter	R	✓	✓	✓	✓	✓	✓	✓	✓	✓	✓	✓	✓	✓	
Steering linkage (2wd)	L		✓		✓		✓		✓		✓		✓		✓
Ball joints	L			✓			✓			✓			✓		
Front wheel bearings	S/I			✓			✓			✓			✓		
Brake linings	S/I			✓			✓			✓			✓		
Air cleaner element	R				✓				✓				✓		
Spark plugs	R				✓				✓				✓		
Auto trans fluid & filter	R					✓					✓				
Auto trans bands	Adj					✓					✓				
Transfer case fluid	R					✓					✓				
Engine coolant ①	R						✓						✓		
Spark plug cables	R								✓						
Accessory drive belt ②	S/I								✓						
PCV valve ②	S/I								✓						

R: Replace   S/I: Service or Inspect   L: Lubricate   Adj: Adjust

① Replace every 36 months, regardless of mileage

② Inspect and replace if necessary.

FREQUENT OPERATION MAINTENANCE (SEVERE SERVICE)

If a vehicle is operated under any of the following conditions it is considered severe service:

- Extremely dusty areas.
- 50% or more of the vehicle operation is in 32°C (90°F) or higher temperatures, or constant operation in temperatures below 0°C (32°F).
- Prolonged idling (vehicle operation in stop and go traffic).
- Frequent short running periods (engine does not warm to normal operating temperatures).
- Police, taxi, delivery usage or trailer towing usage.

Oil & oil filter change: change every 3000 miles.

Air filter/air pump air filter: change every 24,000 miles.

Engine coolant level, hoses & clamps: check every 6,000 miles.

Exhaust system: check every 6000 miles.

Drive belts: check every 18,000 miles; replace every 24,000 miles.

Crankcase inlet air filter (6 & 8 cyl.): clean every 24,000 miles.

Oxygen sensor: replace every 82,500 miles.

Automatic transmission fluid, filter & bands: change & adjust every 12,000 miles.

Steering linkage: lubricate every 6000 miles.

Rear axle fluid: change every 12,000 miles.

93591C47

For Accessory Drive Belt illustrations, see Section 1 of this manual

SCHEDULED MAINTENANCE INTERVALS
2000-02—DODGE DURANGO

TO BE SERVICED	TYPE OF SERVICE	VEHICLE MILEAGE INTERVAL (x1000)													
		7.5	15	22.5	30	37.5	45	52.5	60	67.5	75	82.5	90	97.5	100
Engine oil & filter	R	✓	✓	✓	✓	✓	✓	✓	✓	✓	✓	✓	✓	✓	✓
Front wheel bearings	S/I			✓			✓			✓				✓		
Brake linings	S/I			✓			✓			✓				✓		
Air cleaner element	R				✓				✓					✓		
Spark plugs	R				✓				✓					✓		
Transfer case fluid	R					✓					✓					
Engine coolant ①	R						✓							✓		
Spark plug wires (5.2 & 5.9L)	R								✓							
PCV valve	R								✓							
Drive belt tensioner ②	S/I								✓							
Automatic transmission fluid	R															✓
Automatic transmission bands	Adj.															✓

R: Replace   S/I: Service or Inspect

① Replace every 36 months, regardless of mileage.

② Replace if necessary

FREQUENT OPERATION MAINTENANCE (SEVERE SERVICE)

If a vehicle is operated under any of the following conditions it is considered severe service:

- Extremely dusty areas.

- 50% or more of the vehicle operation is in 32°C (90°F) or higher temperatures, or constant operation in temperatures below 0°C (32°F).

- Prolonged idling (vehicle operation in stop and go traffic).

- Frequent short running periods (engine does not warm to normal operating temperatures).

- Police, taxi, delivery usage or trailer towing usage.

Oil & oil filter change: change every 3000 miles.

Air filter: change every 24,000 miles.

Engine coolant level, hoses & clamps: check every 6,000 miles.

Exhaust system: check every 6000 miles.

Drive belts: check every 18,000 miles; replace every 24,000 miles.

Oxygen sensor: replace every 82,500 miles.

Automatic transmission fluid, filter & bands: change & adjust every 12,000 miles.

Rear axle fluid: change every 12,000 miles.

93591C48

SCHEDULED MAINTENANCE INTERVALS
DAIMLERCHRYSLER CORPORATION
DODGE DURANGO
2000-02

The following should be used as a guide when determining the amount of work required for a particular service.
In estimating how long a particular Scheduled Maintenance Service should take, please observe the following:

- Labor Time is time based on field research and data supplied by the vehicle manufacturer.
- Labor time operations are given in hours and tenths of an hour.
- All labor operations are to be used as a guide.

Mechanic Skill Level Codes:
(A) PRECISION: Highly skilled with multiple certification.
(B) GENERAL: Normally skilled with certification.
(C) MAINTENANCE: Semi-skilled working on certification.

	LABOR TIME
7500 Mile Service (C)	
All Models	.5
15000 Mile Service (C)	
All Models	.6
22500 Mile Service (C)	
All Models	.8
30000 Mile Service (B)	
All Models	1.6
37500 Mile Service (B)	
All Models	1.2

	LABOR TIME
45000 Mile Service (B)	
All Models	1.7
52500 Mile Service (C)	
All Models	.5
60000 Mile Service (B)	
All Models	3.0
67500 Mile Service (C)	
All Models	.8
75000 Mile Service (C)	
All Models	1.2

	LABOR TIME
82500 Mile Service (C)	
All Models	.5
90000 Mile Service (B)	
All Models	1.9
97500 Mile Service (C)	
All Models	.5
100000 Mile Service (B)	
All Models	1.8

93551C48A"

For Tire, Wheel and Ball Joint specifications, see Section 1 of this manual

CHRYSLER CORP.
Jeep Cherokee • Grand Cherokee • Wrangler • Liberty

ENGINE AND VEHICLE IDENTIFICATION

		Engine						Model Year	
Code ①	Liters (cc)	Cu. In.	Cyl.	Fuel Sys.	Engine Type	Eng. Mfg.		Code ②	Year
N	4.7 (4701)	287	8	MFI	SOHC	Chrysler		X	1999
P	2.5 (2458)	150	4	MFI	OHV	Chrysler		Y	2000
S	4.0 (3966)	242	6	MFI	OHV	Chrysler		1	2001
Z	5.9 (5899)	360	8	MFI	OHV	Chrysler		2	2002
K	3.7 (3701)	226	6	MFI	SOHC	Chrysler		3	2003
1	2.4 (2429)	148	4	MFI	DOHC	Chrysler			

MFI: Multi-port Fuel Injection

OHV: Over Head Valve

SOHC: Single Overhead Camshaft

① 8th position of VIN

② 10th position of VIN

93591C49

GENERAL ENGINE SPECIFICATIONS

Year	Model	Engine Displacement Liters (cc)	Engine Series (ID/VIN)	Fuel System	Net Horsepower @ rpm	Net Torque @ rpm (ft. lbs.)	Bore x Stroke (in.)	Compression Ratio	Oil Pressure @ rpm
1999	Cherokee	2.5 (2458)	P	MFI	125@5400	149@3250	3.88x3.19	9.1:1	37@1600
		4.0 (3966)	S	MFI	190@4750	225@4000	3.88x3.44	8.8:1	37@1600
	Grand Cherokee	4.0 (3966)	S	MFI	195@4600	230@3000	3.88x3.44	8.8:1	37@1600
		4.7 (4701)	N	MFI	235@4800	295@3200	3.66x3.40	9.3:1	25@3000
	Wrangler	2.5 (2464)	P	MFI	120@5400	139@3500	3.88x3.19	9.1:1	37@1600
		4.0 (3958)	S	MFI	181@4600	222@2800	3.88x3.44	8.8:1	37@1600
2000	Cherokee	2.5 (2458)	P	MFI	125@5400	149@3250	3.88x3.19	9.1:1	37@1600
		4.0 (3966)	S	MFI	190@4750	225@4000	3.88x3.44	8.8:1	37@1600
	Grand Cherokee	4.0 (3966)	S	MFI	195@4600	230@3000	3.88x3.44	8.8:1	37@1600
		4.7 (4701)	N	MFI	235@4800	295@3200	3.66x3.40	9.3:1	25@3000
	Wrangler	2.5 (2464)	P	MFI	120@5400	139@3500	3.88x3.19	9.1:1	37@1600
		4.0 (3958)	S	MFI	181@4600	222@2800	3.88x3.44	8.8:1	37@1600
2001	Cherokee	2.5 (2458)	P	MFI	125@5400	149@3250	3.88x3.19	9.1:1	37@1600
		4.0 (3966)	S	MFI	190@4750	225@4000	3.88x3.44	8.8:1	37@1600
	Grand Cherokee	4.0 (3966)	S	MFI	195@4600	230@3000	3.88x3.44	8.8:1	37@1600
		4.7 (4701)	N	MFI	235@4800	295@3200	3.66x3.40	9.3:1	25@3000
	Wrangler	2.5 (2464)	P	MFI	120@5400	139@3500	3.88x3.19	9.1:1	37@1600
		4.0 (3958)	S	MFI	181@4600	222@2800	3.88x3.44	8.8:1	37@1600
2002	Grand Cherokee	4.0 (3966)	S	MFI	195@4600	230@3000	3.88x3.44	8.8:1	37@1600
		4.7 (4701)	N	MFI	235@4800	295@3200	3.66x3.40	9.3:1	25@3000
	Wrangler	2.5 (2464)	P	MFI	120@5400	139@3500	3.88x3.19	9.1:1	37@1600
		4.0 (3958)	S	MFI	181@4600	222@2800	3.88x3.44	8.8:1	37@1600
	Liberty	2.4 (2429)	1	MFI	150@5200	165@4000	3.44x3.98	9.5:1	NA
		3.7 (3701)	K	MFI	210@5200	225@4200	3.66x3.40	9.1:1	25-110@3000

MFI: Multi Port Fuel Injection

93591C50

ENGINE TUNE-UP SPECIFICATIONS

Year	Engine Displacement Liters (cc)	Engine ID/VIN	Spark Plug Gap (in.)	Ignition Timing (deg.)	Fuel Pump (psi)	Idle Speed (rpm)	Valve Clearance Intake	Valve Clearance Exhaust
1999	2.5 (2458)	P	0.035	①	47-51	①	HYD	HYD
	4.0 (3966)	S	0.035	①	47-51	①	HYD	HYD
	4.7 (4701)	N	0.040	①	47-51	①	HYD	HYD
2000	2.5 (2458)	P	0.035	①	47-51	①	HYD	HYD
	4.0 (3966)	S	0.035	①	47-51	①	HYD	HYD
	4.7 (4701)	N	0.040	①	47-51	①	HYD	HYD
2001	2.5 (2458)	P	0.035	①	47-51	①	HYD	HYD
	4.0 (3966)	S	0.035	①	47-51	①	HYD	HYD
	4.7 (4701)	N	0.040	①	47-51	①	HYD	HYD
2002	2.4 (2429)	1	0.050	①	44-54	①	HYD	HYD
	2.5 (2458)	P	0.035	①	47-51	①	HYD	HYD
	3.7 (3701)	K	0.042	①	44-54	①	HYD	HYD
	4.0 (3966)	S	0.035	①	47-51	①	HYD	HYD
	4.7 (4701)	N	0.040	①	47-51	①	HYD	HYD

NOTE: The Vehicle Emission Control Information label often reflects specification changes made during production. The label figures must be used if they differ from those in this chart.

HYD: Hydraulic

① Ignition timing and idle speed are controlled by the PCM. No adjustement is necessary.

93591C51

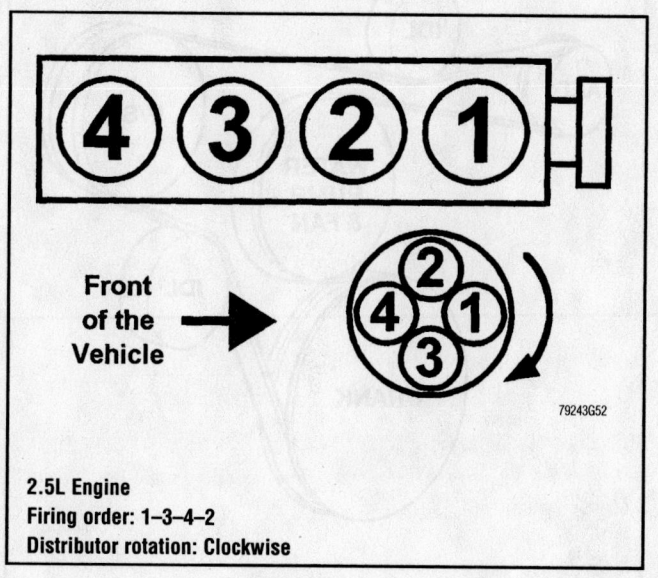

2.5L Engine
Firing order: 1–3–4–2
Distributor rotation: Clockwise

79243G52

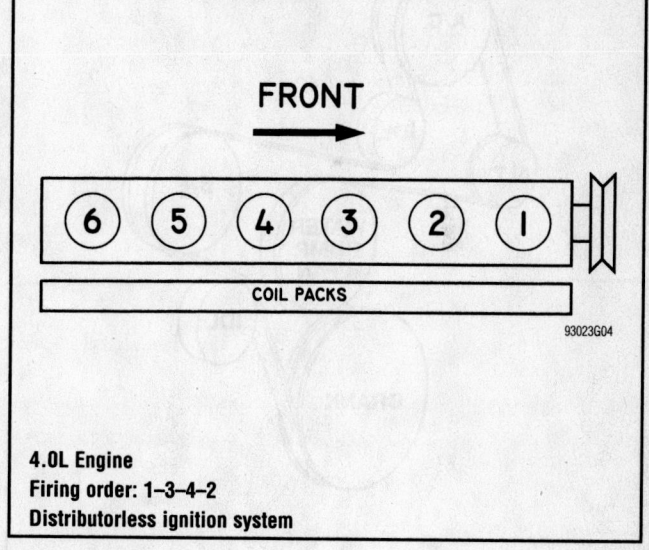

4.0L Engine
Firing order: 1–3–4–2
Distributorless ignition system

93023G04

For Wheel Alignment specifications, see Section 1 of this manual

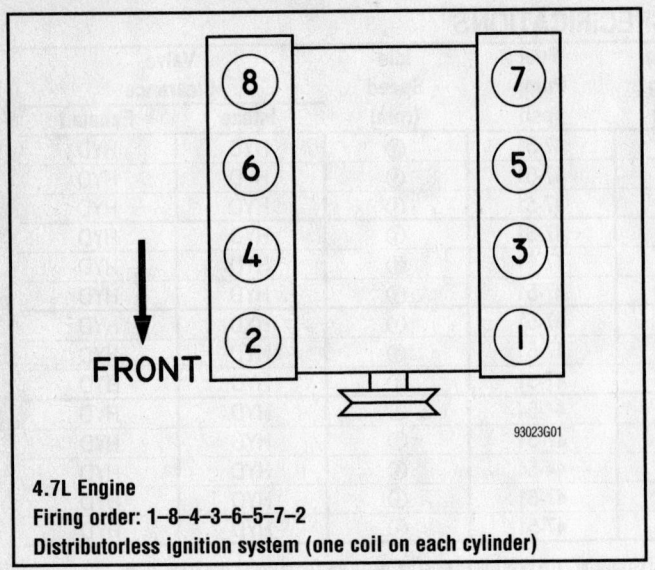

4.7L Engine
Firing order: 1–8–4–3–6–5–7–2
Distributorless ignition system (one coil on each cylinder)

93023G01

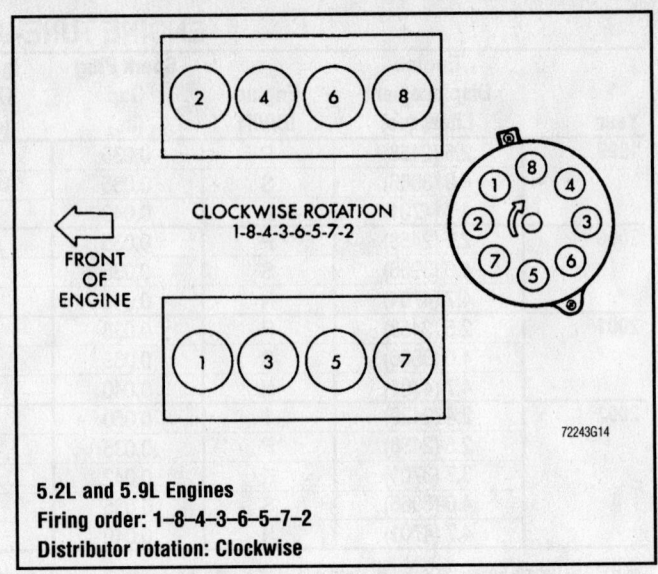

CLOCKWISE ROTATION
1-8-4-3-6-5-7-2

FRONT OF ENGINE

5.2L and 5.9L Engines
Firing order: 1–8–4–3–6–5–7–2
Distributor rotation: Clockwise

72243G14

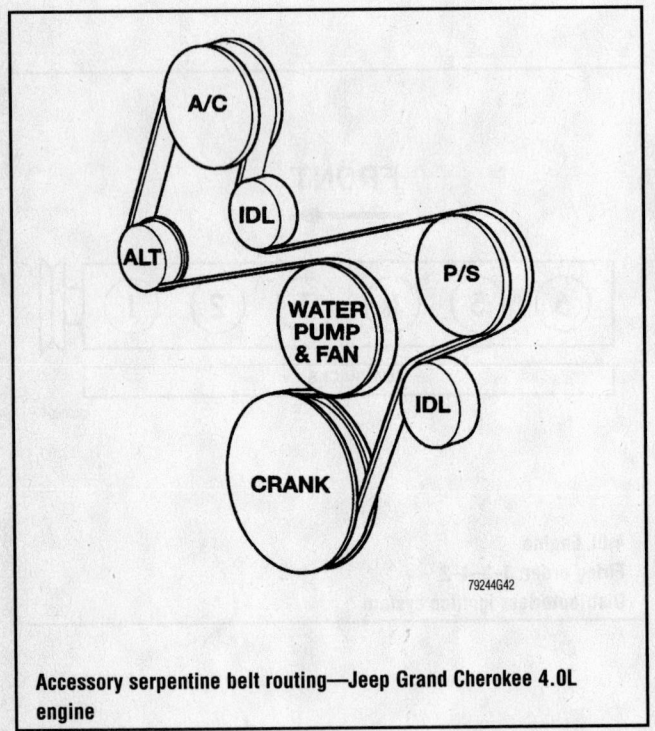

Accessory serpentine belt routing—Jeep Grand Cherokee 4.0L engine

79244G42

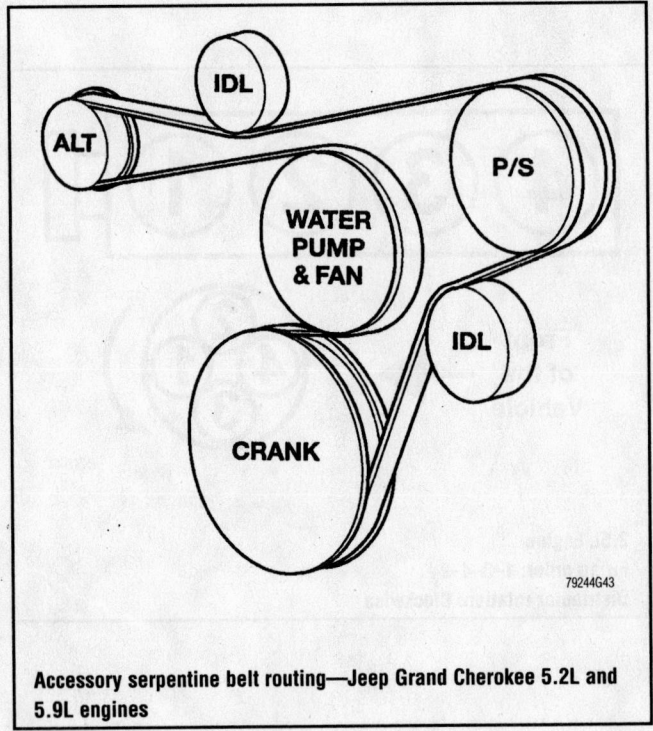

Accessory serpentine belt routing—Jeep Grand Cherokee 5.2L and 5.9L engines

79244G43

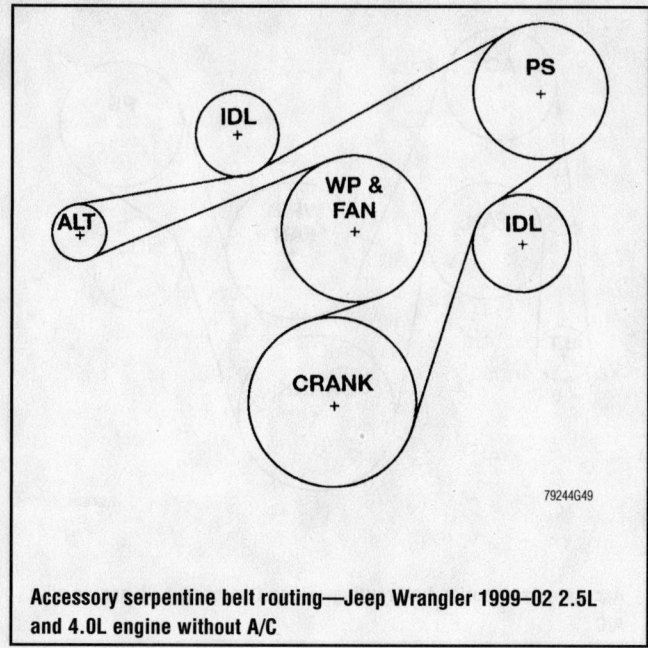

79244G49

Accessory serpentine belt routing—Jeep Wrangler 1999–02 2.5L and 4.0L engine without A/C

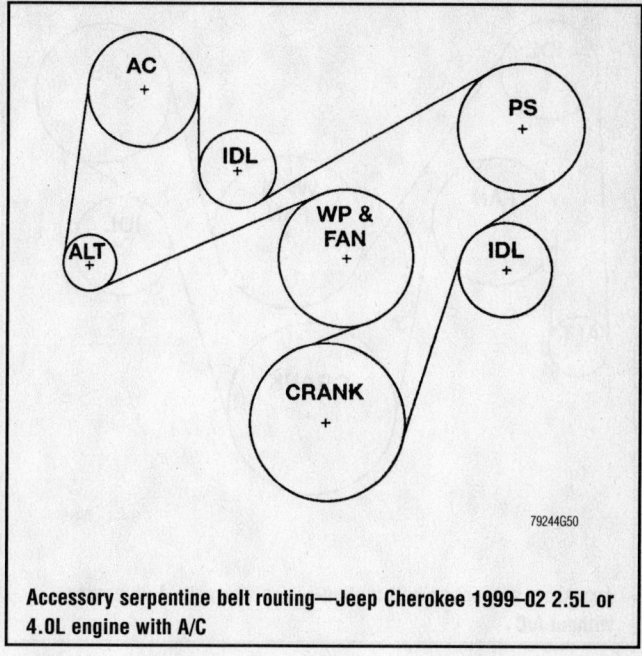

79244G50

Accessory serpentine belt routing—Jeep Cherokee 1999–02 2.5L or 4.0L engine with A/C

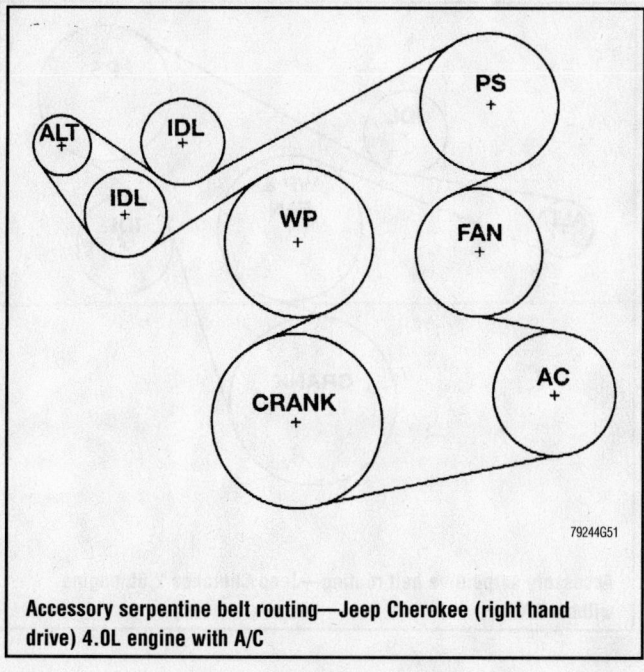

79244G51

Accessory serpentine belt routing—Jeep Cherokee (right hand drive) 4.0L engine with A/C

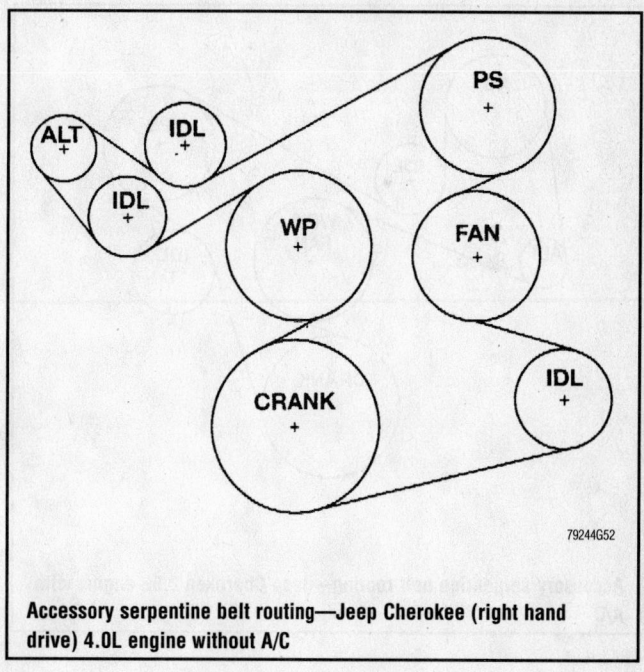

79244G52

Accessory serpentine belt routing—Jeep Cherokee (right hand drive) 4.0L engine without A/C

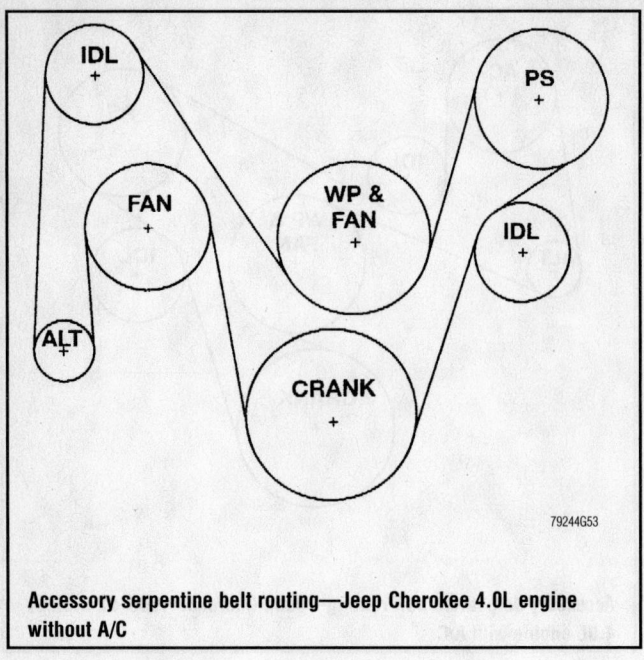

Accessory serpentine belt routing—Jeep Cherokee 4.0L engine without A/C

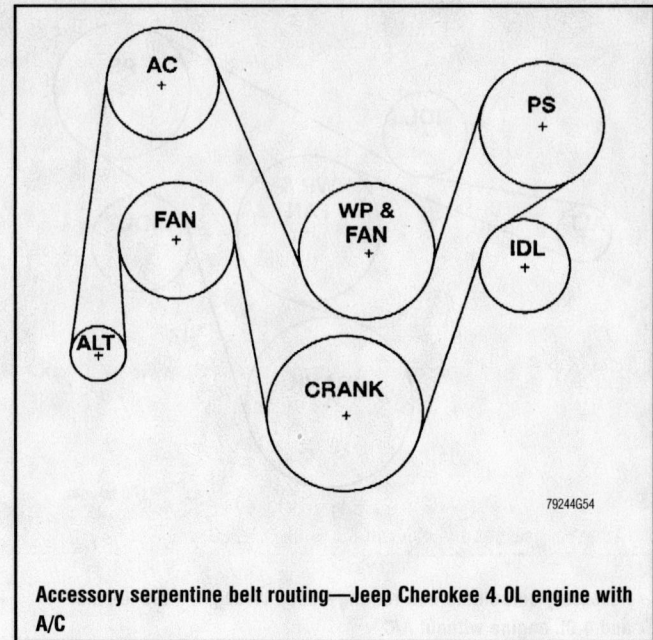

Accessory serpentine belt routing—Jeep Cherokee 4.0L engine with A/C

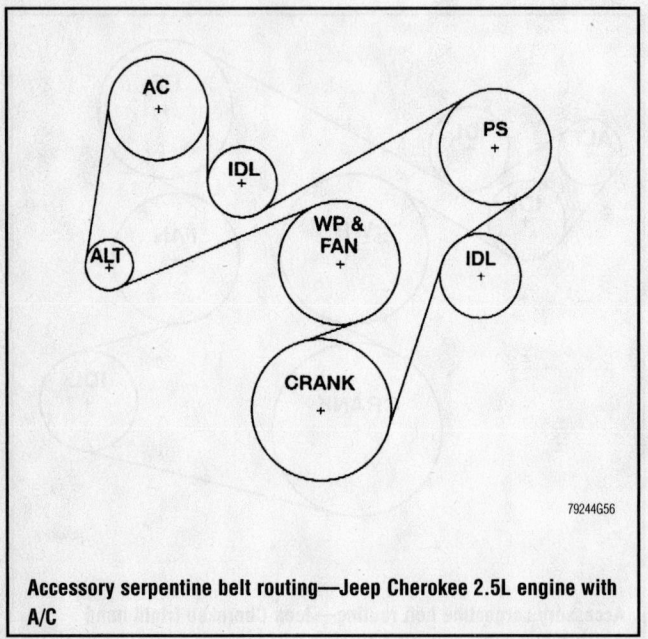

Accessory serpentine belt routing—Jeep Cherokee 2.5L engine with A/C

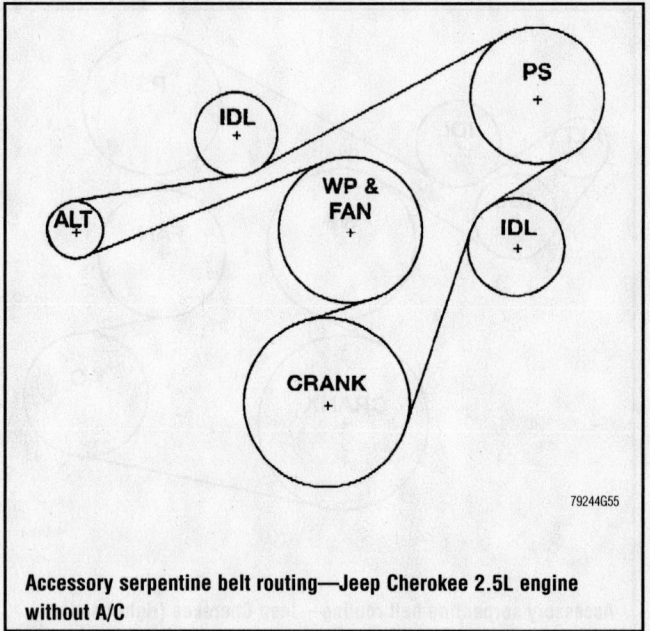

Accessory serpentine belt routing—Jeep Cherokee 2.5L engine without A/C

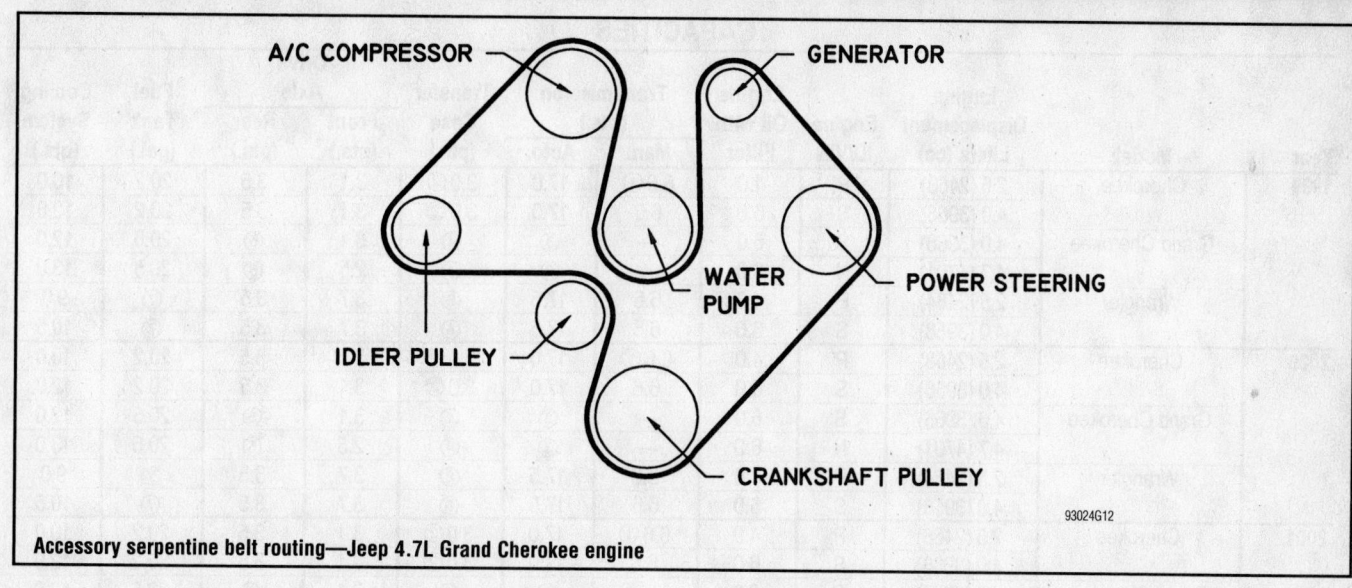

A/C COMPRESSOR — GENERATOR

WATER PUMP

POWER STEERING

IDLER PULLEY

CRANKSHAFT PULLEY

93024G12

Accessory serpentine belt routing—Jeep 4.7L Grand Cherokee engine

1 - GENERATOR PULLEY
2 - ACCESSORY DRIVE BELT
3 - POWER STEERING PUMP PULLEY
4 - CRANKSHAFT PULLEY
5 - IDLER PULLEY
6 - TENSIONER
7 - A/C COMPRESSOR PULLEY
8 - WATER PUMP PULLEY

93591CHH

Accessory serpentine belt routing—Jeep 3.7L and 4.7L engines

CAPACITIES

Year	Model	Engine Displacement Liters (cc)	Engine ID/VIN	Engine Oil with Filter	Transmission (pts.) Man.	Transmission (pts.) Auto.	Transfer Case (pts.)	Drive Axle Front (pts.)	Drive Axle Rear (pts.)	Fuel Tank (gal.)	Cooling System (qts.)
1999	Cherokee	2.5 (2468)	P	4.0	6.6①	17.0	3.0②	3.1	3.5	20.2	10.0
		4.0 (3966)	S	6.0	6.6	17.0	3.0②	3.1	3.5	20.2	12.0
	Grand Cherokee	4.0 (3966)	S	6.0	—	③	④	3.1	⑤	20.5	12.0
		4.7 (4701)	N	6.0	—	③	④	2.5	⑤	20.5	13.0
	Wrangler	2.5 (2464)	P	4.0	6.6	17.5	⑥	3.7	3.5	⑦	9.0
		4.0 (3958)	S	6.0	6.6	17.5	⑥	3.7	3.5	⑦	10.5
2000	Cherokee	2.5 (2468)	P	4.0	6.6①	17.0	3.0②	3.1	3.5	20.2	10.0
		4.0 (3966)	S	6.0	6.6	17.0	3.0②	3.1	3.5	20.2	12.0
	Grand Cherokee	4.0 (3966)	S	6.0	—	③	④	3.1	⑤	20.5	12.0
		4.7 (4701)	N	6.0	—	③	④	2.5	⑤	20.5	13.0
	Wrangler	2.5 (2464)	P	4.0	6.6	17.5	⑥	3.7	3.5	⑦	9.0
		4.0 (3958)	S	6.0	6.6	17.5	⑥	3.7	3.5	⑦	10.5
2001	Cherokee	2.5 (2468)	P	4.0	6.6①	17.0	3.0②	3.1	3.5	20.2	10.0
		4.0 (3966)	S	6.0	6.6	17.0	3.0②	3.1	3.5	20.2	12.0
	Grand Cherokee	4.0 (3966)	S	6.0	—	③	④	3.1	⑤	20.5	12.0
		4.7 (4701)	N	6.0	—	③	④	2.5	⑤	20.5	13.0
	Wrangler	2.5 (2464)	P	4.0	6.6	17.5	⑥	3.7	3.5	⑦	9.0
		4.0 (3958)	S	6.0	6.6	17.5	⑥	3.7	3.5	⑦	10.5
2002	Grand Cherokee	4.0 (3966)	S	6.0	—	③	④	3.1	⑤	20.5	12.0
		4.7 (4701)	N	6.0	—	③	④	2.5	⑤	20.5	13.0
	Wrangler	2.5 (2464)	P	4.0	6.6	17.5	⑥	3.7	3.5	⑦	9.0
		4.0 (3958)	S	6.0	6.6	17.5	⑥	3.7	3.5	⑦	10.5
	Liberty	2.4 (2429)	1	5.0	2.4	—	⑧	2.6	⑨	18.5	10.1
		3.7 (3701)	K	5.0	2.4	10.0	⑧	2.6	⑨	18.5	13.0

① 2WD: 7.0 pts.

② Command-Trac - 2.2 pts.

③ 42RE: 19-22 pts.
44RE: 19-22 pts.
45RFE: 28.0 pts.
46RE: 19.5-28.0 pts.

④ 242 NVG: 3.0 pts.
247 NVG: 2.5 pts.
249 NVG: 2.5 pts.

⑤ 194 RBI: 3.5 pts.
198 RBI: 3.75 pts.
216 RBA: 4.75 pts.
226 RBA: 4.75 pts.

⑥ Command-Trac:
Automatic: 2.2 pts.
Manual: 3.3 pts.

⑦ Standard: 15.0 gals.
Optional: 19.6 gals.

⑧ NV231: 2.2 pts.
NV242: 2.85 pts.

⑨ Model 35: 3.7 pts.
8 1/4 inch: 4.4 pts.

93591C52

VALVE SPECIFICATIONS

Year	Engine Displacement Liters (cc)	Engine ID/VIN	Seat Angle (deg.)	Face Angle (deg.)	Spring Test Pressure (lbs. @ in.)	Spring Installed Height (in.)	Stem-to-Guide Clearance (in.)		Stem Diameter (in.)	
							Intake	Exhaust	Intake	Exhaust
1999	2.5 (2458)	P	44.5	45	184-196@ 1.216	1.640	0.0010-0.0030	0.0010-0.0030	0.3110-0.3120	0.3110-0.3120
	4.0 (3966)	S	44.5	45	184-196@ 1.216	1.640	0.0010-0.0030	0.0010-0.0030	0.3110-0.3120	0.3110-0.3120
	4.7 (4701)	N	44.5-45	45-45.5	176.2-192.4 @1.1532	1.602	0.0011-0.0017	0.0029	0.2728-0.2739	0.2717-0.2728
2000	2.5 (2458)	P	44.5	45	184-196@ 1.216	1.640	0.0010-0.0030	0.0010-0.0030	0.3110-0.3120	0.3110-0.3120
	4.0 (3966)	S	44.5	45	184-196@ 1.216	1.640	0.0010-0.0030	0.0010-0.0030	0.3110-0.3120	0.3110-0.3120
	4.7 (4701)	N	44.5-45	45-45.5	176.2-192.4 @1.1532	1.602	0.0011-0.0017	0.0029	0.2728-0.2739	0.2717-0.2728
2001	2.5 (2458)	P	44.5	45	184-196@ 1.216	1.640	0.0010-0.0030	0.0010-0.0030	0.3110-0.3120	0.3110-0.3120
	4.0 (3966)	S	44.5	45	184-196@ 1.216	1.640	0.0010-0.0030	0.0010-0.0030	0.3110-0.3120	0.3110-0.3120
	4.7 (4701)	N	44.5-45	45-45.5	176.2-192.4 @1.1532	1.602	0.0011-0.0017	0.0029	0.2728-0.2739	0.2717-0.2728
2002	2.4 (2429)	1	NA	NA	NA	NA	NA	NA	NA	NA
	2.5 (2458)	P	44.5	45	184-196@ 1.216	1.640	0.0010-0.0030	0.0010-0.0030	0.3110-0.3120	0.3110-0.3120
	3.7 (3701)	K	44.5-45	45-45.5	221-242@ 1.107	1.619	0.0008-0.0028	0.0019-0.0039	0.2729-0.2739	0.2717-0.2728
	4.0 (3966)	S	44.5	45	184-196@ 1.216	1.640	0.0010-0.0030	0.0010-0.0030	0.3110-0.3120	0.3110-0.3120
	4.7 (4701)	N	44.5-45	45-45.5	176.2-192.4 @1.1532	1.602	0.0011-0.0017	0.0029	0.2728-0.2739	0.2717-0.2728

NA: Information not available

93591C53

CRANKSHAFT AND CONNECTING ROD SPECIFICATIONS
All measurements are given in inches.

Year	Engine Displacement Liters (cc)	Engine ID/VIN	Crankshaft				Connecting Rod		
			Main Brg. Journal Dia.	Main Brg. Oil Clearance	Shaft End-play	Thrust on No.	Journal Diameter	Oil Clearance	Side Clearance
1999	2.5 (2507)	P	2.4996-2.5001	0.0010-0.0025	0.0015-0.0065	2	2.2080-2.2085	0.0015-0.0020	0.0100-0.0190
	4.0 (3966)	S	2.4996-2.5001 ①	0.0010-0.0025	0.0015-0.0065	2	2.0934-2.0955	0.0015-0.0020	0.0100-0.0190
	4.7 (4701)	N	2.4996-2.5005	0.0002-0.0013	0.0021-0.0112	2	2.0076-2.0082	0.0004-0.0019	0.0040-0.0138
2000	2.5 (2507)	P	2.4996-2.5001	0.0010-0.0025	0.0015-0.0065	2	2.2080-2.2085	0.0015-0.0020	0.0100-0.0190
	4.0 (3966)	S	2.4996-2.5001 ①	0.0010-0.0025	0.0015-0.0065	2	2.0934-2.0955	0.0015-0.0020	0.0100-0.0190
	4.7 (4701)	N	2.4996-2.5005	0.0002-0.0013	0.0021-0.0112	2	2.0076-2.0082	0.0004-0.0019	0.0040-0.0138
2001	2.5 (2507)	P	2.4996-2.5001	0.0010-0.0025	0.0015-0.0065	2	2.2080-2.2085	0.0015-0.0020	0.0100-0.0190
	4.0 (3966)	S	2.4996-2.5001 ①	0.0010-0.0025	0.0015-0.0065	2	2.0934-2.0955	0.0015-0.0020	0.0100-0.0190
	4.7 (4701)	N	2.4996-2.5005	0.0002-0.0013	0.0021-0.0112	2	2.0076-2.0082	0.0004-0.0019	0.0040-0.0138
2002	2.4 (2429)	1	NA	NA	NA	NA	NA	NA	NA
	2.5 (2458)	P	2.4996-2.5001	0.0010-0.0025	0.0015-0.0065	2	2.2080-2.2085	0.0015-0.0020	0.0100-0.0190
	3.7 (3701)	K	2.4996-2.5005	0.0020-0.0034	0.0021-0.0112	2	2.2794-2.2797	0.0004-0.0019	0.0040-0.0138
	4.0 (3966)	S	2.4996-2.5001 ①	0.0010-0.0025	0.0015-0.0065	2	2.0934-2.0955	0.0015-0.0020	0.0100-0.0190
	4.7 (4701)	N	2.4996-2.5005	0.0002-0.0013	0.0021-0.0112	2	2.0076-2.0082	0.0004-0.0019	0.0040-0.0138

NA: Information not available
① No 7: 2.4980-2.4995

93591C54

PISTON AND RING SPECIFICATIONS
All measurements are given in inches.

Year	Engine Displacement Liters (cc)	Engine ID/VIN	Piston Clearance	Ring Gap			Ring Side Clearance		
				Top Compression	Bottom Compression	Oil Control	Top Compression	Bottom Compression	Oil Control
1999	2.5 (2507)	P	0.0013-0.0021	0.0090-0.0240	0.0190-0.0380	0.0100-0.0600	0.0017-0.0033	0.0017-0.0033	0.0024-0.0083
	4.0 (3966)	S	0.0008-0.0015	0.0090-0.0240	0.0190-0.0380	0.0100-0.0600	0.0017-0.0033	0.0017-0.0033	0.0024-0.0083
	4.7 (4701)	N	0.0008-0.0020	0.0146-0.0249	0.0146-0.0249	0.0100-0.0500	0.0020-0.0041	0.0016-0.0032	0.0007-0.0091
2000	2.5 (2507)	P	0.0013-0.0021	0.0090-0.0240	0.0190-0.0380	0.0100-0.0600	0.0017-0.0033	0.0017-0.0033	0.0024-0.0083
	4.0 (3966)	S	0.0008-0.0015	0.0090-0.0240	0.0190-0.0380	0.0100-0.0600	0.0017-0.0033	0.0017-0.0033	0.0024-0.0083
	4.7 (4701)	N	0.0008-0.0020	0.0146-0.0249	0.0146-0.0249	0.0100-0.0500	0.0020-0.0041	0.0016-0.0032	0.0007-0.0091
2001	2.5 (2507)	P	0.0013-0.0021	0.0090-0.0240	0.0190-0.0380	0.0100-0.0600	0.0017-0.0033	0.0017-0.0033	0.0024-0.0083
	4.0 (3966)	S	0.0008-0.0015	0.0090-0.0240	0.0190-0.0380	0.0100-0.0600	0.0017-0.0033	0.0017-0.0033	0.0024-0.0083
	4.7 (4701)	N	0.0008-0.0020	0.0146-0.0249	0.0146-0.0249	0.0100-0.0500	0.0020-0.0041	0.0016-0.0032	0.0007-0.0091
2002	2.4 (2429)	1	NA	NA	NA	NA	NA	NA	NA
	2.5 (2507)	P	0.0013-0.0021	0.0090-0.0240	0.0190-0.0380	0.0100-0.0600	0.0017-0.0033	0.0017-0.0033	0.0024-0.0083
	3.7 (3701)	K	0.0014	0.0146-0.0249	0.0146-0.0249	0.0100-0.0300	0.0020-0.0037	0.0016-0.0031	0.0007-0.0091
	4.0 (3966)	S	0.0008-0.0015	0.0090-0.0240	0.0190-0.0380	0.0100-0.0600	0.0017-0.0033	0.0017-0.0033	0.0024-0.0083
	4.7 (4701)	N	0.0008-0.0020	0.0146-0.0249	0.0146-0.0249	0.0100-0.0500	0.0020-0.0041	0.0016-0.0032	0.0007-0.0091

NA: Information not available

93591C55

TORQUE SPECIFICATIONS
All readings in ft. lbs.

Year	Engine Displacement Liters (cc)	Engine ID/VIN	Cylinder Head Bolts	Main Bearing Bolts	Rod Bearing Bolts	Crankshaft Damper Bolts	Flywheel Bolts	Manifold Intake	Manifold Exhaust	Spark Plugs	Lug Nuts
1999	2.5 (2458)	P	①	80	33	80	105	②	②	27	85-115
	4.0 (3966)	S	③	80	33	80	105	④	④	27	85-115
	4.7 (4701)	N	⑤	⑥	15⑦	130	45	9	18	27	85-115
2000	2.5 (2458)	P	①	80	33	80	105	②	②	27	85-115
	4.0 (3966)	S	③	80	33	80	105	④	④	27	85-115
	4.7 (4701)	N	⑤	⑥	15⑦	130	45	9	18	27	85-115
2001	2.5 (2458)	P	①	80	33	80	105	②	②	27	85-115
	4.0 (3966)	S	③	80	33	80	105	④	④	27	85-115
	4.7 (4701)	N	⑤	⑧	15⑦	130	45	9	18	27	85-115
2002	2.4 (2429)	1	NA	NA	NA	NA	NA	NA	NA	NA	85-115
	2.5 (2458)	P	①	80	33	80	105	②	②	27	85-115
	3.7 (3701)	K	⑧	⑨	⑩	130	45	9	18	27	85-115
	4.0 (3966)	S	③	80	33	80	105	④	④	27	85-115
	4.7 (4701)	N	⑤	⑥	15⑦	130	45	9	18	27	85-115

NA: Information not available

① Step 1: 22 ft. lbs.
Step 2: 45 ft. lbs.
Step 3: Bolts 1-6 to 110 ft. lbs.
Step 4: Bolt 7 to 100 ft. lbs.
Step 5: Bolts 8-10 to 110 ft. lbs.

② Bolt 1: 30 ft. lbs.
Bolts 2-7: 23 ft. lbs.

③ Step 1: 22 ft. lbs.
Step 2: 45 ft. lbs.
Step 3: Bolts 1-10 to 110 ft. lbs.
Step 4: Bolt 11 to 100 ft. lbs.
Step 5: Bolts 12-14 to 110 ft. lbs.

④ Bolts 1-5 and 8-11: 24 ft. lbs.
Bolts 6-7: 23 ft. lbs.

⑤ M11 bolts: 60 ft. lbs.
M8 bolts: 250 inch lbs.

⑥ Step 1: Bolts 1-10 to 25 inch lbs.
Step 2: Bolts 1-10 plus 90 degrees
Step 3: Bolts A-K to 40 ft. lbs.
Step 4: Bolts A1-A5 to 20 ft. lbs.

⑦ Plus 110 degrees

⑧ See procedure

⑨ Bed plate torque. Refer to procedure

⑩ 20 ft. lbs. + 90 degrees

93591C56

WHEEL ALIGNMENT

Year	Model		Caster Range (+/-Deg.)	Caster Preferred Setting (Deg.)	Camber Range (+/-Deg.)	Camber Preferred Setting (Deg.)	Toe-in (in.)	Steering Axis Inclination (Deg.)
1999	Cherokee Sport		1.50	+7.00	0.50	-0.25	0.25+/-0.25	—
	Grand Cherokee	F	0.75	+6.75	0.37	-0.37	0.20+/-0.20	—
		R	—	—	0.25	-0.25	0.25+/-0.25	—
	Wrangler	F	1.00	+7.00	0.63	-0.25	0.15+/-0.07	—
		R	—	—	0.25	-0.25	0.25+/-0.25	—
2000	Cherokee Sport		1.50	+7.00	0.50	-0.25	0.25+/-0.25	—
	Grand Cherokee	F	0.75	+6.75	0.37	-0.37	0.20+/-0.20	—
		R	—	—	0.25	-0.25	0.25+/-0.25	—
	Wrangler	F	1.00	+7.00	0.63	-0.25	0.15+/-0.07	—
		R	—	—	0.25	-0.25	0.25+/-0.25	—
2001	Cherokee Sport		1.50	+7.00	0.50	-0.25	0.25+/-0.25	—
	Grand Cherokee	F	0.75	+6.75	0.37	-0.37	0.20+/-0.20	—
		R	—	—	0.25	-0.25	0.25+/-0.25	—
	Wrangler	F	1.00	+7.00	0.63	-0.25	0.15+/-0.07	—
		R	—	—	0.25	-0.25	0.25+/-0.25	—
2002	Liberty	F	0.60	+3.5	0.375	0	0.20+/-0.125	—
		R	—	—	0.375	-0.25	0.25+/-0.41	—
	Grand Cherokee	F	0.75	+6.75	0.37	-0.37	0.20+/-0.20	—
		R	—	—	0.25	-0.25	0.25+/-0.25	—
	Wrangler	F	1.00	+7.00	0.63	-0.25	0.15+/-0.07	—
		R	—	—	0.25	-0.25	0.25+/-0.25	—

93591C57

Timing belt service is covered in Section 3 of this manual

TIRE, WHEEL AND BALL JOINT SPECIFICATIONS

Year	Model	OEM Tires		Tire Pressures (psi)		Wheel Size	Ball Joint Inspection
		Standard	Optional	Front	Rear		
1999	Cherokee	P215/75R15	P225/75R15 P225/70R15	33	33	7-J	①
	Grand Cherokee	P215/75R15	P225/75R15 P235/75R15 P245/70R15 P225/70R16	36	36	7-J	①
	Wrangler	P205/75R15	P215/75R15 P225/75R15 30x9.5R15LT	33	33	7-JJ	①
2000	Cherokee	P215/75R15	P225/75R15 P225/70R15	33	33	7-J	①
	Grand Cherokee	P215/75R15	P225/75R15 P235/75R15 P245/70R15 P225/70R16	33	33	7-J	①
	Wrangler SE	P205/75R15	P215/75R15 P225/75R15	33	33	6JJ/7JJ 7JJ	①
	Wrangler Sport	P215/75R15	P225/75R15 30x9.5R15LT	33 29	33 29	6JJ/7JJ 8JJ	①
	Wrangler Sahara	P225/70R15	30x9.5R15LT	Std: 33 Opt: 29	Std: 33 Opt: 29	7JJ/8JJ	①
2001	Cherokee	P215/75R15	P225/75R15 P225/70R15	33	33	7-J	①
	Grand Cherokee	P215/75R15	P225/75R15 P235/75R15 P245/70R15 P225/70R16	33	33	7-J	①
	Wrangler SE	P205/75R15	P215/75R15 P225/75R15	33	33	6JJ/7JJ 7JJ	①
	Wrangler Sport	P215/75R15	P225/75R15 30x9.5R15LT	33 29	33 29	6JJ/7JJ 8JJ	①
	Wrangler Sahara	P225/70R15	30x9.5R15LT	Std: 33 Opt: 29	Std: 33 Opt: 29	7JJ/8JJ	①
2002	Liberty	P215/75R16	P235/70R16 P225/70R15	NA	NA	NA	①
	Grand Cherokee	P215/75R15	P225/75R15 P235/75R15 P245/70R15 P225/70R16	33	33	7-J	①
	Wrangler SE	P205/75R15	P215/75R15 P225/75R15	33	33	6JJ/7JJ 7JJ	①
	Wrangler Sport	P215/75R15	P225/75R15 30x9.5R15LT	33 29	33 29	6JJ/7JJ 8JJ	①
	Wrangler Sahara	P225/70R15	30x9.5R15LT	Std: 33 Opt: 29	Std: 33 Opt: 29	7JJ/8JJ	①

NA: Information not available

OEM: Original Equipment Manufacturer

PSI: Pounds Per Square Inch

STD: Standard

OPT: Optional

① Replace if any measurable movement is found.

93591C58

BRAKE SPECIFICATIONS
All measurements in inches unless noted

| Year | Model | Brake Disc | | | Brake Drum | | | Minimum Lining Thickness | | Brake Caliper Mounting Bolts (ft. lbs.) |
		Original Thickness	Minimum Thickness	Maximum Run-out	Original Inside Diameter	Max. Wear Limit	Maximum Machine Diameter	Front	Rear	
1999	Cherokee	0.94	0.89	0.005	①	②	③	0.030	④	11
	Grand Cherokee	—	⑤	0.003	—	—	—	0.030	0.030	⑥
	Wrangler	0.94	0.89	0.005	9.00	②	9.06	0.030	④	11
2000	Cherokee	0.94	0.89	0.005	①	②	③	0.030	④	11
	Grand Cherokee	—	⑤	0.003	—	—	—	0.030	0.030	⑥
	Wrangler	0.94	0.89	0.005	9.00	②	9.06	0.030	④	11
2001	Cherokee	0.94	0.89	0.005	①	②	③	0.030	④	11
	Grand Cherokee	—	⑤	0.003	—	—	—	0.030	0.030	⑥
	Wrangler	0.94	0.89	0.005	9.00	②	9.06	0.030	④	11
2002	Liberty	NA	0.8937	0.005	①	②	③	0.030	④	11
	Grand Cherokee	—	⑤	0.003	—	—	—	0.030	0.030	⑥
	Wrangler	0.94	0.89	0.005	9.00	②	9.06	0.030	④	11

NA: Information not available

① Maximum diameter is listed on outside of drum

② Standard: 9.00 in.
Optional 10.00 in.

③ Standard: 9.06 in.
Optional 10.06 in.

④ Riveted brake shoes: 0.030 in.
Bonded brake shoes: 0.060 in.

⑤ Front: 0.965 in.
Rear: 0.335 in.

⑥ Slide pin bolts: 21-30 ft. lbs.
Anchor bolts: 66-85 ft. lbs.

93591C59

Heater Core replacement is covered in Section 2 of this manual

SCHEDULED MAINTENANCE INTERVALS
JEEP Wrangler, Cherokee, Grand Cherokee, Liberty

TO BE SERVICED	TYPE OF SERVICE	VEHICLE MILEAGE INTERVAL (x1000)												
		7.5	15	22.5	30	37.5	45	52.5	60	67.5	75	82.5	90	97.5
Engine oil & filter	R	✓	✓	✓	✓	✓	✓	✓	✓	✓	✓	✓	✓	✓
Brake hoses & linings	S/I	✓	✓	✓	✓	✓	✓	✓	✓	✓	✓	✓	✓	✓
Engine coolant level, hoses & clamps	S/I	✓	✓	✓	✓	✓	✓	✓	✓	✓	✓	✓	✓	✓
Exhaust system	S/I	✓	✓	✓	✓	✓	✓	✓	✓	✓	✓	✓	✓	✓
Lubricate steering linkage (4x2)	S/I	✓	✓	✓	✓	✓	✓	✓	✓	✓	✓	✓	✓	✓
Lubricate steering linkage (4x4)	S/I	✓		✓		✓		✓		✓		✓		✓
Air filter	R				✓				✓				✓	
Automatic transmission fluid & filter	R				✓				✓				✓	
Spark plugs	R				✓				✓				✓	
Transfer case fluid	R				✓				✓				✓	
Drive belts	S/I				✓				✓				✓	
Front & rear axle oil	R				✓				✓				✓	
Prop shaft universal joints	S/I				✓				✓				✓	
Rotate tires	S/I				✓				✓				✓	
Engine coolant	R						✓				✓			
Manual transmission fluid	R					✓					✓			
Distributor cap & rotor	R								✓					
Fuel filter	R								✓					
Ignition cables	R								✓					

R: Replace S/I: Service or Inspect

FREQUENT OPERATION MAINTENANCE (SEVERE SERVICE)

If a vehicle is operated under any of the following conditions it is considered severe service:

- Extremely dusty areas.

- 50% or more of the vehicle operation is in 32°C (90°F) or higher temperatures, or constant operation in temperatures below 0°C (32°F).

- Prolonged idling (vehicle operation in stop and go traffic.

- Frequent short running periods (engine does not warm to normal operating temperatures).

- Police, taxi, delivery usage or trailer towing usage.

Oil & oil filter change: change every 3000 miles.

Automatic transmission fluid, filter & bands: change & adjust every 12,000 miles.

Brake hoses & linings: check every 12,000 miles.

Liubricate steering linkage: check every 3000 miles.

Manual transmission fluid: change every 18,000 miles.

Prop shaft universal joints: lubricate every 3000 miles.

Front & rear axle oil: change every 12,000 miles.

93591C60

SCHEDULED MAINTENANCE INTERVALS
DAIMLERCHRYSLER CORPORATION
JEEP CHEROKEE, GRAND CHEROKEE, LIBERTY, WRANGLER

The following should be used as a guide when determining the amount of work required for a particular service. In estimating how long a particular Scheduled Maintenance Service should take, please observe the following:

● Labor Time is time based on field research and data supplied by the vehicle manufacturer.
● Labor time operations are given in hours and tenths of an hour.
● All labor operations are to be used as a guide.

Mechanic Skill Level Codes:
(A) PRECISION: Highly skilled with multiple certification.
(B) GENERAL: Normally skilled with certification.
(C) MAINTENANCE: Semi-skilled working on certification.

	LABOR TIME
7500 Mile Service (C)	
All Models	.9
w/4WD add	.1
15000 Mile Service (C)	
All Models	.9
22500 Mile Service (C)	
All Models	.9
w/4WD add	.1
30000 Mile Service (B)	
All Models	3.0
w/AT add	.5
w/4WD add	.5
37500 Mile Service (B)	
All Models	1.4
w/4WD add	.1

	LABOR TIME
45000 Mile Service (B)	
All Models	1.8
52500 Mile Service (C)	
All Models	.9
w/4WD add	.1
60000 Mile Service (B)	
All Models	3.8
w/AT add	.5
w/4WD add	.5
67500 Mile Service (C)	
All Models	.9
w/4WD add	.1

	LABOR TIME
75000 Mile Service (B)	
All Models	2.0
82500 Mile Service (C)	
All Models	.9
w/4WD add	.1
90000 Mile Service (B)	
All Models	3.1
w/AT add	.5
w/4WD add	.5
97500 Mile Service (C)	
All Models	.9
w/4WD add	.1

93551C61

Brake service is covered in Section 4 of this manual

FORD MOTOR CO.
Ford Escape

ENGINE AND VEHICLE IDENTIFICATION

	Engine							Model Year	
Code ①	Liters (cc)	Cu. In.	Cyl.	Fuel Sys.	Engine Type	Eng. Mfg.		Code ②	Year
B	2.0 (1998)	121	4	SFI	DOHC	Ford		1	2001
1	3.0 (3049)	182	6	SFI	DOHC	Ford		2	2002
								3	2003

SFI: Multi-port Fuel Injection

DOHC: Double Overhead Camshafts

① 8th digit of VIN

② 10th digit of VIN

93591C62

GENERAL ENGINE SPECIFICATIONS

Year	Model	Engine Displacement Liters (cc)	Engine ID/VIN	Fuel System Type	Net Horsepower @ rpm	Net Torque @ rpm (ft. lbs.)	Bore x Stroke (in.)	Com-pression Ratio	Oil Pressure @ rpm
2001	Escape	2.0 (1998)	B	SFI	135@5500	135@4500	3.34x3.46	9.6:1	54-80 ①
	Escape	3.0 (3049)	1	SFI	200@5500	200@4500	3.50x3.13	10.0:1	45 ①
2002	Escape	2.0 (1998)	B	SFI	135@5500	135@4500	3.34x3.46	9.6:1	54-80 ①
	Escape	3.0 (3049)	1	SFI	200@5500	200@4500	3.50x3.13	10.0:1	45 ①

SFI: Multi-port Fuel Injection

① The manufacturer does not provide an engine speed specification for oil pump pressure.

93591C63

ENGINE TUNE-UP SPECIFICATIONS

Year	Engine Displacement Liters (cc)	Engine ID/VIN	Spark Plug Gap (in.)	Ignition Timing (deg.)		Fuel Pump (psi)	Idle Speed (rpm)		Valve Clearance	
				MT	AT		MT	AT	Intake	Exhaust
2001	2.0 (1998)	B	0.039-0.043	10 BTDC	—	65	①	—	HYD.	HYD.
	3.0 (3049)	1	0.052-0.056	10 BTDC	10 BTDC	65	①	①	HYD.	HYD.
2002	2.0 (1998)	B	0.051	10 BTDC	—	65	①	—	HYD.	HYD.
	3.0 (3049)	1	0.052-0.056	10 BTDC	10 BTDC	65	①	①	HYD.	HYD.

BTDC: Before Top Dead Center

HYD: Hydraulic lash adjusters

① Refer to Vehicle Emission Control Information Label

93591C64

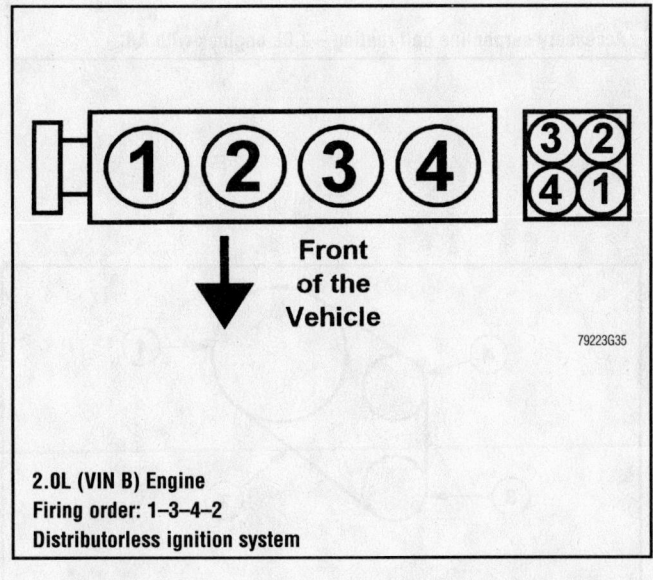

2.0L (VIN B) Engine
Firing order: 1–3–4–2
Distributorless ignition system

79223G35

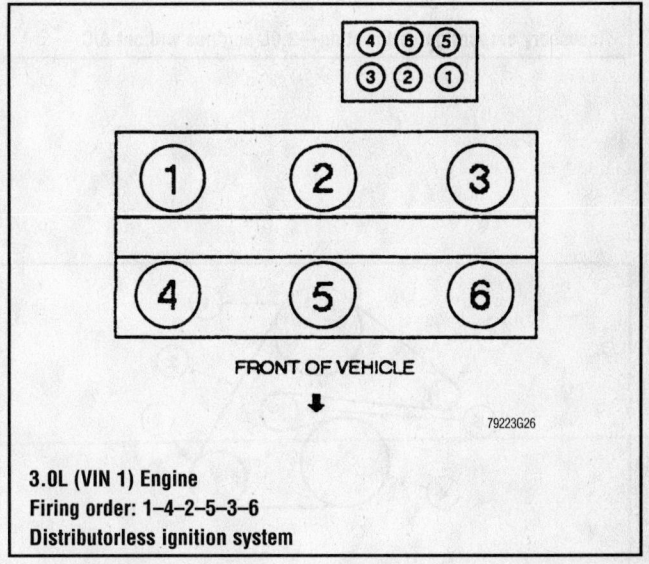

3.0L (VIN 1) Engine
Firing order: 1–4–2–5–3–6
Distributorless ignition system

79223G26

For complete Engine Mechanical specifications, see Section 1 of this manual

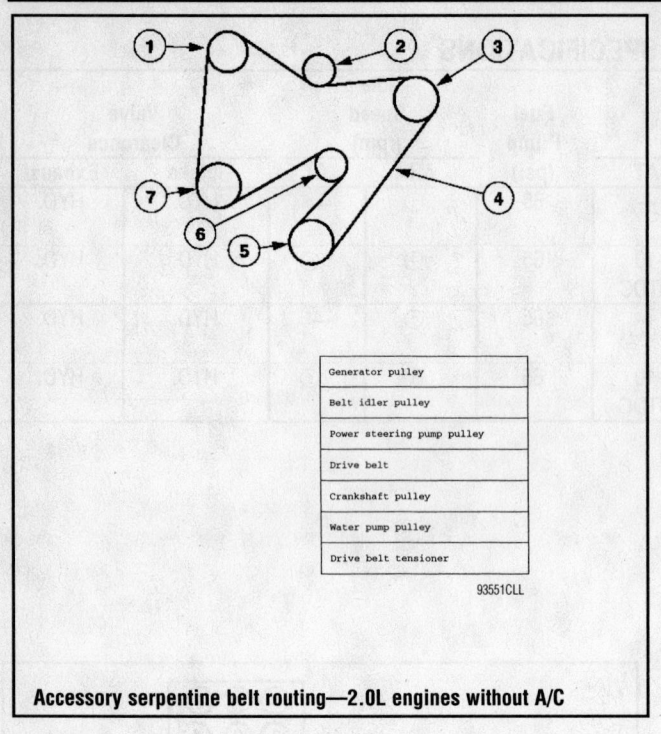

Generator pulley	
Belt idler pulley	
Power steering pump pulley	
Drive belt	
Crankshaft pulley	
Water pump pulley	
Drive belt tensioner	

93551CLL

Accessory serpentine belt routing—2.0L engines without A/C

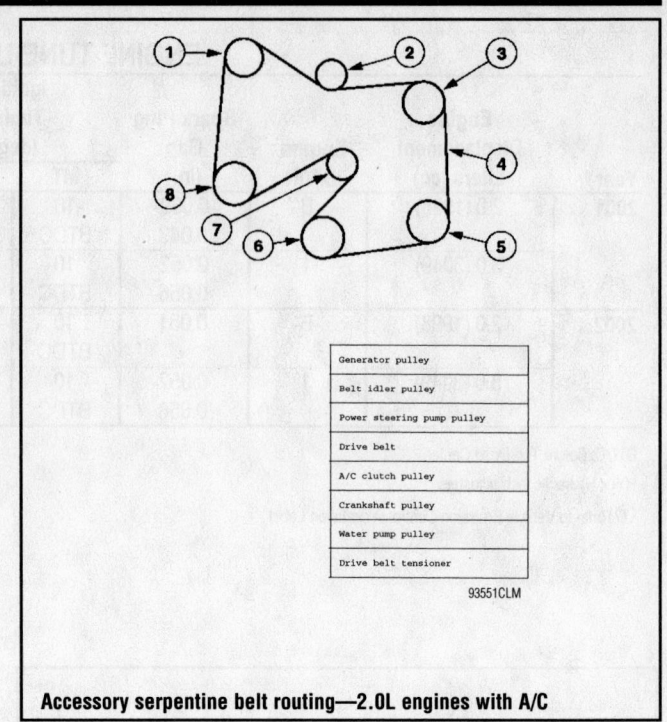

Generator pulley	
Belt idler pulley	
Power steering pump pulley	
Drive belt	
A/C clutch pulley	
Crankshaft pulley	
Water pump pulley	
Drive belt tensioner	

93551CLM

Accessory serpentine belt routing—2.0L engines with A/C

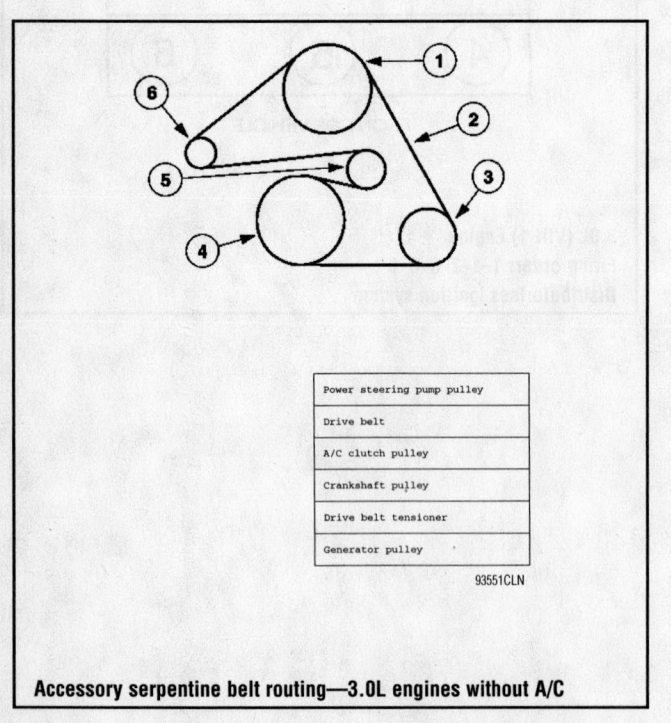

Power steering pump pulley	
Drive belt	
A/C clutch pulley	
Crankshaft pulley	
Drive belt tensioner	
Generator pulley	

93551CLN

Accessory serpentine belt routing—3.0L engines without A/C

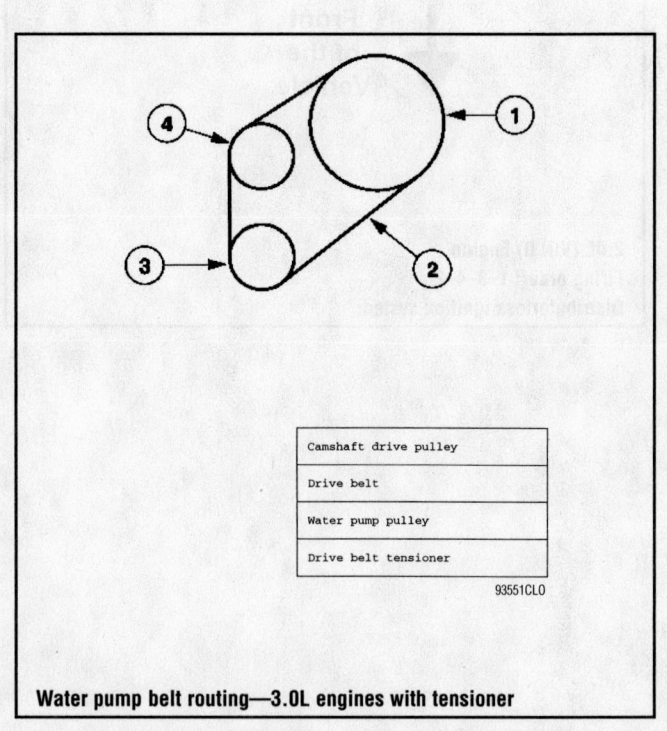

Camshaft drive pulley	
Drive belt	
Water pump pulley	
Drive belt tensioner	

93551CLO

Water pump belt routing—3.0L engines with tensioner

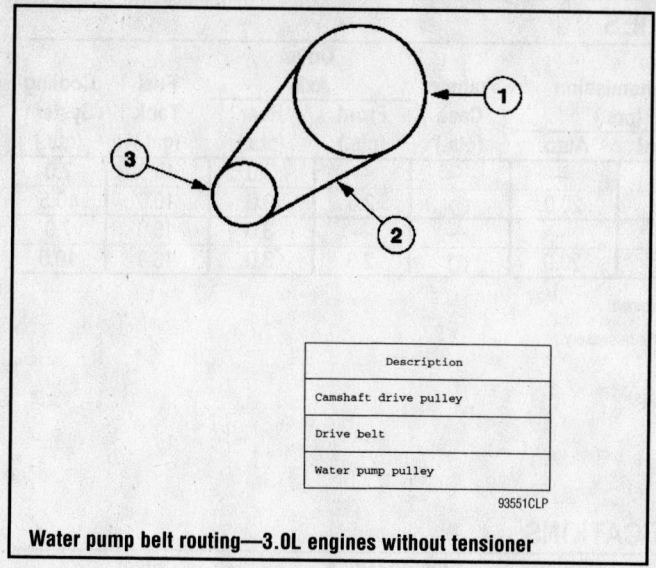

Description
Camshaft drive pulley
Drive belt
Water pump pulley

93551CLP

Water pump belt routing—3.0L engines without tensioner

For Accessory Drive Belt illustrations, see Section 1 of this manual

CAPACITIES

Year	Model	Engine Displacement Liters (cc)	Engine ID/VIN	Engine Oil with Filter (qts.)	Transmission (pts.) Manual	Transmission (pts.) Auto.	Transfer Case (pts.)	Drive Axle Front (pts.)	Drive Axle Rear (pts.)	Fuel Tank (gal.)	Cooling System (qts.)
2001	Escape	2.0 (1998)	B	4.5	5.7	—	—	—	3.0	15.0	7.0
	Escape	3.0 (3049)	1	5.8	5.7	20.0	①	2.6	3.0	16.0	10.5
2002	Escape	2.0 (1998)	B	4.5	5.7	—	—	—	3.0	15.0	7.0
	Escape	3.0 (3049)	1	5.5	5.7	20.0	①	2.6	3.0	16.0	10.5

NOTE: All capacities are approximate. Add fluid gradually and check to be sure a proper fluid level is obtained.

① The transfer case is lubricated for life and is not to be checked unless a leak is suspected or a repair is necessary.

93591C65

VALVE SPECIFICATIONS

Year	Engine Displacement Liters (cc)	Engine ID/VIN	Seat Angle (deg.)	Face Angle (deg.)	Spring Test Pressure (lbs. @ in.)	Spring Installed Height (in.)	Stem-to-Guide Clearance (in.) Intake	Stem-to-Guide Clearance (in.) Exhaust	Stem Diameter (in.) Intake	Stem Diameter (in.) Exhaust
2001	2.0 (1998)	B	45	45	①	1.346	0.0007-0.0025	0.0007-0.0025	0.2374	0.2374
	3.0 (3049)	1	44.75	45.5	153@ 1.18	1.57	0.0008-0.0027	0.0018-0.0037	0.2352-0.2360	0.2343-0.2350
2002	2.0 (1998)	B	45	45	①	1.346	0.0007-0.0025	0.0007-0.0025	0.2374	0.2374
	3.0 (3049)	1	44.75	45.5	153@ 1.18	1.57	0.0008-0.0027	0.0018-0.0037	0.2352-0.2360	0.2343-0.2350

① Intake: 82.1@ 0.988

Exhaust: 95@ 1.0275

93591C66

CRANKSHAFT AND CONNECTING ROD SPECIFICATIONS

All measurements are given in inches.

| Year | Engine Displacement Liters (cc) | Engine ID/VIN | Crankshaft | | | | Connecting Rod | | |
			Main Brg. Journal Dia.	Main Brg. Oil Clearance	Shaft End-play	Thrust on No.	Journal Diameter	Oil Clearance	Side Clearance
2001	2.0 (1998)	B	2.2827-2.2835	①	0.0035-0.0102	3	1.8460-1.8468	0.0006-0.0028	0.0040-0.0110
	3.0 (3049)	1	2.4791-2.4800	0.0010-0.0018	0.0043-0.0091	3	1.9673-1.9681	0.0010-0.0025	0.0039-0.0118
2002	2.0 (1998)	B	2.2827-2.2835	①	0.0035-0.0102	3	1.8460-1.8468	0.0006-0.0028	0.0040-0.0110
	3.0 (3049)	1	2.4791-2.4800	0.0010-0.0018	0.0043-0.0091	3	1.9673-1.9681	0.0011-0.0026	0.0039-0.0118

① Journals 1, 2 and 4: 0.0010 - 0.0017 in.
Journal 3: 0.0012 - 0.0019 in.

93591C67

PISTON AND RING SPECIFICATIONS

All measurements are given in inches.

| Year | Engine Displacement Liters (cc) | Engine ID/VIN | Piston Clearance | Ring Gap | | | Ring Side Clearance | | |
				Top Compression	Bottom Compression	Oil Control	Top Compression	Bottom Compression	Oil Control
2001	2.0 (1998)	B	0.0004-0.0012	0.0100-0.0300	0.0100-0.0300	0.0160-0.0660	0.0015-0.0032	0.0015-0.0035	snug
	3.0 (3049)	1	0.0005-0.0009	0.0039-0.0098	0.0106-0.0165	0.0059-0.0256	0.0016-0.0030	0.0016-0.0033	snug
2002	2.0 (1998)	B	0.0004-0.0012	0.0100-0.0300	0.0100-0.0300	0.0160-0.0660	0.0015-0.0032	0.0015-0.0035	snug
	3.0 (3049)	1	0.0005-0.0009	0.0039-0.0098	0.0106-0.0165	0.0059-0.0256	0.0016-0.0030	0.0016-0.0033	snug

93591C68

For Tire, Wheel and Ball Joint specifications, see Section 1 of this manual

TORQUE SPECIFICATIONS
All readings in ft. lbs.

Year	Engine Displacement Liters (cc)	Engine ID/VIN	Cylinder Head Bolts	Main Bearing Bolts	Rod Bearing Bolts	Crankshaft Damper Bolts	Flywheel Bolts	Manifold Intake	Manifold Exhaust	Spark Plugs	Lug Nuts
2001	2.0 (1998)	B	①	②	③	80-87	83	13	12	11	98
	3.0 (3049)	1	④	⑤	⑥	⑦	59	⑧	15	11	98
2002	2.0 (1998)	B	①	②	③	80-87	83	13	12	11	98
	3.0 (3049)	1	④	⑤	⑥	⑦	59	⑧	15	11	98

① Step 1: 15 ft. lbs. (20 Nm).
   Step 2: 30 ft. lbs. (40 Nm).
   Step 3: Plus an additional 90 degrees.

② Step 1: 18 ft. lbs.
   Step 2: +60 degrees

③ Step 1: 26 ft. lbs.
   Step 2: +90 degrees

④ Step 1: 30 ft. lbs. (40 Nm).
   Step 2: Tighten the bolts 90 degrees.
   Step 3: Loosen the bolts one full turn.
   Step 4: 30 ft. lbs. (40 Nm).
   Step 5: Tighten the bolts 90 degrees.
   Step 6: Tighten the bolts 90 degrees.

⑤ Step 1: Fasteners 1-8: 18 ft. lbs.
   Step 2: Fasteners 9-19: 30 ft. lbs.
   Step 3: Fasteners 1-16: +90 degrees
   Step 4: fasteners 17-22: 18 ft. lbs.

⑥ Step 1: 17 ft. lbs.
   Step 2: 32 ft. lbs.

⑦ Step 1: 89 ft. lbs.
   Step 2: Loosen 1 full turn
   Step 3: 37 ft. lbs.
   Step 4: 66 ft. lbs.

⑧ 89 inch lbs.

93591C69

WHEEL ALIGNMENT

Year	Model		Caster Range (+/-Deg.)	Caster Preferred Setting (Deg.)	Camber Range (+/-Deg.)	Camber Preferred Setting (Deg.)	Toe-in (in.)	Steering Axis Inclination (Deg.)
2001	ALL	F	NA	+1.93	NA	-0.84	0.12+/-0.12	11.40
		R	NA	NA	NA	-0.04	0.09+/-0.11	NA
2002	ALL	F	NA	+1.93	NA	-0.84	0.12+/-0.12	11.40
		R	NA	NA	NA	-0.04	0.09+/-0.11	NA

93591C70

TIRE, WHEEL AND BALL JOINT SPECIFICATIONS

| Year | Model | OEM Tires | | Tire Pressures (psi) | | Wheel Size | Ball Joint Inspection |
		Standard	Optional	Front	Rear		
2001	Escape	P225/70SR15	P235/70R16	NA	NA	NA	0.030 in.
2002	Escape	P215/70R16	P225/70SR15 P235/70R16	NA	NA	NA	0.030 in.

OEM: Original Equipment Manufacturer

PSI: Pounds Per Square Inch

STD: Standard

OPT: Optional

NA: Not Available

93591C71

BRAKE SPECIFICATIONS
All measurements in inches unless noted

| Year | Model | Brake Disc | | | Brake Drum | | | Minimum Lining Thickness | Brake Caliper | |
		Original Thickness	Minimum Thickness	Maximum Run-out	Original Inside Diameter	Max. Wear Limit	Maximum Machine Diameter		Bracket Bolts (ft. lbs.)	Mounting Bolts (ft. lbs.)
2001	Escape	0.940	0.860	0.002	9.06	0.06	8.92	0.039	111	26
2002	Escape	0.940	0.860	0.004	9.06	0.06	8.92	0.039	111	26

93591C72

For Wheel Alignment specifications, see Section 1 of this manual

SCHEDULED MAINTENANCE INTERVALS
Ford Escape

TO BE SERVICED	TYPE OF SERVICE	VEHICLE MILEAGE INTERVAL (x1000)												
		5	10	15	20	25	30	35	40	45	50	55	60	65
Air cleaner filter	R						✓						✓	
Accessory drive belt	S/I												✓	
Brake system ①	S/I			✓			✓			✓			✓	
Clutch pedal operation	S/I						✓						✓	
Cooling fan operation	S/I		✓		✓		✓		✓		✓		✓	
Cooling system hoses and clamps	S/I			✓			✓			✓			✓	
CV-joint boots & axle seals	S/I						✓						✓	
Engine coolant	R	Ten years or 150,000 miles												
Engine oil & filter	R	✓	✓	✓	✓	✓	✓	✓	✓	✓	✓	✓	✓	✓
Exterior Lights	S/I	Check monthly												
PCV valve	S/I												✓	
Exhaust system & heat shields	S/I						✓						✓	
Parking brake system	S/I	Every 6 months												
Power steering fluid	S/I	Every 6 months												
Rotate tires	S/I	✓		✓		✓		✓		✓		✓		✓
Steering linkage	S/I						✓						✓	
Spark plugs	R	Change at 100,000 miles												
Suspension components	S/I						✓						✓	

R: Replace   S/I: Inspect and service, if necessary   L: Lubricate   A: Adjust   C: Clean

① Inspect the reservoir fluid level, rotor and or drum, brake lines, hoses, calipers and or wheel cylinders

FREQUENT OPERATION MAINTENANCE (SEVERE SERVICE)

If a vehicle is operated under any of the following conditions it is considered severe service:

- Extremely dusty areas.

- 50% or more of the vehicle operation is in 32°C (90°F) or higher temperatures, or constant operation in temperatures below 0°C (32°F).

- Prolonged idling (vehicle operation in stop and go traffic).

- Frequent short running periods (engine does not warm to normal operating temperatures).

- Police, taxi, delivery usage or trailer towing usage.

Oil & oil filter change: change every 3000 miles.

Air filter element: change every 15,000 miles.

93591C73

SCHEDULED MAINTENANCE INTERVALS
FORD MOTOR COMPANY
ESCAPE

The following should be used as a guide when determining the amount of work required for a particular service. In estimating how long a particular Scheduled Maintenance Service should take, please observe the following:

- Labor Time is time based on field research and data supplied by the vehicle manufacturer.
- Labor time operations are given in hours and tenths of an hour.
- All labor operations are to be used as a guide.

Mechanic Skill Level Codes:
(A) PRECISION: Highly skilled with multiple certification.
(B) GENERAL: Normally skilled with certification.
(C) MAINTENANCE: Semi-skilled working on certification.

	LABOR TIME		LABOR TIME		LABOR TIME
5000 Mile Service (C)		**25000 Mile Service (C)**		**50000 Mile Service (C)**	
All Models	.9	All Models	.9	All Models	.7
10000 Mile Service (C)		**30000 Mile Service (B)**		**55000 Mile Service (C)**	
All Models	.5	All Models	1.9	All Models	.9
15000 Mile Service (C)		**35000 Mile Service (C)**		**60000 Mile Service (B)**	
All Models	1.6	All Models	.9	All Models	1.9
20000 Mile Service (C)		**40000 Mile Service (C)**		**65000 Mile Service (C)**	
All Models	.5	All Models	.5	All Models	.9
		45000 Mile Service (C)			
		All Models	1.6		

93081C81

For Maintenance Interval recommendations, see Section 1 of this manual

FORD MOTOR CO.
Ford Explorer • Mercury Mountaineer

ENGINE AND VEHICLE IDENTIFICATION

			Engine					Model Year	
Code ①	Liters (cc)	Cu. In.	Cyl.	Fuel Sys.	Type	Eng. Mfg.	Code ②		Year
E	4.0 (4000)	244	6	MFI	SOHC	Ford	X		1999
P	5.0 (4949)	302	8	MFI	OHV	Ford	Y		2000
X	4.0 (3998)	244	6	MFI	OHV	Ford	1		2001
W	4.6 (4601)	281	8	MFI	SOHC	Ford	2		2002
							3		2003

MFI: Multi-port Fuel Injection

OHV: Overhead Valve

SOHC: Single Overhead Camshaft

① 8th digit of the Vehicle Identification Number (VIN)

② 10th digit of the Vehicle Identification Number (VIN)

93591C74

GENERAL ENGINE SPECIFICATIONS

Year	Model	Engine Displacement Liters (cc)	Engine VIN	Fuel System Type	Net Horsepower @ rpm	Net Torque @ rpm (ft. lbs.)	Bore x Stroke (in.)	Compression Ratio	Oil Pressure @ rpm
1999	Explorer Sport	4.0 (4000)	X	MFI	160@4000	225@2500	3.81x3.39	9.0:1	40-60@2000
	Explorer/Mountaineer	4.0 (4000)	E	MFI	160@4000	225@2500	3.81x3.39	9.0:1	40-60@2000
	Explorer/Mountaineer	5.0 (4949)	P	MFI	210@4500	280@3500	4.00x3.00	9.0:1	40-60@2500
2000	Explorer Sport	4.0 (4000)	X	MFI	160@4000	225@2500	3.81x3.39	9.0:1	40-60@2000
	Explorer/Mountaineer	5.0 (4949)	P	MFI	210@4500	280@3500	4.00x3.00	9.0:1	40-60@2500
2001	Explorer/Mountaineer	4.0 (4000)	E	MFI	160@4000	225@2500	3.81x3.39	9.0:1	40-60@2000
	Explorer/Mountaineer	5.0 (4949)	P	MFI	210@4500	280@3500	4.00x3.00	9.0:1	40-60@2500
2002	Explorer/Mountaineer	4.0 (4000)	E	MFI	160@4000	225@2500	3.81x3.39	9.0:1	40-60@2000
	Explorer/Mountaineer	4.6 (4601)	W	MFI	239@4750	282@4000	3.55x3.54	NA	20-45@1500

MFI: Multi-port Fuel Injection

NA: Information not available

93591C75

GASOLINE ENGINE TUNE-UP SPECIFICATIONS

Year	Engine Displacement Liters (cc)	Engine ID/VIN	Spark Plug Gap (in.)	Ignition Timing (deg.)		Fuel Pump (psi)	Idle Speed (rpm)		Valve Clearance	
				MT	AT		MT	AT	In.	Ex.
1999	4.0 (3950)	X	0.052-0.056	10B ①	10B ①	35-45	①	①	HYD	HYD
	4.0 (3950)	E	0.052-0.056	—	10B ①	57-73	—	①	HYD	HYD
	5.0 (4949)	P	0.052-0.056	—	10B ①	35-45	—	①	HYD	HYD
2000	4.0 (3950)	E	0.052-0.056	10B ①	10B ①	57-73	①	①	HYD	HYD
	4.0 (3950)	X	0.052-0.056	10B ①	10B ①	57-73	①	①	HYD	HYD
	5.0 (4949)	P	0.052-0.056	—	10B ①	57-73	—	①	HYD	HYD
2001	4.0 (3950)	E	0.052-0.056	—	10B ①	57-73	—	①	HYD	HYD
	4.0 (3950)	X	0.052-0.056	10B ①	10B ①	57-73	①	①	HYD	HYD
	5.0 (4949)	P	0.052-0.056	—	10B ①	57-73	—	①	HYD	HYD
2002	4.0 (3950)	E	0.052-0.056	10B ①	10B ①	57-73	①	①	HYD	HYD
	4.6 (4601)	W	0.052-0.056	—	10B ①	57-73	—	①	HYD	HYD

NOTE: The Vehicle Emission Control Information label often reflects specification changes changes made during production. The label figures must be used if they differ from those in this chart.

B: Before top dead center

HYD: Hydraulic

① Idle speed and ignition timing are electronically controlled and cannot be adjusted

93591C76

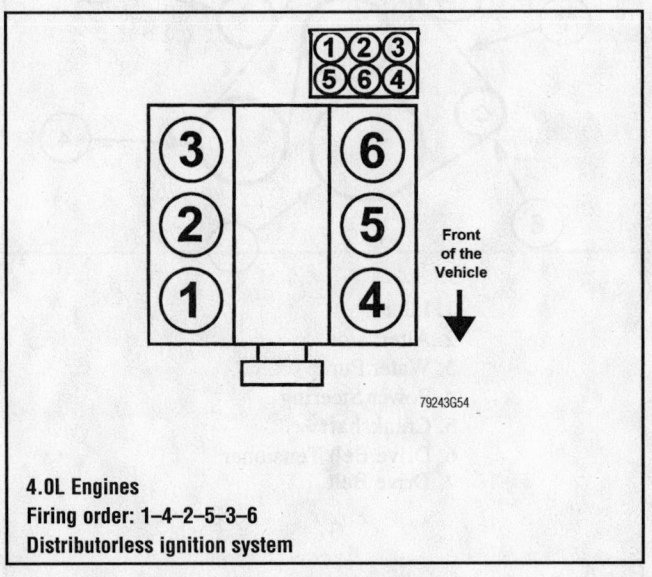

4.0L Engines
Firing order: 1–4–2–5–3–6
Distributorless ignition system

79243G54

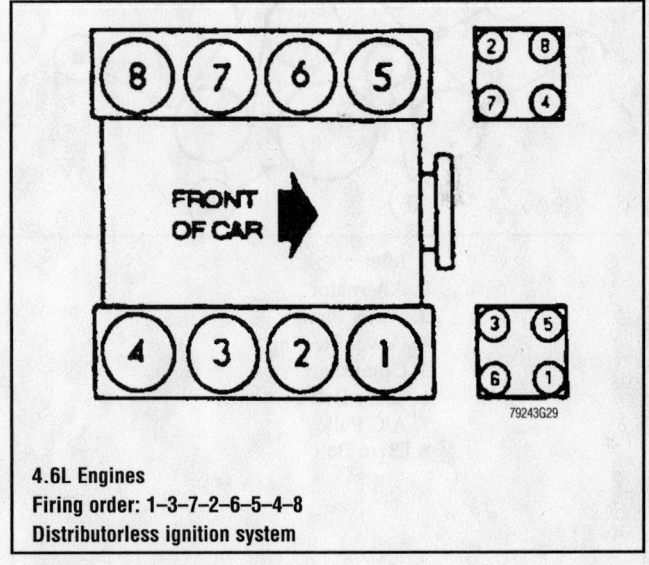

4.6L Engines
Firing order: 1–3–7–2–6–5–4–8
Distributorless ignition system

79243G29

For Tune-up, Capacities and Firing orders, see Section 1 of this manual

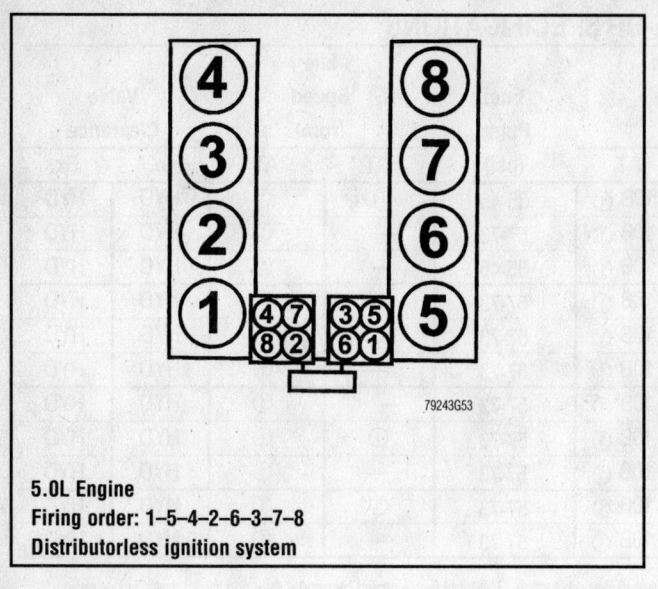

5.0L Engine
Firing order: 1–5–4–2–6–3–7–8
Distributorless ignition system

79243G53

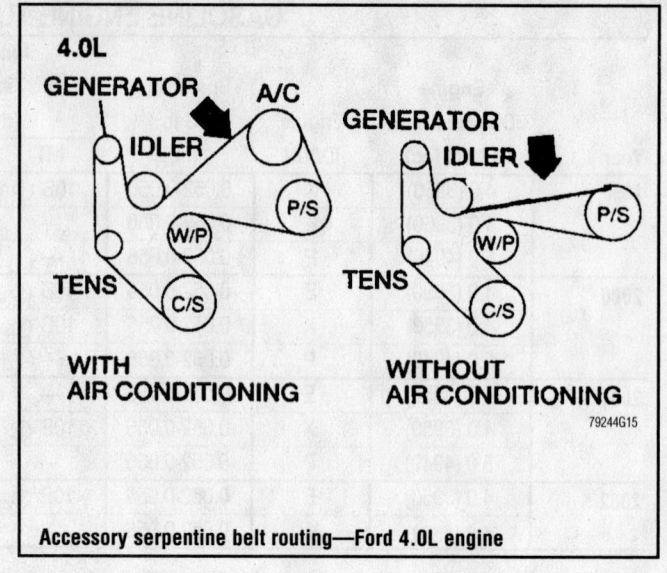

Accessory serpentine belt routing—Ford 4.0L engine

79244G15

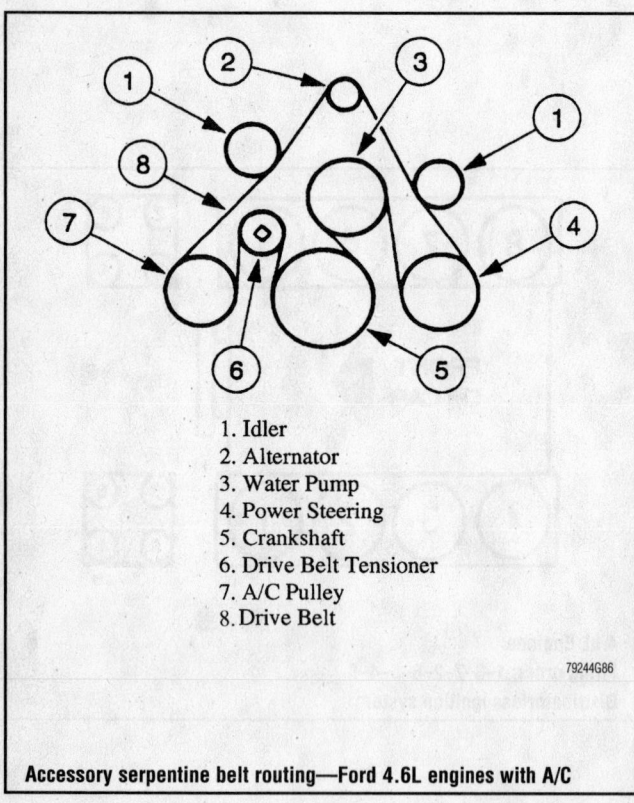

1. Idler
2. Alternator
3. Water Pump
4. Power Steering
5. Crankshaft
6. Drive Belt Tensioner
7. A/C Pulley
8. Drive Belt

79244G86

Accessory serpentine belt routing—Ford 4.6L engines with A/C

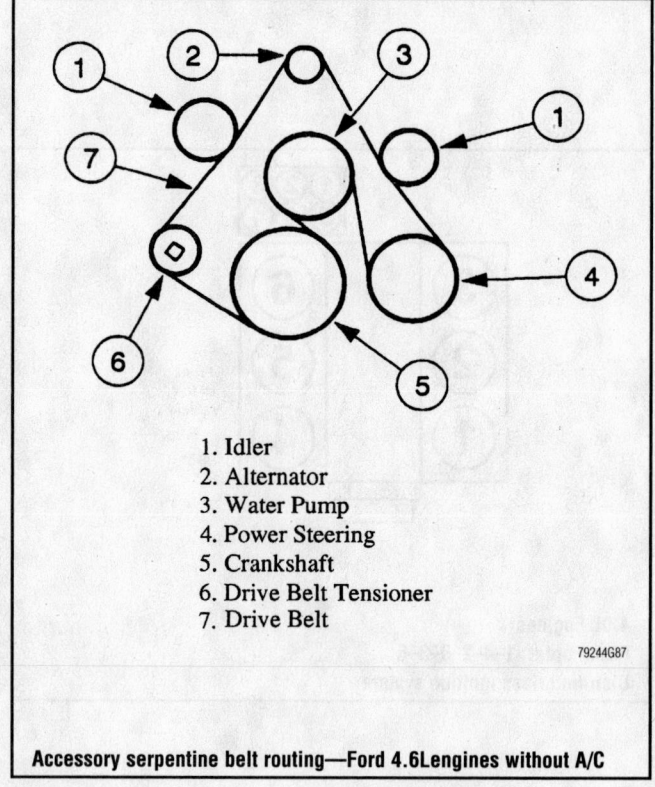

1. Idler
2. Alternator
3. Water Pump
4. Power Steering
5. Crankshaft
6. Drive Belt Tensioner
7. Drive Belt

79244G87

Accessory serpentine belt routing—Ford 4.6Lengines without A/C

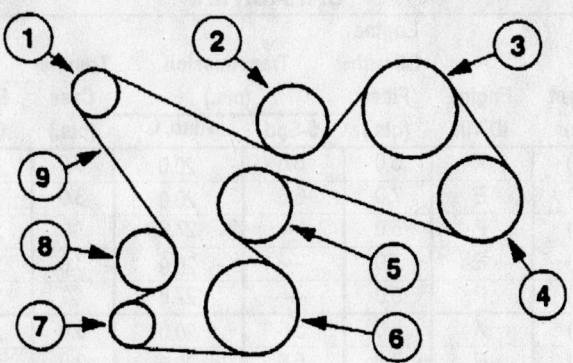

1. Alternator pulley
2. Belt idler pulley
3. Power steering pulley
4. A/C compressor pulley
5. Water pump
6. Crankshaft pulley
7. Belt idler pulley
8. Drive belt tensioner
9. Drive belt

79244G19

Accessory serpentine belt routing—Ford Explorer/Mercury Mountaineer 5.0L engine

CAPACITIES

Year	Model	Engine Displacement Liters (cc)	Engine ID/VIN	Engine Oil with Filter (qts.)	Transmission (pts.) 5-Spd	Transmission (pts.) Auto. ①	Transfer Case (pts.)	Drive Axle Front (pts.)	Drive Axle Rear (pts.)	Fuel Tank (gal.)	Cooling System (qts.)
1999	Explorer Sport	4.0 (3950)	X	5.0	5.6	20.0	3.0	3.25	5.5	17.5	16.0
	Explorer	4.0 (4000)	E	7.0	5.6	20.0	3.0	3.25	5.5	21.0	16.0
	Explorer	5.0 (4949)	P	5.0	—	27.8	②	3.25	5.5	21.0	16.0
	Mountaineer	4.0 (4000)	E	7.0	—	20.0	3.0	3.25	5.5	21.0	16.0
	Mountaineer	5.0 (4949)	P	5.0	—	27.8	②	3.25	5.5	21.0	16.0
2000	Explorer Sport	4.0 (3950)	X	5.0	5.6	20.0	3.0	3.25	5.5	17.5	12.0
	Explorer	4.0 (4000)	E	7.0	5.6	20.0	3.0	3.25	5.5	21.0	14.0
	Explorer	5.0 (4949)	P	5.0	—	27.8	②	3.25	5.5	21.0	15.7
	Mountaineer	4.0 (4000)	E	7.0	—	20.0	3.0	3.25	5.5	21.0	14.0
	Mountaineer	5.0 (4949)	P	5.0	—	27.8	②	3.25	5.5	21.0	15.7
2001	Explorer Sport	4.0 (3950)	X	5.0	5.6	20.0	2.6	3.6	5.5	17.5	12.0
	Explorer	4.0 (4000)	E	7.0	5.6	20.0	3.0	3.25	5.5	③	14.0
	Explorer	5.0 (4949)	P	5.0	—	27.8	②	3.25	5.5	21.0	15.7
	Mountaineer	4.0 (4000)	E	7.0	—	20.0	3.0	3.25	5.5	21.0	14.0
	Mountaineer	5.0 (4949)	P	5.0	—	27.8	②	3.25	5.5	21.0	15.7
2002	Explorer	4.0 (4000)	E	5.0	5.0	25.4	3.0	3.25	2.75	④	⑤
	Explorer	4.6 (4601)	W	6.5	—	25.4	②	3.25	2.75	22.5	⑥
	Mountaineer	4.0 (4000)	E	5.0	5.0	25.4	3.0	3.25	2.75	22.5	⑤
	Mountaineer	4.6 (4601)	W	6.5	—	25.4	②	3.25	2.75	22.5	⑥

NOTE: All capacities are approximate. Add fluid gradually and check to be sure a proper fluid level is obtained.

① Dry fill

② Part time: 3.0
   Full time: 2.6

③ 4-door Explorer: 21.0
   Explorer sport Trac: 20.5

④ 4-door Explorer: 22.5
   Explorer Sport: 17.5
   Explorer Sport Trac: 20.5

⑤ w/o auxiliary heater: 16.3
   w/auxiliary heater: 18.3

⑥ w/o auxiliary heater: 18.6
   w/auxiliary heater: 20.1

93591C77

VALVE SPECIFICATIONS

Year	Engine Displacement Liters (cc)	Engine ID/VIN	Seat Angle (deg.)	Face Angle (deg.)	Spring Test Pressure (lbs. @ in.)	Spring Installed Height (in.)	Stem-to-Guide Clearance (in.)		Stem Diameter (in.)	
							Intake	Exhaust	Intake	Exhaust
1999	4.0 (3950)	X	45	44	138@1.22	1.580-1.610	0.0008-0.0025	0.0018-0.0035	0.3159-0.3167	0.3149-0.3156
	4.0 (4000)	E	45	45	202-225@1.413-1.445	1.569-1.601	0.0010-0.0020	0.0010-0.0020	0.2740-0.2748	0.2730-0.2740
	5.0 (4949)	P	45	44	200@1.20	①	0.0010-0.0027	0.0015-0.0032	0.3415-0.3423	0.3410-0.3418
2000	4.0 (3950)	X	45	44	138@1.22	1.580-1.610	0.0008-0.0025	0.0018-0.0035	0.3159-0.3167	0.3149-0.3156
	4.0 (4000)	E	45	45	202-225@1.413-1.445	1.569-1.601	0.0010-0.0020	0.0010-0.0020	0.2740-0.2748	0.2730-0.2740
	5.0 (4949)	P	45	44	200@1.20	①	0.0010-0.0027	0.0015-0.0032	0.3415-0.3423	0.3410-0.3418
2001	4.0 (3950)	X	45	44	138@1.22	1.580-1.610	0.0008-0.0025	0.0018-0.0035	0.3159-0.3167	0.3149-0.3156
	4.0 (4000)	E	45	45	202-225@1.413-1.445	1.569-1.601	0.0010-0.0020	0.0010-0.0020	0.2740-0.2748	0.2730-0.2740
	5.0 (4949)	P	45	44	200@1.20	①	0.0010-0.0027	0.0015-0.0032	0.3415-0.3423	0.3410-0.3418
2002	4.0 (4000)	E	45	45	202-225@1.413-1.445	1.569-1.601	0.0010-0.0020	0.0010-0.0020	0.2740-0.2748	0.2730-0.2740
	4.6 (4601)	W	44.5-45.1	45.25-45.75	162-180@1.134	1.675	0.0008-0.0027	0.0018-0.0037	0.2746-0.2750	0.2736-0.2740

① Intake: 1.75-1.81 in.
  Exhaust: 1.59 in.

93591C78

CRANKSHAFT AND CONNECTING ROD SPECIFICATIONS

All measurements are given in inches.

Year	Engine Displacement Liters (cc)	Engine ID/VIN	Crankshaft Main Brg. Journal Dia.	Main Brg. Oil Clearance	Shaft End-play	Thrust on No.	Connecting Rod Journal Diameter	Oil Clearance	Side Clearance
1999	4.0 (3950)	X	2.2433-2.2441	0.0008-0.0015	0.0020-0.0120	3	2.1252-2.1260	0.0003-0.0024	0.0002-0.0025
	4.0 (4000)	E	2.2430-2.2440	0.0008-0.0015	0.0020-0.0125	3	2.7252-2.7260	0.0003-0.0024	0.0036-0.0106
	5.0 (4949)	P	2.2482-2.2490	0.0008-0.0015	0.0040-0.0080	3	2.1228-2.1236	0.0008-0.0015	0.0010-0.0020
2000	4.0 (3950)	X	2.2433-2.2441	0.0008-0.0015	0.0020-0.0120	3	2.1252-2.1260	0.0003-0.0024	0.0002-0.0025
	4.0 (4000)	E	2.2430-2.2440	0.0008-0.0015	0.0020-0.0125	3	2.7252-2.7260	0.0003-0.0024	0.0036-0.0106
	5.0 (4949)	P	2.2482-2.2490	0.0008-0.0015	0.0040-0.0080	3	2.1228-2.1236	0.0008-0.0015	0.0010-0.0020
2001	4.0 (3950)	X	2.2433-2.2441	0.0008-0.0015	0.0020-0.0120	3	2.1252-2.1260	0.0003-0.0024	0.0002-0.0025
	4.0 (4000)	E	2.2430-2.2440	0.0008-0.0015	0.0020-0.0125	3	2.7252-2.7260	0.0003-0.0024	0.0036-0.0106
	5.0 (4949)	P	2.2482-2.2490	0.0008-0.0015	0.0040-0.0080	3	2.1228-2.1236	0.0008-0.0015	0.0010-0.0020
2002	4.0 (4000)	E	2.2430-2.2440	0.0008-0.0015	0.0020-0.0125	3	2.7252-2.7260	0.0003-0.0024	0.0036-0.0106
	4.6 (4601)	W	2.6500-2.6570	0.0009-0.0018	0.0051-0.0118	3	2.0867-2.0870	0.0010-0.0027	0.0059 0.0177

93591C79

PISTON AND RING SPECIFICATIONS
All measurements are given in inches.

Year	Engine Displacement Liters (cc)	Engine ID/VIN	Piston Clearance	Ring Gap			Ring Side Clearance		
				Top Compression	Bottom Compression	Oil Control	Top Compression	Bottom Compression	Oil Control
1999	4.0 (3950)	X	0.0008-0.0019	0.015-0.023	0.015-0.023	0.015-0.055	0.0020-0.0033	0.0020-0.0033	SNUG
	4.0 (4000)	E	0.0008-0.0019	0.015-0.023	0.015-0.023	0.015-0.055	0.0010-0.0030	0.0010-0.0030	SNUG
	5.0 (4949)	P	0.0012-0.0020	0.010-0.020	0.018-0.028	0.010-0.040	0.0013-0.0033	0.0013-0.0033	SNUG
2000	4.0 (3950)	X	0.0008-0.0019	0.015-0.023	0.015-0.023	0.015-0.055	0.0020-0.0033	0.0020-0.0033	SNUG
	4.0 (4000)	E	0.0008-0.0019	0.015-0.023	0.015-0.023	0.015-0.055	0.0010-0.0030	0.0010-0.0030	SNUG
	5.0 (4949)	P	0.0012-0.0020	0.010-0.020	0.018-0.028	0.010-0.040	0.0013-0.0033	0.0013-0.0033	SNUG
2001	4.0 (3950)	X	0.0008-0.0019	0.015-0.023	0.015-0.023	0.015-0.055	0.0020-0.0033	0.0020-0.0033	SNUG
	4.0 (4000)	E	0.0008-0.0019	0.008-0.018	0.015-0.024	0.015-0.055	0.0160-0.0300	0.0120-0.0260	SNUG
	5.0 (4949)	P	0.0012-0.0020	0.010-0.020	0.018-0.028	0.010-0.040	0.0013-0.0033	0.0013-0.0033	SNUG
2002	4.0 (4000)	E	0.0008-0.0019	0.015-0.023	0.015-0.023	0.015-0.055	0.0010-0.0030	0.0010-0.0030	SNUG
	4.6 (4601)	W	0.0005-0.0010	0.0394	0.0394	0.0500	0.0012-0.0031	0.0012-0.0031	SNUG

93591C80

TORQUE SPECIFICATIONS
All readings in ft. lbs.

Year	Engine Displacement Liters (cc)	Engine ID/VIN	Cylinder Head Bolts	Main Bearing Bolts	Rod Bearing Bolts	Crankshaft Damper Bolts	Flywheel Bolts	Manifold Intake *	Manifold Exhaust	Spark Plugs	Lug Nut
1999	4.0 (3950)	X	①	72	②	③	52	④	15-18	15	100
	4.0 (4000)	E	⑤	67-74	⑥	⑦	⑧	9-10	15-18	15	100
	5.0 (4949)	P	⑨	61-68	19-24	110-130	75-85	23-25	26-32	15	100
2000	4.0 (3950)	X	①	72	②	③	52	④	15-18	15	100
	4.0 (4000)	E	⑩	72	19-24	⑪	54-64	9-10	15-18	15	100
	5.0 (4949)	P	⑨	61-68	19-24	110-130	75-85	⑫	26-32	15	100
2001	4.0 (3950)	X	①	72	②	③	52	④	15-18	15	100
	4.0 (4000)	E	⑩	67-74	19-24	⑪	54-64	13	15-18	15	100
	5.0 (4949)	P	⑨	61-68	19-24	110-130	75-85	⑫	26-32	15	100
2002	4.0 (4000)	E	⑬	72	②	⑭	75-85	7	16	15	100
	4.6 (4601)	W	⑮	⑯	⑰	⑱	59	18	15	12	100

* NOTE: Applies to Lower Manifold only.

① Step 1: 25 ft. lbs.
Step 2: 53 ft. lbs.
Step 3: Plus 90 degrees

② Step 1: 15 ft. lbs.
Step 2: +90 degrees

③ Step 1: 33 ft. lbs.
Step 2: +90 degrees

④ Step 1: 27 inch lbs.
Step 2: 89 inch lbs.
Step 3: 10 ft. lbs.
Step 4: 12 ft. lbs.

⑤ Step 1: 26 ft. lbs. (35 Nm)
Step 2: +90 degrees
Step 2: +90 degrees
8mm bolts: 24 ft. lbs. (32 Nm)

⑥ Step 1: 15 ft. lbs.(20 Nm)
Step 2: +90 degrees

⑦ Step 1: 44 ft. lbs.
Step 2: +90 degrees

⑧ Step 1: 19-25 ft. lbs
Step 2: +90 degrees

⑨ Step 1: 25-35 ft. lbs.
Step 2: 45-55 ft. lbs.
Step 3: Plus 90 degrees

⑩ Step 1: 28 ft. lbs.
Step 2: Plus 90 degrees
Step 3: Plus 90 degrees

⑪ Step 1: 20-28 ft. lbs.
Step 2: Loosen two turns
Step 3: 20-25 ft. lbs.

⑫ Step 1: 89 inch lbs. (10 Nm)
Step 2: 24 ft. lbs. (32 Nm)

⑬ Step 1: 24 ft. lbs.
Step 2: plus 90 degrees

⑭ Step 1: 37 ft. lbs.
Step 2: plus 90 degrees

⑮ Step 1: 30 ft. lbs.
Step 2: +90 degrees
Step 3: back off 1 full turn
Step 4: 30 ft. lbs.
Step 5: +90 degrees
Step 6: +90 additional degrees

⑯ Bedplate
Bolts 1-20: 106 inch lbs.
Bolts 1-10: 20 ft. lbs.
Bolts 11-20: 30 ft. lbs.
Cross-mounted bolts: Step 1, 30 ft. lbs.; Step 2, plus 90 deg

⑰ Step 1: 17 ft. lbs.
Step 2: 32 ft. lbs.
Step 3: +90-120 degrees

⑱ Step 1: 66 ft. lbs.
Step 2: back off 1 full turn
Step 3: 37 ft. lbs.
Step 4: +90 degrees

93591C81

WHEEL ALIGNMENT

Year	Model		Caster Range (+/-Deg.)	Caster Preferred Setting (Deg.)	Camber Range (+/-Deg.)	Camber Preferred Setting (Deg.)	Toe-in (in.)	Steering Axis Inclination (Deg.)
1999	Explorer		1.00	+4.20	0.50	-0.50	0.12+/-0.25	—
	Explorer Sport		1.00	①	0.70	-0.50	0.12+/-0.25	—
	Mountaineer		1.00	+4.20	0.50	-0.50	0.12+/-0.25	—
2000	Explorer		1.00	+4.20	0.50	-0.50	0.12+/-0.25	—
	Explorer Sport		1.00	①	0.70	-0.50	0.12+/-0.25	—
	Mountaineer		1.00	+4.20	0.50	-0.50	0.12+/-0.25	—
2001	Explorer		1.00	+4.20	0.50	-0.50	0.12+/-0.25	—
	Explorer Sport		1.00	①	0.70	-0.50	0.12+/-0.25	—
	Explorer Sport Trac		1.00	①	0.70	-0.50	0.12+/-0.25	—
	Mountaineer		1.00	+4.20	0.50	-0.50	0.12+/-0.25	—
2002	Explorer	F	1.00	②	0.80	-0.50	0.10+/-0.25	—
		R	—	—	0.80	-0.50	0.10+/-0.25	—
	Explorer Sport		1.00	①	0.70	-0.50	0.12+/-0.25	—
	Explorer Sport Trac		1.00	①	0.70	-0.50	0.12+/-0.25	—
	Mountaineer	F	1.00	②	0.80	-0.50	0.10+/-0.25	—
		R	—	—	0.80	-0.50	0.10+/-0.25	—

① Left side: +3.95
   Right side: +4.45

② Left side: +5.1
   Right side: +5.3

93591C82

Heater Core replacement is covered in Section 2 of this manual

TIRE, WHEEL AND BALL JOINT SPECIFICATIONS

Year	Model	OEM Tires		Tire Pressures (psi)		Wheel Size	Ball Joint Inspection
		Standard	Optional	Front	Rear		
1999	Explorer	P225/70R15	P235/75R15SL	30	35	7-JJ	0.030 in. ①
			P255/70R16	26	26	7-JJ	
	Mountaineer	P225/70R15	P235/75R15SL	30	35	7-JJ	0.030 in. ①
			P255/70R16	26	26	7-JJ	
2000	Explorer	P225/70R15	P235/75R15SL	30	35	7-JJ	0.030 in. ①
			P255/70R16	26	26	7-JJ	
	Mountaineer	P225/70R15	P235/75R15SL	30	35	7-JJ	0.030 in. ①
			P255/70R16	26	26	7-JJ	
2001	Explorer	P225/70R15	P235/70R15	②	②	7-JJ	0.030 in. ①
			P255/70R16	②	②	7-JJ	
	Explorer Sport	P235/70R15	P255/70R16	②	②	7-JJ	0.030 in. ①
	Explorer Sport Trac	P235/70R15	P255/70R16	②	②	7-JJ	0.030 in. ①
	Mountaineer	P225/70R15	P235/70R15	②	②	7-JJ	0.030 in. ①
			P255/70R16	②	②	7-JJ	
2002	Explorer	P225/70R16	P245/70R16	②	②	7-JJ	0.030 in. ①
			P255/70R16	②	②	7-JJ	
			P255/70HR16	②	②	7-JJ	
	Explorer Sport	P235/70R15	P255/70R16	②	②	7-JJ	0.030 in. ①
	Explorer Sport Trac	P235/70R15	P255/70R16	②	②	7-JJ	0.030 in. ①
	Mountaineer	P225/70R16	P245/70R16	②	②	7-JJ	0.030 in. ①
			P255/70R16	②	②	7-JJ	
			P255/70HR16	②	②	7-JJ	

OEM: Original Equipment Manufacturer

PSI: Pounds Per Square Inch

STD: Standard

OPT: Optional

① Both upper and lower

② See placard on vehicle

93591C83

BRAKE SPECIFICATIONS
All measurements in inches unless noted

Year	Model		Brake Disc			Brake Drum Diameter			Minimum Lining Thickness	Brake Caliper	
			Original Thickness	Minimum Thickness	Maximum Runout	Original Inside Diameter	Max. Wear Limit	Maximum Machine Diameter		Bracket Bolts (ft. lbs.)	Mounting Bolts (ft. lbs.)
1999	Explorer	F	1.020	0.980	0.0005	—	—	—	0.100	72-97	21-26
		R	0.480	0.440	0.0024	—	—	—	0.039	80	20
	Explorer Sport	F	1.020	0.980	0.0016	—	—	—	0.100	72-97	21-26
		R	0.480	0.440	0.0024	10.00	10.09	10.06	0.039	80	20
	Mountaineer	F	1.020	0.980	0.0005	—	—	—	0.100	72-97	21-26
		R	0.480	0.440	0.0024	—	—	—	0.039	80	20
2000	Explorer	F	1.020	0.980	0.0005	—	—	—	0.100	72-97	21-26
		R	0.480	0.440	0.0024	—	—	—	0.039	80	20
	Explorer Sport	F	1.020	0.980	0.0016	—	—	—	0.100	72-97	21-26
		R	0.480	0.440	0.0024	10.00	10.09	10.06	0.039	80	20
	Mountaineer	F	1.020	0.980	0.0005	—	—	—	0.100	72-97	21-26
		R	0.480	0.440	0.0024	—	—	—	0.039	80	20
2001	Explorer	F	1.020	0.980	0.0005	—	—	—	0.100	72-97	21-26
		R	0.480	0.440	0.0024	—	—	—	0.039	80	20
	Explorer Sport	F	1.020	0.980	0.0016	—	—	—	0.100	72-97	21-26
		R	0.480	0.440	0.0024	10.00	10.09	10.06	0.039	80	20
	Explorer Sport Trac	F	1.020	0.980	0.0005	—	—	—	0.100	72-97	21-26
		R	0.480	0.440	0.0024	—	—	—	0.039	80	20
	Mountaineer	F	1.020	0.980	0.0005	—	—	—	0.100	72-97	21-26
		R	0.480	0.440	0.0024	—	—	—	0.039	80	20
2002	Explorer	F	1.020	0.980	0.0005	—	—	—	0.100	83	24
		R	0.480	0.440	0.0024	—	—	—	0.039	—	24
	Explorer Sport	F	1.020	0.980	0.0016	—	—	—	0.100	72-97	21-26
		R	0.480	0.440	0.0024	10.00	10.09	10.06	0.039	80	20
	Explorer Sport Trac	F	1.020	0.980	0.0005	—	—	—	0.100	72-97	21-26
		R	0.480	0.440	0.0024	—	—	—	0.039	80	20
	Mountaineer	F	1.020	0.980	0.0005	—	—	—	0.100	72-97	21-26
		R	0.480	0.440	0.0024	—	—	—	0.039	80	20

NOTE: Due to changes made during production, refer to manufacturer's specifications if they differ from those in this chart

93591C84

Brake service is covered in Section 4 of this manual

SCHEDULED MAINTENANCE INTERVALS

1999-00 Ford Explorer, Explorer Sport, Explorer Sport-Trac, Mercury Mountaineer

TO BE SERVICED	TYPE OF SERVICE	5	10	15	20	25	30	35	40	45	50	55	60	65
Engine oil & filter	R	✓	✓	✓	✓	✓	✓	✓	✓	✓	✓	✓	✓	✓
Driveshaft fittings	L	✓	✓	✓	✓	✓	✓	✓	✓	✓	✓	✓	✓	✓
Exhaust system	I	✓	✓	✓	✓	✓	✓	✓	✓	✓	✓	✓	✓	✓
Cooling system hoses	I		✓				✓			✓			✓	
Coolant strength	I		✓				✓			✓			✓	
Brake caliper rails	L		✓				✓			✓			✓	
Air cleaner filter	R						✓						✓	
Front wheel bearings (2wd)	I/L						✓						✓	
Fuel filter ①	R										✓			
PCV valve	R												✓	
Accessory drive belts	S/I												✓	
Spark plugs	R						every 100,000 miles							
Transfer case fluid	R												✓	
Manual trans. fluid	R												✓	
Coolant ②	R										✓			
Differential fluid ③	R						every 100,000 miles							
Clutch reservoir level	I	✓	✓	✓	✓	✓	✓	✓	✓	✓	✓	✓	✓	✓
Brake hoses	I			✓			✓			✓			✓	
Parking brake system	I						✓						✓	

R: Replace   S: Service   I: Inspect   L: Lubricate

① Recommended, but not required in Calif.

② Change at 50,000 miles, then every 30,000 miles or 36 months

③ Except synthetic

FREQUENT OPERATION MAINTENANCE (SEVERE SERVICE)

If a vehicle is operated under any of the following conditions it is considered severe service:

- Towing a trailer or using a camper or car-top carrier.

- Repeated short trips of less than 5 miles in temperatures below freezing, or trips of less than 10 miles in any temperature.

- Extensive idling or low-speed driving for long distance as in heavy commercial use, such as delivery, taxi or police cars.

- Operating on rough, muddy or salt-covered roads.

- Operating on unpaved or dusty roads.

- Driving in extremely hot (over 90°) conditions.

Engine oil & filter: replace every 3000 miles.

Air cleaner filter: service or inspect every 6000 miles.

Exhaust system: check every 6000 miles.

Rotate tires every 9000 miles. (City delivery vehicles & other unique applications that require constant turning may need frequent tire rotation.)

Automatic transmission fluid & filter: change every 21,000 miles.

93591C85

SCHEDULED MAINTENANCE INTERVALS

2001-02 Ford Explorer, Explorer Sport, Explorer Sport-Trac, Mercury Mountaineer

TO BE SERVICED	TYPE OF SERVICE	VEHICLE MILEAGE INTERVAL (x1000)												
		5	10	15	20	25	30	35	40	45	50	55	60	65
Engine oil & filter	R	✓	✓	✓	✓	✓	✓	✓	✓	✓	✓	✓	✓	✓
Tires	Rotate	✓	✓	✓	✓	✓	✓	✓	✓	✓	✓	✓	✓	✓
Auto trans. fluid	I			✓			✓			✓			✓	
Brake pads/shoes	I			✓			✓			✓			✓	
Coolant hoses	S/I			✓			✓			✓			✓	
Steering linkage	I			✓			✓			✓			✓	
Cabin air filter	R			✓			✓			✓			✓	
Ball joints (2wd)	L			✓			✓			✓			✓	
Exhaust system	I						✓						✓	
Engine air filter	R						✓						✓	
Fuel filter ①	R						✓						✓	
Auto trans fluid (4-speed)	R						✓						✓	
Green coolant ②	R									✓				
Wheel bearings (2wd)	L												✓	
Manual trans. fluid	R												✓	
Spark plugs	R	every 100,000 miles											
PCV valve	R	every 100,000 miles											
Orange coolant	R	every 150,000 miles											
Auto trans fluid (5-speed)	R	every 150,000 miles											
Differential fluid	R	every 150,000 miles											
Accessory drive belts	R	every 150,000 miles											
Transfer case fluid	R	every 150,000 miles											

R: Replace   S: Service   I: Inspect   L: Lubricate

① Recommended, but not required in Calif.

② Change every 30,000 miles or 36 months thereafter

FREQUENT OPERATION MAINTENANCE (SEVERE SERVICE)

If a vehicle is operated under any of the following conditions it is considered severe service:

- Towing a trailer or using a camper or car-top carrier.

- Repeated short trips of less than 5 miles in temperatures below freezing, or trips of less than 10 miles in any temperature.

- Extensive idling or low-speed driving for long distance as in heavy commercial use, such as delivery, taxi or police cars.

- Operating on rough, muddy or salt-covered roads.

- Operating on unpaved or dusty roads.

- Driving in extremely hot (over 90°) conditions.

Engine oil & filter: replace every 3000 miles.
Air cleaner filter: service or inspect every 6000 miles.
Exhaust system: check every 6000 miles.
Automatic transmission fluid & filter: change every 30,000 miles.
Transfer case fluid: change every 60,000 miles
Fule filter: change every 15,000 miles
Spark plugs: change every 60,000 miles
2wd front wheel bearings: lubricate every 30,000 miles

93591C86

For complete Engine Mechanical specifications, see Section 1 of this manual

SCHEDULED MAINTENANCE INTERVALS
FORD MOTOR COMPANY
FORD EXPLORER, EXPLORER SPORT (1999-2000)
MERCURY MOUNTAINEER (1999-2000)

The following should be used as a guide when determining the amount of work required for a particular service.
In estimating how long a particular Scheduled Maintenance Service should take, please observe the following:

● Labor Time is time based on field research and data supplied by the vehicle manufacturer.
● Labor time operations are given in hours and tenths of an hour.
● All labor operations are to be used as a guide.

Mechanic Skill Level Codes:
(A) PRECISION: Highly skilled with multiple certification.
(B) GENERAL: Normally skilled with certification.
(C) MAINTENANCE: Semi-skilled working on certification.

	LABOR TIME		LABOR TIME		LABOR TIME
5000 Mile Service (C)		**30000 Mile Service (B)**		**50000 Mile Service (C)**	
All models	.9	All models	1.4	All models	.9
10000 Mile Service (C)		**35000 Mile Service (C)**		**55000 Mile Service (C)**	
All models	.9	All models	.9	All models	.9
15000 Mile Service (C)		**40000 Mile Service (C)**		**60000 Mile Service (B)**	
All models	1.0	All models	.9	All models	2.0
20000 Mile Service (C)		**45000 Mile Service (C)**		**65000 Mile Service (C)**	
All models	.9	All models	1.2	All models	.9
25000 Mile Service (C)					
All models	.9				

93591CEX

SCHEDULED MAINTENANCE INTERVALS
FORD MOTOR COMPANY
FORD EXPLORER, EXPLORER SPORT, EXPLORER SPORT-TRAC (2001-03)
MERCURY MOUNTAINEER (2001-03)

The following should be used as a guide when determining the amount of work required for a particular service. In estimating how long a particular Scheduled Maintenance Service should take, please observe the following:

● Labor Time is time based on field research and data supplied by the vehicle manufacturer.
● Labor time operations are given in hours and tenths of an hour.
● All labor operations are to be used as a guide.

Mechanic Skill Level Codes:
(A) PRECISION: Highly skilled with multiple certification.
(B) GENERAL: Normally skilled with certification.
(C) MAINTENANCE: Semi-skilled working on certification.

	LABOR TIME		LABOR TIME		LABOR TIME
5000 Mile Service (C)		**30000 Mile Service (B)**		**50000 Mile Service (C)**	
All models	.9	All models	1.4	All models	.9
10000 Mile Service (C)		**35000 Mile Service (C)**		**55000 Mile Service (C)**	
All models	.9	All models	.9	All models	.9
15000 Mile Service (C)		**40000 Mile Service (C)**		**60000 Mile Service (B)**	
All models	1.0	All models	.9	All models	2.2
20000 Mile Service (C)		**45000 Mile Service (C)**		**65000 Mile Service (C)**	
All models	.9	All models	1.2	All models	.9
25000 Mile Service (C)					
All models	.9				

93591CEY

For Accessory Drive Belt illustrations, see Section 1 of this manual

FORD MOTOR CO.
Ford Full-Size • Expedition • Excursion • Lincoln Navigator

ENGINE AND VEHICLE IDENTIFICATION

Engine							Model Year	
Code ①	Liters (cc)	Cu. In.	Cyl.	Fuel Sys.	Type	Eng. Mfg.	Code ②	Year
S	6.8 (6802)	415	10	MFI	SOHC	Ford	X	1999
6	4.6 (4588)	280	8	MFI	SOHC	Ford	Y	2000
F	7.3 (7292)	445	8	DI	OHV	Navistar	1	2001
L	5.4 (5409)	330	8	EFI	SOHC	Ford	2	2002
R	5.4 (5409)	330	8	EFI	DOHC	Ford	3	2003
A	5.4 (5409)	330	8	EFI	DOHC	Ford		
W	4.6 (4588)	280	8	MFI	SOHC	Ford		

MFI: Multi-port Fuel Injection

DI: Direct Injection Turbo-Diesel

EFI: Electronic Fuel Injection

OHV: Overhead Valve

SOHC: Single Overhead Camshaft

DOHC: Dual Overhead Camshaft

① 8th digit of the Vehicle Identification Number (VIN)

② 10th digit of the Vehicle Identification Number (VIN)

93591C87

GENERAL ENGINE SPECIFICATIONS

Year	Model	Engine Displacement Liters (cc)	Engine VIN	Fuel System Type	Net Horsepower @ rpm	Net Torque @ rpm (ft. lbs.)	Bore x Stroke (in.)	Compression Ratio	Oil Pressure @ rpm
1999	Expedition	4.6 (4588)	6/W	MFI	210@4400	290@3250	3.55x3.54	9.0:1	20-45@1500
	Expedition	5.4 (5409)	L	MFI	235@4250	330@3000	3.55X4.17	9.0:1	40-70@1500
2000	Excursion	5.4 (5409)	L	MFI	235@4250	330@3000	3.55X4.17	9.0:1	40-70@1500
	Excursion	6.8 (6802)	S	MFI	265@4250	410@2750	4.09X4.17	9.0:1	40-70@1500
	Excursion	7.3 (7292)	F	DI	210@3000	425@2000	4.11x4.18	17.5:1	40-70@3000
	Expedition	4.6 (4588)	6/W	MFI	210@4400	290@3250	3.55x3.54	9.0:1	20-45@1500
	Expedition	5.4 (5409)	L	MFI	235@4250	330@3000	3.55X4.17	9.0:1	40-70@1500
	Navigator	5.4 (5409)	L	EFI	300@5000	335@2750	3.55X4.17	9.0:1	40-70@1500
		5.4 (5409)	R	EFI	300@5000	335@2750	3.55X4.17	9.0:1	40-70@1500
		5.4 (5409)	A	EFI	300@5000	335@2750	3.55X4.17	9.0:1	40-70@1500
2001	Excursion	5.4 (5409)	L	MFI	235@4250	330@3000	3.55X4.17	9.0:1	40-70@1500
	Excursion	6.8 (6802)	S	MFI	265@4250	410@2750	4.09X4.17	9.0:1	40-70@1500
	Excursion	7.3 (7292)	F	DI	210@3000	425@2000	4.11x4.18	17.5:1	40-70@3000
	Expedition	4.6 (4588)	6/W	MFI	210@4400	290@3250	3.55x3.54	9.0:1	20-45@1500
	Expedition	5.4 (5409)	L	MFI	235@4250	330@3000	3.55X4.17	9.0:1	40-70@1500
	Navigator	5.4 (5409)	L	EFI	300@5000	335@2750	3.55X4.17	9.0:1	40-70@1500
		5.4 (5409)	R	EFI	300@5000	335@2750	3.55X4.17	9.0:1	40-70@1500
		5.4 (5409)	A	EFI	300@5000	335@2750	3.55X4.17	9.0:1	40-70@1500
2002-03	Excursion	5.4 (5409)	L	MFI	235@4250	330@3000	3.55X4.17	9.0:1	40-70@1500
	Excursion	6.8 (6802)	5	MFI	265@4250	410@2750	4.09X4.17	9.0:1	40-70@1500
	Excursion	7.3 (7292)	S	DI	210@3000	425@2000	4.11x4.18	17.5:1	40-70@3000
	Expedition	4.6 (4588)	6/W	MFI	210@4400	290@3250	3.55x3.54	9.0:1	20-45@1500
	Expedition	5.4 (5409)	L	MFI	235@4250	330@3000	3.55X4.17	9.0:1	40-70@1500
	Navigator	5.4 (5409)	L	EFI	300@5000	335@2750	3.55X4.17	9.0:1	40-70@1500
		5.4 (5409)	R	EFI	300@5000	335@2750	3.55X4.17	9.0:1	40-70@1500
		5.4 (5409)	A	EFI	300@5000	335@2750	3.55X4.17	9.0:1	40-70@1500

MFI: Multi-port Fuel Injection

EFI:Electronic Fuel Injection

DI: Direct Injection Turbo-Diesel

93591C88

For Tire, Wheel and Ball Joint specifications, see Section 1 of this manual

GASOLINE ENGINE TUNE-UP SPECIFICATIONS

Year	Engine Displacement Liters (cc)	Engine ID/VIN	Spark Plug Gap (in.)	Ignition Timing (deg.) ① MT	Ignition Timing (deg.) ① AT	Fuel Pump (psi) ②	Idle Speed (rpm) MT	Idle Speed (rpm) AT	Valve Clearance In.	Valve Clearance Ex.
1999	4.6 (4588)	W	0.052-0.056	—	10B	30-45	—	①	HYD	HYD
	4.6 (4588)	6	0.052-0.056	—	10B	30-45	—	①	HYD	HYD
	5.4 (5409)	L	0.052-0.056	—	10B	28-45	—	①	HYD	HYD
	6.8 (6802)	5	0.052-0.055	—	10B	28-45	—	①	HYD	HYD
2000	4.6 (4588)	W	0.052-0.056	—	10B	30-45	—	①	HYD	HYD
	4.6 (4588)	6	0.052-0.056	—	10B	30-45	—	①	HYD	HYD
	5.4 (5409)	L	0.052-0.056	—	10B	28-45	—	①	HYD	HYD
	6.8 (6802)	5	0.052-0.055	—	10B	28-45	—	①	HYD	HYD
2001	4.6 (4588)	W	0.052-0.056	—	10B	30-45	—	①	HYD	HYD
	4.6 (4588)	6	0.052-0.056	—	10B	30-45	—	①	HYD	HYD
	5.4 (5409)	L	0.052-0.056	—	10B	28-45	—	①	HYD	HYD
	6.8 (6802)	5	0.052-0.055	—	10B	28-45	—	①	HYD	HYD
2002-03	4.6 (4588)	W	0.052-0.056	—	10B	30-45	—	①	HYD	HYD
	4.6 (4588)	6	0.052-0.056	—	10B	30-45	—	①	HYD	HYD
	5.4 (5409)	L	0.052-0.056	—	10B	28-45	—	①	HYD	HYD
	6.8 (6802)	5	0.052-0.055	—	10B	28-45	—	①	HYD	HYD

NOTE: The Vehicle Emission Control Information label often reflects specification changes changes made during production. The label figures must be used if they differ from those in this chart.

B: Before top dead center

HYD: Hydraulic

NA: Not Available

① Idle speed and timing are electronically controlled and cannot be adjusted

② With engine running

93591C89

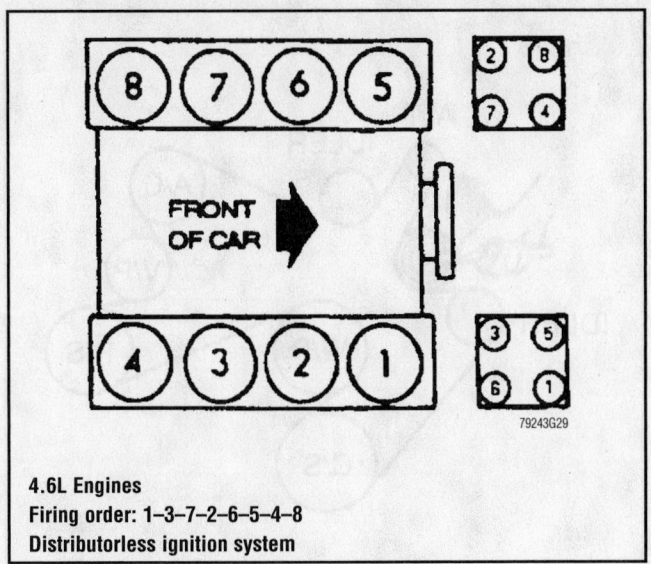

4.6L Engines
Firing order: 1–3–7–2–6–5–4–8
Distributorless ignition system

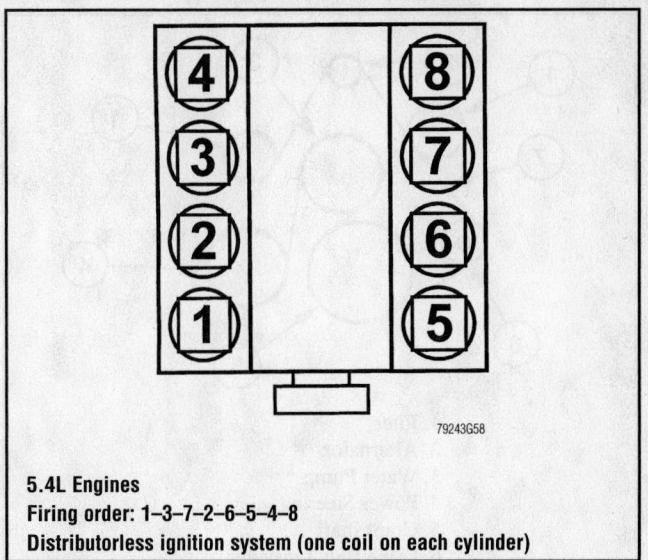

5.4L Engines
Firing order: 1–3–7–2–6–5–4–8
Distributorless ignition system (one coil on each cylinder)

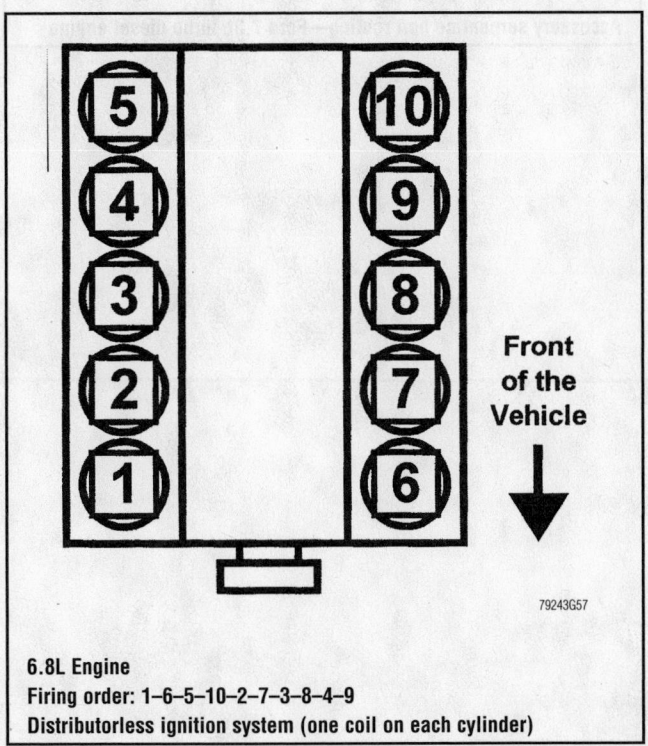

6.8L Engine
Firing order: 1–6–5–10–2–7–3–8–4–9
Distributorless ignition system (one coil on each cylinder)

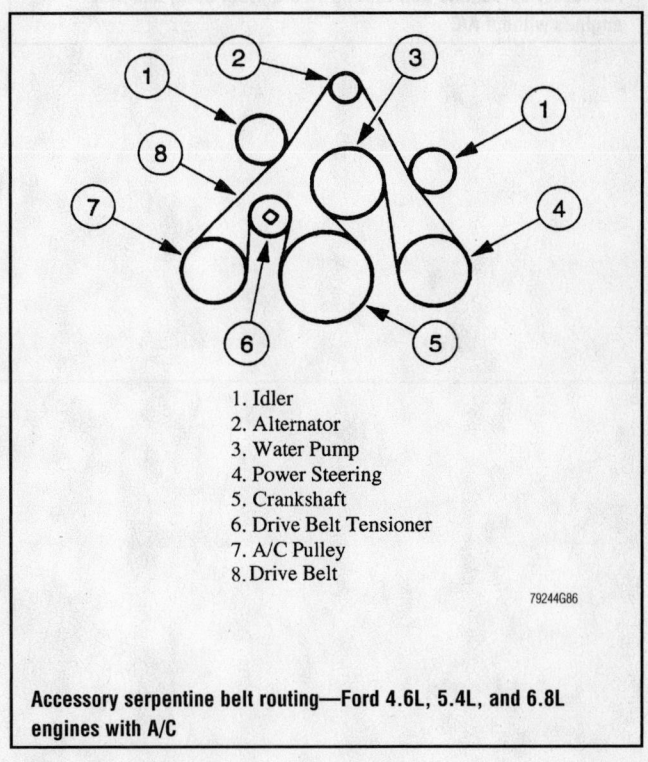

1. Idler
2. Alternator
3. Water Pump
4. Power Steering
5. Crankshaft
6. Drive Belt Tensioner
7. A/C Pulley
8. Drive Belt

Accessory serpentine belt routing—Ford 4.6L, 5.4L, and 6.8L engines with A/C

For Wheel Alignment specifications, see Section 1 of this manual

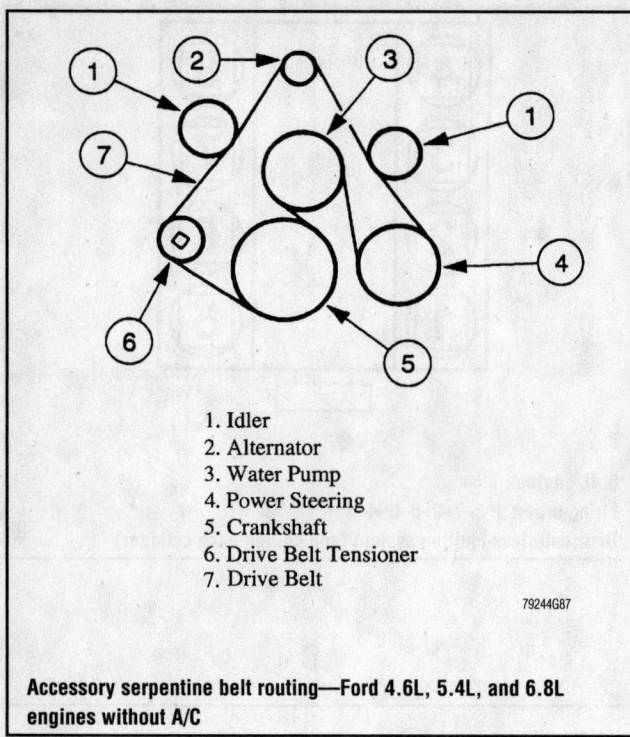

1. Idler
2. Alternator
3. Water Pump
4. Power Steering
5. Crankshaft
6. Drive Belt Tensioner
7. Drive Belt

79244G87

Accessory serpentine belt routing—Ford 4.6L, 5.4L, and 6.8L engines without A/C

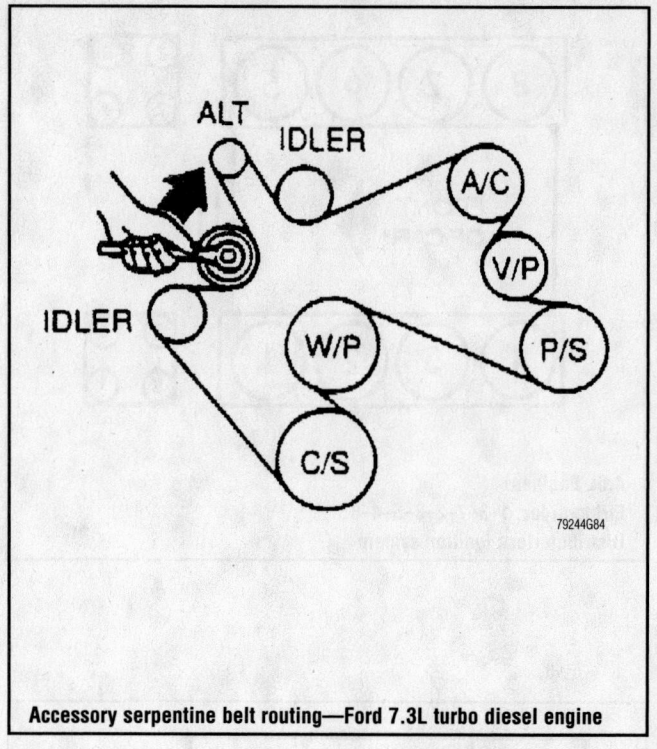

79244G84

Accessory serpentine belt routing—Ford 7.3L turbo diesel engine

DIESEL ENGINE TUNE-UP SPECIFICATIONS

Year	Engine ID/VIN	Engine Displacement cu. in. (cc)	Valve Clearance Intake (in.)	Valve Clearance Exhaust (in.)	Intake Valve Opens (deg.)	Injection Pump Setting (deg.)	Injection Nozzle Pressure (psi) New	Injection Nozzle Pressure (psi) Used	Idle Speed (rpm)	Cranking Compression Pressure (psi)
2000	F	7.3 (7292)	HYD	HYD	—	①	1875	1425	②	③
2001	F	7.3 (7292)	HYD	HYD	—	①	1875	1425	②	③
2002-03	F	7.3 (7292)	HYD	HYD	—	①	1875	1425	②	③

NOTE: The Vehicle Emission Control Information label often reflects specification changes made during production. The label figures must be used if they differ from those in this chart

HYD: Hydraulic

B: Before top dead center

NA: Not Available

① PCM controlled

② See underhood emission label

③ Compression pressure in the lowest cylinder must be at least 75% of the highest cylinder

Minimum pressure: 195 psi

Maximum pressure: 440 psi

93591C91

CAPACITIES

Year	Model	Engine Displacement Liters (cc)	Engine ID/VIN	Engine Oil with Filter (qts.)	Transmission (pts.) 5-Spd	Transmission (pts.) Auto.	Transfer Case (pts.)	Drive Axle Front (pts.)	Drive Axle Rear (pts.)	Fuel Tank (gal.)	Cooling System (qts.)
1999	Expedition	4.6 (4588)	6/W	6.0	—	①	4.0	3.7	5.5	24.5 ②	17.9
	Expedition	5.4 (5409)	L	6.0	—	①	4.0	3.7	5.5	24.5 ②	20.8
	Navigator	5.4 (5409)	L/R/A	6.0	—	①	4.0	3.7	5.5	24.5 ②	20.8
2000	Excursion	5.4 (5409)	L	6.0	—	①	4.0	6.0	6.0	44.0	26.4
	Excursion	6.8 (6802)	S	6.0	—	①	4.0	6.0 ③	6.0 ③	44.0	28.5
	Excursion	7.3 (7292)	F	14.0	—	①	4.0	6.0 ③	6.0 ③	44.0	23.0
	Expedition	4.6 (4588)	6/W	6.0	—	①	4.0	3.7	5.5	24.5 ②	17.9
	Expedition	5.4 (5409)	L	6.0	—	①	4.0	3.7	5.5	24.5 ②	20.8
	Navigator	5.4 (5409)	L/R/A	6.0	—	①	4.0	3.7	5.5	24.5 ②	20.8
2001	Excursion	5.4 (5409)	L	6.0	—	①	4.0	6.0	6.0	44.0	26.4
	Excursion	6.8 (6802)	S	6.0	—	①	4.0	6.0 ③	6.0 ③	44.0	28.5
	Excursion	7.3 (7292)	F	14.0	—	①	4.0	6.0 ③	6.0 ③	44.0	23.0
	Expedition	4.6 (4588)	6/W	6.0	—	①	4.0	3.7	5.5	24.5 ②	17.9
	Expedition	5.4 (5409)	L	6.0	—	①	4.0	3.7	5.5	24.5 ②	20.8
	Navigator	5.4 (5409)	L/R/A	6.0	—	①	4.0	3.7	5.5	30	20.8
2002-03	Excursion	5.4 (5409)	L	6.0	—	①	4.0	6.0	6.0	44.0	26.4
	Excursion	6.8 (6802)	S	6.0	—	①	4.0	6.0 ③	6.0 ③	44.0	28.5
	Excursion	7.3 (7292)	F	14.0	—	①	4.0	6.0 ③	6.0 ③	44.0	23.0
	Expedition	4.6 (4588)	6/W	6.0	—	①	4.0	3.7	5.5	24.5 ②	17.9
	Expedition	5.4 (5409)	L	6.0	—	①	4.0	3.7	5.5	24.5 ②	20.8
	Navigator	5.4 (5409)	L/R/A	6.0	—	①	4.0	3.7	5.5	30	20.8

NOTE: All capacities are approximate. Add fluid gradually and check to be sure a proper fluid level is obtained.

① With 4R70W: 31 pts.
   With E40D: 32.0 pts.
   With 4R100: 32.0 pts.

② Also available with a 30 gallon

③ Heavy duty: 7.5 pts.

93591C92

VALVE SPECIFICATIONS

Year	Engine Displacement Liters (cc)	Engine ID/VIN	Seat Angle (deg.)	Face Angle (deg.)	Spring Test Pressure (lbs. @ in.)	Spring Installed Height (in.)	Stem-to-Guide Clearance (in.)		Stem Diameter (in.)	
							Intake	Exhaust	Intake	Exhaust
1999	4.6 (4588)	6/W	44.75	45.67	NA	1.566-1.637	0.0450-0.0900	0.0015-0.0033	0.2738-0.2728	0.2751-0.2741
	5.4 (5409)	L/R/A	45	45.5	150@1.10	1.570	0.0008-0.0027	0.0018-0.0037	0.275-0.2746	0.274-0.2736
	6.8 (6802)	S	44.50-45.25	45.25-45.75	150@1.10	1.570	0.0008-0.0027	0.0018-0.0037	0.275-0.2746	0.274-0.2735
2000	4.6 (4588)	6/W	44.75	45.67	NA	1.566-1.637	0.0450-0.0900	0.0015-0.0033	0.2738-0.2728	0.2751-0.2741
	5.4 (5409)	L/R/A	45	45.5	150@1.10	1.570	0.0008-0.0027	0.0018-0.0037	0.275-0.2746	0.274-0.2736
	6.8 (6802)	S	44.50-45.25	45.25-45.75	150@1.10	1.570	0.0008-0.0027	0.0018-0.0037	0.275-0.2746	0.274-0.2735
	7.3 (7292)	F	②	②	200@1.38	③	0.0055	0.0055	0.3119-0.3126	0.3119-0.3126
2001	4.6 (4588)	6/W	44.75	45.67	NA	1.566-1.637	0.0450-0.0900	0.0015-0.0033	0.2738-0.2728	0.2751-0.2741
	5.4 (5409)	L/R/A	45	45.5	150@1.10	1.570	0.0008-0.0027	0.0018-0.0037	0.275-0.2746	0.274-0.2736
	6.8 (6802)	S	44.50-45.25	45.25-45.75	150@1.10	1.570	0.0008-0.0027	0.0018-0.0037	0.275-0.2746	0.274-0.2735
	7.3 (7292)	F	②	②	200@1.38	③	0.0055	0.0055	0.3119-0.3126	0.3119-0.3126
2002-03	4.6 (4588)	6/W	44.75	45.67	NA	1.566-1.637	0.0450-0.0900	0.0015-0.0033	0.2738-0.2728	0.2751-0.2741
	5.4 (5409)	L/R/A	45	45.5	150@1.10	1.570	0.0008-0.0027	0.0018-0.0037	0.275-0.2746	0.274-0.2736
	6.8 (6802)	S	44.50-45.25	45.25-45.75	150@1.10	1.570	0.0008-0.0027	0.0018-0.0037	0.275-0.2746	0.274-0.2735
	7.3 (7292)	F	②	②	200@1.38	③	0.0055	0.0055	0.3119-0.3126	0.3119-0.3126

② Intake: 30 degrees
   Exhaust: 37.5 degrees
③ Intake: 1.767 in.
   Exhaust: 1.833 in.

93591C93

For Tune-up, Capacities and Firing orders, see Section 1 of this manual

CRANKSHAFT AND CONNECTING ROD SPECIFICATIONS

All measurements are given in inches.

Year	Engine Displacement Liters (cc)	Engine ID/VIN	Crankshaft				Connecting Rod		
			Main Brg. Journal Dia.	Main Brg. Oil Clearance	Shaft End-play	Thrust on No.	Journal Diameter	Oil Clearance	Side Clearance
1999	4.6 (4588)	6/W	2.6500-2.6570	0.0011-0.0026	0.0051-0.0120	5	2.0870-2.8670	0.0011-0.0026	0.0006-0.0177
	5.4 (5409)	L/R/A	2.6568-2.6576	0.0009-0.0019	0.0015-0.0030	5	2.0859-2.0867	0.0010-0.0025	0.0006-0.0177
	6.8 (6802)	S	2.6568-2.6576	0.0009-0.0019	0.0015-0.0030	5	2.0859-2.0867	0.0010-0.0025	0.0006-0.0177
2000	4.6 (4588)	6/W	2.6500-2.6570	0.0011-0.0026	0.0051-0.0120	5	2.0870-2.8670	0.0011-0.0026	0.0006-0.0177
	5.4 (5409)	L/R/A	2.6568-2.6576	0.0009-0.0019	0.0015-0.0030	5	2.0859-2.0867	0.0010-0.0025	0.0006-0.0177
	6.8 (6802)	S	2.6568-2.6576	0.0009-0.0019	0.0015-0.0030	5	2.0859-2.0867	0.0010-0.0025	0.0006-0.0177
	7.3 (7292)	F	3.1228-3.1236	0.0018-0.0036	0.0025-0.0085	4	2.4980-2.4990	0.0015-0.0045	0.0120-0.0240
2001	4.6 (4588)	6/W	2.6500-2.6570	0.0011-0.0026	0.0051-0.0120	5	2.0870-2.8670	0.0011-0.0026	0.0006-0.0177
	5.4 (5409)	L/R/A	2.6568-2.6576	0.0009-0.0019	0.0015-0.0030	5	2.0859-2.0867	0.0010-0.0025	0.0006-0.0177
	6.8 (6802)	S	2.6568-2.6576	0.0009-0.0019	0.0015-0.0030	5	2.0859-2.0867	0.0010-0.0025	0.0006-0.0177
	7.3 (7292)	F	3.1228-3.1236	0.0018-0.0036	0.0025-0.0085	4	2.4980-2.4990	0.0015-0.0045	0.0120-0.0240
2002-03	4.6 (4588)	6/W	2.6500-2.6570	0.0011-0.0026	0.0051-0.0120	5	2.0870-2.8670	0.0011-0.0026	0.0006-0.0177
	5.4 (5409)	L/R/A	2.6568-2.6576	0.0009-0.0019	0.0015-0.0030	5	2.0859-2.0867	0.0010-0.0025	0.0006-0.0177
	6.8 (6802)	S	2.6568-2.6576	0.0009-0.0019	0.0015-0.0030	5	2.0859-2.0867	0.0010-0.0025	0.0006-0.0177
	7.3 (7292)	F	3.1228-3.1236	0.0018-0.0036	0.0025-0.0085	4	2.4980-2.4990	0.0015-0.0045	0.0120-0.0240

93591C94

PISTON AND RING SPECIFICATIONS
All measurements are given in inches.

Year	Engine Displacement Liters (cc)	Engine ID/VIN	Piston Clearance	Ring Gap			Ring Side Clearance		
				Top Compression	Bottom Compression	Oil Control	Top Compression	Bottom Compression	Oil Control
1999	4.6 (4588)	6/W	0.0005-0.0010	0.010-0.020	0.010-0.020	0.006-0.026	0.0016-0.0031	0.0012-0.0031	SNUG
	5.4 (5409)	L/R/A	0.0000-0.0010	0.005-0.011	0.098-0.015	0.006-0.026	0.0012-0.0037	0.0012-0.0037	SNUG
	6.8 (6802)	S	0.0000-0.0010	0.005-0.011	0.010-0.016	0.006-0.026	0.0012-0.0037	0.0012-0.0037	SNUG
2000	4.6 (4588)	6/W	0.0005-0.0010	0.010-0.020	0.010-0.020	0.006-0.026	0.0016-0.0031	0.0012-0.0031	SNUG
	5.4 (5409)	L/R/A	0.0000-0.0010	0.005-0.011	0.098-0.015	0.006-0.026	0.0012-0.0037	0.0012-0.0037	SNUG
	6.8 (6802)	S	0.0000-0.0010	0.005-0.011	0.010-0.016	0.006-0.026	0.0012-0.0037	0.0012-0.0037	SNUG
	7.3 (7292)	F	0.0044-0.0057	0.014-0.024	0.062-0.072	0.012-0.024	0.0013-0.0033	0.0013-0.0033	SNUG
2001	4.6 (4588)	6/W	0.0005-0.0010	0.010-0.020	0.010-0.020	0.006-0.026	0.0016-0.0031	0.0012-0.0031	SNUG
	5.4 (5409)	L/R/A	0.0000-0.0010	0.005-0.011	0.098-0.015	0.006-0.026	0.0012-0.0037	0.0012-0.0037	SNUG
	6.8 (6802)	S	0.0000-0.0010	0.005-0.011	0.010-0.016	0.006-0.026	0.0012-0.0037	0.0012-0.0037	SNUG
	7.3 (7292)	F	0.0044-0.0057	0.014-0.024	0.062-0.072	0.012-0.024	0.0013-0.0033	0.0013-0.0033	SNUG
2002-03	4.6 (4588)	6/W	0.0005-0.0010	0.010-0.020	0.010-0.020	0.006-0.026	0.0016-0.0031	0.0012-0.0031	SNUG
	5.4 (5409)	L/R/A	0.0000-0.0010	0.005-0.011	0.098-0.015	0.006-0.026	0.0012-0.0037	0.0012-0.0037	SNUG
	6.8 (6802)	S	0.0000-0.0010	0.005-0.011	0.010-0.016	0.006-0.026	0.0012-0.0037	0.0012-0.0037	SNUG
	7.3 (7292)	F	0.0044-0.0057	0.014-0.024	0.062-0.072	0.012-0.024	0.0013-0.0033	0.0013-0.0033	SNUG

93591C95

TORQUE SPECIFICATIONS
All readings in ft. lbs.

	Engine Displacement Liters (cc)	Engine ID/VIN	Cylinder Head Bolts	Main Bearing Bolts	Rod Bearing Bolts	Crankshaft Damper Bolts	Flywheel Bolts	Manifold Intake *	Exhaust	Spark Plugs	Lug Nut
1999	4.6 (4588)	6/W	①	②	29-33	③	54-64	④	18	7-14	83-113
	5.4 (5409)	L/R/A	⑤	⑥	⑦	③	54-64	⑧	18	9-20	100
	6.8 (6802)	S	⑤	⑥	⑥	③	54-64	⑧	17-20	7-14	140
2000	4.6 (4588)	6/W	①	②	29-33	③	54-64	④	18	7-14	83-113
	5.4 (5409)	L/R/A	⑤	⑥	⑥	③	54-64	⑧	18	9-20	100
	6.8 (6802)	S	⑤	⑥	⑥	③	54-64	⑧	17-20	7-14	140
	7.3 (7292)	F	⑨	95	70	90	89	18	45	—	140
2001	4.6 (4588)	6/W	①	②	29-33	③	54-64	④	18	7-14	83-113
	5.4 (5409)	L/R/A	⑤	⑥	⑥	③	54-64	⑧	18	9-20	100
	6.8 (6802)	S	⑤	⑥	⑥	③	54-64	⑧	17-20	7-14	140
	7.3 (7292)	F	⑨	95	70	90	89	18	45	—	140
2002-03	4.6 (4588)	6/W	①	②	29-33	③	54-64	④	18	7-14	83-113
	5.4 (5409)	L/R/A	⑤	⑥	⑥	③	54-64	⑧	18	9-20	100
	6.8 (6802)	S	⑤	⑥	⑥	③	54-64	⑧	17-20	7-14	140
	7.3 (7292)	F	⑨	95	70	90	89	18	45	—	140

* NOTE: Applies to Lower Manifold only.

① Step 1: 30 ft. lbs.
  Step 2: Plus 85-95 degrees
  Step 3: Plus 85-95 degrees

② Jack screws:
  Step 1: 45 inch lbs.
  Step 2: 98 inch lbs.
  Cross-mounted cap bolts:
  Step 1: 24 ft. lbs.
  Step 2: Plus 90 degrees

③ Step 1: 88 ft. lbs.
  Step 2: Loosen bolt
  Step 3: 39 ft. lbs.
  Step 4: Plus 90 degrees

④ Step 1: 18 inch lbs.
  Step 2: 89 inch lbs. lbs.

⑤ Step 1: 27-32 inch lbs.
  Step 2: Plus 90 degrees
  Step 3: Plus 90 degrees

⑥ Step 1: 27-32 ft. lbs.
  Step 2: Plus 90 degrees

⑦ Step 1: 30-33 ft. lbs.
  Step 2: 90-120 degrees

⑧ Step 1: 18 inch lbs.
  Step 2: 71-106 inch lbs.

⑨ Step 1: 65 ft. lbs.
  Step 2: 85 ft. lbs.
  Step 3: 105 ft. lbs.

93591C96

WHEEL ALIGNMENT

Year	Model		Caster Range (+/-Deg.)	Caster Preferred Setting (Deg.)	Camber Range (+/-Deg.)	Camber Preferred Setting (Deg.)	Toe-in (in.)	Steering Axis Inclination (Deg.)
1999	Expedition	2WD base	1.00	+6.10	0.70	-0.30	0.06+/-0.25	—
		2WD air	1.00	+6.10	0.70	-0.30	0.06+/-0.25	—
		4WD base	1.00	+5.10	0.70	-0.20	0.30+/-0.25	—
		4WD air	1.00	+5.00	0.40	-0.40	0.20+/-0.25	—
	Navigator	2WD	1.0	+6.10	0.75	-0.33	0.06+/-0.25	—
		4WD	1.0	+5.10	0.40	-0.40	0.20+/-0.25	—
2000	Excursion	2WD	2.00	+4.00	1.00	+0.62	0.03+/-0.25	—
		4WD	2.00	+3.50	1.00	+0.25	0.03+/-0.25	—
	Expedition	2WD base	1.00	+6.10	0.70	-0.30	0.06+/-0.25	—
		2WD air	1.00	+6.10	0.70	-0.30	0.06+/-0.25	—
		4WD base	1.00	+5.10	0.70	-0.20	0.30+/-0.25	—
		4WD air	1.00	+5.00	0.40	-0.40	0.20+/-0.25	—
	Navigator	2WD	1.0	+6.10	0.75	-0.33	0.06+/-0.25	—
		4WD	1.0	+5.10	0.40	-0.40	0.20+/-0.25	—
2001	Excursion	2WD	2.00	+4.00	1.00	+0.62	0.03+/-0.25	—
		4WD	2.00	+3.50	1.00	+0.25	0.03+/-0.25	—
	Expedition	2WD base	1.00	+6.10	0.70	-0.30	0.06+/-0.25	—
		2WD air	1.00	+6.10	0.70	-0.30	0.06+/-0.25	—
		4WD base	1.00	+5.10	0.70	-0.20	0.30+/-0.25	—
		4WD air	1.00	+5.00	0.40	-0.40	0.20+/-0.25	—
	Navigator	2WD	1.0	+6.10	0.75	-0.33	0.06+/-0.25	—
		4WD	1.0	+5.10	0.40	-0.40	0.20+/-0.25	—
2002-03	Excursion	2WD	2.00	+4.00	1.00	+0.62	0.03+/-0.25	—
		4WD	2.00	+3.50	1.00	+0.25	0.03+/-0.25	—
	Expedition	2WD base	1.00	+6.10	0.70	-0.30	0.06+/-0.25	—
		2WD air	1.00	+6.10	0.70	-0.30	0.06+/-0.25	—
		4WD base	1.00	+5.10	0.70	-0.20	0.30+/-0.25	—
		4WD air	1.00	+5.00	0.40	-0.40	0.20+/-0.25	—
	Navigator	2WD	1.0	+6.10	0.75	-0.33	0.06+/-0.25	—
		4WD	1.0	+5.10	0.40	-0.40	0.20+/-0.25	—

93591C97

TIRE, WHEEL AND BALL JOINT SPECIFICATIONS

| Year | Model | OEM Tires | | Tire Pressures (psi) | | Wheel Size | Ball Joint Inspection |
		Standard	Optional	Front	Rear		
1999	Expedition	P255/70R16	P265/70R17	35	35	7-JJ	0.030 in. ①
	Navigator	P245/75R16	P255/75R17	30	35	7.5	0.030 in. ①
2000	Expedition	P255/70R16	P265/70R17	35	35	7-JJ	0.030 in. ①
	Excursion	P255/70R16	P265/70R17	35	35	7-JJ	0.030 in. ①
	Navigator	P245/75R16	P255/75R17	30	35	7.5	0.030 in. ①
2001	Expedition	P255/70R16	P265/70R17	35	35	7-JJ	0.030 in. ①
	Excursion	P255/70R16	P265/70R17	35	35	7-JJ	0.030 in. ①
	Navigator	P245/75R16	P255/75R17	30	35	7.5	0.030 in. ①
2002-03	Expedition	P255/70R16	P265/70R17	35	35	7-JJ	0.030 in. ①
	Excursion	P255/70R16	P265/70R17	35	35	7-JJ	0.030 in. ①
	Navigator	P245/75R16	P255/75R17	30	35	7.5	0.030 in. ①

OEM: Original Equipment Manufacturer

PSI: Pounds Per Square Inch

STD: Standard

OPT: Optional

① Both upper and lower

93591C98

BRAKE SPECIFICATIONS
All measurements in inches unless noted

Year	Model		Brake Disc Original Thickness	Brake Disc Minimum Thickness	Brake Disc Maximum Runout	Brake Drum Diameter Original Inside Diameter	Brake Drum Diameter Max. Wear Limit	Brake Drum Diameter Maximum Machine Diameter	Minimum Lining Thickness	Brake Caliper Bracket Bolts (ft. lbs.)	Brake Caliper Mounting Bolts (ft. lbs.)
1999	Expedition	F	1.023	0.964	0.0025	—	—	—	0.030	125-168	21-26
		R	0.700	0.657	0.0250	—	—	—	0.030	120	20
	Navigator	F	1.023	0.964	0.0025	—	—	—	0.030	125-168	21-26
		R	0.700	0.657	0.0250	—	—	—	0.030	120	20
2000	Excursion		1.220	1.180	0.0025	12.00	12.09	12.06	0.030	166	42
	Expedition	F	1.023	0.964	0.0025	—	—	—	0.030	125-168	21-26
		R	0.700	0.657	0.0250	—	—	—	0.030	120	20
	Navigator	F	1.023	0.964	0.0025	—	—	—	0.030	125-168	21-26
		R	0.700	0.657	0.0250	—	—	—	0.030	120	20
2001	Excursion		1.220	1.180	0.0025	12.00	12.09	12.06	0.030	166	42
	Expedition	F	1.023	0.964	0.0025	—	—	—	0.030	125-168	21-26
		R	0.700	0.657	0.0250	—	—	—	0.030	120	20
	Navigator	F	1.023	0.964	0.0025	—	—	—	0.030	125-168	21-26
		R	0.700	0.657	0.0250	—	—	—	0.030	120	20
2002-03	Excursion		1.220	1.180	0.0025	12.00	12.09	12.06	0.030	166	42
	Expedition	F	1.023	0.964	0.0025	—	—	—	0.030	125-168	21-26
		R	0.700	0.657	0.0250	—	—	—	0.030	120	20
	Navigator	F	1.023	0.964	0.0025	—	—	—	0.030	125-168	21-26
		R	0.700	0.657	0.0250	—	—	—	0.030	120	20

NOTE: Due to changes made during production, refer to manufacturer's specifications if they differ from those in this chart

F: Front

R: Rear

93591C99

Timing belt service is covered in Section 3 of this manual

SCHEDULED MAINTENANCE INTERVALS
FORD—EXPEDITION & LINCOLN—NAVIGATOR

TO BE SERVICED	TYPE OF SERVICE	5	10	15	20	25	30	35	40	45	50	55	60	65	70	75	80	85	90	95	100	105	110	115	120
Accessory drive belt	S/I												✓												✓
Air cleaner filter ①	R						✓						✓						✓			✓			✓
Automatic transmission fluid	R						✓						✓						✓			✓			✓
Automatic transmission shift linkage	S/I & L	✓	✓	✓	✓	✓	✓	✓	✓	✓	✓	✓	✓	✓	✓	✓	✓	✓	✓	✓	✓	✓	✓	✓	✓
Brake caliper, slide rails	L			✓			✓			✓			✓			✓			✓			✓			✓
Brake system, hoses & lines	S/I			✓			✓			✓			✓			✓			✓			✓			✓
Clutch reservoir fluid level	S/I	✓	✓	✓	✓	✓	✓	✓	✓	✓	✓	✓	✓	✓	✓	✓	✓	✓	✓	✓	✓	✓	✓	✓	✓
Engine coolant ②	R										✓					✓							✓		
Engine cooling system hoses, clamps & coolant	S/I			✓			✓			✓			✓			✓			✓			✓			✓
Engine oil & filter	R	✓	✓	✓	✓	✓	✓	✓	✓	✓	✓	✓	✓	✓	✓	✓	✓	✓	✓	✓	✓	✓	✓	✓	✓
Exhaust system	S/I	✓	✓	✓	✓	✓	✓	✓	✓	✓	✓	✓	✓	✓	✓	✓	✓	✓	✓	✓	✓	✓	✓	✓	✓
Front wheel bearings	S/I & L						✓						✓						✓						✓
Front/rear axle driveshaft slip yoke	L						✓						✓						✓						✓
Front/rear axle fluid ③	R																				✓				
Fuel filter	R			✓			✓			✓			✓			✓			✓			✓			✓
Manual transmission fluid	R												✓												✓
Parking brake system	S/I						✓						✓						✓						✓
PCV valve	R												✓												✓
Rotate tires	S/I	✓	✓	✓	✓	✓	✓	✓	✓	✓	✓	✓	✓	✓	✓	✓	✓	✓	✓	✓	✓	✓	✓	✓	✓

93591CA1

SCHEDULED MAINTENANCE INTERVALS
FORD—EXPEDITION & LINCOLN—NAVIGATOR

TO BE SERVICED	TYPE OF SERVICE	VEHICLE MILEAGE INTERVAL (x1000)																						
		5	10	15	20	25	30	35	40	45	50	55	60	65	70	75	80	85	90	95	100	105	110	115	120
Spark plugs	R																				✓				
Steering linkage, suspension, driveshaft U joints	S/I & L	✓	✓	✓	✓	✓	✓	✓	✓	✓	✓	✓	✓	✓	✓	✓	✓	✓	✓	✓	✓	✓	✓	✓	✓

R: Replace   S/I: Service or Inspect

① Perform this at the mileage shown or every 30 months, whichever occurs first.

② Drain, flush and refill the cooling system initially at 50,000 miles or 48 months, whichever occurs first, then every 30,000 miles or 30 months thereafter.

③ The axle lubricant must be replaced every 100,000 miles or if the axle has been submerged under water. Otherwise the lube should not be checked or changed unless a repair is required.

FREQUENT OPERATION MAINTENANCE (SEVERE SERVICE)

If a vehicle is operated under any of the following conditions it is considered severe service:

- Towing a trailer or using a camper or car-top carrier.
- Repeated short trips of less than 5 miles in temperatures below freezing, or trips of less than 10 miles in any temperature.
- Extensive idling or low-speed driving for long distance as in heavy commercial use, such as delivery, taxi or police cars.
- Operating on rough, muddy or salt-covered roads.
- Operating on unpaved or dusty roads.
- Driving in extremely hot (over 90°) conditions.

Engine oil & filter: replace every 3000 miles.

Tires: rotate and inspect every 6000 miles.

Clutch reservoir fluid level: inspect every 6000 miles.

Automatic transmission shift linkage: lubricate every 6000 miles.

Steering linkage: suspension, U-joints: lubricate every 6000 miles.

Exhaust system: inspect for leaks of damage every 6000 miles.

Fuel filter: replace every 15,000 miles.

Automatic transmission fluid: change every 21,000 miles.

Crankcase emission air filter: replace every 60,000 miles.

PCV valve: replace every 60,000 miles.

Accessory drive belt: inspect every 60,000 miles.

Spark plugs: replace every 99,000 miles.

93591CA2

Heater Core replacement is covered in Section 2 of this manual

SCHEDULED MAINTENANCE INTERVALS
FORD—EXCURSION

TO BE SERVICED	TYPE OF SERVICE	\multicolumn{20}{c}{VEHICLE MILEAGE INTERVAL (x1000)}																		
		5	10	15	20	25	30	35	40	45	50	55	60	65	70	75	80	85	90	95	100
Accessory drive belt	S/I																				✓
Air cleaner filter ①②	R						✓						✓						✓		
Automatic transmission fluid ③	R						✓						✓						✓		
Engine coolant ④⑤	R										✓						✓				
Engine cooling system hoses, clamps & coolant ⑥	S/I			✓			✓			✓			✓			✓			✓		
Engine oil & filter	R	✓	✓	✓	✓	✓	✓	✓	✓	✓	✓	✓	✓	✓	✓	✓	✓	✓	✓	✓	✓
Exhaust system	S/I			✓			✓			✓			✓			✓			✓		
Front wheel bearings	S/I & L																		✓		
Front/rear axle lubricant ⑦	R																				✓
Fuel filter ⑧	R						✓						✓						✓		
PCV valve	R	\multicolumn{20}{c}{Every 120,000 miles}																			
Rotate tires	S/I	✓	✓	✓	✓	✓	✓	✓	✓	✓	✓	✓	✓	✓	✓	✓	✓	✓	✓	✓	✓
Spark plugs	R																				✓
Steering linkage, suspension, driveshaft U joints	S/I & L	✓	✓	✓	✓	✓	✓	✓	✓	✓	✓	✓	✓	✓	✓	✓	✓	✓	✓	✓	✓

R: Replace   S/I: Service or Inspect

① Perform this at the mileage shown or every 30 months, whichever occurs first.

② 7.3L DIT Diesel engine: the air filter should be replaced when the restriction gauge is in the red zone.

③ Except the E40D transmission.

④ Drain, flush and refill the cooling system initially at 50,000 miles or 48 months, whichever occurs first, then every 30,000 miles or 30 months thereafter.

⑤ 7.3L DIT Diesel engine: add 4 pints of FW-15 each time the coolant is replaced.

⑥ 7.3L DIT Diesel engine: add 8-10 oz. of FW-15 to the engine coolant every 15,000 miles.

⑦ The axle lubricant must be replaced every 100,000 miles of if the axle has been submerged under water. Otherwise the lube should not be checked or changed unless a repair is required.

⑧ 7.3L DIT Diesel engine: the fuel filter should be replaced when the restriction lamp is illuminated.

FREQUENT OPERATION MAINTENANCE (SEVERE SERVICE)

If a vehicle is operated under any of the following conditions it is considered severe service:
- Towing a trailer or using a camper or car-top carrier.
- Repeated short trips of less than 5 miles in temperatures below freezing, or trips of less than 10 miles in any temperature.
- Extensive idling or low-speed driving for long distances as in heavy commercial use, such as delivery, taxi or police cars.
- Operating on rough, muddy or salt-covered roads.
- Operating on unpaved or dusty roads.

Engine oil & filter: replace every 3000 miles.
Tires: rotate and inspect every 6000 miles.
Steering linkage, suspension, U-joints: lubricate every 6000 miles.
Exhaust system: inspect for leaks or damage every 12,000 miles.
Fuel filter: replace every 15,000 miles.
Automatic transmission fluid: change ever 21,000 miles.
Front wheel bearings (2WD): inspect and repack every 30,000 miles.
Rear axle lubricant (E-Super Duty only): replace every 30,000 miles.
Spark plugs (except 4.2L engine): replace every 60,000 miles.
PCV valve: replace every 60,000 miles.
Accessory drive belt: inspect every 60,000 miles.
Spark plugs: replace every 99,000 miles.

93591CA3

GENERAL MOTORS
Cadillac Escalade • Chevrolet Suburban • Tahoe • GMC Denali • Yukon

ENGINE AND VEHICLE IDENTIFICATION

Code ①	Liters (cc)	Cu. In.	Cyl.	Fuel Sys.	Engine Type	Eng. Mfg.
J	7.4 (7440)	454	8	MFI	OHV	CPC
R	5.7 (5735)	350	8	MFI	OHV	CPC
S	6.5 (6473)	395	8	DSL	OHV	CPC
T	5.3 (5327)	325	8	MFI	OHV	CPC
U	6.0 (5966)	364	8	MFI	OHV	CPC
V	4.8 (4802)	293	8	MFI	OHV	CPC
G	8.1 (8128)	496	8	MFI	OHV	CPC

Code ②	Year
X	1999
Y	2000
1	2001
2	2002
3	2003

CPC: Chevrolet/Pontiac/Canada

DSL: Diesel

MFI: Multi-port Fuel Injection

① 8th position of VIN

② 10th position of VIN

93591CA4

Brake service is covered in Section 4 of this manual

GENERAL ENGINE SPECIFICATIONS
All measurements are given in inches.

Year	Model	Engine Displacement Liters (cc)	Engine Series (ID/VIN)	Fuel System	Net Horsepower @ rpm	Net Torque @ rpm (ft. lbs.)	Bore x Stroke (in.)	Compression Ratio	Oil Pressure @ rpm
1999	Denali	5.7 (5735)	R	MFI	255@4600	335@2800	4.00x3.48	9.4:1	18@2000
	Denali	6.5 (6374)	S	DSL	180@3400	360@1700	4.06x3.82	21.5:1	30-43@2000
	Escalade	5.7 (5735)	R	MFI	255@4600	335@2800	4.00x3.48	9.4:1	18@2000
	Escalade	6.5 (6374)	S	DSL	180@3400	360@1700	4.06x3.82	21.5:1	30-43@2000
	Suburban	5.7 (5735)	R	MFI	255@4600	335@2800	4.00x3.48	9.4:1	18@2000
	Suburban	7.4 (7440)	J	MFI	290@4200	410@3200	4.25x4.00	9.0:1	40@2000
	Suburban	6.5 (6374)	S	DSL	180@3400	360@1700	4.06x3.82	21.5:1	30-43@2000
	Tahoe/Yukon	5.7 (5735)	R	MFI	255@4600	335@2800	4.00x3.48	9.4:1	18@2000
2000	Denali	5.7 (5735)	R	MFI	255@4600	335@2800	4.00x3.48	9.4:1	18@2000
	Denali	6.5 (6374)	S	DSL	180@3400	360@1700	4.06x3.82	21.5:1	30-43@2000
	Escalade	5.7 (5735)	R	MFI	255@4600	335@2800	4.00x3.48	9.4:1	18@2000
	Escalade	6.5 (6374)	S	DSL	180@3400	360@1700	4.06x3.82	21.5:1	30-43@2000
	Suburban	5.7 (5735)	R	MFI	255@4600	335@2800	4.00x3.48	9.4:1	18@2000
	Suburban	7.4 (7440)	J	MFI	290@4200	410@3200	4.25x4.00	9.0:1	40@2000
	Suburban	6.5 (6374)	S	DSL	180@3400	360@1700	4.06x3.82	21.5:1	30-43@2000
	Tahoe/Yukon	5.7 (5735)	R	MFI	255@4600	335@2800	4.00x3.48	9.4:1	18@2000
2001	Denali	6.0 (5967)	U	MFI	300@4400	360@4000	4.00x3.62	9.4:1	18@2000
	Denali XL	6.0 (5967)	U	MFI	300@4400	360@4000	4.00x3.62	9.4:1	18@2000
	Suburban	5.3 (5327)	T	MFI	285@4000	360@4000	3.78x3.62	9.5:1	18@2000
	Suburban	6.0 (5967)	U	MFI	300@4400	360@4000	4.00x3.62	9.4:1	18@2000
	Suburban	8.1 (8128)	G	MFI	340@4200	455@3200	4.25x4.37	9.1:1	10@2000
	Tahoe	4.8 (4802)	V	MFI	270@5200	285@4000	3.78x3.27	9.5:1	18@2000
	Tahoe	5.3 (5327)	T	MFI	285@4000	360@4000	3.78x3.62	9.5:1	18@2000
	Yukon	4.8 (4802)	V	MFI	270@5200	285@4000	3.78x3.27	9.5:1	18@2000
	Yukon	5.3 (5327)	T	MFI	285@4000	360@4000	3.78x3.62	9.5:1	18@2000
	Yukon XL	5.3 (5327)	T	MFI	285@4000	360@4000	3.78x3.62	9.5:1	18@2000
	Yukon XL	6.0 (5967)	U	MFI	300@4400	360@4000	4.00x3.62	9.4:1	18@2000
	Yukon XL	8.1 (8128)	G	MFI	340@4200	455@3200	4.25x4.37	9.1:1	10@2000
2002-03	Denali	6.0 (5967)	U	MFI	300@4400	360@4000	4.00x3.62	9.4:1	18@2000
	Denali XL	6.0 (5967)	U	MFI	300@4400	360@4000	4.00x3.62	9.4:1	18@2000
	Suburban	5.3 (5327)	T	MFI	285@4000	360@4000	3.78x3.62	9.5:1	18@2000
	Suburban	6.0 (5967)	U	MFI	300@4400	360@4000	4.00x3.62	9.4:1	18@2000
	Suburban	8.1 (8128)	G	MFI	340@4200	455@3200	4.25x4.37	9.1:1	10@2000
	Tahoe	4.8 (4802)	V	MFI	270@5200	285@4000	3.78x3.27	9.5:1	18@2000
	Tahoe	5.3 (5327)	T	MFI	285@4000	360@4000	3.78x3.62	9.5:1	18@2000
	Yukon	4.8 (4802)	V	MFI	270@5200	285@4000	3.78x3.27	9.5:1	18@2000
	Yukon	5.3 (5327)	T	MFI	285@4000	360@4000	3.78x3.62	9.5:1	18@2000
	Yukon XL	5.3 (5327)	T	MFI	285@4000	360@4000	3.78x3.62	9.5:1	18@2000
	Yukon XL	6.0 (5967)	U	MFI	300@4400	360@4000	4.00x3.62	9.4:1	18@2000
	Yukon XL	8.1 (8128)	G	MFI	340@4200	455@3200	4.25x4.37	9.1:1	10@2000

DSL: Diesel

MFI: Multi-port Fuel Injection

93591CA5

GASOLINE ENGINE TUNE-UP SPECIFICATIONS

Year	Engine Displacement Liters (cc)	Engine ID/VIN	Spark Plugs Gap (in.)	Ignition Timing (deg.) MT	Ignition Timing (deg.) AT	Fuel Pump (psi)	Idle Speed (rpm) MT	Idle Speed (rpm) AT	Valve Clearance In.	Valve Clearance Ex.
1999	5.7 (5735)	R	0.060	—	①	60-66 ②	—	525	HYD	HYD
	7.4 (7440)	J	0.060	—	①	60-66 ②	—	675 ③	HYD	HYD
2000	5.7 (5735)	R	0.060	—	①	60-66 ②	—	525	HYD	HYD
	7.4 (7440)	J	0.060	—	①	60-66 ②	—	675 ③	HYD	HYD
2001	4.8 (4802)	V	0.060	—	①	55-62 ②	—	④	HYD	HYD
	5.3 (5327)	T	0.060	—	①	55-62 ②	—	④	HYD	HYD
	6.0 (5966)	U	0.060	—	①	55-62 ②	—	④	HYD	HYD
	8.1 (8128)	G	0.060	—	①	55-62 ②	—	④	HYD	HYD
2002-03	4.8 (4802)	V	0.060	—	①	55-62 ②	—	④	HYD	HYD
	5.3 (5327)	T	0.060	—	①	55-62 ②	—	④	HYD	HYD
	6.0 (5966)	U	0.060	—	①	55-62 ②	—	④	HYD	HYD
	8.1 (8128)	G	0.060	—	①	55-62 ②	—	④	HYD	HYD

NOTE: The Vehicle Emission Control Information label often reflects specification changes made during production. The label figures must be used if they differ from those in this chart.

HYD: Hydraulic

① Ignition timing is preset and cannot be adjusted
② With key ON and engine OFF
③ Over 8500 GVW
④ Idle speed is maintained by the Powertrain Control Module (PCM)

93591CB8

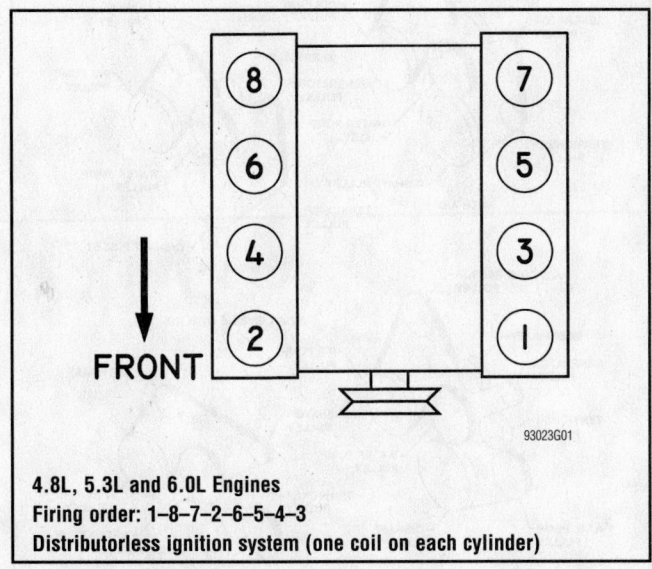

4.8L, 5.3L and 6.0L Engines
Firing order: 1–8–7–2–6–5–4–3
Distributorless ignition system (one coil on each cylinder)

93023G01

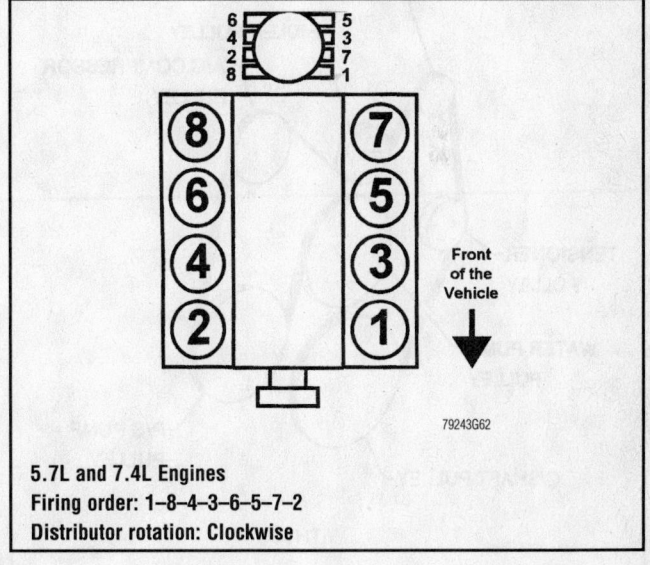

5.7L and 7.4L Engines
Firing order: 1–8–4–3–6–5–7–2
Distributor rotation: Clockwise

79243G62

For complete Engine Mechanical specifications, see Section 1 of this manual

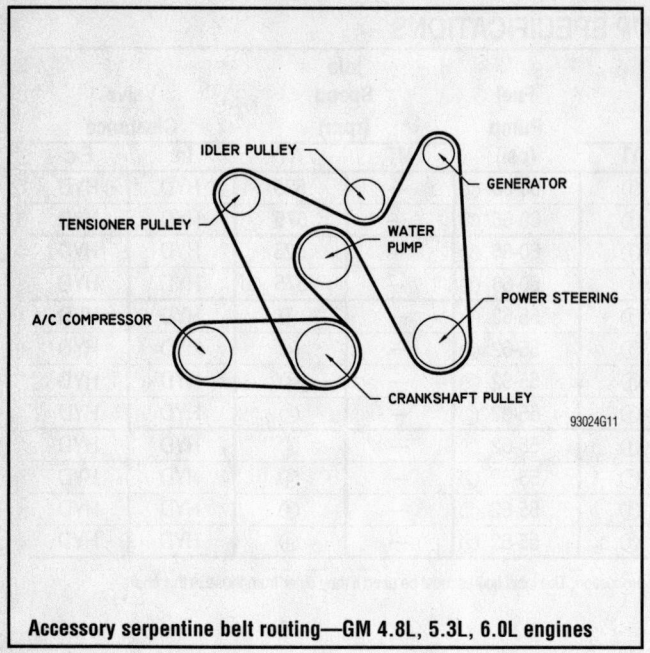

Accessory serpentine belt routing—GM 4.8L, 5.3L, 6.0L engines

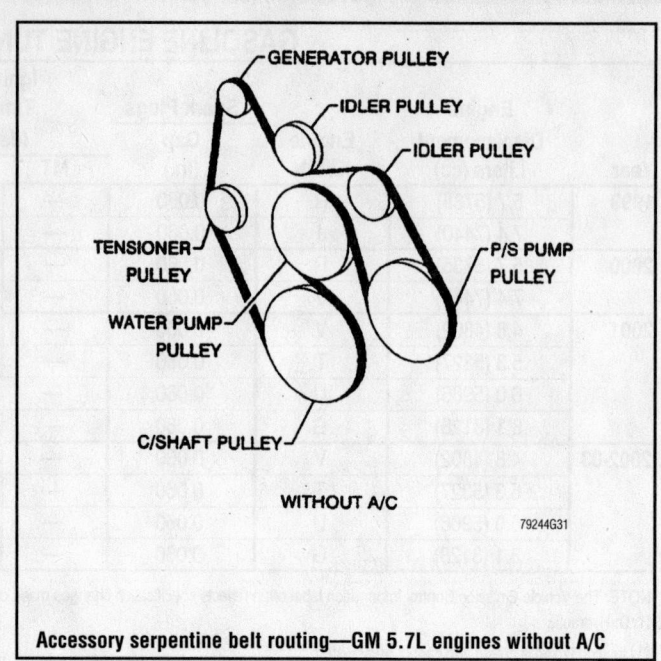

Accessory serpentine belt routing—GM 5.7L engines without A/C

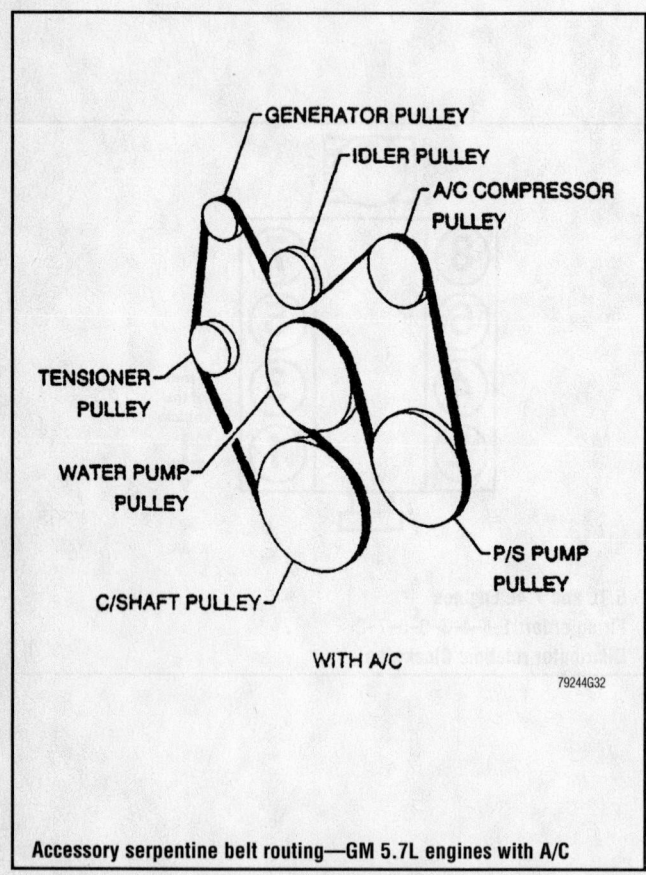

Accessory serpentine belt routing—GM 5.7L engines with A/C

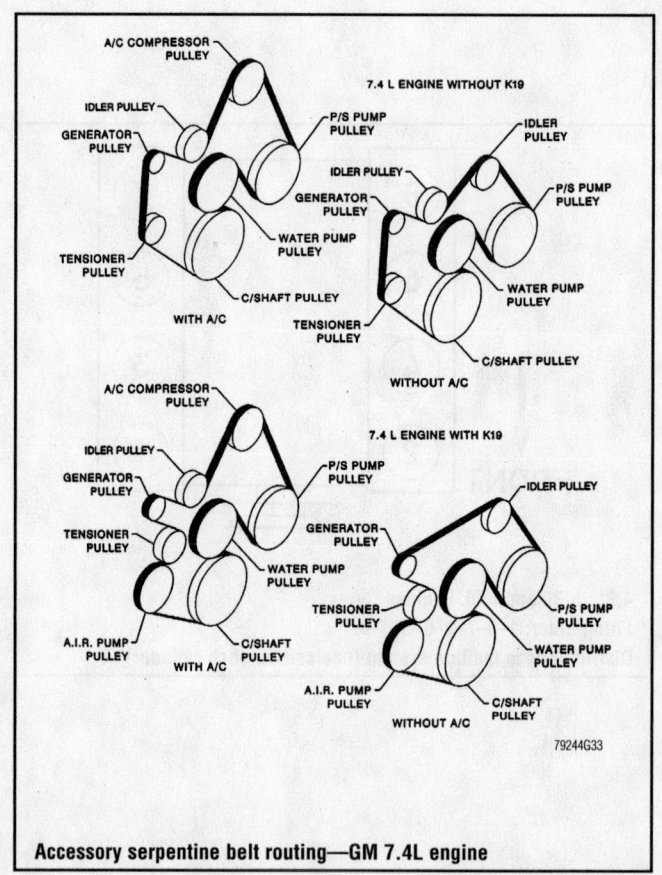

Accessory serpentine belt routing—GM 7.4L engine

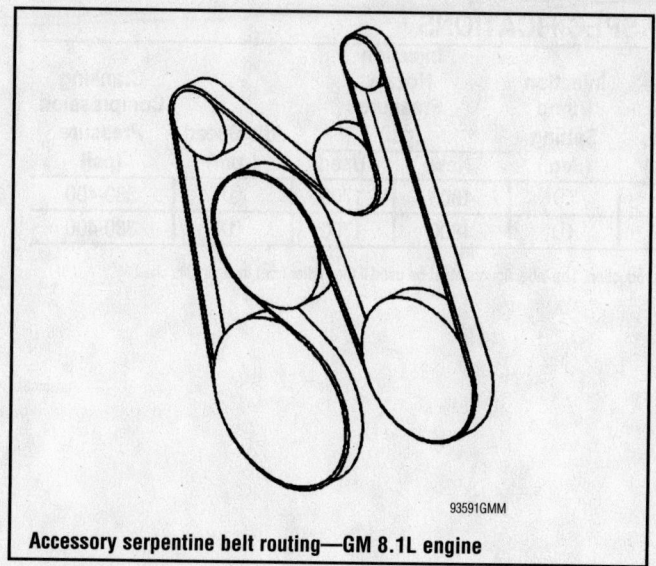

Accessory serpentine belt routing—GM 8.1L engine

93591GMM

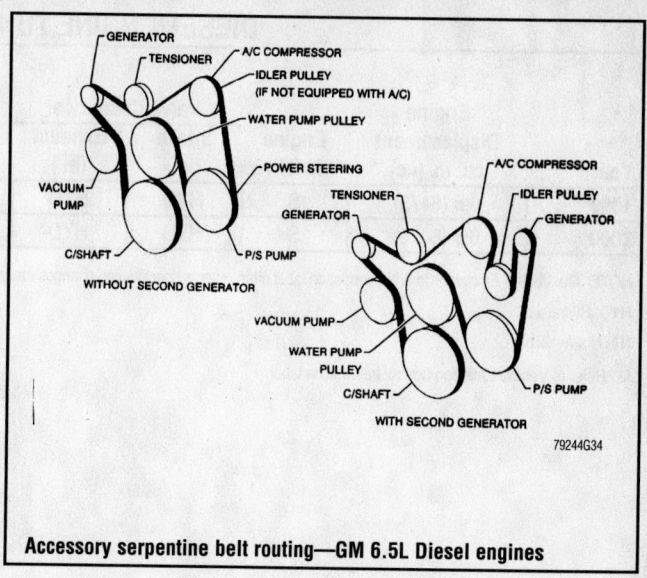

GENERATOR
TENSIONER
A/C COMPRESSOR
IDLER PULLEY
(IF NOT EQUIPPED WITH A/C)
WATER PUMP PULLEY
POWER STEERING
VACUUM PUMP
C/SHAFT
P/S PUMP
WITHOUT SECOND GENERATOR

A/C COMPRESSOR
TENSIONER
IDLER PULLEY
GENERATOR
GENERATOR
VACUUM PUMP
WATER PUMP PULLEY
C/SHAFT
P/S PUMP
WITH SECOND GENERATOR

79244G34

Accessory serpentine belt routing—GM 6.5L Diesel engines

For Accessory Drive Belt illustrations, see Section 1 of this manual

DIESEL ENGINE TUNE-UP SPECIFICATIONS

Year	Engine Displacement cu. in. (cc)	Engine ID/VIN	Valve Clearance		Intake Valve Opens (deg.)	Injection Pump Setting (deg.)	Injection Nozzle Pressure (psi)		Idle Speed (rpm)	Cranking Compression Pressure (psi)
			Intake (in.)	Exhaust (in.)			New	Used		
1999	6.5 (6473)	S	HYD	HYD	①	①	1800	1700	①	380-400
2000	6.5 (6473)	S	HYD	HYD	①	①	1800	1700	①	380-400

NOTE: The Vehicle Emission Control Information label often reflects specification changes made during production. The label figures must be used if they differ from those in this chart.

HYD: Hydraulic

NA: Not Available

① Refer to Vehicle Emission Control Information label

93591CB9

CAPACITIES

Year	Model	Engine Displacement Liters (cc)	Engine ID/VIN	Engine Oil with Filter (qts.)	Transmission (pts.) 5-Spd	Transmission (pts.) Auto.	Transfer Case (pts.)	Drive Axle Front (pts.)	Drive Axle Rear (pts.)	Fuel Tank (gal.)	Cooling System (qts.)
1999	Denali	5.7 (5735)	R	5.0	—	①	②	③	④	⑤	18.0
	Denali	6.5 (6473)	S	7.0	—	①	—	—	④	⑤	23.8
	Escalade	5.7 (5735)	R	5.0	—	①	②	③	④	⑤	18.0
	Escalade	6.5 (6473)	S	7.0	—	①	—	—	④	⑤	23.8
	Suburban	7.4 (7440)	J	6.0	—	①	②	③	④	25.0 ⑥	24.5
	Suburban	5.7 (5735)	R	5.0	—	①	—	—	④	⑤	18.0
	Tahoe	5.7 (5735)	R	5.0	—	①	②	③	④	⑤	18.0
	Tahoe	6.5 (6473)	S	7.0	—	①	—	—	④	⑤	23.8
	Yukon	5.7 (5735)	R	5.0	—	①	②	③	④	⑤	18.0
	Yukon	6.5 (6473)	S	7.0	—	①	—	—	④	⑤	23.8
2000	Denali	5.7 (5735)	R	5.0	—	①	②	③	④	⑤	18.0
	Denali	6.5 (6473)	S	7.0	—	①	—	—	④	⑤	23.8
	Escalade	5.7 (5735)	R	5.0	—	①	②	③	④	⑤	18.0
	Escalade	6.5 (6473)	S	7.0	—	①	—	—	④	⑤	23.8
	Suburban	7.4 (7440)	J	6.0	—	①	②	③	④	25.0 ⑥	24.5
	Suburban	5.7 (5735)	R	5.0	—	①	—	—	④	⑤	18.0
	Tahoe	5.7 (5735)	R	5.0	—	①	②	③	④	⑤	18.0
	Tahoe	6.5 (6473)	S	7.0	—	①	—	—	④	⑤	23.8
	Yukon	5.7 (5735)	R	5.0	—	①	②	③	④	⑤	18.0
	Yukon	6.5 (6473)	S	7.0	—	①	—	—	④	⑤	23.8
2001	Denali	6.0 (5967)	U	6.0	—	⑦	②	⑧	⑨	26.0	14.8 ⑪
	Denali XL	6.0 (5967)	U	6.0	—	⑦	②	⑧	⑨	⑩	14.8 ⑪
	Suburban	5.3 (5327)	T	6.0	—	⑦	②	⑧	⑨	⑩	13.4 ⑫
	Suburban	6.0 (5967)	U	6.0	—	⑦	②	⑧	⑨	⑩	15.8 ⑬
	Suburban	8.1 (8128)	G	6.5	—	⑦	②	⑧	⑨	⑩	20.7
	Tahoe	4.8 (4802)	V	6.0	—	⑦	②	⑧	⑨	26.0	13.4 ⑫
	Tahoe	5.3 (5327)	T	6.0	—	⑦	②	⑧	⑨	26.0	13.4 ⑫
	Yukon	4.8 (4802)	V	6.0	—	⑦	②	⑧	⑨	26.0	13.4 ⑫
	Yukon	5.3 (5327)	T	6.0	—	⑦	②	⑧	⑨	26.0	13.4 ⑫
	Yukon XL	5.3 (5327)	T	6.0	—	⑦	②	⑧	⑨	⑩	13.4 ⑫
	Yukon XL	6.0 (5967)	U	6.0	—	⑦	②	⑧	⑨	⑩	15.8 ⑬
	Yukon XL	8.1 (8128)	G	6.5	—	⑦	②	⑧	⑨	⑩	20.7

93591CB0

For Tire, Wheel and Ball Joint specifications, see Section 1 of this manual

CAPACITIES

Year	Model	Engine Displacement Liters (cc)	Engine ID/VIN	Engine Oil with Filter (qts.)	Transmission (pts.) 5-Spd	Transmission (pts.) Auto.	Transfer Case (pts.)	Drive Axle Front (pts.)	Drive Axle Rear (pts.)	Fuel Tank (gal.)	Cooling System (qts.)
2002-03	Denali	6.0 (5967)	U	6.0	—	⑦	②	⑧	⑨	26.0	14.8 ⑪
	Denali XL	6.0 (5967)	U	6.0	—	⑦	②	⑧	⑨	⑩	14.8 ⑪
	Suburban	5.3 (5327)	T	6.0	—	⑦	②	⑧	⑨	⑩	13.4 ⑫
	Suburban	6.0 (5967)	U	6.0	—	⑦	②	⑧	⑨	⑩	15.8 ⑬
	Suburban	8.1 (8128)	G	6.5	—	⑦	②	⑧	⑨	⑩	20.7
	Tahoe	4.8 (4802)	V	6.0	—	⑦	②	⑧	⑨	26.0	13.4 ⑫
	Tahoe	5.3 (5327)	T	6.0	—	⑦	②	⑧	⑨	26.0	13.4 ⑫
	Yukon	4.8 (4802)	V	6.0	—	⑦	②	⑧	⑨	26.0	13.4 ⑫
	Yukon	5.3 (5327)	T	6.0	—	⑦	②	⑧	⑨	26.0	13.4 ⑫
	Yukon XL	5.3 (5327)	T	6.0	—	⑦	②	⑧	⑨	⑩	13.4 ⑫
	Yukon XL	6.0 (5967)	U	6.0	—	⑦	②	⑧	⑨	⑩	15.8 ⑬
	Yukon XL	8.1 (8128)	G	6.5	—	⑦	②	⑧	⑨	⑩	20.7

NOTE: All capacities are approximate. Add fluid gradually and check to be sure a proper fluid level is obtained.

① 4L60E trans.: 10.0 pts.
4L80E trans.: 14.5 pts.

② NV241 and NV243: 4.5 pts.
4401 and 4470: 6.6 pts.

③ K2 models: 3.5 pts.
K3 models: 4.5 pts.

④ 8.5 in. ring gear: 4.2 pts.
9.5 in. ring gear: 6.5 pts.
9.75 in. ring gear: 6.0 pts.
10.5 in. ring gear: 6.5 pts.

⑤ Std. available with 25 and 34 gallon tanks
Chassis cab available with 22, 30 and 34 gallon tanks

⑥ Optional 31 and 40 gallon tanks

⑦ 4L60E trans.: 10.0 pts.
4L80E trans.: 15.4 pts.

⑧ 8.25 in ring gear: 3.5 pts.
9.25 ring gear: 3.7 pts.

⑨ 8.6 in. ring gear: 4.8 pts.
9.5 & 10.5 in. ring gear: 5.5 pts.
11.5 in. ring gear: 7.7 pts.

⑩ 1500 XL: 32.5 gallon tank
2500 XL: 38.5 gallon tank

⑪ With optional oil cooler: 15.4 qts.

⑫ With optional A/C: 14.9 qts.
With front A/C: 14.4 qts.
With front and rear A/C: 15.8 qts.

⑬ With optional oil cooler: 15.4 qts.

93591CC1

VALVE SPECIFICATIONS

Year	Engine Displacement Liters (cc)	Engine ID/VIN	Seat Angle (deg.)	Face Angle (deg.)	Spring Test Pressure (lbs. @ in.)	Spring Installed Height (in.)	Stem-to-Guide Clearance (in.)		Stem Diameter (in.)	
							Intake	Exhaust	Intake	Exhaust
1999	5.7 (5735)	R	46	45	187-203@1.27	1.69-1.71	0.0010-0.0027	0.0010-0.0027	NA	NA
	6.5 (6473)	S	46	45	230@1.40	1.80	0.0010-0.0027	0.0010-0.0027	NA	NA
	7.4 (7440)	J	46	45	238-262@1.34	1.83	0.0010-0.0029 ①	0.0012-0.0031 ①	NA	NA
2000	5.7 (5735)	R	46	45	187-203@1.27	1.69-1.71	0.0010-0.0027	0.0010-0.0027	NA	NA
	6.5 (6473)	S	46	45	230@1.40	1.80	0.0010-0.0027	0.0010-0.0027	NA	NA
	7.4 (7440)	J	46	45	238-262@1.34	1.83	0.0010-0.0029 ①	0.0012-0.0031 ①	NA	NA
2001	4.8 (4802)	V	46	45	220@1.32	1.80	0.0010-0.0026	0.0010-0.0026	0.3132-0.3140	0.3132-0.3140
	5.3 (5327)	T	46	45	220@1.32	1.80	0.0010-0.0026	0.0010-0.0026	0.3132-0.3140	0.3132-0.3140
	6.0 (5966)	U	46	45	220@1.32	1.80	0.0010-0.0026	0.0010-0.0026	0.3132-0.3140	0.3132-0.3140
	8.1 (8128)	G	46	45	216-236@1.34	1.81-1.84	0.0010-0.0029	0.0012-0.0031	0.3715-0.3722	0.3713-0.3720
2002-03	4.8 (4802)	V	46	45	220@1.32	1.80	0.0010-0.0026	0.0010-0.0026	0.3132-0.3140	0.3132-0.3140
	5.3 (5327)	T	46	45	220@1.32	1.80	0.0010-0.0026	0.0010-0.0026	0.3132-0.3140	0.3132-0.3140
	6.0 (5966)	U	46	45	220@1.32	1.80	0.0010-0.0026	0.0010-0.0026	0.3132-0.3140	0.3132-0.3140
	8.1 (8128)	G	46	45	216-236@1.34	1.81-1.84	0.0010-0.0029	0.0012-0.0031	0.3715-0.3722	0.3713-0.3720

NA: Not Available

① Service limit:
  Intake: 0.0037 MAX
  Exhaust: 0.0049 MAX

93591CC2

For Wheel Alignment specifications, see Section 1 of this manual

CRANKSHAFT AND CONNECTING ROD SPECIFICATIONS
All measurements are given in inches.

Year	Engine Displacement Liters (cc)	Engine ID/VIN	Crankshaft				Connecting Rod		
			Main Brg. Journal Dia.	Main Brg. Oil Clearance	Shaft End-play	Thrust on No.	Journal Diameter	Oil Clearance	Side Clearance
1999	5.7 (5735)	R	①	②	0.0020-0.0080	5	2.0978-2.0998	0.0013-0.0035	0.0060-0.0140
	6.5 (6473)	S	③	④	0.0039-0.0100	3	⑤	0.0018-0.0039	0.0067-0.0248
	7.4 (7440)	J	2.7482-2.7489	⑧	0.0050-0.0110	5	2.1990-2.1996	0.0011-0.0029	0.0013-0.0230
2000	5.7 (5735)	R	①	②	0.0020-0.0080	5	2.0978-2.0998	0.0013-0.0035	0.0060-0.0140
	6.5 (6473)	S	③	④	0.0039-0.0100	3	⑤	0.0018-0.0039	0.0067-0.0248
	7.4 (7440)	J	2.7482-2.7489	⑥	0.0050-0.0110	5	2.1990-2.1996	0.0011-0.0029	0.0013-0.0230
2001	4.8 (4802)	V	2.5580-2.5593	0.0007-0.0021	0.0015-0.0078	5	2.0990-2.1000	0.0006-0.0030	0.0043-0.0200
	5.3 (5327)	T	2.5580-2.5593	0.0007-0.0021	0.0015-0.0078	5	2.0990-2.1000	0.0006-0.0030	0.0043-0.0200
	6.0 (5966)	U	2.5580-2.5593	0.0007-0.0021	0.0015-0.0078	5	2.0990-2.1000	0.0006-0.0030	0.0043-0.0200
	8.1 (8128)	G	2.7482-2.7489	⑦	0.0050-0.0110	NA	2.1990-2.1996	0.0008-0.0025	0.0151-0.0270
2002-03	4.8 (4802)	V	2.5580-2.5593	0.0007-0.0021	0.0015-0.0078	5	2.0990-2.1000	0.0006-0.0030	0.0043-0.0200
	5.3 (5327)	T	2.5580-2.5593	0.0007-0.0021	0.0015-0.0078	5	2.0990-2.1000	0.0006-0.0030	0.0043-0.0200
	6.0 (5966)	U	2.5580-2.5593	0.0007-0.0021	0.0015-0.0078	5	2.0990-2.1000	0.0006-0.0030	0.0043-0.0200
	8.1 (8128)	G	2.7482-2.7489	⑦	0.0050-0.0110	NA	2.1990-2.1996	0.0008-0.0025	0.0151-0.0270

NA: Not Available

① No. 1: 2.4484 in.-2.4493 in.
   Nos. 2, 3, 4: 2.4481 in.-2.4490 in.
   No. 5: 2.4479 in.-2.4488 in.

② No. 1: 0.0007 in.-0.0021 in.
   Nos. 2, 3, 4: 0.0009 in.-0.0024 in.
   No. 5: 0.0010 in.-0.0027 in.

③ No. 1, 2, 3, 4: 2.9517 in.-2.9520 in. (Blue)
   2.9520 in.-2.9524 in. (Orange/Red)
   2.9524 in.-2.9527 in. (White)
   No. 5: 2.9515 in.-2.9518 in. (Blue)
   2.9518 in.-2.9522 in. (Orange/Red)
   2.9522 in.-2.9525 in. (White)

④ No. 1, 2, 3, 4: 0.0018 in.-0.0033 in.
   No. 5: 0.0022 in.-0.0037 in.

⑤ 2.399 in.-2.400 in. (Green)
   2.400 in.-2.401 in. (Yellow)

⑥ No. 1: 0.0017 in.-0.0030 in.
   No. 2, 3, 4: 0.0011 in.-0.0024 in.
   No. 5: 0.0025 in.-0.0038 in.

⑦ No. 1, 2, 3, 4: 0.0008-0.0020 in.
   No. 5: 0.0014-0.0026 in.

93591CC3

PISTON AND RING SPECIFICATIONS
All measurements are given in inches.

Year	Engine Displacement Liters (cc)	Engine ID/VIN	Piston Clearance	Ring Gap			Ring Side Clearance		
				Top Compression	Bottom Compression	Oil Control	Top Compression	Bottom Compression	Oil Control
1999	5.7 (5735)	R	0.0007-0.0021	0.010-0.020	0.018-0.026	0.010-0.030	0.0012-0.0032	0.0012-0.0032	0.0020-0.0070
	6.5 (6473)	S	①	0.010-0.020	0.030-0.039	0.010-0.020	0.0015-0.0031	0.0015-0.0031	0.0016-0.0035
	7.4 (7440)	J	0.0018-0.0030	0.010-0.0180	0.016-0.0240	0.010-0.030	0.0012-0.0029	0.0012-0.0029	0.0050-0.0065
2000	5.7 (5735)	R	0.0007-0.0021	0.010-0.020	0.018-0.026	0.010-0.030	0.0012-0.0032	0.0012-0.0032	0.0020-0.0070
	6.5 (6473)	S	①	0.010-0.020	0.030-0.039	0.010-0.020	0.0015-0.0031	0.0015-0.0031	0.0016-0.0035
	7.4 (7440)	J	0.0018-0.0030	0.010-0.0180	0.016-0.0240	0.010-0.030	0.0012-0.0029	0.0012-0.0029	0.0050-0.0065
2001	4.8 (4802)	V	0.0010-0.0024	0.009-0.015	0.017-0.025	0.007-0.027	0.0016-0.0033	0.0016-0.0031	0.0004-0.0087
	5.3 (5327)	T	0.0010-0.0024	0.009-0.015	0.017-0.025	0.007-0.027	0.0016-0.0033	0.0016-0.0031	0.0004-0.0087
	6.0 (5966)	U	0.0010-0.0024	0.009-0.015	0.017-0.025	0.007-0.027	0.0016-0.0033	0.0016-0.0031	0.0004-0.0087
	8.1 (8128)	G	②	0.012-0.018	0.017-0.025	0.010-0.030	0.0012-0.0029	0.0012-0.0029	0.002-0.008
2002-03	4.8 (4802)	V	0.0010-0.0024	0.009-0.015	0.017-0.025	0.007-0.027	0.0016-0.0033	0.0016-0.0031	0.0004-0.0087
	5.3 (5327)	T	0.0010-0.0024	0.009-0.015	0.017-0.025	0.007-0.027	0.0016-0.0033	0.0016-0.0031	0.0004-0.0087
	6.0 (5966)	U	0.0010-0.0024	0.009-0.015	0.017-0.025	0.007-0.027	0.0016-0.0033	0.0016-0.0031	0.0004-0.0087
	8.1 (8128)	G	②	0.012-0.018	0.017-0.025	0.010-0.030	0.0012-0.0029	0.0012-0.0029	0.002-0.008

① 1-6: 0.0037-0.0047 in.
  7-8: 0.0042-0.0052 in.

② Interference fit (coated piston)

93591CC4

For Maintenance Interval recommendations, see Section 1 of this manual

TORQUE SPECIFICATIONS
All readings in ft. lbs.

Year	Engine Displacement Liters (cc)	Engine ID/VIN	Cylinder Head Bolts	Main Bearing Bolts	Rod Bearing Bolts	Crankshaft Damper Bolts	Flywheel Bolts	Manifold Intake *	Manifold Exhaust	Spark Plugs	Lug Nut
1999	5.7 (5735)	R	①	②	③	74	74	③	④	15	⑤
	6.5 (6473)	S	⑥	⑦	48	200	65	31	26	—	⑤
	7.4 (7440)	J	85	100	45	110	67	30	22	15	⑤
2000	5.7 (5735)	R	⑥	②	⑧	74	74	③	④	15	⑤
	6.5 (6473)	S	⑥	⑦	48	200	65	31	26	—	⑤
	7.4 (7440)	J	85	100	45	110	67	30	22	15	⑤
2001	4.8 (4802)	V	⑨	⑩	⑪	⑫	⑬	⑭	⑮	12	140
	5.3 (5327)	T	⑨	⑩	⑪	⑫	⑬	⑭	⑮	12	140
	6.0 (5966)	U	⑨	⑩	⑪	⑫	⑬	⑭	⑮	12	140
	8.1 (8128)	G	⑯	⑰	⑰	189	⑱	⑲	⑳	15	140
2002-03	4.8 (4802)	V	⑨	⑩	⑪	⑫	⑬	⑭	⑮	12	140
	5.3 (5327)	T	⑨	⑩	⑪	⑫	⑬	⑭	⑮	12	140
	6.0 (5966)	U	⑨	⑩	⑪	⑫	⑬	⑭	⑮	12	140
	8.1 (8128)	G	⑯	⑰	⑰	189	⑱	⑲	⑳	15	140

* NOTE: Applies to Lower Manifold only.

① Step 1: 22 ft. lbs.
Step 2: 90 degrees
Step 3: 90 degrees,
(except medium length bolts at front and rear)
Step 4: Tighten medium length bolts,
at front and rear an additional 50 degrees

② Outer bolts on caps 2-4: 67 ft. lbs.
All others: 74 ft. lbs.

③ Step 1: 22 ft. lbs.
Step 2:
Short bolt: Plus 55 degrees
Medium bolt: Plus 65 degrees
Long bolt: Plus 75 degrees

④ Tighten bolts to 12 ft. lbs.
Retorque to 22 ft. lbs.

⑤ All 5 & 6 stud single rear wheels: 110 ft. lbs.
All 8 stud single rear wheels: 120 ft. lbs.

⑥ Step 1: 20 ft. lbs.
Step 2: 50 ft. lbs.
Step 3: 50 ft. lbs.
Step 4: Plus 90-100 degrees

⑦ Outer bolts: 100 ft. lbs.
Inner bolts: 111 ft. lbs.

⑧ Tighten all bolts to 20 ft. lbs.
Retorque to 50 ft. lbs.

⑨ Step 1: 22 ft. lbs.
Step 2: 90 degrees
Step 3: 90 degrees,
(except medium length bolts at front and rear)
Step 4: Tighten medium length bolts,
at front and rear an additional 50 degrees

⑩ Inner bolts;
Step 1: 15 ft. lbs.

⑪ Step 1: 15 ft. lbs.
Step 2: 60 degrees

⑫ Use a new bolt
Step 1: 37 ft. lbs.
Step 2: 140 degrees

⑬ Step 1: 15 ft. lbs.
Step 2: 37 ft. lbs.
Step 3: 74 ft. lbs.

⑭ Step 1: 44 in. lbs.
Step 2: 89 in. lbs.

⑮ Nuts: 39 ft. lbs.
Stud: 16 ft. lbs.

⑯ 1st pass: 22 ft. lbs.
2nd pass: 22 ft. lbs.,
plus 120 degrees
Final pass:
Short bolt: Plus 60 degrees
Medium bolt: Plus 45 degrees
Long bolt: Plus 30 degrees

⑰ 22 ft. lbs., plus 90 degrees

⑱ 1st pass: 30 ft. lbs.
2nd pass: 59 ft. lbs.
3rd pass: 70 ft. lbs.

⑲ 1st & 2nd pass: 44 inch lbs.
3rd pass: 89 inch lbs.
4th pass: 106 inch lbs.

⑳ Center bolt: 26 ft. lbs.
Nut: 12 ft. lbs.
Stud: 15 ft. lbs.

93591CC5

WHEEL ALIGNMENT

Year	Model	Caster Range (+/-Deg.)	Caster Preferred Setting (Deg.)	Camber Range (+/-Deg.)	Camber Preferred Setting (Deg.)	Toe-in (in.)	Steering Axis Inclination (Deg.)
1999	2WD	1.00	+3.75	0.50	+0.50	0.24+/-0.20	—
	2WD HD	not adjustable	not adjustable	0.50	+1.25	0.12+/-0.12	—
	4WD	1.00	+3.00	0.50	+0.65	0.24+/-0.20	—
	4WD HD	1.00	+3.00	0.50	+0.50	0.24+/-0.20	—
2000	2WD	1.00	+3.75	0.50	+0.50	0.24+/-0.20	—
	2WD HD	not adjustable	not adjustable	0.50	+1.25	0.12+/-0.12	—
	4WD	1.00	+3.00	0.50	+0.65	0.24+/-0.20	—
	4WD HD	1.00	+3.00	0.50	+0.50	0.24+/-0.20	—
2001	2WD	1.00	①	0.50	+0.25	0.10+/-0.20	—
	2WD HD	1.00	②	0.50	+0.25	0.10+/-0.12	—
	4WD	1.00	③	0.50	+0.25	0.10+/-0.20	—
	4WD HD	1.00	②	0.50	+0.25	0.10+/-0.20	—
2002-03	2WD	1.00	①	0.50	+0.25	0.10+/-0.20	—
	2WD HD	1.00	②	0.50	+0.25	0.10+/-0.12	—
	4WD	1.00	③	0.50	+0.25	0.10+/-0.20	—
	4WD HD	1.00	②	0.50	+0.25	0.10+/-0.20	—

① Left side: 3.90
   Right side: 4.70
② Left side: 4.50
   Right side: 4.75
③ Left side: 3.50
   Right side: 4.50

93591CC6

For Tune-up, Capacities and Firing orders, see Section 1 of this manual

TIRE, WHEEL AND BALL JOINT SPECIFICATIONS

| Year | Model | OEM Tires | | Tire Pressures (psi) | | Wheel Size | Ball Joint Inspection |
		Standard	Optional	Front	Rear		
1999	Denali	P265/70R16	None	36	36	7-JJ	U ① L: 0.090 in.
	Tahoe/Yukon, 2wd	P235/75R15	None	36	36	6.5-JJ	U ① L: 0.090 in.
	Tahoe/Yukon, 4wd	P245/75R16	P265/75R16	36	36	7-JJ	②
	1500 Suburban 2wd	P235/75R15XL	LT245/75R16E	36	36	7-JJ	L ③
	1500 Suburban 4wd	P245/75R16C	LT245/75R16E	36	36	6.5-JJ	L ③
	2500 Suburban	LT245/75R16E	None	36	36	6.5-JJ	L ③
2000	Denali	P265/70R16	None	36	36	7-JJ	U ② L: 0.090 in.
	Tahoe/Yukon, 2wd	P235/75R15	None	36	36	6.5-JJ	U ① L: 0.090 in.
	Tahoe/Yukon, 4wd	P245/75R16	P265/75R16	36	36	7-JJ	②
	1500 Suburban 2wd	P235/75R15XL	LT245/75R16E	36	36	7-JJ	L ③
	1500 Suburban 4wd	P245/75R16C	LT245/75R16E	36	36	6.5-JJ	L ③
	2500 Suburban	LT245/75R16E	None	36	36	6.5-JJ	U: 0.125 in. L ③
2001	Denali	P265/70R16	None	36	36	7-JJ	U ② L: 0.090 in.
	Tahoe/Yukon, 2wd	P235/75R15	None	36	36	6.5-JJ	U ① L: 0.090 in.
	Tahoe/Yukon, 4wd	P245/75R16	P265/75R16	36	36	7-JJ	②
	1500 Suburban 2wd	P235/75R15XL	LT245/75R16E	36	36	7-JJ	L ③
	1500 Suburban 4wd	P245/75R16C	LT245/75R16E	36	36	6.5-JJ	L ③
	2500 Suburban	LT245/75R16E	None	36	36	6.5-JJ	U: 0.125 in. L ③
2002-03	Denali	P265/70R16	None	36	36	7-JJ	U ② L: 0.090 in.
	Tahoe/Yukon, 2wd	P235/75R15	None	36	36	6.5-JJ	U ① L: 0.090 in.
	Tahoe/Yukon, 4wd	P245/75R16	P265/75R16	36	36	7-JJ	②
	1500 Suburban 2wd	P235/75R15XL	LT245/75R16E	36	36	7-JJ	L ③
	1500 Suburban 4wd	P245/75R16C	LT245/75R16E	36	36	6.5-JJ	L ③
	2500 Suburban	LT245/75R16E	None	36	36	6.5-JJ	U: 0.125 in. L ③

OEM: Original Equipment Manufacturer

PSI: Pounds Per Square Inch

STD: Standard

OPT: Optional

L: Lower

U: Upper

① Replace if any movement is noted or if stud can be moved by hand

② Ball joint is adjustable, refer to manual for procedure

③ Do not lift truck. Inspect the boss into which the grease fitting is threaded. Replace if the boss is flush or receded below the surface of the ball joint

93591CC7

BRAKE SPECIFICATIONS

All measurements in inches unless noted

Year	Model		Brake Disc Original Thickness	Brake Disc Minimum Thickness	Brake Disc Maximum Runout	Brake Drum Diameter Original Inside Diameter	Brake Drum Diameter Max. Wear Limit	Brake Drum Diameter Maximum Machine Diameter	Minimum Lining Thickness	Brake Caliper Bracket Bolts (ft. lbs.)	Brake Caliper Mounting Bolts (ft. lbs.)
1999	Denali	F	1.500	1.480	0.004	—	—	—	0.030	NA	38
		R	—	—	—	①	②	③	0.030	NA	—
	Envoy	F	1.030	0.965	0.003	—	—	—	0.030	52	④
		R	0.787	0.728	0.004	9.50	9.59	9.56	0.030	NA	—
	Escalade	F	1.500	1.480	0.004	—	—	—	0.030	NA	38
		R	—	—	—	①	②	③	0.030	NA	—
	Suburban	F	1.500	1.480	0.004	—	—	—	0.030	NA	38
		R	—	—	—	①	②	③	0.030	NA	—
	Tahoe	F	1.500	1.480	0.004	—	—	—	0.030	NA	38
		R	—	—	—	①	②	③	0.030	NA	—
	Yukon	F	1.500	1.480	0.004	—	—	—	0.030	NA	38
		R	—	—	—	①	②	③	0.030	NA	—
2000	Denali	F	1.500	1.480	0.004	—	—	—	0.030	NA	38
		R	—	—	—	①	②	③	0.030	NA	—
	Envoy	F	1.030	0.965	0.003	—	—	—	0.030	52	④
		R	0.787	0.728	0.004	9.50	9.59	9.56	0.030	NA	—
	Escalade	F	1.500	1.480	0.004	—	—	—	0.030	NA	38
		R	—	—	—	①	②	③	0.030	NA	—
	Suburban	F	1.500	1.480	0.004	—	—	—	0.030	NA	38
		R	—	—	—	①	②	③	0.030	NA	—
	Tahoe	F	1.500	1.480	0.004	—	—	—	0.030	NA	38
		R	—	—	—	①	②	③	0.030	NA	—
	Yukon	F	1.500	1.480	0.004	—	—	—	0.030	NA	38
		R	—	—	—	①	②	③	0.030	NA	—
2001	Denali	F	⑤	⑥	0.005	—	—	—	0.030	⑦	80
		R	⑧	⑨	0.005	—	—	—	0.030	⑩	⑪
	Denali XL	F	⑤	⑥	0.005	—	—	—	0.030	⑦	80
		R	⑧	⑨	0.005	—	—	—	0.030	⑩	⑪
	Suburban	F	⑤	⑥	0.005	—	—	—	0.030	⑦	80
		R	⑧	⑨	0.005	—	—	—	0.030	⑩	⑪
	Tahoe	F	⑤	⑥	0.005	—	—	—	0.030	⑦	80
		R	⑧	⑨	0.005	—	—	—	0.030	⑩	⑪
	Yukon	F	⑤	⑥	0.005	—	—	—	0.030	⑦	80
		R	⑧	⑨	0.005	—	—	—	0.030	⑩	⑪
	Yukon XL	F	⑤	⑥	0.005	—	—	—	0.030	⑦	80
		R	⑧	⑨	0.005	—	—	—	0.030	⑩	⑪

93591CC8

For complete service labor times, order Nichols' Chilton Labor Guide

BRAKE SPECIFICATIONS
All measurements in inches unless noted

Year	Model		Brake Disc			Brake Drum Diameter			Minimum Lining Thickness	Brake Caliper	
			Original Thickness	Minimum Thickness	Maximum Runout	Original Inside Diameter	Max. Wear Limit	Maximum Machine Diameter		Bracket Bolts (ft. lbs.)	Mounting Bolts (ft. lbs.)
2002-03	Denali	F	⑤	⑥	0.005	—	—	—	0.030	⑦	80
		R	⑧	⑨	0.005	—	—	—	0.030	⑩	⑪
	Denali XL	F	⑤	⑥	0.005	—	—	—	0.030	⑦	80
		R	⑧	⑨	0.005	—	—	—	0.030	⑩	⑪
	Suburban	F	⑤	⑥	0.005	—	—	—	0.030	⑦	80
		R	⑧	⑨	0.005	—	—	—	0.030	⑩	⑪
	Tahoe	F	⑤	⑥	0.005	—	—	—	0.030	⑦	80
		R	⑧	⑨	0.005	—	—	—	0.030	⑩	⑪
	Yukon	F	⑤	⑥	0.005	—	—	—	0.030	⑦	80
		R	⑧	⑨	0.005	—	—	—	0.030	⑩	⑪
	Yukon XL	F	⑤	⑥	0.005	—	—	—	0.030	⑦	80
		R	⑧	⑨	0.005	—	—	—	0.030	⑩	⑪

NA: Not Available

① Available with 1 in., 11.15 in. and 13 in. drums

② 10 in. drum: 10.05
   11.15 in. drum: 11.24
   1 in. drum: 13.09

③ 1 in. drum: 10.09
   11.15 in. drum: 11.21
   1 in. drum: 13.06

④ 2WD: 38 ft. lbs.
   4WD: 77 ft. lbs.

⑤ Vacuum: 1.14 in.
   Hydraulic: 1.50 in.

⑥ Vacuum: 1.10 in.
   Hydraulic: 1.46 in.

⑦ 15 series: 129 ft. lbs.
   25 series: 221 ft. lbs.

⑧ Vacuum: 0.787 in.
   Hydraulic: 1.14 in.

⑨ Vacuum: 0.748 in.
   Hydraulic: 1.10 in.

⑩ Vacuum: 148 ft. lbs.
   Hydraulic (9000 lbs.): 122 ft. lbs.
   Hydraulic (12,000 lbs.): 221 ft. lbs.

⑪ 15 series: 31 ft. lbs.
   25 series: 80 ft. lbs.

93591CC9

SCHEDULED MAINTENANCE INTERVALS
GENERAL MOTORS—DENALI, DENALI XL, SUBURBAN, TAHOE, YUKON & YUKON XL—DIESEL

TO BE SERVICED	TYPE OF SERVICE	5	8	10	15	20	23	25	30	35	38	40	45	50	53	55	60	65	68	70	75	80	83	85	90	95	98
Air intake system	S/I			✓		✓			✓			✓		✓			✓			✓		✓			✓		
Automatic transmission fluid ①	R										Every 50,000 miles																
Brake system	S/I	✓	✓		✓			✓		✓			✓			✓		✓			✓			✓			✓
Chassis & suspension grease points	L	✓		✓	✓	✓		✓	✓	✓		✓	✓	✓	✓	✓	✓		✓	✓	✓		✓	✓	✓		
Cooling fan operation	S/I			✓		✓			✓			✓		✓			✓			✓		✓			✓		
Crankcase depression regular valve system hoses	S/I																✓										
CV-joint boots & axle seals	S/I	✓			✓	✓		✓	✓	✓		✓	✓	✓		✓	✓		✓	✓	✓		✓	✓	✓		
EGR system ②	S/I																✓										
Engine coolant	R										Every 150,000 miles																
Engine cooling system hoses & radiator	S/I & C								Initially at 100,000 miles, then every 50,000 miles																		
Engine oil & filter	R	✓			✓	✓		✓	✓	✓		✓	✓	✓		✓	✓		✓	✓	✓		✓	✓	✓	✓	
Front wheel bearings ③	S/I & L								✓								✓								✓		
Fuel filter	R								✓								✓								✓		
Rear/front axle fluid level	S/I	✓			✓	✓		✓	✓	✓		✓	✓	✓		✓	✓		✓	✓	✓	✓	✓	✓			
Rotate tires	S/I	✓	✓		✓		✓		✓		✓		✓		✓		✓		✓		✓		✓		✓		✓
Shields & underhood insulation ①	S/I			✓		✓						✓					✓			✓		✓			✓		

R: Replace   S/I: Inspect and service, if necessary   L: Lubricate   C: Clean

① Vehicles with a GVWR of 8500 lbs or more only.
② If equipped.
③ 2-wheel drive models only.

FREQUENT OPERATION MAINTENANCE (SEVERE SERVICE)

If a vehicle is operated under any of the following conditions it is considered severe service:

- Towing a trailer or using a camper or car-top carrier.
- Repeated short trips of less than 5 miles in temperatures below freezing, or trips of less than 10 miles in any temperature.
- Extensive idling or low-speed driving for long distances as in heavy commercial use, such as delivery, taxi or police cars.
- Operating on rough, muddy or salt-covered roads.
- Operating on unpaved or dusty roads.
- Driving in extremely hot (over 90°) conditions.

Engine oil & filter: replace every 2500 miles.

Chassis and suspension grease points: lubricate every 2500 miles

Rear/front axle fluid level: inspect initially at 5000 miles, then every 2500 miles thereafter.

Rotate tires: every 7500 miles.

Air cleaner filter: inspect every 15,000 miles.

Front wheel bearings (2-wheel drive only): clean, inspect and repack every 15,000 miles.

93591CC0

SCHEDULED MAINTENANCE INTERVALS
GENERAL MOTORS—DENALI, DENALI XL, SUBURBAN, TAHOE, YUKON & YUKON XL—GASOLINE

TO BE SERVICED	TYPE OF SERVICE	VEHICLE MILEAGE INTERVAL (x1000)															
		7.5	15	22.5	30	37.5	45	52.5	60	67.5	75	82.5	90	97.5	105	112.5	120
Accessory drive belt	S/I								✓								✓
Automatic transmission fluid ①	R	Every 50,000 miles															
Brake system	S/I	✓	✓	✓	✓	✓	✓	✓	✓	✓	✓	✓	✓	✓	✓	✓	✓
Chassis & suspension grease points	L	✓	✓	✓	✓	✓	✓	✓	✓	✓	✓	✓	✓	✓	✓	✓	✓
Cooling fan operation	S/I		✓		✓		✓		✓		✓		✓		✓		✓
CV-joint boots & axle seals	S/I	✓	✓	✓	✓	✓	✓	✓	✓	✓	✓	✓	✓	✓	✓	✓	✓
EGR system	S/I								✓								✓
Engine coolant	R	Every 150,000 miles															
Engine oil & filter	R	✓	✓	✓	✓	✓	✓	✓	✓	✓	✓	✓	✓	✓	✓	✓	✓
EVAP system	S/I								✓								✓
Front wheel bearings ①	S/I & L				✓				✓					✓			✓
Fuel filter	R								✓								✓
Fuel system	S/I								✓								✓
Rear/front axle fluid level	S/I	✓	✓	✓	✓	✓	✓	✓	✓	✓	✓	✓	✓	✓	✓	✓	✓
Rotate tires	S/I	✓	✓	✓	✓	✓	✓	✓	✓	✓	✓	✓	✓	✓	✓	✓	✓
Shields & underhood insulation ②	S/I		✓		✓		✓		✓		✓		✓		✓		✓
Spark plugs	R	Every 100,000 miles															
Spark plug wires	S/I	Every 100,000 miles															

R: Replace   S/I: Inspect and service, if necessary   L: Lubricate
① 2-wheel drive models only.
② Vehicles with a GVWR or 8500 lbs. or more only.

FREQUENT OPERATION MAINTENANCE (SEVERE SERVICE)
If a vehicle is operated under any of the following conditions it is considered severe service:
- Towing a trailer or using a camper or car-top carrier.
- Repeated short trips of less than 5 miles in temperatures below freezing, or trips of less than 10 miles in any temperature.
- Extensive idling or low-speed driving for long distances as in heavy commercial use, such as delivery, taxi or police cars.
- Operating on rough, muddy or salt-covered roads.
- Operating on unpaved or dusty roads.
- Driving in extremely hot (over 90°) conditions.

Engine oil & filter: replace every 3000 miles or 3 months, whichever occurs first.
Chassis and suspension grease points: lubricate every 3000 miles.
Rear/front axle fluid level: inspect every 3000 miles.
Rotate the tires ever 6000 miles.
Brake system components: inspect ever 6000 miles.
Front wheel bearings (2-wheel drive only): clean, inspect and repack every 15,000 miles.
Shields & underhood insulation (vehicles w/GVWR over 8500 lbs. only): inspect every 15,000 miles.
Cooling fan system hoses & connections: inspect every 15,000 miles.
Fuel filter: replace every 30,000 miles.
Air cleaner filter: inspect every 45,000 miles.
Automatic transmission fluid & filter: replace every 50,000 miles.
Accessory drive belt: inspect every 60,000 miles.
Fuel system tank, cap and lines: inspect every 60,000 miles.
EVAP system: inspect every 60,000 miles.
EGR system: inspect every 60,000 miles.
PCV system: inspect every 100,000 miles.
Engine cooling system components: inspect and clean every 150,000 miles.

93591CD1

SCHEDULED MAINTENANCE INTERVALS
GENERAL MOTORS CORPORATION
DENALI, DENALI XL, SUBURBAN, TAHOE, YUKON, YUKON XL (GASOLINE)

The following should be used as a guide when determining the amount of work required for a particular service. In estimating how long a particular Scheduled Maintenance Service should take, please observe the following:

● Labor Time is time based on field research and data supplied by the vehicle manufacturer.
● Labor time operations are given in hours and tenths of an hour.
● All labor operations are to be used as a guide.

Mechanic Skill Level Codes:
(A) PRECISION: Highly skilled with multiple certification.
(B) GENERAL: Normally skilled with certification.
(C) MAINTENANCE: Semi-skilled working on certification.

	LABOR TIME
7500 Mile Service (C)	
All Models	1.5
w/4WD add	.1
15000 Mile Service (C)	
All Models	1.6
w/4WD add	.1
22500 Mile Service (C)	
All Models	1.5
w/4WD add	.1
30000 Mile Service (B)	
All Models	2.1
w/4WD add	.1
37500 Mile Service (C)	
All Models	1.5
w/4WD add	.1

	LABOR TIME
45000 Mile Service (C)	
All Models	1.6
w/4WD add	.1
52500 Mile Service (C)	
All Models	1.5
w/4WD add	.1
60000 Mile Service (B)	
All Models	2.5
w/4WD add	.1
67500 Mile Service (C)	
All Models	1.6
w/4WD add	.1
75000 Mile Service (C)	
All Models	1.5
w/4WD add	.1
82500 Mile Service (C)	
All Models	1.5
w/4WD add	.1

	LABOR TIME
90000 Mile Service (B)	
All Models	2.1
w/4WD add	.1
97500 Mile Service (C)	
All Models	1.6
w/4WD add	.1
105000 Mile Service (B)	
All Models	3.5
w/4WD add	.1
112500 Mile Service (C)	
All Models	1.5
w/4WD add	.1
120000 Mile Service (B)	
All Models	2.5
w/4WD add	.1

93551CD3

Timing belt service is covered in Section 3 of this manual

SCHEDULED MAINTENANCE INTERVALS
GENERAL MOTORS CORPORATION
DENALI, DENALI XL, SUBURBAN, TAHOE, YUKON, YUKON XL (DIESEL)

The following should be used as a guide when determining the amount of work required for a particular service. In estimating how long a particular Scheduled Maintenance Service should take, please observe the following:

● Labor Time is time based on field research and data supplied by the vehicle manufacturer.
● Labor time operations are given in hours and tenths of an hour.
● All labor operations are to be used as a guide.

Mechanic Skill Level Codes:
(A) PRECISION: Highly skilled with multiple certification.
(B) GENERAL: Normally skilled with certification.
(C) MAINTENANCE: Semi-skilled working on certification.

Service	LABOR TIME	Service	LABOR TIME	Service	LABOR TIME
5000 Mile Service (C) All Models	1.5	**35000 Mile Service (C)** All Models	.8	**68000 Mile Service (C)** All Models	1.0
8000 Mile Service (C) All Models	1.0	**38000 Mile Service (C)** All Models	.8	**75000 Mile Service (C)** All Models	1.5
10000 Mile Service (C) All Models	.8	**40000 Mile Service (C)** All Models	.8	**80000 Mile Service (C)** All Models	.8
15000 Mile Service (C) All Models	.8	**45000 Mile Service (C)** All Models	1.5	**83000 Mile Service (C)** All Models	1.0
20000 Mile Service (C) All Models	.8	**50000 Mile Service (C)** All Models	.8	**85000 Mile Service (C)** All Models	.8
23000 Mile Service (C) All Models	1.0	**53000 Mile Service (C)** All Models	1.0	**90000 Mile Service (B)** All Models	2.4
25000 Mile Service (C) All Models	.8	**55000 Mile Service (C)** All Models	.8	**95000 Mile Service (C)** All Models	.8
30000 Mile Service (B) All Models	2.4	**60000 Mile Service (B)** All Models	2.5	**98000 Mile Service (C)** All Models	1.0
		65000 Mile Service (C) All Models	.8		

93551CD4

GENERAL MOTORS
Chevrolet Blazer • Trail Blazer • Xtreme • GMC Envoy • Jimmy • Oldsmobile Bravada

ENGINE AND VEHICLE IDENTIFICATION

Code ①	Liters (cc)	Cu. In.	Cyl.	Fuel Sys.	Engine Type	Eng. Mfg.
S	4.2 (4200)	256	6	MFI	DOHC	CPC
W	4.3 (4293)	263	6	MFI	OHV	CPC
X	4.3 (4293)	263	6	MFI	OHV	CPC

Code ②	Year
X	1999
Y	2000
1	2001
2	2002
3	2003

CPC: Chevrolet/Pontiac/Canada

MFI: Multi-port Fuel Injection

① 8th position of VIN

② 10th position of VIN

93591CT1

Heater Core replacement is covered in Section 2 of this manual

GENERAL ENGINE SPECIFICATIONS
All measurements are given in inches.

Year	Model	Engine Displacement Liters (cc)	Engine Series (ID/VIN)	Fuel System	Net Horsepower @ rpm	Net Torque B rpm (ft. lbs.)	Bore x Stroke (in.)	Com-pression Ratio	Oil Pressure B rpm
1999	Bradava	4.3 (4293)	W	MFI	①	②	4.00x3.48	9.2:1	18@2000
		4.3 (4293)	X	MFI	③	④	4.00x3.48	9.2:1	18@2000
	Envoy	4.3 (4293)	W	MFI	230@4600	285@2800	4.00x3.48	9.4:1	18@2000
		4.3 (4293)	X	MFI	③	④	4.00x3.48	9.2:1	18@2000
	Jimmy	4.3 (4293)	W	MFI	①	②	4.00x3.48	9.2:1	18@2000
		4.3 (4293)	X	MFI	③	④	4.00x3.48	9.2:1	18@2000
	Blazer	4.3 (4293)	W	MFI	230@4600	285@2800	4.00x3.48	9.4:1	18@2000
		4.3 (4293)	X	MFI	③	④	4.00x3.48	9.2:1	18@2000
2000	Bradava	4.3 (4293)	W	MFI	①	②	4.00x3.48	9.2:1	18@2000
		4.3 (4293)	X	MFI	③	④	4.00x3.48	9.2:1	18@2000
	Envoy	4.3 (4293)	W	MFI	230@4600	285@2800	4.00x3.48	9.4:1	18@2000
		4.3 (4293)	X	MFI	③	④	4.00x3.48	9.2:1	18@2000
	Jimmy	4.3 (4293)	W	MFI	①	②	4.00x3.48	9.2:1	18@2000
		4.3 (4293)	X	MFI	③	④	4.00x3.48	9.2:1	18@2000
	Blazer	4.3 (4293)	W	MFI	190@4400	250@2800	4.00x3.48	9.4:1	18@2000
		4.3 (4293)	X	MFI	③	④	4.00x3.48	9.2:1	18@2000
2001	Bradava	4.3 (4293)	W	MFI	①	②	4.00x3.48	9.2:1	18@2000
	Envoy	4.3 (4293)	W	MFI	230@4600	285@2800	4.00x3.48	9.4:1	18@2000
	Jimmy	4.3 (4293)	W	MFI	①	②	4.00x3.48	9.2:1	18@2000
	Xtreme	4.3 (4293)	W	MFI	190@4400	250@2800	4.00x3.48	9.4:1	18@2000
	Blazer	4.3 (4293)	W	MFI	190@4400	250@2800	4.00x3.48	9.4:1	18@2000
2002-03	Bradava	4.2 (4200)	S	MFI	270@6000	275@3600	3.66x4.02	10.0:1	12@1200
	Envoy	4.2 (4200)	S	MFI	270@6000	275@3600	3.66x4.02	10.0:1	12@1200
	TrailBlazer	4.2 (4200)	S	MFI	270@6000	275@3600	3.66x4.02	10.0:1	12@1200
	Xtreme	4.3 (4293)	W	MFI	190@4400	250@2800	4.00x3.48	9.4:1	18@2000
	Blazer	4.3 (4293)	W	MFI	190@4400	250@2800	4.00x3.48	9.4:1	18@2000

MFI: Multi-port Fuel Injection

① 2WD: 245@2800
  4WD: 250@2800

② 2WD: 170B4400
  4WD: 180B4400

③ Below 15,000 GVWR: 180B3400
  Above 15,000 GVWR: 190B3400

④ 2WD: 180B4400
  4WD: 190B4400

93591CT2

GASOLINE ENGINE TUNE-UP SPECIFICATIONS

Year	Engine Displacement Liters (cc)	Engine ID/VIN	Spark Plugs Gap (in.)	Ignition Timing (deg.) MT	Ignition Timing (deg.) AT	Fuel Pump (psi)	Idle Speed (rpm) MT	Idle Speed (rpm) AT	Valve Clearance In.	Valve Clearance Ex.
1999	4.3 (4293)	W	0.060	①	①	58-64 ②	600	625	HYD	HYD
	4.3 (4293)	X	0.045	—	③	41-47	—	④	HYD	HYD
2000	4.3 (4293)	W	0.060	①	①	58-64 ②	600	625	HYD	HYD
	4.3 (4293)	X	0.045	—	③	41-47	—	④	HYD	HYD
2001	4.3 (4293)	W	0.060	①	①	58-64 ②	600	625	HYD	HYD
2002-03	4.2 (4200)	S	0.050	—	③	NA	—	④	HYD	HYD
	4.3 (4293)	W	0.060	①	①	58-64 ②	600	625	HYD	HYD

NOTE: The Vehicle Emission Control Information label often reflects specification changes made during production. The label figures must be used if they differ from those in this chart.

HYD: Hydraulic

① Distributorless ignition, cannot be adjusted

② With key ON and engine OFF

③ Distributorless ignition, cannot be adjusted

④ Idle speed is maintained by the PCM

93591CT3

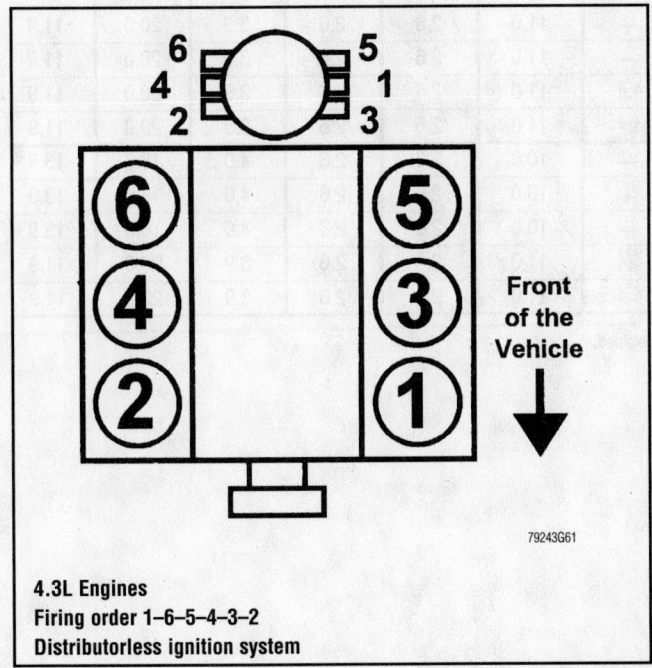

4.3L Engines
Firing order 1–6–5–4–3–2
Distributorless ignition system

79243G61

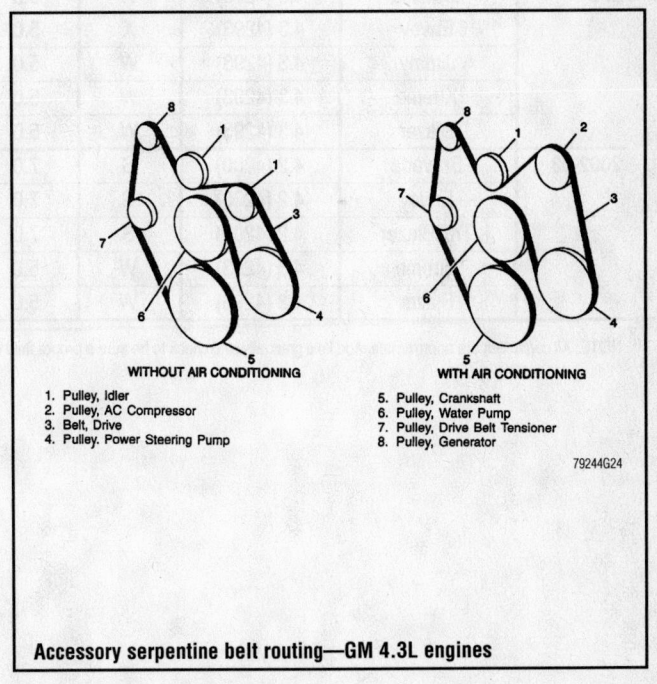

WITHOUT AIR CONDITIONING

1. Pulley, Idler
2. Pulley, AC Compressor
3. Belt, Drive
4. Pulley, Power Steering Pump

WITH AIR CONDITIONING

5. Pulley, Crankshaft
6. Pulley, Water Pump
7. Pulley, Drive Belt Tensioner
8. Pulley, Generator

79244G24

Accessory serpentine belt routing—GM 4.3L engines

Brake service is covered in Section 4 of this manual

CAPACITIES

Year	Model	Engine Displacement Liters (cc)	Engine ID/VIN	Engine Oil with Filter (qts.)	Transmission (pts.) 5-Spd	Transmission (pts.) Auto.	Transfer Case (pts.)	Drive Axle Front (pts.)	Drive Axle Rear (pts.)	Fuel Tank (gal.)	Cooling System (qts.)
1999	Bravada	4.3 (4293)	W	5.0	—	11.0	2.6	2.6	3.9	20.0	11.9
		4.3 (4293)	X	5.0	—	11.0	2.6	2.6	3.9	20.0	11.9
	Envoy	4.3 (4293)	W	5.0	—	11.0	2.6	2.6	3.9	20.0	11.9
		4.3 (4293)	X	5.0	—	11.0	2.6	2.6	3.9	20.0	11.9
	Jimmy	4.3 (4293)	W	5.0	—	11.0	2.6	2.6	3.9	20.0	11.9
		4.3 (4293)	X	5.0	—	11.0	2.6	2.6	3.9	20.0	11.9
	Blazer	4.3 (4293)	W	5.0	4.4	11.0	2.6	2.6	3.9	20.0	11.9
		4.3 (4293)	X	5.0	—	11.0	2.6	2.6	3.9	20.0	11.9
2000	Bravada	4.3 (4293)	W	5.0	—	11.0	2.6	2.6	3.9	18.6	11.9
		4.3 (4293)	X	5.0	—	11.0	2.6	2.6	3.9	18.6	11.9
	Envoy	4.3 (4293)	W	5.0	—	11.0	2.6	2.6	3.9	20.0	11.9
		4.3 (4293)	X	5.0	—	11.0	2.6	2.6	3.9	20.0	11.9
	Jimmy	4.3 (4293)	W	5.0	—	11.0	2.6	2.6	3.9	20.0	11.9
		4.3 (4293)	X	5.0	—	11.0	2.6	2.6	3.9	20.0	11.9
	Blazer	4.3 (4293)	W	5.0	4.4	11.0	2.6	2.6	3.9	20.0	11.9
		4.3 (4293)	X	5.0	—	11.0	2.6	2.6	3.9	20.0	11.9
2001	Bravada	4.3 (4293)	X	5.0	—	11.0	2.6	2.6	3.9	18.6	11.9
	Envoy	4.3 (4293)	X	5.0	—	11.0	2.6	2.6	3.9	20.0	11.9
	Jimmy	4.3 (4293)	W	5.0	—	11.0	2.6	2.6	3.9	20.0	11.9
	Xtreme	4.3 (4293)	W	5.0	4.4	11.0	2.6	2.6	3.9	20.0	11.9
	Blazer	4.3 (4293)	W	5.0	4.4	11.0	2.6	2.6	3.9	20.0	11.9
2002-03	Bravada	4.2 (4200)	S	7.0	—	10.0	2.6	2.6	4.0	18.6	13.9
	Envoy	4.2 (4200)	S	7.0	—	10.0	2.6	2.6	4.0	18.6	13.9
	TrailBlazer	4.2 (4200)	S	7.0	—	10.0	2.6	2.6	4.0	18.6	13.9
	Xtreme	4.3 (4293)	W	5.0	4.4	11.0	2.6	2.6	3.9	20.0	11.9
	Blazer	4.3 (4293)	W	5.0	4.4	11.0	2.6	2.6	3.9	20.0	11.9

NOTE: All capacities are approximate. Add fluid gradually and check to be sure a proper fluid level is obtained.

93591CT4

VALVE SPECIFICATIONS

Year	Engine Displacement Liters (cc)	Engine ID/VIN	Seat Angle (deg.)	Face Angle (deg.)	Spring Test Pressure (lbs. @ in.)	Spring Installed Height (in.)	Stem-to-Guide Clearance (in.)		Stem Diameter (in.)	
							Intake	Exhaust	Intake	Exhaust
1999	4.3 (4293)	W	46	45	187-203@1.27	1.69-1.71	0.0010	0.0020	NA	NA
	4.3 (4293)	X	46	45	187-203@1.27	1.69-1.71	0.0010	0.0020	NA	NA
2000	4.3 (4293)	W	46	45	187-203@1.27	1.69-1.71	0.0010	0.0020	NA	NA
	4.3 (4293)	X	46	45	187-203@1.27	1.69-1.71	0.0010	0.0020	NA	NA
2001	4.3 (4293)	W	46	45	187-203@1.27	1.69-1.71	0.0010	0.0020	NA	NA
2002-03	4.2 (4200)	S	NA	NA	130-142@1.26	NA	0.0011-0.0025	0.0015-0.0030	NA	NA
	4.3 (4293)	W	46	45	187-203@1.27	1.69-1.71	0.0010	0.0020	NA	NA

NA: Not Available

93591CT5

For complete Engine Mechanical specifications, see Section 1 of this manual

CRANKSHAFT AND CONNECTING ROD SPECIFICATIONS

All measurements are given in inches.

| Year | Engine Displacement Liters (cc) | Engine ID/VIN | Crankshaft | | | | Connecting Rod | | |
			Main Brg. Journal Dia.	Main Brg. Oil Clearance	Shaft End-play	Thrust on No.	Journal Diameter	Oil Clearance	Side Clearance
1999	4.3 (4293)	W	①	②	0.0020-0.0070	4	2.2487-2.2497	0.0013-0.0035	0.0060-0.0140
	4.3 (4293)	X	③	②	0.0020-0.0060	4	2.2487-2.2497	0.0013-0.0035	0.0060-0.0140
2000	4.3 (4293)	W	①	②	0.0020-0.0070	4	2.2487-2.2497	0.0013-0.0035	0.0060-0.0140
	4.3 (4293)	X	③	②	0.0020-0.0060	4	2.2487-2.2497	0.0013-0.0035	0.0060-0.0140
2001	4.3 (4293)	W	①	②	0.0020-0.0070	4	2.2487-2.2497	0.0013-0.0035	0.0060-0.0140
2002-03	4.2 (4200)	S	2.7567-2.7574	0.0004-0.0025	0.0044-0.0153	4	2.2337-2.2342	0.0008-0.0025	0.0019-0.0137
	4.3 (4293)	W	①	②	0.0020-0.0070	4	2.2487-2.2497	0.0013-0.0035	0.0060-0.0140

① No. 1: 2.4488-2.4495
  Nos. 2, 3: 2.4485-2.4494
  No. 4: 2.4480-2.4489

② No. 1: 0.0008-0.0020
  Nos. 2, 3: 0.0011-0.0023
  No. 4: 0.0017-0.0032

③ No. 1: 2.4484-2.4493
  Nos. 2, 3: 2.4481-2.4490
  No. 4: 2.4479-2.4488

93591CT6

PISTON AND RING SPECIFICATIONS

All measurements are given in inches.

Year	Engine Displacement Liters (cc)	Engine ID/VIN	Piston Clearance	Ring Gap			Ring Side Clearance		
				Top Compression	Bottom Compression	Oil Control	Top Compression	Bottom Compression	Oil Control
1999	4.3 (4293)	W	0.0007-0.0017	0.010-0.030	0.018-0.026	0.065 Max.	0.0042 Max.	0.0042 Max.	0.0020-0.0070
	4.3 (4293)	X	0.0007-0.0017	0.010-0.030	0.018-0.026	0.065 Max.	0.0042 Max.	0.0042 Max.	0.0020-0.0070
2000	4.3 (4293)	W	0.0007-0.0017	0.010-0.030	0.018-0.026	0.065 Max.	0.0042 Max.	0.0042 Max.	0.0020-0.0070
	4.3 (4293)	X	0.0007-0.0017	0.010-0.030	0.018-0.026	0.065 Max.	0.0042 Max.	0.0042 Max.	0.0020-0.0070
2001	4.3 (4293)	W	0.0007-0.0017	0.010-0.030	0.018-0.026	0.065 Max.	0.0042 Max.	0.0042 Max.	0.0020-0.0070
2002-03	4.2 (4200)	S	0.0004-0.0017	0.0079-0.0157	0.0118-0.0197	0.0098-0.0299	0.0017-0.0037	0.0017-0.0037	0.0023-0.0085
	4.3 (4293)	W	0.0007-0.0017	0.010-0.030	0.018-0.026	0.065 Max.	0.0042 Max.	0.0042 Max.	0.0020-0.0070

93591CT7

For Accessory Drive Belt illustrations, see Section 1 of this manual

TORQUE SPECIFICATIONS
All readings in ft. lbs.

Year	Engine Displacement Liters (cc)	Engine ID/VIN	Cylinder Head Bolts	Main Bearing Bolts	Rod Bearing Bolts	Crankshaft Damper Bolts	Flywheel Bolts	Manifold Intake *	Manifold Exhaust	Spark Plugs	Lug Nut
1999	4.3 (4293)	W	①	77	②	74	74	③	④	11	90
	4.3 (4293)	X	①	77	②	74	74	③	④	11	90
2000	4.3 (4293)	W	①	77	②	74	74	③	④	11	90
	4.3 (4293)	X	①	77	②	74	74	③	④	11	90
2001	4.3 (4293)	X	①	77	②	74	74	③	④	11	90
2002-03	4.2 (4200)	S	⑤	⑥	⑦	⑧	⑨	12	⑩	13	92
	4.3 (4293)	X	①	77	②	74	74	③	④	11	90

* NOTE: Applies to Lower Manifold only.

① 1st pass: 22 ft. lbs.
2nd pass:
Short bolt: Plus 55 degrees
Medium bolt: Plus 65 degrees
Long bolt: Plus 75 degrees

② 20 ft. lbs. plus 70 degrees

③ Lower intake manifold:
1st pass: 27 inch lbs.
2nd pass: 106 inch lbs.
Final pass: 11 ft. lbs.
Upper manifold bolts:
1st pass: 44 inch lbs.
2nd pass: 88 inch lbs.

④ Tighten bolts to 12 ft. lbs.
Retorque to 22 ft. lbs.

⑤ Cylinder head bolts (14)
1st pass: 22 ft. lbs.
2nd pass: Plus 155 degrees
2 short end bolts: 15 ft. lbs.
1 long end bolt: 13 ft. lbs.

⑥ 18 ft. lbs., plus 155 depress

⑦ 18 ft. lbs., plus 110 degrees

⑧ 111 ft. lbs., plus 180 degrees

⑨ 18 ft. lbs., plus 50 degrees

⑩ 1st pass: 18 ft. lbs.
2nd pass: 18 ft. lbs.
3rd pass: 18 ft. lbs.

93591CT8

WHEEL ALIGNMENT

Year	Model		Caster Range (+/-Deg.)	Caster Preferred Setting (Deg.)	Camber Range (+/-Deg.)	Camber Preferred Setting (Deg.)	Toe-in (in.)	Steering Axis Inclination (Deg.)
1999	Blazer/Jimmy/Envoy	2WD	1.00	+3.00	1.00	0	0.10+/-0.05	—
		4WD	1.00	+2.00	1.00	0	0.10+/-0.05	—
	Bravada	2WD	1.00	+3.00	1.00	0	0.10+/-0.11	—
		4WD	1.00	+2.00	1.00	0	0.10+/-0.10	—
2000	Blazer/Jimmy/Envoy	2WD	1.00	+3.00	1.00	0	0.10+/-0.05	—
		4WD	1.00	+2.00	1.00	0	0.10+/-0.05	—
	Bravada	2WD	1.00	+3.00	1.00	0	0.10+/-0.11	—
		4WD	1.00	+2.00	1.00	0	0.10+/-0.10	—
2001	Blazer/Jimmy/ Envoy/Bravada	2WD	1.00	①	1.00	0	0.10+/-0.05	—
		4WD	1.00	①	1.00	0	0.10+/-0.05	—
	Xtreme	2WD	1.00	②	1.00	0	0.10+/-0.11	—
		4WD	1.00	②	1.00	0	0.10+/-0.10	—
2002-03	Bravada/Envoy/ TrailBlazer	2WD	1.00	+3.50	1.00	-0.5	-0.5+/-0.5	—
		4WD	1.00	+3.50	1.00	-0.5	-0.5+/-0.5	—
	Blazer	2WD	1.00	①	1.00	0	0.10+/-0.05	—
		4WD	1.00	①	1.00	0	0.10+/-0.05	—
	Xtreme	2WD	1.00	②	1.00	0	0.10+/-0.11	—
		4WD	1.00	②	1.00	0	0.10+/-0.10	—

① Left Side: +2.80
   Right Side: +3.30

② Left Side: +4.70
   Right Side: +5.20

93591CT9

For Tire, Wheel and Ball Joint specifications, see Section 1 of this manual

TIRE, WHEEL AND BALL JOINT SPECIFICATIONS

| Year | Model | OEM Tires | | Tire Pressures (psi) | | Wheel Size | Ball Joint Inspection |
		Standard	Optional	Front	Rear		
1999	Blazer/Jimmy/Envoy	P205/70R15	P235/70R15 P235/75R15	36	36	6-JJ	U: 0.125 in. L ①
	Bravada	P235/70R15	None	33	35	7-JJ	②
2000	Blazer/Jimmy/Envoy	P205/70R15	P235/70R15 P235/75R15	36	36	6-JJ	L ①
	Bravada	P235/70R15	None	33	35	7-JJ	②
2001	Blazer/Jimmy/ Envoy	P205/70R15	P235/70R15 P235/75R15	36	36	6-JJ	L ①
	Bravada	P235/70R15	None	33	35	7-JJ	②
2002-03	Blazer/Envoy/TrailBlazer	P205/70R15	P235/70R15 P235/75R15	36	36	6-JJ	L ①
	Bravada	P235/70R15	None	33	35	7-JJ	②

OEM: Original Equipment Manufacturer

PSI: Pounds Per Square Inch

STD: Standard

OPT: Optional

L: Lower

U: Upper

① Do not lift truck. Inspect the boss into which the grease fitting is threaded. Replace if the boss is flush or receded below the surface of the ball joint

② Replace if any measurable movement is found.

93591CS1

BRAKE SPECIFICATIONS
All measurements in inches unless noted

Year	Model		Brake Disc Original Thickness	Brake Disc Minimum Thickness	Brake Disc Maximum Runout	Brake Drum Diameter Original Inside Diameter	Brake Drum Diameter Max. Wear Limit	Brake Drum Diameter Maximum Machine Diameter	Minimum Lining Thickness	Brake Caliper Bracket Bolts (ft. lbs.)	Brake Caliper Mounting Bolts (ft. lbs.)
1999	Bravada	F	1.030	0.965	0.003	—	—	—	0.030	NA	38
		R	0.787	0.728	0.004	9.50	9.59	9.56	0.030	NA	—
	Envoy	F	1.030	0.965	0.003	—	—	—	0.030	52	①
		R	0.787	0.728	0.004	9.50	9.59	9.56	0.030	NA	—
	Jimmy	F	1.030	0.965	0.003	—	—	—	0.030	52	①
		R	0.787	0.728	0.004	9.50	9.59	9.56	0.030	NA	—
	Blazer	F	1.030	0.965	0.003	—	—	—	0.030	52	①
		R	0.787	0.728	0.004	9.50	9.59	9.56	0.030	NA	—
2000	Bravada	F	1.030	0.965	0.003	—	—	—	0.030	52	①
		R	0.787	0.728	0.004	9.50	9.59	9.56	0.030	NA	—
	Envoy	F	1.030	0.965	0.003	—	—	—	0.030	52	①
		R	0.787	0.728	0.004	9.50	9.59	9.56	0.030	NA	—
	Jimmy	F	1.030	0.965	0.003	—	—	—	0.030	52	①
		R	0.787	0.728	0.004	9.50	9.59	9.56	0.030	NA	—
	Blazer	F	1.030	0.965	0.003	—	—	—	0.030	52	①
		R	0.787	0.728	0.004	9.50	9.59	9.56	0.030	NA	—
2001	Bravada	F	1.030	0.965	0.003	—	—	—	0.030	52	①
		R	0.787	0.728	0.004	9.50	9.59	9.56	0.030	NA	—
	Envoy	F	1.030	0.965	0.003	—	—	—	0.030	52	①
		R	0.787	0.728	0.004	9.50	9.59	9.56	0.030	NA	—
	Jimmy	F	1.030	0.965	0.003	—	—	—	0.030	52	①
		R	0.787	0.728	0.004	9.50	9.59	9.56	0.030	NA	—
	Blazer	F	1.030	0.965	0.003	—	—	—	0.030	52	①
		R	0.787	0.728	0.004	9.50	9.59	9.56	0.030	NA	—
	Xtreme	F	1.030	0.965	0.003	—	—	—	0.030	52	①
		R	0.787	0.728	0.004	9.50	9.59	9.56	0.030	NA	—
2002-03	Bravada	F	1.030	0.965	0.003	—	—	—	0.030	52	①
		R	0.787	0.728	0.004	9.50	9.59	9.56	0.030	NA	—
	Envoy	F	1.030	0.965	0.003	—	—	—	0.030	52	①
		R	0.787	0.728	0.004	9.50	9.59	9.56	0.030	NA	—
	TrailBlazer	F	1.030	0.965	0.003	—	—	—	0.030	52	①
		R	0.787	0.728	0.004	9.50	9.59	9.56	0.030	NA	—
	Blazer	F	1.030	0.965	0.003	—	—	—	0.030	52	①
		R	0.787	0.728	0.004	9.50	9.59	9.56	0.030	NA	—
	Xtreme	F	1.030	0.965	0.003	—	—	—	0.030	52	①
		R	0.787	0.728	0.004	9.50	9.59	9.56	0.030	NA	—

NA: Not Available

① 2WD: 38 ft. lbs.
4WD: 77 ft. lbs.

93591CS2

For Wheel Alignment specifications, see Section 1 of this manual

SCHEDULED MAINTENANCE INTERVALS
GM—BLAZER, JIMMY, BRAVADA, ENVOY, XTREME & TRAILBLAZER

TO BE SERVICED	TYPE OF SERVICE	VEHICLE MILEAGE INTERVAL (x1000)															
		7.5	15	22.5	30	37.5	45	52.5	60	67.5	75	82.5	90	97.5	105	112.5	120
Accessory drive belt	S/I								✓								✓
Air cleaner filter	R				✓				✓				✓				✓
Automatic transmission fluid	R	Every 50,000 miles															
Brake system ①	S/I	✓	✓	✓	✓	✓	✓	✓	✓	✓	✓	✓	✓	✓	✓	✓	✓
Chassis & suspension grease points	L	✓	✓	✓	✓	✓	✓	✓	✓	✓	✓	✓	✓	✓	✓	✓	✓
CV-joint boots & axle seals	S/I	✓	✓	✓	✓	✓	✓	✓	✓	✓	✓	✓	✓	✓	✓	✓	✓
Engine coolant system ②	S/I	Every 150,000 miles															
Engine oil & filter	R	✓	✓	✓	✓	✓	✓	✓	✓	✓	✓	✓	✓	✓	✓	✓	✓
Front wheel bearings	S/I & L				✓				✓				✓				✓
Fuel filter	R				✓				✓				✓				✓
Fuel tank, cap & lines	S/I								✓								✓
PCV valve	S/I	Every 100,000 miles															
Rear/front axle fluid level	S/I	✓	✓	✓	✓	✓	✓	✓	✓	✓	✓	✓	✓	✓	✓	✓	✓
Rotate tires	S/I	✓	✓	✓	✓	✓	✓	✓	✓	✓	✓	✓	✓	✓	✓	✓	✓
Spark plug wires	S/I	Every 100,000 miles															
Spark plugs	R	Every 100,000 miles															

R: Replace   S/I: Inspect and service, if necessary   L: Lubricate

① This should be performed when the tires are removed for rotation.

② Drain, flush and refill the cooling system, inspect the system hoses, and clean the radiator and condenser.

FREQUENT OPERATION MAINTENANCE (SEVERE SERVICE)

If a vehicle is operated under any of the following conditions it is considered severe service:

- Towing a trailer or using a camper or car-top carrier.

- Repeated short trips of less than 5 miles in temperatures below freezing, or trips of less than 10 miles in any temperature.

- Extensive idling or low-speed driving for long distances as in heavy commercial use, such as delivery, taxi or police cars.

- Operating on rough, muddy or salt-covered roads.

- Operating on unpaved or dusty roads.

- Driving in extremely hot (over 90°) conditions.

Engine oil & filter: replace every 3000 miles or 3 months, whichever occurs first.

Chassis and suspension grease points: lubricate every 3000 miles.

Rear/front axle fluid level: inspect every 3000 miles.

Rotate the tires ever 6000 miles.

Brake system components: inspect ever 6000 miles.

Front wheel bearings (2-wheel drive only): clean, inspect and repack every 15,000 miles.

Air cleaner filter: inspect every 15,000 miles.

Automatic transmission fluid & filter: replace every 15,000 miles.

93591CS3

SCHEDULED MAINTENANCE INTERVALS
GENERAL MOTORS CORPORATION
BLAZER, JIMMY, BRAVADA, ENVOY,
XTREME, TRAIL BLAZER

The following should be used as a guide when determining the amount of work required for a particular service.
In estimating how long a particular Scheduled Maintenance Service should take, please observe the following:

● Labor Time is time based on field research and data supplied by the vehicle manufacturer.
● Labor time operations are given in hours and tenths of an hour.
● All labor operations are to be used as a guide.

Mechanic Skill Level Codes:
(A) PRECISION: Highly skilled with multiple certification.
(B) GENERAL: Normally skilled with certification.
(C) MAINTENANCE: Semi-skilled working on certification.

	LABOR TIME		LABOR TIME		LABOR TIME
7500 Mile Service (C)		**45000 Mile Service (C)**		**90000 Mile Service (B)**	
All models	.7	All models	.7	All models	1.1
w/4WD add	.2	w/4WD add	.2	w/4WD add	.2
15000 Mile Service (C)		w/AT add	.5	w/AT add	.5
All models	.7	**52500 Mile Service (C)**		**97500 Mile Service (C)**	
w/4WD add	.2	All models	.7	All models	.7
w/AT add	.5	w/4WD add	.2	w/4WD add	.2
22500 Mile Service (C)		**60000 Mile Service (B)**		**105000 Mile Service (C)**	
All models	.7	All models	1.3	All models	.7
w/4WD add	.2	w/4WD add	.2	w/4WD add	.2
30000 Mile Service (B)		w/AT add	.5	w/AT add	.5
All models	1.1	**67500 Mile Service (C)**		**112500 Mile Service (C)**	
w/4WD add	.2	All models	.7	All models	.7
w/AT add	.5	w/4WD add	.2	w/4WD add	.2
37500 Mile Service (C)		**75000 Mile Service (C)**		**120000 Mile Service (B)**	
All models	.7	All models	.7	All models	1.3
w/4WD add	.2	w/4WD add	.2	w/4WD add	.2
		w/AT add	.5	w/AT add	.5
		82500 Mile Service (C)			
		All models	.7		
		w/4WD add	.2		

93551CS3

For Maintenance Interval recommendations, see Section 1 of this manual

GENERAL MOTORS
Pontiac Vibe

GENERAL ENGINE SPECIFICATIONS

Year	Model	Engine Displacement Liters (cc)	Engine Series (ID/VIN)	Fuel System	Net Horsepower @ rpm	Net Torque @ rpm (ft. lbs.)	Bore x Stroke (in.)	Compression Ratio	Oil Pressure @ idle
2003	Vibe	1.8 (1794)	1ZZ-FE	EFI	①	②	3.11x3.60	10.0:1	4.2
	Vibe	1.8 (1796)	2ZZ-GE	EFI	180@7600	130@6800	3.23x3.35	11.5:1	2.8

EFI: Electronic Fuel Injection

① 2WD models: 130@6000
   4WD models: 123@6000

② 2WD models: 126@4200
   4WD models: 119@4200

93591CF1

ENGINE TUNE-UP SPECIFICATIONS

Year	Engine Displacement Liters (cc)	Engine ID/VIN	Spark Plug Gap (in.)	Ignition Timing (deg.)	Fuel Pump (psi)	Idle Speed (rpm) MT	Idle Speed (rpm) AT	Valve Clearance In.	Valve Clearance Ex.
2003	1.8 (1794)	1ZZ-FE	0.043	①	44-50	650-750	650-750	0.0059-0.0098	0.0098-0.0138
	1.8 (1796)	2ZZ-GE	0.043	②	44-50	750-850	700-800	0.0031-0.0071	0.0087-0.0126

Note: The Vehicle Emission Control Information label often reflects specification changes made during production. The label figures must be used if they differ from those in this chart.

① With terminal TC and CG of DLC3 connected: 8-12 degrees BTDC
   With terminal TC and CG of DLC3 disconnected: 10-18 degrees BTDC

② With terminal TC and CG of DLC3 connected: 8-12 degrees BTDC
   With terminal TC and CG of DLC3 disconnected:
   A/T: 10-18 degrees BTDC

93591CF2

CAPACITIES

Year	Model	Engine Displacement Liters (cc)	Engine ID/VIN	Engine Oil with Filter	Transmission (pts.) 5-Spd	6-Spd	Auto.	Drive Axle Front (pts.)	Rear (pts.)	Fuel Tank (gal.)	Cooling System (qts.)
2003	Vibe	1.8 (1794)	1ZZ-FE	3.9	4.0	4.0	5.2	3.0	—	13.2	①
	Vibe	1.8 (1796)	2ZZ-GE	4.7	4.8	4.8	6.1	②	—	13.2	6.0

Note: All capacities are approximate. Add fluid gradually and check to be sure a proper fluid level is obtained.

① M/T with Nippodenso radiator: 5.6
  A/T with Nippodenso radiator: 6.2
  M/T with Harrison radiator: 6.3
  A/T with Harrison radiator: 6.2

② Included in transaxle capacity

93591CF3

VALVE SPECIFICATIONS

Year	Engine Displacement Liters (cc)	Engine ID/VIN	Seat Angle (deg.)	Face Angle (deg.)	Spring Test Pressure (lbs. @ in.)	Spring Installed Height (in.)	Stem-to-Guide Clearance (in.) Intake	Exhaust	Stem Diameter (in.) Intake	Exhaust
2003	1.8 (1794)	1ZZ-FE	45	44.5	31.3-34.8 @ 1.252	1.323	0.0010-0.0024	0.0012-0.0026	0.2154-0.2159	0.2152-0.2158
	1.8 (1796)	2ZZ-GE	45	44.5	①	1.516	0.0010-0.0023	0.0012-0.0025	0.2145-0.2156	0.2144-0.2154

① Intake: 49.6-55.5 @ 1.516
  Exhaust: 47.6-52.6 @ 1.516

93591CF4

For Tune-up, Capacities and Firing orders, see Section 1 of this manual

CRANKSHAFT AND CONNECTING ROD SPECIFICATIONS
All measurements are given in inches.

Year	Engine Displacement Liters (cc)	Engine ID/VIN	Crankshaft				Connecting Rod		
			Main Brg. Journal Dia.	Main Brg. Oil Clearance	Shaft End-play	Thrust on No.	Journal Diameter	Oil Clearance	Side Clearance
2003	1.8 (1762)	1ZZ-FE	1.8893-1.8898	0.0006-0.0013	0.0008-0.0087	3	1.7320-1.7323	0.0011-0.0024	0.0063-0.0135
	1.8 (1796)	2ZZ-GE	1.8893-1.8898	0.0006-0.0013	0.0016-0.0094	3	1.7713-1.7717	0.0011-0.0020	0.0063-0.0135

93591CF5

PISTON AND RING SPECIFICATIONS
All measurements are given in inches.

Year	Engine Displacement Liters (cc)	Engine ID/VIN	Piston Clearance	Ring Gap			Ring Side Clearance		
				Top Compression	Bottom Compression	Oil Control	Top Compression	Bottom Compression	Oil Control
2003	1.8 (1762)	1ZZ-FE	0.0026-0.0035	0.0098-0.0138	0.0138-0.0197	0.0059-0.0197	0.0008-0.0028	0.0012-0.0028	0.0012-0.0043
	1.8 (1796)	2ZZ-GE	0.0003-0.0015	0.0098-0.0138	0.0138-0.0197	NA	0.0009-0.0028	0.0012-0.0028	NA

NA - Not available

93591CF6

TORQUE SPECIFICATIONS
All readings in ft. lbs.

Year	Engine Displacement Liters (cc)	Engine ID/VIN	Cylinder Head Bolts	Main Bearing Bolts	Rod Bearing Bolts	Crankshaft Damper Bolts	Flywheel Bolts	Manifold		Spark Plugs	Lug Nuts
								Intake	Exhaust		
2003	1.8 (1794)	1ZZ-FE	①	②	③	102	①	22	27	18	76
	1.8 (1796)	2ZZ-GE	④	⑤	⑥	87	①	⑦	37	13	76

① Step 1: 36 ft. lbs.
　Step 2: 90 degree turn

② 12 pointed bolts:
　Step 1: 33 ft. lbs.
　Step 2: 90 degree turn
　Hex head bolts: 14 ft. lbs.

③ Step 1: 15 ft. lbs.
　Step 2: 90 degree turn

④ Step 1: 26 ft. lbs.
　Step 2: 180 degree turn

⑤ 12 pointed bolts:
　Step 1: 16 ft. lbs.
　Step 2: 32 ft. lbs.
　Step 3: 45 degree turn
　Step 4: 45 degree turn
　Hex head bolts: 13 ft. lbs.

⑥ Step 1: 22 ft. lbs.
　Step 2: 90 degree turn

⑦ Bolt A: 25 ft. lbs.
　Bolt B: 34 ft. lbs.

93591CF7

WHEEL ALIGNMENT

Year	Model		Caster		Camber		Toe-in (in.)	Steering Axis Inclination (Deg.)
			Range (+/-Deg.)	Preferred Setting (Deg.)	Range (+/-Deg.)	Preferred Setting (Deg.)		
2003	Vibe - 2WD	F	0.75	+2.78	0.75	-0.77	0+/-0.08	12.47+/-0.75
		R	—	—	0.50	-1.45	0.11+/-0.11	—
	Vibe - 4WD	F	0.75	+2.77	0.75	-0.48	0+/-0.08	12.22+/-0.75
		R	—	—	0.75	-0.73	0.08+/-0.08	—

93591CF8

For complete service labor times, order Nichols' Chilton Labor Guide

TIRE, WHEEL AND BALL JOINT SPECIFICATIONS

| Year | Model | OEM Tires | | Tire Pressures (psi) | | Wheel Size | Ball Joint Inspection |
		Standard	Optional	Front	Rear		
2003	Vibe	205/55R16	—	33	33	6.5-JJ	9-26 in. ①

OEM: Original Equipment Manufacturer

PSI: Pounds Per Square Inch

STD: Standard

OPT: Optional

① Torque required in inch lbs. to rotate ball joint when removed from the knuckle

93591CF9

BRAKE SPECIFICATIONS
All measurements in inches unless noted

| Year | Model | | Brake Disc | | | Brake Drum Diameter | | | Minimum Lining Thickness | Brake Caliper | |
			Original Thickness	Minimum Thickness	Maximum Runout	Original Inside Diameter	Max. Wear Limit	Maximum Machine Diameter		Bracket Bolts (ft. lbs.)	Mounting Bolts (ft. lbs.)
2003	Vibe	F	0.984	0.906	0.0020	—	—	—	0.039	25	79
		R	0.354	0.295	0.0059	9.00	—	9.04	0.039	—	34

F: Front

R: Rear

93591CF0

SCHEDULED MAINTENANCE INTERVALS
PONTIAC—VIBE

TO BE SERVICED	TYPE OF SERVICE	7.5	15	22.5	30	37.5	45	52.5	60	67.5	75	82.5	90	97.5
Engine oil & filter	R	✓	✓	✓	✓	✓	✓	✓	✓	✓	✓	✓	✓	✓
Drive belts	S/I								✓	✓	✓	✓	✓	✓
Automatic transaxle fluid & filter	S/I		✓		✓		✓		✓		✓		✓	
Ball joints & dust covers	S/I		✓		✓		✓		✓		✓		✓	
Bolts & nuts on body & chassis	S/I		✓		✓		✓		✓		✓		✓	
Brake line pipes & hoses	S/I		✓		✓		✓		✓		✓		✓	
Brake linings & drums	S/I		✓		✓		✓		✓		✓		✓	
Brake pads & discs (front & rear if equipped)	S/I		✓		✓		✓		✓		✓		✓	
Differential oil	S/I		✓		✓		✓		✓		✓		✓	
Drive shaft boots (except Supra)	S/I		✓		✓		✓		✓		✓		✓	
Manual transaxle oil	S/I		✓		✓		✓		✓		✓		✓	
Steering gear housing oil	S/I		✓		✓		✓		✓		✓		✓	
Steering linkage	S/I		✓		✓		✓		✓		✓		✓	
Air filter	R				✓				✓				✓	
Spark plugs	R				✓				✓				✓	
Spark plugs (platinum tip)	R								✓					
Exhaust system	S/I				✓				✓				✓	
Fuel lines & connections	S/I				✓				✓				✓	
Valve clearance	S/I				✓				✓				✓	
Engine coolant	R						✓				✓			
Fuel tank cap gasket	R								✓					
Charcoal canister	S/I								✓					

R: Replace   S/I: Service or Inspect

FREQUENT OPERATION MAINTENANCE (SEVERE SERVICE)

If a vehicle is operated under any of the following conditions it is considered severe service:

- Extremely dusty areas.

- 50% or more of the vehicle operation is in 32°C (90°F) or higher temperatures, or constant operation in temperatures below 0°C (32°F).

- Prolonged idling (vehicle operation in stop and go traffic).

- Frequent short running periods (engine does not warm to normal operating temperatures).

- Police, taxi, delivery usage or trailer towing usage.

Oil & oil filter: change every 6000 miles.

Bolts & nuts on chassis & body: tighten every 7500 miles.

Ball joints & dust covers: service or inspect every 12,000 miles.

Brake linings & drums: service or inspect ever 12,000 miles.

Brake pads & discs (front & rear if equipped): service or inspect every 12,000 miles.

Drive shaft boots & except Supra): service or inspect every 12,000 miles.

Steering linkage: service or inspect every 12,000 miles.

Air filter: service or inspect every 15,000 miles.

Exhaust system: service or inspect every 15,000 miles.

Timing belt: replace every 60,000 miles.

93591CG1

GEO CHEVROLET
Tracker

ENGINE AND VEHICLE IDENTIFICATION CHART

			Engine Code					Model Year	
Code	Liters (cc)	Cu. In.	Cyl.	Fuel Sys.	Engine Type	Eng. Mfg.		Code	Year
6	1.6 (1590)	97	4	MFI	SOHC	Suzuki		X	1999
1	2.5 (2494)	152	6	MFI	DOHC	Suzuki		Y	2000
C	2.0 (1997)	122	4	MFI	DOHC	Suzuki		1	2001
								2	2002
								3	2003

MFI: Multiport Fuel Injection
DOHC: Dual Overhead Cam
SOHC: Single Overhead Cam

93591CI6

GENERAL ENGINE SPECIFICATIONS

Year	Model	Engine Displacement Liters (cc)	Engine ID/VIN	Fuel System Type	Net Horsepower @ rpm	Net Torque @ rpm (ft. lbs.)	Bore x Stroke (in.)	Compression Ratio	Oil Pressure @ rpm
1999	Tracker	1.6 (1590)	6	MFI	95@5600	98@4000	2.95x3.54	9.5:1	47-61@4000
		2.0 (1997)	C	MFI	127@6000	134@3000	3.31x3.54	NA	55-67@4000
2000	Tracker	1.6 (1590)	6	MFI	95@5600	98@4000	2.95x3.54	9.5:1	47-61@4000
		2.0 (1997)	C	MFI	127@6000	134@3000	3.31x3.54	NA	55-67@4000
2001	Tracker	2.0 (1997)	C	MFI	127@6000	134@3000	3.31x3.54	NA	55-67@4000
		2.5 (2494)	1	MFI	140@6500	151@4000	3.31x2.95	9.5:1	55-67@4000
2002	Tracker	2.0 (1997)	C	MFI	127@6000	134@3000	3.31x3.54	NA	55-67@4000
		2.5 (2494)	1	MFI	140@6500	151@4000	3.31x2.95	9.5:1	55-67@4000

MFI: Multi-port Fuel Injection
NA: Not available

93591CI7

ENGINE TUNE-UP SPECIFICATIONS

Year	Engine Displacement Liters (cc)	Engine ID/VIN	Spark Plugs Gap (in.)	Ignition Timing (deg.) MT	AT	Fuel Pump (psi)	Idle Speed (rpm) MT	AT	Valve Clearance In.	Ex.
1999	1.6 (1590)	6	0.030	5B	5B	30-37	800-850	800-850	0.0050-0.0070	0.0050-0.0070
	2.0 (1997)	C	0.041	5B	5B	35-43	700-800	700-800	HYD	HYD
2000	1.6 (1590)	6	0.030	5B	5B	30-37	800-850	800-850	0.0050-0.0070	0.0050-0.0070
	2.0 (1997)	C	0.041	5B	5B	35-43	700-800	700-800	HYD	HYD
2001	2.5 (2494)	1	0.040	5B	5B	30-45	700-800	700-800	HYD	HYD
	2.0 (1997)	C	0.041	5B	5B	35-43	700-800	700-800	HYD	HYD
2002	2.5 (2494)	1	0.040	5B	5B	30-45	700-800	700-800	HYD	HYD
	2.0 (1997)	C	0.041	5B	5B	35-43	700-800	700-800	HYD	HYD

HYD: Hydraulic

93591CI8

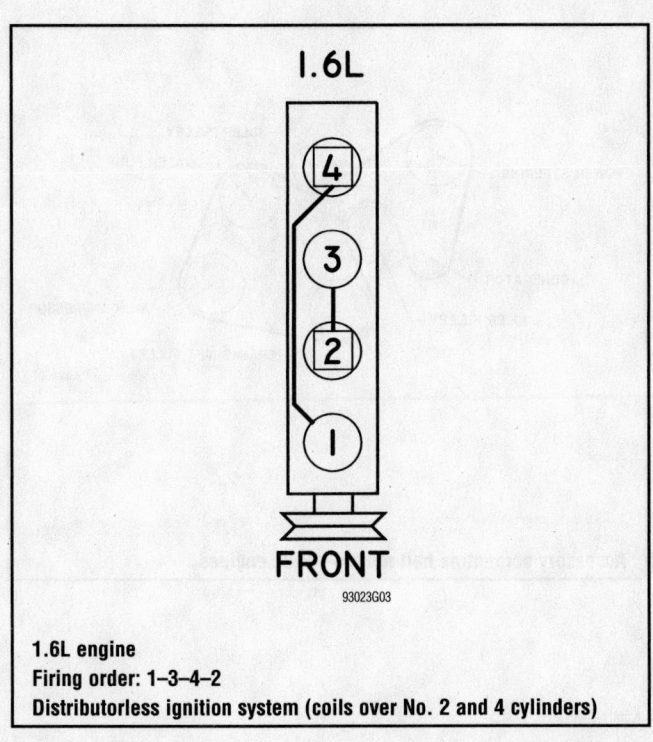

1.6L engine
Firing order: 1–3–4–2
Distributorless ignition system (coils over No. 2 and 4 cylinders)

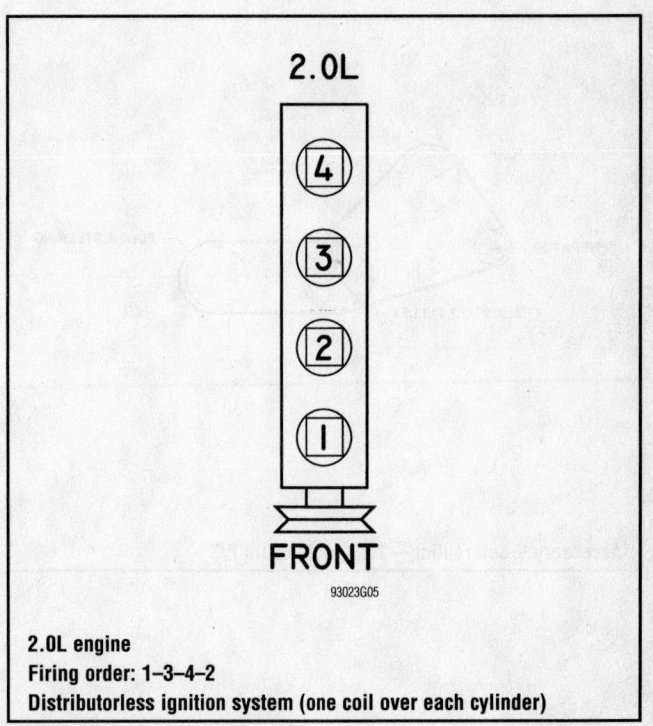

2.0L engine
Firing order: 1–3–4–2
Distributorless ignition system (one coil over each cylinder)

Timing belt service is covered in Section 3 of this manual

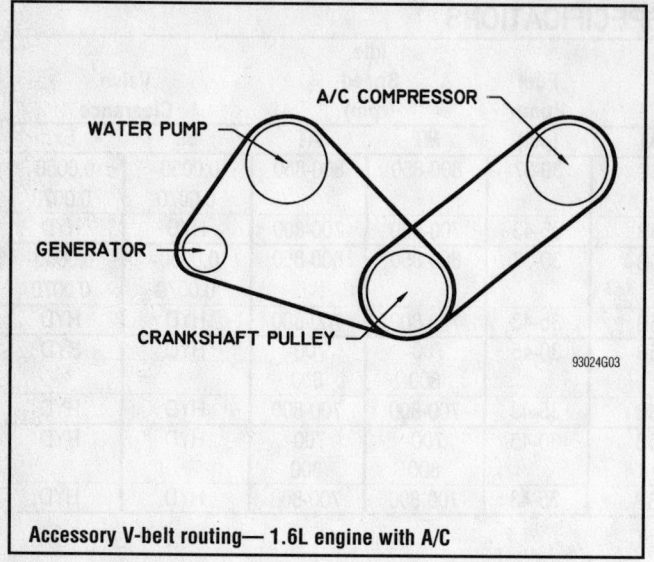

Accessory V-belt routing— 1.6L engine with A/C

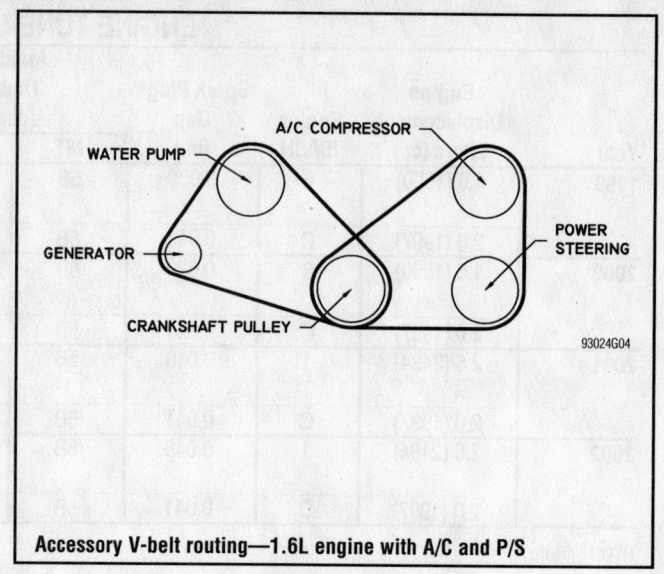

Accessory V-belt routing—1.6L engine with A/C and P/S

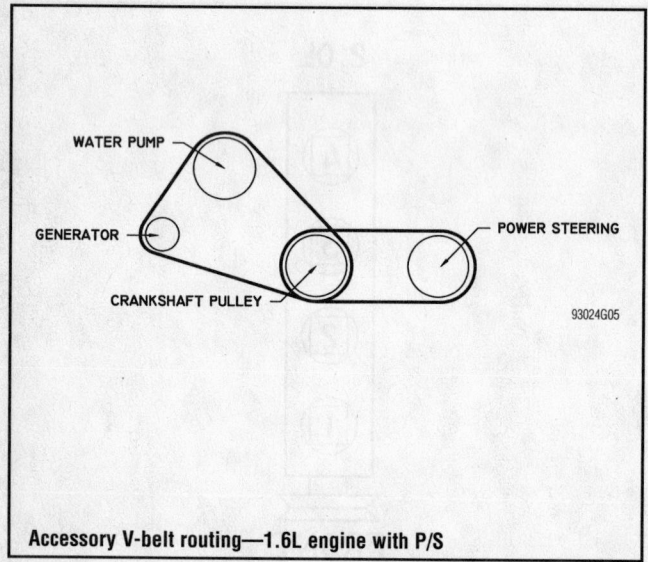

Accessory V-belt routing—1.6L engine with P/S

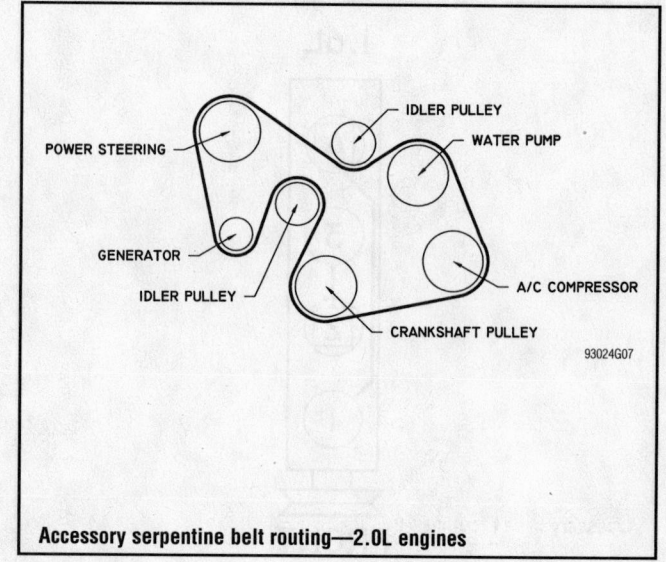

Accessory serpentine belt routing—2.0L engines

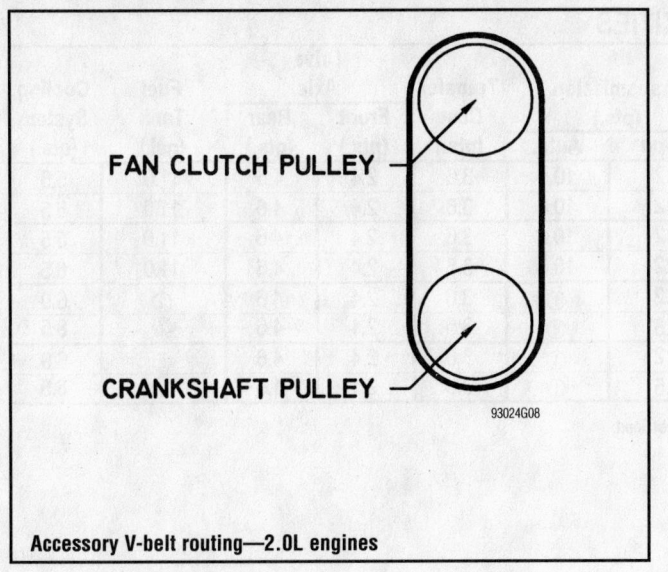

Accessory V-belt routing—2.0L engines

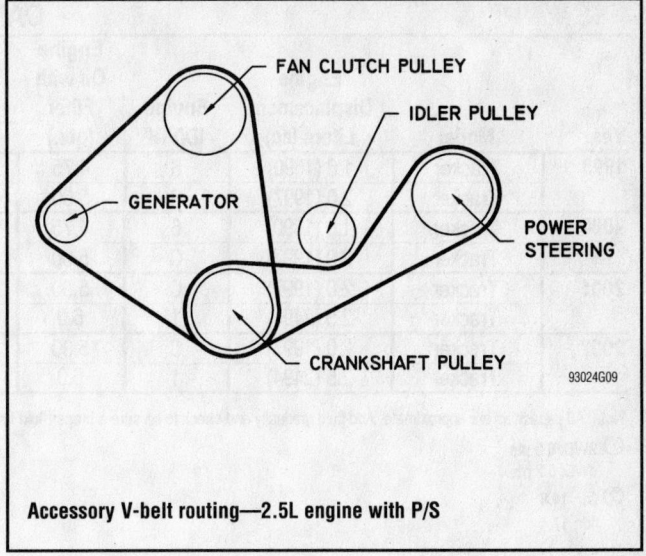

Accessory V-belt routing—2.5L engine with P/S

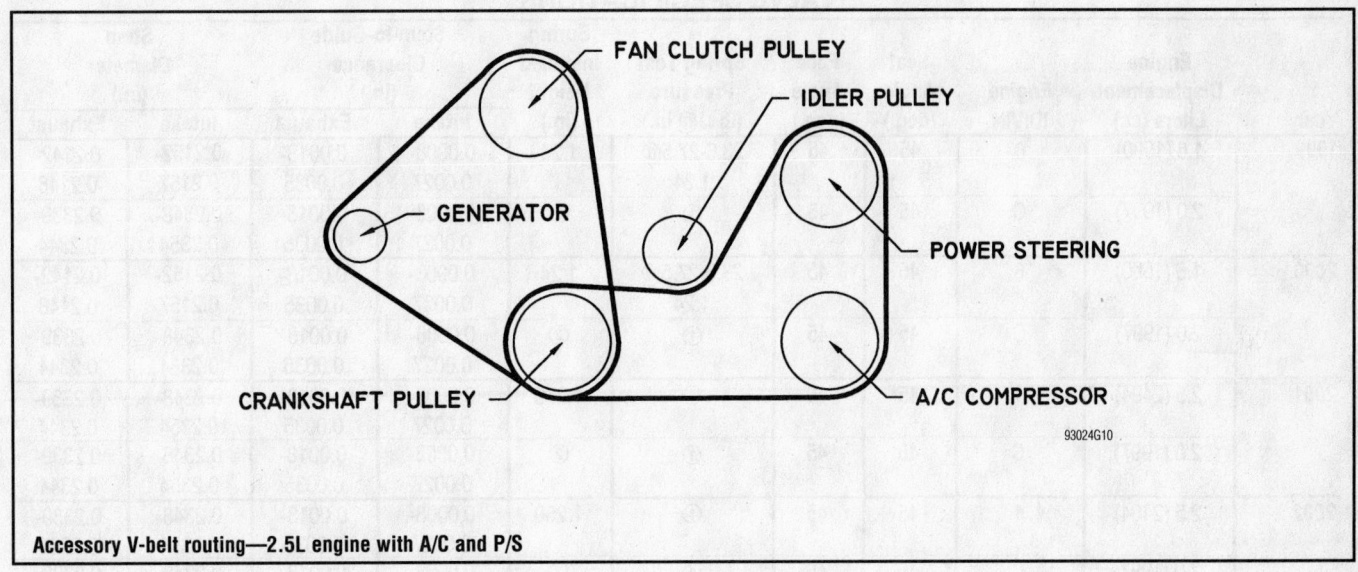

Accessory V-belt routing—2.5L engine with A/C and P/S

Heater Core replacement is covered in Section 2 of this manual

CAPACITIES

Year	Model	Engine Displacement Liters (cc)	Engine ID/VIN	Engine Oil with Filter (qts.)	Transmission (pts.)		Transfer Case (pts.)	Drive Axle		Fuel Tank (gal.)	Cooling System (qts.)
					5-Spd	Auto.		Front (pts.)	Rear (pts.)		
1999	Tracker	1.6 (1590)	6	4.75	3.2	10.6	3.6	2.4	4.6	11.0	5.5
	Tracker	2.0 (1997)	C	5.90	3.2	10.6	3.6	2.4	4.6	11.0	6.5
2000	Tracker	1.6 (1590)	6	4.75	3.2	10.6	3.6	2.4	4.6	11.0	5.5
	Tracker	2.0 (1997)	C	5.90	3.2	10.6	3.6	2.4	4.6	11.0	6.5
2001	Tracker	2.0 (1997)	C	5.90	3.2	①	3.6	2.4	4.6	②	6.9
	Tracker	2.5 (2494)	1	6.0	5.5	①	3.6	2.1	4.6	②	8.5
2002	Tracker	2.0 (1997)	C	5.90	3.2	①	3.6	2.4	4.6	②	6.9
	Tracker	2.5 (2494)	1	6.0	5.5	①	3.6	2.1	4.6	②	8.5

Note: All capacities are approximate. Add fluid gradually and check to be sure a proper fluid level is obtained.

① 2WD: 6.0 pts
   4WD: 5.2 pts.

② 2dr: 14.8
   4dr: 17.4

93591CI9

VALVE SPECIFICATIONS

Year	Engine Displacement Liters (cc)	Engine ID/VIN	Seat Angle (deg.)	Face Angle (deg.)	Spring Test Pressure (lbs. @ in.)	Spring Installed Height (in.)	Stem-to-Guide Clearance (in.)		Stem Diameter (in.)	
							Intake	Exhaust	Intake	Exhaust
1999	1.6 (1590)	6	45	45	23.6-27.5@ 1.24	1.24	0.0008- 0.0027	0.0018- 0.0035	0.2152- 0.2157	0.2142- 0.2148
	2.0 (1997)	C	45	45	①	②	0.0008- 0.0027	0.0018- 0.0035	0.2348- 0.2354	0.2339- 0.2344
2000	1.6 (1590)	6	45	45	23.6-27.5@ 1.24	1.24	0.0008- 0.0027	0.0018- 0.0035	0.2152- 0.2157	0.2142- 0.2148
	2.0 (1997)	C	45	45	①	②	0.0008- 0.0027	0.0018- 0.0035	0.2348- 0.2354	0.2339- 0.2344
2001	2.5 (2494)	1	45	45	①	1.250	0.0008- 0.0027	0.0018- 0.0035	0.2348- 0.2354	0.2339- 0.2344
	2.0 (1997)	C	45	45	①	②	0.0008- 0.0027	0.0018- 0.0035	0.2348- 0.2354	0.2339- 0.2344
2002	2.5 (2494)	1	45	45	①	1.250	0.0008- 0.0027	0.0018- 0.0035	0.2348- 0.2354	0.2339- 0.2344
	2.0 (1997)	C	45	45	①	②	0.0008- 0.0027	0.0018- 0.0035	0.2348- 0.2354	0.2339- 0.2344

① Inner: 13.6-17.4@1.08
   Outer: 30.4-39.2@1.25

② Inner spring: 1.08
   Outer spring: 1.25

93591CI0

CRANKSHAFT AND CONNECTING ROD SPECIFICATIONS
All measurements are given in inches.

Year	Engine Displacement Liters (cc)	Engine ID/VIN	Crankshaft				Connecting Rod		
			Main Brg. Journal Dia.	Main Brg. Oil Clearance	Shaft End-play	Thrust on No.	Journal Diameter	Oil Clearance	Side Clearance
1999	1.6 (1590)	6	2.0465-2.0472	0.0006-0.0023	0.0044-0.0149	3	1.7316-1.7322	0.0008-0.0031	NA
	2.0 (1997)	C	2.2828-2.2834	0.0008-0.0023	0.0039-0.0165	3	1.9678-1.9685	0.0016-0.0031	NA
2000	1.6 (1590)	6	2.0465-2.0472	0.0006-0.0023	0.0044-0.0149	3	1.7316-1.7322	0.0008-0.0031	NA
	2.0 (1997)	C	2.2828-2.2834	0.0008-0.0023	0.0039-0.0165	3	1.9678-1.9685	0.0016-0.0031	NA
2001	2.5 (2494)	1	2.5583-2.5590	0.0008-0.0023	0.0044-0.0149	2	1.9678-1.9685	0.0016-0.0031	NA
	2.0 (1997)	C	2.2828-2.2834	0.0008-0.0023	0.0039-0.0165	3	1.9678-1.9685	0.0016-0.0031	NA
2002	2.5 (2494)	1	2.5583-2.5590	0.0008-0.0023	0.0044-0.0149	2	1.9678-1.9685	0.0016-0.0031	NA
	2.0 (1997)	C	2.2828-2.2834	0.0008-0.0023	0.0039-0.0165	3	1.9678-1.9685	0.0016-0.0031	NA

NA: Not Available

93591CJ1

PISTON AND RING SPECIFICATIONS
All measurements are given in inches.

Year	Engine Displacement Liters (cc)	Engine ID/VIN	Piston Clearance	Ring Gap			Ring Side Clearance		
				Top Compression	Bottom Compression	Oil Control	Top Compression	Bottom Compression	Oil Control
1999	1.6 (1590)	6	0.0008-0.0015	0.0079-0.0275	0.0138-0.0275	0.0039-0.0669	0.0012-0.0027	0.0008-0.0023	NA
	2.0 (1997)	C	0.0008-0.0015	0.0079-0.0276	0.0138-0.0276	0.0079-0.0709	0.0012-0.0027	0.0008-0.0023	NA
2000	1.6 (1590)	6	0.0008-0.0015	0.0079-0.0275	0.0138-0.0275	0.0039-0.0669	0.0012-0.0027	0.0008-0.0023	NA
	2.0 (1997)	C	0.0008-0.0015	0.0079-0.0276	0.0138-0.0276	0.0079-0.0709	0.0012-0.0027	0.0008-0.0023	NA
2001	2.5 (2494)	1	0.0008-0.0015	0.0079-0.0276	0.0138-0.0276	0.0079-0.0709	0.0012-0.0027	0.0008-0.0023	NA
	2.0 (1997)	C	0.0008-0.0015	0.0079-0.0276	0.0138-0.0276	0.0079-0.0709	0.0012-0.0027	0.0008-0.0023	NA
2002	2.5 (2494)	1	0.0008-0.0015	0.0079-0.0276	0.0138-0.0276	0.0079-0.0709	0.0012-0.0027	0.0008-0.0023	NA
	2.0 (1997)	C	0.0008-0.0015	0.0079-0.0276	0.0138-0.0276	0.0079-0.0709	0.0012-0.0027	0.0008-0.0023	NA

NA: Not Available

93591CJ2

Brake service is covered in Section 4 of this manual

TORQUE SPECIFICATIONS
All readings in ft. lbs.

Year	Engine Displacement Liters (cc)	Engine ID/VIN	Cylinder Head Bolts	Main Bearing Bolts	Rod Bearing Bolts	Crankshaft Damper Bolts	Flywheel Bolts	Manifold		Spark Plugs	Lug Nut
								Intake	Exhaust		
1999	1.6 (1590)	6	①	39 ②	25.5	94 ③	58	17	17	21	70
	2.0 (1997)	C	④	⑤	33	109	51	17	17	18	70
2000	1.6 (1590)	6	①	39 ②	25.5	94 ③	58	17	17	21	70
	2.0 (1997)	C	④	⑤	33	109	51	17	17	18	70
2001	2.5 (2494)	1	④	⑤	33	109	51	17	22	18	70
	2.0 (1997)	C	④	⑤	33	109	51	17	17	18	70
2002	2.5 (2494)	1	④	⑤	33	109	51	17	22	18	70
	2.0 (1997)	C	④	⑤	33	109	51	17	17	18	70

① Step 1: 26 ft. lbs.

Step 2: 41 ft. lbs.

Step 3: Loosen in reverse order to 0 ft. lbs.

Step 4: 26 ft. lbs.

Step 5: 52 ft. lbs.

② Use multiple passes to arrive at final torque.

③ Value shown is for crankshaft timing belt sprocket

④ Step 1: 38 ft. lbs.

Step 2: 61 ft. lbs.

Step 3: Loosen in reverse order to 0 ft. lbs.

Step 4: 38 ft. lbs.

Step 5: 76 ft. lbs.

Step 6: Tighten 6mm bolt to 8 ft. lbs.

⑤ 10mm: 43.5 ft. lbs.

8mm: 19.5 ft. lbs.

93591CJ3

WHEEL ALIGNMENT

Year	Model	Caster		Camber		Toe-in (in.)	Steering Axis Inclination (Deg.)
		Range (+/-Deg.)	Preferred Setting (Deg.)	Range (+/-Deg.)	Preferred Setting (Deg.)		
1999	Tracker	1.00	+2.67	1.00	0	0+/-0.16	—
2000	Tracker	1.00	+2.67	1.00	0	0+/-0.16	—
2001	Tracker	1.00	+2.67	1.00	0	0+/-0.16	—
2002	Tracker	1.00	+2.67	1.00	0	0+/-0.16	—

93591CJ4

TIRE, WHEEL AND BALL JOINT SPECIFICATIONS

| Year | Model | OEM Tires | | Tire Pressures (psi) | | Wheel Size | Ball Joint Inspection |
		Standard	Optional	Front	Rear		
1999	Tracker 2wd	P195/75R15	None	23	23	5.5-JJ	①
	Tracker 4wd	P205/75R15	None	23	23	5.5-JJ	①
2000	Tracker 2wd	P195/75R15	None	23	23	5.5-JJ	①
	Tracker 4wd	P205/75R15	None	23	23	5.5-JJ	①
2001	Tracker 2wd	P195/75R15	None	23	23	5.5-JJ	①
	Tracker 4wd	P205/75R15	None	23	23	5.5-JJ	①
2002	Tracker 2wd	P195/75R15	None	23	23	5.5-JJ	①
	Tracker 4wd	P205/75R15	None	23	23	5.5-JJ	①

OEM: Original Equipment Manufacturer

PSI: Pounds Per Square Inch

STD: Standard

OPT: Optional

① Replace if any measurable movement is found.

93591CJ5

BRAKE SPECIFICATIONS
All measurements in inches unless noted

| Year | Model | Brake Disc | | | Brake Drum Diameter | | | Minimum Lining Thickness | | Brake Caliper | |
		Original Thickness	Minimum Thickness	Maximum Runout	Original Inside Diameter	Max. Wear Limit	Maximum Machine Diameter	Front	Rear	Bracket Bolts (ft. lbs.)	Mounting Bolts (ft. lbs.)
1999	Tracker ①	0.670	0.590	0.006	8.66	8.74	8.74	0.08	0.04	51-72	19-21
	Tracker ②	0.670	0.590	0.006	8.66	8.74	8.74	0.08	0.04	51-72	19-21
2000	Tracker ①	0.670	0.590	0.006	8.66	8.74	8.74	0.08	0.04	61.5	19-21
	Tracker ②	0.670	0.590	0.006	8.66	8.74	8.74	0.08	0.04	61.5	19-21
2001	Tracker ①	0.670	0.590	0.006	8.66	8.74	8.74	0.08	0.04	61.5	19-21
	Tracker ②	0.670	0.590	0.006	8.66	8.74	8.74	0.08	0.04	61.5	19-21
2002	Tracker ①	0.670	0.590	0.006	8.66	8.74	8.74	0.08	0.04	61.5	19-21
	Tracker ②	0.670	0.590	0.006	8.66	8.74	8.74	0.08	0.04	61.5	19-21

① 2-door model

② 4-door model

93591CJ6

For complete Engine Mechanical specifications, see Section 1 of this manual

SCHEDULED MAINTENANCE INTERVALS
CHEVROLET—TRACKER

TO BE SERVICED	TYPE OF SERVICE	VEHICLE MILEAGE INTERVAL (x1000)												
		7.5	15	22.5	30	37.5	45	52.5	60	67.5	75	82.5	90	97.5
Engine oil & filter	R	✓	✓	✓	✓	✓	✓	✓	✓	✓	✓	✓	✓	✓
Automatic transmission fluid ①	S/I	✓	✓	✓	✓	✓	✓	✓	✓	✓	✓	✓	✓	✓
Manual transmission oil ②	S/I	✓	✓	✓	✓	✓	✓	✓	✓	✓	✓	✓	✓	✓
Steering system	S/I	✓	✓	✓	✓	✓	✓	✓	✓	✓	✓	✓	✓	✓
Transfer & differential oil ②	S/I	✓	✓	✓	✓	✓	✓	✓	✓	✓	✓	✓	✓	✓
Wheel discs & free wheeling hubs	S/I	✓	✓	✓	✓	✓	✓	✓	✓	✓	✓	✓	✓	✓
Suspension system	S/I	✓	✓	✓	✓	✓	✓	✓	✓	✓	✓	✓	✓	✓
Brake discs & pads (front)	S/I		✓		✓		✓		✓		✓		✓	
Brake drums & shoes (rear)	S/I		✓		✓		✓		✓		✓		✓	
Brake fluid ③	S/I		✓		✓		✓		✓		✓		✓	
Brake hoses & pipes	S/I		✓		✓		✓		✓		✓		✓	
Brake pedal	S/I		✓		✓		✓		✓		✓		✓	
Brake lever & cable	S/I		✓		✓		✓		✓		✓		✓	
Clutch	S/I		✓		✓		✓		✓		✓		✓	
Idle speed	S/I		✓		✓		✓		✓		✓		✓	
Propeller shafts	S/I		✓		✓		✓		✓		✓		✓	
Valve lash (clearance)	S/I		✓		✓		✓		✓		✓		✓	
Wheel bearings	S/I		✓		✓		✓		✓		✓		✓	
Air cleaner filter element	R				✓				✓				✓	
Engine coolant	R				✓				✓				✓	
Fuel filter	R				✓				✓				✓	
Spark plugs	R				✓				✓				✓	
Cooling system hoses	S/I				✓				✓				✓	
Drive belt(s)	S/I				✓				✓				✓	
Exhaust pipes & mountings	S/I				✓				✓				✓	
Fuel lines & connections	S/I				✓				✓				✓	
Camshaft timing belt	R								✓				✓	
Distributor cap & rotor	S/I								✓					
Emission-related hoses	S/I								✓					
Oxygen sensor	S/I											✓		
EVAP canister	R	every 100,000 miles												
PCV valve	R							✓						

93591CJ7

SCHEDULED MAINTENANCE INTERVALS
CHEVROLET—TRACKER

TO BE SERVICED	TYPE OF SERVICE	VEHICLE MILEAGE INTERVAL (x1000)												
		7.5	15	22.5	30	37.5	45	52.5	60	67.5	75	82.5	90	97.5
EGR system	S/I							✓						
Fuel Injectors	S/I	every 100,000 miles												
TWC converter	S/I	every 100,000 miles												

R: Replace   S/I: Service or Inspect

① Replace at 100,000 miles.

② Replace oil every 30,000 miles.

③ Replace every 60,000 miles.

FREQUENT OPERATION MAINTENANCE (SEVERE SERVICE)

If a vehicle is operated under any of the following conditions it is considered severe service:

- Extremely dusty areas.

- 50% or more of the vehicle operation is in 32°C (90°F) or higher temperatures, or constant operation in temperatures below 0°C (32°F).

- Prolonged idling (vehicle operation in stop and go traffic).

- Frequent short running periods (engine does not warm to normal operating temperatures).

- Police, taxi, delivery usage or trailer towing usage.

Oil & oil filter: replace every 3000 miles.

Air cleaner filter element: service or inspect every 3000 miles & replace every 15,000 miles.

Steering wheel free play, gear box oil & linkage: service or inspect every 3000 miles.

Brake & nuts on chassis: tighten every 6000 miles.

Brake discs & pads (front): service or inspect every 6000 miles.

Brake drums & shoes (rear): service or inspect every 6000 miles.

Exhaust pipes & mountings: tighten every 6000 miles.

Propeller shafts: service or inspect every 6000 miles.

Automatic transmission fluid & filter: replace every 15,000 miles.

Distributor cap & ignition wires: service or inspect every 15,000 miles.

Drive belt(s): service or inspect every 15,000 miles.

Manual transmission oil: replace every 15,000 miles.

Transfer & differential oil: replace every 15,000 miles.

93591CJ8

For Accessory Drive Belt illustrations, see Section 1 of this manual

SCHEDULED MAINTENANCE INTERVALS
GENERAL MOTORS CORPORATION
CHEVROLET TRACKER

The following should be used as a guide when determining the amount of work required for a particular service. In estimating how long a particular Scheduled Maintenance Service should take, please observe the following:

- Labor Time is time based on field research and data supplied by the vehicle manufacturer.
- Labor time operations are given in hours and tenths of an hour.
- All labor operations are to be used as a guide.

Mechanic Skill Level Codes:
(A) PRECISION: Highly skilled with multiple certification.
(B) GENERAL: Normally skilled with certification.
(C) MAINTENANCE: Semi-skilled working on certification.

	LABOR TIME		LABOR TIME		LABOR TIME
7500 Mile Service (B)		**37500 Mile Service (B)**		**75000 Mile Service (B)**	
All models	.7	All models	.7	All models	2.2
w/AT add	.1	w/AT add	.1	w/AT add	.1
w/4WD add	.2	w/4WD add	.2	w/4WD add	.2
15000 Mile Service (B)		**45000 Mile Service (B)**		**82500 Mile Service (B)**	
All models	2.2	All models	2.2	All models	.9
w/AT add	.1	w/AT add	.1	w/AT add	.1
w/4WD add	.2	w/4WD add	.2	w/4WD add	.2
22500 Mile Service (B)		**52500 Mile Service (B)**		**90000 Mile Service (B)**	
All models	.7	All models	1.0	All models	4.5
w/AT add	.1	w/AT add	.1	w/AT add	.1
w/4WD add	.2	w/4WD add	.2	w/4WD add	.2
30000 Mile Service (B)		**60000 Mile Service (B)**		**97500 Mile Service (B)**	
All models	3.1	All models	5.3	All models	.7
w/AT add	.1	w/AT add	.1	w/AT add	.1
w/4WD add	.2	w/4WD add	.2	w/4WD add	.2
		67500 Mile Service (B)			
		All models	.7		
		w/AT add	.1		
		w/4WD add	.2		

93591CJ9

HONDA
CR-V

ENGINE AND VEHICLE IDENTIFICATION CHART

		Engine Code					Model Year	
Code	Liters (cc)	Cu. In.	Cyl.	Fuel Sys.	Engine Type	Eng. Mfg.	Code ①	Year
B20B4	2.0 (1973)	120	4	SMFI	DOHC	Honda	X	1999
B20Z2	2.0 (1973)	120	4	SMFI	DOHC	Honda	Y	2000
K24A1	2.4 (NA)	146	4	SMFI	DOHC	Honda	1	2001
							2	2002
							3	2003

DOHC: Double Overhead Cam

SMFI: Sequential Multi-port Fuel Injection

① 10th position of VIN

NA: Not Avalaible

93591CK1

GENERAL ENGINE SPECIFICATIONS

Year	Model	Engine Displacement Liters (cc)	Engine ID/VIN	Fuel System Type	Net Horsepower @ rpm	Net Torque @ rpm (ft. lbs.)	Bore x Stroke (in.)	Com-pression Ratio	Oil Pressure @ rpm
1999	CR-V	2.0 (1973)	B20Z2	SMFI	146@6200	133@4500	3.31x3.50	9.6:1	50@3000
2000	CR-V	2.0 (1973)	B20Z2	SMFI	146@6200	133@4500	3.31x3.50	9.6:1	50@3000
2001	CR-V	2.0 (1973)	B20Z2	SMFI	146@6200	133@4500	3.31x3.50	9.6:1	50@3000
2002	CR-V	2.4 (NA)	K24A1	SMFI	160@6000	162@3600	NA	9.6:1	44@3000

NA: Not Avalaible

SMFI: Sequential Multi-port Fuel Injection

93591CK2

For Tire, Wheel and Ball Joint specifications, see Section 1 of this manual

ENGINE TUNE-UP SPECIFICATIONS

Year	Engine Displacement Liters (cc)	Engine ID/VIN	Spark Plug Gap (in.)	Ignition Timing (deg.)		Fuel Pump (psi)	Idle Speed (rpm)		Valve Clearance (in.)	
				MT	AT		MT	AT	In.	Ex.
1999	2.0 (1973)	B20Z2	0.039-0.043	14-18B	14-18B	38-46	700-800	700-800	0.003-0.005	0.006-0.008
2000	2.0 (1973)	B20Z2	0.039-0.043	14-18B	14-18B	38-46	700-800	700-800	0.003-0.005	0.006-0.008
2001	2.0 (1973)	B20Z2	0.039-0.043	14-18B	14-18B	38-46	700-800	700-800	0.003-0.005	0.006-0.008
2002	2.4 (NA)	K24A1	0.039-0.043	6-10B	6-10B	50	600-700	600-700	NA	NA

NOTE: The Vehicle Emission Control Information label often reflects changes made during production and must be used if they differ from this chart.

NOTE: The fuel pressure readings are given with the vacuum hose connected to the regulator and the engine running

B: Before top dead center

HYD: Hydraulic

93591CK3

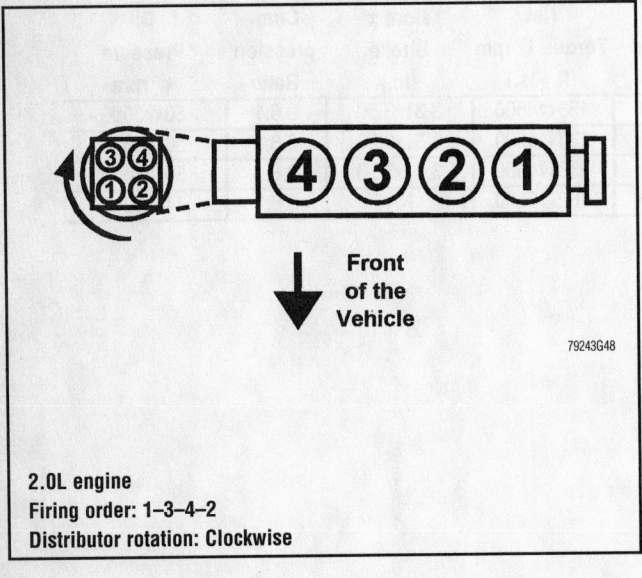

Front of the Vehicle

79243G48

2.0L engine
Firing order: 1–3–4–2
Distributor rotation: Clockwise

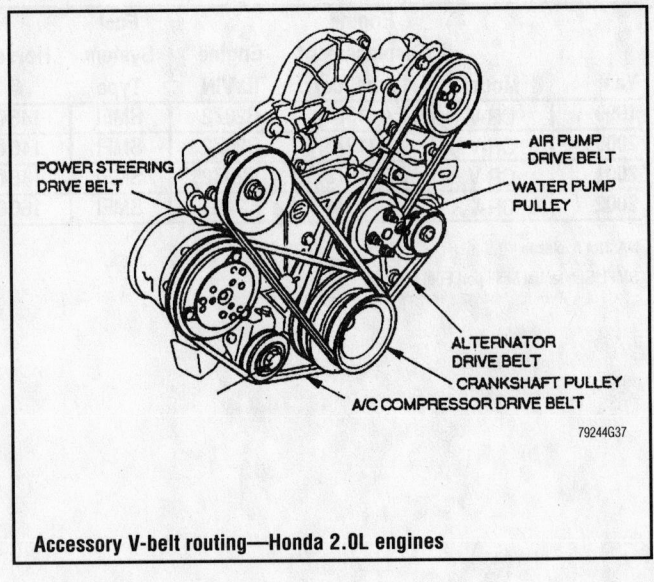

POWER STEERING DRIVE BELT

AIR PUMP DRIVE BELT

WATER PUMP PULLEY

ALTERNATOR DRIVE BELT

CRANKSHAFT PULLEY

A/C COMPRESSOR DRIVE BELT

79244G37

Accessory V-belt routing—Honda 2.0L engines

CAPACITIES

Year	Model	Engine Displacement Liters (cc)	Engine ID/VIN	Engine Oil with Filter (qts.)	Transmission (pts.)		Transfer Case (pts.)	Drive Axle		Fuel Tank (gal.)	Cooling System (qts.)
					5-Spd	Auto.		Front (pts.)	Rear (pts.)		
1999	CR-V	2.0 (1973)	B20Z2	4.0	3.6	①	②	②	2.2	15.3	4.1
2000	CR-V	2.0 (1973)	B20Z2	4.0	3.6	①	②	②	2.2	15.3	4.1
2001	CR-V	2.0 (1973)	B20Z2	4.0	3.6	①	②	②	2.2	15.3	4.1
2002	CR-V	2.4 (NA)	K24A1	4.4	8.0	6.6	②	②	2.2	15.3	5.8

NOTE: All capacities are approximate. Add fluid gradually and check to be sure a proper fluid level is obtained.

① 4WD: 6.2
  2WD: 5.8

② Included in transaxle refill figure

93591CK4

VALVE SPECIFICATIONS

Year	Engine Displacement Liters (cc)	Engine ID/VIN	Seat Angle (deg.)	Face Angle (deg.)	Spring Test Pressure (lbs. @ in.)	Spring Installed Height (in.)	Stem-to-Guide Clearance (in.)		Stem Diameter (in.)	
							Intake	Exhaust	Intake	Exhaust
1999	2.0 (1973)	B20Z2	45	45	NA	①	0.0010-0.0020	0.0020-0.0030	0.2591-0.2594	0.2579-0.2583
2000	2.0 (1973)	B20Z2	45	45	NA	①	0.0010-0.0020	0.0020-0.0030	0.2591-0.2594	0.2579-0.2583
2001	2.0 (1973)	B20Z2	45	45	NA	①	0.0010-0.0020	0.0020-0.0030	0.2591-0.2594	0.2579-0.2583
2002	2.4 (NA)	K24A1	NA	NA	NA	①	0.0012-0.0022	0.0022-0.0031	0.2156-0.2159	0.2146-0.2150

NA: Not Available

① Valve spring free length:
  Intake: 1.668 in.
  Exhaust: 1.745 in.

93591CK5

For Wheel Alignment specifications, see Section 1 of this manual

CRANKSHAFT AND CONNECTING ROD SPECIFICATIONS
All measurements are given in inches

Year	Engine Displacement Liters (cc)	Engine ID/VIN	Crankshaft				Connecting Rod		
			Main Brg. Journal Dia.	Main Brg. Oil Clearance	Shaft End-play	Thrust on No.	Journal Diameter	Oil Clearance	Side Clearance
1999	2.0 (1973)	B20Z2	①	②	0.0040-0.0140	4	1.7707-1.7717	0.0008-0.0015	0.0060-0.0120
2000	2.0 (1973)	B20Z2	①	②	0.0040-0.0140	4	1.7707-1.7717	0.0008-0.0015	0.0060-0.0120
2001	2.0 (1973)	B20Z2	①	②	0.0040-0.0140	4	1.7707-1.7717	0.0008-0.0015	0.0060-0.0120
2002	2.4 (NA)	K24A1	③	④	0.0040-0.0140	3	1.8888-1.8898	0.0008-0.0019	0.016

① Nos. 1, 2, 4 and 5: 2.1644-2.1654
  No. 3: 2.1642-2.1651

② Nos. 1, 2, 4 and 5: 0.0009-0.0017
  No. 3: 0.0012-0.0019

③ Except No. 3: 2.1648-2.1657
  No. 3: 2.1644-2.1654

④ Except No. 3: 0.0007-0.0016
  No. 3: 0.0010-0.0019

93591CK6

PISTON AND RING SPECIFICATIONS
All measurements are given in inches

Year	Engine Displacement Liters (cc)	Engine ID/VIN	Piston Clearance	Ring Gap			Ring Side Clearance		
				Top Compression	Bottom Compression	Oil Control	Top Compression	Bottom Compression	Oil Control
1999	2.0 (1973)	B20Z2	0.0004-0.0016	0.0080-0.0120	0.0160-0.0220	0.0080-0.0200	0.0022-0.0031	0.0014-0.0024	NA
2000	2.0 (1973)	B20Z2	0.0004-0.0016	0.0080-0.0120	0.0160-0.0220	0.0080-0.0200	0.0022-0.0031	0.0014-0.0024	NA
2001	2.0 (1973)	B20Z2	0.0004-0.0016	0.0080-0.0120	0.0160-0.0220	0.0080-0.0200	0.0022-0.0031	0.0014-0.0024	NA
2002	2.4 (NA)	K24A1	0.0008-0.0016	0.0080-0.0140	0.0160-0.0220	0.0080-0.0280	0.0018-0.0028	0.0020-0.0030	NA

NA: Not Applicable

93591CK7

TORQUE SPECIFICATIONS
All readings in ft. lbs.

Year	Engine Displacement Liters (cc)	Engine ID/VIN	Cylinder Head Bolts	Main Bearing Bolts	Rod Bearing Bolts	Crankshaft Damper Bolts	Flywheel Bolts	Manifold		Spark Plugs	Lug Nut
								Intake	Exhaust		
1999	2.0 (1973)	B20Z2	①	②	23	130	54	17	23	13	80
2000	2.0 (1973)	B20Z2	①	②	23	130	54	17	23	13	80
2001	2.0 (1973)	B20Z2	①	②	23	130	54	17	23	13	80
2002	2.4 (NA)	K24A1	③	④	⑤	181	NA	16	33	13	80

NOTE: Dip main bearing bolts and crankshaft damper bolt in clean engine oil prior to tightening.

① Step 1: 22 ft. lbs.
Step 2: 63 ft. lbs.

② Step 1: 18 ft. lbs.
Step 2: 56 ft. lbs.

③ Step 1: 29 ft. lbs.
Step 2: +90 degrees
Step 3: +90 degrees
Step 4: NEW BOLT ONLY +90 degrees

④ 22 ft. lbs. +56 degrees

⑤ 14 ft. lbs. +90 degrees

93591CK8

WHEEL ALIGNMENT

Year	Model		Caster		Camber		Toe-in (in.)	Steering Axis Inclination (Deg.)
			Range (+/-Deg.)	Preferred Setting (Deg.)	Range (+/-Deg.)	Preferred Setting (Deg.)		
1999	CR-V	F	1.00	+2.10	1.00	0	0+/-0.12	—
		R	—	—	1.00	-1.00	0.08+/-0.08	—
2000	CR-V	F	1.00	+2.10	1.00	0	0+/-0.12	—
		R	—	—	1.00	-1.00	0.08+/-0.08	—
2001	CR-V	F	1.00	+2.10	1.00	0	0+/-0.12	—
		R	—	—	1.00	-1.00	0.08+/-0.08	—
2002	CR-V	F	1.00	+1.75	0.75	0	0+/-0.08	—
		R	—	—	0.75	-1.00	0.08+/-0.08	—

93591CK9

For Maintenance Interval recommendations, see Section 1 of this manual

TIRE, WHEEL AND BALL JOINT SPECIFICATIONS

Year	Model	OEM Tires		Tire Pressures (psi)		Wheel Size	Ball Joint Inspection
		Standard	Optional	Front	Rear		
1999	CR-V	P205/70R15	None	26	26	6JJ	NS
2000	CR-V	P205/70R15	None	26	26	6JJ	NS
2001	CR-V	P205/70R15	None	26	26	6JJ	NS
2002	CR-V	P205/70R15	None	26	26	6JJ	NS

OEM: Original Equipment Manufacturer

PSI: Pounds Per Square Inch

NS: Not specified by manufacturer

93591CK0

BRAKE SPECIFICATIONS
HONDA CR-V
All measurements in inches unless noted

Year	Model		Brake Disc			Brake Drum Diameter			Minimum Lining Thickness		Brake Caliper	
			Original Thickness	Minimum Thickness	Maximum Runout	Original Inside Diameter	Max. Wear Limit	Maximum Machine Diameter	Front	Rear	Bracket Bolts (ft. lbs.)	Mounting Bolts (ft. lbs.)
1999	CR-V	F	0.929	0.830	0.004	—	—	—	0.060	—	80	36
		R	—	—	—	8.66	8.70	8.70	—	0.080	—	—
2000	CR-V	F	0.929	0.830	0.004	—	—	—	0.060	—	80	36
		R	—	—	—	8.66	8.70	8.70	—	0.080	—	—
2001	CR-V	F	0.929	0.830	0.004	—	—	—	0.060	—	80	36
		R	—	—	—	8.66	8.70	8.70	—	0.080	—	—
2002	CR-V	F	0.910	0.830	0.004	—	—	—	0.060	—	80	25
		R	0.350	0.280	0.004	—	—	—	0.040	—	41	16

F: Front

R: Rear

93591CL1

SCHEDULED MAINTENANCE INTERVALS
1999-01 HONDA—CRV

TO BE SERVICED	TYPE OF SERVICE	VEHICLE MILEAGE INTERVAL (x1000)															
		7.5	15	22.5	30	37.5	45	52.5	60	67.5	75	82.5	90	97.5	105	112.5	120
Accessory drive belts	I & A				✓				✓				✓				✓
Air cleaner element	R				✓				✓				✓				✓
Air conditioning filter	R				✓				✓				✓				✓
Brake fluid	R						✓						✓				
Brake hoses & lines (including ABS)	I		✓		✓		✓		✓		✓		✓		✓		✓
Cooling system hoses & connections	I		✓		✓		✓		✓		✓		✓		✓		✓
Engine coolant	R						✓						✓				
Engine oil	R	✓	✓	✓	✓	✓	✓	✓	✓	✓	✓	✓	✓	✓	✓	✓	✓
Engine oil and coolant levels	I	Inspect at each fuel stop															
Engine oil filter	R		✓		✓		✓		✓		✓		✓		✓		✓
Exhaust system	I		✓		✓		✓		✓		✓		✓		✓		✓
Fluid levels and condition	I		✓		✓		✓		✓		✓		✓		✓		✓
Front and rear brakes	I		✓		✓		✓		✓		✓		✓		✓		✓
Fuel lines & connection	I		✓		✓		✓		✓		✓		✓		✓		✓
Halfshaft boots	I		✓		✓		✓		✓		✓		✓		✓		✓
Idle speed	I & A												✓				
Parking brake system	I & A		✓		✓		✓		✓		✓		✓		✓		✓
Rear differential fluid	R												✓				
Rotate and inspect tires	I	✓	✓	✓	✓	✓	✓	✓	✓	✓	✓	✓	✓	✓	✓	✓	✓
Spark plugs	R				✓				✓				✓				✓
Supplemental Restrain system (SRS)	I	Inspect the SRS 10 years after production															
Suspension components	I		✓		✓		✓		✓		✓		✓		✓		✓
Tie rod ends, steering gear box & boots	I		✓		✓		✓		✓		✓		✓		✓		✓
Timing balancer belt ①	R														✓		
Timing belt	R														✓		
Transmission fluid	R						✓				✓		✓				
Valve clearance	I				✓				✓				✓				✓
Water pump	S/I														✓		

R: Replace   I: Inspect   A: Adjust

FREQUENT OPERATION MAINTENANCE (SEVERE SERVICE)

If a vehicle is operated under any of the following conditions it is considered severe service:

- Towing a trailer or using a camper or car-top carrier.
- Repeated short trips of less than 5 miles in temperatures below freezing, or trips of less than 10 miles in any temperature.
- Extensive idling or low-speed driving for long distances as in heavy commercial use, such as delivery, taxi or police cars.
- Operating on rough, muddy or salt-covered roads.
- Operating on unpaved or dusty roads.
- Driving in extremely hot (over 90°) conditions.

Air cleaner element: replace every 15,000 miles

Engine oil and filter: replace every 3750 miles or 6 months, whichever occurs first.

Timing belt: replace every 60,000 miles if the vehicle is regularly driven in temperatures above 110°F or below -20°F.

Transmission fluid: replace every 30,000 miles.

Rear differential fluid: replace every 60,000 miles.

Front and rear brakes: inspect every 7500 miles or 6 months, whichever occurs first.

Locks and hinges: lubricate every 15,000 miles.

Tie rods, steering gear box, boots: inspect every 7500 miles or 6 months, whichever occurs first.

Suspension components: inspect every 7500 miles or 6 months, whichever occurs first.

Halfshaft boots: inspect every 7500 miles or 6 months, whichever occurs first.

93591CL2

SCHEDULED MAINTENANCE INTERVALS
2002 HONDA—CRV

TO BE SERVICED	TYPE OF SERVICE	VEHICLE MILEAGE INTERVAL (x1000)											
		10	20	30	40	50	60	70	80	90	100	110	120
Accessory drive belts	I & A			✓			✓			✓			✓
Air cleaner element	R			✓			✓			✓			✓
Air conditioning filter	R			✓			✓			✓			✓
Brake fluid	R											✓	
Brake hoses & lines (including ABS)	I		✓		✓		✓		✓		✓		
Cooling system hoses & connections	I		✓		✓		✓		✓		✓		
Engine coolant	R												✓
Engine oil	R	✓	✓	✓	✓	✓	✓	✓	✓	✓	✓	✓	✓
Engine oil and coolant levels	I	Inspect at each fuel stop											
Engine oil filter	R		✓		✓		✓		✓		✓		
Exhaust system	I		✓		✓		✓		✓		✓		
Fluid levels and condition	I		✓		✓		✓		✓		✓		
Front and rear brakes	I		✓		✓		✓		✓		✓		
Fuel lines & connection	I		✓		✓		✓		✓		✓		
Halfshaft boots	I		✓		✓		✓		✓		✓		
Idle speed	I & A											✓	
Parking brake system	I & A		✓		✓		✓				✓		
Rear differential fluid	R										✓		
Rotate and inspect tires	I	✓	✓	✓	✓	✓	✓	✓	✓	✓	✓	✓	✓
Spark plugs	R											✓	
Suspension components	I		✓		✓		✓		✓		✓		
Tie rod ends, steering gear box & boots	I		✓		✓		✓		✓		✓		
Transmission fluid	R												✓
Valve clearance	I											✓	

R: Replace   I: Inspect   A: Adjust

FREQUENT OPERATION MAINTENANCE (SEVERE SERVICE)

If a vehicle is operated under any of the following conditions it is considered severe service:

- Towing a trailer or using a camper or car-top carrier.
- Repeated short trips of less than 5 miles in temperatures below freezing, or trips of less than 10 miles in any temperature.
- Extensive idling or low-speed driving for long distances as in heavy commercial use, such as delivery, taxi or police cars.
- Operating on rough, muddy or salt-covered roads.
- Operating on unpaved or dusty roads.
- Driving in extremely hot (over 90°) conditions.

Air cleaner element: replace every 15,000 miles

Engine oil and filter: replace every 3750 miles or 6 months, whichever occurs first.

Timing belt: replace every 60,000 miles if the vehicle is regularly driven in temperatures above 110°F or below -20°F.

Transmission fluid: replace every 30,000 miles.

Rear differential fluid: replace every 60,000 miles.

Front and rear brakes: inspect every 7500 miles or 6 months, whichever occurs first.

Locks and hinges: lubricate every 15,000 miles.

Tie rods, steering gear box, boots: inspect every 7500 miles or 6 months, whichever occurs first.

Suspension components: inspect every 7500 miles or 6 months, whichever occurs first.

Halfshaft boots: inspect every 7500 miles or 6 months, whichever occurs first.

93591CHY

SCHEDULED MAINTENANCE INTERVALS
HONDA
CR-V

The following should be used as a guide when determining the amount of work required for a particular service. In estimating how long a particular Scheduled Maintenance Service should take, please observe the following:

- Labor Time is time based on field research and data supplied by the vehicle manufacturer.
- Labor time operations are given in hours and tenths of an hour.
- All labor operations are to be used as a guide.

Mechanic Skill Level Codes:
(A) PRECISION: Highly skilled with multiple certification.
(B) GENERAL: Normally skilled with certification.
(C) MAINTENANCE: Semi-skilled working on certification.

	LABOR TIME		LABOR TIME		LABOR TIME
7500 Mile Service (C)		**45000 Mile Service (C)**		**90000 Mile Service (B)**	
CR-V	1.3	CR-V	1.8	CR-V	5.9
15000 Mile Service (C)		**52500 Mile Service (C)**		**97500 Mile Service (C)**	
CR-V	1.4	CR-V	1.3	CR-V	1.3
22500 Mile Service (C)		**60000 Mile Service (B)**		**105000 Mile Service (C)**	
CR-V	1.3	CR-V	3.9	CR-V	1.8
30000 Mile Service (B)		**67500 Mile Service (C)**		*Replace timing belt add*	
CR-V	4.0	CR-V	1.3	**112500 Mile Service (C)**	
37500 Mile Service (C)		**75000 Mile Service (C)**		CR-V	1.4
CR-V	1.3	CR-V	1.8	**120000 Mile Service (B)**	
		82500 Mile Service (C)		CR-V	4.0
		CR-V	1.3		

93551CL3

HONDA
Passport

ENGINE AND VEHICLE IDENTIFICATION

Code	Liters (cc)	Cu. In.	Cyl.	Fuel Sys.	Engine Type	Eng. Mfg.
W	3.2 (3165)	193	6	MFI	DOHC	Isuzu

MFI: Multi-port Fuel Injection
DOHC: Double Overhead Camshaft
SOHC: Single Overhead Camshaft
① 10th position of VIN

Code ①	Year
X	1999
Y	2000
1	2001
2	2002
3	2003

93591CL4

GENERAL ENGINE SPECIFICATIONS

Year	Model	Engine Displacement Liters (cc)	Engine Series (ID/VIN)	Fuel System	Net Horsepower @ rpm	Net Torque @ rpm (ft. lbs.)	Bore x Stroke (in.)	Compression Ratio	Oil Pressure @ rpm
1999	Passport	3.2 (3165)	W	MFI	205@5400	214@3000	3.68x3.03	9.1:1	57-80@3000
2000	Passport	3.2 (3165)	W	MFI	205@5400	214@3000	3.68x3.03	9.1:1	57-80@3000
2001	Passport	3.2 (3165)	W	MFI	205@5400	214@3000	3.68x3.03	9.1:1	57-80@3000
2002	Passport	3.2 (3165)	W	MFI	205@5400	214@3000	3.68x3.03	9.1:1	57-80@3000

MFI: Multiport fuel injection

93591CL5

ENGINE TUNE-UP SPECIFICATIONS

Year	Engine Displacement Liters (cc)	Engine ID/VIN	Spark Plug Gap (in.)	Ignition Timing (deg.)		Fuel Pump (psi)	Idle Speed (rpm)		Valve Clearance	
				MT	AT		MT	AT	In.	Ex.
1999	3.2 (3165)	W	0.040	①	①	48-55	750	750	0.009-0.013	0.010-0.014
2000	3.2 (3165)	W	0.040	①	①	48-55	750	750	0.009-0.013	0.010-0.014
2001	3.2 (3165)	W	0.040	①	①	48-55	750	750	0.009-0.013	0.010-0.014
2002	3.2 (3165)	W	0.040	①	①	48-55	750	750	0.009-0.013	0.010-0.014

NOTE: The Vehicle Emission Control Information label often reflects specification changes made during production. The label figures must be used if they differ from those in this chart.

B: Before top dead center

HYD: Hydraulic

① Controlled by the PCM

93591CL6

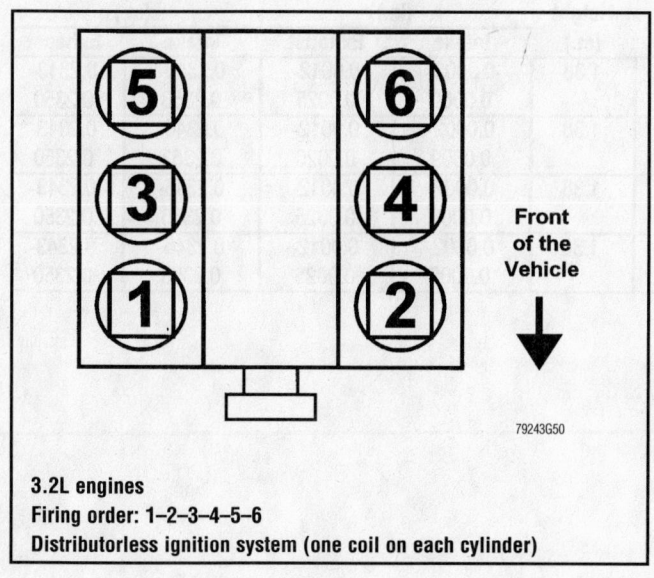

Front of the Vehicle

79243G50

3.2L engines
Firing order: 1–2–3–4–5–6
Distributorless ignition system (one coil on each cylinder)

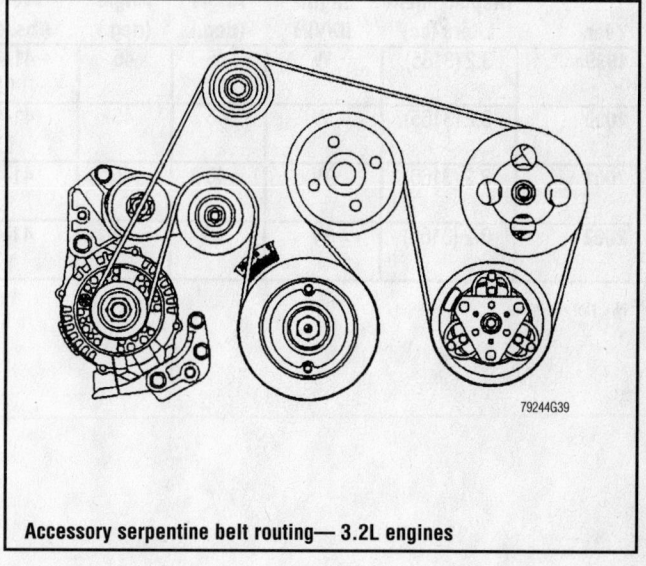

79244G39

Accessory serpentine belt routing— 3.2L engines

CAPACITIES

Year	Model	Engine Displacement Liters (cc)	Engine ID/VIN	Oil with Filter (qts.)	Engine Transmission (pts.) Man.	Engine Transmission (pts.) Auto.	Transfer Case (pts.)	Drive Axle Front (pts.)	Drive Axle Rear (pts.)	Fuel Tank (gal.)	Cooling System (qts.)
1999	Passport	3.2 (3165)	W	5.0	6.2	18.2	3.0	3.0	3.74	21.1	11.2
2000	Passport	3.2 (3165)	W	5.0	6.2	18.2	3.0	2.6	3.74	21.1	11.6
2001	Passport	3.2 (3165)	W	5.0	6.2	18.2	3.0	2.6	3.74	19.5	11.6
2002	Passport	3.2 (3165)	W	5.0	6.2	18.2	3.0	2.6	3.74	19.5	11.6

NOTE: All capacities are approximate. Add fluid gradually and check to ensure a proper level has been reached.

① 4.0 pts. if equipped with Torque On Demand

② A/T: 7.4 qts.
  M/T: 7.0 qts.

93591CL7

VALVE SPECIFICATIONS

Year	Engine Displacement Liters (cc)	Engine ID/VIN	Seat Angle (deg.)	Face Angle (deg.)	Spring Test Pressure (lbs. @ in.)	Spring Installed Height (in.)	Stem-to-Guide Clearance (in.) Intake	Stem-to-Guide Clearance (in.) Exhaust	Stem Diameter (in.) Intake	Stem Diameter (in.) Exhaust
1999	3.2 (3165)	W	45	45	41-44@ 1.38	1.38	0.0002-0.0009	0.0012-0.0025	0.2346-0.2353	0.2343-0.2350
2000	3.2 (3165)	W	45	45	41-44@ 1.38	1.38	0.0002-0.0009	0.0012-0.0025	0.2346-0.2353	0.2343-0.2350
2001	3.2 (3165)	W	45	45	41-44@ 1.38	1.38	0.0002-0.0009	0.0012-0.0025	0.2346-0.2353	0.2343-0.2350
2002	3.2 (3165)	W	45	45	41-44@ 1.38		0.0002-0.0009	0.0012-0.0025	0.2346-0.2353	0.2343-0.2350

NA: Not Available

93591CL8

CRANKSHAFT AND CONNECTING ROD SPECIFICATIONS

All measurements are given in inches.

Year	Engine Displacement Liters (cc)	Engine ID/VIN	Crankshaft				Connecting Rod		
			Main Brg. Journal Dia.	Main Brg. Oil Clearance	Shaft End-play	Thrust on No.	Journal Diameter	Oil Clearance	Side Clearance
1999	3.2 (3165)	W	2.5165-2.5170	0.0007-0.0017	0.0024-0.0094	3	2.1229-2.1235	0.0010-0.0023	0.0050-0.0150
2000	3.2 (3165)	W	2.5165-2.5170	0.0007-0.0017	0.0024-0.0094	3	2.1229-2.1235	0.0010-0.0023	0.0050-0.0150
2001	3.2 (3165)	W	2.5165-2.5170	0.0007-0.0017	0.0024-0.0094	3	2.1229-2.1235	0.0010-0.0023	0.0050-0.0150
2002	3.2 (3165)	W	2.5165-2.5170	0.0007-0.0017	0.0024-0.0094	3	2.1229-2.1235	0.0010-0.0023	0.0050-0.0150

93591CL9

PISTON AND RING SPECIFICATIONS

All measurements are given in inches.

Year	Engine Displacement Liters (cc)	Engine ID/VIN	Piston Clearance	Ring Gap			Ring Side Clearance		
				Top Compression	Bottom Compression	Oil Control	Top Compression	Bottom Compression	Oil Control
1999	3.2 (3165)	W	NA	0.0118-0.0157	0.0177-0.0236	0.006-0.018	0.0006-0.0015	0.0006-0.0015	NA
2000	3.2 (3165)	W	NA	0.0118-0.0157	0.0177-0.0236	0.0060-0.018	0.0006-0.002	0.0006-0.0015	NA
2001	3.2 (3165)	W	NA	0.0118-0.0157	0.0177-0.0236	0.0060-0.018	0.0006-0.002	0.0006-0.0015	NA
2002	3.2 (3165)	W	NA	0.0118-0.0157	0.0177-0.0236	0.0060-0.018	0.0006-0.002	0.0006-0.0015	NA

NA: Not Available

93591CL0

Timing belt service is covered in Section 3 of this manual

TORQUE SPECIFICATIONS
All readings in ft. lbs.

Year	Engine Displacement Liters (cc)	Engine ID/VIN	Cylinder Head Bolts	Main Bearing Bolts	Rod Bearing Bolts	Crankshaft Damper Bolts	Flywheel Bolts	Manifold Intake	Manifold Exhaust	Spark Plugs	Lug Nuts
1999	3.2 (3165)	W	⑦	29	40	123	40	18	42	13	87
2000	3.2 (3165)	W	⑦	29	40	123	40	18	42	13	87
2001	3.2 (3165)	W	⑦	29	40	123	40	18	42	13	87
2002	3.2 (3165)	W	⑦	29	40	123	40	18	42	13	87

① Step 1: 18 ft. lbs.
Step 2: Plus 90 degrees
Step 3: Plus 90 degrees
Step 4: Plus 90 degrees

② Step 1: 37 ft. lbs.
Step 2: Plus 45 degrees
Step 3: Plus 15 degrees

③ Step 1: 25 ft. lbs.
Step 2: Plus 45 degrees
Step 3: Plus 15 degrees

④ Crankshaft sprocket:
Step 1: 94 ft. lbs.
Step 2: Plus 45 degrees
Crankshaft balancer:
Step 1: 14 ft. lbs.
Step 2: Plus 45 degrees

⑤ Step 1: 48 ft. lbs.
Step 2: Plus 30 degrees
Step 3: Plus 15 degrees

⑥ Step 1: 112 inch lbs.
Step 2: 14 ft. lbs.
Step 3: 14 ft. lbs.

⑦ Step 1: 21 ft. lbs.
Step 2: 47 ft. lbs.

⑧ Step 1: 22 ft. lbs.
Step 2: Plus 55-65 degrees
Step 3: Crankcase side bolts to 29 ft. lbs.

93591CM1

WHEEL ALIGNMENT

Year	Model		Caster Range (+/-Deg.)	Caster Preferred Setting (Deg.)	Camber Range (+/-Deg.)	Camber Preferred Setting (Deg.)	Toe-in (in.)	Steering Axis Inclination (Deg.)
1999	Passport	F	0.75	+2.50	0.50	0	0+/-0.08	—
		R	—	—	1.00	0	0+/-0.20	—
2000	Passport	F	1.00	+2.50	0.50	0	0+/-0.08	—
		R	—	—	1.00	0	0+/-0.20	—
2001	Passport	F	1.00	+2.50	0.50	0	0+/-0.08	—
		R	—	—	1.00	0	0+/-0.20	—
2002	Passport	F	1.00	+2.50	0.50	0	0+/-0.08	—
		R	—	—	1.00	0	0+/-0.20	—

93591CM2

TIRE, WHEEL AND BALL JOINT SPECIFICATIONS

Year	Model	OEM Tires		Tire Pressures (psi)		Wheel Size	Ball Joint Inspection
		Standard	Optional	Front	Rear		
1999	Passport LX 2wd	P235/75R15	None	29	29	6.5-JJ	NS
	Passport LX 4wd	P235/75R15	P245/70R16	29	29	6.5J/7J	NS
	Passport EX 2wd	P245/70R16	None	29	29	7J	NS
	Passport EX 4wd	P245/70R16	None	29	29	7-JJ	NS
2000	Passport LX 2wd	P225/75R16	None	29	29	6.5-JJ	NS
	Passport LX 4wd	P245/70R16	None	29	29	7J	NS
	Passport EX 2wd	P245/70R16	None	29	29	7J	NS
	Passport EX 4wd	P245/70R16	None	29	29	7-JJ	NS
2001	Passport LX 2wd	P225/75R16	None	29	29	6.5-JJ	NS
	Passport LX 4wd	P245/70R16	None	29	26	7J	NS
	Passport EX 2wd	P245/70R16	None	29	26	7J	NS
	Passport EX 4wd	P245/70R16	None	29	26	7-JJ	NS
2002	Passport LX 2wd	P225/75R16	None	29	29	6.5-JJ	NS
	Passport LX 4wd	P245/70R16	None	29	26	7J	NS
	Passport EX 2wd	P245/70R16	None	29	26	7J	NS
	Passport EX 4wd	P245/70R16	None	29	26	7-JJ	NS

L: Lower

U: Upper

NS: Not specified by manufacturer

① Torque required in inch lbs. to rotate ball joint when removed from the knuckle

93591CM3

BRAKE SPECIFICATIONS

All measurements in inches unless noted

Year	Model		Brake Disc				Brake Drum Diameter			Minimum Lining Thickness	Brake Caliper	
			Original Thickness	Machine Thickness	Minimum Thickness	Maximum Runout	Original Inside Diameter	Max. Wear Limit	Maximum Machine Diameter		Bracket Bolts (ft. lbs.)	Mounting Bolts (ft. lbs.)
1999	Passport	F	1.020	0.983	0.969	0.005	—	—	—	0.039	115	54
		R	0.710	0.668	0.654	0.005	11.6	11.67	NA	0.039	76	32
2000	Passport	F	1.020	0.983	0.969	0.005	—	—	—	0.039	115	54
		R	0.710	0.668	0.654	0.005	11.6	11.67	NA	0.039	76	32
2001	Passport	F	1.020	0.983	0.969	0.005	—	—	—	0.039	115	54
		R	0.710	0.668	0.654	0.005	—	—	—	0.039	76	32
2002	Passport	F	1.020	0.983	0.969	0.005	—	—	—	0.039	115	54
		R	0.710	0.668	0.654	0.005	—	—	—	0.039	76	32

NA: Not Available

① Heavy duty models: 1.140 in.

② Heavy duty models: 1.100 in.

③ Heavy duty models: 1.080 in.

93591CM4

Heater Core replacement is covered in Section 2 of this manual

SCHEDULED MAINTENANCE INTERVALS
HONDA—PASSPORT

TO BE SERVICED	TYPE OF SERVICE	7.5	15	22.5	30	37.5	45	52.5	60	67.5	75	82.5	90	97.5	105	112.5	120
Accelerator linkage ①	L	✓	✓	✓	✓	✓	✓	✓	✓	✓	✓	✓	✓	✓	✓	✓	✓
Accessory drive belts ②	S/I				✓				✓				✓				✓
Air cleaner filter	R				✓				✓				✓				✓
Auto cruise control linkage & hose ③	S/I		✓		✓		✓		✓		✓		✓		✓		✓
Automatic transmission fluid level ③	S/I	✓		✓		✓		✓		✓				✓		✓	
Battery fluid level ③	S/I	✓	✓	✓	✓	✓	✓	✓	✓	✓	✓	✓	✓	✓	✓	✓	✓
Body and chassis ①	L	✓	✓	✓	✓	✓	✓	✓	✓	✓	✓	✓	✓	✓	✓	✓	✓
Brake fluid level ③	S/I	✓	✓	✓	✓	✓	✓	✓	✓	✓	✓	✓	✓	✓	✓	✓	✓
Brake lines & hoses ③	S/I	✓	✓	✓	✓	✓	✓	✓	✓	✓	✓	✓	✓	✓	✓	✓	✓
Brake pedal play ③	S/I		✓		✓		✓		✓		✓		✓		✓		✓
Clutch fluid level ③	S/I	✓	✓	✓	✓	✓	✓	✓	✓	✓	✓	✓	✓	✓	✓	✓	✓
Clutch lines & hose ③	S/I				✓				✓				✓				✓
Clutch pedal free-play ③	S/I		✓		✓		✓		✓		✓		✓		✓		✓
Clutch pedal spring, bushing and clevis pin ①	S/I		✓		✓		✓		✓		✓		✓		✓		✓
Cooling and heating system hoses ③	S/I		✓		✓		✓		✓		✓		✓		✓		✓
Driveshaft flange torque ③	S/I	✓		✓		✓		✓		✓		✓		✓		✓	
Drum and disc brakes ③	S/I		✓		✓		✓		✓		✓		✓		✓		✓
Engine coolant	R				✓				✓				✓				✓
Engine coolant level ③	S/I	✓	✓	✓	✓	✓	✓	✓	✓	✓	✓	✓	✓	✓	✓	✓	✓
Engine oil & filter ③	R	✓	✓	✓	✓	✓	✓	✓	✓	✓	✓	✓	✓	✓	✓	✓	✓
Exhaust system ③	S/I	✓	✓	✓	✓	✓											✓
Front and rear axle lubricant	R		✓			✓			✓					✓			✓
Front and rear driveshafts ①	S/I	✓	✓	✓	✓	✓	✓	✓	✓	✓	✓	✓	✓	✓	✓	✓	✓
Front wheel bearings	S/I & L				✓				✓				✓				✓
Fuel lines & tank cap ③	S/I								✓								✓
Inspect for fluid leaks ③	S/I	✓	✓	✓	✓	✓	✓	✓	✓	✓	✓	✓	✓	✓	✓	✓	✓
Key lock cylinder ③	L		✓		✓		✓		✓		✓		✓		✓		✓
Manual transmission and transfer case fluid ④	R		✓		✓				✓				✓				✓
Parking brake system ③	S/I		✓		✓		✓		✓		✓		✓		✓		✓
Power steering fluid	R				✓				✓				✓				✓
Radiator core and A/C condenser	S/I & C								✓								✓
Rotate tires	S/I	✓	✓	✓	✓	✓	✓	✓	✓	✓	✓	✓	✓	✓	✓	✓	✓
Shift-on-the-fly system gear fluid ③	S/I		✓		✓		✓		✓		✓		✓		✓		✓
Spark plugs	R	Every 100,000 miles															
Starter safety switch ③	S/I	✓	✓	✓	✓	✓	✓	✓	✓	✓	✓	✓	✓	✓	✓	✓	✓
Steering operation ③	S/I	✓	✓	✓	✓	✓	✓	✓	✓	✓	✓	✓	✓	✓	✓	✓	✓
Suspension & steering ③	S/I	✓	✓	✓	✓	✓	✓	✓	✓	✓	✓	✓	✓	✓	✓	✓	✓
Throttle linkage ③	S/I		✓		✓		✓		✓		✓		✓		✓		✓
Timing belt	R											✓					

93591CM4A

SCHEDULED MAINTENANCE INTERVALS
HONDA—PASSPORT

TO BE SERVICED	TYPE OF SERVICE	VEHICLE MILEAGE INTERVAL (x1000)															
		7.5	15	22.5	30	37.5	45	52.5	60	67.5	75	82.5	90	97.5	105	112.5	120
Tires and wheels ③	S/I	✓	✓	✓	✓	✓	✓	✓	✓	✓	✓	✓	✓	✓	✓	✓	✓
Valve clearance ④	A								✓								✓

R: Replace   S/I: Service or Inspect   L: Lubricate   A: Adjust   C: Clean

① Perform this at the mileage indicated or every 6 months, whichever occurs first.

② Perform this at the mileage indicated or every 24 months, whichever occurs first.

③ Perform this at the mileage indicated or every 12 months, whichever occurs first.

④ 3.2L V6 engine.

FREQUENT OPERATION MAINTENANCE (SEVERE SERVICE)

If a vehicle is operated under any of the following conditions it is considered severe service:

- Towing a trailer or using a camper or car-top carrier.
- Repeated short trips of less than 5 miles in temperatures below freezing.
- Extensive idling or low-speed driving for long distances as in heavy commercial use, such as delivery, taxi or police cars.
- Operating on rough, muddy or salt-covered roads.
- Operating on unpaved or dusty roads.

Air cleaner element: replace every 15,000 miles

Engine oil and filter: replace every 3000 miles or 3 months, whichever occurs first.

Automatic transmission fluid: replace every 20,000 miles.

Rear axle lubricant: replace every 15,000 miles.

93591CM4B

Brake service is covered in Section 4 of this manual

SCHEDULED MAINTENANCE INTERVALS
HONDA
PASSPORT

The following should be used as a guide when determining the amount of work required for a particular service.
In estimating how long a particular Scheduled Maintenance Service should take, please observe the following:

● Labor Time is time based on field research and data supplied by the vehicle manufacturer.
● Labor time operations are given in hours and tenths of an hour.
● All labor operations are to be used as a guide.

Mechanic Skill Level Codes:
(A) PRECISION: Highly skilled with multiple certification.
(B) GENERAL: Normally skilled with certification.
(C) MAINTENANCE: Semi-skilled working on certification.

	LABOR TIME
7500 Mile Service (C)	
Passport	1.8
15000 Mile Service (B)	
Passport	3.6
22500 Mile Service (C)	
Passport	1.8
30000 Mile Service (B)	
Passport	5.3
37500 Mile Service (C)	
Passport	1.8
45000 Mile Service (C)	
Passport	1.9

	LABOR TIME
52500 Mile Service (C)	
Passport	1.8
60000 Mile Service (B)	
Passport	5.3
Adjust valves add	
67500 Mile Service (C)	
Passport	1.8
75000 Mile Service (B)	
Passport	2.7
Replace timing belt add	
82500 Mile Service (C)	
Passport	1.8

	LABOR TIME
90000 Mile Service (B)	
Passport	5.3
97500 Mile Service (C)	
Passport	1.8
105000 Mile Service (B)	
Passport	3.1
112500 Mile Service (C)	
Passport	1.8
120000 Mile Service (B)	
Passport	5.3
Adjust valves add	

93551CM5

HYUNDAI
Sante-Fe

ENGINE AND VEHICLE IDENTIFICATION CHART

		Engine Code					Model Year	
Code	Liters (cc)	Cu. In.	Cyl.	Fuel Sys.	Engine Type	Eng. Mfg.	Code ①	Year
B	2.4 (2351)	120	4	MFI	DOHC	Hyundai	1	2001
D	2.7 (2656)	120	6	MFI	DOHC	Hyundai	2	2002
							3	2003

DOHC: Double Overhead Cam

MFI: Multi-port Fuel Injection

① 8th position of VIN

93591CX1

GENERAL ENGINE SPECIFICATIONS

Year	Model	Engine Displacement Liters (cc)	Engine ID/VIN	Fuel System Type	Net Horsepower @ rpm	Net Torque @ rpm (ft. lbs.)	Bore x Stroke (in.)	Com- pression Ratio	Oil Pressure @ rpm
2001	Santa Fe	2.4 (2351)	B	MFI	149@5500	156@3000	3.41x3.94	10:01	①
		2.7 (2656)	D	MFI	181@6000	177@4000	3.41x2.95	10:01	②
2002	Santa Fe	2.4 (2351)	B	MFI	149@5500	156@3000	3.41x3.94	10:01	①
		2.7 (2656)	D	MFI	181@6000	177@4000	3.41x2.95	10:01	②

MFI: Multi-port Fuel Injection

① 11.6 Psi (80 kPa) @ idle.

② 7.3 Psi (50 kPa) or more @ idle.

93591CX2

For complete Engine Mechanical specifications, see Section 1 of this manual

ENGINE TUNE-UP SPECIFICATIONS

Year	Engine Displacement Liters (cc)	Engine ID/VIN	Spark Plug Gap (in.)	Ignition Timing (deg.)		Fuel Pump (psi)	Idle Speed (rpm)		Valve Clearance (in.)	
				MT	AT		MT	AT	In.	Ex.
2001	2.4 (2351)	B	0.039-0.043	2-12B	2-12B	37	625-825	625-825	NA	NA
	2.7 (2656)	D	0.040-0.043	7-19B	7-19B	37	625-825	625-825	NA	NA
2002	2.4 (2351)	B	0.039-0.043	2-12B	2-12B	37	625-825	625-825	NA	NA
	2.7 (2656)	D	0.040-0.043	7-19B	7-19B	37	625-825	625-825	NA	NA

NOTE: The Vehicle Emission Control Information label often reflects changes made during production and must be used if they differ from this chart.

NOTE: The fuel pressure readings are given with the vacuum hose connected to the regulator and the engine running

B: Before top dead center

HYD: Hydraulic

NA;: Not Availible

93591CX3

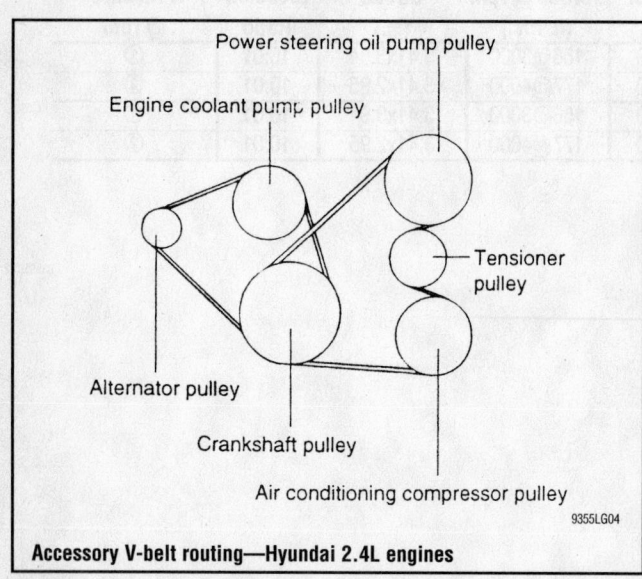

Accessory V-belt routing—Hyundai 2.4L engines

9355LG04

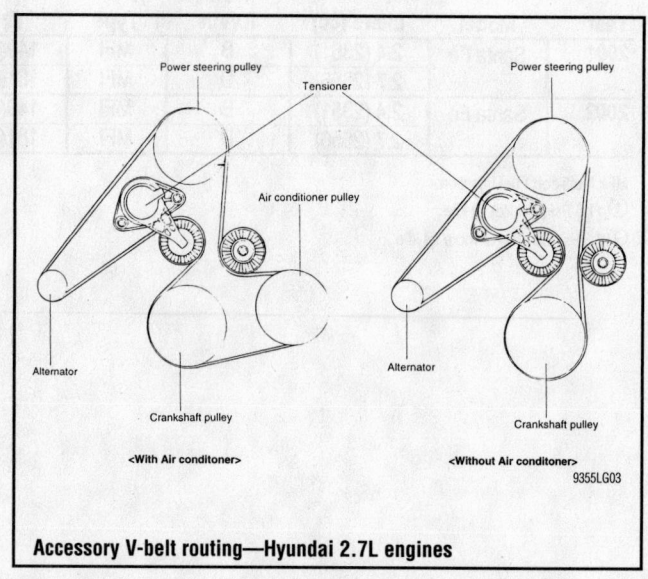

Accessory V-belt routing—Hyundai 2.7L engines

9355LG03

CAPACITIES

Year	Model	Engine Displacement Liters (cc)	Engine ID/VIN	Engine Oil with Filter (qts.)	Transmission (pts.)		Transfer Case (pts.)	Drive Axle		Fuel Tank (gal.)	Cooling System (qts.)
					Man.	Auto.		Front (pts.)	Rear (pts.)		
2001	Santa Fe	2.4 (2351)	B	4.53	2.2	8.2	—	—	2.2	14.3	7.35
		2.7 (2656)	D	4.76	2.2	8.94	—	—	2.2	14.3	8.94
2002	Santa Fe	2.4 (2351)	B	4.53	2.2	8.2	—	—	2.2	14.3	7.35
		2.7 (2656)	D	4.76	2.2	8.94	—	—	2.2	14.3	8.94

NOTE: All capacities are approximate. Add fluid gradually and check to be sure a proper fluid level is obtained.

93591CX4

VALVE SPECIFICATIONS

Year	Engine Displacement Liters (cc)	Engine ID/VIN	Seat Angle (deg.)	Face Angle (deg.)	Spring Test Pressure (lbs. @ in.)	Spring Free Length (in.)	Stem-to-Guide Clearance (in.)		Stem Diameter (in.)	
							Intake	Exhaust	Intake	Exhaust
2001	2.4 (2351)	B	45	45	NA	1.804	0.0008-0.0019	0.0020-0.0030	0.2580-0.2590	0.2571-0.2579
	2.7 (2656)	D	45	45	NA	1.670	0.008-0.0020	0.0014-0.0026	0.2350-0.2354	0.2340-0.2350
2002	2.4 (2351)	B	45	45	NA	1.804	0.0008-0.0019	0.0020-0.0030	0.2580-0.2590	0.2571-0.2579
	2.7 (2656)	D	45	45	NA	1.670	0.008-0.0020	0.0014-0.0026	0.2350-0.2354	0.2340-0.2350

NA: Not Available

93591CX5

For Accessory Drive Belt illustrations, see Section 1 of this manual

CRANKSHAFT AND CONNECTING ROD SPECIFICATIONS

All measurements are given in inches

Year	Engine Displacement Liters (cc)	Engine ID/VIN	Crankshaft				Connecting Rod		
			Main Brg. Journal Dia.	Main Brg. Oil Clearance	Shaft End-play	Thrust on No.	Journal Diameter	Oil Clearance	Side Clearance
2001	2.4 (2351)	B	NA	①	0.0020-0.0098	3	2.2434-2.2411	0.0007-0.0014	0.004-0.0098
	2.7 (2656)	D	NA	0.0002-0.0009	0.0024-0.0094	3	2.2434-2.2411	0.0007-0.0014	0.0039-0.0098
2002	2.4 (2351)	B	NA	①	0.0020-0.0098	3	2.2434-2.2411	0.0007-0.0014	0.004-0.0098
	2.7 (2656)	D	NA	0.0002-0.0009	0.0024-0.0094	3	2.2434-2.2411	0.0007-0.0014	0.0039-0.0098

NA: Not Available

① Nos. 1, 2, 4 and 5: 0.0007-0.0014
No. 3: 0.0009-0.0016

93591CX6

PISTON AND RING SPECIFICATIONS

All measurements are given in inches

Year	Engine Displacement Liters (cc)	Engine ID/VIN	Piston Clearance	Ring Gap			Ring Side Clearance		
				Top Compression	Bottom Compression	Oil Control	Top Compression	Bottom Compression	Oil Control
2001	2.4 (2351)	B	0.0008-0.0016	0.0098-0.0138	0.0157-0.0216	0.0039-0.0157	0.0012-0.0028	0.0008-0.0024	0.0024-0.0059
	2.7 (2656)	D	0.0004-0.0012	0.0079-0.0138	0.0146-0.0205	0.0079-0.0276	0.0016-0.0031	0.0012-0.0028	NA
2002	2.4 (2351)	B	0.0008-0.0016	0.0098-0.0138	0.0157-0.0216	0.0039-0.0157	0.0012-0.0028	0.0008-0.0024	0.0024-0.0059
	2.7 (2656)	D	0.0004-0.0012	0.0079-0.0138	0.0146-0.0205	0.0079-0.0276	0.0016-0.0031	0.0012-0.0028	NA

NA: Not Applicable

93591CX7

TORQUE SPECIFICATIONS
All readings in ft. lbs.

Year	Engine Displacement Liters (cc)	Engine ID/VIN	Cylinder Head Bolts	Main Bearing Bolts	Rod Bearing Bolts	Crankshaft Damper Bolts	Flywheel Bolts	Manifold Intake	Manifold Exhaust	Spark Plugs	Lug Nut
2001	2.4 (2351)	B	①	②	③	14-22	94-101	④	⑤	15-22	66-83
	2.7 (2656)	D	⑥	⑦	⑧	—	53-56	14-15	18-22	15-22	66-83
2002	2.4 (2351)	B	①	②	③	14-22	94-101	④	⑤	15-22	66-83
	2.7 (2656)	D	⑥	⑦	⑧	—	53-56	14-15	18-22	15-22	66-83

NOTE: Dip main bearing bolts and crankshaft damper bolt in clean engine oil prior to tightening.

① If using used parts:
Step 1: 14 ft. lbs. (20 Nm).
Step 2: plus an additional 90 degrees.
Step 3: plus an additional 90 degrees.
If using new parts:
Step 1: 46 ft. lbs. (64 Nm)
Step 2: Release the bolts.
Step 3: 14 ft. lbs. (20 Nm)
Step 4: plus an additional 90 degrees.
Step 5: plus an additional 90 degrees.

② 15 ft. lbs. Plus 90 degrees
③ 14 ft. lbs. Plus 90 degrees

④ Bolt (M8): 11-14 ft. lbs.
Nut: 22-30 ft. lbs.

⑤ Bolt (M8): 18-2 ft. lbs.
Bolt (M10): 25-40

⑥ Step 1: 14 ft. lbs.
Step 2: plus an additional 90 degrees.
Step 3: plus an additional 90 degrees.

⑥ Step 1: 18 ft. lbs.
Step 2: plus an additional 58-62 degrees.
Step 3: plus an additional 43-47 degrees.

⑦ Bolt (M10): 10-12 ft. lbs.
Bolt (M7): 7-9

⑧ 12-15 ft. lbs. Plus 90-94 degrees

93591CX8

WHEEL ALIGNMENT

Year	Model		Caster Range (+/-Deg.)	Caster Preferred Setting (Deg.)	Camber Range (+/-Deg.)	Camber Preferred Setting (Deg.)	Toe-in (in.)	Steering Axis Inclination (Deg.)
2001	Santa Fe	F	2 + or - 30'	2.00	0 + or - 30'	0	0.008	—
		R	—	—	0 + or - 30'	0	0.008	—
2002	Santa Fe	F	2 + or - 30'	2.00	0 + or - 30'	0	0.008	—
		R	—	—	0 + or - 30'	0	0.008	—

F: Front
R: Rear

93591CX9

For Tire, Wheel and Ball Joint specifications, see Section 1 of this manual

TIRE, WHEEL AND BALL JOINT SPECIFICATIONS

Year	Model	OEM Tires		Tire Pressures (psi)		Wheel Size	Ball Joint Inspection
		Standard	Optional	Front	Rear		
2001	Santa Fe	P225/70R16	None	30	30	6.5Jx16	NS
2002	Santa Fe	P225/70R16	None	30	30	6.5Jx16	NS

OEM: Original Equipment Manufacturer

PSI: Pounds Per Square Inch

NS: Not specified by manufacturer

93591CX0

BRAKE SPECIFICATIONS
Hyundai Santa Fe
All measurements in inches unless noted

Year	Model		Brake Disc			Brake Drum Diameter			Minimum Lining Thickness		Brake Caliper	
			Original Thickness	Minimum Thickness	Maximum Runout	Original Inside Diameter	Max. Wear Limit	Maximum Machine Diameter	Pad	Shoe	Bracket Bolts (ft. lbs.)	Mounting Bolts (ft. lbs.)
2001	Santa Fe	F	1.0200	0.960	0.002	—	—	—	0.079	—	58-73	16-24
		R	0.390	0.330	0.040	10.00	10.08	—	0.080	0.590	36-43	16-24
2002	Santa Fe	F	1.0200	0.960	0.002	—	—	—	0.079	—	58-73	16-24
		R	0.390	0.330	0.040	10.00	10.08	—	0.080	0.590	36-43	16-24

F: Front

R: Rear

93591CY1

INFINITI
QX4

ENGINE AND VEHICLE IDENTIFICATION

		Engine						Model Year	
Code ①	Liters (cc)	Cu. In.	Cyl.	Fuel Sys.	Engine Type	Eng. Mfg.	Code ②		Year
VG33E	3.3 (3277)	199.8	6	MFI	SOHC	Nissan	X		1999
VQ35DE	3.5 (3498)	213	6	MFI	DOHC	Nissan	Y		2000
							1		2001
							2		2002
							3		2003

MFI: Multi-port Fuel Injection

SOHC: Single Overhead Camshaft

DOHC: Double Overhead Camshafts

① Located on the timing belt cover

② 10th digit of the Vehicle Identification Number (VIN)

93591CZA

GENERAL ENGINE SPECIFICATIONS

Year	Model	Engine Displacement Liters (cc)	Engine ID/VIN	Fuel System Type	Net Horsepower @ rpm	Net Torque @ rpm (ft. lbs.)	Bore x Stroke (in.)	Com-pression Ratio	Oil Pressure @ rpm
1999	QX4	3.3 (3277)	VG33E	MFI	170@4800	200@2800	3.60x3.27	8.9:1	60-65@2000
2000	QX4	3.3 (3277)	VG33E	MFI	170@4800	200@2800	3.60x3.27	8.9:1	60-65@2000
2001	QX4	3.3 (3277)	VG33E	MFI	170@4800	200@2800	3.60x3.27	8.9:1	60-65@2000
2002	QX4	3.5 (3498)	VQ35DE	MFI	240@6000	265@3200	3.760X3.205	10.0:1	43@2000

MFI: Multi-port Fuel Injection

93591CZB

For Wheel Alignment specifications, see Section 1 of this manual

ENGINE TUNE-UP SPECIFICATIONS

Year	Engine Displacement Liters (cc)	Engine ID/VIN	Spark Plug Gap (in.)	Ignition Timing (deg.)		Fuel Pump (psi) ①	Idle Speed (rpm)		Valve Clearance (in.)	
				MT	AT		MT	AT ②	In.	Ex.
1999	3.3 (3277)	VG33E	0.039-0.043	13-17B	13-17B	34	700-800	700-800	HYD	HYD
2000	3.3 (3277)	VG33E	0.039-0.043	13-17B	13-17B	34	700-800	700-800	HYD	HYD
2001	3.3 (3277)	VG33E	0.039-0.043	13-17B	13-17B	34	700-800	700-800	HYD	HYD
2002	3.5 (3498)	VQ35DE	0.044	15B	15B	35	700-800	700-800	HYD	HYD

NOTE: The Vehicle Emission Control Information label often reflects specification changes made during production. The label figures must be used if they differ from those in this chart.

B: Before top dead center

HYD: Hydraulic

① System pressure at idle with vacuum hose connected
Should increase to 43 psi when disconnected

② Automatic transmission in Neutral

93591CZC

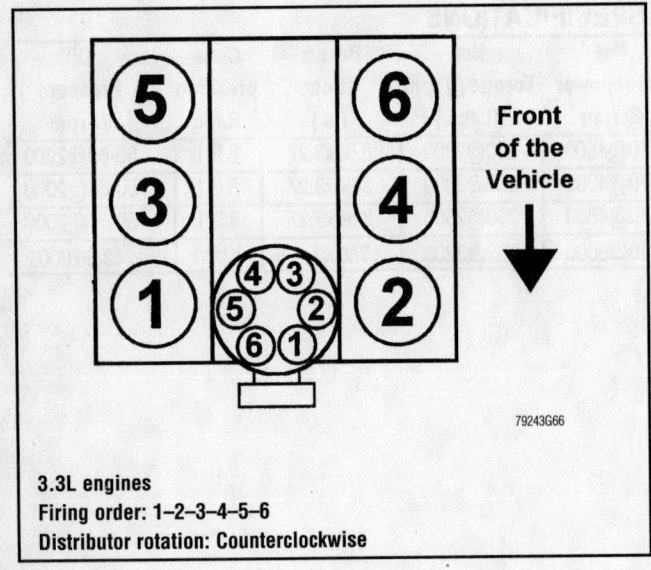

3.3L engines
Firing order: 1–2–3–4–5–6
Distributor rotation: Counterclockwise

79243G66

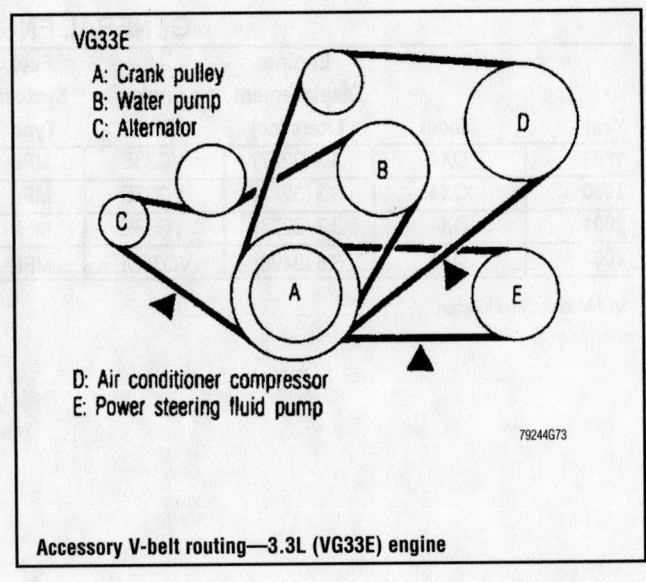

VG33E
A: Crank pulley
B: Water pump
C: Alternator

D: Air conditioner compressor
E: Power steering fluid pump

Accessory V-belt routing—3.3L (VG33E) engine

79244G73

CAPACITIES

Year	Model	Engine Displacement Liters (cc)	Engine ID/VIN	Engine Oil with Filter (qts.)	Transmission (pts.) 5-Spd	Transmission (pts.) Auto.	Transfer Case (pts.)	Drive Axle Front (pts.)	Drive Axle Rear (pts.)	Fuel Tank (gal.)	Cooling System (qts.)
1999	QX4	3.3 (3277)	VG33E	3.8	—	18.0	5.3	4.4	5.9	21.1	11.25
2000	QX4	3.3 (3277)	VG33E	3.8	—	18.0	5.3	4.4	5.9	21.1	11.25
2001	QX4	3.3 (3277)	VG33E	3.8	—	18.0	5.3	4.4	5.9	21.1	11.25
2002	QX4	3.5 (3498)	VQ35DE	5.25	—	18.0	5.3	3.9	5.9	21.1	9.75

NOTE: All capacities are approximate. Add fluid gradually and check to be sure a proper fluid level is obtained.

93591CZD

VALVE SPECIFICATIONS

Year	Engine Displacement Liters (cc)	Engine ID/VIN	Seat Angle (deg.)	Face Angle (deg.)	Spring Test Pressure (lbs. @ in.)	Spring Installed Height (in.)	Stem-to-Guide Clearance (in.) Intake	Stem-to-Guide Clearance (in.) Exhaust	Stem Diameter (in.) Intake	Stem Diameter (in.) Exhaust
1999	3.3 (3277)	VG33E	45	45.25-46.75	①	NA	0.0008-0.0021	0.0016-0.0029	0.2742-0.2748	0.3135-0.3138
2000	3.3 (3277)	VG33E	45	45.25-46.75	①	NA	0.0008-0.0021	0.0016-0.0029	0.2742-0.2748	0.3135-0.3138
2001	3.3 (3277)	VG33E	45	45.25-46.75	①	NA	0.0008-0.0021	0.0016-0.0029	0.2742-0.2748	0.3135-0.3138
2002	3.5 (3498)	VQ35DE	45.15-45.45	45	45.4@1.457	1.457	0.0008-0.0021	0.0016-0.0029	0.2348-0.2354	0.2341-0.2346

NA: Not Available

① Inner: 57.3 @ 0.984
　Outer: 117.7 @ 1.181

93591CDE

For Maintenance Interval recommendations, see Section 1 of this manual

CRANKSHAFT AND CONNECTING ROD SPECIFICATIONS
All measurements are given in inches.

Year	Engine Displacement Liters (cc)	Engine ID/VIN	Crankshaft				Connecting Rod		
			Main Brg. Journal Dia.	Main Brg. Oil Clearance	Shaft End-play	Thrust on No.	Journal Diameter	Oil Clearance	Side Clearance
1999	3.3 (3277)	VG33E	2.4790-2.4793	0.0011-0.0022	0.0020-0.0067	4	1.9967-1.9675	0.0006-0.0021	0.0079-0.0138
2000	3.3 (3277)	VG33E	2.4790-2.4793	0.0011-0.0022	0.0020-0.0067	4	1.9967-1.9675	0.0006-0.0021	0.0079-0.0138
1999	3.3 (3277)	VG33E	2.4790-2.4793	0.0011-0.0022	0.0020-0.0067	4	1.9967-1.9675	0.0006-0.0021	0.0079-0.0138
2000	3.3 (3277)	VG33E	2.4790-2.4793	0.0011-0.0022	0.0020-0.0067	4	1.9967-1.9675	0.0006-0.0021	0.0079-0.0138
2001	3.3 (3277)	VG33E	2.4790-2.4793	0.0011-0.0022	0.0020-0.0067	4	1.9967-1.9675	0.0006-0.0021	0.0079-0.0138
2002	3.5 (3498)	VQ35DE	①	0.0014-0.0018	0.0118	4	②	0.0013-0.0023	0.0079-0.0138

NA - Not Available

① There are 24 different grades, ranging from A (2.3612) to 7 (2.3603)

② Grade 0: 2.0460-2.0462

   Grade 1: 2.0457-2.0460

   Grade 2: 2.0445-2.0457

93591CZF

PISTON AND RING SPECIFICATIONS
All measurements are given in inches.

Year	Engine Displacement Liters (cc)	Engine ID/VIN	Piston Clearance	Ring Gap			Ring Side Clearance		
				Top Compression	Bottom Compression	Oil Control	Top Compression	Bottom Compression	Oil Control
1999	3.3 (3277)	VG33E	①	0.0083-0.0157	0.0197-0.0272	0.0079-0.0272	0.0009-0.0030	0.0012-0.0028	0.0006-0.0073
2000	3.3 (3277)	VG33E	①	0.0083-0.0157	0.0197-0.0272	0.0079-0.0272	0.0009-0.0030	0.0012-0.0028	0.0006-0.0073
2001	3.3 (3277)	VG33E	①	0.0083-0.0157	0.0197-0.0272	0.0079-0.0272	0.0009-0.0030	0.0012-0.0028	0.0006-0.0073
2002	3.5 (3498)	VQ35DE	0.0004-0.0012	0.0091-0.0130	0.0130-0.0189	0.0079-0.0236	0.0016-0.0031	0.0012-0.0028	0.0006-0.0020

① Cylinders 1, 2, 6: 0.0010 - 0.0018 in.

   Cylinders 3 and 4: 0.0006 - 0.0010 in.

   Cylinder 5: 0.0012-0.0016 in.

93591CDFG

TORQUE SPECIFICATIONS
All readings in ft. lbs.

Year	Engine Displacement Liters (cc)	Engine ID/VIN	Cylinder Head Bolts	Main Bearing Bolts	Rod Bearing Bolts	Crankshaft Damper Bolts	Flywheel Bolts	Manifold		Spark Plugs	Lug Nuts
								Intake	Exhaust		
1999	3.3 (3277)	VG33E	①	67-74	②	141-156	61-69	①	21-25	14-22	87-108
2000	3.3 (3277)	VG33E	①	67-74	②	141-156	61-69	①	21-25	14-22	87-108
2001	3.3 (3277)	VG33E	①	67-74	②	141-156	61-69	①	21-25	14-22	87-108
2002	3.5 (3498)	VQ35DE	③	④	⑤	⑥	61-69	⑦	21-24	14-22	87-108

① The cylinder heads and the lower intake manifold are installed together

Step 1: Tighten the cylinder head bolts to 22 ft. lbs.

Step 2: Tighten the cylinder head bolts to 43 ft. lbs.

Step 3: Loosen the cylinder head bolts completely

Step 4: Tighten the cylinder head bolts to 84 inch lbs.

Step 5: Tighten the intake manifold fasteners to 35 inch lbs.

Step 6: Tighten the intake manifold fasteners to 13 ft. lbs.

Step 7: Tighten the intake manifold fasteners to 12-14 ft. lbs.

Step 8: Loosen all intake manifold fasteners completely

Step 9: Tighten the cylinder head bolts to 22 ft. lbs.

Step 10: Tighten the cylinder head bolts 60-65 degrees

Step 11: Tighten the cylinder head sub-bolts to 80-105 inch lbs.

Step 12: Tighten the intake manifold fasteners to 35 inch lbs.

Step 13: Tighten the intake manifold fasteners to 78 inch lbs.

Step 14: Tighten the intake manifold fasteners to 70-84 inch lbs.

② 10-12 ft. lbs. plus 60-65 degrees or 28-33 ft. lbs.

③ Step 1: 72 ft. lbs.

Step 2: Loosen all bolts completely
Step 3: 25-33 ft. lbs.
Step 4: +90 degrees
Step 5: +90 degrees

④ Step 1: 24-28 ft. lbs.

Step 2: +90 degrees

⑤ Step 1: 15 ft. lbs.

Step 2: +90 degrees

⑥ 29-36 ft. lbs. +60-66 degrees

⑦ Step 1: 44-86 inch lbs.

Step 2: 20-23 ft. lbs.

93591CDH

For Tune-up, Capacities and Firing orders, see Section 1 of this manual

WHEEL ALIGNMENT

Year	Model	Caster Range (+/-Deg.)	Caster Preferred Setting (Deg.)	Camber Range (+/-Deg.)	Camber Preferred Setting (Deg.)	Toe-in (in.)	Axis Inclination (Deg.)
1999	QX4	0.75	+3.00	0.75	+0.10	0.08+/-0.04	—
2000	QX4	0.75	+3.00	0.75	+0.10	0.08+/-0.04	—
2001	QX4	0.75	+3.00	0.75	+0.10	0.08+/-0.04	—
2002	QX4 ①	0.75	+3.00	0.75	+0.17	0.08+/-0.04	—

① Assumes P245/65R17 tire

93591CZI

TIRE, WHEEL AND BALL JOINT SPECIFICATIONS

Year	Model	OEM Tires Standard	OEM Tires Optional	Tire Pressures (psi) Front	Tire Pressures (psi) Rear	Wheel Size	Ball Joint Inspection
1999	QX4	P245/70R16	None	35	35	7-JJ	①
2000	QX4	P245/70R16	None	35	35	7-JJ	①
2001	QX4	P245/70R16	None	35	35	7-JJ	①
2002	QX4	P245/70R16	P245/65R17	②	②	Std: 7J/Opt: 8J	③

OEM: Original Equipment Manufacturer

PSI: Pounds Per Square Inch

STD: Standard

OPT: Optional

L: Lower

U: Upper

① Replace if any measurable movement is found.

② See placard on vehicle

③ Turning torque: 4.3-43 inch lbs.

93591CZJ

BRAKE SPECIFICATIONS

All measurements in inches unless noted

| Year | Model | Brake Disc | | | Brake Drum Diameter | | | Minimum Lining Thickness | | Brake Caliper | |
		Original Thickness	Minimum Thickness	Maximum Runout	Original Inside Diameter	Max. Wear Limit	Maximum Machine Diameter	Front	Rear	Bracket Bolts (ft. lbs.)	Mounting Bolts (ft. lbs.)
1999	QX4	1.100	1.024	0.004	11.60	NA	11.67	0.079	0.059	53-72	24-31
2000	QX4	1.100	1.024	0.004	11.60	NA	11.67	0.079	0.059	53-72	24-31
2001	QX4	1.100	1.024	0.004	11.60	NA	11.67	0.079	0.059	53-72	24-31
2002	QX4	1.100	1.024	0.003	11.61	NA	11.67	0.079	0.059	①	①

NA: Not Available

① Torque member mounting bolt: 127-134

93591CFK

SCHEDULED MAINTENANCE INTERVALS
1999-00 Infiniti—QX4

TO BE SERVICED	TYPE OF SERVICE	VEHICLE MILEAGE INTERVAL (x1000)												
		7.5	15	22.5	30	37.5	45	52.5	60	67.5	75	82.5	90	97.5
Engine oil & filter	R	✓	✓	✓	✓	✓	✓	✓	✓	✓	✓	✓	✓	✓
Brake lines & cables	S/I		✓		✓		✓		✓		✓		✓	
Brake pads, discs, drums & linings	S/I		✓		✓		✓		✓		✓		✓	
Driveshaft boots & propeller shaft	S/I				✓								✓	
Front wheel bearings (4x2)	S/I				✓								✓	
Automatic transmission, transfer & differential gear oil ①	S/I		✓		✓		✓		✓		✓		✓	
Front wheel bearings (4x4)	S/I				✓				✓				✓	
Air cleaner filter	R				✓				✓				✓	
Engine coolant	R				✓				✓				✓	
Spark plugs	R				✓				✓				✓	
Drive belt(s)	S/I				✓				✓				✓	
Exhaust system	S/I				✓				✓				✓	
Fuel lines	S/I				✓				✓				✓	
Steering gear (box) & linkage, axle & suspension parts	S/I				✓				✓				✓	
Vapor lines	S/I		✓						✓				✓	
Timing belt ②	R													

R: Replace   S/I: Service or Inspect

① Differential (w/limited-slip differential) oil: replace oil every 30,000 miles.

② Timing belt: replace at 105,000 miles.

FREQUENT OPERATION MAINTENANCE (SEVERE SERVICE)

If a vehicle is operated under any of the following conditions it is considered severe service:

- Extremely dusty areas.

- 50% or more of the vehicle operation is in 32°C (90°F) or higher temperatures, or constant operation in temperatures below 0°C (32°F).

- Prolonged idling (vehicle operation in stop and go traffic).

- Frequent short running periods (engine does not warm to normal operating temperatures).

- Police, taxi, delivery usage or trailer towing usage.

Oil & oil filter: replace every 3750 miles.

Brake pads, discs, drums & linings: service or inspect every 7500 miles.

Driveshaft boots & propeller shaft: service or inspect every 7500 miles.

Exhaust system: service or inspect every 7500 miles.

Steering gear (box) & linkage, (steering damper-4x4), axle & suspension parts: service or inspect every 7500 miles.

Steering linkage ball joints & front suspension ball joints: service or inspect every 7500 miles.

93591CFL

SCHEDULED MAINTENANCE INTERVALS
2001-02 Infiniti—QX4

TO BE SERVICED	TYPE OF SERVICE	VEHICLE MILEAGE INTERVAL (x1000)															
		3.75	7.5	11.3	15	18.8	22.5	26.3	30	33.8	37.5	41.3	45	48.8	52.5	56.3	60
Engine oil & filter	R	✓	✓	✓	✓	✓	✓	✓	✓	✓	✓	✓	✓	✓	✓	✓	✓
Brake lines & cables	S/I				✓				✓				✓				✓
Brake pads, discs, drums & linings	I		✓		✓		✓		✓		✓		✓		✓		✓
Driveshaft boots & propeller shaft	L/I		✓		✓		✓		✓		✓		✓		✓		✓
Front wheel bearings (4x2)	I								✓								✓
Automatic & manual transmission, transfer & differential gear oil ①	I				✓				✓				✓				✓
LSD gear oil	R								✓								✓
Front wheel bearing grease (4x4)	R								✓								✓
Timing belt ②	R																
Air cleaner filter	R								✓								✓
Engine coolant ③	R																✓
Spark plugs	R								platinum tipped plugs every 105,000 miles								
Drive belt(s)	S/I								✓								✓
Cabin air filter	I/R		I		R		I		R		I		R		I		R
Exhaust system	I		✓		✓		✓		✓		✓		✓		✓		✓
Fuel lines	S/I								✓								✓
Steering gear (box) & linkage, axle & suspension parts	I		✓		✓		✓		✓		✓		✓		✓		✓
Vapor lines	S/I								✓								✓

R: Replace   S/I: Service or Inspect   L: Lubricate

① Differential (w/limited-slip differential) oil: replace oil every 30,000 miles.

② Timing belt: replace at 105,000 miles.

③ After 60,000, replace every 30,000

FREQUENT OPERATION MAINTENANCE (SEVERE SERVICE)

If a vehicle is operated under any of the following conditions it is considered severe service:

- Extremely dusty areas.

- 50% or more of the vehicle operation is in 32°C (90°F) or higher temperatures, or constant operation in temperatures below 0°C (32°F).

- Prolonged idling (vehicle operation in stop and go traffic).

- Frequent short running periods (engine does not warm to normal operating temperatures).

- Police, taxi, delivery usage or trailer towing usage.

Oil & oil filter: replace every 3750 miles.

Brake pads, discs, drums & linings: service or inspect every 7500 miles.

Driveshaft boots & propeller shaft: service or inspect every 7500 miles.

Exhaust system: service or inspect every 7500 miles.

Steering gear (box) & linkage, (steering damper-4x4), axle & suspension parts: service or inspect every 7500 miles.

Steering linkage ball joints & front suspension ball joints: service or inspect every 7500 miles.

93591CFM

SCHEDULED MAINTENANCE INTERVALS
INFINITI
QX4

The following should be used as a guide when determining the amount of work required for a particular service.
In estimating how long a particular Scheduled Maintenance Service should take, please observe the following:

● Labor Time is time based on field research and data supplied by the vehicle manufacturer.
● Labor time operations are given in hours and tenths of an hour.
● All labor operations are to be used as a guide.

Mechanic Skill Level Codes:
(A) PRECISION: Highly skilled with multiple certification.
(B) GENERAL: Normally skilled with certification.
(C) MAINTENANCE: Semi-skilled working on certification.

	LABOR TIME		LABOR TIME		LABOR TIM
7500 Mile Service (C)		**45000 Mile Service (C)**		**75000 Mile Service (C)**	
All Models	.6	All Models	1.0	All Models	1.
15000 Mile Service (C)		**52500 Mile Service (C)**		**82500 Mile Service (C)**	
All Models	1.0	All Models	.6	All Models	.
22500 Mile Service (C)		**60000 Mile Service (B)**		**90000 Mile Service (B)**	
All Models	.5	All Models	3.8	All Models	3.
30000 Mile Service (B)		*w/4WD add*	.5	*w/4WD add*	.
All Models	3.0	*Valve clearance add*		**97500 Mile Service (C)**	
w/4WD add	.5	**67500 Mile Service (C)**		All Models	.
37500 Mile Service (C)		All Models	.6	*Valve clearance add*	
All Models	.6				

93591CZN

ISUZU
Amigo • Rodeo • Trooper • VehiCROSS

ENGINE AND VEHICLE IDENTIFICATION

	Engine							Model Year	
Code	Liters (cc)	Cu. In.	Cyl.	Fuel Sys.	Engine Type	Eng. Mfg.	Code ①		Year
D	2.2 (2198)	134	4	MFI	DOHC	Isuzu	X		1999
W	3.2 (3165)	193	6	MFI	DOHC	Isuzu	Y		2000
X	3.5 (3494)	213	6	MFI	DOHC	Isuzu	1		2001
							2		2002
							3		2003

MFI: Multi-port Fuel Injection

DOHC: Double Overhead Camshaft

SOHC: Single Overhead Camshaft

① 10th position of VIN

93591CM7

GENERAL ENGINE SPECIFICATIONS

Year	Model	Engine Displacement Liters (cc)	Engine Series (ID/VIN)	Fuel System	Net Horsepower @ rpm	Net Torque @ rpm (ft. lbs.)	Bore x Stroke (in.)	Compression Ratio	Oil Pressure @ rpm
1999	Amigo	2.2 (2198)	D	MFI	130@5200	144@4000	3.39x3.72	9.6:1	22@800
		3.2 (3165)	W	MFI	205@5400	214@3000	3.68x3.03	9.1:1	60-80@3000
	Rodeo	2.2 (2198)	D	MFI	130@5200	144@4000	3.39x3.72	9.6:1	22@800
		3.2 (3165)	W	MFI	205@5400	214@3000	3.68x3.03	9.1:1	60-80@3000
	Trooper	3.5 (3494)	X	MFI	215@5400	230@3000	3.68x3.35	9.1:1	60-80@3000
	VehiCROSS	3.5 (3494)	X	MFI	215@5400	230@3000	3.68x3.35	9.1:1	60-80@3000
2000	Amigo	2.2 (2198)	D	MFI	130@5200	144@4000	3.39x3.72	9.6:1	22@800
		3.2 (3165)	W	MFI	205@5400	214@3000	3.68x3.03	9.1:1	60-80@3000
	Rodeo	2.2 (2198)	D	MFI	130@5200	144@4000	3.39x3.72	9.6:1	22@800
		3.2 (3165)	W	MFI	205@5400	214@3000	3.68x3.03	9.1:1	60-80@3000
	Trooper	3.5 (3494)	X	MFI	215@5400	230@3000	3.68x3.35	9.1:1	60-80@3000
	VehiCROSS	3.5 (3494)	X	MFI	215@5400	230@3000	3.68x3.35	9.1:1	60-80@3000
2001	Rodeo	2.2 (2198)	D	MFI	130@5200	144@4000	3.39x3.72	10.0:1	22@800
		3.2 (3165)	W	MFI	205@5400	214@3000	3.68x3.03	9.1:1	60-80@3000
	Rodeo Sport	2.2 (2198)	D	MFI	130@5200	144@4000	3.39x3.72	10.0:1	22@800
		3.2 (3165)	W	MFI	205@5400	214@3000	3.68x3.03	9.1:1	60-80@3000
	Trooper	3.5 (3494)	X	MFI	215@5400	230@3000	3.68x3.35	9.1:1	60-80@3000
	VehiCROSS	3.5 (3494)	X	MFI	215@5400	230@3000	3.68x3.35	9.1:1	60-80@3000
2002	Rodeo	2.2 (2198)	D	MFI	130@5200	144@4000	3.39x3.72	10.0:1	22@800
		3.2 (3165)	W	MFI	205@5400	214@3000	3.68x3.03	9.1:1	60-80@3000
	Rodeo Sport	2.2 (2198)	D	MFI	130@5200	144@4000	3.39x3.72	10.0:1	22@800
		3.2 (3165)	W	MFI	205@5400	214@3000	3.68x3.03	9.1:1	60-80@3000
	Trooper	3.5 (3494)	X	MFI	215@5400	230@3000	3.68x3.35	9.1:1	60-80@3000

MFI: Multiport fuel injection

93591CM8

Timing belt service is covered in Section 3 of this manual

ENGINE TUNE-UP SPECIFICATIONS

Year	Engine Displacement Liters (cc)	Engine ID/VIN	Spark Plug Gap (in.)	Ignition Timing (deg.)		Fuel Pump (psi)	Idle Speed (rpm)		Valve Clearance	
				MT	AT		MT	AT	In.	Ex.
1999	2.2 (2198)	D	0.040	①	①	41-55	800	800	HYD	HYD
	3.2 (3165)	W	0.040	①	①	48-55	750	750	0.009-0.013	0.010-0.014
	3.5 (3494)	X	0.040	①	①	48-55	750	750	0.009-0.013	0.010-0.014
2000	2.2 (2198)	D	0.040	①	①	41-55	800	800	HYD	HYD
	3.2 (3165)	W	0.040	①	①	48-55	750	750	0.009-0.013	0.010-0.014
	3.5 (3494)	X	0.040	①	①	48-55	750	750	0.009-0.013	0.010-0.014
2001	2.2 (2198)	D	0.040	①	①	41-55	800	800	HYD	HYD
	3.2 (3165)	W	0.040	①	①	48-55	750	750	0.009-0.013	0.010-0.014
	3.5 (3494)	X	0.040	①	①	48-55	750	750	0.009-0.013	0.010-0.014
2002	2.2 (2198)	D	0.040	①	①	41-55	800	800	HYD	HYD
	3.2 (3165)	W	0.040	①	①	48-55	750	750	0.009-0.013	0.010-0.014
	3.5 (3494)	X	0.040	①	①	48-55	750	750	0.009-0.013	0.010-0.014

NOTE: The Vehicle Emission Control Information label often reflects specification changes made during production. The label figures must be used if they differ from those in this chart.

B: Before top dead center

HYD: Hydraulic

① Controlled by the PCM

93591CM9

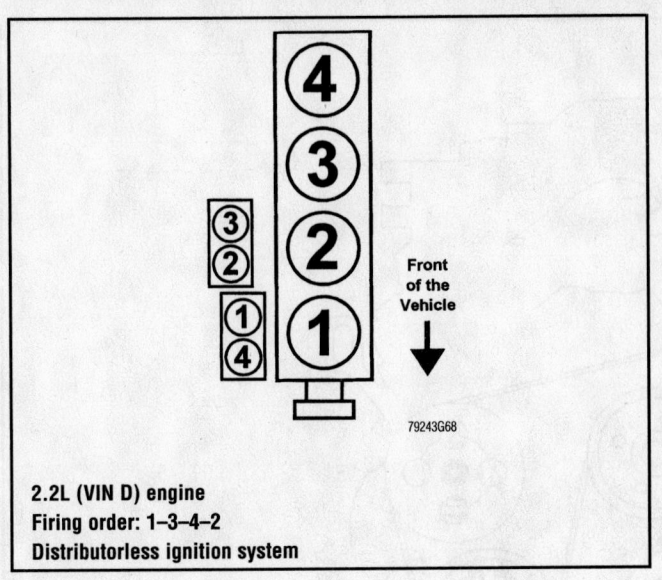

2.2L (VIN D) engine
Firing order: 1–3–4–2
Distributorless ignition system

79243G68

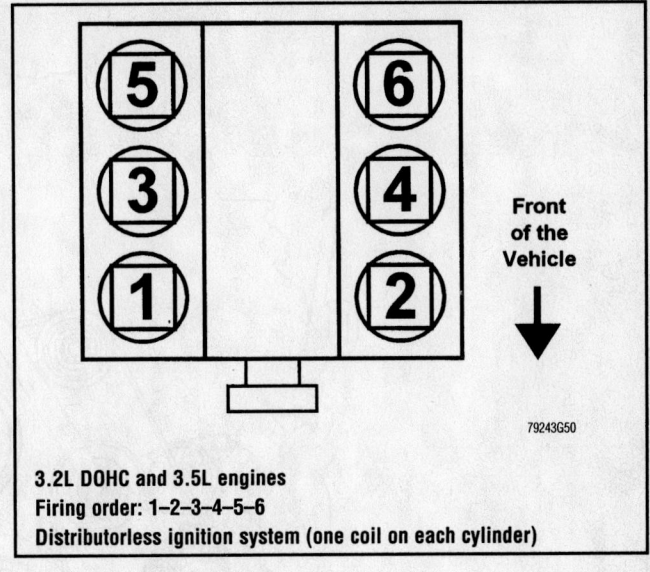

Front
of the
Vehicle

3.2L DOHC and 3.5L engines
Firing order: 1–2–3–4–5–6
Distributorless ignition system (one coil on each cylinder)

79243G50

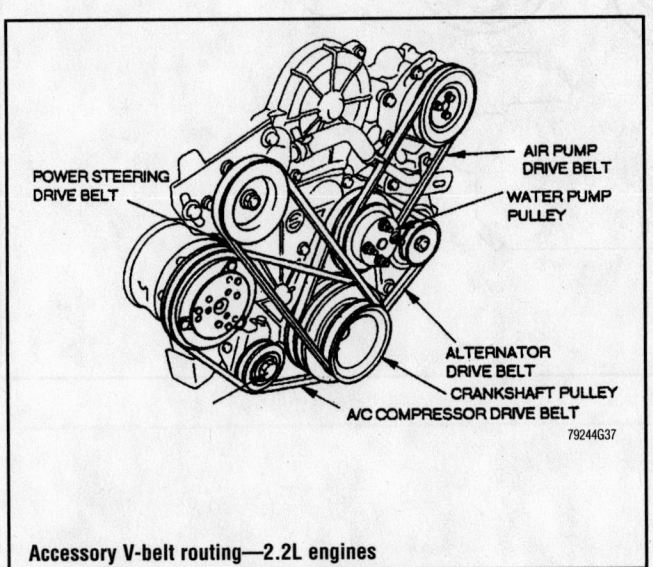

POWER STEERING
DRIVE BELT

AIR PUMP
DRIVE BELT

WATER PUMP
PULLEY

ALTERNATOR
DRIVE BELT

CRANKSHAFT PULLEY

A/C COMPRESSOR DRIVE BELT

79244G37

Accessory V-belt routing—2.2L engines

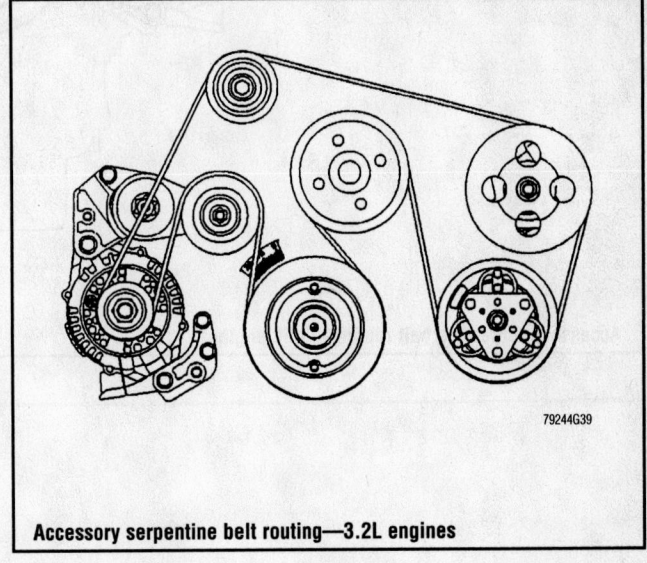

79244G39

Accessory serpentine belt routing—3.2L engines

Heater Core replacement is covered in Section 2 of this manual

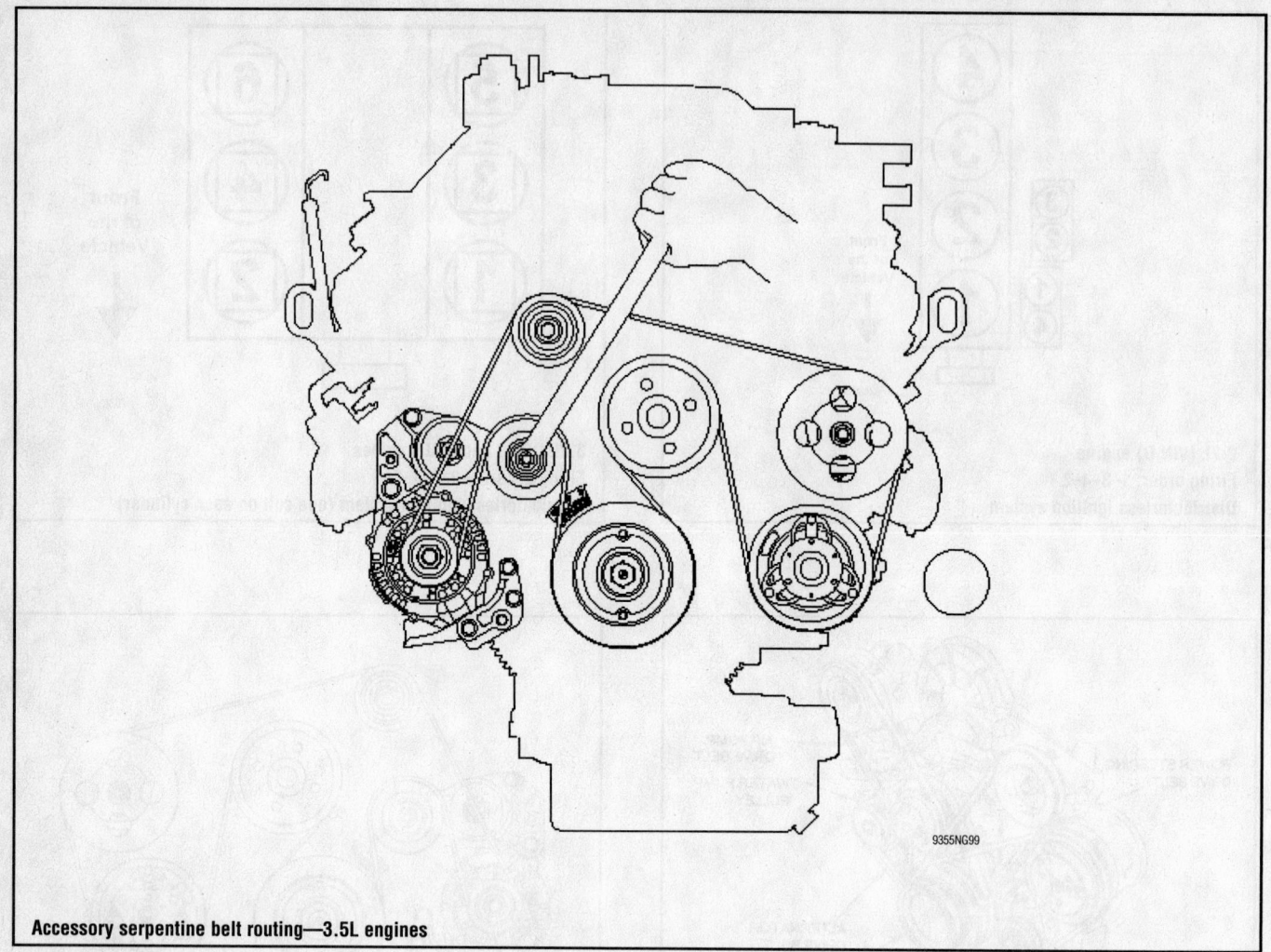

9355NG99

Accessory serpentine belt routing—3.5L engines

CAPACITIES

Year	Model	Engine Displacement Liters (cc)	Engine ID/VIN	Oil with Filter (qts.)	Engine Transmission (pts.) Man.	Auto.	Transfer Case (pts.)	Drive Axle Front (pts.)	Rear (pts.)	Fuel Tank (gal.)	Cooling System (qts.)
1999	Amigo	2.2 (2198)	D	4.8	4.5	—	3.0	3.0	3.7	17.7	7.3
		3.2 (3165)	W	5.0	6.2	—	3.0	3.0	3.7	17.7	11.2
	Rodeo	2.2 (2198)	D	4.8	4.5	18.2	3.0	3.0	3.7	21.1	7.3
		3.2 (3165)	W	5.0	6.2	18.2	3.0	3.0	3.7	21.1	11.2
	VehiCROSS	3.5 (3494)	X	5.0	—	18.2	3.0①	3.0	6.4	22.5	②
	Trooper	3.5 (3494)	X	5.0	5.8	18.2	3.0①	3.0	6.4	22.5	②
2000	Amigo	2.2 (2198)	D	4.8	4.5	—	3.0	3.0	3.7	17.7	7.3
		3.2 (3165)	W	5.0	6.2	—	3.0	3.0	3.7	17.7	11.2
	Rodeo	2.2 (2198)	D	4.8	4.5	18.2	3.0	2.2	3.7	21.1	7.3
		3.2 (3165)	W	5.0	6.2	18.2	3.0	2.6	3.7	21.1	11.6
	Trooper	3.5 (3494)	X	5.0	5.8	18.2	3.0①	3.0	6.4	22.5	②
	VehiCROSS	3.5 (3494)	X	5.0	—	18.2	3.0①	3.0	6.4	22.5	②
2001	Rodeo Sport	2.2 (2198)	D	4.8	6.2	—	3.0	2.2	3.7	17.7	7.3
		3.2 (3165)	W	5.0	6.2	18.2	3.0	2.6	3.7	17.7	11.2
	Rodeo	2.2 (2198)	D	4.8	4.5	—	3.0	2.2	3.7	20.0	7.3
		3.2 (3165)	W	5.0	6.2	18.2	3.0	2.6	3.7	20.0	11.6
	Trooper	3.5 (3494)	X	5.0	5.8	18.2	3.0①	3.0	6.4	22.5	②
	VehiCROSS	3.5 (3494)	X	5.0	—	18.2	4.0	3.0	4.6	22.5	7.4
2002	Rodeo Sport	2.2 (2198)	D	4.8	6.2	—	3.0	2.2	3.7	17.7	7.3
		3.2 (3165)	W	5.0	6.2	18.2	3.0	2.6	3.7	17.7	11.2
	Rodeo	2.2 (2198)	D	4.8	4.5	—	3.0	2.2	3.7	20.0	7.3
		3.2 (3165)	W	5.0	6.2	18.2	3.0	2.6	3.7	20.0	11.6
	Trooper	3.5 (3494)	X	5.0	5.8	18.2	3.0①	3.0	6.4	22.5	②

NOTE: All capacities are approximate. Add fluid gradually and check to ensure a proper level has been reached.

① 4.0 pts. if equipped with Torque On Demand
② A/T: 7.4 qts.
　M/T: 7.0 qts.

93591CM0

Brake service is covered in Section 4 of this manual

VALVE SPECIFICATIONS

Year	Engine Displacement Liters (cc)	Engine ID/VIN	Seat Angle (deg.)	Face Angle (deg.)	Spring Test Pressure (lbs. @ in.)	Spring Installed Height (in.)	Stem-to-Guide Clearance (in.)		Stem Diameter (in.)	
							Intake	Exhaust	Intake	Exhaust
1999	2.2 (2198)	D	NA	NA	NA	NA	0.0012-0.0022	0.0016-0.0026	NA	NA
	3.2 (3165)	W	45	45	41-44@1.38	1.38	0.0002-0.0009	0.0012-0.0025	0.2346-0.2353	0.2343-0.2350
	3.5 (3494)	X	45	45	41-44@1.38	1.38	0.0002-0.0009	0.0012-0.0025	0.2346-0.2353	0.2343-0.2350
2000	2.2 (2198)	D	NA	NA	NA	NA	0.0012-0.0022	0.0016-0.0026	NA	NA
	3.2 (3165)	W	45	45	41-44@1.38	1.38	0.0002-0.0009	0.0012-0.0025	0.2346-0.2353	0.2343-0.2350
	3.5 (3494)	X	45	45	41-44@1.38	1.38	0.0002-0.0009	0.0012-0.0025	0.2346-0.2353	0.2343-0.2350
2001	2.2 (2198)	D	NA	NA	NA	NA	0.0012-0.0022	0.0016-0.0026	NA	NA
	3.2 (3165)	W	45	45	41-44@1.38	1.38	0.0002-0.0009	0.0012-0.0025	0.2346-0.2353	0.2343-0.2350
	3.5 (3494)	X	45	45	41-44@1.38	1.38	0.0002-0.0009	0.0012-0.0025	0.2346-0.2353	0.2343-0.2350
2002	2.2 (2198)	D	NA	NA	NA	NA	0.0012-0.0022	0.0016-0.0026	NA	NA
	3.2 (3165)	W	45	45	41-44@1.38	1.38	0.0002-0.0009	0.0012-0.0025	0.2346-0.2353	0.2343-0.2350
	3.5 (3494)	X	45	45	41-44@1.38	1.38	0.0002-0.0009	0.0012-0.0025	0.2346-0.2353	0.2343-0.2350

NA: Not Available

93591CN1

CRANKSHAFT AND CONNECTING ROD SPECIFICATIONS
All measurements are given in inches.

Year	Engine Displacement Liters (cc)	Engine ID/VIN	Crankshaft				Connecting Rod		
			Main Brg. Journal Dia.	Main Brg. Oil Clearance	Shaft End-play	Thrust on No.	Journal Diameter	Oil Clearance	Side Clearance
1999	2.2 (2198)	D	2.2590-2.2610	0.0007-0.0016	0.0004-0.0008	2	1.9090-1.9100	0.0002-0.0012	0.0050-0.0150
	3.2 (3165)	W	2.5165-2.5170	0.0007-0.0017	0.0024-0.0094	3	2.1229-2.1235	0.0010-0.0023	0.0050-0.0150
	3.5 (3494)	X	2.5165-2.5170	0.0007-0.0017	0.0024-0.0094	3	2.1229-2.1235	0.0010-0.0023	0.0050-0.0150
2000	2.2 (2198)	D	2.2590-2.2610	0.0007-0.0016	0.0004-0.0008	2	1.9090-1.9100	0.0002-0.0012	0.0050-0.0150
	3.2 (3165)	W	2.5165-2.5170	0.0007-0.0017	0.0024-0.0094	3	2.1229-2.1235	0.0010-0.0023	0.0050-0.0150
	3.5 (3494)	X	2.5165-2.5170	0.0007-0.0017	0.0024-0.0094	3	2.1229-2.1235	0.0010-0.0023	0.0050-0.0150
2001	2.2 (2198)	D	2.2590-2.2610	0.0007-0.0016	0.0004-0.0008	2	1.9090-1.9100	0.0002-0.0012	0.0050-0.0150
	3.2 (3165)	W	2.5165-2.5170	0.0007-0.0017	0.0024-0.0094	3	2.1229-2.1235	0.0010-0.0023	0.0050-0.0150
	3.5 (3494)	X	2.5165-2.5170	0.0007-0.0017	0.0024-0.0094	3	2.1229-2.1235	0.0010-0.0023	0.0050-0.0150
2002	2.2 (2198)	D	2.2590-2.2610	0.0007-0.0016	0.0004-0.0008	2	1.9090-1.9100	0.0002-0.0012	0.0050-0.0150
	3.2 (3165)	W	2.5165-2.5170	0.0007-0.0017	0.0024-0.0094	3	2.1229-2.1235	0.0010-0.0023	0.0050-0.0150
	3.5 (3494)	X	2.5165-2.5170	0.0007-0.0017	0.0024-0.0094	3	2.1229-2.1235	0.0010-0.0023	0.0050-0.0150

93591CN2

For complete Engine Mechanical specifications, see Section 1 of this manual

PISTON AND RING SPECIFICATIONS
All measurements are given in inches.

Year	Engine Displacement Liters (cc)	Engine ID/VIN	Piston Clearance	Ring Gap			Ring Side Clearance		
				Top Compression	Bottom Compression	Oil Control	Top Compression	Bottom Compression	Oil Control
1999	2.2 (2198)	D	NA	0.0118-0.0195	0.0118-0.0195	0.016-0.055	0.0008-0.0546	0.0008-0.0546	NA
	3.2 (3165)	W	NA	0.0118-0.0157	0.0177-0.0236	0.006-0.018	0.0006-0.0015	0.0006-0.0015	NA
	3.5 (3494)	X	NA	0.0118-0.0157	0.0177-0.0236	0.006-0.016	0.0006-0.0015	0.0006-0.0015	NA
2000	2.2 (2198)	D	NA	0.0118-0.0195	0.0118-0.0195	0.016-0.055	0.0008-0.0546	0.0008-0.0546	NA
	3.2 (3165)	W	NA	0.0118-0.0157	0.0177-0.0236	0.0060-0.018	0.0006-0.002	0.0006-0.0015	NA
	3.5 (3494)	X	NA	0.0118-0.0157	0.0177-0.0236	0.006-0.018	0.0006-0.0015	0.0006-0.0015	NA
2001	2.2 (2198)	D	NA	0.0118-0.0195	0.0118-0.0195	0.016-0.055	0.0008-0.0546	0.0008-0.0546	NA
	3.2 (3165)	W	NA	0.0118-0.0157	0.0177-0.0236	0.0060-0.018	0.0006-0.002	0.0006-0.0015	NA
	3.5 (3494)	X	NA	0.0118-0.0157	0.0177-0.0236	0.006-0.018	0.0006-0.0015	0.0006-0.0015	NA
2002	2.2 (2198)	D	NA	0.0118-0.0195	0.0118-0.0195	0.016-0.055	0.0008-0.0546	0.0008-0.0546	NA
	3.2 (3165)	W	NA	0.0118-0.0157	0.0177-0.0236	0.0060-0.018	0.0006-0.002	0.0006-0.0015	NA
	3.5 (3494)	X	NA	0.0118-0.0157	0.0177-0.0236	0.006-0.018	0.0006-0.0015	0.0006-0.0015	NA

NA: Not Available

93591CN3

TORQUE SPECIFICATIONS
All readings in ft. lbs.

Year	Engine Displacement Liters (cc)	Engine ID/VIN	Cylinder Head Bolts	Main Bearing Bolts	Rod Bearing Bolts	Crankshaft Damper Bolts	Flywheel Bolts	Manifold		Spark Plugs	Lug Nuts
								Intake	Exhaust		
1999	2.2 (2198)	D	①	②	③	④	⑤	16	⑥	18	87
	3.2 (3165)	W	⑦	29	40	123	40	18	42	13	87
	3.5 (3494)	X	⑦	⑧	40	123	40	18	38	13	87
2000	2.2 (2198)	D	①	②	③	④	⑤	16	⑥	18	87
	3.2 (3165)	W	⑦	29	40	123	40	18	42	13	87
	3.5 (3494)	X	⑦	⑧	40	123	40	18	38	13	87
2001	2.2 (2198)	D	①	②	③	④	⑤	16	⑥	18	87
	3.2 (3165)	W	⑦	29	40	123	40	18	42	13	87
	3.5 (3494)	X	⑦	⑧	40	123	40	18	38	13	87
2002	2.2 (2198)	D	①	②	③	④	⑤	16	⑥	18	87
	3.2 (3165)	W	⑦	29	40	123	40	18	42	13	87
	3.5 (3494)	X	⑦	⑧	40	123	40	18	38	13	87

① Step 1: 18 ft. lbs.
　Step 2: Plus 90 degrees
　Step 3: Plus 90 degrees
　Step 4: Plus 90 degrees

② Step 1: 37 ft. lbs.
　Step 2: Plus 45 degrees
　Step 3: Plus 15 degrees

③ Step 1: 25 ft. lbs.
　Step 2: Plus 45 degrees
　Step 3: Plus 15 degrees

④ Crankshaft sprocket:
　Step 1: 94 ft. lbs.
　Step 2: Plus 45 degrees
　Crankshaft balancer:
　Step 1: 14 ft. lbs.
　Step 2: Plus 45 degrees

⑤ Step 1: 48 ft. lbs.
　Step 2: Plus 30 degrees
　Step 3: Plus 15 degrees

⑥ Step 1: 112 inch lbs.
　Step 2: 14 ft. lbs.
　Step 3: 14 ft. lbs.

⑦ Step 1: 21 ft. lbs.
　Step 2: 47 ft. lbs.

⑧ Step 1: 22 ft. lbs.
　Step 2: Plus 55-65 degrees
　Step 3: Crankcase side bolts to 29 ft. lbs.

93591CN4

For Accessory Drive Belt illustrations, see Section 1 of this manual

WHEEL ALIGNMENT

Year	Model		Caster Range (+/-Deg.)	Caster Preferred Setting (Deg.)	Camber Range (+/-Deg.)	Camber Preferred Setting (Deg.)	Toe-in (in.)	Steering Axis Inclination (Deg.)
1999	Amigo	F	0.75	+2.50	0.30	0	0+/-0.08	—
		R	—	—	1.00	0	0+/-0.20	—
	Rodeo	F	0.75	+2.50	0.50	0	0+/-0.08	—
		R	—	—	1.00	0	0+/-0.20	—
	Trooper		0.75	+2.10	0.50	0	0+/-0.08	—
	VehiCROSS		0.75	+2.10	0.50	0	0+/-0.08	—
2000	Amigo	F	0.75	+2.50	0.30	0	0+/-0.08	—
		R	—	—	1.00	0	0+/-0.20	—
	Rodeo	F	1.00	+2.50	0.50	0	0+/-0.08	—
		R	—	—	1.00	0	0+/-0.20	—
	Trooper		0.75	+2.10	0.50	0	0+/-0.08	—
	VehiCROSS		0.75	+2.10	0.50	0	0+/-0.08	—
2001	Rodeo Sport	F	1.00	+2.50	0.50	0	0+/-0.08	—
		R	—	—	1.00	0	0+/-0.20	—
	Rodeo	F	1.00	+2.50	0.50	0	0+/-0.08	—
		R	—	—	1.00	0	0+/-0.20	—
	Trooper		0.75	+2.10	0.50	0	0+/-0.08	—
	VehiCROSS		0.75	+2.10	0.50	0	0+/-0.08	—
2002	Rodeo Sport	F	1.00	+2.50	0.50	0	0+/-0.08	—
		R	—	—	1.00	0	0+/-0.20	—
	Rodeo	F	1.00	+2.50	0.50	0	0+/-0.08	—
		R	—	—	1.00	0	0+/-0.20	—
	Trooper		0.75	+2.10	0.50	0	0+/-0.08	—

93591CN5

TIRE, WHEEL AND BALL JOINT SPECIFICATIONS

| Year | Model | OEM Tires | | Tire Pressures (psi) | | Wheel Size | Ball Joint Inspection |
		Standard	Optional	Front	Rear		
1999	Amigo 2wd/4wd	P235/75R15	P245/70R16	26	26	6.5-JJ	U: 4-28 ① L: 4-55
	Rodeo S 2wd/4wd	P235/75R15	none	29	29	6.5-JJ	U: 4-28 ① L: 4-55
	Rodeo LS, LE 2wd	P235/75R15	P245/70R16	Std: 29 Opt: 26	Std: 29 Opt: 26	Std: 6.5JJ Opt: 7JJ	U: 4-28 ① L: 4-55
	Rodeo LS, LE 4wd	P245/70R16	none	26	26	7JJ	U: 4-28 ① L: 4-55
	Trooper	P245/70R16	none	30	35	7-JJ	U: 4-28 ① L: 4-55
	VehiCROSS	P245/70R16	none	26	26	7JJ	NS
2000	Amigo	P245/70R16	none	26	26	7JJ	U: 4-28 ① L: 4-55
	Rodeo 2wd/4wd	P245/70R16	none	29	29	7JJ	U: 4-28 ① L: 4-55
	Trooper	P245/70R15	none	26	26	7JJ	U: 4-28 ① L: 4-55
	VehiCROSS	P245/70R16	none	26	26	7JJ	NS
2001	Rodeo Sport	P245/70R16	none	26	26	7JJ	U: 4-28 ① L: 4-55
	Rodeo 2wd/4wd	P225/70R16	P245/70R16	29	29	7JJ	U: 4-28 ① L: 4-55
	Trooper	P245/70R15	none	26	26	7JJ	U: 4-28 ① L: 4-55
	VehiCROSS	P245/60R18	none	26	26	7JJ	NS
2002	Rodeo Sport	P245/70R16	none	26	26	7JJ	U: 4-28 ① L: 4-55
	Rodeo 2wd/4wd	P225/70R16	P245/70R16	29	29	7JJ	U: 4-28 ① L: 4-55
	Trooper	P245/70R15	none	26	26	7JJ	U: 4-28 ① L: 4-55

L: Lower
U: Upper
NS: Not specified by manufacturer
① Torque required in inch lbs. to rotate ball joint when removed from the knuckle

93591CN6

For Tire, Wheel and Ball Joint specifications, see Section 1 of this manual

BRAKE SPECIFICATIONS
All measurements in inches unless noted

Year	Model		Brake Disc Original Thickness	Brake Disc Machine Thickness	Brake Disc Minimum Thickness	Maximum Runout	Brake Drum Diameter Original Inside Diameter	Max. Wear Limit	Maximum Machine Diameter	Minimum Lining Thickness	Brake Caliper Bracket Bolts (ft. lbs.)	Brake Caliper Mounting Bolts (ft. lbs.)
1999	Amigo	F	1.020	0.983	0.969	0.005	—	—	—	0.039	115	54
		R	0.710	0.668	0.654	0.005	11.6	11.67	NA	0.039	76	32
	Rodeo	F	1.020	0.983	0.969	0.005	—	—	—	0.039	115	54
		R	0.710	0.668	0.654	0.005	11.6	11.67	NA	0.039	76	32
	Trooper	F	1.024	0.983	0.969	0.005	—	—	—	0.039	115	54
		R	0.710	0.668	0.654	0.005	—	—	NA	0.039	76	32
	VehiCROSS	F	1.024	0.983	0.969	0.005	—	—	—	0.039	115	54
		R	0.710	0.668	0.654	0.005	—	—	—	0.039	76	32
2000	Amigo	F	1.020	0.983	0.969	0.005	—	—	—	0.039	115	54
		R	0.710	0.668	0.654	0.005	11.6	11.67	NA	0.039	76	32
	Rodeo	F	1.020	0.983	0.969	0.005	—	—	—	0.039	115	54
		R	0.710	0.668	0.654	0.005	11.6	11.67	NA	0.039	76	32
	Trooper	F	1.024	0.983	0.969	0.005	—	—	—	0.039	115	54
		R	0.710	0.668	0.654	0.005	—	—	NA	0.039	76	32
	VehiCROSS	F	1.024	0.983	0.969	0.005	—	—	—	0.039	115	54
		R	0.710	0.668	0.654	0.005	—	—	—	0.039	76	32
2001	Rodeo Sport	F	1.020	0.983	0.969	0.005	—	—	—	0.039	115	54
		R	0.710	0.668	0.654	0.005	11.6	11.67	NA	0.039	76	32
	Rodeo	F	1.020	0.983	0.969	0.005	—	—	—	0.039	115	54
		R	0.710	0.668	0.654	0.005	11.6	11.67	NA	0.039	76	32
	Trooper	F	1.024	0.983	0.969	0.005	—	—	—	0.039	115	54
		R	0.710	0.668	0.654	0.005	—	—	—	0.039	76	32
	VehiCROSS	F	1.024	0.983	0.969	0.005	—	—	—	0.039	115	54
		R	0.710	0.668	0.654	0.005	—	—	—	0.039	76	32
2002	Rodeo Sport	F	1.020	0.983	0.969	0.005	—	—	—	0.039	115	54
		R	0.710	0.668	0.654	0.005	11.6	11.67	NA	0.039	76	32
	Rodeo	F	1.020	0.983	0.969	0.005	—	—	—	0.039	115	54
		R	0.710	0.668	0.654	0.005	11.6	11.67	NA	0.039	76	32
	Trooper	F	1.024	0.983	0.969	0.005	—	—	—	0.039	115	54
		R	0.710	0.668	0.654	0.005	—	—	—	0.039	76	32

NA: Not Available

93591CN7

SCHEDULED MAINTENANCE INTERVALS
Isuzu—Amigo, Rodeo, Rodeo Sport, Trooper & vehiCROSS

TO BE SERVICED	TYPE OF SERVICE	VEHICLE MILEAGE INTERVAL (x1000)															
		7.5	15	22.5	30	37.5	45	52.5	60	67.5	75	82.5	90	97.5	105	112.5	120
Accelerator linkage ①	L	✓	✓	✓	✓	✓	✓	✓	✓	✓	✓	✓	✓	✓	✓	✓	✓
Accessory drive belts ②	S/I				✓				✓				✓				✓
Air cleaner filter	R				✓				✓				✓				✓
Auto cruise control linkage & hose ③	S/I		✓		✓		✓		✓		✓		✓		✓		✓
Automatic transmission fluid level ③	S/I	✓		✓		✓		✓		✓		✓		✓		✓	
Battery fluid level ③	S/I	✓	✓	✓	✓	✓	✓	✓	✓	✓	✓	✓	✓	✓	✓	✓	✓
Body and chassis ①	L	✓	✓	✓	✓	✓	✓	✓	✓	✓	✓	✓	✓	✓	✓	✓	✓
Brake fluid level ③	S/I	✓	✓	✓	✓	✓	✓	✓	✓	✓	✓	✓	✓	✓	✓	✓	✓
Brake lines & hoses ③	S/I	✓	✓	✓	✓	✓	✓	✓	✓	✓	✓	✓	✓	✓	✓	✓	✓
Brake pedal play ③	S/I		✓		✓		✓		✓		✓		✓		✓		✓
Clutch fluid level ③	S/I	✓	✓	✓	✓	✓	✓	✓	✓	✓	✓	✓	✓	✓	✓	✓	✓
Clutch lines & hose ③	S/I				✓				✓				✓				✓
Clutch pedal free-play ③	S/I		✓		✓		✓		✓		✓		✓		✓		✓
Clutch pedal spring, bushing and clevis pin ①	S/I		✓		✓		✓		✓		✓		✓		✓		✓
Cooling and heating system hoses ③	S/I		✓		✓		✓		✓		✓		✓		✓		✓
Driveshaft flange torque ③	S/I	✓		✓		✓		✓		✓		✓		✓		✓	
Drum and disc brakes ③	S/I		✓		✓		✓		✓		✓		✓		✓		✓
Engine coolant	R				✓				✓				✓				✓
Engine coolant level ③	S/I	✓	✓	✓	✓	✓	✓	✓	✓	✓	✓	✓	✓	✓	✓	✓	✓
Engine oil & filter ③	R	✓	✓	✓	✓	✓	✓	✓	✓	✓	✓	✓	✓	✓	✓	✓	✓
Exhaust system ③	S/I	✓	✓	✓	✓	✓	✓	✓	✓	✓	✓	✓	✓	✓	✓	✓	✓
Front and rear axle lubricant	R		✓		✓				✓				✓				✓
Front and rear driveshafts ①	S/I	✓	✓	✓	✓	✓	✓	✓	✓	✓	✓	✓	✓	✓	✓	✓	✓
Front wheel bearings	S/I & L					✓			✓					✓			✓
Fuel lines & tank cap ③	S/I								✓								✓
Inspect for fluid leaks ③	S/I	✓	✓	✓	✓	✓	✓	✓	✓	✓	✓	✓	✓	✓	✓	✓	✓
Key lock cylinder ③	L		✓		✓		✓		✓		✓		✓		✓		✓
Manual transmission and transfer case fluid ④	R		✓		✓				✓				✓				✓
Parking brake system ③	S/I		✓		✓		✓		✓		✓		✓		✓		✓
Power steering fluid	R				✓				✓				✓				✓
Radiator core and A/C condenser	S/I & C								✓								✓
Rotate tires	S/I	✓	✓	✓	✓	✓	✓	✓	✓	✓	✓	✓	✓	✓	✓	✓	✓
Shift-on-the-fly system gear fluid ③	S/I		✓		✓		✓		✓		✓		✓		✓		✓
Spark plug wires ⑤	S/I								✓								✓
Spark plugs	R	Every 100,000 miles															
Starter safety switch ③	S/I	✓	✓	✓	✓	✓	✓	✓	✓	✓	✓	✓	✓	✓	✓	✓	✓
Steering operation ③	S/I	✓	✓	✓	✓	✓	✓	✓	✓	✓	✓	✓	✓	✓	✓	✓	✓
Suspension & steering ③	S/I	✓	✓	✓	✓	✓	✓	✓	✓	✓	✓	✓	✓	✓	✓	✓	✓
Throttle linkage ③	S/I		✓		✓		✓		✓		✓		✓		✓		✓
Timing belt	R										✓						

93591CN8

For Wheel Alignment specifications, see Section 1 of this manual

SCHEDULED MAINTENANCE INTERVALS
Isuzu—Amigo, Rodeo, Rodeo Sport, Trooper & vehiCROSS

TO BE SERVICED	TYPE OF SERVICE	VEHICLE MILEAGE INTERVAL (x1000)															
		7.5	15	22.5	30	37.5	45	52.5	60	67.5	75	82.5	90	97.5	105	112.5	120
Tires and wheels ③	S/I	✓	✓	✓	✓	✓	✓	✓	✓	✓	✓	✓	✓	✓	✓	✓	✓
Valve clearance ④	A								✓								✓

R: Replace   S/I: Service or Inspect   L: Lubricate   A: Adjust   C: Clean

① Perform this at the mileage indicated or every 6 months, whichever occurs first.

② Perform this at the mileage indicated or every 24 months, whichever occurs first.

③ Perform this at the mileage indicated or every 12 months, whichever occurs first.

④ 3.2L V6 engine.

⑤ 2.2L 4 cyl. engine.

FREQUENT OPERATION MAINTENANCE (SEVERE SERVICE)

If a vehicle is operated under any of the following conditions it is considered severe service:

- Towing a trailer or using a camper or car-top carrier.

- Repeated short trips of less than 5 miles in temperatures below freezing.

- Extensive idling or low-speed driving for long distances as in heavy commercial use, such as delivery, taxi or police cars.

- Operating on rough, muddy or salt-covered roads.

- Operating on unpaved or dusty roads.

Air cleaner element: replace every 15,000 miles

Engine oil and filter: replace every 3000 miles or 3 months, whichever occurs first.

Automatic transmission fluid: replace every 20,000 miles.

Rear axle lubricant: replace every 15,000 miles.

93591CN9

SCHEDULED MAINTENANCE INTERVALS
ISUZU
AMIGO, RODEO, RODEO SPORT, TROOPER, vehiCROSS

The following should be used as a guide when determining the amount of work required for a particular service. In estimating how long a particular Scheduled Maintenance Service should take, please observe the following:

● Labor Time is time based on field research and data supplied by the vehicle manufacturer.
● Labor time operations are given in hours and tenths of an hour.
● All labor operations are to be used as a guide.

Mechanic Skill Level Codes:
(A) PRECISION: Highly skilled with multiple certification.
(B) GENERAL: Normally skilled with certification.
(C) MAINTENANCE: Semi-skilled working on certification.

	LABOR TIME
7500 Mile Service (C)	
Amigo	2.8
Rodeo, Rodeo Sport	1.8
Trooper, vehiCROSS	2.8
15000 Mile Service (B)	
Amigo	5.1
Rodeo, Rodeo Sport	5.5
Trooper, vehiCROSS	5.1
22500 Mile Service (C)	
Amigo	2.8
Rodeo, Rodeo Sport	2.8
Trooper, vehiCROSS	2.8
30000 Mile Service (B)	
Amigo	8.3
Rodeo, Rodeo Sport	7.2
Trooper, vehiCROSS	8.3
37500 Mile Service (C)	
Amigo	2.8
Rodeo, Rodeo Sport	2.1
Trooper, vehiCROSS	2.8

	LABOR TIME
45000 Mile Service (B)	
Amigo	5.5
Rodeo, Rodeo Sport	4.5
Trooper, vehiCROSS	5.5
52500 Mile Service (C)	
Amigo	3.1
Rodeo, Rodeo Sport	3.1
Trooper, vehiCROSS	3.1
60000 Mile Service (B)	
Amigo	9.4
Rodeo, Rodeo Sport	9.1
Trooper, vehiCROSS	9.4
67500 Mile Service (C)	
Amigo	2.1
Rodeo, Rodeo Sport	2.1
Trooper, vehiCROSS	2.8
75000 Mile Service (B)	
Amigo	5.1
Rodeo, Rodeo Sport	4.5
Trooper, vehiCROSS	5.1
Replace timing belt add	
82500 Mile Service (C)	
Amigo	2.8
Rodeo, Rodeo Sport	2.1
Trooper, vehiCROSS	2.8

	LABOR TIME
90000 Mile Service (B)	
Amigo	9,4
Rodeo, Rodeo Sport	7.2
Trooper, vehiCROSS	9.4
97500 Mile Service (C)	
Amigo	2.8
Rodeo, Rodeo Sport	2.1
Trooper, vehiCROSS	2.8
105000 Mile Service (B)	
Amigo	5.5
Rodeo, Rodeo Sport	4.5
Trooper, vehiCROSS	5.5
112500 Mile Service (C)	
Amigo	3.1
Rodeo, Rodeo Sport	3.1
Trooper, vehiCROSS	3.1
120000 Mile Service (B)	
Amigo	9.4
Rodeo, Rodeo Sport	9.1
Trooper, vehiCROSS	9.4

93551C01

For Maintenance Interval recommendations, see Section 1 of this manual

KIA
Sportage

ENGINE AND VEHICLE IDENTIFICATION

Code ①	Liters (cc)	Cu. In.	Cyl.	Fuel Sys.	Engine Type	Eng. Mfg.
3	2.0 (1998)	122	4	MFI	DOHC	KIA

MFI: Multi-port Fuel Injection

DOHC: Double Overhead Camshafts

① 8th digit of VIN

② 10th digit of VIN

Code ②	Year
X	1999
Y	2000
1	2001
2	2002
3	2003

Model Year

93591C03

GENERAL ENGINE SPECIFICATIONS

Year	Model	Engine Displacement Liters (cc)	Engine VIN	Fuel System Type	Net Horsepower @ rpm	Net Torque @ rpm (ft. lbs.)	Bore x Stroke (in.)	Compression Ratio	Oil Pressure @ rpm
1999	Sportage	2.0 (1998)	3	MFI	130@5500	127@4000	3.39x3.39	9.2:1	43-57 ①
2000	Sportage	2.0 (1998)	3	MFI	130@5500	127@4000	3.39x3.39	9.2:1	43-57 ①
2001	Sportage	2.0 (1998)	3	MFI	130@5500	127@4000	3.39x3.39	9.2:1	43-57 ①
2002	Sportage	2.0 (1998)	3	MFI	130@5500	127@4000	3.39x3.39	9.2:1	43-57 ①

MFI: Multi-port Fuel Injection

① The manufacturer does not provide an engine speed specification for oil pump pressure.

93591C04

ENGINE TUNE-UP SPECIFICATIONS

Year	Engine Displacement Liters (cc)	Engine VIN	Spark Plug Gap (in.)	Ignition Timing (deg.)		Fuel Pump (psi)	Idle Speed (rpm)		Valve Clearance	
				MT	AT		MT	AT	Intake	Exhaust
1999	2.0 (1998)	3	0.039-0.043	4B	4B	38	750-850	750-850	HYD.	HYD.
2000	2.0 (1998)	3	0.039-0.043	4B	4B	38	750-850	750-850	HYD.	HYD.
2001	2.0 (1998)	3	0.039-0.043	4B	4B	38	750-850	750-850	HYD.	HYD.
2002	2.0 (1998)	3	0.039-0.043	4B	4B	38	750-850	750-850	HYD.	HYD.

B: Before Top Dead Center

HYD: Hydraulic lash adjusters

93591C05

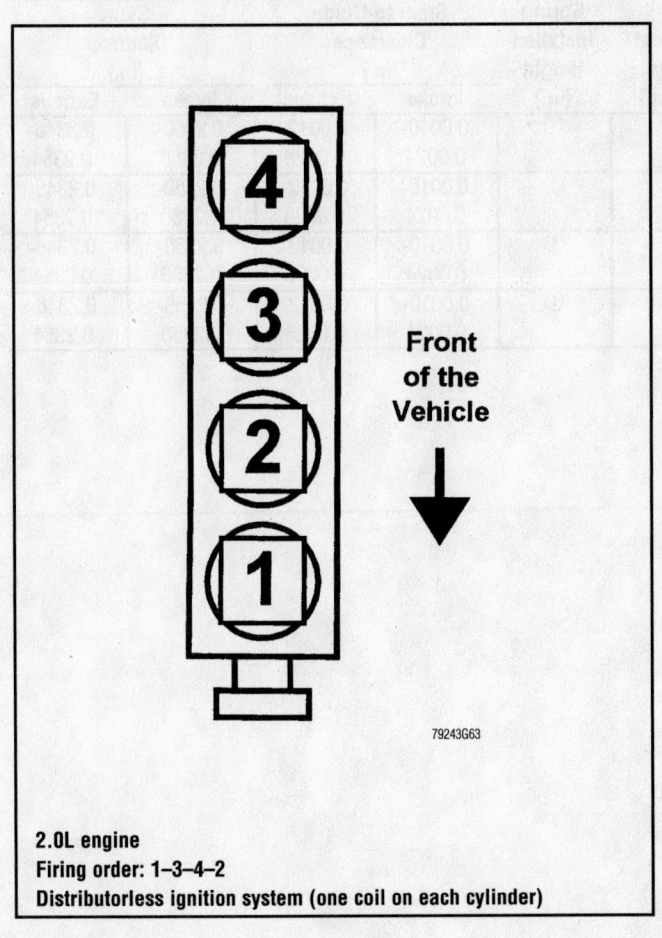

Front of the Vehicle

2.0L engine
Firing order: 1–3–4–2
Distributorless ignition system (one coil on each cylinder)

79243G63

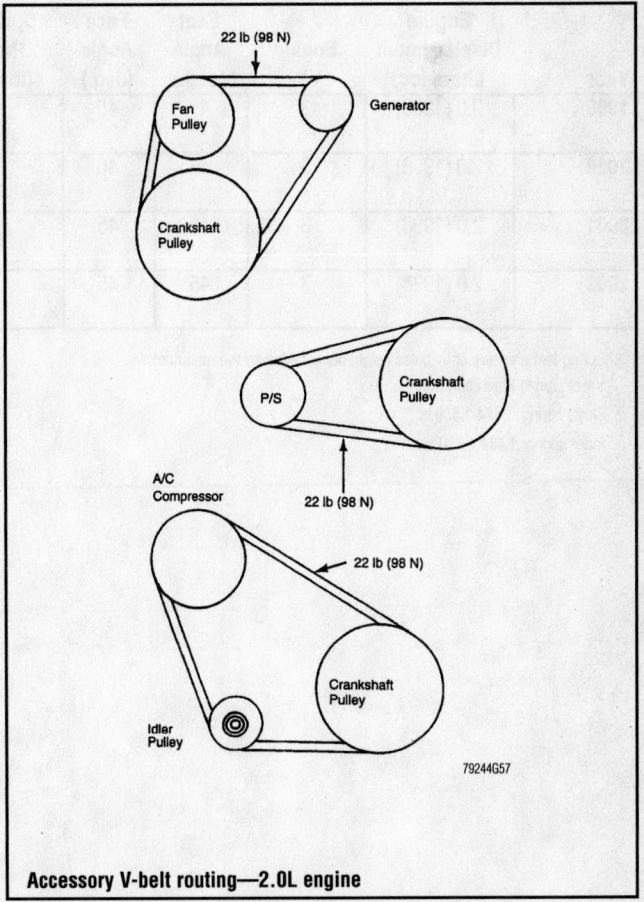

Accessory V-belt routing—2.0L engine

79244G57

For Tune-up, Capacities and Firing orders, see Section 1 of this manual

CAPACITIES

Year	Model	Engine Displacement Liters (cc)	Engine VIN	Engine Oil with Filter (qts.)	Transmission (pts.)		Transfer Case (pts.)	Drive Axle		Fuel Tank (gal.)	Cooling System (qts.)
					Manual	Auto.		Front (pts.)	Rear (pts.)		
1999	Sportage	2.0 (1998)	3	4.4	2.6	5.4	2.8	2.6	2.6	15.8	7.9
2000	Sportage	2.0 (1998)	3	4.4	2.6	5.4	2.8	2.6	2.6	15.8	7.9
2001	Sportage	2.0 (1998)	3	4.4	2.6	5.4	2.8	2.6	2.6	15.8	7.9
2002	Sportage	2.0 (1998)	3	4.4	2.6	5.4	2.8	2.6	2.6	15.8	7.9

NOTE: All capacities are approximate. Add fluid gradually and check to be sure a proper fluid level is obtained.

93591C06

VALVE SPECIFICATIONS

Year	Engine Displacement Liters (cc)	Engine VIN	Seat Angle (deg.)	Face Angle (deg.)	Spring Test Pressure (lbs. @ in.)	Spring Installed Height (in.)	Stem-to-Guide Clearance (in.)		Stem Diameter (in.)	
							Intake	Exhaust	Intake	Exhaust
1999	2.0 (1998)	3	45	45	①	①	0.0010-0.0024	0.0012-0.0026	0.2350-0.2356	0.2348-0.2354
2000	2.0 (1998)	3	45	45	①	①	0.0010-0.0024	0.0012-0.0026	0.2350-0.2356	0.2348-0.2354
2001	2.0 (1998)	3	45	45	①	①	0.0010-0.0024	0.0012-0.0026	0.2350-0.2356	0.2348-0.2354
2002	2.0 (1998)	3	45	45	①	①	0.0010-0.0024	0.0012-0.0026	0.2350-0.2356	0.2348-0.2354

① Spring test pressure or installed height not provided by the manufacturer.
Valve Spring Free Length:
Outer spring: 1.524-1.539 in.
Inner spring: 1.484-1.496 in.

93591C07

CRANKSHAFT AND CONNECTING ROD SPECIFICATIONS

All measurements are given in inches.

Year	Engine Displacement Liters (cc)	Engine VIN	Crankshaft				Connecting Rod		
			Main Brg. Journal Dia.	Main Brg. Oil Clearance	Shaft End-play	Thrust on No.	Journal Diameter	Oil Clearance	Side Clearance
1999	2.0 (1998)	3	2.3597-2.3604	①	0.0031-0.0071	3	2.0055-2.0061	0.0090-0.0021	0.0040-0.0103
2000	2.0 (1998)	3	2.3597-2.3604	①	0.0031-0.0071	3	2.0055-2.0061	0.0090-0.0021	0.0040-0.0103
2001	2.0 (1998)	3	2.3597-2.3604	①	0.0031-0.0071	3	2.0055-2.0061	0.0090-0.0021	0.0040-0.0103
2002	2.0 (1998)	3	2.3597-2.3604	①	0.0031-0.0071	3	2.0055-2.0061	0.0090-0.0021	0.0040-0.0103

① Journals 1, 2 and 4: 0.0010 - 0.0017 in.
Journal 3: 0.0012 - 0.0019 in.

93591C08

PISTON AND RING SPECIFICATIONS

All measurements are given in inches.

Year	Engine Displacement Liters (cc)	Engine VIN	Piston Clearance	Ring Gap			Ring Side Clearance		
				Top Compression	Bottom Compression	Oil Control	Top Compression	Bottom Compression	Oil Control
1999	2.0 (1998)	3	0.0019-0.0024	0.006-0.012	0.008-0.014	0.008-0.028	0.001-0.003	0.001-0.003	SNUG
2000	2.0 (1998)	3	0.0019-0.0024	0.006-0.012	0.008-0.014	0.008-0.028	0.001-0.003	0.001-0.003	SNUG
2001	2.0 (1998)	3	0.0019-0.0024	0.006-0.012	0.008-0.014	0.008-0.028	0.001-0.003	0.001-0.003	SNUG
2002	2.0 (1998)	3	0.0019-0.0024	0.006-0.012	0.008-0.014	0.008-0.028	0.001-0.003	0.001-0.003	SNUG

93591C09

TORQUE SPECIFICATIONS
All readings in ft. lbs.

Year	Engine Displacement Liters (cc)	Engine VIN	Cylinder Head Bolts	Main Bearing Bolts	Rod Bearing Bolts	Crankshaft Damper Bolts	Flywheel Bolts	Manifold Intake	Manifold Exhaust	Spark Plugs	Lug Nuts
1999	2.0 (1998)	3	62	63	50	11	73	16	31	11-17	73
2000	2.0 (1998)	3	62	63	50	11	73	16	31	11-17	73
2001	2.0 (1998)	3	62	63	50	11	73	16	31	11-17	73
2002	2.0 (1998)	3	62	63	50	11	73	16	31	11-17	73

93591C00

WHEEL ALIGNMENT

Year	Model		Caster Range (+/-Deg.)	Caster Preferred Setting (Deg.)	Camber Range (+/-Deg.)	Camber Preferred Setting (Deg.)	Toe-in (in.)	Steering Axis Inclination (Deg.)
1999	Sportage	F	0.75	+3.58	0.75	+0.44	0.10+/-0.10	—
		R	—	—	—	0	0	—
2000	Sportage	F	0.75	+3.58	0.75	+0.44	0.10+/-0.10	—
		R	—	—	—	0	0	—
2001	Sportage	F	0.75	+3.58	0.75	+0.44	0.10+/-0.10	—
		R	—	—	—	0	0	—
2002	Sportage	F	0.75	+3.58	0.75	+0.44	0.10+/-0.10	—
		R	—	—	—	0	0	—

93591CP1

TIRE, WHEEL AND BALL JOINT SPECIFICATIONS

| Year | Model | OEM Tires | | Tire Pressures (psi) | | Wheel Size | Ball Joint Inspection |
		Standard	Optional	Front	Rear		
1999	Sportage	P205/75R15	none	26	26	6-JJ	①
2000	Sportage	P205/75R15	none	26	26	6-JJ	①
2001	Sportage	P205/75R15	none	26	26	6-JJ	①
2002	Sportage	P205/75R15	none	26	26	6-JJ	①

OEM: Original Equipment Manufacturer

PSI: Pounds Per Square Inch

STD: Standard

OPT: Optional

① Replace if any measurable movement is found.

93591CP2

BRAKE SPECIFICATIONS

All measurements in inches unless noted

| Year | Model | Brake Disc | | | Brake Drum | | | Minimum Lining Thickness | | Brake Caliper Mounting Bolts (ft. lbs.) |
		Original Thickness	Minimum Thickness	Maximum Run-out	Original Inside Diameter	Max. Wear Limit	Maximum Machine Diameter	Front	Rear	
1999	Sportage	0.940	0.880	0.004	NA	9.89	NA	0.080	0.060	72
2000	Sportage	0.940	0.880	0.004	NA	9.89	NA	0.080	0.060	72
2001	Sportage	0.940	0.880	0.004	NA	9.89	NA	0.080	0.060	72
2002	Sportage	0.940	0.880	0.004	NA	9.89	NA	0.080	0.060	72

NA: Not Available

93591CP3

SCHEDULED MAINTENANCE INTERVALS
Kia—Sportage

TO BE SERVICED	TYPE OF SERVICE	VEHICLE MILEAGE INTERVAL (x1000)															
		7.5	15	22.5	30	37.5	45	52.5	60	67.5	75	82.5	90	97.5	105	112.5	120
Accessory drive belt	S/I				✓				✓				✓				✓
Air cleaner filter	R				✓				✓				✓				✓
Automatic transmission fluid	R				✓		✓		✓		✓		✓		✓		✓
Ball joints	S/I				✓				✓				✓				✓
Brake lines & connections	S/I				✓				✓				✓				✓
Chassis/body fasteners	S/I				✓				✓				✓				✓
Cooling system	S/I				✓				✓				✓				✓
CV-joint boots	S/I		✓		✓		✓		✓		✓		✓		✓		✓
Disc brakes	S/I		✓		✓		✓		✓		✓		✓		✓		✓
Driveshaft U-joints	L		✓		✓		✓		✓		✓		✓		✓		✓
Drum brakes	S/I				✓				✓				✓				✓
Emission hoses & tubes	S/I								✓								✓
Emission hoses & tubes (Cal)	R															✓	
Engine coolant	R				✓				✓				✓				✓
Engine oil & filter	R	✓	✓	✓	✓	✓	✓	✓	✓	✓	✓	✓	✓	✓	✓	✓	✓
Exhaust system heat shields	S/I				✓				✓				✓				✓
Front differential fluid	R				✓				✓				✓				✓
	S/I	✓	✓	✓	✓	✓	✓	✓	✓	✓	✓	✓	✓	✓	✓	✓	✓
Fuel filter	R				✓				✓				✓				✓
Fuel lines & hoses	S/I				✓				✓				✓				✓
Idle speed	S/I				✓				✓				✓				✓
Locks & hinges	L	✓	✓	✓	✓	✓	✓	✓	✓	✓	✓	✓	✓	✓	✓	✓	✓
Manual transmission fluid	R				✓				✓				✓				✓
PCV valve	S/I								✓								✓
Rear differential fluid	R				✓				✓				✓				✓
	S/I	✓	✓	✓	✓	✓	✓	✓	✓	✓	✓	✓	✓	✓	✓	✓	✓
Spark plug wires	S/I								✓								✓
Spark plugs	R				✓				✓				✓				✓
Steering operation & linkage	S/I				✓				✓				✓				✓
Timing belt	R														✓		
	S/I								✓				✓				
Timing belt (non-California)	R								✓								✓
Transfer case fluid	R				✓				✓				✓				✓

93591CP4

SCHEDULED MAINTENANCE INTERVALS
Kia—Sportage

TO BE SERVICED	TYPE OF SERVICE	VEHICLE MILEAGE INTERVAL (x1000)														
		7.5	15	22.5	30	37.5	45	52.5	60	67.5	75	82.5	90	97.5	105	112.5	120
Transfer case fluid	S/I		✓		✓		✓		✓		✓		✓		✓		✓
Transmission fluid	S/I	✓	✓	✓	✓	✓	✓	✓	✓	✓	✓	✓	✓	✓	✓	✓	✓

R: Replace   S/I: Inspect and service, if needed   L: Lubricate

FREQUENT OPERATION MAINTENANCE (SEVERE SERVICE)

If a vehicle is operated under any of the following conditions it is considered severe service:

- Towing a trailer or using a camper or car-top carrier.
- Repeated short trips of less than 5 miles in temperatures below freezing, or trips of less than 10 miles in any temperature.
- Extensive idling or low-speed driving for long distances as in heavy commercial use, such as delivery, taxi or police cars.
- Operating on rough, muddy or salt-covered roads.
- Operating on unpaved or dusty roads.
- Driving in extremely hot (over 90°) conditions.

Engine oil & filter: replace every 5000 miles or 5 months, whichever occurs first.

Air cleaner filter: inspect and replace if necessary, every 15,000 miles or 15 months, whichever occurs first.

Transfer case fluid: inspect the level every 5000 miles or 5 months, and replace every 15,000 miles or 15 months, whichever occurs first.

Transmission fluid: inspect the level every 5000 miles or 5 months, and replace every 15,000 miles or 15 months, whichever occurs first.

Front differential fluid: inspect the level every 5000 miles or 5 months, and replace every 15,000 miles or 15 months, whichever occurs first.

Rear differential fluid: inspect the level every 5000 miles or 5 months, and replace every 15,000 miles or 15 months, whichever occurs first.

93591CP5

Timing belt service is covered in Section 3 of this manual

SCHEDULED MAINTENANCE INTERVALS
KIA
SPORTAGE

The following should be used as a guide when determining the amount of work required for a particular service.
In estimating how long a particular Scheduled Maintenance Service should take, please observe the following:

● Labor Time is time based on field research and data supplied by the vehicle manufacturer.
● Labor time operations are given in hours and tenths of an hour.
● All labor operations are to be used as a guide.

Mechanic Skill Level Codes:
(A) PRECISION: Highly skilled with multiple certification.
(B) GENERAL: Normally skilled with certification.
(C) MAINTENANCE: Semi-skilled working on certification.

Service	LABOR TIME
7500 Mile Service (C)	
All models	.6
15000 Mile Service (C)	
All models	1.1
22500 Mile Service (C)	
All models	.6
30000 Mile Service (B)	
All models	3.1
37500 Mile Service (C)	
All models	.6

Service	LABOR TIME
45000 Mile Service (C)	
All models	1.1
52500 Mile Service (C)	
All models	.6
60000 Mile Service (B)	
All models	5.3
67500 Mile Service (C)	
All models	.6
75000 Mile Service (C)	
All models	1.1
82500 Mile Service (C)	
All models	.6

Service	LABOR TIME
90000 Mile Service (B)	
All models	5.3
97500 Mile Service (C)	
All models	.6
105000 Mile Service (C)	
All models	1.1
112500 Mile Service (C)	
All models	.6
120000 Mile Service (B)	
All models	5.3

93551CP7Y

LAND ROVER
Discovery • Discovery Series II • Freelander • Range Rover

ENGINE AND VEHICLE IDENTIFICATION

			Engine					Model Year	
Code ①	Liters (cc)	Cu. In.	Cyl.	Fuel Sys.	Engine Type	Eng. Mfg.		Code ②	Year
2	4.0 (3950)	241	8	MFI	OHV	Land Rover		X	1999
J	4.6 (4554)	278	8	MFI	OHV	Land Rover		Y	2000
V	4.0 (3950)	241	8	MFI	OHV	Land Rover		1	2001
G	2.5 (2497)	152	6	MFI	DOHC	Land Rover		2	2002
								3	2003

MFI: Multi-port Fuel Injection

OHV: Overhead Valve

DOHC: Dual Overhead Camshaft

① 8th digit of the Vehicle Identification Number (VIN)

② 10th digit of the Vehicle Identification Number (VIN)

93591CFM

GENERAL ENGINE SPECIFICATIONS

Year	Model	Engine Displacement Liters (cc)	Engine ID/VIN	Fuel System Type	Net Horsepower @ rpm	Net Torque @ rpm (ft. lbs.)	Bore x Stroke (in.)	Com-pression Ratio	Oil Pressure @ rpm
1999	Discovery	4.0 (3950)	2	MFI	188@4750	250@2600	3.7x2.8	9.34:1	35@2400
	Discovery Series II	4.0 (3950)	2	MFI	188@4750	250@2600	3.7x2.8	9.34:1	35@2400
	Range Rover SE	4.0 (3950)	V	MFI	188@4750	250@2600	3.7x2.8	9.34:1	35@2400
	Range Rover HSE	4.6 (4554)	J	MFI	222@4750	300@2600	3.7x3.2	9.34:1	35@2400
2000	Discovery Series II	4.0 (3950)	2	MFI	188@4750	250@2600	3.7x2.8	9.34:1	35@2400
	Range Rover SE	4.0 (3950)	V	MFI	188@4750	250@2600	3.7x2.8	9.34:1	35@2400
	Range Rover Country	4.0 (3950)	V	MFI	188@4750	250@2600	3.7x2.8	9.34:1	35@2400
	Range Rover Vitesse	4.6 (4554)	J	MFI	222@4750	300@2600	3.7x3.2	9.34:1	35@2400
	Range Rover HSE	4.6 (4554)	J	MFI	222@4750	300@2600	3.7x3.2	9.34:1	35@2400
2001	Discovery Series II	4.0 (3950)	2	MFI	188@4750	250@2600	3.7x2.8	9.34:1	35@2400
	Range Rover	4.6 (4554)	J	MFI	222@4750	300@2600	3.7x3.2	9.34:1	35@2400
2002	Discovery Series II	4.0 (3950)	2	MFI	188@4750	250@2600	3.7x2.8	9.34:1	35@2400
	Freelander	2.5 (2497)	G	MFI	175@6250	177@3000	3.2x3.3	10.5:1	44@3000
	Range Rover	4.6 (4554)	J	MFI	222@4750	300@2600	3.7x3.2	9.34:1	35@2400

MFI: Multi-port Fuel Injection

93591CFN

Heater Core replacement is covered in Section 2 of this manual

GASOLINE ENGINE TUNE-UP SPECIFICATIONS

Year	Engine Displacement Liters (cc)	Engine ID/VIN	Spark Plug Gap (in.)	Ignition Timing (deg.)		Fuel Pump (psi)	Idle Speed (rpm)		Valve Clearance	
				MT	AT		MT	AT	Intake	Exhaust
1999	4.0 (3950)	2	0.035	—	①	34-37	—	675-725	HYD	HYD
	4.0 (3950)	V	0.038	—	①	34-37	—	675-725	HYD	HYD
	4.6 (4554)	J	0.038	—	①	34-37	—	680-720	HYD	HYD
2000	4.0 (3950)	2	0.035	—	①	34-37	—	675-725	HYD	HYD
	4.0 (3950)	V	0.038	—	①	34-37	—	675-725	HYD	HYD
	4.6 (4554)	J	0.038	—	①	34-37	—	680-720	HYD	HYD
2001	4.0 (3950)	2	0.035	—	①	34-37	—	675-725	HYD	HYD
	4.6 (4554)	J	0.038	—	①	34-37	—	680-720	HYD	HYD
2002	4.0 (3950)	2	0.035	—	①	34-37	—	675-725	HYD	HYD
	2.5 (2497)	G	0.039	—	①	54.7	—	700-800	HYD	HYD
	4.6 (4554)	J	0.038	—	①	34-37	—	680-720	HYD	HYD

HYD: Hydraulic

① Automatically controlled by the Powertrain Control Module (PCM).

93591CFO

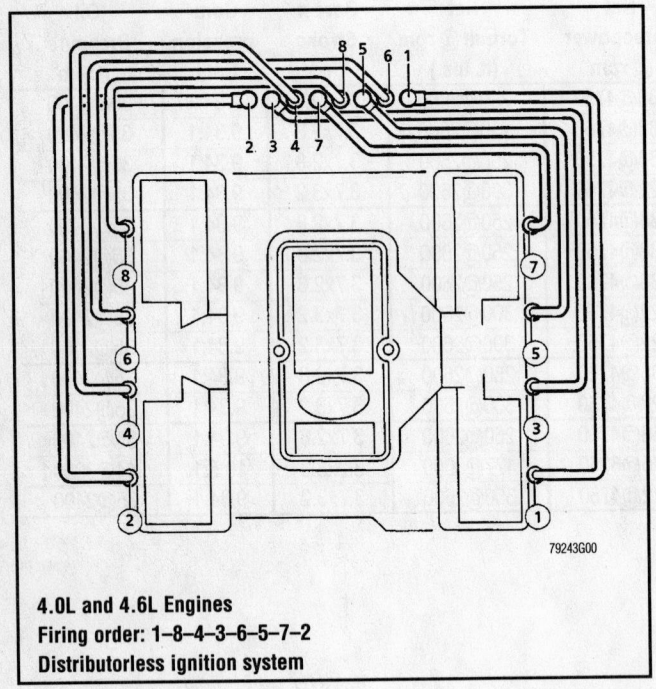

4.0L and 4.6L Engines
Firing order: 1–8–4–3–6–5–7–2
Distributorless ignition system

79243G00

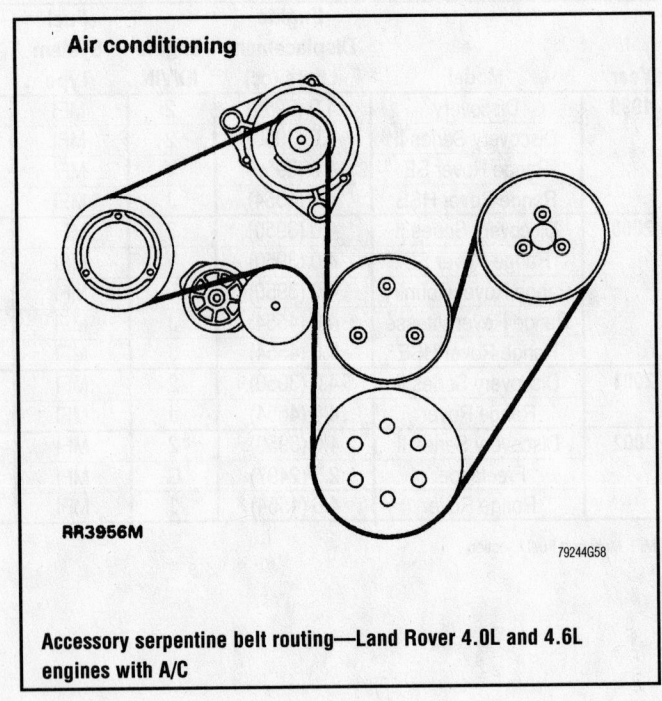

Air conditioning

RR3956M

79244G58

Accessory serpentine belt routing—Land Rover 4.0L and 4.6L engines with A/C

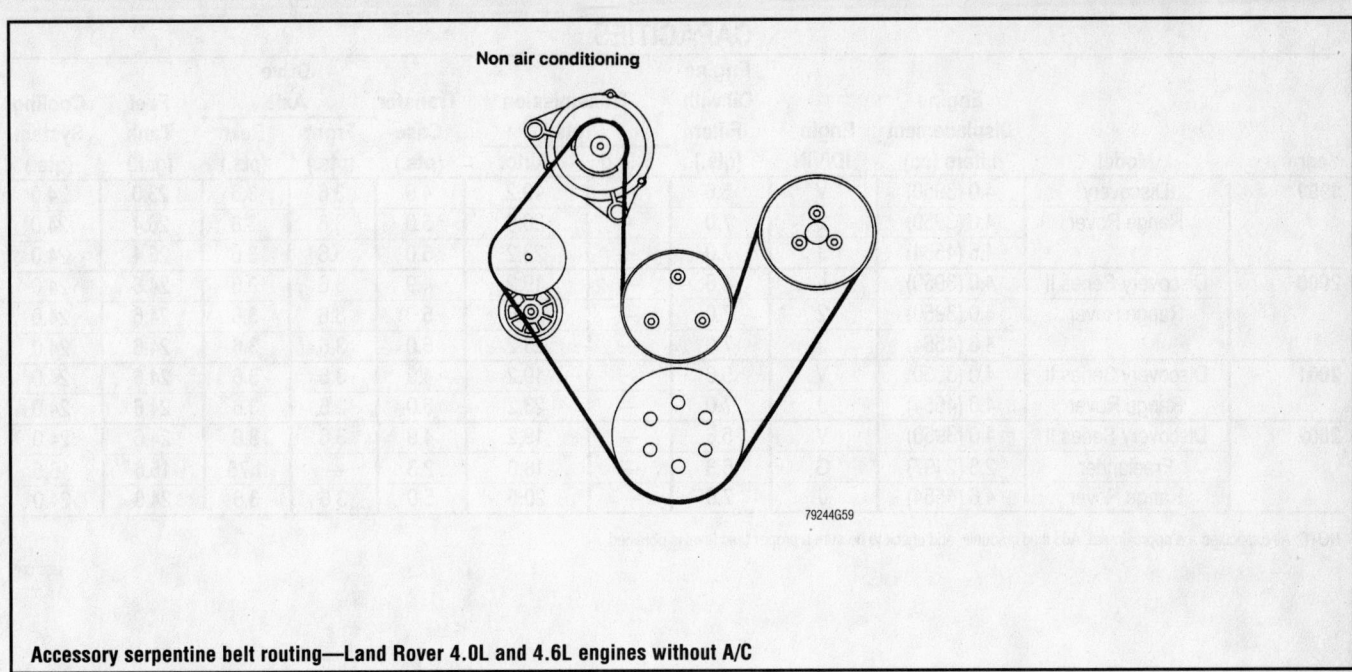

Non air conditioning

79244G59

Accessory serpentine belt routing—Land Rover 4.0L and 4.6L engines without A/C

Brake service is covered in Section 4 of this manual

CAPACITIES

Year	Model	Engine Displacement Liters (cc)	Engine ID/VIN	Engine Oil with Filter (qts.)	Transmission (pts.)		Transfer Case (pts.)	Drive Axle		Fuel Tank (gal.)	Cooling System (qts.)
					5-Spd	Auto.		Front (pts.)	Rear (pts.)		
1999	Discovery	4.0 (3950)	V	5.6	—	19.2	4.9	3.6	3.6	23.0	24.0
	Range Rover	4.0 (3950)	2	7.0	—	20.5	5.0	3.6	3.6	26.4	24.0
		4.6 (4554)	J	7.0	—	23.2	5.0	3.6	3.6	26.4	24.0
2000	Discovery Series II	4.0 (3950)	V	5.6	—	19.2	4.9	3.6	3.6	24.6	24.0
	Range Rover	4.0 (3950)	2	7.0	—	20.5	5.0	3.6	3.6	24.6	24.0
		4.6 (4554)	J	7.0	—	23.2	5.0	3.6	3.6	24.6	24.0
2001	Discovery Series II	4.0 (3950)	V	5.6	—	19.2	4.9	3.6	3.6	24.6	24.0
	Range Rover	4.6 (4554)	J	7.0	—	23.2	5.0	3.6	3.6	24.6	24.0
2002	Discovery Series II	4.0 (3950)	V	5.6	—	19.2	4.9	3.6	3.6	24.6	24.0
	Freelander	2.5 (2497)	G	6.3	—	18.0	2.3	—	1.75	15.6	16.5
	Range Rover	4.6 (4554)	J	7.0	—	20.5	5.0	3.6	3.6	24.6	24.0

NOTE: All capacities are approximate. Add fluid gradually and check to be sure a proper fluid level is obtained.

93591CFP

VALVE SPECIFICATIONS

Year	Engine Displacement Liters (cc)	Engine ID/VIN	Seat Angle (deg.)	Face Angle (deg.)	Spring Test Pressure (lbs. @ in.)	Spring Installed Height (in.)	Stem-to-Guide Clearance (in.)		Stem Diameter (in.)	
							Intake	Exhaust	Intake	Exhaust
1999	4.0 (3950)	2	46	46	76@1.59	1.59	0.0010-0.0026	0.0015-0.0031	0.3411-0.3417	0.3406-0.3412
	4.0 (3950)	V	46	46	76@1.59	1.59	0.0010-0.0026	0.0015-0.0031	0.3411-0.3417	0.3406-0.3412
	4.6 (4554)	J	46	46	76@1.59	1.59	0.0010-0.0026	0.0015-0.0031	0.3411-0.3417	0.3406-0.3412
2000	4.0 (3950)	2	46	46	76@1.59	1.59	0.0010-0.0026	0.0015-0.0031	0.3411-0.3417	0.3406-0.3412
	4.0 (3950)	V	46	46	76@1.59	1.59	0.0010-0.0026	0.0015-0.0031	0.3411-0.3417	0.3406-0.3412
	4.6 (4554)	J	46	46	76@1.59	1.59	0.0010-0.0026	0.0015-0.0031	0.3411-0.3417	0.3406-0.3412
2001	4.0 (3950)	V	46	46	76@1.59	1.59	0.0010-0.0026	0.0015-0.0031	0.3411-0.3417	0.3406-0.3412
	4.6 (4554)	J	46	46	76@1.59	1.59	0.0010-0.0026	0.0015-0.0031	0.3411-0.3417	0.3406-0.3412
2002	4.0 (3950)	V	46	46	76@1.59	1.59	0.0010-0.0026	0.0015-0.0031	0.3411-0.3417	0.3406-0.3412
	2.5 (2497)	G	45	45	155 1.48	1.48	0.0013-0.0025	0.0015-0.0031	0.2343-0.2349	0.2341-0.2347
	4.6 (4554)	J	46	46	76@1.59	1.59	0.0010-0.0026	0.0015-0.0031	0.3411-0.3417	0.3406-0.3412

93591CFQ

For complete Engine Mechanical specifications, see Section 1 of this manual

CRANKSHAFT AND CONNECTING ROD SPECIFICATIONS
All measurements are given in inches.

Year	Engine Displacement Liters (cc)	Engine ID/VIN	Crankshaft				Connecting Rod		
			Main Brg. Journal Dia.	Main Brg. Oil Clearance	Shaft End-play	Thrust on No.	Journal Diameter	Oil Clearance	Side Clearance
1999	4.0 (3950)	2	2.4995-2.5000	0.0004-0.0019	0.004-0.008	3	2.1850-2.1856	0.0006-0.0022	0.006-0.014
	4.0 (3950)	V	2.4995-2.5000	0.0004-0.0019	0.004-0.008	3	2.1850-2.1856	0.0006-0.0022	0.006-0.014
	4.6 (4554)	J	2.4995-2.5000	0.0004-0.0019	0.004-0.008	3	2.1850-2.1856	0.0006-0.0022	0.006-0.014
2000	4.0 (3950)	2	2.4995-2.5000	0.0004-0.0019	0.004-0.008	3	2.1850-2.1856	0.0006-0.0022	0.006-0.014
	4.0 (3950)	V	2.4995-2.5000	0.0004-0.0019	0.004-0.008	3	2.1850-2.1856	0.0006-0.0022	0.006-0.014
	4.6 (4554)	J	2.4995-2.5000	0.0004-0.0019	0.004-0.008	3	2.1850-2.1856	0.0006-0.0022	0.006-0.014
2001	4.0 (3950)	V	2.4995-2.5000	0.0004-0.0019	0.004-0.008	3	2.1850-2.1856	0.0006-0.0022	0.006-0.014
	4.6 (4554)	J	2.4995-2.5000	0.0004-0.0019	0.004-0.008	3	2.1850-2.1856	0.0006-0.0022	0.006-0.014
2002	4.0 (3950)	V	2.4995-2.5000	0.0004-0.0019	0.004-0.008	3	2.1850-2.1856	0.0006-0.0022	0.006-0.014
	2.5 (2497)	G	2.6670-2.6673	0.0008-0.0015	0.004-0.0120	4	2.1279-2.1281	0.0009-0.0016	0.007-0.011
	4.6 (4554)	J	2.4995-2.5000	0.0004-0.0019	0.004-0.008	3	2.1850-2.1856	0.0006-0.0022	0.006-0.014

93591CFR

PISTON AND RING SPECIFICATIONS

All measurements are given in inches.

Year	Engine Displacement Liters (cc)	Engine ID/VIN	Piston Clearance	Ring Gap			Ring Side Clearance		
				Top Compression	Bottom Compression	Oil Control	Top Compression	Bottom Compression	Oil Control
1999	4.0 (3950)	2	0.0010-0.0020	0.010-0.020	0.016-0.030	0.014-0.050	0.0020-0.0040	0.0020-0.0040	SNUG
	4.0 (3950)	V	0.0010-0.0020	0.010-0.020	0.016-0.030	0.014-0.050	0.0020-0.0040	0.0020-0.0040	SNUG
	4.6 (4554)	J	0.0010-0.0020	0.010-0.020	0.016-0.030	0.014-0.050	0.0020-0.0040	0.0020-0.0040	SNUG
2000	4.0 (3950)	2	0.0010-0.0020	0.010-0.020	0.016-0.030	0.014-0.050	0.0020-0.0040	0.0020-0.0040	SNUG
	4.0 (3950)	V	0.0010-0.0020	0.010-0.020	0.016-0.030	0.014-0.050	0.0020-0.0040	0.0020-0.0040	SNUG
	4.6 (4554)	J	0.0010-0.0020	0.010-0.020	0.016-0.030	0.014-0.050	0.0020-0.0040	0.0020-0.0040	SNUG
2001	4.0 (3950)	V	0.0010-0.0020	0.010-0.020	0.016-0.030	0.014-0.050	0.0020-0.0040	0.0020-0.0040	SNUG
	4.6 (4554)	J	0.0010-0.0020	0.010-0.020	0.016-0.030	0.014-0.050	0.0020-0.0040	0.0020-0.0040	SNUG
2002	4.0 (3950)	V	0.0010-0.0020	0.010-0.020	0.016-0.030	0.014-0.050	0.0020-0.0040	0.0020-0.0040	SNUG
	2.5 (2497)	G	0.0013-0.0014	0.008-0.0140	0.011-0.0180	0.010-0.0390	0.0020-0.0031	0.0012-0.0024	0.0004-0.0071
	4.6 (4554)	J	0.0010-0.0020	0.010-0.020	0.016-0.030	0.014-0.050	0.0020-0.0040	0.0020-0.0040	SNUG

93591CFS

For Accessory Drive Belt illustrations, see Section 1 of this manual

TORQUE SPECIFICATIONS
All readings in ft. lbs.

Year	Engine Displacement Liters (cc)	Engine ID/VIN	Cylinder Head Bolts	Main Bearing Bolts	Rod Bearing Bolts	Crankshaft Damper Bolts	Flywheel Bolts	Manifold		Spark Plugs	Lug Nuts
								Intake	Exhaust		
1999	4.0 (3950)	2	①	②	③	200	58	④	40	15	103
	4.0 (3950)	V	①	②	③	200	58	④	40	15	103
	4.6 (4554)	J	①	②	③	200	58	④	40	15	103
2000	4.0 (3950)	2	①	②	③	200	58	④	40	15	103
	4.0 (3950)	V	①	②	③	200	58	④	40	15	103
	4.6 (4554)	J	①	②	③	200	58	④	40	15	103
2001	4.0 (3950)	V	①	②	③	200	58	④	40	15	103
	4.6 (4554)	J	①	②	③	200	58	④	40	15	103
2002	4.0 (3950)	V	①	②	③	200	58	④	40	15	103
	2.5 (2497)	G	⑤	⑥	⑦	118	55	18	18	18	85
	4.6 (4554)	J	①	②	③	200	58	④	40	15	103

① Step 1: 15 ft. lbs.
   Step 2: additional 90 degrees
   Step 3: additional 90 degrees

② Bolts 1-8
   Step 1: 10 ft. lbs.
   Step 2: 53 ft. lbs.
   Bolts 9 & 10
   Step 1: 10 ft. lbs.
   Step 2: 68 ft. lbs.
   Bolts 11-20
   Step 1: 10 ft. lbs.
   Step 2: 33 ft. lbs.

③ Step 1: 15 ft. lbs.
   Step 2: additional 80 degrees

④ Step 1: 84 inch lbs.
   Step 2: 38 ft. lbs.

⑤ Step 1: 18 ft. lbs.
   Step 2: retighten to 18 ft. lbs.
   Step 3: Retighten to 18 ft. lbs.
   Step 4: Tighten all an additional 180 degrees

⑥ Bearing cradle
   Step 1: 15 ft. lbs.
   Step 2: plus 90 degrees

⑦ Step 1: 15 ft. lbs.
   Step 2: plus 70 degrees

93591CFT

WHEEL ALIGNMENT

Year	Model		Caster Range (+/-Deg.)	Caster Preferred Setting (Deg.)	Camber Range (+/-Deg.)	Camber Preferred Setting (Deg.)	Toe-in (in.)	Steering Axis Inclination (Deg.)
1999	Discovery		0.10	+3.70	—	0	0.16+/-0.83	—
	Range Rover		—	+4.00	—	0	0.05+/-0.03	—
2000	Discovery		0.10	+3.70	—	0	0.16+/-0.83	—
	Range Rover		—	+4.00	—	0	0.05+/-0.03	—
2001	Discovery		0.10	+3.70	—	0	0.16+/-0.83	—
	Range Rover		—	+4.00	—	0	0.05+/-0.03	—
2002	Discovery		0.10	+3.70	—	0	0.16+/-0.83	—
	Freelander	F	1.00	+3.42	0.75	-0.25	0.23+/-0.25 (out)	12.3
		R	—	—	0.75	+0.50	0.30+/-0.50	
	Range Rover		—	+4.00	—	0	0.05+/-0.03	—

93591CFU

TIRE, WHEEL AND BALL JOINT SPECIFICATIONS

Year	Model	OEM Tires Standard	OEM Tires Optional	Tire Pressures (psi) Front	Tire Pressures (psi) Rear	Wheel Size	Ball Joint Inspection
1999	Discovery	255/65R16	255/55R18	28	46	8J	①
	Range Rover SE	255/65R16	None	28	38	8J	①
	Range Rover HSE	255/55HR18	None	28	38	8-J	①
2000	Discovery	255/65R16	255/55R18	28	46	8J	①
	Range Rover SE	255/65R16	None	28	38	8J	①
	Range Rover HSE	255/55HR18	None	28	38	8-J	①
2001	Discovery	255/65R16	255/55R18	28	46	8J	①
	Range Rover SE	255/65R16	None	28	38	8J	①
	Range Rover HSE	255/55HR18	None	28	38	8-J	①
2002	Discovery SD	255/65HR16	None	28	46	8J	①
	Discovery SE	P255/55HR18	None	28	46	8J	①
	Freelander S	P215/65R16	None	26 ②	26 ②	6.0J	NA
	Freelander SE	P225/55HR17	None	26 ②	26 ②	7.0J	
	Freelander HSE	P225/55HR17	None	26 ②	26 ②	7.0J	
	Range Rover HSE	P225/65HR18	None	28	38	8-J	①

OEM: Original Equipment Manufacturer

PSI: Pounds Per Square Inch

STD: Standard

OPT: Optional

NA: Not Available

① Replace if any measurable movement is found.

② For towing or max GVW: 30

93591CFV

For Tire, Wheel and Ball Joint specifications, see Section 1 of this manual

BRAKE SPECIFICATIONS

All measurements in inches unless noted

| Year | Model | Front Brake Disc | | | Rear Brake Disc | | | Minimum Lining Thickness | | Brake Caliper | |
		Original Thickness	Minimum Thickness	Maximum Run-out	Original Thickness	Minimum Thickness	Maximum Run-out	Front	Rear	Bracket Bolts (ft. lbs.)	Mounting Bolts (ft. lbs.)
1999	Discovery	0.984	0.866	0.006	0.500	0.461	0.006	0.079	0.079	①	②
	Range Rover	1.000	0.870	0.006	0.500	0.460	0.006	0.080	0.080	③	④
2000	Discovery	0.984	0.866	0.006	0.500	0.461	0.006	0.079	0.079	①	②
	Range Rover	1.000	0.870	0.006	0.500	0.460	0.006	0.080	0.080	③	④
2001	Discovery	0.984	0.866	0.006	0.500	0.461	0.006	0.079	0.079	①	②
	Range Rover	1.000	0.870	0.006	0.500	0.460	0.006	0.080	0.080	③	④
2002	Discovery	0.984	0.866	0.006	0.500	0.461	0.006	0.079	0.079	①	②
	Freelander	0.822	0.708	0.0016	⑤	⑥	⑦	0.118	0.079	20	70
	Range Rover	1.000	0.870	0.006	0.500	0.460	0.006	0.080	0.080	③	④

① Front: 129 ft. lbs.
Rear: 70 ft. lbs.

② Both front and rear calipers: 22 ft. lbs.

③ Front: 122 ft. lbs.
Rear: 74 ft. lbs.

④ Front: 19 ft. lbs.
Rear: 26 ft. lbs.

⑤ Drum original diameter: 10.0 in.

⑥ Drum discard diameter: 10.059 in.

⑦ Drum out of round limit: 0.0005 in.

93591CFW

SCHEDULED MAINTENANCE INTERVALS
Land Rover—Discovery, Range Rover & Freelander

TO BE SERVICED	TYPE OF SERVICE	VEHICLE MILEAGE INTERVAL (x1000)															
		7.5	15	22.5	30	37.5	45	52.5	60	67.5	75	82.5	90	97.5	105	112.5	120
Air cleaner filter	R				✓				✓				✓				✓
Battery fluid level	R				✓				✓				✓				✓
Brake fluid level	S/I		✓				✓				✓				✓		
Brake lines	S/I	✓	✓	✓	✓	✓	✓	✓	✓	✓	✓	✓	✓	✓	✓	✓	✓
Brake pads, calipers & rotors	S/I	✓	✓	✓	✓	✓	✓	✓	✓	✓	✓	✓	✓	✓	✓	✓	✓
Coolant hoses	S/I	✓	✓	✓	✓	✓	✓	✓	✓	✓	✓	✓	✓	✓	✓	✓	✓
Door locks & hinges	L		✓		✓		✓		✓		✓		✓		✓		
Driveshafts & U-joints	L		✓		✓		✓		✓		✓		✓		✓		✓
Engine & transmission mounts	S/I					✓							✓				
Engine coolant	R				✓				✓				✓				✓
Engine oil & filter	R	✓	✓	✓	✓	✓	✓	✓	✓	✓	✓	✓	✓	✓	✓	✓	✓
Exhaust system & heat shields	S/I	✓	✓	✓	✓	✓	✓	✓	✓	✓	✓	✓	✓	✓	✓	✓	✓
Front and rear axle oil	R				✓				✓				✓				✓
Fuel filter	R								✓								✓
Fuel lines	S/I		✓		✓		✓		✓		✓		✓		✓		✓
Oxygen sensors	R											✓					
Parking brake	S/I		✓		✓		✓		✓		✓		✓		✓		✓
Power steering fluid	S/I		✓		✓		✓		✓		✓		✓		✓		✓
Radiator and A/C condenser	S/I		✓		✓		✓		✓		✓		✓		✓		✓
Seat belts	S/I		✓		✓		✓		✓		✓		✓		✓		✓
Serpentine drive belt	R										✓						
Serpentine drive belt	S/I				✓				✓				✓				✓
Shock absorbers	S/I		✓		✓		✓		✓		✓		✓		✓		✓
Spark plugs	R				✓				✓				✓				✓
Steering box	S/I & A						✓						✓				
Steering linkage	S/I		✓		✓		✓		✓		✓		✓		✓		✓
Air bag (SRS)	S/I	Every 10 years															
Suspension links & mountings	S/I		✓		✓		✓		✓		✓		✓		✓		✓
Tires	S/I	✓	✓	✓	✓	✓	✓	✓	✓	✓	✓	✓	✓	✓	✓	✓	✓
Transfer gearbox oil	R				✓				✓				✓				✓
Transmission fluid	R				✓				✓				✓				✓
Transmission fluid filter	R				✓								✓				

93591CFX

For Wheel Alignment specifications, see Section 1 of this manual

SCHEDULED MAINTENANCE INTERVALS
Land Rover—Discovery, Range Rover & Freelander

TO BE SERVICED	TYPE OF SERVICE	VEHICLE MILEAGE INTERVAL (x1000)															
		7.5	15	22.5	30	37.5	45	52.5	60	67.5	75	82.5	90	97.5	105	112.5	120
Wheel speed sensor wiring	S/I		✓		✓		✓		✓		✓		✓		✓		✓
Wiper blades	S/I	✓	✓	✓	✓	✓	✓	✓	✓	✓	✓	✓	✓	✓	✓	✓	✓

R: Replace   S/I: Service or Inspect   L: Lubricate   A: Adjust

FREQUENT OPERATION MAINTENANCE (SEVERE SERVICE)

If a vehicle is operated under any of the following conditions it is considered severe service:

- Towing a trailer or using a camper or car-top carrier.
- Repeated short trips of less than 5 miles in temperatures below freezing, or trips of less than 10 miles in any temperature.
- Extensive idling or low-speed driving for long distances as in heavy commercial use, such as delivery, taxi or police cars.
- Operating on rough, muddy or salt-covered roads.
- Operating on unpaved or dusty roads.
- Frequent operation in temperatures above 90°F.

Air cleaner element: replace every 15,000 miles

Brake fluid: replace every 15,000 miles

Brake fluid level: inspect initially at 7,500 miles, then every 15,000 miles

Brake pads, calipers & rotors: inspect every 3,750 miles

Driveshafts & U-joints: lubricate every 7,500 miles

Engine & transmission mounts: inspect every 22,500 miles

Engine coolant: replace every 15,000 miles

Engine oil & filter: replace every 3,750 miles.

Front & rear axle oil: replace every 15,000 miles.

Fuel filter: replace every 30,000 miles.

Power steering fluid level: inspect every 7,500 miles.

Serpentine drive belt: inspect every 15,000 miles and replace every 30,000 miles.

Shock absorbers: inspect every 7,500 miles.

Spark plugs: replace every 15,000 miles.

Steering rods, joints & dust covers: inspect every 7,500 miles.

Suspension links & mountings: inspect every 7,500 miles.

Tires: inspect every 3,750 miles.

Transfer gearbox oil: replace every 15,000 miles.

Transmission fluid & filter: replace every 15,000 miles.

93591CFY

SCHEDULED MAINTENANCE INTERVALS
LAND ROVER
DISCOVERY,
RANGE ROVER,
FREELANDER

The following should be used as a guide when determining the amount of work required for a particular service. In estimating how long a particular Scheduled Maintenance Service should take, please observe the following:

- Labor Time is time based on field research and data supplied by the vehicle manufacturer.
- Labor time operations are given in hours and tenths of an hour.
- All labor operations are to be used as a guide.

Mechanic Skill Level Codes:
(A) PRECISION: Highly skilled with multiple certification.
(B) GENERAL: Normally skilled with certification.
(C) MAINTENANCE: Semi-skilled working on certification.

	LABOR TIME		LABOR TIME		LAB TIM
7500 Mile Service (C)		**45000 Mile Service (C)**		**90000 Mile Service (B)**	
All Models	2.4	All Models	1.1	All Models	6.
15000 Mile Service (B)		**52500 Mile Service (C)**		**97500 Mile Service (C)**	
All Models	5.0	All Models	2.7	All Models	2.
22500 Mile Service (C)		**60000 Mile Service (B)**		**105000 Mile Service (B)**	
All Models	3.0	All Models	5.9	All Models	5.
30000 Mile Service (B)		**67500 Mile Service (C)**		**112500 Mile Service (C)**	
All Models	5.9	All Models	2.6	All Models	2.
37500 Mile Service (C)		**75000 Mile Service (B)**		**120000 Mile Service (B)**	
All Models	2.4	All Models	5.2	All Models	5.
		82500 Mile Service (C)			
		All Models	2.4		

93591CFZ

LEXUS
RX300

ENGINE AND VEHICLE IDENTIFICATION

Engine							Model Year	
Code ①	Liters (cc)	Cu. In.	Cyl.	Fuel Sys.	Engine Type	Eng. Mfg.	Code ②	Year
1MZ-FE	3.0 (2995)	183	6	SFI	DOHC	Toyota	X	1999
							Y	2000
							1	2001
							2	2002
							3	2003

SFI: Sequential Fuel Injection

MFI: Multi-port Fuel Injection

DOHC: Double Overhead Camshaft

① Stamped on the left side of the engine block

② 10th digit of the Vehicle Identification Number (VIN)

93591CZ0

GENERAL ENGINE SPECIFICATIONS

Year	Model	Engine Displacement Liters (cc)	Engine Series (ID/VIN)	Fuel System	Net Horsepower @ rpm	Net Torque @ rpm (ft. lbs.)	Bore x Stroke (in.)	Compression Ratio	Oil Pressure @ rpm
1999	RX300	3.0 (2995)	1MZ-FE	SFI	220@5800	222@4400	3.44x3.27	10.5:1	43-78@3000
2000	RX300	3.0 (2995)	1MZ-FE	SFI	220@5800	222@4400	3.44x3.27	10.5:1	43-78@3000
2001	RX300	3.0 (2995)	1MZ-FE	SFI	220@5800	222@4400	3.44x3.27	10.5:1	43-78@3000
2002	RX300	3.0 (2995)	1MZ-FE	SFI	220@5800	222@4400	3.44x3.27	10.5:1	43-78@3000

NA: Not Available

SFI: Sequential Fuel Injection

MFI: Multi-port Fuel Injection

93591CZP

ENGINE TUNE-UP SPECIFICATIONS

Year	Engine Displacement Liters (cc)	Engine ID/VIN	Spark Plug Gap (in.)	Ignition Timing (deg.)*	Fuel Pump (psi)	Idle Speed (rpm) MT	Idle Speed (rpm) AT	Valve Clearance Intake	Valve Clearance Exhaust
1999	3.0 (2995)	1MZ-FE	0.043	8-12B	44-50	—	650-750	0.006-0.010	0.010-0.014
2000	3.0 (2995)	1MZ-FE	0.043	8-12B	44-50	—	650-750	0.006-0.010	0.010-0.014
2001	3.0 (2995)	1MZ-FE	0.043	8-12B	44-50	—	650-750	0.006-0.010	0.010-0.014
2002	3.0 (2995)	1MZ-FE	0.039-0.043	8-12B	44-50	—	650-750	0.006-0.010	0.010-0.014

NOTE: The Vehicle Emission Control Information label often reflects specification changes made during production. The label figures must be used if they differ from those in this chart.

B: Before top dead center

* With terminals TC and CG connected to DLC3

93591CZQ

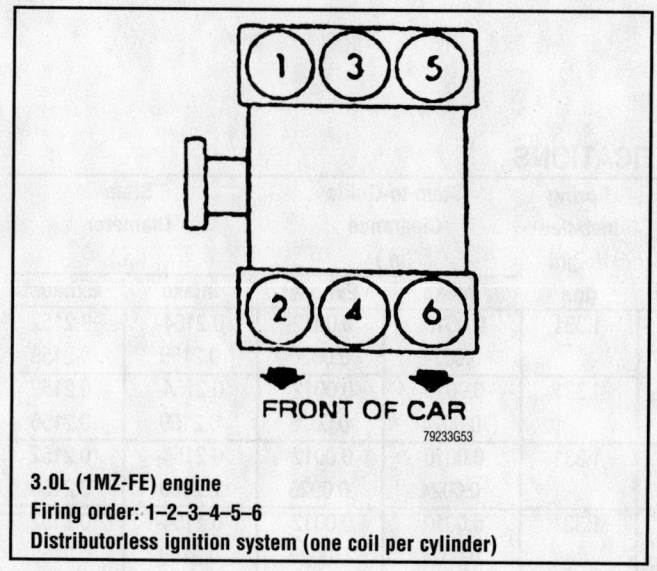

3.0L (1MZ-FE) engine
Firing order: 1–2–3–4–5–6
Distributorless ignition system (one coil per cylinder)

79233G53

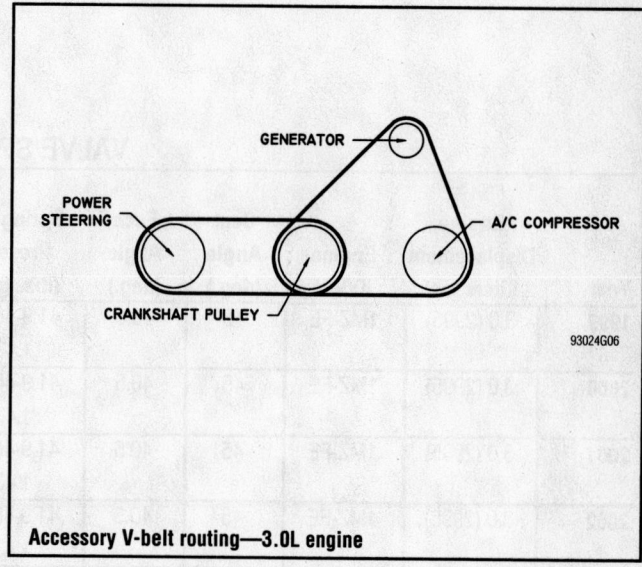

Accessory V-belt routing—3.0L engine

93024G06

For Tune-up, Capacities and Firing orders, see Section 1 of this manual

CAPACITIES

Year	Model	Engine Displacement Liters (cc)	Engine ID/VIN	Engine Oil with Filter (qts.)	Transmission (pts.)		Transfer Case (pts.)	Drive Axle		Fuel Tank (gal.)	Cooling System (qts.)
					5-Spd	Auto.*		Front (pts.)	Rear (pts.)		
1999	RX300	3.0 (2995)	1MZ-FE	5.0	—	①	2.0	1.9	1.9	17.2	9.5
2000	RX300	3.0 (2995)	1MZ-FE	5.0	—	①	2.0	1.9	1.9	17.2	9.5
2001	RX300	3.0 (2995)	1MZ-FE	5.0	—	①	2.0	1.9	1.9	17.2	9.5
2002	RX300	3.0 (2995)	1MZ-FE	5.0	—	①	2.0	1.9	1.9	17.2	9.5

*After draining, add the following amounts, then, fill to the cold full line.

① U140E Transaxle:

Dry Fill: 17.44 pts.

Drain and Refill: 7.4 pts.

U140F Transaxle:

Dry Fill: 19.34 pts.

Drain and Refill: 8.6 pts.

93591CZR

VALVE SPECIFICATIONS

Year	Engine Displacement Liters (cc)	Engine ID/VIN	Seat Angle (deg.)	Face Angle (deg.)	Spring Test Pressure (lbs. @ in.)	Spring Installed Height (in.)	Stem-to-Guide Clearance (in.)		Stem Diameter (in.)	
							Intake	Exhaust	Intake	Exhaust
1999	3.0 (2995)	1MZ-FE	45	40.5	41.9-46.3@ 1.437	1.331	0.0010-0.0024	0.0012-0.0026	0.2154-0.2159	0.2152 0.2156
2000	3.0 (2995)	1MZ-FE	45	40.5	41.9-46.3@ 1.437	1.331	0.0010-0.0024	0.0012-0.0026	0.2154-0.2159	0.2152 0.2156
2001	3.0 (2995)	1MZ-FE	45	40.5	41.9-46.3@ 1.437	1.331	0.0010-0.0024	0.0012-0.0026	0.2154-0.2159	0.2152 0.2156
2002	3.0 (2995)	1MZ-FE	45	40.5	41.9-46.3@ 1.437	1.331	0.0010-0.0024	0.0012-0.0026	0.2154-0.2159	0.2152 0.2156

93591CZS

CRANKSHAFT AND CONNECTING ROD SPECIFICATIONS

All measurements are given in inches.

Year	Engine Displacement Liters (cc)	Engine ID/VIN	Crankshaft				Connecting Rod		
			Main Brg. Journal Dia.	Main Brg. Oil Clearance	Shaft End-play	Thrust on No.	Journal Diameter	Oil Clearance	Side Clearance
1999	3.0 (2995)	1MZ-FE	2.4011-2.4016	①	0.0016-0.0095	2	2.0863-2.0866	0.0015-0.0025	0.0059-0.0188
2000	3.0 (2995)	1MZ-FE	2.4011-2.4016	①	0.0016-0.0095	2	2.0863-2.0866	0.0015-0.0025	0.0059-0.0188
2001	3.0 (2995)	1MZ-FE	2.4011-2.4016	①	0.0016-0.0095	2	2.0863-2.0866	0.0015-0.0025	0.0059-0.0188
2002	3.0 (2995)	1MZ-FE	2.4011-2.4016	①	0.0016-0.0095	2	2.0863-2.0866	0.0015-0.0025	0.0059-0.0188

① Journals 1 and 4: 0.0006 - 0.0013 in.
　Journals 2 and 3: 0.0010 - 0.0018 in.

93591CZT

PISTON AND RING SPECIFICATIONS

All measurements are given in inches.

Year	Engine Displacement Liters (cc)	Engine ID/VIN	Piston Clearance	Ring Gap			Ring Side Clearance		
				Top Compression	Bottom Compression	Oil Control	Top Compression	Bottom Compression	Oil Control
1999	3.0 (2995)	1MZ-FE	0.0033-0.0042	0.0098-0.0138	0.0138-0.0177	0.0059-0.0157	0.0008-0.0028	0.0008-0.0024	SNUG
2000	3.0 (2995)	1MZ-FE	0.0033-0.0042	0.0098-0.0138	0.0138-0.0177	0.0059-0.0157	0.0008-0.0028	0.0008-0.0024	SNUG
2001	3.0 (2995)	1MZ-FE	0.0033-0.0042	0.0098-0.0138	0.0138-0.0177	0.0059-0.0157	0.0008-0.0028	0.0008-0.0024	SNUG
2002	3.0 (2995)	1MZ-FE	0.0033-0.0042	0.0098-0.0138	0.0138-0.0177	0.0059-0.0157	0.0008-0.0028	0.0008-0.0024	SNUG

93591CZU

TORQUE SPECIFICATIONS
All readings in ft. lbs.

Year	Engine Displacement Liters (cc)	Engine ID/VIN	Cylinder Head Bolts	Main Bearing Bolts	Rod Bearing Bolts	Crankshaft Damper Bolts	Flywheel Bolts	Manifold Intake	Manifold Exhaust	Spark Plugs	Lug Nuts
1999	3.0 (2995)	1MZ-FE	①	②	③	159	61	32	36	13	70
2000	3.0 (2995)	1MZ-FE	①	②	③	159	61	32	36	13	70
2001	3.0 (2995)	1MZ-FE	①	②	③	159	61	32	36	13	70
2002	3.0 (2995)	1MZ-FE	①	②	③	159	61	32	36	13	70

① Step 1: 12 point bolts to 40 ft. lbs.
   Step 2: 12 point bolts plus 90 degrees
   Step 3: Hex head recessed bolt to 13 ft. lbs.

② Step 1: 12 point cap bolts to 16 ft. lbs.
   Step 2: 12 point cap bolts plus 90 degrees
   Step 3: Hex head side bolts to 20 ft. lbs.

③ Step 1: 18 ft. lbs.
   Step 2: Plus 90 degrees

93591CZV

WHEEL ALIGNMENT

Year	Model		Caster Range (+/-Deg.)	Caster Preferred Setting (Deg.)	Camber Range (+/-Deg.)	Camber Preferred Setting (Deg.)	Toe-in (in.)	Steering Axis Inclination (Deg.)
1999	RX300	F	0.75	+2.08	0.75	-0.33	0.04+/-0.08	12.16+/-0.75
		2WD R	—	—	0.75	-0.33	0.08+/-0.08	—
		4WD R	—	—	0.75	-0.35	0.12+/-0.08	—
2000	RX300	F	0.75	+2.08	0.75	-0.33	0.04+/-0.08	12.16+/-0.75
		2WD R	—	—	0.75	-0.33	0.08+/-0.08	—
		4WD R	—	—	0.75	-0.35	0.12+/-0.08	—
2001	RX300	F	0.75	+2.08	0.75	-0.33	0.04+/-0.08	12.16+/-0.75
		2WD R	—	—	0.75	-0.33	0.08+/-0.08	—
		4WD R	—	—	0.75	-0.35	0.12+/-0.08	—
2002	RX300	F	0.75	+2.08	0.75	-0.33	0.04+/-0.08	12.16+/-0.75
		2WD R	—	—	0.75	-0.33	0.08+/-0.08	—
		4WD R	—	—	0.75	-0.35	0.12+/-0.08	—

93591CZW

TIRE, WHEEL AND BALL JOINT SPECIFICATIONS

Year	Model	OEM Tires		Tire Pressures (psi)		Wheel Size	Ball Joint Inspection
		Standard	Optional	Front	Rear		
1999	RX300	P255/70HR16	None	30	30	6.5-JJ	①
2000	RX300	P255/70HR16	None	30	30	6.5-JJ	①
2001	RX300	P255/70HR16	None	30	30	6.5-JJ	①
2002	RX300	P255/70HR16	None	30	30	6.5-JJ	①

OEM: Original Equipment Manufacturer

PSI: Pounds Per Square Inch

STD: Standard

OPT: Optional

① Replace if any measurable movement is found.

93591CZX

BRAKE SPECIFICATIONS

All measurements in inches unless noted

Year	Model		Brake Disc			Brake Drum Diameter			Minimum Lining Thickness	Brake Caliper	
			Original Thickness	Minimum Thickness	Maximum Runout	Original Inside Diameter	Max. Wear Limit	Maximum Machine Diameter		Bracket Bolts (ft. lbs.)	Mounting Bolts (ft. lbs.)
1999	RX300	F	1.020	1.024	0.0020	—	—	—	0.039	79	25
		R	0.394	0.354	0.0059	—	—	—	0.039	34	14
2000	RX300	F	1.020	1.024	0.0020	—	—	—	0.039	79	25
		R	0.394	0.354	0.0059	—	—	—	0.039	34	14
2001	RX300	F	1.020	1.024	0.0020	—	—	—	0.039	79	25
		R	0.394	0.354	0.0059	—	—	—	0.039	34	14
2002	RX300	F	1.020	1.024	0.0020	—	—	—	0.039	79	25
		R	0.394	0.354	0.0059	—	—	—	0.039	34	14

F: Front

R: Rear

93591CZY

SCHEDULED MAINTENANCE INTERVALS

LEXUS—RX300

TO BE SERVICED	TYPE OF SERVICE	VEHICLE MILEAGE INTERVAL (x1000)												
		7.5	15	22.5	30	37.5	45	52.5	60	67.5	75	82.5	90	97.5
Engine oil & filter	R	✓	✓	✓	✓	✓	✓	✓	✓	✓	✓	✓	✓	✓
Automatic transmission fluid	S/I		✓		✓		✓		✓		✓		✓	
Ball joints & dust covers	S/I		✓		✓		✓		✓		✓		✓	
Bolts & nuts on chassis & body	S/I		✓		✓		✓		✓		✓		✓	
Brake linings & drums	S/I		✓		✓		✓		✓		✓		✓	
Brake line pipes & hoses	S/I		✓		✓		✓		✓		✓		✓	
Brake pads & discs (front & rear)	S/I		✓		✓		✓		✓		✓		✓	
Propeller shaft grease	S/I		✓		✓		✓		✓		✓		✓	
Steering knuckle & chassis grease	S/I		✓		✓		✓		✓		✓		✓	
Steering linkage	S/I		✓		✓		✓		✓		✓		✓	
Air cleaner filter	R					✓			✓				✓	
Spark plugs	R					✓			✓				✓	
Drive belts	S/I					✓			✓				✓	
Exhaust pipes & mountings	S/I					✓			✓				✓	
Fuel lines & connections	S/I					✓			✓				✓	
Engine coolant	R						✓				✓			
Charcoal canister	R								✓					
Fuel tank cap gasket	R								✓					
Oxygen sensors (exc. Calif.)①	R													

R: Replace   S/I: Service or Inspect

① Heated oxygen sensors (except Calif.): replace every 80,000 miles.

FREQUENT OPERATION MAINTENANCE (SEVERE SERVICE)

If a vehicle is operated under any of the following conditions it is considered severe service:

- Extremely dusty areas.

- 50% or more of the vehicle operation is in 32°C (90°F) or higher temperatures, or constant operation in temperatures below 0°C (32°F).

- Prolonged idling (vehicle operation in stop and go traffic).

- Frequent short running periods (engine does not warm to normal operating temperatures).

- Police, taxi, delivery usage or trailer towing usage.

Air cleaner filter: service or inspect every 3750 miles

Engine oil & filter: replace every 3750 miles.

Ball joints & dust covers: service or inspect every 7500 miles.

Bolts & nuts on chassis & body: service or inspect every 7500 miles.

Brake pads & discs (front & rear): service or inspect every 7500 miles.

Steering knuckle & chassis grease: service or inspect every 7500 miles.

Steering linkage: service or inspect every 7500 miles.

Exhaust pipes & mountings: service or inspect every 15,000 miles.

93591CZZ

SCHEDULED MAINTENANCE INTERVALS
LEXUS
RX300

The following should be used as a guide when determining the amount of work required for a particular service. In estimating how long a particular Scheduled Maintenance Service should take, please observe the following:

● Labor Time is time based on field research and data supplied by the vehicle manufacturer.
● Labor time operations are given in hours and tenths of an hour.
● All labor operations are to be used as a guide.

Mechanic Skill Level Codes:
(A) PRECISION: Highly skilled with multiple certification.
(B) GENERAL: Normally skilled with certification.
(C) MAINTENANCE: Semi-skilled working on certification.

	LABOR TIME		LABOR TIME		LABOR TIME
7500 Mile Service (C)		**37500 Mile Service (C)**		**75000 Mile Service (C)**	
RX300	.4	RX300	.4	RX300	1.6
15000 Mile Service (C)		**45000 Mile Service (B)**		**82500 Mile Service (C)**	
RX300	1.6	RX300	2.1	RX300	.4
22500 Mile Service (C)		**52500 Mile Service (C)**		**90000 Mile Service (B)**	
RX300	.4	RX300	.4	RX300	2.4
30000 Mile Service (B)		**60000 Mile Service (B)**		**97500 Mile Service (C)**	
RX300	2.4	RX300	3.9	RX300	.4
		67500 Mile Service (C)			
		RX300	.4		

93551C001

Timing belt service is covered in Section 3 of this manual

LEXUS
LX470

ENGINE AND VEHICLE IDENTIFICATION

		Engine					Model Year	
Code ①	Liters (cc)	Cu. In.	Cyl.	Fuel Sys.	Engine Type	Eng. Mfg.	Code ②	Year
2UZ-FE	4.7 (4664)	285	8	SFI	DOHC	Toyota	X	1999
							Y	2000
SFI: Sequential Fuel Injection							1	2001
DOHC: Double Overhead Camshaft							2	2002
① Stamped on the left side of the engine block							3	2003
② 10th digit of the Vehicle Identification Number (VIN)								

93591CYA

GENERAL ENGINE SPECIFICATIONS

Year	Model	Engine Displacement Liters (cc)	Engine Series (ID/VIN)	Fuel System	Net Horsepower @ rpm	Net Torque @ rpm (ft. lbs.)	Bore x Stroke (in.)	Com- pression Ratio	Oil Pressure @ rpm
1999	LX470	4.7 (4665)	2UZ-FE	SFI	230@4800	320@3400	3.70x3.31	9.6:1	43-85@3000
2000	LX470	4.7 (4665)	2UZ-FE	SFI	230@4800	320@3400	3.70x3.31	9.6:1	43-85@3000
2001	LX470	4.7 (4665)	2UZ-FE	SFI	245@4800	315@3400	3.70x3.31	9.6:1	45-65@3000
2002	LX470	4.7 (4665)	2UZ-FE	SFI	245@4800	315@3400	3.70x3.31	9.6:1	45-65@3000

SFI: Sequential Fuel Injection

93591CYB

ENGINE TUNE-UP SPECIFICATIONS

Year	Engine Displacement Liters (cc)	Engine ID/VIN	Spark Plug Gap (in.)	Ignition Timing (deg.)*	Fuel Pump (psi)	Idle Speed (rpm)		Valve Clearance	
						MT	AT	Intake	Exhaust
1999	4.7 (4664)	2UZ-FE	0.043	5-15B	38-44	—	650-750	0.006-0.009	0.011-0.014
2000	4.7 (4664)	2UZ-FE	0.043	5-15B	38-44	—	650-750	0.006-0.009	0.011-0.014
2001	4.7 (4664)	2UZ-FE	0.031	5-15B	38-44	—	650-750	0.006-0.010	0.010-0.014
2002	4.7 (4664)	2UZ-FE	0.031	5-15B	38-44	—	650-750	0.006-0.010	0.010-0.014

NOTE: The Vehicle Emission Control Information label often reflects specification changes made during production. The label figures must be used if they differ from those in this chart.

B: Before top dead center

* With terminals TC and E1 connected to DLC1

93591CYC

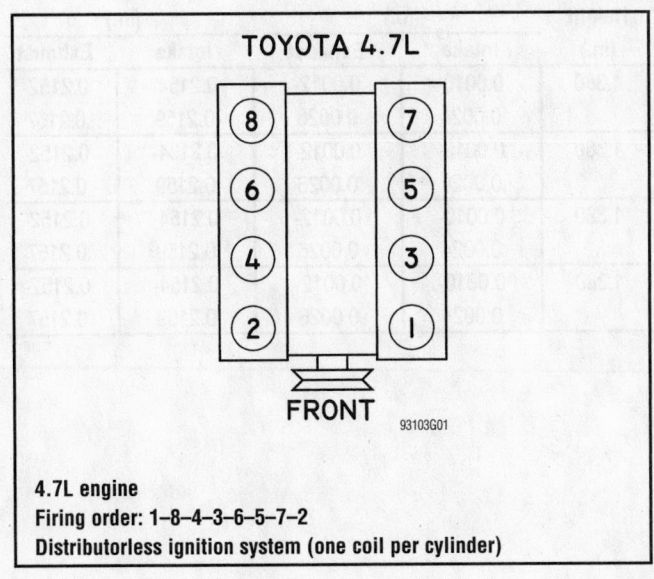

TOYOTA 4.7L

FRONT

93103G01

4.7L engine
Firing order: 1–8–4–3–6–5–7–2
Distributorless ignition system (one coil per cylinder)

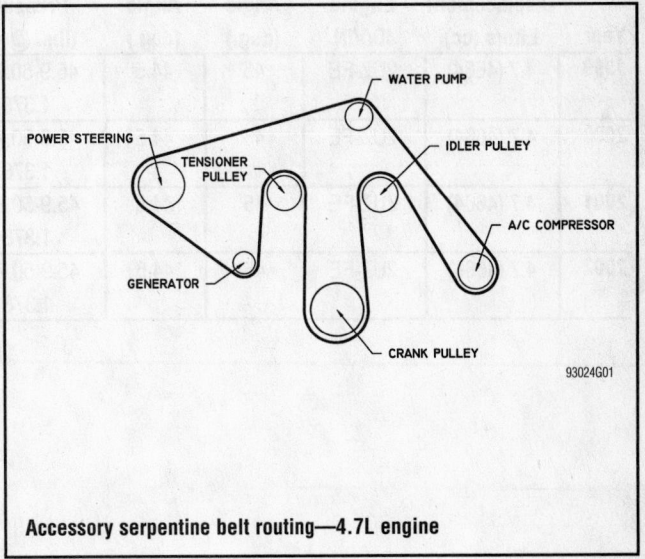

Accessory serpentine belt routing—4.7L engine

93024G01

Heater Core replacement is covered in Section 2 of this manual

CAPACITIES

Year	Model	Engine Displacement Liters (cc)	Engine ID/VIN	Engine Oil with Filter (qts.)	Transmission (pts.) 5-Spd	Transmission (pts.) Auto.*	Transfer Case (pts.)	Drive Axle Front (pts.)	Drive Axle Rear (pts.)	Fuel Tank (gal.)	Cooling System (qts.)
1999	LX470	4.7 (4664)	2UZ-FE	7.2	—	4.0	2.8	3.6	6.8	25.4	①
2000	LX470	4.7 (4664)	2UZ-FE	7.2	—	3.6	2.8	3.6	6.8	25.4	①
2001	LX470	4.7 (4664)	2UZ-FE	7.2	—	3.6	2.8	3.6	6.8	25.4	①
2002	LX470	4.7 (4664)	2UZ-FE	6.6	—	4.2	2.6	2.4	7.7	25.4	12.3

*After draining, add the following amounts, then fill to the cold full line

① With rear heater: 9.5
  Without rear heater: 8.5

93591CYD

VALVE SPECIFICATIONS

Year	Engine Displacement Liters (cc)	Engine ID/VIN	Seat Angle (deg.)	Face Angle (deg.)	Spring Test Pressure (lbs. @ in.)	Spring Installed Height (in.)	Stem-to-Guide Clearance (in.) Intake	Stem-to-Guide Clearance (in.) Exhaust	Stem Diameter (in.) Intake	Stem Diameter (in.) Exhaust
1999	4.7 (4664)	2UZ-FE	45	44.5	45.9-50.7@ 1.378	1.380	0.0010- 0.0024	0.0012- 0.0026	0.2154- 0.2159	0.2152- 0.2157
2000	4.7 (4664)	2UZ-FE	45	44.5	45.9-50.7@ 1.378	1.380	0.0010- 0.0024	0.0012- 0.0026	0.2154- 0.2159	0.2152- 0.2157
2001	4.7 (4664)	2UZ-FE	45	44.5	45.9-50.7@ 1.378	1.380	0.0010- 0.0024	0.0012- 0.0026	0.2154- 0.2159	0.2152- 0.2157
2002	4.7 (4664)	2UZ-FE	45	44.5	45.9-50.7@ 1.378	1.380	0.0010- 0.0024	0.0012- 0.0026	0.2154- 0.2159	0.2152- 0.2157

93591CYE

CRANKSHAFT AND CONNECTING ROD SPECIFICATIONS

All measurements are given in inches.

| Year | Engine Displacement Liters (cc) | Engine ID/VIN | Crankshaft | | | | Connecting Rod | | |
			Main Brg. Journal Dia.	Main Brg. Oil Clearance	Shaft End-play	Thrust on No.	Journal Diameter	Oil Clearance	Side Clearance
1999	4.7 (4665)	2UZ-FE	2.6373-2.6378	0.0016-0.0023	0.0008-0.0087	3	2.0465-2.0472	0.0011-0.0021	0.0063-0.0138
2000	4.7 (4665)	2UZ-FE	2.6373-2.6378	0.0016-0.0023	0.0008-0.0087	3	2.0465-2.0472	0.0011-0.0021	0.0063-0.0138
2001	4.7 (4665)	2UZ-FE	2.6373-2.6378	①	0.0008-0.0087	3	2.0465-2.0472	0.0011-0.0021	0.0063-0.0138
2002	4.7 (4665)	2UZ-FE	2.6373-2.6378	①	0.0008-0.0087	3	2.0465-2.0472	0.0011-0.0021	0.0063-0.0138

① Nos. 1 and 2: 0.0011-0.0018
All others: 0.0016-0.0023

93591CYF

PISTON AND RING SPECIFICATIONS

All measurements are given in inches.

| Year | Engine Displacement Liters (cc) | Engine ID/VIN | Piston Clearance | Ring Gap | | | Ring Side Clearance | | |
				Top Compression	Bottom Compression	Oil Control	Top Compression	Bottom Compression	Oil Control
1999	4.7 (4664)	2UZ-FE	0.0035-0.0044	0.0118-0.0197	0.0157-0.0256	0.0051-0.0189	0.0012-0.0031	0.0012-0.0028	SNUG
2000	4.7 (4664)	2UZ-FE	0.0035-0.0044	0.0118-0.0197	0.0157-0.0256	0.0051-0.0189	0.0012-0.0031	0.0012-0.0028	SNUG
2001	4.7 (4664)	2UZ-FE	0.0035-0.0044	0.0118-0.0157	0.0157-0.0217	0.0051-0.0150	0.0012-0.0031	0.0012-0.0028	SNUG
2002	4.7 (4664)	2UZ-FE	0.0035-0.0044	0.0118-0.0157	0.0157-0.0217	0.0051-0.0150	0.0012-0.0031	0.0012-0.0028	SNUG

93591CYG

Brake service is covered in Section 4 of this manual

TORQUE SPECIFICATIONS
All readings in ft. lbs.

Year	Engine Displacement Liters (cc)	Engine ID/VIN	Cylinder Head Bolts	Main Bearing Bolts	Rod Bearing Bolts	Crankshaft Damper Bolts	Flywheel Bolts	Manifold Intake	Manifold Exhaust	Spark Plugs	Lug Nuts
1999	4.7 (4664)	2UZ-FE	①	②	③	181	④	13	33	13	97
2000	4.7 (4664)	2UZ-FE	①	②	③	181	④	13	33	13	97
2001	4.7 (4664)	2UZ-FE	⑤	②	③	181	④	13	33	13	97
2002	4.7 (4664)	2UZ-FE	⑤	②	③	181	④	13	33	13	97

① Step 1: 24 ft. lbs.
   Step 2: Plus 180 degrees

② Step 1: 20 ft. lbs.
   Step 2: Plus 90 degrees

③ Step 1: 18 ft. lbs.
   Step 2: Plus 90 degrees

④ Step 1: 35 ft. lbs.
   Step 2: Plus 90 degrees

⑤ Step 1: 24
   Step 2: Plus 90 degrees
   Step 3: Plus 90 degrees

93591CYH

WHEEL ALIGNMENT

Year	Model	Caster Range (+/-Deg.)	Caster Preferred Setting (Deg.)	Camber Range (+/-Deg.)	Camber Preferred Setting (Deg.)	Toe-in (in.)	Steering Axis Inclination (Deg.)
1999	LX470	0.75	+3.08	0.75	0	0+/-0.08	12.25+/-0.75
2000	LX470	0.75	+3.08	0.75	0	0+/-0.08	12.25+/-0.75
2001	LX470	0.75	+3.08	0.75	0	0+/-0.08	12.25+/-0.75
2002	LX470	0.75	①	0.75	+0.13	0.05+/-0.08	10.635+/-0.75

Note: All alignment specifications are based on nominal ride height and standard tires

① P245/70R16: +2.95

P265/70R16: +3.00

93591CYI

TIRE, WHEEL AND BALL JOINT SPECIFICATIONS

| Year | Model | OEM Tires | | Tire Pressures (psi) | | Wheel Size | Ball Joint Inspection |
		Standard	Optional	Front	Rear		
1999	LX470	P275/70HR16	None	32	32	8-JJ	①
2000	LX470	P275/70HR16	None	32	32	8-JJ	①
2001	LX470	P275/70HR16	None	32	32	8-JJ	②
2002	LX470	P275/70HR16	None	32	32	8-JJ	②

OEM: Original Equipment Manufacturer

PSI: Pounds Per Square Inch

STD: Standard

OPT: Optional

① Replace if any measurable movement is found.

② Upper ball joint turning torque: 6-39 inch lbs.

    Lower ball joint turning torque: 1-22 inch lbs.

    Lower ball joint excessive play: 0.020 in.

93591CYJ

BRAKE SPECIFICATIONS

All measurements in inches unless noted

| Year | Model | | Brake Disc | | | Brake Drum Diameter | | | Minimum Lining Thickness | Brake Caliper | |
			Original Thickness	Minimum Thickness	Maximum Runout	Original Inside Diameter	Max. Wear Limit	Maximum Machine Diameter		Bracket Bolts (ft. lbs.)	Mounting Bolts (ft. lbs.)
1999	LX470	F	1.260	1.181	0.0028	—	—	—	0.039	76	90
		R	0.709	0.611	0.0040	—	—	—	0.039	—	65
2000	LX470	F	1.260	1.181	0.0028	—	—	—	0.039	76	90
		R	0.709	0.611	0.0040	—	—	—	0.039	—	65
2001	LX470	F	1.260	1.181	0.0028	—	—	—	0.039	76	90
		R	0.709	0.611	0.0040	—	—	—	0.039	—	65
2002	LX470	F	1.260	1.181	0.0028	—	—	—	0.039		90
		R	0.709	0.611	0.0040	—	—	—	0.039	—	20

F: Front

R: Rear

93591CYK

For complete Engine Mechanical specifications, see Section 1 of this manual

SCHEDULED MAINTENANCE INTERVALS
Lexus LX470

TO BE SERVICED	TYPE OF SERVICE	VEHICLE MILEAGE INTERVAL (x1000)												
		7.5	15	22.5	30	37.5	45	52.5	60	67.5	75	82.5	90	97.5
Engine oil & filter	R	✓	✓	✓	✓	✓	✓	✓	✓	✓	✓	✓	✓	✓
Automatic transmission fluid & filter	S/I		✓		✓		✓		✓		✓		✓	
Ball joints & dust covers	S/I		✓		✓		✓		✓		✓		✓	
Bolts & nuts on chassis & body	S/I		✓		✓		✓		✓		✓		✓	
Brake line pipes & hoses	S/I		✓		✓		✓		✓		✓		✓	
Brake pads & discs	S/I		✓		✓		✓		✓		✓		✓	
Propeller shaft grease	S/I		✓		✓		✓		✓		✓		✓	
Steering knuckle & chassis grease	S/I		✓		✓		✓		✓		✓		✓	
Steering linkage	S/I		✓		✓		✓		✓		✓		✓	
Transfer and differential oil	S/I		✓		✓		✓		✓		✓		✓	
Air cleaner filter	R				✓				✓				✓	
Spark plugs ①	R				✓				✓				✓	
Drive belts	S/I				✓				✓				✓	
Exhaust pipes & mountings	S/I				✓				✓				✓	
Fuel lines & connections	S/I				✓				✓				✓	
Engine coolant	R						✓				✓			
Charcoal canister	R								✓					
Fuel tank cap gasket	R								✓					
Heated oxygen sensors (except Calif.) ②	R													

R: Replace   S/I: Service or Inspect

① Platinum plugs, replace every 100,000 miles

② Heated oxygen sensors (except Calif.): replace every 80,000 miles.

FREQUENT OPERATION MAINTENANCE (SEVERE SERVICE)

If a vehicle is operated under any of the following conditions it is considered severe service:

- Extremely dusty areas.

- 50% or more of the vehicle operation is in 32°C (90°F) or higher temperatures, or constant operation in temperatures below 0°C (32°F).

- Prolonged idling (vehicle operation in stop and go traffic).

- Frequent short running periods (engine does not warm to normal operating temperatures).

- Police, taxi, delivery usage or trailer towing usage.

Air cleaner filter: service or inspect every 3750 miles.

Engine oil & filter: replace every 3750 miles.

Ball joints & dust covers: service or inspect every 7500 miles.

Bolts & nuts on chassis & body: service or inspect every 7500 miles.

Brake pads & discs (front & rear): service or inspect every 7500 miles.

Steering knuckle & chassis grease: service or inspect every 7500 miles.

Steering linkage: service or inspect every 7500 miles.

Propeller shaft grease: service or inspect every 7500 miles.

Exhaust pipes & mountings: service or inspect every 15,000 miles.

93591CYL

SCHEDULED MAINTENANCE INTERVALS
LEXUS
LX470

The following should be used as a guide when determining the amount of work required for a particular service.
In estimating how long a particular Scheduled Maintenance Service should take, please observe the following:

● Labor Time is time based on field research and data supplied by the vehicle manufacturer.
● Labor time operations are given in hours and tenths of an hour.
● All labor operations are to be used as a guide.

Mechanic Skill Level Codes:
(A) PRECISION: Highly skilled with multiple certification.
(B) GENERAL: Normally skilled with certification.
(C) MAINTENANCE: Semi-skilled working on certification.

	LABOR TIME		LABOR TIME		LABOR TIME
7500 Mile Service (C)		**37500 Mile Service (C)**		**75000 Mile Service (C)**	
LX470	.4	LX470	.4	LX470	1.7
15000 Mile Service (C)		**45000 Mile Service (B)**		**82500 Mile Service (C)**	
LX470	1.7	LX470	2.2	LX470	.4
22500 Mile Service (C)		**52500 Mile Service (C)**		**90000 Mile Service (B)**	
LX470	.6	LX470	.4	LX470	2.5
30000 Mile Service (B)		**60000 Mile Service (B)**		**97500 Mile Service (C)**	
LX470	2.5	LX470	4.2	LX470	.4
		67500 Mile Service (C)			
		LX470	.4		

93551CC6A

For Accessory Drive Belt illustrations, see Section 1 of this manual

MAZDA
Tribute

ENGINE AND VEHICLE IDENTIFICATION

		Engine							Model Year	
Code ①	Liters (cc)	Cu. In.	Cyl.	Fuel Sys.	Engine Type	Eng. Mfg.		Code ②		Year
B	2.0 (1998)	121	4	SFI	DOHC	Ford		1		2001
1	3.0 (3049)	182	6	SFI	DOHC	Ford		2		2002
								3		2003

SFI: Multi-port Fuel Injection
DOHC: Double Overhead Camshafts
① 8th digit of VIN
② 10th digit of VIN

93591CU2

GENERAL ENGINE SPECIFICATIONS

Year	Model	Engine Displacement Liters (cc)	Engine ID/VIN	Fuel System Type	Net Horsepower @ rpm	Net Torque @ rpm (ft. lbs.)	Bore x Stroke (in.)	Com-pression Ratio	Oil Pressure @ rpm①
2001	Tribute	2.0 (1998)	B	SFI	135@5500	135@4500	3.34x3.46	9.6:1	54-80
	Tribute	3.0 (3049)	1	SFI	200@5500	200@4500	3.50x3.13	10.0:1	45
2002	Tribute	2.0 (1998)	B	SFI	135@5500	135@4500	3.34x3.46	9.6:1	54-80
	Tribute	3.0 (3049)	1	SFI	200@5500	200@4500	3.50x3.13	10.0:1	45

SFI: Multi-port Fuel Injection

① The manufacturer does not provide an engine speed specification for oil pump pressure.

93591CU3

ENGINE TUNE-UP SPECIFICATIONS

Year	Engine Displacement Liters (cc)	Engine ID/VIN	Spark Plug Gap (in.)	Ignition Timing (deg.)		Fuel Pump (psi)	Idle Speed (rpm)		Valve Clearance	
				MT	AT		MT	AT	Intake	Exhaust
2001	2.0 (1998)	B	0.039–0.043	10 BTDC	—	65	①	—	HYD.	HYD.
	3.0 (3049)	1	0.052–0.056	10 BTDC	10 BTDC	65	①	①	HYD.	HYD.
2002	2.0 (1998)	B	0.051	10 BTDC	—	65	①	—	HYD.	HYD.
	3.0 (3049)	1	0.052–0.056	10 BTDC	10 BTDC	65	①	①	HYD.	HYD.

BTDC: Before Top Dead Center

HYD: Hydraulic lash adjusters

① Refer to Vehicle Emission Control Information Label

93591CU4

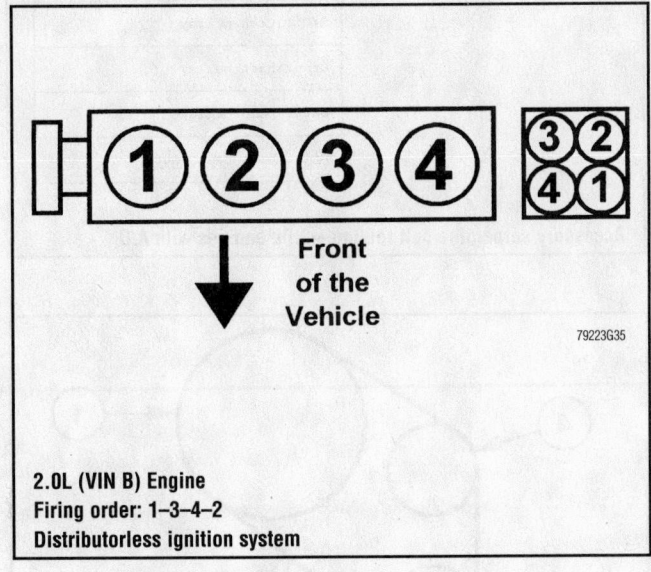

2.0L (VIN B) Engine
Firing order: 1–3–4–2
Distributorless ignition system

79223G35

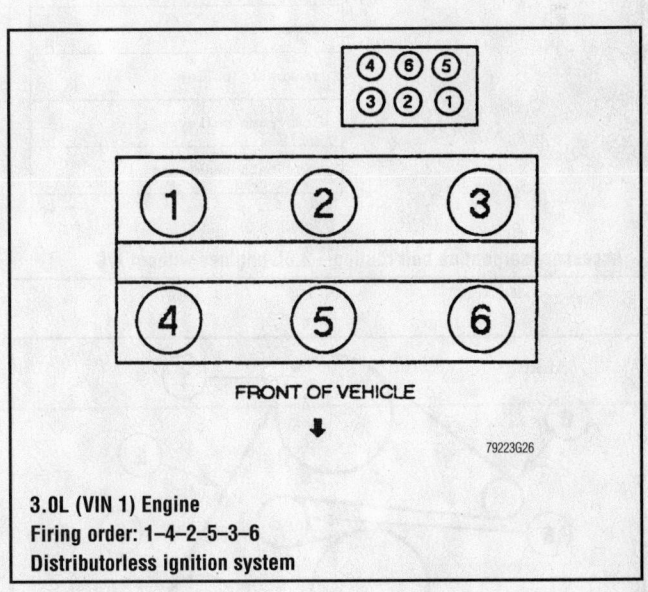

3.0L (VIN 1) Engine
Firing order: 1–4–2–5–3–6
Distributorless ignition system

79223G26

For Tire, Wheel and Ball Joint specifications, see Section 1 of this manual

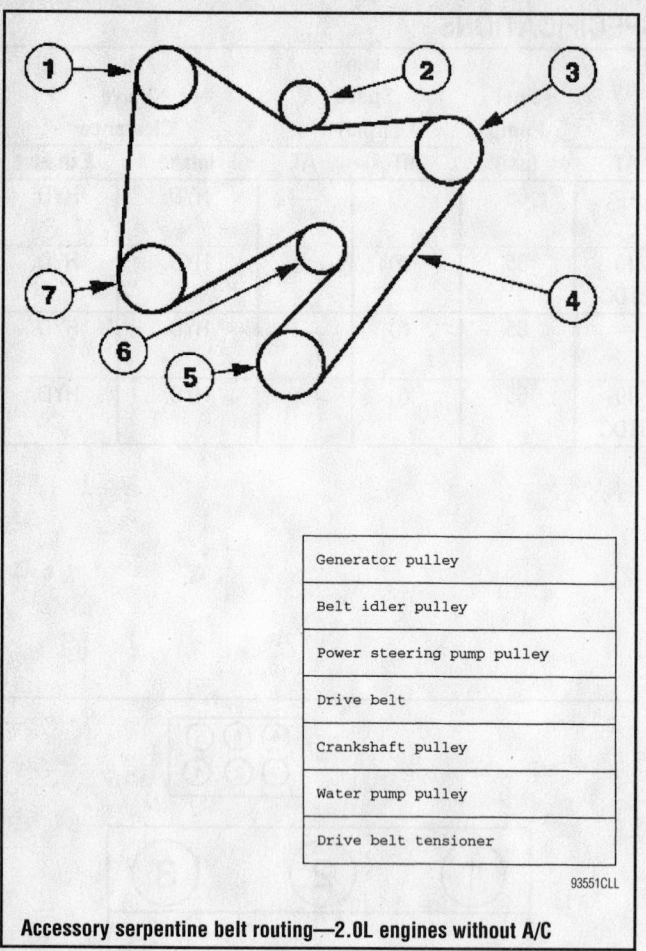

Generator pulley
Belt idler pulley
Power steering pump pulley
Drive belt
Crankshaft pulley
Water pump pulley
Drive belt tensioner

93551CLL

Accessory serpentine belt routing—2.0L engines without A/C

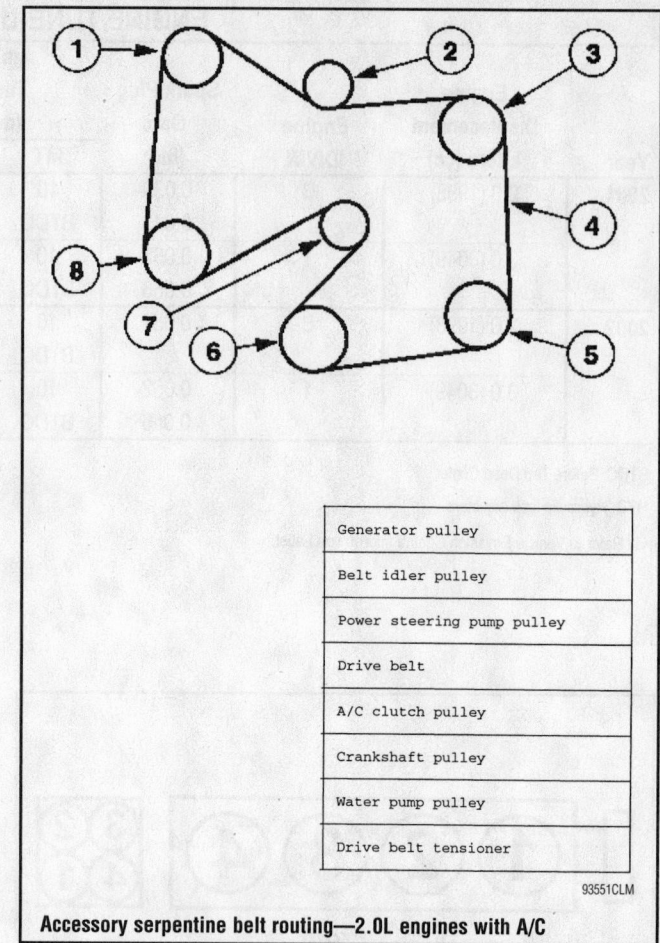

Generator pulley
Belt idler pulley
Power steering pump pulley
Drive belt
A/C clutch pulley
Crankshaft pulley
Water pump pulley
Drive belt tensioner

93551CLM

Accessory serpentine belt routing—2.0L engines with A/C

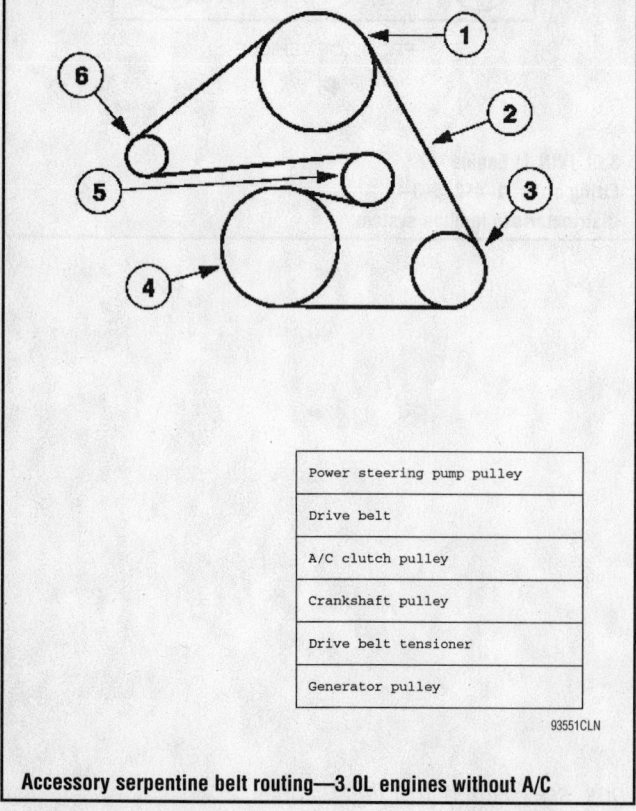

Power steering pump pulley
Drive belt
A/C clutch pulley
Crankshaft pulley
Drive belt tensioner
Generator pulley

93551CLN

Accessory serpentine belt routing—3.0L engines without A/C

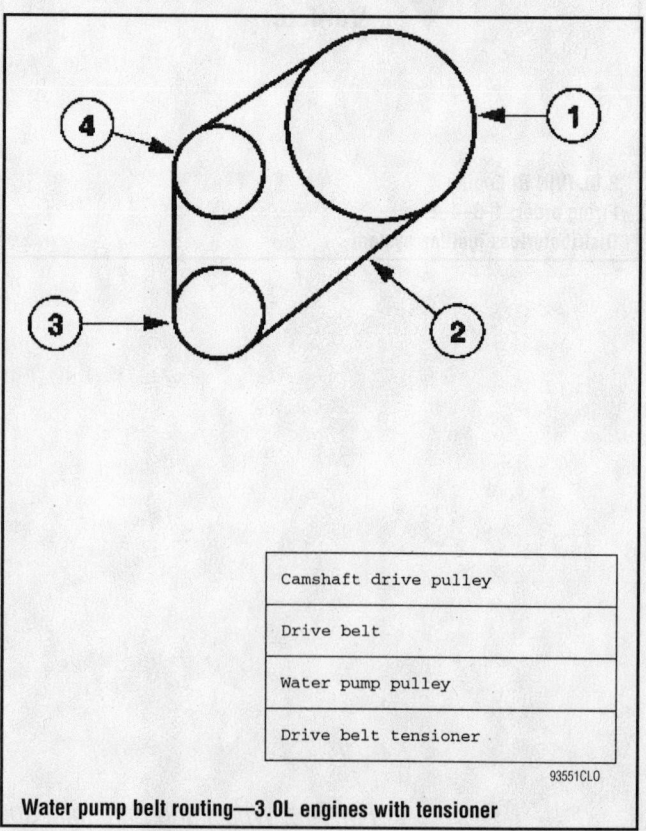

Camshaft drive pulley
Drive belt
Water pump pulley
Drive belt tensioner

93551CLO

Water pump belt routing—3.0L engines with tensioner

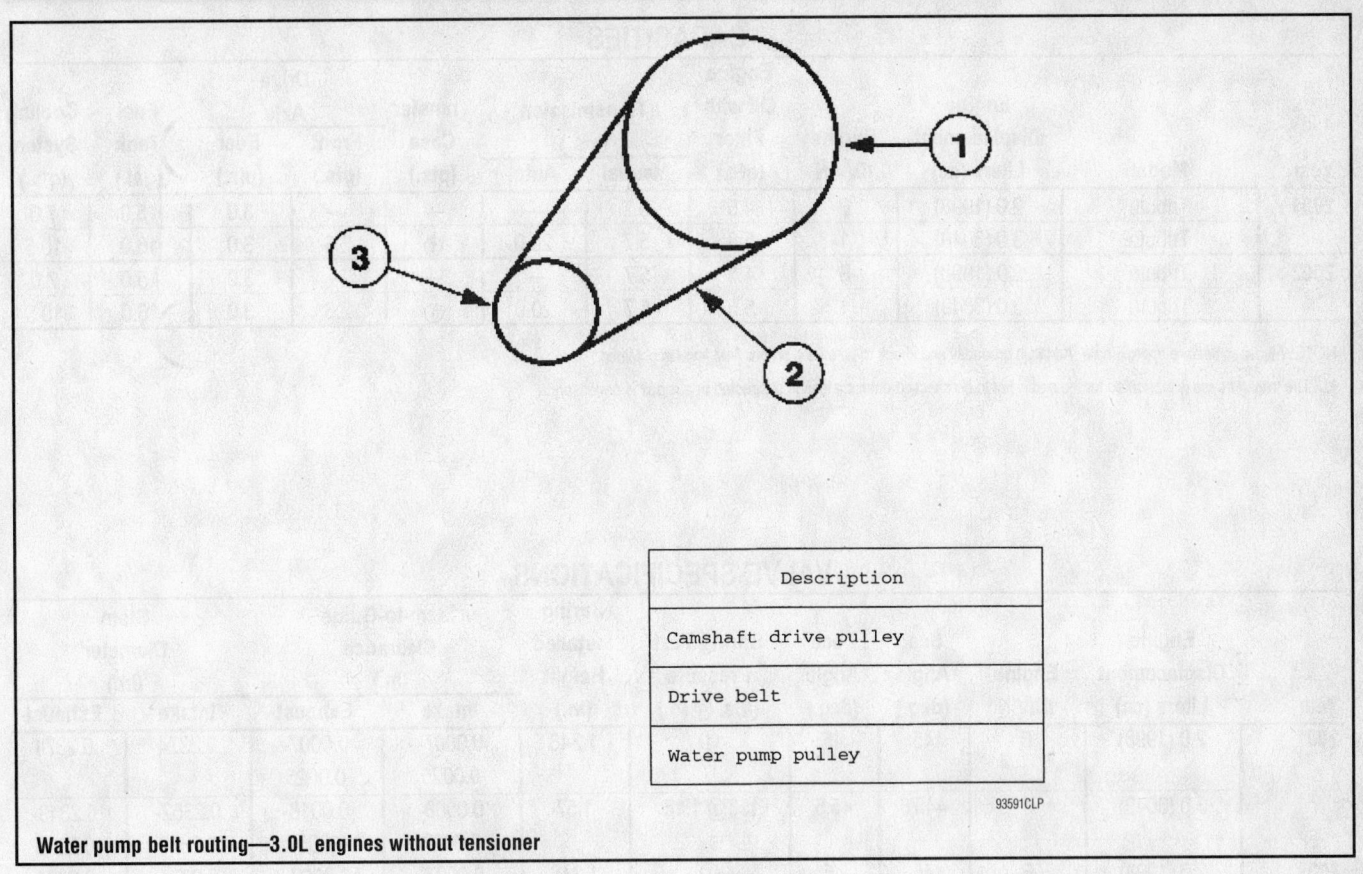

Description
Camshaft drive pulley
Drive belt
Water pump pulley

93591CLP

Water pump belt routing—3.0L engines without tensioner

For Wheel Alignment specifications, see Section 1 of this manual

CAPACITIES

Year	Model	Engine Displacement Liters (cc)	Engine ID/VIN	Engine Oil with Filter (qts.)	Transmission (pts.)		Transfer Case (pts.)	Drive Axle		Fuel Tank (gal.)	Cooling System (qts.)
					Manual	Auto.		Front (pts.)	Rear (pts.)		
2001	Tribute	2.0 (1998)	B	4.5	5.7	—	—	—	3.0	15.0	7.0
	Tribute	3.0 (3049)	1	5.8	5.7	20.0	①	2.6	3.0	16.0	10.5
2002	Tribute	2.0 (1998)	B	4.5	5.7	—	—	—	3.0	15.0	7.0
	Tribute	3.0 (3049)	1	5.5	5.7	20.0	①	2.6	3.0	16.0	10.5

NOTE: All capacities are approximate. Add fluid gradually and check to be sure a proper fluid level is obtained.

① The transfer case is lubricated for life and is not to be checked unless a leak is suspected or a repair is necessary.

93591CU5

VALVE SPECIFICATIONS

Year	Engine Displacement Liters (cc)	Engine ID/VIN	Seat Angle (deg.)	Face Angle (deg.)	Spring Test Pressure (lbs. @ in.)	Spring Installed Height (in.)	Stem-to-Guide Clearance (in.)		Stem Diameter (in.)	
							Intake	Exhaust	Intake	Exhaust
2001	2.0 (1998)	B	45	45	①	1.346	0.0007-0.0025	0.0007-0.0025	0.2374	0.2374
	3.0 (3049)	1	44.75	45.5	153@ 1.18	1.57	0.0008-0.0027	0.0018-0.0037	0.2352-0.2360	0.2343-0.2350
2002	2.0 (1998)	B	45	45	①	1.346	0.0007-0.0025	0.0007-0.0025	0.2374	0.2374
	3.0 (3049)	1	44.75	45.5	153@ 1.18	1.57	0.0008-0.0027	0.0018-0.0037	0.2352-0.2360	0.2343-0.2350

① Intake: 82.1@ 0.988

Exhaust: 95@ 1.0275

93591CU6

CRANKSHAFT AND CONNECTING ROD SPECIFICATIONS

All measurements are given in inches.

| Year | Engine Displacement Liters (cc) | Engine ID/VIN | Crankshaft | | | | Connecting Rod | | |
			Main Brg. Journal Dia.	Main Brg. Oil Clearance	Shaft End-play	Thrust on No.	Journal Diameter	Oil Clearance	Side Clearance
2001	2.0 (1998)	B	2.2827-2.2835	①	0.0035-0.0102	3	1.8460-1.8468	0.0006-0.0028	0.0040-0.0110
	3.0 (3049)	1	2.4791-2.4800	0.0010-0.0018	0.0043-0.0091	3	1.9673-1.9681	0.0010-0.0025	0.0039-0.0118
2002	2.0 (1998)	B	2.2827-2.2835	①	0.0035-0.0102	3	1.8460-1.8468	0.0006-0.0028	0.0040-0.0110
	3.0 (3049)	1	2.4791-2.4800	0.0010-0.0018	0.0043-0.0091	3	1.9673-1.9681	0.0011-0.0026	0.0039-0.0118

① Journals 1, 2 and 4: 0.0010 - 0.0017 in.
Journal 3: 0.0012 - 0.0019 in.

93591CU7

PISTON AND RING SPECIFICATIONS

All measurements are given in inches.

| Year | Engine Displacement Liters (cc) | Engine ID/VIN | Piston Clearance | Ring Gap | | | Ring Side Clearance | | |
				Top Compression	Bottom Compression	Oil Control	Top Compression	Bottom Compression	Oil Control
2001	2.0 (1998)	B	0.0004-0.0012	0.0100-0.0300	0.0100-0.0300	0.0160-0.0660	0.0015-0.0032	0.0015-0.0035	snug
	3.0 (3049)	1	0.0005-0.0009	0.0039-0.0098	0.0106-0.0165	0.0059-0.0256	0.0016-0.0030	0.0016-0.0033	snug
2002	2.0 (1998)	B	0.0004-0.0012	0.0100-0.0300	0.0100-0.0300	0.0160-0.0660	0.0015-0.0032	0.0015-0.0035	snug
	3.0 (3049)	1	0.0005-0.0009	0.0039-0.0098	0.0106-0.0165	0.0059-0.0256	0.0016-0.0030	0.0016-0.0033	snug

93591CU8

TORQUE SPECIFICATIONS
All readings in ft. lbs.

Year	Engine Displacement Liters (cc)	Engine ID/VIN	Cylinder Head Bolts	Main Bearing Bolts	Rod Bearing Bolts	Crankshaft Damper Bolts	Flywheel Bolts	Manifold Intake	Manifold Exhaust	Spark Plugs	Lug Nuts
2001	2.0 (1998)	B	①	②	③	80-87	83	13	12	11	98
	3.0 (3049)	1	④	⑤	⑥	⑦	59	⑧	15	11	98
2002	2.0 (1998)	B	①	②	③	80-87	83	13	12	11	98
	3.0 (3049)	1	④	⑤	⑥	⑦	59	⑧	15	11	98

① Step 1: 15 ft. lbs. (20 Nm).

 Step 2: 30 ft. lbs. (40 Nm).

 Step 3: Plus an additional 90 degrees.

② Step 1: 18 ft. lbs.

 Step 2: +60 degrees

③ Step 1: 26 ft. lbs.

 Step 2: +90 degrees

④ Step 1: 30 ft. lbs. (40 Nm).

 Step 2: Tighten the bolts 90 degrees.

 Step 3: Loosen the bolts one full turn.

 Step 4: 30 ft. lbs. (40 Nm).

 Step 5: Tighten the bolts 90 degrees.

 Step 6: Tighten the bolts 90 degrees.

⑤ Step 1: Fasteners 1-8: 18 ft. lbs.

 Step 2: Fasteners 9-19: 30 ft. lbs.

 Step 3: Fasteners 1-16: +90 degrees

 Step 4: fasteners 17-22: 18 ft. lbs.

⑥ Step 1: 17 ft. lbs.

 Step 2: 32 ft. lbs.

⑦ Step 1: 89 ft. lbs.

 Step 2: Loosen 1 full turn

 Step 3: 37 ft. lbs.

 Step 4: 66 ft. lbs.

⑧ 89 inch lbs.

93591CU9

WHEEL ALIGNMENT

Year	Model		Caster Range (+/-Deg.)	Caster Preferred Setting (Deg.)	Camber Range (+/-Deg.)	Camber Preferred Setting (Deg.)	Toe-in (in.)	Steering Axis Inclination (Deg.)
2001	ALL	F	NA	+1.93	NA	-0.84	0.12+/-0.12	11.40
		R	NA	NA	NA	-0.04	0.09+/-0.11	NA
2002	ALL	F	NA	+1.93	NA	-0.84	0.12+/-0.12	11.40
		R	NA	NA	NA	-0.04	0.09+/-0.11	NA

93591CU0

TIRE, WHEEL AND BALL JOINT SPECIFICATIONS

Year	Model	OEM Tires Standard	OEM Tires Optional	Tire Pressures (psi) Front	Tire Pressures (psi) Rear	Wheel Size	Ball Joint Inspection
2001	Tribute	P225/70SR15	P235/70R16	NA	NA	NA	0.030 in.
2002	Tribute	P215/70R16	P225/70SR15 P235/70R16	NA	NA	NA	0.030 in.

OEM: Original Equipment Manufacturer

PSI: Pounds Per Square Inch

STD: Standard

OPT: Optional

NA: Not Available

93591CV1

For Tune-up, Capacities and Firing orders, see Section 1 of this manual

BRAKE SPECIFICATIONS

All measurements in inches unless noted

Year	Model	Brake Disc			Brake Drum			Minimum Lining Thickness	Brake Caliper	
		Original Thickness	Minimum Thickness	Maximum Run-out	Original Inside Diameter	Max. Wear Limit	Maximum Machine Diameter		Bracket Bolts (ft. lbs.)	Mounting Bolts (ft. lbs.)
2001	Tribute	0.940	0.860	0.002	9.06	0.06	8.92	0.039	111	26
2002	Tribute	0.940	0.860	0.004	9.06	0.06	8.92	0.039	111	26

93591CV2

SCHEDULED MAINTENANCE INTERVALS
Mazda Tribute

TO BE SERVICED	TYPE OF SERVICE	VEHICLE MILEAGE INTERVAL (x1000)												
		5	10	15	20	25	30	35	40	45	50	55	60	65
Air cleaner filter	R						✓						✓	
Accessory drive belt	S/I												✓	
Brake system ①	S/I			✓			✓			✓			✓	
Clutch pedal operation	S/I						✓						✓	
Cooling fan operation	S/I		✓		✓		✓		✓		✓		✓	
Cooling system hoses and clamps	S/I			✓			✓			✓			✓	
CV-joint boots & axle seals	S/I						✓						✓	
Engine coolant	R	Ten years or 150,000 miles												
Engine oil & filter	R	✓	✓	✓	✓	✓	✓	✓	✓	✓	✓	✓	✓	✓
Exterior Lights	S/I	Check monthly												
PCV valve	S/I												✓	
Exhaust system & heat shields	S/I						✓						✓	
Parking brake system	S/I	Every 6 months												
Power steering fluid	S/I	Every 6 months												
Rotate tires	S/I	✓		✓		✓		✓		✓		✓		✓
Steering linkage	S/I						✓						✓	
Spark plugs	R	Change at 100,000 miles												
Suspension components	S/I						✓						✓	

R: Replace   S/I: Inspect and service, if necessary   L: Lubricate   A: Adjust   C: Clean

① Inspect the reservoir fluid level, rotor and or drum, brake lines, hoses, calipers and or wheel cylinders

FREQUENT OPERATION MAINTENANCE (SEVERE SERVICE)

If a vehicle is operated under any of the following conditions it is considered severe service:

- Extremely dusty areas.

- 50% or more of the vehicle operation is in 32°C (90°F) or higher temperatures, or constant operation in temperatures below 0°C (32°F).

- Prolonged idling (vehicle operation in stop and go traffic).

- Frequent short running periods (engine does not warm to normal operating temperatures).

- Police, taxi, delivery usage or trailer towing usage.

Oil & oil filter change: change every 3000 miles.

Air filter element: change every 15,000 miles.

93591CAZ

For complete service labor times, order Nichols' Chilton Labor Guide

SCHEDULED MAINTENANCE INTERVALS
MAZDA
TRIBUTE

The following should be used as a guide when determining the amount of work required for a particular service. In estimating how long a particular Scheduled Maintenance Service should take, please observe the following:

● Labor Time is time based on field research and data supplied by the vehicle manufacturer.
● Labor time operations are given in hours and tenths of an hour.
● All labor operations are to be used as a guide.

Mechanic Skill Level Codes:
(A) PRECISION: Highly skilled with multiple certification.
(B) GENERAL: Normally skilled with certification.
(C) MAINTENANCE: Semi-skilled working on certification.

	LABOR TIME		LABOR TIME		LABOR
5000 Mile Service (C)		**25000 Mile Service (C)**		**50000 Mile Service (C)**	
All Models	.9	All Models	.9	All Models	.7
10000 Mile Service (C)		**30000 Mile Service (B)**		**55000 Mile Service (C)**	
All Models	.5	All Models	1.9	All Models	.9
15000 Mile Service (C)		**35000 Mile Service (C)**		**60000 Mile Service (B)**	
All Models	1.6	All Models	.9	All Models	1.9
20000 Mile Service (C)		**40000 Mile Service (C)**		**65000 Mile Service (C)**	
All Models	.5	All Models	.5	All Models	.9
		45000 Mile Service (C)			
		All Models	1.6		

93081CV2

MITSUBISHI
Montero • Montero Sport

ENGINE AND VEHICLE IDENTIFICATION CHART

	Engine							Model Year	
Code	Liters (cc)	Cu. In.	Cyl.	Fuel Sys.	Engine Type	Eng. Mfg.		Code	Year
G	2.4 (2351)	143.4	4	MFI	SOHC	Mitsubishi		X	1999
M	3.5 (3497)	213.4	6	MFI	SOHC	Mitsubishi		Y	2000
R	3.5 (3497)	213.4	6	MFI	SOHC	Mitsubishi		1	2001
H	3.0 (2972)	181.4	6	MFI	SOHC	Mitsubishi		2	2002
P	3.0 (2972)	181.4	6	MFI	SOHC	Mitsubishi		3	2003

MFI: Multi-port Fuel Injection

93591CV0

GENERAL ENGINE SPECIFICATIONS

Year	Engine Displacement Liters (cc)	Engine VIN	Fuel System Type	Net Horsepower @ rpm	Net Torque @ rpm (ft. lbs.)	Bore x Stroke (in.)	Compression Ratio	Oil Pressure @ rpm
1999	2.4 (2351)	G	MFI	134@5500	148@3000	3.41x3.94	9.5:1	41@2000
	3.5 (3497)	M	MFI	200@5000	228@3500	3.66x3.38	9.0:1	30-80@2000
	3.0 (2972)	P	MFI	173@5500	188@4500	3.59x2.99	9.0:1	30-80@2000
2000	3.5 (3497)	M	MFI	200@5000	228@3500	3.66x3.38	9.0:1	30-80@2000
	3.0 (2972)	P	MFI	173@5500	188@4500	3.59x2.99	9.0:1	30-80@2000
2001	3.5 (3497)	M	MFI	200@5000	228@3500	3.66x3.38	9.0:1	30-80@2000
	3.0 (2972)	P	MFI	173@5500	188@4500	3.59x2.99	9.0:1	30-80@2000
2002	3.5 (3497)	R	MFI	200@5000	235@3500	3.66x3.38	9.0:1	30-80@2000
	3.0 (2972)	P	MFI	173@5500	188@4500	3.59x2.99	9.0:1	30-80@2000

93591CW1

ENGINE TUNE-UP SPECIFICATIONS

Year	Engine Displacement Liters (cc)	Engine VIN	Spark Plugs Gap (in.)	Ignition Timing (deg.) MT	AT	Fuel Pump (psi)	Idle Speed (rpm) MT	AT	Valve Clearance In.	Ex.
1999	2.4 (2351)	G	0.039-0.043	5B	—	38 ①	750	—	HYD	HYD
	3.5 (3479)	M	0.039-0.043	—	5B	38 ①	—	700	HYD	HYD
	3.0 (2972)	P	0.039-0.043	5B	5B	38 ①	750	750	HYD	HYD
2000	3.5 (3479)	M	0.039-0.043	—	5B	38 ①	—	700	HYD	HYD
	3.0 (2972)	P	0.039-0.043	—	5B	38 ①	—	750	HYD	HYD
2001	3.5 (3479)	M	0.039-0.043	—	5B	38 ①	—	700	HYD	HYD
	3.0 (2972)	P	0.039-0.043	—	5B	38 ①	—	750	HYD	HYD
2002	3.5 (3479)	R	0.039-0.043	—	5B	38 ①	—	700	HYD	HYD
	3.0 (2972)	P	0.039-0.043	—	5B	38 ①	—	750	HYD	HYD

B: Before top dead center

HYD: Hydraulic

① With vacuum hose connected

93591CW2

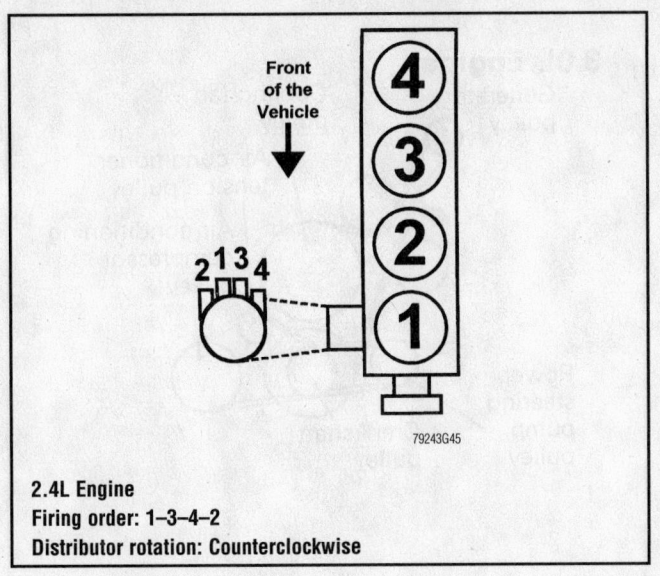

2.4L Engine
Firing order: 1–3–4–2
Distributor rotation: Counterclockwise

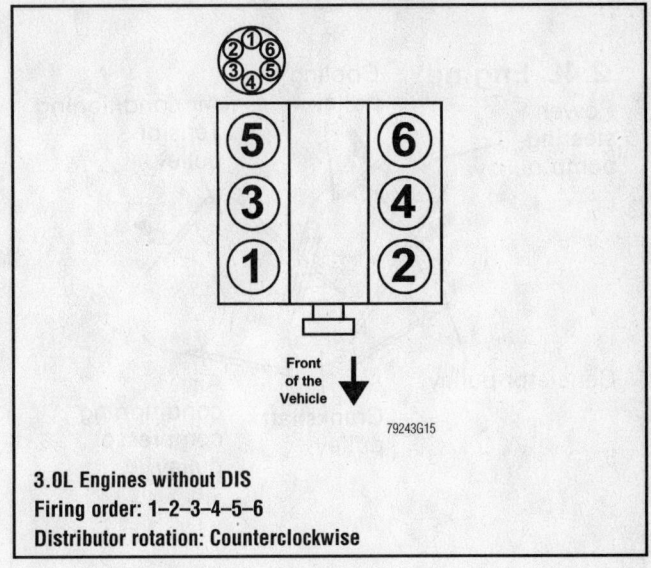

3.0L Engines without DIS
Firing order: 1–2–3–4–5–6
Distributor rotation: Counterclockwise

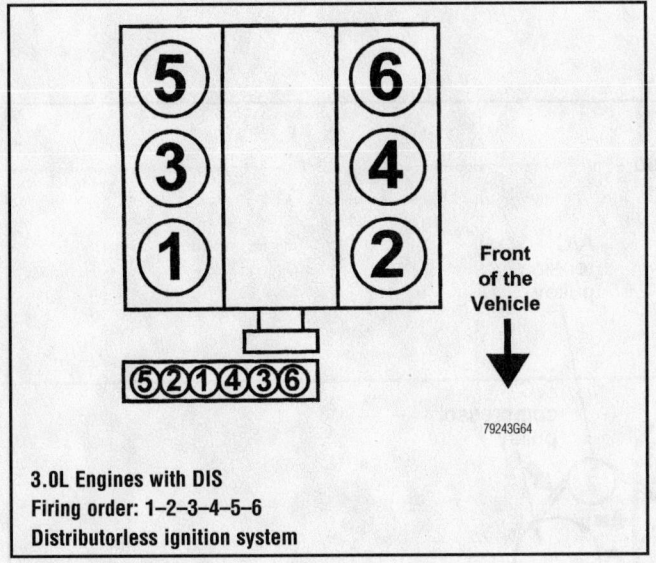

3.0L Engines with DIS
Firing order: 1–2–3–4–5–6
Distributorless ignition system

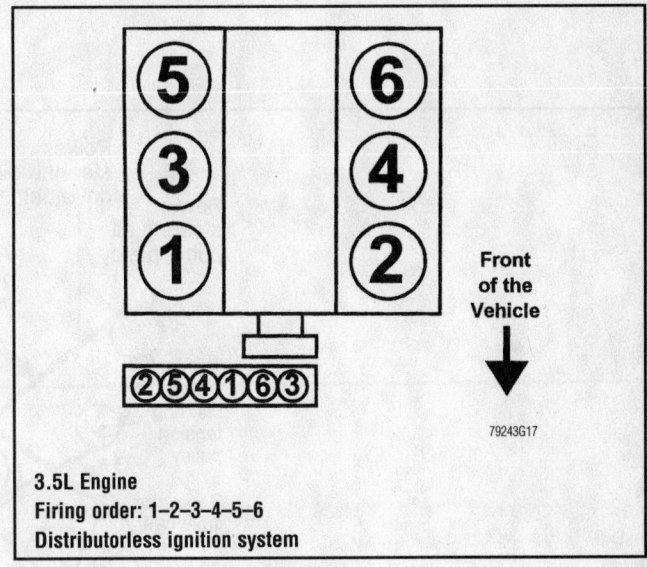

3.5L Engine
Firing order: 1–2–3–4–5–6
Distributorless ignition system

Timing belt service is covered in Section 3 of this manual

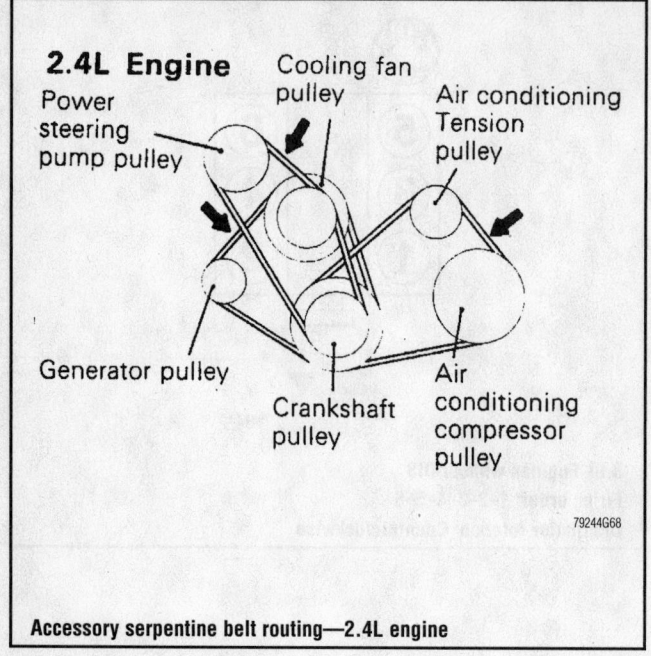

2.4L Engine

- Power steering pump pulley
- Cooling fan pulley
- Air conditioning Tension pulley
- Generator pulley
- Crankshaft pulley
- Air conditioning compressor pulley

79244G68

Accessory serpentine belt routing—2.4L engine

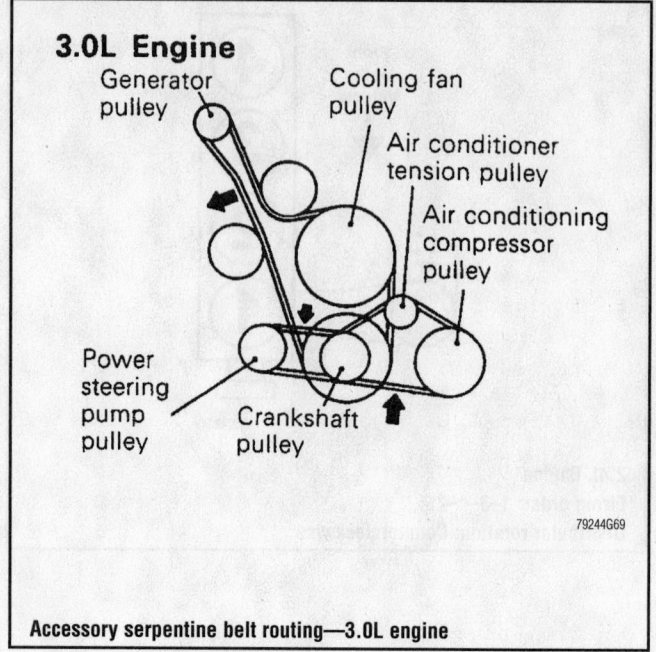

3.0L Engine

- Generator pulley
- Cooling fan pulley
- Air conditioner tension pulley
- Air conditioning compressor pulley
- Power steering pump pulley
- Crankshaft pulley

79244G69

Accessory serpentine belt routing—3.0L engine

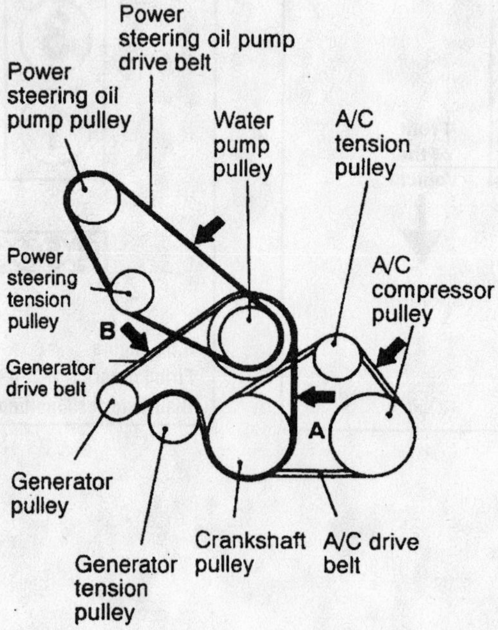

- Power steering oil pump drive belt
- Power steering oil pump pulley
- Water pump pulley
- A/C tension pulley
- Power steering tension pulley
- A/C compressor pulley
- B
- Generator drive belt
- A
- Generator pulley
- Crankshaft pulley
- A/C drive belt
- Generator tension pulley

79244G70

Accessory serpentine belt routing—3.5L engine

CAPACITIES

Year	Model	Engine Displacement Liters (cc)	Engine VIN	Engine Oil with Filter (qts.)	Transmission (pts.) 5-Spd	Transmission (pts.) Auto.	Transfer Case (pts.)	Drive Axle Front (pts.)	Drive Axle Rear (pts.)	Fuel Tank (gal.)	Cooling System (qts.)
1999	Montero	3.5 (3497)	M	5.5	—	17.8	5.2	2.4	6.6	24.3	10.0
	Montero Sport	2.4 (2351)	G	4.5	4.8	—	—	—	3.2	19.5	8.5
		3.0 (2972)	P	5.2	4.8	20.8	4.8	2.4	5.5	19.5	9.5
		3.5 (3497)	M	5.2	—	20.8	4.8	2.4	5.5	19.5	9.5
2000	Montero	3.5 (3497)	M	5.1	—	19.6	6.0	2.6	3.4	23.8	9.5
	Montero Sport	3.5 (3497)	M	4.7	—	20.6	5.2	1.9	5.6	19.5	①
		3.0 (2972)	P	4.6	—	20.6	5.2	1.9	6.8	19.5	①
2001	Montero	3.5 (3497)	M	5.1	—	19.6	6.0	2.6	3.4	23.8	9.5
	Montero Sport	3.5 (3497)	M	4.7	—	20.6	5.2	1.9	5.6	19.5	①
		3.0 (2972)	P	4.6	—	20.6	5.2	1.9	6.8	19.5	①
2002	Montero	3.5 (3497)	R	5.1	—	19.6	6.0	2.6	3.4	23.8	9.5
	Montero Sport	3.5 (3497)	M	4.7	—	20.6	5.2	1.9	5.6	19.5	①
		3.0 (2972)	P	4.6	—	20.6	5.2	1.9	6.8	19.5	①

① without rear heater: 9.5
with rear heater: 10.6

93591CW3

VALVE SPECIFICATIONS

Year	Engine Displacement Liters (cc)	Engine VIN	Seat Angle (deg.)	Face Angle (deg.)	Spring Test Pressure (lbs. @ in.)	Spring Installed Height (in.)	Stem-to-Guide Clearance (in.) Intake	Stem-to-Guide Clearance (in.) Exhaust	Stem Diameter (in.) Intake	Stem Diameter (in.) Exhaust
1999	2.4 (2351)	G	44-44.5	45-45.5	60@1.74	1.740	0.0008-0.0020	0.0012-0.0027	0.240	0.230
	3.5 (3497)	M	44-44.5	45-45.5	60@1.74	1.740	0.0008-0.0020	0.0016-0.0028	0.236	0.236
	3.0 (2972)	P	44-44.5	45-45.5	60@1.74	1.740	0.0008-0.0020	0.0016-0.0028	0.240	0.240
2000	3.5 (3497)	M	44-44.5	45-45.5	60@1.74	1.740	0.0008-0.0020	0.0016-0.0028	0.236	0.236
	3.0 (2972)	P	44-44.5	45-45.5	60@1.74	1.740	0.0008-0.0020	0.0016-0.0028	0.240	0.240
2001	3.5 (3497)	M	44-44.5	45-45.5	60@1.74	1.740	0.0008-0.0020	0.0016-0.0028	0.236	0.236
	3.0 (2972)	P	44-44.5	45-45.5	60@1.74	1.740	0.0008-0.0020	0.0016-0.0028	0.240	0.240
2002	3.5 (3497)	R	44-44.5	45-45.5	60@1.74	1.740	0.0008-0.0020	0.0016-0.0028	0.236	0.236
	3.0 (2972)	P	44-44.5	45-45.5	60@1.74	1.740	0.0008-0.0020	0.0016-0.0028	0.240	0.240

93591CW4

Heater Core replacement is covered in Section 2 of this manual

CRANKSHAFT AND CONNECTING ROD SPECIFICATIONS

All measurements are given in inches.

Year	Engine Displacement Liters (cc)	Engine ID/VIN	Crankshaft				Connecting Rod		
			Main Brg. Journal Dia.	Main Brg. Oil Clearance	Shaft End-play	Thrust on No.	Journal Diameter	Oil Clearance	Side Clearance
1999	2.4 (2351)	G	2.2436-2.2441	0.0008-0.0040	0.0020-0.0098	3	1.7709-1.7717	0.0008-0.0040	0.0039-0.0160
	3.0 (2972)	P	2.3614-2.3622	0.0008-0.0040	0.0020-0.0120	3	2.1646-2.1654	0.0008-0.0040	0.0039-0.0160
	3.5 (3497)	M	2.3614-2.3622	0.0008-0.0040	0.0020-0.0120	3	1.9700	0.0008-0.0040	0.0039-0.0160
2000	3.0 (2972)	P	2.3614-2.3622	0.0008-0.0040	0.0020-0.0120	3	2.1646-2.1654	0.0008-0.0040	0.0039-0.0160
	3.5 (3497)	M	2.3614-2.3622	0.0008-0.0040	0.0020-0.0120	3	1.9700	0.0008-0.0040	0.0039-0.0160
2001	3.0 (2972)	P	2.3614-2.3622	0.0008-0.0040	0.0020-0.0120	3	2.1646-2.1654	0.0008-0.0040	0.0039-0.0160
	3.5 (3497)	M	2.3614-2.3622	0.0008-0.0040	0.0020-0.0120	3	1.9700	0.0008-0.0040	0.0039-0.0160
2002	3.0 (2972)	P	2.3614-2.3622	0.0008-0.0040	0.0020-0.0120	3	2.1646-2.1654	0.0008-0.0040	0.0039-0.0160
	3.5 (3497)	R	2.3614-2.3622	0.0008-0.0040	0.0020-0.0120	3	1.9700	0.0008-0.0040	0.0039-0.0160

93591CW5

PISTON AND RING SPECIFICATIONS

All measurements are given in inches.

Year	Engine Displacement Liters (cc)	Engine ID/VIN	Piston Clearance	Ring Gap			Ring Side Clearance		
				Top Compression	Bottom Compression	Oil Control	Top Compression	Bottom Compression	Oil Control
1999	2.4 (2351)	G	0.0008-0.0016	0.0098-0.0310	0.0157-0.0310	0.0039-0.0390	0.0012-0.0040	0.0012-0.0040	Snug
	3.0 (2972)	P	0.0008-0.0020	0.0118-0.0310	0.0177-0.0310	0.0079-0.0390	0.0012-0.0040	0.0008-0.0040	Snug
	3.5 (3497)	M	0.0008-0.0020	0.0118-0.0310	0.0177-0.0310	0.0079-0.0390	0.0012-0.0040	0.0008-0.0040	Snug
2000	3.0 (2972)	P	0.0008-0.0020	0.0118-0.0310	0.0177-0.0310	0.0079-0.0390	0.0012-0.0040	0.0008-0.0040	Snug
	3.5 (3497)	M	0.0008-0.0020	0.0118-0.0310	0.0177-0.0310	0.0079-0.0390	0.0012-0.0040	0.0008-0.0040	Snug
2001	3.0 (2972)	P	0.0008-0.0020	0.0118-0.0310	0.0177-0.0310	0.0079-0.0390	0.0012-0.0040	0.0008-0.0040	Snug
	3.5 (3497)	M	0.0008-0.0020	0.0118-0.0310	0.0177-0.0310	0.0079-0.0390	0.0012-0.0040	0.0008-0.0040	Snug
2002	3.0 (2972)	P	0.0008-0.0020	0.0118-0.0310	0.0177-0.0310	0.0079-0.0390	0.0012-0.0040	0.0008-0.0040	Snug
	3.5 (3497)	R	0.0008-0.0020	0.0118-0.0310	0.0177-0.0310	0.0079-0.0390	0.0012-0.0040	0.0008-0.0040	Snug

93591CW6

TORQUE SPECIFICATIONS
All readings in ft. lbs.

Year	Engine ID/VIN	Engine Displacement Liters (cc)	Cylinder Head Bolts	Main Bearing Bolts	Rod Bearing Bolts	Crankshaft Damper Bolts	Flywheel Bolts	Manifold Intake *	Manifold Exhaust	Spark Plugs	Lug Nut
1999	G	2.4 (2351)	①	②	③	87	98	14	④	18	100
	M	3.5 (3497)	80	54	38	134	54	16	22	18	100
	P	3.0 (2972)	80	69	37	134	55	16	33	18	100
2000	M	3.5 (3497)	80	54	38	134	54	16	22	18	100
	P	3.0 (2972)	80	69	37	134	55	16	33	18	100
2001	M	3.5 (3497)	80	54	38	134	54	16	22	18	100
	P	3.0 (2972)	80	69	37	134	55	16	33	18	100
2002	R	3.5 (3497)	⑤	54	⑥	134	54	16	33	18	100
	P	3.0 (2972)	⑤	69	38	134	55	16	33	18	100

* NOTE: Applies to Lower Manifold only.

① Step 1: 58 ft. lbs., then, loosen completely
　Step 2: 14 ft. lbs. plus 1/4 turn
　Step 3: Plus an additional 1/4 turn

② Step 1: 18 ft. lbs.
　Step 2: Plus 1/4 turn

③ Step 1: 14 ft. lbs.
　Step 2: Plus 1/4 turn

④ M8 fasteners: 22 ft. lbs.
　M10 fasteners: 36 ft. lbs.

⑤ Step 1: 80 ft. lbs.
　Step 2: back off to 0
　Step 3: 80 ft. lbs.

⑥ Step 1: 25 ft. lbs.
　Step 2: +90 degrees

93591CW7

WHEEL ALIGNMENT

Year	Model		Caster Range (+/-Deg.)	Caster Preferred Setting (Deg.)	Camber Range (+/-Deg.)	Camber Preferred Setting (Deg.)	Toe-in (in.)	Axis Inclination (Deg.)
1999	Montero	F	1.00	+3.00	0.50	+0.66	0.14+/-0.14	—
		R	—	—	1.00	0	0	—
	Montero Sport	F	1.00	+2.66	0.50	+4.00	0.14+/-0.14	—
		R	—	—	—	0	0	—
2000	Montero	F	1.00	+3.00	0.50	+0.66	0.14+/-0.14	—
		R	—	—	1.00	0	0	—
	Montero Sport	F	1.00	+2.66	0.50	+4.00	0.14+/-0.14	—
		R	—	—	—	0	0	—
2001	Montero	F	0.50	+3.83	0.50	0	0.10+/-0.10	—
		R	—	—	0.50	0	0.12+/-0.12	—
	Montero Sport	F	1.00	+2.66	0.50	①	0.14+/-0.14	—
		R	—	—	—	0	0	—
2002	Montero	F	1.00	+3.83	0.50	0	0.10+/-0.10	—
		R	—	—	0.50	0	0.12+/-0.12	—
	Montero Sport	F	1.00	+2.66	0.50	①	0.14+/-0.14	—
		R	—	—	1.00	+2.66	0	—

① Right: 0.42
　Left: 0.92

93591CW8

Brake service is covered in Section 4 of this manual

TIRE, WHEEL AND BALL JOINT SPECIFICATIONS

Year	Model	OEM Tires		Tire Pressures (psi)		Wheel Size	Ball Joint Inspection
		Standard	Optional	Front	Rear		
1999	Montero	P265/70HR15	None	26	26	7-JJ	U: 7-30 in. ① L: 0.010 in.
	Montero Sport	P235/75R15	P265/70R15	26	26	Std: 6-JJ Opt: 7-JJ	U: 7-30 in. ① L: 0.010 in.
2000	Montero	P265/70HR15	None	26	26	7-JJ	U: 7-30 in. ① L: 0.010 in.
	Montero Sport	P235/75R15	P265/70R15	26	26	Std: 6-JJ Opt: 7-JJ	U: 7-30 in. ① L: 0.010 in.
2001	Montero	P265/70HR16	None	29	29	7-JJ	U: 7-30 in. ① L: 0.010 in.
	Montero Sport	P235/75R15	P255/70R16	26	26	Std: 6-JJ Opt: 7-JJ	U: 7-30 in. ① L: 0.010 in.
2002	Montero	P265/70HR16	None	29	29	7-JJ	U: 7-30 in. ① L: 0.010 in.
	Montero Sport	P235/75R15	P255/70R16	26	26	Std: 6-JJ Opt: 7-JJ	U: 7-30 in. ① L: 0.010 in.

OEM: Original Equipment Manufacturer

PSI: Pounds Per Square Inch

STD: Standard

OPT: Optional

① Torque required in inch lbs. to rotate ball joint when removed from the knuckle

93591CW9

BRAKE SPECIFICATIONS
All measurements in inches unless noted

Year	Model		Brake Disc Original Thickness	Brake Disc Minimum Thickness	Brake Disc Maximum Runout	Brake Drum Diameter Original Inside Diameter	Brake Drum Diameter Max. Wear Limit	Brake Drum Diameter Maximum Machine Diameter	Minimum Lining Thickness Front	Minimum Lining Thickness Rear	Brake Caliper Bracket Bolts (ft. lbs.)	Brake Caliper Mounting Bolts (ft. lbs.)
1999	Montero	F	1.060	1.000	0.002	—	—	—	0.079	—	65	54
		R	0.710	0.646	0.003	—	—	—	—	0.040	65	32
	Montero Sport	F	0.940	0.880	0.002	—	—	—	0.079		65	55
		R	0.700	0.650	0.003	10.63	—	10.71	—	①	94	32
2000	Montero	F	1.060	1.000	0.002	—	—	—	0.079	—	65	54
		R	0.710	0.646	0.003	—	—	—	—	0.040	65	32
	Montero Sport	F	0.940	0.880	0.002	—	—	—	0.079	—	65	55
		R	0.700	0.650	0.003	10.63	—	10.71	—	①	94	32
2001	Montero	F	1.023	0.960	0.002	—	—	—	0.079	—	65	54
		R	0.866	0.803	0.003	—	—	—	—	0.079	65	32
	Montero Sport	F	0.940	0.880	0.001	—	—	—	0.079	—	65	55
		R	0.710	0.650	0.003	10.63	—	10.71	—	①	94	32
2002	Montero	F	1.023	0.960	0.002	—	—	—	0.079	—	83	66
		R	0.866	0.803	0.003	—	—	—	—	0.079	65	33
	Montero Sport	F	0.940	0.880	0.001	—	—	—	0.079	—	65	55
		R	0.710	0.650	0.003	10.63	—	10.71	—	①	65	32

① Disc pad: 0.79
Brake shoe: 0.04

93591CW0

For complete Engine Mechanical specifications, see Section 1 of this manual

SCHEDULED MAINTENANCE INTERVALS
Mitsubishi—Montero & Montero Sport

TO BE SERVICED	TYPE OF SERVICE	VEHICLE MILEAGE INTERVAL (x1000)												
		7.5	15	22.5	30	37.5	45	52.5	60	67.5	75	82.5	90	97.5
Engine oil & filter	R	✓	✓	✓	✓	✓	✓	✓	✓	✓	✓	✓	✓	✓
Automatic transmission & transfer oil	S/I		✓		✓		✓		✓		✓		✓	
Brake hoses	S/I		✓		✓		✓		✓		✓		✓	
Disc brake pads & rotors	S/I		✓		✓		✓		✓		✓		✓	
Drive shaft boots	S/I		✓		✓		✓		✓		✓		✓	
Air cleaner filter	R				✓				✓				✓	
Automatic transmission & transfer oil	R				✓				✓				✓	
Engine coolant	R				✓				✓				✓	
Ball joints & steering linkage seals	S/I				✓				✓				✓	
Drive belt(s)	S/I				✓				✓				✓	
Drum brake linings & wheel cylinders	S/I				✓				✓				✓	
Exhaust system	S/I				✓				✓				✓	
Front & rear axle	S/I				✓				✓				✓	
Fuel hoses	S/I				✓				✓				✓	
Manual transmission & transfer oil	S/I				✓				✓				✓	
Propeller shaft joint	S/I				✓				✓				✓	
Ignition cables	R								✓					
Timing belt	R								✓					
Distributor cap & rotor	S/I								✓					
EVAP system (except canister)	S/I								✓					
EGR valve ①	S/I													
EVAP canister ①	S/I													
PCV system ②	S/I													
Spark plugs ③	R													

R: Replace   S/I: Service or Inspect

① Replace at 100,000 miles.

② PCV system (except EVAP canister): service or inspect at 100,000 miles.

③ Iron tips: 30,000 miles
   Platinum tips: 60,000 miles
   Irridium tips: 100,000 miles

FREQUENT OPERATION MAINTENANCE (SEVERE SERVICE)

If a vehicle is operated under any of the following conditions it is considered severe service:

- Extremely dusty areas.

- 50% or more of the vehicle operation is in 32°C (90°F) or higher temperatures, or constant operation in temperatures below 0°C (32°F).

- Prolonged idling (vehicle operation in stop and go traffic).

- Frequent short running periods (engine does not warm to normal operating temperatures).

- Police, taxi, delivery usage or trailer towing usage.

Oil & oil filter: replace every 3000 miles.

Front disc brake pads (dusty or salty conditions): service or inspect every 6000 miles.

Front disc brake pads: service or inspect every 7500 miles.

Air cleaner filter: service or inspect every 15,000 miles.

Rear drum brake linings & rear wheel cylinders: service or inspect every 15,000 miles.

Spark plugs (iron tip): replace every 15,000 miles.

PCV system: service or inspect every 60,000 miles.

93591CHZ

SCHEDULED MAINTENANCE INTERVALS
MITSUBISHI
MONTERO
MONTERO SPORT

The following should be used as a guide when determining the amount of work required for a particular service. In estimating how long a particular Scheduled Maintenance Service should take, please observe the following:

● Labor Time is time based on field research and data supplied by the vehicle manufacturer.
● Labor time operations are given in hours and tenths of an hour.
● All labor operations are to be used as a guide.

Mechanic Skill Level Codes:
(A) PRECISION: Highly skilled with multiple certification.
(B) GENERAL: Normally skilled with certification.
(C) MAINTENANCE: Semi-skilled working on certification.

	LABOR TIME
7500 Mile Service (C)	
All Models	.4
15000 Mile Service (C)	
All Models	.7
22500 Mile Service (C)	
All Models	.4
30000 Mile Service (B)	
All Models	2.2
Renew spark plugs add	
w/4WD add	
w/AT add	

	LABOR TIME
37500 Mile Service (C)	
All Models	.4
45000 Mile Service (C)	
All Models	.7
52500 Mile Service (C)	
All Models	.4
60000 Mile Service (B)	
All Models	5.5
Renew spark plugs add	
w/4WD add	
w/AT add	
67500 Mile Service (C)	
All Models	.4

	LABOR TIME
75000 Mile Service (C)	
All Models	.7
82500 Mile Service (C)	
All Models	.4
90000 Mile Service (B)	
All Models	2.3
Renew spark plugs add	
w/4WD add	
w/AT add	
97500 Mile Service (C)	
All Models	.4

93551CH1A"

For Accessory Drive Belt illustrations, see Section 1 of this manual

NISSAN
Pathfinder • Xterra

ENGINE AND VEHICLE IDENTIFICATION

		Engine					Model Year	
Code ①	Liters (cc)	Cu. In.	Cyl.	Fuel Sys.	Engine Type	Eng. Mfg.	Code ②	Year
VG33E	3.3 (3277)	199.8	6	MFI	SOHC	Nissan	X	1999
VG33ER	3.3 (3277)	199.8	6	MFI	SOHC	Nissan	Y	2000
KA24DE	2.4 (2389)	146	4	MFI	DOHC	Nissan	1	2001
VQ35DE	3.5 (3498)	213	6	MFI	DOHC	Nissan	2	2002
							3	2003

MFI: Multi-port Fuel Injection

SOHC: Single Overhead Camshaft

DOHC: Double Overhead Camshafts

① Located on the timing belt cover

② 10th digit of the Vehicle Identification Number (VIN)

93591CH2

GENERAL ENGINE SPECIFICATIONS

Year	Model	Engine Displacement Liters (cc)	Engine ID/VIN	Fuel System Type	Net Horsepower @ rpm	Net Torque @ rpm (ft. lbs.)	Bore x Stroke (in.)	Compression Ratio	Oil Pressure @ rpm
1999	Pathfinder	3.3 (3277)	VG33E	MFI	170@4800	200@2800	3.60X3.27	8.9:1	60-65@2000
2000	Xterra	3.3 (3277)	VG33E	MFI	170@4800	200@2800	3.60X3.27	8.9:1	60-65@2000
	Pathfinder	3.3 (3277)	VG33E	MFI	170@4800	200@2800	3.60X3.27	8.9:1	60-65@2000
2001	Xterra	2.4 (2398)	KA24DE	MFI	143@5200	154@4000	3.50X3.78	9.2:1	60-70@3000
	Xterra	3.3 (3277)	VG33E	MFI	170@4800	200@2800	3.60X3.27	8.9:1	60-65@2000
	Pathfinder	3.3 (3277)	VG33E	MFI	170@4800	200@2800	3.60X3.27	8.9:1	60-65@2000
2002	Xterra	2.4 (2398)	KA24DE	MFI	143@5200	154@4000	3.50X3.78	9.2:1	60-70@3000
	Xterra	3.3 (3277)	VG33E	MFI	170@4800	200@2800	3.602X3.270	8.9:1	60-65@2000
	Xterra	3.3 (3277)	VG33ER	MFI	210@4800	231@2800	3.602X3.270	8.3:1	60-65@2000
	Pathfinder	3.5 (3498)	VQ35DE	MFI	240@6000	265@3200	3.760X3.205	10.0:1	43@2000

MFI: Multi-port Fuel Injection

93591CH3

ENGINE TUNE-UP SPECIFICATIONS

Year	Engine Displacement Liters (cc)	Engine ID/VIN	Spark Plug Gap (in.)	Ignition Timing (deg.) MT	Ignition Timing (deg.) AT	Fuel Pump (psi) ①	Idle Speed (rpm) MT	Idle Speed (rpm) AT ②	Valve Clearance (in.) In.	Valve Clearance (in.) Ex.
1999	3.3 (3277)	VG33E	0.039-0.043	13-17B	13-17B	34	700-800	700-800	HYD	HYD
2000	3.3 (3277)	VG33E	0.039-0.043	13-17B	13-17B	34	700-800	700-800	HYD	HYD
2001	2.4 (2398)	KA24DE	0.043	18-22B	—	34	750-850	—	0.012-0.015	0.013-0.016
	3.3 (3277)	VG33E	0.039-0.043	13-17B	13-17B	34	700-800	700-800	HYD	HYD
2002	2.4 (2398)	KA24DE	0.043	18-22B	—	34	750-850	—	0.012-0.015	0.013-0.016
	3.3 (3277)	VG33ER	0.043	10B	10B	35	700-800	700-800	HYD	HYD
	3.3 (3277)	VG33E	0.039-0.043	13-17B	13-17B	34	700-800	700-800	HYD	HYD
	3.5 (3498)	VQ35DE	0.044	15B	15B	35	700-800	700-800	HYD	HYD

NOTE: The Vehicle Emission Control Information label often reflects specification changes made during production. The label figures must be used if they differ from those in this chart.

B: Before top dead center

HYD: Hydraulic

① System pressure at idle with vacuum hose connected
　Should increase to 43 psi when disconnected

② Automatic transmission in Neutral

93591CH4

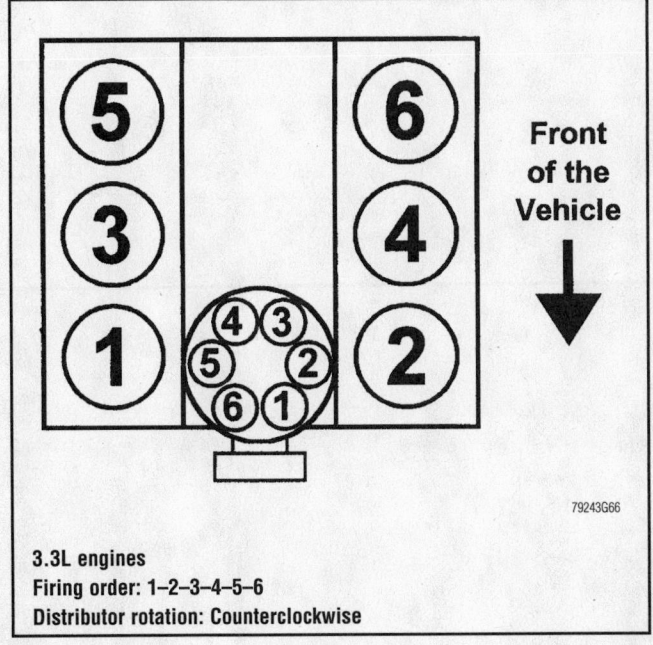

Front of the Vehicle

3.3L engines
Firing order: 1–2–3–4–5–6
Distributor rotation: Counterclockwise

79243G66

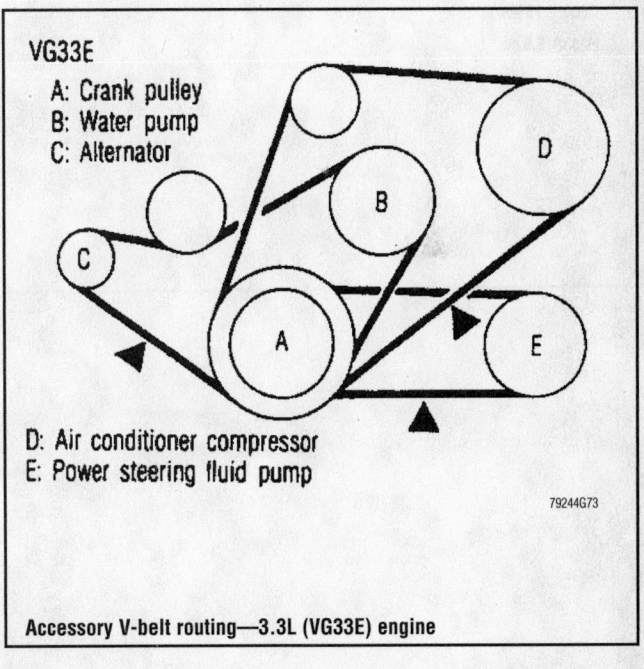

VG33E
A: Crank pulley
B: Water pump
C: Alternator

D: Air conditioner compressor
E: Power steering fluid pump

Accessory V-belt routing—3.3L (VG33E) engine

79244G73

For Tire, Wheel and Ball Joint specifications, see Section 1 of this manual

CAPACITIES

Year	Model	Engine Displacement Liters (cc)	Engine ID/VIN	Engine Oil with Filter (qts.)	Transmission (pts.)		Transfer Case (pts.)	Drive Axle		Fuel Tank (gal.)	Cooling System (qts.)
					5-Spd	Auto.		Front (pts.)	Rear (pts.)		
1999	Pathfinder	3.3 (3277)	VG33E	3.8	①	②	4.8	4.4	4.9	21.1	11.25
2000	Xterra	3.3 (3277)	VG33E	③	①	16.8	4.8	2.8	④	19.4	⑤
	Pathfinder	3.3 (3277)	VG33E	3.8	⑥	②	4.8	4.4	4.9	21.1	11.25
2001	Xterra	2.4 (2398)	KA24DE	3.75	8.5	—	—	—	3.1	19.4	7.75
	Xterra	3.3 (3277)	VG33E	3.8	⑥	②	4.8	4.4	4.9	19.4	11.25
	Pathfinder	3.3 (3277)	VG33E	3.8	⑥	②	4.8	4.4	4.9	21.1	11.25
2002	Xterra	2.4 (2398)	KA24DE	3.75	8.5	—	—	—	3.1	19.4	7.75
	Xterra	3.3 (3277)	VG33E	3.5	⑥	②	4.8	3.1	5.9	19.4	11.6
	Xterra	3.3 (3277)	VG33ER	3.5	⑥	②	4.8	3.1	5.9	19.4	11.6
	Pathfinder	3.5 (3498)	VQ35DE	5.25	10.34	18.0	⑦	3.9	5.9	21.1	9.75

NOTE: All capacities are approximate. Add fluid gradually and check to be sure a proper fluid level is obtained.

① 2WD: 4.25 pts.
   4WD: 10.4 pts.
② 2WD: 17.5 pts.
   4WD: 18 pts.
③ 2WD: 3.75 qts.
   4WD: 4.125 qts.
④ H190A: 3.1 pts.
   C200: 2.75 pts.
   H233B: 5.9 pts.

⑤ 2WD: 8.6 qts.; 4WD: 9.5 qts.
⑥ 2WD: 5.125 pts.
   4WD: 10.75 pts.
⑦ Part time: 2.375; full time: 2.625 pts.

93591CH5

VALVE SPECIFICATIONS

Year	Engine Displacement Liters (cc)	Engine ID/VIN	Seat Angle (deg.)	Face Angle (deg.)	Spring Test Pressure (lbs. @ in.)	Spring Installed Height (in.)	Stem-to-Guide Clearance (in.)		Stem Diameter (in.)	
							Intake	Exhaust	Intake	Exhaust
1999	3.3 (3277)	VG33E	45	45.25-46.75	①	NA	0.0008-0.0021	0.0016-0.0029	0.2742-0.2748	0.3135-0.3138
2000	3.3 (3277)	VG33E	45	45.25-46.75	①	NA	0.0008-0.0021	0.0016-0.0029	0.2742-0.2748	0.3135-0.3138
2001	2.4 (2389)	KA24DE	45.15-45.45	NA	93.9 @ 1.148	NA	0.0008-0.0021	0.0016-0.0029	0.2742-0.2748	0.2734-0.2740
	3.3 (3277)	VG33E	45	45.25-46.75	①	NA	0.0008-0.0021	0.0016-0.0029	0.2742-0.2748	0.3135-0.3138
2002	2.4 (2389)	KA24DE	45.15-45.45	NA	93.9 @ 1.148	NA	0.0008-0.0021	0.0016-0.0029	0.2742-0.2748	0.2734-0.2740
	3.3 (3277)	VG33E	45	45.25-46.75	①	NA	0.0008-0.0021	0.0012-0.0019	0.2742-0.2748	0.3135-0.3138
	3.3 (3277)	VG33ER	45	45.25-46.75	①	NA	0.0008-0.0021	0.0012-0.0019	0.2742-0.2748	0.3135-0.3138
	3.5 (3498)	VQ35DE	45.15-45.45	45	45.4 @ 1.457	1.457	0.0008-0.0021	0.0016-0.0029	0.2348-0.2354	0.2341-0.2346

NA: Not Available

① Inner: 57.3 @ 0.984
  Outer: 117.7 @ 1.181

93591CH6

For Wheel Alignment specifications, see Section 1 of this manual

CRANKSHAFT AND CONNECTING ROD SPECIFICATIONS

All measurements are given in inches.

Year	Engine Displacement Liters (cc)	Engine ID/VIN	Crankshaft				Connecting Rod		
			Main Brg. Journal Dia.	Main Brg. Oil Clearance	Shaft End-play	Thrust on No.	Journal Diameter	Oil Clearance	Side Clearance
1999	3.3 (3277)	VG33E	2.4790-2.4793	0.0011-0.0022	0.0020-0.0067	4	1.9967-1.9675	0.0006-0.0021	0.0079-0.0138
2000	3.3 (3277)	VG33E	2.4790-2.4793	0.0011-0.0022	0.0020-0.0067	4	1.9967-1.9675	0.0006-0.0021	0.0079-0.0138
2001	2.4 (2389)	KA24DE	①	0.0008-0.0019	0.0020-0.0071	NA	2.0866-2.0871	0.0004-0.0014	0.0080-0.0160
	3.3 (3277)	VG33E	2.4790-2.4793	0.0011-0.0022	0.0020-0.0067	4	1.9967-1.9675	0.0006-0.0021	0.0079-0.0138
2002	2.4 (2389)	KA24DE	①	0.0008-0.0019	0.0020-0.0071	NA	2.0866-2.0871	0.0004-0.0014	0.0080-0.0160
	3.3 (3277)	VG33E	②	③	0.0020-0.0067	4	1.9667-1.9675	0.0009-0.0025	0.0079-0.0138
	3.3 (3277)	VG33ER	②	③	0.0020-0.0067	4	1.9667-1.9675	0.0009-0.0025	0.0079-0.0138
	3.5 (3498)	VQ35DE	④	0.0014-0.0018	0.0118	4	⑤	0.0013-0.0023	0.0079-0.0138

NA - Not Available

① Grade No. 1: 2.3609-2.3612

   Grade No. 1: 2.3606-2.3609

   Grade No. 2: 1.9668-1.9670

② Except No. 1

   Grade 0: 2.4790-2.4793

   Grade 1: 2.4787-2,4790

   Grade 2: 2.4784-2.4787

   No. 1

   Grade 3: 2.4683-2.4793

   Grade 4: 2.4789-2.4791

   Grade 5: 2.4786-2.4789

   Grade 6: 2.4784-2.4786

③ No. 1: 0.0012-0.0019

   Nos. 2, 3, 4: 0.0015-0.0026

④ There are 24 different grades, ranging from A (2.3612) to 7 (2.3603)

⑤ Grade 0: 2.0460-2.0462

   Grade 1: 2.0457-2.0460

   Grade 2: 2.0445-2.0457

93591CH7

PISTON AND RING SPECIFICATIONS
All measurements are given in inches.

Year	Engine Displacement Liters (cc)	Engine ID/VIN	Piston Clearance	Ring Gap			Ring Side Clearance		
				Top Compression	Bottom Compression	Oil Control	Top Compression	Bottom Compression	Oil Control
1999	3.3 (3277)	VG33E	①	0.0083-0.0157	0.0197-0.0272	0.0079-0.0272	0.0009-0.0030	0.0012-0.0028	0.0006-0.0073
2000	3.3 (3277)	VG33E	①	0.0083-0.0157	0.0197-0.0272	0.0079-0.0272	0.0009-0.0030	0.0012-0.0028	0.0006-0.0073
2001	2.4 (2389)	KA24DE	0.0008-0.0016	0.0110-0.0205	0.0177-0.0272	0.0079-0.0272	0.0016-0.0031	0.0012-0.0028	0.0026-0.0053
	3.3 (3277)	VG33E	①	0.0083-0.0157	0.0197-0.0272	0.0079-0.0272	0.0009-0.0030	0.0012-0.0028	0.0006-0.0073
2002	2.4 (2389)	KA24DE	0.0008-0.0016	0.0110-0.0205	0.0177-0.0272	0.0079-0.0272	0.0016-0.0031	0.0012-0.0028	0.0026-0.0053
	3.3 (3277)	VG33E	①	0.0083-0.0122	0.0197-0.0236	0.0079-0.0236	0.0016-0.0031	0.0012-0.0028	0.0006-0.0073
	3.3 (3277)	VG33ER	②	0.0083-0.0122	0.0197-0.0236	0.0079-0.0236	0.0016-0.0031	0.0012-0.0028	0.0006-0.0073
	3.5 (3498)	VQ35DE	0.0004-0.0012	0.0091-0.0130	0.0130-0.0189	0.0079-0.0236	0.0016-0.0031	0.0012-0.0028	0.0006-0.0020

① Cylinders 1, 2, 6: 0.0010 - 0.0018 in.
Cylinders 3 and 4: 0.0006 - 0.0010 in.
Cylinder 5: 0.0012-0.0016 in.

② Cylinders 3, 4: 0.0006-0.0010 in.
Cylinders 1, 2, 5, 6: 0.0010-0.0018 in.

93591CH8

For Maintenance Interval recommendations, see Section 1 of this manual

TORQUE SPECIFICATIONS
All readings in ft. lbs.

Year	Engine Displacement Liters (cc)	Engine ID/VIN	Cylinder Head Bolts	Main Bearing Bolts	Rod Bearing Bolts	Crankshaft Damper Bolts	Flywheel Bolts	Manifold		Spark Plugs	Lug Nuts
								Intake	Exhaust		
1999	3.3 (3277)	VG33E	①	67-74	②	141-156	61-69	①	21-25	14-22	87-108
2000	3.3 (3277)	VG33E	①	67-74	②	141-156	61-69	①	21-25	14-22	87-108
2001	2.4 (2389)	KA24DE	③	34-41	④	105-112	105-112	12-14	28-35	14-22	87-108
	3.3 (3277)	VG33E	①	67-74	②	141-156	61-69	①	21-25	14-22	87-108
2002	2.4 (2389)	KA24DE	③	34-41	④	105-112	105-112	12-14	28-35	14-22	87-108
	3.3 (3277)	VG33E	①	67-74	②	141-156	61-69	①	21-25	14-22	87-108
	3.5 (3498)	VQ35DE	⑤	⑥	⑦	⑧	61-69	⑨	21-24	14-22	87-108
	3.3 (3277)	VG33ER	①	67-74	⑩	141-156	61-69	①	21-25	14-22	87-108

① The cylinder heads and the lower intake manifold are installed together

Step 1: Tighten the cylinder head bolts to 22 ft. lbs.

Step 2: Tighten the cylinder head bolts to 43 ft. lbs.

Step 3: Loosen the cylinder head bolts completely

Step 4: Tighten the cylinder head bolts to 84 inch lbs.

Step 5: Tighten the intake manifold fasteners to 35 inch lbs.

Step 6: Tighten the intake manifold fasteners to 13 ft. lbs.

Step 7: Tighten the intake manifold fasteners to 12-14 ft. lbs.

Step 8: Loosen all intake manifold fasteners completely

Step 9: Tighten the cylinder head bolts to 22 ft. lbs.

Step 10: Tighten the cylinder head bolts 60-65 degrees

Step 11: Tighten the cylinder head sub-bolts to 80-105 inch lbs.

Step 12: Tighten the intake manifold fasteners to 35 inch lbs.

Step 13: Tighten the intake manifold fasteners to 78 inch lbs.

Step 14: Tighten the intake manifold fasteners to 70-84 inch lbs.

② 10-12 ft. lbs. plus 60-65 degrees or 28-33 ft. lbs.

③ Step 1: Tighten the cylinder head bolts to 22 ft. lbs.

Step 2: Tighten the cylinder head bolts to 59 ft. lbs.

Step 3: Loosen the cylinder head bolts completely

Step 4: Tighten the cylinder head bolts to 18-25 ft. lbs.

Step 5: Tighten the cylinder head bolts an additional 86-91 degrees

④ Step 1: Tighten the connecting rod bearing bolts to 10-12 ft. lbs.

Step 2: Tighten the connecting rod bearing bolts an additional 60 degrees

⑤ Step 1: 72 ft. lbs.

Step 2: Loosen all bolts completely

Step 3: 25-33 ft. lbs.

Step 4: +90 degrees

Step 5: +90 degrees

⑥ Step 1: 24-28 ft. lbs.

Step 2: +90 degrees

⑦ Step 1: 15 ft. lbs.

Step 2: +90 degrees

⑧ 29-36 ft. lbs. +60-66 degrees

⑨ Step 1: 44-86 inch lbs.

Step 2: 20-23 ft. lbs.

⑩ 10-12 ft. lbs. +60-65 degrees

93591CH9

WHEEL ALIGNMENT

Year	Model	Caster Range (+/-Deg.)	Caster Preferred Setting (Deg.)	Camber Range (+/-Deg.)	Camber Preferred Setting (Deg.)	Toe-in (in.)	Axis Inclination (Deg.)
1999	Pathfinder	0.75	+3.00	0.75	+0.17	0.08+/-0.04	—
2000	Pathfinder	0.75	+3.00	0.75	+0.17	0.08+/-0.04	—
	Xterra	0.75	+0.60	0.50	+0.42	0.12+/-0.04	—
2001	Pathfinder	0.75	+3.00	0.75	+0.17	0.08+/-0.04	—
	Xterra	0.75	+0.60	0.50	+0.42	0.12+/-0.04	—
2002	Pathfinder	0.75	+3.00	0.75	+0.17	0.08+/-0.04	—
	Xterra 2wd	0.50	+2.57	0.50	+0.33	0.16+/-0.04	—
	Xterra 4wd	0.50	+2.10	0.50	+0.60	0.16+/-0.04	—

93591CH0

TIRE, WHEEL AND BALL JOINT SPECIFICATIONS

Year	Model	OEM Tires Standard	OEM Tires Optional	Tire Pressures (psi) Front	Tire Pressures (psi) Rear	Wheel Size	Ball Joint Inspection
1999	Pathfinder	P235/70R15	P265/70R15	30	30	Std: 6.5-JJ Opt: 7-JJ	U: 0.020 in. L: ①
2000	Xterra	P235/70R15	P265/70R15	26	26	7JJ	U: 0.020 in. L: ①
	Pathfinder	P235/70R15	P265/70R15	30	30	Std: 6.5-JJ Opt: 7-JJ	U: 0.020 in. L: ①
2001	Xterra	P235/70R15	P265/70R15	26	26	7JJ	U: 0.020 in. L: ①
	Pathfinder	P235/70R15	P265/70R15	30	30	Std: 6.5-JJ Opt: 7-JJ	U: 0.020 in. L: ①
2002	Xterra SE, SE S/C, and XE S/C	P265/70R16	None	②	②	7JJ	③
	Xterra XE, XE V6	P265/70R15	None	②	②	7JJ	③
	Pathfinder LE	P245/65SR17	none	②	②	8J	④
	Pathfinder SE	P255/65SR16	None	②	②	7JJ	④

OEM: Original Equipment Manufacturer

PSI: Pounds Per Square Inch

STD: Standard

OPT: Optional

L: Lower

U: Upper

① Replace if any measurable movement is found.

② See placard on vehicle

③ Axial play
Upper: 0
Lower: 0.008 in.

④ Turning torque: 4.3-43 inch lbs.

93591CI1

For Tune-up, Capacities and Firing orders, see Section 1 of this manual

BRAKE SPECIFICATIONS
All measurements in inches unless noted

Year	Model	Brake Disc			Brake Drum Diameter			Minimum Lining Thickness		Brake Caliper	
		Original Thickness	Minimum Thickness	Maximum Runout	Original Inside Diameter	Max. Wear Limit	Maximum Machine Diameter	Front	Rear	Bracket Bolts (ft. lbs.)	Mounting Bolts (ft. lbs.)
1999	Pathfinder	1.100	1.024	0.004	11.60	NA	11.67	0.079	0.059	53-72	16-23
2000	Xterra	①	②	0.003	③	NA	④	0.079	0.059	53-72	24-31
	Pathfinder	1.100	1.024	0.004	11.60	NA	11.67	0.079	0.059	53-72	16-23
2001	Xterra	①	②	0.003	③	NA	④	0.079	0.059	53-72	24-31
	Pathfinder	1.100	1.024	0.004	11.60	NA	11.67	0.079	0.059	53-72	16-23
2002	Pathfinder	1.100	1.024	0.003	11.61	NA	11.67	0.079	0.059	⑤	⑤
	Xterra	1.100	1.024	0.003	11.61	NA	11.67	0.079	0.059	⑥	⑥

NA: Not Available

① 2WD: 0.870
   4WD: 1.020

② 2WD: 0.787
   4WD: 0.945

③ 2WD: 10.20
   4WD: 11.60

④ 2WD: 10.30
   4WD: 11.67

⑤ Torque member mounting bolt: 127-134
   Main pin bolt: 24-31

⑥ Torque member mounting bolt: 101-130
   Main pin bolt: 17-22

93591CI2

SCHEDULED MAINTENANCE INTERVALS
1999-00 Nissan—Xterra & Pathfinder

TO BE SERVICED	TYPE OF SERVICE	VEHICLE MILEAGE INTERVAL (x1000)												
		7.5	15	22.5	30	37.5	45	52.5	60	67.5	75	82.5	90	97.5
Engine oil & filter	R	✓	✓	✓	✓	✓	✓	✓	✓	✓	✓	✓	✓	✓
Brake lines & cables	S/I		✓		✓		✓		✓		✓		✓	
Brake pads, discs, drums & linings	S/I		✓		✓		✓		✓		✓		✓	
Driveshaft boots & propeller shaft	S/I				✓				✓				✓	
Front wheel bearings (4x2)	S/I				✓				✓				✓	
Automatic & manual transmission, transfer & differential gear oil ①	S/I		✓		✓		✓		✓		✓		✓	
Front wheel bearings (4x4)	S/I				✓				✓				✓	
Air cleaner filter	R				✓				✓				✓	
Engine coolant	R				✓				✓				✓	
Spark plugs	R				✓				✓				✓	
Drive belt(s)	S/I				✓				✓				✓	
Exhaust system	S/I				✓				✓				✓	
Fuel lines	S/I				✓				✓				✓	
Steering gear (box) & linkage, axle & suspension parts	S/I				✓				✓				✓	
Vapor lines	S/I				✓				✓				✓	
Timing belt ②	R													

R: Replace   S/I: Service or Inspect

① Differential (w/limited-slip differential) oil: replace oil every 30,000 miles.

② Timing belt: replace at 105,000 miles.

FREQUENT OPERATION MAINTENANCE (SEVERE SERVICE)

If a vehicle is operated under any of the following conditions it is considered severe service:

- Extremely dusty areas.

- 50% or more of the vehicle operation is in 32°C (90°F) or higher temperatures, or constant operation in temperatures below 0°C (32°F).

- Prolonged idling (vehicle operation in stop and go traffic).

- Frequent short running periods (engine does not warm to normal operating temperatures).

- Police, taxi, delivery usage or trailer towing usage.

Oil & oil filter: replace every 3750 miles.

Brake pads, discs, drums & linings: service or inspect every 7500 miles.

Driveshaft boots & propeller shaft: service or inspect every 7500 miles.

Exhaust system: service or inspect every 7500 miles.

Steering gear (box) & linkage, (steering damper-4x4), axle & suspension parts: service or inspect every 7500 miles.

Steering linkage ball joints & front suspension ball joints: service or inspect every 7500 miles.

93591CI3

For complete service labor times, order Nichols' Chilton Labor Guide

SCHEDULED MAINTENANCE INTERVALS
2001-02 Nissan—Xterra & Pathfinder

TO BE SERVICED	TYPE OF SERVICE	VEHICLE MILEAGE INTERVAL (x1000)															
		3.75	7.5	11.3	15	18.8	22.5	26.3	30	33.8	37.5	41.3	45	48.8	52.5	56.3	60
Engine oil & filter	R	✓	✓	✓	✓	✓	✓	✓	✓	✓	✓	✓	✓	✓	✓	✓	✓
Brake lines & cables	S/I				✓				✓				✓				✓
Brake pads, discs, drums & linings	I		✓		✓		✓		✓		✓		✓		✓		✓
Driveshaft boots & propeller shaft	L/I		✓		✓		✓		✓		✓		✓		✓		✓
Front wheel bearings (4x2)	I								✓								✓
Automatic & manual transmission, transfer & differential gear oil ①	I					✓			✓				✓				✓
LSD gear oil	R								✓								✓
Front wheel bearing grease (4x4)	R								✓								✓
Timing belt ②	R																
Air cleaner filter	R								✓								✓
Engine coolant ③	R																✓
Spark plugs	R	platinum tipped plugs every 105,000 miles															
Drive belt(s)	S/I								✓								✓
Cabin air filter	I/R		I		R		I		R		I		R		I		R
Exhaust system	I		✓		✓		✓		✓		✓		✓		✓		✓
Fuel lines	S/I								✓								✓
Steering gear (box) & linkage, axle & suspension parts	I		✓		✓		✓		✓		✓		✓		✓		✓
Vapor lines	S/I								✓								✓

R: Replace   S/I: Service or Inspect   L: Lubricate

① Differential (w/limited-slip differential) oil: replace oil every 30,000 miles.

② Timing belt: replace at 105,000 miles.

③ After 60,000, replace every 30,000

FREQUENT OPERATION MAINTENANCE (SEVERE SERVICE)

If a vehicle is operated under any of the following conditions it is considered severe service:

- Extremely dusty areas.

- 50% or more of the vehicle operation is in 32°C (90°F) or higher temperatures, or constant operation in temperatures below 0°C (32°F).

- Prolonged idling (vehicle operation in stop and go traffic).

- Frequent short running periods (engine does not warm to normal operating temperatures).

- Police, taxi, delivery usage or trailer towing usage.

Oil & oil filter: replace every 3750 miles.

Brake pads, discs, drums & linings: service or inspect every 7500 miles.

Driveshaft boots & propeller shaft: service or inspect every 7500 miles.

Exhaust system: service or inspect every 7500 miles.

Steering gear (box) & linkage, (steering damper-4x4), axle & suspension parts: service or inspect every 7500 miles.

Steering linkage ball joints & front suspension ball joints: service or inspect every 7500 miles.

93591CI4

SCHEDULED MAINTENANCE INTERVALS
NISSAN/INFINITI
PATHFINDER, XTERRA

The following should be used as a guide when determining the amount of work required for a particular service. In estimating how long a particular Scheduled Maintenance Service should take, please observe the following:

- Labor Time is time based on field research and data supplied by the vehicle manufacturer.
- Labor time operations are given in hours and tenths of an hour.
- All labor operations are to be used as a guide.

Mechanic Skill Level Codes:
(A) PRECISION: Highly skilled with multiple certification.
(B) GENERAL: Normally skilled with certification.
(C) MAINTENANCE: Semi-skilled working on certification.

	LABOR TIME		LABOR TIME		LABOR TIME
7500 Mile Service (C)		**45000 Mile Service (C)**		**75000 Mile Service (C)**	
All Models	.6	All Models	1.0	All Models	1.0
15000 Mile Service (C)		**52500 Mile Service (C)**		**82500 Mile Service (C)**	
All Models	1.0	All Models	.6	All Models	.5
22500 Mile Service (C)		**60000 Mile Service (B)**		**90000 Mile Service (B)**	
All Models	.5	All Models	3.8	All Models	3.8
30000 Mile Service (B)		*w/4WD add*	.5	*w/4WD add*	.5
All Models	3.0	*Valve clearance add*		**97500 Mile Service (C)**	
w/4WD add	.5	**67500 Mile Service (C)**		All Models	.6
37500 Mile Service (C)		All Models	.6	*Valve clearance add*	
All Models	.6				

93591CI5

SATURN
Vue

ENGINE AND VEHICLE IDENTIFICATION

Engine							Model Year	
Code ①	Liters (cc)	Cu. In.	Cyl.	Fuel Sys.	Engine Type	Eng. Mfg.	Code ②	Year
D	2.2 (2199)	134	4	SFI	DOHC	Saturn	2	2002
B	3.0 (3000)	183	6	SFI	DOHC	Saturn	3	2003

SFI: Sequential Fuel Injection

DOHC: Double Overhead Camshafts

① 8th digit of VIN

② 10th digit of VIN

93591CVA

GENERAL ENGINE SPECIFICATIONS

Year	Model	Engine Displacement Liters (cc)	Engine ID/VIN	Fuel System Type	Net Horsepower @ rpm	Net Torque @ rpm (ft. lbs.)	Bore x Stroke (in.)	Compression Ratio	Oil Pressure @ rpm
2002-03	VUE	2.2 (1901)	D	SFI	137@5800	147@4400	3.38x3.5	9.5:1	50-80@1000
		3.0 (3000)	B	SFI	182@6000	184@3400	3.38x3.50	10.0:1	50-80@1000

SFI: Sequential Fuel Injection

93591CVB

ENGINE TUNE-UP SPECIFICATIONS

Year	Engine Displacement Liters (cc)	Engine ID/VIN	Spark Plug Gap (in.)	Ignition Timing (deg.)		Fuel Pump (psi) ①	Idle Speed (rpm)		Valve Clearance	
				MT	AT		MT ②	AT ②	In.	Ex.
2002-03	2.2 (2199)	D	0.045	③	③	50-60	④	④	HYD	HYD
	3.0 (3000)	B	0.043	③	③	50-60	④	④	HYD	HYD

NOTE: The Vehicle Emission Control Information label often reflects specification changes made during production. The label figures must be used if they differ from those in this chart.

HYD: Hydraulic

① Pressure measured at idle

② Idle speed measured with manual transmission in Neutral; automatic transmission in D (drive)

③ Engines equipped with Distributorless Ignition System (DIS). Ignition timing is not adjustable

④ Refer to the Vehicle Emission Control Information label

93591CVC

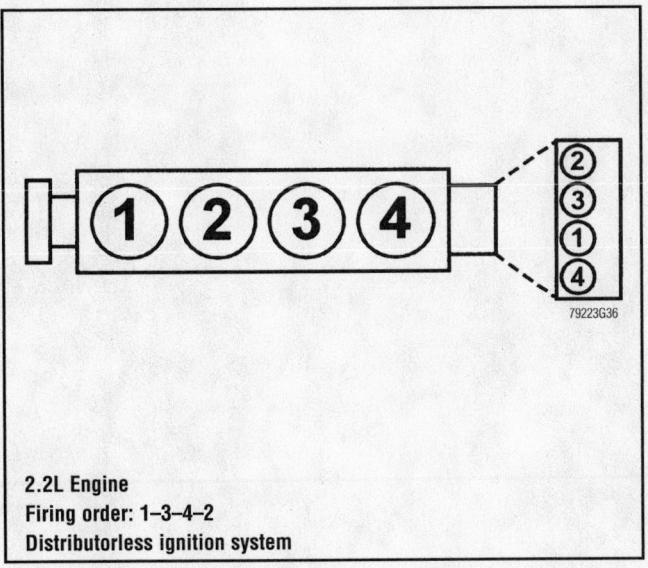

2.2L Engine
Firing order: 1–3–4–2
Distributorless ignition system

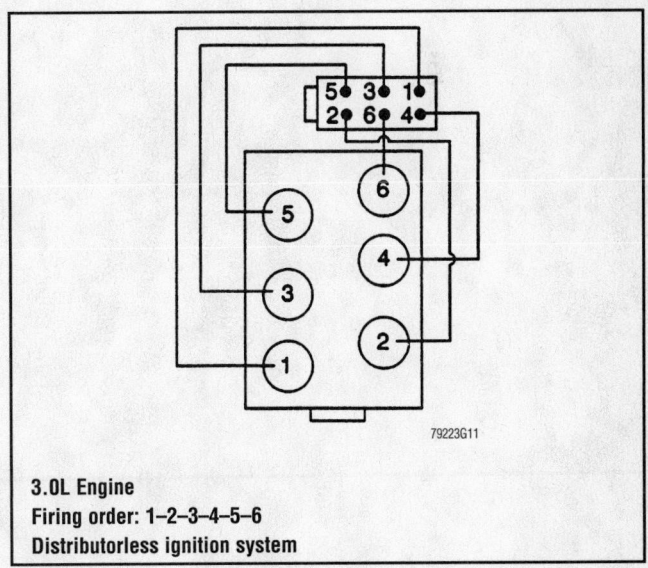

3.0L Engine
Firing order: 1–2–3–4–5–6
Distributorless ignition system

Timing belt service is covered in Section 3 of this manual

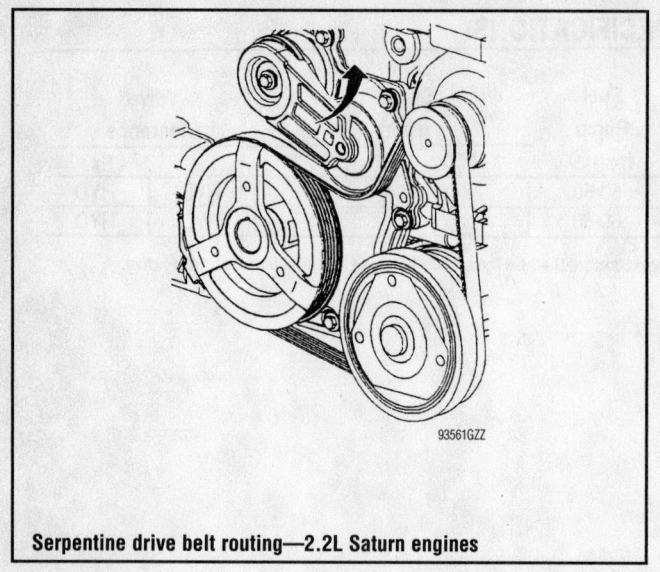

93561GZZ

Serpentine drive belt routing—2.2L Saturn engines

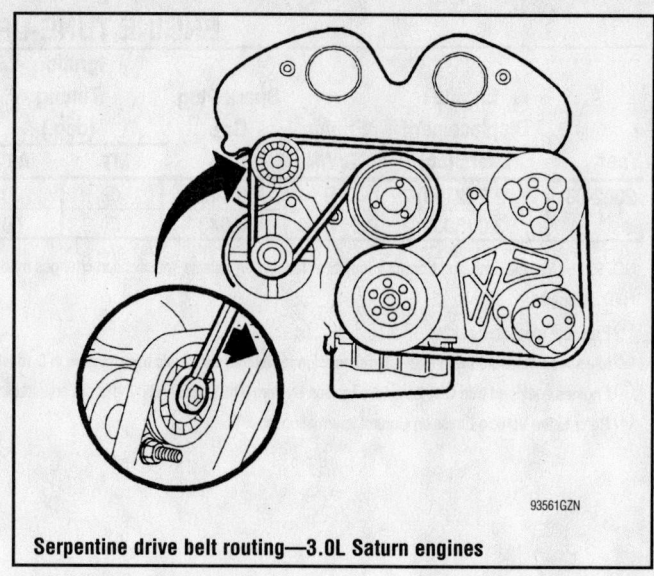

93561GZN

Serpentine drive belt routing—3.0L Saturn engines

CAPACITIES

Year	Model	Engine Displacement Liters (cc)	Engine ID/VIN	Engine Oil with Filter (qts.)	Transaxle (qts.)		Fuel Tank (gal.)	Cooling System (qts.)
					Manual	Auto. ①		
2002-03	VUE	2.2 (2199)	D	5.0	2.0	6.9	15.7	②
		3.0 (3000)	B	5.0	2.0	6.9	15.7	②

NOTE: All capacities are approximate. Add fluid gradually and ensure a proper fluid level is obtained.

① Specification is for overhaul. 8.4 pts. with fluid and filter change

② 2.2L with manual transaxle: 7.4 qts.
2.2L with automatic transaxle: 7.3 qts.
3.0L: 7.8 qts.

93591CVD

VALVE SPECIFICATIONS

Year	Engine Displacement Liters (cc)	Engine ID/VIN	Seat Angle (deg.)	Face Angle (deg.)	Spring Test Pressure (lbs. @ in.)	Spring Free-Length (in.)	Stem-to-Guide Clearance (in.)		Stem Diameter (in.)	
							Intake	Exhaust	Intake	Exhaust
2002-03	2.2 (2199)	D	44.5-45.4	45-45.5	①	1.6100	0.0012	0.0020	0.2344	0.2337
					②		0.0022	0.0026	0.2355	0.2343
	3.0 (3000)	B	45	45	56.6 @ 1.338	NA	0.0012	0.0016	0.2344	0.2341
							0.0022	0.0026	0.2350	0.2346

NA: Not available

① Valve spring load closed: 245-271 N

② Valve spring load open: 525-575 N

93591CVE

Heater Core replacement is covered in Section 2 of this manual

CRANKSHAFT AND CONNECTING ROD SPECIFICATIONS
All measurements are given in inches.

Year	Engine Displacement Liters (cc)	Engine ID/VIN	Crankshaft				Connecting Rod		
			Main Brg. Journal Dia.	Main Brg. Oil Clearance	Shaft End-play	Thrust on No.	Journal Diameter	Oil Clearance	Side Clearance
2002-03	2.2 (2199)	D	2.2045-2.2050	0.0012 0.0026	0.0012-0.0150	3	1.9291-1.9297	0.0001-0.0021	0.0028-0.0146
	3.0 (3000)	B	2.6763-2.6766	0.0060 0.0017	0.0004-0.0300	3	1.927-1.9280	0.0001-0.0021	0.0027-0.0110

93591CVF

PISTON AND RING SPECIFICATIONS
All measurements are given in inches.

Year	Engine Displacement Liters (cc)	Engine ID/VIN	Piston Clearance	Ring Gap			Ring Side Clearance		
				Top Compression	Bottom Compression	Oil Control	Top Compression	Bottom Compression	Oil Control
2002-03	2.2 (2199)	D	0.0004-0.0016	0.008-0.016	0.0014 0.0022	0.0010 0.0030	0.0028-0.0146	0.0005-0.0024	SNUG
	3.0 (3000)	B	0.0010-0.0018	0.0008-0.0015	0.0118 0.0196	0.0157 0.0551	0.0027-0.0110	0.0005-0.0024	SNUG

NA: Not available

① Piston No. 2 and 3: 0.0002-0.0017
Piston No. 1, 4: 0.0003-0.0021

93591CVG

TORQUE SPECIFICATIONS
All readings in ft. lbs.

Year	Engine Displacement Liters (cc)	Engine ID/VIN	Cylinder Head Bolts	Main Bearing Bolts	Rod Bearing Bolts	Crankshaft Damper Bolts	Flywheel Bolts	Manifold		Spark Plugs	Lug Nuts
								Intake	Exhaust		
2002-03	2.2 (2199	D	①	②	③	④	②	⑤	13	15	92
	3.0 (3000)	B	⑥	⑦	26	15	②	15	15	18	92

① Step 1: 22 ft. lbs.
　Step 2: 155 degrees
② 39 ft. lbs. Plus 25 degrees
③ 18 ft. lbs. Plus 100 degrees
④ 74 ft. lbs. Plus 75 degrees
⑤ 89 inch lbs.

⑥ Step 1: 18 ft. lbs.
　Step 2: plus 90 degrees
　Step 3: plus 90 degrees
　Step 4: plus 90 degrees
　Step 5: plus 15 degrees

⑦ Step 1: 37 ft. lbs.
　Step 2: plus 60 degrees
　Step 3: plus 15 degrees

93591CVH

WHEEL ALIGNMENT

Year	Model		Caster		Camber		Toe-in (in.)	Steering Axis Inclination (Deg.)
			Range (+/-Deg.)	Preferred Setting (Deg.)	Range (+/-Deg.)	Preferred Setting (Deg.)		
2002-03	VUE	F	2.60-3.40	3.00	-1.00	0.60	0.20 +/- 0.15	—
		R	—	—	—	-0.05	0.10 +/- 0.10	—

93591CVI

Brake service is covered in Section 4 of this manual

TIRE, WHEEL AND BALL JOINT SPECIFICATIONS

| Year | Model | | OEM Tires | | Tire Pressures (psi) | | Wheel Size | Ball Joint Inspection |
			Standard	Optional	Front	Rear		
2002-03	VUE	F	P235/65R16	P215/70R16	①	①	NS	NS
		R	P235/65R16	P215/70R16	①	①	NS	NS

OEM: Original Equipment Manufacturer

PSI: Pounds Per Square Inch

① Check the placard on the drivers side sill

NS: Not specified by manufacturer

93591CVJ

BRAKE SPECIFICATIONS
All measurements in inches unless noted

| Year | Model | | Brake Disc | | | Brake Drum Diameter | | | Minimum Lining Thickness | Brake Caliper | |
			Original Thickness	Minimum Thickness	Maximum Runout	Original Inside Diameter	Max. Wear Limit	Maximum Machine Diameter		Bracket Bolt (ft. lbs.)	Mounting Bolt (ft. lbs.)
2002-03	VUE	F	1.020	0.960	0.001	—	—	—	0.080	118	24
		R	—	—	—	9.84	9.90	9.90	0.040	63	27

NA: Not Available

F: Front

R: Rear

93591CVK

SCHEDULED MAINTENANCE INTERVALS
SATURN—VUE

TO BE SERVICED	TYPE OF SERVICE	VEHICLE MILEAGE INTERVAL (x1000)												
		3	6	9	12	15	18	21	24	27	30	33	36	39
Engine oil & filter	R		✓		✓		✓		✓		✓		✓	
Lubricate chassis, suspension and steering linkage	S/I		✓		✓		✓		✓		✓		✓	
Lubricate transaxle shift linkage and parking brake cable guides	S/I		✓		✓		✓		✓		✓		✓	
Lubricate underbody contact points & linkage	S/I		✓		✓		✓		✓		✓		✓	
Driveshaft boots, suspension bushings & ball joint seals	S/I		✓				✓		✓		✓		✓	
Exhaust system & throttle linkage	S/I		✓		✓		✓		✓		✓		✓	
Rotate tires	S/I		✓		✓		✓				✓			
Brake hoses & brake lining	S/I		✓				✓						✓	
Accessory drive belt(s)	S/I						✓						✓	
Engine coolant level, hoses & clamps	S/I						✓						✓	
Air filter element	R										✓			
Engine coolant	R												✓	
Manual transaxle oil	R		✓											
Spark plugs ①	R										✓			
Automatic transaxle fluid & filter	S/I										✓			
Ignition cables & fuel systems	S/I										✓			
Vacuum line/hose	S/I										✓			
Fuel filter ②	R													

S/I: Service or Inspect

R: Replace

① Platinum tip spark plugs: replace every 100,000 miles

② Replace every 60,000 miles

FREQUENT OPERATION MAINTENANCE (SEVERE SERVICE)

If a vehicle is operated under any of the following conditions it is considered severe service:

- Extremely dusty areas

- 50% or more of the vehicle operation is in 32°C (90°F) or higher temperatures, or constant operation in temperatures below 0°C (32°F)

#NAME?

- Frequent short running periods (engine does not warm to normal operating temperatures)

- Police, taxi, delivery usage or trailer towing usage

Engine oil & oil filter: change every 3000 miles

93591CVL

For complete Engine Mechanical specifications, see Section 1 of this manual

SUBARU
Forester

ENGINE AND VEHICLE IDENTIFICATION CHART

		Engine Code						Model Year	
Code ①	Liters (cc)	Cu. In.	Cyl.	Fuel Sys.	Type	Eng. Mfg.		Code ②	Year
6	2.5 (2457)	150	4	MFI	DOHC	Subaru		X	1999
								Y	2000
								1	2001
								2	2002
								3	2003

MFI: Multiport Fuel Injection

DOHC: Double Overhead Camshafts

① 6th digit of the VIN.

② 10th digit of the VIN.

93591CY6

GENERAL ENGINE SPECIFICATIONS

Year	Model	Engine Displacement Liters (cc)	Engine ID/VIN	Fuel System Type	Net Horsepower @ rpm	Net Torque @ rpm (ft. lbs.)	Bore x Stroke (in.)	Compression Ratio	Oil Pressure psi @ rpm
1999	Forester	2.5 (2457)	6	MFI	165@5600	162@4000	3.92x3.11	9.7:1	14 @ 800
2000	Forester	2.5 (2457)	6	MFI	165@5600	162@4000	3.92x3.11	9.7:1	14 @ 800
2001	Forester	2.5 (2457)	6	MFI	165@5600	162@4000	3.92x3.11	9.7:1	14 @ 800
2002-03	Forester	2.5 (2457)	6	MFI	165@5600	162@4000	3.92x3.11	9.7:1	14 @ 800

MFI: Multi-port Fuel Injection

93591CY7

ENGINE TUNE-UP SPECIFICATIONS

Year	Engine Displacement Liters (cc)	Engine ID/VIN	Spark Plugs Gap (in.)	Ignition Timing (deg.) ①		Fuel Pump (psi)	Idle Speed (rpm) ②		Valve Clearance ③	
				MT	AT		MT	AT	In.	Ex.
1999	2.5 (2457)	6	0.039-0.043	7-23 BTDC	7-23 BTDC	34-38	600-800	600-800	0.0071-0.0087	0.0090-0.0106
2000	2.5 (2457)	6	0.039-0.043	7-23 BTDC	7-23 BTDC	34-38	600-800	600-800	0.0071-0.0087	0.0090-0.0106
2001	2.5 (2457)	6	0.039-0.043	7-23 BTDC	7-23 BTDC	34-38	600-800	600-800	0.0071-0.0087	0.0090-0.0106
2002-03	2.5 (2457)	6	0.039-0.043	7-23 BTDC	7-23 BTDC	34-38	600-800	600-800	0.0071-0.0087	0.0090-0.0106

BTDC: Before Top Dead Center

① At idle speed.

② With engine under no load.

③ With engine cold.

93591CY8

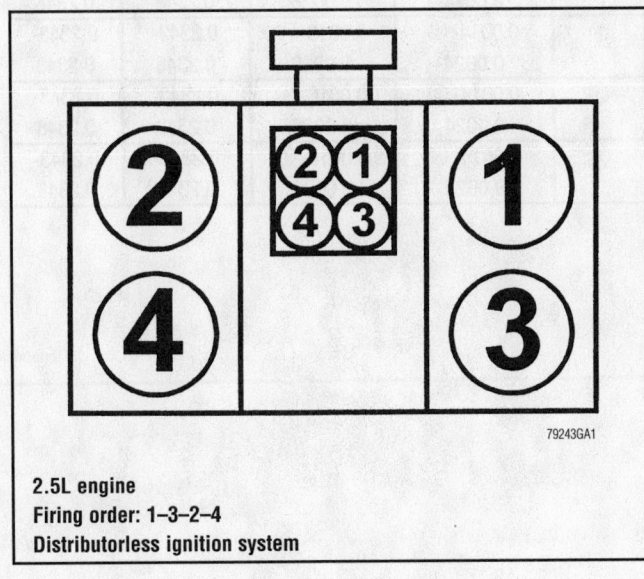

2.5L engine
Firing order: 1–3–2–4
Distributorless ignition system

79243GA1

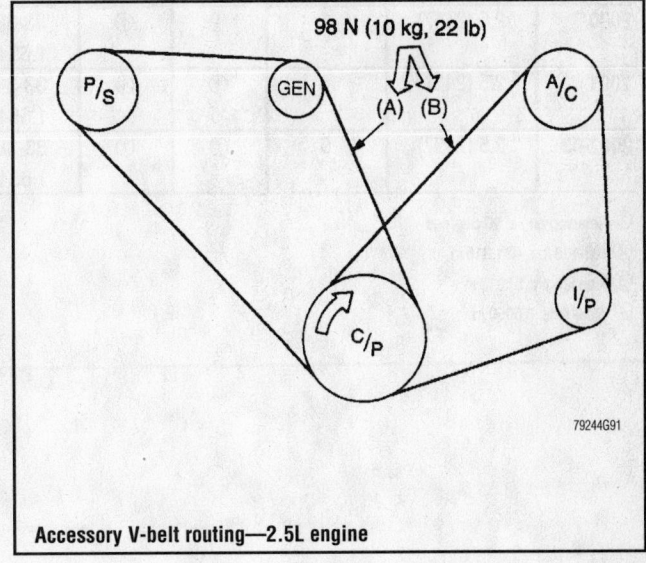

Accessory V-belt routing—2.5L engine

79244G91

For Accessory Drive Belt illustrations, see Section 1 of this manual

CAPACITIES

Year	Model	Engine Displacement Liters (cc)	Engine ID/VIN	Engine Oil with Filter (qts.)	Transmission (pts.)			Transfer Case (pts.)	Drive Axle		Fuel Tank (gal.)	Cooling System (qts.)
					4-Spd	5-Spd	Auto.		Front (pts.)	Rear (pts.)		
1999	Forester	2.5 (2457)	6	4.7	—	7.4	20	—	2.6 ①	1.6	15.9	6.3
2000	Forester	2.5 (2457)	6	4.7	—	7.4	20	—	2.6 ①	1.6	15.9	6.3
2001	Forester	2.5 (2457)	6	4.7	—	7.4	20	—	2.6 ①	1.6	15.9	6.3
2002-03	Forester	2.5 (2457)	6	4.7	—	7.4	20	—	2.6 ①	1.6	15.9	6.3

① A/T differential only

93591CY9

VALVE SPECIFICATIONS

Year	Engine Displacement Liters (cc)	Engine ID/VIN	Seat Angle (deg.)	Face Angle (deg.)	Spring Test Pressure (lbs. @ in.)	Spring Installed Height (in.)	Stem-to-Guide Clearance (in.)		Stem Diameter (in.)	
							Intake	Exhaust	Intake	Exhaust
1999	2.5 (2457)	6	①	①	33-38 @ 1.654 ②	③	0.0014- ③ 0.0024	0.0016- ④ 0.0026	0.2343- 0.2348	0.2343- 0.2348
2000	2.5 (2457)	6	①	①	33-38 @ 1.654 ②	③	0.0014- ③ 0.0024	0.0016- ④ 0.0026	0.2343- 0.2348	0.2343- 0.2348
2001	2.5 (2457)	6	①	①	33-38 @ 1.654 ②	③	0.0014- ③ 0.0024	0.0016- ④ 0.0026	0.2343- 0.2348	0.2343- 0.2348
2002-03	2.5 (2457)	6	①	①	33-38 @ 1.654 ②	③	0.0014- ③ 0.0024	0.0016- ④ 0.0026	0.2343- 0.2348	0.2343- 0.2348

① Refacing angle: 90 degrees
② 102-118 lbs. @ 1.315 in.
③ Free length: 1.8913 in.
④ Wear limit: 0.0059 in.

93591CY0

CRANKSHAFT AND CONNECTING ROD SPECIFICATIONS

All measurements are given in inches.

Year	Engine Displacement Liters (cc)	Engine ID/VIN	Crankshaft				Connecting Rod		
			Main Brg. Journal Dia.	Main Brg. Oil Clearance	Shaft End-play	Thrust on No.	Journal Diameter	Oil Clearance	Side Clearance
1999	2.5 (2457)	6	2.3619-2.3625	①	0.0012-0.0098	3	1.8891-1.8898	0.0004-0.0020	0.0028-0.0160
2000	2.5 (2457)	6	2.3619-2.3625	①	0.0012-0.0098	3	1.8891-1.8898	0.0004-0.0020	0.0028-0.0160
2001	2.5 (2457)	6	2.3619-2.3625	①	0.0012-0.0098	3	1.8891-1.8898	0.0004-0.0020	0.0028-0.0160
2002-03	2.5 (2457)	6	2.3619-2.3625	①	0.0012-0.0098	3	1.8891-1.8898	0.0004-0.0020	0.0028-0.0160

① Journals 1 and 5: 0.0001-0.0016
Journals 2 and 4: 0.0004-0.0018
Journal 3: 0.0004-0.0016

93591CZ1

PISTON AND RING SPECIFICATIONS

All measurements are given in inches.

Year	Engine Displacement Liters (cc)	Engine ID/VIN	Piston Clearance	Ring Gap			Ring Side Clearance		
				Top Compression	Bottom Compression	Oil Control	Top Compression	Bottom Compression	Oil Control
1999	2.5 (2457)	6	0.0004-0.0020	0.0079-0.0390	0.0146-0.0390	0.0079-0.0590	0.0016-0.0059	0.0012-0.0059	NA
2000	2.5 (2457)	6	0.0004-0.0020	0.0079-0.0390	0.0146-0.0390	0.0079-0.0590	0.0016-0.0059	0.0012-0.0059	NA
2001	2.5 (2457)	6	0.0004-0.0020	0.0079-0.0390	0.0146-0.0390	0.0079-0.0590	0.0016-0.0059	0.0012-0.0059	NA
2002-03	2.5 (2457)	6	0.0004-0.0020	0.0079-0.0390	0.0146-0.0390	0.0079-0.0590	0.0016-0.0059	0.0012-0.0059	NA

NA: Not Available

93591CZ2

For Tire, Wheel and Ball Joint specifications, see Section 1 of this manual

TORQUE SPECIFICATIONS
All readings in ft. lbs.

Year	Engine Displacement Liters (cc)	Engine ID/VIN	Cylinder Head Bolts	Main Bearing Bolts	Rod Bearing Bolts	Crankshaft Damper Bolts	Flywheel Bolts	Manifold Intake	Manifold Exhaust	Spark Plugs	Lug Nut
1999	2.5 (2457)	6	①	②	31-34	123-137	51-55	14-17	19-26 ③	13-17	58-72
2000	2.5 (2457)	6	①	②	31-34	123-137	51-55	14-17	19-26 ③	13-17	58-72
2001	2.5 (2457)	6	①	②	31-34	123-137	51-55	14-17	19-26 ③	13-17	58-72
2002-03	2.5 (2457)	6	①	②	31-34	123-137	51-55	14-17	19-26 ③	13-17	58-72

① Step 1: Tighten all bolts, in sequence, to 22 ft. lbs.

Step 2: Tighten all bolts, in sequence, to 51 ft. lbs.

Step 3: Loosen all bolts 180 degrees (one-half turn)

Step 4: Loosen all bolts another 180 degrees (one-half turn)

Step 5: Tighetn bolts A and B, in sequence, to 25 ft. lbs.

Step 6: Tighten bolts C, D, E and F, in sequence, to 11 ft. lbs.

Step 7: Tighten all bolts, in sequence, 80-90 degrees

Step 8: Tighten all bolts, in sequence, another 80-90 degrees

② Split engine case bolts:

10mm bolts: 33-37 ft. lbs.

8mm bolts: A thru G to 17-20 ft. lbs. and H to 5 ft. lbs.

③ No separate exhaust manifold is used, the front pipe bolts directly to the cylinder heads

93591CZ3

WHEEL ALIGNMENT

Year	Model		Caster Range (+/-Deg.)	Caster Preferred Setting (Deg.)	Camber Range (+/-Deg.)	Camber Preferred Setting (Deg.)	Toe-in (in.)	Steering Axis Inclination (Deg.)
1999	Forester	F	0.75	+2.58	0.50	-0.25	0+/-0.12	—
		R	—	—	0.75	-0.58	0.08+/-0.04	—
2000	Forester	F	0.75	+2.58	0.50	-0.25	0+/-0.12	—
		R	—	—	0.75	-0.58	0.08+/-0.04	—
2001	Forester	F	0.75	+2.58	0.50	-0.25	0+/-0.12	—
		R	—	—	0.75	-0.58	0.08+/-0.04	—
2002-03	Forester	F	0.75	+2.58	0.50	-0.25	0+/-0.12	—
		R	—	—	0.75	-0.58	0.08+/-0.04	—

93591CZ4

TIRE, WHEEL AND BALL JOINT SPECIFICATIONS

Year	Model	OEM Tires		Tire Pressures (psi)		Wheel Size	Ball Joint Inspection
		Standard	Optional	Front	Rear		
1999	Forester	P205/70R15 95S	P215/60R16 94H	29	26①	②	0.012 in. ③
2000	Forester	P205/70R15 95S	P215/60R16 94H	29	26①	②	0.012 in. ③
2001	Forester	P205/70R15 95S	P215/60R16 94H	29	26①	②	0.012 in. ③
2002-03	Forester	P205/70R15 95S	P215/60R16 94H	29	26①	②	0.012 in. ③

OEM: Original Equipment Manufacturer

PSI: Pounds Per Square Inch

STD: Standard

OPT: Optional

① Wigh full load: 36 psi.

② With standard tires: 6-JJ

   With optional tires: 6.5-JJ

③ Apply 154 lbs. vertical force

93591CZ5

BRAKE SPECIFICATIONS
All measurements in inches unless noted

Year	Model		Brake Disc			Brake Drum Diameter			Minimum Lining Thickness		Brake Caliper	
			Original Thickness	Minimum Thickness	Maximum Runout	Original Inside Diameter	Max. Wear Limit	Maximum Machine Diameter	Front	Rear	Bracket Bolts (ft. lbs.)	Mounting Bolts (ft. lbs.)
1999	Forester	F	0.940	0.870	0.003	—	—	—	0.059	—	51-65	25-31
		R	0.390	0.340	0.004	9.00①	9.079②	NA	—	0.059	—	25-31
2000	Forester	F	0.940	0.870	0.003	—	—	—	0.059	—	51-65	25-31
		R	0.390	0.340	0.004	9.00①	9.079②	NA	—	0.059	—	25-31
2001	Forester	F	0.940	0.870	0.003	—	—	—	0.059	—	51-65	25-31
		R	0.390	0.340	0.004	9.00①	9.079②	NA	—	0.059	—	25-31
2002-03	Forester	F	0.940	0.870	0.003	—	—	—	0.059	—	51-65	25-31
		R	0.390	0.340	0.004	9.00①	9.079②	NA	—	0.059	—	25-31

NA: Not Available

① Parking brake drum on vehicles with rear disc brakes: 6.69 in.

② Parking brake drum on vehicles with rear disc brakes: 6.73 in.

93591CZ6

For Wheel Alignment specifications, see Section 1 of this manual

SCHEDULED MAINTENANCE INTERVALS
SUBARU—FORESTER

TO BE SERVICED	TYPE OF SERVICE	VEHICLE MILEAGE INTERVAL (x1000)																
		3	7.5	15	22.5	30	37.5	45	52.5	60	67.5	75	82.5	90	97.5	105	112.5	120
Accessory drive belts	R									✓								✓
Accessory drive belts	S/I					✓								✓				
Air cleaner filter	R					✓				✓				✓				✓
Automatic transmission fluid	S/I					✓				✓				✓				✓
Axle shaft joints	S/I			✓		✓		✓		✓		✓		✓		✓		✓
Brake fluid	R					✓				✓				✓				✓
Brake system lines	S/I			✓		✓		✓		✓		✓		✓		✓		✓
Clutch operation	S/I			✓		✓		✓		✓		✓		✓		✓		✓
Disc brake pads & rotors	S/I			✓		✓		✓		✓		✓		✓		✓		✓
Drums brake linings & drums	S/I					✓				✓				✓				✓
Engine coolant	R					✓				✓				✓				✓
Engine cooling system, hoses & connections	S/I					✓				✓				✓				✓
Engine oil & filter	R	✓	✓	✓	✓	✓	✓	✓	✓	✓	✓	✓	✓	✓	✓	✓	✓	✓
Front & rear axle boots	S/I			✓		✓		✓		✓		✓		✓		✓		✓
Front & rear wheel bearings	S/I & L									✓								✓
Fuel filter	R					✓				✓				✓				✓
Parking & service brake systems' operation	S/I			✓		✓		✓		✓		✓		✓		✓		✓
Spark plugs	R									✓								✓
Steering & suspension	S/I			✓		✓		✓		✓		✓		✓		✓		✓
Timing belt	R															✓		
Timing belt	S/I					✓				✓				✓				
Transmission & differential fluid levels	S/I					✓				✓				✓				✓
Valve clearance	S/I															✓		

R: Replace   S/I: Inspect and service, if needed   L: Lubricate

FREQUENT OPERATION MAINTENANCE (SEVERE SERVICE)

If a vehicle is operated under any of the following conditions it is considered severe service:

- Towing a trailer or using a camper or car-top carrier.
- Repeated short trips of less than 5 miles in temperatures below freezing, or trips of less than 10 miles in any temperature.
- Extensive idling or low-speed driving for long distances as in heavy commercial use, such as delivery, taxi or police cars.
- Operating on rough, muddy or salt-covered roads, or extensive mountain driving.
- Operating on unpaved or dusty roads.
- Driving in extremely hot (over 90°) conditions.

Engine oil and filter: replace every 3000 miles or 3 months, whichever occurs first.

Fuel filter: replace every 7500 miles or 7.5 months, whichever occurs first.

Fuel system, hoses & connections: inspect every 7500 miles or 7.5 months, whichever occurs first.

Transmission & differential fluid: replace every 15,000 miles.

Automatic transmission fluid: replace every 15,000 miles.

Brake fluid: replace every 15,000 miles.

Disc brake pads & rotors: inspect every 7500 miles or 7.5 months, whichever occurs first.

Front & rear axle boots: inspect every 7500 miles or 7.5 months, whichever occurs first.

Axle shaft boots: inspect every 7500 miles or 7.5 months, whichever occurs first.

Drum brake linings & drums: inspect every 7500 miles or 7.5 months, whichever occurs first.

Brake lines: inspect every 7500 miles or 7.5 months, whichever occurs first.

Parking & service brake system operation: inspect every 7500 miles or 7.5 months, whichever occurs first.

Clutch operation: inspect every 7500 miles or 7.5 months, whichever occurs first.

93591CZ7

SCHEDULED MAINTENANCE INTERVALS
SUBARU
FORESTER

The following should be used as a guide when determining the amount of work required for a particular service. In estimating how long a particular Scheduled Maintenance Service should take, please observe the following:

- Labor Time is time based on field research and data supplied by the vehicle manufacturer.
- Labor time operations are given in hours and tenths of an hour.
- All labor operations are to be used as a guide.

Mechanic Skill Level Codes:
(A) PRECISION: Highly skilled with multiple certification.
(B) GENERAL: Normally skilled with certification.
(C) MAINTENANCE: Semi-skilled working on certification.

	LABOR TIME		LABOR TIME		LABOR TIME
3000 Mile Service (C)		**45000 Mile Service (C)**		**90000 Mile Service (B)**	
All Models	.4	All Models	1.0	All Models	5.3
7500 Mile Service (C)		**52500 Mile Service (C)**		**97500 Mile Service (C)**	
All Models	.4	All Models	.4	All Models	.4
15000 Mile Service (C)		**60000 Mile Service (B)**		**105000 Mile Service (C)**	
All Models	1.0	All Models	5.5	All Models	1.0
22500 Mile Service (C)		**67500 Mile Service (C)**		*Replace timing belt add*	2.7
All Models	.4	All Models	.4	**112500 Mile Service (C)**	
30000 Mile Service (B)		**75000 Mile Service (C)**		All Models	.4
All Models	5.3	All Models	1.0	**120000 Mile Service (B)**	
37500 Mile Service (C)		**82500 Mile Service (C)**		All Models	5.5
All Models	.4	All Models	.4		

93551CZ7A

For Maintenance Interval recommendations, see Section 1 of this manual

SUZUKI
Vitara • Grand Vitara

ENGINE AND VEHICLE IDENTIFICATION CHART

Engine Code								Model Year	
Code	Liters (cc)	Cu. In.	Cyl.	Fuel Sys.	Engine Type	Eng. Mfg.		Code	Year
0	1.6 (1590)	97	4	MFI	SOHC	Suzuki		X	1999
5	2.0 (1997)	121.8	4	MFI	DOHC	Suzuki		Y	2000
6	2.5 (2494)	152	6	MFI	DOHC	Suzuki		1	2001
								2	2002
								3	2003

MFI: Multiport Fuel Injection

DOHC: Dual Overhead Cam

SOHC: Single Overhead Cam

93591CRA

GENERAL ENGINE SPECIFICATIONS

Year	Model	Engine Displacement Liters (cc)	Engine ID/VIN	Fuel System Type	Net Horsepower @ rpm	Net Torque @ rpm (ft. lbs.)	Bore x Stroke (in.)	Compression Ratio	Oil Pressure @ rpm
1999	Vitara	1.6 (1590)	0	MFI	95@5600	98@4000	2.95x3.54	9.5:1	47-61@3000
		2.0 (1997)	5	MFI	127@6000	134@3000	3.31x3.54	NA	55-67@4000
	Grand Vitara	2.5 (2494)	6	MFI	140@6500	151@4000	3.31x2.95	9.5:1	55-67@4000
2000	Vitara	1.6 (1590)	0	MFI	95@5600	98@4000	2.95x3.54	9.5:1	47-61@3000
		2.0 (1997)	5	MFI	127@6000	134@3000	3.31x3.54	NA	55-67@4000
	Grand Vitara	2.5 (2494)	6	MFI	140@6500	151@4000	3.31x2.95	9.5:1	55-67@4000
2001	Vitara	2.0 (1997)	5	MFI	127@6000	134@3000	3.31x3.54	NA	55-67@4000
	Grand Vitara	2.5 (2494)	6	MFI	140@6500	151@4000	3.31x2.95	9.5:1	55-67@4000
2002	Vitara	2.0 (1997)	5	MFI	127@6000	134@3000	3.31x3.54	NA	55-67@4000
	Grand Vitara	2.5 (2494)	6	MFI	140@6500	151@4000	3.31x2.95	9.5:1	55-67@4000

MFI: Multi-port Fuel Injection

NA: Not available

93591CRB

ENGINE TUNE-UP SPECIFICATIONS

Year	Engine Displacement Liters (cc)	Engine ID/VIN	Spark Plugs Gap (in.)	Ignition Timing (deg.)		Fuel Pump (psi)	Idle Speed (rpm)		Valve Clearance	
				MT	AT		MT	AT	In.	Ex.
1999	1.6 (1590)	0	0.040	5B	5B	30-37	700-800	700-800	0.0050-0.0070	0.0050-0.0070
	2.0 (1997)	5	0.040	5B	5B	30-37	700-800	700-800	HYD	HYD
	2.5 (2494)	6	0.040	5B	·5B	30-45	700-800	700-800	HYD	HYD
2000	1.6 (1590)	0	0.040	5B	5B	30-37	700-800	700-800	0.0050-0.0070	0.0050-0.0070
	2.0 (1997)	5	0.040	5B	5B	30-37	700-800	700-800	HYD	HYD
	2.5 (2494)	6	0.040	5B	5B	30-45	700-800	700-800	HYD	HYD
2001	2.0 (1997)	5	0.040	5B	5B	30-37	700-800	700-800	HYD	HYD
	2.5 (2494)	6	0.040	5B	5B	30-45	700-800	700-800	HYD	HYD
2002	2.0 (1997)	5	0.040	5B	5B	30-37	700-800	700-800	HYD	HYD
	2.5 (2494)	6	0.040	5B	5B	30-45	700-800	700-800	HYD	HYD

HYD: Hydraulic

93591CRC

For Tune-up, Capacities and Firing orders, see Section 1 of this manual

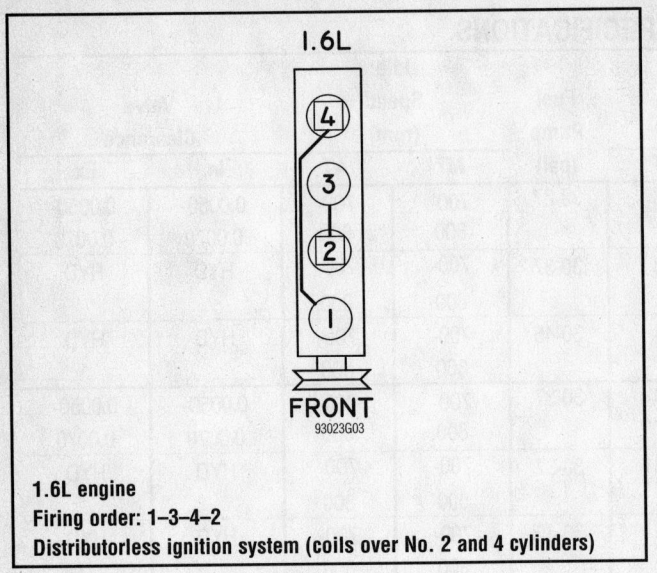

1.6L engine
Firing order: 1–3–4–2
Distributorless ignition system (coils over No. 2 and 4 cylinders)

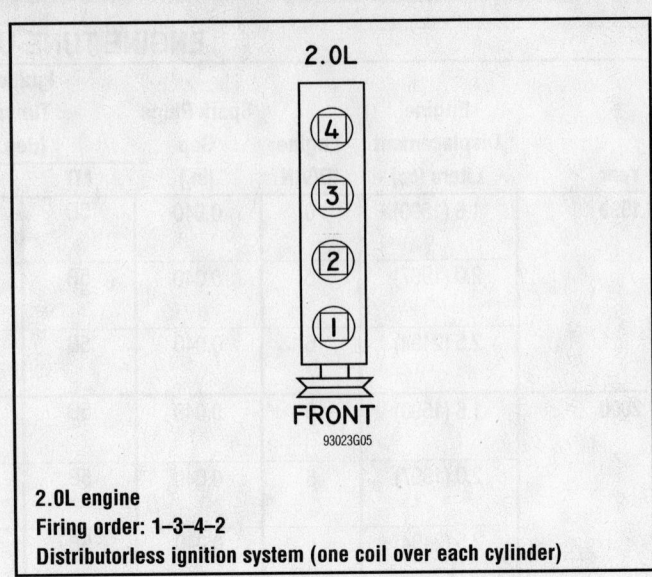

2.0L engine
Firing order: 1–3–4–2
Distributorless ignition system (one coil over each cylinder)

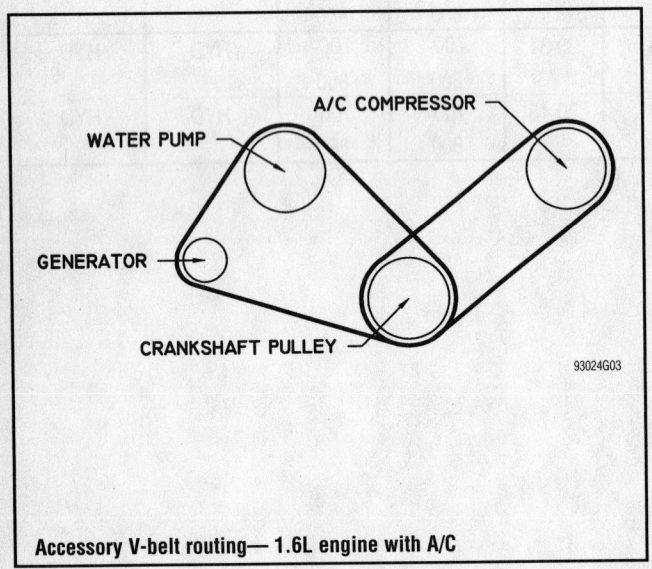

Accessory V-belt routing—1.6L engine with A/C

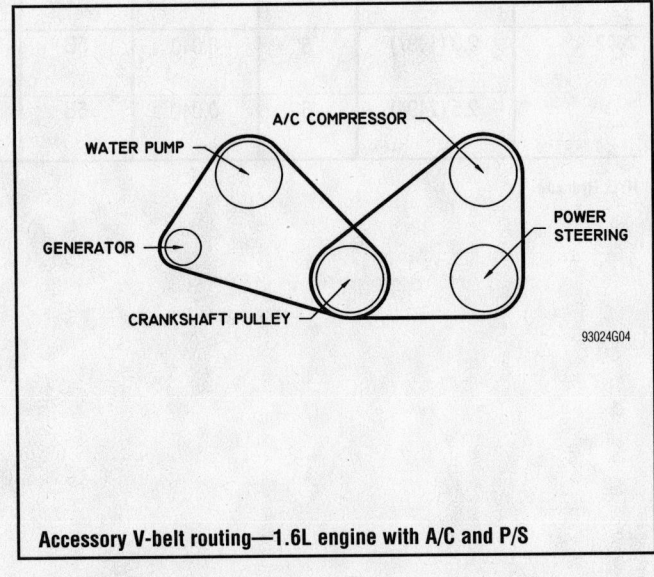

Accessory V-belt routing—1.6L engine with A/C and P/S

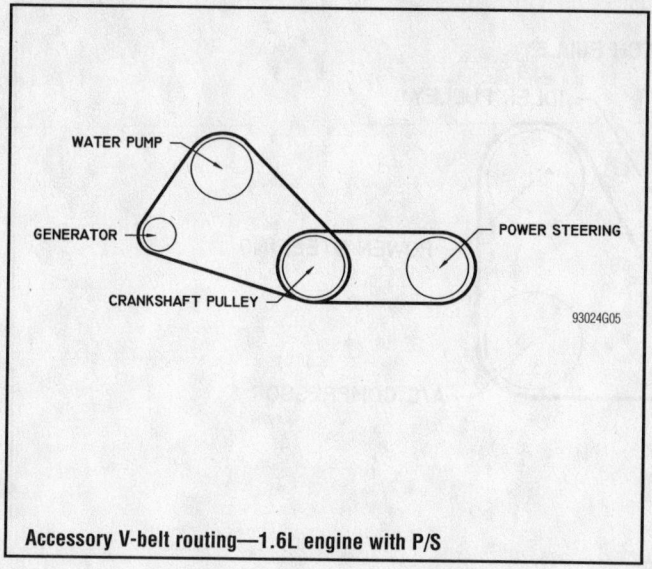

Accessory V-belt routing—1.6L engine with P/S

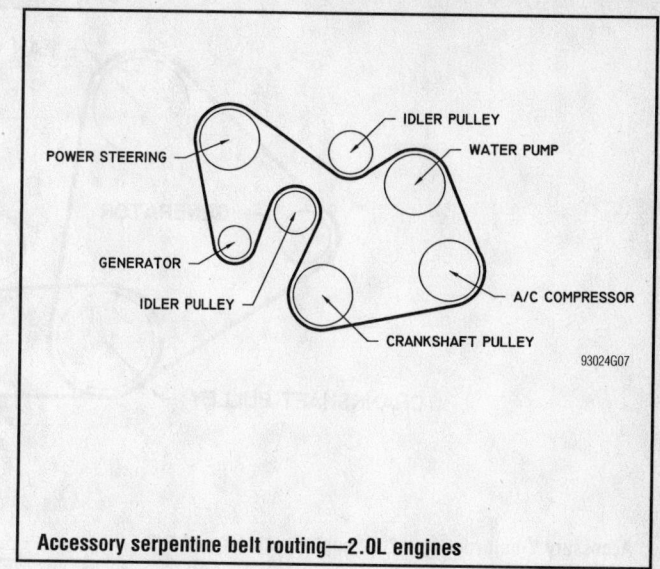

Accessory serpentine belt routing—2.0L engines

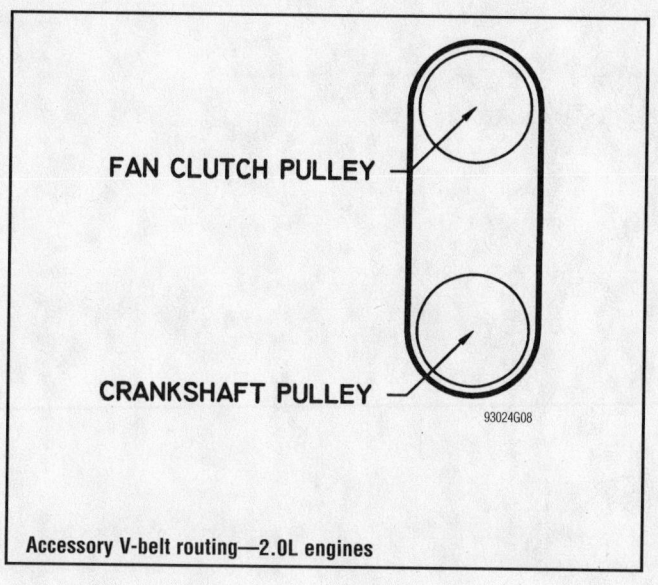

Accessory V-belt routing—2.0L engines

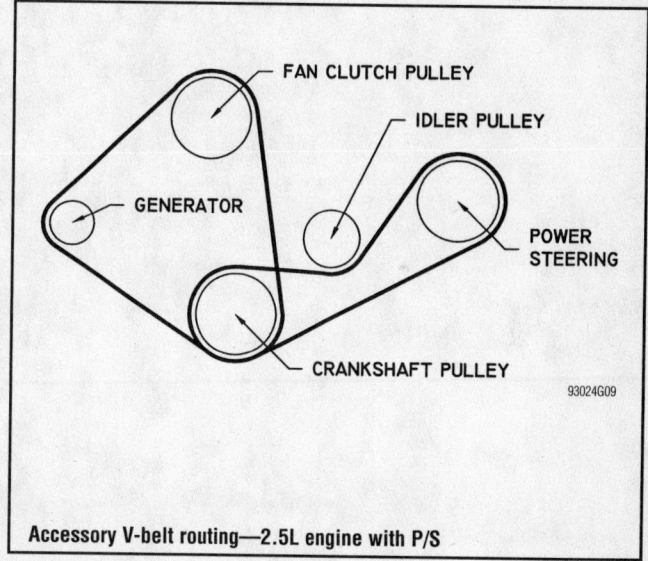

Accessory V-belt routing—2.5L engine with P/S

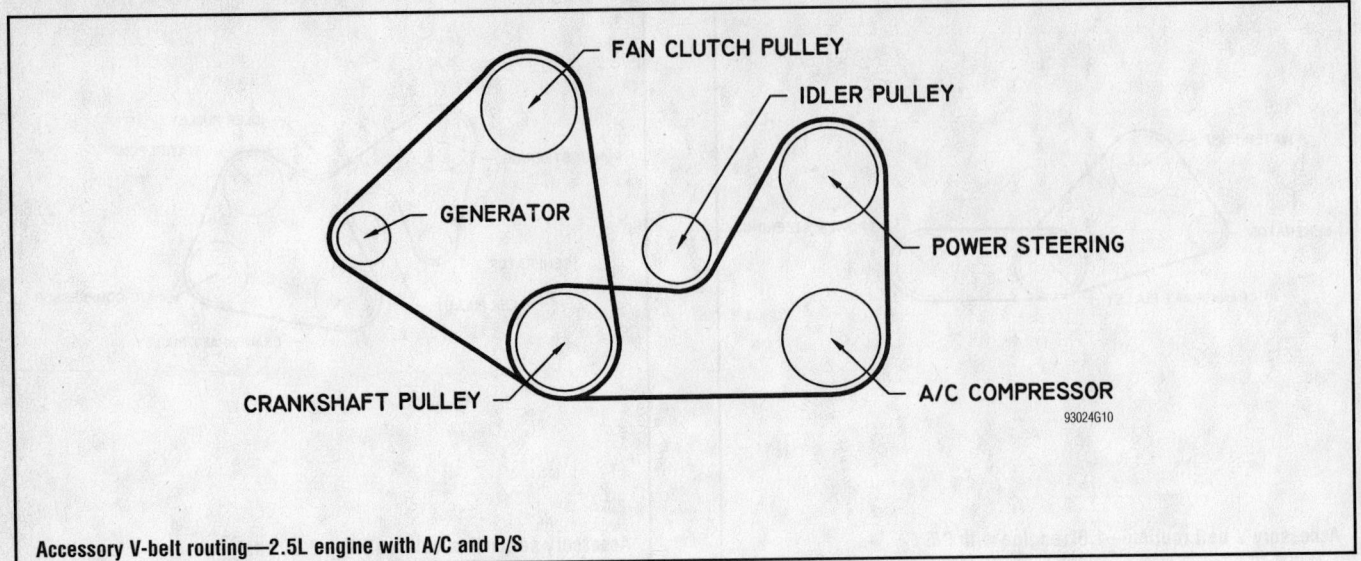

Accessory V-belt routing—2.5L engine with A/C and P/S

CAPACITIES

Year	Model	Engine Displacement Liters (cc)	Engine ID/VIN	Engine Oil with Filter (qts.)	Transmission (pts.) 5-Spd	Transmission (pts.) Auto.	Transfer Case (pts.)	Drive Axle Front (pts.)	Drive Axle Rear (pts.)	Fuel Tank (gal.)	Cooling System (qts.)
1999	Vitara	1.6 (1590)	0	4.75	①	②	3.6	2.1	4.6	11	5.5
		2.0 (1997)	5	5.9	①	②	3.6	2.1	4.6	③	5.5
	Grand Vitara	2.5 (2494)	6	6	5.5	②	3.6	2.1	4.6	18.5	5.5
2000	Vitara	1.6 (1590)	0	4.75	①	②	3.6	2.1	4.6	11	5.5
		2.0 (1997)	5	5.9	①	②	3.6	2.1	4.6	③	5.5
	Grand Vitara	2.5 (2494)	6	6	5.5	②	3.6	2.1	4.6	18.5	5.5
2001	Vitara	2.0 (1997)	5	5.9	①	④	3.6	2.1	4.6	③	6.9
	Grand Vitara	2.5 (2494)	6	6	5.5	④	3.6	2.1	4.6	18.5	8.5
2002	Vitara	2.0 (1997)	5	5.9	①	④	3.6	2.1	4.6	③	6.9
	Grand Vitara	2.5 (2494)	6	6	5.5	④	3.6	2.1	4.6	18.5	8.5

Note: All capacities are approximate. Add fluid gradually and check to be sure a proper fluid level is obtained.

① 2-wheel drive model: 4.0 pts.
   4-wheel drive model: 3.2 pts.

② 3-speed transmission:
   Fluid drain, and filter and pan removal only: 5.9 pts.
   After complete transmission overhaul: 10.8 pts.
   4-speed overdrive transmission:
   Fluid drain, and filter and pan removal only: 5.3 pts.
   After complete transmission overhaul: 14.6 pts.

③ 2-door model: 11 gals.
   4-door model: 14.5 gals.

④ 2WD: 6.0 pts
   4WD: 5.2 pts.

93591CRD

VALVE SPECIFICATIONS

Year	Engine Displacement Liters (cc)	Engine ID/VIN	Seat Angle (deg.)	Face Angle (deg.)	Spring Test Pressure (lbs. @ in.)	Spring Installed Height (in.)	Stem-to-Guide Clearance (in.)		Stem Diameter (in.)	
							Intake	Exhaust	Intake	Exhaust
1999	1.6 (1590)	0	45	45	24-28@1.24	1.245	0.0008-0.0018	0.0018-0.0028	0.2152-0.2157	0.2142-0.2148
	2.0 (1997)	5	45	45	①	1.250	0.0008-0.0027	0.0018-0.0035	0.2348-0.2354	0.2339-0.2344
	2.5 (2494)	6	45	45	①	1.250	0.0008-0.0027	0.0018-0.0035	0.2348-0.2354	0.2339-0.2344
2000	1.6 (1590)	0	45	45	24-28@1.24	1.245	0.0008-0.0018	0.0018-0.0028	0.2152-0.2157	0.2142-0.2148
	2.0 (1997)	5	45	45	①	1.250	0.0008-0.0027	0.0018-0.0035	0.2348-0.2354	0.2339-0.2344
	2.5 (2494)	6	45	45	①	1.250	0.0008-0.0027	0.0018-0.0035	0.2348-0.2354	0.2339-0.2344
2001	2.0 (1997)	5	45	45	①	1.250	0.0008-0.0027	0.0018-0.0035	0.2348-0.2354	0.2339-0.2344
	2.5 (2494)	6	45	45	①	1.250	0.0008-0.0027	0.0018-0.0035	0.2348-0.2354	0.2339-0.2344
2002	2.0 (1997)	5	45	45	①	1.250	0.0008-0.0027	0.0018-0.0035	0.2348-0.2354	0.2339-0.2344
	2.5 (2494)	6	45	45	①	1.250	0.0008-0.0027	0.0018-0.0035	0.2348-0.2354	0.2339-0.2344

① Inner: 13.6-17.4@1.08
  Outer: 30.4-39.2@1.25

93591CRE

CRANKSHAFT AND CONNECTING ROD SPECIFICATIONS
All measurements are given in inches.

Year	Engine Displacement Liters (cc)	Engine ID/VIN	Crankshaft				Connecting Rod		
			Main Brg. Journal Dia.	Main Brg. Oil Clearance	Shaft End-play	Thrust on No.	Journal Diameter	Oil Clearance	Side Clearance
1999	1.6 (1590)	0	2.0465-2.0472	0.0006-0.0023	0.0044-0.0149	3	1.7316-1.7322	0.0008-0.0031	NA
	2.0 (1997)	5	2.2828-2.2834	0.0008-0.0023	0.0039-0.0165	3	1.9660-1.9709	0.0008-0.0031	NA
	2.5 (2494)	6	2.5583-2.5590	0.0008-0.0023	0.0044-0.0149	2	1.9678-1.9685	0.0016-0.0031	NA
2000	1.6 (1590)	0	2.0465-2.0472	0.0006-0.0023	0.0044-0.0149	3	1.7316-1.7322	0.0008-0.0031	NA
	2.0 (1997)	5	2.2828-2.2834	0.0008-0.0023	0.0039-0.0165	3	1.9660-1.9709	0.0008-0.0031	NA
	2.5 (2494)	6	2.5583-2.5590	0.0008-0.0023	0.0044-0.0149	2	1.9678-1.9685	0.0016-0.0031	NA
2001	2.0 (1997)	5	2.2828-2.2834	0.0008-0.0023	0.0039-0.0165	3	1.9660-1.9709	0.0008-0.0031	NA
	2.5 (2494)	6	2.5583-2.5590	0.0008-0.0023	0.0044-0.0149	2	1.9678-1.9685	0.0016-0.0031	NA
2002	2.0 (1997)	5	2.2828-2.2834	0.0008-0.0023	0.0039-0.0165	3	1.9660-1.9709	0.0008-0.0031	NA
	2.5 (2494)	6	2.5583-2.5590	0.0008-0.0023	0.0044-0.0149	2	1.9678-1.9685	0.0016-0.0031	NA

NA: Not Available

93591CRF

Timing belt service is covered in Section 3 of this manual

PISTON AND RING SPECIFICATIONS
All measurements are given in inches.

Year	Engine Displacement Liters (cc)	Engine ID/VIN	Piston Clearance	Ring Gap			Ring Side Clearance		
				Top Compression	Bottom Compression	Oil Control	Top Compression	Bottom Compression	Oil Control
1999	1.6 (1590)	0	0.0008-0.0015	0.0079-0.0275	0.0138-0.0275	0.0039-0.0669	0.0012-0.0027	0.0008-0.0023	NA
	2.0 (1997)	5	0.0008-0.0015	0.0079-0.0276	0.0138-0.0276	0.0079-0.0709	0.0012-0.0027	0.0008-0.0023	NA
	2.5 (2494)	6	0.0008-0.0015	0.0079-0.0276	0.0138-0.0276	0.0079-0.0709	0.0012-0.0027	0.0008-0.0023	NA
2000	1.6 (1590)	0	0.0008-0.0015	0.0079-0.0275	0.0138-0.0275	0.0039-0.0669	0.0012-0.0027	0.0008-0.0023	NA
	2.0 (1997)	5	0.0008-0.0015	0.0079-0.0276	0.0138-0.0276	0.0079-0.0709	0.0012-0.0027	0.0008-0.0023	NA
	2.5 (2494)	6	0.0008-0.0015	0.0079-0.0276	0.0138-0.0276	0.0079-0.0709	0.0012-0.0027	0.0008-0.0023	NA
2001	2.0 (1997)	5	0.0008-0.0015	0.0079-0.0276	0.0138-0.0276	0.0079-0.0709	0.0012-0.0027	0.0008-0.0023	NA
	2.5 (2494)	6	0.0008-0.0015	0.0079-0.0276	0.0138-0.0276	0.0079-0.0709	0.0012-0.0027	0.0008-0.0023	NA
2002	2.0 (1997)	5	0.0008-0.0015	0.0079-0.0276	0.0138-0.0276	0.0079-0.0709	0.0012-0.0027	0.0008-0.0023	NA
	2.5 (2494)	6	0.0008-0.0015	0.0079-0.0276	0.0138-0.0276	0.0079-0.0709	0.0012-0.0027	0.0008-0.0023	NA

NA: Not Available

93591CRG

TORQUE SPECIFICATIONS

All readings in ft. lbs.

Year	Engine Displacement Liters (cc)	Engine ID/VIN	Cylinder Head Bolts	Main Bearing Bolts	Rod Bearing Bolts	Crankshaft Damper Bolts	Flywheel Bolts	Manifold		Spark Plugs	Lug Nut
								Intake	Exhaust		
1999	1.6 (1590)	0	①	36-41	24-26	94 ②	57	13-20	13-20	14-21	58-80
	2.0 (1997)	5	③	④	33	109	51	13-20	13-20	14-21	58-80
	2.5 (2494)	6	③	④	33	109	51	16.5	22.0	18	69
2000	1.6 (1590)	0	①	36-41	24-26	94 ②	57	13-20	13-20	14-21	58-80
	2.0 (1997)	5	③	④	33	109	51	13-20	13-20	14-21	58-80
	2.5 (2494)	6	③	④	33	109	51	16.5	22.0	18	69
2001	2.0 (1997)	5	③	④	33	109	51	13-20	13-20	14-21	70
	2.5 (2494)	6	③	④	33	109	51	17	22	18	70
	2.0 (1997)	C	③	④	33	109	51	17	17	18	70
2002	2.0 (1997)	5	③	④	33	109	51	13-20	13-20	14-21	70
	2.5 (2494)	6	③	④	33	109	51	17	22	18	70

① Step 1: 26 ft. lbs.

   Step 2: 41 ft. lbs.

   Step 3: Loosen in reverse order to 0 ft. lbs.

   Step 4: 26 ft. lbs.

   Step 5: 52 ft. lbs.

② Value shown is for crankshaft timing belt sprocket

③ Step 1: 38 ft. lbs.

   Step 2: 61 ft. lbs.

   Step 3: Loosen in reverse order to 0 ft. lbs.

   Step 4: 38 ft. lbs.

   Step 5: 76 ft. lbs.

   Step 6: Tighten 6mm bolt to 8 ft. lbs.

④ 10mm: 43.5 ft. lbs.

   8mm: 19.5 ft. lbs.

⑤ Use multiple passes to arrive at final torque.

⑥ Value shown is for crankshaft timing belt sprocket

93591CRH

Heater Core replacement is covered in Section 2 of this manual

WHEEL ALIGNMENT

Year	Model	Caster Range (+/-Deg.)	Caster Preferred Setting (Deg.)	Camber Range (+/-Deg.)	Camber Preferred Setting (Deg.)	Toe-in (in.)	Steering Axis Inclination (Deg.)
1999	Vitara	1.00	+2.66	1.00	0	0+/-0.08	—
	Grand Vitara	1.00	+2.66	1.00	0	0+/-0.08	—
2000	Vitara	1.00	+2.66	1.00	0	0+/-0.08	—
	Grand Vitara	1.00	+2.66	1.00	0	0+/-0.08	—
2001	Vitara	1.00	+2.66	1.00	0	0+/-0.08	—
	Grand Vitara	1.00	+2.66	1.00	0	0+/-0.08	—
2002	Vitara	1.00	+2.66	1.00	0	0+/-0.08	—
	Grand Vitara	1.00	+2.66	1.00	0	0+/-0.08	—

93591CRI

TIRE, WHEEL AND BALL JOINT SPECIFICATIONS

Year	Model	OEM Tires Standard	OEM Tires Optional	Tire Pressures (psi) Front	Tire Pressures (psi) Rear	Wheel Size	Ball Joint Inspection
1999	Vitara 2-door 2wd	P195/75SR15	P215/65SR16	26	26	6-JJ	①
	Vitara 2-door 4wd	P205/75R15	none	26	26	6.5JJ	①
	Vitara 4-door	P215/65R16	none	26	26	6.5JJ	①
	Grand Vitara	P235/65SR16	none	26	26	7-JJ	①
2000	Vitara 1.6L 2wd	P195/75SR15	none	26	26	5.5JJ	①
	Vitara 1.6L 4wd	P205/75R15	none	26	26	6.5JJ	①
	Vitara 2.0L	P215/65R16	none	26	26	6.5JJ	①
	Grand Vitara	P235/65SR16	none	26	26	7-JJ	①
2001	Vitara 2.0L	P215/65R16	none	26	26	6.5JJ	①
	Grand Vitara	P235/65SR16	none	26	26	7-JJ	①
2002	Vitara 2.0L	P215/65R16	none	26	26	6.5JJ	①
	Grand Vitara	P235/65SR16	none	26	26	7-JJ	①

OEM: Original Equipment Manufacturer

PSI: Pounds Per Square Inch

STD: Standard

OPT: Optional

① Replace if any measurable movement is found.

93591CRJ

BRAKE SPECIFICATIONS

All measurements in inches unless noted

Year	Model	Brake Disc			Brake Drum Diameter			Minimum Lining Thickness		Brake Caliper	
		Original Thickness	Minimum Thickness	Maximum Runout	Original Inside Diameter	Max. Wear Limit	Maximum Machine Diameter	Front	Rear	Bracket Bolts (ft. lbs.)	Mounting Bolts (ft. lbs.)
1999	Vitara	0.670	0.590	0.006	8.66	8.74	8.74	0.08	0.04	51-72	19-21
	Grand Vitara	0.866	0.787	0.006	8.66	8.74	8.74	0.08	0.04	61.5	①
2000	Vitara	0.670	0.590	0.006	8.66	8.74	8.74	0.08	0.04	61.5	19-21
	Grand Vitara	0.866	0.787	0.006	8.66	8.74	8.74	0.08	0.04	61.5	①
2001	Vitara	0.670	0.590	0.006	8.66	8.74	8.74	0.08	0.04	61.5	19-21
	Grand Vitara	0.866	0.787	0.006	8.66	8.74	8.74	0.08	0.04	61.5	①
2002	Vitara	0.670	0.590	0.006	8.66	8.74	8.74	0.08	0.04	61.5	19-21
	Grand Vitara	0.866	0.787	0.006	8.66	8.74	8.74	0.08	0.04	61.5	①

① 10mm: 37 ft. lbs.
12mm: 62 ft. lbs.

93591CRK

Brake service is covered in Section 4 of this manual

SCHEDULED MAINTENANCE INTERVALS
SUZUKI—VITARA, GRAND VITARA

TO BE SERVICED	TYPE OF SERVICE	VEHICLE MILEAGE INTERVAL (x1000)												
		7.5	15	22.5	30	37.5	45	52.5	60	67.5	75	82.5	90	97.5
Engine oil & filter	R	✓	✓	✓	✓	✓	✓	✓	✓	✓	✓	✓	✓	✓
Automatic transmission fluid ①	S/I	✓	✓	✓	✓	✓	✓	✓	✓	✓	✓	✓	✓	✓
Manual transmission oil ②	S/I	✓	✓	✓	✓	✓	✓	✓	✓	✓	✓	✓	✓	✓
Steering system	S/I	✓	✓	✓	✓	✓	✓	✓	✓	✓	✓	✓	✓	✓
Transfer & differential oil ②	S/I	✓	✓	✓	✓	✓	✓	✓	✓	✓	✓	✓	✓	✓
Wheel discs & free wheeling hubs	S/I	✓	✓	✓	✓	✓	✓	✓	✓	✓	✓	✓	✓	✓
Suspension system	S/I	✓	✓	✓	✓	✓	✓	✓	✓	✓	✓	✓	✓	✓
Brake discs & pads (front)	S/I		✓		✓		✓		✓		✓		✓	
Brake drums & shoes (rear)	S/I		✓		✓		✓		✓		✓		✓	
Brake fluid ③	S/I		✓		✓		✓		✓		✓		✓	
Brake hoses & pipes	S/I		✓		✓		✓		✓		✓		✓	
Brake pedal	S/I		✓		✓		✓		✓		✓		✓	
Brake lever & cable	S/I		✓		✓		✓		✓		✓		✓	
Clutch	S/I		✓		✓		✓		✓		✓		✓	
Idle speed	S/I		✓		✓		✓		✓		✓		✓	
Propeller shafts	S/I		✓		✓		✓		✓		✓		✓	
Valve lash (clearance)	S/I		✓		✓		✓		✓		✓		✓	
Wheel bearings	S/I		✓		✓		✓		✓		✓		✓	
Air cleaner filter element	R				✓				✓				✓	
Engine coolant	R				✓				✓				✓	
Fuel filter	R				✓				✓				✓	
Spark plugs	R				✓				✓				✓	
Cooling system hoses	S/I				✓				✓				✓	
Drive belt(s)	S/I				✓				✓				✓	
Exhaust pipes & mountings	S/I				✓				✓				✓	
Fuel lines & connections	S/I				✓				✓				✓	
Camshaft timing belt	R								✓					✓
Distributor cap & rotor	S/I							✓						
Emission-related hoses & tubes	S/I							✓						
Oxygen sensor	S/I											✓		
EVAP canister	R	every 100,000 miles												
PCV valve	R							✓						
EGR system	S/I							✓						

93591CRL

SCHEDULED MAINTENANCE INTERVALS
SUZUKI—VITARA, GRAND VITARA

TO BE SERVICED	TYPE OF SERVICE	VEHICLE MILEAGE INTERVAL (x1000)												
		7.5	15	22.5	30	37.5	45	52.5	60	67.5	75	82.5	90	97.5
Fuel Injectors	S/I	every 100,000 miles												
TWC converter	S/I	every 100,000 miles												

R: Replace   S/I: Service or Inspect

① Replace at 100,000 miles.

② Replace oil every 30,000 miles.

③ Replace every 60,000 miles.

FREQUENT OPERATION MAINTENANCE (SEVERE SERVICE)

 If a vehicle is operated under any of the following conditions it is considered severe service:

- Extremely dusty areas.

- 50% or more of the vehicle operation is in 32°C (90°F) or higher temperatures, or constant operation in temperatures below 0°C (32°F).

- Prolonged idling (vehicle operation in stop and go traffic).

- Frequent short running periods (engine does not warm to normal operating temperatures).

- Police, taxi, delivery usage or trailer towing usage.

Oil & oil filter: replace every 3000 miles.

Air cleaner filter element: service or inspect every 3000 miles & replace every 15,000 miles.

Steering wheel free play, gear box oil & linkage: service or inspect every 3000 miles.

Brake & nuts on chassis: tighten every 6000 miles.

Brake discs & pads (front): service or inspect every 6000 miles.

Brake drums & shoes (rear): service or inspect every 6000 miles.

Exhaust pipes & mountings: tighten every 6000 miles.

Propeller shafts: service or inspect every 6000 miles.

Automatic transmission fluid & filter: replace every 15,000 miles.

Distributor cap & ignition wires: service or inspect every 15,000 miles.

Drive belt(s): service or inspect every 15,000 miles.

Manual transmission oil: replace every 15,000 miles.

Transfer & differential oil: replace every 15,000 miles.

93591CRM

For complete Engine Mechanical specifications, see Section 1 of this manual

SCHEDULED MAINTENANCE INTERVALS
SUZUKI
VITARA, GRAND VITARA

The following should be used as a guide when determining the amount of work required for a particular service.
In estimating how long a particular Scheduled Maintenance Service should take, please observe the following:

- Labor Time is time based on field research and data supplied by the vehicle manufacturer.
- Labor time operations are given in hours and tenths of an hour.
- All labor operations are to be used as a guide.

Mechanic Skill Level Codes:
(A) PRECISION: Highly skilled with multiple certification.
(B) GENERAL: Normally skilled with certification.
(C) MAINTENANCE: Semi-skilled working on certification.

	LABOR TIME		LABOR TIME		LABOR TIME
7500 Mile Service (B)		**37500 Mile Service (B)**		**75000 Mile Service (B)**	
All models	.7	All models	.7	All models	2.2
w/AT add	.1	*w/AT add*	.1	*w/AT add*	.1
w/4WD add	.2	*w/4WD add*	.2	*w/4WD add*	.2
15000 Mile Service (B)		**45000 Mile Service (B)**		**82500 Mile Service (B)**	
All models	2.2	All models	2.2	All models	.9
w/AT add	.1	*w/AT add*	.1	*w/AT add*	.1
w/4WD add	.2	*w/4WD add*	.2	*w/4WD add*	.2
22500 Mile Service (B)		**52500 Mile Service (B)**		**90000 Mile Service (B)**	
All models	.7	All models	1.0	All models	4.5
w/AT add	.1	*w/AT add*	.1	*w/AT add*	.1
w/4WD add	.2	*w/4WD add*	.2	*w/4WD add*	.2
30000 Mile Service (B)		**60000 Mile Service (B)**		**97500 Mile Service (B)**	
All models	3.1	All models	5.3	All models	.7
w/AT add	.1	*w/AT add*	.1	*w/AT add*	.1
w/4WD add	.2	*w/4WD add*	.2	*w/4WD add*	.2
		67500 Mile Service (B)			
		All models	.7		
		w/AT add	.1		
		w/4WD add	.2		

93591CRN

TOYOTA
4Runner • Highlander • RAV4

ENGINE AND VEHICLE IDENTIFICATION

Code ①			Engine					Code ②	Model Year
	Liters (cc)	Cu. In.	Cyl.	Fuel Sys.	Engine Type	Eng. Mfg.			Year
1AZ-FE	2.0 (1998)	122	4	SFI	DOHC	Toyota		X	1999
1MZ-FE	3.0 (2995)	183	6	SFI	DOHC	Toyota		Y	2000
3S-FE	2.0 (1998)	122	4	MFI	DOHC	Toyota		1	2001
3RZ-FE	2.7 (2693)	164	4	MFI	DOHC	Toyota		2	2002
5VZ-FE	3.4 (3378)	206	6	MFI	DOHC	Toyota		3	2003
2AZ-FE	2.4 (2362)	144	4	SFI	DOHC	Toyota			

SFI: Sequential Fuel Injection

MFI: Multi-port Fuel Injection

DOHC: Double Overhead Camshaft

① Stamped on the left side of the engine block

② 10th digit of the Vehicle Identification Number (VIN)

93591CAN

GENERAL ENGINE SPECIFICATIONS

Year	Model	Engine Displacement Liters (cc)	Engine Series (ID/VIN)	Fuel System	Net Horsepower @ rpm	Net Torque @ rpm (ft. lbs.)	Bore x Stroke (in.)	Compression Ratio	Oil Pressure @ rpm
1999	RAV4	2.0 (1998)	3S-FE	MFI	127@5400	132@4600	3.40x3.40	9.5:1	NA
	4Runner	2.7 (2693)	3RZ-FE	MFI	150@4800	177@4000	3.74x3.74	9.5:1	36-71@3000
		3.4 (3378)	5VZ-FE	MFI	183@4800	217@3600	3.68x3.23	9.6:1	NA
2000	RAV4	2.0 (1998)	3S-FE	MFI	127@5400	132@4600	3.40x3.40	9.5:1	NA
	4Runner	2.7 (2693)	3RZ-FE	MFI	150@4800	177@4000	3.74x3.74	9.5:1	36-71@3000
		3.4 (3378)	5VZ-FE	MFI	183@4800	217@3600	3.68x3.23	9.6:1	NA
2001	RAV4	2.0 (1998)	1AZ-FE	SFI	127@5400	132@4600	3.40x3.40	9.5:1	NA
	4Runner	3.4 (3378)	5VZ-FE	MFI	183@4800	217@3600	3.68x3.23	9.6:1	NA
	Highlander	2.4 (2362)	2AZ-FE	SFI	155@5600	163@4000	3.48x3.78	NA	36@3000
		3.0 (2995)	1MZ-FE	SFI	220@5800	222@4400	3.44x3.27	10.5:1	43-78@3000
2002	RAV4	2.0 (1998)	1AZ-FE	SFI	127@5400	132@4600	3.40x3.40	9.5:1	NA
	4Runner	3.4 (3378)	5VZ-FE	MFI	183@4800	217@3600	3.68x3.23	9.6:1	NA
	Highlander	2.4 (2362)	2AZ-FE	SFI	155@5600	163@4000	3.48x3.78	NA	36@3000
		3.0 (2995)	1MZ-FE	SFI	220@5800	222@4400	3.44x3.27	10.5:1	43-78@3000

NA: Not Available

SFI: Sequential Fuel Injection

MFI: Multi-port Fuel Injection

93591CAO

For Accessory Drive Belt illustrations, see Section 1 of this manual

ENGINE TUNE-UP SPECIFICATIONS

Year	Engine Displacement Liters (cc)	Engine ID/VIN	Spark Plug Gap (in.)	Ignition Timing (deg.)*	Fuel Pump (psi)	Idle Speed (rpm) MT	Idle Speed (rpm) AT	Valve Clearance Intake	Valve Clearance Exhaust
1999	2.0 (1998)	3S-FE	0.043	8-12B	44-50	700-800	700-800	0.007-0.011	0.011-0.015
	3.0 (2995)	1MZ-FE	0.043	8-12B	44-50	—	650-750	0.006-0.010	0.010-0.014
	2.7 (2693)	3RZ-FE	0.031	8-12B	38-44	650-750	650-750	0.006-0.010	0.010-0.014
	3.4 (3378)	5VZ-FE	0.031	8-12B	38-44	650-750	650-750	0.006-0.010	0.010-0.014
2000	2.0 (1998)	3S-FE	0.043	8-12B	44-50	700-800	700-800	0.007-0.011	0.011-0.015
	3.0 (2995)	1MZ-FE	0.043	8-12B	44-50	—	650-750	0.006-0.010	0.010-0.014
	2.7 (2693)	3RZ-FE	0.031	8-12B	38-44	650-750	650-750	0.006-0.010	0.010-0.014
	3.4 (3378)	5VZ-FE	0.031	8-12B	38-44	650-750	650-750	0.006-0.010	0.010-0.014
2001	2.0 (1998)	1AZ-FE	0.043	8-12B	44-50	700-800	700-800	0.007-0.011	0.011-0.015
	2.4 (2362)	2AZ-FE	0.041	8-12B	44-50	—	600-700	0.007-0.011	0.012-0.016
	3.0 (2995)	1MZ-FE	0.043	8-12B	44-50	—	650-750	0.006-0.010	0.010-0.014
	3.4 (3378)	5VZ-FE	0.031	8-12B	38-44	—	650-750	0.006-0.010	0.010-0.014
2002	2.0 (1998)	1AZ-FE	0.043	8-12B	44-50	700-800	700-800	0.008-0.011	0.012-0.016
	2.4 (2362)	2AZ-FE	0.041	8-12B	44-50	—	600-700	0.007-0.011	0.012-0.016
	3.0 (2995)	1MZ-FE	0.039-0.043	8-12B	44-50	—	650-750	0.006-0.010	0.010-0.014
	3.4 (3378)	5VZ-FE	0.039-0.043	8-12B	38-44	—	650-750	0.006-0.009	0.011-0.014

NOTE: The Vehicle Emission Control Information label often reflects specification changes made during production. The label figures must be used if they differ from those in this chart.

B: Before top dead center

* With terminals TC and CG connected to DLC3

93591CAP

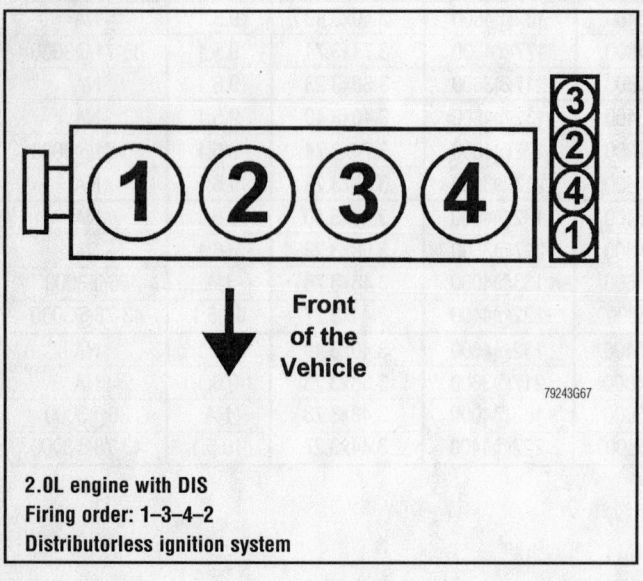

2.0L engine with DIS
Firing order: 1–3–4–2
Distributorless ignition system

79243G67

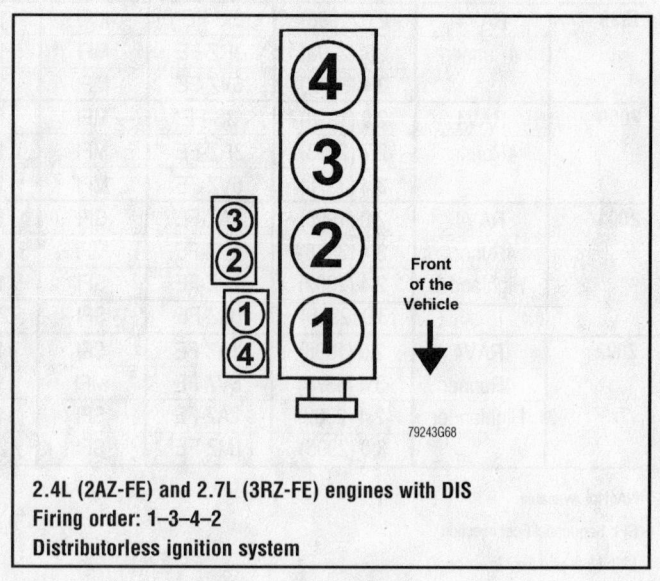

2.4L (2AZ-FE) and 2.7L (3RZ-FE) engines with DIS
Firing order: 1–3–4–2
Distributorless ignition system

79243G68

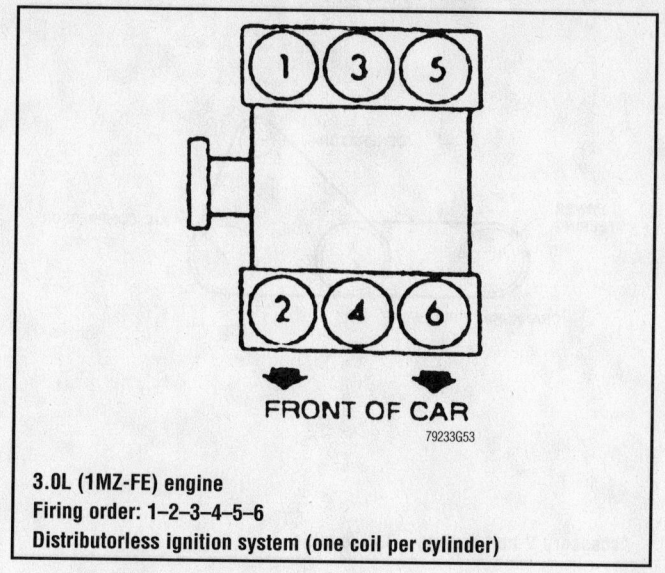

3.0L (1MZ-FE) engine
Firing order: 1–2–3–4–5–6
Distributorless ignition system (one coil per cylinder)

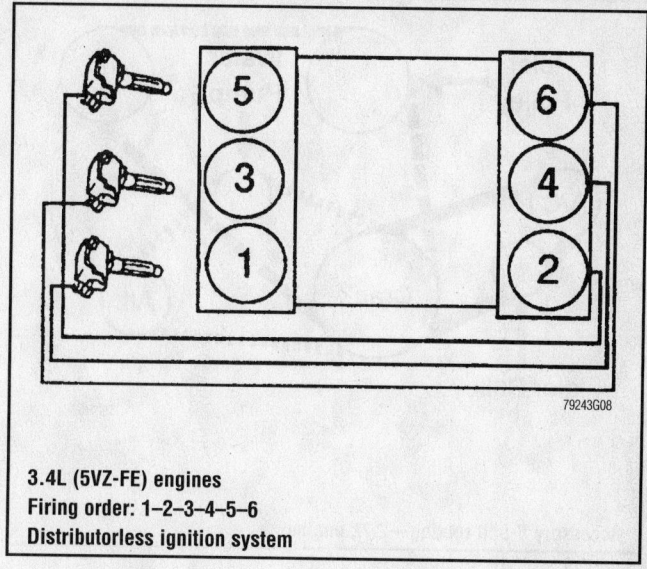

3.4L (5VZ-FE) engines
Firing order: 1–2–3–4–5–6
Distributorless ignition system

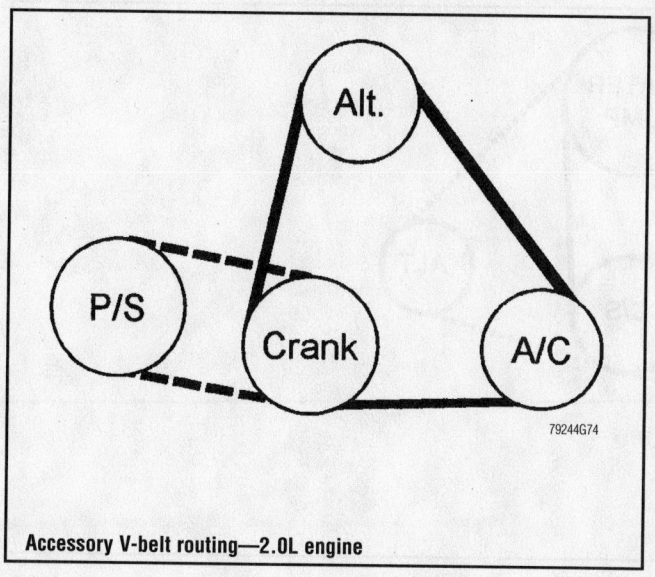

Accessory V-belt routing—2.0L engine

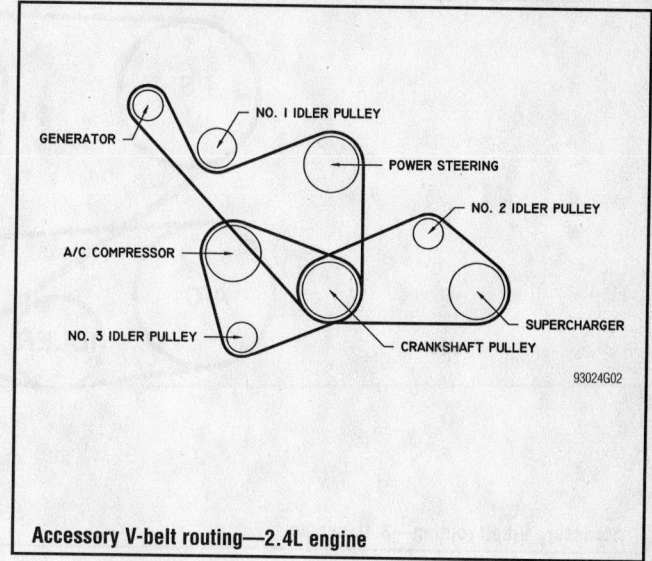

Accessory V-belt routing—2.4L engine

For Tire, Wheel and Ball Joint specifications, see Section 1 of this manual

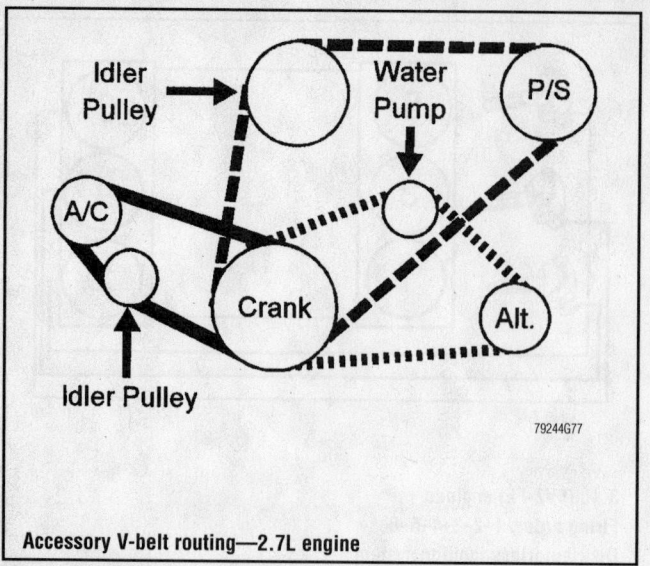

Accessory V-belt routing—2.7L engine

79244G77

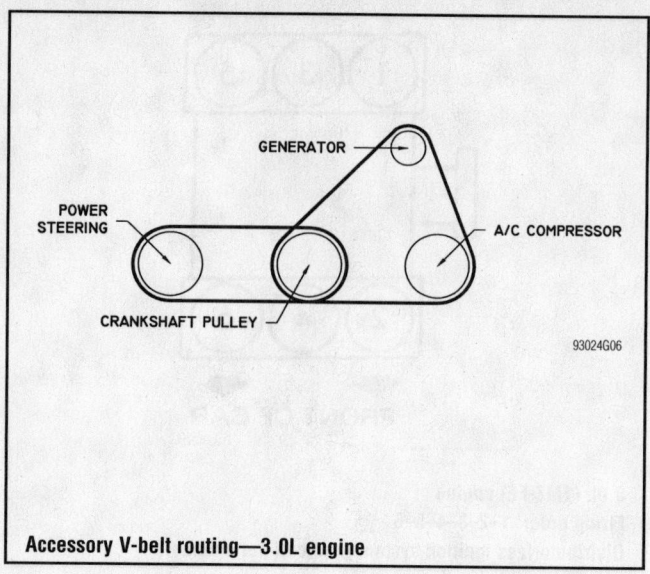

Accessory V-belt routing—3.0L engine

93024G06

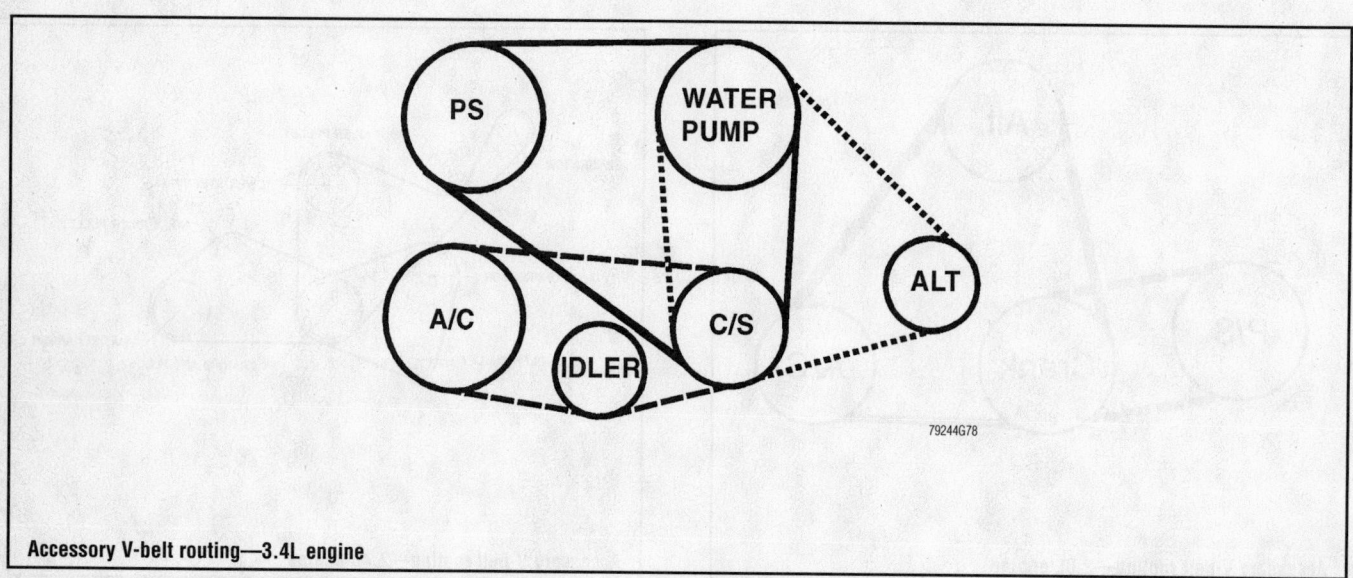

Accessory V-belt routing—3.4L engine

79244G78

CAPACITIES

Year	Model	Engine Displacement Liters (cc)	Engine ID/VIN	Engine Oil with Filter (qts.)	Transmission (pts.) 5-Spd	Transmission (pts.) Auto.*	Transfer Case (pts.)	Drive Axle Front (pts.)	Drive Axle Rear (pts.)	Fuel Tank (gal.)	Cooling System (qts.)
1999	RAV4	2.0 (1998)	3S-FE	4.1	①	7.0	2.0	—	2.0	15.3	②
	4Runner	2.7 (2693)	3RZ-FE	5.8	③	④	2.3	⑤	⑥	18.0	⑦
		3.4 (3378)	5VZ-FE	5.5	③	④	2.4	⑤	⑥	18.0	⑧
2000	RAV4	2.0 (1998)	3S-FE	4.1	①	7.0	2.0	—	2.0	15.3	②
	4Runner	2.7 (2693)	3RZ-FE	5.8	③	④	⑨	⑤	⑥	18.0	⑦
		3.4 (3378)	5VZ-FE	5.5	③	④	⑨	⑤	⑥	18.0	⑧
2001	RAV4	2.0 (1998)	1AZ-FE	4.4	⑩	⑪	2.0	—	2.0	14.8	⑫
	4Runner	3.4 (3378)	5VZ-FE	5.5	—	⑬	2.6	⑤	⑥	18.5	⑧
	Highlander	2.4 (2362)	2AZ-FE	4.0	—	⑪	2.0	—	2.0	19.8	6.8
		3.0 (2995)	1MZ-FE	5.0	—	⑪	2.0	—	2.0	19.8	9.9
2002	RAV4	2.0 (1998)	1AZ-FE	4.6	⑩	⑪	2.0	—	2.0	14.8	⑫
	4Runner	3.4 (3378)	5VZ-FE	5.5	—	⑬	2.6	⑤	⑭	18.5	⑧
	Highlander	2.4 (2362)	2AZ-FE	4.0	—	⑪	2.0	—	2.0	19.8	6.8
		3.0 (2995)	1MZ-FE	5.0	—	⑪	2.0	—	2.0	19.8	9.9

*After draining, add the following amounts, then, fill to the cold full line.

① 2WD: 8.2
   4WD: 10.6

② M/T: 8.5
   A/T: 8.1

③ W59:
   2WD: 5.4
   4WD: 5.2
   R150, R150F:
   2WD: 5.4
   4WD: 4.6

④ A43D: 5.0
   A340E: 3.4
   A340F: 4.2

⑤ Without ADD: 2.32
   With ADD: 2.44

⑥ 2WD: 5.8
   4WD with differential locks: 5.8
   4WD without differential locks: 5.2

⑦ With rear heater: 11.6
   Without rear heater: 10.6
   A340F: 4.2

⑧ With rear heater: 9.5
   Without rear heater: 8.5

⑨ VF2A: 2.2
   VF3AM: 2.6

⑩ 2wd: 5.2
   4wd: 7.2

⑪ 2wd: 7.0
   4wd: 8.2

⑫ MT: 6.7
   AT: 6.6

⑬ 2wd: 3.4
   4wd: 4.2

⑭ 2wd: 5.82
   4wd w/o diff. Lock: 5.18
   4wd w/diff. Lock: 5.82

93591CAQ

For Wheel Alignment specifications, see Section 1 of this manual

VALVE SPECIFICATIONS

Year	Engine Displacement Liters (cc)	Engine ID/VIN	Seat Angle (deg.)	Face Angle (deg.)	Spring Test Pressure (lbs. @ in.)	Spring Installed Height (in.)	Stem-to-Guide Clearance (in.)		Stem Diameter (in.)	
							Intake	Exhaust	Intake	Exhaust
1999	2.0 (1998)	3S-FE	45	44.5	36.8-42.5@ 1.366	1.366	0.0010-0.0024	0.0012-0.0026	0.2350-0.2356	0.2348-0.2354
	3.0 (2995)	1MZ-FE	45	40.5	41.9-46.3@ 1.437	1.331	0.0010-0.0024	0.0012-0.0026	0.2154-0.2159	0.2152-0.2156
	2.7 (2693)	3RZ-FE	45	44.5	40.0-46.0@ 1.406	1.406	0.0010-0.0024	0.0012-0.0026	0.2350-0.2356	0.2348-0.2354
	3.4 (3378)	5VZ-FE	45	44.5	41.9-46.3@ 1.311	1.311	0.0010-0.0024	0.0012-0.0026	0.2350-0.2356	0.2348-0.2354
2000	2.0 (1998)	3S-FE	45	44.5	36.8-42.5@ 1.366	1.366	0.0010-0.0024	0.0012-0.0026	0.2350-0.2356	0.2348-0.2354
	3.0 (2995)	1MZ-FE	45	40.5	41.9-46.3@ 1.437	1.331	0.0010-0.0024	0.0012-0.0026	0.2154-0.2159	0.2152-0.2156
	2.7 (2693)	3RZ-FE	45	44.5	40.0-46.0@ 1.406	1.406	0.0010-0.0024	0.0012-0.0026	0.2350-0.2356	0.2348-0.2354
	3.4 (3378)	5VZ-FE	45	44.5	41.9-46.3@ 1.311	1.311	0.0010-0.0024	0.0012-0.0026	0.2350-0.2356	0.2348-0.2354
2001	2.0 (1998)	1AZ-FE	45	44.5	36.8-42.5@ 1.366	1.366	0.0010-0.0024	0.0012-0.0026	0.2350-0.2356	0.2348-0.2354
	2.4 (2362)	2AZ-FE	45	44.5	NA	NA	0.0010-0.0024	0.0012-0.0026	0.2154-0.2159	0.2152-0.2157
	3.0 (2995)	1MZ-FE	45	40.5	41.9-46.3@ 1.437	1.331	0.0010-0.0024	0.0012-0.0026	0.2154-0.2159	0.2152-0.2156
	3.4 (3378)	5VZ-FE	45	44.5	41.9-46.3@ 1.311	1.311	0.0010-0.0024	0.0012-0.0026	0.2350-0.2356	0.2348-0.2354
2002	2.0 (1998)	1AZ-FE	45	44.5	41.4-45.9@ 1.339	1.339	0.0010-0.0024	0.0012-0.0026	0.2154-0.2159	0.2152-0.2157
	2.4 (2362)	2AZ-FE	45	44.5	NA	NA	0.0010-0.0024	0.0012-0.0026	0.2154-0.2159	0.2152-0.2157
	3.0 (2995)	1MZ-FE	45	40.5	41.9-46.3@ 1.437	1.331	0.0010-0.0024	0.0012-0.0026	0.2154-0.2159	0.2152-0.2156
	3.4 (3378)	5VZ-FE	45	44.5	41.9-46.3@ 1.311	1.311	0.0010-0.0024	0.0012-0.0026	0.2350-0.2356	0.2348-0.2354

93591CAR

CRANKSHAFT AND CONNECTING ROD SPECIFICATIONS

All measurements are given in inches.

Year	Engine Displacement Liters (cc)	Engine ID/VIN	Crankshaft				Connecting Rod		
			Main Brg. Journal Dia.	Main Brg. Oil Clearance	Shaft End-play	Thrust on No.	Journal Diameter	Oil Clearance	Side Clearance
1999	2.0 (1998)	3S-FE	2.1653-2.1655	0.0010-0.0017	0.0008-0.0087	3	2.0466-2.0472	0.0009-0.0022	0.0063-0.0123
	3.0 (2995)	1MZ-FE	2.4011-2.4016	①	0.0016-0.0095	2	2.0863-2.0866	0.0015-0.0025	0.0059-0.0188
	2.7 (2693)	3RZ-FE	2.2615-2.3620	0.0012-0.0022	0.0008-0.0087	3	2.0861-2.0866	0.0009-0.0022	0.0063-0.0123
	3.4 (3378)	5VZ-FE	2.5191-2.5197	0.0008-0.0015	0.0008-0.0087	2	2.1648-2.1654	0.0009-0.0021	0.0059-0.0130
2000	2.0 (1998)	3S-FE	2.1653-2.1655	0.0010-0.0017	0.0008-0.0087	3	2.0466-2.0472	0.0009-0.0022	0.0063-0.0123
	3.0 (2995)	1MZ-FE	2.4011-2.4016	①	0.0016-0.0095	2	2.0863-2.0866	0.0015-0.0025	0.0059-0.0188
	2.7 (2693)	3RZ-FE	2.2615-2.3620	0.0012-0.0022	0.0008-0.0087	3	2.0861-2.0866	0.0009-0.0022	0.0063-0.0123
	3.4 (3378)	5VZ-FE	2.5191-2.5197	0.0008-0.0015	0.0008-0.0087	2	2.1648-2.1654	0.0009-0.0021	0.0059-0.0130
2001	2.0 (1998)	1AZ-FE	2.1653-2.1655	0.0010-0.0017	0.0008-0.0087	3	2.0466-2.0472	0.0009-0.0022	0.0063-0.0123
	2.4 (2362)	2AZ-FE	2.0654-2.1648	0.0009-0.0019	0.0016-0.0094	2	1.8894-1.8898	0.0009-0.0019	0.0063-0.0143
	3.0 (2995)	1MZ-FE	2.4011-2.4016	①	0.0016-0.0095	2	2.0863-2.0866	0.0015-0.0025	0.0059-0.0188
	3.4 (3378)	5VZ-FE	2.5191-2.5197	0.0008-0.0015	0.0008-0.0087	2	2.1648-2.1654	0.0009-0.0021	0.0059-0.0130
2002	2.0 (1998)	1AZ-FE	2.1649-2.1655	0.0010-0.0016	0.0008-0.0087	3	1.8894-1.8898	0.0009-0.0019	0.0063-0.0143
	2.4 (2362)	2AZ-FE	2.0654-2.1648	0.0009-0.0019	0.0016-0.0094	2	1.8894-1.8898	0.0009-0.0019	0.0063-0.0143
	3.0 (2995)	1MZ-FE	2.4011-2.4016	①	0.0016-0.0095	2	2.0863-2.0866	0.0015-0.0025	0.0059-0.0188
	3.4 (3378)	5VZ-FE	2.5191-2.5197	②	0.0008-0.0087	2	2.1648-2.1654	0.0009-0.0021	0.0059-0.0130

① Journals 1 and 4: 0.0006 - 0.0013 in.
  Journals 2 and 3: 0.0010 - 0.0018 in.
② No. 1: 0.0008-0.0015 in.
  All others: 0.0009-0.0017 in.

93591CAS

For Maintenance Interval recommendations, see Section 1 of this manual

PISTON AND RING SPECIFICATIONS

All measurements are given in inches.

Year	Engine Displacement Liters (cc)	Engine ID/VIN	Piston Clearance	Ring Gap			Ring Side Clearance		
				Top Compression	Bottom Compression	Oil Control	Top Compression	Bottom Compression	Oil Control
1999	2.0 (1998)	3S-FE	0.0056-0.0064	0.0106-0.0185	0.0177-0.0256	0.0039-0.0177	0.0012-0.0028	0.0012-0.0028	SNUG
	3.0 (2995)	1MZ-FE	0.0033-0.0042	0.0098-0.0138	0.0138-0.0177	0.0059-0.0157	0.0008-0.0028	0.0008-0.0024	SNUG
	2.7 (2693)	3RZ-FE	0.0019-0.0028	0.0118-0.0157	0.0157-0.0194	0.0051-0.0150	0.0008-0.0028	0.0012-0.0028	SNUG
	3.4 (3378)	5VZ-FE	0.0053-0.0060	0.0118-0.0197	0.0157-0.0236	0.0059-0.0217	0.0016-0.0031	0.0012-0.0028	SNUG
2000	2.0 (1998)	3S-FE	0.0056-0.0064	0.0106-0.0185	0.0177-0.0256	0.0039-0.0177	0.0012-0.0028	0.0012-0.0028	SNUG
	3.0 (2995)	1MZ-FE	0.0033-0.0042	0.0098-0.0138	0.0138-0.0177	0.0059-0.0157	0.0008-0.0028	0.0008-0.0024	SNUG
	2.7 (2693)	3RZ-FE	0.0019-0.0028	0.0118-0.0157	0.0157-0.0194	0.0051-0.0150	0.0008-0.0028	0.0012-0.0028	SNUG
	3.4 (3378)	5VZ-FE	0.0053-0.0060	0.0118-0.0197	0.0157-0.0236	0.0059-0.0217	0.0016-0.0031	0.0012-0.0028	SNUG
2001	2.0 (1998)	1AZ-FE	0.0056-0.0064	0.0106-0.0185	0.0177-0.0256	0.0039-0.0177	0.0012-0.0028	0.0012-0.0028	SNUG
	2.4 (2362)	2AZ-FE	0.0020-0.0029	0.0087-0.0126	0.0197-0.0236	0.0039-0.0138	0.0012-0.0028	0.0012-0.0028	SNUG
	3.0 (2995)	1MZ-FE	0.0033-0.0042	0.0098-0.0138	0.0138-0.0177	0.0059-0.0157	0.0008-0.0028	0.0008-0.0024	SNUG
	3.4 (3378)	5VZ-FE	0.0053-0.0060	0.0118-0.0197	0.0157-0.0236	0.0059-0.0217	0.0016-0.0031	0.0012-0.0028	SNUG
2002	2.0 (1998)	1AZ-FE	0.0025-0.0034	0.0118-0.0157	0.0185-0.0244	0.0039-0.0138	0.0008-0.0028	0.0008-0.0024	0.0028-0.0059
	2.4 (2362)	2AZ-FE	0.0020-0.0029	0.0087-0.0126	0.0197-0.0236	0.0039-0.0138	0.0012-0.0028	0.0012-0.0028	SNUG
	3.0 (2995)	1MZ-FE	0.0033-0.0042	0.0098-0.0138	0.0138-0.0177	0.0059-0.0157	0.0008-0.0028	0.0008-0.0024	SNUG
	3.4 (3378)	5VZ-FE	0.0053-0.0060	0.0118-0.0197	0.0157-0.0236	0.0059-0.0217	0.0016-0.0031	0.0012-0.0028	SNUG

93591CAT

TORQUE SPECIFICATIONS
All readings in ft. lbs.

Year	Engine Displacement Liters (cc)	Engine ID/VIN	Cylinder Head Bolts	Main Bearing Bolts	Rod Bearing Bolts	Crankshaft Damper Bolts	Flywheel Bolts	Manifold Intake	Manifold Exhaust	Spark Plugs	Lug Nuts
1999	2.0 (1998)	3S-FE	①	43	②	80	③	14	36	13	76
	3.0 (2995)	1MZ-FE	④	⑤	②	159	61	32	36	13	70
	2.7 (2693)	3RZ-FE	⑥	⑦	⑧	116	⑨	22	36	14	83
	3.4 (3378)	5VZ-FE	⑩	⑪	②	184	63-67	13	30	13	76
2000	2.0 (1998)	3S-FE	①	43	②	80	③	14	36	13	76
	3.0 (2995)	1MZ-FE	④	⑤	②	159	61	32	36	13	70
	2.7 (2693)	3RZ-FE	⑥	⑦	⑧	116	⑨	22	36	14	83
	3.4 (3378)	5VZ-FE	⑩	⑪	②	184	63-67	13	30	13	76
2001	2.0 (1998)	1AZ-FE	①	43	②	80	③	14	36	13	76
	2.4 (2362)	2AZ-FE	⑫	29	②	125	72	22	27	14	76
	3.0 (2995)	1MZ-FE	④	⑤	②	159	61	32	36	13	70
	3.4 (3378)	5VZ-FE	⑩	⑪	②	184	63-67	13	30	13	76
2002	2.0 (1998)	1AZ-FE	⑫	⑬	②	125	⑭	22	25	15	76
	2.4 (2362)	2AZ-FE	⑫	29	②	125	72	22	27	14	76
	3.0 (2995)	1MZ-FE	④	⑤	②	159	61	32	36	13	70
	3.4 (3378)	5VZ-FE	⑮	⑪	②	213	61	13	30	13	76

① Step 1: 35 ft. lbs.
Step 2: Plus 90 degrees

② Step 1: 18 ft. lbs.
Step 2: Plus 90 degrees

③ Manual transmission: 65 ft. lbs.
Automatic transmission: 61 ft. lbs.

④ Step 1: 12 point bolts to 40 ft. lbs.
Step 2: 12 point bolts plus 90 degrees
Step 3: Hex head recessed bolt to 13 ft. lbs.

⑤ Step 1: 12 point cap bolts to 16 ft. lbs.
Step 2: 12 point cap bolts plus 90 degrees
Step 3: Hex head side bolts to 20 ft. lbs.

⑥ Step 1: 29 ft. lbs.
Step 2: Plus 90 degrees
Step 3: Plus 90 degrees

⑦ Step 1: 29 ft. lbs.
Step 2: Plus 90 degrees

⑧ Step 1: 33 ft. lbs.

⑩ Step 1: 25 ft. lbs.
Step 2: Plus 90 degrees
Recessed head: 13 ft. lbs.

⑪ Step 1: 45 ft. lbs.
Step 2: Plus 90 degrees

⑫ Step 1: 58
Step 2: plus 90 degrees

⑬ Step 1: 15 ft. lbs.
Step 2: 30 ft. lbs.
Step 3: +90 degrees

⑭ MT: 96 ft. lbs.
AT: 72 ft. lbs.

⑮ 12-pointed bolts
Step 1: 25 ft. lbs.
Step 2: +90 degrees
Step 3: +90 degrees
Recessed heads: 13 ft. lbs.

93591CAU

For Tune-up, Capacities and Firing orders, see Section 1 of this manual

WHEEL ALIGNMENT

Year	Model		Caster Range (+/-Deg.)	Caster Preferred Setting (Deg.)	Camber Range (+/-Deg.)	Camber Preferred Setting (Deg.)	Toe-in (in.)	Steering Axis Inclination (Deg.)
1999	RAV4	2WD	0.75	+1.42	0.75	-0.33	0+/-0.08	11+/-0.75
		4WD	0.75	+1.33	0.75	-0.25	0+/-0.08	11+/-0.75
	4Runner	2WD	0.75	+3.25	0.75	-0.25	0.08+/-0.08	11+/-0.75
		4WD	0.75	+3.06	0.75	-0.25	0.08+/-0.08	11+/-0.75
2000	RAV4	2WD	0.75	+1.42	0.75	-0.33	0+/-0.08	11+/-0.75
		4WD	0.75	+1.33	0.75	-0.25	0+/-0.08	11+/-0.75
	4Runner	2WD	0.75	+3.25	0.75	-0.25	0.08+/-0.08	11+/-0.75
		4WD	0.75	+3.06	0.75	-0.25	0.08+/-0.08	11+/-0.75
2001	RAV4	2WD	0.75	+1.42	0.75	-0.33	0+/-0.08	11+/-0.75
		4WD	0.75	+1.33	0.75	-0.25	0+/-0.08	11+/-0.75
	Highlander	2WD F	0.75	+2.75	0.75	-0.67	0+/-0.08	10.75+/-0.75
		4WD F	0.75	+2.75	0.75	-0.58	0+/-0.08	10.58+/-0.75
		2WD R	—	—	0.75	-1.33	0.12+/-0.08	—
		4WD R	—	—	0.75	-0.75	0.12+/-0.08	—
	4Runner	2WD	0.75	+3.25	0.75	-0.25	0.08+/-0.08	11+/-0.75
		4WD	0.75	+3.06	0.75	-0.25	0.08+/-0.08	11+/-0.75
2002	RAV4	2WD	0.75	+2.00	0.75	-0.42	0+/-0.08	11+/-0.75
		4WD	0.75	+1.92	0.75	-0.33	0.04+/-0.08	10.75+/-0.75
	Highlander	2WD F	0.75	+2.75	0.75	-0.67	0+/-0.08	10.75+/-0.75
		4WD F	0.75	+2.75	0.75	-0.58	0+/-0.08	10.58+/-0.75
		2WD R	—	—	0.75	-1.33	0.12+/-0.08	—
		4WD R	—	—	0.75	-0.75	0.12+/-0.08	—
	4Runner	2WD	0.75	+3.25	0.75	-0.25	0.08+/-0.08	11+/-0.75
		4WD	0.75	+3.06	0.75	-0.25	0.08+/-0.08	11+/-0.75

93591CAV

TIRE, WHEEL AND BALL JOINT SPECIFICATIONS

| Year | Model | OEM Tires | | Tire Pressures (psi) | | Wheel Size | Ball Joint Inspection |
		Standard	Optional	Front	Rear		
1999	RAV4	P215/70R16	P235/60HR15	28	28	6.5-JJ	4-30 ②
	4Runner	P225/75R15	P265/70R16	Std: 29 Opt: 32	Std: 29 Opt: 32	7-JJ	6-39 ②
2000	RAV4	P215/70R16	P235/60HR15	28	28	6.5-JJ	4-30 ②
	4Runner	P225/75R15	P265/70R16	Std: 29 Opt: 32	Std: 29 Opt: 32	7-JJ	6-39 ②
2001	RAV4	P215/70R16	P235/60HR16	29	29	③	4-30 ②
	Highlander	P225/70R16	None	30	30	6.5-JJ	①
	4Runner	P225/75R15	P265/70R16	Std: 29 Opt: 32	Std: 29 Opt: 32	7-JJ	6-39 ②
2002	RAV4	P215/70R16	P235/60HR16	29	29	③	9-43 ②
	Highlander	P225/70R16	None	30	30	6.5-JJ	①
	4Runner SR5	P225/75R15	P265/70R16	Std: 29 Opt: 32	Std: 29 Opt: 32	7-JJ	④
	4Runner Limited	P265/70R16	None	32 Opt: 32	32 Opt: 32	7-JJ	④

OEM: Original Equipment Manufacturer

PSI: Pounds Per Square Inch

STD: Standard

OPT: Optional

① Replace if any measurable movement is found.

② Torque required in inch lbs. to rotate ball joint when removed from the knuckle

③ Steel wheel: 6.5J; aluminum wheel: 7JJ

④ Turning torque: upper 6-39 in. lbs.; lower 0.8-21.7 in. lbs.

93591CAX

BRAKE SPECIFICATIONS

All measurements in inches unless noted

Year	Model		Brake Disc			Brake Drum Diameter			Minimum Lining Thickness	Brake Caliper	
			Original Thickness	Minimum Thickness	Maximum Runout	Original Inside Diameter	Max. Wear Limit	Maximum Machine Diameter		Bracket Bolts (ft. lbs.)	Mounting Bolts (ft. lbs.)
1999	RAV4		0.709	0.630	0.0020	9.00	—	9.08	0.039	78	20
	4Runner		0.866	0.787	0.0028	11.61	—	11.69	0.039	—	90
200	RAV4		0.709	0.630	0.0020	9.00	—	9.08	0.039	78	20
	Highlander	F	1.102	1.024	0.0020	—	—	—	0.039	79	25
		R	0.394	0.354	0.0059	—	—	—	0.039	43	25
	4Runner		0.866	0.787	0.0028	11.61	—	11.69	0.039	—	90
2001	RAV4		0.709	0.630	0.0020	9.00	—	9.08	0.039	78	20
	Highlander	F	1.102	1.024	0.0020	—	—	—	0.039	79	25
		R	0.394	0.354	0.0059	—	—	—	0.039	43	25
	4Runner		0.866	0.787	0.0028	11.61	—	11.69	0.039	—	90
2002	RAV4		0.984	0.906	0.0020	9.00	—	9.08	0.039	78	20
	Highlander	F	1.102	1.024	0.0020	—	—	—	0.039	79	25
		R	0.394	0.354	0.0059	—	—	—	0.039	43	25
	4Runner		0.866	0.787	0.0028	11.61	—	11.69	0.039	—	90

F: Front

R: Rear

93591CAY

SCHEDULED MAINTENANCE INTERVALS
TOYOTA—4RUNNER

TO BE SERVICED	TYPE OF SERVICE	VEHICLE MILEAGE INTERVAL (x1000)																		
		5	10	15	20	25	30	35	40	45	50	55	60	65	70	75	80	85	90	95
Automatic transmission and differential fluid	S/I			✓			✓			✓			✓			✓			✓	
Ball joints and boots	S/I			✓			✓			✓			✓			✓			✓	
Brake linings, discs/drums, lines & hoses	S/I			✓			✓			✓			✓			✓			✓	
Charcoal canister	S/I												✓							
Drive belts	S/I						✓						✓						✓	
Driveshaft bushing (4WD)	L						✓						✓						✓	
Engine coolant	R						✓						✓						✓	
Engine oil & filter	R	✓	✓	✓	✓	✓	✓	✓	✓	✓	✓	✓	✓	✓	✓	✓	✓	✓	✓	✓
Exhaust pipes & mounts	S/I			✓			✓			✓			✓			✓			✓	
Fuel tank cap gasket	S/I						✓						✓						✓	
Halfshaft boots & flange bolts	S/I			✓			✓			✓			✓			✓				
Limited slip differential fluid	R						✓						✓						✓	
Manual transmission and differential fluid	S/I						✓						✓						✓	
Non-platinum spark plugs	R						✓						✓						✓	
Platinum spark plugs	R												✓							
Propeller shaft (4WD)	L			✓			✓			✓			✓			✓				
Propeller shaft bolts	S/I			✓			✓			✓			✓			✓			✓	
Rack and pinion assembly	S/I			✓			✓			✓			✓			✓			✓	
Rear wheel bearing	L						✓						✓						✓	
Steering linkage	S/I			✓			✓			✓			✓			✓			✓	
Valves	S/I												✓							

R: Replace   S/I: Service or Inspect   L: Lubricate

FREQUENT OPERATION MAINTENANCE (SEVERE SERVICE)

If a vehicle is operated under any of the following conditions it is considered severe service:

- Towing a trailer or using a camper or car-top carrier.

- Repeated short trips of less than 5 miles in temperatures below freezing.

- Excessive idling or low-speed driving for long distances as in heavy commercial use, such as delivery, taxi or police cars.

- Operating on rough, muddy or salt-covered roads.

- Operating on unpaved or dusty roads.

Oil filter: service or inspect every 5000 miles or 4 months, whichever occurs first.

Brake linings and discs or drums: service or inspect every 5000 miles or 4 months, whichever occurs first.

Steering linkage: service or inspect every 5000 miles or 4 months, whichever occurs first.

Ball joints and boots: service or inspect every 5000 miles or 4 months, whichever occurs first.

Brake discs & pads (front): service or inspect every 6000 miles.

Halfshaft boots: service or inspect every 5000 miles or 4 months. Retighten the flange bolts, whichever occurs first.

Body chassis bolts and nuts: service or inspect every 5000 miles or 4 months, whichever occurs first.

Transmission and differential fluid: replace every 15,000 miles or 12 months, whichever occurs first.

Transfer case and differential fluid: replace every 15,000 miles or 12 months, whichever occurs first.

Timing belt: replace every 60,000 miles or 48 months, whichever occurs first.

93591CAZA

Please visit our web site at www.chiltononline.com

SCHEDULED MAINTENANCE INTERVALS
TOYOTA—HIGHLANDER

TO BE SERVICED	TYPE OF SERVICE	VEHICLE MILEAGE INTERVAL (x1000)												
		7.5	15	22.5	30	37.5	45	52.5	60	67.5	75	82.5	90	97.5
Engine oil & filter	R	✓	✓	✓	✓	✓	✓	✓	✓	✓	✓	✓	✓	✓
Automatic transmission fluid	S/I		✓		✓		✓		✓		✓		✓	
Ball joints & dust covers	S/I		✓		✓		✓		✓		✓		✓	
Bolts & nuts on chassis & body	S/I		✓		✓		✓		✓		✓		✓	
Brake linings & drums	S/I		✓		✓		✓		✓		✓		✓	
Brake line pipes & hoses	S/I		✓		✓		✓		✓		✓		✓	
Brake pads & discs (front & rear)	S/I		✓		✓		✓		✓		✓		✓	
Propeller shaft grease	S/I		✓		✓		✓		✓		✓		✓	
Steering knuckle & chassis grease	S/I		✓		✓		✓		✓		✓		✓	
Steering linkage	S/I		✓		✓		✓		✓		✓		✓	
Air cleaner filter	R				✓				✓				✓	
Spark plugs ①	R				✓				✓				✓	
Drive belts	S/I				✓				✓				✓	
Exhaust pipes & mountings	S/I				✓				✓				✓	
Fuel lines & connections	S/I				✓				✓				✓	
Engine coolant	R					✓					✓			
Charcoal canister	R								✓					
Fuel tank cap gasket	R								✓					
Heated oxygen sensors (except Calif.) ②	R													

R: Replace   S/I: Service or Inspect

① Platinum plugs are replaced at 100,000 mile intervals

② Heated oxygen sensors (except Calif.): replace every 80,000 miles.

FREQUENT OPERATION MAINTENANCE (SEVERE SERVICE)

If a vehicle is operated under any of the following conditions it is considered severe service:

- Extremely dusty areas.

- 50% or more of the vehicle operation is in 32°C (90°F) or higher temperatures, or constant operation in temperatures below 0°C (32°F).

- Prolonged idling (vehicle operation in stop and go traffic).

- Frequent short running periods (engine does not warm to normal operating temperatures).

- Police, taxi, delivery usage or trailer towing usage.

Air cleaner filter: service or inspect every 3750 miles

Engine oil & filter: replace every 3750 miles.

Ball joints & dust covers: service or inspect every 7500 miles.

Bolts & nuts on chassis & body: service or inspect every 7500 miles.

Brake pads & discs (front & rear): service or inspect every 7500 miles.

Steering knuckle & chassis grease: service or inspect every 7500 miles.

Steering linkage: service or inspect every 7500 miles.

Exhaust pipes & mountings: service or inspect every 15,000 miles.

93591CB1

SCHEDULED MAINTENANCE INTERVALS
TOYOTA—RAV4

TO BE SERVICED	TYPE OF SERVICE	VEHICLE MILEAGE INTERVAL (x1000)																		
		5	10	15	20	25	30	35	40	45	50	55	60	65	70	75	80	85	90	95
Automatic transmission and differential fluid	S/I			✓			✓			✓			✓			✓			✓	
Ball joints and boots	S/I			✓			✓			✓			✓			✓			✓	
Brake linings, discs/drums, lines & hoses	S/I			✓			✓			✓			✓			✓			✓	
Charcoal canister	S/I												✓							
Drive belts	S/I						✓						✓						✓	
Driveshaft bushing (4WD)	L						✓						✓						✓	
Engine coolant	R						✓						✓						✓	
Engine oil & filter	R	✓	✓	✓	✓	✓	✓	✓	✓	✓	✓	✓	✓	✓	✓	✓	✓	✓	✓	✓
Exhaust pipes & mounts	S/I			✓			✓			✓			✓			✓			✓	
Fuel tank cap gasket	S/I						✓						✓						✓	
Halfshaft boots & flange bolts	S/I			✓			✓			✓			✓			✓			✓	
Limited slip differential fluid	R						✓						✓						✓	
Manual transmission and differential fluid	S/I						✓						✓						✓	
Non-platinum spark plugs	R						✓						✓						✓	
Platinum spark plugs	R												✓							
Propeller shaft bolts	S/I			✓			✓			✓			✓			✓			✓	
Rack and pinion assembly	S/I			✓			✓			✓			✓			✓			✓	
Steering linkage	S/I			✓			✓			✓			✓			✓			✓	
Transfer case and differential fluid	S/I			✓			✓			✓			✓			✓			✓	
Valves	S/I												✓							

R: Replace   S/I: Service or Inspect   L: Lubricate

FREQUENT OPERATION MAINTENANCE (SEVERE SERVICE)

If a vehicle is operated under any of the following conditions it is considered severe service:

- Towing a trailer or using a camper or car-top carrier.
- Repeated short trips of less than 5 miles in temperatures below freezing.
- Excessive idling or low-speed driving for long distances as in heavy commercial use, such as delivery, taxi or police cars.
- Operating on rough, muddy or salt-covered roads.
- Operating on unpaved or dusty roads.

Oil filter: service or inspect every 5000 miles or 4 months, whichever occurs first.

Brake linings and discs or drums: service or inspect every 5000 miles or 4 months, whichever occurs first.

Steering linkage: service or inspect every 5000 miles or 4 months, whichever occurs first.

Ball joints and boots: service or inspect every 5000 miles or 4 months, whichever occurs first.

Brake discs & pads (front): service or inspect every 6000 miles.

Halfshaft boots: service or inspect every 5000 miles or 4 months. Retighten the flange bolts, whichever occurs first.

Body chassis bolts and nuts: service or inspect every 5000 miles or 4 months, whichever occurs first.

Transmission and differential fluid: replace every 15,000 miles or 12 months, whichever occurs first.

Transfer case and differential fluid: replace every 15,000 miles or 12 months, whichever occurs first.

Timing belt: replace every 60,000 miles or 48 months, whichever occurs first.

93591CB2

Timing belt service is covered in Section 3 of this manual

SCHEDULED MAINTENANCE INTERVALS
TOYOTA
RAV4

The following should be used as a guide when determining the amount of work required for a particular service.
In estimating how long a particular Scheduled Maintenance Service should take, please observe the following:

- Labor Time is time based on field research and data supplied by the vehicle manufacturer.
- Labor time operations are given in hours and tenths of an hour.
- All labor operations are to be used as a guide.

Mechanic Skill Level Codes:
(A) PRECISION: Highly skilled with multiple certification.
(B) GENERAL: Normally skilled with certification.
(C) MAINTENANCE: Semi-skilled working on certification.

	LABOR TIME			LABOR TIME			LABOR TIME
5000 Mile Service (C)			**35000 Mile Service (C)**			**70000 Mile Service (C)**	
RAV4	.4		RAV4	.4		RAV4	.4
10000 Mile Service (C)			**40000 Mile Service (C)**			**75000 Mile Service (B)**	
RAV4	.4		RAV4	.4		RAV4	1.9
15000 Mile Service (B)			**45000 Mile Service (B)**			**80000 Mile Service (C)**	
RAV4	1.9		RAV4	1.9		RAV4	.4
20000 Mile Service (C)			**50000 Mile Service (C)**			**85000 Mile Service (C)**	
RAV4	.4		RAV4	.4		RAV4	.4
25000 Mile Service (C)			**55000 Mile Service (C)**			**90000 Mile Service (B)**	
RAV4	.4		RAV4	.3		RAV4	3.7
30000 Mile Service (B)			**60000 Mile Service (B)**			**95000 Mile Service (C)**	
RAV4	3.7		RAV4	3.7		RAV4	.4
			65000 Mile Service (C)				
			RAV4	.4			

93551C002

SCHEDULED MAINTENANCE INTERVALS
TOYOTA
4RUNNER

The following should be used as a guide when determining the amount of work required for a particular service. In estimating how long a particular Scheduled Maintenance Service should take, please observe the following:

- Labor Time is time based on field research and data supplied by the vehicle manufacturer.
- Labor time operations are given in hours and tenths of an hour.
- All labor operations are to be used as a guide.

Mechanic Skill Level Codes:
(A) PRECISION: Highly skilled with multiple certification.
(B) GENERAL: Normally skilled with certification.
(C) MAINTENANCE: Semi-skilled working on certification.

	LABOR TIME
5000 Mile Service (C)	
4Runner	.4
10000 Mile Service (C)	
4Runner	.4
15000 Mile Service (B)	
4Runner	1.3
20000 Mile Service (C)	
4Runner	.4
25000 Mile Service (C)	
4Runner	.4
30000 Mile Service (B)	
4Runner	3.1

	LABOR TIME
35000 Mile Service (C)	
4Runner	.4
40000 Mile Service (C)	
4Runner	.4
45000 Mile Service (B)	
4Runner	1.3
50000 Mile Service (C)	
4Runner	.4
55000 Mile Service (C)	
4Runner	.3
60000 Mile Service (B)	
4Runner	3.1
65000 Mile Service (C)	
4Runner	.4

	LABOR TIME
70000 Mile Service (C)	
4Runner	.4
75000 Mile Service (B)	
4Runner	1.1
80000 Mile Service (C)	
4Runner	.4
85000 Mile Service (C)	
4Runner	.4
90000 Mile Service (B)	
4Runner	3.1
95000 Mile Service (C)	
4Runner	.4

93551C003

Heater Core replacement is covered in Section 2 of this manual

SCHEDULED MAINTENANCE INTERVALS
TOYOTA
HIGHLANDER

The following should be used as a guide when determining the amount of work required for a particular service.
In estimating how long a particular Scheduled Maintenance Service should take, please observe the following:

● Labor Time is time based on field research and data supplied by the vehicle manufacturer.
● Labor time operations are given in hours and tenths of an hour.
● All labor operations are to be used as a guide.

Mechanic Skill Level Codes:
(A) PRECISION: Highly skilled with multiple certification.
(B) GENERAL: Normally skilled with certification.
(C) MAINTENANCE: Semi-skilled working on certification.

	LABOR TIME		LABOR TIME		LABOR TIME
7500 Mile Service (C)		**37500 Mile Service (C)**		**70000 Mile Service (C)**	
Highlander	.4	Highlander	.4	Highlander	.4
15000 Mile Service (B)		**45000 Mile Service (B)**		**75000 Mile Service (B)**	
Highlander	1.8	Highlander	2.5	Highlander	2.3
20000 Mile Service (C)		**52500 Mile Service (C)**		**82500 Mile Service (C)**	
Highlander	.4	Highlander	.4	Highlander	.4
22500 Mile Service (C)		**60000 Mile Service (B)**		**90000 Mile Service (B)**	
Highlander	.4	Highlander	3.8	Highlander	3.5
30000 Mile Service (B)		**67500 Mile Service (C)**		**97500 Mile Service (C)**	
Highlander	3.5	Highlander	.4	Highlander	.4

93551C004

TOYOTA
Land Cruiser • Sequoia

ENGINE AND VEHICLE IDENTIFICATION

Engine						
Code ①	Liters (cc)	Cu. In.	Cyl.	Fuel Sys.	Engine Type	Eng. Mfg.
2UZ-FE	4.7 (4664)	285	8	SFI	DOHC	Toyota

Model Year	
Code ②	Year
X	1999
Y	2000
1	2001
2	2002
3	2003

SFI: Sequential Fuel Injection

DOHC: Double Overhead Camshaft

① Stamped on the left side of the engine block

② 10th digit of the Vehicle Identification Number (VIN)

93591CB3

GENERAL ENGINE SPECIFICATIONS

Year	Model	Engine Displacement Liters (cc)	Engine Series (ID/VIN)	Fuel System	Net Horsepower @ rpm	Net Torque @ rpm (ft. lbs.)	Bore x Stroke (in.)	Com-pression Ratio	Oil Pressure @ rpm
1999	Land Cruiser	4.7 (4664)	2UZ-FE	SFI	230@4800	320@3400	3.70x3.30	9.6:1	45-65@3000
2000	Land Cruiser	4.7 (4664)	2UZ-FE	SFI	230@4800	320@3400	3.70x3.30	9.6:1	45-65@3000
2001	Land Cruiser	4.7 (4664)	2UZ-FE	SFI	245@4800	315@3400	3.70x3.30	9.6:1	45-65@3000
	Sequoia	4.7 (4664)	2UZ-FE	SFI	245@4800	315@3400	3.70x3.30	9.6:1	45-65@3000
2002	Land Cruiser	4.7 (4664)	2UZ-FE	SFI	245@4800	315@3400	3.70x3.30	9.6:1	45-65@3000
	Sequoia	4.7 (4664)	2UZ-FE	SFI	245@4800	315@3400	3.70x3.30	9.6:1	45-65@3000

SFI: Sequential Fuel Injection

93591CB4

Brake service is covered in Section 4 of this manual

ENGINE TUNE-UP SPECIFICATIONS

Year	Engine Displacement Liters (cc)	Engine ID/VIN	Spark Plug Gap (in.)	Ignition Timing (deg.)*	Fuel Pump (psi)	Idle Speed (rpm)		Valve Clearance	
						MT	AT	Intake	Exhaust
1999	4.7 (4664)	2UZ-FE	0.043	5-15B	38-44	—	650-750	0.006-0.009	0.011-0.014
2000	4.7 (4664)	2UZ-FE	0.043	5-15B	38-44	—	650-750	0.006-0.009	0.011-0.014
2001	4.7 (4664)	2UZ-FE	0.031	5-15B	38-44	—	650-750	0.006-0.010	0.010-0.014
2002	4.7 (4664)	2UZ-FE	0.031	5-15B	38-44	—	650-750	0.006-0.010	0.010-0.014

NOTE: The Vehicle Emission Control Information label often reflects specification changes made during production. The label figures must be used if they differ from those in this chart.

B: Before top dead center

* With terminals TC and E1 connected to DLC1

93591CB5

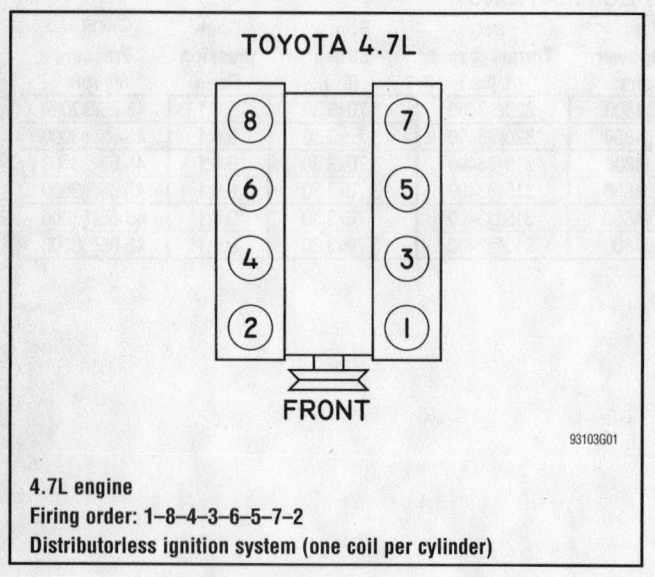

TOYOTA 4.7L

FRONT

93103G01

4.7L engine
Firing order: 1–8–4–3–6–5–7–2
Distributorless ignition system (one coil per cylinder)

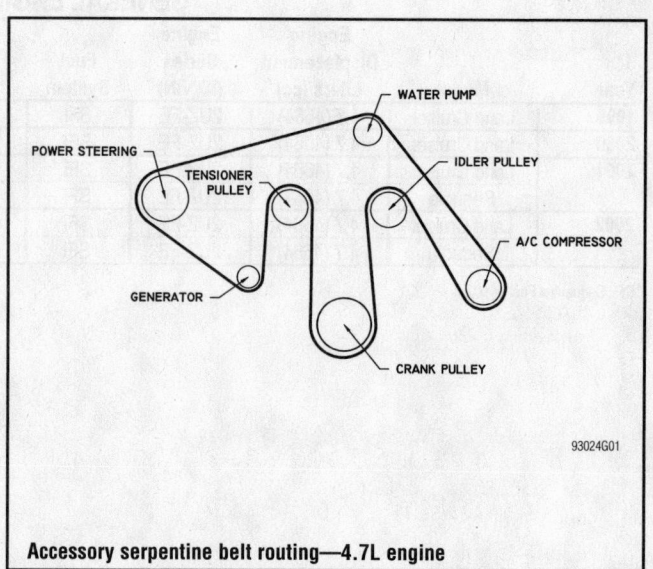

93024G01

Accessory serpentine belt routing—4.7L engine

CAPACITIES

Year	Model	Engine Displacement Liters (cc)	Engine ID/VIN	Engine Oil with Filter (qts.)	Transmission (pts.) 5-Spd	Transmission (pts.) Auto.*	Transfer Case (pts.)	Drive Axle Front (pts.)	Drive Axle Rear (pts.)	Fuel Tank (gal.)	Cooling System (qts.)
1999	Land Cruiser	4.7 (4664)	2UZ-FE	7.2	—	4.0	3.6	3.6	6.8	25.1	①
2000	Land Cruiser	4.7 (4664)	2UZ-FE	7.2	—	3.6	3.6	3.6	6.8	25.1	①
2001	Land Cruiser	4.7 (4664)	2UZ-FE	7.2	—	3.6	3.6	3.6	6.8	25.1	①
	Sequoia	4.7 (4664)	2UZ-FE	5.5	—	3.6	2.4	3.6	6.6	21.6	②
2002	Land Cruiser	4.7 (4664)	2UZ-FE	7.2	—	7.4	2.8	3.6	③	25.4	②
	Sequoia	4.7 (4664)	2UZ-FE	6.6	—	4.2	2.6	2.4	7.7	26.1	12.3

*After draining, add the following amounts, then fill to the cold full line

① With rear heater: 9.5
Without rear heater: 8.5

② With rear heater: 16.2
Without rear heater: 15.6

③ w/o diff. Lock: 6.98
w/diff lock: 6.76

93591CB6

VALVE SPECIFICATIONS

Year	Engine Displacement Liters (cc)	Engine ID/VIN	Seat Angle (deg.)	Face Angle (deg.)	Spring Test Pressure (lbs. @ in.)	Spring Installed Height (in.)	Stem-to-Guide Clearance (in.) Intake	Stem-to-Guide Clearance (in.) Exhaust	Stem Diameter (in.) Intake	Stem Diameter (in.) Exhaust
1999	4.7 (4664)	2UZ-FE	45	44.5	45.9-50.7@ 1.378	1.380	0.0010- 0.0024	0.0012- 0.0026	0.2154- 0.2159	0.2152- 0.2157
2000	4.7 (4664)	2UZ-FE	45	44.5	45.9-50.7@ 1.378	1.380	0.0010- 0.0024	0.0012- 0.0026	0.2154- 0.2159	0.2152- 0.2157
2001	4.7 (4664)	2UZ-FE	45	44.5	45.9-50.7@ 1.378	1.380	0.0010- 0.0024	0.0012- 0.0026	0.2154- 0.2159	0.2152- 0.2157
2002	4.7 (4664)	2UZ-FE	45	44.5	45.9-50.7@ 1.378	1.380	0.0010- 0.0024	0.0012- 0.0026	0.2154- 0.2159	0.2152- 0.2157

93591CB7

For complete Engine Mechanical specifications, see Section 1 of this manual

CRANKSHAFT AND CONNECTING ROD SPECIFICATIONS

All measurements are given in inches.

Year	Engine Displacement Liters (cc)	Engine ID/VIN	Crankshaft				Connecting Rod		
			Main Brg. Journal Dia.	Main Brg. Oil Clearance	Shaft End-play	Thrust on No.	Journal Diameter	Oil Clearance	Side Clearance
1999	4.7 (4665)	2UZ-FE	2.6373-2.6378	0.0016-0.0023	0.0008-0.0087	3	2.0465-2.0472	0.0011-0.0021	0.0063-0.0138
2000	4.7 (4665)	2UZ-FE	2.6373-2.6378	0.0016-0.0023	0.0008-0.0087	3	2.0465-2.0472	0.0011-0.0021	0.0063-0.0138
2001	4.7 (4665)	2UZ-FE	2.6373-2.6378	①	0.0008-0.0087	3	2.0465-2.0472	0.0011-0.0021	0.0063-0.0138
2002	4.7 (4665)	2UZ-FE	2.6373-2.6378	①	0.0008-0.0087	3	2.0465-2.0472	0.0011-0.0021	0.0063-0.0138

① Nos. 1 and 2: 0.0011-0.0018
All others: 0.0016-0.0023

93591CB8A

PISTON AND RING SPECIFICATIONS

All measurements are given in inches.

Year	Engine Displacement Liters (cc)	Engine ID/VIN	Piston Clearance	Ring Gap			Ring Side Clearance		
				Top Compression	Bottom Compression	Oil Control	Top Compression	Bottom Compression	Oil Control
1999	4.7 (4664)	2UZ-FE	0.0035-0.0044	0.0118-0.0197	0.0157-0.0256	0.0051-0.0189	0.0012-0.0031	0.0012-0.0028	SNUG
2000	4.7 (4664)	2UZ-FE	0.0035-0.0044	0.0118-0.0197	0.0157-0.0256	0.0051-0.0189	0.0012-0.0031	0.0012-0.0028	SNUG
2001	4.7 (4664)	2UZ-FE	0.0035-0.0044	0.0118-0.0157	0.0157-0.0217	0.0051-0.0150	0.0012-0.0031	0.0012-0.0028	SNUG
2002	4.7 (4664)	2UZ-FE	0.0035-0.0044	0.0118-0.0157	0.0157-0.0217	0.0051-0.0150	0.0012-0.0031	0.0012-0.0028	SNUG

93591CB9A

TORQUE SPECIFICATIONS
All readings in ft. lbs.

Year	Engine Displacement Liters (cc)	Engine ID/VIN	Cylinder Head Bolts	Main Bearing Bolts	Rod Bearing Bolts	Crankshaft Damper Bolts	Flywheel Bolts	Manifold Intake	Manifold Exhaust	Spark Plugs	Lug Nuts
1999	4.7 (4664)	2UZ-FE	①	②	③	181	④	13	33	13	97
2000	4.7 (4664)	2UZ-FE	①	②	③	181	④	13	33	13	97
2001	4.7 (4664)	2UZ-FE	⑤	②	③	181	④	13	33	13	97
2002	4.7 (4664)	2UZ-FE	⑤	②	③	181	④	13	33	13	97

① Step 1: 24 ft. lbs.
 Step 2: Plus 180 degrees

② Step 1: 20 ft. lbs.
 Step 2: Plus 90 degrees

③ Step 1: 18 ft. lbs.
 Step 2: Plus 90 degrees

④ Step 1: 35 ft. lbs.
 Step 2: Plus 90 degrees

⑤ Step 1: 24
 Step 2: Plus 90 degrees
 Step 3: Plus 90 degrees

93591CB0

For Accessory Drive Belt illustrations, see Section 1 of this manual

WHEEL ALIGNMENT

Year	Model	Caster Range (+/-Deg.)	Caster Preferred Setting (Deg.)	Camber Range (+/-Deg.)	Camber Preferred Setting (Deg.)	Toe-in (in.)	Steering Axis Inclination (Deg.)
1999	Land Cruiser	0.75	+2.50	0.75	+0.08	0.04+/-0.08	12.17+/-0.75
2000	Land Cruiser	0.75	+2.50	0.75	+0.08	0.04+/-0.08	12.17+/-0.75
2001	Land Cruiser	0.75	+2.50	0.75	+0.08	0.04+/-0.08	12.17+/-0.75
	Sequoia	0.75	+2.95	0.75	+0.13	0.05+/-0.08	10.65+/-0.75
2002	Land Cruiser	0.75	+2.50	0.75	+0.08	0.04+/-0.08	12.17+/-0.75
	Sequoia	0.75	①	0.75	+0.13	0.05+/-0.08	10.635+/-0.75

Note: All alignment specifications are based on nominal ride height and standard tires

① P245/70R16: +2.95

　P265/70R16: +3.00

93591CC1A

TIRE, WHEEL AND BALL JOINT SPECIFICATIONS

Year	Model	OEM Tires Standard	OEM Tires Optional	Tire Pressures (psi) Front	Tire Pressures (psi) Rear	Wheel Size	Ball Joint Inspection
1999	Land Cruiser	P275/70R15	None	32	32	8-JJ	①
2000	Land Cruiser	P275/70R15	None	32	32	8-JJ	①
2001	Land Cruiser	P275/70R15	None	32	32	8-JJ	①
	Sequoia	P245/70R16	P265/70R16	32	Std: 35; Opt: 32	7-JJ	②
2002	Land Cruiser	P275/70R16	None	29③	32③	8-JJ	①
	Sequoia	P245/70R16	P265/70R16	32	Std: 35; Opt: 32	7-JJ	②

OEM: Original Equipment Manufacturer

PSI: Pounds Per Square Inch

STD: Standard

OPT: Optional

① Replace if any measurable movement is found.

② Upper ball joint turning torque: 6-39 inch lbs.
　Lower ball joint turning torque: 1-22 inch lbs.
　Lower ball joint excessive play: 0.020 in.

③ Trailer towing: front 32; rear 35

93591CC2A

BRAKE SPECIFICATIONS
All measurements in inches unless noted

Year	Model		Brake Disc			Brake Drum Diameter			Minimum Lining Thickness	Brake Caliper	
			Original Thickness	Minimum Thickness	Maximum Runout	Original Inside Diameter	Max. Wear Limit	Maximum Machine Diameter		Bracket Bolts (ft. lbs.)	Mounting Bolts (ft. lbs.)
1999	Land Cruiser	F	1.260	1.181	0.0059	—	—	—	0.039	—	90
		R	0.709	0.611	—	—	—	—	0.039	65	76
2000	Land Cruiser	F	1.260	1.181	0.0059	—	—	—	0.039	—	90
		R	0.709	0.611	—	—	—	—	0.039	65	76
2001	Land Cruiser	F	1.260	1.181	0.0059	—	—	—	0.039	—	90
		R	0.709	0.611	—	—	—	—	0.039	65	76
	Sequoia	F	1.102	1.024	0.0028	—	—	—	0.039	—	90
		R	0.709	0.611	0.0039	—	—	—	0.039	—	65
2002	Land Cruiser	F	1.260	1.181	0.0028	—	—	—	0.039	—	90
		R	0.709	0.611	—	—	—	—	0.039	—	20
	Sequoia	F	1.102	1.024	0.0028	—	—	—	0.039	—	90
		R	0.709	0.611	0.0039	—	—	—	0.039	—	65

F: Front
R: Rear

93591CC3A

For Tire, Wheel and Ball Joint specifications, see Section 1 of this manual

SCHEDULED MAINTENANCE INTERVALS
Toyota Sequoia

TO BE SERVICED	TYPE OF SERVICE	VEHICLE MILEAGE INTERVAL (x1000)												
		7.5	15	22.5	30	37.5	45	52.5	60	67.5	75	82.5	90	97.5
Engine oil & filter	R	✓	✓	✓	✓	✓	✓	✓	✓	✓	✓	✓	✓	✓
Automatic transmission fluid & filter	S/I		✓		✓		✓		✓		✓		✓	
Ball joints & dust covers	S/I		✓		✓		✓		✓		✓		✓	
Bolts & nuts on chassis & body	S/I		✓		✓		✓		✓		✓		✓	
Brake line pipes & hoses	S/I		✓		✓		✓		✓		✓		✓	
Brake pads & discs	S/I		✓		✓		✓		✓		✓		✓	
Propeller shaft grease	S/I		✓		✓		✓		✓		✓		✓	
Steering knuckle & chassis grease	S/I		✓		✓		✓		✓		✓		✓	
Steering linkage	S/I		✓		✓		✓		✓		✓		✓	
Transfer and differential oil	S/I		✓		✓		✓		✓		✓		✓	
Air cleaner filter	R				✓				✓				✓	
Spark plugs ①	R				✓				✓				✓	
Drive belts	S/I				✓				✓				✓	
Exhaust pipes & mountings	S/I				✓				✓				✓	
Fuel lines & connections	S/I				✓				✓				✓	
Engine coolant	R						✓				✓			
Charcoal canister	R								✓					
Fuel tank cap gasket	R								✓					
Heated oxygen sensors (except Calif.) ②	R													

R: Replace   S/I: Service or Inspect

① Platinum plugs, replace every 100,000 miles

② Heated oxygen sensors (except Calif.): replace every 80,000 miles.

FREQUENT OPERATION MAINTENANCE (SEVERE SERVICE)

If a vehicle is operated under any of the following conditions it is considered severe service:

- Extremely dusty areas.

- 50% or more of the vehicle operation is in 32°C (90°F) or higher temperatures, or constant operation in temperatures below 0°C (32°F).

- Prolonged idling (vehicle operation in stop and go traffic).

- Frequent short running periods (engine does not warm to normal operating temperatures).

- Police, taxi, delivery usage or trailer towing usage.

Air cleaner filter: service or inspect every 3750 miles

Engine oil & filter: replace every 3750 miles.

Ball joints & dust covers: service or inspect every 7500 miles.

Bolts & nuts on chassis & body: service or inspect every 7500 miles.

Brake pads & discs (front & rear): service or inspect every 7500 miles.

Steering knuckle & chassis grease: service or inspect every 7500 miles.

Steering linkage: service or inspect every 7500 miles.

Propeller shaft grease: service or inspect every 7500 miles.

Exhaust pipes & mountings: service or inspect every 15,000 miles.

93591CC4A

SCHEDULED MAINTENANCE INTERVALS
Toyota LAND CRUISER

TO BE SERVICED	TYPE OF SERVICE	VEHICLE MILEAGE INTERVAL (x1000)																		
		5	10	15	20	25	30	35	40	45	50	55	60	65	70	75	80	85	90	95
Automatic transmission and differential fluid	S/I			✓			✓			✓			✓			✓			✓	
Ball joints and boots	S/I			✓			✓			✓			✓			✓			✓	
Brake linings and discs, lines & hoses	S/I			✓			✓			✓			✓			✓			✓	
Charcoal canister	S/I												✓							
Drive belts	S/I						✓						✓						✓	
Driveshaft bushing (4WD)	L						✓						✓						✓	
Engine coolant	R						✓						✓						✓	
Engine oil & filter	R	✓	✓	✓	✓	✓	✓	✓	✓	✓	✓	✓	✓	✓	✓	✓	✓	✓	✓	✓
Exhaust pipes & mounts	S/I			✓			✓			✓			✓			✓			✓	
Fuel lines & connections, fuel tank vapor vent system hoses, fuel tank band	S/I						✓						✓						✓	
Fuel tank cap gasket	S/I						✓						✓						✓	
Halfshaft boots & flange bolts	S/I			✓			✓			✓			✓			✓			✓	
Limited slip differential fluid	R						✓						✓						✓	
Non-platinum spark plugs	R						✓						✓						✓	
Platinum spark plugs	R												✓							
Propeller shaft (4WD)	L			✓			✓			✓			✓			✓			✓	
Propeller shaft bolts	S/I			✓			✓			✓			✓			✓			✓	
Rack and pinion assembly	S/I			✓			✓			✓			✓			✓			✓	
Rear wheel bearing	L						✓						✓						✓	
Steering Knuckle	L			✓			✓			✓			✓			✓			✓	
Steering linkage	S/I			✓			✓			✓			✓			✓			✓	
Valves	S/I												✓							

R: Replace   S/I: Service or Inspect   L: Lubricate

FREQUENT OPERATION MAINTENANCE (SEVERE SERVICE)

If a vehicle is operated under any of the following conditions it is considered severe service:

- Towing a trailer or using a camper or car-top carrier.
- Repeated short trips of less than 5 miles in temperatures below freezing.
- Excessive idling or low-speed driving for long distances as in heavy commercial use, such as delivery, taxi or police cars.
- Operating on rough, muddy or salt-covered roads.
- Operating on unpaved or dusty roads.

Oil filter: service or inspect every 5000 miles or 4 months, whichever occurs first.

Brake linings and discs or drums: service or inspect every 5000 miles or 4 months, whichever occurs first.

Steering linkage: service or inspect every 5000 miles or 4 months, whichever occurs first.

Ball joints and boots: service or inspect every 5000 miles or 4 months, whichever occurs first.

Brake discs & pads (front): service or inspect every 6000 miles.

Halfshaft boots: service or inspect every 5000 miles or 4 months. Retighten the flange bolts, whichever occurs first.

Body chassis bolts and nuts: service or inspect every 5000 miles or 4 months, whichever occurs first.

Transmission and differential fluid: replace every 15,000 miles or 12 months, whichever occurs first.

Transfer case and differential fluid: replace every 15,000 miles or 12 months, whichever occurs first.

Timing belt: replace every 60,000 miles or 48 months, whichever occurs first.

93591CC5A

For Wheel Alignment specifications, see Section 1 of this manual

SCHEDULED MAINTENANCE INTERVALS
TOYOTA
LAND CRUISER

The following should be used as a guide when determining the amount of work required for a particular service.
In estimating how long a particular Scheduled Maintenance Service should take, please observe the following:

● Labor Time is time based on field research and data supplied by the vehicle manufacturer.
● Labor time operations are given in hours and tenths of an hour.
● All labor operations are to be used as a guide.

Mechanic Skill Level Codes:
(A) PRECISION: Highly skilled with multiple certification.
(B) GENERAL: Normally skilled with certification.
(C) MAINTENANCE: Semi-skilled working on certification.

	LABOR TIME			LABOR TIME			LABOR TIME
5000 Mile Service (C)			**35000 Mile Service (C)**			**70000 Mile Service (C)**	
Land Cruiser	.4		Land Cruiser	.4		Land Cruiser	.4
10000 Mile Service (C)			**40000 Mile Service (C)**			**75000 Mile Service (B)**	
Land Cruiser	.4		Land Cruiser	.4		Land Cruiser	2.1
15000 Mile Service (B)			**45000 Mile Service (B)**			**80000 Mile Service (C)**	
Land Cruiser	2.1		Land Cruiser	2.1		Land Cruiser	.4
20000 Mile Service (C)			**50000 Mile Service (C)**			**85000 Mile Service (C)**	
Land Cruiser	.4		Land Cruiser	.4		Land Cruiser	.4
25000 Mile Service (C)			**55000 Mile Service (C)**			**90000 Mile Service (B)**	
Land Cruiser	.4		Land Cruiser	.3		Land Cruiser	3.8
30000 Mile Service (B)			**60000 Mile Service (B)**			**95000 Mile Service (C)**	
Land Cruiser	3.8		Land Cruiser	3.8		Land Cruiser	.4
			65000 Mile Service (C)				
			Land Cruiser	.4			

93551CC7A

SCHEDULED MAINTENANCE INTERVALS
TOYOTA
SEQUOIA

The following should be used as a guide when determining the amount of work required for a particular service.
In estimating how long a particular Scheduled Maintenance Service should take, please observe the following:

● Labor Time is time based on field research and data supplied by the vehicle manufacturer.
● Labor time operations are given in hours and tenths of an hour.
● All labor operations are to be used as a guide.

Mechanic Skill Level Codes:
(A) PRECISION: Highly skilled with multiple certification.
(B) GENERAL: Normally skilled with certification.
(C) MAINTENANCE: Semi-skilled working on certification.

	LABOR TIME		LABOR TIME		LABOR TIME
7500 Mile Service (C)		**37500 Mile Service (C)**		**70000 Mile Service (C)**	
Seqoia	.4	Seqoia	.4	Seqoia	.4
15000 Mile Service (B)		**45000 Mile Service (B)**		**75000 Mile Service (B)**	
Seqoia	2.1	Seqoia	2.1	Seqoia	2.1
20000 Mile Service (C)		**52500 Mile Service (C)**		**82500 Mile Service (C)**	
Seqoia	.4	Seqoia	.4	Seqoia	.4
22500 Mile Service (C)		**60000 Mile Service (B)**		**90000 Mile Service (B)**	
Seqoia	.4	Seqoia	3.8	Seqoia	3.8
30000 Mile Service (B)		**67500 Mile Service (C)**		**97500 Mile Service (C)**	
Seqoia	3.8	Seqoia	.4	Seqoia	.4

93551CC8A

For Maintenance Interval recommendations, see Section 1 of this manual

TOYOTA/LEXUS
Matrix

ENGINE AND VEHICLE IDENTIFICATION

Engine							Model Year	
Code ①	Liters (cc)	Cu. In.	Cyl.	Fuel Sys.	Engine Type	Eng. Mfg.	Code ②	Year
1ZZ-FE	1.8 (1794)	109	4	EFI	DOHC	Toyota	3	2003
2ZZ-GE	1.8 (1796)	109.5	4	EFI	DOHC	Toyota		

EFI: Electronic Fuel Injection

DOHC: Double Overhead Camshaft

① 8th digit of VIN

② 10th digit of VIN

93591CQA

GENERAL ENGINE SPECIFICATIONS

Year	Model	Engine Displacement Liters (cc)	Engine Series (ID/VIN)	Fuel System	Net Horsepower @ rpm	Net Torque @ rpm (ft. lbs.)	Bore x Stroke (in.)	Com-pression Ratio	Oil Pressure @ idle
2003	Matrix	1.8 (1794)	1ZZ-FE	EFI	①	②	3.11x3.60	10.0:1	4.2
	Matrix	1.8 (1796)	2ZZ-GE	EFI	180@7600	130@6800	3.23x3.35	11.5:1	2.8

EFI: Electronic Fuel Injection

① 2WD models: 130@6000
   4WD models: 123@6000

② 2WD models: 126@4200
   4WD models: 119@4200

93591CQB

ENGINE TUNE-UP SPECIFICATIONS

Year	Engine Displacement Liters (cc)	Engine ID/VIN	Spark Plug Gap (in.)	Ignition Timing (deg.)	Fuel Pump (psi)	Idle Speed (rpm)		Valve Clearance	
						MT	AT	In.	Ex.
2003	1.8 (1794)	1ZZ-FE	0.043	①	44-50	650-750	650-750	0.0059-0.0098	0.0098-0.0138
	1.8 (1796)	2ZZ-GE	0.043	②	44-50	750-850	700-800	0.0031-0.0071	0.0087-0.0126

Note: The Vehicle Emission Control Information label often reflects specification changes made during production. The label figures must be used if they differ from those in this chart.

① With terminal TC and CG of DLC3 connected: 8-12 degrees BTDC
With terminal TC and CG of DLC3 disconnected: 10-18 degrees BTDC

② With terminal TC and CG of DLC3 connected: 8-12 degrees BTDC
With terminal TC and CG of DLC3 disconnected:
A/T: 10-18 degrees BTDC
M/T: 4-12 degrees BTDC

93591CQC

CAPACITIES

Year	Model	Engine Displacement Liters (cc)	Engine ID/VIN	Engine Oil with Filter	Transmission (pts.)			Drive Axle		Fuel Tank (gal.)	Cooling System (qts.)
					5-Spd	6-Spd	Auto.	Front (pts.)	Rear (pts.)		
2003	Matrix	1.8 (1794)	1ZZ-FE	3.9	4.0	4.0	5.2	NA	1.04	①	6.9
	Matrix	1.8 (1796)	2ZZ-GE	4.7	4.8	4.8	6.1	NA	—	13.2	7.1

Note: All capacities are approximate. Add fluid gradually and check to be sure a proper fluid level is obtained.

NA - Not available

① 2WD: 13.2 gallons
4WD: 11.9 gallons

93591CQD

For Tune-up, Capacities and Firing orders, see Section 1 of this manual

VALVE SPECIFICATIONS

Year	Engine Displacement Liters (cc)	Engine ID/VIN	Seat Angle (deg.)	Face Angle (deg.)	Spring Test Pressure (lbs. @ in.)	Spring Installed Height (in.)	Stem-to-Guide Clearance (in.)		Stem Diameter (in.)	
							Intake	Exhaust	Intake	Exhaust
2003	1.8 (1794)	1ZZ-FE	45	44.5	31.3-34.8@ 1.252	1.323	0.0010-0.0024	0.0012-0.0026	0.2154-0.2159	0.2152-0.2158
	1.8 (1796)	2ZZ-GE	45	44.5	①	1.516	0.0010-0.0023	0.0012-0.0025	0.2145-0.2156	0.2144-0.2154

① Intake: 49.6-55.5@1.516
  Exhaust: 47.6-52.6@1.516

93591CQE

CRANKSHAFT AND CONNECTING ROD SPECIFICATIONS
All measurements are given in inches.

Year	Engine Displacement Liters (cc)	Engine ID/VIN	Crankshaft				Connecting Rod		
			Main Brg. Journal Dia.	Main Brg. Oil Clearance	Shaft End-play	Thrust on No.	Journal Diameter	Oil Clearance	Side Clearance
2003	1.8 (1762)	1ZZ-FE	1.8893-1.8898	0.0006-0.0013	0.0008-0.0087	3	1.7320-1.7323	0.0011-0.0024	0.0063-0.0135
	1.8 (1796)	2ZZ-GE	1.8893-1.8898	0.0006-0.0013	0.0016-0.0094	3	1.7713-1.7717	0.0011-0.0020	0.0063-0.0135

93591CQF

PISTON AND RING SPECIFICATIONS
All measurements are given in inches.

| Year | Engine Displacement Liters (cc) | Engine ID/VIN | Piston Clearance | Ring Gap | | | Ring Side Clearance | | |
				Top Compression	Bottom Compression	Oil Control	Top Compression	Bottom Compression	Oil Control
2003	1.8 (1762)	1ZZ-FE	0.0026-0.0035	0.0098-0.0138	0.0138-0.0197	0.0059-0.0197	0.0008-0.0028	0.0012-0.0028	0.0012-0.0043
	1.8 (1796)	2ZZ-GE	0.0003-0.0015	0.0098-0.0138	0.0138-0.0197	NA	0.0009-0.0028	0.0012-0.0028	NA

NA - Not available

93591CQG

TORQUE SPECIFICATIONS
All readings in ft. lbs.

| Year | Engine Displacement Liters (cc) | Engine ID/VIN | Cylinder Head Bolts | Main Bearing Bolts | Rod Bearing Bolts | Crankshaft Damper Bolts | Flywheel Bolts | Manifold | | Spark Plugs | Lug Nuts |
								Intake	Exhaust		
2003	1.8 (1794)	1ZZ-FE	①	②	③	102	①	22	27	18	76
	1.8 (1796)	2ZZ-GE	④	⑤	⑥	87	①	⑦	37	13	76

① Step 1: 36 ft. lbs.
  Step 2: 90 degree turn
② 12 pointed bolts:
  Step 1: 33 ft. lbs.
  Step 2: 90 degree turn
  Hex head bolts: 14 ft. lbs.
③ Step 1: 15 ft. lbs.
  Step 2: 90 degree turn
④ Step 1: 26 ft. lbs.
  Step 2: 180 degree turn

⑤ 12 pointed bolts:
  Step 1: 16 ft. lbs.
  Step 2: 32 ft. lbs.
  Step 3: 45 degree turn
  Step 4: 45 degree turn
  Hex head bolts: 13 ft. lbs.
⑥ Step 1: 22 ft. lbs.
  Step 2: 90 degree turn
⑦ Bolt A: 25 ft. lbs.
  Bolt B: 34 ft. lbs.

93591CQH

For complete service labor times, order Nichols' Chilton Labor Guide

WHEEL ALIGNMENT

Year	Model		Caster Range (+/-Deg.)	Caster Preferred Setting (Deg.)	Camber Range (+/-Deg.)	Camber Preferred Setting (Deg.)	Toe-in (in.)	Steering Axis Inclination (Deg.)
2003	Matrix - 2WD	F	0.75	+2.78	0.75	-0.77	0+/-0.08	12.47+/-0.75
		R	—	—	0.50	-1.45	0.11+/-0.11	—
	Matrix - 4WD	F	0.75	+2.77	0.75	-0.48	0+/-0.08	12.22+/-0.75
		R	—	—	0.75	-0.73	0.08+/-0.08	—

93591CQI

TIRE, WHEEL AND BALL JOINT SPECIFICATIONS

Year	Model	OEM Tires Standard	OEM Tires Optional	Tire Pressures (psi) Front	Tire Pressures (psi) Rear	Wheel Size	Ball Joint Inspection
2003	Matrix	205/55R16	—	33	33	6.5-JJ	9-26 in. ①

OEM: Original Equipment Manufacturer

PSI: Pounds Per Square Inch

STD: Standard

OPT: Optional

① Torque required in inch lbs. to rotate ball joint when removed from the knuckle

93591CQJ

BRAKE SPECIFICATIONS
All measurements in inches unless noted

Year	Model			Brake Disc			Brake Drum Diameter			Minimum Lining Thickness	Brake Caliper	
			Original Thickness	Minimum Thickness	Maximum Runout	Original Inside Diameter	Max. Wear Limit	Maximum Machine Diameter		Bracket Bolts (ft. lbs.)	Mounting Bolts (ft. lbs.)
2003	Matrix	F	0.984	0.906	0.0020	—	—	—	0.039	25	79
		R	0.354	0.295	0.0059	9.00	—	9.04	0.039	—	34

F: Front

R: Rear

93591CQK

SCHEDULED MAINTENANCE INTERVALS
TOYOTA—MATRIX

TO BE SERVICED	TYPE OF SERVICE	VEHICLE MILEAGE INTERVAL (x1000)												
		7.5	15	22.5	30	37.5	45	52.5	60	67.5	75	82.5	90	97.5
Engine oil & filter	R	✓	✓	✓	✓	✓	✓	✓	✓	✓	✓	✓	✓	✓
Drive belts	S/I								✓	✓	✓	✓	✓	✓
Automatic transaxle fluid & filter	S/I		✓		✓		✓		✓		✓		✓	
Ball joints & dust covers	S/I		✓		✓		✓		✓		✓		✓	
Bolts & nuts on body & chassis	S/I		✓		✓		✓		✓		✓		✓	
Brake line pipes & hoses	S/I		✓		✓		✓		✓		✓		✓	
Brake linings & drums	S/I		✓		✓		✓		✓		✓		✓	
Brake pads & discs (front & rear if equipped)	S/I		✓		✓		✓		✓		✓		✓	
Differential oil	S/I		✓		✓		✓		✓		✓		✓	
Drive shaft boots (except Supra)	S/I		✓		✓		✓		✓		✓		✓	
Manual transaxle oil	S/I		✓		✓		✓		✓		✓		✓	
Steering gear housing oil	S/I		✓		✓		✓		✓		✓		✓	
Steering linkage	S/I		✓		✓		✓		✓		✓		✓	
Air filter	R				✓				✓				✓	
Spark plugs	R				✓				✓				✓	
Spark plugs (platinum tip)	R								✓					
Exhaust system	S/I				✓				✓				✓	
Fuel lines & connections	S/I				✓				✓				✓	
Valve clearance	S/I				✓				✓				✓	
Engine coolant	R						✓				✓			
Fuel tank cap gasket	R								✓					
Charcoal canister	S/I								✓					

R: Replace S/I: Service or Inspect

FREQUENT OPERATION MAINTENANCE (SEVERE SERVICE)
If a vehicle is operated under any of the following conditions it is considered severe service:

- Extremely dusty areas.

- 50% or more of the vehicle operation is in 32°C (90°F) or higher temperatures, or constant operation in temperatures below 0°C (32°F).

- Prolonged idling (vehicle operation in stop and go traffic).

- Frequent short running periods (engine does not warm to normal operating temperatures).

- Police, taxi, delivery usage or trailer towing usage.

Oil & oil filter: change every 6000 miles.

Bolts & nuts on chassis & body: tighten every 7500 miles.

Ball joints & dust covers: service or inspect every 12,000 miles.

Brake linings & drums: service or inspect ever 12,000 miles.

Brake pads & discs (front & rear if equipped): service or inspect every 12,000 miles.

Drive shaft boots & except Supra): service or inspect every 12,000 miles.

Steering linkage: service or inspect every 12,000 miles.

Air filter: service or inspect every 15,000 miles.

Exhaust system: service or inspect every 15,000 miles.

Timing belt: replace every 60,000 miles.

93591CQL

HEATER CORES

2

ACURA

MDX

REMOVAL & INSTALLATION

1. Disconnect the negative battery cable.

2. Drain the cooling system.

3. Recover the refrigerant using approved equipment.

4. Disconnect the heater valve cable from the valve arm. Turn the valve arm to the fully opened position.

5. Disconnect the heater hoses from the heater unit.

6. Remove the mounting nut from the heater unit. Be careful not to bend or damage fuel or brake lines.

7. Remove the dashboard as follows:

   a. Remove the center console by unlatching the clips.

   b. Remove the dashboard lower cover screw, gently pull down on the cover to disengage the clips and disconnect the electrical connections.

   c. Remove the dashboard side cover by gently pulling and turning to unfasten the clips.

   d. While holding the glove box, remove the box stop from each side, then disconnect the lock from the damper.

   e. Remove the glove ox bolts and the glove box.

   f. Remove the shift lever assembly.

   g. Remove the front door trim, lick panel and A-pillar trim from both sides.

   h. Remove the cap from the front pillar corner trim. Unfasten the screw, slide the trim upward along the pillar and

Fastener Locations

A ▶ : Bolt, 2   B ▶ : Bolt, 5   C ▶ : Bolt, 3   D ▶ : Bolt, 2   E ▶ : Bolt, 2   F ▶ : Bolt, 1

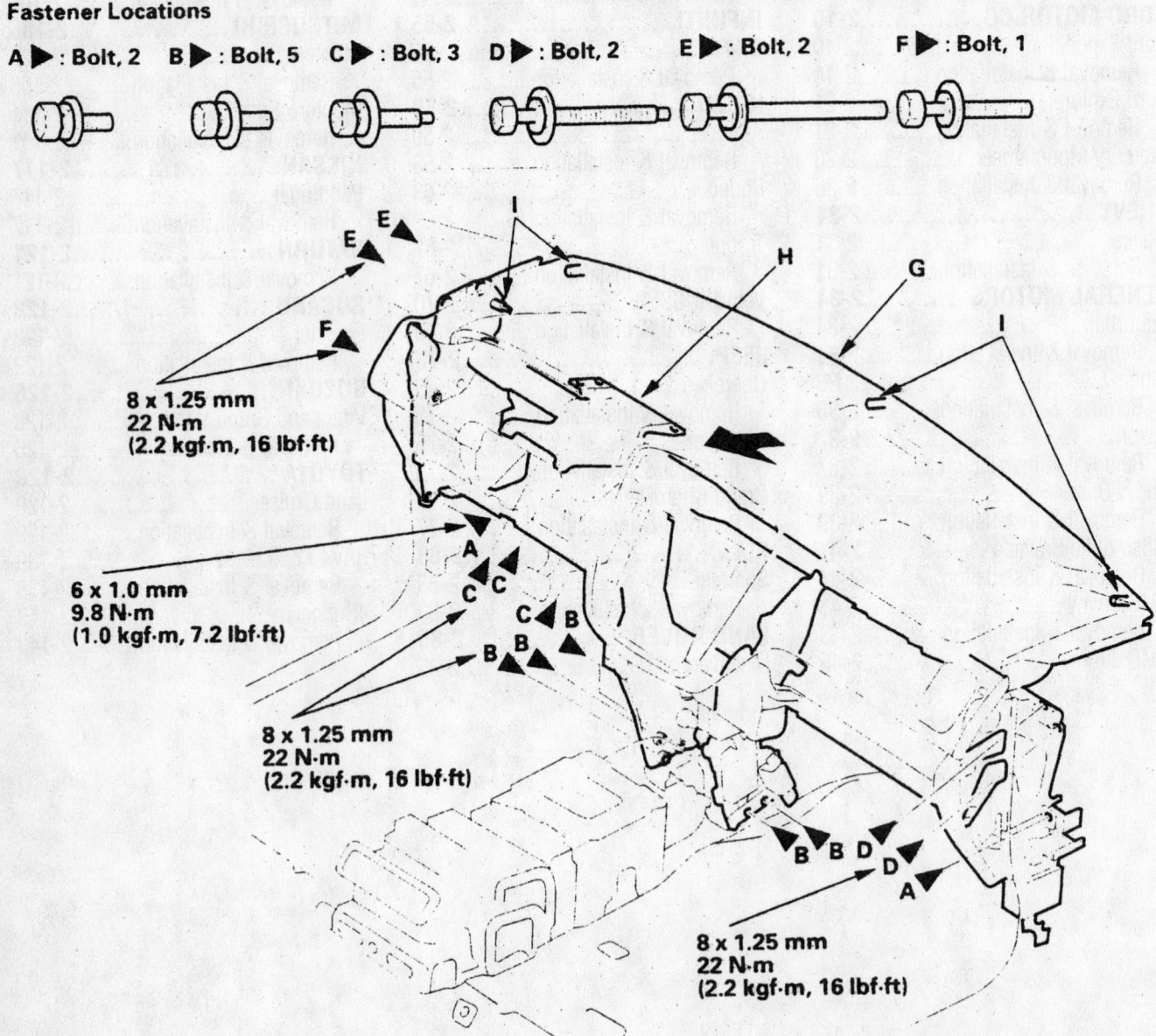

8 x 1.25 mm
22 N·m
(2.2 kgf·m, 16 lbf·ft)

6 x 1.0 mm
9.8 N·m
(1.0 kgf·m, 7.2 lbf·ft)

8 x 1.25 mm
22 N·m
(2.2 kgf·m, 16 lbf·ft)

8 x 1.25 mm
22 N·m
(2.2 kgf·m, 16 lbf·ft)

93552G91

Exploded view of the dashboard mounting—Acura MDX

6 x 1.0 mm
9.8 N·m (1.0 kgf·m, 7.2 lbf·ft)

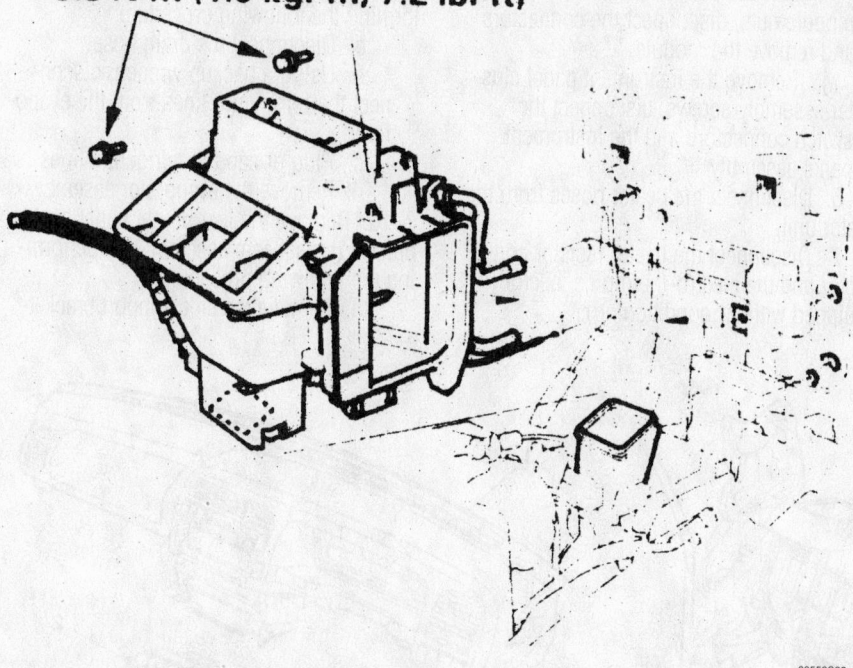

93552G92

Exploded view of the evaporator mounting—Acura MDX

remove it. Remove the remaining clips from the body.

i. On the drivers side, remove the fuel/relay box nut and pull out the box.

j. Remove the steering column

k. On the passenger side remove the fuse/relay bolt and pull out the box.

l. Disconnect all electrical connections from the dashboard.

m. If equipped with a navigation system, pull back the carpet, remove the harness cushions and then pull out the GPS harness.

n. Remove all harness and connector clips.

o. Remove all the bolts and lift up on the dashboard to release the dashboard and steering hanger beam from the guide pins.

p. Remove the dashboard through the door.

8. Remove the evaporator as follows:

a. Disconnect the receiver and suction lines from the evaporator.

b. Remove the mounting nuts and plug the lines to avoid system contamination.

c. Remove the plastic brace and glove box frame.

d. Disconnect the wire harness and evaporator temperature sensor connector.

e. Remove the self-tapping screws, the nuts and the evaporator.

9. Remove the mounting bolts and the heater unit.

10. Remove the self-tapping screws and the clamp, then pull the heater core from the case being careful not to bend the pipes.

To install:

11. Install the heater core in the case.

12. Install the clamp and the screws.

13. Install the heater unit and tighten the bolts to 7 ft. lbs. (10 Nm).

14. Install the evaporator in the reverse order of removal. Tighten all the retainers to 7 ft. lbs. (10 Nm) .

15. Install the dashboard in the reverse order of removal keeping in mind the following points:

a. Make sure the dashboard is seated properly and that the wiring harness and steering hanger beam wire harness are not pinched.

b. Refering to the accompanying illustration, tighten bolts **(A)** to 7 ft. lbs. (10 Nm). Tighten all the other bolts to 16 ft. lbs. (22 Nm). Apply thread lock to the **B** bolts before installation.

c. Ensure that all electrical connectors are properly connected.

16. Install the mounting nut to the heater unit and tighten to 9 ft. lbs. (13 Nm).

17. Connect the heater hoses.

18. Connect the heater valve cable and adjust as follows:

19. Fill the cooling system

20. Connect the battery cable.

SLX

REMOVAL & INSTALLATION

✳✳ CAUTION

The vehicle is equipped with a driver's side and a passenger's side air bag. Before starting service procedures on components, especially under the instrument panel and/or near the steering column, disable the air bag systems. There is sufficient voltage in the system to cause a deployment for up to 15 seconds after the battery has been disconnected, the ignition turned OFF or fuse C-21 is removed from the fuse panel.

1. If equipped with an air bag, perform the following procedures:

a. Disconnect the negative battery cable, then disconnect the positive battery cable.

b. Disconnect the yellow 2-pin connector located at the base of the steering column.

c. Remove the glove box and disconnect the yellow 2-pin connector located behind the glove box.

2. Disconnect the negative battery cable.

3. Drain the cooling system.

4. If equipped with air conditioning, discharge and recover the refrigerant.

5. Remove the instrument panel assembly by performing the following procedure:

a. At the front console assembly, disconnect the switch connectors; then, remove the console-to-chassis screws and the console.

b. At the lower cluster assembly, remove the cluster-to-instrument panel screws, disconnect the cigarette lighter and light connectors and remove the lower cluster.

c. Remove the glove box and the instrument panel lower cover and the passenger knee bolster reinforcement.

d. At the left side, remove the instrument panel lower cover and the knee bolster assembly.

e. At the top of the instrument panel, pry the 8 claws on the front side toward you, raise the defroster grille and remove it.

f. At the SRS adjust bracket and cross beam, under the passenger air bag module, remove the 2 attaching bolts and remove the instrument panel assembly.

g. Disconnect the air conditioning control cables from the unit.

h. Remove the instrument harness connectors (5 on the driver's side and 3 on the passenger's side), the passenger air bag module connector, the radio antenna plug and the center bracket ground cable bolt.

i. Remove the passenger's air bag module nuts, disconnect the connectors and remove the module.

j. Remove the instrument panel cluster assembly screws, disconnect the switch connectors and the instrument panel assembly.

6. Disconnect the heater hoses from the heater unit.

7. Disconnect the heater resistor connector and the electro-thermo connector (if equipped with air conditioning).

8. Remove the heater duct.

9. If equipped with air conditioning, remove the evaporator assembly by performing the following procedure:

a. Disconnect the drain hose.

b. Using a backup wrench, disconnect the refrigerant lines from the evaporator.

c. Plug or cap the refrigerant lines.

d. Remove the evaporator assembly.

10. Remove the instrument panel center bracket (crossbeam assembly) by performing the following procedure:

a. Remove the side support bracket

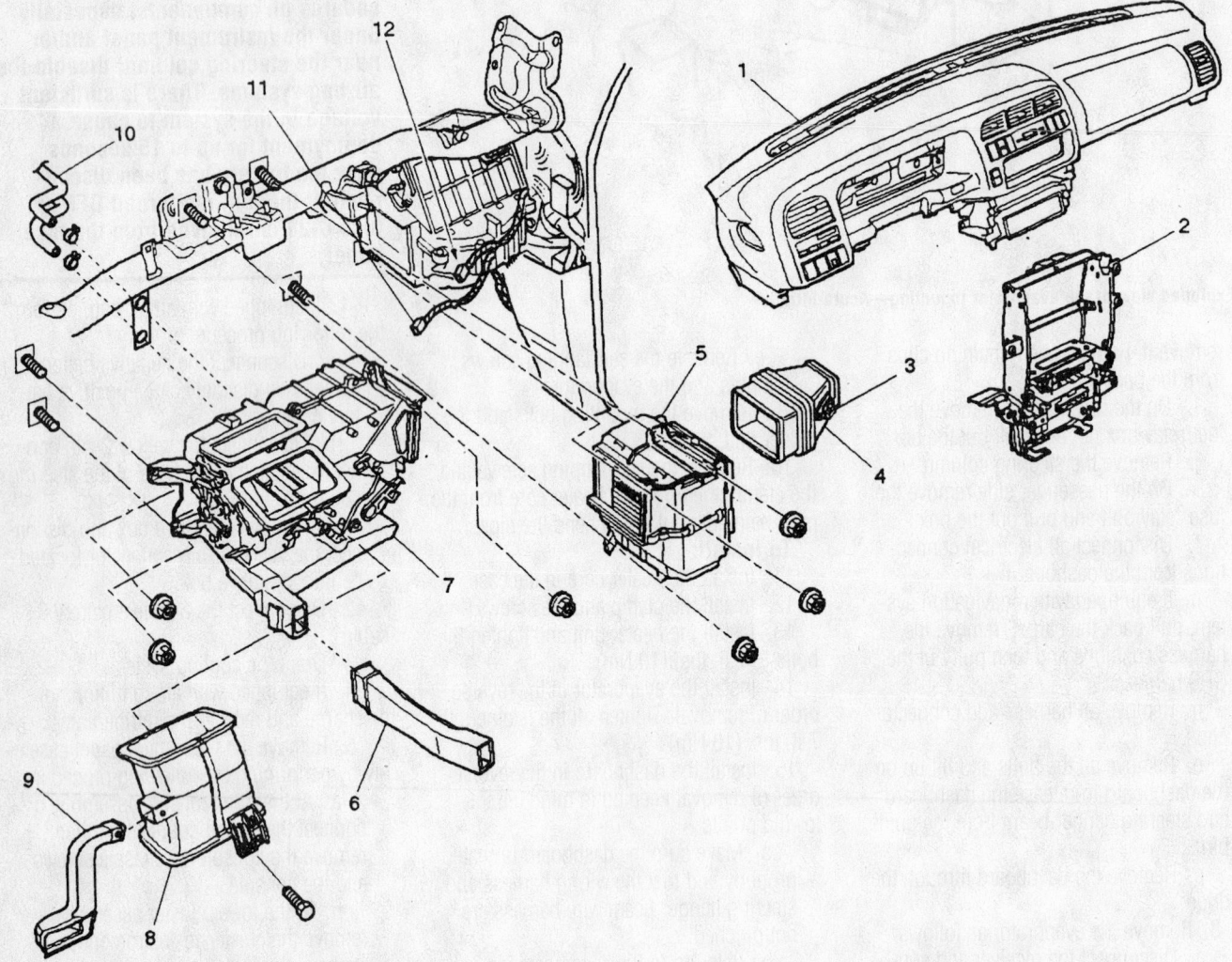

(1) Instrument Panel Assembly	(7) Heater Unit Assembly
(2) Instrument Panel Center Bracket	(8) Center Ventilation Lower Duct
(3) Resistor	(9) Driver Lap Vent Nozzle
(4) Duct	(10) Water Hose
(5) Evaporator Assembly (A/C only)	(11) Electro Thermo Connector (With A/C)
(6) Rear Heater Duct	(12) Resistor Connector

93113G01

Exploded view of the heater unit and related components—Acura SLX

bolts and brackets from both sides of the vehicle.

b. Remove the crossbeam center bracket nuts, disconnect the electrical connectors and the center bracket.

11. Remove the rear heater duct and heater assembly.

12. Disassemble the heater unit assembly by performing the following procedure:

a. Remove the lower air duct; do not remove the link unit.

b. Remove the temperature control case screws and lift the case from the heater unit.

c. Remove the heater core.

To install:

13. Assemble the heater unit assembly by performing the following procedure:

a. Install the heater core into the heater unit.

b. Install the temperature control case onto the unit and secure with screws.

c. Install the lower air duct.

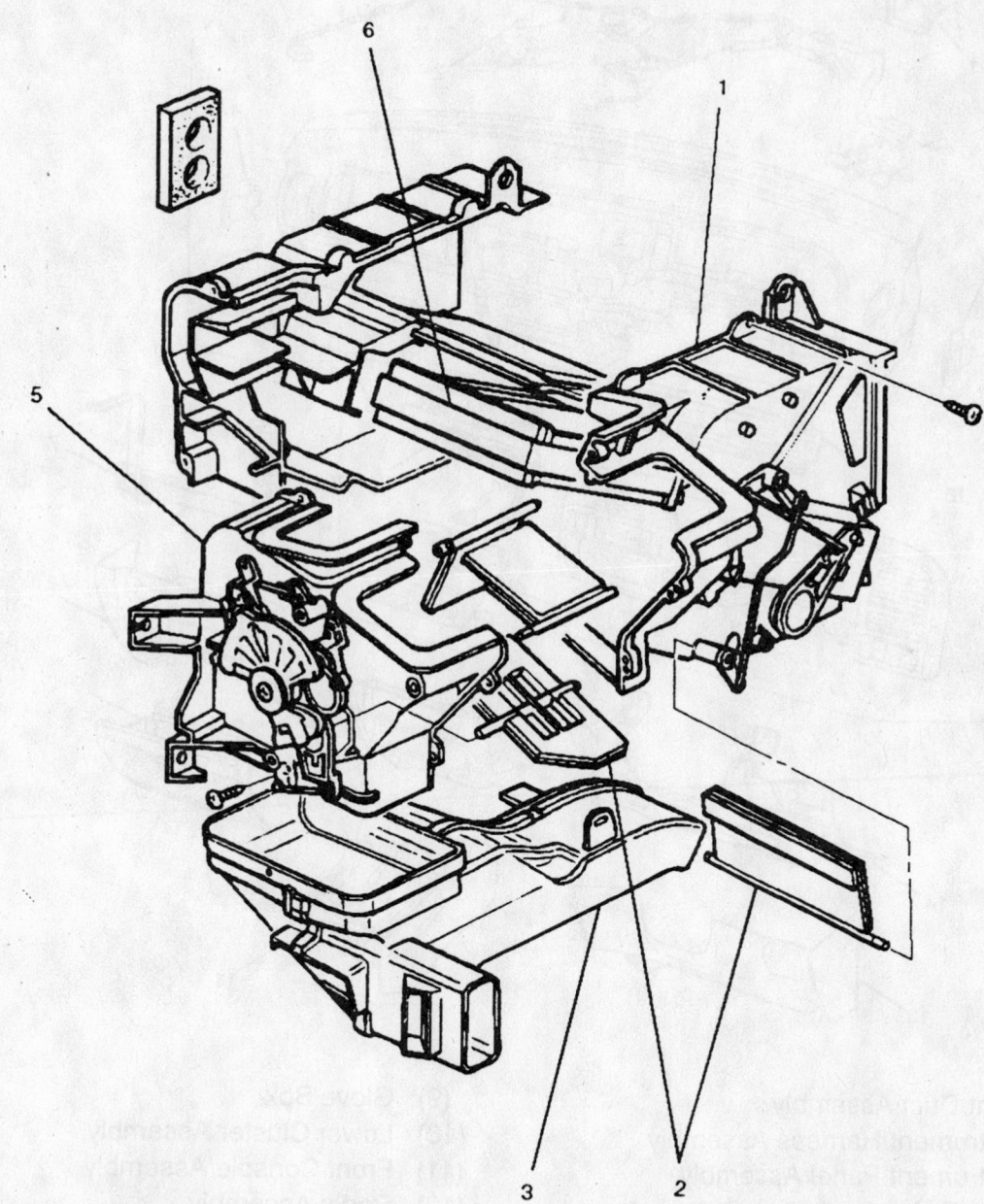

(1) Case (Temperature Control)
(2) Mode Door
(3) Duct
(5) Case (Mode Control)
(6) Heater Core

Exploded view of the heater unit—Acura SLX

93113G02

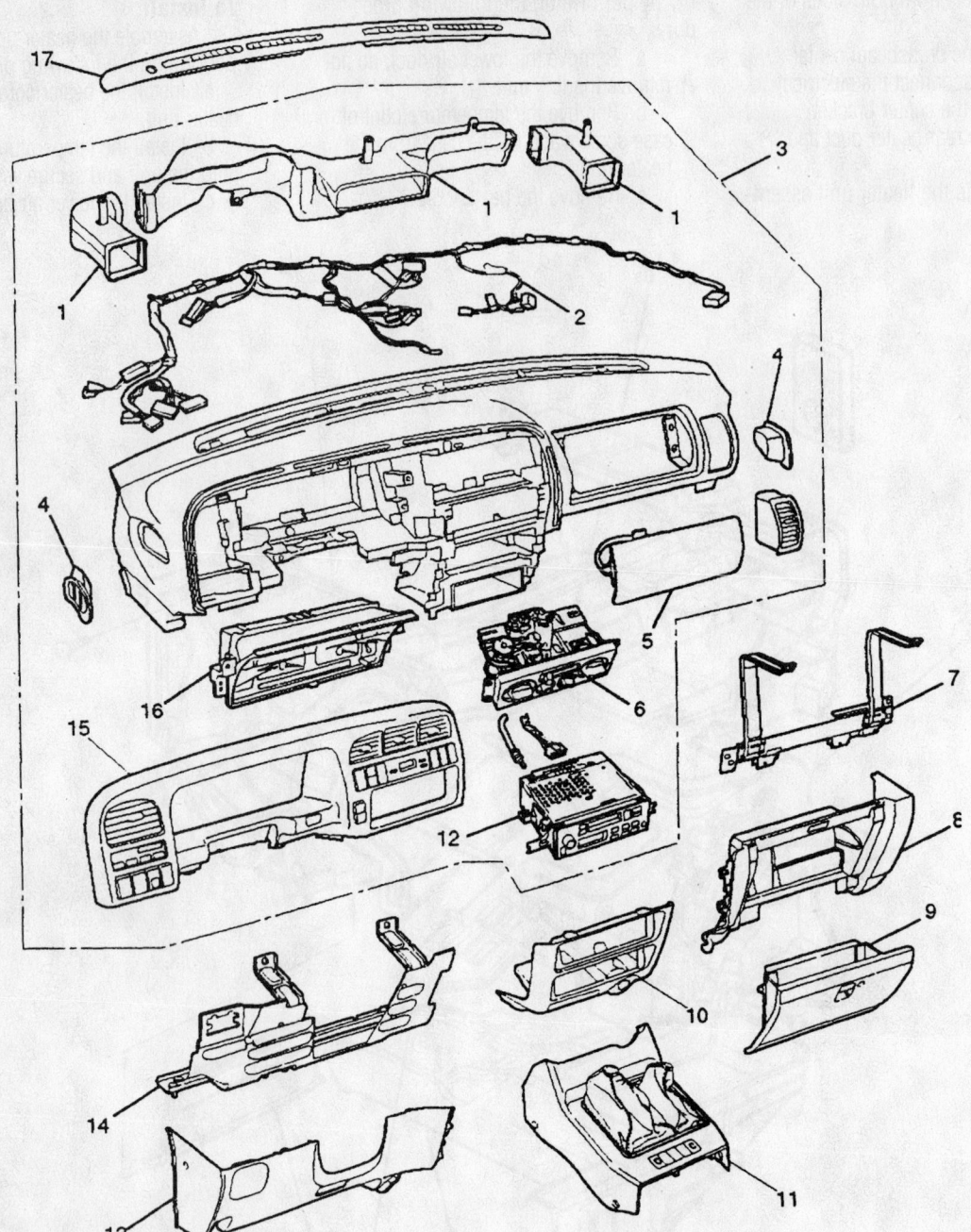

(1) Vent Duct Assembly
(2) Instrument Harness Assembly
(3) Instrument Panel Assembly
(4) Side Defroster Grille
(5) Passenger Inflator Module
(6) Control Lever Assembly
(7) Passenger Knee Bolster Reinforcement Assembly
(8) Instrument Panel Passenger Lower Cover Assembly
(9) Glove Box
(10) Lower Cluster Assembly
(11) Front Console Assembly
(12) Radio Assembly
(13) Instrument Panel Driver Lower Cover Assembly
(14) Driver Knee Bolster Assembly
(15) Instrument Panel Cluster Assembly
(16) Meter Assembly
(17) Front Defroster Grille

93113G03

Exploded view of the instrument panel and accessories—Acura SLX

14. Install the heater unit assembly into the vehicle.

15. Install the rear heater duct.

16. Install the instrument panel cross beam assembly by reversing the removal procedures.

17. If equipped with air conditioning, install the evaporator assembly by performing the following procedures:

a. If installing a new evaporator assembly, add 1.7 fl. oz. (50mL) of refrigerant oil to the evaporator.

b. Using new O-rings and a backup wrench, install the refrigerant lines and torque the outlet line to 18 ft. lbs. (25 Nm) and the inlet line to 11 ft. lbs. (15 Nm).

18. Install the heater duct.

19. Connect the heater resistor connec-tor and the electro-thermo connector (if equipped with air conditioning).

20. Connect the heater hoses to the heater unit.

21. Install the instrument panel assembly by reversing the removal procedures.

22. If equipped with air conditioning, evacuate, charge and leak test the system.

23. Refill the cooling system.

24. Connect the negative battery cable.

✳✳ CAUTION

Never use an air bag assembly from another vehicle and/or different model year. Starting in 1999, the air bag assemblies are equipped with identification colors on the bar code label as follows: YELLOW for the driver's air bag assembly, WHITE for the passenger's air bag assembly.

25. Enable the air bags by performing the following procedure:

a. Connect the passenger's side air bag yellow 2-pin connector.

b. Install the glove box.

c. At the base of the steering column, connect the yellow 2-pin connector.

d. Install the air bag fuse C-21 (if removed) or connect the negative battery cable.

e. Turn the ignition switch ON and verify that the AIR BAG warning light flashes 7 times and then turns OFF.

26. Run the engine to normal operating temperatures and check for leaks. Check the systems for correct operation.

CHRYSLER CORPORATION

Dodge Durango

REMOVAL & INSTALLATION

1. Disconnect the negative battery cable.

✳✳ CAUTION

After disconnecting the negative battery cable, wait 2 minutes for the driver's/passenger's air bag system capacitor to discharge before attempting to do any work around the steering column or instrument

2. Remove the instrument panel by performing the following procedure:

a. Remove the trim from the right and left door sills.

b. Remove the trim from the right and left cowl side inner panels.

c. Remove the steering column opening cover from the instrument panel.

d. Remove the 2 hood latch release handle-to-instrument panel lower reinforcement screws and lower the release handle to the floor.

e. Disconnect the driver's side air bag module wire harness connector.

f. If equipped, disconnect the overdrive lockout switch harness connector.

g. Do not disassemble the steering column but remove the assembly from the vehicle.

h. From under the driver's side of the instrument panel, disconnect or remove the following items:

- The screw from the center of the headlight/dash-to-instrument panel bulkhead wire harness connector and disconnect the connector.
- The 2 body wire harness connectors from the 2 instrument panel wire harness connector that secure the outboard side of the instrument panel bulkhead connector.
- The 3 wire harness connectors from the 3 junction block connector receptacles located closest to the dash panel; 1 from the body wire harness and 2 from the headlight/dash wire harness.
- The plastic park brake release linkage rod-to-rear parking brake release handle lever. Disengage the linkage rod from the handle lever.
- The stoplight switch connector.
- The vacuum harness connector from the left side of the heater/air conditioning housing assembly.

i. Remove the instrument panel center support bracket.

j. Remove the instrument panel wire harness ground screw located on the left side of the air bag control module (ACM) mount on the floor panel transmission tunnel.

k. Disconnect the instrument panel wiring harness-to-ACM connector receptacle.

l. Remove the glove box.

m. Working through the glove box opening, disconnect or remove the following items:

- The 2 halves of the radio antenna coaxial cable connector near the center of the lower instrument panel.
- The antenna half of the radio antenna coaxial cable from the retainer clip near the outboard side of the lower instrument panel.
- The blower motor wire harness connector located near the heater/air conditioning housing assembly support brace.

n. From the passenger's side, disconnect or remove the following items:

- The 2 instrument panel wire harness connectors from the infinity speaker amplifier connector receptacles on the right cowl.
- The instrument panel wire harness radio ground eyelet-to-stud nut on the right cowl.

o. Loosen the right and left instrument panel cowl side roll-down bracket screws.

p. Remove the 5 top instrument panel-to-dash panel screws.

q. Pull the instrument panel rearward until the right and left cowl side roll-down bracket screws are in the roll-down slot position of both brackets.

r. Roll down the instrument panel and install a temporary hook in the center hole on top of the instrument panel; secure the other end to the top of the dash panel. The hook is to support the instrument panel in its rolled down position.

s. With the instrument panel in the rolled-down position, disconnect or remove the following items:

- The 2 instrument panel wire harness connector from the door jumper wire harness connectors located on the right bracket.
- The instrument panel wire harness connector from the blower

motor resistor connector receptacle.
- The temperature control cable flag retainer from the top of the heater/air conditioning housing assembly. Pull the cable core

adjuster clip off the blend-air door lever.
- The demister duct flexible hose from the adapter on the top of the heater/air conditioning housing assembly.

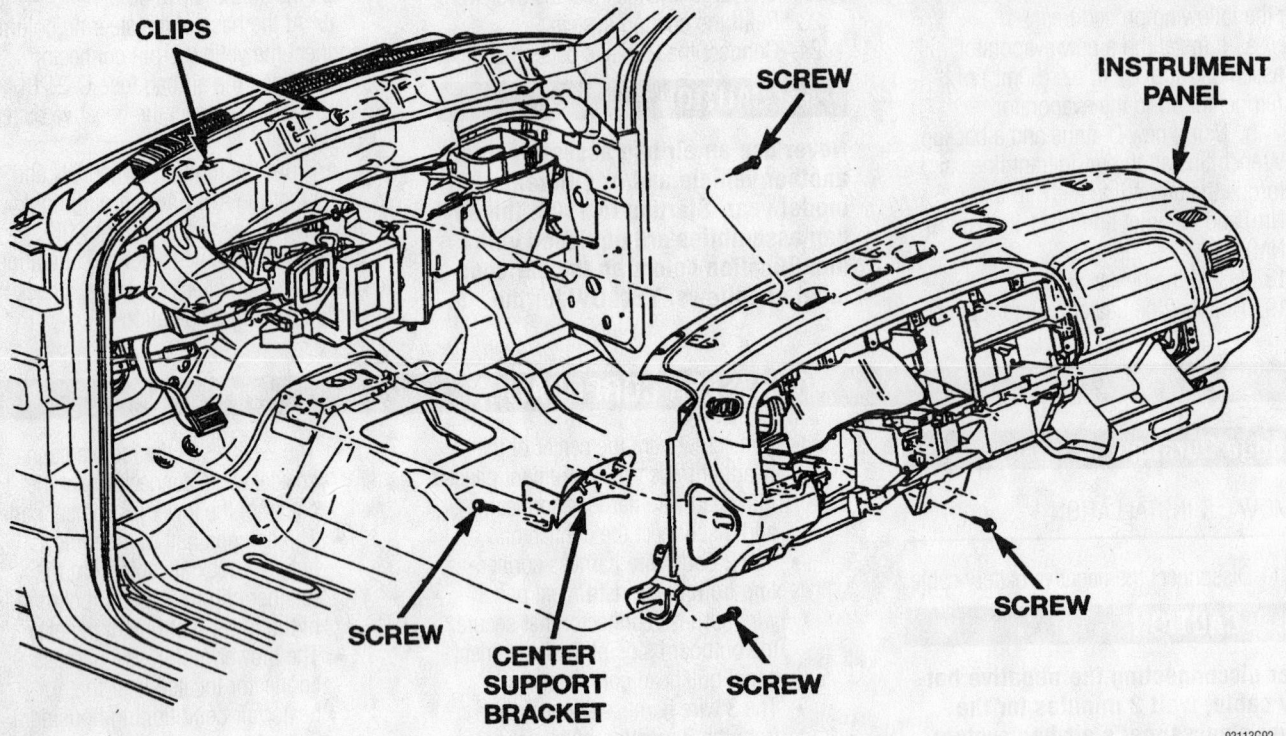

CLIPS **SCREW** **INSTRUMENT PANEL**

SCREW **CENTER SUPPORT BRACKET** **SCREW** **SCREW**

93113G92

View of the instrument panel assembly—Dodge Durango

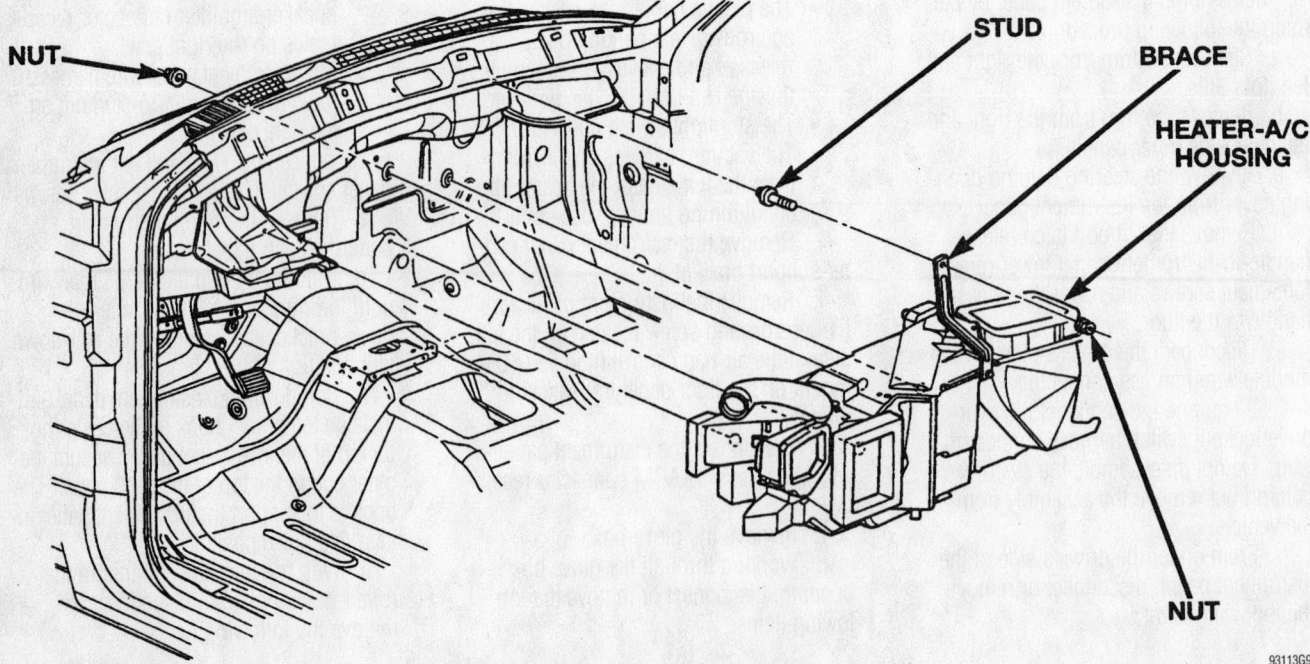

NUT **STUD** **BRACE**

HEATER-A/C HOUSING

NUT

93113G93

View of the heater/air conditioning assembly—Dodge Durango

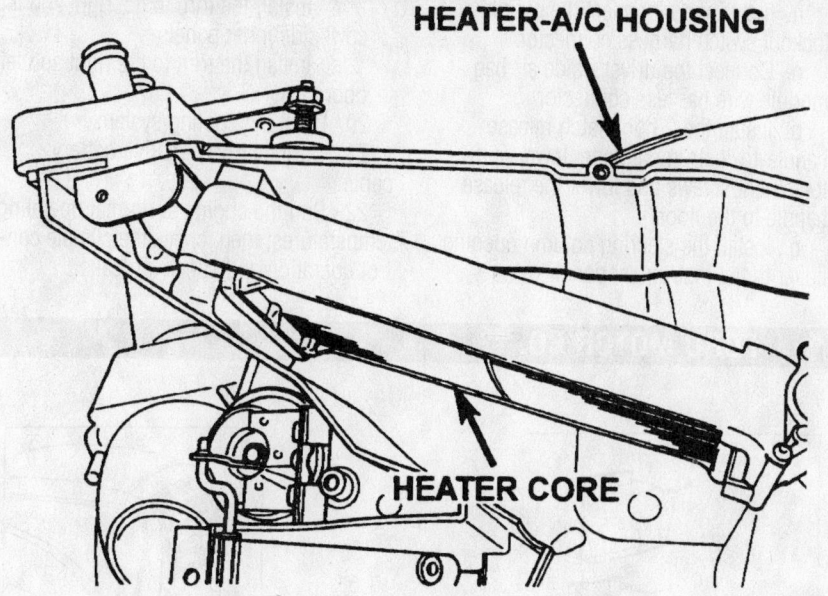

HEATER-A/C HOUSING

HEATER CORE

93113G94

View of the heater core—Dodge Durango

t. With aid of an assistant, lift the instrument panel from the vehicle.

3. If equipped with air conditioning, perform the following procedure:

a. Discharge and recover the air conditioning system refrigerant.

b. Disconnect the refrigerant line from the evaporator inlet and outlet tubes. Plug the openings to prevent contamination.

4. Drain the cooling system into a clean container for reuse.

5. Disconnect the heater hoses from the heater core. Plug the openings.

6. Remove the 4 heater/air conditioning housing assembly-to-chassis nuts.

7. Remove the heater/air conditioning housing assembly-to-mounting brace nut; the nut is located on the passenger side of the vehicle.

8. Pull the heater/air conditioning housing assembly rearward for the studs and drain tube to clear the dash panel hole.

9. Remove the heater/air conditioning housing assembly from the vehicle.

10. Remove the heater housing cover screws and the cover.

11. Remove the heater core from the heater/air conditioning housing assembly.

To install:

12. Install the heater core to the heater/air conditioning housing assembly.

13. Install the heater housing cover and the cover screws.

14. Install the heater/air conditioning housing assembly.

15. Install the heater/air conditioning

housing assembly-to-mounting brace nut; the nut is located on the passenger side of the vehicle.

16. Install the 4 heater/air conditioning housing assembly-to-chassis nuts.

17. Connect the heater hoses to the heater core.

18. If equipped with air conditioning, perform the following procedure:

a. Using new gaskets, connect the refrigerant line to the evaporator inlet and outlet tubes.

b. Evacuate and charge the air conditioning system.

19. Install the instrument panel by performing the following procedures:

a. With aid of an assistant, install the instrument panel into the vehicle.

b. With the instrument panel in the rolled-down position, connect or install the following items:

- The demister duct flexible hose to the adapter on the top of the heater/air conditioning housing assembly.
- The temperature control cable flag retainer to the top of the heater/air conditioning housing assembly.
- The instrument panel wire harness connector to the blower motor resistor connector receptacle.
- The 2 instrument panel wire harness connector to the door jumper wire harness connectors located on the right bracket.

c. Move the instrument panel forward

until the right and left cowl side roll-down bracket screws are in the roll-down slot position of both brackets.

d. Install the 5 top instrument panel-to-dash panel screws.

e. Install the right and left instrument panel cowl side roll-down bracket screws.

f. At the passenger's side, connect or install the following items:

- The instrument panel wire harness radio ground eyelet-to-stud nut on the right cowl.
- The 2 instrument panel wire harness connectors to the infinity speaker amplifier connector receptacles on the right cowl.

g. Working through the glove box opening, connect or install the following items:

- The blower motor wire harness connector located near the heater/air conditioning housing assembly support brace.
- The antenna half of the radio antenna coaxial cable to the retainer clip near the outboard side of the lower instrument panel.
- The 2 halves of the radio antenna coaxial cable connector near the center of the lower instrument panel.

h. Install the glove box.

i. Connect the instrument panel wiring harness-to-ACM connector receptacle.

j. Install the instrument panel wire harness ground screw located on the left side of the air bag control module (ACM) mount on the floor panel transmission tunnel.

k. Install the instrument panel center support bracket.

l. Under the driver's side of the instrument panel, connect or install the following items:

- The vacuum harness connector to the left side of the heater/air conditioning housing assembly.
- The stoplight switch connector.
- Engage the linkage rod to the handle lever. Install the plastic park brake release linkage rod-to-rear parking brake release handle lever.
- The 3 wire harness connectors to the 3 junction block connector receptacles located closest to the dash panel; 1 at the body wire harness and 2 at the headlight/dash wire harness.
- The 2 body wire harness connec-

Heater Core replacement is covered in Section 2 of this manual

tors to the 2 instrument panel wire harness connector that secure the outboard side of the instrument panel bulkhead connector.

• Connect the connector. Install the screw at the center of the headlight/dash-to-instrument panel bulkhead wire harness connector.

m. Install the steering column assembly.

n. If equipped, connect the overdrive lockout switch harness connector.

o. Connect the driver's side air bag module wire harness connector.

p. Install the 2 hood latch release handle-to-instrument panel lower reinforcement screws and lower the release handle to the floor.

q. Install the steering column opening cover to the instrument panel.

r. Install the trim to the right and left cowl side inner panels.

s. Install the trim to the right and left door sills.

20. Refill the cooling system.

21. Connect the negative battery cable.

22. Run the engine to normal operating temperatures; then, check the climate control operation and check for leaks.

FORD MOTOR CO.

Ford Expedition

REMOVAL & INSTALLATION

1. Disconnect the negative battery cable.

✳✳ CAUTION

After disconnecting the negative battery cable, wait 1 minute for the SRS module to deplete its energy.

2. Drain the cooling system into a clean container for reuse.

3. Remove the instrument panel by performing the following procedure:

a. If equipped, remove the floor console assembly.

b. Remove the lower steering column cover bolts and the cover.

c. Remove both front door scuff plates.

d. Remove both side cowl trim panels.

e. Disconnect the electrical connector from the Brake Pedal Position (BPP) switch.

f. Remove the radio ground and the GEM/CTM ground bolts.

g. Disconnect the left side instrument panel main wiring harness connector.

h. In the engine compartment, remove the bulkhead wiring harness connector bolts and disconnect the wiring connectors.

i. In the driver's compartment, release the 6 locking tabs and remove the bulkhead electrical connector from the instrument panel.

j. Disconnect the air bag diagnostic monitor electrical connector.

k. Disconnect the inertia fuel shutoff switch electrical connector.

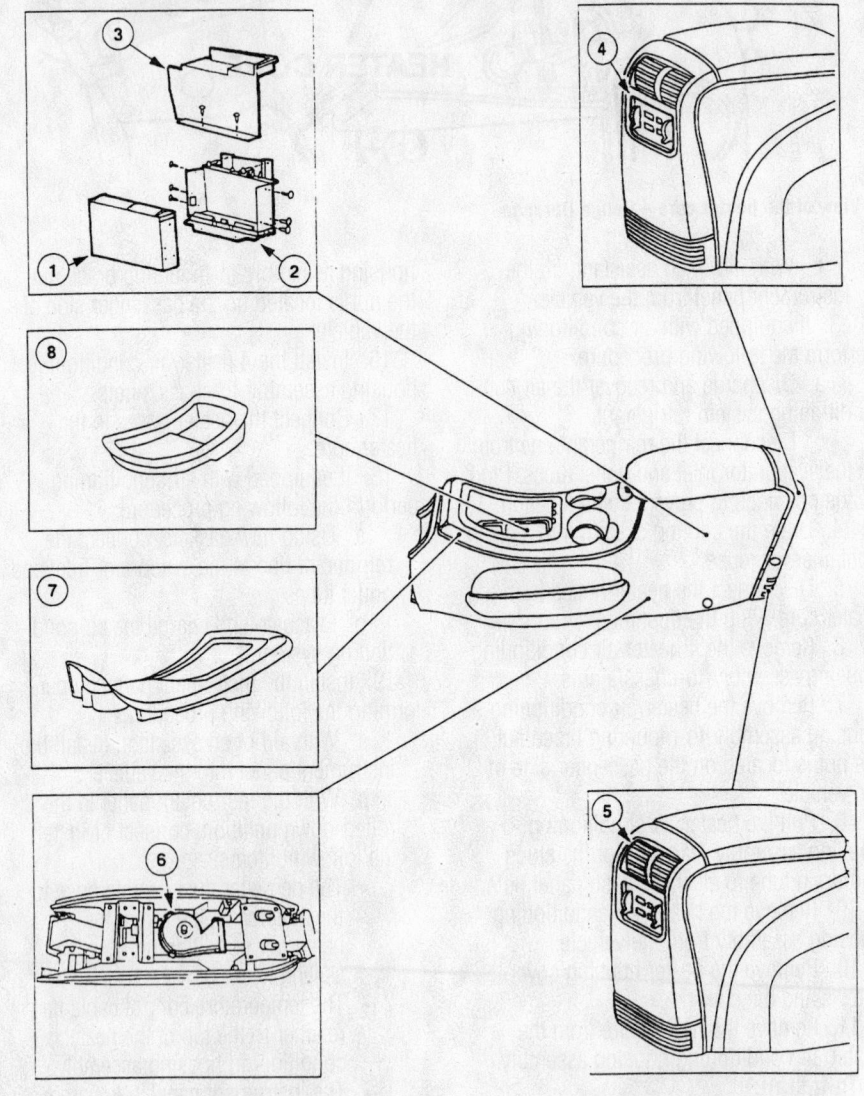

1. Digital audio compact disc player
2. Compact disc player mounting bracket
3. Compact disc player compartment trim panel
4. Radio and A/C integral control assembly
5. A/C register (upper)
6. Blower assembly
7. Center console finish panel
8. Console finish panel mat

93113GM2

Exploded view of the instrument panel components—Ford Expedition

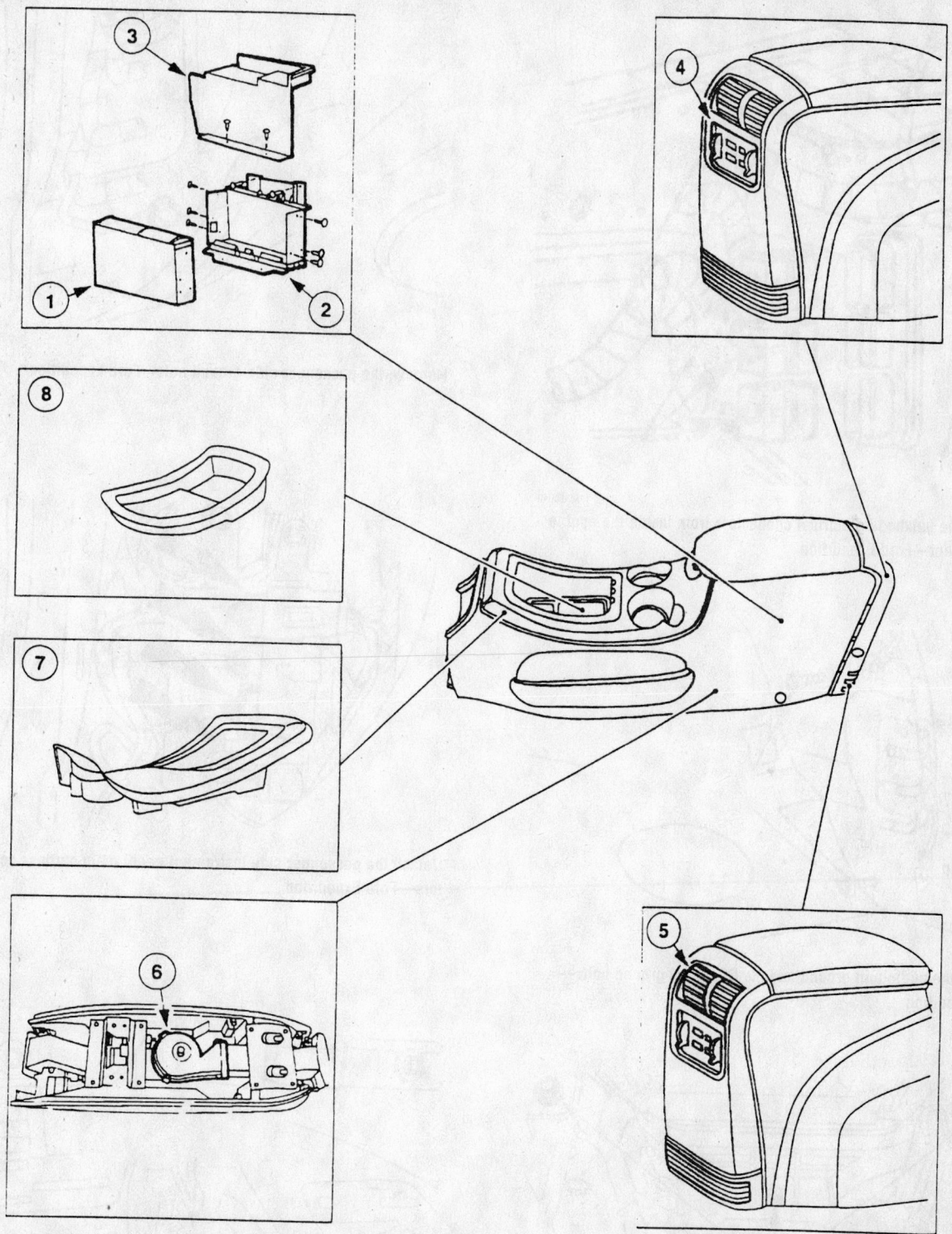

1. Digital audio compact disc player
2. Compact disc player mounting bracket
3. Compact disc player compartment trim panel
4. Radio and A/C integral control assembly
5. A/C register (upper)
6. Blower assembly
7. Center console finish panel
8. Console finish panel mat

93113GM3

Exploded view of the floor console components—Ford Expedition

Brake service is covered in Section 4 of this manual

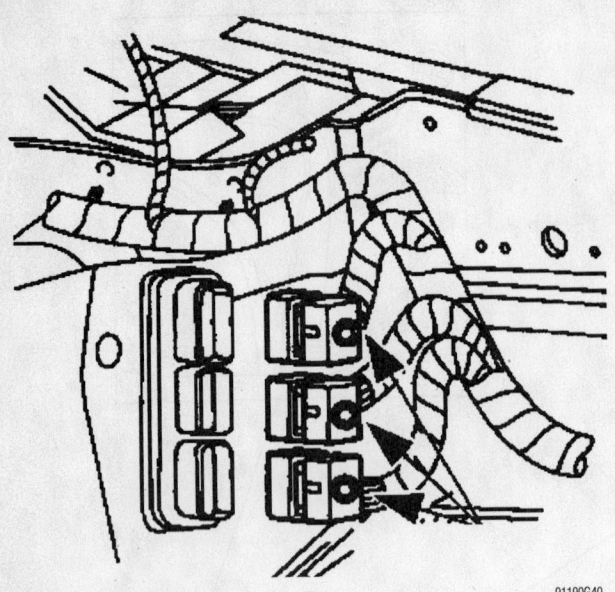

91190G40

Remove the bulkhead electrical connectors from inside the engine compartment—Ford Expedition

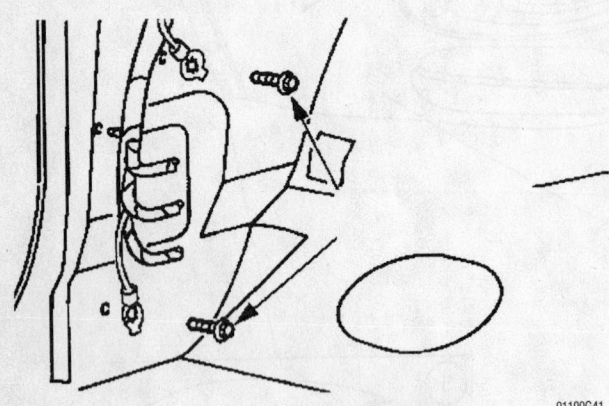

91190G41

Remove the audio unit ground and the GEM/CTM ground bolts—Ford Expedition

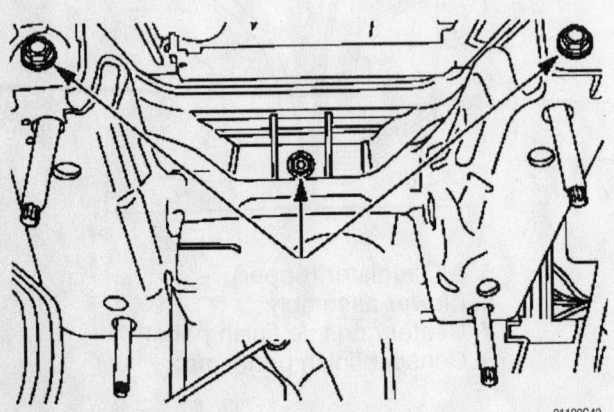

91190G42

Remove the instrument panel bolts through the steering column opening—Ford Expedition

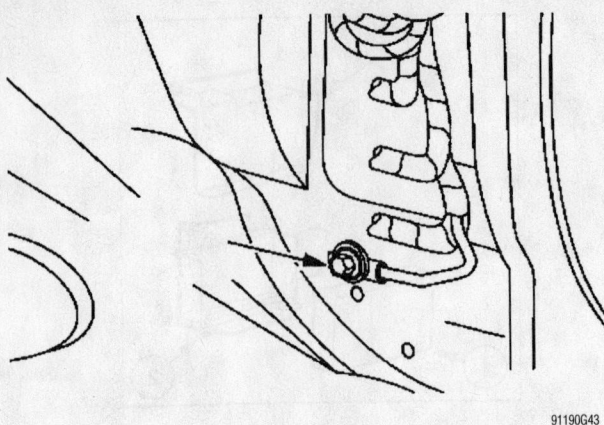

91190G43

Remove the passenger side ground bolt—Ford Expedition

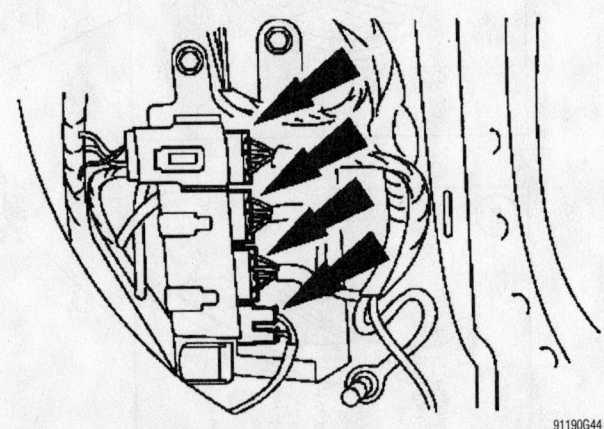

91190G44

Detach the passenger side instrument panel main harness connectors—Ford Expedition

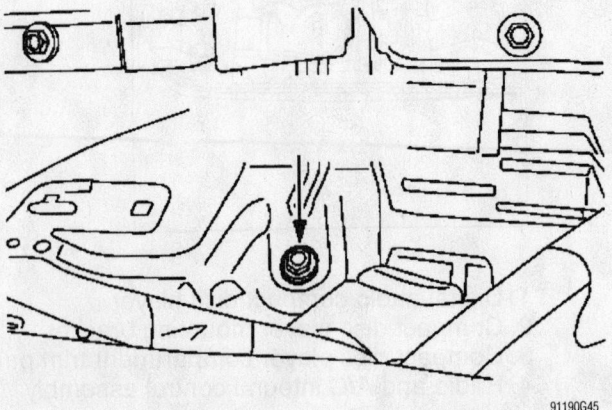

91190G45

Remove the instrument panel bolt on the relay bracket—Ford Expedition

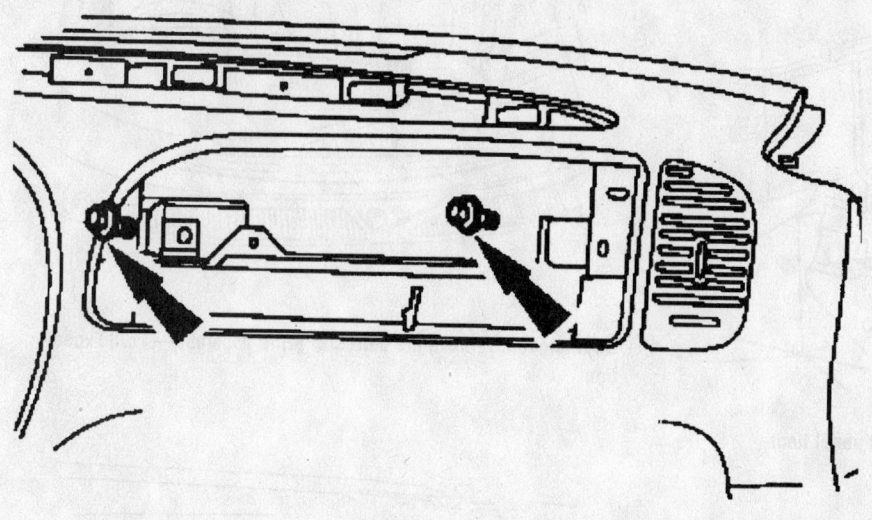

Remove the instrument panel bolts through the passenger side air bag module opening—Ford Expedition

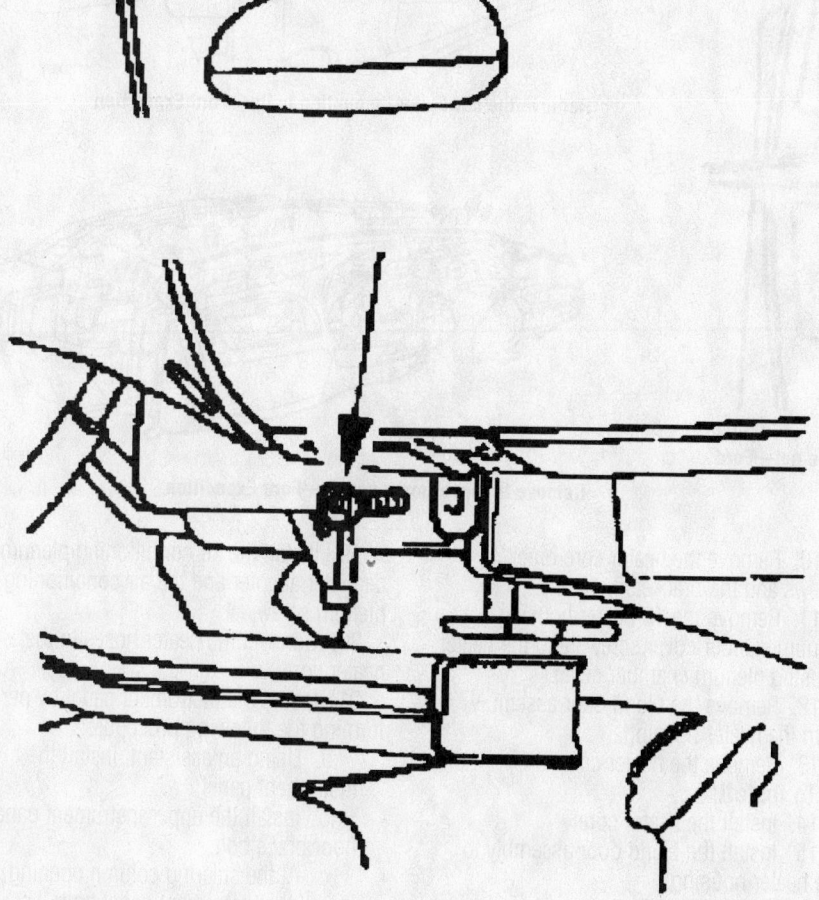

Remove the instrument panel reinforcement bolt below the driver's side corner of the glove compartment—Ford Expedition

l. Remove the right side ground bolts.

m. Disconnect the right side instrument panel wiring harness connectors.

n. Disconnect the electronic blend door actuator electrical connector.

o. Disconnect the climate control head vacuum harness connector.

p. Remove the steering column opening cover reinforcement nuts and the cover reinforcement.

q. At the base of the steering column, disconnect the air bag sliding contact and the anti-theft sensor electrical connectors.

r. At the steering column, disconnect the remaining electrical connectors.

s. If equipped with a transmission range indicator, remove the bolt and disconnect the cable.

t. Remove the steering column-to-instrument panel nuts and lower the steering column.

u. Remove the right side front fender splash shield screws and move the shield away from the panel.

v. Disconnect the antenna cable from the antenna base.

w. Remove the instrument panel relay cover and disconnect the autolamp sensor electrical connector and/or the sunload sensor connector.

x. Remove the glove box.

y. At the passenger's air bag module, remove the screws, disconnect the electrical connector and remove the air bag module.

Place the air bag module in a safe place with the front facing upward.

a. Remove the right side assist handle screw covers, the screws and the handle.

b. At both doors, pull back the weatherstrip seals and remove the windshield garnish moldings.

c. Remove the instrument panel reinforcement bolt below the left side corner of the glove box.

d. Through the air bag module opening, remove the instrument panel bolts.

e. Remove the upper instrument panel cowl covers and bolts.

f. At the relay bracket, remove the instrument panel bolt.

g. At the lower left side of the cigar lighter, remove the instrument panel bolt.

h. At the both sides, remove the instrument panel-to-cowl side nuts.

i. At the steering column opening, remove the instrument panel bolts.

For complete Engine Mechanical specifications, see Section 1 of this manual

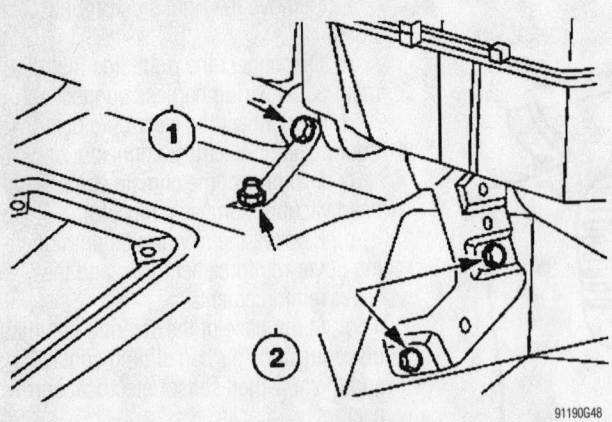

Position the carpet aside and loosen the instrument panel floor brace—Ford Expedition

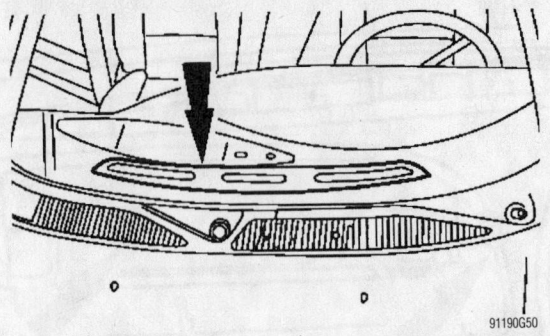

On Navigator, remove the defroster grille assembly—Ford Expedition

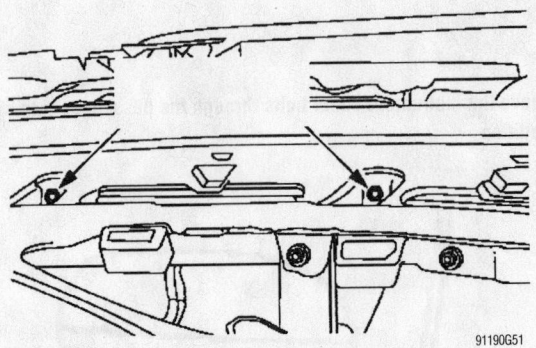

Remove the cowl panel mounting bolts—Ford Expedition

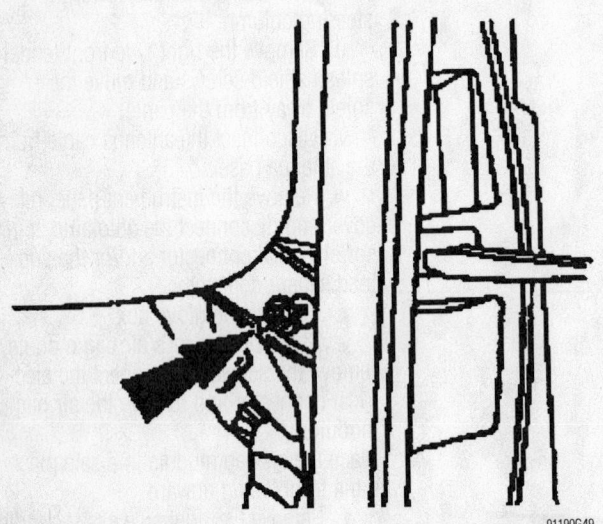

Remove the passenger side instrument panel cowl side nut—Ford Expedition

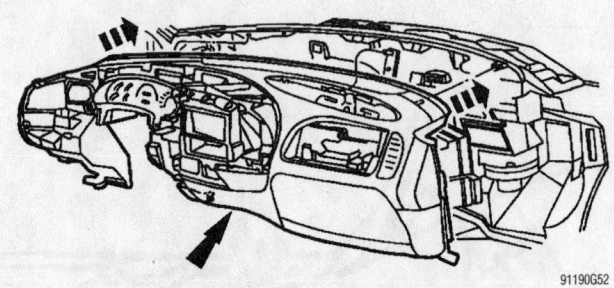

Remove the instrument panel —Ford Expedition

j. Remove the upper instrument panel floor brace bolt.

k. Using an assistant, remove the instrument panel.

4. If equipped with the 5.4L 4V engine, remove the junction block splash shield.

5. If equipped with the 5.4L 4V engine, remove the bolts and disconnect the cable ends from the starter relay.

6. If equipped with the 5.4L 4V engine, remove the junction block bracket.

7. Compress the holding tabs and disconnect the heater hoses from the heater core.

8. Remove the air conditioning plenum screw and the air conditioning plenum demister adapter.

9. Disconnect the vacuum line.

10. Remove the heater core bracket screws and the bracket.

11. Remove the 13 heater housing plenum camber cover screws and the heater housing plenum chamber cover.

12. Remove the blend door assembly from the heater housing.

13. Remove the heater core.

To install:

14. Install the heater core.

15. Install the blend door assembly to the heater housing.

16. Install the 13 heater housing plenum camber cover and the heater housing plenum chamber cover screws.

17. Install the heater core bracket and the bracket screws.

18. Connect the vacuum line.

19. Install the air conditioning plenum demister adapter and the air conditioning plenum screw.

20. Connect the heater hoses to the heater core.

21. Install the instrument panel by performing the following procedure:

a. Using an assistant, install the instrument panel.

b. Install the upper instrument panel floor brace bolt.

c. At the steering column opening, install the instrument panel bolts.

d. At the both sides, install the instrument panel-to-cowl side nuts.

e. At the lower left side of the cigar lighter, install the instrument panel bolt.

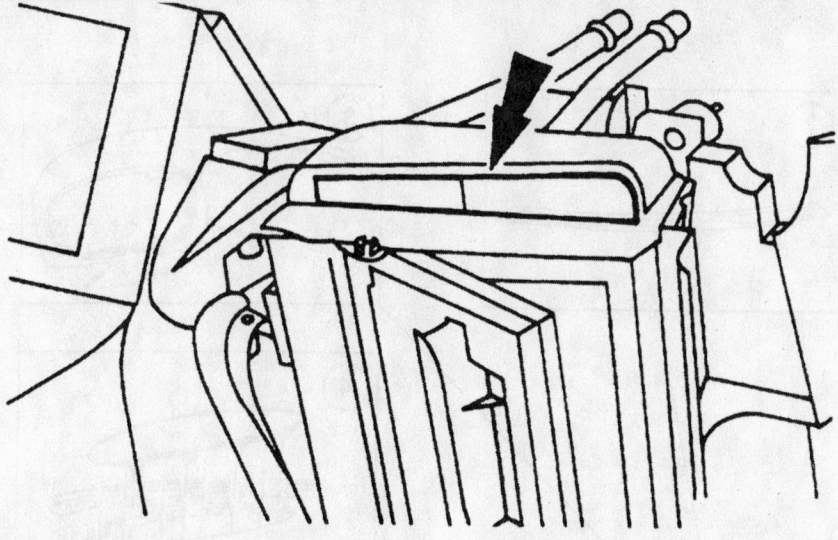

View of the heater core—Ford Expedition

93113GM1

f. At the relay bracket, install the instrument panel bolt.

g. Install the upper instrument panel cowl bolts and covers.

h. Through the air bag module opening, install the instrument panel bolts.

i. Install the instrument panel reinforcement bolt below the left side corner of the glove box.

j. At both doors, install the windshield garnish moldings and the weatherstrip seals.

k. Install the right side assist handle, the screws and the handle screw covers.

l. At the passenger's air bag module, install the air bag module, connect the electrical connector and install the air bag module screws.

m. Install the glove box.

n. Connect the autolamp sensor electrical connector and/or the sunload sensor connector; then, install the instrument panel relay cover.

o. Connect the antenna cable to the antenna base.

p. Install the right side front fender splash shield and screws.

q. Install the steering column and the steering column-to-instrument panel nuts.

r. If equipped with a transmission range indicator, connect the cable and install the bolt.

s. At the steering column, connect the remaining electrical connectors.

t. At the base of the steering column, connect the air bag sliding contact and the anti-theft sensor electrical connectors.

u. Install the steering column opening cover reinforcement and the cover reinforcement nuts.

v. Connect the climate control head vacuum harness connector.

w. Connect the electronic blend door actuator electrical connector.

x. Connect the right side instrument panel wiring harness connectors.

y. Install the right side ground bolts.

z. Connect the inertia fuel shutoff switch electrical connector.

aa. Connect the air bag diagnostic monitor electrical connector.

bb. In the driver's compartment, install the bulkhead electrical connector to the instrument panel.

cc. In the engine compartment, connect the bulkhead wiring harness connectors and the install wiring connector bolts.

dd. Connect the left side instrument panel main wiring harness connector.

ee. Install the radio ground and the GEM/CTM ground bolts.

ff. Connect the electrical connector to the Brake Pedal Position (BPP) switch.

gg. Install both side cowl trim panels.

hh. Install both front door scuff plates.

ii. Install the lower steering column cover and the cover bolts.

jj. If equipped, install the floor console assembly.

22. Refill the cooling system.

23. Connect the negative battery cable.

24. Run the engine to normal operating temperatures; then, check the climate control operation and check for leaks.

Ford Excursion

REMOVAL & INSTALLATION

1. Drain and recycle the engine coolant.

✳✳ CAUTION

Never open, service or drain the radiator or cooling system when hot; serious burns can occur from the steam and hot coolant. Also, when draining engine coolant, keep in mind that cats and dogs are attracted to ethylene glycol antifreeze and could drink any that is left in an uncovered container or in puddles on the ground. This will prove fatal in sufficient quantities. Always drain coolant into a sealable container. Coolant should be reused unless it is contaminated or is several years old.

2. Disconnect the heater water hoses from the heater core.

3. Disengage the stops and lower the glove compartment door.

4. Remove the electronic blend door actuator and bracket assembly.

✳✳ WARNING

>The heater core cover must be raised vertically before removal to avoid damage to the heater core housing.

5. Remove the heater core cover screws and remove the cover.

6. Remove the heater core from the housing.
To install:

✳✳ WARNING

Position the temperature blend door manually to properly align the actuator and the door. Do not power the actuator electrically. If it is not engaged with the temperature blend door, damage to the actuator may occur.

➡**Add gasket between housing and cover before installing cover.**

7. The installation is the reverse of the removal.

Lincoln Navigator

REMOVAL & INSTALLATION

1. Disconnect the negative battery cable.

For Accessory Drive Belt illustrations, see Section 1 of this manual

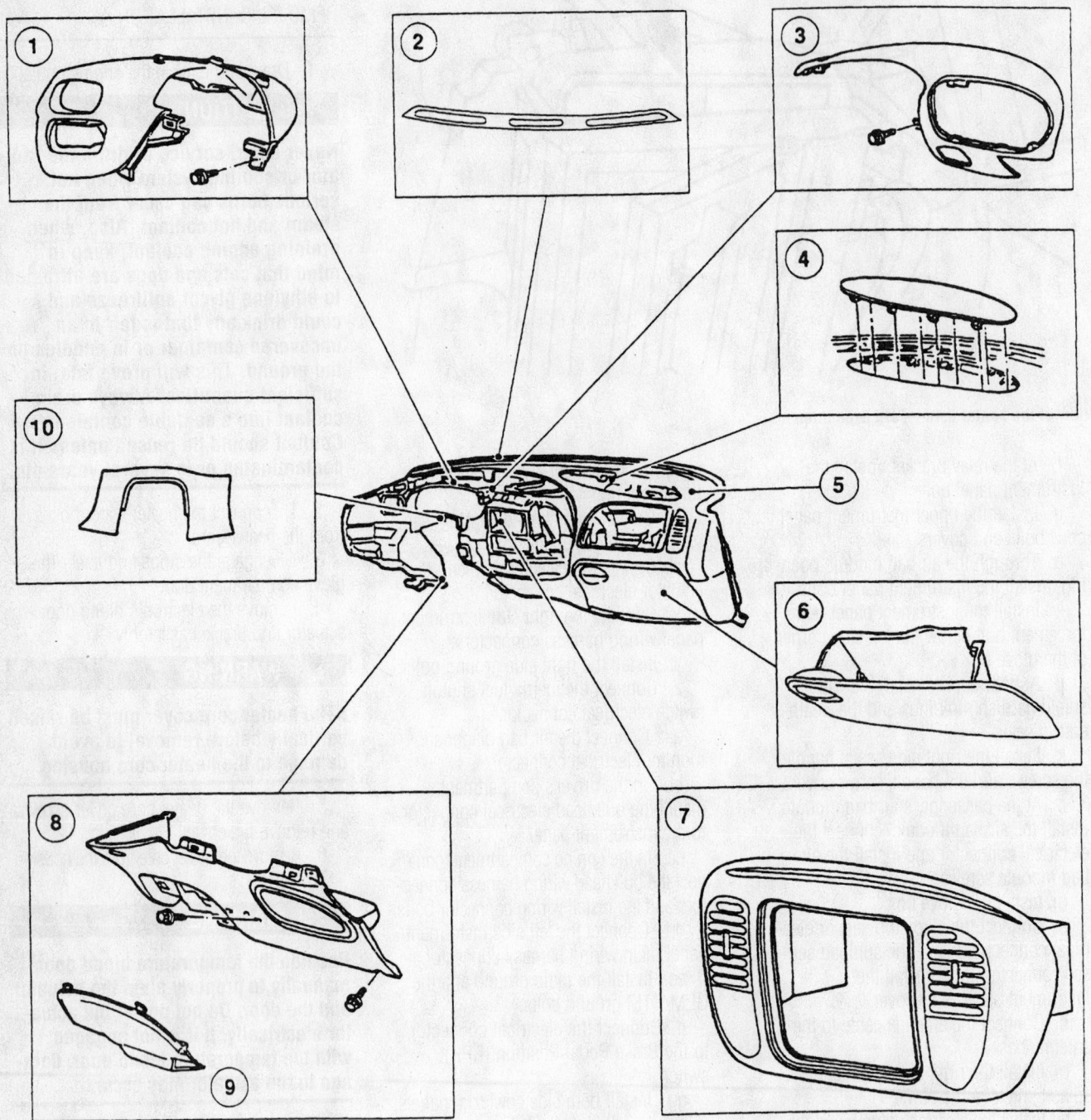

1. Instrument panel finish panel
2. Instrument panel defroster opening grille assembly
3. Instrument cluster panel
4. Instrument panel relay cover
5. Instrument panel
6. Glove compartment
7. Center instrument panel finish panel
8. Instrument panel steering column cover
9. Instrument panel fuse door
10. Steering column opening cover

93113GM2

Exploded view of the instrument panel components—Lincoln Navigator

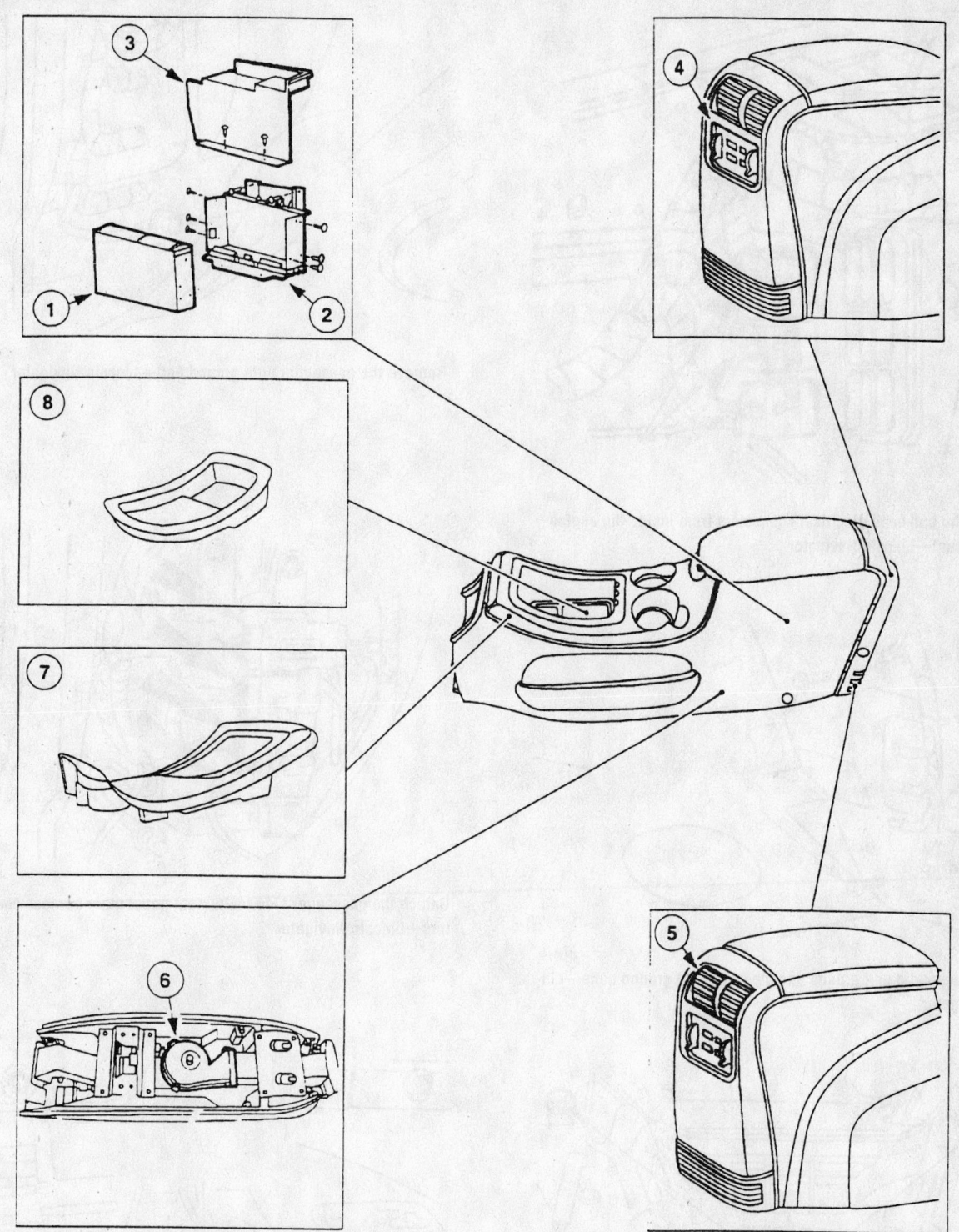

1. Digital audio compact disc player
2. Compact disc player mounting bracket
3. Compact disc player compartment trim panel
4. Radio and A/C integral control assembly
5. A/C register (upper)
6. Blower assembly
7. Center console finish panel
8. Console finish panel mat

93113GM3

Exploded view of the floor console components—Lincoln Navigator

For Tire, Wheel and Ball Joint specifications, see Section 1 of this manual

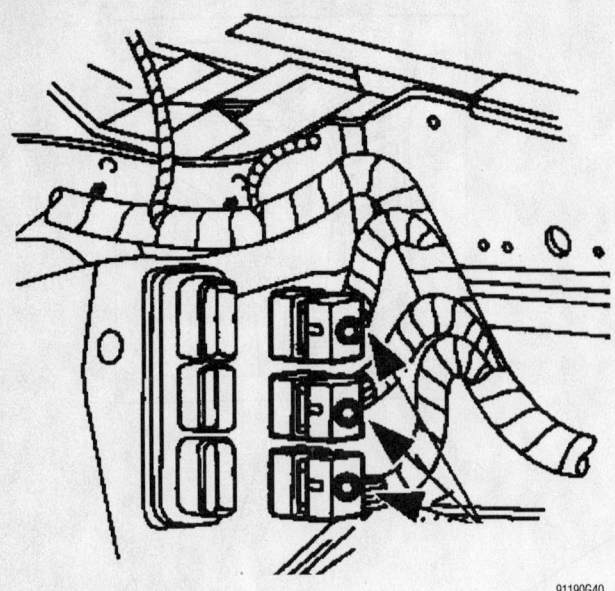

91190G40

Remove the bulkhead electrical connectors from inside the engine compartment—Lincoln Navigator

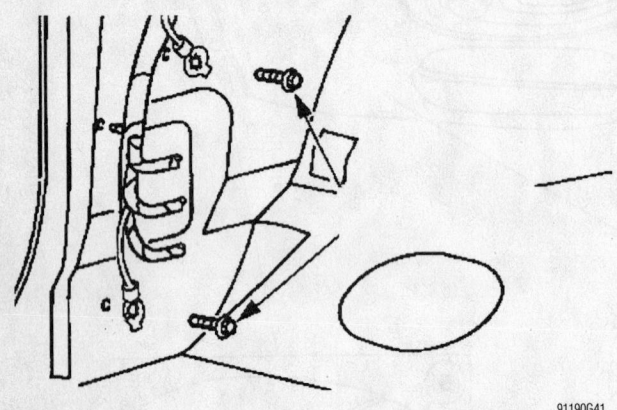

91190G41

Remove the audio unit ground and the GEM/CTM ground bolts—Lincoln Navigator

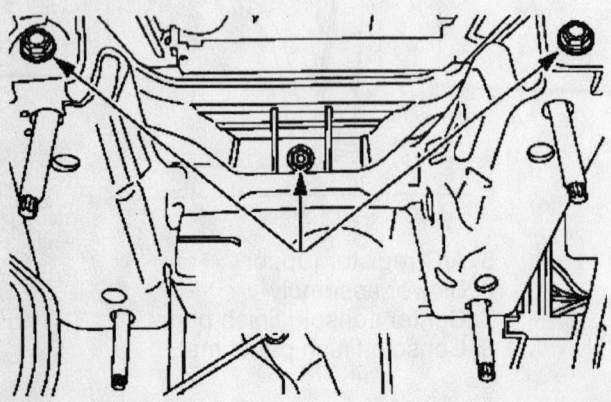

91190G42

Remove the instrument panel bolts through the steering column opening—Lincoln Navigator

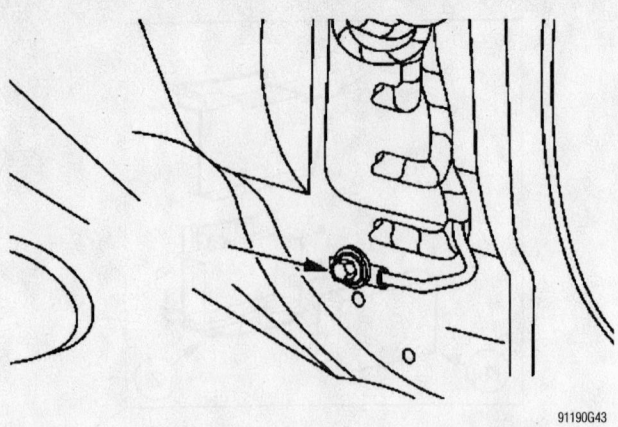

91190G43

Remove the passenger side ground bolt—Lincoln Navigator

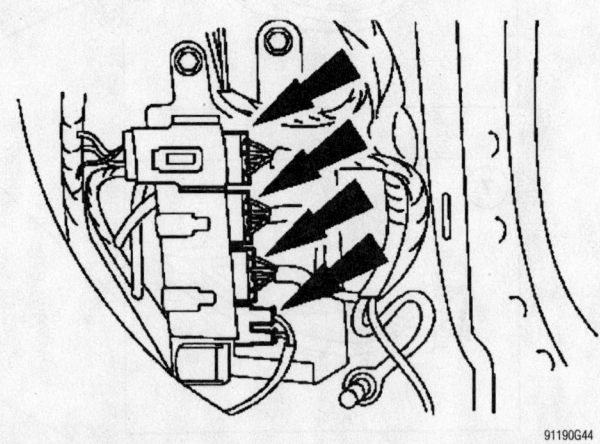

91190G44

Detach the passenger side instrument panel main harness connectors—Lincoln Navigator

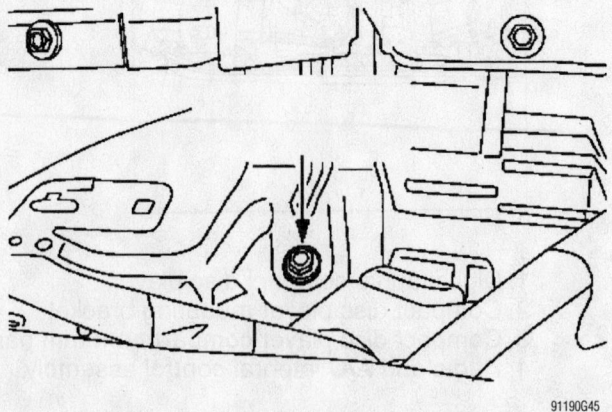

91190G45

Remove the instrument panel bolt on the relay bracket—Lincoln Navigator

After disconnecting the negative battery cable, wait 1 minute for the SRS module to deplete its energy.

2. Drain the cooling system into a clean container for reuse.

3. Remove the instrument panel by performing the following procedure:

a. If equipped, remove the floor console assembly.

b. Remove the lower steering column cover bolts and the cover.

c. Remove both front door scuff plates.

d. Remove both side cowl trim panels.

e. Disconnect the electrical connector from the Brake Pedal Position (BPP) switch.

f. Remove the radio ground and the GEM/CTM ground bolts.

g. Disconnect the left side instrument panel main wiring harness connector.

h. In the engine compartment, remove the bulkhead wiring harness connector bolts and disconnect the wiring connectors.

i. In the driver's compartment, release the 6 locking tabs and remove the bulkhead electrical connector from the instrument panel.

j. Disconnect the air bag diagnostic monitor electrical connector.

k. Disconnect the inertia fuel shutoff switch electrical connector.

l. Remove the right side ground bolts.

m. Disconnect the right side instrument panel wiring harness connectors.

n. Disconnect the electronic blend door actuator electrical connector.

o. Disconnect the climate control head vacuum harness connector.

p. Remove the steering column opening cover reinforcement nuts and the cover reinforcement.

q. At the base of the steering column, disconnect the air bag sliding contact and the anti-theft sensor electrical connectors.

r. At the steering column, disconnect the remaining electrical connectors.

s. If equipped with a transmission range indicator, remove the bolt and disconnect the cable.

t. Remove the steering column-to-

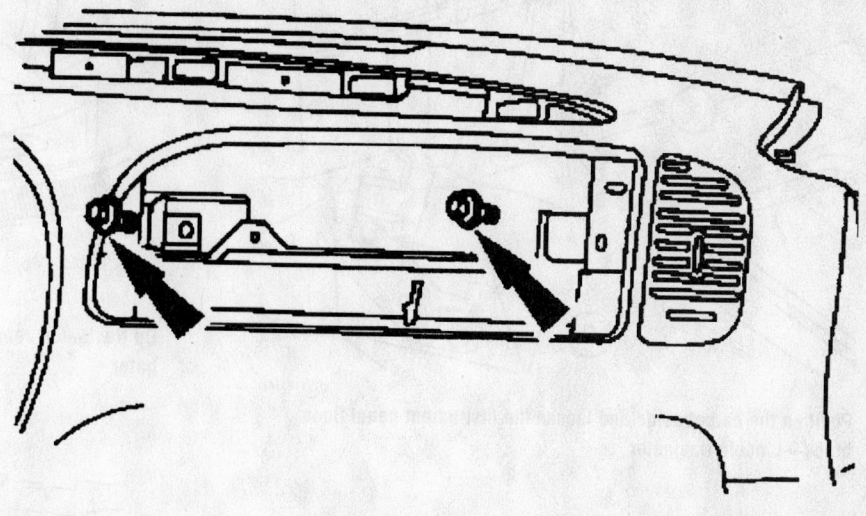

91190G46

Remove the instrument panel bolts through the passenger side air bag module opening—Lincoln Navigator

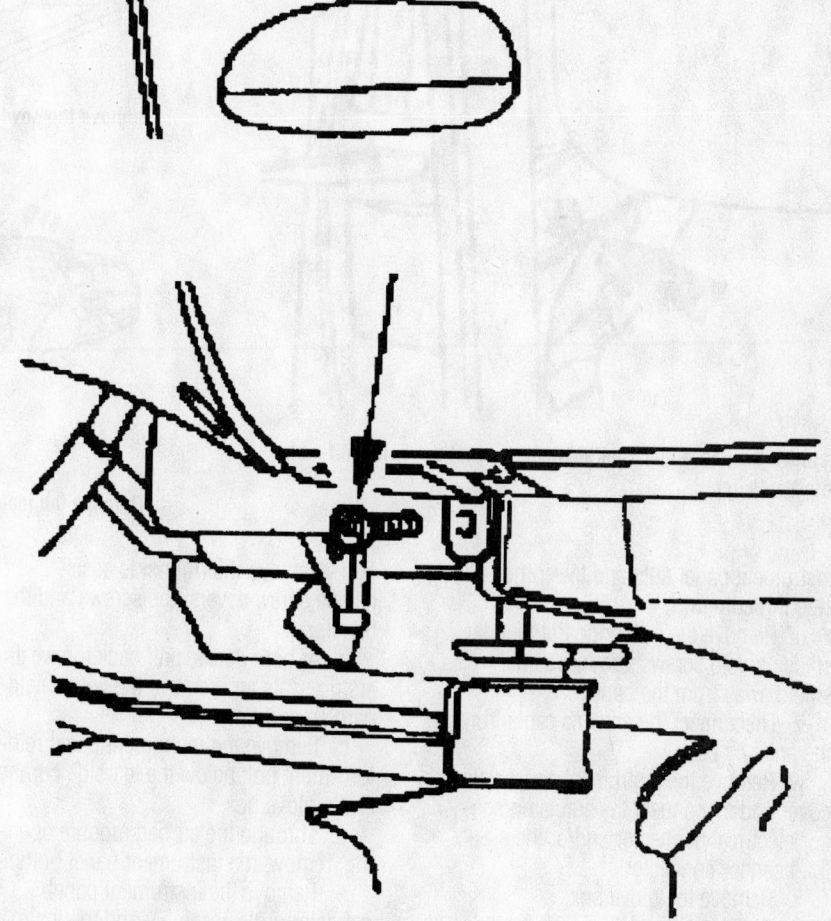

91190G47

Remove the instrument panel reinforcement bolt below the driver's side corner of the glove compartment—Lincoln Navigator

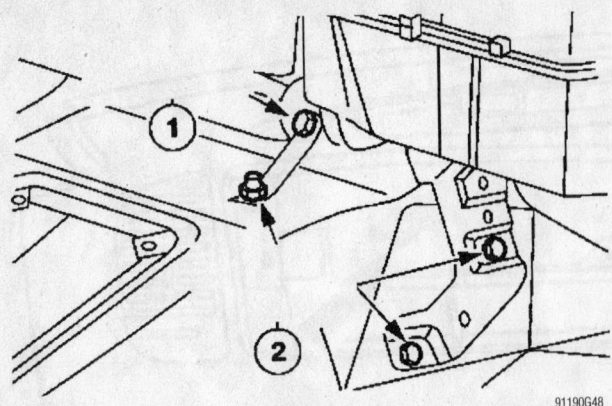

Position the carpet aside and loosen the instrument panel floor brace—Lincoln Navigator

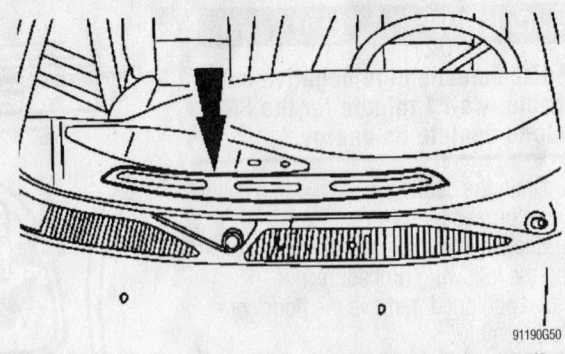

On Navigator, remove the defroster grille assembly—Lincoln Navigator

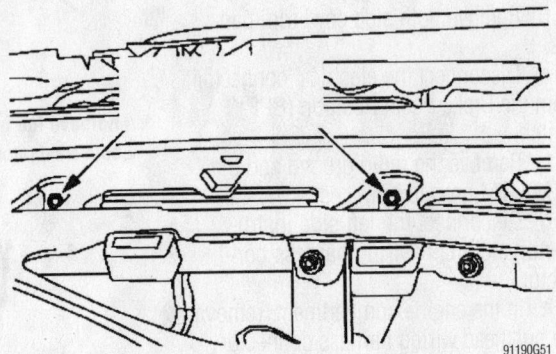

Remove the cowl panel mounting bolts—Lincoln Navigator

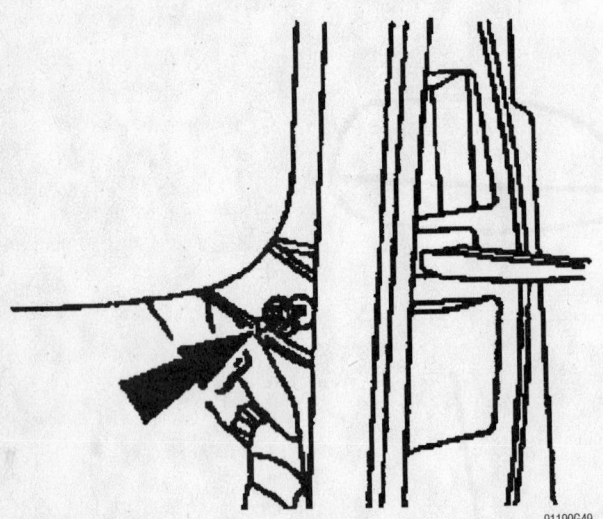

Remove the passenger side instrument panel cowl side nut—Lincoln Navigator

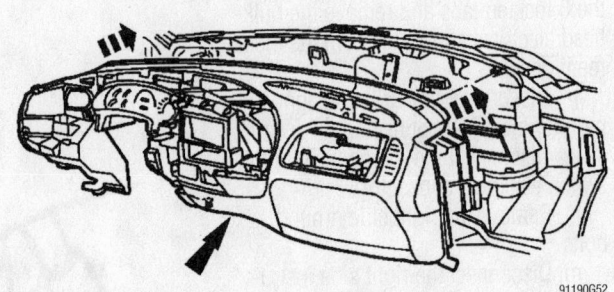

Remove the instrument panel —Lincoln Navigator

instrument panel nuts and lower the steering column.

u. Remove the right side front fender splash shield screws and move the shield away from the panel.

v. Disconnect the antenna cable from the antenna base.

w. Remove the instrument panel relay cover and disconnect the autolamp sensor electrical connector and/or the sunload sensor connector.

x. Remove the glove box.

y. At the passenger's air bag module, remove the screws, disconnect the electrical connector and remove the air bag module.

Place the air bag module in a safe place with the front facing upward.

a. Remove the right side assist handle screw covers, the screws and the handle.

b. At both doors, pull back the weatherstrip seals and remove the windshield garnish moldings.

c. Remove the instrument panel reinforcement bolt below the left side corner of the glove box.

d. Through the air bag module opening, remove the instrument panel bolts.

e. Remove the instrument panel defroster grille assembly and the instrument panel cowl top bolts.

f. At the relay bracket, remove the instrument panel bolt.

g. At the lower left side of the cigar lighter, remove the instrument panel bolt.

h. At the both sides, remove the instrument panel-to-cowl side nuts.

i. At the steering column opening, remove the instrument panel bolts.

j. Remove the upper instrument panel floor brace bolt.

k. Using an assistant, remove the instrument panel.

4. If equipped with the 5.4L 4V engine, remove the junction block splash shield.

5. If equipped with the 5.4L 4V engine, remove the bolts and disconnect the cable ends from the starter relay.

6. If equipped with the 5.4L 4V engine, remove the junction block bracket.

7. Compress the holding tabs and disconnect the heater hoses from the heater core.

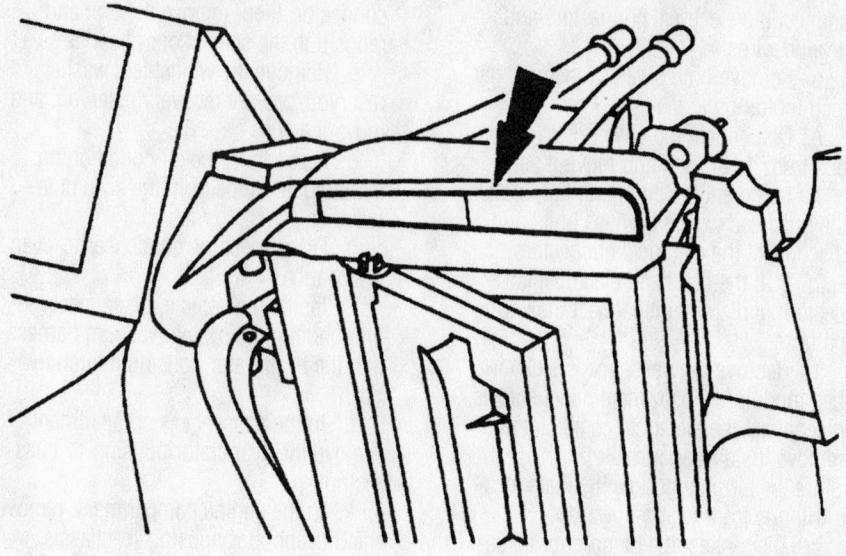

View of the heater core—Lincoln Navigator

93113GM1

8. Remove the air conditioning plenum screw and the air conditioning plenum demister adapter.

9. Disconnect the vacuum line.

10. Remove the heater core bracket screws and the bracket.

11. Remove the 13 heater housing plenum camber cover screws and the heater housing plenum chamber cover.

12. Remove the blend door assembly from the heater housing.

13. Remove the heater core.

To install:

14. Install the heater core.

15. Install the blend door assembly to the heater housing.

16. Install the 13 heater housing plenum camber cover and the heater housing plenum chamber cover screws.

17. Install the heater core bracket and the bracket screws.

18. Connect the vacuum line.

19. Install the air conditioning plenum demister adapter and the air conditioning plenum screw.

20. Connect the heater hoses to the heater core.

21. Install the instrument panel by performing the following procedure:

a. Using an assistant, install the instrument panel.

b. Install the upper instrument panel floor brace bolt.

c. At the steering column opening, install the instrument panel bolts.

d. At the both sides, install the instrument panel-to-cowl side nuts.

e. At the lower left side of the cigar lighter, install the instrument panel bolt.

f. At the relay bracket, install the instrument panel bolt.

g. Install the instrument panel cowl top bolts and the instrument panel defroster grille assembly.

h. Through the air bag module opening, install the instrument panel bolts.

i. Install the instrument panel reinforcement bolt below the left side corner of the glove box.

j. At both doors, install the windshield garnish moldings and the weatherstrip seals.

k. Install the right side assist handle, the screws and the handle screw covers.

l. At the passenger's air bag module, install the air bag module, connect the electrical connector and install the air bag module screws.

m. Install the glove box.

n. Connect the autolamp sensor electrical connector and/or the sunload sensor connector; then, install the instrument panel relay cover.

o. Connect the antenna cable to the antenna base.

p. Install the right side front fender splash shield and screws.

q. Install the steering column and the steering column-to-instrument panel nuts.

r. If equipped with a transmission range indicator, connect the cable and install the bolt.

s. At the steering column, connect the remaining electrical connectors.

t. At the base of the steering column, connect the air bag sliding contact and the anti-theft sensor electrical connectors.

u. Install the steering column opening cover reinforcement and the cover reinforcement nuts.

v. Connect the climate control head vacuum harness connector.

w. Connect the electronic blend door actuator electrical connector.

x. Connect the right side instrument panel wiring harness connectors.

y. Install the right side ground bolts.

z. Connect the inertia fuel shutoff switch electrical connector.

aa. Connect the air bag diagnostic monitor electrical connector.

bb. In the driver's compartment, install the bulkhead electrical connector to the instrument panel.

cc. In the engine compartment, connect the bulkhead wiring harness connectors and the install wiring connector bolts.

dd. Connect the left side instrument panel main wiring harness connector.

ee. Install the radio ground and the GEM/CTM ground bolts.

ff. Connect the electrical connector to the Brake Pedal Position (BPP) switch.

gg. Install both side cowl trim panels.

hh. Install both front door scuff plates.

ii. Install the lower steering column cover and the cover bolts.

jj. If equipped, install the floor console assembly.

22. Refill the cooling system.

23. Connect the negative battery cable.

24. Run the engine to normal operating temperatures; then, check the climate control operation and check for leaks.

Ford Explorer

REMOVAL & INSTALLATION

1999–00

1. Disconnect the negative battery cable.

❈❈ **CAUTION**

After disconnecting the negative battery cable, wait for 1 minute for the SRS module to deplete its energy.

2. Drain the cooling system into a clean container for reuse.

3. Disconnect the heater hoses from the heater core.

4. Remove the steering column by performing the following procedure:

a. Position the front wheels in the straight-ahead direction.

b. At the both sides of the steering wheel, remove the cover plugs, the steering wheel-to-air bag module screws, disconnect the air bag electrical connector and carefully remove the air bag module.

✳✳ CAUTION

Safely store the air bag module with the front side facing upward.

c. Remove the steering wheel-to-steering column nut.

d. Using a steering wheel puller, press the steering wheel from the steering column.

e. Remove the parking brake release handle screws and move the release handle aside.

f. Remove the hood release screws and move the hood release aside.

g. Remove the 2 instrument panel-to-steering column cover screws and the cover.

h. Remove the instrument panel steering column opening reinforcement bolts and the reinforcement.

i. Remove the ignition switch bolt and disconnect the ignition switch electrical connector.

j. At the base of the steering column, disconnect the electrical connectors.

k. If equipped with an automatic transmission, remove the transmission range indicator bolt and the cable.

l. If equipped with an automatic transmission, disconnect the shift cable from the steering column shift tube lever and the steering column bracket.

m. Disconnect the brake shift interlock solenoid electrical connector.

n. Remove the air bag sliding contact.

o. Remove the upper intermediate steering shaft-to-column shaft bolt and discard the bolt.

p. Remove the lower steering column-to-instrument panel nuts and the steering column.

5. Remove the instrument panel by performing the following procedure:

a. Disconnect the Brake Pedal Position (BPP) switch electrical connector.

b. Remove the push pins and remove both cowl side trim panels.

c. At the right side cowl panel, disconnect the electrical connectors and ground wires.

d. Remove both sides windshield garnish moldings.

e. Disconnect the power distribution box from its bracket and move it aside.

f. In the engine compartment, loosen the bulkhead wiring harness bolts and disconnect the electrical connectors.

g. Pull the bulkhead electrical connector handle and disconnect the wiring harness.

h. Remove the passenger's side air bag module-to-instrument panel screws, disconnect the electrical connector and remove the air bag module.

Store the air bag module in a safe location with the front facing upward.

a. Disconnect the blend door actuator's electrical connector.

b. Disconnect the climate control vacuum harness connector.

c. Disconnect the radio's antenna connector.

d. Remove the glove compartment.

e. Remove the instrument panel defroster grille.

f. Remove the upper instrument panel bolts.

g. If equipped, remove the upper series floor console.

h. Under the steering column, remove the instrument panel brace bolt.

i. At both sides, remove the windshield side garnish moldings.

j. Remove both the right and left instrument panel-to-cowl bolts.

k. Remove the fuse panel door.

l. Pull the instrument panel away from the dash.

m. Loosen the instrument panel-to-body harness bolt and disconnect the harness.

n. Using an assistant, remove the instrument panel.

6. At the PCM, disconnect the electrical connector; then, remove the 2 PCM cover nuts, the cover and the PCM.

➡ **The PCM is located at the right side of the instrument panel.**

7. Remove the evaporator core by perform the following procedure:

a. Discharge and recover the air conditioning system refrigerant.

b. Remove the refrigerant lines from the evaporator core. Discard the O-rings.

c. Disconnect the air conditioning cycling switch.

d. Disconnect the blower motor electrical connectors.

e. Disconnect the speed control servo connector; then, remove the bolt and reposition the speed control servo.

f. Remove the windshield washer reservoir/coolant recovery reservoir, and move it aside.

g. Disconnect the air conditioning manifold and tube from the accumulator/drier.

h. Disconnect the condenser-to-evaporator tube.

i. Inside the vehicle, disconnect the air conditioning system vacuum harness and the evaporator housing mounting nut.

j. In the passenger's compartment, remove the evaporator housing-to-chassis nut.

k. In the engine compartment, remove the 3 evaporator housing-to-chassis nuts.

l. Remove the air conditioning accumulator from the evaporator core.

m. If equipped with a 5.0L engine, remove the evaporator housing heat shield screw, clips and the shield.

n. Remove the evaporator housing cover screws and the cover.

o. Remove the evaporator core from the housing.

8. Remove the PCM ground strap screw and the heat sink.

9. In the engine compartment, remove the heater housing air plenum nuts.

10. Remove the heater core-to-air plenum cover screws and the cover.

11. Remove the heater core.

To install:

12. Install the heater core.

13. Install the heater core-to-air plenum cover and the cover screws.

14. In the engine compartment, install the heater housing air plenum nuts.

15. Install the PCM ground strap screw and the heat sink.

16. Install the evaporator core by perform the following procedure:

a. Install the evaporator core to the housing.

b. Install the evaporator housing cover and the cover screws.

c. If equipped with a 5.0L engine, install the evaporator housing heat shield, clips and the shield screw.

d. Install the air conditioning accumulator to the evaporator core.

e. In the engine compartment, install the 3 evaporator housing-to-chassis nuts.

f. In the passenger's compartment, install the evaporator housing-to-chassis nut.

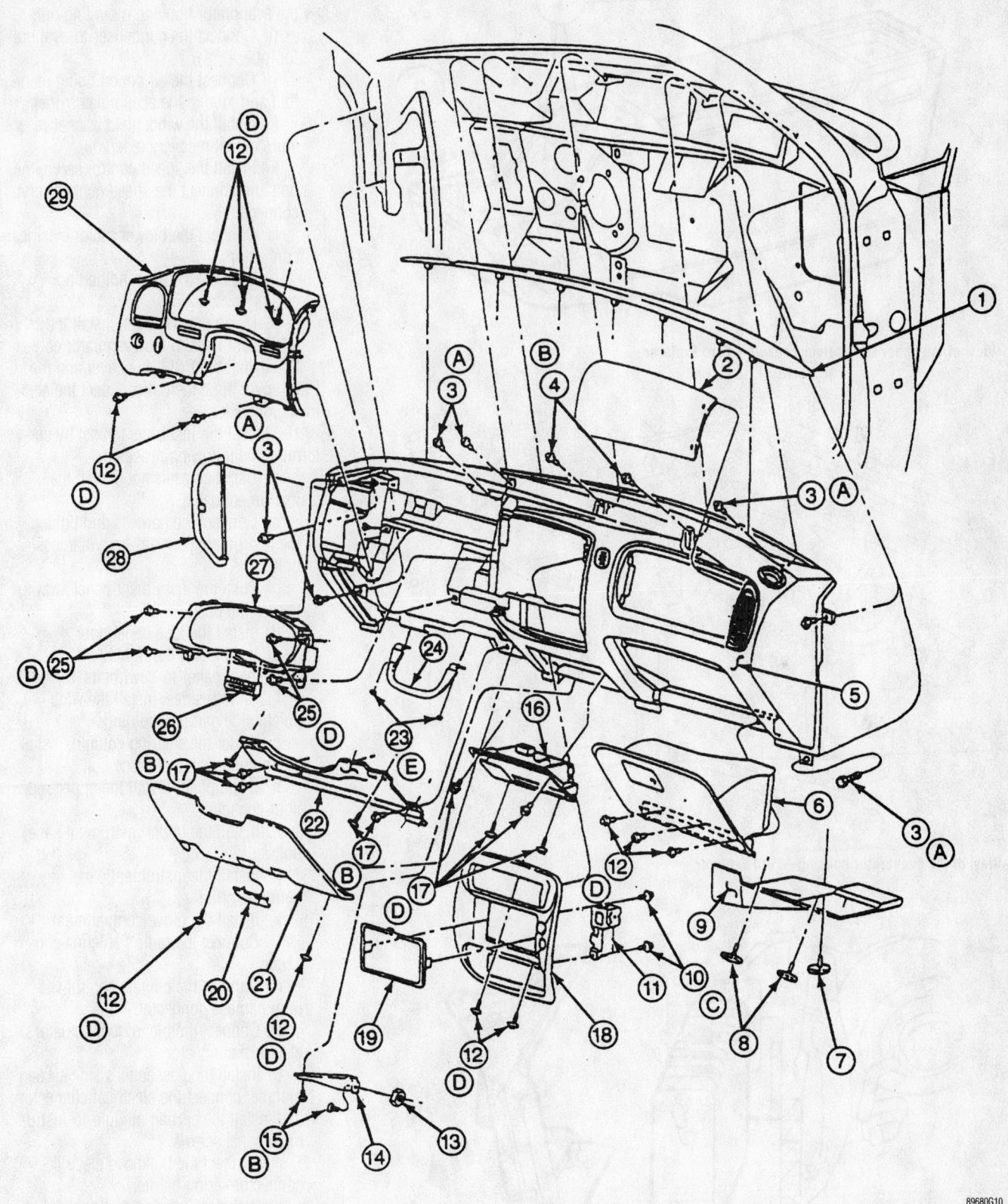

Exploded view of the instrument panel assembly—Ford Explorer

89680G10

For Tune-up, Capacities and Firing orders, see Section 1 of this manual

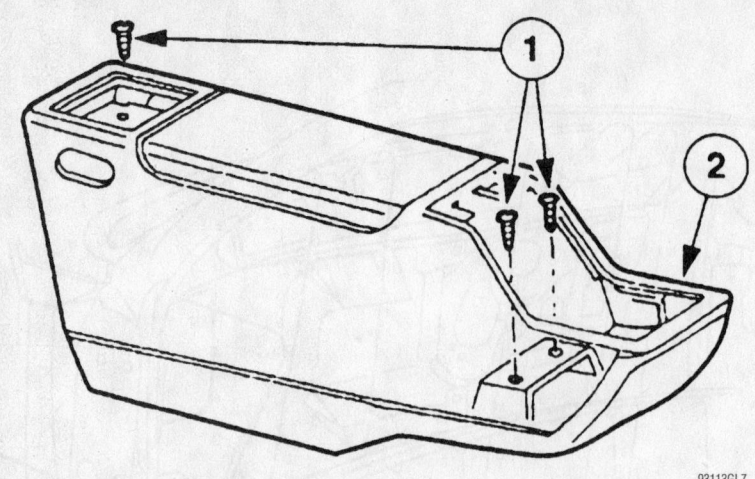

View of the upper series floor console—Ford Explorer

93113GL7

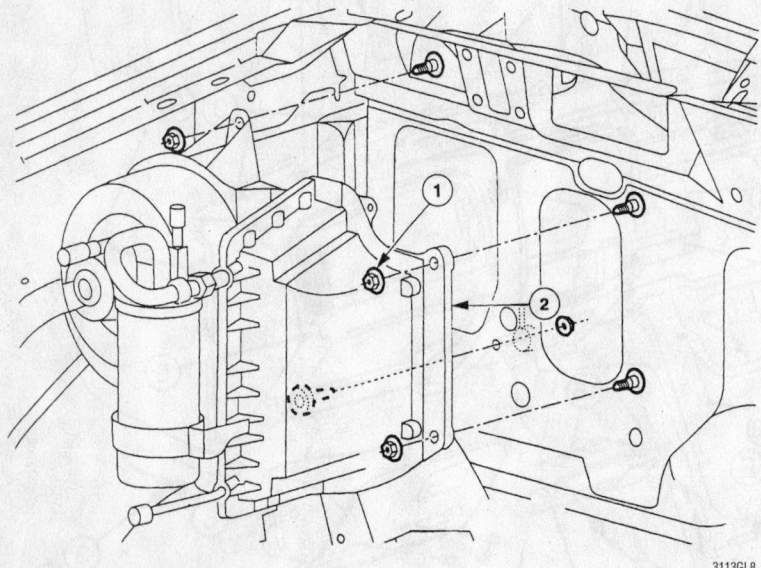

View of the evaporator housing—Ford Explorer

3113GL8

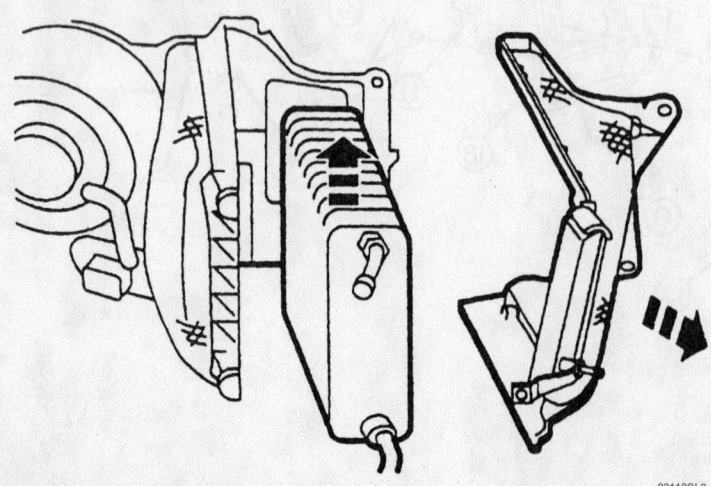

View of the evaporator core—Ford Explorer

93113GL9

g. Inside the vehicle, connect the air conditioning system vacuum harness and the evaporator housing mounting nut.

h. Connect the condenser-to-evaporator tube.

i. Connect the air conditioning manifold and tube to the accumulator/drier.

j. Install the windshield washer reservoir/coolant recovery reservoir.

k. Install the speed control servo, the bolt and connect the speed control servo connector.

l. Connect the blower motor electrical connectors.

m. Connect the air conditioning cycling switch.

n. Using new O-rings, install the refrigerant lines to the evaporator core.

17. Install the PCM, the cover and the PCM cover nuts; then, disconnect the electrical connector.

18. Install the instrument panel by performing the following procedure:

a. Using an assistant, install the instrument panel.

b. Connect the harness and tighten the instrument panel-to-body harness bolt.

c. Push the instrument panel away toward the dash.

d. Install the fuse panel door.

e. Install both the right and left instrument panel-to-cowl bolts.

f. At both sides, install the windshield side garnish moldings.

g. Under the steering column, install the instrument panel brace bolt.

h. If equipped, install the upper series floor console.

i. Install the upper instrument panel bolts.

j. Install the instrument panel defroster grille.

k. Install the glove compartment.

l. Connect the radio's antenna connector.

m. Connect the climate control vacuum harness connector.

n. Connect the blend door actuator's electrical connector.

o. Install the passenger's side air bag module, connect the electrical connector and install the air bag module-to-instrument panel screws.

p. Connect the bulkhead electrical connector wiring harness.

q. In the engine compartment, connect the electrical connectors and tighten the bulkhead wiring harness bolts.

r. Connect the power distribution box to its bracket.

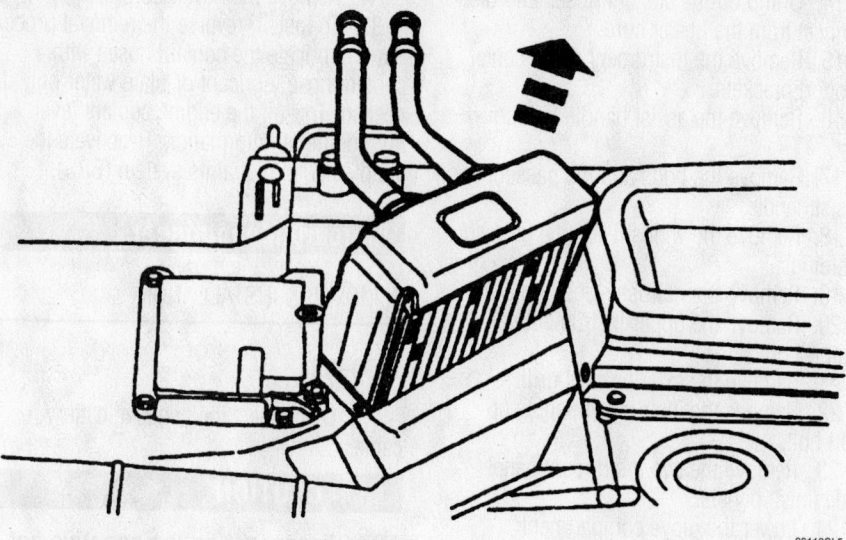

93113GL5

View of the heater core—Ford Explorer

s. Install both sides windshield garnish moldings.

t. At the right side cowl panel, connect the electrical connectors and ground wires.

u. Install both cowl side trim panels and the push pins.

v. Connect the Brake Pedal Position (BPP) switch electrical connector.

19. Install the steering column by performing the following procedure:

a. Install the lower steering column and the steering column-to-instrument panel nuts; then, torque the nuts to 10–13 ft. lbs. (13–17 Nm).

b. Using a new bolt, install the upper intermediate steering shaft-to-column shaft bolt and torque to 19–25 ft. lbs. (26–34 Nm).

c. Install the air bag sliding contact.

d. Connect the brake shift interlock solenoid electrical connector.

e. If equipped with an automatic transmission, connect the shift cable from the steering column shift tube lever and the steering column bracket.

f. If equipped with an automatic transmission, install the transmission range indicator cable and bolt.

g. At the base of the steering column, connect the electrical connectors.

h. Connect the ignition switch electrical connector and install the ignition switch bolt.

i. Install the instrument panel steering column opening reinforcement and the reinforcement bolts.

j. Install the instrument panel-to-steering column cover and the 2 cover screws.

k. Install the hood release and the hood release screws.

l. Install the parking brake release handle and the release handle screws.

m. Install the steering wheel to the steering column.

n. Install the steering wheel-to-steering column nut and torque the nut to 25–34 ft. lbs. (34–46 Nm).

o. At the both sides of the steering wheel, install the air bag module, connect the air bag electrical connector, install the steering wheel-to-air bag module screws and the cover plugs.

20. Connect the heater hoses to the heater core.

21. Refill the cooling system.

22. Connect the negative battery cable.

23. Evacuate and charge the air conditioning system.

24. Run the engine to normal operating temperatures; then, check the climate control operation and check for leaks.

2001

1. Drain the radiator.
2. Remove the heater water hoses.

❊❊ WARNING

Electronic modules are sensitive to static electrical charges. If exposed to these charges, damage may result.

3. Remove the steering column.

4. Disconnect the brake pedal position (BPP) switch electrical connector.

5. If equipped, disconnect the clutch pedal position (CPP) switch electrical connector.

6. Remove the LH and RH cowl side trim panels.

7. Disconnect the electrical connectors and the ground wires on the RH cowl panel.

8. Disconnect the power distribution box from the bracket and position aside.

9. Disconnect the bulkhead wiring harness connectors from inside the engine compartment.

10. Remove the bulkhead connector insulator.

11. Unclip the bulkhead electrical connectors from the dash panel.

12. Remove the passenger side air bag module.

13. Disconnect the blend door actuator electrical connector.

14. Disconnect the climate control vacuum harness connector.

15. Disconnect the radio antenna cable in-line connector.

16. Raise the glove compartment.

17. Remove the instrument panel defroster opening grille.

18. Remove the instrument panel cowl top bolts.

19. If equipped, remove the upper series floor console.

20. Remove the instrument panel brace bolt from under the steering column opening.

21. Remove the windshield side garnish mouldings.

22. Remove the RH instrument panel cowl side bolt.

23. Remove the instrument panel fuse panel door.

24. Remove the LH instrument panel cowl side bolts.

25. Position the instrument panel away from the dash panel.

26. Disconnect the instrument panel to body harness.

➡Two technicians are required to carry out this step.

27. Remove the instrument panel.
28. Recover the refrigerant.
29. Disconnect the A/C cycling switch.
30. Disconnect the blower electrical connectors.
31. Position the speed control servo aside.

32. Remove the nuts and the screws from the windshield washer reservoir/coolant recovery reservoir. Set the reservoir aside.

33. Disconnect the A/C manifold and tube from the suction accumulator/drier.

34. Disconnect the condenser to evaporator tube.

35. Disconnect the A/C system vacuum harness and the A/C evaporator housing mounting nut inside the vehicle.

36. Remove the A/C evaporator housing.

37. Remove the powertrain control module (PCM).

38. Remove the PCM heat sink.

39. Remove the heater air plenum nuts from the engine side of the dash panel.

40. Remove the heater core cover to air plenum screws.

41. Lift off the cover.

42. Remove the heater core.

43. To install, reverse the removal procedure. Be sure to install a new oval foam seal around the heater core inlet and outlet tubes.

2002

1. Deactivate the supplemental restraints system (SRS).

2. Release the two floor console front clips (one each side).

3. On vehicles with manual transmission, remove the gearshift lever handle.

4. Remove the console finish panel mat.

5. On vehicles with automatic transmission:

   a. If equipped, remove the ashtray assembly.

   b. Disconnect the electrical connector.

   c. Remove the console finish panel screw.

6. Remove the console finish panel by lifting up at the rear and sliding it rearward. Disconnect the electrical connector(s).

7. Remove the floor console front screws.

8. If equipped, disconnect the electrical connector and release the wiring harness locators.

9. Remove the floor console center bolts.

10. Remove the two floor console access covers (one each side).

11. Remove the two floor console rear bolts (one each side).

12. Remove the floor console by lifting up at the rear of the console and sliding it rearward.

13. Remove the screws and remove the snow shield.

14. Crimp off the coolant hoses and disconnect from the heater core.

15. Remove the instrument panel center support brackets.

16. Remove the assist handle bolt covers.

17. Remove the bolts and the passenger assist handle.

18. Remove the windshield side garnish moldings.

19. Remove the defroster opening grill.

20. Remove the upper instrument panel support bolts.

21. Remove the exterior cowl grill.

22. Remove the instrument panel support bolt.

23. Remove the two instrument panel side finish panels.

24. Lower the glove compartment.

25. Loosen the LH instrument panel side support bolts until half the threads are exposed.

✳✳ WARNING

Be sure the instrument panel is properly supported to avoid possible damage to the instrument panel wiring or components. To avoid damage to the instrument panel wiring or components, do not use excessive force when moving the instrument panel away from the dash panel. The instrument panel may be supported by installing threaded rods or equivalent, with the same diameter and thread pitch, in place of the instrument panel support bolts.

26. Properly support the instrument panel and remove the RH instrument panel support bolts to allow the instrument panel to be pulled away from the dash panel.

27. Remove the center console floor duct.

28. Remove the screws and remove the RH floor duct.

29. Remove the screws and remove the LH floor duct.

30. Disconnect the electrical connectors and detach the pin-type retainers.

31. Remove the screws and remove the heater core tube cover.

32. Remove the RH heater core cover screws.

33. Remove the LH heater core cover screws and remove the heater core cover.

➡**The instrument panel will need to be moved to facilitate removal of the heater core.**

34. Remove the heater core.

35. To install, reverse the removal procedure. Lubricate the coolant hoses with coolant hose lubricant or plain water only if needed. Top-off the engine coolant level. For additional information. Reactivate the supplemental restraints system (SRS).

Mercury Mountaineer

REMOVAL & INSTALLATION

1999–00

1. Disconnect the negative battery cable.

✳✳ CAUTION

After disconnecting the negative battery cable, wait for 1 minute for the SRS module to deplete its energy.

2. Drain the cooling system into a clean container for reuse.

3. Disconnect the heater hoses from the heater core.

4. Remove the steering column by performing the following procedure:

   a. Position the front wheels in the straight-ahead direction.

   b. At the both sides of the steering wheel, remove the cover plugs, the steering wheel-to-air bag module screws, disconnect the air bag electrical connector and carefully remove the air bag module.

✳✳ CAUTION

Safely store the air bag module with the front side facing upward.

   c. Remove the steering wheel-to-steering column nut.

   d. Using a steering wheel puller, press the steering wheel from the steering column.

   e. Remove the parking brake release handle screws and move the release handle aside.

   f. Remove the hood release screws and move the hood release aside.

   g. Remove the 2 instrument panel-to-steering column cover screws and the cover.

   h. Remove the instrument panel steering column opening reinforcement bolts and the reinforcement.

   i. Remove the ignition switch bolt and disconnect the ignition switch electrical connector.

   j. At the base of the steering column, disconnect the electrical connectors.

k. If equipped with an automatic transmission, remove the transmission range indicator bolt and the cable.

l. If equipped with an automatic transmission, disconnect the shift cable from the steering column shift tube lever and the steering column bracket.

m. Disconnect the brake shift interlock solenoid electrical connector.

n. Remove the air bag sliding contact.

o. Remove the upper intermediate steering shaft-to-column shaft bolt and discard the bolt.

p. Remove the lower steering column-to-instrument panel nuts and the steering column.

5. Remove the instrument panel by performing the following procedure:

a. Disconnect the Brake Pedal Position (BPP) switch electrical connector.

b. Remove the push pins and remove both cowl side trim panels.

c. At the right side cowl panel, disconnect the electrical connectors and ground wires.

d. Remove both sides windshield garnish moldings.

e. Disconnect the power distribution box from its bracket and move it aside.

f. In the engine compartment, loosen the bulkhead wiring harness bolts and disconnect the electrical connectors.

g. Pull the bulkhead electrical connector handle and disconnect the wiring harness.

h. Remove the passenger's side air bag module-to-instrument panel screws, disconnect the electrical connector and remove the air bag module. Store the air bag module in a safe location with the front facing upward.

a. Disconnect the blend door actuator's electrical connector.

b. Disconnect the climate control vacuum harness connector.

c. Disconnect the radio's antenna connector.

d. Remove the glove compartment.

e. Remove the instrument panel defroster grille.

f. Remove the upper instrument panel bolts.

g. If equipped, remove the upper series floor console.

h. Under the steering column, remove the instrument panel brace bolt.

i. At both sides, remove the windshield side garnish moldings.

j. Remove both the right and left instrument panel-to-cowl bolts.

k. Remove the fuse panel door.

l. Pull the instrument panel away from the dash.

m. Loosen the instrument panel-to-body harness bolt and disconnect the harness.

n. Using an assistant, remove the instrument panel.

6. At the PCM, disconnect the electrical connector; then, remove the 2 PCM cover nuts, the cover and the PCM.

➡ **The PCM is located at the right side of the instrument panel.**

7. Remove the evaporator core by perform the following procedure:

a. Discharge and recover the air conditioning system refrigerant.

b. Remove the refrigerant lines from the evaporator core. Discard the O-rings.

c. Disconnect the air conditioning cycling switch.

d. Disconnect the blower motor electrical connectors.

e. Disconnect the speed control servo connector; then, remove the bolt and reposition the speed control servo.

f. Remove the windshield washer reservoir/coolant recovery reservoir, and move it aside.

g. Disconnect the air conditioning manifold and tube from the accumulator/drier.

h. Disconnect the condenser-to-evaporator tube.

i. Inside the vehicle, disconnect the air conditioning system vacuum harness and the evaporator housing mounting nut.

j. In the passenger's compartment, remove the evaporator housing-to-chassis nut.

k. In the engine compartment, remove the 3 evaporator housing-to-chassis nuts.

l. Remove the air conditioning accumulator from the evaporator core.

m. If equipped with a 5.0L engine, remove the evaporator housing heat shield screw, clips and the shield.

n. Remove the evaporator housing cover screws and the cover.

o. Remove the evaporator core from the housing.

8. Remove the PCM ground strap screw and the heat sink.

9. In the engine compartment, remove the heater housing air plenum nuts.

10. Remove the heater core-to-air plenum cover screws and the cover.

11. Remove the heater core.

To install:

12. Install the heater core.

13. Install the heater core-to-air plenum cover and the cover screws.

14. In the engine compartment, install the heater housing air plenum nuts.

15. Install the PCM ground strap screw and the heat sink.

16. Install the evaporator core by perform the following procedure:

a. Install the evaporator core to the housing.

b. Install the evaporator housing cover and the cover screws.

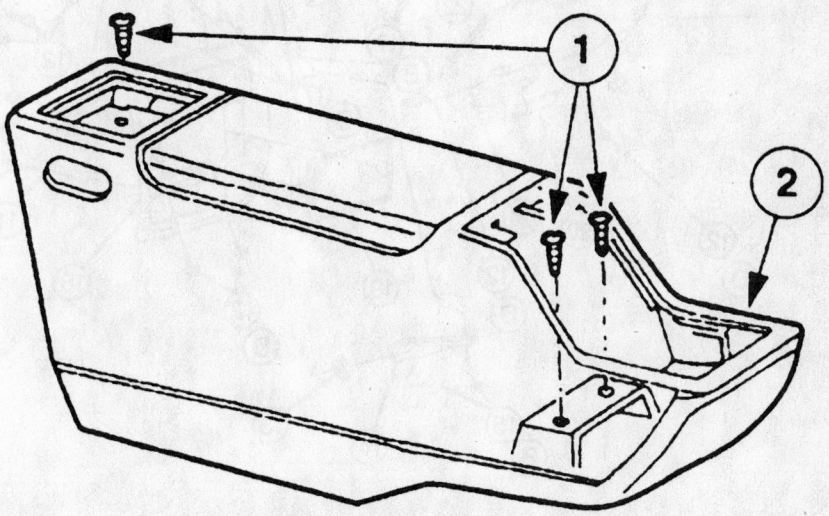

View of the upper series floor console—Mercury Mountaineer

93113GL7

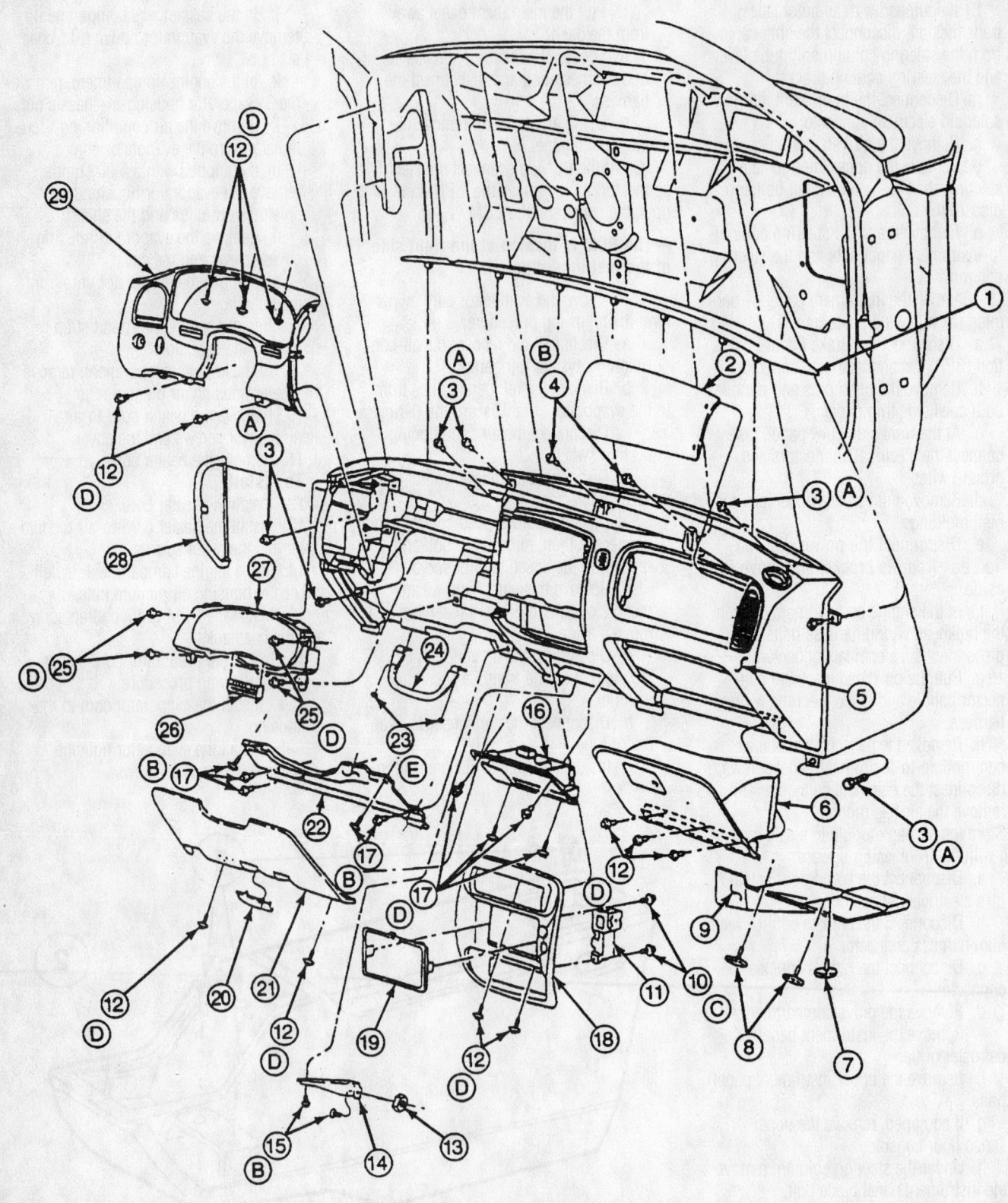

Exploded view of the instrument panel assembly—Mercury Mountaineer

89680G10

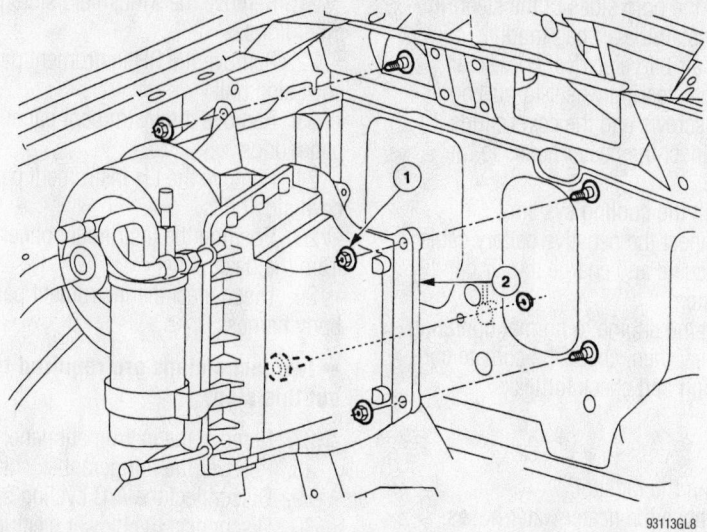

View of the evaporator housing—Mercury Mountaineer

93113GL8

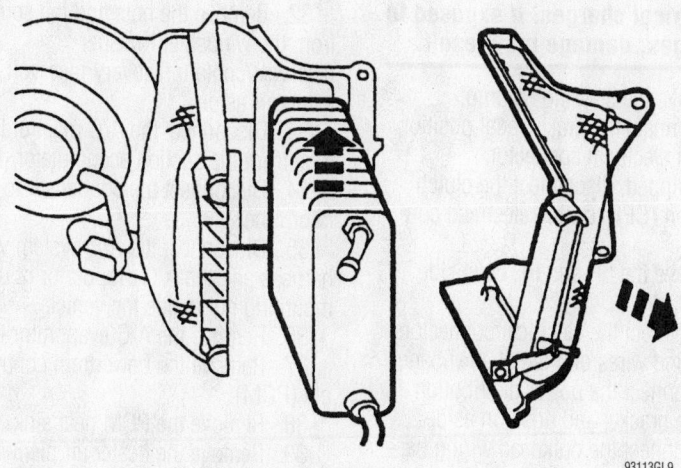

View of the evaporator core—Mercury Mountaineer

93113GL9

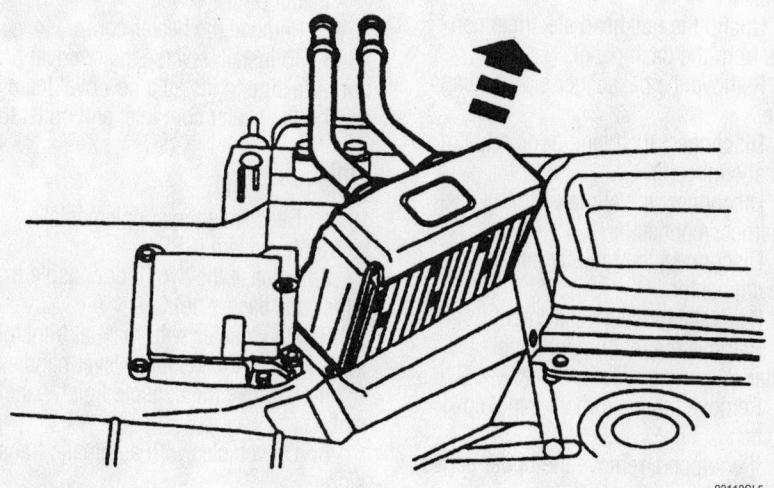

View of the heater core—Mercury Mountaineer

93113GL5

c. If equipped with a 5.0L engine, install the evaporator housing heat shield, clips and the shield screw.

d. Install the air conditioning accumulator to the evaporator core.

e. In the engine compartment, install the 3 evaporator housing-to-chassis nuts.

f. In the passenger's compartment, install the evaporator housing-to-chassis nut.

g. Inside the vehicle, connect the air conditioning system vacuum harness and the evaporator housing mounting nut.

h. Connect the condenser-to-evaporator tube.

i. Connect the air conditioning manifold and tube to the accumulator/drier.

j. Install the windshield washer reservoir/coolant recovery reservoir.

k. Install the speed control servo, the bolt and connect the speed control servo connector.

l. Connect the blower motor electrical connectors.

m. Connect the air conditioning cycling switch.

n. Using new O-rings, install the refrigerant lines to the evaporator core.

17. Install the PCM, the cover and the PCM cover nuts; then, disconnect the electrical connector.

18. Install the instrument panel by performing the following procedure:

a. Using an assistant, install the instrument panel.

b. Connect the harness and tighten the instrument panel-to-body harness bolt.

c. Push the instrument panel away toward the dash.

d. Install the fuse panel door.

e. Install both the right and left instrument panel-to-cowl bolts.

f. At both sides, install the windshield side garnish moldings.

g. Under the steering column, install the instrument panel brace bolt.

h. If equipped, install the upper series floor console.

i. Install the upper instrument panel bolts.

j. Install the instrument panel defroster grille.

k. Install the glove compartment.

l. Connect the radio's antenna connector.

m. Connect the climate control vacuum harness connector.

Timing belt service is covered in Section 3 of this manual

n. Connect the blend door actuator's electrical connector.

o. Install the passenger's side air bag module, connect the electrical connector and install the air bag module-to-instrument panel screws.

p. Connect the bulkhead electrical connector wiring harness.

q. In the engine compartment, connect the electrical connectors and tighten the bulkhead wiring harness bolts.

r. Connect the power distribution box to its bracket.

s. Install both sides windshield garnish moldings.

t. At the right side cowl panel, connect the electrical connectors and ground wires.

u. Install both cowl side trim panels and the push pins.

v. Connect the Brake Pedal Position (BPP) switch electrical connector.

19. Install the steering column by performing the following procedure:

a. Install the lower steering column and the steering column-to-instrument panel nuts; then, torque the nuts to 10–13 ft. lbs. (13–17 Nm).

b. Using a new bolt, install the upper intermediate steering shaft-to-column shaft bolt and torque to 19–25 ft. lbs. (26–34 Nm).

c. Install the air bag sliding contact.

d. Connect the brake shift interlock solenoid electrical connector.

e. If equipped with an automatic transmission, connect the shift cable from the steering column shift tube lever and the steering column bracket.

f. If equipped with an automatic transmission, install the transmission range indicator cable and bolt.

g. At the base of the steering column, connect the electrical connectors.

h. Connect the ignition switch electrical connector and install the ignition switch bolt.

i. Install the instrument panel steering column opening reinforcement and the reinforcement bolts.

j. Install the instrument panel-to-steering column cover and the 2 cover screws.

k. Install the hood release and the hood release screws.

l. Install the parking brake release handle and the release handle screws.

m. Install the steering wheel to the steering column.

n. Install the steering wheel-to-steering column nut and torque the nut to 25–34 ft. lbs. (34–46 Nm).

o. At the both sides of the steering wheel, install the air bag module, connect the air bag electrical connector, install the steering wheel-to-air bag module screws and the cover plugs.

20. Connect the heater hoses to the heater core.

21. Refill the cooling system.

22. Connect the negative battery cable.

23. Evacuate and charge the air conditioning system.

24. Run the engine to normal operating temperatures; then, check the climate control operation and check for leaks.

2001

1. Drain the radiator.
2. Remove the heater water hoses.

✳✳ WARNING

Electronic modules are sensitive to static electrical charges. If exposed to these charges, damage may result.

3. Remove the steering column.
4. Disconnect the brake pedal position (BPP) switch electrical connector.
5. If equipped, disconnect the clutch pedal position (CPP) switch electrical connector.
6. Remove the LH and RH cowl side trim panels.
7. Disconnect the electrical connectors and the ground wires on the RH cowl panel.
8. Disconnect the power distribution box from the bracket and position aside.
9. Disconnect the bulkhead wiring harness connectors from inside the engine compartment.
10. Remove the bulkhead connector insulator.
11. Unclip the bulkhead electrical connectors from the dash panel.
12. Remove the passenger side air bag module.
13. Disconnect the blend door actuator electrical connector.
14. Disconnect the climate control vacuum harness connector.
15. Disconnect the radio antenna cable in-line connector.
16. Raise the glove compartment.
17. Remove the instrument panel defroster opening grille.
18. Remove the instrument panel cowl top bolts.
19. If equipped, remove the upper series floor console.
20. Remove the instrument panel brace bolt from under the steering column opening.

21. Remove the windshield side garnish mouldings.
22. Remove the RH instrument panel cowl side bolt.
23. Remove the instrument panel fuse panel door.
24. Remove the LH instrument panel cowl side bolts.
25. Position the instrument panel away from the dash panel.
26. Disconnect the instrument panel to body harness.

➡**Two technicians are required to carry out this step.**

27. Remove the instrument panel.
28. Recover the refrigerant.
29. Disconnect the A/C cycling switch.
30. Disconnect the blower electrical connectors.
31. Position the speed control servo aside.
32. Remove the nuts and the screws from the windshield washer reservoir/coolant recovery reservoir. Set the reservoir aside.
33. Disconnect the A/C manifold and tube from the suction accumulator/drier.
34. Disconnect the condenser to evaporator tube.
35. Disconnect the A/C system vacuum harness and the A/C evaporator housing mounting nut inside the vehicle.
36. Remove the A/C evaporator housing.
37. Remove the powertrain control module (PCM).
38. Remove the PCM heat sink.
39. Remove the heater air plenum nuts from the engine side of the dash panel.
40. Remove the heater core cover to air plenum screws.
41. Lift off the cover.
42. Remove the heater core.
43. To install, reverse the removal procedure. Be sure to install a new oval foam seal around the heater core inlet and outlet tubes.

2002

1. Deactivate the supplemental restraints system (SRS).
2. Release the two floor console front clips (one each side).
3. On vehicles with manual transmission, remove the gearshift lever handle.
4. Remove the console finish panel mat.
5. On vehicles with automatic transmission:

a. If equipped, remove the ashtray assembly.

b. Disconnect the electrical connector.

c. Remove the console finish panel screw.

6. Remove the console finish panel by lifting up at the rear and sliding it rearward. Disconnect the electrical connector(s).

7. Remove the floor console front screws.

8. If equipped, disconnect the electrical connector and release the wiring harness locators.

9. Remove the floor console center bolts.

10. Remove the two floor console access covers (one each side).

11. Remove the two floor console rear bolts (one each side).

12. Remove the floor console by lifting up at the rear of the console and sliding it rearward.

13. Remove the screws and remove the snow shield.

14. Crimp off the coolant hoses and disconnect from the heater core.

15. Remove the instrument panel center support brackets.

16. Remove the assist handle bolt covers.

17. Remove the bolts and the passenger assist handle.

18. Remove the windshield side garnish moldings.

19. Remove the defroster opening grill.

20. Remove the upper instrument panel support bolts.

21. Remove the exterior cowl grill.

22. Remove the instrument panel support bolt.

23. Remove the two instrument panel side finish panels.

24. Lower the glove compartment.

25. Loosen the LH instrument panel side support bolts until half the threads are exposed.

✳✳ WARNING

Be sure the instrument panel is properly supported to avoid possible damage to the instrument panel wiring or components. To avoid damage to the instrument panel wiring or components, do not use excessive force when moving the instrument panel away from the dash panel. The instrument panel may be supported by installing threaded rods or equivalent, with the same diameter and thread pitch, in place of the instrument panel support bolts.

26. Properly support the instrument panel and remove the RH instrument panel support bolts to allow the instrument panel to be pulled away from the dash panel.

27. Remove the center console floor duct.

28. Remove the screws and remove the RH floor duct.

29. Remove the screws and remove the LH floor duct.

30. Disconnect the electrical connectors and detach the pin-type retainers.

31. Remove the screws and remove the heater core tube cover.

32. Remove the RH heater core cover screws.

33. Remove the LH heater core cover screws and remove the heater core cover.

➡ **The instrument panel will need to be moved to facilitate removal of the heater core.**

34. Remove the heater core.

35. To install, reverse the removal procedure. Lubricate the coolant hoses with coolant hose lubricant or plain water only if needed. Top-off the engine coolant level. For additional information. Reactivate the supplemental restraints system (SRS).

CHEVROLET

Tracker

REMOVAL & INSTALLATION

1. Disconnect the negative battery cable.

2. Disable the SIR by performing the following procedure:

a. From the fuse box, located near the base of the steering column, remove the AIR BAG fuse.

b. Remove the steering wheel side cap, disconnect the Connector Positive Assurance (CPA) and the yellow 2-way driver's inflator module connectors.

c. Pull the instrument panel compartment out by pushing the right-side and left-side stoppers (located on both sides) inward.

d. Disconnect the CPA and the yellow 4-way passenger's inflator module connectors.

➡ **With the AIR BAG fuse removed and the ignition switch turned ON, the AIR**

BAG warning light will be ON; this is normal operation and does not indicate a SIR system malfunction.

3. Drain the cooling system into a clean container for reuse.

4. Remove the instrument panel as follows:

a. Remove the center console.

b. Remove the lower steering column cover by loosening the mounting screws.

c. Remove the glove box.

d. Detach the wiring harness connectors from the heater unit and the blower motor assembly.

e. Detach the wiring harness connectors from the ignition switch, contact coil and combination switch.

f. Open the hood.

g. Remove the steering column shaft joint bolt, then separate the steering column shaft from the lower steering shaft.

h. Loosen all of the steering column-to-firewall and instrument panel brace bolts.

i. If equipped, remove the shift (key)

interlock cable screw. Disconnect the cable from the ignition switch.

j. Remove the steering column from the vehicle.

✳✳ WARNING

Do not rest the steering column assembly on the steering wheel with the air bag module facing downward and the column vertical—personal injury may be the result.

k. Disconnect the speedometer cable from the speedometer, then remove the instrument cluster.

l. Remove the hood latch handle.

m. Remove the radio, the heater control panel and the heater control cables from the instrument panel.

n. Disconnect and label all wiring harness connectors from the instrument panel.

o. Remove the instrument panel mounting screws and bolts. Remove the side cover plates and the instrument panel mounting fasteners from the side of the assembly. Then, remove the upper cover plates and loosen the remaining mounting fasteners

p. Have an assistant help you carefully lift the instrument panel up and out of the vehicle. When separating the instrument panel from the firewall, ensure that all of the cables, wires and hoses are disconnected form the instrument panel.

5. Remove the 2 bolts and the right-side instrument panel center support.

6. If equipped with air conditioning, remove the evaporator.

7. Remove the 2 screws securing the SIR harness clip on the Sensing and Diagnostic Module (SDM) bracket.

8. Disconnect the SDM electrical connector.

9. Remove the 4 screws and the SDM bracket from the vehicle.

10. Remove the speedometer cable and antenna cable (if equipped) from the heater case.

11. Remove the floor duct from the heater case.

12. If equipped with air conditioning, disconnect the electrical jumper harness for the air conditioning amplifier.

13. Remove the relay bracket screws and the relay bracket.

14. From the engine compartment, remove the 2 heater assembly-to-chassis nuts and the 2 bolts.

15. Remove the heater case from the vehicle.

16. Remove the dampers and linkages from the heater case.

17. Remove the heater core bracket screw and the bracket.

18. Remove the heater core from the heater case.

To install:

19. Install the heater core to the heater case.

20. Install the heater core bracket and the bracket screw.

21. Install the dampers and linkages to the heater case.

22. Install the heater case to the vehicle.

23. In the engine compartment, install the 2 heater assembly-to-chassis nuts and the 2 bolts. Torque the nuts/bolts to 89 inch lbs. (10 Nm).

24. Install the relay bracket and the relay bracket screws.

25. If equipped with air conditioning, connect the electrical jumper harness for the air conditioning amplifier.

26. Install the floor duct to the heater case.

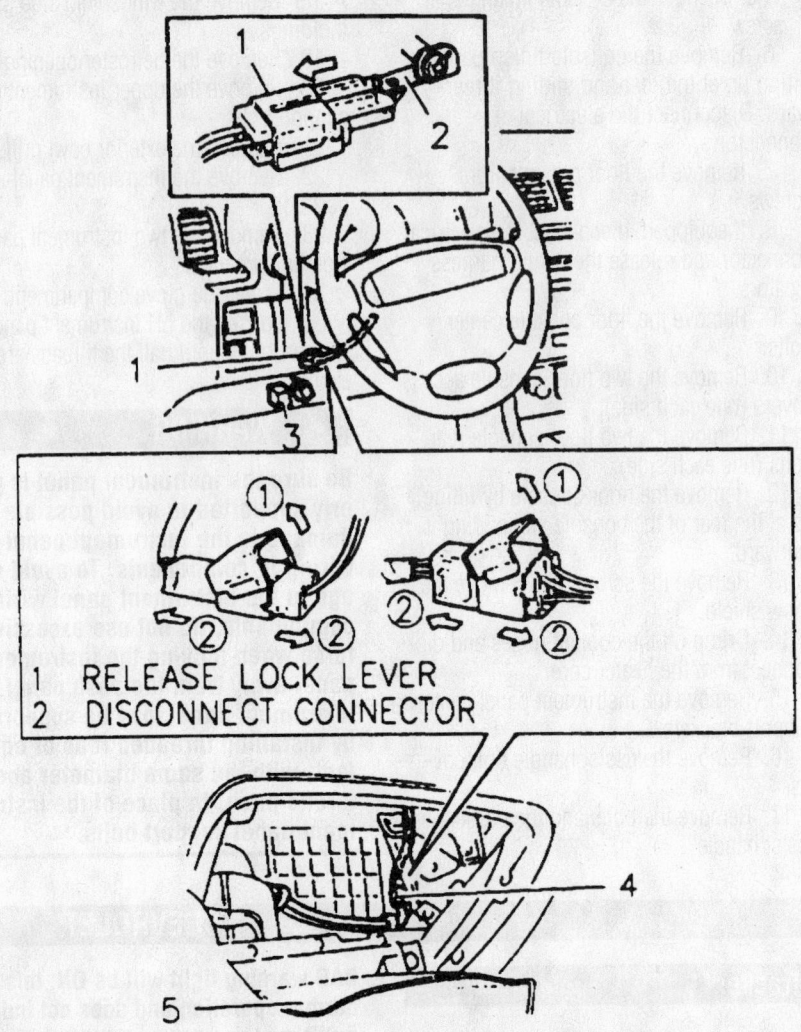

1 RELEASE LOCK LEVER
2 DISCONNECT CONNECTOR

1 YELLOW 2-WAY SIR CONNECTOR (DRIVER)
2 CONNECTOR POSITION ASSURANCE (CPA)
3 AIR BAG FUSE
4 YELLOW 4-WAY SIR CONNECTOR (PASSENGER)
5 GLOVE BOX

93113G90

Disabling the air bag system—Tracker

27. Install the speedometer cable and antenna cable (if equipped) to the heater case.

28. Install the SDM bracket and the 4 screws to the vehicle. Torque the screws to 49 inch lbs. (5.5 Nm).

29. Connect the SDM electrical connector.

30. Install the 2 screws securing the SIR harness clip on the Sensing and Diagnostic Module (SDM) bracket. Torque the screws to 49 inch lbs. (5.5 Nm)

31. If equipped with air conditioning, install the evaporator.

32. Install the 2 bolts and the right-side instrument panel center support.

33. Install the instrument panel as follows:

a. Have an assistant help you position the instrument panel in the vehicle. When installing the instrument panel on the firewall, ensure that all of the cables, wires and hoses are routed properly.

b. Install and tighten the instrument panel mounting screws and bolts.

c. Reattach all wiring harness connectors to the instrument panel.

d. Install the radio, the heater control panel and the heater control cables. Be sure to adjust the heater control cables.

e. Install the hood latch handle.

f. Install the instrument cluster, then connect the cable to the speedometer.

g. Install the steering column in the vehicle.

h. If equipped, connect the cable from the ignition switch, then install the shift (key) interlock cable screw.

i. Install and tighten all of the steering column-to-firewall and instrument panel brace bolts to 221 inch lbs. (25 Nm).

j. Install and tighten the steering col-umn shaft joint bolt to 221 inch lbs. (25 Nm).

k. Reattach the wiring harness connectors to the ignition switch, contact coil and combination switch.

l. Reattach the wiring harness con-

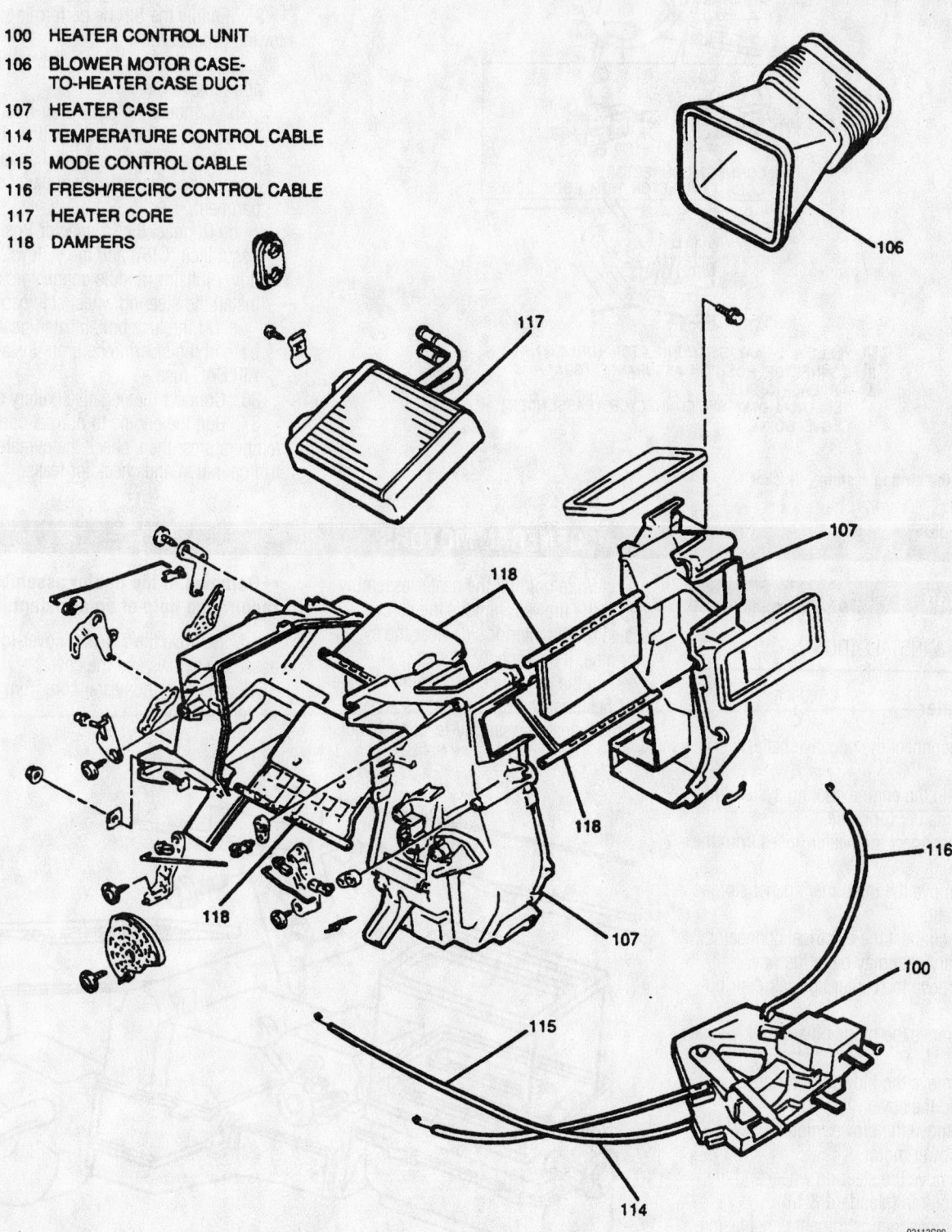

100 HEATER CONTROL UNIT
106 BLOWER MOTOR CASE-TO-HEATER CASE DUCT
107 HEATER CASE
114 TEMPERATURE CONTROL CABLE
115 MODE CONTROL CABLE
116 FRESH/RECIRC CONTROL CABLE
117 HEATER CORE
118 DAMPERS

93113G89

Exploded view of the heater case assembly and related components—Tracker

Brake service is covered in Section 4 of this manual

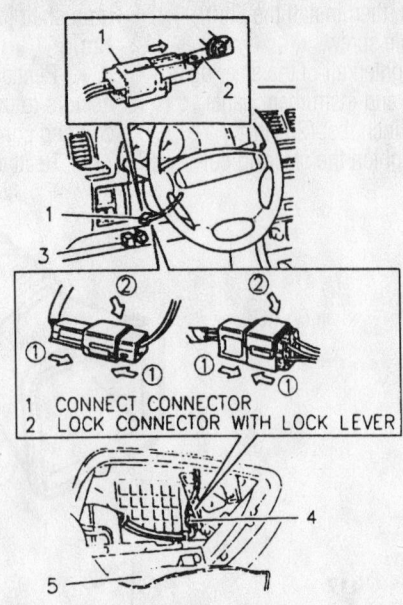

1 YELLOW 2-WAY SIR CONNECTOR (DRIVER)
2 CONNECTOR POSITION ASSURANCE (CPA)
3 AIR BAG-IG FUSE
4 YELLOW 4-WAY SIR CONNECTOR (PASSENGER)
5 GLOVE BOX

93113G91

Enabling the air bag system—Tracker

nectors to the heater unit and the blower motor assembly.

　m. Install the glove box.

　n. Install the lower steering column cover.

　o. Install the center console.

34. Refill the cooling system.

35. Enable the SIR by performing the following procedure:

　a. Turn the ignition switch to LOCK and remove the key.

　b. Connect the Connector Positive Assurance (CPA) and the yellow 4-way passenger's inflator module connectors.

　c. Close the instrument panel compartment.

　d. Connect the Connector Positive Assurance (CPA) and the yellow 2-way driver Inflator module connectors and install the steering wheel side cap.

　e. At the fuse box, located near the base of the steering column, install the AIR BAG fuse.

36. Connect the negative battery cable.

37. Run the engine to normal operating temperatures; then, check the climate control operation and check for leaks.

GENERAL MOTORS

Suburban

REMOVAL & INSTALLATION

Front Heater

1. Disconnect the negative battery cable.

2. Drain the engine cooling system into a clean container for reuse.

3. Disconnect the heater hoses from the heater core.

4. Remove the instrument panel storage compartment.

5. Disconnect the electrical connectors, as necessary, that may be in the way.

6. Remove the center floor air distribution duct.

7. Remove the hinge pillar trim kick panels.

8. Remove the blower motor cover screws and the cover.

9. Remove the blower motor screws and the blower motor.

10. Remove the steering wheel and the steering column (standard & tilt).

11. Remove the instrument panel fasteners and pull the instrument panel back far enough to gain access to the heater assembly.

12. While holding the heater assembly against the firewall, remove the screw located on the interior side near the evaporator pipe, if equipped.

13. In the engine compartment, remove the 4 heater assembly-to-chassis screws and the 2 heater assembly-to-chassis nuts.

➡**Removal of the heater assembly may require the help of an assistant.**

14. Remove the 7 heater cover-to-heater assembly screws and the cover.

15. Remove the heater core from the heater assembly.

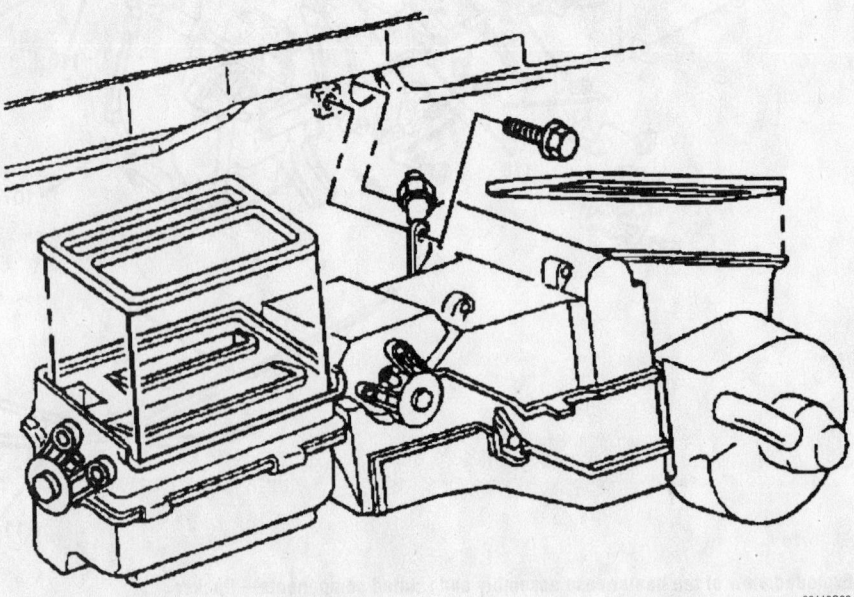

93113G80

View of the front heater assembly—Suburban

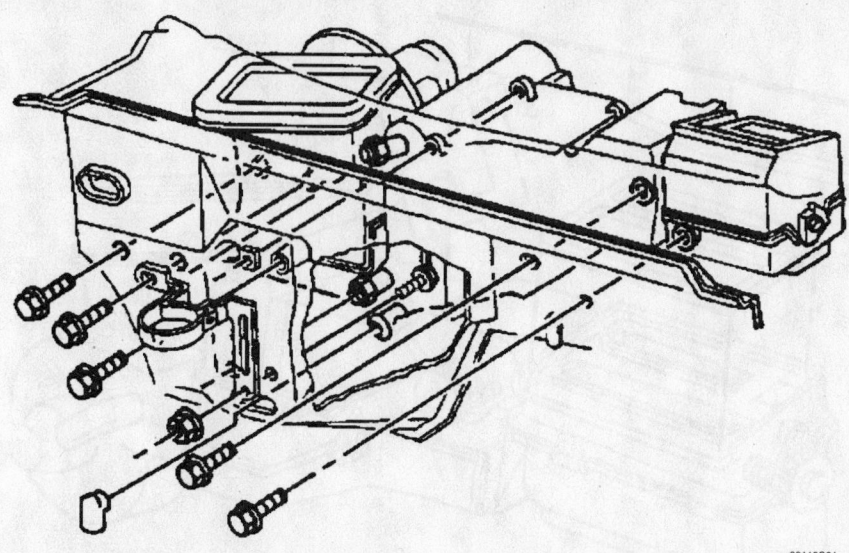

Location of the front heater assembly-to-chassis fasteners—Suburban

93113G81

To install:

16. Install the heater core to the heater assembly.

17. Install the heater cover and the 7 heater cover-to-heater assembly screws.

➡**Installation of the heater assembly may require the help of an assistant.**

18. In the engine compartment, install the 4 heater assembly-to-chassis screws and the 2 heater assembly-to-chassis nuts. Torque the screws to 17 inch lbs. (1.9 Nm) and the nuts to 25 inch lbs. (2.8 Nm).

19. While holding the heater assembly against the firewall, install the screw located on the interior side near the evaporator pipe, if equipped. Torque the screw to 97 inch lbs. (11 Nm).

20. Install the instrument panel and the instrument panel fasteners.

21. Install the steering column (standard & tilt) and the steering wheel.

22. Install the blower motor and the blower motor screws.

23. Install the blower motor the cover and the cover screws.

24. Install the hinge pillar trim kick panels.

25. Install the center floor air distribution duct.

26. Connect the electrical connectors that were disconnected.

27. Install the instrument panel storage compartment.

28. Disconnect the heater hoses to the heater core.

29. Refill the engine cooling system.

30. Connect the negative battery cable.

31. Run the engine to normal operating temperatures; then, check the climate control operation and check for leaks.

Rear Auxiliary Heater

1. Disconnect the negative battery cable.

2. Drain the engine cooling system into a clean container for reuse.

3. Remove the rear quarter trim panel, as necessary.

4. Remove the right rear quarter trim panel.

5. Remove the right rear wheelhouse.

6. Disconnect the heater hoses from the rear auxiliary heater core.

7. Disconnect the electrical connectors, as necessary.

8. Remove the drain valve.

9. Remove the rear auxiliary heater assembly-to-chassis nuts and bolts.

10. Remove the rear auxiliary heater assembly.

11. If necessary, remove the blower motor from the heater assembly.

12. Remove the rear auxiliary heater assembly cover.

13. Remove the heater core from the rear auxiliary assembly.

To install:

14. Install the heater core to the rear auxiliary assembly.

15. Install the rear auxiliary heater assembly cover.

16. If removed, install the blower motor to the heater assembly.

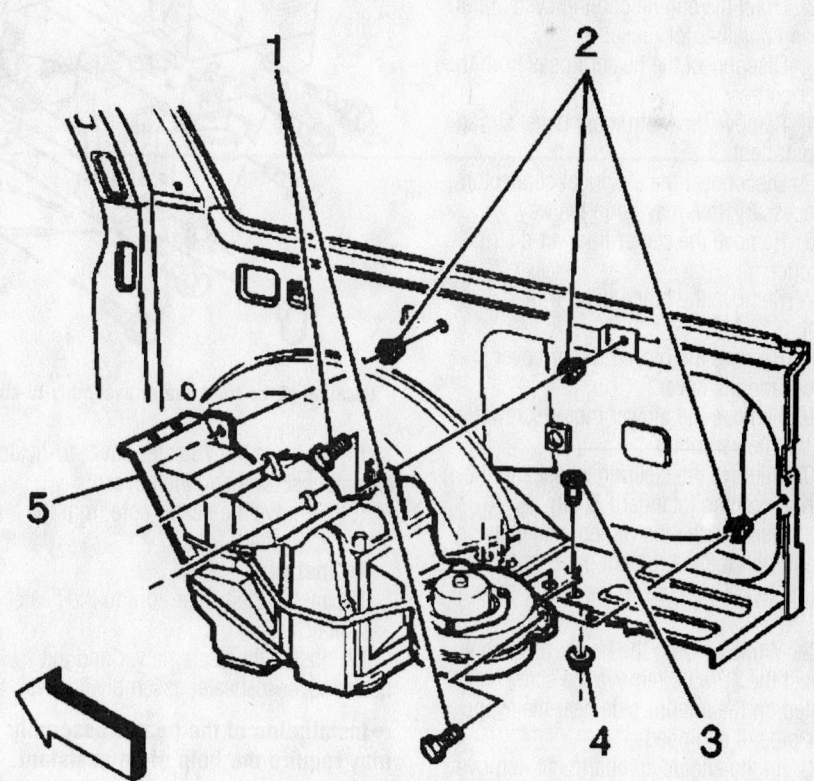

View of the rear auxiliary heater assembly—Suburban

93113G82

For complete Engine Mechanical specifications, see Section 1 of this manual

17. Install the rear auxiliary heater assembly.

18. Install the rear auxiliary heater assembly-to-chassis nuts and bolts. Torque the bolts to 13 inch lbs. (1.5 Nm) and the nuts to 89 inch lbs. (10 Nm).

19. Install the drain valve.

20. Connect the electrical connectors, as necessary.

21. Connect the heater hoses from the rear auxiliary heater core.

22. Install the right rear wheelhouse.

23. Install the right rear quarter trim panel.

24. Install the rear quarter trim panel, as necessary.

25. Refill the engine cooling system.

26. Connect the negative battery cable.

27. Run the engine to normal operating temperatures; then, check the climate control operation and check for leaks.

Tahoe

REMOVAL & INSTALLATION

Front Heater

1. Disconnect the negative battery cable.

2. Drain the engine cooling system into a clean container for reuse.

3. Disconnect the heater hoses from the heater core.

4. Remove the instrument panel storage compartment.

5. Disconnect the electrical connectors, as necessary, that may be in the way.

6. Remove the center floor air distribution duct.

7. Remove the hinge pillar trim kick panels.

8. Remove the blower motor cover screws and the cover.

9. Remove the blower motor screws and the blower motor.

10. Remove the steering wheel and the steering column (standard & tilt).

11. Remove the instrument panel fasteners and pull the instrument panel back far enough to gain access to the heater assembly.

12. While holding the heater assembly against the firewall, remove the screw located on the interior side near the evaporator pipe, if equipped.

13. In the engine compartment, remove the 4 heater assembly-to-chassis screws and the 2 heater assembly-to-chassis nuts.

➡**Removal of the heater assembly may require the help of an assistant.**

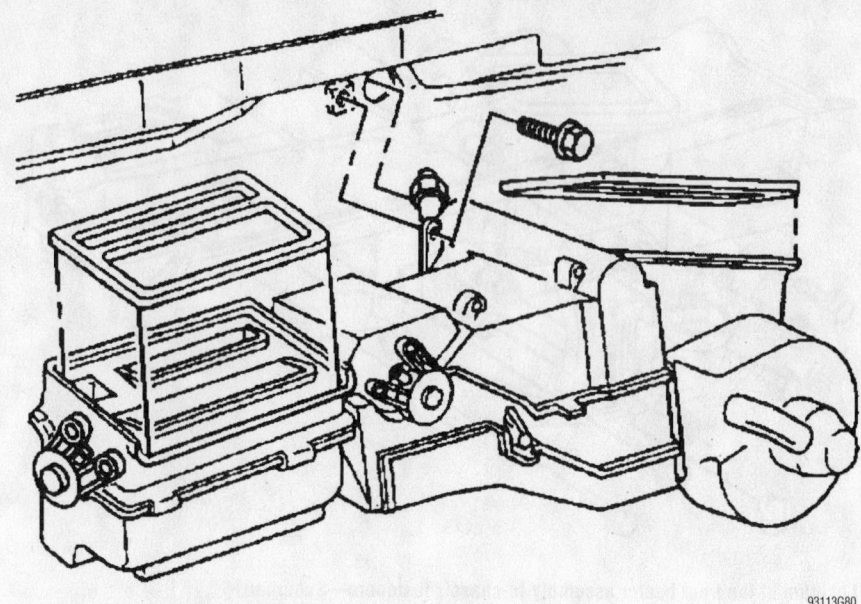

View of the front heater assembly—Tahoe

93113G80

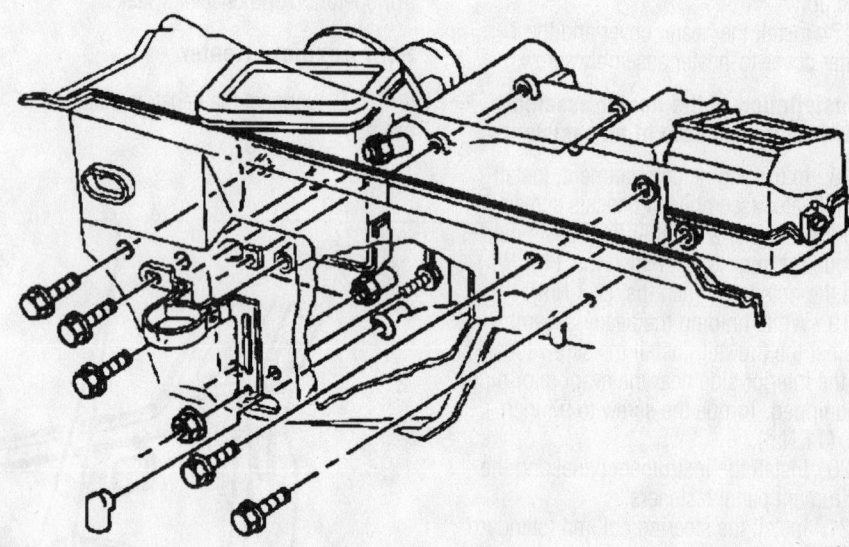

Location of the front heater assembly-to-chassis fasteners—Tahoe

93113G81

14. Remove the 7 heater cover-to-heater assembly screws and the cover.

15. Remove the heater core from the heater assembly.

To install:

16. Install the heater core to the heater assembly.

17. Install the heater cover and the 7 heater cover-to-heater assembly screws.

➡**Installation of the heater assembly may require the help of an assistant.**

18. In the engine compartment, install the 4 heater assembly-to-chassis screws and the 2 heater assembly-to-chassis nuts. Torque the screws to 17 inch lbs.

(1.9 Nm) and the nuts to 25 inch lbs. (2.8 Nm).

19. While holding the heater assembly against the firewall, install the screw located on the interior side near the evaporator pipe, if equipped. Torque the screw to 97 inch lbs. (11 Nm).

20. Install the instrument panel and the instrument panel fasteners.

21. Install the steering column (standard & tilt) and the steering wheel.

22. Install the blower motor and the blower motor screws.

23. Install the blower motor the cover and the cover screws.

24. Install the hinge pillar trim kick panels.

25. Install the center floor air distribution duct.
26. Connect the electrical connectors that were disconnected.
27. Install the instrument panel storage compartment.
28. Disconnect the heater hoses to the heater core.
29. Refill the engine cooling system.
30. Connect the negative battery cable.
31. Run the engine to normal operating temperatures; then, check the climate control operation and check for leaks.

Rear Auxiliary Heater

1. Disconnect the negative battery cable.
2. Drain the engine cooling system into a clean container for reuse.
3. Remove the rear quarter trim panel, as necessary.
4. Remove the right rear quarter trim panel.
5. Remove the right rear wheelhouse.
6. Disconnect the heater hoses from the rear auxiliary heater core.
7. Disconnect the electrical connectors, as necessary.
8. Remove the drain valve.

9. Remove the rear auxiliary heater assembly-to-chassis nuts and bolts.
10. Remove the rear auxiliary heater assembly.
11. If necessary, remove the blower motor from the heater assembly.
12. Remove the rear auxiliary heater assembly cover.
13. Remove the heater core from the rear auxiliary assembly.
To install:
14. Install the heater core to the rear auxiliary assembly.
15. Install the rear auxiliary heater assembly cover.
16. If removed, install the blower motor to the heater assembly.
17. Install the rear auxiliary heater assembly.
18. Install the rear auxiliary heater assembly-to-chassis nuts and bolts. Torque the bolts to 13 inch lbs. (1.5 Nm) and the nuts to 89 inch lbs. (10 Nm).
19. Install the drain valve.
20. Connect the electrical connectors, as necessary.
21. Connect the heater hoses from the rear auxiliary heater core.

22. Install the right rear wheelhouse.
23. Install the right rear quarter trim panel.
24. Install the rear quarter trim panel, as necessary.
25. Refill the engine cooling system.
26. Connect the negative battery cable.
27. Run the engine to normal operating temperatures; then, check the climate control operation and check for leaks.

Yukon

REMOVAL & INSTALLATION

Front Heater

1. Disconnect the negative battery cable.
2. Drain the engine cooling system into a clean container for reuse.
3. Disconnect the heater hoses from the heater core.
4. Remove the instrument panel storage compartment.
5. Disconnect the electrical connectors, as necessary, that may be in the way.
6. Remove the center floor air distribution duct.
7. Remove the hinge pillar trim kick panels.
8. Remove the blower motor cover screws and the cover.
9. Remove the blower motor screws and the blower motor.
10. Remove the steering wheel and the steering column (standard & tilt).
11. Remove the instrument panel fasteners and pull the instrument panel back far enough to gain access to the heater assembly.
12. While holding the heater assembly against the firewall, remove the screw located on the interior side near the evaporator pipe, if equipped.
13. In the engine compartment, remove the 4 heater assembly-to-chassis screws and the 2 heater assembly-to-chassis nuts.

➡**Removal of the heater assembly may require the help of an assistant.**

14. Remove the 7 heater cover-to-heater assembly screws and the cover.
15. Remove the heater core from the heater assembly.
To install:
16. Install the heater core to the heater assembly.

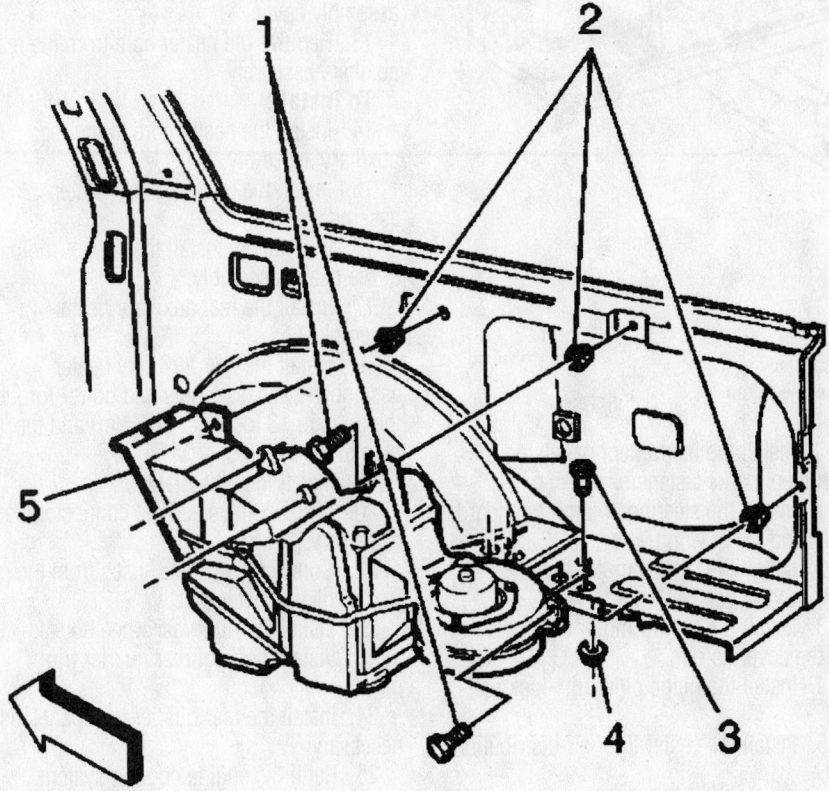

View of the rear auxiliary heater assembly—Tahoe

93113G82

For Accessory Drive Belt illustrations, see Section 1 of this manual

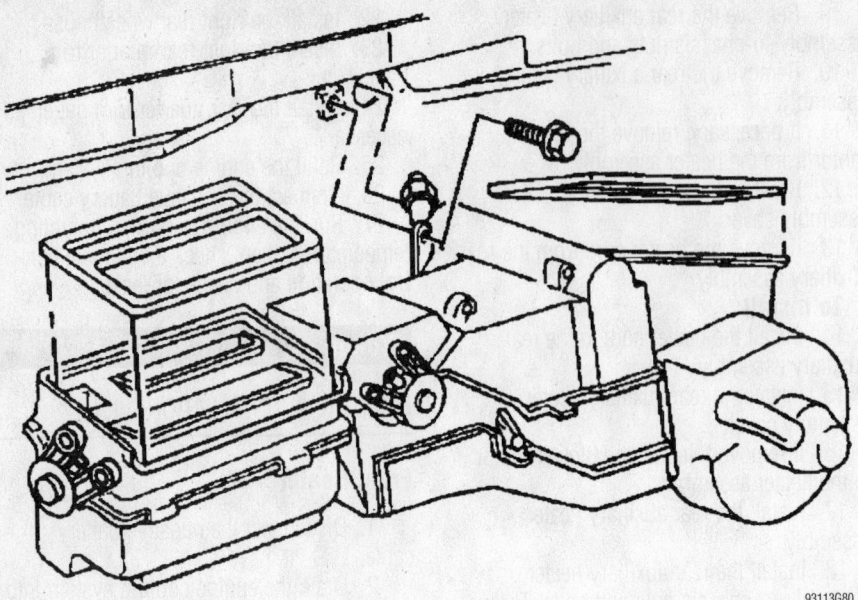

View of the front heater assembly—Yukon

93113G80

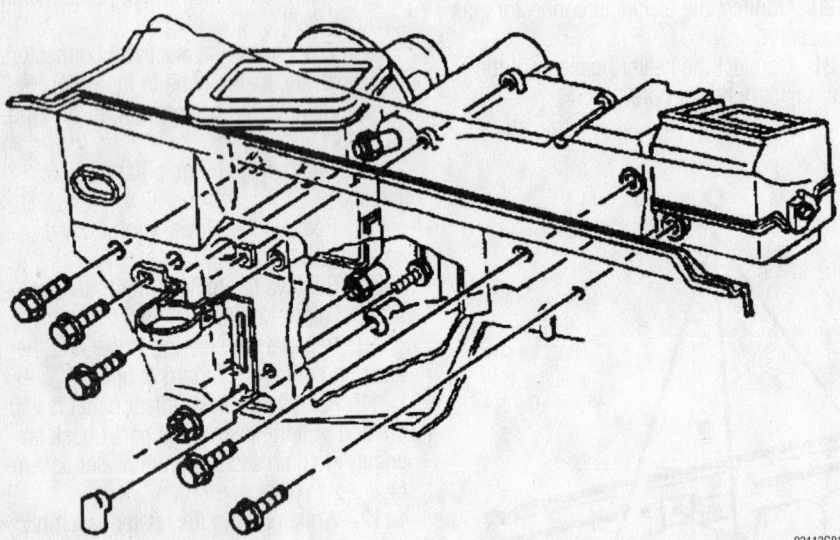

Location of the front heater assembly-to-chassis fasteners—Yukon

93113G81

17. Install the heater cover and the 7 heater cover-to-heater assembly screws.

➡**Installation of the heater assembly may require the help of an assistant.**

18. In the engine compartment, install the 4 heater assembly-to-chassis screws and the 2 heater assembly-to-chassis nuts. Torque the screws to 17 inch lbs. (1.9 Nm) and the nuts to 25 inch lbs. (2.8 Nm).

19. While holding the heater assembly against the firewall, install the screw located on the interior side near the evaporator pipe, if equipped. Torque the screw to 97 inch lbs. (11 Nm).

20. Install the instrument panel and the instrument panel fasteners.

21. Install the steering column (standard & tilt) and the steering wheel.

22. Install the blower motor and the blower motor screws.

23. Install the blower motor the cover and the cover screws.

24. Install the hinge pillar trim kick panels.

25. Install the center floor air distribution duct.

26. Connect the electrical connectors that were disconnected.

27. Install the instrument panel storage compartment.

28. Disconnect the heater hoses to the heater core.

29. Refill the engine cooling system.

30. Connect the negative battery cable.

31. Run the engine to normal operating temperatures; then, check the climate control operation and check for leaks.

Rear Auxiliary Heater

1. Disconnect the negative battery cable.

2. Drain the engine cooling system into a clean container for reuse.

3. Remove the rear quarter trim panel, as necessary.

4. Remove the right rear quarter trim panel.

5. Remove the right rear wheelhouse.

6. Disconnect the heater hoses from the rear auxiliary heater core.

7. Disconnect the electrical connectors, as necessary.

8. Remove the drain valve.

9. Remove the rear auxiliary heater assembly-to-chassis nuts and bolts.

10. Remove the rear auxiliary heater assembly.

11. If necessary, remove the blower motor from the heater assembly.

12. Remove the rear auxiliary heater assembly cover.

13. Remove the heater core from the rear auxiliary assembly.

To install:

14. Install the heater core to the rear auxiliary assembly.

15. Install the rear auxiliary heater assembly cover.

16. If removed, install the blower motor to the heater assembly.

17. Install the rear auxiliary heater assembly.

18. Install the rear auxiliary heater assembly-to-chassis nuts and bolts. Torque the bolts to 13 inch lbs. (1.5 Nm) and the nuts to 89 inch lbs. (10 Nm).

19. Install the drain valve.

20. Connect the electrical connectors, as necessary.

21. Connect the heater hoses from the rear auxiliary heater core.

22. Install the right rear wheelhouse.

23. Install the right rear quarter trim panel.

24. Install the rear quarter trim panel, as necessary.

25. Refill the engine cooling system.

26. Connect the negative battery cable.

27. Run the engine to normal operating temperatures; then, check the climate control operation and check for leaks.

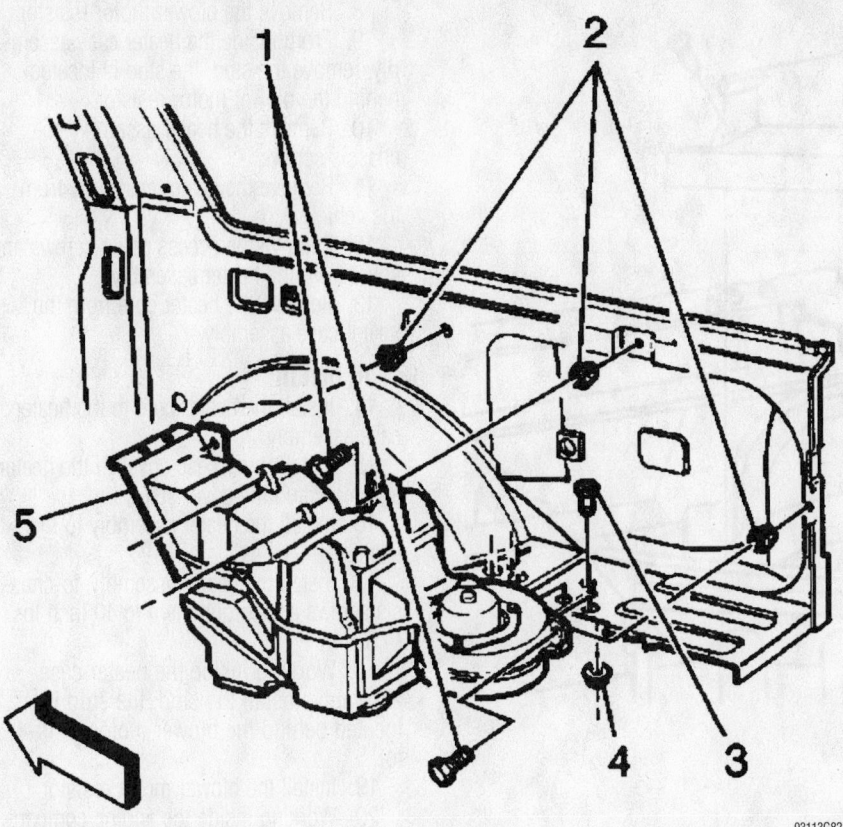

View of the rear auxiliary heater assembly—Yukon

Chevrolet Blazer

REMOVAL & INSTALLATION

1. Disconnect the negative battery cable.

2. Drain the cooling system into a clean container for reuse.

3. Remove the heater hoses from the heater core.

4. Remove the instrument panel as follows:

   a. Disable the air bag system.

   b. Set the parking brake and block the wheels.

   c. Disconnect the parking brake release cable from the parking brake lever.

   d. Unfasten the screws that retain the DLC instrument panel left side sound insulator. Feed the DLC through the hole in the sound insulator.

   e. Unfasten the right side sound insulator panel screws and remove the panel.

   f. Unfasten the screws that attach the instrument panel left side sound insulator to the knee bolster and cowl panel.

   g. Unfasten the nut that attaches the left side sound insulator to the accelerator pedal bracket.

   h. Unplug the remote control door lock receiver module electrical connector.

   i. Remove the door lock receiver module from the left side sound insulator. Remove the left side sound insulator.

   j. Unfasten the screws that attach the instrument panel center sound insulator to the knee bolster, instrument panel, heater assembly and floor duct.

   k. Remove the center sound insulator.

   l. Unfasten the screws that attach the courtesy lamp to the knee bolster.

   m. Unfasten the screws that attach the knee bolster to the instrument panel.

   n. Disconnect the lap cooler duct from the knee bolster.

   o. Unplug the lighter electrical connection and remove the knee bolster.

   p. Unfasten the steering column-to-instrument panel nuts and lower the column.

   q. Unfasten the screws that attach the instrument panel accessory trim plate to the instrument panel.

   r. Remove the trim plate and unplug all necessary electrical connection.

   s. Remove the heater and/or air conditioning control assembly.

   t. Remove the radio and the storage compartment assembly (if equipped).

   u. If necessary, remove the instrument cluster.

   v. Unfasten the left and right instrument panel pivot bolts and the panel lower support bolt.

   w. Unfasten the speaker grilles retaining screws and remove the speaker grilles.

   x. Remove the windshield defroster

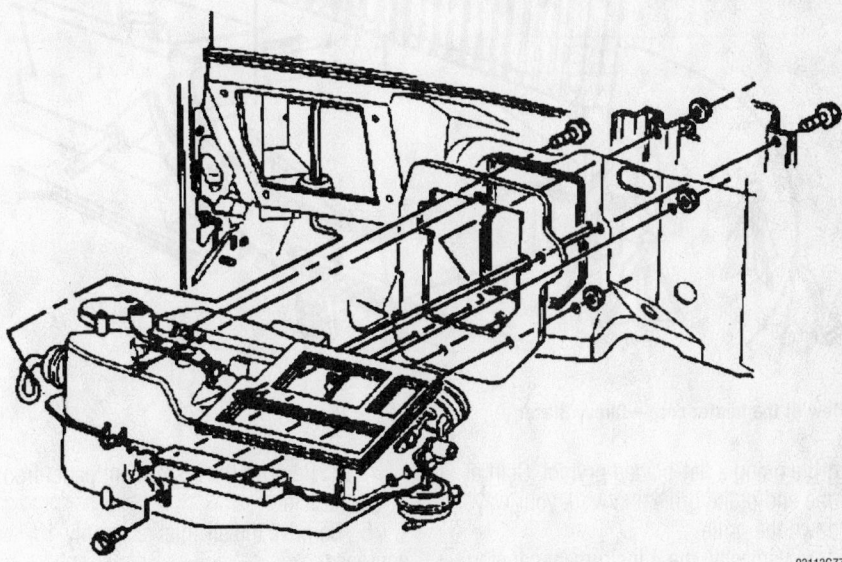

View of the heater case assembly—Chevy Blazer

For Tire, Wheel and Ball Joint specifications, see Section 1 of this manual

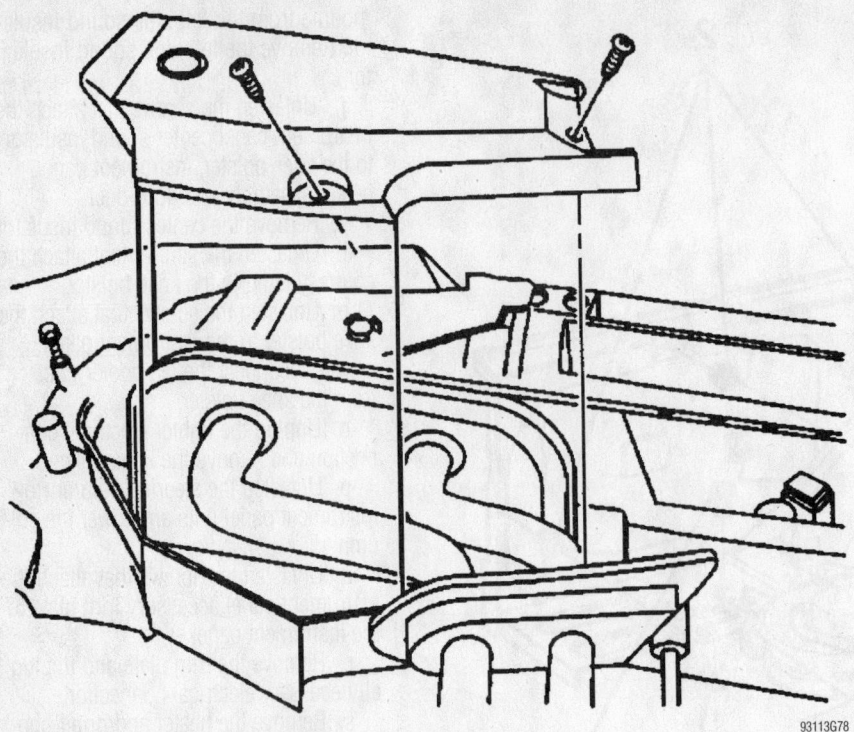

View of the heater case cover—Chevy Blazer

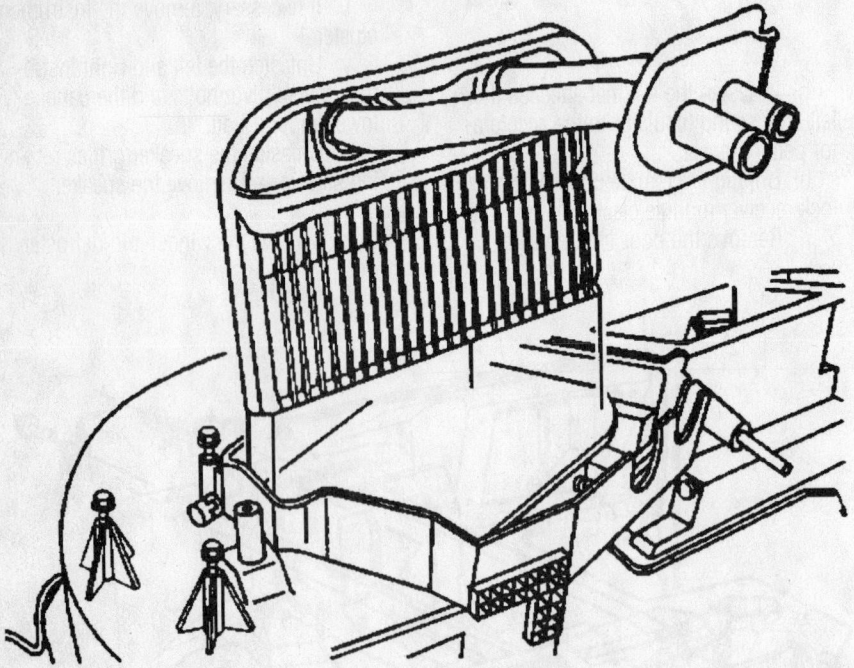

View of the heater core—Chevy Blazer

grille using a flat-bladed prytool. Start at one end of the grille and work your way down the grille.

    y. Unfasten the 4 instrument panel upper support screws.

    z. Tag and unplug all necessary electrical connections.

    aa. Remove the instrument panel from the vehicle.

  5. Remove the air inlet assembly, if equipped.

  6. Remove the vacuum hoses.

  7. From inside the engine compartment, remove the heater assembly studs.

  8. Remove the blower motor resistor.

  9. From inside the heater case assembly, remove the stud; the stud is located behind the blower motor resistor.

  10. Remove the heater assembly-to-chassis screws.

  11. Remove the heater assembly from the vehicle.

  12. Remove the access cover screws and cover from the heater assembly.

  13. Remove the heater core from the heater case assembly.

To install:

  14. Install the heater core to the heater case assembly.

  15. Install the access cover to the heater assembly and the cover screws.

  16. Install the heater assembly to the vehicle.

  17. Install the heater assembly-to-chassis screws and torque them to 40 inch lbs. (4.5 Nm).

  18. Working inside the heater case assembly, install the stud; the stud is located behind the blower motor resistor.

  19. Install the blower motor resistor.

  20. Working inside the engine compartment, install the heater assembly studs and torque them to 17 inch lbs. (1.9 Nm).

  21. Install the vacuum hoses.

  22. Install the air inlet assembly, if equipped.

  23. Install the instrument panel as follows:

    a. Rest the instrument panel on the lower pivot studs.

    b. Attach the electrical connections.

    c. Install but do not tighten the 4 upper instrument panel support screws.

    d. Install the left and right panel pivot bolts. Tighten the bolts to 102 inch lbs. (11.5 Nm).

    e. Install the panel lower support bolt. Tighten the bolt to 102 inch lbs. (11.5 Nm).

    f. Tighten the upper support screws to 17 inch lbs. (1.9 Nm).

    g. Install the windshield defroster grille and the speaker grilles.

    h. Install the radio and storage compartment assembly (if equipped).

    i. If removed, install the instrument cluster.

    j. Install the heater and/or air conditioning control assembly.

    k. attach the electrical connections to the instrument panel accessory trim plate.

    l. Place the trim plate in position and

install its retaining screws. Tighten the screws to 17 inch lbs. (1.9 Nm).

m. Place the steering column into position and install it retaining nuts. Tighten the nuts to 22 ft. lbs. (30 Nm).

n. Attach the lighter electrical connection ;and the lap cooler duct to the knee bolster.

o. Place the knee bolster into position and install its retaining screws. Tighten the Torx® head screws to 80 inch lbs. (9 Nm) and the hex head screws to 17 inch lbs. (1.9 Nm).

p. Place the courtesy lamp in position and install its screws. Tighten the screws to 17 inch lbs. (1.9 Nm).

q. Place the instrument panel center sound insulator in position. Install the screws that attach the center sound insulator to the knee bolster, instrument panel and the floor duct. Tighten the screws to 17 inch lbs. (1.9 Nm).

r. Install the screw that attaches the center sound insulator to the heater assembly. Tighten the screw to 13 inch lbs. (1.5 Nm).

s. >Install the remote control door lock receiver module to the instrument panel left side sound insulator.

t. >Attach the door lock receiver electrical connection.

u. Install the nut that attaches the left side sound insulator to the accelerator pedal bracket. Tighten the nut to 35 inch lbs. (4 Nm).

v. Install the screw that attaches the left side sound insulator to cowl panel. Tighten the screw to 13 inch lbs. (1.5 Nm).

w. Install the screws that attach the left side sound insulator to knee bolster. Tighten the screw to 17 inch lbs. (1.9 Nm).

x. Feed the DLC through the hole in the sound insulator, place the DLC in position and install its retaining screws. Tighten the screws to 21 inch lbs. (2.4 Nm).

y. Install the right side sound insulator and tighten the screws

z. Connect the parking brake release cable to the lever.

aa. Enable the air bag system.

24. Install the heater hoses to the heater core.

25. Refill the cooling system.

26. >Connect the negative battery cable.

27. Run the engine to normal operating temperatures; then, check the climate control operation and check for leaks.

Oldsmobile Bravada

REMOVAL & INSTALLATION

1. Disconnect the negative battery cable.

2. >Drain the cooling system into a clean container for reuse.

3. Remove the heater hoses from the heater core.

4. Remove the instrument panel as follows:

a. Disable the air bag system.

b. Set the parking brake and block the wheels.

c. Disconnect the parking brake release cable from the parking brake lever.

d. Unfasten the screws that retain the DLC instrument panel left side sound insulator. Feed the DLC through the hole in the sound insulator.

e. Unfasten the right side sound insulator panel screws and remove the panel.

f. Unfasten the screws that attach the instrument panel left side sound insulator to the knee bolster and cowl panel.

g. Unfasten the nut that attaches the left side sound insulator to the accelerator pedal bracket.

h. Unplug the remote control door lock receiver module electrical connector.

i. Remove the door lock receiver module from the left side sound insulator. Remove the left side sound insulator.

j. Unfasten the screws that attach the instrument panel center sound insulator to the knee bolster, instrument panel, heater assembly and floor duct.

k. Remove the center sound insulator.

l. Unfasten the screws that attach the courtesy lamp to the knee bolster.

m. Unfasten the screws that attach the knee bolster to the instrument panel.

n. Disconnect the lap cooler duct from the knee bolster.

o. Unplug the lighter electrical connection and remove the knee bolster.

p. Unfasten the steering column-to-instrument panel nuts and lower the column.

q. Unfasten the screws that attach the instrument panel accessory trim plate to the instrument panel.

r. Remove the trim plate and unplug all necessary electrical connection.

s. Remove the heater and/or air conditioning control assembly.

t. Remove the radio and the storage compartment assembly (if equipped).

u. If necessary, remove the instrument cluster.

v. Unfasten the left and right instrument panel pivot bolts and the panel lower support bolt.

w. Unfasten the speaker grilles retaining screws and remove the speaker grilles.

x. Remove the windshield defroster grille using a flat-bladed prytool. Start at one end of the grille and work your way down the grille.

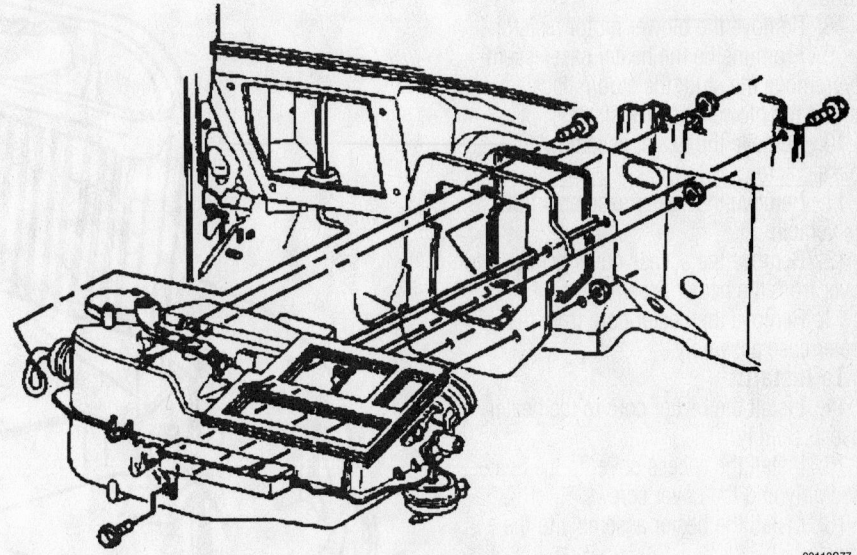

93113G77

View of the heater case assembly—Oldsmobile Bravada

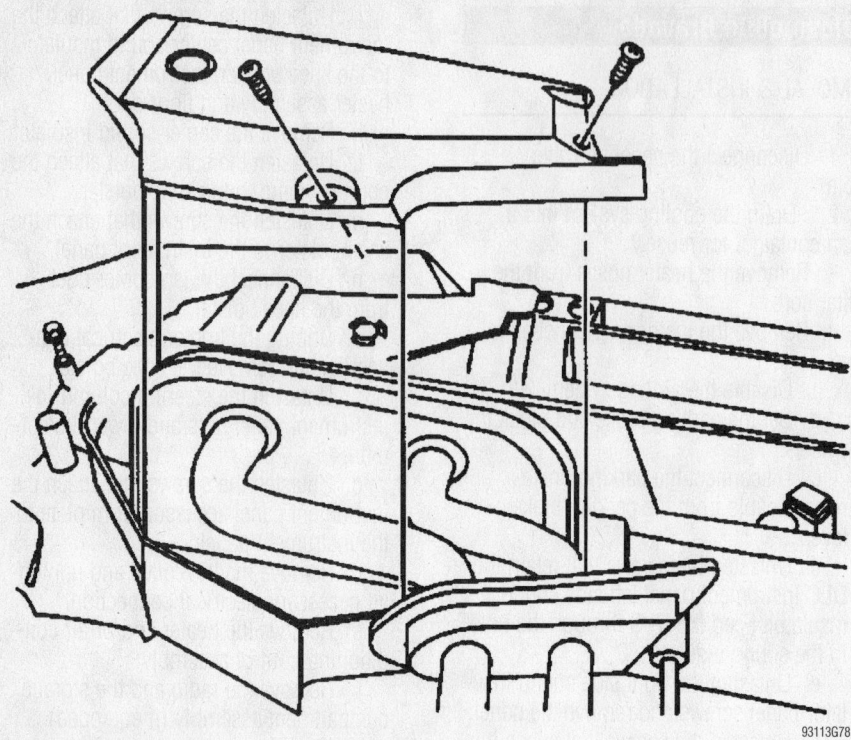

View of the heater case cover—Oldsmobile Bravada

y. Unfasten the 4 instrument panel upper support screws.

z. Tag and unplug all necessary electrical connections.

aa. Remove the instrument panel from the vehicle.

5. Remove the air inlet assembly, if equipped.

6. Remove the vacuum hoses.

7. From inside the engine compartment, remove the heater assembly studs.

8. Remove the blower motor resistor.

9. From inside the heater case assembly, remove the stud; the stud is located behind the blower motor resistor.

10. Remove the heater assembly-to-chassis screws.

11. Remove the heater assembly from the vehicle.

12. Remove the access cover screws and cover from the heater assembly.

13. Remove the heater core from the heater case assembly.

To install:

14. Install the heater core to the heater case assembly.

15. Install the access cover to the heater assembly and the cover screws.

16. Install the heater assembly to the vehicle.

17. Install the heater assembly-to-chassis screws and torque them to 40 inch lbs. (4.5 Nm).

18. Working inside the heater case assembly, install the stud; the stud is located behind the blower motor resistor.

19. Install the blower motor resistor.

20. Working inside the engine compartment, install the heater assembly studs and torque them to 17 inch lbs. (1.9 Nm).

21. Install the vacuum hoses.

22. Install the air inlet assembly, if equipped.

23. Install the instrument panel as foll the accelerator pedal bracket. Tighten the nut to 35 inch lbs. (4 Nm).

v. Install the screw that attaches the left side sound insulator to cowl panel. Tighten the screw to 13 inch lbs. (1.5 Nm).

w. Install the screws that attach the left side sound insulator to knee bolster. Tighten the screw to 17 inch lbs. (1.9 Nm).

x. Feed the DLC through the hole in the sound insulator, place the DLC in position and install its retaining screws. Tighten the screws to 21 inch lbs. (2.4 Nm).

y. Install the right side sound insulator and tighten the screws

z. Connect the parking brake release cable to the lever.

aa. Enable the air bag system.

24. Install the heater hoses to the heater core.

25. Refill the cooling system.

26. Connect the negative battery cable.

27. Run the engine to normal operating temperatures; then, check the climate control operation and check for leaks.

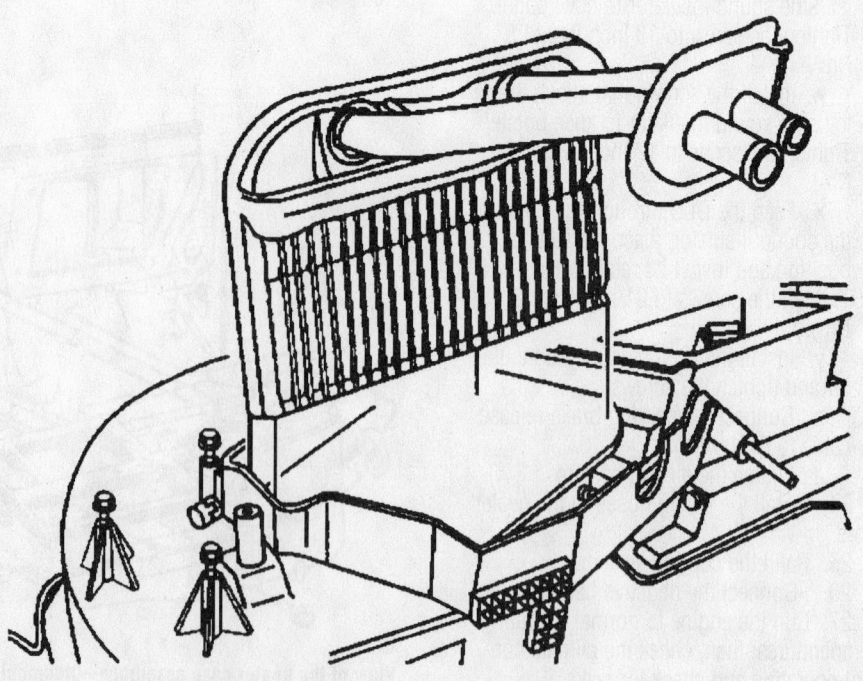

View of the heater core—Oldsmobile Bravada

GMC Envoy

REMOVAL & INSTALLATION

1. Disconnect the negative battery cable.

2. Drain the cooling system into a clean container for reuse.

3. Remove the heater hoses from the heater core.

4. Remove the instrument panel as follows:

a. Disable the air bag system.

b. Set the parking brake and block the wheels.

c. Disconnect the parking brake release cable from the parking brake lever.

d. Unfasten the screws that retain the DLC instrument panel left side sound insulator. Feed the DLC through the hole in the sound insulator.

e. Unfasten the right side sound insulator panel screws and remove the panel.

f. Unfasten the screws that attach the instrument panel left side sound insulator to the knee bolster and cowl panel.

g. Unfasten the nut that attaches the left side sound insulator to the accelerator pedal bracket.

h. Unplug the remote control door lock receiver module electrical connector.

i. Remove the door lock receiver module from the left side sound insulator. Remove the left side sound insulator.

j. Unfasten the screws that attach the instrument panel center sound insulator to the knee bolster, instrument panel, heater assembly and floor duct.

k. Remove the center sound insulator.

l. Unfasten the screws that attach the courtesy lamp to the knee bolster.

m. Unfasten the screws that attach the knee bolster to the instrument panel.

n. Disconnect the lap cooler duct from the knee bolster.

o. Unplug the lighter electrical connection and remove the knee bolster.

p. Unfasten the steering column-to-instrument panel nuts and lower the column.

q. Unfasten the screws that attach the instrument panel accessory trim plate to the instrument panel.

r. Remove the trim plate and unplug all necessary electrical connection.

s. Remove the heater and/or air conditioning control assembly.

t. Remove the radio and the storage compartment assembly (if equipped).

u. If necessary, remove the instrument cluster.

v. Unfasten the left and right instrument panel pivot bolts and the panel lower support bolt.

w. Unfasten the speaker grilles retaining screws and remove the speaker grilles.

x. Remove the windshield defroster grille using a flat-bladed prytool. Start at one end of the grille and work your way down the grille.

y. Unfasten the 4 instrument panel upper support screws.

z. Tag and unplug all necessary electrical connections.

aa. Remove the instrument panel from the vehicle.

5. Remove the air inlet assembly, if equipped.

6. Remove the vacuum hoses.

7. From inside the engine compartment, remove the heater assembly studs.

8. Remove the blower motor resistor.

9. From inside the heater case assembly, remove the stud; the stud is located behind the blower motor resistor.

10. Remove the heater assembly-to-chassis screws.

11. Remove the heater assembly from the vehicle.

12. Remove the access cover screws and cover from the heater assembly.

13. Remove the heater core from the heater case assembly.

To install:

14. Install the heater core to the heater case assembly.

15. Install the access cover to the heater assembly and the cover screws.

16. Install the heater assembly to the vehicle.

17. Install the heater assembly-to-chassis screws and torque them to 40 inch lbs. (4.5 Nm).

18. Working inside the heater case assembly, install the stud; the stud is located behind the blower motor resistor.

19. Install the blower motor resistor.

20. Working inside the engine compartment, install the heater assembly studs and torque them to 17 inch lbs. (1.9 Nm).

21. Install the vacuum hoses.

22. Install the air inlet assembly, if equipped.

23. Install the instrument panel as follows:

a. Rest the instrument panel on the lower pivot studs.

b. Attach the electrical connections.

c. Install but do not tighten the 4 upper instrument panel support screws.

d. Install the left and right panel pivot bolts. Tighten the bolts to 102 inch lbs. (11.5 Nm).

e. Install the panel lower support bolt. Tighten the bolt to 102 inch lbs. (11.5 Nm).

f. Tighten the upper support screws to 17 inch lbs. (1.9 Nm).

g. Install the windshield defroster grille and the speaker grilles.

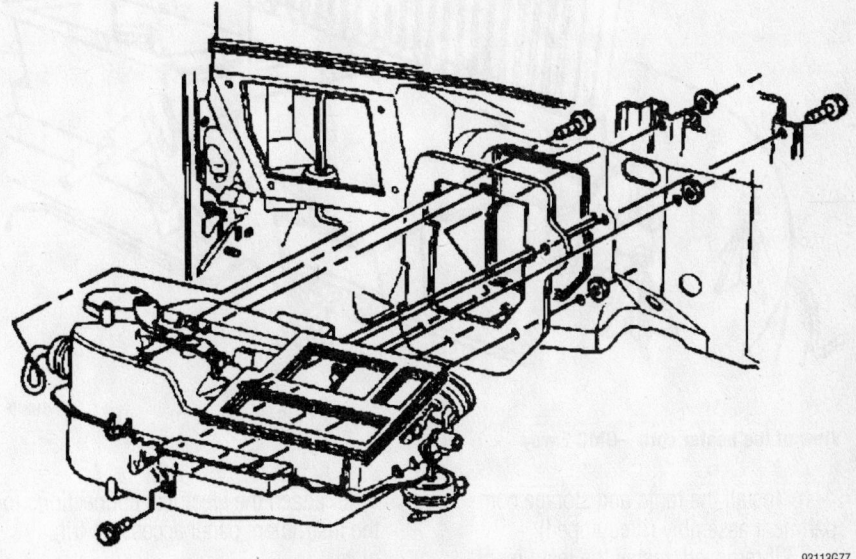

View of the heater case assembly—GMC Envoy

93113G77

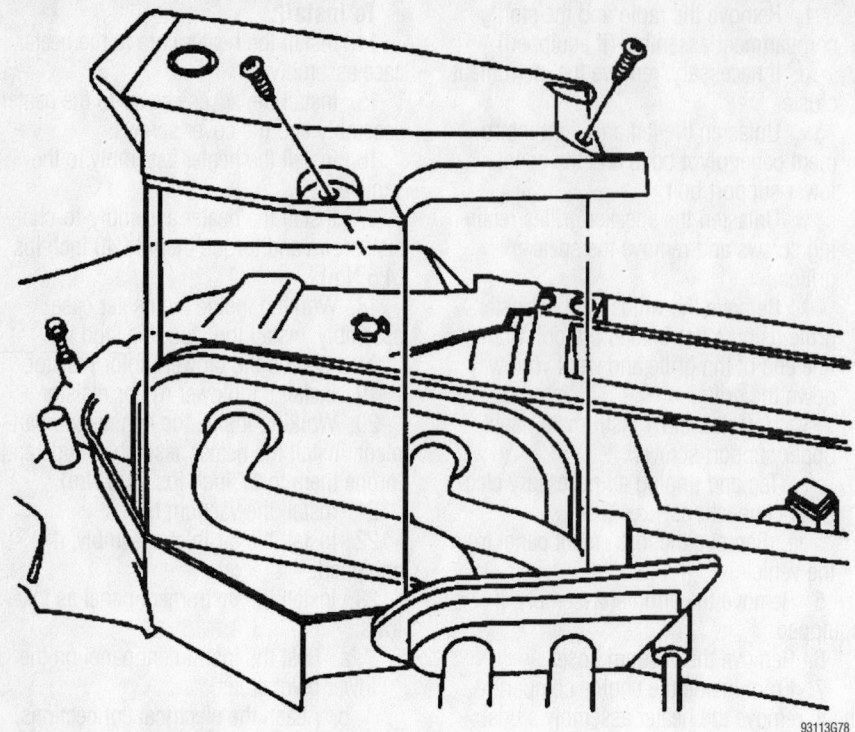

View of the heater case cover—GMC Envoy

93113G78

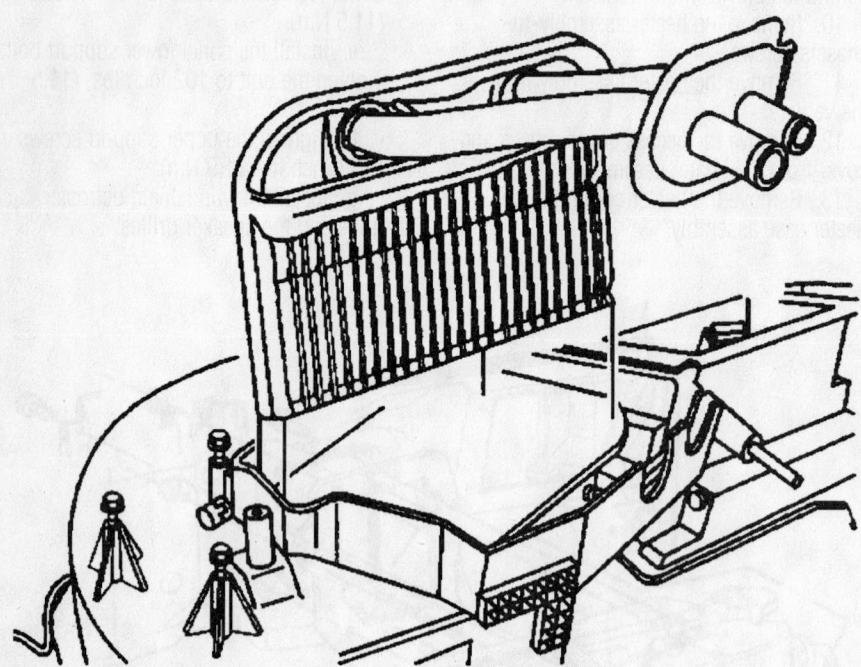

View of the heater core—GMC Envoy

93113G79

h. Install the radio and storage compartment assembly (if equipped).

i. If removed, install the instrument cluster.

j. Install the heater and/or air conditioning control assembly.

k. attach the electrical connections to the instrument panel accessory trim plate.

l. Place the trim plate in position and install its retaining screws. Tighten the screws to 17 inch lbs. (1.9 Nm).

m. Place the steering column into position and install its retaining nuts. Tighten the nuts to 22 ft. lbs. (30 Nm).

n. Attach the lighter electrical connection and the lap cooler duct to the knee bolster.

o. Place the knee bolster into position and install its retaining screws. Tighten the Torx® head screws to 80 inch lbs. (9 Nm) and the hex head screws to 17 inch lbs. (1.9 Nm).

p. Place the courtesy lamp in position and install its screws. Tighten the screws to 17 inch lbs. (1.9 Nm).

q. Place the instrument panel center sound insulator in position. Install the screws that attach the center sound insulator to the knee bolster, instrument panel and the floor duct. Tighten the screws to 17 inch lbs. (1.9 Nm).

r. Install the screw that attaches the center sound insulator to the heater assembly. Tighten the screw to 13 inch lbs. (1.5 Nm).

s. Install the remote control door lock receiver module to the instrument panel left side sound insulator.

t. Attach the door lock receiver electrical connection.

u. Install the nut that attaches the left side sound insulator to the accelerator pedal bracket. Tighten the nut to 35 inch lbs. (4 Nm).

v. Install the screw that attaches the left side sound insulator to cowl panel. Tighten the screw to 13 inch lbs. (1.5 Nm).

w. Install the screws that attach the left side sound insulator to knee bolster. Tighten the screw to 17 inch lbs. (1.9 Nm).

x. Feed the DLC through the hole in the sound insulator, place the DLC in position and install its retaining screws. Tighten the screws to 21 inch lbs. (2.4 Nm).

y. Install the right side sound insulator and tighten the screws

z. Connect the parking brake release cable to the lever.

aa. Enable the air bag system.

24. Install the heater hoses to the heater core.

25. Refill the cooling system.

26. Connect the negative battery cable.

27. Run the engine to normal operating temperatures; then, check the climate control operation and check for leaks.

GMC Jimmy

REMOVAL & INSTALLATION

1. Disconnect the negative battery cable.

2. Drain the cooling system into a clean container for reuse.

3. Remove the heater hoses from the heater core.

4. Remove the instrument panel as follows:

   a. Disable the air bag system.

   b. Set the parking brake and block the wheels.

   c. Disconnect the parking brake release cable from the parking brake lever.

   d. Unfasten the screws that retain the DLC instrument panel left side sound insulator. Feed the DLC through the hole in the sound insulator.

   e. Unfasten the right side sound insulator panel screws and remove the panel.

   f. Unfasten the screws that attach the instrument panel left side sound insulator to the knee bolster and cowl panel.

   g. Unfasten the nut that attaches the left side sound insulator to the accelerator pedal bracket.

   h. Unplug the remote control door lock receiver module electrical connector.

   i. Remove the door lock receiver module from the left side sound insulator. Remove the left side sound insulator.

   j. Unfasten the screws that attach the instrument panel center sound insulator to the knee bolster, instrument panel, heater assembly and floor duct.

   k. Remove the center sound insulator.

   l. Unfasten the screws that attach the courtesy lamp to the knee bolster.

   m. Unfasten the screws that attach the knee bolster to the instrument panel.

   n. Disconnect the lap cooler duct from the knee bolster.

   o. Unplug the lighter electrical connection and remove the knee bolster.

   p. Unfasten the steering column-to-instrument panel nuts and lower the column.

   q. Unfasten the screws that attach the instrument panel accessory trim plate to the instrument panel.

   r. Remove the trim plate and unplug all necessary electrical connection.

   s. Remove the heater and/or air conditioning control assembly.

   t. Remove the radio and the storage compartment assembly (if equipped).

   u. If necessary, remove the instrument cluster.

   v. Unfasten the left and right instrument panel pivot bolts and the panel lower support bolt.

   w. Unfasten the speaker grilles retaining screws and remove the speaker grilles.

   x. Remove the windshield defroster grille using a flat-bladed prytool. Start at one end of the grille and work your way down the grille.

   y. Unfasten the 4 instrument panel upper support screws.

   z. Tag and unplug all necessary electrical connections.

   aa. Remove the instrument panel from the vehicle.

5. Remove the air inlet assembly, if equipped.

6. Remove the vacuum hoses.

7. From inside the engine compartment, remove the heater assembly studs.

8. Remove the blower motor resistor.

9. From inside the heater case assembly, remove the stud; the stud is located behind the blower motor resistor.

10. Remove the heater assembly-to-chassis screws.

11. Remove the heater assembly from the vehicle.

12. Remove the access cover screws and cover from the heater assembly.

13. Remove the heater core from the heater case assembly.

To install:

14. Install the heater core to the heater case assembly.

15. Install the access cover to the heater assembly and the cover screws.

16. Install the heater assembly to the vehicle.

17. Install the heater assembly-to-chassis screws and torque them to 40 inch lbs. (4.5 Nm).

18. Working inside the heater case assembly, install the stud; the stud is located behind the blower motor resistor.

19. Install the blower motor resistor.

20. Working inside the engine compartment, install the heater assembly studs and torque them to 17 inch lbs. (1.9 Nm).

21. Install the vacuum hoses.

22. Install the air inlet assembly, if equipped.

23. Install the instrument panel as follows:

   a. Rest the instrument panel on the lower pivot studs.

   b. Attach the electrical connections.

   c. Install but do not tighten the 4 upper instrument panel su to 13 inch lbs. (1.5 Nm).

   d. Install the left and right panel pivot bolts. Tighten the bolts to 102 inch lbs. (11.5 Nm).

   e. Install the panel lower support bolt. Tighten the bolt to 102 inch lbs. (11.5 Nm).

   f. Tighten the upper support screws to 17 inch lbs. (1.9 Nm).

   g. Install the windshield defroster grille and the speaker grilles.

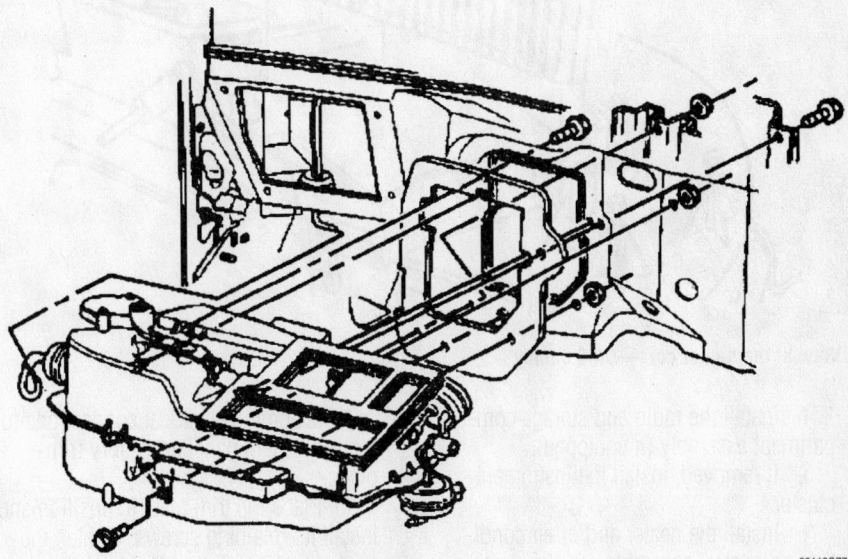

View of the heater case assembly—GMC Jimmy

93113G77

For Tune-up, Capacities and Firing orders, see Section 1 of this manual

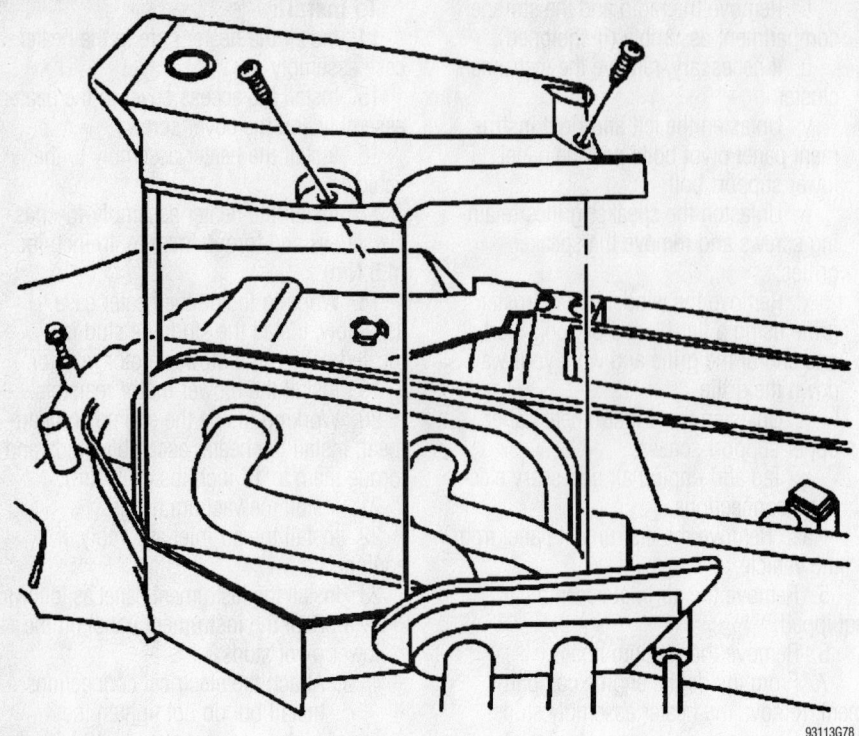

93113G78

View of the heater case cover—GMC Jimmy

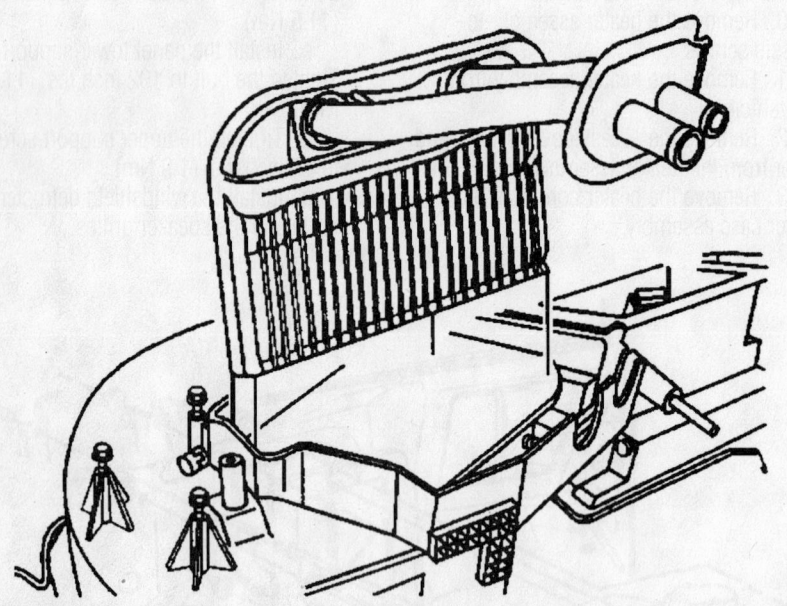

93113G79

View of the heater core—GMC Jimmy

h. Install the radio and storage compartment assembly (if equipped).

i. If removed, install the instrument cluster.

j. Install the heater and/or air conditioning control assembly.

k. attach the electrical connections to the instrument panel accessory trim plate.

l. Place the trim plate in position and install its retaining screws. Tighten the screws to 17 inch lbs. (1.9 Nm).

m. Place the steering column into position and install its retaining nuts. Tighten the nuts to 22 ft. lbs. (30 Nm).

n. Attach the lighter electrical connection and the lap cooler duct to the knee bolster.

o. Place the knee bolster into position and install its retaining screws. Tighten the Torx® head screws to 80 inch lbs. (9 Nm) and the hex head screws to 17 inch lbs. (1.9 Nm).

p. Place the courtesy lamp in position and install its screws. Tighten the screws to 17 inch lbs. (1.9 Nm).

q. Place the instrument panel center sound insulator in position. Install the screws that attach the center sound insulator to the knee bolster, instrument panel and the floor duct. Tighten the screws to 17 inch lbs. (1.9 Nm).

r. Install the screw that attaches the center sound insulator to the heater assembly. Tighten the screw to 13 inch lbs. (1.5 Nm).

s. Install the remote control door lock receiver module to the instrument panel left side sound insulator.

t. Attach the door lock receiver electrical connection.

u. Install the nut that attaches the left side sound insulator to the accelerator pedal bracket. Tighten the nut to 35 inch lbs. (4 Nm).

v. Install the screw that attaches the left side sound insulator to cowl panel. Tighten the screw to 13 inch lbs. (1.5 Nm).

w. Install the screws that attach the left side sound insulator to knee bolster. Tighten the screw to 17 inch lbs. (1.9 Nm).

x. Feed the DLC through the hole in the sound insulator, place the DLC in position and install its retaining screws. Tighten the screws to 21 inch lbs. (2.4 Nm).

y. Install the right side sound insulator and tighten the screws

z. Connect the parking brake release cable to the lever.

aa. Enable the air bag system.

24. Install the heater hoses to the heater core.

25. Refill the cooling system.

26. Connect the negative battery cable.

27. Run the engine to normal operating temperatures; then, check the climate control operation and check for leaks.

HONDA

CR-V

REMOVAL & INSTALLATION

1. Disconnect the negative battery cable.

2. Drain the cooling system into a clean container for reuse.

3. In the engine compartment, open the heater valve cable clamp and disconnect the cable from the heater valve arm. Then, turn the heater valve to the fully opened position.

4. Disconnect the heater hoses from the heater core.

5. Remove the heater housing-to-chassis nut.

➡**When removing the heater housing nut, be careful not to damage or bend the fuel lines, the brake lines, etc.**

6. Remove the instrument panel by performing the following procedure:

　a. Remove the driver's side lower instrument panel cover screws, disengage the clips and remove the lower cover.

　b. Remove the knee bolster bolts and the knee bolster.

　c. Remove the glove box stops from each side of the glove box.

　d. Remove the glove box-to-instrument panel bolts and the glove box.

　e. Remove the lower console cover by disengaging the 4 clips and removing the cover.

　f. Remove the 6 center pocket-to-instrument panel screws; then, insert a flat tipped screwdriver at the upper right side corner of the center pocket, push down on the top of the hook and remove the center pocket/beverage holder assembly.

　g. Remove the center instrument panel lower cover screws and disengage the clips on the upper left side; then, disconnect the electrical connectors and remove the cover.

　h. Gently, push the power window switch from the instrument panel's lower cover opening by hand. Disconnect the electrical connectors and remove the power window switch.

　i. Close the driver's side air vent; then, gently, push out the clips and pull out the vent. Disconnect the electrical connectors and remove the vent.

　j. Gently, push out the driver's side

defogger trim; then, disconnect the electrical connector and remove the side defogger trim.

　k. At the base of the steering wheel, remove the access panel and disconnect the air bag electrical connector.

　l. Remove the steering column covers screws and the covers.

　m. Remove the steering column-to-instrument panel nuts/bolts and lower the steering column.

　n. Remove the instrument panel side covers.

　o. Disconnect the wiring harness connector and remove the nuts.

　p. Move the under-dash fuse/relay box.

　q. Disconnect the antenna connector and the harness clips.

　r. Remove the connector holder from the instrument panel frame.

　s. Remove the control unit/relay bracket from behind the center of the instrument panel.

　t. Remove the passenger's side lower instrument panel cover.

　u. Disconnect the connectors and the harness clips.

　v. Remove the instrument panel-to-chassis bolts.

　w. Using an assistant, remove the instrument panel.

7. Remove the evaporator housing by performing the following procedure:

　a. Discharge and recover the air conditioning system refrigerant.

　b. In the engine compartment, remove the refrigerant lines-to-evaporator housing bolts.

　c. Separate the lines, discard the grommets and plug the openings to prevent contamination.

　d. Disconnect the evaporator housing's temperature sensor connector.

　e. Remove the evaporator housing-to-chassis screws/nut and remove the evaporator housing.

8. Disconnect the mode control motor and the air mix control motor electrical connectors and remove the wiring harness clips and the wiring harness from the heater housing.

9. Remove the heater duct clip, the heater housing-to-chassis nuts and the heater housing.

10. Remove the heater core cover screws and the cover.

11. Remove the heater core pipe clamp screws and the clamp.

12. Remove the heater core from the heater housing.

To install:

13. Install the heater core in the heater housing.

14. Install the heater core pipe clamp and the clamp screws.

15. Install the heater core cover and the cover screws.

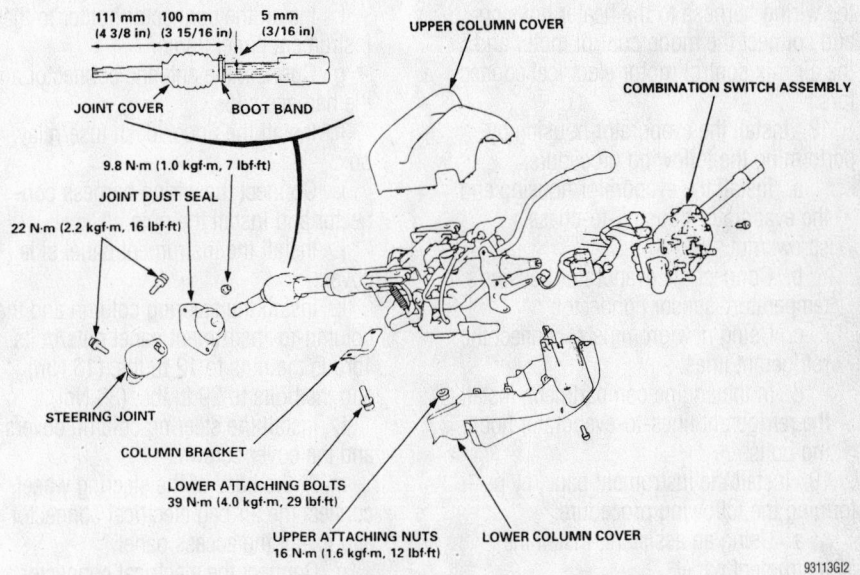

111 mm　100 mm　5 mm
(4 3/8 in)　(3 15/16 in)　(3/16 in)

UPPER COLUMN COVER

JOINT COVER　　BOOT BAND

COMBINATION SWITCH ASSEMBLY

9.8 N·m (1.0 kgf·m, 7 lbf·ft)

JOINT DUST SEAL

22 N·m (2.2 kgf·m, 16 lbf·ft)

STEERING JOINT

COLUMN BRACKET

LOWER ATTACHING BOLTS
39 N·m (4.0 kgf·m, 29 lbf·ft)

UPPER ATTACHING NUTS
16 N·m (1.6 kgf·m, 12 lbf·ft)

LOWER COLUMN COVER

93113GI2

Exploded view of the steering column and related components—Honda CR-V

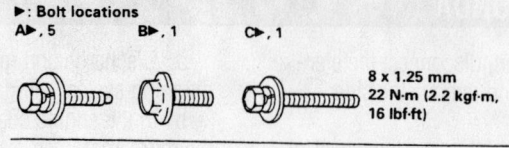

► : Bolt locations
A► , 5   B► , 1   C► , 1

8 x 1.25 mm
22 N·m (2.2 kgf·m,
16 lbf·ft)

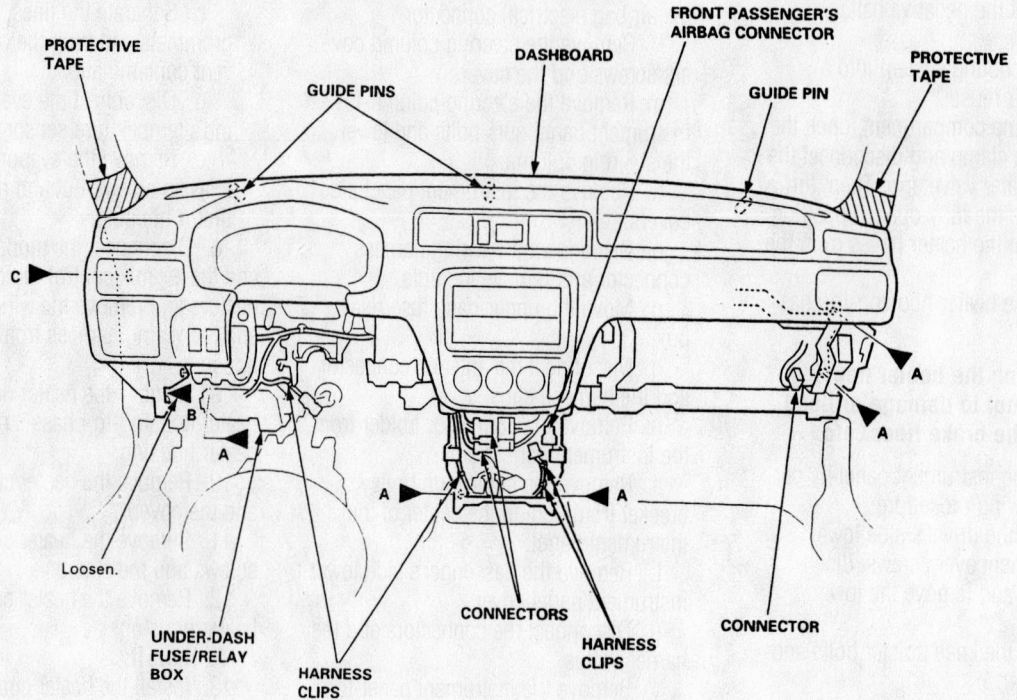

PROTECTIVE TAPE

GUIDE PINS

DASHBOARD

FRONT PASSENGER'S AIRBAG CONNECTOR

GUIDE PIN

PROTECTIVE TAPE

C ►

A

B

A

Loosen.

A   A

UNDER-DASH FUSE/RELAY BOX

HARNESS CLIPS

CONNECTORS

HARNESS CLIPS

CONNECTOR

93113GI3

Exploded view of the instrument panel and related components—Honda CR-V

16. Install the heater housing, the heater housing-to-chassis nuts and the heater duct clip.

17. Install the wiring harness clips and the wiring harness to the heater housing and connect the mode control motor and the air mix control motor electrical connectors.

18. Install the evaporator housing by performing the following procedure:

a. Install the evaporator housing and the evaporator housing-to-chassis screws/nut.

b. Connect the evaporator housing's temperature sensor connector.

c. Using new grommets, connect the refrigerant lines.

d. In the engine compartment, install the refrigerant lines-to-evaporator housing bolts.

19. Install the instrument panel by performing the following procedure:

a. Using an assistant, install the instrument panel.

b. Install the instrument panel-to-chassis bolts.

c. Connect the connectors and the harness clips.

d. Install the passenger's side lower instrument panel cover.

e. Install the control unit/relay bracket to the center of the instrument panel.

f. Install the connector holder to the instrument panel frame.

g. Connect the antenna connector and the harness clips.

h. Install the under-dash fuse/relay box.

i. Connect the wiring harness connector and install the nuts.

j. Install the instrument panel side covers.

k. Install the steering column and the column-to-instrument panel nuts/bolts. Torque the nuts to 12 ft. lbs. (16 Nm) and the bolts to 29 ft. lbs. (39 Nm).

l. Install the steering column covers and the cover screws.

m. At the base of the steering wheel, connect the air bag electrical connector and install the access panel.

n. Connect the electrical connector and install the driver's side defogger trim.

o. Connect the electrical connectors and install driver's side air vent.

p. Connect the electrical connectors

and install the power window switch to the instrument panel's lower cover opening.

q. Install the center instrument panel lower cover and engage the clips on the upper left side; then, connect the electrical connectors and install the cover screws.

r. Install the center pocket/beverage holder assembly and the 6 center pocket-to-instrument panel screws.

s. Install the lower console cover by engaging the 4 clips.

t. Install the glove box and the glove box-to-instrument panel bolts.

u. Install the glove box stops to each side of the glove box.

v. Install the knee bolster and the knee bolster bolts.

w. Install the driver's side lower instrument panel cover, engage the clips and install the lower cover screws.

➡ **When installing the heater housing nut, be careful not to damage or bend the fuel lines, the brake lines or etc.**

20. Install the heater housing-to-chassis nut.

21. Connect the heater hoses to the heater core.

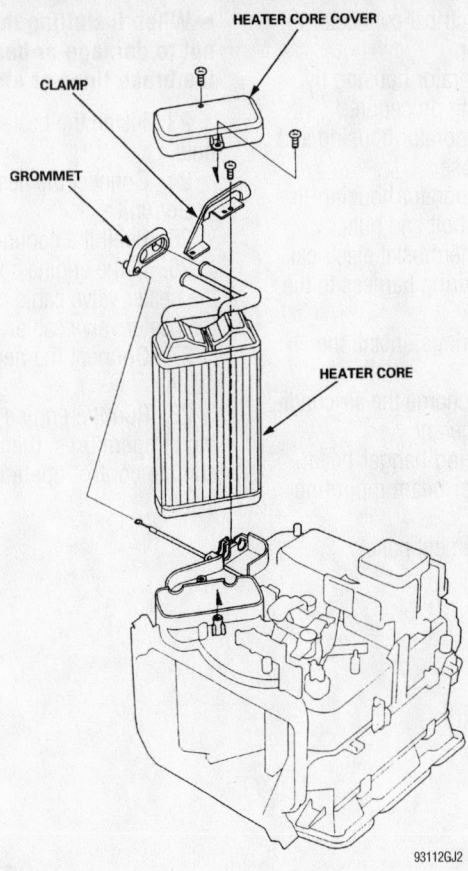

HEATER CORE COVER

CLAMP

GROMMET

HEATER CORE

93112GJ2

Exploded view of the heater core and housing—Honda CR-V

22. In the engine compartment, connect the cable to the heater valve arm and close the heater valve cable clamp.

23. Refill the cooling system.

24. Connect the negative battery cable.

25. Evacuate and charge and leak test the air conditioning system refrigerant.

26. Run the engine to normal operating temperatures; then, check the climate control operation and check for leaks.

Odyssey

REMOVAL & INSTALLATION

➥Make sure to acquire the anti-theft code for the radio and write down the frequencies for the radio's preset buttons.

1. Disconnect the negative battery cable.

✳✳ CAUTION

Wait at least 3 minutes for the air bag to deplete its energy before

working on the steering wheel or instrument panel.

2. In the engine compartment, remove the heater valve cable clamp; then, disconnect the heater valve cable and rotate the heater valve to the fully open position.

3. Drain the engine coolant into a clean container for reuse.

4. Disconnect the heater hoses from the heater unit.

5. Remove the heater housing-to-chassis nuts.

6. Remove the center console.

7. Remove the instrument panel.

8. Remove the steering hanger beam mounting bolts and the steering hanger beam.

9. Remove the evaporator housing by performing the following procedure:

   a. Discharge and recover the air conditioning system refrigerant.

   b. Remove the refrigerant lines. Discard the O-rings. Plug the openings to prevent contamination.

   c. Disconnect the thermostat electrical connector and the wiring harness from the evaporator.

   d. Remove the evaporator housing-to-chassis screws, the bolt and nuts.

   e. Disconnect the drain hose and remove the evaporator housing.

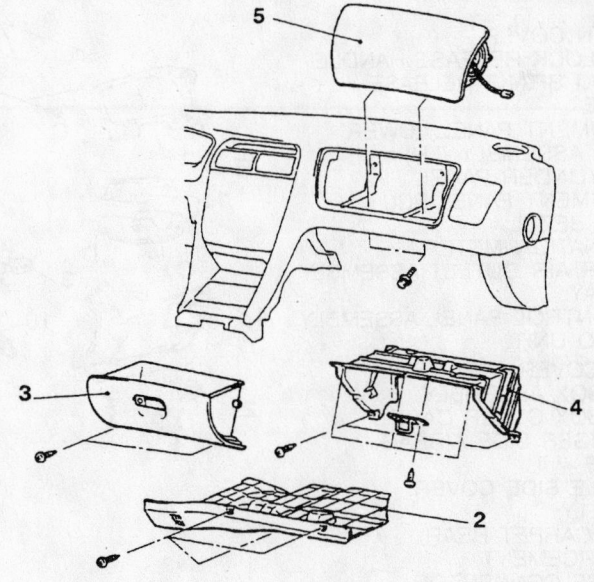

2. UNDERCOVER
3. GLOVE BOX ASSEMBLY
4. GLOVE BOX CASE
5. AIR BAG MODULE

93112GG1

View of the heater housing, evaporator housing and related components—Honda Odyssey

10. Disconnect the electrical connector from the mode control motor.

11. Remove the wiring harness clips from the heater housing.

12. Remove the heater housing-to-chassis nuts and the heater housing.

13. Remove the heater housing screws and separate the housings.

14. Remove the heater core from the heater housing.

To install:

15. Install the heater core to the heater housing.

16. Assemble the housings and install the heater housing screws.

17. Install the heater housing and the heater housing-to-chassis nuts.

18. Install the wiring harness clips to the heater housing.

19. Connect the electrical connector to the mode control motor.

20. Install the evaporator housing by performing the following procedure:

a. Install the evaporator housing and connect the drain hose.

b. Install the evaporator housing-to-chassis screws, the bolt and nuts.

c. Connect the thermostat electrical connector and the wiring harness to the evaporator.

d. Using new O-rings, install the refrigerant lines.

e. Evacuate and charge the air conditioning system refrigerant.

21. Install the steering hanger beam and the steering hanger beam mounting bolts.

22. Install the instrument panel.

➡**When installing the nuts, be careful not to damage or bend the fuel lines, the brake lines or etc.**

23. Install the heater housing-to-chassis nuts.

24. Connect the heater hoses to the heater unit.

25. Refill the cooling system.

26. In the engine compartment, Install the heater valve cable clamp; then, connect the heater valve cable.

27. Connect the negative battery cable.

28. Run the engine to normal operating temperatures; then, check the climate control operation and check for leaks.

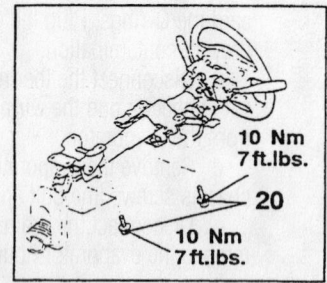

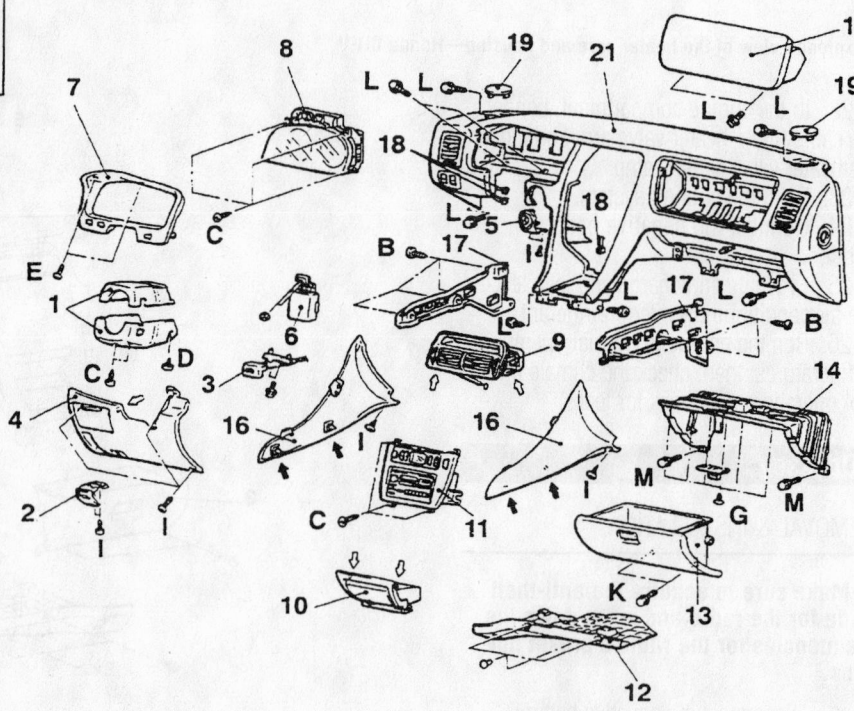

1. COLUMN COVER
2. HOOD LOCK RELEASE HANDLE
3. PARKING BRAKE RELEASE HANDLE
4. INSTRUMENT PANEL LOWER COVER ASSEMBLY (LH)
5. KEY CYLINDER PANEL
6. INSTRUMENT PANEL ECU
7. METER BEZEL
8. COMBINATION METER
9. CENTER AIR OUTLET ASSEMBLY
10. ASHTRAY
11. AIR CONTROL PANEL ASSEMBLY & AUDIO UNIT
12. UNDERCOVER ASSEMBLY
13. GLOVEBOX ASSEMBLY
14. GLOVEBOX OUTER CASE
15. PASSENGER SIDE AIRBAG MODULE
16. CONSOLE SIDE COVER ASSEMBLY
17. FLOOR CARPET REAR REINFORCEMENT
18. HARNESS CONNECTOR
19. PLUG
20. STEERING COLUMN MOUNTIN BOLT
21. INSTRUMENT PANEL

NOTE
(1) ⇦ : metal clip position
(2) ◀ : plastic clip position

93112GG2

View of the steering hanger beam and related components—Honda Odyssey

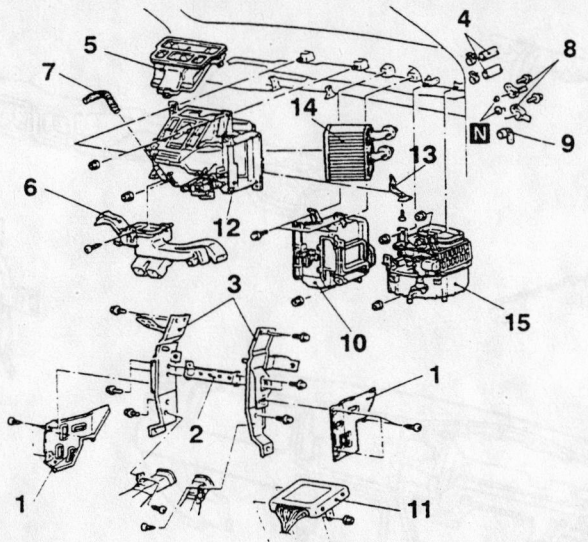

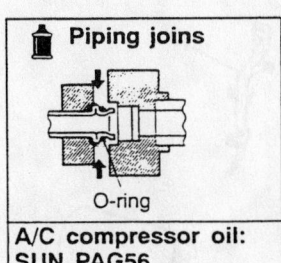

Piping joins

O-ring

A/C compressor oil:
SUN PAG56

1. FLOOR CARPET FRONT REINFORCEMENT
2. ECU BRACKET
3. CENTER STAY ASSEMBLY
4. HEATER HOSE CONNECTION
5. CENTER DUCT ASSEMBLY
6. FOOT DISTRIBUTION DUCT
7. BREATHER HOSE
8. SUCTION PIPE, LIQUID PIPE B AND COOLING UNIT CONNECTION
10. DRAIN HOSE
11. EVAPORATOR
12. ENGINE CONTROL MODULE
13. HEATER UNIT
14. HEATER CORE SUPPORT
15. HEATER CORE

93112GG3

Exploded view of the heater core, the heater housing and related components—Honda Odyssey

Passport

REMOVAL & INSTALLATION

1. If equipped with an air bag, perform the following procedure:

a. Turn the ignition to the LOCK position and remove the key.

b. From the lower left dash side fuse block, remove the SRS-1 fuse.

c. Disconnect the 2-pin yellow connector located at the base of the steering column.

d. Remove the glove box assembly.

e. Disconnect the 2-pin yellow connector located behind the glove box.

2. Disconnect the negative battery cable.

3. If equipped, discharge and recover the air conditioning system refrigerant.

4. Remove the evaporator lines at the firewall. Plug the air conditioning lines to minimize contamination.

5. Disconnect the cooling system hoses and drain the coolant into a clean container for reuse. Plug the cooling system hoses.

6. Remove the instrument panel by performing the following procedure:

a. Remove the lower center cover screw and pull it out at the clip positions; then, disconnect the cigarette lighter connector.

b. Remove both the rear and front console.

c. Remove the dash side trim panel sill plates and the panels.

d. Remove the 2 glove box screws and the glove box.

e. Remove the 2 hood release screws, the 6 instrument panel driver's lower cover assembly screws and the cover assembly.

f. Remove the 5 instrument cluster screws and the 2 clips. Disconnect the 8 switch connectors and remove the instrument cluster assembly.

g. Remove the 6 driver's knee bolster assembly bolts and screws and the knee bolster assembly.

h. Remove the 4 control lever assembly bolts; then, disconnect the 3 control cables (unit side) and the 3 harness connectors.

i. Remove the 4 radio/audio sub box assembly screws and the radio/audio sub box assembly.

j. Disconnect or remove the following instrument panel harness connectors or items:

- The 6 driver's side connectors
- The 3 passenger's side connectors
- The 2 center connectors
- Passenger's inflator module connector
- Radio antenna cable plug
- Ground cable bolt on the left dash side panel
- The 8 instrument panel-to-chassis bolts and the 3 nuts.

k. Remove the instrument panel assembly.

7. Remove the instrument panel bracket by performing the following procedure:

a. Remove the 2 passenger's inflator module bolts and 4 nuts.

b. Remove the 4 meter assembly screws. Then, disconnect the meter wiring harness connectors and remove the meter assembly.

c. Remove the 5 vent duct assembly screws and the assembly.

d. Remove the 3 lower passenger's bracket screws and the bracket.

e. Remove the 9 passenger's knee bolster reinforcement screws and the reinforcement.

f. Remove the 6 instrument panel center reinforcement screws and the reinforcement.

g. Remove the instrument panel wiring harness assembly clips and the wiring harness.

h. Remove the 2 instrument panel bracket nuts and 2 bolts for each bracket; then, remove the bracket(s).

8. Remove the 5 cross beam assembly nuts, 2 bolts and the 6 lower bolts; then, remove the crossbeam.

9. Disconnect the resistor wiring connector.

10. Remove the duct from the heater assembly.

11. If equipped with air conditioning, remove the evaporator assembly.

12. Remove the driver's lap vent.

13. Remove the lower ventilation duct.

14. Remove the footrest, the carpet, the 3 clips and the rear heater duct.

15. Remove the heater assembly.

16. Remove the mode control case-to-

Timing belt service is covered in Section 3 of this manual

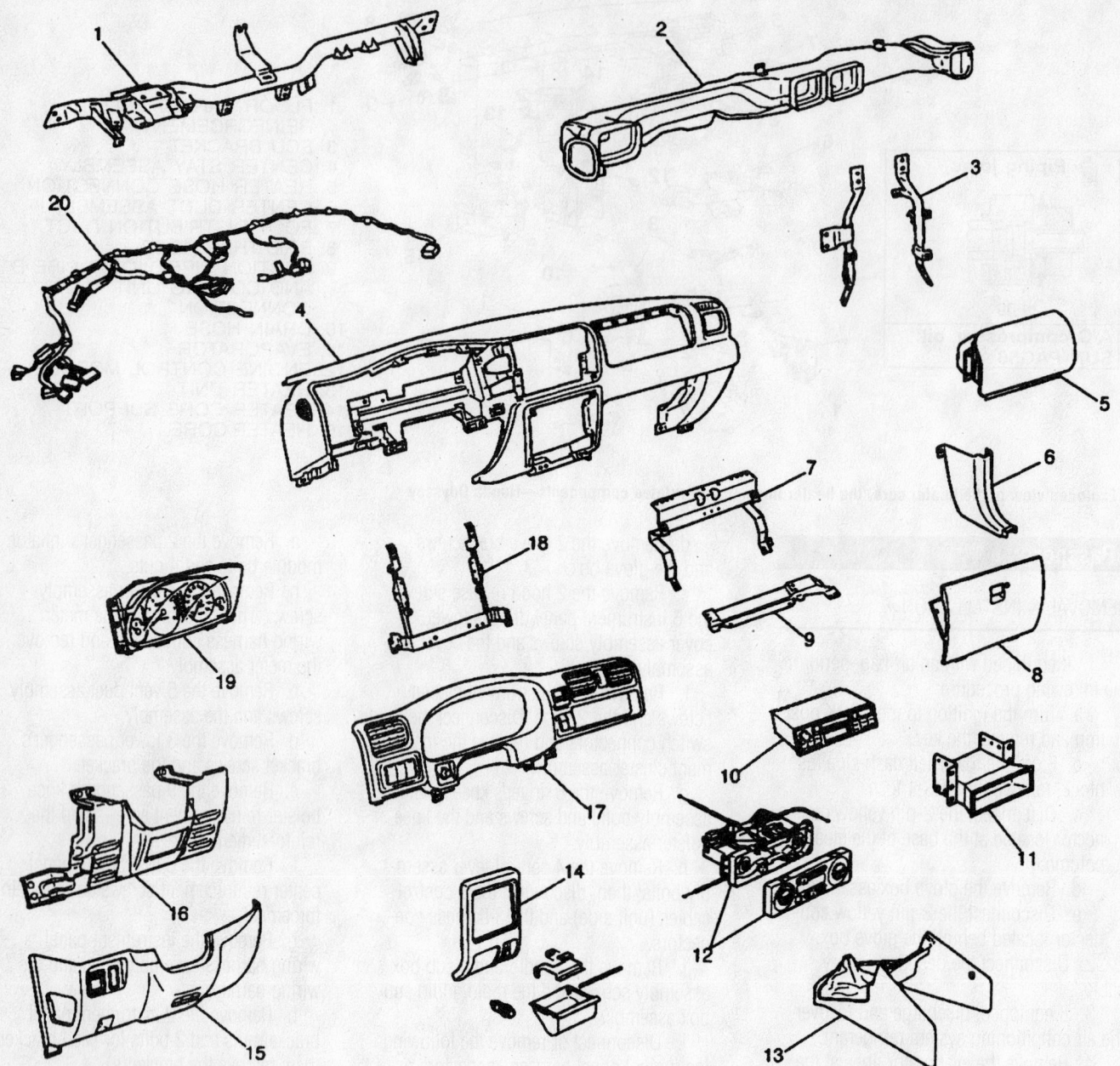

1 Cross Beam
2 Vent Duct Assembly
3 Instrument Panel Bracket
4 Instrument Panel Assembly
5 Passenger Inflator Module
6 Dash Side Trim Panel
7 Passenger Knee Bolster Reinforcement Assembly
8 Glove Box
9 Passenger Lower Bracket
10 Radio Assembly

11 Audio Sub Box
12 Control Lever Assembly
13 Front Console Assembly
14 Lower Center Cover
15 Instrument Panel Driver Lower Cover Assembly
16 Driver Knee Bolster Assembly
17 Meter Cluster Assembly
18 Instrument Panel Center Reinforcement
19 Meter Assembly
20 Instrument Harness Assembly

93113GB8

Exploded view of the instrument panel—Honda Passport

temperature control case screws and remove the mode control case; do not remove the link unit.

17. Remove the temperature control case screws and separate the cases.

18. Remove the heater core from the case.

To install:

19. Install the heater core to the case.

20. Assemble the temperature control cases and install the case screws.

21. Install the mode control case and the mode control case-to-temperature control case screws.

22. Install the heater assembly.

23. Install the rear heater duct, the footrest, the carpet, and the 3 clips.

24. Install the lower ventilation duct.

25. Install the driver's lap vent.

26. If equipped with air conditioning, install the evaporator assembly.

27. Install the duct to the heater assembly.

28. Connect the resistor wiring connector.

29. Install the crossbeam, the 5 cross beam assembly nuts, 2 bolts and the 6 lower bolts.

30. Install the instrument panel bracket by performing the following procedure:

a. Install the instrument panel bracket and the 2 nuts and 2 bolts for each bracket.

b. Install the instrument panel wiring harness assembly and the wiring harness clips.

c. Install the instrument panel center

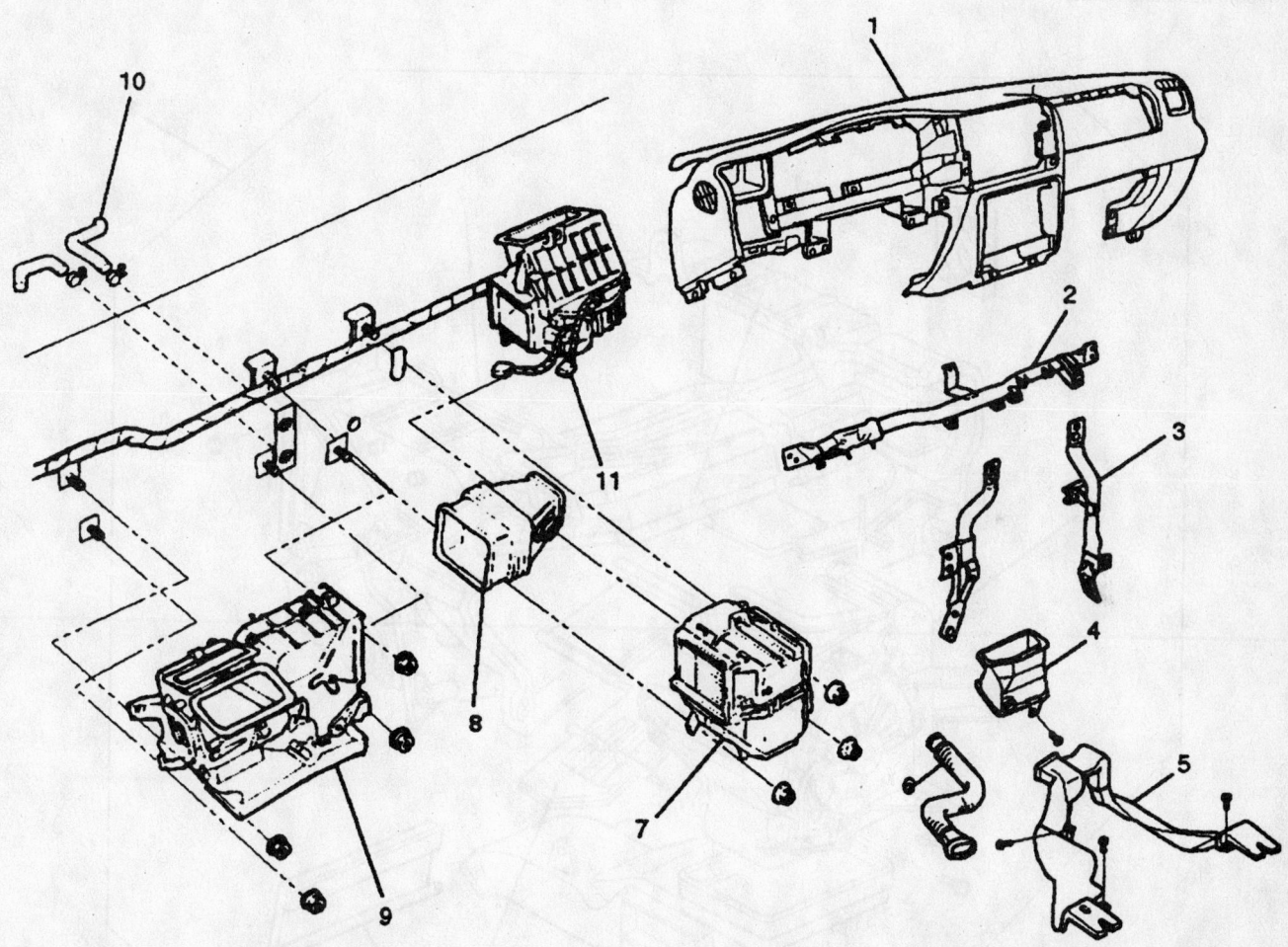

1	Instrument Panel Assembly	6	Driver Lap Vent Duct
2	Cross Beam Assembly	7	Evaporator Assembly (A/C only)
3	Instrument Panel Bracket	8	Duct
4	Ventilation Lower Duct	9	Heater Unit Assembly
5	Rear Heater Duct	10	Heater Hose
		11	Resistor Connector

93113GB9

View of the heater and air conditioning housing assemblies and related components—Honda Passport

Heater Core replacement is covered in Section 2 of this manual

reinforcement and the 6 reinforcement screws.

d. Install the passenger knee bolster reinforcement and the 9 reinforcement screws.

e. Install the lower passenger bracket and the 3 bracket screws.

f. Install the vent duct assembly and the 5 vent duct assembly screws.

g. Install the meter assembly and the 4 meter assembly screws; then, connect the meter wiring harness connectors.

h. Install the 2 passenger's inflator module bolts and 4 nuts.

31. Install the instrument panel by performing the following procedure:

a. Install the instrument panel assembly.

b. Connect or install the following instrument panel harness connectors or items:

- The 6 driver's side connectors
- The 3 passenger's side connectors
- The 2 center connectors
- Passenger's inflator module connector
- Radio antenna cable plug

- Ground cable bolt on the left dash side panel
- The 8 instrument panel-to-chassis bolts and the 3 nuts.

c. Install the radio/audio sub box assembly and the 4 radio/audio sub box assembly screws.

d. Connect the 3 control cables (unit side) and the 3 harness connectors. Install the 4 control lever assembly bolts.

e. Install the knee bolster assembly and the 6 driver's knee bolster assembly bolts and screws.

f. Install the instrument cluster

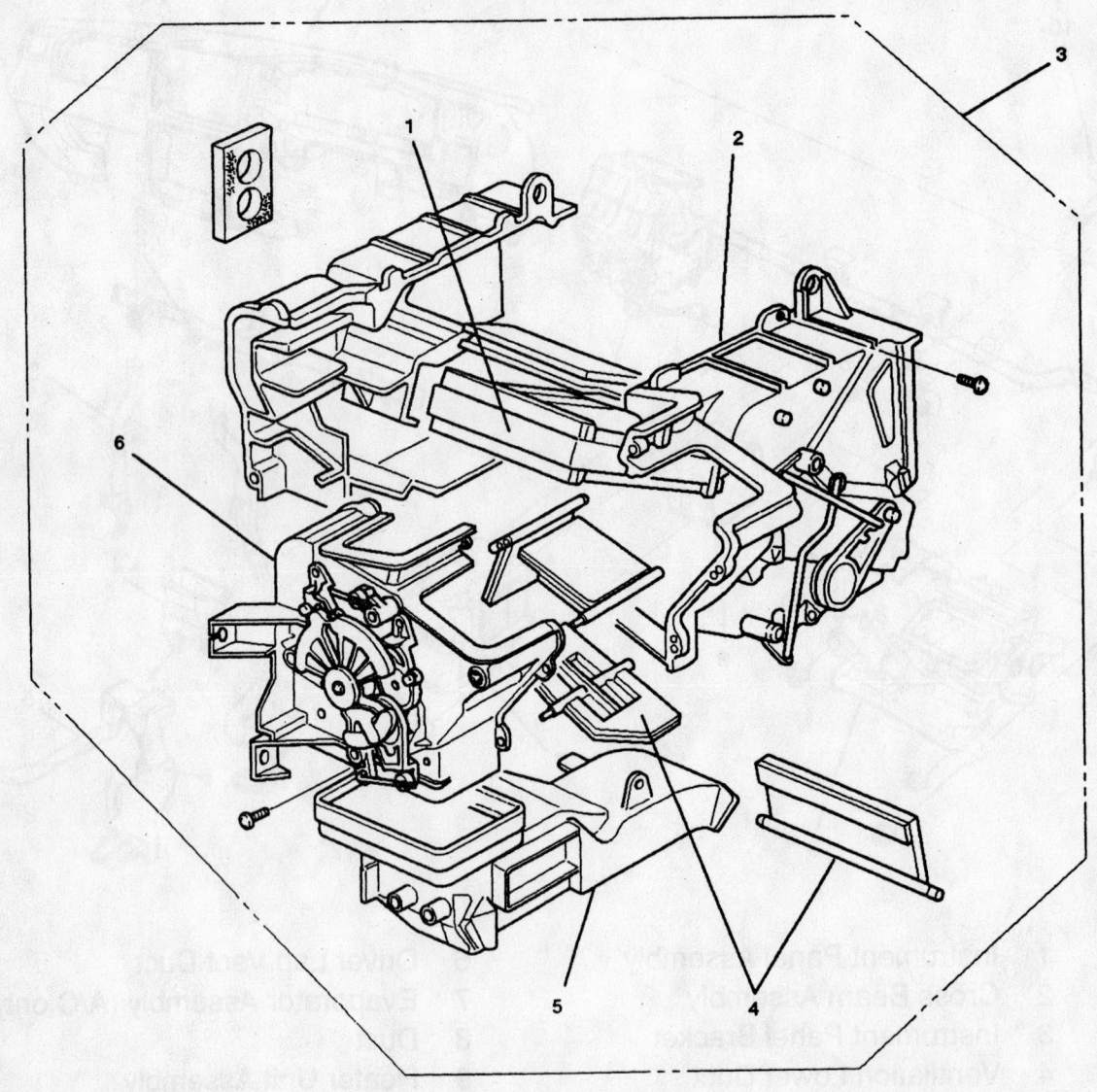

1	Heater Core	4	Mode Door
2	Case (Temperature Control)	5	Duct
3	Heater Unit	6	Case (Mode Control)

93113GB0

Exploded view of the heater housing assembly—Honda Passport

assembly. Connect the 8 switch connectors. Install 5 instrument cluster screws and the 2 clips.

g. Install the 2 hood release screws, the 6 instrument panel driver's lower cover assembly screws and the cover assembly.

h. Install the glove box and the 2 glove box screws.

i. Install the dash side trim panel sill plates and the panels.

j. Install both the rear and front console.

k. Connect the cigarette lighter connector and install the lower center cover screw.

32. Connect the cooling system hoses.

33. Refill the cooling system.

34. Install the evaporator lines at the firewall.

35. If equipped, evacuate and charge the air conditioning system.

36. Connect the negative battery cable.

37. If equipped with an air bag, perform the following procedure:

a. Turn the ignition to the LOCK position and remove the key.

b. Connect the 2-pin yellow connector located behind the glove box.

c. Install the glove box assembly.

d. Connect the 2-pin yellow connector located at the base of the steering column.

e. At the lower left dash side fuse block, install the SRS-1 fuse.

f. Turn the ignition switch to ON and verify that the AIR BAG warning light flashes 7 times and turns OFF.

38. Run the engine to normal operating temperatures; then, check the climate control operation and check for leaks.

INFINITI

QX4

REMOVAL & INSTALLATION

1. Disconnect the negative battery cable.

⁕⁕ CAUTION

After disconnecting the negative battery cable, wait for at least 3 minutes before working on the steering column or instrument panel.

2. Drain the cooling system into a clean container for reuse.

3. Disconnect the heater hoses from the heater core.

4. Remove the driver's side air bag and steering wheel by performing the following procedure:

a. Place the front wheels in the straight-ahead position.

b. Remove the lower lid from the steering wheel and disconnect the air bag module connector.

c. Remove the side lids from both sides of the steering wheel.

d. Using the Tamper Resistant Torx® tool T50, remove the left and right Torx® bolts.

e. Carefully, remove the air bag module.

⁕⁕ CAUTION

Place the air bag module in safe place with the front facing upward.

f. Remove the steering wheel nut.

g. Using a steering wheel puller, press the steering wheel from the steering column.

5. Remove the passenger's side air bag by performing the following procedure:

a. Remove the glove box clips and disconnect the passenger's side air bag module connector.

b. Remove the lower panel screws; then, disconnect the harness connector and remove the air bag module bracket.

c. Using the Tamper Resistant Torx® tool T50, remove the passenger's side air bag module bolts.

d. Carefully, remove the air bag module.

⁕⁕ CAUTION

Place the air bag module in safe place with the front facing upward.

6. Remove the instrument panel by performing the following procedure:

a. Remove the steering column cover and the combination switch.

b. Remove the instrument panel side lower finisher.

c. At the driver's side, remove the lower panel screws, disconnect the electrical harness connectors and remove the panel.

d. Remove the cluster lid **"A"** screws and the cluster lid **"A"**.

e. Remove the combination meter screws, disconnect the electrical harness connectors and remove the combination meter.

f. Remove the cluster lid **"C"** screws, disconnect the electrical harness connectors and remove the cluster lid **"C"**.

g. Remove the audio assembly screws and the audio assembly.

h. Remove the air conditioning control unit screws, disconnect the electrical harness connectors and the air conditioning control unit.

i. Remove the ashtray.

j. Remove the shifter (automatic transmission) or shift lever boot (manual transmission); then, remove the screw and disconnect the harness connector.

k. Remove the console box; then, remove the screw and disconnect the harness connector.

l. Remove the lower instrument center panel screws and the lower instrument center panel.

m. Remove the defroster grille.

n. At both sides, remove the pillar garnishes.

o. Remove the instrument panel and pads nuts and bolts.

p. Using an assistant, remove the instrument panel.

7. Remove the defroster nozzle and the heater nozzle from the heater housing.

8. Disconnect the electrical connector and/or control cable from the heater housing.

9. Remove the heater housing-to-chassis fasteners and remove the heater housing.

10. Separate the heater core from the heater housing and remove the heater core.

To install:

11. Install the heater core and assemble the heater housing.

12. Install the heater housing and the heater housing-to-chassis fasteners.

13. Connect the electrical connector and/or control cable to the heater housing.

14. Install the defroster nozzle and the heater nozzle to the heater housing.

15. Install the passenger's side air bag by performing the following procedure:

a. Carefully, install the air bag module.

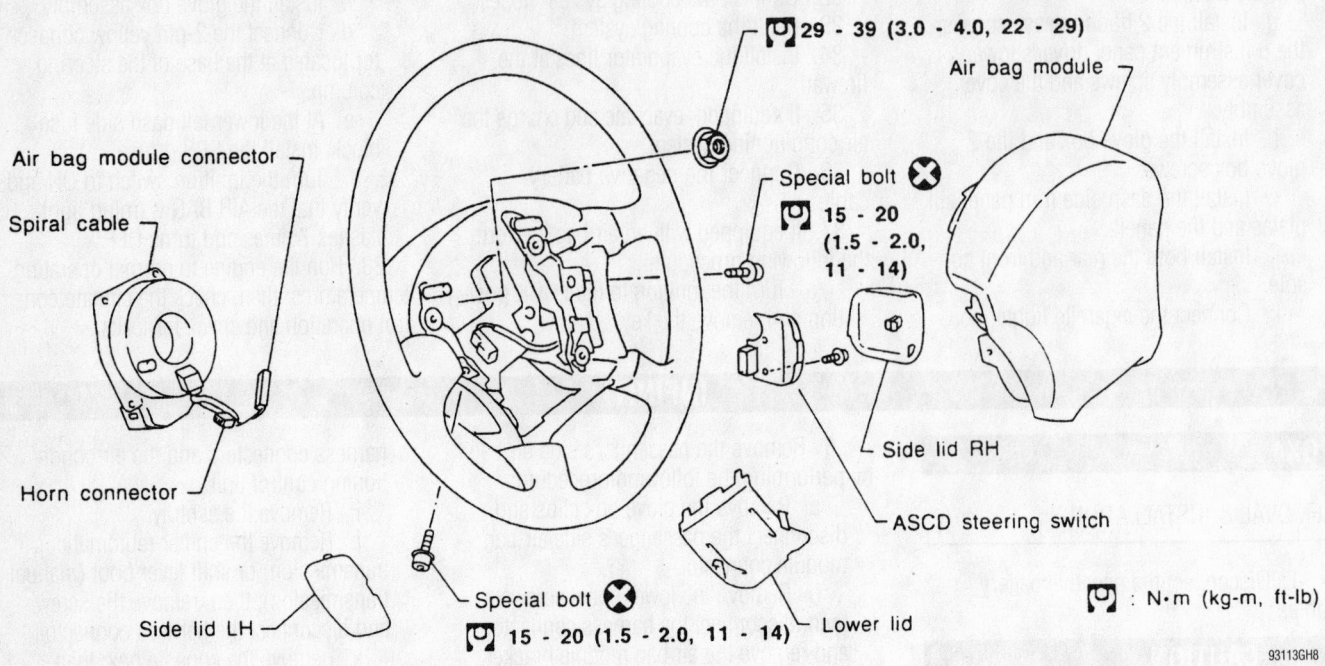

29 - 39 (3.0 - 4.0, 22 - 29)

Air bag module

Air bag module connector

Spiral cable

Special bolt ⊗

15 - 20 (1.5 - 2.0, 11 - 14)

Horn connector

Side lid RH

ASCD steering switch

: N•m (kg-m, ft-lb)

Side lid LH

Special bolt ⊗

15 - 20 (1.5 - 2.0, 11 - 14)

Lower lid

93113GH8

Exploded view of the driver's side air bag module and steering wheel—Infiniti QX4

b. Using the Tamper Resistant Torx® tool T50, install the passenger's side air bag module bolts. Torque the bolts to 11–18 ft. lbs. (15–25 Nm).

c. Connect the harness connector and install the air bag module bracket; then, install the lower panel screws.

d. Connect the passenger's side air bag module connector and install the glove box clips.

16. Install the instrument panel by performing the following procedure:

a. Using an assistant, install the instrument panel.

b. Install the instrument panel and pads nuts and bolts.

c. At both sides, install the pillar garnishes.

d. Install the defroster grille.

e. Install the lower instrument center panel and the lower instrument center panel screws.

f. Install the console box; then, install the screw and connect the harness connector.

g. Connect the harness connector and install the screw; then, install the shifter (automatic transmission) or shift lever boot (manual transmission).

h. Install the ashtray.

i. Install the air conditioning control

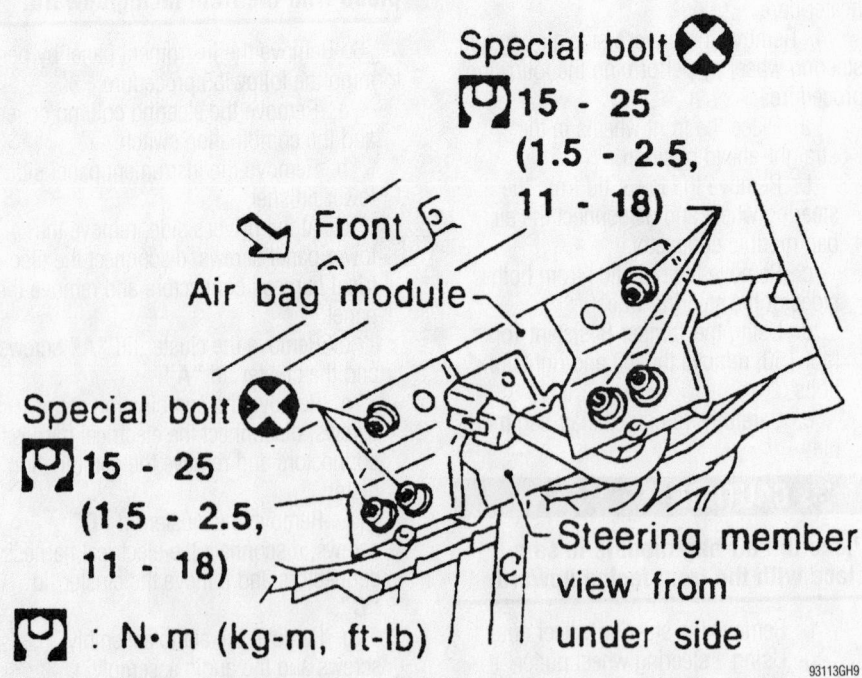

Special bolt ⊗

15 - 25 (1.5 - 2.5, 11 - 18)

Front

Air bag module

Special bolt ⊗

15 - 25 (1.5 - 2.5, 11 - 18)

Steering member view from under side

: N•m (kg-m, ft-lb)

93113GH9

Exploded view of the passenger's side air bag module—Infiniti QX4

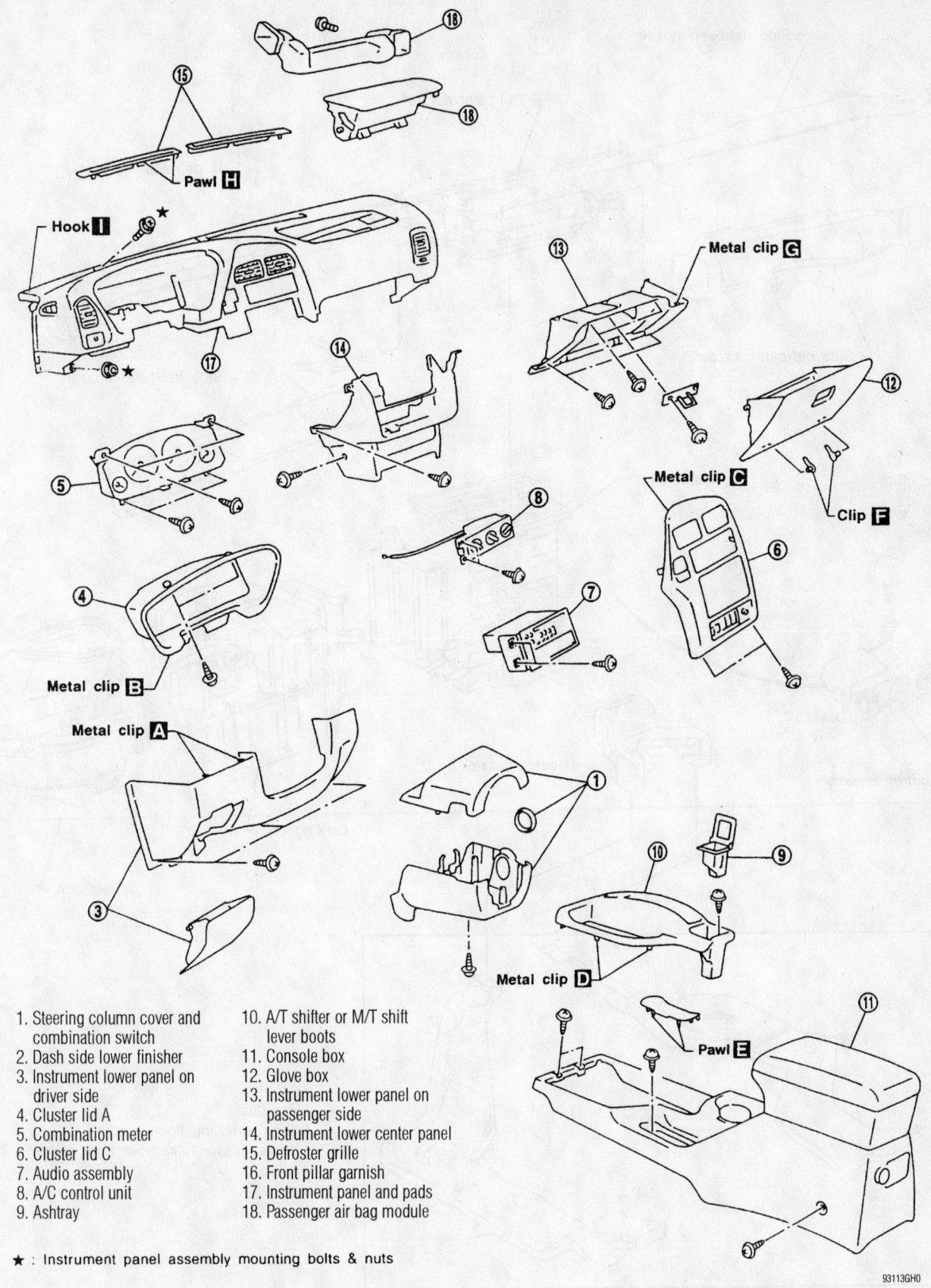

1. Steering column cover and combination switch
2. Dash side lower finisher
3. Instrument lower panel on driver side
4. Cluster lid A
5. Combination meter
6. Cluster lid C
7. Audio assembly
8. A/C control unit
9. Ashtray
10. A/T shifter or M/T shift lever boots
11. Console box
12. Glove box
13. Instrument lower panel on passenger side
14. Instrument lower center panel
15. Defroster grille
16. Front pillar garnish
17. Instrument panel and pads
18. Passenger air bag module

★ : Instrument panel assembly mounting bolts & nuts

93113GH0

Exploded view of the instrument panel and related accessories—Infiniti QX4

For complete Engine Mechanical specifications, see Section 1 of this manual

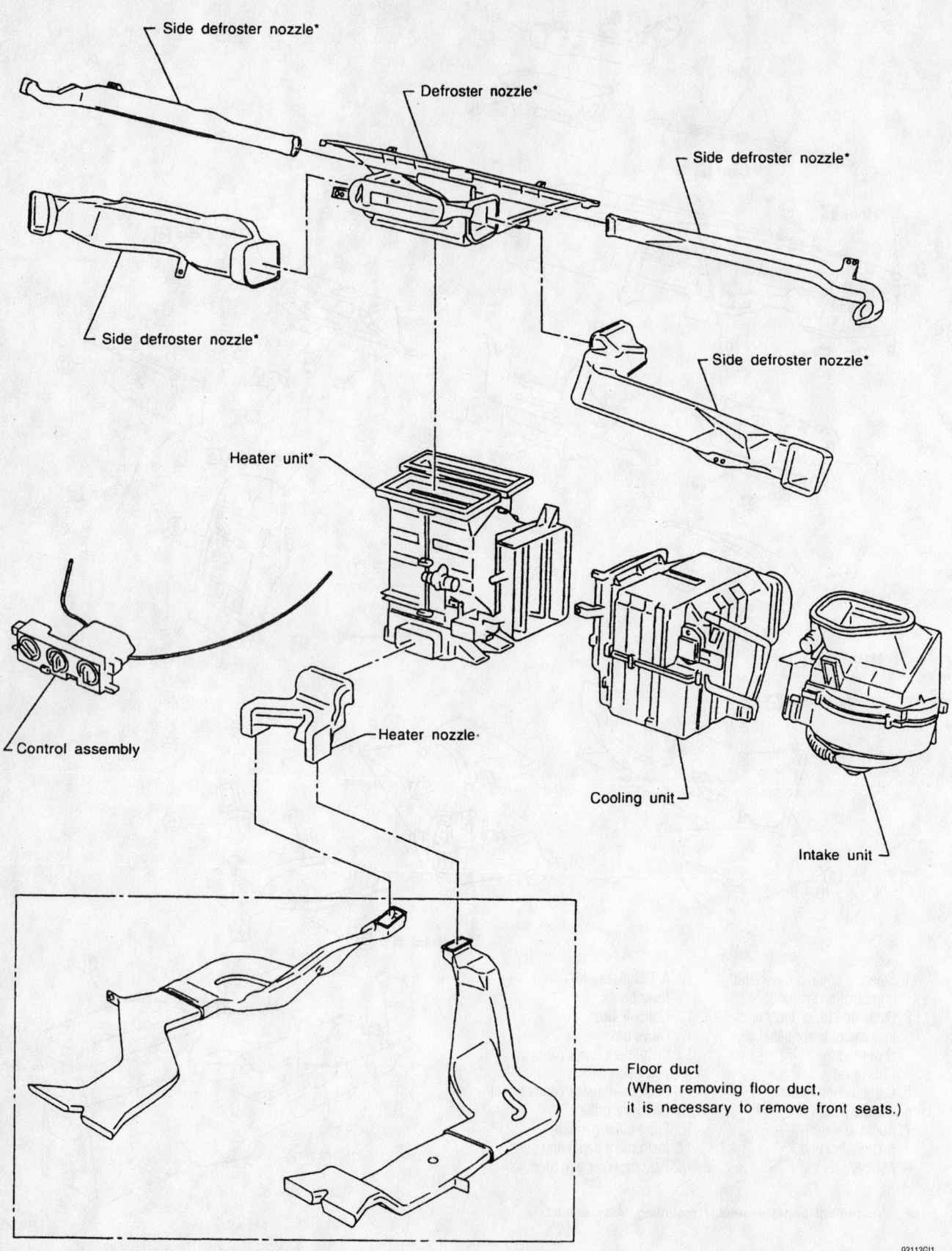

- Side defroster nozzle*
- Defroster nozzle*
- Side defroster nozzle*
- Side defroster nozzle*
- Side defroster nozzle*
- Heater unit*
- Control assembly
- Heater nozzle*
- Cooling unit
- Intake unit
- Floor duct
(When removing floor duct,
it is necessary to remove front seats.)

93113GI1

Exploded view of the heater housing, the evaporator housing, the ventilation dusts and related accessories—Infiniti QX4

unit, connect the electrical harness connectors and the air conditioning control unit screws.

j. Install the audio assembly and the audio assembly screws.

k. Install the cluster lid **"C"**, connect the electrical harness connectors and install the cluster lid **"C"** screws.

l. Install the combination meter, connect the electrical harness connectors and install the combination meter screws.

m. Install the cluster lid **"A"** and the cluster lid **"A"** screws.

n. At the driver's side, install the lower panel, connect the electrical har-

ness connectors and install the panel screws.

o. Install the instrument panel side lower finisher.

p. Install the combination switch and the steering column cover.

17. Install the driver's side air bag and steering wheel by performing the following procedure:

a. Install the steering wheel to the steering column.

b. Install the steering wheel nut. Torque the nut to 22–29 ft. lbs. (29–39 Nm).

c. Carefully, install the air bag module.

d. Using the Tamper Resistant Torx® tool T50, install the left and right Torx® bolts. Torque the bolts to 11–14 ft. lbs. (15–20 Nm).

e. Install the side lids to both sides of the steering wheel.

f. Connect the air bag module connector and install the lower lid to the steering wheel.

18. Connect the heater hoses to the heater core.

19. Refill the cooling system.

20. Connect the negative battery cable.

21. Run the engine to normal operating temperatures; then, check the climate control operation and check for leaks.

ISUZU

Amigo

REMOVAL & INSTALLATION

1. If equipped with an air bag, perform the following procedure:

a. Turn the ignition to the LOCK position and remove the key.

b. From the lower left dash side fuse block, remove the SRS-1 fuse.

c. Disconnect the 2-pin yellow connector located at the base of the steering column.

d. Remove the glove box assembly.

e. Disconnect the 2-pin yellow connector located behind the glove box.

2. Disconnect the negative battery cable.

3. If equipped, discharge and recover the air conditioning system refrigerant.

4. Remove the evaporator lines at the firewall. Plug the air conditioning lines to minimize contamination.

5. Disconnect the cooling system hoses and drain the coolant into a clean container for reuse. Plug the cooling system hoses.

6. Remove the instrument panel by performing the following procedure:

a. Remove the lower center cover screw and pull it out at the clip positions; then, disconnect the cigarette lighter connector.

b. Remove both the rear and front console.

c. Remove the dash side trim panel sill plates and the panels.

d. Remove the 2 glove box screws and the glove box.

e. Remove the 2 hood release screws, the 6 instrument panel driver's lower cover assembly screws and the cover assembly.

f. Remove 5 instrument cluster screws, the 2 clips; then, disconnect the 8 switch connectors and remove the instrument cluster assembly.

g. Remove the 6 driver's knee bolster assembly bolts and screws and the knee bolster assembly.

h. Remove the 4 control lever assembly bolts; then, disconnect the 3 control cables (unit side) and the 3 harness connectors.

i. Remove the 4 radio/audio sub box assembly screws and the radio/audio sub box assembly.

j. Disconnect or remove the following instrument panel harness connectors or items:

- The 6 driver's side connectors
- The 3 passenger's side connectors
- The 2 center connectors
- Passenger's inflator module connector
- Radio antenna cable plug
- Ground cable bolt on the left dash side panel
- The 8 instrument panel-to-chassis bolts and the 3 nuts.

k. Remove the instrument panel assembly.

7. Remove the instrument panel bracket by performing the following procedure:

a. Remove the 2 passenger's inflator module bolts and 4 nuts.

b. Remove the 4 meter assembly screws; then, disconnect the meter wiring

harness connectors and remove the meter assembly.

c. Remove the 5 vent duct assembly screws and the assembly.

d. Remove the 3 lower passenger bracket screws and the bracket.

e. Remove the 9 passenger knee bolster reinforcement screws and the reinforcement.

f. Remove the 6 instrument panel center reinforcement screws and the reinforcement.

g. Remove the instrument panel wiring harness assembly clips and the wiring harness.

h. Remove the 2 instrument panel bracket nuts and 2 bolts for each bracket; then, remove the bracket(s).

8. Remove the 5 cross beam assembly nuts, 2 bolts and the 6 lower bolts; then, remove the crossbeam.

9. Disconnect the resistor wiring connector.

10. Remove the duct from the heater assembly.

11. If equipped with air conditioning, remove the evaporator assembly.

12. Remove the driver's lap vent.

13. Remove the lower ventilation duct.

14. Remove the footrest, the carpet, the 3 clips and the rear heater duct.

15. Remove the heater assembly.

16. Remove the mode control case-to-temperature control case screws and remove the mode control case; do not remove the link unit.

17. Remove the temperature control case screws and separate the cases.

18. Remove the heater core from the case.

For Accessory Drive Belt illustrations, see Section 1 of this manual

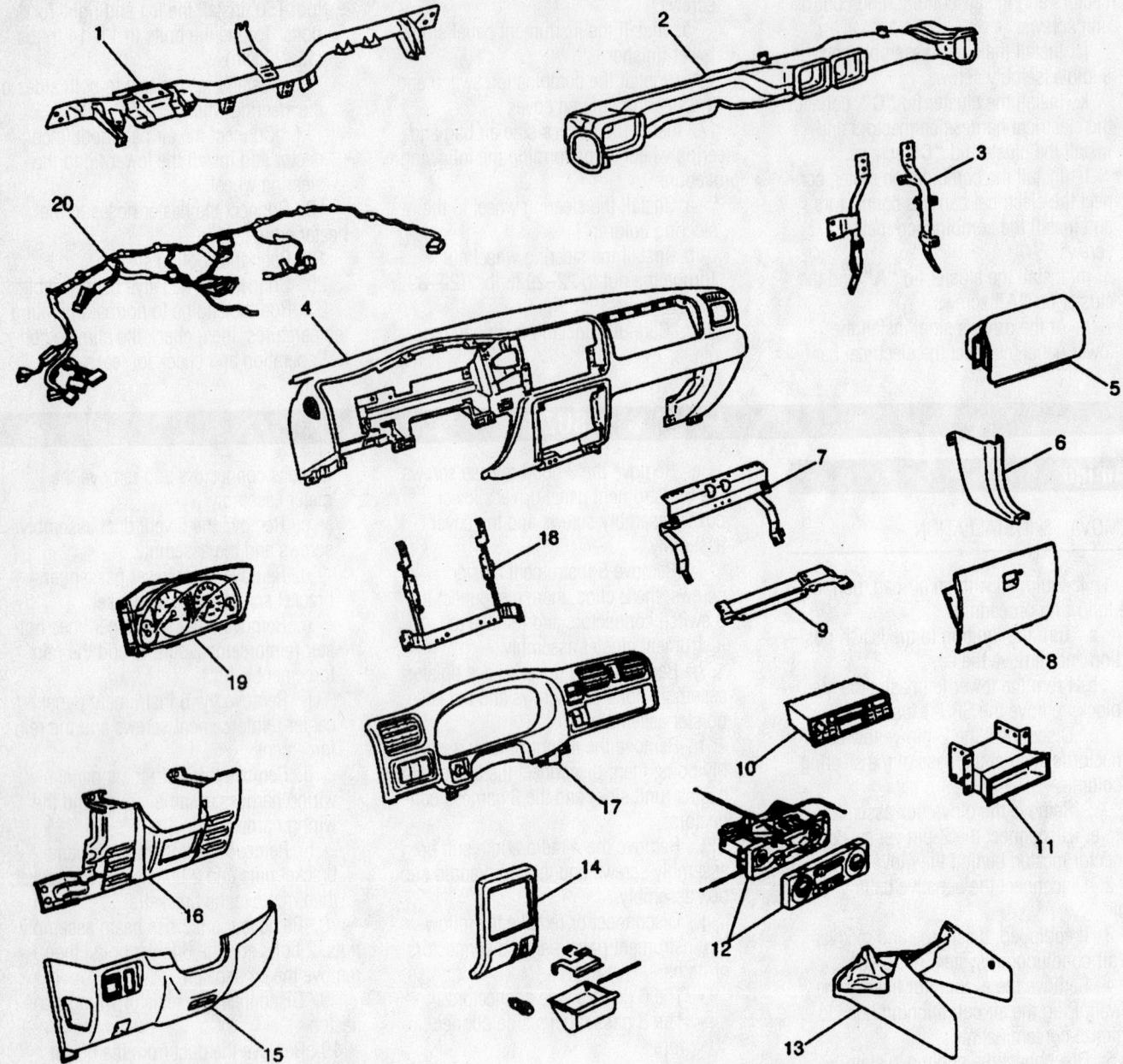

1	Cross Beam	11	Audio Sub Box
2	Vent Duct Assembly	12	Control Lever Assembly
3	Instrument Panel Bracket	13	Front Console Assembly
4	Instrument Panel Assembly	14	Lower Center Cover
5	Passenger Inflator Module	15	Instrument Panel Driver Lower Cover Assembly
6	Dash Side Trim Panel	16	Driver Knee Bolster Assembly
7	Passenger Knee Bolster Reinforcement Assembly	17	Meter Cluster Assembly
8	Glove Box	18	Instrument Panel Center Reinforcement
9	Passenger Lower Bracket	19	Meter Assembly
10	Radio Assembly	20	Instrument Harness Assembly

93113GB8

Exploded view of the instrument panel—Isuzu Amigo

To install:

19. Install the heater core to the case.

20. Assemble the temperature control cases and install the case screws.

21. Install the mode control case and the mode control case-to-temperature control case screws.

22. Install the heater assembly.

23. Install the rear heater duct, the footrest, the carpet, and the 3 clips.

24. Install the lower ventilation duct.

25. Install the driver's lap vent.

26. If equipped with air conditioning, install the evaporator assembly.

27. Install the duct to the heater assembly.

28. Connect the resistor wiring connector.

29. Install the crossbeam, the 5 crossbeam assembly nuts, 2 bolts and the 6 lower bolts.

30. Install the instrument panel bracket by performing the following procedure:

a. Install the instrument panel bracket and the 2 nuts and 2 bolts for each bracket.

b. Install the instrument panel wiring harness assembly and the wiring harness clips.

c. Install the instrument panel center reinforcement and the 6 reinforcement screws.

d. Install the passenger knee bolster reinforcement and the 9 reinforcement screws.

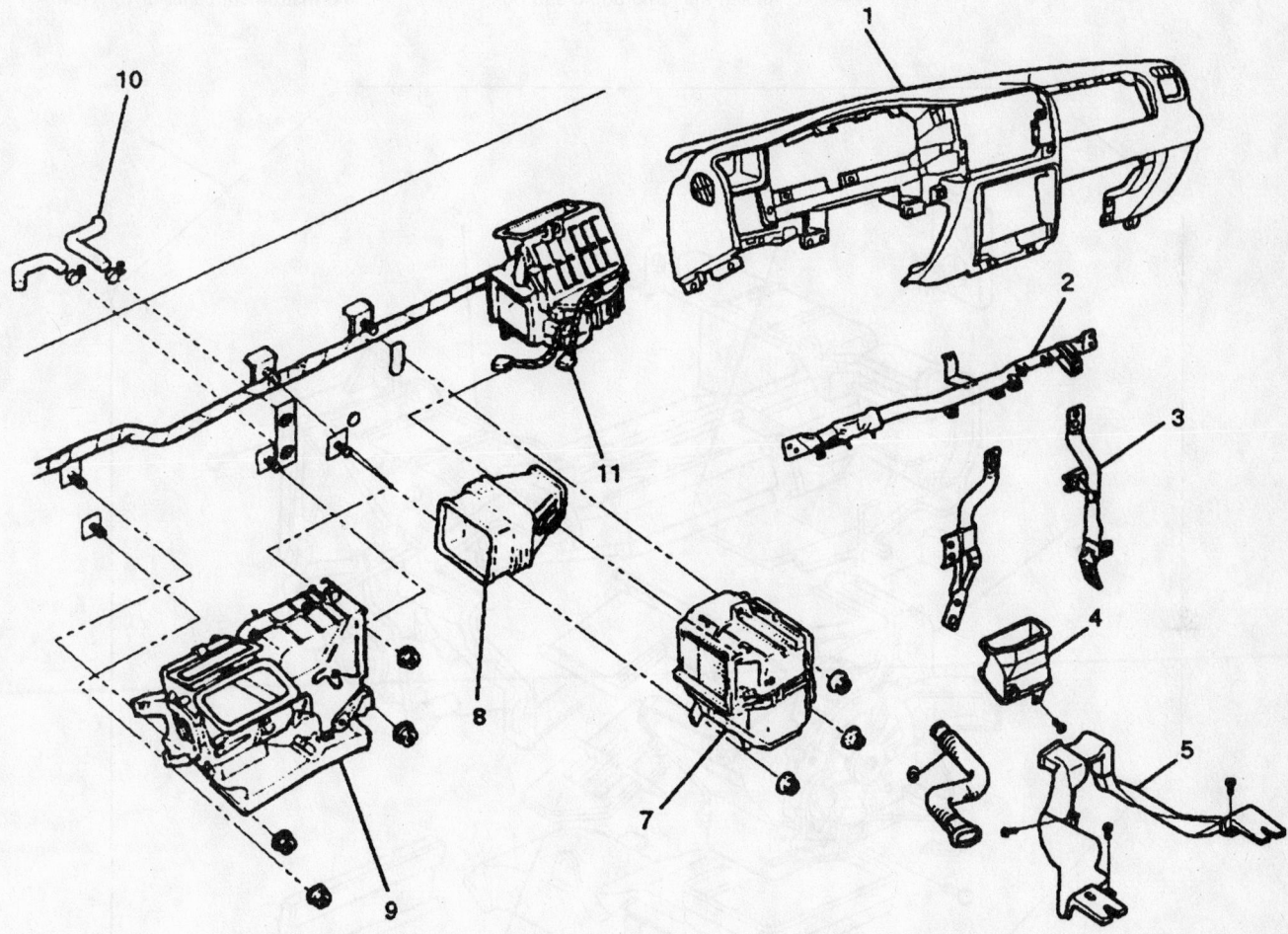

1	Instrument Panel Assembly	6	Driver Lap Vent Duct
2	Cross Beam Assembly	7	Evaporator Assembly (A/C only)
3	Instrument Panel Bracket	8	Duct
4	Ventilation Lower Duct	9	Heater Unit Assembly
5	Rear Heater Duct	10	Heater Hose
		11	Resistor Connector

93113GB9

View of the heater and air conditioning housing assemblies and related components—Isuzu Amigo

For Tire, Wheel and Ball Joint specifications, see Section 1 of this manual

e. Install the lower passenger bracket and the 3 bracket screws.

f. Install the vent duct assembly and the 5 vent duct assembly screws.

g. Install the meter assembly and the 4 meter assembly screws; then, connect the meter wiring harness connectors.

h. Install the 2 passenger's inflator module bolts and 4 nuts.

31. Install the instrument panel by performing the following procedure:

a. Install the instrument panel assembly.

b. Connect or install the following instrument panel harness connectors or items:

- The 6 driver's side connectors
- The 3 passenger's side connectors
- The 2 center connectors
- Passenger's inflator module connector
- Radio antenna cable plug
- Ground cable bolt on the left dash side panel
- The 8 instrument panel-to-chassis bolts and the 3 nuts.

c. Install the radio/audio sub box assembly and the 4 radio/audio sub box assembly screws.

d. Connect the 3 control cables (unit side) and the 3 harness connectors. Install the 4 control lever assembly bolts.

e. Install the knee bolster assembly and the 6 driver's knee bolster assembly bolts and screws.

f. Install the instrument cluster assembly. Connect the 8 switch connectors. Install 5 instrument cluster screws and the 2 clips.

g. Install the 2 hood release screws, the 6 instrument panel driver's lower

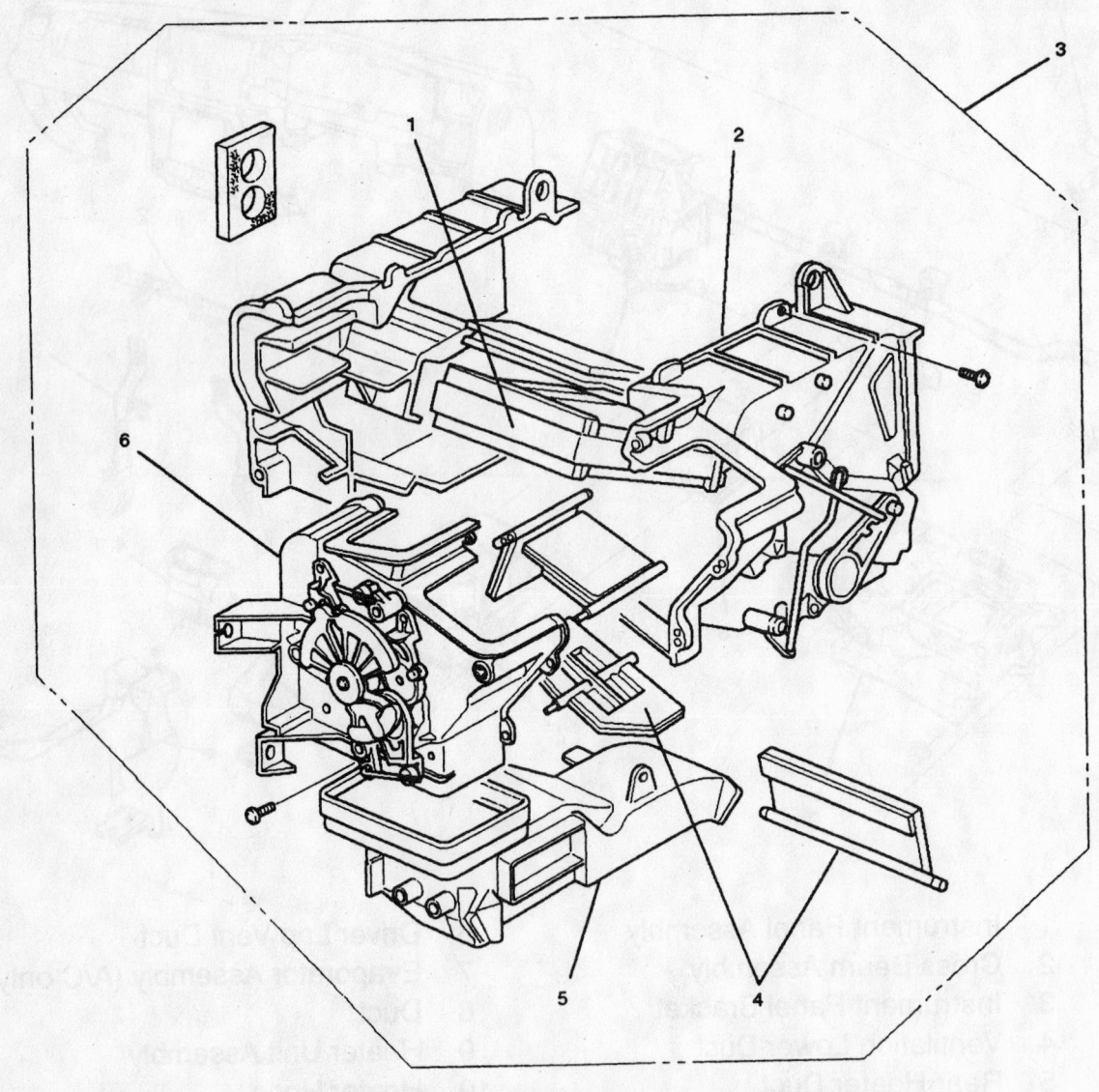

1	Heater Core	4	Mode Door
2	Case (Temperature Control)	5	Duct
3	Heater Unit	6	Case (Mode Control)

93113GB0

Exploded view of the heater housing assembly—Isuzu Amigo

cover assembly screws and the cover assembly.

h. Install the glove box and the 2 glove box screws.

i. Install the dash side trim panel sill plates and the panels.

j. Install both the rear and front console.

k. Connect the cigarette lighter connector and install the lower center cover screw.

32. Connect the cooling system hoses.

33. Refill the cooling system.

34. Install the evaporator lines at the firewall.

35. If equipped, evacuate and charge the air conditioning system.

36. Connect the negative battery cable.

37. If equipped with an air bag, perform the following procedure:

a. Turn the ignition to the LOCK position and remove the key.

b. Connect the 2-pin yellow connector located behind the glove box.

c. Install the glove box assembly.

d. Connect the 2-pin yellow connector located at the base of the steering column.

e. At the lower left dash side fuse block, install the SRS-1 fuse.

f. Turn the ignition switch to ON and verify that the AIR BAG warning light flashes 7 times and turns OFF.

38. Run the engine to normal operating temperatures; then, check the climate control operation and check for leaks.

Rodeo

REMOVAL & INSTALLATION

1. If equipped with an air bag, perform the following procedure:

a. Turn the ignition to the LOCK position and remove the key.

b. From the lower left dash side fuse block, remove the SRS-1 fuse.

c. Disconnect the 2-pin yellow connector located at the base of the steering column.

d. Remove the glove box assembly.

e. Disconnect the 2-pin yellow connector located behind the glove box.

2. Disconnect the negative battery cable.

3. If equipped, discharge and recover the air conditioning system refrigerant.

4. Remove the evaporator lines at the firewall. Plug the air conditioning lines to minimize contamination.

5. Disconnect the cooling system hoses and drain the coolant into a clean container for reuse. Plug the cooling system hoses.

6. Remove the instrument panel by performing the following procedure:

a. Remove the lower center cover screw and pull it out at the clip positions; then, disconnect the cigarette lighter connector.

b. Remove both the rear and front console.

c. Remove the dash side trim panel sill plates and the panels.

d. Remove the 2 glove box screws and the glove box.

e. Remove the 2 hood release screws, the 6 instrument panel driver's lower cover assembly screws and the cover assembly.

f. Remove 5 instrument cluster screws and the 2 clips. Then, disconnect the 8 switch connectors and remove the instrument cluster assembly.

g. Remove the 6 driver's knee bolster assembly bolts and screws and the knee bolster assembly.

h. Remove the 4 control lever assembly bolts; then, disconnect the 3 control cables (unit side) and the 3 harness connectors.

i. Remove the 4 radio/audio sub box assembly screws and the radio/audio sub box assembly.

j. Disconnect or remove the following instrument panel harness connectors or items:

- The 6 driver's side connectors
- The 3 passenger's side connectors
- The 2 center connectors
- Passenger's inflator module connector
- Radio antenna cable plug
- Ground cable bolt on the left dash side panel
- The 8 instrument panel-to-chassis bolts and the 3 nuts.

k. Remove the instrument panel assembly.

7. Remove the instrument panel bracket by performing the following procedure:

a. Remove the 2 passenger's inflator module bolts and 4 nuts.

b. Remove the 4 meter assembly screws. Then, disconnect the meter wiring harness connectors and remove the meter assembly.

c. Remove the 5 vent duct assembly screws and the assembly.

d. Remove the 3 lower passenger bracket screws and the bracket.

e. Remove the 9 passenger knee bolster reinforcement screws and the reinforcement.

f. Remove the 6 instrument panel center reinforcement screws and the reinforcement.

g. Remove the instrument panel wiring harness assembly clips and the wiring harness.

h. Remove the 2 instrument panel bracket nuts and 2 bolts for each bracket; then, remove the bracket(s).

8. Remove the 5 cross beam assembly nuts, 2 bolts and the 6 lower bolts; then, remove the crossbeam.

9. Disconnect the resistor wiring connector.

10. Remove the duct from the heater assembly.

11. If equipped with air conditioning, remove the evaporator assembly.

12. Remove the driver's lap vent.

13. Remove the lower ventilation duct.

14. Remove the footrest, the carpet, the 3 clips and the rear heater duct.

15. Remove the heater assembly.

16. Remove the mode control case-to-temperature control case screws and remove the mode control case; do not remove the link unit.

17. Remove the temperature control case screws and separate the cases.

18. Remove the heater core from the case.

To install:

19. Install the heater core to the case.

20. Assemble the temperature control cases and install the case screws.

21. Install the mode control case and the mode control case-to-temperature control case screws.

22. Install the heater assembly.

23. Install the rear heater duct, the footrest, the carpet, and the 3 clips.

24. Install the lower ventilation duct.

25. Install the driver's lap vent.

26. If equipped with air conditioning, install the evaporator assembly.

27. Install the duct to the heater assembly.

28. Connect the resistor wiring connector.

29. Install the crossbeam, the 5 crossbeam assembly nuts, 2 bolts and the 6 lower bolts.

30. Install the instrument panel bracket by performing the following procedure:

1	Cross Beam	11	Audio Sub Box
2	Vent Duct Assembly	12	Control Lever Assembly
3	Instrument Panel Bracket	13	Front Console Assembly
4	Instrument Panel Assembly	14	Lower Center Cover
5	Passenger Inflator Module	15	Instrument Panel Driver Lower Cover Assembly
6	Dash Side Trim Panel	16	Driver Knee Bolster Assembly
7	Passenger Knee Bolster Reinforcement Assembly	17	Meter Cluster Assembly
8	Glove Box	18	Instrument Panel Center Reinforcement
9	Passenger Lower Bracket	19	Meter Assembly
10	Radio Assembly	20	Instrument Harness Assembly

93113GB8

Exploded view of the instrument panel—Isuzu Rodeo

a. Install the instrument panel bracket and the 2 nuts and 2 bolts for each bracket.

b. Install the instrument panel wiring harness assembly and the wiring harness clips.

c. Install the instrument panel center reinforcement and the 6 reinforcement screws.

d. Install the passenger knee bolster reinforcement and the 9 reinforcement screws.

e. Install the lower passenger bracket and the 3 bracket screws.

f. Install the vent duct assembly and the 5 vent duct assembly screws.

g. Install the meter assembly and the 4 meter assembly screws. Then, connect the meter wiring harness connectors.

h. Install the 2 passenger's inflator module bolts and 4 nuts.

31. Install the instrument panel by performing the following procedure:

a. Install the instrument panel assembly.

b. Connect or install the following instrument panel harness connectors or items:

- The 6 driver's side connectors
- The 3 passenger's side connectors
- The 2 center connectors
- Passenger's inflator module connector
- Radio antenna cable plug

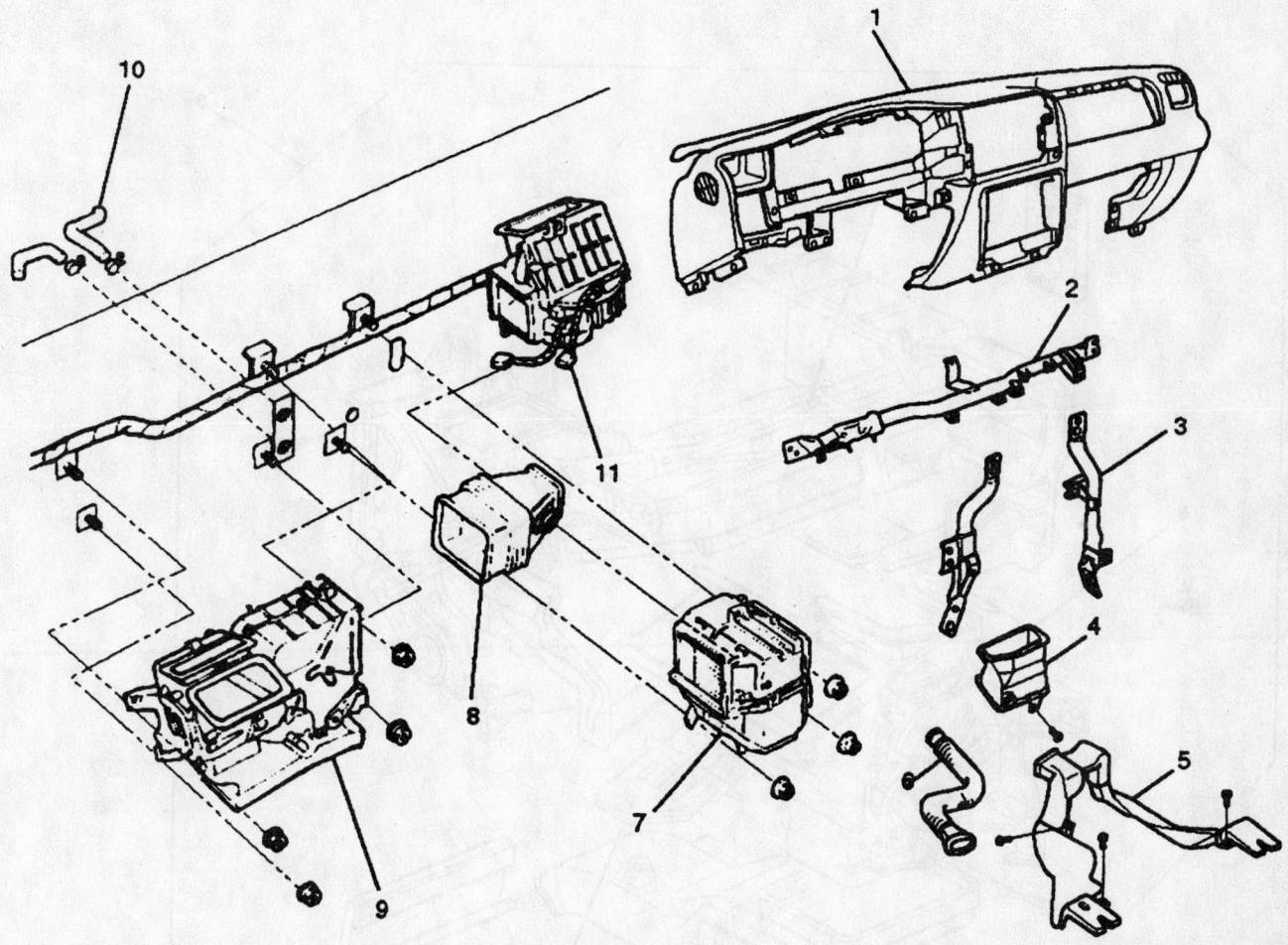

1	Instrument Panel Assembly	6	Driver Lap Vent Duct
2	Cross Beam Assembly	7	Evaporator Assembly (A/C only)
3	Instrument Panel Bracket	8	Duct
4	Ventilation Lower Duct	9	Heater Unit Assembly
5	Rear Heater Duct	10	Heater Hose
		11	Resistor Connector

93113GB9

View of the heater and air conditioning housing assemblies and related components—Isuzu Rodeo

For Maintenance Interval recommendations, see Section 1 of this manual

- Ground cable bolt on the left dash side panel
- The 8 instrument panel-to-chassis bolts and the 3 nuts.

c. Install the radio/audio sub box assembly and the 4 radio/audio sub box assembly screws.

d. Connect the 3 control cables (unit side) and the 3 harness connectors. Install the 4 control lever assembly bolts.

e. Install the knee bolster assembly and the 6 driver's knee bolster assembly bolts and screws.

f. Install the instrument cluster assembly. Connect the 8 switch connectors. Install 5 instrument cluster screws and the 2 clips.

g. Install the 2 hood release screws, the 6 instrument panel driver's lower cover assembly screws and the cover assembly.

h. Install the glove box and the 2 glove box screws.

i. Install the dash side trim panel sill plates and the panels.

j. Install both the rear and front console.

k. Connect the cigarette lighter connector and install the lower center cover screw.

32. Connect the cooling system hoses.

33. Refill the cooling system.

34. Install the evaporator lines at the firewall.

35. If equipped, evacuate and charge the air conditioning system.

36. Connect the negative battery cable.

37. If equipped with an air bag, perform the following procedure:

a. Turn the ignition to the LOCK position and remove the key.

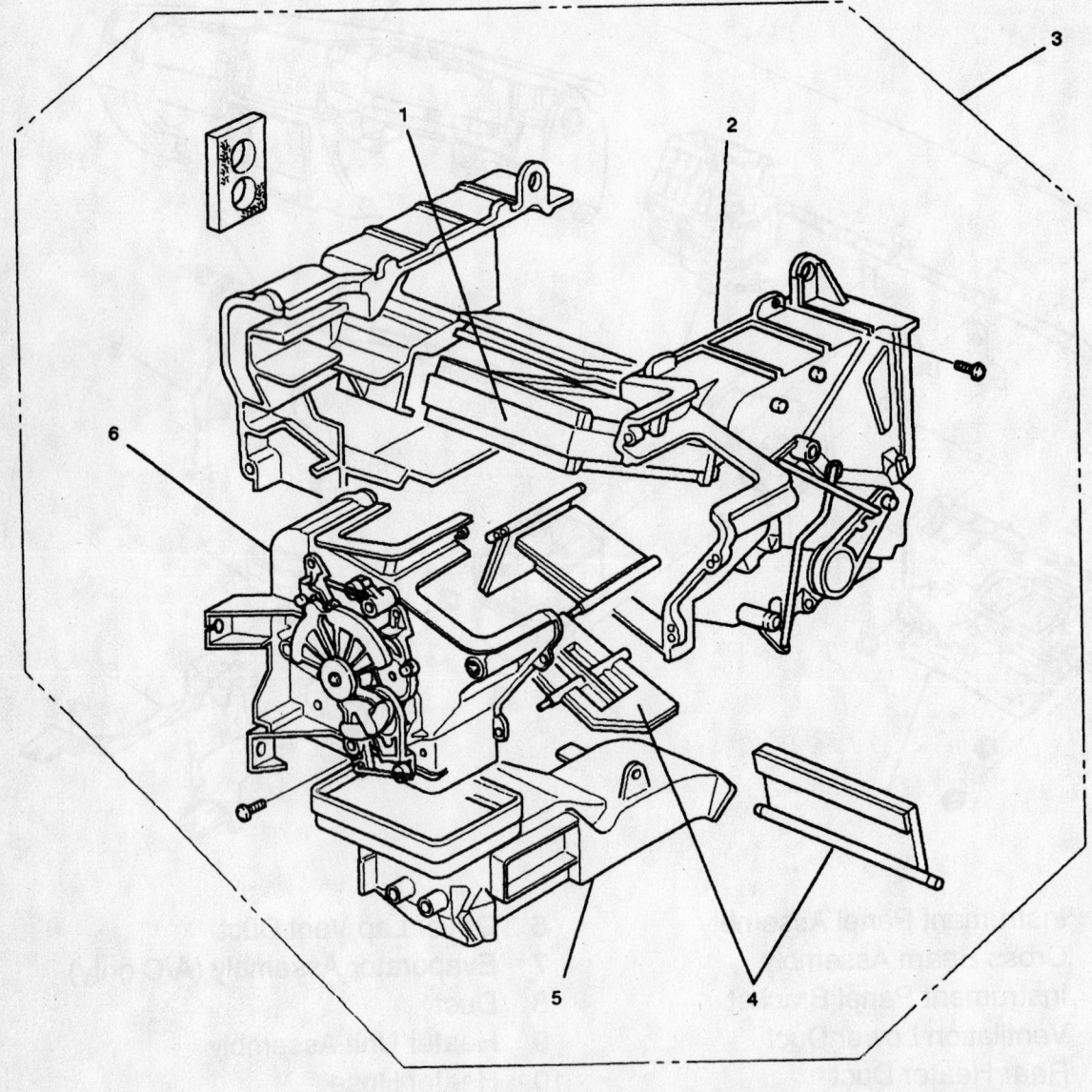

| 1 | Heater Core | 4 | Mode Door |
| 2 | Case (Temperature Control) | 5 | Duct |

Exploded view of the heater housing assembly—Isuzu Rodeo

93113GB0

b. Connect the 2-pin yellow connector located behind the glove box.

c. Install the glove box assembly.

d. Connect the 2-pin yellow connector located at the base of the steering column.

e. At the lower left dash side fuse block, install the SRS-1 fuse.

f. Turn the ignition switch to ON and verify that the AIR BAG warning light flashes 7 times and turns OFF.

38. Run the engine to normal operating temperatures; then, check the climate control operation and check for leaks.

Trooper

REMOVAL & INSTALLATION

✳✳ CAUTION

The vehicle is equipped with a driver's side and a passenger's side air bag. Before starting service procedures on components, especially under the instrument panel and/or near the steering column, disable the air bag systems. There is sufficient voltage in the system to cause a deployment for up to 15 seconds after the battery has been disconnected, the ignition turned OFF or fuse C-21 is removed from the fuse panel.

1. If equipped with an air bag, perform the following procedures:

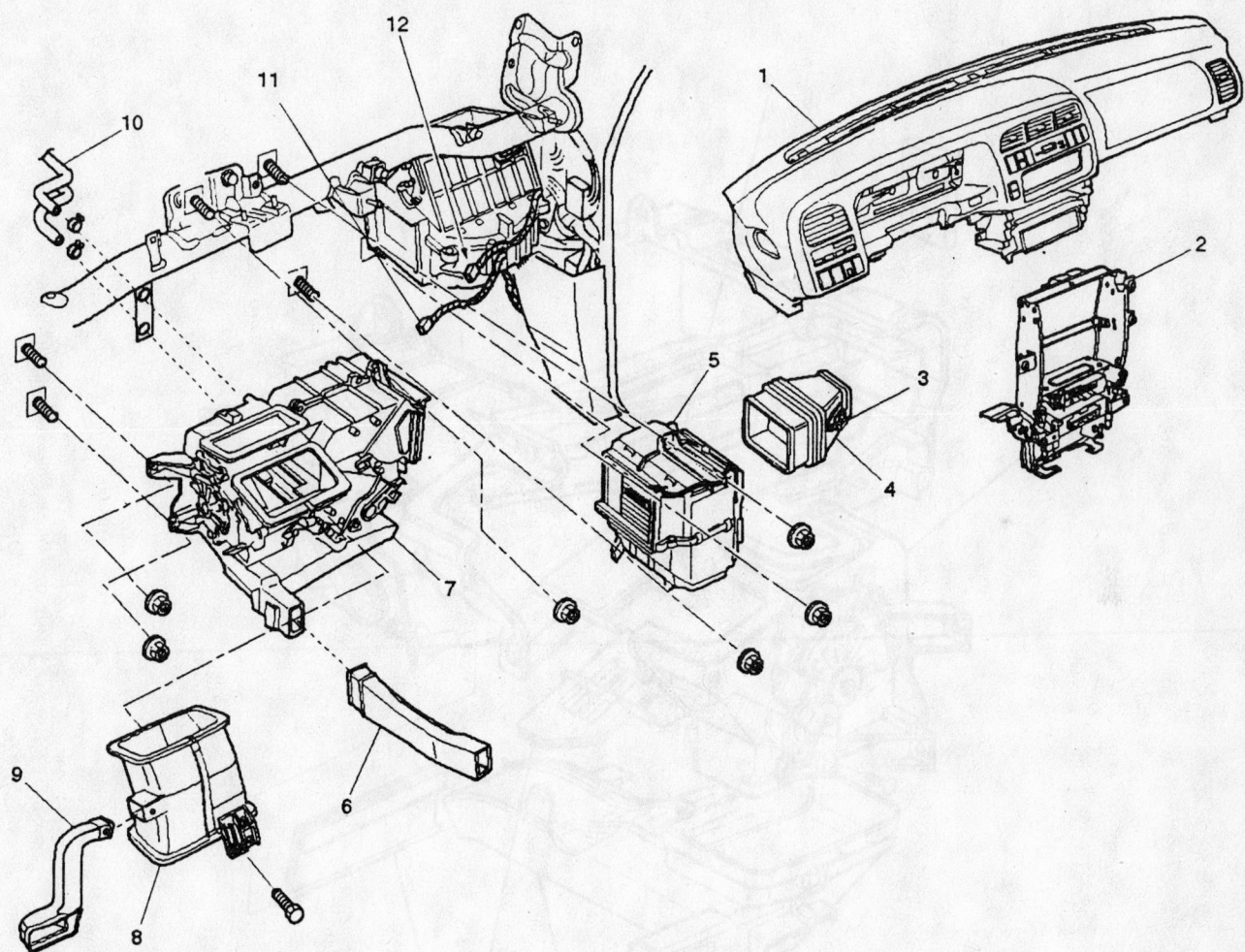

(1) Instrument Panel Assembly
(2) Instrument Panel Center Bracket
(3) Resistor
(4) Duct
(5) Evaporator Assembly (A/C only)
(6) Rear Heater Duct
(7) Heater Unit Assembly
(8) Center Ventilation Lower Duct
(9) Driver Lap Vent Nozzle
(10) Water Hose
(11) Electro Thermo Connector (With A/C)
(12) Resistor Connector

93113G01

Exploded view of the heater unit and related components—Isuzu Trooper

For Tune-up, Capacities and Firing orders, see Section 1 of this manual

a. Disconnect the negative battery cable, then disconnect the positive battery cable.

b. Disconnect the yellow 2-pin connector located at the base of the steering column.

c. Remove the glove box and disconnect the yellow 2-pin connector located behind the glove box.

2. Disconnect the negative battery cable.

3. Drain the cooling system.

4. If equipped with air conditioning, discharge and recover the refrigerant.

5. Remove the instrument panel assembly by performing the following procedure:

a. At the front console assembly, disconnect the switch connectors; then, remove the console-to-chassis screws and the console.

b. At the lower cluster assembly, remove the cluster-to-instrument panel screws, disconnect the cigarette lighter and light connectors and remove the lower cluster.

c. Remove the glove box and the instrument panel lower cover and the passenger knee bolster reinforcement.

d. At the left side, remove the instrument panel lower cover and the knee bolster assembly.

e. At the top of the instrument panel, pry the 8 claws on the front side toward you, raise the defroster grille and remove it.

f. At the SRS adjust bracket and cross beam under the passenger air bag module, remove the 2 fixing bolts and remove the instrument panel assembly.

g. Disconnect the air conditioning control cables from the unit.

h. Remove the instrument panel har-

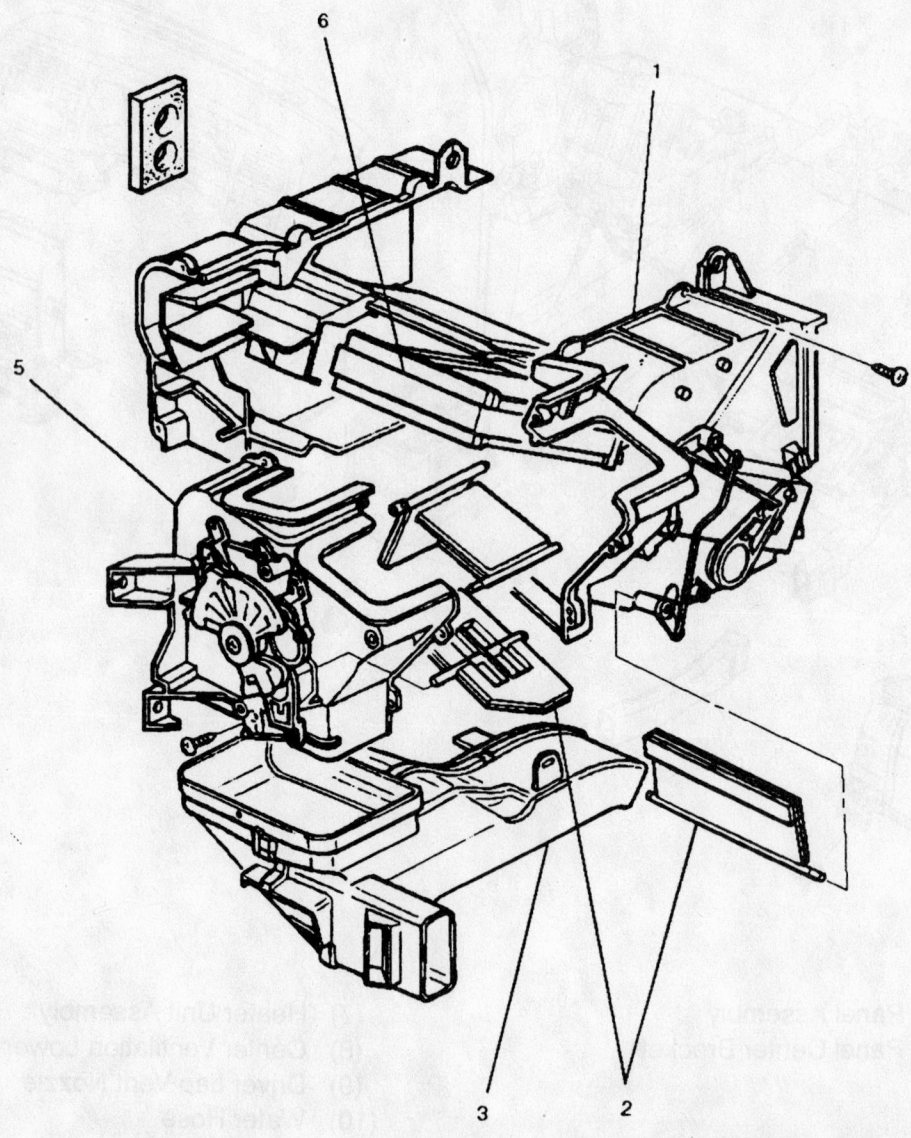

(1) Case (Temperature Control)
(2) Mode Door
(3) Duct
(5) Case (Mode Control)
(6) Heater Core

93113G02

Exploded view of the heater unit—Isuzu Trooper

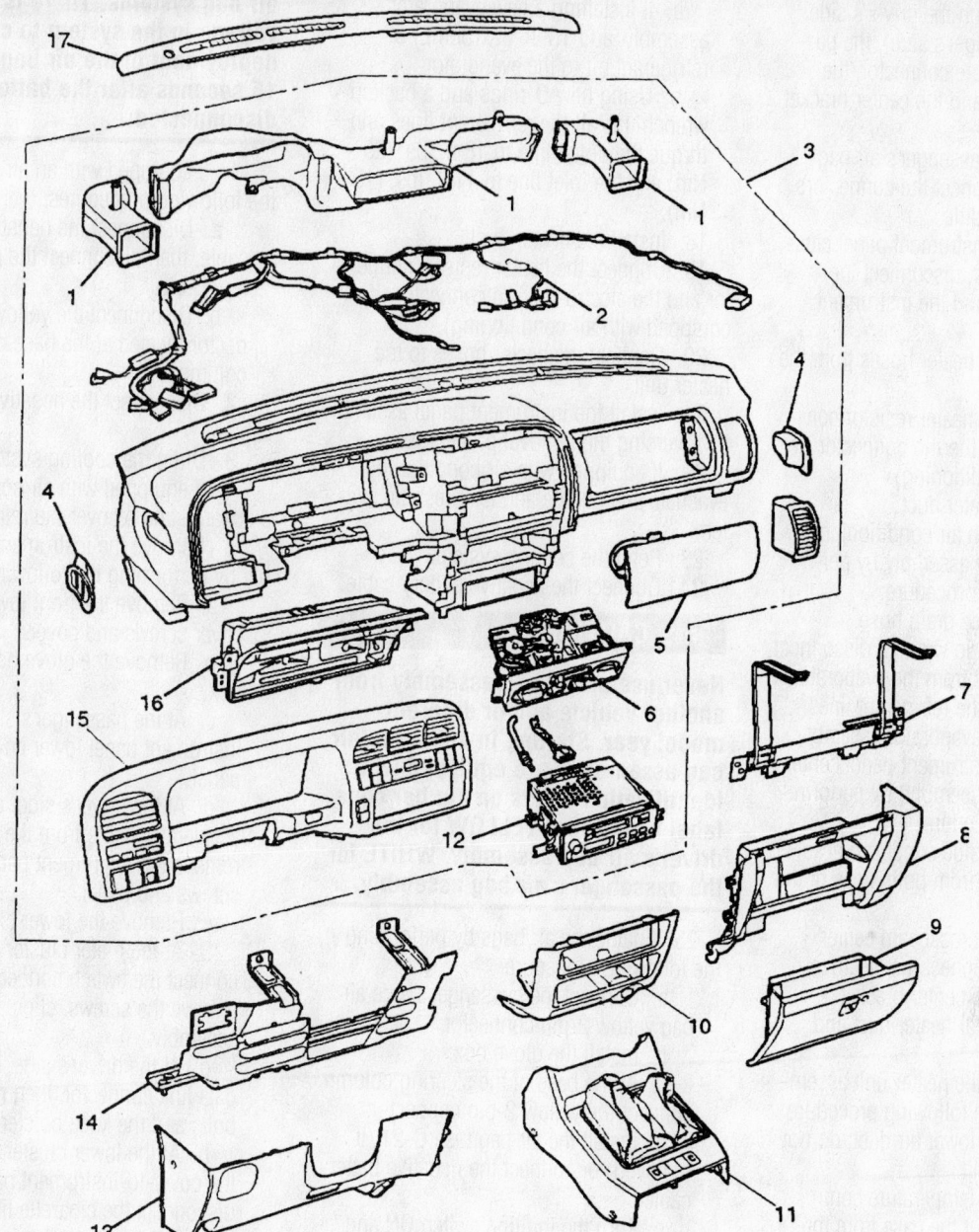

(1) Vent Duct Assembly
(2) Instrument Harness Assembly
(3) Instrument Panel Assembly
(4) Side Defroster Grille
(5) Passenger Inflator Module
(6) Control Lever Assembly
(7) Passenger Knee Bolster Reinforcement Assembly
(8) Instrument Panel Passenger Lower Cover Assembly

(9) Glove Box
(10) Lower Cluster Assembly
(11) Front Console Assembly
(12) Radio Assembly
(13) Instrument Panel Driver Lower Cover Assembly
(14) Driver Knee Bolster Assembly
(15) Instrument Panel Cluster Assembly
(16) Meter Assembly
(17) Front Defroster Grille

93113G03

Exploded view of the instrument panel and accessories—Isuzu Trooper

For complete service labor times, order Nichols' Chilton Labor Guide

ness connectors (5 on the driver's side and 3 on the passenger's side), the passenger air bag module connector, the radio antenna plug and the center bracket ground cable bolt.

i. Remove the passenger's air bag module nuts, disconnect the connectors and remove the module.

j. Remove the instrument panel cluster assembly screws, disconnect the switch connectors and the instrument panel assembly.

6. Disconnect the heater hoses from the heater unit.

7. Disconnect the heater resistor connector and the electro thermo connector (if equipped with air conditioning).

8. Remove the heater duct.

9. If equipped with air conditioning, remove the evaporator assembly by performing the following procedure:

a. Disconnect the drain hose.

b. Using a backup wrench, disconnect the refrigerant lines from the evaporator.

c. Plug or cap the refrigerant lines.

d. Remove the evaporator assembly.

10. Remove the instrument panel center bracket (crossbeam assembly) by performing the following procedure:

a. Remove the side support bracket bolts and brackets from both sides of the vehicle.

b. Remove the crossbeam center bracket nuts, disconnect the electrical connectors and the center bracket.

11. Remove the rear heater duct and heater assembly.

12. Disassemble the heater unit assembly by performing the following procedure:

a. Remove the lower air duct; do not remove the link unit.

b. Remove the temperature control case screws and lift the case from the heater unit.

c. Remove the heater core.

To install:

13. Assemble the heater unit assembly by performing the following procedure:

a. Install the heater core into the heater unit.

b. Install the temperature control case onto the unit and secure with screws.

c. Install the lower air duct.

14. Install the heater unit assembly into the vehicle.

15. Install the rear heater duct.

16. Install the instrument panel cross beam assembly by reversing the removal procedures.

17. If equipped with air conditioning, install the evaporator assembly by performing the following procedures:

a. If installing a new evaporator assembly, add 1.7 fl. oz. (50mL) of refrigerant oil to the evaporator.

b. Using new O-rings and a backup wrench, install the refrigerant lines and torque the outlet line to 18 ft. lbs. (25 Nm) and the inlet line to 11 ft. lbs. (15 Nm).

18. Install the heater duct.

19. Connect the heater resistor connector and the electro-thermo connector (if equipped with air conditioning).

20. Connect the heater hoses to the heater unit.

21. Install the instrument panel assembly by reversing the removal procedures.

22. If equipped with air conditioning, evacuate and charge and leak-test the system.

23. Refill the cooling system.

24. Connect the negative battery cable.

✳✳ CAUTION

Never use an air bag assembly from another vehicle and/or different model year. Starting in 1999, the air bag assemblies are equipped with identification colors on the bar code label as follows: YELLOW for the driver's air bag assembly, WHITE for the passenger's air bag assembly.

25. Enable the air bags by performing the following procedure:

a. Connect the passenger's side air bag yellow 2-pin connector.

b. Install the glove box.

c. At the base of the steering column, connect the yellow 2-pin connector.

d. Install the air bag fuse C-21 (if removed) or connect the negative battery cable.

e. Turn the ignition switch ON and verify that the AIR BAG warning light flashes 7 times and then turns OFF.

26. Run the engine to normal operating temperatures and check for leaks. Check the systems for correct operation.

VehiCROSS

REMOVAL & INSTALLATION

✳✳ CAUTION

The vehicle is equipped with a driver's side and a passenger's side air bag. Before starting service procedures on components, especially under the instrument panel and/or near the steering column, disable the air bag systems. There is sufficient voltage in the system to cause a deployment of the air bags for up to 15 seconds after the battery has been disconnected.

1. If equipped with an air bag, perform the following procedures:

a. Disconnect the negative battery cable, then disconnect the positive battery cable.

b. Disconnect the yellow 3-pin connector located at the base of the steering column.

2. Disconnect the negative battery cable.

3. Drain the cooling system.

4. If equipped with air conditioning, discharge and recover the refrigerant.

5. Remove the instrument panel assembly by performing the following procedure:

a. Remove the front lower console cover screws and cover.

b. Remove the glove box door and box.

c. At the passenger's side, remove the instrument panel lower cover screws and panel.

d. At the driver's side, disconnect the accelerator cable from the pedal and remove the instrument panel lower cover screws and panel.

e. Remove the lower cluster.

f. At the meter cluster assembly, disconnect the switch connectors, then remove the screws, clips and the meter assembly.

g. At the driver's side, disconnect the data link connector, then remove the bolts and the knee bolster.

h. At the lower cluster cover, remove the cover-to-instrument panel screws, disconnect the cigarette lighter and remove the lower cover.

i. Disconnect the air conditioning control cables from the unit.

j. Remove the instrument panel cluster assembly screws, disconnect the switch connectors and the instrument panel assembly.

k. Remove the instrument harness connectors, the radio antenna plug.

l. Remove the side defroster grille. Remove the instrument panel nuts, bolts and screws and the instrument panel.

6. Remove the passenger's air bag reinforcement screws and the reinforcement.

7. At the meter assembly, disconnect the electrical connector, then remove the screws and the meter assembly.

8. Remove the radio and the vent duct assembly.

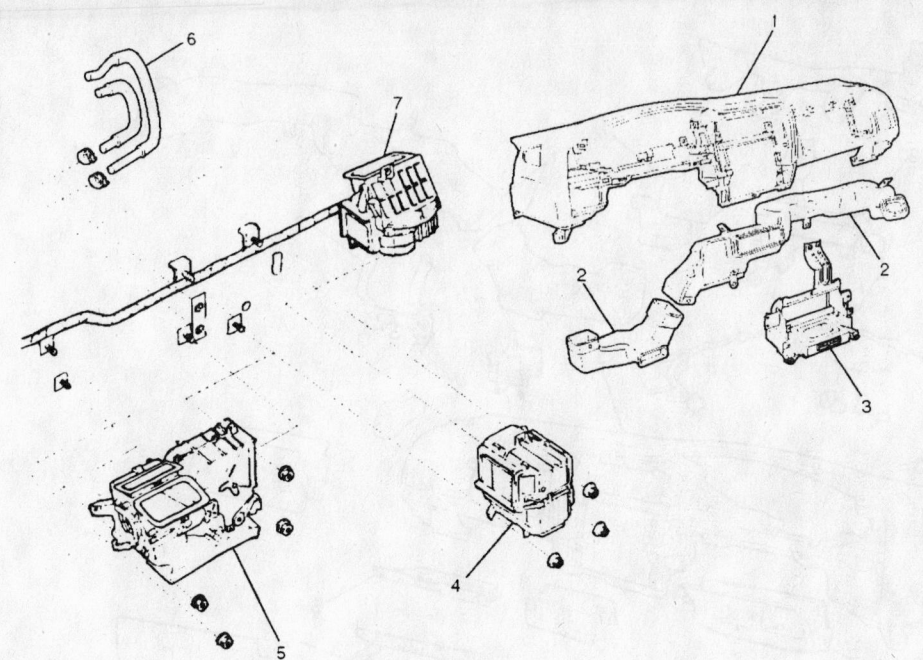

1 Instrument Panel Assembly

2 Center & Lower Vent Duct

3 Instrument Panel Center Bracket

4 Evaporator Assembly

5 Heater Unit Assembly

6 Heater Hose

7 Blower Unit

93113G04

Exploded view of the heater unit and related components—Isuzu VehiCROSS

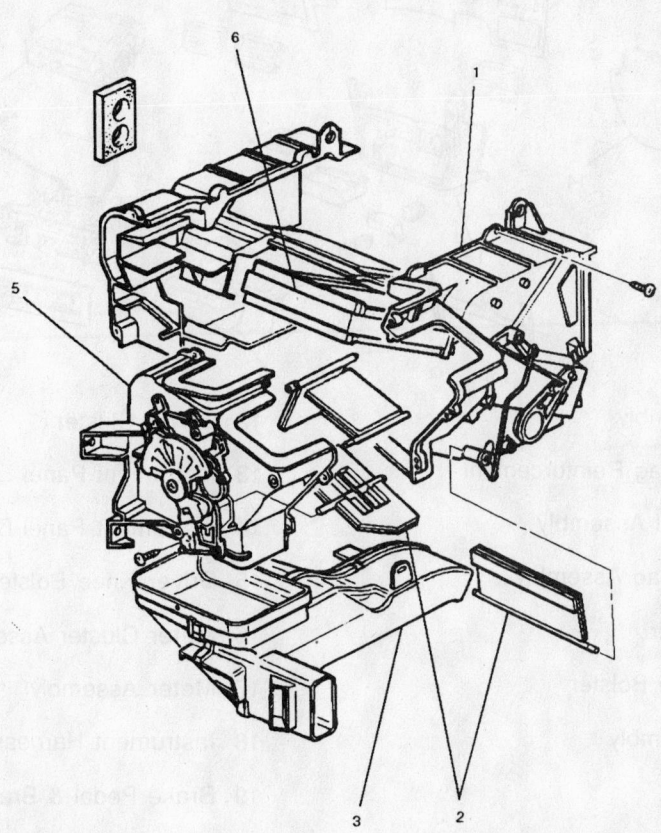

(1) Case (Temperature Control)   (5) Case (Mode Control)
(2) Mode Door   (6) Heater Core
(3) Duct

93113G02

Exploded view of the heater unit—Isuzu VehiCROSS

Please visit our web site at www.chiltononline.com

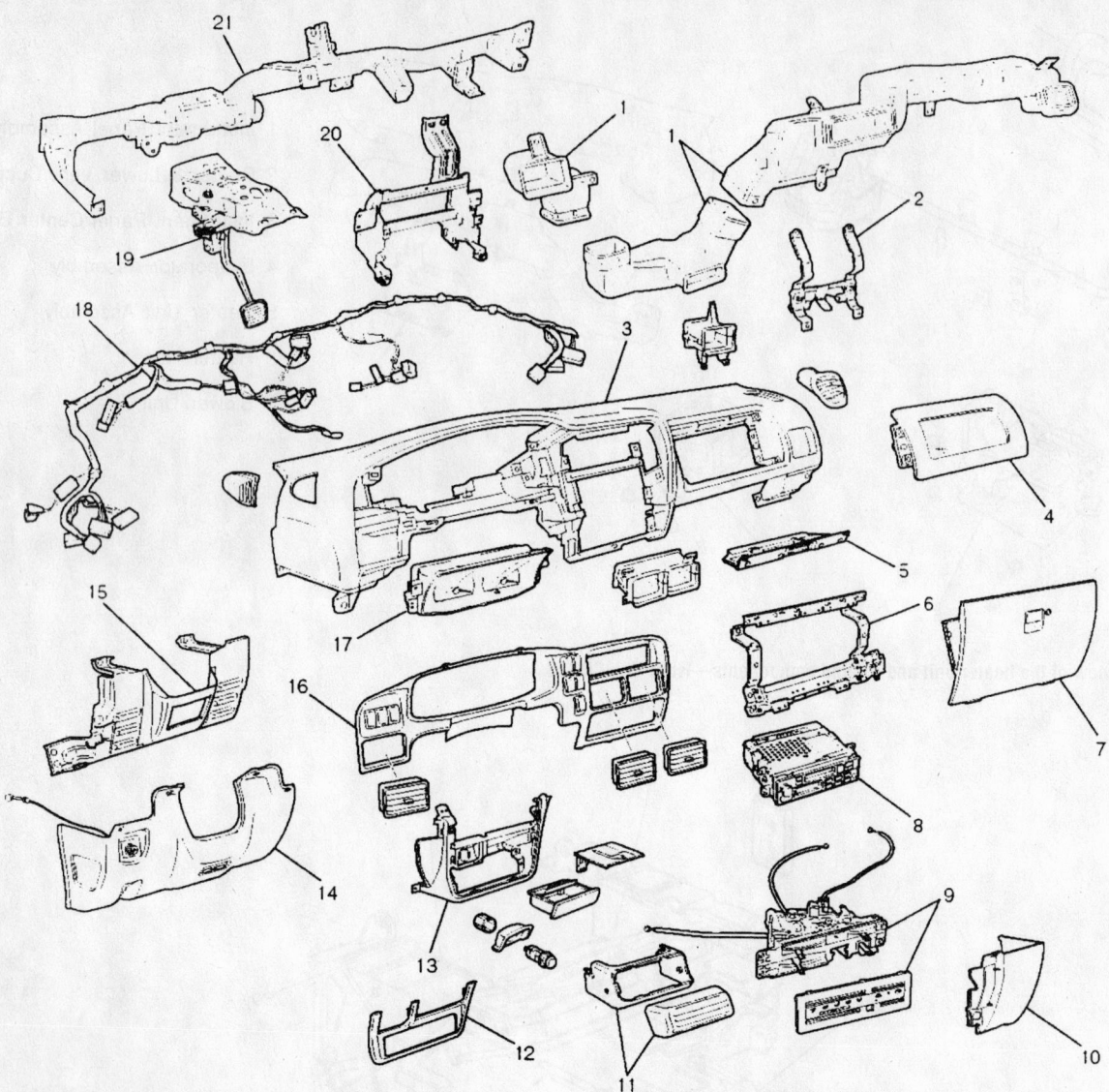

1. Vent Duct Assembly
2. Passenger Air Bag Reinforcement
3. Instrument Panel Assembly
4. Passenger Air Bag Assembly
5. Glove Box Cover
6. Passenger Knee Bolster
7. Glove Box Assembly
8. Radio Assembly
9. Air Conditioner Control Lever Assembly
10. Instrument Panel Passenger Lower Cover
11. Front Lower Console Cover
12. Lower Cluster
13. Instrument Panel Lower Center Cover
14. Instrument Panel Driver Lower Cover
15. Driver Knee Bolster
16. Meter Cluster Assembly
17. Meter Assembly
18. Instrument Harness Assembly
19. Brake Pedal & Bracket Assembly
20. Instrument Panel Center Bracket
21. Cross Beam Assembly

93113G05

Exploded view of the instrument panel and accessories—Isuzu VehiCROSS

9. Remove the passenger's knee bolster screws and bolster.

10. Remove the instrument panel center bracket bolts, nuts and bracket.

11. Remove the heater resistor connectors and the electro thermo connector (if equipped with air conditioning).

12. Remove the blower motor assembly.

13. If equipped with air conditioning, remove the evaporator assembly by performing the following procedure:

a. Disconnect the drain hose.

b. Using a backup wrench, disconnect the refrigerant lines from the evaporator.

c. Plug or cap the refrigerant lines.

d. Remove the evaporator assembly.

14. Remove the driver's side lap vent duct.

15. Remove the center and lower vent ducts.

16. Remove the heater assembly.

17. Disassemble the heater unit assembly by performing the following procedure:

a. Remove the lower air duct; do not remove the link unit.

b. Remove the temperature control case screws and lift the case from the heater unit.

c. Remove the heater core.

To install:

18. Assemble the heater unit assembly by performing the following procedure:

a. Install the heater core into the heater unit.

b. Install the temperature control case onto the unit and secure with screws.

c. Install the lower air duct.

19. Install the heater unit assembly into the vehicle.

20. Install the center and lower vent ducts.

21. Install the driver's side lap vent duct.

22. Install the instrument panel cross beam assembly by reversing the removal procedures.

23. If equipped with air conditioning, install the evaporator assembly by performing the following procedures:

a. If installing a new evaporator assembly, add 1.7 fl. oz. (50mL) of refrigerant oil to the evaporator.

b. Using new O-rings and a backup wrench, install the refrigerant lines and torque the outlet line to 18 ft. lbs. (25 Nm) and the inlet line to 11 ft. lbs. (15 Nm).

24. Install the blower motor.

25. Connect the heater resistor connectors and the electro-thermo connector (if equipped with air conditioning).

26. Connect the heater hoses to the heater unit.

27. Install the instrument panel assembly by reversing the removal procedures.

28. If equipped with air conditioning, evacuate and recharge the system.

29. Refill the cooling system.

30. Connect the negative battery cable.

✳✳ CAUTION

Never use an air bag assembly from another vehicle and/or different model year.

31. Enable the air bag by performing the following procedure:

a. At the base of the steering column, connect the yellow 3-pin connector.

b. Connect the negative battery cable.

c. Turn the ignition switch ON and verify that the AIR BAG warning light flashes 7 times and then turns OFF.

32. Run the engine to normal operating temperatures and check for leaks. Check the systems for correct operation.

JEEP

Cherokee

REMOVAL & INSTALLATION

1. Disconnect and remove the negative battery.

✳✳ CAUTION

After disconnecting the negative battery cable, wait 2 minutes for the driver's/passenger's air bag system capacitor to discharge before attempting to do any work around the steering column or instrument.

2. Drain the cooling system into a clean container for reuse.

3. Remove the instrument panel by performing the following procedure:

a. Turn the steering wheel in the straight-ahead position.

b. Remove the knee blocker from the instrument panel.

c. Remove the steering column; do not remove the air bag module, the steering wheel or switches from the steering column.

d. From under the driver's side of the instrument panel, disconnect the following items:

- Instrument panel wiring harness connector from the 100-way wiring harness connector at the left side of the inner panel.
- Side window demister hose at the heater/air conditioning housing demister/defroster duct on the driver's side.

e. Remove the glove box.

f. Reaching through the glove box opening, disconnect the following items:

- Two halves of the heater/air conditioning system vacuum harness connector.
- Instrument panel wiring harness connector from the heater/air conditioning system wiring harness connector.
- Instrument panel wiring harness connector from the passenger's side air bag module wiring harness connector.

- Side window demister hose at the heater/air conditioning housing demister/defroster duct (passenger's side).
- Two halves of the radio antenna coaxial cable connector.
- Two instrument panel wiring harness connectors from the passenger air bag ON/OFF switch wiring harness connector.
- Passenger's side air bag ON/OFF switch wiring harness from the retainer clip on the plenum bracket that supports the heater/air conditioning housing just inboard of the fuse block module.
- Two lower passenger's side air bag module bracket-to-dash panel nuts.

g. Remove the upper cover from the instrument panel.

h. Remove the 3 instrument panel-to-door hinge pillar screws.

i. Remove the 4 upper instrument panel-to-dash nuts.

j. Using an assistant, remove the instrument panel from the vehicle.

Timing belt service is covered in Section 3 of this manual

4. If equipped with air conditioning, discharge and recover the air conditioning system refrigerant.

5. Disconnect the refrigerant lines from the evaporator. Plug the refrigerant openings to prevent evaporation.

6. Disconnect the heater hoses from the heater core tubes.

7. Disconnect the heater/air conditioning system vacuum supply line connector from the T-fitting near the heater core tubes.

8. In the engine compartment, remove the 5 heater/air conditioning housing-to-chassis nuts. If necessary, loosen the battery hold-downs and reposition the battery for access.

9. Remove the cowl plenum drain tube from the heater/air conditioning housing stud; it's located behind the cylinder head on the cowl.

10. From the bottom of the heater/air conditioning housing, remove the floor duct.

11. On the passenger side, remove the heater/air conditioning housing-to-plenum bracket screw.

12. Pull the heater/air conditioning housing down far enough to clear the defrost/demist and fresh air ducts, then, rearward far enough to clear the mounting studs and the evaporator drain tube to clear the dash panel holes.

13. Remove the heater/air conditioning housing assembly from the vehicle.

14. Remove the heater/air conditioning housing upper case.

15. Lift the heater core from the lower half of the heater/air conditioning housing.

To install:

16. Assemble the heater core into the lower half of the heater/air conditioning housing.

17. Install the heater/air conditioning housing upper case.

18. Install the heater/air conditioning housing assembly to the vehicle.

19. On the passenger's side, install the heater/air conditioning housing-to-plenum bracket screw.

20. At the bottom of the heater/air conditioning housing, install the floor duct.

21. Install the cowl plenum drain tube to the heater/air conditioning housing stud; it's located behind the cylinder head on the cowl.

22. In the engine compartment, install the 5 heater/air conditioning housing-to-chassis nuts.

23. Connect the heater/air conditioning system vacuum supply line connector to the T-fitting near the heater core tubes.

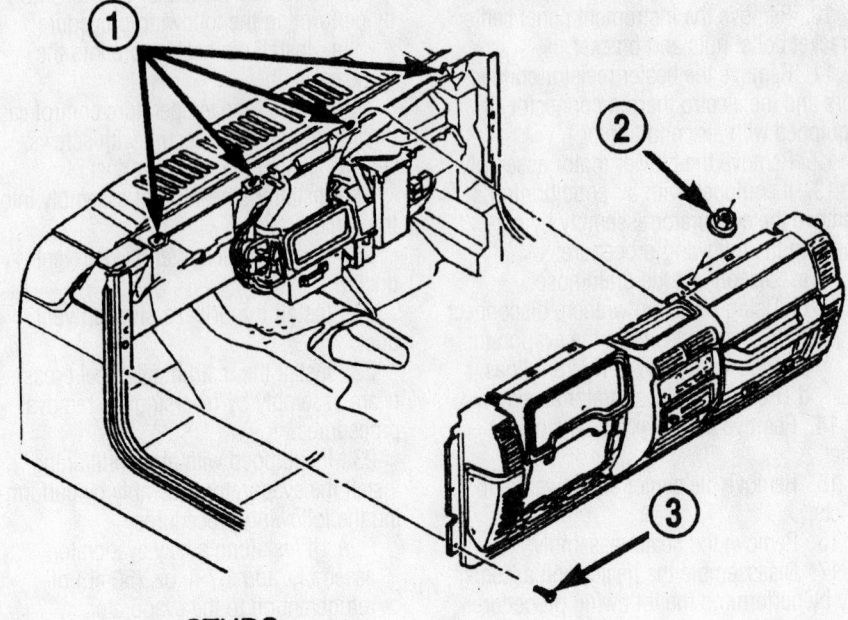

1 – STUDS
2 – NUT
3 – SCREW

93113GA6

View of the instrument panel and fasteners—Jeep Cherokee

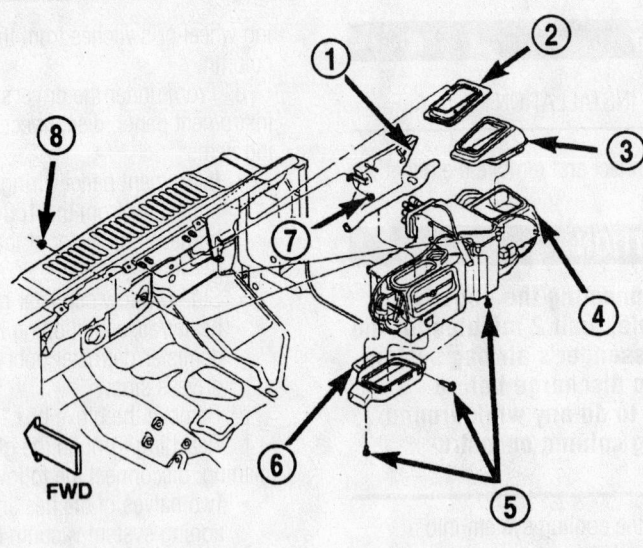

1 – DEFROST/DEMIST DUCT
2 – COLLAR
3 – FRESH AIR DUCT
4 – HEATER-A/C HOUSING
5 – SCREWS
6 – FLOOR DUCT
7 – NUT
8 – NUT

93113GA7

Exploded view of the heater core assembly—Jeep Cherokee

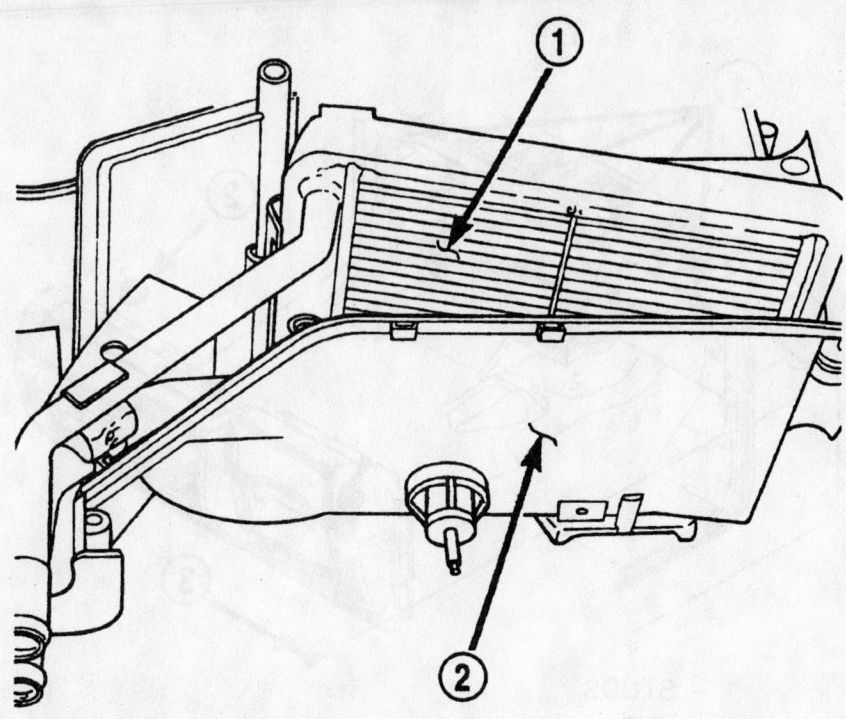

1 – HEATER CORE
2 – LOWER HEATER-A/C HOUSING

93113GA8

View of the heater core—Jeep Cherokee

24. Connect the heater hoses to the heater core tubes.

25. Connect the refrigerant lines to the evaporator.

26. If equipped with air conditioning, evacuate and charge the air conditioning system refrigerant.

27. Install the instrument panel by performing the following procedure:

   a. Using an assistant, install the instrument panel to the vehicle.

   b. Install the 4 upper instrument panel-to-dash nuts.

   c. Install the 3 instrument panel-to-door hinge pillar screws.

   d. Install the upper cover to the instrument panel.

   e. Reaching through the glove box opening, connect the following items.

   • Two lower passenger's side air bag module bracket-to-dash panel nuts.

   • Passenger's side air bag ON/OFF switch wiring harness to the retainer clip on the plenum bracket that supports the heater/air conditioning housing just inboard of the fuse block module.

   • Two instrument panel wiring harness connectors to the passenger air bag ON/OFF switch wiring harness connector.

   • Two halves of the radio antenna coaxial cable connector.

   • Side window demister hose at the heater/air conditioning housing demister/defroster duct (passenger's side).

   • Instrument panel wiring harness connector to the passenger's side air bag module wiring harness connector.

   • Instrument panel wiring harness connector to the heater/air conditioning system wiring harness connector.

   • Two halves of the heater/air conditioning system vacuum harness connector.

   f. Install the glove box.

   g. Under the driver's side of the instrument panel, connect the following items:

   • Side window demister hose at the heater/air conditioning housing demister/defroster duct on the driver's side.

   • Instrument panel wiring harness connector to the 100-way wiring harness connector at the left side of the inner panel.

   h. Install the steering column.

   i. Install the knee blocker to the instrument panel.

28. Connect and remove the negative battery.

29. Refill the cooling system.

30. Run the engine to normal operating temperatures; then, check the climate control operation and check for leaks.

Wrangler

REMOVAL & INSTALLATION

1. Disconnect and remove the negative battery.

✳✳ CAUTION

After disconnecting the negative battery cable, wait 2 minutes for the driver's/passenger's air bag system capacitor to discharge before attempting to do any work around the steering column or instrument.

2. Drain the cooling system into a clean container for reuse.

3. Remove the instrument panel by performing the following procedure:

   a. Turn the steering wheel in the straight-ahead position.

   b. Remove the knee blocker from the instrument panel.

   c. Remove the steering column; do not remove the air bag module, the steering wheel or switches from the steering column.

   d. From under the driver's side of the instrument panel, disconnect the following items:

   • Instrument panel wiring harness connector from the 100-way wiring harness connector at the left side of the inner panel.

   • Side window demister hose at the heater/air conditioning housing demister/defroster duct on the driver's side.

   e. Remove the glove box.

   f. Reaching through the glove box opening, disconnect the following items:

   • Two halves of the heater/air conditioning system vacuum harness connector.

   • Instrument panel wiring harness

Heater Core replacement is covered in Section 2 of this manual

connector from the heater/air conditioning system wiring harness connector.

- Instrument panel wiring harness connector from the passenger's side air bag module wiring harness connector.
- Side window demister hose at the heater/air conditioning housing demister/defroster duct (passenger's side).
- Two halves of the radio antenna coaxial cable connector.
- Two instrument panel wiring harness connectors from the passenger air bag ON/OFF switch wiring harness connector.
- Passenger's side air bag ON/OFF switch wiring harness from the retainer clip on the plenum bracket that supports the heater/air conditioning housing just inboard of the fuse block module.
- Two lower passenger's side air bag module bracket-to-dash panel nuts.

g. Remove the upper cover from the instrument panel.

h. Remove the 3 instrument panel-to-door hinge pillar screws.

i. Remove the 4 upper instrument panel-to-dash nuts.

j. Using an assistant, remove the instrument panel from the vehicle.

4. If equipped with air conditioning, discharge and recover the air conditioning system refrigerant.

5. Disconnect the refrigerant lines from the evaporator. Plug the refrigerant openings to prevent evaporation.

6. Disconnect the heater hoses from the heater core tubes.

7. Disconnect the heater/air conditioning system vacuum supply line connector from the T-fitting near the heater core tubes.

8. In the engine compartment, remove the 5 heater/air conditioning housing-to-chassis nuts. If necessary, loosen the battery hold-downs and reposition the battery for access.

9. Remove the cowl plenum drain tube from the heater/air conditioning housing stud; it's located behind the cylinder head on the cowl.

10. From the bottom of the heater/air conditioning housing, remove the floor duct.

11. On the passenger side, remove the heater/air conditioning housing-to-plenum bracket screw.

12. Pull the heater/air conditioning housing down far enough to clear the defrost/demist and fresh air ducts, then,

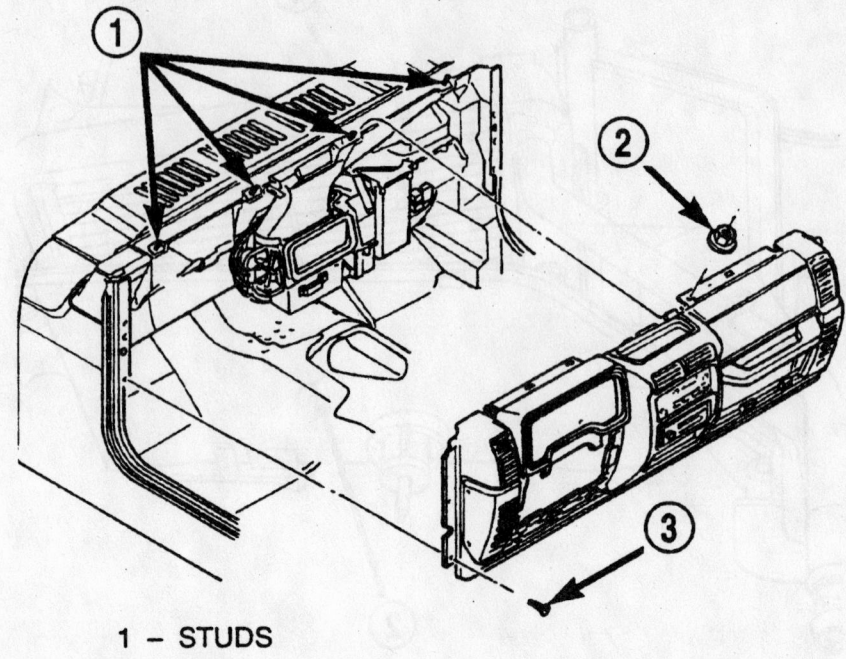

1 – STUDS
2 – NUT
3 – SCREW

93113GA6

View of the instrument panel and fasteners—Jeep Wrangler

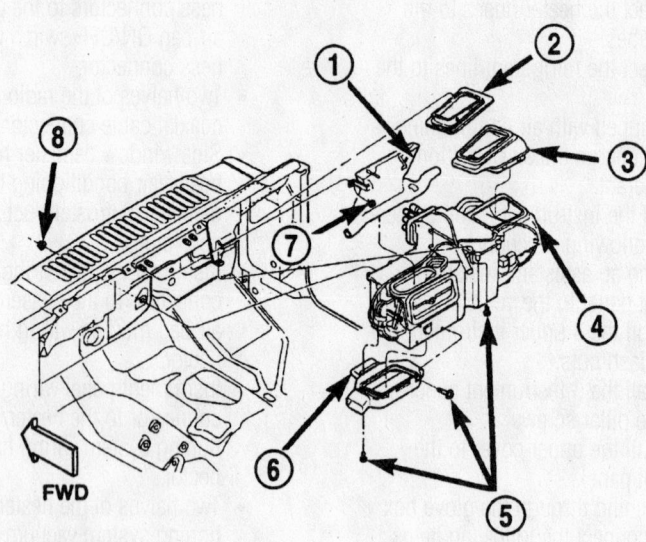

1 – DEFROST/DEMIST DUCT
2 – COLLAR
3 – FRESH AIR DUCT
4 – HEATER-A/C HOUSING
5 – SCREWS
6 – FLOOR DUCT
7 – NUT
8 – NUT

93113GA7

Exploded view of the heater core assembly—Jeep Wrangler

rearward far enough to clear the mounting studs and the evaporator drain tube to clear the dash panel holes.

13. Remove the heater/air conditioning housing assembly from the vehicle.

14. Remove the heater/air conditioning housing upper case.

15. Lift the heater core from the lower half of the heater/air conditioning housing.

To install:

16. Assemble the heater core into the lower half of the heater/air conditioning housing.

17. Install the heater/air conditioning housing upper case.

18. Install the heater/air conditioning housing assembly to the vehicle.

19. On the passenger's side, install the heater/air conditioning housing-to-plenum bracket screw.

20. At the bottom of the heater/air conditioning housing, install the floor duct.

21. Install the cowl plenum drain tube to the heater/air conditioning housing stud; it's located behind the cylinder head on the cowl.

22. In the engine compartment, install the 5 heater/air conditioning housing-to-chassis nuts.

23. Connect the heater/air conditioning system vacuum supply line connector to the T-fitting near the heater core tubes.

24. Connect the heater hoses to the heater core tubes.

25. Connect the refrigerant lines to the evaporator.

26. If equipped with air conditioning, evacuate and charge the air conditioning system refrigerant.

27. Install the instrument panel by performing the following procedure:

a. Using an assistant, install the instrument panel to the vehicle.

b. Install the 4 upper instrument panel-to-dash nuts.

c. Install the 3 instrument panel-to-door hinge pillar screws.

d. Install the upper cover to the instrument panel.

e. Reaching through the glove box opening, connect the following items.

- Two lower passenger's side air bag module bracket-to-dash panel nuts.
- Passenger's side air bag ON/OFF switch wiring harness to the retainer clip on the plenum bracket that supports the heater/air condi-

tioning housing just inboard of the fuse block module.

- Two instrument panel wiring harness connectors to the passenger air bag ON/OFF switch wiring harness connector.
- Two halves of the radio antenna coaxial cable connector.
- Side window demister hose at the heater/air conditioning housing demister/defroster duct (passenger's side).
- Instrument panel wiring harness connector to the passenger's side air bag module wiring harness connector.
- Instrument panel wiring harness connector to the heater/air conditioning system wiring harness connector.
- Two halves of the heater/air conditioning system vacuum harness connector.

f. Install the glove box.

g. Under the driver's side of the instrument panel, connect the following items:

- Side window demister hose at the heater/air conditioning housing demister/defroster duct on the driver's side.
- Instrument panel wiring harness connector to the 100-way wiring harness connector at the left side of the inner panel.

h. Install the steering column.

i. Install the knee blocker to the instrument panel.

28. Connect and remove the negative battery.

29. Refill the cooling system.

30. Run the engine to normal operating temperatures; then, check the climate control operation and check for leaks.

Grand Cherokee

REMOVAL & INSTALLATION

1. Disconnect and remove the negative battery.

※※ CAUTION

After disconnecting the negative battery cable, wait 2 minutes for the driver's/passenger's air bag system capacitor to discharge before attempting to do any work around the steering column or instrument.

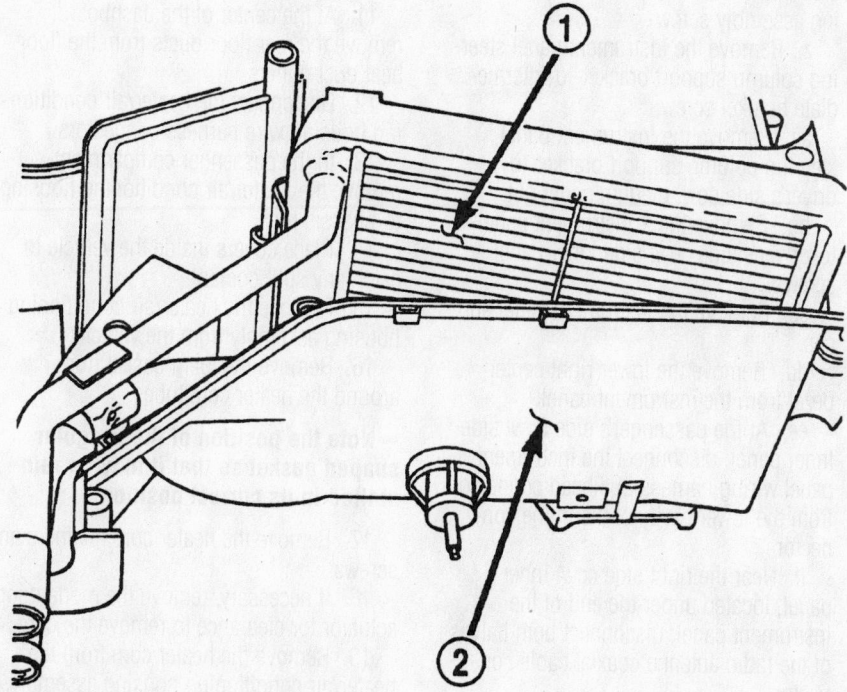

1 – HEATER CORE
2 – LOWER HEATER-A/C HOUSING

93113GA8

View of the heater core—Jeep Cherokee Wrangler

Brake service is covered in Section 4 of this manual

2. Drain the cooling system into a clean container for reuse.

3. Remove the instrument panel by performing the following procedure:

a. Turn the steering wheel in the straight-ahead position.

b. Remove the A-pillar trim from both sides of the vehicle.

c. Remove the top cover from the instrument panel.

d. Near the windshield line, remove the 4 instrument panel-to-chassis nuts.

e. Remove the scuff plates from both front door sills.

f. Remove the trim panels from both sides of the inner cowl.

g. Remove the floor console.

h. Remove the fuse cover from the junction box.

i. Remove the instrument panel cluster bezel.

j. Remove the steering column opening cover from the instrument panel.

k. Remove the steering column bracket from the instrument panel column support bracket.

l. Remove the lower steering column shroud cover-to-multifunction switch screw; then, unsnap both halves of the shroud cover from the steering column.

m. Disconnect the instrument panel wiring harness connectors from the following steering column components:

- Both lower clockspring connector receptacles
- Left multifunction switch receptacle
- Right multifunction switch receptacle
- Both ignition switch receptacles
- Shifter interlock solenoid receptacle
- Sentry Key Immobilizer Module (SKIM) receptacle, if equipped

n. Turn the ignition switch to ON position; then, release and remove the shifter interlock cable connector from the ignition lock housing receptacle.

o. Turn the ignition switch to OFF position; this will prevent the steering wheel from turning and the loss of the clockspring centering following steering column removal.

p. Remove the 4 steering column-to-instrument panel steering column bracket nuts.

q. Remove the steering column from the instrument panel.

r. Disconnect both side body wiring harness bulkhead connectors, the Ignition Off Draw (IOD) wiring harness con-

nector and the fused wiring harness connector from the junction block connector receptacles.

s. Disconnect the instrument panel wiring harness-to-floor console component connectors:

- Air bag control module connector receptacle
- Parking brake switch terminal
- Transmission shifter connector receptacle

t. Remove the 2 instrument panel wiring harness-to-floor console ground terminals located behind the air bag control module.

u. Disconnect the instrument panel wiring harness-to-floor console retainers.

v. Remove the instrument panel-to-floor console bracket screws and the bracket.

w. Remove the driver's side floor duct-to-heater/air conditioning housing assembly screw and remove the duct.

x. If equipped with a manual heating-air conditioning system, disconnect the vacuum harness connector from behind the driver's side floor duct.

y. Remove the instrument panel steering column support bracket-to-driver's side of the heater/air conditioning housing assembly screw

z. Remove the instrument panel steering column support bracket-to-intermediate bracket screw.

aa. Remove the instrument panel steering column support bracket-to-driver's side cowl plenum panel nut.

bb. Remove the 2 instrument panel-to-driver's side cowl side inner panel screws.

cc. Remove the instrument panel end cap.

dd. Remove the lower right center bezel from the instrument panel.

ee. At the passenger's side cowl side inner panel, disconnect the instrument panel wiring harness bulkhead connector from the lower cavity of the inline connector.

ff. Near the right side cowl inner panel, located under the end of the instrument panel, disconnect both halves of the radio antenna coaxial cable connector.

gg. Disconnect the 2 instrument panel-to-heater/air conditioning assembly wiring harness connectors.

hh. At the passenger's side, remove the 2 instrument panel structural duct-to-

heater/air conditioning housing assembly screws.

ii. At the passenger's side cowl side inner panel, remove the 2 instrument panel-to-passenger's side cowl side inner panel screws.

jj. With the help of an assistant, lift the instrument panel from the vehicle.

4. Discharge and recover the air conditioning system refrigerant.

5. Disconnect the air conditioning system lines at the evaporator. Plug the openings to prevent contamination.

6. Disconnect the heater hoses from the heater core. Plug the openings to prevent coolant loss.

7. If equipped with a manual temperature control system, unplug the heater/air conditioning system vacuum supply line connector from the T-fitting located near the heater core tubes.

8. From the passenger's side inner fender shield, remove the coolant reservoir/overflow bottle.

9. From the passenger's side in the engine compartment dash panel, remove the PCM; DO NOT unplug it, just move it aside.

10. In the engine compartment, remove the heater/air conditioning housing-to-chassis nuts.

11. At the center of the dashboard, remove the rear floor ducts from the floor heat duct outlets.

12. Disconnect the heater/air conditioning housing wire harness connectors.

13. In the passenger compartment, remove the heater/air conditioning housing-to-chassis nuts.

14. Place covers inside the vehicle to catch any spilt coolant.

15. Remove the heater/air conditioning housing assembly from the vehicle.

16. Remove the foam gasket from around the heater core tubes.

➡ **Note the position of the irregular shaped gasket so that it may be reinstalled in its correct position.**

17. Remove the heater core retainers and screws.

18. If necessary, remove the mode door actuator for clearance to remove the core.

19. Remove the heater core from the heater/air conditioning housing assembly.

To install:

20. Install the heater core tom the heater/air conditioning housing assembly.

21. If removed, install the mode door actuator.

1 – COLUMN MOUNTING NUTS
2 – COUPLER BOLT

93113GA1

View of the steering column mounting nuts—Jeep Grand Cherokee

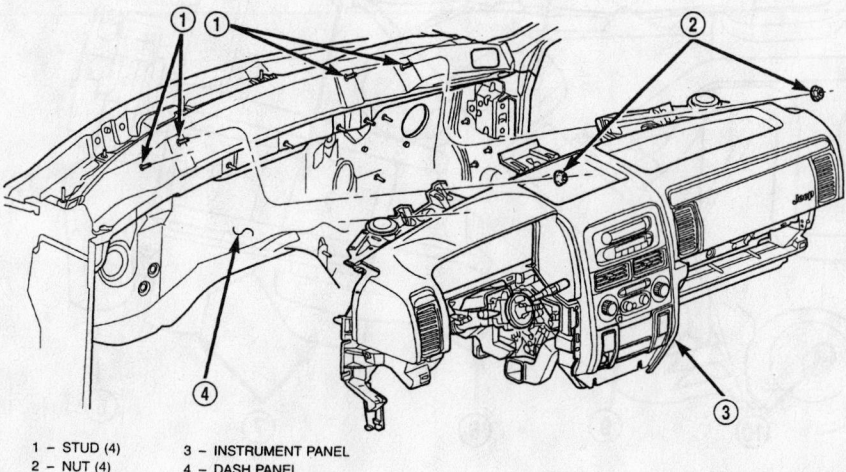

1 – STUD (4)   3 – INSTRUMENT PANEL
2 – NUT (4)   4 – DASH PANEL

93113GA2

View of the instrument panel assembly— Jeep Grand Cherokee

22. Install the heater core retainers and screws.

23. Install the foam gasket around the heater core tubes.

24. Install the heater/air conditioning housing assembly to the vehicle.

25. In the passenger compartment, install the heater/air conditioning housing-to-chassis nuts.

26. Connect the heater/air conditioning housing wire harness connectors.

27. At the center of the dashboard, install the rear floor ducts to the floor heat duct outlets.

28. In the engine compartment, install the heater/air conditioning housing-to-chassis nuts.

29. At the passenger's side in the engine compartment dash panel, install the PCM.

30. At the passenger's side inner fender shield, install the coolant reservoir/overflow bottle.

31. If equipped with a manual temperature control system, plug the heater/air conditioning system vacuum supply line connector to the T-fitting located near the heater core tubes.

32. Connect the heater hoses to the heater core.

33. Connect the air conditioning system lines to the evaporator.

34. Evacuate and charge the air conditioning system refrigerant.

35. Remove the instrument panel by performing the following procedure:

   a. Turn the steering wheel in the straight-ahead position.

   b. Install the A-pillar trim to both sides of the vehicle.

   c. Install the top cover to the instrument panel.

   d. Near the windshield line, install the 4 instrument panel-to-chassis nuts.

   e. Install the scuff plates to both front door sills.

   f. Install the trim panels to both sides of the inner cowl.

   g. Install the floor console.

   h. Install the fuse cover to the junction box.

   i. Install the instrument panel cluster bezel.

   j. Install the steering column opening cover to the instrument panel.

   k. Install the steering column bracket to the instrument panel column support bracket.

   l. Install the lower steering column shroud cover-to-multifunction switch

For complete Engine Mechanical specifications, see Section 1 of this manual

screw; then, snap both halves of the shroud cover to the steering column.

m. Connect the instrument panel wiring harness connectors to the following steering column components:

- Sentry Key Immobilizer Module (SKIM) receptacle, if equipped
- Shifter interlock solenoid receptacle
- Both ignition switch receptacles
- Right multifunction switch receptacle
- Left multifunction switch receptacle
- Both lower clockspring connector receptacles

n. Turn the ignition switch to ON position; then, release and install the shifter interlock cable connector to the ignition lock housing receptacle.

o. Install the 4 steering column-to-instrument panel steering column bracket nuts.

p. Install the steering column to the instrument panel.

q. Connect both side body wiring harness bulkhead connectors, the Ignition Off Draw (IOD) wiring harness connector and the fused wiring harness connector to the junction block connector receptacles.

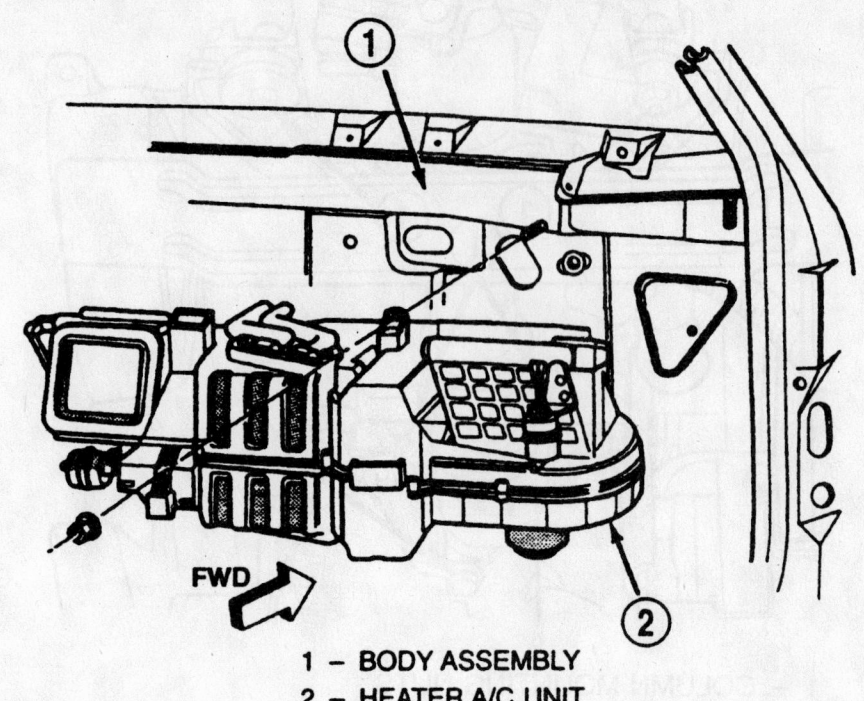

1 – BODY ASSEMBLY
2 – HEATER A/C UNIT

93113GA3

View of the heater/air conditioning housing assembly—Jeep Grand Cherokee

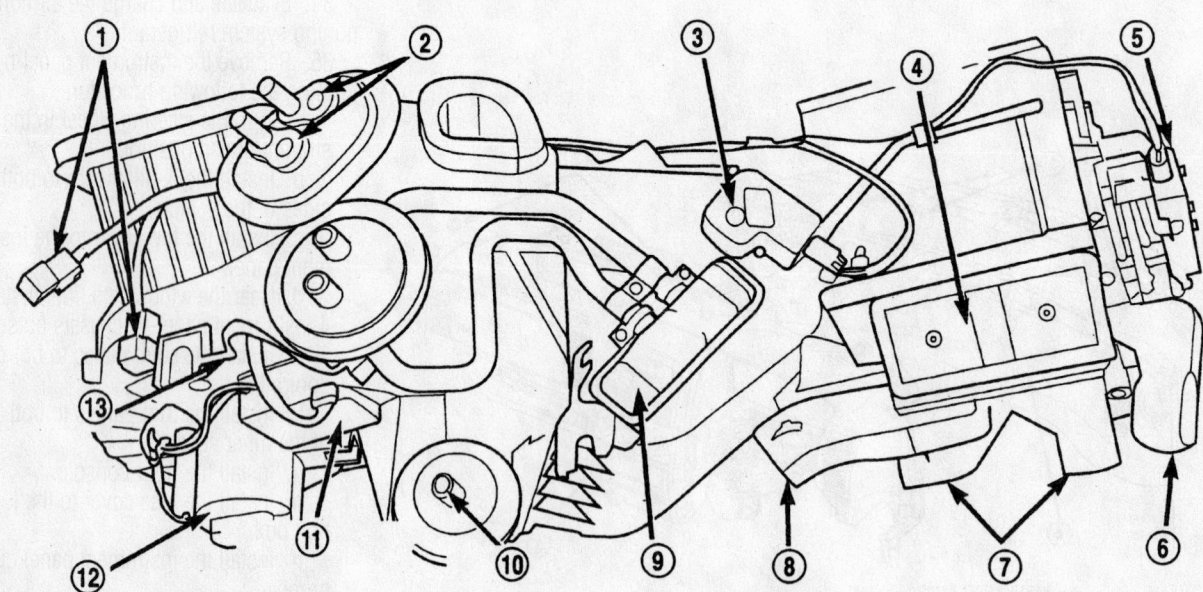

1 – ELECTRICAL CONNECTORS	8 – FLOOR DUCT
2 – EVAPORATOR FITTINGS (CAPPED)	9 – HEATER CORE AND TUBES
3 – ELECTRIC ACTUATOR	10 – HOUSING DRAIN
4 – OUTLET TO DEFROSTER DUCTS	11 – BLOWER MOTOR CONTROLLER/POWER MODULE
5 – ELECTRIC ACTUATOR	12 – BLOWER MOTOR
6 – FLOOR DUCT	13 – GROUND STRAP
7 – TO REAR PASSENGER FLOOR AIR DUCTS	

93113GA4

View of the heater core, the heater/air conditioning housing and related components—Jeep Grand Cherokee

r. Disconnect the instrument panel wiring harness-to-floor console component connectors:

- Transmission shifter connector receptacle
- Parking brake switch terminal
- Air bag control module connector receptacle

s. Install the 2 instrument panel wiring harness-to-floor console ground terminals located behind the air bag control module.

t. Connect the instrument panel wiring harness-to-floor console retainers.

u. Install the instrument panel-to-floor console bracket and the screws.

v. Install the driver's side floor duct and the duct-to-heater/air conditioning housing assembly screw.

w. If equipped with a manual heating-air conditioning system, connect the vacuum harness connector behind the driver's side floor duct.

x. Install the instrument panel steering column support bracket-to-driver's side of the heater/air conditioning housing assembly screw.

y. Install the instrument panel steering column support bracket-to-intermediate bracket screw.

z. Install the instrument panel steering column support bracket-to-driver's side cowl plenum panel nut.

aa. Install the 2 instrument panel-to-driver's side cowl side inner panel screws.

bb. Install the instrument panel end cap.

cc. Install the lower right center bezel to the instrument panel.

dd. At the passenger's side cowl side inner panel, connect the instrument panel wiring harness bulkhead connector to the lower cavity of the inline connector.

ee. Near the right cowl side inner panel located under the end of the instrument panel, connect both halves of the radio antenna coaxial cable connector.

ff. Connect the 2 instrument panel-to-heater/air conditioning assembly wiring harness connectors.

gg. At the passenger's side, install the 2 instrument panel structural duct-to-heater/air conditioning housing assembly screws.

hh. At the passenger's side cowl side inner panel, install the 2 instrument

panel-to-passenger's side cowl side inner panel screws.

ii. With the help of an assistant, lift the instrument panel into the vehicle.

36. Refill the cooling system.

37. Connect the negative battery.

38. Run the engine to normal operating temperatures. Check the climate control operation and check for leaks.

Liberty

REMOVAL & INSTALLATION

1. Disconnect and remove the negative battery.

✳✳ CAUTION

After disconnecting the negative battery cable, wait 2 minutes for the driver's/passenger's air bag system capacitor to discharge before attempting to do any work around the steering column or instrument.

2. Drain the cooling system into a clean container for reuse.

3. Remove the instrument panel by performing the following procedure:

a. Turn the steering wheel in the straight-ahead position.

b. Remove the A-pillar trim from both sides of the vehicle.

c. Remove the top cover from the instrument panel.

d. Remove the speakers.

e. Remove the floor console.

f. Remove the radio.

g. Remove the center support bracket.

h. Remove the trim panels from both sides of the inner cowl.

i. Remove the fuse cover from the junction box.

j. Remove the instrument panel cluster bezel.

k. Remove the steering column opening cover from the instrument panel.

l. Remove the steering column bracket from the instrument panel column support bracket.

m. Remove the lower steering column shroud cover-to-multifunction switch screw; then, unsnap both halves of the shroud cover from the steering column.

n. Disconnect the instrument panel wiring harness connectors from the following steering column components:

- Both lower clockspring connector receptacles

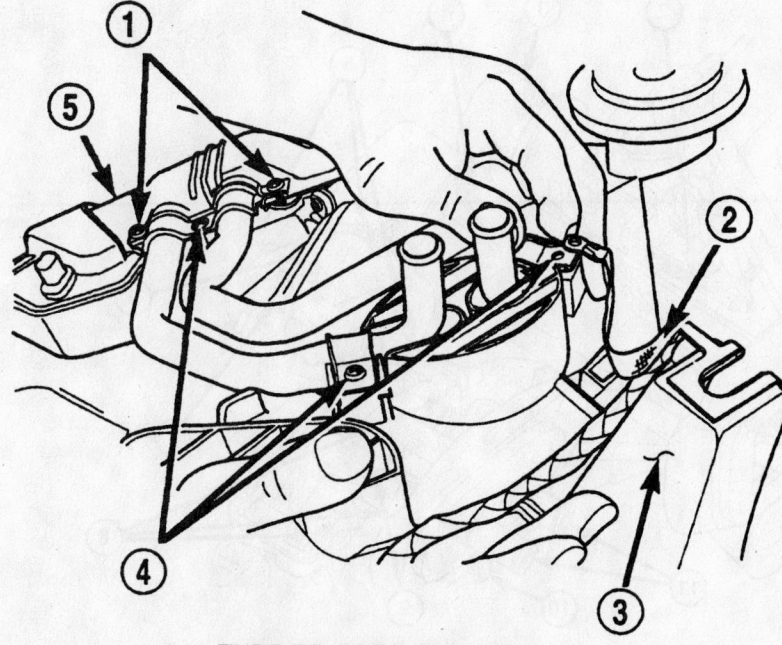

1 – TUBE-TO-CORE CLAMPS
2 – GROUND STRAP
3 – HVAC HOUSING
4 – TUBE RETAINERS AND SCREWS
5 – HEATER CORE

93113GA5

View of the heater core screws, gasket and retainers—Jeep Grand Cherokee

For Accessory Drive Belt illustrations, see Section 1 of this manual

- Left multifunction switch receptacle
- Right multifunction switch receptacle
- Both ignition switch receptacles
- Shifter interlock solenoid receptacle
- Sentry Key Immobilizer Module (SKIM) receptacle, if equipped

o. Turn the ignition switch to ON position; then, release and remove the shifter interlock cable connector from the ignition lock housing receptacle.

p. Turn the ignition switch to OFF position; this will prevent the steering wheel from turning and the loss of the clockspring centering following steering column removal.

q. Remove the 4 steering column-to-instrument panel steering column bracket nuts.

r. Remove the steering column from the instrument panel.

4. Remove the driver's side cowl trim cover.

5. Disconnect the green and light blue wire harness bulk connectors at the junction block.

6. Disconnect the electrical connector at the inner side of the pedal support bracket.

7. Remove the 2 bolts from the front and the 2 from the side of the pedal support bracket.

8. Remove the glove box.

9. Remove the 2 HVAC mounting bolts behind the center trim.

10. Remove the passenger trim bezel.

11. Remove the HVAC mount bolt above the glove box.

12. Remove the HVAC bolt at the lower outside glove box opening.

13. Remove the passenger trim cover, disconnect the blower resistor, remove the rolldown brackets ta the right cowl side panel.

14. Disconnect the vacuum check valve and the vacuum reservoir.

15. Disconnect the blower connectors.

16. Remove the 4 top bolts connecting the instrument panel to the cowl.

17. Roll the instrument panel rearward and disconnect the wiring.

18. Remove the panel.

19. Discharge and recover the air conditioning system refrigerant.

20. Disconnect the air conditioning system lines at the evaporator. Plug the openings to prevent contamination.

21. Disconnect the heater hoses from the heater core. Plug the openings to prevent coolant loss.

22. If equipped with a manual temperature control system, unplug the heater/air conditioning system vacuum supply line connector from the T-fitting located near the heater core tubes.

23. Remove all remaining fasteners and connections and remove the HVAC unit.

24. Disconnect all remaining hoses and wires.

25. Remove the blower motor.

26. Pop out the grommet on the vacuum supply line and slide hole.

27. Remove the foam gasket from around the heater core tubes.

28. Pry off the 4 snap clips that hold the halves of the unit together and separate the unit halves.

29. Installation is the reverse of removal.

30. Refill the cooling system.

31. Connect the negative battery.

32. Evacuate, charge and leak test the system.

33. Run the engine to normal operating temperatures. Check the climate control operation and check for leaks.

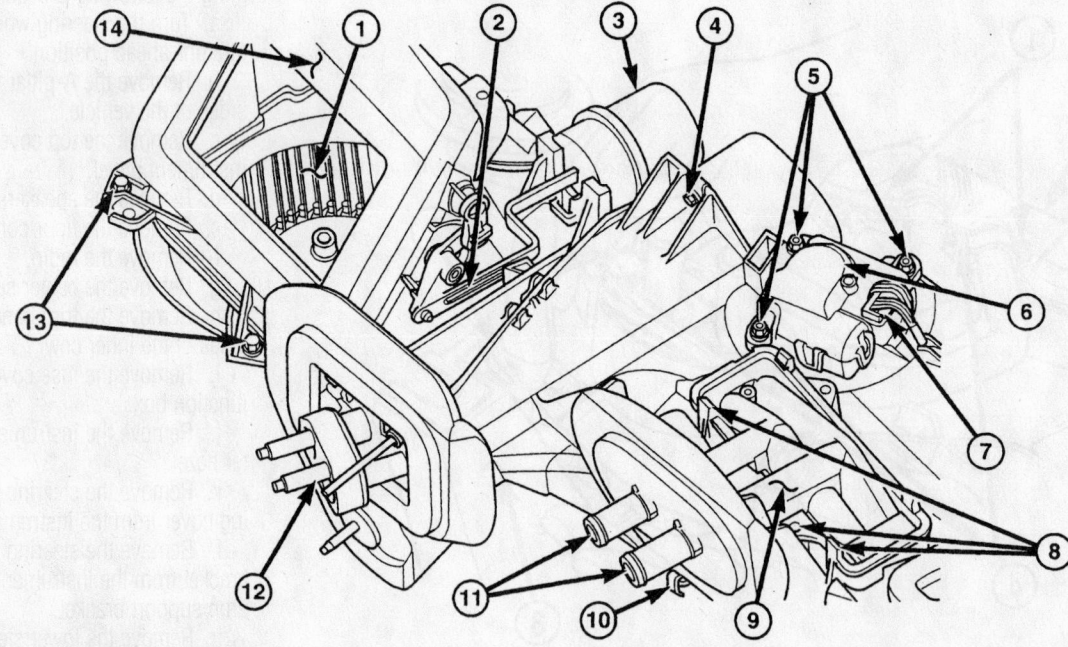

1 - BLOWER MOTOR AND CAGE
2 - RECIRCULATION DOOR ACTUATOR LINKAGE
3 - RECIRCULATION DOOR VACUUM ACTUATOR
4 - CASE RETAINER SCREW
5 - BLEND DOOR ACTUATOR MOUNTING SCREWS
6 - ELECTRIC BLEND DOOR ACTUATOR
7 - ELECTRICAL CONNECTOR FOR BLEND DOOR ACTUATOR
8 - HEATER CORE RETAINER TABS (4) AND SCREWS (2)
9 - HEATER CORE
10 - HVAC CASE RETAINER CLIP
11 - HEATER CORE INPUT AND OUTPUT CONNECTIONS
12 - EVAPORATOR CONNECTION FLANGE
13 - HVAC CASE RETAINER SCREWS
14 - HVAC HOUSING

HVAC case components—Liberty

9355PG99

KIA

Sportage

REMOVAL & INSTALLATION

1. Disconnect the negative battery cable.

> ❊❊ **CAUTION**
>
> **After disconnecting the negative battery cable, wait for at least 10 minutes for the air bag module to deplete its stored energy.**

2. Remove the driver's side air bag and steering wheel by performing the following procedure:

   a. Position the front wheels in the straight-ahead position.

   b. Remove the 4 steering wheel-to-air bag module bolts.

   c. Carefully, lift the air bag module and disconnect the electrical connector.

> ❊❊ **CAUTION**
>
> **Place the air bag module in a safe location with the front facing upward.**

   d. Remove the steering wheel-to-steering column nut.

➡ **If may be necessary to mark the steering wheel to steering column alignment.**

   e. Using a steering wheel puller, press the steering wheel from the steering column.

3. Drain the cooling system into a clean container for reuse.

4. Discharge and recover the air conditioning system refrigerant.

5. Remove the instrument panel by performing the following procedure:

   a. Remove both the rear and front consoles.

   b. Remove the knee bolster assembly.

   c. Remove the "T" bar section.

   d. Remove the relay bracket.

   e. Remove the turn signal assembly and the upper/lower steering column covers.

   f. Remove the hood release handle lockscrew, the hood release handle and the cable assembly nut.

   g. Remove the left side front pillar trim and the lower left side cover.

   h. Remove the 2 left side of the "T" bar-to-chassis bolts.

   i. At the left side of the instrument panel, remove the 3 instrument panel-to-chassis bolts.

   j. Remove the ashtray.

   k. Remove the center panel trim.

   l. Remove the ventilation control panel.

   m. At the center of the windshield next to the windshield, remove the cap and the mounting bolt.

   n. Remove the right side front pillar trim and the lower right side cover.

   o. Remove the 2 right side of the "T" bar-to-chassis bolts.

   p. At the right side of the instrument panel, remove the 3 instrument panel-to-chassis bolts.

   q. Remove the steering column-to-instrument panel bolts and lower the steering column.

   r. Disconnect the instrument panel electrical connectors.

   s. Remove the instrument panel.

6. Remove the blower/evaporator housing by performing the following procedure:

   a. Disconnect the air conditioning

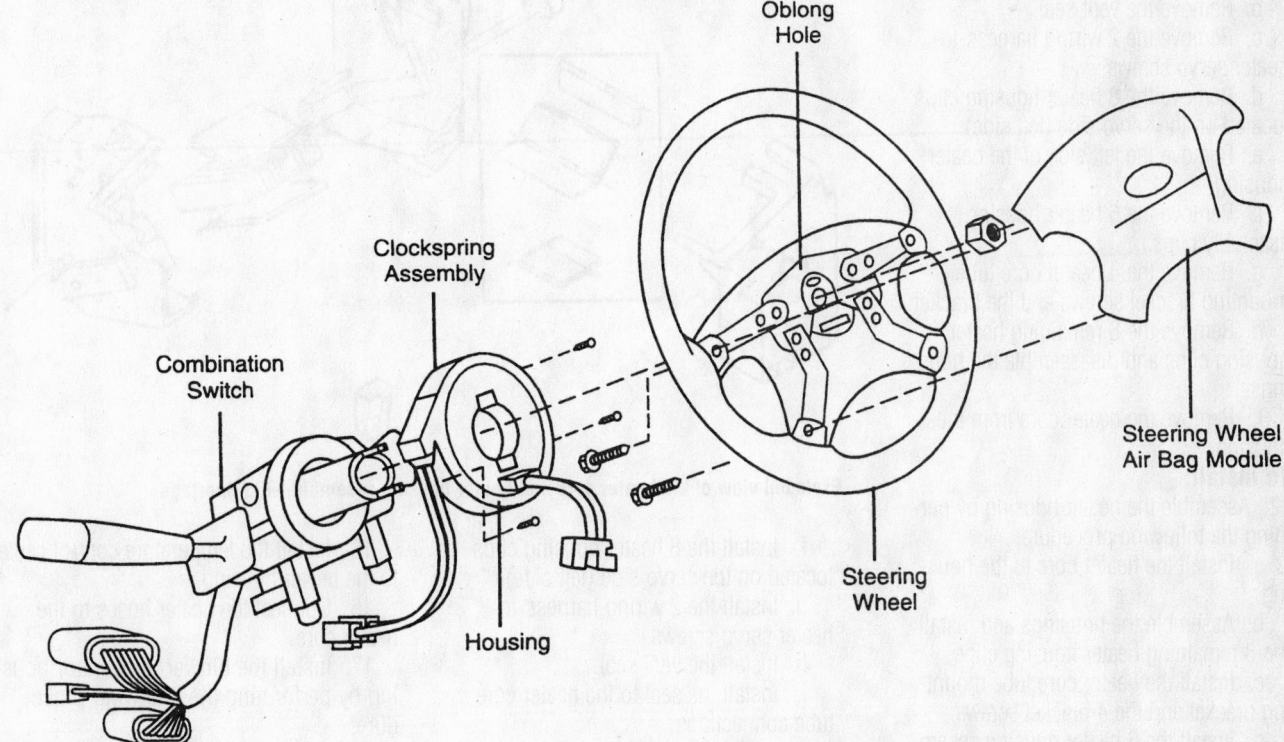

Exploded view of the steering wheel and air bag module assembly—Kia Sportage

93113GI4

For Tire, Wheel and Ball Joint specifications, see Section 1 of this manual

refrigerant lines from the evaporator core and discard the gaskets. Plug the openings to prevent contamination.

b. Disconnect the fresh air control cable from the blower/evaporator housing inlet duct.

c. Disconnect the 5 connectors from the bottom of the blower/evaporator housing.

d. Move the carpeting from the bulkhead to gain access to the hole cover plate.

e. Remove the 4 hole cover plate nuts and the plate.

f. Remove the 2 upper blower/evaporator housing bolts.

g. Remove the 2 lower blower/evaporator housing-to-bulkhead nuts.

h. Remove the blower/evaporator housing.

7. Disconnect the heater hoses from the heater core.

8. Remove the temperature control cable from the heater housing.

9. Remove the 2 lower heater housing nuts and the upper heater housing-to-bulkhead nut.

10. Remove the heater housing.

11. Disassemble the heater housing by performing the following procedure:

a. Remove the seal from the heater core tube connections.

b. Remove the vent seal.

c. Remove the 2 wiring harness-to-heater servo screws.

d. Remove the 8 heater housing clips located on the servo side (left side).

e. Remove the left side of the heater housing.

f. Remove the 6 heater housing assembly clips.

g. Remove the 4 heater core tube mounting bracket screws and the bracket.

h. Remove the 8 remaining heater housing clips and disassemble the housings.

i. Remove the heater core from the housing.

To install:

12. Assemble the heater housing by performing the following procedure:

a. Install the heater core to the housing.

b. Assemble the housings and install the 8 remaining heater housing clips.

c. Install the heater core tube mounting bracket and the 4 bracket screws.

d. Install the 6 heater housing assembly clips.

e. Install the left side of the heater housing.

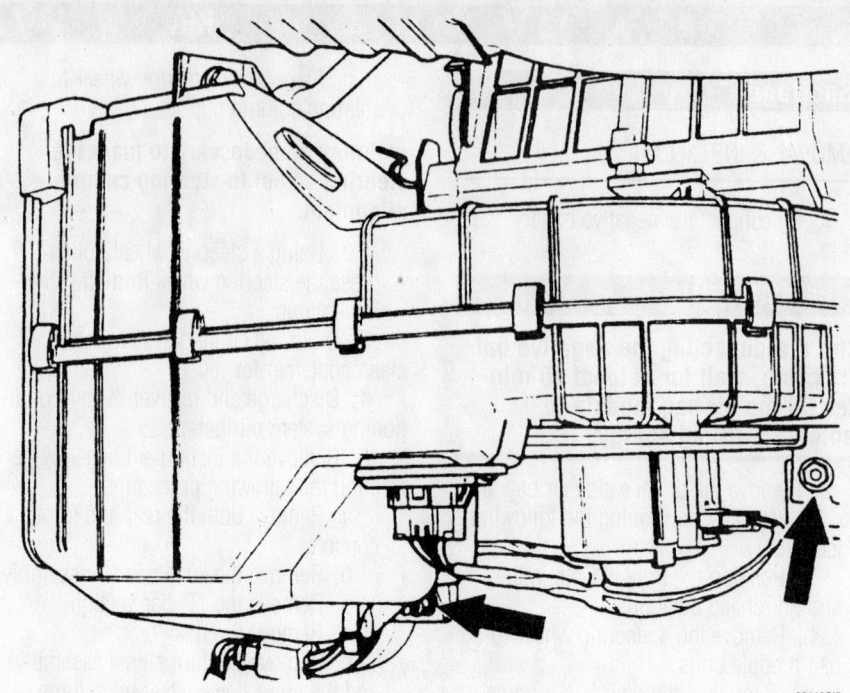

View of the blower/evaporator housing assembly—Kia Sportage

93113GI5

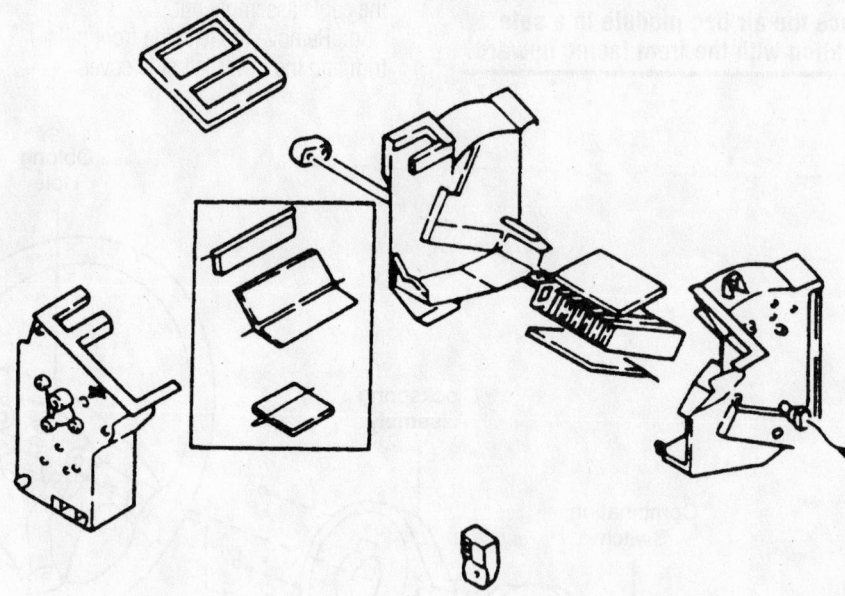

Exploded view of the heater core and heater housing assembly—Kia Sportage

93113GI6

f. Install the 8 heater housing clips located on the servo side (left side).

g. Install the 2 wiring harness-to-heater servo screws.

h. Install the vent seal.

i. Install the seal to the heater core tube connections.

13. Install the heater housing.

14. Install the 2 lower heater housing nuts and the upper heater housing-to-bulkhead nut.

15. Install the temperature control cable to the heater housing.

16. Connect the heater hoses to the heater core.

17. Install the blower/evaporator housing by performing the following procedure:

a. Install the blower/evaporator housing.

b. Install the 2 lower blower/evaporator housing-to-bulkhead nuts.

c. Install the 2 upper blower/evaporator housing bolts.

d. Install the hole cover plate and the 4 plate nuts.

e. Move the carpeting over the bulkhead.

f. Connect the 5 connectors to the bottom of the blower/evaporator housing.

g. Connect the fresh air control cable to the blower/evaporator housing inlet duct.

h. Using new gaskets, connect the air conditioning refrigerant lines to the evaporator core.

18. Install the instrument panel by performing the following procedure:

a. Install the instrument panel.

b. Connect the instrument panel electrical connectors.

c. Install the steering column and lower the steering column-to-instrument panel bolts. Torque the bolts to 15 ft. lbs. (20 Nm).

d. At the right side of the instrument panel, install the 3 instrument panel-to-chassis bolts.

e. Install the 2 right side of the "T" bar-to-chassis bolts.

f. Install the right side front pillar trim and the lower right side cover.

g. At the center of the windshield next to the windshield, install the cap and the mounting bolt.

h. Install the ventilation control panel.

i. Install the center panel trim.

j. Install the ashtray.

k. At the left side of the instrument panel, install the 3 instrument panel-to-chassis bolts.

l. Install the 2 left side of the "T" bar-to-chassis bolts.

m. Install the lower left side cover and the left side front pillar trim.

n. Install the cable assembly nut, the hood release handle and the hood release handle lockscrew.

o. Install the turn signal assembly and the upper/lower steering column covers.

p. Install the relay bracket.

q. Install the "T" bar section.

r. Install the knee bolster assembly.

s. Install both the rear and front consoles.

19. Evacuate and charge the air conditioning system refrigerant.

20. Refill the cooling system.

21. Install the driver's side air bag and steering wheel by performing the following procedure:

a. Install the steering wheel to the steering column.

b. Install the steering wheel-to-steering column nut and torque the nut to 33 ft. lbs. (45 Nm).

c. Carefully, install the air bag module and connect the electrical connector.

d. Install the 4 steering wheel-to-air bag module bolts and torque to 72–106 inch lbs. (8–12 Nm).

22. Connect the negative battery cable.

23. Run the engine to normal operating temperatures; then, check the climate control operation and check for leaks.

LAND ROVER

Discovery

REMOVAL & INSTALLATION

1. Turn the steering wheel 90° from horizontal.

2. Turn the ignition switch OFF.

3. Disconnect the negative (()) battery cable and then the positive (+) battery cable.

➡**Wait at least 20 minutes for the air bag(s) to discharge before performing any work on the system(s).**

4. Drain the cooling system into a clean container for reuse.

5. Discharge and recover the air conditioning system refrigerant.

6. Remove the dash panel assembly by performing the following procedure:

a. Move the seats rearward.

b. Working under the dash panel assembly, disconnect the air bag multi-plug electrical connectors.

c. Remove the glove box.

d. Remove the center console assembly.

e. Remove the driver's SRS module, by disconnecting or removing following items:

- Under the steering wheel, release the lower dash panel cover turnbuckles and remove the lower dash panel.

- Disconnect the air bag harness connector from the yellow air bag column harness.

- Using a special socket, remove the 2 tamper-proof resistor air bag module-to-steering wheel screws.

- Remove the air bag module from the steering wheel.

✳✳ CAUTION

Do not allow the air bag module to hang by electrical harness.

- Disconnect the air bag harness connector.

- Remove the air bag module.

f. Remove the passenger's SRS module, by disconnecting or removing following items:

- Open the glove box door and disconnect the air bag module electrical connector.

- Using a special socket and a long extension, remove both front air bag module-to-dash panel screws.

- Using a Torx® socket, remove both rear air bag module-to-dash panel screws.

- Remove the air bag module from the dash panel.

✳✳ CAUTION

Do not allow the air bag module to hang by electrical harness.

- Disconnect the air bag harness connector.

g. Release the clamp and lower the steering column.

h. Remove the steering wheel and the steering column switch.

i. Remove the instrument housing.

j. Disconnect the electrical connectors and remove the radio.

k. Remove the exterior mirrors switch panel and the coin tray.

l. Remove the switch panel and the clock.

m. Remove the passenger's side relay assembly mounting bracket screw and move the assembly aside.

n. Turn the heater controls fully clockwise. Noting the position of the levers, disconnect the heater control cables from the levers and outer cable from the retaining clips.

o. Remove the 4 dash panel-to-center lower mounting bracket bolts.

p. Remove the 4 dash panel-to-side lower mounting bracket bolts.

q. Working below the steering column, remove the 4 driver's knee bolster pads screws and both knee bolster pads.

   r. Remove the 4 instrument bracket-to-dash panel nuts.

   s. Using an assistant, partially maneuver the dash panel rearward.

   t. At the driver's side, disconnect the 6 dash harness-to-main harness multi-plugs connectors.

   u. Disconnect the 3 dash harness-to-fusebox multi-plug connectors.

   v. Using an assistant, remove the dash panel assembly from the vehicle.

   7. In the engine compartment, disconnect the heater hoses from the heater tubes.

   8. Release the P-clip securing both the high and low air conditioning pressure tubes.

   9. Remove both the high and low pressure tubes-to-evaporator bolts; then, separate the tubes from the evaporator and discard the O-rings.

   10. At the lower right side of the heater/air conditioning housing, disconnect the heater-to-blower motor multi-plug connector.

   11. Remove the 3 blower motor housing screws and remove the blower motor housing.

   12. Remove the 5 heater/air conditioning housing-to-chassis screws.

   13. Remove the 2 center console front mounting bracket-to-chassis bolts and remove the center console front mounting bracket.

   14. Disconnect both drain tubes.

   15. Carefully, remove the heater/air conditioning housing from the vehicle.

   16. Remove both right side footwell outlet-to-heater/air conditioning housing screws and the outlet.

   17. Remove the heater hoses-to-heater assembly clip.

   18. Slide the heater core from the heater/air conditioning housing.

To install:

   19. If installing a new heater core, transfer the heater hoses to the new heater core.

   20. Slide the heater core into the heater/air conditioning housing.

   21. Install the heater hoses-to-heater assembly clip.

   22. Install the right side footwell outlet and both outlet-to-heater/air conditioning housing screws.

   23. Carefully, install the heater/air conditioning housing to the vehicle.

   24. Connect both drain tubes.

   25. Install the center console front mounting bracket and the 2 center console front mounting bracket-to-chassis bolts.

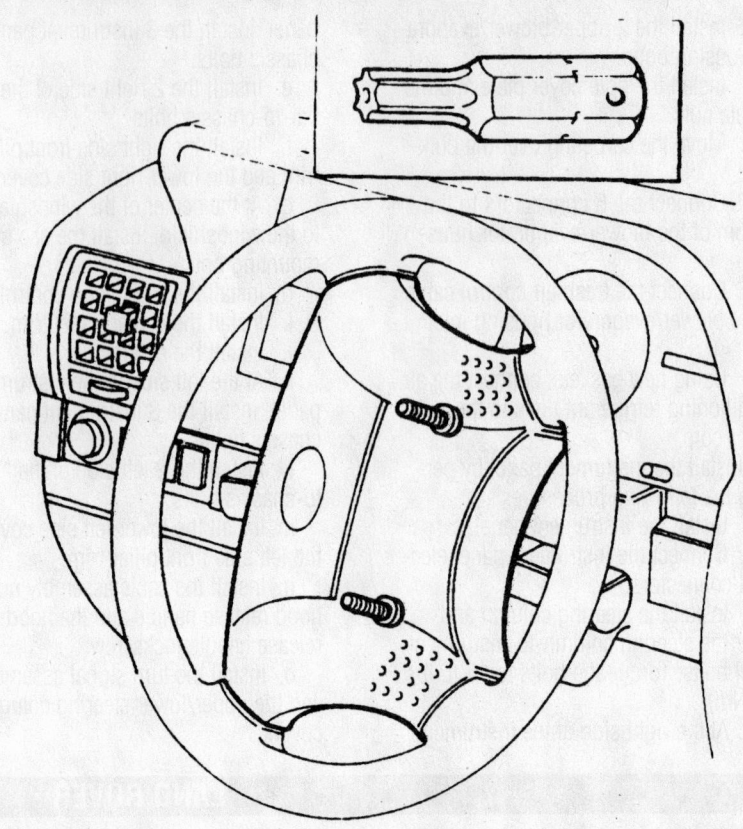

93113GC8

View of the steering wheel, SRS module and special tool—Land Rover Discovery

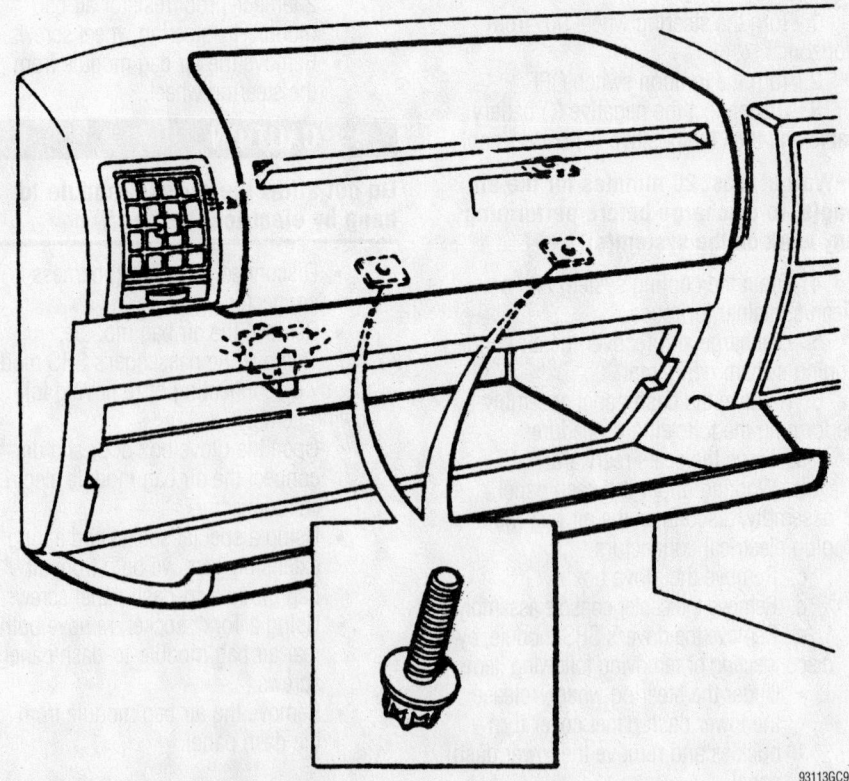

93113GC9

View of the passenger's side SRS module and special screws—Land Rover Discovery

26. Install the 5 heater/air conditioning housing-to-chassis screws.

27. Install the blower motor housing and the 3 blower motor housing screws.

28. At the lower right side of the heater/air conditioning housing, connect the heater-to-blower motor multi-plug connector.

29. Using new O-rings, assemble the tubes to the evaporator and install both the high and low pressure tubes-to-evaporator bolts.

30. Install the P-clip securing both the high and low air conditioning pressure tubes.

31. In the engine compartment, connect the heater hoses to the heater tubes.

32. Install the dash panel assembly by performing the following procedure:

   a. Using an assistant, install the dash panel assembly to the vehicle.

   b. Connect the 3 dash harness-to-fuse box multi-plug connectors.

   c. At the driver's side, connect the 6 dash harness-to-main harness multi-plugs connectors.

   d. Using an assistant, partially maneuver the dash panel forward.

   e. Install the 4 instrument bracket-to-dash panel nuts.

   f. Working below the steering column, install both knee bolster pads and the 4 driver's knee bolster pads screws.

   g. Install the 4 dash panel-to-side lower mounting bracket bolts.

   h. Install the 4 dash panel-to-center lower mounting bracket bolts.

   i. Noting the position of the levers, connect the heater control cables to the levers and outer cable to the retaining clips.

   j. Install the passenger's side relay assembly mounting bracket and the assembly screw.

   k. Install the switch panel and the clock.

   l. Install the exterior mirrors switch panel and the coin tray.

   m. Connect the electrical connectors and install the radio.

   n. Install the instrument housing.

   o. Install the steering column switch and the steering wheel.

   p. Raise the steering column and install the clamp.

   q. Install the passenger's SRS module, by connecting or installing following items:

   • Connect the air bag harness connector.

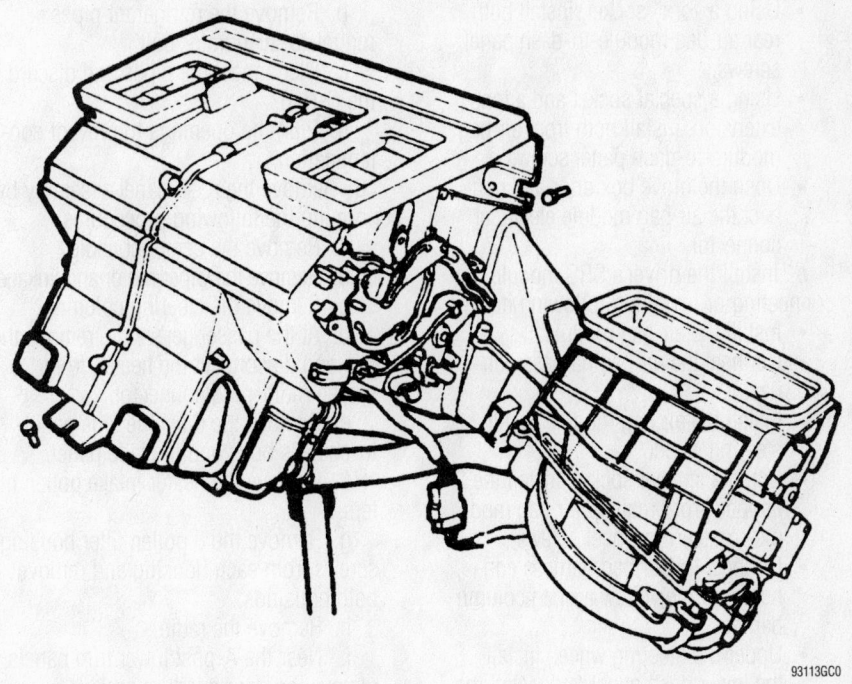

93113GC0

View of the heater/air conditioning housing assembly—Land Rover Discovery

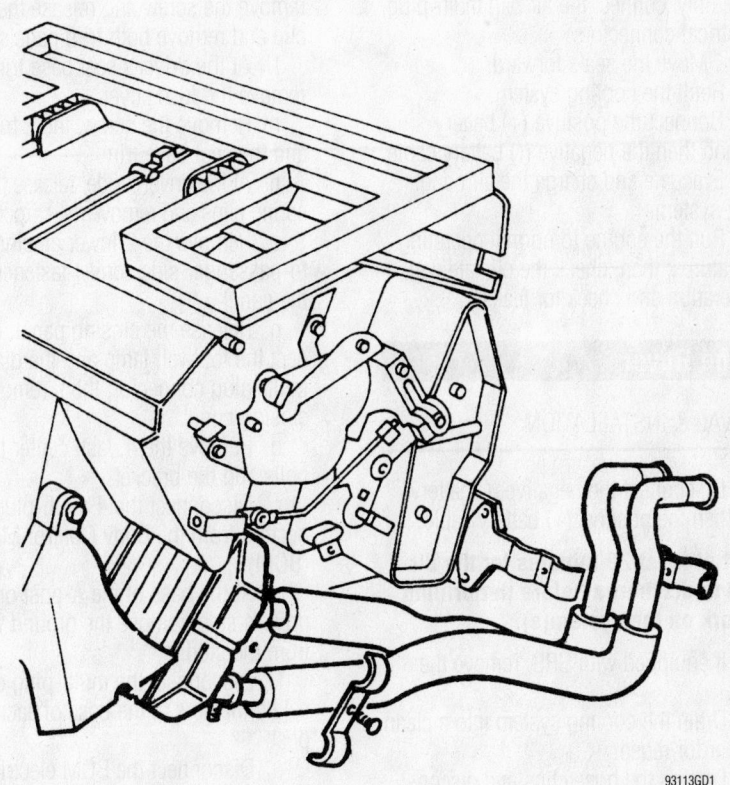

93113GD1

View of the heater core and related components—Land Rover Discovery

For Maintenance Interval recommendations, see Section 1 of this manual

- Install the air bag module to the dash panel.
- Using a Torx® socket, install both rear air bag module-to-dash panel screws.
- Using a special socket and a long extension, install both front air bag module-to-dash panel screws.
- Open the glove box door and connect the air bag module electrical connector.

r. Install the driver's SRS module, by connecting or installing following items:

- Install the air bag module.
- Connect the air bag harness connector.
- Install the air bag module to the steering wheel.
- Using a special socket, install the 2 tamper-proof resistor air bag module-to-steering wheel screws.
- Connect the air bag harness connector to the yellow air bag column harness.
- Under the steering wheel, install the lower dash panel and install the lower dash panel cover turnbuckles.

s. Install the center console assembly.

t. Install the glove box.

u. Working under the dash panel assembly, connect the air bag multi-plug electrical connectors.

v. Move the seats forward.

33. Refill the cooling system.

34. Connect the positive (+) battery cable and then the negative (() battery cable.

35. Evacuate and charge the air conditioning system.

36. Run the engine to normal operating temperatures; then, check the climate control operation and check for leaks.

Range Rover

REMOVAL & INSTALLATION

1. Disconnect the negative (() battery cable; then the positive (+) battery cable.

➡**Wait at least 20 minutes for the air bag(s) to discharge before performing any work on the system(s).**

2. If equipped with SRS, remove the battery.

3. Drain the cooling system into a clean container for reuse.

4. Loosen the hose clips and disconnect the heater hoses from the heater tubes.

5. If equipped with air conditioning, perform the following procedure:

a. Discharge and recover the air conditioning refrigerant.

b. Remove the refrigerant pipes mount-to-evaporator bolt

c. Disconnect the pipes and discard the O-rings.

d. Plug the openings to prevent contamination.

6. Remove the dash panel assembly by performing the following procedures:

a. Remove the center console.

b. Remove the wiper motor and linkage.

c. Remove the steering column.

d. At the passenger's side, remove the clip and disconnect the heated front screen multi-plug connector.

e. Remove the 6 scuttle side panel-to-chassis bolts and the side panel.

f. Remove the heater intake pollen filters.

g. Remove the 8 pollen filter housing screws from each housing and remove both housings.

h. Remove the radio.

i. Near the A-post lower trim panels, remove the door aperture seal.

j. If equipped with a footrest on the driver's side, remove the 3 foot rest-to-A-post lower trim bolts and remove the foot rest.

k. At each A-post's lower trim panel, remove the screw and release the spring clip and remove both trim panels.

l. At the driver's seat base trim, remove the fuse cover.

m. Remove the screw, the 2 trim studs and the seat base trim.

n. At the driver's side, release the 4 spring clips and remove the carpet retainer.

o. Remove the 2 lower closing panel-to-passenger side scrivet fasteners and the panel.

p. Release the closing panel, disconnect the footwell lamp and the diagnostic multi-plug connector; then, remove the closing panel.

q. Remove the 4 dash center bracket bolts and the bracket.

r. Disconnect the 4 multi-plug connectors from the Body Control Module (BCM).

s. At the base of the A-post on the driver's side, remove the ground wires from the stud.

t. Disconnect the multi-plug electrical connectors at the base of each A-post.

u. Disconnect the BCM electrical harness from the sill and move it into the dash panel so it will not hamper removal of the dash panel.

v. At the brake and clutch switches, disconnect the multi-plug electrical connectors and vacuum hose.

w. If equipped with SRS, disconnect the following items:

- The SRS harness connector from the main wiring harness
- The SRS harness connector from the control module

x. Remove both front wheel arch liners.

y. At the left wheel arch, remove the 2 air cleaner baffle scrivet fasteners and the baffle.

z. If equipped with SRS, disconnect both SRS crash sensor electrical connectors.

aa. Remove the 4 battery tray bolts and the 2 air cleaner-to-valance bolts.

bb. Raise the air cleaner and battery tray to access the crash sensor harness clips.

cc. Disconnect the crash sensor harness-to-valance clips; then, move the harnesses into the wheel arches.

dd. Disconnect the 3 crash sensor harness-to-underside wheel arches; then, move the harness through the bulkhead and into the dash.

ee. At the top of the dash panel, remove the 4 dash panel-to-scuttle panel tube bolts.

ff. Remove the dash panel-to-chassis bolts. Pull the panel rearward and support it on 50mm deep wooden blocks.

7. With the dash panel supported on 50mm deep wooden blocks, remove the face level vent ducts-to-dash screws; there is a vent duct located on both sides of the dash.

8. Remove the face level vent ducts inserts from the heater unit.

9. Remove the passenger's side blower duct.

10. Remove the 4 heater control panel-to-dash panel screws and remove the panel.

11. Disconnect the 4 multi-plug connectors from the heater control panel and remove the control panel.

12. Remove the 5 center switch assembly-to-dash screws and remove the center switch assembly.

13. Disconnect the multi-plug connectors from the solar sensor and the alarm LED; then push the leads into the dash ducting.

14. Disconnect the harness-to-dash ducting clip; then, position the solar sensor/LED harness aside.

15. Disconnect the water temperature sensor-to-heater core inlet pipe clip and position the sensor aside.

16. Disconnect the multi-plug connector from the evaporator sensor.

17. Disconnect the 2 harness-to-heater base clips.

18. Remove the 4 heater housing-to-dash frame bolts.

19. Using an assistant, hold the harness away and remove the heater housing.

20. Remove the right side duct-to-heater/air conditioning housing screws and the duct.

21. Remove the heater pipe bracket screw.

22. Remove the 2 right side servo-to-heater/air conditioning housing screws and the servo.

23. Remove the heater core/pipe assembly-to-heater/air conditioning housing clips and the heater core/pipe assembly.

24. If installing a new heater core, remove the 2 heater core-to-heater pipe assembly screws and separate the heater pipe assembly. Discard the O-rings.

To install:

25. If installing a new heater core, install new O-rings, the heater pipe assembly and the 2 heater core-to-heater pipe assembly screws.

26. Install the heater core/pipe assembly and the heater core/pipe assembly-to-heater/air conditioning housing clips.

27. Install the right side servo and the 2 servo-to-heater/air conditioning housing screws.

28. Install the heater pipe bracket screw.

29. Install the right side duct and the duct-to-heater/air conditioning housing screws.

30. Using an assistant, install the heater housing.

31. Install the 4 heater housing-to-dash frame bolts.

32. Connect the 2 harness-to-heater base clips.

33. Connect the multi-plug connector to the evaporator sensor.

34. Connect the water temperature sensor-to-heater core inlet pipe clip.

35. Connect the harness-to-dash ducting clip.

36. Connect the multi-plug connectors to the solar sensor and the alarm LED.

37. Install the center switch assembly and the 5 center switch assembly-to-dash screws.

38. Install the control panel and connect the 4 multi-plug connectors to the heater control panel.

39. Install the heater control panel and the 4 panel-to-dash panel screws.

40. Install the passenger's side blower duct.

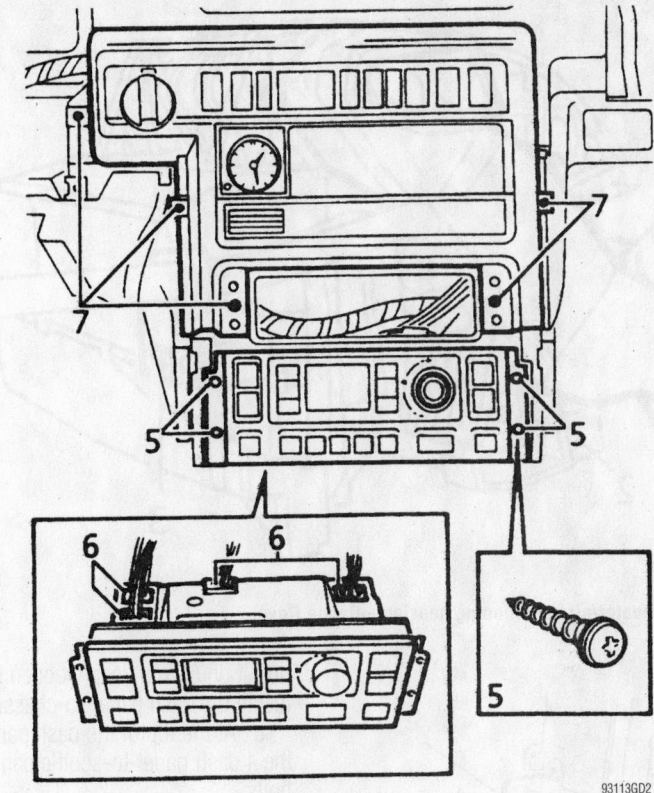

View of the center switch assembly—Range Rover

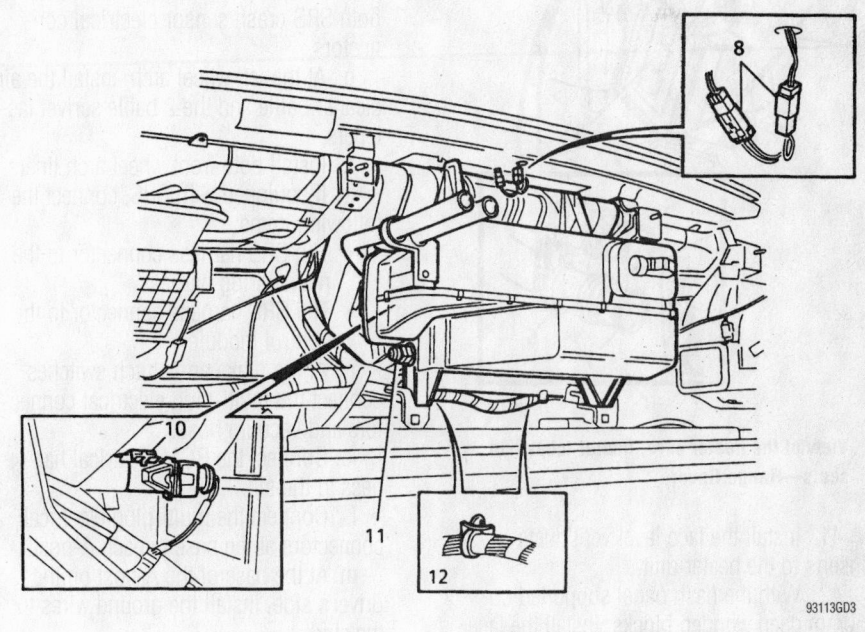

View of the heater/air conditioning housing assembly and related components—Range Rover

For Tune-up, Capacities and Firing orders, see Section 1 of this manual

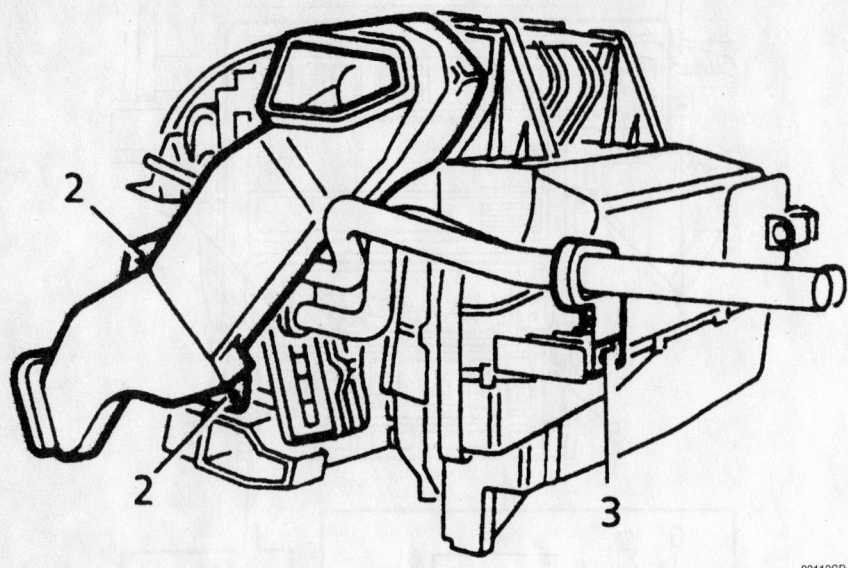

93113GD4

View of the heater/air conditioning housing—Range Rover

View of the heater core, heater tubes and seals—Range Rover

93113GD5

View of the heater core, heater tubes and seals—Range Rover

41. Install the face level vent ducts inserts to the heater unit.

42. With the dash panel supported on 50mm deep wooden blocks, install the face level vent ducts-to-dash screws; there is a vent duct located on both sides of the dash.

43. Install the dash panel assembly by performing the following procedures:
   a. Push the panel forward and sup-

port it on 50mm deep wooden blocks. Install the dash panel-to-chassis bolts.
   b. At the top of the dash panel, install the 4 dash panel-to-scuttle panel tube bolts.
   c. Connect the 3 crash sensor harness-to-underside wheel arches.
   d. Connect the crash sensor harness-to-valance clips.
   e. Install the 4 battery tray bolts and the 2 air cleaner-to-valance bolts.
   f. If equipped with SRS, connect both SRS crash sensor electrical connectors.
   g. At the left wheel arch, install the air cleaner baffle and the 2 baffle scrivet fasteners.
   h. Install both front wheel arch liners.
   i. If equipped with SRS, connect the following items:
   • The SRS harness connector to the main wiring harness
   • The SRS harness connector to the control module
   j. At the brake and clutch switches, connect the multi-plug electrical connectors and vacuum hose.
   k. Connect the BCM electrical harness to the sill.
   l. Connect the multi-plug electrical connectors at the base of each A-post.
   m. At the base of the A-post on the driver's side, install the ground wires to the stud.
   n. Connect the 4 multi-plug connectors to the Body Control Module (BCM).
   o. Install the dash center bracket and the 4 bracket bolts.
   p. Install the closing panel; then, con-

nect the footwell lamp and the diagnostic multi-plug connector.
   q. Install the lower closing panel and the 2 panel-to-passenger side scrivet fasteners.
   r. At the driver's side, install the carpet retainer and the 4 spring clips.
   s. Install the seat base trim, the 2 trim studs and the screw.
   t. At the driver's seat base trim, install the fuse cover.
   u. At each A-post's lower trim panel, install both trim panels, the screw and the spring clip.
   v. If equipped with a foot rest on the driver's side, install the foot rest and the 3 foot rest-to-A-post lower trim bolts.
   w. Near the A-post lower trim panels, install the door aperture seal.
   x. Install the radio.
   y. Install the both pollen filter housings and the 8 housing screws to each housing.
   z. Install the heater intake pollen filters.
   aa. Install the scuttle side panel and the 6 side panel-to-chassis bolts.
   bb. At the passenger's side, connect the heated front screen multi-plug connector and install the clip.
   cc. Install the steering column.
   dd. Install the wiper motor and linkage.
   ee. Install the center console.

44. If equipped with air conditioning, perform the following procedure:
   a. Lubricate and install new O-rings and connect the pipes.
   b. Install the refrigerant pipes mount-to-evaporator bolt.

45. Connect the heater hoses to the heater tubes and install the hose clips.

46. Refill the cooling system.

47. If the battery was removed, install it.

48. Connect the positive (+) battery cable; then, the negative (() battery cable.

49. Evacuate and charge the air conditioning system.

50. Run the engine to normal operating temperatures; then, check the climate control operation and check for leaks.

Freelander

REMOVAL & INSTALLATION

1. Drain cooling system.
2. Release 2 clips securing heater hoses to heater and disconnect hoses.
3. Remove fascia.
4. Remove 2 bolts from console sup-

port bracket, release 2 relays and remove bracket.

5. Disconnect 4 multiplugs from heater controls.

6. Disconnect multiplug from heater and release diagnostic socket.

7. Models with A/C: Disconnect multiplug from evaporator.

8. Remove bolt from duct of left side outer face level vent and release duct from heater.

9. Remove bolt from duct of right side outer face level vent and release duct from heater.

10. Release duct of right side demister vent.

11. Release duct of left side demister vent.

12. Release air inlet connector hose.

13. Release harness from 2 clips on heater.

14. Remove 2 nuts and 1 bolt securing heater to body and remove heater.

➡**Do not carry out further dismantling if component is removed for access only.**

15. Disconnect 2 multiplugs from heater controls.

16. Release air blend control cable from lever and abutment.

17. Release air distribution cable from lever and abutment.

18. Remove heater controls.

19. Remove heater assembly.

20. Remove screw securing pipe clamp to heater casing and remove clamp.

21. Remove 2 screws from core cover and remove cover.

22. Remove heater core.

To install:

23. Fit heater core to heater assembly.

24. Fit core cover and secure with screws.

25. Fit pipe clamp and secure with screw.

26. Fit heater assembly.

27. Position heater controls to replacement heater.

28. Connect air distribution cable to lever and abutment.

29. Connect air blend cable to lever and abutment.

30. Connect multiplugs to heater controls.

31. Fit heater and secure with nuts and bolt.

32. Position harness and connect to clips on heater.

33. Fit air inlet connector hose.

34. Fit right side demister vent duct to heater.

35. Fit right side outer face level vent duct to heater and secure to body with bolt.

36. Fit left side demister vent duct to heater.

37. Fit left side outer face level vent duct to heater and secure to body with bolt.

38. Models with A/C: Connect multiplug to evaporator.

39. Connect multiplug to heater and secure diagnostic socket.

40. Connect multiplugs to heater controls.

41. Fit console support bracket and secure with bolts.

42. Fit relays to console support bracket.

43. Fit fascia.

44. Connect heater hoses and secure with clips.

45. Refill cooling system.

LEXUS

LX 470

REMOVAL & INSTALLATION

Front Heater

1. Disconnect the negative battery cable.

2. Drain the cooling system into a clean container for reuse.

3. Disconnect the heater hoses from the heater core.

4. Remove the steering wheel by performing the following procedure:

   a. Position the front wheels facing straight-ahead.

   b. Remove the steering wheel side covers.

   c. Using a Torx® wrench, loosen the 2 screws located at each side of the steering wheel until the screw's circumference groove catches on the screw case.

   d. Pull the air bag module from the steering wheel and disconnect the electrical connector.

✳✳ **CAUTION**

Place the air bag module in a safe place with the front side facing upward.

   e. Remove the steering wheel nut.

   f. Place alignment marks on the steering wheel and the main shaft.

   g. Using a steering wheel puller, press the steering wheel from the steering column.

5. Remove the instrument panel and reinforcement by performing the following procedure:

   a. Remove the front door scuff plates, the cowl side trim and the front door opening trim.

   b. At the driver's side, remove the 2 assist grip plugs, the 2 screws and assist grip and the front pillar garnish.

   c. At the passenger's side, remove the 4 assist grip plugs, the 4 screws, the 2 assist grips and the front pillar garnish.

   d. Remove the instrument cluster finish panel.

   e. Remove the 2 screws and the hood lock control cable.

   f. Remove the 2 screws and the fuel lid control cable lever.

   g. Remove the lower No. 1 panel screw and the panel.

   h. Remove the lower left side panel.

   i. Remove the 3 steering column cover screws and the covers.

   j. At the steering column, disconnect the electrical connectors; then, remove the clamp, the 3 screws and the combination switch.

   k. Remove the No. 2 heater-to-register duct screw and the duct.

   l. Remove the steering column-to-instrument panel bolts and the steering column.

   m. At the combination meter, disconnect the electrical connectors; then, remove the 4 screws and the combination meter.

   n. Remove the glove compartment door stoppers, the 2 screws and the glove box door.

   o. At the passenger's side air bag module, remove the No. 1 undercover, pull the air bag connector up from the undercover and disconnect it; then, remove the air bag.

✳✳ **CAUTION**

Place the air bag module in a safe place with the front side facing upward.

   p. Remove the 3 lower No. 2 panel screws and the panel.

q. Remove the center cluster; then, pry the center cluster from the dash by prying the 8 clips in the following order:
- Left side
- Right side
- Top left side
- Top right side

r. Remove the 4 radio screws, pull the radio outward, disconnect the electrical connectors and remove the radio.

s. At the rear console panel, remove the transfer shift lever knob; then, pry the panel upward disengaging the 4 clips (2 on each side) and remove the panel.

t. At the rear of the console, remove the 2 rear end panel-to-console screws; then, pry the end panel rearward disengaging the 2 clips and remove the panel.

u. If not equipped with a rear air conditioning system, disconnect the connector and control cable; then, remove the 3 rear heater control panel screws and the panel.

v. Remove the 4 rear console box-to-chassis screws/bolts and the console box.

w. Remove the center lower cluster finish panel by prying panel rearward disengaging the 5 clips; then, disconnect the electrical connector.

x. Remove the 2 front console-to-chassis bolts/screws, disengage the 2 clips and remove the console.

y. At the instrument panel, disconnect the junction connectors (the connectors can be disconnected by loosening the bolts), the instrument panel-to-chassis 8 bolts and 2 nuts. Using an assistant, remove the instrument panel.

z. Disconnect the electrical connector and remove the ECM.

aa. Remove the No. 3 and No. 4 heater-to-register ducts.

bb. Remove the floor brace, the No. 1 brace and the reinforcement.

6. Remove the evaporator housing by performing the following procedure:

a. Discharge and recover the air conditioning system refrigerant.

b. Remove the air conditioning liquid line clamp.

c. Remove the air conditioning suction line clamp.

d. Disconnect both air conditioning lines and plug the openings to prevent contamination. Discard the 4 O-rings.

e. Remove the antenna relay electrical connector, the 2 screws and the relay.

f. Remove the evaporator housing-to-chassis 4 screws/2 nuts and the housing.

7. Remove the heater housing by performing the following procedure:

a. Remove the defroster nozzle.

b. Disconnect the electrical connector.

c. Remove the 4 nuts and the heater housing.

8. Remove the heater core-to-heater housing packing, the screw, the bracket, the clamp and the heater core.

To install:

9. Install the heater core, the clamp, the bracket, the screw and the heater core-to-heater housing packing.

10. Install the heater housing by performing the following procedure:

a. Install the heater housing and the 4 nuts.

b. Connect the electrical connector.

c. Install the defroster nozzle.

11. Install the evaporator housing by performing the following procedure:

a. Install the evaporator housing and the housing-to-chassis 4 screws and 2 nuts.

b. Install the antenna relay, the 2 screws and the electrical connector.

c. Using new O-rings, connect both air conditioning lines.

d. Install the air conditioning liquid line and suction line clamp.

12. Install the instrument panel and reinforcement by performing the following procedure:

a. Install the reinforcement, the No. 1 brace and the floor brace.

b. Install the No. 3 and No. 4 heater-to-register ducts.

c. Install the ECM and connect the electrical connector.

d. Using an assistant, install the instrument panel, connect the junction connectors, the instrument panel-to-chassis 8 bolts and 2 nuts.

e. Install the front the console, engage the 2 clips and install the 2 console-to-chassis bolts/screws.

f. Connect the electrical connector; then, install the center lower cluster finish panel by engaging the 5 clips.

g. Install the console box and the 4 rear console box-to-chassis screws/bolts.

h. If not equipped with a rear air conditioning system, install rear heater control panel, the 3 panel screws; then, connect the connector and control cable.

i. Install the rear of the console and engage the 2 clips; then, install the 2 rear end panel-to-console screws.

j. Install the rear console panel and engage the 4 clips (2 on each side); then, install the transfer shift lever knob.

k. Install the radio, connect the electrical connectors and the 4 radio screws.

l. Install the center cluster and engage the 8 center cluster clips.

m. Install the lower No. 2 panel and the 3 panel screws.

n. Install the passenger's side air bag module, connect it and install the No. 1 undercover.

o. Install the glove box door, the 2 screws and the glove compartment door stoppers.

p. Install the combination meter and the 4 screws; then, connect the electrical connectors.

q. Install the steering column and the

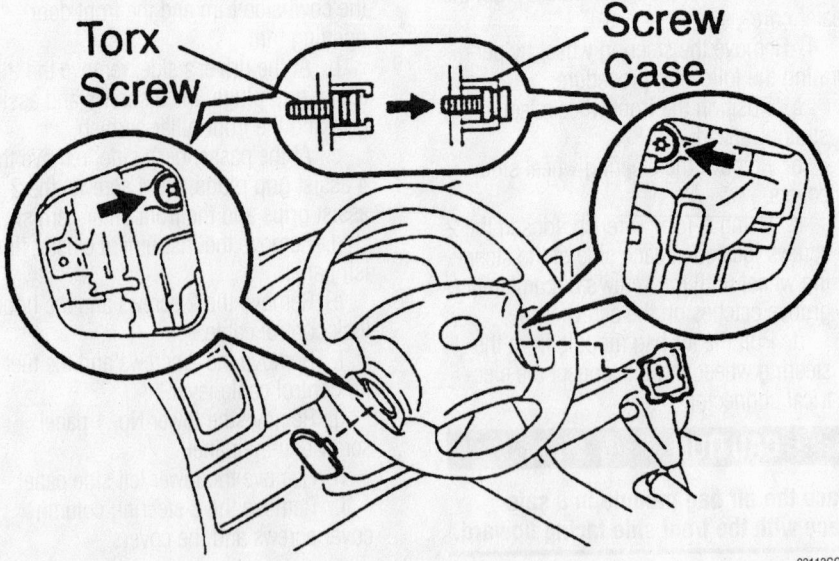

View the steering wheel's Torx® bolts—Lexus LX 470

93113GG4

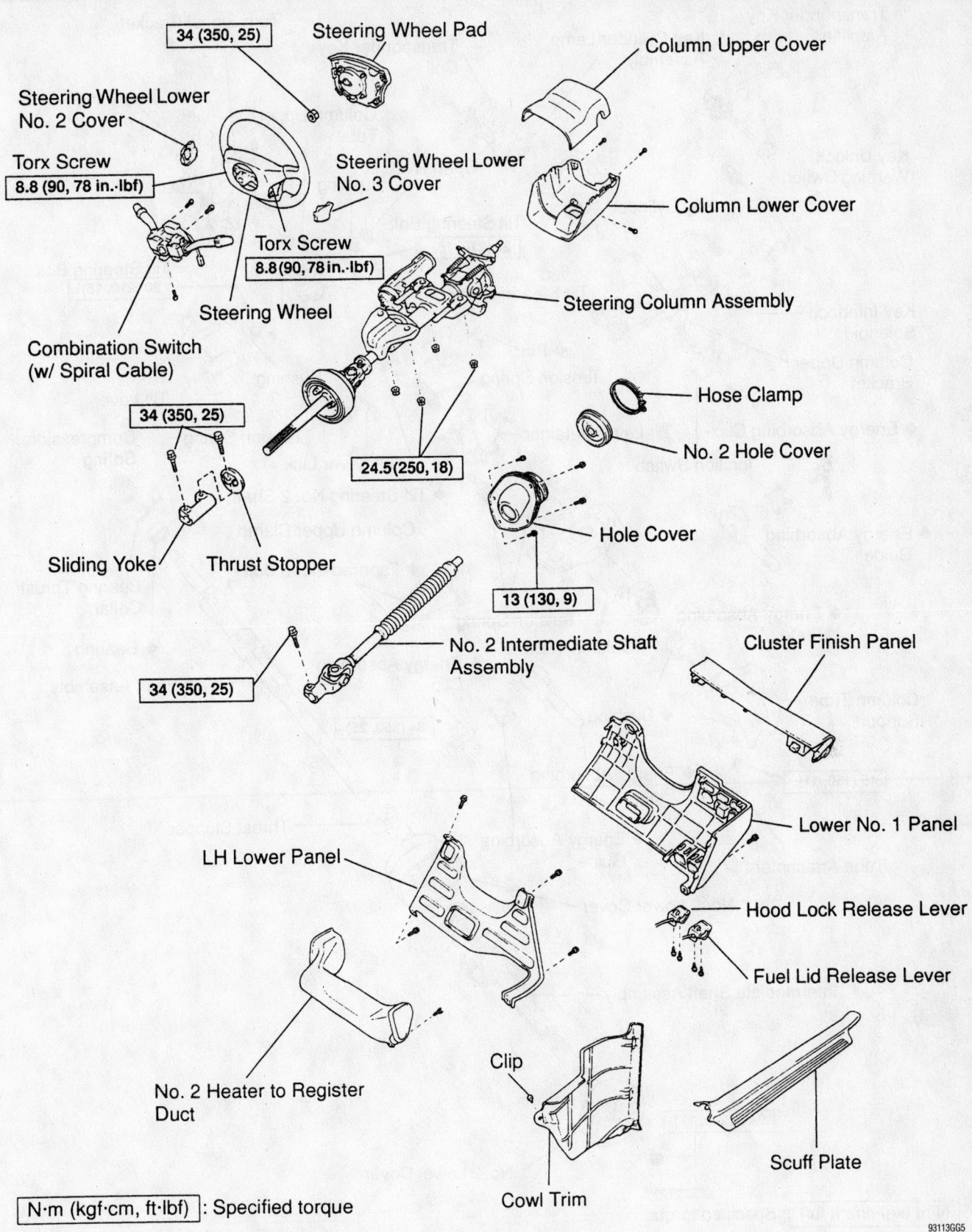

34 (350, 25)
Steering Wheel Pad
Column Upper Cover

Steering Wheel Lower
No. 2 Cover

Column Lower Cover

Torx Screw
8.8 (90, 78 in.·lbf)

Steering Wheel Lower
No. 3 Cover

Torx Screw
8.8 (90, 78 in.·lbf)

Steering Column Assembly

Steering Wheel

Combination Switch
(w/ Spiral Cable)

Hose Clamp

No. 2 Hole Cover

34 (350, 25)

24.5 (250, 18)

Hole Cover

Sliding Yoke   Thrust Stopper

13 (130, 9)

34 (350, 25)

No. 2 Intermediate Shaft
Assembly

Cluster Finish Panel

Lower No. 1 Panel

LH Lower Panel

Hood Lock Release Lever

Fuel Lid Release Lever

Clip

No. 2 Heater to Register
Duct

Scuff Plate

Cowl Trim

N·m (kgf·cm, ft·lbf) : Specified torque

93113GG5

Exploded view the steering column—Lexus LX 470 (Part 1 of 2)

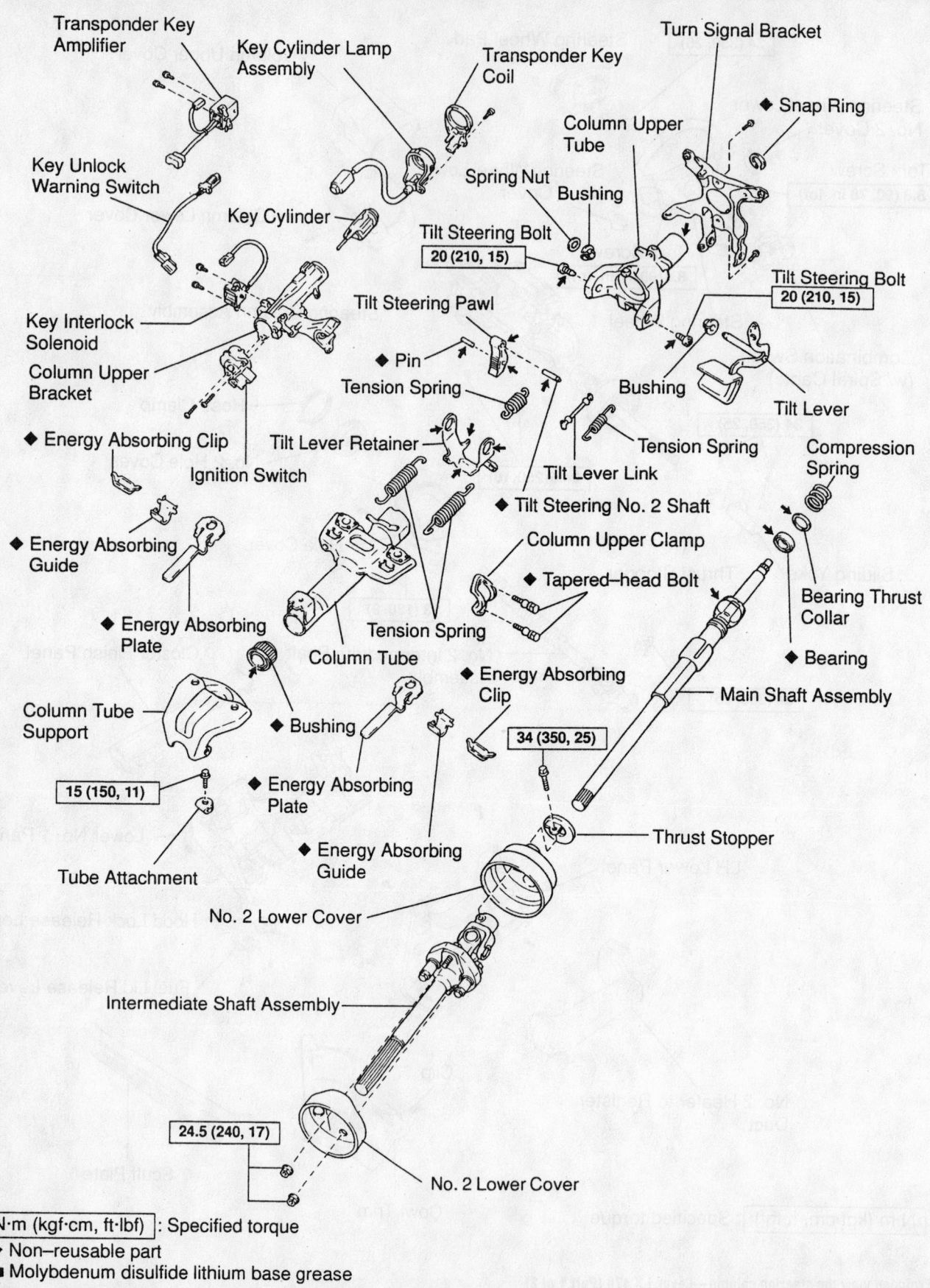

Transponder Key Amplifier

Key Cylinder Lamp Assembly

Transponder Key Coil

Turn Signal Bracket

◆ Snap Ring

Column Upper Tube

Key Unlock Warning Switch

Spring Nut

Bushing

Tilt Steering Bolt
20 (210, 15)

Key Cylinder

Tilt Steering Bolt
20 (210, 15)

Key Interlock Solenoid

Tilt Steering Pawl

◆ Pin

Tension Spring

Bushing

Tilt Lever

Column Upper Bracket

◆ Energy Absorbing Clip

Tilt Lever Retainer

Ignition Switch

Tension Spring

Tilt Lever Link

Tilt Steering No. 2 Shaft

Compression Spring

◆ Energy Absorbing Guide

Column Upper Clamp

◆ Tapered–head Bolt

Bearing Thrust Collar

◆ Energy Absorbing Plate

◆ Bearing

◆ Energy Absorbing Clip

Tension Spring

Column Tube

Main Shaft Assembly

Column Tube Support

◆ Bushing

34 (350, 25)

15 (150, 11)

◆ Energy Absorbing Plate

Thrust Stopper

◆ Energy Absorbing Guide

Tube Attachment

No. 2 Lower Cover

Intermediate Shaft Assembly

24.5 (240, 17)

No. 2 Lower Cover

N·m (kgf·cm, ft·lbf) : Specified torque

◆ Non–reusable part

N Molybdenum disulfide lithium base grease

Exploded view the steering column—Lexus LX 470 (Part 2 of 2)

93113GG6

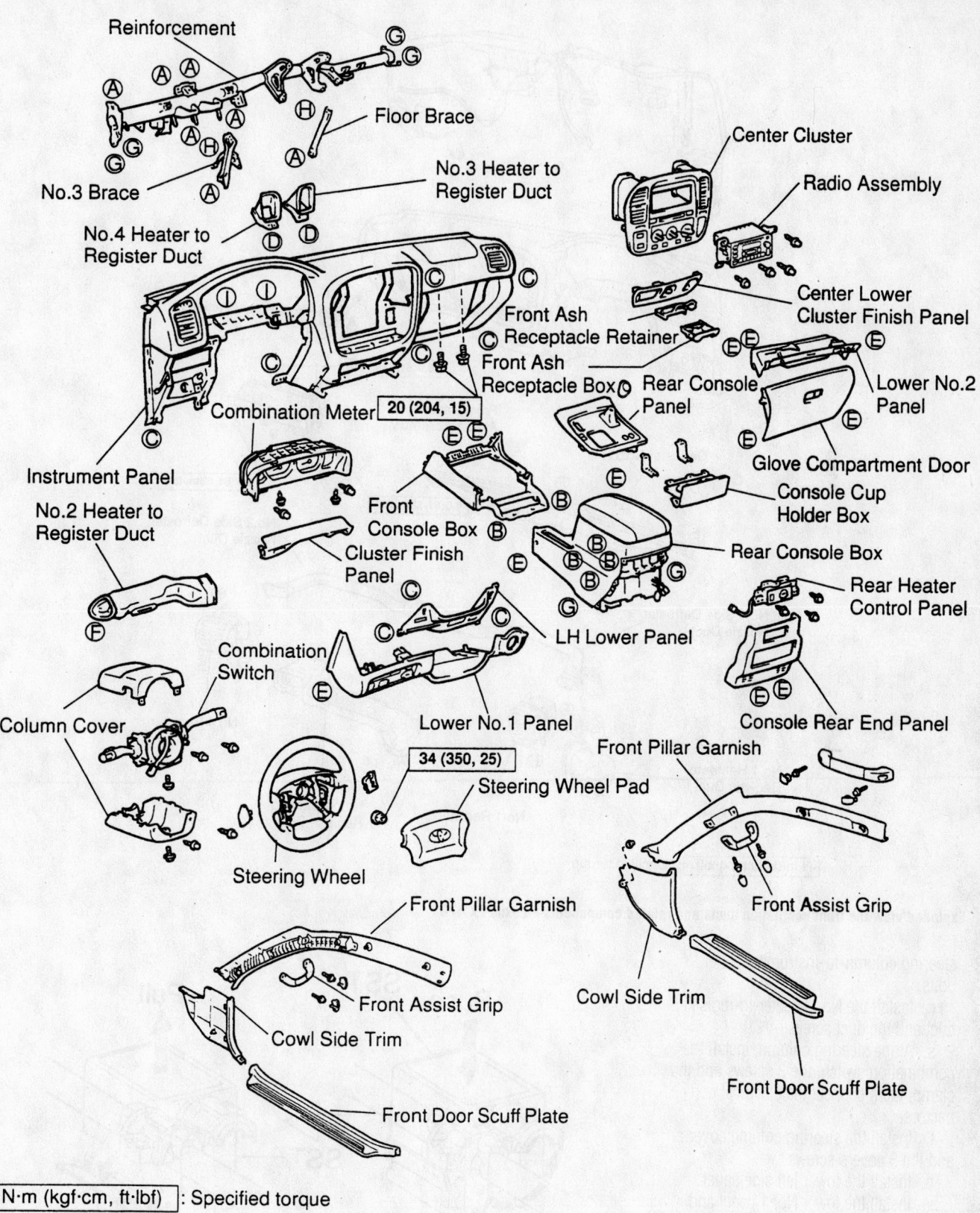

Reinforcement

Floor Brace

No.3 Brace

No.3 Heater to Register Duct

No.4 Heater to Register Duct

Center Cluster

Radio Assembly

Center Lower Cluster Finish Panel

Front Ash Receptacle Retainer

Front Ash Receptacle Box

Rear Console Panel

Lower No.2 Panel

Combination Meter **20 (204, 15)**

Glove Compartment Door

Instrument Panel

Console Cup Holder Box

No.2 Heater to Register Duct

Front Console Box

Cluster Finish Panel

Rear Console Box

Rear Heater Control Panel

LH Lower Panel

Combination Switch

Column Cover

Lower No.1 Panel

Console Rear End Panel

34 (350, 25)

Steering Wheel Pad

Front Pillar Garnish

Steering Wheel

Front Assist Grip

Front Pillar Garnish

Front Assist Grip

Cowl Side Trim

Cowl Side Trim

Front Door Scuff Plate

Front Door Scuff Plate

N·m (kgf·cm, ft·lbf) : Specified torque

93113GG7

Exploded view the instrument panel and related components—LX 470

Timing belt service is covered in Section 3 of this manual

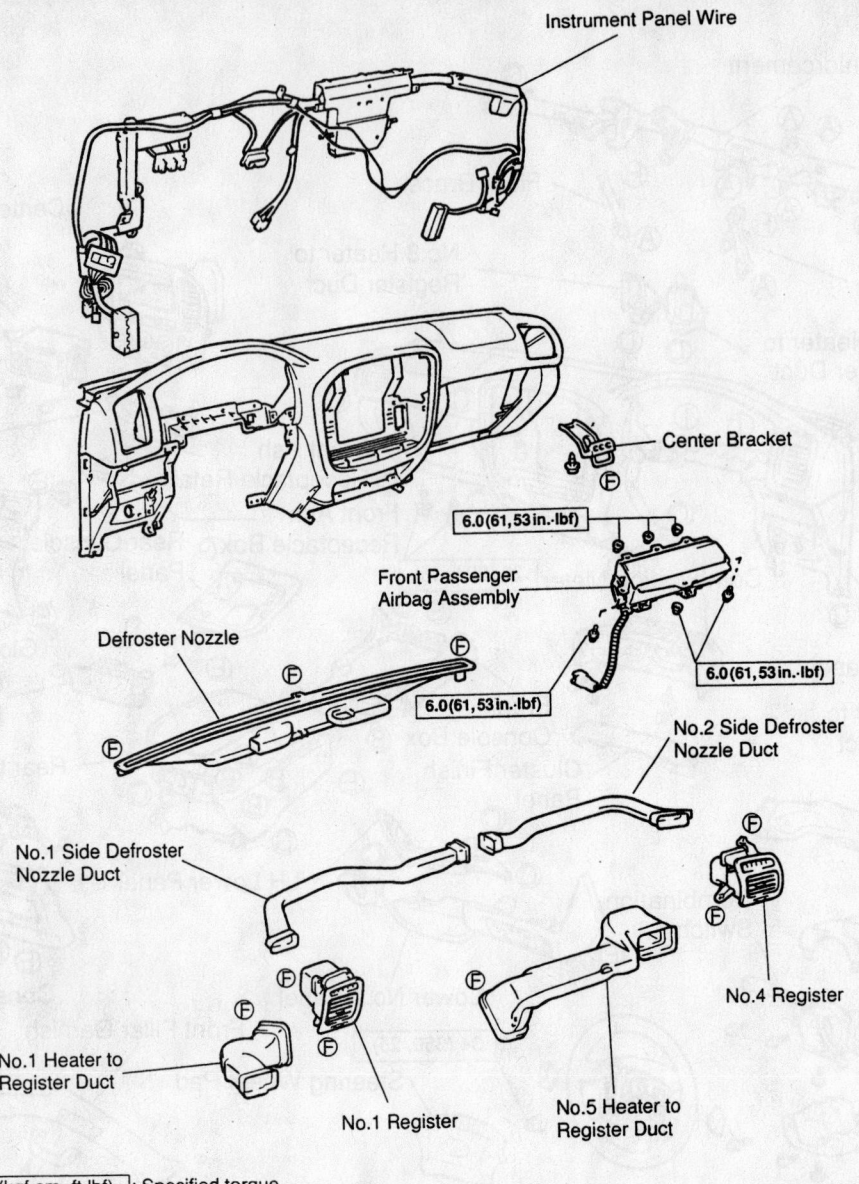

Exploded view the front ventilation ducts and related components—Lexus LX 470

steering column-to-instrument panel bolts.

r. Install the No. 2 heater-to-register duct and the duct screw.

s. At the steering column, install the combination switch, the 3 screws and the clamp; then, connect the electrical connectors.

t. Install the steering column covers and the 3 covers screws.

u. Install the lower left side panel.

v. Install the lower No. 1 panel and the panel screw.

w. Install the fuel lid control cable lever and the 2 screws.

x. Install the hood lock control cable and the 2 screws.

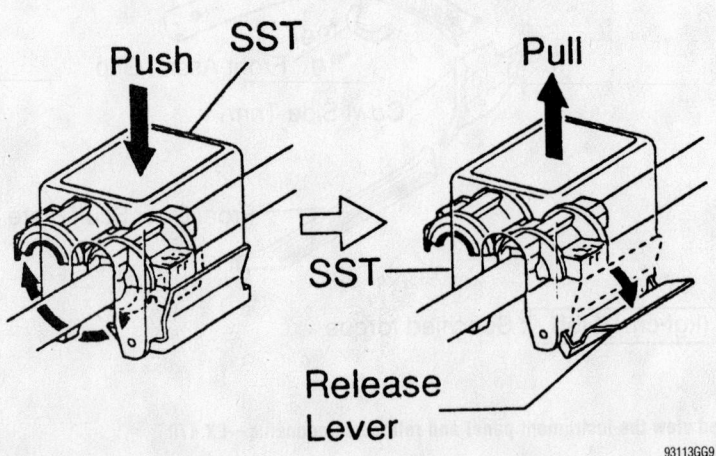

View the air conditioning line clamp removal tool—Lexus LX 470

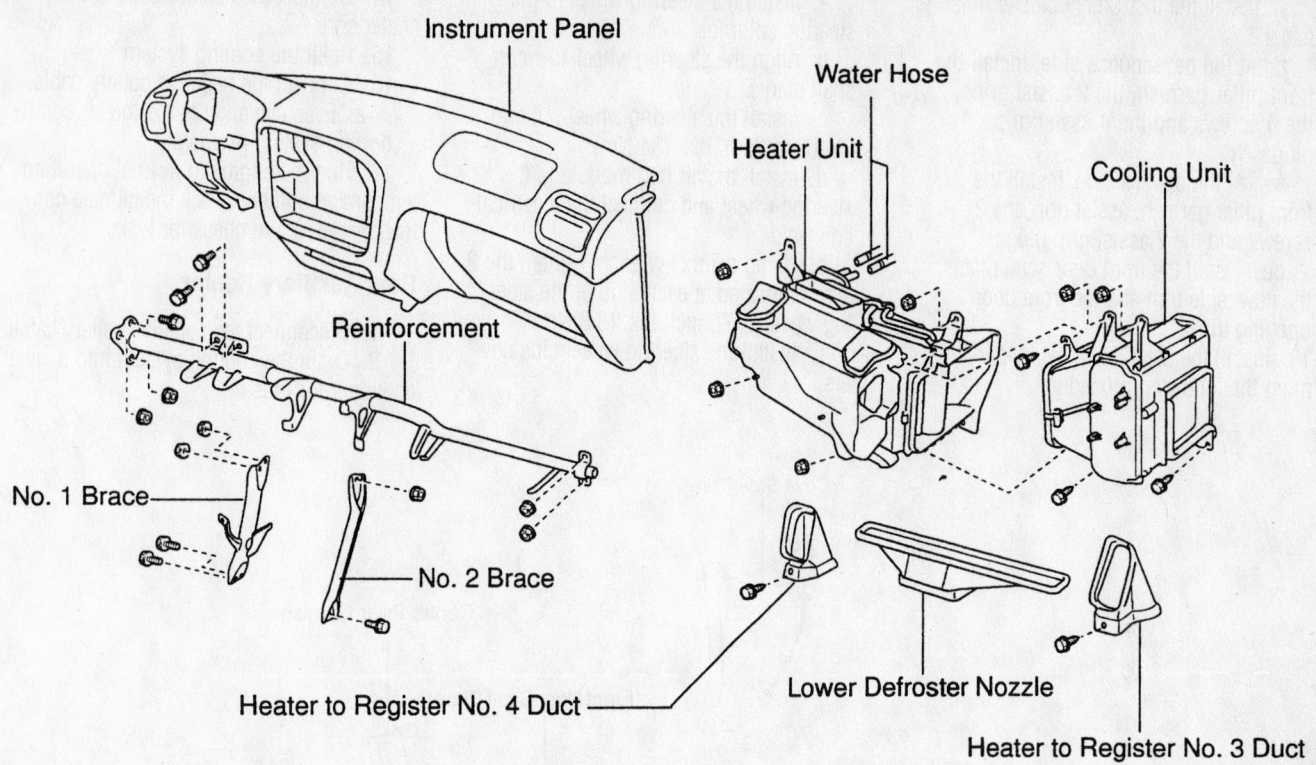

Instrument Panel

Water Hose

Heater Unit

Cooling Unit

Reinforcement

No. 1 Brace

No. 2 Brace

Heater to Register No. 4 Duct

Lower Defroster Nozzle

Heater to Register No. 3 Duct

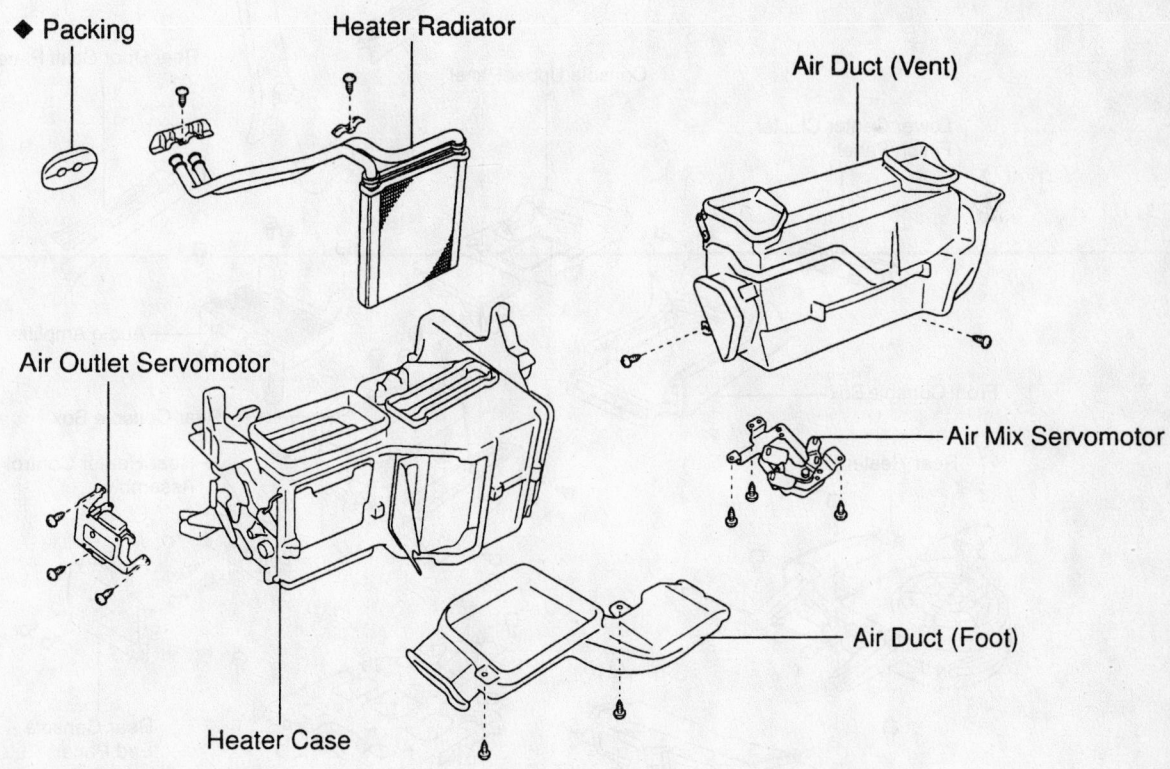

◆ Packing

Heater Radiator

Air Duct (Vent)

Air Outlet Servomotor

Air Mix Servomotor

Air Duct (Foot)

Heater Case

◆ Non–reusable part

93113GG0

Exploded view the front heater core, heater housing, evaporator housing and related components—Lexus LX 470

Heater Core replacement is covered in Section 2 of this manual

y. Install the instrument cluster finish panel.

z. At the passenger's side, install the front pillar garnish, the 2 assist grips, the 4 screws and the 4 assist grip plugs.

aa. At the driver's side, install the front pillar garnish, assist grip, the 2 screws and the 2 assist grip plugs.

bb. Install the front door scuff plates, the cowl side trim and the front door opening trim.

13. Install the steering wheel by performing the following procedure:

a. Install the steering wheel to the steering column.

b. Align the steering wheel-to-main shaft marks.

c. Install the steering wheel nut and torque to 25 ft. lbs. (34 Nm).

d. Install the air bag module to the steering wheel and connect the electrical connector.

e. Using a Torx® wrench, tighten the 2 screws located at each side of the steering wheel to 78 inch lbs. (8.8 Nm).

f. Install the steering wheel side covers.

14. Connect the heater hoses to the heater core.

15. Refill the cooling system.

16. Connect the negative battery cable.

a. Evacuate and charge the air conditioning system refrigerant.

17. Run the engine to normal operating temperatures; then, check the climate control operation and check for leaks.

Rear Auxiliary Heater

1. Disconnect the negative battery cable.

2. Drain the cooling system into a clean container for reuse.

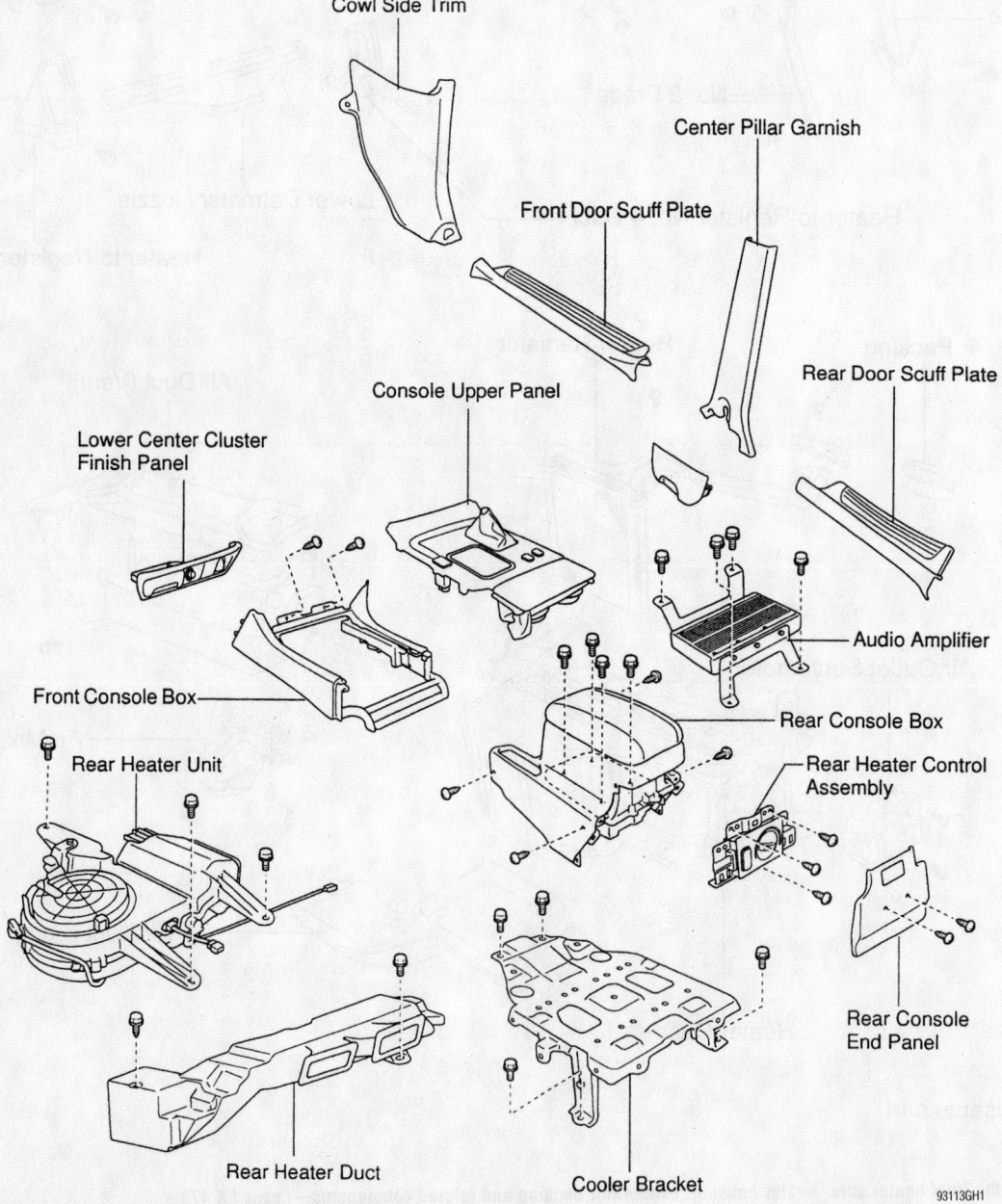

Exploded view of the rear heater housing and related components—Lexus LX 470

93113GH1

3. Disconnect the heater hoses from the rear heater core.

4. Remove the front seats.

5. Remove the rear heater control assembly.

6. Remove the rear console box.

7. Remove the front console box cover.

8. Remove the lower center cluster finish panel.

9. Remove the front door scuff plates.

10. Remove the cowl side trim.

11. Remove the rear door scuff plates.

12. Remove the center pillar garnishes.

13. Slide the carpet rearward.

14. Remove the cooler bracket bolts and the bracket.

15. Remove the rear heater duct bolt/screw and the duct.

16. Disconnect the rear heater housing electrical connector.

17. Remove the 3 rear heater housing-to-chassis bolts and the heater housing.

18. Remove the heater core-to-heater housing 3 screws and 2 clamps.

19. Remove the heater core from the heater housing.

To install:

20. Install the heater core to the heater housing.

21. Install the heater core-to-heater housing 3 screws and 2 clamps.

22. Install the heater housing and the 3 rear heater housing-to-chassis bolts.

23. Connect the rear heater housing electrical connector.

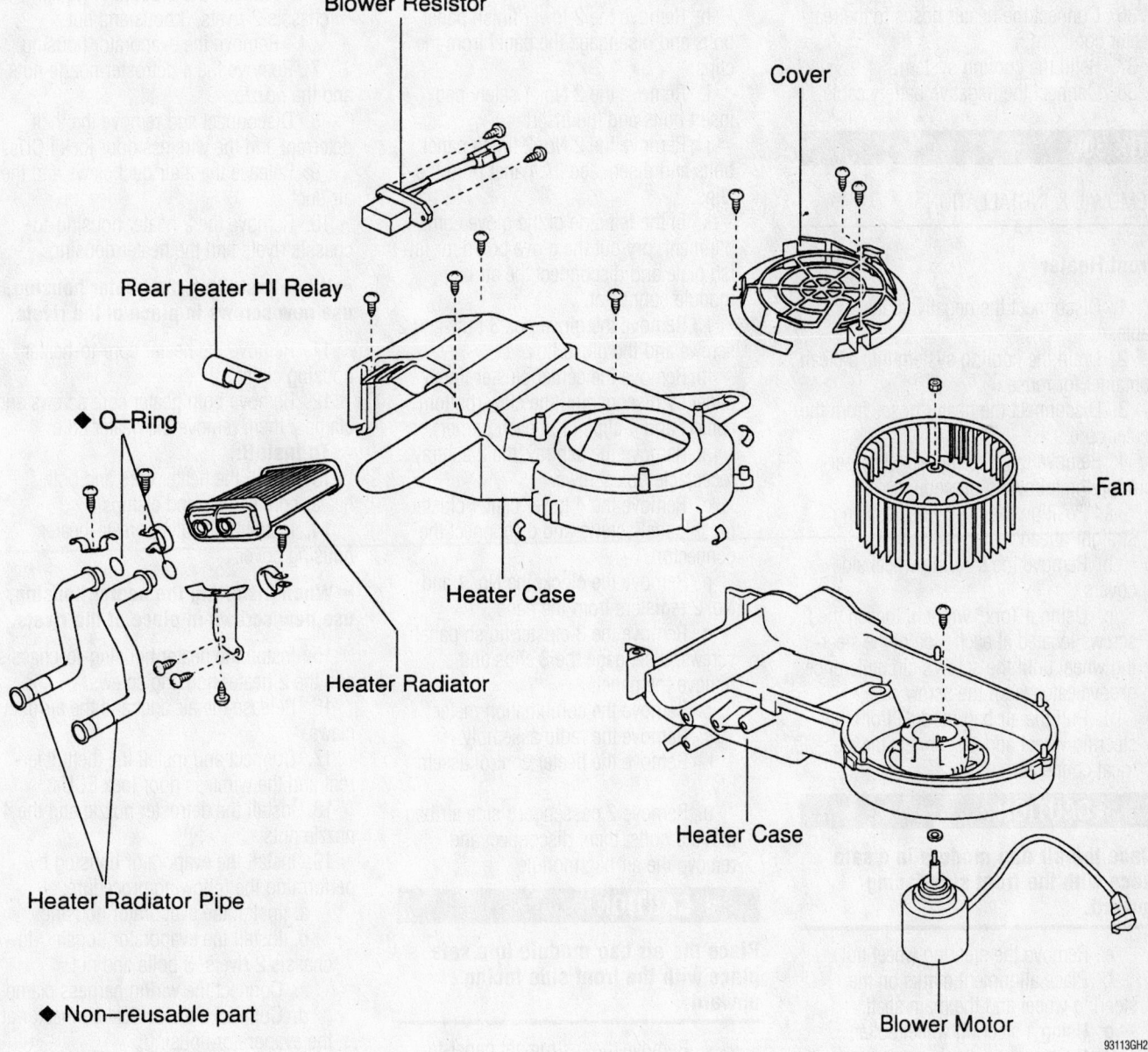

Blower Resistor

Cover

Rear Heater HI Relay

◆ O–Ring

Heater Radiator

Heater Case

Fan

Heater Case

Heater Radiator Pipe

Blower Motor

◆ Non–reusable part

93113GH2

Exploded view of the rear heater core, heater housing and related components—Lexus LX 470

Brake service is covered in Section 4 of this manual

24. Install the rear heater duct and the duct bolt/screw.

25. Install the cooler bracket and the bracket bolts.

26. Slide the carpet rearward.

27. Install the center pillar garnishes.

28. Install the rear door scuff plates.

29. Install the cowl side trim.

30. Install the front door scuff plates.

31. Install the lower center cluster finish panel.

32. Install the front console box cover.

33. Install the rear console box.

34. Install the rear heater control assembly.

35. Install the front seats.

36. Connect the heater hoses to the rear heater core.

37. Refill the cooling system.

38. Connect the negative battery cable.

RX 300

REMOVAL & INSTALLATION

Front Heater

1. Disconnect the negative battery cable.

2. Drain the cooling system into a clean container for reuse.

3. Disconnect the heater hoses from the heater core.

4. Remove the steering wheel by performing the following procedure:

   a. Position the front wheels facing straight-ahead.

   b. Remove the steering wheel side covers.

   c. Using a Torx® wrench, loosen the 2 screws located at each side of the steering wheel until the screw's circumference groove catches on the screw case.

   d. Pull the air bag module from the steering wheel and disconnect the electrical connector.

✳✳ CAUTION

Place the air bag module in a safe place with the front side facing upward.

   e. Remove the steering wheel nut.

   f. Place alignment marks on the steering wheel and the main shaft.

   g. Using a steering wheel puller, press the steering wheel from the steering column.

5. Remove the instrument panel and reinforcement by performing the following procedure:

   a. Remove the front door scuff plates.

   b. Remove the cowl side boards.

   c. Remove the front door trim covers.

   d. Remove the front pillar garnish by disengaging the 5 clips. If equipped with a tweeter speaker, disconnect the electrical connector.

   e. Remove the steering column covers-to-steering column screws and the covers.

   f. Remove the combination switch-to-steering column screws, disconnect the electrical connector(s) and remove the combination switch.

   g. Remove the 2 hood open lever screws and the hood open lever.

   h. Remove the 2 lower finish panel bolts and disengage the panel from the 3 clips.

   i. Remove the 2 No. 1 safety pad insert bolts and the insert.

   j. Remove the 2 No. 2 finish panel bolts and disengage the panel from the 4 clips.

   k. In the left side of the glove compartment, pry out the glove box door finish plate and disconnect the air bag module connector.

   l. Remove the glove box 3 nuts and 2 screws and the glove box.

   m. Remove the center cluster finish panel by disengaging the claw (bottom center) and 4 clips (1 at each corner).

   n. Remove the ashtray, the 2 ashtray receptacle box screws.

   o. Remove the 4 lower center cluster finish panel screws and disconnect the connector.

   p. Remove the clock, the No. 1 and No. 2 registers from the panel.

   q. Remove the 3 cluster finish panel screws, disengage the 8 clips and remove the panel.

   r. Remove the combination meter.

   s. Remove the radio assembly.

   t. Remove the heater control assembly.

   u. Remove 2 passenger's side air bag module bolts; then, disconnect and remove the air bag module.

✳✳ CAUTION

Place the air bag module in a safe place with the front side facing upward.

   v. Remove the instrument panel-to-chassis 5 bolts and nut.

   w. Remove the audio amplifier.

   x. Remove the No. 1 and No. 2 braces.

   y. Remove the No. 2 cowl brace.

   z. Remove the instrument panel reinforcement.

6. Remove the evaporator housing by performing the following procedure:

   a. Discharge and recover the air conditioning system refrigerant.

   b. In the engine compartment, remove the refrigerant lines-to-cowl connector bolts; then, disconnect the lines and discard the O-rings.

   c. Disconnect the electrical connector at the evaporator housing.

   d. Disconnect the wiring harness clamp.

   e. Remove the evaporator housing-to-chassis 2 rivets, 3 bolts and nut.

   f. Remove the evaporator housing.

7. Remove the 4 defroster nozzle nuts and the nozzle.

8. Disconnect and remove the theft deterrent and the wireless door lock ECUs.

9. Release the 2 air duct claws and the air duct.

10. Remove the 2 heater housing-to-chassis rivets and the heater housing.

➡**When installing the heater housing, use new screws in place of the rivets.**

11. Remove the heater core-to-heater housing cover.

12. Remove both heater core screws and clamps; then, remove the heater core.

To install:

13. Install the heater core and both heater core screws and clamps.

14. Install the heater core-to-heater housing cover.

➡**When installing the heater housing, use new screws in place of the rivets.**

15. Install the heater housing-to-chassis and the 2 heater housing screws.

16. Release the air duct and the air duct claws.

17. Connect and install the theft deterrent and the wireless door lock ECUs.

18. Install the defroster nozzle and the 4 nozzle nuts.

19. Install the evaporator housing by performing the following procedure:

   a. Install the evaporator housing.

   b. Install the evaporator housing-to-chassis 2 rivets, 3 bolts and nut.

   c. Connect the wiring harness clamp.

   d. Connect the electrical connector at the evaporator housing.

   e. In the engine compartment, use new O-rings and install the refrigerant lines-to-cowl connector and install the bolts.

20. Install the instrument panel and rein-

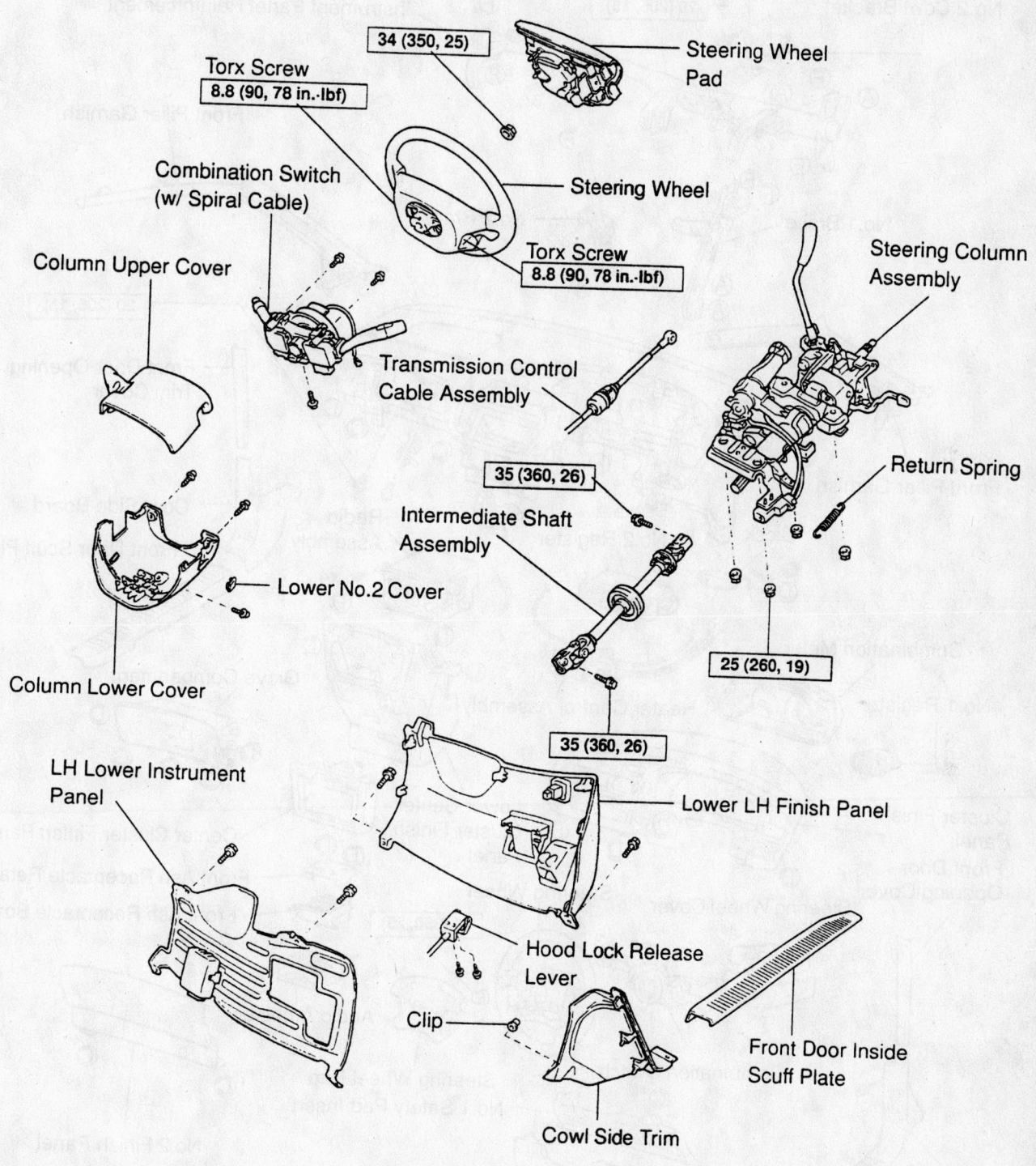

34 (350, 25)

Torx Screw
8.8 (90, 78 in.·lbf)

Steering Wheel Pad

Combination Switch
(w/ Spiral Cable)

Steering Wheel

Column Upper Cover

Torx Screw
8.8 (90, 78 in.·lbf)

Steering Column Assembly

Transmission Control Cable Assembly

35 (360, 26)

Return Spring

Intermediate Shaft Assembly

Lower No.2 Cover

25 (260, 19)

Column Lower Cover

35 (360, 26)

LH Lower Instrument Panel

Lower LH Finish Panel

Hood Lock Release Lever

Clip

Front Door Inside Scuff Plate

Cowl Side Trim

N·m (kgf·cm, ft·lbf) : Specified torque

93113GH3

Exploded view of the steering wheel, steering column and related components—Lexus RX 300

For complete Engine Mechanical specifications, see Section 1 of this manual

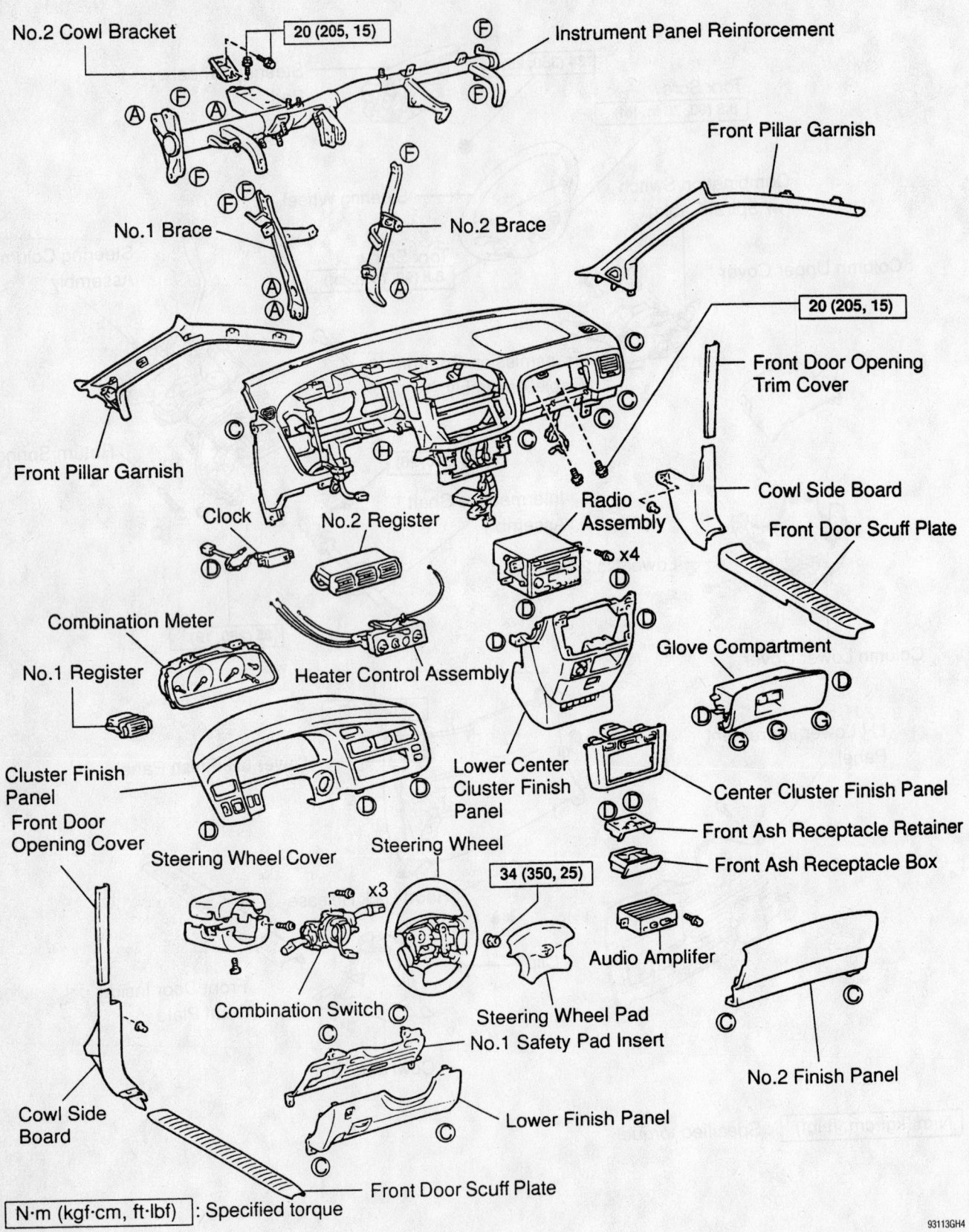

No.2 Cowl Bracket

20 (205, 15)

Instrument Panel Reinforcement

Front Pillar Garnish

No.1 Brace

No.2 Brace

20 (205, 15)

Front Door Opening Trim Cover

Front Pillar Garnish

Cowl Side Board

Front Door Scuff Plate

Clock

No.2 Register

Radio Assembly

Combination Meter

No.1 Register

Heater Control Assembly

Glove Compartment

Cluster Finish Panel

Lower Center Cluster Finish Panel

Center Cluster Finish Panel

Front Door Opening Cover

Steering Wheel Cover

Steering Wheel

34 (350, 25)

Front Ash Receptacle Retainer

Front Ash Receptacle Box

Audio Amplifer

Combination Switch

Steering Wheel Pad

No.1 Safety Pad Insert

No.2 Finish Panel

Cowl Side Board

Lower Finish Panel

Front Door Scuff Plate

N·m (kgf·cm, ft·lbf) : Specified torque

93113GH4

Exploded view of the instrument panel and related components—Lexus RX 300

forcement by performing the following procedure:

a. Install the instrument panel reinforcement.

b. Install the No. 2 cowl brace.

c. Install the No. 1 and No. 2 braces.

d. Install the audio amplifier.

e. Install the instrument panel-to-chassis 5 bolts and nut.

f. Connect and install the air bag module and the 2 passenger's side air bag module bolts.

g. Install the heater control assembly.

h. Install the radio assembly.

i. Install the combination meter.

j. Install the cluster finish panel, engage the 8 clips and install the panel screws.

k. Install the No. 1 and No. 2 registers and the clock to the panel.

l. Connect the lower center cluster finish panel connector and install the 4 lower center cluster finish panel screws.

m. Install the 2 ashtray receptacle box screws and the ashtray.

n. Install the center cluster finish panel by engaging the 4 clips (1 at each corner) and the claw (bottom center).

o. Install the glove box and the glove box 3 nuts and 2 screws.

p. In the left side of the glove compartment, connect the air bag module connector and install the glove box door finish plate.

q. Install the No. 2 finish panel, engage the 4 panel clips and install the 3 panel bolts.

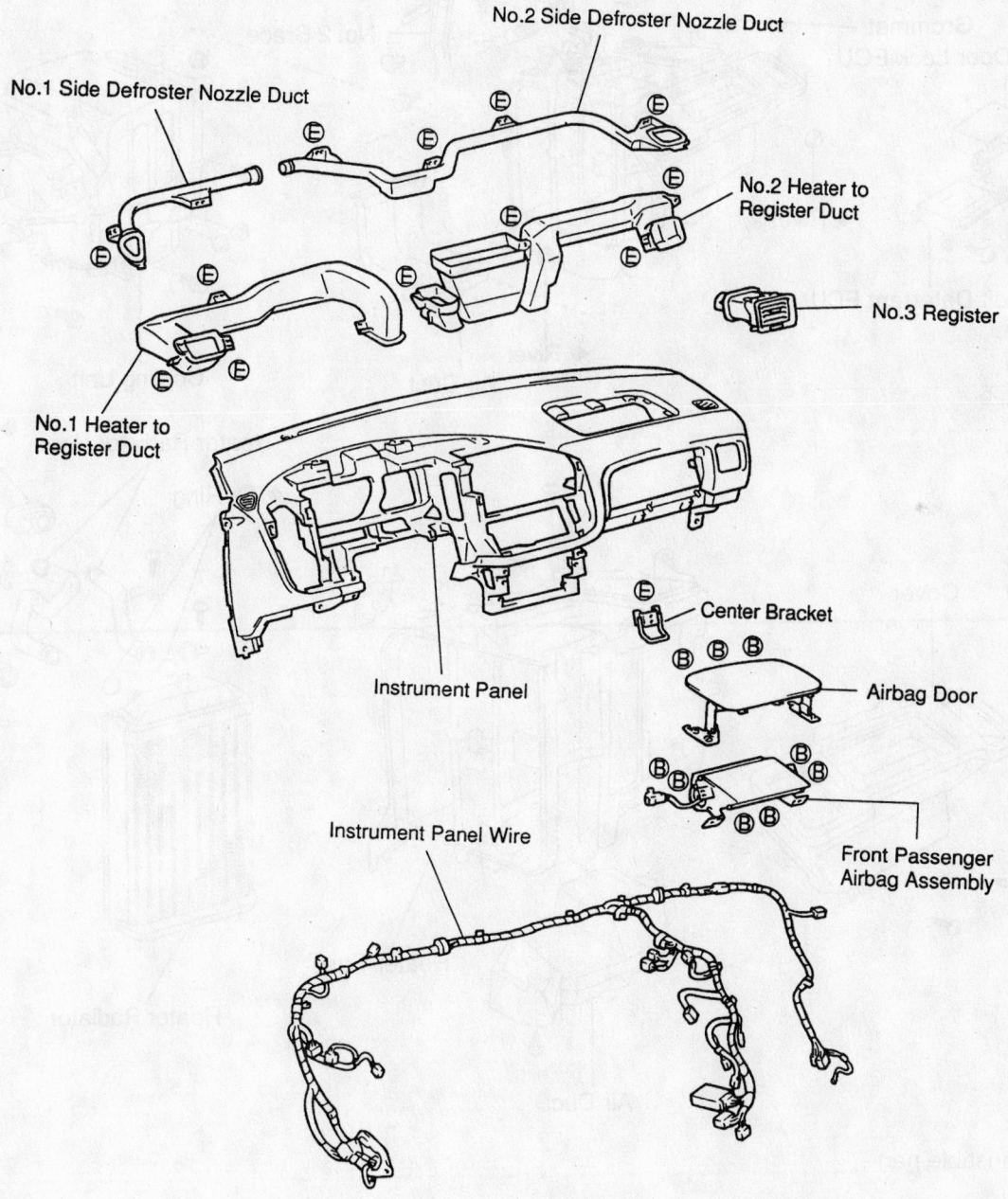

Exploded view of the ventilation system and related components—Lexus RX 300

93113GH5

For Accessory Drive Belt illustrations, see Section 1 of this manual

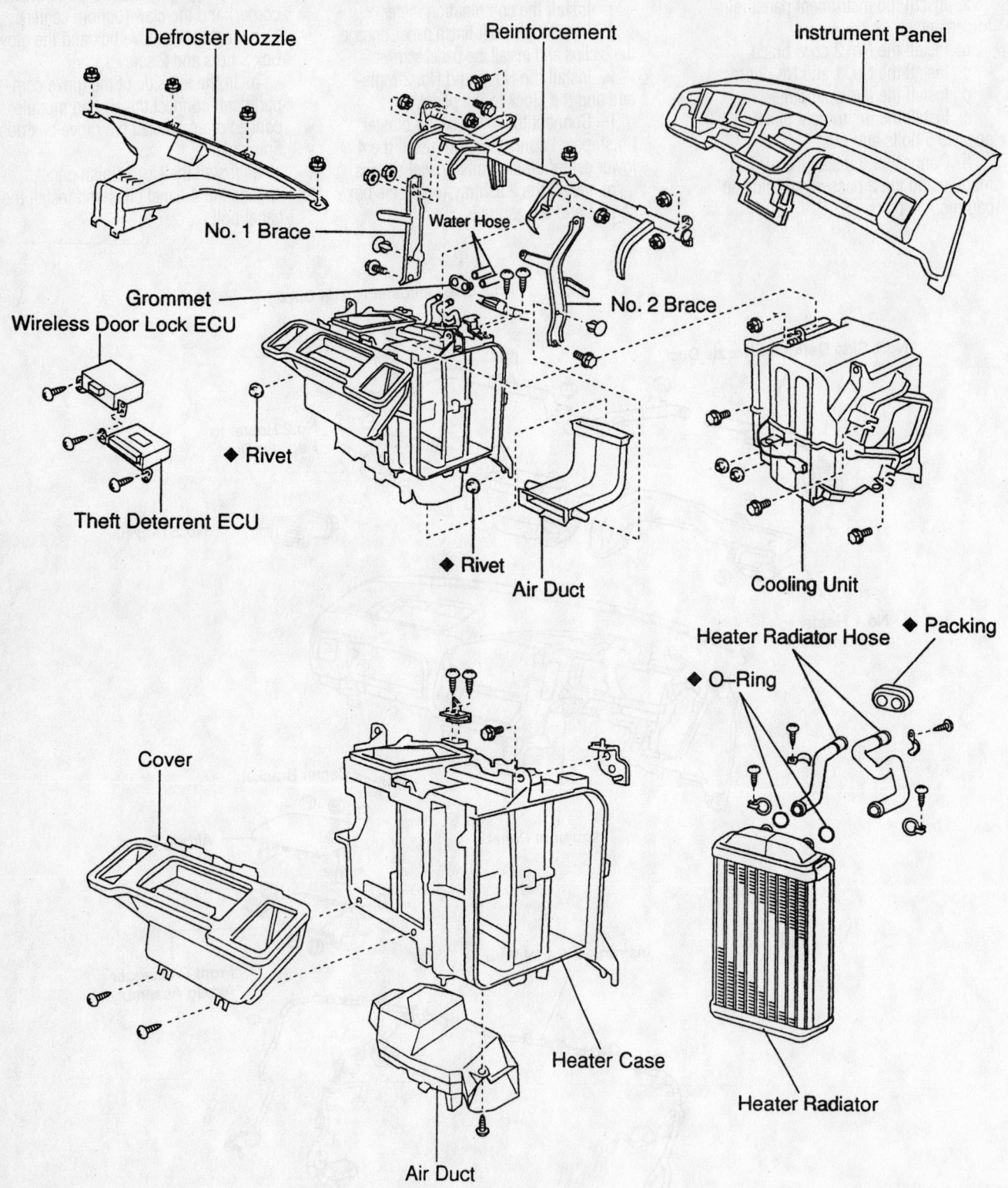

Defroster Nozzle

Reinforcement

Instrument Panel

No. 1 Brace

Water Hose

Grommet

Wireless Door Lock ECU

No. 2 Brace

◆ Rivet

Theft Deterrent ECU

◆ Rivet

Air Duct

Cooling Unit

Heater Radiator Hose

◆ Packing

◆ O-Ring

Cover

Heater Case

Heater Radiator

Air Duct

◆ Non-reusable part

93113GH6

Exploded view of the heater core, heater housing, evaporator housing and related components—Lexus RX 300

r. Install the No. 1 safety pad insert and the 2 insert bolts.

s. Install the finish panel, engage the 3 finish panel clips and install 2 lower finish panel bolts.

t. Install the hood open lever and the 2 hood open lever screws.

u. Install the combination switch, connect the electrical connector(s) and install the combination switch-to-steering column screws.

v. Install the steering column covers and the covers-to-steering column screws.

w. Install the front pillar garnish by engaging the 5 clips. If equipped with a tweeter speaker, connect the electrical connector.

x. Install the front door trim covers.

y. Install the cowl side boards.

z. Install the front door scuff plates.

21. Install the steering wheel by performing the following procedure:

a. Install the steering wheel to the steering column.

b. Align the steering wheel-to-main shaft marks.

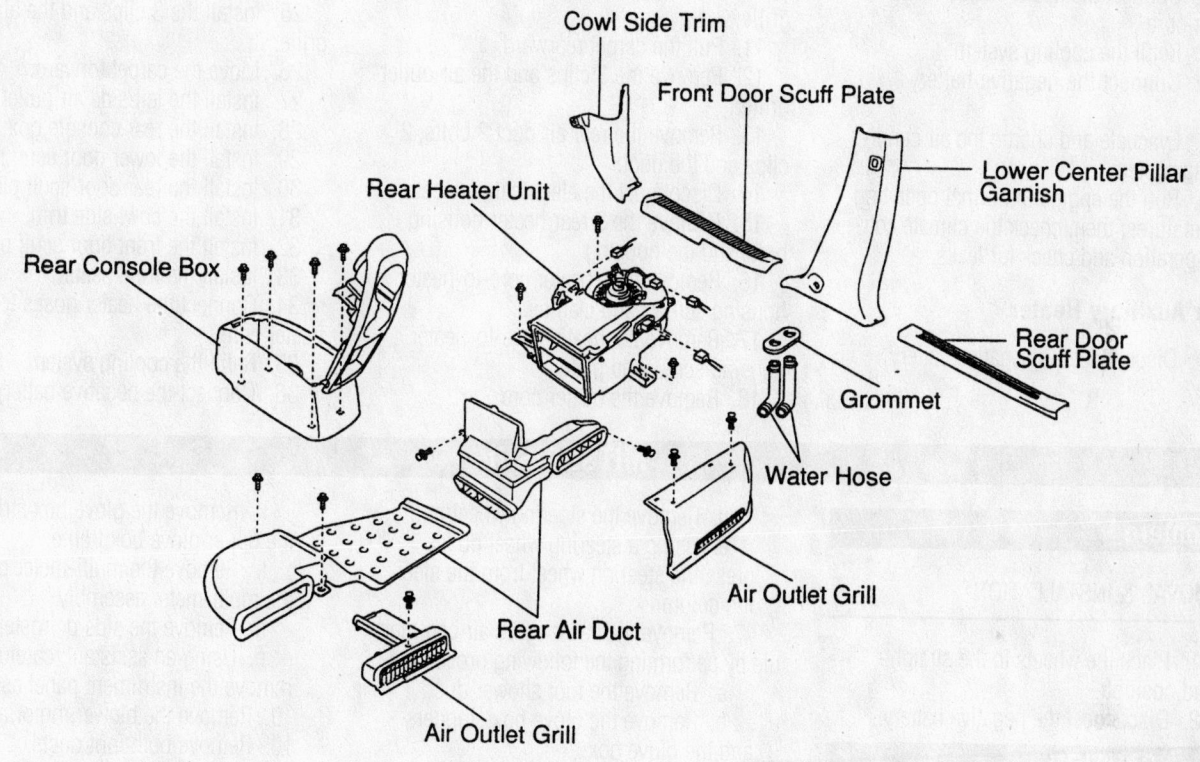

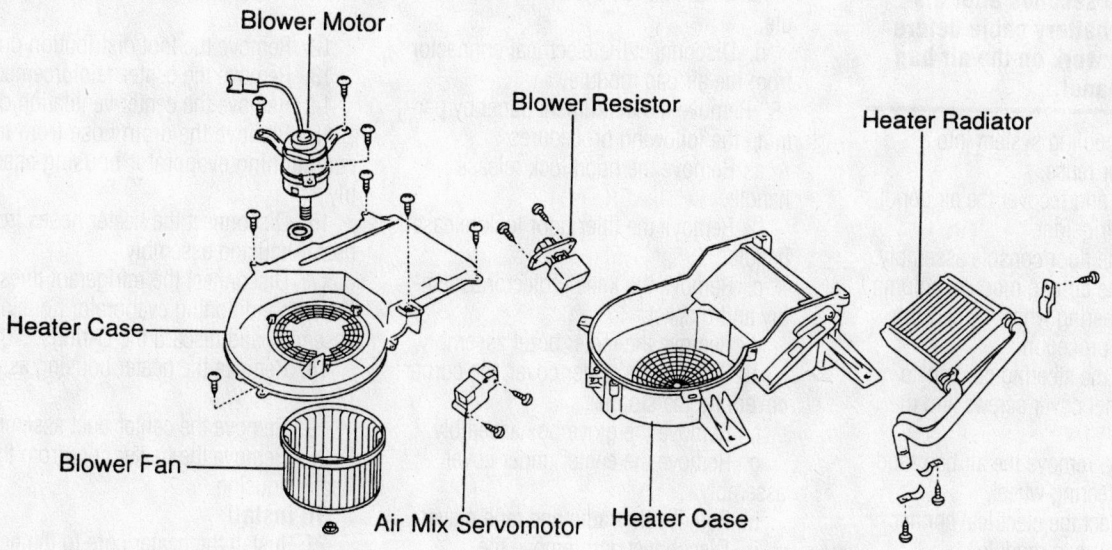

Exploded view of the rear heater core, the rear heater housing and related components—Lexus RX 300

For Tire, Wheel and Ball Joint specifications, see Section 1 of this manual

c. Install the steering wheel nut and torque the nut to 25 ft. lbs. (34 Nm).

d. Install the air bag module to the steering wheel and connect the electrical connector.

e. Using a Torx® wrench, tighten the steering wheel screws to 78 inch lbs. (8.8 Nm).

f. Install the steering wheel side covers.

22. Connect the heater hoses to the heater core.

23. Refill the cooling system.

24. Connect the negative battery cable.

25. Evacuate and charge the air conditioning system.

26. Run the engine to normal operating temperatures; then, check the climate control operation and check for leaks.

Rear Auxiliary Heater

1. Disconnect the negative battery cable.

2. Drain the cooling system into a clean container for reuse.

3. Disconnect the heater hoses from the rear heater core.

4. Remove the front seats.

5. Remove the front door scuff plates.

6. Remove the cowl side trim.

7. Remove the rear door scuff plates.

8. Remove the lower door scuff plates.

9. Remove the rear console box.

10. Remove the left side air outlet grille.

11. Pull the carpet rearward.

12. Remove the 3 clips and the air outlet grille.

13. Remove the rear air duct 2 bolts, 2 clips and the duct.

14. Disconnect the electrical connectors.

15. Remove the 3 rear heater housing bolts and the housing.

16. Remove both heater core-to-heater housing screws and clamps.

17. Remove the heater core-to-heater housing screw and plate.

18. Remove the heater core.

To install:

19. Install the heater core.

20. Install the heater core-to-heater housing screw and plate.

21. Install both heater core-to-heater housing screws and clamps.

22. Install the rear heater housing and the 3 housing bolts.

23. Connect the electrical connectors.

24. Install the rear air duct and the duct 2 bolts and 2 clips.

25. Install the 3 clips and the air outlet grille.

26. Move the carpet forward.

27. Install the left side air outlet grille.

28. Install the rear console box.

29. Install the lower door scuff plates.

30. Install the rear door scuff plates.

31. Install the cowl side trim.

32. Install the front door scuff plates.

33. Install the front seats.

34. Connect the heater hoses to the rear heater core.

35. Refill the cooling system.

36. Connect the negative battery cable.

MITSUBISHI

Montero

REMOVAL & INSTALLATION

1. Place the wheels in the straight-ahead position.

2. Disconnect the negative battery.

❄❄ CAUTION

Wait at least 60 seconds after disconnecting the battery cable before performing any work on the air bag or instrument panel.

3. Drain the cooling system into a clean container for reuse.

4. Discharge and recover the air conditioning system refrigerant.

5. Remove the floor console assembly.

6. Remove the air bag module, column covers and the steering wheel by performing the following procedure:

a. Remove the steering column-to-instrument panel cover screws and the cover.

b. Carefully, remove the air bag module from the steering wheel.

c. Disconnect the electrical connectors from the air bag module.

❄❄ CAUTION

Store the air bag module facing up.

d. Remove the steering wheel nut.

e. Using a steering wheel puller, press the steering wheel from the steering column.

7. Remove the passenger's air bag module by performing the following procedure:

a. Remove the foot shower duct.

b. Remove the glove box stoppers and the glove box.

c. Remove the passenger's air bag module nut and remove the air bag module.

d. Disconnect the electrical connector from the air bag module.

8. Remove the instrument panel by performing the following procedures:

a. Remove the hood lock release handle.

b. Remove the filler door lock release handle.

c. Remove the knee protector assembly and bracket.

d. Remove the meter bezel assembly.

e. Remove the under cover, the corner cover and the stopper.

f. Remove the glove box assembly.

g. Remove the center under cover assembly.

h. Remove the radio and tape player.

i. Disconnect and remove the heater/air conditioning control assembly.

j. Remove the combination meter and the speaker.

k. Remove the glove box striker and the upper glove box frame.

l. Remove the multi-meter panel and the multi-meter assembly.

m. Remove the side defroster grille.

n. Using an assistant, carefully, remove the instrument panel assembly.

9. Remove the blower motor assembly.

10. Remove both foot ducts.

11. Remove the joint duct from the air conditioning evaporator housing assembly.

12. Remove the foot distribution duct.

13. Remove the center reinforcement.

14. Remove the center ventilation duct.

15. Remove the drain hose from the air conditioning evaporator housing assembly.

16. Disconnect the heater hoses from the heater housing assembly.

17. Disconnect the refrigerant lines from the air conditioning evaporator housing assembly and discard the O-rings.

18. Remove the heater housing assembly.

19. Remove the center duct assembly.

20. Remove the heater core from the heater housing.

To install:

21. Install the heater core to the heater housing.

22. Install the center duct assembly.

23. Install the heater housing assembly.

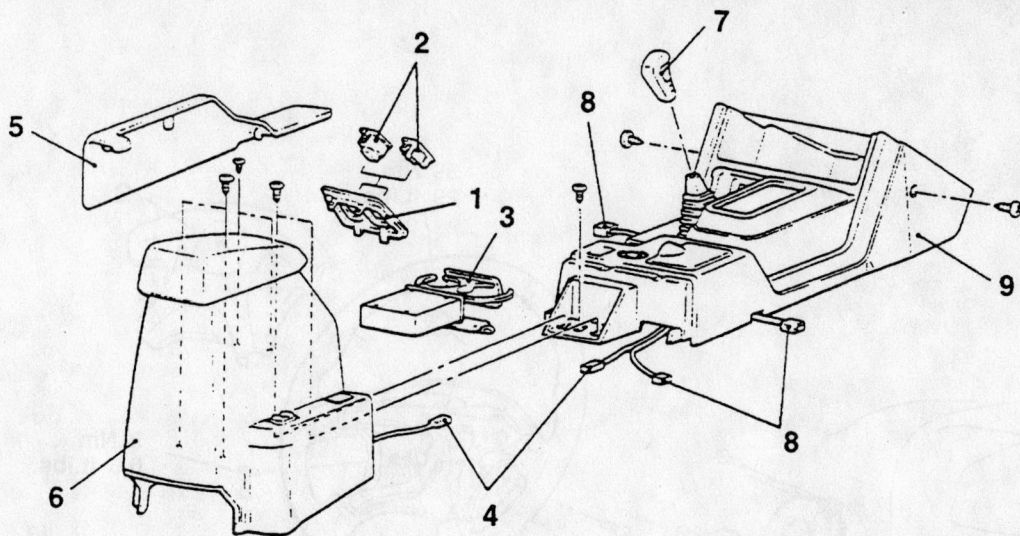

1. Switch panel
2. Suspension control switch or hole cover
3. Cup holder assembly
4. Rear console harness connector
5. Side panel A
6. Rear console assembly
7. Transfer shift lever knob
8. Floor console harness connector
9. Front console assembly

93113GE2

Exploded view of the floor console and related components—Mitsubishi Montero

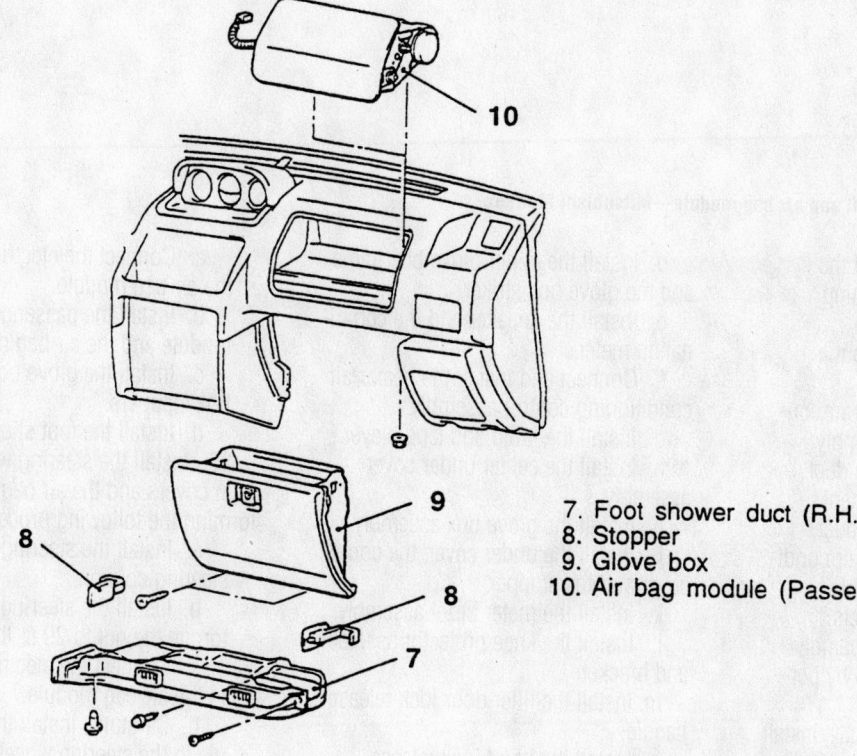

7. Foot shower duct (R.H.)
8. Stopper
9. Glove box
10. Air bag module (Passenger's side)

93113GE3

Exploded view of the passenger's side air bag module—Mitsubishi Montero

For Wheel Alignment specifications, see Section 1 of this manual

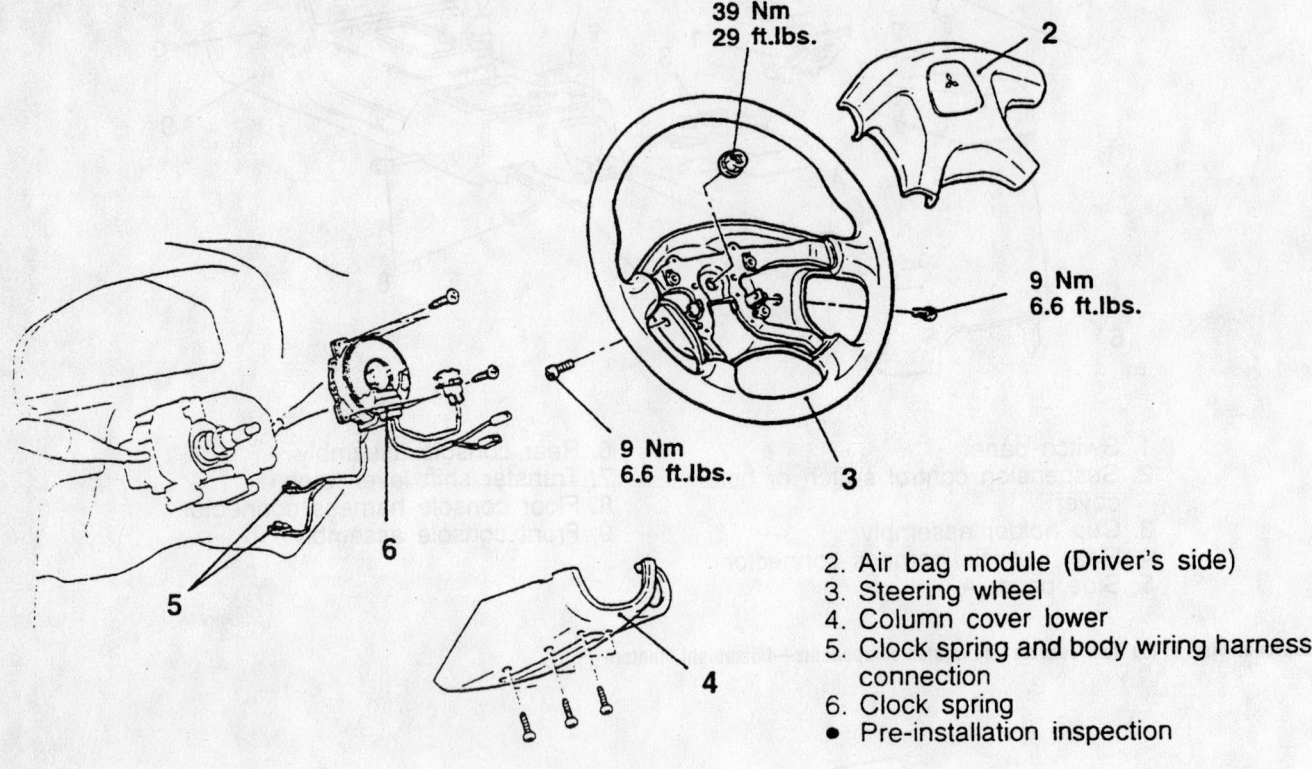

39 Nm
29 ft.lbs.

9 Nm
6.6 ft.lbs.

9 Nm
6.6 ft.lbs.

2. Air bag module (Driver's side)
3. Steering wheel
4. Column cover lower
5. Clock spring and body wiring harness connection
6. Clock spring
● Pre-installation inspection

93113GE4

Exploded view of the steering wheel and air bag module—Mitsubishi Montero

24. Using new O-rings, connect the refrigerant lines to the air conditioning evaporator housing assembly.

25. Connect the heater hoses to the heater housing assembly.

26. Install the drain hose to the air conditioning evaporator housing assembly.

27. Install the center ventilation duct.

28. Install the center reinforcement.

29. Install the foot distribution duct.

30. Install the joint duct to the air conditioning evaporator housing assembly.

31. Install both foot shower ducts.

32. Install the blower motor assembly.

33. Install the instrument panel by performing the following procedures:

  a. Using an assistant, carefully, install the instrument panel assembly.

  b. Install the side defroster grille.

  c. Install the multi-meter assembly and the multi-meter panel.

  d. Install the upper glove box frame and the glove box striker.

  e. Install the speaker and the combination meter.

  f. Connect and install the heater/air conditioning control assembly.

  g. Install the radio and tape player.

  h. Install the center under cover assembly.

  i. Install the glove box assembly.

  j. Install the under cover, the corner cover and the stopper.

  k. Install the meter bezel assembly.

  l. Install the knee protector assembly and bracket.

  m. Install the filler door lock release handle.

  n. Install the hood lock release handle.

34. Install the passenger's air bag module by performing the following procedure:

  a. Connect the electrical connector to the air bag module.

  b. Install the passenger's air bag module and the air bag module nut.

  c. Install the glove box and the glove box stoppers.

  d. Install the foot shower duct.

35. Install the steering wheel, the column covers and the air bag module by performing the following procedure:

  a. Install the steering wheel to the steering column.

  b. Install the steering wheel nut and torque the nut to 29 ft. lbs. (39 Nm).

  c. Connect the electrical connectors to the air bag module.

  d. Carefully, install the air bag module to the steering wheel.

  e. Install the steering column-to-instrument panel cover and the cover screws.

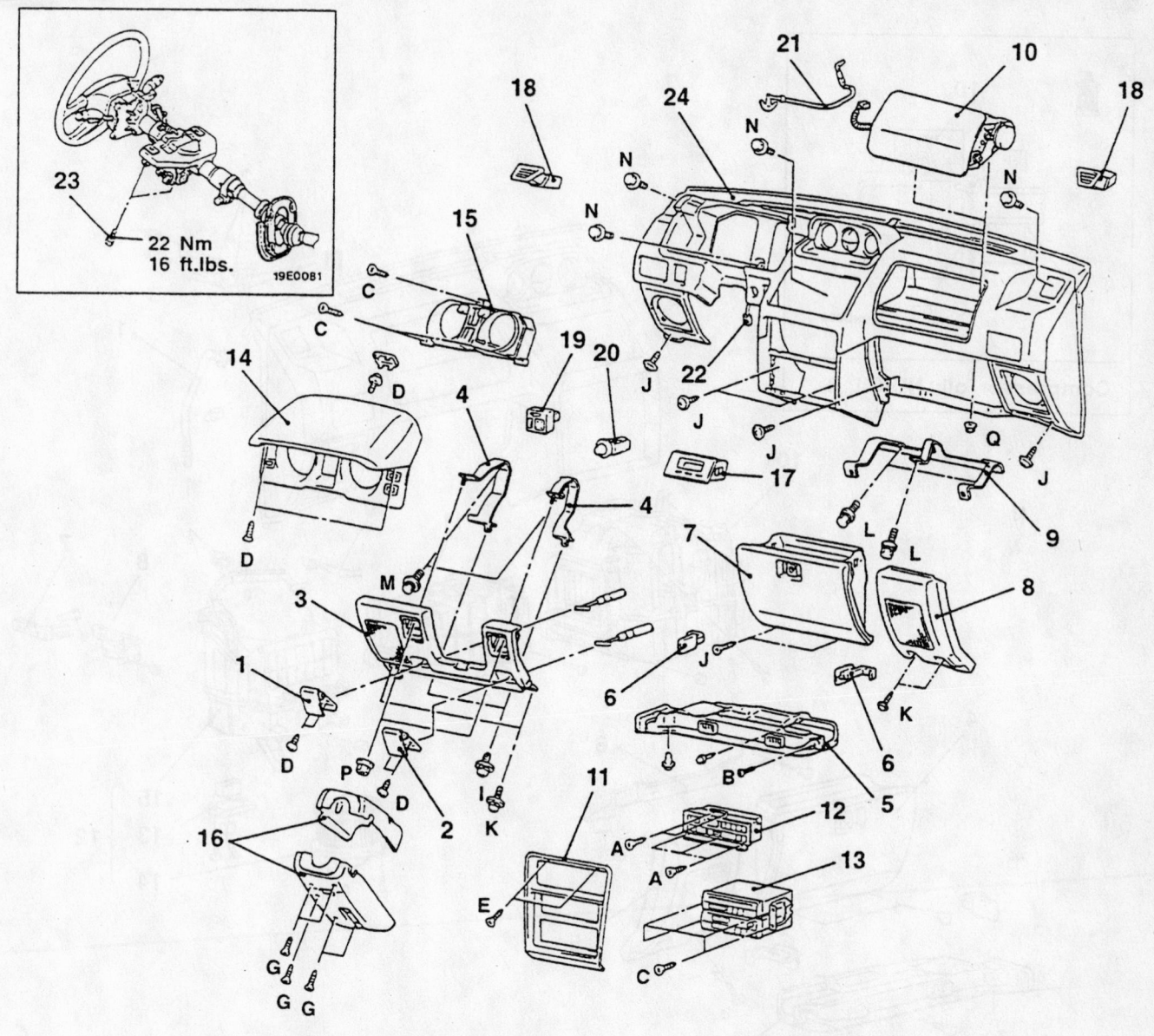

22 Nm
16 ft.lbs.
19E0081

Exploded view of the instrument panel and related components—Mitsubishi Montero

1. Hood lock release handle
2. Fuel filler door lock release handle
3. Knee protector
4. Stay
5. Foot shower duct (R.H.)
6. Glove box stopper
7. Glove box assembly
8. Corner cover
9. Stay
10. Passenger-side air bag module assembly
11. Center panel
12. Heater control assembly
13. Radio and tape player
14. Meter bezel assembly
15. Combination meter
16. Column cover
17. Clock
18. Side defroster garnish
19. Door mirror control switch
20. Rheostat
21. Ventilation control wire
22. Harness connector
23. Steering column installation bolts
24. Instrument panel assembly

93113GE5

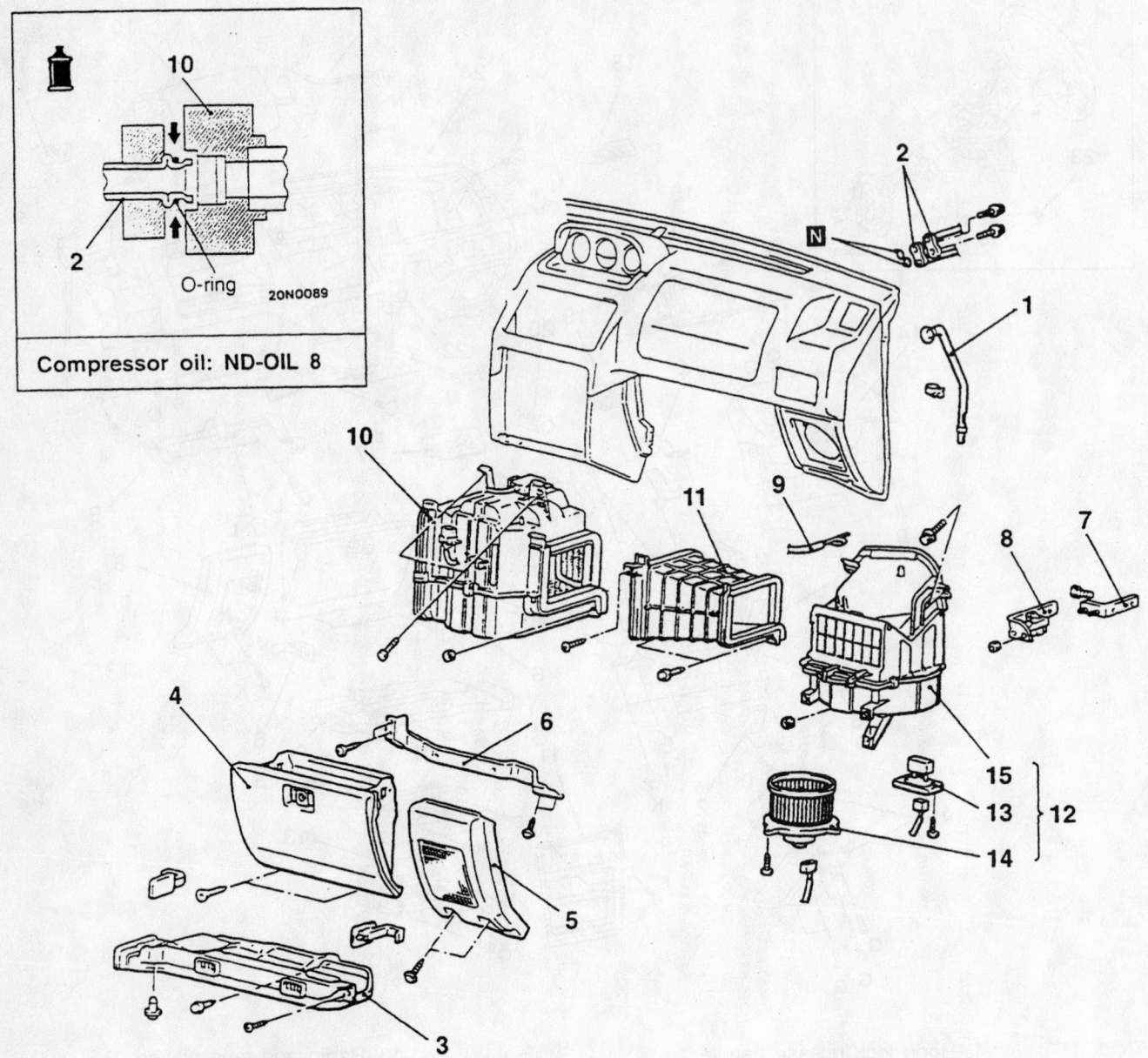

Compressor oil: ND-OIL 8

20N0089

O-ring

1. Drain hose
2. Liquid pipe and suction hose connection
3. Foot shower duct (R.H.)
4. Glove box
5. Corner cover
6. Lower frame
7. Engine control relay assembly
8. Bracket
9. Air selection control wire connection
10. Evaporator

11. Duct joint
12. Blower assembly
13. Resistor
14. Blower motor assembly
15. Blower case assembly

93113GE6

Exploded view of the air conditioning evaporator housing, blower motor assembly and related components—Mitsubishi Montero

36. Install the floor console assembly.
37. Refill the cooling system.
38. Connect the negative battery.
39. Evacuate and charge the air conditioning system refrigerant.
40. Run the engine to normal operating temperatures; then, check the climate control operation and check for leaks.

Montero Sport

REMOVAL & INSTALLATION

Front Heater System

1. Place the wheels in the straight-ahead position.
2. Disconnect the negative battery.

※※ CAUTION

Wait at least 60 seconds after disconnecting the battery cable before performing any work on the air bag or instrument panel.

3. Drain the cooling system into a clean container for reuse.
4. Discharge and recover the air conditioning system refrigerant.
5. Remove the floor console assembly.
6. Remove the air bag module, column covers and the steering wheel by performing the following procedure:
   a. Remove the steering column-to-instrument panel cover screws and the cover.

b. Carefully, remove the air bag module from the steering wheel.
c. Disconnect the electrical connectors from the air bag module.

※※ CAUTION

Store the air bag module facing up.

d. Remove the steering wheel nut.
e. Using a steering wheel puller, press the steering wheel from the steering column.

7. Remove the passenger's air bag module by performing the following procedure:
   a. Remove the glove box stoppers and lower the glove box.
   b. Remove the passenger's air bag

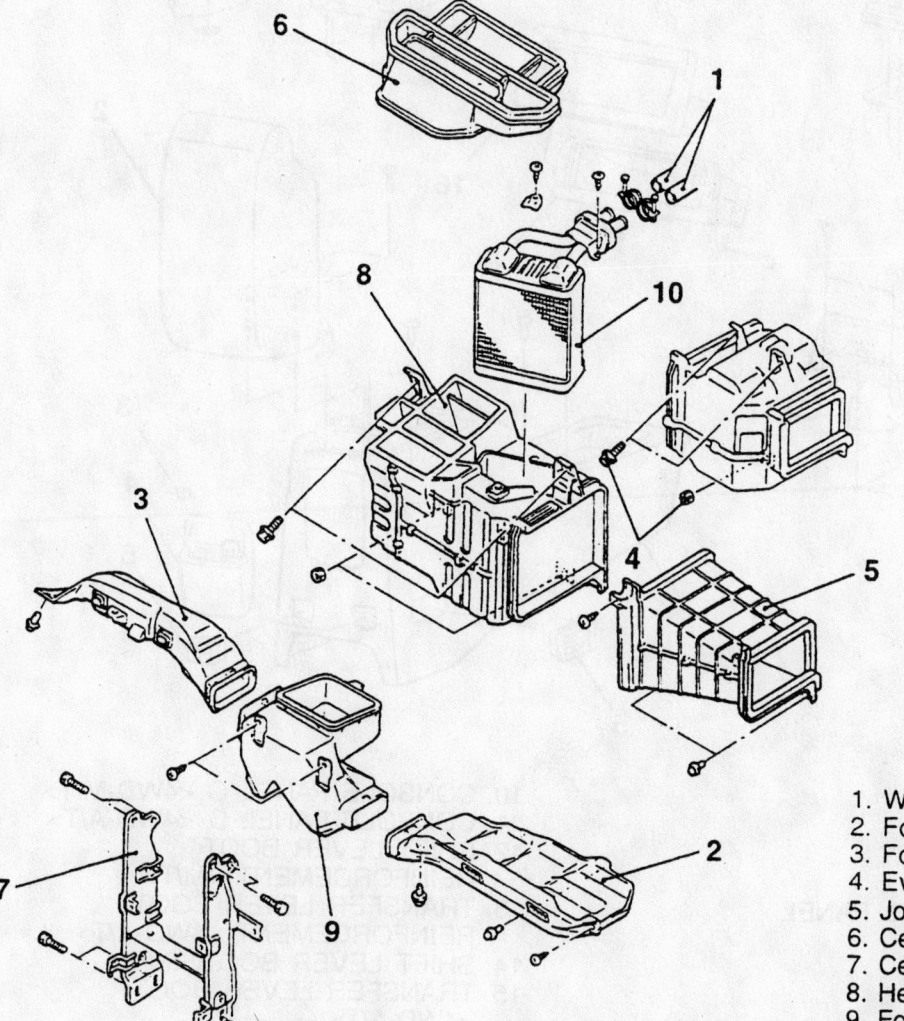

1. Water hoses connection
2. Foot shower duct (RH)
3. Foot shower duct (LH)
4. Evaporator mounting bolt and nut
5. Joint duct
6. Center duct assembly
7. Center reinforcement
8. Heater unit
9. Foot distribution duct
10. Heater core

93113GE7

Exploded view of the heater housing, air conditioning evaporator housing and related components—Mitsubishi Montero

For Tune-up, Capacities and Firing orders, see Section 1 of this manual

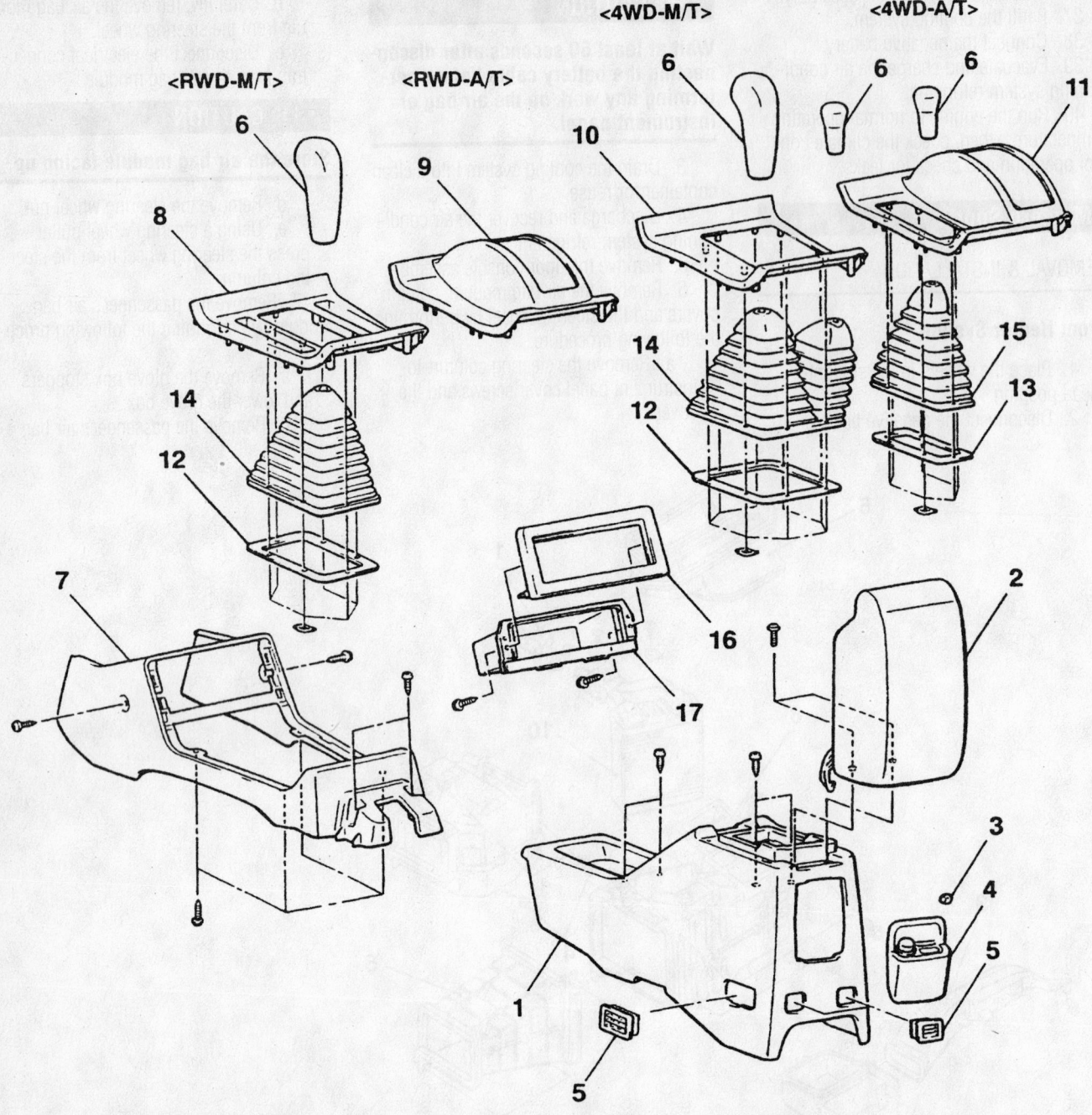

<RWD-M/T> <RWD-A/T> <4WD-M/T> <4WD-A/T>

1. REAR FLOOR CONSOLE
   ASSEMBLY
2. CONSOLE LID ASSEMBLY
3. KNOB
4. REAR HEATER CONTROL PANEL
   ASSEMBLY
5. FOOT GRILL
6. SHIFT LEVER KNOB
7. FRONT FLOOR CONSOLE
   ASSEMBLY
8. CONSOLE PANEL A <RWD-M/T>
9. CONSOLE PANEL B <RWD-A/T>

10. CONSOLE PANEL C <4WD-M/T>
11. CONSOLE PANEL D <4WD-A/T>
12. SHIFT LEVER BOOT
    REINFORCEMENT <M/T>
13. TRANSFER LEVER BOOT
    REINFORCEMENT <4WD-A/T>
14. SHIFT LEVER BOOT <M/T>
15. TRANSFER LEVER BOOT
    <4WD-A/T>
16. CONSOLE PANEL
17. BOX

93113GD6

Exploded view of the floor console and related components—Mitsubishi Montero Sport

module bolts and remove the air bag module.

   c. Disconnect the electrical connector from the air bag module.

  8. Remove the instrument panel by performing the following procedures:

   a. Remove the hood lock release handle.

   b. Remove the knee protector assembly and bracket.

   c. Remove the meter bezel assembly.

   d. Remove the under cover, the corner cover and the stopper.

   e. Remove the glove box assembly and the ashtray.

   f. Remove the center under cover assembly and the cup holder assembly.

   g. Remove the radio and tape player.

   h. Disconnect and remove the heater/air conditioning control assembly.

   i. Remove the combination meter and the speaker.

   j. Remove the glove box striker and the upper glove box frame.

   k. Remove the multi-meter panel and the multimeter assembly.

   l. Remove the side defroster grille.

   m. Using an assistant, carefully, remove the instrument panel assembly.

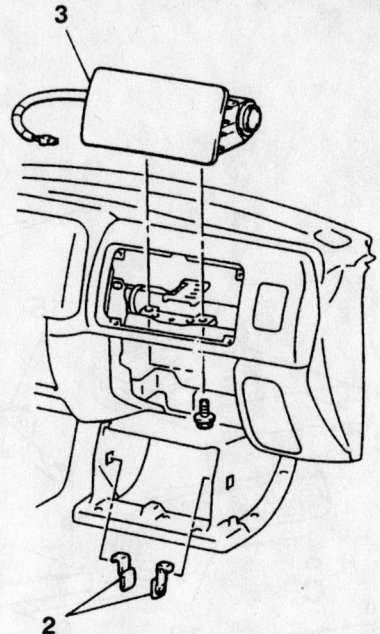

1. NEGATIVE (–) BATTERY CABLE CONNECTION
2. STOPPER
3. AIR BAG MODULE
• PRE-INSTALLATION INSPECTION

93113GD7

Exploded view of the passenger's side air bag module—Mitsubishi Montero Sport

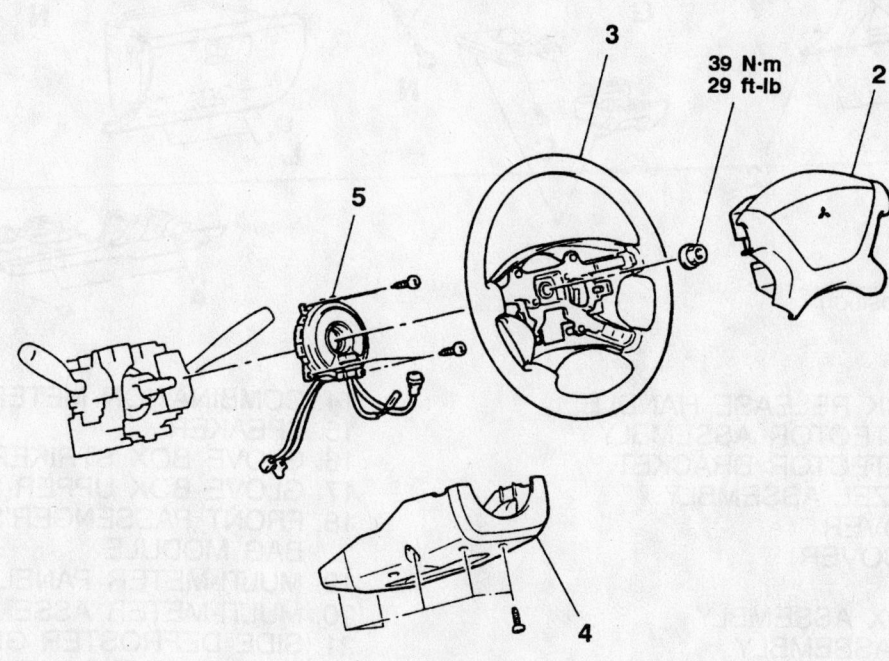

39 N·m
29 ft-lb

2. AIR BAG MODULE
3. STEERING WHEEL
4. COLUMN COVER LOWER
5. CLOCK SPRING

93113GD8

Exploded view of the steering wheel and air bag module—Mitsubishi Montero Sport

For complete service labor times, order Nichols' Chilton Labor Guide

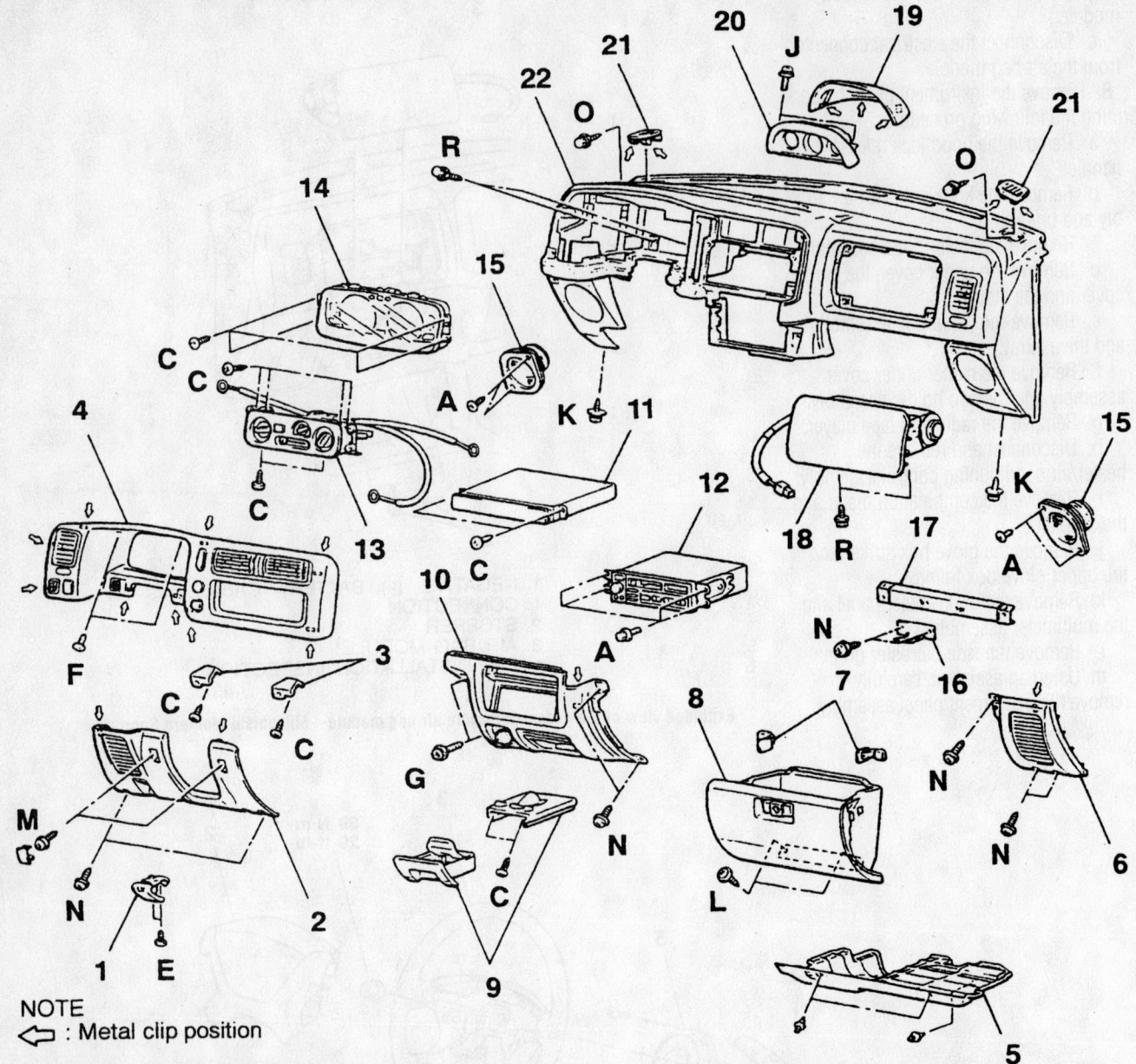

NOTE
◁ : Metal clip position

1. HOOD LOCK RELEASE HANDLE
2. KNEE PROTECTOR ASSEMBLY
3. KNEE PROTECTOR BRACKET
4. METER BEZEL ASSEMBLY
5. UNDER COVER
6. CORNER COVER
7. STOPPER
8. GLOVE BOX ASSEMBLY
9. ASHTRAY ASSEMBLY
10. CENTER UNDER COVER
    ASSEMBLY
11. CUP HOLDER ASSEMBLY
12. RADIO AND TAPE PLAYER
13. HEATER CONTROL ASSEMBLY

14. COMBINATION METER
15. SPEAKER
16. GLOVE BOX STRIKER
17. GLOVE BOX UPPER FRAME
18. FRONT PASSENGER'S SIDE AIR
    BAG MODULE
19. MULTI-METER PANEL
20. MULTI-METER ASSEMBLY
21. SIDE DEFROSTER GRILL
22. INSTRUMENT PANEL ASSEMBLY

93113GD9

Exploded view of the instrument panel and related components—Mitsubishi Montero Sport

9. Remove the blower motor assembly.

10. Remove the joint duct from the air conditioning evaporator housing assembly.

11. Remove the center reinforcement.

12. Remove the center ventilation duct.

13. Remove the drain hose from the air conditioning evaporator housing assembly.

14. Disconnect the heater hoses from the heater housing assembly.

15. Disconnect the refrigerant lines from the air conditioning evaporator housing assembly and discard the O-rings.

16. Remove the heater housing assembly.

17. Remove the heater core from the heater housing.

To install:

18. Install the heater core to the heater housing.

19. Install the heater housing assembly.

20. Using new O-rings, connect the refrigerant lines to the air conditioning evaporator housing assembly.

21. Connect the heater hoses to the heater housing assembly.

22. Install the drain hose to the air conditioning evaporator housing assembly.

23. Install the center ventilation duct.

24. Install the center reinforcement.

25. Install the joint duct to the air conditioning evaporator housing assembly.

26. Install the blower motor assembly.

27. Install the instrument panel by performing the following procedures:

a. Using an assistant, carefully, install the instrument panel assembly.

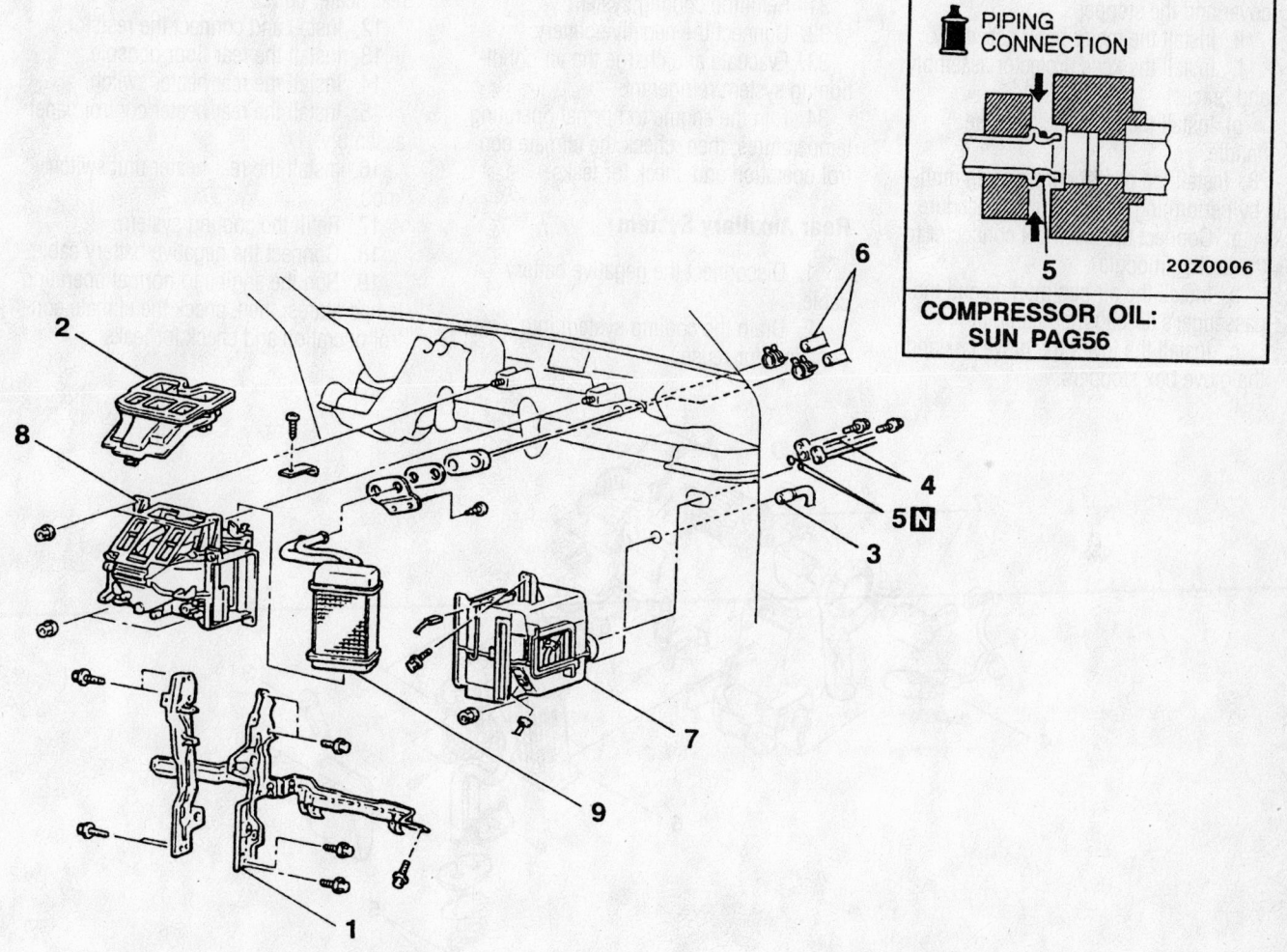

PIPING CONNECTION

5

20Z0006

COMPRESSOR OIL: SUN PAG56

1. CENTER REINFORCEMENT
2. CENTER VENTILATION DUCT
3. DRAIN HOSE <VEHICLES WITH A/C>
4. SUCTION PIPE OR HOSE AND DISCHARGE PIPE CONNECTION <VEHICLES WITH A/C>
5. O-RING
6. HEATER HOSE CONNECTION
7. EVAPORATOR <VEHICLES WITH A/C>
8. HEATER UNIT
9. HEATER CORE

93113GD0

Exploded view of the heater housing, air conditioning evaporator housing and related components—Mitsubishi Montero Sport

b. Install the side defroster grille.

c. Install the multi-meter assembly and the multimeter panel.

d. Install the upper glove box frame and the glove box striker.

e. Install the speaker and the combination meter.

f. Connect and install the heater/air conditioning control assembly.

g. Install the radio and tape player.

h. Install the center under cover assembly and the cup holder assembly.

i. Install the glove box assembly and the ashtray.

j. Install the under cover, the corner cover and the stopper.

k. Install the meter bezel assembly.

l. Install the knee protector assembly and bracket.

m. Install the hood lock release handle.

28. Install the passenger's air bag module by performing the following procedure:

a. Connect the electrical connector to the air bag module.

b. Install the air bag module and the passenger's air bag module bolts.

c. Install the lower the glove box and the glove box stoppers.

29. Install the steering wheel, the column covers and the air bag module by performing the following procedure:

a. Install the steering wheel to the steering column.

b. Install the steering wheel nut and torque the nut to 29 ft. lbs. (39 Nm).

c. Connect the electrical connectors to the air bag module.

d. Carefully, install the air bag module to the steering wheel.

e. Install the steering column-to-instrument panel cover and the cover screws.

30. Install the floor console assembly.

31. Refill the cooling system.

32. Connect the negative battery.

33. Evacuate and charge the air conditioning system refrigerant.

34. Run the engine to normal operating temperatures; then, check the climate control operation and check for leaks.

Rear Auxiliary System

1. Disconnect the negative battery cable.

2. Drain the cooling system into a clean container for reuse.

3. Remove the rear heater unit switch knob.

4. Remove the rear heater control panel assembly.

5. Remove the rear heater switch.

6. Remove the rear floor console.

7. Disconnect and remove the resistor.

8. Disconnect the rear heater hoses from the rear heater core.

9. Remove the rear heater core from the rear heater housing.

To install:

10. Install the rear heater core to the rear heater housing.

11. Connect the rear heater hoses to the rear heater core.

12. Install and connect the resistor.

13. Install the rear floor console.

14. Install the rear heater switch.

15. Install the rear heater control panel assembly.

16. Install the rear heater unit switch knob.

17. Refill the cooling system.

18. Connect the negative battery cable.

19. Run the engine to normal operating temperatures; then, check the climate control operation and check for leaks.

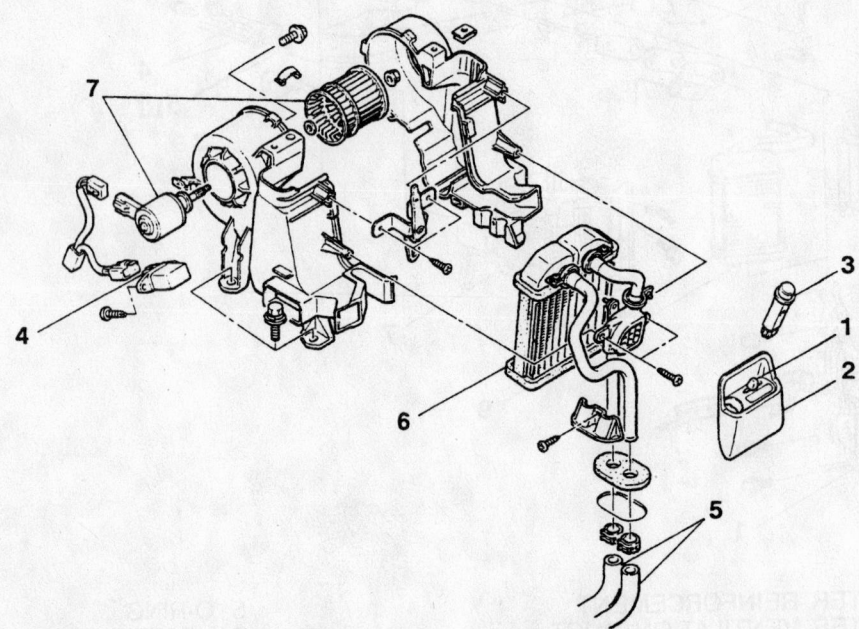

1. KNOB
2. REAR HEATER CONTROL PANEL ASSEMBLY
3. REAR HEATER SWITCH
4. RESISTOR
● DRAINING AND SUPPLYING OF COOLANT
5. REAR HEATER HOSE CONNECTION
6. REAR HEATER CORE ASSEMBLY
7. REAR BLOWER MOTOR ASSEMBLY

93113GE1

Exploded view of the rear heater core and related components—Mitsubishi Montero Sport

NISSAN

Pathfinder

REMOVAL & INSTALLATION

1. Disconnect the negative battery cable.

> ※※ **CAUTION**
>
> **After disconnecting the negative battery cable, wait for at least 3 minutes before working on the steering column or instrument panel.**

2. Drain the cooling system into a clean container for reuse.

3. Disconnect the heater hoses from the heater core.

4. Remove the driver's side air bag and steering wheel by performing the following procedure:

   a. Place the front wheels in the straight-ahead position.

   b. Remove the lower lid from the steering wheel and disconnect the air bag module connector.

   c. Remove the side lids from both sides of the steering wheel.

   d. Using the Tamper Resistant Torx® tool T50, remove the left and right Torx® bolts.

   e. Carefully, remove the air bag module.

> ※※ **CAUTION**
>
> **Place the air bag module in safe place with the front facing upward.**

   f. Remove the steering wheel nut.

   g. Using a steering wheel puller, press the steering wheel from the steering column.

5. Remove the passenger's side air bag by performing the following procedure:

   a. Remove the glove box clips and disconnect the passenger's side air bag module connector.

   b. Remove the lower panel screws; then, disconnect the harness connector and remove the air bag module bracket.

   c. Using the Tamper Resistant Torx®

tool T50, remove the passenger's side air bag module bolts.

   d. Carefully, remove the air bag module.

> ※※ **CAUTION**
>
> **Place the air bag module in safe place with the front facing upward.**

6. Remove the instrument panel by performing the following procedure:

   a. Remove the steering column cover and the combination switch.

   b. Remove the instrument panel side lower finisher.

   c. At the driver's side, remove the lower panel screws, disconnect the electrical harness connectors and remove the panel.

   d. Remove the cluster lid "A" screws and the cluster lid "A".

   e. Remove the combination meter screws, disconnect the electrical harness connectors and remove the combination meter.

   f. Remove the cluster lid "C" screws,

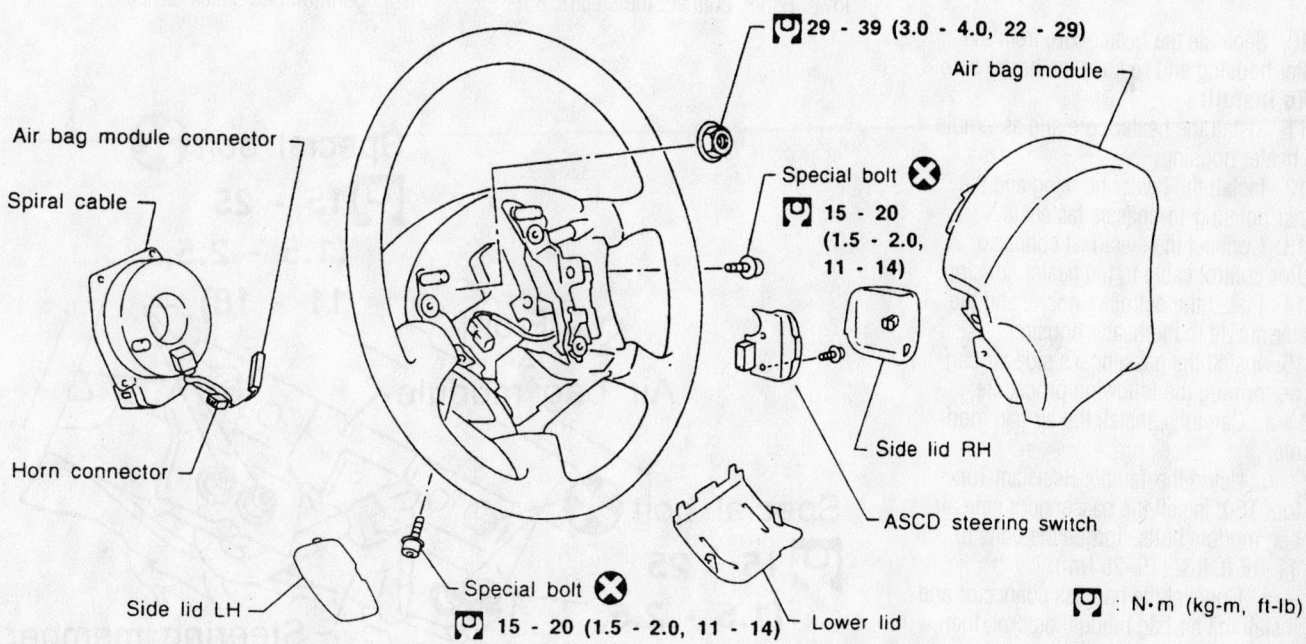

Exploded view of the driver's side air bag module and steering wheel—Nissan Pathfinder

Timing belt service is covered in Section 3 of this manual

disconnect the electrical harness connectors and remove the cluster lid "C".

g. Remove the audio assembly screws and the audio assembly.

h. Remove the air conditioning control unit screws, disconnect the electrical harness connectors and the air conditioning control unit.

i. Remove the ashtray.

j. Remove the shifter (automatic transmission) or shift lever boot (manual transmission); then, remove the screw and disconnect the harness connector.

k. Remove the console box; then, remove the screw and disconnect the harness connector.

l. Remove the lower instrument center panel screws and the lower instrument center panel.

m. Remove the defroster grille.

n. At both sides, remove the pillar garnishes.

o. Remove the instrument panel and pads nuts and bolts.

p. Using an assistant, remove the instrument panel.

7. Remove the defroster nozzle and the heater nozzle from the heater housing.

8. Disconnect the electrical connector and/or control cable from the heater housing.

9. Remove the heater housing-to-chassis fasteners and remove the heater housing.

10. Separate the heater core from the heater housing and remove the heater core.

To install:

11. Install the heater core and assemble the heater housing.

12. Install the heater housing and the heater housing-to-chassis fasteners.

13. Connect the electrical connector and/or control cable to the heater housing.

14. Install the defroster nozzle and the heater nozzle to the heater housing.

15. Install the passenger's side air bag by performing the following procedure:

a. Carefully, install the air bag module.

b. Using the Tamper Resistant Torx® tool T50, install the passenger's side air bag module bolts. Torque the bolts to 11–18 ft. lbs. (15–25 Nm).

c. Connect the harness connector and install the air bag module bracket; then, install the lower panel screws.

d. Connect the passenger's side air bag module connector and install the glove box clips.

16. Install the instrument panel by performing the following procedure:

a. Using an assistant, position the instrument panel.

b. Install the instrument pads, nuts and bolts.

c. At both sides, install the pillar garnishes.

d. Install the defroster grille.

e. Install the lower instrument center panel and the lower instrument center panel screws.

f. Install the console box; then, install the screw and connect the harness connector.

g. Connect the harness connector and install the screw; then, install the shifter (automatic transmission) or shift lever boot (manual transmission).

h. Install the ashtray.

i. Install the air conditioning control unit, connect the electrical harness connectors and the air conditioning control unit screws.

j. Install the audio assembly and the audio assembly screws.

k. Install the cluster lid "C", connect the electrical harness connectors and install the cluster lid "C" screws.

l. Install the combination meter, connect the electrical harness connectors and install the combination meter screws.

m. Install the cluster lid "A" and the cluster lid "A" screws.

n. At the driver's side, install the lower panel, connect the electrical harness connectors and install the panel screws.

o. Install the instrument panel side lower finisher.

p. Install the combination switch and the steering column cover.

17. Install the driver's side air bag and steering wheel by performing the following procedure:

a. Install the steering wheel to the steering column.

b. Install the steering wheel nut. Torque the nut to 22–29 ft. lbs. (29–39 Nm).

c. Carefully, install the air bag module.

d. Using the Tamper Resistant Torx® tool T50, install the left and right Torx® bolts. Torque the bolts to 11–14 ft. lbs. (15–20 Nm).

e. Install the side lids to both sides of the steering wheel.

f. Connect the air bag module connector and install the lower lid to the steering wheel.

18. Connect the heater hoses to the heater core.

19. Refill the cooling system.

20. Connect the negative battery cable.

21. Run the engine to normal operating temperatures; then, check the climate control operation and check for leaks.

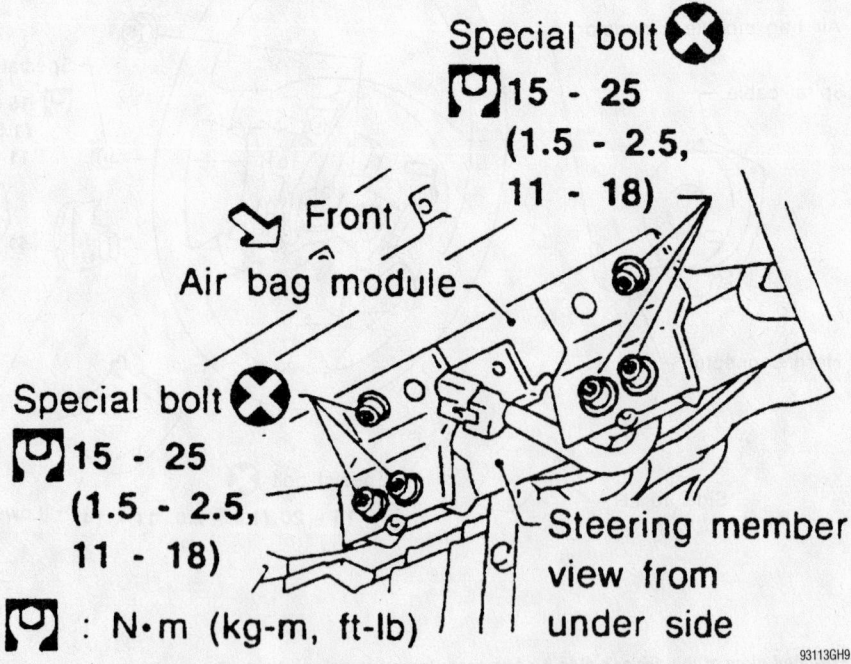

Special bolt ⊗
15 - 25
(1.5 - 2.5, 11 - 18)

Front
Air bag module

Special bolt ⊗
15 - 25
(1.5 - 2.5, 11 - 18)

: N•m (kg-m, ft-lb)

Steering member view from under side

93113GH9

Exploded view of the passenger's side air bag module—Nissan Pathfinder

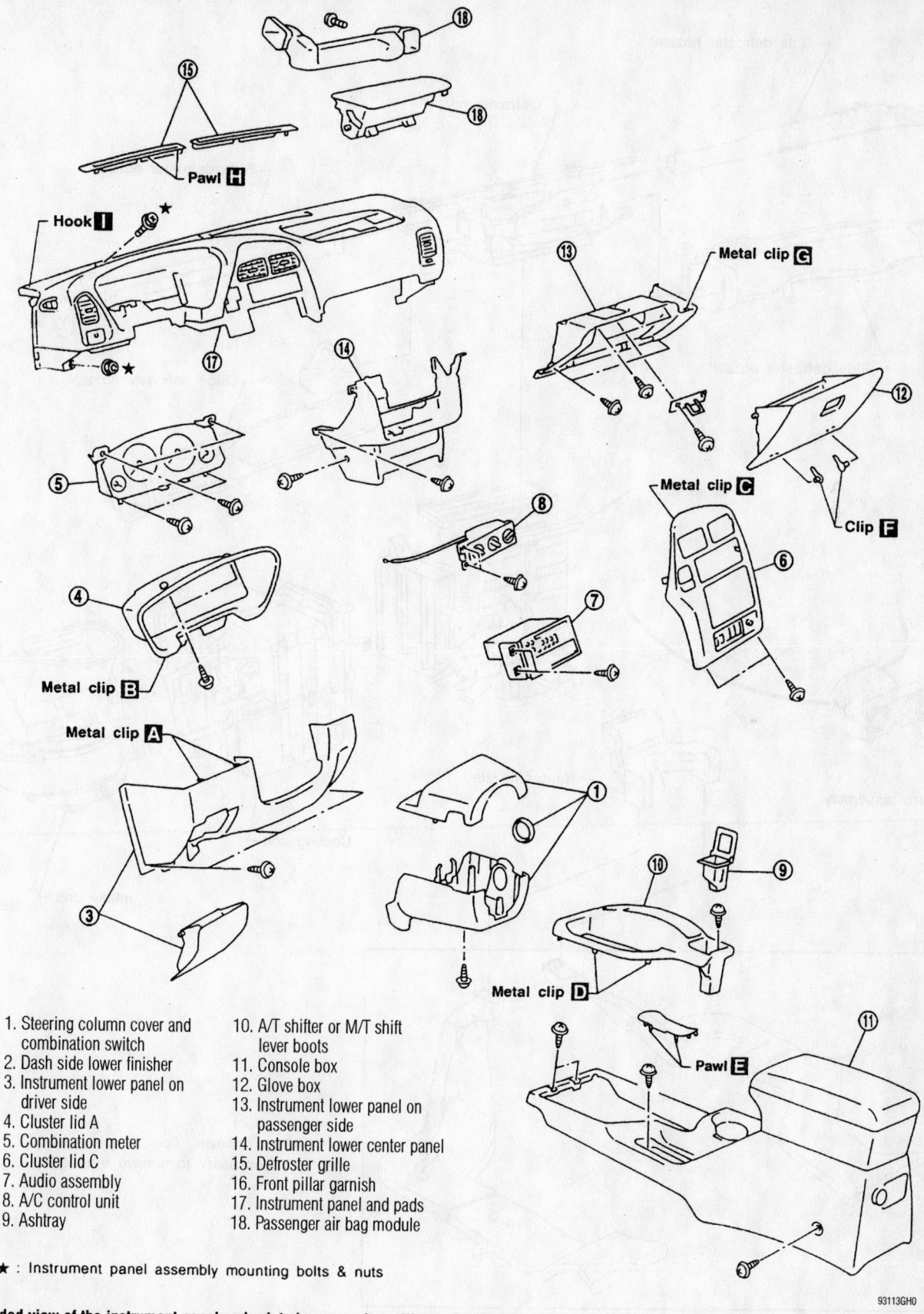

1. Steering column cover and combination switch
2. Dash side lower finisher
3. Instrument lower panel on driver side
4. Cluster lid A
5. Combination meter
6. Cluster lid C
7. Audio assembly
8. A/C control unit
9. Ashtray
10. A/T shifter or M/T shift lever boots
11. Console box
12. Glove box
13. Instrument lower panel on passenger side
14. Instrument lower center panel
15. Defroster grille
16. Front pillar garnish
17. Instrument panel and pads
18. Passenger air bag module

★ : Instrument panel assembly mounting bolts & nuts

93113GH0

Exploded view of the instrument panel and related accessories—Nissan Pathfinder

Heater Core replacement is covered in Section 2 of this manual

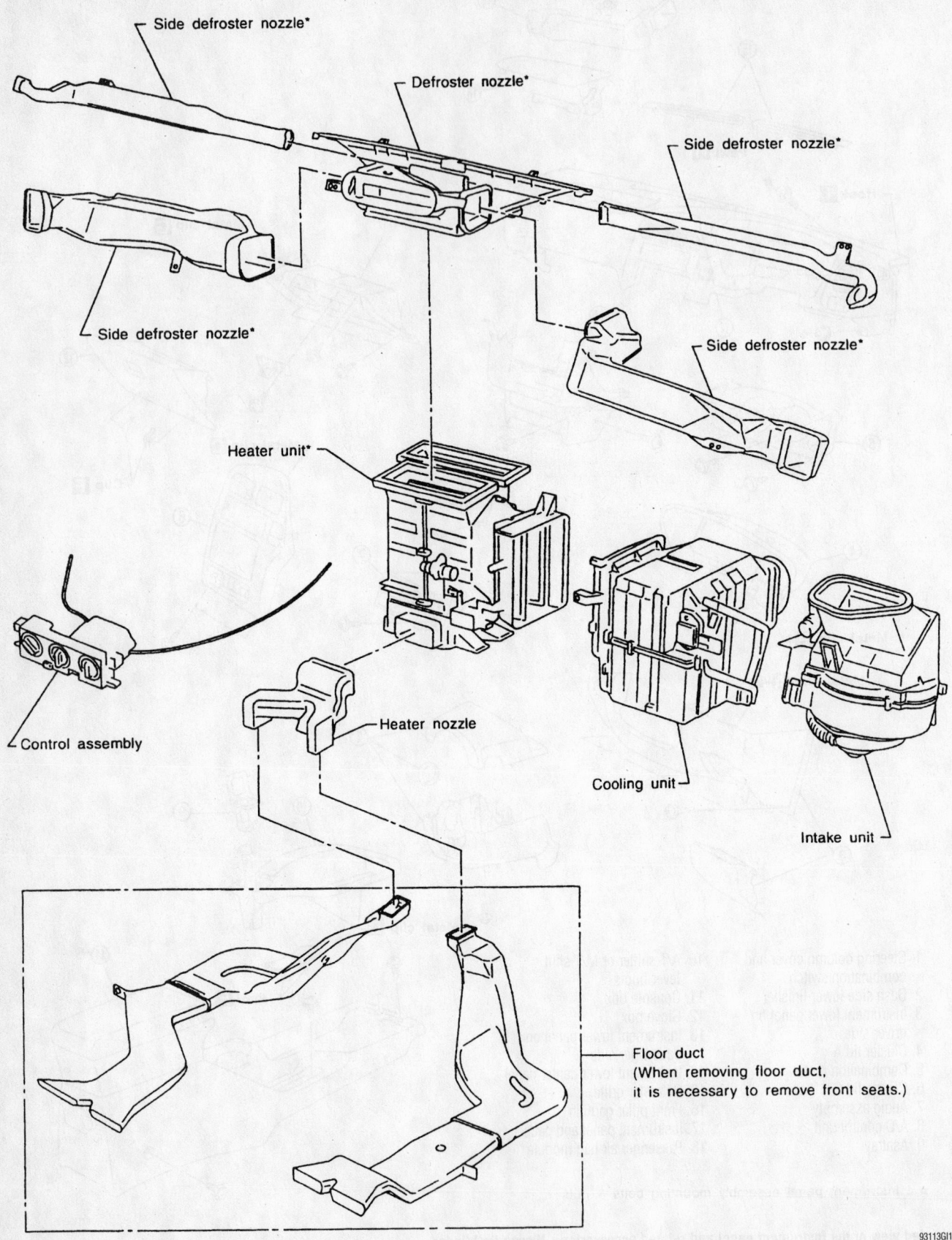

Side defroster nozzle*

Defroster nozzle*

Side defroster nozzle*

Side defroster nozzle*

Side defroster nozzle*

Heater unit*

Control assembly

Heater nozzle

Cooling unit

Intake unit

Floor duct
(When removing floor duct,
it is necessary to remove front seats.)

93113GI1

Exploded view of the heater housing, the evaporator housing, the ventilation dusts and related accessories—Nissan Pathfinder

SATURN

REMOVAL & INSTALLATION

Saturn VUE

1. Disable the air bag system.
2. Record all preset radio stations.
3. Disconnect the negative battery cable.
4. Drain and recycle the engine coolant.
5. Recover the A/C system refrigerant using approved equipment.
6. Remove or disconnect the following:
   - Suction line from the Thermal Expansion Valve (TXV) and cap the TXV and the line
   - TXV thermister connection
   - Heater core outlet and inlet hose's from the core and plug the lines
   - Instrument Panel (IP) right end panel by gently tugging to disengage the clip
   - IP knee bolster panel
   - IP left end panel by gently tugging to disengage the clip
   - Door jamb switch electrical connections on both sides
7. Place the shifter in neutral.
   - Horse shoe bezel at the shifter by first pulling up at the rear to disengage the clips and slide it up and over the shifter to access the electrical connections. Disengage all electrical connections.
   - Glove box-to-IP screws and upper glove box-to-radio bezel screws
   - Glove box
   - Radio bezel by pulling the lower edge forward first and then the top to disengage the clips
   - Temperature cable, blower switch and IP 20-way connections from the controller
   - Air bag telltale, foglamp switch, dimmer switch and hazard switch connections
   - Upper and lower steering column shrouds
   - Cluster bezel screws and the bezel
   - A-pillar garnish moldings
   - Right IP deflector assembly-to-intermediate duct screw
   - IP cover-to-cross car beam bolt covers and the cover-to-beam bolts
   - IP cover-to-IP retaining screws at the lower edge of the cover and the radio opening
   - IP cover by lifting it then moving it

rearwards passenger side first and walk the cluster opening around the steering wheel and out the drivers side door
   - Cluster-to-IP screws and the cluster
   - Radio
   - Passenger side air bag
   - Center, right and left shifter close-out panels
   - Right and left intermediate ducts
   - Cluster connector from the retainer
   - IP fuse block from the bracket
   - IP ground wire from the H brace
   - Brake Control Module (BCM) from the retainer
   - Data Link Controller (DLC) from the retainer
   - Right door sill plate trim
   - IP retainer fasteners and retainer
   - Heater duct
   - Heater core cover
   - Heater core pipe cover and pipe foam seal
8. Grab the heater core at the end tanks and remove the core. If the core sticks, spray the perimeter of the core seal and the pips at the front of the dash with soapy water can aid in removal.

To install:

9. Spary the dash seal at the core pipe openings and seal with soapy water to aid in installation.
10. Install or connect the following:
   - Heater core
   - Pipe seal and cover. Tighten the cover retainers to 9 inch lbs. (1 Nm).
   - Heater core cover and tighten the cover retainers to 9 inch lbs. (1 Nm)

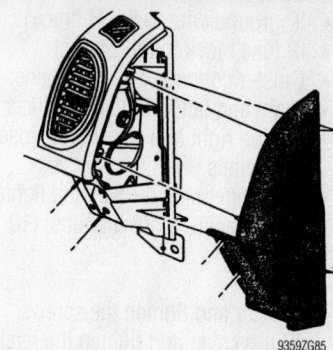

Remove the instrument Panel right end panel—Saturn Vue

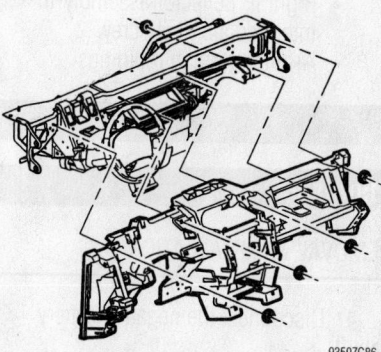

Remove the IP retainer—Saturn Vue

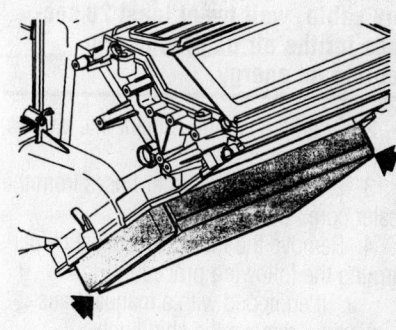

Grab the heater core at the end tanks and remove the core—Saturn Vue

   - Heater duct
   - IP retainer by aligning the 4-way locator (tapered boss) and outboard locators with the corresponding holes or slots in the beam, then tighten the fasteners starting from the center to 88 inch lbs. (10 Nm)
   - Right door sill plate trim

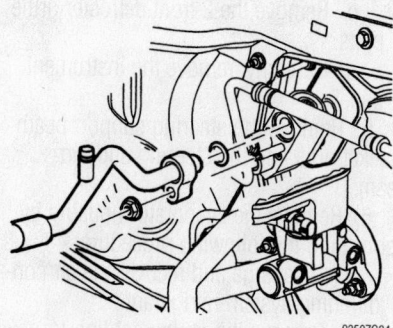

Disconnect the suction line from the Thermal Expansion Valve (TXV)—Saturn Vue

Brake service is covered in Section 4 of this manual

- DLC to the retainer
- BCM to the retainer
- IP ground wire to the H brace
- IP fuse block to the bracket
- Cluster connector to the retainer
- Right and left intermediate ducts
- Center, right and left shifter close-out panels
- Passenger side air bag and tighten the fasteners to 88 inch lbs. (10 Nm)
- Radio
- Cluster and tighten the screws
- IP top cover and tighten the retainers to 22 inch lbs. (2.5 Nm)
- IP top cover-to-beam bolts and tighten to 88 inch lbs. (10 Nm)
- Ip top cover bolt covers
- Right IP deflector assembly-to-intermediate duct screw
- A-pillar garnish moldings

- Cluster bezel and screws
- Upper and lower steering column shrouds

11. Center the temperature knob by aligning the controller housing alignment tab with the slot in the shaft.

12. Align the air temperature cable lug with the detent spring point.

- Temperature cable to the control head by aligning the retention tabs and the knob shaft and snap into place
- IP harness to the blower switch and controller
- Air bag telltale, foglamp switch, dimmer switch and hazard switch connections
- Radio bezel
- Glove box

13. Place the shifter in neutral.

- Horse shoe bezel electrical connections

- Horse shoe bezel and place the shifter in park
- Door jamb switch electrical connections on both sides
- IP left end panel
- IP knee bolster panel
- IP right end panel
- Heater core outlet and inlet hose's to the core and position the clamp at 9 o'clock
- TXV thermister connection
- Suction line to the TXV using new seal washer and tighten to 12 ft. lbs. (16 Nm) and cap the TXV and the line

14. Recharge the A/C system refrigerant using approved equipment.

15. Refill the engine cooling system.

16. Connect the negative battery cable.

17. Reset all preset radio stations.

18. Enable the air bag system.

SUBARU

Forester

REMOVAL & INSTALLATION

1. Disconnect the negative battery cable.

※※ CAUTION

After disconnecting the negative battery cable, wait for at least 20 seconds for the air bag module to deplete its energy.

2. Drain the engine coolant into a clean container for reuse.

3. Disconnect the heater hoses from the heater core.

4. Remove the instrument panel by performing the following procedure:

   a. If equipped with a manual transmission, remove the shift knob.

   b. Remove both the front and rear console covers.

   c. Remove the console box-to-chassis screws and the console box.

   d. Remove the 3 lower left side cover assembly screws, disengage the 3 upper clips, remove the cover assembly.

   e. Using a screwdriver, disconnect the data link connector from the lower cover.

   f. Remove the knee panel.

   g. At the glove box, remove the right side cover screw, the clip and the side cover.

   h. Remove the glove box screws and remove the glove box.

   i. Remove the center panel bezel.

   j. Remove the audio assembly screws, disconnect the electrical connectors and remove the audio assembly.

   k. Remove the 2 steering column-to-instrument panel bolts and lower the steering column.

   l. Move the temperature control switch to FULL HOT, the mode selector switch to DEF and the recirculation switch to FRESH positions.

   m. Disconnect the temperature control cable and the mode control cable from the heater housing.; then, the recirculation control cable from the intake housing.

   n. Disconnect the electrical harness connectors.

   o. Remove the instrument panel-to-chassis bolts.

   p. Remove the 2 front defroster grille bolts.

   q. Carefully, remove the instrument panel.

5. Remove the steering support beam bracket nuts and the steering support beam.

6. Remove the evaporator housing by performing the following procedure:

   a. Discharge and recover the air conditioning system refrigerant.

   b. Remove the refrigerant line-to-cowl connector bolt, separate the lines, discard the O-rings and plug the openings to prevent contamination.

   c. Disconnect the electrical harness connector from the evaporator housing.

   d. Disconnect the drain hose.

   e. Remove the evaporator housing nut/bolts and the evaporator housing.

7. Remove the heater housing-to-chassis bolts and the heater housing.

8. Remove the heater core from the heater housing.

To install:

9. Install the heater core to the heater housing.

10. Install the heater housing and the heater housing-to-chassis bolts.

11. Install the steering support beam and the steering support beam bracket nuts.

12. Install the evaporator housing by performing the following procedure:

   a. Install the evaporator housing and the evaporator housing nut/bolts.

   b. Connect the drain hose.

   c. Connect the electrical harness connector to the evaporator housing.

   d. Using new O-rings, assemble the refrigerant lines and install the refrigerant line-to-cowl connector bolt.

13. Install the instrument panel by performing the following procedure:

   a. Carefully, install the instrument panel.

   b. Install the 2 front defroster grille bolts.

   c. Install the instrument panel-to-chassis bolts.

   d. Connect the electrical harness connectors.

   e. Connect the temperature control cable and the mode control cable to

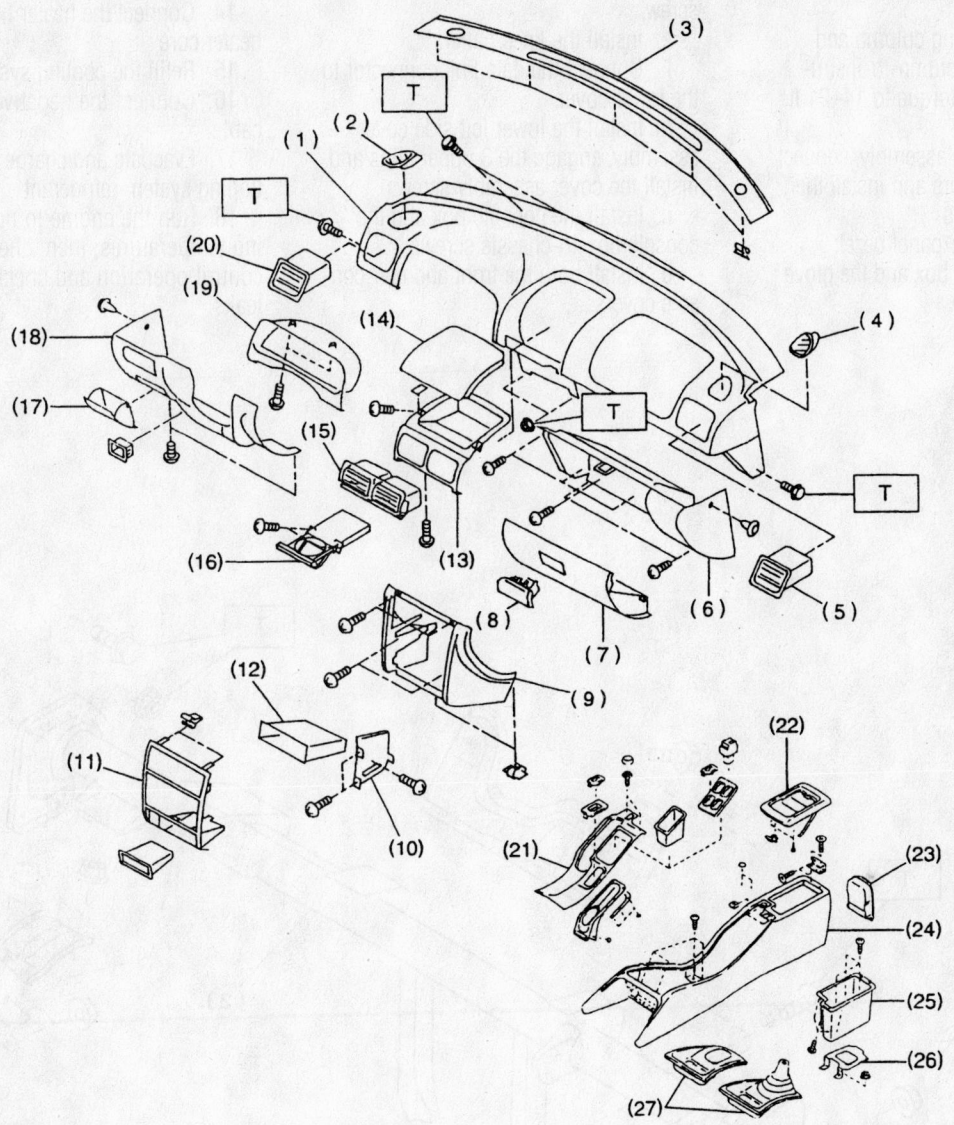

1	Pad & frame	12	Pocket
2	Grille side (D)	13	Panel center
3	Front def. grille	14	Center pocket lid
4	Grille side (P)	15	Grille center
5	Grille vent (P)	16	Cup holder
6	Glove box panel	17	Side pocket
7	Glove box lid	18	Lower cover ASSY
8	Knob	19	Meter visor
9	Instrument panel center console	20	Grille vent (D)
10	BRKT (Radio)	21	Console cover
11	Center console cover	22	Console lid

23	Rear cup holder
24	Console box
25	Console pocket
26	Rear console BRKT
27	Front cover

Tightening torque: N·m (kg-m, ft-lb)
T: 7±1 (0.7±0.1, 5.1±0.7)

93113GI8

Exploded view of the instrument panel assembly—Subaru Forester

For complete Engine Mechanical specifications, see Section 1 of this manual

the heater housing. Then, the recirculation control cable to the intake housing.

f. Install the steering column and lower the 2 steering column-to-instrument panel bolts and torque to 14–21 ft. lbs. (20–30 Nm).

g. Install the audio assembly, connect the electrical connectors and install the audio assembly screws.

h. Install the center panel bezel.

i. Install the glove box and the glove box screws.

j. At the glove box, install the right side cover, the clip and the side cover screw.

k. Install the knee panel.

l. Connect the data link connector to the lower cover.

m. Install the lower left side cover assembly, engage the 3 upper clips and install the cover assembly screws.

n. Install the console box and the console box-to-chassis screws.

o. Install both the front and rear console covers.

p. If equipped with a manual transmission, install the shift knob.

14. Connect the heater hoses to the heater core.

15. Refill the cooling system.

16. Connect the negative battery cable.

17. Evacuate and charge the air conditioning system refrigerant.

18. Run the engine to normal operating temperatures; then, check the climate control operation and check for leaks.

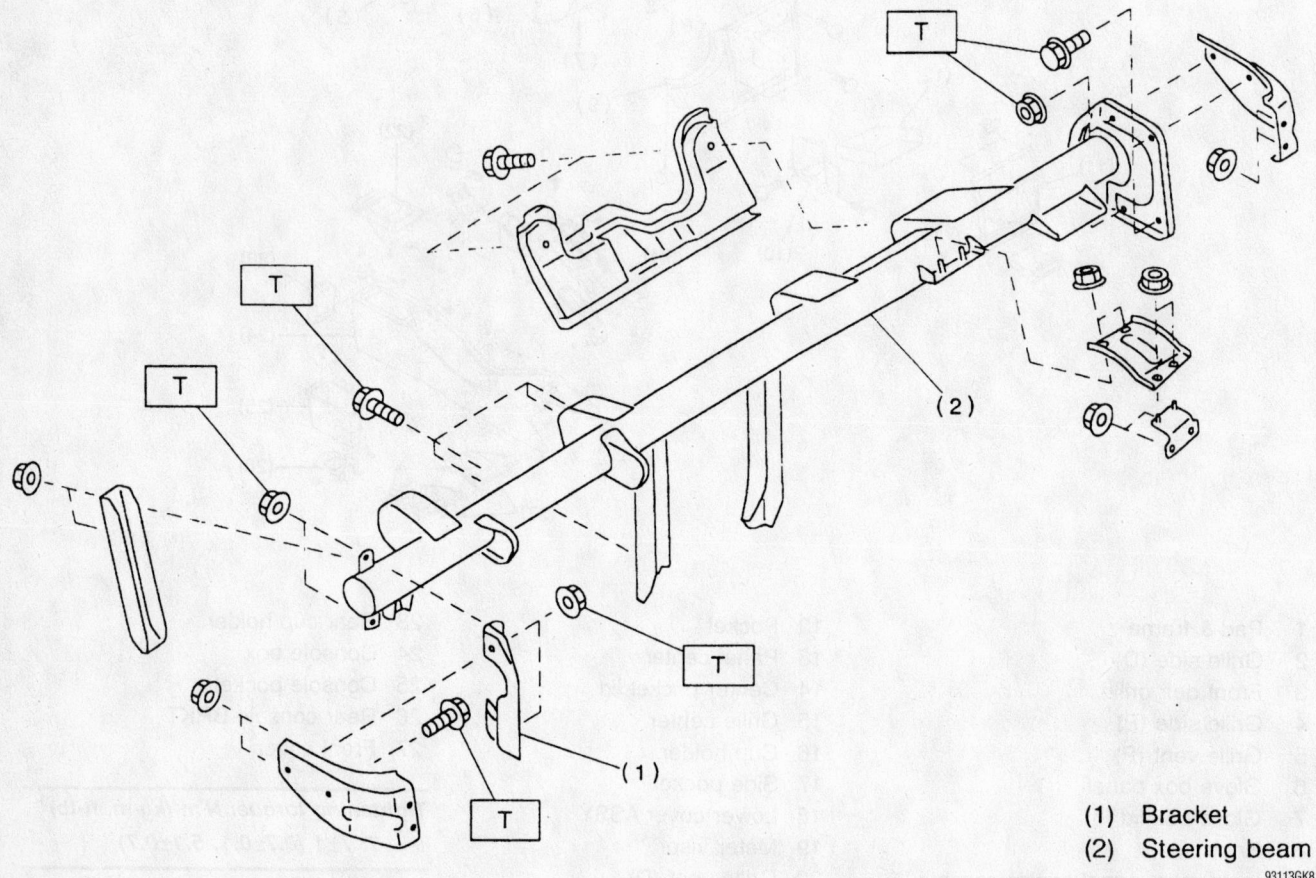

(1) Bracket
(2) Steering beam

93113GK8

Exploded view of the steering support beam assembly—Subaru Forester

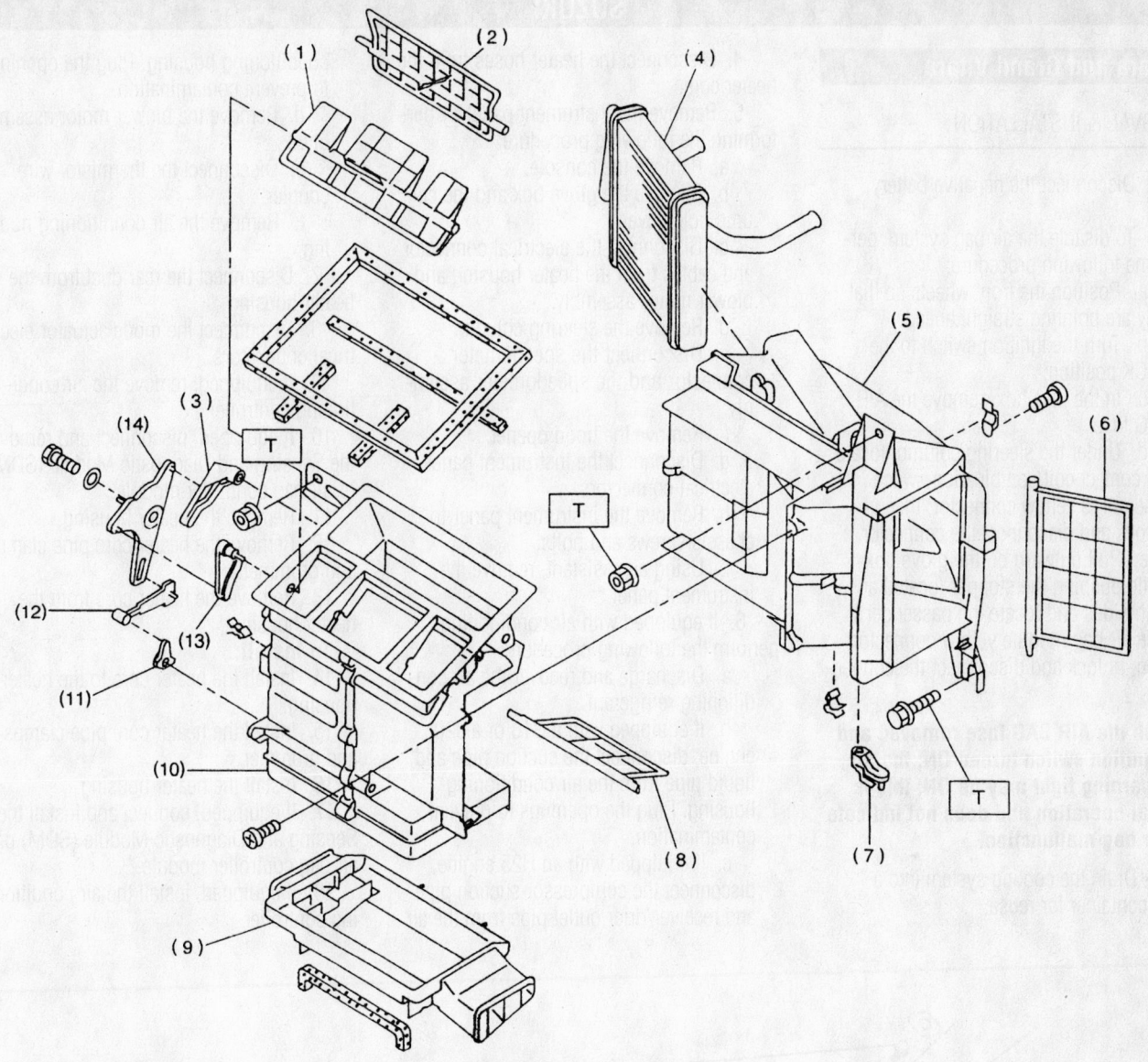

1	Vent door	7	**Mix lever**	13	Vent lever
2	DEF door	8	Foot door	14	Side link
3	DEF lever	9	Foot duct		
4	Heater core	10	Heater case REAR		
5	Heater case FRONT	11	Foot lever lower		
6	Mix door	12	Foot lever upper		

Tightening torque: N·m (kg-m, ft-lb)
T: 7.35±1.96
(0.750±0.200, 5.421±1.446)

93113GI7

Exploded view of the heater core, heater housing and related components—Subaru Forester

For Accessory Drive Belt illustrations, see Section 1 of this manual

SUZUKI

Vitara and Grand Vitara

REMOVAL & INSTALLATION

1. Disconnect the negative battery cable.

2. To disable the air bag system, perform the following procedure:

   a. Position the front wheels so that they are pointing straight ahead.

   b. Turn the ignition switch to the LOCK position.

   c. In the fuse box, remove the AIR BAG fuse.

   d. Under the steering column, locate the contact coil/combination switch assembly's yellow connector; then, unlock and disconnect the connector.

   e. Pull outward on the glove box while pushing the stopper located at both sides and locate the passenger's side air bag module yellow connector; then, unlock and disconnect the connector.

➡ **With the AIR BAG fuse removed and the ignition switch turned ON; the air bag warning light may be ON; this is normal operation and does not indicate an air bag malfunction.**

3. Drain the cooling system into a clean container for reuse.

4. Disconnect the heater hoses from the heater core.

5. Remove the instrument panel by performing the following procedure:

   a. Remove the console.

   b. Remove the glove box and the column hole cover.

   c. Disconnect the electrical connector and cables from the heater housing and blower motor assembly.

   d. Remove the steering column.

   e. Disconnect the speedometer connector and the speedometer assembly.

   f. Remove the hood opener.

   g. Disconnect the instrument panel electrical connectors.

   h. Remove the instrument panel-to-chassis screws and bolts.

   i. Using an assistant, remove the instrument panel.

6. If equipped with air conditioning, perform the following procedure:

   a. Discharge and recover the air conditioning refrigerant.

   b. If equipped with a G16 or a J20 engine, disconnect the suction pipe and liquid pipe from the air conditioning housing. Plug the openings to prevent contamination.

   c. If equipped with an H25 engine, disconnect the compressor suction pipe and receiver/drier outlet pipe from the air

conditioning housing. Plug the openings to prevent contamination.

   d. Remove the blower motor assembly.

   e. Disconnect the thermistor wire coupler.

   f. Remove the air conditioning housing.

7. Disconnect the rear duct from the heater housing.

8. Disconnect the mode actuator electrical connectors.

9. If equipped, remove the air conditioning controller.

10. If equipped, disconnect and remove the Sensing and Diagnostic Module (SDM) or air bag controller module.

11. Remove the heater housing.

12. Remove the heater core pipe clamps and grommet.

13. Remove the heater core from the heater housing.

To install:

14. Install the heater core to the heater housing.

15. Install the heater core pipe clamps and grommet.

16. Install the heater housing.

17. If equipped, connect and install the Sensing and Diagnostic Module (SDM) or air bag controller module.

18. If equipped, install the air conditioning controller.

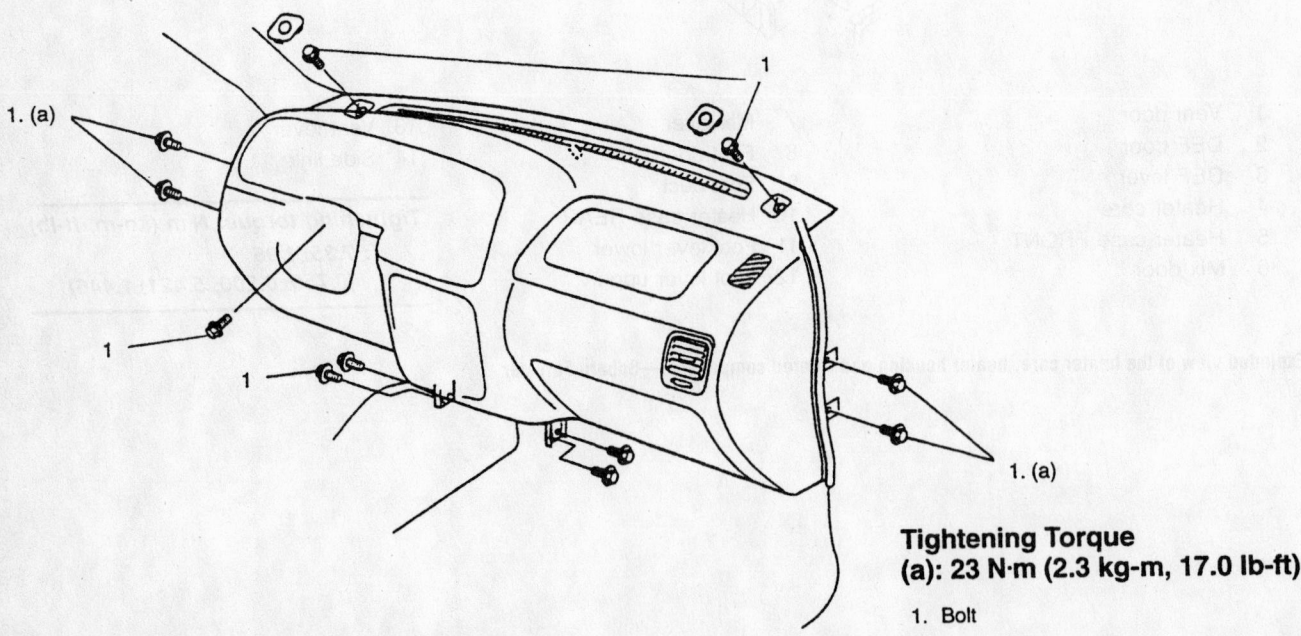

**Tightening Torque
(a): 23 N·m (2.3 kg-m, 17.0 lb-ft)**

1. Bolt

93113GE8

View of the instrument panel and fasteners—Suzuki Vitara

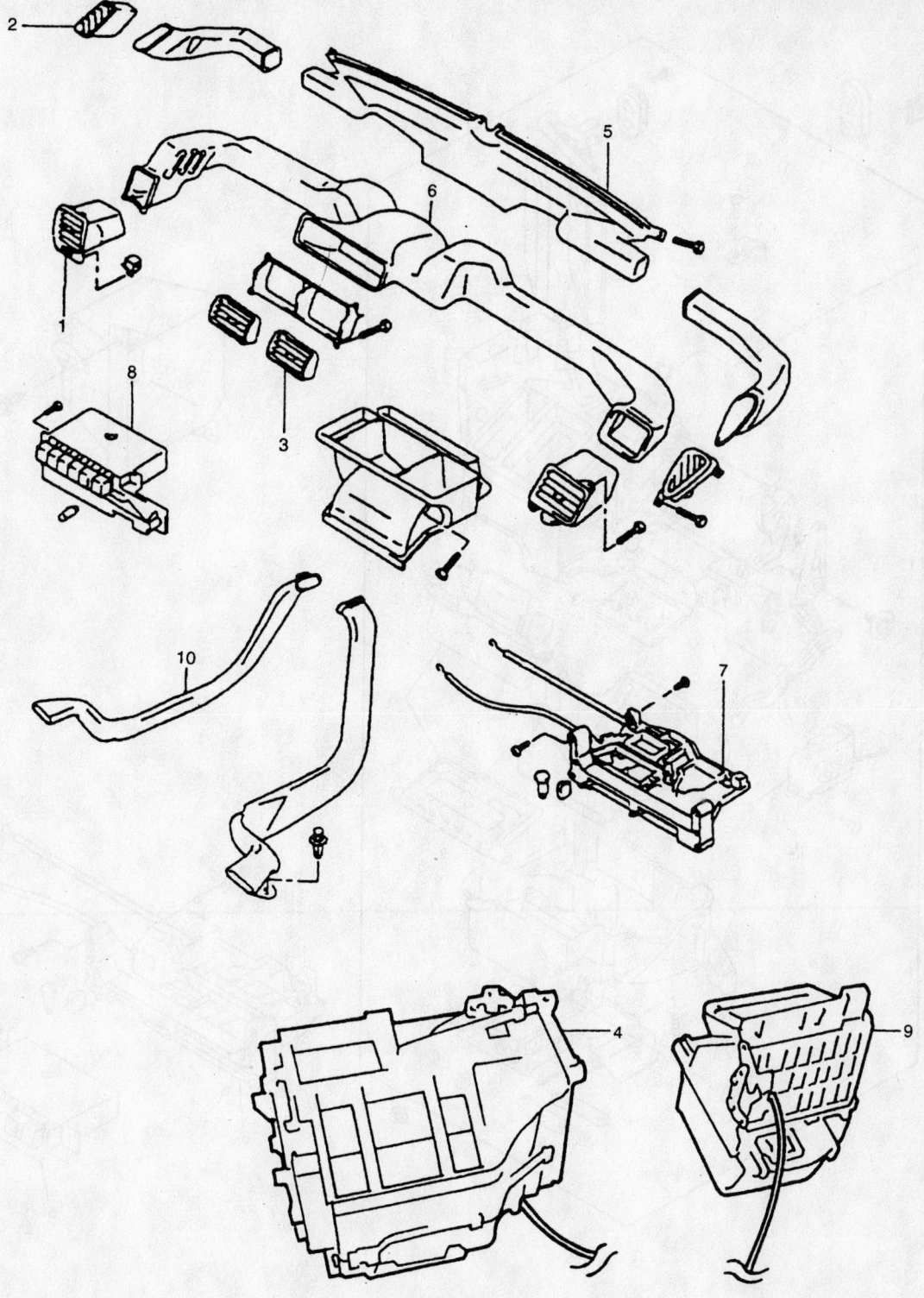

1. Side ventilator outlet
2. Side defroster outlet
3. Center ventilatior outlet
4. Heater unit
5. Defroster duct
6. Ventilator duct
7. Control lever
8. Mode control switch
9. Blower unit
10. Rear duct

93113GE9

Exploded view of the heater housing and ventilation ducts—Suzuki Vitara

For Tire, Wheel and Ball Joint specifications, see Section 1 of this manual

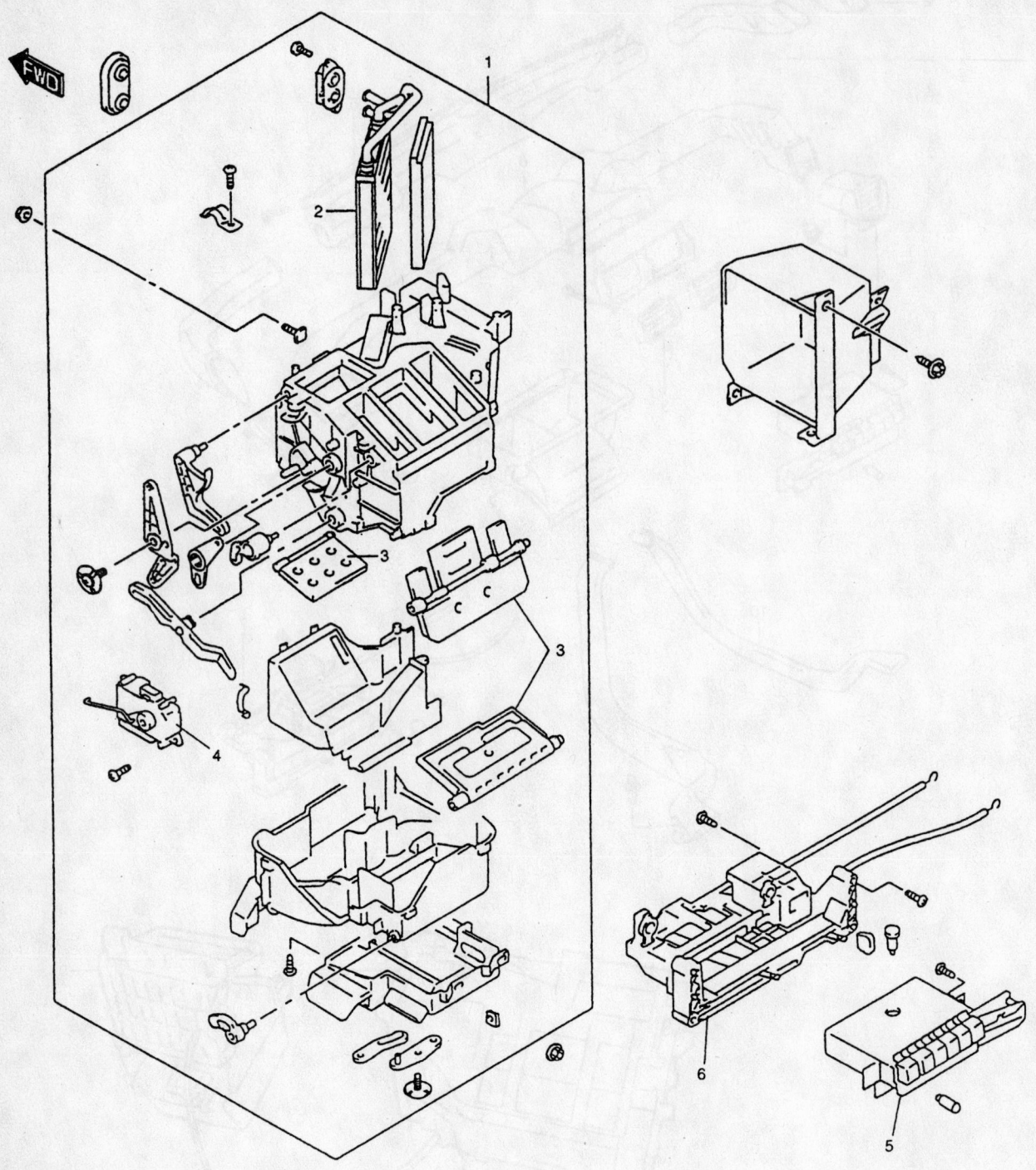

1. Heater assembly
2. Heater core
3. Damper
4. Mode actuator
5. Mode control switch
6. Control lever assembly

93113GE0

Exploded view of the heater core, heater housing and related components—Suzuki Vitara

19. Connect the mode actuator electrical connectors.

20. Connect the rear duct from the heater housing.

21. If equipped with air conditioning, perform the following procedure:

a. Install the air conditioning housing.

b. Connect the thermistor wire coupler.

c. Install the blower motor assembly.

d. If equipped with an H25 engine, connect the compressor suction pipe and receiver/drier outlet pipe to the air conditioning housing.

e. If equipped with a G16 or a J20 engine, connect the suction pipe and liquid pipe to the air conditioning housing.

f. Evacuate and charge the air conditioning system.

22. Install the instrument panel by performing the following procedure:

a. Using an assistant, install the instrument panel.

b. Install the instrument panel-to-chassis screws and bolts.

c. Connect the instrument panel electrical connectors.

d. Install the hood opener.

e. Connect the speedometer connector and the speedometer assembly.

f. Install the steering column.

g. Connect the electrical connector and cables to the heater housing and blower motor assembly.

h. Install the glove box and the column hole cover.

i. Install the console.

23. Connect the heater hoses to the heater core.

24. Refill the cooling system.

25. To enable the air bag system, perform the following procedure:

a. Push inward on the glove box while pushing the stopper located at both sides and connect the passenger's side air bag module yellow connector and lock it.

b. Under the steering column, connect the contact coil/combination switch assembly's yellow connector.

c. In the fuse box, install the AIR BAG fuse.

d. Turn the ignition switch ON and verify that the AIR BAG warning light flashes 7 times and turns OFF; if the system does not operate as described, perform the Air Bad Diagnostic System Check.

26. Connect the negative battery cable.

27. Run the engine to normal operating temperatures; then, check the climate control operation and check for leaks.

TOYOTA

Land Cruiser

REMOVAL & INSTALLATION

Front Heater

1. Disconnect the negative battery cable.

2. Drain the cooling system into a clean container for reuse.

3. Disconnect the heater hoses from the heater core.

4. Remove the steering wheel by performing the following procedure:

a. Position the front wheels facing straight-ahead.

b. Remove the steering wheel side covers.

c. Using a Torx® wrench, loosen the 2 screws located at each side of the steering wheel until the screw's circumference groove catches on the screw case.

d. Pull the air bag module from the steering wheel and disconnect the electrical connector.

❋❋ CAUTION

Place the air bag module in a safe place with the front side facing upward.

e. Remove the steering wheel nut.

f. Place alignment marks on the steering wheel and the main shaft.

g. Using a steering wheel puller, press the steering wheel from the steering column.

5. Remove the instrument panel and reinforcement by performing the following procedure:

a. Remove the front door scuff plates, the cowl side trim and the front door opening trim.

b. At the driver's side, remove the 2 assist grip plugs, the 2 screws and assist grip and the front pillar garnish.

c. At the passenger's side, remove the 4 assist grip plugs, the 4 screws, the 2 assist grips and the front pillar garnish.

d. Remove the instrument cluster finish panel.

e. Remove the 2 screws and the hood lock control cable.

f. Remove the 2 screws and the fuel lid control cable lever.

g. Remove the lower No. 1 panel screw and the panel.

h. Remove the lower left side panel.

i. Remove the 3 steering column cover screws and the covers.

j. At the steering column, disconnect the electrical connectors; then, remove the clamp, the 3 screws and the combination switch.

k. Remove the No. 2 heater-to-register duct screw and the duct.

l. Remove the steering column-to-

instrument panel bolts and the steering column.

m. At the combination meter, disconnect the electrical connectors; then, remove the 4 screws and the combination meter.

n. Remove the glove compartment door stoppers, the 2 screws and the glove box door.

o. At the passenger's side air bag module, remove the No. 1 undercover, pull the air bag connector up from the undercover and disconnect it; then, remove the air bag.

❋❋ CAUTION

Place the air bag module in a safe place with the front side facing upward.

p. Remove the 3 lower No. 2 panel screws and the panel.

q. Remove the center cluster; then, pry the center cluster from the dash by prying the 8 clips in the following order:
- Left side
- Right side
- Top left side
- Top right side

r. Remove the 4 radio screws, pull the radio outward, disconnect the electrical connectors and remove the radio.

s. At the rear console panel, remove the transfer shift lever knob. Pry the

For Wheel Alignment specifications, see Section 1 of this manual

panel upward disengaging the 4 clips (2 on each side) and remove the panel.

t. At the rear of the console, remove the 2 rear end panel-to-console screws; then, pry the end panel rearward disengaging the 2 clips and remove the panel.

u. If not equipped with a rear air conditioning system, disconnect the connector and control cable; then, remove the 3 rear heater control panel screws and the panel.

v. Remove the 4 rear console box-to-chassis screws/bolts and the console box.

w. Remove the center lower cluster finish panel by prying panel rearward disengaging the 5 clips; then, disconnect the electrical connector.

x. Remove the 2 front console-to-chassis bolts/screws, disengage the 2 clips and remove the console.

y. At the instrument panel, disconnect the junction connectors (the connectors can be disconnected by loosening the bolts), the instrument panel-to-chassis 8 bolts and 2 nuts. Using an assistant, remove the instrument panel.

z. Disconnect the electrical connector and remove the ECM.

aa. Remove the No. 3 and No. 4 heater-to-register ducts.

bb. Remove the floor brace, the No. 1 brace and the reinforcement.

6. Remove the evaporator housing by performing the following procedure:

a. Discharge and recover the air conditioning system refrigerant.

b. Remove the air conditioning liquid line clamp.

c. Remove the air conditioning suction line clamp.

d. Disconnect both air conditioning lines and plug the openings to prevent contamination. Discard the 4 O-rings.

e. Remove the antenna relay electrical connector, the 2 screws and the relay.

f. Remove the evaporator housing-to-chassis 4 screws/2 nuts and the housing.

7. Remove the heater housing by performing the following procedure:

a. Remove the defroster nozzle.

b. Disconnect the electrical connector.

c. Remove the 4 nuts and the heater housing.

8. Remove the heater core-to-heater housing packing, the screw, the bracket, the clamp and the heater core.

To install:

9. Install the heater core, the clamp, the bracket, the screw and the heater core-to-heater housing packing.

10. Install the heater housing by performing the following procedure:

a. Install the heater housing and the 4 nuts.

b. Connect the electrical connector.

c. Install the defroster nozzle.

11. Install the evaporator housing by performing the following procedure:

a. Install the evaporator housing and the housing-to-chassis 4 screws and 2 nuts.

b. Install the antenna relay, the 2 screws and the electrical connector.

c. Using new O-rings, connect both air conditioning lines.

d. Install the air conditioning liquid line and suction line clamp.

12. Install the instrument panel and reinforcement by performing the following procedure:

a. Install the reinforcement, the No. 1 brace and the floor brace.

b. Install the No. 3 and No. 4 heater-to-register ducts.

c. Install the ECM and connect the electrical connector.

d. Using an assistant, install the instrument panel, connect the junction connectors, the instrument panel-to-chassis 8 bolts and 2 nuts.

e. Install the front the console, engage the 2 clips and install the 2 console-to-chassis bolts/screws.

f. Connect the electrical connector; then, install the center lower cluster finish panel by engaging the 5 clips.

g. Install the console box and the 4 rear console box-to-chassis screws/bolts.

h. If not equipped with a rear air conditioning system, install rear heater control panel, the 3 panel screws; then, connect the connector and control cable.

i. Install the rear of the console and engage the 2 clips; then, install the 2 rear end panel-to-console screws.

j. Install the rear console panel and engage the 4 clips (2 on each side); then, install the transfer shift lever knob.

k. Install the radio, connect the electrical connectors and the 4 radio screws.

l. Install the center cluster and engage the 8 center cluster clips.

m. Install the lower No. 2 panel and the 3 panel screws.

n. Install the passenger's side air bag module, connect it and install the No. 1 undercover.

o. Install the glove box door, the 2 screws and the glove compartment door stoppers.

p. Install the combination meter and the 4 screws; then, connect the electrical connectors.

q. Install the steering column and the steering column-to-instrument panel bolts.

r. Install the No. 2 heater-to-register duct and the duct screw.

s. At the steering column, install the combination switch, the 3 screws and the clamp; then, connect the electrical connectors.

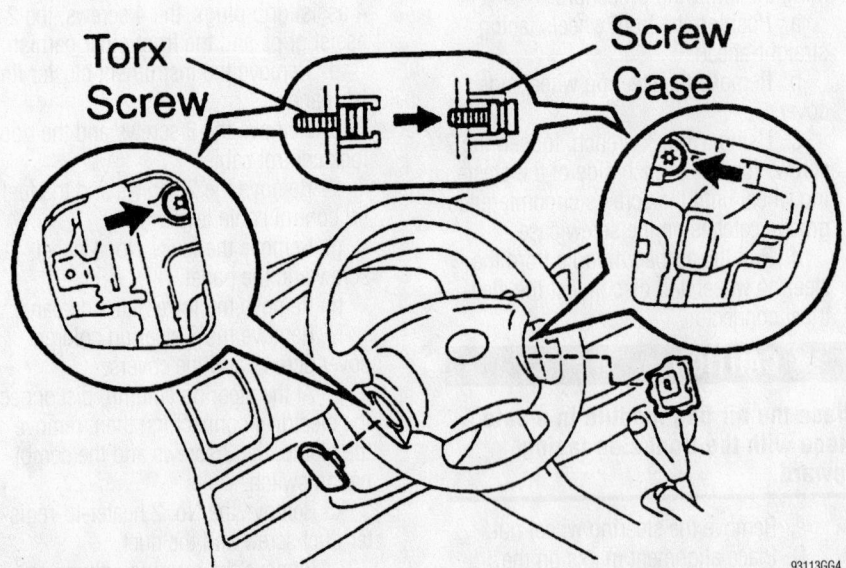

View the steering wheel's Torx® bolts—Toyota Land Cruiser

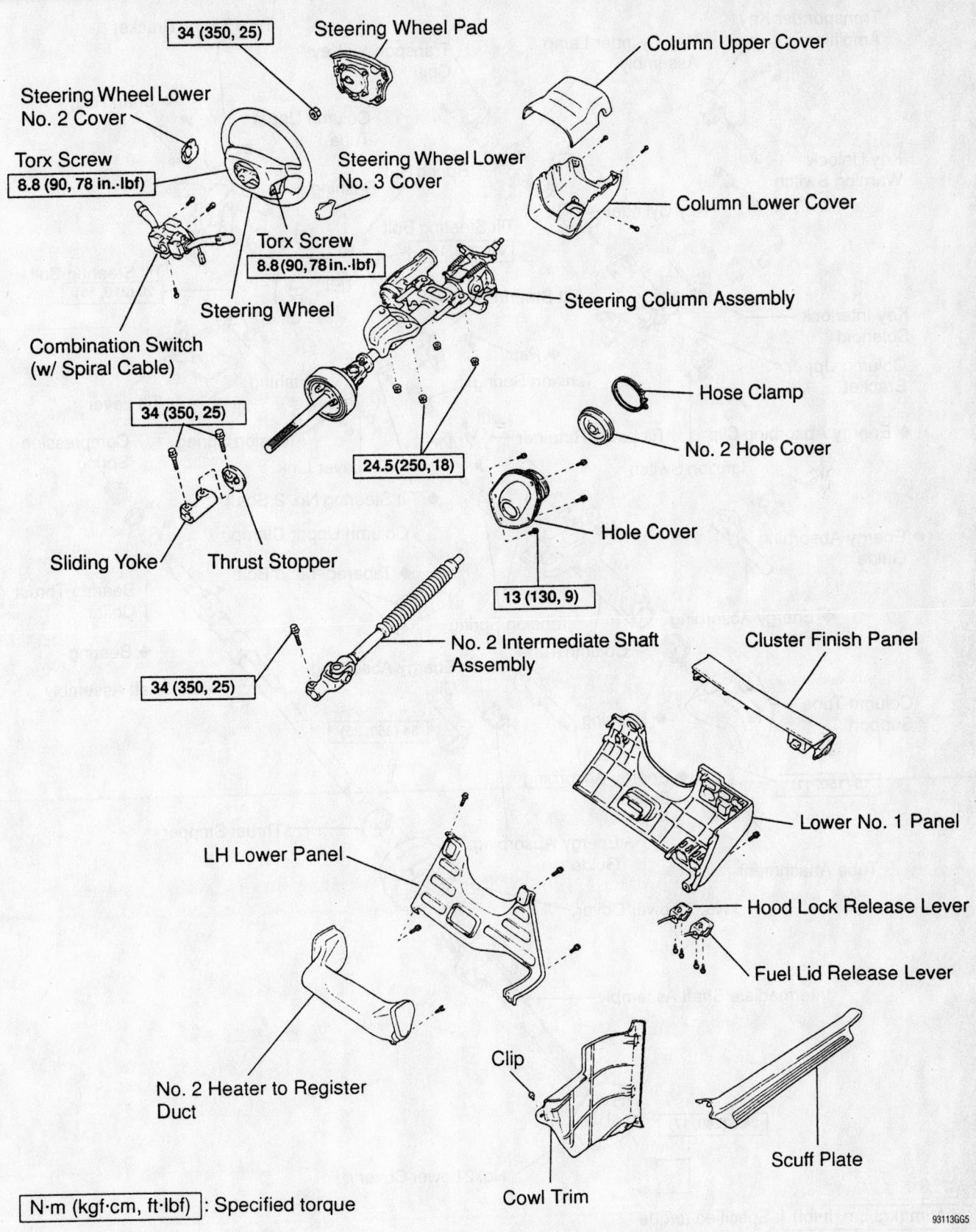

34 (350, 25)

Steering Wheel Pad

Column Upper Cover

Steering Wheel Lower No. 2 Cover

Torx Screw
8.8 (90, 78 in.·lbf)

Column Lower Cover

Steering Wheel Lower No. 3 Cover

Torx Screw
8.8 (90, 78 in.·lbf)

Steering Wheel

Steering Column Assembly

Combination Switch (w/ Spiral Cable)

Hose Clamp

No. 2 Hole Cover

34 (350, 25)

24.5 (250, 18)

Hole Cover

13 (130, 9)

Sliding Yoke

Thrust Stopper

34 (350, 25)

No. 2 Intermediate Shaft Assembly

Cluster Finish Panel

Lower No. 1 Panel

LH Lower Panel

Hood Lock Release Lever

Fuel Lid Release Lever

No. 2 Heater to Register Duct

Clip

Cowl Trim

Scuff Plate

N·m (kgf·cm, ft·lbf) : Specified torque

93113GG5

Exploded view the steering column—Toyota Land Cruiser (Part 1 of 2)

For Maintenance Interval recommendations, see Section 1 of this manual

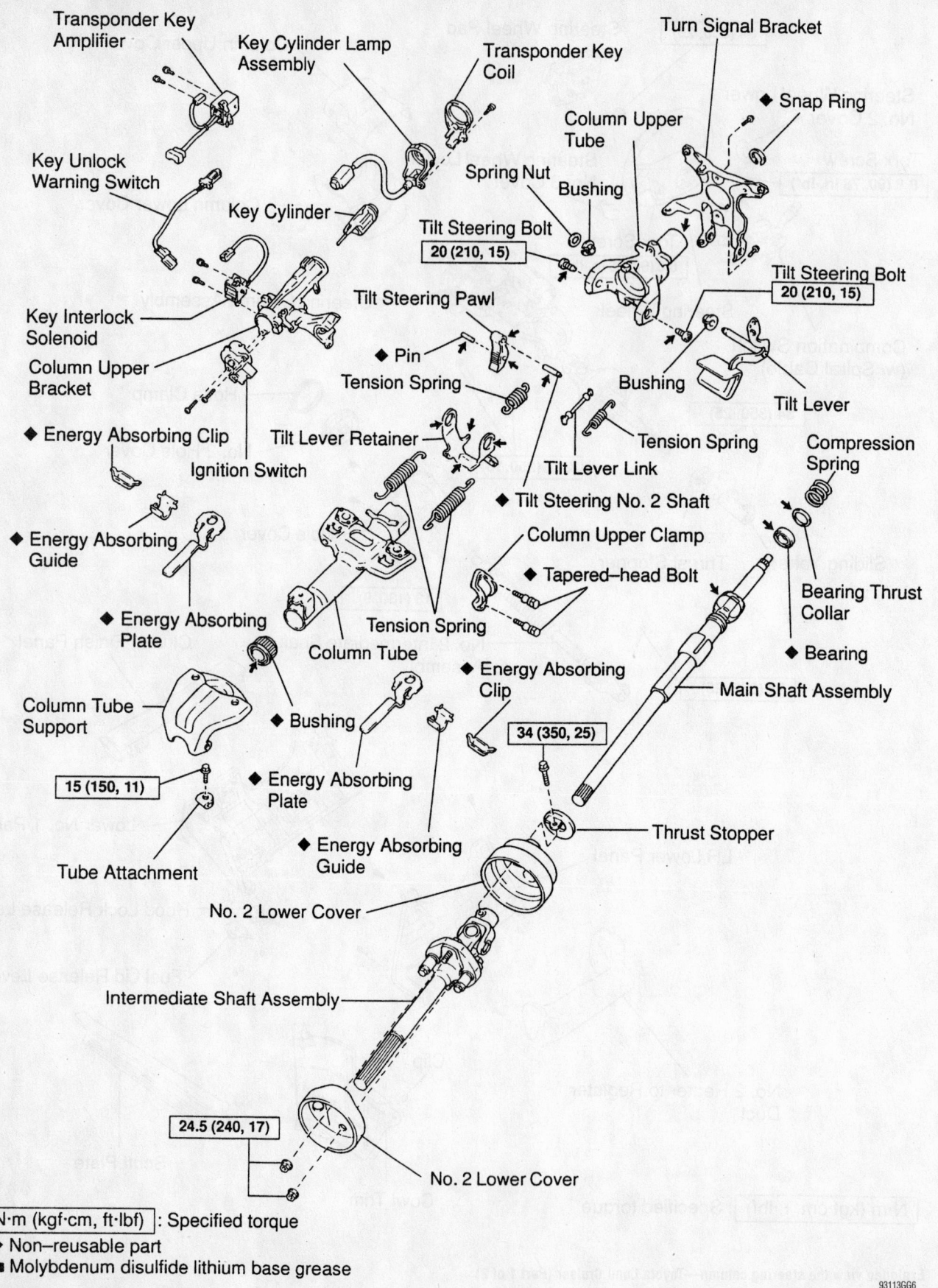

Transponder Key Amplifier

Key Cylinder Lamp Assembly

Transponder Key Coil

Turn Signal Bracket

◆ Snap Ring

Key Unlock Warning Switch

Column Upper Tube

Spring Nut

Bushing

Key Cylinder

Tilt Steering Bolt
20 (210, 15)

Tilt Steering Bolt
20 (210, 15)

Key Interlock Solenoid

Tilt Steering Pawl

◆ Pin

Column Upper Bracket

Tension Spring

Bushing

Tilt Lever

◆ Energy Absorbing Clip

Tilt Lever Retainer

Ignition Switch

Tension Spring

Tilt Lever Link

Compression Spring

◆ Energy Absorbing Guide

◆ Tilt Steering No. 2 Shaft

Column Upper Clamp

Bearing Thrust Collar

◆ Tapered–head Bolt

◆ Energy Absorbing Plate

◆ Bearing

◆ Energy Absorbing Clip

Main Shaft Assembly

Column Tube Support

Tension Spring

Column Tube

34 (350, 25)

◆ Bushing

15 (150, 11)

◆ Energy Absorbing Plate

Thrust Stopper

Tube Attachment

◆ Energy Absorbing Guide

No. 2 Lower Cover

Intermediate Shaft Assembly

24.5 (240, 17)

No. 2 Lower Cover

N·m (kgf·cm, ft·lbf) : Specified torque

◆ Non–reusable part

◀ Molybdenum disulfide lithium base grease

Exploded view the steering column—Toyota Land Cruiser (Part 2 of 2)

93113GG6

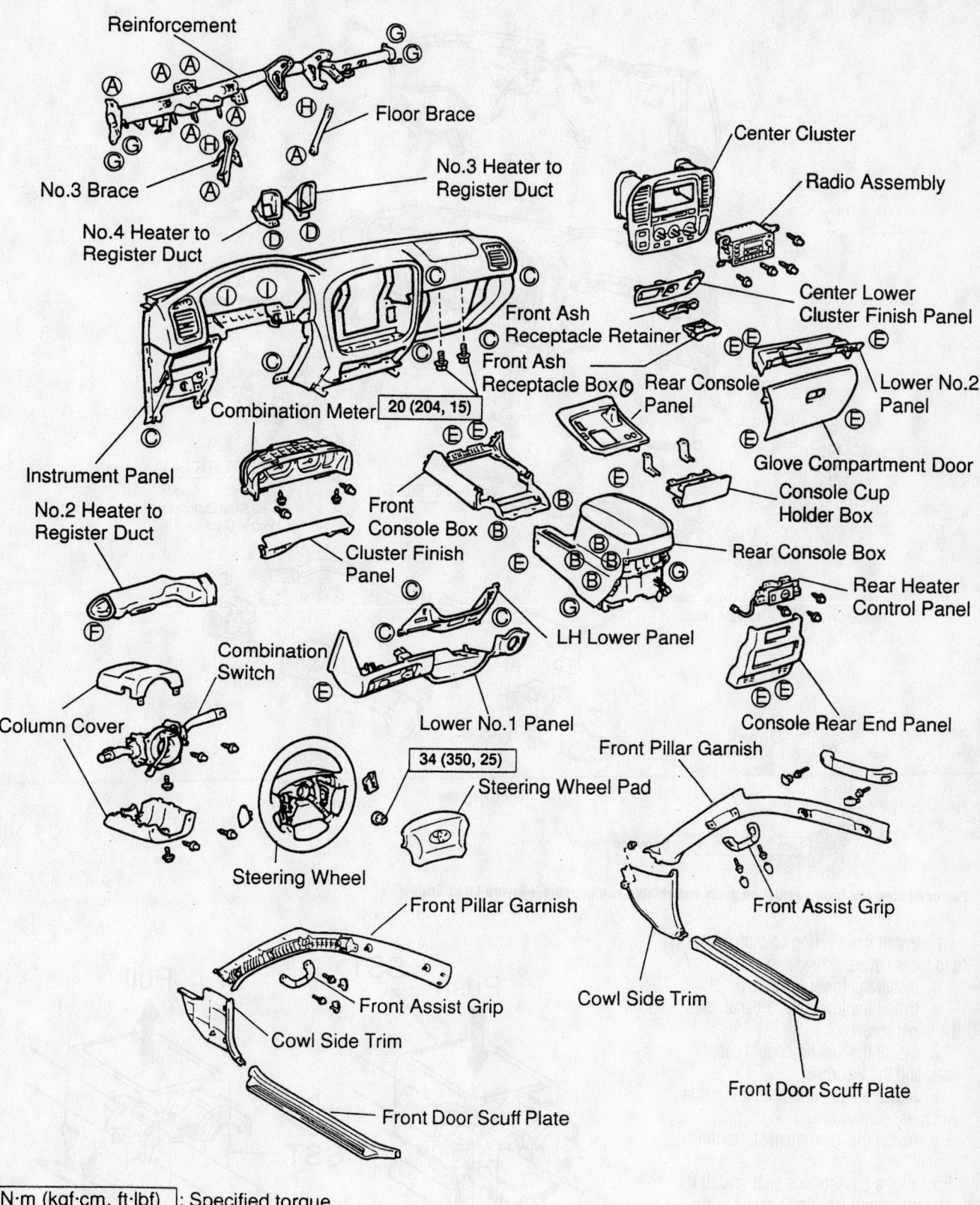

Reinforcement

Floor Brace

No.3 Brace

No.4 Heater to Register Duct

No.3 Heater to Register Duct

Center Cluster

Radio Assembly

Center Lower Cluster Finish Panel

Front Ash Receptacle Retainer

Front Ash Receptacle Box

Rear Console Panel

Lower No.2 Panel

Glove Compartment Door

Instrument Panel

No.2 Heater to Register Duct

Combination Meter   20 (204, 15)

Front Console Box

Cluster Finish Panel

Console Cup Holder Box

Rear Console Box

LH Lower Panel

Rear Heater Control Panel

Combination Switch

Column Cover

Lower No.1 Panel

Console Rear End Panel

Steering Wheel

34 (350, 25)

Steering Wheel Pad

Front Pillar Garnish

Front Pillar Garnish

Front Assist Grip

Cowl Side Trim

Front Door Scuff Plate

Front Pillar Garnish

Front Assist Grip

Cowl Side Trim

Front Door Scuff Plate

N·m (kgf·cm, ft·lbf) : Specified torque

93113GG7

Exploded view the instrument panel and related components—Toyota Land Cruiser

For Tune-up, Capacities and Firing orders, see Section 1 of this manual

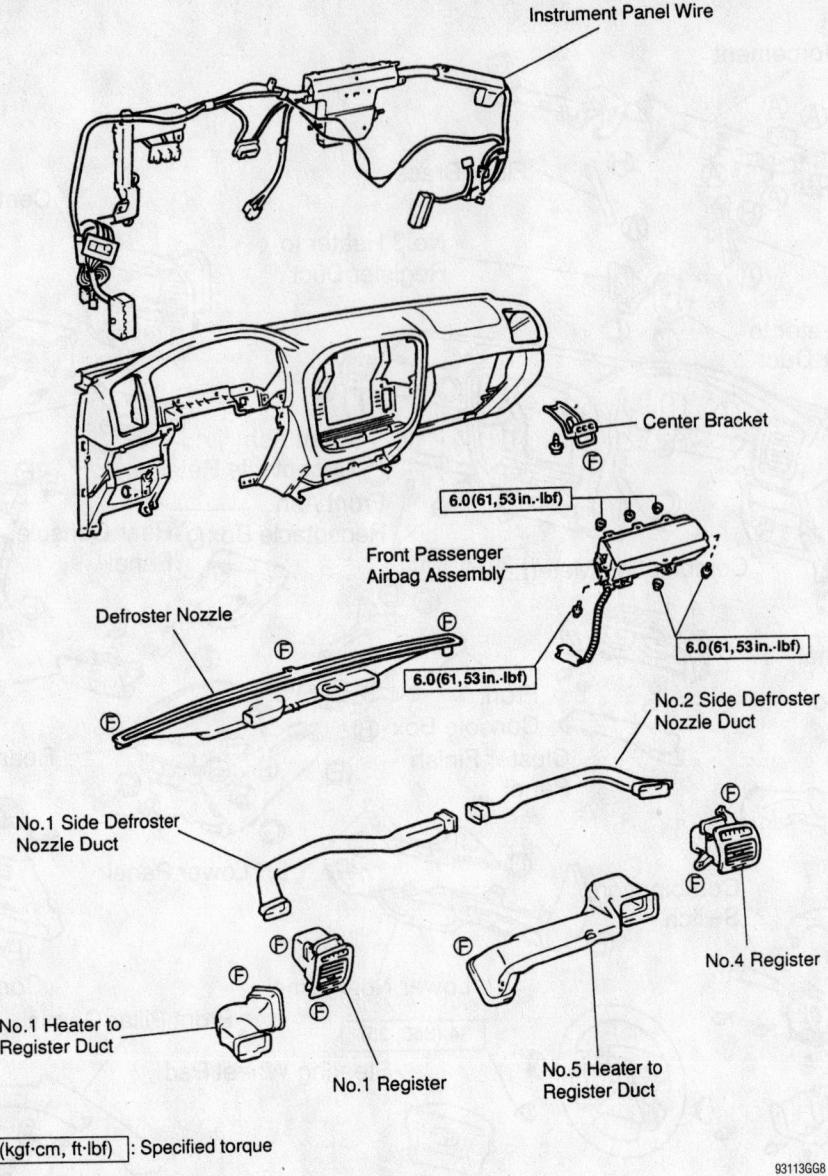

Instrument Panel Wire

Center Bracket

6.0 (61, 53 in.-lbf)

Front Passenger
Airbag Assembly

6.0 (61, 53 in.-lbf)

Defroster Nozzle

6.0 (61, 53 in.-lbf)

No.2 Side Defroster
Nozzle Duct

No.1 Side Defroster
Nozzle Duct

No.4 Register

No.1 Heater to
Register Duct

No.1 Register

No.5 Heater to
Register Duct

N·m (kgf·cm, ft·lbf) : Specified torque

93113GG8

Exploded view the front ventilation ducts and related components—Toyota Land Cruiser

t. Install the steering column covers and the 3 covers screws.

u. Install the lower left side panel.

v. Install the lower No. 1 panel and the panel screw.

w. Install the fuel lid control cable lever and the 2 screws.

x. Install the hood lock control cable and the 2 screws.

y. Install the instrument cluster finish panel.

z. At the passenger's side, install the front pillar garnish, the 2 assist grips, the 4 screws and the 4 assist grip plugs.

aa. At the driver's side, install the front pillar garnish, assist grip, the 2 screws and the 2 assist grip plugs.

bb. Install the front door scuff plates,

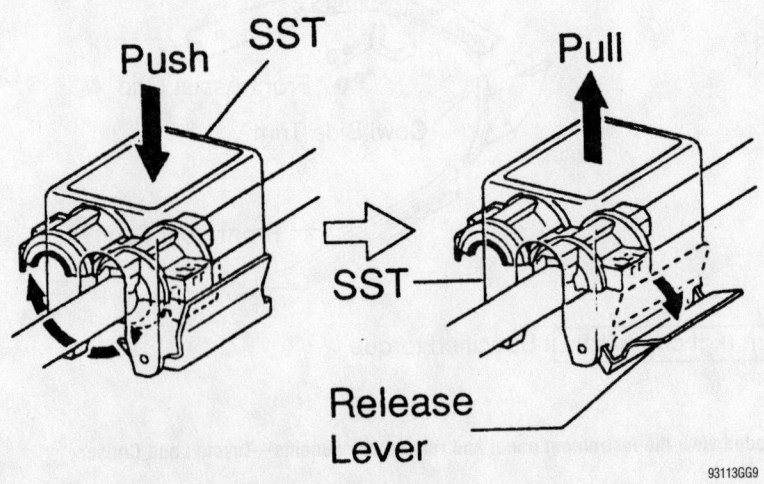

Push   SST   Pull

SST

Release
Lever

93113GG9

View the air conditioning line clamp removal tool—Toyota Land Cruiser

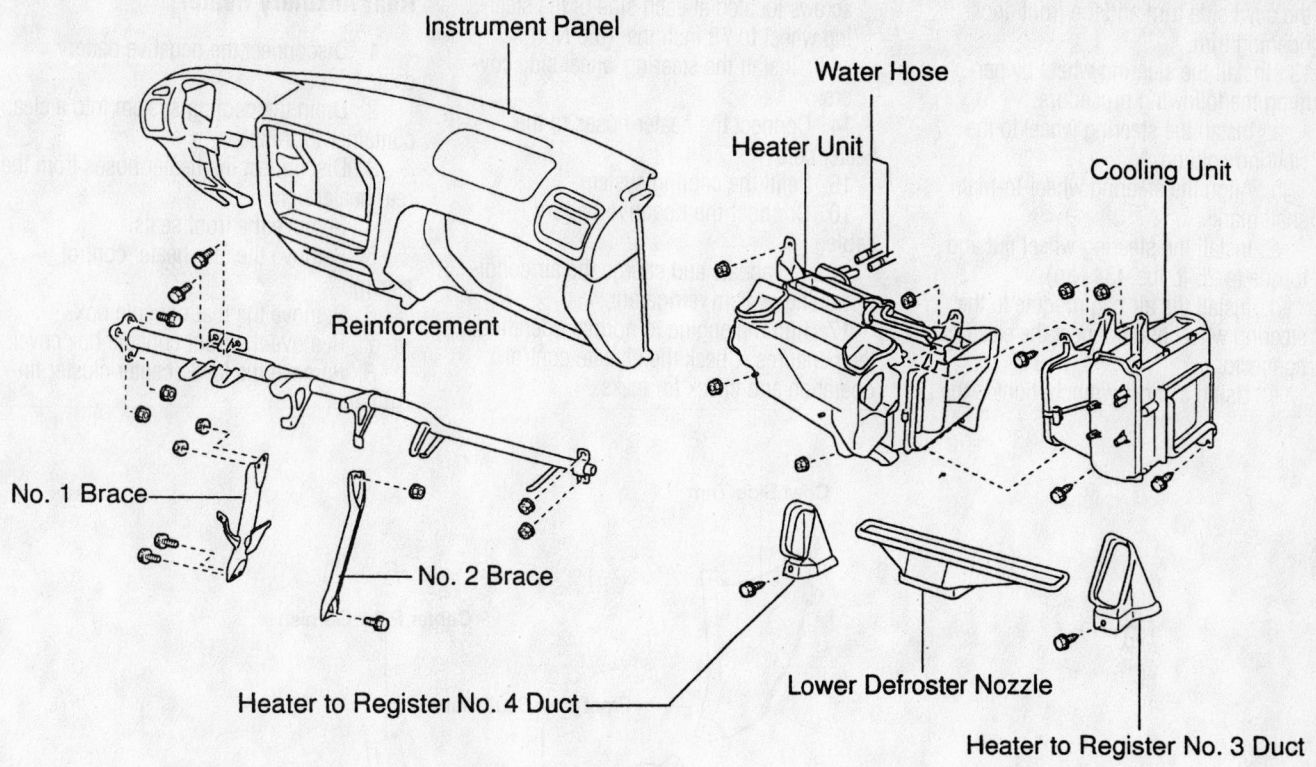

Instrument Panel

Water Hose

Heater Unit

Cooling Unit

Reinforcement

No. 1 Brace

No. 2 Brace

Heater to Register No. 4 Duct

Lower Defroster Nozzle

Heater to Register No. 3 Duct

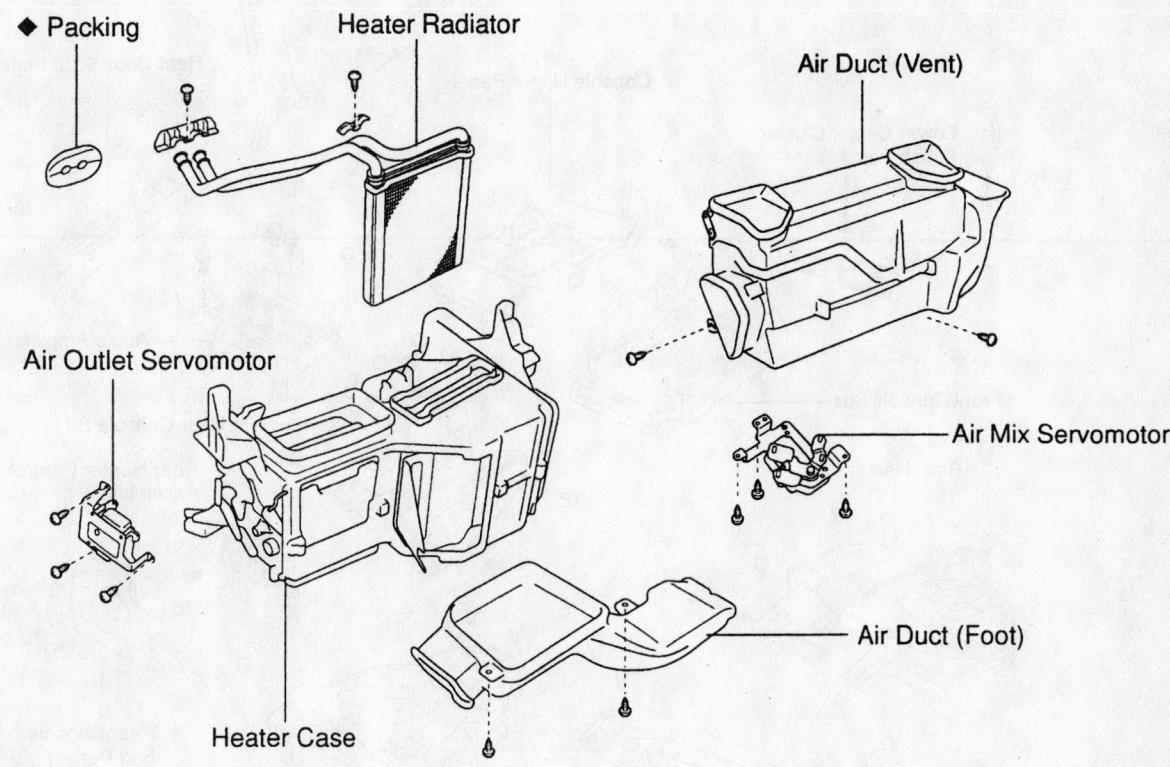

◆ Packing

Heater Radiator

Air Duct (Vent)

Air Outlet Servomotor

Air Mix Servomotor

Air Duct (Foot)

Heater Case

◆ Non–reusable part

93113GG0

Exploded view of the front heater core, heater housing, evaporator housing and related components—Toyota Land Cruiser

For complete service labor times, order Nichols' Chilton Labor Guide

the cowl side trim and the front door opening trim.

13. Install the steering wheel by performing the following procedure:

    a. Install the steering wheel to the steering column.

    b. Align the steering wheel-to-main shaft marks.

    c. Install the steering wheel nut and torque to 25 ft. lbs. (34 Nm).

    d. Install the air bag module to the steering wheel and connect the electrical connector.

    e. Using a Torx® wrench, tighten the 2 screws located at each side of the steering wheel to 78 inch lbs. (8.8 Nm).

    f. Install the steering wheel side covers.

14. Connect the heater hoses to the heater core.

15. Refill the cooling system.

16. Connect the negative battery cable.

    a. Evacuate and charge the air conditioning system refrigerant.

17. Run the engine to normal operating temperatures. Check the climate control operation and check for leaks.

Rear Auxiliary Heater

1. Disconnect the negative battery cable.

2. Drain the cooling system into a clean container for reuse.

3. Disconnect the heater hoses from the rear heater core.

4. Remove the front seats.

5. Remove the rear heater control assembly.

6. Remove the rear console box.

7. Remove the front console box cover.

8. Remove the lower center cluster finish panel.

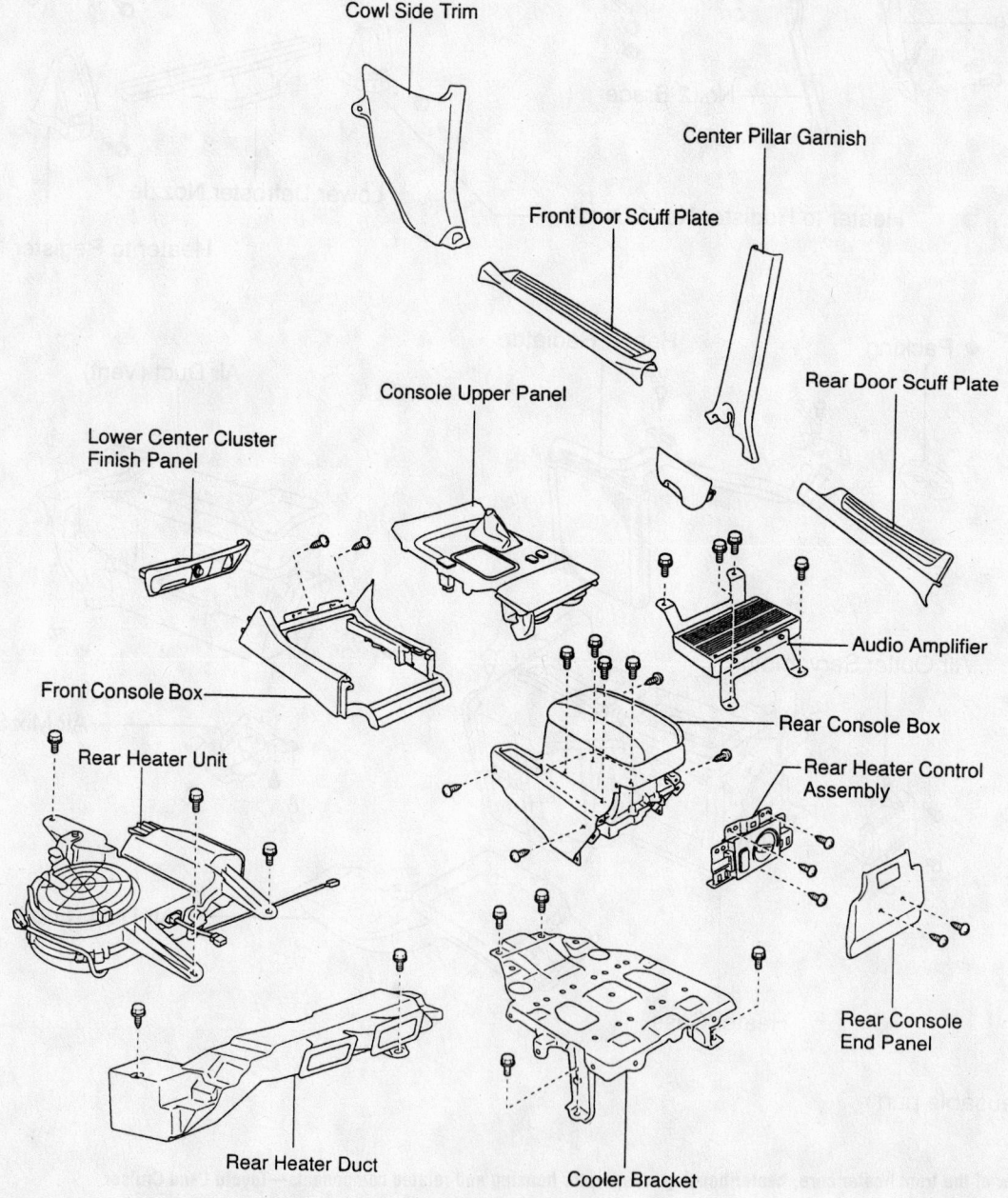

Cowl Side Trim

Center Pillar Garnish

Front Door Scuff Plate

Rear Door Scuff Plate

Console Upper Panel

Lower Center Cluster Finish Panel

Audio Amplifier

Front Console Box

Rear Console Box

Rear Heater Unit

Rear Heater Control Assembly

Rear Console End Panel

Rear Heater Duct

Cooler Bracket

93113GH1

Exploded view of the rear heater housing and related components—Toyota Land Cruiser

9. Remove the front door scuff plates.
10. Remove the cowl side trim.
11. Remove the rear door scuff plates.
12. Remove the center pillar garnishes.
13. Slide the carpet rearward.
14. Remove the cooler bracket bolts and the bracket.
15. Remove the rear heater duct bolt/screw and the duct.
16. Disconnect the rear heater housing electrical connector.
17. Remove the 3 rear heater housing-to-chassis bolts and the heater housing.

18. Remove the heater core-to-heater housing 3 screws and 2 clamps.
19. Remove the heater core from the heater housing.

To install:
20. Install the heater core to the heater housing.
21. Install the heater core-to-heater housing 3 screws and 2 clamps.
22. Install the heater housing and the 3 rear heater housing-to-chassis bolts.
23. Connect the rear heater housing electrical connector.

24. Install the rear heater duct and the duct bolt/screw.
25. Install the cooler bracket and the bracket bolts.
26. Slide the carpet rearward.
27. Install the center pillar garnishes.
28. Install the rear door scuff plates.
29. Install the cowl side trim.
30. Install the front door scuff plates.
31. Install the lower center cluster finish panel.
32. Install the front console box cover.
33. Install the rear console box.

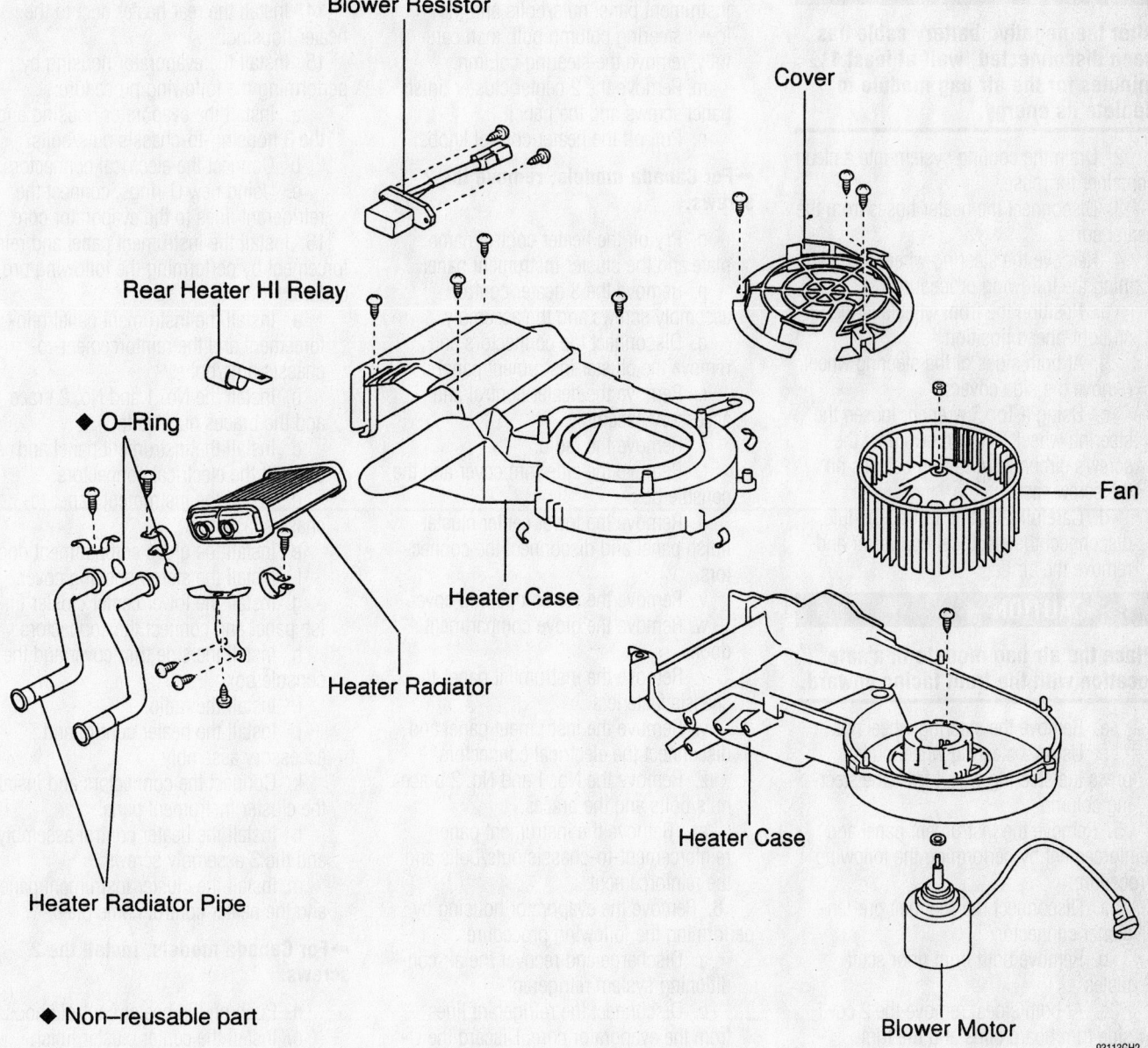

Blower Resistor
Rear Heater HI Relay
◆ O–Ring
Heater Radiator Pipe
Heater Radiator
Heater Case
Cover
Fan
Heater Case
Blower Motor
◆ Non–reusable part

93113GH2

Exploded view of the rear heater core, heater housing and related components—Toyota Land Cruiser

34. Install the rear heater control assembly.

35. Install the front seats.

36. Connect the heater hoses to the rear heater core.

37. Refill the cooling system.

38. Connect the negative battery cable.

RAV4

REMOVAL & INSTALLATION

1. Disconnect the negative battery cable.

✳✳ CAUTION

After the negative battery cable has been disconnected, wait at least 1½ minutes for the air bag module to deplete its energy.

2. Drain the cooling system into a clean container for reuse.

3. Disconnect the heater hoses from the heater core.

4. Remove the steering wheel by performing the following procedure:

   a. Position the front wheels in the straight-ahead position.

   b. At both sides of the steering wheel, remove the side covers.

   c. Using a Torx® wrench, loosen the steering wheel Torx® screws until the screw's circumference ring catches on the screw case.

   d. Carefully, lift the air bag module, disconnect the electrical connector and remove the air bag.

✳✳ CAUTION

Place the air bag module in a safe location with the front facing upward.

   e. Remove the steering wheel nut.

   f. Using a steering wheel puller, press the steering wheel from the steering column.

5. Remove the instrument panel and reinforcement by performing the following procedure:

   a. Disconnect the seat belt pre-tensioner connector.

   b. Remove both front door scuff plates.

   c. At both sides, remove the 2 cowl side trim board clips and the trim boards.

   d. Remove the combination switch-to-steering column screws, disconnect the electrical connectors and remove the combination switch.

   e. Remove the 4 steering column cover screws and the cover.

   f. Remove the cluster finish panel screw and the panel.

   g. Remove the 4 combination meter screws, disconnect the electrical connectors and the meter.

   h. Remove the hood lock release lever.

   i. Remove the 2 lower finish panel screws and the panel.

   j. For USA models, remove the lower panel insert.

   k. Remove the No. 2 heater-to-register duct.

   l. Remove the steering column-to-instrument panel nuts/bolts and the lower steering column bolt; then carefully, remove the steering column.

   m. Remove the 2 center cluster finish panel screws and the panel.

   n. Pull off the heater control knobs.

➡ **For Canada models, remove the 2 screws.**

   o. Pry off the heater control name plate and the cluster instrument panel.

   p. Remove the 3 heater control assembly screws and the assembly.

   q. Disconnect the connectors and remove the cluster instrument panel.

   r. Remove the heater control and accessory assembly.

   s. Remove the radio.

   t. Remove the side trim cover and the console box.

   u. Remove the lower center cluster finish panel and disconnect the connectors.

   v. Remove the stereo opening cover.

   w. Remove the glove compartment door.

   x. Remove the instrument panel-to-chassis fasteners.

   y. Remove the instrument panel and disconnect the electrical connectors.

   z. Remove the No. 1 and No. 2 brace nuts/bolts and the braces.

   aa. Remove the instrument panel reinforcement-to-chassis nuts/bolts and the reinforcement.

6. Remove the evaporator housing by performing the following procedure:

   a. Discharge and recover the air conditioning system refrigerant.

   b. Disconnect the refrigerant lines from the evaporator core. Discard the O-rings and plug the openings to prevent contamination.

   c. Disconnect the electrical connectors.

   d. Remove the 3 evaporator housing-to-chassis nuts/bolts and the housing.

7. Remove the rear heater duct from the heater housing.

8. Remove the heater housing-to-chassis nuts and the housing.

9. Remove the 2 defroster nozzle-to-heater housing screws and the nozzle.

10. Remove the 2 heater core-to-heater housing screws, clamps and the heater core.

To install:

11. Install the heater core and the 2 heater core-to-heater housing screws and clamps.

12. Install the defroster nozzle and the 2 nozzle-to-heater housing screws.

13. Install the heater housing and the housing-to-chassis nuts.

14. Install the rear heater duct to the heater housing.

15. Install the evaporator housing by performing the following procedure:

   a. Install the evaporator housing and the 3 housing-to-chassis nuts/bolts.

   b. Connect the electrical connectors.

   c. Using new O-rings, connect the refrigerant lines to the evaporator core.

16. Install the instrument panel and reinforcement by performing the following procedure:

   a. Install the instrument panel reinforcement and the reinforcement-to-chassis nuts/bolts.

   b. Install the No. 1 and No. 2 brace and the braces nuts/bolts.

   c. Install the instrument panel and connect the electrical connectors.

   d. Install the instrument panel-to-chassis fasteners.

   e. Install the glove compartment door.

   f. Install the stereo opening cover.

   g. Install the lower center cluster finish panel and connect the connectors.

   h. Install the side trim cover and the console box.

   i. Install the radio.

   j. Install the heater control and accessory assembly.

   k. Connect the connectors and install the cluster instrument panel.

   l. Install the heater control assembly and the 3 assembly screws.

   m. Install the cluster instrument panel and the heater control name plate.

➡ **For Canada models, install the 2 screws.**

   n. Push on the heater control knobs.

   o. Install the center cluster finish panel and the 2 panel screws.

   p. Carefully, install the steering column. Then, install the steering column-to-instrument panel nuts/bolts and torque the nuts/bolts 19 ft. lbs. (25 Nm) and the lower steering column bolt to 26 ft. lbs. (5 Nm).

q. Install the No. 2 heater-to-register duct.

r. For USA models, install the lower panel insert.

s. Install the lower finish panel and the 2 panel screws.

t. Install the hood lock release lever.

u. Install the combination meter, connect the electrical connectors and the 4 meter screws.

v. Install the cluster finish panel and the panel screw.

w. Install the steering column cover and the 4 cover screws.

x. Install the combination switch, connect the electrical connectors and install the combination switch-to-steering column screws.

y. At both sides, install the cowl side trim board and the 2 trim boards clips.

z. Install both front door scuff plates.

aa. Connect the seat belt pre-tensioner connector.

17. Install the steering wheel by performing the following procedure:

a. Install the steering wheel to the steering column.

b. Install the steering wheel nut and torque the nut to 25 ft. lbs. (34 Nm).

c. Connect the electrical connector and carefully, install the air bag module.

d. Using a Torx® wrench, tighten the

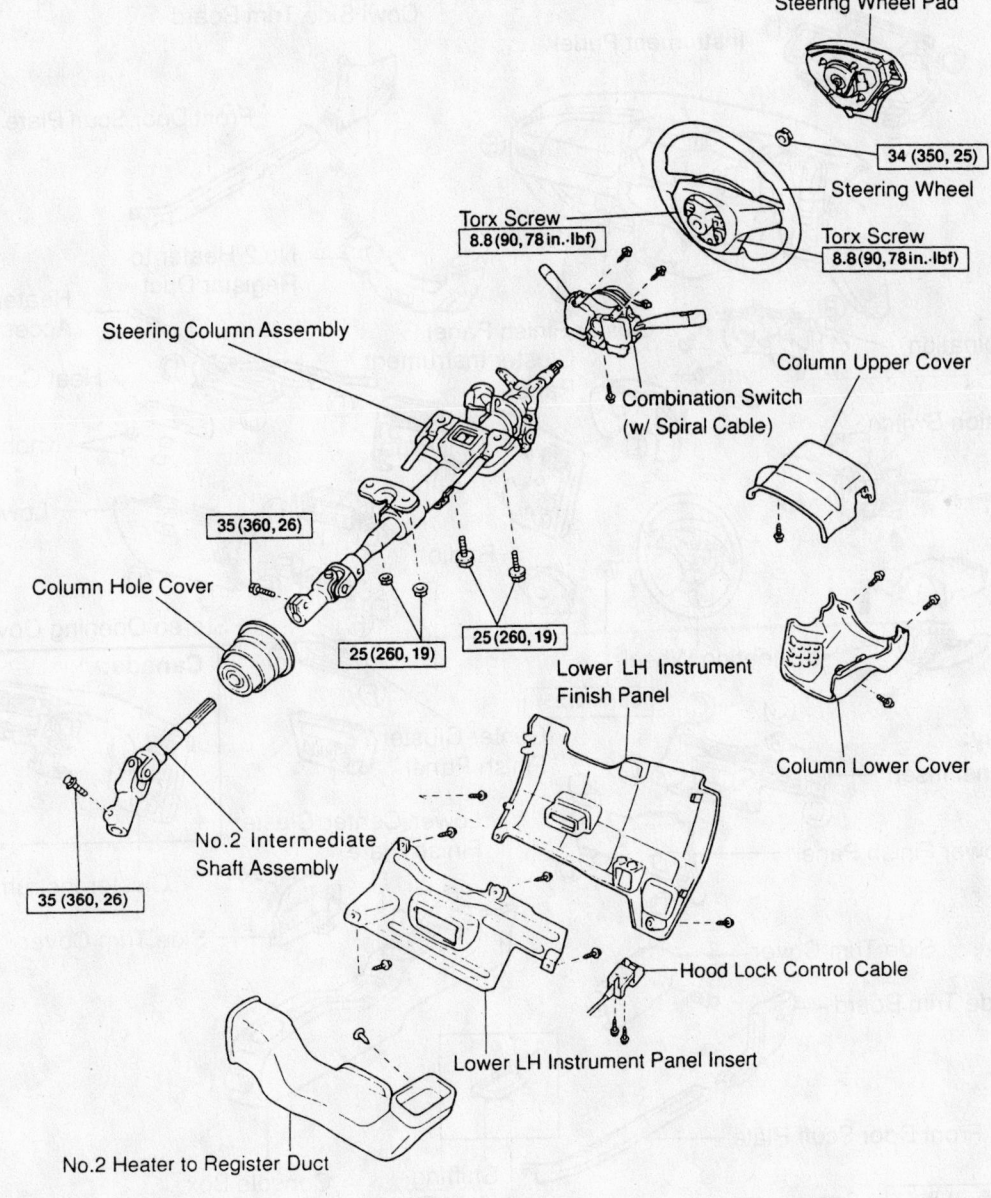

Steering Wheel Pad

34 (350, 25)

Steering Wheel

Torx Screw
8.8 (90, 78 in.·lbf)

Torx Screw
8.8 (90, 78 in.·lbf)

Steering Column Assembly

Column Upper Cover

Combination Switch
(w/ Spiral Cable)

35 (360, 26)

Column Hole Cover

25 (260, 19)

25 (260, 19)

Lower LH Instrument
Finish Panel

Column Lower Cover

No.2 Intermediate
Shaft Assembly

35 (360, 26)

Hood Lock Control Cable

Lower LH Instrument Panel Insert

No.2 Heater to Register Duct

N·m (kgf·cm, ft·lbf) : Specified torque

93113GK4

Exploded view of the steering wheel, air bag module, steering column and related components—Toyota RAV4

Timing belt service is covered in Section 3 of this manual

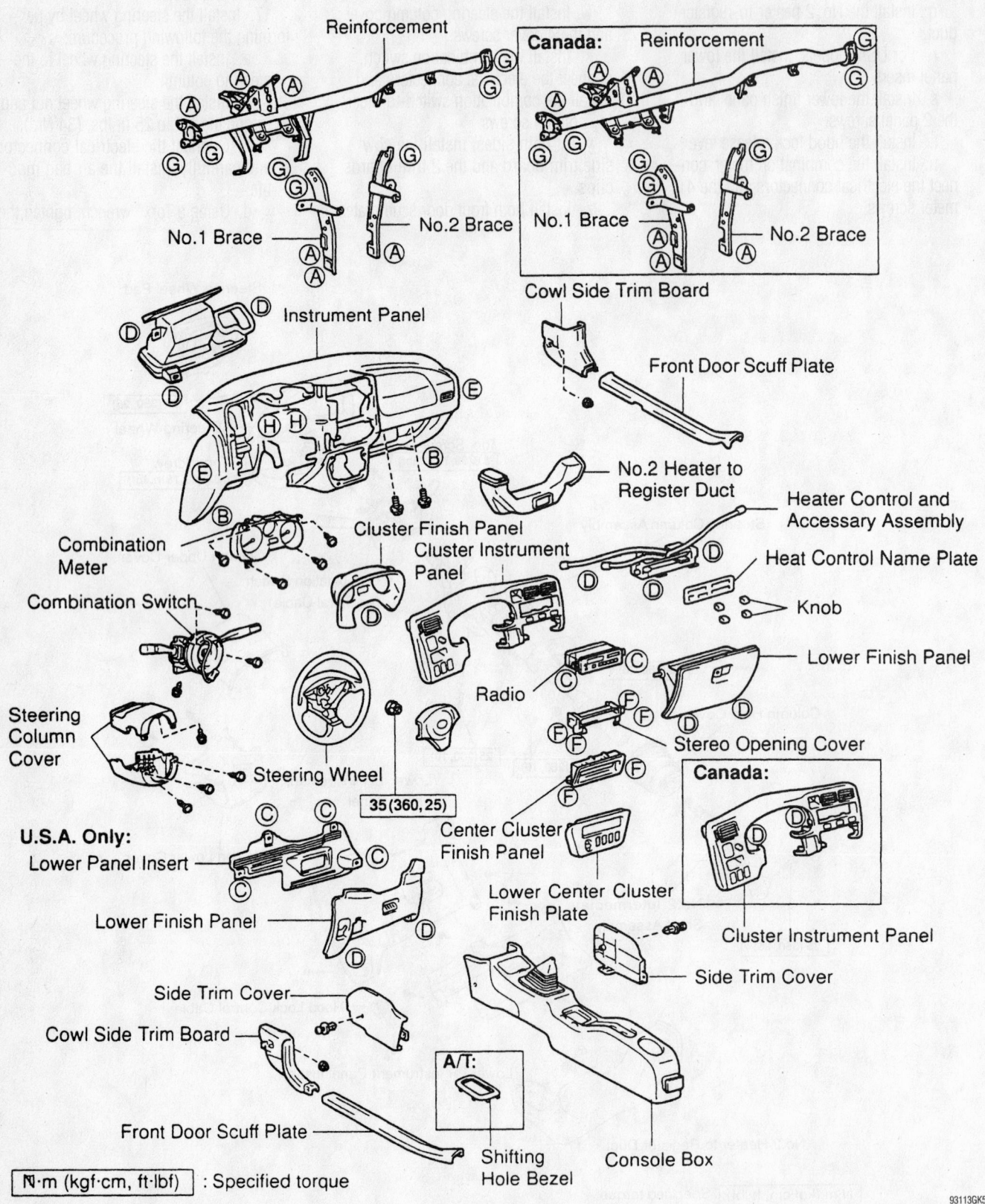

Reinforcement

Canada: Reinforcement

No.1 Brace

No.2 Brace

Canada:
No.1 Brace

No.2 Brace

Cowl Side Trim Board

Instrument Panel

Front Door Scuff Plate

No.2 Heater to Register Duct

Heater Control and Accessary Assembly

Heat Control Name Plate

Knob

Combination Meter

Cluster Finish Panel
Cluster Instrument Panel

Combination Switch

Radio

Lower Finish Panel

Steering Column Cover

Steering Wheel

35 (360, 25)

Stereo Opening Cover

Canada:

U.S.A. Only:
Lower Panel Insert

Center Cluster Finish Panel

Lower Finish Panel

Lower Center Cluster Finish Plate

Cluster Instrument Panel

Side Trim Cover

Side Trim Cover

Cowl Side Trim Board

A/T:

Front Door Scuff Plate

Shifting Hole Bezel

Console Box

N·m (kgf·cm, ft·lbf) : Specified torque

93113GK5

Exploded view of the instrument panel and related components—Toyota RAV4

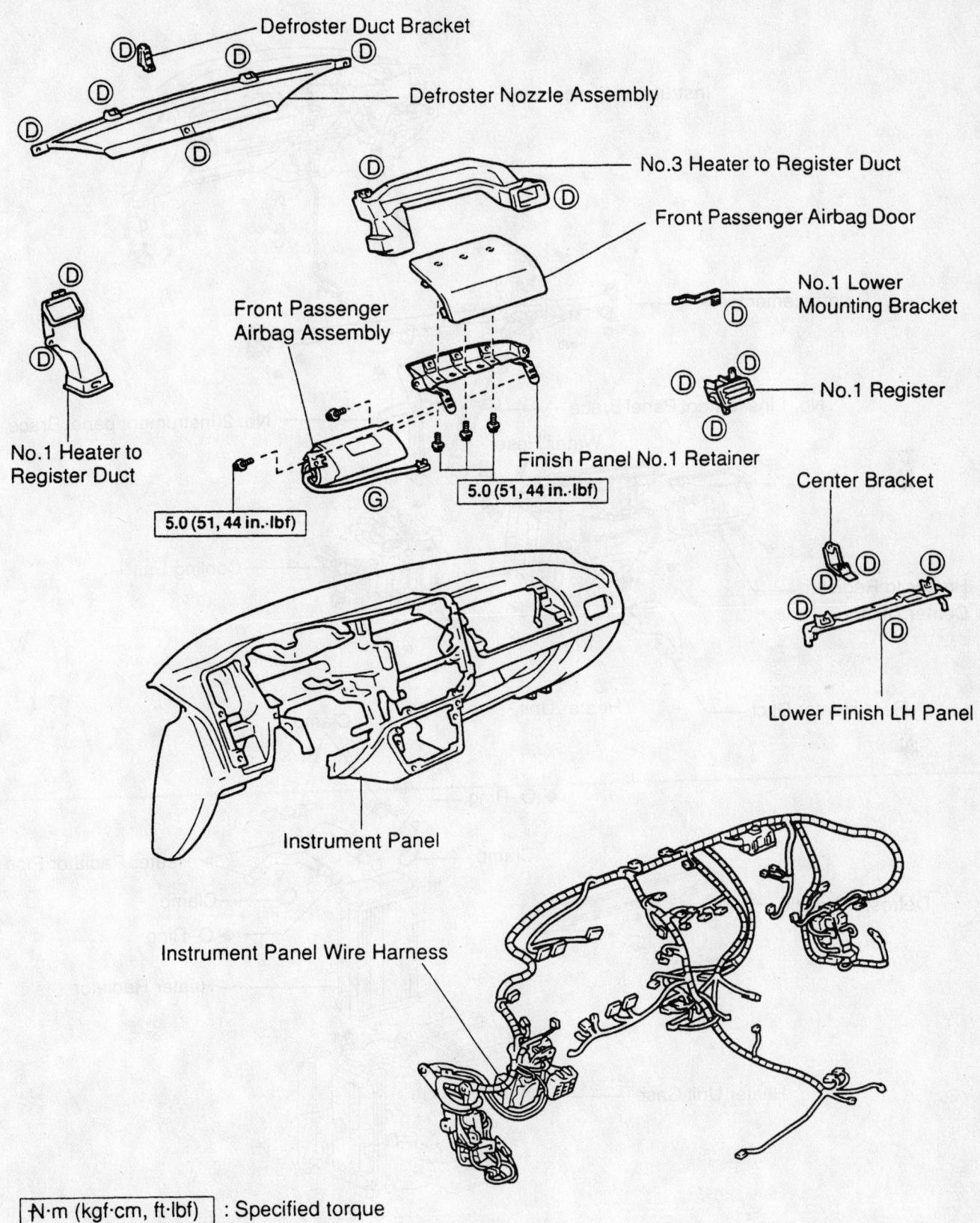

Defroster Duct Bracket

Defroster Nozzle Assembly

No.3 Heater to Register Duct

Front Passenger Airbag Door

No.1 Lower Mounting Bracket

Front Passenger Airbag Assembly

No.1 Register

No.1 Heater to Register Duct

Finish Panel No.1 Retainer

Center Bracket

5.0 (51, 44 in.·lbf)

5.0 (51, 44 in.·lbf)

Lower Finish LH Panel

Instrument Panel

Instrument Panel Wire Harness

N·m (kgf·cm, ft·lbf) : Specified torque

93113GK6

Exploded view of the instrument panel air bag module, ventilation components and wiring harness—Toyota RAV4

Heater Core replacement is covered in Section 2 of this manual

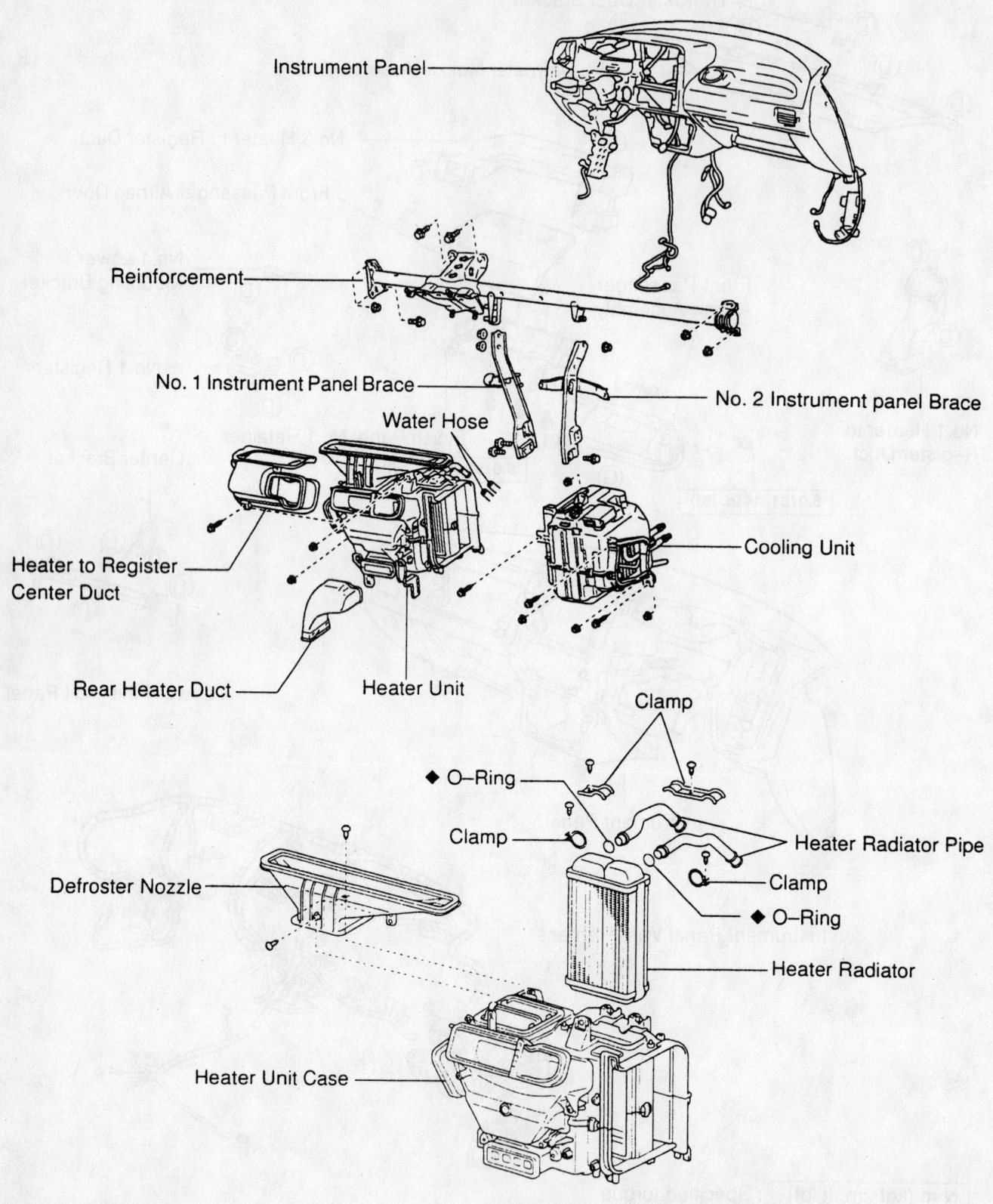

Instrument Panel

Reinforcement

No. 1 Instrument Panel Brace

No. 2 Instrument panel Brace

Water Hose

Heater to Register Center Duct

Cooling Unit

Rear Heater Duct

Heater Unit

Clamp

◆ O–Ring

Clamp

Heater Radiator Pipe

Clamp

◆ O–Ring

Defroster Nozzle

Heater Radiator

Heater Unit Case

◆ Non–reusable part

93113GK7

Exploded view of the heater core, heater housing, evaporator housing and related components—Toyota RAV4

steering wheel screws to 78 inch lbs. (8.8 Nm).

e. At both sides of the steering wheel, install the side covers.

18. Connect the heater hoses to the heater core.

19. Refill the cooling system.

20. Connect the negative battery cable.

21. Evacuate and charge the air conditioning system.

22. Run the engine to normal operating temperatures. Check the climate control operation and check for leaks.

4Runner

REMOVAL & INSTALLATION

Front Heater

1. Disconnect the negative battery cable.

❊❊ CAUTION

After the negative battery cable has been disconnected, wait at least 1½ minutes for the air bag module to deplete its energy.

2. Drain the cooling system into a clean container for reuse.

3. Disconnect the heater hoses from the heater core.

4. Remove the steering wheel by performing the following procedure:

a. Position the front wheels in the straight-ahead position.

b. At both sides of the steering wheel, remove the side covers.

c. Using a Torx® wrench, loosen the steering wheel screws until the screw's circumference ring catches on the screw case.

d. Carefully, lift the air bag module, disconnect the electrical connector and remove the air bag.

❊❊ CAUTION

Place the air bag module in a safe location with the front facing upward.

e. Remove the steering wheel nut.

f. Using a steering wheel puller, press the steering wheel from the steering column.

5. Remove the instrument panel and reinforcement by performing the following procedure:

a. Remove both front door scuff plates.

b. Remove both cowl side trims.

c. Remove the 2 hood lock release lever screws and the hood lock release lever.

d. Remove the 2 fuel lid release lever screws and the fuel lid release lever.

e. Remove the 4 lower finish panel bolts and the panel.

f. Remove the No. 1 and No. 2 heater-to-register duct screw and the ducts.

g. Pry out the starter switch bezel.

h. Remove the steering column cover screws and the covers.

i. Remove the combination switch-to-steering column screws, disconnect the electrical connector and the combination switch.

j. Remove the steering column-to-instrument panel nuts/bolts and the lower steering column bolt; then carefully, remove the steering column.

k. Remove the 4 cluster finish panel screws and the panel.

l. Remove the 4 combination meter screws, disconnect the electrical connectors and remove the combination meter.

m. Pry out the parking brake hole cover.

n. Pry out the upper console panel.

o. Disengage the 7 center cluster finish panel clips and remove the panel.

➡**Remove the center cluster finish panel clips by starting at the bottom and working toward the top.**

p. Remove the heater control knobs.

q. Remove the 2 rear console box bolts/screws and the rear console box.

r. Remove the upper console panel garnish.

s. Remove the 2 glove compartment door screws and the door.

t. Disconnect the passenger's side air bag module electrical connector.

u. Remove the glove box light.

v. Remove the 3 lower No. 2 finish panel bolts and the panel.

w. Remove the 3 glove compartment door reinforcement bolts and the reinforcement.

x. Remove the No. 4 heater-to-register duct.

y. Remove the radio assembly.

z. Remove the side bracket bolt and the bracket.

aa. If equipped with manual air conditioning, remove the heater control assembly.

bb. If equipped with automatic air conditioning, remove the air conditioning control assembly.

cc. Remove the instrument panel-to-chassis nut and 2 bolts; then, remove the instrument panel.

dd. Remove the instrument panel reinforcement-to-chassis nuts/bolts and the reinforcement.

6. Remove the defroster nozzle and heater-to-register duct.

7. Remove the evaporator housing by performing the following procedure:

a. Discharge and recover the air conditioning system refrigerant.

b. Disconnect the refrigerant lines from the evaporator core. Discard the O-rings and plug the openings to prevent contamination.

c. Disconnect the electrical connectors.

d. Remove the 3 evaporator housing-to-chassis screws and the housing.

8. Disconnect the mode control servomotor connector.

9. Disconnect the aspirator hose from the room temperature sensor.

10. Disconnect the heater valve control cable.

11. Remove the heater housing-to-chassis nuts and the heater housing.

12. Remove the 3 heater core-to-heater housing screws, the 2 clips and clamp.

13. Remove the heater core from the heater housing.

To install:

14. Install the 3 heater core-to-heater housing screws, the 2 clips and clamp.

15. Install the heater housing and the heater housing-to-chassis nuts.

16. Connect the heater valve control cable.

17. Connect the aspirator hose to the room temperature sensor.

18. Connect the mode control servomotor connector.

19. Install the defroster nozzle and heater-to-register duct.

20. Install the evaporator housing by performing the following procedure:

a. Install the evaporator housing and the 3 housing-to-chassis screws.

b. Connect the electrical connectors.

c. Using new O-rings, connect the refrigerant lines to the evaporator core.

21. Install the instrument panel and reinforcement by performing the following procedure:

a. Install the instrument panel reinforcement and the reinforcement-to-chassis nuts/bolts.

Brake service is covered in Section 4 of this manual

b. Install the instrument panel and the instrument panel-to-chassis nut and 2 bolts.

c. If equipped with automatic air conditioning, install the air conditioning control assembly.

d. If equipped with manual air conditioning, install the heater control assembly.

e. Install the side bracket and the bracket bolt.

f. Install the radio assembly.

g. Install the No. 4 heater-to-register duct.

h. Install the glove compartment door reinforcement and the 3 reinforcement bolts.

i. Install the lower No. 2 finish panel and the 3 panel bolts.

j. Install the glove box light.

k. Connect the passenger's side air bag module electrical connector.

l. Install the glove compartment door and the 2 door screws.

m. Install the upper console panel garnish.

n. Install the rear console box and the 2 rear console box bolts/screws.

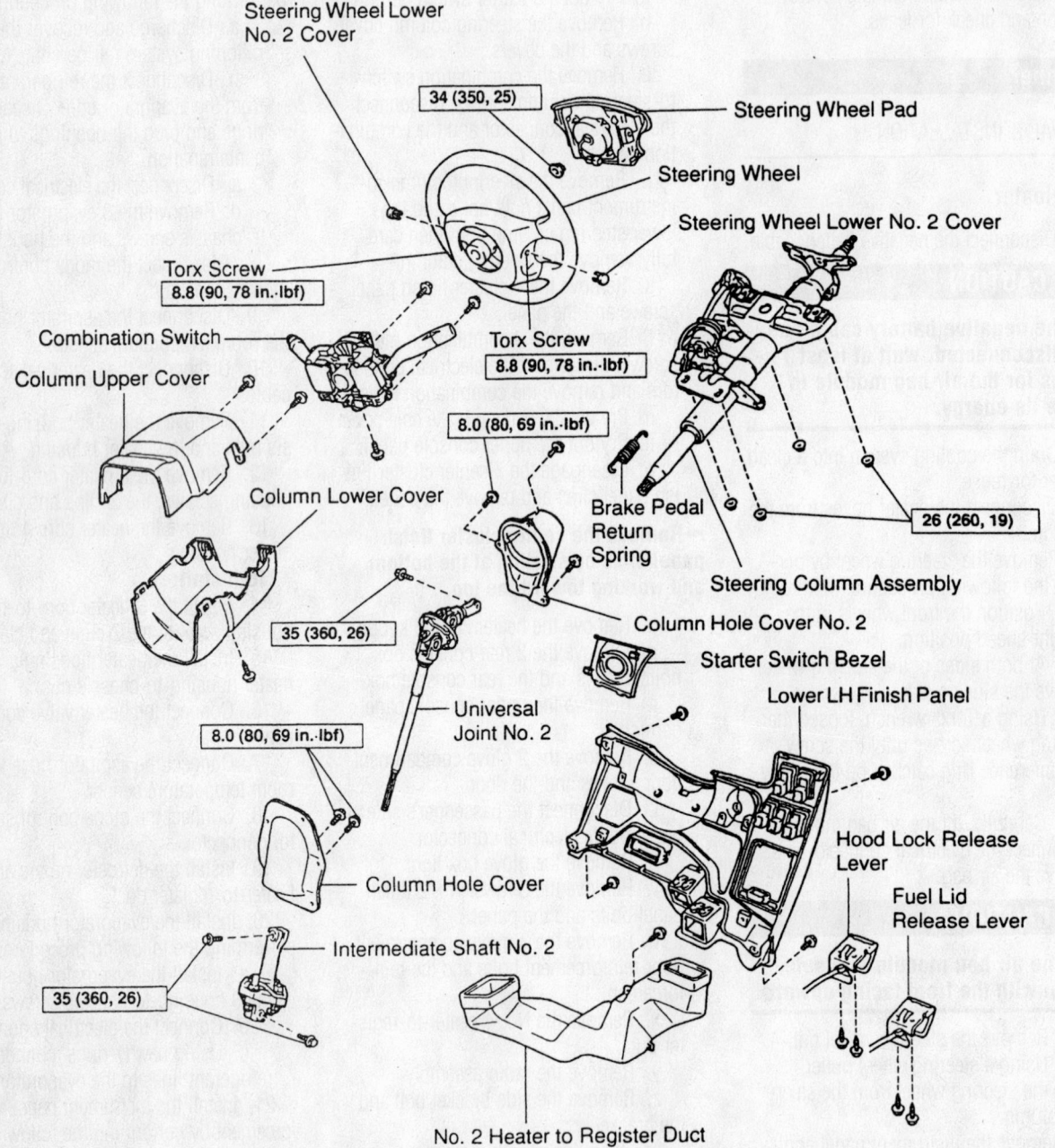

Exploded view of the steering wheel, air bag module, steering column and related components—Toyota 4Runner

93113GJ9

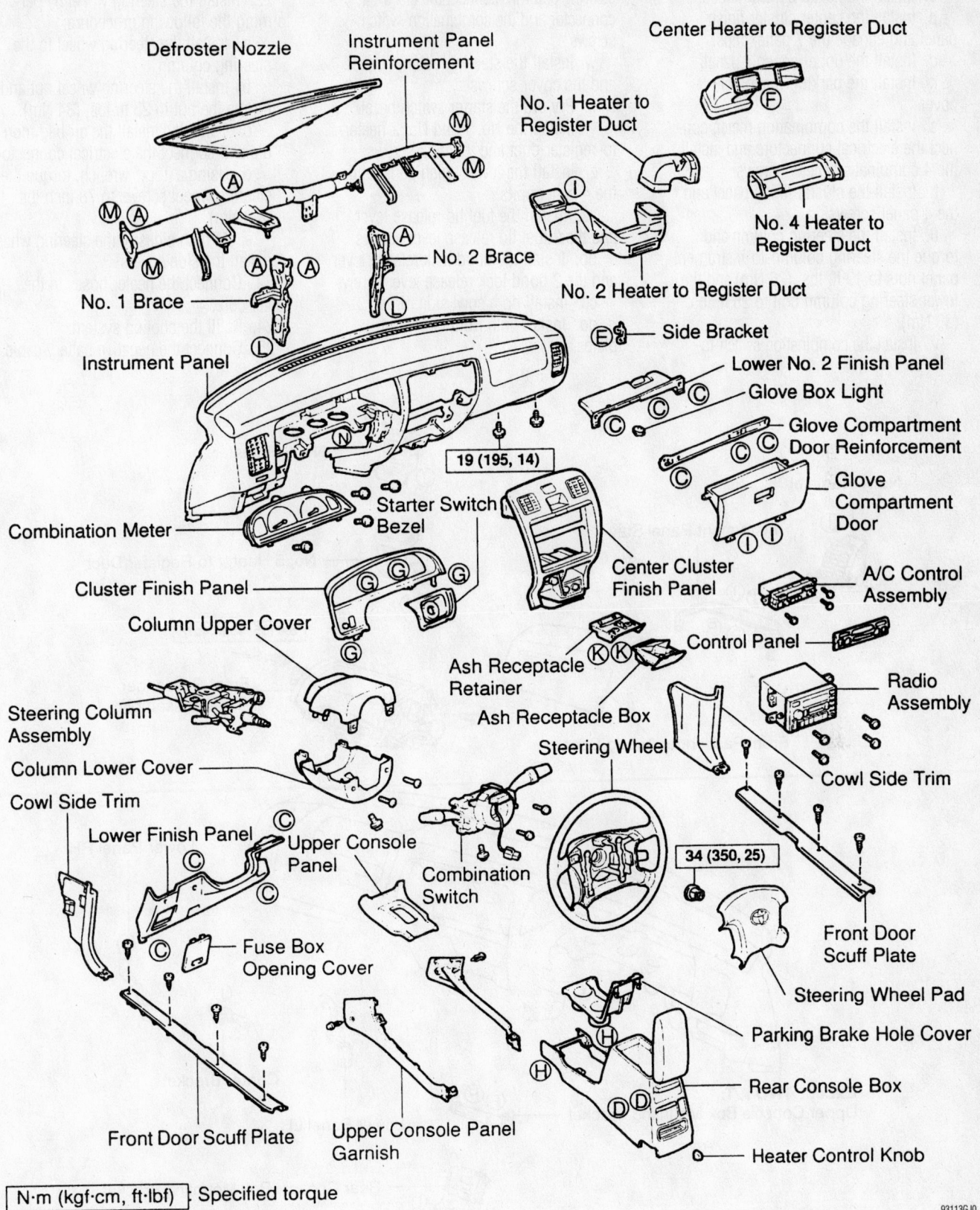

Defroster Nozzle

Instrument Panel Reinforcement

Center Heater to Register Duct

No. 1 Heater to Register Duct

No. 4 Heater to Register Duct

No. 2 Heater to Register Duct

No. 2 Brace

No. 1 Brace

Instrument Panel

Side Bracket

Lower No. 2 Finish Panel

Glove Box Light

Glove Compartment Door Reinforcement

Glove Compartment Door

19 (195, 14)

Combination Meter

Starter Switch Bezel

Center Cluster Finish Panel

A/C Control Assembly

Control Panel

Cluster Finish Panel

Column Upper Cover

Radio Assembly

Steering Column Assembly

Ash Receptacle Retainer

Ash Receptacle Box

Steering Wheel

Column Lower Cover

Cowl Side Trim

Cowl Side Trim

Lower Finish Panel

Upper Console Panel

Combination Switch

34 (350, 25)

Fuse Box Opening Cover

Front Door Scuff Plate

Steering Wheel Pad

Parking Brake Hole Cover

Rear Console Box

Front Door Scuff Plate

Upper Console Panel Garnish

Heater Control Knob

N·m (kgf·cm, ft·lbf) : Specified torque

93113GJ0

Exploded view of the instrument panel and related components—Toyota 4Runner

For complete Engine Mechanical specifications, see Section 1 of this manual

o. Install the heater control knobs.

p. Install the center cluster finish panel and engage the 7 panel clips.

q. Install the upper console panel.

r. Install the parking brake hole cover.

s. Install the combination meter, connect the electrical connectors and install the 4 combination meter screws.

t. Install the cluster finish panel and the 4 panel screws.

u. Install the steering column and torque the steering column-to-instrument panel nuts to 19 ft. lbs. (26 Nm) and the lower steering column bolt to 26 ft. lbs. (35 Nm).

v. Install the combination switch-to-steering column, connect the electrical connector and the combination switch screws.

w. Install the steering column cover and the cover screws.

x. Pry out the starter switch bezel.

y. Install the No. 1 and No. 2 heater-to-register duct and the duct screws.

z. Install the lower finish panel and the 4 panel bolts.

aa. Install the fuel lid release lever and the 2 fuel lid release lever screws.

bb. Install the hood lock release lever and the 2 hood lock release lever screws.

cc. Install both cowl side trims.

dd. Install both front door scuff plates.

22. Install the steering wheel by performing the following procedure:

a. Install the steering wheel to the steering column.

b. Install the steering wheel nut and torque the nut to 25 ft. lbs. (34 Nm).

c. Carefully, install the air bag module and connect the electrical connector.

d. Using a Torx® wrench, torque the steering wheel screws to 78 inch lbs. (8.8 Nm).

e. At both sides of the steering wheel, install the side covers.

23. Connect the heater hoses to the heater core.

24. Refill the cooling system.

25. Connect the negative battery cable.

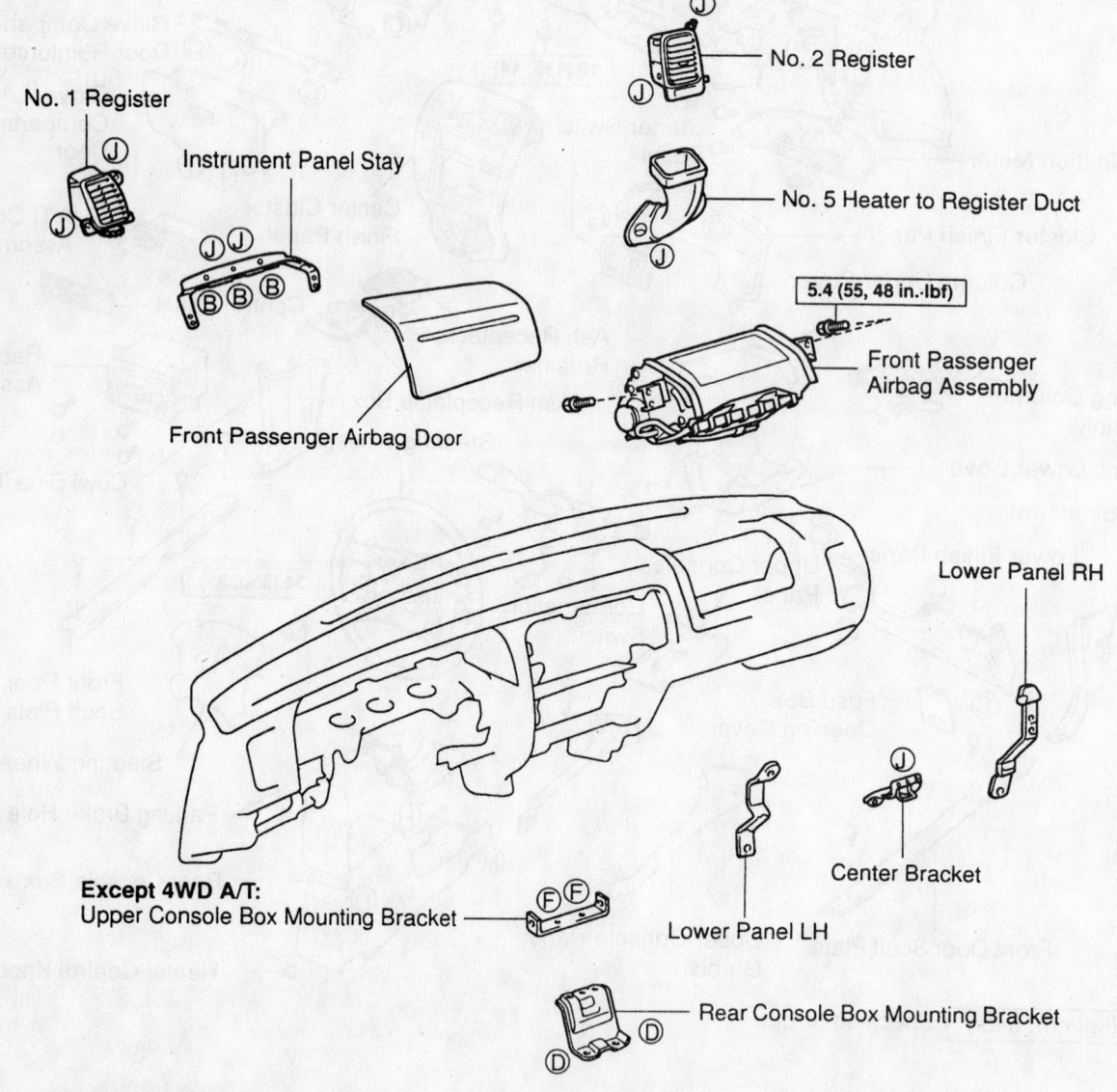

No. 1 Register

No. 2 Register

Instrument Panel Stay

No. 5 Heater to Register Duct

5.4 (55, 48 in.·lbf)

Front Passenger Airbag Assembly

Front Passenger Airbag Door

Lower Panel RH

Lower Panel LH

Center Bracket

Except 4WD A/T:
Upper Console Box Mounting Bracket

Rear Console Box Mounting Bracket

N·m (kgf·cm, ft·lbf) : Specified torque

93113GK1

Exploded view of the instrument panel air bag module, ventilation components and brackets—Toyota 4Runner

26. Evacuate and charge the air conditioning system.

27. Run the engine to normal operating temperatures; then, check the climate control operation and check for leaks.

Rear Auxiliary Heater

1. Disconnect the negative battery cable.
2. Drain the cooling system into a clean container for reuse.
3. Remove the front seats.
4. Remove the center console box.
5. Move the floor carpet backward.
6. Disconnect the rear heater hoses from the rear heater core.
7. Remove the rear heater duct bolt, screw and duct.
8. Remove the rear heater control assembly.
9. Disconnect the electrical connectors.
10. Remove the 3 rear heater housing-to-chassis screws and the housing.
11. Remove the blower resistor, the rear heater relay and with wiring harness from the rear heater housing.
12. Remove the rear heater housing case screws and separate the cases.
13. Remove the rear heater core.

To install:
14. Install the rear heater core.
15. Assemble the rear heater housing case and install the case screws.
16. Install the blower resistor, the rear heater relay and wiring harness to the rear heater housing.

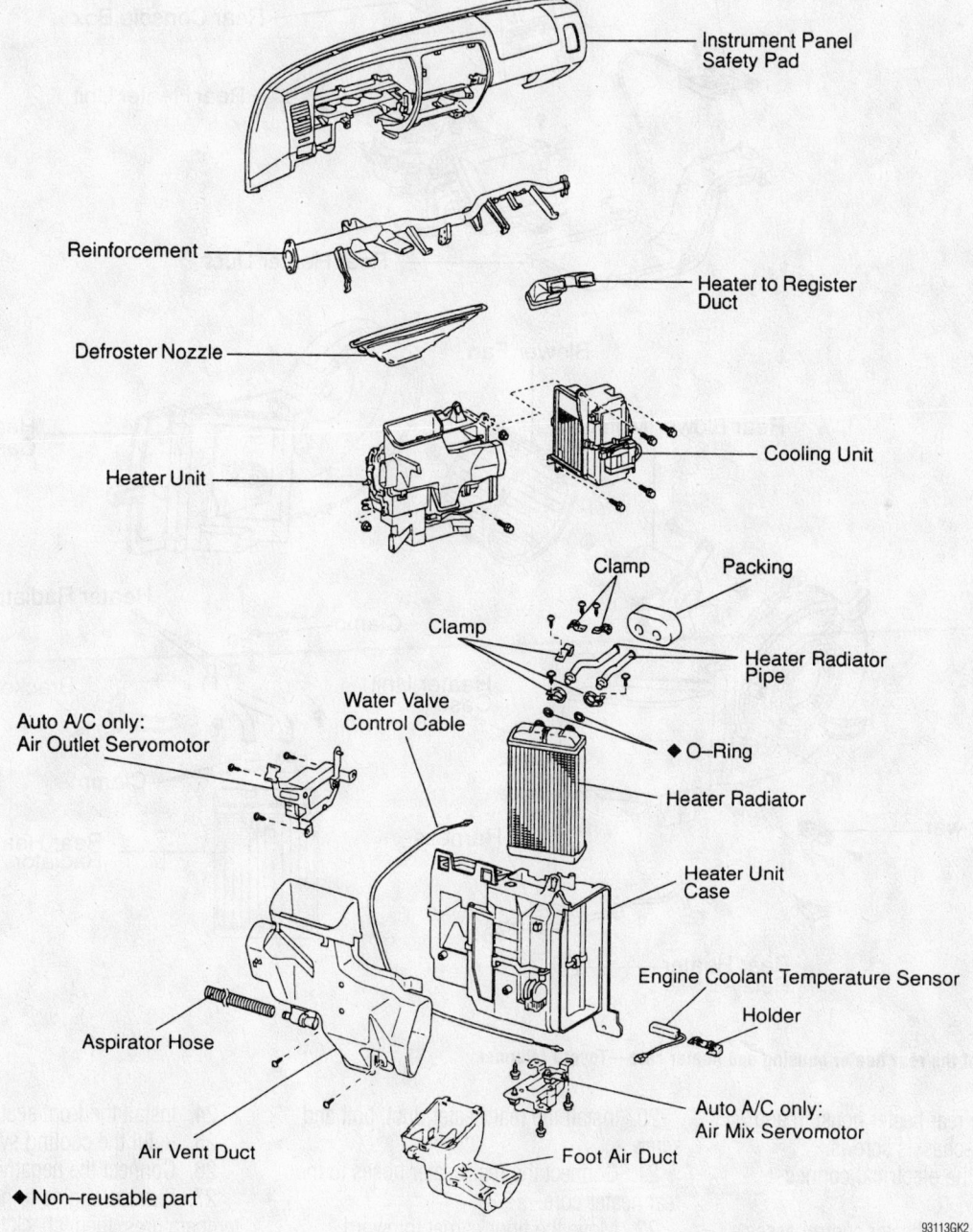

Exploded view of the front heater core, heater housing, evaporator housing and related components—Toyota 4Runner

For Accessory Drive Belt illustrations, see Section 1 of this manual

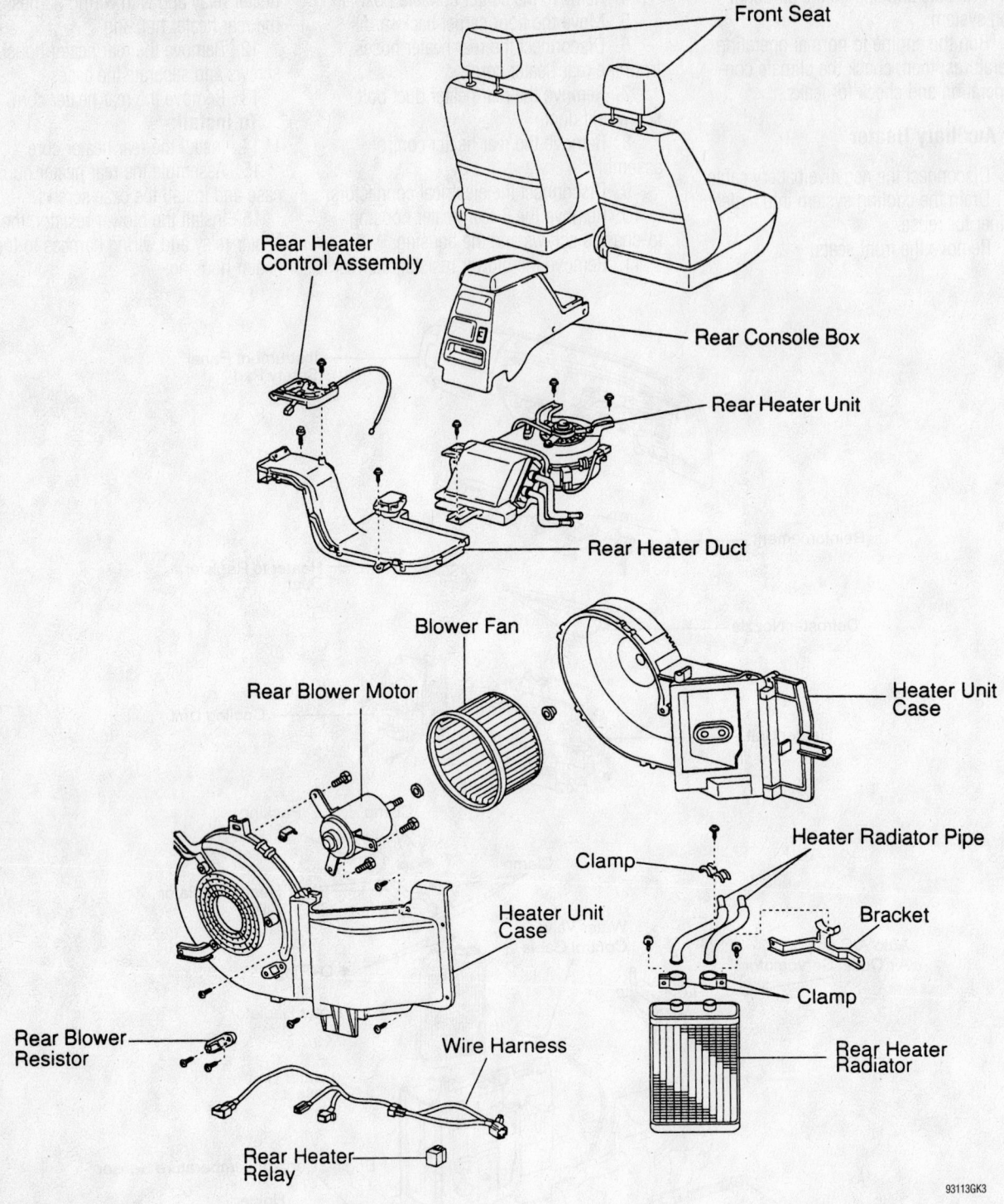

Front Seat

Rear Heater Control Assembly

Rear Console Box

Rear Heater Unit

Rear Heater Duct

Blower Fan

Rear Blower Motor

Heater Unit Case

Heater Unit Case

Heater Radiator Pipe

Clamp

Bracket

Clamp

Rear Blower Resistor

Wire Harness

Rear Heater Radiator

Rear Heater Relay

93113GK3

Exploded view of the rear heater housing and heater core—Toyota 4Runner

17. Install the rear heater housing and the 3 housing-to-chassis screws.
18. Connect the electrical connectors.
19. Install the rear heater control assembly.

20. Install the rear heater duct, bolt and screw.
21. Connect the rear heater hoses to the rear heater core.
22. Move the floor carpet foreword.
23. Install the center console box.

24. Install the front seats.
25. Refill the cooling system.
26. Connect the negative battery cable.
27. Run the engine to normal operating temperatures; then, check the climate control operation and check for leaks.

TIMING BELTS

3

TIMING BELTS

General Information

Timing belts are typically only used on overhead camshaft engines. Timing belts are used to synchronize the crankshaft with the camshaft, similar to a timing chain on an overhead valve (pushrod) engine. Unlike a timing belt, a timing chain will normally last the life of the engine without needing service or replacement. Timing belts use raised teeth to mesh with sprockets to operate the valve train of an overhead camshaft engine.

Whenever a vehicle with an unknown service history comes into your repair facility or is recently purchased, here are some points that should be asked to help prevent costly engine damage:

• Does the owner know if, or when the belt was replaced?

• If the vehicle purchased is used, or the condition and mileage of the last timing belt replacement are unknown, it is recommended to inspect, replace, or at least inform the owner that the vehicle is equipped with a timing belt.

• Note the mileage of the vehicle. The average replacement interval for a timing belt is approximately 60,000 miles (96,000 km).

Interference engines

Engines, chain- or belt-driven, can be classified as either free-running or interference, depending on what would happen if the piston-to-valve timing is disrupted. A free-running engine is designed with enough clearance between the pistons and valves to allow the crankshaft to rotate (pistons still moving) while the camshaft stays in one position (several valves fully open). If this condition occurs normally, no internal engine damage will result. In an interference engine, there is not enough clearance between the pistons and valves to allow the crankshaft to turn without the camshaft being in time.

An interference engine can suffer extensive internal damage if a timing belt fails. The piston design does not allow clearance for the valve to be fully open and the piston to be at the top of its stroke. If the belt fails, the piston will collide with the valve and will bend or break the valve, damage the piston, and/or bend a connecting rod. When this type of failure occurs, the engine will need to be replaced or disassembled for further internal inspection; either choice costing many times that of replacing the timing belt.

Timing Belt Service

INSPECTION

➡ For manufacturer's recommended service interval, refer to the maintenance interval chart located in this manual.

The average replacement interval for a timing belt is approximately 60,000 miles (96,000 km). If, however, the timing belt is inspected earlier or more frequently than suggested, and shows signs of wear or defects, the belt should be replaced at that time.

❈❈ WARNING

Never allow antifreeze, oil or solvents to come into with a timing belt. If this occurs immediately wash the solution from the timing belt. Also, never excessive bend or twist the timing belt; this can damage the belt so that its lifetime is severely shortened.

Inspect both sides of the timing belt. Replace the belt with a new one if any of the following conditions exist:

Never bend or twist a timing belt excessively, and do not allow solvents, antifreeze, gasoline, acid or oil to come into contact with the belt

Clean the timing belt before inspection so that imperfections or defects are easier to recognize

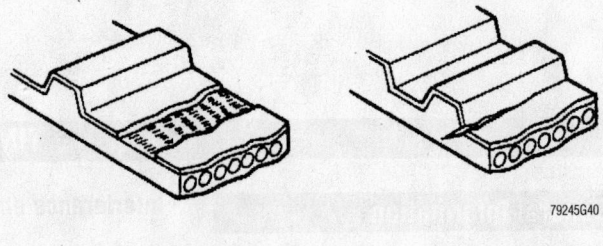

Inspect the timing belt for damage, such as a broken or missing tooth, which may be due to a damaged pulley

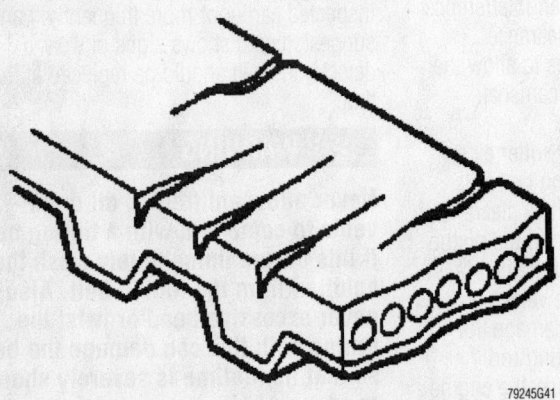

Back surface worn or cracked from a possible overheated engine or interference with the belt cover

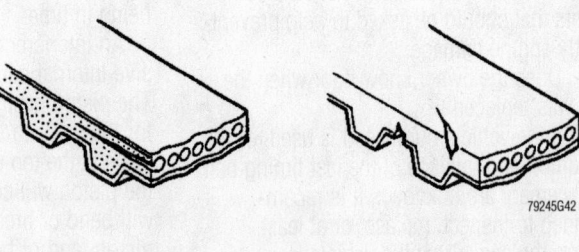

Side wear from improper installation or a defective pulley plate

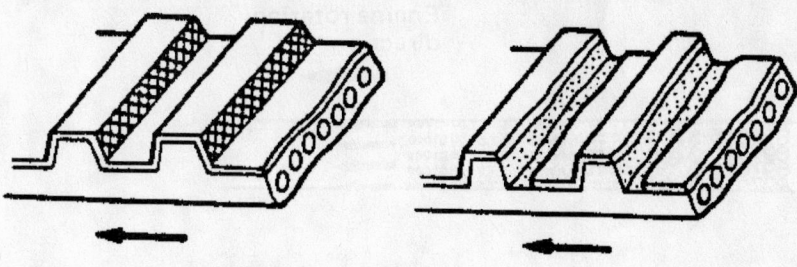

Rotating direction

79245G43

Worn teeth from excessive belt tension, camshaft or distributor not turning properly, or fluid leaking on the belt

- Hardening of the rubber — back side is glossy without resilience and leaves no indentation when pressed with a fingernail
- Cracks on the rubber backing
- Cracks or peeling of the canvas backing
- Cracks on rib root
- Cracks on belt sides
- Missing teeth or chunks of teeth
- Abnormal wear of belt sides — the sides are normal if they are sharp, as if cut by a knife.

If none of these conditions exist, the belt does not need replacement unless it is at the recommended interval. The belt MUST be replaced at the recommended interval.

❊❊ WARNING

On interference engines, it is very important to replace the timing belt at the recommended intervals, otherwise expensive engine damage will likely result if the belt fails.

REMOVAL & INSTALLATION

Acura SLX

3.2L AND 3.5L ENGINES

1. Disconnect the negative battery cable.
2. Drain the engine coolant into a sealable container.
3. Remove the air cleaner assembly and intake air duct.
4. Disconnect the upper radiator hose from the coolant inlet.
5. Remove the upper fan shroud from the radiator.
6. Remove the 4 nuts retaining the cooling fan assembly. Remove the cooling fan from the fan pulley.

7. Loosen and remove the drive belts.
8. Remove the upper timing belt covers.
9. Remove the fan pulley assembly.
10. Rotate the crankshaft to align the camshaft timing marks with the pointer dots on the back covers. Verify that the pointer on the crankshaft aligns with the mark on the lower timing cover.

➡**When the timing marks are aligned, the No. 2 piston is at Top Dead Center (TDC) of the compression.**

❊❊ WARNING

Align the camshaft and crankshaft sprockets with their alignment marks before removing the timing belt. Failure to align the belt and sprocket marks may result in valve damage.

11. Use tool No. J-8614-01 or a suitable pulley holding tool to remove the crankshaft pulley center bolt. Remove the crankshaft pulley.
12. If present, disconnect the 2 oil cooler hose bracket bolts on the timing cover. Move the oil cooler hoses and bracket off of the lower timing cover.
13. Remove the lower timing belt cover.
14. Remove the pusher assembly (tensioner) from below the belt tensioner pulley. The pusher rod must always face upward to prevent oil leakage. Depress the pusher rod, and insert a wire pin into the hole to keep the pusher rod retracted.
15. Remove the timing belt.
16. Inspect the water pump and replace it if there is any doubt about its condition.
17. Repair any oil or coolant leaks before installing a new timing belt. If the timing belt has been contaminated with oil or coolant, or is damaged, it must be replaced.

To install:
18. Verify that the sprocket timing marks are still aligned and that the groove and the keyway on the crankshaft timing sprocket align with the mark on the oil pump. The white pointers on the camshaft timing sprockets should align with the dots on the front plate.
19. Install the timing belt. Use clips to secure the belt onto each sprocket until the installation is complete. Align the dotted marks on the timing belt with the timing mark opposite the groove on the crankshaft sprocket.

➡**The arrows on the timing belt must follow the belt's direction of rotation. The manufacturer's trademark on the belt's spine should be readable left-to-right when the belt is installed.**

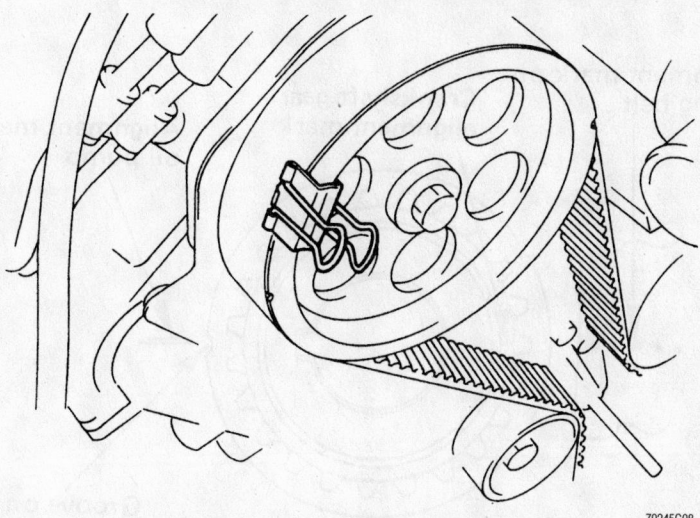

Using a double clip to hold the belt in place—Acura 3.2L and 3.5L engines

79245G08

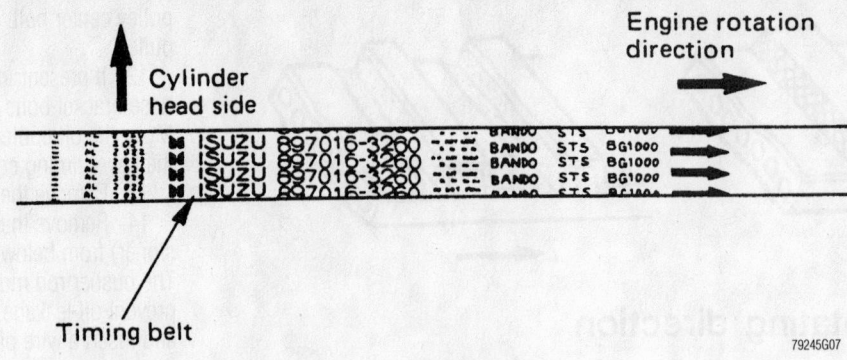

For maximum timing belt life, install the belt as shown—Acura 3.2L and 3.5L engines

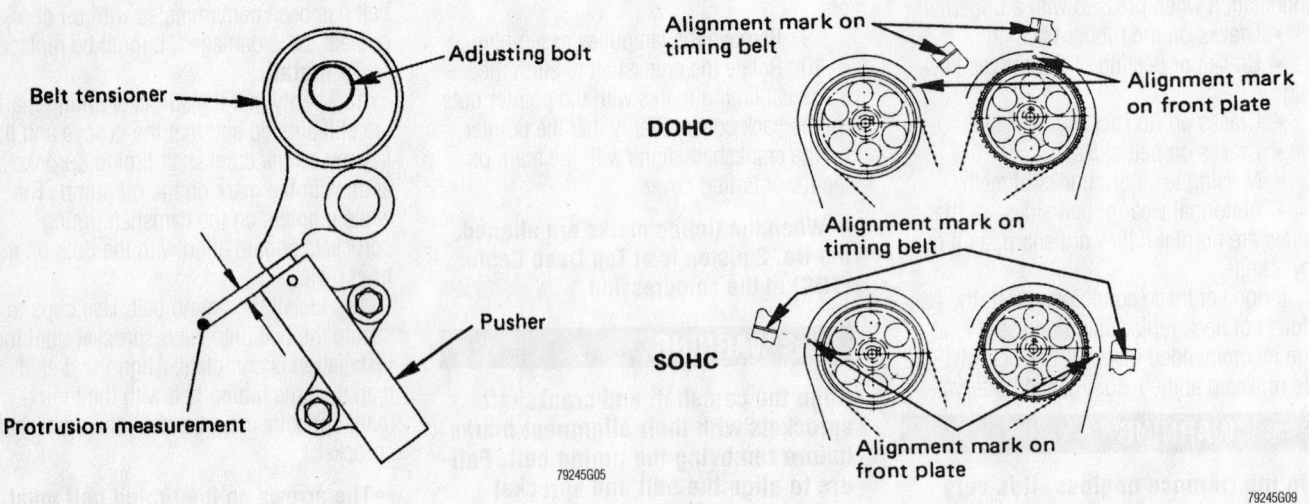

View of timing belt tensioner and pusher—Acura 3.2L and 3.5L engines

Proper camshaft alignment marks for timing belt installation—Acura 3.2L and 3.5L engines

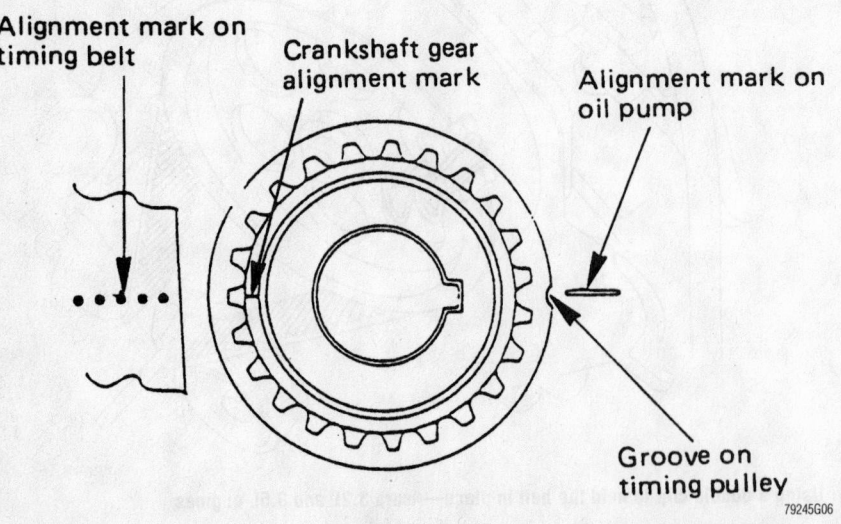

Proper crankshaft alignment marks for timing belt installation—Acura 3.2L and 3.5L engines

20. Align the white line on the timing belt with the alignment mark on the right bank camshaft timing pulley. Secure the belt with a clip.

✷✷ WARNING

If any binding is felt when adjusting the timing belt tension by turning the crankshaft, STOP turning the engine, because the pistons may be hitting the valves.

21. Rotate the crankshaft counterclockwise to remove the slack between the crankshaft sprocket and the right camshaft timing belt sprocket.
22. Install the belt around the water pump pulley.
23. Install the belt on the idler pulley.
24. Align the white alignment mark on

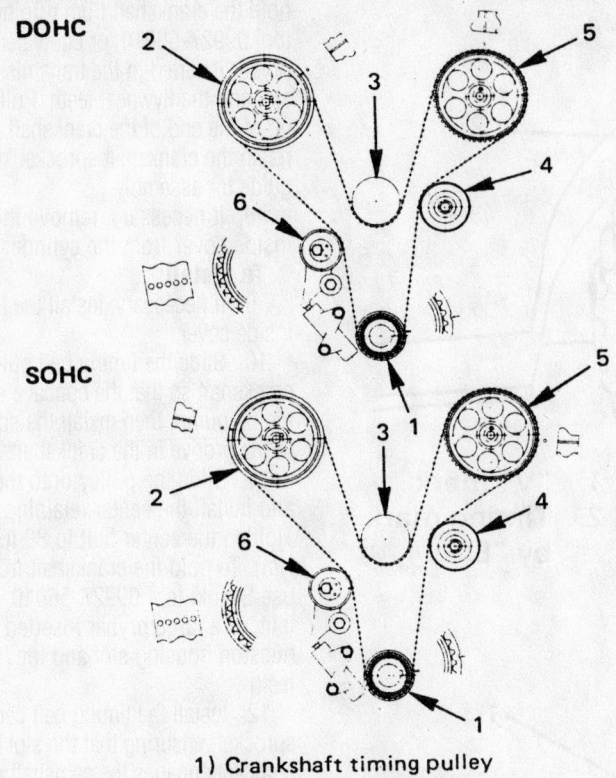

1) Crankshaft timing pulley
2) RH bank timing pulley
3) Water pump pulley
4) Idler pulley
5) LH bank timing pulley
6) Tension pulley

79245G10

Timing belt routing—Acura 3.2L and 3.5L engines

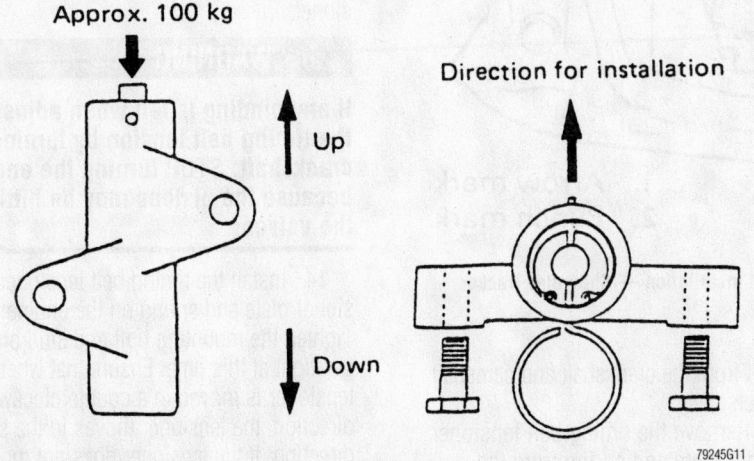

Approx. 100 kg

Up

Down

Direction for installation

79245G11

View of timing belt tensioner pusher—Acura 3.2L and 3.5L engines

the timing belt with the alignment mark on the left bank camshaft timing belt sprocket.

25. Install the crankshaft pulley and tighten the center bolt by hand. Rotate the crankshaft pulley clockwise to give slack between the crankshaft timing belt pulley and the right bank camshaft timing belt pulley.

26. Insert a 1.4mm piece of wire through the hole in the pusher to hold the rod in. Install the pusher assembly while pushing the tension pulley toward the belt.

27. Pull the pin out from the pusher to release the rod.

28. Remove the clamps from the sprockets. Rotate the crankshaft pulley clockwise 2 turns. Measure the rod protrusion to ensure it is between 0.16 – 0.24 in. (4 – 6mm).

29. If the tensioner pulley bracket pivot bolt was removed, tighten it to 31 ft. lbs. (42 Nm).

30. Tighten the pusher bolts to 14 ft. lbs. (19 Nm).

31. Remove the crankshaft pulley. Install the lower and upper timing belt covers and tighten their bolts to 12 ft. lbs. (17 Nm).

32. Fit the oil cooler hose onto the timing cover and tighten its mounting bracket bolts to 16 ft. lbs. (22 Nm).

33. Install the crankshaft pulley and tighten the pulley bolt to 123 ft. lbs. (167 Nm).

34. Install fan pulley assembly and tighten the bolts to 16 ft. lbs. (22 Nm).

35. Install and adjust the accessory drive belts.

36. Install the cooling fan assembly and tighten the bolts to 72 inch lbs. (8 Nm).

37. Install the upper fan shroud.

38. Install the air cleaner assembly and intake air duct.

39. Connect the upper radiator hose to the coolant inlet.

40. Refill and bleed the cooling system.

41. Connect the negative battery cable.

Chevrolet Tracker

1.6L 16-VALVE ENGINE

The 1.6L 16-valve engine is known as an interference motor, because it is fabricated with such close tolerances between the pistons and valves that, if the timing belt is incorrectly positioned, jumps teeth on one of the sprockets or breaks, the valve and pistons will come into contact. This can cause severe internal engine damage

➡ **Do not rotate the crankshaft counterclockwise or attempt to rotate the crankshaft by turning the camshaft sprocket.**

1. Remove the timing belt cover.

2. If the timing belt is not already marked with a directional arrow, use white paint, a grease pencil or correction fluid to do so.

3. Rotate the crankshaft clockwise until the timing mark on the camshaft sprocket and the V mark on the timing belt inside cover are aligned, and the punch mark on

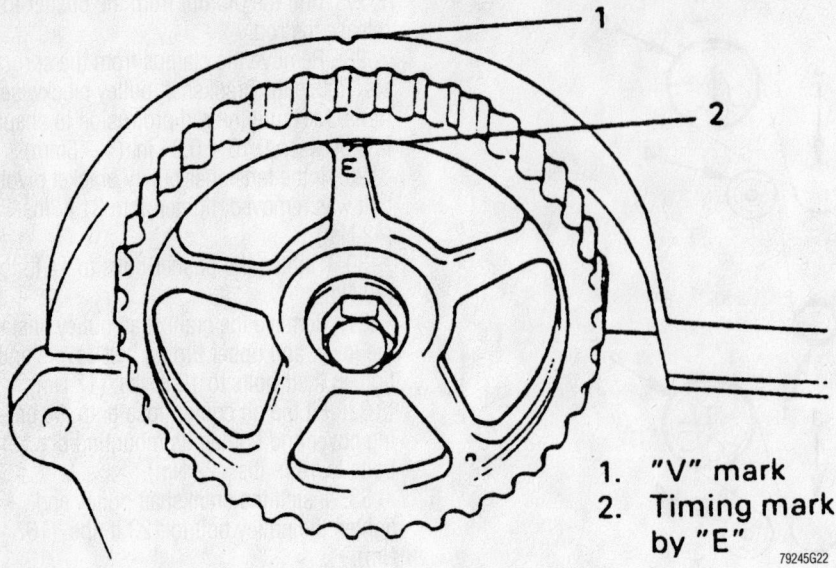

1. "V" mark
2. Timing mark by "E"

Camshaft timing marks — Chevrolet Tracker 1.6L 16-valve engine

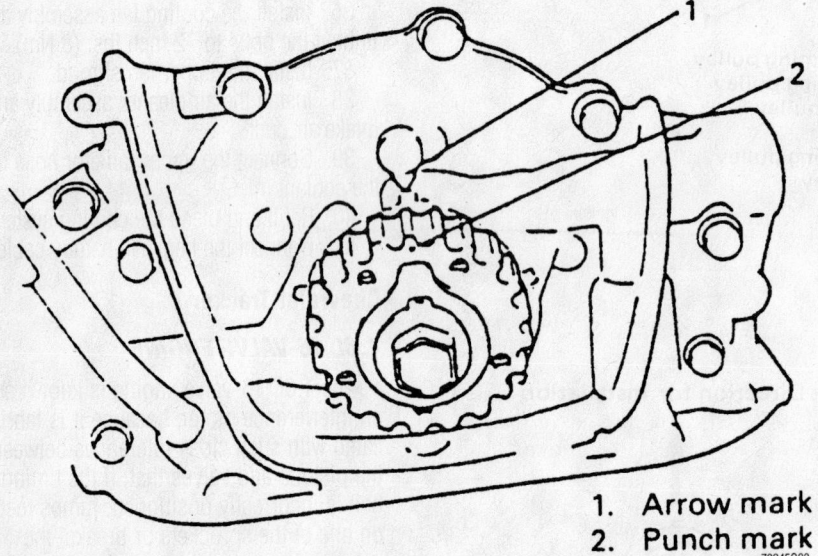

1. Arrow mark
2. Punch mark

Align the punch mark with the arrow for proper timing belt installation — Chevrolet Tracker 1.6L 16-valve engine

the crankshaft sprocket is aligned with the mark on the engine.

> ❊❊ **WARNING**
>
> **Do not rotate the crankshaft or camshaft once the timing belt is removed, because the valves and pistons can come into contact, which may cause internal engine damage.**

4. Disconnect one end of the tensioner spring. Loosen the timing belt tensioner bolt and stud, then, using your finger, press the tensioner plate up and remove the timing belt from the crankshaft and camshaft sprockets.

5. Remove the timing belt tensioner, tensioner plate and spring from the engine.

6. Install Suzuki tool 09917-68220, or equivalent, onto the camshaft sprocket to hold the camshaft from rotating. Loosen the camshaft sprocket retaining bolt, then pull the camshaft sprocket off of the end of the camshaft.

7. Remove the crankshaft timing belt sprocket by loosening the center bolt, while preventing the crankshaft from rotating. To hold the crankshaft from turning, use Suzuki tool 09927-56010, or equivalent, or a large prybar inserted in the transmission housing slot and the flywheel teeth. Pull the sprocket off of the end of the crankshaft. Be sure to retain the crankshaft sprocket key and belt guide for assembly.

8. If necessary, remove the timing belt inside cover from the cylinder head.

To install:

9. If necessary, install the timing belt inside cover.

10. Slide the timing belt guide on the crankshaft so that the concave side faces the oil pump, then install the sprocket key in the groove in the crankshaft.

11. Slide the pulley onto the crankshaft, and install the center retaining bolt. Tighten the center bolt to 80 ft. lbs. (110 Nm). To hold the crankshaft from turning, use Suzuki tool 09927-56010, or equivalent, or a large prybar inserted in the transmission housing slot and the flywheel teeth.

12. Install the timing belt camshaft sprocket, ensuring that the slot in the sprocket engages the camshaft (pulley) pin; this ensures that the sprocket is properly positioned on the end of the camshaft. Secure the camshaft with the holding tool used during removal, then tighten the sprocket bolt to 44 ft. lbs. (60 Nm).

13. Assemble the timing belt tensioner plate and the tensioner, making sure that the lug of the tensioner plate engages the tensioner.

> ❊❊ **WARNING**
>
> **If any binding is felt when adjusting the timing belt tension by turning the crankshaft, STOP turning the engine, because the pistons may be hitting the valves.**

14. Install the timing belt tensioner, tensioner plate and spring on the engine. Tighten the mounting bolt and stud only finger-tight at this time. Ensure that when the tensioner is moved in a counterclockwise direction, the tensioner moves in the same direction. If the tensioner does not move, remove it and the tensioner plate to reassemble them properly.

15. Loosen all rocker arm valve lash locknuts and adjusting screws. This will permit movement of the camshaft without any rocker arm associated drag, which is essential for proper timing belt tensioning. If the camshaft does not rotate freely (free of rocker arm drag), the belt will not be properly tensioned.

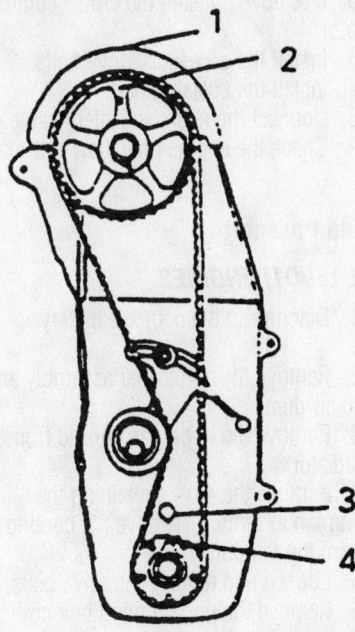

1. "V" mark on cylinder head cover
2. Timing mark by "E" on camshaft timing belt pulley
3. Arrow mark on oil pump case
4. Punch mark on crankshaft timing belt pulley

79245G47

Rotate the crankshaft clockwise until the camshaft and crankshaft timing marks are aligned — Chevrolet Tracker 1.6L 16-valve engine

16. Rotate the camshaft sprocket clockwise until the timing mark on the sprocket and the V mark on the timing belt inside cover are aligned.

17. Using a wrench, or socket and breaker bar, on the crankshaft sprocket center bolt, turn the crankshaft clockwise until the punch mark on the sprocket is aligned with the arrow mark on the oil pump.

18. With the camshaft and crankshaft marks properly aligned, push the tensioner up with your finger and install the timing belt on the 2 sprockets, ensuring that the drive side of the belt is free of all slack. Release your finger from the tensioner. Be sure to install the timing belt so that the directional arrow is pointing in the appropriate direction.

➡ In this position, the No. 4 cylinder is at Top Dead Center (TDC) on the compression stroke.

19. Rotate the crankshaft clockwise 2 full revolutions, then tighten the tensioner stud to 97 inch lbs. (11 Nm). Then, tighten the tensioner bolt to 18 ft. lbs. (24 Nm).

20. Ensure that all 4 timing marks are still aligned as before; if they are not, remove the timing belt, and install and tension it again.

21. Install the timing belt cover and all related components.

Honda CR-V

2.0L (B20B4) ENGINE

1. Disconnect the negative battery cable.

2. Position crankshaft so that No. 1 piston is at Top Dead Center (TDC).

3. Remove the splash guard.

4. Remove the accessory drive belts.

5. If equipped, remove the cruise control actuator.

6. Place a piece of wood between the oil pan and the jack, support the engine with a jack.

7. Remove upper engine bracket.

8. Remove the valve cover.

9. Remove the timing belt covers.

10. Loosen the adjusting bolt 180 degrees. Release the tension from the belt by pushing on the tensioner, then retighten the adjusting bolt.

11. Remove the timing belt.

To install:

12. Be sure the timing marks are properly aligned.

13. Install the timing belt on the pulleys following this sequence:

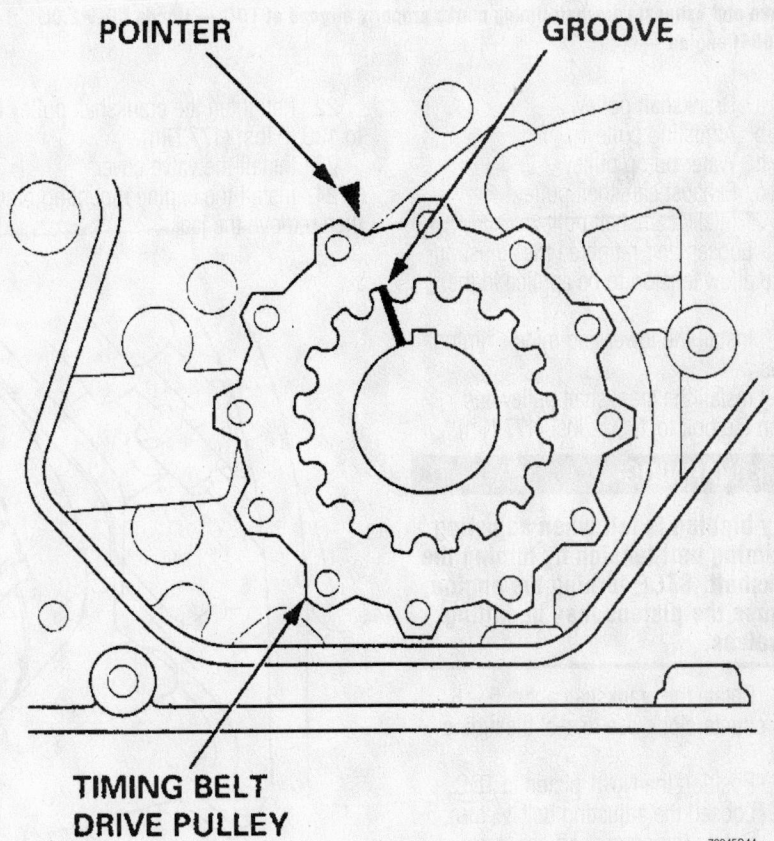

POINTER   GROOVE

TIMING BELT DRIVE PULLEY

79245G44

Crankshaft timing mark will be easier to verify when clean — Honda CR-V 2.0L (B20B4) engine

Timing belt service is covered in Section 3 of this manual

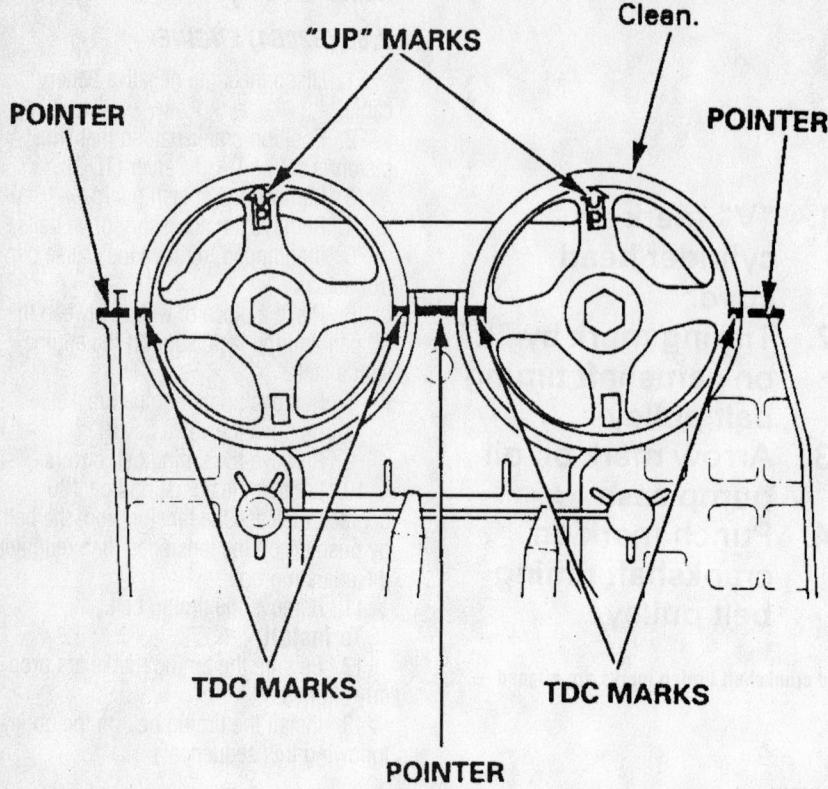

Intake and exhaust camshaft timing marks properly aligned at TDC — Honda CR-V 2.0L (B20B4) engine

a. Crankshaft pulley.
b. Adjusting pulley.
c. Water pump pulley.
d. Exhaust camshaft pulley.
e. Intake camshaft pulley.

14. Loosen and retighten the adjusting bolt to allow tension to be applied to the belt.

15. Install the lower and middle timing covers.

16. Install the crankshaft pulley and tighten the bolt to 130 ft. lbs. (177 Nm).

✳✳ WARNING

If any binding is felt when adjusting the timing belt tension by turning the crankshaft, STOP turning the engine, because the pistons may be hitting the valves.

17. Rotate the crankshaft about 5 – 6 times counterclockwise to seat the timing belt.

18. Position the No. 1 piston to TDC.

19. Loosen the adjusting bolt ½ turn.

20. Rotate the crankshaft counterclockwise 3 teeth on the camshaft pulley.

21. Tighten the adjusting bolt to 40 ft. lbs. (54Nm).

22. Retighten the crankshaft pulley bolt to 130 ft. lbs. (177 Nm).

23. Install the valve cover.

24. Install the engine mounting bracket, then remove the jack.

25. If removed, install the cruise control actuator.

26. Install the accessory drive belts.

27. Install the splash guard.

28. Connect the negative battery cable.

29. Check the engine operation and road test.

Honda Passport

3.2L (6VD1) ENGINES

1. Disconnect the negative battery cable.

2. Remove the air cleaner assembly and intake air duct.

3. Remove the upper fan shroud from the radiator.

4. Remove the 4 nuts retaining the cooling fan assembly. Remove the cooling fan from the fan pulley.

5. Loosen and remove the drive belts.

6. Remove the upper timing belt covers.

7. Remove the fan pulley assembly.

8. Rotate the crankshaft to align the camshaft timing marks with the pointer dots on the back covers. Verify that the pointer on the crankshaft aligns with the mark on the lower timing cover.

➡**When the timing marks are aligned, the No. 2 piston is at Top Dead Center (TDC) of the compression.**

✳✳ WARNING

Align the camshaft and crankshaft sprockets with their alignment

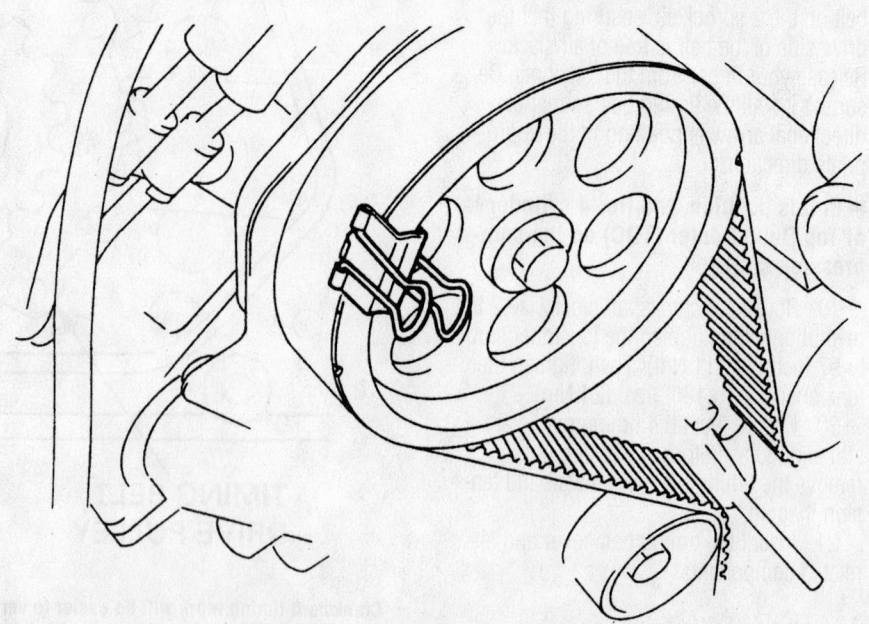

Using a double clip to hold the belt in place — Honda Passport 3.2L engine

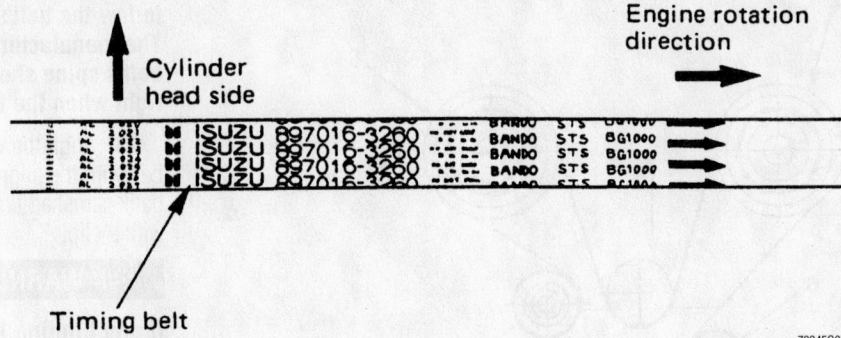

For maximum timing belt life, install the belt as shown — Honda Passport 3.2L engine

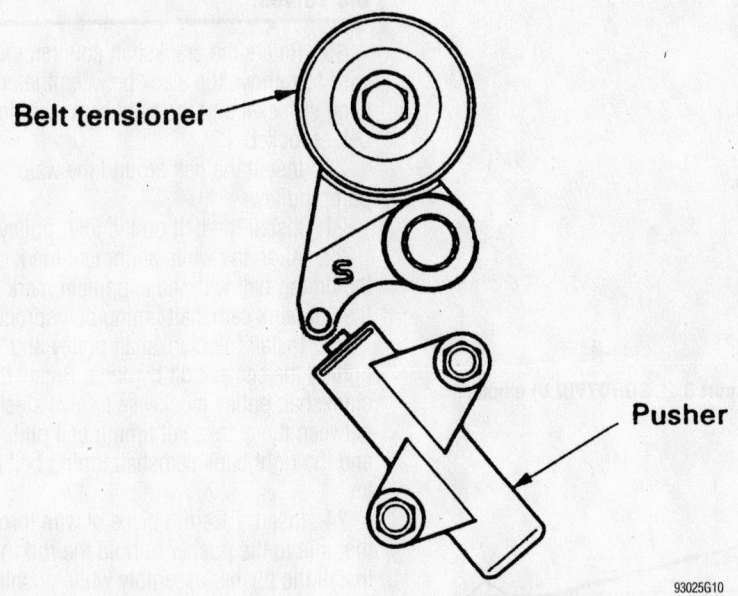

View of timing belt tensioner and pusher — Honda Passport 3.2L engine

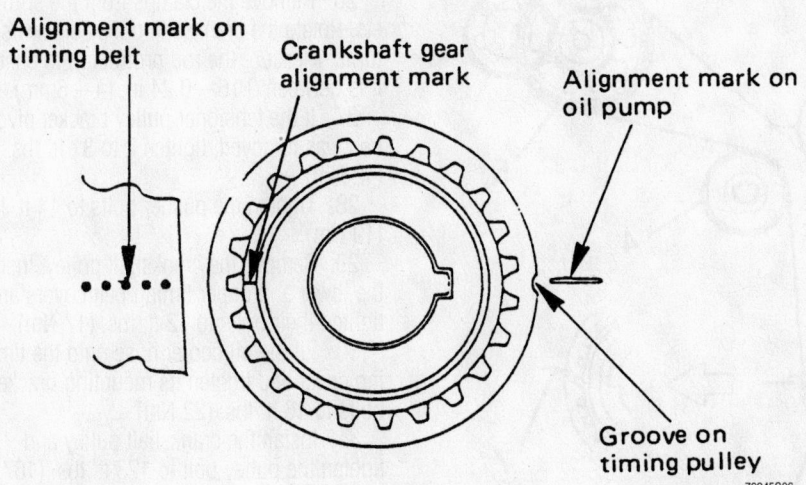

Proper crankshaft alignment marks for timing belt installation — Honda Passport 3.2L SOHC (VIN V) engine

marks before removing the timing belt. Failure to align the belt and sprocket marks may result in valve damage.

9. Use tool No. J-8614-01 or a suitable pulley holding tool to remove the crankshaft pulley center bolt. Remove the crankshaft pulley.

10. If present, disconnect the 2 oil cooler hose bracket bolts on the timing cover. Move the oil cooler hoses and bracket off of the lower timing cover.

11. Remove the lower timing belt cover.

12. Remove the pusher assembly (tensioner) from below the belt tensioner pulley. The pusher rod must always face upward to prevent oil leakage. Depress the pusher rod and insert a wire pin into the hole to keep the pusher rod retracted.

13. Remove the timing belt.

14. Inspect the water pump and replace it if there is any doubt about its condition.

15. Repair any oil or coolant leaks before installing a new timing belt. If the timing belt has been contaminated with oil or coolant, or is damaged, it must be replaced.

To install:

16. Verify that the sprocket timing marks are still aligned and that the groove and the keyway on the crankshaft timing sprocket align with the mark on the oil pump. The white pointers on the camshaft timing sprockets should align with the dots on the front plate.

17. Install the timing belt. Use clips to secure the belt onto each sprocket until the installation is complete. Align the dotted marks on the timing belt with the timing mark opposite the groove on the crankshaft sprocket.

Heater Core replacement is covered in Section 2 of this manual

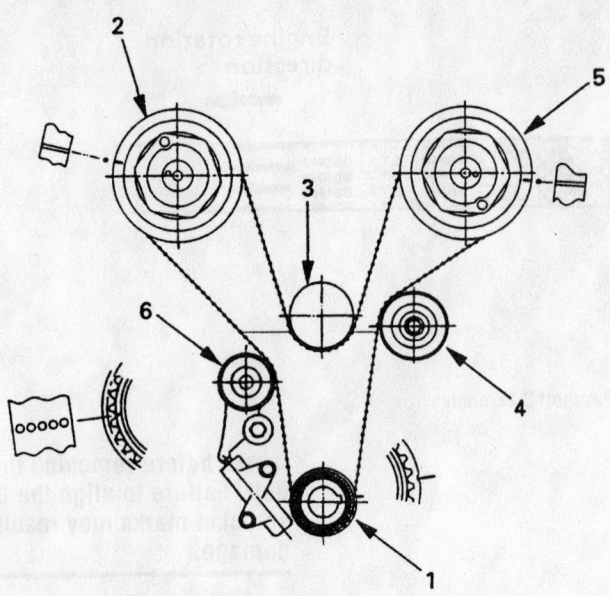

1) Crankshaft timing pulley
2) RH bank timing pulley
3) Water pump pulley
4) Idler pulley
5) LH bank timing pulley
6) Tension pulley

93025G09

Timing mark alignment and timing belt routing — Honda Passport 3.2L SOHC (VIN V) engine

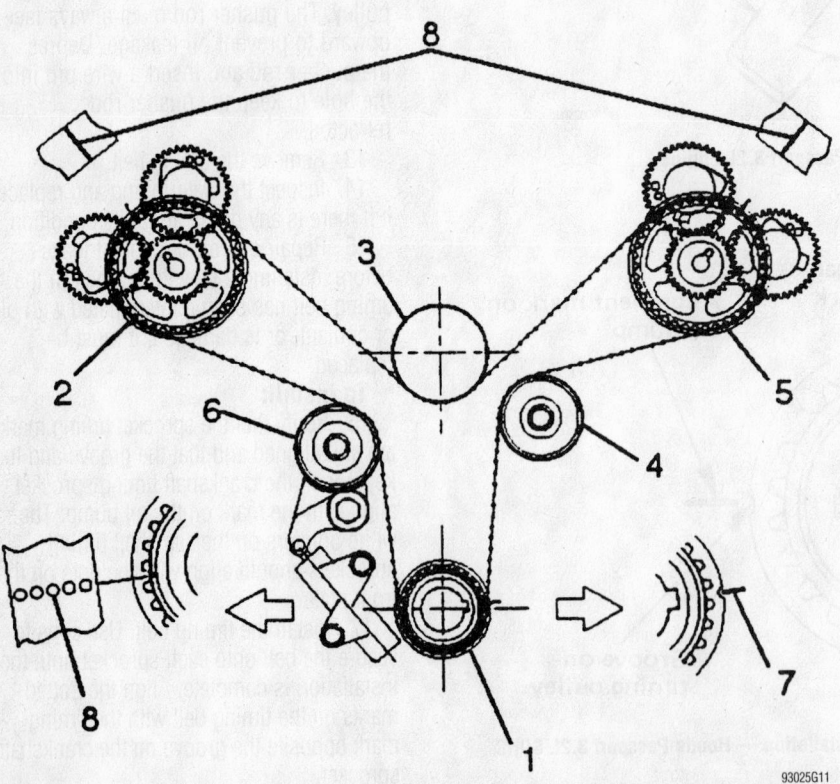

93025G11

Timing mark alignment and timing belt routing — Honda Passport 3.2L DOHC (VIN W) engine

➡The arrows on the timing belt must follow the belt's direction of rotation. The manufacturer's trademark on the belt's spine should be readable left-to-right when the belt is installed.

18. Align the white line on the timing belt with the alignment mark on the right bank camshaft timing pulley. Secure the belt with a clip.

✳✳ WARNING

If any binding is felt when adjusting the timing belt tension by turning the crankshaft, STOP turning the engine, because the pistons may be hitting the valves.

19. Rotate the crankshaft counterclockwise to remove the slack between the crankshaft sprocket and the right camshaft timing belt sprocket.

20. Install the belt around the water pump pulley.

21. Install the belt on the idler pulley.

22. Align the white alignment mark on the timing belt with the alignment mark on the left bank camshaft timing belt sprocket.

23. Install the crankshaft pulley and tighten the center bolt by hand. Rotate the crankshaft pulley clockwise to give slack between the crankshaft timing belt pulley and the right bank camshaft timing belt pulley.

24. Insert a 1.4mm piece of wire through the hole in the pusher to hold the rod in. Install the pusher assembly while pushing the tension pulley toward the belt.

25. Pull the pin out from the pusher to release the rod.

26. Remove the clamps from the sprockets. Rotate the crankshaft pulley clockwise 2 turns. Measure the rod protrusion to ensure it is between 0.16 – 0.24 in. (4 – 6mm).

27. If the tensioner pulley bracket pivot bolt was removed, tighten it to 31 ft. lbs. (42 Nm).

28. Tighten the pusher bolts to 14 ft. lbs. (19 Nm).

29. Remove the crankshaft pulley. Install the lower and upper timing belt covers and tighten their bolts to 12 ft. lbs. (17 Nm).

30. Fit the oil cooler hose onto the timing cover and tighten its mounting bracket bolts to 16 ft. lbs. (22 Nm).

31. Install the crankshaft pulley and tighten the pulley bolt to 123 ft. lbs. (167 Nm).

32. Install fan pulley assembly and tighten the bolts to 16 ft. lbs. (22 Nm).

33. Install and adjust the accessory drive belts.

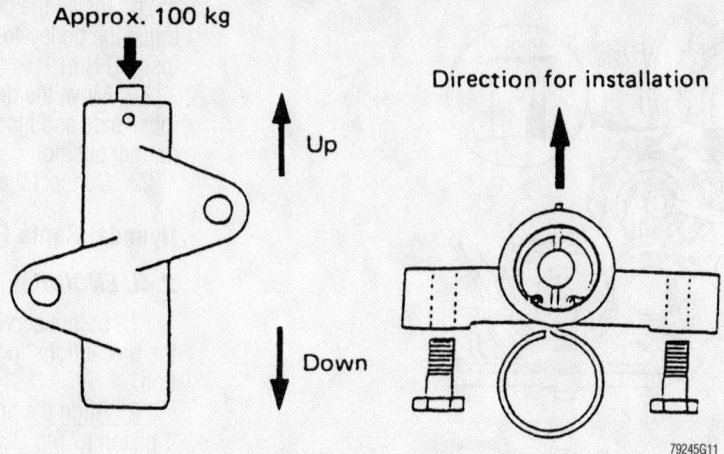

Approx. 100 kg

Up

Down

Direction for installation

79245G11

View of timing belt tensioner pusher — Honda Passport 3.2L engine

34. Install the cooling fan assembly and tighten the bolts to 72 inch lbs. (8 Nm).

35. Install the upper fan shroud.

36. Install the air cleaner assembly and intake air duct.

37. Connect the negative battery cable.

2.2L (X22SE/D) ENGINE

1. Disconnect the negative battery cable.

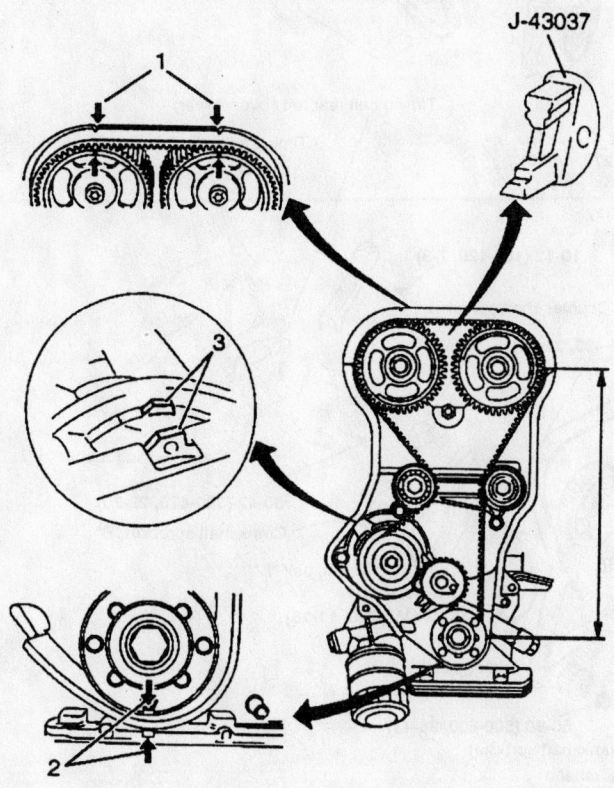

J-43037

93025G12

Aligning the timing marks and installing the timing belt — Honda Passport 2.2L (X22SE/D) engine

2. Using a box-end wrench on the drive belt adjuster, turn the adjuster clockwise and remove the drive belt.

3. From the left rear of the engine compartment, disconnect the 3 electrical connectors from the chassis harness.

4. Remove the crankshaft pulley-to-crankshaft bolts and remove the pulley.

5. From the front of the engine, remove the nut and the engine harness cover.

6. Remove the timing belt cover.

7. Rotate the crankshaft to position the timing marks at Top Dead Center (TDC) of the No. 1 cylinder's compression stroke.

➡**Mark the rotational direction of the timing belt for reinstallation purposes.**

8. Remove the timing belt tensioner adjusting bolt and the tensioner from the engine.

9. Remove the timing belt.

To install:

10. Install the timing belt tensioner and finger-tighten the tensioner bolt.

11. Inspect the timing marks to be sure that the engine is positioned at TDC of the No. 1 cylinder's compression stroke.

12. Using tool J-43037, or equivalent, place it between the intake and exhaust sprockets to prevent the camshaft gear from moving during the timing belt installation.

13. Install the timing belt.

14. Position the timing belt to ensure that the tension side of the belt is taut and move the timing belt tension adjusting lever clockwise until the tensioner pointer is flowing.

15. If installing a used timing belt (used over 60 min. from new), the pointer should be positioned approximately 0.16 in. (4mm) to the left of the "V" notch when viewed from the front of the engine.

16. If installing a new timing belt, the pointer should be positioned at the center of the "V" notch when viewed from the front of the engine.

17. Torque the timing belt tensioner adjusting bolt to 18 ft. lbs. (25 Nm).

18. Install the timing belt front cover and torque the bolts to 53 inch lbs. (6 Nm).

19. Install the engine harness connectors.

Brake service is covered in Section 4 of this manual

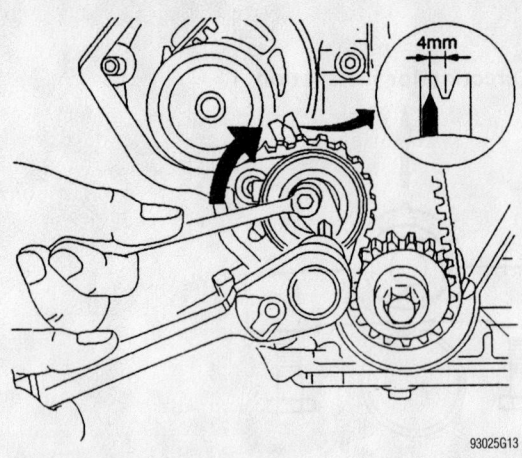

Tensioning the timing belt for a used timing belt — Honda Passport 2.2L (X22SE/D) engine

20. Install the crankshaft pulley and toque the pulley-to-crankshaft bolts to 14 ft. lbs. (20 Nm).

21. Move the drive belt tensioner to the loose side and install the drive belt to its normal position.

22. Connect the negative battery cable.

Hyundai Santa Fe

2.4L ENGINE

1. Before servicing the vehicle, refer to the precautions in the beginning of this section.

2. Align the timing marks to set the No. 1 piston to Top Dead Center (TDC) by rotating the crankshaft clockwise. The timing

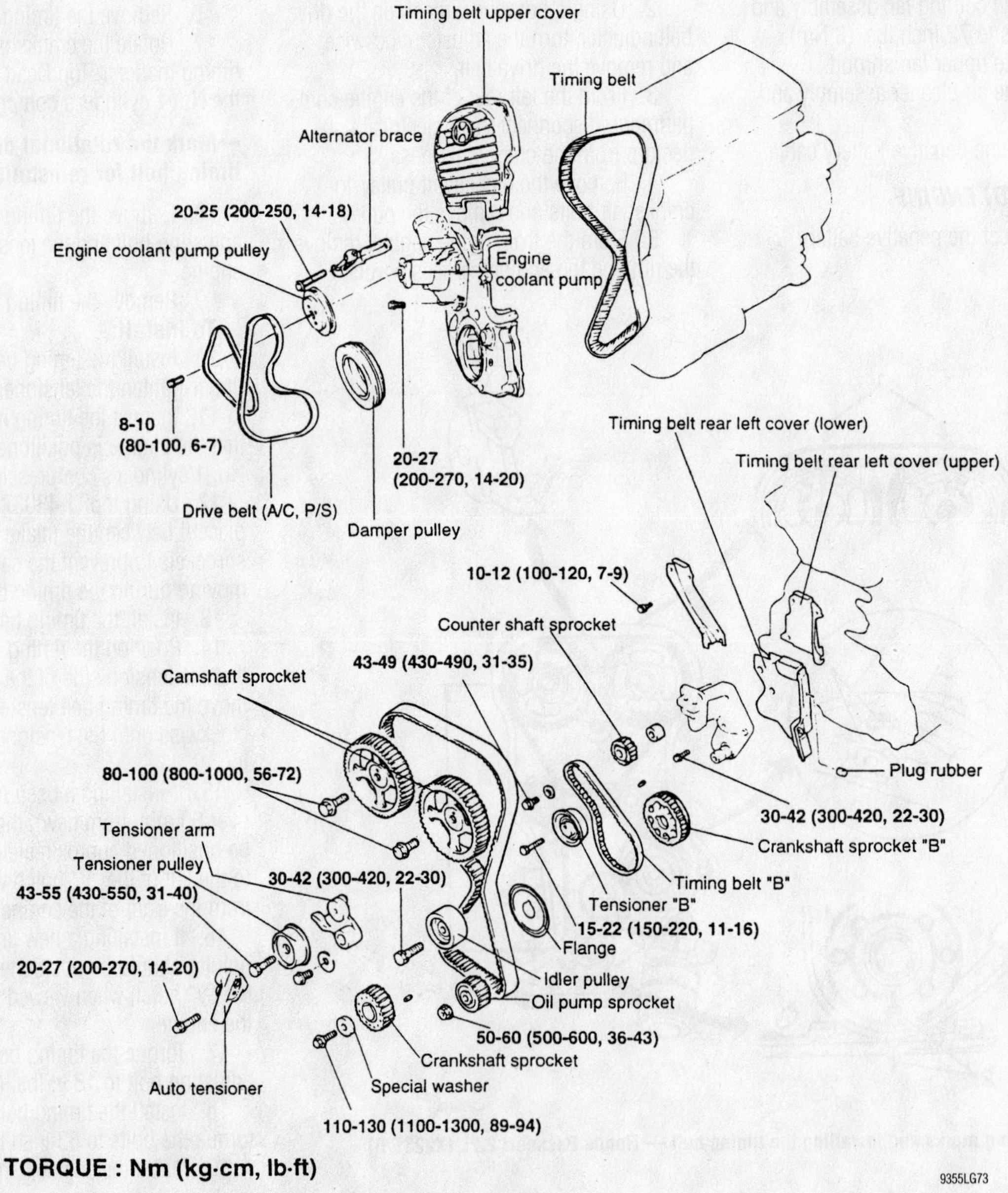

Timing belt upper cover

Timing belt

Alternator brace

20-25 (200-250, 14-18)

Engine coolant pump pulley

Engine coolant pump

8-10 (80-100, 6-7)

20-27 (200-270, 14-20)

Drive belt (A/C, P/S)

Damper pulley

Timing belt rear left cover (lower)

Timing belt rear left cover (upper)

10-12 (100-120, 7-9)

Counter shaft sprocket

43-49 (430-490, 31-35)

Camshaft sprocket

80-100 (800-1000, 56-72)

Tensioner arm

Tensioner pulley

43-55 (430-550, 31-40)

30-42 (300-420, 22-30)

20-27 (200-270, 14-20)

Auto tensioner

Special washer

Crankshaft sprocket

110-130 (1100-1300, 89-94)

Plug rubber

30-42 (300-420, 22-30)

Crankshaft sprocket "B"

Timing belt "B"

Tensioner "B"

15-22 (150-220, 11-16)

Flange

Idler pulley

Oil pump sprocket

50-60 (500-600, 36-43)

TORQUE : Nm (kg·cm, lb·ft)

Exploded view of the timing belt assembly and related components — 2.4L engine

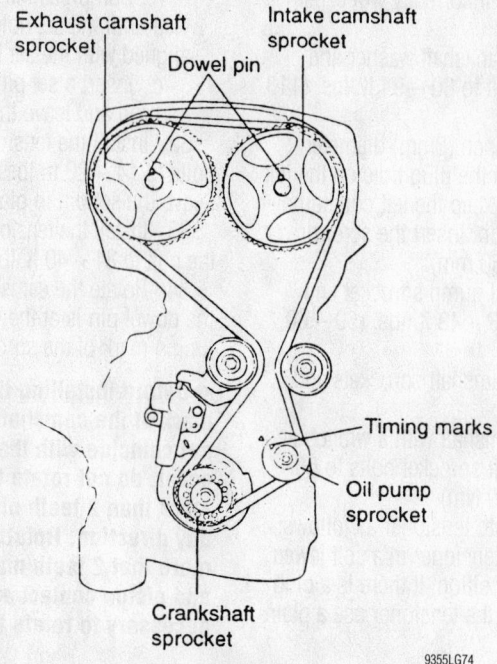

Correct sprocket alignment when the belt is installed — 2.4L engine

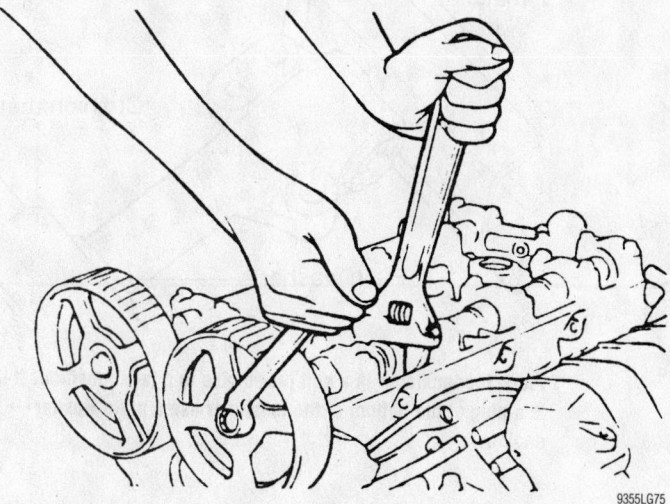

Hold the camshaft with a wrench and loosen the camshaft sprocket bolts — 2.4L engine

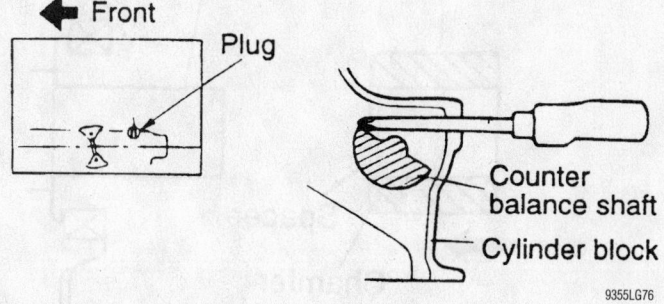

Insert a 0.3 inch (8mm) diameter screwdriver to keep the left counterbalance shaft in position — 2.4L engine

marks of the camshaft sprocket and the cylinder head cover should be aligned and the dowel pin of the camshaft sprocket should be at the upper side.

3. Remove the crankshaft pulley, water pump pulley and drive Belt.

4. Remove the timing belt cover.

5. Remove the auto tensioner.

6. Mark the timing belt is being reused, mark an arrow on the belt noting the direction of rotation or the front of the engine to make sure the belt is reinstalled in its original position.

7. Remove the timing belt.

8. Hold the camshaft with a wrench and loosen the camshaft sprocket bolts.

9. Remove the sprockets.

10. When removing the oil pump socket nut, first remove the plug at the side of the block and insert a 0.3 inch (8mm) diameter screwdriver to keep the left counterbalance shaft in position. Insert the screwdriver at least 2.36 inch (60 mm).

11. Remove the oil pump sprocket nut and the sprocket.

12. Loosen the right counterbalance shaft sprocket bolt until you can loosen it by hand.

13. Remove the tensioner **B** and timing belt **B**. Refer to the accompanying illustration for tensioner and belt identification.

✳✳ CAUTION

Do not attempt to loosen bolts while holding the sprocket with pliers or any tool after removing timing belt B

14. Remove the crankshaft sprocket **B** from the crankshaft.
   To install:
   15. Install the crankshaft sprocket **B** to the crankshaft.

✳✳ CAUTION

Pay attention to the direction of the flange. If it is installed in the wrong direction, the belt will break.

16. Apply engine oil to the outer surface of the spacer lightly and install the spacer to the right counterbalance shaft. Be sure to install spacer correctly.

17. Install the counterbalance shaft sprocket onto the right counterbalance shaft and then tighten the flange bolt by hand until it is tight.

18. Align the timing mark on each sprocket with its corresponding timing mark on the front case.

For complete Engine Mechanical specifications, see Section 1 of this manual

19. Install timing belt **B** and make sure there is no slack.

20. Install tensioner **B** so that the center of the pulley is located on the left side of the mounting bolt and the pulley flange faces the front of the engine.

21. Align the timing mark on the right counterbalance shaft sprocket with the timing mark on the front case.

22. Lift the tensioner **B** to tighten tensioner **B** so that its tension side is pulled tight. Tighten the bolt on tensioner **B**. As the bolt is being tightened, make sure the shaft does not turn. If the shaft turns, the belt will be overtightened.

23. Make sure the timing marks are aligned.

24. Check the belt tension by depressing the center of the belt span with an index finger. The deflection should be 0.20 – 0.28 inch (5 – 7mm).

25. Install the flange and crankshaft sprocket making sure it is installed properly.

Installing the flange incorrectly will cause the belt to break.

26. Install the crankshaft washer and bolt. Tighten the bolt to 80 – 94 ft. lbs. (110 – 130 Nm).

27. Insert a 0.3 inch (8mm) diameter screwdriver through the plug hole on the left side of the block to keep the left counterbalance shaft in position. Insert the screwdriver at least 2.36 inch (60 mm).

28. Install the oil pump sprocket and tighten the nut to 36 – 43 ft. lbs. (50 – 60 Nm)

29. Install the camshaft sprockets and the bolts.

30. Hold the camshaft with a wrench and tighten the camshaft sprocket bolts to 56 – 72 ft. lbs. (80 – 100 Nm).

31. Reset the auto tensioner as follows:

   a. Place the tensioner in a soft jawed vice in a level position. If there is a plug at the bottom of the tensioner use a plain washer.

   b. Compress the rod slowly using the vice until the set hole in the rod is aligned with the set hole on the cylinder.

   c. Insert a set pin through the body and rod and leave the pin installed.

32. Install the tensioner and tighten the bolts to 14 – 20 ft. lbs. (20 – 27 Nm). Leave the set pin in place.

33. Install the tensioner pulley and tighten the bolt to 31 – 40 ft. lbs. (43 – 55 Nm).

34. Rotate the camshaft sprockets so that the dowel pin is at the upper side. Set the timing mark of the sprocket correctly.

➡ **Before installing the belt, the timing mark of the camshaft sprocket doers not coincide with that of the rocker cover, do not rotate the cam sprocket more than 2 teeth of the sprocket in any direction. Rotating the sprocket more that 2 teeth may make the valve and piston contact each other. If it is necessary to rotate the sprocket more**

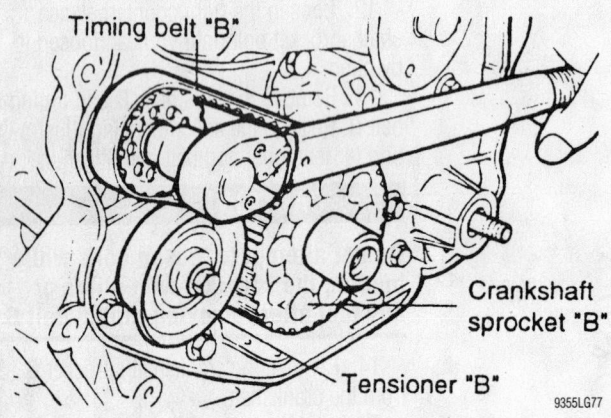

Remove tensioner B and timing belt B — 2.4L engine

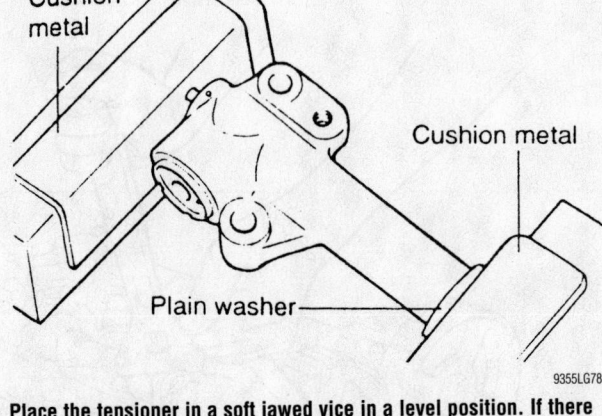

Place the tensioner in a soft jawed vice in a level position. If there is a plug at the bottom of the tensioner use a plain washer — 2.4L engine

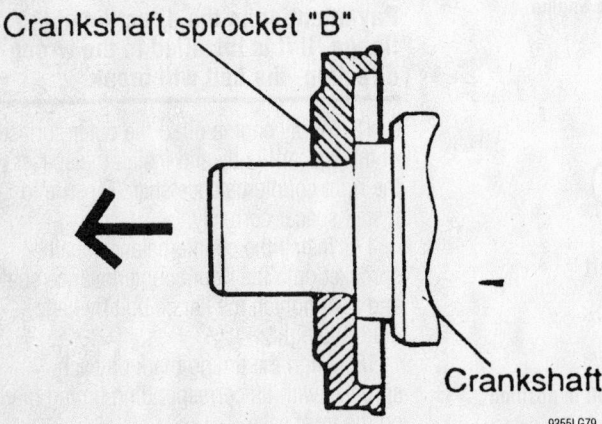

Pay attention to the direction of the flange, if it is installed in the wrong direction, the belt will break — 2.4L engine

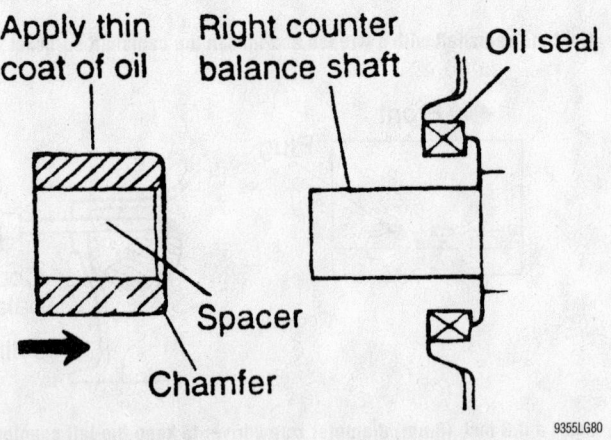

Apply engine oil to the outer surface of the spacer lightly and install the spacer to the right counterbalance shaft. Be sure to install spacer correctly — 2.4L engine

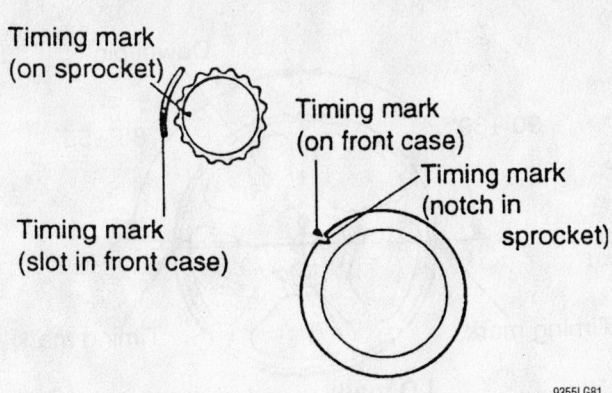

Align the timing mark on each sprocket with its corresponding timing mark on the front case — 2.4L engine

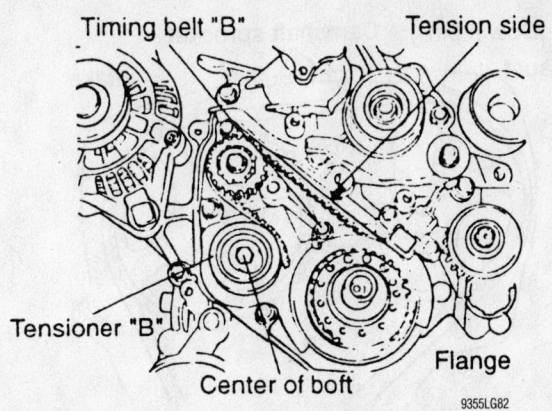

Align the timing mark on the right counterbalance shaft sprocket with the timing mark on the front case — 2.4L engine

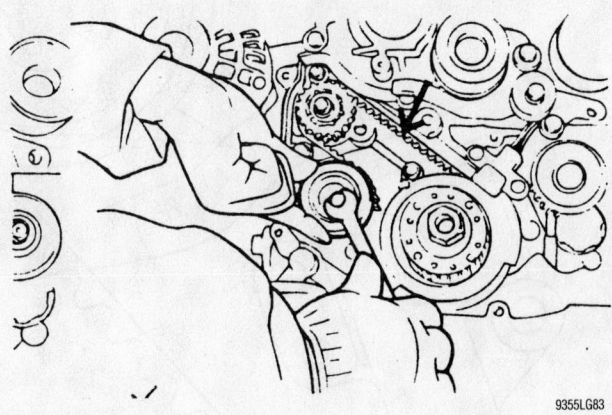

Lift the tensioner B to tighten tensioner B so that its tension side is pulled tight — 2.4L engine

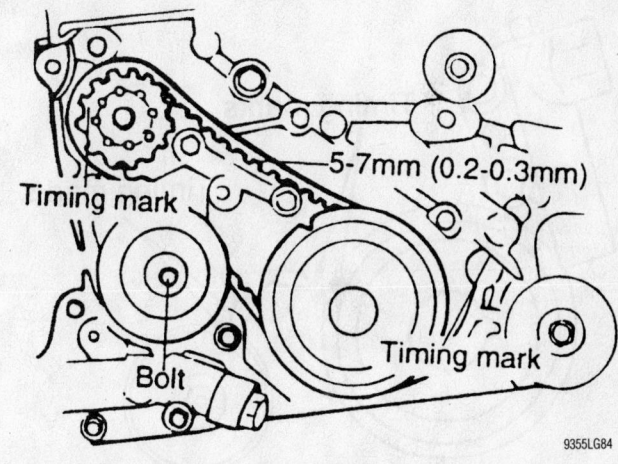

Make sure the timing belt B marks are aligned and the belt tensioner is correct — 2.4L engine

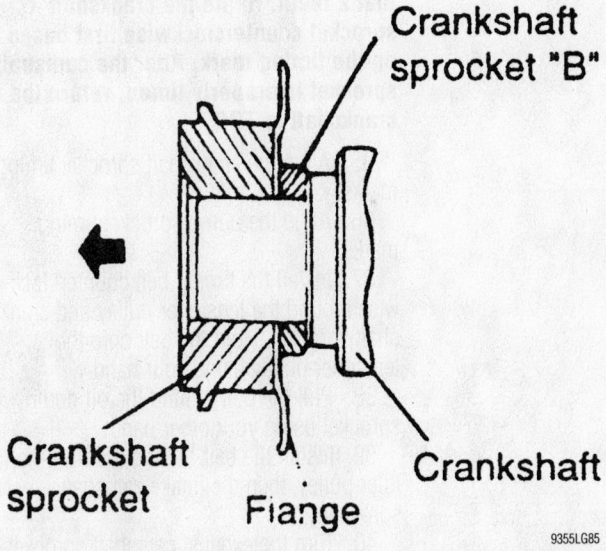

Install the flange and crankshaft sprocket making sure it is installed properly — 2.4L engine

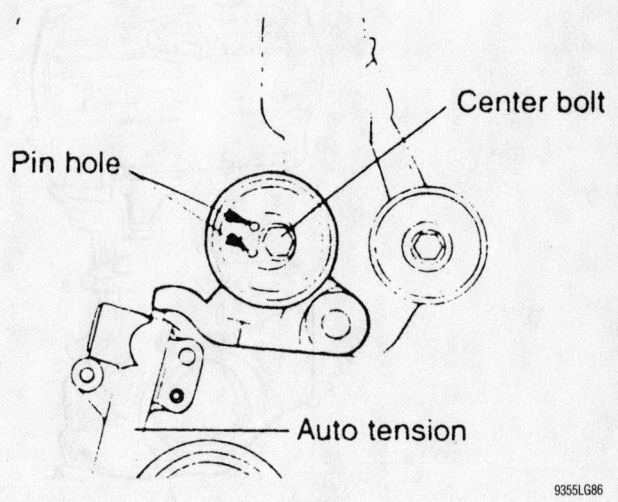

Install the tensioner pulley — 2.4L engine

For Accessory Drive Belt illustrations, see Section 1 of this manual

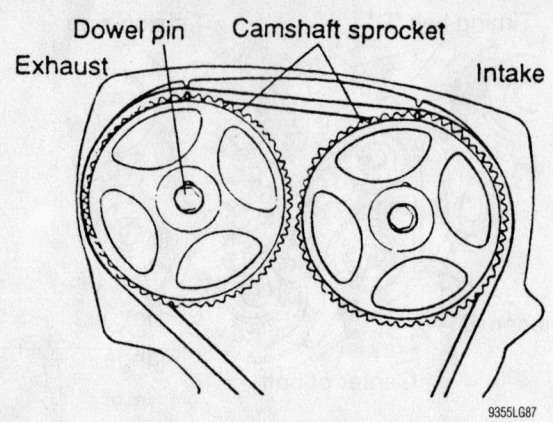

Rotate the camshaft sprockets so that the dowel pin is at the upper side, set the timing mark of the sprocket correctly — 2.4L engine

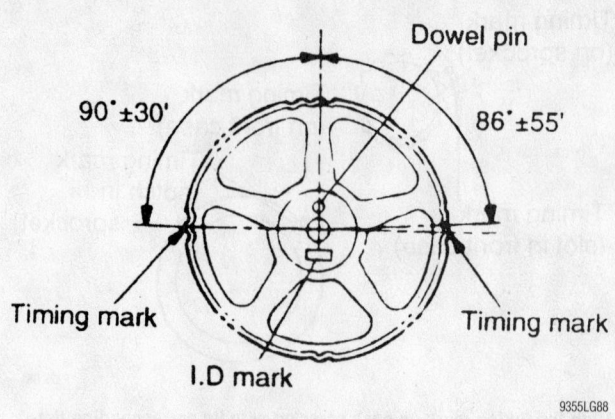

Camshaft sprocket alignment — 2.4L engine

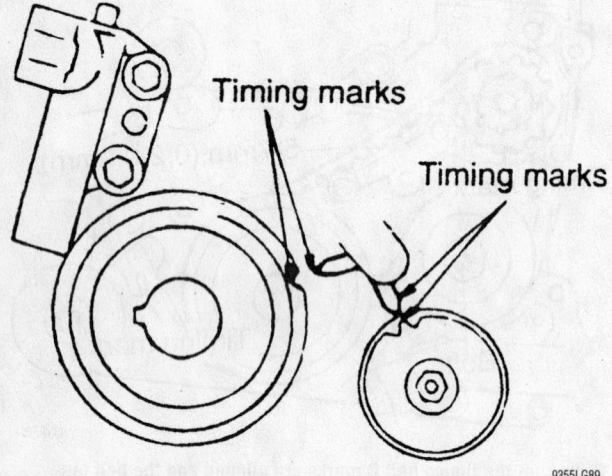

Align the oil pump sprocket timing marks — 2.4L engine

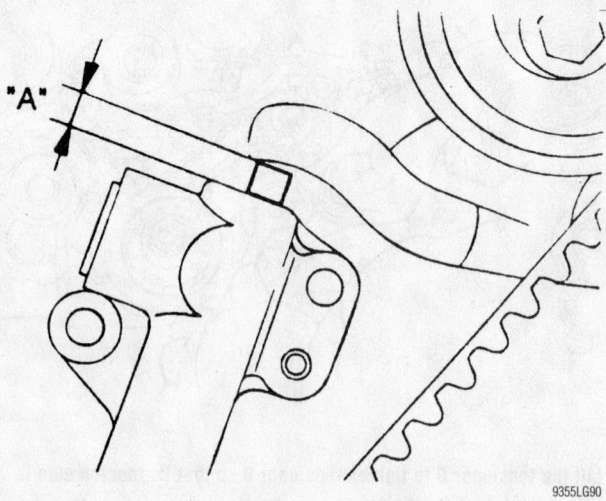

Measure the auto tensioner protrusion A — 2.4L engine

that 2 teeth, rotate the crankshaft sprocket counterclockwise first based on the timing mark. After the camshaft sprocket is properly timed, return the crankshaft to TDC.

35. Align the crankshaft sprocket timing marks.

36. Align the pump sprocket timing marks.

37. Install the timing belt counterclockwise around the tensioner pulley and crankshaft sprocket. Hold the belt onto the tensioner pulley using your hand.

38. Pull the belt around the oil pump sprocket using your other hand.

39. Install the belt around the right-hand idler pulley, then the intake camshaft sprocket.

40. Turn the exhaust camshaft sprocket one tooth clockwise to align its timing mark with the cylinder top surface, then pull the belt around the exhaust camshaft sprocket.

41. Raise the tensioner pulley gently so

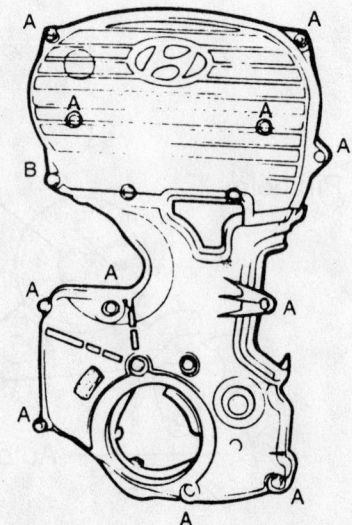

A:8-10 N.m (80-100 kg.cm, 6-7 lb.ft)
B:10-12 N.m (100-120 kg.cm, 7-9 lb.ft)

Timing cover bolt location and torque specifications A — 2.4L engine

that the belt does not sag and temporarily tighten the pulley center bolt.

42. Recheck that all timing marks are correct.

43. Remove the set pin from the auto tensioner.

44. Rotate the crankshaft two turns clockwise and let it sit for around 15 minutes. After 15 minutes, measure the auto tensioner protrusion **A** (the distance between the tensioner arm and tensioner) as shown in the accompanying illustration. The specification should be 0.24 – 0.35 inch (6 – 9mm).

45. Install the timing covers. Tighten the

bolts as shown in the accompanying illustration to specification as shown in the accompanying illustration.

2.7L ENGINE

1. Before servicing the vehicle, refer to the precautions in the beginning of this section.

2. Remove the engine cover.

3. Using a 16mm wrench, rotate the tensioner arm clockwise about 14 degrees and remove the drive belt.

4. Remove the power steering pump pulley, idler pulley, tensioner pulley and crankshaft pulley.

5. Remove the upper and lower timing covers.

6. Remove the timing belt tensioner.

7. Rotate the crankshaft clockwise and align the timing mark to set the No. 1 cylinder to Top Dead Center (TDC). Make sure the timing marks of the camshaft sprocket and cylinder head cover should align with each other.

➡**If reusing the belt, mark the direction of rotation on the belt to ensure proper belt installation.**

8. Unbolt the tensioner and remove the belt.

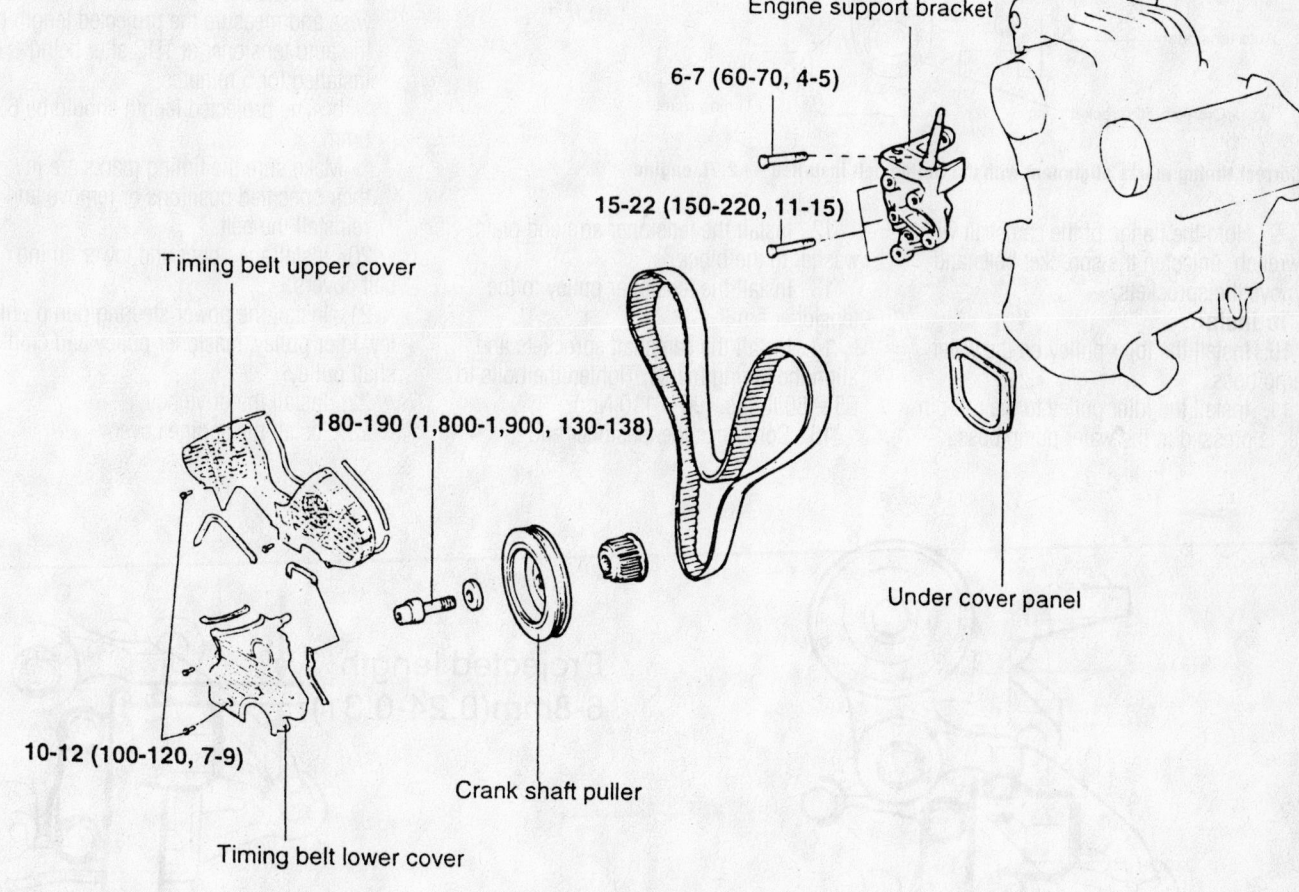

Engine support bracket

6-7 (60-70, 4-5)

15-22 (150-220, 11-15)

Timing belt upper cover

180-190 (1,800-1,900, 130-138)

Under cover panel

10-12 (100-120, 7-9)

Crank shaft puller

Timing belt lower cover

TORQUE : Nm (kg.cm, lb.ft)

9355LG92

Exploded view of the timing belt assembly — 2.7L engine

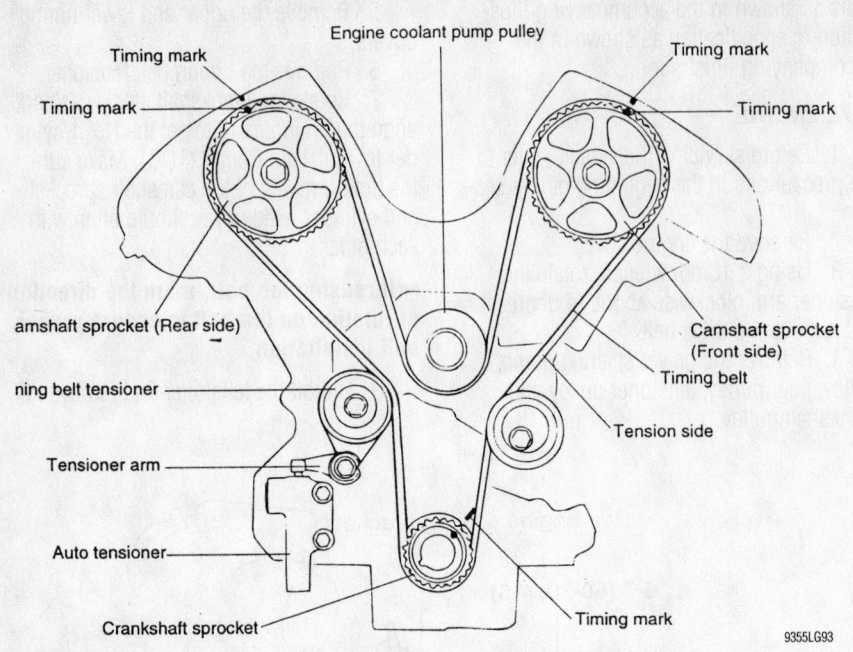

Timing mark
Engine coolant pump pulley
Timing mark
Timing mark
Timing mark
amshaft sprocket (Rear side)
Camshaft sprocket (Front side)
ning belt tensioner
Timing belt
Tensioner arm
Tension side
Auto tensioner
Crankshaft sprocket
Timing mark

9355LG93

Correct timing marks alignment with the timing belt installed — 2.7L engine

9. Hold the flange of the camshaft with a wrench, unfasten the sprocket bolts and remove the sprockets.

To install:

10. Install the idler pulley on the water pump boss.

11. Install the idler pulley to the roll pin that is pressed in the water pump boss.

12. Install the tensioner arm and plain washer to the block.

13. Install the tensioner pulley to the tensioner arm.

14. Install the camshaft sprockets and align the timing marks. Tighten the bolts to 65 – 80 ft. lbs. (90 – 110 Nm).

15. Compress the tensioner and

install a set pin to keep the plunger in position.

16. Install the tensioner and tighten the bolt to 14 – 20 ft. lbs. (20 – 27 Nm).

17. Align the sprocket timing marks and install the belt in the following order:

- Crankshaft sprocket
- Idler pulley
- Camshaft sprocket on the left hand side
- Water pump pulley
- Camshaft sprocket on the right hand side
- Tensioner pulley

18. Remove the tensioner set pin.

19. Adjust the timing belt tension as follows:

a. Rotate the crankshaft 2 turns clockwise and measure the projected length of the auto tensioner at TDC after being installed for 5 minutes.

b. The projected length should be 6 – 8mm.

Make sure the timing marks are in their specified positions or remove and reinstall the belt.

20. Install the upper and lower timing belt covers.

21. Install the power steering pump pulley, idler pulley, tensioner pulley and crankshaft pulley.

22. Install the drive belt.

23. Install the engine cover.

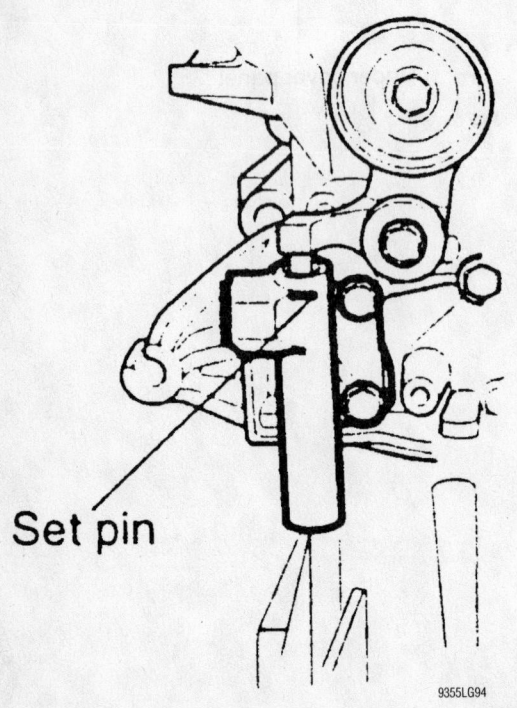

Set pin

9355LG94

Remove the set pin from the auto tensioner — 2.7L engine

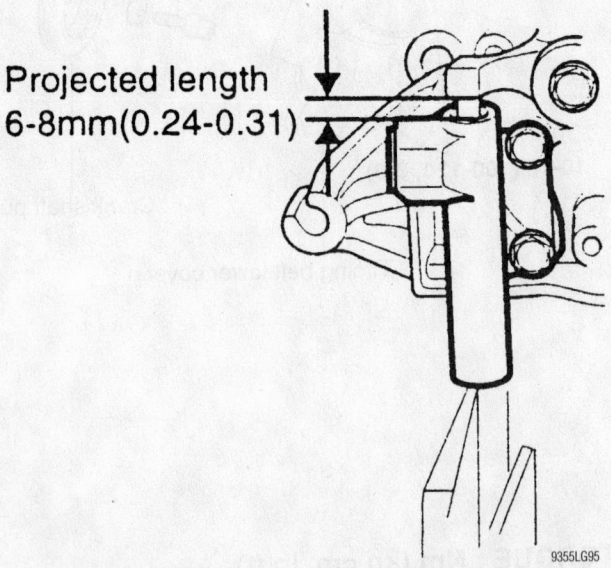

Projected length 6-8mm(0.24-0.31)

9355LG95

The projected length of the auto tensioner at Top Dead Center (TDC) should be 6 – 8mm — 2.7L engine

Infiniti QX4

3.3L (VG33E) ENGINE

1. Remove the engine undercover.
2. Remove the radiator shroud, the fan and the pulleys.
3. Drain the coolant from the radiator and remove the water pump hose.

✳✳ CAUTION

When draining the coolant, keep in mind that cats and dogs are attracted by the ethylene glycol antifreeze, and are quite likely to drink any that is left in an uncovered container or in puddles on the ground. This will prove fatal in sufficient quantity. Always drain the coolant into a sealable container. Coolant should be reused unless it is contaminated or several years old.

4. Remove the radiator.
5. Remove the power steering, air conditioning compressor and alternator drive belts.
6. Remove the spark plugs.
7. Remove the distributor protector (dust shield).
8. Remove the air conditioning compressor drive belt idler pulley and bracket.
9. Remove the fresh air intake tube at the cylinder head cover.
10. Disconnect the radiator hose at the thermostat housing.

11. Remove the crankshaft pulley bolt, then pull off the pulley with a suitable puller.
12. Remove the bolts, then remove the front upper and lower timing belt covers.
13. Set the No. 1 piston at Top Dead Center (TDC) of its compression stroke. Align the punchmark on the left camshaft sprocket with the punchmark on the timing belt upper rear cover. Align the punchmark on the crankshaft sprocket with the notch on the oil pump housing. Temporarily install the crank pulley bolt so the crankshaft can be rotated if necessary.
14. Loosen the timing belt tensioner and return spring, then remove the timing belt.

To install:

✳✳ CAUTION

Before installing the timing belt, confirm that the No. 1 cylinder is set at the TDC of the compression stroke.

15. Remove both cylinder head covers and loosen all rocker arm shaft retaining bolts.

➡**The rocker arm shaft bolts MUST be loosened so that the correct belt tension can be obtained.**

16. Install the tensioner and the return spring. Using a hexagon wrench, turn the tensioner clockwise and temporarily tighten the locknut.
17. Be sure that the timing belt is clean and free from oil or water.

18. When installing the timing belt, align the white lines on the belt with the punchmarks on the camshaft and crankshaft sprockets. Have the arrow on the timing belt pointing toward the front belt covers.

➡**A good way (although rather tedious!) to check for proper timing belt installation is to count the number of belt teeth between the timing marks. There are 133 teeth on the belt; there should be 40 teeth between the timing marks on the left and right side camshaft sprockets, and 43 teeth between the timing marks on the left side camshaft sprocket and the crankshaft sprocket.**

19. While keeping the tensioner steady, loosen the locknut with a hex wrench.
20. Turn the tensioner approximately 70 – 80 degrees clockwise with the wrench, then tighten the locknut.

✳✳ WARNING

If any binding is felt when adjusting the timing belt tension by turning the crankshaft, STOP turning the engine, because the pistons may be hitting the valves.

21. Turn the crankshaft in a clockwise direction several times, then **slowly** set the No. 1 piston to TDC of the compression stroke.
22. Apply 22 lbs. (10 kg) of pressure (push it in!) to the center span of the timing belt between the right side camshaft sprocket and the tensioner pulley, then loosen the tensioner locknut.
23. Using a 0.0138 in. (0.35mm) thick feeler gauge (the actual width of the blade **must** be ½ in. or 13mm!), turn the crankshaft clockwise (**slowly!**). The timing belt should move approximately 2½ teeth. Tighten the tensioner locknut, turn the crankshaft slightly and remove the feeler gauge.
24. Slowly rotate the crankshaft clockwise several more times, then set the No. 1 piston to TDC of the compression stroke.
25. Position the 2 timing covers on the block, then tighten the mounting bolts to 24 ft. lbs. (35 Nm).
26. Press the crankshaft pulley onto the shaft, then tighten the bolt to 90 – 98 ft. lbs. (123 – 132 Nm).
27. Connect the radiator hose to the thermostat housing.

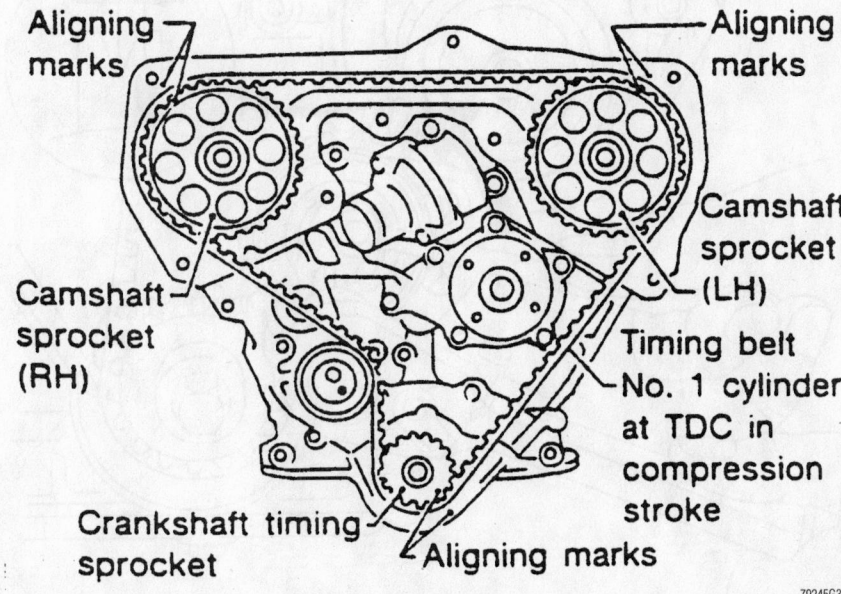

Aligning marks

Aligning marks

Camshaft sprocket (LH)

Camshaft sprocket (RH)

Timing belt No. 1 cylinder at TDC in compression stroke

Crankshaft timing sprocket

Aligning marks

79245G35

Timing belt alignment mark locations — Infiniti QX4 3.3L engine

For Wheel Alignment specifications, see Section 1 of this manual

28. Reconnect the fresh air intake tube at the cylinder head cover.

29. Install the air conditioning compressor drive belt idler pulley and bracket.

30. Install the distributor protector (dust shield).

31. Install the spark plugs.

32. Install the power steering, air conditioning compressor and alternator drive belts.

33. Install the radiator.

34. Reconnect the water pump hose and fill the engine with coolant. Install the fan shroud and pulleys.

35. Install the engine undercover.

36. Start the engine and check for any leaks.

Isuzu Amigo

2.2L (X22SE/D) ENGINE

1. Disconnect the negative battery cable.

2. Using a box-end wrench on the drive belt adjuster, turn the adjuster clockwise and remove the drive belt.

3. From the left rear of the engine compartment, disconnect the 3 electrical connectors from the chassis harness.

4. Remove the crankshaft pulley-to-crankshaft bolts and remove the pulley.

5. From the front of the engine, remove the nut and the engine harness cover.

6. Remove the timing belt cover.

7. Rotate the crankshaft to position the timing marks at Top Dead Center (TDC) of the No. 1 cylinder's compression stroke.

➡**Mark the rotational direction of the timing belt for reinstallation purposes.**

8. Remove the timing belt tensioner adjusting bolt and the tensioner from the engine.

9. Remove the timing belt.

To install:

10. Install the timing belt tensioner and finger-tighten the tensioner bolt.

11. Inspect the timing marks to be sure that the engine is positioned at TDC of the No. 1 cylinder's compression stroke.

12. Using tool J-43037, or equivalent, place it between the intake and exhaust sprockets to prevent the camshaft gear from moving during the timing belt installation.

13. Install the timing belt.

14. Position the timing belt to ensure that the tension side of the belt is taut and move the timing belt tension adjusting lever clockwise until the tensioner pointer is flowing.

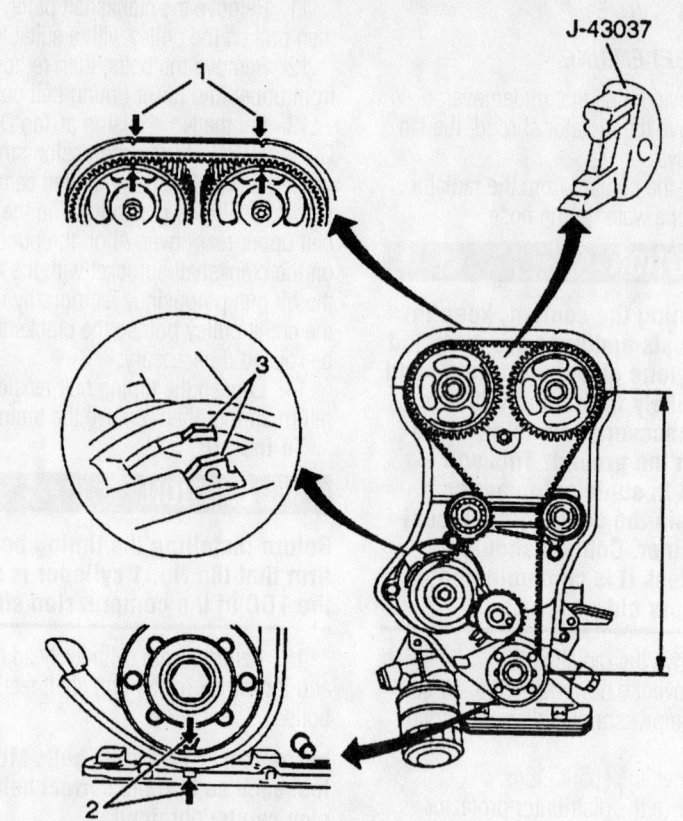

Aligning the timing marks and installing the timing belt — Isuzu Amigo and Rodeo 2.2L (X22SE/D) engine

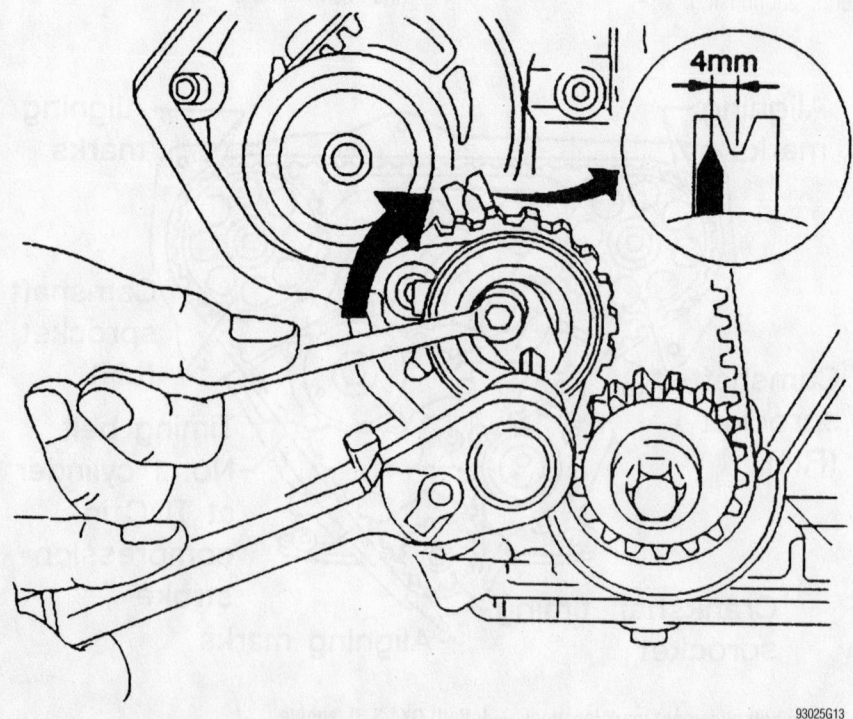

Tensioning the timing belt for a used timing belt — Isuzu Amigo and Rodeo 2.2L (X22SE/D) engine

15. If installing a used timing belt (used over 60 min. from new), the pointer should be positioned approximately 0.16 in. (4mm) to the left of the "V" notch when viewed from the front of the engine.

16. If installing a new timing belt, the pointer should be positioned at the center of the "V" notch when viewed from the front of the engine.

17. Torque the timing belt tensioner adjusting bolt to 18 ft. lbs. (25 Nm).

18. Install the timing belt front cover and torque the bolts to 53 inch lbs. (6 Nm).

19. Install the engine harness connectors.

20. Install the crankshaft pulley and toque the pulley-to-crankshaft bolts to 14 ft. lbs. (20 Nm).

21. Move the drive belt tensioner to the loose side and install the drive belt to its normal position.

22. Connect the negative battery cable.

3.2L AND 3.5L (6VE1) ENGINES

1. Disconnect the negative battery cable.

2. Remove the air cleaner assembly and intake air duct.

3. Remove the upper fan shroud from the radiator.

4. Remove the 4 nuts retaining the cooling fan assembly. Remove the cooling fan from the fan pulley.

5. Loosen and remove the drive belts.

6. Remove the upper timing belt covers.

7. Remove the fan pulley assembly.

8. Rotate the crankshaft to align the camshaft timing marks with the pointer dots on the back covers. Verify that the pointer on the crankshaft aligns with the mark on the lower timing cover.

➡When the timing marks are aligned, the No. 2 piston is at Top Dead Center (TDC) compression.

❊❊ WARNING

Align the camshaft and crankshaft sprockets with their alignment marks before removing the timing belt. Failure to align the belt and sprocket marks may result in valve damage.

9. Use tool No. J-8614-01, or a suitable pulley holding tool to remove the crankshaft pulley center bolt. Remove the crankshaft pulley.

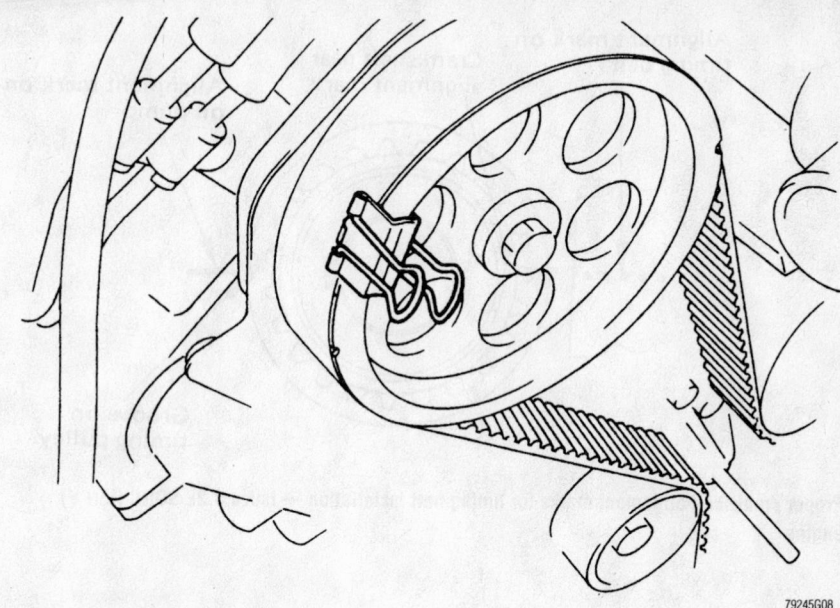

79245G08

Using a double clip to hold the belt in place — Isuzu 3.2L and 3.5L engines

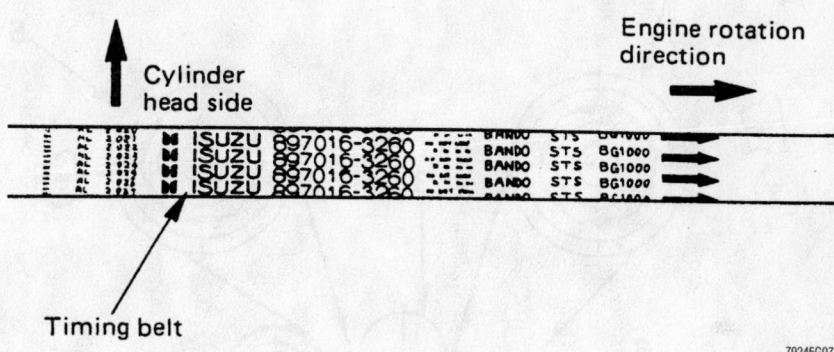

79245G07

For maximum timing belt life, install the belt as shown — Isuzu 3.2L and 3.5L engines

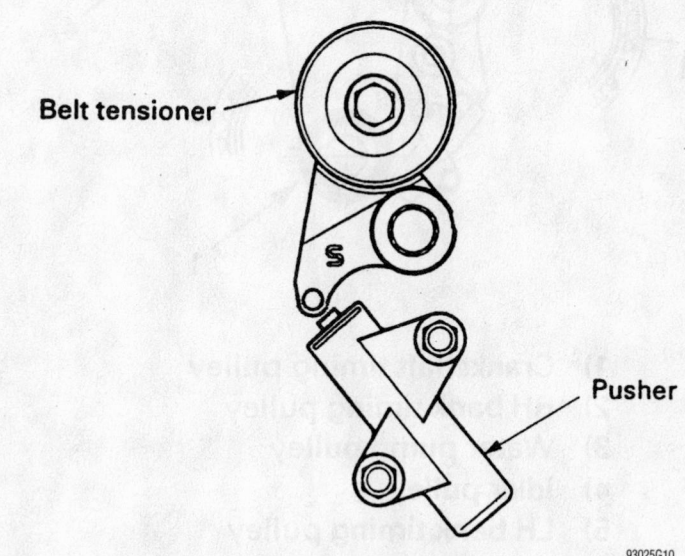

93025G10

View of timing belt tensioner and pusher — Isuzu 3.2L and 3.5L engines

For Maintenance Interval recommendations, see Section 1 of this manual

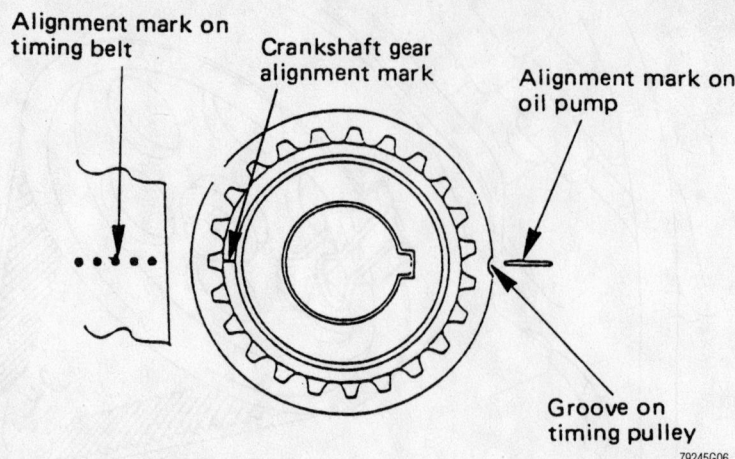

Alignment mark on timing belt

Crankshaft gear alignment mark

Alignment mark on oil pump

Groove on timing pulley

79245G06

Proper crankshaft alignment marks for timing belt installation — Isuzu 3.2L SOHC (VIN V) engine

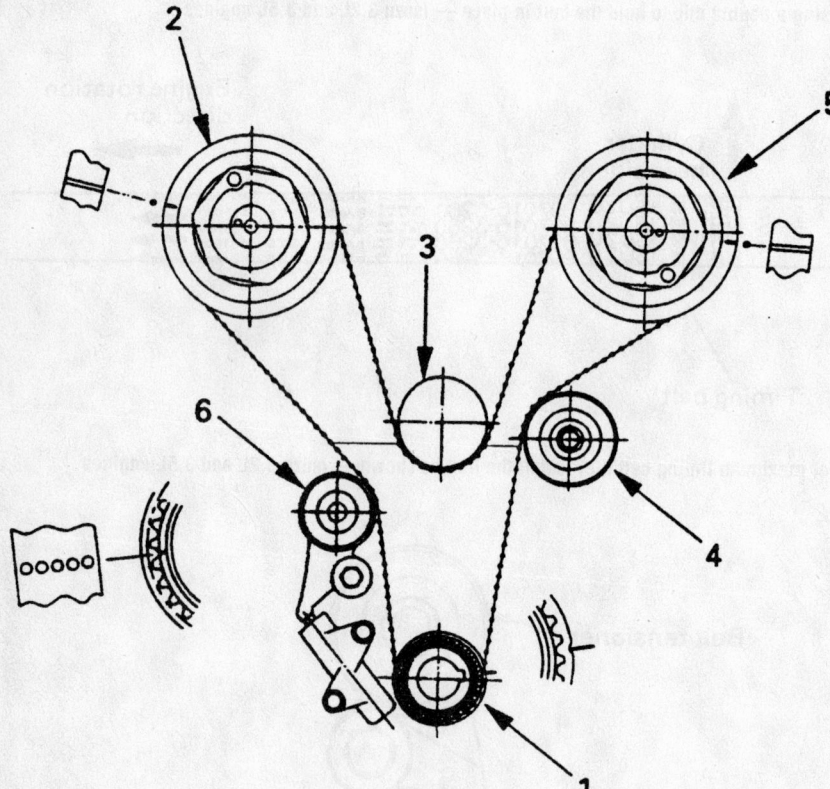

1) **Crankshaft timing pulley**
2) **RH bank timing pulley**
3) **Water pump pulley**
4) **Idler pulley**
5) **LH bank timing pulley**
6) **Tension pulley**

93025G09

Timing mark alignment and timing belt routing — Isuzu 3.2L SOHC (VIN V) engine

10. If present, disconnect the 2 oil cooler hose bracket bolts on the timing cover. Move the oil cooler hoses and bracket off of the lower timing cover.

11. Remove the lower timing belt cover.

12. Remove the pusher assembly (tensioner) from below the belt tensioner pulley. The pusher rod must always face upward to prevent oil leakage. Depress the pusher rod, and insert a wire pin into the hole to keep the pusher rod retracted.

13. Remove the timing belt.

14. Inspect the water pump and replace it if there is any doubt about its condition.

15. Repair any oil or coolant leaks before installing a new timing belt. If the timing belt has been contaminated with oil or coolant, or is damaged, it must be replaced.

To install:

16. Verify that the sprocket timing marks are still aligned and that the groove and the keyway on the crankshaft timing sprocket align with the mark on the oil pump. The white pointers on the camshaft timing sprockets should align with the dots on the front plate.

17. Install the timing belt. Use clips to secure the belt onto each sprocket until the installation is complete. Align the dotted marks on the timing belt with the timing mark opposite the groove on the crankshaft sprocket.

➡The arrows on the timing belt must follow the belt's direction of rotation. The manufacturer's trademark on the belt's spine should be readable left-to-right when the belt is installed.

18. Align the white line on the timing belt with the alignment mark on the right bank camshaft timing pulley. Secure the belt with a clip.

✳✳ WARNING

If any binding is felt when adjusting the timing belt tension by turning the crankshaft, STOP turning the engine, because the pistons may be hitting the valves.

19. Rotate the crankshaft counterclockwise to remove the slack between the crankshaft sprocket and the right camshaft timing belt sprocket.

20. Install the belt around the water pump pulley.

21. Install the belt on the idler pulley.

22. Align the white alignment mark on the timing belt with the alignment mark on the left bank camshaft timing belt sprocket.

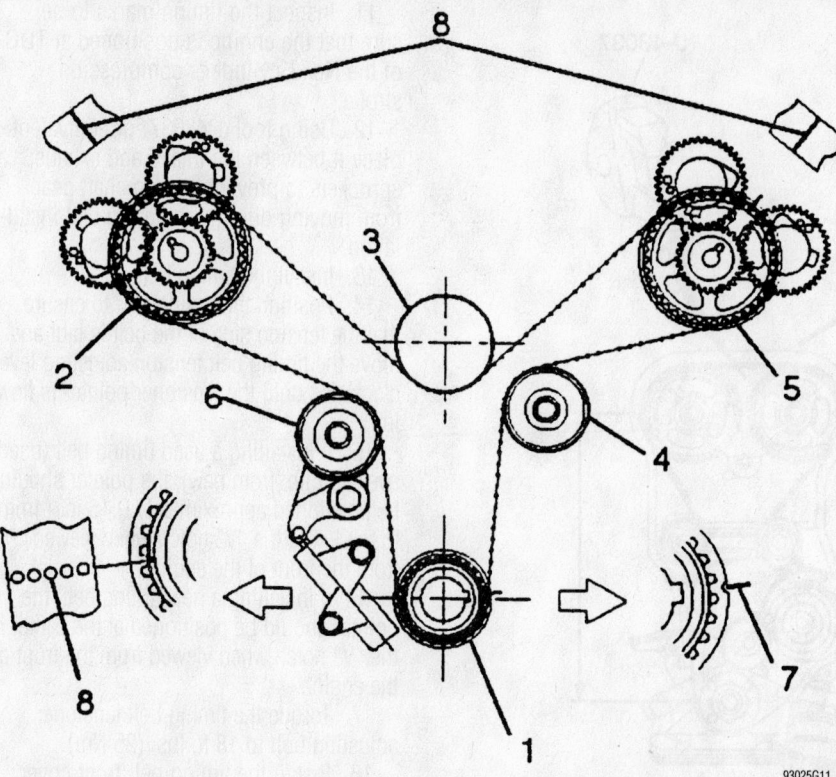

Timing mark alignment and timing belt routing — Isuzu 3.2L (VIN W) and 3.5L DOHC engines

93025G11

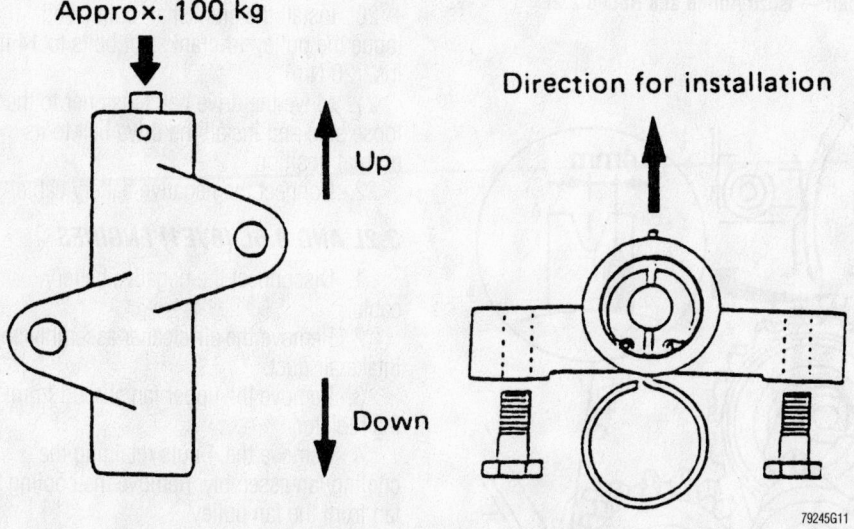

Approx. 100 kg

Up

Down

Direction for installation

79245G11

View of timing belt tensioner pusher — Isuzu 3.2L and 3.5L engines

23. Install the crankshaft pulley and tighten the center bolt by hand. Rotate the crankshaft pulley clockwise to give slack between the crankshaft timing belt pulley and the right bank camshaft timing belt pulley.

24. Insert a 1.4mm piece of wire through the hole in the pusher to hold the

rod in. Install the pusher assembly while pushing the tension pulley toward the belt.

25. Pull the pin out from the pusher to release the rod.

26. Remove the clamps from the sprockets. Rotate the crankshaft pulley clockwise 2 turns. Measure the rod protru-

sion to ensure it is between 0.16 – 0.24 in. (4–6mm).

27. If the tensioner pulley bracket pivot bolt was removed, tighten it to 31 ft. lbs. (42 Nm).

28. Tighten the pusher bolts to 14 ft. lbs. (19 Nm).

29. Remove the crankshaft pulley. Install the lower and upper timing belt covers and tighten their bolts to 12 ft. lbs. (17 Nm).

30. Fit the oil cooler hose onto the timing cover and tighten its mounting bracket bolts to 16 ft. lbs. (22 Nm).

31. Install the crankshaft pulley and tighten the pulley bolt to 123 ft. lbs. (167 Nm).

32. Install fan pulley assembly and tighten the bolts to 16 ft. lbs. (22 Nm).

33. Install and adjust the accessory drive belts.

34. Install the cooling fan assembly and tighten the bolts to 72 inch lbs. (8 Nm).

35. Install the upper fan shroud.

36. Install the air cleaner assembly and intake air duct.

37. Connect the negative battery cable.

Isuzu Rodeo

2.2L (X22SE/D) ENGINE

1. Disconnect the negative battery cable.

2. Using a box-end wrench on the drive belt adjuster, turn the adjuster clockwise and remove the drive belt.

3. From the left rear of the engine compartment, disconnect the 3 electrical connectors from the chassis harness.

4. Remove the crankshaft pulley-to-crankshaft bolts and remove the pulley.

5. From the front of the engine, remove the nut and the engine harness cover.

6. Remove the timing belt cover.

7. Rotate the crankshaft to position the timing marks at Top Dead Center (TDC) of the No. 1 cylinder's compression stroke.

➡**Mark the rotational direction of the timing belt for reinstallation purposes.**

8. Remove the timing belt tensioner adjusting bolt and the tensioner from the engine.

9. Remove the timing belt.
   To install:

10. Install the timing belt tensioner and finger-tighten the tensioner bolt.

For Tune-up, Capacities and Firing orders, see Section 1 of this manual

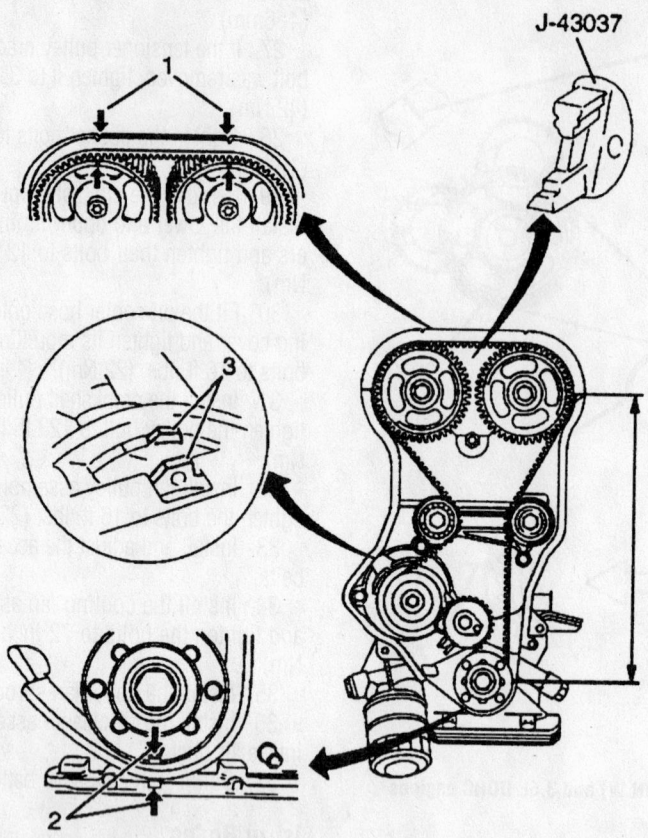

J-43037

Aligning the timing marks and installing the timing belt — Isuzu Amigo and Rodeo 2.2L (X22SE/D) engine

93025G12

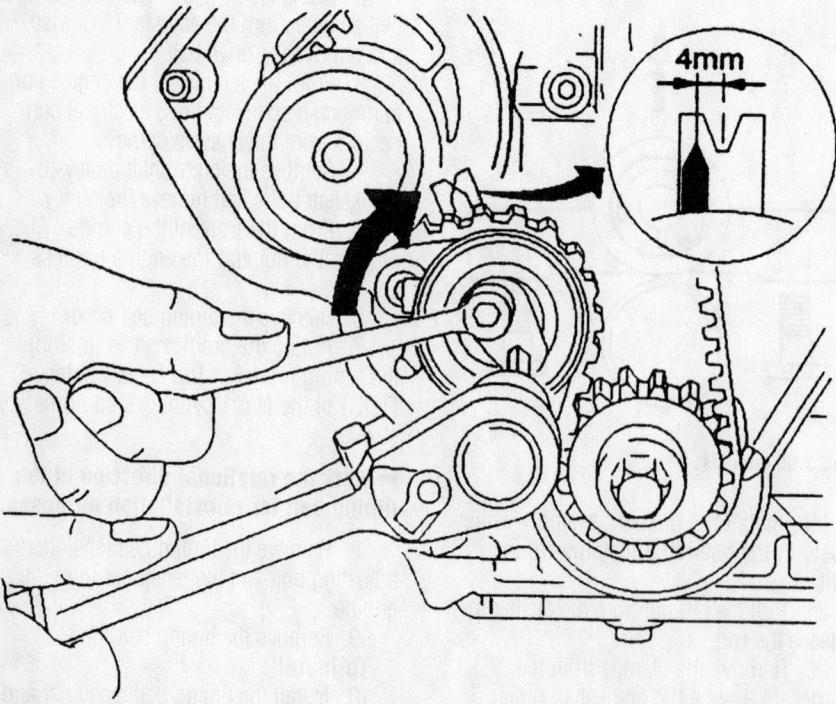

4mm

Tensioning the timing belt for a used timing belt — Isuzu Amigo and Rodeo 2.2L (X22SE/D) engine

93025G13

11. Inspect the timing marks to be sure that the engine is positioned at TDC of the No. 1 cylinder's compression stroke.

12. Using tool J-43037, or equivalent, place it between the intake and exhaust sprockets to prevent the camshaft gear from moving during the timing belt installation.

13. Install the timing belt.

14. Position the timing belt to ensure that the tension side of the belt is taut and move the timing belt tension adjusting lever clockwise until the tensioner pointer is flowing.

15. If installing a used timing belt (used over 60 min. from new), the pointer should be positioned approximately 0.16 in. (4mm) to the left of the "V" notch when viewed from the front of the engine.

16. If installing a new timing belt, the pointer should be positioned at the center of the "V" notch when viewed from the front of the engine.

17. Torque the timing belt tensioner adjusting bolt to 18 ft. lbs. (25 Nm).

18. Install the timing belt front cover and torque the bolts to 53 inch lbs. (6 Nm).

19. Install the engine harness connectors.

20. Install the crankshaft pulley and toque the pulley-to-crankshaft bolts to 14 ft. lbs. (20 Nm).

21. Move the drive belt tensioner to the loose side and install the drive belt to its normal position.

22. Connect the negative battery cable.

3.2L AND 3.5L (6VE1) ENGINES

1. Disconnect the negative battery cable.

2. Remove the air cleaner assembly and intake air duct.

3. Remove the upper fan shroud from the radiator.

4. Remove the 4 nuts retaining the cooling fan assembly. Remove the cooling fan from the fan pulley.

5. Loosen and remove the drive belts.

6. Remove the upper timing belt covers.

7. Remove the fan pulley assembly.

8. Rotate the crankshaft to align the camshaft timing marks with the pointer dots on the back covers. Verify that the pointer on the crankshaft aligns with the mark on the lower timing cover.

➡When the timing marks are aligned, the No. 2 piston is at Top Dead Center (TDC) compression.

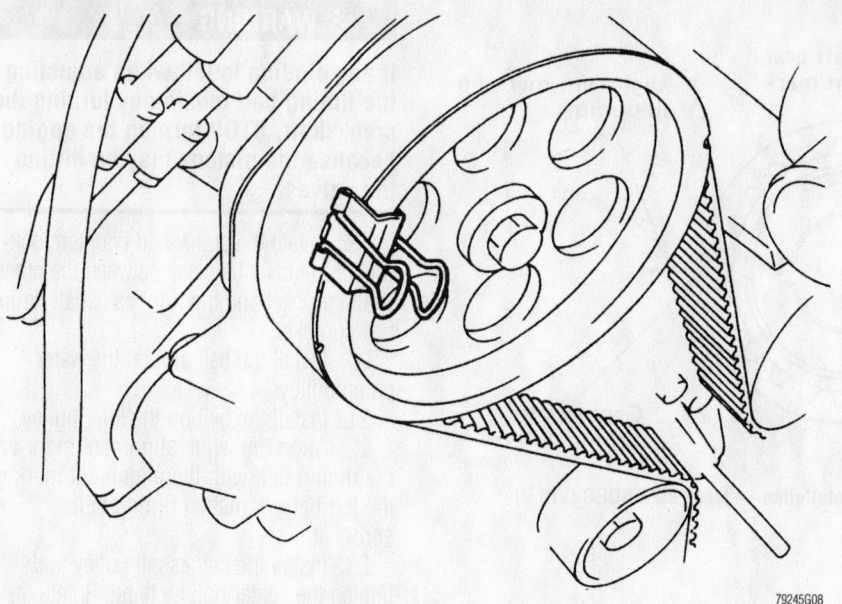

Using a double clip to hold the belt in place — Isuzu 3.2L and 3.5L engines

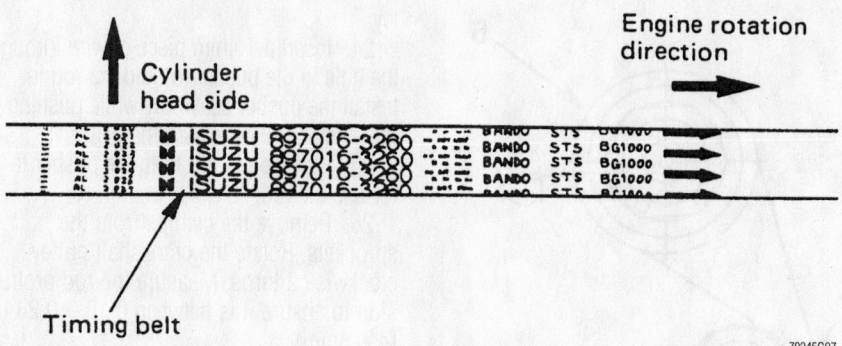

For maximum timing belt life, install the belt as shown — Isuzu 3.2L and 3.5L engines

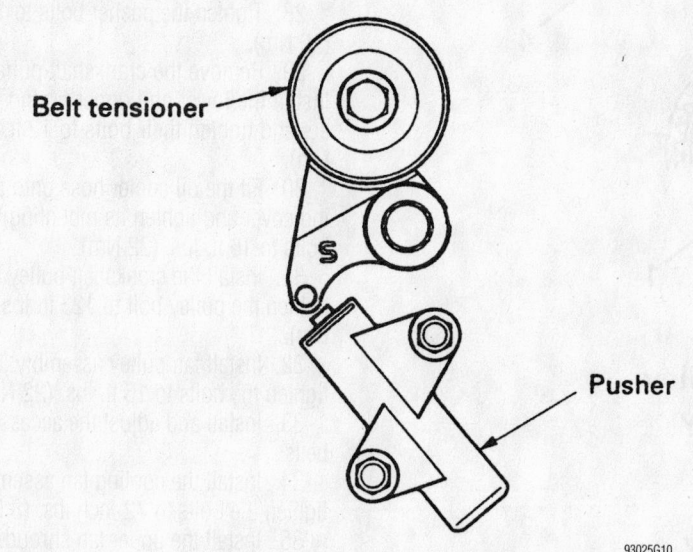

View of timing belt tensioner and pusher — Isuzu 3.2L and 3.5L engines

※※ WARNING

Align the camshaft and crankshaft sprockets with their alignment marks before removing the timing belt. Failure to align the belt and sprocket marks may result in valve damage.

9. Use tool No. J-8614-01, or a suitable pulley holding tool to remove the crankshaft pulley center bolt. Remove the crankshaft pulley.

10. If present, disconnect the 2 oil cooler hose bracket bolts on the timing cover. Move the oil cooler hoses and bracket off of the lower timing cover.

11. Remove the lower timing belt cover.

12. Remove the pusher assembly (tensioner) from below the belt tensioner pulley. The pusher rod must always face upward to prevent oil leakage. Depress the pusher rod, and insert a wire pin into the hole to keep the pusher rod retracted.

13. Remove the timing belt.

14. Inspect the water pump and replace it if there is any doubt about its condition.

15. Repair any oil or coolant leaks before installing a new timing belt. If the timing belt has been contaminated with oil or coolant, or is damaged, it must be replaced.

To install:

16. Verify that the sprocket timing marks are still aligned and that the groove and the keyway on the crankshaft timing sprocket align with the mark on the oil pump. The white pointers on the camshaft timing sprockets should align with the dots on the front plate.

17. Install the timing belt. Use clips to secure the belt onto each sprocket until the installation is complete. Align the dotted marks on the timing belt with the timing mark opposite the groove on the crankshaft sprocket.

➡ The arrows on the timing belt must follow the belt's direction of rotation. The manufacturer's trademark on the belt's spine should be readable left-to-right when the belt is installed.

18. Align the white line on the timing belt with the alignment mark on the right bank camshaft timing pulley. Secure the belt with a clip.

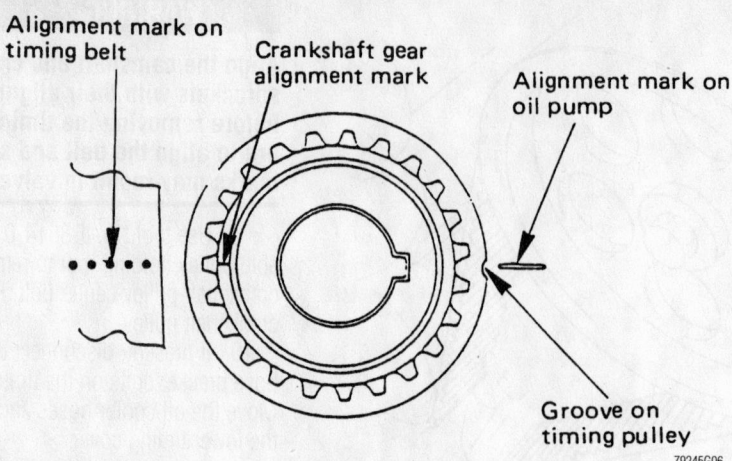

Proper crankshaft alignment marks for timing belt installation — Isuzu 3.2L SOHC (VIN V) engine

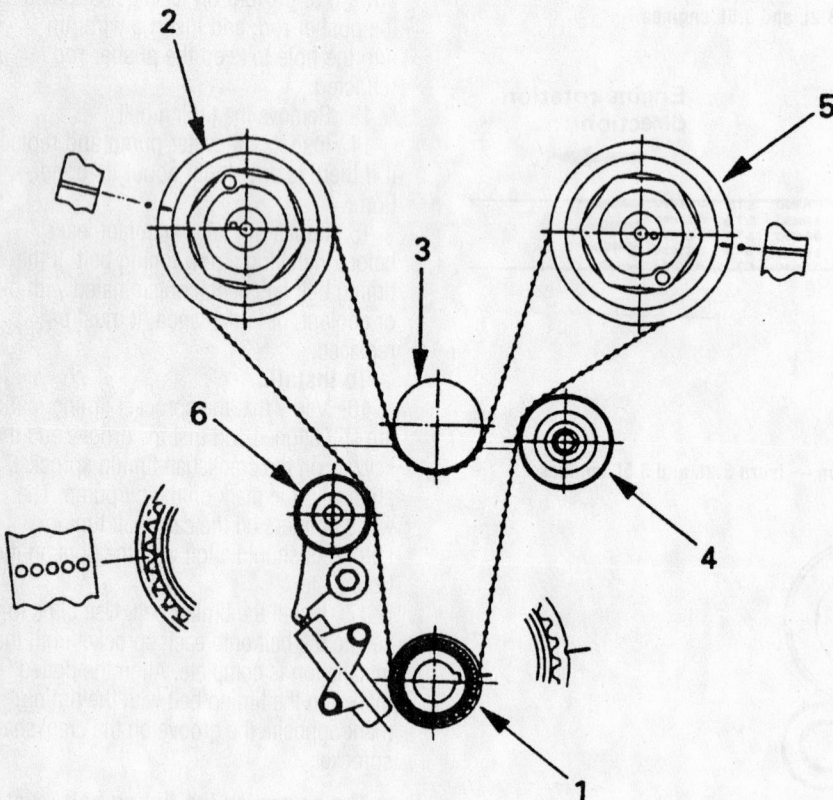

1) Crankshaft timing pulley
2) RH bank timing pulley
3) Water pump pulley
4) Idler pulley
5) LH bank timing pulley
6) Tension pulley

Timing mark alignment and timing belt routing — Isuzu 3.2L SOHC (VIN V) engine

✲✲ WARNING

If any binding is felt when adjusting the timing belt tension by turning the crankshaft, STOP turning the engine, because the pistons may be hitting the valves.

19. Rotate the crankshaft counterclockwise to remove the slack between the crankshaft sprocket and the right camshaft timing belt sprocket.

20. Install the belt around the water pump pulley.

21. Install the belt on the idler pulley.

22. Align the white alignment mark on the timing belt with the alignment mark on the left bank camshaft timing belt sprocket.

23. Install the crankshaft pulley and tighten the center bolt by hand. Rotate the crankshaft pulley clockwise to give slack between the crankshaft timing belt pulley and the right bank camshaft timing belt pulley.

24. Insert a 1.4mm piece of wire through the hole in the pusher to hold the rod in. Install the pusher assembly while pushing the tension pulley toward the belt.

25. Pull the pin out from the pusher to release the rod.

26. Remove the clamps from the sprockets. Rotate the crankshaft pulley clockwise 2 turns. Measure the rod protrusion to ensure it is between 0.16 – 0.24 in. (4 – 6mm).

27. If the tensioner pulley bracket pivot bolt was removed, tighten it to 31 ft. lbs. (42 Nm).

28. Tighten the pusher bolts to 14 ft. lbs. (19 Nm).

29. Remove the crankshaft pulley. Install the lower and upper timing belt covers and tighten their bolts to 12 ft. lbs. (17 Nm).

30. Fit the oil cooler hose onto the timing cover and tighten its mounting bracket bolts to 16 ft. lbs. (22 Nm).

31. Install the crankshaft pulley and tighten the pulley bolt to 123 ft. lbs. (167 Nm).

32. Install fan pulley assembly and tighten the bolts to 16 ft. lbs. (22 Nm).

33. Install and adjust the accessory drive belts.

34. Install the cooling fan assembly and tighten the bolts to 72 inch lbs. (8 Nm).

35. Install the upper fan shroud.

36. Install the air cleaner assembly and intake air duct.

37. Connect the negative battery cable.

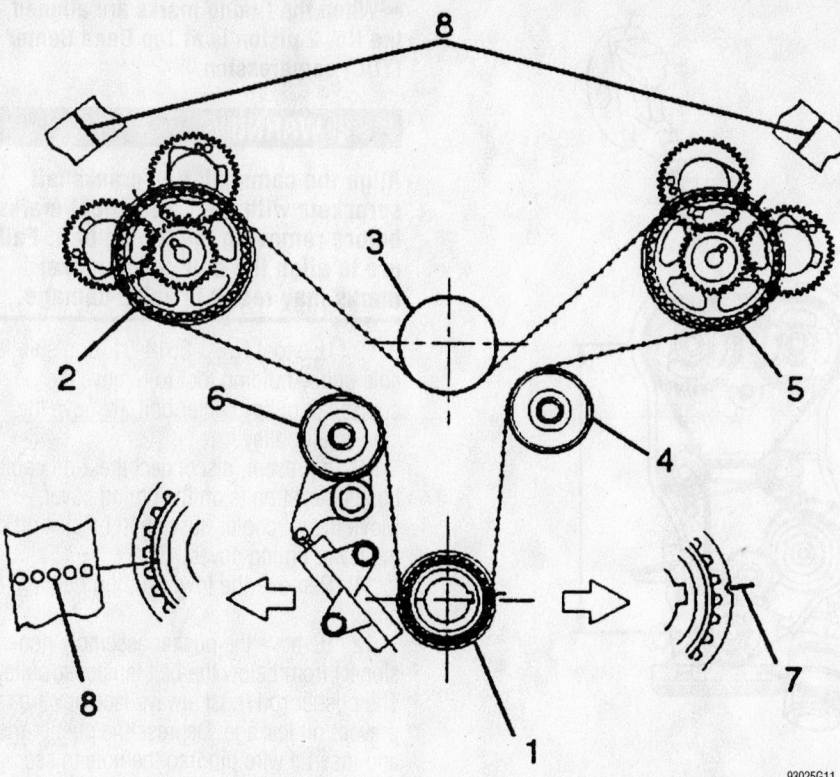

Timing mark alignment and timing belt routing — Isuzu 3.2L (VIN W) and 3.5L DOHC engines

93025G11

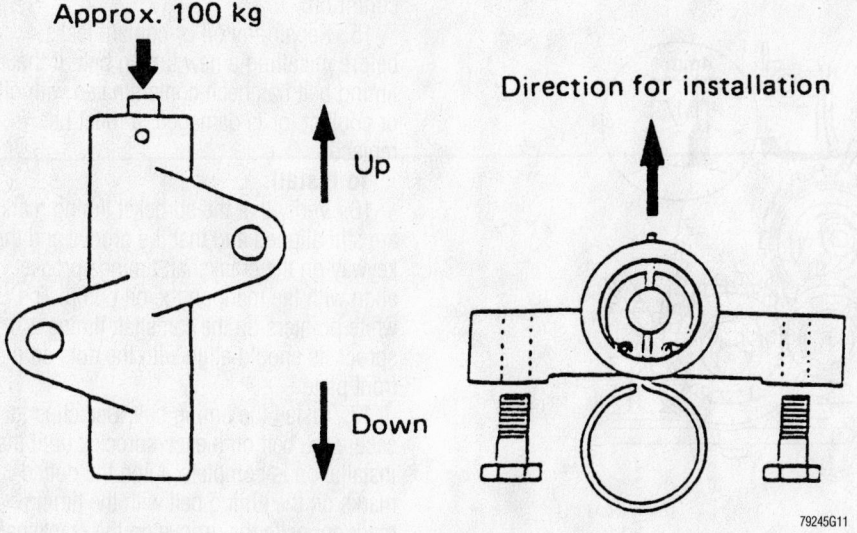

Approx. 100 kg

Up

Down

Direction for installation

79245G11

View of timing belt tensioner pusher — Isuzu 3.2L and 3.5L engines

Isuzu Trooper

2.2L (X22SE/D) ENGINE

1. Disconnect the negative battery cable.
2. Using a box-end wrench on the drive belt adjuster, turn the adjuster clockwise and remove the drive belt.
3. From the left rear of the engine compartment, disconnect the 3 electrical connectors from the chassis harness.

4. Remove the crankshaft pulley-to-crankshaft bolts and remove the pulley.
5. From the front of the engine, remove the nut and the engine harness cover.
6. Remove the timing belt cover.
7. Rotate the crankshaft to position the timing marks at Top Dead Center (TDC) of the No. 1 cylinder's compression stroke.

➡**Mark the rotational direction of the timing belt for reinstallation purposes.**

8. Remove the timing belt tensioner adjusting bolt and the tensioner from the engine.
9. Remove the timing belt.

To install:

10. Install the timing belt tensioner and finger-tighten the tensioner bolt.
11. Inspect the timing marks to be sure that the engine is positioned at TDC of the No. 1 cylinder's compression stroke.
12. Using tool J-43037, or equivalent, place it between the intake and exhaust sprockets to prevent the camshaft gear from moving during the timing belt installation.
13. Install the timing belt.
14. Position the timing belt to ensure that the tension side of the belt is taut and move the timing belt tension adjusting lever clockwise until the tensioner pointer is flowing.
15. If installing a used timing belt (used over 60 min. from new), the pointer should be positioned approximately 0.16 in. (4mm) to the left of the "V" notch when viewed from the front of the engine.
16. If installing a new timing belt, the pointer should be positioned at the center of the "V" notch when viewed from the front of the engine.
17. Torque the timing belt tensioner adjusting bolt to 18 ft. lbs. (25 Nm).
18. Install the timing belt front cover and torque the bolts to 53 inch lbs. (6 Nm).
19. Install the engine harness connectors.
20. Install the crankshaft pulley and toque the pulley-to-crankshaft bolts to 14 ft. lbs. (20 Nm).
21. Move the drive belt tensioner to the loose side and install the drive belt to its normal position.
22. Connect the negative battery cable.

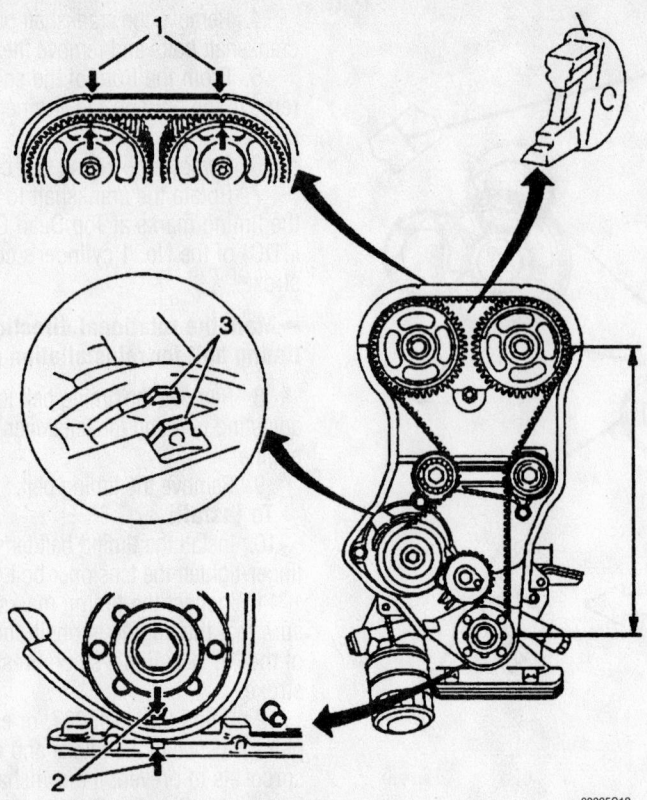

93025G12

Aligning the timing marks and installing the timing belt — Isuzu Amigo and Rodeo 2.2L (X22SE/D) engine

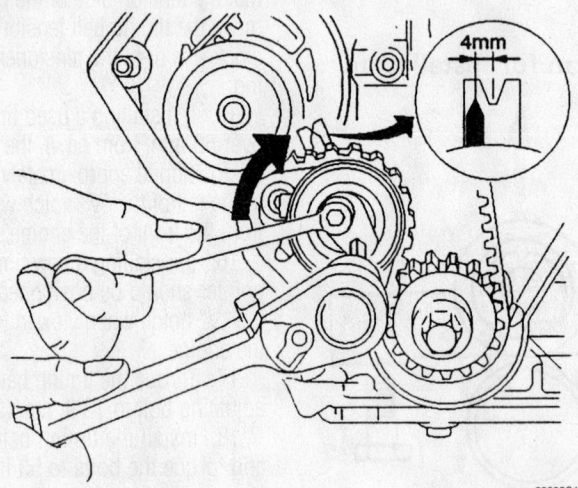

93025G13

Tensioning the timing belt for a used timing belt — Isuzu Amigo and Rodeo 2.2L (X22SE/D)

3.2L AND 3.5L (6VE1) ENGINES

1. Disconnect the negative battery cable.

2. Remove the air cleaner assembly and intake air duct.

3. Remove the upper fan shroud from the radiator.

4. Remove the 4 nuts retaining the cooling fan assembly. Remove the cooling fan from the fan pulley.

5. Loosen and remove the drive belts.

6. Remove the upper timing belt covers.

7. Remove the fan pulley assembly.

8. Rotate the crankshaft to align the camshaft timing marks with the pointer dots on the back covers. Verify that the pointer on the crankshaft aligns with the mark on the lower timing cover.

➡When the timing marks are aligned, the No. 2 piston is at Top Dead Center (TDC) compression.

✳✳ WARNING

Align the camshaft and crankshaft sprockets with their alignment marks before removing the timing belt. Failure to align the belt and sprocket marks may result in valve damage.

9. Use tool No. J-8614-01, or a suitable pulley holding tool to remove the crankshaft pulley center bolt. Remove the crankshaft pulley.

10. If present, disconnect the 2 oil cooler hose bracket bolts on the timing cover. Move the oil cooler hoses and bracket off of the lower timing cover.

11. Remove the lower timing belt cover.

12. Remove the pusher assembly (tensioner) from below the belt tensioner pulley. The pusher rod must always face upward to prevent oil leakage. Depress the pusher rod, and insert a wire pin into the hole to keep the pusher rod retracted.

13. Remove the timing belt.

14. Inspect the water pump and replace it if there is any doubt about its condition.

15. Repair any oil or coolant leaks before installing a new timing belt. If the timing belt has been contaminated with oil or coolant, or is damaged, it must be replaced.

To install:

16. Verify that the sprocket timing marks are still aligned and that the groove and the keyway on the crankshaft timing sprocket align with the mark on the oil pump. The white pointers on the camshaft timing sprockets should align with the dots on the front plate.

17. Install the timing belt. Use clips to secure the belt onto each sprocket until the installation is complete. Align the dotted marks on the timing belt with the timing mark opposite the groove on the crankshaft sprocket.

➡The arrows on the timing belt must follow the belt's direction of rotation. The manufacturer's trademark on the belt's spine should be readable left-to-right when the belt is installed.

18. Align the white line on the timing belt with the alignment mark on the right bank camshaft timing pulley. Secure the belt with a clip.

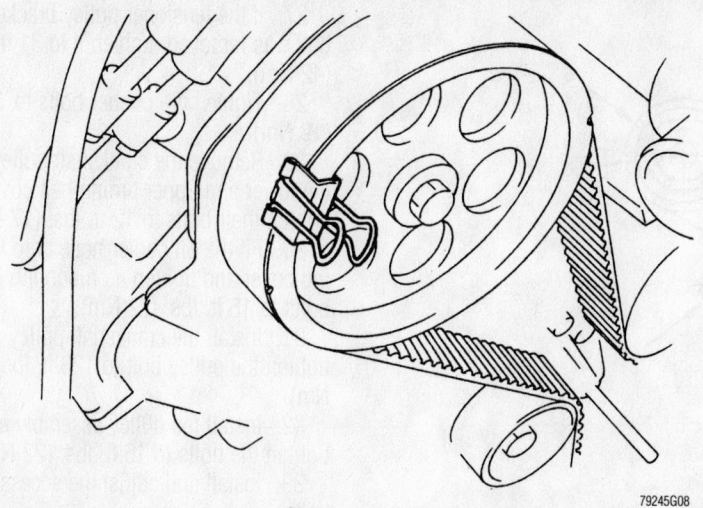

Using a double clip to hold the belt in place — Isuzu 3.2L and 3.5L engines

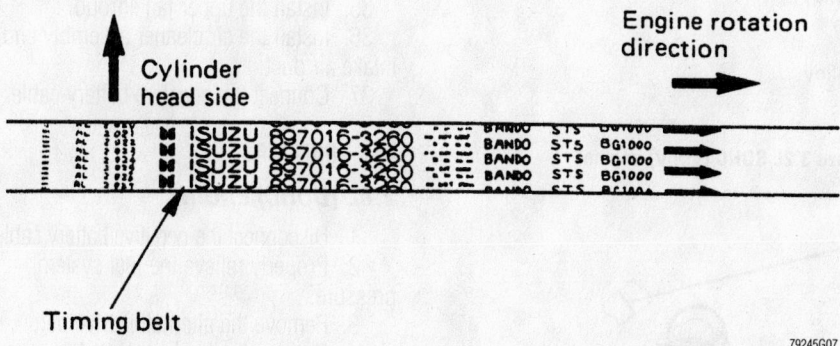

Engine rotation direction

Cylinder head side

Timing belt

For maximum timing belt life, install the belt as shown — Isuzu 3.2L and 3.5L engines

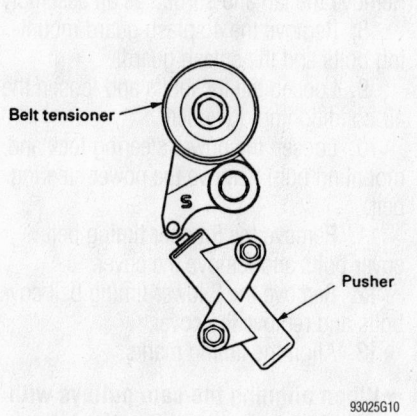

Belt tensioner

Pusher

S

View of timing belt tensioner and pusher — Isuzu 3.2L and 3.5L engines

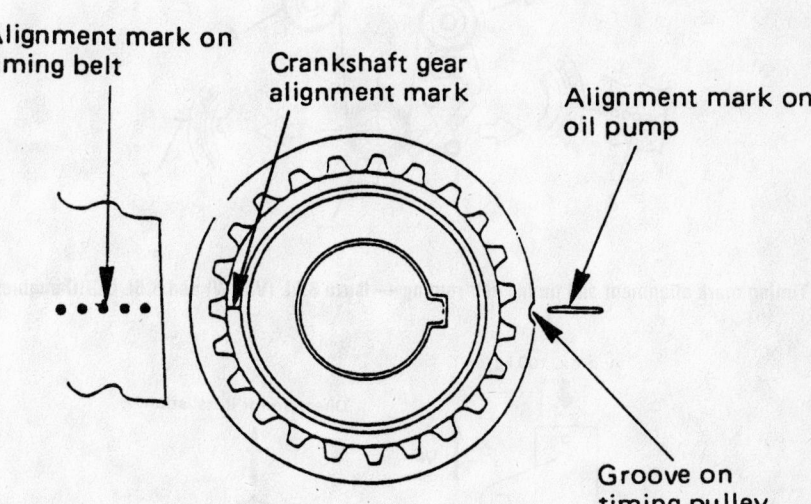

Alignment mark on timing belt

Crankshaft gear alignment mark

Alignment mark on oil pump

Groove on timing pulley

Proper crankshaft alignment marks for timing belt installation — Isuzu 3.2L SOHC (VIN V) engine

✳✳ WARNING

If any binding is felt when adjusting the timing belt tension by turning the crankshaft, STOP turning the engine, because the pistons may be hitting the valves.

19. Rotate the crankshaft counterclockwise to remove the slack between the crankshaft sprocket and the right camshaft timing belt sprocket.

20. Install the belt around the water pump pulley.

21. Install the belt on the idler pulley.

22. Align the white alignment mark on the timing belt with the alignment mark on the left bank camshaft timing belt sprocket.

23. Install the crankshaft pulley and tighten the center bolt by hand. Rotate the crankshaft pulley clockwise to give slack between the crankshaft timing belt pulley and the right bank camshaft timing belt pulley.

24. Insert a 1.4mm piece of wire through the hole in the pusher to hold the rod in. Install the pusher assembly while pushing the tension pulley toward the belt.

25. Pull the pin out from the pusher to release the rod.

26. Remove the clamps from the sprockets. Rotate the crankshaft pulley clockwise 2 turns. Measure the rod protrusion to ensure it is between 0.16 – 0.24 in. (4 – 6mm).

Timing belt service is covered in Section 3 of this manual

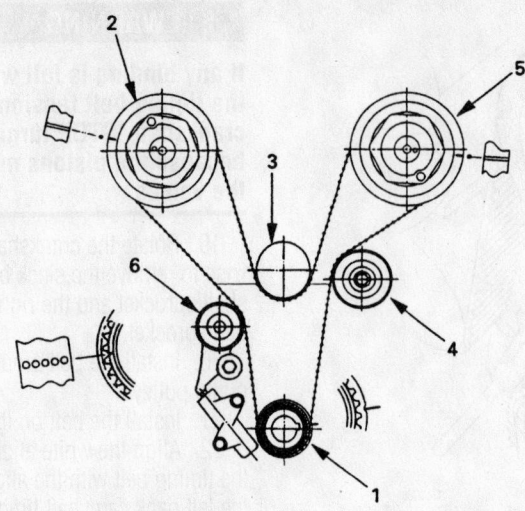

1) Crankshaft timing pulley
2) RH bank timing pulley
3) Water pump pulley
4) Idler pulley
5) LH bank timing pulley
6) Tension pulley

93025G09

Timing mark alignment and timing belt routing — Isuzu 3.2L SOHC (VIN V) engine

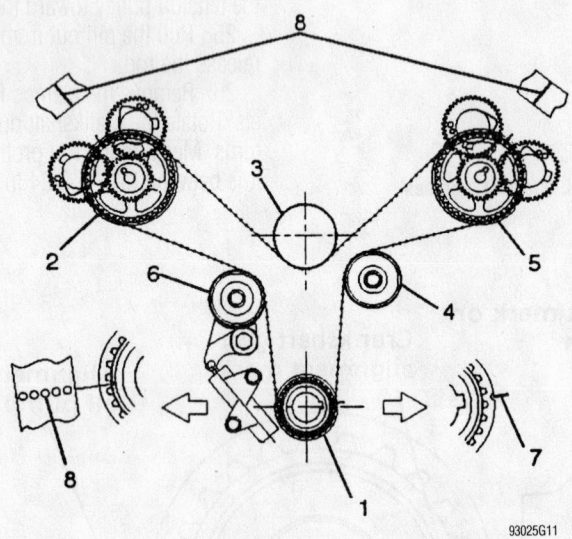

93025G11

Timing mark alignment and timing belt routing — Isuzu 3.2L (VIN W) and 3.5L DOHC engines

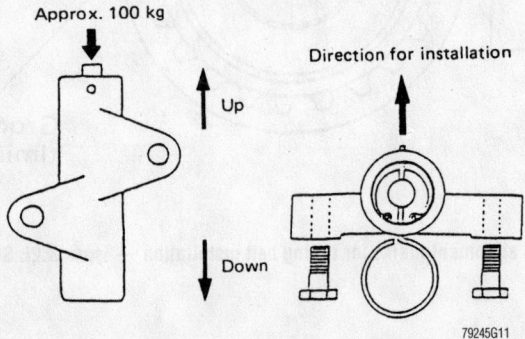

79245G11

View of timing belt tensioner pusher — Isuzu 3.2L and 3.5L engines

27. If the tensioner pulley bracket pivot bolt was removed, tighten it to 31 ft. lbs. (42 Nm).

28. Tighten the pusher bolts to 14 ft. lbs. (19 Nm).

29. Remove the crankshaft pulley. Install the lower and upper timing belt covers and tighten their bolts to 12 ft. lbs. (17 Nm).

30. Fit the oil cooler hose onto the timing cover and tighten its mounting bracket bolts to 16 ft. lbs. (22 Nm).

31. Install the crankshaft pulley and tighten the pulley bolt to 123 ft. lbs. (167 Nm).

32. Install fan pulley assembly and tighten the bolts to 16 ft. lbs. (22 Nm).

33. Install and adjust the accessory drive belts.

34. Install the cooling fan assembly and tighten the bolts to 72 inch lbs. (8 Nm).

35. Install the upper fan shroud.

36. Install the air cleaner assembly and intake air duct.

37. Connect the negative battery cable.

KIA Sportage

2.0L (DOHC) ENGINE

1. Disconnect the negative battery cable.

2. Properly relieve the fuel system pressure.

3. Remove the alternator drive belt.

4. Remove the fresh air duct from the top of the radiator.

5. Remove the upper radiator hose.

6. Remove the 4 attaching nuts to the clutch fan.

7. Remove the 5 fan shroud bolts. Remove the fan and shroud as an assembly.

8. Remove the 4 splash guard mounting bolts and the splash guard.

9. Loosen the lockbolts and loosen the air conditioning drive belt.

10. Loosen the power steering lock and mounting bolt. Remove the power steering belt.

11. Remove the 5 upper timing belt cover bolts and remove the cover.

12. Remove the 2 lower timing belt cover bolts and remove the cover.

13. Align the timing marks.

➡When aligning the cam pulleys with the seal plate marks, align the left cam pulley I mark and the right cam pulley on the E mark.

❊❊ **WARNING**

When aligning the timing marks, do not turn the timing gear counterclockwise. Damage to the engine will occur.

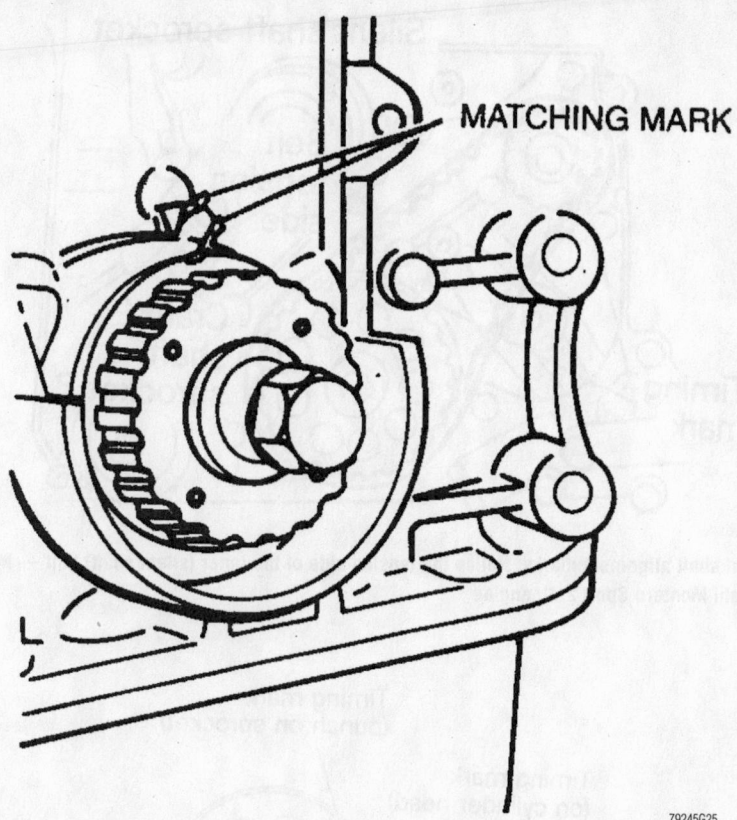

79245G25

Align the crankshaft marks before removing the timing belt—KIA Sportage 2.0L (DOHC) engine

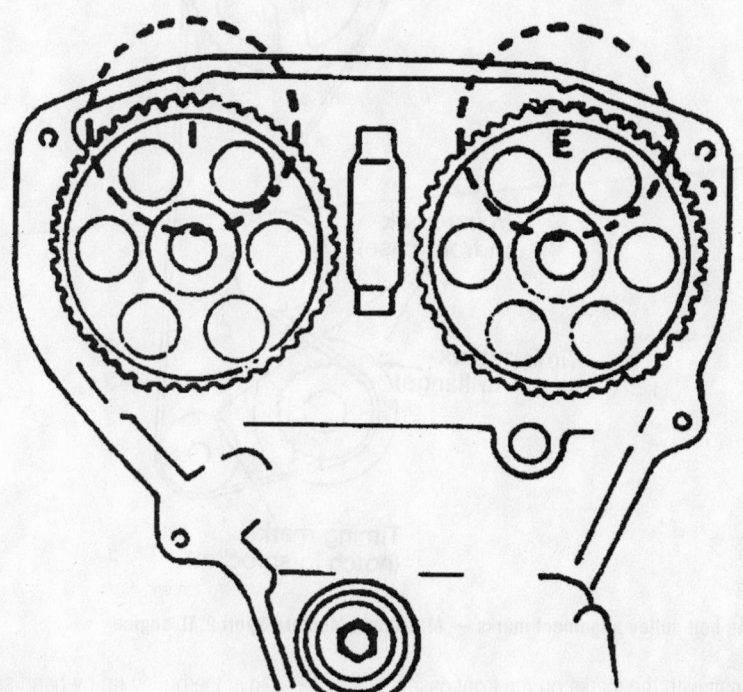

79245G28

Proper alignment of the intake and exhaust camshaft pulley timing marks—KIA Sportage 2.0L (DOHC) engine

14. Loosen the tensioner bolt. Pry the tensioner away from the belt. Tighten the tensioner bolt to relieve the pressure against the timing belt.

15. Remove the timing belt.

16. Remove the camshaft pulley attaching bolts. Use a driver placed through one of the holes in the pulley to prevent it from moving when the attaching bolt is removed. Remove and mark the pulleys.

17. Remove the lower timing belt pulley and locking bolt.

To install:

18. Install the camshaft pulleys. Tighten the bolts to 35 – 48 ft. lbs. (47 – 65 Nm).

19. Install the lower timing belt pulley and locking bolt. Tighten the bolt to 120 ft. lbs. (162 Nm).

20. If necessary, align the timing marks.

➡When aligning the cam pulleys with the seal plate marks, align the left cam pulley "I" mark and the right cam pulley on the "E" mark.

✳✳ WARNING

When aligning the timing marks, do not turn the timing gear counterclockwise. Damage to the engine will occur.

21. Loosen the tensioner bolt. Pry the tensioner away from the belt. Tighten tensioner bolt to relieve the pressure against the timing belt.

22. Install the timing belt.

✳✳ WARNING

If any binding is felt when adjusting the timing belt tension by turning the crankshaft, STOP turning the engine, because the pistons may be hitting the valves.

23. Loosen the tensioner bolt and allow the tensioner to tighten the timing belt. Tighten the tensioner bolt 27 – 38 ft. lbs. (37 – 52 Nm).

24. Check the timing belt deflection. If there is more than 0.30 – 0.33 in. (7.5 – 8.5mm) replace the tensioner spring.

25. Install the 2 lower timing belt cover bolts to the cover.

26. Install the 5 upper timing belt cover bolts to the cover.

27. Install and adjust the air conditioning and power steering drive belts.

28. Install the splash guard.

29. Install and tighten the alternator belt.

Heater Core replacement is covered in Section 2 of this manual

30. Install the upper radiator hose.

31. Install the fan and shroud as an assembly.

32. Install the 4 attaching nuts to the clutch fan.

33. Install the 5 fan shroud bolts.

34. Install the fresh air duct to the top of the radiator.

35. Properly fill the cooling system.

36. Connect the negative battery cable.

37. Start the engine and check for leaks.

38. Road test the vehicle.

Mitsubishi Montero

2.4L (VIN G) ENGINE

1. Be sure that the engine's No. 1 piston is at Top Dead Center (TDC) in the compression stroke.

✷✷ CAUTION

Wait at least 90 seconds after the negative battery cable is disconnected to prevent possible deployment of the air bag.

2. Disconnect the negative battery cable.

3. Remove the spark plug wires from the tree on the upper cover.

4. Drain the cooling system.

5. Remove the shroud, fan and accessory drive belts.

6. Remove the radiator as required.

7. Remove the power steering pump, alternator, air conditioning compressor, tension pulley and accompanying brackets, as required.

8. Remove the upper front timing belt cover.

9. Remove the water pump pulley and the crankshaft pulley(s).

10. Remove the lower timing belt cover mounting screws and remove the cover.

11. If the belt(s) are to be reused, mark the direction of rotation on the belt.

12. Remove the timing (outer) belt tensioner and remove the belt. Unbolt the tensioner from the block and remove.

13. Remove the outer crankshaft sprocket and flange.

14. Remove the silent shaft (inner) belt tensioner and remove the inner belt. Unbolt the tensioner from the block and remove it.

15. To remove the camshaft sprockets, use SST MB990767-01 and MIT308239 or their equivalents.

To install:

16. Install the camshaft sprockets and tighten the center bolt to 65 ft. lbs. (90 Nm).

17. Align the timing mark of the silent shaft belt sprockets on the crankshaft and

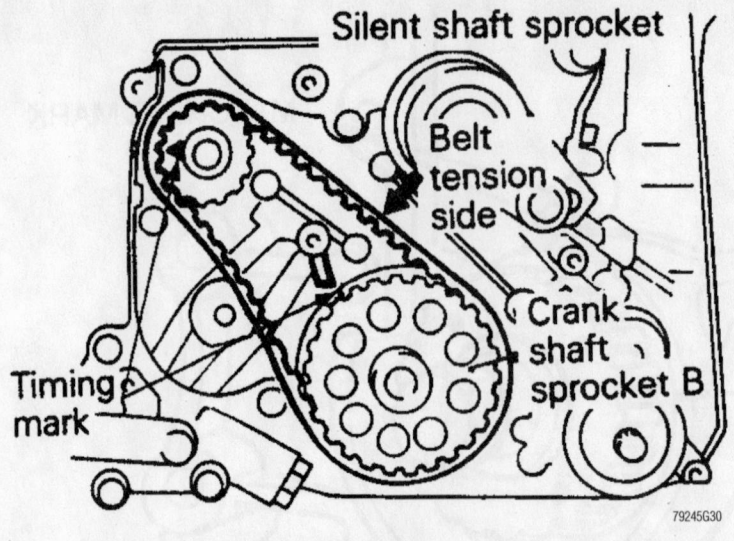

Silent shaft alignment marks. Notice the tension side of the inner (silent shaft) belt — Mitsubishi Montero Sport 2.4L engine

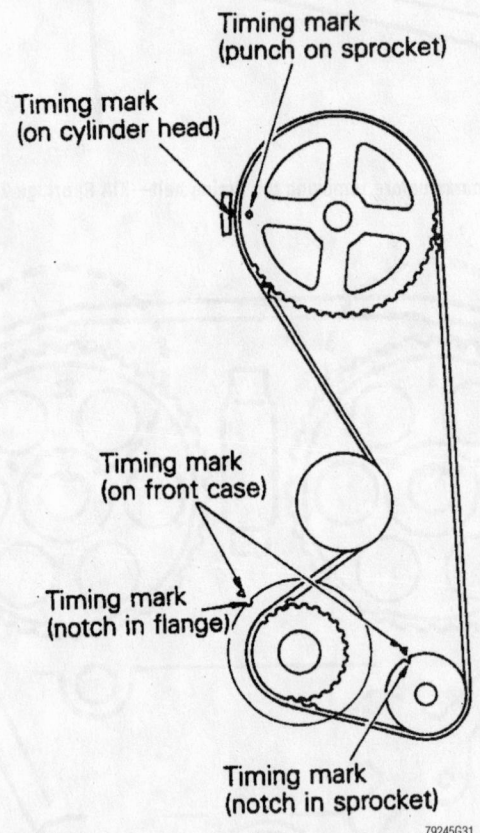

Timing belt pulley alignment marks — Mitsubishi Montero Sport 2.4L engine

silent shaft with the marks on the front case. Wrap the silent shaft belt around the sprockets so there is no slack in the upper span of the belt and the timing marks are aligned.

18. Install the tensioner initially so the actual center of the pulley is above and to the left of the installation bolt.

19. Move the pulley up by hand so the center span of the long side of the belt deflects about ¼ in. (6mm).

20. Hold the pulley tightly so it does not rotate when the bolt is tightened. Tighten the bolt to 15 ft. lbs. (20 Nm). If the pulley has moved, the belt will be too tight.

21. Install the timing belt tensioner fully toward the water pump and temporarily tighten the bolts. Place the upper end of the spring against the water pump body. Align the timing marks of the cam, crankshaft and oil pump sprockets with the corresponding marks on the front case or head.

➡**If the following steps are not followed exactly, there is a chance that the silent shaft alignment will be 180 degrees off. This will cause a noticeable vibration in the engine and the entire procedure will have to be repeated.**

22. Before installing the timing belt, ensure that the left side silent shaft is in the correct position.

➡**It is possible to align the timing marks on the camshaft sprocket, crankshaft sprocket and the oil pump sprocket with the left balance shaft out of alignment.**

23. With the timing mark on the oil pump pulley aligned with the mark on the front case, check the alignment of the left balance shaft to assure correct shaft timing.

   a. Remove the plug located on the left side of the block in the area of the starter.

   b. Insert a tool having a shaft diameter of 0.3 in. (8mm) into the hole.

   c. With the timing marks still aligned, the tool must be able to go in at least 2⅓ in. If it can only go in about 1 inch, turn the oil pump sprocket one complete revolution.

   d. Recheck the position of the balance shaft with the timing marks realigned. Leave the tool in place to hold the silent shaft while continuing.

24. Install the belt to the crankshaft sprocket, oil pump sprocket and the camshaft sprocket, in that order. While doing so, be sure there is no slack between the sprockets except where the tensioner will take it up when released.

25. Recheck the timing marks' alignment.

26. If all are aligned, loosen the tensioner mounting bolt and allow the tensioner to apply tension to the belt.

27. Remove the tool that is holding the silent shaft in place and turn the crankshaft clockwise a distance equal to 2 teeth of the camshaft sprocket. This will allow the tensioner to automatically tension the belt the proper amount.

⁂ **WARNING**

Do not manually apply pressure to the tensioner. This will over tighten the belt and will cause a howling noise.

28. First tighten the lower mounting bolt and then tighten the upper spacer bolt.

⁂ **WARNING**

If any binding is felt when adjusting the timing belt tension by turning the crankshaft, STOP turning the engine, because the pistons may be hitting the valves.

29. To verify that belt tension is correct, check that the deflection of the longest span (between the camshaft and oil pump sprockets) is ½ in. (13mm).

30. Install the lower timing belt cover. Be sure the packing is properly positioned in the inner grooves of the covers when installing.

31. Install the water pump pulley and the crankshaft pulley(s).

32. Install the upper front timing belt cover.

33. Install the power steering pump, alternator, air conditioning compressor, tension pulley and accompanying brackets, as required.

34. Install the radiator, shroud, fan and accessory drive belts.

35. Install the spark plug wires to the tree on the upper cover.

36. Refill the cooling system.

37. Connect the negative battery cable. Start the engine and check for leaks.

3.0L (VIN H) AND 3.5L (VIN M AND R) ENGINES

1. Disconnect the negative battery cable.

2. Drain the engine coolant and store it for reinstallation. Remove the upper radiator hose.

3. Remove the cooling fan shroud assembly.

4. Remove the cooling fan-to-clutch bolts and the fan.

5. Remove the cooling fan clutch-to-water pump nuts and the clutch assembly.

6. Remove the drive belts for the alternator, power steering pump and air conditioning compressor.

7. Disconnect the electrical connectors from the alternator.

8. Remove the alternator-to-engine bolts and the alternator bracket-to-engine bolts; then, remove the alternator and bracket from the engine.

9. Remove the power steering pump cover. Remove the power steering pump-to-engine bolts and move the pump aside with the hoses and electrical connector attached.

10. Remove the air conditioning compressor-to-bracket bolts and move the compressor aside with the lines and electrical connector attached.

11. Remove the air conditioning compressor bracket-to-engine bolts and the bracket.

12. Remove the timing indicator bracket (near crankshaft pulley) bolts and the bracket.

13. Remove the accessory mount assembly-to-engine bolts and the mount assembly.

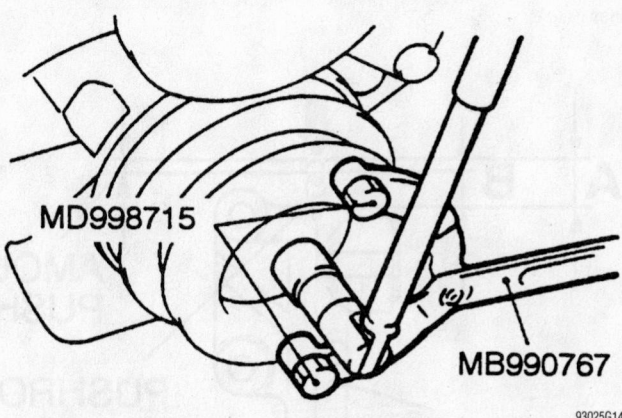

MD998715

MB990767

93025G14

Removing or installing the crankshaft pulley bolt — Mitsubishi 1999 – 01 3.0L (VIN H) and 3.5L (VIN M and R) engines

Brake service is covered in Section 4 of this manual

14. Remove the upper timing belt cover assembly.

15. Using the End Yoke Holder tool MD990767 and 2 Crankshaft Pulley Holder Pin tools MD998715, or equivalent to hold the crankshaft pulley, and a socket wrench, remove the crankshaft pulley bolt and the pulley.

16. Remove the lower timing belt cover.

17. Rotate the crankshaft clockwise to align the timing marks to position the No. 1 cylinder at the Top Dead Center (TDC) of its compression stroke.

18. Use chalk to mark the rotating (clockwise) direction of the timing belt for reinstallation purposes.

19. Loosen the auto-tensioner pulley center bolt and remove the timing bolt.

20. Remove the auto-tensioner pulley and the auto-tensioner arm assembly.

To install:

21. Press the end of the auto-tensioner inward with 72 – 145 ft. lbs. (98 – 196 Nm) of force and measure the distance that the pushrod is pushed in. If the standard distance is not 0.04 in. (1mm), replace the auto-tensioner.

22. Position the auto-tensioner in a soft-jawed vise and SLOWLY compress the pushrod until the pushrod and housing holes align; then, install a setting pin to secure the auto-tensioner in the retracted position.

23. Align the camshaft and crankshaft TDC timing marks.

24. Install the timing belt (noting its rotational direction) so that there is no deflection between the sprockets and pulleys in the following manner:
- Crankshaft sprocket
- Idler pulley
- Left camshaft sprocket
- Water pump pulley
- Right camshaft sprocket
- Tension pulley

25. Turn the camshaft sprocket counterclockwise until the tension side of the timing belt is firmly stretched, then, recheck the timing marks.

26. Using the Tension Pulley Socket Wrench tool MD998767, or equivalent, push the tensioner pulley into the timing belt and secure the center bolt.

27. Using the Crankshaft Pulley Spacer tool MD998769, or equivalent, rotate the crankshaft ¼ turn counterclockwise, then, turn it again clockwise to align the timing marks.

28. Loosen the timing belt tensioner center bolt. Using the Tension Pulley Socket Wrench tool MD998767, or equivalent, and a torque wrench, apply 39 inch lbs. (4.4 Nm) pressure on the timing belt. Torque the tensioner pulley center bolt to 35 ft. lbs. (48 Nm).

29. Remove the setting pin from the auto-tensioner.

30. Rotate the crankshaft 2 complete revolutions and realign the timing marks. Then, wait for 5 minutes until the auto-tensioner pushrod extends to its standard value. If the standard value is not 0.15 – 0.20 in. (3.8 – 5.0mm), repeat the adjustment procedure. If

View of the timing belt alignment marks — Mitsubishi 1999 – 01 3.0L (VIN H) and 3.5L (VIN M and R) engines

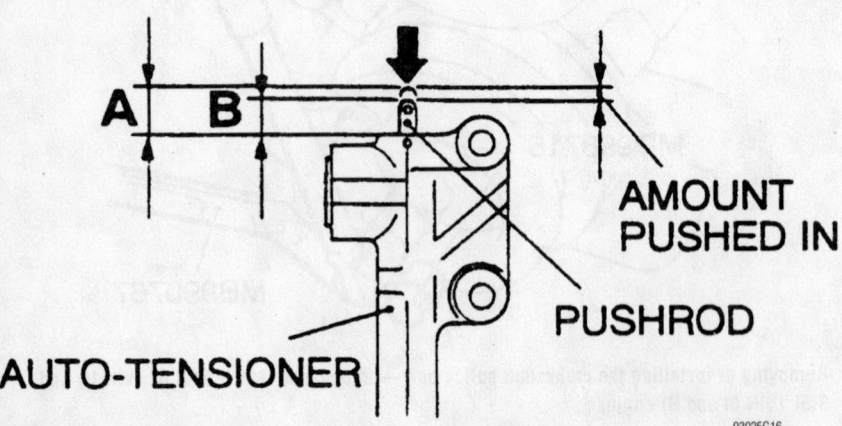

Inspecting the auto-tensioner movement—Mitsubishi 1999–01 3.0L (VIN H) and 3.5L (VIN M and R) engines

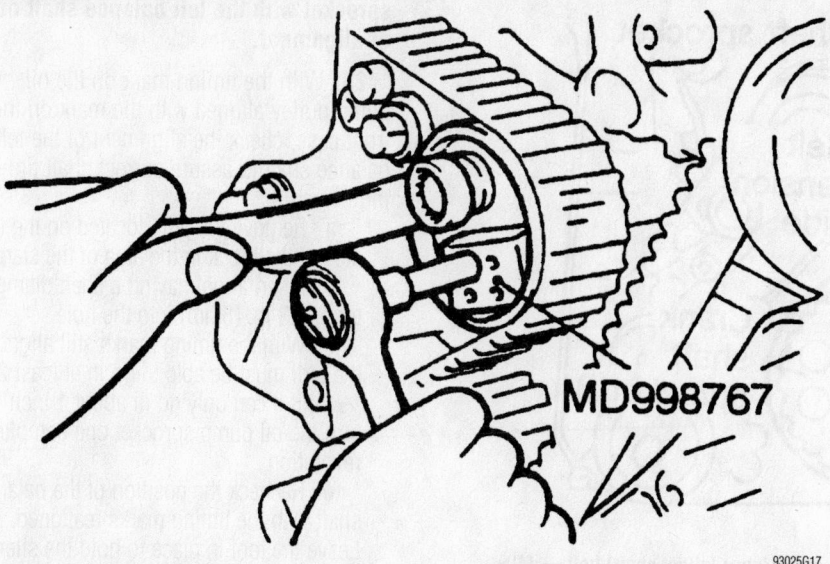

Adjusting the timing belt tensioner pulley—Mitsubishi 1999–01 3.0L (VIN H) and 3.5L (VIN M and R) engines

93025G17

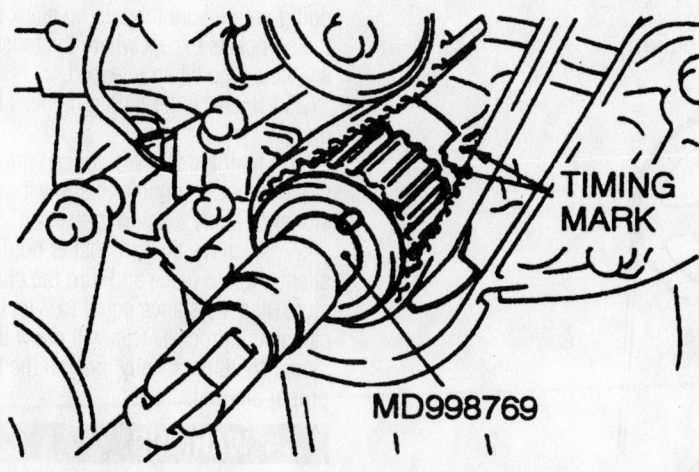

Using crankshaft spacer tool to rotate the crankshaft—Mitsubishi 1999–01 3.0L (VIN H) and 3.5L (VIN M and R) engines

93025G18

the standard value is still not achieved, replace the auto-tensioner.

31. Install the lower timing belt cover and crankshaft pulley.

32. Using the End Yoke Holder tool MD990767 and 2 Crankshaft Pulley Holder Pin tools MD998715, or equivalent to hold the crankshaft pulley, and a socket torque wrench, torque the crankshaft pulley bolt to 134 ft. lbs. (181 Nm).

33. Install the upper timing belt cover assembly.

34. Install the remaining items by reversing the removal procedures.

35. Refill the cooling system.

36. Connect the negative battery cable.

Mitsubishi Montero Sport

2.4L (VIN G) ENGINE

1. Be sure that the engine's No. 1 piston is at Top Dead Center (TDC) in the compression stroke.

✳✳ CAUTION

Wait at least 90 seconds after the negative battery cable is disconnected to prevent possible deployment of the air bag.

2. Disconnect the negative battery cable.

3. Remove the spark plug wires from the tree on the upper cover.

4. Drain the cooling system.

5. Remove the shroud, fan and accessory drive belts.

6. Remove the radiator as required.

7. Remove the power steering pump, alternator, air conditioning compressor, tension pulley and accompanying brackets, as required.

8. Remove the upper front timing belt cover.

9. Remove the water pump pulley and the crankshaft pulley(s).

10. Remove the lower timing belt cover mounting screws and remove the cover.

11. If the belt(s) are to be reused, mark the direction of rotation on the belt.

12. Remove the timing (outer) belt tensioner and remove the belt. Unbolt the tensioner from the block and remove.

13. Remove the outer crankshaft sprocket and flange.

14. Remove the silent shaft (inner) belt tensioner and remove the inner belt. Unbolt the tensioner from the block and remove it.

15. To remove the camshaft sprockets, use SST MB990767-01 and MIT308239 or their equivalents.

To install:

16. Install the camshaft sprockets and tighten the center bolt to 65 ft. lbs. (90 Nm).

17. Align the timing mark of the silent shaft belt sprockets on the crankshaft and silent shaft with the marks on the front case. Wrap the silent shaft belt around the sprockets so there is no slack in the upper span of the belt and the timing marks are aligned.

18. Install the tensioner initially so the actual center of the pulley is above and to the left of the installation bolt.

19. Move the pulley up by hand so the center span of the long side of the belt deflects about ¼ in. (6mm).

20. Hold the pulley tightly so it does not rotate when the bolt is tightened. Tighten the bolt to 15 ft. lbs. (20 Nm). If the pulley has moved, the belt will be too tight.

21. Install the timing belt tensioner fully toward the water pump and temporarily tighten the bolts. Place the upper end of the spring against the water pump body. Align the timing marks of the cam, crankshaft and oil pump sprockets with the corresponding marks on the front case or head.

For complete Engine Mechanical specifications, see Section 1 of this manual

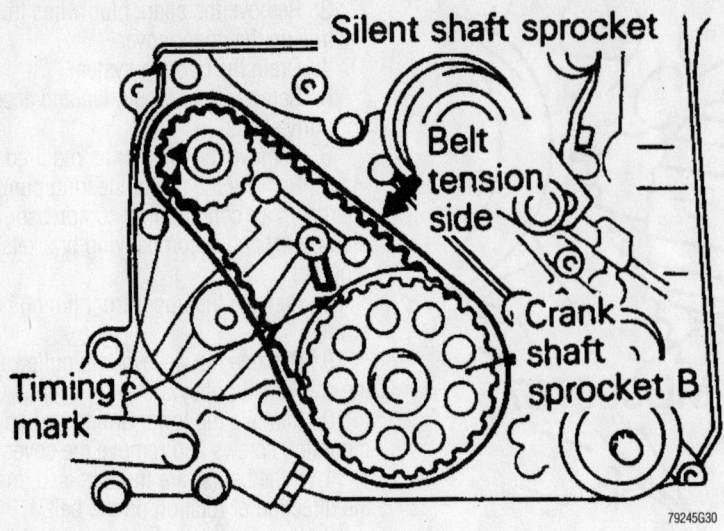

Silent shaft alignment marks. Notice the tension side of the inner (silent shaft) belt — Mitsubishi Montero Sport 2.4L engine

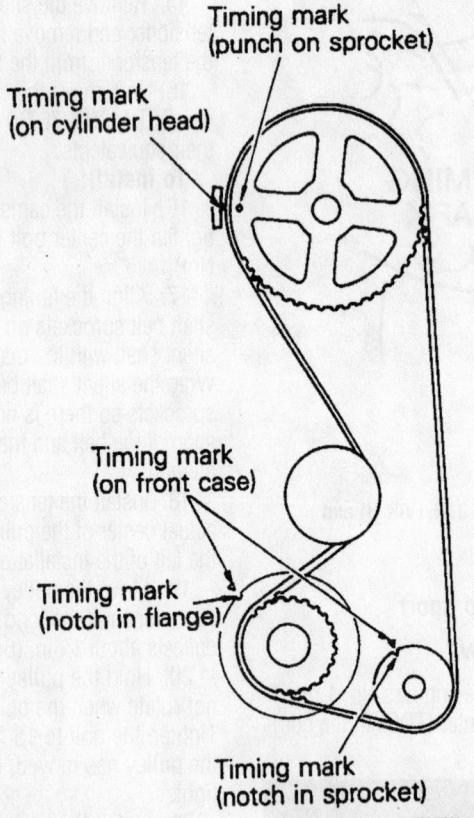

Timing belt pulley alignment marks — Mitsubishi Montero Sport 2.4L engine

➡ **If the following steps are not followed exactly, there is a chance that the silent shaft alignment will be 180 degrees off. This will cause a noticeable vibration in the engine and the entire procedure will have to be repeated.**

22. Before installing the timing belt, ensure that the left side silent shaft is in the correct position.

➡ **It is possible to align the timing marks on the camshaft sprocket, crankshaft sprocket and the oil pump**

sprocket with the left balance shaft out of alignment.

23. With the timing mark on the oil pump pulley aligned with the mark on the front case, check the alignment of the left balance shaft to assure correct shaft timing.

  a. Remove the plug located on the left side of the block in the area of the starter.

  b. Insert a tool having a shaft diameter of 0.3 in. (8mm) into the hole.

  c. With the timing marks still aligned, the tool must be able to go in at least 2 1/3 in. If it can only go in about 1 inch, turn the oil pump sprocket one complete revolution.

  d. Recheck the position of the balance shaft with the timing marks realigned. Leave the tool in place to hold the silent shaft while continuing.

24. Install the belt to the crankshaft sprocket, oil pump sprocket and the camshaft sprocket, in that order. While doing so, be sure there is no slack between the sprockets except where the tensioner will take it up when released.

25. Recheck the timing marks' alignment.

26. If all are aligned, loosen the tensioner mounting bolt and allow the tensioner to apply tension to the belt.

27. Remove the tool that is holding the silent shaft in place and turn the crankshaft clockwise a distance equal to 2 teeth of the camshaft sprocket. This will allow the tensioner to automatically tension the belt the proper amount.

✳✳ WARNING

Do not manually apply pressure to the tensioner. This will over tighten the belt and will cause a howling noise.

28. First tighten the lower mounting bolt and then tighten the upper spacer bolt.

✳✳ WARNING

If any binding is felt when adjusting the timing belt tension by turning the crankshaft, STOP turning the engine, because the pistons may be hitting the valves.

29. To verify that belt tension is correct, check that the deflection of the longest span (between the camshaft and oil pump sprockets) is 1/2 in. (13mm).

30. Install the lower timing belt cover. Be sure the packing is properly positioned in

the inner grooves of the covers when installing.

31. Install the water pump pulley and the crankshaft pulley(s).

32. Install the upper front timing belt cover.

33. Install the power steering pump, alternator, air conditioning compressor, tension pulley and accompanying brackets, as required.

34. Install the radiator, shroud, fan and accessory drive belts.

35. Install the spark plug wires to the tree on the upper cover.

36. Refill the cooling system.

37. Connect the negative battery cable. Start the engine and check for leaks.

3.0L (VIN H) AND 3.5L (VIN M AND R) ENGINES

1. Disconnect the negative battery cable.

2. Drain the engine coolant and store it for reinstallation. Remove the upper radiator hose.

3. Remove the cooling fan shroud assembly.

4. Remove the cooling fan-to-clutch bolts and the fan.

5. Remove the cooling fan clutch-to-water pump nuts and the clutch assembly.

6. Remove the drive belts for the alternator, power steering pump and air conditioning compressor.

7. Disconnect the electrical connectors from the alternator.

8. Remove the alternator-to-engine bolts and the alternator bracket-to-engine bolts; then, remove the alternator and bracket from the engine.

9. Remove the power steering pump cover. Remove the power steering pump-to-engine bolts and move the pump aside with the hoses and electrical connector attached.

10. Remove the air conditioning compressor-to-bracket bolts and move the compressor aside with the lines and electrical connector attached.

11. Remove the air conditioning compressor bracket-to-engine bolts and the bracket.

12. Remove the timing indicator bracket (near crankshaft pulley) bolts and the bracket.

13. Remove the accessory mount assembly-to-engine bolts and the mount assembly.

14. Remove the upper timing belt cover assembly.

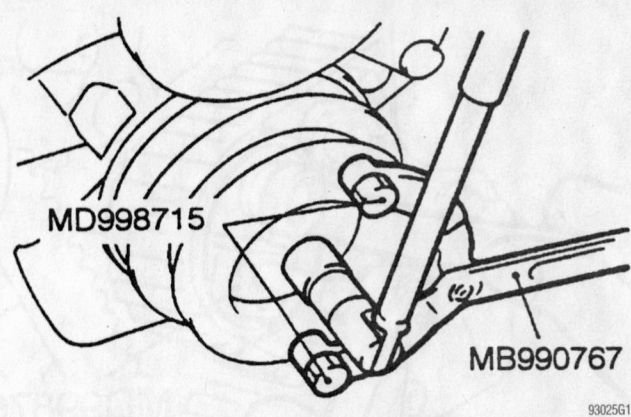

Removing or installing the crankshaft pulley bolt — Mitsubishi 1999 – 01 3.0L (VIN H) and 3.5L (VIN M and R) engines

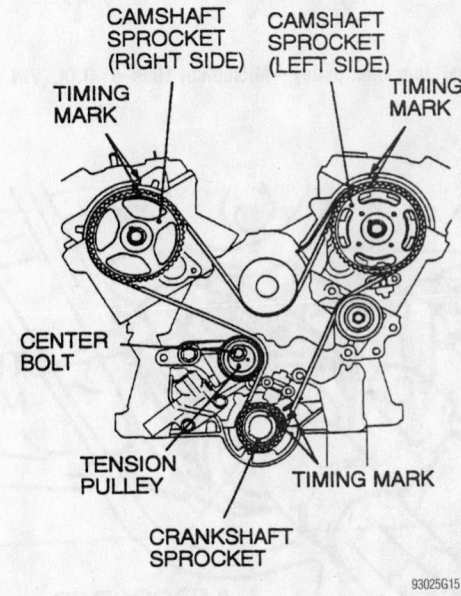

View of the timing belt alignment marks — Mitsubishi 1999 – 01 3.0L (VIN H) and 3.5L (VIN M and R) engines

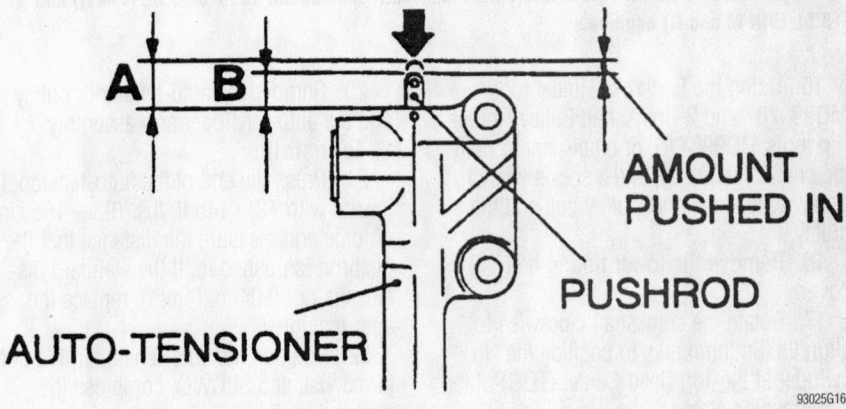

Inspecting the auto-tensioner movement—Mitsubishi 1999–01 3.0L (VIN H) and 3.5L (VIN M and R) engines

For Accessory Drive Belt illustrations, see Section 1 of this manual

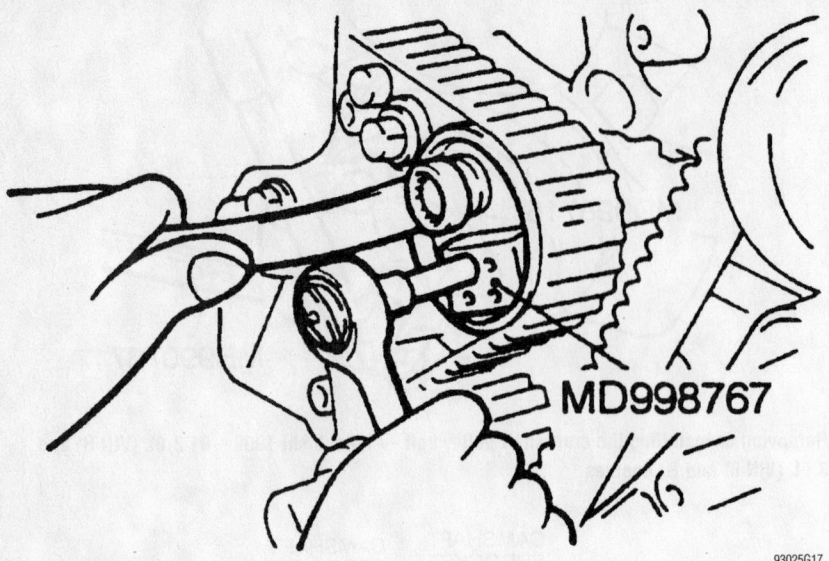

Adjusting the timing belt tensioner pulley—Mitsubishi 1999–01 3.0L (VIN H) and 3.5L (VIN M and R) engines

93025G17

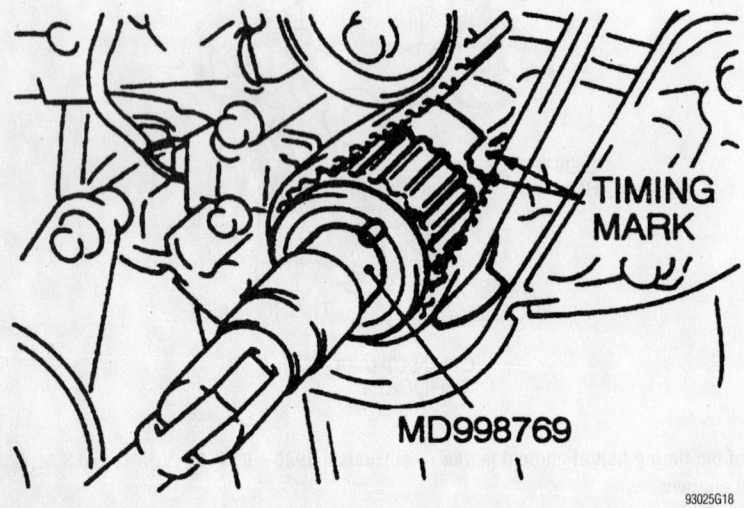

Using crankshaft spacer tool to rotate the crankshaft—Mitsubishi 1999–01 3.0L (VIN H) and 3.5L (VIN M and R) engines

93025G18

15. Using the End Yoke Holder tool MD990767 and 2 Crankshaft Pulley Holder Pin tools MD998715, or equivalent to hold the crankshaft pulley, and a socket wrench, remove the crankshaft pulley bolt and the pulley.

16. Remove the lower timing belt cover.

17. Rotate the crankshaft clockwise to align the timing marks to position the No. 1 cylinder at the Top Dead Center (TDC) of its compression stroke.

18. Use chalk to mark the rotating (clockwise) direction of the timing belt for reinstallation purposes.

19. Loosen the auto-tensioner pulley center bolt and remove the timing bolt.

20. Remove the auto-tensioner pulley and the auto-tensioner arm assembly.

To install:

21. Press the end of the auto-tensioner inward with 72 – 145 ft. lbs. (98 – 196 Nm) of force and measure the distance that the pushrod is pushed in. If the standard distance is not 0.04 in. (1mm), replace the auto-tensioner.

22. Position the auto-tensioner in a soft-jawed vise and SLOWLY compress the pushrod until the pushrod and housing holes align; then, install a setting pin to secure the auto-tensioner in the retracted position.

23. Align the camshaft and crankshaft TDC timing marks.

24. Install the timing belt (noting its rotational direction) so that there is no deflection between the sprockets and pulleys in the following manner:
- Crankshaft sprocket
- Idler pulley
- Left camshaft sprocket
- Water pump pulley
- Right camshaft sprocket
- Tension pulley

25. Turn the camshaft sprocket counterclockwise until the tension side of the timing belt is firmly stretched, then, recheck the timing marks.

26. Using the Tension Pulley Socket Wrench tool MD998767, or equivalent, push the tensioner pulley into the timing belt and secure the center bolt.

27. Using the Crankshaft Pulley Spacer tool MD998769, or equivalent, rotate the crankshaft ¼ turn counterclockwise, then, turn it again clockwise to align the timing marks.

28. Loosen the timing belt tensioner center bolt. Using the Tension Pulley Socket Wrench tool MD998767, or equivalent, and a torque wrench, apply 39 inch lbs. (4.4 Nm) pressure on the timing belt. Torque the tensioner pulley center bolt to 35 ft. lbs. (48 Nm).

29. Remove the setting pin from the auto-tensioner.

30. Rotate the crankshaft 2 complete revolutions and realign the timing marks. Then, wait for 5 minutes until the auto-tensioner pushrod extends to its standard value. If the standard value is not 0.15 – 0.20 in. (3.8 – 5.0mm), repeat the adjustment procedure. If the standard value is still not achieved, replace the auto-tensioner.

31. Install the lower timing belt cover and crankshaft pulley.

32. Using the End Yoke Holder tool MD990767 and 2 Crankshaft Pulley Holder Pin tools MD998715, or equivalent to hold the crankshaft pulley, and a socket torque wrench, torque the crankshaft pulley bolt to 134 ft. lbs. (181 Nm).

33. Install the upper timing belt cover assembly.

34. Install the remaining items by reversing the removal procedures.

35. Refill the cooling system.

36. Connect the negative battery cable.

Nissan Pathfinder

3.0L (VG30E) AND 3.3L (VG33E) ENGINES

1. Remove the engine undercover.

2. Remove the radiator shroud, the fan and the pulleys.

3. Drain the coolant from the radiator and remove the water pump hose.

✳✳ CAUTION

When draining the coolant, keep in mind that cats and dogs are attracted by the ethylene glycol antifreeze, and are quite likely to drink any that is left in an uncovered container or in puddles on the ground. This will prove fatal in sufficient quantity. Always drain the coolant into a sealable container. Coolant should be reused unless it is contaminated or several years old.

4. Remove the radiator.
5. Remove the power steering, air conditioning compressor and alternator drive belts.
6. Remove the spark plugs.
7. Remove the distributor protector (dust shield).
8. Remove the air conditioning compressor drive belt idler pulley and bracket.
9. Remove the fresh air intake tube at the cylinder head cover.
10. Disconnect the radiator hose at the thermostat housing.
11. Remove the crankshaft pulley bolt, then pull off the pulley with a suitable puller.
12. Remove the bolts, then remove the front upper and lower timing belt covers.
13. Set the No. 1 piston at Top Dead Center (TDC) of its compression stroke. Align the punchmark on the left camshaft sprocket with the punchmark on the timing belt upper rear cover. Align the punchmark on the crankshaft sprocket with the notch on the oil pump housing. Temporarily install the crank pulley bolt so the crankshaft can be rotated if necessary.
14. Loosen the timing belt tensioner and return spring, then remove the timing belt.
To install:

✳✳ CAUTION

Before installing the timing belt, confirm that the No. 1 cylinder is set at the TDC of the compression stroke.

15. Remove both cylinder head covers and loosen all rocker arm shaft retaining bolts.

➡**The rocker arm shaft bolts MUST be loosened so that the correct belt tension can be obtained.**

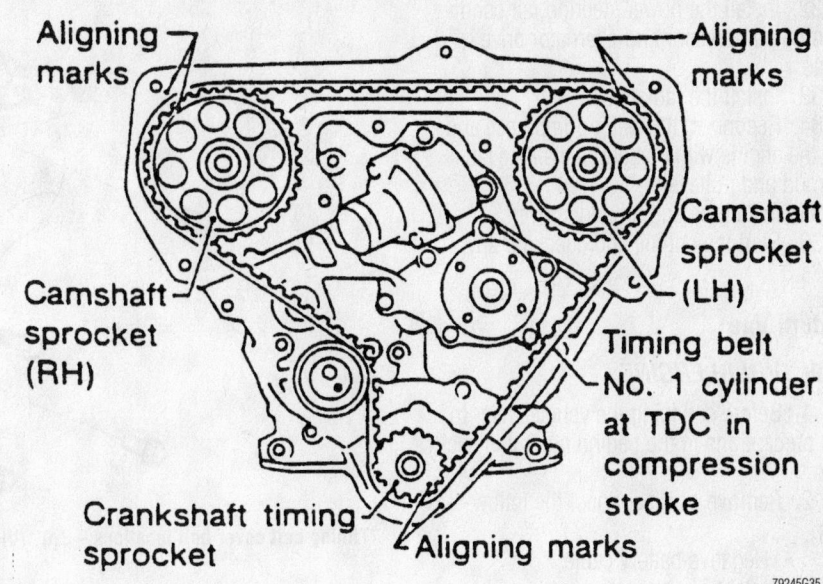

Timing belt alignment mark locations — Nissan Pathfinder 3.0L (VG30E) and 3.3L (VG33E) engines

16. Install the tensioner and the return spring. Using a hexagon wrench, turn the tensioner clockwise and temporarily tighten the locknut.
17. Be sure that the timing belt is clean and free from oil or water.
18. When installing the timing belt, align the white lines on the belt with the punchmarks on the camshaft and crankshaft sprockets. Have the arrow on the timing belt pointing toward the front belt covers.

➡**A good way (although rather tedious!) to check for proper timing belt installation is to count the number of belt teeth between the timing marks. There are 133 teeth on the belt; there should be 40 teeth between the timing marks on the left and right side camshaft sprockets, and 43 teeth between the timing marks on the left side camshaft sprocket and the crankshaft sprocket.**

19. While keeping the tensioner steady, loosen the locknut with a hex wrench.
20. Turn the tensioner approximately 70 – 80 degrees clockwise with the wrench, then tighten the locknut.

✳✳ WARNING

If any binding is felt when adjusting the timing belt tension by turning the crankshaft, STOP turning the engine, because the pistons may be hitting the valves.

21. Turn the crankshaft in a clockwise direction several times, then **slowly** set the No. 1 piston to TDC of the compression stroke.
22. Apply 22 lbs. (10 kg) of pressure (push it in!) to the center span of the timing belt between the right side camshaft sprocket and the tensioner pulley, then loosen the tensioner locknut.
23. Using a 0.0138 in. (0.35mm) thick feeler gauge (the actual width of the blade **must** be ½ in. or 13mm!), turn the crankshaft clockwise (**slowly!**). The timing belt should move approximately 2½ teeth. Tighten the tensioner locknut, turn the crankshaft slightly and remove the feeler gauge.
24. Slowly rotate the crankshaft clockwise several more times, then set the No. 1 piston to TDC of the compression stroke.
25. Position the 2 timing covers on the block, then tighten the mounting bolts to 24 ft. lbs. (35 Nm).
26. Press the crankshaft pulley onto the shaft, then tighten the bolt to 90 – 98 ft. lbs. (123 – 132 Nm).
27. Connect the radiator hose to the thermostat housing.
28. Reconnect the fresh air intake tube at the cylinder head cover.
29. Install the air conditioning compressor drive belt idler pulley and bracket.
30. Install the distributor protector (dust shield).
31. Install the spark plugs.

32. Install the power steering, air conditioning compressor and alternator drive belts.

33. Install the radiator.

34. Reconnect the water pump hose and fill the engine with coolant. Install the fan shroud and pulleys.

35. Install the engine undercover.

36. Start the engine and check for any leaks.

Saturn Vue

3.0L (VIN B) ENGINE

1. Before servicing the vehicle, refer to the precautions in the beginning of this section.

2. Remove or disconnect the following:

- Negative battery cable
- Intake air resonator
- Right front wheel and splash shield
- Water pump and idler bolts, loosen only
- Accessory drive belt
- Right engine mount
- Water pump pulley
- Idler pulley
- Electrical harness connectors at the harness channel
- Electrical harness channel from the front cover and position aside
- Drive belt tensioner
- Front timing belt cover

3. Rotate the crankshaft clockwise to 60 degrees Before Top Dead Center (BTDC) using crank hub Torx socket J42098.

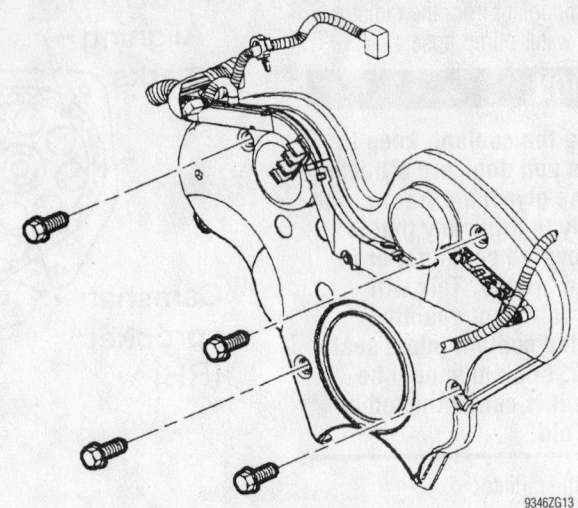

Timing belt cover bolt locations – 3.0L (VIN B) engine

Rotate the crankshaft clockwise to 60 degrees Before Top Dead Center (BTDC) using crank hub Torx socket J42098 — GM 3.0L (VIN B) engine

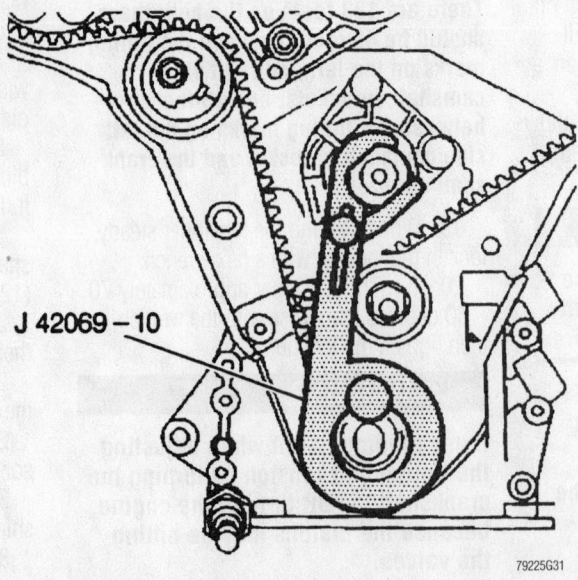

J 42069 – 10

Rotate the crankshaft in a clockwise direction using tool J42098 until the number one cylinder is at Top dead center (TDC) and tighten the lever arm to the water pump pulley flange — GM 3.0L (VIN B) engine

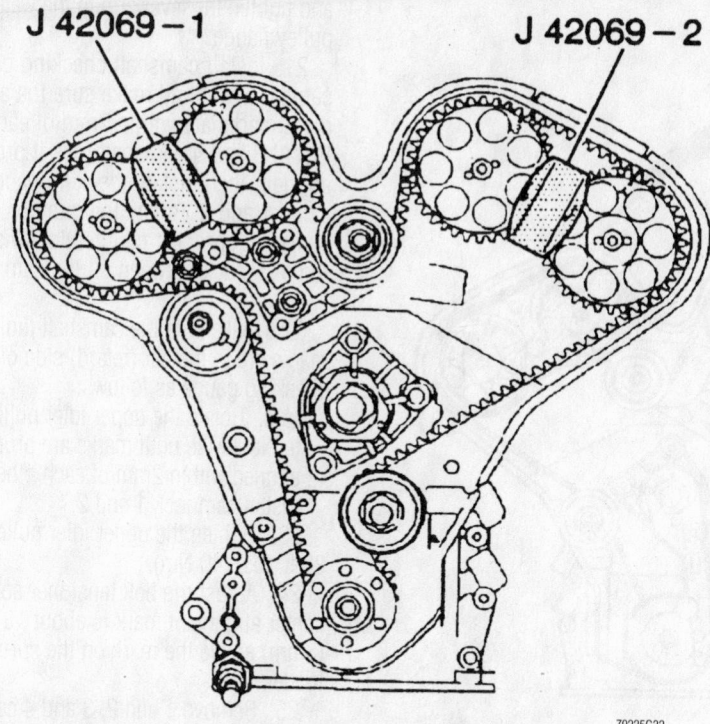

Locking the camshaft — GM 3.0L (VIN B) engine

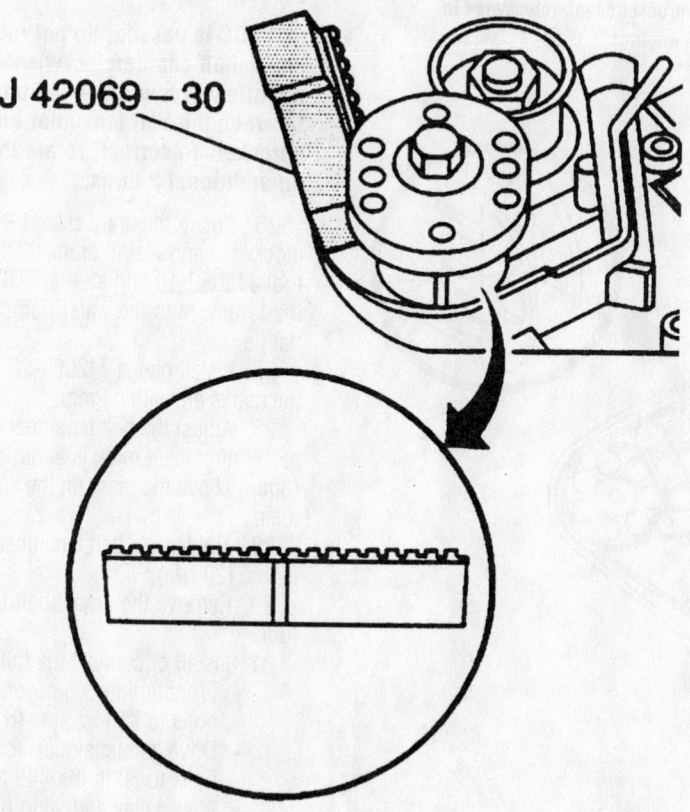

Install wedge tool J42069-30 to lock the belt onto the crankshaft hub sprocket — GM 3.0L (VIN B) engine

4. Install crankshaft locking tool J42069-10.

5. Rotate the crankshaft in a clockwise direction using tool J42098 until the number one cylinder is at Top dead center (TDC) and tighten the lever arm to the water pump pulley flange.

➡**Make sure the alignment of the crankshaft is not 180 degrees off. The alignment mark must align with the corresponding marks on the rear timing belt cover.**

6. Install 1 – 2 and 3 – 4 camshaft locks, timing belt alignment kit J42069-01 and J42069-2.

✳✳ CAUTION

Do not rotate the crankshaft unless crankshaft is at 60 degrees BTDC or the valve could hit the crankshaft.

7. Loosen the timing belt tensioner and idler pulleys.

8. Remove the timing belt.

To install:

9. Position the crank at 60 degrees BTDC.

10. Install the lower idler pulley and tighten the bolt to 30 ft. lbs. (40 Nm).

11. Install the upper idle pulley and hand tighten the bolt.

12. Install the timing belt tensioner and hand tighten the nut.

13. With the camshaft locks installed use the green test belt from the alignment kit J42069 to make sure the distance between the camshaft sprockets.

14. Start at the number 1 and 2 sprockets, route the belt over the sprockets while aligning the belt marks to marks on the sprocket and rear cover.

15. Install the belt around the tensioner and upper idler pulley.

16. Route the belt over the 3 and 4 sprockets while aligning the belt marks to marks on the sprocket and rear cover.

17. Install the belt around the crankshaft hub sprockets while aligning marks on the belt to marks on the sprocket.

18. Install wedge tool J42069-30 to lock the belt onto the crankshaft hub sprocket.

19. Carefully rotate the crankshaft counterclockwise to get belt slack on the lower idler pulley, then route the belt around the lower idler pulley and remove the wedge tool.

20. Install the crankshaft locking tool

For Wheel Alignment specifications, see Section 1 of this manual

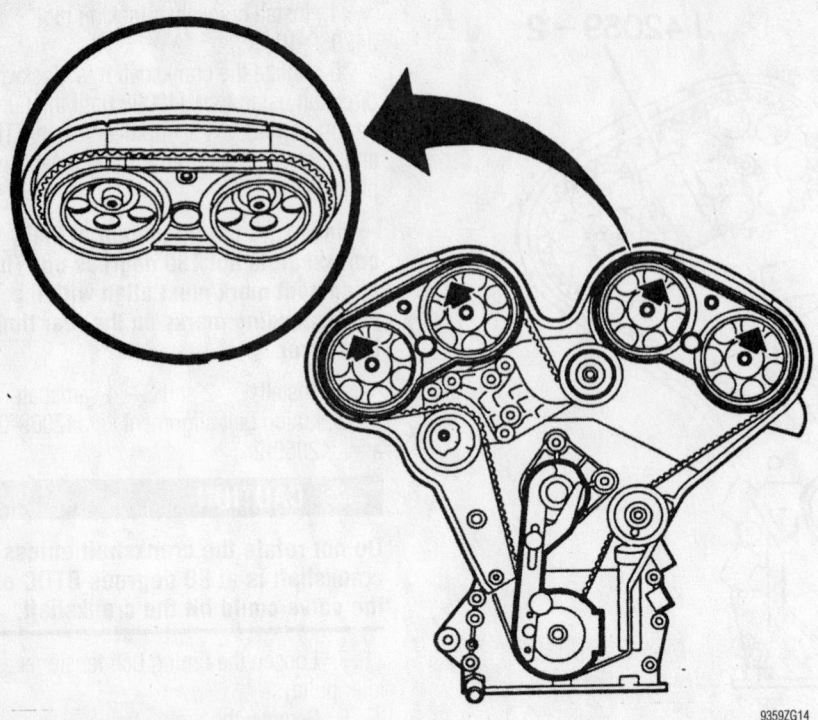

9359ZG14

Install camshaft checking gauge to cams 3 and 4 and make sure the alignment marks are within 2mm of each other. Install 3 and 4 camshaft lock if properly aligned, then install checking gauge on cams 1 and 2. Rotate the number 1 camshaft sprocket counterclockwise to remove slack between all the cam sprockets — 3.0L (VIN B) engines

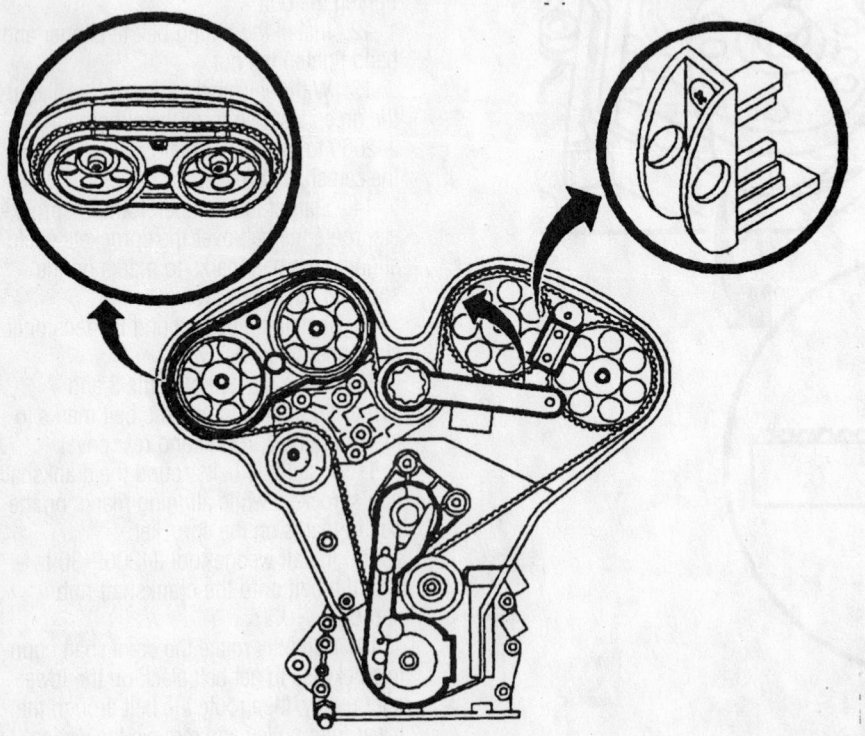

9359ZG15

Make sure the camshaft timing marks are to the left (retard) side of the checking gauge by rotating the upper idler pulley counterclockwise until marks are properly aligned within 2mm of each other and install cam lock 1 and 2 — 3.0L (VIN B) engines

and tighten the lever arm to the water pump pulley flange.

21. Install camshaft checking gauge to cams 3 and 4 and make sure the alignment marks are within 2mm of each other. Install 3 and 4 camshaft lock if properly aligned, then install checking gauge on cams 1 and 2. Rotate the number 1 camshaft sprocket counterclockwise to remove slack between all the cam sprockets.

22. Make sure the camshaft timing marks are to the left (retard) side of the checking gauge as follows:

a. Rotate the upper idler pulley counterclockwise until marks are properly aligned within 2mm of each other and install cam lock 1 and 2.

23. Tighten the upper idler pulley bolt to 30 ft. lbs. (40 Nm).

24. Adjust the belt tensioner so that the center alignment mark is about ⅛ inch (3mm) above the mark on the spring loader.

25. Remove 1 and 2, 3 and 4 camshaft locks, belt alignment kit J42069-1 and J42069-2 and camshaft locking tool J42069-10.

➡**If TDC is passed, do not rotate the crankshaft counterclockwise. this will not allow proper slack to be taken up between the belt tensioner and crank sprocket. To correct, rotate the crank an additional 2 turns.**

26. Rotate the crankshaft 1 ¾ turns clockwise and install crankshaft locking tool J42069-10 and stop at TDC. Tighten the lever arm to the water pump pulley flange.

27. Install gauge J42069-20 and verify the marks are within 2mm.

28. Adjust the belt tensioner so that the center alignment mark is about ⅛ inch (3mm) above the mark on the spring loaded idler.

29. Tighten the belt tensioner nut to 15 ft. lbs. (20 Nm).

30. Remove the crankshaft locking tool.

31. Install or connect the following:
  • Front timing belt cover. Torque the bolts to 71 inch lbs. (8 Nm).
  • Drive belt tensioner. Torque the bolts to 30 ft. lbs. (40 Nm).
  • Idler pulley and snug the bolt.
  • Electrical harness.
  • Water pump pulley
  • Accessory drive belt

32. Tighten the water pump pulley bolts to 71 inch lbs. (8 Nm0 and the idler puller bolt to 15 ft. lbs. (20 Nm).

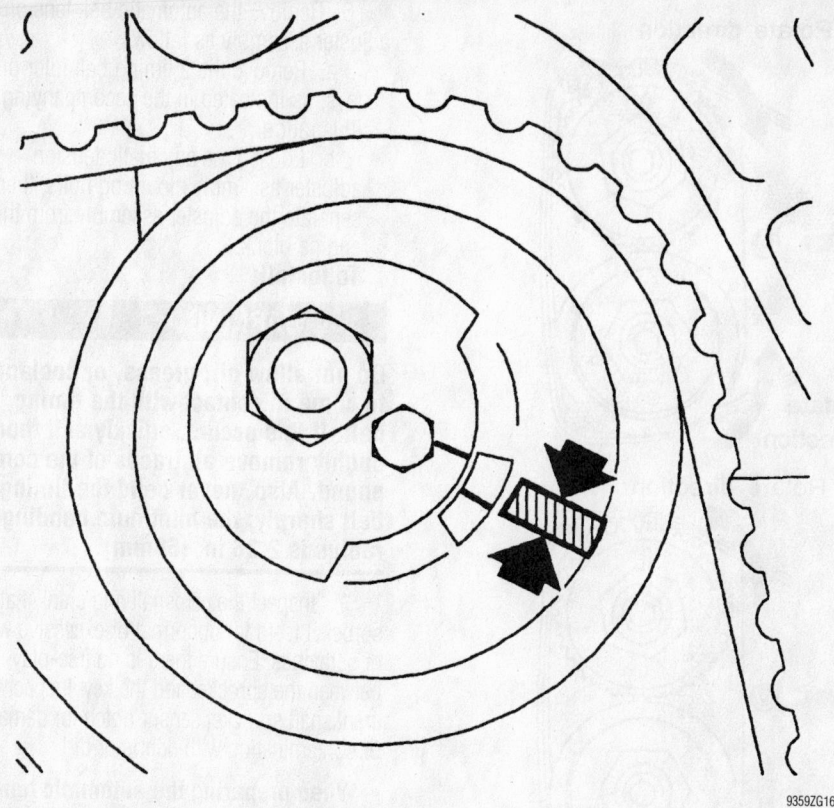

9359ZG16

Adjust the belt tensioner so that the center alignment mark is about ⅛ inch (3mm) above the mark on the spring loader — 3.0L (VIN B) engines

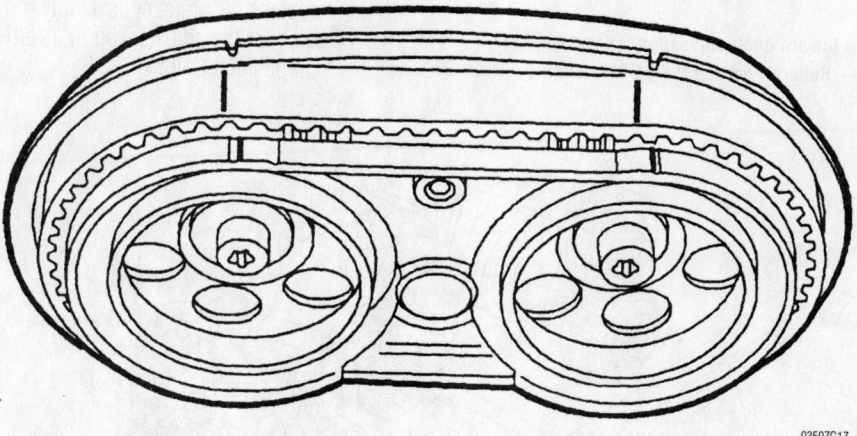

9359ZG17

Install gauge J42069-20 and verify the marks are within 2mm — 3.0L (VIN B) engines

- Engine mount bracket and tighten the bolt to 41 ft. lbs. (55 Nm)
- Engine mount-to-frame rail and engine mount bracket, tighten the fasteners to 37 ft. lbs. (50 Nm)

33. Remove the engine support.
- Splash shield
- Wheel

- Intake air resonator
- Negative battery cable

Subaru Forester

2.5L DOHC ENGINE

When servicing the timing belt, note the following:

- The intake and exhaust camshafts can be rotated independently when the timing

belt is removed. If the intake and exhaust valves are lifted off of their seats simultaneously, their heads will contact each other, possibly causing damage.

- When the timing belt is removed, the camshafts are positioned so that none of the valves are lifted off of their seats, resulting in a "zero-lift" position.
- The left-hand cylinder head camshafts must be rotated from the "zero-lift" position as little as possible when orienting it for timing belt installation, otherwise possible valve head interference may occur.
- Never allow the camshafts to rotate in the direction shown in the accompanying illustration, which would cause both the intake and exhaust valves to lift simultaneously, causing interference.

1. Remove all necessary components to gain access to the timing belt.

2. If equipped with manual transmissions, loosen the 2 timing belt guide mounting bolts, then separate the guide from the engine block.

3. If the directional arrow and alignment marks on the timing belt are faded, and the belt is to be reused, remark the belt with white paint or a grease pencil as follows:

a. Using a Subaru tool No. ST-499987500 Crankshaft Socket, or equivalent, installed on the crankshaft sprocket, rotate the crankshaft until the crankshaft sprocket, left-hand exhaust camshaft sprocket, left-hand intake camshaft sprocket, right-hand intake camshaft sprocket and right-hand exhaust camshaft sprocket timing mark notches are aligned with the respective marks on the belt cover and engine block.

b. Make alignment and/or arrow marks on the timing belt in relation to the sprockets as indicated in the accompanying illustration.
- Z1: 54.5 tooth length
- Z2: 51 tooth length
- Z3: 28 tooth length

4. Loosen the center bolt from the timing belt idler pulley, then remove the idler pulley from the engine block.

✳✳ WARNING

After removing the timing belt, DO NOT rotate the camshafts. Damage to the valves may occur.

5. Carefully remove the timing belt from all of the sprockets.

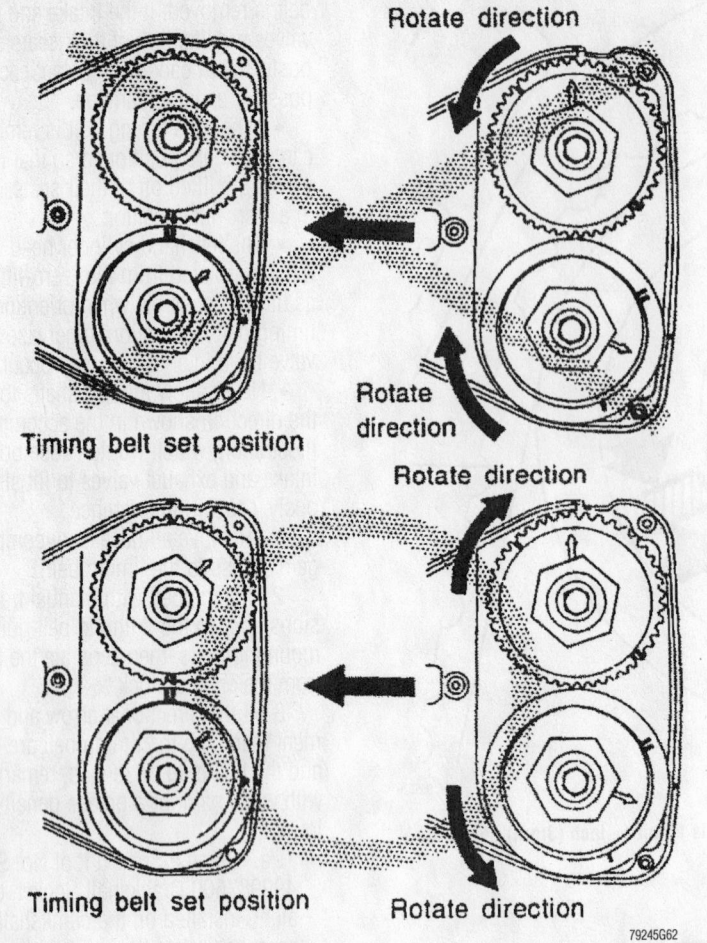

If the camshafts must be rotated, do not turn them in toward each other (upper diagram); only rotate them away from each other (lower diagram) — Subaru Forester 2.5L DOHC engine

6. Remove the automatic belt tension adjuster assembly as follows:

a. Remove the 2 timing belt idler pulleys, as indicated in the accompanying illustration.

b. Loosen the automatic tension adjuster assembly mounting bolts, then separate the adjuster assembly from the engine block.

To install:

✳✳ WARNING

Do not allow oil, grease, or coolant to come in contact with the timing belt. If this occurs, quickly and thoroughly remove all traces of the compound. Also, never bend the timing belt sharply; the minimum bending radius is 2.36 in. (60mm).

7. Inspect the camshaft and crankshaft sprocket teeth for abnormal or excessive wear or scratches. Ensure there is no free-play between the sprocket and the key. Inspect the crankshaft sprocket sensor notch for damage or contamination with debris or dirt.

➡When preparing the automatic tension adjuster assembly for installation, adhere to the following points:

- Always use a vertical press, rather than a horizontal press or vise, to depress the adjuster assembly rod
- Depress the adjuster rod in a vertical position ONLY

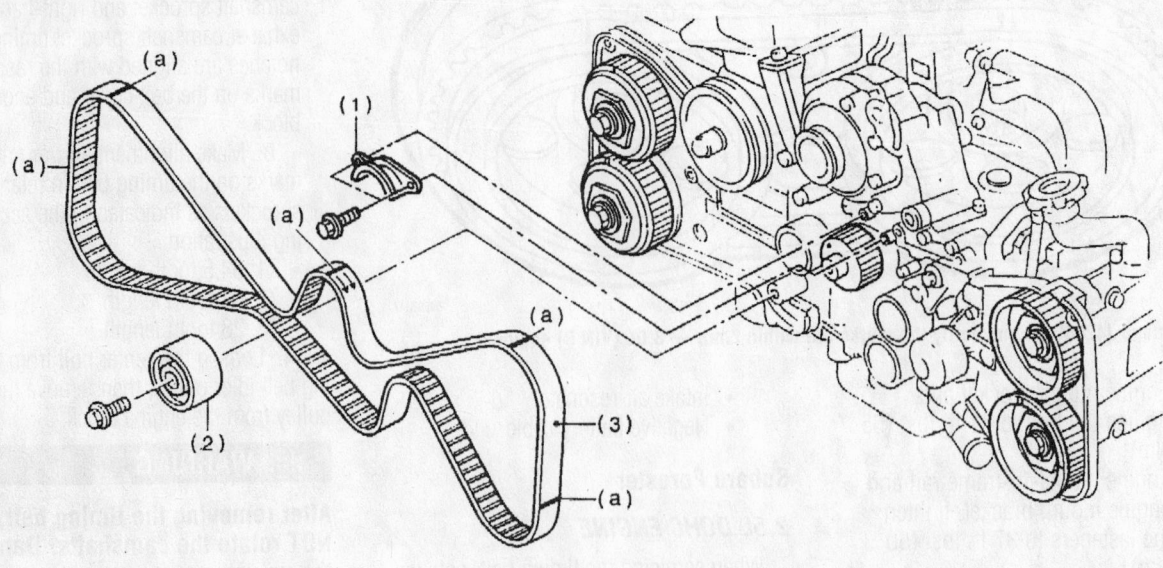

Timing belt guide (MT vehicles only)

(2) Belt idler
(3) Timing belt

(4) Alignment marks

Timing belt routing and timing belt guide (manual transmission equipped vehicles only) location — Subaru Forester 2.5L DOHC engine

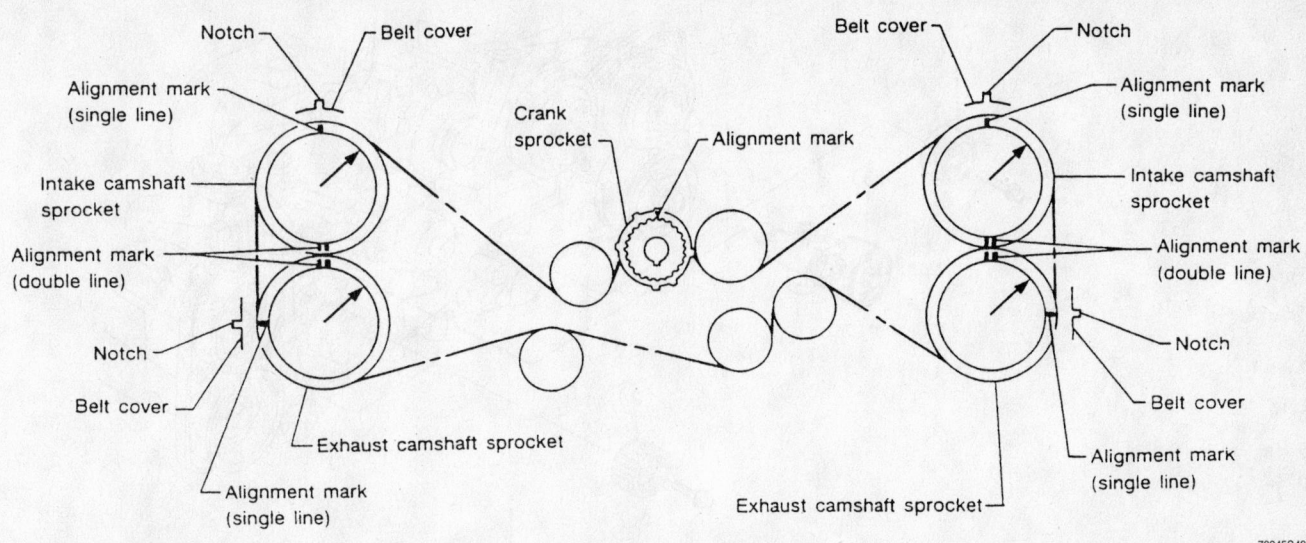

79245G49

Before removing the timing belt, turn the crankshaft sprocket until all of the alignment marks are aligned as indicated — Subaru Forester 2.5L DOHC engine

79245G50

On models equipped with a manual transmission, loosen the 2 timing belt guide bolts and separate the guide from the engine block — Subaru Forester 2.5L DOHC engine

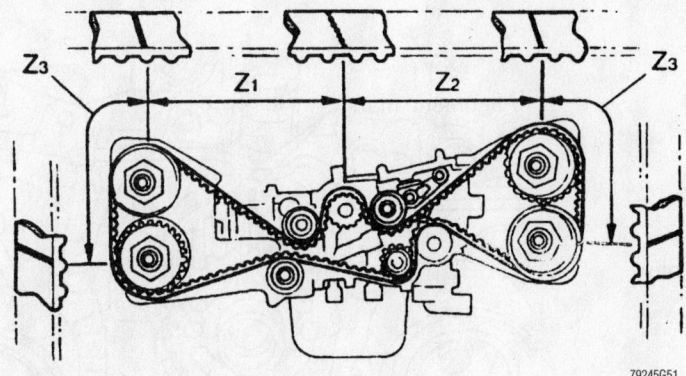

79245G51

If the original marks on the timing belt are worn or faded, make new alignment marks in the positions indicated — Subaru Forester 2.5L DOHC engine

• Depress the adjuster rod slowly (taking more than 3 minutes) with a force of 66 lbs. (30 kg)

• Do not allow the press force to exceed 2205 lbs. (1000 kg)

• Press the adjuster rod in as far as the end surface of the cylinder — do not press the rod into the cylinder, which may cause damage to the assembly

• Do not release the press force from the rod until the stopper pin is completely inserted in the cylinder

8. Prepare the automatic timing belt tension adjuster assembly for installation as follows:

a. Position the adjuster assembly in a vertical press.

b. Slowly depress the adjuster rod with a force of 66 lbs. (30 kg) until the hole in the rod is aligned with the hole in the adjuster cylinder housing.

c. Insert a 0.08 in. (2mm) diameter stopper pin or Allen wrench through the hole in the cylinder housing and rod, then slowly release the press force from the adjuster rod.

9. Install the adjuster assembly onto the engine block.

10. Install timing belt idler pulley No. 2 on the engine block.

11. Install the timing belt idler pulley No. 1 on the engine block.

12. If the camshaft and crankshaft timing marks are no longer aligned, perform the following:

For Tune-up, Capacities and Firing orders, see Section 1 of this manual

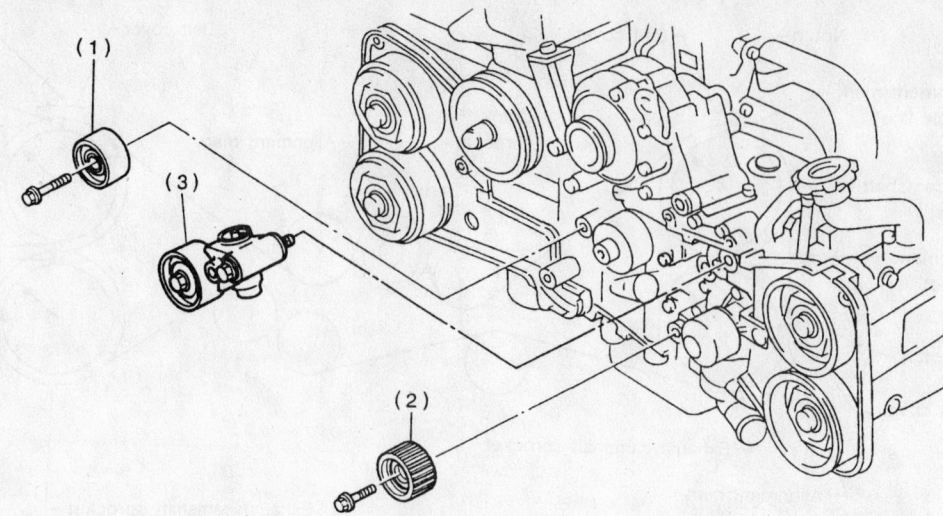

(1) Belt idler
(2) Belt idler No. 2
(3) Automatic belt tension adjuster ASSY

79245G52

It is necessary to remove the automatic adjuster assembly and reset the pushrod for timing belt installation — Subaru Forester 2.5L DOHC engine

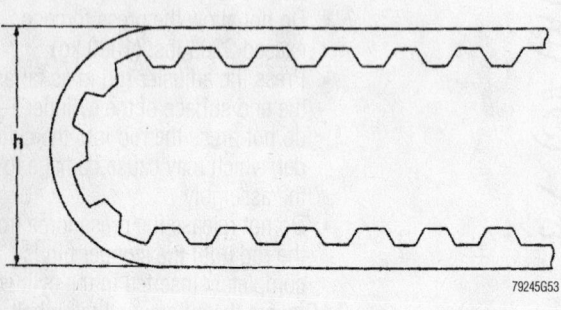

79245G53

Never bend the timing belt into a radius tighter than 2.36 in./60mm (h), otherwise it will be damaged beyond use — Subaru Forester 2.5L DOHC engine

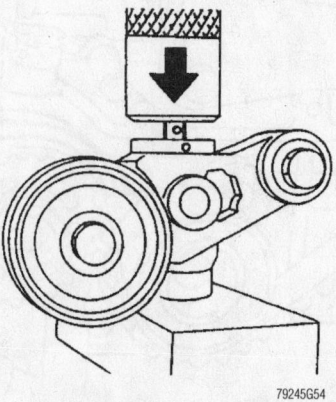

79245G54

Use a vertical press to push the adjuster rod into its housing until it is flush with the assembly's outer surface . . . — Subaru Forester 2.5L DOHC engine

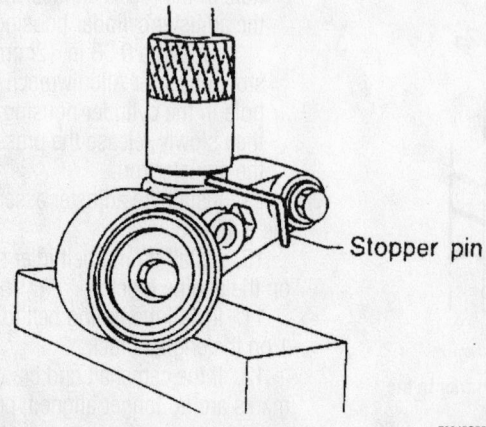

Stopper pin

79245G55

. . . then insert a 0.08 in. (2mm) diameter pin or Allen wrench into the housing and rod holes to hold it in position — Subaru Forester 2.5L DOHC engine

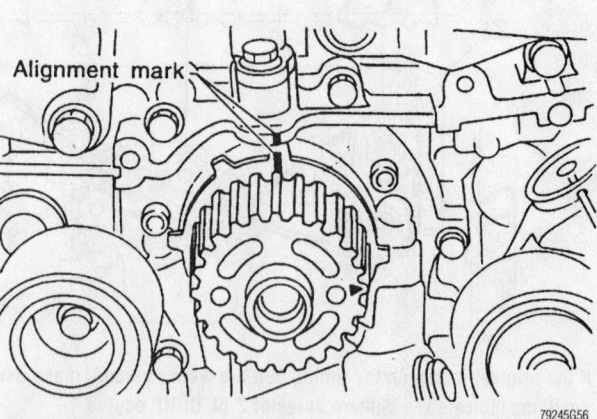

Alignment mark

79245G56

If the camshaft sprockets are no longer aligned, rotate the crankshaft sprocket until the marks are aligned . . . — Subaru Forester 2.5L DOHC engine

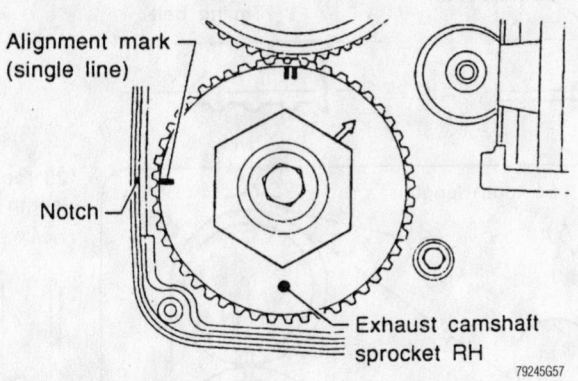

79245G57

. . . then turn the right-hand exhaust camshaft until the single line mark is aligned with the notch in the belt cover—Subaru Forester 2.5L DOHC engine

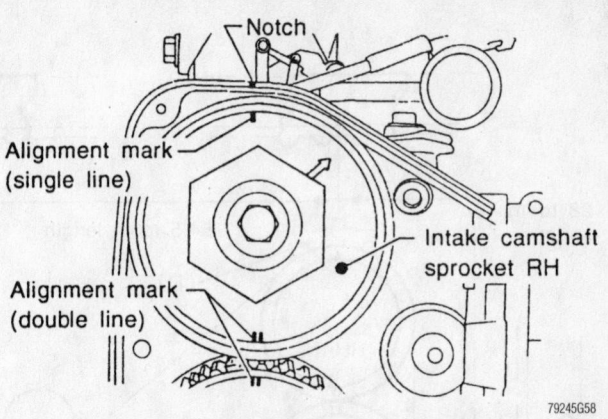

79245G58

Spin the right-hand intake camshaft sprocket so that the single line mark is aligned with the notch in the belt cover—at this point, the double line marks on both right-hand camshaft sprockets must be aligned—Subaru Forester 2.5L DOHC engine

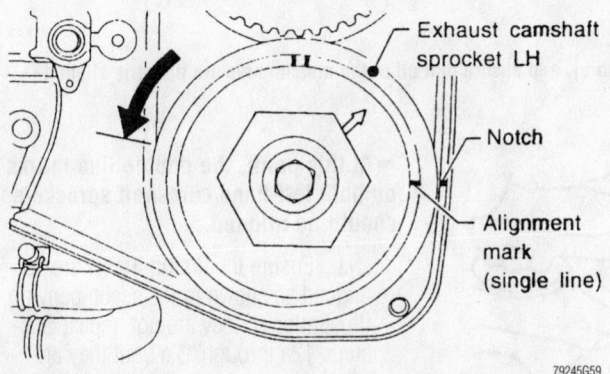

79245G59

Rotate the left-hand exhaust camshaft until the single line mark is aligned with the notch in the belt cover . . .—Subaru Forester 2.5L DOHC engine

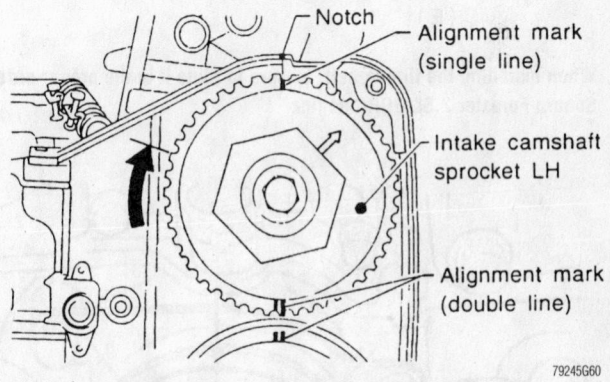

79245G60

. . . then align the single line mark on the left-hand intake sprocket with the belt cover notch—Subaru Forester 2.5L DOHC engine

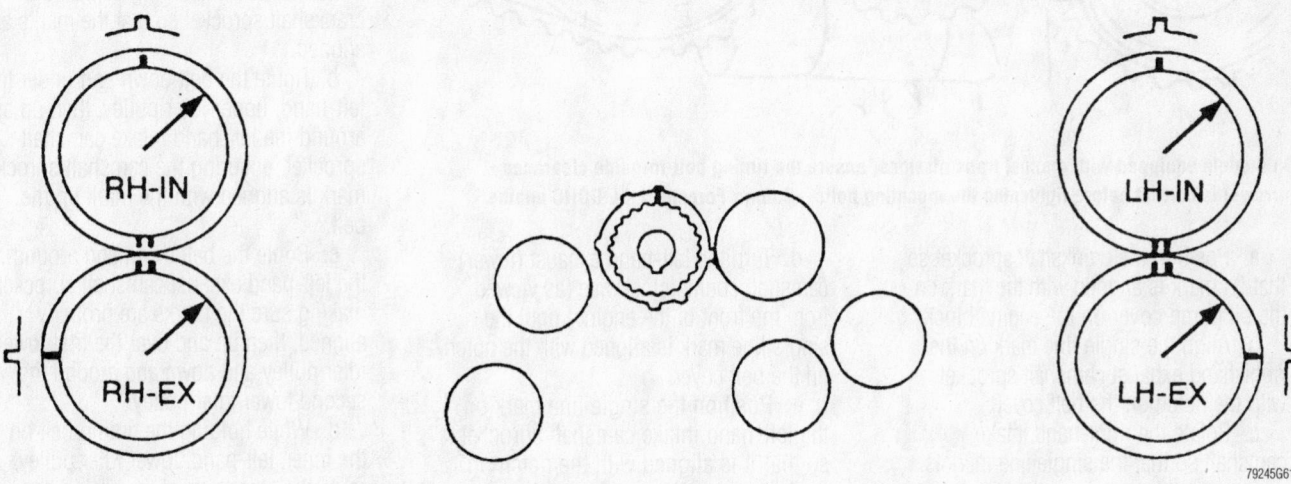

79245G61

After orienting all 5 sprockets, the alignment marks should be positioned as shown—Subaru Forester 2.5L DOHC engine

For complete service labor times, order Nichols' Chilton Labor Guide

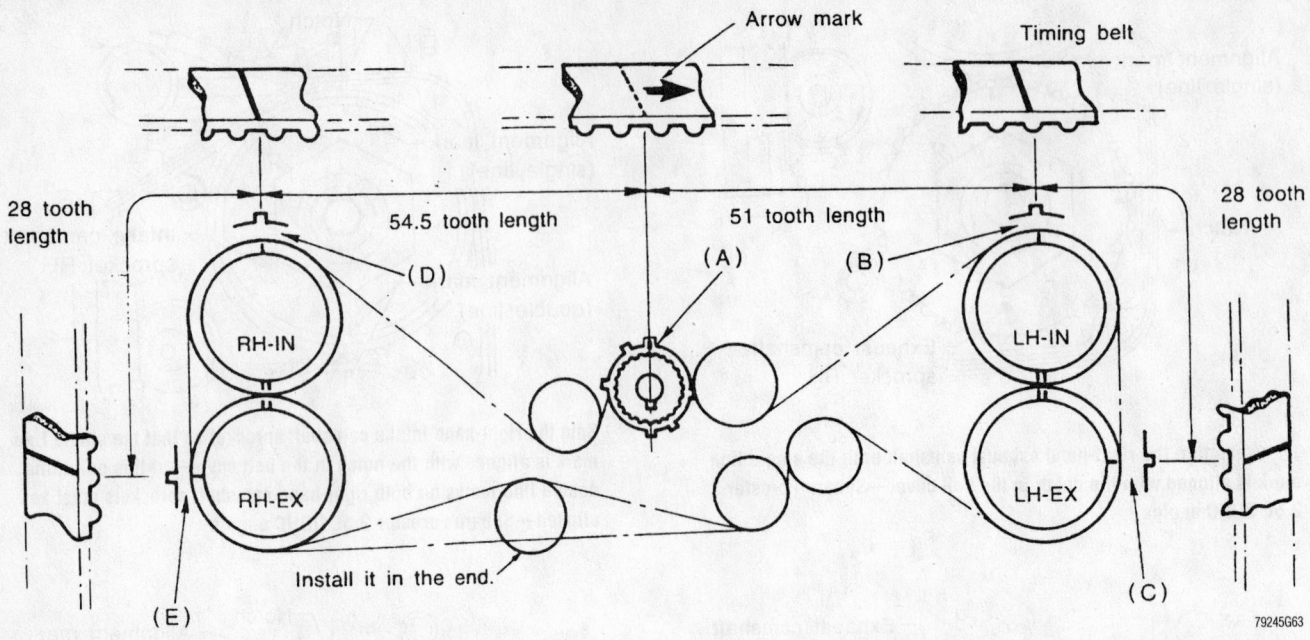

When installing the timing belt, be sure to route it in the proper order (a through e), and ensure that all of the matchmarks are properly aligned—Subaru Forester 2.5L DOHC engine

On models equipped with manual transmissions, ensure the timing belt-to-guide clearance (arrows) is correct before tightening the mounting bolts—Subaru Forester 2.5L DOHC engine

a. Position the crankshaft sprocket so that its mark is aligned with the mark on the oil pump cover on the engine block.

b. Align the single line mark on the right-hand exhaust camshaft sprocket with the notch on the belt cover.

c. Rotate the right-hand intake camshaft so that the single line mark is aligned with the notch on the belt cover.

➡️ At this point, the double line marks on both right-hand camshaft sprockets should be aligned.

d. Turn the left-hand exhaust (lower) camshaft counterclockwise (as viewed from the front of the engine) until the single line mark is aligned with the notch on the belt cover.

e. Position the single line mark on the left-hand intake camshaft sprocket so that it is aligned with the notch on the belt cover. When rotating the camshaft, do so only in a clockwise direction (as viewed from the front of the engine).

➡️ At this point, the double line marks on both left-hand camshaft sprockets should be aligned.

f. Ensure the timing marks are aligned as shown in the accompanying illustration. If they are not, repeat Substeps 12a through 12e until they are properly aligned.

13. Install the timing belt around the camshaft, crankshaft and idler pulleys so that the positioning marks on the timing belt are aligned with the marks on the sprockets as follows:

a. Position the timing belt on the crankshaft sprocket so that the marks are aligned.

b. Route the belt down and under the left-hand, upper idler pulley, then up and around the left-hand intake camshaft sprocket, ensuring the camshaft sprocket mark is aligned with the mark on the belt.

c. Route the belt down and around the left-hand exhaust camshaft sprocket, making sure the marks are properly aligned, then up and over the first lower idler pulley and down and around the second lower idler pulley.

d. While holding the timing belt on the inner, left-hand, lower idler pulley, route the other side of the timing belt (from the crankshaft sprocket) down and under the right-hand upper idler pulley.

e. Route the timing belt up and around the right-hand intake camshaft sprocket so that the belt and sprocket marks are aligned.

    f. Position the belt down and around the right-hand exhaust camshaft sprocket, ensuring the positioning marks are aligned.

14. Install the right-hand lower idler pulley so that the timing belt is routed over the top side of it.

➡️**Once the belt is completely installed on all of the pulleys and sprockets, ensure that the positioning marks are still all aligned.**

15. After ensuring all of the marks are still aligned, use a pair of pliers to withdraw the stopper pin or Allen wrench from the adjuster assembly housing.

16. On models with manual transmissions, perform the following:

    a. Install the timing belt guide by temporarily tightening the mounting bolts.

    b. Position the timing belt guide so that there is 0.019 – 0.059 in. (0.5 – 1.5mm) clearance between the timing belt and the belt guide.

    c. Tighten the guide mounting bolts securely, then double check the guide clearance.

17. Install the timing belt covers and all remaining engine components.

Suzuki Vitara

➡️**During these procedures, identify all components removed from the engine so that they may be reinstalled in their original positions. If discarding the old components so that new components can be installed, identifying the old items is not necessary.**

1.6L 16-VALVE ENGINE

The 1.6L 16-valve engine is known as an interference motor, because it is fabricated with such close tolerances between the pistons and valves that, if the timing belt is incorrectly positioned, jumps teeth on one of the sprockets or breaks, the valve and pistons will come into contact. This can cause severe internal engine damage.

✳✳ WARNING

Do not rotate the crankshaft counter-clockwise or attempt to rotate the crankshaft by turning the camshaft sprocket.

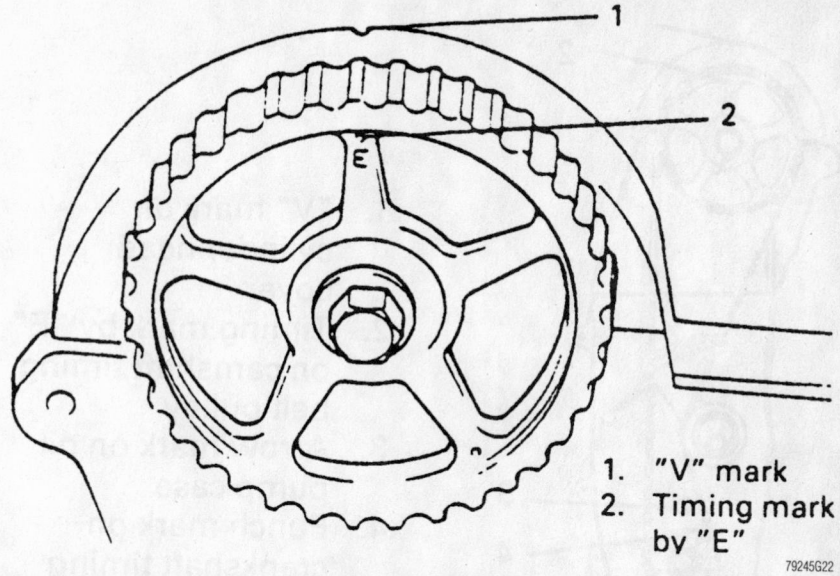

1. "V" mark
2. Timing mark by "E"

79245G22

Camshaft timing marks — Suzuki Vitara 1.6L 16-valve engine

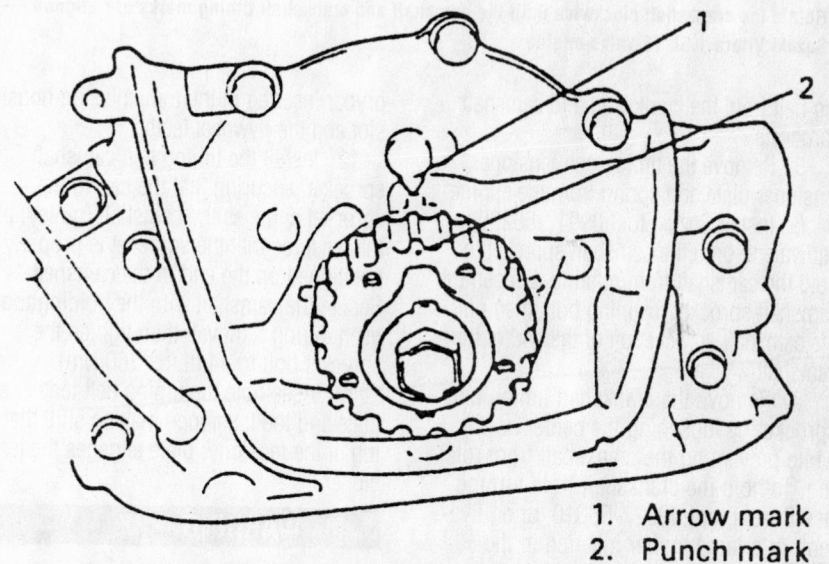

1. Arrow mark
2. Punch mark

79245G23

Align the punch mark with the arrow for proper timing belt installation — Suzuki Vitara 1.6L 16-valve engine

1. Remove the timing belt cover.
2. If the timing belt is not already marked with a directional arrow, use white paint, a grease pencil or correction fluid to do so.
3. Rotate the crankshaft clockwise until the timing mark on the camshaft sprocket and the "V" mark on the timing belt inside cover are aligned, and the punch mark on the crankshaft sprocket is aligned with the mark on the engine.

✳✳ WARNING

Do not rotate the crankshaft or camshaft once the timing belt is removed, because the valves and pistons can come into contact, which may cause internal engine damage.

4. Disconnect one end of the tensioner spring. Loosen the timing belt tensioner bolt and stud, then, using your finger, press the tensioner plate up and remove the tim-

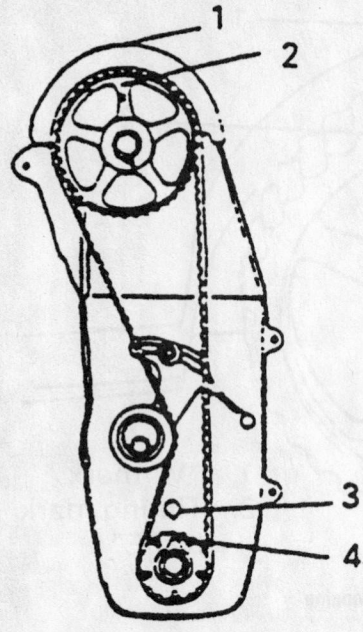

1. "V" mark on cylinder head cover
2. Timing mark by "E" on camshaft timing belt pulley
3. Arrow mark on oil pump case
4. Punch mark on crankshaft timing belt pulley

79245G47

Rotate the crankshaft clockwise until the camshaft and crankshaft timing marks are aligned — Suzuki Vitara 1.6L 16-valve engine

ing belt from the crankshaft and camshaft sprockets.

5. Remove the timing belt tensioner, tensioner plate and spring from the engine.

6. Install Suzuki tool 09917-68220, or equivalent, onto the camshaft sprocket to hold the camshaft from rotating. Loosen the camshaft sprocket retaining bolt, then pull the camshaft sprocket off of the end of the camshaft.

7. Remove the crankshaft timing belt sprocket by loosening the center bolt, while preventing the crankshaft from rotating. To hold the crankshaft from turning, use Suzuki tool 09927-56010, or equivalent, or a large prybar inserted in the transmission housing slot and the flywheel teeth. Pull the sprocket off of the end of the crankshaft. Be sure to retain the crankshaft sprocket key and belt guide for assembly.

8. If necessary, remove the timing belt inside cover from the cylinder head.

To install:

9. If necessary, install the timing belt inside cover.

10. Slide the timing belt guide on the crankshaft so that the concave side faces the oil pump, then install the sprocket key in the groove in the crankshaft.

11. Slide the pulley onto the crankshaft, and install the center retaining bolt. Tighten the center bolt to 80 ft. lbs. (110 Nm). To hold the crankshaft from turning, use Suzuki tool 09927-56010, or equivalent, or a large

prybar inserted in the transmission housing slot and the flywheel teeth.

12. Install the timing belt camshaft sprocket, ensuring that the slot in the sprocket engages the camshaft (pulley) pin; this ensures that the sprocket is properly positioned on the end of the camshaft. Secure the camshaft with the holding tool used during removal, then tighten the sprocket bolt to 44 ft. lbs. (60 Nm).

13. Assemble the timing belt tensioner plate and the tensioner, making sure that the lug of the tensioner plate engages the tensioner.

❊❊ WARNING

If any binding is felt when adjusting the timing belt tension by turning the crankshaft, STOP turning the engine, because the pistons may be hitting the valves.

14. Install the timing belt tensioner, tensioner plate and spring on the engine. Tighten the mounting bolt and stud only finger-tight at this time. Ensure that when the tensioner is moved in a counterclockwise direction, the tensioner moves in the same direction. If the tensioner does not move, remove it and the tensioner plate to reassemble them properly.

15. Loosen all rocker arm valve lash locknuts and adjusting screws. This will permit movement of the camshaft without any rocker arm associated drag, which is

essential for proper timing belt tensioning. If the camshaft does not rotate freely (free of rocker arm drag), the belt will not be properly tensioned.

16. Rotate the camshaft sprocket clockwise until the timing mark on the sprocket and the "V" mark on the timing belt inside cover are aligned.

17. Using a wrench, or socket and breaker bar, on the crankshaft sprocket center bolt, turn the crankshaft clockwise until the punch mark on the sprocket is aligned with the arrow mark on the oil pump.

18. With the camshaft and crankshaft marks properly aligned, push the tensioner up with your finger and install the timing belt on the 2 sprockets, ensuring that the drive side of the belt is free of all slack. Release your finger from the tensioner. Be sure to install the timing belt so that the directional arrow is pointing in the appropriate direction.

➡ In this position, the No. 4 cylinder is at Top Dead Center (TDC) on the compression stroke.

19. Rotate the crankshaft clockwise 2 full revolutions, then tighten the tensioner stud to 97 inch lbs. (11 Nm). Then, tighten the tensioner bolt to 18 ft. lbs. (24 Nm).

20. Ensure that all 4 timing marks are still aligned as before; if they are not, remove the timing belt, and install and tension it again.

21. Install the timing belt cover and all related components.

Toyota 4Runner

3.4L (5VZ-FE) ENGINE

1. Disconnect the negative battery cable.

❊❊ CAUTION

Wait 90 seconds from the time the key is turned to LOCK and the negative battery cable is disconnected to begin work. This allows the SRS capacitor to discharge and prevent deployment of the air bag(s).

2. Raise and safely support the vehicle.
3. Remove the engine undercover.
4. Drain the engine coolant.

❊❊ CAUTION

Never open, service or drain the radiator or cooling system when hot; serious burns can occur from the steam and hot coolant. Also, when draining engine coolant, keep in mind that cats and dogs are attracted

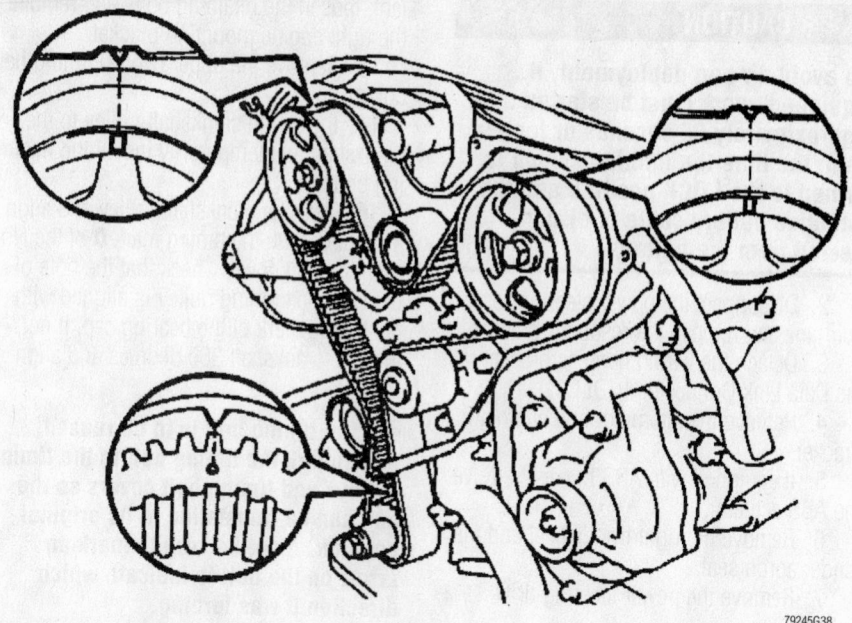

Crankshaft and camshaft timing mark locations — Toyota 4Runner 3.4L (5VZ-FE) engine

79245G38

to ethylene glycol antifreeze and could drink any that is left in an uncovered container or in puddles on the ground. This will prove fatal in sufficient quantities. Always drain coolant into a sealable container. Coolant should be reused unless it is contaminated or is several years old.

5. Disconnect the upper radiator hose from the engine.

6. Remove the power steering drive belt.

7. Remove the air conditioning drive belt by loosening the idler pulley nut and the adjusting bolt.

8. If equipped with air conditioning, disconnect the compressor from the engine and set aside. Do not disconnect the lines from the compressor.

9. If equipped with air conditioning, disconnect the air conditioning bracket.

10. Remove the fan with the fluid coupling and fan pulleys.

11. Loosen the lockbolt, pivot bolt, and the adjusting bolt and the alternator drive belt.

12. Remove the No. 2 fan shroud by removing the 2 clips.

13. Disconnect the power steering pump from the engine and set aside. Do not disconnect the lines from the pump.

14. Remove the oil dipstick and the guide.

15. Remove the No. 2 timing belt cover as follows:

a. Detach the camshaft position sen-

sor connector from the No. 2 timing belt cover.

b. Disconnect the 4 spark plug wire clamps from the No. 2 timing belt cover.

c. Remove the 6 bolts and remove the timing belt cover.

16. Remove the fan bracket as follows:

a. Remove the power steering adjusting strut by removing the nut.

b. Remove the fan bracket by removing the bolt and nut.

17. Using SST 09213-54015, or equivalent, remove the crankshaft pulley.

18. Remove the starter wire bracket and the No. 1 timing belt cover.

19. Remove the timing belt guide.

20. Set the No. 1 cylinder at Top Dead Center (TDC) of the compression stroke, as follows:

a. Temporarily install the crankshaft pulley bolt to the crankshaft.

b. Turn the crankshaft and align the timing marks of the crankshaft timing pulley and the oil pump body.

c. Check that the timing marks of the camshaft timing pulleys and the No. 3 timing belt cover are aligned. If not, turn the crankshaft pulley one revolution (360 degrees).

➡ If reusing the timing belt, be sure that you can still read the installation marks. If not, place new installation marks on the timing belt to match the timing marks of the camshaft timing pulleys.

21. Remove the timing belt tensioner by alternately loosening the 2 bolts.

22. Remove the right and left camshaft pulleys.

23. Remove the No. 2 idler pulley.

24. Using a 10mm hex wrench, remove the pivot bolt, No. 1 idler pulley and the plate washer.

25. Remove the timing belt guide and remove the timing belt.

26. Remove the crankshaft timing pulley.

To install:

27. Install the crankshaft timing belt pulley, as follows:

a. Align the timing belt pulley set key with the key groove of the timing pulley and slide on the timing pulley.

b. Slide on the timing belt pulley with the flange side facing inward.

28. Install the plate washer and the No. 1 idler pulley with the pivot bolt and tighten it to 26 ft. lbs. (35 Nm). Check that the pulley bracket moves smoothly.

29. Install the No. 2 timing belt idler with the bolt. Tighten the bolt to 30 ft. lbs. (40 Nm). Check that the pulley bracket moves smoothly.

30. Install the left and right camshaft timing pulleys.

31. Set the No. 1 cylinder to TDC of the compression stroke, as follows:

a. Using the crankshaft pulley bolt, turn the crankshaft and align the timing marks of the crankshaft timing pulley and the oil pump body.

b. Using SST 09960-10010, or equivalent, to turn the camshaft pulley to align the marks of the camshaft timing belt pulley and the No. 3 timing belt cover.

32. Install the timing belt, as follows:

➡ **The engine should be cold.**

a. Face the front mark on the timing belt forward.

b. Align the installation mark on the timing belt with the timing mark of the crankshaft timing pulley.

c. Align the installation marks on the timing belt with the timing marks of the camshaft pulleys.

33. Install the timing belt in the following order:

- Left camshaft pulley
- No. 2 idler pulley
- Right camshaft pulley
- Water pump pulley
- Crankshaft pulley
- No. 1 idler pulley

Timing belt service is covered in Section 3 of this manual

❊❊ WARNING

If any binding is felt when adjusting the timing belt tension by turning the crankshaft, STOP turning the engine, because the pistons may be hitting the valves.

34. Set the timing belt tensioner as follows:

 a. Using a press, slowly press in the pushrod using 220 – 2205 lbs. (981 – 9807 N) of force.

 b. Align the holes of the pushrod and housing, pass a 1.27mm wrench through the holes to keep the setting position of the pushrod.

 c. Release the press and install the dust boot to the tensioner.

35. Install the timing belt tensioner and alternately tighten the bolts to 20 ft. lbs. (27 Nm). Using pliers, remove the 1.27mm wrench from the belt tensioner.

36. Check the valve timing, as follows:

 a. Slowly turn the crankshaft and align the timing marks of the crankshaft timing pulley and the oil pump body. Always turn the crankshaft pulley clockwise.

 b. Check that the timing marks of the right and left timing pulleys align with the timing marks of the No. 3 timing belt cover. If the marks do not align, remove the timing belt and reinstall it.

37. Install the timing belt guide with the cup side facing outward.

38. Install the No. 1 timing belt cover and starter wire bracket. Tighten the timing belt cover fasteners to 80 inch lbs. (9 Nm).

39. Install the crankshaft pulley, as follows:

 a. Align the pulley set key with the key groove of the pulley and slide the pulley.

 b. Using SST 09213-54014, or equivalent, tighten the bolt to 184 ft. lbs. (250 Nm).

40. Install the fan bracket with the bolt and nut.

41. Install the No. 2 timing belt cover, and tighten the bolts to 80 inch lbs. (9 Nm). Install the remaining components.

42. Fill the cooling system with coolant.

43. Connect the negative battery cable.

44. Start the engine and check for leaks.

45. Check the ignition timing.

Toyota RAV4

2.0L (3S-FE) ENGINE

The timing belt is not adjustable.

1. Disconnect the negative battery cable.

❊❊ CAUTION

To avoid air bag deployment, if equipped, work must be started after approximately 90 seconds or longer from the time the ignition switch is turned to the LOCK position and the negative battery cable is disconnected from the battery.

2. Disconnect the power steering reservoir tank and remove the reservoir bracket.

3. Detach the wiring harness bracket for the Data Link Connector 1 (DLC1).

4. Remove the alternator and alternator bracket.

5. If equipped with ABS brakes, remove the ABS actuator.

6. Remove the right front wheel and the fender apron seal.

7. Remove the power steering drive belt.

8. Slightly raise the engine using a block of wood and floor jack under the oil pan to prevent damage.

9. Remove the 4 bolts, 2 nuts, and right-hand mounting bracket.

10. Remove the spark plugs.

11. Using SST 09213-54015, or equivalent, loosen the crankshaft pulley bolt and remove it by pulling it straight off the crankshaft.

12. Using SST 09249-63010, or equivalent, loosen the retaining bolts and remove the right engine mounting bracket.

13. Remove the upper (No. 2) timing belt cover.

14. Install the crankshaft pulley to the crankshaft and temporarily install the retaining bolt.

15. Turn the crankshaft pulley and align its groove with the timing mark **0** of the No. 1 timing belt cover. Check that the hole of the camshaft timing pulley is aligned with the timing mark of the bearing cap. If not, turn the crankshaft 360 degrees and align the marks.

➡ **If the timing belt is to be reused, matchmark the timing belt to the timing pulleys and timing belt covers so the belt can be reinstalled in its original position. Also, be sure to mark an arrow on the belt to indicate which direction it was turning.**

16. Remove the timing belt from the camshaft timing pulley.

17. Hold the camshaft sprocket with a spanner wrench and remove the mounting bolt. Remove the camshaft pulley.

18. Remove the crankshaft pulley bolt and remove the crankshaft pulley.

19. Remove the No. 1 timing belt cover.

20. Remove the timing belt guide and the timing belt.

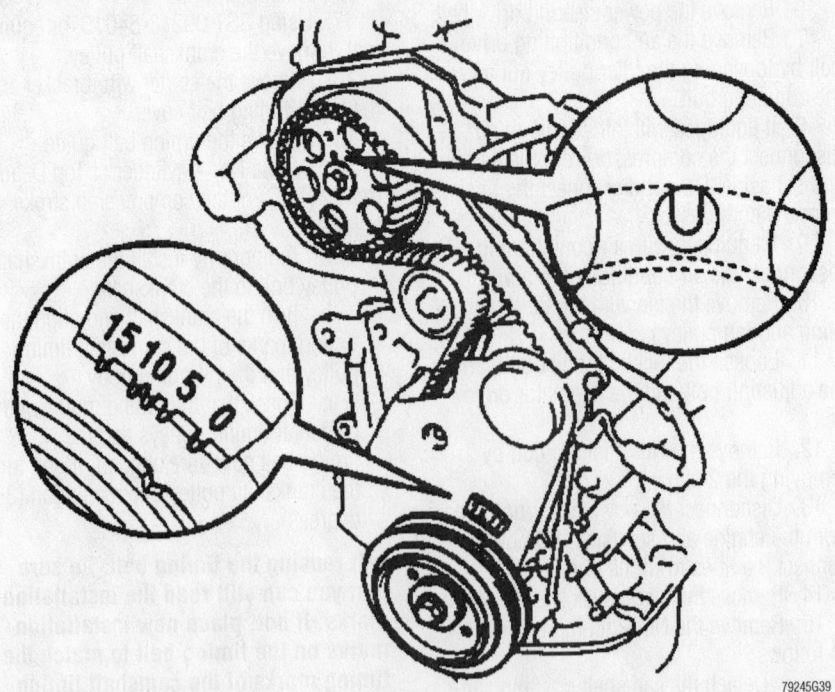

It is necessary to align the timing reference indicators prior to removing the timing belt — Toyota RAV4 2.0L (3S-FE) engine

21. Remove the No. 1 idler pulley and tension spring.

22. Remove the No. 2 idler pulley.

23. Remove the crankshaft timing pulley.

24. Support the oil pump sprocket with a spanner wrench, then remove the mounting bolt and remove the sprocket.

To install:

25. Install the oil pump pulley. Tighten the nut to 18 ft. lbs. (24 Nm).

26. Install the crankshaft timing pulley. Align the pulley set key with the key groove of the pulley. Slide on the pulley facing the flange side inward.

27. Install the No. 2 idler pulley and tighten the mounting bolt to 31 ft. lbs. (42 Nm). Be sure that the pulley moves smoothly.

28. Install the No. 1 idler pulley with the bolt and the tension spring. Pry the pulley toward the left as far as it will go and tighten the bolt. Make sure that the pulley moves smoothly.

29. Temporarily install the timing belt. Using the crankshaft pulley bolt, turn the crankshaft and position the key groove of the crankshaft timing pulley upward. If reusing the timing belt, align the points marked during removal.

30. Install the timing belt on the crankshaft timing pulley, oil pump pulley, No. 1 idler pulley, water pump pulley and the No. 2 idler pulley.

31. Install the timing belt guide.

➡ **If the old timing belt is being reinstalled, be sure the directional arrow is facing in the original direction and that the belt and sprocket/cover matchmarks are properly aligned.**

32. Install the lower (No. 1) timing belt cover and new gasket with the 4 bolts.

33. Align the crankshaft pulley set key with the pulley key groove. Temporarily install the crankshaft pulley and bolt.

34. Align the camshaft knock pin with the groove of the pulley, and slide the timing pulley onto the camshaft with the plate washer and set bolt.

35. Tighten the pulley set bolt to 40 ft. lbs. (54 Nm).

✳✳ WARNING

If any binding is felt when adjusting the timing belt tension by turning the crankshaft, STOP turning the engine, because the pistons may be hitting the valves.

36. Turn the crankshaft pulley and align the **0** mark on the lower (No. 1) timing belt cover.

37. Finish installing the timing belt and check the valve timing, as follows:

   a. If reusing the old timing belt, align the matchmarks made previously and install the timing belt onto the camshaft pulley.

   b. Align the marks on the timing belt with the marks on the camshaft pulley.

   c. Loosen the No. 1 idler pulley set bolt ½ turn.

   d. Turn the crankshaft pulley 2 complete revolutions TDC to TDC. ALWAYS turn the crankshaft CLOCKWISE. Check that the pulleys are still in alignment with the timing marks.

   e. If the No. 1 idler pulley uses a green tension spring, slowly turn the crankshaft pulley 1⅞ revolutions, and align its groove with the mark at 45 degrees BTDC (for the No. 1 cylinder) of the No. 1 timing belt cover.

   f. Tighten the No. 1 idler pulley set bolt to 31 ft. lbs. (42 Nm).

   g. Be sure there is belt tension between the crankshaft and camshaft timing pulleys.

38. Place the right-hand engine mounting bracket in position but do not install the bolts.

39. Install the upper (No. 2) timing cover with a new gasket(s).

40. Remove the engine crankshaft pulley bolt and pulley.

41. Using SST 09249-63010, or equivalent, install the mounting bolts for the right-hand mounting bracket. Tighten the mounting bolts to 38 ft. lbs. (52 Nm).

42. Align the crankshaft pulley set key with the pulley key groove. Install the pulley. Tighten the pulley bolt to 80 ft. lbs. (108 Nm).

43. Install the spark plugs.

44. Install the right-hand mounting insulator, as follows:

   a. Attach the mounting insulator to the body and mounting bracket with the 4 bolts and 2 nuts.

   b. Tighten the 3 bolts to hold the mounting insulator to the body. Tighten the bolts to 47 ft. lbs. (64 Nm).

   c. Tighten the 2 nuts and bolt to hold the mounting insulator to the mounting bracket. Tighten the bolt to 27 ft. lbs. (37 Nm) and the nut to 38 ft. lbs. (52 Nm).

45. Install and adjust the power steering pump drive belt.

46. Install the right-hand engine undercover.

47. Install the right front wheel.

48. Lower the engine.

49. If equipped, install the ABS actuator.

50. Install the alternator and alternator bracket.

51. Install the wiring harness bracket for the DLC1.

52. Install the power steering reservoir bracket and reservoir.

53. Connect the negative battery cable.

54. Start the engine and check the timing.

Toyota Land Cruiser

4.7L (2UZ-FE) ENGINE

1. Disconnect the negative battery cable.

2. Raise and safely support the vehicle.

3. Remove the oil pan protector and the engine under cover.

4. Drain the cooling system and store the coolant for refilling purposes.

5. Lower the vehicle and remove the battery clamp cover.

6. From the top of the engine, remove the fuel return hose, the engine cover nuts/bolts and the cover.

7. Remove the air cleaner and the intake air connector assembly.

8. Remove the cooling fan pulley by performing the following procedures:

   a. Loosen the 4 fan clutch-to-fan pulley nuts.

   b. Using a box-end wrench on the serpentine drive belt tensioner bolt, rotate the tensioner counterclockwise and remove the drive belt.

➡ **The serpentine drive belt tensioner bolt is a left-hand thread.**

   c. Remove the fan clutch-to-fan pulley nuts, the fan, the clutch assembly and the fan pulley.

9. Remove the radiator by performing the following procedures:

   a. Disconnect the upper, lower and reservoir hoses from the radiator.

   b. Disconnect and plug the automatic transmission oil cooler at the radiator. Disconnect the automatic transmission oil cooler hoses from the fan shroud clamp.

   c. Remove the radiator reservoir tank.

   d. Remove the fan shroud-to-radiator bolts and the shroud.

   e. Remove the 2 upper radiator-to-chassis nuts.

Heater Core replacement is covered in Section 2 of this manual

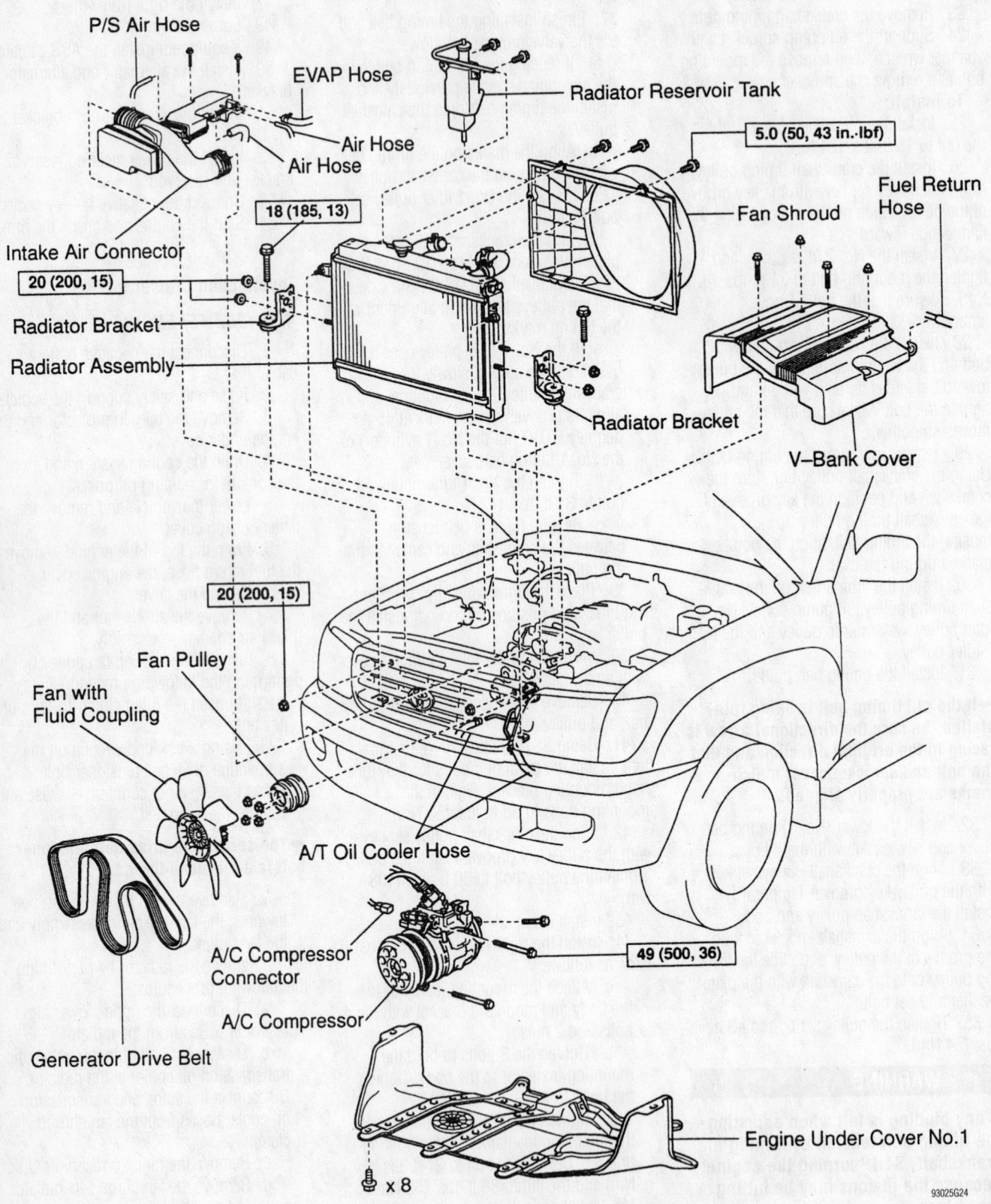

P/S Air Hose

EVAP Hose

Radiator Reservoir Tank

Air Hose

Air Hose

5.0 (50, 43 in.-lbf)

18 (185, 13)

Fan Shroud

Fuel Return Hose

Intake Air Connector

20 (200, 15)

Radiator Bracket

Radiator Assembly

Radiator Bracket

V–Bank Cover

20 (200, 15)

Fan Pulley

Fan with Fluid Coupling

A/T Oil Cooler Hose

A/C Compressor Connector

49 (500, 36)

A/C Compressor

Generator Drive Belt

Engine Under Cover No.1

x 8

93025G24

Exploded view of vehicle components for timing belt replacement—Toyota Land Cruiser 4.7L (2UZ-FE) engine

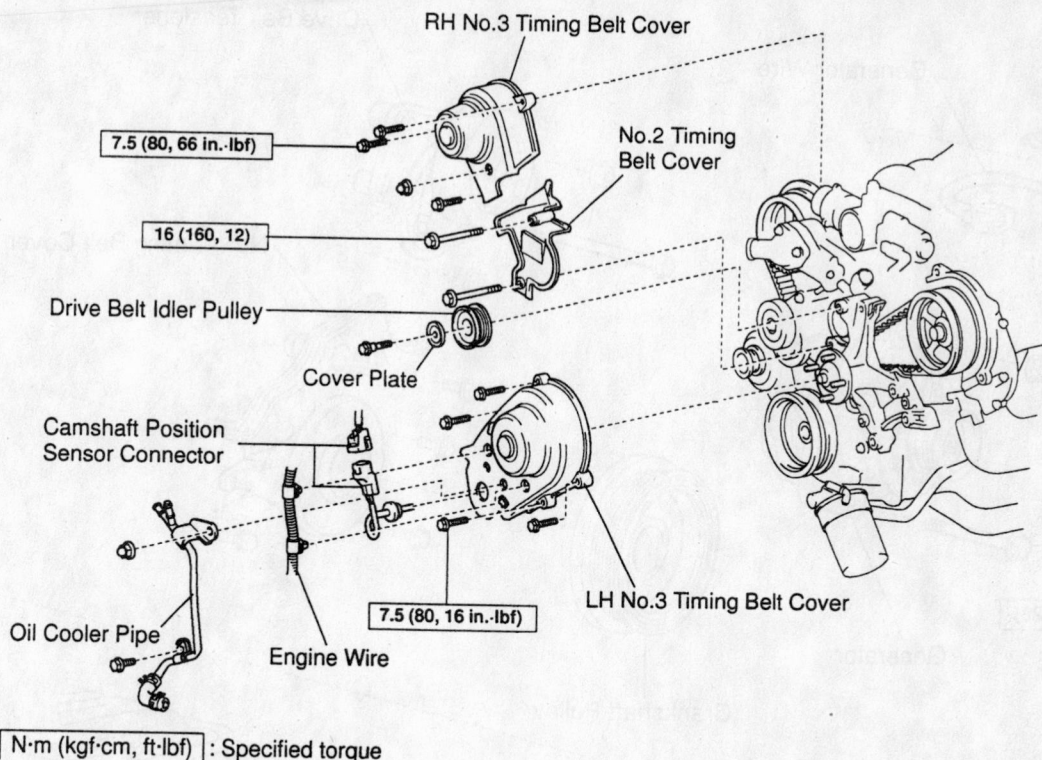

RH No.3 Timing Belt Cover

No.2 Timing Belt Cover

7.5 (80, 66 in.·lbf)

16 (160, 12)

Drive Belt Idler Pulley

Cover Plate

Camshaft Position Sensor Connector

7.5 (80, 16 in.·lbf)

LH No.3 Timing Belt Cover

Oil Cooler Pipe

Engine Wire

N·m (kgf·cm, ft·lbf) : Specified torque

93025G25

Exploded view of upper timing belt covers—Toyota Land Cruiser 4.7L (2UZ-FE) engine

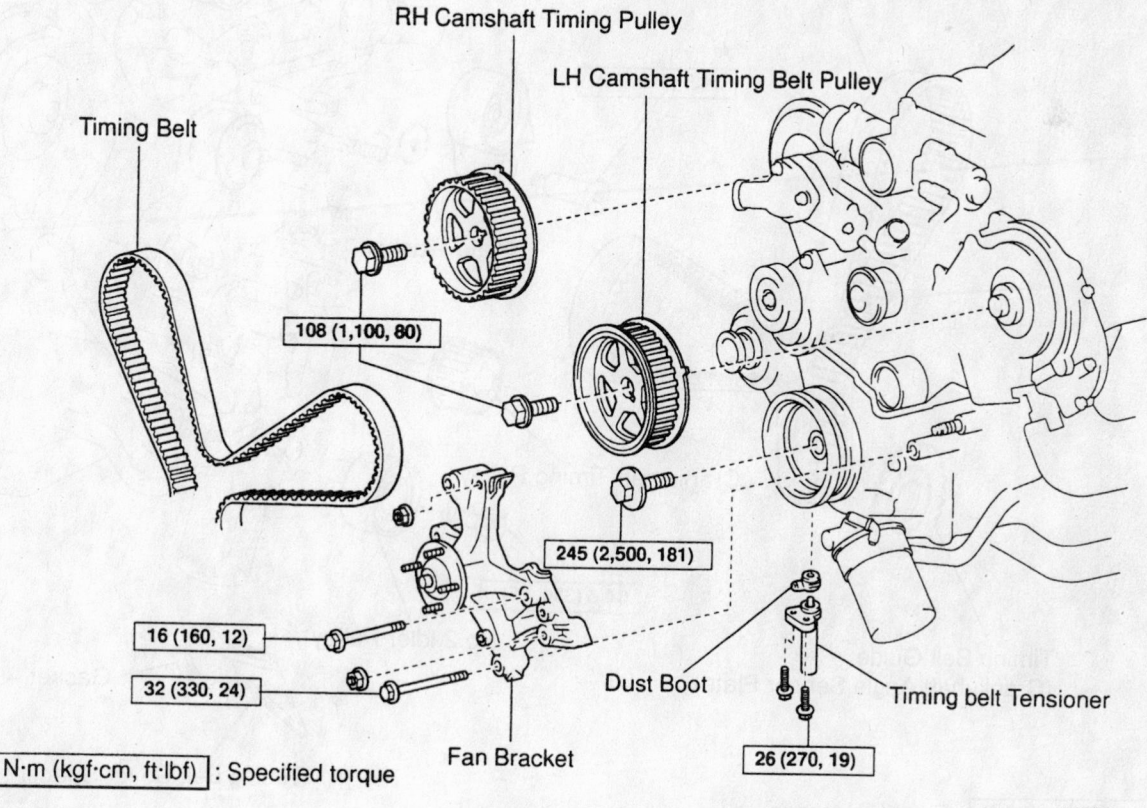

RH Camshaft Timing Pulley

LH Camshaft Timing Belt Pulley

Timing Belt

108 (1,100, 80)

245 (2,500, 181)

16 (160, 12)

32 (330, 24)

Dust Boot

Timing belt Tensioner

Fan Bracket

26 (270, 19)

N·m (kgf·cm, ft·lbf) : Specified torque

93025G26

Exploded view of upper timing sprockets and components—Toyota Land Cruiser 4.7L (2UZ-FE) engine

Brake service is covered in Section 4 of this manual

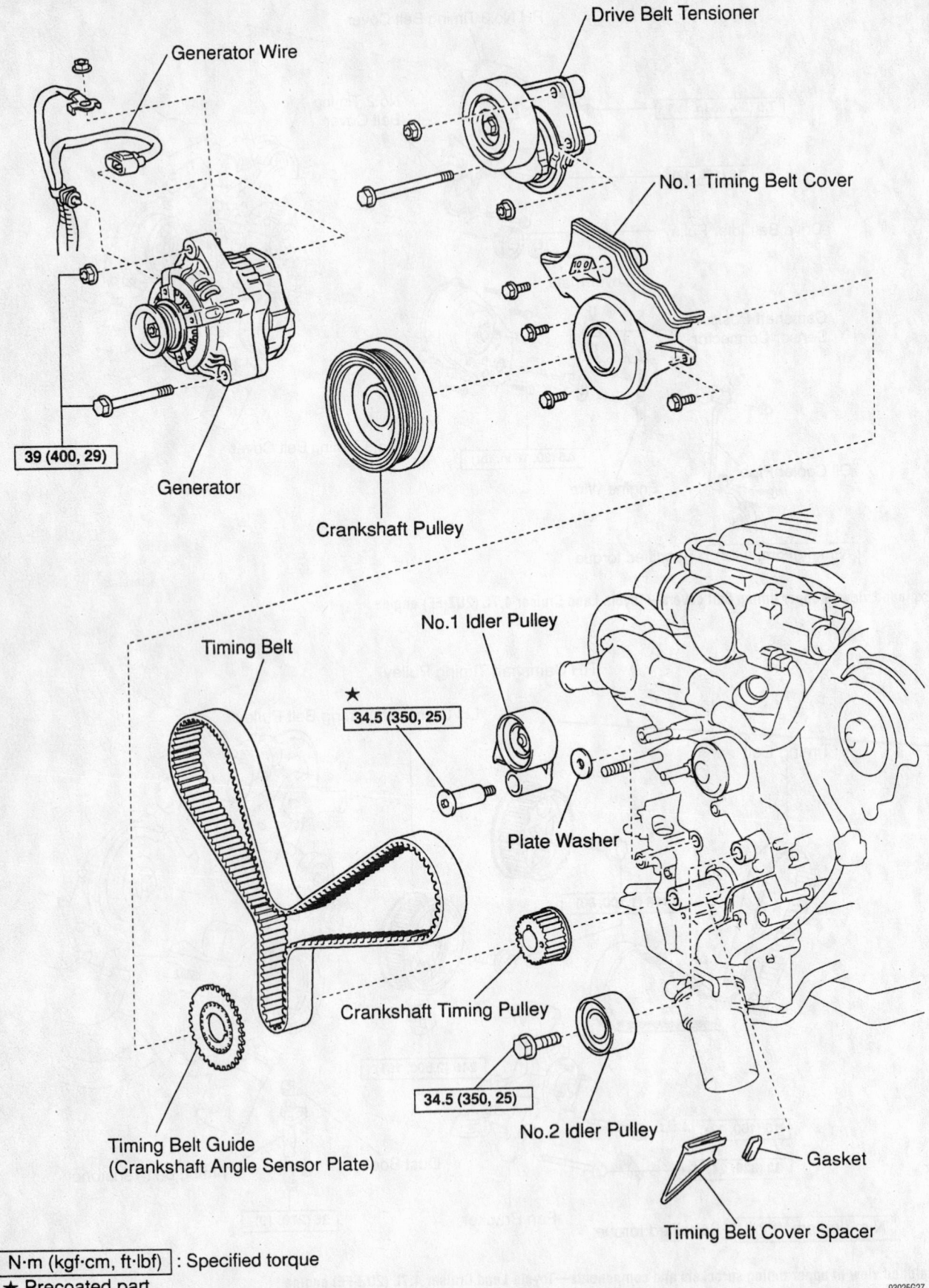

Generator Wire

Drive Belt Tensioner

No.1 Timing Belt Cover

39 (400, 29)

Generator

Crankshaft Pulley

Timing Belt

No.1 Idler Pulley

★ 34.5 (350, 25)

Plate Washer

Crankshaft Timing Pulley

Timing Belt Guide
(Crankshaft Angle Sensor Plate)

34.5 (350, 25)

No.2 Idler Pulley

Gasket

Timing Belt Cover Spacer

N·m (kgf·cm, ft·lbf) : Specified torque
★ Precoated part

93025G27

Exploded view of lower timing belt cover, sprockets and components—Toyota Land Cruiser 4.7L (2UZ-FE) engine

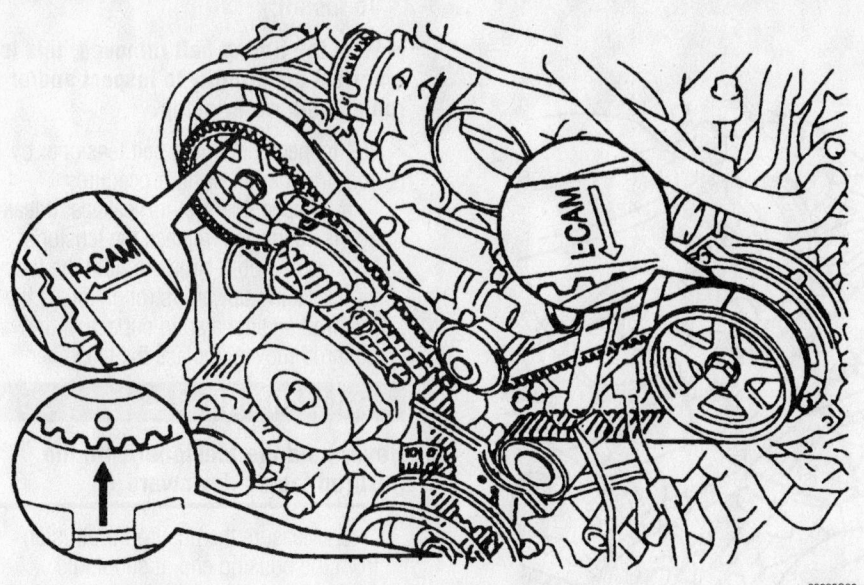

Alignment of timing belt with the timing sprockets—Toyota Land Cruiser 4.7L (2UZ-FE) engine

f. Remove the middle radiator-to-chassis nut/bolts and brackets.

g. Carefully, lift the radiator from the vehicle.

10. Remove the serpentine drive belt idler pulley bolt, cover plate and pulley.

11. Remove the right side (No. 3) timing belt cover.

12. Remove the left side (No. 3) timing belt cover by performing the following procedures:

a. Disconnect the engine wire from both wire clamps.

b. Disconnect the camshaft position sensor wire from the wire clamp on the left-side (No.3) timing belt cover.

c. Disconnect the sensor connector from the connector bracket.

d. Disconnect the sensor connector.

e. Remove the wire grommet from the left-side (No. 3) timing belt cover.

f. Remove the oil cooler tube bolts and tube.

13. Remove the middle (No. 2) timing belt cover bolts and cover.

14. Remove the cooling fan bracket nuts/bolts and bracket.

➡ **If reusing the timing belt, make sure that there are 3 installation marks on the belt; if there are none, install them.**

15. Using the Crankshaft Pulley Holding tool 09213-70010, Bolt tool 90105-08076 and Companion Flange Holding tool 09330-00021, or equivalent, loosen the crankshaft pulley bolt.

16. Position the No. 1 cylinder to approximately 50 degrees After Top Dead Center (ATDC) of the compression stroke by performing the following procedures:

a. Rotate the crankshaft pulley (CLOCKWISE) to align its groove with the timing mark "0" on the lower (No. 1) timing belt cover.

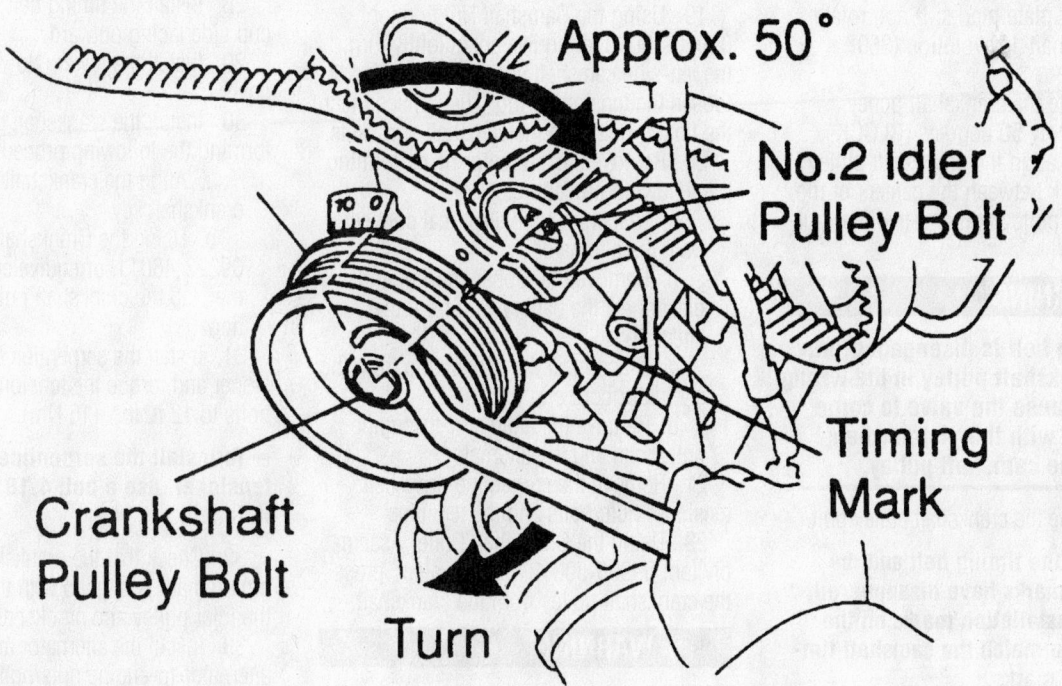

Aligning of crankshaft pulley timing mark with the center line of the crankshaft pulley bolt and the idler pulley bolt—Toyota Land Cruiser 4.7L (2UZ-FE) engine

For complete Engine Mechanical specifications, see Section 1 of this manual

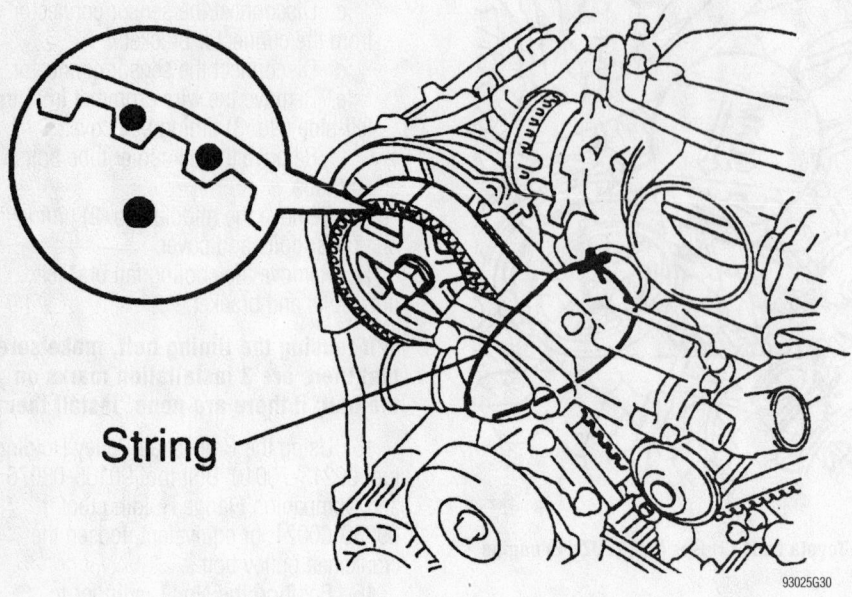

String

Securing the timing belt with string and matchmarking the camshaft with the timing belt—Toyota Land Cruiser 4.7L (2UZ-FE) engine

b. Check that the camshaft sprocket timing marks are aligned with the rear timing belt plate marks; if not, rotate the crankshaft 1 revolution (360 degrees).

c. Rotate the crankshaft pulley approximately 50 degrees (CLOCKWISE) and align the crankshaft pulley timing mark between the centers of the crankshaft pulley bolt and the idler pulley bolt.

✳✳ WARNING

If the timing belt is disengaged, having the crankshaft pulley in the wrong angle can cause the valve to come into contact with the piston when removing the camshaft pulley.

17. Remove the crankshaft pulley bolt.

➡If reusing the timing belt and the installation marks have disappeared, place new installation marks on the timing belt to match the camshaft timing sprocket marks.

➡To avoid meshing the timing sprocket and the timing belt, secure one with a string; then, place matchmarks on the timing belt and the right-side camshaft timing sprocket.

18. Remove the timing belt tensioner bolts and the tensioner.

19. Using the Camshaft Holding tool 09960-10010, or equivalent, slightly turn the left-side camshaft sprocket clockwise to loosen the tension spring. Then, disconnect the timing belt from the camshaft sprockets.

20. Remove the alternator by performing the following procedures:

a. Disconnect the electrical connector from the alternator.

b. Remove the rubber cap/nut and disconnect the battery wire from the alternator.

c. Disconnect the wire clamp from the alternator cord clip.

d. Remove the alternator-to-engine nuts/bolts and the alternator.

21. Remove the serpentine drive belt tensioner nuts/bolts and the tensioner.

22. Using the Crankshaft Puller Assembly tool 09950-50012, or equivalent, press the crankshaft pulley from the crankshaft.

✳✳ WARNING

DO NOT rotate the crankshaft pulley.

23. Remove the lower (No. 1) timing belt cover bolts and the cover.

24. Remove the timing belt guide, spacer and the timing belt.

To install:

➡With the timing belt removed, this is a perfect opportunity to inspect and/or replace the water pump.

25. Inspect the timing belt tensioner by performing the following procedures:

a. Inspect the seal for leakage; if leakage is suspected, replace the tensioner.

b. Using both hands to hold the tensioner facing upward, strongly press the pushrod against a solid surface. If the pushrod moves, replace the tensioner.

✳✳ WARNING

Never hold the tensioner with the pushrod facing downward.

c. Measure the pushrod protrusion from the housing end, it should be 0.413–0.453 in. (10.5–11.5mm). If the protrusion is not as specified, replace the tensioner.

26. Temporarily install the timing belt by performing the following procedures:

a. Align the timing belt's installation mark with the crankshaft timing sprocket.

b. Install the timing belt on the crankshaft timing sprocket, the No. 1 idler pulley and the No. 2 idler pulley.

27. Install the gasket to the timing belt cover spacer and install the cover spacer.

28. Install the timing belt guide with the cup side facing outward.

29. Install the lower (No. 1) timing belt cover.

30. Install the crankshaft pulley by performing the following procedures:

a. Align the crankshaft pulley with the crankshaft key.

b. Using the Crankshaft Installer tool 09223-46011, or equivalent, and a hammer, tap the crankshaft pulley into position.

31. Install the serpentine drive belt tensioner and torque the tensioner-to-engine bolts to 12 ft. lbs. (16 Nm).

➡To install the serpentine drive belt tensioner, use a bolt 4.18 in. (106mm) in length.

32. Check that the crankshaft pulley's timing mark is aligned with the centers of the idler pulley and crankshaft pulley bolts.

33. Install the alternator and torque the alternator-to-engine nuts/bolts to 29 ft. lbs. (39 Nm). Connect the alternator's electrical connectors and clip.

34. Install the timing belt to the left-side camshaft by performing the following procedures:

a. Rotate the left-side camshaft pulley

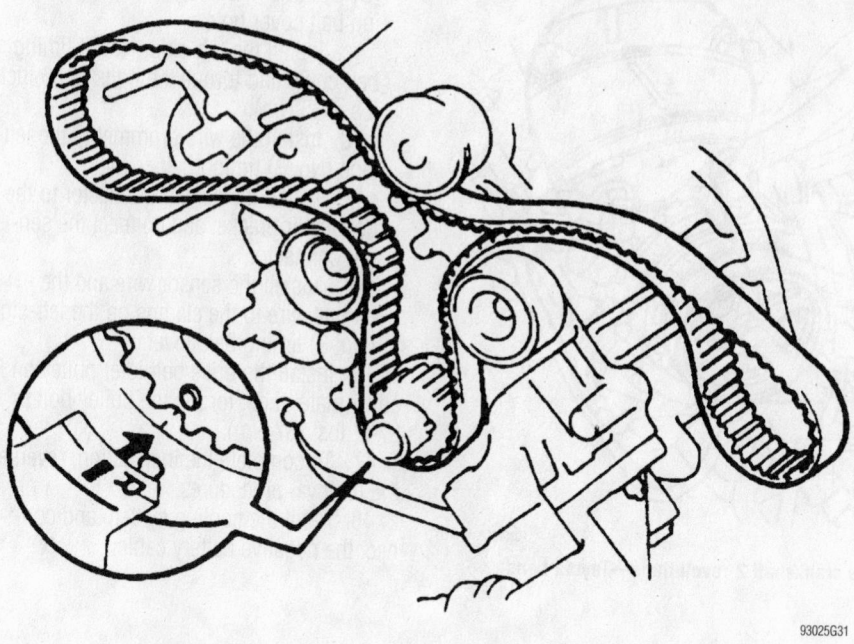

Installing the timing belt on the crankshaft sprocket—Toyota Land Cruiser 4.7L (2UZ-FE) engine

93025G31

to align the timing belt installation mark with the camshaft sprocket's timing mark and slide the belt onto the camshaft timing sprocket.

　b. Using the Camshaft Holding tool 09960-10010, or equivalent, slightly turn the left-side camshaft sprocket counterclockwise to place tension on the timing belt between the crankshaft sprocket and the camshaft sprocket.

35. Rotate the right-side camshaft pulley to align the timing belt installation mark with the camshaft sprocket's timing mark and slide the belt onto the camshaft timing sprocket.

36. Using a vertical press, slowly press the pushrod into the housing using 200–2205 lbs. (981–9807 N) until the holes align, then, install a 1.27mm Allen® wrench to secure the pushrod and release the press. Install the dust boot on the tensioner housing.

37. Install the timing belt tensioner and torque the bolts to 19 ft. lbs. (26 Nm).

38. Using a pair of pliers, remove the Allen® wrench from the tensioner housing.

39. Check the valve timing by performing the following procedure:

　a. Temporarily install the crankshaft pulley bolt.

　b. Slowly, rotate the crankshaft pulley 2 revolutions (CLOCKWISE) and realign the TDC marks.

➡ **If the pulley/sprocket timing marks do not realign, remove the timing belt and reinstall it.**

40. Using the Crankshaft Pulley Holding tool 09213-70010, Bolt tool 90105-08076 and Companion Flange Holding tool 09330-00021; or equivalent, torque the crankshaft pulley bolt to 181 ft. lbs. (245 Nm).

41. Install the cooling fan bracket and torque the 12mm (head size) bolt to 12 ft. lbs. (16 Nm) and the 14mm (head size) bolt to 24 ft. lbs. (32 Nm).

42. Install the air conditioning compressor.

43. Install the middle (No. 2) timing belt cover and torque the bolts to 12 ft. lbs. (16 Nm).

44. Install the upper right-side (No. 3) timing belt cover and torque the bolts to 66 inch lbs. (7.5 Nm).

45. Install the upper left-side (No. 3) timing belt cover by performing the following procedures:

　a. Install the oil cooler tube and bolt.

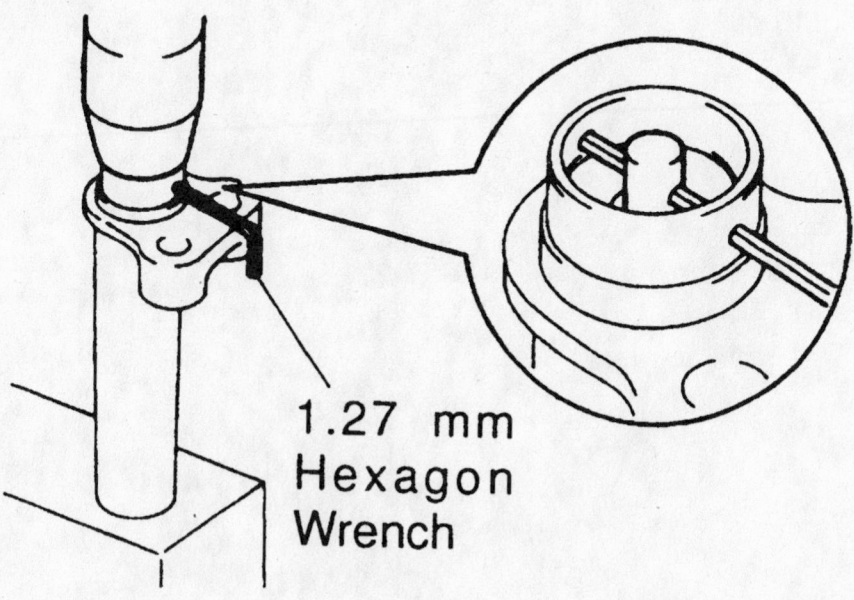

1.27 mm Hexagon Wrench

93025G32

Securing the timing belt tensioner pushrod—Toyota Land Cruiser 4.7L (2UZ-FE) engine

For Accessory Drive Belt illustrations, see Section 1 of this manual

Checking the TDC alignment marks after rotating the crankshaft 2 revolutions—Toyota Land Cruiser 4.7L (2UZ-FE) engine

b. Feed the Camshaft Position Sensor (CPS) through the left-side (No. 3) timing belt cover hole.

c. Install the left-side (No. 3) timing belt cover and torque the bolts to 66 inch lbs. (7.5 Nm).

d. Install the wire grommet to the left-side (No. 3) timing belt cover.

e. Install the sensor connector to the connector bracket and connect the sensor connector.

f. Install the sensor wire and the engine wire to the clamps on the left-side (No. 3) timing belt cover.

46. Install the drive belt idler pulley and cover plate; then, torque the pulley bolt to 27 ft. lbs. (37 Nm).

47. To complete the installation, reverse the removal procedures.

48. Refill the cooling system and connect the negative battery cable.

BRAKES

4

ACURA

Brake Caliper

REMOVAL & INSTALLATION

SLX

FRONT

1. Raise and safely support the vehicle.
2. Remove some brake fluid from the reservoir.
3. Remove the front wheels.
4. Disconnect the brake fluid line from the caliper. Plug the line to prevent fluid loss.
5. Loosen the brake caliper mounting bolt and guide bolt. Remove the caliper from the mount.
6. Remove the brake pads and clips from the caliper. Inspect the brake pads for wear and replace them if necessary.

To install:

7. Fill the brake caliper with clean brake fluid and connect the fluid line to the caliper using new washers. Tighten the brake line banjo fitting to 26 ft. lbs. (35 Nm). Install the brake pads and clips onto the caliper.
8. Install the caliper onto the mounting bracket. Lubricate the caliper bolts and their boots. Then, install the caliper mounting bolts and tighten them to 54 ft. lbs. (74 Nm).
9. Refill and bleed the brake system.
10. Install the front wheels and lower the vehicle.

REAR

1. Raise and safely support the vehicle.
2. Remove some brake fluid from the reservoir.
3. Remove the rear wheels.
4. Disconnect the brake fluid line from the caliper. Plug the line to prevent fluid loss.
5. Loosen the brake caliper mounting bolt and guide bolt. Remove the caliper from the mount bracket.
6. Remove the brake pads and clips from the caliper. Inspect the brake pads for wear; replace them if necessary.
7. If necessary for servicing, unbolt the caliper mounting bracket from the backing plate.

To install:

8. If removed, install the caliper mounting bracket and tighten its bolts to 76 ft. lbs. (103 Nm).
9. Fill the brake caliper with clean brake fluid and connect the fluid line to the caliper

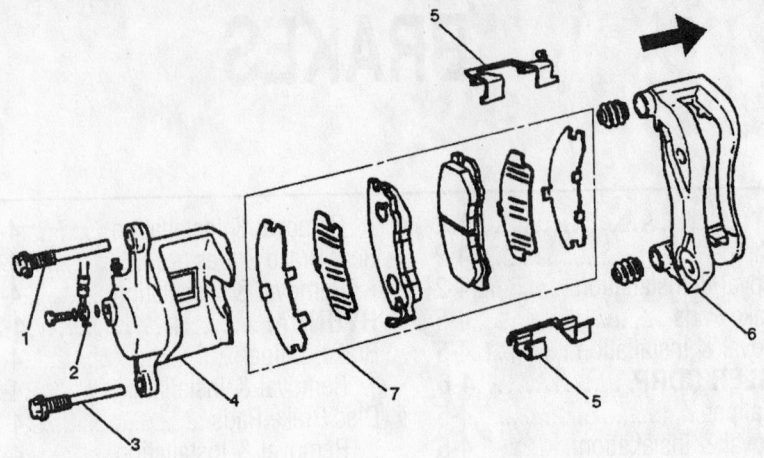

(1) Guide Bolt
(2) Brake Flexible Hose
(3) Lock Bolt
(4) Caliper Assembly
(5) Clip
(6) Support Bracket with Pad Assembly
(7) Pad Assembly

93026G02

Front caliper assembly—SLX

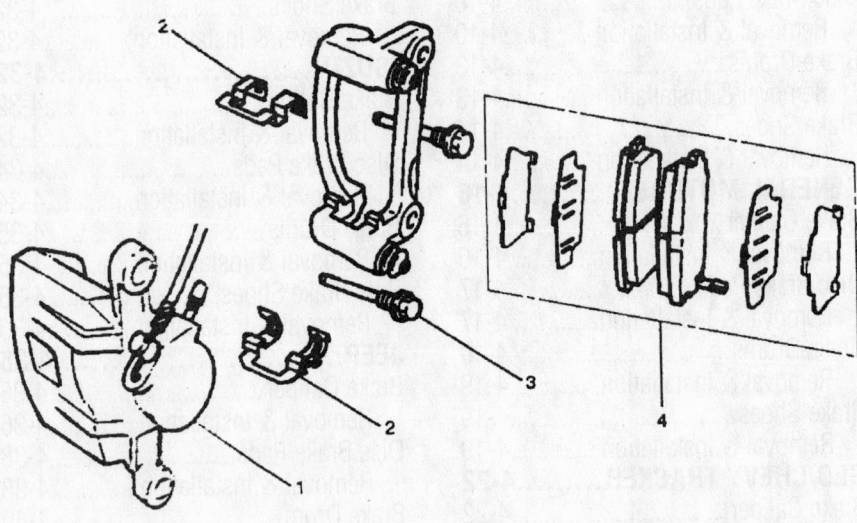

(1) Caliper Assembly
(2) Clip
(3) Lock Bolt
(4) Pad Assembly

93026G01

Rear caliper assembly—SLX

using new washers. Tighten the brake line banjo fitting to 26 ft. lbs. (35 Nm). Install the brake pads and clips onto the caliper.
10. Install the caliper on the mounting bracket. Lubricate the caliper bolts and their

boots. Then, install the caliper mounting bolts. Tighten them to 32 ft. lbs. (44 Nm).
11. Refill and bleed the brake system.
12. Install the rear wheels and lower the vehicle.

MDX

FRONT

1. Remove some fluid from the reservoir with a suction pump.
2. Raise and safely support the vehicle.
3. Remove the front wheels.

4. Remove the banjo bolt and disconnect the brake hose from the caliper. Plug the hose to prevent fluid loss and contamination.

5. Remove the mounting bolts and remove the caliper from its mounting bracket.

To Install:

6. Fit the caliper over the pads and onto its mounting bracket.
7. Torque both caliper bolts to 27 ft. lbs. (36 Nm).
8. Reconnect the brake hose to the caliper using new sealing washers. Care-

Front Brake Caliper Overhaul

⚠CAUTION

Frequent inhalation of brake pad dust, regardless of material composition, could be hazardous to your health.
- Avoid breathing dust particles.
- Never use an air hose or brush to clean brake assemblies. Use an OSHA-approved vacuum cleaner.

Remove, disassemble, inspect, reassemble, and install the caliper, and note these items:

- Do not spill brake fluid on the vehicle; it may damage the paint; if brake fluid gets on the paint, wash it off immediately with water.
- To prevent dripping brake fluid, cover disconnected hose joints with rags or shop towels.
- Clean all parts in brake fluid and air dry; blow out all passages with compressed air.
- Before reassembling, check that all parts are free of dirt and other foreign particles.
- Replace parts with new ones as specified in the illustration.
- Make sure no dirt or other foreign matter gets in the brake fluid.
- Make sure no grease or oil gets on the brake discs or pads.
- When reusing pads, always reinstall them in their original positions to prevent loss of braking efficiency.
- Do not reuse drained brake fluid. Use only clean Genuine Honda DOT 3 Brake Fluid. Non-Honda brake fluid can cause corrosion and shorten the life of the system.
- Coat the piston, piston seal groove, and caliper bore with clean brake fluid.
- Replace all rubber parts with new ones.
- After installing the caliper, check the brake hose and line for leaks, interference, and twisting.

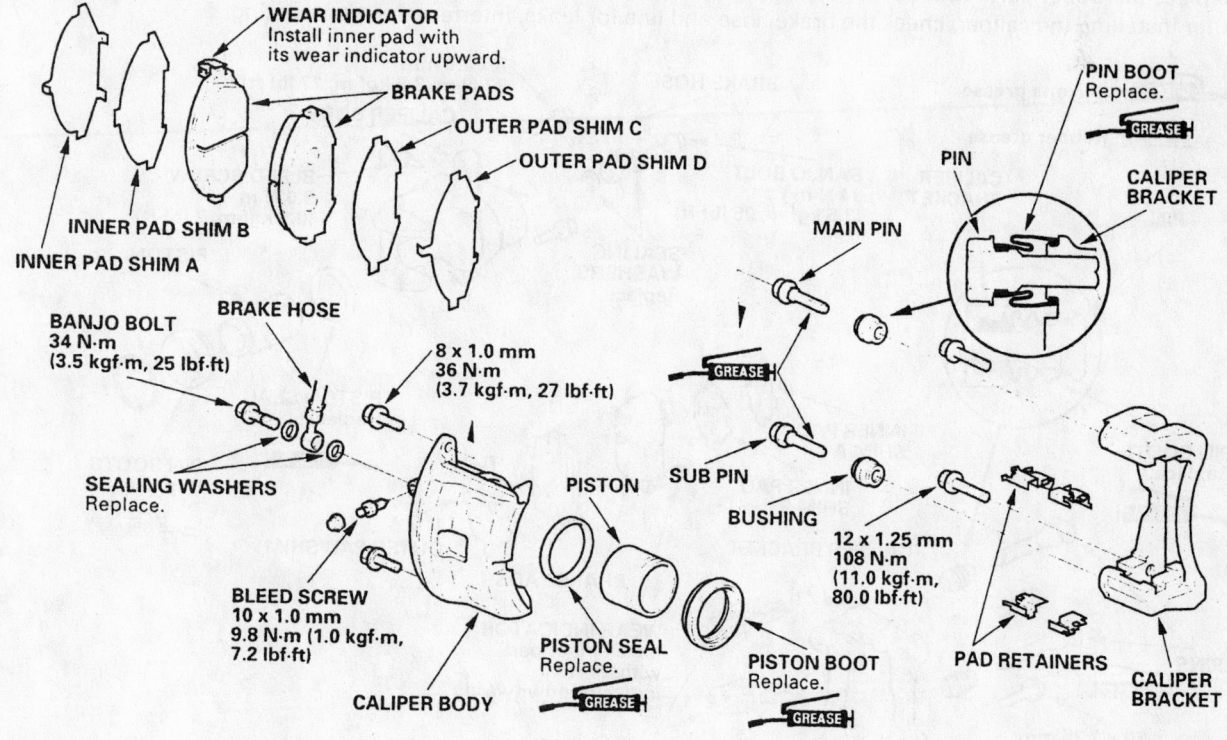

Exploded view of the front caliper components—MDX

fully torque the banjo bolt to 25 ft. lbs. (34 Nm).

9. Fill the reservoir with fluid and bleed the brakes.

10. Install the front wheels and lower the vehicle.

REAR

1. Remove some fluid from the reservoir with a suction pump.
2. Raise and safely support the vehicle.
3. Remove the rear wheels.

4. Remove the banjo bolt and disconnect the brake hose from the caliper. Plug the hose to prevent fluid loss and contamination.

5. Remove the 2 caliper mounting bolts. Remove the caliper from its mounting bracket.

Rear Brake Caliper Overhaul

⚠ CAUTION

Frequent inhalation of brake pad dust, regardless of material composition, could be hazardous to your health.
- Avoid breathing dust particles.
- Never use an air hose or brush to clean brake assemblies. Use an OSHA-approved vacuum cleaner.

Remove, disassemble, inspect, reassemble, and install the caliper, and note these items:

- Do not spill brake fluid on the vehicle; It may damage the paint; If brake fluid gets on the paint, wash it off immediately with water.
- To prevent dripping brake fluid, cover disconnected hose joints with rags or shop towels.
- Clean all parts in brake fluid and air dry; blow out all passages with compressed air.
- Before reassembling, check that all parts are free of dirt and other foreign particles.
- Replace parts with new ones as specified in the illustration.
- Make sure no dirt or other foreign matter gets into the brake fluid.
- Make sure no grease or oil gets on the brake discs or pads.
- When reusing pads, always reinstall them in their original positions to prevent loss of braking efficiency.
- Do not reuse drained brake fluid. Use only clean Genuine Honda DOT 3 Brake Fluid. Non-Honda brake fluid can cause corrosion and shorten the life of the system.
- Coat the piston, piston seal groove, and caliper bore with clean brake fluid.
- Replace all rubber parts with new ones.
- After installing the caliper, check the brake hose and line for leaks, interference, and twisting.

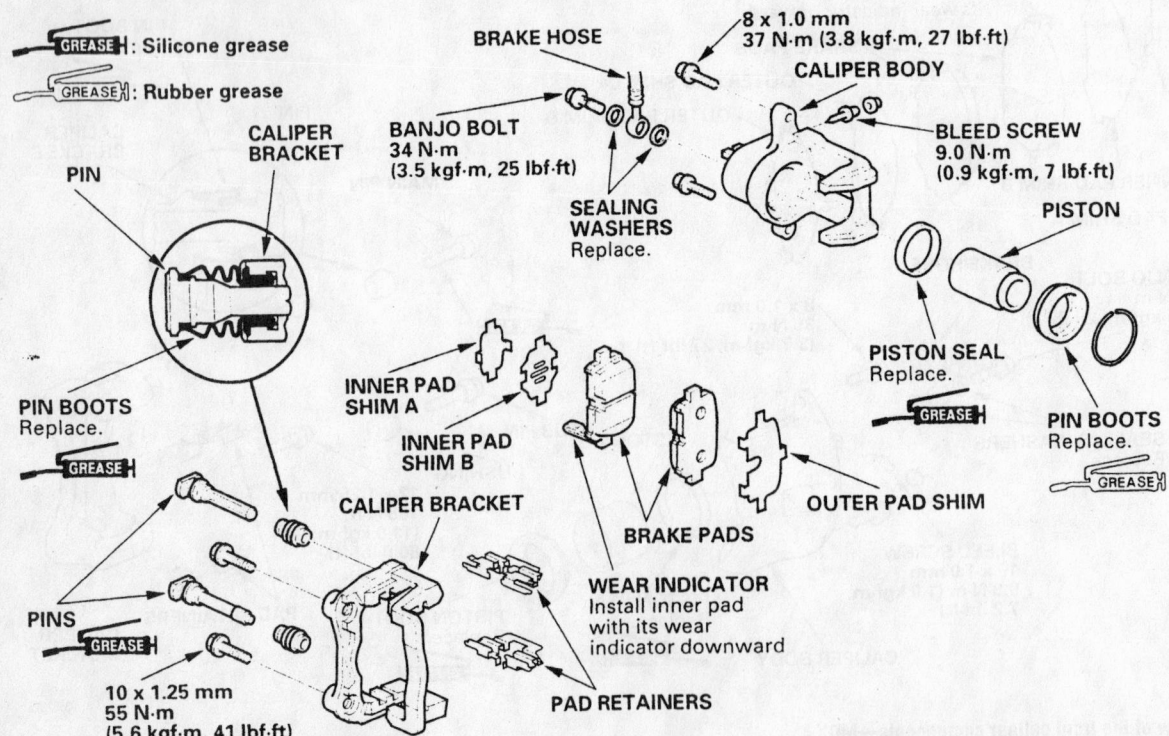

Exploded view of the rear caliper components—MDX

93352GZB

To Install:

6. Fit the caliper over the pads and onto its mounting bracket.

7. Tighten the caliper bolts to 27 ft. lbs. (37 Nm).

8. Reconnect the brake hose with new sealing washers. Tighten the banjo bolt to 25 ft. lbs. (34 Nm).

9. Fill the reservoir with fluid and bleed the brake system. Adjust the parking brake if necessary.

10. Install the rear wheels and lower the vehicle.

Disc Brake Pads

REMOVAL & INSTALLATION

SLX

FRONT

1. Remove about ½ of the brake fluid from the master cylinder reservoir to prevent overflow when the caliper piston is compressed.

2. Raise and safely support the vehicle.

3. Remove the front wheels.

4. Remove the brake caliper from the caliper bracket without disconnecting the brake line. Support the caliper with a length of wire. Do not let the caliper hang from the brake hose.

5. Remove the brake pads and shims. Inspect the brake rotor and machine or replace as necessary. Check the minimum thickness (specification is cast into the rotor) before machining.

To install:

6. Use a large C-clamp or brake piston tool to push the caliper piston into its bore.

7. Apply a thin coat of brake grease to both sides of both inner shims. Assemble the pads and shims, then install them into the caliper. The wear indicator on the inner pad must face down.

8. Install the calipers. Clean and lubricate the caliper mounting bolts and lubricate the mounting bolt boots. Install the mounting bolts and tighten them to 54 ft. lbs. (74 Nm).

9. Install the front wheels and lower vehicle.

10. Apply the brakes several times to seat the pads before moving the vehicle. Check the fluid level in the master cylinder reservoir and add as necessary.

REAR

1. Use a vacuum pump to remove some brake fluid from the master cylinder reservoir to prevent overflow when the caliper piston is compressed.

2. Raise and safely support the vehicle.

3. Remove the rear wheels.

4. Remove the brake caliper from the caliper bracket without disconnecting the brake line. Support the caliper with a length of wire. Do not let the caliper hang from the brake hose.

5. Remove the brake pads and shims. Inspect the brake rotor and machine or replace as necessary. Check the minimum thickness (specification is cast into the rotor) before machining.

To install:

6. Use a large C-clamp or brake piston tool to push the caliper piston into its bore.

7. Apply a thin coat of brake grease to both sides of both inner shims. Assemble the pads and shims, then install them into the caliper. The wear indicator on the inner pad must face down.

8. Install the calipers. Clean and lubricate the caliper mounting bolts and lubricate the mounting bolt boots. Install the mounting bolts and tighten them to 32 ft. lbs. (44 Nm).

9. Install the rear wheels and lower vehicle.

10. Apply the brakes several times to seat the pads before moving the vehicle. Check the fluid level in the master cylinder reservoir and add as necessary.

MDX

FRONT

1. Raise and support the vehicle safely.

2. Remove the front wheels.

3. Remove a small amount of brake fluid from the reservoir using a suction pump.

4. Unbolt the brake hose clamp from the knuckle by removing the retaining bolts.

5. Remove the lower caliper retaining bolt and pivot the caliper upward, off of the pads.

6. Remove the pad shim and pad retainers. Remove the disc brake pads from the caliper.

To install:

7. Clean the caliper thoroughly; remove any rust from the lip of the disc or rotor. Check the brake rotor for grooves or cracks. If any heavy scoring is present, the rotor must be replaced.

8. Install the pad retainers. Apply molyb-

denum brake grease to both surfaces of the shims and the back of the disc brake pads.

9. Install the pads and shims. The pad with the wear indicator goes in the inboard position.

10. Push in the caliper piston so the caliper will fit over the pads. This is most easily accomplished with a pad spreader or large C-clamp.

11. Pivot the caliper down into position and tighten the mounting bolt to 27 ft. lbs. (37 Nm).

12. Connect the brake hose to the knuckle, if removed.

13. Install the wheel and lower the vehicle to the ground.

14. Add brake fluid to the master cylinder reservoir and install the cap.

15. Depress the brake pedal several times and make sure that the movement feels normal. The first brake pedal application may result in a very long pedal action due to the pistons being retracted. Always make several brake applications before starting the vehicle. Bleed the system if necessary.

REAR

1. Raise and safely support the vehicle.

2. Remove a small amount of brake fluid from the reservoir using a suction pump.

3. Remove the rear wheels.

4. Remove the 2 caliper mounting bolts and remove the caliper from the bracket.

5. Remove the pads, shims, and pad retainers.

To install:

6. Clean the caliper thoroughly; remove any dirt or dust. Check the brake rotor for grooves or cracks and machine or replace, as necessary.

7. Install the pad retainers. Apply molybdenum brake grease to both surfaces of the shims and the back of the disc brake pads.

8. Install the pads and shims. The wear retainer on the inboard pad faces down.

9. Use a suitable tool to push caliper piston into its bore and enable the caliper to fit over the pads. Lubricate the piston boot with silicon grease. Avoid twisting the boot.

10. Install the brake caliper. Tighten the mounting bolts to 27 ft. lbs. (37 Nm).

11. Install the rear wheels. Lower the vehicle.

12. Add brake fluid to the master cylinder reservoir. Depress the brake pedal several times to seat the pads. Bleed the brakes if necessary.

CHRYSLER CORP.

Brake Caliper

REMOVAL & INSTALLATION

Durango

1. Raise and support the front end on jackstands.
2. Remove the wheels.
3. Disconnect the rubber brake hose from the tubing at the frame mount. If the pistons are to be removed from the caliper, leave the brake hose connected to the caliper. Check the rubber hose for cracks or chafed spots.
4. Plug the brake line to prevent loss of fluid.
5. Remove the caliper slide pins.
6. Remove the caliper and brake pads from the rotor adapter.

To install:
7. Position the outboard shoe in the caliper. The shoe should not rattle in the caliper. If it does, or if any movement is obvious, bend the shoe tabs over the caliper to tighten the fit.
8. Slide the caliper into position on the adapter and over the rotor.
9. Align the caliper and start the pins in by hand.
10. Tighten the pins to 22 ft. lbs. (30 Nm).
11. Connect the brake hose to the caliper. Use new washers to attach the hose fitting if the original washers are scored, worn or damaged.
12. Fill and bleed the brake system.
13. Install the wheels.
14. Lower the vehicle.

Disc Brake Pads

REMOVAL & INSTALLATION

Durango

1. Raise and support the front end on jackstands.
2. Remove the wheels.
3. Press the caliper piston back into the bore with a suitable prytool. Use a large C-clamp to drive the piston into the bore of additional force is required.
4. Remove the caliper mounting bolts with a ⅜ in. hex wrench or socket.
5. Rotate the caliper rearward off the rotor and out from its mount.
6. Set the caliper on a crate or sturdy box, then remove the inboard and outboard

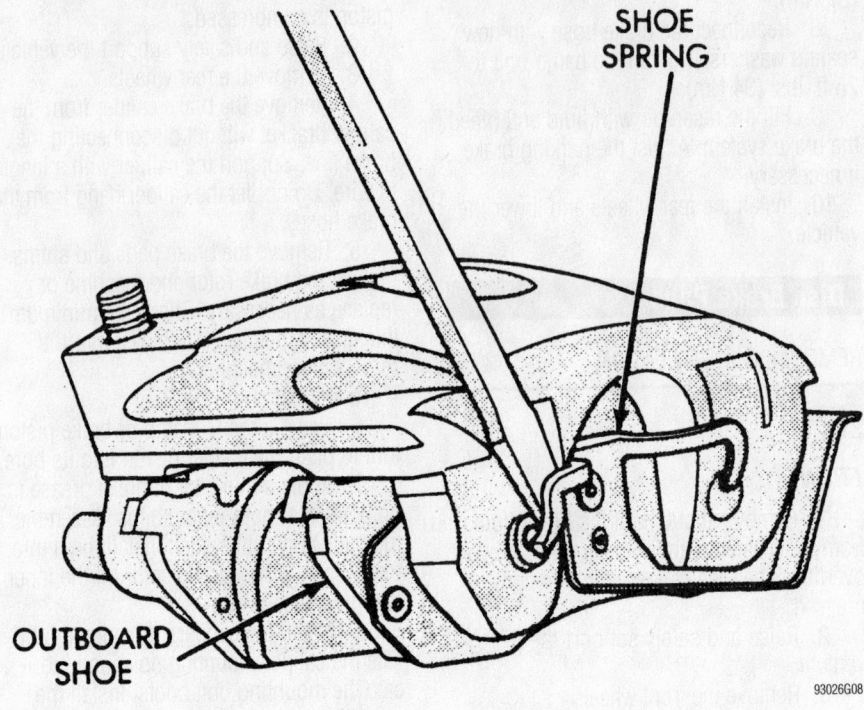

Prying the disc brake from the 4WD front brake caliper assembly—1999–02 Dakota

brake pads. The inboard pad has a spring clip that holds it in the caliper. Tilt this pad out at the top to unseat the clip. The outboard pad has a retaining spring that secures it in the caliper. Unseat 1 spring end and rotate the pad out of the caliper.
7. Secure the caliper to a chassis or suspension component with a sturdy wire. Do not let it hang from the hose.

To install:
8. Clean the caliper and steering knuckle sliding surfaces with a wire brush. Then, apply a coat of Mopar® multi-mileage grease or equivalent.
9. Clean the caliper slide pins with brake cleaner or brake fluid. Then apply a light coating of silicone grease to the pins.

➡**If there is minor rust or corrosion on the pins, first polish them with a crocus cloth. If they are severely rusted, replace them.**

10. Install the inboard pad and its spring clip.
11. Install the outboard brake pad.
12. Install the caliper over the rotor and seat it in its original position until flush.

13. Final tighten the caliper slide pins to 18–26 ft. lbs. (25–35 Nm).
14. Install the wheels.
15. Lower the vehicle.
16. Pump the brakes several times to seat the pads.

Brake Drums

REMOVAL & INSTALLATION

Durango

CHRYSLER SERVO TYPE WITH SINGLE ANCHOR

1. Raise and safely support the truck.
2. Remove the plug from the brake adjustment access hole.
3. Insert a thin bladed screwdriver through the adjusting hole and hold the adjusting lever away from the starwheel.
4. Release the brake by prying down against the starwheel with a brake spoon.
5. Remove the rear wheel and clips from the wheel studs. Remove the brake drum.

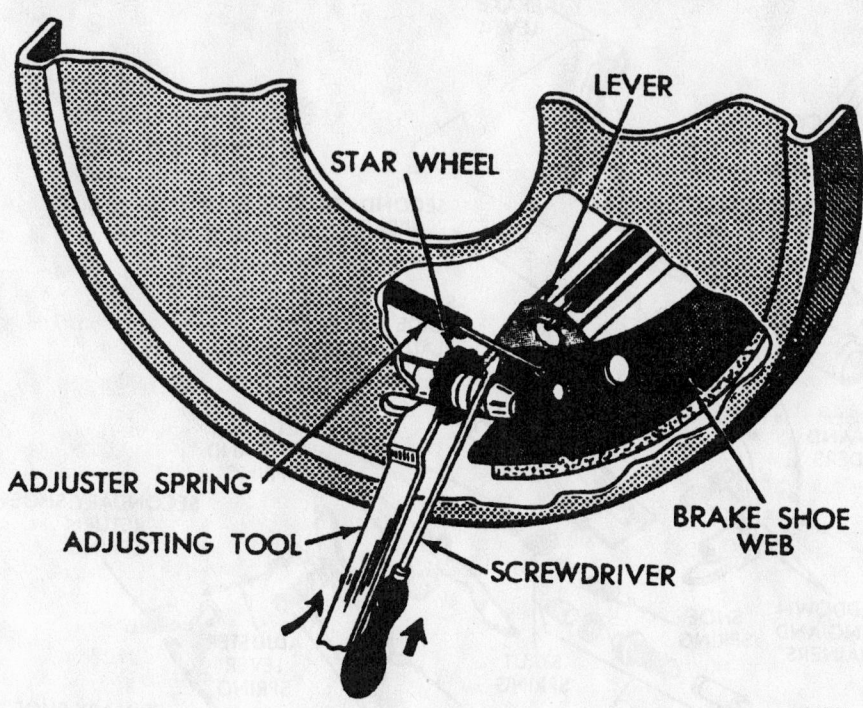

Use a lever releasing tool to depress the adjuster lever while turning the starwheel with a pry-tool—Bendix brakes

6. Installation is the reverse of removal. Adjust the brakes.

BENDIX DUO-SERVO TYPE

1. Raise and safely support the vehicle.
2. Remove the rear wheel and tire.
3. Remove the axle shaft nuts, washers and cones. If the cones do not readily release, rap the axle shaft sharply in the center.
4. Remove the axle shaft.
5. Remove the outer hub nut.
6. Straighten the lockwasher tab and remove it along with the inner nut and bearing.
7. Carefully remove the drum.

To install:

8. Position the drum on the axle housing.
9. Install the bearing and inner nut. While rotating the wheel and tire, tighten the adjusting nut until a slight drag is felt.
10. Back off the adjusting nut ⅙ turn so that the wheel rotates freely without excessive end-play.
11. Install the lockrings and nut. Place a new gasket on the hub and install the axle shaft, cones, lockwashers and nuts.

12. Install the wheel and tire.
13. Road-test the vehicle.

Brake Shoes

REMOVAL & INSTALLATION

Durango

SERVO TYPE WITH SINGLE ANCHOR

1. Raise and support the vehicle.
2. Remove the rear wheel, drum retaining clips and the brake drum.
3. Remove the brake shoe return springs, noting how the secondary spring overlaps the primary spring.
4. Remove the brake shoe retainer, springs and nails.
5. Disconnect the automatic adjuster cable from the anchor and unhook it from the lever. Remove the cable, cable guide, and anchor plate.
6. Remove the spring and lever from the shoe web.
7. Spread the anchor ends of the primary and secondary shoes and remove the parking brake spring and strut.

8. Disconnect the parking brake cable and remove the brake assembly.
9. Remove the primary and secondary brake shoe assemblies and the star adjuster as an assembly. Block the wheel cylinders to retain the pistons.

To install:

10. Measure the drum as described in this section.
11. Apply a thin coat of lubricant to the support platforms.
12. Attach the parking brake lever to the rear of the secondary shoe.
13. Place the primary and secondary shoes in their relative positions on a workbench.
14. Lubricate the adjuster screw threads. Install it between the primary and secondary shoes with the star wheel next to the secondary shoe. The star wheels are stamped with an **L** (left) and **R** (right).
15. Overlap the ends of the primary and second brake shoes and install the adjusting spring and lever at the anchor end.
16. Hold the shoes in position and install the parking brake cable into the lever.
17. Install the parking brake strut and spring between the parking brake lever and primary shoe.
18. Place the brake shoes on the support and install the retainer nails and springs.
19. Install the anchor pin plate.
20. Install the eye of the adjusting cable over the anchor pin and install the return spring between the anchor pin and primary shoe.
21. Install the cable guide in the secondary shoe and install the secondary return spring. Be sure that the primary spring overlaps the secondary spring.
22. Position the adjusting cable in the groove of the cable guide and engage the hook of the cable in the adjusting lever.
23. Install the brake drum and retaining clips. Install the wheel and tire.
24. Adjust the brakes and road-test the truck.

BENDIX DUO-SERVO TYPE

1. Unhook and remove the adjusting lever return spring.
2. Remove the lever from the lever pivot pin.
3. Unhook the adjuster lever from the adjuster cable.
4. Unhook the upper shoe-to-shoe spring.
5. Unhook and remove the shoe holddown springs.

Timing belt service is covered in Section 3 of this manual

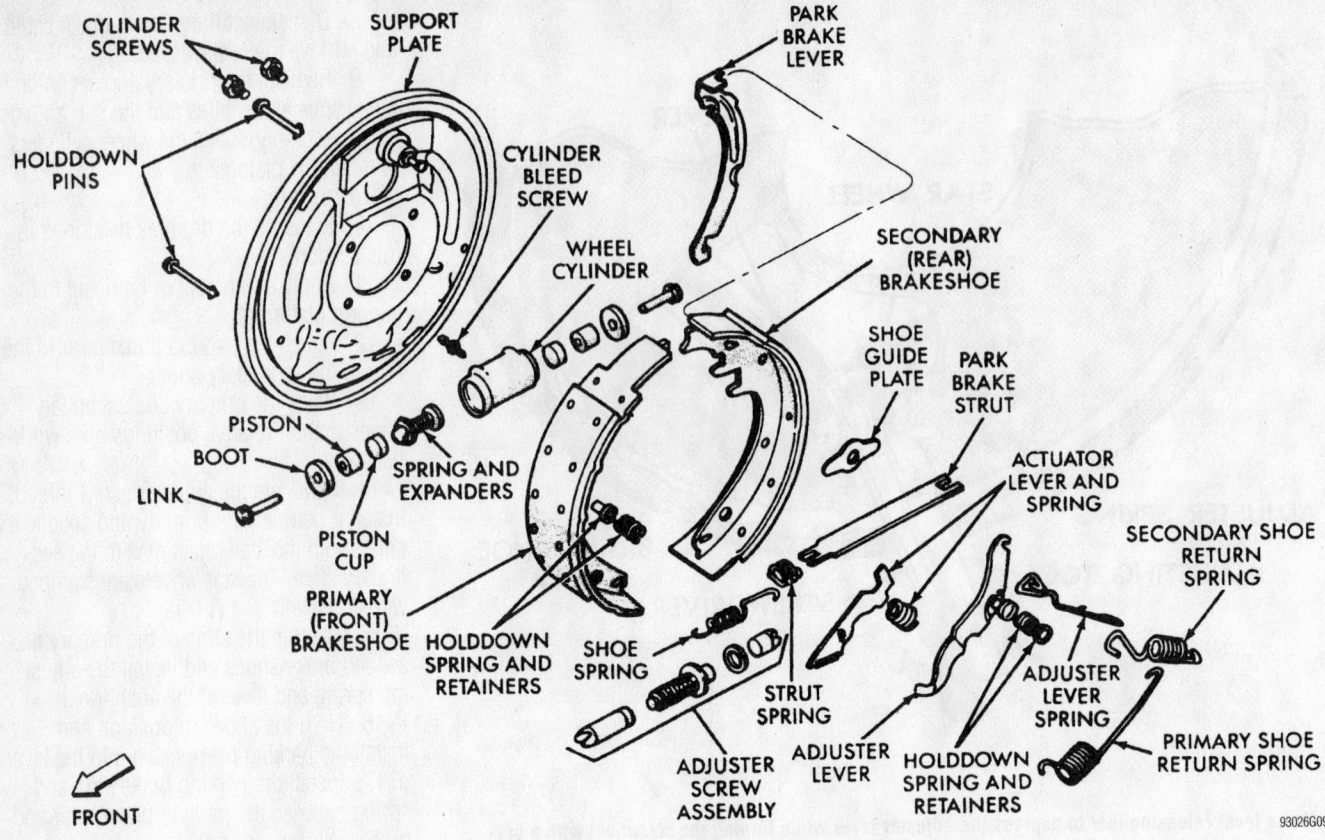

CYLINDER SCREWS

SUPPORT PLATE

PARK BRAKE LEVER

HOLDDOWN PINS

CYLINDER BLEED SCREW

WHEEL CYLINDER

SECONDARY (REAR) BRAKESHOE

SHOE GUIDE PLATE

PARK BRAKE STRUT

PISTON

BOOT

LINK

SPRING AND EXPANDERS

PISTON CUP

PRIMARY (FRONT) BRAKESHOE

HOLDDOWN SPRING AND RETAINERS

SHOE SPRING

ACTUATOR LEVER AND SPRING

SECONDARY SHOE RETURN SPRING

STRUT SPRING

ADJUSTER SCREW ASSEMBLY

ADJUSTER LEVER

HOLDDOWN SPRING AND RETAINERS

ADJUSTER LEVER SPRING

PRIMARY SHOE RETURN SPRING

FRONT

93026G09

Exploded view of the rear brake components—B-Series Van, Dakota, Durango and Ram Pick-Up

6. Disconnect the parking brake cable from the parking brake lever.

7. Remove the shoes with the lower shoe-to-shoe spring and star wheel as an assembly.

To install:

8. The pivot screw and adjusting nut on the left side have left-hand threads and right-hand threads on the right side.

9. Lubricate and assemble the star wheel assembly. Lubricate the guide pads on the support plates.

10. Assemble the star wheel, lower shoe-to-shoe spring, and the primary and secondary shoes. Position this assembly on the support plate.

11. Install and hook the hold-down springs.

12. Install the upper shoe-to-shoe spring.

13. Install the cable and retaining clip.

14. Position the adjuster lever return spring on the pivot (green springs on left brakes and red springs on right brakes).

15. Install the adjuster lever. Route the adjuster cable and connect it to the adjuster.

16. Install the brake drum and adjust the brakes.

FORD

Brake Caliper

REMOVAL & INSTALLATION

Explorer and Mountaineer

FRONT

1. Siphon part of the brake fluid out of the master cylinder to avoid overflow when the caliper piston is pressed into the caliper bore.

2. Raise the vehicle and support it safely. Remove the wheel and tire assembly.

3. Position an 8 in. (20cm) C-clamp on the caliper and tighten the clamp to move the caliper piston into the bore approximately 1/8 in. (3mm). Avoid clamp contact with the outer shoe spring clip. Remove the clamp.

➡️**Do not pry the piston away from the rotor.**

4. Clean excess dirt from the pin tab area.

5. Using a 1/4 in. drive socket and a light hammer, tap the upper caliper pin towards the outboard side until the pin tabs pass the spindle face.

6. Compress the inboard pin tab, if equipped, with pliers and, with a hammer, drive the pin out until the tab slips into the spindle groove.

7. Place an end of a 7/16 in. (11mm) diameter punch against the end of the caliper pin and tap the pin out of the caliper slide groove.

8. Repeat Steps 5, 6 and 7 to remove the lower pin.

9. Disconnect and plug the brake hose at the caliper. Remove the caliper from the rotor.

To install:

10. Make sure the caliper mounting surfaces are free of dirt. Lubricate the caliper grooves with disc brake caliper grease and install the caliper.

11. From the caliper outboard side, position the pin between the caliper and spindle grooves. The pin must be positioned so the tabs will be installed against the spindle outer face.

12. Tap the pin on the outboard end with a hammer until the retention tabs on the sides of the pin contact the spindle face.

13. Repeat Steps 11 and 12 for the lower pin.

➡️**During installation, do not allow the tabs of the caliper pin to be tapped too far into the spindle groove. If this happens, it will be necessary to tap the other end of the caliper pin until the tabs snap in place. The tabs on each end of the pin must be free to catch on the spindle face.**

14. Connect the brake hose to the caliper. Bleed the brake system.

15. Install the wheel and tire assembly and lower the vehicle. Check the brake fluid level and check the brakes for proper operation.

REAR

1. Siphon part of the brake fluid out of the master cylinder to avoid overflow when the caliper piston is pressed into the caliper bore.

2. Raise the vehicle and support it safely. Remove the wheel and tire assembly.

3. Position an 8 in. (20cm) C-clamp on the caliper and tighten the clamp to move the caliper piston into the bore approximately ⅛ in. (3mm). Remove the clamp.

➡️**Do not pry the piston away from the rotor.**

4. Clean excess dirt from the retainer bolt area.

5. Using a Torx® socket, remove the 2 retainer bolts securing the caliper to the bracket and adapter plate.

6. Disconnect and plug the brake hose at the caliper. Remove the caliper from the rotor.

To install:

7. Make sure the caliper mounting surfaces are free of dirt. Lubricate the caliper grooves with disc brake caliper grease and install the caliper.

8. Position the caliper to the bracket and secure in place with the retainer bolts. Tighten the bolts to 20 ft. lbs. (27 Nm).

9. Install the caliper brake hose using new washers. Tighten the bolt to 29 ft. lbs. (40 Nm).

10. Fill and bleed the brake system.

11. Install the wheel and tire assembly and lower the vehicle. Check the brake fluid level and check the brakes for proper operation.

Expedition and Navigator

FRONT RAIL SLIDER TYPE

1. Raise and safely support the vehicle and remove the wheels.

2. Place an 8 in. (20cm) C-clamp on the caliper and tighten the clamp to bottom the caliper piston in the cylinder bore. Bear the clamp on the outer pad; never press directly on the piston! Remove the C-clamp.

3. Clean the excess dirt from around the caliper pin tabs.

4. Drive the upper caliper pin inward until the tabs on the pin touch the spindle.

5. Insert a small prybar into the slot provided behind the pin tabs on the inboard side of the pin.

6. Using needle-nose pliers, compress the outboard end of the pin while, at the same time, prying with the prybar until the tabs slip into the groove in the spindle.

7. Place the end of a ⁷⁄₁₆ in. (11mm) punch against the end of the caliper pin and drive the pin out of the caliper slide groove.

8. Repeat this procedure for the lower pin.

9. Lift the caliper off of the rotor.

10. Remove the brake pads from the caliper.

11. Disconnect the brake hose from the caliper, then plug the brake line to prohibit contamination of the brake fluid by water or dirt.

To install:

12. Connect the brake hose to the caliper. When connecting the brake fluid hose to the caliper, it is recommended that a new copper washer be used at the connection of the brake hose and caliper.

13. Thoroughly clean the areas of the caliper and spindle assembly which contact each other during the sliding action of the caliper.

14. Install the brake pads onto the caliper.

15. Position the caliper on the spindle assembly. Lightly lubricate the caliper sliding grooves with caliper pin grease.

16. Position a new upper pin with the retention tabs next to the spindle groove.

➡️**Don't use the bolt and nut with the new pin.**

17. Carefully drive the pin, at the outboard end, inward until the tabs contact the spindle face.

18. Repeat the procedure for the lower pin.

✴✴ WARNING

Don't drive the pins in too far or it will be necessary to drive them back out until the tabs snap into place.

The tabs on each end of the pin must be free to catch on the spindle sides!

19. Install the wheels.

FRONT PIN SLIDER TYPE

1. Break the front wheel lug nuts loose, then raise and support the front of the vehicle safely on jackstands.

2. Remove the front wheels.

3. Remove the front brake hose bolt, then remove the copper washers and plug the front brake hose.

4. Remove the 2 front disc brake caliper slide pins, then lift the caliper off of the front caliper anchor plate.

To install:

5. Install the front disc brake caliper onto the caliper anchor plate. Install the 2 slide pins. Tighten the slider pins/bolts to the following values:
- All except F- and E-250/350: 21–26 ft. lbs. (28–36 Nm)
- F-250, F-350, E-250 and E-350: 141–190 ft. lbs. (191–259 Nm)

6. Using new copper washers, attach the front brake hose to the brake caliper. Install and tighten the retaining bolt to 23–29 ft. lbs. (30–40 Nm)

7. Bleed the brake system.

8. Clean the wheel hub mounting surface.

9. Install the front wheels and snug the lug nuts to fully seat the wheel against the hub.

10. Lower the vehicle until some of the vehicle's weight rests on the front tires, then tighten the lug nuts to 83–112 ft. lbs. (113–153 Nm).

11. Lower the vehicle completely.

12. Make sure that the brakes are operating correctly.

REAR

1. Remove enough brake fluid from the brake master cylinder reservoir until it is ½ full.

2. Raise and safely support the vehicle.

3. Remove the wheel and tire assembly.

4. Remove the hollow bolt connecting the brake hose to the disc brake caliper and plug the brake hose. Discard 2 copper sealing washers.

5. Remove 2 brake caliper slide pins and lift the caliper off the anchor plate.

To install:

6. Retract the disc brake caliper pistons fully in the piston bores using an old brake pad or block of wood and a C-clamp or equivalent.

Heater Core replacement is covered in Section 2 of this manual

7. Place the disc brake caliper above the rotor and install it with a rotating motion. Make sure the inner and outer pads are properly positioned and the anti-rattle clips are correctly installed. The brake caliper bleed screw should be positioned on top of the caliper when assembled on the vehicle.

8. Lubricate the locating pins and the inside of the insulators with silicone grease. Install the locating pins through the caliper insulators and hand-start the threads into the steering knuckle attaching holes. Tighten the locating pins to 25 ft. lbs. (34 Nm).

9. Remove the plug and install the brake hose to the disc brake caliper using 2 new copper sealing washers. Tighten the hollow bolt to 29 ft. lbs. (40 Nm).

10. Bleed the brake system, filling the brake master cylinder reservoir as required.

11. Install the wheel and tire assembly. Tighten the lug nuts in a star pattern to 83–112 ft. lbs. (113–153 Nm).

12. Lower the vehicle.

13. Pump the brake pedal several times to position the brake pads prior to moving the vehicle.

14. Road-test the vehicle and check for proper brake system operation.

Disc Brake Pads

REMOVAL & INSTALLATION

Explorer and Mountaineer

FRONT

1. Siphon part of the brake fluid out of the master cylinder to avoid overflow when the caliper piston is pressed into the caliper bore.

2. Raise the vehicle and support it safely. Remove the wheel and tire assembly.

3. Remove the brake caliper, but do not disconnect the brake hose. Secure the caliper aside with mechanic's wire.

4. Compress the anti-rattle clip and remove the inner brake pad from the caliper.

5. Press each ear of the outer brake pad away from the caliper and slide the torque buttons out of the retention notches.

To install:

6. Bottom out the caliper piston in the caliper bore using an 8 in. (20cm) C-clamp or equivalent and a worn out inner brake pad or block of wood to push against the piston. Do not attempt to bottom out the piston with the outer brake pad installed.

7. Place a new anti-rattle clip on the lower end of the inner brake pad. Make sure

the tabs on the clip are properly positioned and the clip is fully seated.

8. Position the inner brake pad and anti-rattle clip in the pad abutment with the ant-rattle clip tab against the pad abutment and the loop-type spring away from the rotor. Compress the anti-rattle clip and slide the upper end of the pad in position.

9. Install the outer pad, making sure the torque buttons on the pad are seated solidly in the matching holes in the caliper.

10. Install the caliper on the spindle.

11. Install the wheel and tire assembly and lower the vehicle. Apply the brakes several times before moving the vehicle to seat the pads.

12. Check the brake fluid level. Check the brakes for proper operation.

REAR

1. Siphon part of the brake fluid out of the master cylinder to avoid overflow when the caliper piston is pressed into the caliper bore.

2. Raise the vehicle and support it safely. Remove the wheel and tire assembly.

3. Remove the brake caliper, but do not disconnect the brake hose. Secure the caliper aside with mechanic's wire.

4. Remove the inner and outer brake pad from the caliper.

To install:

5. Bottom out the caliper piston in the caliper bore using an 8 in. (20cm) C-clamp or equivalent and a worn out inner brake pad or block of wood to push against the piston. Do not attempt to bottom out the piston with the outer brake pad installed.

6. Position the inboard brake pad in the caliper and press the retainer spring fully into the caliper piston.

7. Start one end of the outboard brake shoe and lining on the caliper and rotate it down until the locating lugs and the retainer spring are fully seated.

8. Install new shoe slippers on the rear wheel disc brake adapter.

9. Install the caliper on the spindle.

10. Install the wheel and tire assembly and lower the vehicle. Apply the brakes several times before moving the vehicle to seat the pads.

11. Check the brake fluid level. Check the brakes for proper operation.

Expedition and Navigator

FRONT

1. Break the front wheel lug nuts loose, then raise and support the front of the vehicle safely on jackstands.

2. Remove the front wheels.

3. Remove the front disc brake calipers.

4. Note the position and orientation of the brake pads and anti-rattle clip. Remove the brake shoes and lings from the front disc brake caliper anchor plate, then remove the anti-rattle clips.

To install:

5. Thoroughly clean the caliper and spindle sliding areas.

6. Place a new anti-rattle clip on the lower end of the inboard shoe. Make sure the tabs on the clip are positioned correctly and the loop-type spring is away from the rotor.

7. Place the lower end of the inner brake pad in the spindle assembly pad abutment, against the anti-rattle clip and slide the upper end of the pad into position. Be sure the clip is still in position.

8. Check and make sure the caliper piston is fully bottomed in the cylinder bore. Use a large C-clamp, bearing on a piece of wood, to bottom the piston, if necessary.

9. Position the outer brake pad on the caliper and press the pad tabs into place with your fingers. If the pad cannot be pressed into place by hand, use a C-clamp. Be careful not to damage the lining with the clamp. Bend the tabs to prevent rattling.

10. Lightly lubricate the caliper sliding grooves with caliper pin grease.

11. Install the brake caliper onto the anchor plate.

12. Bleed the brake system.

13. Clean the wheel hub mounting surface.

14. Install the front wheels and snug the lug nuts to fully seat the wheel against the hub.

15. Lower the vehicle until some of the vehicle's weight rests on the front tires, then tighten the lug nuts to 83–112 ft. lbs. (113–153 Nm).

16. Lower the vehicle completely.

17. Make sure that the brakes are operating correctly.

REAR

1. Remove the rear brake caliper from the rear hub without disconnecting the brake hose.

2. Remove the brake pads and anti-rattle spring.

3. Thoroughly clean the areas of the caliper and caliper support assembly which contact each other during the sliding action of the caliper.

To install:

4. Position the brake shoes and anti-rattle clips on the disc brake caliper support bracket.

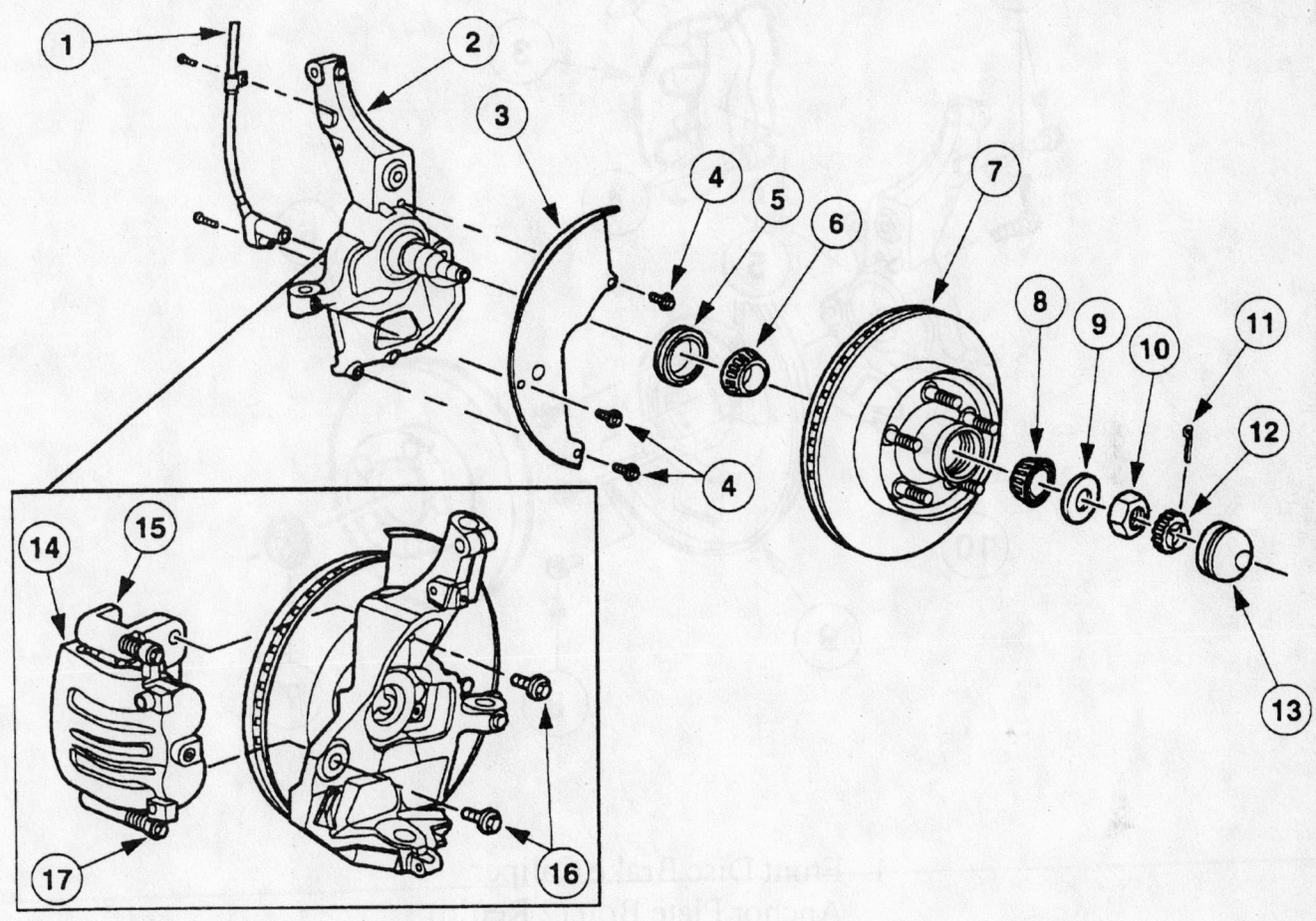

1 Front Brake Anti-Lock Sensor
2 Front Wheel Spindle
3 Front Disc Brake Rotor Shield
4 Rotor Shield Bolt
5 Grease Seal
6 Front Wheel Bearing

7 Front Disc Brake Hub and Rotor
8 Front Wheel Bearing
9 Front Wheel Outer Bearing Retainer Washer
10 Hub Spindle Nut
11 Cotter Pin

12 Nut Retainer
13 Hub Grease Cap
14 Disc Brake Caliper
15 Front Disc Brake Caliper Anchor Plate
16 Caliper Anchor Plate Bolts
17 Disc Brake Caliper Bolt

93026G22

Exploded view of the 2WD front disc brake assembly—1999–02 Explorer and Mountaineer

Brake service is covered in Section 4 of this manual

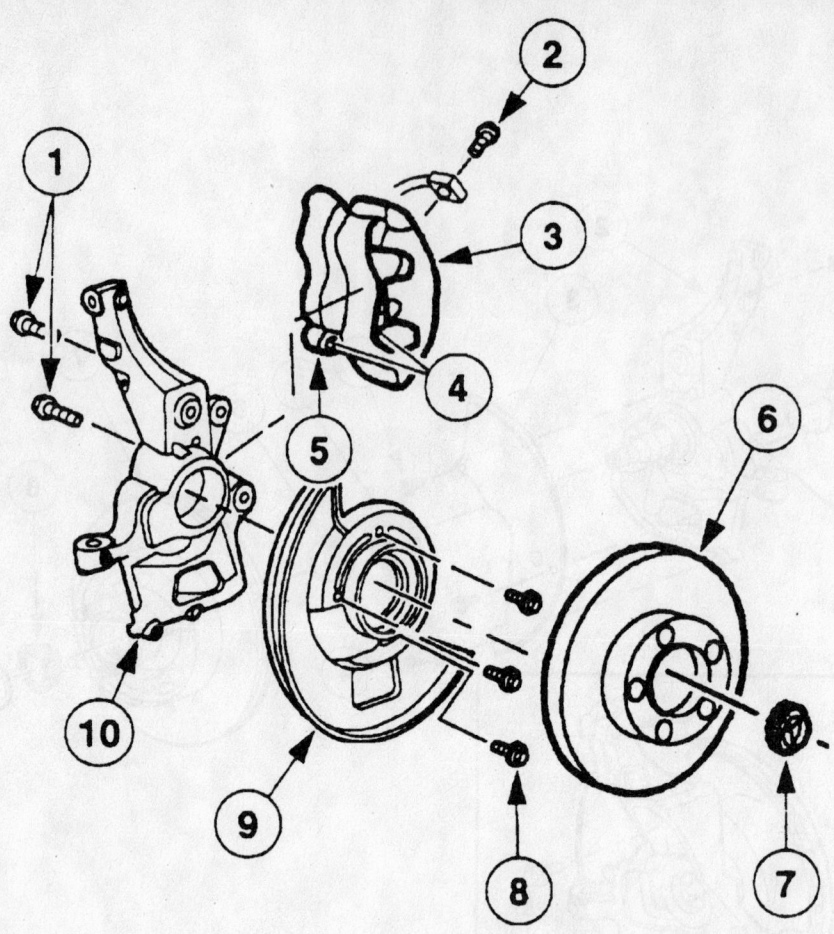

1 Front Disc Brake Caliper
   Anchor Plate Bolt (2 Req'd)

2 Front Brake Hose Bolt

3 Disc Brake Caliper

4 Pads

5 Front Disc Brake Caliper
   Anchor Plate

6 Front Disc Brake Rotor

7 Front Axle Wheel Hub
   Retainer

8 Front Disc Brake Rotor Shield
   Bolt (3 Req'd)

9 Front Disc Brake Rotor Shield

10 Front Wheel Knuckle

93026G23

Exploded view of the 4WD front disc brake assembly—1999–02 Explorer and Mountaineer

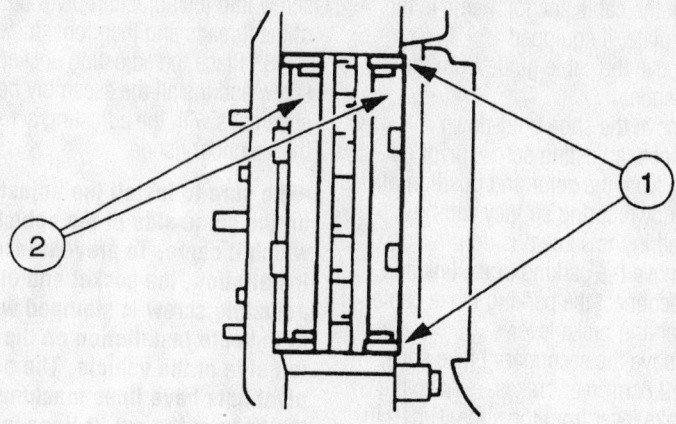

1 stainless slippers

2 pads

93026G24

Position of the front disc brake components—1999–02 Explorer and Mountaineer

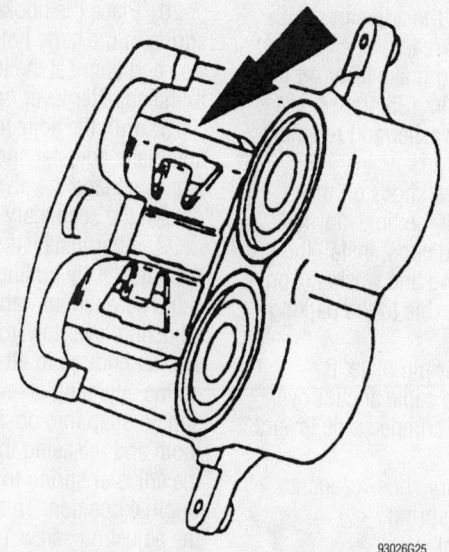

93026G25

View of the front disc brake anti-rattle spring—1999–02 Explorer and Mountaineer

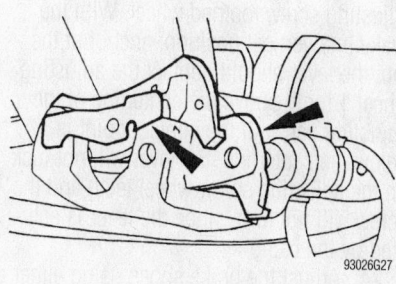

93026G27

Installing the rear disc brake pads—Explorer and Mountaineer

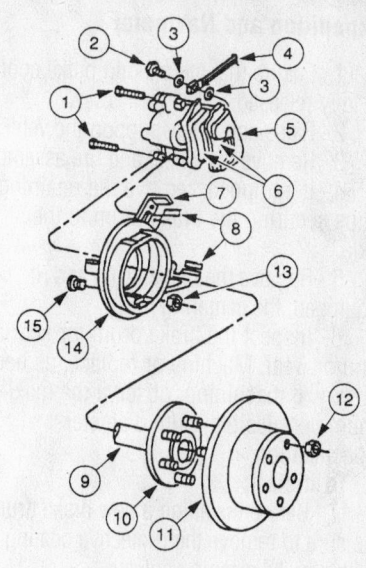

1 Brake Caliper Bolt
2 Flow Bolt
3 Copper Washer (2 Req'd)
4 Rear Wheel Brake Hose
5 Rear Disc Brake Caliper
6 Brake Pads
7 Rear Disc Brake Caliper Anchor Plate
8 Shoe Slippers
9 Axle Shaft
10 Lug Bolt (5 Req'd)
11 Rear Disc Brake Rotor
12 Keeper Nut
13 Rear Wheel Disc Brake Adapter Nut
14 Rear Wheel Disc Brake Adapter
15 Rear Wheel Disc Brake Adapter Bolt

93026G26

Exploded view of the rear disc brake assembly—Explorer and Mountaineer

Brake Drums

REMOVAL & INSTALLATION

Explorer and Mountaineer

1. Raise and safely support the vehicle. Remove the wheel and tire assembly.

2. Remove the retaining nuts, if equipped, and remove the brake drum.

3. Inspect the brake drum surface for wear, scoring and runout. Machine or replace, as necessary.

To install:

4. Install the brake drum and secure in place with the retainer nuts, if equipped.

5. Adjust the rear brakes.

6. Install the wheel. Lower the vehicle.

5. Install the rear disc brake caliper onto the rear support bracket.

6. Bleed the brake system.

7. Clean the wheel hub mounting surface.

8. Install the front wheels and snug the lug nuts to fully seat the wheel against the hub.

9. Lower the vehicle until some of the vehicle's weight rests on the front tires, then tighten the lug nuts to 83–112 ft. lbs. (113–153 Nm).

10. Lower the vehicle completely.

11. Make sure that the brakes are operating correctly.

For complete Engine Mechanical specifications, see Section 1 of this manual

Expedition and Navigator

1. Ensure that the parking brake control is fully released.

2. Raise and safely support the vehicle.

3. Remove the wheel and tire assembly.

4. If equipped, remove the retaining clips securing the brake drum to the axle.

5. Remove the brake drum and, if equipped, the centering ring.

6. Inspect the brake drum for scoring and/or wear. Machine or replace, as necessary. If machining, observe the maximum permissible drum diameter specification.

To install:

7. Before installing a new brake drum, be sure to remove the protective coating with a brake cleaning solvent.

8. If needed, adjust the rear brake shoes to fit the drum using Brake Adjustment Gauge D81L-1103-C or equivalent.

9. Position the brake drum on the axle hub and if equipped, the centering ring.

10. Install the wheel and tire assembly. Tighten the lug nuts in a star pattern to 84–112 ft. lbs. (113–153 Nm).

11. Lower the vehicle.

12. Pump the brake pedal several times to position the rear brake shoes before moving the vehicle.

13. Road-test the vehicle and check for proper brake system operation.

Brake Shoes

REMOVAL & INSTALLATION

Explorer and Mountaineer

1. Raise and safely support the vehicle. Remove the wheel and tire assembly and the brake drum.

2. Pull backward on the adjusting lever cable to disengage the adjusting lever from the adjusting screw. Move the outboard side of the adjusting screw upward and back off the pivot nut as far as it will go.

3. Pull the adjusting lever, cable and automatic adjuster spring down and toward the rear to unhook the pivot hook from the large hole in the secondary shoe web. Do not pry the pivot hook from the hole.

4. Remove the automatic adjuster spring and adjusting lever.

5. Remove the secondary shoe-to-anchor spring using a suitable brake spring removal/installation tool. Using the tool, remove the primary shoe-to-anchor spring

and unhook the cable anchor. Remove the anchor pin plate, if equipped.

6. Remove the cable guide from the secondary shoe.

7. Remove the shoe hold-down springs, shoes, adjusting screw, pivot nut and socket. Note the color and position of each hold-down spring so they can be reassembled in the same position.

8. Remove the parking brake link and spring. Disconnect the parking brake cable from the parking brake lever.

9. Remove the secondary brake shoe. On 9 in. (22.8cm) rear brakes, remove the parking brake lever from the shoe. On 10 in. (25.4cm) rear brakes, remove the retainer clip and spring washer and remove the parking brake lever.

To install:

10. Clean the backing plate ledge pads and sand lightly. Apply a light coating of high temperature lithium grease to the points where the brake shoes touch the backing plate. Lubricate the adjusting cable eye and the anchor pin area.

11. Install the parking brake lever on the secondary shoe. On 10 in. (25.4cm) brakes, secure with the spring washer and retaining clip.

12. Position the brake shoes on the backing plate and install the hold-down spring pins, springs and cups. Install the parking brake link, spring and washer. Connect the parking brake cable to the parking brake lever.

13. Install the anchor pin plate, if equipped, and place the cable anchor over the anchor pin with the crimped side toward the backing plate.

14. Install the primary shoe-to-anchor spring using the brake spring removal/installation tool.

15. Install the cable guide on the secondary shoe with the flanged hole fitted into the hole in the secondary shoe. Thread the cable around the cable guide groove.

➡ **Make sure the cable is positioned in the groove and not between the guide and shoe web.**

16. Install the secondary shoe-to-anchor (long) spring.

➡ **Make sure the cable end is not cocked or binding on the anchor pin when installed. All parts should be flat on the anchor pin.**

17. Apply high temperature lithium grease to the threads and the socket end of the adjusting screw. Turn the adjusting

screw into the adjusting pivot nut to the end of the threads and then loosen, ½ turn.

18. Place the adjusting socket on the screw and install the assembly between the shoe ends with the adjusting screw nearest the secondary shoe.

➡ **Be sure to install the adjusting screw on the same side of the vehicle from which it came. To prevent incorrect installation, the socket end of each adjusting screw is stamped with R or L, to indicate installation on the right or left side of the vehicle. The adjusting pivot nuts have lines machined around the body of the nut, 2 lines indicating the right side nut and 1 line indicating the left side nut.**

19. Hook the cable hook into the hole in the adjusting lever from the outboard plate side. The adjusting levers are also stamped with an **R** or **L** to indicate right or left side installation.

20. Place the hooked end of the adjuster spring in the large hole in the primary shoe web and connect the loop end of the spring to the adjuster lever hole.

21. Pull the adjuster lever, cable and automatic adjuster spring down toward the rear to engage the pivot hook in the large hole in the secondary shoe web.

22. After installation, check the action of the adjuster by pulling the section of the cable between the cable guide and the adjusting lever toward the secondary shoe web far enough to lift the lever past a tooth on the adjusting screw wheel. The lever should snap into position behind the next tooth and releasing the cable should cause the adjuster spring to return the lever to its original position. This return action will turn the adjusting screw 1 tooth.

23. If pulling the cable does not produce the action described previously, or if lever action is sluggish instead of positive and sharp, check the position of the lever on the adjusting screw toothed wheel. With the brake in a vertical position, anchor at the top, the lever should contact the adjusting wheel 1 tooth above the centerline of the adjusting screw. If the contact point is below the centerline, the lever will not lock on the adjusting screw wheel teeth and the screw will not turn, since the lever is actuated by the cable.

24. Adjust the brake shoes using either a brake adjustment gauge or manually with the drums installed.

25. Install the wheels, and lower the vehicle.

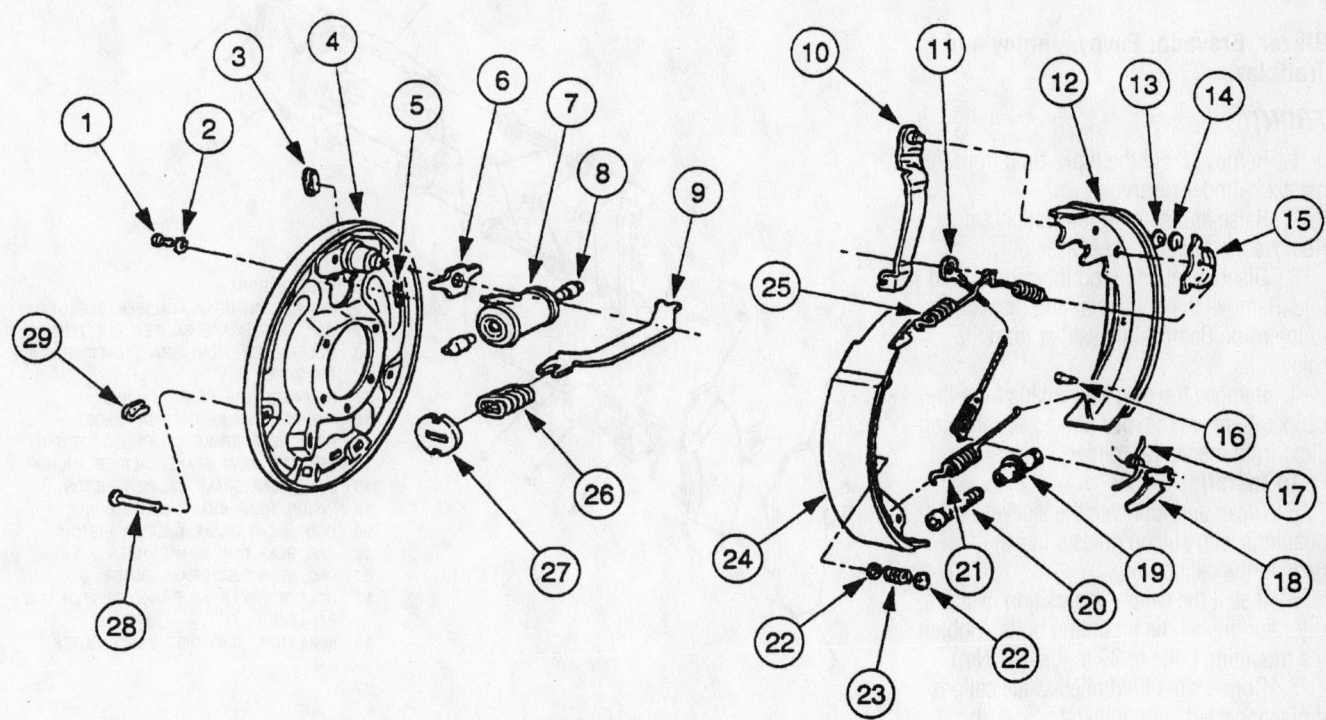

1 Wheel Cylinder-to-Backing Plate Bolt (2 Req'd)
2 Washer
3 Inspection Hole Cover
4 Brake Backing Plate
5 Lining Inspection Hole
6 Anchor Pin Guide Plate
7 Rear Wheel Cylinder
8 Wheel Cylinder Brake Shoe Link
9 Parking Brake Strut
10 Parking Brake Lever
11 Brake Shoe Adjusting Lever Cable

12 Rear Brake Shoe and Lining, Secondary
13 Washer
14 Parking Brake Lever Pin Retainer
15 Cable Guide
16 Adjusting Lever Pin
17 Adjusting Lever Return Spring
18 Brake Shoe Adjusting Lever
19 Brake Shoe Adjusting Screw Nut
20 Brake Adjuster Screw
21 Brake Shoe Adjusting Screw Spring

22 Brake Shoe Hold-Down Spring Cup
23 Brake Shoe Hold-Down Spring
24 Rear Brake Shoe and Lining, Primary
25 Brake Shoe Retracting Spring, Short
26 Parking Brake Link Spring
27 Parking Brake Spring Retainer
28 Brake Shoe Hold-Down Spring Pin
29 Brake Adjusting Hole Cover

93026G21

Exploded view of the rear brake shoes and components—Explorer and Mountaineer

For Accessory Drive Belt illustrations, see Section 1 of this manual

Brake Caliper

REMOVAL & INSTALLATION

Blazer, Bravada, Envoy, Jimmy and Trailblazer

FRONT

1. Remove ⅔ of the brake fluid from the master cylinder reservoir.

2. Raise and support the vehicle safely. Remove the tire and wheel assembly.

3. Disconnect and plug the caliper fluid line. Remove the bolts retaining the caliper to the rotor. Remove the caliper from the rotor.

4. Remove the disc brake pads from the caliper. Remove the disc brake pad retaining clips from inside the caliper.

To install:

5. Clean and lubricate the sleeves and bushings with silicon grease. Install the pads in the caliper.

6. Install the caliper in position over the rotor and install the mounting bolts. Tighten the mounting bolts to 38 ft. lbs. (51 Nm).

7. Connect the fluid lines to the caliper, if disconnected, and tighten to 33 ft. lbs. (45 Nm).

8. Install the wheel and tire assembly.

9. Lower the vehicle and refill the master cylinder to the correct level. Bleed the brake system if the fluid lines were disconnected from the caliper.

REAR

1. Raise and safely support the vehicle.
2. Remove rear wheels.
3. Remove brake hose and cap line.
4. Remove retainers from caliper and remove caliper.

To install:

5. Install brake pads if removed.
6. Install caliper over rotor, and onto mounts.
7. Install retainers, and tighten to 23 ft. lbs. or (31 Nm).
8. Install brake hose, and tighten to 20 ft. lbs. (27 Nm).
9. Bleed brake system.
10. Install tires.
11. Lower the vehicle.
12. Lower the vehicle, refill the master cylinder and pump pedal to attain full brake pedal before Road-testing the vehicle.

Denali, Escalade, Suburban, Tahoe and Yukon

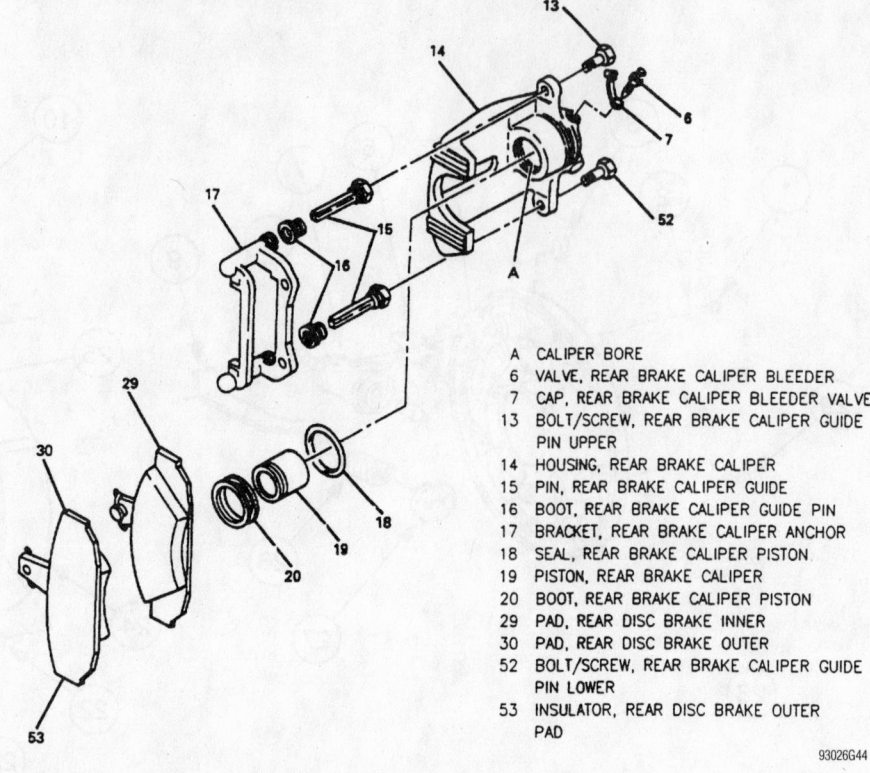

A	CALIPER BORE
6	VALVE, REAR BRAKE CALIPER BLEEDER
7	CAP, REAR BRAKE CALIPER BLEEDER VALVE
13	BOLT/SCREW, REAR BRAKE CALIPER GUIDE PIN UPPER
14	HOUSING, REAR BRAKE CALIPER
15	PIN, REAR BRAKE CALIPER GUIDE
16	BOOT, REAR BRAKE CALIPER GUIDE PIN
17	BRACKET, REAR BRAKE CALIPER ANCHOR
18	SEAL, REAR BRAKE CALIPER PISTON
19	PISTON, REAR BRAKE CALIPER
20	BOOT, REAR BRAKE CALIPER PISTON
29	PAD, REAR DISC BRAKE INNER
30	PAD, REAR DISC BRAKE OUTER
52	BOLT/SCREW, REAR BRAKE CALIPER GUIDE PIN LOWER
53	INSULATOR, REAR DISC BRAKE OUTER PAD

93026G44

Rear Brake Caliper—Bravada

➡ **There are 2 caliper designs and they can be identified by the method used to secure the assembly to the spindle bracket. The Delco caliper is secured by a bolt and sleeve combination. The Bendix caliper assembly is secured by a slider, spring and bolt.**

1. Remove the cover on the master cylinder and siphon enough fluid out of the reservoirs to bring the level to ⅓ full. This step prevents spilling fluid when the piston is pushed back.

2. Raise and support the vehicle safely. Remove the front wheels and tires.

3. Position a C-clamp around the outside pad and caliper; tighten the C-clamp until the caliper piston bottoms in its bore.

4. Remove the brake hose from the caliper by removing the inlet fitting.

5. Remove the bolt and sleeve or bolt and slider assemblies that hold the caliper and then lift the caliper off the rotor.

6. Remove the inboard and outboard pad.

To install:

7. Install the pads onto the caliper.
8. Position the caliper onto the knuckle/rotor assembly and secure the assembly with the mounting bolts or sliders.

9. Reconnect the brake line to the caliper.

10. Bleed the brakes

11. Pump the brake pedal and verify there is minimal brake pedal travel.

12. Check the brake fluid level. Install the tire and wheel assembly.

13. Lower the vehicle.

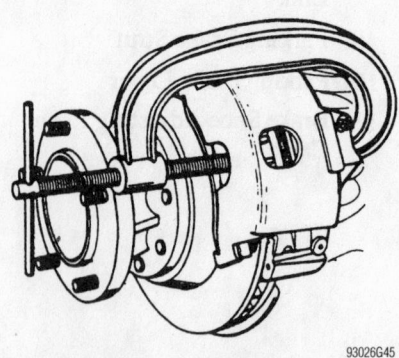

93026G45

Compressing the caliper piston—Denali, Escalade, Suburban, Tahoe and Yukon

Disc Brake Pads

REMOVAL & INSTALLATION

Blazer, Bravada, Envoy, Jimmy and Trailblazer

FRONT

1. Remove ⅔ of the brake fluid from the master cylinder.
2. Raise and safely support the vehicle.
3. Place a C-clamp around the outer pad and caliper; tighten the C-clamp until the piston is fully compressed in the caliper. Remove the brake caliper pads.
4. Remove the inboard pad and retaining spring from the caliper.
5. Remove the outboard pad from the caliper.
6. Remove the sleeves and bushings.

To install:

7. Clean and lubricate the sleeves and bushing with silicone lubricant and install them in the caliper.
8. Clip the retaining spring onto the inboard pad and install the pad in the caliper.
9. Install the outboard pad into the caliper.
10. Install the caliper in position over the rotor and install the mounting bolts. Bend the tabs, on the outboard brake pad, over the caliper.
11. Install the wheel and tire assemblies.
12. Lower the vehicle, refill the master cylinder and pump pedal to attain full brake pedal before Road-testing the vehicle.

REAR

1. Remove ⅔ of the brake fluid from the master cylinder.
2. Raise and safely support the vehicle.
3. Remove wheels
4. Place a C-clamp around the outer pad and caliper; tighten the C-clamp until the piston is fully compressed in the caliper. Remove top caliper retainer, and rotate caliper away from rotor.
5. Remove the inboard pad and retaining spring from the caliper.
6. Remove the outboard pad from the caliper.

To install:

7. Clean and lubricate the sleeves and bushing with silicone lubricant and install them in the caliper.
8. Clip the retaining spring onto the inboard pad and install the pad in the caliper.

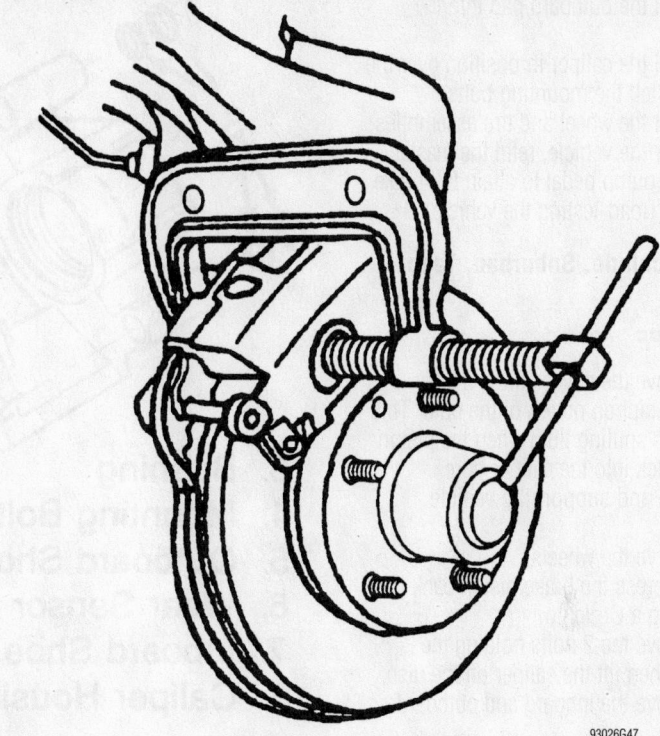

Compressing the caliper piston with a C-clamp—Blazer, Bravada, Envoy, Jimmy and Trailblazer

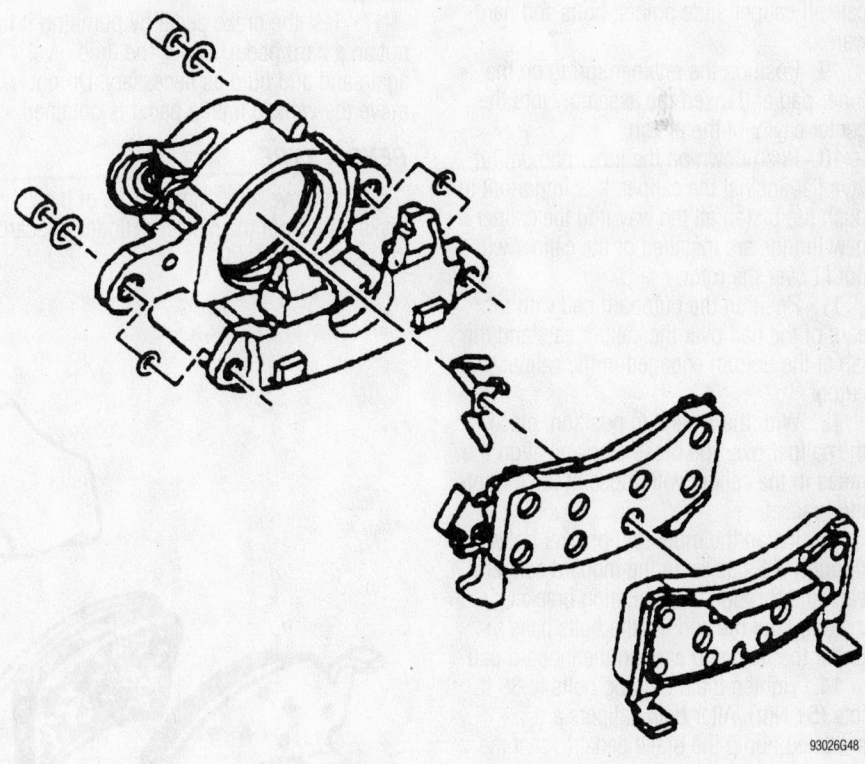

Exploded view of the disc brake assembly—Astro, Blazer, Bravada, Envoy, Jimmy, Safari, S10/S15 Pick-Up and Sonoma

For Tire, Wheel and Ball Joint specifications, see Section 1 of this manual

9. Install the outboard pad into the caliper.

10. Install the caliper in position over the rotor and install the mounting bolts.

11. Install the wheel and tire assemblies.

12. Lower the vehicle, refill the master cylinder and pump pedal to attain full brake pedal before Road-testing the vehicle.

Denali, Escalade, Suburban, Tahoe and Yukon

DELCO TYPE

1. Remove the cover on the master cylinder and siphon out ⅔ of the fluid. This step prevents spilling fluid when the piston is pushed back into the caliper bore.

2. Raise and support the vehicle safely.

3. Remove the wheels.

4. Compress the brake piston back into its bore using a C-clamp.

5. Remove the 2 bolts holding the caliper and then lift the caliper off the disc.

6. Remove the inboard and outboard shoe.

7. Remove the pad support spring from the piston, if equipped.

To install:

8. Thoroughly inspect, clean and lubricate all caliper slide points, bolts and hardware.

9. Position the retainer spring on the inner pad and insert the assembly into the center cavity of the piston.

10. Push down on the inner pad until it lays flat against the caliper. It is important to push the piston all the way into the caliper if new linings are installed or the caliper will not fit over the rotor.

11. Position the outboard pad with the ears of the pad over the caliper ears and the tab at the bottom engaged in the caliper cutout.

12. With the 2 pads in position, place the caliper over the brake disc and align the holes in the caliper with those of the mounting bracket.

13. Install the mounting bracket bolts through the sleeves in the inboard caliper ears and through the mounting bracket, making sure the ends of the bolts pass under the retaining ears on the inboard pad.

14. Tighten the mounting bolts to 38 ft. lbs. (51 Nm). After both calipers are mounted pump the brake pedal to seat the pad against the rotor. Use a pair of channel lock pliers to bend over the upper ears of the outer pad so it isn't loose.

15. Install the wheels and lower the vehicle.

16. Add fluid to the master cylinder

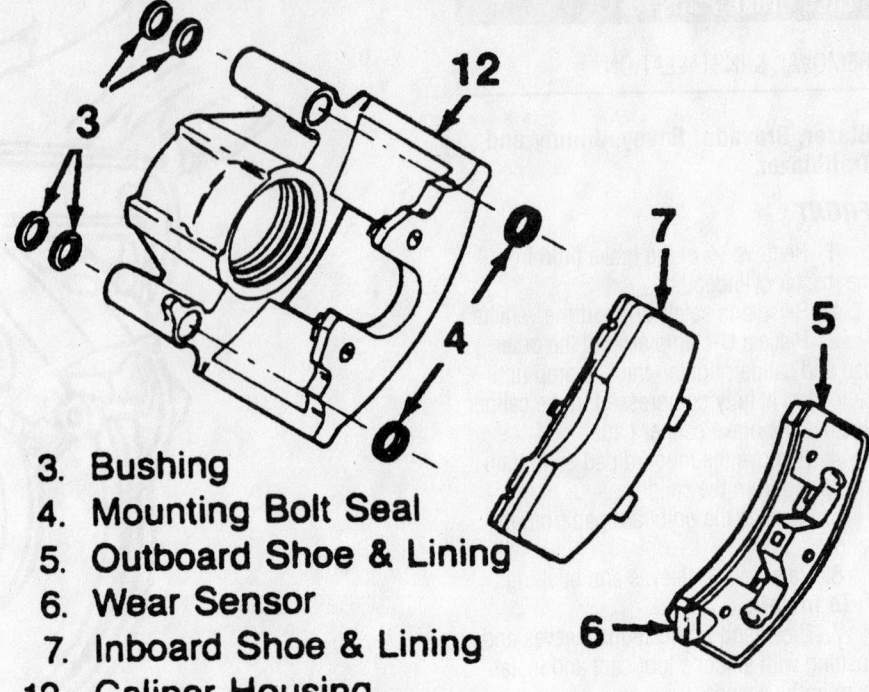

3. **Bushing**
4. **Mounting Bolt Seal**
5. **Outboard Shoe & Lining**
6. **Wear Sensor**
7. **Inboard Shoe & Lining**
12. **Caliper Housing**

93026G49

Replacing the disc brake pads—Delco type

reservoirs so they are ¼ in. (6.35mm) from the top.

17. Test the brake pedal by pumping it to obtain a hard pedal. Check the fluid level again and add fluid as necessary. Do not move the vehicle until a pedal is obtained.

BENDIX TYPE

1. Remove approximately ⅓ of the brake fluid from the master cylinder. Discard the used brake fluid.

2. Raise and support the vehicle safely and remove the wheel.

3. Push the piston back into its bore. This can be done by using a C-clamp.

4. Remove the bolt at the caliper slider. Use a brass drift pin to remove the slider and spring.

5. Rotate the caliper up and forward from the bottom and lift it off the caliper support.

6. Tie the caliper out of the way with a

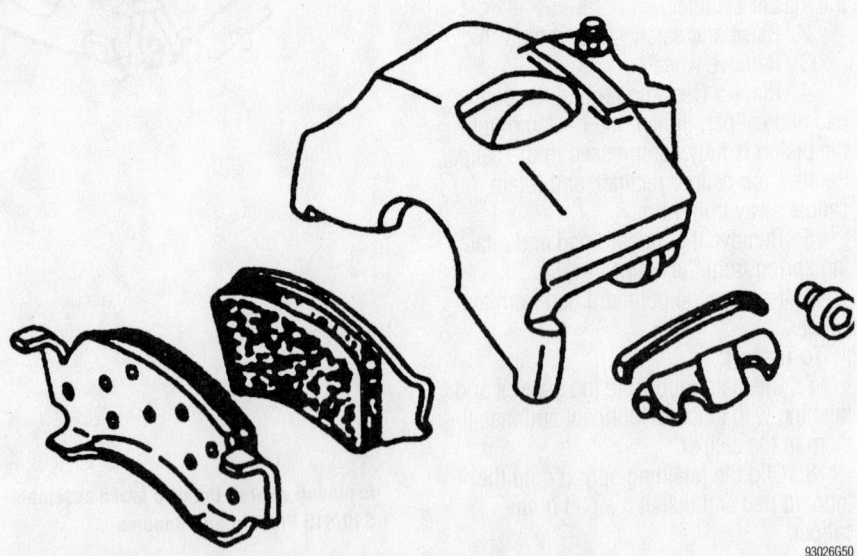

93026G50

Replacing the disc brake pads—Bendix type

piece of wire. Be careful not to damage the brake line.

7. Remove the inner shoe from the caliper support. Discard the inner shoe clip.

8. Remove the outer shoe from the caliper.

To install:

9. Thoroughly clean, inspect and lubricate the caliper, slider and spring with silicone.

10. Install a new inboard shoe clip on the shoe.

11. Install the lower end of the inboard shoe into the groove provided in the support. Slide the upper end of the shoe into position. Be sure the clip remains in position.

12. Position the outboard shoe in the caliper, with the ears at the top of the shoe over the caliper ears and the tab at the bottom of the shoe engaged in the caliper cutout. If assembly is difficult, a C-clamp may be used. Be careful not to damage the lining.

13. Position the caliper over the brake disc, top edge first. Rotate the caliper downward onto the support.

14. Place the spring over the caliper support key, install the assembly between the support and lower caliper groove. Tap into place until the key retaining screw can be installed.

15. Install the screw and torque to 15 ft. lbs. (20 Nm). The boss must fit fully into the circular cutout in the key.

16. Install the wheel and add brake fluid as necessary.

Brake Drums

REMOVAL & INSTALLATION

Blazer, Bravada, Envoy, Jimmy and Trailblazer

1. Raise and safely support the vehicle.
2. Remove the wheel and tire assembly.
3. Remove the brake drum. If the drum will not pull of the axle, use a rubber mallet and tap it around the edge.

To install:

4. Install the drum on the axle and install the wheel and tire assembly.
5. Lower the vehicle.
6. Refill the master cylinder and pump pedal to attain full brake pedal before road-testing the vehicle.

Denali, Escalade, Suburban, Tahoe and Yukon

W/SEMI-FLOATING AXLES

1. Raise and support the vehicle safely.
2. Mark the relationship of the wheel to the hub and remove the wheel.
3. Mark the relationship of the drum to the hub and pull the drum from the brake assembly. If the brake drums have been scored from worn linings, the brake adjuster must be backed off so the brake shoes will retract from the drum. The adjuster can be backed off by inserting a brake adjusting tool through the access hole provided. In some cases the access hole is provided in the brake drum. A metal cover plate is over the hole. This may be removed by using a hammer and chisel.

To install:

4. Align the mark on the drum to mark on hub and install drum
5. Align the mark on the wheel to mark on drum and install wheel
6. Adjust brake lining as needed. Pump brakes

W/FULL FLOATING AXLES

To remove the drums from full floating rear axles, the axle shaft will have to be removed. Full-floating rear axles can be identified by a bearing housing that protrudes through the center of the wheel.

1. Raise and support the vehicle safely.
2. Remove the wheel.
3. Remove the axle shaft.
4. Remove the retaining ring, key and adjusting nut.
5. Remove the hub and drum.

To install:

6. Install the hub and drum to the tube.
7. Install the adjusting nut and torque to specification.
8. Install the key and retaining ring.
9. Install the axle shaft and wheel.

Brake Shoes

REMOVAL & INSTALLATION

Blazer, Bravada, Envoy, Jimmy and Trailblazer

1. Raise and safely support the vehicle.
2. Remove the wheel and tire assembly.
3. Remove the brake drum.
4. Remove the return springs from the brake shoes. Remove the shoe guide.
5. Remove the hold-down springs and pins. Remove the actuator lever and pivot.
6. Remove the lever return spring. Remove the actuator link.

7. Remove the parking brake strut and spring. Remove the parking brake lever.

8. Remove the brake shoes and the adjuster assembly.

To install:

9. Lubricate the contact points on the backing plate and the adjuster with lithium grease.

10. Install the parking brake lever, adjusting screw and spring assembly.

11. Install the shoe assembly onto the backing plate.

12. Install the parking brake lever, strut and strut spring.

13. Install the actuator lever and lever pivot. Install the actuator link.

14. Install the lever spring, the hold-down pins and springs.

15. Install the shoe guide. Install the return springs and install the brake drum in position.

16. Adjust the brakes as follows:

   a. Remove the knockout area in the backing plate, behind the adjuster assembly.

   b. Ensure the parking brake system is adjusted properly with no tension on the cables or parking brake lever. The tops of the shoes should be firmly seated against the upper spring retaining anchor, if not as specified, loosen the parking brake cables.

   c. Install the drum and turn the brake adjuster until the wheels can just be turned by hand.

   d. Then, back the adjuster off 24 notches. No brake drag should be felt after 12 notches.

   e. Install an adjusting hole plug in the backing plate to prevent dirt and moisture from entering.

   f. Readjust the parking brake cable as necessary.

17. Install the wheel and tire assemblies.

18. Lower the vehicle, refill the master cylinder and pump pedal to attain full brake pedal before Road-testing the vehicle.

Denali, Escalade, Suburban, Tahoe and Yukon

LEADING/TRAILING BRAKES

1. Raise the vehicle and support it safely.
2. Remove the tire and wheel assembly.
3. Remove the brake drums.
4. Raise the lever arm of the actuator until the upper end is clear of the slot in the adjuster screw.

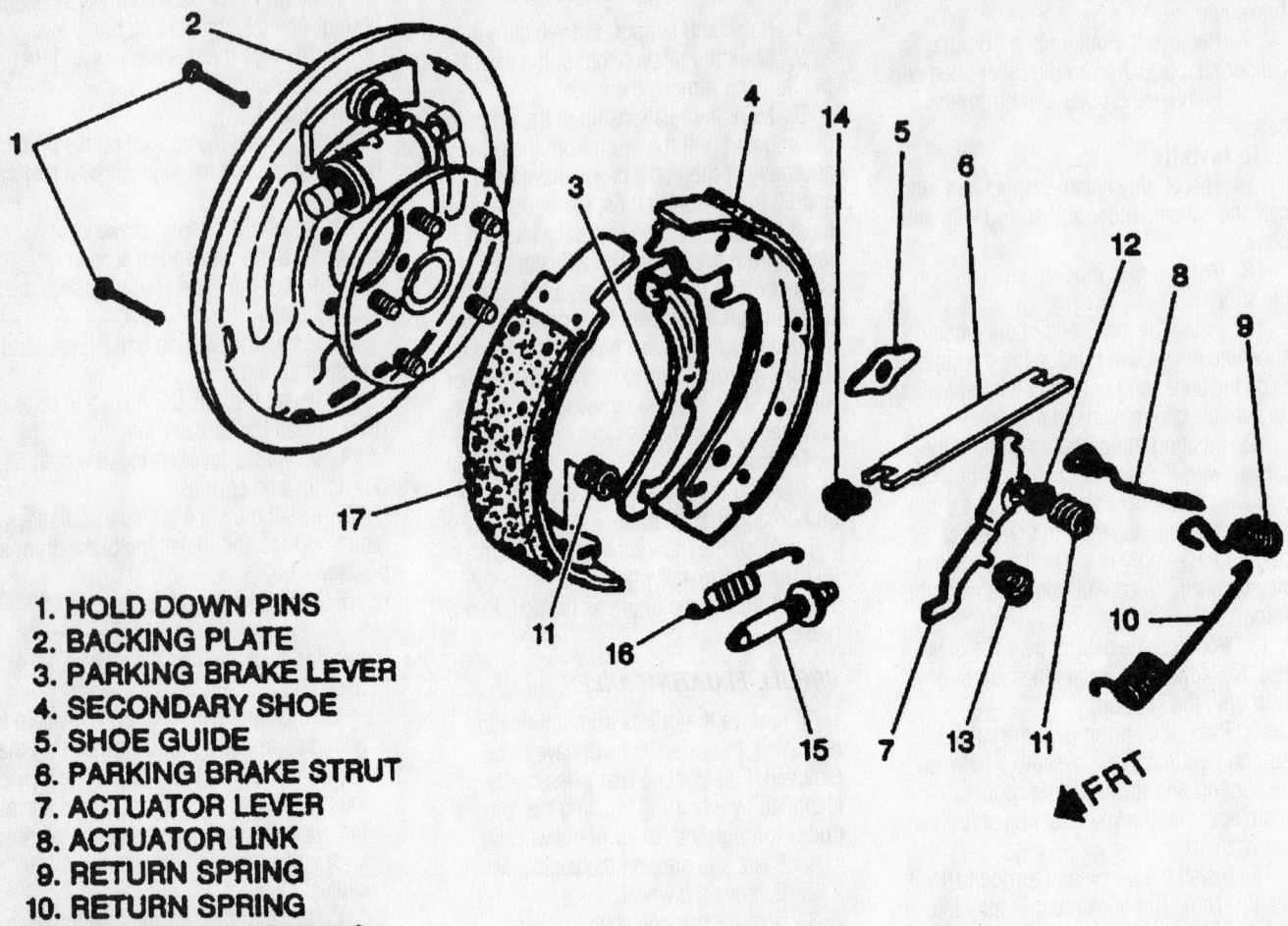

1. HOLD DOWN PINS
2. BACKING PLATE
3. PARKING BRAKE LEVER
4. SECONDARY SHOE
5. SHOE GUIDE
6. PARKING BRAKE STRUT
7. ACTUATOR LEVER
8. ACTUATOR LINK
9. RETURN SPRING
10. RETURN SPRING
11. HOLD DOWN SPRING
12. LEVER PIVOT
13. LEVER RETURN SPRING
14. STRUT SPRING
15. ADJUSTING SCREW ASSEMBLY
16. ADJUSTING SCREW SPRING
17. PRIMARY SHOE

93026G51

Exploded view of the drum brake components— Blazer, Bravada, Envoy, Jimmy and Trailblazer

5. Slide the actuator off the adjuster pin. Disconnect the actuator spring from the shoe.

6. Remove the hold-down spring assemblies and pins.

7. Pull the bottom ends of the shoes apart and lift the lower return spring over the anchor plate. Allow the shoe ends to come together and remove the spring.

8. Remove the shoe assembly, along with the upper return spring and the adjusting screw assembly.

9. Remove the upper return spring and the adjusting screw assembly from the shoes.

10. Remove the retaining ring, pin, spring washer, and parking brake lever.

To install:

11. Clean adjuster wheel and the backing plates with a suitable cleaner. Lubricate the backing plate contact points, levers and adjuster with a suitable lubricant.

12. Assemble the parking lever, spring washer (concave side facing the brake lever), pin, and retaining ring onto the rearward shoe.

13. Install the adjuster pin in the forward shoe with the pin projecting 0.276 in. (7mm) from the side of the shoe web where the adjuster actuator is installed.

14. With the brake shoes resting on a flat surface (the shoe with the parking lever to the rear of the vehicle), install the upper return spring.

15. Install the adjuster screw assembly with the spring clip facing the backing plate.

16. Place the shoes in position on the backing plate. Do not place the lower shoe webs under the anchor plate.

17. Install the lower return spring, spread the bottom of the shoes and position the shoe against the backing plate.

18. Install the hold-down pins and spring assemblies.

19. Install the adjuster actuator over the end of the adjuster pin so the top leg engages the notch in the adjuster screw.

20. Install the actuator spring, being careful not to over-stretch it more than 3.27 in. (83mm).

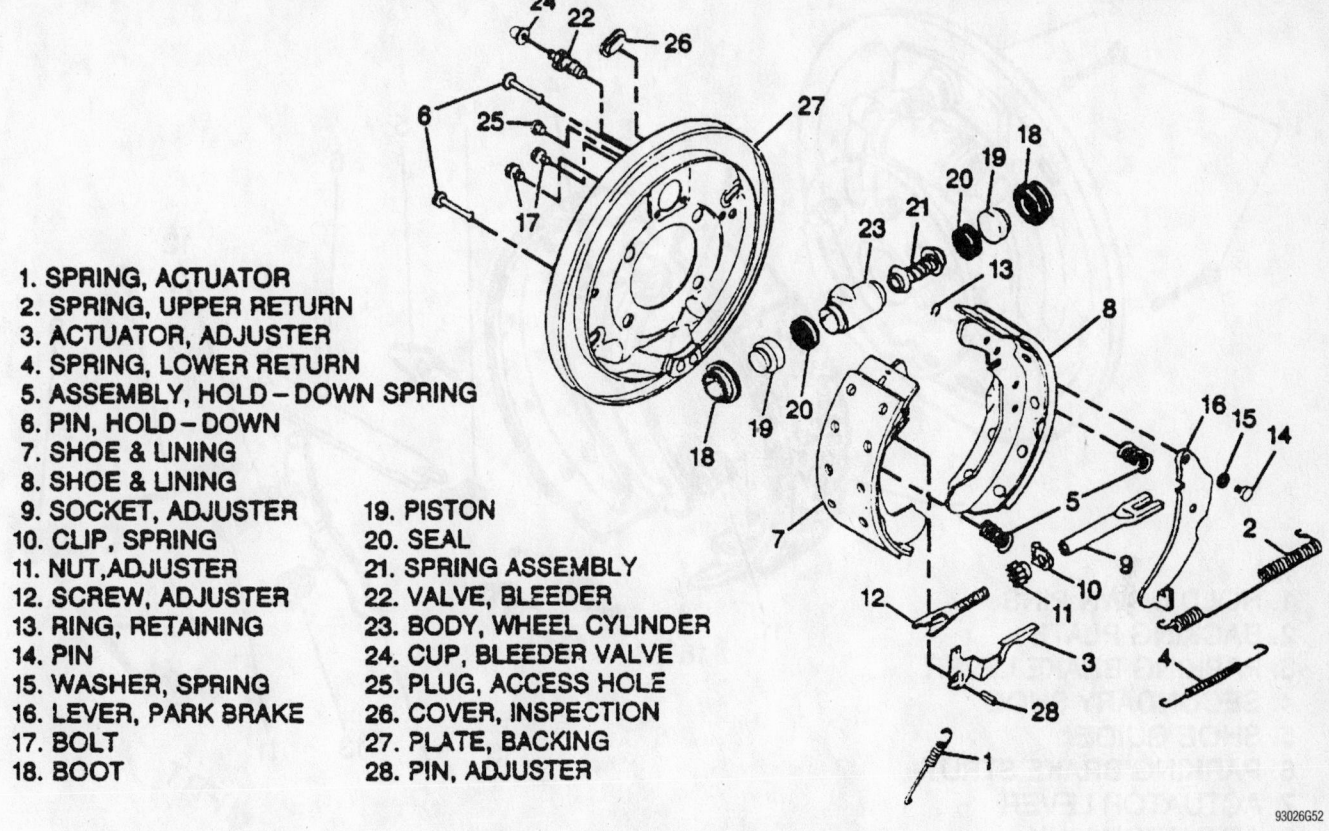

1. SPRING, ACTUATOR
2. SPRING, UPPER RETURN
3. ACTUATOR, ADJUSTER
4. SPRING, LOWER RETURN
5. ASSEMBLY, HOLD – DOWN SPRING
6. PIN, HOLD – DOWN
7. SHOE & LINING
8. SHOE & LINING
9. SOCKET, ADJUSTER
10. CLIP, SPRING
11. NUT, ADJUSTER
12. SCREW, ADJUSTER
13. RING, RETAINING
14. PIN
15. WASHER, SPRING
16. LEVER, PARK BRAKE
17. BOLT
18. BOOT

19. PISTON
20. SEAL
21. SPRING ASSEMBLY
22. VALVE, BLEEDER
23. BODY, WHEEL CYLINDER
24. CUP, BLEEDER VALVE
25. PLUG, ACCESS HOLE
26. COVER, INSPECTION
27. PLATE, BACKING
28. PIN, ADJUSTER

93026G52

Exploded view of the rear drum brake assembly—Denali, Escalade, Suburban, Tahoe and Yukon with Leading/Trailing type

21. Install the parking brake cable to the lever.

22. Adjust the parking brake if the shoes will not totally retract.

23. Install the drum, tire and wheel assembly. Adjust the rear brakes and lower the vehicle.

DUO-SERVO BRAKES

1. Raise the vehicle and support it safely.

2. Remove the tire and wheel assembly.

3. Remove the brake drums.

4. Using a brake tool, remove the shoe return springs.

5. Remove the shoe guide.

6. Remove the hold-down springs and pins.

7. Remove the actuator lever and pivot.

8. Remove the lever return spring.

9. Remove the actuator link, parking brake strut, spring retaining ring.

10. Remove the parking brake lever and washer.

11. Remove the shoe assemblies.

12. Remove the adjuster screw and spring from the shoe assembly.

To install:

13. Use a brake cleaning fluid to remove dirt from the brake drum. Check the drums for scoring, cracks and for out-of-round; service the drums as necessary.

14. Check the wheel cylinders by carefully pulling the lower edges of the wheel cylinder boots away from the cylinders. If there is excessive leakage, the inside of the cylinder will drip fluid; repair or replace as necessary.

15. Check the flange plate, which is located around the axle, for leakage of differential lubricant.

16. Lightly lubricate the parking brake cable, parking brake lever where it enters the shoe and the backing plate-to-shoe contact points. Use high temperature, waterproof, grease or special brake lube.

17. Install the parking brake lever into the secondary shoe with the attaching bolt, spring washer, lockwasher, and nut. It is important that the lever move freely before the shoe is attached. Move the assembly and check for proper action.

18. Lubricate the adjusting screw and make sure it works freely.

19. Connect the adjuster screw and spring to the bottom portion of both shoes. Ensure the spring does not interfere with the adjuster rotation when installed. The primary (smaller shoe pad area) to the front and secondary shoe (larger shoe pad area) to the rear of the vehicle.

20. Install the shoe assembly. Ensuring the shoe webs are positioned correctly against the wheel cylinder.

21. Install the parking brake cable.

22. Secure the primary shoes with the hold-down pin and spring.

23. Install the parking brake strut and the strut spring.

24. Install the actuator lever and pivot, securing the assembly with the hold-down pin and spring. Install the actuator link and spring.

25. Install the return springs.

26. Check the operation of the self-adjusting mechanism by moving the actuating lever by hand.

27. Adjust the brakes and install the drum.

28. Adjust the parking brake.

29. Install the tire and wheel assembly.

30. Lower the vehicle.

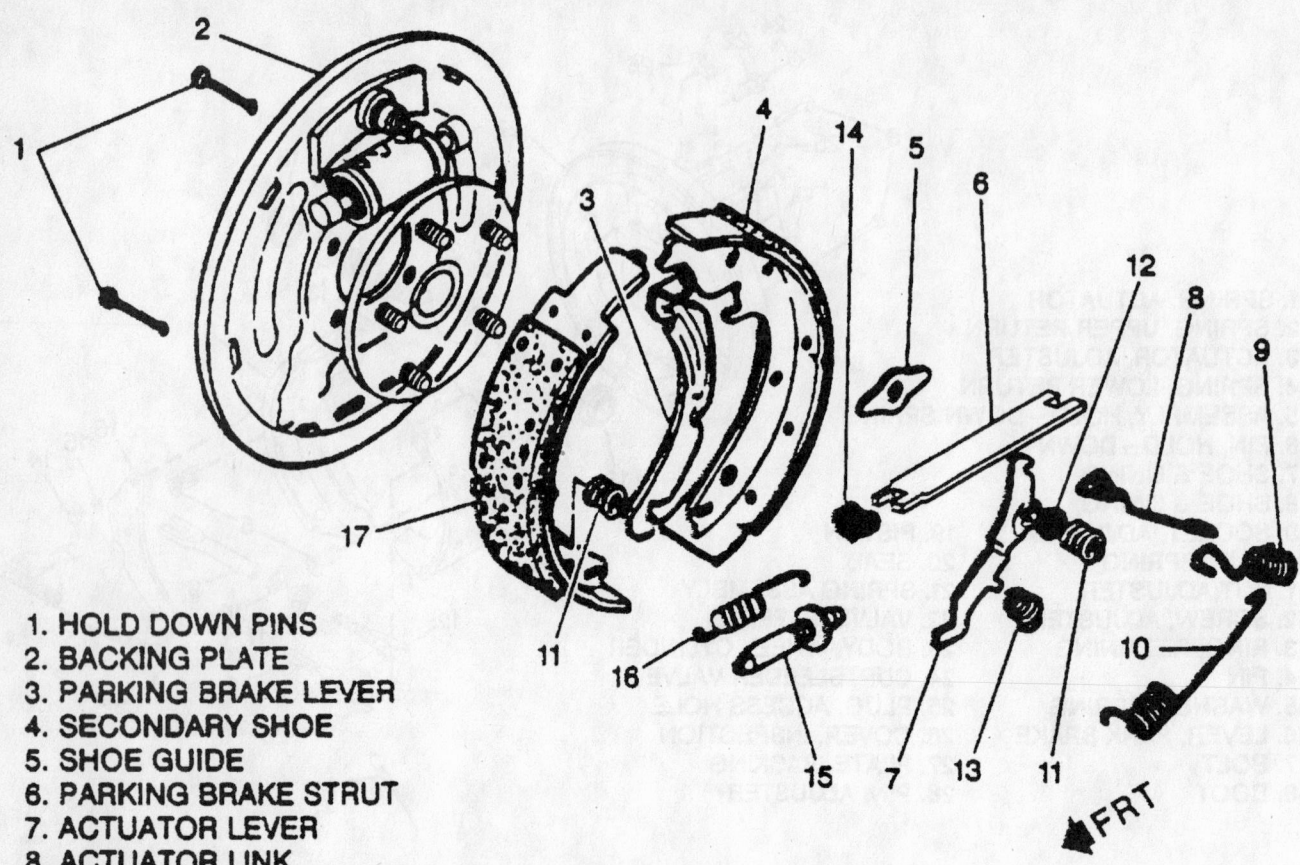

1. HOLD DOWN PINS
2. BACKING PLATE
3. PARKING BRAKE LEVER
4. SECONDARY SHOE
5. SHOE GUIDE
6. PARKING BRAKE STRUT
7. ACTUATOR LEVER
8. ACTUATOR LINK
9. RETURN SPRING
10. RETURN SPRING
11. HOLD DOWN SPRING
12. LEVER PIVOT
13. LEVER RETURN SPRING
14. STRUT SPRING
15. ADJUSTING SCREW ASSEMBLY
16. ADJUSTING SCREW SPRING
17. PRIMARY SHOE

93026G53

Exploded view of the rear drum brake assembly—Denali, Escalade, Suburban, Tahoe and Yukon with Duo-Servo type

GEO/CHEVY TRACKER

Brake Caliper

REMOVAL & INSTALLATION

1. Raise and safely support the vehicle.
2. Remove the wheels.
3. Disconnect and plug the brake line.
4. Remove the caliper mounting bolts (guide pins) and remove the caliper from the vehicle.
   To install:
5. Install the caliper on the vehicle. Tighten the mounting bolts to 20 ft. lbs. (27 Nm).

6. Connect the hydraulic brake line, using 2 new washers. Torque the union bolt to 17 ft. lbs. (23 Nm).
7. Replace the front wheels.
8. Lower the vehicle.
9. Fill the brake reservoir and bleed the hydraulic brake system.

Disc Brake Pads

REMOVAL & INSTALLATION

1. Siphon about ⅔ of the fluid out of the master cylinder.

2. Raise and safely support the vehicle.
3. Remove the wheels.
4. Remove the brake caliper mounting bolts and remove the caliper from the mounting bracket.
5. Support the caliper with a wire.
6. Using a large pair of plies or a C-clamp compress the caliper piston back into the bore.
7. Remove the disc brake pads and any shims from the caliper mounting bracket.
   To install:
8. Install the brake pads and any shims removed from the caliper mounting bracket.
9. Install the caliper on the mounting

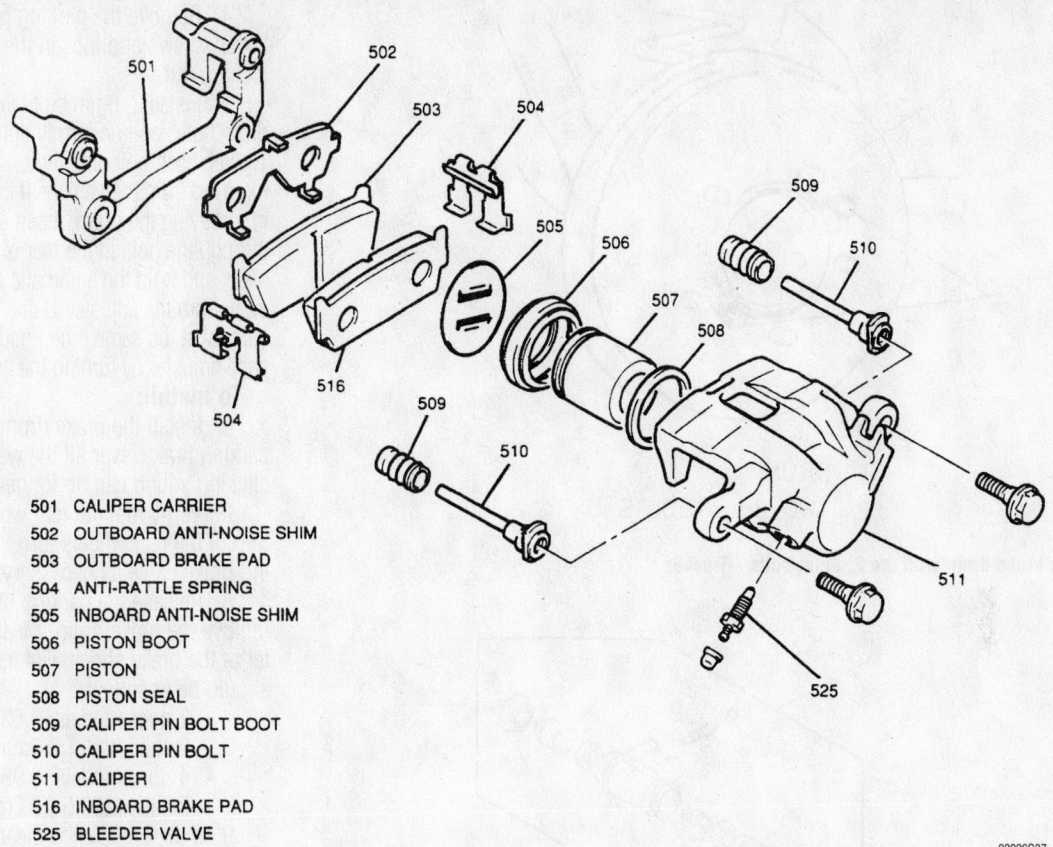

501	CALIPER CARRIER
502	OUTBOARD ANTI-NOISE SHIM
503	OUTBOARD BRAKE PAD
504	ANTI-RATTLE SPRING
505	INBOARD ANTI-NOISE SHIM
506	PISTON BOOT
507	PISTON
508	PISTON SEAL
509	CALIPER PIN BOLT BOOT
510	CALIPER PIN BOLT
511	CALIPER
516	INBOARD BRAKE PAD
525	BLEEDER VALVE

93026G37

Front disc brake components—Tracker

bracket and install the mounting bolts. Tighten the mounting bolts to 20 ft. lbs. (27 Nm).

10. Install the front wheels and lower the vehicle.

❊❊ CAUTION

Do not attempt to drive the vehicle until after the following step is performed.

11. Depress the brake pedal repeatedly until a firm pedal is obtained. Do not attempt to drive the vehicle unless a firm pedal is obtained.

12. Check the fluid level in the master cylinder. Add fresh brake fluid, as necessary.

13. Road-test the vehicle.

Brake Drums

REMOVAL & INSTALLATION

1. Raise and safely support the vehicle.
2. Remove the rear wheel(s).
3. Release the parking brake.

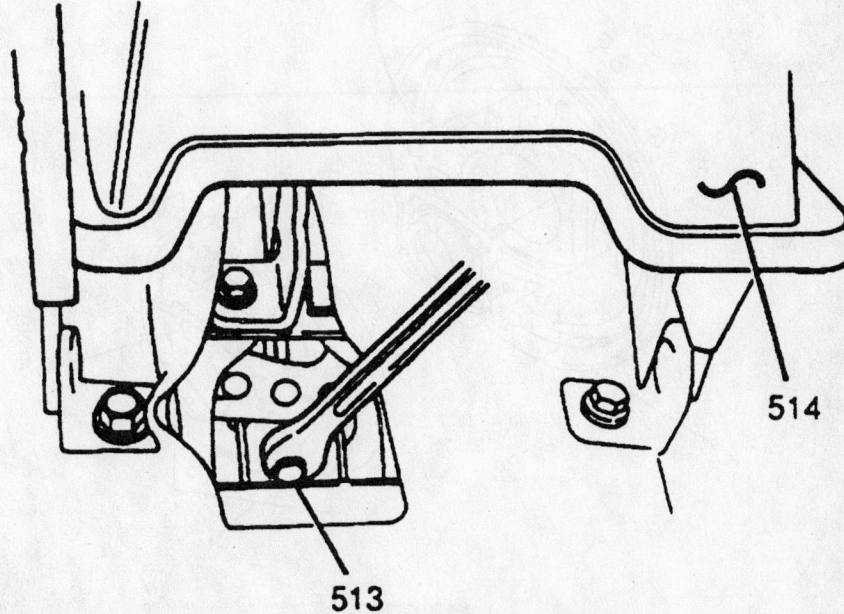

513 PARKING BRAKE CABLE LOCKNUT

514 PARKING BRAKE LEVER COVER

93026G38

Reducing the adjuster to remove the brake drum—Tracker

For Tune-up, Capacities and Firing orders, see Section 1 of this manual

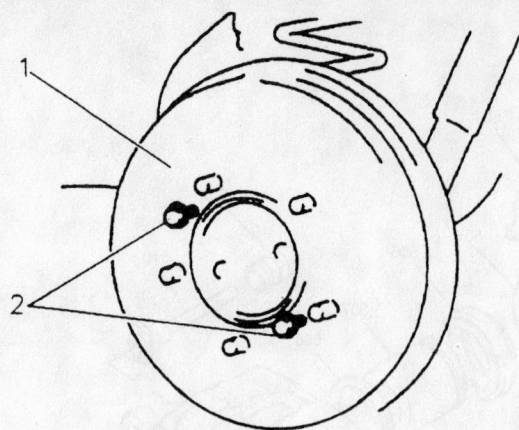

1 DRUM
2 TWO 8mm BOLTS

93026G39

Removing the brake drum with the 2, 8mm bolts—Tracker

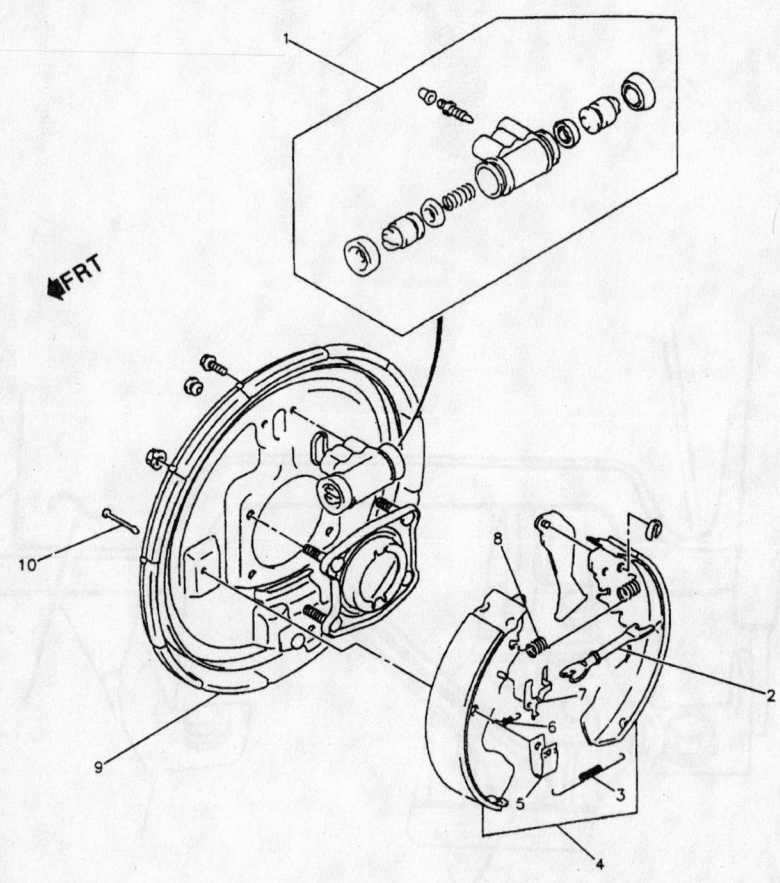

1 WHEEL CYLINDER
2 ADJUSTER
3 SHOE RETURN LOWER SPRING
4 BRAKE SHOES
5 SHOE HOLD DOWN SPRING
6 ADJUSTER SPRING
7 PAWL LEVER
8 SHOE RETURN UPPER SPRING
9 BACKING PLATE
10 SHOE HOLD DOWN PIN

93026G40

Exploded view of the rear brake components—Tracker

4. Remove the parking brake lever cover screws and loosen the brake cable locking nut.

5. Install 2, 8mm bolts into the brake drum holes and uniformly tighten each bolt. Tighten each bolt until the brake drum is removed from the vehicle. If there is difficulty in removing the drum, insert a small tool through the hole in the rear of the backing plate, and hold the automatic adjusting lever away from the adjuster. Using another narrow, flat tool at the same time, reduce the brake shoe adjuster by turning the adjusting wheel.

To install:

6. Install the brake drum and pull the parking brake lever all the way up until a clicking sound can no longer be heard.

7. Verify that the rear wheels will not turn. If the rear wheels turn, adjust the parking brake cable as necessary.

8. Release the parking brake and remove the brake drum. Measure the diameter of the brake shoes. Outer diameter should be as follows:
  - 2 door models: 8.638 (0.0012 inches (219 (0.3mm)
  - 4 door models: 9.980 (0.0079 inches (253.5 (0.2mm)

9. If the brake shoe clearance is not correct, adjust the brake shoes until the clearance is correct.

10. Reinstall the brake drum, replace the wheel(s), and safely lower the vehicle.

11. Adjust the parking brake and install the cover with the 2 screws.

12. Road-test the vehicle for proper brake operation.

Brake Shoes

REMOVAL & INSTALLATION

1. Raise and safely support the vehicle.
2. Remove the rear wheel(s).
3. Remove the brake drum.
4. Using a suitable tool, remove the brake shoe return spring.
5. Using a brake spring hold-down tool, disengage the hold-down spring and retainers from the front shoe. Remove the hold-down retainer pinch
6. Disconnect the anchor spring from the front shoe and remove the front shoe.
7. Remove the anchor spring from the rear shoe. Using a brake spring hold-down tool, disengage the hold-down spring and retainers from the rear shoe. Remove the hold-down pinch
8. Disengage the parking brake lever from the parking brake cable and remove the rear shoe.

9. Remove the C-washer, the automatic adjuster lever and spring, the C-washer, and the parking brake lever from the rear shoe.

10. Thoroughly clean the backing plate and brake hardware with brake cleaning solvent. Apply high temperature grease to the backing plate shoe contact points, anchor plate and shoe contact points, adjusting bolt, and adjuster and brake shoe contact points.

To install:

11. Reinstall the automatic adjuster lever and the parking brake lever to the rear shoe using new C-washers.

12. Connect the parking brake lever to the parking brake cable. Set the adjuster and spring to the rear shoe.

13. Set the rear brake shoe in place, install the hold-down pin and install the hold-down spring and retainers. Make sure that the shoe is inserted in the wheel cylinder and that the other end is in the anchor plate.

14. Install the anchor spring to the rear shoe.

15. Install the front shoe to the other end of the anchor spring and set the front shoe in place. Make sure that the front shoe engages the wheel cylinder, adjuster mechanism and spring, and the anchor plate.

16. Reinstall the front brake shoe hold-down pin and secure with the hold-down spring and retainers using a suitable tool.

17. Install the return spring.

18. Install the brake drum and pull the parking brake lever all the way up until a clicking sound can no longer be heard.

19. Verify that the rear wheels will not turn. If the rear wheels turn, adjust the parking brake cable as necessary.

20. Release the parking brake and remove the brake drum. Measure the diameter of the brake shoes. Brake diameter should be as follows:
- 2 door models: 8.638 (0.0012 inches (219 (0.3mm)
- 4 door models: 9.980 (0.0079 inches (253.5 (0.2mm)

21. If the brake shoe clearance is not correct, adjust the brake shoes until the clearance is correct.

22. Reinstall the brake drum, replace the wheel(s), and safely lower the vehicle.

23. Road-test the vehicle for proper brake operation.

HONDA

Brake Caliper

REMOVAL & INSTALLATION

Passport and 1999–01 CR-V

FRONT

1. Raise and safely support the vehicle. Remove the front wheels.

2. Remove some brake fluid from the master cylinder reservoir.

3. Disconnect and plug the brake fluid line from the caliper.

4. Remove the brake caliper mounting bolt and guide bolt and remove the caliper from the mount. The brackets can be remove for additional work space.

5. Remove the brake pads and clips from the caliper. Inspect the brake pads for wear; replace them, if necessary.

To install:

6. Install the brake pads and clips onto the caliper.

7. If the caliper bracket was removed, tighten the bolts to 103–126 ft. lbs. (139–171 Nm).

8. Install the caliper on the mounting bracket. Torque the caliper-to-mounting bracket bolts to:
- 4-cylinder models: 20–27 ft. lbs. (27–37 Nm)
- 6-cylinder models: 54 ft. lbs. (74 Nm)

9. Connect the fluid line to the caliper using new washers. Torque the brake line banjo fitting to 26 ft. lbs. (35 Nm).

✳✳ WARNING

Be sure the hook end of the flexible brake line is positioned in the anti-rotation cavity.

10. Refill the master cylinder reservoir and bleed the brake system.

11. Install the front wheels and lower the vehicle.

REAR

1. Raise and safely support the vehicle. Remove the rear wheels.

2. Remove some fluid from the master cylinder reservoir.

3. Disconnect and plug the brake fluid line from the caliper.

➡**Discard the parking brake cable mounting pin after removal.**

4. If equipped with caliper actuated parking brakes; remove the mounting pin from the parking brake cable and disconnect the parking cable from the disc caliper.

5. Remove the brake caliper mounting bolt and guide bolt and remove the caliper from the mount.

6. Remove the brake pads and clips from the caliper. Inspect the brake pads for wear; replace them, if necessary.

To install:

7. Install the brake pads and clips onto the caliper.

8. If the mounting bracket was removed, tighten the bolts to 69–84 ft. lbs. (93–114 Nm).

9. Install the caliper on the mounting bracket. Torque the caliper-to-mounting bracket bolts to 12–17 ft. lbs. (16–24 Nm), or 32 ft. lbs. (43 Nm) on vehicles with shoe-type parking brakes.

10. Connect the parking brake cable to the caliper and install a new mounting pin.

11. Connect the fluid line to the caliper using new washers. Torque the brake line banjo fitting to 26 ft. lbs. (35 Nm).

12. Refill the master cylinder reservoir and bleed the brake system.

13. Install the rear wheels and lower the vehicle.

2002 CR-V

FRONT

1. Remove the upper and lower bolts.

2. Lift off the caliper.

3. Remove the pad springs.

4. Remove the pads and shims.

5. Remove the pad retainers.

6. Installation is the reverse of removal. Coat both sides of the shims and the backs of the pads with brake grease. Torque the bolts to 25 ft. lbs. (34 Nm). If the hose was disconnected, torque the banjo bolt to 25 ft. lbs. (34 Nm).

REAR

1. Remove the caliper pin bolts.

2. Lift off the caliper and suspend it safely.

3. Remove the pads and shims.

4. Remove the pad retainers.

5. Installation is the reverse of removal.

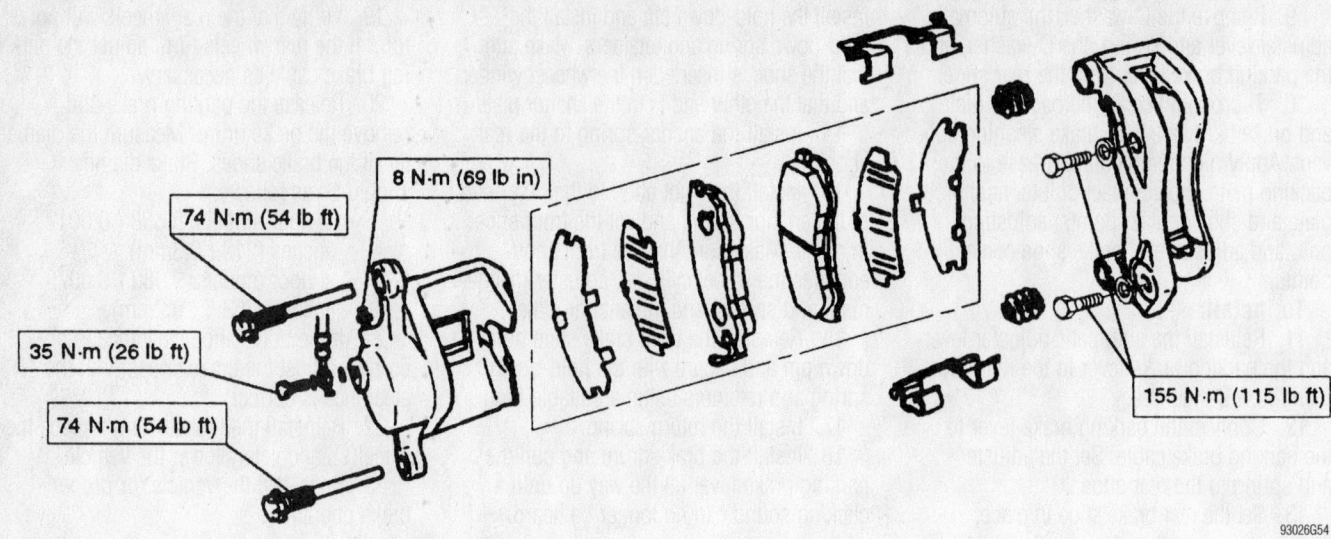

Exploded view of the front caliper components—Passport

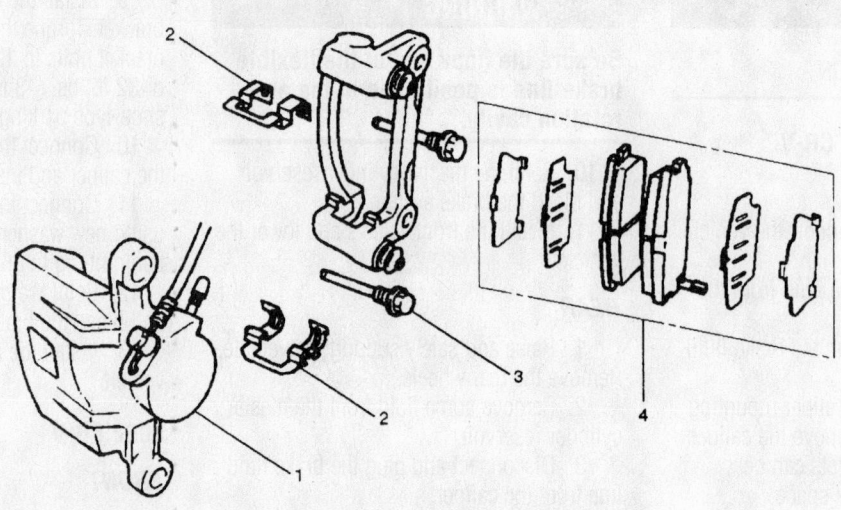

(1) Caliper Assembly
(2) Clip
(3) Lock Bolt
(4) Pad Assembly

93026G55

Exploded view of the rear caliper components—Passport

Coat both sides of the shims and the backs of the pads with brake grease. Torque the bolts to 16 ft. lbs. (22 Nm). If the hose was disconnected, torque the banjo bolt to 16 ft. lbs. (22 Nm).

Disc Brake Pads

REMOVAL & INSTALLATION

Passport and 1999–01 CR-V

FRONT

Most disc brake pads are equipped with wear indicators. If a squealing noise occurs from the brakes while driving, check the pad wear indicator plate. If there is evidence of the indicator plate contacting the brake disc, the brake pad should be replaced.

1. Remove ½ of the volume of brake fluid from the master cylinder to prevent overflow when the caliper piston is compressed.
2. Raise and safely support the vehicle.
3. Remove the wheel and tire assemblies.
4. Remove the brake caliper without disconnecting the brake line. Support the caliper with a length of wire. Do not let the caliper hang from the brake hose.

➡On some disc brake systems it is not necessary to remove the caliper when installing new brake pads. Remove the lower slide bolt and rotate the caliper upward to remove the pads.

5. Remove the brake pads and shims. Inspect the brake rotor and machine or replace as necessary. Check the minimum thickness (specification is cast into the rotor) before machining.

To install:
6. Use a suitable tool to push the caliper piston into its bore.
7. Apply a thin coat of grease to the rear face of the brake pad and install the shim. Install the brake pads.
8. Install the calipers. Lubricate the caliper bolts and boots. If equipped with a

4-cylinder engine, tighten the caliper mounting bolts to 24 ft. lbs. (33 Nm). If equipped with a 6-cylinder engine, tighten the caliper mounting bolts to 54 ft. lbs. (74 Nm).

9. Install the wheel and tire assemblies and lower the vehicle.

10. Apply the brakes several times to seat the pads before moving the vehicle. Check the fluid in the master cylinder and add as necessary.

REAR

1. Raise and safely support the vehicle. Remove the rear wheels.

2. Remove the brake caliper mounting bolts and remove the caliper without disconnecting the brake fluid line. Support the caliper so it does not hang on the brake line.

3. Remove the brake pads and retaining clips from the caliper.

4. If equipped with caliper-activated parking brakes; use tool J-37617 or equivalent, to rotate the piston clockwise until it retracts into the bore. Align the notches of the piston face so the centerline of the notches is perpendicular to the centerline of the mounting bosses.

To install:

5. Install the new brake pads and clips in the caliper and install the caliper in the mounting bracket.

6. Tighten the caliper mounting bolts to 12–17 ft. lbs. (16–24 Nm), or 32 ft. lbs. (43 Nm) on vehicles with shoe-type parking brakes.

7. Install the rear wheels. Check the brake fluid level.

8. Pump the brake pedal until pressure is felt before moving the vehicle.

2002 CR-V

FRONT

1. Remove the lower bolt.
2. Pivot the caliper up and hold the pads.
3. Remove the pad springs.
4. Remove the pads and shims.
5. Remove the pad retainers.
6. Installation is the reverse of removal. Coat both sides of the shims and the backs of the pads with brake grease. Torque the lower bolt to 25 ft. lbs. (34 Nm).

REAR

1. Remove the caliper pin bolts.
2. Lift off the caliper and suspend it safely.

3. Remove the pads and shims.
4. Remove the pad retainers.
5. Installation is the reverse of removal. Coat both sides of the shims and the backs of the pads with brake grease. Torque the bolts to 16 ft. lbs. (22 Nm).

Brake Drums

REMOVAL & INSTALLATION

CR-V and Passport

1. Raise and safely support the vehicle. Release the parking brake.
2. Remove the rear wheels.
3. Use chalk to mark the brake drum to one of the wheel studs as an index mark for reinstallation.
4. Remove the retaining screw that holds the brake drum to the axle flange.
5. Pull the brake drum from the axle flange.

To install:

6. Align the index mark and install the brake drum to the axle flange.
7. Install the retaining screw to secure the brake drum to the axle flange.
8. Install the rear wheels.

Rear Brake Shoes

REMOVAL & INSTALLATION

CR-V and Passport

1. Raise and safely support the vehicle.
2. Remove the rear wheels.
3. Remove the brake drums.
4. Remove the brake return springs.
5. Remove the leading shoe holding pin and spring, and then the leading shoe.
6. Remove the self-adjuster and the adjuster lever.
7. Remove the trailing shoe holding pin and spring.
8. Disconnect the parking brake cable from the trailing shoe and remove the trailing shoe. Remove the parking brake lever from the trailing shoe.

To install:

9. Attach the parking brake lever to the trailing shoe.
10. Connect the parking brake cable to the parking brake lever.
11. Apply a thin coat of high temperature grease to the shoe contact points on the brake backing plate contact surface (B), and self-adjuster (D).

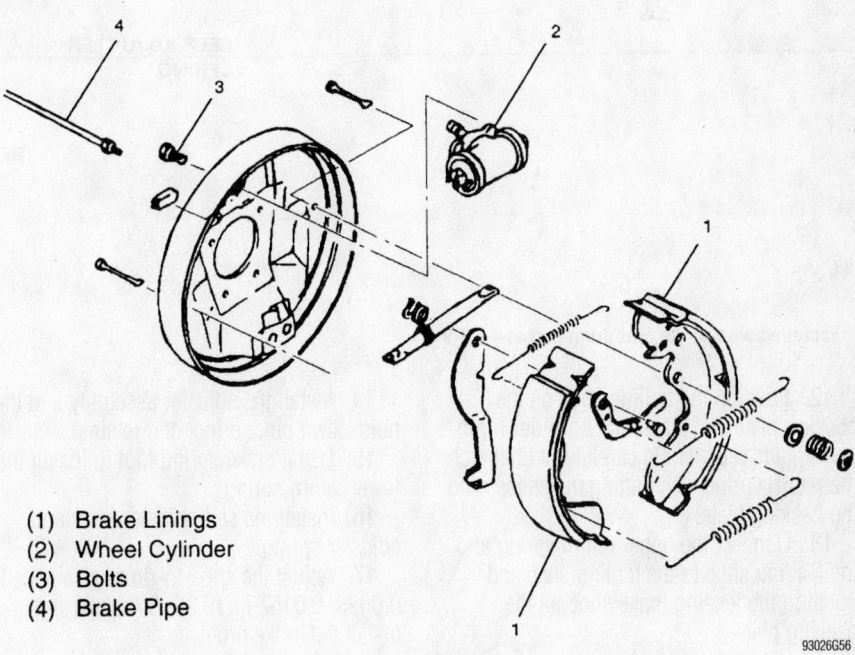

(1) Brake Linings
(2) Wheel Cylinder
(3) Bolts
(4) Brake Pipe

Exploded view of the rear drum brakes—Passport

93026G56

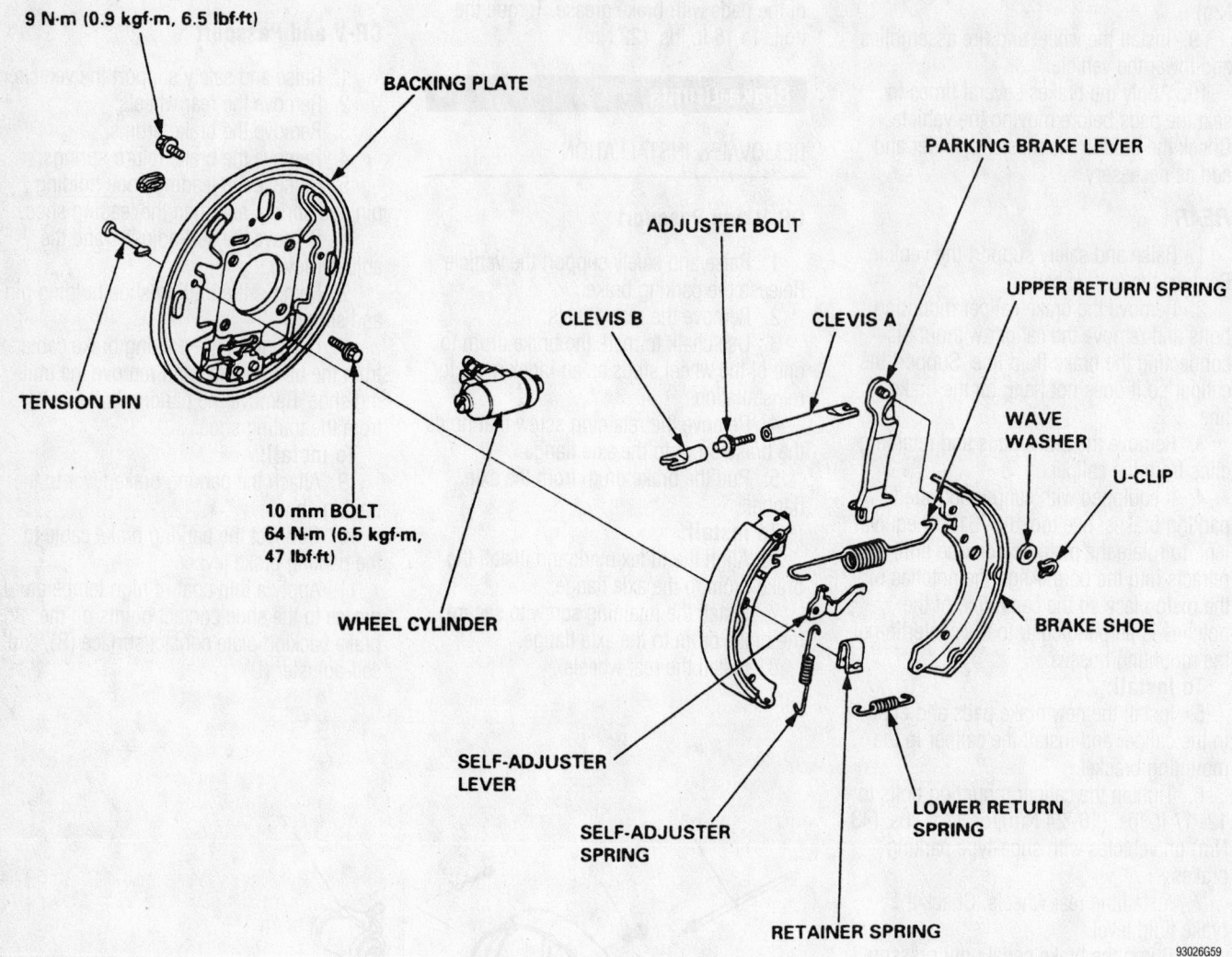

9 N·m (0.9 kgf·m, 6.5 lbf·ft)

BACKING PLATE

TENSION PIN

10 mm BOLT
64 N·m (6.5 kgf·m,
47 lbf·ft)

WHEEL CYLINDER

SELF-ADJUSTER
LEVER

SELF-ADJUSTER
SPRING

ADJUSTER BOLT

CLEVIS B

CLEVIS A

PARKING BRAKE LEVER

UPPER RETURN SPRING

WAVE
WASHER

U-CLIP

BRAKE SHOE

LOWER RETURN
SPRING

RETAINER SPRING

93026G59

Exploded view of the rear drum brakes—CR-V

12. Position the trailing shoe on the backing plate and install the hold-down pin, spring, and retainer. Be careful not to stretch the return spring when fitting the shoes onto the backing plate.

13. Connect the upper return spring and the leading shoe to the trailing shoe and position the leading brake shoe on the backing plate.

14. Install the adjuster assembly and the hold-down pin, spring, and retainer.

15. Use a brake spring tool to install the lower return spring.

16. Install the self-adjuster lever and adjuster spring.

17. Adjust the shoe-to-drum clearance to 0.0098–0.0157 in. (0.25–0.40mm) and install the brake drum.

18. Check the brake drum for scoring or other wear. Machine or replace as necessary. Check the maximum brake drum diameter specification when machining.

19. Install the rear wheels. Lower the vehicle.

20. Road-test the vehicle.

HYUNDAI

Brake Caliper

REMOVAL & INSTALLATION

Santa Fe

1. Before servicing the vehicle, refer to the precautions in the beginning of this section.
2. Remove the wheel.
3. Remove the brake hose from the caliper.
4. Remove the caliper mounting bolts.
5. Remove the caliper.

To install:

6. Install the caliper.
7. Install the caliper mounting bolts and tighten to 58–73 ft. lbs. (80–100 Nm).
8. Install the brake hose to the caliper and tighten the fitting too 18–22 ft. lbs. (25–30 Nm).
9. Install the wheel.
10. Bleed the brake system.

Disc Brake Pads

REMOVAL & INSTALLATION

Santa Fe

1. Before servicing the vehicle, refer to the precautions in the beginning of this section.
2. Remove the wheel.
3. Remove the caliper mounting bolts.
4. Remove the caliper and support aside with wire. Do not let the caliper hang by the hose.
5. Remove the pads and shims.

To install:

6. Install the pads, clips and shims.
7. Bottom the caliper piston using tool 09581-11000 or a C-clamp
8. Install the caliper mounting bolts and tighten to 16–24 ft. lbs. (22–32Nm).
9. Install the wheel.

Brake Shoes

REMOVAL & INSTALLATION

Santa Fe

1. Before servicing the vehicle, refer to the precautions in the beginning of this section.
2. Remove the wheel.
3. Remove the drum.
4. Remove the brake shoe hold-down spring.
5. Remove the brake strut.
6. Remove the brake shoe return spring.
7. Remove the brake shoes.

To install:

8. Install the brake shoes.
9. Install the brake shoe return spring.
10. Install the brake strut.
11. Install the brake shoe hold-down spring.
12. Install the brake drum.
13. Install the wheel.
14. Adjust the brake shoes.

INFINITI

Brake Caliper

REMOVAL & INSTALLATION

QX4

FRONT

1. Raise the vehicle and support safely.
2. Remove the appropriate tire and wheel assembly.
3. Remove the bolt attaching the brake hose to the caliper. Plug the brake hose to prevent brake fluid loss.
4. Remove the caliper support mounting bolts and lift the caliper assembly from the knuckle.

To install

5. Position caliper assembly onto the knuckle and install the bolts. Make sure the rotor fits between the brake pads. Torque the bolts to 53–72 ft. lbs. (72–97 Nm).
6. Use new copper washers and connect the brake hose to the caliper. Torque the brake hose attaching bolt to 12–14 ft. lbs. (17–20 Nm).
7. Bleed the brake system.
8. Apply the brake pedal and inspect the system. Ensure proper operation and no leakage.

9. Install tire and wheel assembly. Lower the vehicle and road-test.

REAR

➡**Unlike most rear disc brake designs, this system does not incorporate the parking brake system into the rear brake caliper. The rear brake system is serviced in the same manner as the front system.**

1. Raise the vehicle and support safely.
2. Remove the appropriate tire and wheel assembly.
3. Remove the caliper support mounting bolts and lift the caliper assembly from the baffle plate.
4. Loosen the brake fluid hose with a wrench and turn the caliper to disconnect it from the brake hose. Plug the brake hose to prevent brake fluid loss.

To install:

5. Use a new copper washer and connect the brake hose to the caliper. Torque the hose fitting to 11 ft. lbs. (15 Nm).
6. Position the caliper assembly over the baffle plate and install the bolts. Make sure the rotor fits between the brake pads. Torque the bolts to 28–38 ft. lbs. (38–52 Nm).

7. Bleed the brake system.
8. Apply the brake pedal and inspect the system. Ensure proper operation and no leakage.
9. Install tire and wheel assembly. Lower the vehicle and road-test.

Disc Brake Pads

REMOVAL & INSTALLATION

QX4

➡**Both the front and rear disc brake pads can be serviced using the same procedure.**

1. Using a syringe, siphon brake fluid from the reservoir, leaving reservoir approximately ½ full.
2. Raise and properly support the vehicle.
3. Remove the wheel assemblies.
4. Remove the lower pin bolt from the brake caliper.
5. Swivel the caliper up and away from the torque member. Tie the caliper to a suspension member so that it is out of the way.
6. Lift the 2 brake pads out of the torque member.

Timing belt service is covered in Section 3 of this manual

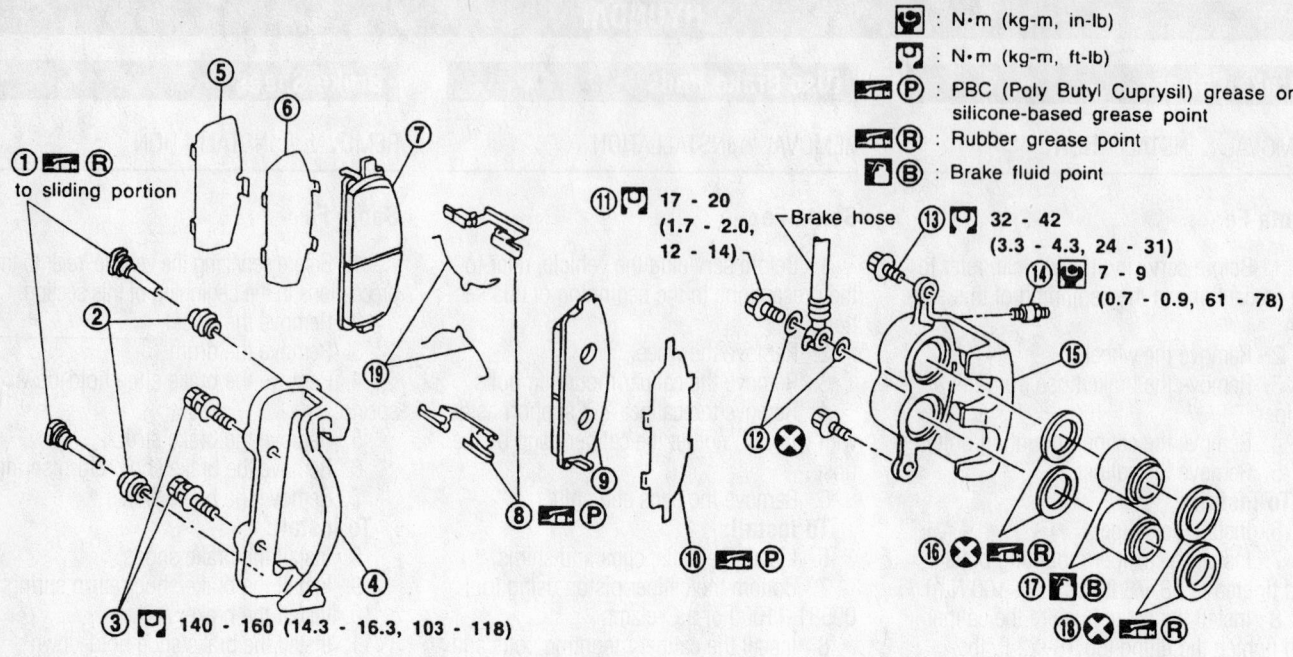

: N·m (kg-m, in-lb)

: N·m (kg-m, ft-lb)

P : PBC (Poly Butyl Cuprysil) grease or silicone-based grease point

R : Rubber grease point

B : Brake fluid point

①	Main pin	⑧	Pad retainer
②	Pin boot	⑨	Outer pad
③	Torque member fixing bolt	⑩	Outer shim
④	Torque member	⑪	Connecting bolt
⑤	Shim cover	⑫	Copper washer
⑥	Inner shim	⑬	Main pin bolt
⑦	Inner pad		

⑭	Bleed valve		
⑮	Cylinder body		
⑯	Piston seal		
⑰	Piston		
⑱	Piston boot		
⑲	Pad spring		

93026G60

Exploded view of the dual piston caliper front brake components—QX4

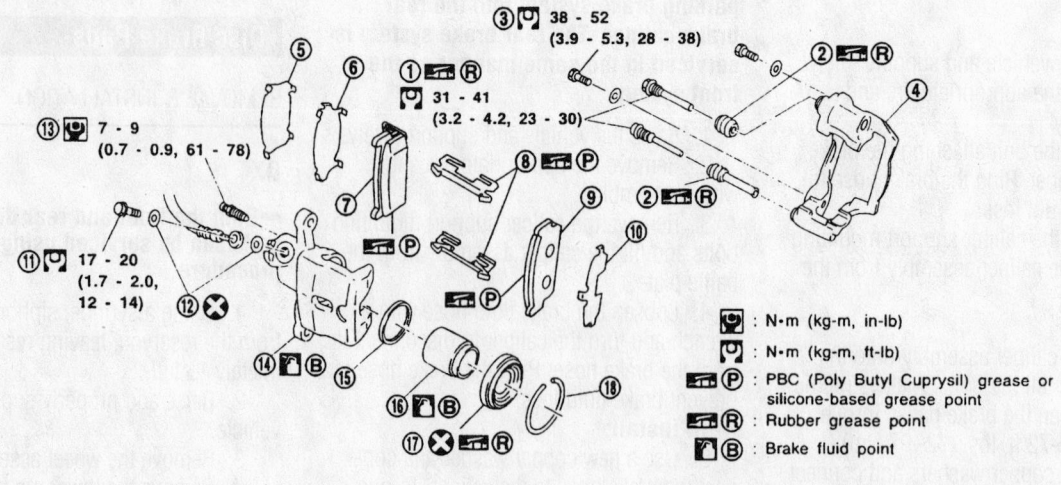

: N·m (kg-m, in-lb)

: N·m (kg-m, ft-lb)

P : PBC (Poly Butyl Cuprysil) grease or silicone-based grease point

R : Rubber grease point

B : Brake fluid point

①	Main pin bolt	⑦	Inner pad
②	Pin boot	⑧	Pad retainer
③	Torque member fixing bolt	⑨	Outer pad
④	Torque member	⑩	Outer shim
⑤	Shim cover	⑪	Connecting bolt
⑥	Inner shim	⑫	Copper washer

⑬	Bleed valve	
⑭	Cylinder body	
⑮	Piston seal	
⑯	Piston	
⑰	Piston boot	
⑱	Retainer	

93026G61

Exploded view of the rear disc brake components—QX4

7. Remove the inner and outer shims. Remove the 2 pad retainers if they are not still attached to the pads.

8. Check the pad thickness and replace the pads if they are less than 0.079 in. (2mm) thick.

To install:

9. Install the inner and outer shims into the torque member.

10. Install a pad retainer to the bottom of each pad.

11. Install the pads into the torque member.

12. Use a C-clamp or hammer handle and press the caliper piston(s) back into the housing.

13. Untie the caliper and swivel it back into position over the torque plate so that the dust boot is not pinched. Install the pin

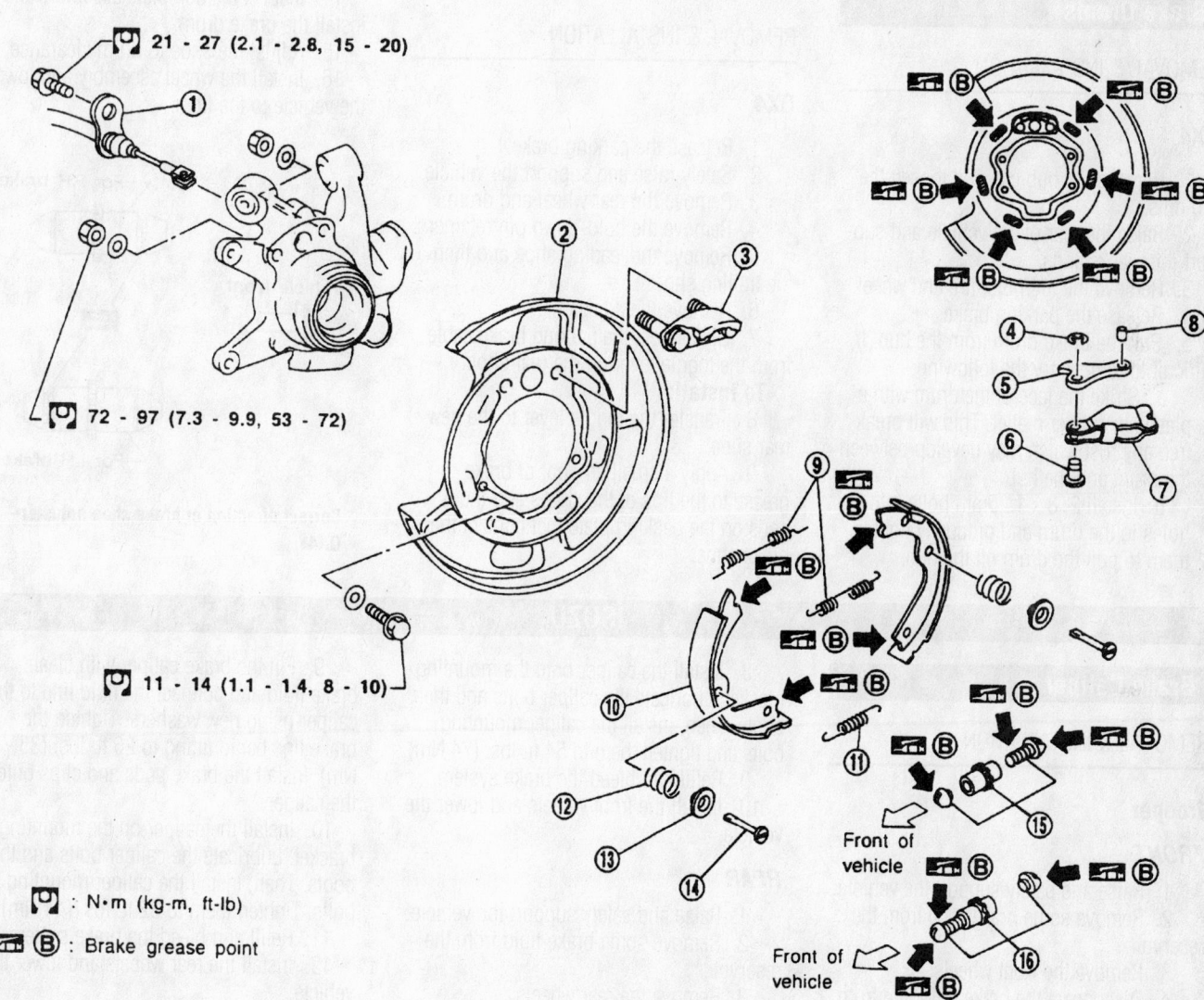

21 - 27 (2.1 - 2.8, 15 - 20)

72 - 97 (7.3 - 9.9, 53 - 72)

11 - 14 (1.1 - 1.4, 8 - 10)

: N·m (kg-m, ft-lb)

B : Brake grease point

Front of vehicle

Front of vehicle

①	Parking brake cable	⑦	Toggle lever	⑫	Anti-rattle spring
②	Back plate	⑧	Stopper pin	⑬	Retainer
③	Anchor block	⑨	Return spring	⑭	Anti-rattle pin
④	E-ring	⑩	Shoe	⑮	Adjuster assembly LH
⑤	Lever	⑪	Adjuster spring	⑯	Adjuster assembly RH
⑥	Pin				

93026G62

Exploded view of the rear disc brake drum assembly—QX4

Heater Core replacement is covered in Section 2 of this manual

bolt and torque it to 16–23 ft. lbs. (22–31 Nm).

14. Check the condition of the pin boot. Gently pull on it to expel any trapped air.

15. Install the wheel and lower the vehicle.

16. Pump the brakes until the pedal is firm and check the level of brake fluid. Road-test the vehicle.

Brake Drums

REMOVAL & INSTALLATION

QX4

1. Remove the hub cap and loosen the lug nuts.

2. Raise the rear of the vehicle and support it on jackstands.

3. Remove the lug nuts, tire and wheel.

4. Release the parking brake.

5. Pull the brake drum from the hub. If difficult to remove try the following:

   a. Strike the face of the drum with a plastic or rubber mallet. This will break free any rust which may develop between the drum and the hub.

   b. Install 2, 8 x 1.25mm bolts into the holes in the drum and gradually tighten them to pull the drum off the hub.

To install:

6. Install the brake drum to the hub.

7. Install the tire, wheel and lug nuts.

8. Remove the jackstands and lower the vehicle.

9. Road-test the vehicle to ensure that the brakes are working properly.

Brake Shoes

REMOVAL & INSTALLATION

QX4

1. Release the parking brake.

2. Safely raise and support the vehicle.

3. Remove the rear wheel and drum.

4. Remove the hold-down pin retainers.

5. Remove the leading shoe and then the trailing shoe.

6. Remove the adjuster.

7. Disconnect the parking brake cable from the toggle lever on the rear shoe.

To install:

8. Transfer the toggle lever to the new rear shoe.

9. Apply a small amount of brake grease to the tips of the shoes and the 6 pads on the backing plate that contact the brake shoe.

10. Shorten the adjuster by turning it.

11. Connect the parking brake cable to the toggle lever on the rear shoe.

12. Install the lower return spring to both shoes and install the shoes on the backing plate with the hold down pins and retainers.

13. Install the adjuster and the remaining springs. Pay attention to the direction of the adjuster assembly.

14. Inspect the complete assembly and install the brake drum.

15. Adjust the shoe to drum clearance.

16. Install the wheel assembly and lower the vehicle to the floor.

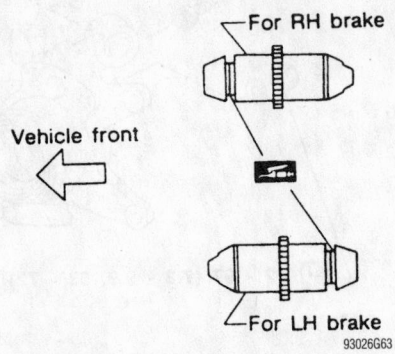

Correct direction of brake shoe adjuster— QX4

ISUZU

Brake Caliper

REMOVAL & INSTALLATION

Trooper

FRONT

1. Raise and safely support the vehicle.

2. Remove some brake fluid from the reservoir.

3. Remove the front wheels.

4. Disconnect the brake fluid line from the caliper. Plug the line to prevent fluid loss.

5. Loosen the brake caliper mounting bolt and guide bolt. Remove the caliper from the mount.

6. Remove the brake pads and clips from the caliper. Inspect the brake pads for wear and replace them if necessary.

To install:

7. Fill the brake caliper with clean brake fluid and connect the fluid line to the caliper using new washers. Tighten the brake line banjo fitting to 26 ft. lbs. (35 Nm). Install the brake pads and clips onto the caliper.

8. Install the caliper onto the mounting bracket. Lubricate the caliper bolts and their boots. Then, install the caliper mounting bolts and tighten them to 54 ft. lbs. (74 Nm).

9. Refill and bleed the brake system.

10. Install the front wheels and lower the vehicle.

REAR

1. Raise and safely support the vehicle.

2. Remove some brake fluid from the reservoir.

3. Remove the rear wheels.

4. Disconnect the brake fluid line from the caliper. Plug the line to prevent fluid loss.

5. Loosen the brake caliper mounting bolt and guide bolt. Remove the caliper from the mount bracket.

6. Remove the brake pads and clips from the caliper. Inspect the brake pads for wear; replace them if necessary.

7. If necessary for servicing, unbolt the caliper mounting bracket from the backing plate.

To install:

8. If removed, install the caliper mounting bracket and tighten its bolts to 76 ft. lbs. (103 Nm).

9. Fill the brake caliper with clean brake fluid and connect the fluid line to the caliper using new washers. Tighten the brake line banjo fitting to 26 ft. lbs. (35 Nm). Install the brake pads and clips onto the caliper.

10. Install the caliper on the mounting bracket. Lubricate the caliper bolts and their boots. Then, install the caliper mounting bolts. Tighten them to 32 ft. lbs. (44 Nm).

11. Refill and bleed the brake system.

12. Install the rear wheels and lower the vehicle.

Amigo and Rodeo

FRONT

1. Raise and safely support the vehicle. Remove the front wheels.

2. Remove some brake fluid from the master cylinder reservoir.

3. Disconnect and plug the brake fluid line from the caliper.

4. Remove the brake caliper mounting bolt and guide bolt and remove the caliper from the mount. The brackets can be remove for additional work space.

5. Remove the brake pads and clips

from the caliper. Inspect the brake pads for wear; replace them, if necessary.

To install:

6. Install the brake pads and clips onto the caliper.

7. If the caliper bracket was removed, tighten the bolts to 103–126 ft. lbs. (139–171 Nm).

8. Install the caliper on the mounting bracket. Torque the caliper-to-mounting bracket bolts to:
- 4-cylinder models: 20–27 ft. lbs. (27–37 Nm)
- 6-cylinder models: 54 ft. lbs. (74 Nm)

9. Connect the fluid line to the caliper using new washers. Torque the brake line banjo fitting to 26 ft. lbs. (35 Nm).

❈❈ WARNING

Be sure the hook end of the flexible brake line is positioned in the anti-rotation cavity.

10. Refill the master cylinder reservoir and bleed the brake system.

11. Install the front wheels and lower the vehicle.

REAR

1. Raise and safely support the vehicle. Remove the rear wheels.

2. Remove some fluid from the master cylinder reservoir.

3. Disconnect and plug the brake fluid line from the caliper.

➡ **Discard the parking brake cable mounting pin after removal.**

4. If equipped with caliper actuated parking brakes; remove the mounting pin from the parking brake cable and disconnect the parking cable from the disc caliper.

5. Remove the brake caliper mounting bolt and guide bolt and remove the caliper from the mount.

6. Remove the brake pads and clips from the caliper. Inspect the brake pads for wear; replace them, if necessary.

To install:

7. Install the brake pads and clips onto the caliper.

8. If the mounting bracket was removed, tighten the bolts to 69–84 ft. lbs. (93–114 Nm).

9. Install the caliper on the mounting bracket. Torque the caliper-to-mounting bracket bolts to 12–17 ft. lbs. (16–24 Nm),

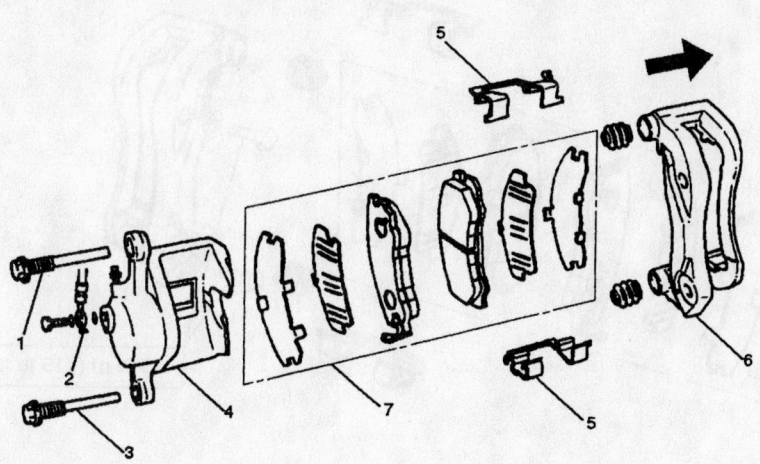

(1) Guide Bolt	(4) Caliper Assembly
(2) Brake Flexible Hose	(5) Clip
(3) Lock Bolt	(6) Support Bracket with Pad Assembly
	(7) Pad Assembly

93026G02

Front caliper assembly—Trooper

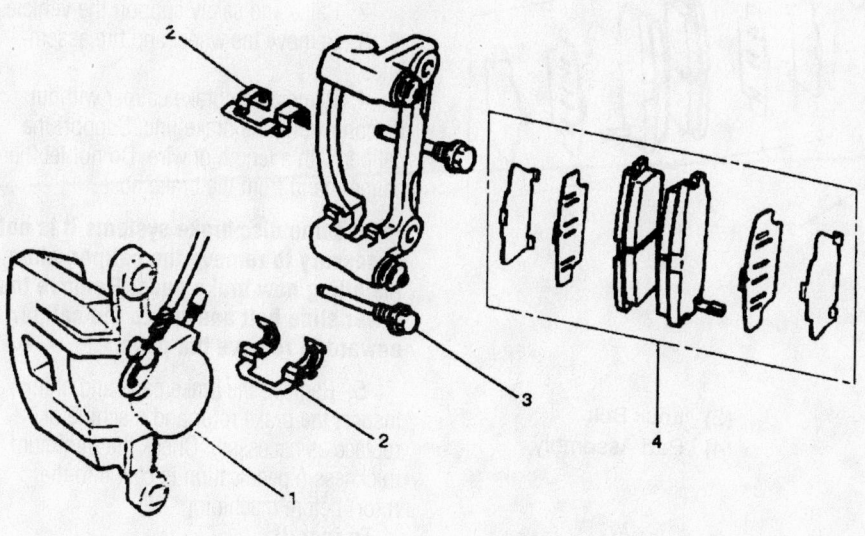

(1) Caliper Assembly	(3) Lock Bolt
(2) Clip	(4) Pad Assembly

93026G01

Rear caliper assembly—Trooper

Brake service is covered in Section 4 of this manual

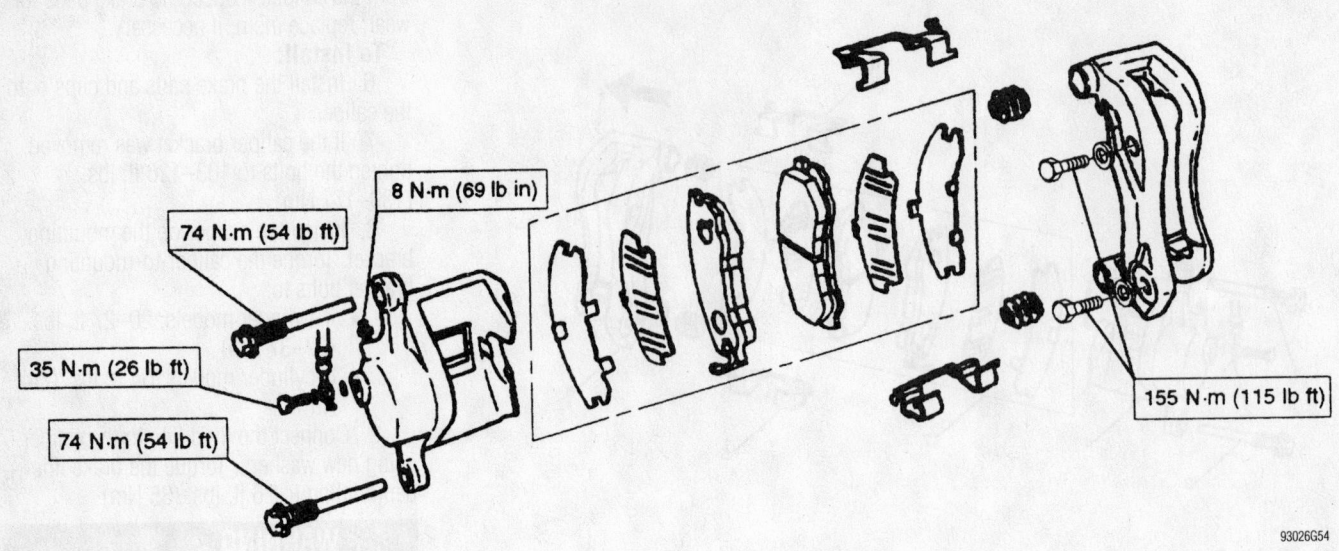

8 N·m (69 lb in)

74 N·m (54 lb ft)

35 N·m (26 lb ft)

74 N·m (54 lb ft)

155 N·m (115 lb ft)

93026G54

Exploded view of the front caliper components—Rodeo

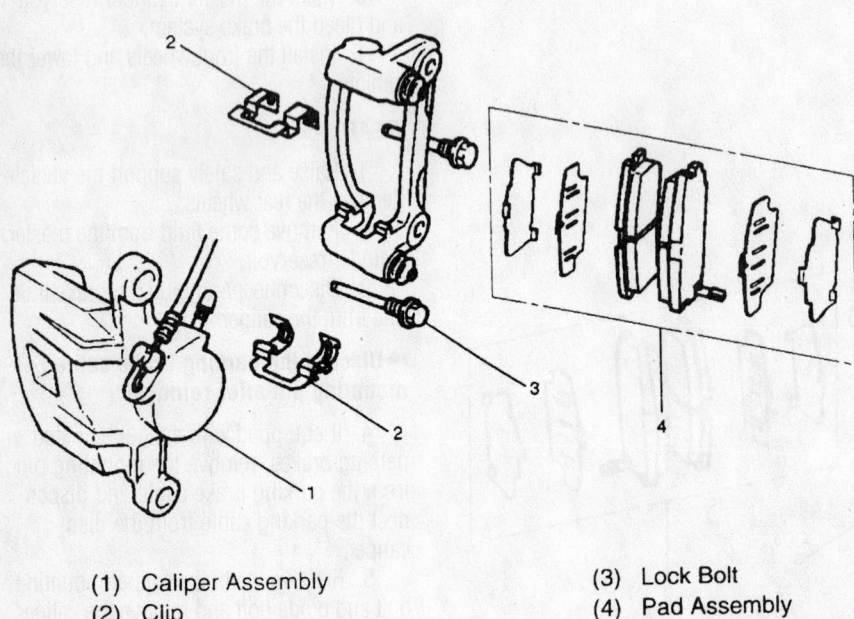

(1) Caliper Assembly
(2) Clip
(3) Lock Bolt
(4) Pad Assembly

93026G55

Exploded view of the rear caliper components—Rodeo

or 32 ft. lbs. (43 Nm) on vehicles with shoe-type parking brakes.

10. Connect the parking brake cable to the caliper and install a new mounting pin.

11. Connect the fluid line to the caliper using new washers. Torque the brake line banjo fitting to 26 ft. lbs. (35 Nm).

12. Refill the master cylinder reservoir and bleed the brake system.

13. Install the rear wheels and lower the vehicle.

Disc Brake Pads

REMOVAL & INSTALLATION

Amigo and Rodeo

FRONT

Most disc brake pads are equipped with wear indicators. If a squealing noise occurs from the brakes while driving, check the pad wear indicator plate. If there is evidence of the indicator plate contacting the brake disc, the brake pad should be replaced.

1. Remove ½ of the volume of brake fluid from the master cylinder to prevent overflow when the caliper piston is compressed.

2. Raise and safely support the vehicle.

3. Remove the wheel and tire assemblies.

4. Remove the brake caliper without disconnecting the brake line. Support the caliper with a length of wire. Do not let the caliper hang from the brake hose.

➡On some disc brake systems it is not necessary to remove the caliper when installing new brake pads. Remove the lower slide bolt and rotate the caliper upward to remove the pads.

5. Remove the brake pads and shims. Inspect the brake rotor and machine or replace as necessary. Check the minimum thickness (specification is cast into the rotor) before machining.

To install:

6. Use a suitable tool to push the caliper piston into its bore.

7. Apply a thin coat of grease to the rear face of the brake pad and install the shim. Install the brake pads.

8. Install the calipers. Lubricate the caliper bolts and boots. If equipped with a 4-cylinder engine, tighten the caliper mounting bolts to 24 ft. lbs. (33 Nm). If equipped with a 6-cylinder engine, tighten the caliper mounting bolts to 54 ft. lbs. (74 Nm).

9. Install the wheel and tire assemblies and lower the vehicle.

10. Apply the brakes several times to seat the pads before moving the vehicle. Check the fluid in the master cylinder and add as necessary.

REAR

1. Raise and safely support the vehicle. Remove the rear wheels.

2. Remove the brake caliper mounting bolts and remove the caliper without disconnecting the brake fluid line. Support the caliper so it does not hang on the brake line.

3. Remove the brake pads and retaining clips from the caliper.

4. If equipped with caliper-activated parking brakes; use tool J-37617 or equivalent to rotate the piston clockwise until it retracts into the bore. Align the notches of the piston face so the centerline of the notches is perpendicular to the centerline of the mounting bosses.

To install:

5. Install the new brake pads and clips in the caliper and install the caliper in the mounting bracket.

6. Tighten the caliper mounting bolts to 12–17 ft. lbs. (16–24 Nm), or 32 ft. lbs. (43 Nm) on vehicles with shoe-type parking brakes.

7. Install the rear wheels. Check the brake fluid level.

8. Pump the brake pedal until pressure is felt before moving the vehicle.

Trooper

FRONT

1. Remove about ½ of the brake fluid from the master cylinder reservoir to prevent overflow when the caliper piston is compressed.

2. Raise and safely support the vehicle.

3. Remove the front wheels.

4. Remove the brake caliper from the caliper bracket without disconnecting the brake line. Support the caliper with a length of wire. Do not let the caliper hang from the brake hose.

5. Remove the brake pads and shims. Inspect the brake rotor and machine or replace as necessary. Check the minimum thickness (specification is cast into the rotor) before machining.

To install:

6. Use a large C-clamp or brake piston tool to push the caliper piston into its bore.

7. Apply a thin coat of brake grease to both sides of both inner shims. Assemble

the pads and shims, then install them into the caliper. The wear indicator on the inner pad must face down.

8. Install the calipers. Clean and lubricate the caliper mounting bolts and lubricate the mounting bolt boots. Install the mounting bolts and tighten them to 54 ft. lbs. (74 Nm).

9. Install the front wheels and lower the vehicle.

10. Apply the brakes several times to seat the pads before moving the vehicle. Check the fluid level in the master cylinder reservoir and add as necessary.

REAR

1. Use a vacuum pump to remove some brake fluid from the master cylinder reservoir to prevent overflow when the caliper piston is compressed.

2. Raise and safely support the vehicle.

3. Remove the rear wheels.

4. Remove the brake caliper from the caliper bracket without disconnecting the brake line. Support the caliper with a length of wire. Do not let the caliper hang from the brake hose.

5. Remove the brake pads and shims. Inspect the brake rotor and machine or replace as necessary. Check the minimum thickness (specification is cast into the rotor) before machining.

To install:

6. Use a large C-clamp or brake piston tool to push the caliper piston into its bore.

7. Apply a thin coat of brake grease to both sides of both inner shims. Assemble the pads and shims, then install them into the caliper. The wear indicator on the inner pad must face down.

8. Install the calipers. Clean and lubricate the caliper mounting bolts and lubricate the mounting bolt boots. Install the mounting bolts and tighten them to 32 ft. lbs. (44 Nm).

9. Install the rear wheels and lower the vehicle.

10. Apply the brakes several times to seat the pads before moving the vehicle. Check the fluid level in the master cylinder reservoir and add as necessary.

Brake Drums

REMOVAL & INSTALLATION

Amigo and Rodeo

1. Raise and safely support the vehicle. Release the parking brake.

2. Remove the rear wheels.

3. Use chalk to mark the brake drum to one of the wheel studs as an index mark for reinstallation.

4. Remove the retaining screw that holds the brake drum to the axle flange.

5. Pull the brake drum from the axle flange.

To install:

6. Align the index mark and install the brake drum to the axle flange.

7. Install the retaining screw to secure the brake drum to the axle flange.

8. Install the rear wheels.

Rear Brake Shoes

REMOVAL & INSTALLATION

Amigo and Rodeo

1. Raise and safely support the vehicle.

2. Remove the rear wheels.

3. Remove the brake drums.

4. Remove the brake return springs.

5. Remove the leading shoe holding pin and spring, and then the leading shoe.

6. Remove the self-adjuster and the adjuster lever.

7. Remove the trailing shoe holding pin and spring.

8. Disconnect the parking brake cable from the trailing shoe and remove the trailing shoe. Remove the parking brake lever from the trailing shoe.

To install:

9. Attach the parking brake lever to the trailing shoe.

10. Connect the parking brake cable to the parking brake lever.

11. Apply a thin coat of high temperature grease to the shoe contact points on the brake backing plate (locations A and C in the accompanying illustration), piston contact surface (B), and self-adjuster (D).

12. Position the trailing shoe on the backing plate and install the hold-down pin, spring, and retainer. Don't stretch the return spring when fitting the shoes onto the backing plate.

13. Connect the upper return spring and the leading shoe to the trailing shoe and position the leading brake shoe on the backing plate.

14. Install the adjuster assembly and the hold-down pin, spring, and retainer.

For complete Engine Mechanical specifications, see Section 1 of this manual

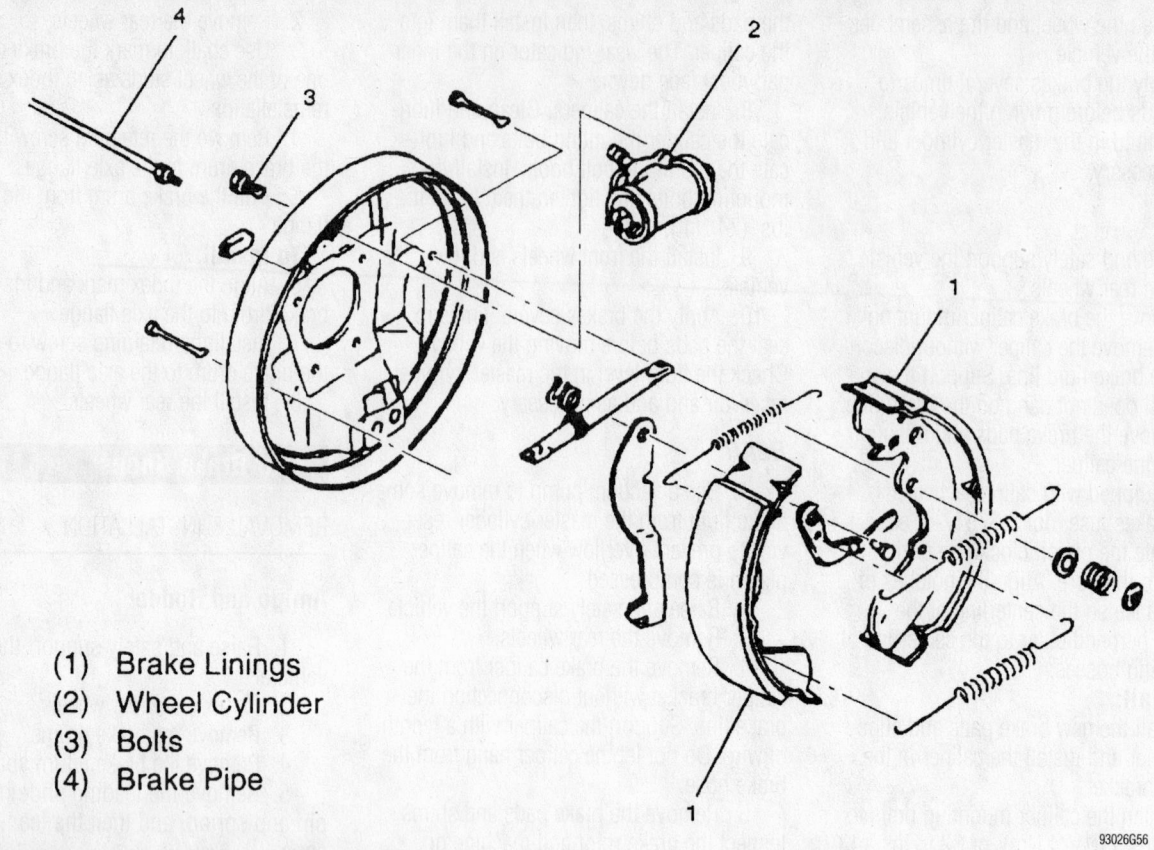

(1) Brake Linings
(2) Wheel Cylinder
(3) Bolts
(4) Brake Pipe

93026G56

Exploded view of the rear drum brakes—Amigo and Rodeo

15. Use a brake spring tool to install the lower return spring.

16. Install the self-adjuster lever and adjuster spring.

17. Adjust the shoe-to-drum clearance to 0.0098–0.0157 in. (0.25–0.40mm) and install the brake drum.

18. Check the brake drum for scoring or other wear. Machine or replace as necessary. Check the maximum brake drum diameter specification when machining.

19. Install the rear wheels. Lower the vehicle.

20. Road-test the vehicle.

JEEP

Brake Caliper

REMOVAL & INSTALLATION

Cherokee

1. Drain ⅔ of the brake fluid from the front reservoir. Use the bleeder screw at the front outlet port to drain the fluid. If equipped with anti-lock brakes, relieve the system pressure.

2. Raise and safely support the vehicle.

3. Remove the wheels.

4. Place a C-clamp on the caliper so the solid end contacts the back of the caliper and screw end contacts the metal part of the outboard brake pad.

5. Tighten the clamp until the caliper moves far enough to force the piston to the bottom of the piston bore. This will back the brake pads off of the rotor surface to facili-tate the removal and installation of the caliper assembly.

6. Remove the C-clamp.

7. Remove both of the mounting bolts and lift the caliper off the rotor.

8. If the caliper is being removed, it is necessary to disconnect the brake fluid hose. Clean the brake fluid hose-to-caliper connection thoroughly. Remove the hose-to-caliper bolt. Cap or tape the open ends to keep dirt out. Discard the copper gaskets.

To install:

9. Connect the brake line to the caliper with new sealing washers and fitting bolt. Hand-tighten the fitting bolt.

10. Position the caliper into place over the rotor.

11. Coat the caliper mounting bolt with silicone grease and torque them to 84–180 inch lbs. (10–20 Nm).

12. Position the brake line clear of all chassis components, untwisted and free of kinks. Torque the fitting bolt to 23 ft. lbs. (31 Nm).

13. Install the wheels.

14. Fill the master cylinder with fluid and bleed the brake system.

15. Before driving the vehicle, pump the brakes several times to seat the pads.

Grand Cherokee

FRONT

1. Drain ⅔ of the brake fluid from the front reservoir. Use the bleeder screw at the front outlet port to drain the fluid. If equipped with anti-lock brakes, relieve the system pressure.

2. Raise and safely support the vehicle.

3. Remove the wheels.

4. Insert a small prybar through the caliper opening and pry the caliper (using the outboard brake pad) to bottom the pistons in the caliper bore.

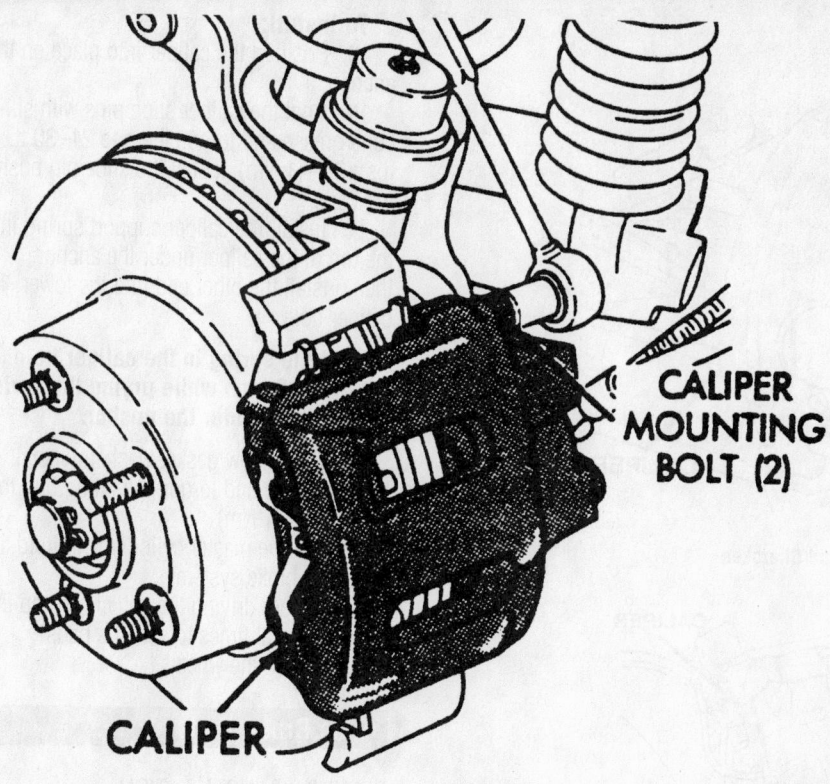

Front caliper mounting—Cherokee

➡ **This will back the brake pads off of the rotor surface to facilitate the removal and installation of the caliper assembly.**

5. Remove the brake hose-to-caliper bolt, hose and washers.

6. Pry the caliper support spring out of the caliper.

7. Remove both caliper slide pin bushing caps and slide pins.

8. Lift the caliper from the anchor.

To install:

9. Position the caliper into place on the anchor.

10. Coat the caliper slide pins with silicone grease and torque them to 21–30 ft. lbs. (29–41 Nm). Install the slide pin bushing caps.

11. Install the caliper support spring in the top of the caliper under the anchor; then, install the other end into the lower caliper hole.

➡ **Hold the spring in the caliper hole with your thumb while prying the spring end out and under the anchor.**

12. Using new gasket washers, install the brake line and torque the fitting bolt to 23 ft. lbs. (31 Nm).

13. Fill the master cylinder with fluid and bleed the brake system.

14. Before driving the vehicle, pump the brakes several times to seat the pads.

15. Install the wheels.

REAR

1. Drain ⅔ of the brake fluid from the front reservoir. Use the bleeder screw at the front outlet port to drain the fluid. If equipped with anti-lock brakes, relieve the system pressure.

2. Raise and safely support the vehicle.

3. Remove the wheels.

4. Insert a small prybar through the caliper opening and pry the caliper (using the outboard brake pad) to bottom the piston in the caliper bore.

➡ **This will back the brake pads off of the rotor surface to facilitate the removal and installation of the caliper assembly.**

5. Remove the brake hose-to-caliper bolt, hose and washers.

6. Pry the caliper support spring out of the caliper.

7. Remove both caliper slide pin bushing caps and slide pins.

8. Lift the caliper from the anchor.

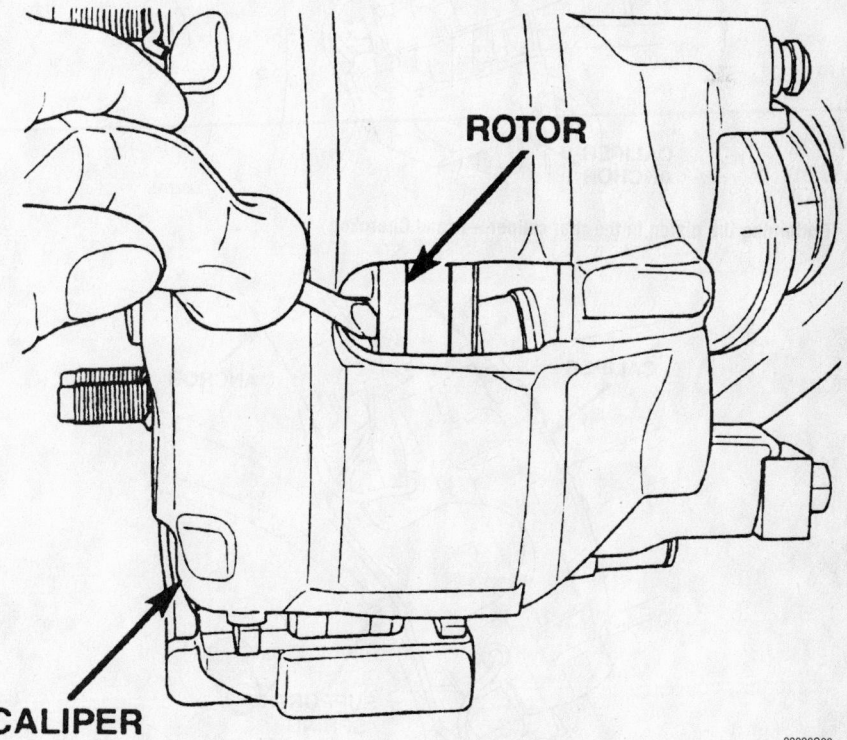

Bottoming the pistons in the front caliper—Grand Cherokee

For Accessory Drive Belt illustrations, see Section 1 of this manual

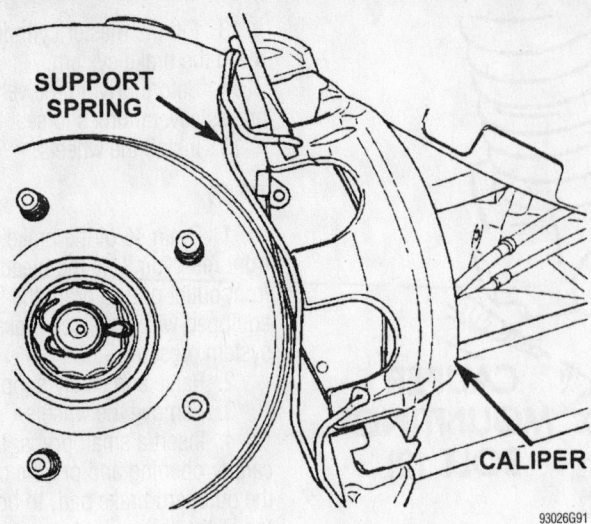

View of the support springs on the front caliper—Grand Cherokee

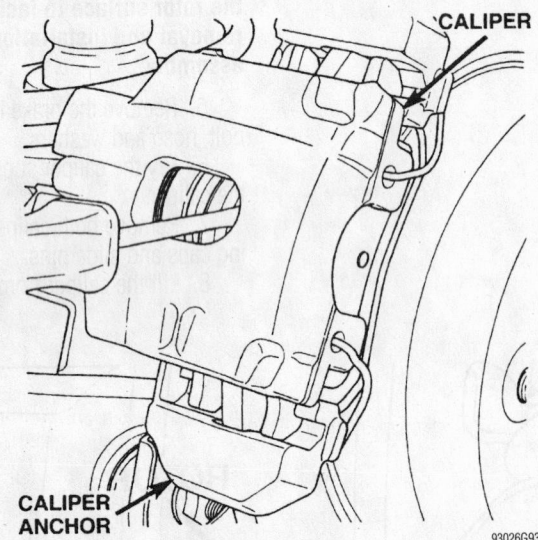

Bottoming the piston in the rear caliper—Grand Cherokee

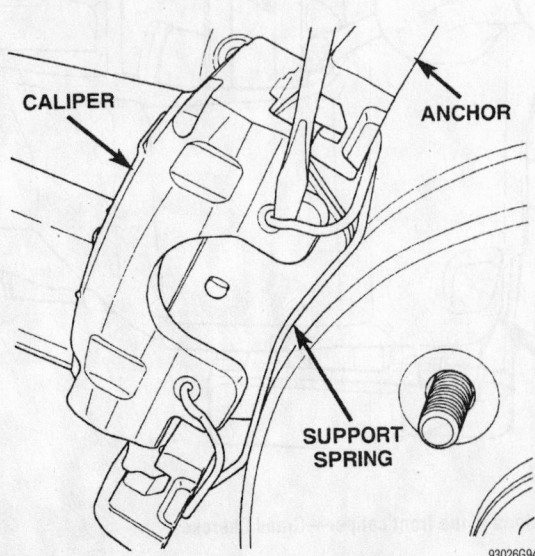

View of the support springs on the rear caliper—Grand Cherokee

To install:

9. Position the caliper into place on the anchor.

10. Coat the caliper slide pins with silicone grease and torque them to 21–30 ft. lbs. (29–41 Nm). Install the slide pin bushing caps.

11. Install the caliper support spring in the top of the caliper under the anchor; then, install the other end into the lower caliper hole.

➡ **Hold the spring in the caliper hole with your thumb while prying the spring end out and under the anchor.**

12. Using new gasket washers, install the brake line and torque the fitting bolt to 23 ft. lbs. (31 Nm).

13. Fill the master cylinder with fluid and bleed the brake system.

14. Before driving the vehicle, pump the brakes several times to seat the pads.

15. Install the wheels.

Disc Brake Pads

REMOVAL & INSTALLATION

Cherokee

1. Raise and safely support the vehicle.

2. Drain ⅔ of the brake fluid from the front reservoir. Use the bleeder screw at the front outlet port to drain the fluid.

3. Raise and support the vehicle safely.

4. Remove the wheels.

5. Remove the brake caliper. Use a suitable tool to compress the caliper piston into the bore.

6. Hold the anti-rattle clip against the caliper anchor plate and remove the outboard brake pad.

7. Remove the inboard pad and its anti-rattle clip.

To install:

8. Clean all the mounting holes and bushing grooves in the caliper ears. Clean the mounting bolts. Replace the bolts if they are corroded or if the threads are damaged. Wipe the inside of the caliper clean, including the exterior of the dust boot. Inspect the dust boot for cuts or cracks and for proper seating in the piston bore. If evidence of fluid leakage is noted, the caliper should be rebuilt.

➡ **Do not use abrasives on the bolts. This will destroy their protective plating.**

9. Install the inboard anti-rattle clip on the trailing end of the anchor plate. The split

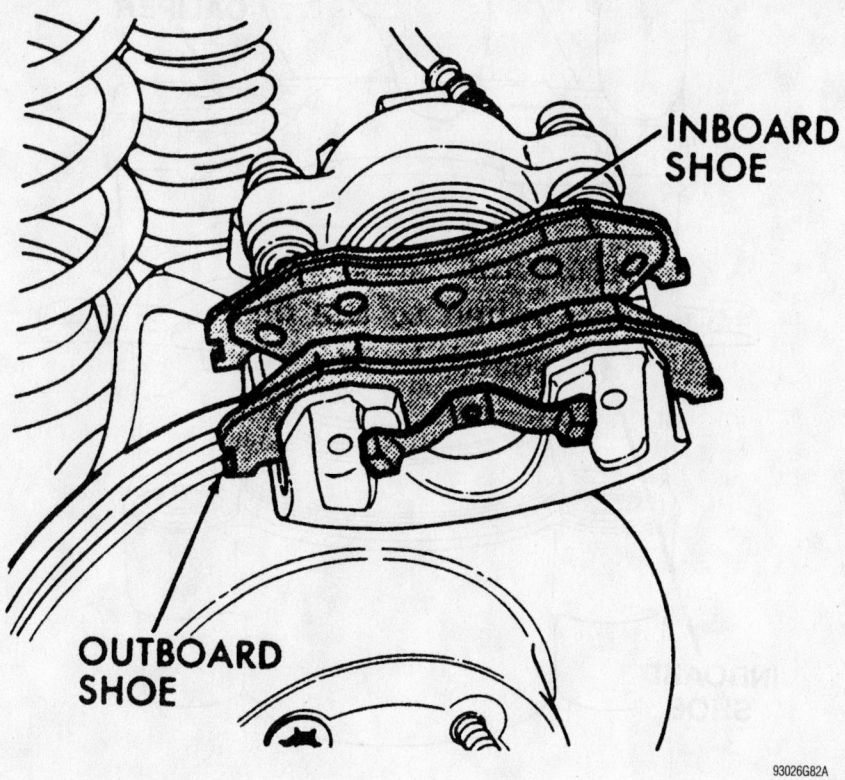

Front disc brake pad installation—Cherokee

front outlet port to drain the fluid. If equipped with anti-lock brakes, relieve the system pressure.

2. Raise and safely support the vehicle.

3. Remove the wheels.

4. Insert a small prybar through the caliper opening and pry the caliper (using the outboard brake pad) to bottom the pistons in the caliper bore.

➡️**This will back the brake pads off of the rotor surface to facilitate the removal and installation of the caliper assembly.**

5. Pry the caliper support spring out of the caliper.

6. Remove both caliper slide pin bushing caps and slide pins.

7. Lift the caliper from the anchor.

8. Using a piece of mechanics wire, support the caliper so there is not tension on the brake hose.

9. Remove the brake pads from the caliper.

To install:

10. Position the brake pads onto the caliper.

11. Position the caliper into place on the anchor.

end of the clip must face away from the rotor.

10. Install the inboard pad in the caliper. The pad must lay flat against the piston.

11. Install the outboard pad in the caliper while holding the anti-rattle clip.

12. With the pads installed, position the caliper over the rotor. Line up the mounting holes in the caliper and the support bracket and insert the mounting bolts. Make sure the bolts pass under the retaining ears on the inboard shoes. Push the bolts through until they engage the holes of the outboard pad and caliper ears. Thread the bolts into the support bracket and tighten them to 11 ft. lbs. (15 Nm).

13. Fill the master cylinder with brake fluid and pump the brake pedal to seat the pads.

14. Install the wheel assembly and lower the vehicle. Check the level of the brake fluid in the master cylinder and fill as necessary. Test the operation of the brakes before taking the vehicle onto the road.

Grand Cherokee

FRONT

1. Drain ⅔ of the brake fluid from the front reservoir. Use the bleeder screw at the

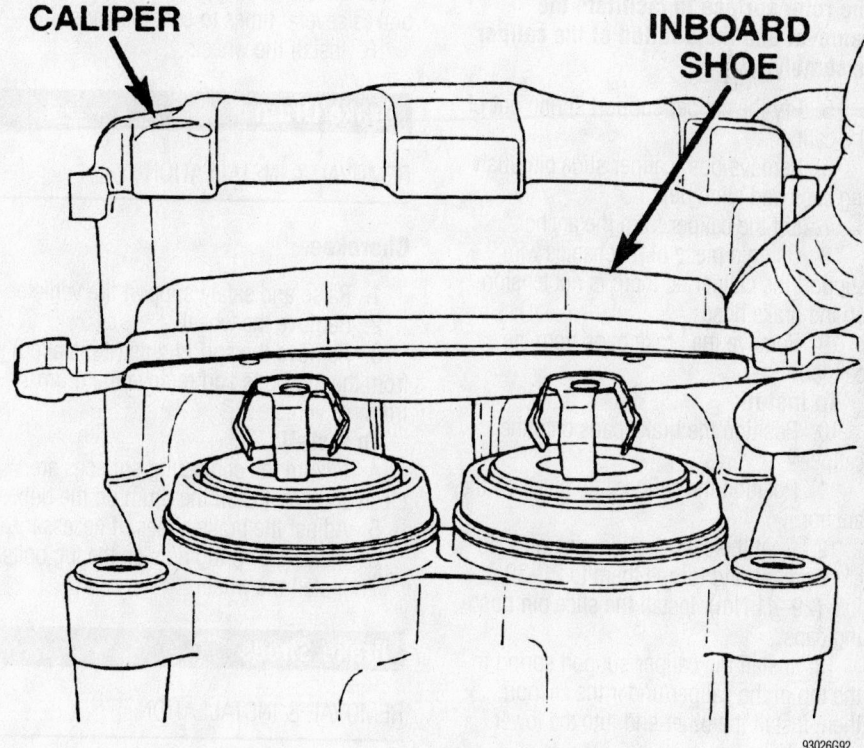

Installing the inward brake pad on the front caliper pistons—Grand Cherokee

For Tire, Wheel and Ball Joint specifications, see Section 1 of this manual

12. Coat the caliper slide pins with silicone grease and torque them to 21–30 ft. lbs. (29–41 Nm). Install the slide pin bushing caps.

13. Install the caliper support spring in the top of the caliper under the anchor; then, install the other end into the lower caliper hole.

➡ **Hold the spring in the caliper hole with your thumb while prying the spring end out and under the anchor.**

14. Fill the master cylinder with fluid and bleed the brake system.

15. Before driving the vehicle, pump the brakes several times to seat the pads.

16. Install the wheels.

REAR

1. Drain ⅔ of the brake fluid from the front reservoir. Use the bleeder screw at the front outlet port to drain the fluid. If equipped with anti-lock brakes, relieve the system pressure.

2. Raise and safely support the vehicle.

3. Remove the wheels.

4. Insert a small prybar through the caliper opening and pry the caliper (using the outboard brake pad) to bottom the piston in the caliper bore.

➡ **This will back the brake pads off of the rotor surface to facilitate the removal and installation of the caliper assembly.**

5. Pry the caliper support spring out of the caliper.

6. Remove both caliper slide pin bushing caps and slide pins.

7. Lift the caliper from the anchor.

8. Using a piece of mechanics wire, support the caliper so there is not tension on the brake hose.

9. Remove the brake pads from the caliper.

To install:

10. Position the brake pads onto the caliper.

11. Position the caliper into place on the anchor.

12. Coat the caliper slide pins with silicone grease and torque them to 21–30 ft. lbs. (29–41 Nm). Install the slide pin bushing caps.

13. Install the caliper support spring in the top of the caliper under the anchor; then, install the other end into the lower caliper hole.

➡ **Hold the spring in the caliper hole with your thumb while prying the spring end out and under the anchor.**

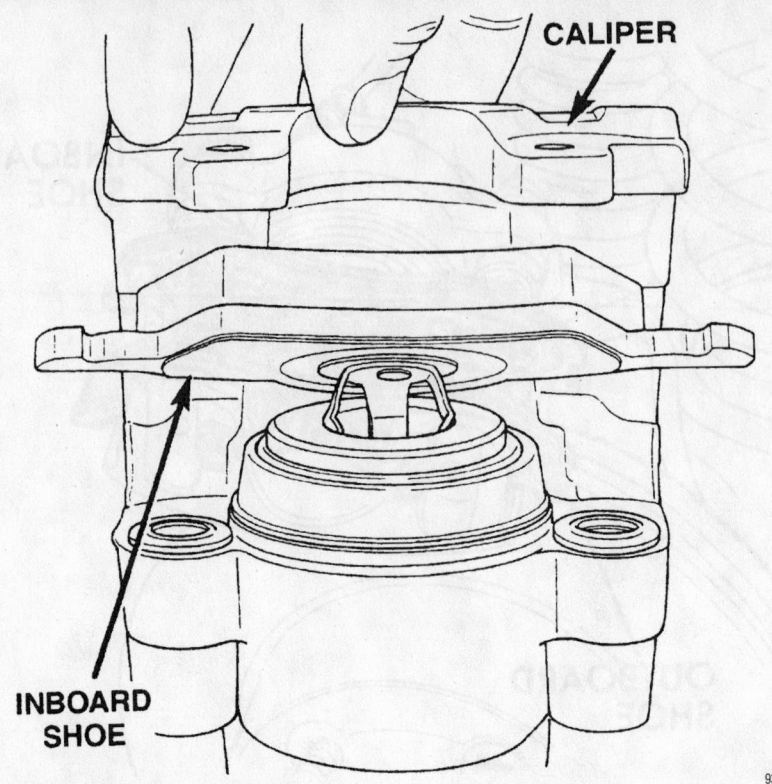

Installing the inward brake pad on the rear caliper piston—Grand Cherokee

14. Fill the master cylinder with fluid and bleed the brake system.

15. Before driving the vehicle, pump the brakes several times to seat the pads.

16. Install the wheels.

Brake Drum

REMOVAL & INSTALLATION

Cherokee

1. Raise and safely support the vehicle.

2. Remove the wheel.

3. Remove the spring nuts (if installed) from the lug bolts and remove the drum from the vehicle.

To install:

4. Ensure the contacting surfaces are clean and flat. Install the drum on the hub.

5. Adjust the brake shoes, if necessary.

6. Install the spring nuts on the lug bolts.

7. Install the wheel.

Brake Shoes

REMOVAL & INSTALLATION

Cherokee

1. Raise and safely support the vehicle.

2. Remove the wheel and brake drum.

3. Remove the U-clip and washer securing the adjuster cable to the parking brake lever.

4. Remove the primary and secondary return springs from the anchor pin.

5. Remove the hold-down springs, retainers and pins.

6. Install spring clamps on the wheel cylinders to hold the pistons in place.

7. Remove the adjuster lever, adjuster screw and spring.

8. Remove the adjuster cable and cable guide.

9. Remove the brake shoes and parking brake strut.

10. Disconnect the cable from the parking brake lever and remove the lever.

To install:

11. Clean the support plate with brake cleaner.

12. Apply multi-purpose grease to the brake shoe contact surfaces on the backing plate.

13. Lubricate the adjuster screw threads.

14. Attach the parking brake lever to the secondary brake shoe. Use a new washer and U-clip.

15. Remove the wheel cylinder clamps.

16. Attach the parking brake cable to the lever.

17. Install the brake shoes on the sup-

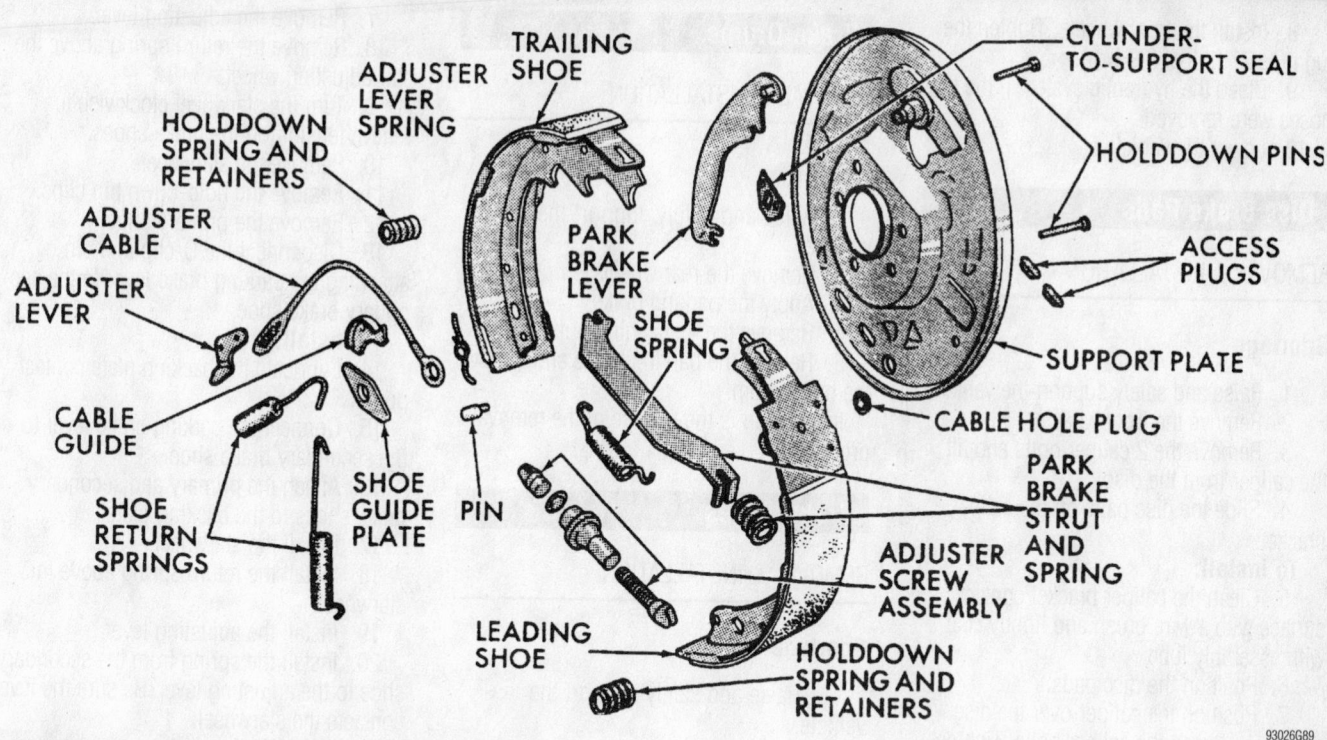

TRAILING SHOE

ADJUSTER LEVER SPRING

HOLDDOWN SPRING AND RETAINERS

ADJUSTER CABLE

ADJUSTER LEVER

CABLE GUIDE

SHOE RETURN SPRINGS

SHOE GUIDE PLATE

PIN

LEADING SHOE

PARK BRAKE LEVER

SHOE SPRING

ADJUSTER SCREW ASSEMBLY

HOLDDOWN SPRING AND RETAINERS

CYLINDER-TO-SUPPORT SEAL

HOLDDOWN PINS

ACCESS PLUGS

SUPPORT PLATE

CABLE HOLE PLUG

PARK BRAKE STRUT AND SPRING

93026G89

Exploded view of the rear drum brake components—Cherokee

port plate. Secure the shoes with new hold-down springs, pins and retainers.

18. Install the parking brake strut and spring.

19. Install the guide plate and adjuster cable to the anchor pin.

20. Install the return springs.

21. Install the adjuster cable guide on the secondary shoe.

22. Install the adjuster screw, spring and lever. Connect to the adjuster cable.

23. Adjust the shoes to the drum. Install the drum.

24. Install the wheel/tire assemblies and lower the vehicle.

25. Verify a firm brake pedal before moving the vehicle.

KIA

Brake Caliper

REMOVAL & INSTALLATION

Sportage

1. Raise and safely support the vehicle.

2. Remove the front wheels.

3. Remove the 2 caliper bolts and lift the caliper from the disc.

4. Disconnect the brake fluid flex line by removing the retaining bolt if the caliper is to be replaced.

To install:

5. Seat the caliper piston using a C-clamp.

6. Connect the brake fluid flex line bolt to the caliper. Tighten the bolt to 17 ft. lbs. (23.5 Nm).

7. Position the caliper over the disc assembly. Install the caliper bolts. Tighten to 72 ft. lbs. (98 Nm).

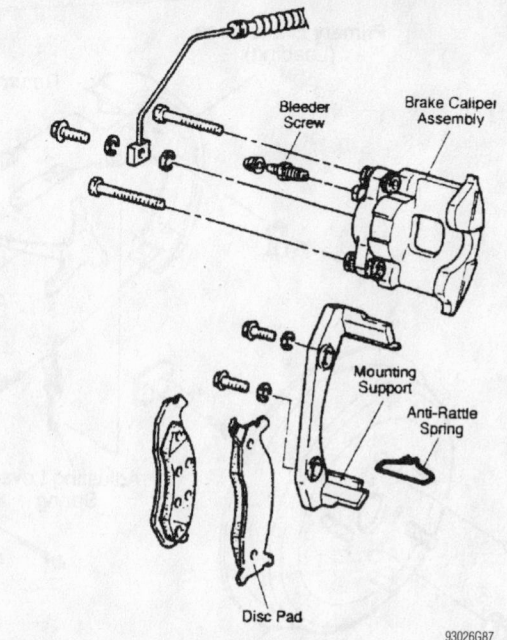

Bleeder Screw

Brake Caliper Assembly

Mounting Support

Anti-Rattle Spring

Disc Pad

93026G87

Exploded view of the front disc brake assembly—Sportage

For Wheel Alignment specifications, see Section 1 of this manual

8. Install the front wheels. Tighten the lug nuts 77 ft. lbs. (99 Nm).

9. Bleed the hydraulic system if the flex hoses were removed.

10. Lower the vehicle.

Disc Brake Pads

REMOVAL & INSTALLATION

Sportage

1. Raise and safely support the vehicle.
2. Remove the front wheels.
3. Remove the 2 caliper bolts and lift the caliper from the disc.
4. Slide the disc pads off the caliper bracket.

To install:

5. Clean the caliper bracket contact surface with a wire brush and lightly coat with assembly lube.

6. Position the disc pads.

7. Position the caliper over the disc assembly. Install the caliper bolts. Tighten to 72 ft. lbs. (98 Nm).

8. Install the front wheels. Tighten the lug nuts 77 ft. lbs. (99 Nm).

9. Bleed the hydraulic system if the flex hoses were removed.

10. Lower the vehicle.

Brake Drums

REMOVAL & INSTALLATION

Sportage

1. Raise and safely support the vehicle.
2. Remove the rear wheels.
3. Apply the parking brake.
4. Remove the 4 attaching nuts.
5. Release the parking brake and remove the brake drum.

Installation is the reverse of the removal procedure.

Brake Shoes

REMOVAL & INSTALLATION

Sportage

1. Raise and safely support the vehicle.
2. Remove the rear wheels.
3. Apply the parking brake.
4. Remove the 4 attaching nuts.
5. Release the parking brake and remove the brake drum.
6. Remove the spring from the secondary shoe to the adjusting lever.

7. Remove the adjusting lever.

8. Remove the return spring above the star adjusting wheel.

9. Turn the starwheel clockwise to relieve tension on the brake shoes.

10. Remove the starwheel.

11. Remove the hold-down pin clips.

12. Remove the primary shoe.

13. Disconnect the C-clip and pin attaching the parking brake lever to the secondary brake shoe.

To install:

14. Lubricate the backing plate contact points.

15. Connect the parking brake lever to the secondary brake shoe.

16. Attach the primary and secondary brake shoes to the backing plate.

17. Install the starwheel.

18. Install the return spring above the starwheel.

19. Install the adjusting lever.

20. Install the spring from the secondary shoe to the adjusting lever. Be sure the lever contacts the starwheel.

21. Install the brake drum and retaining nuts.

22. Adjust the rear brakes through the slot in the rear of the backing plate.

23. Install the wheels. Tighten the wheel lugs to 77 ft. lbs. (99 Nm).

24. Lower the vehicle.

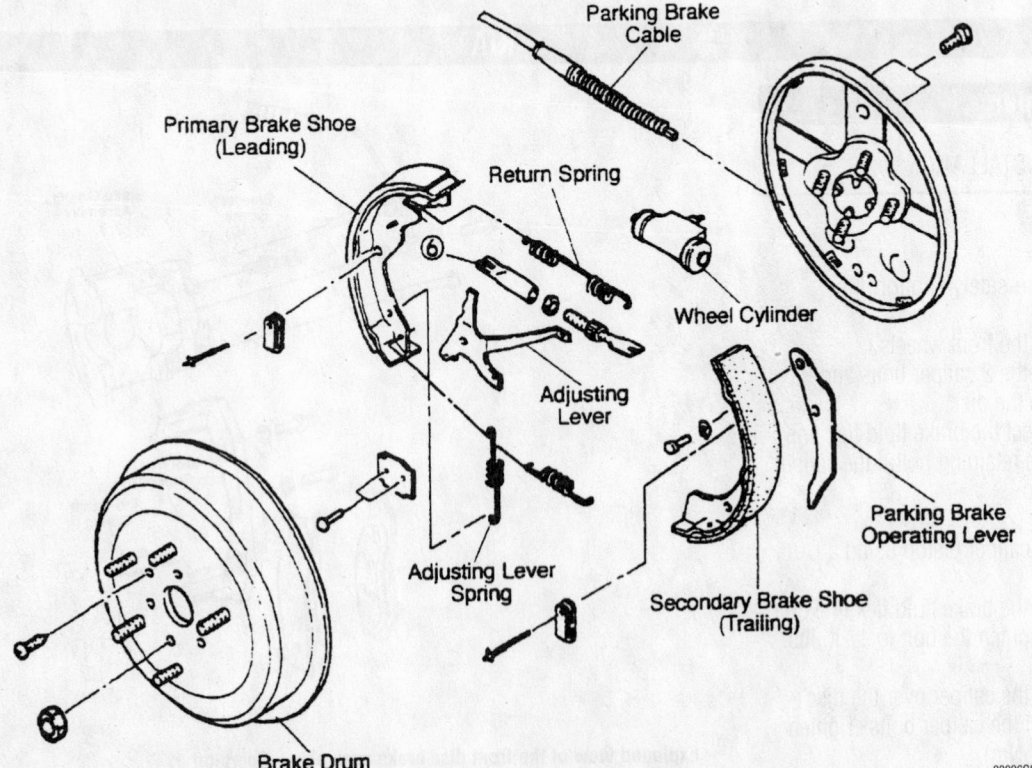

Exploded view of the rear drum brake assembly—Sportage

93026G88

LEXUS

Brake Caliper

REMOVAL & INSTALLATION

LX470

FRONT

1. Disconnect the negative battery cable from the battery.
2. Raise and support the vehicle safely.
3. Remove the wheels.
4. Disconnect the brake hose from the caliper by removing the union bolt and 2 gaskets. Plug the end of the hose to prevent loss of fluid.
5. Remove the bolts that attach the caliper to the torque plate.
6. Lift the bottom of the caliper up and remove the caliper assembly.

To install:

7. Grease the caliper slides and bolts with lithium grease or equivalent. Install the caliper and secure with the bolts. Torque the bolts to 90 ft. lbs. (123 Nm).
8. Connect the brake hose to the caliper, using 2 new washers. Make sure the flexible hose lock is securely in the lock hole of the caliper. Torque the union bolt to 22 ft. lbs. (30 Nm).
9. Fill the brake system to the proper level and bleed the brake system.
10. Install the tire and wheel assembly.
11. Top off the brake fluid level in the master cylinder. Check for leaks and proper brake operation.
12. Connect the negative battery cable to the battery.

REAR

1. Remove the brake line from the caliper.
2. Hold the siding pin and remove the 2 bolts.
3. Remove the caliper from the torque plate.
4. Remove the pads and shims.
5. Remove the pad support plates.
6. Installation is the reverse of removal. Torque the caliper bolts to 20 ft. lbs. (26 Nm). Torque the brake line union bolt to 22 ft. lbs. (30 Nm).

RX300

FRONT

1. Disconnect the negative battery cable from the battery.

2. Raise and support the vehicle safely.
3. Remove the wheels.
4. Disconnect the brake hose from the caliper by removing the union bolt and 2 gaskets. Plug the end of the hose to prevent loss of fluid.
5. Remove the caliper mounting bolts.
6. Remove the caliper, pads, shims and support plates.

To install:

7. Grease the caliper slides and bolts with lithium grease or equivalent. Install the support plates, shims, pads and caliper and secure with the bolts. Torque the bolts to 25 ft. lbs. (34 Nm).
8. Connect the brake hose to the caliper, using 2 new washers. Make sure the flexible hose lock is securely in the lock hole of the caliper. Torque the union bolt to 21 ft. lbs. (29 Nm).
9. Fill the brake system to the proper level and bleed the brake system.
10. Install the tire and wheel assembly.
11. Top off the brake fluid level in the master cylinder. Check for leaks and proper brake operation.

12. Connect the negative battery cable to the battery.

REAR

1. Disconnect the brake line from the caliper.
2. Remove the caliper mounting bolt.
3. Remove the caliper, pads, shims and support plates.
4. Remove the main pin.
5. Installation is the reverse of removal. Torque the main pin to 20 ft. lbs. (26 Nm), the caliper bolt to 14 ft. lbs. (20 Nm), and the union bolt to 21 ft. lbs. (29 Nm).

Disc Brake Pads

REMOVAL & INSTALLATION

LX470

FRONT

1. Raise the vehicle and support it safely.
2. Remove the wheels.

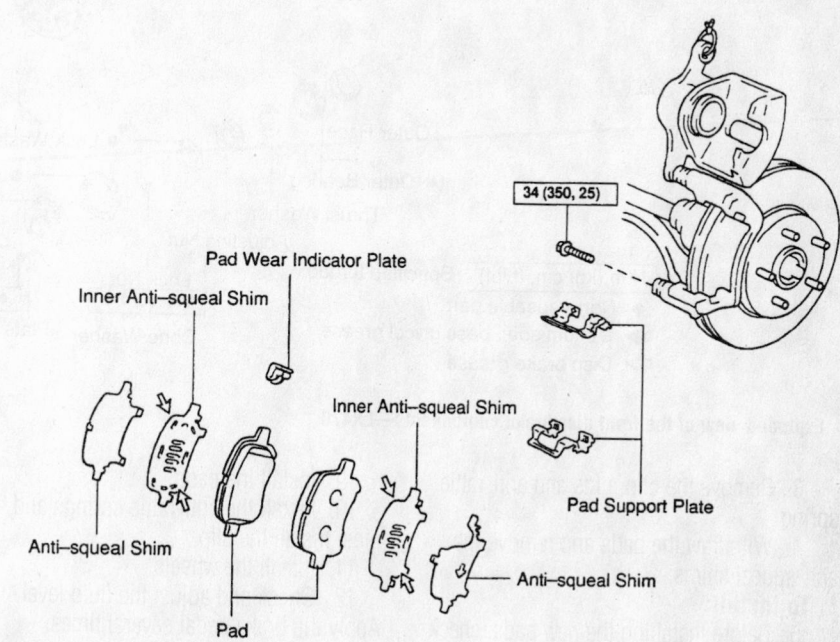

N·m (kgf·cm, ft·lbf) : Specified torque

⬅ Disc brake grease

Exploded view of the front disc brake components—RX300

93026G86

For Maintenance Interval recommendations, see Section 1 of this manual

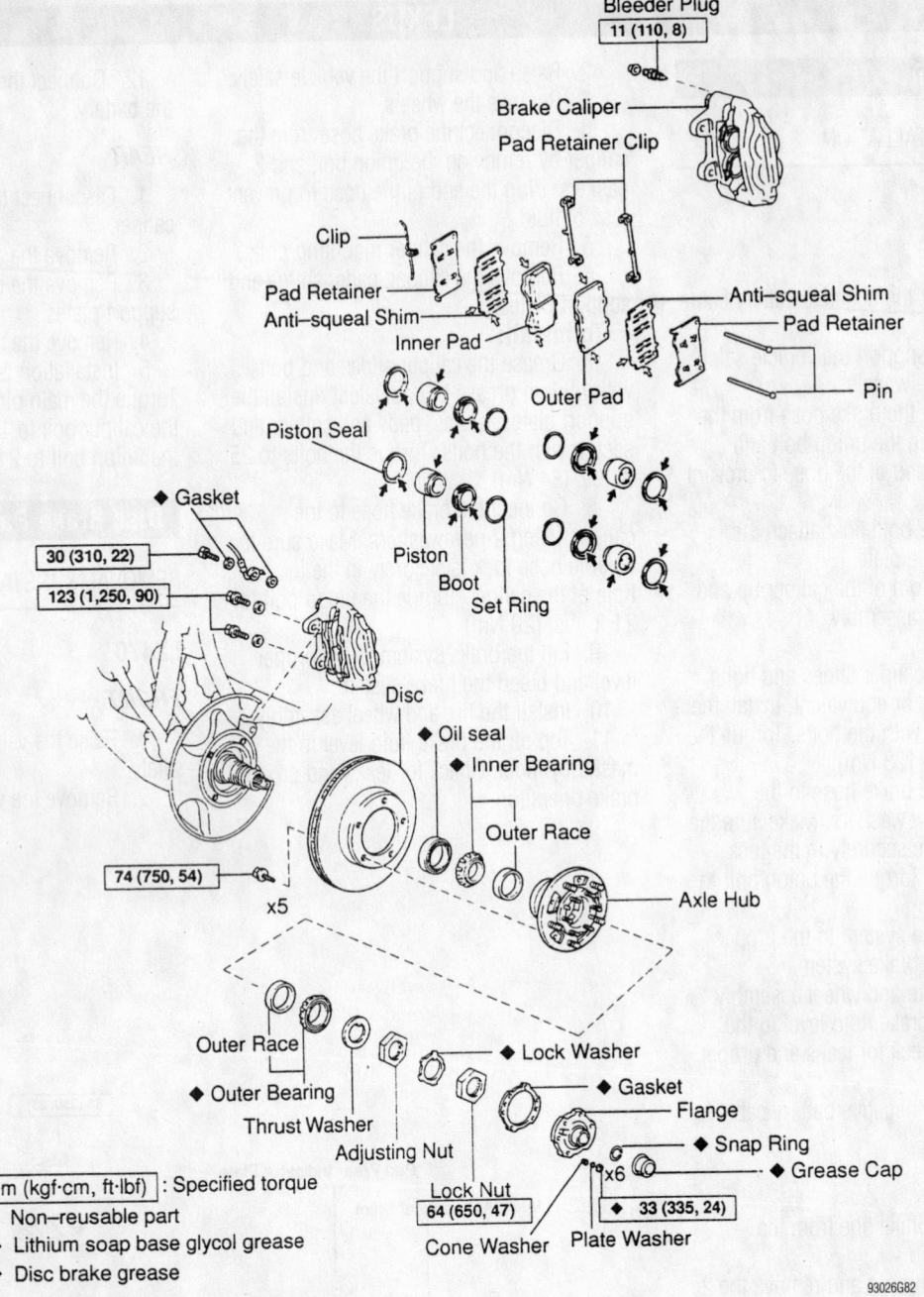

Bleeder Plug
11 (110, 8)

Brake Caliper
Pad Retainer Clip

Clip
Pad Retainer
Anti–squeal Shim
Inner Pad
Outer Pad

Anti–squeal Shim
Pad Retainer
Pin

Piston Seal

Gasket
30 (310, 22)
123 (1,250, 90)

Piston
Boot
Set Ring

Disc

Oil seal
Inner Bearing
Outer Race

74 (750, 54)
x5

Axle Hub

Outer Race
Outer Bearing
Thrust Washer
Adjusting Nut

Lock Washer
Gasket
Flange
Snap Ring
Grease Cap

N·m (kgf·cm, ft·lbf) : Specified torque
◆ Non–reusable part
➤ Lithium soap base glycol grease
⇨ Disc brake grease

Lock Nut
64 (650, 47)
Cone Washer

Plate Washer
33 (335, 24)

x6

93026G82

Exploded view of the front disc brake components—LX470

3. Remove the clip, pins and anti-rattle spring.

4. Withdraw the pads and remove the anti-squeal shims.

To install:

5. Before installing the new pads, check the disc thickness and disc runout.

6. Siphon out a small amount of brake fluid from the reservoir.

7. Press in the pistons with a hammer handle or equivalent.

8. Apply disc brake grease to both sides of the inner anti-squeal shim. Install the anti-squeal shims to the new pads.

9. Install the pads.

10. Install the anti-rattle springs and pins. Install the clip.

11. Install the wheels.

12. Check and adjust the fluid level. Apply the brake pedal several times.

13. Road-test the vehicle for proper operation.

REAR

1. Raise the vehicle and support it safely.

2. Remove the wheels.

3. Remove the brake caliper and suspend it so the hose is not stretched.

4. Remove the brake pads, anti-squeal shim, pad support plates and wear indicators.

To install:

5. Before installing the new pads, check the disc thickness and disc runout.

6. Install the pad support plates.

7. Install the pad wear indicator plate to each pads.

8. Install the anti-squeal shim to the outer pad. Install the pads so the wear indicator plate is facing upward.

9. Install the brake caliper.

10. Install the wheels.

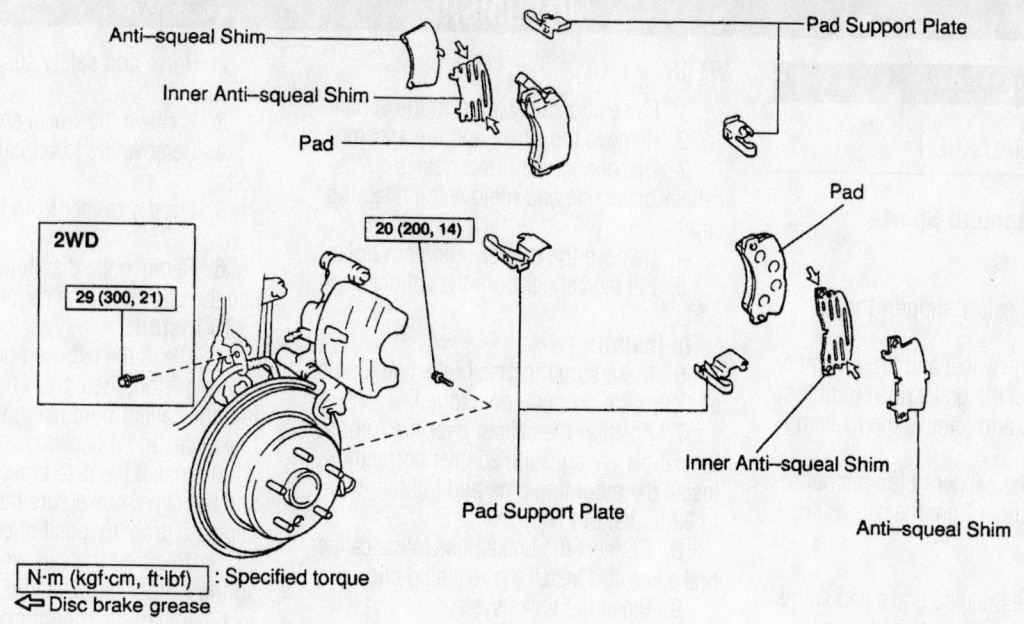

N·m (kgf·cm, ft·lbf) : Specified torque
⇐ Disc brake grease

Exploded view of the rear disc brake components—RX300

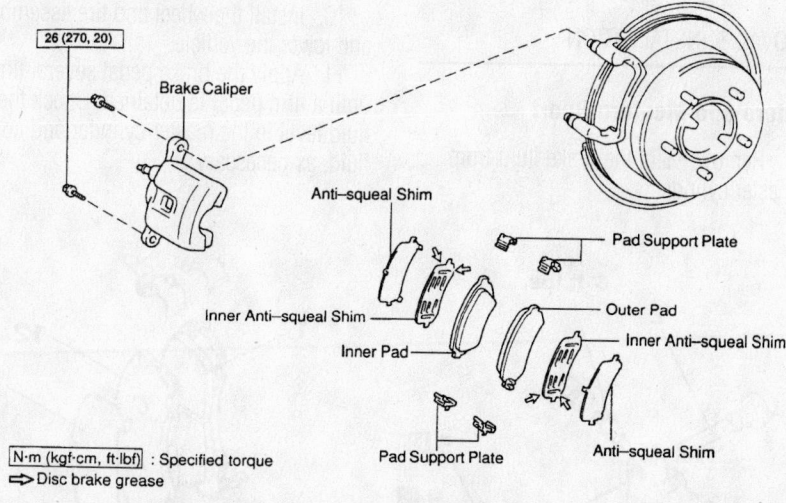

N·m (kgf·cm, ft·lbf) : Specified torque
⇒ Disc brake grease

Exploded view of the rear disc brake components—LX470

11. Apply the brake pedal several times.
12. Road-test the vehicle for proper operation.

RX300

FRONT

1. Hold the sliding pin and remove the lower bolt.
2. Lift the caliper up and secure it.
3. Remove the pads, 4 shims and wear indicator plate. Remove the 2 pad support plates.

➡The support plates can be reused, provided they have sufficient rebound, are not deformed or cracked, show no signs of wear and are cleaned of all rust and debris.

To install:

➡Always use new shims and wear indicators, even when re-installing the original pads.

4. Install a wear indicator plate on the inner pad.
5. Apply disc brake grease to both sides of the inner anti-squeal shims and install the shims.
6. Install the inner pad with the wear indicator plate facing upwards.

7. Install the outer pad.
8. Install the caliper. Torque the bolt to 25 ft. lbs.

REAR

1. On 2WD, unbolt the brake hose from the shock absorber.
2. Remove the caliper installation bolt from the torque plate.
3. Lift the caliper up and secure it.
4. Remove the pads, 4 shims and 4 support plates.

➡The support plates can be reused, provided they have sufficient rebound, are not deformed or cracked, show no signs of wear and are cleaned of all rust and debris.

To install:

➡Always use new shims and wear indicators, even when re-installing the original pads.

5. Install the support plates.
6. Apply disc brake grease to both sides of the anti-squeal shims.
7. Install the anti-squeal shims.
8. Install the inner pad with the wear indicator plate facing upwards.
9. Install the outer pad.
10. Install the caliper. Torque the bolt to 14 ft. lbs. (20 Nm).
11. On 2WD, connect the line to the shock absorber. Torque the bolt to 21 ft. lbs. (29 Nm).

For Tune-up, Capacities and Firing orders, see Section 1 of this manual

Brake Caliper

REMOVAL & INSTALLATION

Montero and Montero Sport

FRONT

1. Raise and safely support the vehicle.
2. Remove the wheel and tire assembly.
3. Disconnect the brake hose from the caliper brake line and remove the retaining clip.
4. Remove the caliper guide pin bolts.
5. Lift the caliper from the caliper support.

To install:

6. Make sure the disc brake pad shims and clips are properly positioned.
7. Position the caliper over the rotor so the caliper engages the adapter correctly. Install the mounting pins and tighten to 54 ft. lbs. (74 Nm) .
8. Connect the brake hose to the caliper brake line and install the retaining clip.
9. Bleed the brake system.
10. Install the wheel and tire assembly.

REAR

1. Raise and safely support the vehicle.
2. Remove the wheel and tire assembly.
3. Disconnect the brake hose from the caliper brake line and remove the retaining clip.
4. Remove the caliper guide pin bolts.
5. Lift the caliper from the caliper support.

To install:

6. Make sure the disc brake pad shims and clips are properly positioned.
7. Position the caliper over the rotor so the caliper engages the adapter correctly. Install the mounting pins and tighten them to 32 ft. lbs. (44 Nm).
8. Connect the brake hose to the caliper brake line and install the retaining clip.
9. Bleed the brake system.
10. Install the wheel and tire assembly.

Disc Brake Pads

REMOVAL & INSTALLATION

Montero and Montero Sport

1. Remove ½ of the brake fluid from the master cylinder.

2. Raise and safely support the vehicle.
3. Remove the wheel and tire assembly.
4. Remove the lower caliper guide pin bolt.
5. Lift the caliper from the caliper support.
6. Remove the disc brake pads, shims, and the clips from the caliper support.

To install:

7. Clean the exposed portion of the caliper piston, then press the piston back into the caliper bore using the old inner brake pad and a C-clamp.
8. Install the disc brake pads, shims, and the clips. Make sure the shims and clips are properly positioned.
9. Position the caliper over the rotor so the caliper engages the adapter correctly. Install the mounting pin(s) and tighten the front caliper to 54 ft. lbs. (74 Nm) and the rear caliper to 32 ft. lbs. (44 Nm).
10. Install the wheel and tire assembly and lower the vehicle.
11. Apply the brake pedal several times until a firm pedal is obtained. Check the fluid level in the master cylinder and add fluid, as necessary.

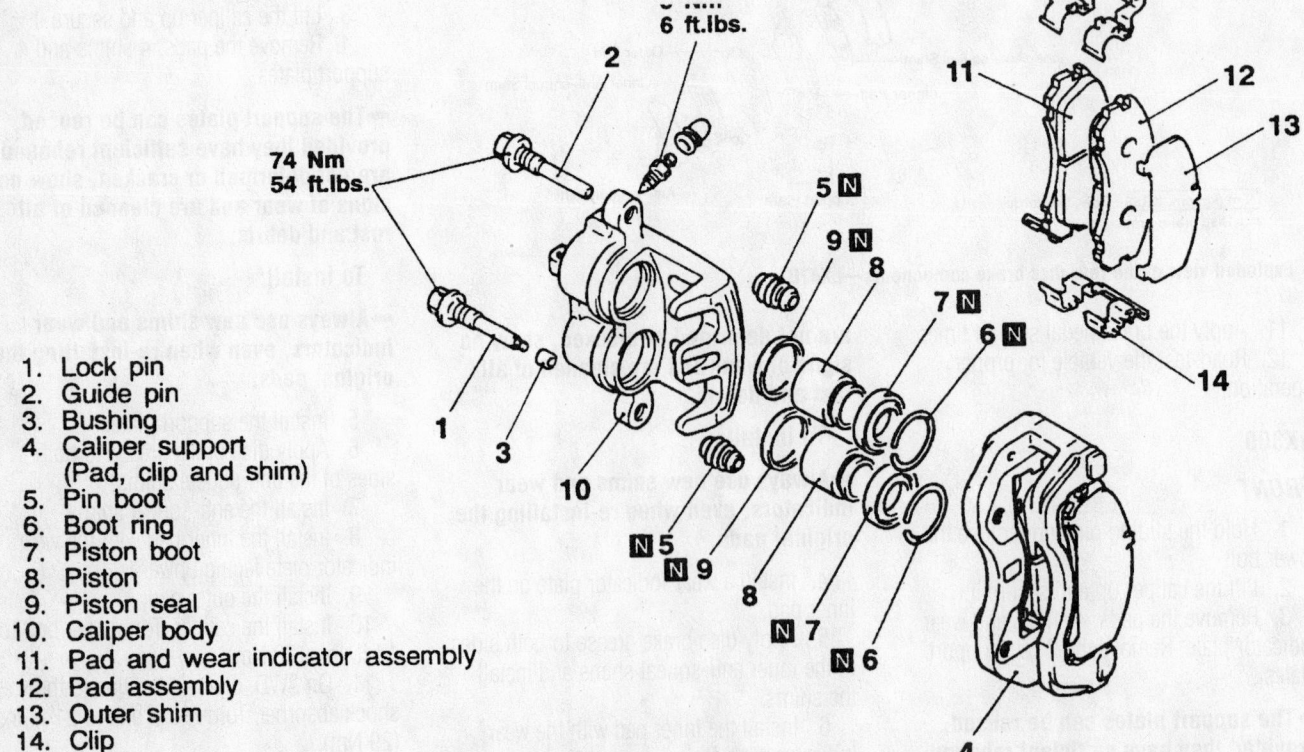

1. Lock pin
2. Guide pin
3. Bushing
4. Caliper support
   (Pad, clip and shim)
5. Pin boot
6. Boot ring
7. Piston boot
8. Piston
9. Piston seal
10. Caliper body
11. Pad and wear indicator assembly
12. Pad assembly
13. Outer shim
14. Clip

Exploded view of the front disc brake assembly—Montero and Montero Sport

93026G99

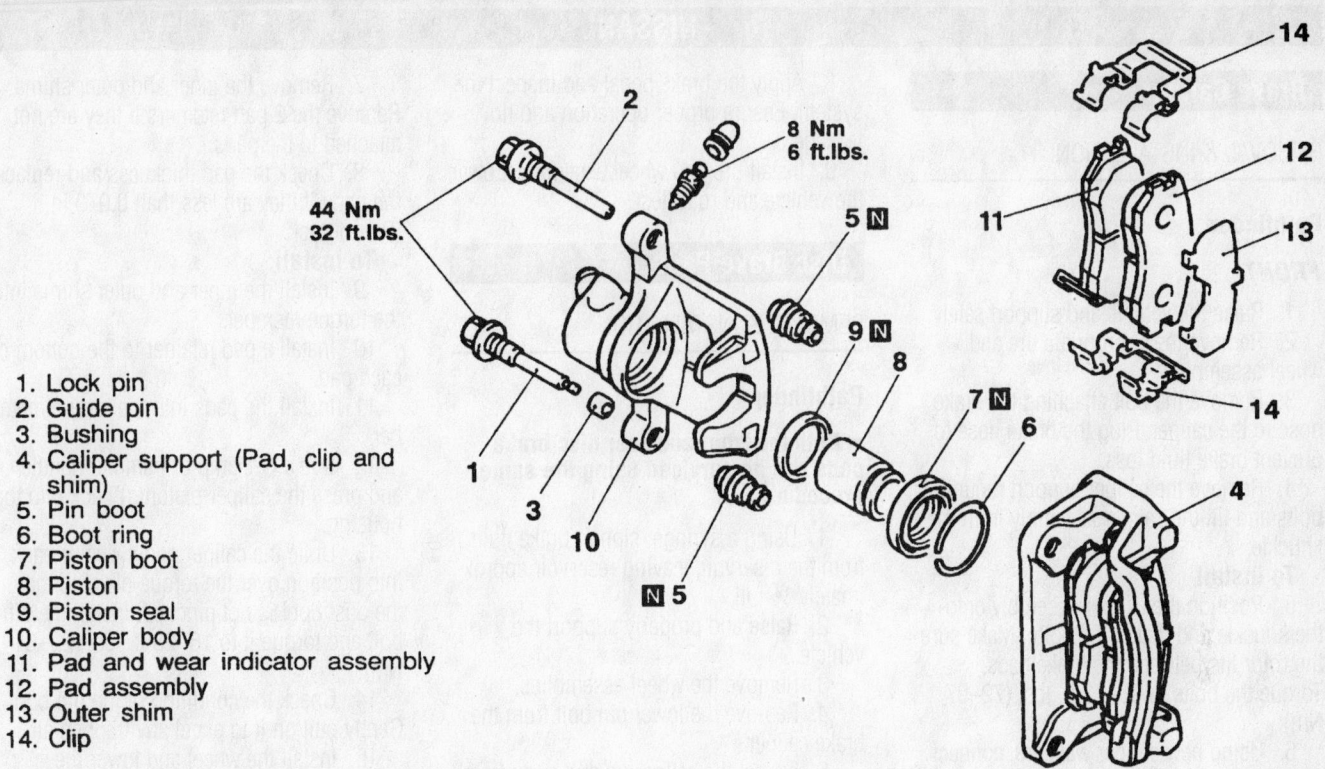

1. Lock pin
2. Guide pin
3. Bushing
4. Caliper support (Pad, clip and shim)
5. Pin boot
6. Boot ring
7. Piston boot
8. Piston
9. Piston seal
10. Caliper body
11. Pad and wear indicator assembly
12. Pad assembly
13. Outer shim
14. Clip

Exploded view of the rear disc brake assembly—Montero

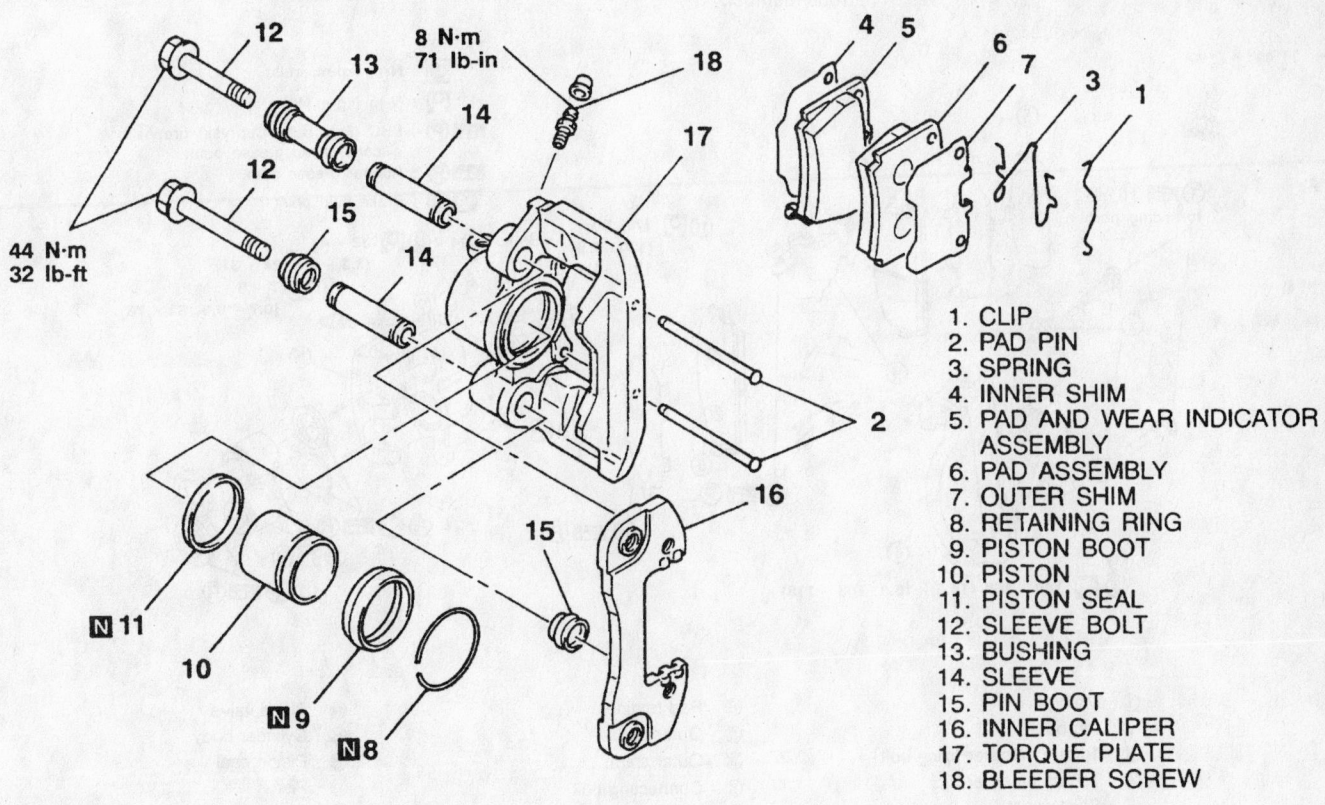

1. CLIP
2. PAD PIN
3. SPRING
4. INNER SHIM
5. PAD AND WEAR INDICATOR ASSEMBLY
6. PAD ASSEMBLY
7. OUTER SHIM
8. RETAINING RING
9. PISTON BOOT
10. PISTON
11. PISTON SEAL
12. SLEEVE BOLT
13. BUSHING
14. SLEEVE
15. PIN BOOT
16. INNER CALIPER
17. TORQUE PLATE
18. BLEEDER SCREW

Exploded view of the rear disc brake assembly—Montero Sport

For complete service labor times, order Nichols' Chilton Labor Guide

NISSAN

Brake Caliper

REMOVAL & INSTALLATION

Pathfinder

FRONT

1. Raise the vehicle and support safely.
2. Remove the appropriate tire and wheel assembly.
3. Remove the bolt attaching the brake hose to the caliper. Plug the brake hose to prevent brake fluid loss.
4. Remove the caliper support mounting bolts and lift the caliper assembly from the knuckle.

To install

5. Position the caliper assembly onto the knuckle and install the bolts. Make sure the rotor fits between the brake pads. Torque the bolts to 53–72 ft. lbs. (72–97 Nm).
6. Using new copper washers, connect the brake hose to the caliper. Torque the brake hose attaching bolt to 12–14 ft. lbs. (17–20 Nm).
7. Bleed the brake system.

8. Apply the brake pedal and inspect the system. Ensure proper operation and no leakage.
9. Install tire and wheel assembly. Lower the vehicle and road-test.

Disc Brake Pads

REMOVAL & INSTALLATION

Pathfinder

➡ **Both the front and rear disc brake pads can be serviced using the same procedure.**

1. Using a syringe, siphon brake fluid from the reservoir, leaving reservoir approximately ½ full.
2. Raise and properly support the vehicle.
3. Remove the wheel assemblies.
4. Remove the lower pin bolt from the brake caliper.
5. Swivel the caliper up and away from the torque member. Tie the caliper to a suspension member so that it is out of the way.
6. Lift the 2 brake pads out of the torque member.

7. Remove the inner and outer shims. Remove the 2 pad retainers if they are not attached to the pads.
8. Check the pad thickness and replace the pads if they are less than 0.079 in. (2mm) thick.

To install:

9. Install the inner and outer shims into the torque member.
10. Install a pad retainer to the bottom of each pad.
11. Install the pads into the torque member.
12. Use a C-clamp or hammer handle and press the caliper piston(s) back into the housing.
13. Untie the caliper and swivel it back into position over the torque plate so that the dust boot is not pinched. Install the pin bolt and torque it to 16–23 ft. lbs. (22–31 Nm).
14. Check the condition of the pin boot. Gently pull on it to expel any trapped air.
15. Install the wheel and lower the vehicle.
16. Pump the brakes until the pedal is firm and check the level of brake fluid. Road-test the vehicle.

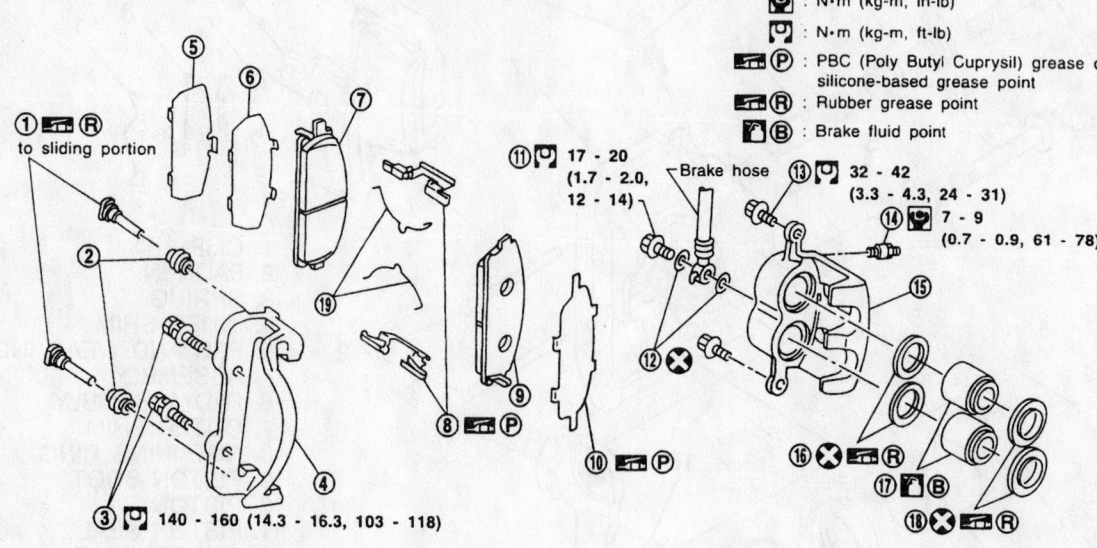

① Main pin
② Pin boot
③ Torque member fixing bolt
④ Torque member
⑤ Shim cover
⑥ Inner shim
⑦ Inner pad
⑧ Pad retainer
⑨ Outer pad
⑩ Outer shim
⑪ Connecting bolt
⑫ Copper washer
⑬ Main pin bolt
⑭ Bleed valve
⑮ Cylinder body
⑯ Piston seal
⑰ Piston
⑱ Piston boot
⑲ Pad spring

Exploded view of the dual piston caliper front brake components—Pathfinder

93026G10

Brake Drums

REMOVAL & INSTALLATION

Pathfinder

1. Remove the hub cap and loosen the lug nuts.

2. Raise the rear of the vehicle and support it on jackstands.

3. Remove the lug nuts, tire and wheel.

4. Release the parking brake.

5. Pull the brake drum from the hub. If difficult to remove try the following:

    a. Strike the face of the drum with a plastic or rubber mallet. This will break free any rust that may develop between the drum and the hub.

    b. Install 2, M8x1.25mm bolts into the holes in the drum and gradually tighten them to pull the drum off the hub.

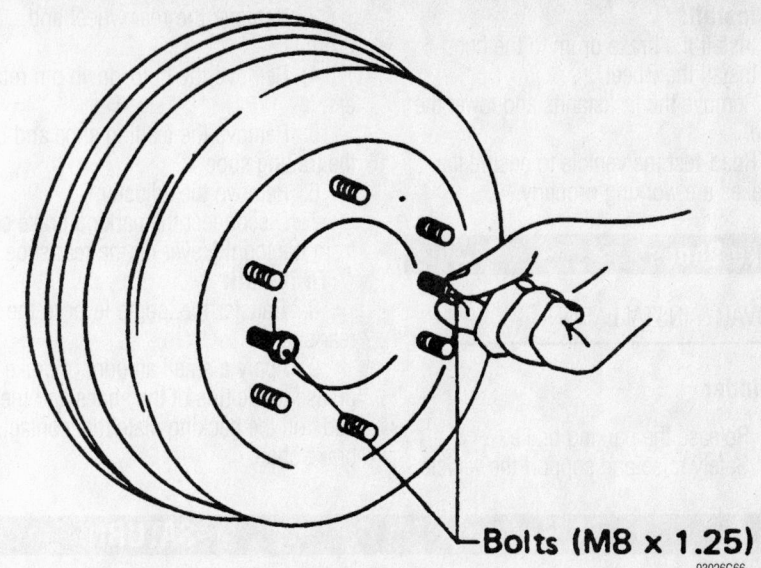

Bolts (M8 x 1.25)

93026G66

Install and tighten 2 bolts to remove a stubborn brake drum—Pathfinder

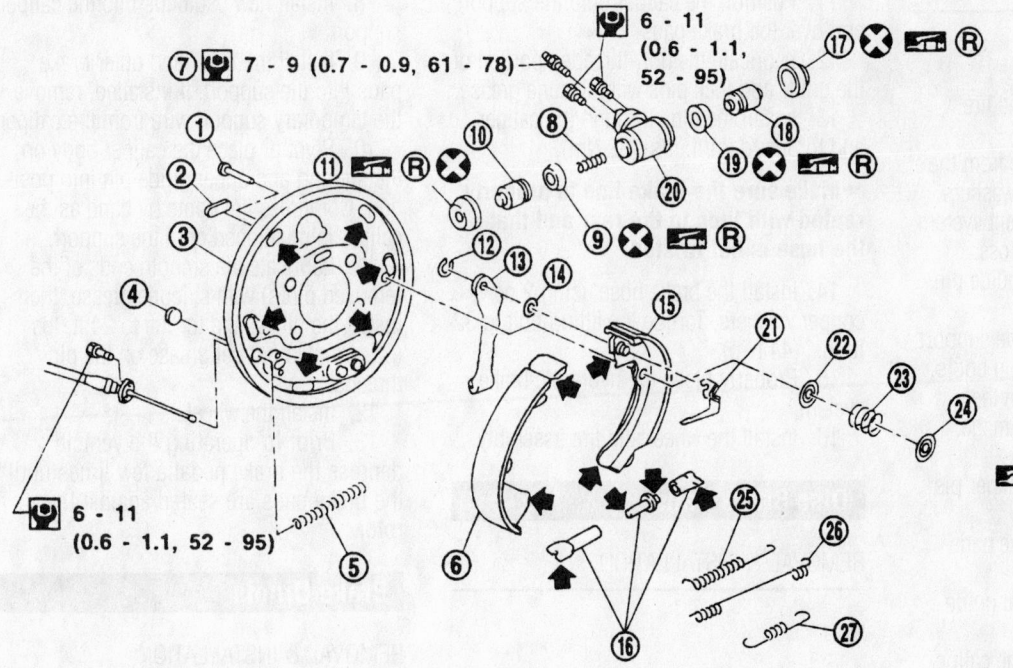

: Rubber grease point

: Brake grease point

: N•m (kg-m, ft-lb)

: N•m (kg-m, in-lb)

93026G69

1. Shoe hold pin	10. Piston	19. Piston cup
2. Plug	11. Boot	20. Wheel cylinder
3. Back plate	12. Retainer ring	21. Adjuster lever
4. Check plug	13. Toggle lever	22. Spring seat
5. Spring	14. Wave washer	23. Shoe hold spring
6. Shoe (leading side)	15. Shoe (trailing side)	24. Retainer
7. Air bleeder	16. Adjuster	25. Adjuster spring
8. Spring	17. Boot	26. Return spring (upper)
9. Piston cup	18. Piston	27. Return spring (lower)

Drum brake assembly exploded view—Pathfinder w/LT30C system

To install:

6. Install the brake drum to the hub.

7. Install the wheel.

8. Remove the jackstands and lower the vehicle.

9. Road-test the vehicle to ensure that the brakes are working properly.

Brake Shoes

REMOVAL & INSTALLATION

Pathfinder

1. Release the parking brake.
2. Safely raise and support the vehicle.

3. Remove the rear wheel and drum.

4. Remove the hold-down pin retainers.

5. Remove the leading shoe and then the trailing shoe.

6. Remove the adjuster.

7. Disconnect the parking brake cable from the toggle lever on the rear shoe.

To install:

8. Transfer the toggle lever to the new rear shoe.

9. Apply a small amount of brake grease to the tips of the shoes and the 6 pads on the backing plate that contact the brake shoe.

10. Shorten the adjuster by turning it.

11. Connect the parking brake cable to the toggle lever on the rear shoe.

12. Install the lower return spring to both shoes and install the shoes on the backing plate with the hold down pins and retainers.

13. Install the adjuster and the remaining springs. Pay attention to the direction of the adjuster assembly.

14. Inspect the complete assembly and install the brake drum.

15. Adjust the shoe to drum clearance.

16. Install the wheel assembly and lower the vehicle to the floor.

SATURN

Brake Caliper

REMOVAL & INSTALLATION

Vue

1. Remove the front wheel and tire assembly.

2. Disconnect the brake hose from the caliper and discard the 2 copper washers.

3. Plug the openings to prevent system contamination or excessive fluid loss.

4. Remove the lock pin and guide pin from the caliper.

5. Remove the caliper from the support, being careful not to damage the pin boots.

6. Remove the pin boots from the caliper support and inspect for damage.

To install:

7. If necessary, bottom the caliper piston by using a C-clamp.

8. If removed, install the brake pads and clips to the caliper support.

9. Lubricate the pin boots and guide pins with silicone grease.

10. Install the pin boots into the caliper

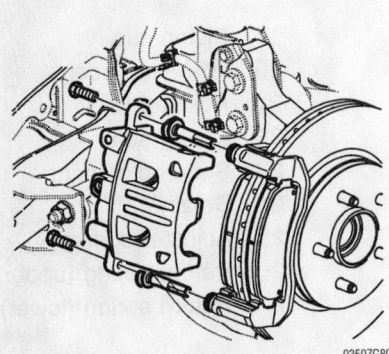

Exploded view of the caliper mounting— Saturn Vue

support, using the pin to assure that the boot passes all the way through the support.

11. Position the caliper onto the support and over the brake pads.

12. Lubricate the non-threaded portion of the guide and lock pins with silicone grease.

13. Install the pins through the caliper and torque to 24 ft. lbs. (32 Nm).

➡**Make sure the brake line is properly routed with loop to the rear and that the hose is not twisted.**

14. Install the brake hose using 2 new copper washers. Torque the fitting bolt to 32 ft. lbs. (44 Nm).

15. Properly bleed the hydraulic brake system.

16. Install the wheel and tire assembly.

Disc Brake Pads

REMOVAL & INSTALLATION

Vue

1. Remove the front wheels.

2. Remove the caliper lower lock pin.

3. Either pivot the caliper up on the guide pin or remove the upper guide pin and support the caliper from the strut using a coat hanger or length of wire.

4. Remove the 2 brake pads and the pad clips from the caliper support. Discard the old pad clips.

5. Check the caliper pins, pin boots and the piston boot for deterioration or damage.

To install:

6. Using a C-clamp, bottom the piston all the way into the caliper bore.

7. Carefully lift the inner edge of the piston boot by hand to release any trapped air.

8. Install new pad clips into the caliper support.

9. Install the inner and outer brake pads into the support. If installed, remove the temporary support wire from the caliper.

10. Pivot or place the caliper body on the support and upper guide pin into position. Compress the boots by hand as the caliper is positioned onto the support.

11. Lubricate the smooth ends of the removed pin(s) with silicone grease, then install the pin(s) and torque to 24 ft. lbs. (32 Nm). Do not get grease on the pin threads.

12. Install the wheels.

13. Prior to operating the vehicle, depress the brake pedal a few times until the brake pads are seated against the rotor.

Brake Drums

REMOVAL & INSTALLATION

Vue

1. Remove the rear wheel and tire assembly.

2. Remove the brake drum.

3. If necessary, turn the starwheel of the brake adjuster assembly to loosen the brake shoes and allow for drum removal.

4. Remove the drum.

To install:

5. Install brake drum over brake shoes and onto hub.

6. Install tire and wheel assembly. Torque to the proper specification.

7. Adjust the brakes.

8. Road test for braking operation.

Brake Shoes

REMOVAL & INSTALLATION

Vue

1. Remove the wheels and brake drums.

2. Pull the park brake sure towards the front of the vehicle and release the cable from the lever.

3. Remove the upper adjuster spring and lower return spring.

4. Remove the trailing (rear) shoe hold-down spring and pin.

5. Remove the trailing (rear) shoe and adjuster assembly.

6. Remove the leading (front) shoe hold-down spring and pin.

7. Remove the parking brake lever retainer, washer and the lever from the shoe.

8. Disassemble the brake adjuster socket, screw and nut, then clean the components in denatured alcohol. Inspect the assembly, making sure the screw threads smoothly into the adjusting nut over the full threaded length.

9. Inspect the wheel cylinder for signs of leakage and for cut or damaged boots. Do not attempt to repair a damaged cylinder, the assembly must be replaced.

To install:

10. Lubricate the adjuster assembly, the 6 backing plate raised shoe contact pads, the brake lever pin and surfaces which contact brake shoe webs with brake lubricant.

11. Install the leading (front) shoe, hold-down pin and spring onto the backing plate.

12. Attach the parking brake lever onto the trailing (rear) shoe and secure it by installing then crimping the washer.

13. Install the trailing (rear) shoe, hold-down pin and spring onto the backing plate.

14. Install the adjuster assembly and lever.

15. Make sure that both shoes are properly seated to the anchor and wheel cylinder accordingly.

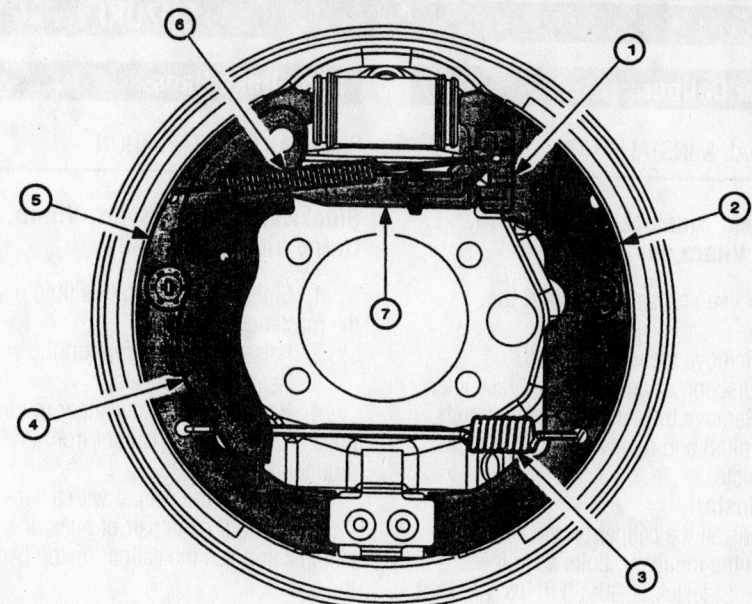

① Adjuster lever

② Leading (front) brake shoe

③ Lower return spring

④ Park brake lever

⑤ Trailing (rear) brake shoe

⑥ Upper return spring

⑦ Adjuster assembly

9359ZG81

View of the installed drum brake assembly components—Saturn Vue

16. Make sure the adjuster notches properly engage the shoe notches.

17. Install the lower return spring and then the upper adjuster spring.

18. Make sure the adjuster lever is engaged with the teeth on the adjuster assembly.

19. Pull the park brake sure towards the front of the vehicle and attach the cable to the lever.

holes of the shoes.

20. Verify the correct location of all brake components, if necessary, use the other side brake assembly for comparison.

21. Using a suitable drum clearance gauge, measure the inner diameter of the brake drum and adjust the outside diameter of the brake shoes to 0.02 inch (0.50mm) less than the inner diameter of the drum.

22. Repeat the procedure for the opposite brake shoes and install the brake drums.

23. If the wheel cylinders have been replaced, bleed the hydraulic brake system.

24. Install the rear wheels.

25. Apply and release the brake pedal 20 times to allow the adjuster to properly position the brake shoes.

26. Check and adjust the parking brake cable, as necessary.

Timing belt service is covered in Section 3 of this manual

SUZUKI

Brake Caliper

REMOVAL & INSTALLATION

Sidekick, Sidekick Sport, Vitara, Grand Vitara and X-90

1. Raise and safely support the vehicle.
2. Remove the wheels.
3. Disconnect and plug the brake line.
4. Remove the caliper mounting bolts (guide pins) and remove the caliper from the vehicle.

To install:

5. Install the caliper on the vehicle. Tighten the mounting bolts as follows:
   - Sidekick, X-90: 20 ft. lbs. (27 Nm).
   - Sidekick Sport: bottom bolt to 37 ft. lbs.; top bolt to 42 ft. lbs. (58 Nm)
6. Connect the hydraulic brake line, using 2 new washers. Torque the union bolt to 17 ft. lbs. (23 Nm).
7. Replace the front wheels.
8. Lower the vehicle.
9. Fill the brake reservoir and bleed the hydraulic brake system.

Disc Brake Pads

REMOVAL & INSTALLATION

Sidekick, Sidekick Sport, Vitara, Grand Vitara and X-90

1. Siphon about ⅔ of the fluid out of the master cylinder.
2. Raise and safely support the vehicle.
3. Remove the wheels.
4. Remove the brake caliper mounting bolts and remove the caliper from the mounting bracket.
5. Support the caliper with a wire.
6. Using a large pair of plies or a C-clamp compress the caliper piston back into the bore.
7. Remove the disc brake pads and any shims from the caliper mounting bracket.

To install:

8. Install the brake pads and any shims removed from the caliper mounting bracket.
9. Install the caliper on the mounting bracket and install the mounting bolts. Tighten the mounting bolts to 20 ft. lbs. (27 Nm).
10. Install the front wheels and lower the vehicle.

✳✳ CAUTION

Do not attempt to drive the vehicle until after the following step is performed.

11. Depress the brake pedal repeatedly until a firm pedal is obtained. Do not attempt to drive the vehicle unless a firm pedal is obtained.
12. Check the fluid level in the master cylinder. Add fresh brake fluid, as necessary.
13. Road-test the vehicle.

Brake Drums

REMOVAL & INSTALLATION

Sidekick

1. Raise and safely support the vehicle.
2. Remove the rear wheel(s).
3. Release the parking brake.
4. Remove the parking brake lever cover screws and loosen the brake cable locking nut.
5. Install 2, 8mm bolts into the brake

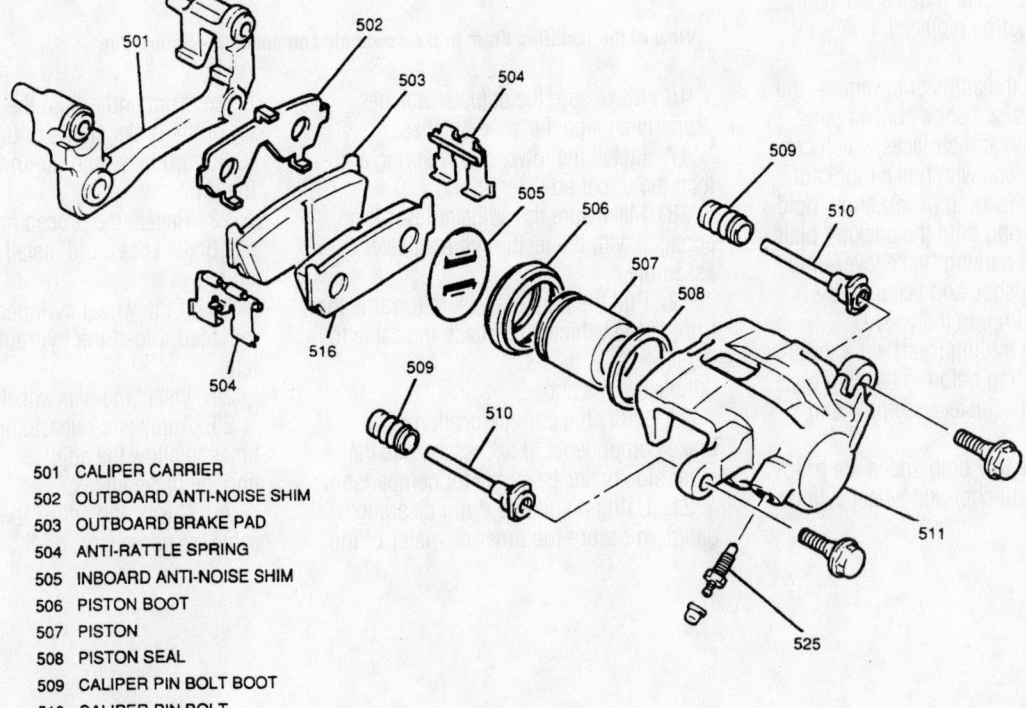

501	CALIPER CARRIER
502	OUTBOARD ANTI-NOISE SHIM
503	OUTBOARD BRAKE PAD
504	ANTI-RATTLE SPRING
505	INBOARD ANTI-NOISE SHIM
506	PISTON BOOT
507	PISTON
508	PISTON SEAL
509	CALIPER PIN BOLT BOOT
510	CALIPER PIN BOLT
511	CALIPER
516	INBOARD BRAKE PAD
525	BLEEDER VALVE

93026G37

Front disc brake components—Sidekick, Sidekick Sport and X-90 models

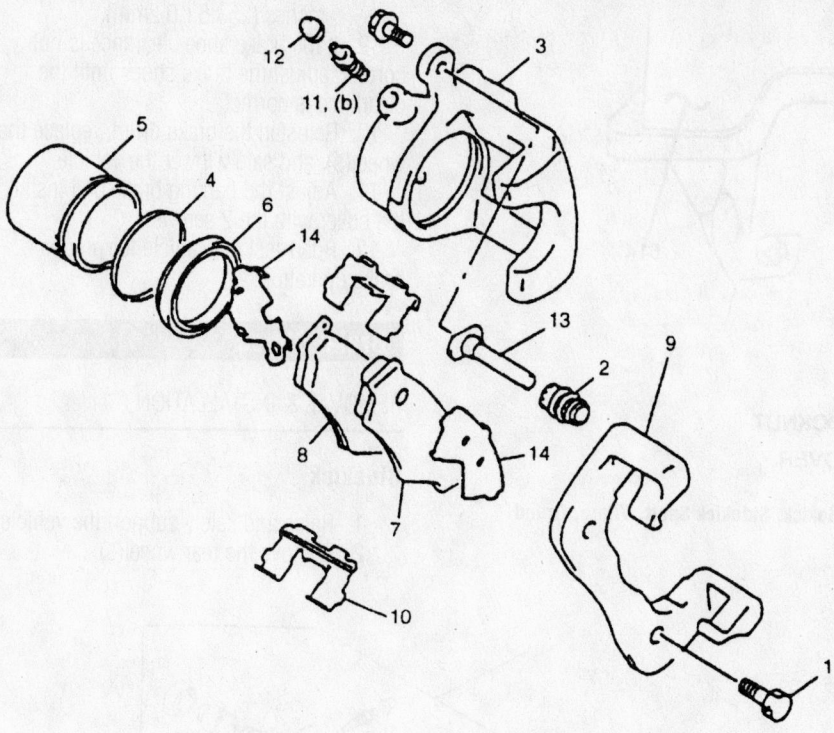

Front disc brake components—Vitara

1. Caliper (slide) pin bolt
2. Boot
3. Disc brake caliper (disc brake cylinder)
4. Piston seal
5. Disc brake piston
6. Cylinder boot
7. Disc brake inner pad
8. Disc brake outer pad
9. Brake caliper carrier
10. Pad spring
11. Bleeder plug
12. Bleeder plug cap
13. Caliper pin
14. Anti noise shim
15. Inner shim

Tightening torque
(a): 8.0 N·m (0.80 kg-m, 6.0 lb-ft)
(b): 8.5 N·m (0.85 kg-m, 6.5 lb-ft)

93026G41

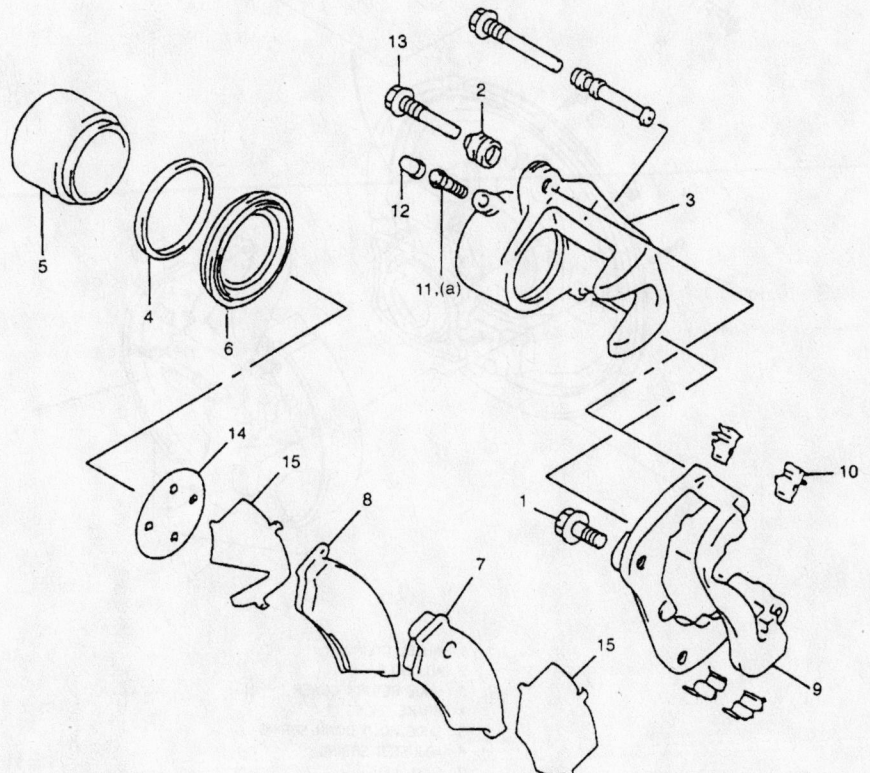

Front disc brake components—Grand Vitara

1. Caliper (slide) pin bolt
2. Boot
3. Disc brake caliper (disc brake cylinder)
4. Piston seal
5. Disc brake piston
6. Cylinder boot
7. Disc brake inner pad
8. Disc brake outer pad
9. Brake caliper carrier
10. Pad spring
11. Bleeder plug
12. Bleeder plug cap
13. Caliper pin
14. Anti noise shim
15. Inner shim

Tightening torque
(a): 8.0 N·m (0.80 kg-m, 6.0 lb-ft)

93026G42

Heater Core replacement is covered in Section 2 of this manual

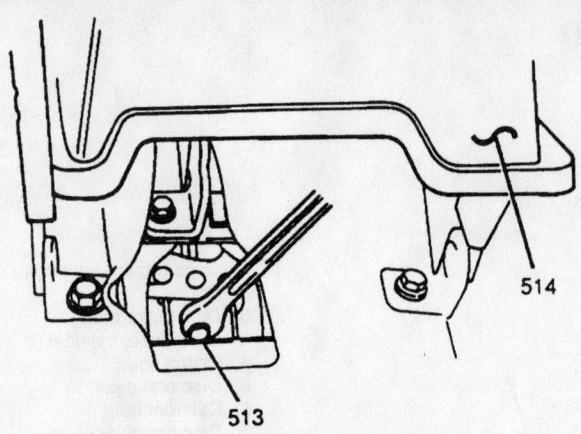

513 PARKING BRAKE CABLE LOCKNUT
514 PARKING BRAKE LEVER COVER

93026G38

Reducing the adjuster to remove the brake drum— Sidekick, Sidekick Sport, Vitara, Grand Vitara and X-90

drum holes and uniformly tighten each bolt. Tighten each bolt until the brake drum is removed from the vehicle. If there is difficulty in removing the drum, insert a small tool through the hole in the rear of the backing plate, and hold the automatic adjusting lever away from the adjuster. Using another narrow, flat tool at the same time, reduce the brake shoe adjuster by turning the adjusting wheel.

To install:

6. Install the brake drum and pull the parking brake lever all the way up until a clicking sound can no longer be heard.

7. Verify that the rear wheels will not turn. If the rear wheels turn, adjust the parking brake cable as necessary.

8. Release the parking brake and remove the brake drum. Measure the diameter of the brake shoes. Outer diameter should be as follows:

- For 2 door models: 8.638 (0.0012 inches (219 (0.3mm)

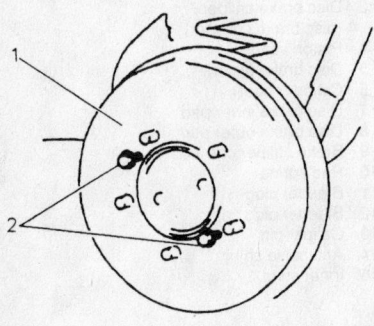

1 DRUM
2 TWO 8mm BOLTS

93026G39

Removing the brake drum with the 2, 8mm bolts— Sidekick, Sidekick Sport, Vitara, Grand Vitara and SUZ-90

- For 4 door models: 9.980 (0.0079 inches (253.5 (0.2mm)

9. If the brake shoe clearance is not correct, adjust the brake shoes until the clearance is correct.

10. Reinstall the brake drum, replace the wheel(s), and safely lower the vehicle.

11. Adjust the parking brake and install the cover with the 2 screws.

12. Road-test the vehicle for proper brake operation.

Brake Shoes

REMOVAL & INSTALLATION

Sidekick

1. Raise and safely support the vehicle.
2. Remove the rear wheel(s).

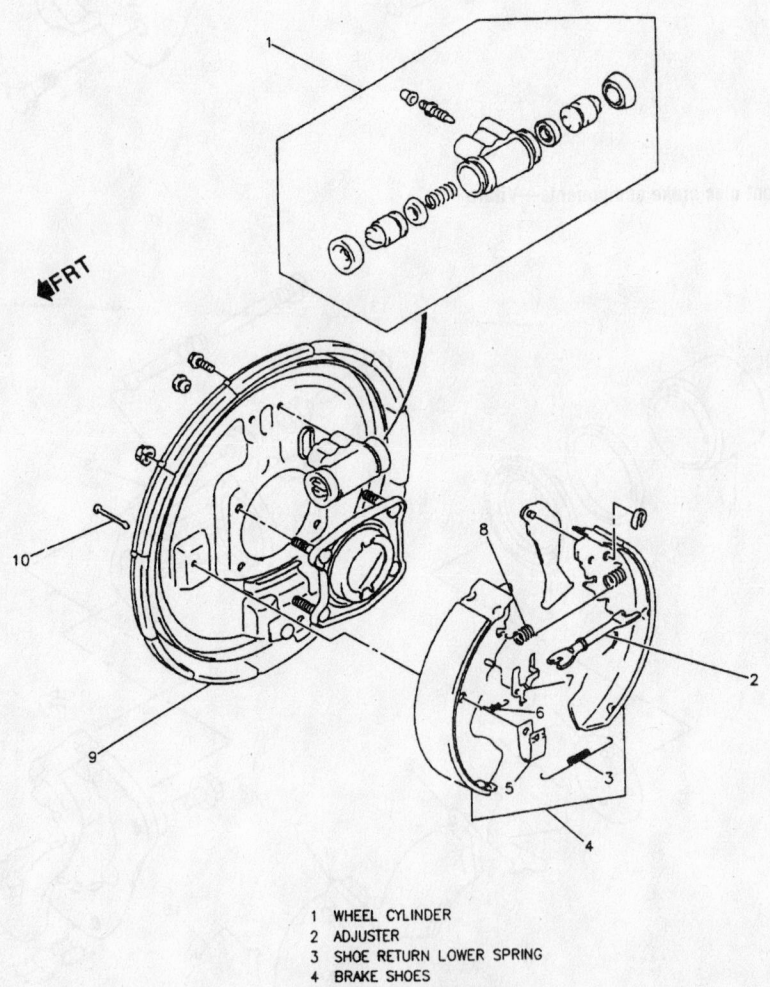

1   WHEEL CYLINDER
2   ADJUSTER
3   SHOE RETURN LOWER SPRING
4   BRAKE SHOES
5   SHOE HOLD DOWN SPRING
6   ADJUSTER SPRING
7   PAWL LEVER
8   SHOE RETURN UPPER SPRING
9   BACKING PLATE
10 SHOE HOLD DOWN PIN

93026G40

Exploded view of the rear brake components—Sidekick, Sidekick Sport and X-90 models

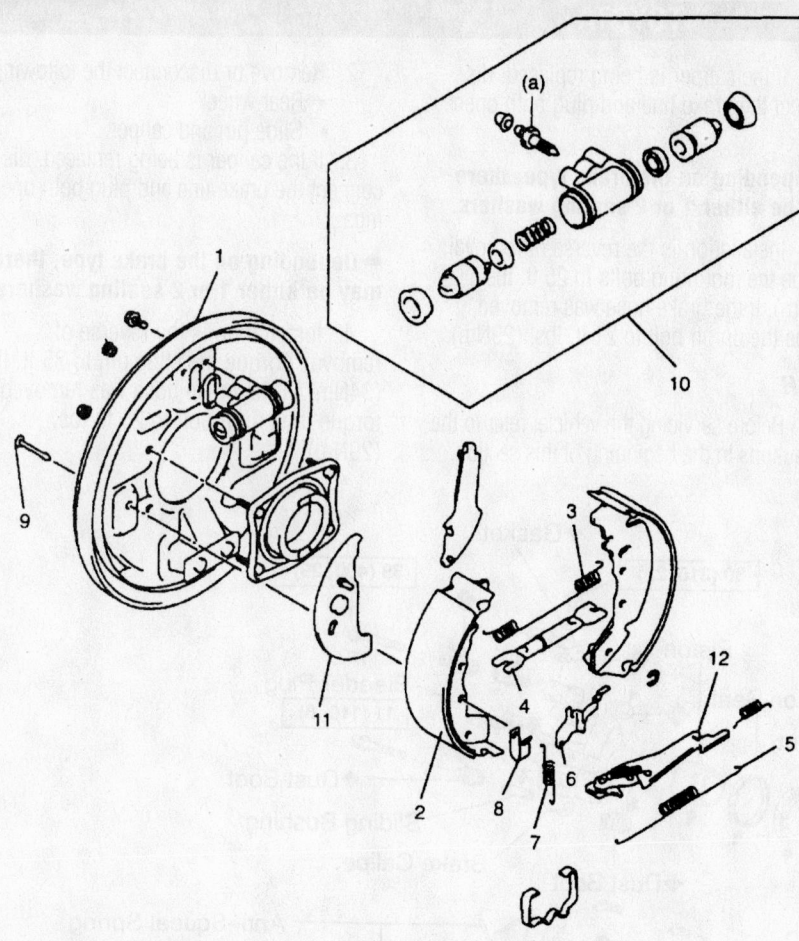

1. Brake back plate
2. Brake shoe
3. Shoe return upper spring
4. Adjuster
5. Shoe return lower spring
6. Adjuster lever
7. Adjuster spring
8. Shoe hold down spring
9. Shoe hold down pin
10. Wheel cylinder
11. Link
12. Brake strut

Tightening torque
(a): 7.5 N·m (0.75 kg-m, 5.5 lb-ft)

93026G43

Exploded view of the rear brake components—Vitara and Grand Vitara

3. Remove the brake drum.

4. Using a suitable tool, remove the brake shoe return spring.

5. Using a brake spring hold-down tool, disengage the hold-down spring and retainers from the front shoe. Remove the hold-down retainer pinch

6. Disconnect the anchor spring from the front shoe and remove the front shoe.

7. Remove the anchor spring from the rear shoe. Using a brake spring hold-down tool, disengage the hold-down spring and retainers from the rear shoe. Remove the hold-down pinch

8. Disengage the parking brake lever from the parking brake cable and remove the rear shoe.

9. Remove the C-washer, the automatic adjuster lever and spring, the C-washer, and the parking brake lever from the rear shoe.

10. Thoroughly clean the backing plate and brake hardware with brake cleaning solvent. Apply high temperature grease to the backing plate shoe contact points, anchor plate and shoe contact points, adjusting bolt, and adjuster and brake shoe contact points.

To install:

11. Reinstall the automatic adjuster lever and the parking brake lever to the rear shoe using new C-washers.

12. Connect the parking brake lever to the parking brake cable. Set the adjuster and spring to the rear shoe.

13. Set the rear brake shoe in place, install the hold-down pin and install the hold-down spring and retainers. Make sure that the shoe is inserted in the wheel cylinder and that the other end is in the anchor plate.

14. Install the anchor spring to the rear shoe.

15. Install the front shoe to the other end of the anchor spring and set the front shoe in place. Make sure that the front shoe engages the wheel cylinder, adjuster mechanism and spring, and the anchor plate.

16. Reinstall the front brake shoe hold-

down pin and secure with the hold-down spring and retainers using a suitable tool.

17. Install the return spring.

18. Install the brake drum and pull the parking brake lever all the way up until a clicking sound can no longer be heard.

19. Verify that the rear wheels will not turn. If the rear wheels turn, adjust the parking brake cable as necessary.

20. Release the parking brake and remove the brake drum. Measure the diameter of the brake shoes. Brake diameter should be as follows:

- For 2 door models: 8.638 (0.0012 inches (219 (0.3mm)
- For 4 door models: 9.980 (0.0079 inches (253.5 (0.2mm)

21. If the brake shoe clearance is not correct, adjust the brake shoes until the clearance is correct.

22. Reinstall the brake drum, replace the wheel(s), and safely lower the vehicle.

23. Road-test the vehicle for proper brake operation.

Brake service is covered in Section 4 of this manual

Brake Caliper

REMOVAL & INSTALLATION

Highlander

FRONT

1. Before servicing the vehicle, refer to the precautions in the beginning of this section.
2. Remove or disconnect the following:
   • Front wheel
   • 2 mounting bolts and caliper

3. If the caliper is being replaced, disconnect the brake line and plug both openings.

➡ **Depending on the brake type, there may be either 1 or 2 sealing washers.**

4. Installation is the reverse of removal. Torque the mounting bolts to 25 ft. lbs. (34Nm). If the brake hose was removed, torque the union bolt to 21 ft. lbs. (29Nm).

REAR

1. Before servicing the vehicle, refer to the precautions in the beginning of this section.

2. Remove or disconnect the following:
   • Rear wheel
   • Slide pin and caliper

3. If the caliper is being replaced, disconnect the brake line and plug both openings.

➡ **Depending on the brake type, there may be either 1 or 2 sealing washers.**

4. Installation is the reverse of removal. Torque the slide pin to 25 ft. lbs. (34Nm). If the brake hose was removed, torque the union bolt to 21 ft. lbs. (29Nm).

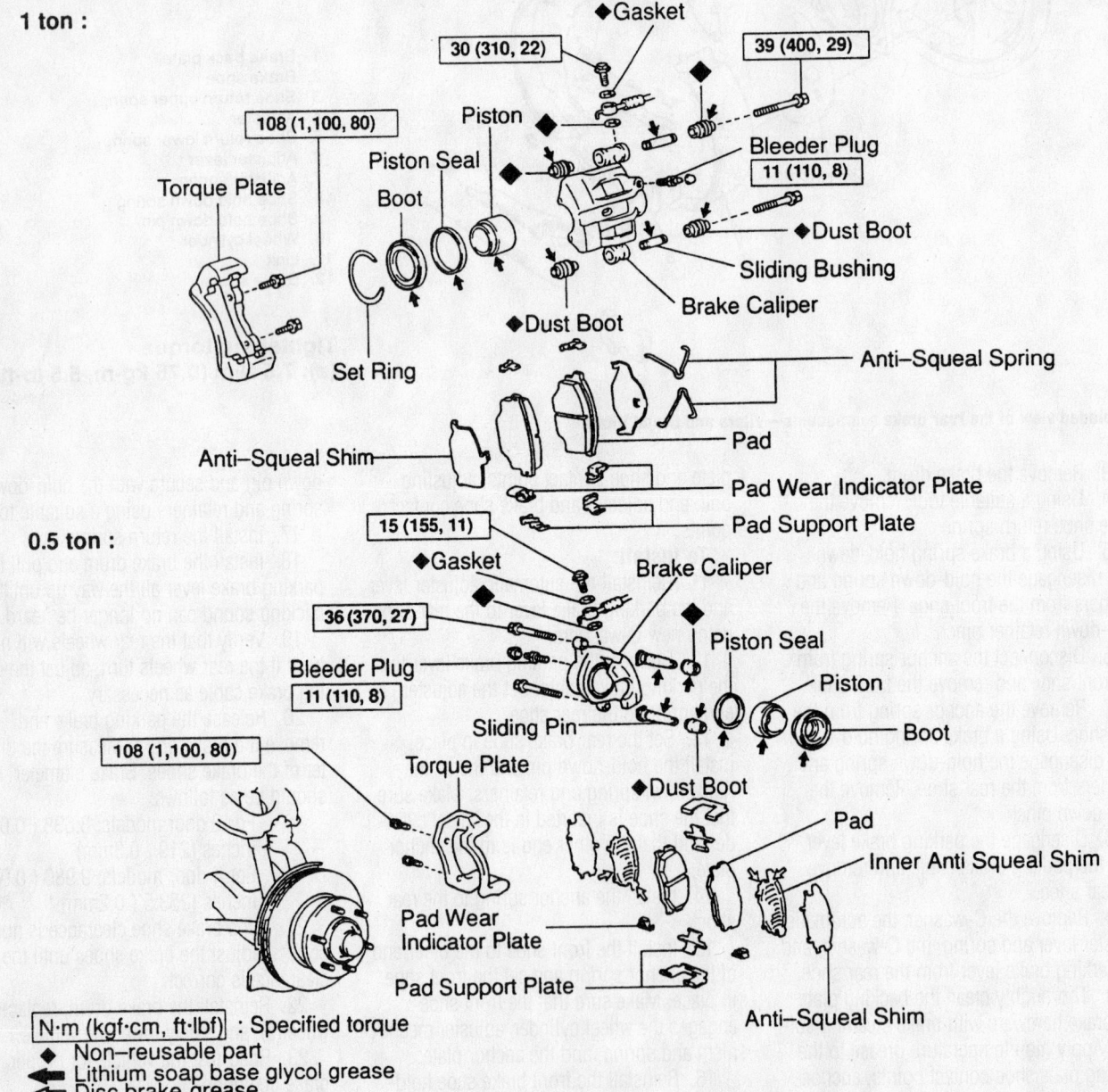

1 ton :

◆Gasket
30 (310, 22)
39 (400, 29)
108 (1,100, 80)
Piston
Bleeder Plug
11 (110, 8)
Torque Plate
Piston Seal
Boot
◆Dust Boot
Sliding Bushing
Brake Caliper
Set Ring
◆Dust Boot
Anti–Squeal Spring
Anti–Squeal Shim
Pad
Pad Wear Indicator Plate
Pad Support Plate

0.5 ton :

15 (155, 11)
◆Gasket
Brake Caliper
36 (370, 27)
Piston Seal
Bleeder Plug
11 (110, 8)
Piston
Sliding Pin
Boot
108 (1,100, 80)
Torque Plate
◆Dust Boot
Pad
Inner Anti Squeal Shim
Pad Wear Indicator Plate
Pad Support Plate
Anti–Squeal Shim

N·m (kgf·cm, ft·lbf) : Specified torque
◆ Non–reusable part
◄ Lithium soap base glycol grease
◄ Disc brake grease

93026G70

Single piston type caliper assembly—4Runner

4Runner

1. Disconnect the negative battery cable from the battery.
2. Raise and support the vehicle safely.
3. Remove the wheels.
4. Disconnect the brake hose from the caliper by removing the union bolt and 2 gaskets. Plug the end of the hose to prevent loss of fluid.
5. Remove the bolts that attach the caliper to the torque plate.
6. Lift the bottom of the caliper up and remove the caliper assembly.

To install:

7. Grease the caliper slides and bolts with lithium grease or equivalent. Install the caliper and secure with the bolts. Torque the bolts to 90 ft. lbs. (123 Nm).
8. Connect the brake hose to the caliper, using 2 new washers. Make sure the flexible hose lock is securely in the lock hole of the caliper. Torque the union bolt to 11 ft. lbs. (15 Nm).
9. Fill the brake system to the proper level and bleed the brake system.
10. Install the tire and wheel assembly.

11. Top off the brake fluid level in the master cylinder. Check for leaks and proper brake operation.
12. Connect the negative battery cable to the battery.

RAV4

1. Raise and safely support the vehicle.
2. Remove the wheel(s).
3. Remove the union bolt and 2 gaskets and remove the flexible brake hose from the caliper. Use a suitable container to catch the brake fluid as it drains out.
4. Hold the sliding pin and loosen the 2 caliper mounting bolts. Remove the bolts and remove the caliper from the torque plate.
5. Remove the brake pads and brake hardware.

To install:

6. Install the brake pads and brake hardware.
7. Install the caliper to the torque plate with the 2 mounting bolts. Torque the bolts to 25 ft. lbs. (34 Nm).
8. Reconnect the flexible brake hose to

the caliper with the 2 gaskets and the union bolt. Torque the union bolt to 22 ft. lbs. (30 Nm).
9. Refill the master cylinder with brake fluid and bleed the brake system.
10. Check for proper operation and make sure there are no leaks.

Land Cruiser

1. Disconnect the negative battery cable from the battery.
2. Raise and support the vehicle safely.
3. Remove the wheels.
4. Disconnect the brake hose from the caliper by removing the union bolt and 2 gaskets. Plug the end of the hose to prevent loss of fluid.
5. Remove the bolts that attach the caliper to the torque plate.
6. Lift the bottom of the caliper up and remove the caliper assembly.

To install:

7. Grease the caliper slides and bolts with lithium grease or equivalent. Install the caliper and secure with the bolts. Torque the bolts to 90 ft. lbs. (123 Nm).

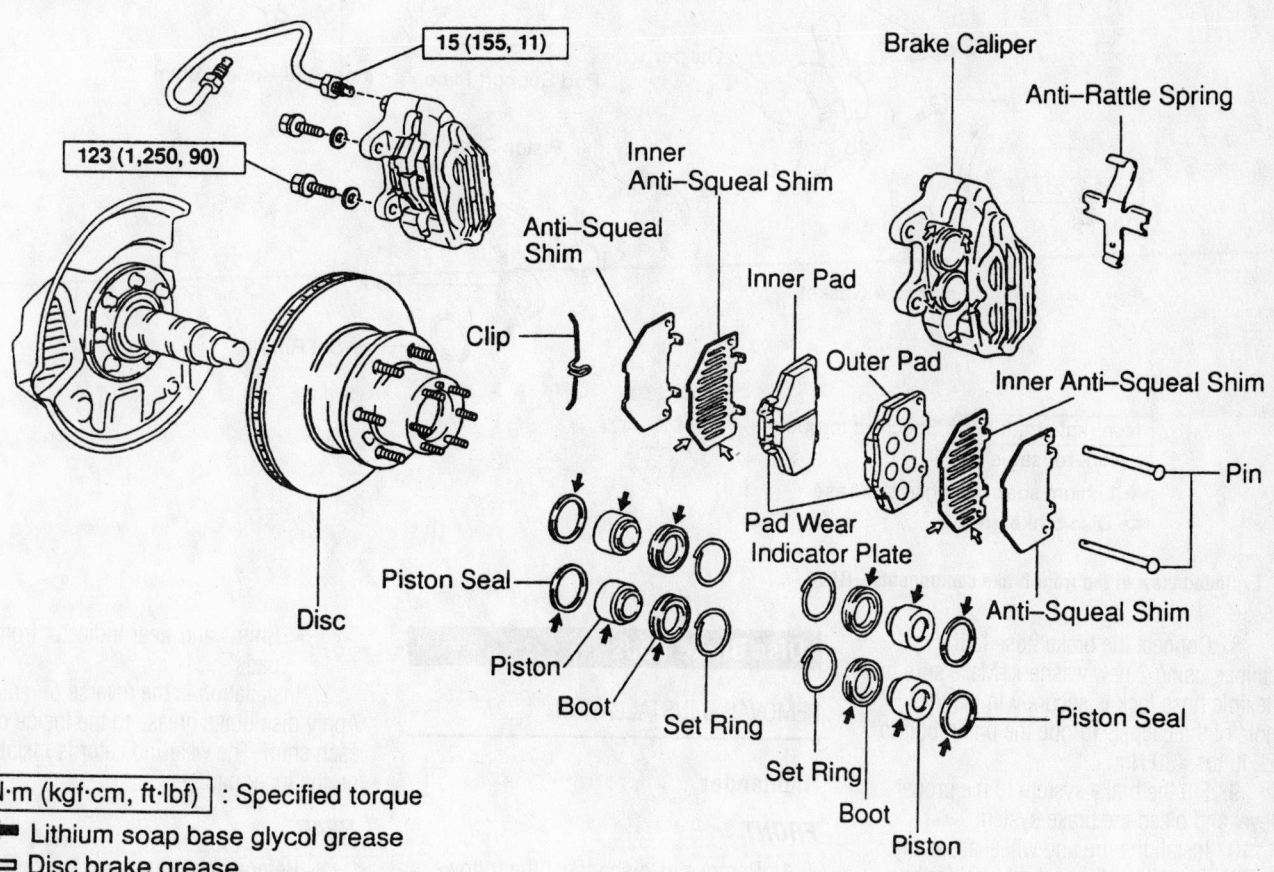

Dual piston type caliper assembly—4Runner

For complete Engine Mechanical specifications, see Section 1 of this manual

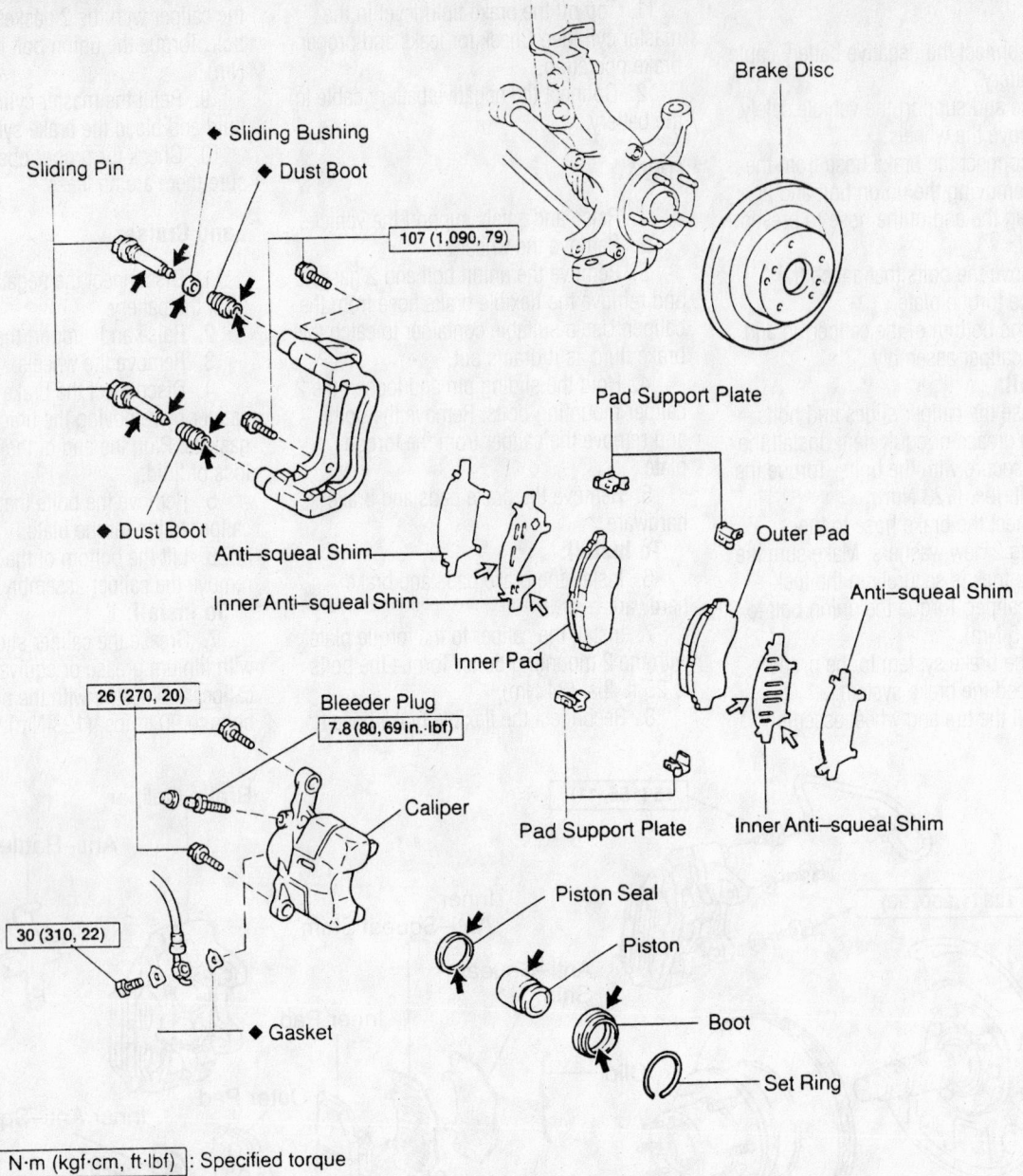

Exploded view of the front brake components—RAV4

8. Connect the brake hose to the caliper, using 2 new washers. Make sure the flexible hose lock is securely in the lock hole of the caliper. Torque the union bolt to 22 ft. lbs. (30 Nm).

9. Fill the brake system to the proper level and bleed the brake system.

10. Install the tire and wheel assembly.

11. Top off the brake fluid level in the master cylinder. Check for leaks and proper brake operation.

12. Connect the negative battery cable to the battery.

Disc Brake Pads

REMOVAL & INSTALLATION

Highlander

FRONT

1. Remove or disconnect the following:

- Front wheel
- 2 mounting bolts and caliper
- Pads and shims

- Shims and wear indicator from the pads

2. Installation is the reverse of removal. Apply disc brake grease to the inside of each shim. The wear indicator is installed on the inner pad.

REAR

1. Remove or disconnect the following:

- Rear wheel
- Caliper
- Pads and shims

- Shims and wear indicator from the pads

2. Installation is the reverse of removal. Apply disc brake grease to the inside of each shim. The wear indicator is installed on the inner pad.

4Runner

FRONT W/2-WHEEL DRIVE

1. Raise the vehicle and support it safely.
2. Remove the wheel and tire assembly.
3. When servicing the front pads, loosen the brake caliper upper side mounting bolt. Loosen and remove the lower side mounting bolt. Lift the cylinder and suspend it so the hose is not stretched.
4. If equipped, remove the anti-squeal spring.
5. Remove the brake pads.

To install:

6. Siphon a small amount of brake fluid from the reservoir. Press in the brake caliper piston with a hammer handle or equivalent.
7. Before installing the new pads, check the disc thickness and disc runout.
8. Install the pad support plates.
9. Install the anti-squeal shims to each pad.

➡**Apply disc brake grease to both sides of the inner anti-squeal shims.**

10. Install the disc pads so the wear indicator plate is facing downward.
11. If removed, install the anti-squeal springs.
12. Carefully install the brake caliper so the boot is not wedged. Torque the caliper mounting bolts, as follows:
- 2-Wheel drive w/PD60 type disc: 29 ft. lbs. (39 Nm)
- 2-wheel drive w/FS17 type disc: 65 ft. lbs. (88 Nm)
13. Install the wheel and tire assembly.
14. Check and adjust the fluid level. Apply the brake pedal several times.
15. Road-test the vehicle for proper operation.

FRONT W/4-WHEEL DRIVE

1. Raise the vehicle and support it safely.
2. Remove the wheel and tire assembly.
3. Remove the clip, pins, and the anti-rattle spring.
4. Remove the pads and the anti-squeal shims.
5. Remove the caliper, but do not disconnect the brake hose.

To install:

6. Before installing the new pads, check the disc thickness and disc runout.
7. Siphon out a small amount of brake fluid from the reservoir.
8. Temporarily install the old inner brake pad. Press in the pistons with a C-clamp or equivalent. Remove the old inner brake pad.
9. Apply disc brake grease to both sides of the inner anti-squeal shim. Install the anti-squeal shims to the new pads.
10. Install the pads.
11. Install the anti-rattle springs and pins. Install the clip.
12. Install the caliper and the mounting bolts. Torque the mounting bolts to 90 ft. lbs. (123 Nm).
13. Install the wheel and tire assembly.
14. Check and adjust the fluid level. Apply the brake pedal several times.
15. Road-test the vehicle for proper operation.

REAR

1. Raise the vehicle and support it safely.
2. Remove the wheel and tire assembly.
3. Remove the brake caliper and suspend it with a wire so the hose is not stretched or stressed.
4. Remove the brake pads, anti-squeal shim, pad support plates and wear indicators.

To install:

5. Before installing the new pads, check the disc thickness and disc runout.
6. Temporarily install the old inner brake pad. Press in the piston with a C-clamp or equivalent. Remove the old inner brake pad.
7. Install the pad support plates.
8. Install the pad wear indicator plate to each pads.
9. Install the anti-squeal shim to the outer pad. Install the pads so the wear indicator plate is facing upward.
10. Install the brake caliper. Torque the main sliding pin and the sub pin to 65 ft. lbs. (88 Nm).
11. Install the wheel and tire assembly.
12. Apply the brake pedal several times.
13. Road-test the vehicle for proper operation.

RAV4

1. Raise and safely support the vehicle.
2. Remove the wheel(s).
3. Temporarily install 2 wheel stud nuts to hold the brake rotor in place.

4. If necessary, siphon a sufficient quantity of brake fluid from the master cylinder reservoir to prevent any brake fluid from overflowing the master cylinder when removing or installing new pads. This may be necessary, as the piston must be forced into the caliper bore to provide sufficient clearance when installing the pads.
5. Grasp the caliper from behind and carefully pull it towards you. This will start to seat the piston(s) in its bore. Using a C-clamp or other suitable tool, press the piston the remaining way into the caliper. Be careful not to cock the piston in the bore. Also, do not force the piston or the caliper and piston may be damaged.
6. Hold the sliding pin and loosen the 2 caliper mounting bolts. Remove the bolts and remove the caliper from the torque plate.
7. Secure the caliper assembly out of the way with a wire; so as not to stress the flexible hose.
8. Slide out the old brake pads along with any anti-squeal shims, springs, pad wear indicators and pad support plates. Make sure to note the position of all assorted pad hardware.

To install:

9. Check the brake disc (rotor) for thickness and run-out. Inspect the caliper and piston assembly for breaks, cracks, fluid seepage or other damage. Overhaul or replace as necessary.
10. Install the pad support plates into the torque plate.
11. Install the pad wear indicators onto the pads. Be sure the arrow on the indicator plate is pointing in the direction of rotation.
12. Install the anti-squeal shims on the outside of each pad and then install the pad assemblies into the torque plate.
13. Install the caliper to the torque plate with the 2 mounting bolts. Torque the bolts to 20 ft. lbs. (26 Nm).
14. Remove the 2 temporary wheel stud nuts and check that the rotor turns freely.
15. Reinstall the wheel(s), safely lower the vehicle, and road-test for proper brake operation.
16. Be sure to pump the brakes several times prior to moving the vehicle.

Land Cruiser

FRONT

1. Raise the vehicle and support it safely.
2. Remove the wheels.

3. Remove the clip, pins and anti-rattle spring.

4. Withdraw the pads and remove the anti-squeal shims.

To install:

5. Before installing the new pads, check the disc thickness and disc runout.

6. Siphon out a small amount of brake fluid from the reservoir.

7. Press in the pistons with a hammer handle or equivalent.

8. Apply disc brake grease to both sides of the inner anti-squeal shim. Install the anti-squeal shims to the new pads.

9. Install the pads.

10. Install the anti-rattle springs and pins. Install the clip.

11. Install the wheels.

12. Check and adjust the fluid level. Apply the brake pedal several times.

13. Road-test the vehicle for proper operation.

REAR

1. Raise the vehicle and support it safely.

2. Remove the wheels.

3. Remove the brake caliper and suspend it so the hose is not stretched.

4. Remove the brake pads, anti-squeal shim, pad support plates and wear indicators.

To install:

5. Before installing the new pads, check the disc thickness and disc runout.

6. Install the pad support plates.

7. Install the pad wear indicator plates on each pad.

8. Install the anti-squeal shim to the outer pad. Install the pads so the wear indicator plate is facing upward.

9. Install the brake caliper.

10. Install the wheels.

11. Apply the brake pedal several times.

12. Road-test the vehicle for proper operation.

Brake Drums

REMOVAL & INSTALLATION

4Runner

1. Raise and safely support the vehicle.

2. Remove the rear wheel(s).

3. Remove the brake drum from the axle hub. If there is difficulty in removing the drum, insert a suitable tool through the hole in the rear of the backing plate, and hold the automatic adjusting lever away from the adjuster. Using another suitable tool at the same time, reduce the brake shoe adjuster by turning the adjusting wheel.

To install:

4. Install the brake drum and pull the parking brake lever all the way up until a clicking sound can no longer be heard.

5. Verify that the rear wheels will not

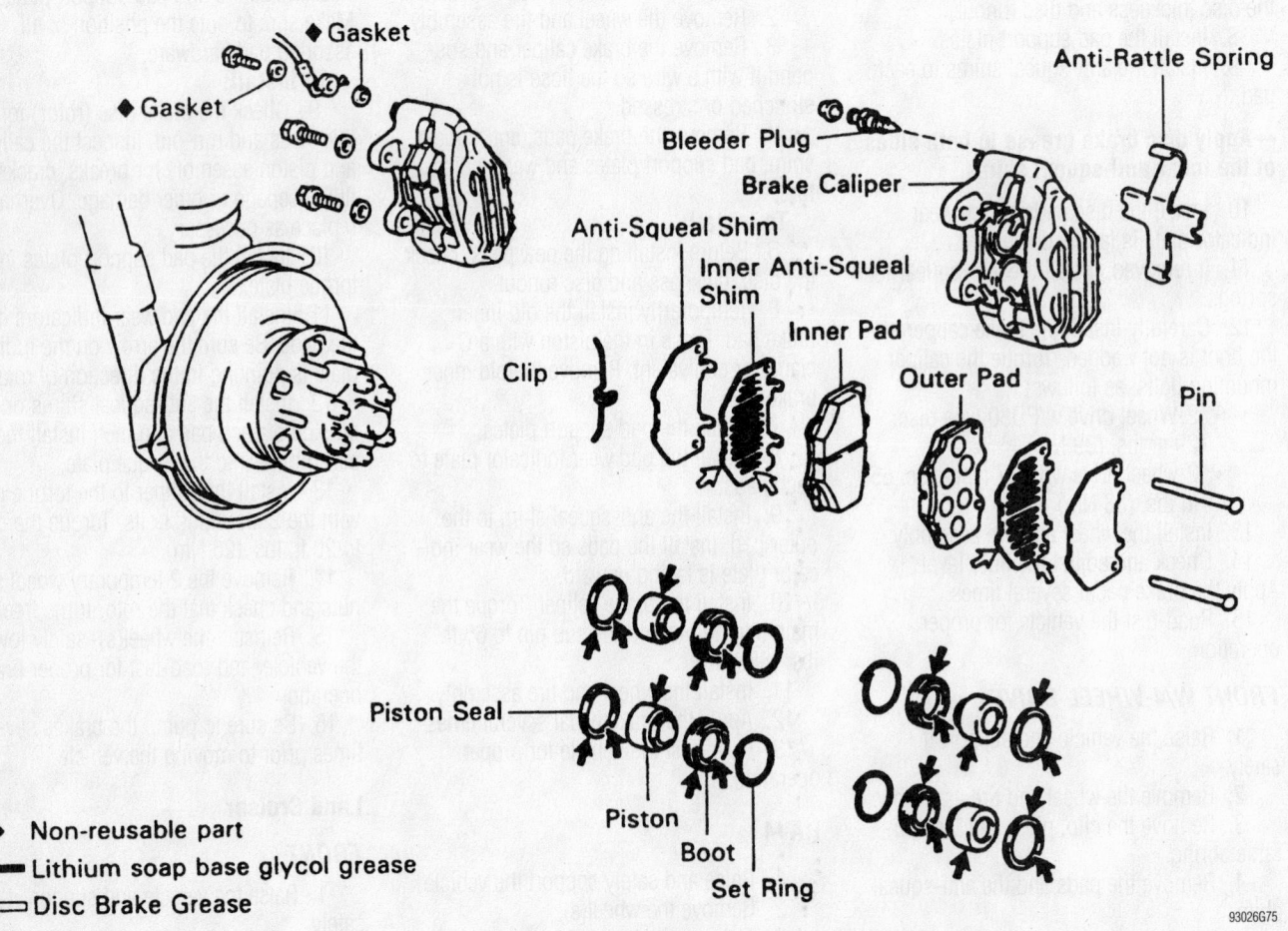

◆ Non-reusable part

◀ Lithium soap base glycol grease

⇦ Disc Brake Grease

Exploded view of the front disc brake components—Land Cruiser

93026G75

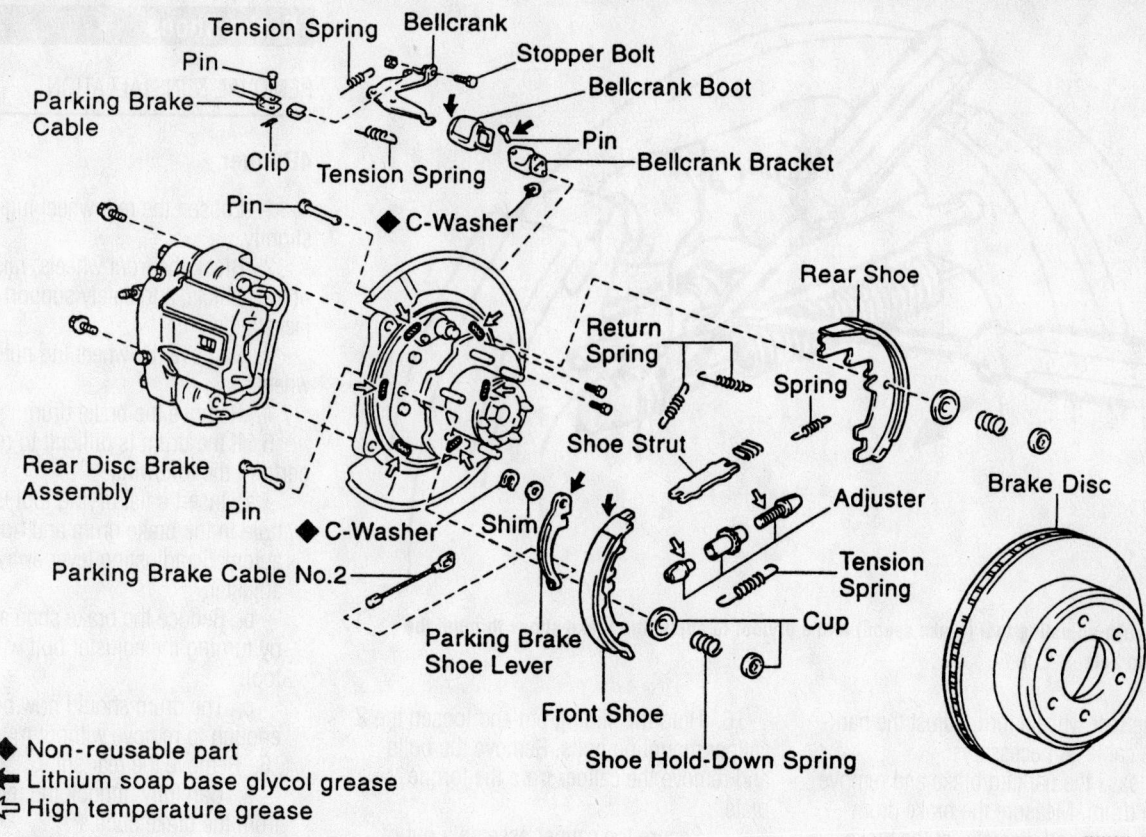

Non-reusable part

◆ Non-reusable part
← Lithium soap base glycol grease
⇐ High temperature grease

93026G76

Exploded view of the rear disc brake components—Land Cruiser

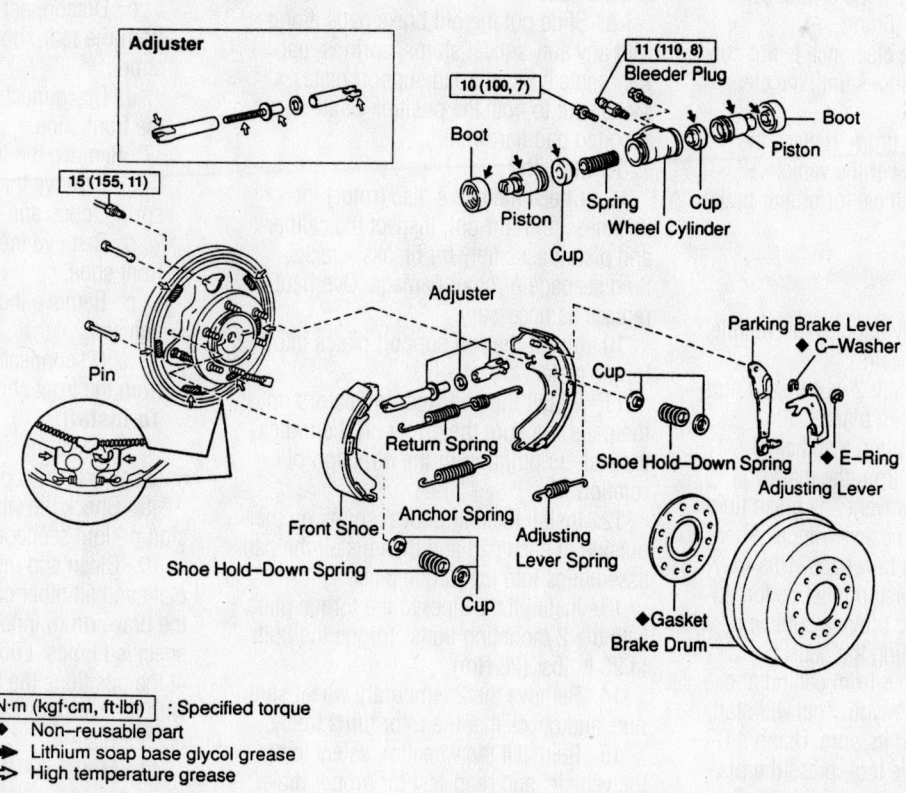

N·m (kgf·cm, ft·lbf) : Specified torque
◆ Non-reusable part
➤ Lithium soap base glycol grease
⇒ High temperature grease

93026G77

Exploded view of the rear brake drums components

For Tire, Wheel and Ball Joint specifications, see Section 1 of this manual

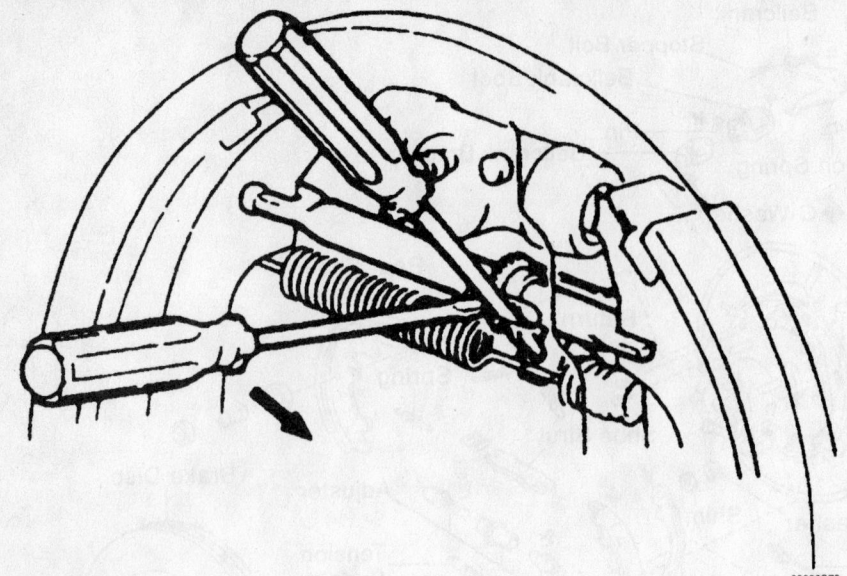

Use a brake adjusting tool (brake spoon) and a prytool to adjust the brake shoes through the adjusting hole

turn. If the rear wheels turn, adjust the parking brake cable as necessary.

6. Release the parking brake and remove the brake drum. Measure the brake drum inside diameter and diameter of the brake shoes. Check that the difference between the diameters is the correct shoe clearance. Clearance is 0.024 in. (6mm).

7. If the brake shoe clearance is not correct, adjust the brake shoes until the clearance is correct.

8. Install the brake drum, replace the wheel(s); and safely lower the vehicle.

9. Road-test the vehicle for proper brake operation.

RAV4

1. Raise and safely support the vehicle.
2. Remove the wheel(s).
3. Temporarily install 2 wheel stud nuts to hold the brake rotor in place.
4. If necessary, siphon a sufficient quantity of brake fluid from the master cylinder reservoir to prevent any brake fluid from overflowing the master cylinder when removing or installing new pads. This may be necessary, as the piston must be forced into the caliper bore to provide sufficient clearance when installing the pads.
5. Grasp the caliper from behind and carefully pull it towards you. This will start to seat the piston(s) in its bore. Using a C-clamp or other suitable tool, press the piston the remaining way into the caliper. Be careful not to cock the piston in the bore. Also, do not force the piston or the caliper and piston may be damaged.

6. Hold the sliding pin and loosen the 2 caliper mounting bolts. Remove the bolts and remove the caliper from the torque plate.

7. Secure the caliper assembly out of the way with a wire, to avoid stressing the flexible hose.

8. Slide out the old brake pads along with any anti-squeal shims, springs, pad wear indicators and pad support plates. Make sure to note the position of all assorted pad hardware.

To install:

9. Check the brake disc (rotor) for thickness and run-out. Inspect the caliper and piston assembly for breaks, cracks, fluid seepage or other damage. Overhaul or replace as necessary.

10. Install the pad support plates into the torque plate.

11. Install the pad wear indicators onto the pads. Be sure the arrow on the indicator plate is pointing in the direction of rotation.

12. Install the anti-squeal shims on the outside of each pad and then install the pad assemblies into the torque plate.

13. Install the caliper to the torque plate with the 2 mounting bolts. Torque the bolts to 20 ft. lbs. (26 Nm).

14. Remove the 2 temporary wheel stud nuts and check that the rotor turns freely.

15. Reinstall the wheel(s), safely lower the vehicle, and road-test for proper brake operation.

16. Be sure to pump the brakes several times prior to moving the vehicle.

REMOVAL & INSTALLATION

4Runner

1. Loosen the rear wheel lug nuts slightly.

2. Block the front wheels, raise the rear of the vehicle, and safely support it with jackstands.

3. Remove the wheel lug nuts and the wheel.

4. Remove the brake drum.

5. If the drum is difficult to remove, perform the following:

a. Insert a flat prying tool through the hole in the brake drum and hold the automatic adjusting lever away from the adjuster.

b. Reduce the brake shoe adjustment by turning the adjuster bolt with a brake tool.

c. The drum should now be loose enough to remove without much effort.

6. Remove the rear shoe.

a. Carefully unhook the return spring from the brake shoe.

b. Remove the shoe hold-down spring, cups and the pin.

c. Disconnect the anchor spring from the rear shoe and remove the rear shoe.

d. Disconnect the anchor spring from the front shoe.

7. Remove the front shoe.

a. Remove the shoe hold-down spring, cups and pin.

b. Remove the return spring from the front shoe.

c. Remove the front shoe with the adjuster.

d. Disconnect the parking brake cable from the front shoe.

To install:

8. Inspect the shoes for signs of unusual wear or scoring.

9. Check the wheel cylinder for any sign of fluid seepage or frozen pistons.

10. Clean and inspect the brake backing plate and all other components. Check that the brake drum inner diameter is within specified limits. Lubricate the backing plate at the positions the brakes come in contact with the backing plate. Also lubricate the anchor plate.

11. Mount the automatic adjuster assembly onto a new rear brake shoe.

12. Install the front shoe.

a. Install the parking brake cable to the front shoe.

b. Install the front shoe with the adjuster.

c. Install the return spring to the front shoe.

d. Install the shoe hold-down spring, cups and pin.

13. Install the rear shoe.

a. Install the anchor spring to the front shoe.

b. Install the anchor spring to the rear shoe and install the rear shoe.

c. Install the shoe hold-down spring, cups and the pin.

d. Hook the return spring to the brake shoe.

14. Install the brake drum.

15. Adjust the brake shoes until a slight drag is felt when the drum is spun by hand.

16. Remove the brake drum and check the clearance between brake shoes and brake drum. Adjust the clearance to specification.

17. Pull the parking lever all the way up until a clicking sound can no longer be heard. Verify that the drum doesn't turn. If the drum turns, adjust the parking brake cable.

18. Install the rear wheels, tighten the wheel lug nuts and lower the vehicle.

19. Retighten the wheel lug nuts and pump the brake pedal a few times before moving the vehicle. Adjust the rear brakes again if necessary.

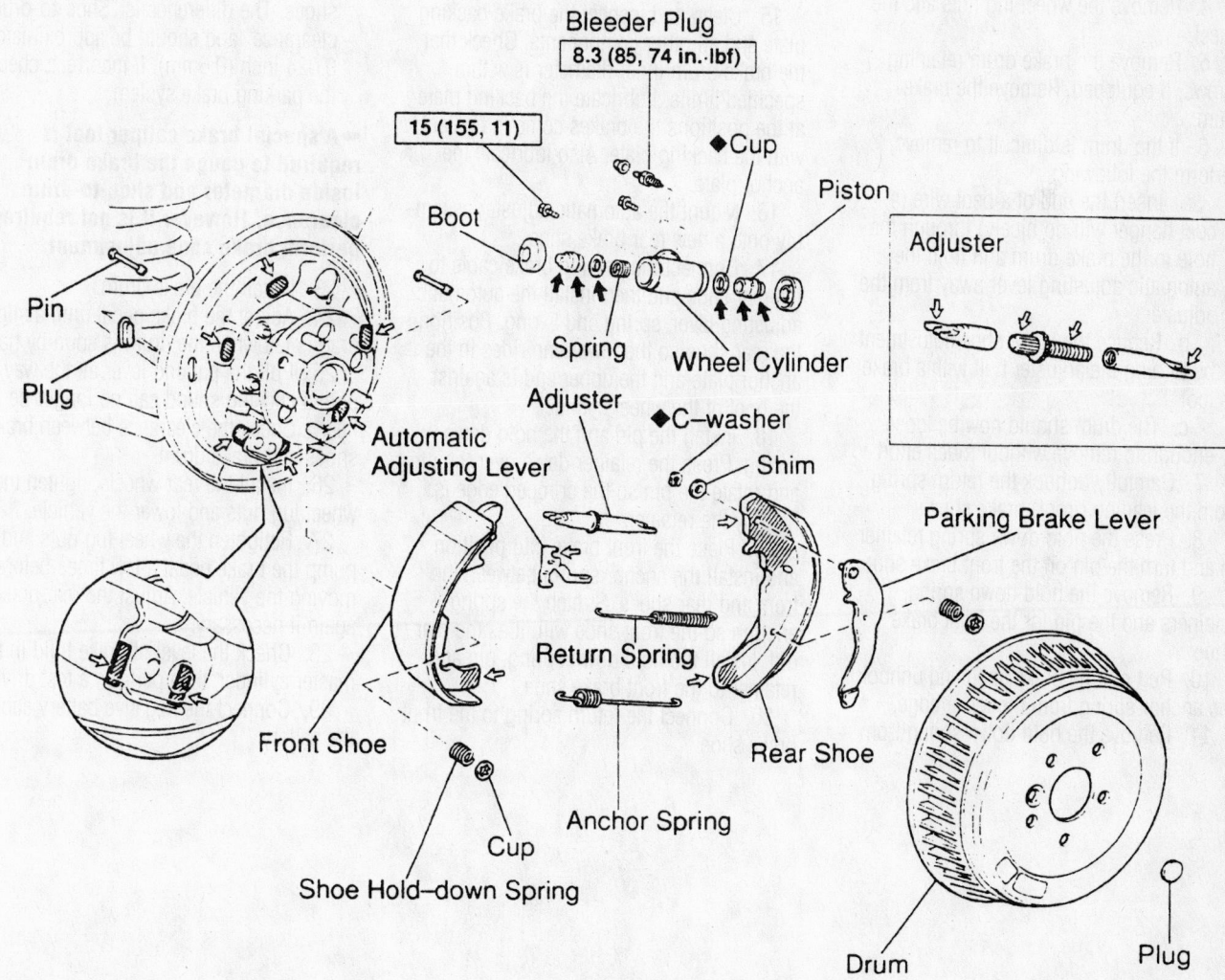

Bleeder Plug
8.3 (85, 74 in.·lbf)

15 (155, 11)

Cup

Piston

Boot

Spring

Wheel Cylinder

Adjuster

Pin

Plug

Automatic Adjusting Lever

Adjuster

◆C-washer

Shim

Parking Brake Lever

Front Shoe

Return Spring

Rear Shoe

Cup

Anchor Spring

Shoe Hold-down Spring

Drum

Plug

N·m (kgf·cm, ft·lbf) : Specified torque

◆ Non-reusable part

➡ Lithium soap base glycol grease

⇨ High temperature grease

93026G80

Exploded view of the rear drum brake components—RAV4

For Wheel Alignment specifications, see Section 1 of this manual

20. Check the level of brake fluid in the master cylinder, then perform a test drive.

21. Connect the negative battery cable to the battery.

RAV4

1. Disconnect the negative battery cable from the battery.

2. Loosen the rear wheel lug nuts slightly. Release the parking brake.

3. Block the front wheels, raise the rear of the vehicle, and safely support it with jackstands.

4. Remove the wheel lug nuts and the wheel.

5. Remove the brake drum retaining screws, if equipped. Remove the brake drum.

6. If the drum is difficult to remove, perform the following:

 a. Insert the end of a bent wire (a coat hanger will do nicely) through the hole in the brake drum and hold the automatic adjusting lever away from the adjuster.

 b. Reduce the brake shoe adjustment by turning the adjuster bolt with a brake tool.

 c. The drum should now be loose enough to remove without much effort.

7. Carefully unhook the return spring from the leading (front) brake shoe.

8. Press the hold down spring retainer in and turn the pin on the front brake shoe.

9. Remove the hold down spring, retainers and the pin for the front brake shoe.

10. Pull out the brake shoe and unhook the anchor spring from the lower edge.

11. Remove the hold down spring from the trailing (rear) shoe. Pull the shoe out with the adjuster, automatic adjuster assembly and springs attached. Disconnect the parking brake cable. Remove the tension/return and anchor springs from the rear shoe.

12. Unhook the adjusting lever spring from the rear shoe and then remove the automatic adjuster assembly.

To install:

13. Inspect the shoes for signs of unusual wear or scoring.

14. Check the wheel cylinder for any sign of fluid seepage or frozen pistons.

15. Clean and inspect the brake backing plate and all other components. Check that the brake drum inner diameter is within specified limits. Lubricate the backing plate at the positions the brakes come in contact with the backing plate. Also lubricate the anchor plate.

16. Mount the automatic adjuster assembly onto a new rear brake shoe.

17. Connect the parking brake cable to the rear shoe and then install the automatic adjusting lever, spring and E-ring. Position the rear shoe so the lower end rides in the anchor plate and the upper end is against the boot of the wheel cylinder.

18. Install the pin and the hold down spring. Press the retainer down over the pin and rotate the pin so the crimped edge is held by the retainer.

19. Place the front brake into position and install the anchor spring between the front and rear shoes. Stretch the spring enough so the front shoe will fit as the rear did. Install the hold down spring, pin and retainer to the front brake shoe.

20. Connect the return spring to the front brake shoe.

21. Check the operation of the automatic adjuster mechanism:

 a. Apply the parking brake lever and verifying the adjusting bolt turns.

 b. Adjust the strut to where it is the shortest possible length.

 c. Install the brake drum.

 d. Apply the parking brake lever until the clicking sound can no longer be heard.

22. Check the clearance between the brake shoes and drum:

 a. Remove the brake drum.

 b. Measure the brake drum inside diameter and diameter of the brake shoes. The difference is "Shoe-to-drum clearance" and should be approximately 0.024 inch (0.6mm). If incorrect, check the parking brake system.

➡A special brake caliper tool is required to gauge the brake drum inside diameter and shoe-to-drum clearance. However it is not required to perform brake shoe adjustment.

23. Install the brake drum.

24. Adjust the brake pedal until a slight drag is felt when the drum is spun by hand.

25. Pull the parking lever all the way up until a clicking sound can no longer be heard. Check the clearance between brake shoes and brake drum.

26. Install the rear wheels, tighten the wheel lug nuts and lower the vehicle.

27. Retighten the wheel lug nuts and pump the brake pedal a few times before moving the vehicle. Adjust the rear brakes again if necessary.

28. Check the level of brake fluid in the master cylinder, then perform a test drive.

29. Connect the negative battery cable to the battery.

ACURA

MDX

5

PRECAUTIONS

Before servicing any vehicle, please be sure to read all of the following precautions, which deal with personal safety, prevention of component damage, and important points to take into consideration when servicing a motor vehicle:

• Never open, service or drain the radiator or cooling system when the engine is hot; serious burns can occur from the steam and hot coolant.

• Observe all applicable safety precautions when working around fuel. Whenever servicing the fuel system, always work in a well-ventilated area. Do not allow fuel spray or vapors to come in contact with a spark, open flame, or excessive heat (a hot drop light, for example). Keep a dry chemical fire extinguisher near the work area. Always keep fuel in a container specifically designed for fuel storage; also, always properly seal fuel containers to avoid the possibility of fire or explosion. Refer to the additional fuel system precautions later in this section.

• Fuel injection systems often remain pressurized, even after the engine has been turned **OFF**. The fuel system pressure must be relieved before disconnecting any fuel lines. Failure to do so may result in fire and/or personal injury.

• Brake fluid often contains polyglycol ethers and polyglycols. Avoid contact with the eyes and wash your hands thoroughly after handling brake fluid. If you do get brake fluid in your eyes, flush your eyes with clean, running water for 15 minutes. If

eye irritation persists, or if you have taken brake fluid internally, seek medical assistance IMMEDIATELY.

• The EPA warns that prolonged contact with used engine oil may cause a number of skin disorders, including cancer. You should make every effort to minimize your exposure to used engine oil. Protective gloves should be worn when changing oil. Wash your hands and any other exposed skin areas as soon as possible after exposure to used engine oil. Soap and water, or waterless hand cleaner should be used.

• All new vehicles are now equipped with an air bag system. The system must be disabled before performing service on or around system components, steering column, instrument panel components, wiring and sensors. Failure to follow safety and disabling procedures could result in accidental air bag deployment, possible personal injury and unnecessary system repairs.

• Always wear safety goggles when working with, or around, the air bag system. When carrying a non-deployed air bag, be sure the bag and trim cover are pointed away from your body. When placing a non-deployed air bag on a work surface, always face the bag and trim cover upward, away from the surface. This will reduce the motion of the module if it is accidentally deployed. Refer to the additional air bag system precautions later in this section.

• Clean, high quality brake fluid from a sealed container is essential to the safe and proper operation of the brake system. You

should always buy the correct type of brake fluid for your vehicle. If the brake fluid becomes contaminated, completely flush the system with new fluid. Never reuse any brake fluid. Any brake fluid that is removed from the system should be discarded. Also, do not allow any brake fluid to come in contact with a painted surface; it will damage the paint.

• Never operate the engine without the proper amount and type of engine oil; doing so WILL result in severe engine damage.

• Timing belt maintenance is extremely important. Many models utilize an interference-type, non-freewheeling engine. If the timing belt breaks, the valves in the cylinder head may strike the pistons, causing potentially serious (also time-consuming and expensive) engine damage. Refer to the maintenance interval charts in the front of this manual for the recommended replacement interval for the timing belt, and to the timing belt section for belt replacement and inspection.

• Disconnecting the negative battery cable on some vehicles may interfere with the functions of the on-board computer system(s) and may require the computer to undergo a relearning process once the negative battery cable is reconnected.

• When servicing drum brakes, only disassemble and assemble one side at a time, leaving the remaining side intact for reference.

• Only an MVAC-trained, EPA-certified automotive technician should service the air conditioning system or its components.

ENGINE REPAIR

➡**Disconnecting the negative battery cable on some vehicles may interfere with the functions of the on board computer system. The computer may undergo a relearning process once the negative battery cable is reconnected.**

Distributor

The MDX is equipped with a Distributorless Ignition System (DIS).

Alternator

REMOVAL

1. Before servicing the vehicle, refer to the precautions in the beginning of this section.

2. Remove or disconnect the following:
   • Negative battery cable
   • Intake manifold and ignition coil covers
   • Accessory drive belt
   • Alternator wiring harness connectors
   • Alternator mounting bolts
   • Wiring harness clamp
   • Alternator

INSTALLATION

1. Install or connect the following:
   • Alternator
   • Wiring harness clamp. Tighten the bolt to 105 inch lbs. (12 Nm).
   • Alternator mounting bolts. Tighten the 10mm bolt to 33 ft. lbs. (44

Nm) and the 8mm bolt to 16 ft. lbs. (22 Nm).
   • Alternator wiring harness connectors. Tighten the battery terminal nut to 105 inch lbs. (12 Nm).
   • Accessory drive belt
   • Intake manifold and ignition coil covers
   • Negative battery cable

Ignition Timing

ADJUSTMENT

The MDX is equipped with a Distributorless Ignition System (DIS). The ignition timing is controlled by the Powertrain Control module (PCM). No adjustment is necessary.

Engine Assembly

REMOVAL & INSTALLATION

➡️**The engine and transaxle are removed from the vehicle as a unit.**

1. Before servicing the vehicle, refer to the precautions in the beginning of this section.
2. Drain the cooling system.
3. Drain the power steering system.
4. Drain the transaxle fluid.
5. Drain the engine oil.
6. Relieve fuel system pressure.
7. Remove or disconnect the following:
   - Negative battery cable
   - Battery and tray
   - Intake and ignition coil covers
   - Air intake duct
   - Left engine wire harness connectors
   - Relay bracket
   - Starter cable and harness clamp
   - Accelerator cable
   - Cruise control cable
   - Fuel lines
   - EVAP canister hose
8. Remove the drivers side center console lower panel and pull back the cover to access steering joint cover.
   - Steering joint bolt
   - Powertrain Control Module (PCM) connectors
   - Heated Oxygen (HO_2S) sensor connector and grommet. Pull the PCM harness through the firewall.
   - Brake booster vacuum line
   - Power steering hose's clamps and clips
   - Fuse/Relay box battery cable
   - Accessory drive belts
   - Front wheels
   - Splash shield
   - Front sub-frame stiffener
   - Exhaust front pipe
   - Propeller shaft
   - Shift control cable
   - Transfer assembly
   - Ball joints
   - Stabilizer bar links
   - Power steering hose and pressure switch connector
   - Transaxle lower front mount
   - Transaxle lower rear mount
   - A/C compressor
   - Heater hoses
   - Radiator hoses
   - Ground cable

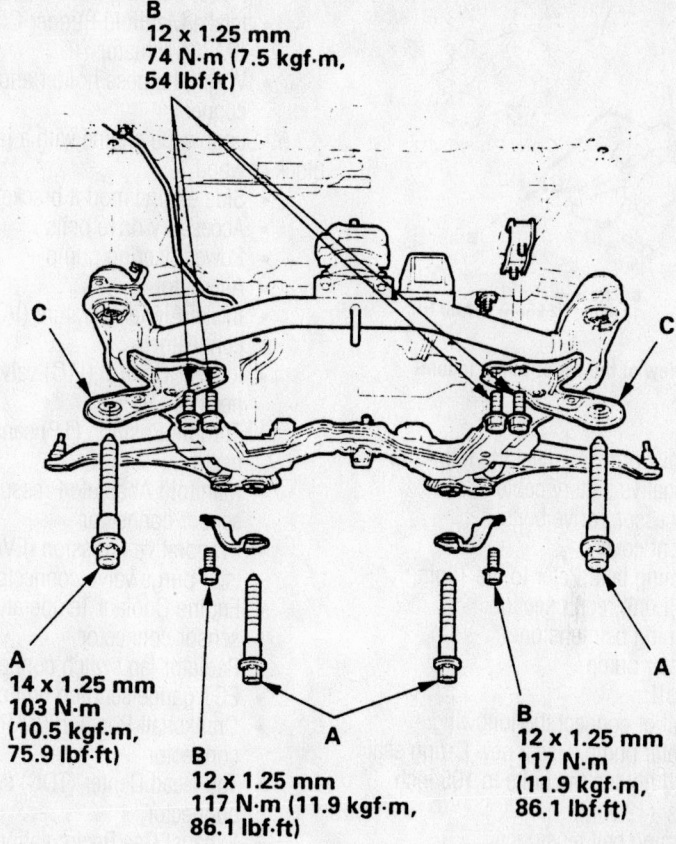

Sub-frame fastener locations and tightening torque—MDX

   - Transaxle oil cooler lines
   - Radiator
9. Attach a hoist to the engine lifting eyes and support the powertrain weight.
10. Remove or disconnect the following:
    - Side engine mount bracket
    - Front mount bracket support nut
11. Matchmark the front subframe to the mounting points.
12. Remove or disconnect the following:
    - Front subframe
    - All remaining hoses and electrical connections
13. Lower the powertrain away from the vehicle.

To install:

14. Raise the powertrain into position.
15. Installation is the reverse of removal but please note the following steps:
    - A/C compressor bolts to 16 ft. lbs. (22 Nm)
    - Front subframe. Use new bolts and tighten the 14mm bolts to 76 ft. lbs. (103 Nm). Tighten the front brace bolts to 54 ft. lbs. (74 Nm) and the rear brace bolts to 86 ft. lbs. (117 Nm).

   - Transaxle lower front mount nuts to 28 ft. lbs. (38 Nm)
   - Transaxle lower rear mount bolts to 28 ft. lbs. (38 Nm)
   - Front mount bracket support nut to 40 ft. lbs. (54 Nm)
   - Side engine mount bracket bolts to 33 ft. lbs. (44 Nm) and the through bolt to 40 ft. lbs. (54 Nm)
16. Fill the engine crankcase to the correct level.
17. Fill the transaxle to the correct level.
18. Fill the cooling system.
19. Fill the power steering system.
20. Start the engine and check for leaks.
21. Check the wheel alignment and adjust as necessary.

Water Pump

REMOVAL & INSTALLATION

1. Before servicing the vehicle, refer to the precautions in the beginning of this section.
2. Drain the cooling system.

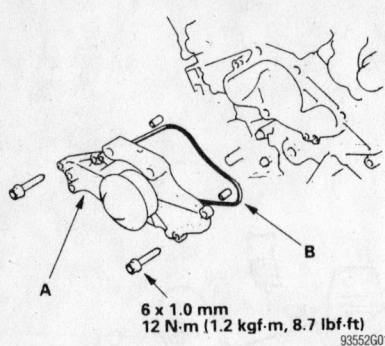

6 x 1.0 mm
12 N·m (1.2 kgf·m, 8.7 lbf·ft)
93552G01

Exploded view of the water pump mounting

3. Remove or disconnect the following:
   - Negative battery cable
   - Accessory drive belts
   - Front cover
   - Timing belt. Refer to the Timing Belt unit repair section.
   - Timing belt tensioner
   - Water pump

To install:

4. Install or connect the following:
   - Water pump. Use a new O-ring seal and tighten the bolts to 105 inch lbs. (12 Nm).
   - Timing belt tensioner
   - Timing belt
   - Front cover
   - Accessory drive belts
   - Negative battery cable
5. Fill the cooling system.
6. Start the engine and check for leaks.

Cylinder Head

REMOVAL & INSTALLATION

1. Before servicing the vehicle, refer to the precautions in the beginning of this section.
2. Drain the cooling system.
3. Relieve the fuel system pressure.
4. Remove or disconnect the following:
   - Negative battery cable
   - Ignition coil covers
   - Intake manifold cover
   - Air intake tube
   - Accelerator cable
   - Cruise control cable
   - EVAP canister, breather, water bypass hoses and the bypass pipe bolt
   - Fuel lines
   - Brake booster vacuum line
   - Intake manifold vacuum line
   - Positive Crankcase Ventilation (PCV) valve and hose

- Power steering hose clamp
- Intake Manifold Runner Control (IMRC) actuator
- Wiring harness holder and joint connector

5. Support the engine with a jack and a block of wood.
   - Side engine mount bracket
   - Accessory drive belts
   - Power steering pump
   - Alternator
   - Intake Air Temperature (IAT) sensor connector
   - Idle Air Control (IAC) valve connector
   - Throttle Position (TP) sensor connector
   - Manifold Absolute Pressure (MAP) sensor connector
   - Evaporative Emission (EVAP) canister purge valve connector
   - Engine Coolant Temperature (ECT) sensor connector
   - Radiator fan switch connectors
   - ECT gauge sending unit connector
   - Crankshaft Position (CKP) sensor connector
   - Top Dead Center (TDC) sensor connector
   - Exhaust Gas Recirculation (EGR) connector
   - Variable Valve Timing and Valve Lift Electronic Control (VTEC) solenoid valve connector
   - VTEC oil pressure switch connector
   - Oil pressure switch connector
   - Ignition coils
   - Intake manifold
   - Fuel injector connectors
   - Fuel supply manifold
   - Fuel injection air control valve vacuum lines
   - Front cover
   - Timing belt. Refer to the Timing Belt unit repair section.
   - Radiator hoses

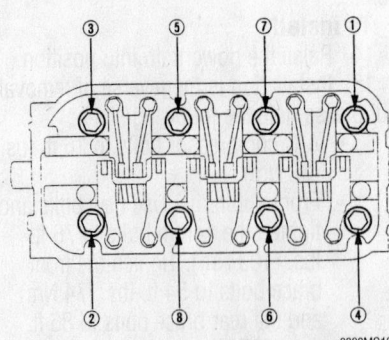

Cylinder head bolt loosening sequence—MDX

9308MG12

- Heater hoses
- Front and rear exhaust manifolds
- Coolant cross-over pipe
- Valve covers

6. Loosen the cylinder head bolts in sequence and ⅓ turns until all bolts are loose.
7. Remove the cylinder head.

To install:

8. Align the crankshaft and camshaft sprocket TDC marks as shown.
9. Install the cylinder heads with new gaskets.

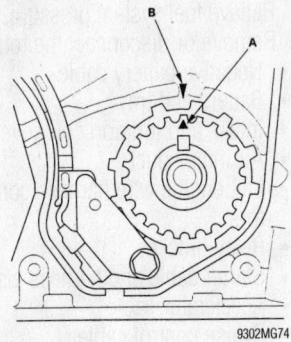

9302MG74

Crankshaft timing belt sprocket TDC marks. Align sprocket mark (A) with pointer (B)—MDX

FRONT:

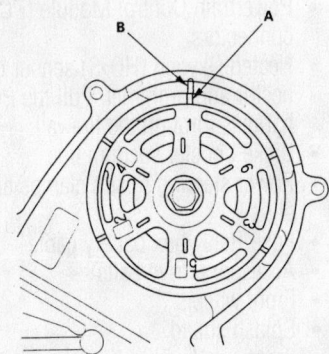

REAR:

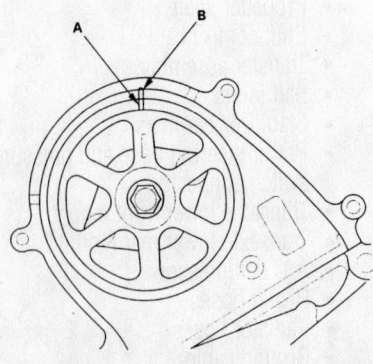

9302MG85

Camshaft TDC marks. Align sprocket mark (A) with the back cover pointer (B)—MDX

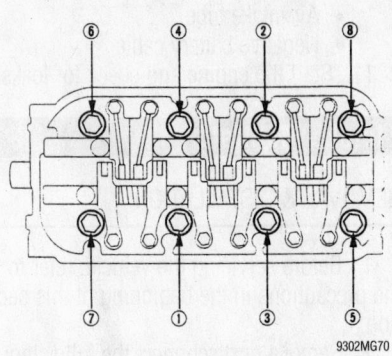

Cylinder head bolt tightening sequence— MDX

10. Apply clean engine oil to the cylinder head bolt threads and flanges.

11. Tighten the cylinder head bolts in sequence as follows:
   a. Step 1: 29 ft. lbs. (39 Nm)
   b. Step 2: 51 ft. lbs. (69 Nm)
   c. Step 3: 72 ft. lbs. (98 Nm)

12. Install or connect the following:
   - Valve covers
   - Coolant cross-over pipe
   - Front and rear exhaust manifolds
   - Heater hoses
   - Radiator hoses
   - Timing belt
   - Front cover
   - Fuel injection air control valve vacuum lines
   - Fuel supply manifold
   - Fuel injector connectors
   - Intake manifold
   - Ignition coils
   - Oil pressure switch connector
   - VTEC oil pressure switch connector
   - VTEC solenoid valve connector
   - EGR connector
   - TDC sensor connector
   - CKP sensor connector
   - ECT gauge sending unit connector
   - Radiator fan switch connectors
   - ECT sensor connector
   - MAP sensor connector
   - TP sensor connector
   - IAC valve connector
   - IAT sensor connector
   - Alternator
   - Power steering hose clamp
   - Power steering pump
   - Side engine mount bracket
   - PCV valve and hose
   - Intake manifold vacuum line
   - Brake booster vacuum line
   - Fuel lines
   - Cruise control cable
   - Accelerator cable

- Intake manifold cover
- Ignition coil covers
- Accessory drive belts
- Air intake tube
- EVAP control canister hose and vacuum hose
- Negative battery cable

13. Fill the cooling system.

14. Start the engine and check for leaks.

Rocker Arms/Shafts

REMOVAL & INSTALLATION

1. Before servicing the vehicle, refer to the precautions in the beginning of this section.

2. Remove or disconnect the following:
   - Negative battery cable
   - Air intake tube
   - Ignition coil covers
   - Intake manifold cover
   - Intake manifold
   - Valve cover

3. Loosen the valve adjuster locknuts

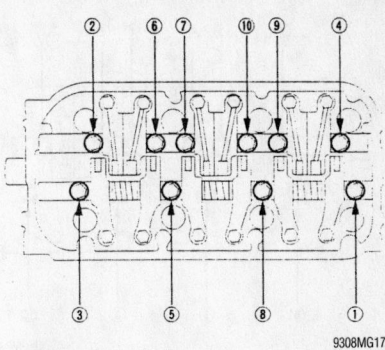

Rocker arm shaft loosening sequence— MDX

and screws so that all valves are closed.

4. Loosen the rocker arm shaft bolts evenly in sequence.

5. Remove the rocker arms and shafts from the vehicle as an assembly.

➡ **Keep all valvetrain components in order for assembly.**

6. Remove the rocker arms and springs from the rocker arm shafts.

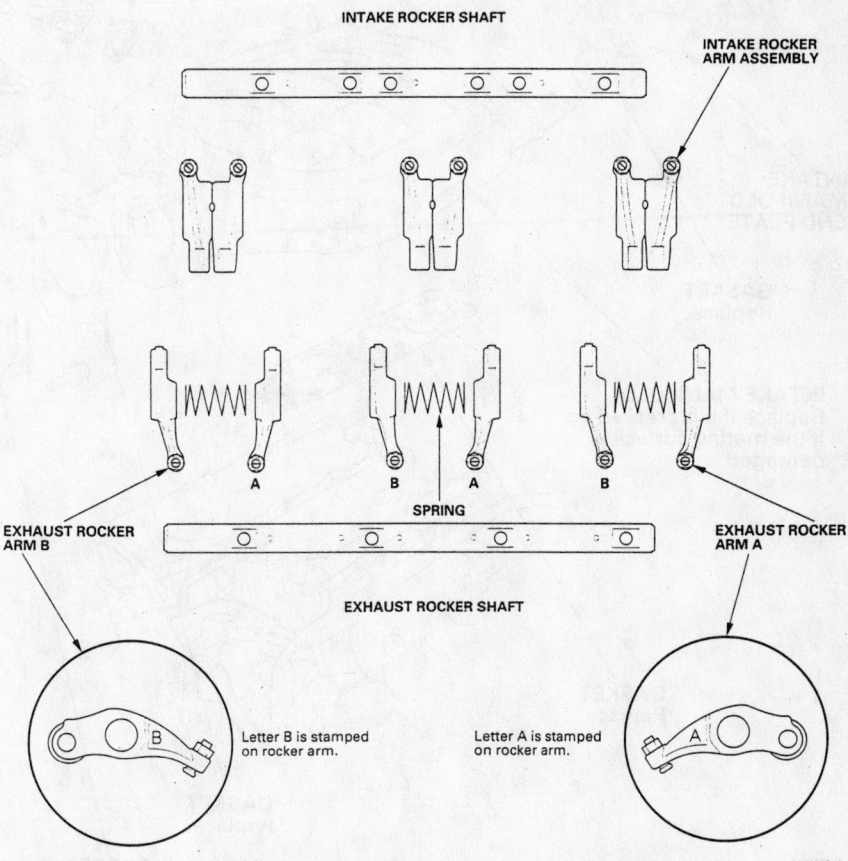

Exploded view of the rocker arms and shafts—MDX

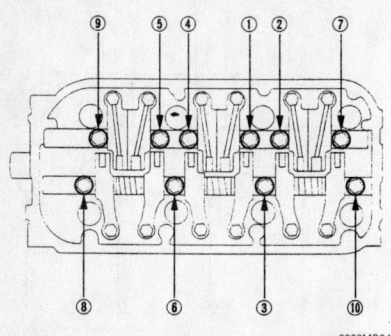

Rocker shaft tightening sequence—MDX

9302MG84

To install:

7. Assemble the rocker arms and springs to the rocker arm shafts in their original positions.

8. Install the rocker arm assemblies. Tighten the bolts in sequence and in multiple passes to 17 ft. lbs. (24 Nm).

9. Adjust the valve clearance.

10. Install or connect the following:
- Valve covers
- Intake manifold
- Intake manifold cover
- Ignition coil covers
- Air intake tube
- Negative battery cable

11. Start the engine and check for leaks.

Intake Manifold

REMOVAL & INSTALLATION

1. Before servicing the vehicle, refer to the precautions in the beginning of this section.

2. Remove or disconnect the following:
- Negative battery cable

INTAKE MANIFOLD END PLATE

**6 x 1.0 mm
12 N·m (1.2 kgf·m, 8.7 lbf·ft)**

GASKET
Replace.

**8 x 1.25 mm
22 N·m (2.2 kgf·m, 16 lbf·ft)**

GASKET
Replace.

BOOST PLATE

**6 x 1.0 mm
12 N·m (1.2 kgf·m, 8.7 lbf·ft)**

**INTAKE AIR TEMPERATURE (IAT)SENSOR
18 N·m (1.8 kgf·m, 13 lbf·ft)**

O-RING
Replace.

**12 x 1.5 mm
26 N·m (2.7 kgf·m, 20 lbf·ft)**

DOWEL PIN

INTAKE MANIFOLD END PLATE

GASKET
Replace.

INTAKE MANIFOLD
Replace if it is cracked or if the mating surface is damaged.

GASKET
Replace.

GASKET
Replace.

SPACER

GASKET
Replace.

**8 x 1.25 mm
22 N·m (2.2 kgf·m, 16 lbf·ft)**

THROTTLE BODY

9308MG20

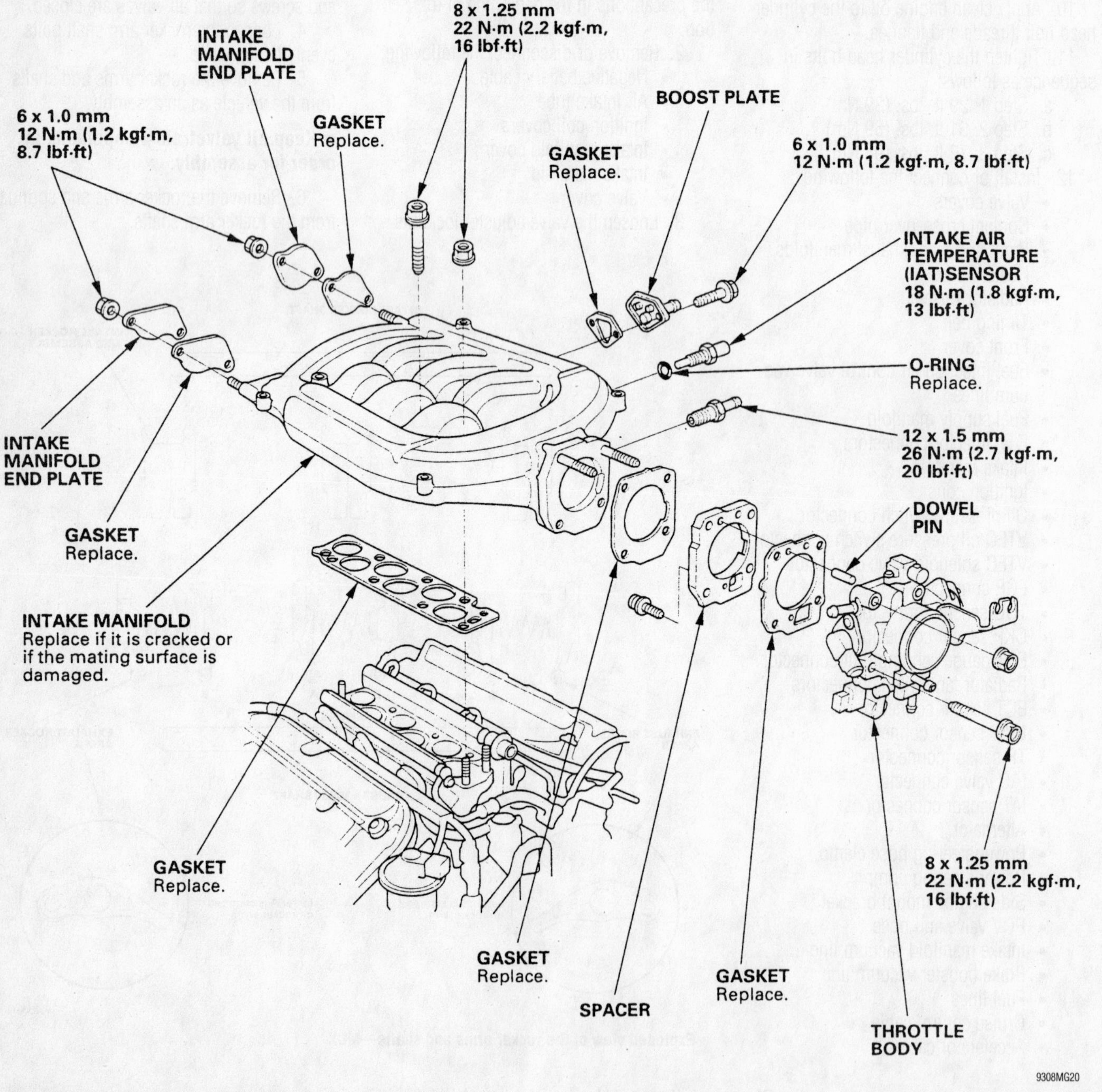

Exploded view of the intake manifold—MDX

- Evaporative Emissions (EVAP) control canister hose and vacuum hose
- Air intake tube
- Intake manifold cover
- Accelerator cable
- Cruise control cable
- Brake booster vacuum line
- Intake manifold vacuum line
- Positive Crankcase Ventilation (PCV) valve and hose
- Intake Air Temperature (IAT) sensor connector
- Idle Air Control (IAC) valve connector
- Throttle Position (TP) sensor connector
- Manifold Absolute Pressure (MAP) sensor connector
- Intake manifold

To install:

3. Install or connect the following:
- New intake manifold gasket
- Intake manifold. Tighten the fasteners in sequence and in several passes to 16 ft. lbs. (22 Nm).
- MAP sensor connector
- TP sensor connector
- IAC valve connector
- IAT sensor connector
- PCV valve and hose
- Intake manifold vacuum line
- Brake booster vacuum line
- Cruise control cable

- Accelerator cable
- Intake manifold cover
- Air intake tube
- EVAP control canister hose and vacuum hose
- Negative battery cable

4. Start the engine and check for proper operation.

Exhaust Manifold

REMOVAL & INSTALLATION

1. Before servicing the vehicle, refer to the precautions in the beginning of this section.
2. Remove or disconnect the following:
- Negative battery cable
- Exhaust manifold heat shield
- Heated Oxygen (HO2S) sensor connector
- Exhaust front pipe
- Exhaust manifold bracket, if equipped
- Exhaust manifold

To install:

3. Install or connect the following:
- Exhaust manifold. Tighten the fasteners to 23 ft. lbs. (31 Nm).
- Exhaust manifold bracket, if equipped. Tighten the bolts to 33 ft. lbs. (44 Nm).

- Exhaust front pipe. Tighten the nuts to 40 ft. lbs. (55 Nm).
- Heated Oxygen (HO2S) sensor connector
- Exhaust manifold heat shield
- Negative battery cable

Front Crankshaft Seal

REMOVAL & INSTALLATION

1. Before servicing the vehicle, refer to the precautions in the beginning of this section.
2. Remove or disconnect the following:
- Negative battery cable
- Accessory drive belts
- Side engine mount
- Valve cover
- Crankshaft pulley
- Front cover
- Balance shaft belt, if equipped
- Timing belt. Refer to the Timing Belt unit repair section.
- Top Dead Center (TDC) sensor, if equipped
- Crankshaft timing sprocket
- Front crankshaft seal

To install:

3. Lubricate the crankshaft seal lip with grease prior to installation.
4. Install the front crankshaft seal so that it is flush with the surface of the oil pump housing.
5. Install or connect the following:
- Crankshaft timing sprocket
- Top Dead Center (TDC) sensor, if equipped
- Timing belt. Refer to the Timing Belt unit repair section.
- Balance shaft belt, if equipped
- Front cover
- Crankshaft pulley. Tighten the bolt to 181 ft. lbs. (245 Nm).

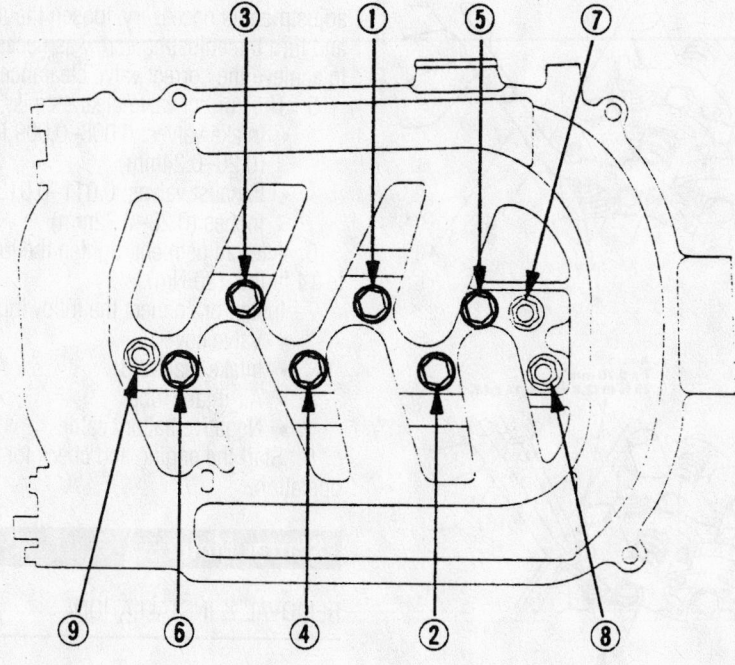

Intake manifold torque sequence—MDX

9302MG71

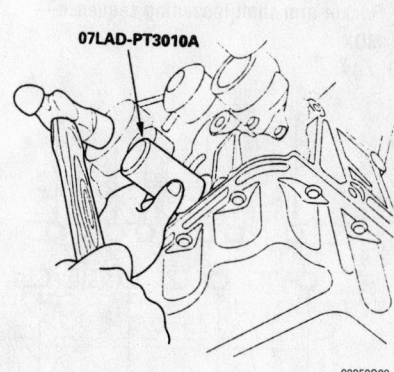

07LAD-PT3010A

Front crankshaft seal installation

93352G02

Timing belt service is covered in Section 3 of this manual

- Valve cover
- Side engine mount
- Accessory drive belts
- Negative battery cable

6. Check the engine oil level and add if necessary.

7. Start the engine and check for leaks.

Camshaft

REMOVAL & INSTALLATION

1. Before servicing the vehicle, refer to the precautions in the beginning of this section.

2. Remove or disconnect the following:
- Negative battery cable
- Air intake tube
- Accessory drive belts
- Front cover
- Timing belt. Refer to the Timing Belt unit repair section.
- Camshaft sprockets
- Timing belt rear covers
- Ignition coil covers
- Intake manifold cover
- Intake manifold
- Valve cover
- Rocker arms and shaft assembly
- Camshaft thrust cover
- Camshaft

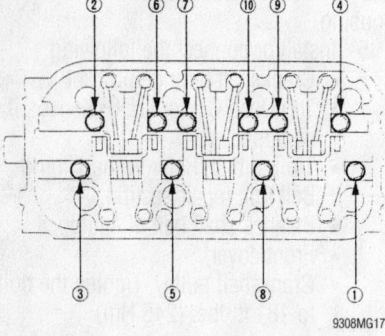

Rocker arm shaft loosening sequence—MDX

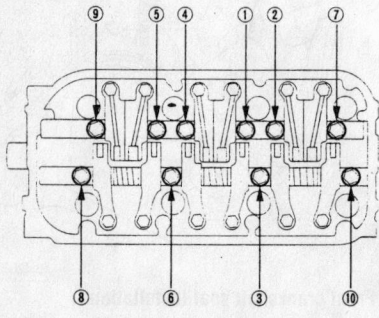

Rocker shaft tightening sequence—MDX

To install:

➡ **Use new O-rings, seals and gaskets when installing the camshaft.**

3. Install or connect the following:
- Camshaft
- Camshaft thrust cover. Tighten the bolts to 16 ft. lbs. (22 Nm).
- Rocker arms and shaft assembly
- Valve cover
- Intake manifold
- Intake manifold cover
- Ignition coil covers
- Timing belt rear covers
- Camshaft sprockets. Tighten the bolts to 67 ft. lbs. (90 Nm).
- Timing belt
- Front cover
- Accessory drive belts
- Air intake tube
- Negative battery cable

4. Start the engine and check for leaks.

Valve Lash

ADJUSTMENT

Adjust the valves only when the cylinder head temperature is less than 100°F (38°C).

1. Before servicing the vehicle, refer to the precautions in the beginning of this section.

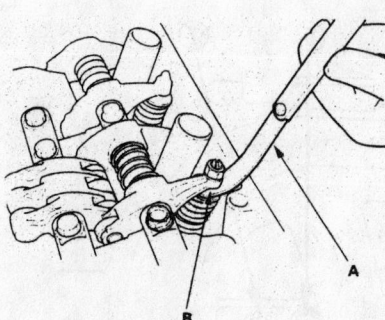

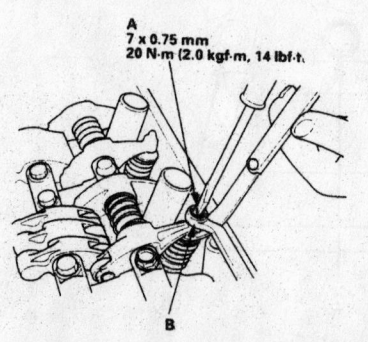

A
7 x 0.75 mm
20 N·m (2.0 kgf·m, 14 lbf·t.

Inspect the valve clearance, adjust to specification and tighten the retainer to specification

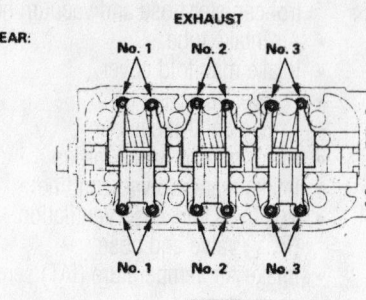

Adjusting screw locations:

Valve adjusting retainer locations

2. Remove or disconnect the following:
- Negative battery cable
- Air intake tube
- Intake manifold
- Valve cover

3. Rotate the crankshaft so that the valves to be adjusted are closed and the rocker arm is contacting the camshaft lobe base circle.

4. Measure the valve clearance. If adjustment is necessary, loosen the locknut and turn the adjusting screw as necessary to achieve the correct valve clearance.

5. The correct valve clearance is:
- Intake valves: 0.008–0.009 inches (0.20–0.24mm)
- Exhaust valves: 0.011–0.013 inches (0.28–0.32mm)

6. After adjustment, tighten the locknuts to 14 ft. lbs. (20 Nm).

7. Install or connect the following:
- Valve cover
- Intake manifold
- Air intake tube
- Negative battery cable

8. Start the engine and check for proper operation.

Starter Motor

REMOVAL & INSTALLATION

1. Before servicing the vehicle, refer to the precautions in the beginning of this section.

2. Remove or disconnect the following:

- Negative battery cable and wait at least 3 minutes
- Transmission fluid cooler line clamp
- Starter wiring harness connectors
- Starter motor

To install:

3. Install or connect the following:
- Starter motor. Tighten the 10mm bolt to 33 ft. lbs. (44 Nm) and the 12mm bolt to 47 ft. lbs. (64 Nm).
- Starter wiring harness connectors. Tighten the battery cable nut to 79 inch lbs. (9 Nm).
- Transmission fluid cooler line clamp
- Negative battery cable

Oil Pan

REMOVAL & INSTALLATION

1. Before servicing the vehicle, refer to the precautions in the beginning of this section.
2. Drain the engine oil.
3. Remove or disconnect the following:
- Negative battery cable
- Front splash shield
- Heated Oxygen (HO$_2$S) sensor connector
- Exhaust front pipe
- Torque converter cover, if equipped with an automatic transaxle
- Oil pan

To install:

4. Install the oil pan. Apply liquid gasket as shown.
5. Tighten the bolts in sequence to 105 inch lbs. (12 Nm).

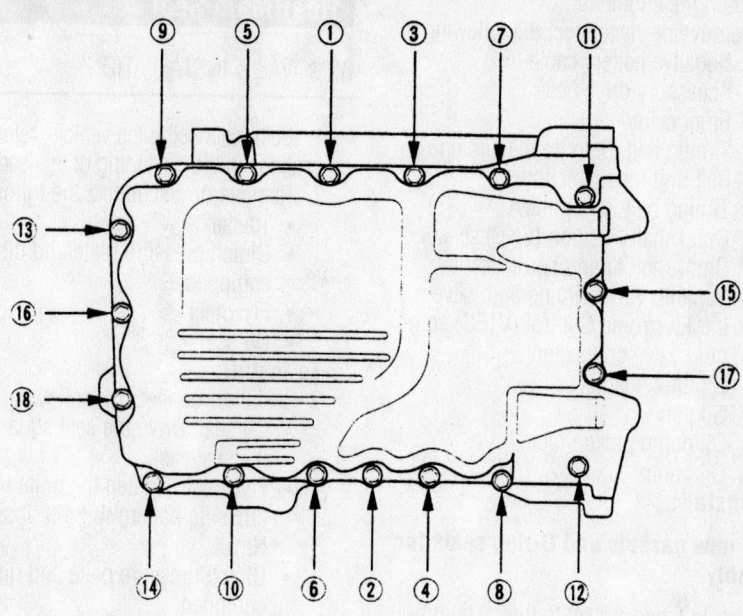

Oil pan tightening sequence—MDX

6. Install or connect the following:
- Torque converter cover, if removed
- Exhaust front pipe
- Subframe center beam, if removed
- HO$_2$S sensor connector
- Front splash shield
- Negative battery cable

Oil Pump

REMOVAL & INSTALLATION

1. Before servicing the vehicle, refer to the precautions in the beginning of this section.

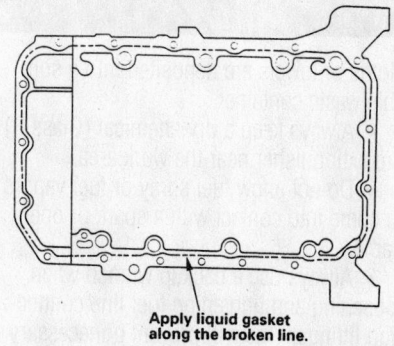

Apply liquid gasket along the broken line.

9302MG75

Apply liquid gasket to the inner threads of the bolt holes and the engine block along the area indicated by the broken line— MDX

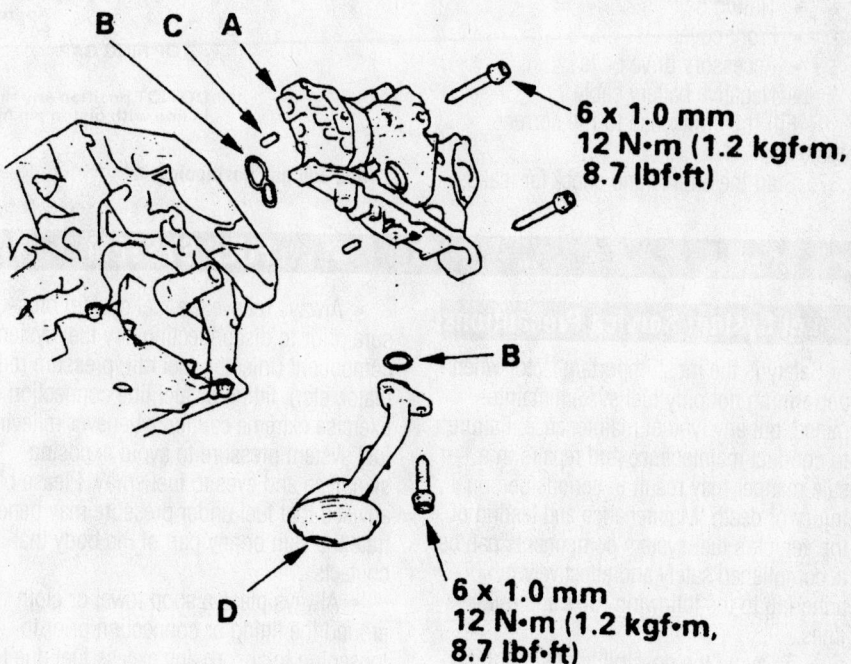

**6 x 1.0 mm
12 N•m (1.2 kgf•m, 8.7 lbf•ft)**

**6 x 1.0 mm
12 N•m (1.2 kgf•m, 8.7 lbf•ft)**

93552G05

Exploded view of the oil pump assembly

Heater Core replacement is covered in Section 2 of this manual

2. Drain the engine oil.
3. Remove or disconnect the following:
   - Negative battery cable
   - Accessory drive belts
   - Front cover
   - Timing belt. Refer to the Timing Belt unit repair section.
   - Timing belt idler pulley
   - Crankshaft Position (CKP) sensor
   - Crankshaft timing sprocket
   - Variable Valve Timing and Valve Lift Electronic Control (VTEC) solenoid valve connector
   - Oil filter adapter
   - Oil pan
   - Oil pump pickup tube
   - Oil pump

To install:

➡**Use new gaskets and O-ring seals for assembly.**

4. Apply liquid gasket to the oil pump and to the bolt hole threads.
5. Install or connect the following:
   - Oil pump. Tighten the bolts to 105 inch lbs. (12 Nm).
   - Oil pump pickup tube. Tighten the bolts to 105 inch lbs. (12 Nm).
   - Oil pan
   - Oil filter adapter
   - VTEC solenoid valve connector
   - Crankshaft timing sprocket
   - CKP sensor
   - Timing belt idler pulley
   - Timing belt
   - Front cover
   - Accessory drive belts
   - Negative battery cable
6. Fill the crankcase to the correct level.
7. Start the engine and check for leaks.

Rear Main Seal

REMOVAL & INSTALLATION

1. Before servicing the vehicle, refer to the precautions in the beginning of this section.
2. Remove or disconnect the following:
   - Transaxle
   - Clutch pressure plate and disc, if equipped
   - Flywheel
   - Oil seal

To install:
3. Install or connect the following:
   - Oil seal. Drive the seal square into the seal case.
   - Flywheel. Tighten the bolts in a crossing pattern to 54 ft. lbs. (73 Nm).
   - Clutch pressure plate and disc, if equipped
   - Transaxle
4. Check the fluid levels.
5. Start the engine and check for leaks.

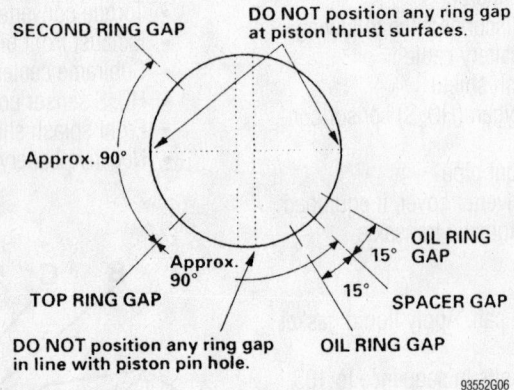

Ring end gap positioning

Piston and Ring

POSITIONING

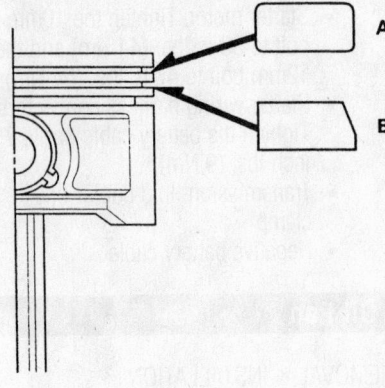

9302AG06

Compression ring identification—3.5L engine

FUEL SYSTEM

Fuel System Service Precautions

Safety is the most important factor when performing not only fuel system maintenance, but any type of maintenance. Failure to conduct maintenance and repairs in a safe manner may result in serious personal injury or death. Maintenance and testing of the vehicle's fuel system components can be accomplished safely and effectively by adhering to the following rules and guidelines:

- To avoid the possibility of fire and personal injury, always disconnect the negative battery cable unless the repair or test procedure requires that battery voltage be applied.

- Always relieve the fuel system pressure prior to disconnecting any fuel system component (injector, fuel rail, pressure regulator, etc.), fitting or fuel line connection. Exercise extreme caution whenever relieving fuel system pressure to avoid exposing skin, face and eyes to fuel spray. Please be advised that fuel under pressure may penetrate the skin or any part of the body that it contacts.

- Always place a shop towel or cloth around the fitting or connection prior to loosening to absorb any excess fuel due to spillage. Ensure that all fuel spillage (should it occur) is quickly removed from engine surfaces. Ensure that all fuel soaked cloths or towels are deposited into a suitable waste container.

- Always keep a dry chemical (Class B) fire extinguisher near the work area.

- Do not allow fuel spray or fuel vapors to come into contact with a spark or open flame.

- Always use a backup wrench when loosening and tightening fuel line connection fittings. This will prevent unnecessary stress and torsion to fuel line piping. Always follow the proper torque specifications.

- Always replace worn fuel fitting O-rings with new. Do not substitute fuel hose or equivalent, where fuel pipe is installed.

Fuel System Pressure

RELIEVING

1. Before servicing the vehicle, refer to the precautions in the beginning of this section.
2. Disconnect the negative battery cable.
3. Remove the fuel filler cap.
4. Place a shop towel over the fuel pulsation damper.
5. Loosen the fuel pulsation damper 1 turn.
6. When service is completed, replace the sealing washer and tighten the pulsation damper to 16 ft. lbs. (22 Nm).
7. Replace the fuel filler cap.
8. Connect the negative battery cable.
9. Start the engine and check for leaks.

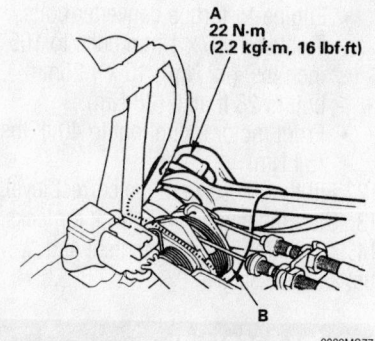

A
22 N·m
(2.2 kgf·m, 16 lbf·ft)

B

9302MG77

Use a wrench on the fuel pulsation damper (A). Place a rag over the damper (B) when relieving residual fuel pressure—MDX

Fuel Filter

REMOVAL & INSTALLATION

1. Before servicing the vehicle, refer to the precautions in the beginning of this section.
2. Relieve the fuel system pressure.
3. Remove or disconnect the following:
- Negative battery cable
- Driver's side second row seat and cut the carpet along the dotted line. Be careful not to cut the wiring harness under the carpet.
- Access panel
- Fuel pump module

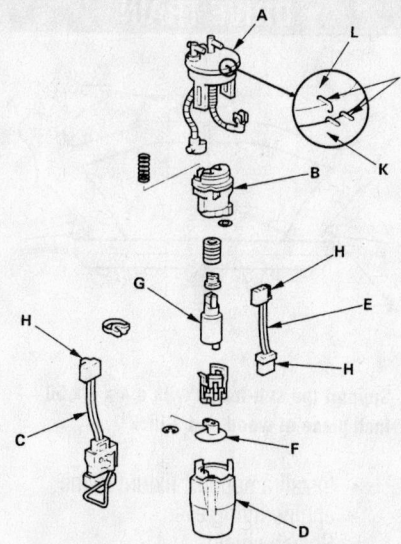

A. Bracket
B. Fuel filter
C. Fuel gauge sender
D. Case
E. Wire harness
F. Suction filter
G. Fuel pump
H. Connectors
J. Alignment marks
K. Fuel tank
L. Fuel pump module

9308MG26

Exploded view of the fuel pump module—MDX

4. Disassemble the fuel pump module and remove the fuel filter.
To install:
5. Install the fuel filter and assemble the fuel pump module.
6. Install or connect the following:
- Fuel pump module
- Access panel
- Carpet and seat
- Negative battery cable
7. Start the engine and check for leaks.

Fuel Pump

REMOVAL & INSTALLATION

1. Before servicing the vehicle, refer to the precautions in the beginning of this section.
2. Relieve the fuel system pressure.
3. Remove or disconnect the following:
- Negative battery cable
- Driver's side second row seat and cut the carpet along the dotted line. Be careful not to cut the wiring harness under the carpet.

- Access panel
- Fuel pump module wiring connector
- Fuel supply and return lines
- Fuel pump locknut
- Fuel pump module
To install:
4. Install or connect the following:
- Fuel pump module. Use a new seal and align the matchmarks.
- Fuel pump locknut
- Fuel supply and return lines
- Fuel pump module wiring connector
- Access panel
- Carpet and seat
- Negative battery cable
5. Start the engine and check for leaks.

Fuel Injector

REMOVAL & INSTALLATION

1. Before servicing the vehicle, refer to the precautions in the beginning of this section.
2. Relieve the fuel system pressure.
3. Remove or disconnect the following:
- Negative battery cable
- Intake manifold
- Fuel lines
- Fuel injector connectors
- Fuel pressure regulator vacuum line
- Fuel supply manifold
4. Separate the fuel injectors from the fuel supply manifold.
To install:
5. Install the fuel injectors to the fuel supply manifold with new cushion rings and O-rings.
6. Install new seal rings to the intake manifold.
7. Install or connect the following:
- Fuel supply manifold and injector assembly. Tighten the bolts to 86 inch lbs. (10 Nm).
- Fuel pressure regulator vacuum line
- Fuel injector connectors
- Fuel lines
- Intake manifold
- Negative battery cable
8. Start the engine and check for leaks.

Brake service is covered in Section 4 of this manual

DRIVE TRAIN

Transaxle Assembly

REMOVAL & INSTALLATION

1. Before servicing the vehicle, refer to the precautions in the beginning of this section.
2. Drain the transaxle.
3. Drain the power steering system.
4. Remove the engine appearance covers.
5. Remove the drivers side center console lower panel and pull back the cover to access steering joint cover.
6. Remove or disconnect the following:
   - Steering joint bolt
   - Steering joint from the steering gearbox pinion shaft
   - Air intake assembly
   - Battery
   - Battery tray
   - Power steering pump hose and the clamp bolt
   - Transmission breather tube
   - Cooler hose from the clamp on the starter
   - Transaxle oil cooler lines
   - Starter motor
   - Shift control solenoid valve connectors
   - Transaxle ground cable
   - 8P connector from the bracket and the connector
   - Clutch pressure switch connectors
   - Joint connector and transmission range switch connector from the brackets
   - Countershaft speed sensor connector
   - Heated Oxygen (HO2S) sensor connectors
   - Transmission housing mounting bolts
   - Nut from the front mount and the ground cable from the engine
   - Bulkhead cover, windshield wiper arms, cowl cover sealing and cover

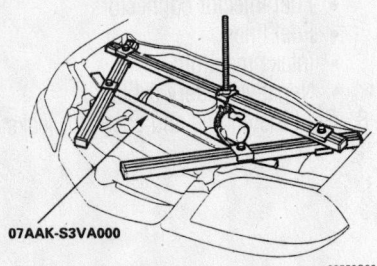

07AAK-S3VA000

93552G08

Support the engine while removing the transaxle—MDX

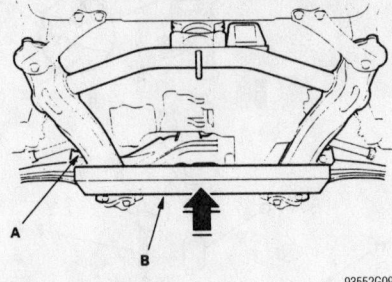

A

B

93552G09

Support the sub-frame with a 4 x 4 x 50 inch piece of wood and a jack

- Install a support fixture to the engine lifting eyes.
- Splash shield
- Front sub-frame stiffener
- Exhaust front pipe
- Lower control arms from the knuckle
- Stabilizer bar links
- Tie rod ends from the knuckle
- Left driveshaft from the differential
- Right driveshaft from the intermediate shaft
- Propeller shaft from the companion flange
- Shift cable cover and holder
- Shift control cable and lever

7. Install a 6 x 1 x 14mm bolt and nut on the cable cover, then reinstall the cable cover to the torque converter housing. If this is not done, the bolt head of the cable cover may prevent torque converter removal.
   - Transfer assembly
   - Engine-to-torque converter bolts
   - Power steering pressure switch connection
   - Power steering hose clamp, then the hose from the pipe at the sub-frame
   - Transmission lower mount nuts

8. Matchmark the front subframe to the vehicle body.
   - Rear mount bracket bolts

9. Support the sub-frame with a 4 x 4 x 50 inch piece of wood and a jack.
   - Sub-frame
   - Transaxle lower mounts
   - Driveshafts from the differential and intermediate shaft
   - Intermediate shaft
   - Transmission front mount bracket
   - Tranmission flange bolts
   - Transmission

To install:

➡**Use new circlips, split pins and self-locking nuts for assembly.**

10. Installation is the reverse of removal. Please note the following specifications:
   - Transmission housing bolts and harness clamp bolts to 47 ft. lbs. (64 Nm)
   - Transmission housing bolts to 40 ft. lbs. (54 Nm)
   - Front mount bracket bolts to 28 ft. lbs. (38 Nm)
   - Intermediate shaft bolts to 29 ft. lbs. (39 Nm)

11. Raise the subframe into position and align the matchmarks. Tighten the subframe bolts to 76 ft. lbs. (103 Nm). Tighten the front subframe bracket bolts to 54 ft. lbs. (74 Nm) and the rear bracket bolts to 86 ft. lbs. (117 Nm).
   - Rear engine mount bolts to 28 ft. lbs. (38 Nm)
   - Engine-to-torque converter bolts. Tighten the 6 x 1 mm bolts to 105 inch lbs. (12 Nm), 10 x 1.25mm bolt to 28 ft. lbs. (38 Nm).
   - Front motor mount nut to 40 ft. lbs. (54 Nm)

12. Fill the transaxle to the correct level.
13. Start the engine and check for leaks.
14. Check the wheel alignment and adjust as necessary.

Transfer Assembly

REMOVAL & INSTALLATION

1. Before servicing the vehicle, refer to the precautions in the beginning of this section.
2. Remove or disconnect the following:
   - Negative battery cable
   - Heated Oxygen (HO2S) sensor connectors
   - Front sub-frame stiffener
   - Exhaust front pipe
   - Propeller shaft from the transfer assembly
   - Transfer assembly bolts and the assembly

To install:

3. Install or connect the following:
   - New O-ring on the transfer cover
   - Dowel pin on the assembly
   - Transfer assembly and tighten the bolts to 33 ft. lbs. (44 Nm)
   - Propeller shaft
   - Exhaust front pipe
   - Front sub-frame stiffener and tighten the bolts to 40 ft. lbs. (54 Nm)
   - Heated Oxygen (HO2S) sensor connectors
   - Negative battery cable

Halfshaft

REMOVAL & INSTALLATION

1. Before servicing the vehicle, refer to the precautions in the beginning of this section.
2. Drain the transaxle.
3. Remove or disconnect the following:
   - Negative battery cable
   - Front wheels
   - Spindle nut
   - Stabilizer bar link
   - Lower ball joint
4. Pry the inboard joint from the transaxle or intermediate shaft.

5. Remove the outer CV-joint stub shaft from the hub by tapping the stub shaft with a plastic hammer.

To install:

➡**Use new circlips, split pins and self-locking nuts for assembly.**

6. Install the outer CV-joint stub shaft into the hub.
7. Install the inner CV-joint to the transaxle or intermediate shaft until the circlip locks in the retaining groove.
8. Install or connect the following:
   - Lower ball joint. Tighten the nut to 43–51 ft. lbs. (59–69 Nm).
   - Stabilizer bar link. Tighten the nut to 58 ft. lbs. (78 Nm).

   - Spindle nut. Tighten the nut to 181 ft. lbs. (245 Nm).
   - Front wheels
   - Negative battery cable
9. Fill the transaxle to the correct level and check for leaks.

CV-Joint

OVERHAUL

Front

OUTBOARD JOINT

1. Before servicing the vehicle, refer to the precautions in the beginning of this section.

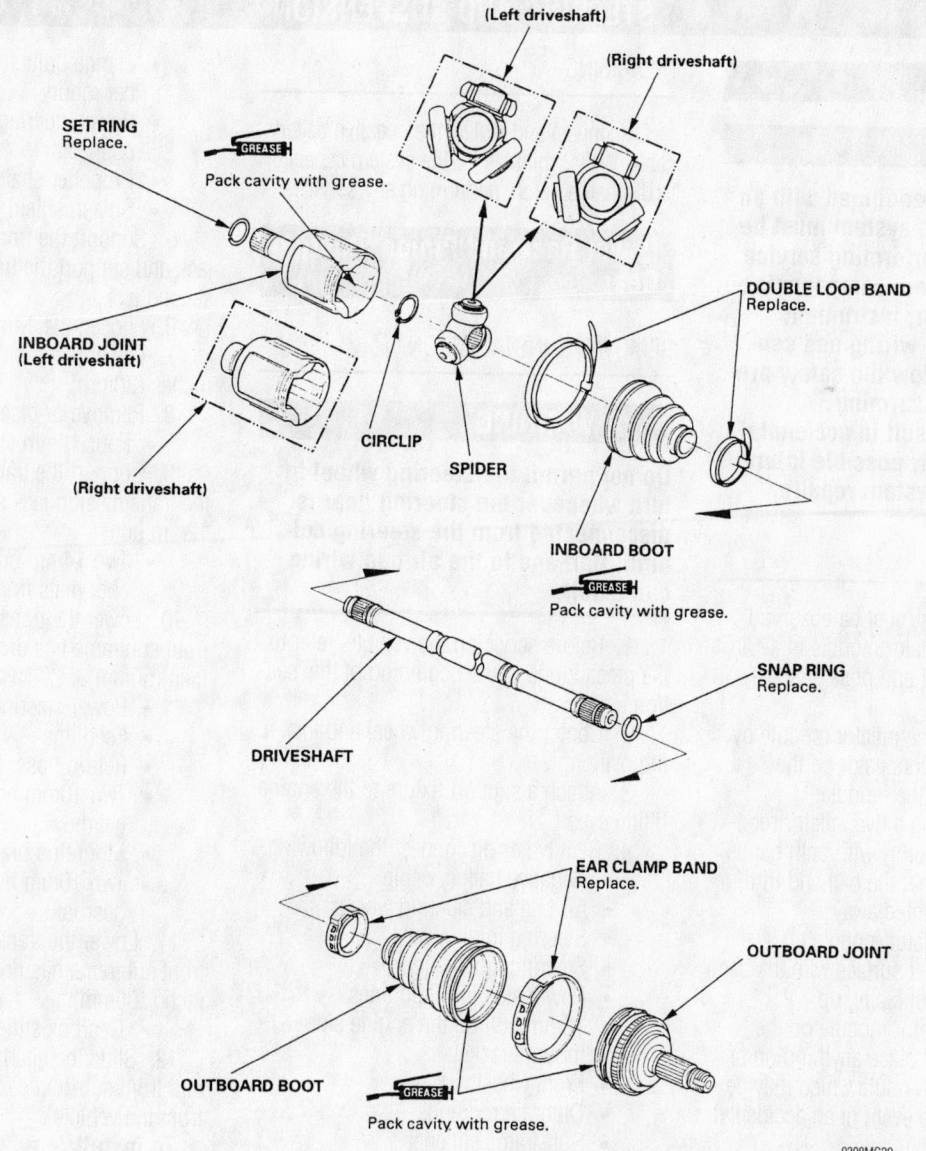

SET RING
Replace.

GREASE
Pack cavity with grease.

(Left driveshaft)

(Right driveshaft)

INBOARD JOINT
(Left driveshaft)

(Right driveshaft)

CIRCLIP

SPIDER

DOUBLE LOOP BAND
Replace.

INBOARD BOOT

GREASE
Pack cavity with grease.

DRIVESHAFT

SNAP RING
Replace.

EAR CLAMP BAND
Replace.

OUTBOARD JOINT

OUTBOARD BOOT

GREASE
Pack cavity with grease.

9308MG29

Front axle exploded view—MDX

For complete Engine Mechanical specifications, see Section 1 of this manual

2. Remove or disconnect the following:
- Axle halfshaft from the vehicle and place it in a vise
- Outboard joint boot clamps and push the boot back
- Outboard joint by driving it off the axle shaft with a brass drift and hammer
- Outboard joint boot

To install:

➡**Use new circlips and boot clamps for assembly.**

3. Install the outboard joint boot and clamps to the axle shaft.

4. Fill the outboard joint with grease. Install the outboard joint to the axle shaft.

Tap the stub shaft with a brass hammer to seat the circlip.

5. Fill the outboard joint boot with grease and install the boot clamps.

6. Install the axle halfshaft to the vehicle.

INBOARD JOINT

1. Before servicing the vehicle, refer to the precautions in the beginning of this section.

2. Remove or disconnect the following:
- Axle halfshaft from the vehicle.
- Inboard joint boot clamps and push the boot back
- Inboard joint housing from the axle
- Rollers from the spider

- Snapring and the spider from the axle shaft
- Inboard joint boot

To install:

➡**Use new circlips and boot clamps for assembly.**

3. Install or connect the following:
- Inboard joint boot and clamps to the axle shaft
- Spider with a new snapring
- Rollers to the spider

4. Fill the joint housing with grease and install it.

5. Fill the inboard joint boot with grease and install the boot clamps.

6. Install the axle halfshaft to the vehicle.

STEERING AND SUSPENSION

Air Bag

❋❋ CAUTION

Some vehicles are equipped with an air bag system. The system must be disarmed before performing service on, or around, system components, the steering column, instrument panel components, wiring and sensors. Failure to follow the safety precautions and the disarming procedure could result in accidental air bag deployment, possible injury and unnecessary system repairs.

PRECAUTIONS

Several precautions must be observed when handling the inflator module to avoid accidental deployment and possible personal injury.

- Never carry the inflator module by the wires or connector on the underside of the module.
- When carrying a live inflator module, hold securely with both hands, and ensure that the bag and trim cover are pointed away.
- Place the inflator module on a bench or other surface with the bag and trim cover facing up.
- With the inflator module on the bench, never place anything on or close to the module which may be thrown in the event of an accidental deployment.

Before servicing the vehicle, also make sure to refer to the precautions in the beginning of this section as well.

DISARMING

Disconnect and isolate the negative battery cable. Wait 3 minutes for the system capacitor to discharge before performing any service.

Power Rack and Pinion Steering Gear

REMOVAL & INSTALLATION

❋❋ WARNING

Do not permit the steering wheel to turn whenever the steering gear is disconnected from the steering column. Damage to the air bag wiring can result.

1. Before servicing the vehicle, refer to the precautions in the beginning of this section.

2. Center the steering wheel and lock it in position.

3. Attach a support fixture to the engine lifting eyes.

4. Remove or disconnect the following:
- Negative battery cable
- Air bag and steering wheel
- Steering joint cover
- Steering flexible joint
- Power steering fluid lines
- 10mm bolt on the engine side mount bracket
- Front wheels
- Outer tie rod ends
- Sub-frame stiffener
- Heated Oxygen (HO_2S) sensor connectors
- 3 way catalytic converter from the mufflers

- Flange bolts from the exhaust rubber mount
- Power steering pressure switch connector
- Propeller shaft protector
- Splash shield

5. Support the front subframe with a jack and support the transmission with a second jack.

6. Loosen the 14mm subframe bolts.

7. Lower the subframe about 1 3/16 inches (30mm).

8. Remove or disconnect the following:
- Four 12mm stiffener plate bolts

9. Support the transfer case by raising the transmission jack and remove the two 12mm bolts.
- Two 14mm bolts and the rear stiffener plats from the sub-frame

10. Lower the transmission jack until the front subframe has dropped about 1 15/16 inch (50mm).
- Power steering line brackets
- Feed line
- Return hose
- Two 10mm bolts from the right side gearbox
- Mounting bracket and cushion
- Two 10mm bolts from the left side gearbox

11. Lower the transmission jack until the front subframe has dropped about 3 15/16 inch (100mm).
- Gearbox stiffener bracket

12. Slide the gearbox between the body and front sub-frame towards the left and from the vehicle.

To install:

13. Position the steering gear in the vehicle.

14. Install or connect the following:

- Left steering gear mounting bolts. Tighten the bolts to 43 ft. lbs. (58 Nm).
- Right steering gear mounting bracket. Tighten the bolts to 29 ft. lbs. (39 Nm).
- Return hose
- Feed line
- Power steering line mounting brackets and tighten the bolts to 7 ft. lbs. (10 Nm)

15. Raise the subframe into position. Tighten the 14mm bolts to 76 ft. lbs. (103 Nm) and the 12mm bolts to 86 ft. lbs. (117 Nm).

16. Install or connect the following:
- Front stiffener plates. Tighten the 14mm bolts to 76 ft. lbs. (103 Nm) and the 12mm bolts to 54 ft. lbs. (74 Nm).
- Splash shield
- Propeller shaft protector
- Power steering pressure switch
- 3 way catalytic converter and mufflers. Tighten the nuts to 25 ft. lbs. (33 Nm)
- Rubber exhaust mount and tighten the bolts to 28 ft. lbs. (38 Nm)
- HO2S sensor connectors
- Sub-frame stiffener plate
- 10mm flange bolts on the engine side mount bracket to 33 ft. lbs. (44 Nm)
- Power steering hoses
- Outer tie rod ends
- Front wheels. Position the wheels straight-ahead.
- Steering flexible joint. Tighten the pinch bolts to 16 ft. lbs. (22 Nm).
- Steering joint cover
- Negative battery cable

17. Fill the power steering system.

18. Check the wheel alignment and adjust as necessary.

Strut

REMOVAL & INSTALLATION

Front

1. Before servicing the vehicle, refer to the precautions in the beginning of this section.

2. Remove or disconnect the following:
- Front wheel
- Wheel speed sensor wiring bracket
- Brake hose bracket
- Stabilizer bar link

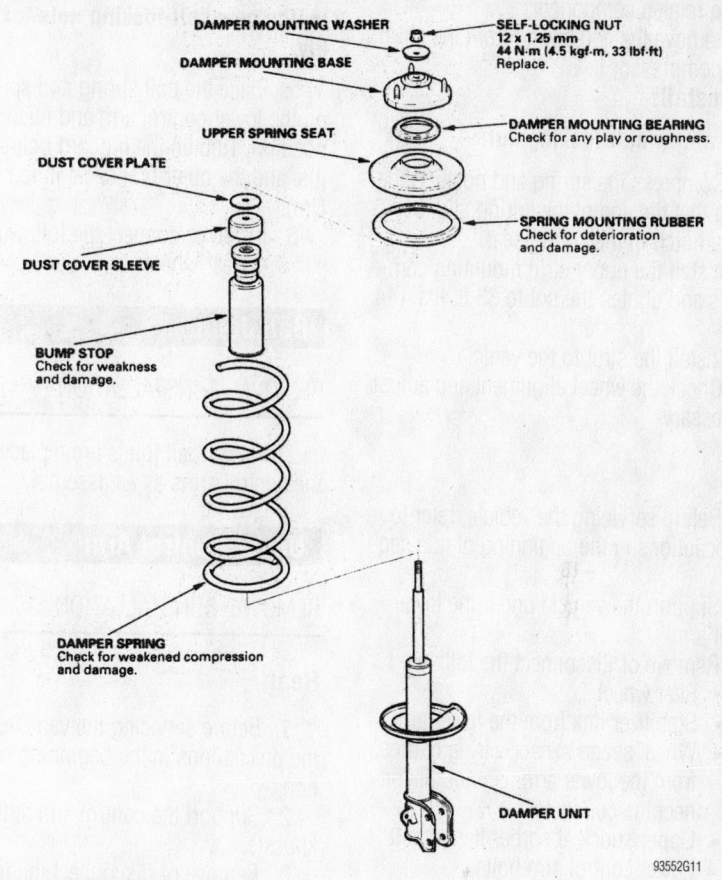

Exploded view of the front strut assembly

- Strut pinch bolts
- Upper mount nuts
- Strut

To install:

3. Install or connect the following:
- Strut. Tighten the upper mount nuts to 43 ft. lbs. (59 Nm).
- Strut pinch bolts. Tighten the nuts to 116 ft. lbs. (157 Nm).
- Stabilizer bar link. Tighten the nut to 58 ft. lbs. (78 Nm).
- Brake hose bracket
- Wheel speed sensor wiring bracket
- Front wheel

4. Check the wheel alignment and adjust as necessary.

Shock Absorber

REMOVAL & INSTALLATION

Rear

1. Before servicing the vehicle, refer to the precautions in the beginning of this section.

2. Support the vehicle under the lower control arm.

3. Remove or disconnect the following:
- Rear wheel
- Upper shock absorber flange bolt
- Lower shock absorber nut
- Shock absorber

To install:

4. Install or connect the following:
- Shock absorber. Tighten the fasteners to 47 ft. lbs. (64 Nm).
- Rear wheel

Coil Spring

REMOVAL & INSTALLATION

Front

1. Before servicing the vehicle, refer to the precautions in the beginning of this section.

2. Remove the strut from the vehicle and install in a strut spring compressor. Compress the spring until the end of the spring comes away from the spring seat.

For Accessory Drive Belt illustrations, see Section 1 of this manual

3. Remove the upper strut mount, spring seat and related components.

4. Remove the coil spring from the strut spring compressor.

To install:

➡**Use a new self-locking nut.**

5. Compress the spring and position the strut so that the end of the spring aligns with the notch in the spring seat.

6. Install the upper strut mounting components and tighten the nut to 33 ft. lbs. (44 Nm).

7. Install the strut to the vehicle.

8. Check the wheel alignment and adjust as necessary.

Rear

1. Before servicing the vehicle, refer to the precautions in the beginning of this section.

2. Support the vehicle under the lower control arm.

3. Remove or disconnect the following:
- Rear wheel
- Stabilizer link from the lower arm
- Wheel speed sensor wiring harness from the lower arm. Do not disconnect the connector.
- Upper shock absorber flange bolt
- Lower control arm bolts

4. Lower the floor jack and remove the coil spring and spring seats.

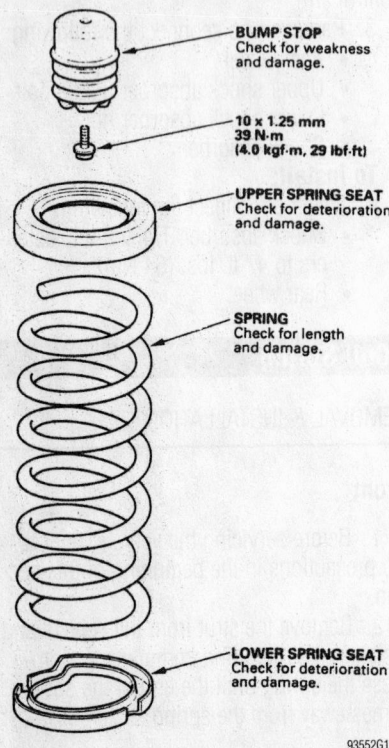

Exploded view of the rear spring assembly

To install:

➡**Use new self-locking nuts for assembly.**

5. Place the coil spring and spring seats on the lower control arm and raise into position. Tighten the inboard bolt to 61 ft. lbs. and the outer bolt to 54 ft. lbs. (74 Nm).

6. Install or connect the following:
- Rear wheel

Ball Joint

REMOVAL & INSTALLATION

The lower ball joints are replaced with the control arms as an assembly.

Upper Control Arm

REMOVAL & INSTALLATION

Rear

1. Before servicing the vehicle, refer to the precautions in the beginning of this section.

2. Support the control arm at the knuckle.

3. Remove or disconnect the following:
- Wheel
- Upper ball joint from the knuckle
- Upper arm bolt and the arm

4. Installation is the reverse of removal. Tighten the arm bolt to 47 ft. lbs. (64 Nm) and the ball joint nut to 36–43 ft. lbs. (49–59 Nm).

Lower Control Arm

REMOVAL & INSTALLATION

Front

1. Before servicing the vehicle, refer to the precautions in the beginning of this section.

2. Remove or disconnect the following:
- Front wheel
- Lower ball joint
- Front inner flange bolt
- Rear inner flange bolt
- Lower control arm

To install:

➡**Use a new split pin for assembly.**

3. Install or connect the following:
- Lower control arm. Tighten the inner flange bolts to 69 ft. lbs. (93 Nm).

- Lower ball joint. Tighten the nut to 43–51 ft. lbs. (59–69 Nm).
- Front wheel

4. Check the wheel alignment and adjust as necessary.

Rear

LOWER ARM (A)

1. Before servicing the vehicle, refer to the precautions in the beginning of this section.

2. Remove or disconnect the following:
- Lower arm mounting bolt and nut
- Lower arm

3. Installation is the reverse of removal. Tighten the bolt to 103 ft. lbs. (140 Nm) and the nut to 47 ft. lbs. (64 Nm).

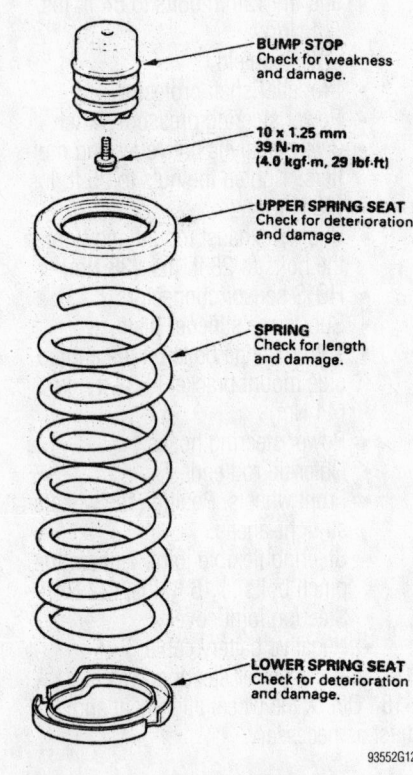

Rear lower arm (A) mounting

LOWER ARM (B)

1. Before servicing the vehicle, refer to the precautions in the beginning of this section.

2. Support the control arm with a jack.

3. Remove or disconnect the following:
- Wheel
- Stabilizer link from the lower arm
- Wheel speed sensor wiring harness from the lower arm. Do not disconnect the connector.
- Flange bolts that attaches the lower arm to the knuckle

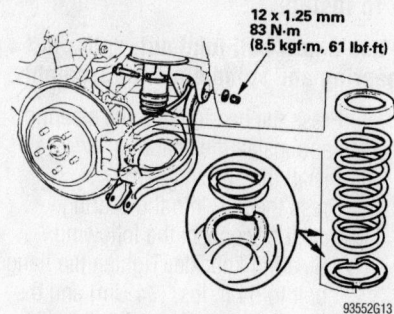

Rear lower arm (B) mounting

93552G13

4. Spring assembly
   • Inner nuts and bolts and the arm
5. Install or connect the following:
   • Arm, inner bolt and loosely install the nut
   • Spring assembly
6. Raise the arm into position and install the flange bolt.
7. Raise the rear suspension with a floor jack to load the vehicle weight.
   • Tighten the flange bolt to 54 ft. lbs. (74 Nm) and the inner nut and bolt to 61 ft. lbs. (83 Nm).
   • Wheel speed sensor harness
   • Wheel
8. Check the vehicle alignment.

CONTROL ARM BUSHING REPLACEMENT

The control arm bushings are serviced with the control arms as an assembly.

Wheel Bearings

ADJUSTMENT

The wheel bearings are sealed units and are not adjustable.

REMOVAL & INSTALLATION

Front

1. Before servicing the vehicle, refer to the precautions in the beginning of this section.
2. Remove or disconnect the following:
   • Front wheel
   • Brake hose mounting bolt
   • Brake caliper
   • Wheel speed sensor
   • Spindle nut
   • Brake rotor
   • Outer tie rod end
   • Lower ball joint
   • Steering knuckle

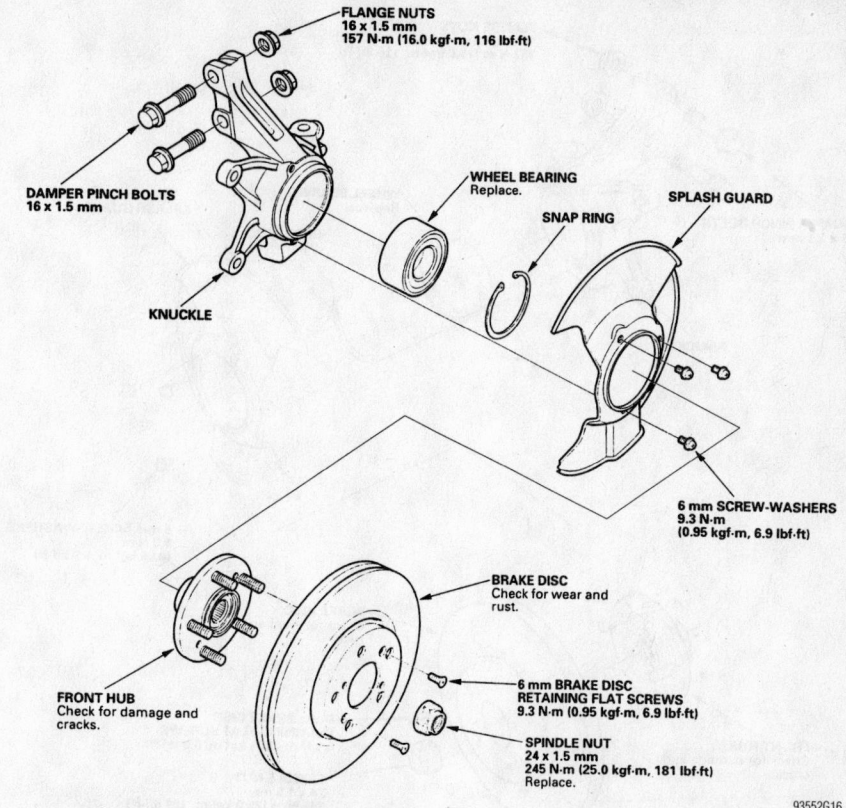

Exploded view of the front wheel bearing assembly

93552G16

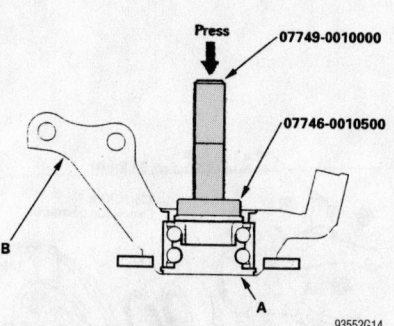

Press the wheel bearing out of the knuckle

93552G14

3. Press the hub out of the wheel bearing.
4. Remove the splash guard.
5. Remove the snapring and press the wheel bearing out of the steering knuckle.
6. If necessary, press the inner bearing race off of the hub.

To install:

➡**Use a new ball joint nut, split pin, snapring and spindle nut for assembly.**

7. Press the bearing into the steering knuckle and install the snapring.
8. Install the splash guard.
9. Press the hub into the bearing.
10. Install or connect the following:

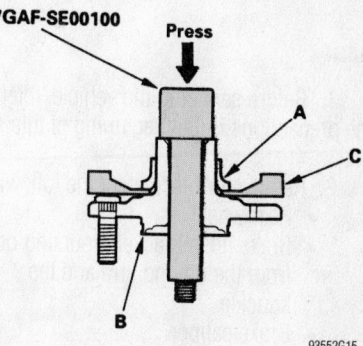

Press the wheel bearing inner race from the hub

93552G15

   • Steering knuckle. Tighten the ball joint nut to 43–51 ft. lbs. (59–69 Nm) and the damper flange bolts to 116 ft. lbs. (157 Nm).
   • Outer tie rod end. Tighten the nut to 40 ft. lbs. (54 Nm).
   • Wheel speed sensor, if equipped
   • Brake caliper and rotor
   • Brake hose
   • Spindle nut. Tighten the nut to 181 ft. lbs. (245 Nm).
   • Front wheel
11. Check the wheel alignment and adjust as necessary.

For Tire, Wheel and Ball Joint specifications, see Section 1 of this manual

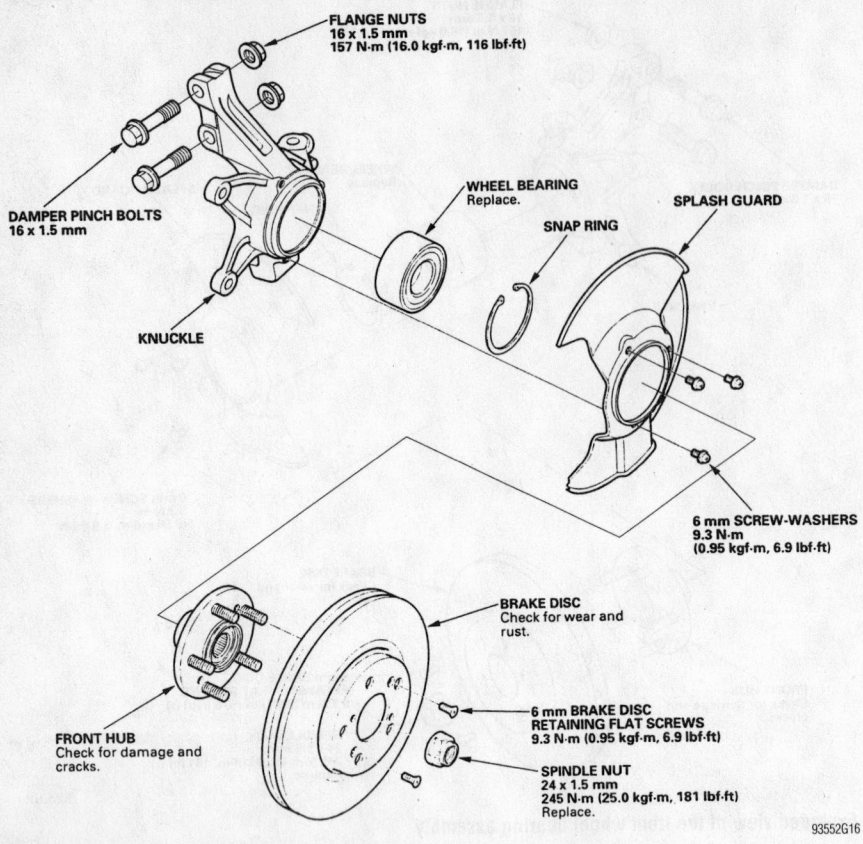

FLANGE NUTS
16 x 1.5 mm
157 N·m (16.0 kgf·m, 116 lbf·ft)

DAMPER PINCH BOLTS
16 x 1.5 mm

KNUCKLE

WHEEL BEARING
Replace.

SNAP RING

SPLASH GUARD

6 mm SCREW-WASHERS
9.3 N·m
(0.95 kgf·m, 6.9 lbf·ft)

BRAKE DISC
Check for wear and
rust.

FRONT HUB
Check for damage and
cracks.

**6 mm BRAKE DISC
RETAINING FLAT SCREWS**
9.3 N·m (0.95 kgf·m, 6.9 lbf·ft)

SPINDLE NUT
24 x 1.5 mm
245 N·m (25.0 kgf·m, 181 lbf·ft)
Replace.

93552G16

Press the wheel bearing into the knuckle

Rear

1. Before servicing the vehicle, refer to the precautions in the beginning of this section.

2. Remove or disconnect the following:
   - Rear wheel
   - Brake hose bracket mounting bolts from the trailing arm and the knuckle
   - Brake caliper
   - Wheel speed sensor
   - Spindle nut
   - Brake rotor
   - Upper ball joint
   - Lower arm (A)
   - Lower arm (B) from the trailing arm

3. Support the lower arm (B)
   - Steering knuckle

4. Press the hub out of the wheel bearing.

5. Remove the splash guard.

6. Remove the snapring and press the wheel bearing out of the steering knuckle.

7. If necessary, press the inner bearing race off of the hub.

To install:

➡ **Use a new ball joint nut, split pin, snapring and spindle nut for assembly.**

8. Press the bearing into the steering knuckle and install the snapring.

9. Install the splash guard.

10. Press the hub into the bearing.

11. Install or connect the following:
    - Steering knuckle. Tighten the flange bolt to 54 ft. lbs. (74 Nm) and the lower shock nut to 47 ft. lbs. (64 Nm)
    - Lower arm (B) to the trailing arm and tighten the bolts to 47 ft. lbs. (64 Nm)
    - Lower arm (A)
    - Upper ball joint and tighten the nut to 40 ft. lbs. (54 Nm)
    - Brake rotor and tighten the screws to 7 ft. lbs. (9 Nm)
    - Spindle nut and tighten to 181 ft, lbs. (245 Nm)
    - Wheel speed sensor
    - Brake caliper and tighten the bolts to 41 ft. lbs. (55 Nm)
    - Brake hose bracket mounting bolts to the knuckle and trailing arm
    - Rear wheel

12. Check the wheel alignment and adjust as necessary.

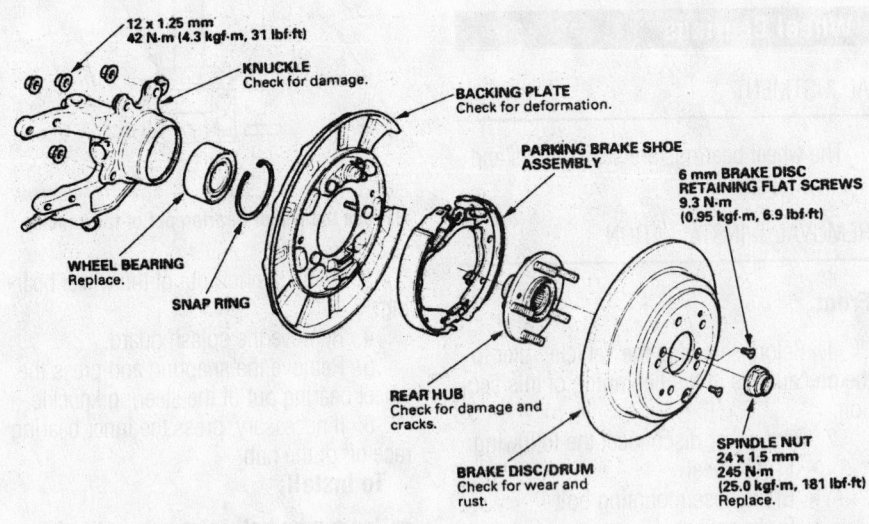

12 x 1.25 mm
42 N·m (4.3 kgf·m, 31 lbf·ft)

KNUCKLE
Check for damage.

BACKING PLATE
Check for deformation.

**PARKING BRAKE SHOE
ASSEMBLY**

**6 mm BRAKE DISC
RETAINING FLAT SCREWS**
9.3 N·m
(0.95 kgf·m, 6.9 lbf·ft)

WHEEL BEARING
Replace.

SNAP RING

REAR HUB
Check for damage and
cracks.

BRAKE DISC/DRUM
Check for wear and
rust.

SPINDLE NUT
24 x 1.5 mm
245 N·m
(25.0 kgf·m, 181 lbf·ft)
Replace.

93552G17

Exploded view of the rear wheel bearing assembly

ACURA

SLX

PRECAUTIONS

Before servicing any vehicle, please be sure to read all of the following precautions, which deal with personal safety, prevention of component damage and important points to take into consideration when servicing a motor vehicle:

• Never open, service or drain the radiator or cooling system when the engine is hot; serious burns can occur from the steam and hot coolant.

• Observe all applicable safety precautions when working around fuel. Whenever servicing the fuel system, always work in a well-ventilated area. Do not allow fuel spray or vapors to come in contact with a spark, open flame, or excessive heat (a hot drop light, for example). Keep a dry chemical fire extinguisher near the work area. Always keep fuel in a container specifically designed for fuel storage; also, always properly seal fuel containers to avoid the possibility of fire or explosion. Refer to the additional fuel system precautions later in this section.

• Fuel injection systems often remain pressurized, even after the engine has been turned **OFF**. The fuel system pressure must be relieved before disconnecting any fuel lines. Failure to do so may result in fire and/or personal injury.

• Brake fluid often contains polyglycol ethers and polyglycols. Avoid contact with the eyes and wash your hands thoroughly after handling brake fluid. If you do get brake fluid in your eyes, flush your eyes with clean, running water for 15 minutes. If eye irritation persists, or if you have taken brake fluid internally, seek medical assistance IMMEDIATELY.

• The EPA warns that prolonged contact with used engine oil may cause a number of skin disorders, including cancer. You should make every effort to minimize your exposure to used engine oil. Protective gloves should be worn when changing oil. Wash your hands and any other exposed skin areas as soon as possible after exposure to used engine oil. Soap and water, or waterless hand cleaner should be used.

• All new vehicles are now equipped with an air bag system, often referred to as a Supplemental Restraint System (SRS) or Supplemental Inflatable Restraint (SIR) system. The system must be disabled before performing service on or around system components, steering column, instrument panel components, wiring and sensors. Failure to follow safety and disabling procedures could result in accidental air bag deployment, possible personal injury and unnecessary system repairs.

• Always wear safety goggles when working with, or around, the air bag system. When carrying a non-deployed air bag, be sure the bag and trim cover are pointed away from your body. When placing a non-deployed air bag on a work surface, always face the bag and trim cover upward, away from the surface. This will reduce the motion of the module if it is accidentally deployed. Refer to the additional air bag system precautions later in this section.

• Clean, high quality brake fluid from a sealed container is essential to the safe and proper operation of the brake system. You should always buy the correct type of brake fluid for your vehicle. If the brake fluid becomes contaminated, completely flush the system with new fluid. Never reuse any brake fluid. Any brake fluid that is removed from the system should be discarded. Also, do not allow any brake fluid to come in contact with a painted surface; it will damage the paint.

• Never operate the engine without the proper amount and type of engine oil; doing so WILL result in severe engine damage.

• Timing belt maintenance is extremely important. Many models utilize an interference-type, non-freewheeling engine. If the timing belt breaks, the valves in the cylinder head may strike the pistons, causing potentially serious (also time-consuming and expensive) engine damage. Refer to the maintenance interval charts in the front of this manual for the recommended replacement interval for the timing belt and to the timing belt section for belt replacement and inspection.

• Disconnecting the negative battery cable on some vehicles may interfere with the functions of the on-board computer system(s) and may require the computer to undergo a relearning process once the negative battery cable is reconnected.

• When servicing drum brakes, only disassemble and assemble one side at a time, leaving the remaining side intact for reference.

• Only an MVAC-trained, EPA-certified automotive technician should service the air conditioning system or its components.

ENGINE REPAIR

➡**Disconnecting the negative battery cable on some vehicles may interfere with the functions of the on board computer system. The computer may undergo a relearning process once the negative battery cable is reconnected.**

Alternator

REMOVAL

1. Before servicing the vehicle, refer to the precautions in the beginning of this section.
2. Remove or disconnect the following:
   • Negative battery cable

• Accessory drive belt
• Alternator wiring connectors
• Alternator

INSTALLATION

1. Before servicing the vehicle, refer to the precautions in the beginning of this section.
2. Install or connect the following:
   • Alternator. Tighten the 10mm bolts to 30 ft. lbs. (41 Nm) and the 8mm bolts to 15 ft. lbs. (21 Nm).
   • Alternator wiring connectors
   • Accessory drive belt
   • Negative battery cable

Ignition Timing

ADJUSTMENT

Ignition timing is controlled by the Powertrain Control Module (PCM) and is not adjustable.

Engine Assembly

REMOVAL & INSTALLATION

1. Before servicing the vehicle, refer to the precautions in the beginning of this section.

2. Drain the cooling system.
3. Remove or disconnect the following:
- Battery
- Hood
- Air cleaner assembly
- Accelerator cable
- Cruise control cable
- Canister vacuum line
- Brake booster vacuum line
- Engine wiring harness connectors
- Transmission harness connectors and bracket
- Engine ground cable
- Starter harness connector
- Alternator harness connector
- Coolant reservoir tank hose
- Radiator hoses
- Heater hoses
- Upper fan shroud
- Radiator
- Cooling fan
- Accessory drive belt
- Power steering pump
- A/C compressor
- Heated Oxygen Sensor (HO2S) connectors
- Left and right exhaust front pipes
- Fuel lines
- Flywheel dust cover
- Transmission
- Left and right engine mounts
- Engine

To install:
4. Install or connect the following:
- Engine
- Left and right engine mounts. Tighten the bolts to 30 ft. lbs. (41 Nm).
- Transmission
- Flywheel dust cover
- Fuel lines
- Left and right exhaust front pipes
- HO2S connectors
- A/C compressor
- Power steering pump
- Accessory drive belt
- Cooling fan. Tighten the nuts to 16 ft. lbs. (22 Nm).
- Radiator
- Upper fan shroud
- Heater hoses
- Radiator hoses
- Coolant reservoir tank hose
- Alternator harness connector
- Starter harness connector
- Engine ground cable
- Transmission harness connectors and bracket
- Engine wiring harness connectors

- Brake booster vacuum line
- Canister vacuum line
- Cruise control cable
- Cruise control cable
- Accelerator cable
- ir cleaner assembly
- Hood
- Battery

5. Fill the cooling system. Check all fluid levels and adjust as necessary.
6. Start the engine and check for leaks.

Water Pump

REMOVAL & INSTALLATION

1. Before servicing the vehicle, refer to the precautions in the beginning of this section.
2. Drain the cooling system.
3. Remove or disconnect the following:
- Negative battery cable
- Upper radiator hose
- Timing belt
- Idler pulley
- Water pump

To install:

➡ **Apply Loctite® 262 to bolt number 3 prior to installation.**

4. Install or connect the following:
- Water pump. Tighten the bolts in two passes, in sequence, to 18 ft. lbs. (25 Nm).
- Idler pulley
- Timing belt
- Upper radiator hose
- Negative battery cable
5. Fill the cooling system.
6. Start the engine and check for leaks.

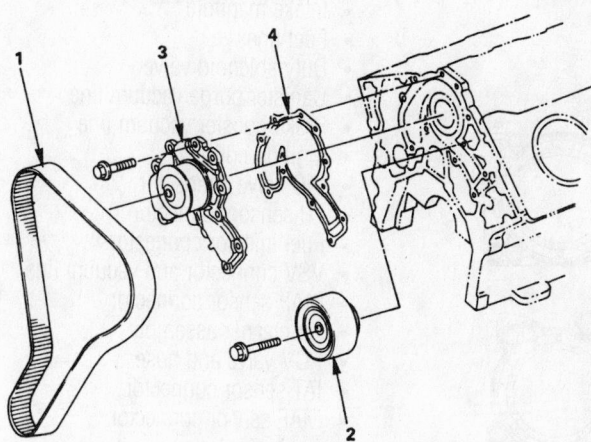

Exploded view of the water pump mounting

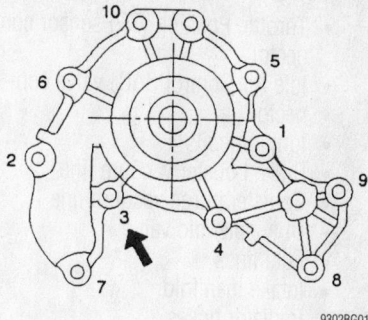

Water pump torque sequence. Apply LOC-TITE® 262 to bolt number 3 (arrow)

Cylinder Head

REMOVAL & INSTALLATION

1. Before servicing the vehicle, refer to the precautions in the beginning of this section.
2. Drain the cooling system.
3. Remove or disconnect the following:
- Negative battery cable
- Hood
- Engine cover
- Mass Air Flow (MAF) sensor connector
- Intake Air Temperature (IAT) sensor connector
- Positive Crankcase Ventilation (PCV) valve and hose
- Air cleaner assembly
- Manifold Absolute Pressure (MAP) sensor connector
- Vacuum Switching Valve (VSV) connector and vacuum line
- Fuel injector connectors

1. Timing belt
2. Idle pulley
3. Water pump assembly
4. Gasket

- Throttle Position (TP) sensor connector
- Idle Air Control (IAC) valve connector
- Ignition coils
- Brake booster vacuum line
- Canister purge vacuum line
- Duty solenoid valve
- Fuel lines
- Intake manifold
- Radiator hoses
- Engine coolant manifold
- Upper fan shroud
- Accessory drive belt and tensioner
- Cooling fan and pulley
- Alternator
- Idler pulley
- Power steering pump and bracket
- A/C compressor
- Crankshaft pulley
- Oil cooler hoses
- Timing belt cover
- Valve covers
- Timing belt
- Left and right exhaust front pipes
- Oil dipstick tube
- Cylinder heads

To install:

➡**Use new head bolts when installing the cylinder head. Do not apply oil to the head bolt threads.**

➡**The left and right cylinder head gaskets are not interchangeable.**

4. Install the cylinder heads with new gaskets. Tighten the bolts to 47 ft. lbs. (64 Nm).
5. Install or connect the following:
   - Oil dipstick tube
   - Left and right exhaust front pipes
   - Timing belt

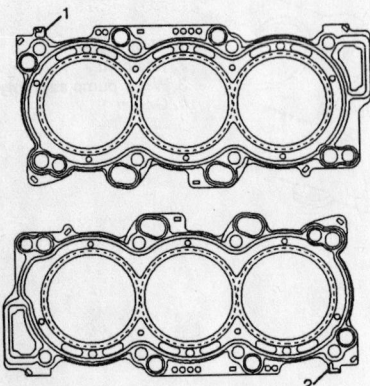

Right (1) and left (2) head gasket identification mark locations

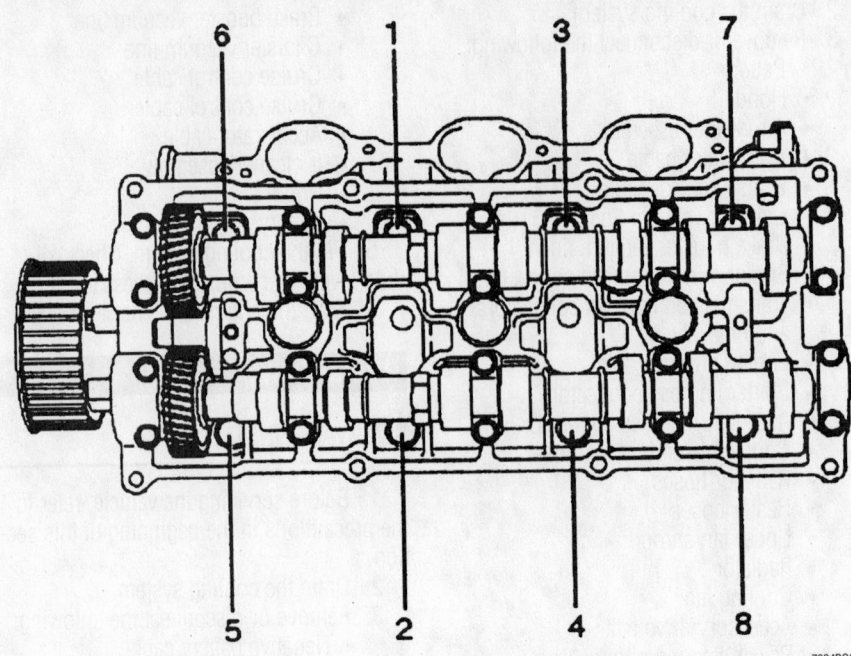

Cylinder head torque sequence

- Valve covers
- Timing belt cover
- Oil cooler hoses
- Crankshaft pulley. Tighten the pulley bolt to 123 ft. lbs. (167 Nm).
- A/C compressor
- Power steering pump and bracket. Tighten the bolts to 34 ft. lbs. (46 Nm).
- Idler pulley
- Alternator
- Cooling fan and pulley
- Accessory drive belt and tensioner
- Upper fan shroud
- Engine coolant manifold
- Radiator hoses
- Intake manifold
- Fuel lines
- Duty solenoid valve
- Canister purge vacuum line
- Brake booster vacuum line
- Ignition coils
- IAC valve connector
- TP sensor connector
- Fuel injector connectors
- VSV connector and vacuum line
- MAP sensor connector
- Air cleaner assembly
- PCV valve and hose
- IAT sensor connector
- MAF sensor connector
- Engine cover
- Hood
- Negative battery cable

6. Fill the cooling system.
7. Start the engine. Check for leaks and proper operation.

Rocker Arms/Shafts

REMOVAL & INSTALLATION

The 3.5L engine is not equipped with rocker arms. The camshaft lobes act directly on the valve shims.

Intake Manifold

REMOVAL & INSTALLATION

1. Before servicing the vehicle, refer to the precautions in the beginning of this section.
2. Remove or disconnect the following:
   - Negative battery cable
   - Engine cover
   - Air cleaner assembly
   - Accelerator cable
   - Cruise control cable
   - Brake booster vacuum line
   - Manifold Absolute Pressure (MAP) sensor connector
   - Idle Air Control (IAC) valve connector
   - Throttle Position (TP) sensor connector
   - Canister purge solenoid connector
   - Electronic Vacuum Sensing Valve (EVSV) connector and vacuum line
   - Exhaust Gas Recirculation (EGR) valve
   - Positive Crankcase Ventilation (PCV) valve and hose
   - Pressure regulator vacuum line

- Ventilation hose
- Throttle body
- Fuel lines
- Fuel injector connectors
- Intake manifold

To install:

3. Install or connect the following:
- Intake manifold. Tighten the fasteners to 18 ft. lbs. (25 Nm).
- Fuel injector connectors
- Fuel lines
- Throttle body. Tighten the bolts to 88 inch lbs. (10 Nm).
- Ventilation hose
- Pressure regulator vacuum line
- PCV valve and hose
- EGR valve
- EVSV connector and vacuum line
- Canister purge solenoid connector
- TP sensor connector
- IAC valve connector
- MAP sensor connector
- Brake booster vacuum line
- Cruise control cable
- Accelerator cable
- Air cleaner assembly
- Engine cover
- Negative battery cable

4. Start the engine and check for proper operation.

Exhaust Manifold

REMOVAL & INSTALLATION

1. Before servicing the vehicle, refer to the precautions in the beginning of this section.
2. Remove or disconnect the following:
- Negative battery cable
- Air cleaner assembly
- Heated Oxygen Sensor (HO2S) connectors
- Right torsion bar
- Exhaust Gas Recirculation (EGR) pipe and bracket
- Left and right exhaust front pipes
- Heat shields
- Accessory drive belt
- A/C compressor and bracket
- Exhaust manifolds

To install:

3. Install or connect the following:
- Exhaust manifolds. Tighten the bolts to 38 ft. lbs. (52 Nm).
- A/C compressor and bracket
- Accessory drive belt
- Heat shields
- Left and right exhaust front pipes

- EGR pipe and bracket
- Right torsion bar
- HO2S connectors
- Air cleaner assembly
- Negative battery cable

4. Start the engine and check for leaks.

Front Crankshaft Seal

REMOVAL & INSTALLATION

1. Before servicing the vehicle, refer to the precautions in the beginning of this section.
2. Remove or disconnect the following:
- Negative battery cable
- Air cleaner assembly
- Upper fan shroud
- Accessory drive belt and tensioner
- Cooling fan and pulley
- Idler pulley
- Power steering pump and move it aside
- Crankshaft pulley
- Timing belt cover
- Timing belt
- Crankshaft timing sprocket
- Oil seal

To install:

3. Install or connect the following:
- Oil seal so that it is flush with the oil pump housing
- Crankshaft timing sprocket
- Timing belt
- Timing belt cover
- Crankshaft pulley. Tighten the bolt to 123 ft. lbs. (167 Nm).
- Power steering pump

- Idler pulley
- Cooling fan and pulley
- Accessory drive belt and tensioner
- Upper fan shroud
- Air cleaner assembly
- Negative battery cable

4. Start the engine and check for leaks.

Camshaft and Valve Lifters

REMOVAL & INSTALLATION

1. Before servicing the vehicle, refer to the precautions in the beginning of this section.
2. Remove or disconnect the following:
- Negative battery cable
- Air cleaner assembly
- Upper fan shroud
- Accessory drive belt and tensioner
- Cooling fan and pulley
- Idler pulley
- Power steering pump and move it aside
- Crankshaft pulley
- Timing belt cover
- Timing belt
- Ignition coils
- Valve covers
- Camshafts
- Valve shims and tappets

➡**Keep the valve shims and tappets in order for installation.**

To install:

3. Install the valve tappets and shims in their original locations.
4. Using Gear Spring Lever J-42686,

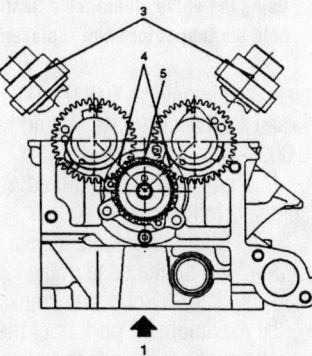

1   **Right Bank**
2   **Left Bank**

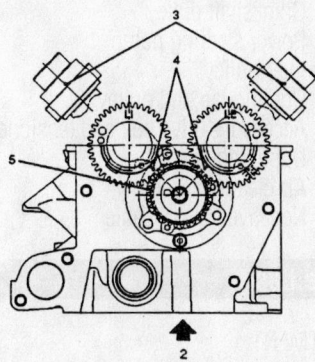

3   **Alignment Mark on Camshaft Drive Gear**
4   **Alignment Mark on Camshaft**
5   **Alignment Mark on Retainer**

7924BG11

Camshaft alignment marks for the left and right cylinder heads

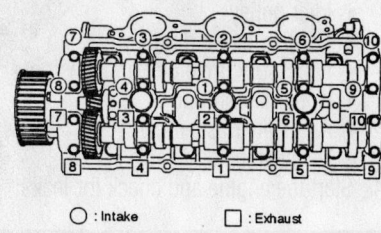

○ : Intake   □ : Exhaust

7924BG12

Camshaft retaining bracket tightening sequence

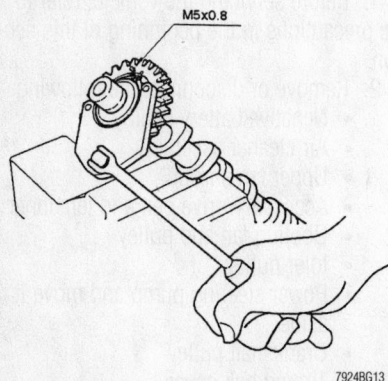

7924BG13

Aligning the sub gear with the Gear Spring Lever J-42686

turn the sub gear clockwise to align the 5mm bolt holes in the sub gear and the camshaft driven gear. Tighten the 5mm bolt.

5. Install the camshafts. Align the timing marks as shown. Tighten the bolts in sequence to 89 inch lbs. (10 Nm).

6. Install or connect the following:
- Valve covers
- Ignition coils
- Timing belt
- Timing belt cover
- Crankshaft pulley
- Power steering pump
- Idler pulley
- Cooling fan and pulley
- Accessory drive belt and tensioner
- Upper fan shroud
- Air cleaner assembly
- Negative battery cable

Valve Lash

ADJUSTMENT

➡ **Measure valve clearance with the engine cold.**

1. Before servicing the vehicle, refer to the precautions in the beginning of this section.

2. Remove the valve covers.

3. Check the valve clearance with the camshafts positioned as shown. Intake valve

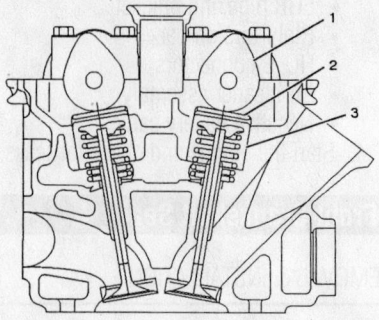

9302BG02

Cross section of the cylinder head. Note the position of the camshaft lobe (1), adjustment shim (2) and the tappet (3)

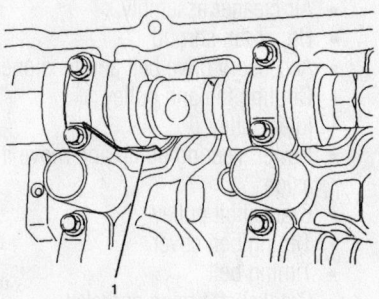

9302BG03

Valve clearance adjusting tool J–42689 (1)

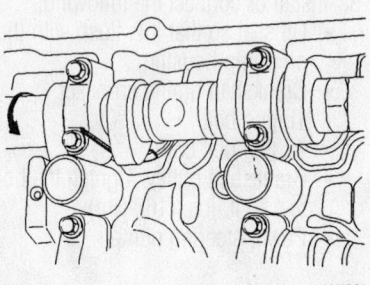

9302BG04

Using the valve clearance adjusting tool to hold the tappet for shim replacement

clearance should be 0.0091–0.0130 inches. Exhaust valve clearance should be 0.0098–0.0138 inches.

4. If adjustment is required, replace the shims as follows:

a. Step 1: Position special tool J–42689 on the edge of the tappet.

b. Step 2: Rotate the crankshaft until the maximum lift portion of the camshaft lobe contacts the upper edge of the special tool and presses the tappet down to create enough clearance between the adjustment shim and the camshaft for the shim to be removed.

c. Step 3: Replace shims as necessary to achieve correct valve clearance.

d. Step 4: Repeat for each valve to be adjusted.

5. Replace the valve covers. Tighten the bolts to 80 inch lbs. (9 Nm).

Starter Motor

REMOVAL & INSTALLATION

1. Before servicing the vehicle, refer to the precautions in the beginning of this section.

2. Remove or disconnect the following:
- Negative battery cable
- Heated Oxygen Sensor (HO2S) connectors
- Exhaust front pipe
- Heat shield
- Starter wiring connectors
- Starter motor

To install:

3. Install or connect the following:
- Starter motor. Tighten the bolts to 30 ft. lbs. (40 Nm).
- Starter wiring connectors
- Heat shield
- Exhaust front pipe
- HO2S connectors
- Negative battery cable

Oil Pan

REMOVAL & INSTALLATION

1. Before servicing the vehicle, refer to the precautions in the beginning of this section.

2. Drain the engine oil.

3. Remove or disconnect the following:
- Negative battery cable
- Front wheels
- Oil level dipstick
- Stone guard
- Radiator under fan shroud
- Suspension crossmember
- Flywheel dust cover
- Pitman arm
- Idler arm

4. If equipped with 4 wheel drive, unbolt and lower the front axle housing assembly for clearance.
- Oil pan
- Lower crankcase

To install:

5. Apply a bead of silicone sealant to the crankcase flange and install the crankcase. Tighten the fasteners in sequence to 89 inch lbs. (10 Nm).

6. Apply a bead of silicone sealant to the oil pan flange and install the oil pan. Tighten the fasteners to 89 inch lbs. (10 Nm).

7. If equipped, raise the axle housing

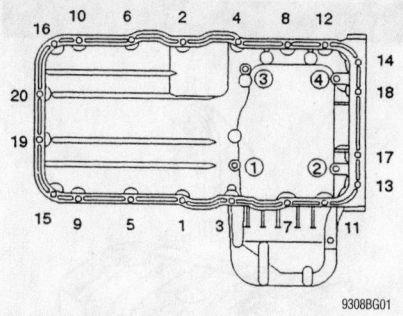

Oil pan torque sequence

assembly into position. Tighten the axle case bolts to 61 ft. lbs. (82 Nm) and the mounting bolts to 112 ft. lbs. (152 Nm).

8. Install or connect the following:
- Pitman arm. Tighten the nut to 159 ft. lbs. (216 Nm).
- Idler arm. Tighten the bolt to 33 ft. lbs. (44 Nm).
- Flywheel dust cover
- Suspension crossmember. Tighten the bolts to 58 ft. lbs. (78 Nm).
- Radiator under fan shroud
- Stone guard
- Oil level dipstick
- Front wheels
- Negative battery cable
9. Fill the crankcase with engine oil.
10. Start the engine and check for leaks.

Oil Pump

REMOVAL & INSTALLATION

1. Before servicing the vehicle, refer to the precautions in the beginning of this section.
2. Remove or disconnect the following:
- Timing belt
- Oil pan
- Oil pick-up tube
- Oil filter adapter
- Oil pump

To install:

3. Apply silicone sealant to the oil pump mounting surface and install the oil pump. Tighten the bolts in sequence to 18 ft. lbs. (25 Nm).
4. Install or connect the following:
- Oil filter adapter

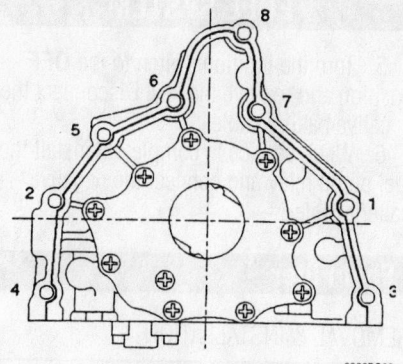

Oil pump torque sequence

- Oil pickup tube
- Oil pan
- Timing belt

Rear Main Seal

REMOVAL & INSTALLATION

1. Before servicing the vehicle, refer to the precautions in the beginning of this section.
2. Remove or disconnect the following:
- Negative battery cable
- Transmission
- Flywheel by loosening the flywheel bolts in a 2-step crisscross sequence
- Rear main seal, using a seal puller

➡**Do not damage the crankshaft sealing surface.**

To install:

3. Install or connect the following:
- New rear main seal, by lubricating it with engine oil
- Flywheel, using new flywheel bolts. Tighten the bolts, in a 2-step crisscross pattern, to 40 ft. lbs. (54 Nm).
- Transmission
- Negative battery cable
4. Check the oil and refill as necessary.

Piston and Ring

POSITIONING

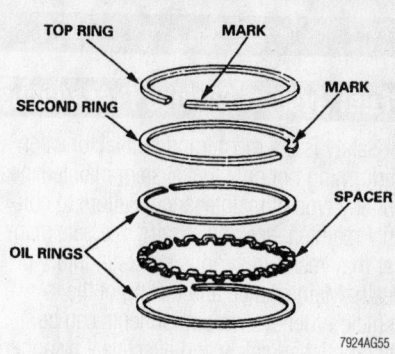

Piston ring top mark locations

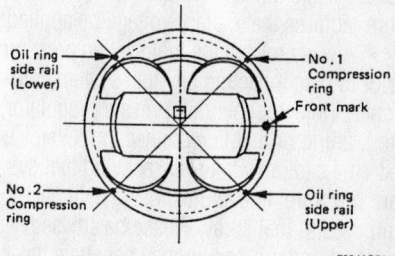

Piston ring end-gap spacing

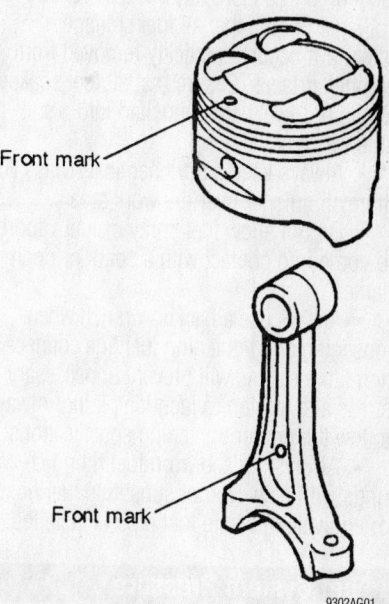

Piston and rod front marks

Timing belt service is covered in Section 3 of this manual

FUEL SYSTEM

Fuel System Service Precautions

Safety is the most important factor when performing not only fuel system maintenance but any type of maintenance. Failure to conduct maintenance and repairs in a safe manner may result in serious personal injury or death. Maintenance and testing of the vehicle's fuel system components can be accomplished safely and effectively by adhering to the following rules and guidelines:

- To avoid the possibility of fire and personal injury, always disconnect the negative battery cable unless the repair or test procedure requires that battery voltage be applied.
- Always relieve the fuel system pressure prior to disconnecting any fuel system component (injector, fuel rail, pressure regulator, etc.), fitting or fuel line connection. Exercise extreme caution whenever relieving fuel system pressure, to avoid exposing skin, face and eyes to fuel spray. Please be advised that fuel under pressure may penetrate the skin or any part of the body that it contacts.
- Always place a shop towel or cloth around the fitting or connection prior to loosening to absorb any excess fuel due to spillage. Ensure that all fuel spillage (should it occur) is quickly removed from engine surfaces. Ensure that all fuel soaked cloths or towels are deposited into a suitable waste container.
- Always keep a dry chemical (Class B) fire extinguisher near the work area.
- Do not allow fuel spray or fuel vapors to come into contact with a spark or open flame.
- Always use a backup wrench when loosening and tightening fuel line connection fittings. This will prevent unnecessary stress and torsion to fuel line piping. Always follow the proper tightening specifications.
- Always replace worn fuel fitting O-rings with new. Do not substitute fuel hose or equivalent, where fuel pipe is installed.

Fuel System Pressure

RELIEVING

1. Before servicing the vehicle, refer to the precautions in the beginning of this section.
2. Remove the fuel filler cap.
3. Remove the fuel pump relay from the underhood relay box.
4. Start the engine and let it run until it stalls, then crank the engine for an additional 30 seconds.
5. Turn the ignition switch to the **OFF** position and remove the key. Disconnect the negative battery cable.
6. When service is completed, install the fuel pump relay and connect the negative battery cable.

Fuel Filter

REMOVAL & INSTALLATION

1. Before servicing the vehicle, refer to the precautions in the beginning of this section.
2. Relieve the fuel system pressure.
3. Remove or disconnect the following:
   - Fuel lines from the fuel filter
   - Fuel filter

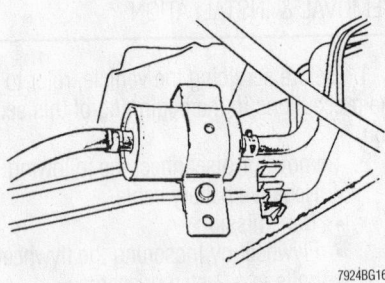

Fuel filter mounting location

To install:

4. Install or connect the following:
   - Fuel filter and tighten the bracket bolt. Note the fuel flow directional arrow on the filter.
   - Fuel lines to the fuel filter
   - Negative battery cable
5. Start the engine and inspect the fuel filter connections for leaks.

Fuel Pump

REMOVAL & INSTALLATION

1. Before servicing the vehicle, refer to the precautions in the beginning of this section.
2. Relieve fuel system pressure.
3. Drain the fuel tank.
4. Remove or disconnect the following:
   - Negative battery cable
   - Right rear inner fender liner
   - Fuel filler and vent hoses
   - Fuel tank skid plate
   - Fuel tank wiring connectors
   - Fuel supply and return lines

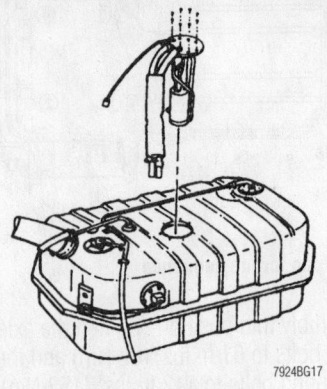

Fuel pump assembly mounting

- Fuel tank
- Fuel pump assembly

To install:

5. Install or connect the following:
   - Fuel pump assembly
   - Fuel tank. Tighten the bolts to 27 ft. lbs. (36 Nm).
   - Fuel supply and return lines
   - Fuel tank wiring connectors
   - Fuel tank skid plate
   - Fuel filler and vent hoses
   - Right rear inner fender liner
   - Negative battery cable
6. Start the engine and check for leaks.

Fuel Injector

REMOVAL & INSTALLATION

1. Before servicing the vehicle, refer to the precautions in the beginning of this section.
2. Relieve fuel system pressure.
3. Remove or disconnect the following:
   - Negative battery cable
   - Engine cover
   - Fuel injector wiring connectors
   - Fuel lines
   - Fuel supply manifold with injectors attached
   - Clips
   - Injectors from the supply manifold

To install:

4. Install or connect the following:
   - New O-rings on the fuel injectors
   - Fuel injectors
   - Fuel supply manifold with injectors attached. Tighten the bolts to 60 inch lbs. (6.5 Nm).
   - Fuel lines
   - Fuel injector wiring connectors
   - Engine cover
   - Negative battery cable
5. Start the engine and check for leaks.

DRIVE TRAIN

Transmission Assembly

REMOVAL & INSTALLATION

2 WHEEL DRIVE

1. Before servicing the vehicle, refer to the precautions in the beginning of this section.
2. Remove the hood.
3. Install a support fixture to the engine lifting eyes.
4. Remove or disconnect the following:
   - Negative battery cable
   - Front console assembly and wiring connectors
   - Shift lock cable
   - Shift control rod
   - Selector lever assembly
   - Driveshaft
   - Wiring harness heat shield
   - Transmission mount and crossmember
   - Heated Oxygen Sensor (HO2S) connectors
   - Left and right exhaust front pipes
   - Transmission oil cooler lines
   - Starter motor
   - Fuel line bracket
   - Transmission harness connectors
   - Flywheel under covers
   - Torque converter
   - Transmission flange bolts
   - Transmission

To install:

➡**Use new torque converter bolts.**

5. Install or connect the following:
   - Transmission. Tighten the large bolts to 56 ft. lbs. (76 Nm) and the small bolts to 69 inch lbs. (8 Nm).
   - Torque converter. Tighten the bolts to 40 ft. lbs. (54 Nm).
   - Flywheel under covers
   - Transmission harness connectors
   - Fuel line bracket
   - Starter motor. Tighten the bolts to 30 ft. lbs. (40 Nm).
   - Transmission oil cooler lines
   - Left and right exhaust front pipes. Tighten the manifold flange fasteners to 49 ft. lbs. (67 Nm), and the exhaust flange bolts to 32 ft. lbs. (43 Nm).
   - HO2S connectors
   - Crossmember. Tighten the bolts to 37 ft. lbs. (50 Nm).

   - Transmission mount. Tighten the bolts to 30 ft. lbs. (41 Nm).
   - Wiring harness heat shield
   - Driveshaft. Tighten the flange bolts to 46 ft. lbs. (63 Nm).
   - Selector lever assembly
   - Shift control rod
   - Shift lock cable
   - Front console assembly and wiring connectors
   - Negative battery cable
6. Remove the engine support fixture and install the hood.

4 WHEEL DRIVE

1. Before servicing the vehicle, refer to the precautions in the beginning of this section.
2. Remove the hood.
3. Install a support fixture to the engine lifting eyes.
4. Remove or disconnect the following:
   - Negative battery cable
   - Transfer case shift lever knob
   - Front console assembly and wiring connectors
   - Shift lock cable
   - Shift control rod
   - Selector lever assembly
   - Transfer case shift lever
   - Transfer case skid plate
   - Front and rear driveshafts
   - Wiring harness heat shield
   - Transmission mount and crossmember
   - Right torsion bar, if equipped with Torque On Demand (TOD) system
   - Heated Oxygen Sensor (HO2S) connectors
   - Left and right exhaust front pipes
   - Transmission oil cooler lines
   - Starter motor
   - Fuel line bracket
   - Transmission harness connectors
   - Flywheel under covers
   - Torque converter
   - Transmission flange bolts
   - Transmission

To install:

➡**Use new torque converter bolts.**

5. Install or connect the following:
   - Transmission. Tighten the large bolts to 56 ft. lbs. (76 Nm) and the small bolts to 69 inch lbs. (8 Nm).
   - Torque converter. Tighten the bolts to 40 ft. lbs. (54 Nm).

   - Flywheel under covers
   - Transmission harness connectors
   - Fuel line bracket
   - Starter motor. Tighten the bolts to 30 ft. lbs. (40 Nm).
   - Transmission oil cooler lines
   - Left and right exhaust front pipes. Tighten the manifold flange fasteners to 49 ft. lbs. (67 Nm), and the exhaust flange bolts to 32 ft. lbs. (43 Nm).
   - HO2S connectors
   - Right torsion bar, if removed
   - Crossmember. Tighten the bolts to 37 ft. lbs. (50 Nm).
   - Transmission mount. Tighten the bolts to 30 ft. lbs. (41 Nm).
   - Wiring harness heat shield
   - Front and rear driveshafts. Tighten the flange bolts to 46 ft. lbs. (63 Nm).
   - Transfer case skid plate. Tighten the bolts to 27 ft. lbs. (37 Nm).
   - Transfer case shift lever
   - Selector lever assembly
   - Shift control rod
   - Shift lock cable
   - Front console assembly and wiring connectors
   - Transfer case shift lever knob
   - Negative battery cable
6. Remove the engine support fixture and install the hood.

Transfer Case Assembly

REMOVAL & INSTALLATION

1. Before servicing the vehicle, refer to the precautions in the beginning of this section.
2. Remove or disconnect the following:
   - Negative battery cable
   - Transfer case skid plate
   - Front and rear driveshafts
   - Heated Oxygen Sensor (HO2S) connectors
   - Left and right exhaust front pipes
   - Transfer case control lever knob
   - Selector lever assembly
   - Transfer case control lever
   - Vehicle Speed Sensor (VSS) connector
   - 4 wheel drive switch connector
   - 4 wheel drive actuator connector
   - Transfer case flange fasteners
   - Transfer case

Heater Core replacement is covered in Section 2 of this manual

To install:

3. Install or connect the following:

- Transfer case. Tighten the flange fasteners to 34 ft. lbs. (46 Nm).
- 4 wheel drive actuator connector
- 4 wheel drive switch connector
- VSS connector
- Transfer case control lever
- Selector lever assembly
- Transfer case control lever knob
- Left and right exhaust front pipes
- HO2S connectors
- Front and rear driveshafts. Tighten the bolts to 46 ft. lbs. (63 Nm).
- Transfer case skid plate. Tighten the bolts to 27 ft. lbs. (37 Nm).
- Negative battery cable

Halfshaft

REMOVAL & INSTALLATION

1. Before servicing the vehicle, refer to the precautions in the beginning of this section.

2. Remove or disconnect the following:

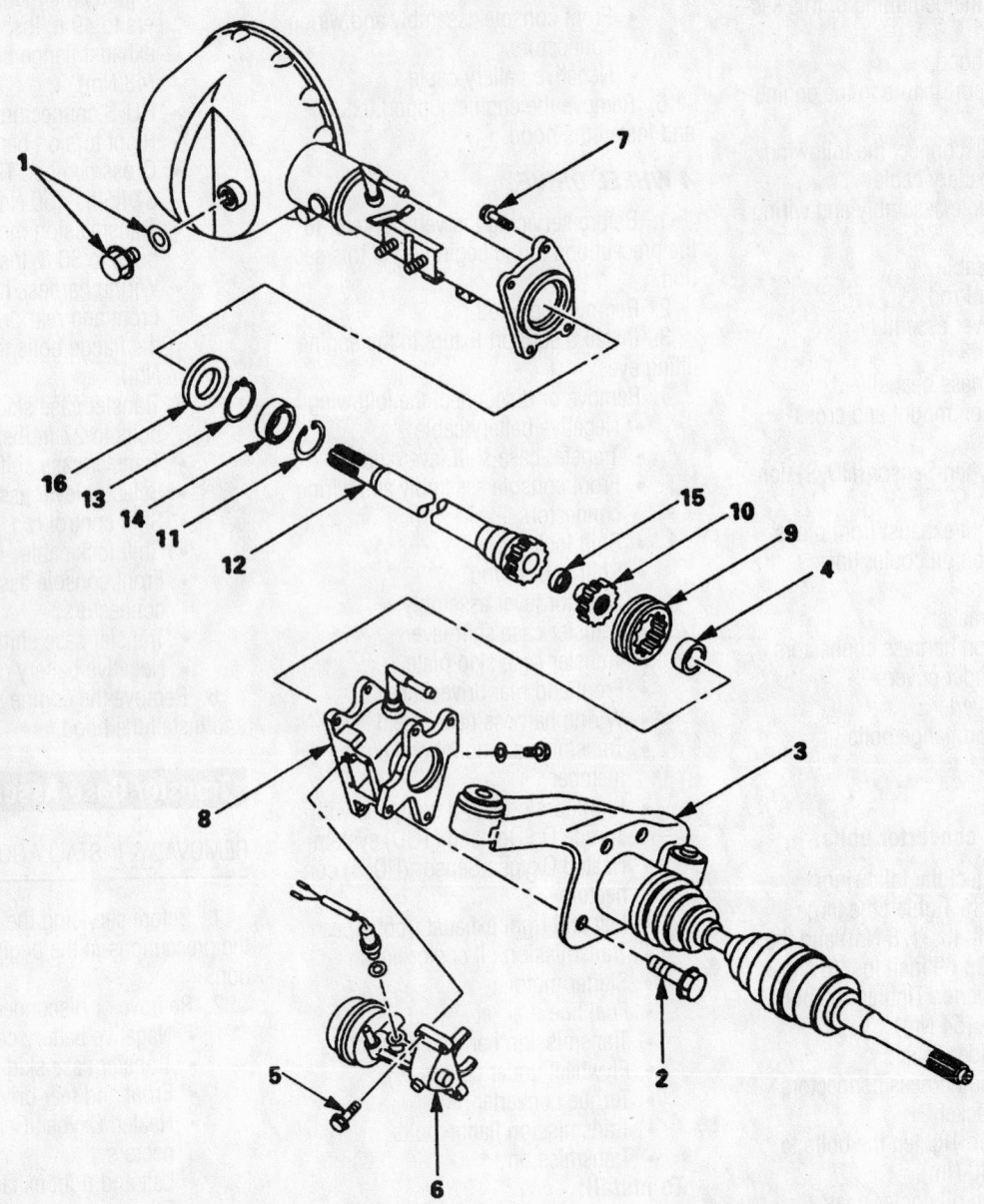

1. Filler plug
2. Bolt
3. Front axle drive shaft (LH side)
4. Spacer
5. Bolt
6. Actuator assembly
7. Bolt
8. Housing
9. Sleeve
10. Clutch gear
11. Snap ring
12. Inner shaft
13. Snap ring
14. Inner shaft bearing
15. Needle bearing
16. Oil seal

7924BG26

Exploded view of the left halfshaft, axle shaft, and axle disconnect

- Negative battery cable
- Front wheel
- Radiator skid plate
- Transfer case skid plate
- Brake calipers and mounting bracket
- Brake rotor
- Wheel speed sensor
- Steering knuckle

3. Support the axle housing with a jack.

Unbolt the axle mounting bracket and remove the halfshaft/bracket assembly.

To install:

4. Install or connect the following:
- Axle/bracket assembly. Tighten the bracket flange bolts to 85 ft. lbs. (116 Nm) and the bracket mounting bolts to 112 ft. lbs. (152 Nm).
- Steering knuckle
- Wheel speed sensor

- Brake rotor
- Brake caliper and mounting bracket. Tighten the bracket bolts to 115 ft. lbs. (155 Nm).
- Transfer case skid plate. Tighten the bolts to 27 ft. lbs. (37 Nm).
- Radiator skid plate. Tighten the bolts to 58 ft. lbs. (78 Nm).
- Front wheel
- Negative battery cable

5. Check the wheel alignment and adjust as necessary.

CV-Joints

OVERHAUL

Outer CV-Joint

The outer CV-joint is serviced with the axle shaft as an assembly. The outer CV-joint boot can be serviced by removing the inner CV-joint.

Inner CV-Joint

1. Before servicing the vehicle, refer to the precautions in the beginning of this section.
2. Remove or disconnect the following:
- Halfshaft from the vehicle
- Snapring and bearing
- Snapring and oil seal
- Mounting bracket
- CV-joint boot
- Circlip and inner joint housing
- Snapring and spacer
- Inner joint balls
- Snapring and inner CV-joint

To install:

3. Install or connect the following:
- Inner CV-joint and snapring
- Inner joint balls
- Spacer and snapring
- Inner joint housing and circlip. Add 150 grams CV-joint grease.

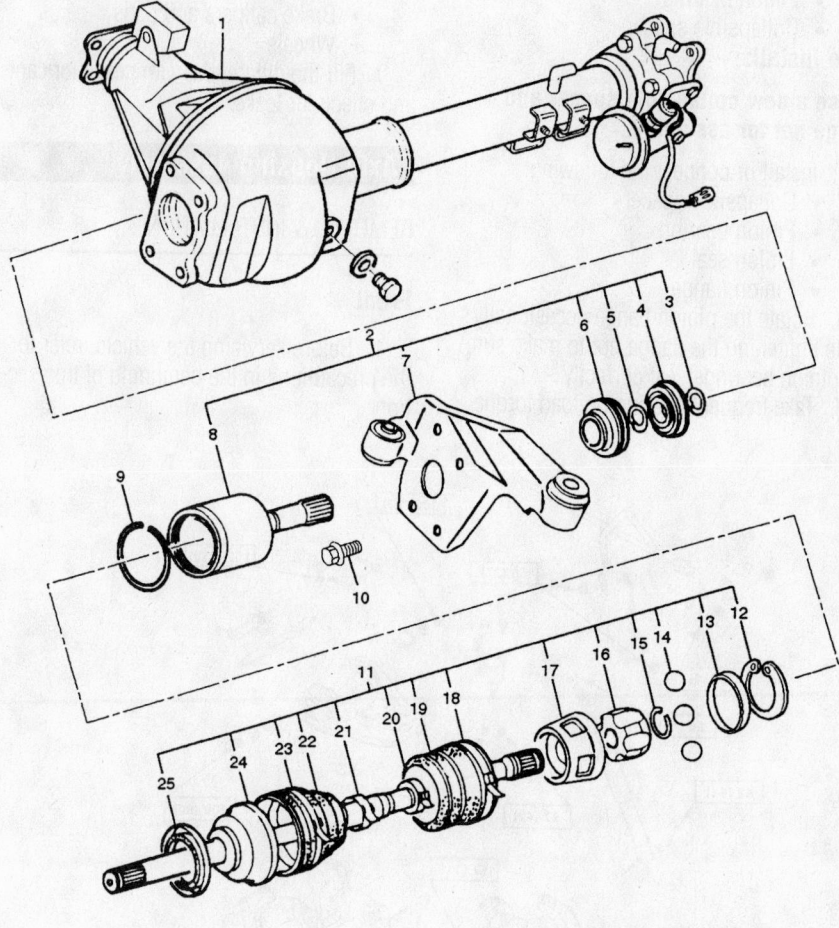

1	Axle Case and Differential
2	DOJ Case Assembly
3	Snap Ring
4	Bearing
5	Snap Ring
6	Oil Seal
7	Bracket
8	DOJ Case
9	Circlip
10	Bolt
11	Drive Shaft Joint Assembly
12	Snap Ring
13	Spacer
14	Ball
15	Snap Ring
16	Ball Retainer
17	Ball Guide
18	Band
19	Bellows
20	Band
21	Band
22	Bellows
23	Band
24	BJ Shaft
25	Dust Seal

9308BG03

Exploded view of the right halfshaft and mounting bracket

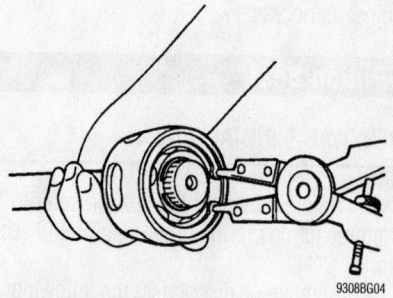

9308BG04

CV-joint spacer snapring—Inner CV-Joint

Brake service is covered in Section 4 of this manual

- CV-joint boot
- Mounting bracket
- Oil seal and snapring
- Bearing and snapring

4. Install the halfshaft and mounting bracket to the vehicle.

5. Check the wheel alignment and adjust as necessary.

Axle Shaft, Bearing and Seal

REMOVAL & INSTALLATION

1. Before servicing the vehicle, refer to the precautions in the beginning of this section.

2. Remove or disconnect the following:
- Rear wheel
- Disc brake caliper and bracket
- Disc brake rotor
- Wheel speed sensor bracket
- Parking brake cable and bracket
- Parking brake shoes
- Axle shaft
- Snapring and discard it
- Bearing by pressing it off the axle shaft with the bearing holder and oil seal

To install:

3. Install or connect the following:
- New oil seal into the bearing housing
- Bearing housing onto the axle shaft
- Bearing by pressing it onto the axle shaft
- New snapring
- Axle shaft. Use new lock washers and tighten the bearing holder nuts to 54 ft. lbs. (74 Nm).
- Parking brake shoes
- Parking brake cable and bracket
- Wheel speed sensor bracket
- Disc brake rotor
- Disc brake caliper and bracket. Tighten the bracket bolts to 76 ft. lbs. (103 Nm).
- Rear wheel

4. Check the rear axle oil level and adjust as necessary.

Pinion Seal

REMOVAL & INSTALLATION

1. Before servicing the vehicle, refer to the precautions in the beginning of this section.

2. Remove or disconnect the following:
- Driveshaft
- Wheels
- Brake calipers and pads

➡ **The brake calipers and pads must be removed so that there is no additional drag when measuring pinion bearing preload.**

3. Use an inch lb. torque wrench and measure and record the amount of torque required to maintain pinion rotation through several revolutions.

4. Remove or disconnect the following:
- Pinion flange
- Pinion seal
- Pinion bearing
- Collapsible spacer

To install:

➡ **Use a new collapsible spacer and flange nut for assembly.**

5. Install or connect the following:
- Collapsible spacer
- Pinion bearing
- Pinion seal
- Pinion flange

6. Rotate the pinion flange occasionally while tightening the flange nut to make sure the pinion bearings seat correctly.

7. Take frequent bearing preload torque readings. Tighten the flange nut to achieve the preload torque readings originally recorded.

❊❊ CAUTION

Never loosen the pinion nut to reduce bearing preload. If it is necessary to reduce bearing preload, install a new collapsible spacer and pinion nut.

8. Install or connect the following:
- Driveshaft
- Brake calipers and pads
- Wheels

9. Fill the differential with gear lubricant and check for leaks.

Axle Housing Assembly

REMOVAL & INSTALLATION

Front

1. Before servicing the vehicle, refer to the precautions in the beginning of this section.

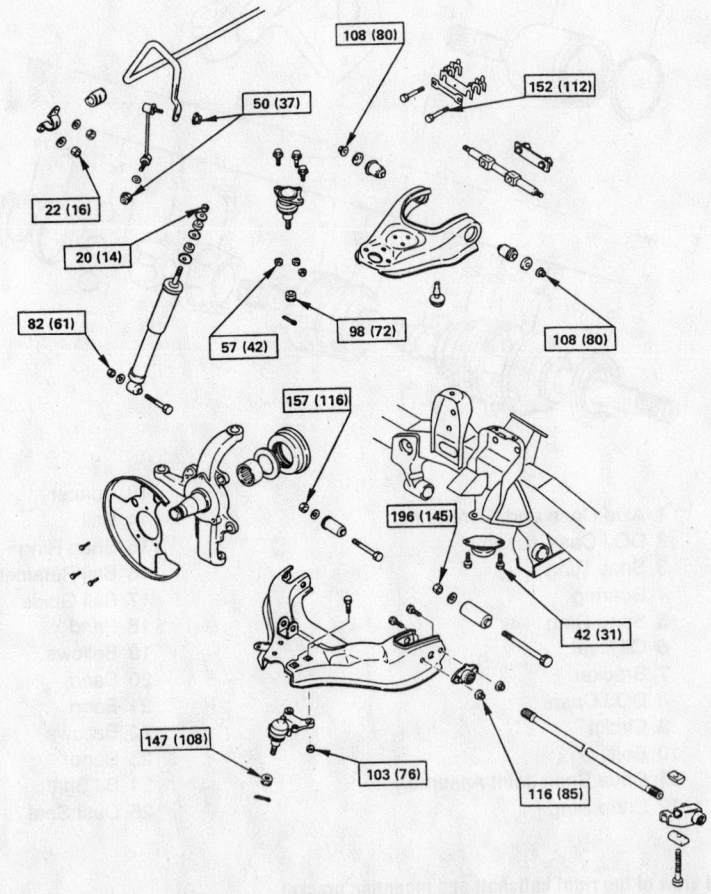

N•m (ft • lbs)

7924BG48

Exploded view of the front suspension, showing the tightening specifications

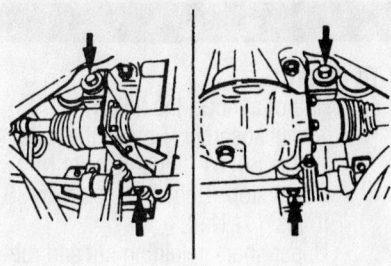

Axle assembly mounting bracket bolt locations

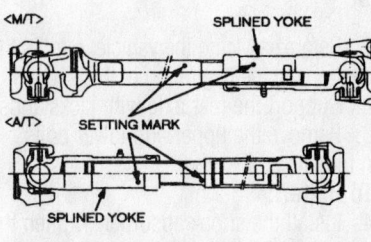

Driveshaft alignment mark locations

2. Remove or disconnect the following:
- Negative battery cable
- Front wheels
- Radiator skid plate
- Transfer case skid plate
- Brake calipers and mounting brackets
- Brake rotors
- Wheel speed sensors
- Axle disconnect actuator
- Vacuum Switching Valve (VSV)
- Steering knuckles
- Idler arm
- Pitman arm
- Front suspension crossmember
- Front driveshaft
- Front axle bracket mounting bolts
- Right halfshaft and mounting bracket from the axle
- Axle assembly from the vehicle with the left halfshaft and bracket attached
- Left halfshaft and bracket from the axle disconnect housing

To install:

➡ **Use new nuts, bolts and snaprings for assembly.**

3. Install or connect the following:
- Left halfshaft and mounting bracket. Tighten the bracket flange bolts to 85 ft. lbs. (116 Nm).
- Axle housing. Tighten the bracket mounting bolts to 112 ft. lbs. (152 Nm).

- Right halfshaft and mounting bracket. Tighten the bracket flange bolts to 85 ft. lbs. (116 Nm).
- Front driveshaft. Tighten the bolts to 46 ft. lbs. (63 Nm).
- Front suspension crossmember. Tighten the bolts to 58 ft. lbs. (78 Nm).
- Pitman arm
- Idler arm
- Steering knuckles
- VSV valve
- Axle disconnect actuator
- Wheel speed sensors
- Brake rotors
- Brake calipers and mounting brackets. Tighten the bracket bolts to 115 ft. lbs. (155 Nm).
- Transfer case skid plate. Tighten the bolts to 27 ft. lbs. (37 Nm).
- Radiator skid plate. Tighten the bolts to 58 ft. lbs. (78 Nm).
- Front wheels
- Negative battery cable

4. Check the wheel alignment and adjust as necessary.

Rear

1. Before servicing the vehicle, refer to the precautions in the beginning of this section.

2. Support the rear axle with a jack.
3. Remove or disconnect the following:
- Rear wheels
- Rear driveshaft
- Parking brake cables
- Axle breather hose
- Wheel speed sensor connectors and bracket
- Brake fluid hose
- Shock absorbers
- Coil springs
- Stabilizer bar linkage
- Lateral rod
- Center link
- Trailing links
- Axle housing from the vehicle

To install:

4. Install or connect the following:
- Axle housing, raise it into position
- Trailing links
- Center link
- Lateral rod
- Stabilizer bar linkage
- Coil springs
- Shock absorbers
- Brake fluid hose
- Wheel speed sensor connectors and bracket
- Axle breather hose
- Parking brake cables
- Rear driveshaft
- Rear wheels

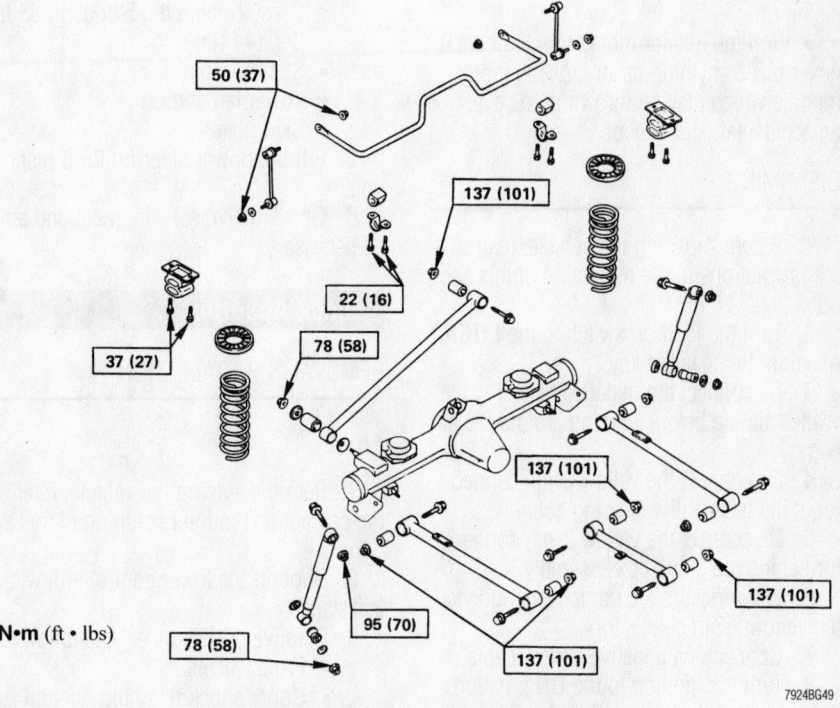

N•m (ft • lbs)

Exploded view of the rear suspension, showing the tightening specifications

For complete Engine Mechanical specifications, see Section 1 of this manual

STEERING AND SUSPENSION

Air Bag

❊❊ CAUTION

Some vehicles are equipped with an air bag system. The system must be disarmed before performing service on, or around, system components, the steering column, instrument panel components, wiring and sensors. Failure to follow the safety precautions and the disarming procedure could result in accidental air bag deployment, possible injury and unnecessary system repairs.

PRECAUTIONS

Several precautions must be observed when handling the inflator module to avoid accidental deployment and possible personal injury.

• Never carry the inflator module by the wires or connector on the underside of the module.

• When carrying a live inflator module, hold securely with both hands, and ensure that the bag and trim cover are pointed away from you.

• Place the inflator module on a bench or other surface with the bag and trim cover facing up.

• With the inflator module on the bench, never place anything on or close to the module which may be thrown in the event of an accidental deployment.

DISARMING

1. Before servicing the vehicle, refer to the precautions in the beginning of this section.
2. Turn the ignition switch to the **LOCK** position. Remove the key.
3. Disconnect the negative battery cable. Wait 1 minute before working around the air bags.
4. Disconnect the yellow 2-pin connector at the base of the steering column.
5. Disconnect the yellow 2-pin connector behind the glove box assembly.
6. When repairs are completed, connect the yellow 2-pin connectors.
7. Connect the negative battery cable.
8. Turn the ignition to the **ON** position, but don't start the engine. The AIR BAG warning light should turn on and flash on and off 7 times, and then turn off. This light sequence indicates that the SRS system is functioning normally. If the AIR BAG light doesn't come on, or stays on longer than 7 seconds, the system must be diagnosed.

Recirculating Ball Steering Gear

REMOVAL & INSTALLATION

1. Before servicing the vehicle, refer to the precautions in the beginning of this section.
2. Disable the air bag system.
3. Remove or disconnect the following:
   • Skid plates
   • Lower fan shroud
   • Stabilizer bar
   • Power steering pressure and return lines
   • Pitman arm
   • Steering column intermediate shaft
   • Steering gear

To install:
4. Install or connect the following:
   • Steering gear. Tighten the bolts to 33 ft. lbs. (44 Nm).
   • Steering column intermediate shaft. Tighten the pinch bolt to 18 ft. lbs. (25 Nm).
   • Pitman arm. Tighten the nut to 159 ft. lbs. (216 Nm).
   • Power steering pressure and return lines. Tighten the fittings to 33 ft. lbs. (44 Nm).
   • Stabilizer bar
   • Lower fan shroud
   • Skid plates
5. Fill the power steering fluid reservoir.
6. Check the wheel alignment and adjust as necessary.

Shock Absorber

REMOVAL & INSTALLATION

Front

1. Before servicing the vehicle, refer to the precautions in the beginning of this section.
2. Support the lower control arm with a jackstand.
3. Remove or disconnect the following:
   • Front wheels
   • Upper shock retaining nut and rubber bushing
   • Suspension bump stops
   • Shock absorber

To install:
4. Install or connect the following:
   • Shock absorber. Tighten the lower bolt to 60–61 ft. lbs. (82–84 Nm).
   • Bump stop. Tighten the bolts to 30 ft. lbs. (41 Nm).
   • Upper shock retaining nut and rubber bushing. Tighten the nut to 14–15 ft. lbs. (19–20 Nm).
   • Front wheels

Rear

1. Before servicing the vehicle, refer to the precautions in the beginning of this section.
2. Support the rear axle with jackstands.
3. Remove the upper and lower bolts, then the shock absorber.

To install:
4. Install the shock absorber. Tighten the upper bolt to 70 ft. lbs. (95 Nm). Tighten the lower bolt to 58 ft. lbs. (78 Nm).
5. Remove the jackstands.

Coil Spring

REMOVAL & INSTALLATION

1. Before servicing the vehicle, refer to the precautions in the beginning of this section.
2. Support the vehicle under the frame.
3. Support the rear axle with a jack.
4. Remove or disconnect the following:
   • Rear wheels
   • Stabilizer bar links
   • Parking brake cable brackets
   • Shock absorbers
5. Lower the rear axle with the jack to release the coil spring tension. Remove the coil springs and insulators.

To install:
6. Place the coil springs on the axle assembly and the insulators on top of the springs.
7. Raise the axle assembly into position.
8. Install or connect the following:
   • Shock absorbers
   • Parking brake cable brackets
   • Stabilizer bar links. Tighten the nuts to 37 ft. lbs. (50 Nm).
   • Rear wheels

Torsion Bar

REMOVAL & INSTALLATION

1. Before servicing the vehicle, refer to the precautions in the beginning of this section.

2. Matchmark the adjusting bolt and end piece, then remove the bolt, end piece, and seat.

3. Matchmark the height control arm to the torsion bar, then remove the height control arm.

4. Matchmark the torsion bar to the lower control arm, then remove the torsion bar.

To install:

5. Apply grease to the torsion bar splines.

6. Apply grease to the contact points of the height control arm, adjusting bolt end piece and seat.

7. Align the matchmarks and install the torsion bar.

8. Align the matchmarks and install the height control arm.

9. Install the adjusting bolt, seat and end piece.

10. Tighten the adjusting bolt to align the matchmarks.

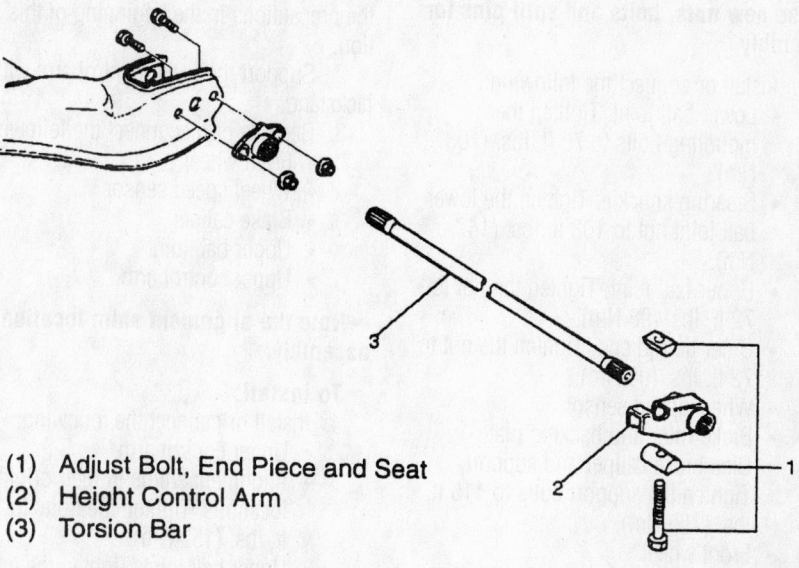

(1) Adjust Bolt, End Piece and Seat
(2) Height Control Arm
(3) Torsion Bar

Exploded view of the torsion bar assembly

Upper Ball Joint

REMOVAL & INSTALLATION

1. Before servicing the vehicle, refer to the precautions in the beginning of this section.

2. Support the lower control arm with a floor jack.

3. Remove or disconnect the following:
- Front wheel
- Wheel speed sensor
- Upper ball joint

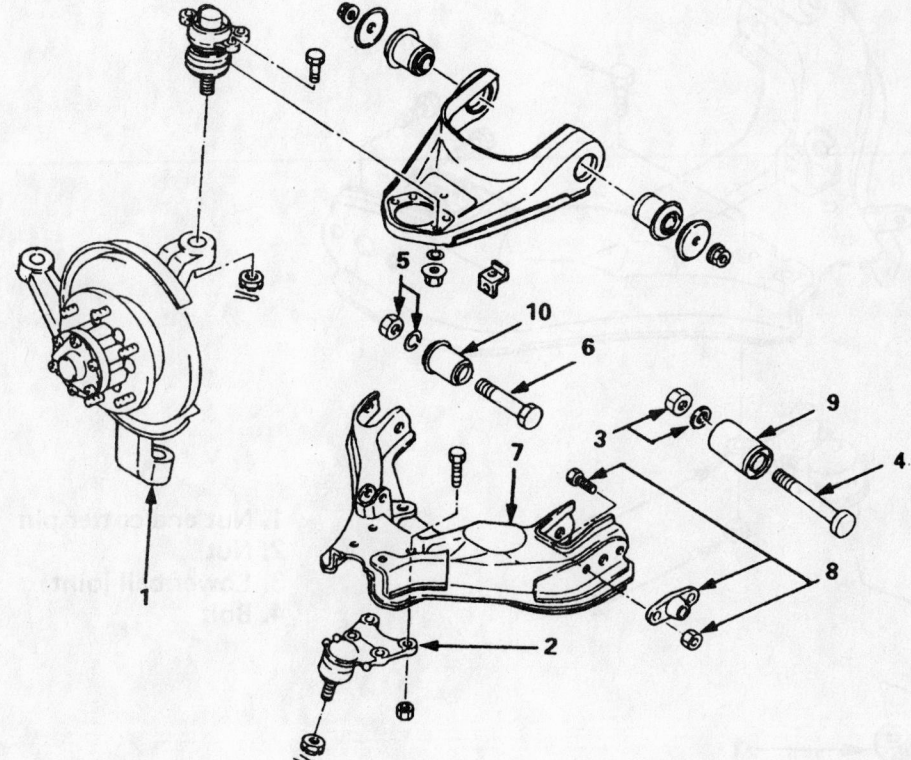

1. Knuckle
2. Lower end
3. Nut and washer, rear
4. Bolt, rear
5. Nut and washer, front
6. Bolt, front
7. Lower control arm assembly
8. Torsion bar arm bracket
9. Bushing, rear
10. Bushing, front

Exploded view of the control arm and ball joint components

For Accessory Drive Belt illustrations, see Section 1 of this manual

To install:

➡**Use new nuts, bolts and split pins for assembly.**

4. Install or connect the following:
- Upper ball joint. Tighten the mounting bolts to 42 ft. lbs. (57 Nm) and the nut to 72 ft. lbs. (96 Nm).
- Wheel speed sensor
- Front wheel

Lower Ball Joint

REMOVAL & INSTALLATION

1. Before servicing the vehicle, refer to the precautions in the beginning of this section.

2. Support the lower control arm with a jackstand.

3. Remove or disconnect the following:
- Front wheel
- Disc brake caliper and support
- Brake rotor and backing plate
- Wheel speed sensor

- Outer tie rod end
- Upper ball joint
- Steering knuckle
- Lower ball joint

To install:

➡**Use new nuts, bolts and split pins for assembly.**

4. Install or connect the following:
- Lower ball joint. Tighten the mounting bolts to 76 ft. lbs. (103 Nm).
- Steering knuckle. Tighten the lower ball joint nut to 108 ft. lbs. (147 Nm).
- Upper ball joint. Tighten the nut to 72 ft. lbs. (98 Nm).
- Outer tie rod end. Tighten the nut to 72 ft. lbs. (98 Nm).
- Wheel speed sensor
- Brake rotor and backing plate
- Disc brake caliper and support. Tighten the support bolts to 115 ft. lbs. (155 Nm).
- Front wheel

5. Check the wheel alignment and adjust as necessary.

Upper Control Arm

REMOVAL & INSTALLATION

1. Before servicing the vehicle, refer to the precautions in the beginning of this section.

2. Support the lower control arm with a jackstand.

3. Remove or disconnect the following:
- Front wheel
- Wheel speed sensor
- Brake caliper
- Upper ball joint
- Upper control arm

➡**Note the alignment shim location for assembly.**

To install:

4. Install or connect the following:
- Upper control arm
- Alignment shims in their original locations. Tighten the bolts to 112 ft. lbs. (152 Nm).
- Upper ball joint. Tighten the nut to 72 ft. lbs. (98 Nm).

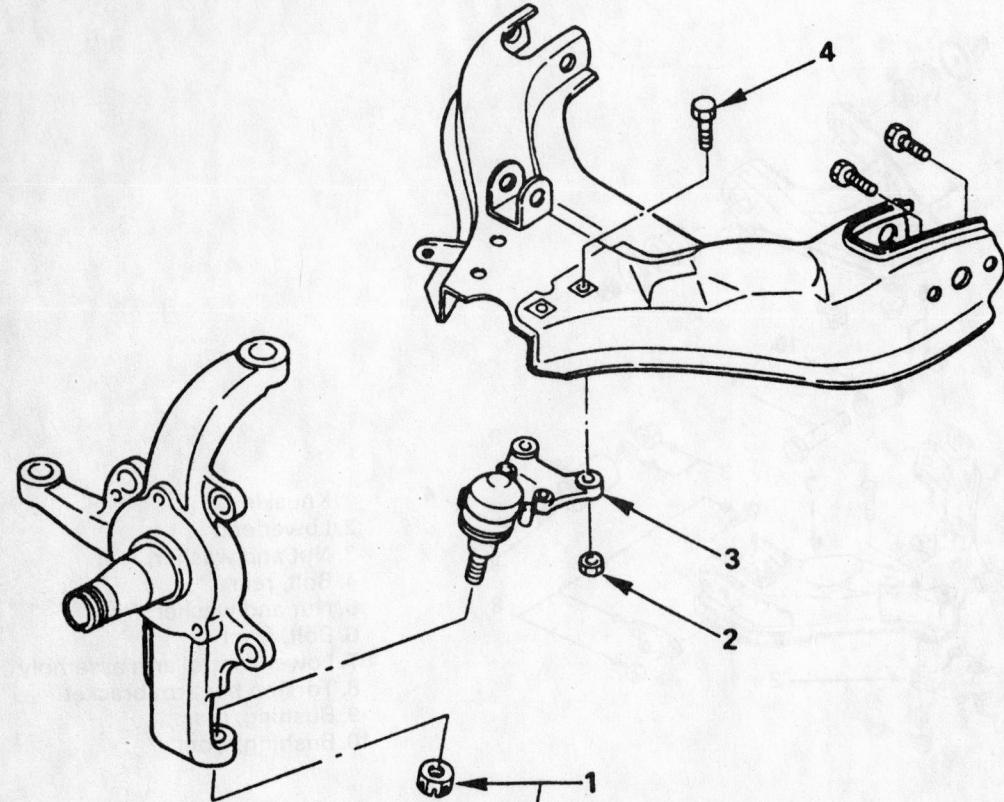

1. Nut and cotter pin
2. Nut
3. Lower ball joint
4. Bolt

7924BG35

Exploded view of the lower ball joint mounting and related components

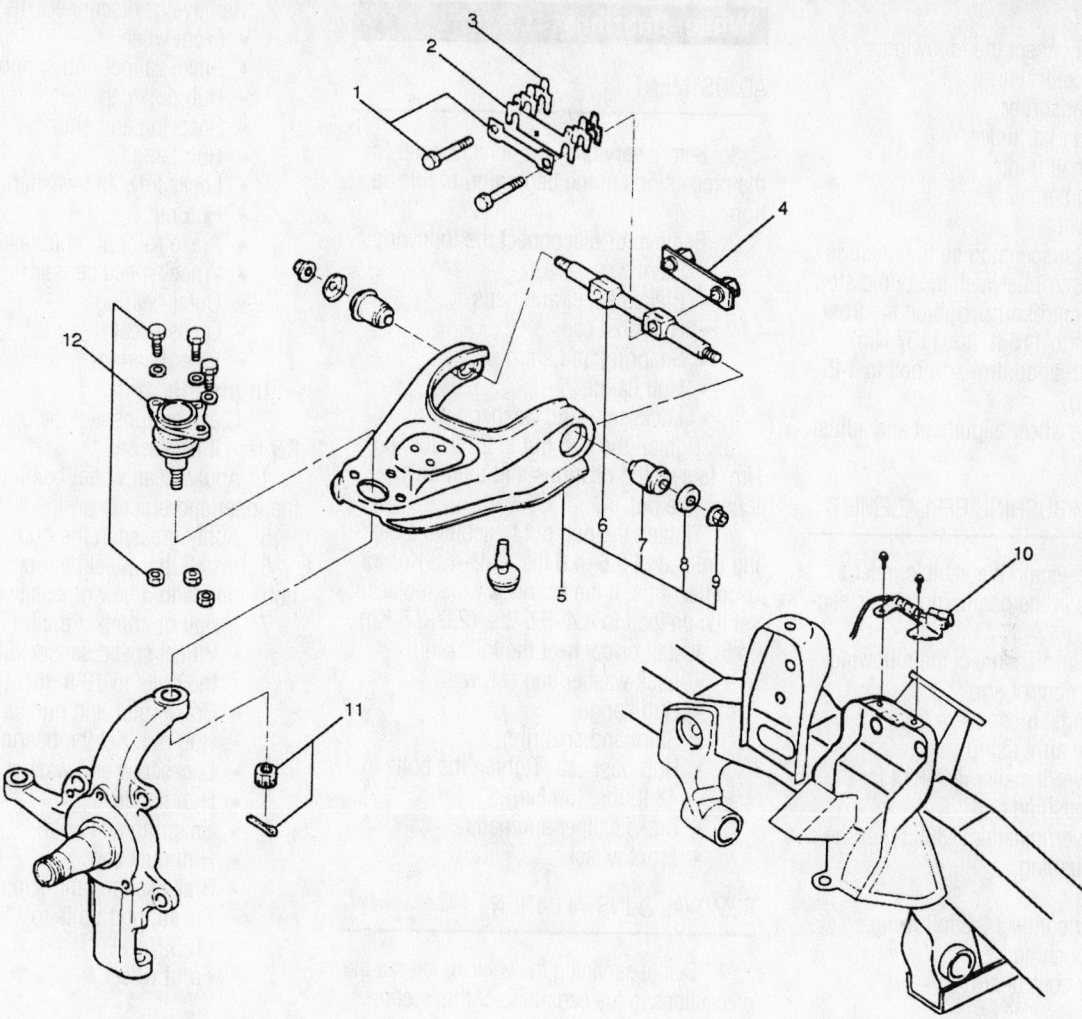

1   Bolt and Plate
2   Camber Shims
3   Caster Shims
4   Nut Assembly
5   Upper Control Arm Assembly
6   Fulcrum Pin

7   Bushing
8   Plate
9   Nut
10 Speed Sensor Cable
11 Nut and Cotter Pin
12 Upper Ball Joint

9308BG06

Upper control arm and related parts

- Brake caliper
- Wheel speed sensor
- Front wheel

5. Check the wheel alignment and adjust as necessary.

CONTROL ARM BUSHING REPLACEMENT

1. Before servicing the vehicle, refer to the precautions in the beginning of this section.

2. Remove the upper control arm.

3. Remove the nuts and washers from the fulcrum pin.

4. Press the bushings out of the control arm.

To install:

5. Press the bushings into the control arm.

6. Install the fulcrum pin washers and nuts.

7. Install the upper control arm.

8. Raise the suspension so that there is 0.79 inches (20mm) between the bump stop and the lower control arm. Tighten the fulcrum pin nuts to 80 ft. lbs. (108 Nm).

9. Check the wheel alignment and adjust as necessary.

Lower Control Arm

REMOVAL & INSTALLATION

1. Before servicing the vehicle, refer to the precautions in the beginning of this section.

2. Remove or disconnect the following:

- Front wheel
- Torsion bar
- Lower ball joint
- Stabilizer bar link
- Shock absorber
- Lower control arm

For Tire, Wheel and Ball Joint specifications, see Section 1 of this manual

To install:

3. Install or connect the following:
- Lower control arm
- Shock absorber
- Stabilizer bar link
- Lower ball joint
- Torsion bar
- Front wheel

4. Raise the suspension so that there is 0.79 inches (20mm) between the bump stop and the lower control arm. Tighten the front control arm bolt to 116 ft. lbs. (157 Nm) and tighten the rear control arm bolt to 145 ft. lbs. (196 Nm).

5. Check the wheel alignment and adjust as necessary.

CONTROL ARM BUSHING REPLACEMENT

1. Before servicing the vehicle, refer to the precautions in the beginning of this section.

2. Remove or disconnect the following:
- Lower control arm
- Bushings, press them from the control arm, using Remover/Installer J-36833 for the front bushing and Remover/Installer J-36834 for the rear bushing.

To install:

3. Install or connect the following:
- New bushings
- Lower control arm

Wheel Bearings

ADJUSTMENT

1. Before servicing the vehicle, refer to the precautions in the beginning of this section.

2. Remove or disconnect the following:
- Front wheel
- Brake caliper and pads
- Hub dust cap
- Snapring and shim
- Hub flange
- Lockscrew and washer

3. Tighten the hub nut to 22 ft. lbs. (29 Nm) to seat the bearings and then fully loosen the nut.

4. Tighten the hub nut to achieve a bearing preload of 2.6–4.0 lbs. (1.2–1.8 Kg) for used bearings. If the bearings were replaced, set the preload to 4.4–5.5 lbs. (2.0–2.5 Kg).

5. Install or connect the following:
- Lock washer and screw
- Hub flange
- Shim and snapring
- Hub dust cap. Tighten the bolts to 43 ft. lbs. (59 Nm).
- Brake caliper and pads
- Front wheel

REMOVAL & INSTALLATION

1. Before servicing the vehicle, refer to the precautions in the beginning of this section.

2. Remove or disconnect the following:
- Front wheel
- Brake caliper and support
- Hub dust cap
- Snapring and shim
- Hub flange
- Lockscrew and washer
- Hub nut
- Brake rotor and hub assembly
- Wheel speed sensor ring
- Outer bearing
- Grease seal
- Inner bearing

To install:

3. Clean and inspect the bearings. Replace if necessary.

4. Apply clean wheel bearing grease to the inner and outer bearings.

5. Apply grease in the hub.

6. Install the wheel bearings into the hub along with a new grease seal.

7. Install or connect the following:
- Wheel speed sensor ring. Tighten the bolts to 13 ft. lbs. (18 Nm).
- Brake rotor and hub assembly
- Hub nut. Set the bearing preload.
- Lockscrew and washer
- Hub flange
- Snapring and shim
- Hub dust cap
- Brake caliper and support. Tighten the support bolts to 115 ft. lbs. (155 Nm).
- Front wheel

PRECAUTIONS

Before servicing any vehicle, please be sure to read all of the following precautions, which deal with personal safety, prevention of component damage and important points to take into consideration when servicing a motor vehicle:

• Never open, service or drain the radiator or cooling system when the engine is hot; serious burns can occur from the steam and hot coolant.

• Observe all applicable safety precautions when working around fuel. Whenever servicing the fuel system, always work in a well-ventilated area. Do not allow fuel spray or vapors to come in contact with a spark, open flame, or excessive heat (a hot drop light, for example). Keep a dry chemical fire extinguisher near the work area. Always keep fuel in a container specifically designed for fuel storage; also, always properly seal fuel containers to avoid the possibility of fire or explosion. Refer to the additional fuel system precautions later in this section.

• Fuel injection systems often remain pressurized, even after the engine has been turned **OFF**. The fuel system pressure must be relieved before disconnecting any fuel lines. Failure to do so may result in fire and/or personal injury.

• Brake fluid often contains polyglycol ethers and polyglycols. Avoid contact with the eyes and wash your hands thoroughly after handling brake fluid. If you do get brake fluid in your eyes, flush your eyes with clean, running water for 15 minutes. If eye irritation persists, or if you have taken brake fluid internally, seek medical assistance IMMEDIATELY.

• The EPA warns that prolonged contact with used engine oil may cause a number of skin disorders, including cancer! You should make every effort to minimize your exposure to used engine oil. Protective gloves should be worn when changing oil. Wash your hands and any other exposed skin areas as soon as possible after exposure to used engine oil. Soap and water, or waterless hand cleaner should be used.

• All new vehicles are now equipped with an air bag system. The system must be disabled before performing service on or around system components, steering column, instrument panel components, wiring and sensors. Failure to follow safety and disabling procedures could result in accidental air bag deployment, possible personal injury and unnecessary system repairs.

• Always wear safety goggles when working with, or around, the air bag system. When carrying a non-deployed air bag, be sure the bag and trim cover are pointed away from your body. When placing a non-deployed air bag on a work surface, always face the bag and trim cover upward, away from the surface. This will reduce the motion of the module if it is accidentally deployed. Refer to the additional air bag system precautions later in this section.

• Clean, high quality brake fluid from a sealed container is essential to the safe and proper operation of the brake system. You should always buy the correct type of brake fluid for your vehicle. If the brake fluid becomes contaminated, completely flush the system with new fluid. Never reuse any brake fluid. Any brake fluid that is removed from the system should be discarded. Also, do not allow any brake fluid to come in contact with a painted surface; it will damage the paint.

• Never operate the engine without the proper amount and type of engine oil; doing so WILL result in severe engine damage.

• Timing belt maintenance is extremely important! Many models utilize an interference-type, non-freewheeling engine. If the timing belt breaks, the valves in the cylinder head may strike the pistons, causing potentially serious (also time-consuming and expensive) engine damage. Refer to the maintenance interval charts in the front of this manual for the recommended replacement interval for the timing belt and to the timing belt section for belt replacement and inspection.

• Disconnecting the negative battery cable on some vehicles may interfere with the functions of the on-board computer system(s) and may require the computer to undergo a relearning process once the negative battery cable is reconnected.

• When servicing drum brakes, only disassemble and assemble one side at a time, leaving the remaining side intact for reference.

• Only an MVAC-trained, EPA-certified automotive technician should service the air conditioning system or its components.

ENGINE REPAIR

➡**Disconnecting the negative battery cable on some vehicles may interfere with the functions of the on-board computer systems and may require the computer to undergo a relearning process.**

Ignition Timing

ADJUSTMENT

The Digital Motor Electronics (DME) control, unit controls all ignition and fuel injection functions. Ignition timing is fully electronically controlled; there is no vacuum advance or manual adjustment. Ignition functions are calculated from internal maps and from the same sensors used for the fuel injection system. On vehicles with an automatic transmission, the control unit will retard ignition timing briefly when the transmission is about to shift up or down. For this reason, there is a data link between the DME control unit and the transmission control unit.

Since the ignition timing is controlled by the DME, checking and adjusting the timing is impossible. There is no method of setting dynamic or static timing.

Alternator

REMOVAL & INSTALLATION

➡**When the battery is disconnected the radio code, on-board computer and clock settings will be lost. The radio code should be obtained before disconnecting the battery or radio. Once the battery has been reconnected, the radio will not function unless the code is keyed in.**

1. Before servicing the vehicle, refer to the precautions in the beginning of this section.
2. Drain the cooling system.
3. Remove or disconnect the following:
   • Negative battery cable
   • Drive belt
   • Fan cowling
   • Alternator bolts
   • Alternator electrical connectors
   • Alternator roller by releasing the screw
   • Alternator

To install:
4. Replace the sealing ring for the alternator.

5. Install or connect the following:
- Alternator. Torque the bolts to 31 ft. lbs. (43 Nm).
- Alternator roller
- Alternator electrical connectors
- Fan cowling
- Drive belt
- Negative battery cable

6. Fill the cooling system to the proper level.

7. Start the vehicle and check for leaks, repair if necessary.

Engine Assembly

REMOVAL & INSTALLATION

1. Fully open the hood and properly secure it in place.

2. Properly relieve the fuel system pressure.

3. Evacuate and recover the A/C system.

4. Drain the cooling system.

5. Drain the engine oil.

6. Drain the power steering fluid.

7. Before servicing the vehicle, refer to the precautions in the beginning of this section.

8. Remove or disconnect the following:
- Negative battery cable
- Engine cover
- Throttle cable from the intake filter housing
- Intake filter housing
- Mass Air Flow (MAF) sensor
- Windshield washer reservoir
- Brake booster hose from the suction jet pump
- Fuel feed line from the injection pipe
- Engine splash shield and reinforcement plate
- Drive belt
- Power steering pump
- A/C system lines between the compressor and condenser
- A/C suction line
- Transmission
- Starter electrical connectors and heat shield
- Starter
- Oil lines to the transmission on the heat exchanger
- Radiator
- Coolant hoses on the alternator and thermostat housing
- Coolant hoses from the coolant manifold

- Heating valve and hoses
- Fuel tank vent hose
- Engine wire harness from the control unit box
- Transmission wire harness from the control unit box
- Oxygen (O_2S) sensor wiring and place all wires on top of the engine
- Expansion tank
- Supply reservoir from the carrier
- Ground strap from the oil filter housing
- Left and right swivel bearings
- Output shafts
- Propeller shaft
- Partition wall
- Ground tape from the right side engine support
- Upper nuts from the left and right side engine mounts and install an engine removal tool to the locating lugs
- Engine from the vehicle

To install:

9. Carefully lower the engine into the engine compartment.

10. Install or connect the following:
- Engine mounts. Torque the bolts to 32 ft. lbs. (45 Nm).
- Ground tape
- Partition wall
- Propeller shafts
- Output shafts
- Left and right swivel bearings
- Ground strap to the oil filter housing
- Supply reservoir
- Expansion tank
- O_2S sensor electrical connector
- Engine and transmission wire harness to the control unit box
- Fuel tank vent hose
- Heater valve and hoses
- Coolant hoses to the manifold and thermostat housing
- Radiator
- Transmission
- Oil lines to the transmission. Torque the nuts to 25 ft. lbs. (34 Nm).
- Starter and electrical connectors
- A/C lines
- Power steering pump
- Drive belt
- Engine splash shield and reinforcement plate
- Fuel feed line to the injection pipe
- Brake booster vacuum hose
- Windshield washer reservoir

- MAF sensor
- Intake filter housing
- Throttle cable to the filter housing
- Engine cover
- Negative battery cable

11. Fill and bleed the power steering system.

12. Fill and bleed the coolant system.

13. Recharge the A/C system.

14. Fill the engine with clean oil.

15. Start the vehicle and check for leaks, repair if necessary.

Water Pump

REMOVAL & INSTALLATION

1. Before servicing the vehicle, refer to the precautions in the beginning of this section.

2. Drain the cooling system.

3. Remove or disconnect the following:
- Negative battery cable
- Vibration damper
- Thermostat housing
- Water pump pulley
- Coolant hoses
- Water pump and discard the seal

To install:

4. Clean and remove any residual debris or gasket material from the engine mounting surface for the water pump.

5. Install the water pump with a new gasket. Torque the bolts as follows:
   a. M6 bolts: 78 inch lbs. (9 Nm).
   b. M7 bolts: 89 inch lbs. (10 Nm).
   c. M8 bolts: 16 ft. lbs. (22 Nm).

6. Install or connect the following:
- Coolant hoses to the water pump
- Water pump pulley
- Thermostat housing
- Vibration damper
- Negative battery cable

7. Fill and bleed the cooling system.

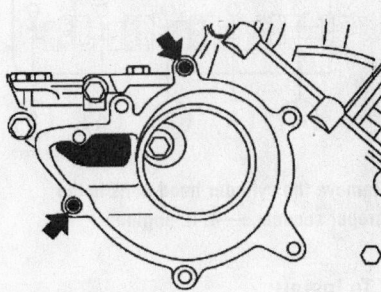

9308KG01

Exploded view of the water pump—4.4L engine

8. Start the vehicle, check for leaks and repair as necessary.

Cylinder Head

REMOVAL & INSTALLATION

Left Side

1. Before servicing the vehicle, refer to the precautions in the beginning of this section.

2. Properly relieve the fuel system pressure.

3. Drain the cooling system.

4. Remove or disconnect the following:

- Negative battery cable
- Left side exhaust manifold
- Cylinder head cover
- Spark plugs
- Intake manifold
- Coolant manifold
- Left side camshaft adjustment unit

5. Install special tool 11–5–180 and pull back until the flywheel is no longer secured.

6. Lift the timing chain and hold it under tension.

7. Crank the engine at the central bolt against the direction of rotation to 45 degrees Before Top Dead Center (BTDC).

8. Remove the special tools.

9. Remove or disconnect the following:

- Guide rail from the cylinder head
- Cylinder head bolts in the proper sequence
- Cylinder head

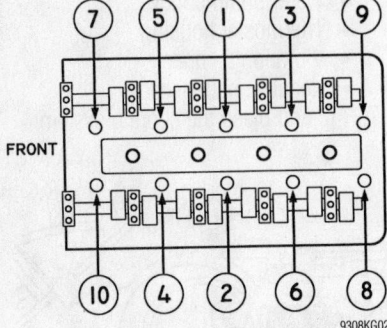

Remove the cylinder head bolts in the proper sequence—4.4L engine

To install:

10. Thoroughly clean all mounting surfaces and check the head for warpage. Take care not to drop any pieces of gasket or

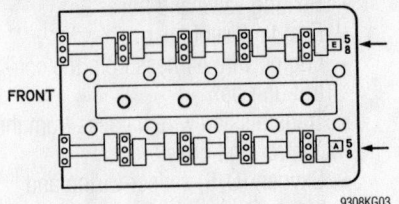

Rotate the camshafts until the markings face upward—4.4L engine

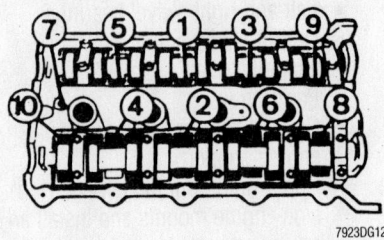

Cylinder head bolt torque sequence—4.4L engine

debris into the oil or coolant passages. Check the condition of the head locating dowel sleeves and clean out the bolt threads with a tap.

11. Mount the cylinder head on the block and use new bolts. Do not remove the coating on the head bolts, apply oil to the threads and torque the bolts in the following sequence:

  a. Step 1: 22 ft. lbs. (30 Nm).
  b. Step 2: 80 degrees.
  c. Step 3: 80 degrees.

12. Install or connect the following:

- Guide rail screw and rotate the camshafts until the markings face upward
- Camshaft sprockets and the timing chain tensioner. Torque the bolts to 11 ft. lbs. (15 Nm).
- Left side camshaft adjustment unit
- Coolant manifold
- Spark plugs
- Cylinder head cover. Torque the bolts to 10 ft. lbs. (15 Nm).
- Left side exhaust manifold
- Negative battery cable

13. Fill and bleed the cooling system.

14. Change the engine oil and filter.

15. Start the vehicle, check for leaks and repair as necessary.

Right Side

1. Before servicing the vehicle, refer to the precautions in the beginning of this section.

2. Properly relieve the fuel system pressure.

3. Drain the cooling system.

4. Remove or disconnect the following:

- Negative battery cable
- Right side exhaust manifold
- Cylinder head cover
- Spark plugs
- Fan clutch and impeller
- Intake manifold
- Coolant manifold
- Right side cam adjustment unit

5. Install special tool 11–5–180 and pull back until the flywheel is no longer secured.

6. Lift the timing chain and hold it under tension.

7. Crank the engine at the central bolt against the direction of rotation to 45 degrees Before Top Dead Center (BTDC).

8. Remove the special tools.

9. Remove or disconnect the following:

- Guide rail from the cylinder head
- Cylinder head bolts in the proper sequence
- Cylinder head

To install:

10. Thoroughly clean all mounting surfaces and check the head for warpage. Take care not to drop any pieces of gasket or debris into the oil or coolant passages. Check the condition of the head locating dowel sleeves and clean out the bolt threads with a tap.

11. Mount the cylinder head on the block and use new bolts. Do not remove the coating on the head bolts, apply oil to the threads and torque the bolts in the following sequence:

  a. Step 1: 22 ft. lbs. (30 Nm).
  b. Step 2: 80 degrees.
  c. Step 3: 80 degrees.

12. Install or connect the following:

- Guide rail screw and rotate the camshafts until the markings face upward
- Camshaft sprockets and the timing chain tensioner. Torque the bolts to 11 ft. lbs. (15 Nm).
- Left side camshaft adjustment unit
- Coolant manifold
- Spark plugs
- Cylinder head cover. Torque the bolts to 10 ft. lbs. (15 Nm).
- Left side exhaust manifold
- Negative battery cable

13. Fill and bleed the cooling system.

14. Change the engine oil and filter.

- Start the vehicle and check for leaks, repair if necessary.

Intake Manifold

REMOVAL & INSTALLATION

1. Before servicing the vehicle, refer to the precautions in the beginning of this section.
2. Properly relieve the fuel system pressure.
3. Remove or disconnect the following:
   - Both battery cables
   - Acoustic cover
   - Ignition coil electrical connectors
   - Throttle bellows
   - Intake filter housing
   - Mass Air Flow (MAF) sensor
   - Wiring harness from the intake manifold
   - Air injection vacuum control hoses
   - Throttle body vacuum hose
   - Fuel feed line from the injection pipe
   - Engine ventilation hose from the cylinder head cover
   - Engine ventilation hose from the oil separator
   - Brake booster vacuum hose
   - Oil separator from the cover
   - Decoupling elements from under the intake manifold
   - Oil drain hose from the rear cover after raising the intake manifold slightly
   - Intake manifold

➡ **The intake manifold is vibrationally separated from the cylinder head by decoupling elements and gaskets.**

To install:

4. Install or connect the following:
   - Decoupling elements to the cylinder head
   - Intake manifold
   - Remaining components of the decoupling elements. Torque the M6 nuts to 89 inch lbs. (10 Nm), the M7 nuts to 10 ft. lbs. (15 Nm) and the M8 nuts to 16 ft. lbs. (22 Nm).
   - Oil drain hose to the rear cover
   - Oil separator
   - Brake booster vacuum hose
   - Engine vent hose to the oil separator and the cylinder head
   - Fuel feed line to the injection pipe
   - Throttle body vacuum hose
   - Air injection vacuum control hoses
   - Wiring harness to the intake manifold
   - MAF sensor
   - Intake filter housing
   - Throttle bellows
   - Ignition coil electrical connectors
   - Acoustic cover
   - Both battery cables

Exhaust Manifold

REMOVAL & INSTALLATION

1. Before servicing the vehicle, refer to the precautions in the beginning of this section.
2. Remove or disconnect the following:
   - Negative battery cable
   - Exhaust system
   - Reinforcement plate
   - Propeller shaft, left side manifold only
   - Screw connection at the exhaust manifold
   - Exhaust manifold downward and discard the gaskets

To install:

3. Remove the old gasket off of the cylinder head and exhaust manifold and replace the gasket. The gasket beads face the exhaust manifolds.
4. Install or connect the following:
   - Exhaust manifold with new gaskets. Torque the bolts to 10 ft. lbs. (15 Nm).
   - Screw connection at the exhaust manifold
   - Propeller shaft, left side manifold only
   - Reinforcement plate
   - Exhaust system
   - Negative battery cable

Camshaft and Valve Lifters

REMOVAL & INSTALLATION

Left Camshaft (Cylinder Bank 5–8)

1. Before servicing the vehicle, refer to the precautions in the beginning of this section.
2. Remove or disconnect the following:
   - Negative battery cable
   - Splash guard
   - Left and right cylinder head covers

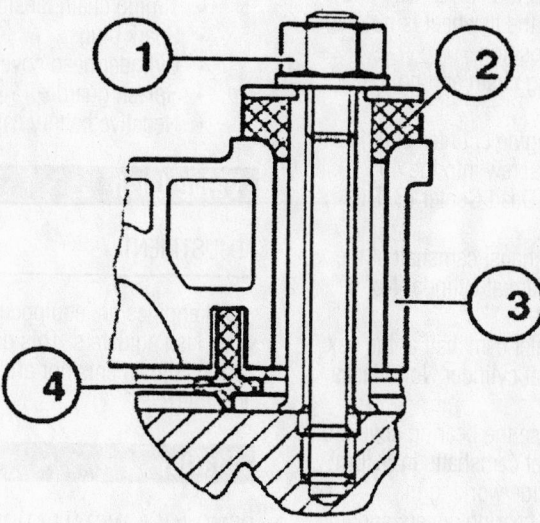

(1)   Nut

(2)   Decoupling element

(3)   Intake air manifold

(4)   Seal

9308KG04

Separating the intake manifold from the cylinder head

- Spark plugs
- Timing chain tensioning piston
- Top left timing case cover
- Left camshaft adjustment unit and distributor
- Oil lines to the left and right cylinder head

3. Install special tool 11–2–300 and pull back on it until the flywheel is no longer secured in position.

4. Lift the timing chain and hold it under tension.

5. Crank the engine counter-engine wise on the central screw into the 45 degrees Before Top Dead Center (BTDC) position.

6. Rotate the exhaust camshaft at the hex head until the cam at cylinder No. 6 faces upward.

7. Rotate the inlet camshaft at the hex head until the cam at cylinder No. 8 faces upward.

8. Evenly release the bearing covers on the exhaust and inlet camshafts in ½turn steps from the outside working in.

9. Remove the bearing covers and remove the camshafts.

10. Remove the hydraulic valve lifters. To remove, use tool No. 11-3-250 to pull them out of the cylinder head. Be sure that no damage occurs to the guides in the head. Inspect the bearing surfaces of the tappets for wear and scoring.

To install:

11. If the lifters were removed, install them with tool No. 11-32-250.

12. Lubricate and install the camshafts in their correct position.

13. Install or connect the following:
- Exhaust and inlet camshafts and rotate them until exhaust cam at cylinder No. 6 faces up and inlet cam at cylinder No. 8 faces up
- Bearing caps. The exhaust camshaft bearing covers are marked A1 to A5 and the inlet covers are marked E1 to E5. Evenly tighten the bearing covers in ½ turn steps from the outside working in.

14. Torque the camshaft bearing covers as follows:
   a. M6 to 89 inch lbs. (10 Nm).
   b. M7 to 9 ft. lbs. (14 Nm).
   c. M8 to 15 ft. lbs. (20 Nm).

15. Rotate the camshafts until the markings face upwards.

16. Using special tools 11–2–446 and 11–2–442 align the inlet and exhaust camshafts with an open-end wrench so the tools rest without a gap on the cylinder head.

17. Crank the engine from the 45 degrees BTDC position in the direction of rotation up to the TDC position.

18. Hold the crankshaft and install the distributor on the camshaft adjustment unit.

19. Install or connect the following:
- Camshaft adjustment unit
- Left timing case cover
- Timing chain tensioning piston
- Spark plugs
- Cylinder head covers
- Splash guard
- Negative battery cable

Right Camshaft (Cylinder Bank 1–4)

1. Before servicing the vehicle, refer to the precautions in the beginning of this section.

2. Remove or disconnect the following:
- Negative battery cable
- Splash guard
- Left and right cylinder head covers
- Spark plugs
- Timing chain tensioning piston
- Top right timing case cover
- Right camshaft adjustment unit and distributor
- Oil lines to the left and right cylinder head

3. Install special tool 11–2–300 and pull back on it until the flywheel is no longer secured in position.

4. Lift the timing chain and hold it under tension.

5. Crank the engine counter-engine wise on the central screw into the 45 degrees Before Top Dead Center (BTDC) position.

6. Rotate the exhaust camshaft at the hex head until the cam at cylinder No. 6 faces upward.

7. Rotate the inlet camshaft at the hex head until the cam at cylinder No. 8 faces upward.

8. Evenly release the bearing covers on the exhaust and inlet camshafts in ½turn steps from the outside working in.

9. Remove the bearing covers and remove the camshafts.

10. Remove the hydraulic valve lifters. To remove, use tool No. 11-3-250 to pull them out of the cylinder head. Be sure that no damage occurs to the guides in the head. Inspect the bearing surfaces of the tappets for wear and scoring.

To install:

11. If the lifters were removed, install them with tool No. 11-32-250.

12. Lubricate and install the camshafts in their correct position.

13. Install or connect the following:
- Exhaust and inlet camshafts and

rotate them until exhaust cam at cylinder No. 6 faces up and inlet cam at cylinder No. 8 faces up
- Bearing caps. The exhaust camshaft bearing covers are marked A1 to A5 and the inlet covers are marked E1 to E5. Evenly tighten the bearing covers in ½ turn steps from the outside working in.

14. Torque the camshaft bearing covers as follows:
   a. M6 to 89 inch lbs. (10 Nm).
   b. M7 to 9 ft. lbs. (14 Nm).
   c. M8 to 15 ft. lbs. (20 Nm).

15. Rotate the camshafts until the markings face upwards.

16. Using special tools 11–2–446 and 11–2–442 align the inlet and exhaust camshafts with an open-end wrench so the tools rest without a gap on the cylinder head.

17. Crank the engine from the 45 degrees BTDC position in the direction of rotation up to the TDC position.

18. Hold the crankshaft and install the distributor on the camshaft adjustment unit.

19. Install or connect the following:
- Camshaft adjustment unit
- Both timing case covers
- Timing chain tensioning piston
- Spark plugs
- Cylinder head covers
- Splash guard
- Negative battery cable

Valve Lash

ADJUSTMENT

All engines are equipped with hydraulic valve lash adjusters. This design does not permit adjustments nor are adjustments possible.

Starter

REMOVAL & INSTALLATION

➡**When the battery is disconnected, the radio code, on-board computer and clock settings will be lost. The radio code should be obtained before disconnecting the battery or radio. Once the battery has been reconnected, the radio will not function unless the code is keyed in.**

1. If needed, read the stored fault memories from the control module.

2. Relieve the fuel system pressure.

3. Set the ignition switch to the **OFF** position.

4. Before servicing the vehicle, refer to the precautions in the beginning of this section.

5. Remove or disconnect the following:
- Negative battery cable
- Reinforcement plate
- Positive battery cable from the starter
- Starter electrical connectors
- Heat shield
- Starter from the transmission mount
- Starter

To install:

6. Install or connect the following:
- Starter to the transmission mount
- Heat shield. Torque the bolts to 38 ft. lbs. (47 Nm).
- Starter electrical connectors
- Positive battery cable to the starter
- Reinforcement plate
- Negative battery cable

Oil Pan

REMOVAL & INSTALLATION

1. Before servicing the vehicle, refer to the precautions in the beginning of this section.
2. Drain the engine oil.
3. Remove or disconnect the following:
- Negative battery cable
- Reinforcement plate
- Oil level switch plug
- Cable guide clips
- Lower oil pan section

➡**To remove the upper section of the oil pan, proceed with the following steps.**

4. Remove or disconnect the following:
- Upper nuts on the left and right engine mounts
- Front splash guard
- Positive battery cable from the starter
- Left and right swivel bearings
- Output shafts
- Bearing pedestal from the right output shaft
- Propeller shaft
- Steering spindle from the steering gear and support the front axle
- Front axle support from the engine carrier and slightly lower the axle support

- Drive belt
- Vane pump from the oil pan
- Adjustable plate from the oil pan after releasing the tension from the A/C compressor belt
- Guide tube for the oil dipstick
- Oil return line from the oil separator to the oil pan
- Oil pump snorkel
- Cable guide for the positive lead
- Upper oil pan section towards the rear of the vehicle

To install

5. Clean the mounting surfaces.
6. Check the seals on the oil pipes and replace it if necessary. Lubricate the seals with oil.
7. Install or connect the following:
- Install upper oil pan. Torque the bolts to 89 inch lbs. (10 Nm) and lower the engine
- Cable guide for the positive lead
- Banjo bolt for the oil return pipe from the oil filter at the oil pan
- Drive belt
- Left and right engine mounts Torque the bottom bolts to 32 ft. lbs. (43 Nm).
- Lower oil pan with a new gasket. Torque the bolts, beginning in the middle and working to the outside to 89 inch lbs. (10 Nm).
- Plug for the level switch, making sure to replace the O-ring
- Steering spindle to the steering gear
- Propeller shaft
- Bearing pedestal
- Output shafts
- Positive battery cable
- Engine splash guards
- Oil dipstick guide tube, making sure to replace the O-ring
- Reinforcement plate
- Negative battery cable
8. Fill the engine with clean oil.
9. Start the vehicle and check for leaks, repair if necessary.

Oil Pump

REMOVAL & INSTALLATION

1. Before servicing the vehicle, refer to the precautions in the beginning of this section.
2. Drain the engine oil.

3. Remove or disconnect the following:
- Negative battery cable
- Oil pan
- Oil pump sprocket wheel and chain
- Oil pump

To install

4. Check the seals on the oil pipes and replace it if necessary. Lubricate the seals with oil.
5. Check the seal in the oil pump and replace it if necessary. Screw the hexagon adapter back into the oil pump until it stops.
6. Install or connect the following:
- Oil pump. Torque the bolts to 17 ft. lbs. (22 Nm).
- Oil pump sprocket wheel and chain. Torque the nut to 35 ft. lbs. (47 Nm).
- Oil pan
- Negative battery cable
7. Fill the engine with clean oil.
8. Start the vehicle and check for leaks, repair if necessary.

Rear Main Seal

REMOVAL & INSTALLATION

The rear main bearing oil seal can be replaced after the transmission and flywheel has been removed from the engine.

1. Before servicing the vehicle, refer to the precautions in the beginning of this section.
2. Drain the transmission fluid.
3. Remove or disconnect the following:
- Negative battery cable
- Transmission
- Flywheel assembly
- Oil seal, using a suitable seal removal tool

To install:

4. Coat the sealing lips of the new seal with oil.
5. Install or connect the following:
- New seal into the end cover housing with a suitable seal installation tool
- Flywheel
- Transmission
- Negative battery cable
6. Fill the transmission with new fluid.
7. Start the engine and check that oil pressure is present; if the oil pressure lamp does not extinguish within 5–7 seconds of starting the engine, turn the engine **OFF**.
8. Check and top off all fluid levels.

Timing belt service is covered in Section 3 of this manual

Timing Chain, Sprockets, Front Cover and Seal

REMOVAL & INSTALLATION

1. Before servicing the vehicle, refer to the precautions in the beginning of this section.

2. Drain the engine oil.

3. Remove or disconnect the following:
   - Negative battery cable
   - Spark plugs
   - Cylinder head covers

➡ **In the Top Dead Center (TDC) firing position, the inlet camshaft twists in the splines of the camshaft adjustment unit.**

4. Remove or disconnect the following:
   - Oil lines from the cylinder head
   - Vibration damper and rotate the engine at the central bolt so that the first cylinder is at the TDC position.
   - Timing chain tensioning piston
   - Both top timing case covers
   - Left hand threaded nut from the sensor gear on cylinder bank 1–4
   - Left hand threaded nut from the sensor gear on cylinder bank 5–8

5. Slacken the screw connection for the exhaust camshaft on cylinder bank 5–8 by ½ turn.

6. Slacken the screw connection for the exhaust camshaft on cylinder bank 1–4 by ½ turn.

7. Slacken the screw connection for the inlet camshaft on cylinder bank 5–8 by ½ turn.

8. Slacken the screw connection for the inlet camshaft on cylinder bank 1–4 by ½ turn.

9. Align the camshafts and install special tool 11–2–445/441 to the camshafts on cylinder back 1–4.

10. Align the camshafts and install special tool 11–2–446/442 to the camshafts on cylinder back 5–8.

11. Remove or disconnect the following:
    - Central bolt and hub from the vibration damper
    - Oil pump
    - Water pump and thermostat housing
    - Bottom timing case cover
    - Tensioning rail and oil guide
    - Timing chain from the camshaft adjustment unit

To install:

12. Install or connect the following:
    - Timing chain over the reversing rail, camshaft adjustment unit and crankshaft sprocket wheel for cylinder bank 5–8
    - Timing chain inside screw-in pin
    - Timing chain onto the camshaft adjustment unit for cylinder bank 1–4

13. Raise the timing chain slightly by the guide rail and slide the rail over the pin until the retaining lug can be heard snapping into place on the lower guide pin.

14. Align the timing chain to the guide rail.

15. Install the oil guide for the bow cover in the pivot rail.

16. Install the tensioning rail screw. Press the cover against the timing chain and secure the cover with the plastic strap.

17. Install or connect the following:
    - Upper oil pan section and secure the crankshaft in the TDC position with special tool 11–5–180
    - Special tool 11–7–380 to the right side cylinder bank and install special tool 11–4–230
    - Adjustment screw into the tensioning rail and hand tighten
    - Special tool 11–6–440 to the camshaft adjustment unit on cylinder bank 5–8 and move it 31 ft. lbs. (40 Nm) to the left hand stop.

18. Tighten the screw connection on the inlet camshaft on cylinder bank 5–8 to 10 ft. lbs. (15 Nm) and back off by ¼ turn.

19. Tighten the screw connection on the exhaust camshaft on cylinder bank 5–8 to 10 ft. lbs. (15 Nm) and back off by ¼ turn.

20. Special tool 11–6–440 to the camshaft adjustment unit on cylinder bank 1–4 and move it 31 ft. lbs. (40 Nm) to the left hand stop.

21. Tighten the screw connection on the inlet camshaft on cylinder bank 1–4 to 10 ft. lbs. (15 Nm) and back off by ¼ turn.

22. Tighten the screw connection on the exhaust camshaft on cylinder bank 1–4 to 10 ft. lbs. (15 Nm) and back off by ¼ turn.

23. Tighten the tensioning rail by turning the adjusting screw on special tool 11–4–230.

➡ **When the timing chain is pretensioned, the camshaft adjustment unit moves and must be reset to the left hand stop.**

24. Tighten the inlet camshaft screw connection on cylinder bank 5–8 to 85 ft. lbs. (110 Nm).

25. Tighten the exhaust camshaft screw connection on cylinder bank 5–8 to 85 ft. lbs. (110 Nm).

26. Install special tool 11–6–451 to the camshaft adjustment unit on cylinder bank 1–4 and move it 31 ft. lbs. (40 Nm) to the left hand stop.

27. Tighten the inlet camshaft screw connection on cylinder bank 1–4 to 85 ft. lbs. (110 Nm).

28. Tighten the exhaust camshaft screw connection on cylinder bank 1–4 to 85 ft. lbs. (110 Nm).

29. Install the sensor gear to cylinder bank 1–4 and hand tighten the nut.

30. Align the locating bore on the sensor gear to the positioning pin on special tool 11–6–451. Press the tool downward and align it to the cylinder head. Torque the sensor gear screw to 30 ft. lbs. (40 Nm). Remove the special tool.

31. Install special tool 11–6–452 to the camshaft adjustment unit on cylinder bank 5–8 and move it 31 ft. lbs. (40 Nm) to the left hand stop.

32. Tighten the inlet camshaft screw connection on cylinder bank 5–8 to 85 ft. lbs. (110 Nm).

33. Tighten the exhaust camshaft screw connection on cylinder bank 5–8 to 85 ft. lbs. (110 Nm).

34. Install the sensor gear to cylinder bank 5–8 and hand tighten the nut.

35. Align the locating bore on the sensor gear to the positioning pin on special tool 11–6–452. Press the tool downward and align it to the cylinder head. Torque the sensor gear screw to 30 ft. lbs. (40 Nm). Remove the special tool.

36. Check for the correct seating of the dowel sleeves. Clean the sealing surfaces thoroughly, and then place a new gasket on the lower cover.

37. Trim the protruding ends of the gasket, making sure the cutting tool is level. Do not allow the pieces to fall into the engine.

38. Position the lower cover and install the mounting bolts with an even distribution of pressure. Tighten the 6mm bolts to 84 inch lbs. (10 Nm), 8mm bolts to 16 ft. lbs. (22 Nm) and 10mm bolts to 35 ft. lbs. (47 Nm).

39. Install the oil seal in the timing case cover using tool No. 11–1–220. Make sure the seal is flush with the cover.

40. Install the vibration damper hub and install the mounting bolt. Tighten the hub bolt to:
    a. Step 1: 74 ft. lbs. (100 Nm).
    b. Step 2: turn an additional 60 degrees.
    c. Step 3: turn an additional 60 degrees.
    d. Step 4: turn an additional 30 degrees.

41. Position the vibration damper pul-

leys and install the mounting bolts for the damper.

42. Install or connect the following:
- Water pump pulley
- Drive belt and the cooling fan. Rotate the fan counterclockwise to install.
- Intake hose between the throttle body and the air volume meter
- Battery positive cable for the alternator and install the protective tube mounting fasteners and connect the remaining wires to the alternator
- Oil filter housing and the return pipe and replace the housing cover
- Alternator and cylinder head cover
- Negative battery cable

43. Fill the engine with clean oil.

44. Start the vehicle and check for leaks, repair if necessary

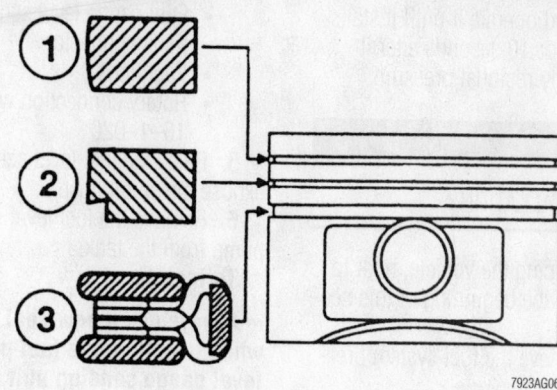

BMW engines compression and oil control ring locations

Piston and Ring Positioning

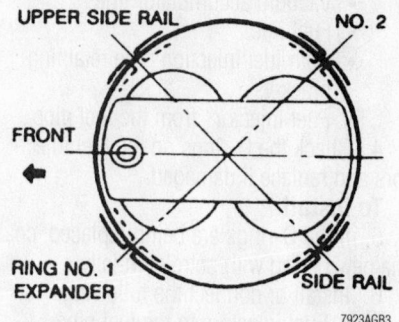

BMW engines piston ring end-gap spacing

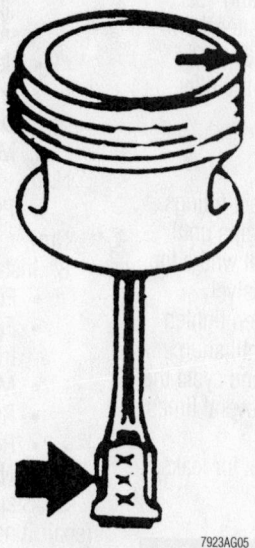

BMW engines connecting rod-to-piston positioning

FUEL SYSTEM

Fuel System Service Precautions

Safety is the most important factor when performing not only fuel system maintenance but also any type of maintenance. Failure to conduct maintenance and repairs in a safe manner may result in serious personal injury or death. Maintenance and testing of the vehicle's fuel system components can be accomplished safely and effectively by adhering to the following rules and guidelines.

1. To avoid the possibility of fire and personal injury, always disconnect the negative battery cable unless the repair or test procedure requires that battery voltage be applied.

2. Always relieve the fuel system pressure prior to disconnecting any fuel system component (injector, fuel rail, pressure reg-ulator, etc.), fitting or fuel line connection. Exercise extreme caution whenever relieving fuel system pressure, to avoid exposing skin, face and eyes to fuel spray. Fuel under pressure may penetrate the skin or any part of the body that it contacts.

3. Always place a shop towel or cloth around the fitting or connection prior to loosening to absorb any excess fuel due to spillage. Ensure that all fuel spillage (should it occur) is quickly removed from engine surfaces. Ensure that all fuel soaked cloths or towels are deposited into a suitable waste container.

4. Always keep a dry chemical (Class B) fire extinguisher near the work area.

5. Do not allow fuel spray or fuel vapors to come into contact with a spark or open flame.

6. Always use a back-up wrench when loosening and tightening fuel line connection fittings. This will prevent unnecessary stress and torsion to fuel line piping. Always follow the proper torque specifications.

7. Always replace worn fuel fitting O-rings with new. Do not substitute fuel hose or equivalent where fuel pipe is installed.

Fuel System Pressure

RELIEVING

To relieve the pressure in the system, locate fuel pump relay located on the cowl. The relay can sometimes be distinguished by the orange color of the housing. Unplug and remove the relay, and place it in a safe location. With the fuel pump relay removed,

Heater Core replacement is covered in Section 2 of this manual

start the engine and operate it until it stalls. Crank the engine for 10 seconds after it stalls to remove any residual pressure.

Fuel Filter

REMOVAL & INSTALLATION

1. Before servicing the vehicle, refer to the precautions in the beginning of this section.
2. Properly relieve the fuel system pressure.
3. Remove or disconnect the following:
   - Negative battery cable
   - Fuel pressure regulator and seal the fuel line before and after the filter with special tool 13–3–010
   - Clips and fuel line from the filter
   - Fuel filter

To install:
   - New fuel filter
   - Fuel lines onto the correct fittings. Tighten the fuel line clamps until tight, but not to the point where the fuel lines become excessively pinched or damaged, then tighten the mounting bracket until snug.
   - Negative battery cable and cycle the ignition **ON** and **OFF** several times to build fuel pressure
4. Start the vehicle and check for leaks, repair if necessary.

Fuel Pump

REMOVAL & INSTALLATION

1. Before servicing the vehicle, refer to the precautions in the beginning of this section.
2. Drain the fuel tank.
3. Properly relieve the fuel system pressure.
4. Remove or disconnect the following:
   - Negative battery cable
   - Rear seat bench
   - Rubber plug above the sender unit and fold the rubber mat back
   - Metal cover

- Fuel gauge level sending unit electrical connector
- Fuel lines
- Rotary connection with special tool 16–1–020

5. Raise the fuel level sensor and expose the spiral hose.
6. Remove the fuel level sensor and fuel pump from the tank.

To install:

➡**Always use a new seal or gasket when installing the fuel pump or fuel level gauge sending unit assembly.**

7. Install or connect the following:
   - Fuel pump into the fuel tank taking care not to bend or damage the fuel sending unit assembly
   - New seal and torque the sealing ring using tool No. 16-1-020 as follows:
     a. Metal sealing rings: 26 ft. lbs. (35 Nm).
     b. Plastic sealing rings: 41 ft. lbs. (55 Nm).
8. Install or connect the following:
   - Fuel lines
   - Fuel gauge level sending unit electrical connector
   - Metal cover
   - Rubber plug above the sender unit
   - Rear seat bench
   - Negative battery cable
9. Start the vehicle and check for leaks, repair if necessary.

Fuel Injector(s)

REMOVAL & INSTALLATION

1. Before servicing the vehicle, refer to the precautions in the beginning of this section.
2. Properly relieve the fuel system pressure.
3. Remove or disconnect the following:
   - Negative battery cable
   - Acoustic cover
   - Knock Sensor (KS) electrical connector from the cable strips

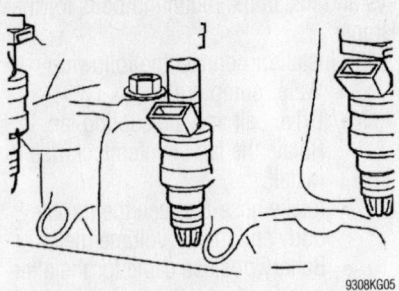

9308KG05

Remove the fuel injector from the injection pipe—4.4L engine

   - Changeover valve electrical connector
   - Camshaft (CMP) sensor electrical connector
   - Left side cylinder head ignition coil cover
   - Ignition coil electrical connectors
   - Ignition coil cover
   - Cable strip from the fuel injectors
   - Vacuum accumulator lines
   - Fuel line
   - Both fuel injection pipe retaining brackets
   - Fuel injectors from the fuel pipe
4. Check the O-rings on the fuel injectors and replace if damaged.
To install:
5. If the O-rings are being replaced, coat the new O-ring with petroleum jelly.
6. Install or connect the following:
   - Fuel injectors to the fuel pipe
   - Fuel line
   - Vacuum lines to the vacuum accumulator
   - Cable strip to the fuel injectors
   - Ignition coil electrical connectors
   - Ignition coil cover
   - CMP sensor electrical connector
   - Changeover valve electrical connector
   - KS electrical connector
   - Acoustic cover
   - Negative battery cable
7. Start the vehicle and check for leaks, repair if necessary.

DRIVE TRAIN

Transmission Assembly

REMOVAL & INSTALLATION

Automatic Transmission

1. Before servicing the vehicle, refer to the precautions in the beginning of this section.
2. Drain the transmission fluid.
3. Remove or disconnect the following:
   - Negative battery cable
   - Exhaust system
   - Front splash guard
   - Reinforcement plate
   - Heat shields and unclip the Oxygen (O_2S) sensor
   - Stabilizer bar and slide it forward
   - Rear heat shield
   - Front propeller shaft and unclip the vent line from the transmission
   - Retaining plate and brace the clamping bush
   - Bracket for the oil lines at the oil pan
   - Power steering pump oil line bracket
   - Oil line and banjo bolt
   - Union screw on the oil return line and properly support the transmission and transfer case assembly
   - O_2S cable from the transmission crossmember
   - Transmission crossmember
   - Nuts for the center bearing after bracing the propeller shaft

➡**Do not allow the propeller shaft to damage the CV-joints.**

   - Transmission output flange by bending the propeller shaft downward at the center bearing
   - Cable connector from the transmission case
   - Impulse sensor electrical connector
   - Torque converter retaining screws
4. Support the engine at the front housing and turn the front wheel to the right lock position.
5. Remove the remaining screws and remove the transmission and transfer case assembly.

To install:
6. Align the transmission/transfer case as an assembly.

7. Torque the transmission to engine screws as follows:
   a. M8 Hex screws: 18 ft. lbs. (24 Nm).
   b. M10 Hex screws: 33 ft. lbs. (45 Nm).
   c. M12 Hex screws: 60 ft. lbs. (82 Nm).
   d. M8 Torx bolts: 16 ft. lbs. (21 Nm).
   e. M10 Torx bolts: 31 ft. lbs. (42 Nm).
   f. M12 Torx bolts: 54 ft. lbs. (72 Nm).
8. Install or connect the following:
   - Torque converter. Torque the bolts to 30 ft. lbs. (40 Nm).
   - Impulse sensor electrical connector
   - Transmission case cable connector
   - Transmission output flange
   - Center bearing. Torque the bolts to 16 ft. lbs. (21 Nm).
   - Transmission crossmember. Torque the bolts to 16 ft. lbs. (21 Nm).
   - O_2S cable to the crossmember
   - Union screw to the oil return line. Torque the screw to 21 ft. lbs. (28 Nm).
   - Oil line and banjo bolt. Torque the bolt to 21 ft. lbs. (28 Nm).
   - Oil line bracket for the power steering pump
   - Retaining plate
   - Front propeller shaft. Torque the bolts to 47 ft. lbs. (64 Nm) and clip the vent line to the transmission.
   - Rear heat shield
   - Stabilizer bar. Torque the bolts to 16 ft. lbs. (22 Nm).
   - Front heat shields
   - O_2S connector
   - Reinforcement plate
   - Front splash guard
   - Exhaust system
   - Negative battery cable
9. Fill the transmission with the proper fluid to the proper level.
10. Start the vehicle and check for leaks, repair if necessary.

Transfer Case

REMOVAL & INSTALLATION

1. Before servicing the vehicle, refer to the precautions in the beginning of this section.

2. Drain the transmission fluid.
3. Remove or disconnect the following:
   - Negative battery cable
   - Exhaust system
   - Front splash guard
   - Reinforcement plate
   - Heat shields and unclip the Oxygen (O_2S) sensor
   - Stabilizer bar and slide it forward
   - Rear heat shield
   - Front propeller shaft and unclip the vent line from the transmission. Properly support the transmission.
   - O_2S cable from the transmission crossmember
   - Transmission crossmember
   - O_2S cable from the transfer case
   - Center bearing after bracing the properller shaft. Do not allow the propeller shaft to damage the CV-joints.
   - Transmission output flange by bending the propeller shaft downward at the center bearing
   - Transfer case from the transmission

To install:
4. Connect the transfer case to the transmission. Torque the bolts as follows:
   a. M8 Hex screws: 18 ft. lbs. (24 Nm).
   b. M10 Hex screws: 33 ft. lbs. (45 Nm).
   c. M12 Hex screws: 60 ft. lbs. (82 Nm).
   d. M8 Torx bolts: 16 ft. lbs. (21 Nm).
   e. M10 Torx bolts: 31 ft. lbs. (42 Nm).
   f. M12 Torx bolts: 54 ft. lbs. (72 Nm).
5. Install or connect the following:
   - Transmission output flange
   - Center bearing. Torque the bolts to 16 ft. lbs. (21 Nm).
   - Transmission crossmember. Torque the bolts to 16 ft. lbs. (21 Nm).
   - O_2S cable to the crossmember
   - Union screw to the oil return line. Torque the screw to 21 ft. lbs. (28 Nm).
   - Front propeller shaft. Torque the bolts to 47 ft. lbs. (64 Nm) and clip the vent line to the transmission.
   - Rear heat shield
   - Stabilizer bar. Torque the bolts to 16 ft. lbs. (22 Nm).
   - Front heat shields

Brake service is covered in Section 4 of this manual

- O₂S connector
- Reinforcement plate
- Front splash guard
- Exhaust system
- Negative battery cable

6. Fill the transmission with the proper fluid to the proper level.

7. Start the vehicle and check for leaks, repair if necessary.

Halfshafts

REMOVAL & INSTALLATION

Front

1. Before servicing the vehicle, refer to the precautions in the beginning of this section.

2. Remove or disconnect the following:
- Negative battery cable
- Front wheel
- Reinforcement plate
- Front splash guard
- Swivel bearing
- Output shaft from the differential by pressing it out with special tool 31–5–110

To install:

3. Install or connect the following:
- New output shaft radial seal
- New snap ring on the output shaft

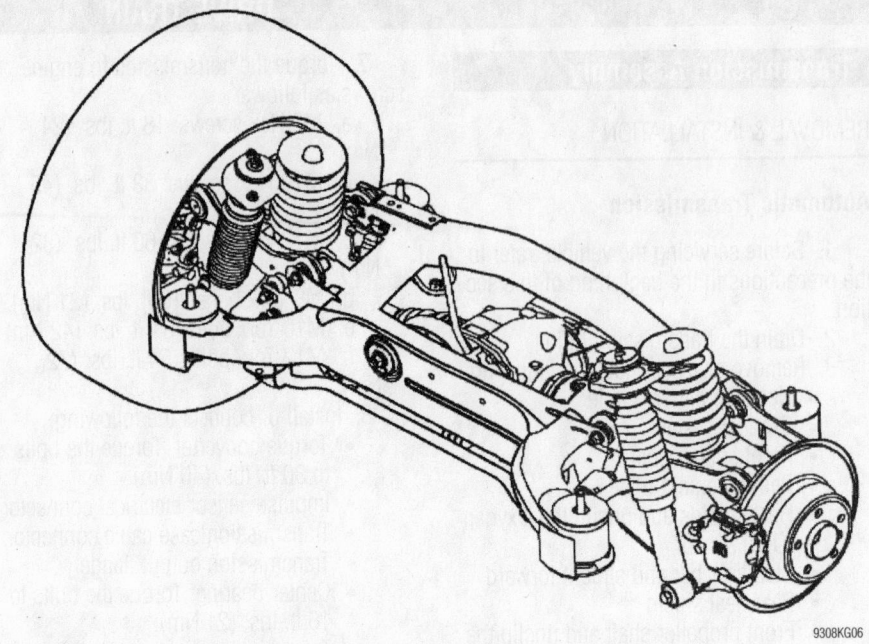

Exploded view of the rear halfshaft and suspension system—X5 Series

- Press the output shaft in by pushing it in over the resistance of the snap ring
- Swivel bearing
- Front splash guard
- Reinforcement plate
- Front wheel
- Negatice battery cable

Rear

1. Before servicing the vehicle, refer to the precautions in the beginning of this section.

2. Remove or disconnect the following:
- Negative battery cable
- Rear tire and wheel assembly
- Collar nut
- Brake disc
- ABS sensor
- Retaining nut from the output flange. Remove the drive flange hub

➡**The wheel bearing will be destroyed when the flange is removed. The wheel bearing must be replaced.**

- Halfshaft from the vehicle by removing the shaft from the final drive output flange and by pressing the halfshaft out of the drive flange hub using tool Nos. 33-2-116, 201, 202 and 203

3. Press out the drive flange hub.

4. Pull out the seal with a suitable tool.

5. If the bearing inner race is damaged, pull it off of the drive flange hub with tool No. 33–3–240.

To install:

6. Using an appropriate bearing installer, install the wheel bearing assembly, install the seal, then insert the snapring, and install the drive flange hub.

7. Install or connect the following:

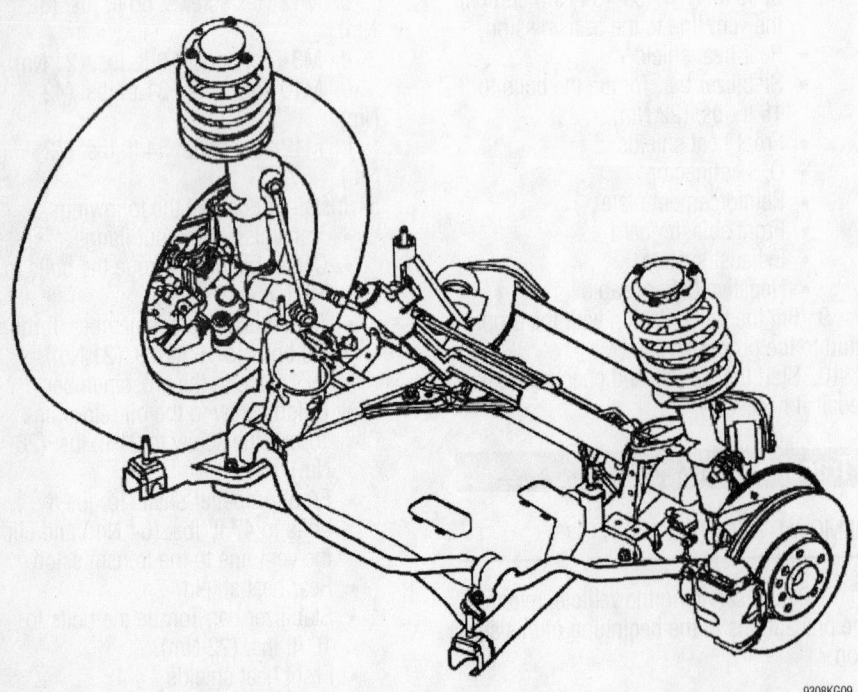

Exploded view of the front axle assembly—X5 Series

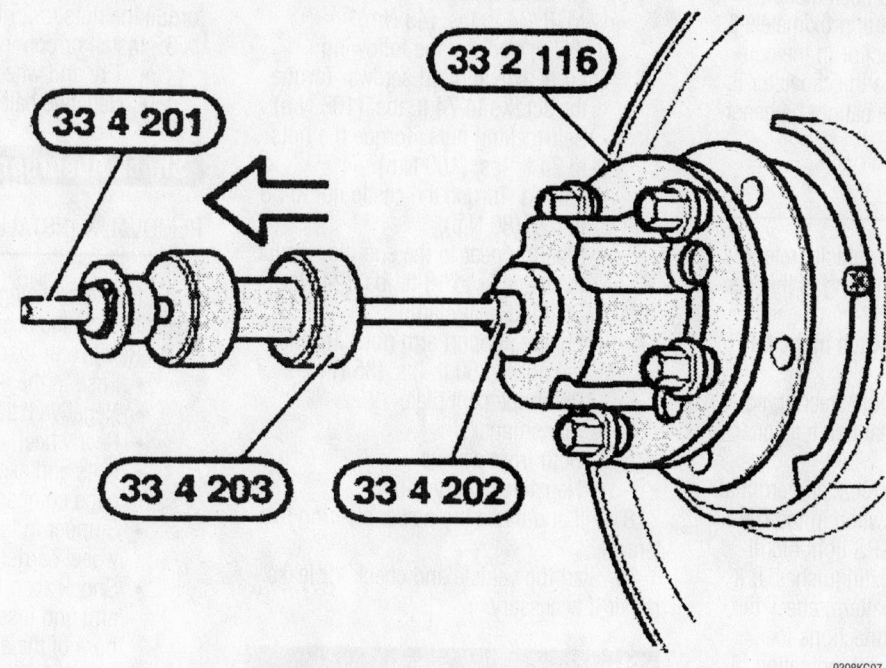

Halfshaft removal tools Nos. 33-2-116, 33-4-201, 201 and 203

9308KG07

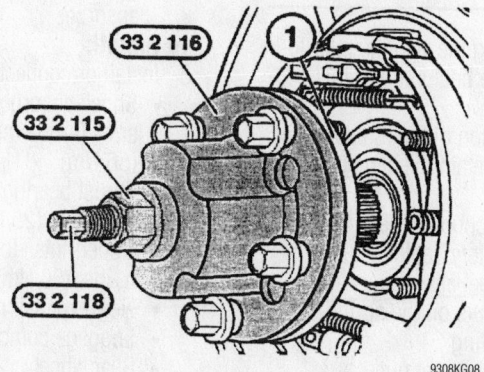

Drive flange hub tool Nos. 33-2-115, 116 and 118 used for drive flange installation

9308KG08

- Axle shaft seal
- Output shaft, using tool Nos. 33-2-115, 116 and 118
- Output shaft to the final drive. Torque the bolts to 61 ft. lbs. (83 Nm).
- Outer nut with bearing surface lightly oiled. Torque the nut to 310 ft. lbs. (420 Nm).
- Brake disc
- ABS sensor
- Rear tire and wheel assembly
- Negative battery cable

STEERING AND SUSPENSION

Air Bag

✳✳ CAUTION

The vehicles are equipped with an air bag system. The system must be disarmed before performing service on, or around, system components, the steering column, instrument panel components, wiring and sensors. Failure to follow the safety precautions and the disarming procedure could result in accidental air bag deployment, possible personal injury and unnecessary system repairs.

PRECAUTIONS

Several precautions must be observed when handling the inflator module to avoid accidental deployment and possible personal injury.

1. Never carry the inflator module by the wires or connector on the underside of the module.

2. When carrying a live inflator module, hold securely with both hands, and ensure that the bag and trim cover are pointed away.

3. Place the inflator module on a bench or other surface with the bag and trim cover facing up.

4. With the inflator module on the bench, never place anything on or close to the module that may be thrown in the event of an accidental deployment.

DISARMING

1. Before servicing the vehicle, refer to the precautions in the beginning of this section.

2. Place the ignition switch in the **OFF** position.

3. Disconnect the negative battery terminal and cover the battery terminal to prevent accidental contact.

For complete Engine Mechanical specifications, see Section 1 of this manual

4. Once the battery has been disconnected, wait for a period of approximately 5 seconds allowing the capacitor in the control unit to discharge. Once the capacitor is discharged, a trigger pulse cannot be generated inadvertently.

REARMING

1. Before servicing the vehicle, refer to the precautions in the beginning of this section.

2. Place the ignition switch in the **OFF** position.

3. Attach the sensors, the steering column connector and the seat belt tensioner connectors.

4. Connect the negative battery terminal.

5. Place the ignition switch in the **ON** position. Check that the SRS light illuminates for 6 seconds and extinguishes. If it illuminates in any other pattern, check the components and their connections for proper operation and recheck operation of the warning light.

Power Rack and Pinion Steering

REMOVAL & INSTALLATION

1. Before servicing the vehicle, refer to the precautions in the beginning of this section.

2. Set the steering gear in the straight ahead position by aligning the marks on the steering gear and spindle.

3. Drain the power steering fluid.

4. Remove or disconnect the following:
   - Negative battery cable
   - Front wheels
   - Reinforcement plate
   - Nuts from the left and right engine support arms and raise the engine slightly
   - Lower clamping screw
   - Steering gear clamps
   - Tie rod by pressing it off with special tool 32–3–090
   - Self-locking nuts and brace the front axle support
   - Banjo bolts and slide the steering gear out through the left wheel opening

To install:

5. Install the steering gear through the left side wheel opening

6. Install new sealing rings and banjo bolts. Torque the bolts as follows:
   a. M10: 7 ft. lbs. (12 Nm).
   b. M14: 25 ft. lbs. (35 Nm).
   c. M16: 29 ft. lbs. (40 Nm).
   d. M18: 34 ft. lbs. (45 Nm).

7. Install or connect the following:
   - Front axle support screws. Torque the screws to 74 ft. lbs. (100 Nm).
   - Self-locking nuts. Torque the nuts to 74 ft. lbs. (100 Nm).
   - Tie rod. Torque the castle nut to 58 ft. lbs. (80 Nm).
   - Steering gear to the spindle. Torque the fastener to 18 ft. lbs. (24 Nm).
   - Steering gear clamp
   - Engine support arm nuts. Torque the nuts to 60 ft. lbs. (85 Nm).
   - Reinforcement plate
   - Splash guard
   - Both front wheels
   - Negative battery cable

8. Fill and bleed the power steering system.

9. Start the vehicle and check for leaks, repair if necessary.

Strut

REMOVAL & INSTALLATION

1. Before servicing the vehicle, refer to the precautions in the beginning of this section.

2. Mark the position of the threaded pin to the wheel arch to retain the camber setting when installed.

3. Remove or disconnect the following:
   - Negative battery cable
   - Tire and wheel assembly
   - Two of the nuts on the spring strut support bearing
   - Center strut bracket nut
   - Speed sensor/brake wear cable and disconnect the plug housing
   - Swivel bearing and tie it aside
   - Remaining nut on the spring strut support bearing
   - Strut assembly

To install:

4. Install or connect the following:
   - Strut assembly
   - One nut to the spring strut support bearing and hand tighten at this time
   - Swivel bearing. Torque the new self-locking nut to 176 ft. lbs. (250 Nm).
   - Speed sensor/brake wear cable and connect the housing plug
   - Center strut bracket. Torque the nut to 74 ft. lbs. (100 Nm).

5. Align the three upper spring strut support bearing nuts and match the threaded pin with the mark made during the removal procedure. When aligned properly torque the nuts to 25 ft. lbs. (34 Nm).

6. Install or connect the following:
   - Tire and wheel assembly
   - Negative battery cable

Shock Absorber

REMOVAL & INSTALLATION

1. Before servicing the vehicle, refer to the precautions in the beginning of this section.

2. Remove or disconnect the following:
   - Fuse in the air supply system
   - Negative battery cable
   - Rear wheel
   - Nuts and expansion rivets and luggage compartment trim
   - 3 upper nuts after supporting the wheel carrier
   - Shock absorber from the swinging arm and insert a bushing into the bore of the arm
   - Thrust bearing after bracing the piston rod with a ring spanner
   - Upper nut and remove the shock absorber

To install:

3. Install or connect the following:
   - Shock absorber to the swinging arm. Torque the bolt to 41 ft. lbs. (56 Nm).
   - Thrust bearing. Torque the nut to 18 ft. lbs. (25 Nm).
   - Upper nuts. Torque the nuts to 41 ft. lbs. (56 Nm).
   - New expansion rivets and nuts
   - Luggage compartment trim
   - Rear wheel
   - Negative battery cable
   - Fuse for the air supply system

Coil Spring

REMOVAL & INSTALLATION

Front

✳✳ CAUTION

This procedure calls for the spring to be compressed. A compressed spring has high potential energy and if released suddenly can cause severe damage and personal injury.

1. Before servicing the vehicle, refer to the precautions in the beginning of this section.

2. disconnect the negative battery cable.

3. Remove the strut from the vehicle and mount in a vise using a strut holder. This will prevent damage to the strut tube

4. Using a proper spring compressor, compress the spring until the stress on the thrust bearing is released.

5. Remove the top nut of the strut mount. Counterhold the strut rod during removal.

6. Pull the strut mount off the strut rod. Note the positioning of the spacers and washer for replacement.

7. Pull the spring off the strut and place aside in a safe area.

8. slowly release the compression of the spring.

To install:

9. Install or connect the following:
- Spring in the compressor and compress
- Spring and strut mount with all the spacers and washers in their original positions. Torque the new strut rod nut: 47 ft. lbs. (65 Nm).

10. Release the spring slowly and check that it seats in the spring holders. Install the strut in the vehicle.

11. Connect the negative battery cable.

Lower Ball Joint

REMOVAL & INSTALLATION

1. Before servicing the vehicle, refer to the precautions in the beginning of this section.

2. Remove or disconnect the following:
- Negative battery cable
- Push rod/integral link assembly and properly support the wheel carrier
- Shock absorber from the swinging arm
- Circlip

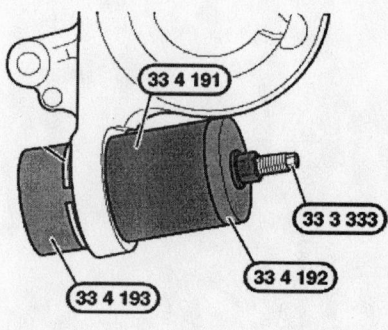

Remove the lower ball joint from the steering knuckle—X5 Series

3. Using special tool 33–4–191, 192, 193 and 33–3–333 pull the ball joint out of the steering knuckle.

To install:

4. Install or connect the following:
- Ball joint into the steering knuckle with special tools 33–4–191, 192, 194 and 33–3–333
- New circlip
- Shock absorber and remove the support from the wheel carrier
- Push rod/integral link
- Negative battery cable

Lower Control Arm

REMOVAL & INSTALLATION

1. Before servicing the vehicle, refer to the precautions in the beginning of this section.

2. Remove or disconnect the following:
- Negative battery cable
- Front wheel
- Control arm from the front axle support and loosen the nut from the control arm to swivel bearing
- Control arm from the swivel bearing by pressing it off with special tool 31–2–240

To install:

3. Install or connect the following:
- Lower control arm to the swivel bearing. Torque the nut to 58 ft. lbs. (80 Nm).
- Lower control arm to the front axle support. Torque the nut to 74 ft. lbs. (100 Nm) plus an additional 90 degrees.
- Front wheel
- Negative battery cable

4. Check and adjust the front end alignment as needed.

BUSHING REPLACEMENT

1. Before servicing the vehicle, refer to the precautions in the beginning of this section.

2. Remove or disconnect the following:
- Negative battery cable
- Control arm and tie it back to prevent damage to the ball joint
- Control arm bushing by installing special tools 31–1–051, 052, 33–3–051, 052, 054 and 310

To install:

3. Install or connect the following:
- Control arm bushing by installing

special tools 31–1–051, 052, 33–3–051, 052, 054 and 310
- Control arm
- Negative battery cable

Upper Control Arm

REMOVAL & INSTALLATION

1. Before servicing the vehicle, refer to the precautions in the beginning of this section.

2. Remove or disconnect the following:
- Negative battery cable
- Wheel assembly
- Fuse for the air supply system, if equipped with air suspension and loosen the pipes on the distributor block
- Upper control arm from the steering knuckle
- Plastic shim and unhook the lines
- Upper control arm

To install:

3. Install or connect the following:
- Upper control arm. Torque the bolt to 74 ft. lbs. (100 Nm).
- Plastic shim and connect the lines
- Upper control arm to the steering knuckle. Torque the bolt to 122 ft. lbs. (165 Nm).
- Fuse for the air supply system and tighten the pipes on the distributor block
- Wheel assembly
- Negative battery cable

Wheel Bearings

ADJUSTMENT

Wheel bearings can not be adjusted and must be replaced as a unit and never be reused once removed.

REMOVAL & INSTALLATION

Front

➡**The wheel bearings are only removed if they are worn. They cannot be removed without destroying them (due to side thrust created by the bearing puller). They cannot be disassembled, repacked or adjusted.**

1. Before servicing the vehicle, refer to the precautions in the beginning of this section.

For Accessory Drive Belt illustrations, see Section 1 of this manual

2. Remove or disconnect the following:
   - Negative battery cable
   - Swivel bearing and clamp it in a vise
   - Drive flange by installing special tools 33–2–116, 150 and 33–4–200
   - Bearing inner race from the flange
   - Circlip
   - Snap ring
   - Bearing by installing special tools 31–2–113, 33–3–261, 262 and 266

To install:

3. Install or connect the following:
   - Wheel bearing with the wider camfer facing the swivel bearing to the drive flange with special tools

33–2–261, 264, 268 and 31–2–113
   - Snap ring and circlip
   - inner race to the drive flange
   - Drive flange to the swivel bearing by using special tool 33–3–261, 266, 268 and 31–2–113
   - Swivel bearing
   - Negative battery cable

Rear

1. Before servicing the vehicle, refer to the precautions in the beginning of this section.

2. Remove or disconnect the following:
   - Negative battery cable
   - Wheel assembly

   - Collar nut
   - Brake disc
   - Drive flange by installing special tool 33–2–116, 33–4–201, 202 and 203
   - Inner race from the drive flange
   - Wheel bearing

To install:

3. Install or connect the following:
   - Wheel bearing
   - Inner race to the drive flange. Torque the bolts to 74 ft. lbs. (100 Nm).
   - Drive flange to the axle shaft. Torque the collar nut to 310 ft. lbs. (420 Nm).
   - Brake disc
   - Wheel assembly
   - Negative battery cable

CHRYSLER CORP.

Dodge-Durango

8

PRECAUTIONS

Before servicing any vehicle, please be sure to read all of the following precautions, which deal with personal safety, prevention of component damage, and important points to take into consideration when servicing a motor vehicle:

• Never open, service or drain the radiator or cooling system when the engine is hot; serious burns can occur from the steam and hot coolant.

• Observe all applicable safety precautions when working around fuel. Whenever servicing the fuel system, always work in a well-ventilated area. Do not allow fuel spray or vapors to come in contact with a spark, open flame, or excessive heat (a hot drop light, for example). Keep a dry chemical fire extinguisher near the work area. Always keep fuel in a container specifically designed for fuel storage; also, always properly seal fuel containers to avoid the possibility of fire or explosion. Refer to the additional fuel system precautions later in this section.

• Fuel injection systems often remain pressurized, even after the engine has been turned **OFF**. The fuel system pressure must be relieved before disconnecting any fuel lines. Failure to do so may result in fire and/or personal injury.

• Brake fluid often contains polyglycol ethers and polyglycols. Avoid contact with the eyes and wash your hands thoroughly after handling brake fluid. If you do get brake fluid in your eyes, flush your eyes with clean, running water for 15 minutes. If eye irritation persists, or if you have taken

brake fluid internally, IMMEDIATELY seek medical assistance.

• The EPA warns that prolonged contact with used engine oil may cause a number of skin disorders, including cancer! You should make every effort to minimize your exposure to used engine oil. Protective gloves should be worn when changing oil. Wash your hands and any other exposed skin areas as soon as possible after exposure to used engine oil. Soap and water, or waterless hand cleaner should be used.

• All new vehicles are now equipped with an air bag system, often referred to as a Supplemental Restraint System (SRS) or Supplemental Inflatable Restraint (SIR) system. The system must be disabled before performing service on or around system components, steering column, instrument panel components, wiring and sensors. Failure to follow safety and disabling procedures could result in accidental air bag deployment, possible personal injury and unnecessary system repairs.

• Always wear safety goggles when working with, or around, the air bag system. When carrying a non-deployed air bag, be sure the bag and trim cover are pointed away from your body. When placing a non-deployed air bag on a work surface, always face the bag and trim cover upward, away from the surface. This will reduce the motion of the module if it is accidentally deployed. Refer to the additional air bag system precautions later in this section.

• Clean, high quality brake fluid from a sealed container is essential to the safe and

proper operation of the brake system. You should always buy the correct type of brake fluid for your vehicle. If the brake fluid becomes contaminated, completely flush the system with new fluid. Never reuse any brake fluid. Any brake fluid that is removed from the system should be discarded. Also, do not allow any brake fluid to come in contact with a painted surface; it will damage the paint.

• Never operate the engine without the proper amount and type of engine oil; doing so WILL result in severe engine damage.

• Timing belt maintenance is extremely important! Many models utilize an interference-type, non-freewheeling engine. If the timing belt breaks, the valves in the cylinder head may strike the pistons, causing potentially serious (also time-consuming and expensive) engine damage. Refer to the maintenance interval charts in the front of this manual for the recommended replacement interval for the timing belt, and to the timing belt section for belt replacement and inspection.

• Disconnecting the negative battery cable on some vehicles may interfere with the functions of the on-board computer system(s) and may require the computer to undergo a relearning process once the negative battery cable is reconnected.

• When servicing drum brakes, only disassemble and assemble one side at a time, leaving the remaining side intact for reference.

• Only an MVAC-trained, EPA-certified automotive technician should service the air conditioning system or its components.

GASOLINE ENGINE REPAIR

➡**Disconnecting the negative battery cable on some vehicles may interfere with the functions of the on board computer system. The computer may undergo a relearning process once the negative battery cable is reconnected.**

Distributor

REMOVAL

3.9L, 5.2L and 5.9L Engines

1. Before servicing the vehicle, refer to the precautions in the beginning of this section.
2. Remove or disconnect the following:

• Negative battery cable
• Air cleaner tube
• Distributor cap
• Camshaft Position (CMP) sensor connector

3. Matchmark the distributor housing and the rotor.
4. Matchmark the distributor housing and the intake manifold.
5. Remove the distributor.

INSTALLATION

Timing Not Disturbed

3.9L, 5.2L AND 5.9L ENGINES

1. Before servicing the vehicle, refer to the precautions in the beginning of this section.

➡**The rotor will rotate clockwise as the gears engage.**

2. Position the rotor slightly counterclockwise of the matchmark made during removal.
3. Install the distributor.
4. Align the distributor housing and intake manifold matchmarks and check that the distributor housing matchmark and rotor are also aligned.
5. Install or connect the following:

• Distributor housing clamp and bolt. Tighten the bolt to 17 ft. lbs. (23 Nm).
• CMP sensor connector
• Distributor cap
• Air cleaner tube
• Negative battery cable

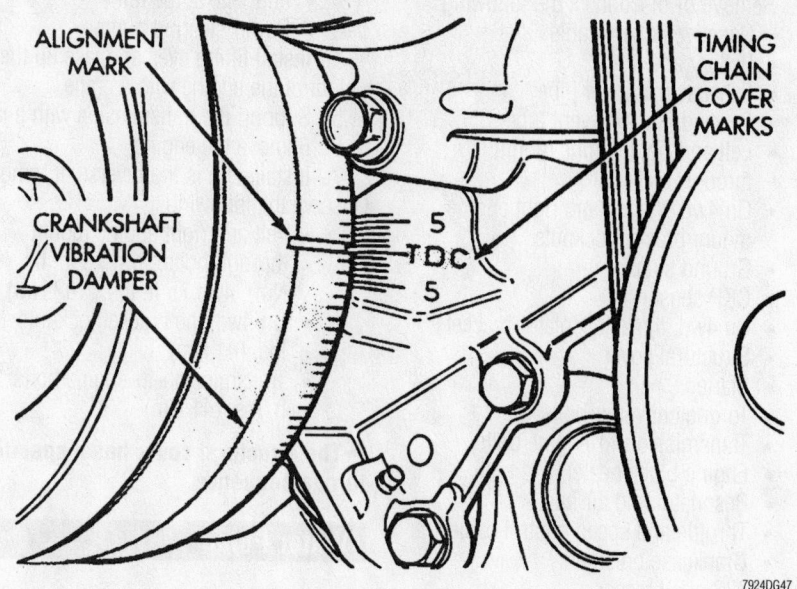

Crankshaft pulley and timing chain cover marks aligned at Top Dead Center (TDC)

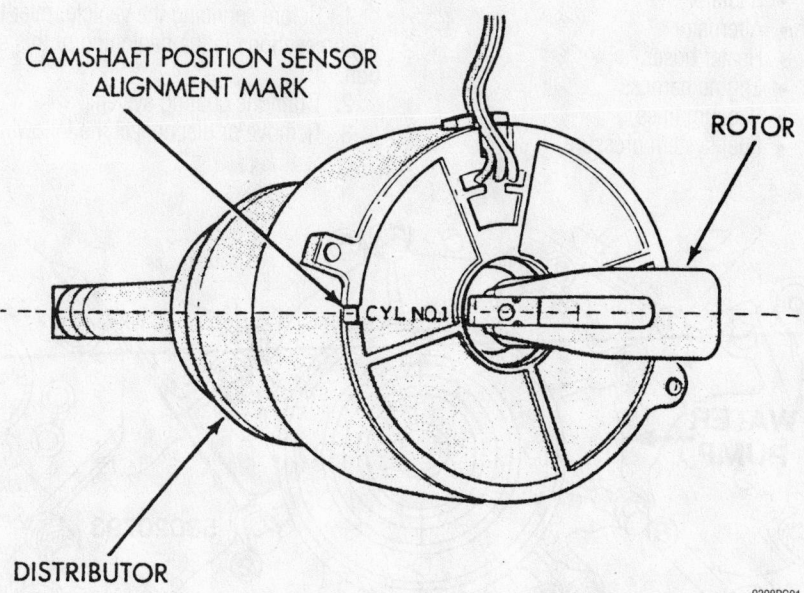

Distributor rotor alignment—3.9L, 5.2L and 5.9L engines

Timing Disturbed

3.9L, 5.2L AND 5.9L ENGINES

1. Before servicing the vehicle, refer to the precautions in the beginning of this section.

2. Set the engine at Top Dead Center (TDC) of the No. 1 cylinder compression stroke.

3. Install the distributor, clamp and bolt.

4. Rotate the distributor so that the rotor aligns with the No. 1 cylinder mark on the Camshaft Position (CMP) sensor. Tighten the clamp bolt to 17 ft. lbs. (23 Nm).

5. Install or connect the following:
- CMP sensor connector
- Distributor cap
- Air cleaner tube
- Negative battery cable

6. The final distributor position must be set with a scan tool. Follow the instructions supplied by the scan tool manufacturer.

Alternator

REMOVAL & INSTALLATION

1. Before servicing the vehicle, refer to the precautions in the beginning of this section.

2. Remove or disconnect the following:
- Negative battery cable
- Accessory drive belt
- Alternator harness connectors
- Alternator

3. Installation is the reverse of removal. Tighten the bolts to the following specifications:

   a. 4.7L engine: Vertical bolt 41 ft. lbs. (56 Nm), long horizontal bolt to 55 ft. lbs. (74 Nm) and short horizontal bolt to 55 ft. lbs. (74 Nm).

   b. 3.9L, 5.2L, 5.9L engines: 30 ft. lbs. (41 Nm).

Ignition Timing

ADJUSTMENT

The ignition timing is controlled by the Powertrain Control Module (PCM) and is not adjustable.

Engine Assembly

REMOVAL & INSTALLATION

3.9L, 5.2L & 5.9L Engines

1. Before servicing the vehicle, refer to the precautions in the beginning of this section.

2. Drain the cooling system.

3. Drain the engine oil.

4. Relieve the fuel system pressure.

5. Remove or disconnect the following:
- Negative battery cable
- Hood
- Upper crossmember and top core support
- Radiator hoses
- Cooling fan and shroud
- Radiator
- Accelerator cable
- Cruise control cable, if equipped
- Transmission cable, if equipped
- Heater hoses
- Intake manifold vacuum lines
- Accessory drive belt
- Power steering pump
- A/C compressor, if equipped

- Engine control sensor harness connectors
- Engine block heater, if equipped
- Fuel line
- Exhaust front pipe
- Starter motor
- Torque converter, if equipped
- Transmission oil cooler lines, if equipped
- Engine mounts
- Transmission flange bolts. Support the transmission.
- Engine

※※ WARNING

Do not lift the engine by the intake manifold. Damage to the manifold may result.

To install:

6. Install or connect the following:
   - Engine. Tighten the engine mount bolts to 70 ft. lbs. (95 Nm).
   - Transmission flange bolts. Tighten the bolts to 40–45 ft. lbs. (54–61 Nm).
   - Transmission oil cooler lines, if equipped
   - Torque converter, if equipped. Tighten the bolts to 23 ft. lbs. (31 Nm).
   - Starter motor
   - Exhaust front pipe
   - Fuel line
   - Engine block heater, if equipped
   - Engine control sensor harness connectors
   - A/C compressor, if equipped
   - Power steering pump
   - Accessory drive belt
   - Intake manifold vacuum lines
   - Heater hoses
   - Transmission cable, if equipped
   - Cruise control cable, if equipped
   - Accelerator cable
   - Radiator
   - Cooling fan and shroud
   - Radiator hoses
   - Upper crossmember and top core support
   - Hood
   - Negative battery cable

7. Fill the crankcase to the correct level.
8. Fill the cooling system.
9. Start the engine and check for leaks.

4.7L Engine

1. Before servicing the vehicle, refer to the precautions in the beginning of this section.
2. Drain the cooling system and engine oil.

3. Remove or disconnect the following:
   - Negative battery cable
   - Battery and tray
   - Exhaust crossover pipe
   - On 4wd, the axle vent tube
   - Left and right engine mount through bolts
   - On 4wd, the left and right engine mount bracket locknuts
   - Ground straps
   - CKP sensor
   - On 4wd, the axle isolator bracket
   - Structural cover
   - Starter
   - Torque converter bolts
   - Transmission-to-engine bolts
   - Engine block heater
   - Resonator and air inlet
   - Throttle and speed control cables
   - Crankcase breathers
   - A/C compressor
   - Shroud and fan assemblies
   - Transmission cooler lines
   - Radiator hoses
   - Radiator
   - Alternator
   - Heater hoses
   - Engine harness
   - Vacuum lines
   - Fuel system pressure

   - Fuel line at the rail
   - Power steering pump

4. Install lifting eyes and take up the weight of the engine with a crane.
5. Support the transmission with a jack.
6. Remove the engine.
7. Installation is the reverse of removal. Observe the following:
   - Left and right engine mount through bolts: 2wd 70 ft. lbs. (95 Nm); 4wd 75 ft. lbs. (102 Nm)
   - On 4wd, the bracket locknuts: 30 ft. lbs. (41 Nm)
   - Transmission-to-engine bolts: 30 ft. lbs. (41 Nm)

➡ The structural cover has a specific torque sequence.

Water Pump

REMOVAL & INSTALLATION

4.7L Engine

1. Before servicing the vehicle, refer to the precautions in the beginning of this section.
2. Drain the cooling system.
3. Remove or disconnect the following:

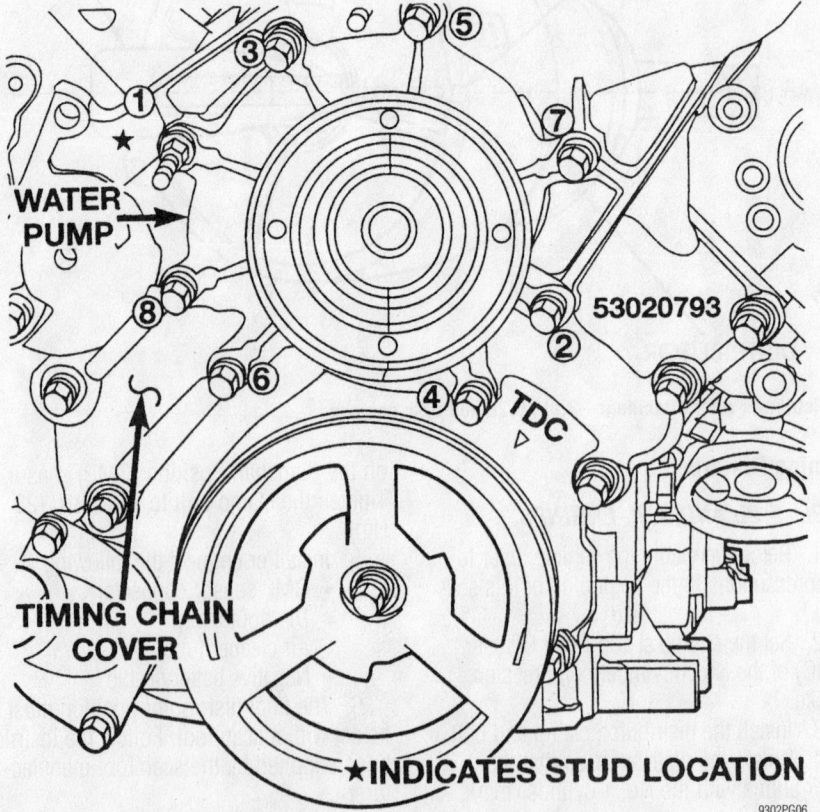

Water pump torque sequence—4.7L engine

- Negative battery cable
- Fan and fan drive assembly from the pump. Don't attempt to remove it from the vehicle, yet.

➡**If a new pump is being installed, don't separate the fan from the drive.**

- Shroud and fan

✳✳ WARNING

Keep the fan upright to avoid fluid loss from the drive.

- Accessory drive belt
- Lower radiator hose
- Water pump
4. Installation is the reverse of removal. tighten the bolts in sequence to 40 ft. lbs. (54 Nm).

3.9L, 5.2L and 5.9L Engines

1. Before servicing the vehicle, refer to the precautions in the beginning of this section.
2. Drain the cooling system.
3. Remove or disconnect the following:
   - Negative battery cable
   - Engine cooling fan and shroud

➡**Do not store the fan clutch assembly horizontally, silicone may leak into the bearing grease and cause contamination.**

- Accessory drive belt
- Water pump pulley

WATER PUMP MOUNTING BOLTS

7924DG03

Water pump mounting bolt locations—3.9L, 5.2L and 5.9L engines

- Lower radiator hose
- Heater hose and tube
- Bypass hose
- Water pump

To install:
4. Install or connect the following:
   - Water pump, using a new gasket. Tighten the bolts to 30 ft. lbs. (40 Nm).
   - Bypass hose
   - Heater hose and tube. Use a new O-ring seal.
   - Lower radiator hose
   - Water pump pulley. Tighten the bolts to 20 ft. lbs. (27 Nm).
   - Accessory drive belt
   - Engine cooling fan and shroud
   - Negative battery cable
5. Fill the cooling system.
6. Start the engine and check for leaks.

Cylinder Head

REMOVAL & INSTALLATION

3.9L Engine

1. Before servicing the vehicle, refer to the precautions in the beginning of this section.
2. Relieve the fuel pressure.
3. Drain the cooling system.
4. Remove or disconnect the following:
   - Negative battery cable

- Accessory drive belt
- Alternator
- A/C compressor, if equipped
- Alternator and A/C compressor bracket
- Air injection pump, if equipped
- Closed Crankcase Ventilation (CCV) system
- Air cleaner and hose
- Fuel line
- Accelerator linkage
- Cruise control cable, if equipped
- Transmission cable, if equipped
- Spark plug wires
- Distributor
- Ignition coil harness connectors
- Engine Coolant Temperature (ECT) sensor connector
- Heater hoses
- Bypass hose
- Intake manifold vacuum lines
- Fuel injector harness connectors
- Valve covers
- Intake manifold
- Exhaust front pipe
- Exhaust manifolds

➡**Keep all valvetrain components in order for assembly.**

- Rocker arms
- Pushrods
- Cylinder heads

To install:

✳✳ WARNING

Position the crankshaft so that no piston is at Top Dead Center (TDC) prior to installing the cylinder heads. Do not rotate the crankshaft during or immediately after rocker arm installation. Wait 5 minutes for the hydraulic lash adjusters to bleed down.

5. Install the cylinder heads with new gaskets. Tighten the bolts in sequence as follows:
   a. Step 1: 50 ft. lbs. (68 Nm).
   b. Step 2: 105 ft. lbs. (143 Nm).
   c. Step 3: 105 ft. lbs. (143 Nm).
6. Install or connect the following:
   - Pushrods in their original locations
   - Rocker arms in their original locations. Tighten the bolts to 21 ft. lbs. (28 Nm).
   - Exhaust manifolds
   - Exhaust front pipe
   - Intake manifold
   - Valve covers

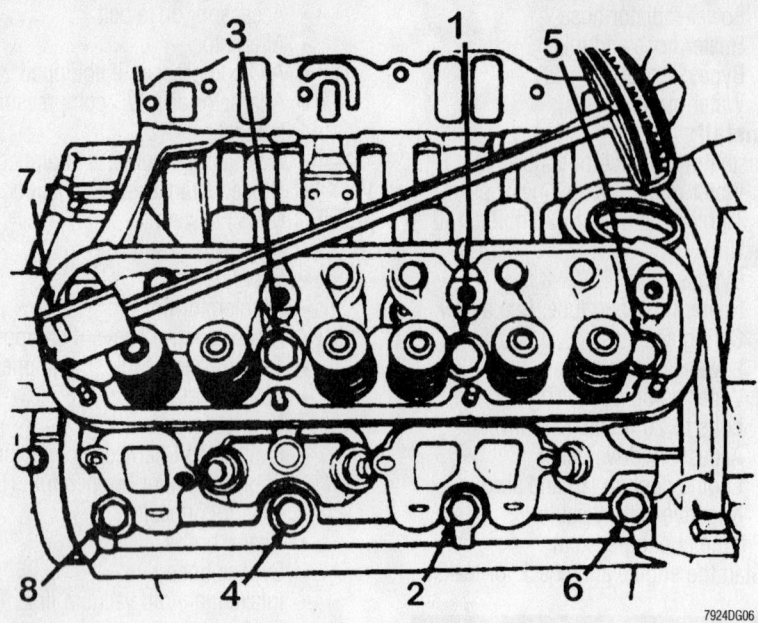

Cylinder head torque sequence—3.9L engine

- Fuel injector harness connectors
- Intake manifold vacuum lines
- Bypass hose
- Heater hoses
- ECT sensor connector
- Ignition coil harness connectors
- Distributor
- Spark plug wires
- Transmission cable, if equipped
- Cruise control cable, if equipped
- Accelerator linkage
- Fuel line
- Air cleaner and hose
- CCV system
- Air injection pump, if equipped
- Alternator and A/C compressor bracket
- Alternator
- A/C compressor
- Accessory drive belt
- Negative battery cable
7. Fill the cooling system.
8. Start the engine and check for leaks.

4.7L Engine

1999

1. Before servicing the vehicle, refer to the precautions in the beginning of this section.
2. Drain the cooling system.
3. Remove or disconnect the following:
- Negative battery cable
- Exhaust Y-pipe
- Intake manifold
- Valve covers
- Engine cooling fan and shroud
- Accessory drive belt

- Oil fill housing
- Power steering pump
- Rocker arms
4. Rotate the crankshaft so that the crankshaft timing mark aligns with the Top Dead Center (TDC) mark on the front cover,

and the **V8** marks on the camshaft sprockets are at 12 o'clock as shown.
- Crankshaft damper
- Front cover
5. Lock the secondary timing chains to the idler sprocket with Timing Chain Locking tool 8515.
6. Matchmark the secondary timing chains to the camshaft sprockets.
7. Remove or disconnect the following:
- Secondary timing chain tensioners
- Cylinder head access plugs
- Secondary timing chain guides
- Camshaft sprockets
- Cylinder heads

➡ **Each cylinder head is retained by ten 11mm bolts and four 8mm bolts.**

To install:

8. Check the cylinder head bolts for signs of stretching and replace as necessary.
9. Lubricate the threads of the 11mm bolts with clean engine oil.
10. Coat the threads of the 8mm bolts with Mopar® Lock and Seal Adhesive.
11. Install the cylinder heads. Use new gaskets and tighten the bolts, in sequence, as follows:

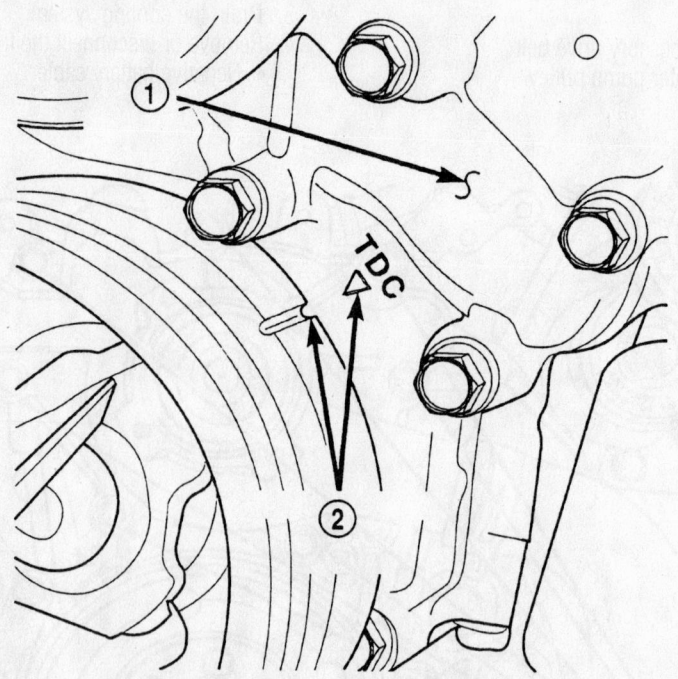

1 – TIMING CHAIN COVER
2 – CRANKSHAFT TIMING MARKS

Crankshaft timing marks—4.7L engine

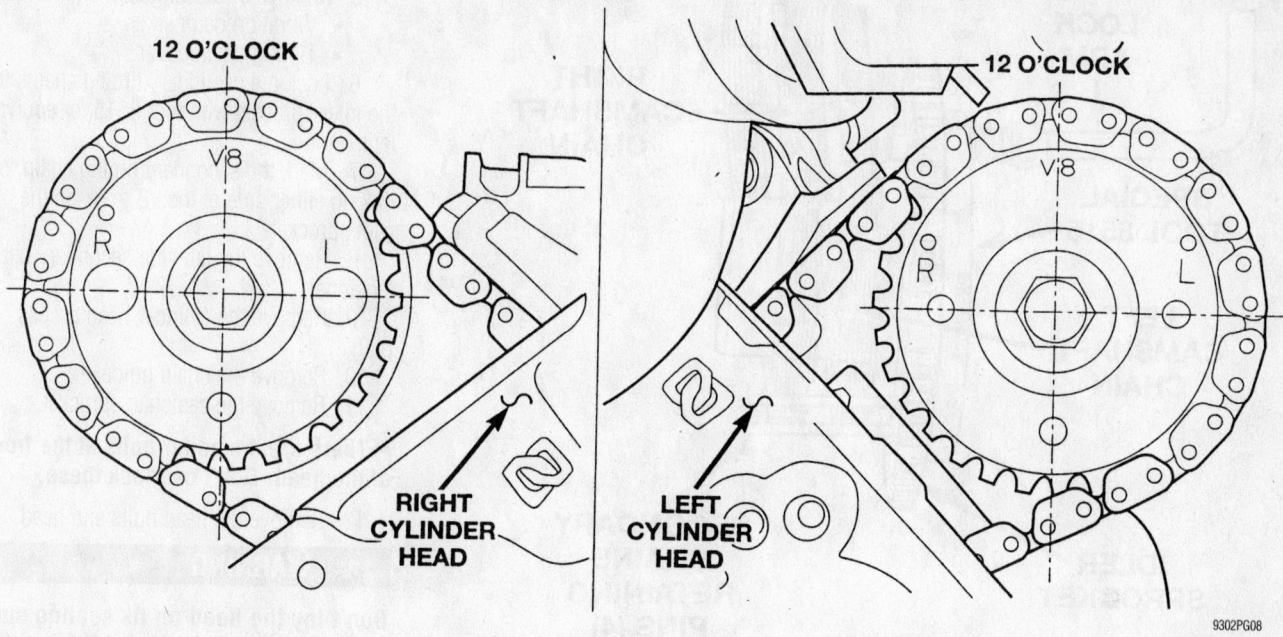

Camshaft positioning—4.7L engine

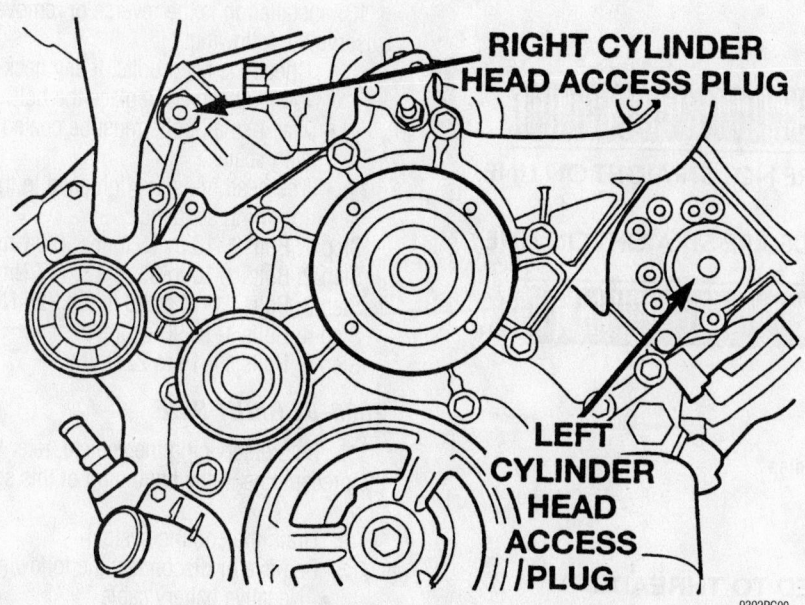

Cylinder head access plug locations—4.7L engine

a. Step 1: Bolts 1–10 to 15 ft. lbs. (20 Nm).

b. Step 2: Bolts 1–10 to 35 ft. lbs. (47 Nm).

c. Step 3: Bolts 11–14 to 18 ft. lbs. (25 Nm).

d. Step 4: Bolts 1–10 plus ¼ (90 degree) turn.

e. Step 5: Bolts 11–14 to 22 ft. lbs. (30 Nm).

12. Install or connect the following:
- Camshaft sprockets. Align the secondary chain matchmarks and tighten the bolts to 90 ft. lbs. (122 Nm).
- Secondary timing chain guides
- Cylinder head access plugs
- Secondary timing chain tensioners. Refer to the timing chain procedure in this section.

13. Remove the Timing Chain Locking tool 8515.

14. Install or connect the following:
- Front cover
- Crankshaft damper. Tighten the bolt to 130 ft. lbs. (175 Nm).
- Rocker arms
- Power steering pump
- Oil fill housing
- Accessory drive belt
- Engine cooling fan and shroud
- Valve covers
- Intake manifold
- Exhaust Y-pipe
- Negative battery cable

15. Fill the cooling system.

16. Start the engine and check for leaks.

2000–02, LEFT SIDE

1. Before servicing the vehicle, refer to the precautions in the beginning of this section.

2. Drain the cooling system.

3. Remove or disconnect the following:
- Negative battery cable
- Exhaust pipe
- Intake manifold
- Cylinder head cover
- Fan shroud and fan
- Accessory drive belt
- Power steering pump

4. Rotate the crankshaft until the damper mark is aligned with the TDC mark. Verify that the V8 mark on the camshaft sprocket is at the 12 o'clock position.

Timing belt service is covered in Section 3 of this manual

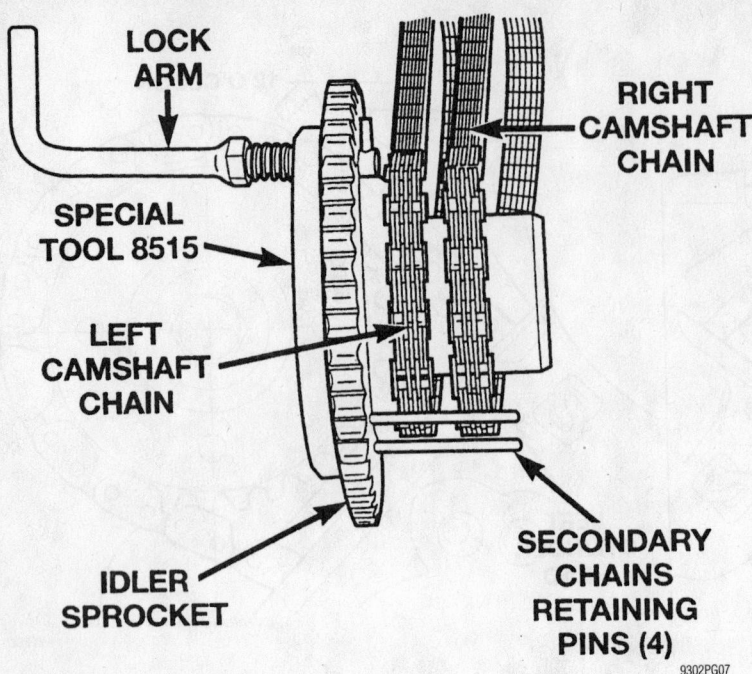

Use the special tool to lock the timing chains on the idler gear—4.7L engine

9302PG07

LOCK ARM

SPECIAL TOOL 8515

LEFT CAMSHAFT CHAIN

IDLER SPROCKET

RIGHT CAMSHAFT CHAIN

SECONDARY CHAINS RETAINING PINS (4)

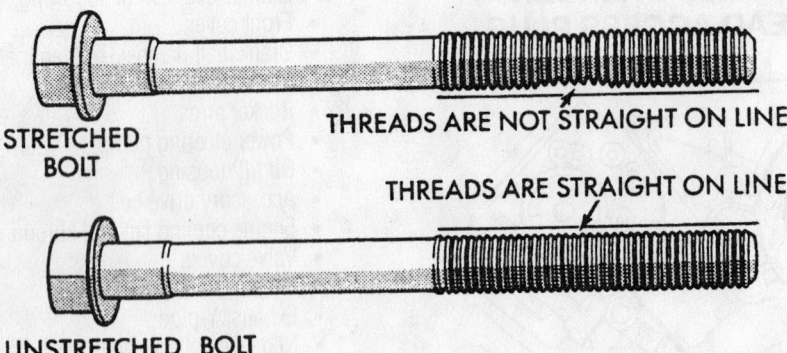

STRETCHED BOLT

THREADS ARE NOT STRAIGHT ON LINE

THREADS ARE STRAIGHT ON LINE

UNSTRETCHED BOLT

9302PG10

Examine the head bolts for signs of stretching—4.7L engine

◆ INDICATES SEALER APPLIED TO THREADS

9302PG11

Cylinder head torque sequence—4.7L engine, right cylinder head shown

5. Remove or disconnect the following:
  • Vibration damper
  • Timing chain cover

6. Lock the secondary timing chains to the idler sprocket with tool 8515, or equivalent.

7. Mark the secondary timing chain, on link on either side of the V8 mark on the cam sprocket.

8. Remove the left side secondary chain tensioner.

9. Remove the cylinder head access plug.

10. Remove the chain guide.

11. Remove the camshaft sprocket.

➡ **There are 4 smaller bolts at the front of the head. Don't overlook these.**

12. Remove the head bolts and head.

❈❈ WARNING

Don't lay the head on its sealing surface. Due to the design of the head gasket, any distortion to the head sealing surface will result in leaks.

13. Installation is the reverse of removal. Observe the following:
  • Check the head bolts. If any necking is observed, replace the bolt.
  • The 4 small bolts must be coated with sealer.
  • The head bolts are tightened in the following sequence:
Step 1: Bolts 1-10 to 15 ft. lbs. (20 Nm)
Step 2: Bolts 1-10 to 35 ft. lbs. (47 Nm)
Step 3: Bolts 11-14 to 18 ft. lbs. (25 Nm)
Step 4: Bolts 1-10 90 degrees
Step 5: Bolts 11-14 to 22 ft. lbs.

2000–02 RIGHT SIDE

1. Before servicing the vehicle, refer to the precautions in the beginning of this section.

2. Drain the cooling system.

3. Remove or disconnect the following:
  • Negative battery cable
  • Exhaust pipe
  • Intake manifold
  • Cylinder head cover
  • Fan shroud and fan
  • Oil filler housing
  • Accessory drive belt

4. Rotate the crankshaft until the damper mark is aligned with the TDC mark. Verify that the V8 mark on the camshaft sprocket is at the 12 o'clock position.

5. Remove or disconnect the following:
  • Vibration damper
  • Timing chain cover

6. Lock the secondary timing chains to

◆ INDICATES SEALER APPLIED TO THREADS

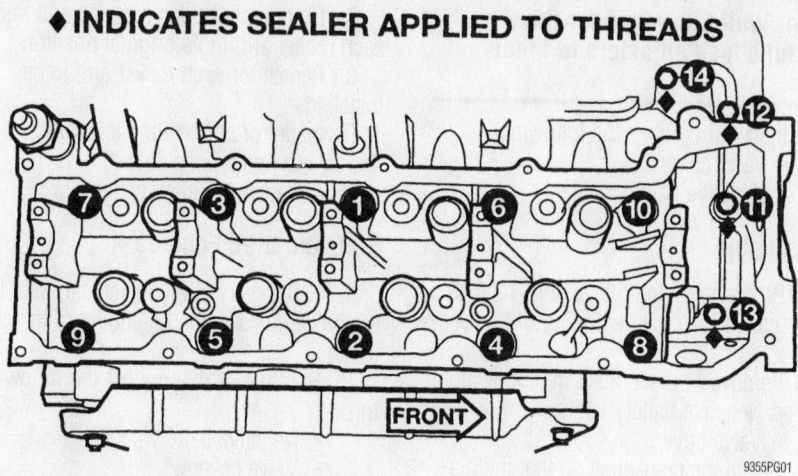

Cylinder head tightening sequence—2000–02 4.7L

the idler sprocket with tool 8515, or equivalent.

7. Mark the secondary timing chain, on link on either side of the V8 mark on the cam sprocket.

8. Remove the left side secondary chain tensioner.

9. Remove the cylinder head access plug.

10. Remove the chain guide.

11. Remove the camshaft sprocket.

✳✳ WARNING

Do not pry on the target wheel for any reason!

➡ **There are 4 smaller bolts at the front of the head. Don't overlook these.**

12. Remove the head bolts and head.

✳✳ WARNING

Don't lay the head on its sealing surface. Due to the design of the head gasket, any distortion to the head sealing surface will result in leaks.

13. Installation is the reverse of removal. Observe the following:
- Check the head bolts. If any necking is observed, replace the bolt.
- The 4 small bolts must be coated with sealer.
- The head bolts are tightened in the following sequence:

Step 1: Bolts 1-10 to 15 ft. lbs. (20 Nm)
Step 2: Bolts 1-10 to 35 ft. lbs. (47 Nm)
Step 3: Bolts 11-14 to 18 ft. lbs. (25 Nm)
Step 4: Bolts 1-10 90 degrees
Step 5: Bolts 11-14 to 22 ft. lbs.

5.2L and 5.9L Engines

1. Before servicing the vehicle, refer to the precautions in the beginning of this section.

2. Drain the cooling system.

3. Remove or disconnect the following:
- Negative battery cable
- Accessory drive belt
- Alternator
- A/C compressor, if equipped
- Air injection pump, if equipped
- Air cleaner assembly
- Closed Crankcase Ventilation (CCV) system
- Evaporative emissions control system
- Fuel line

- Accelerator linkage
- Cruise control cable
- Transmission cable
- Distributor cap and wires
- Ignition coil wiring
- Engine Coolant Temperature (ECT) sensor connector
- Heater hoses
- Bypass hose
- Upper radiator hose
- Intake manifold
- Valve covers

➡ **Keep valvetrain components in order for reassembly.**

- Rocker arms
- Pushrods
- Exhaust manifolds
- Spark plugs
- Cylinder heads

To install:

✳✳ WARNING

Position the crankshaft so that no piston is at Top Dead Center (TDC) prior to installing the cylinder heads. Do not rotate the crankshaft during or immediately after rocker arm installation. Wait 5 minutes for the hydraulic lash adjusters to bleed down.

4. Install the cylinder heads. Use new gaskets and tighten the bolts in sequence as follows:

a. Step 1: 50 ft. lbs. (68 Nm).

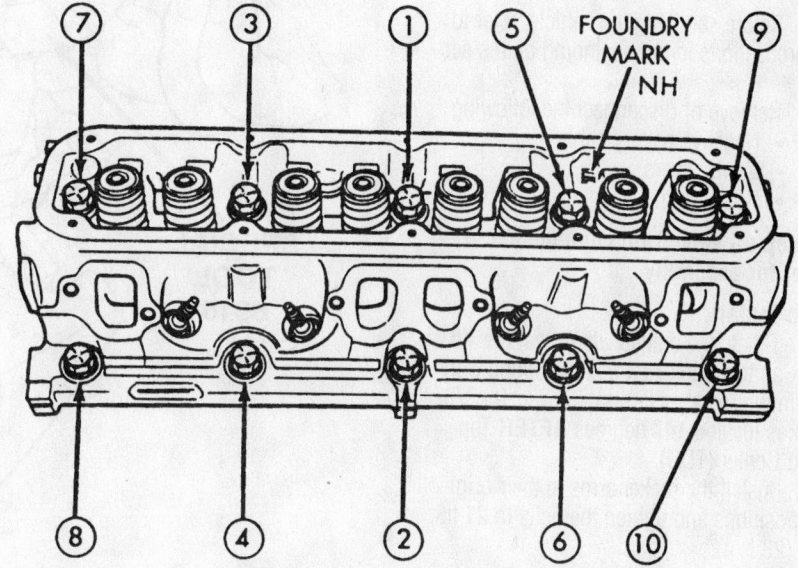

Cylinder head torque sequence—5.2L and 5.9L engines

Heater Core replacement is covered in Section 2 of this manual

b. Step 2: 105 ft. lbs. (143 Nm).

c. Step 3: 105 ft. lbs. (143 Nm).

5. Install or connect the following:
- Spark plugs
- Exhaust manifolds
- Pushrods and rocker arms in their original positions
- Valve covers
- Intake manifold
- Upper radiator hose
- Bypass hose
- Heater hoses
- ECT sensor connector
- Ignition coil wiring
- Distributor cap and wires
- Transmission cable
- Cruise control cable
- Accelerator linkage
- Fuel line
- Evaporative emissions control system
- CCV system
- Air cleaner assembly
- A/C compressor, if equipped
- Air injection pump, if equipped
- Alternator
- Accessory drive belt
- Negative battery cable

6. Fill the cooling system.

7. Start the engine and check for leaks.

Rocker Arms/Shafts

REMOVAL & INSTALLATION

3.9L Engine

1. Before servicing the vehicle, refer to the precautions in the beginning of this section.

2. Remove or disconnect the following:
- Negative battery cable
- Valve covers
- Rocker arms

➤Keep all valvetrain components in order for assembly.

To install:

3. Rotate the crankshaft so that the **V6** mark on the crankshaft damper aligns with the timing mark on the front cover. The **V6** mark is located 147 degrees **AFTER** Top Dead Center (TDC).

4. Install the rocker arms in their original positions and tighten the bolts to 21 ft. lbs. (28 Nm).

✳✳ CAUTION

Do not rotate the crankshaft during or immediately after rocker arm instal-

lation. **Wait 5 minutes for the hydraulic lash adjusters to bleed down.**

5. Install or connect the following:
- Valve covers
- Negative battery cable

4.7L Engine

1. Before servicing the vehicle, refer to the precautions in the beginning of this section.

2. Remove or disconnect the following:
- Negative battery cable
- Valve covers

3. Rotate the crankshaft so that the piston of the cylinder to be serviced is at Bottom Dead Center (BDC) and both valves are closed.

4. Use special tool 8516 to depress the valve and remove the rocker arm.

5. Repeat for each rocker arm to be serviced.

➤Keep valvetrain components in order for reassembly.

To install:

6. Rotate the crankshaft so that the piston of the cylinder to be serviced is at BDC.

7. Compress the valve spring and install each rocker arm in its original position.

8. Repeat for each rocker arm to be installed.

9. Install or connect the following:
- Cylinder head cover
- Negative battery cable

5.2L and 5.9L Engines

1. Before servicing the vehicle, refer to the precautions in the beginning of this section.

2. Remove or disconnect the following:
- Negative battery cable
- Valve covers
- Rocker arms

➤Keep valvetrain components in order for reassembly.

To install:

3. Rotate the crankshaft so that the **V8** mark on the crankshaft damper aligns with the timing mark on the front cover. The **V8** mark is located 147 degrees **AFTER** Top Dead Center (TDC).

4. Install the rocker arms in their original positions and tighten the bolts to 21 ft. lbs. (28 Nm).

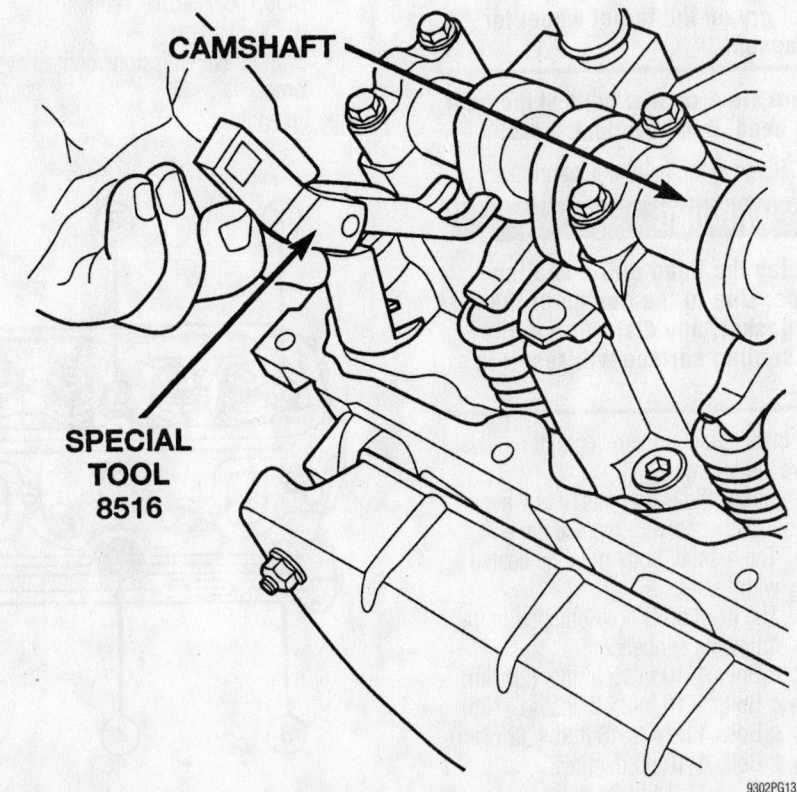

Rocker arm service—4.7L engine

9302PG13

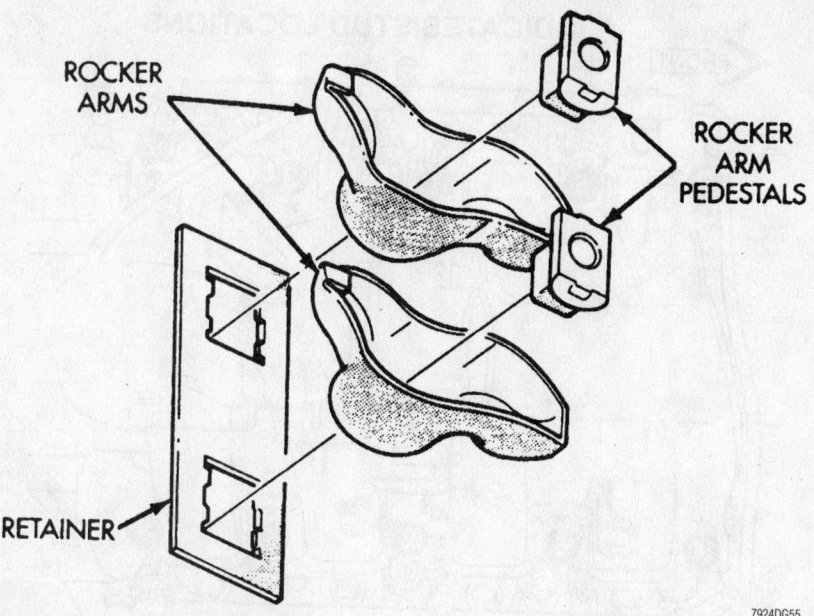

Exploded view of the rocker arm assembly—8.0L engine

7924DG55

> ❊❊ **CAUTION**
>
> **Do not rotate the crankshaft during or immediately after rocker arm installation. Wait 5 minutes for the hydraulic lash adjusters to bleed down.**

5. Install or connect the following:
  • Valve covers
  • Negative battery cable

Intake Manifold

REMOVAL & INSTALLATION

3.9L Engine

1. Before servicing the vehicle, refer to the precautions in the beginning of this section.
2. Drain the cooling system.
3. Remove or disconnect the following:
  • Negative battery cable
  • Accessory drive belt
  • Alternator
  • A/C compressor
  • Alternator and A/C compressor bracket
  • Air cleaner assembly
  • Fuel line
  • Fuel supply manifold
  • Accelerator cable
  • Transmission cable
  • Cruise control cable

  • Distributor cap and wires
  • Ignition coil wiring
  • Engine Coolant Temperature (ECT) sensor connector
  • Heater hose
  • Upper radiator hose

  • Bypass hose
  • Closed Crankcase Ventilation (CCV) system
  • Evaporative emissions system
  • Intake manifold

To install:

4. Install the intake manifold. Use a new gasket and tighten the bolts in sequence as follows:

  a. Step 1: Bolts 1–2 to 72 inch lbs. (8 Nm) using 12 inch lb. (1.4 Nm) increments.

  b. Step 2: Bolts 3–12 to 72 inch lbs. (8 Nm).

  c. Step 3: Bolts 1–12 to 72 inch lbs. (8 Nm).

  d. Step 4: Bolts 1–12 to 12 ft. lbs. (16 Nm).

  e. Step 5: Bolts 1–12 to 12 ft. lbs. (16 Nm).

5. Install or connect the following:
  • Evaporative emissions system
  • CCV system
  • Bypass hose
  • Upper radiator hose
  • Heater hose
  • ECT sensor connector
  • Ignition coil wiring
  • Distributor cap and wires
  • Cruise control cable
  • Transmission cable

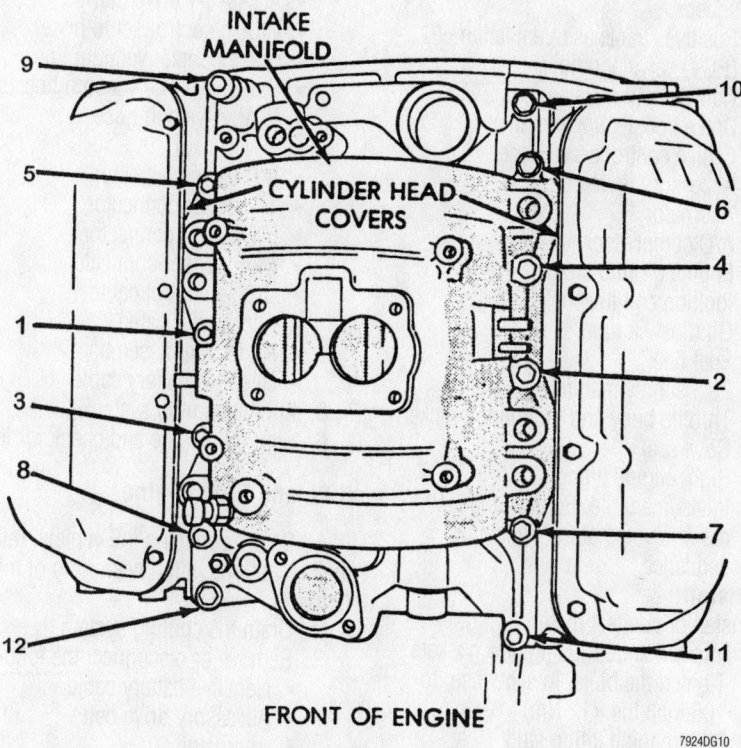

Intake manifold torque sequence—3.9L engine

7924DG10

Brake service is covered in Section 4 of this manual

- Accelerator cable
- Fuel supply manifold
- Fuel line
- Air cleaner assembly
- Alternator and A/C compressor bracket
- A/C compressor
- Alternator
- Accessory drive belt
- Negative battery cable
6. Fill the cooling system.
7. Start the engine and check for leaks.

4.7L Engine

1. Before servicing the vehicle, refer to the precautions in the beginning of this section.
2. Drain the cooling system.
3. Remove or disconnect the following:
- Negative battery cable
- Air cleaner assembly
- Accelerator cable
- Cruise control cable
- Manifold Absolute Pressure (MAP) sensor connector
- Intake Air Temperature (IAT) sensor connector
- Throttle Position (TP) sensor connector
- Idle Air Control (IAC) valve connector
- Engine Coolant Temperature (ECT) sensor
- Positive Crankcase Ventilation (PCV) valve and hose
- Canister purge vacuum line
- Brake booster vacuum line
- Cruise control servo hose
- Accessory drive belt
- Alternator
- A/C compressor
- Engine ground straps
- Ignition coil towers
- Oil dipstick tube
- Fuel line
- Fuel supply manifold
- Throttle body and mounting bracket
- Cowl seal
- Right engine lifting stud
- Intake manifold. Remove the fasteners in reverse of the tightening sequence.

To install:
4. Install or connect the following:
- Intake manifold using new gaskets. Tighten the bolts, in sequence, to 105 inch lbs. (12 Nm).
- Right engine lifting stud
- Cowl seal
- Throttle body and mounting bracket
- Fuel supply manifold

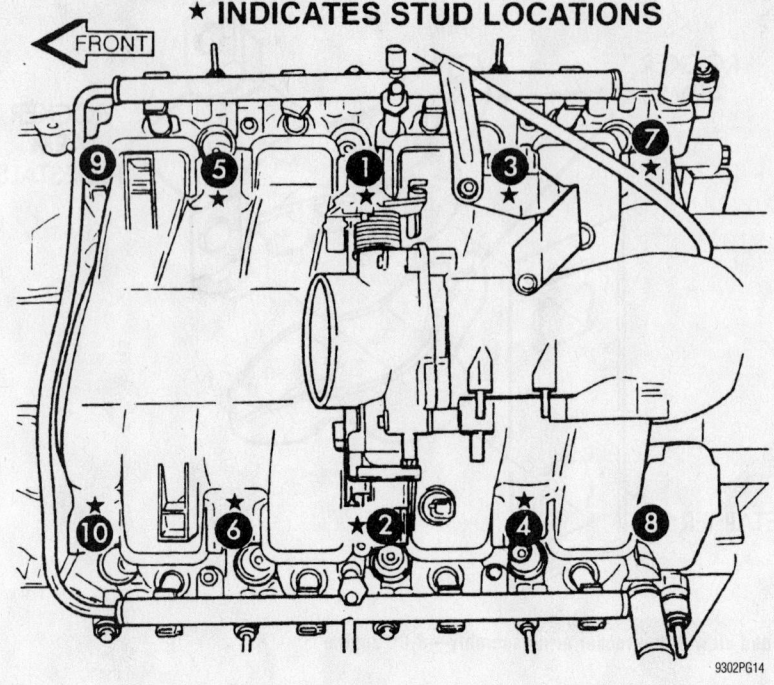

★ **INDICATES STUD LOCATIONS**

FRONT

Intake manifold torque sequence—4.7L engine

9302PG14

- Fuel line
- Oil dipstick tube
- Ignition coil towers
- Engine ground straps
- A/C compressor
- Alternator
- Accessory drive belt
- Cruise control servo hose
- Brake booster vacuum line
- Canister purge vacuum line
- PCV valve and hose
- ECT sensor
- IAC valve connector
- TP sensor connector
- IAT sensor connector
- MAP sensor connector
- Cruise control cable
- Accelerator cable
- Air cleaner assembly
- Negative battery cable
5. Fill the cooling system.
6. Start the engine and check for leaks.

5.2L and 5.9L Engines

1. Before servicing the vehicle, refer to the precautions in the beginning of this section.
2. Drain the cooling system.
3. Remove or disconnect the following:
- Negative battery cable
- Accessory drive belt
- Alternator
- A/C compressor
- Alternator and A/C compressor bracket

- Air cleaner assembly
- Fuel line
- Fuel supply manifold
- Accelerator cable
- Transmission cable
- Cruise control cable
- Distributor cap and wires
- Ignition coil wiring
- Engine Coolant Temperature (ECT) sensor connector
- Heater hose
- Upper radiator hose
- Bypass hose
- Closed Crankcase Ventilation (CCV) system
- Evaporative emissions system
- Intake manifold

To install:
4. Install the intake manifold. Use a new gasket and tighten the bolts in sequence as follows:

a. Step 1: Bolts 1–4 to 72 inch lbs. (8 Nm) using 12 inch lb. (1.4 Nm) increments.

b. Step 2: Bolts 5–12 to 72 inch lbs. (8 Nm).

c. Step 3: Bolts 1–12 to 72 inch lbs. (8 Nm).

d. Step 4: Bolts 1–12 to 12 ft. lbs. (16 Nm).

e. Step 5: Bolts 1–12 to 12 ft. lbs. (16 Nm).
5. Install or connect the following:
- Evaporative emissions system
- CCV system

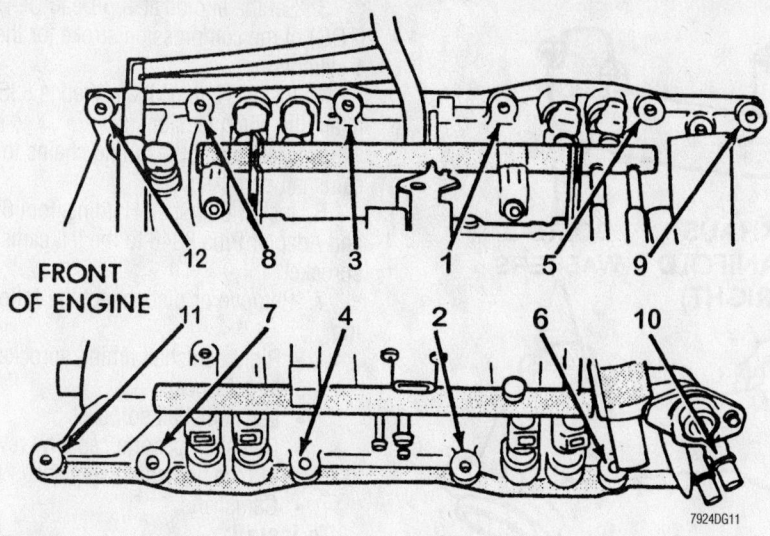

Intake manifold torque sequence—5.2L and 5.9L engines

- Bypass hose
- Upper radiator hose
- Heater hose
- ECT sensor connector
- Ignition coil wiring
- Distributor cap and wires
- Cruise control cable
- Transmission cable
- Accelerator cable
- Fuel supply manifold
- Fuel line
- Air cleaner assembly
- Alternator and A/C compressor bracket
- A/C compressor
- Alternator
- Accessory drive belt
- Negative battery cable

6. Fill the cooling system.
7. Start the engine and check for leaks.

Exhaust Manifold

REMOVAL & INSTALLATION

3.9L Engine

1. Before servicing the vehicle, refer to the precautions in the beginning of this section.
2. Remove or disconnect the following:
   - Negative battery cable
   - Heated Oxygen (HO2S) sensor connectors
   - Exhaust manifold heat shields
   - Exhaust front pipe
   - Exhaust manifolds

To install:

➡ **If the exhaust manifold studs came out with the nuts when removing the exhaust manifolds, replace them with new studs.**

3. Install or connect the following:
   - Exhaust manifolds. Tighten the fasteners to 25 ft. lbs. (34 Nm), starting with the center fasteners and working out to the ends.
   - Exhaust front pipe
   - Exhaust manifold heat shields
   - HO2S sensor connectors

   - Negative battery cable
4. Start the engine and check for leaks.

4.7L Engine

1. Before servicing the vehicle, refer to the precautions in the beginning of this section.
2. Drain the cooling system.
3. Remove or disconnect the following:
   - Battery
   - Power distribution center
   - Battery tray
   - Windshield washer fluid bottle
   - Air cleaner assembly
   - Accessory drive belt
   - A/C compressor
   - A/C accumulator bracket
   - Heater hoses
   - Exhaust manifold heat shields
   - Exhaust Y-pipe
   - Starter motor
   - Exhaust manifolds

To install:
4. Install or connect the following:
   - Exhaust manifolds, using new gaskets. Tighten the bolts to 18 ft. lbs. (25 Nm), starting with the inner bolts and work out to the ends.
   - Starter motor
   - Exhaust Y-pipe
   - Exhaust manifold heat shields
   - Heater hoses
   - A/C accumulator bracket
   - A/C compressor

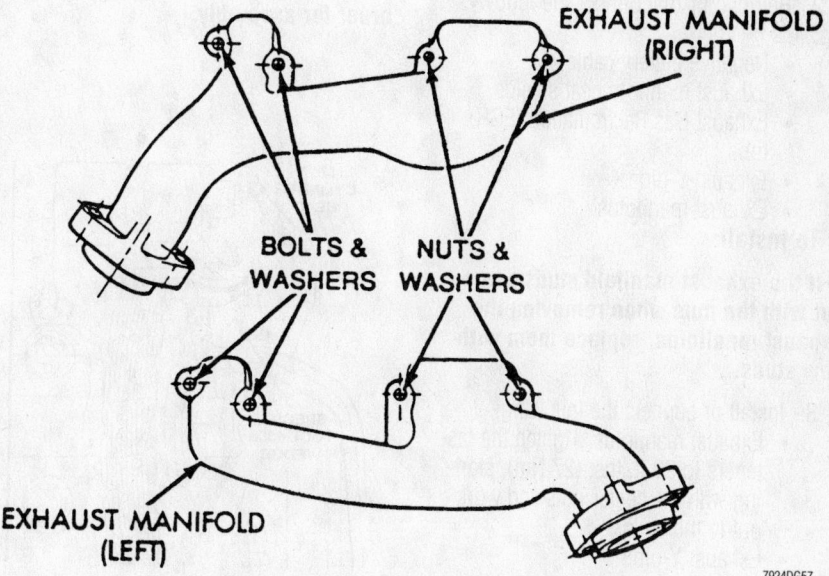

Stud and bolt locations—3.9L engine

For complete Engine Mechanical specifications, see Section 1 of this manual

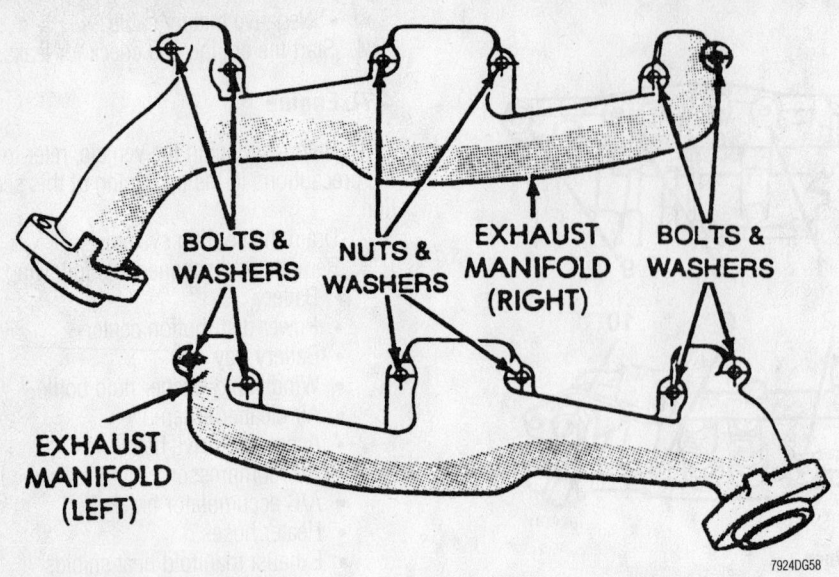

Exhaust manifold fastener locations—5.2L and 5.9L engines

- Accessory drive belt
- Air cleaner assembly
- Windshield washer fluid bottle
- Battery tray
- Power distribution center
- Battery

5. Fill the cooling system.

6. Start the engine and check for leaks.

5.2L and 5.9L Engines

1. Before servicing the vehicle, refer to the precautions in the beginning of this section.

2. Remove or disconnect the following:

- Negative battery cable
- Exhaust manifold heat shields
- Exhaust Gas Recirculation (EGR) tube
- Exhaust Y-pipe
- Exhaust manifolds

To install:

➡**If the exhaust manifold studs came out with the nuts when removing the exhaust manifolds, replace them with new studs.**

3. Install or connect the following:

- Exhaust manifolds. Tighten the fasteners to 20 ft. lbs. (27 Nm), starting with the center nuts and work out to the ends.
- Exhaust Y-pipe
- EGR tube
- Exhaust manifold heat shields
- Negative battery cable

4. Fill the cooling system.

5. Start the engine and check for leaks.

Camshaft and Valve Lifters

REMOVAL & INSTALLATION

4.7L Engine

1. Before servicing the vehicle, refer to the precautions in the beginning of this section.

2. Remove or disconnect the following:

- Negative battery cable
- Cylinder head covers
- Rocker arms
- Hydraulic lash adjusters

➡**Keep all valvetrain components in order for assembly.**

3. Set the engine at Top Dead Center (TDC) of the compression stroke for the No. 1 cylinder.

4. Install Timing Chain Wedge 8350 to retain the chain tensioners.

5. Matchmark the timing chains to the camshaft sprockets.

6. Install Camshaft Holding Tool 6958 and Adapter Pins 8346 to the left camshaft sprocket.

7. Remove or disconnect the following:

- Right camshaft timing sprocket and target wheel
- Left camshaft sprocket
- Camshaft bearing caps, by reversing the tightening sequence
- Camshafts

To install:

8. Install or connect the following:

- Camshafts. Tighten the bearing cap bolts in ½ turn increments, in sequence, to 100 inch lbs. (11 Nm).
- Target wheel to the right camshaft
- Camshaft timing sprockets and chains, by aligning the matchmarks

9. Remove the tensioner wedges and tighten the camshaft sprocket bolts to 90 ft. lbs. (122 Nm).

10. Install or connect the following:

- Hydraulic lash adjusters in their original locations
- Rocker arms in their original locations
- Cylinder head covers
- Negative battery cable

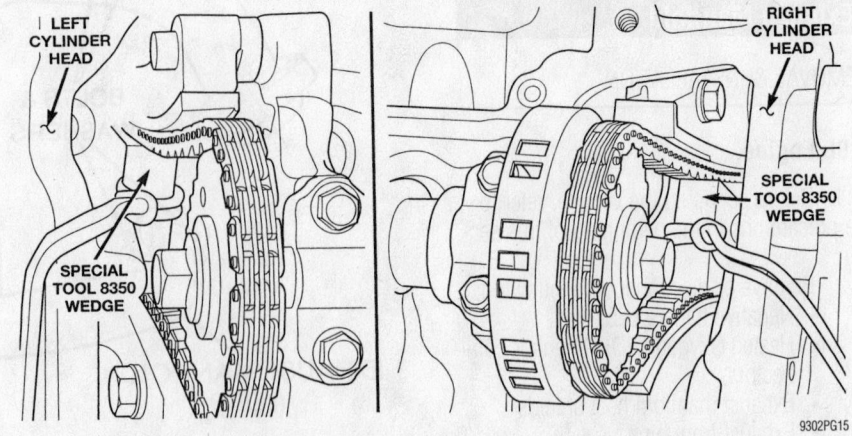

Chain Tensioner Retaining Wedges—4.7L engine

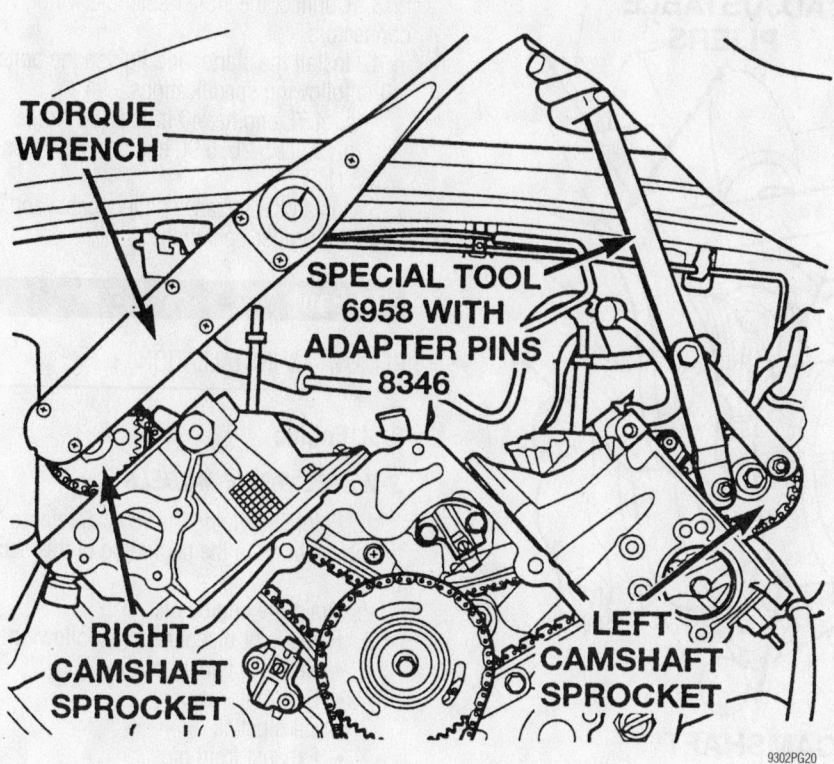

Hold the left camshaft sprocket with a spanner wrench while removing or installing the camshaft sprocket bolts—4.7L engine

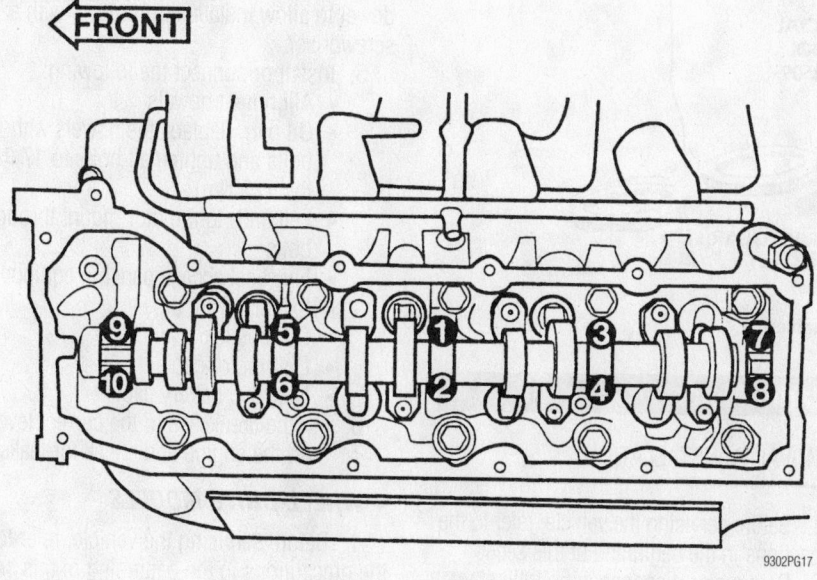

Camshaft bearing cap bolt tightening sequence—4.7L engine

3.9L, 5.2L and 5.9L Engines

1. Before servicing the vehicle, refer to the precautions in the beginning of this section.
2. Drain the cooling system.
3. Recover the A/C refrigerant, if equipped with air conditioning.
4. Set the crankshaft to Top Dead Center (TDC) of the compression stroke for the No. 1 cylinder.
5. Remove or disconnect the following:
   - Negative battery cable
   - Accessory drive belt
   - Power steering pump
   - Water pump
   - Radiator
   - A/C condenser
   - Grille
   - Crankshaft damper
   - Front cover
   - Valve covers
   - Distributor
   - Intake manifold

➡ **Keep all valvetrain components in order for assembly.**

   - Rocker arms and pushrods
   - Hydraulic lifters
   - Timing chain and sprockets
   - Camshaft thrust plate and chain oil tab
   - Camshaft

To install:

6. Install or connect the following:
   - Camshaft
   - Camshaft Holding Tool C-3509
   - Camshaft thrust plate and chain oil tab. Tighten the bolts to 18 ft. lbs. (24 Nm).
   - Timing chain and sprockets. Tighten the camshaft sprocket bolt to 50 ft. lbs. (68 Nm).
7. Remove the camshaft holding tool.
8. Install or connect the following:
   - Hydraulic lifters in their original positions
   - Rocker arms and pushrods in their original positions
   - Intake manifold
   - Distributor
   - Valve covers
   - Front cover
   - Crankshaft damper
   - Grille
   - A/C condenser
   - Radiator
   - Water pump
   - Power steering pump
   - Accessory drive belt
   - Negative battery cable

For Accessory Drive Belt illustrations, see Section 1 of this manual

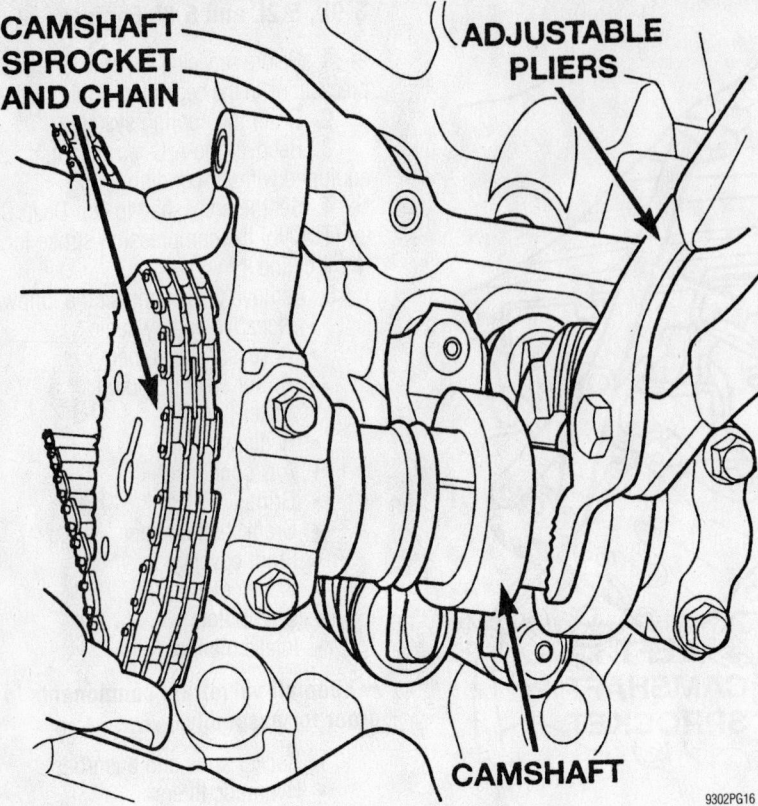

Turn the camshaft with pliers, if needed, to align the dowel in the sprocket—4.7L engine

Labels on image: CAMSHAFT SPROCKET AND CHAIN, ADJUSTABLE PLIERS, CAMSHAFT, 9302PG16

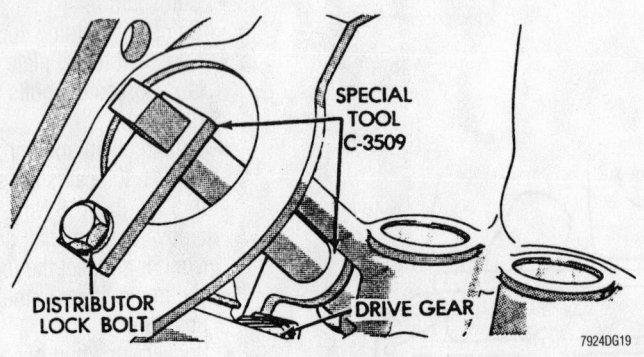

Camshaft holding tool C-3509—3.9L, 5.2L and 5.9L engines

Labels on image: SPECIAL TOOL C-3509, DISTRIBUTOR LOCK BOLT, DRIVE GEAR, 7924DG19

9. Fill the cooling system.
10. Recharge the A/C system, if equipped.
11. Start the engine and check for leaks.

Valve Lash

ADJUSTMENT

All gasoline engines covered in this section use hydraulic lifters. No maintenance or periodic adjustment is required.

Starter Motor

REMOVAL & INSTALLATION

1. Before servicing the vehicle, refer to the precautions in the beginning of this section.
2. Remove or disconnect the following:
   - Negative battery cable
   - Starter mounting bolts
   - Starter solenoid harness connections
   - Starter

To install:
3. Connect the starter solenoid wiring connectors.
4. Install the starter and tighten the bolts to the following specifications:
   a. 4.7L engine: 40 ft. lbs. (54 Nm).
   b. 3.9L, 5.2L, 5.9L engines: 50 ft. lbs. (68 Nm).
5. Install the negative battery cable and check for proper operation.

Oil Pan

REMOVAL & INSTALLATION

3.9L Engine

2-WHEEL DRIVE MODELS

1. Before servicing the vehicle, refer to the precautions in the beginning of this section.
2. Drain the engine oil.
3. Remove or disconnect the following:
   - Negative battery cable
   - Distributor cap
   - Oil dipstick
   - Exhaust front pipe
   - Flywheel access panel, if equipped
   - Left and right motor mount through bolts
   - Oil pan. Raise the engine as necessary for clearance.

To install:
4. Fabricate 4 alignment dowels from 1½ inch x ¼ inch bolts. Cut the heads off the bolts and cut a slot into the top of the dowel to allow installation/removal with a screwdriver.
5. Install or connect the following:
   - Alignment dowels
   - Oil pan. Replace the dowels with bolts and tighten all bolts to 17 ft. lbs. (23 Nm).
   - Left and right motor mount through bolts
   - Flywheel access panel, if equipped
   - Exhaust front pipe
   - Oil dipstick
   - Distributor cap
   - Negative battery cable
6. Fill the crankcase to the correct level.
7. Start the engine and check for leaks.

4-WHEEL DRIVE MODELS

1. Before servicing the vehicle, refer to the precautions in the beginning of this section.
2. Drain the engine oil.
3. Remove or disconnect the following:
   - Negative battery cable
   - Engine oil dipstick

- Front axle. Support the engine
- Exhaust front pipe
- Flywheel access panel, if equipped
- Oil pan

To install:

4. Fabricate 4 alignment dowels from 1½ inch x ¼ inch bolts. Cut the heads off the bolts and cut a slot into the top of the dowel to allow installation/removal with a screwdriver.

5. Install or connect the following:
- Alignment dowels
- Oil pan. Replace the dowels with bolts and tighten all bolts to 17 ft. lbs. (23 Nm).
- Flywheel access panel, if equipped
- Exhaust front pipe
- Front axle
- Engine oil dipstick
- Negative battery cable

6. Fill the crankcase to the correct level.

7. Start the engine and check for leaks.

4.7L Engine

1. Before servicing the vehicle, refer to the precautions in the beginning of this section.

2. Drain the engine oil.

3. Remove or disconnect the following:
- Negative battery cable
- Structural cover
- Exhaust Y-pipe
- Starter motor
- Transmission oil cooler lines

- Oil pan
- Oil pump pickup tube
- Oil pan gasket

To install:

4. Install or connect the following:
- Oil pan gasket
- Oil pump pickup tube, using a new

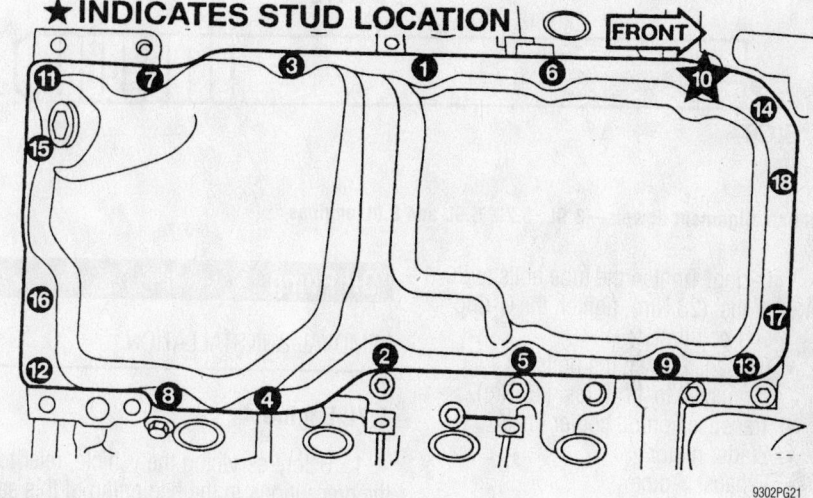

Oil pan mounting bolt tightening sequence—4.7L engine

9302PG21

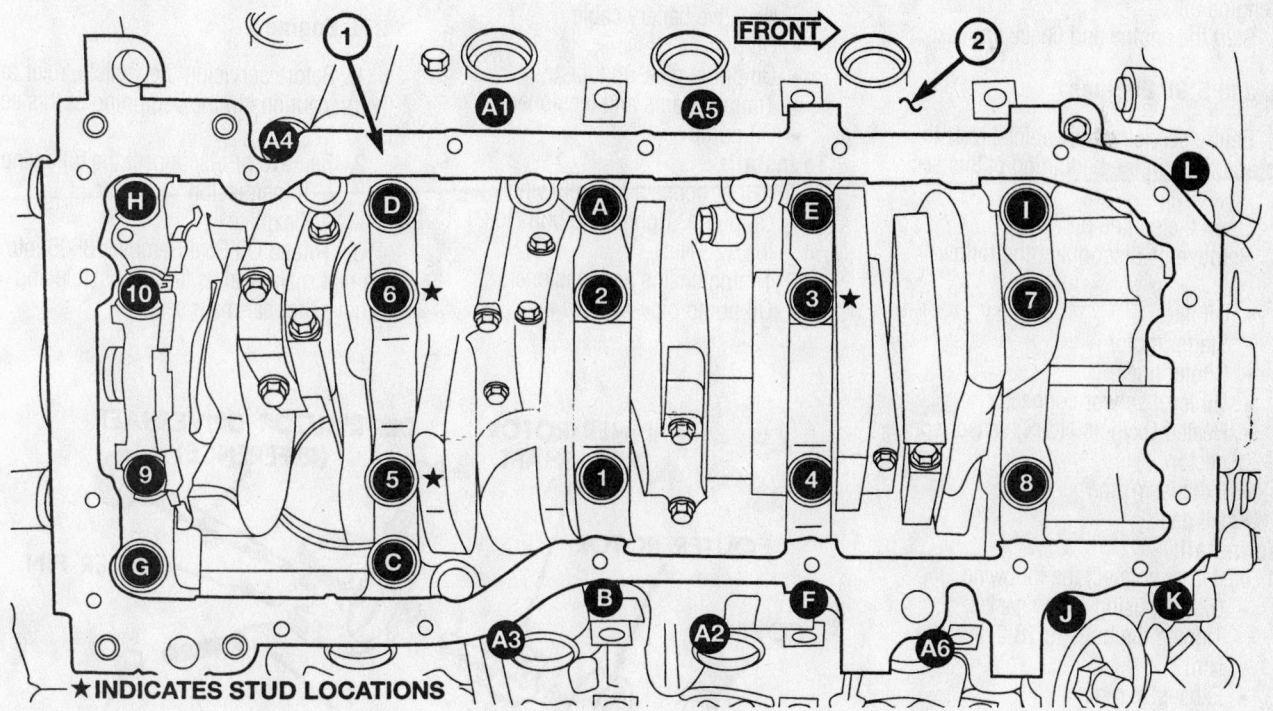

1 - BEDPLATE
2 - CYLINDER BLOCK

Bedplate bolt tightening sequence—4.7L engine

9355PG03

For Tire, Wheel and Ball Joint specifications, see Section 1 of this manual

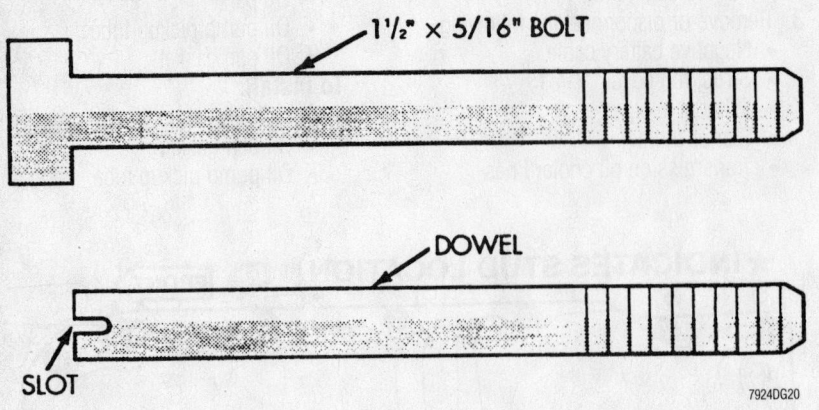

Oil pan alignment dowels—3.9L, 5.2L, 5.9L and 8.0L engines

O-ring. Tighten the tube bolts to 20 ft. lbs. (28 Nm); tighten the O-ring end bolt first.
- Oil pan. Tighten the bolts, in sequence, to 11 ft. lbs. (15 Nm).
- Transmission oil cooler lines
- Starter motor
- Exhaust Y-pipe
- Structural cover
- Negative battery cable

5. Fill the crankcase to the proper level with engine oil.

6. Start the engine and check for leaks.

5.2L and 5.9L Engines

1. Before servicing the vehicle, refer to the precautions in the beginning of this section.

2. Drain the engine oil.

3. Remove or disconnect the following:
- Oil filter
- Starter motor
- Cooler lines
- Oil level sensor connector
- Heated Oxygen (HO₂S) sensor connector
- Exhaust Y-pipe
- Oil pan

To install:

4. Install or connect the following:
- Oil pan, using a new gasket. Tighten the bolts to 18 ft. lbs. (24 Nm).
- Exhaust Y-pipe
- Heated Oxygen (HO₂S) sensor connector
- Oil level sensor connector
- Cooler lines
- Starter motor
- Oil filter

5. Fill the crankcase.

6. Start the engine and check for leaks.

Oil Pump

REMOVAL & INSTALLATION

4.7L Engine

1. Before servicing the vehicle, refer to the precautions in the beginning of this section.

2. Drain the engine oil.

3. Remove or disconnect the following:
- Negative battery cable
- Oil pan
- Oil pump pick-up tube
- Timing chains and tensioners
- Oil pump

To install:

4. Install or connect the following:
- Oil pump. Tighten the bolts to 21 ft. lbs. (28 Nm).
- Timing chains and tensioners
- Oil pump pick-up tube

- Oil pan
- Negative battery cable

5. Fill the crankcase to the correct level.

6. Start the engine and check for leaks.

3.9L, 5.2L and 5.9L Engines

1. Before servicing the vehicle, refer to the precautions in the beginning of this section.

2. Drain the engine oil.

3. Remove or disconnect the following:
- Negative battery cable
- Oil pan
- Oil pump pick-up tube
- Oil pump

To install:

4. Install or connect the following:
- Oil pump. Tighten the bolts to 30 ft. lbs. (41 Nm).
- Oil pump pick-up tube
- Oil pan
- Negative battery cable

5. Fill the crankcase to the correct level.

6. Start the engine and check for leaks.

Rear Main Seal

REMOVAL & INSTALLATION

4.7L Engine

1. Before servicing the vehicle, refer to the precautions in the beginning of this section.

2. Remove or disconnect the following:
- Transmission
- Flexplate

3. Thread Oil Seal Remover 8506 into the rear main seal as far as possible and remove the rear main seal.

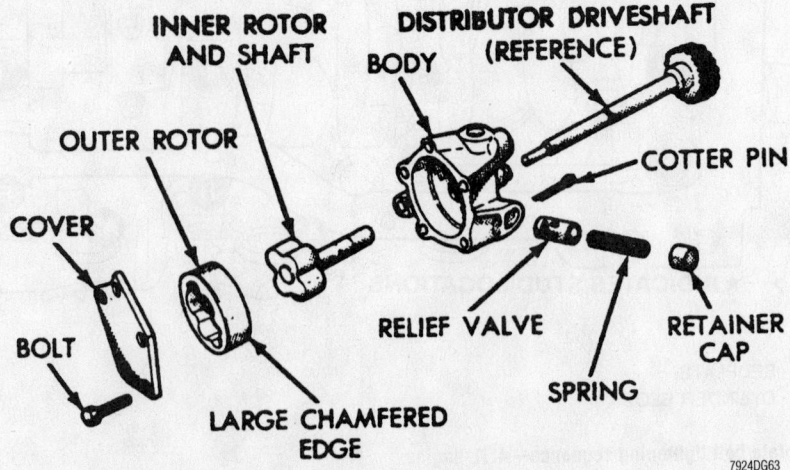

Exploded view of the oil pump assembly—3.9L, 5.2L and 5.9L engines

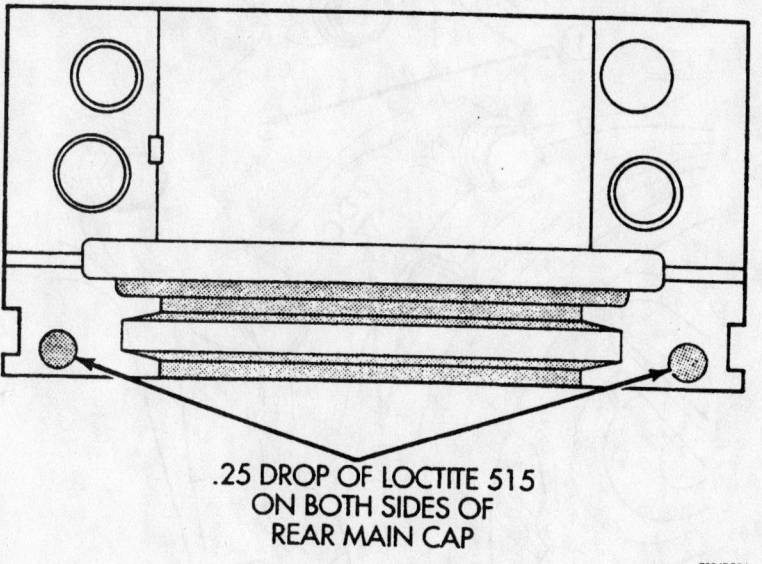

.25 DROP OF LOCTITE 515
ON BOTH SIDES OF
REAR MAIN CAP

7924DG24

Sealant application locations —3.9L, 5.2L and 5.9L engines

To install:

4. Install or connect the following:
- Seal Guide 8349-2 onto the crankshaft
- Rear main seal on the seal guide
- Rear main seal, using the Crankshaft Rear Oil Seal Installer 8349 and Driver Handle C-4171; tap it into place until the installer is flush with the cylinder block
- Flexplate. Tighten the bolts to 45 ft. lbs. (60 Nm).
- Transmission

5. Start the engine and check for leaks.

3.9L, 5.2L and 5.9L Engines

1. Before servicing the vehicle, refer to the precautions in the beginning of this section.
2. Drain the engine oil.
3. Remove or disconnect the following:
- Oil pan
- Oil pump
- Rear main bearing cap

4. Loosen the other main bearing cap bolts for clearance and remove the rear main seal halves.

To install:

5. Install or connect the following:
- New upper seal half to the cylinder block
- New lower seal half to the bearing cap

6. Apply sealant to the rear main bearing cap.

7. Install or connect the following:
- Rear main bearing cap. Tighten **all** main bearing cap bolts to 85 ft. lbs. (115 Nm).
- Oil pump and oil pan

8. Fill the engine.

9. Start the engine and check for leaks.

Timing Chain, Sprockets, Front Cover and Seal

REMOVAL & INSTALLATION

3.9L Engine

1. Before servicing the vehicle, refer to the precautions in the beginning of this section.
2. Drain the cooling system.
3. Remove or disconnect the following:
- Negative battery cable
- Accessory drive belt
- Radiator
- Cooling fan
- Water pump
- Crankshaft pulley
- Front crankshaft seal
- Front cover
- Timing chain and gears

To install:

4. Install or connect the following:
- Timing chain and gears. Align the timing marks and tighten the camshaft sprocket bolt to 35 ft. lbs. (47 Nm).
- Front cover. Tighten the bolts to 30 ft. lbs. (41 Nm).
- Front crankshaft seal
- Crankshaft pulley. Tighten the bolt to 135 ft. lbs. (183 Nm).

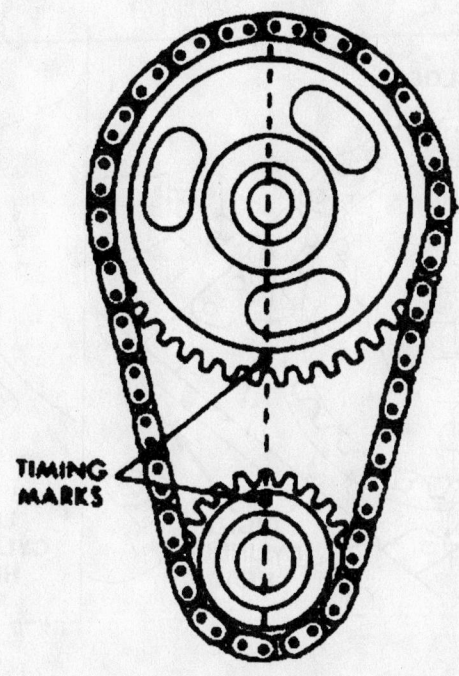

TIMING MARKS

7924DG67

Timing chain alignment marks—3.9L, 5.2L, and 5.9L engines

For Wheel Alignment specifications, see Section 1 of this manual

- Water pump
- Cooling fan
- Radiator
- Accessory drive belt
- Negative battery cable

5. Fill the cooling system.
6. Start the engine and check for leaks.

4.7L Engine

1. Before servicing the vehicle, refer to the precautions in the beginning of this section.

2. Drain the cooling system.

3. Remove or disconnect the following:
- Negative battery cable
- Valve covers
- Camshaft Position (CMP) sensor
- Engine cooling fan and shroud
- Accessory drive belt
- Heater hoses
- Lower radiator hose
- Power steering pump

4. Rotate the crankshaft so that the crankshaft timing mark aligns with the Top Dead Center (TDC) mark on the front cover, and the **V8** marks on the camshaft sprockets are at 12 o'clock.

5. Remove or disconnect the following:
- Crankshaft damper
- Oil fill housing

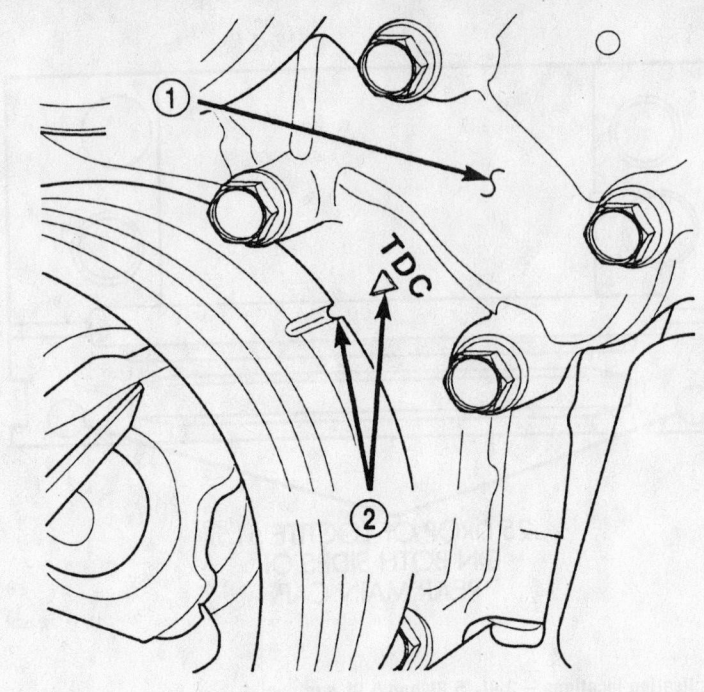

1 – TIMING CHAIN COVER
2 – CRANKSHAFT TIMING MARKS

9308PG04

Crankshaft timing marks—4.7L engine

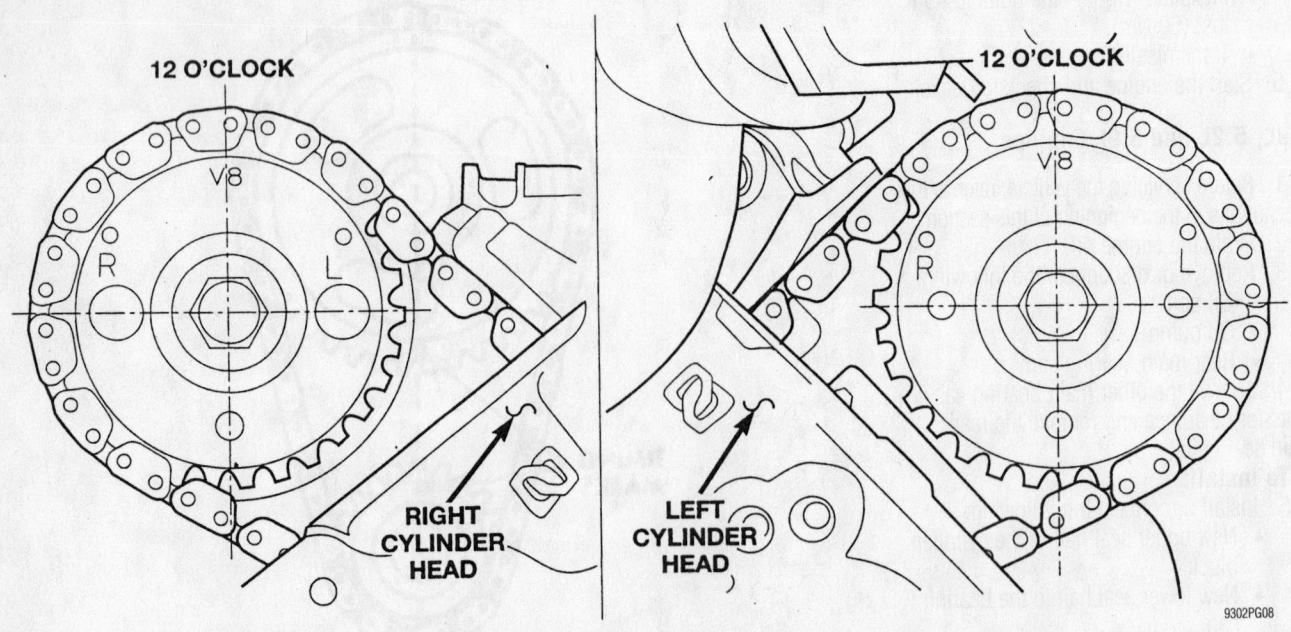

Camshaft positioning—4.7L engine

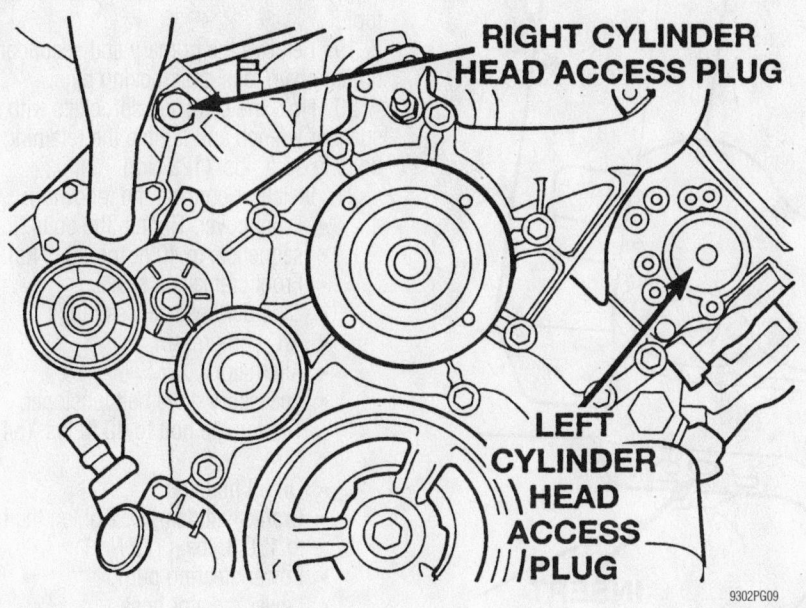

Cylinder head access plug locations—4.7L engine

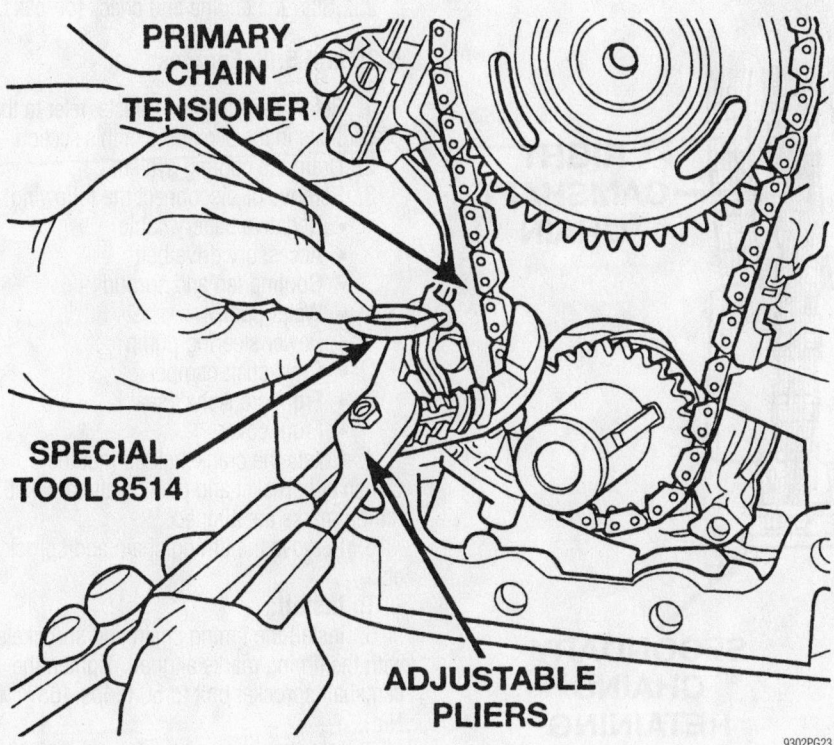

Compress and lock the primary chain tensioner—4.7L engine

- Accessory drive belt tensioner
- Alternator
- A/C compressor
- Front cover
- Front crankshaft seal
- Cylinder head access plugs
- Secondary timing chain guides

6. Compress the primary timing chain tensioner and install a lockpin.

7. Remove the secondary timing chain tensioners.

8. Hold the left camshaft with adjustable pliers and remove the sprocket and chain. Rotate the **left** camshaft 15 degrees **clockwise** to the neutral position.

9. Hold the right camshaft with adjustable pliers and remove the camshaft sprocket. Rotate the **right** camshaft 45 degrees **counterclockwise** to the neutral position.

10. Remove the primary timing chain and sprockets.

To install:

11. Use a small prytool to hold the ratchet pawl and compress the secondary timing chain tensioners in a vise and install locking pins.

➡ **The black bolts fasten the guide to the engine block and the silver bolts fasten the guide to the cylinder head.**

12. Install or connect the following:
- Secondary timing chain guides. Tighten the bolts to 21 ft. lbs. (28 Nm).
- Secondary timing chains to the idler sprocket so that the double plated links on each chain are visible through the slots in the primary idler sprocket

13. Lock the secondary timing chains to the idler sprocket with Timing Chain Locking tool 8515 as shown.

14. Align the primary chain double plated links with the idler sprocket timing mark and the single plated link with the crankshaft sprocket timing mark.

15. Install the primary chain and sprockets. Tighten the idler sprocket bolt to 25 ft. lbs. (34 Nm).

16. Align the secondary chain single plated links with the timing marks on the secondary sprockets. Align the dot at the **L** mark on the left sprocket with the plated link on the left chain and the dot at the **R** mark on the right sprocket with the plated link on the right chain.

17. Rotate the camshafts back from the

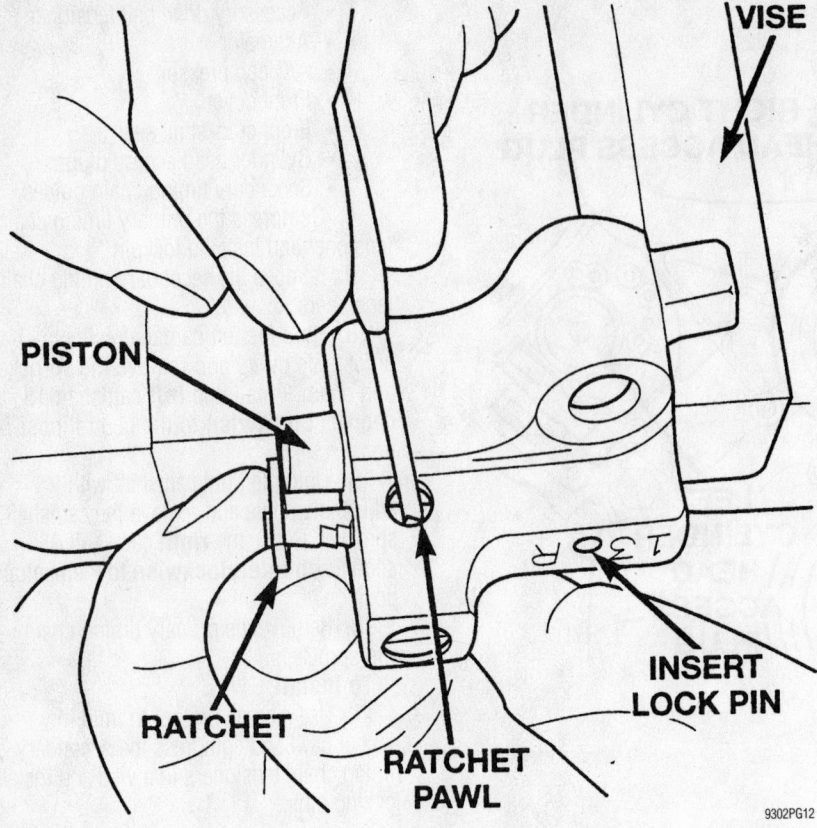

Secondary timing chain tensioner preparation—4.7L engine

9302PG12

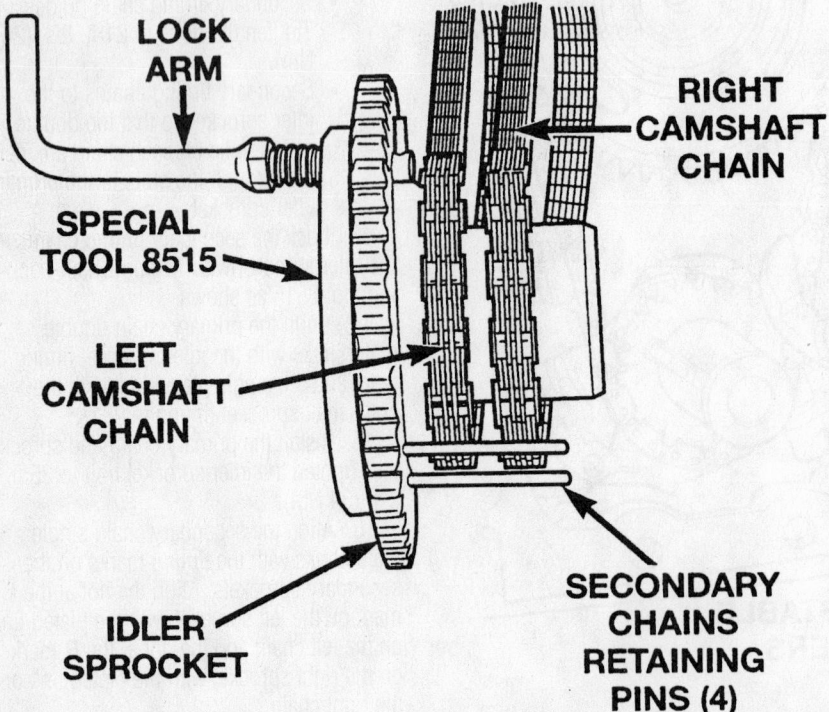

Use the Timing Chain Locking tool to lock the timing chains on the idler gear—4.7L engine

9302PG07

neutral position and install the camshaft sprockets.

18. Remove the secondary chain locking tool.

19. Remove the primary and secondary timing chain tensioner locking pins.

20. Hold the camshaft sprockets with a spanner wrench and tighten the retaining bolts to 90 ft. lbs. (122 Nm).

21. Install or connect the following:
- Front cover. Tighten the bolts, in sequence, to 40 ft. lbs. (54 Nm).
- Front crankshaft seal
- Cylinder head access plugs
- A/C compressor
- Alternator
- Accessory drive belt tensioner. Tighten the bolt to 40 ft. lbs. (54 Nm).
- Oil fill housing
- Crankshaft damper. Tighten the bolt to 130 ft. lbs. (175 Nm).
- Power steering pump
- Lower radiator hose
- Heater hoses
- Accessory drive belt
- Engine cooling fan and shroud
- Camshaft Position (CMP) sensor
- Valve covers
- Negative battery cable

22. Fill the cooling system.

23. Start the engine and check for leaks.

5.2L and 5.9L Engines

1. Before servicing the vehicle, refer to the precautions in the beginning of this section.

2. Drain the cooling system.

3. Remove or disconnect the following:
- Negative battery cable
- Accessory drive belt
- Cooling fan and shroud
- Water pump
- Power steering pump
- Crankshaft damper
- Front crankshaft seal
- Front cover

4. Rotate the crankshaft so that the camshaft sprocket and crankshaft sprocket timing marks are aligned.

5. Remove the timing chain and sprockets.

To install:

6. Install the timing chain and sprockets with the timing marks aligned. Tighten the camshaft sprocket bolt to 50 ft. lbs. (68 Nm).

7. Install or connect the following:
- Front cover. Tighten the cover bolts to 30 ft. lbs. (41 Nm) and the oil pan bolts to 18 ft. lbs. (24 Nm).
- Front crankshaft seal

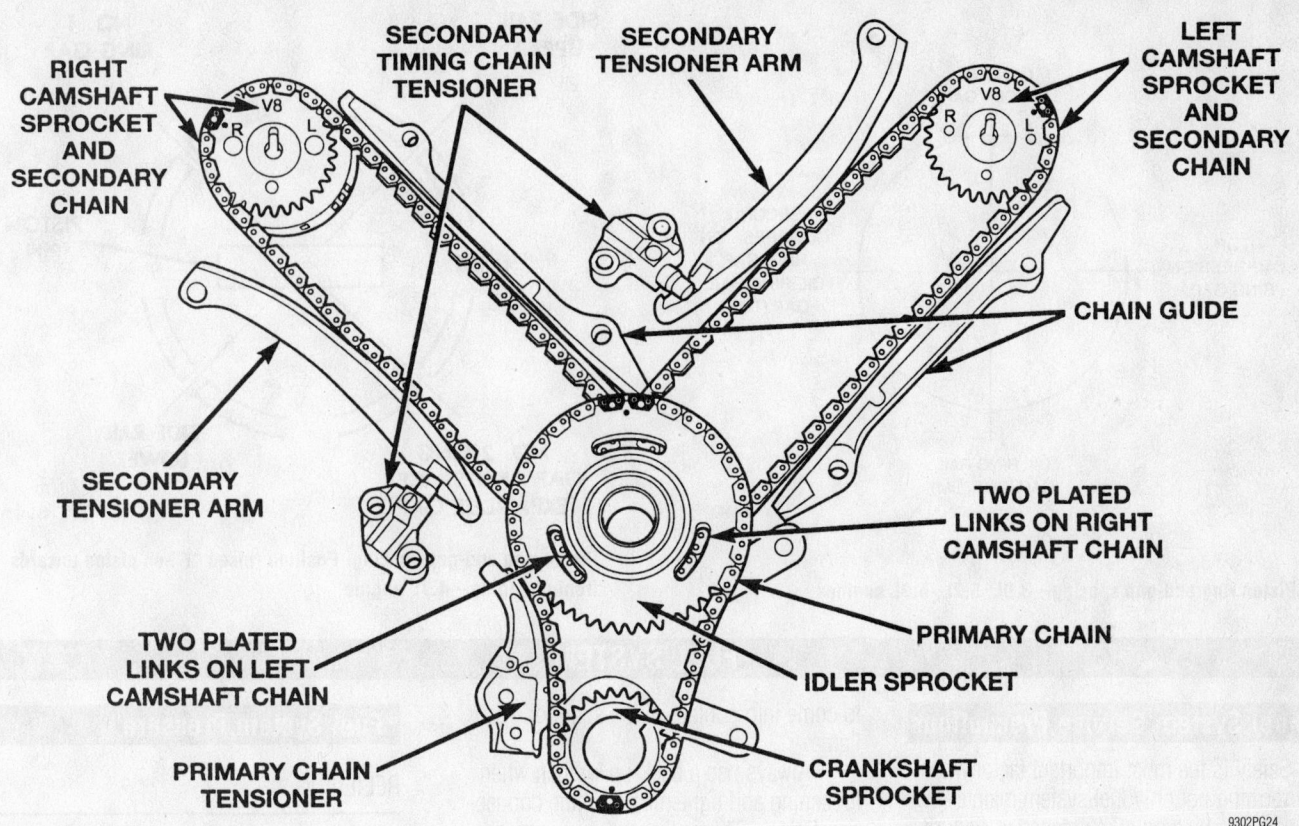

Timing chain system and alignment marks—4.7L engine

★ **INDICATES STUD LOCATIONS**

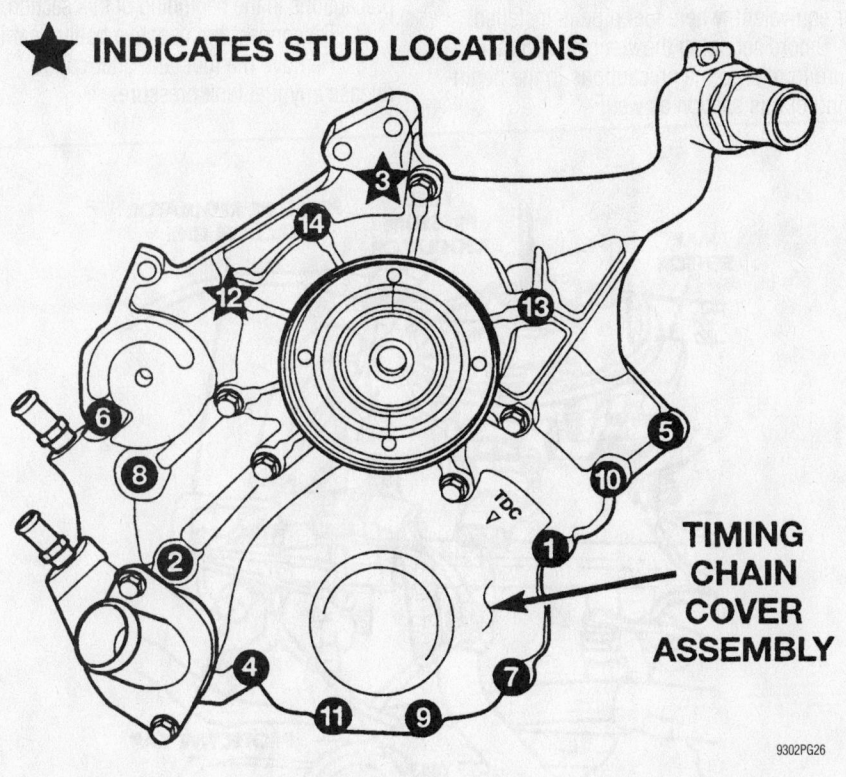

TIMING CHAIN COVER ASSEMBLY

Timing chain cover bolt torque sequence—4.7L engine

- Crankshaft damper. Tighten the bolt to 135 ft. lbs. (183 Nm).
- Power steering pump
- Water pump
- Cooling fan and shroud
- Accessory drive belt
- Negative battery cable
8. Fill the cooling system.
9. Start the engine and check for leaks.

Piston and Ring

POSITIONING

Piston to engine positioning—3.9L, 5.2L, 5.9L engines

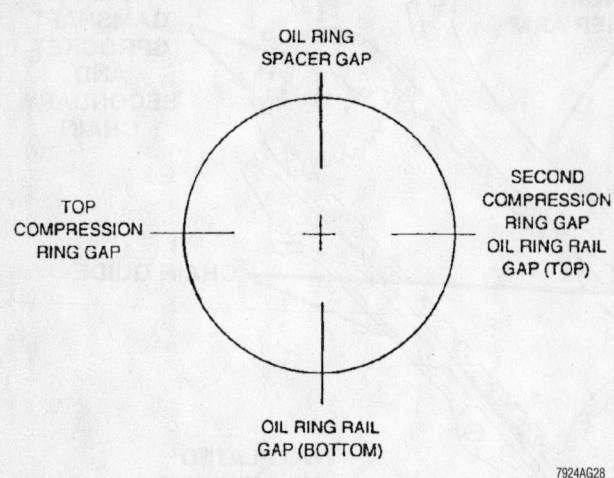

Piston ring end-gap spacing—3.9L, 5.2L, 5.9L engines

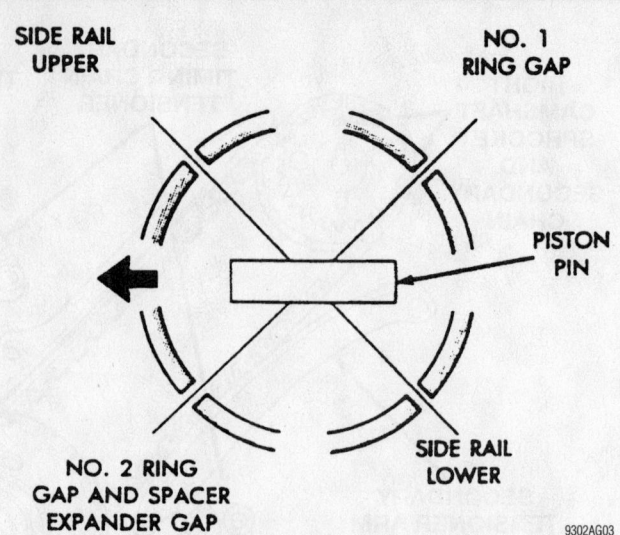

Piston ring end-gap spacing. Position raised "F" on piston towards front of engine—4.7L engine

FUEL SYSTEM

Fuel System Service Precautions

Safety is the most important factor when performing not only fuel system maintenance but any type of maintenance. Failure to conduct maintenance and repairs in a safe manner may result in serious personal injury or death. Maintenance and testing of the vehicle's fuel system components can be accomplished safely and effectively by adhering to the following rules and guidelines.

• To avoid the possibility of fire and personal injury, always disconnect the negative battery cable unless the repair or test procedure requires that battery voltage be applied.

• Always relieve the fuel system pressure prior to detaching any fuel system component (injector, fuel rail, pressure regulator, etc.), fitting or fuel line connection. Exercise extreme caution whenever relieving fuel system pressure to avoid exposing skin, face and eyes to fuel spray. Please be advised that fuel under pressure may penetrate the skin or any part of the body that it contacts.

• Always place a shop towel or cloth around the fitting or connection prior to loosening to absorb any excess fuel due to spillage. Ensure that all fuel spillage (should it occur) is quickly removed from engine surfaces. Ensure that all fuel soaked cloths or towels are deposited into a suitable waste container.

• Always keep a dry chemical (Class B) fire extinguisher near the work area.

• Do not allow fuel spray or fuel vapors to come into contact with a spark or open flame.

• Always use a back-up wrench when loosening and tightening fuel line connection fittings. This will prevent unnecessary stress and torsion to fuel line piping.

• Always replace worn fuel fitting O-rings with new. Do not substitute fuel hose or equivalent, where fuel pipe is installed.

Before servicing the vehicle, also make sure to refer to the precautions in the beginning of this section as well.

Fuel System Pressure

RELIEVING

1999

1. Before servicing the vehicle, refer to the precautions in the beginning of this section.
2. Disconnect the negative battery cable.
3. Remove the fuel tank filler cap to release any fuel tank pressure.

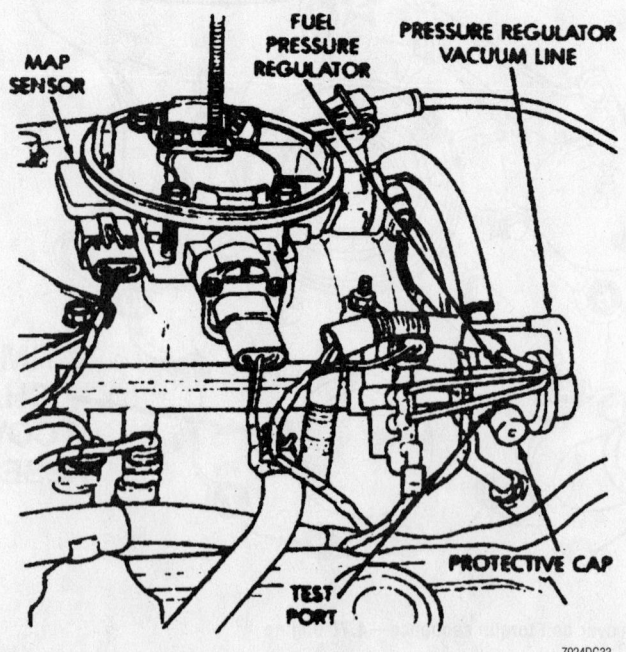

Fuel pressure test port—3.9L engine

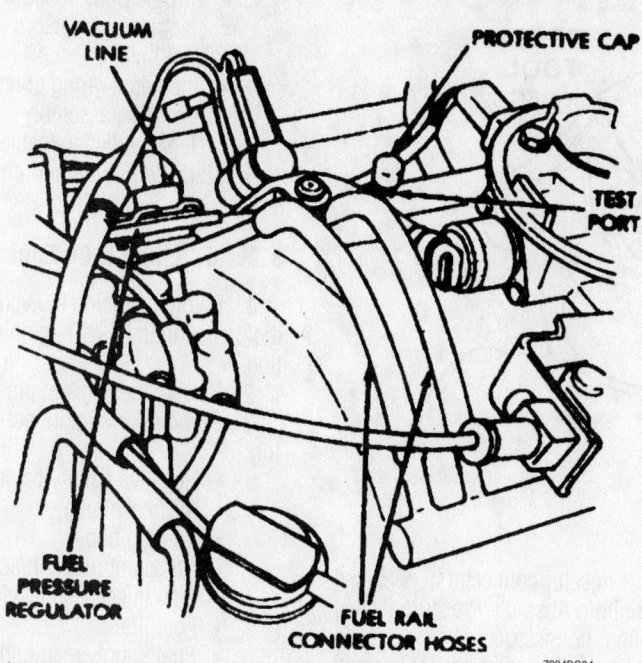

VACUUM LINE

PROTECTIVE CAP

TEST PORT

FUEL PRESSURE REGULATOR

FUEL RAIL CONNECTOR HOSES

7924DG34

Fuel pressure test port—5.2L engine, 5.9L engine is similar

4. Unscrew the plastic cap from the pressure test port on the fuel rail.

5. Obtain a fuel pressure gauge/hose from a fuel pressure gauge tool set No. 5069, or equivalent. Remove the gauge, then place the gauge end of the hose into a suitable gasoline container.

6. Place a shop towel under the test port.

7. Screw the other end of the hose onto the fuel pressure port to relieve the pressure.

8. When the pressure has been relieved, remove the hose and cap the port.

2000–02

1. Remove the fuel filler cap.
2. Remove the fuel pump relay from the power distribution center.
3. Run the engine until it stalls.
4. Turn the key to **OFF**.

Fuel Filter

REMOVAL & INSTALLATION

These engines are equipped with a fuel filter/pressure regulator. This unit is not a regularly maintained unit and is serviced only when a DTC indicates a fault.

1. Before servicing the vehicle, refer to the precautions in the beginning of this section.

2. Relieve the fuel system pressure.
3. Remove or disconnect the following:

   • Negative battery cable
   • Fuel tank

4. Pull the filter/regulator out of the rubber grommet. Cut the hose clamp and remove the fuel line.

To install:

5. Install the filter/regulator with a new clamp and push it into the rubber grommet.
6. Install or connect the following:

   • Fuel tank
   • Negative battery cable

7. Start the engine and check for leaks.

Fuel Pump

REMOVAL & INSTALLATION

1. Before servicing the vehicle, refer to the precautions in the beginning of this section.
2. Relieve the fuel system pressure.
3. Remove or disconnect the following:

   • Negative battery cable
   • Fuel pump module harness connector
   • Fuel line
   • Fuel tank
   • Fuel pump module locknut
   • Fuel pump module

To install:

4. Install or connect the following:

   • Fuel pump module
   • Fuel pump module locknut
   • Fuel tank
   • Fuel line
   • Fuel pump module harness connector
   • Negative battery cable

5. Start the engine and check for leaks.

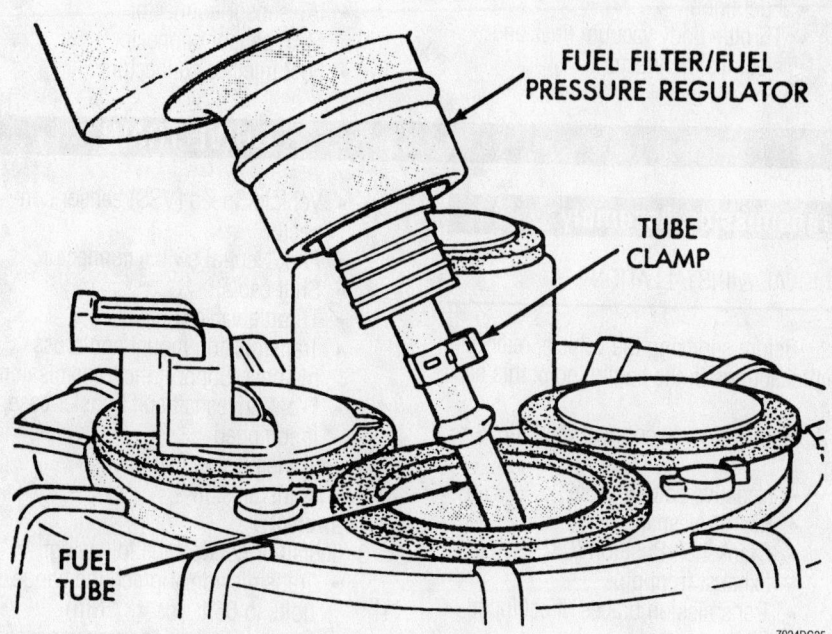

FUEL FILTER/FUEL PRESSURE REGULATOR

TUBE CLAMP

FUEL TUBE

7924DG35

Pull and twist the filter/regulator to remove it from the top of the fuel pump module

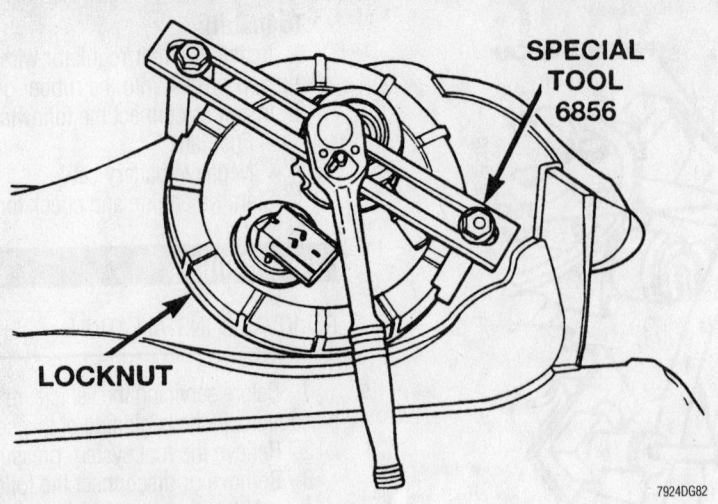

SPECIAL TOOL 6856

LOCKNUT

7924DG82

Fuel pump module locknut removal

Fuel Injector

REMOVAL & INSTALLATION

4.7L Engine

1. Before servicing the vehicle, refer to the precautions in the beginning of this section.
2. Relieve fuel system pressure.
3. Remove or disconnect the following:

- Negative battery cable
- Air intake assembly
- Alternator wiring connectors
- Fuel line
- Throttle body vacuum lines and electrical connectors
- Fuel injector connectors
- Manifold Absolute Pressure (MAP) sensor connector
- Intake Air Temperature (IAT) sensor connector
- Ignition coils
- Fuel supply manifold with injectors attached
- Fuel injectors

To install:

4. Install or connect the following:

- Fuel injectors, using new O-rings
- Fuel supply manifold with injectors attached. Tighten the bolts to 20 ft. lbs. (27 Nm).
- Ignition coils
- IAT sensor connector
- MAP sensor connector
- Fuel injector connectors

- Throttle body vacuum lines and electrical connectors
- Fuel line
- Alternator wiring connectors
- Air intake assembly
- Negative battery cable

5. Start the engine and check for leaks.

3.9L, 5.2L and 5.9L Engines

1. Before servicing the vehicle, refer to the precautions in the beginning of this section.
2. Relieve fuel system pressure.
3. Remove or disconnect the following:

- Negative battery cable
- Air intake tube
- Throttle body
- A/C compressor bracket
- Fuel injector connectors
- Fuel line
- Fuel supply manifold with injectors
- Fuel injectors

To install:

4. Install or connect the following:

- Fuel injectors with new O-ring seals
- Fuel supply manifold with injectors. Tighten the bolts to 17 ft. lbs. (23 Nm).
- Fuel line
- Fuel injector connectors
- A/C compressor bracket
- Throttle body
- Air intake tube
- Negative battery cable

5. Start the engine and check for leaks.

DRIVE TRAIN

Transmission Assembly

REMOVAL & INSTALLATION

1. Before servicing the vehicle, refer to the precautions in the beginning of this section.
2. Remove or disconnect the following:

- Negative battery cable
- Rear driveshaft
- Crankshaft Position (CKP) sensor
- Exhaust front pipe
- Transmission braces, if equipped
- Starter motor
- Transmission oil cooler lines
- Torque converter access cover
- Torque converter
- Transmission oil dipstick tube

- Vehicle Speed (VSS) sensor connector
- Park/Neutral switch connector
- Shift cable
- Throttle valve cable
- Transmission mount and crossmember. Support the transmission.
- Front driveshaft and transfer case, if equipped
- Transmission flange bolts
- Transmission

To install:

3. Install or connect the following:

- Transmission. Tighten the flange bolts to 65 ft. lbs. (87 Nm).
- Front driveshaft and transfer case, if equipped
- Transmission mount and crossmember

- Throttle valve cable
- Shift cable
- Park/Neutral switch connector
- VSS sensor connector
- Transmission oil dipstick tube
- Torque converter. Tighten the bolts to 23 ft. lbs. (31 Nm) for 10.75 inch converters and to 35 ft. lbs. (47 Nm) for 12.2 inch converters.
- Torque converter access cover
- Transmission oil cooler lines
- Starter motor
- Transmission braces, if equipped. Tighten the bolts to 30 ft. lbs. (41 Nm).
- Exhaust front pipe
- CKP sensor
- Rear driveshaft
- Negative battery cable

Transfer Case Assembly

REMOVAL & INSTALLATION

1. Before servicing the vehicle, refer to the precautions in the beginning of this section.
2. Shift the transfer case into **N**.
3. Remove or disconnect the following:
   - Front and rear driveshafts
   - Transmission mount and cross-member. Support the transmission.
   - Vehicle Speed (VSS) sensor connector
   - Shift linkage
   - Vent hose
   - Vacuum hose
   - Indicator switch connector
   - Transfer case attaching nuts
   - Transfer case

To install:

4. Install or connect the following:
   - Transfer case. Tighten the nuts to 26 ft. lbs. (35 Nm).
   - Indicator switch connector
   - Vacuum hose
   - Vent hose
   - Shift linkage
   - VSS sensor connector
   - Transmission mount and cross-member
   - Front and rear driveshafts

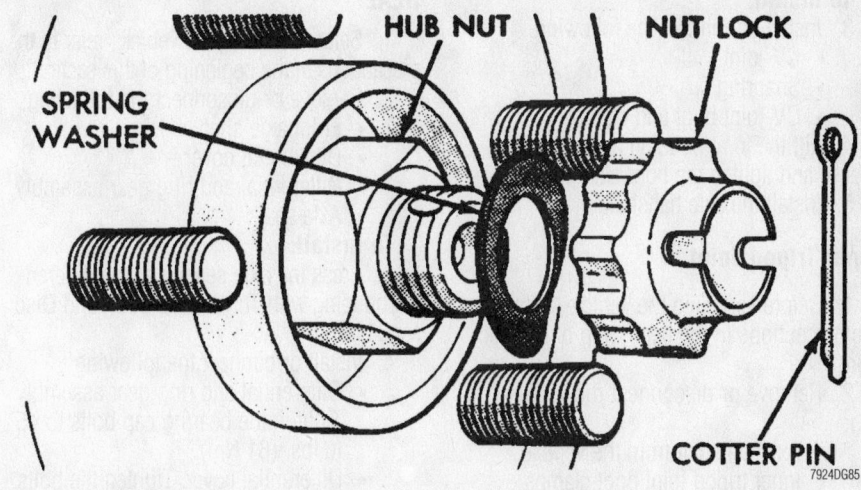

To separate the halfshaft from the hub, remove the cotter pin, nut lock and spring washer from the axle shaft

Halfshaft

REMOVAL & INSTALLATION

1. Before servicing the vehicle, refer to the precautions in the beginning of this section.
2. Remove or disconnect the following:
   - Skid plate, if equipped
   - Front wheel
   - Split pin

   - Nut lock
   - Spring washer
   - Hub nut
   - Brake caliper and rotor
   - Wheel speed sensor, if equipped
   - Wheel bearing and hub assembly
3. Pry the inner tripod joint out of the differential and remove the axle halfshaft.

To install:

4. Install the axle halfshaft so that the snapring is felt to seat in the joint housing groove.
5. Install or connect the following:
   - Wheel bearing and hub assembly
   - Wheel speed sensor, if equipped
   - Brake caliper and rotor
   - Hub nut. Tighten the nut to 180 ft. lbs. (244 Nm).
   - Spring washer
   - Nut lock
   - Split pin
   - Front wheel
   - Skid plate, if equipped

CV-Joints

OVERHAUL

Outer CV-Joint

1. Before servicing the vehicle, refer to the precautions in the beginning of this section.
2. Remove or disconnect the following:
   - Axle halfshaft from the vehicle
   - CV-joint boot and clamps
   - Snapring
   - CV-joint

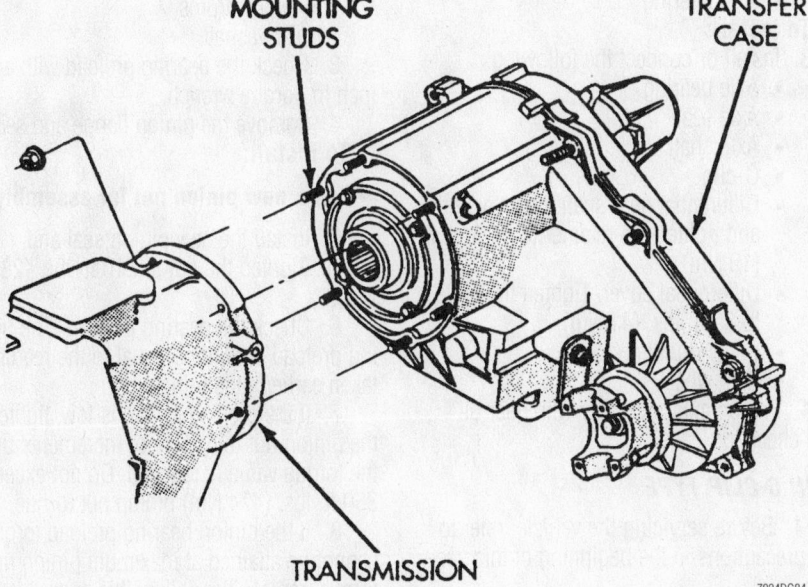

Typical transfer case mounting

To install:

3. Install or connect the following:
- CV-joint
- Snapring
- CV-joint boot and clamps

4. Fill the joint housing and boot with grease and tighten the boot clamps.

5. Install the axle halfshaft.

Inner Tripod Joint

1. Before servicing the vehicle, refer to the precautions in the beginning of this section.

2. Remove or disconnect the following:
- Axle halfshaft from the vehicle
- Inner tripod joint boot clamps
- Tripod joint housing
- Snapring
- Circlip
- Tripod joint

To install:

➡ **Use new snaprings, clips, and boot clamps for assembly.**

3. Install or connect the following:
- Tripod joint
- Circlip
- Snapring
- Tripod joint housing

4. Fill the tripod joint housing and boot with grease and tighten the boot clamps.

5. Install the axle halfshaft.

Axle Shaft, Bearing and Seal

REMOVAL & INSTALLATION

Front

AXLE SHAFT

1. Before servicing the vehicle, refer to the precautions in the beginning of this section.

2. Remove or disconnect the following:
- Front wheel
- Brake caliper and rotor
- Wheel speed sensor, if equipped
- Axle hub nut
- Wheel bearing and hub assembly
- Axle shaft

To install:

3. Install or connect the following:
- Axle shaft
- Wheel bearing and hub assembly
- Axle hub nut. Tighten the nut to 175 ft. lbs. (237 Nm).
- Wheel speed sensor, if equipped
- Brake caliper and rotor
- Front wheel

SEAL

1. Before servicing the vehicle, refer to the precautions in the beginning of this section.

2. Remove or disconnect the following:
- Front axle shafts
- Differential cover
- Differential and ring gear assembly
- Axle seals

To install:

3. Press the axle seals into the differential housing with Turnbuckle 6797 and Disc set 8110.

4. Install or connect the following:
- Differential and ring gear assembly. Tighten the bearing cap bolts to 45 ft. lbs. (61 Nm).
- Differential cover. Tighten the bolts to 30 ft. lbs. (41 Nm).
- Front axle shafts

5. Fill the axle assembly with gear oil and check for leaks.

Rear

C-CLIP TYPE

1. Before servicing the vehicle, refer to the precautions in the beginning of this section.

2. Remove or disconnect the following:
- Rear wheel
- Brake drum
- Differential cover
- Differential gear shaft retainer
- Differential gear shaft
- C-clip
- Axle shaft
- Axle seal
- Axle bearing

To install:

3. Install or connect the following:
- Axle bearing
- Axle seal
- Axle shaft
- C-clip
- Differential gear shaft. Use Loctite® and tighten the retainer to 14 ft. lbs. (19 Nm).
- Differential cover. Tighten the bolts to 30 ft. lbs. (41 Nm).
- Brake drum
- Rear wheel

4. Fill the axle assembly with gear oil and check for leaks.

NON C-CLIP TYPE

1. Before servicing the vehicle, refer to the precautions in the beginning of this section.

2. Remove or disconnect the following:
- Rear wheel
- Brake caliper and rotor, if equipped
- Brake drum, if equipped

- Axle retainer nuts
- Axle shaft, seal and bearing assembly

3. Split the bearing retainer with a chisel and remove the retainer ring.

4. Press the bearing off the axle shaft.

5. Remove the axle seal and retaining plate.

To install:

6. Install the retaining plate and axle seal onto the axle shaft.

7. Pack the wheel bearing with axle grease and press the bearing on to the axle shaft.

8. Press the retaining ring onto the axle shaft.

9. Install or connect the following:
- Axle shaft, seal and bearing assembly. Tighten the nuts to 45 ft. lbs. (61 Nm).
- Brake caliper and rotor, if equipped
- Brake drum, if equipped
- Rear wheel

10. Fill the axle assembly with gear oil and check for leaks.

Pinion Seal

REMOVAL & INSTALLATION

C-Clip Type

1. Before servicing the vehicle, refer to the precautions in the beginning of this section.

2. Remove or disconnect the following:
- Wheels
- Brake drums
- Driveshaft

3. Check the bearing preload with an inch lb. torque wrench.

4. Remove the pinion flange and seal.

To install:

➡ **Use a new pinion nut for assembly.**

5. Install the new pinion seal and flange. Tighten the nut to 210 ft. lbs. (285 Nm).

6. Check the bearing preload. The bearing preload should be equal to the reading taken earlier, plus 5 inch lbs.

7. If the preload torque is low, tighten the pinion nut in 5 inch lb. increments until the torque value is reached. Do not exceed 350 ft. lbs. (474 Nm) pinion nut torque.

8. If the pinion bearing preload torque cannot be attained at maximum pinion nut torque, replace the collapsible spacer.

9. Install or connect the following:
- Driveshaft
- Brake drums
- Wheels

10. Fill the axle assembly with gear oil and check for leaks.

Non C-Clip Type

FRONT

1. Before servicing the vehicle, refer to the precautions in the beginning of this section.

2. Remove or disconnect the following:
- Wheels
- Brake rotors
- Driveshaft

3. Check the bearing preload with an inch lb. torque wrench.

4. Remove the pinion flange and seal.

To install:

➡ **Use a new pinion nut for assembly.**

5. Install the new pinion seal and flange. Tighten the nut to 160 ft. lbs. (217 Nm).

6. Check the bearing preload. The bearing preload should be equal to the reading taken earlier, plus 5 inch lbs.

7. If the preload torque is low, tighten the pinion nut in 5 inch lb. increments until the torque value is reached. Do not exceed 260 ft. lbs. (353 Nm) pinion nut torque.

8. If the pinion bearing preload torque can not be attained at maximum pinion nut torque, replace the collapsible spacer.

9. Install or connect the following:
- Driveshaft
- Brake rotors
- Wheels

10. Fill the axle assembly with gear oil and check for leaks.

REAR

1. Before servicing the vehicle, refer to the precautions in the beginning of this section.

2. Remove or disconnect the following:
- Wheels
- Brake rotors or drums
- Driveshaft

3. Check the bearing preload with an inch lb. torque wrench.

4. Remove the pinion flange and seal.

To install:

➡ **Use a new pinion nut for assembly.**

5. Install the new pinion seal and flange. Tighten the nut to 160 ft. lbs. (217 Nm).

6. Check the bearing preload. The bearing preload should be equal to the reading taken earlier, plus 5 inch lbs.

7. If the preload torque is low, tighten the pinion nut in 5 inch lb. increments until the torque value is reached. Do not exceed 260 ft. lbs. (353 Nm) pinion nut torque.

8. If the pinion bearing preload torque can not be attained at maximum pinion nut torque, remove one or more pinion preload shims.

9. Install or connect the following:
- Driveshaft
- Brake rotors or drums
- Wheels

10. Fill the axle assembly with gear oil and check for leaks.

STEERING AND SUSPENSION

Air Bag

✳✳ CAUTION

Some vehicles are equipped with an air bag system. The system must be disarmed before performing service on, or around, system components, the steering column, instrument panel components, wiring and sensors. Failure to follow the safety precautions and the disarming procedure could result in accidental air bag deployment, possible injury and unnecessary system repairs.

PRECAUTIONS

Several precautions must be observed when handling the inflator module to avoid accidental deployment and possible personal injury.

- Never carry the inflator module by the wires or connector on the underside of the module.
- When carrying a live inflator module, hold securely with both hands, and ensure that the bag and trim cover are pointed away.
- Place the inflator module on a bench or other surface with the bag and trim cover facing up.

- With the inflator module on the bench, never place anything on or close to the module which may be thrown in the event of an accidental deployment.

Before servicing the vehicle, also make sure to refer to the precautions in the beginning of this section as well.

DISARMING

1. Disconnect and isolate the negative battery cable. Wait 2 minutes for the system capacitor to discharge before performing any service.

2. When repairs are completed, connect the negative battery cable.

Recirculating Ball Power Steering Gear

REMOVAL & INSTALLATION

1. Before servicing the vehicle, refer to the precautions in the beginning of this section.

2. Remove or disconnect the following:
- Negative battery cable
- Power steering pressure and return lines
- Intermediate shaft
- Pitman arm

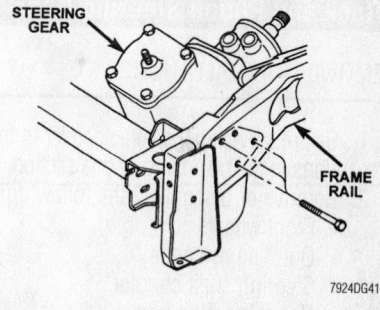

Typical recirculating ball power steering gear mounting

- Steering gear

To install:

3. Install or connect the following:
- Steering gear. Tighten the bolts to 100 ft. lbs. (136 Nm).
- Pitman arm. Tighten the nut to 175 ft. lbs. (237 Nm).
- Intermediate shaft. Tighten the pinch bolt to 36 ft. lbs. (49 Nm).
- Power steering pressure and return lines
- Negative battery cable

4. Fill the power steering fluid reservoir.

5. Start the engine and check for leaks.

Timing belt service is covered in Section 3 of this manual

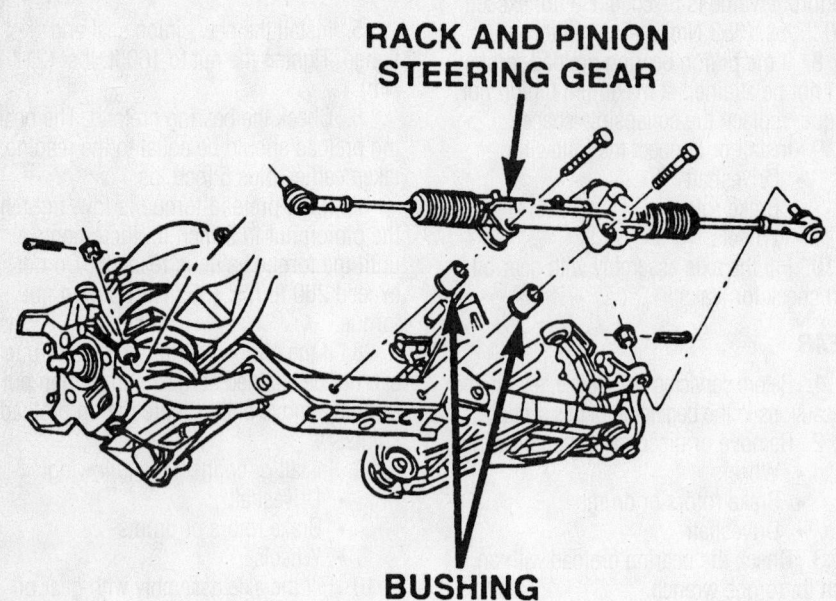

RACK AND PINION STEERING GEAR

BUSHING

7924DG42

Rack and pinion steering gear mounting used on the 2WD models

Rack and Pinion Steering Gear

REMOVAL & INSTALLATION

1. Before servicing the vehicle, refer to the precautions in the beginning of this section.
2. Remove or disconnect the following:
   - Front wheels
   - Outer tie rod ends
   - Steering shaft coupler
   - Power steering hoses
   - Steering gear

To install:

3. Install or connect the following:
   - Steering gear. Tighten the bolts to 190 ft. lbs. (258 Nm).
   - Power steering hoses. Tighten the fittings to 25 ft. lbs. (35 Nm).
   - Steering shaft coupler. Tighten the bolt to 36 ft. lbs. (49 Nm).
   - Outer tie rod ends. Tighten the nuts to 65 ft. lbs. (88 Nm).
   - Front wheels

Shock Absorber

REMOVAL & INSTALLATION

Front

1. Before servicing the vehicle, refer to the precautions in the beginning of this section.

2. Remove or disconnect the following:
3. Install or connect the following:
   - Front wheel
   - Upper mount nut
   - Lower mount bolt
   - Shock absorber

To install:

4. Install or connect the following:
   - Shock absorber. Tighten the lower bolt to 100 ft. lbs. (136 Nm) and the upper nut to 30 ft. lbs. (41 Nm).
   - Front wheel

Rear

1. Before servicing the vehicle, refer to the precautions in the beginning of this section.

2. Support the axle.
3. Remove or disconnect the following:
   - Upper bolt
   - Lower bolt
   - Shock absorber

To install:

4. Install the bolts through the brackets and shock and tighten them as follows: Tighten the lower bolt and nut to 60 ft. lbs. (81 Nm) and the upper bracket nuts to 20 ft. lbs. (27 Nm).

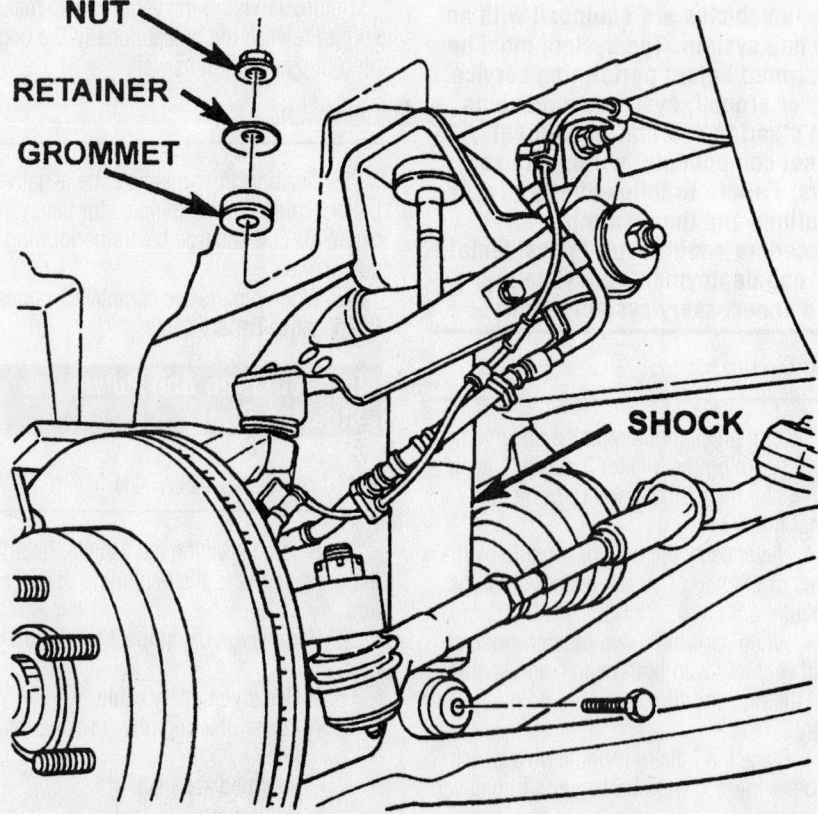

NUT

RETAINER

GROMMET

SHOCK

7924DG86

Front shock absorber mounting—4WD models

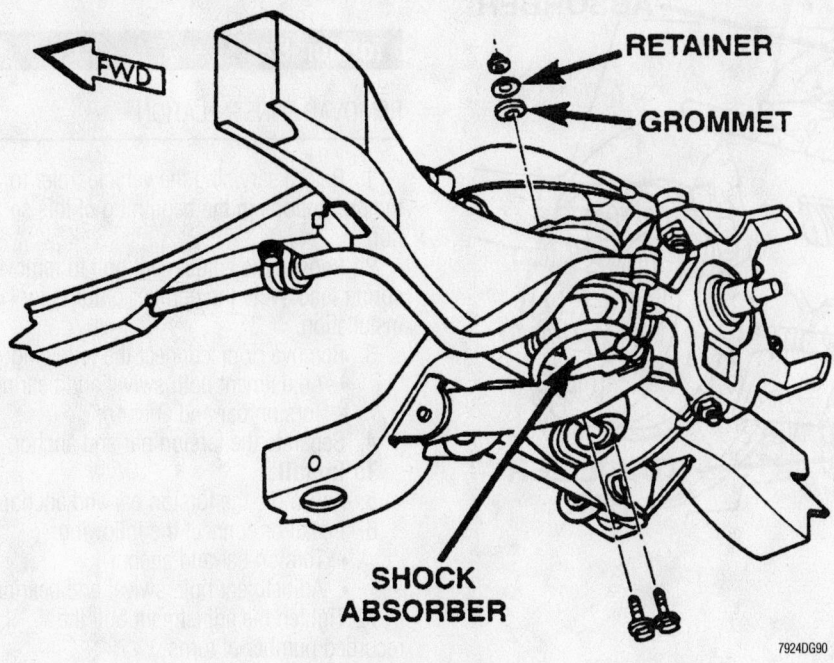

Front shock absorber mounting—2WD models

REMOVAL & INSTALLATION

1. Before servicing the vehicle, refer to the precautions in the beginning of this section.

2. Support the lower control arm on a floor jack.

3. Remove or disconnect the following:

- Front wheel
- Shock absorber
- Stabilizer bar link
- Lower ball joint.

4. Lower the jack and remove the coil spring.

To install:

5. Install the coil spring and raise the control arm into position.

6. Install or connect the following:

- Lower ball joint. Tighten the nut to 135 ft. lbs. (183 Nm).
- Stabilizer bar link
- Shock absorber
- Front wheel

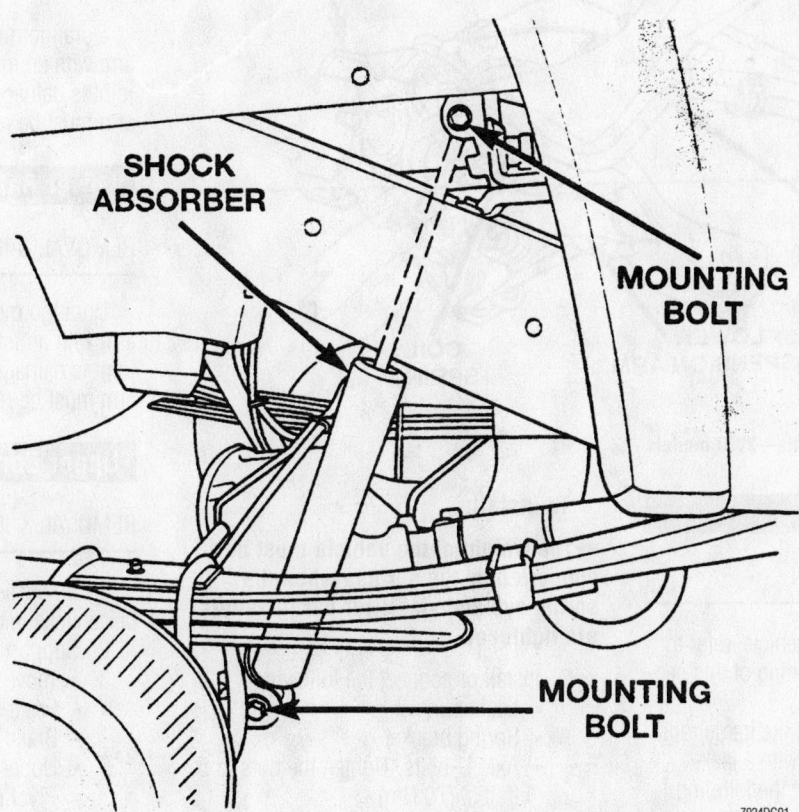

Rear shock absorber mounting

Heater Core replacement is covered in Section 2 of this manual

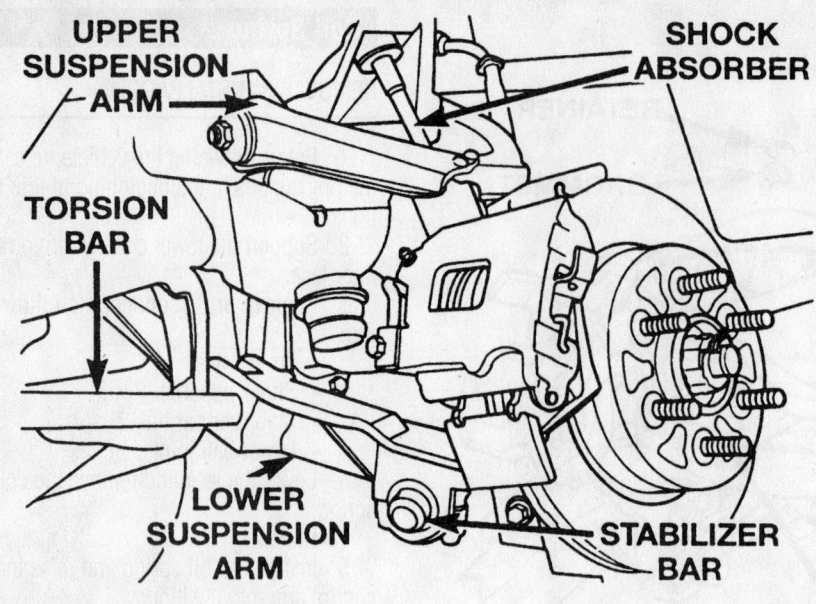

Front suspension components—4WD models

7924DG45

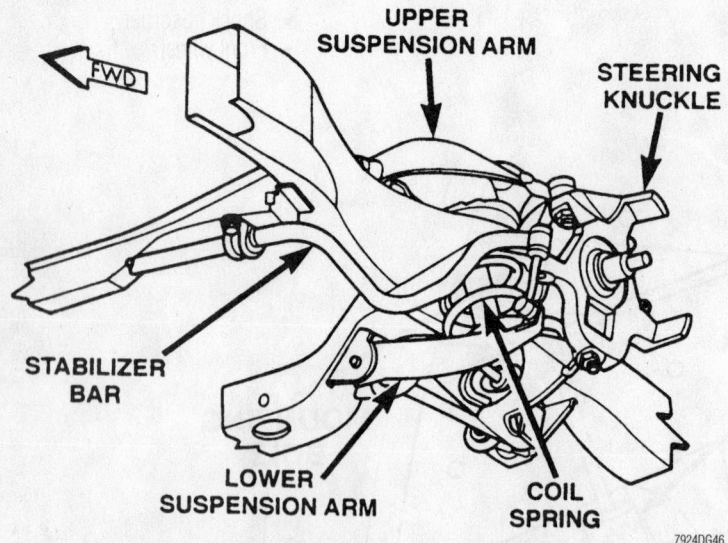

Front suspension components—2WD models

7924DG46

Leaf Spring

REMOVAL & INSTALLATION

1. Before servicing the vehicle, refer to the precautions in the beginning of this section.
2. Support the vehicle at the frame rails.
3. Support the rear axle with a jack.
4. Remove or disconnect the folloling:
   - Rear wheel
   - Stabilizer bar link
   - Axle U-bolts
   - Spring bracket
   - Leaf spring

To install:

→**The weight of the vehicle must be supported by the springs when the spring eye and stabilizer bar fasteners are tightened.**

5. Install or connect the following:
   - Leaf spring
   - Spring bracket
   - Axle U-bolts. Tighten the nuts to 52 ft. lbs. (70 Nm).
   - Stabilizer bar link
   - Rear wheel
6. Tighten the front spring eye bolt and nut to 115 ft. lbs. (156 Nm). Tighten the rear spring eye bolt and nut to 80 ft. lbs.

(108 Nm). Tighten the stabilizer bar nuts 55 ft. lbs. (74 Nm).

Torsion Bar

REMOVAL & INSTALLATION

1. Before servicing the vehicle, refer to the precautions in the beginning of this section.
2. Loosen the adjustment bolt to remove spring load. Note the number of turns for installation.
3. Remove or disconnect the following:
   - Adjustment bolt, swivel and bearing
   - Torsion bar and anchor
4. Separate the torsion bar and anchor.
To install:
5. Assemble the torsion bar and anchor.
6. Install or connect the following:
   - Torsion bar and anchor
   - Adjustment bolt, swivel and bearing
7. Tighten the adjustment bolt the recorded number of turns.

Upper Ball Joint

REMOVAL & INSTALLATION

Durango models utilize an upper control arm with an integral ball joint. If the ball joint is damaged or worn, the upper control arm must be replaced.

Lower Ball Joint

REMOVAL & INSTALLATION

Durango models utilize a lower control arm with an integral ball joint. If the ball joint is damaged or worn, the upper control arm must be replaced.

Upper Control Arm

REMOVAL & INSTALLATION

1. Before servicing the vehicle, refer to the precautions in the beginning of this section.
2. Support the lower control arm.
3. Remove or disconnect the following:
   - Front wheel
   - Brake hose brackets
   - Upper ball joint
   - Pivot mounting nuts
   - Upper control arm
To install:
4. Install or connect the following:
   - Upper control arm. Tighten the pivot nuts to 155 ft. lbs. (210 Nm).

- Upper ball joint. Tighten the nut to 60 ft. lbs. (81 Nm).
- Brake hose brackets
- Front wheel

5. Check the wheel alignment and adjust as necessary.

CONTROL ARM BUSHING REPLACEMENT

The control arm bushings are serviced with the control arm as an assembly.

Lower Control Arm

REMOVAL & INSTALLATION

2 Wheel Drive

1. Before servicing the vehicle, refer to the precautions in the beginning of this section.
2. Remove or disconnect the following:
   - Front wheel
   - Shock absorber
   - Brake caliper and rotor
   - Stabilizer bar link
   - Coil spring
   - Inner mounting bolts
   - Lower control arm

To install:
3. Install or connect the following:
   - Lower control arm. Tighten the front bolt to 130 ft. lbs. (175 Nm) and the rear bolt to 80 ft. lbs. (108 Nm).
   - Coil spring. Tighten the lower ball joint nut to 94 ft. lbs. (127 Nm).
   - Stabilizer bar link
   - Brake caliper and rotor
   - Shock absorber
   - Front wheel

4 Wheel Drive

1. Before servicing the vehicle, refer to the precautions in the beginning of this section.
2. Remove or disconnect the following:
   - Front wheel
   - Front driveshaft
   - Torsion bar
   - Shock absorber
   - Stabilizer bar
   - Lower ball joint
   - Pivot bolts
   - Lower control arm

To install:
3. Install or connect the following:
   - Lower control arm. Tighten the front pivot bolt to 80 ft. lbs. (108

Nm) and the rear bolt to 140 ft. lbs. (190 Nm).
   - Lower ball joint. Tighten the nut to 135 ft. lbs. (183 Nm).
   - Stabilizer bar
   - Shock absorber
   - Torsion bar
   - Front driveshaft
   - Front wheel

CONTROL ARM BUSHING REPLACEMENT

The control arm bushings are serviced with the control arm as an assembly.

Wheel Bearings

ADJUSTMENT

Durango models utilize a hub/bearing assembly which is not adjustable.

REMOVAL & INSTALLATION

2 Wheel Drive

1. Before servicing the vehicle, refer to the precautions in the beginning of this section.
2. Remove or disconnect the following:

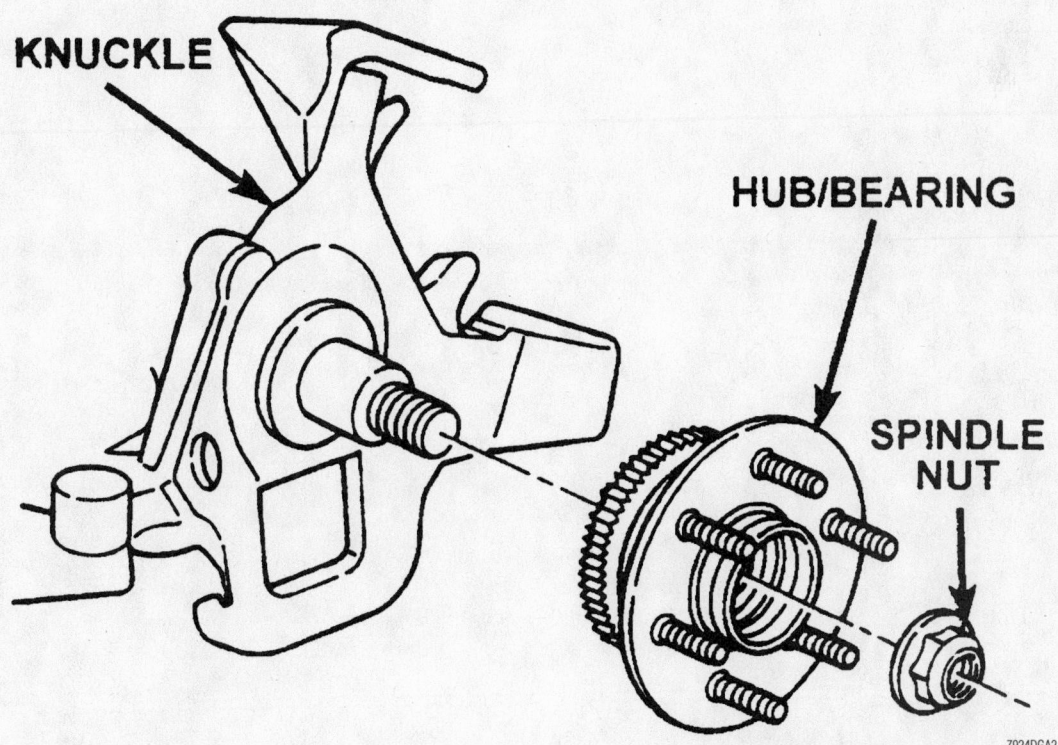

KNUCKLE

HUB/BEARING

SPINDLE NUT

7924DGA2

Hub/bearing assembly—2 wheel drive

Brake service is covered in Section 4 of this manual

- Front wheel
- Brake caliper and rotor
- Spindle nut
- Hub and bearing assembly

To install:

3. Install or connect the following:
- Hub and bearing assembly
- Spindle nut. Tighten the nut to 185 ft. lbs. (251 Nm).
- Brake caliper and rotor
- Front wheel

4 Wheel Drive

1. Before servicing the vehicle, refer to the precautions in the beginning of this section.

2. Remove or disconnect the following:

- Front wheel
- Brake caliper and rotor
- Hub retainer nut
- Hub and bearing assembly

To install:

3. Install or connect the following:
- Hub and bearing assembly. Tighten the bolts to 123 ft. lbs. (166 Nm).
- Hub retainer nut. Tighten the nut to 173 ft. lbs. (235 Nm).
- Brake caliper and rotor
- Front wheel

FORD/MAZDA

Escape • Tribute

9

PRECAUTIONS

Before servicing any vehicle, please be sure to read all of the following precautions, which deal with personal safety, prevention of component damage, and important points to take into consideration when servicing a motor vehicle:

• Never open, service or drain the radiator or cooling system when the engine is hot; serious burns can occur from the steam and hot coolant.

• Observe all applicable safety precautions when working around fuel. Whenever servicing the fuel system, always work in a well-ventilated area. Do not allow fuel spray or vapors to come in contact with a spark, open flame, or excessive heat (a hot drop light, for example). Keep a dry chemical fire extinguisher near the work area. Always keep fuel in a container specifically designed for fuel storage; also, always properly seal fuel containers to avoid the possibility of fire or explosion. Refer to the additional fuel system precautions later in this section.

• Fuel injection systems often remain pressurized, even after the engine has been turned **OFF**. The fuel system pressure must be relieved before disconnecting any fuel lines. Failure to do so may result in fire and/or personal injury.

• Brake fluid often contains polyglycol ethers and polyglycols. Avoid contact with the eyes and wash your hands thoroughly after handling brake fluid. If you do get brake fluid in your eyes, flush your eyes with clean, running water for 15 minutes. If eye irritation persists, or if you have taken brake fluid internally, IMMEDIATELY seek medical assistance.

• The EPA warns that prolonged contact with used engine oil may cause a number of skin disorders, including cancer! You should make every effort to minimize your exposure to used engine oil. Protective gloves should be worn when changing oil. Wash your hands and any other exposed skin areas as soon as possible after exposure to used engine oil. Soap and water, or waterless hand cleaner should be used.

• All new vehicles are now equipped with an air bag system, often referred to as a Supplemental Restraint System (SRS) or Supplemental Inflatable Restraint (SIR) system. The system must be disabled before performing service on or around system components, steering column, instrument panel components, wiring and sensors. Failure to follow safety and disabling procedures could result in accidental air bag deployment, possible personal injury and unnecessary system repairs.

• Always wear safety goggles when working with, or around, the air bag system. When carrying a non-deployed air bag, be sure the bag and trim cover are pointed away from your body. When placing a non-deployed air bag on a work surface, always face the bag and trim cover upward, away from the surface. This will reduce the motion of the module if it is accidentally deployed. Refer to the additional air bag system precautions later in this section.

• Clean, high quality brake fluid from a sealed container is essential to the safe and proper operation of the brake system. You should always buy the correct type of brake fluid for your vehicle. If the brake fluid becomes contaminated, completely flush the system with new fluid. Never reuse any brake fluid. Any brake fluid that is removed from the system should be discarded. Also, do not allow any brake fluid to come in contact with a painted surface; it will damage the paint.

• Never operate the engine without the proper amount and type of engine oil; doing so WILL result in severe engine damage.

• Timing belt maintenance is extremely important! Many models utilize an interference-type, non-freewheeling engine. If the timing belt breaks, the valves in the cylinder head may strike the pistons, causing potentially serious (also time-consuming and expensive) engine damage. Refer to the maintenance interval charts in the front of this manual for the recommended replacement interval for the timing belt, and to the timing belt section for belt replacement and inspection.

• Disconnecting the negative battery cable on some vehicles may interfere with the functions of the on-board computer system(s) and may require the computer to undergo a relearning process once the negative battery cable is reconnected.

• When servicing drum brakes, only disassemble and assemble one side at a time, leaving the remaining side intact for reference.

• Only an MVAC-trained, EPA-certified automotive technician should service the air conditioning system or its components.

ENGINE REPAIR

Distributor

The Escape uses a Direct Ignition System (DIS). No distributor is used.

Alternator

REMOVAL

2.0L Engine

Remove or disconnect the following:
• Negative battery cable
• Drive belt
• Alternator electrical connectors and loosen the upper alternator bolt

while moving the alternator to the rear of the engine
• Alternator. Torque the mounting and adjusting bolts to 33 ft. lbs. (45Nm).

3.0L Engine

Remove or disconnect the following:
• Negative battery cable
• Right side intermediate axle shaft
• Right side splash shield and retainers
• Drive belt
• Alternator electrical connectors
• Alternator. Torque the mounting and adjusting bolts to 35 ft. lbs. (48Nm).

INSTALLATION

2.0L Engine

Install or connect the following:
• Alternator with the upper bolt in the alternator before installation. Torque the bolts to 18 ft. lbs. (25 Nm).
• Alternator electrical connectors
• Drive belt
• Negative battery cable

3.0L Engine

Install or connect the following:
• Alternator. Torque the bolts to 35 ft. lbs. (48 Nm).

- Alternator electrical connectors
- Drive belt
- Negative battery cable

Ignition Timing

ADJUSTMENT

Ignition timing is controlled by the Powertrain Control Module (PCM). No adjustment is necessary or possible.

Engine Assembly

REMOVAL & INSTALLATION

2.0L Engine

MANUAL TRANSMISSION

1. Before servicing the vehicle, refer to the precautions in the beginning of this section.
2. Properly recover the air conditioning system refrigerant.
3. Properly relieve the fuel system pressure.
4. Drain the cooling system.
5. Drain the engine oil.
6. Remove or disconnect the following:
   - Hood
   - Battery and battery tray
   - Air cleaner housing
   - Fuel lines
   - Throttle cable and speed control cable, if equipped
   - Exhaust Gas Recirculation (EGR) vacuum valve regulator
   - EGR electrical connectors and vacuum hoses
   - Brake booster vacuum hose
   - Powertrain Control Module (PCM) wire harness and ground

- Wire harness connector
- Power distribution board electrical connectors
- Evaporative emissions (EVAP) canister vacuum lines
- Upper radiator hose
- Power steering line bracket
- Upper power steering pump bolts
- Coolant hose
- Heater hoses
- Speed control unit, if equipped
- Catalytic converter
- A/C compressor
- Both halfshafts
- Shifter linkages
- Block heater electrical connector, if equipped
- Front transmission through bolt
- Engine-to-transmission bolts
- Lower radiator hose
- Power steering pump
- Clutch slave cylinder line from the bracket and move it aside
- Rear transmission mount
- Left side transmission mount
- Lower ground cable
- Engine mount upper bracket
- Engine and transmission as an assembly by using a proper lifting device
- Alternator electrical connectors
- Knock Sensor (KS) electrical connector
- Oil pressure sender electrical connector
- Starter electrical connector
- Vehicle Speed Sensor (VSS) electrical connector
- Park Neutral Position (PNP) electrical connector
- Fuel charging wire harness electrical connector
- PCM wire harness from the bracket

- PCM ground wire
- Back up lamp switch electrical connector
- Wire harness
- Differential Pressure Feedback (DPFEE) EGR sensor
7. Separate the engine from the transmission.
8. Lock the flywheel to the engine.
9. Clutch pressure plate and disc.
10. Flywheel and rear cover plates.

To install:

11. Install or connect the following:
    - Flywheel. Torque the bolts to 83 ft. lbs. (112 Nm).
    - Clutch disc to the flywheel
    - Pressure plate to the flywheel. Torque the bolts in the proper sequence to 18 ft. lbs. (25 Nm).
    - Transmission to the engine. Torque the bolts to 33 ft. lbs. (45 Nm).
    - Starter. Torque the bolts to 18 ft. lbs. (25 Nm).
    - Wire harness and attach it to the powertrain assembly
    - DPFEE sensor electrical connector
    - Reverse lamp switch electrical connector
    - Ground wire. Torque the bolt to 80 inch lbs. (9 Nm).
    - PCM wire harness to the bracket
    - Fuel charging wire harness electrical connector
    - PNP switch electrical connector
    - VSS electrical connector
    - KS, Oil pressure sender and starter electrical connector. Torque the fasteners to 9 ft. lbs. (12 Nm).
    - Alternator electrical connectors. Torque the fasteners to 71 inch lbs. (8 Nm).
    - Powertrain assembly in the vehicle
    - Left side transmission mount. Torque the side bolts to 41 ft. lbs. (55 Nm) and the center bolt to 66 ft. lbs. (90 Nm).
    - Engine mount upper bracket. Torque the side bolts to 72 ft. lbs. (98 Nm) and the center bolt to 57 ft. lbs. (77 Nm).
    - Ground wire. Torque the bolt to 25 ft. lbs. (34 Nm).
    - Rear transmission mount. Torque the bolts to 41 ft. lbs. (55 Nm).
    - Speed control unit, if equipped. Torque the bolts to 89 inch lbs. (10 Nm).
    - Power steering pump and hand tighten the bolts

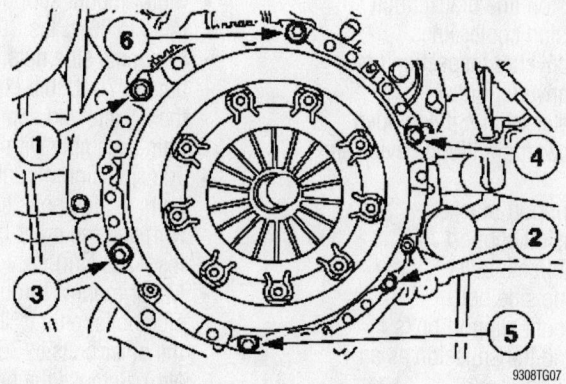

9308TG07

Tighten the pressure plate bolts in the proper sequence—2.0L engine

- Lower power steering line bracket. Torque the bolt to 89 inch lbs. (10 Nm).
- Upper power steering line bolt. Torque the bolt to 15 ft. lbs. (20 Nm).
- Slave cylinder line and clip. Torque the bolts to 16 ft. lbs. (22 Nm).
- Power steering lines. Torque the retaining bolts to 89 inch lbs. (10 Nm). Torque the power steering pump bolts to 18 ft. lbs. (25 Nm).
- Lower radiator hose
- Engine-to-transmission bolts. Torque the bolts to 33 ft. lbs. (45 Nm).
- Front transmission through bolt. Torque the bolt to 66 ft. lbs. (90 Nm).
- Block heater electrical connector
- Shifter linkages. Torque the upper bolt to 33 ft. lbs. (45 Nm) and the lower bolt to 15 ft. lbs. (20 Nm).
- Coolant hose
- Catalytic converter
- Heater hoses
- Upper radiator hose
- EVAP canister vacuum lines
- Power distribution box electrical connector. Torque the fastener to 9 ft. lbs. (12 Nm).
- Wire harness electrical connector
- Ground wires
- PCM wire harness and ground
- Brake booster vacuum supply hose to the intake manifold
- EGR vacuum regulator valve hoses and electrical connector
- Throttle cable and speed control cable, if equipped
- Fuel lines
- Battery tray and battery
- Air cleaner
- Hood

12. Fill the engine with clean oil.
13. Fill and bleed the cooling system.
14. Recharge the A/C system.
15. Start the vehicle, check for leaks and repair if necessary.

3.0L Engine

1. Before servicing the vehicle, refer to the precautions in the beginning of this section.
2. Properly recover the air conditioning system refrigerant.
3. Properly relieve the fuel system pressure.
4. Drain the cooling system.
5. Drain the engine oil.
6. Remove or disconnect the following:

- Hood
- Battery and battery tray
- Air cleaner outlet tube and housing
- Lower radiator air deflectors
- Fuel lines
- Water pump drive belt
- Accelerator cable and speed control cable, if equipped
- Vapor Management Valve (VMV)
- Powertrain Control Module (PCM)
- PCM ground wire
- Thermostat housing and hose assembly and move them aside
- Power distribution box electrical connector
- Power distribution box cover
- Nuts and cables from inside the power distribution box
- Transmission linkage
- Brake booster vacuum hose
- Heater hoses
- Power steering return line
- Power Steering Pressure (PSP) switch electrical connector
- Power steering supply line
- Oil level indicator
- Catalytic converter
- A/C compressor
- Both front wheels
- Intermediate drive shaft, if equipped

7. Separate both side ball joints.
8. Separate both side tie rod ends from the steering knuckles.
9. Separate both sway bar links from the strut mounts.
10. Separate the struts from the steering knuckles.
11. Remove or disconnect the following:

- Both wheel speed sensors, if equipped
- Brake calipers from the steering knuckles and properly support the struts
- Steering shaft from the rack
- Transmission line bracket bolt
- Transmission cooler lines
- Torque converter inspection cover
- Torque converter nuts
- Block heater wiring, if equipped

12. Install a powertrain lifting devise and raise the vehicle.

- Engine support bracket
- Transmission support
- 2 rear subframe bolts
- 2 subframe side bolts
- Motor mount support bolts
- Engine and transmission as an assembly
- Heated Oxygen (HO2S) sensor
- Transmission Range (TR) sensor
- Transmission harness electronic control switch
- Transmission control harness from the bracket
- Starter and wire harness
- Knock Sensor (KS) electrical connector
- Output Shaft Speed (OSS) sensor electrical connector
- HO2S sensor and Exhaust Gas Recirculation (EGR) tube from the exhaust manifold
- Alternator and electrical connectors
- Right side exhaust manifold and gasket
- Halfshaft support bracket and move it aside

13. Separate the engine from the transmission assembly

To install:

14. Install or connect the following:

- Powertrain assembly on the subframe
- Transmission-to-engine bolts. Torque the bolts to 30 ft. lbs. (40 Nm).
- Halfshaft bracket. Torque the bolts to 18 ft. lbs. (25 Nm).
- Right side exhaust manifold and new gasket. Torque the bolts to 15 ft. lbs. (25 Nm).
- Alternator. Torque the larger bolts to 18 ft. lbs. (25 Nm) and smaller bolt to 89 inch lbs. (10 Nm).
- EGR tube and HO2S sensor electrical connectors
- OSS sensor electrical connector
- KS jumper electrical connector
- Starter. Torque the bolts to 18 ft. lbs. (25 Nm).
- Transmission control harness to the bracket. Torque the bolt to 18 ft. lbs. (25 Nm).
- Transmission harness
- Transmission range sensor
- Powertrain assembly
- Motor mount support. Torque the bolts to 66 ft. lbs. (90 Nm).
- Subframe side nuts. Torque the nuts to 76 ft. lbs. (103 Nm). Raise the vehicle and support the powertrain assembly with a lifting device.
- Transmission mount. Torque the bolts to side bolts to 66 ft. lbs. (90 Nm) and the other bolts to 76 ft. lbs. (103 Nm).
- Motor mount. Torque the bolts to side bolts to 66 ft. lbs. (90 Nm) and the other bolts to 76 ft. lbs. (103 Nm). Remove the powertrain lift.
- Block heater electrical connector, if equipped
- Torque converter. Torque the nuts to 27 ft. lbs. (37 Nm).

- Transmission cover plate and plug
- Transmission cooler lines
- Transmission cooler line bracket. Torque the bolt to 15 ft. lbs. (20 Nm).
- Steering shaft to the rack. Torque the bolt to 18 ft. lbs. (25 Nm).
- Struts to the steering knuckles. Torque the bolts to 75 ft. lbs. (102 Nm).
- Brake calipers to the steering knuckles
- Wheel speed sensors, if equipped. Torque the bolts to 89 inch lbs. (10 Nm).
- Sway bar links to the strut mount. Torque the bolts to 41 ft. lbs. (55 Nm).
- Tie rods to the steering knuckles. Torque the bolts to 41 ft. lbs. (55 Nm).
- Ball joints. Torque the bolts to 52 ft. lbs. (70 Nm).
- Intermediate drive shaft, if equipped
- Both front wheels
- A/C compressor
- Lower radiator air deflectors
- Catalytic converter
- Oil level indicator dipstick tube
- Power steering line and bracket. Torque the bolt to 13 ft. lbs. (17 Nm).
- PSP switch electrical connector
- Power steering return line
- Heater hoses
- Vacuum lines
- Transmission linkage
- Wire harness cables and nuts to the power distribution box. Torque the nuts to 89 inch lbs. (10 Nm).
- Power distribution box wire harness
- Thermostat housing and connect the hoses
- Ground wire. Torque the bolt to 89 inch lbs. (10 Nm).
- PCM electrical connector
- VMV electrical connector
- Accelerator cable and speed control cable, if equipped
- Air cleaner assembly
- Water pump drive belt
- Battery and tray

15. Fill and bleed the cooling system.
16. Fill the engine with clean oil.
17. Recharge the A/C system.
18. Inspect and top off the power steering fluid.
19. Start the vehicle, check for leaks and repair if necessary.

Water Pump

REMOVAL & INSTALLATION

2.0L Engine

1. Before servicing the vehicle, refer to the precautions in the beginning of this section.
2. Drain the cooling system.
3. Remove or disconnect the following:
   - Negative battery cable
   - Right front wheel
   - Splash shield
   - Drive belt
   - Water pump pulley
   - Water pump

To install:
4. Install or connect the following:
   - Water pump. Torque the bolts to 89 inch lbs. (10 Nm).
   - Water pump pulley. Torque the bolts to 89 inch lbs. (10 Nm).
   - Drive belt
   - Splash shield
   - Right front wheel
   - Negative battery cable
5. Refill the cooling system.
6. Start the vehicle and check for leaks, repair if necessary.

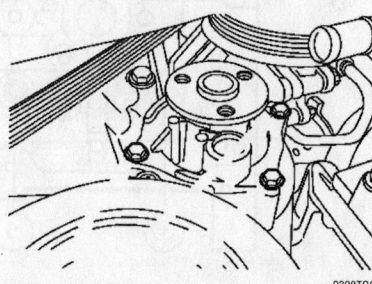

9308TG08

Exploded view of the water pump—2.0L engine

3.0L Engine

1. Before servicing the vehicle, refer to the precautions in the beginning of this section.
2. Drain the cooling system.
3. Remove or disconnect the following:
   - Negative battery cable
   - Air cleaner outlet tube
   - Water pump belt tensioner
   - Coolant hoses
   - Water pump
   - Water pump from the housing

To install:
- Water pump to the housing. Torque the bolts to 89 inch lbs. (10 Nm).
- Water pump. Torque the bolts to 89 inch lbs. (10 Nm).
- Coolant hoses
- Water pump belt tensioner
- Air cleaner outlet tube
- Negative battery cable
4. Refill the cooling system.
5. Start the vehicle and check for leaks, repair if necessary.

Cylinder Head

REMOVAL & INSTALLATION

2.0L Engine

1. Before servicing the vehicle, refer to the precautions in the beginning of this section.
2. Properly relieve the fuel system pressure.
3. Drain the engine oil.
4. Remove or disconnect the following:

   - Negative battery cable
   - Ignition coil bracket
   - Thermostat housing
   - Positive Crankcase Ventilation (PCV) tube
   - Intake manifold
   - Exhaust manifold
   - Power steering bracket and move it aside
   - Valve tappets
   - Engine mount lower bracket
   - Engine mount upper bracket
   - Cylinder head bolts in the proper sequence and discard the gasket

To install:
5. Install a new head gasket and the cylinder head.
6. Lubricate the cylinder head bolt threads.
7. Torque the cylinder head bolts in the proper sequence as follows:
   a. Step 1: 15 ft. lbs. (20 Nm).
   b. Step 2: 30 ft. lbs. (40 Nm).
   c. Step 3: Plus an additional 90 degrees.
8. Install or connect the following:
   - Engine mount upper bracket. Torque the 2 upper bolts to 72 ft. lbs. (98 Nm) and the center bolt to 57 ft. lbs. (77 Nm).
   - Engine mount lower bracket.

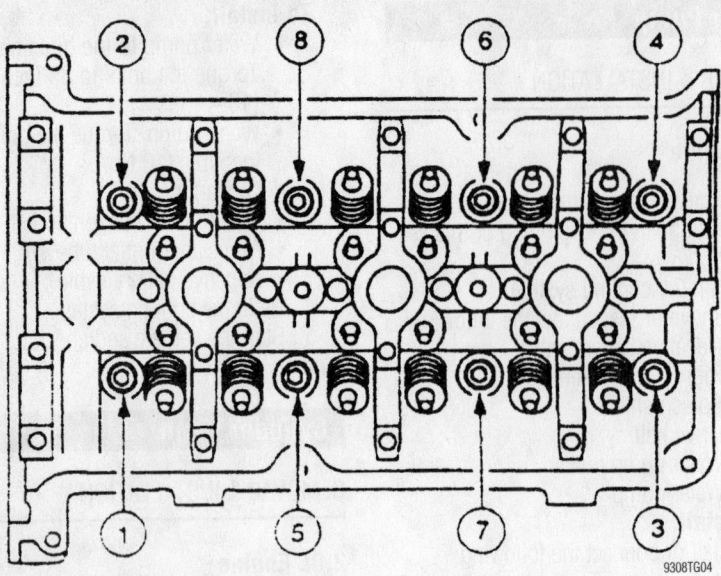

Cylinder head bolt torque sequence 2.0L engine

Torque the bolts to 37 ft. lbs. (50 Nm).

- Valve tappets
- Power steering pump bracket. Torque the bolts to 20 ft. lbs. (28 Nm).
- Exhaust manifold
- Intake manifold
- PCV tube
- Thermostat housing
- Ignition coil bracket
- Negative battery cable

9. Fill the engine with clean oil and replace the filter.

10. Start the vehicle and check for leaks, repair if necessary.

3.0L Engine

The procedure for the left side cylinder head and right side are similar. Changes in the procedure will be noted for either side cylinder head.

1. Before servicing the vehicle, refer to the precautions in the beginning of this section.

2. Properly relieve the fuel system pressure.

- Drain the cooling system.

3. Remove or disconnect the following:

- Negative battery cable
- Camshaft
- Exhaust Gas Recirculation (EGR) tube, right side only
- Exhaust manifold
- Camshaft followers
- Hydraulic lash adjusters and matchmark them for proper installation
- Cylinder head bolts in sequence and discard them

- Cylinder head and discard the gasket

To install:

4. Install a new head gasket and the cylinder head.

5. Lubricate the cylinder head bolt threads.

6. Torque the cylinder head bolts in the proper sequence as follows:

- a. Step 1: 30 ft. lbs. (40 Nm).
- b. Step 2: Additional 90 degrees.
- c. Step 3: Loosen the bolts one full turn.
- d. Step 4: 30 ft. lbs. (40 Nm).
- e. Step 5: Plus an additional 90 degrees.
- f. Step 6: Plus an additional 90 degrees.

7. Install or connect the following:

- Hydraulic lash adjusters
- Camshaft followers
- Camshaft
- Exhaust manifold. Torque the bolts

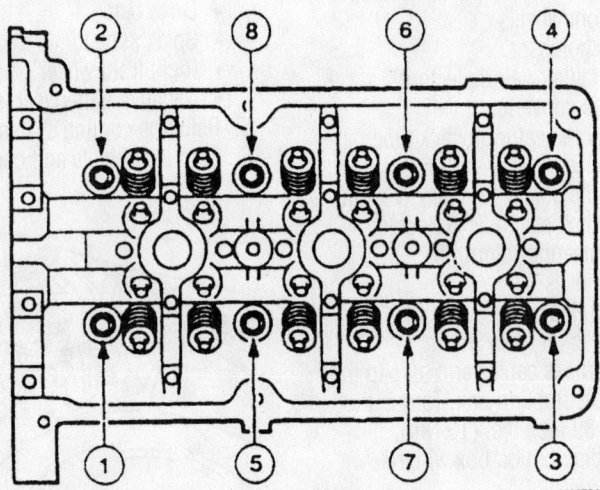

Left side cylinder head bolt torque sequence 3.0L engine

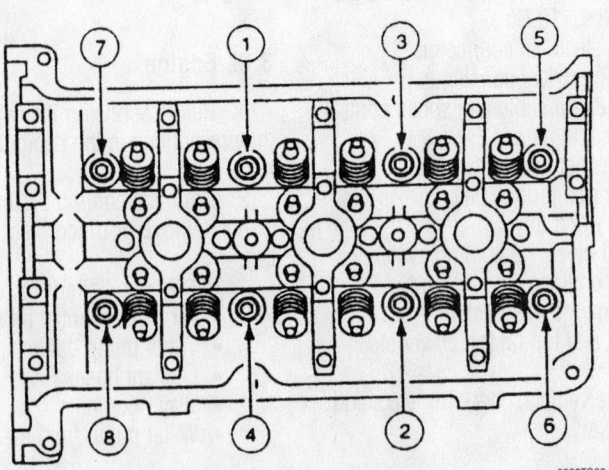

Right side cylinder head bolt torque sequence 3.0L engine

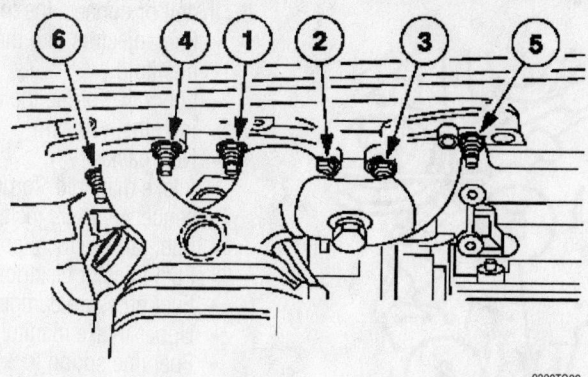

Right side exhaust manifold bolt torque sequence—3.0L engine

in sequence to 15 ft. lbs. (20 Nm), right side only
• EGR tube, right side only
• Coolant bypass tube
• Negative battery cable
8. Fill the coolant to the proper level.
9. Start the vehicle and check for leaks, repair if necessary.

Intake Manifold

REMOVAL & INSTALLATION

2.0L Engine

1. Before servicing the vehicle, refer to the precautions in the beginning of this section.
2. Properly relieve the fuel system pressure.

3. Remove or disconnect the following:
• Negative battery cable
• Fuel injection supply manifold
• Throttle Position (TP) sensor electrical connector
• Idle Air Control (IAC) electrical connector and unclip the harness from the bracket
• Main engine control sensor wiring
• Connector from the bracket
• Powertrain Control Module (PCM) wire harness from the bracket
• Brake booster vacuum hose
• 4 additional vacuum lines
• Positive Crankcase Ventilation (PCV) hose from the intake manifold
• Knock Sensor (KS) electrical connector
• Alternator
• Intake manifold and discard the gasket

4. Clean the mating surfaces.
To install:
5. Install or connect the following:
• New gasket
• Intake manifold. Torque the bolts, in sequence, to 13 ft. lbs. (18 Nm).
• Alternator
• KS electrical connector
• PCV vacuum line
• 4 vacuum lines
• Brake booster vacuum supply hose
• PCM wire harness to the bracket
• Main engine control sensor wiring
• IAC valve electrical connector and attach the harness to the bracket
• TP sensor electrical connector
• Fuel injection supply manifold
• Negative battery cable
6. Start the vehicle and check for leaks, repair if necessary.

3.0L Engine

UPPER

1. Before servicing the vehicle, refer to the precautions in the beginning of this section.
2. Properly relieve the fuel system pressure.
3. Drain the coolant system.
4. Remove or disconnect the following:
• Negative battery cable
• Air cleaner outlet tube
• Engine appearance cover
• Throttle cable
• Speed control cable, if equipped
• Throttle cable bracket
• Throttle Position (TP) sensor electrical connector
• Idle Air Control (IAC) valve electrical connector
• Exhaust Gas Recirculation (EGR) valve vacuum hose and tube
• EGR vacuum regulator valve electrical connector and hose
• Chassis vacuum hose
• Engine vacuum hose
• Positive Crankcase Ventilation (PCV) hose
• Vapor Management Valve (VMV) vacuum hose
• Electrical connectors from the left side of the upper intake manifold
• Power Steering Pressure (PSP) sensor electrical connector
• Upper intake manifold and discard the gasket
5. Clean the mating surfaces.

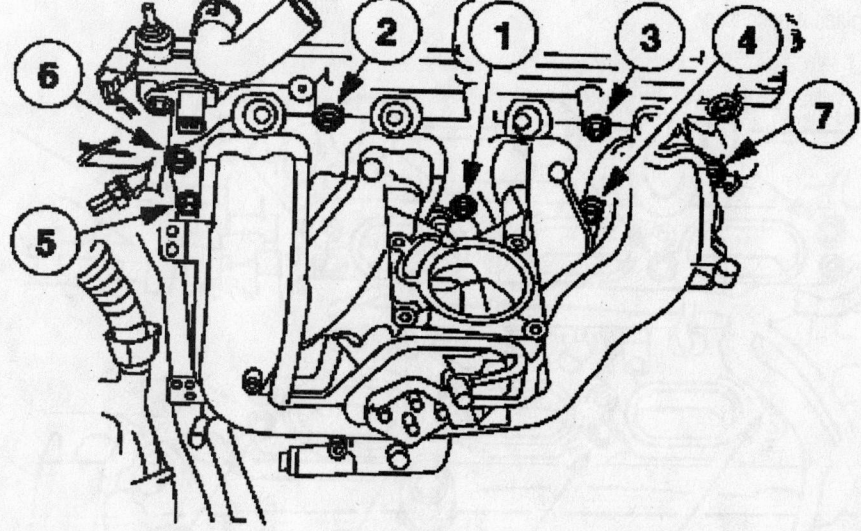

Tighten the intake manifold bolts in the sequence shown—2.0L engine

Timing belt service is covered in Section 3 of this manual

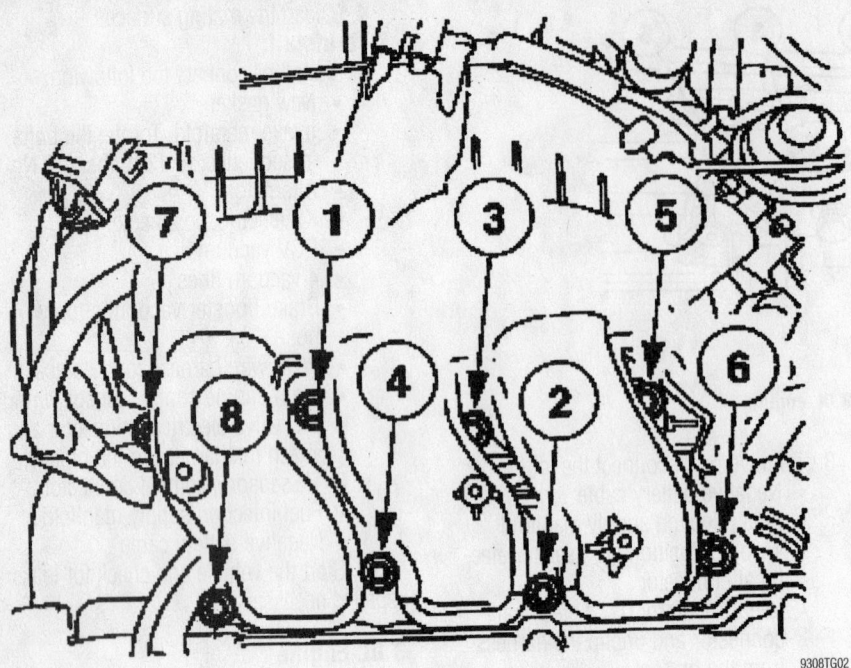

Tighten the upper intake manifold bolts in the sequence shown—3.0L engine

To install:

6. Install or connect the following:
- New gasket
- Intake manifold. Torque the bolts, in sequence, to 89 inch lbs. (10 Nm).
- PSP electrical connector
- Electrical connectors on the left side of the upper intake manifold
- VMV vacuum hose
- Chassis, engine and PCV hoses
- EGR valve vacuum regulator
- EGR valve vacuum hose and tube. Torque the nut to 30 ft. lbs. (40 Nm).
- TP sensor electrical connector
- IAC valve electrical connector
- Throttle cable and speed control cable, if equipped. Torque the bracket bolts to 89 inch lbs. (10 Nm).
- Air cleaner outlet tube
- Engine appearance cover. Torque the bolts to 53 inch lbs. (6 Nm).
- Negative battery cable

7. Fill the coolant system to the proper level.

8. Start the vehicle and check for leaks, repair if necessary.

LOWER

1. Before servicing the vehicle, refer to the precautions in the beginning of this section.

2. Properly relieve the fuel system pressure.

3. Remove or disconnect the following:
- Negative battery cable

- Fuel line spring lock coupling
- Upper intake manifold
- Fuel rail
- Fuel injector electrical connectors
- Fuel pressure damper vacuum line
- Lower intake manifold
- Lower intake manifold from the fuel rail
- Fuel injectors from the manifold and discard the gasket

4. Clean the mating surfaces.

To install:

5. Inspect the fuel injector O-rings and replace if necessary.

6. Install or connect the following:
- Fuel injectors into the lower intake manifold
- Fuel rail. Torque the bolts to 89 inch lbs. (10 Nm).
- New gasket
- Intake manifold. Torque the bolts, in sequence, to 89 inch lbs. (10 Nm).
- Fuel rail electrical connectors
- Fuel injector electrical connectors
- Fuel pressure damper vacuum line
- Upper intake manifold
- Fuel line spring lock coupling
- Negative battery cable

7. Start the vehicle and check for leaks, repair if necessary.

Exhaust Manifold

REMOVAL & INSTALLATION

2.0L Engine

1. Before servicing the vehicle, refer to the precautions in the beginning of this section.

2. Remove or disconnect the following:
- Negative battery cable
- Catalytic converter
- Oil level indicator tube and bracket
- Exhaust manifold and discard the gasket

To install:

3. Clean the sealing surfaces of any old gasket material.

4. Install or connect the following:
- Exhaust manifold and new gasket. Torque the bolts to 12 ft. lbs. (16 Nm).

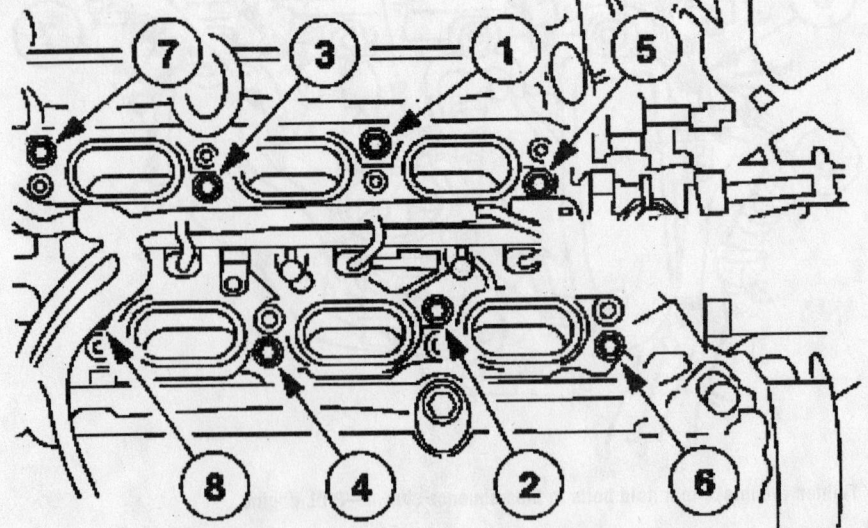

Tighten the lower intake manifold bolts in the sequence shown—3.0L engine

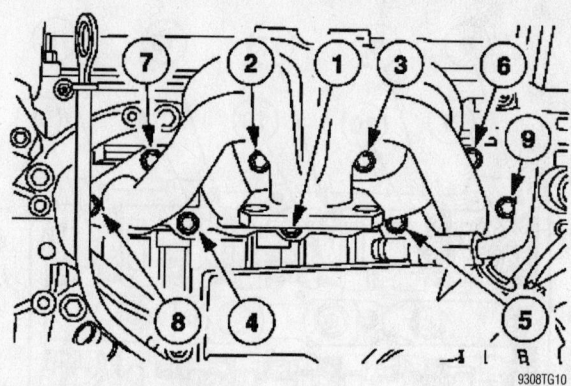

Exhaust manifold bolt torque sequence—2.0L engine

- Oil level indicator tube and bracket. Torque the bolt to 89 inch lbs. (10 Nm).
- Catalytic converter
- Negative battery cable

5. Start the vehicle and check for leaks, repair if necessary.

3.0L Engine

LEFT SIDE

1. Before servicing the vehicle, refer to the precautions in the beginning of this section.

2. Remove or disconnect the following:
- Negative battery cable
- Heated Oxygen (HO2S) sensor and catalyst monitor
- Splash shield
- Exhaust crossover pipe
- Drive belt
- A/C compressor and move it aside
- Exhaust manifold and discard the gasket

To install:

3. Clean the sealing surfaces of any old gasket material.

4. Install or connect the following:
- Exhaust manifold and new gasket. Torque the bolts to 15 ft. lbs. (20 Nm).
- A/C compressor. Torque the bolts to 18 ft. lbs. (20 Nm).
- Drive belt
- Exhaust crossover pipe. Torque the bolts to 30 ft. lbs. (40 Nm).
- Splash shield. Torque the bolts to 80 inch lbs. (9 Nm).
- Left side HO2S sensor and catalyst monitor
- Negative battery cable

5. Start the vehicle and check for leaks, repair if necessary.

RIGHT SIDE

1. Before servicing the vehicle, refer to the precautions in the beginning of this section.

2. Remove or disconnect the following:
- Negative battery cable
- Exhaust Gas Recirculation (EGR) tube
- Alternator
- Right side Heated Oxygen (HO2S) sensor

- Right side exhaust manifold and discard the gasket

To install:

3. Clean the sealing surfaces of any old gasket material.

4. Install or connect the following:
- Exhaust manifold and new gasket. Torque the bolts to 15 ft. lbs. (20 Nm).
- Right side HO2S sensor
- Alternator
- EGR tube
- Negative battery cable

5. Start the vehicle and check for leaks, repair if necessary.

Front Crankshaft Seal

REMOVAL & INSTALLATION

2.0L Engine

1. Before servicing the vehicle, refer to the precautions in the beginning of this section.

2. Remove or disconnect the following:
- Negative battery cable

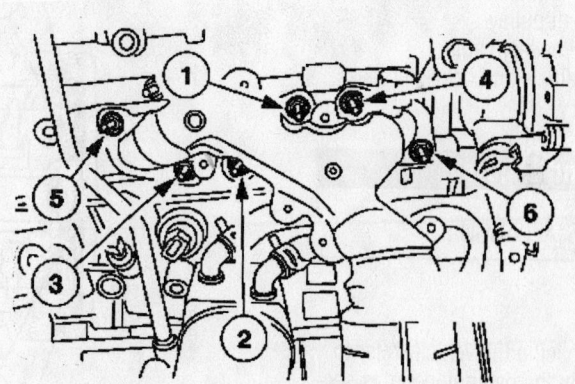

Left side exhaust manifold bolt torque sequence—3.0L engine

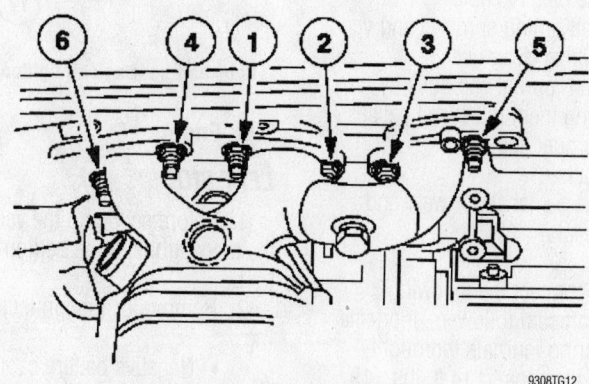

Right side exhaust manifold bolt torque sequence—3.0L engine

Heater Core replacement is covered in Section 2 of this manual

- Timing belt
- Crankshaft sprocket and timing belt guide
- Crankshaft oil seal

➡ **Be careful not to damage the seal surface of the cover.**

To install:

3. Install or connect the following:
- New front crankshaft oil seal
- Timing belt guide and crankshaft sprocket
- Timing belt
- Negative battery cable

4. Start the vehicle and check for leaks, repair if necessary.

3.0L Engine

1. Before servicing the vehicle, refer to the precautions in the beginning of this section.

2. Remove or disconnect the following:
- Negative battery cable
- Crankshaft pulley
- Front oil seal

To install:

3. Install or connect the following:
- New front crankshaft oil seal
- Crankshaft pulley
- Negative battery cable

4. Start the vehicle and check for leaks, repair if necessary.

Camshaft and Lifters

REMOVAL & INSTALLATION

2.0L Engine

1. Before servicing the vehicle, refer to the precautions in the beginning of this section.

2. Remove or disconnect the following:
- Negative battery cable
- Camshaft timing sprocket and verify the valve clearance
- Camshaft journal cap bolts by loosening them in several passes in the proper sequence
- Camshafts

3. Inspect the camshaft for wear and discard the oil seals

To install:

4. Install or connect the following:
- Camshaft cam followers, lubricate the bearing journals thoroughly. Torque the caps to 14 ft. lbs. (19 Nm).
- Exhaust camshaft oil seal
- Camshaft timing sprocket
- Negative battery cable

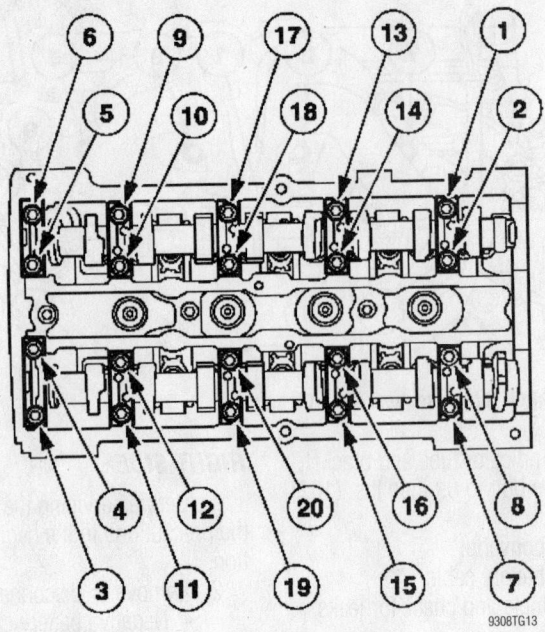

Remove the camshaft bearing caps in sequence—2.0L engine

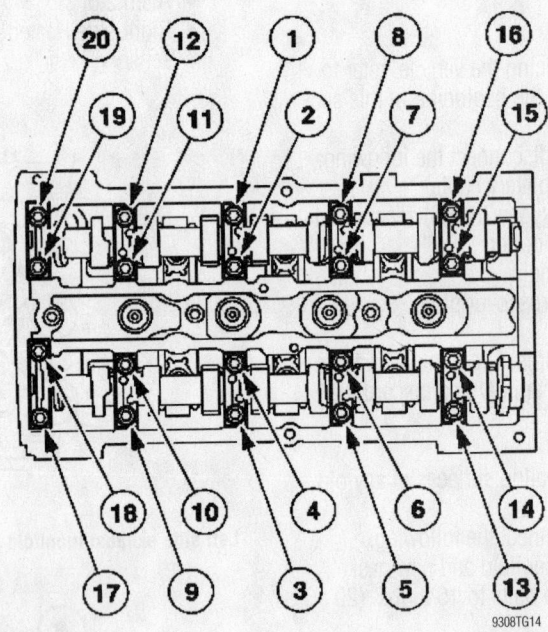

Camshaft bearing cap tightening sequence—2.0L engine

3.0L Engine

LEFT SIDE

1. Before servicing the vehicle, refer to the precautions in the beginning of this section.

2. Remove or disconnect the following:
- Negative battery cable
- Water pump belt
- Timing drive components
- Camshaft oil seal
- Camshaft oil seal retainer

- Camshaft cap bolts by loosening them in sequence
- Camshafts

To install:

3. Install or connect the following:
- Camshaft bearing caps in their original position
- Align the camshafts
- Bearing thrust caps and hand tighten the bolts. When aligned properly, torque the bolts to 89 inch lbs. (10 Nm).
- Timing drive components

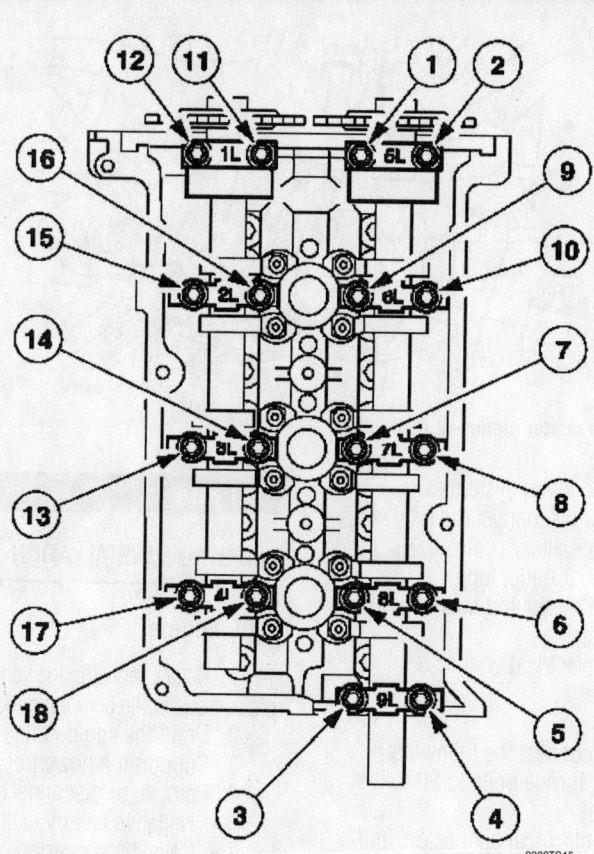

Remove and install the left side camshaft bearing caps in sequence—3.0L engine

- Camshaft oil seal retainer
- Crankshaft oil seal
- Water pump drive pulley
- Water pump belt
- Negative battery cable

RIGHT SIDE

1. Before servicing the vehicle, refer to the precautions in the beginning of this section.

2. Remove or disconnect the following:
   - Negative battery cable
   - Timing drive components
   - Camshaft cap bolts by loosening them in sequence
   - Camshafts caps
   - Camshafts

To install:

3. Install or connect the following:
   - Camshaft bearing caps in their original position
   - Align the camshafts
   - Bearing caps and hand tighten the bolts
   - Bearing thrust caps and hand tighten the bolts. When aligned properly, torque the bolts to 89 inch lbs. (10 Nm).

- Timing drive components
- Negative battery cable

Valve Lash

ADJUSTMENT

2.0L Engine

1. Before servicing the vehicle, refer to the precautions in the beginning of this section.

2. Remove or disconnect the following:
   - Negative battery cable
   - Timing belt

3. Measure each valve's clearance at the base circle with the lobe facing away from the tappet.

4. Use a feeler gauge to measure and record each valve's clearance

5. Remove or disconnect the following:
   - Camshafts
   - Valve tappets from the cylinder head

6. A mid range clearance is recommended as follows:
   a. Intake: 0.006 inch (0.15mm).
   b. Exhaust: 0.012 inch (0.3mm).

To install:

7. Install or connect the following:

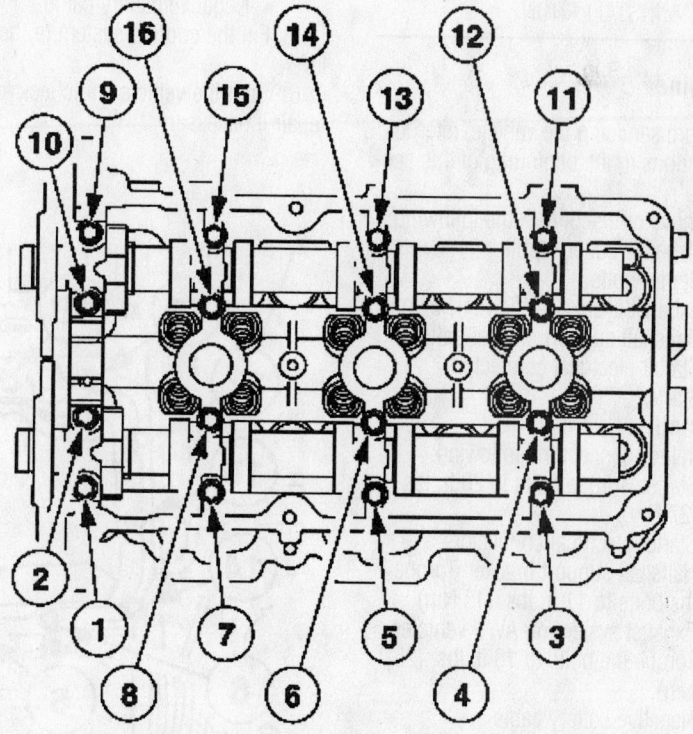

Remove and install the right side camshaft bearing caps in sequence—3.0L engine

Brake service is covered in Section 4 of this manual

- Valve tappets after lubricating them with clean engine oil
- Camshafts and verify each valve's clearance at the base circle with the lobe facing away from the tappet
- Timing belt
- Negative battery cable

3.0L Engine

1. Before servicing the vehicle, refer to the precautions in the beginning of this section.
2. Remove or disconnect the following:
   - Negative battery cable
   - Camshaft followers
   - Hydraulic lash adjusters

➡ **Mark the position of the hydraulic lash adjusters to assure they are assembled in their original position**

3. Inspect the adjusters for scoring or uneven wear in the bore and replace them as required.
   To install:
4. Install or connect the following:
   - Hydraulic lash adjusters after lubricating them with clean engine oil
   - Camshaft followers
   - Negative battery cable

Starter Motor

REMOVAL & INSTALLATION

2.0L Engine

1. Before servicing the vehicle, refer to the precautions in the beginning of this section.
2. Remove or disconnect the following:
   - Negative battery cable
   - Starter bolts
   - Exhaust system, AWD vehicles only
   - Halfshaft support bracket bolts
   - Starter electrical connectors
   - Starter
   To install:
3. Install or connect the following:
   - Starter. Torque bolts to 20 ft. lbs. (27 Nm).
   - Starter electrical connectors
   - Halfshaft support bracket. Torque the bolts to 11 ft. lbs. (15 Nm).
   - Exhaust system on AWD vehicles. Torque the bolts to 18 ft. lbs. (25 Nm).
   - Negative battery cable

3.0L Engine

1. Before servicing the vehicle, refer to the precautions in the beginning of this section.

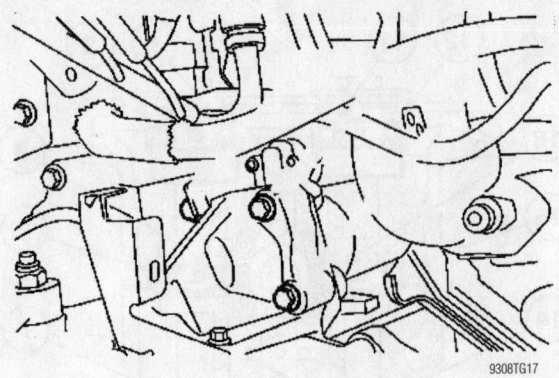

9308TG17

Removal of the starter motor—2.0L engine

2. Drain the cooling system.
3. Remove or disconnect the following:
   - Negative battery cable
   - Air cleaner outlet tube
   - Coolant hoses and move the thermostat aside
   - Starter electrical connectors
   - Starter
   To install:
4. Install or connect the following:
   - Starter. Torque bolts to 20 ft. lbs. (27 Nm).
   - Starter electrical connectors and reposition the thermostat
   - Connect the 4 coolant hoses
   - Air cleaner outlet tube
   - Negative battery cable
5. Fill the cooling system to the proper level.
6. Start the vehicle and check for leaks, repair if necessary.

Oil Pan

REMOVAL & INSTALLATION

2.0L Engine

1. Before servicing the vehicle, refer to the precautions in the beginning of this section.
2. Drain the engine oil.
3. Support the powertrain assembly.
4. Remove or disconnect the following:
   - Negative battery cable
   - Catalytic converter
   - Oil pan and gasket
5. Thoroughly clean the gasket mating surfaces.
   To install:
6. Apply silicone sealer to the oil pan.
7. Install a new gasket on the oil pan.
8. Oil pan. Torque the bolts in sequence to:

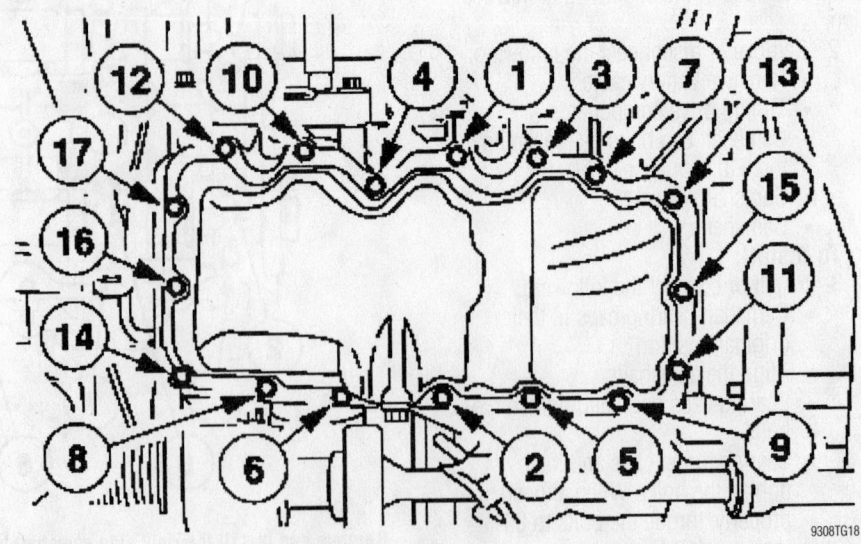

9308TG18

Tighten the oil pan bolts in sequence—2.0L engine

a. Step 1: 53 inch lbs. (6 Nm).
b. Step 2: 106 in lbs. (12 Nm).
9. Install or connect the following:
  • Catalytic converter
  • Negative battery cable
10. Fill the engine with clean oil.
11. Start the engine and check for leaks,
repair if necessary.

3.0L Engine

1. Before servicing the vehicle, refer to the precautions in the beginning of this section.
2. Drain the engine oil.
3. Remove or disconnect the following:
  • Negative battery cable
  • Flexible exhaust pipe
  • Downstream catalyst monitor sensor
  • Oil pan and gasket
4. Thoroughly clean the gasket mating surfaces.
  To install:
5. Apply silicone sealer to the oil pan.
6. Install or connect the following:
  • New gasket on the oil pan
  • Oil pan. Torque the bolts in sequence to 18 ft. lbs. (25 Nm).
  • Flexible exhaust pipe
  • Downstream catalyst monitor sensor
  • Negative battery cable
7. Fill the engine with clean oil.
8. Start the vehicle and check for leaks, repair if necessary.

Oil Pump

REMOVAL & INSTALLATION

2.0L Engine

1. Before servicing the vehicle, refer to the precautions in the beginning of this section.
2. Drain the engine oil.
3. Remove or disconnect the following:
  • Negative battery cable
  • Oil pan
  • Oil pump screen cover and tube
  • Oil pump and discard the gasket
4. Thoroughly clean the gasket mating surfaces.
  To install:
5. Install or connect the following:
  • Oil pump screen cover and tube with a new gasket. Torque the bolts to 89 inch lbs. (10 Nm).
  • Oil pump to the oil pan
  • Oil pan
  • Negative battery cable
6. Refill the engine with clean oil.

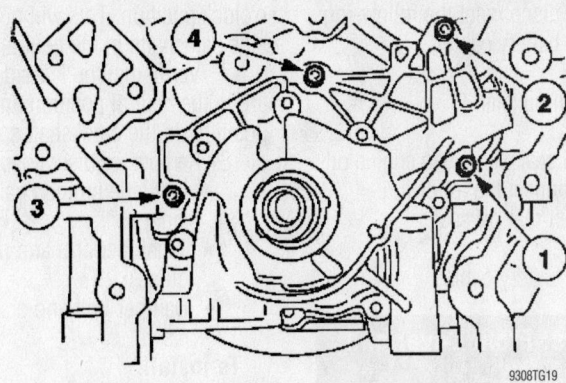

Remove the oil pump bolts in the proper sequence—3.0L engine

Install the oil pump bolts in the proper sequence—3.0L engine

7. Start the engine and check for leaks; repair if necessary.

3.0L Engine

1. Before servicing the vehicle, refer to the precautions in the beginning of this section.
2. Drain the engine oil.
3. Remove or disconnect the following:
  • Negative battery cable
  • Timing drive components
  • Oil pump screen cover and tube
  • Damper bolt and crankshaft sprockets
  • Oil pump bolts in the proper sequence
4. Thoroughly clean the gasket mating surfaces.
  To install:
5. Install or connect the following:
  • Oil pump and bolts in the proper sequence. Torque the bolts to 89 inch lbs. (10 Nm).
  • Crankshaft sprockets
  • Oil pump screen cover and tube
  • Timing drive components
  • Negative battery cable
6. Refill the engine with clean oil.

7. Start the engine and check for leaks; repair if necessary.

Rear Main Seal

REMOVAL & INSTALLATION

2.0L Engine

1. Before servicing the vehicle, refer to the precautions in the beginning of this section.
2. Remove or disconnect the following:
  • Negative battery cable
  • Flywheel
  • Rear main seal
  To install:
3. Coat the oil seal with clean engine oil.
4. Install or connect the following:
  • Crankshaft rear oil seal
  • Flywheel
  • Negative battery cable

3.0L Engine

1. Before servicing the vehicle, refer to the precautions in the beginning of this section.

2. Remove or disconnect the follow-ing:
- Negative battery cable
- Flexplate
- Rear main oil seal

To install:

3. Coat the oil seal with clean engine oil.
4. Install or connect the following:
- Crankshaft rear oil seal
- Flywheel
- Negative battery cable

Timing Gears, Front Cover and Seal

REMOVAL & INSTALLATION

3.0L Engine

1. Before servicing the vehicle, refer to the precautions in the beginning of this section.
2. Remove or disconnect the following:
- Negative battery cable
- Engine front cover
- Ignition pulse wheel and install a damper bolt
- Spark plugs

3. Rotate the crankshaft clockwise to position the keyway at the 11 o'clock position and the camshafts in the correct positions. The No. 1 cylinder will be at Top Dead Center (TDC).
4. Rotate the crankshaft clockwise 120 degrees to the 3 o'clock position to locate the right side camshafts in the neutral position.
5. Remove or disconnect the following:
- Right side timing chain and ten-sioner
- Tensioner arm and timing chain guide

6. Rotate the crankshaft clockwise 2 times to position the keyway at the 11

o'clock position. This will position the left side camshafts in the neutral position.

7. Verify that the left side crankshafts are in the neutral position and mark the link position on the crankshaft sprocket.

8. Remove or disconnect the following:
- Left side timing chain and ten-sioner
- Tensioner arm and timing chain guide
- Damper bolt and crankshaft sprock-ets

To install:

9. Install the crankshaft sprockets.
10. Position the timing chain tensioner in a soft jaw vise. Hold the ratchet lock mechanism away from the ratchet stem and slowly compress the timing chain tensioner
11. If the timing marks on the chain are not visible, use a permanent marker to mark the left and right side timing chains. Mark the timing chains in the following sequence:
  a. Mark any link to use as the crank-shaft timing mark.
  b. Count 29 links from the crankshaft timing mark and mark the link as the exhaust cam sprocket timing mark.
  c. Continue counting to 42 and mark the link as the intake sprocket timing mark
12. Install the guide. Torque the bolts to 18 ft. lbs. (25 Nm).
13. Install the left side timing chain and align the chain in the following sequence:
  a. Mark any link to use as the crank-shaft timing mark.
  b. Count 29 links from the crankshaft timing mark and mark the link as the exhaust cam sprocket timing mark.
  c. Continue counting to 42 and mark the link as the intake sprocket timing mark
14. Install or connect the following:
- Left side timing chain and ten-

sioner arm. Torque the bolts to 18 ft. lbs. (25 Nm).
- Crankshaft damper bolt and rotate the keyway to the 3 o'clock position
15. Verify that the right side camshafts are properly positioned and install the right side timing chain and guide. Torque the bolts to 18 ft. lbs. (25 Nm).
16. Make certain that the timing chain aligns with the marks on the camshaft and crankshaft sprockets

✳✳ CAUTION

Install the pulse wheel with the key-way in the slot stamped 20–25–34Y–30M (Color Blur).

17. Install or connect the following:
- Right side timing chain tensioner and arm. Torque the bolts to 18 ft. lbs. (25 Nm) and remove the damper bolt
- Ignition pulse wheel
- Spark plugs
- Engine front cover
- Negative battery cable

Piston and Ring

POSITIONING

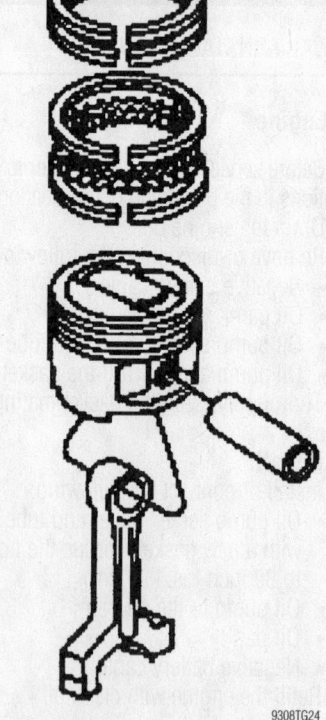

2.0L engine piston and connecting rod positioning ring end-gap spacing

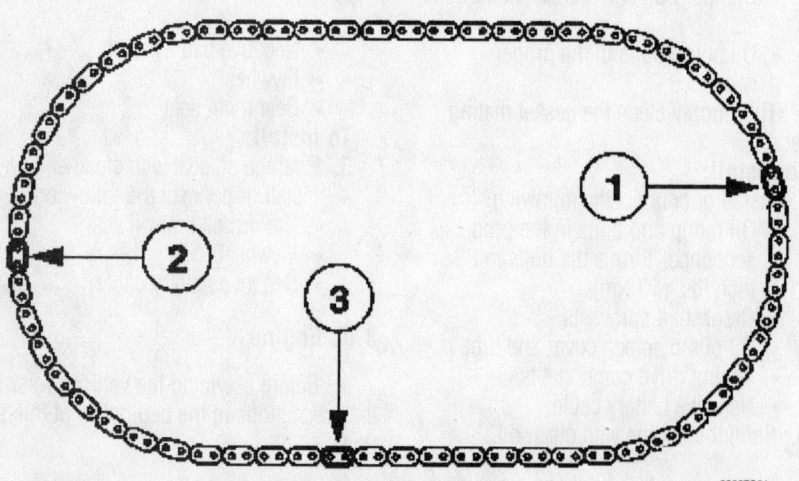

Mark the timing chain in the proper sequence—3.0L engine

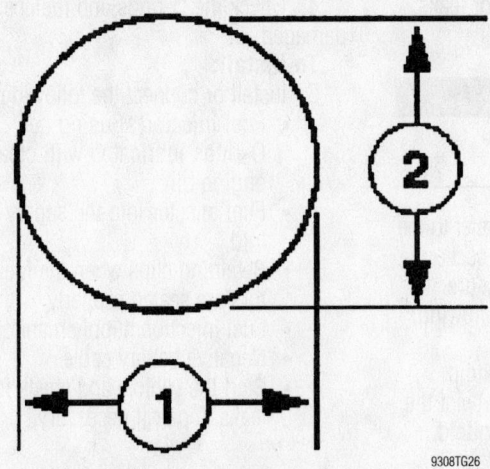

2.0L (VIN B) engine —piston ring end-gap spacing

3.0L (VIN 1) engine—piston ring end-gap spacing

FUEL SYSTEM

Fuel System Service Precautions

Safety is the most important factor when performing not only fuel system maintenance but any type of maintenance. Failure to conduct maintenance and repairs in a safe manner may result in serious personal injury or death. Maintenance and testing of the vehicle's fuel system components can be accomplished safely and effectively by adhering to the following rules and guidelines.

1. To avoid the possibility of fire and personal injury, always disconnect the negative battery cable unless the repair or test procedure requires that battery voltage be applied.

2. Always relieve the fuel system pressure prior to disconnecting any fuel system component (injector, fuel rail, pressure regulator, etc.), fitting or fuel line connection. Exercise extreme caution whenever relieving fuel system pressure, to avoid exposing skin, face and eyes to fuel spray. Please be advised that fuel under pressure may penetrate the skin or any part of the body that it contacts.

3. Always place a shop towel or cloth around the fitting or connection prior to loosening to absorb any excess fuel due to spillage. Ensure that all fuel spillage (should it occur) is quickly removed from engine surfaces. Ensure that all fuel soaked cloths or towels are deposited into a suitable waste container.

4. Always keep a dry chemical (Class B) fire extinguisher near the work area.

5. Do not allow fuel spray or fuel vapors to come into contact with a spark or open flame.

6. Always use a backup wrench when loosening and tightening fuel line connection fittings. This will prevent unnecessary stress and torsion to fuel line piping.

7. Always replace worn fuel fitting O-rings with new. Do not substitute fuel hose or equivalent, where fuel pipe is installed.

Before servicing the vehicle, make sure to refer to the precautions in the beginning of this section as well.

Fuel System Pressure

RELIEVING

2.0L Engine

1. Before servicing the vehicle, refer to the precautions in the beginning of this section.

2. Remove or disconnect the following:

3. Remove the fuel pump relay and start the engine.

4. After the engine stalls, crank the engine 2 more times to be certain that all fuel pressure has been relieved.

5. Turn the ignition switch to the **OFF**-position.

6. Install the fuel pump relay.

3.0L Engine

1. Before servicing the vehicle, refer to the precautions in the beginning of this section.

2. Remove or disconnect the following:

3. Remove the schrader valve cap at the end of the fuel injection supply manifold and attach a fuel pressure gauge.

4. Open the manual valve slowly and drain the fuel into a suitable container.

5. Continue draining the fuel system to relieve fuel pressure.

Fuel Filter

REMOVAL & INSTALLATION

1. Before servicing the vehicle, refer to the precautions in the beginning of this section.

2. Properly relieve the fuel system pressure.

3. Remove or disconnect the following:
   - Negative battery cable
   - Fuel line to the fuel filter

4. Loosen the clamp and remove the filter

To install:

5. Install or connect the following:
   - New clips to the fuel lines
   - Fuel filter and tighten the clamp
   - Fuel lines to the fuel filter
   - Negative battery cable

6. Start the vehicle and check for leaks, repair if necessary.

Fuel Pump

REMOVAL & INSTALLATION

1. Before servicing the vehicle, refer to the precautions in the beginning of this section.

2. Properly relieve the fuel system pressure.

For Accessory Drive Belt illustrations, see Section 1 of this manual

3. Remove or disconnect the following:
- Negative battery cable
- Gas cap to relieve any additional fuel pressure
- Left rear seat cushion and lift the access cover on the scuff plate
- Pin type retainers and move the carpet aside
- Screws from the fuel pump module access cover
- Fuel pump module electrical connectors
- Fuel and vapor lines from the fuel tank
- Fuel pump module and discard the gasket

To install:

4. Install or connect the following:
- New fuel pump module gasket
- Fuel pump module. Torque the module to 60 ft. lbs. (81 Nm).
- Fuel and vapor lines to the fuel tank
- Fuel pump module electrical connectors
- Fuel pump module access cover and tighten the screws securely
- Pin type retainers and reposition the carpet
- Left rear seat cushion
- Gas cap
- Negative battery cable

5. Start the engine and check for leaks, repair if necessary.

Fuel Injectors

REMOVAL & INSTALLATION

1. Before servicing the vehicle, refer to the precautions in the beginning of this section.
2. Release the fuel system pressure.
3. Remove or disconnect the following:
- Negative battery cable
- Fuel injection supply manifold
- Retaining clips and gently twist the fuel injector out of the manifold

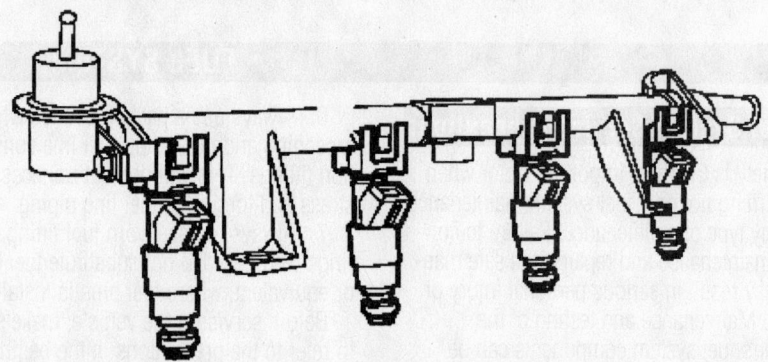

Remove the fuel injectors from the fuel supply manifold—2.0L engine

9308TG22

4. Check the O-rings and replace if damaged.

To install:

5. Install or connect the following:
- Fuel injector(s) using new O-rings lubricated with clean engine oil
- Fuel injector into the supply manifold
- Retaining clips when the fuel injectors are seated properly
- Fuel injection supply manifold
- Negative battery cable
- Start the vehicle and check for leaks, repair if necessary.

DRIVE TRAIN

Transmission Assembly

REMOVAL & INSTALLATION

Manual Transmission

1. Before servicing the vehicle, refer to the precautions in the beginning of this section.
2. Drain the transmission fluid.
3. Remove or disconnect the following:
- Battery cables
- Battery and tray
- Mass Air Flow (MAF) sensor electrical connector
- Accelerator cable from the air cleaner outlet tube
- Emission management tube and hose
- Crankcase ventilation hose
- Air cleaner outlet tube
- Air cleaner housing
- Back-up lamp switch electrical connector
- Front wire harness bracket and move it aside
- Front wire harness bracket spacer

- Wire harness from the rear harness bracket
- Park Neutral Position (PNP) electrical connector
- Rear wire harness bracket and move it aside
- Vehicle Speed Sensor (VSS) electrical connector
- Clutch slave cylinder line from the bracket and move it aside while properly supporting the engine
- Left side transmission support insulator and bracket
- Rear transmission support insulator
- Front transmission support insulator and bracket
- Starter and move it aside
- Top transmission flywheel housing bolts
- Front transmission flywheel housing bolts
- Transfer case, if equipped
- Left side halfshaft
- Rear transmission support insulator bracket
- Shifter linkage and stabilizer bar
- Transverse crossmember

- Front to aft crossmember
- Left side splash shield and properly support the transmission
- Remaining transmission flywheel housing bolts
- Transmission and separate the right side halfshaft from the transmission

To install:

4. Align the right side half shaft to the transmission and position the transmission to the engine.
5. Install or connect the following:
- Transmission flywheel housing bolts. Torque the bolts to 33 ft. lbs. (45 Nm) and remove the transmission support
- Left side splash shield
- Front-to-aft crossmember. Torque the bolts to 66 ft. lbs. (90 Nm).
- Transverse crossmember. Torque the bolts to 85 ft. lbs. (115 Nm).
- Shifter linkage. Torque the bolt to 15 ft. lbs. (20 Nm).
- Stabilizer bar. Torque the bolt to 30 ft. lbs. (40 Nm).
- Rear transmission support bracket.

Torque the bolts to 66 ft. lbs. (90 Nm).
- Left side halfshaft
- Transfer case, if equipped
- Front transmission flywheel housing bolts. Torque the bolts to 33 ft. lbs. (45 Nm).
- Top transmission flywheel housing bolts. Torque the bolts to 33 ft. lbs. (45 Nm).
- Starter. Torque the bolts to 33 ft. lbs. (45 Nm).
- Front transmission support insulator and bracket. Torque the lower bolt to 66 ft. lbs. (90 Nm) and the 3 upper bolts to 41 ft. lbs. (55 Nm).
- Rear transmission support insulator bolt. Torque the bolt to 66 ft. lbs. (90 Nm).
- Left side transmission support insulator bracket. Torque the bolts to 66 ft. lbs. (90 Nm).
- Left side transmission support insulator. Torque the large bolt to 66 ft. lbs. (90 Nm) and the 3 remaining bolts to 41 ft. lbs. (55 Nm).
- Clutch slave cylinder. Torque the bolt to 15 ft. lbs. (20 Nm).
- Clutch slave cylinder line to the bracket and install the retaining clip
- VSS electrical connector
- Rear wire harness bracket. Torque the bolts to 80 inch lbs. (9 Nm).
- PNP switch electrical connector
- Wire harness to the rear bracket
- Front wire harness bracket spacer and bracket. Torque the bolt to 9 ft. lbs. (12 Nm).
- Back-up lamp switch electrical connector
- Air cleaner housing
- MAF sensor electrical connector
- Air cleaner outlet tube
- Crankcase ventilation hose
- Emission management tube and hose
- Accelerator cable to the air cleaner outlet tube
- Battery and tray
- Both battery cables

6. Fill the transmission to the proper level.

7. Start the vehicle and check for leaks, repair if necessary.

Automatic Transmission

1. Before servicing the vehicle, refer to the precautions in the beginning of this section.

2. Remove or disconnect the following:
- Battery cables
- Battery and tray
- Breather tube
- Mass Air Flow (MAF) sensor
- Intake tube and air cleaner cover
- Air cleaner assembly
- Transmission Range (TR) sensor
- Heated Oxygen (H02S) sensors
- Transmission harness connector and bracket
- Wire harness bracket spacer and move the bracket aside
- Shift cable
- Shift cable bracket and move the bracket aside
- Starter electrical connectors
- Starter
- Electrical connectors from the valve cover and install an engine support bar
- Upper transmission retaining bolts
- Left side upper transmission mounting plate
- Rear transmission mount
- Right side engine mount bolt and slightly raise the engine
- Both front wheels and splash shields
- Right side halfshaft and intermediate shaft assembly after matchmarking them
- Cross brace
- Center exhaust pipe and rubber hanger
- Front exhaust pipe and flange
- Rear exhaust pipe flange
- Driveshaft
- PTU vent tube
- Lower transmission bracket
- Access cover
- Flexplate nuts
- Output Shaft Speed (OSS) sensor
- Turbine Shaft Speed (TSS) sensor
- Fluid cooler tube and move it aside
- Fluid cooler line and install a transmission jack
- Bolts from the PTU unit
- Transmission with the PTU unit attached

To install:
3. Install or connect the following:
- Transmission with the PTU unit. Torque the engine-to-transmission mounting bolts to 30 ft. lbs. (40 Nm).
- Fluid cooler line. Torque the fastener to 17 ft. lbs. (23 Nm) and remove the transmission jack.

- Fluid cooler tube. Torque the bolt to 17 ft. lbs. (23 Nm).
- OSS sensor
- TSS sensor
- Flexplate nuts. Torque the nuts to 27 ft. lbs. (36 Nm).
- Access cover
- Cross brace. Torque the bolts to 96 ft. lbs. (130 Nm).
- Transmission bracket. Torque the bolts to 30 ft. lbs. (40 Nm).
- PTU vent tube
- Driveshaft. Torque the bolts to 15 ft. lbs. (20 Nm).
- Exhaust pipe and flange. Torque the bolts to 21 ft. lbs. (29 Nm).
- Exhaust pipe and rubber hanger. Torque the bolts to 21 ft. lbs. (29 Nm).
- Left side halfshaft assembly
- Right side halfshaft and intermediate shaft assembly by aligning the matchmarks
- Splash shields
- Both front wheels and lower the engine on to the right side engine mount
- Right side engine mount bolt. Torque the bolt to 89 ft. lbs. (120 Nm).
- Rear transmission mount. Torque the upper bolt to 89 ft. lbs. (120 Nm) and the lower bolts to 35 ft. lbs. (45 Nm).
- Transmission mount assemble. Torque the bolts to 30 ft. lbs. (40 Nm) and remove the engine support bar.
- Electrical connectors to the valve cover
- Starter. Torque the bolts to 20 ft. lbs. (27 Nm).
- Starter electrical connectors
- Shifter cable and bracket. Torque the bolt to 14 ft. lbs. (19 Nm) and connect the shifter cable.
- Wire harness and install the harness bracket spacer
- Wire harness bracket. Torque the bolt to 89 inch lbs. (10 Nm).
- H02S sensor
- TR sensor and make certain it is properly aligned
- Air cleaner assembly
- Intake tube and air cleaner cover
- Breather tube
- MAF sensor
- Battery tray
- Battery and cables

For Tire, Wheel and Ball Joint specifications, see Section 1 of this manual

4. Fill the transmission with clean fluid to the proper level.

5. Start the vehicle and check for leaks, repair if necessary

Clutch

ADJUSTMENTS

The clutch is hydraulically driven and therefore no adjustment is required.

REMOVAL & INSTALLATION

1. Before servicing the vehicle, refer to the precautions in the beginning of this section.

2. Remove or disconnect the following:
   • Negative battery cable
   • Transmission and lock the flywheel to the engine with special tool 303–103
   • Pressure plate bolts by loosening them evenly
   • Clutch pressure plate and disc

3. Clean the pressure plate and inspect it for burn marks, scores, flatness or ridges, replace if damaged.

4. Inspect the pressure plate diaphragm finger for wear, replace if damaged.

5. Measure the depth of the rivet heads. Minimum depth is 0.012 inch (0.3mm).

6. Inspect the clutch disc for signs of wear and replace if needed.

7. Check the clutch disc runout. Replace the disc if not with specification: 0.027 inch (0.7mm).

To install:

8. Install or connect the following:
   • Clutch disc to the flywheel
   • Pressure plate to the flywheel. Torque the bolts in sequence to 21 ft. lbs. (29 Nm).
   • Transmission
   • Negative battery cable

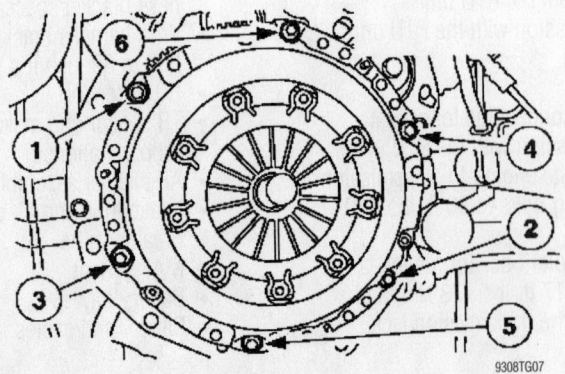

Torque the pressure plate bolts in the proper sequence

9308TG07

9. Check the transmission fluid level and top off if necessary.

Hydraulic Clutch System

BLEEDING

The following procedure is recommended for bleeding the clutch hydraulic system installed on the vehicle. It is recommended that the original clutch tube, with quick-connect fitting be replaced when servicing the hydraulic system, because air can be trapped in the quick-connect fitting and prevent complete bleeding of the system.

1. Before servicing the vehicle, refer to the precautions in the beginning of this section.

2. Clean the dirt and grease from the dust cap.

3. Remove the cap and diaphragm and fill the reservoir ¾ of the way with approved brake fluid C6AZ-19542-AB or DOT 3 equivalent fluid (ESA-M6C25-A).

4. Loosen the bleeder screw cover from the slave cylinder and attach a hose to the screw.

5. Place the hose in a container and slowly pump the clutch pedal several times.

6. With the clutch pedal depressed, loosen the bleeder screw to release the fluid and air.

7. Remove the hose and tighten the bleeder screw.

8. Repeat this procedure until all the air is removed from the hydraulic system.

Transfer Case

REMOVAL & INSTALLATION

3.0L Engine

1. Before servicing the vehicle, refer to the precautions in the beginning of this section.

2. Place the transmission in the NEUTRAL position.

3. Remove or disconnect the following:
   • Negative battery cable
   • Driveshaft
   • Intermediate shaft
   • Exhaust crossover pipe
   • Right side exhaust manifold
   • Heat shield
   • Lower bracket
   • Crossmember brace
   • Transfer case bolts and vent tube
   • Transfer case

To install:

4. Install or connect the following:
   • Transfer case with a new driven gear seal. Torque the bolts to 33 ft. lbs. (45 Nm).
   • Transfer case vent tube
   • Crossmember brace. Torque the bolts to 30 ft. lbs. (40 Nm).
   • Lower bracket. Torque the bolts to 30 ft. lbs. (40 Nm).
   • Exhaust manifold
   • Heat shield. Torque the bolts to 10 ft. lbs. (14 Nm).
   • Exhaust crossover pipe
   • Intermediate shaft
   • Driveshaft
   • Negative battery cable

5. Check the transfer case fluid level and top off if necessary.

6. Start the vehicle and check for leaks, repair if necessary.

Halfshaft

REMOVAL & INSTALLATION

1. Before servicing the vehicle, refer to the precautions in the beginning of this section.

2. Place the transmission in the PARK position.

3. Remove or disconnect the following:
   • Negative battery cable
   • Front wheel
   • Front brake disc
   • Front axle wheel hub nut and discard the nut
   • Tie rod end and separate the lower ball from the steering knuckle
   • Halfshaft from the steering knuckle
   • Halfshaft

To install:

4. When seated properly, the halfshaft bearing retainer circlip will snap into the differential side gear groove.

5. Position the halfshaft and joint so that the splines align with differential side gear splines. Push the halfshaft into side gear.

6. Install or connect the following:
- Halfshaft into the steering knuckle
- Lower ball joint to steering knuckle. Torque the pinch bolt to 52 ft. lbs. (70 Nm).
- Tie rod end. Torque the nut to 41 ft. lbs. (55 Nm).
- New front axle wheel hub nut. Torque the nut to 214 ft. lbs. (290 Nm).
- Front brake disc
- Front wheel
- Negative battery cable

7. Check the fluid level and adjust as needed.

CV-Joints

OVERHAUL

1. Before servicing the vehicle, refer to the precautions in the beginning of this section.

2. Remove or disconnect the following:
- Negative battery cable
- Halfshaft and secure it in a soft-jawed vise
- Inboard halfshaft boot clamp
- Boot from the inboard CV-joint housing

- Tripod joint from the CV-joint housing and matchmark the tripod joint to the halfshaft
- Snapring and boot from the half-shaft
- Outboard halfshaft boot clamps
- Outboard boot back to expose the CV-joint and matchmark the joint to the halfshaft
- Outboard CV-joint from the halfshaft
- Halfshaft retainer circlip and discard it
- Boot from the halfshaft

To install:

3. Install or connect the following:
- Outboard CV-joint and boot
- New halfshaft bearing circlip
- Inboard CV-joint to the halfshaft
- Outboard halfshaft boot forward on to the outboard CV-joint
- New outboard halfshaft boot clamps
- Inboard halfshaft boot
- Tripod joint on the halfshaft by aligning the matchmarks
- New snapring to the tripod joint and lubricate the needle bearings while filling the housing with CV-joint grease, E43Z–19590–A
- Inboard halfshaft boot with new clamps

- Halfshaft
- Negative battery cable

Axle Housing

REMOVAL & INSTALLATION

1. Before servicing the vehicle, refer to the precautions in the beginning of this section.

2. Remove or disconnect the following:
- Negative battery cable
- Rotary blade coupling
- Rear halfshafts
- Axle assembly-to-front bracket bolts
- Rear axle-to-side bracket bolts
- Axle assembly

To install:

3. Install or connect the following:
- Axle assembly
- Rear axle-to-side-bearing bolts. Torque the bolts to 59 ft. lbs. (80 Nm).
- Axle assembly-to-front bracket bolts. Torque the bolts to 59 ft. lbs. (80 Nm).
- Rotary blade coupling
- Negative battery cable

STEERING AND SUSPENSION

Air Bag

✳✳ CAUTION

Some vehicles are equipped with an air bag system. The system MUST BE disabled before performing service on or around system components, steering column, instrument panel components, wiring and sensors. Failure to follow safety and disabling procedures could result in accidental air bag deployment, possible personal injury and unnecessary system repairs.

PRECAUTIONS

Several precautions must be observed when handling the inflator module to avoid accidental deployment and possible personal injury:

1. Never carry the inflator module by the

wires or connector on the underside of the module.

2. When carrying a live inflator module, hold securely with both hands and ensure that the bag and trim cover are pointed away.

3. Place the inflator module on a bench or other surface with the bag and trim cover facing up.

4. With the inflator module on the bench, never place anything on or close to the module, which may be thrown in the event of an accidental deployment.

DISARMING

✳✳ CAUTION

The Supplemental Inflatable Restraint (SIR) system must be disarmed before performing service around SIR system components or SIR system wiring. Failure to do so may cause accidental deployment of

the air bag, resulting in unnecessary SIR system repairs and/or personal injury.

The positive battery cable must be disconnected for a minimum of 1 minute before beginning any air bag work to de-energize the back-up power supply. It is a good idea to disengage both the positive and negative battery cables to ensure that the Air Bag system is definitely discharged.

ARMING THE SYSTEM

✳✳ WARNING

If the air bag simulators have been used, the air bag simulators must be removed and the air bags reconnected when the system is reactivated to avoid non-deployment in a collision resulting in possible personal injury.

1. Disconnect the positive battery cable.
2. Wait 1 minute, this is required for the back-up power supply in the air bag diagnostic monitor to deplete its stored energy.
3. Remove the air bag simulator from the air bag sliding contact connector at the top of the steering column. Reconnect the driver's side air bag module assembly. Position the driver's air bag module on the steering wheel and secure with the 2 bolts and washers. Tighten the bolt and washer assembly to 8–10 ft. lbs. (10–14 Nm).
4. Connect the positive battery cable.
5. Turn the ignition switch from the **OFF** to **RUN** and visually monitor the air bag warning indicator. The light will illuminate continuously for approximately 6 seconds and then turn off. If a fault occurs, the air bag indicator will either fail to light, remain lighted continuously or flash. The flashing may not occur until approximately 30 seconds after the ignition switch has been turned from **OFF** to **RUN**. This is the time needed for the air bag diagnostic monitor to complete testing the system. If the air bag indicator is inoperative, an air bag system fault exists, a tone will sound in a pattern of 5 sets of 5 beeps. If this occurs, the air bag indicator will need to be serviced before further diagnostics can be done.

Steering Gear

REMOVAL & INSTALLATION

1. Before servicing the vehicle, refer to the precautions in the beginning of this section.
2. Place the steering wheel in the straight-ahead position. Lock the steering wheel in place, using a steering wheel holder.

➡**Locking the steering wheel keeps the clockspring in alignment position.**

3. Drain the power steering fluid.
4. Remove or disconnect the following:
- Negative battery cable
- Rear transmission insulator
- Rear transmission insulator bracket, if equipped with an automatic transmission
- Both front wheels
- Rear transmission insulator bracket, if equipped with a manual transmission
- Tie rod end cotter pin and nut
- Tie rod end from the steering knuckle and record the number of turns required to remove the tie rod end

- Steering gear coupling pinch bolt
- Power steering pressure and return lines and bracket
- Steering gear mounting bolts
- Steering gear and separate the steering coupling from the steering gear shaft
- Steering gear

To install:

5. Slide the steering gear rearward to connect the steering coupling to the steering gear shaft
6. Install or connect the following:
- Steering gear mounting bolts. Torque the bolts to 93 ft. lbs. (126 Nm).
- Pressure and return lines and bracket. Torque the bracket bolts to 89 inch lbs. (10 Nm).
- Power steering pressure and return lines to the steering gear. Torque the bolt to 18 ft. lbs. (25 Nm).
- Steering gear pinch bolt and reposition the boot. Torque the bolt to 18 ft. lbs. (25 Nm).
- Tie rod end to the tie rod using the number of turns required to remove the tie rod end
- Jam nuts. Torque the nuts to 35 ft. lbs. (47 Nm).
- Tie rod end to the steering knuckle. Torque the nut to 41 ft. lbs. (57 Nm) and install a new cotter pin
- Rear transmission insulator bracket. Torque the bolts to 66 ft. lbs. (90 Nm).
- Both front wheels
- Rear transmission insulator bracket. Torque the bolts to 66 ft. lbs. (90 Nm).
- Rear transmission insulator. Torque the bolts to 66 ft. lbs. (90 Nm).
- Negative battery cable
7. Fill and bleed the power steering system.
8. Start the vehicle and check for leaks, repair if necessary.
9. Check and adjust the front end alignment.

Strut

REMOVAL & INSTALLATION

1. Before servicing the vehicle, refer to the precautions in the beginning of this section.
2. Install or connect the following:
- Negative battery cable
- Front wheel

- Brake hose grommet from the bracket
- Antilock Brake System (ABS) harness from the strut assembly and move the brake hose bracket aside
- Stabilizer bar link nut and move the bar aside
- Strut to steering knuckle bolts and support the strut assembly
- Upper strut nuts
- Strut and coil spring assembly

To install:

3. Install or connect the following:
- Strut and spring assembly. Torque the upper nuts to 59 ft. lbs. (80 Nm).
- Lower strut assembly to the steering knuckle. Torque the lower bolts to 85 ft. lbs. (115 Nm).
- Stabilizer bar into position. Torque the bolts to 35 ft. lbs. (48 Nm).
- Brake hose bracket. Torque the bolts to 14 ft. lbs. (18 Nm).
- ABS harness to the strut assembly, if equipped
- Brake hose grommet to the bracket
- Front wheel
- Negative battery cable

Shock Absorber

REMOVAL & INSTALLATION

1. Before servicing the vehicle, refer to the precautions in the beginning of this section.
2. Remove or disconnect the following:
- Negative battery cable
- Rear quarter trim panel
- Upper shock absorber nut and raise the vehicle enough to relax the suspension
- Lower shock absorber nut
- Shock absorber

To install:

3. Install or connect the following:
- Shock absorber. Torque the lower nut to 85 ft. lbs. (115 Nm).
- Upper shock absorber nut. Torque the nut to 13 ft. lbs. (18 Nm).
- Rear quarter trim panel
- Negative battery cable

Coil Spring

REMOVAL & INSTALLATION

Front

1. Before servicing the vehicle, refer to the precautions in the beginning of this section.

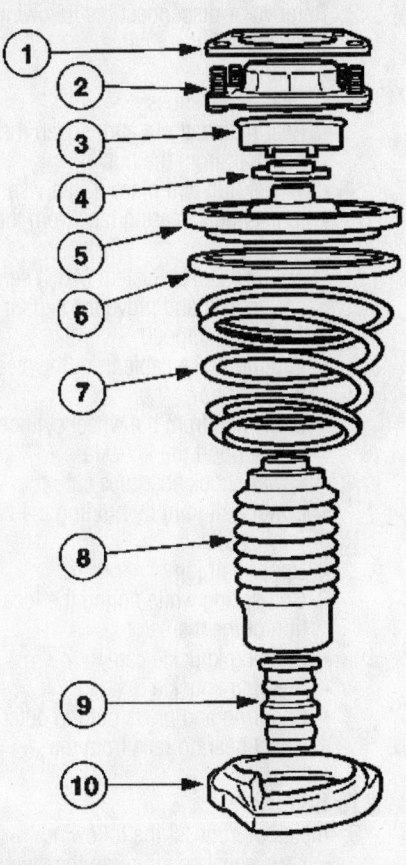

1	Metal sheet plate
2	Upper strut mount
3	Thrust bearing plate
4	Thrust bearing
5	Upper spring seat
6	Upper spring seat isolator
7	Spring
8	Dust boot
9	Rubber bump stopper
10	Lower spring seat

9308TG23

Disassemble the strut assembly in the proper sequence

2. Install or connect the following:
- Negative battery cable
- Front wheel
- Strut and spring assembly and mount the strut assembly in a holding fixture and compress the coil spring using a suitable tool
- Strut piston rod nut

3. Coil spring by disassembling the strut in the following sequence:
   a. Step 1: Metal sheet plate.
   b. Step 2: Upper strut mount.
   c. Step 3: Thrust bearing plate.
   d. Step 4: Thrust bearing.
   e. Step 5: Upper spring seat.
   f. Step 6: Upper spring seat isolator.
   g. Step 7: Coil spring.
   h. Step 8: Dust boot.
   i. Step 9: Rubber bump stopper.
   j. Step 10: Lower spring seat.

To install:

4. Assemble the strut assembly in the reverse order of the removal procedure

5. Install or connect the following:
- Strut piston rod nut. Torque the nut to 76 ft. lbs. (103 Nm) and remove the assembly from the holding fixture
- Strut and spring assembly
- Front wheel
- Negative battery cable

Rear

1. Before servicing the vehicle, refer to the precautions in the beginning of this section.

2. Remove or disconnect the following:
- Wheel and install 1 lug nut to retain the brake drum
- Brake line from the wheel cylinder
- Brake line bracket
- Bolts from the Antilock Braking System (ABS) sensor bracket and move the sensor aside, if equipped
- Rear knuckle and loosen the inside upper and lower arm bolts
- Shock absorber lower nut
- Spring

To install:

3. Install or connect the following:
- Spring to the shock absorber
- Lower shock absorber nut. Torque the nut to 85 ft. lbs. (115 Nm).
- Inside upper and lower arm bolts. Torque the bolts to 85 ft. lbs. (115 Nm).
- ABS sensor bracket, if equipped. Torque the bolts to 80 inch lbs. (9 Nm).
- Brake line bracket. Torque the bolt to 15 ft. lbs. (20 Nm).
- Brake line to the wheel cylinder. Torque the fastener to 11 ft. lbs. (15 Nm) and remove the lug nut
- Wheel
- Negative battery cable

Upper Control Arm

REMOVAL & INSTALLATION

The upper and lower ball joints are an integral part of the control arms and are not a serviceable components. Replacement of the ball joint requires replacing the appropriate control arm.

1. Before servicing the vehicle, refer to the precautions in the beginning of this section.

2. Remove or disconnect the following:
- Negative battery cable
- Rear wheel
- Upper control arm from the knuckle while holding the ball joint stud from turning
- Upper ball joint nut
- Upper control arm
- Upper control arm inner bolt

To install:

3. Install or connect the following:
- Upper control arm inner bolt
- Upper control arm. Torque the bolts to 85 ft. lbs. (115 Nm).
- Upper ball joint nut
- Upper control arm to the knuckle. Torque the ball joint nut to 85 ft. lbs. (115 Nm).
- Rear wheel
- Negative battery cable

Lower Control Arm

REMOVAL & INSTALLATION

Front

1. Remove or disconnect the following:
- Negative battery cable
- Front wheel
- Lower ball joint from the knuckle and support the subframe
- Lower control arm

To install:

2. Install or connect the following:
   - Lower control arm bolts and hand tighten them
   - Pinch bolt to the wheel knuckle. Torque the nut to 52 ft. lbs. (70 Nm) and remove the subframe support
   - Front wheel and jounce the vehicle
3. Torque the inner lower control arm bolt to 148 ft. lbs. (200 Nm) and outer bolt 85 ft. lbs. (115 Nm).

Rear

1. Remove or disconnect the following:
   - Negative battery cable
   - Front wheel
   - Lower ball joint from the knuckle while holding the ball joint stud from moving
   - Lower ball joint nut
   - Lower control arm
   - Lower control arm inner bolt

To install:

2. Install or connect the following:
   - Lower control arm inner bolt
   - Lower control arm. Torque the bolts to 85 ft. lbs. (115 Nm).
   - Lower ball joint nut
   - Lower ball joint the knuckle. Torque the ball joint nut to 85 ft. lbs. (115 Nm).
   - Rear wheel
   - Negative battery cable

Wheel Bearings

REMOVAL & INSTALLATION

Front

1. Before servicing the vehicle, refer to the precautions in the beginning of this section.
2. Remove or disconnect the following:
   - Negative battery cable
   - Front wheel
   - Brake disc
   - Wheel hub nut
   - Tie rod end cotter pin and nut
   - Tie rod end from the knuckle
   - Antilock Brake System (ABS) sensor bolt and move the sensor aside, if equipped
   - Lower ball joint from the knuckle
   - Halfshaft from the knuckle and properly support the halfshaft
   - Steering knuckle

3. Press the hub from the wheel bearing and knuckle
4. Press the inner wheel bearing race from the knuckle and remove the snapring
5. Press the outer wheel bearing race from the knuckle

To install:

6. Install or connect the following:
   - Wheel bearing into the steering knuckle
   - Snapring
   - Wheel hub into the wheel bearing by using a press
   - Steering knuckle. Torque the bolts to 85 ft. lbs. (115 Nm).
   - Halfshaft into the wheel hub
   - Pinch bolt to knuckle. Torque the nut to 52 ft. lbs. (70 Nm).
   - Ball joint stud into the knuckle
   - ABS sensor. Torque the bolt to 80 inch lbs. (9 Nm), if equipped
   - Tie rod end to the knuckle. Torque the nut to 41 ft. lbs. (55 Nm).
   - New cotter pin to the tie rod end nut
   - Wheel hub. Torque the nut to 214 ft. lbs. (290 Nm).
   - Brake disc
   - Front wheel
   - Negative battery cable

Rear

2WD VEHICLES

1. Before servicing the vehicle, refer to the precautions in the beginning of this section.
2. Remove or disconnect the following:
   - Negative battery cable
   - Rear wheel
   - Rear brake drum
   - Wheel hub nut
   - Wheel hub
   - Inner wheel bearing race from the hub
   - Snapring
   - Wheel bearing outer race from the knuckle

To install:

3. Install or connect the following:
   - Wheel bearing in to the knuckle
   - Snapring
   - Wheel hub into the wheel bearing
   - Wheel hub nut. Torque the nut to 214 ft. lbs. (290 Nm).
   - Brake drum
   - Rear wheel
   - Negative battery cable

4WD VEHICLES

1. Before servicing the vehicle, refer to the precautions in the beginning of this section.

2. Remove or disconnect the following:
   - Negative battery cable
   - Rear wheel
   - Rear brake shoes
   - Rear halfshaft nut and loosen the halfshaft from the hub
   - Wheel hub and place it in a vise
   - Inner wheel bearing race from the hub
   - Antilock Brake System (ABS) sensor bracket and move the sensor aside, if equipped
   - Parking brake cable from the steering knuckle
   - Brake line from the wheel cylinder and support the knuckle
   - Lower shock absorber nut
   - Lower ball joint by holding the ball joint stud
   - Upper ball joint
   - Coil spring while noting the location of the insulator
   - Steering knuckle cam
   - Steering knuckle
   - Snapring and press out the outer wheel bearing race from the knuckle

To install:

3. Install or connect the following:
   - New wheel bearing into the steering knuckle
   - Snapring to the knuckle
   - Wheel hub
   - Steering knuckle cam and hand tighten the bolt
   - Coil spring
   - Shock absorber lower nut. Torque the nut to 85 ft. lbs. (115 Nm).
   - Upper ball joint. Torque the nut to 85 ft. lbs. (115 Nm).
   - Lower ball joint. Torque the nut to 85 ft. lbs. (115 Nm). Align the steering knuckle cam and torque the bolt to 85 ft. lbs. (115 Nm).
   - Brake line to the wheel cylinder. Torque the brake line bracket bolt to 15 ft. lbs. (20 Nm) and the brake line fastener to 11 ft. lbs. (15 Nm).
   - Parking brake cable to the backing plate. Torque the bolt to 16 ft. lbs. (22 Nm).
   - ABS sensor bracket. Torque the bolt to 80 inch lbs. (9 Nm), if equipped
   - Halfshaft nut. Torque the nut to 214 ft. lbs. (290 Nm).
   - Brake shoes
   - Rear wheel
   - Negative battery cable
4. Fill and bleed the brake system.
5. Check and adjust the wheel alignment as needed.

FORD MOTOR CO.

Ford-Explorer • Explorer Sport • Explorer Sport Trac • **Mercury-**Mountaineer

PRECAUTIONS

Before servicing any vehicle, please be sure to read all of the following precautions, which deal with personal safety, prevention of component damage, and important points to take into consideration when servicing a motor vehicle:

• Never open, service or drain the radiator or cooling system when the engine is hot; serious burns can occur from the steam and hot coolant.

• Observe all applicable safety precautions when working around fuel. Whenever servicing the fuel system, always work in a well-ventilated area. Do not allow fuel spray or vapors to come in contact with a spark, open flame, or excessive heat (a hot drop light, for example). Keep a dry chemical fire extinguisher near the work area. Always keep fuel in a container specifically designed for fuel storage; also, always properly seal fuel containers to avoid the possibility of fire or explosion. Refer to the additional fuel system precautions later in this section.

• Fuel injection systems often remain pressurized, even after the engine has been turned **OFF**. The fuel system pressure must be relieved before disconnecting any fuel lines. Failure to do so may result in fire and/or personal injury.

• Brake fluid often contains polyglycol ethers and polyglycols. Avoid contact with the eyes and wash your hands thoroughly after handling brake fluid. If you do get brake fluid in your eyes, flush your eyes with clean, running water for 15 minutes. If

eye irritation persists, or if you have taken brake fluid internally, IMMEDIATELY seek medical assistance.

• The EPA warns that prolonged contact with used engine oil may cause a number of skin disorders, including cancer! You should make every effort to minimize your exposure to used engine oil. Protective gloves should be worn when changing oil. Wash your hands and any other exposed skin areas as soon as possible after exposure to used engine oil. Soap and water, or waterless hand cleaner should be used.

• All new vehicles are now equipped with an air bag system, often referred to as a Supplemental Restraint System (SRS) or Supplemental Inflatable Restraint (SIR) system. The system must be disabled before performing service on or around system components, steering column, instrument panel components, wiring and sensors. Failure to follow safety and disabling procedures could result in accidental air bag deployment, possible personal injury and unnecessary system repairs.

• Always wear safety goggles when working with, or around, the air bag system. When carrying a non-deployed air bag, be sure the bag and trim cover are pointed away from your body. When placing a non-deployed air bag on a work surface, always face the bag and trim cover upward, away from the surface. This will reduce the motion of the module if it is accidentally deployed. Refer to the additional air bag system precautions later in this section.

• Clean, high quality brake fluid from a sealed container is essential to the safe and proper operation of the brake system. You should always buy the correct type of brake fluid for your vehicle. If the brake fluid becomes contaminated, completely flush the system with new fluid. Never reuse any brake fluid. Any brake fluid that is removed from the system should be discarded. Also, do not allow any brake fluid to come in contact with a painted surface; it will damage the paint.

• Never operate the engine without the proper amount and type of engine oil; doing so WILL result in severe engine damage.

• Timing belt maintenance is extremely important! Many models utilize an interference-type, non-freewheeling engine. If the timing belt breaks, the valves in the cylinder head may strike the pistons, causing potentially serious (also time-consuming and expensive) engine damage. Refer to the maintenance interval charts in the front of this manual for the recommended replacement interval for the timing belt, and to the timing belt section for belt replacement and inspection.

• Disconnecting the negative battery cable on some vehicles may interfere with the functions of the on-board computer system(s) and may require the computer to undergo a relearning process once the negative battery cable is reconnected.

• When servicing drum brakes, only disassemble and assemble one side at a time, leaving the remaining side intact for reference.

ENGINE REPAIR

Alternator

REMOVAL & INSTALLATION

4.0L OHV and 5.0L Engines

1. Before servicing the vehicle, refer to the precautions in the beginning of this section.

2. Remove or disconnect the following:
   • Negative battery cable
   • Air cleaner outlet tube
   • Drive belt
   • Electrical connectors from the alternator
   • A/C manifold and tube bracket aside, 5.0L engine only
   • Wiring harness to alternator push pin
   • Alternator

To install:

3. Install or connect the following:
   • Alternator. Torque the bolts to 40 ft. lbs. (55 Nm).
   • Push pin for the alternator wiring harness
   • A/C manifold and tube bracket, 5.0L engine only. Torque the bolt to 106 inch lbs. (12 Nm).
   • Electrical connectors to the alternator
   • Drive belt
   • Air cleaner outlet tube
   • Negative battery cable

4.0L OHC Engine

1. Before servicing the vehicle, refer to the precautions in the beginning of this section.

2. Remove or disconnect the following:

   • Negative battery cable
   • Air cleaner outlet tube
   • Accessory drive belt
   • Electrical connectors
   • Wiring harness-to-generator pin-type retainer
   • Stud bolt, the bolts and the alternator

3. To install, reverse the removal procedure. Torque all mounting bolts to 35 ft. lbs. (47Nm).

4.6L Engine

1. Before servicing the vehicle, refer to the precautions in the beginning of this section.

2. Remove or disconnect the following:
   • Negative battery cable
   • Drive belt
   • Electrical connectors

- Stud bolts, the bolts and the generator bracket
- Bolts and remove the alternator

3. To install, reverse the removal procedure. Torque the mounting bolts to 18 ft. lbs. (25Nm).

Ignition Timing

ADJUSTMENT

The ignition timing is preset to 10 degrees Before Top Dead Center (BTDC) and is not adjustable.

Engine Assembly

REMOVAL & INSTALLATION

4.0L OHV Engines

1. Before servicing the vehicle, refer to the precautions in the beginning of this section.
2. Relieve the fuel system pressure.
3. Drain the cooling system.
4. Drain the engine oil.
5. Properly discharge the A/C system.
6. Remove or disconnect the following:
   - Both battery cables
   - Mass Air Flow (MAF) sensor
   - Air cleaner outlet tube
   - Drive belt
   - Accelerator control splash shield
   - Upper and lower radiator hoses
   - Fan guard/shroud
   - Radiator overflow tube
   - Radiator
   - Transmission cooler lines, if equipped
   - Heater hoses
   - Alternator electrical connectors
   - Vacuum reservoir connection
   - Throttle body heater hose
   - A/C cycling switch
   - Powertrain Control Module (PCM) connector
   - Ground wire from the PCM
   - Power steering cut-out switch
   - Peanut fitting from the A/C condenser core
   - A/C high pressure cut-out switch
   - A/C manifold hose
   - Accelerator and speed control cables
   - Brake booster vacuum hose and tube from the intake manifold
   - Vacuum reservoir line

- Fuel lines
- Power steering pressure and return hoses
- Block heater, if equipped
- Engine ground cable
- Automatic transmission harness connectors, if equipped
- Heated Oxygen (HO2S) sensor, if equipped
- Starter
- Catalytic converter
- Torque converter bolts, if equipped and properly support the transmission
- Differential pressure feedback sensor and support the engine with a floor crane
- Exhaust Gas Recirculation (EGR) transducer
- Upper transmission-to-engine bolts
- Engine from the vehicle
- Clutch/ flywheel, if equipped

To install:

7. Install or connect the following:
   - Flywheel. Torque the bolts to 64 ft. lbs. (87 Nm).
   - Engine. Torque the mounting bolts to 85 ft. lbs. (115 Nm).
   - HO2S sensor
   - Upper transmission-to-engine bolts. Torque the bolts to 38 ft. lbs. (51 Nm).
   - EGR transducer and remove the floor jack from the transmission
   - Torque converter-to-engine bolts. Torque the bolts to 38 ft. lbs. (51 Nm).
   - Catalytic converter
   - Transmission wiring harness connectors
   - Engine ground cable. Torque the bolt to 106 inch lbs. (12 Nm).
   - Block heater, if equipped
   - A/C high pressure cut out switch
   - A/C manifold to the evaporator core
   - Power steering cut-out switch
   - Power steering pressure and return lines
   - Engine sensor control wiring harness to the A/C compressor
   - Fuel lines
   - 42 pin connector
   - Vacuum reservoir vacuum line
   - Brake booster vacuum hose and tube
   - Accelerator cable
   - Ground strap. Torque the bolt to 106 inch lbs. (12 Nm).
   - A/C manifold hose

- PCM ground strap. Torque the bolt to 106 inch lbs. (12 Nm).
- PCM wire harness bracket bolt. Torque the bolt to 61 inch lbs. (7 Nm).
- A/C low pressure cut-out switch
- Throttle body heater hose
- Fan clutch and water pump pulley. Torque the bolt 17 ft. lbs. (23 Nm).
- Drive belt
- Alternator electrical connectors
- Inlet and outlet heater hoses
- Fuel charging wiring
- Radiator and fan shroud
- Transmission cooler lines, if equipped
- Upper, lower and overflow hoses to the radiator
- Accelerator control splash shield. Torque the bolts to 89 inch lbs. 10 Nm).
- Drive belt
- Air cleaner outlet tube
- MAF sensor
- Both battery cables

8. Recharge the A/C system.
9. Fill the cooling system.
10. Fill the engine with clean oil.
11. Run the engine and check for leaks and proper operation.
12. Check and adjust the front end alignment.

4.0L OHC Engines

✳✳ CAUTION

If the fuel supply manifold is used as a leverage device, damage may occur to the supply manifold. Care must be taken when working around the fuel supply manifold.

1. Remove or disconnect the following:
   - Accelerator cable from engine
   - Speed control cable from engine
   - Radiator, the fan blade, and the fan shroud
   - Accessory bracket bolts and position bracket aside
   - Alternator wiring
   - Wiring harness retainer and position generator wiring away from engine
   - Engine electrical connector
   - PCM connector
   - PCM ground wire
   - Engine ground wire
   - Brake booster vacuum hose
   - A/C high pressure switch electrical connector

- Bolt and position the A/C lines aside

➤**Heater hose will be removed with engine.**

- Heater hoses
- Fuel line
- Starter motor
- Engine oil
- Oil drain plug
- Transmission portion of wiring harness
- RH and LH heated oxygen sensor connectors
- Transmission control connector
- Output shaft speed sensor connector
- Digital transmission range sensor connector
- Catalyst monitor sensor electrical connector
- Transmission/transfer case portion of the wiring harness from any routing clips or pushpins. Route transmission/transfer case portion of the wiring harness to top of engine.
- Bolt, and position the transmission cooling line bracket aside
- A/C line bracket nut and position it aside
- Power steering return hose
- Power steering pressure hose
- Vapor management valve hose connector
- Eight bolts and the LH and the RH engine support insulator nuts

➤**The lifting eyes should be installed on the exhaust manifold studs for number three and number four cylinders.**

2. Install the lifting eyes.
3. Install the spreader bar to the lifting eyes.
4. Attach a floor crane to the spreader bar and remove the engine.
5. Installation is the reverse of removal. Observe the following torques:
- Left and right engine insulator nuts: 81 ft. lbs. (110 Nm)
- Engine mount nuts: 59 ft. lbs. (80Nm)
- Transmission-to-engine bolts: 35 ft. lbs. (47Nm)
- Torque converter nuts: 35 ft. lbs. (47Nm)

4.6L Engine

1. Before servicing the vehicle, refer to the precautions in the beginning of this section.

2. Relieve the fuel system pressure.
3. Drain the cooling system.
4. Drain the engine oil.
5. Properly discharge the A/C system.
6. Remove or disconnect the following:
- Battery ground cable
- Hood
- Accelerator cable from the throttle body and all clips
- Speed control cable from the throttle body and all clips
- Air cleaner and the air cleaner outlet pipe
- Radiator and transmission cooler
- A/C condenser core

➤**Access the heater control valve bracket through the wheel well.**

- Heater control valve bracket bolt
- Coolant hoses
- All vacuum hoses. Position the heater control valve assembly aside.
- A/C tubes from the A/C compressor
- Fuel charging wiring harness connections
- Fuel charging wiring harness from the clips on the bulkhead. Release the fuel charging wiring harness pin-type retainer.
- All connections from the evaporative emissions (EVAP) canister purge solenoid
- Brake booster vacuum hose
- Fuel tube
- Alternator connections and the wiring anchors
- Wiring harness side of the left and right heated exhaust gas oxygen (HEGO) sensor connectors
- Power steering return hose. Plug the power steering reservoir.
- Power steering pressure tube from the power steering pump
- Brackets
- Oil pressure sender electrical connector
- CKP sensor electrical connector and the wiring anchor
- A/C compressor electrical connector and the wiring anchor
- Wiring harness routing clip and wiring anchor
- Starter
- On 4x4 vehicles, the bolt from the right side axle housing bushing bolt
- Right side and the left side stabilizer bar bracket nuts
- Left side lower axle housing bushing bolt

- Left side upper axle housing bushing bolt
- Right lower mount nut and washer
- Left lower mount nut and washer
- Transmission inspection cover, torque converter bolts and the transmission-to-engine block bolts

7. Install the right side Lifting Eye.
8. Install the left side Lifting Eye.
9. Install the Spreader Bar to the Lifting Eyes.
10. Attach a floor crane to the Spreader Bar and remove the engine.

➤**If engine disassembly is to be carried out, remove the rear crankshaft oil seal slinger and the crankshaft rear oil seal.**

11. Installation is the reverse of removal. Observe the following torques:
- Lower mount nuts: 59 ft. lbs. (81Nm)
- Left side upper axle housing bushing bolt: 49 ft. lbs. (66Nm)
- Left side axle housing bushing bolt: 49 ft. lbs. (66Nm)
- Stabilizer bar bracket nuts: 30 ft. lbs. (40Nm)
- Right side axle housing bushing bolt: 49 ft. lbs. (66Nm)

5.0L Engine

1. Before servicing the vehicle, refer to the precautions in the beginning of this section.
2. Relieve the fuel system pressure.
3. Drain the cooling system.
4. Drain the engine oil.
5. Properly discharge the A/C system.
6. Remove or disconnect the following:
- Battery
- Drive belt
- Fan shroud
- Air cleaner outlet tube
- A/C condenser core
- Upper radiator hose
- Power steering reservoir and move it aside
- Power steering pump
- Spark plug wire bracket from the A/C compressor
- A/C compressor and bracket
- Alternator
- Wide-open A/C cut-off switch electrical connector
- Vapor Management Valve (VMV) hose
- Lower steering column shaft and move it aside
- Left and right hand side vacuum connections

- Accelerator and speed control cables
- Accelerator cable bracket
- Powertrain Control Module (PCM) connector
- PCM ground connector
- Engine bulkhead connector
- Heater hoses
- Transmission fill tube
- Power steering cooler and move it aside
- Ground cable from the engine front cover
- Ground strap from the lower intake manifold
- Transmission cooler lines from the retainer on the right engine mount
- Starter
- Transmission inspection cover
- Torque converter nuts
- Transmission
- Left and right side Heated Oxygen (HO$_2$S) sensor electrical connectors
- Transmission bulkhead connector
- Brake booster vacuum supply line at the left upper intake manifold
- Low oil level sensor electrical connector
- Oil bypass filter
- Lower radiator hose
- Exhaust manifolds
- Left and right side motor mounts and install the lifting brackets to the exhaust manifold studbolts on No. 1–No. 8 cylinders
- Engine from the vehicle
- Flywheel, if equipped

To install:

7. Install or connect the following:
- Flywheel. Torque the new bolts to 85 ft. lbs. (115 Nm).
- Engine. Torque the motor mount bolts to 109 ft. lbs. (148 Nm).
- Exhaust manifolds
- Lower radiator hose
- Oil bypass filter
- Brake booster vacuum supply line to the left side upper intake connection
- Transmission bulkhead connector
- Left and right HO$_2$S sensor connectors
- Transmission
- Torque converter bolts. Torque the bolts to 38 ft. lbs. (51 Nm).
- Transmission inspection cover
- Starter

- Transmission cooler lines to the retainer by the right side motor mount
- Battery-to-starter relay cable
- Ground strap to the engine front cover
- Ground strap to the rear of the lower intake manifold
- Ground cable to the engine front cover
- Power steering cooler
- transmission fill tube
- Heater hoses
- Engine bulkhead connector. Torque the bolt to 89 inch lbs. (10 Nm).
- PCM and body ground connectors. Torque the bolt to 89 inch lbs. (10 Nm).
- Accelerator and speed control cables to the clip
- Accelerator cable bracket. Torque the upper bolts to 15 ft. lbs. (20 Nm) and lower the bolt to 80 inch lbs. (9 Nm).
- Accelerator and sped control cables to the throttle linkage
- Accelerator control shield. Torque the bolt to 106 inch lbs. (12 Nm).
- Left and right side vacuum connections
- Lower power steering column shaft. Torque the bolt to 40 ft. lbs. (55 Nm).
- VMV hose
- Fuel lines
- Wide-open A/C cutoff switch electrical connector
- Alternator and electrical connectors
- A/C compressor
- Spark plug wire bracket. Torque the nut to 89 inch lbs. (10 Nm).
- Power steering pump. Torque the bolts to 21 ft. lbs. (29 Nm).
- Power steering reservoir. Torque the bolts to 106 inch lbs. (12 Nm).
- Upper radiator hose
- Fan blade
- Drive belt
- A/C condenser core
- Air cleaner outlet tube
- Battery and cables

8. Recharge the A/C system.
9. Fill the cooling system.
10. Fill the engine with clean oil.
11. Run the engine and check for leaks and proper operation.
12. Check and adjust the front end alignment.

Water Pump

REMOVAL & INSTALLATION

4.0L OHV Engines

1. Before servicing the vehicle, refer to the precautions in the beginning of this section.
2. Drain the cooling system.
3. Remove or disconnect the following:
- Negative battery cable
- Air cleaner outlet tube
- Fan and radiator shroud
- Water bypass tube
- Drive belt
- Heater hose
- Water pump pulley
- Lower radiator hose
- A/C compressor and bracket assembly and move them aside
- Water pump

To install:

4. Clean the mating surfaces where the water pump attaches to the engine.
5. Install or connect the following:
- Water pump. Torque the bolts to 106 in lbs. (12 Nm).
- A/C compressor mounting bracket. Torque the bolts to 44 ft. lbs. (61 Nm).
- Water pump pulley. Torque the bolts to 20 ft. lbs. (28 Nm).
- Drive belt
- Heater hose
- Lower radiator hose
- Fan and shroud
- Air cleaner outlet tube
- Negative battery cable

6. Fill the cooling system.
7. Start the vehicle and check for leaks, repair if necessary.

4.0L OHC Engine

1. Before servicing the vehicle, refer to the precautions in the beginning of this section.
2. Drain the cooling system.
3. Remove or disconnect the following:
- Fan shroud
- Accessory drive belt
- Idler pulley
- Water bypass hose
- Heater hose
- Lower radiator hose
- Water pump pulley
- Water pump

✳✳ WARNING

Use care when scraping the water pump-to-engine block mating surfaces. Gouges in the aluminum could form leak paths.

4. Clean all the sealing surfaces.

5. To install, reverse the removal procedure. Torque the water pump bolts to 89 inch lbs. (10Nm). Torque the pulley bolts to 18 ft. lbs. (25Nm).

4.6L Engine

1. Before servicing the vehicle, refer to the precautions in the beginning of this section.

2. Drain the cooling system.

3. Remove or disconnect the following:
- Engine cooling fan
- Drive belt
- Water pump pulley
- Water pump

4. Discard the O-ring seal.

5. To install, reverse the removal procedure. Install a new O-ring seal and lubricate with engine coolant.

6. Torque the pump and pulley bolts to 18 ft. lbs. (25Nm).

5.0L Engines

1. Before servicing the vehicle, refer to the precautions in the beginning of this section.

2. Drain the cooling system.

3. Remove or disconnect the following:
- Negative battery cable
- Fan shroud
- Drive belt
- Water bypass hose
- Heater hose from the water pump
- Engine control sensor wiring and move it aside
- Water pump pulley
- Water pump from the engine front cover
- Water pump inlet hose
- Water pump

To install:

4. Clean the mounting surfaces of the pump and front cover thoroughly. Remove all traces of gasket material.

5. Install or connect the following:
- Apply adhesive gasket sealer to both sides of a new gasket and place the gasket on the pump.
- Inlet hose to the water pump. Torque the clamps to 11 ft. lbs. (15 Nm).
- Water pump. Torque the bolts to 20 ft. lbs. (28 Nm).

- Water pump pulley. Torque the bolts to 20 ft. lbs. (28 Nm).
- Engine control sensor wiring
- Heater hose
- Water bypass tube. Torque the clamps to 11 ft. lbs. (15 Nm).
- Dive belt
- Fan shroud
- Negative battery cable

6. Fill the cooling system.

7. Start the vehicle and check for leaks, repair if necessary.

Cylinder Head

REMOVAL & INSTALLATION

4.0L OHV Engine

➡**New cylinder head bolts must be used when installing the cylinder head on the 4.0L OHV engine.**

1. Before servicing the vehicle, refer to the precautions in the beginning of this section.

2. Relieve the fuel system pressure.

3. Evacuate the A/C system.

4. Drain the cooling system.

5. Drain the engine oil.

6. Remove or disconnect the following:
- Negative battery cable
- Drive belt
- Separate the A/C manifold tube from the A/C compressor
- A/C compressor electrical connectors
- A/C compressor mounting bracket and move it aside

- Alternator electrical connectors
- Heater hose and move it aside
- Alternator mounting bracket
- Lower intake manifold
- Both exhaust manifolds

7. Gradually loosen the rocker arm shafts and remove them.
- Matchmark the position of the push rods and remove them
- Cylinder head and gasket

To install:

8. Clean the mating surface where the cylinder head attaches to the engine.

9. Install a new cylinder head gasket and the cylinder head to the engine.

10. Torque the new cylinder head bolts in sequence, as follows:
- a. Step 1: 25 ft. lbs. (34 Nm).
- b. Step 2: 53 ft. lbs. (72 Nm).
- c. Step 3: Plus an additional 90 degrees.

11. Install or connect the following:
- Push rods
- Rocker arm shafts gradually. Torque the bolts to 24 ft lbs. (33 Nm) plus an additional 90 degrees
- Exhaust manifolds
- Lower intake manifold
- Alternator bracket. Torque the bolts to 35 ft. lbs. (47 Nm).
- Heater hose retaining clip
- Alternator electrical connectors
- A/C compressor mounting bracket. Torque the bolts to 35 ft. lbs. (47 Nm).
- A/C compressor electrical connectors
- A/C manifold tube to the A/C compressor

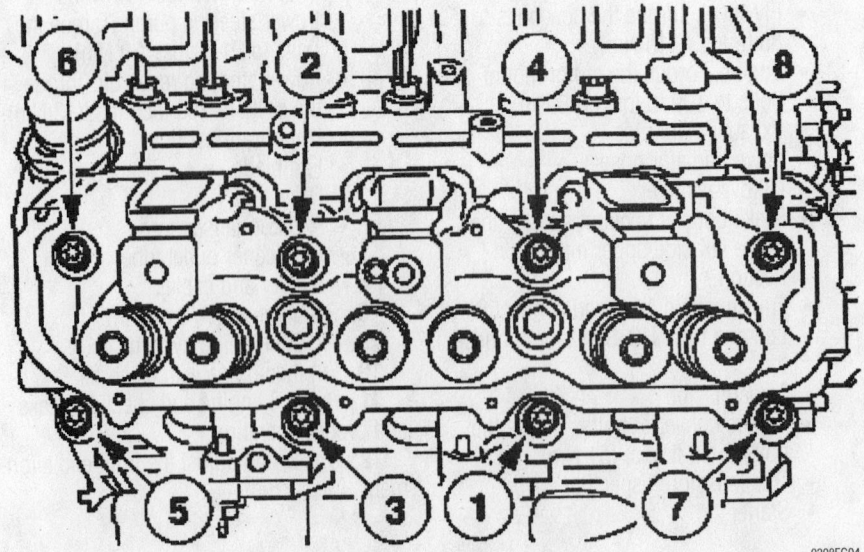

Cylinder head bolt torque sequence 4.0L OHV engine

9308EG04

- Drive belt
- Negative battery cable

12. Fill the cooling system.

13. Fill the engine with clean oil. A filter replacement is also recommended.

14. Recharge the A/C system.

➡When the battery has been disconnected and reconnected, some abnormal drive symptoms may occur while the Powertrain Control Module (PCM) relearns its adaptive strategy. The vehicle may need to be driven about 10 miles (16 km) or more to relearn the strategy.

15. Start the engine and check for leaks.

4.0L SOHC Engine

➡If only one cylinder head is to be removed, only follow the procedures that apply. The following tools, or their equivalents are absolutely necessary to properly perform this procedure:

1. Remove or disconnect the following:
   - Cam Chain Tensioner tool T97T-6K254-A
   - Cam Gear Removal tool T97T-6256-F
   - Cam Gear Torque adapter T97T-6256-G
   - Camshaft Gear Positioning/Holding tool T97T-6256-B
   - Camshaft Gear Positioning/Holding tool adapter T97T-6256-A
   - Camshaft holding tool T97T-6256-C
   - Crankshaft holding tool T97T-6303-A
   - Camshaft holding tool adapter T97T-6256-D

2. Before servicing the vehicle, refer to the precautions in the beginning of this section.

3. Properly relieve the fuel system pressure.

4. Drain the cooling system.

5. Remove or disconnect the following:
   - Negative battery cable
   - Lower intake manifold
   - Fan blade and shroud
   - Valve cover
   - Roller followers, if equipped
   - Drive belt
   - Upper radiator hose and tube
   - Alternator electrical connectors
   - Alternator mounting bracket
   - Engine accessory bracket and move it aside
   - Camshaft Position (CMP) electrical connector
   - Crankshaft Position (CKP) sensor electrical connector
   - Engine Coolant Temperature (ECT) sensor electrical connector
   - Coil pack electrical connector
   - Exhaust Gas Recirculation (EGR) valve electrical connector
   - EGR valve bracket and move it aside
   - Heater hoses
   - Fuel injector electrical connectors
   - Water bypass hose
   - Thermostat housing
   - Spark plug wires
   - Fuel injection supply manifold
   - Fuel injectors
   - Crankcase vent separator spring
   - Oil dipstick housing
   - Exhaust manifold
   - Hydraulic chain tensioner
   - Cassette retaining bolt
   - Camshaft sprocket
   - Cylinder head and discard the gasket

To install:

6. Thoroughly clean all gasket mating surfaces. Remove all traces of old gasket material, oil, grease or dirt.

7. Insure that the rubber band is holding the right-hand chain to the cassette.

8. Install a new head gasket and the cylinder head.

9. Torque the new cylinder head bolts in sequence as follows:

1999
   - Step 1: 26 ft. lbs. (34 Nm).
   - Step 2: Plus 90 degrees.
   - Step 3: Plus an additional 90 degrees.

2000–01
   - Step 1: 28 ft. lbs. (38 Nm).
   - Step 2: Plus 90 degrees.
   - Step 3: Plus an additional 90 degrees.

2002
   - Step 1: 24 ft. lbs. (32 Nm).
   - Step 2: Plus 90 degrees.

10. Install or connect the following:
   - Camshaft sprocket in the cassette and make certain that the camshaft sprocket turns freely on the camshaft
   - Cassette retaining bolt. Torque the bolt to 89 inch lbs. (10 Nm).
   - Exhaust manifold
   - Oil level indicator tube. Torque the bolt to 18 ft. lbs. (25 Nm).
   - Crankcase vent separator and spring
   - Thermostat housing. Torque the bolts to 8 ft. lbs. (11 Nm).
   - Water bypass hose
   - Heater hoses
   - EGR bracket. Torque the bolt to 89 inch lbs. (10 Nm).
   - EGR tube. Torque the nut to 30 ft. lbs. (40 Nm).
   - ECT sensor electrical connector
   - Electrical harness retainer. Torque the bolt to 89 inch lbs. (10 Nm).
   - CKP and CMP electrical connectors

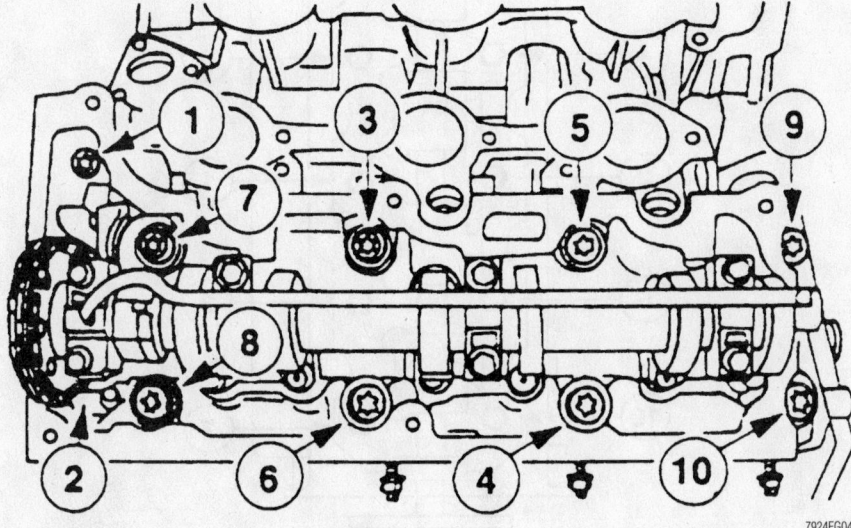

7924EG04

Cylinder head bolt torque sequence 4.0L SOHC engine

Timing belt service is covered in Section 3 of this manual

- Accessory bracket. Torque the bolts to 31 ft. lbs. (42 Nm).
- Alternator mounting bracket. Torque the bolts to 31 ft. lbs. (42 Nm).
- Alternator and electrical connectors
- Drive belt
- Fan shroud
- Roller followers
- Valve cover
- Lower intake manifold
- Negative battery cable

11. Change the engine oil and filter.

12. Refill the cooling system.

13. Start the engine and check for leaks, repair if necessary.

4.6L Engine

1. Before servicing the vehicle, refer to the precautions in the beginning of this section.

2. Drain the cooling system.

3. Remove or disconnect the following:
- Negative battery cable
- Intake manifold
- Timing chains
- Exhaust manifolds
- Water heater tube back and remove. Discard the O-ring seal.

➥The cylinder head bolts must be discarded with new bolts installed. They are tighten-to-yield designed and cannot be reused. Do not use metal scrapers, wire brushes, power abrasive discs or other abrasive means to clean the sealing surfaces. These tools cause scratches and gouges that make leak paths. Use a plastic scraping tool to remove all traces of the head gasket.

- Bolts and the RH cylinder head
- Bolts and the LH cylinder head

4. Installation is the reverse of removal. Observe the following notes and torques:

a. Install the cylinder head on the head gasket and loosely install the bolts.

b. Make sure to tighten the bolts in sequence in six stages.
- Step 1: Tighten to 40 Nm (30 lb-ft).
- Step 2: Tighten an additional 85-95 degrees.
- Step 3: Back out all bolts one full turn (360 degrees).
- Step 4: Tighten to 40 Nm (30 lb-ft).
- Step 5: Tighten an additional 85-95 degrees.
- Step 6: Tighten an additional 85-95 degrees.

c. Install a new O-ring seal and position the water heater tube forward. Lubri-

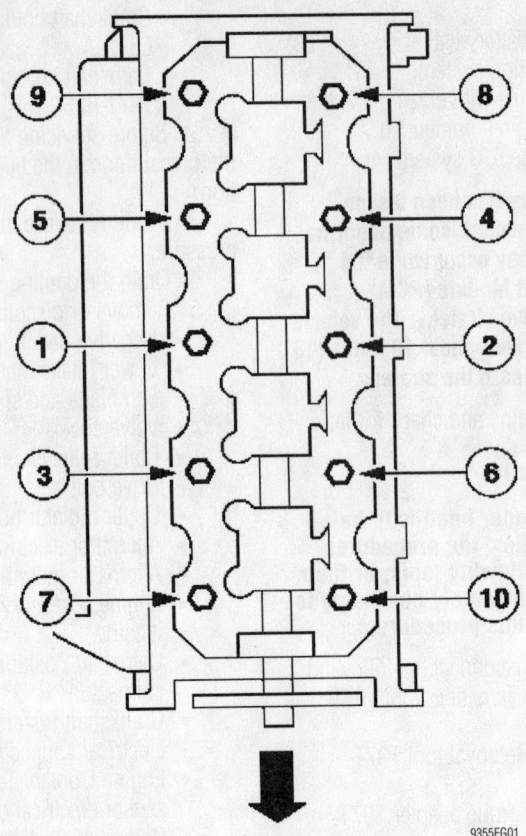

9355EG01

Left cylinder head bolt torque sequence 4.6L engine

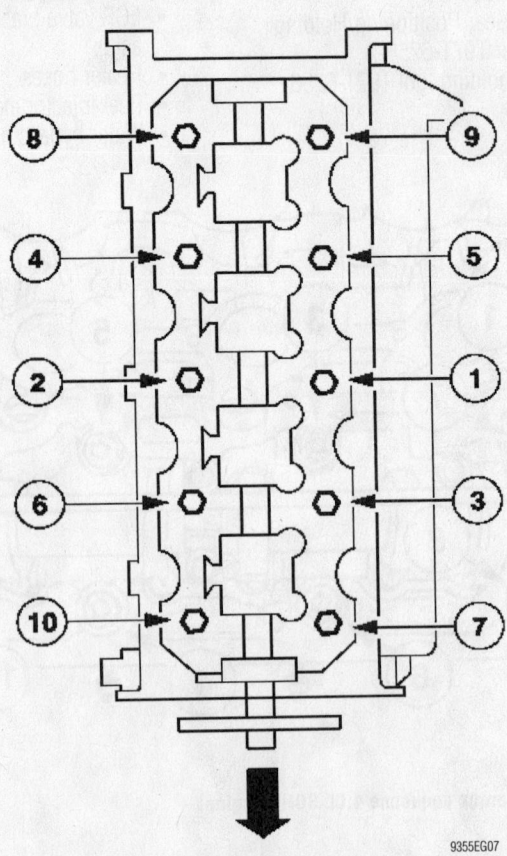

9355EG07

Right cylinder head bolt torque sequence 4.6L engine

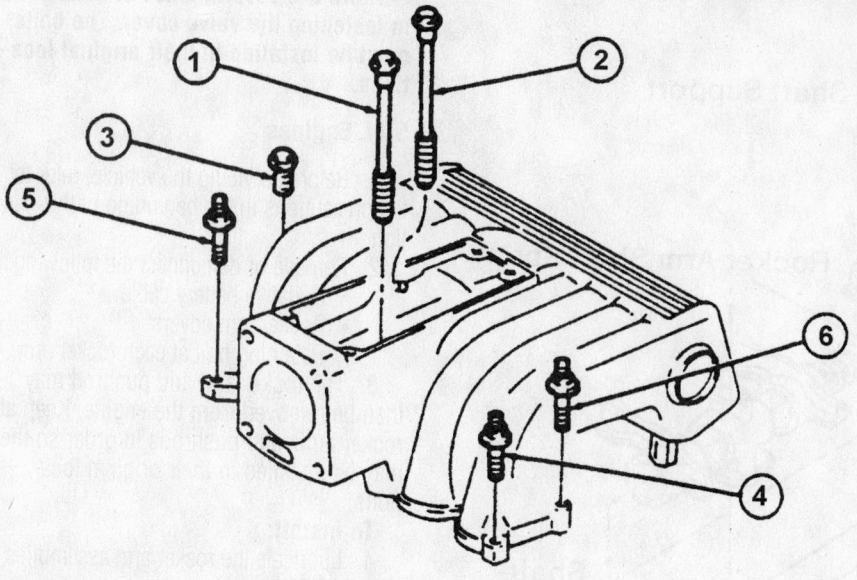

Tighten valve cover bolts in the sequence shown—4.6L engine

7924EG14

cate the O-ring seal with premium engine coolant.

5.0L Engine

1. Before servicing the vehicle, refer to the precautions in the beginning of this section.

2. Drain the cooling system.

3. Remove or disconnect the following:

- Negative battery cable
- Lower intake manifold
- Valve cover
- Matchmark the rocker arms and push rods to ease installation and remove the rocker arm fulcrums
- Rocker arms and fulcrum guides
- Exhaust manifold
- A/C compressor and power steer-

ing bracket, left side cylinder head only

- Alternator electrical connectors, right side cylinder head only
- Transmission oil fill tube from the rear of the right side cylinder head
- Push rods and make certain to matchmark them
- Cylinder head and discard the gasket

To install:

4. Thoroughly clean all the gasket mating surfaces.

5. Position a new cylinder head gasket to the engine block, then install the cylinder head.

6. Install new cylinder head bolts, and torque in 3 steps:

a. Step 1: 30 ft. lbs. (40 Nm).
b. Step 2: 50 ft. lbs. (68 Nm).
c. Step 3: Plus an additional 90 degrees.

7. Install or connect the following:

- Lubricate and install the push rods in their original positions
- Rocker arms and fulcrum guides
- Rocker arm fulcrums. Torque the bolts to 25 ft. lbs. (34 Nm).
- Valve cover
- Transmission oil fill tube. Torque the bolt to 25 ft. lbs. (34 Nm).
- Alternator bracket, if removed. Torque the bolts to 48 ft. lbs. (65 Nm).
- Alternator electrical connectors, if removed
- Exhaust manifold
- Lower intake manifold
- Negative battery cable

8. Fill the cooling system

9. Fill the engine with clean oil and replace the filter.

10. Start the engine and check for leaks, repair if necessary.

Rocker Arms/Roller Followers

REMOVAL & INSTALLATION

4.0L SOHC Engines

➤**A special tool is required to compress the valve spring.**

1. Before servicing the vehicle, refer to the precautions in the beginning of this section.

2. Disconnect the negative battery cable.

3. Remove the valve cover.

4. Rotate the camshaft so that the base circle of the cam is against the cam follower you intend to remove.

➤**If removing more than one cam follower, label them so they can be returned to their original position.**

5. Using special tool T97T-6565-A depress the valve spring. Slide the cam follower over the lash adjuster and out from under the camshaft.

To install:

6. Compress the valve spring and slide the roller follower into position.

7. Release the tension from the spring.

8. Install the valve cover and connect the negative battery cable.

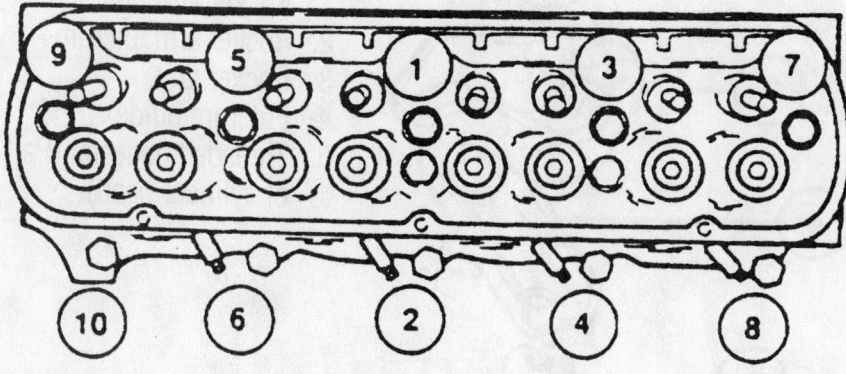

Cylinder head bolt torque sequence 5.0L engine

7924EG05

Heater Core replacement is covered in Section 2 of this manual

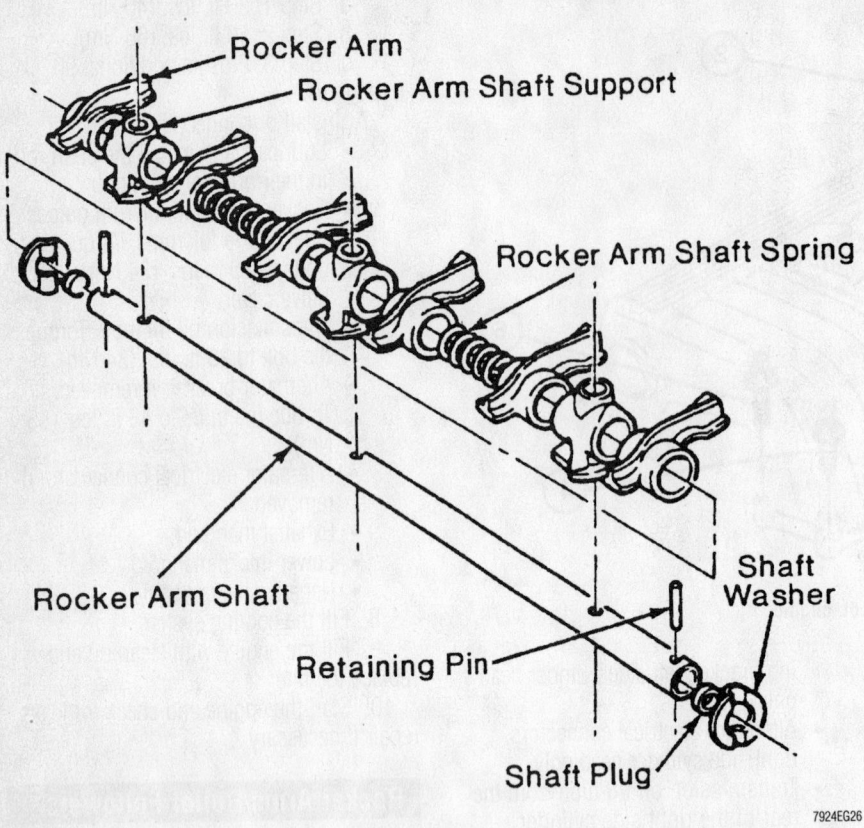

Rocker Arm
Rocker Arm Shaft Support
Rocker Arm Shaft Spring
Rocker Arm Shaft
Retaining Pin
Shaft Washer
Shaft Plug

7924EG26

Rocker arm and shaft assembly—4.0L SOHC engine

4.6L Engine

1. Before servicing the vehicle, refer to the precautions in the beginning of this section.
2. Remove the valve cover.
3. Position the piston of the cylinder being repaired at the bottom of the stroke.
4. Install the special tool between the valve spring coils to prevent valve stem seal damage.

➡ The roller followers are positional. Mark the followers for installation in their original locations.

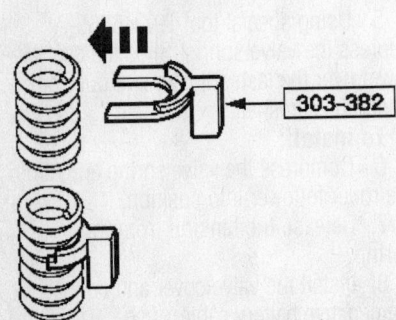

303-382

9355EG06

Install this tool to prevent seal damage—4.6L engine

- Compress the valve springs and remove the camshaft roller followers.
5. Installation is the reverse of removal.

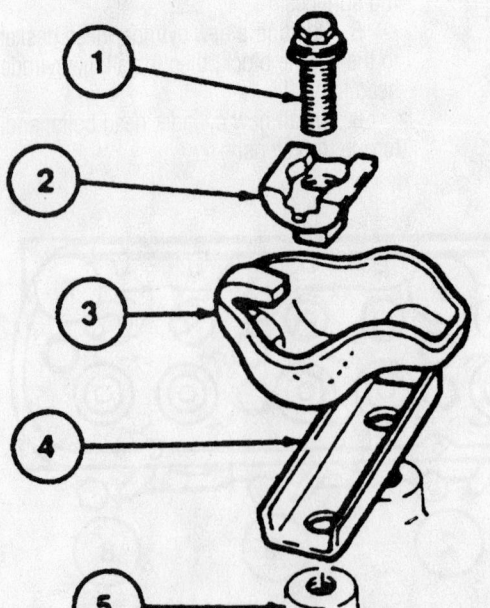

7924EG28

Exploded view of the rocker arm assembly. Notice the fulcrum guide between the pedestals used for extra stability—5.0L engine

➡There are several kinds of bolts used in fastening the valve cover. The bolts must be installed in their original locations.

5.0L Engines

1. Before servicing the vehicle, refer to the precautions in the beginning of this section.
2. Remove or disconnect the following:
- Negative battery cable
- Rocker arm covers
- Retaining bolt at each rocker arm
3. The rocker arm and pushrod may then be removed from the engine. Keep all rocker arms and pushrods in order so they may be installed in their original locations.

To install:
4. Lubricate the rocker arm assemblies with SAE 50W engine oil.
5. Ensure that the fulcrums are properly seated into the cylinder head (3.0L engines) or the fulcrum guide (5.0L engines). Torque the rocker arm fulcrum bolts to 19 ft. lbs. (26 Nm).
6. Install the rocker arm covers and connect the negative battery cable.

4.0L OHV Engine

1. Before servicing the vehicle, refer to the precautions in the beginning of this section.
2. Remove or disconnect the following:

1. Rocker arm bolt
2. Rocker arm fulcrum
3. Rocker arm
4. Fulcrum guide
5. Threaded pedestal (Part of cylinder head)

BOLT
E800544
(6 PLACES)
62-70 N·m
(46-52 FT-LB)

PUSH ROD
6565
(12 PLACES)

NOTE: DRAW BOLTS DOWN EVENLY AND TORQUE IN STAGES

7924EG29

Rocker arm and shaft assembly—4.0L OHV engine

- Negative battery cable
- Rocker arm covers
- Rocker arm shaft stand attaching bolts by loosening the bolts 2 turns at a time, in sequence (from the end of the shaft to the middle of the shaft)
- Rocker arm and shaft assembly

To install:

3. If equipped, loosen the valve lash adjusting screws a few turns. Apply engine oil to the assembly to provide initial lubrication.

4. Install or connect the following:
- Rocker arm shaft assembly to the cylinder head and guide adjusting screws on to the pushrods
- Rocker arm stand. Torque the bolts to 46–52 ft. lbs. (62–70 Nm), 2 turns at a time, in sequence (from

middle of shaft to the end of the shaft).
- Rocker arm covers
- Negative battery cable

Intake Manifold

REMOVAL & INSTALLATION

4.0L OHV Engine

UPPER

The intake manifold is a 4-piece assembly, consisting of the upper intake manifold, the throttle body, the fuel supply manifold, and the lower intake manifold.

1. Before servicing the vehicle, refer to the precautions in the beginning of this section.

2. Remove or disconnect the following:
- Negative battery cable
- Air cleaner outlet tube
- Accelerator cable splash shield
- Spark plug wires from the ignition coil
- Ignition coil and radio interference capacitor electrical connectors
- Throttle Position (TP) sensor electrical connector
- Idle Air Control (IAC) valve electrical connector
- Brake booster vacuum hose
- Positive Crankcase Ventilation (PCV) hose
- Canister purge line from the throttle body

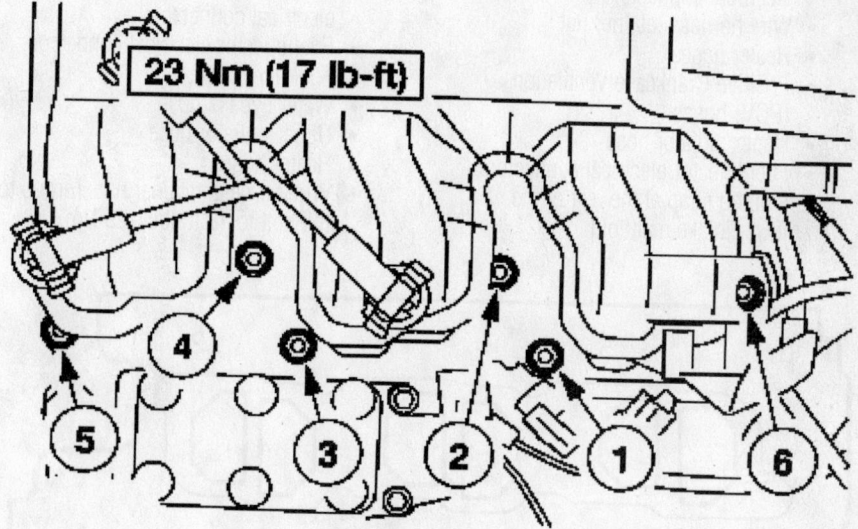

23 Nm (17 lb-ft)

9355EG02

Tighten the upper manifold bolts in the sequence shown—4.0L OHV engine

---

Brake service is covered in Section 4 of this manual

- Fuel line bracket
- Upper intake manifold and discard the gasket

To install:

3. Install or connect the following:
- New gasket and the upper manifold. Torque the nuts to 18 ft. lbs. (25 Nm).
- Fuel line bracket
- Canister purge line to the throttle body
- Brake booster vacuum hose
- IAC valve electrical connector
- TP sensor electrical connector
- Ignition coil and radio interference capacitor electrical connectors
- Spark plug wires to the proper spark plug
- Accelerator cable and speed control cable, if equipped
- Accelerator cable splash shield
- Air cleaner outlet tube
- Negative battery cable

4. Start the vehicle and check for leaks, repair if necessary.

LOWER

1. Before servicing the vehicle, refer to the precautions in the beginning of this section.
2. Drain the cooling system.
3. Remove or disconnect the following:
- Negative battery cable
- Radiator overflow hose and set it aside
- Upper intake manifold
- Water bypass hose
- Engine Coolant Temperature (ECT) electrical connector
- Wire harness retainer nut
- Heater hoses
- Positive Crankcase Ventilation (PCV) hoses
- Upper radiator hose
- Fuel injector electrical connectors
- Ground strap at the rear of the lower intake manifold

- Water temperature indicator sender electrical connector
- Camshaft Position (CMP) sensor and the camshaft synchronizer
- Bolts from the lower intake manifold
- Lower intake manifold and discard the gaskets

To install:

4. Ensure that all of the gasket mating surfaces are clean and free of grease, oil or dirt. Also ensure that the EGR passages in the manifolds and heads are clear.
5. Apply a 1/16 in. (1.6mm) bead of silicone sealer to the points where the cylinder block rails meet the cylinder heads.
6. Position new seals on the cylinder block and new gaskets on the cylinder heads with the gaskets interlocked with the seal tabs. Make sure the holes in the gaskets are aligned with the holes in the cylinder heads.
7. Apply a 1/16 in. (1.6mm) bead of sealer to the outer end of each intake manifold seal for the full width of the seal. Make sure the silicone sealer will not fall into the engine and possibly block oil passages.
8. Install the lower intake manifold and tighten the bolts as follows:
- 27 inch lbs. (3 Nm)
- 89 inch lbs. (10 Nm)
- 10 ft. lbs. (13 Nm)
- 12 ft. lbs. (16 Nm)
9. Install or connect the following:
- CMP sensor and camshaft synchronizer
- Water temperature indicator sender electrical connector
- Fuel injector electrical connectors
- Fuel line
- Water bypass hose
- Upper radiator hose
- Heater hoses
- Wire harness retainer nut. Torque to the bolt to 18 ft. lbs. (25 Nm).

- ECT electrical connector
- Water heater bypass hose to the heater tube
- PCV hoses to the heater tube
- Upper intake manifold
- Radiator overflow hose
- Negative battery cable

10. Fill the cooling system.
11. Start the vehicle and check for leaks, repair if necessary.

4.0L SOHC Engine

1999–01

1. Before servicing the vehicle, refer to the precautions in the beginning of this section.
2. Remove or disconnect the following:
- Negative battery cable
- Air cleaner-to-intake tube
- Accelerator splash shield
- Accelerator and, if equipped with cruise control, speed control cables from the throttle control cam
- Accelerator cable retaining bracket from the upper intake manifold
- Label and disengage all vacuum and electrical connections on the intake manifold.
- Upper intake manifold attaching bolts
- Lift up on the manifold and remove both fuel Vapor Management Valve (VMV) hoses
- Intake manifold and discard the gasket

To install:

➡ **Ford does not specify a sequence for either upper or lower intake manifolds, but it is recommended that you start tightening in the middle and work your way out to the ends. Repeat the tightening sequence several times until the bolts will no longer turn at the specified torque.**

3. Install the lower intake manifold. Torque the bolts to 10 ft. lbs. (13 Nm) for 1999–00, and 13 ft. lbs. (18 Nm) for 2001.
4. Position the upper manifold on the lower manifold.
5. Install or connect the following:
- Attach both VMV hoses to the manifold
- Upper manifold attaching bolts. Torque the bolts to 62 inch lbs. (7 Nm).
- Attach any vacuum and electrical connections that were removed
- Accelerator cable bracket to the

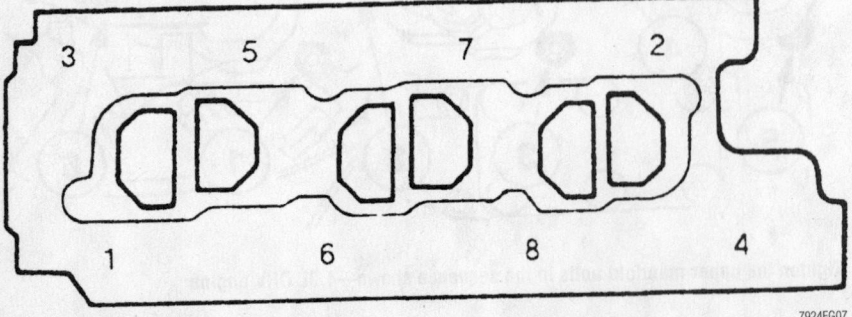

Tighten the lower manifold bolts in the sequence shown—4.0L OHV engine

7924EG07

intake and the cable (or cables if equipped with cruise control) to the throttle cam
- Accelerator splash shield
- Air cleaner-to-intake supply tube
- Negative battery cable

6. Start the vehicle and check for leaks, repair if necessary.

2002

1. Before servicing the vehicle, refer to the precautions in the beginning of this section.
2. Remove or disconnect the following:
- Negative battery cable
- Splash shield
- Air cleaner outlet pipe
- Accelerator and speed control cables from the throttle body
- Accelerator cable and the speed control cable from the bracket. Position the cables away from the intake manifold.
- All the vacuum hoses
- KS electrical connector
- IAC valve electrical connector
- TP sensor electrical connector
- EGR transducer electrical connector
- Wiring anchor
- EGR tube from the EGR valve
- Wiring harness pin-type retainer
- Spark plug wires
- Intake manifold retainers, then slightly reposition the manifold to access the EGR vacuum regulator (EVR) solenoid.
- EVR solenoid vacuum and electrical connections
- Intake manifold
- To install, reverse the removal procedure. Torque the bolts to 89 inch lbs. (10 Nm). There is no particular sequence.

4.6L Engine

1. Before servicing the vehicle, refer to the precautions in the beginning of this section.
2. Remove or disconnect the following:
- Negative battery cable
- Throttle body
- Ignition coils
- Fuel injection supply manifold
- Upper radiator hose and the heater hose
- Alternator
- Accelerator cable bracket
- Throttle body adapter

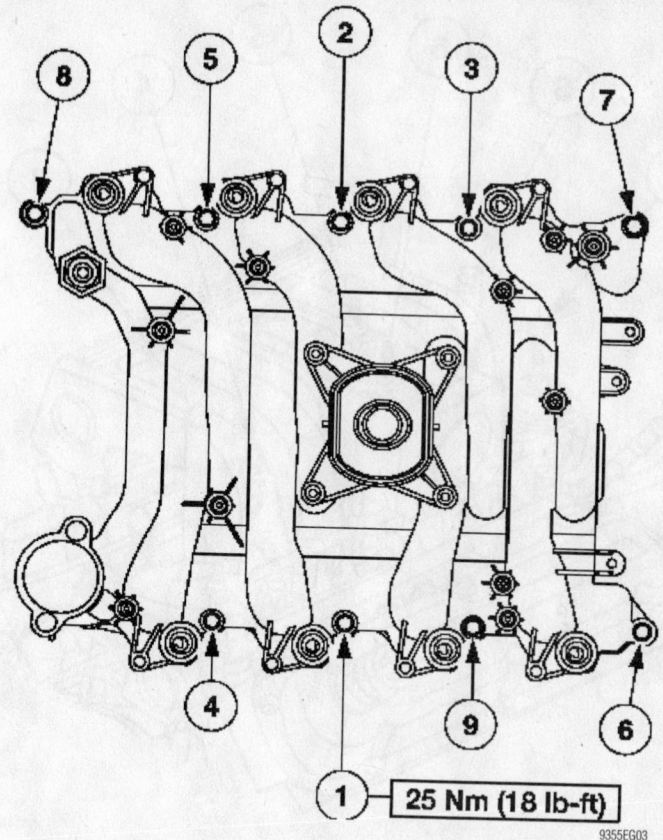

Tighten the manifold bolts in the sequence shown—4.6L engine

9355EG03

- Thermostat housing
- Intake manifold bolts
- Intake manifold. Tilt the manifold to slide it out from underneath the exhaust gas recirculation (EGR) tube.
- To install, reverse the removal procedure. Tighten retaining bolts is sequence shown.

5.0L Engine

UPPER

1. Before servicing the vehicle, refer to the precautions in the beginning of this section.
2. Remove or disconnect the following:
- Negative battery cable
- Air cleaner outlet tube
- Idle Air Control (IAC) valve electrical connector
- Accelerator control splash shield
- Throttle Position (TP) sensor electrical connector
- Accelerator cable and if equipped, speed control cable from the throttle linkage
- Accelerator cable bracket

- Fuel pressure regulator vacuum connection
- Pressure transducer hoses
- Upper Exhaust Gas Recirculation (EGR) valve-to-exhaust manifold tubing
- Engine Vacuum Regulator (EVR) electrical connector
- EVR vacuum connector
- EGR back pressure electrical connector
- Ignition coil bracket
- Accelerator cable from the upper intake manifold clips
- Intake cover plate
- Vacuum connections from the front of the manifold
- Vapor Management Valve (VMV) purge line
- Brake booster vacuum supply line
- Positive Crankcase Ventilation (PCV) hose
- PCV heater hoses
- Upper intake manifold and discard the gasket

To install:

3. Ensure that all of the gasket mating surfaces are clean and free of grease, oil or

---

For complete Engine Mechanical specifications, see Section 1 of this manual

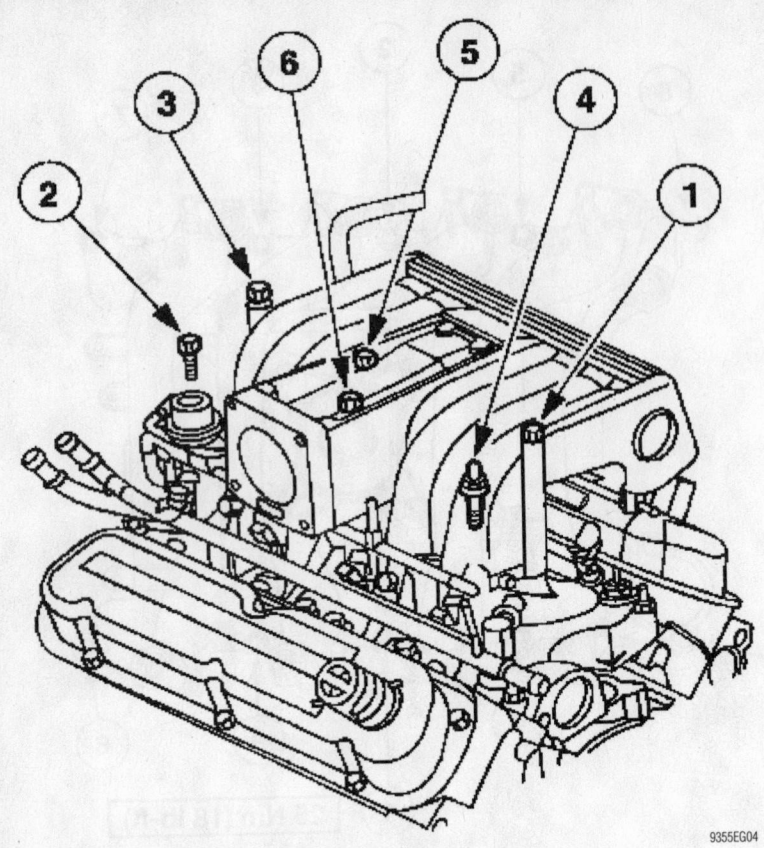

9355EG04

Loosen the upper manifold bolts in the sequence shown—5.0L engine

dirt. Also ensure that the EGR passages in the manifolds and heads are clear.

4. Apply a 1/16 in. (1.6mm) bead of silicone sealer to the points where the cylinder block rails meet the cylinder heads.

5. Position new seals on the cylinder block and new gaskets on the cylinder heads with the gaskets interlocked with the seal tabs. Make sure the holes in the gaskets are aligned with the holes in the cylinder heads.

6. Apply a 1/16 in. (1.6mm) bead of sealer to the outer end of each intake manifold seal for the full width of the seal. Make

sure the silicone sealer will not fall into the engine and possibly block oil passages.

7. Using guide pins to ease installation, carefully lower the intake manifold into position on the cylinder block and cylinder heads. Also, ensure that the water pump bypass hose is installed at the same time.

8. Install or connect the following:
- Intake manifold and new gasket and hand tighten the bolts
- Upper intake vacuum connections
- PCV tube and heater hoses
- Brake booster vacuum supply line
- Torque the intake manifold bolts to 18 ft. lbs. (25 Nm).
- Intake manifold cover plate. Torque the bolts to 15 ft. lbs. (20 Nm).
- Accelerator cable bracket. Torque the upper bolt to 15 ft. lbs. (20 Nm) and lower bolt to 80 inch lbs. (9 Nm).
- Accelerator cable and speed control cable to the throttle linkage, if equipped
- Ignition coils and bracket
- EVR electrical connector
- EVR vacuum connector
- EGR valve vacuum connector
- Fuel pressure regulator vacuum connector
- Upper EGR valve-to-exhaust manifold connector. Torque the fastener to 25 ft. lbs. (34 Nm).
- EGR back pressure transducer electrical connector
- TP sensor electrical connector
- IAC valve electrical connector
- Accelerator control splash shield. Torque the bolt to 89 inch lbs. (10 Nm).
- Air cleaner outlet tube
- Negative battery cable

LOWER

1. Before servicing the vehicle, refer to the precautions in the beginning of this section.

2. Drain the cooling system.

3. Remove or disconnect the following:
- Negative battery cable
- Radiator overflow hose and set it aside
- Upper intake manifold
- Water bypass hose
- Engine Coolant Temperature (ECT) electrical connector
- Wire harness retainer nut
- Heater hoses
- Positive Crankcase Ventilation (PCV) hoses
- Upper radiator hose
- Fuel injector electrical connectors

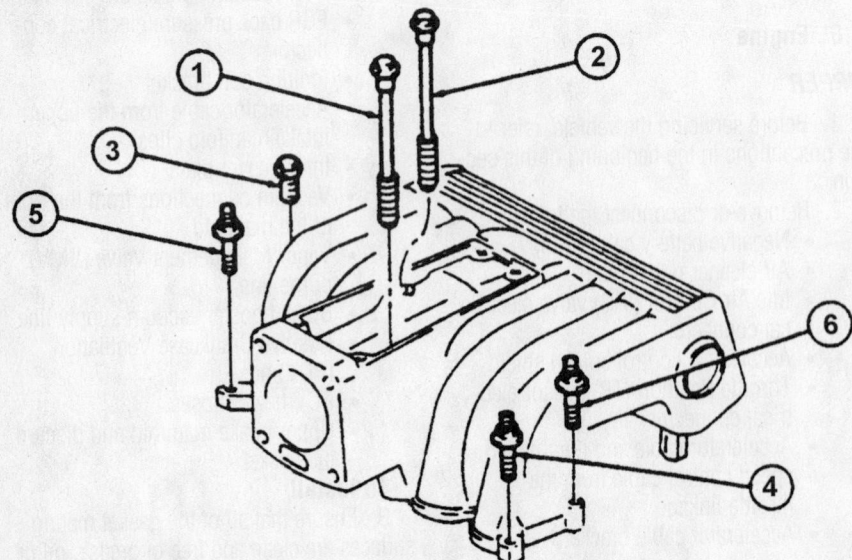

7924EG14

Tighten the upper manifold bolts in the sequence shown—5.0L engine

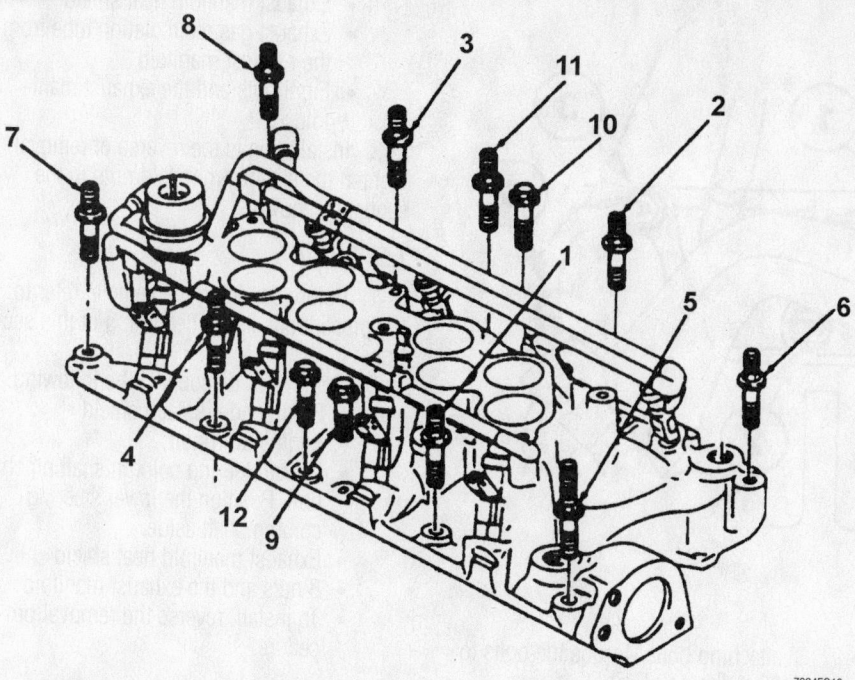

Tighten the lower manifold bolts in the sequence shown—5.0L engine

7924EG13

- Ground strap at the rear of the lower intake manifold
- Water temperature indicator sender electrical connector
- Camshaft Position (CMP) sensor and the camshaft synchronizer
- Bolts from the lower intake manifold
- Lower intake manifold and discard the gaskets

To install:

4. Ensure that all of the gasket mating surfaces are clean and free of grease, oil or dirt. Also ensure that the EGR passages in the manifolds and heads are clear.

5. Apply a ⅟₁₆ in. (1.6mm) bead of silicone sealer to the points where the cylinder block rails meet the cylinder heads.

6. Position new seals on the cylinder block and new gaskets on the cylinder heads with the gaskets interlocked with the seal tabs. Make sure the holes in the gaskets are aligned with the holes in the cylinder heads.

7. Apply a ⅟₁₆ in. (1.6mm) bead of sealer to the outer end of each intake manifold seal for the full width of the seal. Make sure the silicone sealer will not fall into the engine and possibly block oil passages.

8. Install the lower intake manifold and tighten the bolts in 2 steps as follows:
   a. 89 inch lbs. (10 Nm).
   b. 24 ft. lbs. (32 Nm).

9. Install or connect the following:
- CMP sensor and camshaft synchronizer
- Water temperature indicator sender electrical connector
- Fuel injector electrical connectors
- Fuel line
- Water bypass hose
- Upper radiator hose
- Heater hoses

- Wire harness retainer nut. Torque to the bolt to 18 ft. lbs. (25 Nm).
- ECT electrical connector
- Water heater bypass hose to the heater tube
- PCV hoses to the heater tube
- Upper intake manifold
- Radiator overflow hose
- Negative battery cable

10. Fill the cooling system.

11. Start the vehicle and check for leaks, repair if necessary.

Exhaust Manifold

REMOVAL & INSTALLATION

4.0L OHV Engine

1. Before servicing the vehicle, refer to the precautions in the beginning of this section.

2. Remove or disconnect the following:
- Negative battery cable
- Oil level indicator tube and bracket
- Exhaust pipe-to-manifold bolts
- Power steering pump hoses, if removing the left-hand manifold
- Hot air intake shroud that is bolted around the manifold, if removing the right-hand manifold
- Exhaust manifold and gasket

To install:

3. Clean all mating surfaces for the exhaust manifold and cylinder head.

4. Install or connect the following:

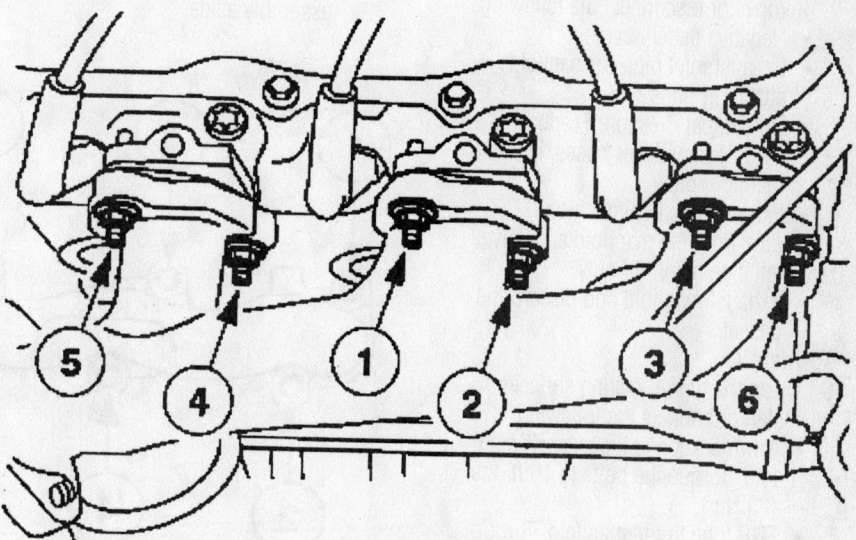

Tighten the right side exhaust manifold in sequence

9308EG10

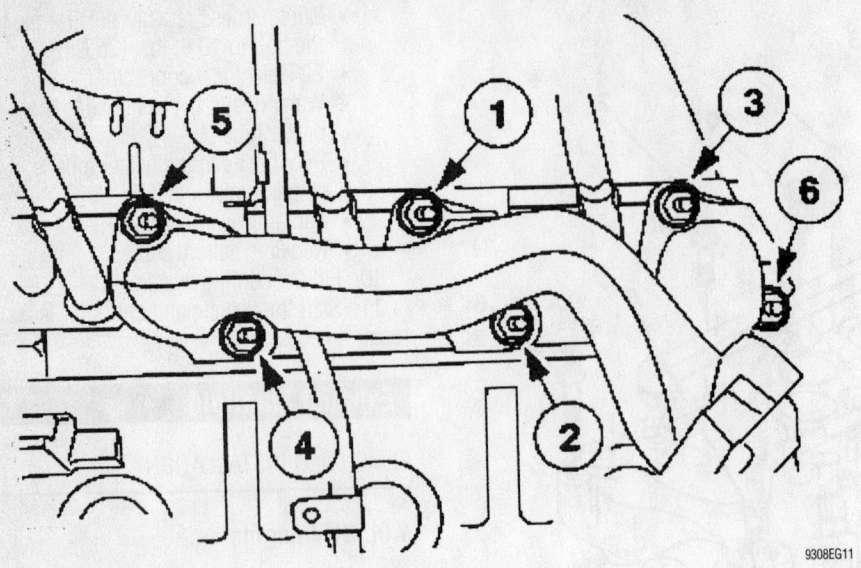

9308EG11

Tighten the left side exhaust manifold in sequence

- New gasket and the exhaust manifold. Torque the right side bolts to 18 ft. lbs. (25 Nm) and the left side bolts to 16 ft. lbs. (22 Nm).
- Exhaust pipe to the manifold
- Oil level bracket. Torque the bolt to 17 ft. lbs. (23 Nm).
- Negative battery cable

5. Start the vehicle and check for leaks, repair if necessary.

4.0L SOHC Engine

1. Before servicing the vehicle, refer to the precautions in the beginning of this section.

2. Remove or disconnect the following:
- Negative battery cable
- Exhaust inlet pipe-to-manifold attaching bolts
- Differential Pressure Feedback EGR (DPFE) transducer hoses, left side manifold only
- Exhaust Gas Recirculation (EGR) tube from the manifold and valve, left side manifold only
- Exhaust manifold and discard the gasket

To install:

3. Clean the gasket mating surfaces.

4. Install or connect the following:
- New gasket and the exhaust manifold. Torque the bolts to 16 ft. lbs. (22 Nm).
- EGR tube to the manifold. Torque the fastener to 30 ft. lbs. (40 Nm) left side manifold only
- DPFE transducer hoses, left side manifold only
- Exhaust inlet pipe-to-manifold

attaching bolts. Torque the bolts to 30 ft. lbs. (40 Nm).
- Negative battery cable

5. Start the vehicle and check for leaks, repair if necessary.

4.6L engine

RIGHT SIDE

1. Before servicing the vehicle, refer to the precautions in the beginning of this section.

2. Remove or disconnect the following:
- Front fender splash shield
- Heater control valve bracket bolt
- Vacuum hose and heater hose. Position the heater control valve assembly aside.

- Exhaust manifold heat shield
- Exhaust gas recirculation tube from the exhaust manifold
- Eight nuts and the exhaust manifold

3. Installation is the reverse of removal. Tighten the exhaust manifold nuts in the sequence shown.

LEFT SIDE

1. Before servicing the vehicle, refer to the precautions in the beginning of this section.

2. Remove or disconnect the following:
- Front fender splash shield
- Engine oil runoff
- Lower steering column shaft pinch bolt. Position the lower steering column shaft aside.
- Exhaust manifold heat shield
- 8 nuts and the exhaust manifold
- To install, reverse the removal procedure.

5.0L Engine

LEFT SIDE

1. Before servicing the vehicle, refer to the precautions in the beginning of this section.

2. Discharge and recover the A/C system.

3. Remove or disconnect the following:
- Negative battery cable
- Spark plug wires and retaining brackets
- A/C manifold and tube
- Condenser to evaporator tube from the A/C condenser core

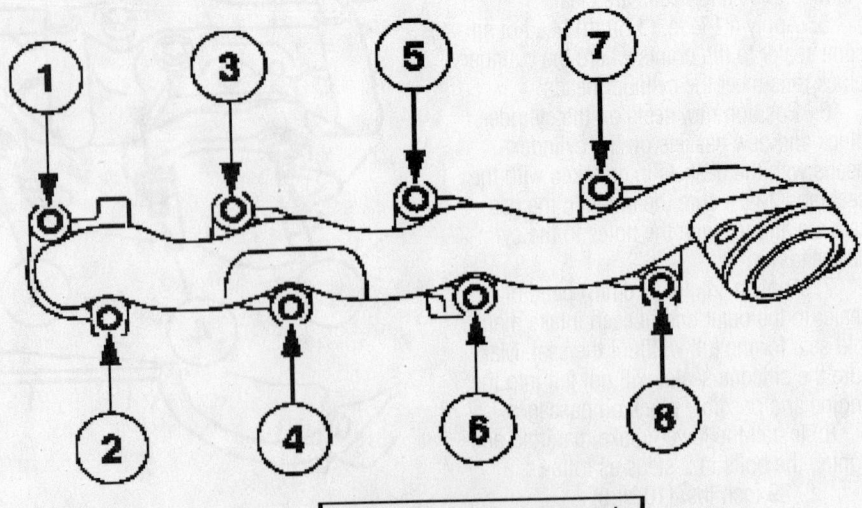

20 Nm (15 lb-ft)

9355EG14

Exhaust manifold torque sequence—4.6L engine

- Oil level indicator tube
- Left wheel
- Wheel well apron pin retainers
- Exhaust manifold and discard the gasket

To install:

4. Clean the gasket mating surfaces.
5. Install or connect the following:
   - New gasket and the exhaust manifold. Torque the bolts to 30 ft. lbs. (40 Nm).
   - Wheel well pin retainers
   - Left wheel
   - Oil level indicator tube. Torque the bolt to 18 ft. lbs. (25 Nm).
   - Condenser to evaporator tube
   - A/C manifold and tube
   - Spark plug wires and brackets
   - Negative battery cable
6. Recharge the A/C system.
7. Start the vehicle and check for leaks, repair if necessary.

RIGHT SIDE

1. Before servicing the vehicle, refer to the precautions in the beginning of this section.
2. Remove or disconnect the following:
   - Negative battery cable
   - Air cleaner outlet tube
   - Drive belt tensioner
   - Alternator electrical connectors
   - Alternator and bracket
   - Exhaust flange
   - Right front wheel
   - Wheel well apron pin retainers
   - Exhaust Gas Recirculation (EGR) valve from the manifold tube
   - Spark plug wires and retaining brackets
   - Exhaust manifold heat shield
   - Exhaust manifold and discard the gasket

To install:

3. Clean the gasket mating surfaces.
4. Install or connect the following:
   - New gasket and the exhaust manifold. Torque the bolts to 30 ft. lbs. (40 Nm).
   - Exhaust manifold heat shield. Torque the bolts to 61 inch lbs. (7 Nm).
   - Spark plug wires and retaining brackets
   - EGR valve to the manifold tube
   - Wheel well apron pin retainers
   - Right front wheel
   - Exhaust flange. Torque the nuts to 26 ft. lbs. (34 Nm).

5. Install the alternator and bracket. Torque the bolts in sequence as follows:
   a. Bolts NO. 1–2 to 26 ft. lbs. (34 Nm).
   b. Bolt NO. 3 to 48 ft. lbs. (65 Nm).
   c. Torque the remaining bolt to 26 ft. lbs. (34 Nm).
6. Install or connect the following:
   - Drive belt tensioner
   - Air cleaner outlet tube
   - Negative battery cable
7. Start the vehicle and check for leaks, repair if necessary.

Camshaft and Valve Lifters

➡**Although Ford suggests that this component is removable while the engine is installed in the vehicle, depending on the particular options with which your truck is equipped, working clearance may be extremely tight and this procedure may be much easier to perform with the engine removed. Before commencing, read through this procedure and make certain enough clearance, or working room, exists with the engine in the vehicle; if there is not enough space, the engine should be removed.**

REMOVAL & INSTALLATION

4.0L OHV Engine

➡**It is necessary to replace the oil pan gasket when removing and installing the engine front cover. It will also be necessary to remove the transmission to properly reseal the oil pan.**

1. Before servicing the vehicle, refer to the precautions in the beginning of this section.
2. Drain the engine oil.
3. Drain the cooling system.
4. Evacuate the A/C system.
5. Relieve fuel system pressure.
6. Remove or disconnect the following:
   - Negative battery cable
   - Radiator
   - A/C compressor. Do not disconnect the lines
   - A/C condenser
   - Fan, spacer and shroud
   - Air cleaner hoses
   - Spark plug wires
   - Ignition coil and bracket
   - Crankshaft pulley and damper
   - Oil pump drive
   - Alternator

   - Fuel lines at the supply manifold
   - Upper and lower intake manifold
   - Rocker arm covers
   - Rocker arm shafts
   - Pushrods and identify them for installation
   - Tappets and identify them for installation
   - Oil pan
   - Engine front cover
   - Water pump

7. Turn the engine by hand until the timing marks align at Top dead Center (TDC) of the power stroke on No.1 piston.
8. Place the timing chain tensioner in the retracted position and install the retaining clip.
9. Check the camshaft end-play. If excessive, you'll have to replace the thrust plate.
10. Remove the camshaft gear attaching bolt and washer, then slide the gear off the camshaft.
11. Remove the camshaft thrust plate.
12. Carefully slide the camshaft out of the engine block, using caution to avoid any damage to the camshaft bearings.

To install:

13. Lubricate the camshaft using a good assembly lubricant.
14. Install or connect the following:
    - Camshaft using caution to avoid any damage to the camshaft bearings
    - Thrust plate. Make sure that it covers the main oil gallery. Torque the screws to 84–120 inch lbs. (9–13 Nm).
15. Rotate the camshaft and crankshaft, as necessary, to align the timing marks.
    - Camshaft gear and chain. Torque the bolt to 44–50 ft. lbs. (60–68 Nm).
16. Remove the clip from the chain tensioner
    - Engine front cover and water pump
    - Crankshaft damper/pulley. Torque the bolt to 46–49 ft. lbs. +90 degrees.
    - Oil pan
17. Coat the tappets with 50W engine oil and place them in their original locations.
18. Apply 50W engine oil to both ends of the pushrods. Install the pushrods in their original locations.
    - Tappets
    - Rocker arm shafts and covers
    - Upper and lower intake manifolds

- Fuel lines to the fuel supply manifold
- Alternator and electrical connectors
- Oil pump drive
- Ignition coil and bracket
- Spark plug wires
- Air cleaner hoses
- Fan, spacer and shroud
- A/C condenser
- A/C compressor
- Radiator
- Negative battery cable

19. Fill the cooling system.
20. Recharge the A/C system.
21. Replace the oil filter and refill the engine with clean oil.
22. Start the engine and check the ignition timing and idle speed; adjust if necessary. Run the engine at fast idle and check for coolant, fuel, vacuum or oil leaks.

4.0L SOHC Engine

1. Before servicing the vehicle, refer to the precautions in the beginning of this section.
2. Remove or disconnect the following:
   - Negative battery cable for safety
   - Valve cover
   - Hydraulic camshaft tensioner

➡**The right-hand camshaft sprocket bolt uses left-hand threads.**

3. For the right-hand camshaft use the Cam Gear Torque Adapter tool T97T-6256-F, to remove the camshaft sprocket bolt.
4. For the left-hand camshaft, remove the sprocket bolt.

➡**When removing the followers, label them so that they may be returned to their original positions.**

5. Using the Valve Spring Compressor tool ST1330-A, remove the camshaft roller followers.
6. Install or connect the following:
   - Camshaft bearing cap bolts and the oil rail
   - Camshaft

To install:
7. Lubricate all of the moving parts with SAE 50W engine oil.
8. Install camshaft onto the cylinder head.
9. Position the oil rail and install the bearing caps and bolts. Torque the bolts in 2 steps:
   a. Step 1—53.5 inch lbs. (6 Nm).
   b. Step 2—11–12.5 ft. lbs. (15–17 Nm).
10. Install or connect the following:
   - Camshaft followers

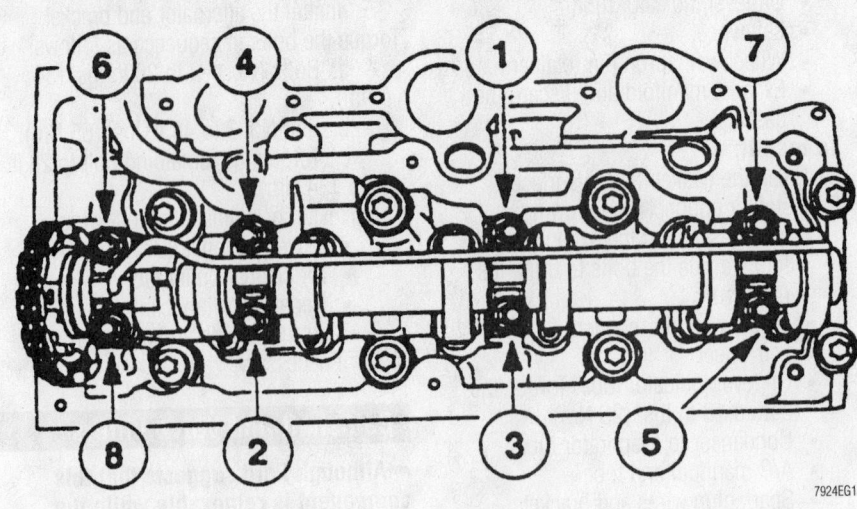

Use the proper sequence to prevent damage to the camshaft both when installing and removing the bearing caps—4.0L SOHC engine

- Camshaft sprocket bolt and hand tighten the bolt
- Camshaft Chain Tensioner T97T-6K254-A in the hole that the hydraulic chain tensioner was in

11. Turn the crankshaft one revolution clockwise until No. 1 piston is Top Dead Center (TDC).
12. Install or connect the following:
   - Crankshaft Holding tool T97T-6303-A on the crankshaft to keep it from turning
   - Position the timing slot on the rear of the camshaft to fit Camshaft Holding tool T97T-6256-C and install the holding tool on the rear of the head
   - Camshaft Gear Holding tool T97T-6256-B and Camshaft Gear Holding tool T97T-6256-A on the front of the cylinder head to securely hold the camshaft gear
   - Tighten the camshaft sprocket bolt to 63 ft. lbs. (85 Nm).

13. Remove the Camshaft Chain Tensioner tool and install the hydraulic chain tensioner, tighten the tensioner to 35–39 ft. lbs. (47–53 Nm).
14. Remove the special tools from engine.
15. Install or connect the following:
   - Valve cover
   - Negative battery cable
16. Start the engine check for leaks and repair if necessary.

4.6L Engine

1. Before servicing the vehicle, refer to the precautions in the beginning of this section.

2. Remove or disconnect the following:

✳✳ WARNING

At no time, when the timing chains are removed and the cylinder heads are installed may the crankshaft or camshaft be rotated. Severe piston and valve damage will occur.

- Timing chains
- Camshaft roller followers
- Camshaft sprocket
- The 13 camshaft bearing cap bolts
- Camshaft bearing caps ladders
- Camshaft from the cylinder head

➡**The valve tappets are positional. Mark each valve tappet for installation in its original location.**

- Valve tappets

To install:
3. Lubricate the camshaft journals with clean engine oil.
4. Install the camshaft onto the cylinder head.
5. Lubricate the camshaft bearing caps with clean engine oil.
6. Install the camshaft bearing caps and loosely install the bolts.
7. Tighten the bolts in the sequence shown.
8. Install the camshaft sprocket. Tighten the sprocket bolt in two stages.
   - Step 1: Tighten to 40 Nm (30 ft. lbs.)
   - Step 2: Tighten an additional 90 degrees.
9. Install the roller followers
10. Install the timing chains

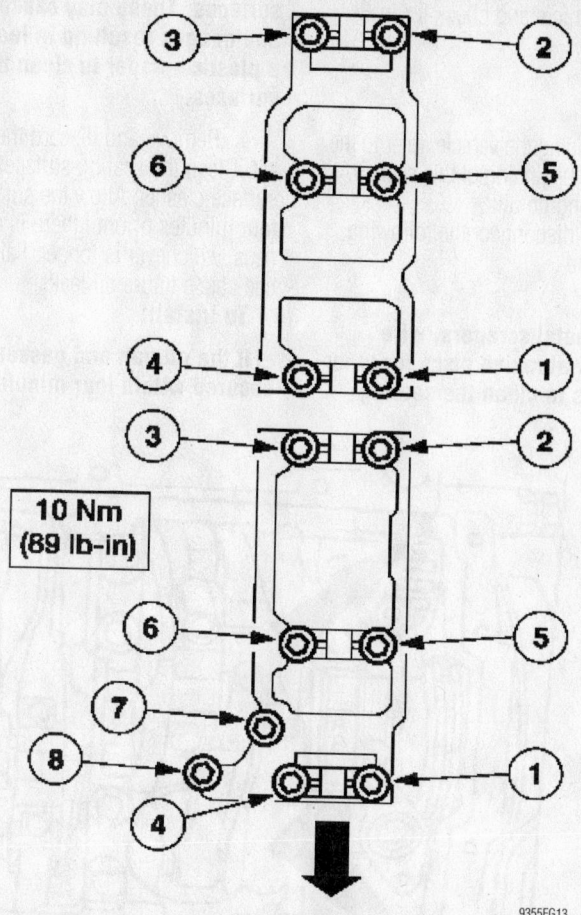

**10 Nm
(89 lb-in)**

9355EG13

Camshaft bearing bolt torque sequence—4.6L engine

5.0L Engine

1. Before servicing the vehicle, refer to the precautions in the beginning of this section.
2. Remove or disconnect the following:
   - Negative battery cable
   - Timing chain cover
   - Camshaft sprocket and chain assembly
   - Upper and lower intake manifolds
   - Both valve covers
   - Loosen the rocker arm bolts and rotate the rocker arms to the side
   - Pushrods in sequence so that they may be installed to their original positions
   - Lifters
   - Camshaft thrust plate bolts and the plate
   - Camshaft from the engine, taking care not to damage the bearings, lobes or journals

To install:

3. Apply SAE 50W engine oil to the camshaft lobes and journals.

4. Install or connect the following:
   - Camshaft

5. Apply SAE 50W engine oil to the camshaft thrust plate. Position the thrust plate with the groove toward the block and install the retaining bolts. Torque to 10–12 ft. lbs. (13–16 Nm).

6. Apply SAE 50W engine oil to the valve tappets and install them. If reusing the old lifters, place them in their original positions.

7. Install or connect the following:
   - Pushrods to their original positions
   - Rocker arms
   - Valve covers
   - Lower and upper intake manifolds
   - Camshaft sprocket and chain assembly. Ensure that the timing marks on the cam and crankshaft sprockets are aligned.
   - Timing chain cover
   - Negative battery cable

8. Start the engine, check for leaks and repair if necessary.

Oil Pan

REMOVAL & INSTALLATION

4.0L OHV Engine

➡Review the complete service procedure before starting this repair.

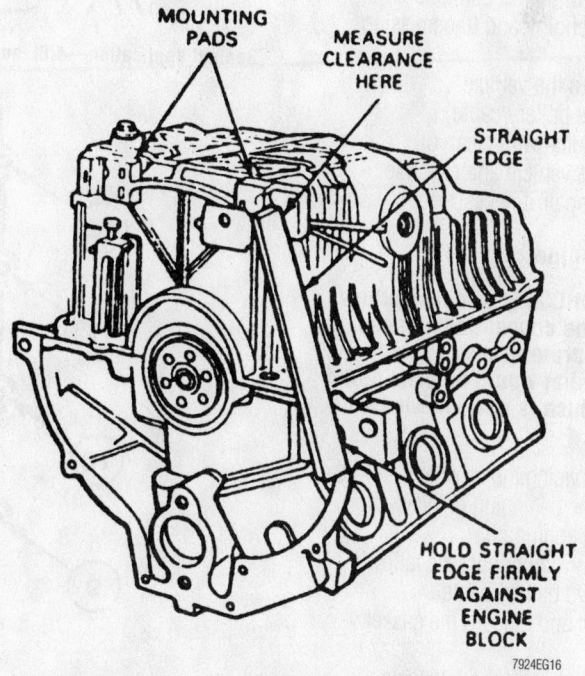

MOUNTING PADS

MEASURE CLEARANCE HERE

STRAIGHT EDGE

HOLD STRAIGHT EDGE FIRMLY AGAINST ENGINE BLOCK

7924EG16

The correct spacer must be used to extend the mounting surface of the oil pan so it is flush with the mounting surface of the engine block—4.0L OHV engine

For Wheel Alignment specifications, see Section 1 of this manual

1. Before servicing the vehicle, refer to the precautions in the beginning of this section.
2. Drain the engine oil.
3. Remove or disconnect the following:
   - Negative battery cable
   - Engine from the vehicle and mount the engine on a suitable engine stand with the oil pan facing up
   - Oil pan attaching bolts (note location of 2 spacers) and remove the pan from the engine block
   - Oil pan gasket and crankshaft rear main bearing cap wedge seal
4. Clean all gasket surfaces on the engine and oil pan. Remove all traces of old gasket and/or sealer.

To install:

5. Install or connect the following:
   - New crankshaft rear main bearing cap wedge seal. The seal should fit snugly into the sides of the rear main bearing cap
   - New oil pan gasket to the engine block and place the oil pan in correct position on the 4 locating studs. Torque the bolts EVENLY to 60–84 inch lbs. (7–10 Nm).
   - Transmission bolts to the engine and oil pan. There are 2 spacers on the rear of the oil pan to allow proper mating of the transmission and oil pan.
   - Spacers to the mounting pads on the rear of the oil pan before bolting the engine and transmission together
   - Engine to the vehicle
   - Negative battery cable
6. Fill the engine with clean oil.
   - Start the vehicle and check for leaks, repair if necessary.

4.0L SOHC Engine

➡The 4.0L SOHC engine does not use an oil pan in the conventional sense. There is a separate access panel that unbolts from what would be considered the oil pan (which is now known as the ladder frame).

1. Before servicing the vehicle, refer to the precautions in the beginning of this section.
2. Drain the engine oil.
3. Remove or disconnect the following:
   - Negative battery cable
   - Oil pan and discard the gasket

To install:

4. Install or connect the following:
   - New gasket and oil pan. Torque the bolts to 80 inch lbs. (9 Nm).
   - Negative battery cable
5. Fill the engine with clean oil.

6. Start the vehicle and check for leaks, repair if necessary.

4.6L Engine

1. Before servicing the vehicle, refer to the precautions in the beginning of this section.
2. Drain the engine oil.
3. Remove or disconnect the following:
   - Front axle
   - Oil pan

➡Do not use metal scrapers, wire brushes, power abrasive discs, or other abrasive means to clean the sealing surfaces. These may cause scratches and gouges resulting in leak paths. Use a plastic scraper to clean the sealing surfaces.

4. Remove and discard the oil pan gasket. Clean the sealing surfaces with metal surface cleaner. Allow the surfaces to dry for four minutes or until there is no sign of wetness, whichever is longer. Failure to do so can cause future oil leaks.

To install:

➡If the oil pan and gasket are not secured within four minutes of sealer

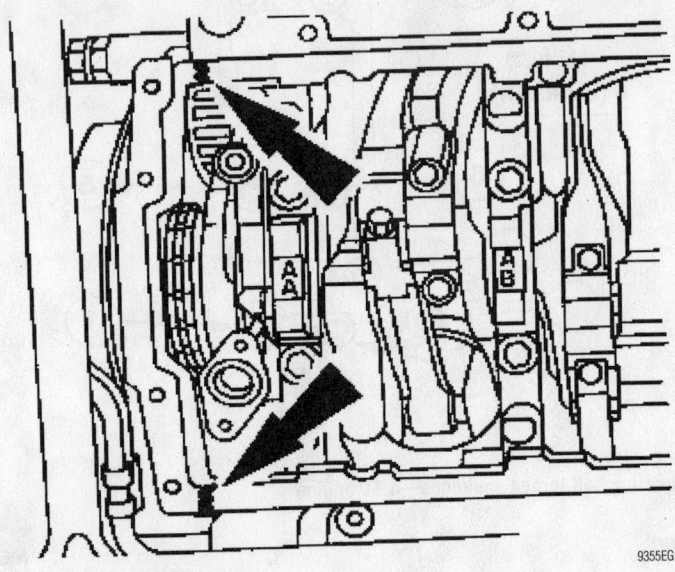

Sealant application—4.6L engine

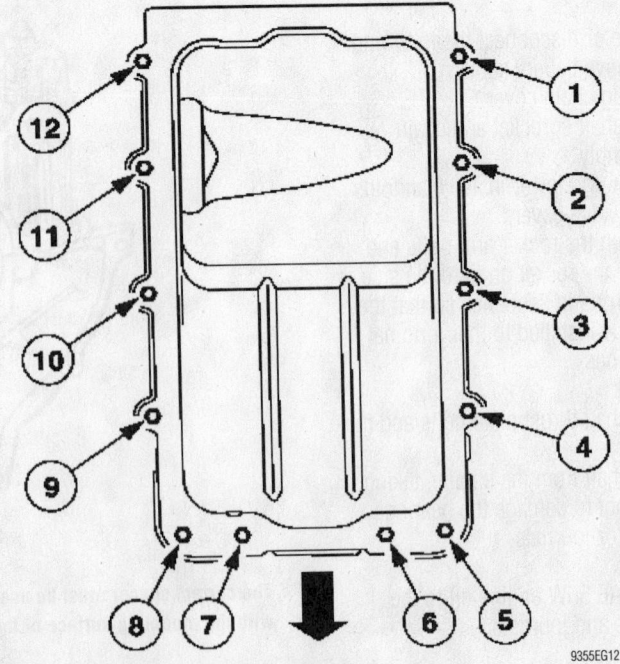

Oil pan bolt torque sequence—4.6L engine

application, the sealant must be removed and the sealing surfaces cleaned with metal surface cleaner.

5. Apply silicone gasket and sealant in the two places shown.

6. Position the new oil pan gasket and the oil pan and loosely install the bolts.

7. Tighten the bolts in three stages in the sequence shown.
- Step 1: Tighten to 2 Nm (18 inch lbs.)
- Step 2: Tighten to 20 Nm (15 ft. lbs.)
- Step 3: Tighten an additional 60 degrees.

8. Install the front axle.

5.0L Engine

➡**The oil pan cannot be removed with the engine in the vehicle.**

1. Before servicing the vehicle, refer to the precautions in the beginning of this section.

2. Drain the engine oil.

3. Remove or disconnect the following:
- Negative battery cable
- Engine from the vehicle
- Oil pan and discard the gasket

To install:

4. Install or connect the following:
- Oil pan with a new gasket. Torque the bolts to 18 ft. lbs. (25 Nm).
- Engine to the vehicle
- Negative battery cable

5. Fill the engine with clean oil.

6. Start the vehicle and check for leaks, repair if necessary.

Oil Pump

REMOVAL & INSTALLATION

4.0L OHV Engines

1. Before servicing the vehicle, refer to the precautions in the beginning of this section.

2. Drain the engine oil.

3. Remove or disconnect the following:
- Negative battery cable
- Oil pan
- Oil pick-up and tube assembly from the pump
- Oil pump retainer bolts and the oil pump

To install:

4. Prime the oil pump with clean engine oil by filling either the inlet or outlet port. Rotate the pump shaft to distribute the oil within the pump body.

5. Install the oil pump and tighten the mounting bolts to 13–15 ft. lbs. (18–20 Nm).

✳✳ WARNING

Do not force the oil pump if it does not seat readily. The oil pump drive-shaft may be misaligned with the distributor or shaft assembly. If the pump is tightened down with the driveshaft misaligned, damage to the pump could occur. To align, rotate the intermediate driveshaft into a new position.

---

6. Install or connect the following:
- Oil pick-up and tube assembly
- Oil pan

7. Fill the engine with clean oil.

8. Start the vehicle and check for leaks, repair if necessary.

4.0L SOHC and 5.0L Engines

➡**The oil pump cannot be removed with the engine in the vehicle.**

1. Before servicing the vehicle, refer to the precautions in the beginning of this section.

2. Drain the engine oil.

3. Remove or disconnect the following:
- Engine from the vehicle
- Oil pan
- Unbolt the oil pick-up tube

4. On the 4.0L engine, perform the following:

   a. Remove the 8 ladder frame bolts that were under the oil pan.

   b. Remove the 2 rear outer ladder frame bolts.

   c. Remove the 7 left-hand and the 8 right-hand ladder frame bolts.

   d. Lift the ladder frame from the engine.

5. Remove the 2 oil pump attaching bolts and the pump.

To install:

6. Submerge the pump in clean engine oil to prime it.

7. On the 4.0L engine, do the following:

   a. Position the ladder frame on the engine.

   b. Install the 8 right-hand and 7 left-hand ladder frame bolts.

   c. Install the 2 rear outer and the 8 frame bolts under the pan.

8. Install the oil pump. Torque the bolts to:

   a. 4.0L: 13–15 ft. lbs. (17–21 Nm).

   b. 5.0L: 23–31 ft. lbs. (30–43 Nm).

9. Install or connect the following:
- Oil pick-up tube
- Oil pan
- Engine to the vehicle
- Negative battery cable

---

10. Fill the engine with clean oil.

11. Start the vehicle and check for leaks, repair if necessary.

4.6L Engines

1. Before servicing the vehicle, refer to the precautions in the beginning of this section.

2. Drain the engine oil.

3. Remove or disconnect the following:
- Negative battery cable
- Timing chains
- Oil pan
- Three bolts and the oil pump screen cover and tube
- Oil pump

To install:

➡**Lubricate the new O-ring seal with clean engine oil.**

4. Clean and inspect the mating surfaces. Install a new O-ring seal.

5. Position the oil pump.

6. Loosely install the bolts.

7. Tighten the bolts to 89 inch lbs. (10Nm).

8. Install the three oil pump screen and cover bolts. Torque the bolts to 18 ft. lbs. (25Nm).

9. Install the timing chains.

10. Install the oil pan.

Rear Main Seal

REMOVAL & INSTALLATION

4.0L OHV and 5.0L Engines

1. Remove the flexplate or flywheel.

✳✳ WARNING

Use care to avoid scratching or damaging the oil seal surface or leakage may occur.

---

2. Using a sharp awl, punch one hole into the crankshaft rear oil seal metal surface between the seal lip and the cylinder block.

3. Screw the threaded end of the special tool into the oil seal. Use the special tool to remove the crankshaft rear oil seal.

To install:

4. Lubricate the outer lips and the inner seal on the crankshaft rear oil seal with clean engine oil.

5. Using the special tool, install the crankshaft rear oil seal. Alternate bolt tightening to correctly seat the crankshaft rear oil seal.

6. Install the flexplate or flywheel.

For Maintenance Interval recommendations, see Section 1 of this manual

4.0L OHC Engine

1. Remove the flexplate or flywheel.

> ⁂ **WARNING**
>
> **Avoid scratching or damaging the oil crankshaft seal running surface during removal of the crankshaft rear oil seal.**

2. Using the special tool, remove the crankshaft rear oil seal.

To install:

➡ **Be sure the crankshaft rear sealing surface is clean and free of any rust or corrosion. To clean the crankshaft rear sealing surface, use extra-fine emery cloth or extra-fine 0000 steel wool with metal surface cleaner.**

3. Lubricate the crankshaft rear oil seal with clean engine oil and install on the special tool.

4. Using the special tool, install the crankshaft rear oil seal.

5. Install the flexplate or flywheel.

4.6L Engine

1. Before servicing the vehicle, refer to the precautions in the beginning of this section.

2. Remove or disconnect the following:
   - Flexplate
   - Crankshaft rear oil seal slinger with a slide hammer
   - Rear oil seal with a slide hammer

3. Installation is the reverse of removal. Note the following:
   - Lubricate the inner lip of the rear crankshaft seal with clean engine oil.
   - Use the two Crankshaft Rear Oil Seal Installers to install the rear oil seal.
   - Using the two Crankshaft Rear Oil Seal Installers and the Crankshaft Rear Oil Slinger Installer, install the crankshaft rear oil slinger.

Timing Chain, Sprockets, Front Cover and Seal

REMOVAL & INSTALLATION

4.0L OHV Engines

1. Before servicing the vehicle, refer to the precautions in the beginning of this section.

2. Remove or disconnect the following:
   - Negative battery cable
   - Engine front cover
   - Rotate the crankshaft and align the timing marks

   - Timing chain tensioner
   - Sprocket bolt
   - Timing chain, camshaft sprocket and crankshaft sprocket as an assembly

To install:

3. Install or connect the following:
   - Timing chain, camshaft and crankshaft sprockets as an assembly

4. Align the timing marks.

5. Install or connect the following:
   - Timing chain tensioner
   - Sprocket bolt. Torque the bolt to 51 ft. lbs. (70 Nm).
   - Engine front cover
   - Negative battery cable

4.0L OHC Engine

1. Before servicing the vehicle, refer to the precautions in the beginning of this section.

2. Drain the engine oil.

3. Remove or disconnect the following:
   - Negative battery cable
   - Engine from the vehicle
   - Oil pan
   - Engine front cover
   - Cylinder heads

4. Lock the jackshaft tensioner by installing a pin.
   - Jackshaft sprocket and chain assembly
   - Left front cassette retaining bolt
   - Cassette chain and tensioner assembly

   - Rear jackshaft plug from the engine
   - Right rear cassette retaining bolt and spacer
   - Right rear cassette chain and tensioner
   - Timing chain (s)

To install:

5. Install or connect the following:
   - Timing chain(s)
   - Right rear cassette chain, tensioner and sprocket
   - Jackshaft sprocket and chain on the engine and remove the tensioner pin

6. Torque the jackshaft sprocket bolt in 2 stages:
   a. 32–35 ft. lbs. (43–47 Nm).
   b. Turn an additional 65 degrees.

7. Install or connect the following:
   - Cylinder heads
   - Front cover
   - Oil pan
   - Engine to the vehicle
   - Negative battery cable

8. Fill the engine with clean oil.

9. Start the vehicle, check for leaks and repair if necessary.

4.6L Engine

SEAL ONLY

1. Before servicing the vehicle, refer to the precautions in the beginning of this section.

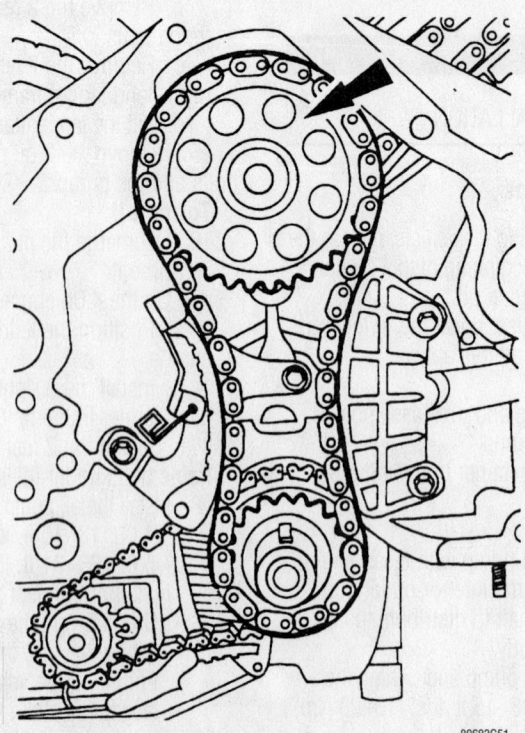

Remove the jackshaft sprocket–4.0L SOHC Engine

89683G51

2. Remove or disconnect the following:
  • Crankshaft pulley
  • Front oil seal with a puller

To install:

3. Lubricate the front cover and the front oil seal inner lip with clean engine oil.

4. Use a seal driver to install the front oil seal into the engine front cover.

5. Install the crankshaft pulley.

FRONT COVER

1. Before servicing the vehicle, refer to the precautions in the beginning of this section.

2. Drain the engine oil and coolant.

3. Remove or disconnect the following:
  • Battery ground cable
  • Both valve covers
  • Radiator and the transmission cooler
  • Drive belt
  • Water pump
  • CMP sensor
  • Upper radiator hose bracket
  • EGR vacuum regulator solenoid
  • Drive belt tensioner
  • Idler pulleys
  • Crankshaft pulley

  • Coolant hose clamps
  • CKP sensor
  • Oil pan to front cover bolts
  • Crankshaft front oil seal
  • Timing cover bolts and studs
  • Front cover from the cylinder block

To install:

➡**If the engine front cover is not secured within four minutes, the sealant must be removed and the sealing area cleaned with metal surface cleaner.**

4. Apply silicone gasket and sealant along the cylinder head to block surface and the oil pan to cylinder block surface.

5. Install the engine front cover on the front cover to cylinder block dowel and loosely install the bolts.

6. Tighten the front cover bolts and stud bolts in the sequence shown.

7. Install a new crankshaft front oil seal.

8. Loosely install the pan-to-cover bolts, then, tighten the bolts in two stages, inner bolts first.
  • Step 1: Tighten to 20 Nm (15 ft. lbs.)
  • Step 2: Tighten an additional 90 degrees

9. The remainder of installation is the reverse of removal.

TIMING CHAINS

1. Before servicing the vehicle, refer to the precautions in the beginning of this section.

2. Remove or disconnect the following:

✳✳ WARNING

Since the engine is not free-wheeling, timing procedures must be followed exactly or piston and valve damage may occur.

  • Front cover
  • Crankshaft sensor ring from the crankshaft

3. Rotate the crankshaft until both camshaft key ways are 90 degrees from the valve cover surface. Make sure the copper links line up with the dots on the camshaft sprocket.

4. Install the special tools 303-380 and 303-413 on the camshaft.

5. Remove or disconnect the following:
  • LH timing chain tensioner
  • RH timing chain tensioner

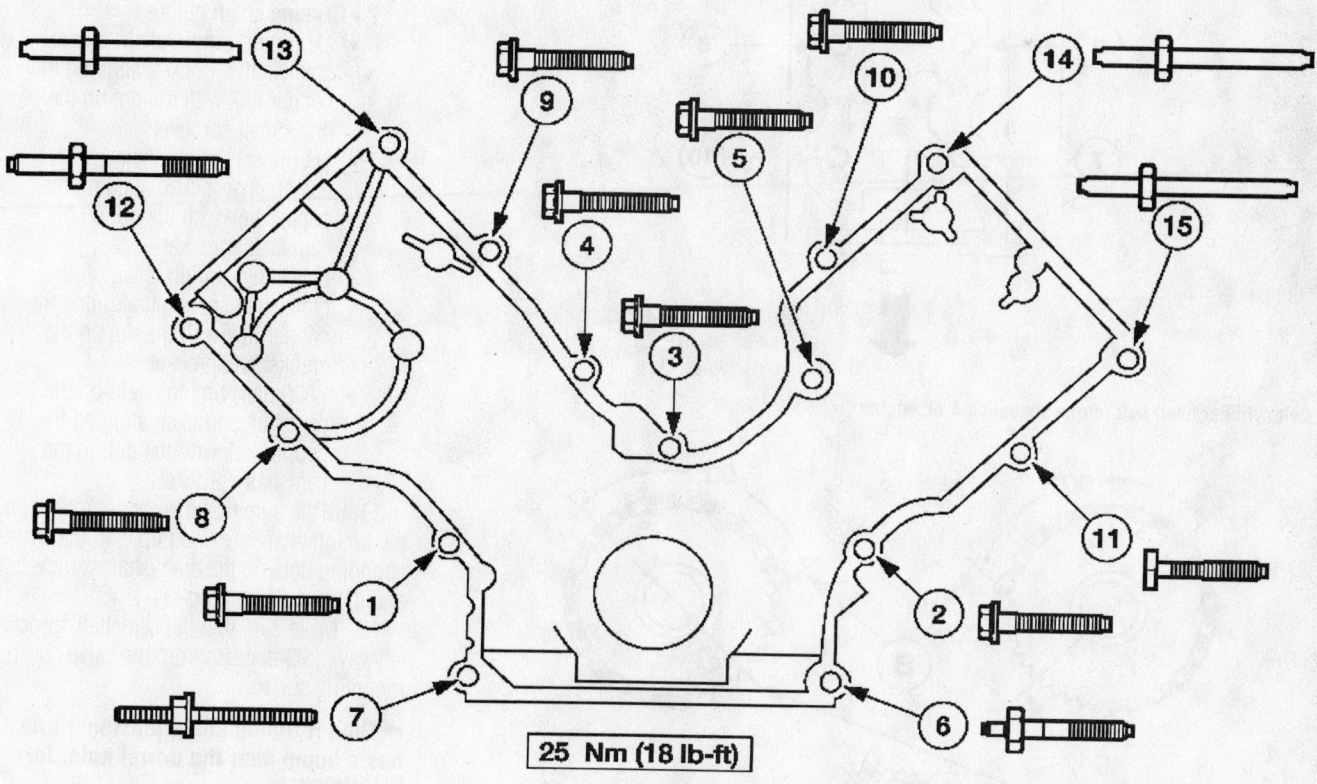

25 Nm (18 lb-ft)

9355EG10

Front cover bolt torque sequence—4.6L engine

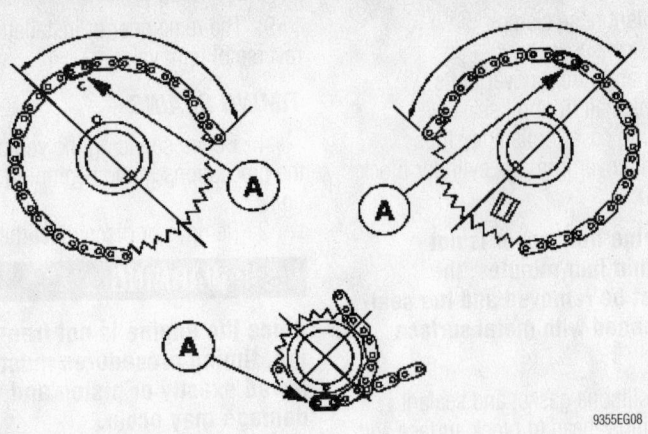

Copper link alignment—4.6L engine

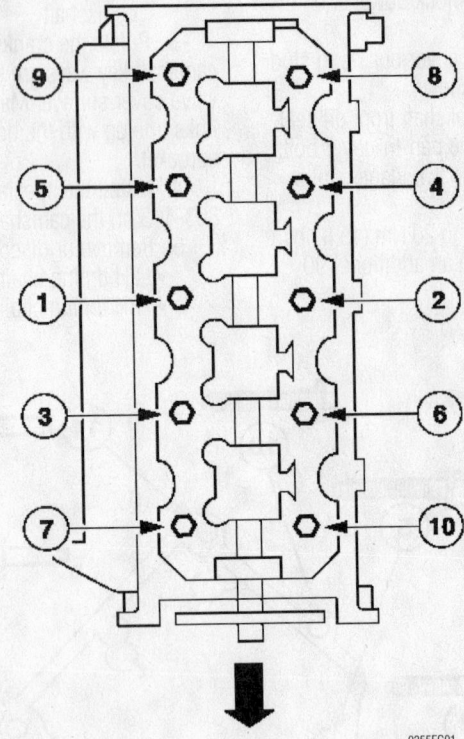

Left cylinder head bolt torque sequence 4.6L engine

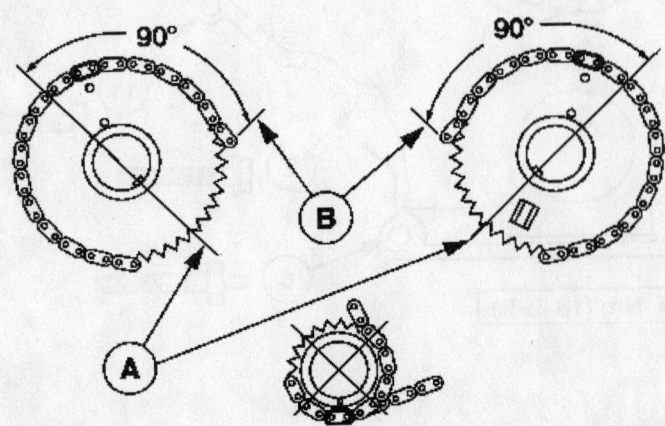

Sprocket alignment—4.6L engine

- LH and RH timing chain tensioner arm from the dowel pins
- Timing chains and crankshaft sprocket
- Timing chain guides

To install:

> ※ **WARNING**
>
> **Do not compress the ratchet assembly. This will damage the ratchet assembly.**

6. Compress the tensioner plunger, using an edge of a vise.

7. Using a small screwdriver or pick, push back and hold the ratchet mechanism.

8. While holding the ratchet mechanism, push the ratchet arm back into the tensioner housing.

9. Install a paper clip into the hole in the tensioner housing to hold the ratchet assembly and plunger in during installation.

➡ **If the copper links are not visible, mark one link on one end and one link on the other end, and use as timing marks.**

10. Install or connect the following:
- Crankshaft sprocket, making sure the timing mark faces forward
- Timing chain guides
- LH (inner) timing chain on the crankshaft sprocket, aligning the copper link with the dot on the crankshaft sprocket
- LH (inner) timing chain on the camshaft sprocket, aligning the copper link with the dot on the camshaft sprocket
- RH (outer) timing chain on the crankshaft sprocket, aligning the copper link with the dot on the crankshaft sprocket
- RH (outer) timing chain on the camshaft sprocket, aligning the copper link with the dot on the camshaft sprocket

11. Make sure that the copper marks on the timing chain are lined up with the corresponding dots on the crankshaft sprockets and the camshaft sprockets.

12. Make sure that the camshaft sprocket keyway is 90 degrees from the valve cover mounting surface.

➡ **The LH timing chain tensioner arm has a bump near the dowel hole, for identification.**

13. Position the LH and RH timing chain tensioner arm on the dowel pins.

14. Position the RH timing chain tensioner and install the bolts.

15. Position the LH timing chain tensioner and install the bolts.

16. Remove both the RH and LH retaining pins from the timing chain tensioner.

17. Remove the special tools from the camshaft.

18. Install the crankshaft sensor ring on the crankshaft.

19. Install the engine front cover.

5.0L Engine

1. Before servicing the vehicle, refer to the precautions in the beginning of this section.

2. Drain the engine oil.

3. Remove or disconnect the following:
   - Negative battery cable
   - Engine front cover

4. Rotate the crankshaft and align the timing marks.

5. Remove or disconnect the following:
   - Camshaft sprocket bolt
   - Timing chain, camshaft sprocket and crankshaft sprocket as an assembly

To install:

6. Install or connect the following:
   - Timing chain, camshaft and crankshaft sprockets as an assembly

7. Align the timing marks.
   - Camshaft sprocket bolt. Torque the bolt to 45 ft. lbs. (61 Nm).
   - Engine front cover
   - Negative battery cable

Piston and Ring

POSITIONING

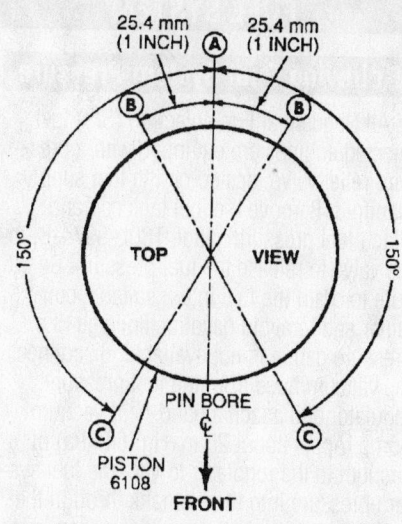

Piston ring end gap spacing

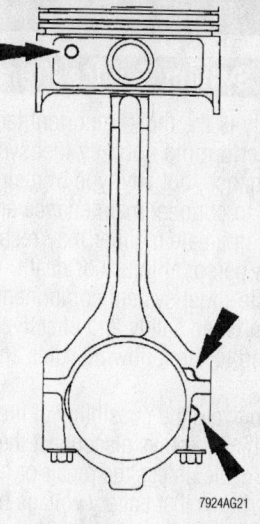

Piston and connecting rod positioning on 4.6L

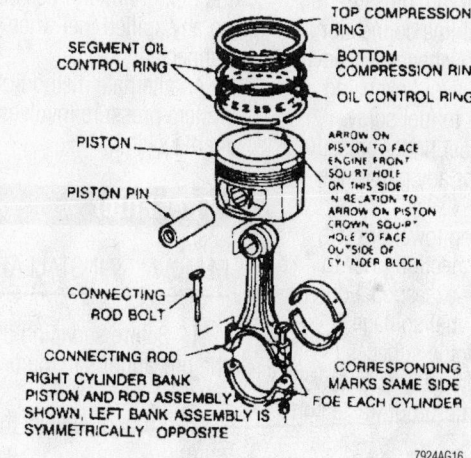

Piston and connecting rod positioning on 4.0L

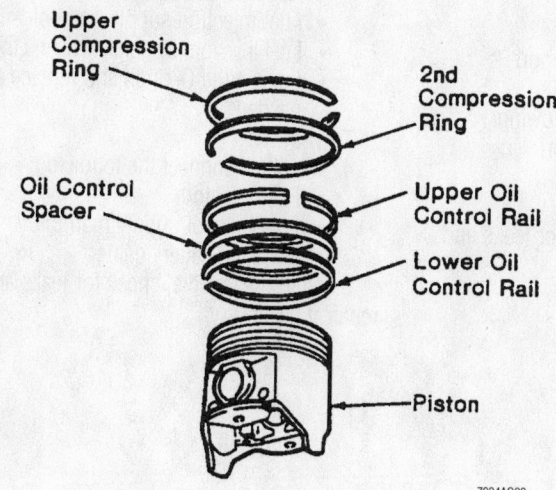

Piston ring positioning

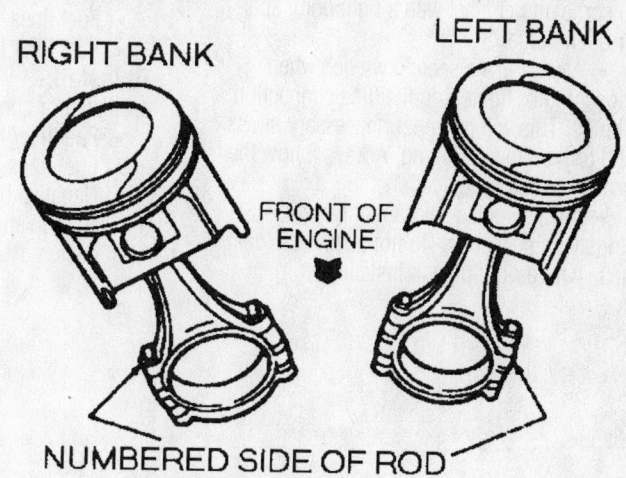

Piston and connecting rod positioning on 5.0L

FUEL SYSTEM

Fuel System Service Precautions

Safety is the most important factor when performing not only fuel system maintenance, but any type of maintenance. Failure to conduct maintenance and repairs in a safe manner may result in serious personal injury or death. Work on a vehicle's fuel system components can be accomplished safely and effectively by adhering to the following rules and guidelines.

• To avoid the possibility of fire and personal injury, always disconnect the negative battery cable unless the repair or test procedure requires that battery voltage be applied.

• Always relieve the fuel system pressure prior to disconnecting any fuel system component (injector, fuel rail, pressure regulator, etc.) fitting or fuel line connection. Exercise extreme caution whenever relieving fuel system pressure, to avoid exposing your skin, face and eyes to fuel spray. Please be advised that fuel under pressure may penetrate the skin or any part of the body that it contacts.

• Always place a shop towel or cloth around the fitting or connection prior to loosening to absorb any excess fuel due to spillage. Ensure that all fuel spillage is quickly remove from engine surfaces. Ensure that all fuel-soaked cloths or towels are deposited into a flame-proof waste container with a lid.

• Always keep a dry chemical (Class B) fire extinguisher near the work area.

• Do not allow fuel spray or fuel vapors to come into contact with a light bulb, spark or open flame.

• Always use a second wrench when loosening or tightening fuel line connection fittings. This will prevent unnecessary stress and torsion to fuel piping. Always follow the proper torque specifications.

• Always replace worn fuel fitting O-rings with new ones. Do not substitute fuel hose where rigid pipe is installed.

Relieving Fuel System Pressure

All Sequential Fuel Injection (SFI) fuel injected engines are equipped with a pressure relief valve located on the fuel supply manifold. Remove the fuel tank cap and attach fuel pressure gauge T80L-9974-B, to the valve to release the fuel pressure. Be sure to drain the fuel into a suitable container and to avoid gasoline spillage. If a pressure gauge is not available, disconnect the vacuum hose from the fuel pressure regulator and attach a hand-held vacuum pump. Apply about 25 in. Hg (84 kPa) of vacuum to the regulator to vent the fuel system pressure into the fuel tank through the fuel return hose. Note that this procedure will remove the fuel pressure from the lines, but not the fuel. Take precautions to avoid the risk of fire and use clean rags to soak up any spilled fuel when the lines are disconnected.

An alternate method of relieving the fuel system pressure involves disconnecting the inertia switch.

Fuel Filter

REMOVAL & INSTALLATION

1. Before servicing the vehicle, refer to the precautions in the beginning of this section.
2. Properly relieve the fuel system pressure.
3. Remove or disconnect the following:

   • Negative battery cable
   • Fuel lines
   • Fuel filter from the support

To install:
4. Install or connect the following:
   • Fuel filter to the support
   • Fuel lines
   • Negative battery cable
5. Start the vehicle, check for leaks and repair if necessary.

Fuel Pump

REMOVAL & INSTALLATION

1. Before servicing the vehicle, refer to the precautions in the beginning of this section.
2. Properly relieve the fuel system pressure.
3. Remove or disconnect the following:
   • Negative battery cable
   • Fuel tank
   • Fuel pressure transducer electrical connector
   • Fuel pump assembly

To install:
4. Install or connect the following:
   • Fuel pump and align the arrow on the flange with the dimple on the fuel tank. Torque the bolts to 80 inch lbs. (9 Nm).
   • Fuel pressure transducer electrical connector
   • Fuel tank
   • Negative battery cable
5. Start the vehicle, check for leaks and repair if necessary.

Fuel Injectors

REMOVAL & INSTALLATION

1. Before servicing the vehicle, refer to the precautions in the beginning of this section.
2. Properly relieve the fuel system pressure.
3. Remove or disconnect the following:
   • Negative battery cable
   • Fuel injection supply manifold
   • Fuel injectors by gently twisting them
   • Inspect the O-rings and replace as needed

To install:
4. Install or connect the following:
   • Fuel injectors
   • Fuel injector supply manifold
   • Negative battery cable
5. Start the vehicle, check for leaks and repair if necessary.

DRIVE TRAIN

Transmission Assembly

REMOVAL & INSTALLATION

Manual Transmission

1999–01

1. Before servicing the vehicle, refer to the precautions in the beginning of this section.
2. Drain the transmission fluid.
3. Place the transmission in **Neutral**.
4. Remove or disconnect the following:
   - Negative battery cable
   - Gearshift lever assembly from the control housing
5. On 2WD vehicles, matchmark the driveshaft to the rear axle flange. Position a drain pan under the rear of the transmission. Remove the driveshaft-to-rear axle flange fasteners and pull the driveshaft rearward to disengage it from the transmission.
   - Heated Oxygen (HO2S) sensor
   - Back-up lamp switch electrical connector
   - Fan shroud
   - Clutch hydraulic line at the clutch housing
   - Speedometer from the transfer case/extension housing and place a wood block on a service jack and position the jack under the engine oil pan
   - Transfer case on 4WD vehicles
   - Starter
6. Position a transmission jack under the transmission.
   - Transmission-to-engine retaining bolts and washers
   - Transmission mount and damper to the crossmember
   - Crossmember and lower the engine jack slightly to angle the transmission assembly. Work the clutch housing off the locating dowels and slide the clutch housing and the transmission rearward until the input shaft clears the clutch disc
   - Exhaust inlet cross over pipe
   - Transmission from the vehicle

To install:

7. Check that the mating surfaces of the clutch housing, engine rear and dowel holes are free of burrs, dirt and paint.
8. Place the transmission on the transmission jack. Position the transmission

under the vehicle, then raise it into position. Align the input shaft splines with the clutch disc splines and work the transmission forward onto the locating dowels.

9. Install or connect the following:
   - Transmission-to-engine retaining bolts and washers. Tighten the retaining bolts to 30–41 ft. lbs. (40–55 Nm).
   - Exhaust inlet cross over pipe and remove the transmission jack
   - Right side transmission mount. Torque the bolt to 81 ft. lbs. (110 Nm).
   - Rear crossmember. Torque the bolts to 53 ft. lbs. (72 Nm).
   - Starter motor and tighten the attaching nuts
   - Transfer case on 4WD vehicles. Torque the bolts to 87 ft. lbs. (119 Nm).
   - Rear driveshaft
   - Starter motor, back-up lamp switch connectors
   - Hydraulic clutch line and bleed the system
   - Speedometer cable
   - Gearshift lever assembly
   - Negative battery cable
10. Fill the transmission fluid to the proper level.
11. Check for proper shifting and operation of the transmission.

2002

1. Before servicing the vehicle, refer to the precautions in the beginning of this section.
2. Disconnect the negative battery cable

➡**Do not remove the gearshift lever knob.**

3. Remove the screw and lift the console finish panel to access the gearshift lever nut. Remove the gearshift lever nut, then remove the upper gearshift lever and console finish panel as an assembly.
4. Install the nut on the front of the lever, then tighten the nut to remove the eccentric stud out of the gearshift lever.
5. place the transmission in NEUTRAL.
6. If the transmission is being disassembled, drain the transmission fluid.
7. Remove or disconnect the following:
   - Front support
   - Starter

- Rear driveshaft and position it aside
- Transfer case

8. Using a suitable high lift jack, support the transmission. Securely strap the jack to the transmission.
9. Remove or disconnect the following:
   - RH crossmember cover, then the four bolts
   - LH crossmember bolts
   - Heat shields from the crossmember
   - Nuts, then lower the crossmember from the vehicle
   - Hydraulic line from the clutch slave cylinder
   - Output shaft speed (OSS) sensor electrical connector
   - Reverse lamp switch electrical connector
   - Wiring harness brackets from the transmission, then disconnect the heated oxygen sensor (HO2S) electrical connectors
   - The two rear three way catalytic converter (TWC) bolts
   - Front TWC bolts and the TWC
   - The eight transmission-to-engine bolts
   - Transmission.

To install:

10. Raise and position the transmission to the engine and clutch.
11. Install the eight transmission-to-engine bolts. Torque the bolts to 44 ft. lbs. (50Nm).
12. Position the front three way catalytic converter (TWC) in the vehicle and install the bolts. Do not tighten the bolts at this time.
13. Using a new gasket and nuts, install the two rear three way catalytic converter (TWC) bolts. Tighten the rear bolts, then the front bolts at this time. Tighten the bolts to 30 ft. lbs. (40Nm).
14. Install or connect the following:
    - Heated oxygen sensor (HO2S) electrical connectors
    - RH and LH crossmember bolts and the RH crossmember cover. Torque the crossmember nuts to 46 ft. lbs. (63Nm).
    - Transmission mount nuts. Torque the nuts to 72 ft. lbs. (98Nm).
    - Heat shields to the crossmember
    - Wiring harness brackets and attach the wiring harness
    - Front support

- Hydraulic line to the clutch slave cylinder
- Starter
- Transfer case
- Driveshaft

15. Position the upper gearshift lever, the outer gearshift lever boot and console finish panel into the console, then install the nut.

16. Connect the battery ground cable.

17. Bleed the clutch hydraulic system.

Automatic Transmission

1999–01

1. Before servicing the vehicle, refer to the precautions in the beginning of this section.

2. Drain the transmission fluid.

3. Place the transmission in **Neutral**.

4. Remove or disconnect the following:
- Negative battery cable
- Fluid level indicator
- Fan shroud
- Transfer case on 4WD vehicles
- Rear driveshaft after matchmarking the yoke and axle flange
- Starter
- Torque converter access cover from the lower right side of the converter housing on the 3.0L engine
- Torque converter access cover from the bottom of the engine oil pan on the 2.3L and 2.5L engines
- Access cover and adapter plate bolts from the lower left side of the converter housing, on all other applications
- Flywheel-to-converter attaching nuts. Use a socket and breaker bar on the crankshaft pulley attaching bolt. Rotate the pulley clockwise as viewed from the front to gain access to each of the nuts.
- Shifter cable
- Transmission wire harness
- Heated Oxygen (HO2S) sensor connector
- Transmission connector
- Digital Transmission Range (TR) sensor connector
- Catalytic converter
- Speedometer cable and/or vehicle speed sensor from the transfer case (4WD) or extension housing (2WD)
- Transmission cooler lines
- Engine rear support-to-crossmember bolts and the crossmember-to-frame side support attaching nuts and bolts
- Crossmember
- Transmission mount
- Transmission upper fill tube

- Vent tube on 4WD vehicles
- Converter housing-to-engine bolts

5. Move the transmission to the rear so it disengages from the dowel pins and the converter is disengaged from the flywheel. Lower the transmission from the vehicle.

6. Remove the torque converter from the transmission, if necessary.

To install:

7. Install the converter on the transmission.

✳✳ WARNING

Before installing an automatic transmission, always check that the torque converter is fully seated into the transmission. Typically, the converter has notches or tangs on the hub that must engage the transmission fluid pump. If they are not engaged in the pump, the transmission will not mate to the engine properly, as the converter will be holding it away. Severe damage to the pump, converter or transmission casting can occur if the transmission-to-engine bolts are tightened to force the transmission to mate to the engine.

Proper installation of the converter requires full engagement of the converter hub in the pump gear. To accomplish this, the converter must be pushed and at the same time rotated through what feels like 2 notches or bumps. When fully installed, rotation of the converter will usually result in a clicking noise heard, caused by the converter surface touching the housing to case bolts.

For reference, a properly installed converter will have a distance from the converter pilot nose from face-to-converter housing outer face of $^{13}/_{32}$–$^{9}/_{16}$ in. (10.5–14.5mm).

8. Rotate the converter so that the drive studs are in alignment with the holes in the flywheel.

9. Move the converter and transmission assembly forward into position, being careful not to damage the flywheel and converter pilot. The converter housing is piloted into position by the dowels in the rear of the engine block.

➡**During this move, to avoid damage, do not allow the transmission to get into a nose down position as this will cause the converter to move forward and disengage from the pump gear.**

10. Install or connect the following:
- Converter housing to engine.

Torque the bolts to 30–41 ft. lbs. (40–55 Nm). The 2 longer bolts are located at the dowel holes.
- Upper fluid filler tube and bracket. Torque the bolt to 41 ft. lbs. (55 Nm).
- Transfer case, 4WD vehicles
- Exhaust bracket. Torque the bolts to 64 ft. lbs. (87 Nm).
- Crossmember. Torque the bolts to 87 ft. lbs. (118 Nm).
- Transmission mount. Torque the bolts to 81 ft. lbs. (110 Nm).
- Transmission cooler lines. Torque the bolts to 23 ft. lbs. (31 Nm).
- Starter. Torque the bolts to 30 ft. lbs. (40 Nm).
- Transmission wire harness
- HO2S sensor
- Transmission connector
- TR sensor connector
- Shift cable and bracket
- Driveshaft. Torque the bolts to 95 ft. lbs. (129 Nm).
- Shroud. Torque the bolts to 71 inch lbs. (8 Nm).
- Fluid level indicator
- Negative battery cable

11. Fill the transmission to the proper level.

12. Start the vehicle and check for leaks, repair if necessary.

2002

1. Before servicing the vehicle, refer to the precautions in the beginning of this section.

2. Place the transmission in **Neutral**.

3. Remove or disconnect the following:
- Negative battery cable
- Transfer case
- Three-way catalytic converter

4. If transmission disassembly is necessary, drain the transmission fluid.

5. Support the transmission with a transmission jack.

6. Remove or disconnect the following:
- RH heat shield
- LH heat shield bolt
- Plastic cover from the right side of the frame rail
- Plastic shield on the right side of the crossmember near the fuel tank
- The two upper crossmember bolts (two on each side)
- RH side lower crossmember bolts
- LH side lower crossmember bolts
- Transmission support insulator nuts and remove the crossmember
- The two bolts and remove the transmission support insulator
- Shift cable and bracket

- Turbine shaft speed (TSS) sensor, output shaft speed (OSS) sensor and intermediate shaft speed (ISS) sensor electrical connectors

➡Clean the area around connector to prevent contamination of the solenoid body connector.

7. Remove or disconnect the following:
- Screw from the solenoid body connector and disconnect the connector
- Harness retainers
- Left side heated oxygen sensor (HO2S) connector from the bracket
- Digital transmission range (TR) sensor connector
- Starter

✳✳ WARNING

Do not damage the cooler tubes

- Transmission cooler tubes
- Access cover

➡Make an identifying mark on the nut, stud, and adapter plate to allow for correct installation.

- The 4 torque converter nuts
- The 7 engine-to-transmission retaining bolts

8. Lower the transmission from the vehicle.
To install:
9. Raise and position the transmission.
10. Align the flexplate to converter marks made at removal.
11. Install seven engine to transmission retaining bolts. Torque to 35 ft. lbs. (48Nm).
12. Install the four torque converter nuts. Torque to 28 ft. lbs. (48Nm).
13. Install or connect the following:
- Access cover
- Transmission cooler tubes
- Starter
- Digital transmission range (TR) sensor connector
- Left side heated oxygen sensor (HO2S) to the bracket
- Wire harness, and install the retainers

✳✳ WARNING

Damage will occur to the solenoid body assembly if the screw is tightened above 44 inch lbs. (5Nm). Always install new O-ring seals on vehicle harness connector. Clean the area around connector to prevent

contamination of the solenoid body connector. Use petroleum jelly to lubricate the O-ring seals to aid in the installation process.

14. Install and lubricate new O-ring seals on the transmission connector and connect the connector.
15. Install or connect the following:
- Turbine shaft speed (TSS) sensor, output shaft speed (OSS) sensor, and intermediate shaft speed (ISS) sensor electrical connectors.
- Shift cable and bracket
- Three-way catalytic converter
- Transmission support insulator. Torque to 68 ft. lbs. (90Nm).
- Crossmember in place and loosely install the two nuts to hold up the crossmember.
- Crossmember bolts. Torque to 52 ft. lbs. (70Nm).
- Rear transmission support nuts. Torque to 52 ft. lbs. (70Nm).
- Heat shields
- Plastic cover to the right side of the frame rail
- Plastic shield on the right side of the crossmember near the fuel tank
- Transfer case

➡If the vehicle was not equipped with a fluid filter, install a Fluid Filter Service Kit (XC3Z-7B155-AA). Follow the instructions supplied in the kit. If the

vehicle was equipped with a fluid filter, install a new filter (XC3Z-7B155-AB).

16. Connect the battery ground cable.
17. Verify that the shift cable is correctly adjusted.

Clutch

REMOVAL & INSTALLATION

1999–00

1. Before servicing the vehicle, refer to the precautions in the beginning of this section.
2. Remove or disconnect the following:
- Negative battery cable
- Clutch hydraulic system master cylinder from the clutch pedal
- Starter
- Hydraulic coupling at the transmission
- Transmission from the vehicle
3. Mark the assembled position of the pressure plate in relation to the flywheel to aid during reassemble.
- Loosen the pressure plate and cover attaching bolts evenly until the pressure plate springs are expanded
- Pressure plate and cover assembly and the clutch disc from the flywheel
- Pilot bearing, only if replacing

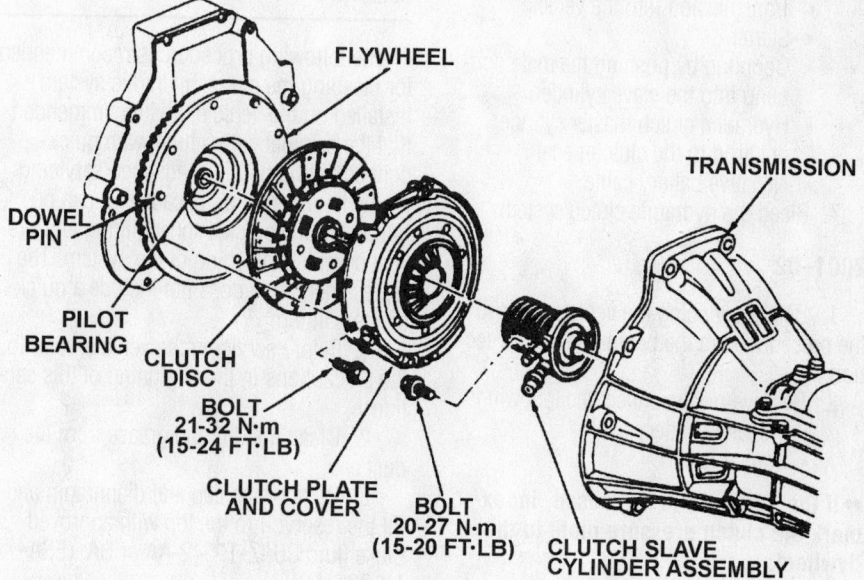

FLYWHEEL
TRANSMISSION
DOWEL PIN
PILOT BEARING
CLUTCH DISC
BOLT 21-32 N·m (15-24 FT·LB)
CLUTCH PLATE AND COVER
BOLT 20-27 N·m (15-20 FT·LB)
CLUTCH SLAVE CYLINDER ASSEMBLY

7924EG17

Clutch disc, pressure plate and bearing assembly

Timing belt service is covered in Section 3 of this manual

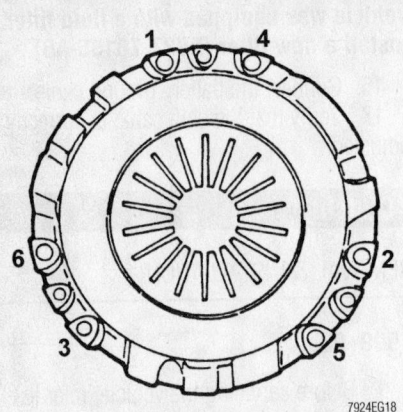

Tighten the bolts gradually in the correct sequence to avoid warping the pressure plate

To install:

4. Position the clutch disc on the flywheel so that the Clutch Alignment Shaft tool T74P-7137-K can enter the clutch pilot bearing and align the disc.

5. When reinstalling the original pressure plate and cover assembly, align the assembly and flywheel according to the marks made during the removal operations.

6. Install or connect the following:

- Pressure plate and cover assembly on the flywheel, align the pressure plate and disc, and install the retaining bolts that fasten the assembly to the flywheel. Torque the bolts to 15–25 ft. lbs. (21–35 Nm) in the proper sequence. Remove the clutch disc pilot tool
- Transmission into the vehicle
- Starter
- Coupling by pushing the male coupling into the slave cylinder
- Hydraulic clutch master cylinder pushrod to the clutch pedal
- Negative battery cable

7. Bleed the hydraulic clutch system.

2001–02

1. Before servicing the vehicle, refer to the precautions in the beginning of this section.

2. Remove or disconnect the following:
- Negative battery cable
- Transmission

➡**If the parts are to be reused, index-mark the clutch pressure plate to the flywheel.**

- Bolts, clutch pressure plate and the clutch disc

To install:

3. Lubricate the transmission input shaft pilot bearing with grease.

4. Adjust the clutch pressure plate:
   a. Using a suitable press, press downward on the fingers until the adjusting ring moves freely.
   b. Rotate the adjusting ring counterclockwise to compress the tension springs. Hold the adjusting ring in this position.
   c. Release the pressure on the fingers. The adjusting ring will stay in the reset position.

5. Position the clutch disc on the flywheel.

6. Align the clutch disc and the clutch pressure plate. Install the bolts and tighten in a star pattern sequence. Torque to 21 ft. lbs. (28Nm).

➡**An "L" is stamped by three bolt holes on the clutch pressure plate. Tighten these to specification first to ensure correct clutch alignment.**

7. Install the transmission.

ADJUSTMENT

Because the clutch is hydraulically driven, there is no adjustment required.

In the event the clutch pedal develops a squeak or uneven feel when depressing, spray the pedal bushing assembly with penetrating oil and work the pedal back-and-forth.

Hydraulic Clutch System

BLEEDING

The following procedure is recommended for bleeding the clutch hydraulic system installed on the vehicle. It is recommended that the original clutch tube, with quick-connect fitting be replaced when servicing the hydraulic system, because air can be trapped in the quick-connect fitting and prevent complete bleeding of the system. The replacement tube does not include a quick-connect fitting.

1. Before servicing the vehicle, refer to the precautions in the beginning of this section.

2. Clean the dirt and grease from the dust cap.

3. Remove the cap and diaphragm and fill the reservoir to the top with approved brake fluid C6AZ-19542-AA or BA, (ESA-M6C25-A).

➡**To keep brake fluid from entering the clutch housing, route a suitable rubber tube of appropriate inside diameter from the bleed screw to a container.**

4. Loosen the bleed screw, located in the slave cylinder body, next to the inlet connection. Fluid will now begin to move from the master cylinder down the tube to the slave cylinder.

➡**The reservoir must be kept full at all times during the bleeding operation, to ensure no additional air enters the system.**

5. Observe the bleed screw outlet. When the slave cylinder is full, a steady stream of fluid will flow from the outlet port. Tighten the bleed screw.

6. Depress the clutch pedal to the floor and hold for 1–2 seconds. Release the pedal as rapidly as possible. The pedal must be released completely. Pause for 1–2 seconds. Repeat 10 times.

7. Check the fluid level in the reservoir. The fluid should be level with the step when the diaphragm is removed.

8. Hold the pedal to the floor, slightly open the bleed screw to allow any additional air to escape. Close the bleed screw, then release the pedal.

9. Check the fluid in the reservoir. The hydraulic system should now be fully bled, and should actuate the clutch.

10. Check the vehicle by starting, pushing the clutch pedal to the floor and selecting reverse gear. There should be no grating of gears. If there is, and the hydraulic system still contains air; repeat the bleeding procedure.

Transfer Case Assembly

REMOVAL & INSTALLATION

1999–00

1. Before servicing the vehicle, refer to the precautions in the beginning of this section.

2. Place the transmission in **Neutral**.

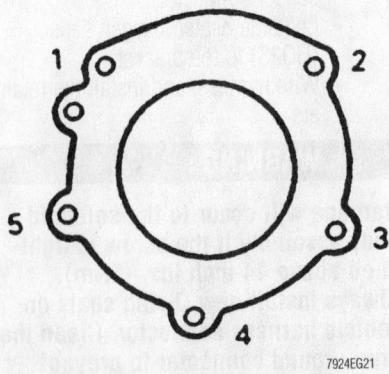

Transfer case-to-extension bolt torque sequence

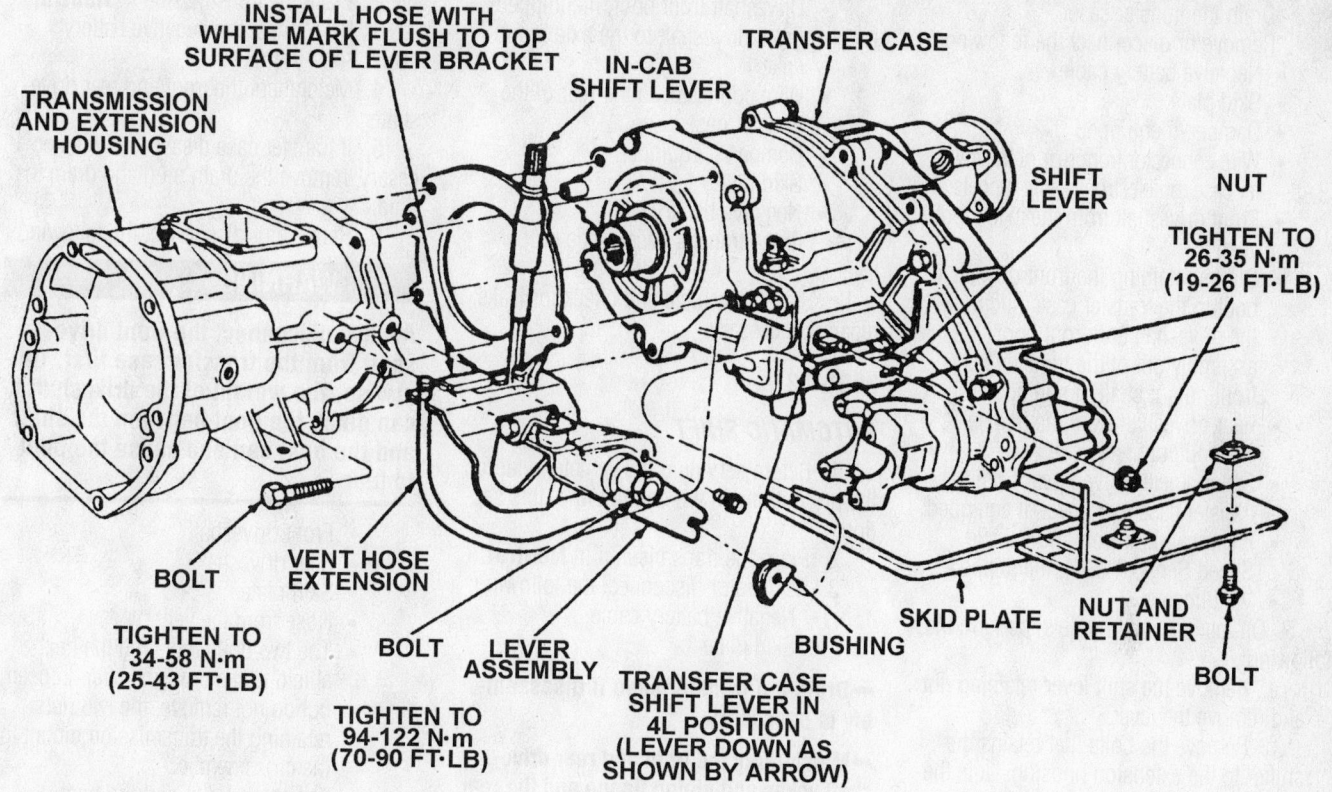

Exploded view of the 13-54 mechanical shift transfer case-to-transmission mounting

7924EG19

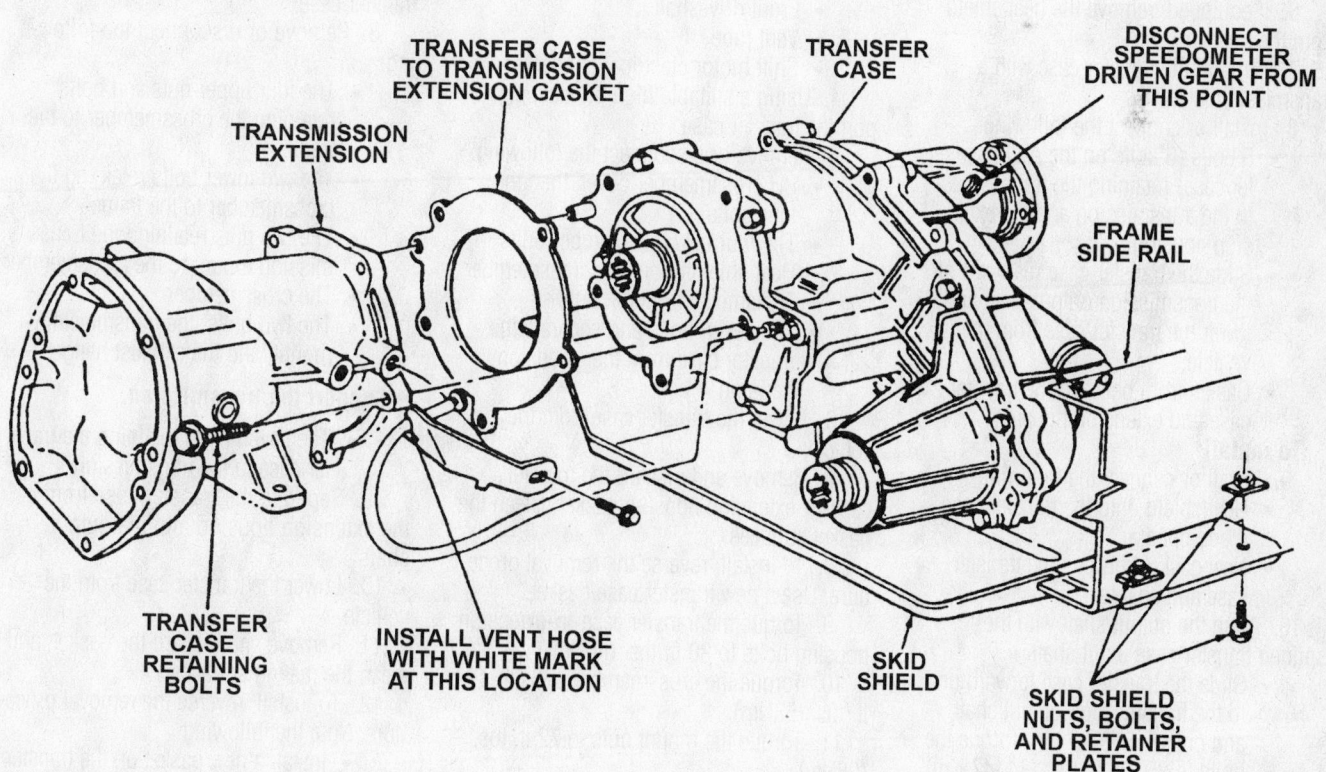

Exploded view of the 13-54 electronic shift transfer case-to-transmission mounting—44-05 model is similar

7924EG20

Heater Core replacement is covered in Section 2 of this manual

3. Drain the transfer case.

4. Remove or disconnect the following:
- Negative battery cable
- Skid plate
- Damper, if equipped
- Wire connector from rear of the transfer case, on electronic-shift models
- Front driveshaft from the axle input yoke
- Clamp retaining the front driveshaft boot to the transfer case, and pull the driveshaft and front boot assembly out of the transfer case front output shaft, if equipped
- Rear driveshaft from the transfer case output shaft yoke
- Speedometer driven gear from the transfer case rear cover, if equipped
- Electrical plug from the Vehicle Speed Sensor (VSS), if equipped
- Vent hose

5. On manual-shift models, perform the following:

a. Remove the shift lever retaining nut and remove the lever.

b. Remove the bolts that retains the shifter to the extension housing. Note the size and location of the bolts to aid during installation. Remove the lever assembly and bushing.

6. If equipped, remove the heat shield from the transfer case.

7. Support the transfer case with a transmission jack.

8. Install or connect the following:
- 5 bolts (6 bolts on the AWD transfer case) retaining the transfer case to the transmission and the extension housing
- Slide the transfer case rearward off the transmission output shaft and lower the transfer case from the vehicle.
- Gasket from between the transfer case and extension housing

To install:

9. Install or connect the following:
- Heat shield onto the transfer case, if equipped
- New gasket between the transfer case and adapter

10. Align the output shaft with the splined transfer case input shaft.
- Slide the transfer case forward on to the transmission output shaft and onto the dowel pin. Torque the bolts to 25–35 ft. lbs. (34–47 Nm).
- Shifter to the extension housing
- Vent tube
- VSS plug, if equipped
- Speedometer driven gear to the rear cover, if equipped

- Driveshaft front boot, if equipped
- Front driveshaft to the axle input yoke
- Wire connector to the rear of the transfer case
- Damper, if equipped
- Skid plate
- Negative battery cable

11. Fill the transfer case to the proper level.

12. Start the vehicle and check for leaks, repair if necessary.

2001–02

AUTOMATIC SHIFT

1. Before servicing the vehicle, refer to the precautions in the beginning of this section.

2. Place the transmission in **Neutral**.

3. Remove or disconnect the following:
- Negative battery cable
- Skid plate

➡ **Drain the transfer case if disassembly is necessary.**

➡ **Match-mark the front and rear driveshaft yokes and pinion flange and the rear driveshaft yoke and rear output flange.**

- Rear driveshaft
- Front driveshaft
- Vent tube
- Shift motor electrical connector

4. Using a suitable high lift jack, support the transfer case.

5. Remove or disconnect the following:
- RH crossmember cover, then the four bolts
- The four LH crossmember bolts
- Heat shields from the crossmember
- Transmission mount nuts
- The seven bolts and separate the transfer case from the extension housing

6. Lower the transfer case from the vehicle.

7. Remove and discard the transfer case-to-extension housing gasket. Clean the gasket surfaces.

8. To install, reverse the removal procedure. Use a new transfer case gasket.

9. Torque the transfer case-to-extension housing bolts to 30 ft. lbs. (40Nm).

10. Torque the crossmember bolts to 46 ft. lbs. (63Nm).

11. Torque the mount nuts to 72 ft. lbs. (98Nm).

ALL WHEEL DRIVE

1. Before servicing the vehicle, refer to the precautions in the beginning of this section.

2. Place the transmission in **Neutral**.

3. Disconnect the negative battery cable.

4. Matchmark the front and rear driveshafts.

5. If transfer case disassembly is necessary, remove the drain plug and drain the fluid.

6. Remove or disconnect the following:

✳✳ WARNING

Always disconnect the front driveshaft from the transfer case first. Otherwise, the weight of the driveshaft can pinch the boot between the shaft and the boot can and cause the boot to tear.

- Front driveshaft
- Rear driveshaft
- Skid plate
- Hose from the vent
- The two bolts retaining the heat shield to the crossmember. Loosen, but do not remove, the two nuts retaining the transmission mount to the crossmember.
- The three bolts and the heat shield

7. Position a high lift jack under the transfer case.

8. Remove or disconnect the following:
- The four upper nuts and bolts retaining the crossmember to the frame
- The two lower bolts retaining the crossmember to the frame
- The two nuts retaining the transmission mount to the crossmember
- The crossmember
- The two bolts, the transmission mount, and the exhaust hanger

➡ **Support the transmission.**

- The seven bolts retaining the transfer case to the transmission

9. Separate the transfer case from the extension housing and the output shaft.

10. Lower the transfer case from the vehicle.

11. Remove and discard the gasket, and clean the mating surfaces.

12. To install, reverse the removal procedure. Note the following:
- Install a new gasket on the transfer case.
- Watch for obstructions when installing the transfer case in the vehicle.

- Install the original transfer case to transmission bolts. If new bolts are required, be sure to install only the specific Ford transfer case to transmission bolts.
- Transfer case-to-extension: 30 ft. lbs. (40Nm)
- Transmission mount bolts: 66 ft. lbs. (90Nm)
- Crossmember bolts/nuts: 52 ft. lbs. (70Nm)
- Heat shield: 18 ft. lbs. (25Nm)

Halfshaft

REMOVAL & INSTALLATION

1999–00

1. Before servicing the vehicle, refer to the precautions in the beginning of this section.
2. Remove or disconnect the following:
   - Negative battery cable
   - Front wheel
   - Manual locking hub assembly from the rotor by pulling straight outward
   - Plastic stub shaft retainer
   - Front disc brake caliper from the anchor and move it aside. Support the front suspension lower arm.
   - Upper ball joint from the steering knuckle
   - Outboard front driveshaft from the hub
   - Halfshaft from the axle housing
   - Driveshaft and joint from the axle housing

To install:

3. Install or connect the following:
   - Halfshaft and joint to the axle housing
   - Outboard front driveshaft to the hub
   - Upper ball joint to the steering knuckle
   - Brake caliper to the anchor and remove the front suspension support
   - New stub shaft retainer
   - Locking hub to the brake rotor
   - Front wheel
   - Negative battery cable

2001–02

FRONT

1. Before servicing the vehicle, refer to the precautions in the beginning of this section.

2. Remove or disconnect the following:
3. Loosen the front axle wheel hub retainer.
   - Wheel and tire assembly
   - Hub retainer and the washer. Discard the front axle wheel hub retainer.
   - The two bolts and position the disc brake caliper aside
   - Tie-rod end from the knuckle. Discard the nut.
   - Stabilizer bar link. Discard the nut.

✳✳ WARNING

Do not allow the knuckle to hang freely. It is possible to overextend and internally separate each inner CV joint from its housing.

   - Upper ball joint from the knuckle

✳✳ WARNING

Do not use a hammer to separate the outboard CV joint from the hub. Damage to the threads and internal CV joint components may result.

4. Press the outboard CV joint until it is loose in the hub.
5. Remove the outboard CV joint from the hub.

✳✳ WARNING

Do not damage the axle shaft oil seal or the machined sealing surface on the inboard CV joint housing.

➡ **A circlip retains the inboard CV joint housing to the differential side gear in the axle.**

6. On the left side, pry the LH inboard CV joint housing from the differential side gear.
7. On the right side, disengage the RH inboard CV joint housing from the axle tube.
8. Pull the halfshaft and the axle shaft away from the axle tube, and separate the inboard CV joint housing from the axle shaft.
9. Remove the halfshaft assembly from the vehicle.

✳✳ WARNING

Do not damage the axle shaft oil seal, the machined sealing surface on the inboard CV joint housing, or the axle shaft splines.

10. To install, reverse the removal procedure.
11. Always install the halfshaft with a new retainer circlip and a new front axle wheel hub retainer.
12. On the right side, check the retainer circlip engagement after reseating the axle shaft and after installing the halfshaft in the axle. On the LH side, check the retainer circlip engagement after installing the halfshaft in the axle. When seated, the retainer circlip will lock the axle shaft and the inboard CV joint housing to the axle.

✳✳ WARNING

Never use power tools to tighten the front axle wheel hub retainer. Torque the retainer to 184 ft. lbs. (250 Nm).

➡ **It may be necessary to support the front suspension lower arm to be able to connect the upper ball joint to the knuckle.**

REAR

1. Before servicing the vehicle, refer to the precautions in the beginning of this section.

2. Remove or disconnect the following:

✳✳ WARNING

Do not loosen the rear axle wheel hub retainer until after the wheel and tire assembly are removed from the vehicle. Wheel bearing damage will occur if the wheel bearing is unloaded with the weight of the vehicle applied.

   - Wheel and tire assembly

➡ **Have an assistant press the brake pedal to keep the axle from rotating.**

   - Hub retainer and the washer
   - Caliper. Position the disc brake caliper out of the way.
   - Brake disc
   - Bolt retaining the parking brake cable bracket to the frame

✳✳ WARNING

Using a rubber hose approximately 37.5 mm (1.5 in) long, cover the stabilizer link bolt threads and nut to prevent boot damage when removing the halfshaft assembly from the vehicle.

Brake service is covered in Section 4 of this manual

✳✳ WARNING

The bolt that retains the upper ball joint to the knuckle is longer and it has fewer threads than the bolt that retains the toe link to the knuckle. Switching these bolts during installation will prevent the pinch arms on the knuckle from correctly retaining toe link to the knuckle. This may cause the toe link to separate from the knuckle during vehicle operation. Failure to follow these instructions may result in personal injury.

- Pinch bolts and disconnect the toe link and upper ball joint from the knuckle

✳✳ WARNING

Using a wood stick, approximately 450 mm (18 in) long and 25 mm (1 in) wide, support the rear suspension upper arm to prevent boot damage when removing the halfshaft assembly from the vehicle.

➡Do not use a hammer to separate the outboard CV joint from the hub. Damage to the threads and internal CV joint components may result. Once the outboard CV joint separates from the hub the knuckle will continue to pivot until the brake backing plate presses against the suspension lower arm. To prevent damage to the brake backing plate, immediately after separating the outboard CV joint from the hub, rest the knuckle on a cushioned support that is tall enough to keep the backing plate from pressing against the suspension lower arm.

3. Press the outboard CV joint until it is loose in the hub.

4. Separate the outboard CV joint from the hub.

5. Rest the knuckle on a cushioned support.

✳✳ WARNING

Do not damage the axle shaft oil seal or the machined sealing surface on the inboard CV joint housing.

➡A circlip retains the inboard CV joint housing to the differential side gear in the axle.

6. Using the special tool, disengage the inboard CV joint housing from the differential side gear.

✳✳ WARNING

To prevent damage to the axle shaft oil seal, install the special tool 205-461 (seal protector) before removing the inboard CV joint housing from the axle.

7. Remove the halfshaft assembly from the vehicle.

8. To install, reverse the removal procedure.

9. To prevent damage to the axle shaft oil seal, install the seal protector before positioning the inboard CV joint housing in the axle.

10. Always install the halfshaft with a new retainer circlip and a new rear axle wheel hub retainer.

➡Never use power tools to tighten the rear axle wheel hub retainer. Torque the retainer to 251 ft. lbs. (340Nm).

CV-Joints

OVERHAUL

1999–00

1. Before servicing the vehicle, refer to the precautions in the beginning of this section.

2. Remove or disconnect the following:
- Negative battery cable
- Halfshaft and place it in a vice with the inboard joint lower than the outboard joint

3. Cut the inner boot clamps with side cutters and remove the clamp from the boot.
- Larger boot end off the joint
- Inboard CV-joint bolts and separate the spacer and grease cap
- Snap-ring retaining the interconnecting shaft end to the CV-joint cage
- CV-joint and discard the washer

➡The outboard CV-joint is non-serviceable other than to replace the boot.

To install:

4. Install or connect the following:
- Slide the boot over the shaft

5. Fill the CV-joint area with grease.
- Assemble the outer boot to the outboard CV-joint and interconnecting shaft. Make certain that the boot is seated in the grooves on the outer race and on the shaft
- New clamps to the boot
- New inner boot to the shaft
- New washer to the end of the shaft

- Assemble the inboard CV-joint to the interconnecting shaft spline until it rests on the washer
- Snap-ring

6. Fill the CV-joint area with grease.
- Boot into position and make certain that it is seated in the grooves on the boot adapter and the shaft
- New clamps and tighten the clamps with crimping pliers
- Spacer to the CV-joint end pilot. Torque the bolts to 25 ft. lbs. (34 Nm).
- Halfshaft
- Negative battery cable

2001–02

FRONT SHAFT

1. Before servicing the vehicle, refer to the precautions in the beginning of this section.

2. Remove or disconnect the following:
- Halfshaft
- The two inboard boot clamps
- Inboard halfshaft boot off the inboard CV joint housing

3. Separate the CV joint from the CV joint housing.

4. Mark the shaft and the inboard CV joint to ease alignment during assembly.

5. Remove the snap ring.

6. Remove the CV joint.

7. Inspect the stop ring for wear or damage. Install a new stop ring as necessary.

8. Remove the inboard halfshaft boot from the shaft assembly.

9. Remove the two outboard boot clamps.

10. Remove the outboard halfshaft boot.

➡If grease is contaminated, clean and inspect the joint for wear. If worn or damaged, install a new outboard CV joint and shaft assembly.

To assemble:

11. Pack the outboard CV joint with grease.

➡Clean the halfshaft boot mounting surfaces of access grease before positioning the halfshaft boot into place.

12. Position the outboard halfshaft boot and outboard boot clamps.

13. Tighten the through-bolt until the installer is in the closed position.

14. Use the CV Boot Clamp Installer to install the outboard CV joint boot clamps.

15. Position the boot clamp on the halfshaft.

16. Position the inboard halfshaft boot.

17. Install a new stop ring, if necessary.

18. Install the CV joint on the halfshaft.

19. Install the snap-ring.

20. Lubricate the three CV joint needle bearings with grease.

21. Fill the inboard CV joint housing with 235 grams of grease.

22. Position the CV joint housing onto the CV joint.

➡**Remove any excess grease from the inboard halfshaft boot mating surface before positioning into place.**

23. Position the inboard halfshaft boot and boot clamp.

24. Insert a wood wand to relieve built-up air pressure in the halfshaft boot.

25. Use the CV Boot Clamp Installer to install the inboard boot clamps.

26. Install the front wheel halfshaft.

REAR SHAFT

1. Before servicing the vehicle, refer to the precautions in the beginning of this section.

2. Remove or disconnect the following:

3. Remove the halfshaft.

4. For the inboard CV joint:

a. Remove and discard the boot clamps.

b. Remove the inboard CV joint housing.

c. Remove and discard the retainer circlip.

d. Slide the boot away from the CV joint.

e. Using a suitable 3-jaw puller, remove the CV joint.

f. Remove and discard the tri-lobe insert and the boot.

5. For the outboard CV joint:

a. Remove and discard the boot clamps.

b. Remove and discard the boot.

➡**Do not disassemble the side shaft assembly. Install a new halfshaft assembly, if the components are worn/damaged.**

6. Inspect the grease packed in the inboard CV joint and the outboard CV joint for contamination. Rub some of the grease from each joint between two fingers. Any gritty feeling indicates contamination. Wash all of the grease from the inboard CV joint, the inboard CV joint housing, the outboard CV joint, and the interconnecting shaft. Thoroughly dry all of the components, and inspect them for wear and damage. Discard the assembly, if necessary. Proceed as follows only if not discarding the assembly.

7. If necessary, remove and discard the outboard dust seal. Tap uniformly around the seal to separate it from the joint.

8. On the inboard end:

a. Remove and discard the retainer circlip.

b. If necessary, remove and discard the inboard dust seal. Tap uniformly around the seal to separate it from the housing.

To assemble:

9. For the outboard CV joint:

a. Slide the boot on the interconnecting shaft.

b. Pack the outboard CV joint with 225 grams (5.29 ounces) of grease.

c. Spread any remaining grease evenly inside the boot.

d. Clean any excess grease from the boot mounting surfaces before installing the boot.

e. Install the boot by seating it in the groove in the CV joint housing.

f. Tighten the through-bolt until the special tool is in the closed position.

g. Using the special tool, install both boot clamps.

h. If removed, use the special tools to install the dust seal.

✳✳ WARNING

Do not overexpand or twist the circlip during installation.

i. Install the retainer circlip.

10. Install the halfshaft in a soft jaw vise.

11. For the inboard CV joint:

a. Position the clamp on the interconnecting shaft.

b. Position the boot on the interconnecting shaft.

➡**The lip on the end of the tri-lobe insert must seat against the end of the boot.**

c. Install the tri-lobe insert.

➡**One side of the inboard CV joint has a champher cut in the edge of joint at the inner diameter near the splines. Install the inboard CV joint so that the champher faces the outboard end of the halfshaft.**

d. Install the CV joint.

e. Install the retainer circlip.

f. Pack the inboard CV joint housing with 250 grams (5.88 ounces) of grease.

g. Spread any remaining grease evenly inside the boot and on the CV joint.

h. Clean any excess grease from the boot mounting surfaces before installing the boot.

i. Install the inboard CV joint housing, seating the boot in the groove in the housing.

12. Set the halfshaft assembled length to 28.44 inches (722.4mm).

13. Hold the inner joint to prevent the assembled length from changing, and insert a wood wand between the boot and the joint to equalize the pressure.

14. Tighten the through-bolt until the special tool is in the closed position.

15. Install both boot clamps.

16. If removed, install the dust seal.

17. Install the halfshaft.

Locking Hubs

REMOVAL & INSTALLATION

The Mountaineer and Explorer models use a locking mechanism mounted in the differential. This system is called a vacuum disconnect axle lock.

Front Axle Tube Bearing

REMOVAL & INSTALLATION

1. Before servicing the vehicle, refer to the precautions in the beginning of this section.

2. Remove or disconnect the following:
- Right-hand halfshaft
- Right-hand axle shaft
- Axle seal, with a slide hammer
- Axle tube bearing, with a slide hammer

3. Clean the bearing and seal surfaces of any foreign debris.

To install:

4. Use an axle bearing replacer and the handle to replace the RH axle tube bearing.

5. Check the bearing depth as shown.

6. Use an axle seal replacer and the handle to replace the axle tube seal.

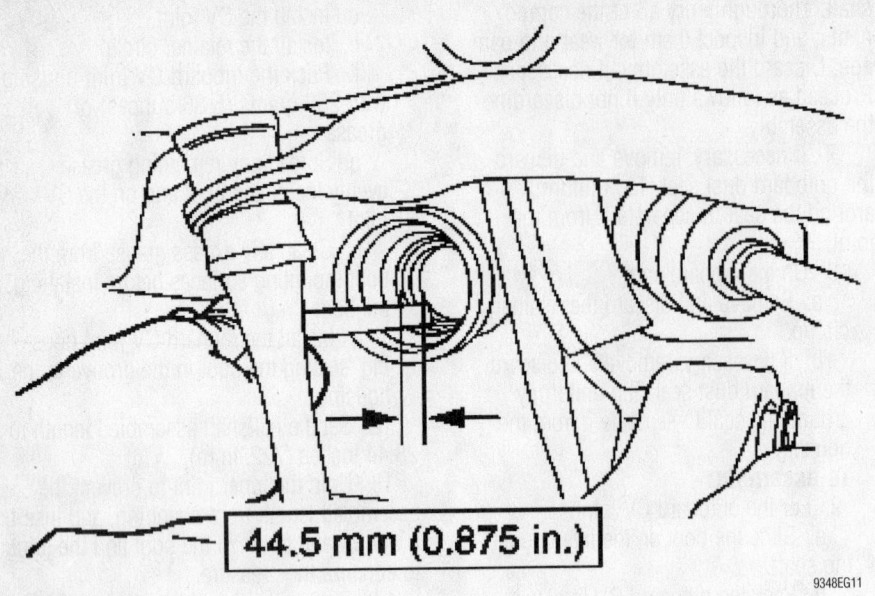

44.5 mm (0.875 in.)

9348EG11

Front axle tube bearing depth

➡ **Care should be taken not to damage the axle seal surface.**

7. Install the axle shaft.
8. Refill the front drive axle to proper level using SAE 80W90.
9. Install the RH halfshaft.

Rear Axle Shaft, Bearing and Seal

REMOVAL & INSTALLATION

Solid Axle

1. Before servicing the vehicle, refer to the precautions in the beginning of this section.
2. Drain the axle housing fluid.
3. Remove or disconnect the following:
   - Negative battery cable
   - Rear wheel
   - Brake drum
   - Wheel speed sensor, if equipped
   - Axle housing cover
   - Bearing retainer nuts
   - Axle shaft and bearing
   - Axle shaft inner oil seal
4. If equipped with ABS, grind a flat spot on the wheel speed sensor tone ring, then split the ring with a chisel.
5. Press the wheel bearing off the axle shaft.
6. Remove the bearing retainer and the outer oil seal.
   To install:
7. Install or connect the following:
   - Outer oil seal to the bearing retainer

   - Bearing retainer to the axle shaft
   - Bearing and retainer ring pressed onto the axle shaft
   - Wheel speed sensor tone ring pressed onto the axle shaft, if equipped
   - Axle shaft inner oil seal
   - Axle shaft and bearing
   - Bearing retainer nuts. Tighten them to 17 ft. lbs. (23 Nm).
   - Wheel speed sensor, if equipped
   - Brake drum
   - Rear wheel
   - Negative battery cable
8. Fill the rear differential to the correct level.

Pinion Seal

REMOVAL & INSTALLATION

Solid Rear Axle

1. Before servicing the vehicle, refer to the precautions in the beginning of this section.
2. Drain the axle housing fluid.
3. Remove or disconnect the following:
   - Negative battery cable
   - Rear wheels
   - Driveshaft
   - Brake calipers and pads or brake drum

➡ **The brake calipers and pads or brake drum must be removed so that there is no additional drag when measuring pinion bearing preload.**

4. Use an inch lb. torque wrench and measure and record the amount of torque required to maintain pinion rotation through several revolutions.
5. Remove or disconnect the following:
   - Pinion flange
   - Pinion seal
   - Pinion bearing
   - Collapsible spacer
   To install:

➡ **Use a new collapsible spacer and flange nut for assembly.**

6. Install or connect the following:
   - Collapsible spacer
   - Pinion bearing
   - Pinion seal
   - Pinion flange
7. Rotate the pinion flange occasionally while tightening the flange nut to make sure the pinion bearings seat correctly.
8. Take frequent bearing preload torque readings. Tighten the flange nut to achieve the preload torque readings originally recorded.

✳✳ CAUTION

Never loosen the pinion nut to reduce bearing preload. If it is necessary to reduce bearing preload, install a new collapsible spacer and pinion nut.

9. Install or connect the following:
   - Driveshaft
   - Brake calipers and pads or brake drum
   - Wheels
   - Negative battery cable
10. Fill the differential with gear lubricant and check for leaks.

Halfshaft Rear Axle

1. Before servicing the vehicle, refer to the precautions in the beginning of this section.
2. Drain the axle housing fluid.
3. Remove or disconnect the following:
   - Rear wheel and tire assemblies
   - Brake caliper and support bracket from the knuckle as an assembly. Wire the caliper and support bracket assembly out of the way.

➡ **Matchmark the driveshaft flange and rear axle pinion flange to maintain initial balance during installation.**

4. Disconnect and position the driveshaft out of the way.
5. Install an inch/pound torque wrench on the nut and record the torque necessary to maintain rotation of the drive pinion gear through several revolutions.

6. Remove and discard the pinion flange nut.

➡ **Matchmark the rear axle pinion flange and drive pinion gear stem to maintain initial balance during installation.**

7. Remove the rear axle pinion flange.

8. Force up on the metal flange of the rear axle drive pinion seal. Install gripping pliers and strike with a hammer until the rear axle drive pinion seal is removed.

To install:

9. Lubricate the new rear drive pinion seal with grease.

➡ **If the rear axle drive pinion seal becomes misaligned during installation, remove the rear axle drive pinion seal and install a new seal.**

10. Drive in the rear axle drive pinion seal.

11. Inspect the rear axle pinion flange seal journal for rust, nicks and scratches prior to installing the flange. Polish the seal journal with fine crocus cloth, if necessary.

12. Lubricate the rear axle pinion flange splines.

13. Install the rear axle pinion flange.

✲✲ WARNING

Do not under any circumstance loosen the nut to reduce preload. If it is necessary to reduce preload, install a new differential drive pinion collapsible spacer and nut.

14. Rotate the pinion occasionally to make sure the pinion bearings seat correctly. Take frequent pinion bearing torque preload readings by rotating the drive pinion gear with an inch/pound torque wrench.

15. If the preload recorded prior to disassembly is lower than the specification for used bearings, then tighten the nut to specification. If the preload recorded prior to disassembly is higher than the specification for used bearings, then tighten the nut to the original reading as recorded.

Pinion bearing preload used bearings: 8–14 inch lbs. (0.9–1.16 Nm); new bearings: 16–29 inch lbs. (1.8–3.2 Nm).

16. Connect the driveshaft. Torque the bolts to 83 ft. lbs. (112 Nm).

17. Install the rear brake calipers.

18. Install the rear wheel and tire assemblies.

Front Axle

➡ **This operation disturbs the differential pinion bearing preload. Carefully reset the preload during assembly.**

✲✲ CAUTION

The electrical power to the air suspension system must be shut off prior to hoisting, jacking or towing an air suspension vehicle. This can be accomplished by turning off the air suspension switch located in the rear jack storage area. Failure to do so can result in unexpected inflation or deflation of the air springs, which can result in shifting of the vehicle during these operations.

1. Before servicing the vehicle, refer to the precautions in the beginning of this section.

2. Index-mark the front driveshaft and pinion flange.

3. Remove or disconnect the following:

   • Front driveshaft from the pinion flange, and position it aside

➡ **Do not allow the driveshaft to hang unsupported.**

4. Using a Nm (inch-pound) torque wrench, measure the torque required to maintain pinion rotation. Record the measurement.

5. Index-mark the pinion flange and the pinion stem.

6. Hold the pinion flange while removing the nut.

7. Place a drain pan under the differential housing.

8. Using a puller, remove the pinion flange.

9. Inspect the pinion flange for burrs and damage. Inspect the end of the pinion flange that contacts the bearing cone, the nut counterbore, and the seal surface for nicks. Discard the pinion flange as necessary.

10. Using a seal remover and impact slide hammer, remove the pinion seal.

11. Remove the front axle drive pinion shaft oil slinger and the differential pinion bearing.

12. Remove and discard the collapsible spacer.

To install:

13. Verify that the splines on the pinion stem are free of burrs. If burrs are evident, remove them with a fine crocus cloth. Work in a rotating motion to wipe the pinion clean.

14. Clean the pinion seal bore.

15. Install a new collapsible spacer.

16. Install the original differential pinion bearing and the front axle drive pinion shaft oil slinger.

17. Lubricate the pinion seal. Use Motorcraft SAE 80W90 Thermally Stable 4x4 Axle Lubricant meeting Ford specification WSP-M2C197-A.

18. Install the pinion seal.

19. Lubricate the pinion flange splines. Use Motorcraft SAE 80W90 Thermally Stable 4x4 Axle Lubricant meeting Ford specification WSP-M2C197-A.

➡ **Never use a metal hammer on the pinion flange or install the flange with power tools. If necessary, use a plastic hammer to tap on a tight fitting flange.**

   • Align the index marks and install the pinion flange.
   • Install the new nut hand-tight.

➡ **Do not loosen the nut to reduce preload. Install a new collapsible spacer and nut if preload reduction is necessary.**

20. Use the special tool to hold the pinion flange while tightening the nut to set the preload.

21. Tighten the nut, rotating the pinion occasionally to ensure the differential pinion bearings are seating correctly. Take frequent differential pinion bearing preload readings by rotating the pinion with a Nm (inch-pound) torque wrench. The final reading must be 0.56 Nm (5 inch lbs.) more than the initial reading taken during removal.

22. Align the index marks and position the front driveshaft.

23. Install the universal joint spider retainers and bolts.

24. Check the fluid level and, if necessary, fill the axle to specification. Use Motorcraft SAE 80W90 Thermally Stable 4x4 Axle Lubricant meeting Ford specification WSP-M2C197-A.

25. Lower the vehicle.

26. If so equipped, reactivate the air suspension.

For Accessory Drive Belt illustrations, see Section 1 of this manual

STEERING AND SUSPENSION

Air Bag

PRECAUTIONS

• Always wear safety glasses when servicing an air bag vehicle, and when handling an air bag.

• Never attempt to service the steering wheel or steering column on an air bag-equipped vehicle without first properly disarming the air bag system. The air bag system should be properly disarmed whenever ANY service procedure in this manual indicates that you should do so.

• When carrying a live air bag module, always make sure the bag and trim cover are pointed away from your body. In the unlikely event of an accidental deployment, the bag will then deploy with minimal chance of injury.

• When placing a live air bag on a bench or other surface, always face the bag and trim cover up, away from the surface. This will reduce the motion of the air bag if it is accidentally deployed.

• If you should come in contact with a deployed air bag, be advised that the air bag surface may contain deposits of sodium hydroxide, which is a product of the gas combustion and is irritating to the skin. Always wear gloves and safety glasses when handling a deployed air bag, and wash your hands with mild soap and water afterwards.

DISARMING THE SYSTEM

1. Before servicing the vehicle, refer to the precautions in the beginning of this section.
2. Disconnect the negative battery cable from the battery.
3. Disconnect the positive battery cable from the battery.
4. Wait 1 minute. This time is required for the back-up power supply in the air bag diagnostic monitor to completely drain. The system is now disarmed.

ARMING THE SYSTEM

1. Before servicing the vehicle, refer to the precautions in the beginning of this section.
2. Connect the positive battery cable.
3. Connect the negative battery cable.
4. Stand outside the vehicle and carefully turn the ignition to the **RUN** position. Be sure that no part of your body is in front of the air bag module on the steering wheel, to prevent injury in case of an accidental air bag deployment.
5. Ensure the air bag indicator light turns off after approximately 6 seconds. If the light does not illuminate at all, does not turn off, or starts to flash, test the system.

Power Rack and Pinion Steering Gear

REMOVAL & INSTALLATION

1999–01

> ☀ **WARNING**
>
> **If equipped, always turn off the Automatic Ride Control (ARC) service switch before lifting the vehicle off of the ground. Failure to do so could damage the ARC system components.**

1. Before servicing the vehicle, refer to the precautions in the beginning of this section.
2. Raise and safely support the front of the vehicle, block the rear wheels and apply the parking brake.
3. Start the engine then rotate the steering wheel from lock-to-lock and record the number of rotations.
4. Divide the number of rotations by 2. This gives the number of rotations to achieve true center of the steering. Turn the wheel in one direction to the full lock.
5. Turn the wheel in the opposite direction the number of turns equal to true steering (lock-to-lock number divided by 2).

> ☀ **WARNING**
>
> **Do not rotate the steering wheel when the shaft is disconnected from the steering gear as damage to the clock spring could occur.**

6. Drain the power steering fluid reservoir.
7. Remove or disconnect the following:
   • Negative battery cable
   • Bolt retaining the lower steering column shaft to the steering gear input shaft
   • Stabilizer bar
   • Quick-connect fittings for the power steering pressure and return hoses at the steering gear housing
   • Nuts securing the power steering cooler and remove the cooler
   • Outer tie rod ends
   • Nuts, bolts and washer assemblies retaining the steering gear housing to the front crossmember
   • Steering gear from the vehicle

To install:

8. Install or connect the following:
   • Position the steering gear to the front crossmember and install the nuts, bolts and washer assemblies. Torque to 94–127 ft. lbs. (128–172 Nm).
   • Power steering cooler retaining bolts
   • Power steering lines to the steering gear housing and torque the fittings to 20–25 ft. lbs. (27–34 Nm).
   • Outer tie rod ends and ensure that the steering shaft or gear input shaft has not been rotated
   • Intermediate shaft-to-steering input shaft retaining (pinch) bolt and torque the bolt to 30–42 ft. lbs. (41–56 Nm)
   • Negative battery cable

9. Fill the power steering pump reservoir.
10. Bleed the air from the power steering system.
11. Ensure that there are no leaks and the fluid is maintained at the proper level.
12. Check the alignment.

2002

1. Before servicing the vehicle, refer to the precautions in the beginning of this section.
2. Place front wheels in the straight-ahead position and the ignition switch in the **OFF** position.
3. Remove the front wheels.
4. Remove the power steering fluid cooler.
5. With a 4.6L engine, remove the shield.
6. Loosen the outer tie-rod end jam nuts.
7. Separate the tie-rod ends from the wheel knuckles. Discard the nuts.
8. Remove the tie-rods, counting the turns for reference during installation.

> ☀ **WARNING**
>
> **Do not allow the intermediate shaft to rotate while it is disconnected from the gear or damage to the clock-spring can occur. If there is evidence that the intermediate shaft has rotated, the clockspring must be removed and re-centered.**

9. Remove the bolt and disconnect the intermediate shaft from the gear.

10. Remove the nut and disconnect the lines from the gear. Drain the fluid into a suitable container.

11. Remove and discard the O-ring seals.

12. Remove the gear mounting nuts. Discard them.

13. With 4wd:

a. Remove the nut and the LH stabilizer bar link. Discard the nut.

b. Remove the LH shock absorber lower mounting bolt and flag nut. Discard the flag nut.

c. Remove the LH lower arm rearward mounting nuts. Discard the nuts.

d. Remove the LH lower arm forward mounting nut and flag bolt. Discard the nut.

14. Remove the lower arm inboard mounting nuts. Discard the nuts.

15. Remove the bolts and detach the bracket from the gear.

16. Remove the steering gear.

To install:

17. Position the gear, attach the bracket and install the bolts. Torque the bolts to 52 ft. lbs. (70Nm).

18. With 4wd:

❊❊ WARNING

Do not tighten the lower arm inboard mounting nuts until the installation procedure is complete and the vehicle is at curb ride height. Make sure to tighten the forward flag bolt and nut before tightening the rearward nuts.

a. Install the flag bolt and the nut. Snug the nut for now. Torque will be 295 ft. lbs. (400 Nm).

b. Install the lower arm inboard mounting nuts. Snug the nuts for now. Torque will be 111 ft. lbs. (150 Nm).

c. Install the bolt and the flag nut. Torque the bolt to 258 ft. lbs. (350 Nm).

d. Install the stabilizer bar link and tighten the nut. Torque the nut to 18 ft. lbs. (25 Nm).

19. Install the gear mounting nuts. Torque them to 148 ft. lbs. (200 Nm).

20. Connect the lines to the gear and tighten the nut to 18 ft. lbs. (25 Nm).

❊❊ WARNING

Do not allow the intermediate shaft to rotate while it is disconnected

from the gear or damage to the clockspring can occur. If there is evidence that the intermediate shaft has rotated, the clockspring must be removed and re-centered.

21. Connect the intermediate shaft to the gear and tighten the bolt to 35 ft. lbs. (48 Nm).

22. With a 4.6L engine, install the shield and the bolts.

23. Install the outer tie-rods, nuts and the cotter pins. Torque the nuts to 52 ft. lbs. (70Nm).

24. Tighten the tie-rod jam nuts to 59 ft. lbs. (80Nm).

25. Install the power steering fluid cooler.

26. Install the wheels.

27. Fill, purge and leak test the system.

28. Check and, if necessary, align the front end.

Shock Absorber

REMOVAL & INSTALLATION

➡**Low pressure gas shocks are charged with Nitrogen gas. Do not attempt to open, puncture or apply heat to them. Prior to installing a new shock absorber, hold it upright and extend it fully. Invert it and fully compress and extend it at least 3 times. This will bleed trapped air.**

Front

1999–01

1. Before servicing the vehicle, refer to the precautions in the beginning of this section.

2. Remove or disconnect the following:
   • Negative battery cable
   • Upper shock-to-frame attaching nut, washer and insulator assembly
   • Lower shock-to-control arm attaching nuts
   • Slightly compress the shock absorber by hand and remove it from the vehicle

To install:

3. Install or connect the following:
   • Position the lower washer and insulator on the shock absorber rod and position the shock absorber to the upper frame bracket mount
   • Position the upper insulator and washer on the shock absorber rod and install the attaching nut loosely.

• Position the lower shock absorber mounting studs into the control arm and install the attaching nuts loosely.

• Torque the lower shock attaching nuts to 15–21 ft. lbs. (21–29 Nm), and the upper shock attaching bolts to 30–40 ft. lbs. (40–55 Nm).

• Negative battery cable

2002

1. Before servicing the vehicle, refer to the precautions in the beginning of this section.

2. Remove or disconnect the following:
   • Wheel
   • Upper shock mounting nuts. Discard the nuts.
   • Nut and the stabilizer bar link. Discard the nut.
   • Bolt, flagnut and the spring and shock absorber as an assembly. Discard the flag nut.

3. Using a suitable spring compressor, compress the spring until the tension is released from the shock absorber.

4. While holding the flats of the washer, remove the nut.

5. Remove the shock absorber. Discard the nut, remove the washer, bushing and the upper mount.

6. Remove the insulator.

7. Remove the dust shield.

8. To install, reverse the removal procedure. Observe the following torques:
   • Center nut: 37 ft. lbs. (50 Nm)
   • Lower shock bolt: 258 ft. lbs. (350 Nm).
   • Sway bar link nut: 18 ft. lbs. (25 Nm)
   • Upper shock nuts: 22 ft. lbs. (30 Nm)

Rear

1999–01

1. Before servicing the vehicle, refer to the precautions in the beginning of this section.

2. Remove or disconnect the following:
   • Upper shock-to-frame attaching nut
   • Lower shock nut
   • Slightly compress the shock absorber by hand and remove it from the vehicle

To install:

3. Install or connect the following:
   • Shock absorber upper end and nut
   • Shock absorber lower end and nut

For Tire, Wheel and Ball Joint specifications, see Section 1 of this manual

- Torque the upper and lower shock attaching nuts to 53 ft. lbs. (72Nm)

2002

1. Before servicing the vehicle, refer to the precautions in the beginning of this section.
2. Remove or disconnect the following:
   - Wheels
   - Upper shock mounting nuts. Discard the nuts.
   - Nut and the stabilizer bar link. Discard the nut.
   - Ball joint pinch bolt. Discard the nut.
   - Bolt, flag nut and the shock absorber and spring as an assembly. Discard the flag nut.
3. Using a suitable spring compressor, compress the spring until the tension is released from the shock absorber.
4. While holding the flats of the washer, remove the nut.
5. Remove the shock absorber. Discard the nut.
6. Remove the washer, bushing (18A103) and the upper mount (18178).
7. Remove the insulator (5536).
8. Remove the dust shield.
9. To install, reverse the removal procedure. Note the following torques:
   - Center nut: 37 ft. lbs. (50 Nm)
   - Lower mounting bolt: 184 ft. lbs. (250 Nm)
   - Pinch bolt: 111 ft. lbs. (150 Nm)
   - Sway bar nut: 18 ft. lbs. (25 Nm)
   - Upper shock mounting nuts: 22 ft. lbs. (30 Nm)

Coil Spring

REMOVAL & INSTALLATION

For 2002, the front and rear coil springs are mounted on the shock absorbers. See Shock absorber Removal and Installation.

Leaf Springs

REMOVAL & INSTALLATION

1. Before servicing the vehicle, refer to the precautions in the beginning of this section.
2. Turn the air suspension switch off, if equipped.
3. Remove or disconnect the following:
   - Negative battery cable
   - Rear wheels and support the rear axle
   - Separate the rear spring from the axle and position the spring plate aside
   - Rear spring

To install:
4. Install or connect the following:
   - Rear spring. Torque the dual bolts to 85 ft. lbs. (115 Nm) and the single bolt to 66 ft. lbs. (90 Nm).
   - Properly position the spring plate and install the U-bolts. Torque the bolts to 87 ft. lbs. (117 Nm).
   - Rear wheels and remove the rear axle support
   - Negative battery cable
5. Turn the air suspension switch on, if equipped.

Torsion Bar

REMOVAL & INSTALLATION

1. Before servicing the vehicle, refer to the precautions in the beginning of this section.
2. Remove or disconnect the following:
   - Negative battery cable
   - Torsion bar cover plate and measure the length of the torsion bar adjustment bolt
   - Relieve torsion bar tension
   - Torsion bar adjustment bolt
   - Torsion bar and insulator

To install:
3. Install or connect the following:
   - Torsion bar in the front suspension lower arm
   - Torsion bar adjuster and position the insulator
4. Preload the torsion bar and install a **NEW** adjuster bolt. Turn the bolt until it reaches the measurement made during the removal procedure.
   - Torsion bar cover plate. Torque the bolts to 46 ft. lbs. (63 Nm).
   - Negative battery cable

Upper Ball Joint

REMOVAL & INSTALLATION

The ball joints on the Mountaineer and Explorer are integral with the control arm. If the ball joint is defective, the entire control arm must be replaced.

Lower Ball Joint

REMOVAL & INSTALLATION

The ball joints on the Mountaineer and Explorer are integral with the control arm. If the ball joint is defective, the entire control arm must be replaced.

Upper Control Arm

REMOVAL & INSTALLATION

1999–01

1. Before servicing the vehicle, refer to the precautions in the beginning of this section.
2. Turn off the air suspension switch, if equipped.
3. Remove or disconnect the following:
   - Negative battery cable
   - Front wheel
   - Pinch bolt and nut from the spindle
   - Upper control arm

To install:
4. Install or connect the following:
   - Upper control arm. Torque the bolts to 112 ft. lbs. (153 Nm).
   - Pinch bolt and nut to the front spindle. Torque the bolt to 46 ft. lbs. (63 Nm).
   - Front wheel
   - Negative battery cable
   - Turn on the air suspension switch, if equipped.

2002

1. Before servicing the vehicle, refer to the precautions in the beginning of this section.
2. Remove or disconnect the following:
   - Wheel
   - Upper ball joint from the wheel knuckle. Remove and discard the nut.
   - Arm-to-frame nuts and shims. Discard the nuts.
   - Upper arm
3. To install, reverse the removal procedure. Torque the upper arm nuts to 111 ft. lbs. (150 Nm). Torque the ball joint nut to 41 ft. lbs. (55 Nm).
4. Check and, if necessary, align the front end.

UPPER CONTROL ARM BUSHING REPLACEMENT

The control arm bushings are not serviceable. If they require service, the upper or lower arm must be replaced.

Lower Control Arm

REMOVAL AND & INSTALLATION

1999–01

1. Before servicing the vehicle, refer to the precautions in the beginning of this section.

2. Turn off the air suspension switch, if equipped.

3. Remove or disconnect the following:
- Negative battery cable
- Front wheel
- Brake rotor shield
- Shock absorber
- Torsion bar
- Lower ball joint from the spindle
- Lower control arm

To install:

4. Install or connect the following:
- Lower control arm assembly to the crossmember and hand tighten the bolts at this time
- Lower ball joint to the spindle. Torque the new castle nut to 21 ft. lbs. (29 Nm).
- Torsion bar. Torque the bolts to 21 ft. lbs. (29 Nm). Check and adjust the ride height.
- Shock absorber. Torque the bolts to 21 ft. lbs. (29 Nm).
- Brake rotor shield
- Torque the lower control arm bolts to 148 ft. lbs. (200 Nm)
- Front wheel
- Negative battery cable

5. Turn on the air suspension switch, if equipped.

6. Check and adjust the front end alignment as needed.

2002

1. Before servicing the vehicle, refer to the precautions in the beginning of this section.

2. Remove or disconnect the following:

➡**For reference during the installation procedure, measure the distance from** the lip of the fender to the center of the wheel hub with the vehicle in a level static ground position.

- Wheel
- Upper shock absorber mounting nuts. Discard the nuts.
- Stabilizer bar connecting link. Discard the nut.
- Bolt, flag nut and the spring and shock assembly. Discard the flag nut.
- Separate the ball joint from the wheel knuckle. Discard the nut.
- Arm-to-frame nuts. Discard the nuts.
- Arm-to-knuckle bolt. Discard the nut.

➡**On 4x4 vehicles, make sure that the crimped area of the outer CV boot clamp is not positioned downward or it will interfere with the removal of the arm.**

- Lower arm

➡**Using a suitable jack stand, raise the suspension until the distance between the lip of the fender and the center of the wheel hub is equal to the measurement taken in the removal procedure before tightening the inboard lower arm mountings. Make sure that the forward mounting is tightened first.**

3. To install, reverse the removal procedure. Observe the following torques:
- Control arm-to-frame flag bolt: 295 ft. lbs. (400 Nm)
- Control arm-to-frame nuts: 111 ft. lbs. (150 Nm)
- Ball joint nut: 129 ft. lbs. (175 Nm)

- Shock lower flag bolt: 258 ft. lbs. (350 Nm)
- Sway bar link nut: 18 ft. lbs. (25 Nm)
- Upper shock mounting nuts: 22 ft. lbs. (30 Nm)

4. Check and, if necessary, align the front end.

LOWER CONTROL ARM BUSHING REPLACEMENT

The control arm bushings are not serviceable. If they require service, the upper or lower arm must be replaced.

Front Wheel Bearings

ADJUSTMENT

The wheel bearings are not adjustable.

REMOVAL & INSTALLATION

1999–01

✴✴ WARNING

If equipped, always turn off the Automatic Ride Control (ARC) service switch before lifting the vehicle off of the ground. Failure to do so could damage the ARC system components.

1. Before servicing the vehicle, refer to the precautions in the beginning of this section.

2. Remove or disconnect the following:
- Negative battery cable
- Front wheels

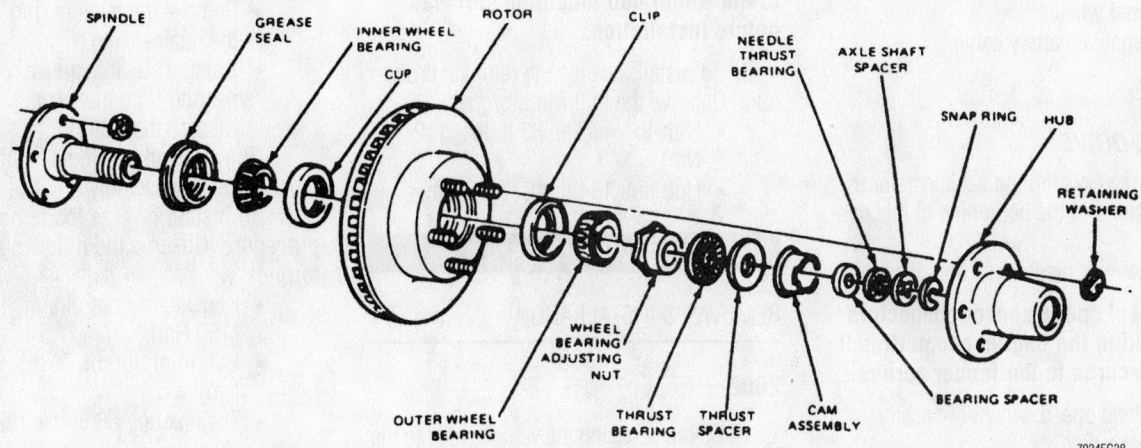

Exploded view of the wheel bearing and automatic locking hub assembly

7924EG38

- Front disc brake caliper, bracket and rotor
- Brake rotor splash shield
- Antilock Brake System (ABS) sensor and wire harness from the steering knuckle, if equipped
- Front wheel hub nut and washer
- Wheel hub/bearing to steering knuckle retaining bolts
- Hub and bearing assembly

✳✳ WARNING

Never reuse the wheel hub nut and washer. This nut is a torque prevailing design and cannot be reused.

➡**The hub shaft is a slip fit into the wheel hub and bearing; a press is not normally required.**

3. Ensure that the wheel hub shaft can be pushed inwards. If not, assemble a press to the front wheel studs and press the wheel hub shaft inwards slightly to break it loose.

To install:

4. Install or connect the following:
- ABS sensor to the wheel hub then position the hub to the front axle shaft and steering knuckle
- Retaining bolts. Torque the bolts to 74–96 ft. lbs. (100–130 Nm).
- Position the hub on the driveshaft and into the steering knuckle. Torque the bolts to 89 inch lbs. (10 Nm).
- Brake rotor splash shield
- ABS sensor. Torque the bolt to 89 inch lbs. (10 Nm).
- Hub washer and nut and tighten to 157–213 ft. lbs. (212–288 Nm)
- Front brake rotor, bracket and caliper
- Front wheels
- Negative battery cable

2002

2-WHEEL DRIVE

1. Before servicing the vehicle, refer to the precautions in the beginning of this section.
2. Remove or disconnect the following:

➡**The wheel speed sensor connectors are located in the engine compartment and are secured to the fender aprons.**

- Wheel speed sensor connector
- Brake disc
- Wiring harness from the retainers

- Bolts, wheel hub and sensor as an assembly. Discard the bolts.
3. To install, reverse the removal procedure.

➡**Apply a thin coat of silicone sealant to the wheel hub mounting surfaces before installation.**

4. Torque the hub-to-knuckle bolts to 83 ft. lbs. (112 Nm)

4-WHEEL DRIVE

1. Before servicing the vehicle, refer to the precautions in the beginning of this section.
2. Remove or disconnect the following:
- Hub nut. Discard the nut.

➡**The wheel speed sensor connectors are located in the engine compartment and are secured to the fender aprons.**

- Wheel speed sensor connector
- Brake disc

✳✳ WARNING

Do not use a hammer to separate the outboard CV joint from the hub. Damage to the threads and internal CV joint components may result.

3. Press the outboard CV joint until it is loose in the hub.
4. Detach the harness from the retainers.

✳✳ WARNING

Do not overextend the CV joint and boots when removing the wheel hub.

5. Remove the bolts, wheel hub and sensor as an assembly. Discard the bolts.

➡**Apply a thin coat of silicone sealant to the wheel hub mounting surfaces before installation.**

6. To install, reverse the removal procedure. Observe the following torques:
- Hub-to-knuckle: 83 ft. lbs. 112 Nm)
- Hub nut: 184 ft. lbs. (250 Nm)

Rear Wheel Bearings

REMOVAL & INSTALLATION

2002

1. Before servicing the vehicle, refer to the precautions in the beginning of this section.

2. Remove or disconnect the following:

➡**Do not loosen the axle wheel hub retainer until the wheel and tire are removed from the vehicle. Wheel bearing damage will occur if the wheel bearing is unloaded with the weight of the vehicle applied.**

- Wheel
- Hub nut and washer. Discard the nut.
- Brake shield
- Nut and bolt and separate the toe link from the wheel knuckle. Discard the nut.
- Nut and bolt and separate the upper ball joint from the wheel knuckle. Discard the nut.

➡**Do not use a hammer to separate the outboard CV joint from the hub. Damage to the threads and internal CV joint components may result.**

3. Press the outboard CV joint until it is loose from the hub.
4. Remove the nut and bolt and the wheel knuckle, hub and bearing as an assembly. Discard the nut.
5. Remove the hub and bearing.

✳✳ WARNING

Make sure that the press adapter outside diameter is slightly smaller than the hub outside diameter or damage to the knuckle will result.

6. Using a suitable press, remove the hub from the bearing. Discard the hub.

➡**The retainer ring is tapered and must be installed flat side down.**

- Remove the retainer ring. Discard the retainer ring.
- Using a suitable press, remove the bearing from the wheel knuckle. Discard the bearing.
7. The hub and bearing cannot be reused after disassembly.
8. To install, reverse the removal procedure. Observe the following torques:
- Knuckle-to-control arm: 111 ft. lbs. (150 Nm)
- Ball joint nut: 66 ft. lbs. (90 Nm)
- Toe link nut: 66 ft. lbs. (90 Nm)
- Hub nut: 258 ft. lbs. (350 Nm)

FORD MOTOR CO.

Ford-Excursion • Expedition • **Lincoln**-Navigator

11

PRECAUTIONS

PRECAUTIONS

Before servicing any vehicle, please be sure to read all of the following precautions, which deal with personal safety, prevention of component damage and important points to take into consideration when servicing a motor vehicle:

• Never open, service or drain the radiator or cooling system when the engine is hot; serious burns can occur from the steam and hot coolant.

• Observe all applicable safety precautions when working around fuel. Whenever servicing the fuel system, always work in a well-ventilated area. Do not allow fuel spray or vapors to come in contact with a spark, open flame, or excessive heat (a hot drop light, for example). Keep a dry chemical fire extinguisher near the work area. Always keep fuel in a container specifically designed for fuel storage; also, always properly seal fuel containers to avoid the possibility of fire or explosion.

• Fuel injection systems often remain pressurized, even after the engine has been turned **OFF**. The fuel system pressure must be relieved before disconnecting any fuel lines. Failure to do so may result in fire and/or personal injury.

• Brake fluid often contains polyglycol ethers and polyglycols. Avoid contact with the eyes and wash your hands thoroughly after handling brake fluid. If you do get brake fluid in your eyes, flush your eyes with clean, running water for 15 minutes. If

eye irritation persists, or if you have taken brake fluid internally, seek medical assistance IMMEDIATELY.

• The EPA warns that prolonged contact with used engine oil may cause a number of skin disorders, including cancer! You should make every effort to minimize your exposure to used engine oil. Protective gloves should be worn when changing oil. Wash your hands and any other exposed skin areas as soon as possible after exposure to used engine oil. Soap and water, or waterless hand cleaner should be used.

• All new vehicles are now equipped with an air bag system, often referred to as a Supplemental Restraint System (SRS) or Supplemental Inflatable Restraint (SIR) system. The system must be disabled before performing service on or around system components, steering column, instrument panel components, wiring and sensors. Failure to follow safety and disabling procedures could result in accidental air bag deployment, possible personal injury and unnecessary system repairs.

• Always wear safety goggles when working with, or around, the air bag system. When carrying a non-deployed air bag, be sure the bag and trim cover are pointed away from your body. When placing a non-deployed air bag on a work surface, always face the bag and trim cover upward, away from the surface. This will reduce the motion of the module if it is accidentally deployed.

• Clean, high quality brake fluid from a

sealed container is essential to the safe and proper operation of the brake system. You should always buy the correct type of brake fluid for your vehicle. If the brake fluid becomes contaminated, completely flush the system with new fluid. Never reuse any brake fluid. Any brake fluid that is removed from the system should be discarded. Also, do not allow any brake fluid to come in contact with a painted surface; it will damage the paint.

• Never operate the engine without the proper amount and type of engine oil; doing so WILL result in severe engine damage.

• Timing belt maintenance is extremely important! Many models utilize an interference-type, non-freewheeling engine. If the timing belt breaks, the valves in the cylinder head may strike the pistons, causing potentially serious (also time-consuming and expensive) engine damage. Refer to the maintenance interval charts in the front of this manual for the recommended replacement interval for the timing belt and to the timing belt section for belt replacement and inspection.

• Disconnecting the negative battery cable on some vehicles may interfere with the functions of the on-board computer system(s) and may require the computer to undergo a relearning process once the negative battery cable is reconnected.

• When servicing drum brakes, only disassemble and assemble one side at a time, leaving the remaining side intact for reference.

• Only an MVAC-trained, EPA-certified automotive technician should service the air conditioning system or its components.

GASOLINE ENGINE REPAIR

➡Disconnecting the negative battery cable on some vehicles may interfere with the functions of the on board computer systems and may require the computer to undergo a relearning process, once the negative battery cable is reconnected.

Distributor

REMOVAL

1. Before servicing the vehicle, refer to the precautions in the beginning of this section.
2. Remove or disconnect the following:
   • Primary wiring connector from the distributor
3. Mark the position of the cap's No. 1 terminal on the distributor base.
   • Cap and adapter
   • Rotor
   • Thick Film Injector (TFI) connector
4. Matchmark the distributor base and engine for installation reference.
   • Hold-down bolt and lift out the distributor

INSTALLATION

Timing Not Disturbed

1. Before servicing the vehicle, refer to the precautions in the beginning of this section.
2. Visually inspect the distributor. The O-ring should fit tightly onto the housing and be free of cuts. The drive gear should be free of nicks, cracks or excessive wear. The distributor shaft should rotate freely, without any binding.
3. Lubricate the distributor gear teeth with a coating of engine oil meeting.
4. Align the locating boss and fully seat the distributor rotor on the distributor shaft, if removed.
5. Rotate the distributor shaft so that the distributor rotor blade points toward the marked position on the distributor base adapter.
6. Install or connect the following:
   • Distributor assembly into the engine block with a slight side-to-side twist.

➡If the vane and vane switch assembly cannot be kept on the leading edge after installation, remove the distributor from the cylinder block by pulling upward enough for the distributor gear to disengage the distributor gear from the camshaft gear. Rotate the distributor rotor enough so that the gear will align on the next tooth of the camshaft gear.

   • Distributor hold-down clamp and bolt; leave it snug
   • Adapter base in place
   • Base attaching bolts
   • Electrical connector to the distributor
   • Distributor cap
   • Spark plug wires
   • Negative battery cable
7. Check the initial timing according to the proper procedure.
8. Adjust the timing, as necessary, then tighten the distributor hold-down bolt to 17–25 ft. lbs. (23–34 Nm).

Timing Disturbed

1. Disconnect the No. 1 spark plug wire and remove the No. 1 spark plug.

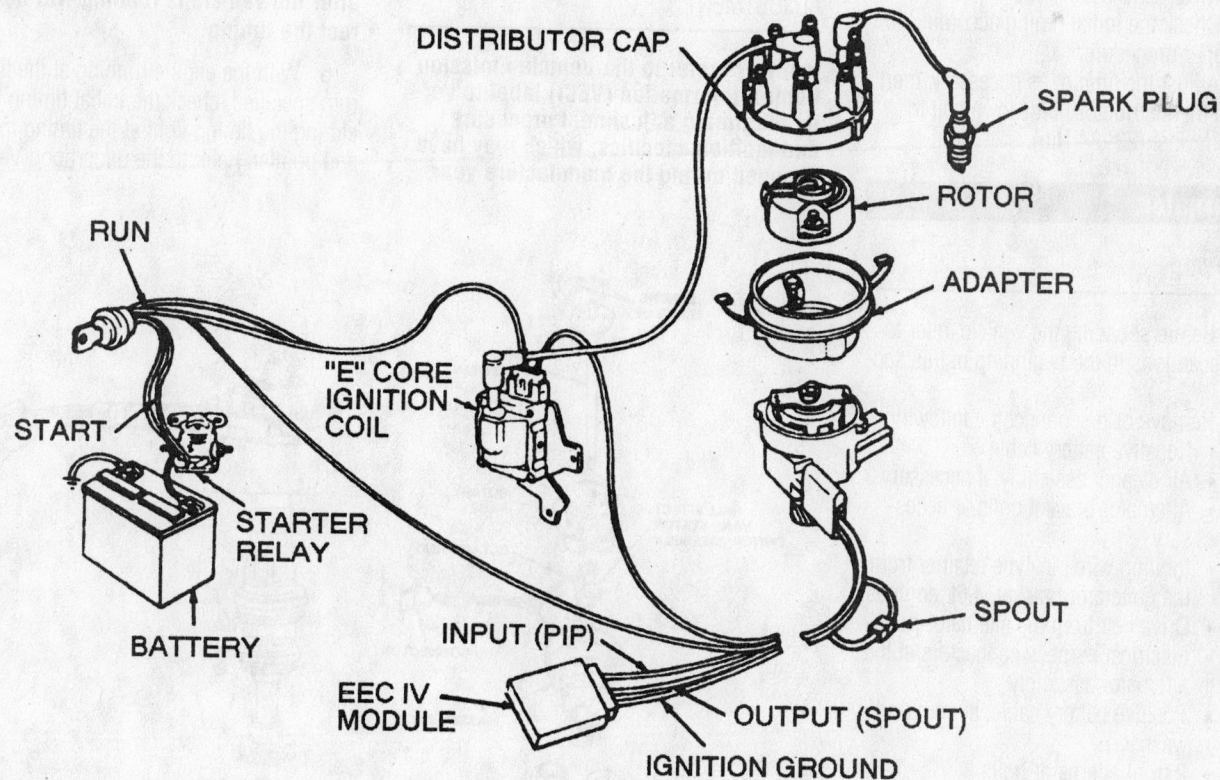

Exploded view of the TFI ignition system with universal distributor

7924FG34

2. Place a finger over the spark plug hole and crank the engine slowly until compression is felt.

3. Align the Top Dead center (TDC) mark on the crankshaft pulley with the pointer on the timing cover. This places the No. 1 cylinder at TDC on the compression stroke.

4. Install or connect the following:
- Distributor assembly into the engine block with a slight side-to-side twist.

➡ **If the vane and vane switch assembly cannot be kept on the leading edge after installation, remove the distributor from the cylinder block by pulling upward enough for the distributor gear to disengage the distributor gear from the camshaft gear. Rotate the distributor rotor enough so that the gear will align on the next tooth of the camshaft gear.**

- Distributor hold-down clamp and bolt; leave it snug
- Adapter base in place
- Base attaching bolts
- Electrical connector to the distributor
- Distributor cap
- Spark plug wires
- Negative battery cable

5. Check the initial timing according to the proper procedure.

6. Adjust the timing, as necessary, then tighten the distributor hold-down bolt to 17–25 ft. lbs. (23–34 Nm).

Alternator

REMOVAL

1. Before servicing the vehicle, refer to the precautions in the beginning of this section.

2. Remove or disconnect the following:
- Negative battery cable
- Air cleaner assembly, if necessary
- Alternator bracket bolts, if necessary
- Ignition wire pin-type retainer from the generator bracket, 4.6L engines
- Drive belt from the alternator pulley
- Electrical harness connectors at the alternator assembly
- Positive battery cable, the nut and washer
- 2 front alternator bolts
- Rear alternator support bracket retaining bolts and the support bracket
- Alternator from the vehicle

INSTALLATION

1. Before servicing the vehicle, refer to the precautions in the beginning of this section.

2. Install or connect the following:
- Alternator in position
- 2 front alternator retaining bolts, loosely
- Alternator bracket and 3 alternator bracket bolts. Tighten to 84 inch lbs. (10 Nm).
- Tighten 2 front alternator retaining bolts to 19 ft. lbs. (26 Nm)
- 2 electrical harness connectors to the alternator assembly
- Positive battery cable, the nut and washer. Tighten the nut to 72 inch lbs. (8 Nm).
- Drive belt on the alternator pulley
- Ignition wire pin-type retainer from the generator bracket, 4.6L engines
- Alternator bracket bolts, if removed
- Air cleaner assembly, if removed
- Negative battery cable

3. Start the engine and check for proper charging system operation.

Ignition Timing

ADJUSTMENT

➡ **Always refer to the Vehicle Emission Control Information (VECI) label to verify the timing adjustment procedure and ignition specifics, which may have changed during the manufacture year.**

Distributorless Ignition System

Base timing for 5.4L and 6.8L distributorless ignition engines is set at the factory at 10 degrees Before Top Dead Center (BTDC) and is not adjustable.

Distributor Ignition System

1. Before servicing the vehicle, refer to the precautions in the beginning of this section.

2. Place transmission in **P**. The air conditioning and heater controls should be in the OFF position.

3. Connect a suitable inductive timing light and a tachometer according to the manufacturer's instructions.

4. Disengage the single wire inline spout connector or remove the shorting bar from the double wire spout connector.

5. Start the engine and bring it up to normal operating temperature.

➡ **To set timing correctly, a remote starter should not be used. Use the ignition key only to start the vehicle. Disconnecting the start wire at the starter relay will cause the Ignition Control Module (ICM) to revert to "start mode timing" after the vehicle is started. Reconnecting the start wire after the vehicle is running will not correct the timing.**

6. With the engine running at the timing rpm specified, check the initial timing by aiming the timing light at the timing marks and pointer. Refer to the underhood Vehicle

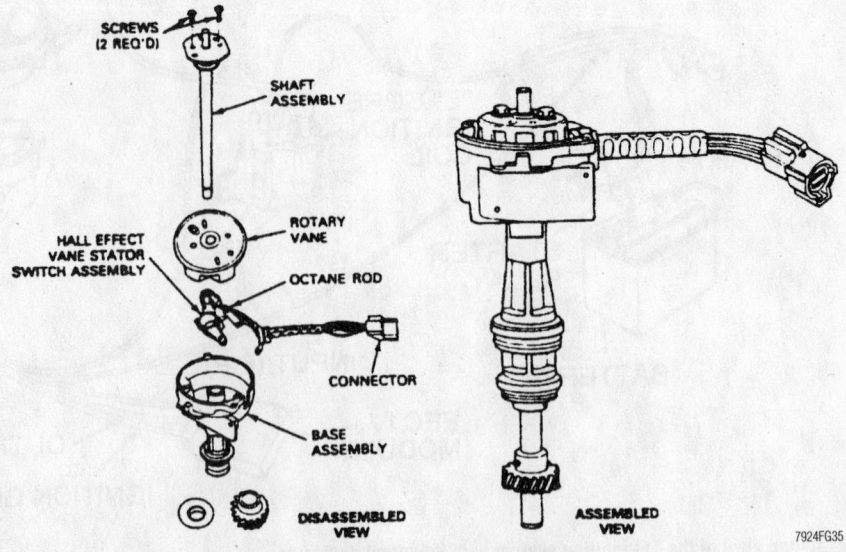

Exploded view of the closed bowl distributor

7924FG35

Emission Control Information (VECI) label for specific specifications.

7. If the marks do not align, shut the engine **OFF** and loosen the distributor hold-down clamp bolt. Start the engine, and while watching the timing marks with the timing light, turn the distributor until the marks are correctly aligned. Shut the engine **OFF**, then tighten the distributor hold-down clamp bolt to 17–25 ft. lbs. (23–34 Nm).

8. Reattach the single wire inline spout connector or reinstall the shorting bar on the double wire spout connector. Check the timing advance to verify the distributor is advancing beyond the initial setting.

9. Remove the timing light and tachometer.

Engine Assembly

REMOVAL & INSTALLATION

✳✳ CAUTION

Fuel injection systems remain under pressure, even after the engine has been turned OFF. The fuel system pressure must be relieved before disconnecting any fuel lines. Failure to do so may result in fire and/or personal injury.

1. Before servicing the vehicle, refer to the precautions in the beginning of this section.

2. Remove or disconnect the following:
   - Both battery cables, negative cable first
   - Hood
   - Coolant
   - Refrigerant, using approved equipment

3. Relieve the fuel system pressure.
   - Engine cooling fan, shroud and radiator
   - Accessory drive belt
   - Engine air cleaner outlet tube
   - Intake manifold assembly
   - Bulkhead connector cover
   - Bulkhead connector
   - The 3 power steering reservoir bracket retaining bolts and move the reservoir aside
   - The 2 Differential Pressure Feedback (DPFE) transducer hoses
   - Upper and lower Exhaust Gas Recirculation (EGR) valve to exhaust manifold tube fittings and the tube

   - Heater water hose
   - Ignition coil, radio capacitor and Camshaft Position (CMP) sensor electrical harness connectors
   - Both ignition coils and mounting bracket bolts, the coil and bracket assemblies
   - Starter motor
   - The 3 lower radiator air deflector screws
   - 5 clips and remove the air deflector
   - Air conditioning compressor electrical harness connector
   - Air conditioning manifold-to-compressor bolt, the manifold and tube assembly
   - The 3 air conditioning compressor retaining bolts and the air conditioning compressor
   - Fluid cooler hoses from the block mounted clip
   - Inspection cover, torque converter bolts and transmission-to-engine retaining bolts, on vehicles with automatic transmissions
   - Upper and lower power steering pump bolts and move the power steering pump aside
   - Exhaust system from the exhaust manifolds and support with wire hung from the crossmember
   - Right-hand and left-hand engine support insulator (mount) through-bolts

4. Install a suitable engine lifting bracket and connect suitable engine lifting equipment to the lifting brackets.

5. Carefully raise the engine out of the engine compartment and place on a work stand. Remove the engine lifting equipment.

To install:

6. Install the engine lifting brackets. Support the engine using a suitable floor crane installed to the lifting equipment and remove the engine from the work stand.

7. Carefully lower the engine into the engine compartment. Start the converter pilot into the flywheel and align the paint marks on the flywheel and torque converter. Be sure the studs on the torque converter align with the holes in the flywheel.

8. Fully engage the engine to the transmission and lower onto the engine support insulators.

9. Remove the engine lifting equipment and brackets.

10. Raise and safely support the vehicle.

11. Install or connect the following:
    - The 6 engine-to-transmission

    retaining bolts and tighten to 30–44 ft. lbs. (40–60 Nm)
    - The engine support insulator through-bolts and tighten to 15–22 ft. lbs. (20–30 Nm)
    - The 4 torque converter retaining nuts and tighten to 22–25 ft. lbs. (20–30 Nm)
    - Transmission housing cover to the cylinder block
    - Exhaust pipes to the exhaust manifolds and tighten to 30 ft. lbs. (41 Nm)
    - Power steering pump in position on the cylinder block and install 4 retaining nuts. Tighten to 15–20 ft. lbs. (20–30 Nm).
    - Starter motor
    - Transmission fluid cooler hoses into the cylinder block mounted clip
    - Air conditioning compressor in position and install the 3 retaining bolts. Tighten the bolts to 15–22 ft. lbs. (20–30 Nm).
    - Air conditioning manifold and tube assembly on the compressor and install the retaining bolt. Tighten the bolt to 14–18 ft. lbs. (18–24 Nm).
    - Air conditioning compressor clutch electrical harness connector
    - Lower radiator air deflector
    - Ignition coil and bracket assemblies. Tighten the retaining nuts to 15–23 ft. lbs. (20–30 Nm).
    - Ignition coil, radio capacitor and CMP sensor electrical harness connectors
    - Rear heater water hose and compress and slide the clamp in position
    - EGR valve-to-exhaust manifold tube and tighten the upper and lower fittings to 26–33 ft. lbs. (35–45 Nm)
    - DPFE transducer hoses
    - Power steering pump reservoir in position and install the 3 retaining bolts. Tighten the bolts to 71–107 inch lbs. (8–12 Nm).
    - Engine bulkhead connector and install the retaining bolt. Tighten the bolt to 36–50 inch lbs. (4–6 Nm).
    - Bulkhead connector cover
    - Intake manifold assembly
    - Accessory drive belt
    - Radiator, cooling fan and shroud

- Engine air cleaner outlet tube

12. Fill the engine with the correct amount and type of engine oil.

13. Fill and bleed the engine cooling system.

14. Connect both battery cables, negative cable last.

15. Start the engine and allow to reach normal operating temperature while checking for leaks.

16. Check all fluid levels.

17. Properly evacuate and recharge the air conditioning system using approved equipment.

18. Install the hood, aligning the marks that were made during removal.

19. Road test the vehicle and check the engine and transmission for proper operation.

Water Pump

REMOVAL & INSTALLATION

1. Before servicing the vehicle, refer to the precautions in the beginning of this section.

2. Remove or disconnect the following:
- Negative battery cable
- Radiator, fan blade assembly and fan shroud
- Accessory drive belt
- Water pump pulley
- Heater hose from the water pump
- Water pump bolts and nuts. Note the locations of the bolts if different lengths.
- Water pump stud bolt, the water pump and the water pump housing gasket. Discard the water pump housing gasket.

To install:

3. Before installing the water pump, be sure to completely clean the water pump mounting surfaces of all dirt, grime and old gasket material.

➡ **All water pump housing bolts, nuts and studs are tightened to 15–22 ft. lbs. (20–30 Nm).**

4. Install or connect the following:
- Water pump onto the engine with a new gasket. Install the water pump stud bolt temporarily finger-tight.
- Water pump mounting nuts and bolts temporarily finger-tight, then tighten all water pump housing fasteners to 15–22 ft. lbs. (20–30 Nm).
- Water outlet tube for the heater, if equipped
- Water pump pulley and accessory drive belt
- Fan shroud, fan blade assembly and the radiator
- Coolant
- Negative battery cable

5. Start the engine and check for any fluid leaks.

6. If necessary, bleed the cooling system.

Cylinder Head

REMOVAL & INSTALLATION

✳✳ CAUTION

Fuel injection systems remain under pressure, even after the engine has been turned OFF. The fuel system pressure must be relieved before disconnecting any fuel lines. Failure to

do so may result in fire and/or personal injury.

➡ **To correctly tighten the cylinder head bolts, an angle torque wrench is needed.**

1. Before servicing the vehicle, refer to the precautions in the beginning of this section.

2. Remove or disconnect the following:
- Air conditioning system, using approved equipment
- Negative battery cable
- Cylinder head covers
- Intake manifold
- Timing chains from the engine
- Exhaust manifolds
- The 2 heater hose retaining bolts, then compress and slide the hose clamp back to remove the heater water hose.
- Any remaining hoses, electrical connections or cables
- Cylinder head bolts, then lift the cylinder head from the engine block. Discard the cylinder head gasket and clean the engine block surface.

To install:

✳✳ WARNING

Cylinder head bolts must be replaced with new ones. They are torque-to-yield designed and cannot be reused.

3. Turn the crankshaft to position the keyway at the 12 o'clock position.

4. Clean and inspect the cylinder head for damage or warpage. Install the cylinder head gasket over the dowel pins. Then, install the cylinder head onto the engine block. Loosely install NEW cylinder head bolts.

➡ **Be sure to tighten the head bolts in 3 steps.**

5. Tighten the cylinder head bolts in the correct sequence using 3 steps, as follows:
   a. Step 1—30 ft. lbs. (40 Nm).
   b. Step 2—tighten an additional 85–95 degrees.
   c. Step 3—tighten another 85–95 degrees.

6. Install or connect the following:
- Heater hose
- Exhaust manifolds
- Timing chains
- Intake manifold
- Cylinder head covers
- Electrical connections and cables
- Refrigerant
- Engine oil

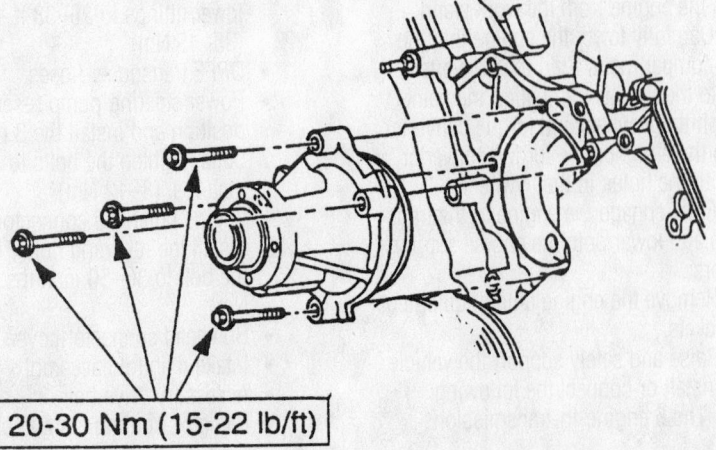

20-30 Nm (15-22 lb/ft)

7924FG02

Exploded view of the water pump mounting—4.6L, 5.4L and 6.8L engines

- Coolant
- Negative battery cable

7. Start the engine and check for any fluid or vacuum leaks.

Rocker Arms

REMOVAL & INSTALLATION

4.6L and 5.4L Engines

1. Before servicing the vehicle, refer to the precautions in the beginning of this section.
2. Disconnect the negative battery cable.
3. Remove the camshaft covers.
4. Position the piston of the cylinder being serviced at the bottom of its travel.

➡ **Two different valve spring compressor tools are used for this procedure. Valve Spring Compressor (T91P-6565-A) is used on the exhaust camshaft and Valve Spring Compressor (T93P-6565-A) is used on the intake camshaft.**

5. Compress the valve spring and remove the rocker arm.

To install:

6. Position the piston of the cylinder being serviced at the bottom of its travel.
7. Apply clean engine oil to the rocker arm, valve stem tip and tappet bore.

➡ **Valve tappet should have no more than 1/16 inch (1.5mm) of travel before installing the rocker arm.**

8. Compress the valve spring using the correct tool and install the rocker arm.

6.8L Engine

1. Before servicing the vehicle, refer to the precautions in the beginning of this section.
2. Disconnect the negative battery cable.
3. Remove the camshaft covers.
4. Position the base circle of the camshaft lobe on the rocker arm to be serviced. Also, be sure the piston is not at the top of its travel near the valve.
5. Compress the valve spring and remove the rocker arm.

To install:

6. Position the base circle of the camshaft lobe over the place where the rocker arm is to be installed.
7. Apply clean engine oil to the rocker arm, valve stem tip and tappet bore.
8. Compress the valve using the special tool and install the rocker arm.
9. Install the rocker arm covers and the remaining components.

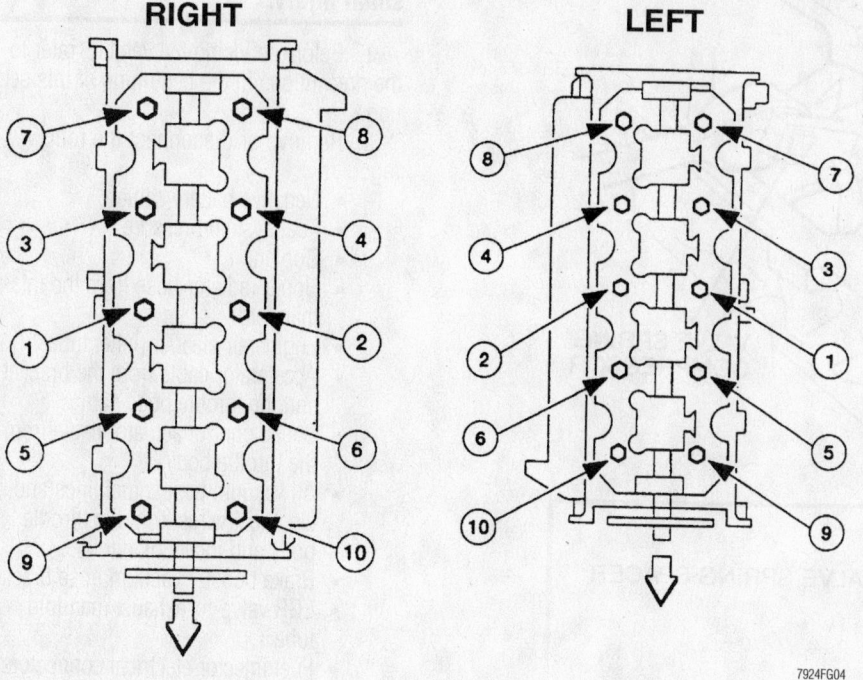

Tighten the cylinder head bolts using 3 steps in this sequence—4.6L and 5.4L engine

7924FG04

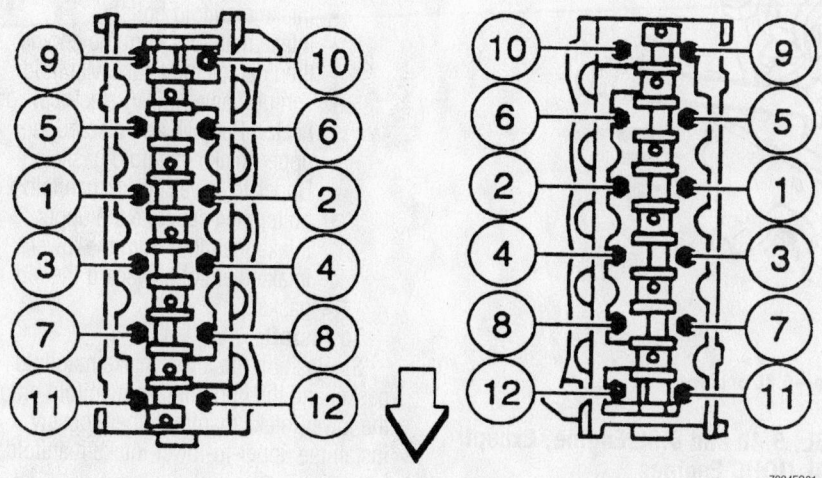

Tighten the cylinder head bolts using 3 steps in this sequence—6.8L engine

7924FG81

Timing belt service is covered in Section 3 of this manual

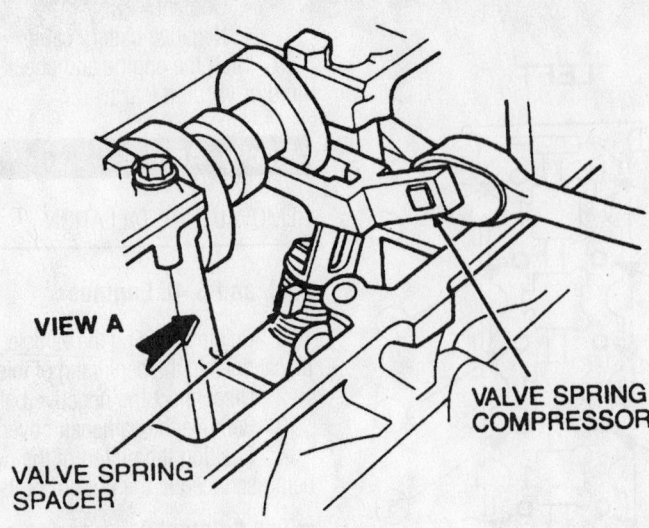

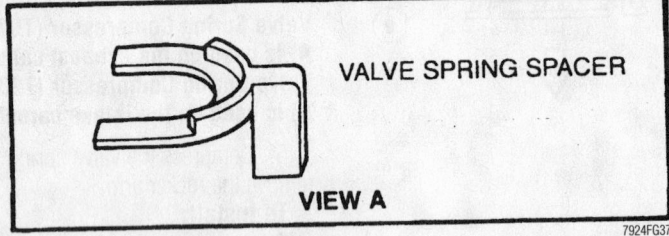

Using the proper tool, compress the valve spring and remove the rocker arm—4.6L and 5.4L engine

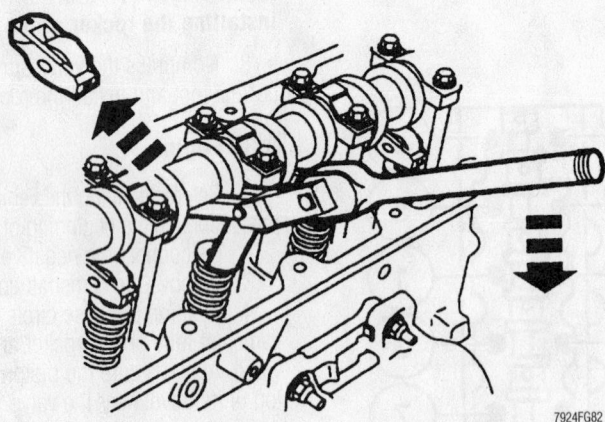

Compress the valve spring and remove the rocker arm—6.8L engine

Intake Manifold

REMOVAL & INSTALLATION

➡When the battery is disconnected and reconnected, some abnormal drive symptoms may occur while the vehicle relearns its adaptive strategy. The vehicle may need to be driven 10 miles (16 km) or more to relearn the strategy.

4.6L, 5.4L and 6.8L Engine, Except 5.4L DOHC Engines

1999–00 MODELS

❄❄ CAUTION

Fuel injection systems remain under pressure, even after the engine has been turned OFF. The fuel system pressure must be relieved before disconnecting any fuel lines. Failure to do so may result in fire and/or personal injury.

1. Before servicing the vehicle, refer to the precautions in the beginning of this section.

2. Remove or disconnect the following:

- Negative battery cable
- Fuel system pressure
- Coolant
- Upper radiator hose from the intake manifold
- Engine air cleaner outlet tube
- Accelerator cable from the bracket and the throttle body cam
- Speed control actuator cable from the throttle body
- All vacuum hoses, fuel lines and electrical wires from the throttle body and intake manifold
- Brake booster vacuum hose bracket
- EGR valve-to-exhaust manifold tube
- Fuel injector electrical connectors
- On the 6.8L engine: the radio interference capacitors from the left side of the intake manifold
- Spark plug wires, or ignition coils, if necessary
- Accessory drive belt
- Alternator
- Power steering oil reservoir bracket and set aside.
- Heater hose from the intake manifold
- Intake manifold bolts
- Intake manifold from the engine, then detach the Intake Manifold Tuning Valve (IMTV) electrical connector. Remove and discard the upper intake manifold gaskets.
- Upper-to-lower intake manifold bolts, then separate the upper intake manifold from the lower intake manifold. Discard the old gasket.

To install:

3. Position the lower intake manifold gasket and the upper intake manifold onto the lower intake manifold, then loosely install the upper-to-lower intake manifold bolts.

➡Be sure to tighten the lower-to-upper manifold bolts in 2 steps.

4. Tighten the 8 lower-to-upper intake manifold bolts in 2 steps following the tightening sequence as follows:

- 18 inch lbs. (2 Nm).
- 72–96 inch lbs. (8–12 Nm).

5. Position the 2 upper intake manifold gaskets on the cylinder heads. Set the upper intake manifold in place on the engine, then loosely install the 9 intake manifold-to-cylinder head bolts.

6. Attach the IMTV electrical connector.

➡ **Check that the thermostat housing is in the correct position before the thermostat housing is installed.**

7. Install the thermostat housing and start the 2 housing bolts.

➡ **Make certain to tighten the intake manifold in 2 steps.**

8. Tighten the intake manifold bolts using the sequence shown, in 2 steps.
- 18 inch lbs. (2 Nm)
- 15–22 ft. lbs. (20–30 Nm)

9. Install or connect the following:
- Heater water hose
- Power steering bracket and install the power steering pump bracket bolts to 71–107 inch lbs. (8–12 Nm)
- All electrical connections, fuel lines, vacuum tubes and coolant hoses to the intake manifold, fuel injectors and throttle body assembly.
- Alternator and the accessory drive belt
- Spark plug wires
- EGR valve-to-exhaust manifold tube. The tube fittings should be tightened to 26–33 ft. lbs. (35–45 Nm).
- Speed actuator cable, if equipped, and the accelerator cable to the throttle body.
- Engine air cleaner outlet tube
- Heater hose
- Coolant
- Negative battery cable. Start the engine and check for fuel, vacuum or coolant leaks.

2001–03 MODELS

❋ CAUTION

Fuel injection systems remain under pressure, even after the engine has been turned OFF. The fuel system pressure must be relieved before disconnecting any fuel lines. Failure to do so may result in fire and/or personal injury.

1. Before servicing the vehicle, refer to the precautions in the beginning of this section.

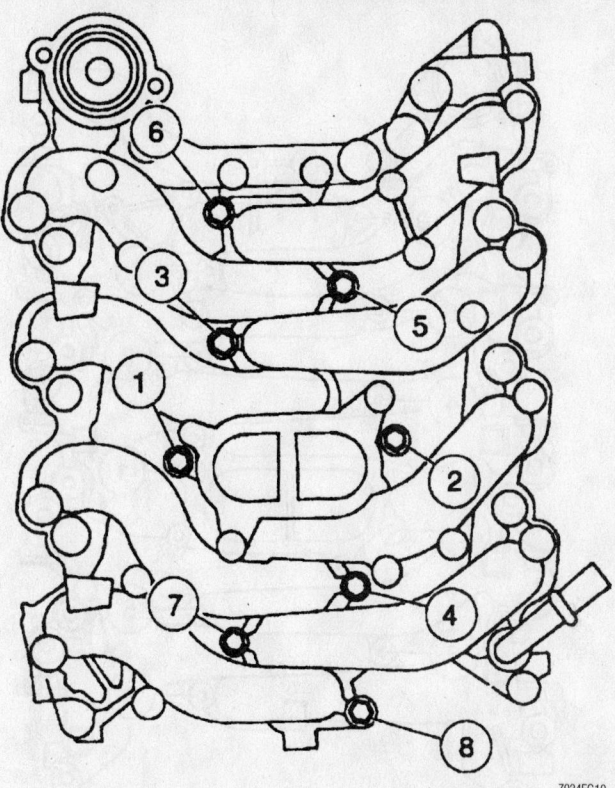

Tighten the lower intake manifold bolts in 2 steps using this sequence—1999–00 4.6L shown, 5.4L SOHC engine similar

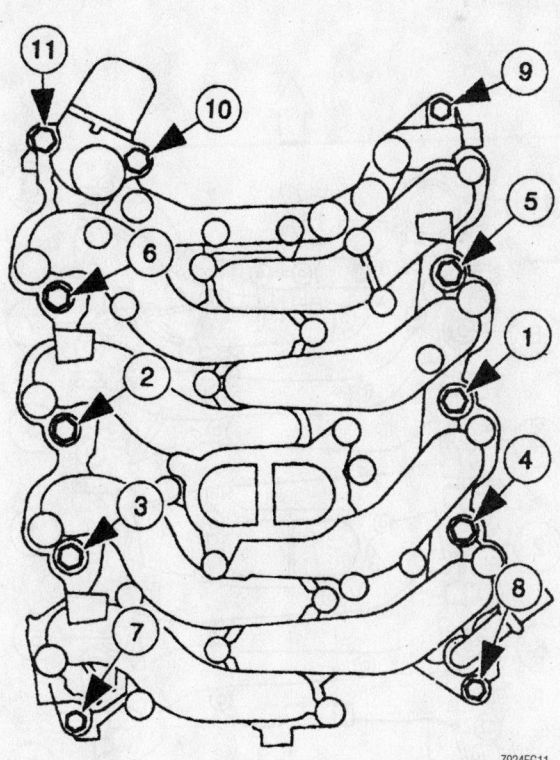

Tighten the upper intake manifold bolts using this sequence—1999–00 4.6L shown, 5.4L SOHC engine similar

Heater Core replacement is covered in Section 2 of this manual

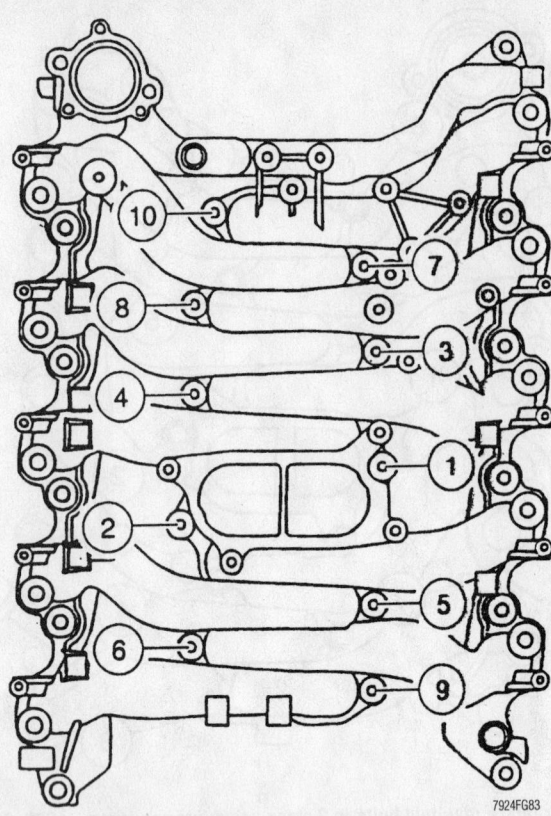

Tighten the upper-to-lower intake manifold bolts using this sequence—1999–00 6.8L engines

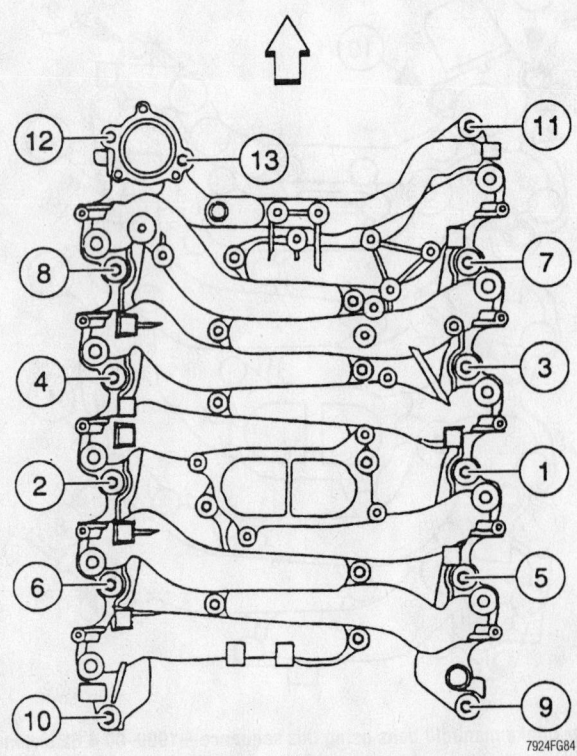

Tighten the upper intake manifold-to-cylinder head bolts using this sequence—1999–00 6.8L engines

2. Remove or disconnect the following:

- Negative battery cable
- Engine cover, if equipped
- Fuel system pressure
- Coolant
- Upper radiator hose from the intake manifold
- Accelerator cable from the bracket and the throttle body cam
- Speed control actuator cable from the throttle body
- Throttle cable return spring
- All vacuum hoses, fuel lines and electrical wires from the throttle body and intake manifold
- Accelerator bracket from the throttle body
- EGR valve-to-exhaust manifold tube
- Brake booster vacuum hose bracket
- Two evaporative emission canister purge valve nuts and position the valve aside
- Four bolts, and the throttle body and adapter as an assembly
- Fuel injector electrical connectors
- Bolts and the eight ignition coils
- On the 6.8L engine: the radio interference capacitors from the left side of the intake manifold
- Alternator upper support bracket
- Heater hose from the intake manifold
- Water thermostat housing and thermostat. Discard the O-ring seal.
- Intake manifold bolts
- Intake manifold from the engine, then detach the Intake Manifold Tuning Valve (IMTV) electrical connector. Remove and discard the upper intake manifold gaskets.
- Upper-to-lower intake manifold bolts, then separate the upper intake manifold from the lower intake manifold. Discard the old gasket.

To install:

3. Position the lower intake manifold gasket and the upper intake manifold onto the lower intake manifold, then loosely install the upper-to-lower intake manifold bolts.

➡**Be sure to tighten the lower-to-upper manifold bolts in 2 steps.**

4. Tighten the 8 lower-to-upper intake manifold bolts in 2 steps following the tightening sequence as follows:

- 18 inch lbs. (2 Nm)
- 18 ft. lbs. (25 Nm)

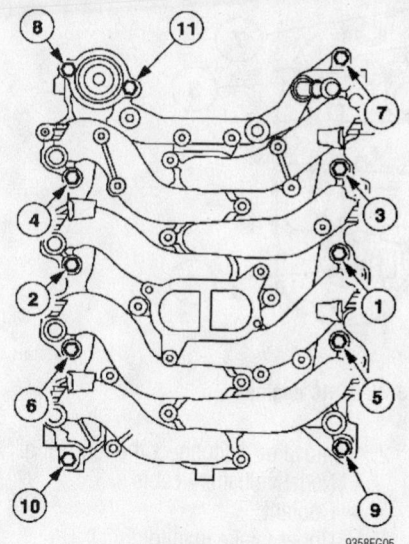

Tighten the upper intake manifold bolts in 2 steps using this sequence—2001–03 4.6L shown, 5.4L SOHC engine similar

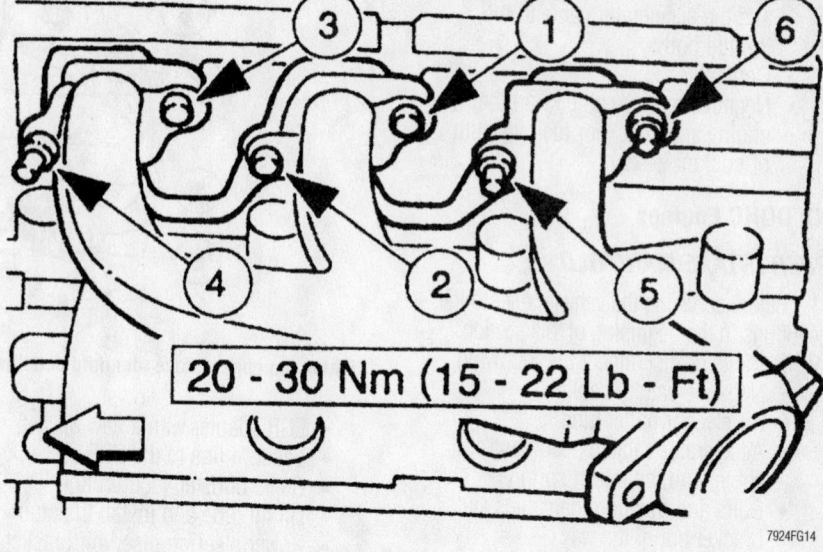

Tighten the lower intake manifold bolts in 2 steps using this sequence—2001–03 6.8L engine

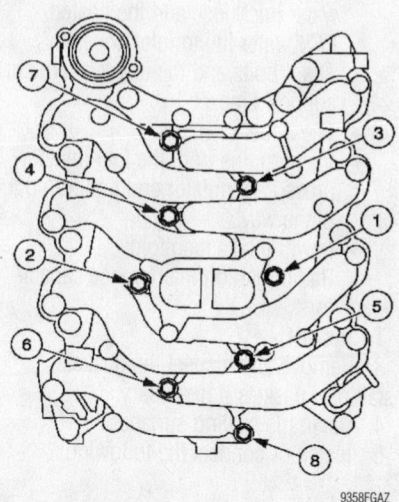

Tighten the lower intake manifold bolts in 2 steps using this sequence—2001–03 4.6L shown, 5.4L SOHC engine similar

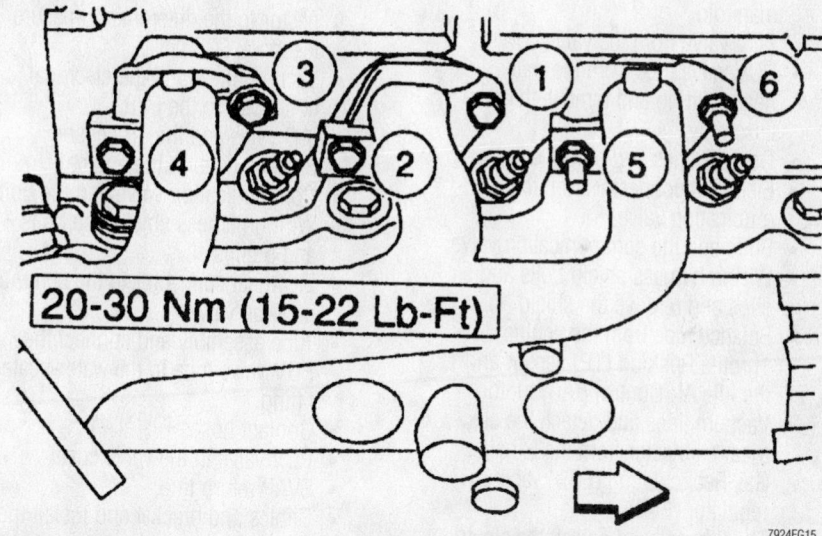

Tighten the upper intake manifold bolts in 2 steps using this sequence—2002–03 6.8L engine

5. Position the 2 upper intake manifold gaskets on the cylinder heads. Set the upper intake manifold in place on the engine, then loosely install the 9 intake manifold-to-cylinder head bolts.

6. Attach the IMTV electrical connector.

➡Check that the thermostat housing is in the correct position before the thermostat housing is installed.

7. Install the thermostat housing and start the 2 housing bolts.

➡Make certain to tighten the intake manifold in 2 steps.

8. Tighten the intake manifold bolts using the sequence shown, in 2 steps on all except 6.8L engines.
- 18 inch lbs. (2 Nm)
- 15–22 ft. lbs. (20–30 Nm)

9. Tighten the upper and lower intake manifold bolts using the sequence shown, in 2 steps on 6.8L engines.
- 18 inch lbs. (2 Nm)
- 89 inch lbs. (10 Nm)

10. Install or connect the following:
- Heater water hose
- All electrical connections, fuel lines,

vacuum tubes and coolant hoses to the intake manifold, fuel injectors and throttle body assembly.
- Alternator support bracket
- Ignition coils
- New throttle body adapter gasket and the throttle body adapter assembly. Tighten the bolts to 89 inch lbs. (10 Nm).
- All remaining hoses and electrical connections
- EGR valve-to-exhaust manifold tube. The tube fittings should be tightened to 26–33 ft. lbs. (35–45 Nm).

Brake service is covered in Section 4 of this manual

- Speed actuator cable, if equipped, and the accelerator cable to the throttle body.
- Coolant
- Negative battery cable. Start the engine and check for fuel, vacuum or coolant leaks.

5.4L DOHC Engines

UPPER INTAKE MANIFOLD

1. Before servicing the vehicle, refer to the precautions in the beginning of this section.
2. Remove or disconnect the following:
   - Negative battery cable
   - Air cleaner outlet tube
   - Accelerator cable, speed control cable, and the return spring
   - Bolts and position the cables and bracket out of the way
   - Evaporative emission return line
   - Positive Crankcase Ventilation (PCV) tube from the upper intake manifold
   - PCV valve from the valve cover
   - PCV valve tube from the water heated fitting and remove the tube assembly
   - Coolant lines and plug the lines.
   - Electrical connector from the communication valve
   - Bolts and the communication valve
   - Wiring harness shield bolts and 5 clips and remove the shield
   - Balance tube from the engine
   - Throttle Position (TP) sensor and the Idle Air Control (IAC) motor
   - Vacuum lines and detach the electrical connector from the Exhaust Gas Recirculation (EGR) vacuum regulator (EVR).
   - The 2 hoses and detach the electrical connector from the differential pressure feedback EGR.
   - The bolts from the power steering reservoir bracket and position it aside.
   - The stud and position the oil fill tube aside.
   - Brake booster vacuum line
   - Vacuum line from the EGR valve
   - Bolts from the EGR adapter
   - Retaining bolts and remove the upper intake manifold.

To install:

3. Clean and inspect the sealing surfaces.
4. Install or connect the following:
   - Upper intake manifold: Tighten the bolts to 89 inch lbs. (10 Nm) and then an additional 90 degrees in the sequence shown.

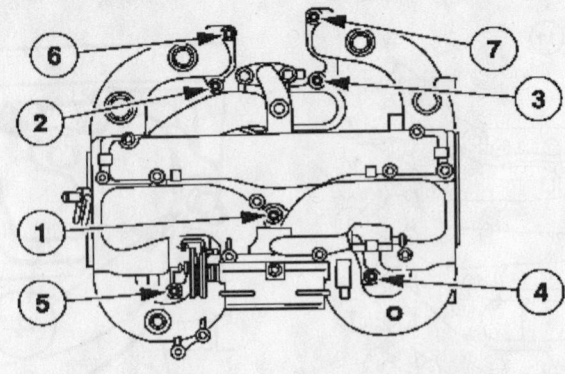

Tighten the upper intake manifold bolts as shown—5.4L DOHC engines

- EGR adapter with a new gasket
- Vacuum line to the EGR valve
- Brake booster vacuum line
- Oil fill tube and install the stud
- Power steering reservoir bracket and install the bolts.
- EGR valve at the exhaust manifold
- The 2 hoses and the electrical connector to the differential pressure feedback EGR
- Vacuum lines and the electrical connector to the EVR
- IAC valve and the TP sensor
- Balance tube on the engine
- Communication valve and the bolts
- Wiring harness shield, the bolts and 5 clips
- Electrical connector to the communication valve
- Tube assembly and connect the PCV valve tube to the water-heated fitting.
- Coolant hoses
- PCV valve in the valve cover
- EVAP return line
- Cables and bracket and install the bolts
- Accelerator cable, speed control cable, and the return spring
- Engine appearance cover and the 3 bolts
- Engine air cleaner and outlet tube

LOWER INTAKE MANIFOLD

✳✳ CAUTION

Fuel injection systems remain under pressure, even after the engine has been turned OFF. The fuel system pressure must be relieved before disconnecting any fuel lines. Failure to do so may result in fire and/or personal injury.

1. Before servicing the vehicle, refer to the precautions in the beginning of this section.

2. Remove or disconnect the following:
   - Negative battery cable
   - Coolant
   - Upper intake manifold
   - Fuel system pressure
   - Fuel lines
   - Engine water bypass hose
   - Electrical connector from the water temperature indicator sender
   - Upper radiator hose, the heater water inlet hose and the heated PCV water fitting inlet hose.
   - The 4 bolts and the upper alternator support bracket
   - The 8 fuel injectors
   - Vacuum line from the fuel injector pressure regulator and position out of the way.
   - Lower intake manifold
   - Radio ignition interference capacitors

To install:

3. Remove and inspect the gaskets, install new gaskets if necessary.
4. Clean the sealing surfaces.
5. Install or connect the following:

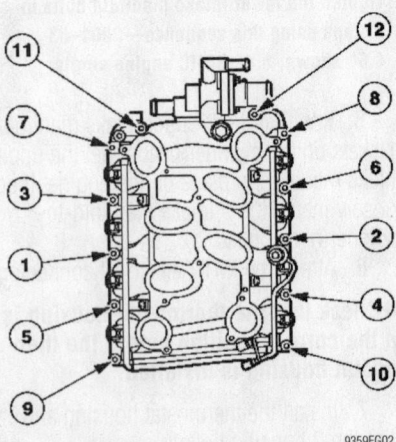

Tighten the lower intake manifold bolts as shown—5.4L DOHC engines

- Radio ignition interference capacitors

➡**Bolts should be hand-started, positions 7–12 first then 1–6.**

- Lower intake manifold: Tighten the bolts to 89 inch lbs. (10 Nm) and then an additional 90 degrees in the sequence shown.
- Water temperature indicator sensor
- Vacuum harness and connect the vacuum line to the fuel injector pressure regulator.
- Fuel injectors
- Upper generator support bracket and the 4 bolts
- Heater water inlet hose, the upper radiator hose and the water heated fitting inlet hose
- Engine water bypass return hose
- Fuel lines
- Upper intake manifold
- Coolant

Exhaust Manifold

REMOVAL & INSTALLATION

1. Before servicing the vehicle, refer to the precautions in the beginning of this section.
2. Remove or disconnect the following:
   - Front fender splash shield

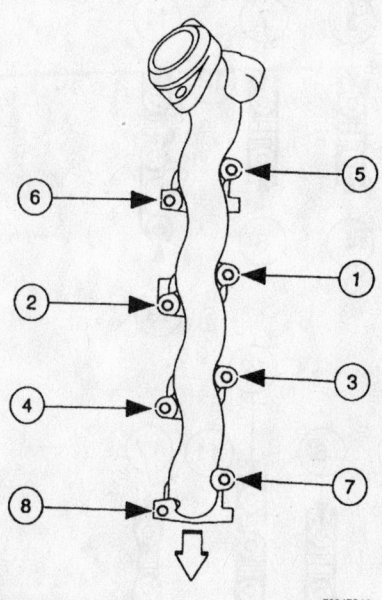

Tighten the exhaust manifold bolts in the sequence shown—4.6L engine shown, 5.4L engine similar

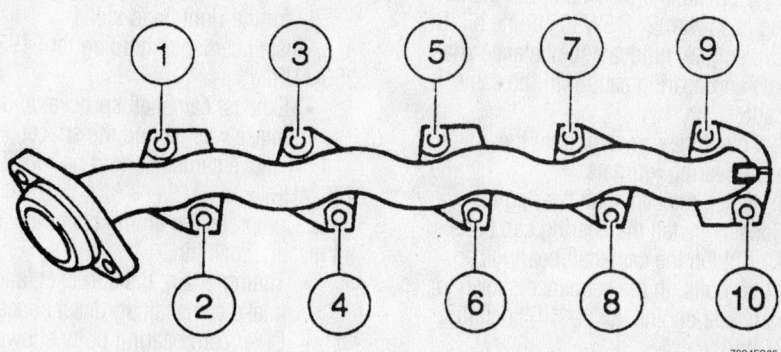

23-27 Nm (17-20 lb/ft)

7924FG85

Tighten the exhaust manifold bolts in the sequence shown—right side of 6.8L engine shown

- For the left-hand exhaust manifold: the Exhaust Gas Recirculation (EGR) valve-to-exhaust manifold tube and if equipped, the DPFE gas recirculation transducer hoses.
- On the 4.6L and 5.4L engines: the catalytic converter-to-exhaust manifold bolts
- On the 6.8L engine: the front exhaust pipe from the manifold
- The exhaust manifold mounting nuts, then remove the exhaust manifold itself. Remove and discard the old gasket.

3. Clean and inspect the exhaust manifold for damage.

To install:

4. Position a new gasket and the exhaust manifold onto the engine block.
5. Install the mounting nuts and tighten following the sequence shown.
   - 4.6L and 5.4L engines: 18 ft. lbs. (25 Nm)
   - 6.8L engines: 17–20 ft. lbs. (23–27 Nm)

6. On the 6.8L engine, tighten the exhaust manifold-to-front pipe fasteners to 27–34 ft. lbs. (34–46 Nm).

7. On the 4.6L and 5.4L engines, attach the catalytic converter to the exhaust manifold, install the catalytic converter-to-exhaust manifold bolts and tighten to 25–34 ft. lbs. (34–46 Nm).

8. For the left-hand exhaust manifold, install the DPFE transducer hoses if equipped, and the EGR valve-to-exhaust manifold tube. Tighten the upper and lower fittings to 26–33 ft. lbs. (35–45 Nm).

9. Install the front fender splash shield.
10. Lower the vehicle to the ground.

Camshaft and Valve Lifters

REMOVAL & INSTALLATION

SOHC Engines

1. Before servicing the vehicle, refer to the precautions in the beginning of this section.
2. Remove the cylinder head covers from the engine.
3. Remove the timing chain.

✳✳ WARNING

At no time, when the timing chains are removed and the cylinder heads are installed may the crankshaft or camshaft be rotated. Severe piston and valve damage will occur.

4. On the 6.8L engine, remove the 6 bolts securing the balance shaft to the cylinder head and remove the shaft.
5. Remove the camshaft roller lifters.
6. On VIN W engines, remove the timing chain camshaft gear by removing the gear retaining bolt.

➡**Keep the bearing caps in order so they can be installed in the same position.**

7. Remove the camshaft bearing cap bolts, then lift the camshaft bearing caps off of the cylinder head.
8. Lift the camshaft from the cylinder head.
9. Remove the rocker arms and pull the lash adjusters out of their bores. Keep all the parts in order. They must be installed in their original positions.

For complete Engine Mechanical specifications, see Section 1 of this manual

To install:

10. Install the lash adjusters and rocker arms in their original positions.

11. Lubricate the camshaft journals and bearing caps with SAE 5W30 engine oil. On the 6.8L engine, lubricate the balance shaft journals and bearing caps with the same lubricant.

12. Lower the camshaft onto the camshaft bearing journals.

13. Install the camshaft bearing caps, then loosely install the bearing cap bolts.

14. Tighten the camshaft bearing cap mounting bolts, in the sequence shown for the particular engine, to 71–107 inch lbs. (8–12 Nm).

15. On the 6.8L engine, align the timing marks and position the balance shaft on the journals, then install the bearing caps. Tighten the bolts in sequence to 89 inch lbs. (10 Nm).

16. On engines equipped with bolt on sprockets, install the camshaft timing chain gear by tightening the retaining bolts as follows:

   a. M10 Step 1: 30 ft. lbs. (40 Nm).
   b. M10 Step 2: Tighten 90 degrees.
   c. M12 Step 1: 90 ft. lbs. (120 Nm).

17. Install the valve lifters.

18. Install the timing chain and sprockets, if applicable.

19. Install the cylinder head covers.

DOHC Engines

1. Before servicing the vehicle, refer to the precautions in the beginning of this section.

2. Remove or disconnect the following:
- Roller followers
- Left hand timing chain for the left hand side and both timing chains for the right hand side

3. Install camshaft holding tool T93P-6256-AHR.
- Exhaust camshaft sprocket and the intake washer and the spacer

4. Remove camshaft holding tool T93P-6256-AHR.

5. Compress the timing chain tensioner and install a lock pin.
- Timing chain, the sprocket, and intake camshaft sprocket spacer
- Outer cam bearing bolts shown in the accompanying illustration

❊❊ WARNING

The outer bolts on the outer cam bearing cap (exhaust) are longer and must be returned to the same location or engine damage may occur.

6. Make sure to identify the camshaft to cylinder head location. Caps are not interchangeable.
- Bolts and the camshaft bearing cap assemblies
- Camshafts

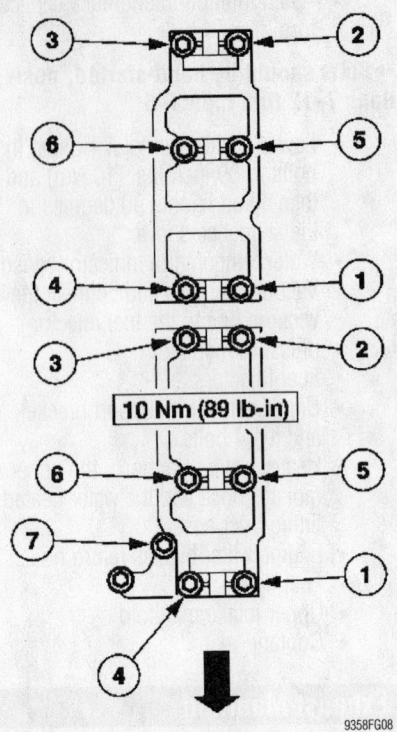

10 Nm (89 lb-in)

Tighten the bearing caps in the sequence shown—Romeo 4.6L engine shown, 5.4L engine similar

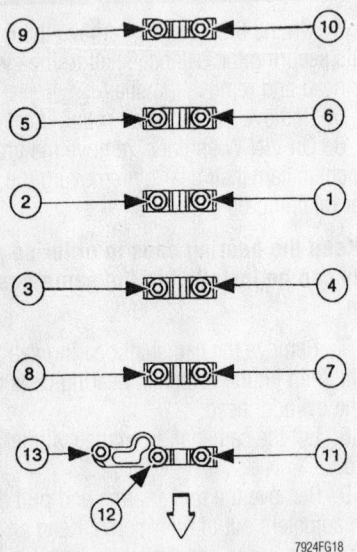

Tighten the bearing caps in the sequence shown—Windsor 4.6L engine shown, 5.4L engine similar

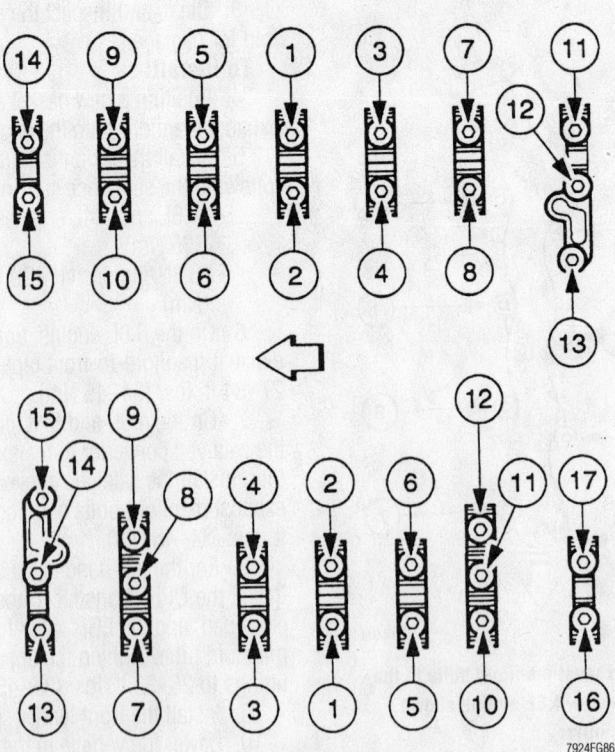

8-12 Nm (71-106 lb/in)

Camshaft bearing cap bolt tightening sequence—6.8L engine

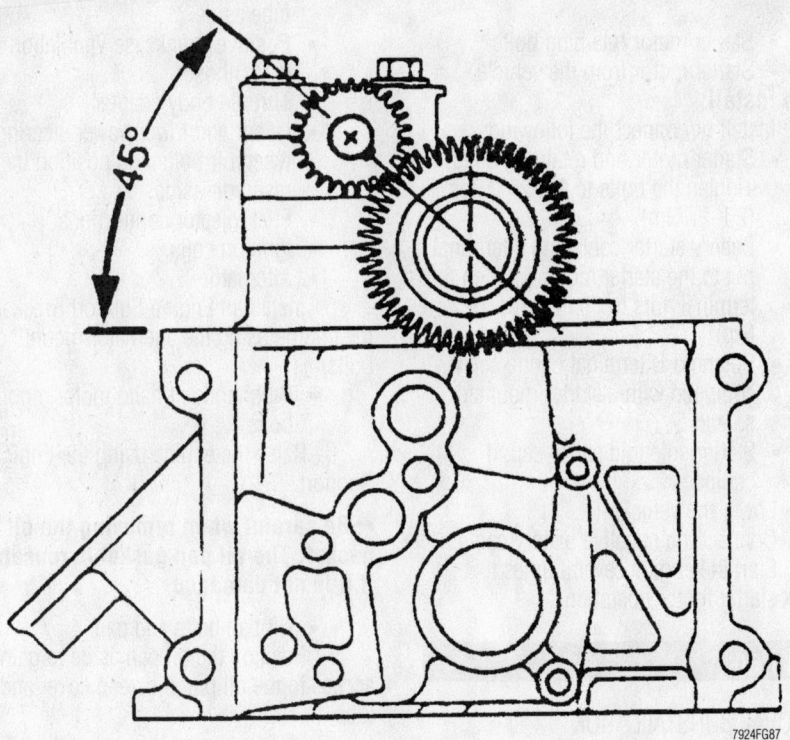

Be sure to align the balance shaft timing mark with the mark on the camshaft gear—6.8L engine

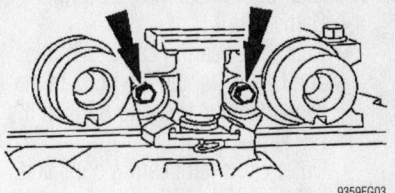

Remove the bolts shown before removing any other bearing cap bolts—DOHC models

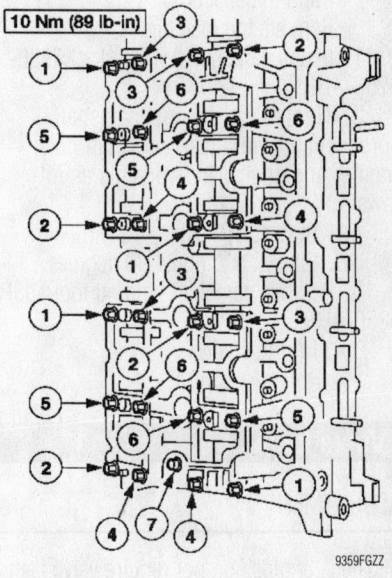

Tighten the camshaft cap bolts as shown to specification

8-12 Nm (71-106 lb/in)

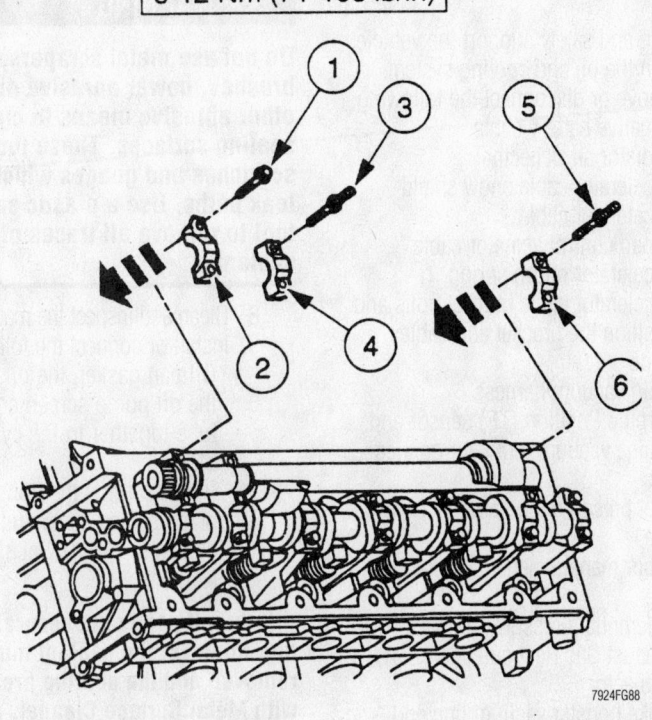

Tighten the balance shaft bearing cap bolts in the sequence shown—6.8L engine

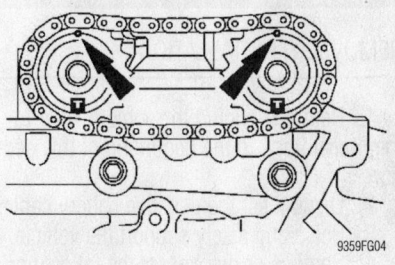

Make sure the timing marks are at 12 o'clock and index at 6 o'clock —DOHC models

To install:

7. Make sure the timing marks are at 12 o'clock and index at 6 o'clock. Refer to the illustration for proper timing mark alignment.

8. Lubricate the camshafts. Use Super Premium SAE 5W-30.

For Accessory Drive Belt illustrations, see Section 1 of this manual

9. Install or connect the following:
- Camshafts
- Camshaft bearing cap assemblies and bolts. Tighten the bolts in the sequence illustrated to 89 inch lbs. (10 Nm).
- Outer camshaft bearing bolts and tighten to 71–106 inch lbs. (8–12 Nm).
- Timing chain, the sprocket, and intake camshaft sprocket spacer as an assembly
- Camshaft sprocket bolt, the washer and the spacer

10. Remove the lock pin.

11. Install camshaft holding tool T93P-6256-AHR.

12. Tighten the exhaust camshaft sprocket and the intake camshaft bolt washer and spacer in two steps as follows:

a. Step 1: 30 ft. lbs. (40 Nm).
b. Step 2: additional 90 degrees.

13. Remove camshaft holding tool T93P-6256-AHR.
- Timing chains
- Roller followers

Valve Lash

ADJUSTMENT

These engines do not require valve lash adjusting, because they utilize hydraulic lash components in their valve actuation systems.

Starter Motor

REMOVAL & INSTALLATION

1. Before servicing the vehicle, refer to the precautions in the beginning of this section.

2. Disconnect the negative battery cable.

3. Raise and safely support the vehicle.

4. Remove or disconnect the following:
- Starter terminal cover
- Terminal nut and separate the battery starter cable from the starter motor
- Solenoid **S** terminal connector, if equipped with a starter mounted solenoid

➡ **To disconnect the hard-shell connector from the solenoid S terminal, grasp the plastic shell and pull off; do not pull on the wire. Pull straight off to prevent damage to the connector and S terminal.**

5. Remove or disconnect the following:
- Starter motor retaining bolts
- Starter motor from the vehicle

To install:

6. Install or connect the following:
- Starter motor and retaining bolts. Tighten the bolts to 15–20 ft. lbs. (20–27 Nm).
- Battery starter cable and a terminal nut to the starter motor. Tighten the terminal nuts to 79 inch lbs. (9 Nm).
- Solenoid **S** terminal connector, if equipped with a starter mounted solenoid
- Starter solenoid safety cap, if equipped

7. Lower the vehicle.

8. Connect the negative battery cable.

9. Start the engine several times to check starter motor operation.

Oil Pan

REMOVAL & INSTALLATION

4X2 Models

SOHC ENGINES

1. Before servicing the vehicle, refer to the precautions in the beginning of this section.

2. Raise and safely support the vehicle.

3. Drain the oil and cooling system,

4. Remove or disconnect the following:
- Negative battery cable
- Radiator air deflector
- Accelerator cable snow shield
- Accelerator cable
- Speed control actuator cable
- Accelerator return spring
- Accelerator cable bracket bolts and position the bracket and cables aside
- Main vacuum harness
- Throttle Position (TP) sensor and engine vacuum regulator connectors
- Fuel pressure regulator vacuum line
- Vapor management valve vacuum line
- Differential pressure feedback Exhaust Gas Recirculation (EGR) connector
- Brake booster vacuum line and bracket
- Fuel lines
- Idle Air Control (IAC) motor electrical connector

- EGR valve-to-exhaust manifold tube aside
- Positive Crankcase Ventilation (PCV) hose
- Throttle body adapter
- Upper and lower power steering reservoir bolts and position the reservoir aside
- Fuel injector connections
- Ignition coils
- Alternator

5. Install an Engine Support Bracket on the engine using the alternator mounting bolts.
- Right and left hand motor mount bolts

6. Raise the engine using the Engine Support.

➡ **Be careful when removing the oil pan gasket. The oil pan gasket is reusable if it is not damaged.**

- Oil pan bolts and pan

7. Position the oil pan aside to gain access to the oil pump screen cover and tube .
- Bolt securing the rear of the oil pump screen cover and tube
- Bolts securing the front of the oil pump screen cover and tube

To install:

✳✳ WARNING

Do not use metal scrapers, wire brushes, power abrasive discs or other abrasive means to clean the sealing surfaces. These tools cause scratches and gouges which make leak paths. Use a plastic scraping tool to remove all traces of old sealant.

8. Clean and inspect the mating surfaces.

9. Install or connect the following:
- Oil pan gasket, the oil pan and the oil pump screen cover and tube together to the cylinder block
- Bolts securing the front of the oil pump screen cover and tube and tighten to 71–107 inch lbs. (8–12 Nm)

➡ **If the oil pan is not secured within four minutes, the sealant must be removed and the sealing area cleaned with Metal Surface Cleaner. Allow to dry until there is no sign of wetness, or four minutes, whichever is longer. Failure to follow this procedure may result in future oil leakage.**

10. Apply the silicone Gasket and Sealant F7AZ-19554-EA at the rear oil seal retainer-to-cylinder block sealing surface.

11. Install the oil pan gasket and the oil pan and loosely install the 16 bolts. Tighten the bolts in three steps, in the sequence illustrated as follows:

    a. Step 1: 18 inch lbs. (2 Nm).
    b. Step 2: 15 ft. lbs. (20 Nm).
    c. Step 3: an additional 60 degrees.

12. Remove the Engine Support.

- Right and left hand motor mount bolts and tighten to 50–68 ft. lbs. (68–92 Nm)
- Alternator
- Fan shroud
- Throttle body adapter and the four bolts. Tighten the bolts to 89 inch lbs. (10 Nm).
- EGR valve-to-exhaust manifold tube
- All vacuum hoses, fuel lines and electrical connections
- Accelerator cable bracket and the bolts
- Accelerator cable, speed control actuator cable and the throttle return spring
- Accelerator cable snow shield and the bolts
- Upper radiator hose and position the clamp
- Air cleaner assembly
- Radiator air deflector
- Negative battery cable

13. Refill and bleed the cooling system.

14. Refill the engine crankcase with clean engine oil.

DOHC ENGINES

1. Before servicing the vehicle, refer to the precautions in the beginning of this section.

2. Raise and safely support the vehicle.

3. Drain the oil.

4. Remove or disconnect the following:
- Intake manifold
- Fan shroud and the engine cooling fan
- Alternator

5. Install an engine support device.
- Two bolts from the engine mounts

6. Raise the engine 2.5 inches (63.5mm).

➡ **Be careful when removing oil pan gasket. It is re-usable if not damaged.**

- Bolts, oil pan and gasket

7. Inspect the oil pan gasket for damage.

To install:

⚹⚹ **WARNING**

Do not use metal scrapers, wire brushes, power abrasive discs or other abrasive means to clean the sealing surfaces. These tools cause scratches and gouges which make leak paths. Use a plastic scraping tool to remove all traces of old sealant.

8. Clean the oil pan mating surfaces.

➡ **If the oil pan and the oil pan gasket are not secured within four minutes, the sealant must be removed and the sealing area cleaned with Metal Surface Cleaner F4AZ-19A536-RA. Allow to dry until there is no sign of wetness, or four minutes, whichever is longer. Failure to follow this procedure can cause future oil leakage.**

9. Apply the silicone Gasket and Sealant F7AZ-19554-EA at the rear oil seal retainer-to-cylinder block sealing surface.

10. Install the oil pan gasket and the oil pan and loosely install the 16 bolts. Tighten the bolts in three steps, in the sequence illustrated as follows:

    a. Step 1: 18 inch lbs. (2 Nm).
    b. Step 2: 15 ft. lbs. (20 Nm).
    c. Step 3: an additional 60 degrees.

11. Lower the vehicle, then the engine.

12. Remove the engine support

13. Install or connect the following:

Tighten the oil pan bolts in the sequence shown

9359FG08

- Two bolts for the engine mounts and tighten to 59 ft. lbs. (80 Nm)
- Alternator
- Cooling fan
- Lower intake manifold

14. Refill the engine crankcase with clean engine oil.

4X4 Models

SOHC ENGINES

1. Before servicing the vehicle, refer to the precautions in the beginning of this section.

2. Raise and safely support the vehicle.

3. Drain the oil.

⚹⚹ **CAUTION**

The air suspension switch must be turned off prior to raising the vehicle. Failure to do so can result in unexpected inflation or deflation of the air springs, which could result in shifting of the vehicle during the repair operation.

4. Turn the air suspension switch off.

5. Support the front axle housing with a jackstand.

6. Remove or disconnect the following:
- Negative battery cable
- Front axle mount bolt
- Front axle support bolts and the support

⚹⚹ **CAUTION**

Use care when lowering the front axle housing, or the vacuum lines to the axle solenoid may become disconnected or damaged.

- Right and front axle housing mount bolts and lower the axle to allow clearance for the oil pan to be removed
- Oil pan and gasket

To install:

⚹⚹ **WARNING**

Do not use metal scrapers, wire brushes, power abrasive discs or other abrasive means to clean the sealing surfaces. These tools cause scratches and gouges which make leak paths. Use a plastic scraping tool to remove all traces of old sealant.

7. Clean the oil pan mating surfaces.

For Tire, Wheel and Ball Joint specifications, see Section 1 of this manual

➡If the oil pan and the oil pan gasket are not secured within four minutes, the sealant must be removed and the sealing area cleaned with Metal Surface Cleaner F4AZ-19A536-RA. Allow to dry until there is no sign of wetness, or four minutes, whichever is longer. Failure to follow this procedure can cause future oil leakage.

8. Apply the silicone Gasket and Sealant F7AZ-19554-EA at the rear oil seal retainer-to-cylinder block sealing surface.

9. Install or connect the following:
- New oil pan gasket and the oil pan and loosely install the bolts

10. Tighten the bolts in three steps in the sequence illustrated as follows:
   a. Step 1: 18 inch lbs. (2 Nm).
   b. Step 2: 15 ft. lbs. (20 Nm).
   c. Step 3: an additional 60 degrees.

11. Install or connect the following:
- Front axle housing and loosely install the bolts
- Axle support and tighten the bolts to 56 ft. lbs. (90 Nm)

12. Refill the engine crankcase with clean engine oil.

13. Turn on the air suspension switch.

DOHC ENGINES

1. Before servicing the vehicle, refer to the precautions in the beginning of this section.

2. Raise and safely support the vehicle.

3. Drain the oil.

4. Remove or disconnect the following:
- Front drive axle assembly

➡Be careful when removing oil pan gasket. It is re-usable if not damaged.

- Bolts, oil pan and gasket

5. Inspect the oil pan gasket for damage.

To install:

✳✳ WARNING

Do not use metal scrapers, wire brushes, power abrasive discs or other abrasive means to clean the sealing surfaces. These tools cause scratches and gouges which make leak paths. Use a plastic scraping tool to remove all traces of old sealant.

6. Clean the oil pan mating surfaces.

➡If the oil pan and the oil pan gasket are not secured within four minutes, the sealant must be removed and the sealing area cleaned with Metal Surface Cleaner F4AZ-19A536-RA. Allow to

dry until there is no sign of wetness, or four minutes, whichever is longer. Failure to follow this procedure can cause future oil leakage.

7. Apply the silicone Gasket and Sealant F7AZ-19554-EA at the rear oil seal retainer-to-cylinder block sealing surface.

8. Install the oil pan gasket and the oil pan and loosely install the 16 bolts. Tighten the bolts in three steps, in the sequence illustrated as follows:
   a. Step 1: 18 inch lbs. (2 Nm).
   b. Step 2: 15 ft. lbs. (20 Nm).
   c. Step 3: an additional 60 degrees.

9. Install the front drive axle assembly.

10. Refill the engine crankcase with clean engine oil.

6.8L Engine

✳✳ CAUTION

Fuel injection systems remain under pressure, even after the engine has been turned OFF. The fuel system pressure must be relieved before disconnecting any fuel lines. Failure to do so may result in fire and/or personal injury.

1. Before servicing the vehicle, refer to the precautions in the beginning of this section.

2. Remove or disconnect the following:
- Negative battery cable
- Fuel system pressure
- Coolant
- Upper radiator hose from the intake manifold
- Air cleaner outlet tube
- Accelerator cable from the bracket and the throttle body cam
- Speed control actuator cable from the throttle body
- All vacuum hoses, fuel lines and electrical wires from the throttle body and intake manifold
- Brake booster vacuum hose bracket
- EGR valve-to-exhaust manifold tube and disconnect the vacuum line
- Connector and vacuum line from the Engine Vacuum Regulator (EVR) solenoid
- The 4 bolts and the throttle body adapter
- Upper fan shroud mounting screws and position the shroud toward the engine.
- The alternator and install the Mod-

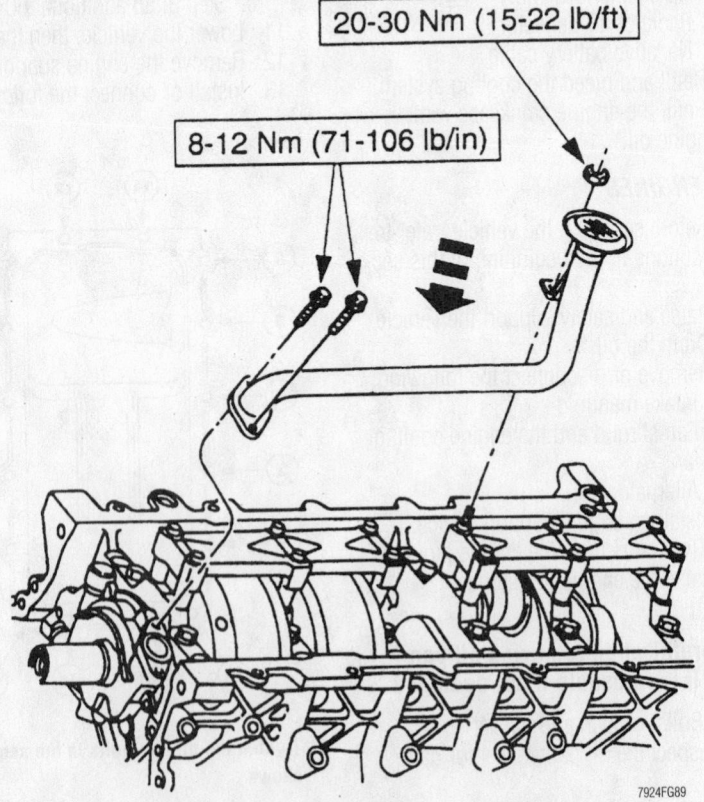

20-30 Nm (15-22 lb/ft)

8-12 Nm (71-106 lb/in)

Oil pump pick-up tube and screen assembly—6.8L engine

7924FG89

ular Engine Support Bracket on the engine using the mounting holes.

- Lower engine mount-to-frame nuts
- Turbine Shaft Speed (TSS) and Output Shaft Speed (OSS) sensors from the transmission. Plug the openings.

3. Lower the vehicle to the floor and raise the engine using a hoist attached to the support bracket.

4. Install an engine support fixture with a J hook to keep the engine raised, then remove the hoist.

5. Remove or disconnect the following:
- Engine oil and filter
- Dual converter Y-pipe and the flywheel inspection cover
- Driveshaft and the 2 transmission mounting nuts
- Transmission. Be sure to support the transmission along the rails of the pan to avoid damage.
- Oil pan mounting bolts and partially lower the pan.
- Oil pump pick-up tube and screen assembly and allow it to drop into the pan.
- Oil pan towards the rear of the vehicle

To install:

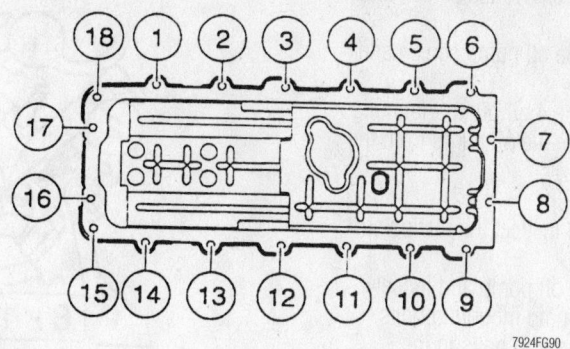

To prevent leaks, tighten the oil pan bolts in the order shown—1999–00 6.8L engine

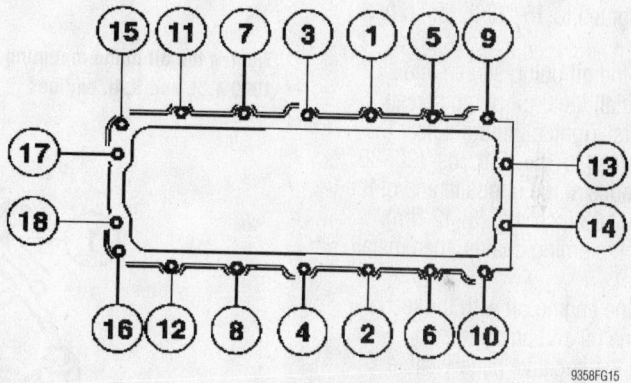

To prevent leaks, tighten the oil pan bolts in the order shown—2001–03 6.8L engine

❄❄ WARNING

To prevent possible oil leaks, use only a plastic scraper to clean the oil pan mounting surface.

6. Clean the oil pan-to-engine mounting surface.

7. Place the oil pump pick-up tube and screen assembly in the oil pan, then position the pan and gasket near the engine.

8. Install the oil pump pick-up tube and screen assembly.
   a. Tighten the nut to 15–22 ft. lbs. (20–30 Nm).
   b. Tighten the 2 bolts to 71–106 inch lbs. (8–12 Nm).

9. Apply a bead of silicone sealant to the areas where the front cover and rear bearing cap meet the engine block.

10. Install the oil pan. Tighten the bolts in sequence using 3 steps as follows:
- 18 inch lbs. (2 Nm)
- 15 ft. lbs. (20 Nm)
- Plus an additional 90°

11. Lower the transmission and install the 2 mounting nuts. Tighten the nuts to 60–80 ft. lbs. (81–108 Nm).

12. Install the driveshaft and the TSS and OSS sensors.

13. Install the flywheel cover and dual converter Y-pipe.

14. Install the oil bypass filter.

15. Lower the vehicle and remove the engine support fixture.

16. Install the engine mounting nuts. Tighten the nuts to 66 ft. lbs. (90 Nm).

17. Remove the modular engine support bracket and install the alternator.

18. Install the fan shroud.

19. Use a new gasket and install the throttle body adapter.

20. Tighten the bolts in 2 steps:
- 71–88 inch lbs. (8–10 Nm).
- 85–95 degrees.

21. Connect the EVR solenoid harness and vacuum line.

22. Attach the vacuum line to the EGR valve.

23. Install the EGR valve-to-exhaust manifold tube. Tighten the fittings to 41 ft. lbs. (55 Nm).

24. Install the EGR transducer.

25. Install all remaining components in the reverse of the removal.

❄❄ WARNING

Operating the engine without the proper amount and type of engine oil will result in severe engine damage.

26. Fill the engine with SAE 5W30 oil.
27. Fill and bleed the cooling system.

Oil Pump

REMOVAL & INSTALLATION

4.6L and 5.4L Engines

1999 Models

1. Before servicing the vehicle, refer to the precautions in the beginning of this section.

2. Disconnect the negative battery cable.

3. Remove the timing chain.

4. Drain the engine oil into a suitable container.

5. Remove the oil pan.

6. Remove the 3 oil pump screen and

For Wheel Alignment specifications, see Section 1 of this manual

cover bolts, then remove the screen and cover.

7. Remove the oil pump screen and cover spacer.

8. Remove the 4 oil pump mounting bolts, then remove the oil pump from the engine.

To install:

9. Clean and inspect the mating surfaces.

10. Install the oil pump and loosely install the 4 oil pump mounting bolts. Tighten the 4 oil pump bolts in the sequence shown to 71–106 inch lbs. (8–12 Nm).

11. Install the oil pump screen and cover spacer, then tighten to 15–22 ft. lbs. (20–30 Nm).

12. Install the oil pump screen and cover, then install the 3 oil pump screen and cover bolts. Tighten the bolts near the oil pick-up screen to 15–22 ft. lbs. (20–30 Nm) and the bolts at the opposite end of the pick-up to 70–106 inch lbs. (8–12 Nm).

13. Install the timing chains, then install the oil pan.

14. Refill the engine oil with the recommended engine oil and amount.

15. Install the negative battery cable.

2000–03 MODELS

1. Before servicing the vehicle, refer to the precautions in the beginning of this section.

2. Disconnect the negative battery cable.

3. Remove or disconnect the following:
- Crankshaft sprocket
- Oil pan
- Bolts and the oil pump

To install:

4. Clean and inspect the mating surfaces.

5. Install or connect the following:
- Oil pump, and the bolts loosely. Tighten the bolts in the sequence illustrated to 89 inch lbs. (10 Nm).
- Oil pan
- Crankshaft sprocket

6.8L Engine

1. Before servicing the vehicle, refer to the precautions in the beginning of this section.

2. Disconnect the negative battery cable.

3. Remove the engine front cover and crankshaft sprocket.

4. Drain the engine oil into a suitable container.

5. Remove the oil pan.

6. Remove the 3 oil pump mounting bolts, then remove the oil pump from the engine.

Tighten the oil pump mounting bolts to 71–106 inch lbs. (8–12 Nm) in the sequence shown—1999 4.6L and 5.4L engines

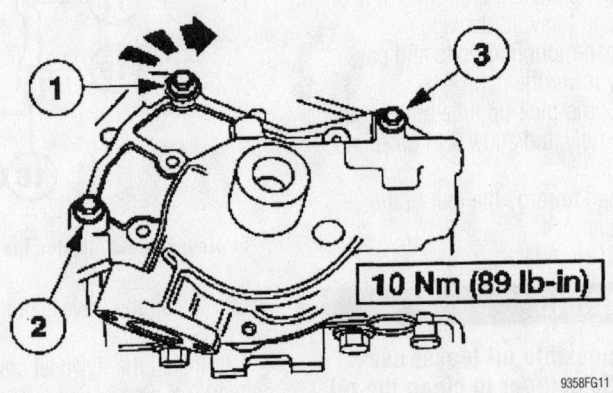

Tighten the oil pump mounting bolts in the sequence shown—4.6L and 5.4L engines

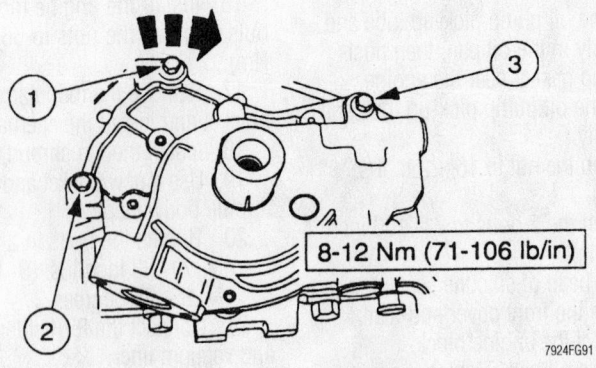

Be sure to tighten the oil pump mounting bolts in the sequence shown—6.8L engine

To install:

7. Clean and inspect the mating surfaces.

8. Install the oil pump and loosely install the oil pump mounting bolts. Tighten the bolts in the sequence shown to 71–106 inch lbs. (8–12 Nm).

9. Install the oil pan.

10. Install the crankshaft sprocket and timing chains.

11. Install the front cover.

12. Refill the engine oil with the recommended engine oil and amount.

13. Install the negative battery cable.

Rear Main Seal

REMOVAL & INSTALLATION

If the crankshaft rear oil seal replacement is the only operation being performed, it can be done in the vehicle as detailed in the following procedure. If the oil seal is being replaced in conjunction with a rear main bearing replacement, the engine must be removed from the vehicle and installed on a work stand.

1. Before servicing the vehicle, refer to the precautions in the beginning of this section.
2. Remove or disconnect the following:
   - Negative battery cable.
   - Transmission
   - Flywheel/flexplate
   - Crankshaft oil slinger from the crankshaft, if equipped
3. Use an awl to punch 2 holes in the crankshaft rear oil seal. Punch the holes on opposite sides of the crankshaft and just above the bearing cap-to-cylinder block split line.
4. Install a sheet metal screw in each hole. Use 2 small prybars to pry against both screws at the same time to remove the crankshaft rear oil seal. It may be necessary to place small blocks of wood against the cylinder block to provide a fulcrum point for the prybars. Use caution throughout this procedure to avoid scratching or otherwise damaging the crankshaft oil seal surface.
5. Clean the oil seal recess in the cylinder block and main bearing cap.

To install:

6. Clean, inspect and polish the rear oil seal rubbing surface on the crankshaft.
7. Coat the new oil seal and the crankshaft with a light film of engine oil.
8. Start the seal in the recess with the seal lip facing forward and install it with a seal driver. Keep the tool straight with the centerline of the crankshaft and install the seal until the tool contacts the cylinder block surface. Remove the tool and inspect the seal to be sure it was not damaged during installation.
9. Install or connect the following:
   - Crankshaft oil slinger, if equipped
   - Flywheel on the crankshaft flange. Coat the threads of the flywheel attaching bolts with Loctite® and install the bolts. Tighten the bolts in sequence across from each other to 75–85 ft. lbs. (102–115 Nm).
   - Transmission
   - Negative battery cable

Timing Chain, Sprockets and Front Cover

REMOVAL & INSTALLATION

4.6L Engine

1. Before servicing the vehicle, refer to the precautions in the beginning of this section.
2. Remove or disconnect the following:
   - Negative battery cable
   - Radiator, fan blade and fan shroud assembly
   - Accessory drive belt
   - Water pump pulley
   - Electrical harness connectors from both ignition coils
   - Both ignition coils with their brackets attached
   - Left-hand and right-hand cylinder head covers
   - The 2 upper power steering pump retaining bolts
   - The 2 lower power steering pump retaining bolts and move the pump aside
   - Crankshaft Position (CKP) sensor electrical harness connector
   - CKP sensor retaining bolt and the sensor
   - Engine oil
   - The 4 oil pan-to-engine front cover retaining bolts
   - Crankshaft damper retaining bolt and washer from the crankshaft
   - Damper from the crankshaft
   - Camshaft Position (CMP) sensor retaining bolt and the sensor
   - Idler pulley bolt and the pulley
   - The 3 belt tensioner retaining bolts and the tensioner
   - The 8 engine front cover retaining bolts and the 7 nuts. Swing the top of the cover out off the dowel pins and remove the cover.
   - Sensor ring from the crankshaft
3. Use Camshaft Positioning Tool T91P-6256-A and Camshaft Positioning Adapters T92P-6256-A to position the camshaft.
4. Rotate the crankshaft until both camshaft keyways are 90 degrees from the cam cover surface. Be sure the copper links line up with the dots on the camshaft sprockets.

✳✳ WARNING

At no time, when the timing chains are removed and the cylinder heads are installed may the crankshaft or the camshaft be rotated. Severe piston and valve damage will occur.

5. Remove or disconnect the following:
   - The 2 left-hand and right-hand tensioner bolts and the timing chain tensioners

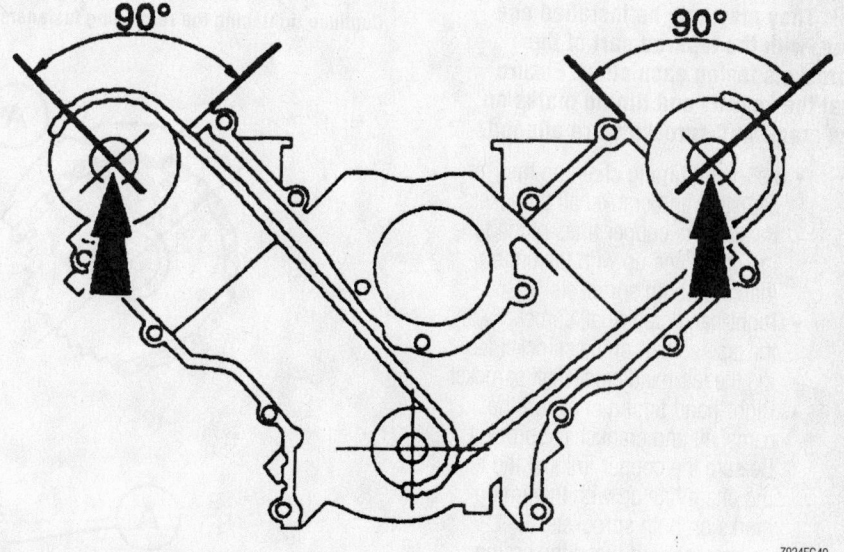

When removing the timing chains, rotate the crankshaft so that the camshaft keyways are positioned as shown—4.6L engines

7924FG49

- The left-hand and right-hand tensioner guides off the dowel pins
- The right-hand timing chain from the camshaft sprocket
- The left-hand timing chain from the camshaft sprocket
- The left-hand and right-hand timing chain guide bolts and the timing chain guides
- Camshaft gear bolt and the camshaft gear, if necessary

To install:

6. Examine the timing chains, looking for the copper links. If the copper links are not visible, lay the chain on a flat surface and pull the chain taught until the opposite sides of the chain contact one another. Mark the links at each end of the chain and use these marks in place of the copper links.

➡ **If the engine jumped time, damage has been done to valves and possibly pistons and/or connecting rods. Any damage must be corrected before installing the timing chains.**

7. Install or connect the following:
- Camshaft gears and tighten the retaining bolt to 81–95 ft. lbs. (110–130 Nm)
- Left-hand and right-hand timing chain guides and retaining bolts. Tighten the retaining bolts to 71–106 inch lbs. (8–12 Nm).
- Left-hand crankshaft sprocket with the tapered part of the sprocket facing away from the engine block

➡ **The crankshaft sprockets are identical. They may only be installed one way, with the tapered part of the sprockets facing each other. Ensure that the keyway and timing marks on the crankshaft sprockets are aligned.**

- Left-hand timing chain on the camshaft and crankshaft sprockets. Be sure the copper links of the timing chain line up with the timing marks on both sprockets.
- Right-hand crankshaft sprocket with the tapered part of the sprocket facing the left-hand crankshaft sprocket
- Right-hand timing chain on the camshaft and crankshaft sprockets. Be sure the copper links of the timing chain line up with the timing marks on both sprockets.

8. It is necessary to bleed the timing chain tensioners before installation. Proceed as follows:
   a. Place the timing chain tensioner in a soft-jawed vise.

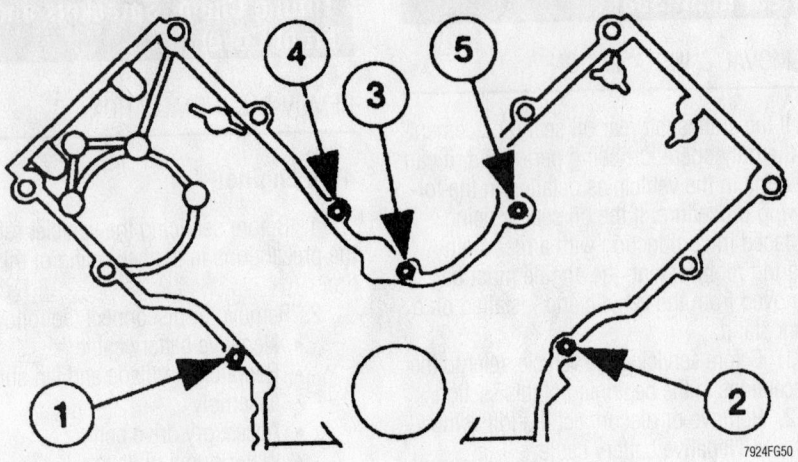

Tighten the first 5 front cover fasteners in the sequence shown—4.6L engine

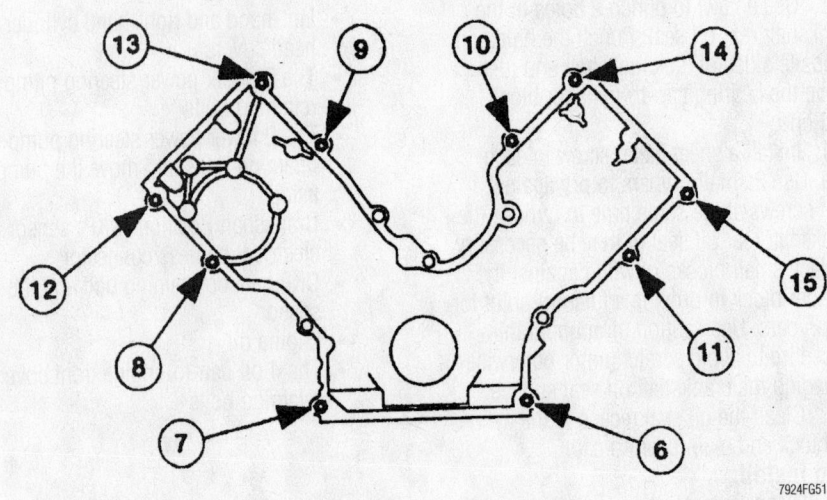

Continue tightening the remaining fasteners in the sequence shown here—4.6L engine

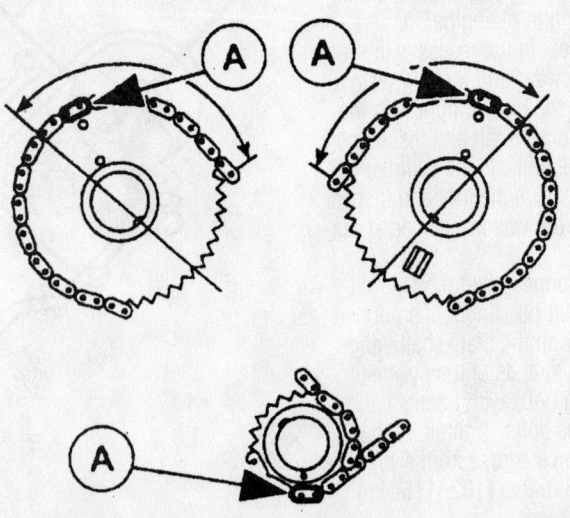

When installing the timing chains, make certain that the copper colored links (A) are aligned with the timing marks—4.6L engine

b. Using a small pick or similar tool, hold the ratchet lock mechanism away from the ratchet stem and slowly compress the tensioner plunger by rotating the vise handle.

✳✳ WARNING

The tensioner must be compressed slowly or damage to the internal seals will result.

c. Once the tensioner plunger bottoms in the tensioner bore, continue to hold the ratchet lock mechanism and push down on the ratchet stem until flush with the tensioner face.

d. While holding the ratchet stem flush to the tensioner face, release the ratchet lock mechanism and install a paper clip or similar tool in the tensioner body to lock the tensioner in the collapsed position.

e. The paper clip must not be removed until the timing chain, tensioner, tensioner arm and timing chain guide are completely installed on the engine.

9. Install or connect the following:
- Left-hand and right-hand timing chain tensioner guides on the dowel pins
- Left-hand and right-hand timing chain tensioners in position and install the retaining bolts. Tighten the bolts to 15–22 ft. lbs. (20–30 Nm).

10. Remove the retaining pins from the timing chain tensioners.

11. Remove Camshaft Positioning Tool T91P-6256-A and Camshaft Positioning Adapters T92P-6256-A from the camshaft.

12. Install or connect the following:
- Crankshaft sensor ring on the crankshaft

13. Apply a bead of silicone sealer along the cylinder head-to-cylinder block and the oil pan-to-cylinder block sealing surfaces.

- Engine front cover carefully onto the dowel pins. Tighten the engine front cover bolts, in sequence, to 15–22 ft. lbs. (20–30 Nm).
- Idler pulley in position and the retaining bolt. Tighten the bolt to 15–22 ft. lbs. (20–30 Nm).
- Drive belt tensioner and the 3 retaining bolts. Tighten the bolts to 15–22 ft. lbs. (20–30 Nm).
- CMP sensor and the retaining bolt. Tighten the bolt to 106 inch lbs. (12 Nm).

- The 4 oil pan-to-engine front cover bolts and tighten, in sequence, in 2 steps: Tighten the bolts to 15 ft. lbs. (20 Nm). Rotate the bolts an additional 60 degrees.
- CKP sensor and the retaining bolt. Tighten the bolt to 106 inch lbs. (12 Nm).
- CKP sensor electrical harness connector
- Power steering pump and the 2 upper and 2 lower retaining bolts. Tighten the bolts to 15–20 ft. lbs. (20–30 Nm).
- Ignition coil and brackets on the engine front cover and the bracket bolts. Tighten the bolts to 15–22 ft. lbs. (20–30 Nm).
- Ignition coil and capacitor electrical harness connectors
- CMP electrical harness connector
- Water pump pulley and tighten the bolts to 15–22 ft. lbs. (20–30 Nm).
- Radiator, fan blade and fan shroud assembly
- Accessory drive belt

14. Refill the engine oil with the recommended type of engine oil and the correct amount.

15. Connect the negative battery cable.

16. Start the engine and check for leaks.

17. Road test the vehicle and check for proper engine operation.

5.4L SOHC and 6.8L Engines

1. Before servicing the vehicle, refer to the precautions in the beginning of this section.

2. Remove or disconnect the following:
- Negative battery cable
- Radiator, fan blade and fan shroud assembly
- Accessory drive belt
- Water pump pulley
- Electrical harness connectors from both ignition coils
- Both ignition coils with their brackets attached
- The 2 upper power steering pump retaining bolts
- The 2 lower power steering pump retaining bolts and move the pump aside
- Crankshaft Position (CKP) sensor electrical harness connector
- Retaining bolt and the CKP sensor.

3. Drain the engine oil.
- The 4 oil pan-to-engine front cover retaining bolts
- The crankshaft damper retaining bolt and washer from the crankshaft
- Crankshaft damper from the crankshaft
- Camshaft Position (CMP) sensor retaining bolt and remove the CMP sensor
- Idler pulley bolt and remove the pulley

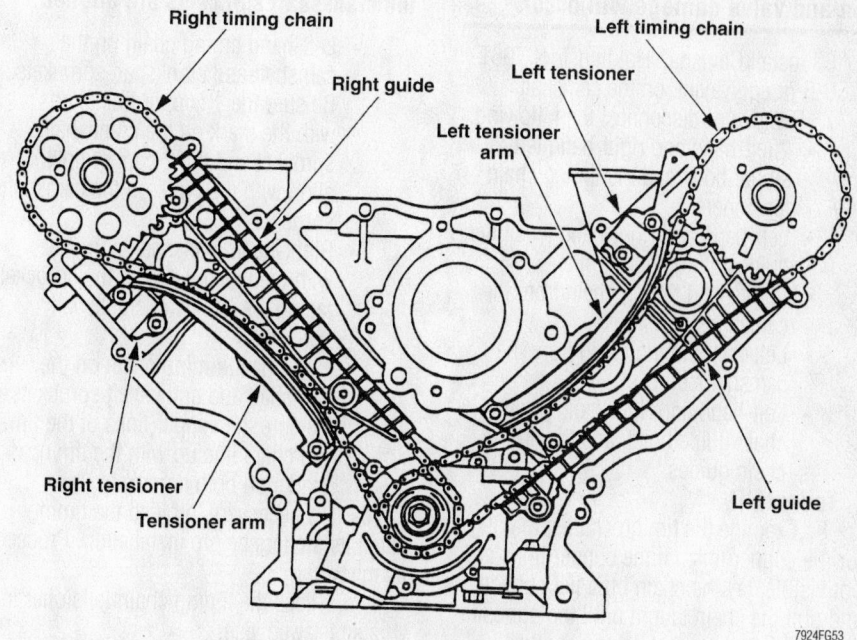

Timing chains and related components—5.4L and 6.8L engines

7924FG53

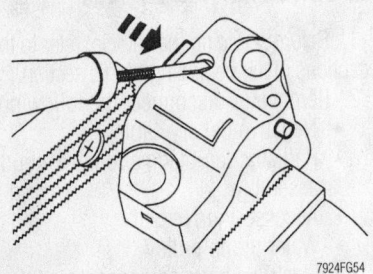

Compress the tensioner while holding the ratchet mechanism with a suitable tool—5.4L and 6.8L engines

- The 3 belt tensioner bolts and remove the tensioner
- The 8 engine front cover retaining bolts and the 7 nuts. Swing the top of the cover out off the dowel pins and remove the cover.
- Sensor ring from the crankshaft

4. Use Crankshaft Holding Tool 303-448 (T93P-6303-A), to position the camshaft.

5. Rotate the crankshaft until both camshaft keyways are 90 degrees from the cam cover surface. Be sure the copper links line up with the dots on the camshaft sprockets.

✳✳ WARNING

At no time, when the timing chains are removed and the cylinder heads are installed may the crankshaft or the camshaft be rotated. Severe piston and valve damage will occur.

6. Install Camshaft Holding Tool T96T-6256-B or equivalent, on the camshaft.

7. Remove or disconnect the following:
- 2 left-hand and right-hand tensioner bolts and the timing chain tensioners
- Left-hand and right-hand tensioner guides off the dowel pins
- Right-hand timing chain from the camshaft sprocket
- Left-hand timing chain from the camshaft sprocket
- Left-hand and right-hand timing chain guide bolts, and the timing chain guides

To install:

8. Examine the timing chains, looking for the copper links. If the copper links are not visible, lay the chain on a flat surface and pull the chain taught until the opposite sides of the chain contact one another. Mark the links at each end of the chain and use these marks in place of the copper links.

If the copper links are not visible, mark one link on one end of the chain and 2 links on the opposite end of the chain—5.4L and 6.8L engines

➡ **If the engine jumped time, damage has been done to valves and possibly pistons and/or connecting rods. Any damage must be corrected before installing the timing chains.**

9. Install or connect the following:
- Left-hand and right-hand timing chain guides and retaining bolts. Tighten the retaining bolts to 89 inch lbs. (10 Nm).
- Left-hand crankshaft sprocket with the tapered part of the sprocket facing away from the engine block.

➡ **The crankshaft sprockets are identical. They may only be installed one way, with the tapered part of the sprockets facing each other. Ensure that the keyway and timing marks on the crankshaft sprockets are aligned.**

- Left-hand timing chain on the camshaft and crankshaft sprockets. Be sure the 1 copper link aligns with the mark on the crankshaft sprocket and the 2 copper links align with the mark on the camshaft sprocket.
- Right-hand crankshaft sprocket with the tapered part of the sprocket facing the left-hand crankshaft sprocket
- Right-hand timing chain on the camshaft and crankshaft sprockets. Be sure the copper links of the timing chain line up with the timing marks on both sprockets.

10. It is necessary to bleed the timing chain tensioners before installation. Proceed as follows:
   a. Place the timing chain tensioner in a soft-jawed vise.
   b. Using a small pick or similar tool, hold the ratchet lock mechanism away from the ratchet stem and slowly compress the tensioner plunger by rotating the vise handle.

✳✳ WARNING

The tensioner must be compressed slowly or damage to the internal seals will result.

   c. Once the tensioner plunger bottoms in the tensioner bore, continue to hold the ratchet lock mechanism and push down on the ratchet stem until flush with the tensioner face.
   d. While holding the ratchet stem flush to the tensioner face, release the ratchet lock mechanism and install a paper clip or similar tool in the tensioner body to lock the tensioner in the collapsed position.
   e. The paper clip must not be removed until the timing chain, tensioner, tensioner arm and timing chain guide are completely installed on the engine.

11. Install the left-hand and right-hand timing chain tensioner guides on the dowel pins.

12. Place the left-hand and right-hand timing chain tensioners in position and install the retaining bolts. Tighten the bolts to 15–22 ft. lbs. (20–30 Nm).

13. Remove the retaining pins from the timing chain tensioners.

14. Remove Cam Holding Tool T96T-6256-B or equivalent, from the camshaft.

15. Install the crankshaft sensor ring on the crankshaft.

16. Apply silicone gasket along the cylinder head-to-cylinder block and engine oil pan-to-cylinder block sealing surfaces.

17. Install or connect the following:
- Engine front cover carefully onto the dowel pins
- Engine front cover bolts, in sequence, in the following manner: bolts 1 through 7: 15–22 ft. lbs.

(20–30 Nm); bolts 6 through 15: 29–40 ft. lbs. (40–55 Nm)

- Idler pulley in position and install the retaining bolt. Tighten the bolt to 15–22 ft. lbs. (20–30 Nm).
- Drive belt tensioner in position and install 3 retaining bolts. Tighten the bolts to 15–22 ft. lbs. (20–30 Nm).
- CMP sensor in position and install the retaining bolt. Tighten the bolt to 106 inch lbs. (12 Nm).
- Damper on the crankshaft. Ensure the crankshaft key and keyway are aligned.
- 4 front oil pan-to-engine front cover retaining bolts and tighten in sequence in 2 steps: Tighten the bolts to 15 ft. lbs. (20 Nm); Rotate the bolts an additional 60 degrees.
- CKP sensor in position and install the retaining bolt. Tighten the bolt to 106 inch lbs. (12 Nm).
- CKP sensor electrical harness connector
- Power steering pump with 2 upper and 2 lower retaining bolts. Tighten the bolts to 15–20 ft. lbs. (20–30 Nm).
- Ignition coil and brackets on the engine front cover and install the bracket bolts. Tighten the bolts to 15–22 ft. lbs. (20–30 Nm).
- Ignition coil and capacitor electrical harness connectors
- CMP electrical harness connector
- Water pump pulley and tighten the bolts to 15–22 ft. lbs. (20–30 Nm)
- Radiator
- Negative battery cable

18. Refill the engine oil with the recommended type of engine oil and the correct amount.

19. Start the engine and check for leaks.

20. Road test the vehicle and check for proper engine operation.

5.4L DOHC Engines

1. Before servicing the vehicle, refer to the precautions in the beginning of this section.

2. Drain the oil.

3. Remove or disconnect the following:
- Negative battery cable
- Valve covers
- Cooling fan
- Water pump pulley
- Crankshaft pulley using Crankshaft Damper Remover 303-009 (T58P-6316-D)

- Nuts and the hose support, if equipped
- Power steering pump and position aside
- Air Conditioner (A/C) muffler bolt and position out of the way
- Crankshaft Position (CKP) sensor
- Crankshaft front oil seal using Front Cover Seal Remover 303-107 (T74P-6700-A)
- Camshaft Position (CMP) sensor
- Belt idler pulley
- Engine front cover bolts, studs, cover and the gaskets
- Sensor ring from the crankshaft

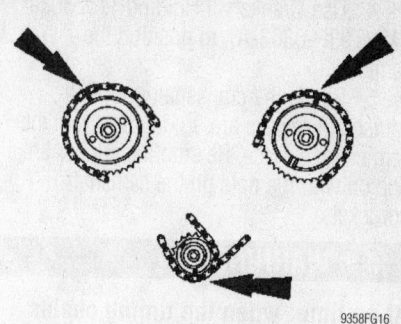

9358FG16

Check the alignment of the timing marks as shown—5.4L and 6.8L engines

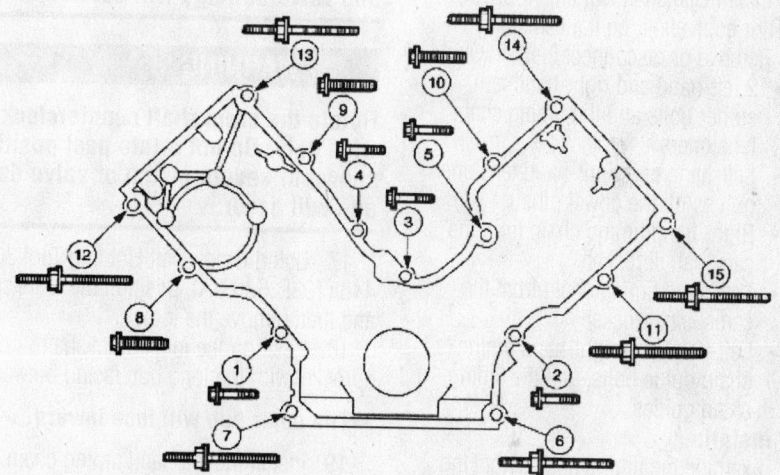

Item	Description
1	Bolt, Hex Flange Head Pilot, M8 x 1.25 x 53
2	Bolt, Hex Flange Head Pilot, M8 x 1.25 x 53
3	Bolt, Hex Flange Head Pilot, M8 x 1.25 x 53
4	Bolt, Hex Flange Head Pilot, M8 x 1.25 x 53
5	Bolt, Hex Flange Head Pilot, M8 x 1.25 x 53
6	Stud, Hex-Head Pilot, M10 x 1.5 x 1.5 x 103.1
7	Stud, Hex-Head Pilot, M10 x 1.5 x 1.5 x 103.1
8	Screw and Washer, Hex Pilot, M10 x 1.5 x 57.5
9	Screw and Washer, Hex Pilot, M10 x 1.5 x 57.5
10	Screw and Washer, Hex Pilot, M10 x 1.5 x 57.5
11	Stud and Washer, Hex-Head Pilot, M10 x 1.5 x M8 x 1.25 x 109.6
12	Stud and Washer, Hex-Head Pilot, M10 x 1.5 x M8 x 1.25 x 109.6
13	Stud and Washer, Hex-Head Pilot, M10 x 1.5 x M8 x 1.25 x 109.6
14	Stud and Washer, Hex-Head Pilot, M10 x 1.5 x M8 x 1.25 x 109.6
15	Stud and Washer, Hex-Head Pilot, M10 x 1.5 x M8 x 1.25 x 109.6

9358FG17

Install the timing cover and tighten the bolts to specification——5.4L and 6.8L engines

4. Use Crankshaft Holding Tool 303-448 (T93P-6303-A), to position the camshaft.

5. Rotate the crankshaft until both camshaft keyways are 90 degrees from the cam cover surface. Be sure the copper links line up with the dots on the camshaft sprockets.

✳✳ WARNING

At no time, when the timing chains are removed and the cylinder heads are installed may the crankshaft or the camshaft be rotated. Severe piston and valve damage will occur.

6. Install Camshaft Holding Tool T96T-6256-B or equivalent, on the camshaft.

7. Remove or disconnect the following:
- 2 left-hand and right-hand tensioner bolts and the timing chain tensioners
- Left-hand and right-hand tensioner guides off the dowel pins
- Right-hand timing chain from the camshaft sprocket
- Left-hand timing chain from the camshaft sprocket
- Left-hand and right-hand timing chain guide bolts, and the timing chain guides

To install:

8. Examine the timing chains, looking for the copper links. If the copper links are not visible, lay the chain on a flat surface and pull the chain taught until the opposite sides of the chain contact one another. Mark the links at each end of the chain and use these marks in place of the copper links.

➡**If the engine jumped time, damage has been done to valves and possibly pistons and/or connecting rods. Any damage must be corrected before installing the timing chains.**

✳✳ WARNING

Do not compress the ratchet assembly. This will damage the ratchet assembly.

9. Compress the tensioner plunger, using the edge of a soft-jawed vise.

10. Use a small screwdriver or pick, push back and hold the ratchet mechanism.

11. While holding the ratchet mechanism, push the ratchet arm back into the tensioner housing.

12. Install a paper clip into the hole in the tensioner housing to hold the ratchet assembly and plunger in during installation.

13. Install the timing chain guides.

14. Remove tool T93P-6256-A.

15. Rotate the left hand camshaft sprocket until the timing mark is approximately at the 12 o' clock position. Rotate the right hand camshaft timing sprocket until the timing mark is approximately in the 11 o' clock position.

16. Install tool T93P-6256-A.

✳✳ WARNING

Unless otherwise instructed, at no time when the timing chains are removed and the cylinder heads are installed is the crankshaft or the camshaft to be rotated. Severe piston and valve damage will occur.

✳✳ WARNING

Rotate the crankshaft counterclockwise only. Do not rotate past position shown or severe piston or valve damage will occur.

17. Using Crankshaft Holding Tool 303-448 (T93P-6303-A), position the crankshaft, and then remove the tool.

18. Position the inner crankshaft sprocket with the long hub facing outward.

➡**The outer hub will face inward.**

19. Install the left hand timing chain onto the crankshaft sprocket, aligning the one copper link on the timing chain with the slot on the crankshaft sprocket.

20. Verify the camshaft sprocket to copper link alignment.

21. Position the tensioner arms and tensioners, and install the bolts. Tighten the bolts to 18 ft. lbs. (25 Nm).

22. Remove the paper clips used to retain the ratchet assemblies.

23. Remove the camshaft holding tool.

24. Position the crankshaft sensor ring on the crankshaft.

25. Apply silicone gasket along the cylinder head-to-cylinder block and engine oil pan-to-cylinder block sealing surfaces.

26. Install or connect the following:
- Engine front cover carefully onto the dowel pins
- Engine front cover bolts, in sequence, in the following manner: bolts 1 through 5: 18 ft. lbs. (25 Nm); bolts 6 through 7: 37 ft. lbs. (50 Nm), bolts 1 through 15 to 18 ft. lbs. (25 Nm)
- Idler pulley in position and install the retaining bolt. Tighten the bolt to 18 ft. lbs. (25 Nm).
- CMP sensor in position and install the retaining bolt. Tighten the bolt to 106 inch lbs. (12 Nm).
- Damper on the crankshaft. Ensure the crankshaft key and keyway are aligned.
- 4 front oil pan-to-engine front cover retaining bolts and tighten in sequence in 3 steps: Tighten the bolts to 18 ft. lbs. (25 Nm), then to 15 ft. lbs. (20 Nm); then rotate the bolts an additional 60 degrees.
- Crankshaft front oil seal using Front Crankshaft Seal Installer 303-635 and Crankshaft Seal Installer/Aligner 303-335 (T88T-6701-A)
- CKP sensor in position and install the retaining bolt. Tighten the bolt to 106 inch lbs. (12 Nm).
- CKP sensor electrical harness connector
- A/C muffler and tighten the nut to 18 ft. lbs. (25 Nm).
- Power steering pump with the upper and lower retaining bolts. Tighten the bolts to 18 ft. lbs. (25 Nm).
- Crankshaft pulley, bolt and washer

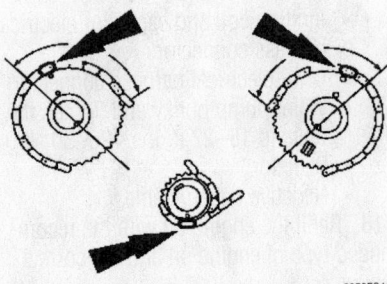

9359FG16

Check the alignment of the timing marks as shown—5.4L DOHC engines

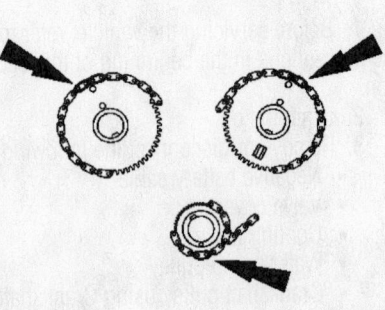

9359FG17

Verify the camshaft sprocket to copper link alignment—5.4L DOHC engines

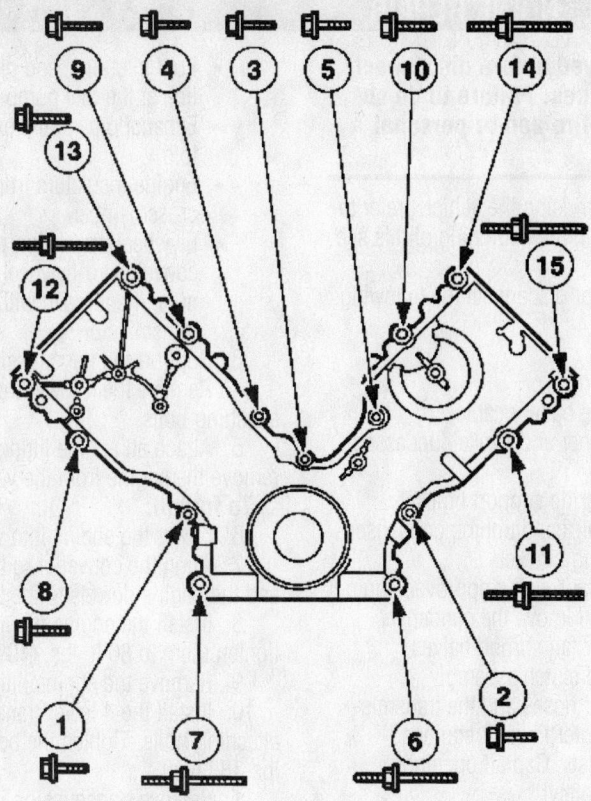

Item	Description
1	Bolt, Hex Flange Head Pilot, M8 x 1.25 x 50
2	Bolt, Hex Flange Head Pilot, M8 x 1.25 x 50
3	Bolt, Hex Flange Head Pilot, M8 x 1.25 x 50
4	Bolt, Hex Flange Head Pilot, M8 x 1.25 x 50
5	Bolt, Hex Flange Head Pilot, M8 x 1.25 x 50
6	Stud, Hex-Head Pilot, M10 x 1.5 x 30
7	Stud, Hex-Head Pilot, M10 x 1.5 x 30
8	Bolt, Hex-Head Flange, M8 x 1.25 x 50
9	Bolt, Hex-Head Flange, M8 x 1.25 x 50
10	Bolt, Hex-Head Flange, M8 x 1.25 x 50
11	Stud Hex Shoulder Pilot, M8 x 1.25 x 65 - M8 x 1.25 x
12	Stud Hex Shoulder Pilot, M8 x 1.25 x 65 - M8 x 1.25 x
13	Bolt Hex Shoulder Pilot, M8 x 1.25 x 65
14	Stud Hex Shoulder Pilot, M8 x 1.25 x 65 - M8 x 1.25 x
15	Bolt Hex Shoulder Pilot, M8 x 1.25 x 65

9359FG18

Install the timing cover and tighten the bolts to specification——5.4L DOHC engines

using Crankshaft Damper Replacer 303-102 (T74P-6316-B) and tighten the bolt in 4 steps as follows:
a. Step 1: 66 ft. lbs. (90 Nm).
b. Step 2: Loosen one full turn.
c. Step 3: 37 ft. lbs. (50 Nm).
d. Step 4: an additional 85-95 degrees.
27. Install or connect the following:
• Water pump pulley and tighten the bolts to 18 ft. lbs. (25 Nm)

• Cooling fan
• Valve covers
• Negative battery cable
28. Refill the engine oil with the recommended type of engine oil and the correct amount.
29. Start the engine and check for leaks.
30. Road test the vehicle and check for proper engine operation.

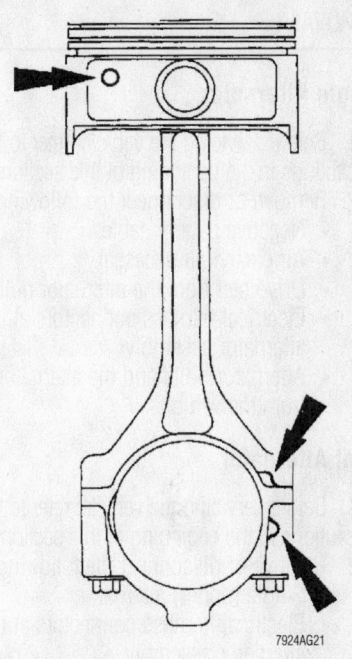

4.6L engines—piston and connecting rod front mark locations

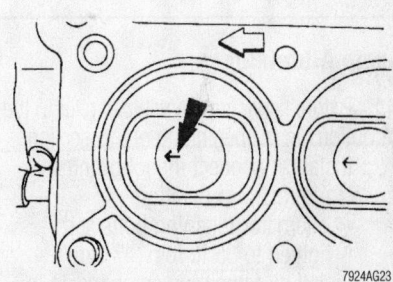

4.6L engines—piston-to-engine orientation

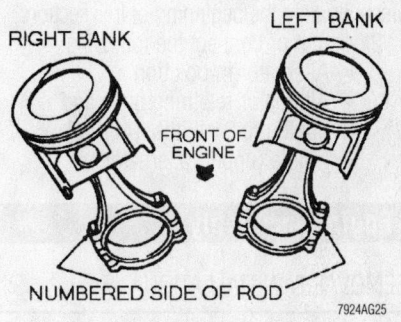

RIGHT BANK | LEFT BANK
FRONT OF ENGINE
NUMBERED SIDE OF ROD
7924AG25

5.4L, 5.8L and 6.8L engines—piston and connecting rod assembly positioning

DIESEL ENGINE REPAIR

Alternator

REMOVAL

Single Alternator

1. Before servicing the vehicle, refer to the precautions in the beginning of this section.
2. Remove or disconnect the following:
   - Negative battery cable
   - Air cleaner, if necessary
   - Drive belt from the alternator pulley
   - Electrical harness connectors at the alternator assembly
   - Alternator bolts and the alternator from the vehicle

Dual Alternator

1. Before servicing the vehicle, refer to the precautions in the beginning of this section.
2. Remove or disconnect the following:
   - Single (upper) alternator
   - Electrical harness connectors at the alternator assembly
   - Alternator bolts and the alternator from the vehicle

INSTALLATION

Single Alternator

1. Before servicing the vehicle, refer to the precautions in the beginning of this section.
2. Install or connect the following:
   - Alternator in position
   - Alternator retaining bolts and tighten to 35 ft. lbs. (47 Nm).
3. Install all remaining components in the reverse order of removal.

Dual Alternator

1. Before servicing the vehicle, refer to the precautions in the beginning of this section.
2. Install or connect the following:
   - Alternator in position
   - Alternator retaining bolts and tighten to 35 ft. lbs. (47 Nm).
   - Single (upper) alternator

Engine Assembly

REMOVAL & INSTALLATION

✳✳ CAUTION

The fuel system remains under pressure, even after the engine has been turned OFF. The fuel system pressure must be relieved before disconnecting any fuel lines. Failure to do so may result in fire and/or personal injury.

1. Before servicing the vehicle, refer to the precautions in the beginning of this section.
2. Remove or disconnect the following:
   - Hood
   - Coolant
   - Engine oil
   - Negative battery cable
   - Air cleaner and intake duct assembly
   - Upper grille support bracket
   - Upper air conditioning condenser mounting bracket
   - Refrigerant, using approved equipment to remove the condenser
   - Radiator fan shroud halves
   - Fan and clutch assembly
   - Radiator hoses and the transmission cooler lines, if equipped
   - Condenser. Cap all openings immediately!
   - Radiator
   - Power steering pump and position it out of the way
   - Fuel supply line heater and alternator wires at the alternator
   - Oil pressure sending unit wire at the sending unit
   - Sender from the firewall and lay it on the engine
   - Accelerator cable and the speed control cable, if equipped, from the injection pump
   - Accelerator cable bracket with the cables attached, from the intake manifold and position it out of the way
   - Transmission kickdown rod from the injection pump, if equipped
   - Main wiring harness connector from the right side of the engine and the ground strap from the rear of the engine
   - Fuel system pressure
   - Fuel return hose from the left rear of the engine
   - The 2 upper transmission-to-engine attaching bolts
   - Heater hoses
   - Water temperature sender wire
   - Overheat light switch wire and position the wire out of the way
   - Battery ground cables from the front of the engine and the cables from the starter
   - Fuel inlet line, and plug the fuel line at the fuel pump
   - Exhaust pipe at the exhaust manifold
   - Engine insulators from the no. 1 crossmember
   - Flywheel inspection plate and the 4 converter-to-flywheel attaching nuts, if equipped with automatic transmission
3. Support the transmission on a jack.
4. Remove the 4 lower transmission attaching bolts.
5. Attach an engine lifting sling and remove the engine from the vehicle.

To install:

6. Lower the engine into vehicle.
7. Align the converter to the flexplate and the engine dowels to the transmission.
8. Install the engine mount bolts and tighten them to 80 ft. lbs. (109 Nm).
9. Remove the engine lifting sling.
10. Install the 4 lower transmission attaching bolts. Tighten the bolts to 65 ft. lbs. (88 Nm).
11. Remove transmission jack.
12. Raise and support the front end.
13. Install or connect the following:
   - 4 converter-to-flywheel attaching nuts, if equipped with an automatic transmission. Tighten the nuts to 34 ft. lbs. (47 Nm).
   - Flywheel inspection plate. Tighten the bolts to 60–90 inch lbs. (6.7–10 Nm).
   - Exhaust pipe at the exhaust manifold
   - Fuel inlet line
   - Battery ground cables to the front of the engine
   - Starter cables at the starter
   - Overheat light switch wire
   - Water temperature sender wire
   - Heater hoses
   - The 2 upper transmission-to-engine attaching bolts. Tighten the bolts to 65 ft. lbs. (88 Nm).
   - Fuel return hose at the left rear of the engine
   - Main wiring harness connector at the right side of the engine and the ground strap from the rear of the engine.
   - Transmission kickdown rod at the injection pump, if equipped
   - Accelerator cable and the speed control cable, if equipped, at the injection pump.
   - Cable bracket with the cables attached, to the intake manifold

- Oil pressure sending unit
- Oil pressure sending unit wire at the sending unit
- Fuel supply line heater and alternator wires at the alternator
- Power steering pump
- Radiator
- Condenser
- Radiator hoses and the transmission cooler lines, if equipped.
- Fan and clutch assembly
- Radiator fan shroud halves
- Upper grille support bracket and upper air conditioning condenser mounting bracket.
- Air cleaner and intake duct assembly

14. Refill the engine oil with the recommended type of engine oil and the correct amount.
15. Connect the negative battery cable.
16. If equipped with air conditioning, charge the system.
17. Fill the cooling system.
18. Install the hood.

Water Pump

REMOVAL & INSTALLATION

1. Disconnect both battery ground cables.
2. Drain the cooling system.
3. Remove or disconnect the following:
   - Radiator shroud, fan clutch and fan

➡ **The fan clutch bolts are right-hand thread. Remove them by turning counter-clockwise.**

- Water pump pulley bolts, loosen only at this time
- Drive belt
- Water pump pulley bolts and the pulley
- Engine Coolant Temperature (ECT) sensor connector
- Heater hose from the water pump
- Bolts attaching the water pump to the front cover and lift off the pump

To install:

4. Thoroughly clean the mating surfaces of the pump and front cover.
5. Install or connect the following:
   - New gasket
   - Water pump over the dowel pins and into place on the front cover
   - Water pump attaching bolts.

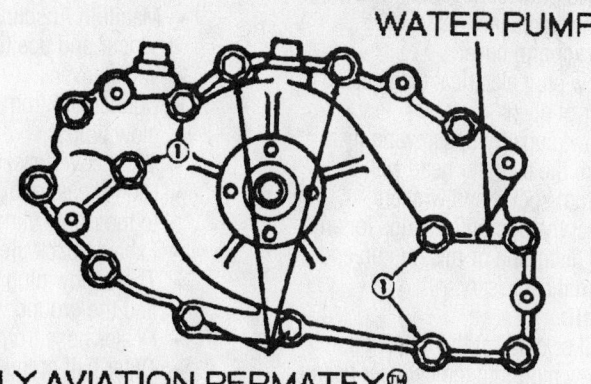

WATER PUMP

APPLY AVIATION PERMATEX® NO. 3 OR EQUIVALENT TO THESE BOLTS

① **THESE BOLTS 2 3/4 IN. LONG ALL OTHERS ARE 1 1/2 IN. LONG**

7924FG56

Apply RTV sealant to the bolts indicated—7.3L diesel engine

Tighten the bolts to 15 ft. lbs. (20 Nm).
- Heater hose to the pump
- ECT sensor connector
- Water pump pulley and start the water pump pulley bolts
- Drive belt
- Tighten the pulley retaining bolts to 12–18 ft. lbs. (16–24 Nm)
- Fan and fan shroud assembly
6. Fill and bleed the cooling system.
7. Connect the battery ground cables.
8. Start the engine and check for leaks.

Glow Plugs

REMOVAL & INSTALLATION

✳✳ CAUTION

The red-striped wiring harness carries 115v direct current. Severe electrical shock may be received. DO NOT pierce.

1. Before servicing the vehicle, refer to the precautions in the beginning of this section.

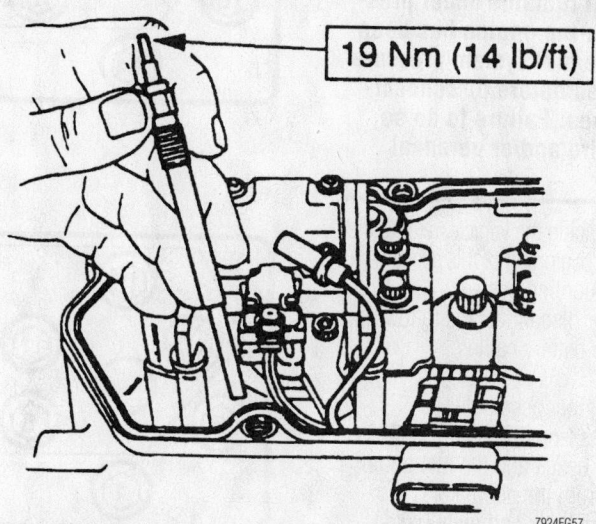

19 Nm (14 lb/ft)

7924FG57

Tighten the glow plugs to 14 ft. lbs. (19 Nm) and attach the connector—7.3L engine

Timing belt service is covered in Section 3 of this manual

2. Remove or disconnect the following:
- Negative battery cable
- Rocker arm cover
- Glow plug electrical leads using a pair of pliers
- Glow plugs by unscrewing them from the cylinder head with a 10mm socket and wrench

3. Inspect the tips of the plugs for any evidence of distortion or missing tip ends; replace them if necessary.

To install:

4. Install or connect the following:
- Glow plug into the cylinder head. Tighten the glow plugs to 14 ft. lbs. (19 Nm).
- Glow plug electrical connector. Be sure that the glow plug wiring is routed to avoid moving components in the engine bay.
- Rocker arm cover

➡When the battery is disengaged and reconnected, some abnormal drive symptoms may occur while the vehicle relearns its adaptive strategy. The vehicle may need to be driven 10 miles (16 km) or more to relearn this strategy.

- Negative battery cable

Cylinder Head

REMOVAL & INSTALLATION

Right Cylinder Head

❊❊ CAUTION

The fuel system remains under pressure, even after the engine has been turned OFF. The fuel system pressure must be relieved before disconnecting any fuel lines. Failure to do so may result in fire and/or personal injury.

1. Before servicing the vehicle, refer to the precautions in the beginning of this section.
2. Drain the cooling system.
3. Remove or disconnect the following:
- Negative battery cables
- Radiator
- Turbocharger assembly
- Fuel lines from the rear of both cylinder heads and the fuel pump
- Wiring from the alternator
- Adjusting bolts and pivot bolts from the alternator and the vacuum pump and remove both units
- Alternator and its bracket

- Engine oil dipstick tube
- Manifold Absolute Pressure (MAP) sensor and position it aside
- Valve cover
- Connectors from the injectors and glow plugs
- Valve cover gasket
- High pressure oil pump supply line to the right cylinder head
- Exhaust back pressure line
- Three glow plug relay bracket nuts and the ground wire
- Heater hose from the cylinder head
- Outer half of the heater distribution box
- Four outboard fuel injector hold-down bolts, retaining screws and four oil deflectors

❊❊ WARNING

Remove the oil drain plugs prior to removing the injectors or oil could enter the combustion chamber which could result in hydrostatic lock and severe engine damage.

- Oil rail drain plugs
- Fuel injectors using Injector Remover No. T94T–9000–AH1, or equivalent. Position the tool's fulcrum beneath the fuel injector hold-down plate and over the edge of the cylinder head. Install the remover screw in the threaded hole of the fuel injector plate. Tighten the screw to lift out the injector from its bore. Place the injector in a suitable protective sleeve such as Rotunda Injector Protective Sleeve, No. 014–00933–2, and set the injector in a suitable holding rack.

4. Use a suitable vacuum tool, such as Rotunda Vacuum Pump, No. 021–00037, or equivalent to remove the oil and fuel left over in the injector bores.

5. Remove or disconnect the following:
- Rocker arms and pushrods, KEEP EVERYTHING IN ORDER
- Four glow plugs
- Right turbo exhaust inlet pipe
- Ground strap from the rear of the cylinder head
- Fuel return line at the front of the cylinder head
- Four inboard fuel injector shoulder bolts
- Cylinder head bolts and attach a Rotunda Cylinder Head Lifting Bracket, 014–00932–2, or equivalent

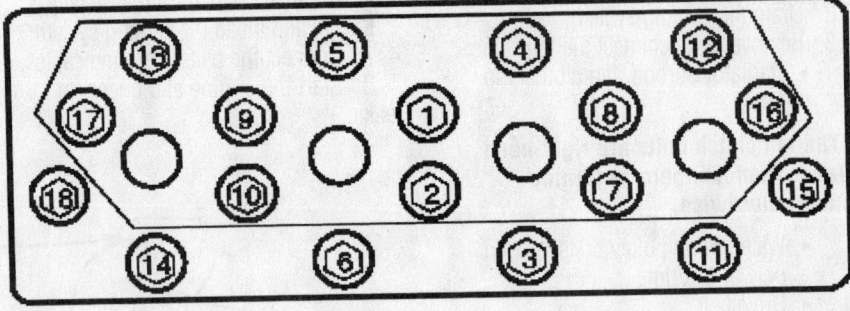

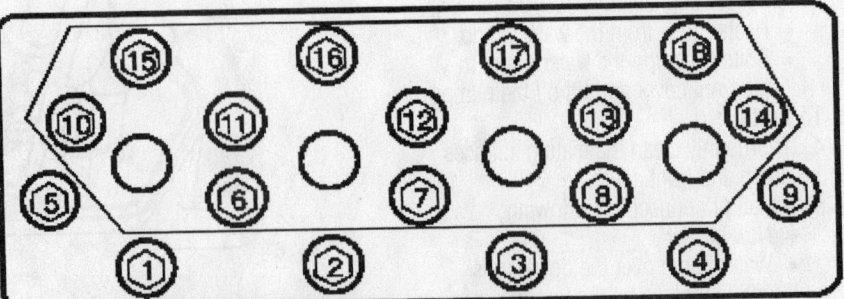

Cylinder head torque sequence—diesel engines

9355FG01

6. Carefully lift the cylinder head out of the engine compartment and remove the head gaskets.

To install:

➡ **To prepare a good seat for the fuel injector O-rings, use a suitable injector sleeve brush to clean any debris from the bore.**

7. Carefully clean the cylinder block and head mating surfaces.

8. Install or connect the following:
- Cylinder head gasket on the engine block and carefully lower the cylinder head in place
- Cylinder head bolt and torque in 3 steps using the sequence shown in the illustration as follows:
  a. Step 1: 65 ft. lbs. (88 Nm).
  b. Step 2: 85 ft. lbs. (115 Nm).
  c. Step 3: 95 ft. lbs. (129 Nm).

➡ **Lubricate the threads and the mating surfaces of the bolt heads and washers with engine oil.**

- Fuel return line to the cylinder head
- Four inboard injector shoulder bolts. Tighten them to 9 ft. lbs. (12 Nm).

9. Install the fuel injectors using special tools as follows:
  a. Lubricate the injectors with clean engine oil. Using new copper washers, carefully push the injectors square into the bore using hand pressure only to seat the O-rings.
  b. Position the open end of Injector Replacer, No. T94T–9000–AH2, or equivalent between the fuel injector body and injector hold-down plate, while positioning the opposite end of the tool over the edge of the cylinder head.
  c. Align the hole in the tool with the threaded hole in the cylinder head and install the bolt from the tool kit. Tighten the bolt to fully seat the injector, then remove the bolt and tool.

10. Install or connect the following:
- Four outboard fuel injector hold-down bolts, the four oil deflectors and retaining screws. Tighten them to 9 ft. lbs. (12 Nm).

11. Dip the pushrod ends in clean engine oil and install the pushrods with the copper colored ends toward the rocker arms, making sure the pushrods are fully seated in the tappet pushrod seats.
- Rocker arms and posts in their original positions. Apply

Lubriplate® grease to the valve stem tips. Turn the engine over by hand until the timing mark is at the 11 o'clock position as viewed from the front.
- Rocker arm posts, bolts and tighten to 27 ft. lbs. (37 Nm)
- Valve covers
- Fuel rail drain plugs, tightening them to 8 ft. lbs. (11 Nm)
- Oil rail drain plugs, tightening them to 53 inch lbs. (6 Nm)
- Heater distribution box
- Heater hose to the cylinder head
- Glow plug relay bracket and ground wire
- Exhaust back pressure line
- Oil supply line to the cylinder head, tightening it to 19 ft. lbs. (26 Nm)
- Dipstick tube
- MAP sensor and screws
- Valve cover gasket
- Wiring to the fuel injectors and glow plugs
- Valve cover, tightening the bolts to 8 ft. lbs. (11 Nm)
- Injector wiring harness to the valve cover gasket
- Alternator, tightening the bracket bolts to 40–55 ft. lbs. (54–75 Nm)
- Alternator wiring
- Drive belt

12. The remainder of the installation is the reverse of the removal. Tighten the fuel pump-to-fuel line banjo bolt to 40 ft. lbs. (54 Nm).

13. Connect both negative battery cables.

14. Refill and bleed the cooling system.

15. Run the engine and check for fuel, coolant and exhaust leaks.

Left Cylinder head

✳✳ CAUTION

The fuel system remains under pressure, even after the engine has been turned OFF. The fuel system pressure must be relieved before disconnecting any fuel lines. Failure to do so may result in fire and/or personal injury.

1. Before servicing the vehicle, refer to the precautions in the beginning of this section.

2. Drain the cooling system.

3. Using approved equipment, recover the refrigerant from the A/C system.

4. Remove or disconnect the following:
- Negative battery cables
- Radiator
- Turbocharger assembly
- Two crankcase breather screws and the breather
- Wiring from the air conditioning compressor
- Four left accessory bracket bolts
- Vacuum hose at the brake vacuum pump
- A/C lines from the compressor
- Power steering lines from the pump
- Left accessory bracket and accessories as an assembly
- Valve cover
- Fuel line assembly between the cylinder heads and fuel pump
- Fuel line nut from the intake manifold stud
- Fuel return line from the cylinder head
- High pressure oil pump supply line from the cylinder head

5. Raise the vehicle and support it safely on jackstands.
- Left turbo exhaust pipe from the manifold

6. Lower the vehicle.
- Oil rail drain plugs
- Four outboard fuel injector hold-down bolts, retaining screws and four oil deflectors

✳✳ WARNING

Remove the oil drain plugs prior to removing the injectors or oil could enter the combustion chamber which could result in hydrostatic lock and severe engine damage.

- Oil rail drain plugs

7. Remove the fuel injectors using Injector Remover No. T94T–9000–AH1, or equivalent. Position the tool's fulcrum beneath the fuel injector hold-down plate and over the edge of the cylinder head. Install the remover screw in the threaded hole of the fuel injector plate. Tighten the screw to lift out the injector from its bore. Place the injector in a suitable protective sleeve such as Rotunda Injector Protective Sleeve, No. 014–00933–2, and set the injector in a suitable holding rack.

8. Use a suitable vacuum tool, such as Rotunda Vacuum Pump, No. 021–00037, or equivalent to remove the oil and fuel left over in the injector bores.

9. Remove or disconnect the following:

Heater Core replacement is covered in Section 2 of this manual

- Rocker arms and pushrods, KEEP EVERYTHING IN ORDER
- Four glow plugs
- Left turbo exhaust inlet pipe
- Main engine harness connectors in the left fender well and position the harness aside
- Four inboard fuel injector shoulder bolts
- Cylinder head bolts and attach a Rotunda Cylinder Head Lifting Bracket, 014–00932–2, or equivalent

10. Carefully lift the cylinder head out of the engine compartment and remove the head gaskets.

To install:

➡ **To prepare a good seat for the fuel injector O-rings, use a suitable injector sleeve brush to clean any debris from the bore.**

11. Carefully clean the cylinder block and head mating surfaces.

12. Install or connect the following:
- Cylinder head gasket on the engine block and carefully lower the cylinder head in place
- Cylinder head bolt and torque in 3 steps using the sequence shown in the illustration as follows:
  a. Step 1: 65 ft. lbs. (88 Nm).
  b. Step 2: 85 ft. lbs. (115 Nm).
  c. Step 3: 95 ft. lbs. (129 Nm).

➡ **Lubricate the threads and the mating surfaces of the bolt heads and washers with engine oil.**

13. Apply anti-seize paste and install the glow plugs, tightening them to 14 ft. lbs. (19 Nm).

14. Install or connect the following:
- Four outboard fuel injector hold-down bolts, the four oil deflectors and retaining screws. Tighten them to 9 ft. lbs. (12 Nm).

15. Dip the pushrod ends in clean engine oil and install the pushrods with the copper colored ends toward the rocker arms, making sure the pushrods are fully seated in the tappet pushrod seats.
- Rocker arms and posts in their original positions. Apply Lubriplate® grease to the valve stem tips. Turn the engine over by hand until the timing mark is at the 11 o'clock position as viewed from the front.
- Rocker arm posts, bolts and tighten to 27 ft. lbs. (37 Nm)
- Valve covers
- Four inboard injector shoulder

bolts. Tighten them to 9 ft. lbs. (12 Nm).

16. Install the fuel injectors using special tools as follows:

a. Lubricate the injectors with clean engine oil. Using new copper washers, carefully push the injectors square into the bore using hand pressure only to seat the O-rings.

b. Position the open end of Injector Replacer, No. T94T–9000–AH2, or equivalent between the fuel injector body and injector hold-down plate, while positioning the opposite end of the tool over the edge of the cylinder head.

c. Align the hole in the tool with the threaded hole in the cylinder head and install the bolt from the tool kit. Tighten the bolt to fully seat the injector, then remove the bolt and tool.

17. Install or connect the following:
- Fuel rail drain plugs, tightening them to 8 ft. lbs. (11 Nm)
- Oil rail drain plugs, tightening them to 53 inch lbs. (6 Nm)
- Heater distribution box
- Valve cover gasket
- Wiring to the fuel injectors and glow plugs
- Valve cover, tightening the bolts to 8 ft. lbs. (11 Nm)
- Engine wiring harness

18. Raise and safely support the vehicle on jackstands.

19. Loosely install the turbo exhaust pipe to the manifold.

20. Lower the vehicle.

21. The remainder of the installation is the reverse of the removal. Tighten the fuel pump-to-fuel line banjo bolt to 40 ft. lbs. (54 Nm).

22. Connect both negative battery cables.

23. Refill and bleed the cooling system.

24. Run the engine and check for fuel, coolant and exhaust leaks.

Rocker Arms

REMOVAL & INSTALLATION

1. Before servicing the vehicle, refer to the precautions in the beginning of this section.

2. Disconnect the ground cables from both batteries.

3. Remove the valve cover attaching screws and remove both valve covers.

4. Remove the valve rocker arm post mounting bolts. Remove the rocker arms

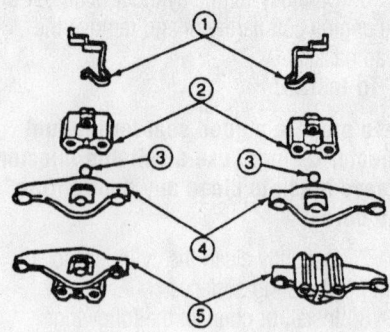

Item	Description
1	Snap Retaining Clip
2	Rocker Arm Pedestal
3	Rocker Arm Ball
4	Rocker Arm
5	Rocker Arm Assembly

7924FG58

Exploded view of the rocker arm assembly—diesel engines

and posts in order and mark them with tape so they can be installed in their original positions.

5. If the cylinder heads are to be removed, then the pushrods can now be removed. Make a holder for the pushrods out of a piece of wood or cardboard and remove the pushrods in order. It is very important that the pushrods be reinstalled in their original order. The pushrods can remain in position if no further disassembly is required.

To install:

6. If the pushrods were removed, install them in their original locations. Make sure they are fully seated in the tappet seats.

➡ **The copper colored end of the pushrod goes toward the rocker arm.**

7. Apply a polyethylene grease to the valve stem tips. Install the rocker arms and posts in their original positions.

8. Turn the engine over by hand until the valve timing mark is at the 11 o'clock position, as viewed from the front of the engine. Install all of the rocker arm post attaching bolts and tighten to 20 ft. lbs. (27 Nm).

9. Install new valve cover gaskets and install the valve cover.

10. Install the battery cables, start the engine and check for leaks.

Turbocharger

REMOVAL & INSTALLATION

1. Before servicing the vehicle, refer to the precautions in the beginning of this section.
2. Remove or disconnect the following:
   - Negative battery cable
   - 2 air intake tube assembly bolts
   - Clamps at the turbocharger
   - Crankcase breather assembly
   - Engine air cleaner, air intake tube and hoses
   - Exhaust outlet clamp from the turbocharger
   - Engine charge exhaust pipe bolt from the transmission, if so equipped
   - Bolts and nuts from the catalytic converter-to-engine charge exhaust pipe, if so equipped
   - 2 bolts retaining the turbocharger exhaust inlet pipe to the left exhaust manifold, loosen only
   - Bolts retaining the left turbocharger exhaust inlet pipe to the turbocharger exhaust inlet adapter, on automatic transmissions
   - 2 bolts retaining the turbocharger exhaust inlet pipe to the right exhaust manifold, loosen only
   - Lower bolt retaining the right turbocharger exhaust inlet pipe to the turbocharger exhaust inlet adapter
   - Upper bolts retaining the right and left turbocharger exhaust inlet pipes to the turbocharger exhaust inlet adapter, on automatic transmissions
   - Right engine lift hook and bolt
   - Air inlet hose clamp at the turbocharger, hose and lay aside
   - Compressor manifold
   - 4 bolts retaining the turbocharger pedestal assembly to the cylinder block
   - Turbocharger assembly and all electrical connectors from it

➡ **If the turbocharger is not being removed for service, install the Fuel/Oil turbo Protector Cap Set T94T-9395-AH.**

   - Oil gallery O-rings

To install:

3. Install or connect the following:
   - Oil gallery O-rings
   - Turbocharger electrical connectors and the turbocharger assembly

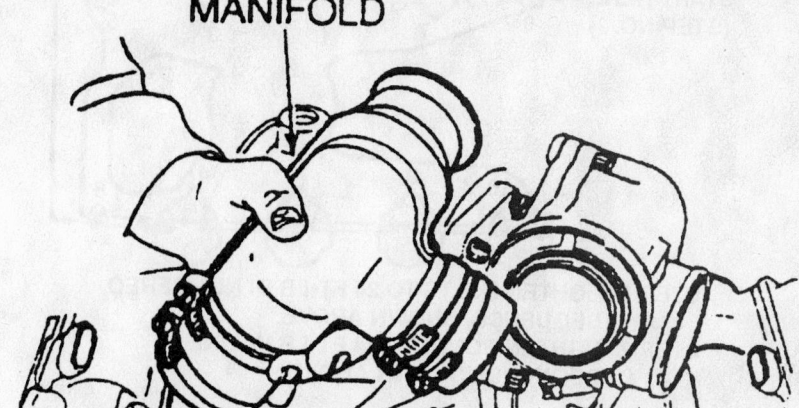

COMPRESSOR MANIFOLD

7924FG59

After loosening the clamps, the compressor manifold can be removed—diesel engines

   - 4 bolts retaining the turbocharger pedestal assembly to the engine block. Tighten the bolts to 18 ft. lbs. (25 Nm).
   - 4 bolts retaining the right and left turbocharger exhaust inlet pipes to the turbocharger exhaust inlet adapter.
   - Compressor manifold, intake manifold hoses and clamps. Be sure the compressor outlet seal is in position.
   - Right engine lift hook and bolt
   - 2 right and left lower bolts (retaining the turbocharger exhaust inlet pipes to the turbocharger exhaust inlet adapter) to 36 ft. lbs. (49 Nm)
   - 4 right and left bolts and nuts (retaining the turbocharger exhaust inlet pipes to the exhaust manifolds) to 36 ft. lbs. (49 Nm)
   - Catalytic converter to the engine charge exhaust pipe bolts and nuts
   - 2 right and left upper bolts (retaining the turbocharger exhaust inlet pipes to the turbocharger exhaust inlet adapter) to 36 ft. lbs. (49 Nm)
   - Exhaust outlet clamp to the turbocharger
   - Air intake tube and hose assembly
   - Negative battery cable

Intake Manifold

REMOVAL & INSTALLATION

✳✳ CAUTION

The fuel system remains under pressure, even after the engine has been turned OFF. The fuel system pressure must be relieved before disconnecting any fuel lines. Failure to do so may result in fire and/or personal injury.

1. Before servicing the vehicle, refer to the precautions in the beginning of this section.
2. Remove or disconnect the following:
   - Both battery ground cables
   - Air cleaner and install clean rags into the air intake of the intake manifold. It is important that no dirt or foreign objects get into the intake.
   - Fuel system pressure
   - Injection pump
   - Fuel return hose from No. 7 and No. 8 rear nozzles
   - Return hose to the fuel tank
   - Engine wiring harness from the engine

Brake service is covered in Section 4 of this manual

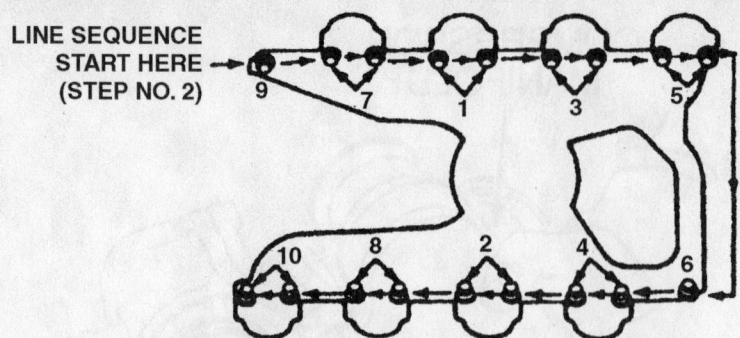

LINE SEQUENCE
START HERE →
(STEP NO. 2)

STEP 1. TIGHTEN BOLTS TO 24 FT•LB IN NUMBERED
SEQUENCE SHOWN ABOVE.
STEP 2. TIGHTEN BOLTS TO 24 FT•LB IN LINE
SEQUENCE SHOWN ABOVE.

7924FG25

Intake manifold bolt torque sequence—diesel engines

➡**The engine harness ground cables must be removed from the back of the left cylinder head.**

- Bolts attaching the intake manifold to the cylinder heads and the manifold
- Crankcase Depression Regulator (CDR) valve tube grommet from the valley pan
- Bolts attaching the valley pan strap to the front of the engine block and the strap
- Valley pan drain plug and the valley pan

To install:

3. Apply a ⅛ inch (3mm) bead of RTV sealer to each end of the cylinder block.

➡**The RTV sealer should be applied immediately prior to the valley pan installation.**

4. Install or connect the following:
- Valley pan drain plug, CDR valve tube and new grommet into the valley pan
- New O-ring and new back-up ring on the CDR valve
- Valley pan strap on the front of the valley pan
- Intake manifold and tighten the bolts to 24 ft. lbs. (33 Nm) using the tightening sequence
- Engine wiring harness and the engine ground wire located to the rear of the left cylinder head
- Injection pump
- No. 7 and No. 8 fuel return hoses and the fuel tank return hose
- Air cleaner
- Battery ground cables to both batteries

➡**If necessary, purge the nozzle high pressure lines of air by loosening the connector ½ to 1 turn and cranking the engine until solid stream of fuel, devoid of any bubbles, flows from the connection.**

5. Run the engine and check for oil and fuel leaks.

※ CAUTION

Keep eyes and hands away from the nozzle spray. Fuel spraying from the nozzle under high pressure can penetrate the skin.

6. Check and adjust the injection pump timing.

Exhaust Manifold

REMOVAL & INSTALLATION

1. Before servicing the vehicle, refer to the precautions in the beginning of this section.
2. Disconnect the ground cables from both batteries.
3. Raise the vehicle and safely support it.
4. Disconnect the muffler inlet pipe from the exhaust manifolds.
5. If removing the right manifold, lower the vehicle. When removing the left manifold, raise the vehicle and remove the manifold from underneath. Bend the tabs on the manifold attaching bolts, then remove the bolts and manifold.

To install:

6. Before installing, clean all mounting surfaces on the cylinder heads and the manifold. Apply an anti-seize compound on the manifold both threads and install the left manifold, using a new gasket and new locking tabs.
7. Tighten the bolts to 45 ft. lbs. (61 Nm) and bend the tabs over the flats on the bolt heads to prevent the bolts from loosening.
8. Raise the vehicle to install the right manifold. Install the right manifold.
9. Connect the inlet pipes to the manifold and tighten. Lower the vehicle, connect the batteries and run the engine to check for exhaust leaks.

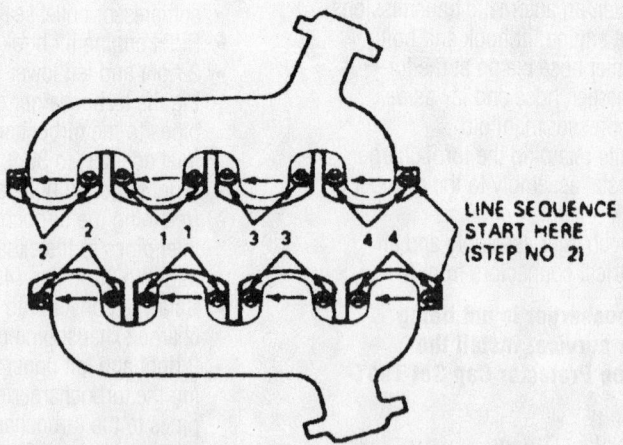

LINE SEQUENCE
START HERE
(STEP NO 2)

STEP 1. TIGHTEN BOLTS TO 35 FT.LB. IN
NUMBERED SEQUENCE SHOWN ABOVE
STEP 2. TIGHTEN BOLTS TO 35 FT.LB. IN
LINE SEQUENCE SHOWN ABOVE

7924FG26

Tighten the manifold bolts in the proper sequence—diesel engines

Camshaft and Valve Lifters

REMOVAL & INSTALLATION

➡**Ford recommends removing the diesel engine from the vehicle for camshaft removal.**

1. Before servicing the vehicle, refer to the precautions in the beginning of this section.
2. Remove or disconnect the following:
   - Intake manifold and valley pan, if equipped
   - Rocker covers
   - Either the rocker arm shafts or loosen the rockers on their pivots and remove the pushrods. The pushrods must be reinstalled in their original positions.
   - Valve lifters in sequence with a magnet. They must be replaced in their original positions.
   - Timing gear cover, timing gear and sprockets

➡**A camshaft removal tool, Ford part no. T65L-6250-a and adapter 14-0314 is needed to remove the diesel camshaft.**

   - Camshafts
   To install:
3. liberally coat the camshaft with oil before installing it. Slide the camshaft into the engine very carefully so as not to scratch the bearing bores with the camshaft lobes. Install the camshaft thrust plate and tighten the attaching screws to 10–12 ft. lbs. (13–16 Nm). Measure the camshaft end-play. If the end-play is more than 0.009 inch (0.228mm), replace the thrust plate. Assemble the remaining components in the reverse order of removal.
4. Install or connect the following:
   - Timing gear and front cover
   - Valve lifters. They must be replaced in their original positions.
   - Pushrods, the rocker arms and the rocker arm covers
   - Intake manifold and valley pan, if equipped

Valve Lash

ADJUSTMENT

Valve lash on the 7.3L diesel engine is not adjustable.

Starter Motor

REMOVAL & INSTALLATION

1. Before servicing the vehicle, refer to the precautions in the beginning of this section.
2. Remove or disconnect the following:
   - Negative battery cable
3. Raise the front of the truck and install jackstands beneath the frame. Firmly apply the parking brake and place blocks in back of the rear wheels.
4. Remove or disconnect the following:
   - Wiring from the starter motor terminals
   - Starter motor retaining bolts, loosen
   - Starter retaining bolts while supporting the starter motor
   - Starter from the vehicle
   To install:
5. The installation is the reverse of removal. Tighten the starter retaining bolts to 15–20 ft. lbs. (20–27 Nm).

Oil Pan

REMOVAL & INSTALLATION

1. Before servicing the vehicle, refer to the precautions in the beginning of this section.
2. Remove the engine.
3. Remove the oil pan bolts.
4. Lower the oil pan.

➡**The oil pan is sealed to the crankcase with RTV silicone sealant in place of a gasket. It may be necessary to separate the pan from the crankcase with a utility knife. Also, the crankshaft may have to be turned to allow the pan to clear the crankshaft throws.**

5. Clean the pan and crankcase mating surfaces thoroughly.
   To install:
6. Apply a ⅛ in. (3mm) bead of RTV silicone sealant to the pan mating surfaces, and a ¼ in. (6mm) bead on the front and rear covers and in the corners; you have 15 minutes within which to install the pan!
7. Install the locating dowels into position.
8. Position the pan on the engine and install the pan bolts loosely.
9. Remove the dowels.
10. Tighten the pan bolts to:
    - ¼ in.-20 bolts: 84 inch lbs. (10 Nm)
    - ⁵⁄₁₆ in.-18 bolts: 14 ft. lbs. (19 Nm)
    - ⅜ in.-16 bolts: 24 ft. lbs. (33 Nm)
11. Install the engine.

Oil Pump

REMOVAL & INSTALLATION

1. Before servicing the vehicle, refer to the precautions in the beginning of this section.
2. Remove or disconnect the following:

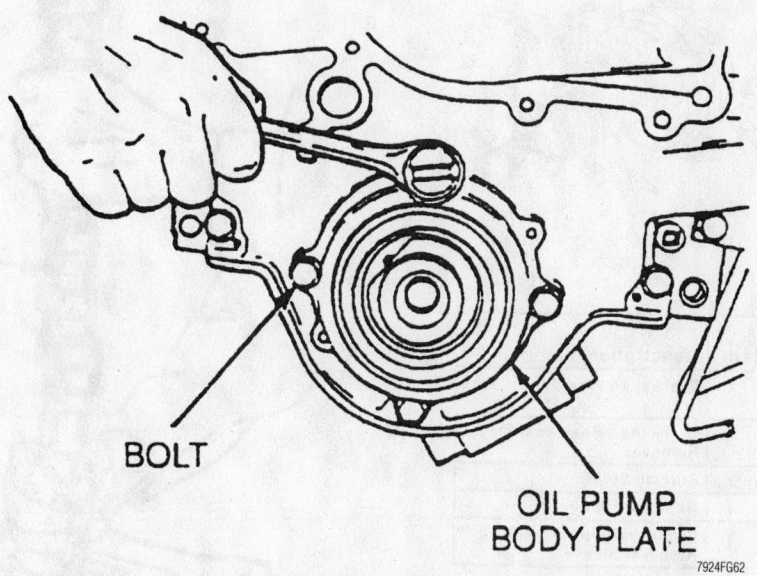

BOLT

OIL PUMP
BODY PLATE

7924FG62

The oil pump is mounted on the cylinder block with 4 bolts—diesel engine

For complete Engine Mechanical specifications, see Section 1 of this manual

- Negative battery cables
- Oil pan
- Oil pick-up tube from the pump
- Bolts and the oil pump

To install:

3. Assemble the pick-up tube and pump. Use a new gasket.

4. Install or connect the following:
- Oil pump and tighten the bolts to 14 ft. lbs. (19 Nm)
- Oil pick up tube
- Oil pan
- Negative battery cables

Rear Main Seal

REMOVAL & INSTALLATION

1. Before servicing the vehicle, refer to the precautions in the beginning of this section.

2. Remove or disconnect the following:
- Transmission
- Flywheel
- Crankshaft rear oil seal bolts and the seal

3. Clean the seal mating surfaces.

4. If installing the old seal, inspect it for damage.

5. Using crankshaft wear ring removal tool T94T-6701-AH1, forcing screw T84T-

7025-B, remover tube T77J-7025-B and wear ring remover sleeve T94T-6701-AH2 (refer to the illustration), remove the wear ring.

To install:

6. Apply RTV silicone sealant to the seal retaining ring and the seal retaining bolts.

7. Using seal replacers T94T-6701-AH3 and T94T-AH4, driver sleeve T79T-6316-A4 (part of T79T-6316-A) and guide pins T94P-7000-P install the wear ring and oil seal.

8. Install or connect the following:
- Seal retaining bolts
- Flywheel
- Transmission

Timing Gears, Front Cover and Seal

REMOVAL & INSTALLATION

➡ The crankshaft gear sprocket is not serviced separately from the crankshaft. Do not try to remove the sprocket or you will damage the crankshaft.

1. Before servicing the vehicle, refer to the precautions in the beginning of this section.

2. Remove or disconnect the following:
- Negative battery cables
- Camshaft

- Sprocket from the camshaft using a press
- Thrust plate and sprocket key

3. Inspect the camshaft and related parts for wear and damage.

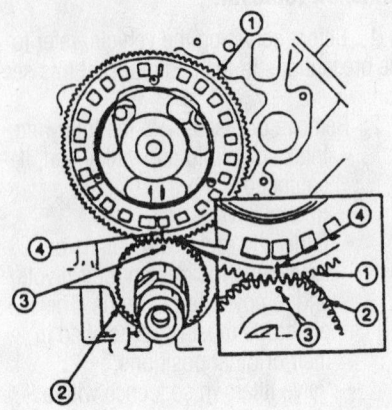

Item	Description
1	Camshaft Sprocket
2	Crankshaft Sprocket
3	Crankshaft Sprocket Timing Mark
4	Camshaft Sprocket Timing Mark

7924FG64

Be sure the timing marks are aligned as illustrated—diesel engines

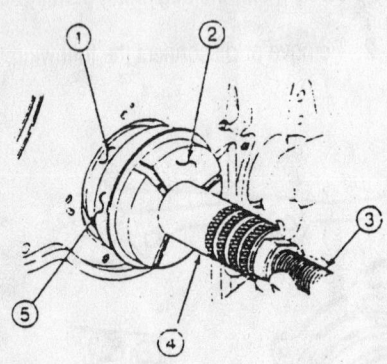

Item	Description
1	Crankshaft Wear Ring
2	Crankshaft Rear Wear Ring Remover
3	Forcing Screw
4	Remover Tube
5	Crankshaft Rear Wear Ring Remover Sleeve

7924FG61

Assemble the seal and wear ring removal tools, then remove the wear ring—diesel engine

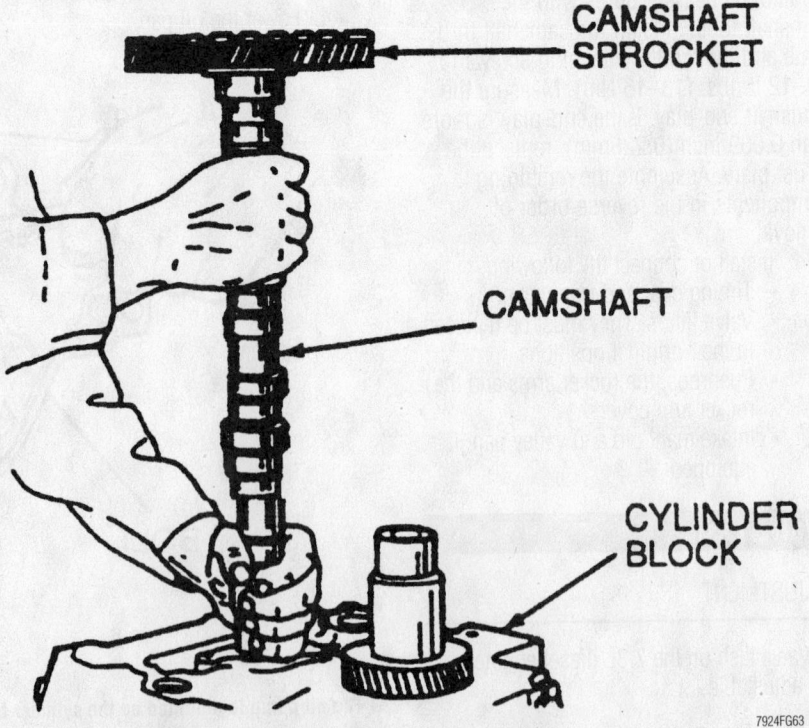

CAMSHAFT SPROCKET

CAMSHAFT

CYLINDER BLOCK

7924FG63

The camshaft gear is removed with the camshaft, then pressed off—diesel engines

To install:

4. Clean the nose of the camshaft.

5. Install or connect the following:
   - Thrust plate
   - Key in the keyway on the camshaft

6. Heat the sprocket in an oven to 500°F (260°C).

7. Remove the sprocket from the oven, align the sprocket keyway with the camshaft key and install the sprocket on the camshaft until it is fully seated. Allow the camshaft assembly to cool before installation
   - Camshaft in the engine and align the timing marks on the gears
   - Negative battery cables

Piston and Ring Positioning

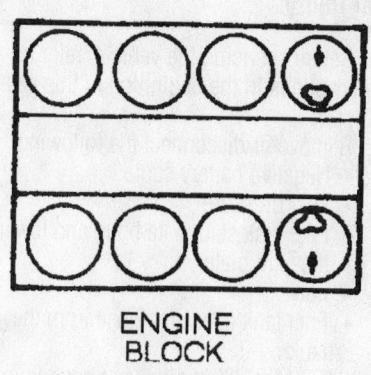

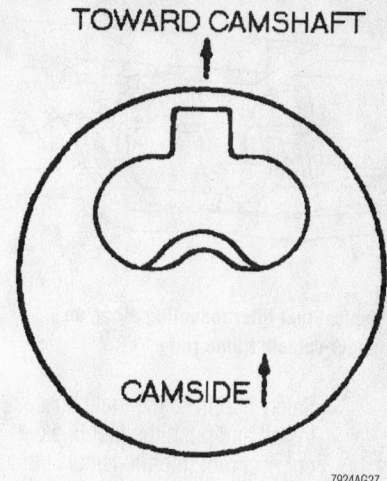

7.3L Diesel engines—piston-to-engine orientation

GASOLINE FUEL SYSTEM

Fuel System Service Precautions

Safety is the most important factor when performing not only fuel system maintenance, but any type of maintenance. Failure to conduct maintenance and repairs in a safe manner may result in serious personal injury or death. Maintenance and testing of the vehicle's fuel system components can be accomplished safely and effectively by adhering to the following rules and guidelines.

- To avoid the possibility of fire and personal injury, always disconnect the negative battery cable unless the repair or test procedure requires that battery voltage be applied.

- Always relieve the fuel system pressure prior to disconnecting any fuel system component (injector, fuel rail, pressure regulator, etc.), fitting or fuel line connection. Exercise extreme caution whenever relieving fuel system pressure, to avoid exposing skin, face and eyes to fuel spray. Please be advised that fuel under pressure may penetrate the skin or any part of the body that it contacts.

- Always place a shop towel or cloth around the fitting or connection prior to loosening to absorb any excess fuel due to spillage. Ensure that all fuel spillage (should it occur) is quickly removed from engine surfaces. Ensure that all fuel soaked cloths or towels are deposited into a suitable waste container.

- Always keep a dry chemical (Class B) fire extinguisher near the work area.

- Do not allow fuel spray or fuel vapors to come into contact with a spark or open flame.

- Always use a back-up wrench when loosening and tightening fuel line connection fittings. This will prevent unnecessary stress and torsion to fuel line piping. Always follow the proper torque specifications.

- Always replace worn fuel fitting O-rings with new. Do not substitute fuel hose or equivalent where fuel pipe is installed.

Fuel System Pressure

RELIEVING

➡**A fuel pressure gauge is needed to correctly perform this procedure.**

✳✳ CAUTION

Fuel injection systems remain under pressure, even after the engine has been turned OFF. The fuel system pressure must be relieved before disconnecting any fuel lines. Failure to do so may result in fire and/or personal injury.

1. Before servicing the vehicle, refer to the precautions in the beginning of this section.

2. Disconnect the negative battery cable and remove the fuel filler cap.

3. Remove the cap from the pressure relief valve on the fuel supply manifold. Install a fuel pressure gauge to the pressure relief valve.

4. Direct the gauge drain hose into a suitable container and depress the pressure relief button.

5. Remove the gauge and replace the cap on the pressure relief valve.

➡**As an alternate method, disconnect the inertia switch and crank the engine for 15–20 seconds until the pressure is relieved.**

Fuel Filter

REMOVAL & INSTALLATION

➡**A fuel line disconnect tool is needed for this procedure.**

✳✳ CAUTION

Fuel injection systems remain under pressure, even after the engine has been turned OFF. The fuel system pressure must be relieved before disconnecting any fuel lines. Failure to do so may result in fire and/or personal injury.

1. Before servicing the vehicle, refer to the precautions in the beginning of this section.

2. Remove or disconnect the following:
   - Negative battery cable

3. Relieve the fuel system pressure.
   - Fuel lines from the fuel filter. Have a drain pan handy to catch any residual fuel once the lines are separated.

4. On newer models, disconnect the fuel lines from the filter as follows:

For Accessory Drive Belt illustrations, see Section 1 of this manual

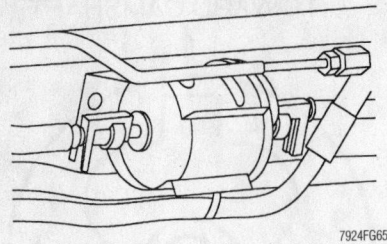

Typical fuel filter mounting along an under-vehicle frame rail

- Safety clip from the male hose
- Install and push the fuel line disconnect tool into the female fitting.
5. Remove or disconnect the following:
   - Male and female fittings from the filter

➡ Inspect the fuel lines for any damage after the fuel is finished draining.

   - Fuel filter from the bracket and the retainer, if equipped. Note the direction of the flow arrow so the replacement filter can be installed correctly.

To install:

6. Install or connect the following:
   - Fuel filter into the mounting bracket with the flow arrow pointing in the correct direction
   - Fuel lines to the fuel filter. Align and push the male tube into the female fitting until a click is heard. Pull on the fitting to ensure that it is fully engaged, then install the safety clip.
7. Lower the vehicle to the ground.

➡ When the battery has been disconnected and reconnected, some abnormal drive symptoms may occur while the Powertrain Control Module (PCM) relearns its adaptive strategy. The vehicle may need to be driven 10 miles (16 km) or more to relearn the strategy.

8. Connect the negative battery cable.

Fuel Pump

REMOVAL & INSTALLATION

Except Excursion

❊❊ **CAUTION**

Fuel injection systems remain under pressure, even after the engine has been turned OFF. The fuel system pressure must be relieved before dis-

connecting any fuel lines. Failure to do so may result in fire and/or personal injury.

1. Before servicing the vehicle, refer to the precautions in the beginning of this section.
2. Remove or disconnect the following:
   - Negative battery cable
   - Fuel pressure
   - Fuel tank skid plate bolts and lower the skid plate
   - Fuel
   - Fuel tank filler pipe hose from the tank
   - Fuel tank filler pipe vent hose from the tank
   - Fuel lines from the fuel pump
   - Front fuel tank connections
   - Rear Evaporative Emissions (EVAP) hose clamp and the hose
   - Electrical connector from the fuel pump
3. Support the fuel tank with a jack.
   - Fuel tank support strap bolts and the fuel tank straps
   - Fuel tank
   - Fuel pump lock ring
   - Fuel pump

To install:

4. Install or connect the following:
   - Fuel pump
   - Fuel pump lock ring and tighten to 66 ft. lbs. lbs. (89 Nm)
   - Fuel tank
5. Tighten the fuel tank strap bolts to 41 ft. lbs. (55 Nm).
6. Tighten the skid plate bolts to 41 ft. lbs. (55 Nm).
7. Connect the negative battery.

Excursion

❊❊ **CAUTION**

Fuel injection systems remain under pressure, even after the engine has been turned OFF. The fuel system pressure must be relieved before disconnecting any fuel lines. Failure to do so may result in fire and/or personal injury.

1. Before servicing the vehicle, refer to the precautions in the beginning of this section.
2. Remove or disconnect the following:
   - Negative battery cable
   - Fuel pressure
   - Fuel tank
   - Fuel pump bolts
   - Fuel pump assembly

3. Squeeze the locking tabs while pushing down.
   - Fuel pump mounting gasket.

To install:

4. Install or connect the following:
   - Fuel pump mounting gasket.
   - Fuel pump assembly
   - Fuel pump bolts and tighten to 80–107 inch lbs. (9–12 Nm)
   - Fuel tank
   - Negative battery cable

Fuel Injector

REMOVAL & INSTALLATION

1999 Models

❊❊ **CAUTION**

Fuel injection systems remain under pressure, even after the engine has been turned OFF. The fuel system pressure must be relieved before disconnecting any fuel lines. Failure to do so may result in fire and/or personal injury.

1. Before servicing the vehicle, refer to the precautions in the beginning of this section.
2. Disconnect the negative battery cable.
3. Partially drain the engine cooling system.
4. Remove the engine air cleaner outlet tube from the throttle body.
5. Properly relieve the fuel system pressure.
6. Remove or disconnect the following:
   - 3 power steering reservoir retaining bolts and the reservoir bracket
   - Accelerator splash shield
   - Vacuum line at the fuel pressure regulator
   - Fuel supply and return lines at the fuel injection supply manifold
   - Electrical harness connectors from each fuel injectors
   - Heater water inlet tube hose and position aside
   - Brake booster bracket nut
   - Brake booster tube
   - Positive Crankcase Ventilation (PCV) hose
   - Exhaust Gas Recirculation (EGR) Differential Pressure Feedback (DPFE) transducer hoses from the EGR tube
   - Upper and lower EGR tube fittings and remove the EGR valve to exhaust manifold tube

- Vapor management valve hose
- 4 fuel injection supply manifold retaining bolts
- Fuel injection supply manifold from the lower intake manifold

7. Remove the fuel injectors from the fuel injection supply manifold as follows:

a. Grasp the injector body and pull while gently rocking the injector from side-to-side to remove the injector from the fuel injection supply manifold. Repeat the removal procedure for fuel injectors left in the intake manifold.

8. Inspect the fuel injector end cap, body and washer for signs of dirt and deterioration.

9. Discard the fuel injector O-rings.

To install:

10. Lubricate new O-rings with clean engine oil and install 2 O-rings on each injector. Do not use silicone grease as it will clog the injectors.

11. Install the fuel injectors into the fuel injection supply manifold using a light, twisting, pushing motion.

12. Install or connect the following:

- Fuel injection supply manifold, pushing it down to ensure all fuel injector O-rings are fully seated in the fuel rail cups and the intake manifold
- 4 fuel injection supply manifold retaining bolts while holding the manifold down. Tighten the retaining bolts to 71–106 inch lbs. (8–12 Nm).
- Fuel supply and return lines to the fuel injection supply manifold
- Vacuum line to the fuel pressure regulator

13. With the fuel injector wiring disconnected, temporarily connect the negative battery cable and turn the ignition switch to the **RUN** position to allow the fuel pump to pressurize the fuel system.

14. Check for fuel leaks.

15. Disconnect the negative battery cable.

16. Install or connect the following:

- Electrical harness connectors to each fuel injector
- Vapor management valve hose
- EGR valve to exhaust manifold tube. Tighten the fittings to 30 ft. lbs. (41 Nm).
- EGR DPFE hoses to the EGR tube
- PCV hose
- Brake booster tube and position 2 clamps. Tighten the retaining nut to 108 inch lbs. (12 Nm).
- Heater water inlet tube hose and properly position clamp
- Accelerator splash shield and tighten 3 retaining bolts to 3 ft. lbs. (4 Nm)
- 3 power steering reservoir bracket bolts
- Engine air cleaner outlet tube to the throttle body
- Negative battery cable

17. Start the engine and allow to idle for several minutes while checking for leaks.

18. Turn the engine **OFF** and recheck for leaks.

19. Road test the vehicle and check for proper engine operation.

2001–03 Models

> **✳✳ CAUTION**
>
> **Fuel injection systems remain under pressure, even after the engine has been turned OFF. The fuel system pressure must be relieved before disconnecting any fuel lines. Failure to do so may result in fire and/or personal injury.**

1. Before servicing the vehicle, refer to the precautions in the beginning of this section.

2. Remove or disconnect the following:

- Negative battery cable
- Engine cover, if equipped
- Air cleaner assembly

3. Relieve the fuel pressure.

- Upper intake manifold
- Fuel pressure regulator valve vacuum hose
- Fuel lines
- Fuel injector electrical connectors
- Fuel injector electrical harness from the fuel injector manifold
- Fuel injection supply manifold
- Fuel rail bolts and the rail

> **✳✳ CAUTION**
>
> **The fuel injectors are deposit-resistant. Do not clean. Use O-rings made of special fuel resistant material. Use of ordinary O-rings can cause the fuel system to leak.**

- Fuel injector retaining clips and the fuel injectors. Inspect the fuel injector O-rings and if necessary.

To install:

➡ **Make sure the injector clips must engage the upper groove on the injectors or fuel leaks may occur.**

4. Lubricate the new O-rings with clean engine oil to aid installation.

5. Install or connect the following:

- Fuel injectors and the rail. Tighten the rail bolts to 89 inch lbs. (10 Nm).
- Fuel injection supply manifold
- Fuel injector electrical harness from the fuel injector manifold
- Fuel injector electrical connectors
- Fuel lines
- Fuel pressure regulator valve vacuum hose
- Upper intake manifold
- Air cleaner assembly
- Engine cover, if equipped
- Negative battery cable

DIESEL FUEL SYSTEM

Fuel System Service Precautions

Safety is the most important factor when performing not only fuel system maintenance but any type of maintenance. Failure to conduct maintenance and repairs in a safe manner may result in serious personal injury or death. Maintenance and testing of the vehicle's fuel system components can be accomplished safely and effectively by adhering to the following rules and guidelines.

- To avoid the possibility of fire and personal injury, always disconnect the negative battery cable unless the repair or test procedure requires that battery voltage be applied.
- Always relieve the fuel system pressure prior to disconnecting any fuel system component (injector, fuel rail, pressure regulator, etc.), fitting or fuel line connection. Exercise extreme caution whenever relieving fuel system pressure, to avoid exposing skin, face and eyes to fuel spray. Please be advised that fuel under pressure may penetrate the skin or any part of the body that it contacts.
- Always place a shop towel or cloth around the fitting or connection prior to loosening to absorb any excess fuel due to

spillage. Ensure that all fuel spillage (should it occur) is quickly removed from engine surfaces. Ensure that all fuel soaked cloths or towels are deposited into a suitable waste container.

• Always keep a dry chemical (Class B) fire extinguisher near the work area.

• Do not allow fuel spray or fuel vapors to come into contact with a spark or open flame.

• Always use a back-up wrench when loosening and tightening fuel line connection fittings. This will prevent unnecessary stress and torsion to fuel line piping. Always follow the proper torque specifications.

• Always replace worn fuel fitting O-rings with new. Do not substitute fuel hose or equivalent where fuel pipe is installed.

Fuel System Pressure

RELIEVING

✳✳ CAUTION

Before removing the fuel tank filler cap, turn the fuel tank filler cap ¼ to ¾ turn counterclockwise and wait for the tank pressure to be relieved. Personal injury may result if the fuel tank filler cap is removed without the pressure fully relieved.

1. Before servicing the vehicle, refer to the precautions in the beginning of this section.

2. Remove the fuel tank filler cap to relieve any pressure in the fuel tank.

3. When servicing the fuel lines, loosen the fuel fitting to allow any residual fuel line pressure to be relieved.

Idle Speed

ADJUSTMENT

1. Before servicing the vehicle, refer to the precautions in the beginning of this section.

2. Place the transmission in **P**.

3. Bring the engine up to normal operating temperature.

➡**Idle speed is measured with the transmission in D.**

4. Ensure that the curb idle adjusting screw is against the stop. If not, correct the vehicle linkage.

5. Check curb idle speed. Curb idle speed is specified on the Vehicle Emissions

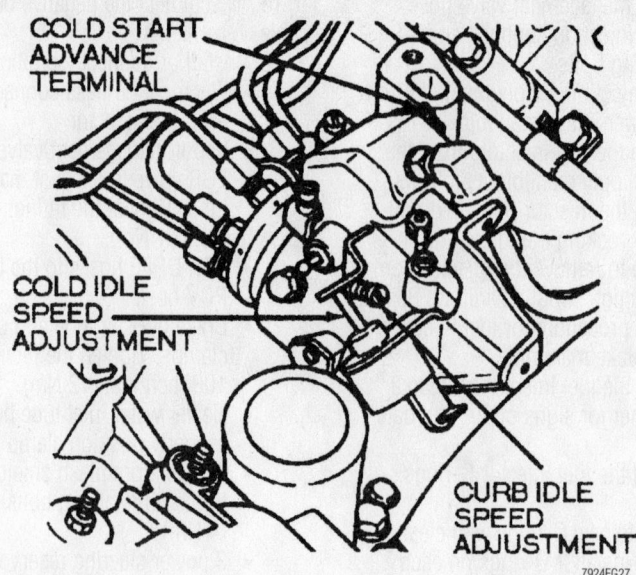

Raise or lower the curb idle speed by turning the curb idle speed adjusting screw—diesel engines

Control Information (VECI) decal on the underside of the vehicle's hood. Adjust the idle speed to specification using the idle speed adjusting screw.

6. Place the transmission in **P**. Rev the engine momentarily, then place the transmission in the specified gear and recheck the idle speed. Adjust again if necessary.

7. Remove the tachometer and close the hood.

Fuel Filter/Water Separator

DRAINING WATER

➡**Drain water from the water separator manual drain valve whenever the warning light comes ON or every 5000 miles (8000km). The "Water in Fuel" light will glow when approximately 3.5 oz. (103.5ml) of water accumulates in the separator.**

✳✳ CAUTION

The fuel system remains under pressure, even after the engine has been turned OFF. The fuel system pressure must be relieved before disconnecting any fuel lines. Failure to do so may result in fire and/or personal injury.

The diesel engines are equipped with a fuel/water separator in the fuel supply line. A "Water in Fuel" indicator light is provided on the instrument panel to alert the driver. The light should glow when the ignition switch is in the **start** position to indicate proper light and water sensor function. If the light glows continuously while the engine is running, the water must be drained from the separator as soon as possible to prevent damage to the fuel injection system.

1. Before servicing the vehicle, refer to the precautions in the beginning of this section.

2. Properly relieve the fuel system pressure.

3. Shut the engine **OFF**. Failure to shut the engine **OFF** before draining the separator will cause air to enter the system.

4. Unscrew the vent on the top center of the separator unit 2 ½ –3 turns.

5. Unscrew the drain screw on the bottom of the separator 1 ½ –2 turns and drain the water into an appropriate container.

6. After the water is completely drained, close the water drain finger-tight.

7. Tighten the vent until snug, then turn it an additional ¼ turn.

8. Start the engine and check the "Water in Fuel" indicator light; it should not be lit. If it is lit and continues to stay so, there is a problem somewhere else in the fuel system.

REMOVAL & INSTALLATION

✳✳ CAUTION

The fuel system remains under pressure, even after the engine has been turned OFF. The fuel system pressure must be relieved before disconnecting any fuel lines. Failure to do so may result in fire and/or personal injury.

1. Before servicing the vehicle, refer to the precautions in the beginning of this section.

2. Remove or disconnect the following:
   - Negative battery cable
   - Turbocharger assembly
   - Baffle and the air inlet crossover manifold

3. Place a suitable container under the drain hose and open the filter drain.
   - The 2 capscrews securing the fuel filter base to the crankcase
   - Water drain hose from the filter

4. Relieve the fuel system pressure.
   - Fuel outlet hose, located between the fuel and filter housing, and the fuel return hose from the fuel pressure regulator valve
   - The 2 fuel supply hoses that connect the regulator block to the cylinder head fuel rails
   - Clamp at the fuel pump end of the hose (loosen only), which connects the fuel filter to the inlet of high pressure stage at the fuel pump.
   - Wiring harness from the right side of the filter housing
   - Electrical connections from the Water In Fuel (WIF) sensor and the fuel heater

   - Fuel filter

5. Use a prybar to remove the fuel filter cap and the filter element will come out with the cap.

6. Depress the element locking tabs and remove the element from the cap.

To install:

7. Clean the mating surfaces and install the filter element onto the cap, making sure the tabs engage.

8. Install or connect the following:
   - Filter gap and press down firmly, but gently, to engage it
   - Wiring connections to the fuel heater and WIF sensor
   - Wiring harness to the filter housing
   - Tighten the clamp at the fuel pump end of the hose, which connects the fuel filter to the inlet of high pressure stage at the fuel pump
   - 2 fuel supply hoses that connect the regulator block to the cylinder head fuel rails
   - Fuel outlet hose, located between the fuel and filter housing, and the fuel return hose from the fuel pressure regulator valve.
   - Water drain hose to the filter
   - The 2 capscrews securing the fuel filter base to the crankcase

   - Air inlet crossover manifold and baffle
   - Turbocharger assembly
   - Negative battery cable

Injection Pump

REMOVAL & INSTALLATION

✳✳ CAUTION

The fuel system remains under pressure, even after the engine has been turned OFF. The fuel system pressure must be relieved before disconnecting any fuel lines. Failure to do so may result in fire and/or personal injury.

1. Before servicing the vehicle, refer to the precautions in the beginning of this section.

2. Remove or disconnect the following:
   - Negative battery cable
   - Turbocharger assembly
   - Fuel system pressure
   - Fuel line banjo bolt at the pump
   - Fuel line fittings at the rear of the cylinder heads

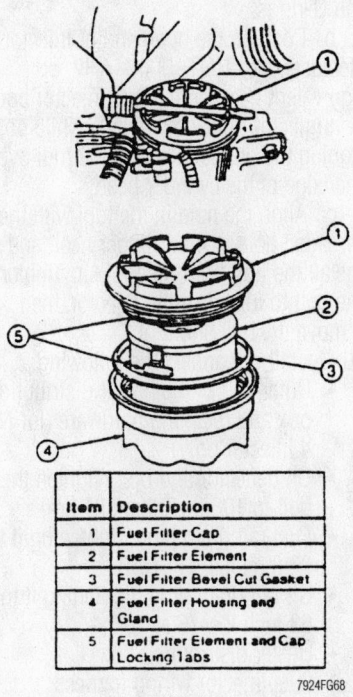

Item	Description
1	Fuel Filter Cap
2	Fuel Filter Element
3	Fuel Filter Bevel Cut Gasket
4	Fuel Filter Housing and Gland
5	Fuel Filter Element and Cap Locking Tabs

7924FG68

Use a prytool to remove the fuel filter cap to gain access to the filter element-diesel fuel systems

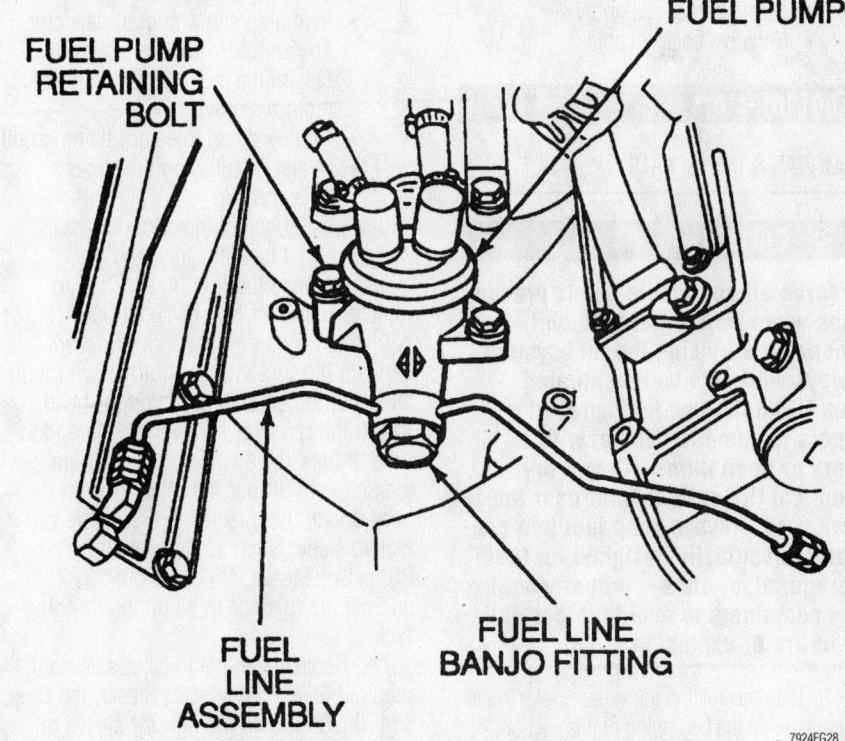

FUEL PUMP RETAINING BOLT

FUEL PUMP

FUEL LINE ASSEMBLY

FUEL LINE BANJO FITTING

7924FG28

Injection pump assembly components—diesel engines

For Wheel Alignment specifications, see Section 1 of this manual

- Fuel lines
- The 3 hose clamps at the injection pump fittings
- Water drain hose at the fuel filter
- Filter and position it forward
- Injection pump retaining bolts, then lift the pump out of the crankcase bore
- Injection pump tappet from the crankcase bore

To install:

3. Rotate the engine so the injection pump eccentric is on the base circle.

4. Install or connect the following:

- Injection pump tappet in the base of the injection pump
- O-ring on the injection pump base
- Injection pump and tighten the bolts to 19–27 ft. lbs. (26–37 Nm)
- Fuel filter and connect the water drain hose
- The 3 fuel hoses at the front of the injection pump
- The fuel line clamps and the fuel filter retaining bolts
- Fuel line assembly and new seal rings at the rear of the pump
- Fuel line fittings at the rear of the cylinder heads
- Fuel line banjo fitting at the pump. Tighten the bolt to 18 ft. lbs. (24 Nm).
- Turbocharger assembly
- Negative battery cable

Fuel Injectors

REMOVAL & INSTALLATION

✳✳ CAUTION

Observe all applicable safety precautions when working around fuel. Whenever servicing the fuel system, always work in a well ventilated area. Do not allow fuel spray or vapors to come in contact with a spark or open flame. Keep a dry chemical fire extinguisher near the work area. Always keep fuel in a container specifically designed for fuel storage; also, always properly seal fuel containers to avoid the possibility of fire or explosion.

1. Before servicing the vehicle, refer to the precautions in the beginning of this section.

✳✳ CAUTION

The red-striped wires on the DI Turbo carry 115 volts DC. A severe electrical shock may be given. Do not pierce the wires.

✳✳ WARNING

Do not pierce the wires or damage to the harness could occur.

Special tools required:

- Slide Hammer, No. T50T–100–A, or equivalent
- Injector Remover, No. T94T–9000–AH1, or equivalent
- Injector Replacer, No. T94T–9000–AH2, or equivalent

2. Remove or disconnect the following:

- Valve cover
- Fuel injector electrical connector

✳✳ WARNING

Remove the oil drain plugs prior to removing the injectors or oil could enter the combustion chamber which could result in hydrostatic lock and severe engine damage.

- Oil rail drain plugs
- Retaining screw and oil deflector. The shoulder bolt on the inboard side of the fuel injector does not require removal.
- Outboard fuel injector retaining bolt
- Heater distribution box screws, nuts and clip
- Outer half of the case (to service No. 4 fuel injector only)

3. Remove the fuel injector using Injector Remover No. T94T–9000–AH1, or equivalent. Position the tool's fulcrum beneath the fuel injector hold-down plate and over the edge of the cylinder head. Install the remover screw in the threaded hole of the fuel injector plate. Tighten the screw to lift out the injector from its bore. Place the injector in a suitable protective sleeve such as Rotunda Injector Protective Sleeve, No. 014–00933–2, and set the injector in a suitable holding rack.

4. Remove the fuel injector sleeves, if required. Insert the Injector Sleeve Tap Plug, 014–00934–3 into the injector sleeve to prevent debris from entering the combustion chamber. Insert Injector Sleeve Tap Pilot into the fuel injector sleeve and tighten 1–1 ½ turns. Attach Slide Hammer T50T–100–A to the Injector Sleeve Tap 014–00934–1 and 014–00934–2 Injector Sleeve Tap Pilot and remove the fuel injector sleeve from the bore.

5. Use Rotunda Injector Sleeve Brush 104–00934–A, or equivalent to clean the injector bore of any sealant residue. Make sure to remove any debris.

To install:

6. If removed, install the fuel injector sleeves using Rotunda Sleeve Replacer, No. 014–00934–4, or equivalent. Apply Thread-lock, No. 262–E2FZ–19554–B, or equivalent to the fuel injector sleeves. Using a rubber mallet, tap on the tool to seat the injector bore. Remove the tool and remove any residue sealant.

7. Clean the fuel injector sleeve using a suitable sleeve brush set. Clean any debris from the sleeve.

8. Clean the injector bore with a lint-free shop towel.

9. Install the fuel injectors using special tools as follows:

   a. Lubricate the injectors with clean engine oil. Using new copper washers, carefully push the injectors square into the bore using hand pressure only to seat the O-rings.

   b. Position the open end of Injector Replacer, No. T94T–9000–AH2, or equivalent between the fuel injector body and injector hold-down plate, while positioning the opposite end of the tool over the edge of the cylinder head.

   c. Align the hole in the tool with the threaded hole in the cylinder head and install the bolt from the tool kit. Tighten the bolt to fully seat the injector, then remove the bolt and tool.

10. Install or connect the following:

- Outer half of the heater distribution box and retaining hardware (for No. 4 injector only)
- Oil deflector and bolt. Tighten the bolt to 108 inch lbs. (12 Nm).
- Fuel rail drain plug, tightening it to 96 inch lbs. (11 Nm)
- Oil rail drain plug, tightening it to 53 inch lbs. (6 Nm)
- Heater distribution box
- Fuel injector wiring harness
- Valve cover

DRIVE TRAIN

Transmission Assembly

REMOVAL & INSTALLATION

Except Excursion

4R100 TRANSMISSION

1. Before servicing the vehicle, refer to the precautions in the beginning of this section.

2. Place the transmission range selector lever in the NEUTRAL position.

➡ All gasoline vehicles will have new adaptive shift strategies. Whenever the vehicle's battery has been disconnected for any type of service or repair the strategy parameters that are stored in the Keep Alive Memory (KAM) will be lost. The strategy will start to relearn once the battery is reconnected and the vehicle is driven. This is a temporary condition and will return to normal operating condition once the Powertrain Control Module (PCM) relearns all the parameters from the driving conditions. There is no set time frame for this process. If this concern is present during downshifts or converter clutch apply, it is not the fault of the shift strategy and will require diagnosis as outlined in the workshop manual. The customer needs to be notified that they may experience slightly different upshifts either soft or firm and that this is a temporary condition and will eventually return to normal operating condition.

3. Remove or disconnect the following:
- Negative battery cable
- Transmission fluid filler tube
- Transmission fluid cooler tubes at the cooler bypass valve. Catch the fluid in a suitable container.
- Transfer case assembly, if equipped

4. Mark the driveshaft yoke and axle flange so they may be installed in their original position.
- Rear driveshaft and install an appropriate plug in the transmission to prevent fluid leakage
- Shift cable from the transmission
- Shift cable from the manual lever
- Cable housing from the bracket and position aside
- Transmission Range (TR) sensor connector
- Shift cable at the transmission and position it aside

- Solenoid body connector
- Turbine Shaft Speed (TSS) sensor and the Output Shaft Speed (OSS) sensor
- Starter motor
- Cylinder block opening cover
- Torque converter-to-flexplate retaining nuts and discard

✳✳ CAUTION

Be sure not to raise the back of the transmission too high. If it makes contact with the underbody, damage to the TSS sensor can occur.

5. Install a high lift transmission jack with the special jack adapter 014-0763 to the transmission.
- Exhaust and crossmember
- Six transmission-to-engine mounting bolts

6. Gently rock the transmission side-to-side to disengage it from the locator dowels.

7. Move the transmission and the transmission jack rearward to clear the engine flexplate.
- Transmission from the vehicle

To install:

8. Installation is the reverse of removal, please note the following specs:

9. Transmission-to-engine bolts to 45 ft. lbs. (61 Nm).

10. Torque converter-to-flexplate nuts to 26 ft. lbs. (35 Nm).

11. Flexplate inspection cover nuts to 25 ft. lbs. (34 Nm) on gasoline engines and 15 ft. lbs. (20 Nm) on diesel engines.

12. Fluid filler tube into the short fluid inlet tube. Tighten the nut to 15 ft. lbs. (22 Nm).

13. Refill all transmissions with the correct amount of Motorcraft MERCON® automatic transmission fluid.

14. Connect the negative battery cable.

4R70W TRANSMISSION

1. Before servicing the vehicle, refer to the precautions in the beginning of this section.

2. Place the transmission range selector lever in the NEUTRAL position.

➡ All gasoline vehicles will have new adaptive shift strategies. Whenever the vehicle's battery has been disconnected for any type of service or repair the strategy parameters that are stored in the Keep Alive Memory (KAM) will be lost. The strategy will start to

relearn once the battery is reconnected and the vehicle is driven. This is a temporary condition and will return to normal operating condition once the Powertrain Control Module (PCM) relearns all the parameters from the driving conditions. There is no set time frame for this process. If this concern is present during downshifts or converter clutch apply, it is not the fault of the shift strategy and will require diagnosis as outlined in the workshop manual. The customer needs to be notified that they may experience slightly different upshifts either soft or firm and that this is a temporary condition and will eventually return to normal operating condition.

3. Remove or disconnect the following:
- Negative battery cable
- Transmission fluid filler tube
- Transmission fluid cooler tubes at the cooler bypass valve. Catch the fluid in a suitable container.
- Transfer case assembly, if equipped

4. Mark the driveshaft yoke and axle flange so they may be installed in their original position.
- Starter motor
- Torque converter-to-flexplate retaining nuts
- Shift linkage
- Transmission cooler lines
- Transmission electrical connectors.
- Transmission Range (TR) sensor
- Output Shaft Speed (OSS) sensor
- Solenoid body assembly electrical connector

5. Install a high lift transmission jack with the special jack adapter 014-0763 to the transmission.

6. Remove or disconnect the following:
- Three Way Converters (TWC) and crossmember
- Fuel line bracket bolt
- Transmission fill tube bolt from the back of the right hand cylinder head
- Transmission fluid fill tube
- 6 transmission to engine bolts

✳✳ CAUTION

The torque converter is heavy and may result in injury if it falls out of the transmission. Secure the torque converter in the transmission.

7. Install a Torque Converter Holding Tool.

8. Lower the transmission from the vehicle.

To install:

9. Installation is the reverse of removal, please note the following specs:

10. Transmission-to-engine bolts to 41 ft. lbs. (55 Nm).

11. Fluid filler tube into the short fluid inlet tube. Tighten the nut to 16 ft. lbs. (23 Nm).

12. Refill all transmissions with the correct amount of Motorcraft MERCON® automatic transmission fluid.

13. Connect the negative battery cable.

Excursion

1. Before servicing the vehicle, refer to the precautions in the beginning of this section.

2. Remove or disconnect the following:
- Negative battery cable
- Front driveshaft and transfer case, on 4-wheel drive vehicles
- All cables, connectors and fluid lines that may interfere with transmission removal. Tag them if helpful for installation.
- The torque converter from the flexplate
- Shift linkage

3. Drain the transmission fluid.

4. Position a transmission jack under the transmission and safety-chain the case to the jack.

5. Remove or disconnect the following:
- Driveshaft
- Transmission rear mount
- Crossmember
- The transmission-to-engine block bolts

❄❄ CAUTION

The torque converter will fall out of the transmission if it is tilted forward. Keep a hand on it while lowering the transmission out of the vehicle.

6. Roll the transmission rearward until the input shaft clears, lower the jack and remove the transmission.

To install:

7. Carefully raise the transmission to the engine or bell housing.

8. Roll the transmission forward and into position.

9. Install or connect the following:
- The transmission-to-engine block bolts. Tighten the bolts to 65 ft. lbs.

(87 Nm) for the diesel or to 50 ft. lbs. (67 Nm) for gasoline engines.
- Crossmember and tighten the bolts to 55 ft. lbs. (74 Nm)
- Transmission rear insulator and lower retainer. Tighten the bolts to 60 ft. lbs. (81 Nm).

10. The rest of the installation is the reverse of removal.

11. Refill all transmissions with the correct amount of Motorcraft MERCON® Automatic Transmission Fluid (ATF).

12. Connect the negative battery cable.

Transfer Case Assembly

REMOVAL & INSTALLATION

❄❄ CAUTION

The catalytic converter is located beside the transfer case. Due to the extreme high temperatures generated by the converter, be careful when removing the transfer case, or personal injury may result.

1. Before servicing the vehicle, refer to the precautions in the beginning of this section.

2. Remove or disconnect the following:
- Negative battery cable
- Fluid from the transfer case
- 4WD indicator switch wire connector at the transfer case
- Skid plate from the frame, if equipped
- Front driveshaft from the front output yoke
- Rear driveshaft from the rear output shaft yoke
- Speedometer driven gear from the transfer case rear bearing retainer
- Retaining rings and shift rod from the transfer case shift lever
- Vent hose from the transfer case
- Heat shield from the frame

3. Support the transfer case with a transmission jack.

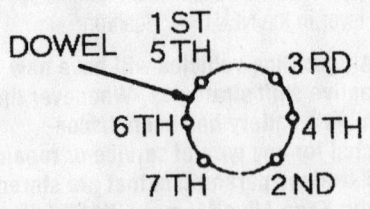

Transfer case-to-adapter bolt torque sequence—Borg-Warner model 13–45

7924FG30

- Bolts retaining the transfer case to the transmission adapter
- Transfer case from the vehicle

To install:

4. When installing place a new gasket between the transfer case and the adapter.

5. Raise the transfer case with the transmission jack so the transmission output shaft aligns with the splined transfer case input shaft. Install the bolts retaining the transfer case to the adapter.

6. Remove the transmission jack from the transfer case.

7. Install or connect the following:
- Rear driveshaft to the rear output shaft yoke. Tighten the bolts to 15 ft. lbs. (20 Nm).
- Shift lever to the transfer case and tighten the retaining nut
- Speedometer driven gear to the transfer case
- 4WD indicator switch wire connector at the transfer case
- Front driveshaft to the front output yoke. Tighten the bolts to 15 ft. lbs. (20 Nm).
- Heat shield to the frame crossmember and the mounting lug on the transfer case. Install and tighten the retaining bolts.
- Skid plate to the frame
- Drain plug

8. Remove the filler plug and install 6 pints. (2.8L) of Dexron®II transmission.

9. Connect the negative battery cable.

Halfshaft

REMOVAL & INSTALLATION

1. Before servicing the vehicle, refer to the precautions in the beginning of this section.

2. Remove or disconnect the following:
- Front wheels
- Front hub cotter pin, retainer and nut

3. Using a floor hydraulic jack, support the lower suspension arm.
- Upper ball joint cotter pin and castle nut
- Knuckle from the front suspension upper arm

4. Lower the lower suspension arm and steering knuckle slightly to facilitate easier halfshaft removal.

5. Remove the 2 disc caliper mounting bolts, then lift the front disc caliper off of the front disc brake caliper anchor plate and position aside. Do not allow the caliper to hang by the brake hose; suspend

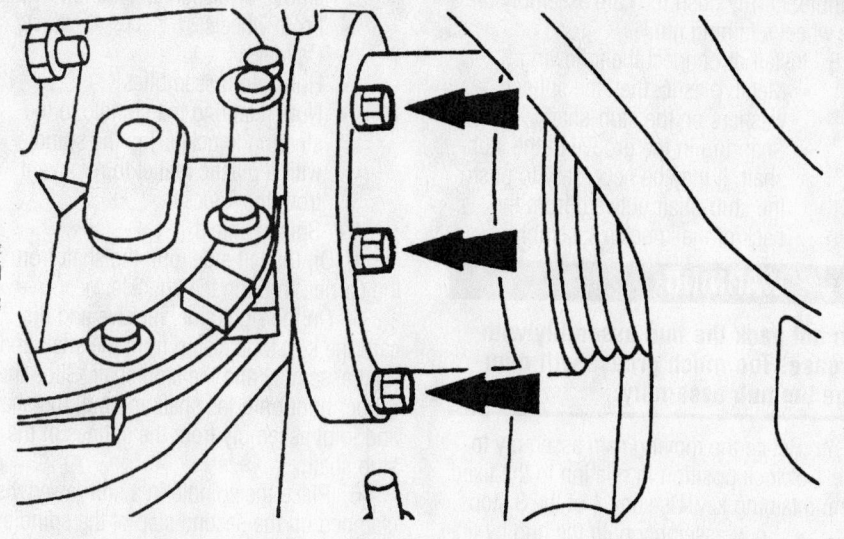

Halfshaft-to-differential mounting bolts (3 of the 6 bolts shown)

7924FG70

it from the vehicle's frame with strong cord or wire.

6. Remove the 6 front halfshaft-to-differential bolts.

✳✳ WARNING

Use care to avoid damaging the hub seal when removing the front halfshaft.

- Inboard end of the halfshaft from the differential case or extension axle case. Separate the front halfshaft and joints from the hub, then remove the halfshaft and joints from the vehicle.

To install:

7. Slide the halfshaft outboard end into the hub, making sure that the splines engage.

8. Situate the inboard end of the halfshaft against the front differential flange and install the 6 halfshaft-to-differential bolts. Tighten the halfshaft bolts to 60 ft. lbs. (82 Nm).

9. Install the front disc brake caliper onto the rotor and anchor plate, then install and tighten the 2 caliper mounting bolts to 21–26 ft. lbs. (28–36 Nm).

10. Lift the lower suspension arm and steering knuckle up until the upper ball joint stud is inserted into the steering knuckle. Install the upper ball joint castle nut and tighten to 57–76 ft. lbs. (77–104 Nm). Install a new cotter pin.

11. Install the hub nut onto the halfshaft and tighten the hub nut to 221 ft. lbs. (300 Nm).

12. Install the hub nut retainer and a new cotter pin.

13. Install the front wheels and tighten the lug nuts in a star-shaped sequence to 83–112 ft. lbs. (113–153 Nm).

14. Lower the vehicle to the ground.

CV-Joint

OVERHAUL

➡**Before continuing with this procedure, make sure to have available a new CV-joint boot kit, for each CV-joint being serviced. The outer CV-joint cannot be disassembled, only the boot can be replaced.**

Inner CV-Joint And Boot

1. Before servicing the vehicle, refer to the precautions in the beginning of this section.

2. Remove the halfshaft assembly from the vehicle.

3. Clamp the halfshaft in a vise equipped with jaw caps to prevent damage to machined surfaces. Do not allow the vise jaws to contact the boot or its clamp.

4. Slide 2 inboard clamp protectors off the boot clamps.

5. Carefully remove 2 boot clamps, and slide the boot off the inner CV-joint and housing.

6. Remove the CV-joint retaining ring and remove the housing.

7. Mark the inner race and the ball cage for assembly.

8. Remove 6 cage balls.

9. Remove the snapring.

10. Remove the inner race and ball cage.

11. Clean all parts in suitable parts cleaning solvent and inspect for wear.

To install:

12. Place the boot and the small boot clamp and protector on the shaft.

13. Place the ball cage on the shaft with the tapered end toward the outer CV-joint.

➡**Line up the marks made at disassembly.**

14. Position the inner race on the driveshaft in the position marked on disassembly.

15. Install the snapring.

16. Lubricate and position 6 balls with suitable CV-joint grease.

17. Place the boot protector and boot clamp on the CV-joint housing. Fill the housing with 8.29 ounces of suitable CV-joint grease.

18. Place the housing to the cage and bearings and install the retaining ring.

19. Remove any excess grease from the mating surface and position the boot and clamp.

20. Adjust the CV-joint to boot spacing to 16.43 in. (417.25mm).

21. After adjusting the CV-joint to boot spacing, insert a dull bladed screwdriver blade to relieve built up air pressure in the boot.

22. Use CV Boot Clamp Installer T95P-3514-A or equivalent, to install the boot clamps.

23. Place the clamp protectors over the boot clamps.

24. Install the halfshaft into the vehicle.

25. Road test the vehicle and check for proper operation.

Outer Boot

1. Remove the inner CV-joint and housing from the halfshaft.

2. Remove the inner boot from the halfshaft.

3. Remove the outer boot clamp protectors and carefully remove the boot clamps.

4. Remove the outer boot from the halfshaft and inspect the grease for contamination.

5. If the grease is contaminated, clean and inspect the joint for wear. Replace the joint and shaft if worn or damaged.

To install:

6. Place the new CV-joint boot on the shaft.

For Tune-up, Capacities and Firing orders, see Section 1 of this manual

7. Using 5.82 ounces (165 grams) of suitable CV-joint grease, pack the outer CV-joint with grease, then spread the remaining grease inside the boot.

8. Clean the boot mounting surface and position the boot in the joint grooves.

9. Place the clamps in position and use CV Boot Clamp Installer T95P-3514-A or equivalent, to install the boot clamps.

10. Place the clamp protectors over the boot clamps.

11. Install the inner boot on the half-shaft.

12. Install the inner CV-joint and housing on the halfshaft.

13. Install the halfshaft in the vehicle.

14. Road test the vehicle and check for proper operation.

Locking Hubs

REMOVAL & INSTALLATION

1. Before servicing the vehicle, refer to the precautions in the beginning of this section.

2. Raise and safely support the vehicle.

3. Remove or disconnect the following:
- Tire
- 3 screws and separate the cap from the body
- Lockring seated in the groove of the hub assembly
- Body assembly from the brake rotor/hub
- Snapring from the groove in the stub-shaft
- 3 thrust washers from the stub-shaft

4. Pull the cam assembly to remove it.

To install:

5. Align the fixed cam retaining key on the cam assembly with the keyway on the spindle. Firmly push the cam assembly on the wheel retaining nut.

6. Install or connect the following:
- Metal, plastic, then the splined washers on the stub-shaft
- Snapring in the groove of the stub-shaft. It may be necessary to push the stub-shaft outward from the back of the knuckle assembly.

✳✳ WARNING

Do not pack the hub assembly with grease. Too much grease will damage the hub assembly.

7. Rotate the moving cam assembly to the 1 o'clock position in relation to the fixed cam retaining key. Use any 1 of the 3 stops.
- Body assembly onto the hub by lining up the 3 legs with the 3 pockets in the cam assembly. Be sure the assembly is in far enough to see the groove in the hub.
- Large lockring in the groove on the hub. Ensure the lockring is seated completely.
- Cap using the 3 screws. Tighten the screws to 35–53 inch lbs. (4–6 Nm).
- Tire and lower the vehicle to the floor

Axle Shaft, Bearing and Seal

REMOVAL & INSTALLATION

Dana 50 Independent Front Suspension

1. Before servicing the vehicle, refer to the precautions in the beginning of this section.

2. Raise and support the front end on jackstands.

3. Remove or disconnect the following:
- Front wheels
- Calipers
- Hub/rotor assemblies
- Nuts retaining the spindle to the steering knuckle. Tap the spindle with a plastic mallet to remove it from the knuckle.
- Splash shield

4. On the left side, pull the shaft from the carrier, through the knuckle.

5. On the right side, remove and discard the keystone clamp from the shaft and joint assembly and the stub shaft. Slide the rubber boot onto the shaft and pull the shaft and joint assembly from the splines of the stub shaft.

6. Place the spindle in a soft-jawed vise clamped on the second step of the spindle.

7. Using a slide hammer and bearing puller, remove the needle bearing from the spindle.

8. Inspect all parts. If the spindle is excessively corroded or pitted it must be replaced. If the U-joints are excessively loose or don't move freely, they must be replaced. If any shaft is bent, it must be replaced.

To install:

9. Clean all dirt and grease from the spindle bearing bore. The bore must be free of nicks and burrs.

10. Insert a new spindle bearing in its bore with the printing facing outward. Drive it into place with drive T80T–4000–S for F-150 and F-250, or T80T–4000–R for the F-350, or their equivalents. Install a new bearing seal with the lip facing away from the bearing.

11. Pack the bearing and hub seal with grease. Install the hub seal with a driver.

12. Install or connect the following:
- Thrust washer on the axle shaft

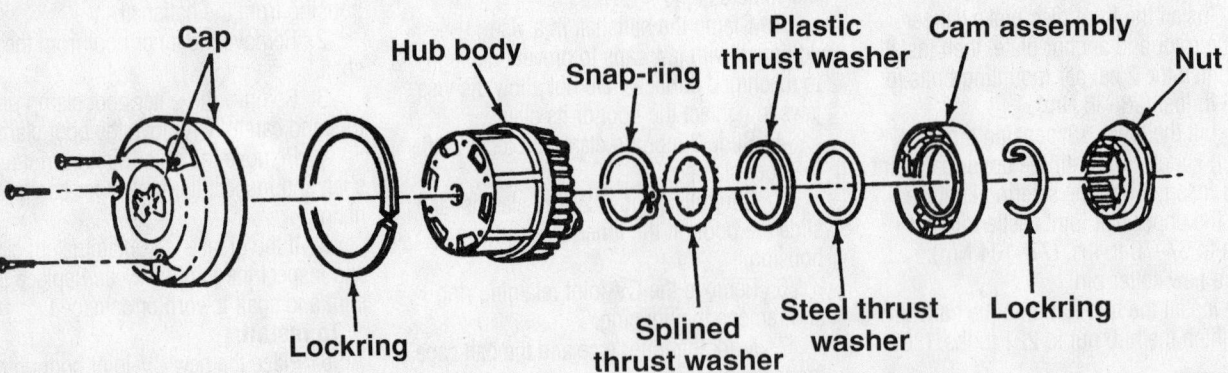

Exploded view of the typical automatic locking hub assembly

- New slinger on the axle shaft
- Rubber V-seal on the slinger. The seal lip should face the spindle.
- Plastic spacer on the axle shaft. The chamfered side of the spacer should be inboard against the axle shaft.

13. Pack the thrust face of the seal in the spindle bore and the V-seal on the axle shaft with heavy duty, high temperature, waterproof wheel bearing grease.

14. On the right side, install the rubber boot and new keystone clamps on the stub shaft and slip yoke. The splines permit only one way of meshing so you'll have to properly align the missing spline in the slip yoke with the gapless male spline on the shaft. Slide the right shaft and joint assembly into the slip yoke, making sure that the splines are fully engaged. Slide the boot over the assembly and crimp the keystone clamp.

15. On the left side, slide the shaft and joint assembly through the knuckle and engage the splines in the carrier.

16. Install or connect the following:
- Splash shield and spindle on the knuckle. Tighten the spindle nuts to 60 ft. lbs. (81 Nm).
- Rotor on the spindle
- Outer wheel bearing into the cup. Make sure that the grease seal lip totally encircles the spindle.
- Wheel bearing, locknut, thrust bearing, snapring and locking hubs.
- Caliper

Spindle and Front Axle Shaft

REMOVAL & INSTALLATION

Dana 60 Monobeam

1. Before servicing the vehicle, refer to the precautions in the beginning of this section.

2. Raise and support the front end on jackstands.

3. Remove or disconnect the following:
- Caliper from the knuckle and wire it out of the way
- Free-running hub
- Front wheel bearings
- Hub and rotor assembly
- Spindle-to-knuckle bolts. Tap the spindle from the knuckle using a plastic mallet.
- Splash shield and caliper support

4. Pull the axle shaft out through the knuckle.

5. Using a slide hammer and bearing cup puller, remove the needle bearing from the spindle.

6. Clean the spindle bore thoroughly and make sure that it is free of nicks and burrs. If the bore is excessively pitted or scored, the spindle must be replaced.

To install:

7. Insert a new spindle bearing in its bore with the printing facing outward. Drive it into place with driver T80T–4000–R, or its equivalent. Install a new bearing seal with the lip facing away from the bearing.

8. Pack the bearing with waterproof wheel bearing grease.

9. Pack the thrust face of the seal in the spindle bore and the V-seal on the axle shaft with waterproof wheel bearing grease.

10. Carefully guide the axle shaft through the knuckle and into the housing. Align the splines and fully seat the shaft.

11. Install or connect the following:
- Bronze spacer on the shaft. The chamfered side of the spacer must be inboard.
- Splash shield and caliper support
- Spindle on the knuckle and install the bolts. Tighten the bolts to 50–60 ft. lbs. (68–81 Nm).
- Hub/rotor assembly on the spindle
- Wheel bearings
- Free-running hub

Right Side Slip Yoke and Stub Shaft, Carrier, Carrier Oil Seal and Bearing

REMOVAL & INSTALLATION

Dana 50 Independent Front Axle

1. Before servicing the vehicle, refer to the precautions in the beginning of this section.

➡ **This procedure requires the use of special tools.**

2. Raise and support the front end on jackstands.

3. Disconnect the front driveshaft from the carrier and wire it up out of the way.

4. Remove the left and right axle shafts and both spindles.

5. Support the carrier with a floor jack and unbolt the carrier from the support arm.

6. Place a drain pan under the carrier, separate the carrier from the support arm and drain the carrier.

7. Remove the carrier from the truck.

8. Place the carrier in holding fixture T57L–500–B with adapters T80T–4000–B.

9. Rotate the slip yoke and shaft assembly from the carrier.

10. Using a slide hammer/puller remove the caged needle bearing and oil seal as a unit. Discard the oil seal and bearing.

To install:

11. Clean the bearing bore thoroughly and make sure that it is free of nicks and burrs.

12. Insert a new bearing in its bore with the printing facing outward. Drive it into place with driver T83T–1244–A, or its equivalent. Install a new bearing seal with the lip facing away from the bearing. Coat the bearing and seal with waterproof wheel bearing grease.

13. Install the slip yoke and shaft assembly into the carrier so that the groove in the shaft is visible in the differential case.

14. Install the snapring in the groove in the shaft. It may be necessary to force the snapring into place with a small prybar. Don't strike the snapring!

15. Remove the carrier from the holding fixture.

16. Clean all traces of sealant from the carrier and support arm. Make sure the mating surfaces are clean. Apply a ¼ in. (6mm) wide bead of RTV sealant to the mating surface of the carrier. The bead must be continuous and should not pass through or outside of the holes. Install the carrier with 5 minutes of applying the sealer.

17. Position the carrier on the jack and raise it into position using guide pins to align it if you'd like. Install and hand-tighten the bolts. Tighten the bolts in a circular pattern to 30–40 ft. lbs. (41–54 Nm).

18. Install the support arm tab bolts and tighten them to 85–100 ft. lbs. (115–136 Nm).

19. Install all other parts in reverse order of removal.

Pinion Seal

REMOVAL & INSTALLATION

Independent Front Axle

1. Before servicing the vehicle, refer to the precautions in the beginning of this section.

➡ **A torque wrench capable of at least 225 ft. lbs. (305 Nm) is required for pinion seal installation.**

2. Raise and safely support the vehicle

with jackstands under the frame rails. Allow the axle to drop to rebound position for working clearance.

3. Mark the companion flanges and U-joints for correct reinstallation position.

4. Remove the driveshaft. Use a suitable tool to hold the companion flange. Remove the pinion nut and companion flange.

5. Use a slide hammer and hook or sheet metal screw to remove the oil seal.

To install:

6. Install a new pinion seal after lubricating the sealing surfaces. Use a suitable seal driver. Install the companion flange and pinion nut. Tighten the nut to 200–220 ft. lbs. (271–298 Nm).

Monobeam Front Axle

1. Before servicing the vehicle, refer to the precautions in the beginning of this section.

➡ **A torque wrench capable of at least 300 ft. lbs. (407 Nm) is required for pinion seal installation.**

2. Raise and support the truck on jackstands.

3. Allow the axle to hang freely.

4. Matchmark and disconnect the driveshaft from the front axle.

5. Using a tool such as T75T–4851–B, or equivalent, hold the pinion flange while removing the pinion nut.

6. Using a puller, remove the pinion flange.

7. Use a puller to remove the seal, or punch the seal out using a pin punch.

To install:

8. Thoroughly clean the seal bore and make sure that it is not damaged in any way. Coat the sealing edge of the new seal with a small amount of 80W/90 oil and drive the seal into the housing using a seal driver.

9. Coat the inside of the pinion flange with clean 80W/90 oil and install the flange onto the pinion shaft.

10. Install the nut on the pinion shaft and tighten it to 250–300 ft. lbs. (339–407 Nm).

11. Connect the driveshaft.

STEERING AND SUSPENSION

Air Bag

❊❊ CAUTION

Some vehicles are equipped with an air bag system. The system must be disabled before performing service on or around system components, steering column, instrument panel components, wiring and sensors. Failure to follow safety and disabling procedures could result in accidental air bag deployment, possible personal injury and unnecessary system repairs.

PRECAUTIONS

Several precautions must be observed when handling the inflator module to avoid accidental deployment and possible personal injury:

• Never carry the inflator module by the wires or connector on the underside of the module.

• When carrying a live inflator module, hold securely with both hands and ensure that the bag and trim cover are pointed away.

• Place the inflator module on a bench or other surface with the bag and trim cover facing up.

• With the inflator module on the bench, never place anything on or close to the module, which may be thrown in the event of an accidental deployment.

DISARMING

❊❊ CAUTION

The Supplemental Inflatable Restraint (SIR) system must be dis-

armed before performing service around SIR system components or SIR system wiring. Failure to do so may cause accidental deployment of the air bag, resulting in unnecessary SIR system repairs and/or personal injury.

The positive battery cable must be disconnected for a minimum of 1 minute before beginning any air bag work to de-energize the back-up power supply. It is a good idea to disengage both the positive and negative battery cables to ensure that the Air Bag system is definitely discharged.

ARMING THE SYSTEM

❊❊ WARNING

If the air bag simulators have been used, the air bag simulators must be removed and the air bags reconnected when the system is reactivated to avoid non-deployment in a collision resulting in possible personal injury.

1. Disconnect the positive battery cable.

2. Wait 1 minute, this is required for the back-up power supply in the air bag diagnostic monitor to deplete its stored energy.

3. Remove the air bag simulator from the air bag sliding contact connector at the top of the steering column. Reconnect the driver's side air bag module assembly. Position the driver's air bag module on the steering wheel and secure with the 2 bolts and washers. Tighten the bolt and washer assembly to 8–10 ft. lbs. (10–14 Nm).

4. Connect the positive battery cable.

5. Turn the ignition switch from the **OFF**

to **RUN** and visually monitor the air bag warning indicator. The light will illuminate continuously for approximately 6 seconds and then turn off. If a fault occurs, the air bag indicator will either fail to light, remain lighted continuously or flash. The flashing may not occur until approximately 30 seconds after the ignition switch has been turned from **OFF** to **RUN**. This is the time needed for the air bag diagnostic monitor to complete testing the system. If the air bag indicator is inoperative, an air bag system fault exists, a tone will sound in a pattern of 5 sets of 5 beeps. If this occurs, the air bag indicator will need to be serviced before further diagnostics can be done.

Steering Gear

REMOVAL & INSTALLATION

Excursion

1. Before servicing the vehicle, refer to the precautions in the beginning of this section.

2. Place the wheels in the straight-ahead position.

3. Remove or disconnect the following:

• Pressure and return lines. Cap the openings.

• Splash shield from the flex coupling

• Flex coupling at the gear

• Pitman arm from the sector shaft

• Steering gear mounting bolts

• Steering gear. It may be necessary to work it free of the flex coupling.

To install:

4. Place the splash shield on the steering gear lugs.

5. Slide the flex coupling into place on the steering shaft. Be sure the steering wheel spokes are still horizontal.

6. Center the steering gear input shaft with the indexing flat facing downward.

7. Slide the steering gear input shaft into the flex coupling and into place on the frame side rail. Install the flex coupling bolt and tighten it to 30 ft. lbs. (41 Nm).

8. Install or connect the following:
- Steering gear mounting bolts and tighten them to 65 ft. lbs. (88 Nm)
- Pitman arm. Tighten the nut to 230 ft. lbs. (312 Nm).
- Pressure, then, the return lines. Tighten the pressure line to 25 ft. lbs. (34 Nm).
- Flex coupling shield

9. Fill the steering reservoir.

10. Run the engine and turn the steering wheel lock-to-lock several times to expel air. Check for leaks.

Expedition & Navigator

1. Before servicing the vehicle, refer to the precautions in the beginning of this section.

2. Remove or disconnect the following:
- Skid plate
- Lower radiator air deflector
- Pitman arm cotter pin and castellated nut from the drag link
- Pitman arm from the drag link
- Dust cover from the steering shaft valve housing
- Intermediate shaft pinch bolt and slide the shaft off the steering gear input shaft
- Power steering pressure hoses at the steering gear
- Steering gear-to-frame rail retaining bolts and the steering gear

3. If replacing or servicing the steering gear, matchmark the sector shaft arm to the sector shaft and remove the steering gear sector shaft arm retaining nut and lockwasher. Remove the sector shaft arm.

To install:

4. If removed, install the steering gear sector shaft arm to the sector shaft aligning the matchmarks made during removal. Install the retaining nut and lockwasher and tighten to 170–228 ft. lbs. (234–316 Nm).

5. Install or connect the following:
- Steering gear in position. Install the 3 retaining bolts and tighten them to 50–68 ft. lbs. (68–92 Nm).
- Power steering pressure hoses to the steering gear using new seals, if necessary.

- Intermediate shaft on the steering gear input shaft and install the shaft pinch bolt. Tighten the pinch bolt to 30–42 ft. lbs. (41–57 Nm).
- Dust cover over the steering shaft valve housing
- Pitman arm to the drag link. Install the castellated nut and tighten to 57–76 ft. lbs. (77–104 Nm). Install a new cotter pin.
- Radiator air deflector and secure with the retaining screws and push clips
- Skid plate and secure it with the retaining bolts

6. Lower the vehicle.

7. Fill and bleed the power steering system.

8. Road test the vehicle and check the steering system for proper operation.

Shock Absorber

REMOVAL & INSTALLATION

Front

EXCEPT EXCURSION

1. Before servicing the vehicle, refer to the precautions in the beginning of this section.

2. If equipped with 2-wheel drive, perform the following:

a. Hold the shock absorber stem and remove the nut from the top of the shock.

b. Raise and safely support the vehicle.

c. Remove the 2 lower retaining nuts and remove the shock absorber.

3. If equipped with 4-wheel drive, perform the following:

a. Hold the shock absorber stem and remove the nut, washer and bushing from the top of the shock absorber stud.

b. Raise and support the vehicle.

c. Remove the lower retaining nut and bolt.

d. Remove the shock absorber from the vehicle.

To install:

4. On 2-wheel drive models, perform the following:

a. Install the washer and bushing to the top stem of the shock absorber.

b. Place the shock absorber up through the coil spring.

c. Install the 2 lower retaining nuts. Tighten the nuts to 22–29 ft. lbs. (30–40 Nm).

d. Lower the vehicle.

e. Install the bushing, washer and retaining nut to the top of the shock absorber stud. Tighten the nut to 34–46 ft. lbs. (47–63 Nm).

5. On 4-wheel drive models, perform the following:

a. Install the washer and bushing to the top stem of the shock absorber.

b. Place the shock absorber up through the coil spring.

c. Install the lower retaining nut and bolt. Tighten to 57–76 ft. lbs. (77–104 Nm).

d. Lower the vehicle.

e. Install the shock absorber upper bushing, washer and retaining nut.

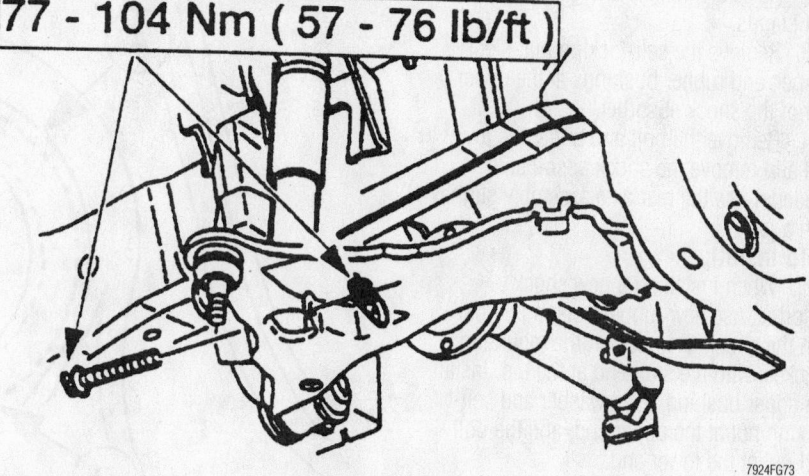

77 - 104 Nm (57 - 76 lb/ft)

7924FG73

Lower shock absorber mounting—late models with torsion bar suspension

Tighten the nut to 22–29 ft. lbs. (30–40 Nm).

6. Road test the vehicle and check for proper operation.

EXCURSION—4X2 MODELS

1. Before servicing the vehicle, refer to the precautions in the beginning of this section.

2. Remove the upper shock absorber retaining nut and upper shock absorber insulator.

3. Raise and support the vehicle.

4. Remove the lower shock absorber retaining nut and remove the shock absorber.

To install:

5. Installation is the reverse of removal. Using new fasteners, tighten the lower nut to 60 ft. lbs. (80 Nm) and the upper nut to 30 ft. lbs. (40 Nm).

EXCURSION—4X4 MODELS

1. Before servicing the vehicle, refer to the precautions in the beginning of this section.

2. Remove the lower shock absorber retaining nut and remove the shock absorber.

3. Remove the upper shock absorber retaining nut and upper shock absorber insulator.

To install:

4. Installation is the reverse of removal. Using new fasteners, tighten the lower and upper nuts to 76 ft. lbs. (103 Nm).

Rear

1. Before servicing the vehicle, refer to the precautions in the beginning of this section.

2. Raise the vehicle and secure on support stands.

3. Remove the self-locking nut, steel washer, and rubber bushings at the upper end of the shock absorber.

4. Remove the bolt and nut at the lower end and remove the shock absorber. If needed, raise the rear axle assembly slightly with a jack.

To install:

5. When installing a new shock absorber, use new rubber bushings. Position the shock absorber on the mounting brackets with the stud end at the top. Install the upper bushing, steel washer and self-locking nut at the upper end, and the bolt and nut at the lower end.

6. Tighten the upper mounting studs to 63–84 ft. lbs. (85–115 Nm) and the lower mounting nuts to 63–84 ft. lbs. (85–115 Nm).

Coil Spring

REMOVAL & INSTALLATION

Front

This procedure applies to 2-wheel drive vehicles only. In order to remove the coil spring, the lower control arm must also be removed.

1. Before servicing the vehicle, refer to the precautions in the beginning of this section.

2. Remove or disconnect the following:
- Wheels
- Disc brake caliper and support aside with wire
- Disc brake adapter
- Rotor
- Rotor splash shield
- Shock absorber
- Bracket supporting the brake hose
- Sway bar link retaining nut and bushing from the lower control arm. Separate the sway bar link from the lower control arm.

3. Install a coil spring compressor and compress the coil spring enough to relieve the tension of the spring between the upper and lower control arms.
- Cotter pin and castellated nut from the lower ball joint
- Lower ball joint from the wheel spindle

4. Matchmark the lower control arm alignment cams for installation reference.
- Lower control arm retaining nuts and bolts
- Lower control arm and the compressed coil spring as an assembly

5. Loosen the coil spring compressor and remove the coil spring from the lower control arm.

6. Inspect the coil spring and replace as needed.

To install:

7. Install or connect the following:
- Coil spring correctly in the saddle of the lower control arm
- Coil spring compressor and compress the coil spring. The end of the coil spring **A**, must cover the hole designated **B** and be visible in the second hole designated as **C**, for proper installation.
- Lower control arm to the frame
- Retaining bolts, adjusting cams, and nuts

8. Align the matchmarks on the adjusting cams. The forward nut must be tightened first while the control arm is held at the curb position height. Tighten the nuts to 197–241 ft. lbs. (270–330 Nm).
- Lower ball joint stud into the wheel spindle
- Castellated nut and tighten to 83–113 ft. lbs. (113–153 Nm)
- New cotter pin
- Sway bar link to the lower control arm
- Bushing and retaining nut. Tighten to 15–21 ft. lbs. (21–29 Nm).

9. Remove the coil spring compressor.
- Shock absorber. Tighten the lower bolts to 22 ft. lbs. (32 Nm) and the top nut to 45 ft. lbs. (61 Nm).
- Brake hose bracket
- Brake rotor splash shield

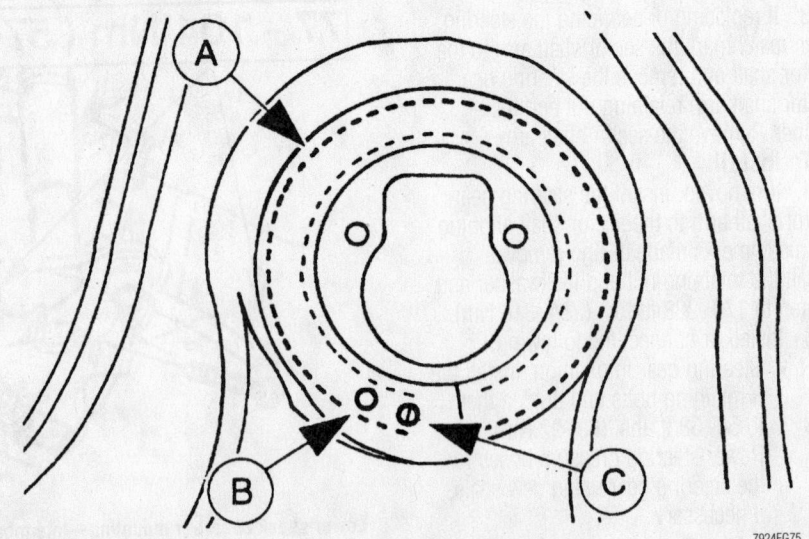

Be sure the coil spring is mounted correctly in the lower control arm

7924FG75

- 3 retaining bolts and tighten them to 90–107 inch lbs. (10–14 Nm)
- Disc brake rotor and caliper assemblies
  - Wheel and tire assembly
10. Lower the vehicle.
11. Pump the brake pedal several times to position the brake pads prior to moving the vehicle.
12. Check the alignment and adjust if out of specification.
13. Road test the vehicle and check for proper operation.

Rear

1. Before servicing the vehicle, refer to the precautions in the beginning of this section.
2. Raise and support the vehicle.
   - Wheel and tire assembly
   - Rear driveshaft
   - Rear Anti-lock Brake System (ABS) sensor electrical connector

➡ **Make sure the parking brake control is fully released.**

3. Pull the front parking brake cable and conduit.
4. Insert a suitable retainer into the parking brake control.
5. Remove or disconnect the following:
   - Rear parking brake cable and conduit from the rear axle
   - Rear parking brake cable and conduit from the rear axle bracket assembly
   - Cable from the caliper
6. Repeat for the other side.
   - Disc brake caliper bolts
7. Repeat for the other side.
   - Axle vent tube
   - Stabilizer bar from the rear axle
   - Bolts from each stabilizer bar retainer
   - Retainers
8. Use a suitable jack to support the rear axle.
   - Lower shock absorber nut and bolt
9. Repeat for the other side.
   - Rear axle-to-trackbar assembly bolt
   - Rear suspension lower arm assembly-to-axle bolt
10. Repeat for the other side.
    - Rear suspension upper arm assembly-to-rear axle bolt
11. Repeat for the other side.
12. Lower the rear axle and remove the coil spring.

To install:

➡ **Do not tighten any nuts or bolts until the rear suspension is raised so that the rear suspension lower arms are parallel to the ground. Once in position, tighten all the nuts and bolts to specification.**

13. Install or connect the following:
    - Rear coil spring
    - Rear suspension upper arm assemblies to the axle and the bolts
    - Rear suspension lower arm assemblies to the axle and the bolts
    - Trackbar assembly to the axle and the bolts
    - Shock absorbers to the axle and the bolts
    - Stabilizer bar to the axle and the stabilizer bar retainers and bolts
    - Disc brake caliper and the bolts
    - Parking brake cables and conduit to the axle
    - Parking brake cable and conduit to the lever
    - Parking brake cable and conduit to the rear axle bracket assembly
    - Tire and wheel assembly
    - Lug nuts and lower the vehicle
    - Wheel cover.
14. Remove the pin from the parking brake control assembly.
    - Rear brake anti-lock sensor (ABS) electrical connector
    - Axle vent tube
    - Driveshaft
15. Lower the vehicle.

Leaf Spring

REMOVAL & INSTALLATION

Excursion

1. Before servicing the vehicle, refer to the precautions in the beginning of this section.
2. Remove or disconnect the following:
3. Raise and support the vehicle.
   - Wheel and tire assembly
4. Support the rear axle with a suitable jack.
   - U-bolt retaining nuts and the U-bolts.
   - Rear spring upper plate.
   - Nut and bolt from the rear spring front hanger bracket.
   - Lower nut and bolt from the rear spring shackle bracket

- Rear spring assembly.
- Nut and bolt from the rear spring shackle assembly and the rear spring shackle

To install:

5. Installation is the reverse of removal, please use new fasteners and note the following torques:
   a. Nut and bolt for the rear spring shackle assembly to 185 ft. lbs. (285 Nm).
   b. Lower nut and bolt for the rear spring shackle bracket to 185 ft. lbs. (285 Nm).
   c. Nut and bolt for the rear spring front hanger bracket to 185 ft. lbs. (285 Nm).
   d. U-bolt retaining nuts and the U-bolts to 185 ft. lbs. (285 Nm).

Track Bar

REMOVAL & INSTALLATION

1. Before servicing the vehicle, refer to the precautions in the beginning of this section.

✳✳ CAUTION

The electrical power to the air suspension system must be shut off prior to hoisting, jacking or towing an air suspension vehicle. This can be accomplished by turning off the air suspension switch located in the PASSENGER SIDE kick panel area. Failure to do so can result in unexpected inflation or deflation of the air springs, which can result in shifting of the vehicle during these operations.

2. Raise and support the vehicle.

✳✳ WARNING

The air suspension height sensor has a plastic harness retainer to suspension that must be unclipped prior to removal.

3. Remove or disconnect the following:
   - Air suspension height sensor electrical connector
   - Metal retaining tabs and remove the air suspension height sensor from the ball studs
   - Passenger side trackbar bolt

- Driver's side trackbar nut and bolt
- Trackbar from the vehicle

To install:

4. The installation is the reverse of the removal.

Torsion Bars

REMOVAL & INSTALLATION

1. Before servicing the vehicle, refer to the precautions in the beginning of this section.

2. Raise and safely support the vehicle.

3. Matchmark and measure the length of the torsion bar being removed.

➡**Torsion bars are marked for left-hand or right-hand installation. If removing both torsion bars, make sure to reinstall in the correct position. If replacing a torsion bar, make sure that it is the correct torsion bar for the side being replaced.**

4. Install Torsion Bar Adjuster T95T-5310-A and Adapter Plates T96T-5310-A, or equivalents.

5. Tighten the adjuster tool until it touches the torsion bar adjuster.

6. Remove the torsion bar adjuster bolt and nut.

7. Remove 6 torsion bar crossmember retaining bolts.

8. Remove the crossmember support.

9. Remove the torsion bar from the vehicle.

To install:

10. Place the torsion bar in position ensuring that it is the correct torsion bar for the side being installed.

11. Install the crossmember support and 6 retaining bolts. Tighten the retaining bolts to 40–50 ft. lbs. (53–72 Nm).

12. Install the torsion bar adjuster bolt and nut, then remove the torsion bar adjuster tool and adapters.

13. Tighten the adjusting bolt until the matchmarks are in alignment.

14. Lower the vehicle.

15. Measure the ride height and adjust if necessary.

16. Check the alignment and adjust if not within specification.

17. Road test the vehicle and check for proper operation.

ADJUSTMENT

1. Place the vehicle on a level surface.

2. Bounce the front end to normalize the ride height.

3. Measure the ride height on both sides of the vehicle from the frame to the ground.

4. Loosen the lock nut and add or delete turns until the vehicle is level and at the correct curb height.

5. Tighten the lock nut.

6. Road test the vehicle and recheck the measurements.

Upper Ball Joint

REMOVAL & INSTALLATION

The upper ball joint is an integral part of the upper control arm and is not a serviceable component. Replacement of the ball joint requires replacing the upper control arm assembly.

Lower Ball Joint

REMOVAL & INSTALLATION

The lower ball joint is an integral part of the lower control arm and is not a serviceable component. Replacement of the ball joint requires replacing the lower control arm.

Upper Control Arm

REMOVAL & INSTALLATION

1. Before servicing the vehicle, refer to the precautions in the beginning of this section.

2. Raise and safely support the vehicle.

3. Remove the wheel and tire assembly.

4. Support the lower control arm with a transmission jack or equivalent adjustable jack.

5. Matchmark the upper control arm adjusting cams.

6. If equipped with 4-wheel Anti-lock Brake System (ABS), remove the speed sensor harness bracket bolt and position the harness aside.

7. Remove the cotter pin and castellated nut from the upper ball joint. Discard the cotter pin.

8. Using Pitman Arm Puller T64P-3590-F or equivalent, separate the ball joint from the wheel spindle.

9. Remove 2 nuts and bolts securing the upper control arm to the body brackets.

10. Remove the upper control arm and ball joint assembly from the vehicle.

To install:

11. Align the matchmarks on the upper control arm adjusting cams.

12. Place the upper control arm in position to the body brackets and install 2 retaining bolts and nuts. Do not tighten at this time.

13. Install the ball joint stud to the wheel spindle and install the castellated nut. Tighten the nut to 56–77 ft. lbs. (76–104 Nm). Install a new cotter pin.

14. Raise the lower control arm to position the control arms at normal curb height and tighten the front upper control arm retaining nut first, to 84–112 ft. lbs. (113–153 Nm).

15. With the vehicle still at curb height, tighten the rear upper control arm retaining nut to 84–112 ft. lbs. (113–153 Nm).

16. Remove the transmission jack or equivalent from under the lower control arm.

17. If equipped with 4-wheel ABS brake system, position the speed sensor harness and install the bracket bolt. Tighten the speed sensor bracket bolt to 62–80 inch lbs. (7–9 Nm).

18. Install the wheel and tire assembly. Torque the lug nuts to 83–112 ft. lbs. (113–153 Nm).

19. Lower the vehicle.

20. Check the alignment and adjust if not within specification.

21. Road test the vehicle and check for proper operation.

Lower Control Arm

REMOVAL & INSTALLATION

1. Before servicing the vehicle, refer to the precautions in the beginning of this section.

➡**These vehicles require that the torsion bar tension be relieved before removing the lower control arm. If either the lower control arm or lower ball joint require replacement, the lower control arm and ball joint must be replaced as a unit.**

2. Raise and safely support the vehicle.

3. Matchmark and measure the length of the torsion bar being removed.

➡**Torsion bars are marked for left-hand or right-hand installation. If removing both torsion bars, make sure to reinstall in the correct position. If replacing a torsion bar, make sure that it is the correct torsion bar for the side being replaced.**

4. Install Torsion Bar Adjuster T95T-5310-A and Adapter Plates T96T-5310-A, or equivalents.

5. Tighten the adjuster tool until it touches the torsion bar adjuster.

6. Remove the torsion bar adjuster bolt and nut.

7. Remove 6 torsion bar crossmember retaining bolts.

8. Remove the crossmember support.

9. Remove the torsion bar from the vehicle.

10. If equipped with 4-wheel Anti-lock Brake System (ABS), disconnect the sensor wire bracket bolt.

11. Remove the lower shock absorber retaining nut and bolt.

12. Disconnect the lower control arm castle nut and cotter pin and separate the ball joint from the steering knuckle.

13. Disconnect the stabilizer bar link from the lower control arm.

14. Remove the 2 lower control arm nuts and bolts.

15. Remove the lower arm and torsion bar.

To install:

16. Install the lower control arm and torsion bar. Tighten the bolts to 150 ft. lbs. (200 Nm).

17. Connect the lower ball joint and tighten the castle nut to 80–110 ft. lbs. (113–153 Nm).

18. Connect the stabilizer bar link to the lower control arm. tighten the nut to 16–21 ft. lbs. (22–28 Nm).

19. Place the torsion bar in position ensuring that it is the correct torsion bar for the side being installed.

20. Install the lower shock absorber retaining nut and bolt. Tighten to 55–74 ft. lbs. (76–103 Nm).

21. Secure the ABS sensor bracket, if equipped.

22. Install the crossmember support and 6 retaining bolts. Tighten the retaining bolts to 40–50 ft. lbs. (53–72 Nm).

23. Install the torsion bar adjuster bolt and nut, then remove the torsion bar adjuster tool and adapters.

24. Tighten the adjusting bolt until the matchmarks are in alignment.

25. Lower the vehicle.

26. Measure the ride height and adjust if necessary.

27. Check the alignment and adjust if not within specification.

28. Road test the vehicle and check for proper operation.

Front Wheel Bearings

ADJUSTMENT

✳✳ CAUTION

If equipped with the automatic air suspension system, the service switch near the right kick panel must be turned OFF before raising the vehicle for service.

➥**On 4-wheel drive vehicles, the front wheel bearings are not adjustable.**

1. Before servicing the vehicle, refer to the precautions in the beginning of this section.

2. Raise and safely support the vehicle.

3. Support the front end.

4. Remove the wheel cover, if equipped.

5. Remove the grease cap.

➥**Check the wheel bearings for sufficient grease.**

6. Remove the cotter pin and retaining washer. Back off the spindle nut. Discard the cotter pin.

7. Adjust the wheel bearings on Navigator and Expedition models as follows:

 a. Tighten the spindle nut to 30 ft. lbs. (40 Nm) while rotating the wheel and tire assembly to seat the wheel bearings.

 b. Back off the spindle nut 2 turns.

 c. Tighten the spindle nut to 17–24 ft. lbs. (23–34 Nm) while rotating the rotor clockwise.

 d. Loosen the nut 175 degrees.

 e. Tighten the nut to 17 inch lbs. (2 Nm).

8. Adjust the wheel bearings on Excursion models as follows:

 a. Tighten the spindle nut to 21 ft. lbs. (28 Nm) while rotating the wheel and tire assembly to seat the wheel bearings.

 b. Back off the spindle nut 120–180 degrees.

 c. Tighten the nut to 17 inch lbs. (2 Nm).

9. Install the retaining washer so the castellations are aligned with the cotter pin hole. Install a new cotter pin.

10. Check the wheel and tire assembly for proper rotation, then install the grease cap. If the wheel still does not rotate properly, inspect and clean or replace the wheel bearings and cups.

11. Install the wheel cover, if equipped.

12. Lower the vehicle.

13. Road test the vehicle and check for proper operation.

REMOVAL & INSTALLATION

2-Wheel Drive

The hub is part of the disc brake rotor and cannot be serviced separately. The inner and outer wheel bearing and races are serviced individually. Be sure to have a new hub grease seal when servicing the wheel bearings.

1. Before servicing the vehicle, refer to the precautions in the beginning of this section.

2. Remove or disconnect the following:
 - Wheels
 - Caliper
 - Brake pads and anti-rattle clips
 - Anchor plate
 - Hub grease cap, cotter pin, retainer washer and the spindle nut
 - Wheel bearing retainer washer and the outer wheel bearing
 - Brake hub and rotor assembly
 - Grease seal
 - Inner wheel bearing

3. Clean and inspect the wheel bearings and races for unusual wear or damage. Replace parts as necessary.

4. Inspect the hub and brake rotor assembly. If required, the hub and brake rotor assembly must be replaced as a unit.

To install:

5. If needed, pack the wheel bearing with a suitable high temperature wheel bearing grease before assembly.

6. Install or connect the following:
 - Inner wheel bearing in the hub and brake rotor assembly
 - New grease seal
 - Hub and rotor assembly on the wheel spindle
 - Outer wheel bearing
 - Retainer washer and the spindle nut

7. Adjust the wheel bearings (2WD only).

8. Install or connect the following:
 - Retaining washer, so the castellations are aligned with the cotter pin hole. Install a new cotter pin.
 - Anchor plate, and the 2 retaining bolts. Tighten the bolts to 125–168 ft. lbs. (170–230 Nm).
 - Anti-rattle clips, and the disc brake pads
 - Caliper
 - Wheels. Tighten the lug nuts to 83–112 ft. lbs. (113–153 Nm).

Heater Core replacement is covered in Section 2 of this manual

9. Check the wheel and tire assembly for proper rotation, then install the grease cap.

10. Lower the vehicle.

11. Road test the vehicle and check for proper operation.

4-Wheel Drive

EXCEPT EXCURSION

The wheel bearings are of the cartridge design and are an integral part of the hub assembly. The bearings are permanently lubricated and require no maintenance or adjustments. If required, a new hub assembly must be installed.

1. Before servicing the vehicle, refer to the precautions in the beginning of this section.

2. Remove or disconnect the following:
- Wheels
- Caliper
- Brake pads and anti-rattle clips
- Anchor plate
- Brake rotor
- Rotor shield
- Speed sensor retaining bolt and move the speed sensor and harness aside, if equipped with 4-wheel Anti-lock Brake System (ABS)
- Hub nut cotter pin, retainer and the hub nut. Discard the cotter pin.
- The 3 hub assembly retaining bolts from the inside of the steering knuckle
- Hub assembly

➡If necessary, use a suitable puller to separate the hub assembly from the CV-joint. Use care not to over extend the CV-joint and boot when removing the hub assembly.

- Grease seal from the steering knuckle

To install:

3. Install or connect the following:
- New grease seal using Bearing Cup Replacer 205-147 (T80T-4000-P), Knuckle Seal Replacer 205-361 (T96T-1175-A) and Threaded Drawbar 204-029 (T77F-1176-A)
- Hub assembly to the steering knuckle and secure with the 3

retaining bolts. Tighten the bolts to 110–145 ft. lbs. (149–201 Nm).
- CV-joint into the hub assembly
- Speed sensor and secure with the retaining bolt, if equipped with 4-wheel ABS. Tighten the bolt to 60–84 inch lbs. (7–9 Nm).
- Brake rotor shield and the 3 retaining bolts. Tighten the bolts to 80–107 inch lbs. (9–12 Nm).
- Hub nut and tighten to 188–254 ft. lbs. (255–345 Nm)
- Hub nut retainer and a new cotter pin
- Brake rotor
- Anchor plate in position and install the 2 retaining bolts. Tighten the bolts to 125–168 ft. lbs. (170–230 Nm).
- Anti-rattle clips and the disc brake pads
- Caliper
- Wheels. Tighten the lug nuts to 83–112 ft. lbs. (113–153 Nm).

4. Lower the vehicle.

5. Pump the brake pedal several times to position the brake pads prior to moving the vehicle.

6. Road test the vehicle and check for proper operation.

EXCURSION

1. Before servicing the vehicle, refer to the precautions in the beginning of this section.

2. Remove or disconnect the following:
- Front brake rotor
- Hub lock by pulling out the retainer ring. And then the hub lock
- Snap ring and the axle shaft thrust washers
- ABS wheel sensor harness and routing clips, if equipped

➡The wheel hub and bearing is a slip fit design and should not require a puller to remove it.

- Four lock nuts
- Wheel hub and bearing
- Brake rotor shield
- Bolt and the Anti-lock Brake System (ABS) sensor, if equipped

3. Place the hub in a soft-jawed vise.

4. Install two nuts on the studs and use the inner nut to remove the studs.

5. Remove and discard the O-ring.

To install:

➡Any time the wheel hub is removed for any reason, a new O-ring seal must be installed. Failure to do so can cause a vacuum leak and loss of four wheel drive operations.

6. Install a new O-ring.

7. Place the hub in a soft-jawed vise.

8. Install two nuts on the studs and use the outer nut to install the studs.

9. Install or connect the following:
- ABS sensor and the bolt. Tighten the bolt to 13 ft. lbs. (18 Nm).
- Brake rotor shield

10. Apply a coat of High Temperature 4x4 Front Axle and Wheel Bearing Grease E8TZ-19590-A to the O-ring area of the wheel hub and bearing before installing the hub and bearing.
- Wheel hub and bearing
- Four lock nuts and tighten to 133 ft. lbs. (180 Nm)
- ABS sensor harness and routing clips, if equipped

✳✳ WARNING

The non-metallic thrust washer must be installed between the two metal thrust washers. Failure to do so will cause severe wear to the non-metallic thrust washer, allowing the axle shaft to travel further in and out during torque thrust causing damage to the wheel hub and bearing, the axle shaft end seal and the axle shaft.

- Three thrust washers onto the axle shaft
- Snap ring

➡Any time the hub lock is removed, a new O-ring seal must be installed. Failure to do so can cause a vacuum leak and loss of four wheel drive functions.

- New O-ring seal
- Hub lock and the retainer ring
- Brake rotor

11. Perform a wheel-end vacuum leak test.

CHEVROLET/SUZUKI

Chevrolet-Tracker • **Suzuki**-Vitara • Grand Vitara

12

PRECAUTIONS

Before servicing any vehicle, please be sure to read all of the following precautions, which deal with personal safety, prevention of component damage and important points to take into consideration when servicing a motor vehicle:

• Never open, service or drain the radiator or cooling system when the engine is hot; serious burns can occur from the steam and hot coolant.

• Observe all applicable safety precautions when working around fuel. Whenever servicing the fuel system, always work in a well-ventilated area. Do not allow fuel spray or vapors to come in contact with a spark, open flame, or excessive heat (a hot drop light, for example). Keep a dry chemical fire extinguisher near the work area. Always keep fuel in a container specifically designed for fuel storage; also, always properly seal fuel containers to avoid the possibility of fire or explosion. Refer to the additional fuel system precautions later in this section.

• Fuel injection systems often remain pressurized, even after the engine has been turned **OFF**. The fuel system pressure must be relieved before disconnecting any fuel lines. Failure to do so may result in fire and/or personal injury.

• Brake fluid often contains polyglycol ethers and polyglycols. Avoid contact with the eyes and wash your hands thoroughly after handling brake fluid. If you do get brake fluid in your eyes, flush your eyes with clean, running water for 15 minutes. If eye irritation persists, or if you have taken brake fluid internally, IMMEDIATELY seek medical assistance.

• The EPA warns that prolonged contact with used engine oil may cause a number of skin disorders, including cancer. You should make every effort to minimize your exposure to used engine oil. Protective gloves should be worn when changing oil. Wash your hands and any other exposed skin areas as soon as possible after exposure to used engine oil. Soap and water, or waterless hand cleaner should be used.

• All new vehicles are now equipped with an air bag system, often referred to as a Supplemental Restraint System (SRS) or Supplemental Inflatable Restraint (SIR) system. The system must be disabled before performing service on or around system components, steering column, instrument panel components, wiring and sensors. Failure to follow safety and disabling procedures could result in accidental air bag deployment, possible personal injury, and unnecessary system repairs.

• Always wear safety goggles when working with, or around, the air bag system. When carrying a non-deployed air bag, be sure the bag and trim cover are pointed away from your body. When placing a non-deployed air bag on a work surface, always face the bag and trim cover upward, away from the surface. This will reduce the motion of the module if it is accidentally deployed. Refer to the additional air bag system precautions later in this section.

• Clean, high quality brake fluid from a sealed container is essential to the safe and proper operation of the brake system. You should always buy the correct type of brake fluid for your vehicle. If the brake fluid becomes contaminated, completely flush the system with new fluid. Never reuse any brake fluid. Any brake fluid that is removed from the system should be discarded. Also, do not allow any brake fluid to come in contact with a painted surface; it will damage the paint.

• Never operate the engine without the proper amount and type of engine oil; doing so WILL result in severe engine damage.

• Timing belt maintenance is extremely important. Many models utilize an interference-type, non-freewheeling engine. If the timing belt breaks, the valves in the cylinder head may strike the pistons, causing potentially serious (also time-consuming and expensive) engine damage. Refer to the maintenance interval charts in the front of this manual for the recommended replacement interval for the timing belt, and to the timing belt section for belt replacement and inspection.

• Disconnecting the negative battery cable on some vehicles may interfere with the functions of the on-board computer system(s) and may require the computer to undergo a relearning process once the negative battery cable is reconnected.

• When servicing drum brakes, only disassemble and assemble one side at a time, leaving the remaining side intact for reference.

ENGINE REPAIR

➡**Disconnecting the negative battery cable on some vehicles may interfere with the functions of the on board computer system. The computer may undergo a relearning process once the negative battery cable is reconnected.**

Distributor

REMOVAL

All engines are equipped with a Distributorless Ignition System (DIS).

Alternator

REMOVAL

1. Before servicing the vehicle, refer to the precautions in the beginning of this section.

2. Remove or disconnect the following:
• Negative battery cable
• Evaporative Emission (EVAP) canister
• Accessory drive belt
• Alternator harness connectors
• Alternator mounting bracket
• Alternator

INSTALLATION

Install or connect the following:
• Alternator
• Alternator mounting bracket. Tighten the bolts to 20 ft. lbs. (27 Nm).
• Alternator harness connectors
• Accessory drive belt. Tighten the alternator bolts to 24 ft. lbs. (33 Nm).
• EVAP canister
• Negative battery cable

Ignition Timing

ADJUSTMENT

1.6L Engine

This engine is equipped with a Distributorless Ignition System (DIS). All timing functions are controlled by the Powertrain Control Module (PCM). No adjustment is possible.

2.0L and 2.5L Engines

➡**The 2.0L and 2.5L engines use a Camshaft Position (CMP) sensor that is rotated to set base timing.**

➡**Check and adjust the ignition timing with the engine at normal operating temperature, all electrical accessories OFF and transmission in P, N for auto-**

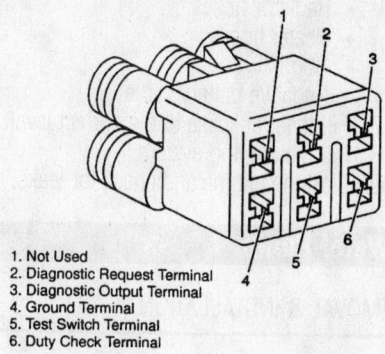

1. Not Used
2. Diagnostic Request Terminal
3. Diagnostic Output Terminal
4. Ground Terminal
5. Test Switch Terminal
6. Duty Check Terminal

7924HG01

Duty Check Data Link Connector terminal identification for ignition timing

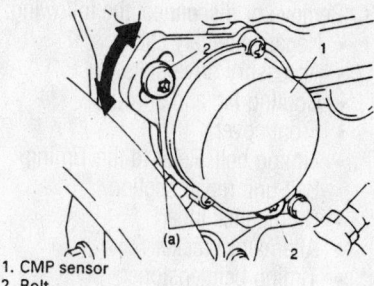

1. CMP sensor
2. Bolt

7924HG82

Camshaft position sensor

matic transmission or neutral for manual transmission.

1. Before servicing the vehicle, refer to the precautions in the beginning of this section.

2. With the engine **OFF**, connect a jumper wire between terminals **4** and **5** of the Data Link Connector (DLC) for Tracker or between terminals **D** and **E** of the DLC for all others.

3. Connect a timing light to the No. 1 spark plug wire and start the engine.

4. Ignition timing at idle should be 4–6 degrees Before Top Dead Center (BTDC).

5. Adjust the timing as necessary, then turn the engine **OFF**.

6. Remove the jumper wire from the DLC and remove the timing light.

Engine Assembly

REMOVAL & INSTALLATION

1.6L Engine

1. Before servicing the vehicle, refer to the precautions in the beginning of this section.

2. Relieve the fuel system pressure.
3. Drain the cooling system and engine oil.
4. Remove or disconnect the following:
   - Negative battery cable
   - Hood
   - Strut tower bar, if equipped
   - Cooling fan and shroud
   - Heater hoses
   - Radiator hoses
   - Bypass hose
   - Radiator
   - Air intake tube
   - Accelerator cable
   - Transmission cable, if equipped
   - Positive Crankcase Ventilation (PCV) valve and hose
   - Exhaust Gas Recirculation (EGR) valve and temperature sensor
   - EGR vacuum valve connector
   - EGR bypass valve connector
   - Idle Air Control (IAC) valve hoses and connector
   - Fuel lines
   - Main engine control wiring harness connectors at the firewall
   - Throttle Position (TP) sensor connector
   - Heated Oxygen (HO2S) sensor connectors
   - Engine Coolant Temperature (ECT) sensor connector
   - Temperature gauge sender connector
   - A/C temperature switch, if equipped
   - Injector harness connectors
   - Alternator wiring connectors
   - Manifold Absolute Pressure (MAP) sensor connector and vacuum line
   - Brake booster vacuum line
   - Evaporative Emission (EVAP) canister and hoses
   - Distributor, if equipped
   - Front skidplate, if equipped
   - Power steering hoses
   - A/C compressor, if equipped
   - Flywheel access cover
   - Torque converter, if equipped
   - Clutch cable, if equipped
   - Transmission cooler lines, if equipped
   - Exhaust front pipe
   - Starter motor
   - Transmission flange fasteners and support the transmission
   - Left and right engine mounts
   - Engine

To install:
5. Install or connect the following:
   - Engine. Tighten the mount bolts to 40 ft. lbs. (54 Nm)
   - Transmission flange fasteners. Tighten them to 62 ft. lbs. (85 Nm).
   - Starter motor
   - Exhaust front pipe
   - Transmission cooler lines, if equipped
   - Clutch cable, if equipped
   - Torque converter, if equipped. Tighten the bolts to 40 ft. lbs. (54 Nm).
   - Flywheel access cover
   - A/C compressor, if equipped
   - Power steering hoses
   - Front skidplate, if equipped
   - Distributor, if equipped
   - EVAP canister and hoses
   - Brake booster vacuum line
   - MAP sensor connector and vacuum line
   - Alternator wiring connectors
   - Injector harness connectors
   - A/C temperature switch, if equipped
   - Temperature gauge sender connector
   - ECT sensor connector
   - HO2S sensor connectors
   - TP sensor connector
   - Main engine control wiring harness connectors at the firewall
   - Fuel lines
   - IAC valve hoses and connector
   - EGR bypass valve connector
   - EGR vacuum valve connector
   - EGR valve and temperature sensor
   - PCV valve and hose
   - Transmission cable, if equipped
   - Accelerator cable
   - Air intake tube
   - Radiator
   - Bypass hose
   - Radiator hoses
   - Heater hoses
   - Cooling fan and shroud
   - Strut tower bar, if equipped
   - Hood
   - Negative battery cable
6. Fill the crankcase to the correct level.
7. Fill the cooling system.
8. Start the engine and check for leaks.

2.0L and 2.5L Engines

1. Before servicing the vehicle, refer to the precautions in the beginning of this section.
2. Relieve the fuel system pressure.
3. Drain the cooling system.
4. Drain the engine oil.

5. Remove or disconnect the following:
- Negative battery cable
- Hood
- Heater hoses
- Radiator hoses
- Cooling fan and shroud
- Radiator overflow tank
- Radiator
- Accelerator cable
- Transmission cable, if equipped
- Strut tower bar, if equipped
- Air intake assembly
- Engine oil dipstick tube
- Transmission oil dipstick tube, if equipped
- Ignition coil covers
- Ignition coil connectors
- Injector connectors
- Camshaft Position (CMP) sensor connector
- Crankshaft Position (CKP) sensor connector
- Throttle Position (TP) sensor connector
- Mass Air Flow (MAF) sensor connector
- Idle Air Control (IAC) valve
- Intake manifold ground cable
- Evaporative Emission (EVAP) canister purge valve
- Exhaust Gas Recirculation (EGR) valve connector
- Heated Oxygen (HO2S) sensor connectors
- Engine Coolant Temperature (ECT) sensor connector
- Alternator wiring connectors
- Oil pressure gauge sender connector
- Power Steering Pressure (PSP) switch connector
- Alternator bracket ground cable
- Brake booster vacuum line
- Tank pressure control vacuum valve hose
- Fuel lines
- EVAP canister
- Power steering pump
- A/C compressor
- Steering shaft lower assembly
- Front differential housing, if equipped
- Exhaust front pipe and bracket
- Transmission oil cooler lines, if equipped
- Transmission stiffener brackets, if equipped
- Flywheel access cover
- Torque converter, if equipped
- Starter motor
- Transmission flange fasteners and support the transmission
- Left and right engine mounts
- Engine

To install:

6. Install or connect the following:
- Engine
- Left and right engine mounts. Tighten the nuts to 36 ft. lbs. (50 Nm).
- Transmission flange fasteners. Tighten them to 58 ft. lbs. (80 Nm).
- Starter motor
- Torque converter. Tighten the bolts to 47 ft. lbs. (65 Nm).
- Flywheel access cover
- Transmission stiffener brackets. Tighten the bolts to 36 ft. lbs. (50 Nm).
- Transmission oil cooler lines, if equipped
- Exhaust front pipe and bracket
- Front differential housing, if equipped
- Steering shaft lower assembly
- A/C compressor
- Power steering pump
- EVAP canister
- Fuel lines
- Tank pressure control vacuum valve hose
- Brake booster vacuum line
- Alternator bracket ground cable
- PSP switch connector
- Oil pressure gauge sender connector
- Alternator wiring connectors
- ECT sensor connector
- HO2S sensor connectors
- EGR valve connector
- EVAP canister purge valve
- Intake manifold ground cable
- IAC valve
- MAF sensor connector
- TP sensor connector
- CKP sensor connector
- CMP sensor connector
- Injector connectors
- Ignition coil connectors
- Ignition coil covers
- Transmission oil dipstick tube, if equipped
- Engine oil dipstick tube
- Air intake assembly
- Strut tower bar, if equipped
- Transmission cable, if equipped
- Accelerator cable
- Radiator
- Radiator overflow tank
- Cooling fan and shroud
- Radiator hoses
- Heater hoses
- Hood
- Negative battery cable

7. Fill the crankcase to the correct level.
8. Fill the cooling system.
9. Start the engine and check for leaks.

Water Pump

REMOVAL & INSTALLATION

1.6L Engines

1. Before servicing the vehicle, refer to the precautions in the beginning of this section.
2. Drain the cooling system.
3. Remove or disconnect the following:
- Negative battery cable
- Accessory drive belts
- Cooling fan and shroud
- Front cover
- Timing belt. Refer to the Timing Belt unit repair section.
- Oil dipstick tube
- Alternator bracket
- Timing belt tensioner
- Water pump

To install:

4. Install or connect the following:
- Water pump with a new gasket. Tighten the bolts to 106 inch lbs. (12 Nm).
- Timing belt tensioner
- Alternator bracket
- Oil dipstick tube. Tighten the bolt to 97 inch lbs. (11 Nm).
- Timing belt. Refer to the Timing Belt unit repair section.
- Front cover
- Cooling fan and shroud
- Accessory drive belts
- Negative battery cable

5. Fill the cooling system.
6. Start the engine and check for leaks.

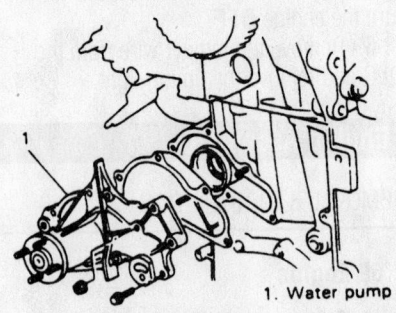

1. Water pump

7924HG04

Exploded view of the water pump mounting—1.6L engines

2.0L Engines

1. Before servicing the vehicle, refer to the precautions in the beginning of this section.
2. Drain the cooling system.
3. Remove or disconnect the following:
   - Negative battery cable
   - Radiator hose at the thermostat housing
   - Heater outlet pipe bolt
   - Alternator belt
   - Water pump

To install:

➡**Use new water pump bolts for assembly.**

4. Install or connect the following:
   - Water pump with a new O-ring seal. Tighten the bolts to 19 ft. lbs. (27 Nm).
   - Alternator belt
   - Heater outlet pipe bolt
   - Radiator hose at the thermostat housing
   - Negative battery cable
5. Fill the cooling system.
6. Start the engine and check for leaks.

2.5L Engine

1. Before servicing the vehicle, refer to the precautions in the beginning of this section.
2. Drain the cooling system.
3. Remove or disconnect the following:
   - Negative battery cable
   - Accessory drive belts
   - Front cover
   - Water pump

To install:

4. Install or connect the following:
   - Water pump with a new O-ring seal. Tighten the bolts to 19 ft. lbs. (27 Nm).
   - Front cover
   - Accessory drive belts
   - Negative battery cable
5. Fill the cooling system.
6. Start the engine and check for leaks.

Cylinder Head

REMOVAL & INSTALLATION

1.6L Engine

1. Before servicing the vehicle, refer to the precautions in the beginning of this section.

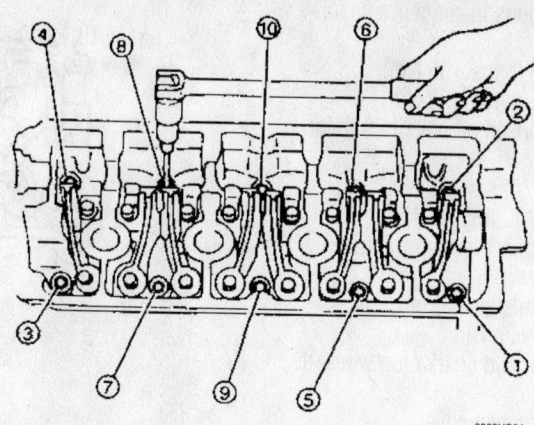

9308HG01

Cylinder head loosening sequence—1.6L engine

2. Relieve the fuel system pressure.
3. Drain the cooling system.
4. Remove or disconnect the following:
   - Negative battery cable
   - Accessory drive belts
   - Air intake pipe
   - Fuel lines
   - Upper radiator hose
   - Coolant bypass hose
   - Alternator bracket
   - Intake manifold brackets
   - Intake manifold
   - Heated Oxygen (HO_2S) sensor connectors
   - Exhaust manifold heat shield
   - Exhaust manifold bracket
   - Exhaust front pipe
   - Exhaust manifold
   - A/C compressor
   - Power steering pump
   - Front cover
   - Timing belt. Refer to the Timing Belt unit repair section.
   - Camshaft timing sprocket
   - Rear timing belt cover
   - Distributor and spark plug wires, if equipped
   - Ignition coils and wiring, if equipped with Distributorless Ignition System (DIS)
   - Valve cover
   - Cylinder head. Loosen the bolts in the sequence shown.

To install:

5. Install the cylinder head with a new gasket.

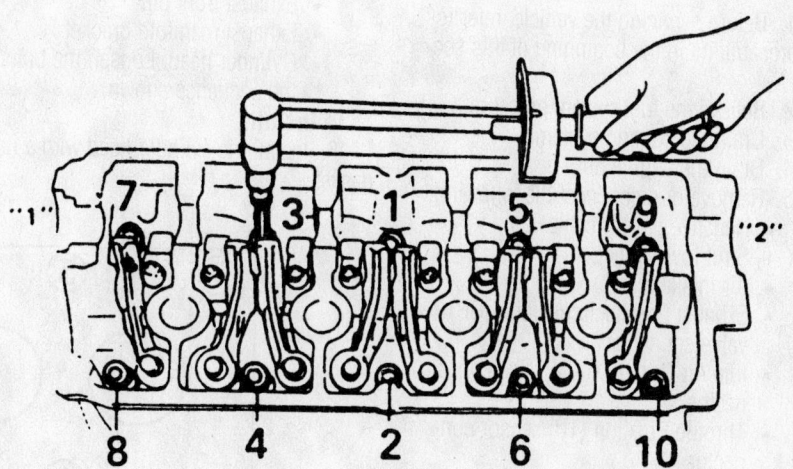

"1": Camshaft pulley side
"2": Distributor side

7924HG08

Cylinder head torque sequence—1.6L engine

6. Tighten the bolts in sequence as follows:

  a. Step 1: 26 ft. lbs. (35 Nm)
  b. Step 2: 41 ft. lbs. (55 Nm)
  c. Step 3: Loosen all bolts to 0 ft. lbs. (0 Nm)
  d. Step 4: 26 ft. lbs. (35 Nm)
  e. Step 5: 52 ft. lbs. (70 Nm)

7. Install or connect the following:

- Valve cover
- Ignition coils and wiring, if equipped with DIS
- Distributor and spark plug wires, if equipped
- Rear timing belt cover
- Camshaft timing sprocket
- Timing belt
- Front cover
- Power steering pump
- A/C compressor
- Exhaust manifold
- Exhaust front pipe
- Exhaust manifold bracket
- Exhaust manifold heat shield
- HO2S sensor connectors
- Intake manifold
- Intake manifold brackets
- Alternator bracket
- Coolant bypass hose
- Upper radiator hose
- Fuel lines
- Air intake pipe
- Accessory drive belts
- Negative battery cable

8. Fill the cooling system.
9. Start the engine and check for leaks.

2.0L Engines

1. Before servicing the vehicle, refer to the precautions in the beginning of this section.
2. Relieve the fuel system pressure.
3. Drain the cooling system.
4. Drain the engine oil.
5. Remove or disconnect the following:

- Negative battery cable
- Strut tower brace
- Air intake tube
- Exhaust Gas Recirculation (EGR) valve connector
- Idle Air Control (IAC) valve connector
- Throttle Position (TP) sensor connector
- Evaporative Emission (EVAP) canister purge valve connector and hose
- Intake manifold ground cable
- Heated Oxygen (HO2S) sensor connectors
- Camshaft Position (CMP) sensor connector

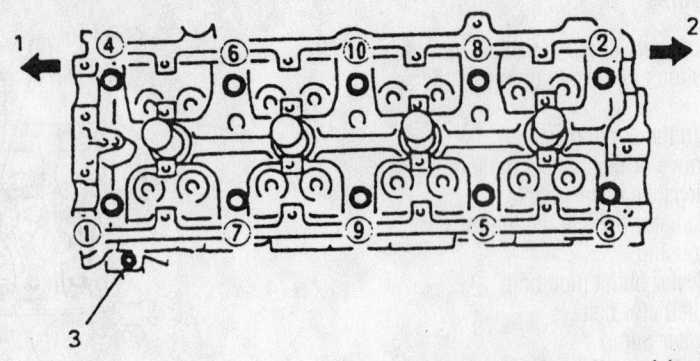

1. Crankshaft pulley side
2. Flywheel side
3. Bolt (M6)

7924HG09

Cylinder head loosening sequence—2.0L engines

- Engine Coolant Temperature (ECT) sensor connector
- Fuel injector connectors
- Ignition coils
- Accelerator cable
- Transmission cable, if equipped
- Brake booster vacuum hose
- Radiator hose
- Bypass hose
- Heater hose
- Fuel lines
- Intake manifold bracket
- Water pipe
- Valve cover
- Accessory drive belts
- Oil pan
- Front cover
- Timing chains
- Camshafts
- Exhaust front pipe
- Exhaust manifold bracket
- Cylinder head. Loosen the bolts in the sequence shown.

To install:

6. Install the cylinder head with a new gasket.

7. Tighten the bolts in sequence as follows:

  a. Step 1: 38 ft. lbs. (52 Nm)
  b. Step 2: 61 ft. lbs. (84 Nm)
  c. Step 3: Loosen all bolts to 0 ft. lbs. (0 Nm)
  d. Step 4: 38 ft. lbs. (52 Nm)
  e. Step 5: 76 ft. lbs. (105 Nm)
  f. Step 6: 6mm bolt to 96 inch lbs. (8 Nm)

8. Install or connect the following:

- Exhaust manifold bracket
- Exhaust front pipe
- Camshafts
- Timing chains
- Front cover
- Oil pan
- Accessory drive belts
- Valve cover
- Water pipe
- Intake manifold bracket
- Fuel lines
- Heater hose
- Bypass hose
- Radiator hose
- Brake booster vacuum hose

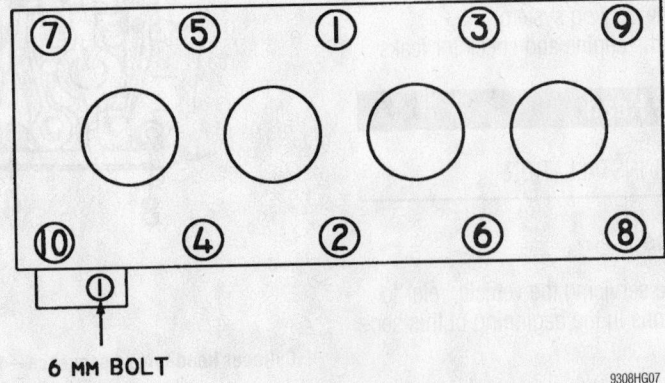

6 MM BOLT

9308HG07

Cylinder head torque sequence—2.0L engines

- Transmission cable, if equipped
- Accelerator cable
- Ignition coils
- Fuel injector connectors
- ECT sensor connector
- CMP sensor connector
- HO2S sensor connectors
- Intake manifold ground cable
- EVAP canister purge valve connector and hose
- TP sensor connector
- IAC valve connector
- EGR valve connector
- Air intake tube
- Strut tower brace
- Negative battery cable

9. Fill the crankcase to the correct level.
10. Fill the cooling system.
11. Start the engine and check for leaks.

2.5L Engine

1. Before servicing the vehicle, refer to the precautions in the beginning of this section.
2. Relieve the fuel system pressure.

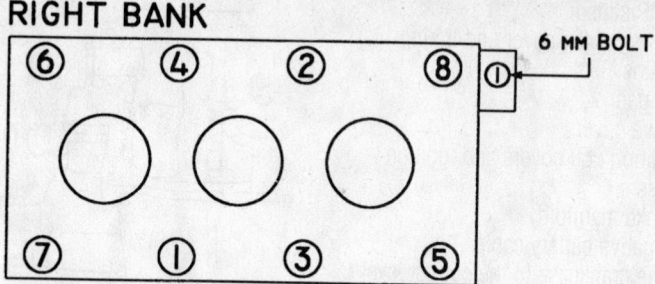

Cylinder head torque sequence—2.5L engine

9308HG08

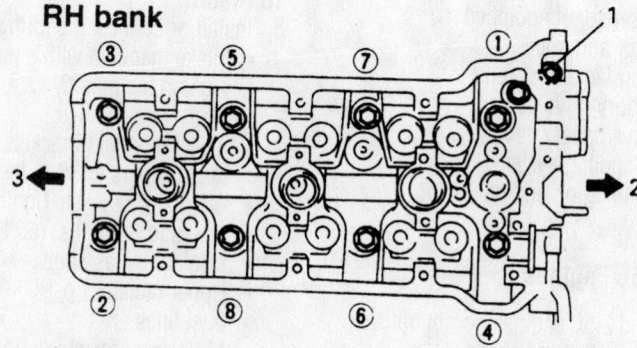

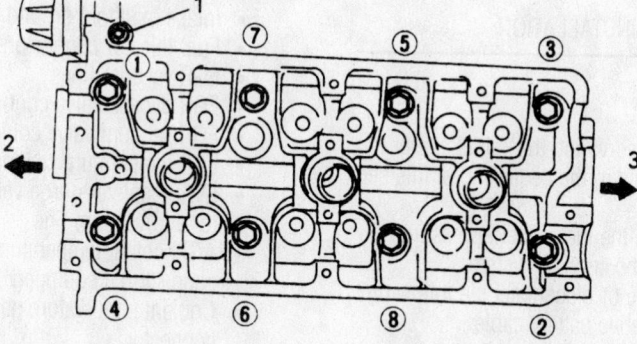

1. Hex hole bolt
2. Timing chain side
3. Flywheel side

9302HG01

Cylinder head loosening sequence—2.5L engine

3. Drain the cooling system and engine oil.
4. Remove or disconnect the following:
   - Negative battery cable
   - Intake manifold
   - Ignition coil covers and ignition coils
   - Valve covers
   - Oil pan
   - Timing chain cover and timing chains
   - Camshaft Position (CMP) sensor
   - Camshafts
   - Exhaust manifolds
   - Water outlet caps
   - Cylinder heads. Loosen the bolts in the sequence shown.

To install:

5. Install the cylinder heads with new gaskets. Tighten the bolts in sequence as follows:
   a. Step 1: 38 ft. lbs. (52 Nm)
   b. Step 2: 61 ft. lbs. (84 Nm)
   c. Step 3: Loosen all bolts to 0 ft. lbs. (0 Nm)
   d. Step 4: 38 ft. lbs. (52 Nm)
   e. Step 5: 76 ft. lbs. (105 Nm)
   f. Step 6: 6mm bolt to 96 inch lbs. (8 Nm)
6. Install or connect the following:
   - Water outlet caps
   - Exhaust manifolds
   - Camshafts

Timing belt service is covered in Section 3 of this manual

- CMP sensor
- Timing chain cover and timing chains
- Oil pan
- Valve covers
- Ignition coil covers and ignition coils
- Intake manifold
- Negative battery cable

7. Fill the crankcase to the correct level
8. Fill the cooling system.
9. Start the engine and check for leaks.

Rocker Arms/Shafts

REMOVAL & INSTALLATION

1.6L Engine

1. Before servicing the vehicle, refer to the precautions in the beginning of this section.
2. Drain the cooling system.
3. Remove or disconnect the following:
   - Negative battery cable
   - Accessory drive belts
   - Cooling fan and shroud
   - Radiator and hoses
   - Distributor, if equipped
   - Camshaft Position (CMP) sensor, if equipped
   - Valve cover
   - Front cover
   - Timing belt. Refer to the Timing Belt unit repair section.
   - Camshaft
   - Rocker arm shaft plug
   - Rear timing belt cover

4. Loosen the rocker arm locknuts and back the valve adjusters off until all rocker arms move freely with no tension.

➡ **Keep all valvetrain components in order for assembly.**

5. Remove or disconnect the following:
   - Intake rocker arms and clips
   - Rocker arm shaft bolts
6. Push the rocker arm shaft towards the rear of the cylinder head and remove the rocker arm shaft O-ring.
7. Remove the exhaust rocker arms and springs by pulling the rocker arm shaft out of the front of the cylinder head.

To install:
8. Insert the rocker arm shaft into the front of the cylinder head, while installing the exhaust rocker arms and springs in their original positions.
9. Push the end of the rocker arm shaft out of the rear of the cylinder head and install a new O-ring.

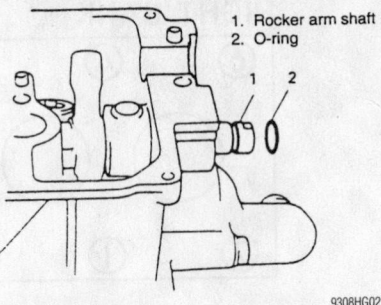

1. Rocker arm shaft
2. O-ring

9308HG02

Rocker arm shaft and O-ring—1.6L engine

10. Install or connect the following:
    - Rocker arm shaft bolts. Tighten them to 96 inch lbs. (8 Nm).
    - Intake rocker arms and clips in their original positions
    - Rear timing belt cover
    - Rocker arm shaft plug. Tighten it to 24 ft. lbs. (33 Nm).
    - Camshaft
    - Timing belt and adjust the valve clearance
    - Valve cover
    - Front cover
    - Distributor, if equipped
    - CMP sensor, if equipped
    - Radiator and hoses
    - Cooling fan and shroud
    - Accessory drive belts
    - Negative battery cable
11. Fill the cooling system.
12. Start the engine and check for leaks.

2.0L and 2.5L Engines

The 2.0L and 2.5L engines do not utilize rocker arms or rocker arm shafts.

Intake Manifold

REMOVAL & INSTALLATION

1.6L Engine

1. Before servicing the vehicle, refer to the precautions in the beginning of this section.
2. Relieve the fuel system pressure.
3. Drain the cooling system.
4. Remove or disconnect the following:
   - Negative battery cable
   - Air intake pipe
   - Accelerator cable and bracket
   - Transmission cable, if equipped
   - Throttle Position (TP) sensor connector
   - Idle Air Control (IAC) valve connector and hoses

- Engine Coolant Temperature (ECT) sensor connector
- Coolant temperature gauge sender connector
- A/C coolant temperature switch connector, if equipped
- Evaporative Emission (EVAP) canister purge valve connector and vacuum line
- Exhaust Gas Recirculation (EGR) temperature sensor connector
- EGR vacuum valve connector
- EGR bypass valve connector
- EGR vacuum lines
- Fuel injector connectors
- Intake manifold ground cable
- Transmission vacuum line, if equipped
- Brake booster vacuum line
- Manifold Absolute Pressure (MAP) sensor vacuum line
- Fuel lines
- Upper radiator hose
- Coolant bypass hose
- Alternator bracket
- Intake manifold brackets
- Intake manifold

To install:
5. Install or connect the following:
   - Intake manifold with a new gasket. Tighten the nuts to 17 ft. lbs. (23 Nm).
   - Intake manifold brackets. Tighten the fasteners to 36 ft. lbs. (50 Nm).
   - Alternator bracket. Tighten the fasteners to 36 ft. lbs. (50 Nm).
   - Coolant bypass hose
   - Upper radiator hose
   - Fuel lines
   - MAP sensor vacuum line
   - Brake booster vacuum line
   - Transmission vacuum line, if equipped
   - Intake manifold ground cable
   - Fuel injector connectors
   - EGR vacuum lines
   - EGR bypass valve connector
   - EGR vacuum valve connector
   - EGR temperature sensor connector
   - EVAP canister purge valve connector and vacuum line
   - A/C coolant temperature switch connector, if equipped
   - Coolant temperature gauge sender connector
   - ECT sensor connector
   - IAC valve connector and hoses
   - TP sensor connector
   - Transmission cable, if equipped
   - Accelerator cable and bracket
   - Air intake pipe
   - Negative battery cable

6. Fill the cooling system.
7. Start the engine and check for leaks.

2.0L Engines

1. Before servicing the vehicle, refer to the precautions in the beginning of this section.
2. Relieve the fuel system pressure.
3. Drain the cooling system.
4. Remove or disconnect the following:
   • Negative battery cable
   • Air intake tube
   • Exhaust Gas Recirculation (EGR) valve connector
   • Idle Air Control (IAC) valve connector
   • Throttle Position (TP) sensor connector
   • Evaporative Emissions (EVAP) canister purge valve connector and hose
   • Intake manifold ground cable
   • Manifold Absolute Pressure (MAP) sensor connector
   • Accelerator cable
   • Transmission cable, if equipped
   • Brake booster vacuum hose
   • Positive Crankcase Ventilation (PCV) valve and hose
   • Fuel pressure regulator vacuum hose
   • Intake manifold vacuum hose
   • Throttle body coolant hoses
   • Water bypass pipe
   • Fuel lines
   • Fuel supply manifold with injectors attached
   • Intake manifold support brackets
   • Intake manifold water pipe
   • Intake manifold

To install:
5. Install or connect the following:
   • Intake manifold with a new gasket. Tighten the fasteners to 17 ft. lbs. (23 Nm).
   • Intake manifold water pipe
   • Intake manifold front support bracket. Tighten the bolts to 36 ft. lbs. (50 Nm).
   • Intake manifold rear support bracket. Tighten the bolts to 18 ft. lbs. (25 Nm).
   • Fuel supply manifold with injectors attached
   • Fuel lines
   • Water bypass pipe
   • Throttle body coolant hoses
   • Intake manifold vacuum hose
   • Fuel pressure regulator vacuum hose

   • PCV valve and hose
   • Brake booster vacuum hose
   • Transmission cable, if equipped
   • Accelerator cable
   • MAP sensor connector
   • Intake manifold ground cable
   • EVAP canister purge valve connector and hose
   • TP sensor connector
   • IAC valve connector
   • EGR valve connector
   • Air intake tube
   • Negative battery cable
6. Fill the cooling system.
7. Start the engine and check for leaks.

2.5L Engine

1. Before servicing the vehicle, refer to the precautions in the beginning of this section.
2. Relieve the fuel system pressure.
3. Drain the cooling system.
4. Remove or disconnect the following:
   • Negative battery cable
   • Strut tower bar
   • Intake Air Temperature (IAT) sensor connector
   • Surge tank cover
   • Air intake assembly
   • Accelerator cable
   • Transmission cable, if equipped
   • Throttle body coolant hoses
   • Fuel injector connectors
   • Throttle Position (TP) sensor connector
   • Mass Air Flow (MAF) sensor connector
   • Idle Air Control (IAC) valve connector
   • Intake manifold ground cables
   • Brake booster vacuum hose
   • Evaporative Emissions (EVAP) canister purge valve connector and hoses
   • Exhaust Gas Recirculation (EGR) valve connector
   • Positive Crankcase Ventilation (PCV) valve and hose
   • Heater hoses
   • EGR pipe
   • Fuel lines
   • Throttle body and intake collector
   • Intake manifold

To install:
5. Install or connect the following:
   • Intake manifold with new gaskets. Tighten the fasteners to 16 ft. lbs. (23 Nm).
   • Throttle body and intake collector

with new gaskets. Tighten the fasteners to 102 inch lbs. (12 Nm).
   • Fuel lines
   • EGR pipe
   • Heater hoses
   • PCV valve and hose
   • EGR valve connector
   • EVAP canister purge valve connector and hoses
   • Brake booster vacuum hose
   • Intake manifold ground cables
   • IAC valve connector
   • MAF sensor connector
   • TP sensor connector
   • Fuel injector connectors
   • Throttle body coolant hoses
   • Transmission cable, if equipped
   • Accelerator cable
   • Air intake assembly
   • Surge tank cover
   • IAT sensor connector
   • Strut tower bar
   • Negative battery cable
6. Fill the cooling system.
7. Start the engine and check for leaks.

Exhaust Manifold

REMOVAL & INSTALLATION

1.6L and 2.0L Engines

1. Before servicing the vehicle, refer to the precautions in the beginning of this section.
2. Remove or disconnect the following:
   • Negative battery cable
   • Strut tower bar, if equipped
   • Air intake assembly and bracket
   • Heated Oxygen (HO2S) sensor connector
   • Exhaust front pipe
   • Exhaust manifold heat shield
   • Exhaust manifold bracket, if equipped
   • Exhaust manifold

To install:
3. Install or connect the following:
   • Exhaust manifold with a new gasket. Tighten the fasteners to 13–20 ft. lbs. (18–28 Nm).
   • Exhaust manifold bracket, if equipped. Tighten the bolts to 36–43 ft. lbs. (50–60 Nm).
   • Exhaust manifold heat shield
   • Exhaust front pipe. Tighten the fasteners to 29–43 ft. lbs. (40–60 Nm).
   • HO2S sensor connector

Heater Core replacement is covered in Section 2 of this manual

- Air intake assembly and bracket
- Strut tower bar, if equipped. Tighten the fasteners to 66 ft. lbs. (90 Nm).
- Negative battery cable

4. Start the engine and check for leaks.

2.5L Engine

1. Before servicing the vehicle, refer to the precautions in the beginning of this section.
2. Remove or disconnect the following:
   - Negative battery cable
   - Strut tower bar
   - Air intake assembly
   - Heated Oxygen (HO2S) sensor connectors
   - Oil dipstick tube
   - Exhaust Gas Recirculation (EGR) pipe
   - Exhaust manifold heat shields
   - Evaporative Emissions (EVAP) canister
   - Front driveshaft, if equipped
   - Exhaust front pipe
   - Exhaust manifold brace
   - Exhaust manifolds

To install:

3. Install or connect the following:
   - Exhaust manifolds with new gaskets. Tighten the nuts to 21 ft. lbs. (30 Nm).
   - Exhaust manifold brace
   - Exhaust front pipe. Tighten the fasteners to 37 ft. lbs. (50 Nm).
   - Front driveshaft, if equipped
   - EVAP canister
   - Exhaust manifold heat shields
   - EGR pipe
   - Oil dipstick tube
   - HO2S sensor connectors
   - Air intake assembly
   - Strut tower bar
   - Negative battery cable

4. Start the engine and check for leaks.

Front Crankshaft Seal

REMOVAL & INSTALLATION

1.6L Engine

1. Before servicing the vehicle, refer to the precautions in the beginning of this section.
2. Drain the cooling system.
3. Remove or disconnect the following:
   - Negative battery cable
   - Accessory drive belts
   - Cooling fan and shroud
   - Water pump pulley

1. Oil seal
2. Oil pump case

7924HG13

Install the new oil pump seal flush with the oil pump housing—1.6L engine

- Crankshaft pulley
- Front cover
- Timing belt. Refer to the Timing Belt unit repair section.
- Crankshaft timing sprocket
- Front crankshaft seal

To install:

4. Install or connect the following:
   - Front crankshaft seal flush with the oil pump housing
   - Crankshaft timing sprocket. Tighten the bolt to 94 ft. lbs. (128 Nm).
   - Timing belt
   - Front cover
   - Crankshaft pulley. Tighten the bolts to 12 ft. lbs. (16 Nm).
   - Water pump pulley
   - Cooling fan and shroud
   - Accessory drive belts
   - Negative battery cable

5. Start the engine and check for leaks.

Camshaft and Valve Lifters

REMOVAL & INSTALLATION

1.6L Engine

1. Before servicing the vehicle, refer to the precautions in the beginning of this section.

2. Drain the cooling system.
3. Remove or disconnect the following:
   - Negative battery cable
   - Radiator
   - Accessory drive belts
   - Crankshaft pulley
   - Front cover
   - Timing belt. Refer to the Timing Belt unit repair section.
   - Camshaft sprocket
   - Valve cover
   - Distributor and case, if equipped
   - Camshaft Position (CMP) sensor and case, if equipped

4. Loosen the rocker arm locknuts and back the valve adjusters off until all rocker arms move freely with no tension.
5. Remove or disconnect the following:
   - Camshaft bearing caps. Loosen the bolts in reverse of the tightening sequence.
   - Camshaft

To install:

6. Install or connect the following:
   - Camshaft
   - Camshaft bearing caps. Tighten the bolts in sequence to 96 inch lbs. (11 Nm).
   - CMP sensor and case, if equipped
   - Distributor and case, if equipped
   - Camshaft sprocket. Tighten the bolt to 44 ft. lbs. (60 Nm).
   - Timing belt and adjust the valve clearance
   - Valve cover
   - Front cover
   - Crankshaft pulley. Tighten the bolts to 12 ft. lbs. (16 Nm).
   - Accessory drive belts
   - Radiator
   - Negative battery cable

7. Fill the cooling system.
8. Start the engine and check for leaks.

7924HG25

Camshaft housing torque sequence—1.6L engine

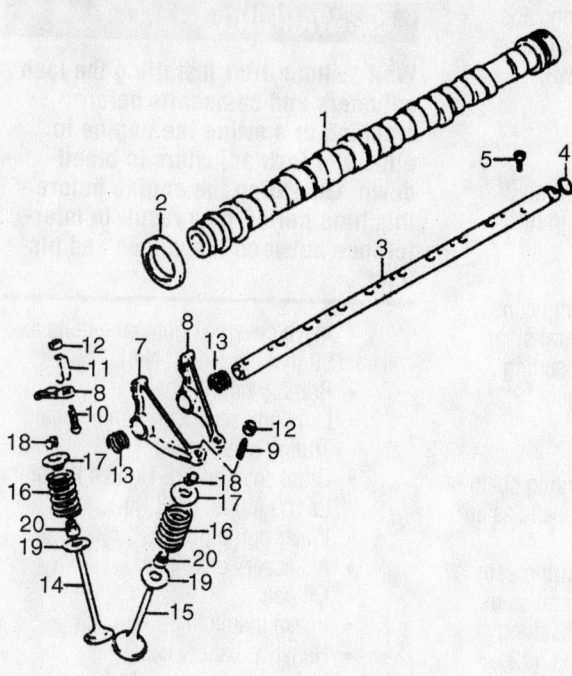

1. Camshaft
2. Camshaft oil seal
3. Rocker arm shaft
4. O ring
5. Rocker shaft bolt
6. Rocker arm (IN)
7. Rocker arm No. 1 (EX)
8. Rocker arm No. 2 (EX)
9. Valve adjusting screw
10. Valve adjusting screw
11. Clip
12. Lock nut
13. Rocker arm spring
14. Intake valve
15. Exhaust valve
16. Valve spring
17. Valve spring retainer
18. Valve cotter
19. Valve spring seat
20. Valve stem seal

7924HG24

Exploded view of the valve train components—1.6L engine

2.0L Engines

1. Before servicing the vehicle, refer to the precautions in the beginning of this section.
2. Drain the engine oil.
3. Drain the cooling system.
4. Remove or disconnect the following:
   - Negative battery cable
   - Oil pan
   - Valve cover
   - Accessory drive belts
   - Crankshaft pulley
   - Front cover
   - Secondary timing chain
   - Camshaft Position (CMP) sensor

➡**Keep all valvetrain components in order for installation.**

   - Camshaft bearing caps. Loosen the bolts in several steps in reverse of the tightening sequence.
   - Camshafts
   - Hydraulic lash adjusters

To install:

5. Install or connect the following:
   - Hydraulic lash adjusters in their original positions
   - Camshafts
   - Camshaft bearing caps in their original positions. Tighten the bolts in several steps in sequence to 96 inch lbs. (11 Nm).

   - CMP sensor
   - Secondary timing chain
   - Front cover
   - Crankshaft pulley. Tighten the bolt to 109 ft. lbs. (148 Nm).
   - Accessory drive belts
   - Valve cover
   - Oil pan
   - Negative battery cable
6. Fill the crankcase to the correct level.
7. Fill the cooling system.
8. Start the engine and check for leaks.

❈❈ WARNING

Wait ½ hour after installing the lash adjusters and camshafts before cranking or starting the engine to allow the lash adjusters to bleed down. Operating the engine before this time period may result in interference between the valves and pistons.

2.5L Engine

1. Before servicing the vehicle, refer to the precautions in the beginning of this section.
2. Drain the engine oil.
3. Drain the cooling system.
4. Remove or disconnect the following:
   - Negative battery cable
   - Intake manifold
   - Oil pan
   - Accessory drive belts
   - Water pump pulley

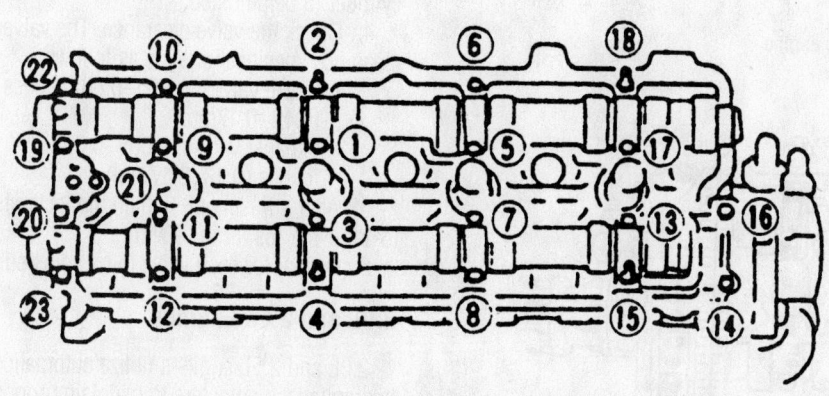

7924HG26

Camshaft housing torque sequence—2.0L engines

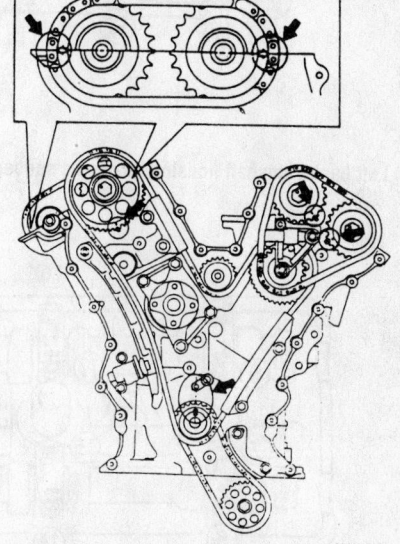

9302HG02

Timing mark alignment—2.5L engine

Brake service is covered in Section 4 of this manual

- Crankshaft pulley
- Timing chain cover

5. Align the timing marks as shown.

❉❉ WARNING

Do not allow the crankshaft or camshafts to rotate once the timing chains have been removed. Valve or piston damage could result.

6. Remove or disconnect the following:
   - Left bank secondary timing chain
   - Primary timing chain
   - Valve covers

➡**Keep all valvetrain components in order for assembly.**

- Right bank camshaft bearing caps. Loosen the bolts in several steps and in the sequence shown.
- Right bank secondary timing chain, exhaust and intake camshafts as an assembly
- Camshaft Position (CMP) sensor

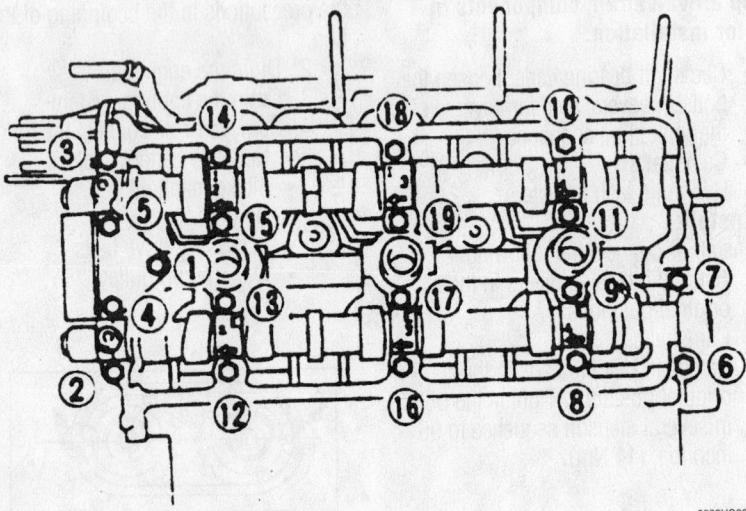

Left bank camshaft housing loosening sequence—2.5L engine

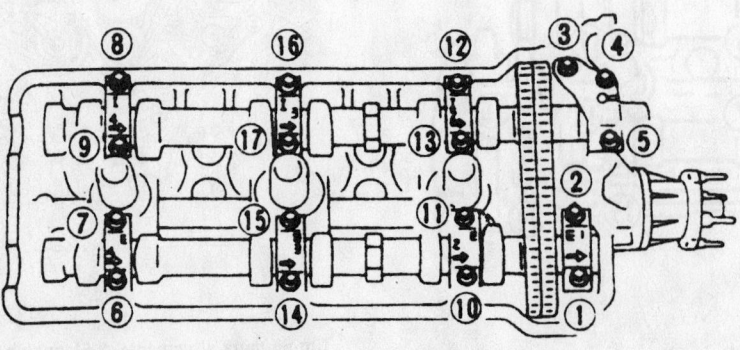

Right bank camshaft housing loosening sequence—2.5L engine

- Left bank camshaft bearing caps. Loosen the bolts in several steps and in the sequence shown.
- Left bank camshafts
- Hydraulic lash adjusters

To install:

7. Install or connect the following:
   - Hydraulic lash adjusters in their original positions
   - Left bank camshafts
   - Left bank camshaft bearing caps. Tighten the bolts in several steps and in reverse of the loosening sequence to 102 inch lbs. (12 Nm).
   - CMP sensor
   - Right bank secondary timing chain, exhaust and intake camshafts as an assembly
   - Right bank camshaft bearing caps. Tighten the bolts in several steps and in reverse of the loosening sequence to 102 inch lbs. (12 Nm).

❉❉ WARNING

Wait ½ hour after installing the lash adjusters and camshafts before cranking or starting the engine to allow the lash adjusters to bleed down. Operating the engine before this time period may result in interference between the valves and pistons.

- Valve covers. Tighten the bolts to 90 inch lbs. (10.5 Nm).
- Primary timing chain
- Left bank secondary timing chain
- Timing chain cover
- Crankshaft pulley. Tighten the bolt to 109 ft. lbs. (148 Nm).
- Water pump pulley
- Accessory drive belts
- Oil pan
- Intake manifold
- Negative battery cable

8. Fill the crankcase to the correct level.

9. Fill the cooling system.

10. Start the engine and check for leaks.

Valve Lash

ADJUSTMENT

1.6L Engines

➡**Measure valve clearance with the engine cold.**

1. Before servicing the vehicle, refer to the precautions in the beginning of this section.

2. Remove the valve cover.

3. Set the engine to Top Dead Center (TDC) of the compression stroke for the cylinder to be adjusted.

4. Check the valve clearance. The valve clearance specifications are as follows:
   - Intake valves: 0.005–0.007 inches (0.13–0.17mm)
   - Exhaust valves: 0.005–0.007 inches (0.13–0.17mm)

5. After adjustment, tighten the locknuts to 11–14 ft. lbs. (15–19 Nm).

6. Repeat for each valve to be adjusted.

2.0L and 2.5L Engines

2.0L and 2.5L engines utilize automatic hydraulic lash adjusters to maintain proper valve lash at all times. Periodic valve lash inspection and adjustment is not necessary or possible.

Starter Motor

REMOVAL & INSTALLATION

1. Before servicing the vehicle, refer to the precautions in the beginning of this section.
2. Remove or disconnect the following:
   - Negative battery cable
   - Starter motor wiring connectors
   - Starter motor

To install:

3. Install or connect the following:
   - Starter motor. Tighten the bolts to 22 ft. lbs. (30 Nm).
   - Starter motor wiring connectors. Tighten the solenoid nut to 11 ft. lbs. (15 Nm).
   - Negative battery cable

Oil Pan

REMOVAL & INSTALLATION

1.6L Engines

1. Before servicing the vehicle, refer to the precautions in the beginning of this section.
2. Drain the engine oil.
3. Remove or disconnect the following:
   - Negative battery cable
   - Front skidplate, if equipped
   - Front differential, if equipped
   - Crankshaft Position (CKP) sensor
   - Left transmission stiffener bracket, if equipped
   - Flywheel access panel
   - Oil pan and oil pump pickup tube

To install:

4. Apply a bead of silicone sealant to the oil pan flange.
5. Install or connect the following:
   - New oil pump pickup tube O-ring seal
   - Oil pan and oil pump pickup tube. Tighten the fasteners to 97 inch lbs. (11 Nm).
   - Flywheel access panel
   - Left transmission stiffener bracket, if equipped
   - CKP sensor
   - Front differential, if equipped
   - Front skidplate, if equipped. Tighten the bolts to 40 ft. lbs. (55 Nm).
   - Negative battery cable
6. Fill the crankcase to the correct level.
7. Start the engine and check for leaks.

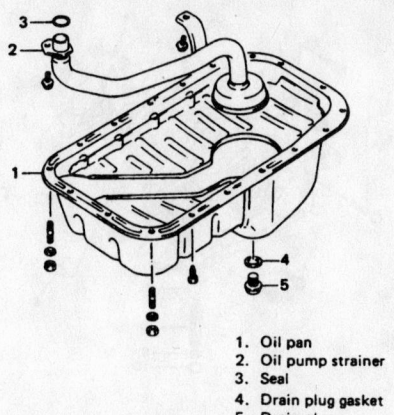

1. Oil pan
2. Oil pump strainer
3. Seal
4. Drain plug gasket
5. Drain plug

7924HG11

Exploded view of the oil pan and pump pickup mounting—1.6L engine

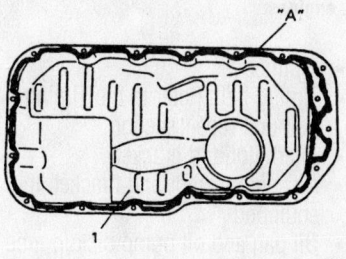

1. Oil pan
A. Sealant

7924HG79

Before installing the oil pan, apply a continuous bead of silicone sealant to the oil pan mating flange—all engines

2.0L Engines

1. Before servicing the vehicle, refer to the precautions in the beginning of this section.
2. Drain the engine oil.
3. Remove or disconnect the following:
   - Negative battery cable
   - Oil dipstick tube
   - Front wheels
   - Front skidplate, if equipped
   - Steering gear
   - Front differential, if equipped
   - Left transmission stiffener bracket, if equipped
   - Flywheel access panel
   - Exhaust front pipe
   - Left and right motor mounts and raise the engine about 1 inch (25mm) for clearance
   - Oil pan and oil pump pickup tube

To install:

4. Apply a bead of silicone sealant to the oil pan flange. Install new oil pump pickup tube O-ring seals.

5. Install or connect the following:
   - Oil pan and oil pump pickup tube. Tighten the fasteners to 97 inch lbs. (11 Nm).
   - Left and right engine mounts. Tighten the nuts to 36 ft. lbs. (50 Nm).
   - Exhaust front pipe
   - Flywheel access panel
   - Left transmission stiffener bracket, if equipped
   - Front differential, if equipped
   - Steering gear
   - Front skidplate, if equipped
   - Front wheels
   - Oil dipstick tube
   - Negative battery cable
6. Fill the crankcase to the correct level.
7. Start the engine and check for leaks.

2.5L Engine

1. Before servicing the vehicle, refer to the precautions in the beginning of this section.
2. Drain the engine oil.
3. Remove or disconnect the following:
   - Negative battery cable
   - Oil dipstick tube
   - Front wheels
   - Front skidplate, if equipped
   - Steering gear
   - Front differential, if equipped
   - Lower oil pan
   - Oil pickup tube bracket
   - Radiator outlet pipe
   - Upper oil pan and oil pickup tube

To install:

4. Install a new O-ring to the lower crankcase.
5. Apply a bead of silicone sealant to the upper oil pan flange.
6. Install or connect the following:
   - New oil pump pickup tube O-ring seals
   - Upper oil pan and oil pump pickup tube and tighten the fasteners as shown
   - Radiator outlet pipe
   - Oil pickup tube bracket
   - Lower oil pan. Tighten the bolts to 97 inch lbs. (11 Nm).
   - Front differential, if equipped
   - Steering gear
   - Front skidplate, if equipped
   - Front wheels
   - Oil dipstick tube
   - Negative battery cable
7. Fill the crankcase to the correct level.
8. Start the engine and check for leaks.

For complete Engine Mechanical specifications, see Section 1 of this manual

1. O-ring

Lower crankcase O-ring seal

9302HG05

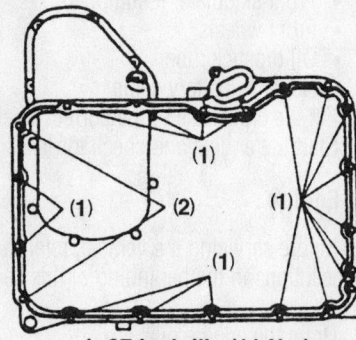

1: 97 Inch ilb. (11 Nm)
2: 20 Ft. lbs. (27 Nm)

9308HG03

Upper oil pan bolt torque values—2.5L engine

Oil Pump

REMOVAL & INSTALLATION

1.6L Engines

1. Before servicing the vehicle, refer to the precautions in the beginning of this section.
2. Drain the engine oil.
3. Drain the cooling system.
4. Remove or disconnect the following:
   • Negative battery cable
   • Accessory drive belts
   • Crankshaft pulley

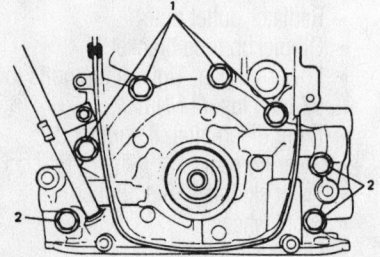

1. No. 1 bolts (short)
2. No. 2 bolts (long)

7924HG14

Oil pump housing short (1) and long bolt (2) locations—1.6L engines

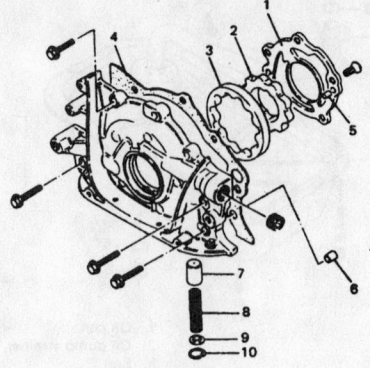

1. Rotor plate
2. Inner rotor
3. Outer rotor
4. Gasket
5. Pin
6. Pin
7. Relief valve
8. Spring
9. Retainer
10. Retainer ring

7924HG12

Exploded view of the oil pump housing—1.6L engines

• Front cover
• Timing belt. Refer to the Timing Belt unit repair section.
• Alternator and bracket
• A/C compressor and bracket, if equipped
• Oil pan and oil pump pickup tube
• Crankshaft timing sprocket
• Oil pump

➡ The oil pump bolts are different lengths. Note their location for assembly.

To install:
5. Install or connect the following:
   • Oil pump with a new gasket. Tighten the bolts to 97 inch lbs. (11 Nm).
   • Crankshaft timing sprocket. Tighten the bolt to 94 ft. lbs. (130 Nm).
   • Oil pan and oil pump pickup tube
   • A/C compressor and bracket, if equipped
   • Alternator and bracket
   • Timing belt
   • Front cover
   • Crankshaft pulley. Tighten the bolts to 12 ft. lbs. (16 Nm).
   • Accessory drive belts
   • Negative battery cable
6. Fill the crankcase to the correct level.
7. Start the engine and check for leaks.

2.0L Engines

1. Before servicing the vehicle, refer to the precautions in the beginning of this section.
2. Drain the engine oil.
3. Remove or disconnect the following:
   • Negative battery cable
   • Oil pan and pickup tube

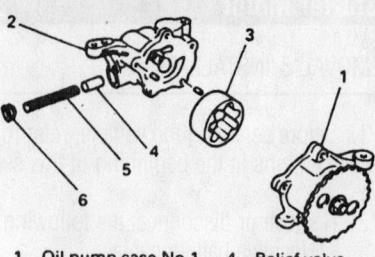

1. Oil pump case No.1
2. Oil pump case No.2
3. Outer rotor
4. Relief valve
5. Relief spring
6. Retainer

7924HG15

Exploded view of oil pump—2.0L and 2.5L engines

• Oil pump sprocket cover
• Oil pump

✳✳ WARNING

Do not remove the sprocket from the oil pump. Damage to the oil pump center shaft and abnormal pump operation may result.

To install:
4. Install or connect the following:
   • Oil pump. Tighten the bolts to 20 ft. lbs. (27 Nm).
   • Oil pump sprocket cover. Tighten the bolts to 108 inch lbs. (12 Nm).
   • Oil pan and pickup tube
   • Negative battery cable
5. Fill the crankcase to the correct level.
6. Start the engine and check for leaks.

2.5L Engine

1. Before servicing the vehicle, refer to the precautions in the beginning of this section.
2. Drain the cooling system.
3. Drain the engine oil.
4. Remove or disconnect the following:
   • Negative battery cable
   • Accessory drive belts
   • Intake manifold
   • Oil pan and oil pickup tube
   • Front cover
   • Oil pump chain guide
   • Oil pump

✳✳ WARNING

Do not remove the sprocket from the oil pump. Damage to the oil pump center shaft and abnormal pump operation may result.

To install:
5. Install or connect the following:
   • Oil pump. Tighten the bolts to 20 ft. lbs. (27 Nm).

- Oil pump chain guide. Tighten the bolts to 97 inch lbs. (11 Nm).
- Front cover
- Oil pan and oil pickup tube
- Intake manifold
- Accessory drive belts
- Negative battery cable
6. Fill the crankcase to the correct level.
7. Fill the cooling system.
8. Start the engine and check for leaks.

Rear Main Seal

REMOVAL & INSTALLATION

1. Before servicing the vehicle, refer to the precautions in the beginning of this section.
2. Remove or disconnect the following:
- Negative battery cable
- Transmission
- Clutch assembly, if equipped
- Flywheel
- Rear main seal

To install:

3. Install or connect the following:
- Rear main seal flush with the cylinder block
- Flywheel. Tighten the bolts in a crossing pattern to 58 ft. lbs. (79 Nm) for 1.6L engines or to 51 ft. lbs. (69 Nm) for all other engines.
- Clutch assembly, if equipped
- Transmission
- Negative battery cable

Timing Chain, Sprockets, Front Cover and Seal

REMOVAL & INSTALLATION

2.0L Engines

1. Before servicing the vehicle, refer to the precautions in the beginning of this section.
2. Drain the cooling system.
3. Drain the engine oil.
4. Remove or disconnect the following:
- Negative battery cable
- Oil pan and pickup tube
- Valve cover
- Bypass pipe and hose
- Accessory drive belts
- Cooling fan and shroud
- Water pump pulley
- Alternator belt tensioner and idler pulleys

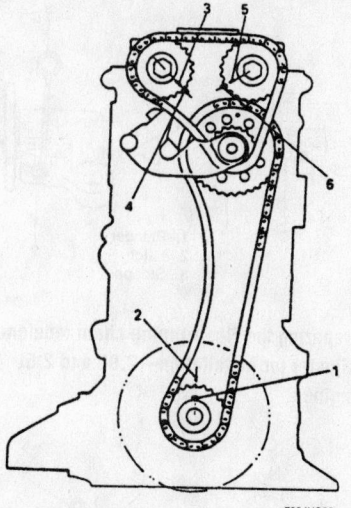

Timing mark alignment—2.0L engines

- Upper radiator hose
- A/C compressor and bracket, if equipped
- Crankshaft pulley
- Front crankshaft seal
- Front cover

5. Rotate the crankshaft to align the timing marks as shown.

✳✳ WARNING

Do not allow the crankshaft or camshafts to rotate once the timing chains have been removed. Valve or piston damage could result.

6. Remove or disconnect the following:
- Second timing chain tensioner
- Camshaft sprockets and second timing chain
- First timing chain tensioner
- Timing chain idler sprocket and first timing chain

To install:

7. Prepare the timing chain tensioners for installation by releasing the latches,

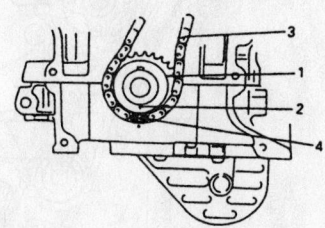

1. Crankshaft timing sprocket   3. 1st timing chain
2. Match mark   4. Yellow plate

Crankshaft and first timing chain alignment—2.0L engines

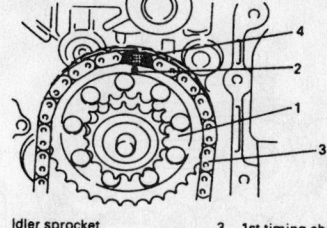

1. Idler sprocket   3. 1st timing chain
2. Match mark on idler sprocket 4. Dark blue plate

Idler sprocket and first timing chain alignment—2.0L engines

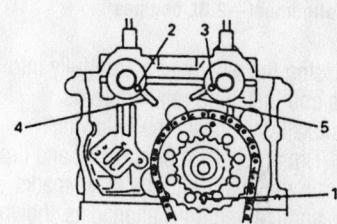

1. Arrow mark on idler sprocket
2. Knock pin of intake camshaft
3. Knock pin of exhaust camsaft
4. Timing mark of intake side
5. Timing mark of exhaust side

Idler sprocket and camshaft alignment—2.0L engines

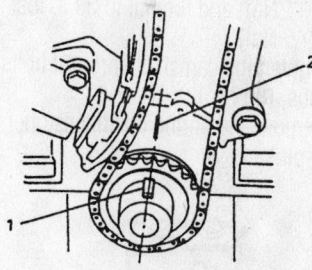

1. Crank timing sprocket key
2. Timing mark

Crankshaft sprocket alignment—2.0L engines

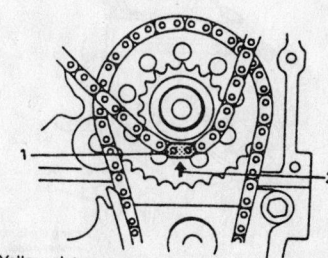

1. Yellow plate
2. Match mark of 2nd timing chain (Arrow mark)

Idler sprocket and second timing chain alignment—2.0L engines

For Accessory Drive Belt illustrations, see Section 1 of this manual

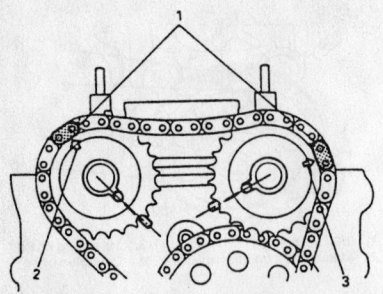

1. Dark blue
2. Arrow mark on intake camshaft timing sprocket
3. Arrow mark on exhaust camshaft timing sprocket

7924HG21

Camshaft sprocket and second timing chain alignment—2.0L engines

compressing the tensioner piston fully into the bore and installing retaining pins.

8. Install or connect the following:
- Timing chain idler sprocket and first timing chain with the matchmarks and colored links aligned as shown
- First timing chain tensioner. Tighten the bolts to 97 inch lbs. (11 Nm).
- Camshaft sprockets and second timing chain with the matchmarks and colored links aligned as shown
- Second timing chain tensioner. Tighten the bolts to 97 inch lbs. (11 Nm) and the nut to 33 ft. lbs. (45 Nm).

9. Tighten the camshaft sprocket bolts to 59 ft. lbs. (80 Nm).

10. Remove the timing chain tensioner retaining pins.

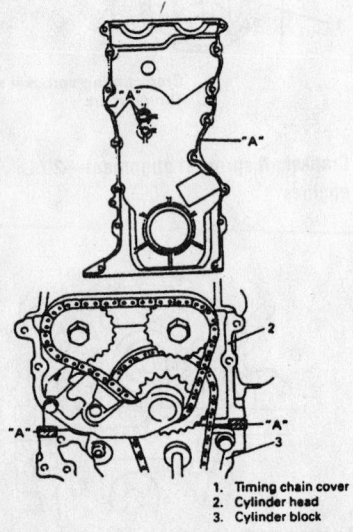

1. Timing chain cover
2. Cylinder head
3. Cylinder block

7924HG10

Prior to installing the timing chain cover on the engine block and cylinder head, apply silicone sealant to the cover as indicated (areas marked A)—2.0L and 2.5L engines

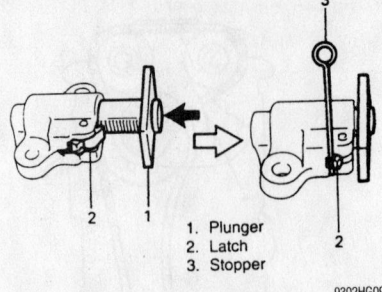

1. Plunger
2. Latch
3. Stopper

9302HG09

Preparing the No. 1 timing chain tensioner adjuster for installation—2.0L and 2.5L engines

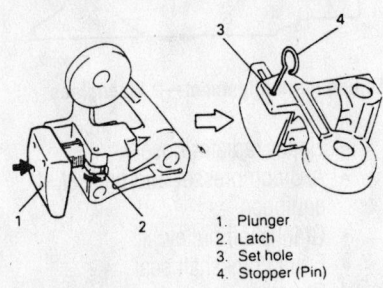

1. Plunger
2. Latch
3. Set hole
4. Stopper (Pin)

9302HG13

No. 2 timing chain tensioner—left bank tensioner shown—2.0L and 2.5L engines

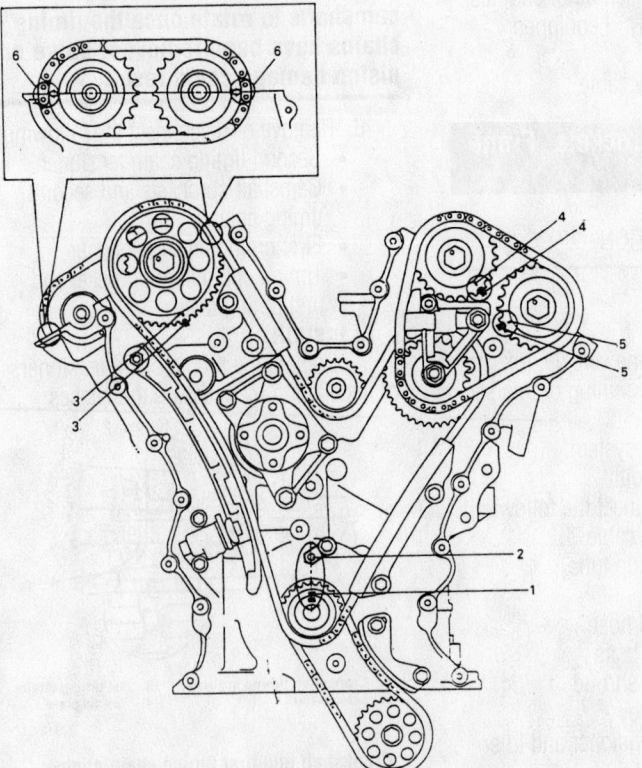

1. Crankshaft Keyway
2. Oil Jet
3. Right Bank No. 1 Chain Marks
4. Left Bank Intake Cam Marks
5. Left Bank Exhaust Cam Marks
6. Right Bank Intake and Exhaust Cam Marks

9302HG07

Timing chain alignment marks—2.5L engine

11. Rotate the crankshaft two complete turns and check that the timing marks align.

12. Install or connect the following:
- Front cover. Apply sealant as shown.
- Front crankshaft seal
- Crankshaft pulley. Tighten the bolt to 109 ft. lbs. (130 Nm).
- A/C compressor and bracket, if equipped
- Upper radiator hose
- Alternator belt tensioner and idler pulleys
- Water pump pulley
- Cooling fan and shroud
- Accessory drive belts
- Bypass pipe and hose
- Valve cover
- Oil pan and pickup tube
- Negative battery cable

13. Fill the crankcase to the correct level.
14. Fill the cooling system.
15. Start the engine and check for leaks.

2.5L Engine

1. Before servicing the vehicle, refer to the precautions in the beginning of this section.

2. Drain the cooling system.

3. Drain the engine oil.
4. Remove or disconnect the following:
- Negative battery cable
- Intake manifold
- Ignition coils
- Valve covers
- Accessory drive belts
- Cooling fan and shroud
- Water pump pulley
- Radiator
- Power steering pump and brackets
- Oil pan and pickup tube

- Crankshaft pulley
- Front crankshaft seal
- Crankshaft Position (CKP) sensor
- Front cover

5. Rotate the crankshaft so that the timing marks are aligned as shown.

❋❋ WARNING

Do not allow the crankshaft or camshafts to rotate once the timing chains have been removed. Valve or piston damage could result.

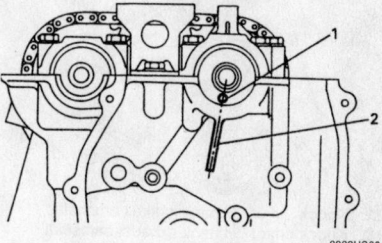

1. Knock pin of intake camshaft
2. Match mark

9302HG08

Right bank camshaft timing marks—2.5L engine

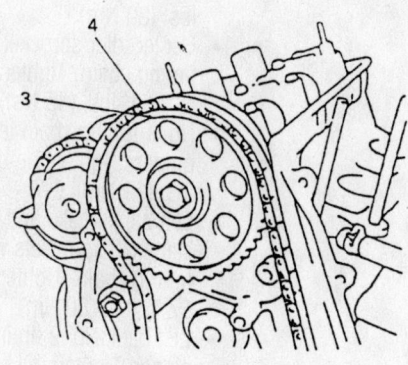

3. Match mark of RH bank 1st timing chain sprocket
4. Silver plate (LH) of 1st timing chain

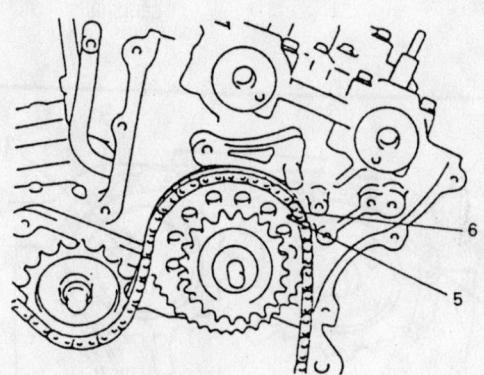

5. Match mark of idler sprocket No.2
6. Silver plate (RH) of 1st timing chain

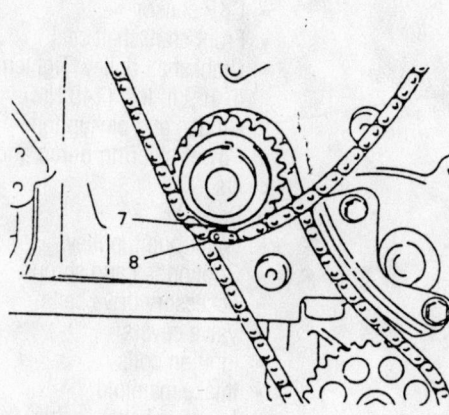

7. Match mark of crankshaft timing sprocket
8. Gold or Yellow plate of 1st timing chain

1. Crank timing pulley key
2. Oil jet

9302HG10

No. 1 timing chain alignment—2.5L engine

For Tire, Wheel and Ball Joint specifications, see Section 1 of this manual

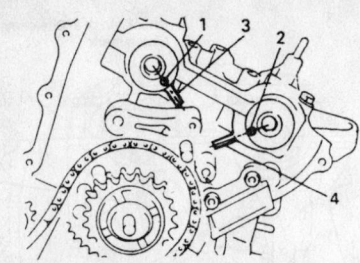

1. Knock pin of LH bank intake camshaft
2. Knock pin of LH bank exhaust camshaft
3. Match mark of intake side
4. Match mark of exhaust side

9302HG11

Left bank camshaft alignment—2.5L engine

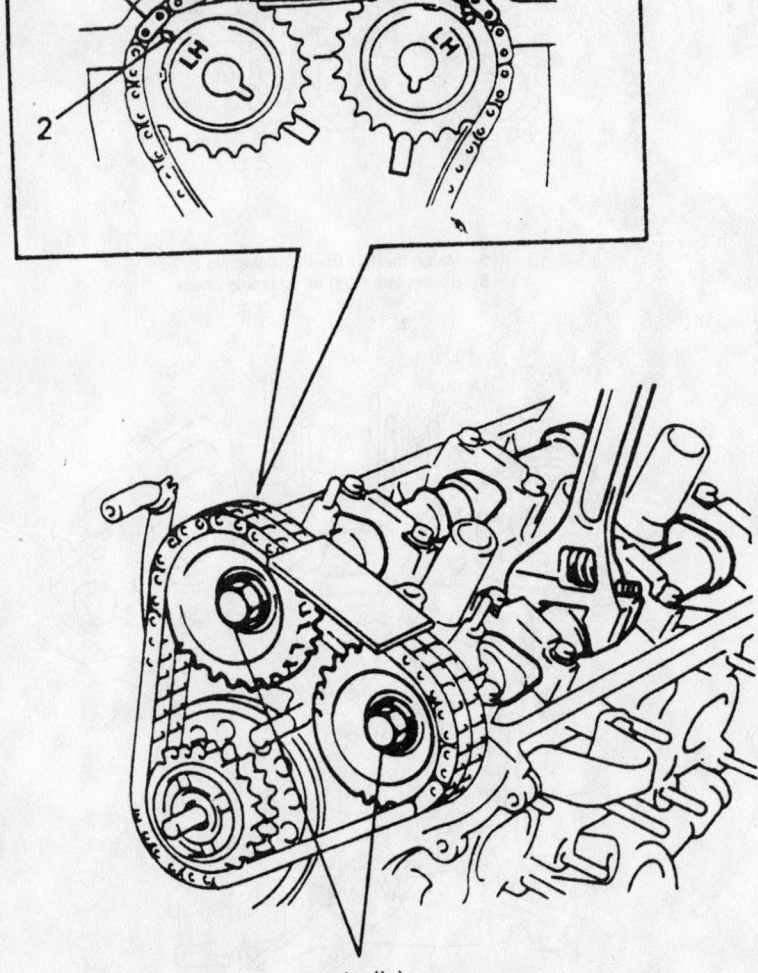

1. **Silver plate**
2. **Arrow mark on intake camshaft timing sprocket**
3. **Arrow mark on exhaust camshaft timing sprocket**
4. **Sprocket bolt**

4, (b)

9302HG12

Align the left bank No. 2 chain silver links—2.5L engine

6. Remove or disconnect the following:
   - Left bank No. 2 timing chain tensioner
   - Left bank intake and exhaust camshaft sprockets with the No. 2 timing chain
   - No. 1 timing chain guides
   - No. 1 timing chain tensioner
   - Center idler sprocket and the No. 1 timing chain
   - Right bank No. 1 timing chain sprocket

7. The right bank No. 2 timing chain is removed with the intake and exhaust camshafts.

To install:

8. Prepare the timing chain tensioners for installation by releasing the latches, compressing the tensioner piston fully into the bore and installing retaining pins.

9. Align the timing chain sprocket matchmarks and colored chain links as shown during assembly.

10. Install or connect the following:
    - Right bank intake and exhaust camshafts with the No. 2 timing chain
    - Right bank No. 1 timing chain sprocket. Tighten the bolt to 58 ft. lbs. (80 Nm).
    - Center idler sprocket and the No. 1 timing chain. Tighten the fastener to 32 ft. lbs. (45 Nm).
    - No. 1 timing chain tensioner and guides. Tighten the bolts to 97 inch lbs. (11 Nm).
    - Left bank intake and exhaust camshaft sprockets with the No. 2 timing chain. Tighten the bolts to 57 ft. lbs. (80 Nm).
    - Left bank No. 2 timing chain tensioner. Tighten the bolts to 97 inch lbs. (11 Nm).

11. Remove the retaining pins from the timing chain tensioners.

12. Rotate the crankshaft two complete turns and check that the timing marks align.

13. Install or connect the following:
    - Front cover. Tighten the bolts to 97 inch lbs. (11 Nm).
    - CKP sensor
    - Front crankshaft seal
    - Crankshaft pulley. Tighten the bolt to 109 ft. lbs. (148 Nm).
    - Oil pan and pickup tube
    - Power steering pump and brackets
    - Radiator
    - Water pump pulley
    - Cooling fan and shroud
    - Accessory drive belts
    - Valve covers
    - Ignition coils
    - Intake manifold
    - Negative battery cable

14. Fill the crankcase to the correct level.

15. Fill the cooling system.

16. Start the engine and check for leaks.

Piston and Ring

POSITIONING

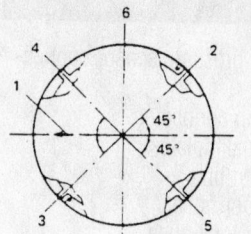

1. Arrow mark
2. 1st ring end gap
3. 2nd ring end gap and oil ring spacer gap
4. Oil ring upper rail gap
5. Oil ring lower rail gap
6. Intake side
7. Exhaust side

7924AG67

Piston ring end-gap spacing—All engines

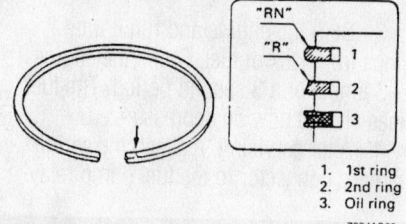

1. 1st ring
2. 2nd ring
3. Oil ring

7924AG68

Compression ring identification marks— All engines

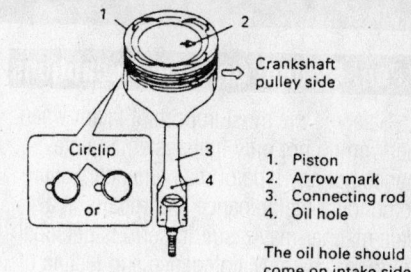

1. Piston
2. Arrow mark
3. Connecting rod
4. Oil hole

The oil hole should come on intake side

7924AG69

Piston and connecting rod positioning— 1.6L engine

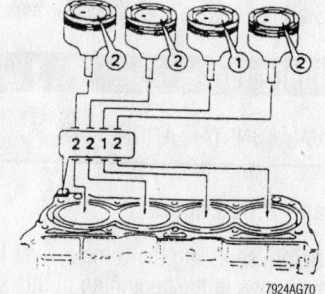

2212

7924AG70

Piston installation—1.6L engine

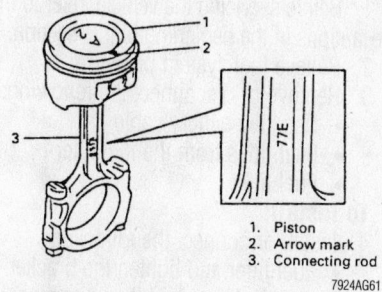

1. Piston
2. Arrow mark
3. Connecting rod

7924AG61

Piston and connecting rod positioning— 2.0L and 2.5L engines

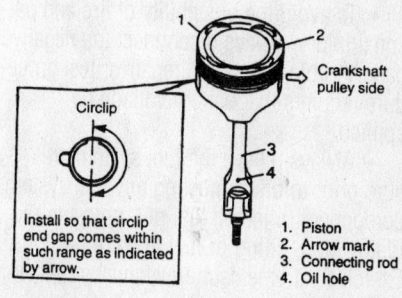

Install so that circlip end gap comes within such range as indicated by arrow.

1. Piston
2. Arrow mark
3. Connecting rod
4. Oil hole

9302AG15

Piston pin circlip installation—2.5L engine

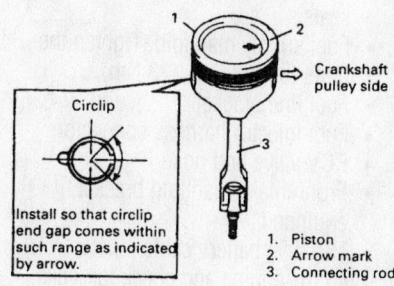

Install so that circlip end gap comes within such range as indicated by arrow.

1. Piston
2. Arrow mark
3. Connecting rod

7924AG62

Piston pin circlip installation—2.0L engines

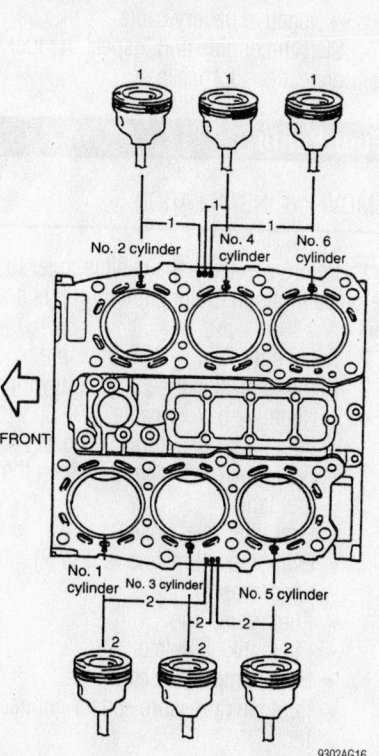

No. 2 cylinder   No. 4 cylinder   No. 6 cylinder

FRONT

No. 1 cylinder   No. 3 cylinder   No. 5 cylinder

9302AG16

Piston identification—2.5L engine

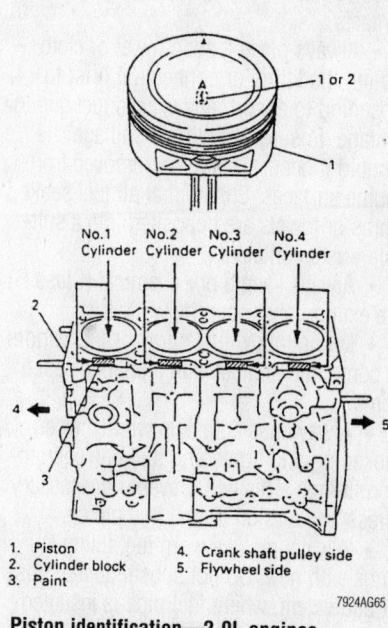

No.1 Cylinder   No.2 Cylinder   No.3 Cylinder   No.4 Cylinder

1. Piston
2. Cylinder block
3. Paint
4. Crank shaft pulley side
5. Flywheel side

7924AG65

Piston identification—2.0L engines
Match pistons with "1" indicators to red cylinder paint marks
Match pistons with "2" indicators to blue cylinder paint marks

For Wheel Alignment specifications, see Section 1 of this manual

FUEL SYSTEM

Fuel System Service Precautions

Safety is the most important factor when performing not only fuel system maintenance but any type of maintenance. Failure to conduct maintenance and repairs in a safe manner may result in serious personal injury or death. Maintenance and testing of the vehicle fuel system components can be accomplished safely and effectively by adhering to the following rules and guidelines.

• To avoid the possibility of fire and personal injury, always disconnect the negative battery cable unless the repair or test procedure requires that battery voltage be applied.

• Always relieve the fuel system pressure prior to disconnecting any fuel system component (injector, fuel rail, pressure regulator, etc.), fitting or fuel line connection. Exercise extreme caution whenever relieving fuel system pressure to avoid exposing skin, face and eyes to fuel spray. Please be advised that fuel under pressure may penetrate the skin or any part of the body that it contacts.

• Always place a shop towel or cloth around the fitting or connection prior to loosening to absorb any excess fuel due to spillage. Ensure that all fuel spillage (should it occur) is quickly removed from engine surfaces. Ensure that all fuel soaked cloths or towels are deposited into a suitable waste container.

• Always keep a dry chemical (Class B) fire extinguisher near the work area.

• Do not allow fuel spray or fuel vapors to come into contact with a spark or open flame.

• Always use a backup wrench when loosening and tightening fuel line connection fittings. This will prevent unnecessary stress and torsion to fuel line piping.

• Always replace worn fuel fitting O-rings with new. Do not substitute fuel hose or equivalent, where fuel pipe is installed.

Fuel System Pressure

RELIEVING

1. Before servicing the vehicle, refer to the precautions in the beginning of this section.
2. Detach the wiring harness connector from the fuel pump relay, located under the left-hand side of the instrument panel near the ECM.

3. Start the engine and run it until it stops from lack of fuel. Crank the engine 2–3 times for a 3 second period. The fuel lines should now be depressurized.
4. After servicing, reattach the wiring harness connector to the fuel pump relay.

Fuel Filter

REMOVAL & INSTALLATION

1. Before servicing the vehicle, refer to the precautions in the beginning of this section.
2. Relieve fuel system pressure.
3. Remove or disconnect the following:
   • Negative battery cable
   • Fuel lines from the fuel filter
   • Fuel filter

To install:
4. Install or connect the following:
   • Fuel filter and tighten the bracket bolt. Note the fuel flow directional arrow.
   • Fuel lines to the fuel filter.
   • Negative battery cable
5. Start the engine and inspect the fuel filter connections for leaks.

Fuel Pump

REMOVAL & INSTALLATION

1. Before servicing the vehicle, refer to the precautions in the beginning of this section.
2. Relieve the fuel system pressure.
3. Remove or disconnect the following:
   • Negative battery cable
   • Fuel filler hose and vent hose
   • Fuel tank inlet valve and drain the fuel tank
   • Fuel filter inlet hose
   • Evaporative Emissions (EVAP) vapor hose
   • Fuel return line
   • Fuel tank skidplate
   • Fuel pump connector
   • Fuel tank pressure sensor connector
   • Fuel tank
   • Fuel pump module

To install:
4. Install or connect the following:
   • Fuel pump module with a new seal. Tighten the bolts to 44 inch lbs. (5 Nm).
   • Fuel tank. Tighten the strap bolts to 37 ft. lbs. (50 Nm).

   • Fuel tank pressure sensor connector
   • Fuel pump connector
   • Fuel tank skidplate
   • Fuel return line
   • EVAP vapor hose
   • Fuel filter inlet hose
   • Fuel tank inlet valve and drain the fuel tank
   • Fuel filler hose and vent hose
   • Negative battery cable
5. Fill the fuel tank.
6. Start the engine and check for leaks.

Fuel Injector

REMOVAL & INSTALLATION

1.6L and 2.0L Engines

1. Before servicing the vehicle, refer to the precautions in the beginning of this section.
2. Relieve the fuel system pressure.
3. Remove or disconnect the following:
   • Negative battery cable
   • Front intake manifold bracket, if equipped
   • Positive Crankcase Ventilation (PCV) valve and hose
   • Fuel injector harness connectors
   • Fuel line bracket
   • Fuel supply manifold
   • Fuel injectors

To install:
4. Install or connect the following:
   • Fuel injectors with new O-ring seals
   • Fuel supply manifold. Tighten the bolts to 17 ft. lbs. (23 Nm).
   • Fuel line bracket
   • Fuel injector harness connectors
   • PCV valve and hose
   • Front intake manifold bracket, if equipped
   • Negative battery cable
5. Start the engine and check for leaks.

2.5L Engine

1. Before servicing the vehicle, refer to the precautions in the beginning of this section.
2. Relieve the fuel system pressure.
3. Remove or disconnect the following:
   • Negative battery cable
   • Air intake tube
   • Throttle body intake collector
   • Fuel lines

- Fuel pressure regulator vacuum line
- Fuel injector harness connectors
- Fuel supply manifold connect pipe
- Fuel supply manifolds
- Fuel injectors

To install:

4. Install or connect the following:

- Fuel injectors with new O-ring seals
- Fuel supply manifolds. Tighten the bolts to 17 ft. lbs. (23 Nm).
- Fuel supply manifold connect pipe. Tighten the bolts to 22 ft. lbs. (30 Nm).
- Fuel injector harness connectors

- Fuel pressure regulator vacuum line
- Fuel lines
- Throttle body intake collector
- Air intake tube
- Negative battery cable

5. Start the engine and check for leaks.

DRIVE TRAIN

Transmission Assembly

REMOVAL & INSTALLATION

Manual Transmission

1. Before servicing the vehicle, refer to the precautions in the beginning of this section.
2. Drain the transmission fluid.
3. Drain the transfer case fluid, if equipped.
4. Remove or disconnect the following:

- Negative battery cable
- Shift lever boots
- Gear shift lever
- Transfer case shift lever, if equipped
- 4WD switch connector, if equipped
- Reverse light switch connector
- Starter motor
- Front driveshaft, if equipped
- Rear driveshaft
- Speedometer cable, if equipped
- Vehicle Speed (VSS) sensor, if equipped
- Clutch slave cylinder or cable
- Flywheel access cover
- Transmission flange bolts and nuts
- Transmission braces, if equipped

A   WOOD BLOCK
H   200 mm (8.0")
T   45 mm (1.8")
W   100–150 mm (4.0–6.0")
7017 DISTRIBUTOR CAP
7018 BULKHEAD

7924HG37

Support the engine with a wooden block between the cylinder head and the firewall—All models

5. Support the transmission with a jack and remove the transmission mount and crossmember.
6. Place a wooden block at the rear of the cylinder head as shown to support the engine when the transmission is removed.
7. Lower the transmission away from the vehicle.

To install:

8. Install or connect the following:

- Transmission. Tighten the flange fasteners to 62–72 ft. lbs. (85–98 Nm).
- Transmission mount and crossmember. Tighten the fasteners to 29–43 ft. lbs. (40–60 Nm).
- Transmission braces, if equipped. Tighten the bolts to 62–72 ft. lbs. (85–98 Nm).
- Flywheel access cover
- Clutch slave cylinder or cable
- VSS sensor, if equipped
- Speedometer cable, if equipped
- Rear driveshaft
- Front driveshaft, if equipped
- Starter motor
- Reverse light switch connector
- 4WD switch connector, if equipped
- Transfer case shift lever, if equipped
- Gear shift lever
- Shift lever boots
- Negative battery cable

9. Fill the transmission to the correct level.
10. Fill the transfer case, if equipped.

Automatic Transmission

1. Before servicing the vehicle, refer to the precautions in the beginning of this section.
2. Drain the transfer case oil, if equipped.
3. Remove or disconnect the following:

- Negative battery cable
- Center console and transfer case shift lever, if equipped
- Transmission dipstick tube
- Transmission wiring harness connectors

- Starter motor
- Front driveshaft, if equipped
- Rear driveshaft
- Gear select cable and bracket
- Throttle Valve (TV) cable, if equipped
- Exhaust front pipe
- Transmission oil cooler lines
- Transmission brace
- Flywheel access cover
- Torque converter
- Speedometer cable, if equipped
- Vehicle Speed (VSS) sensor connector, if equipped
- Transmission flange bolts and nuts
- Transmission braces, if equipped

4. Support the transmission with a jack and remove the transmission mount and crossmember.
5. Place a wooden block at the rear of the cylinder head as shown to support the engine when the transmission is removed.
6. Lower the transmission away from the vehicle.

To install:

7. Install or connect the following:

- Transmission. Tighten the flange fasteners to 62–72 ft. lbs. (85–98 Nm).
- Transmission mount and crossmember. Tighten the fasteners to 29–43 ft. lbs. (40–60 Nm).
- Transmission braces, if equipped. Tighten the bolts to 62–72 ft. lbs. (85–98 Nm).
- VSS sensor connector, if equipped
- Speedometer cable, if equipped
- Torque converter. Tighten the bolts to 47 ft. lbs. (65 Nm).
- Flywheel access cover
- Transmission brace
- Transmission oil cooler lines
- Exhaust front pipe
- TV cable, if equipped
- Gear select cable and bracket
- Rear driveshaft
- Front driveshaft, if equipped
- Starter motor
- Transmission wiring harness connectors

- Transmission dipstick tube
- Center console and transfer case shift lever, if equipped
- Negative battery cable

8. Fill the transmission to the correct level.

9. Fill the transfer case, if equipped.

Clutch

ADJUSTMENTS

These vehicles are equipped with a hydraulic clutch system. No adjustment is necessary.

REMOVAL & INSTALLATION

1. Before servicing the vehicle, refer to the precautions in the beginning of this section.
2. Remove the transmission.
3. Loosen the pressure plate mounting bolts in a 2-step crisscross sequence until the spring tension is relieved.
4. Remove the pressure plate and the clutch disc.

To install:

5. Using a clutch alignment tool, assemble the clutch disc and pressure plate onto the flywheel.
6. Tighten the pressure plate bolts in multiple passes to 17 ft. lbs. (23 Nm).
7. Install the transmission.
8. Check for proper clutch operation.

Hydraulic Clutch System

BLEEDING

1. Before servicing the vehicle, refer to the precautions in the beginning of this section.
2. Fill the master cylinder reservoir to the MAX line with clean brake fluid and keep it at least half full throughout the bleeding procedure.
3. From beneath the vehicle, remove the bleeder plug cap, then attach a clear vinyl tube to the slave cylinder bleeder plug. Insert the open end of the hose into a container.
4. Have an assistant depress the clutch pedal. Open the bleeder after the pedal is depressed.
5. Close the bleeder before releasing the clutch pedal.
6. Repeat until all air bubbles are gone from the hydraulic fluid.
7. Install the bleeder plug cap.
8. Fill the clutch master cylinder fluid reservoir to the specified full level.

Transfer Case Assembly

REMOVAL & INSTALLATION

1. Before servicing the vehicle, refer to the precautions in the beginning of this section.
2. Drain the transfer case oil.
3. Remove or disconnect the following:
   - Negative battery cable
   - Distributor or Camshaft Position (CMP) sensor, if equipped
   - Center console
   - Transmission shift lever and case, if equipped with a manual transmission
   - Transfer case shift lever
   - Front and rear driveshafts
   - Exhaust center pipe
   - Speedometer cable or Vehicle Speed (VSS) sensor, as equipped
   - Vent hose
   - 4WD switch connector

4. Support the transmission with a jack and remove the transmission mount and crossmember.
5. Place a wooden block at the rear of the cylinder head as shown to support the engine when the transfer case is removed.
6. Lower the transfer case away from the vehicle.

To install:

7. Install or connect the following:
   - Transfer case. Tighten the bolts to 30 ft. lbs. (41 Nm).
   - 4WD switch connector
   - Vent hose
   - Speedometer cable or VSS sensor, as equipped
   - Exhaust center pipe
   - Front and rear driveshafts. Tighten the bolts to 36 ft. lbs. (50 Nm).
   - Transfer case shift lever
   - Transmission shift lever and case, if equipped with a manual transmission
   - Center console
   - Distributor or CMP sensor, if equipped
   - Negative battery cable

8. Fill the transfer case.

Halfshaft

REMOVAL & INSTALLATION

Left

1. Before servicing the vehicle, refer to the precautions in the beginning of this section.
2. Remove or disconnect the following:
   - Front wheel
   - Hub drive flange or locking hub, as equipped
   - Snapring
   - Thrust washer
   - Halfshaft flange fasteners
   - Halfshaft

To install:

3. Install or connect the following:
   - Halfshaft. Tighten the flange bolts to 37 ft. lbs. (50 Nm).
   - Thrust washer
   - Snapring
   - Locking hub, if equipped. Tighten the bolts to 24 ft. lbs. (33 Nm).

1. Drive shaft oil seal
2. Double off-set joint (DOJ)
3. Joint circlip
4. DOJ boot
5. Ball joint boot
6. Ball joint assembly (RH side)
7. Drive shaft assembly (LH side)
8. Left drive shaft
9. Drive shaft bearing circlip
10. Drive shaft bearing

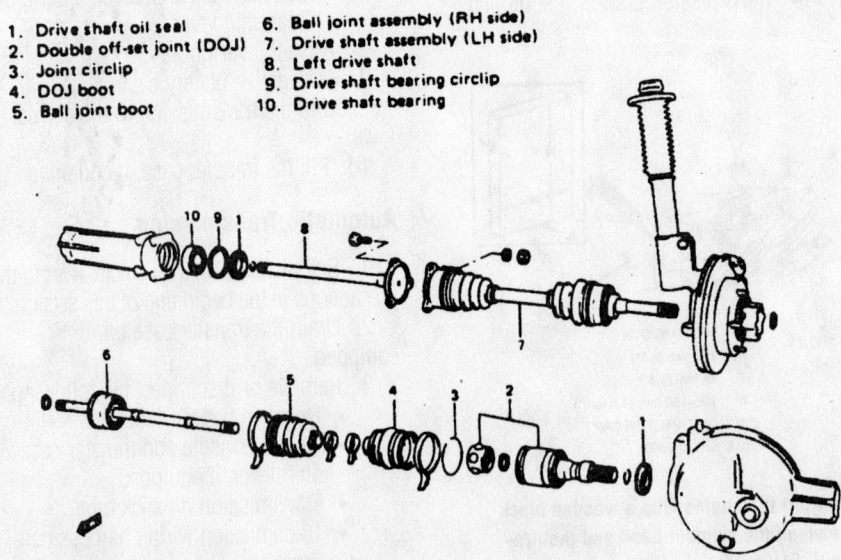

Exploded view of the left- and right-hand halfshaft assemblies

7924HG31

- Hub drive flange, if equipped. Tighten the bolts to 35 ft. lbs. (48 Nm).
- Front wheel

Right

1. Before servicing the vehicle, refer to the precautions in the beginning of this section.
2. Remove or disconnect the following:
   - Front wheel
   - Hub drive flange or locking hub, as equipped
   - Snapring
   - Thrust washer
   - Brake caliper
   - Wheel speed sensor, if equipped
   - Brake rotor
   - Stabilizer bar link
   - Outer tie rod end
   - Lower ball joint
   - Strut bracket bolts
   - Steering knuckle and wheel hub
3. Pry the inboard joint out of the differential and remove the halfshaft.

To install:

4. Insert the inboard joint into the differential until the circlip is felt to seat.
5. Install or connect the following:
   - Steering knuckle and wheel hub
   - Strut bracket bolts. Tighten them to 70 ft. lbs. (95 Nm).
   - Lower ball joint. Tighten the nut to 40 ft. lbs. (55 Nm).
   - Outer tie rod end. Tighten the nut to 35 ft. lbs. (48 Nm).
   - Stabilizer bar link. Tighten the nut to 21 ft. lbs. (29 Nm).
   - Brake rotor
   - Wheel speed sensor, if equipped
   - Brake caliper
   - Thrust washer
   - Snapring
   - Locking hub, if equipped. Tighten the bolts to 24 ft. lbs. (33 Nm).
   - Hub drive flange, if equipped. Tighten the bolts to 35 ft. lbs. (48 Nm).
   - Front wheel
6. Check the wheel alignment and adjust as necessary.

CV-Joints

OVERHAUL

Outer CV-Joint

The outer CV-joint is serviced with the axle shaft as an assembly. The outer CV-

joint boot can be serviced by removing the inner CV-joint.

Inner CV-Joint

1. Before servicing the vehicle, refer to the precautions in the beginning of this section.
2. Remove or disconnect the following:
   - Halfshaft from the vehicle
   - Grease boot clamps
   - Outer race snapring
   - Outer race
   - Shaft snapring
   - Inner race, cage and balls

To install:

3. Install or connect the following:
   - Inner race, cage and balls
   - Shaft snapring
   - Outer race
   - Outer race snapring
4. Fill the outer race and the grease boot with CV-joint grease and tighten the boot clamps.
5. Install the axle halfshaft.

Manual Locking Hubs

REMOVAL & INSTALLATION

1. Before servicing the vehicle, refer to the precautions in the beginning of this section.
2. Set the selector knob to the **FREE** position.
3. Remove or disconnect the following:

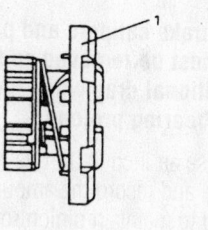

1. Cover

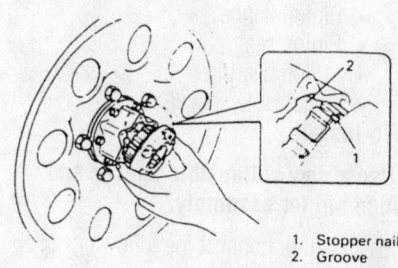

1. Stopper nail
2. Groove

9308HG06

Manual hub alignment—All models

- Hub cover assembly
- Hub body assembly

To install:

4. Install the hub body. Tighten the bolts to 18 ft. lbs. (25 Nm).
5. Align the hub cover stopper nail with the groove in the hub body and install the hub cover. Tighten the bolts to 90 inch lbs. (10 Nm).
6. Check for proper hub operation.

Automatic Locking Hubs

REMOVAL & INSTALLATION

1. Before servicing the vehicle, refer to the precautions in the beginning of this section.
2. Unlock the hub by setting the transfer case in the 2H position and driving backwards at least 6.5 feet (2 meters).
3. Remove or disconnect the following:
   - Hub sub assembly
   - Hub brake assembly

To install:

4. Align the brake assembly key with the slot in the spindle and install the brake assembly.
5. Align the matchmark on the sub assembly with the mark on the brake assembly and install the sub assembly. Tighten the hub bolts to 24 ft. lbs. (33 Nm).
6. Check for proper hub operation.

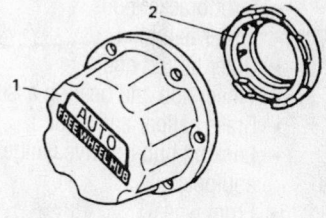

1. Free wheeling hub sub assembly
2. Free wheeling hub brake assembly

9308HG04

Automatic hub—All models

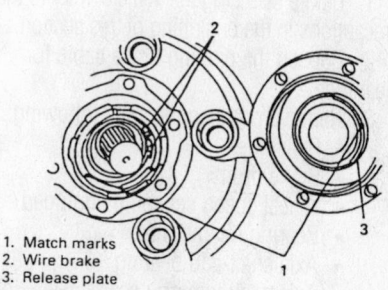

1. Match marks
2. Wire brake
3. Release plate

9308HG05

Automatic hub matchmarks—All models

Spindle Bearings

REMOVAL, PACKING & INSTALLATION

1. Before servicing the vehicle, refer to the precautions in the beginning of this section.

2. Support the control arm with a stand or floor jack.

3. Remove or disconnect the following:
- Front wheel
- Locking hub or drive flange, as equipped
- Brake caliper and rotor
- Wheel hub and bearing assembly
- Outer tie rod end
- Lower ball joint
- Strut bracket bolts
- Wheel spindle and steering knuckle assembly
- Inner oil seal
- Spindle bearing

To install:

4. Fill the recess in the wheel spindle with lithium grease.

5. Coat the spindle bearing and wheel spindle mating surfaces with sealant.

6. Press or drive the spindle bearing into the wheel spindle.

7. Install or connect the following:
- Inner oil seal
- Wheel spindle and steering knuckle assembly
- Strut bracket bolts
- Lower ball joint
- Outer tie rod end
- Wheel hub and bearing assembly
- Brake caliper and rotor
- Locking hub or drive flange, as equipped
- Front wheel

Axle Shaft, Bearing and Seal

REMOVAL & INSTALLATION

1. Before servicing the vehicle, refer to the precautions in the beginning of this section.

2. Loosen the parking brake cable for clearance.

3. Remove or disconnect the following:
- Rear wheel
- Brake drum
- Wheel speed sensor, if equipped
- Bearing retainer nuts
- Axle shaft and bearing
- Axle shaft inner oil seal

4. If equipped with ABS, grind a flat spot on the wheel speed sensor tone ring, then split the ring with a chisel.

5. Grind flat spots on the bearing retainer and split it with a chisel.

6. Press the wheel bearing off the axle shaft.

7. Remove the bearing retainer and the outer oil seal.

To install:

8. Install or connect the following:
- Outer oil seal to the bearing retainer
- Bearing retainer to the axle shaft
- Bearing and retainer ring pressed onto the axle shaft
- Wheel speed sensor tone ring pressed onto the axle shaft, if equipped
- Axle shaft inner oil seal
- Axle shaft and bearing
- Bearing retainer nuts. Tighten them to 17 ft. lbs. (23 Nm).
- Wheel speed sensor, if equipped
- Brake drum
- Rear wheel

9. Fill the rear differential to the correct level.

Pinion Seal

REMOVAL & INSTALLATION

1. Before servicing the vehicle, refer to the precautions in the beginning of this section.

2. Remove or disconnect the following:
- Driveshaft
- Wheels
- Brake calipers and pads or brake drum

➡**The brake calipers and pads or brake drum must be removed so that there is no additional drag when measuring pinion bearing preload.**

3. Use an inch lb. torque wrench and measure and record the amount of torque required to maintain pinion rotation through several revolutions.

4. Remove or disconnect the following:
- Pinion flange
- Pinion seal
- Pinion bearing
- Collapsible spacer

To install:

➡**Use a new collapsible spacer and flange nut for assembly.**

5. Install or connect the following:
- Collapsible spacer
- Pinion bearing
- Pinion seal
- Pinion flange

6. Rotate the pinion flange occasionally while tightening the flange nut to make sure the pinion bearings seat correctly.

7. Take frequent bearing preload torque readings. Tighten the flange nut to achieve the preload torque readings originally recorded.

❊❊ CAUTION

Never loosen the pinion nut to reduce bearing preload. If it is necessary to reduce bearing preload, install a new collapsible spacer and pinion nut.

8. Install or connect the following:
- Driveshaft
- Brake calipers and pads or brake drum
- Wheels

9. Fill the differential with gear lubricant and check for leaks.

Axle Housing Assembly

REMOVAL & INSTALLATION

1. Before servicing the vehicle, refer to the precautions in the beginning of this section.

2. Drain the gear oil.

3. Support the vehicle at the frame with a hoist or jackstands.

4. Support the rear axle with a floor jack.

5. Remove or disconnect the following:
- Rear wheels
- Rear brake drums
- Rear axle shafts
- Load sensing proportioning valve linkage, if equipped
- Brake fluid hose
- Brake backing plates
- Wheel speed sensor connector, if equipped
- Axle vent tube
- Rear driveshaft
- Differential carrier assembly
- Shock absorber lower bolts
- Coil springs
- Upper rods
- Lower rods
- Lateral rod
- Axle housing

To install:

6. Install or connect the following:
- Axle housing
- Upper rods
- Lower rods
- Coil springs
- Lateral rod
- Shock absorber lower bolts

- Differential carrier assembly. Tighten the nuts to 40 ft. lbs. (55 Nm).
- Rear driveshaft
- Axle vent tube
- Wheel speed sensor connector, if equipped

- Brake backing plates
- Brake fluid hose
- Load sensing proportioning valve linkage, if equipped
- Rear axle shafts
- Rear brake drums
- Rear wheels

7. Fill the rear axle to the correct level.
8. Lower the vehicle so that the rear suspension is at curb height.
9. Tighten the upper, lower and lateral rod fasteners to 65 ft. lbs. (90 Nm).
10. Tighten the lower shock absorber fasteners to 62 ft. lbs. (85 Nm).

STEERING AND SUSPENSION

Air Bag

✳✳ CAUTION

Some vehicles are equipped with an air bag system. The system must be disarmed before performing service on, or around, system components, the steering column, instrument panel components, wiring and sensors. Failure to follow the safety precautions and the disarming procedure could result in accidental air bag deployment, possible injury and unnecessary system repairs.

PRECAUTIONS

Several precautions must be observed when handling the inflator module to avoid accidental deployment and possible personal injury.

- Never carry the inflator module by the wires or connector on the underside of the module.
- When carrying a live inflator module, hold securely with both hands and ensure that the bag/trim cover are pointed away.
- Place the inflator module on a bench or other surface with the bag and trim cover facing up.
- With the inflator module on the bench, never place anything on or close to the module which may be thrown in the event of an accidental deployment.
- Never use air bag component parts from another vehicle.
- If there is a chance of electrical shock to any of the air bag components, remove the air bag module before servicing the vehicle.

DISARMING

1. Before servicing the vehicle, refer to the precautions in the beginning of this section.

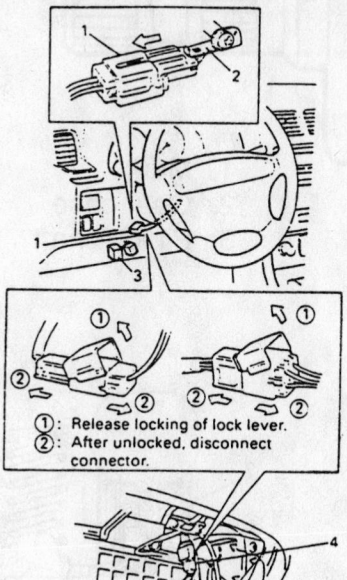

①: Release locking of lock lever.
②: After unlocked, disconnect connector.

Yellow connector of driver air bag (inflator) module
Connector stay
Air bag fuse box
Yellow connector of passenger air bag (inflator) module
Glove box

7924HG36

Air bag component location and identification—All models

2. Remove or disconnect the following:
- Negative battery cable
- AIR BAG fuse
- Driver air bag connector
- Glove box
- Passenger air bag connector

ARMING

When repairs are complete, install or connect the following:
- Passenger air bag connector
- Glove box
- Driver air bag connector
- AIR BAG fuse
- Negative battery cable

Recirculating Ball Steering Gear

REMOVAL & INSTALLATION

1. Before servicing the vehicle, refer to the precautions in the beginning of this section.
2. Remove or disconnect the following:
- Skidplate, if equipped
- Coolant overflow tank
- Intermediate shaft pinch bolt
- Power steering hoses
- Pitman arm center link joint
- Steering gearbox
- Pitman arm

To install:
3. Install or connect the following:
- Pitman arm. Tighten the nut to 102 ft. lbs. (140 Nm).
- Steering gearbox. Tighten the bolts to 62 ft. lbs. (85 Nm).
- Pitman arm center link joint. Tighten the nut to 37 ft. lbs. (50 Nm).
- Power steering hoses
- Intermediate shaft pinch bolt. Tighten it to 18 ft. lbs. (25 Nm).
- Coolant overflow tank
- Skidplate, if equipped
4. Fill the power steering system.
5. Start the engine and check for leaks.
6. Check the wheel alignment and adjust as necessary.

Power Rack and Pinion Steering Gear

REMOVAL & INSTALLATION

1. Before servicing the vehicle, refer to the precautions in the beginning of this section.
2. Remove or disconnect the following:
- Power steering hoses
- Intermediate shaft pinch bolt
- Front wheels
- Outer tie rod ends
- Steering gear

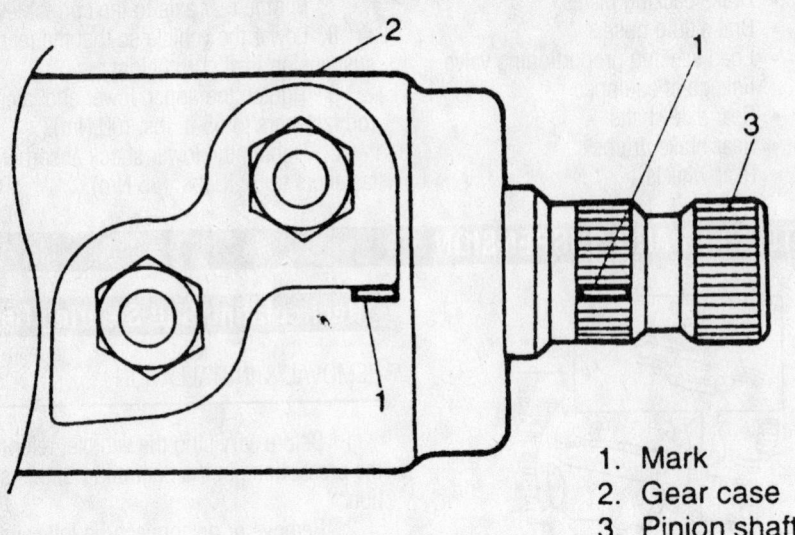

Steering gear centering marks

1. Mark
2. Gear case
3. Pinion shaft

9302HG14

To install:

3. Install or connect the following:
   - Steering gear. Tighten the bolts to 40 ft. lbs. (55 Nm).
   - Outer tie rod ends. Tighten the nuts to 32 ft. lbs. (43 Nm).
   - Front wheels
   - Intermediate shaft pinch bolt. Tighten it to 18 ft. lbs. (25 Nm).
   - Power steering hoses
4. Fill the power steering system.
5. Start the engine and check for leaks.
6. Check the wheel alignment and adjust as necessary.

Strut

REMOVAL & INSTALLATION

1. Before servicing the vehicle, refer to the precautions in the beginning of this section.
2. Support the control arm with a stand or floor jack.
3. Remove or disconnect the following:
   - Front wheel
   - Brake hose bracket
   - Strut bracket bolts
   - Upper strut mount nuts
   - Strut

To install:

4. Install or connect the following:
   - Strut. Tighten the upper mount nuts to 40 ft. lbs. (55 Nm) and the bracket bolts to 70 ft. lbs. (95 Nm).
   - Brake hose bracket
   - Front wheel
5. Check the wheel alignment and adjust as necessary.

Shock Absorber

REMOVAL & INSTALLATION

1. Before servicing the vehicle, refer to the precautions in the beginning of this section.
2. Support the rear axle housing with a hydraulic jack or stand.
3. Remove or disconnect the following:
   - Shock absorber upper locknut and retaining nut
   - Lower shock absorber mounting nut and bolt
   - Rear shock absorber

To install:

4. Install or connect the following:
   - Rear shock absorber
   - Lower mounting nut and bolt
   - Upper retaining nut and locknut
5. Torque the upper mounting nuts to 21

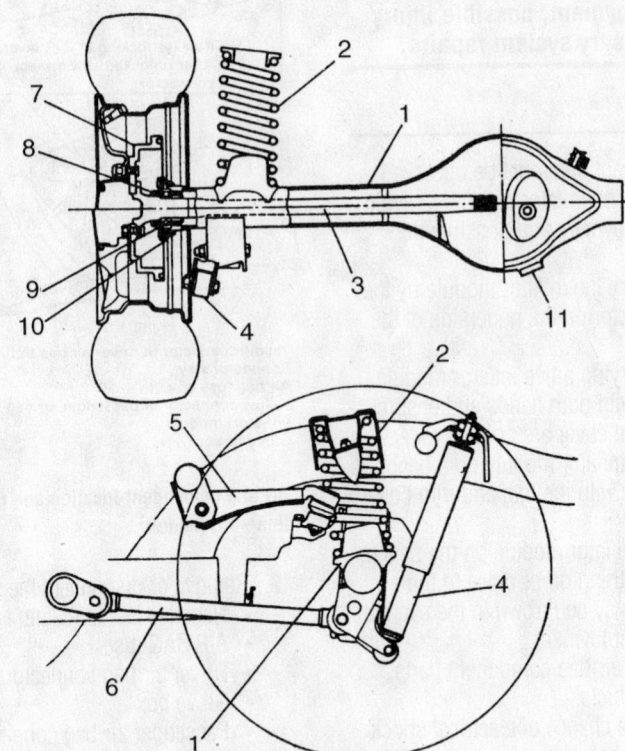

1. Rear axle housing
2. Coil spring
3. Axle shaft
4. Shock absorber
5. Upper arm
6. Trailing rod
7. Brake drum
8. Wheel bearing retainer
9. Rear wheel bearing
10. Brake back plate
11. Oil drain plug

7924HG32

Rear suspension component identification

ft. lbs. (22–35 Nm) and the lower mounting nut/bolt to 62 ft. lbs. (85 Nm) for 1999–01 models; 74 ft. lbs. (100 Nm) for 2002 models.

6. Remove the jack or stand from the rear axle assembly.

Coil Spring

REMOVAL & INSTALLATION

Front

1. Before servicing the vehicle, refer to the precautions in the beginning of this section.
2. Support the vehicle at the frame with a hoist or jackstand.
3. Support the control arm with a floor jack.
4. Remove or disconnect the following:
   - Front wheel
   - Brake caliper and rotor
   - Locking hub or drive flange, if equipped
   - Axle shaft snapring and thrust washer, if equipped
   - Wheel speed sensor, if equipped
   - Stabilizer bar link
   - Lower ball joint
   - Strut bracket bolts
5. Lower the floor jack and remove the coil spring.

To install:

➡**The bottom of the spring has a larger diameter than the top.**

6. Install the coil spring onto the control arm and raise the floor jack.
7. Install or connect the following:
   - Strut bracket bolts. Tighten them to 70 ft. lbs. (95 Nm).
   - Lower ball joint. Tighten the nut to 40 ft. lbs. (55 Nm).
   - Stabilizer bar link. Tighten the nut to 21 ft. lbs. (29 Nm).
   - Wheel speed sensor, if equipped
   - Axle shaft snapring and thrust washer, if equipped
   - Locking hub or drive flange, if equipped
   - Brake caliper and rotor
   - Front wheel
8. Check the wheel alignment and adjust as necessary.

Rear

1. Before servicing the vehicle, refer to the precautions in the beginning of this section.
2. Support the vehicle at the frame with a hoist or jackstand.

3. Support the rear axle housing with a floor jack.
4. Remove or disconnect the following:
   - Rear wheels
   - Parking brake cable hanger
   - Shock absorber lower mounting bolts
   - Wheel speed sensor harness clamps, if equipped
   - Brake pipe E-ring
   - Axle vent hose
5. Lower the floor jack and remove the coil springs.

To install:

6. Install the coil springs onto the axle spring seats and raise the floor jack.
7. Install or connect the following:
   - Axle vent hose
   - Brake pipe E-ring
   - Wheel speed sensor harness clamps, if equipped
   - Shock absorber lower mounting bolts. Tighten them to 62 ft. lbs. (85 Nm) for 1999–01 models; 74 ft. lbs. (100 Nm) for 2002 models.
   - Parking brake cable hanger
   - Rear wheels

Lower Ball Joint

REMOVAL & INSTALLATION

The lower ball joint is serviced with the lower control arm as an assembly.

Lower Control Arm

REMOVAL & INSTALLATION

1. Before servicing the vehicle, refer to the precautions in the beginning of this section.
2. Support the vehicle at the frame with a hoist or jackstand.
3. Support the control arm with a floor jack.
4. Remove or disconnect the following:
   - Front wheel
   - Brake caliper and rotor
   - Locking hub or drive flange, if equipped
   - Axle shaft snapring and thrust washer, if equipped
   - Wheel speed sensor, if equipped
   - Stabilizer bar link
   - Lower ball joint
   - Strut bracket bolts

5. Lower the floor jack and remove the coil spring.
6. Remove the inner control arm bolts and remove the control arm.

To install:

7. Install the inner control arm bolts.
8. Install the coil spring onto the control arm and raise the floor jack.
9. Install or connect the following:
   - Strut bracket bolts. Tighten them to 70 ft. lbs. (95 Nm).
   - Lower ball joint. Tighten the nut to 40 ft. lbs. (55 Nm).
   - Stabilizer bar link. Tighten the nut to 21 ft. lbs. (29 Nm).
   - Wheel speed sensor, if equipped
   - Axle shaft snapring and thrust washer, if equipped
   - Locking hub or drive flange, if equipped
   - Brake caliper and rotor
   - Front wheel
10. Lower the vehicle so that the front suspension is at curb height.
11. Tighten the front inner bolt to 62 ft. lbs. (85 Nm) and the rear inner bolt to 92 ft. lbs. (127 Nm).
12. Check the wheel alignment and adjust as necessary.

CONTROL ARM BUSHING REPLACEMENT

1. Before servicing the vehicle, refer to the precautions in the beginning of this section.
2. Remove the control arm from the vehicle.
3. Remove the control arm bushings with a hydraulic press.

To install:

4. Lubricate the control arm bushings with liquid soap.
5. Press the bushings into the control arm until the bushing flange contacts the housing edge of the control arm.
6. Install the control arm to the vehicle.
7. Check the wheel alignment and adjust as necessary.

Wheel Bearings

ADJUSTMENT

The wheel bearings are not adjustable.

REMOVAL & INSTALLATION

1. Before servicing the vehicle, refer to the precautions in the beginning of this section.
2. Remove or disconnect the following:

- Front wheel
- Brake caliper and rotor
- Locking hub or hub drive flange, if equipped
- Hub grease cap, if equipped
- Wheel speed sensor, if equipped
- Wheel bearing lockwasher
- Wheel bearing locknut and inner washer
- Wheel hub and bearing assembly
- Wheel hub oil seal
- Wheel bearing oil seal

- Snapring

3. Press the wheel bearing and race out of the hub.

To install:

4. Press the wheel bearing and race into the hub so that the race is fully seated in the hub bore.

5. Install or connect the following:

- Snapring
- Wheel bearing oil seal
- Wheel hub oil seal
- Wheel hub and bearing assembly

- Wheel bearing locknut and inner washer. Tighten the nut to 157 ft. lbs. (216 Nm).
- Wheel bearing lockwasher. Tighten the retaining screws to 13 inch lbs. (1.5 Nm).
- Wheel speed sensor, if equipped
- Hub grease cap, if equipped
- Locking hub or hub drive flange, if equipped
- Brake caliper and rotor
- Front wheel

GENERAL MOTORS CORP.

13

Chevrolet-Blazer • Xtreme • Trailblazer • Oldsmobile-Bravada • GMC-Envoy • Jimmy

PRECAUTIONS

Before servicing any vehicle, please be sure to read all of the following precautions, which deal with personal safety, prevention of component damage, and important points to take into consideration when servicing a motor vehicle:

• Never open, service or drain the radiator or cooling system when the engine is hot; serious burns can occur from the steam and hot coolant.

• Observe all applicable safety precautions when working around fuel. Whenever servicing the fuel system, always work in a well-ventilated area. Do not allow fuel spray or vapors to come in contact with a spark, open flame, or excessive heat (a hot drop light, for example). Keep a dry chemical fire extinguisher near the work area. Always keep fuel in a container specifically designed for fuel storage; also, always properly seal fuel containers to avoid the possibility of fire or explosion. Refer to the additional fuel system precautions later in this section.

• Fuel injection systems often remain pressurized, even after the engine has been turned **OFF**. The fuel system pressure must be relieved before disconnecting any fuel lines. Failure to do so may result in fire and/or personal injury.

• Brake fluid often contains polyglycol ethers and polyglycols. Avoid contact with the eyes and wash your hands thoroughly after handling brake fluid. If you do get brake fluid in your eyes, flush your eyes with clean, running water for 15 minutes. If eye irritation persists, or if you have taken brake fluid internally, IMMEDIATELY seek medical assistance.

• The EPA warns that prolonged contact with used engine oil may cause a number of skin disorders, including cancer! You should make every effort to minimize your exposure to used engine oil. Protective gloves should be worn when changing oil. Wash your hands and any other exposed skin areas as soon as possible after exposure to used engine oil. Soap and water, or waterless hand cleaner should be used.

• All new vehicles are now equipped with an air bag system. The system must be disabled before performing service on or around system components, steering column, instrument panel components, wiring and sensors. Failure to follow safety and disabling procedures could result in accidental air bag deployment, possible personal injury and unnecessary system repairs.

• Always wear safety goggles when working with, or around, the air bag system. When carrying a non-deployed air bag, be sure the bag and trim cover are pointed away from your body. When placing a non-deployed air bag on a work surface, always face the bag and trim cover upward, away from the surface. This will reduce the motion of the module if it is accidentally deployed. Refer to the additional air bag system precautions later in this section.

• Clean, high quality brake fluid from a sealed container is essential to the safe and proper operation of the brake system. You should always buy the correct type of brake fluid for your vehicle. If the brake fluid becomes contaminated, completely flush the system with new fluid. Never reuse any brake fluid. Any brake fluid that is removed from the system should be discarded. Also, do not allow any brake fluid to come in contact with a painted surface; it will damage the paint.

• Never operate the engine without the proper amount and type of engine oil; doing so WILL result in severe engine damage.

• Timing belt maintenance is extremely important! Many models utilize an interference-type, non-freewheeling engine. If the timing belt breaks, the valves in the cylinder head may strike the pistons, causing potentially serious (also time-consuming and expensive) engine damage. Refer to the maintenance interval charts in the front of this manual for the recommended replacement interval for the timing belt, and to the timing belt section for belt replacement and inspection.

• Disconnecting the negative battery cable on some vehicles may interfere with the functions of the on-board computer system(s) and may require the computer to undergo a relearning process once the negative battery cable is reconnected.

• When servicing drum brakes, only disassemble and assemble one side at a time, leaving the remaining side intact for reference.

ENGINE REPAIR

Distributor

REMOVAL

1. Before servicing the vehicle, refer to the precautions in the beginning of this section.

2. Remove or disconnect the following:

• Negative battery cable
• Spark plug wires and the coil leads from the distributor
• Electrical connector from the distributor
• Distributor cap fasteners and the cap

3. Using a marker, matchmark the rotor-to-housing and housing-to-intake manifold positions so that they can be matched during installation.

• Distributor hold-down bolt
• Distributor from the engine

4. As the distributor is being removed from the engine the rotor will move in a counterclockwise direction about 42°. This will appear as slightly more than one clock position.

5. Place a second mark on the distributor to mark the position of the rotor segment. This will help to ensure the correct rotor alignment when installing the distributor.

INSTALLATION

Engine Not Disturbed

1. If installing a new distributor, place two marks on the new distributor housing in the same position as the marks on the old distributor housing.

2. Align the rotor with the second mark made on the distributor.

3. Install the distributor in the engine making sure that the mounting hole in the distributor hold-down base is aligned over the mounting hole in the intake manifold.

4. As you are installing the distributor, watch the rotor move in a clockwise direction about 42°.

5. Once the distributor is fully seated, the rotor should be aligned with the first mark made on the distributor housing. If the rotor is not aligned with the first mark made on the housing, the distributor and camshaft teeth have meshed one or more teeth out of alignment. If this is the case, remove the

distributor and reinstall it so that all the marks are aligned.

6. Install or connect the following:
- Hold-down bolt and tighten the bolt to 18 ft. lbs. (25 Nm)
- Distributor cap and engage the electrical connector to the distributor
- Spark plug wires and coil leads
- Negative battery cable

Engine Disturbed

1. Remove the No. 1 cylinder spark plug. Turn the engine using a socket wrench on the large bolt on the front of the crankshaft pulley. Place a finger near the No. 1 spark plug hole and turn the crankshaft until the piston reaches Top Dead Center (TDC). As the engine approaches TDC, you will feel air being expelled by the No. 1 cylinder. If the position is not being met, turn the engine another full turn (360 degree). Once the engine position is correct, install the spark plug.

2. Align the cast arrow in the distributor housing, the driven gear roll pin and the pre-drilled indent hole in the distributor driven gear. If the driven gear is installed correctly, the dimple will be approximately 180ø opposite the rotor segment when it is installed in the distributor.

➡**Installing the distributor 180° out of alignment, or locating the rotor in the wrong holes, may cause a no start condition or can cause premature engine damage and wear.**

3. Make sure the rotor is pointing to the cap hold-down mount nearest the flat side of the housing.

4. Using a long screwdriver, align the oil pump driveshaft in the engine in the mating drive tab in the distributor.

5. Install the distributor in the engine. Make sure the spark plug towers are perpendicular to the centerline of the engine.

6. When the distributor is fully seated, the rotor segment should be aligned with the pointer cast in the distributor base. The pointer will have a "6" cast into it indicating a 6 cylinder engine. If the rotor segment is not within a few degrees of the pointer, the distributor gear may be off a tooth or more. If this is the case repeat the process until the rotor aligns with the pointer.

7. Install the cap and fasten the mounting screws.

8. Tighten the distributor mounting bolt to 18 ft. lbs. (25 Nm).

9. Engage the electrical connections and the spark plug wires.

Alternator

REMOVAL

4.2L Engine

1. Before servicing the vehicle, refer to the precautions in the beginning of this section.

2. Remove or disconnect the following:
- Negative battery cable
- Accessory belt
- Positive battery cable nut from the generator
- A/C line mounting bracket bolt at the engine lift hook
- Right engine lift hook bolts
- Engine lift hook
- Mounting bolts
- Alternator

4.3L Engine

1. Before servicing the vehicle, refer to the precautions in the beginning of this section.

2. Remove or disconnect the following:
- Negative battery cable
- Air inlet duct, if necessary
- Accessory belt
- Heater hose brace
- Wires
- Mounting bolts
- Alternator

INSTALLATION

4.2L Engine

1. Install or connect the following:
- Alternator and loosely install the mounting blots
- Tighten the alternator mounting bolts to 37 ft. lbs. (50 Nm)
- Positive battery cable and secure with the nut; tighten the nut to 80 inch lbs. (9 Nm)
- Engine lift hook and bolts; tighten the bolts to 37 ft. lbs. (50 Nm)
- A/C line bracket to the lift hook, then tighten the retaining bolt to 89 inch lbs. (10 Nm)
- Accessory belt
- Negative battery cable

4.3L Engine

1. Install or connect the following:
- Alternator and loosely install the mounting bolts

- Tighten the rear bolt to 37 ft. lbs. (50 Nm) and the front bolt to 18 ft. lbs. (25 Nm)
- Tighten the brace-to-alternator and brace-to-intake retainers to 18 ft. lbs. (25 Nm). Tighten the brace-to-engine stud nut to 37 ft. lbs. (50 Nm).
- Wires and the battery feed wire nut
- Heater hose bracket
- Accessory belt
- Negative battery cable

Ignition Timing

ADJUSTMENT

The ignition timing is preset and cannot be adjusted.

Engine Assembly

REMOVAL & INSTALLATION

4.2L Engine

1. Before servicing the vehicle, refer to the precautions in the beginning of this section.

2. Drain the engine cooling system

➡**Keep the oil drain plug removed during the engine removal and installation.**

3. Drain the engine oil. Install a suitable plug into the oil pan to prevent oil leakage during the remainder of the procedure.

4. Using the proper equipment, discharge and recover the refrigerant from the A/C system, if equipped.

5. Remove or disconnect the following:

- Hood
- Negative battery cable
- Fuel system pressure
- Air cleaner assembly
- Throttle body
- Manifold Absolute Pressure (MAP) sensor
- Windshield washer solvent container
- Air intake baffle
- Grille
- Headlight housing
- Radiator support brace
- Hood
- A/C lines from the condenser

- Transmission cooler lines from the engine, not the radiator

6. Remove the cooling fan and shroud, tilting the radiator forward, and the cooling fan and shroud rearward for clearance.

- Accessory belt
- Power steering pump bolts; position the pump aside
- Heater hoses from the heater core
- Transmission filler tube bracket nut from the Air Injector Reactor (AIR) adapter
- AIR adapter

7. Install a suitable lift hook to the AIR adapter

8. Remove or disconnect the following:

- Oxygen (O_2) sensor connector
- A/C line from the accumulator
- Front axle actuator electrical connector
- Camshaft phaser actuator valve electrical connector
- Transmission cooler lines from the clips on the right side of the engine block
- Ignition coil harness connectors
- Harness retainer from the clips
- Power brake hose from the booster
- Powertrain Control Module (PCM)
- Fuel lines from the fuel pressure regulator. Cap the lines to avoid excessive fuel leakage.
- All harnesses from the engine harness bracket
- Engine harness bracket bolt and bracket
- Starter electrical connections
- A/C pressure sensor and clutch electrical connector
- Alternator electrical connector and battery lead
- Knock Sensor (KS), Crankshaft Position (CKP) and Camshaft Position (CMP) sensor electrical connectors
- 4 ground on the left side of the block

9. Raise and safely support the vehicle.

- Left and right side driveshafts
- Propeller shaft from the front axle pinion yoke
- Engine protection shield
- Exhaust pipe from the exhaust manifold. Slide the exhaust pipe backward slightly.
- Fuel tank shield, if equipped
- Torque converter access cover and bolts

10. Place a jack on the transmission fluid pan for support.

11. Remove the transmission support.

12. Lower the transmission enough to reach the top bell housing bolts.

13. Remove the top 4 bell housing bolts, there may be 2 harness clips that will need to be removed in order to have access to 2 of the top bolts.

14. Raise the transmission.

15. Reinstall the transmission support using only 2 through bolts.

16. Remove or disconnect the following:

- Remaining bell housing bolts (11 total)
- Left and right engine lower mount nuts
- Oil level sensor electrical connector
- Oil pressure switch electrical connector

17. Carefully lower the vehicle.

18. Remove the left, then the right upper engine mount nut.

19. Install a suitable engine hoist.

20. Raise the engine out of the compartment slowly, keeping the transmission supported.

21. Remove both engine mounts for clearance.

22. Continue raising the engine out of the vehicle.

23. Place the engine on a suitable engine stand.

To install:

24. Remove the engine from the engine stand.

25. Slowly install the engine into the engine compartment, aligning the engine mounts with the brackets.

26. When the engine mounts are aligned, install the engine mounts, putting the mount up through the engine mount brackets before inserting into the chassis mount brackets.

27. Lower the engine onto the mounts and install the upper engine mounting nuts. Tighten the nuts to 51 ft. lbs. (71 Nm).

28. Remove the engine hoist.

29. Lay the radiator into the radiator support, but do not install the radiator completely.

30. Raise and safely support the vehicle.

31. Install the lower bell housing bolts, except the top four.

32. Remove the 2 through bolts secure the transmission support, then lower the transmission.

33. Install the top 4 bell housing bolts and tighten all 11 bolts to 37 ft. lbs. (50 Nm).

34. Raise the transmission.

35. Install or connect the following:

- Transmission support
- 3 torque converter bolts and tighten to 44 ft. lbs. (60 Nm)
- Torque converter bolt cover
- Fuel tank shield, if equipped
- Engine protection shield
- Propeller shaft to the front axle pinion yoke
- Exhaust pipe to the manifold and tighten the bolts to 37 ft. lbs. (50 Nm)
- Oil level switch and oil pressure sender electrical connectors
- Oil pan drain plug and tighten to 19 ft. lbs. (26 Nm)
- Lower radiator hose
- Left and right wheel driveshafts

36. Lower the vehicle.

37. Install or connect the following:

- 4 grounds on the left side of the block
- CMP, CKP and knock sensor electrical connectors
- Alternator and starter electrical connectors and battery leads. Torque the nuts to 80 inch lbs. (9 Nm).
- Fuel lines at the fuel pressure regulator
- Engine harness bracket and bolt. Torque the bolt to 37 ft. lbs. (50 Nm).
- Front differential vent hose, to the engine harness bracket
- PCM
- Power brake hose to the booster
- Harness retainer to its original location
- Ignition coil harness connectors
- Transmission cooler lines to clips on right side of engine block
- Camshaft phaser actuator valve electrical connector
- Front axle actuator electrical connector
- A/C line at the accumulator

38. Remove the lift hook.

- AIR adapter and secure with the studs. Tighten to 18 ft. lbs. (25 Nm).
- Transmission filler tube bracket to AIR adapter stud and secure the bracket with the nut. Torque the nut to 89 inch lbs. (10 Nm).
- Heater hoses to the heater core
- Power steering pump and tighten the bolts to 18 ft. lbs. (25 Nm).

39. The remainder of installation is the

reverse of removal, but please note the following important steps:

40. Connect the negative battery cable
41. Check all powertrain fluid levels and add, as necessary.
42. Refill the engine crankcase.
43. Refill the engine cooling system.
44. Perform the CKP System Variation Learn Procedure, as follows:

a. Install a suitable scan tool and check for Diagnostic Trouble Codes (DTCs). If any DTCs, other than P1336 are set, resolve those codes first, before proceeding with this procedure.

b. With the scan tool, select the crankshaft position variation learn procedure.

c. Observe the fuel cut-off for the 4.2L engine.

d. The scan tool will instruct you to perform certain steps, make sure you follow all directions given by the scan tool exactly.

e. Enable the crankshaft position system variation learn procedure.

➡**While the learn procedure is in progress, release the throttle immediately when the engine started to decelerate. The engine control is returned to the operator and the engine responds to throttle position after the learn procedure is complete.**

f. Slowly increase the engine speed to the RPM that you observed.

g. Immediately release the throttle when fuel cut-out is reached.

h. The scan tool displays: Learn Status: Learned this ignition. If the scan tool does NOT display this message and not other DTCs set, you must perform further troubleshooting.

i. Turn the ignition **OFF** for 30 seconds after the learn procedure has been completed successfully.

45. Start and run the engine, then check for leaks.

4.3L Engine

1. Before servicing the vehicle, refer to the precautions in the beginning of this section.
2. Drain the engine cooling system
3. Drain the engine oil.
4. Remove or disconnect the following:

- Negative battery cable
- Fuel system pressure
- Vacuum reservoir and/or the under-

hood light from the hood, as equipped
- Outer cowl vent grilles
- Hood
- Oxygen Sensor (O_2S) and/or wiring
- Exhaust pipes at the manifolds and loosen the hanger at the catalytic converter. This is necessary to remove the rear catalytic converter cushion mounts for removal of the exhaust assembly.
- Skid plate, if equipped
- Engine-to-transmission pencil braces
- Slave cylinder and position aside, if equipped
- Line clamp at the bell housing
- Wiring from the starter
- Starter
- Transfer case on 1999–01 models, if equipped
- Oil filter
- Engine mount through bolts
- Rear engine mount crossbar, nut and washer
- Bell housing bolts, except the upper left.
- Battery ground (negative) cable from the engine
- Front drive axle bolts and roll the axle downward, on 4WD vehicles
- Air cleaner assembly
- Upper radiator shroud
- Fan assembly
- Drive belt assembly
- Water pump pulley
- Upper radiator hose
- Air conditioning compressor, if equipped, and position aside with the lines intact
- Lower radiator hose
- Oil cooler and overflow lines from the radiator, plug the openings to prevent system contamination or excessive fluid loss
- Radiator and lower radiator shroud
- Power steering hoses from the steering gear, then cap the openings to prevent system contamination or excessive fluid loss
- Heater hoses from the intake manifold and the water pump
- Wiring harness and vacuum lines from the engine
- Throttle cables
- Distributor cap
- Remaining bell housing bolt
- Fuel lines and the bracket

- Ground strap(s) from the rear of the cylinder head
- Front body mount bolts, on 4WD vehicles

5. Support the transmission.
6. Install a lifting device and lift the engine.

To install:

7. Install or connect the following:

- Engine into the vehicle
- Front body mount bolts, on 4WD vehicles
- Ground strap(s) to the rear of the cylinder head
- Fuel lines and the bracket
- Upper left bell-housing bolt
- Distributor, cap and wires
- Throttle cables
- Vacuum lines and wiring harness connectors
- Heater hoses
- Power steering hoses
- Lower shroud and radiator
- Oil cooler lines to the radiator and overflow hose
- Lower radiator hose
- Air conditioning compressor to the engine, if equipped
- Upper radiator hose
- Water pump pulley
- Drive belt assembly
- Fan assembly
- Upper radiator shroud
- Air cleaner assembly
- Front drive axle, for 4WD vehicles
- Battery ground strap to the engine block
- Remaining bell housing bolts
- Engine mount through-bolts. Torque them to 49 ft. lbs. (66 Nm).
- Rear engine mount crossbar nut and washer. Tighten the nut to 33 ft. lbs. (45 Nm).
- Oil filter
- Starter motor
- Flywheel cover
- Clutch slave cylinder, if equipped
- Pencil brace and the skid plate, as equipped
- Catalytic converter Y-pipe assembly and hangers
- Hood
- Outer cowl vent grilles
- Vacuum reservoir and/or the underhood light to the hood, as equipped
- Negative battery cable

8. Check all powertrain fluid levels and add, as necessary.
9. Refill the engine crankcase.

10. Refill the engine cooling system.

11. Start and run the engine, then check for leaks.

Water Pump

REMOVAL & INSTALLATION

1. Before servicing the vehicle, refer to the precautions in the beginning of this section.

2. Disconnect the negative battery cable.

3. Drain the engine cooling system.

4. Relieve the belt tension and remove the accessory drive belts or the serpentine drive belt, as applicable.

5. Remove or disconnect the following:
- Upper fan shroud
- Fan or fan and clutch assembly, as applicable
- Water pump pulley; use a suitable tool to hold the pulley while removing the bolts
- Coolant hose(s) from the water pump

➡ For the hoses on some engines, removal may be easier if the hose is left attached until the pump is free from the block. Once the pump is removed from the engine, the pump may be pulled (giving a better grip and greater leverage) from the tight hose connection.

- Water pump retainers
- Water pump from the engine

❊❊ WARNING

Note the positions of all retainers as some engines will utilize different length fasteners in different locations and/or bolts and studs in different locations.

To install:

6. Clean the gasket mounting surfaces.

➡ The water pumps on some of the engines covered may have been installed using sealer only, no gasket, at the factory. If a gasket is supplied with the replacement part, it should be used. Otherwise, a ⅛ in. (3mm) bead of RTV sealer should be used around the sealing surface of the pump.

7. Apply sealant to the water pump retainer threads.

8. Install or connect the following:
- Water pump using a new gasket. Tighten the water pump retainers to 89 inch lbs. (10 Nm) for 4.2L

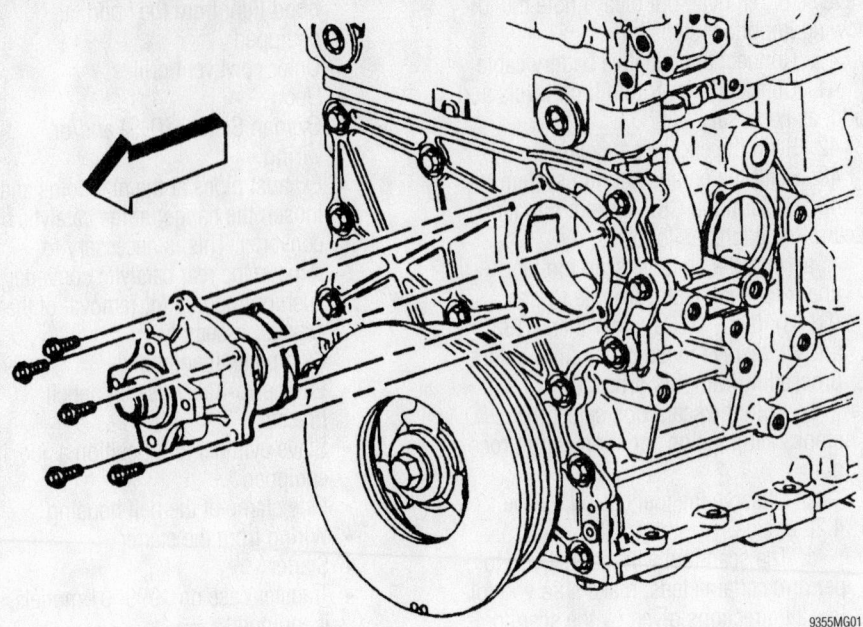

Exploded view of the water pump assembly mounting—4.2L engine

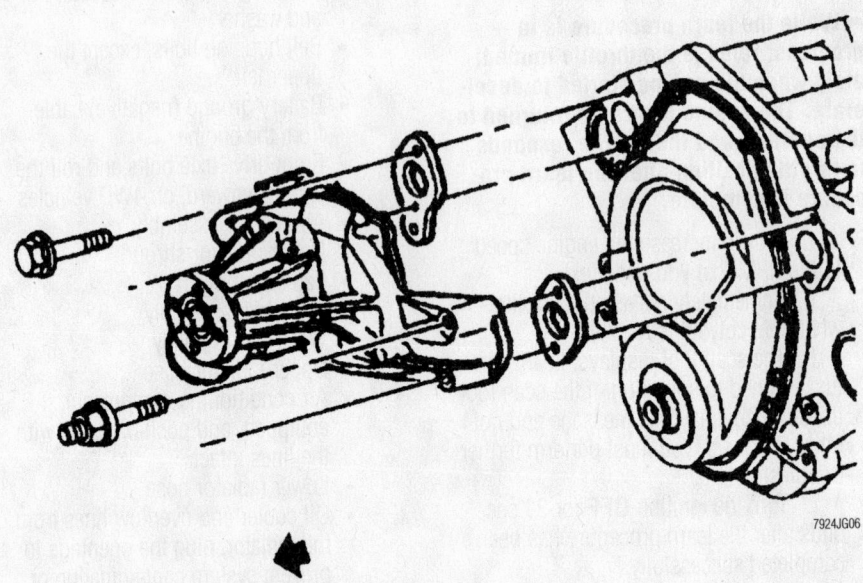

Exploded view of the water pump assembly mounting—4.3L engine

engines, or to 30 ft. lbs. (41 Nm) for 4.3L engines.
- Coolant hose(s)
- Water pump pulley. Tighten the pulley bolts to 18 ft. lbs. (25 Nm).
- Fan or fan and clutch assembly
- Serpentine drive belt (if equipped) by positioning the belt over the pulleys and carefully allow the tensioner back into contact with the belt.
- V-belts (if equipped) and adjust the tension
- Upper fan shroud
- Negative battery cable

9. Refill the engine cooling system.

10. Run the engine and check for leaks.

Cylinder Head

REMOVAL & INSTALLATION

4.2L Engine

1. Before servicing the vehicle, refer to the precautions in the beginning of this section.

2. Disconnect the negative battery cable.

3. Drain the engine cooling system.

4. Remove or disconnect the following:

- Camshaft cover
- Exhaust manifold
- Front cover
- Cylinder head access hole plugs
- Timing chain tensioner shoe bolt and shoe
- Timing chain tensioner guide bolts and guide
- Timing chain and sprockets

5. Unfasten the cylinder head bolts by loosening them in the reverse of the torque sequence, then carefully remove the cylinder head.

6. Remove the cylinder head gasket.

To install:

7. Carefully clean and inspect the cylinder head and the gasket mounting surfaces.

➡The gasket surfaces on both the head and block must be clean of any foreign matter and free of nicks or heavy scratches. The cylinder bolt threads in the block and thread on the bolts must be cleaned (dirt will affect the bolt torque).

➡DO NOT apply sealer to composition steel-asbestos gaskets.

⁂ WARNING

Make sure the number 1 cylinder is at Top Dead Center (TDC).

8. If using a steel only gasket, apply a thin and even coat of sealer to both sides of the gaskets.

9. Place a new gasket over the dowel pins with the bead or the words "This Side Up" facing upwards (as applicable), then carefully lower the cylinder head into position over the gasket and dowels.

10. Apply a coating of 12345493 or equivalent sealer to the threads of the cylinder head bolts, then thread the bolts into position until finger-tight.

11. Tighten the cylinder head bolts in sequence as follows:

a. Tighten the long bolts (1-14), in sequence, to 30 ft. lbs. (40 Nm).

b. Tighten the long bolts, in sequence, an additional 90 degrees.

c. Tighten the long bolts, in sequence, an additional 60 degrees.

d. Tighten the 2 long end bolts to 15 ft. lbs. (20 Nm).

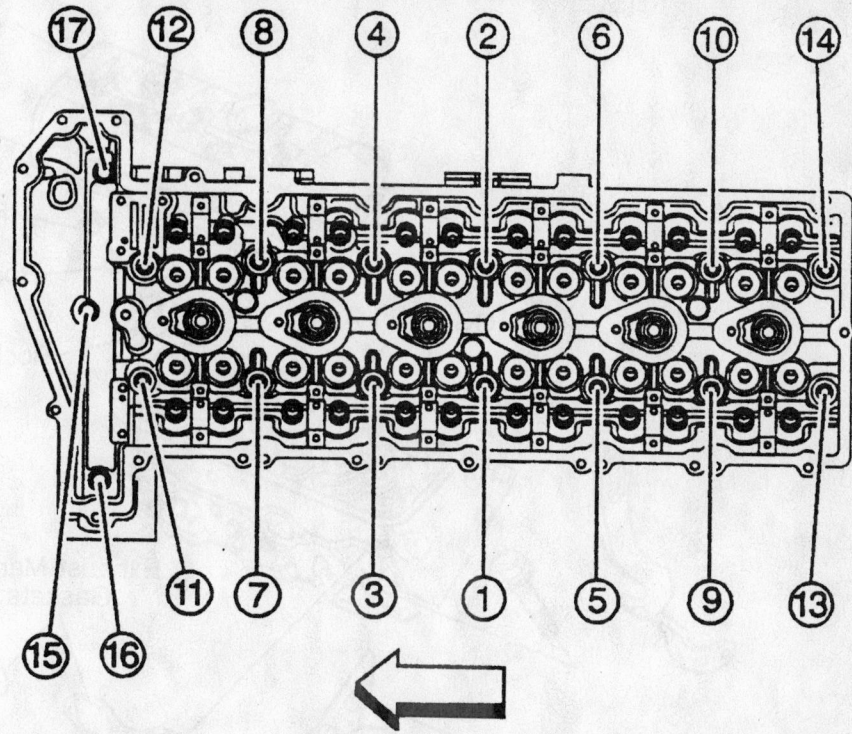

Cylinder head bolt tightening sequence—4.2L engine

9355MG02

e. Tighten the 1 short end bolt to 13 ft. lbs. (18 Nm).

12. Install or connect the following:

- Cylinder head access hole plugs and tighten to 44 inch lbs. (5 Nm)
- Timing chain and sprockets
- Front cover
- Camshaft cover
- Exhaust manifold
- Negative battery cable

13. Properly refill the engine cooling system.

14. Run the engine to check for leaks.

4.3L Engine

1. Before servicing the vehicle, refer to the precautions in the beginning of this section.

2. Properly relieve the fuel system pressure, then disconnect the negative battery cable.

3. Drain the engine cooling system.

4. Remove or disconnect the following:

- Intake manifold
- Exhaust manifold
- Alternator and bracket, if removing the right cylinder head
- Cooling fan assembly, on 1999–01 models
- Air conditioning compressor (posi-

tion it aside with the refrigerant lines attached), on 1999–01 models equipped

- Air pipe bracket and nut from the rear of the power steering pump if removing the left cylinder head on equipped 1999–01 models
- Engine accessory bracket with power steering pump (position the pump aside with the lines attached) and brackets, if removing the left cylinder head
- Wiring harness and clip from the rear of the cylinder head
- Coolant sensor wire
- Wiring from the spark plugs
- Spark plugs, if necessary
- Ground wires and if necessary, the fuel line bracket from the rear of the cylinder head, on 1999–01 models
- Rocker arm cover

5. Loosen the rocker arms and remove the pushrods.

➡If valve train components, such as the rocker arms or pushrods, are to be reused, they must be tagged or arranged to insure installation in their original locations.

6. Unfasten the cylinder head bolts by loosening them in the reverse of the torque

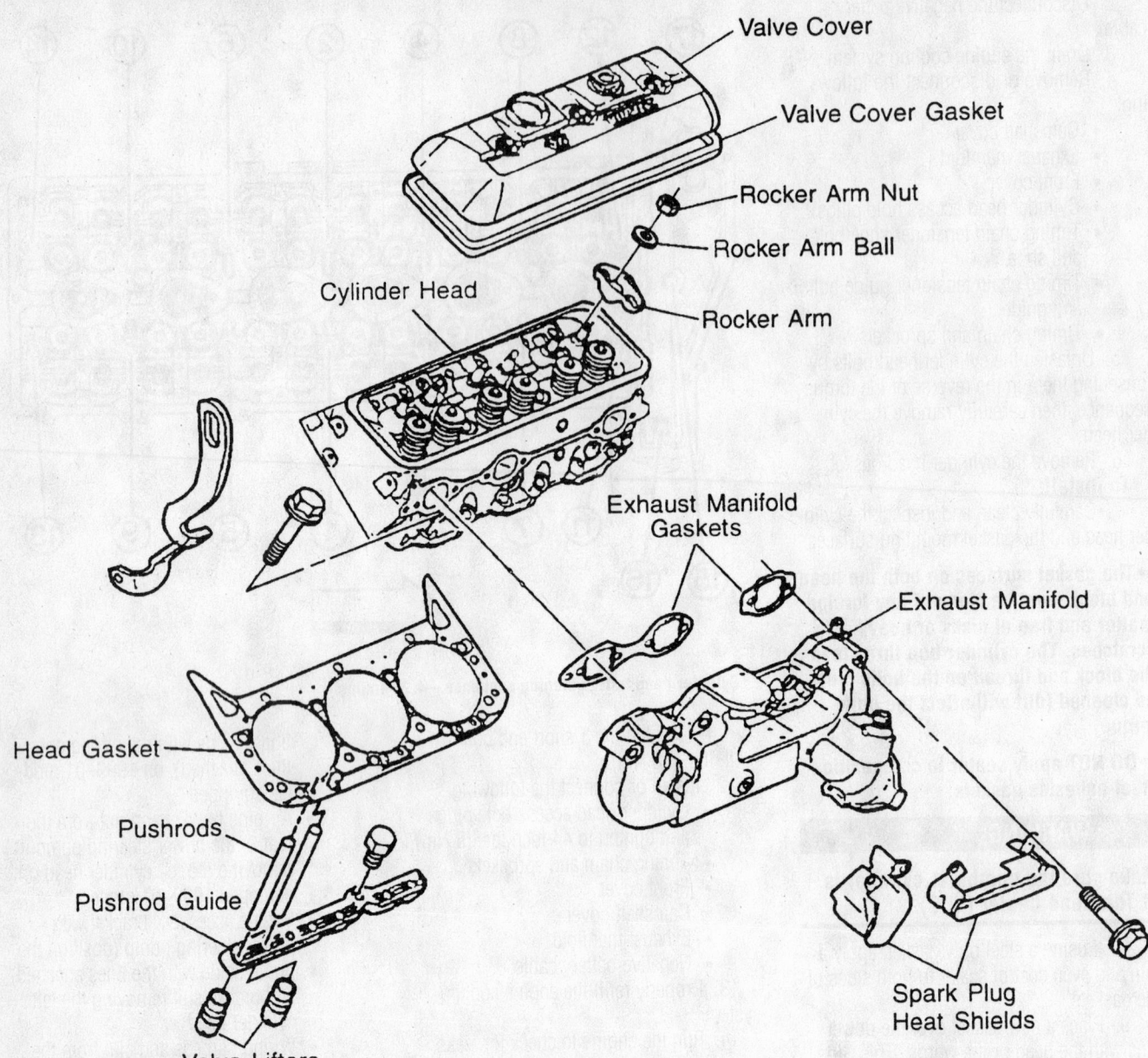

Valve Cover
Valve Cover Gasket
Rocker Arm Nut
Rocker Arm Ball
Rocker Arm
Cylinder Head
Exhaust Manifold Gaskets
Exhaust Manifold
Head Gasket
Pushrods
Pushrod Guide
Spark Plug Heat Shields
Valve Lifters

7924JG25

Cylinder head and related components—4.3L engine

sequence, then carefully remove the cylinder head.

To install:

7. Carefully clean and inspect the cylinder head and the gasket mounting surfaces.

➡**The gasket surfaces on both the head and block must be clean of any foreign matter and free of nicks or heavy scratches. The cylinder bolt threads in the block and thread on the bolts must be cleaned (dirt will affect the bolt torque).**

➡**DO NOT apply sealer to composition steel-asbestos gaskets.**

8. If using a steel only gasket, apply a thin and even coat of sealer to both sides of the gaskets.

9. Place a new gasket over the dowel pins with the bead or the words "This Side Up" facing upwards (as applicable), then carefully lower the cylinder head into position over the gasket and dowels.

10. Apply a coating of 12346004 or equivalent sealer to the threads of the cylinder head bolts, then thread the bolts into position until finger-tight.

11. Install the bolts in sequence to 22 ft. lbs. (30 Nm). The bolts must then be tightened again in sequence in the following order:

a. Short length bolts: (11, 7, 3, 2, 6, 10) 55 degrees.

b. Medium length bolts: (12, 13) 65 degrees.

c. Long length bolts: (1, 4, 8, 5, 9) 75 degrees.

12. Install or connect the following:

• Pushrods, secure the rocker arms and adjust the valves
• Rocker arm cover
• Spark plugs, if removed
• Spark plug wires
• Attach the fuel line bracket (if removed) and ground wires to the rear of the head and tighten the

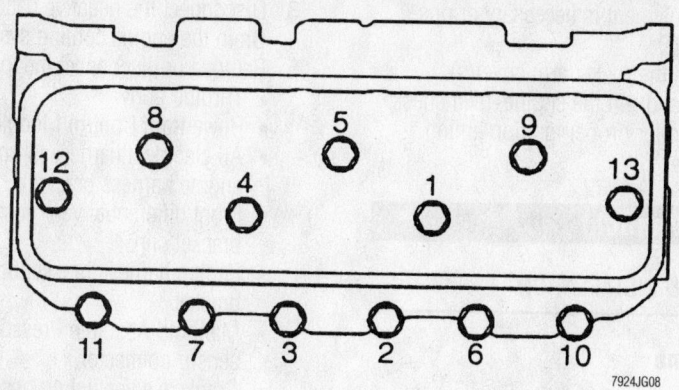

Cylinder head bolt torque sequence—4.3L engine

bolts to 22 ft. lbs. (30 Nm) on 1999–01 models
- Air conditioning compressor and bracket, if the left cylinder head was removed
- Alternator and bracket, if the right cylinder head was removed
- Engine accessory bracket with power steering pump. if the left cylinder head was removed
- Air pipe bracket and nut to the rear of the power steering pump (if equipped), if the left cylinder head was removed on 1999–01 models. Tighten the nut to 30 ft. lbs. (41 Nm).
- A/C compressor, if the left cylinder head was removed
- Cooling fan assembly, if the left cylinder head was removed on 1999–01 models
- Wiring harness and clip to the rear of the cylinder head
- Coolant sensor wire
- Exhaust manifold
- Intake manifold
- Negative battery cable

13. Properly refill the engine cooling system.

14. Run the engine to check for leaks.

Rocker Arms/Shafts

REMOVAL & INSTALLATION

4.2L Engine

1. Before servicing the vehicle, refer to the precautions in the beginning of this section.

2. Remove or disconnect the following:
- Camshaft cover

➡ **Make sure to place the camshaft caps in a rack to keep them in order, so**

they may be installed in their original locations.

- Camshaft cap bolts and caps
- Camshafts

➡ **If valve train components, such as the rocker arms or lash adjusters, are to be reused, they must be tagged or arranged to insure installation in their original locations.**

- Rocker arms
- Valve lash adjusters

To install:

3. Lubricate and fill the valve lash adjusters and the rocker arm roller with engine oil.

4. Install or connect the following
- Valve lash adjusters, in their original locations
- Rocker arm rollers in their original positions
- Camshafts
- Camshaft cap bolts
- Camshaft cover

4.3L Engine

1. Before servicing the vehicle, refer to the precautions in the beginning of this section.

2. Remove or disconnect the following:
- Rocker arm cover(s)
- Rocker arm nut, rocker arm and ball washer

➡ **If only the pushrod is to be removed, loosen the rocker arm nut, swing the rocker arm to the side and remove the pushrod.**

- Pushrod(s)

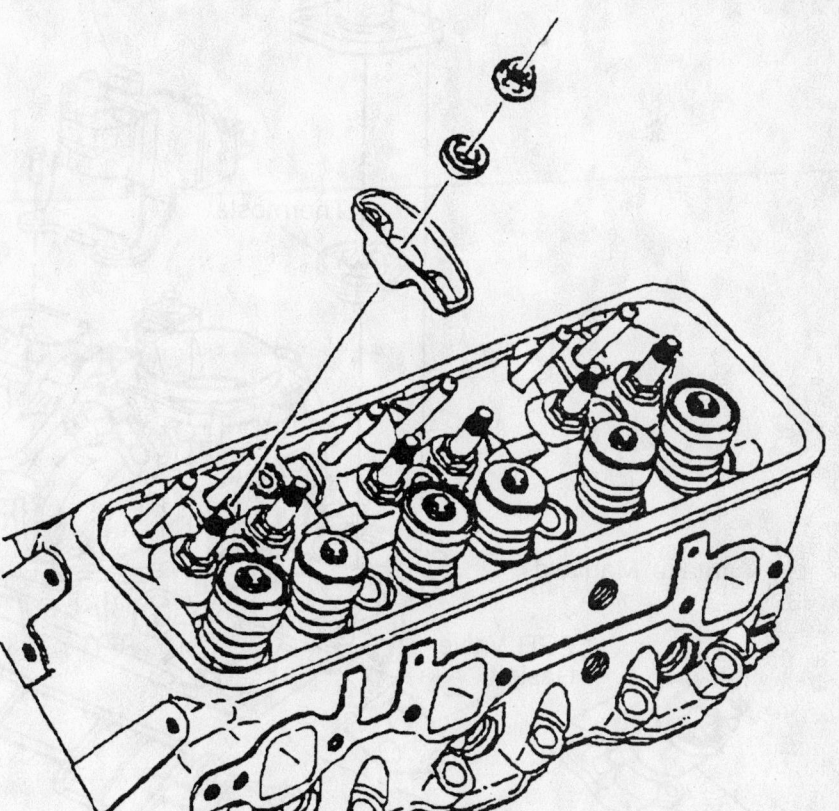

Exploded view of the rocker arm assembly

Heater Core replacement is covered in Section 2 of this manual

To install:

3. Inspect and replace components if worn or damaged.

4. Coat the bearing surfaces of the rocker arms and the rocker arm ball washers with Molykote® or equivalent pre-lube.

5. Install or connect the following:
- Pushrods making sure they seat properly in the lifter
- Rocker arms, ball washers and the nuts

➡ The 4.3L engines are equipped with screw-in rocker arm studs with positive stop shoulders.

- Rocker arm adjusting nuts. Tighten them against the stop shoulders to 18 ft. lbs. (24 Nm). No further

adjustment is necessary or possible.

6. Install the rocker arm cover(s).

7. Start and run the engine, then check for leaks and for proper ignition timing adjustment.

Intake Manifold

REMOVAL & INSTALLATION

4.2L Engine

1. Before servicing the vehicle, refer to the precautions in the beginning of this section.

2. Properly relieve the fuel system pressure.

3. Disconnect the negative battery cable.

4. Drain the engine cooling system.

5. Remove or disconnect the following:
- Throttle body
- Powertrain Control Module (PCM)
- All electrical harnesses from the engine harness bracket
- Front differential vent hose from the bracket clip
- Engine harness bracket bolt and bracket
- Manifold Absolute Pressure (MAP) sensor connector
- Crankcase ventilation hose
- Brake hose from the booster
- Alternator
- Intake manifold bolts and manifold.
- Manifold gasket

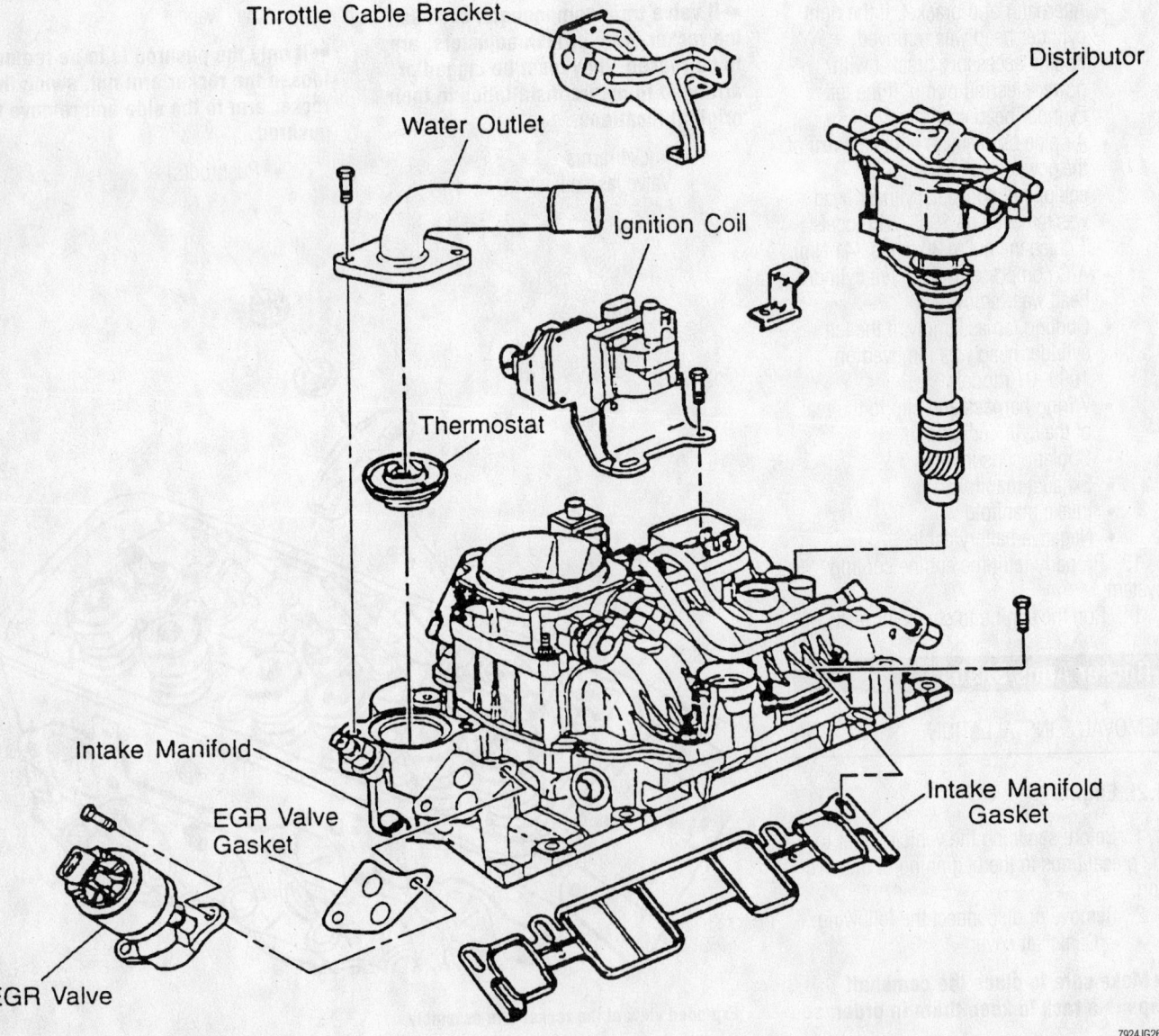

Intake manifold and related components

7924JG26

To install:

6. Clean the gasket mounting surfaces. Be sure to inspect the manifold for warpage and/or cracks. If necessary, replace it.

7. Properly position a new intake manifold gasket.

8. Install or connect the following:
- Intake manifold and bolts. Torque the bolts to 16 ft. lbs. (22 Nm).
- Alternator
- Brake hose to the booster
- Crankcase ventilation hose, lubricating the inner diameter first with 12345884, or equivalent lubricant
- MAP sensor electrical connector
- Engine harness bracket. Tighten the retaining bolt to 37 ft. lbs. (50 Nm).
- Front differential vent hose to the engine harness bracket clip
- All harnesses to their original locations onto the engine harness bracket
- PCM
- Throttle body
- Negative battery cable

9. Refill the engine cooling system.

4.3L Engine

➡️If only the upper intake manifold is being removed, the fuel system pressure does not need to be released. ALWAYS release the pressure before disconnecting any fuel lines.

1. Before servicing the vehicle, refer to the precautions in the beginning of this section.

2. Remove the engine cover, if equipped

3. Properly relieve the fuel system pressure.

4. Disconnect the negative battery cable.

5. Drain the engine cooling system.

6. Remove or disconnect the following:
- Air cleaner and air inlet duct
- Wiring harness connectors and brackets
- Throttle linkage from the upper intake manifold
- Ignition coil
- Fuel lines and bracket from the rear of the lower intake manifold
- Brake booster vacuum hose at the upper intake manifold
- Positive Crankcase Ventilation (PCV) hose at the rear of the upper intake manifold
- Vacuum hoses from both the front and rear of the upper intake

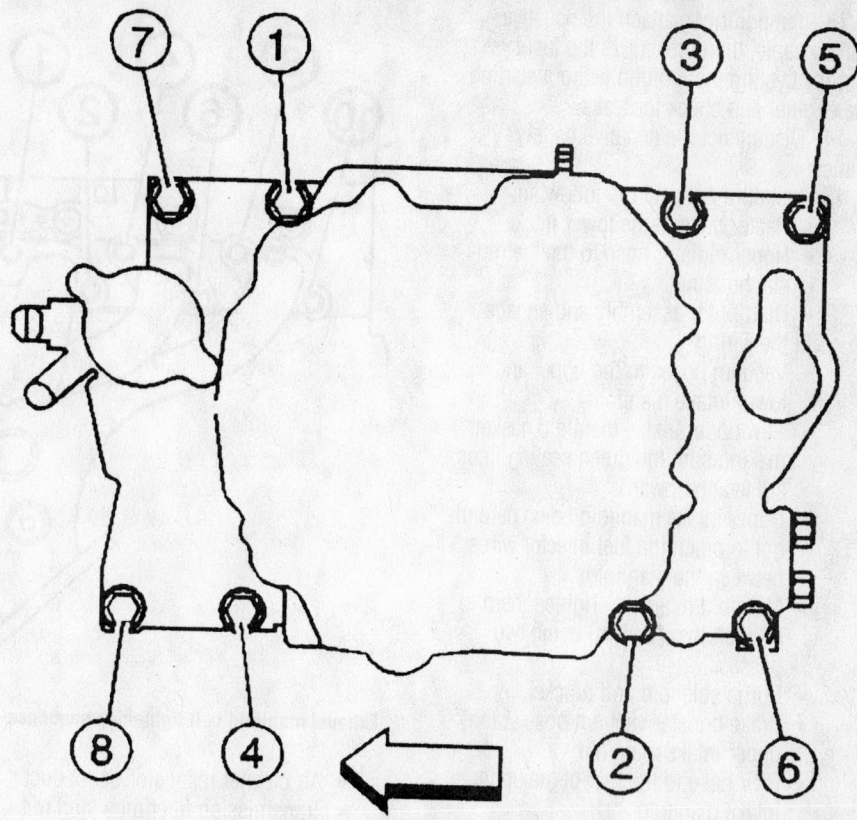

Intake manifold bolt tightening sequence—4.3L engine

9355MG12

- Purge solenoid and bracket
- Upper intake manifold
- Distributor or High Voltage Switch (HVS) assembly
- Upper radiator hose at the thermostat housing
- Heater hose at the lower intake manifold
- Wiring harnesses and brackets
- Automatic transmission dipstick tube
- Exhaust Gas Recirculation (EGR) tube, clamp and tube
- Air conditioning compressor bracket-to-lower intake manifold pencil brace
- Alternator bracket bolts near the thermostat housing
- Lower intake manifold

7. Insert clean rags into the openings in the cylinder head to prevent dirt and debris from entering the engine.

8. Clean the gasket mounting surfaces. Be sure to inspect the manifold for warpage and/or cracks. If necessary, replace it.

To install:

9. Remove the rags from the cylinder heads.

10. Position the gaskets on the cylinder head with the port blocking plates to the rear and the **this side up** stamps facing upward. Then apply a 3⁄16 in. (5mm) bead of RTV sealant on the front and rear of the engine block at the block-to-manifold mating surface. Extend the bead ½ in. (13mm) up each cylinder head to seal and retain the gaskets.

11. Install the lower intake manifold. Tighten the bolts in sequence and in 3 steps, as follows:
   a. Step 1: 26 inch lbs. (3 Nm).
   b. Step 2: 106 inch lbs. (12 Nm).
   c. Step 3: 11 ft. lbs. (15 Nm).

12. Install or connect the following:
- Alternator bracket bolt near the thermostat housing
- EGR tube, clamp and bolt
- Wiring harness to the lower manifold components, including the injector, EGR valve and ECT sensor
- Air conditioning compressor bracket-to-the lower intake manifold pencil braces
- Transmission oil dipstick tube, if necessary
- Fuel supply and return lines to the rear of the lower intake

Brake service is covered in Section 4 of this manual

13. Temporarily reattach the negative battery cable, then pressurize the fuel system (by cycling the ignition without starting the engine) and check for leaks.

14. Disconnect the negative battery cable.

15. Install or connect the following:
- Heater hose to the lower intake
- Upper radiator hose to the thermostat housing
- Distributor assembly and engage the wiring
- Vacuum hoses to the upper and lower intake manifold
- New upper intake manifold gasket, making sure the green sealing lines are facing upward
- Upper intake manifold being careful not to pinch the fuel injector wires between the manifolds
- Manifold retainers. Tighten them to 88 inch lbs. (10 Nm) using two passes.
- Purge solenoid and bracket
- Brake booster vacuum hose at the upper intake manifold
- PCV hose to the rear of the upper intake manifold
- Vacuum hoses to both the front and rear of the manifold assembly
- Throttle linkage to the upper intake
- Ignition coil
- Wiring to the upper intake components including the TP sensor, IAC motor, MAP sensor and the IMTV
- Plastic cover
- Air cleaner and air inlet duct
- Negative battery cable

16. Refill the engine cooling system.

Exhaust Manifold

REMOVAL & INSTALLATION

4.2L Engine

1. Before servicing the vehicle, refer to the precautions in the beginning of this section.

2. Remove or disconnect the following:
- Negative battery cable

➡ It will be easier if the vehicle is only supported to a height where underhood access is still possible, the vehicle may be left in position for the entire procedure. If the vehicle is raised too high for underhood access, it will have to lowered, raised and lowered again during the procedure.

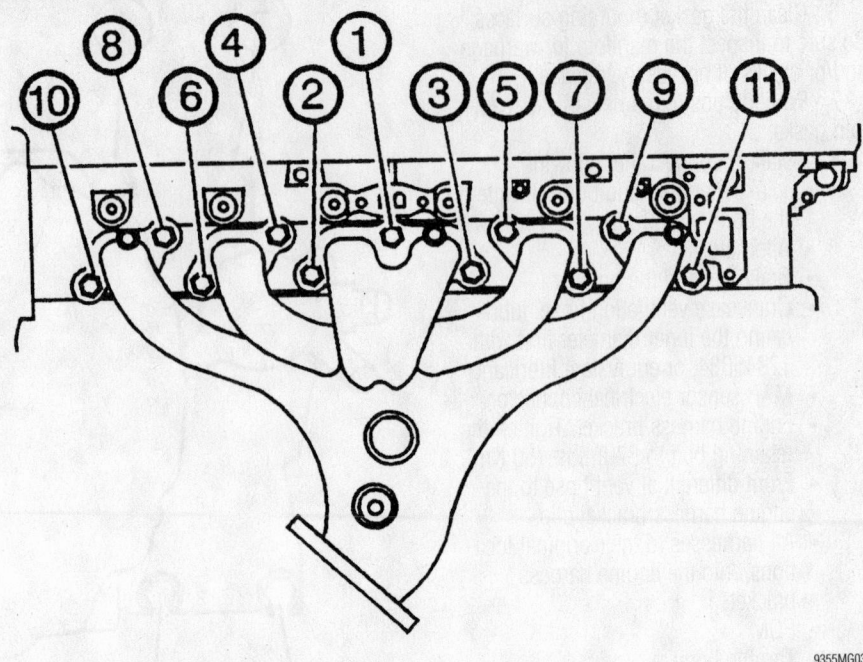

Exhaust manifold bolt tightening sequence—4.2L engine

9355MG03

- Air cleaner resonator outlet duct
- Transmission filler tube stud nut from the Air Injector Reactor (AIR) adapter and move the tube aside
- Oil level indicator tube
- Oxygen (O_2) sensor from the exhaust manifold
- 4 manifold heat shield nuts and shield
- Exhaust pipe bolts from the exhaust manifold
- Exhaust manifold bolts, and manifold
- Old gaskets and discard

To install:

3. Using a putty knife, clean the gasket mounting surfaces. Inspect the exhaust manifold for distortion, cracks or damage; replace if necessary.

4. Apply a threadlock such as GM 12345493 to the threads of the manifold retainers prior to installation.

5. Install or connect the following:
- Exhaust manifold to the cylinder using a new gasket, then tighten the bolts, in 3 passes, in sequence, to 18 ft. lbs. (25 Nm)
- Heat shield studs, if necessary, and tighten to 89 inch lbs. (10 Nm)
- O_2 sensor
- Exhaust manifold heat shield

➡ Apply a suitable anti-seize compound to the exhaust manifold heat shield nuts prior to installation.

- Heat shield nuts and tighten to 44 inch lbs. (5 Nm)
- Exhaust pipe to the manifold with seal and retaining nuts. Tighten the nuts to 37 ft. lbs. (50 Nm).
- Oil level indicator tube
- Transmission filler tube back onto the AIR adapter block stud and secure with the nut. Tighten the bracket nut to 89 inch lbs. (10 Nm).
- Air cleaner resonator outlet duct
- Negative battery cable.

4.3L Engine

1. Before servicing the vehicle, refer to the precautions in the beginning of this section.

2. Remove or disconnect the following:
- Negative battery cable

➡ It will be easier if the vehicle is only supported to a height where underhood access is still possible, the vehicle may be left in position for the entire procedure. If the vehicle is raised too high for underhood access, it will have to lowered, raised and lowered again during the procedure.

- Exhaust pipe from the exhaust manifold. It may be necessary to remove the tires to gain access to the rear manifold bolts.
- Engine oil dipstick tube bolt, if removing the right side manifold on 1999–01 models

- Exhaust Gas Recirculation (EGR) inlet pipe from the left side manifold, if necessary.
- Engine Coolant Temperature (ECT) sensor electrical connection, on 1999–01 models
- Upper radiator support hose and nut, on 1999–01 models
- Steering intermediate shaft, if removing the left side manifold on 1999–01 models
- Wheel house extension, if removing the right side manifold on 1999–01 models
- Spark plugs wires from the plugs
- Nuts attaching the secondary air injection pipe to the manifold, on 1999–01 models
- Air injection pipe and gasket, on 1999–01 models
- Locktangs (unbend), the exhaust manifold retaining bolts, washers and tab washers
- Heat shields
- Exhaust manifold
- Old gaskets and discard

To install:

3. Using a putty knife, clean the gasket mounting surfaces. Inspect the exhaust manifold for distortion, cracks or damage; replace if necessary.

4. On 1999–01 models, apply a threadlock such as GM 12345493 to the threads of the manifold retainers prior to installation.

5. Install or connect the following:
- Exhaust manifold to the cylinder using a new gasket, then tighten the center bolts to 11 ft. lbs. (15 Nm) and the front and rear manifold bolts to 22 ft. lbs. (30 Nm). Once the bolts are tightened, bend the tabs on the washers back over the heads of all bolts in order to lock them in position.
- Spark plug wires to the plugs
- Fender wheelhouse extension and the tire assembly, if removed on 1999–01 models
- Secondary air injection pipe with a NEW gasket to the manifold and tighten the nuts to 18 ft. lbs. (25 Nm), if removed on 1999–01 models
- EGR inlet pipe, if removed
- ECT sensor electrical connection, if removed
- Upper radiator hose support and nut, if removed

- Steering intermediate shaft., if removed on 1999–01 models (left side manifold only)
- Engine oil dipstick tube bolt to 106 inch lbs. (12 Nm), if removed
- Exhaust pipe to the manifold
- Negative battery cable

Camshaft and Valve Lifters

REMOVAL & INSTALLATION

4.2L Engine

1. Before servicing the vehicle, refer to the precautions in the beginning of this section.

2. Disconnect the negative battery cable.

3. Discharge and recover the refrigerant from the air conditioning system, using the proper equipment.

4. Remove or disconnect the following:
- Intake manifold
- A/C line from the oil level indicator tube
- A/C line from the accumulator
- A/C bracket bolt from the engine lift hook
- Engine lift bracket
- Ignition control module electrical connectors
- Ignition control module bolts and module

✳✳ WARNING

Be careful not to damage the clips that hold the harness housing in place.

- Engine electrical harness housing from the camshaft cover
- Fuel injection harness electrical connector
- Camshaft cover bolts and cover
- Exhaust and intake sprocket bolts

5. Install a suitable sprocket holding tool onto the cylinder head and adjust the horizontal bolts into the camshaft sprockets to maintain timing chain tension and avoid disturbing the timing chain components.

6. Carefully move the sprockets with the timing chain off of the camshafts.

➡**Make sure to place the camshaft caps in a rack to keep them in order, so they may be installed in their original locations.**

7. Remove or disconnect the following:
- Camshaft cap bolts and caps
- Camshafts

To install:

8. Coat the camshaft journals with engine oil.
- Camshafts, in their original position
- Camshaft caps, in their original locations. Tighten the bolts to 106 inch lbs. (12 Nm).

9. Carefully place the camshaft sprockets back onto the camshafts and remove the holding tool.

10. Install or connect the following:
- Intake camshaft sprocket washer and bolt and the exhaust camshaft actuator bolt. Tighten the intake camshaft sprocket bolt to 22 ft. lbs. (30 Nm), plus an additional 135 degrees and the exhaust camshaft actuator bolt to 18 ft. lbs. (25 Nm), plus an additional 135 degrees.
- New camshaft cover seal
- New rubber ignition control module seals
- Camshaft cover and bolts. Tighten the bolts to 89 inch lbs. (10 Nm).
- Ignition control module. Tighten the bolts to 89 inch lbs. (10 Nm).
- Ignition control module electrical connectors
- Fuel injector electrical connectors
- Engine electrical harness housing
- A/C line bracket to the oil level indicator tube stud and secure with the nut. Tighten the nut to 62 inch lbs. (7 Nm).
- Engine lift bracket and secure the lift hook with the bolts. Tighten the bolts to 37 ft. lbs. (50 Nm).
- A/C line bracket to the engine lift bracket. Tighten the bolt to 89 inch lbs. (10 Nm).
- Intake manifold

11. Using the proper equipment, recharge the A/C system.

4.3L Engine

1. Before servicing the vehicle, refer to the precautions in the beginning of this section.

2. Properly relieve the fuel system pressure.

3. Disconnect the negative battery cable.

4. Drain the engine cooling system.

5. Discharge and recover the refrigerant from the air conditioning system.

6. Remove or disconnect the following:
- Radiator
- Air conditioning condenser
- Rocker arm covers
- Intake manifold assembly
- Rocker arms, pushrods and lifters
- Crankshaft pulley and hub
- Engine front (timing) cover

7. Align the timing marks on the crankshaft and camshaft sprockets.
- Camshaft sprocket and timing chain
- Balance shaft drive gear, if equipped
- Camshaft thrust plate
- Camshaft by installing the sprocket bolts or longer bolts the camshaft end to act as a handle; then, remove the camshaft while turning slightly from side to side, as necessary.

➡ **Take care not to damage the camshaft bearings when removing the camshaft.**

To install:

8. Lubricate the camshaft journals with clean engine oil or a suitable pre-lube.

9. Install or connect the following:
- Camshaft being extremely careful not to contact the bearings with the cam lobes

- Thrust plate. Torque the bolts to 106 inch lbs. (12 Nm).
- Balance shaft drive gear, if equipped
- Timing chain and camshaft sprocket
- Engine front (timing) cover
- Crankshaft pulley and hub
- Valve lifters, pushrods and rocker arms. Adjust the valve clearance.
- Intake manifold assembly
- Rocker arm covers
- Radiator
- Negative battery cable

10. Refill the engine cooling system.

Valve Lash

ADJUSTMENT

The 4.2L engines do not require a periodic valve lash adjustment.

The 4.3L engines are equipped with screw-in rocker arm studs with positive stop shoulders. Because the shoulders that allow the rocker arms to be tightened into proper position, no adjustments are necessary or possible. If a valvetrain problem is suspected, check that the rocker arm nuts are tightened to 18 ft. lbs. (24 Nm). When valve lash falls out of specification (valve tap is heard), replace the rocker arm, pushrod and hydraulic lifter on the offending cylinder.

Starter Motor

REMOVAL & INSTALLATION

4.2L Engine

1. Before servicing the vehicle, refer to the precautions in the beginning of this section.
2. Disconnect the negative battery cable.
3. Raise and safely support the vehicle.
4. Remove the left front tire and wheel assembly.
5. Working in the left fender area, disconnect the positive battery lead from the solenoid.
6. Remove or disconnect the following:
- Starter mount bolt and nut
- Starter motor

To install:
7. Install or connect the following:
- Starter motor
- Starter mounting bolt and nut. Tighten to 37 ft. lbs. (50 Nm).
- Positive battery cable to the starter. Tighten the nut to 80 inch lbs. (9 Nm).
- Left front tire and wheel assembly
8. Carefully lower the vehicle, then connect the negative battery cable.

4.3L Engine

2WD MODELS

1. Before servicing the vehicle, refer to the precautions in the beginning of this section.
2. Remove or disconnect the following:
- Negative battery cable
- Wires from the starter solenoid
- Starter motor mounting bolts
- Starter motor and if equipped, the shims

To install:
3. Install or connect the following:
- Starter motor into position
- Starter motor inboard bolt but do not tighten it at this time
- Starter motor shims, if equipped
- Outboard starter motor bolt. Tighten the bolts to 32 ft. lbs. (43 Nm).
- Wires to the solenoid
- Negative battery cable

4WD MODELS

1. Before servicing the vehicle, refer to the precautions in the beginning of this section.

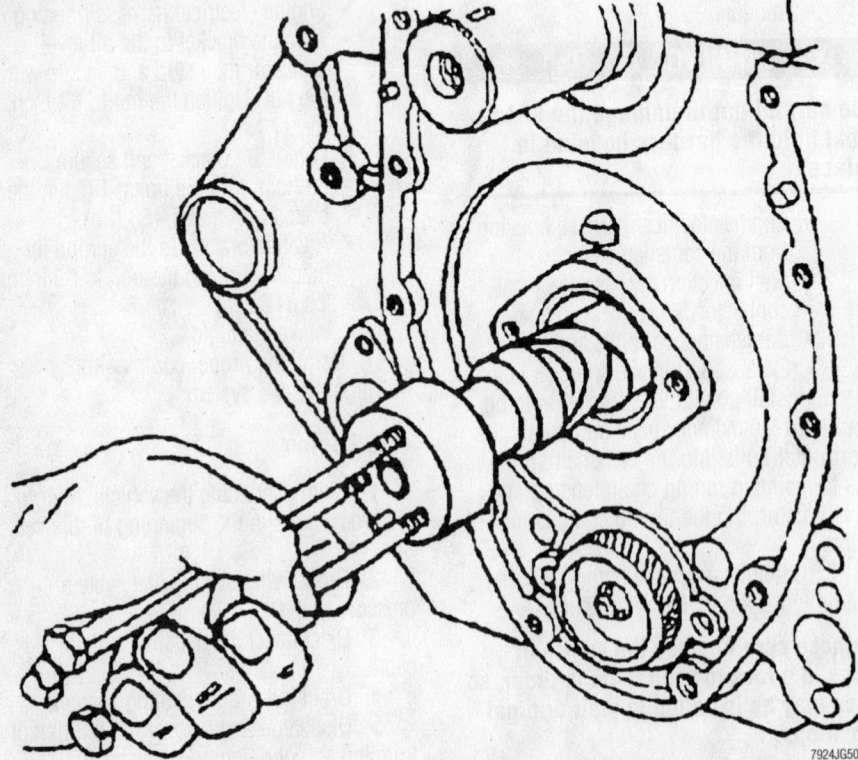

7924JG50

Thread 3 long bolts into the camshaft to use as a handle, then withdraw it from the engine

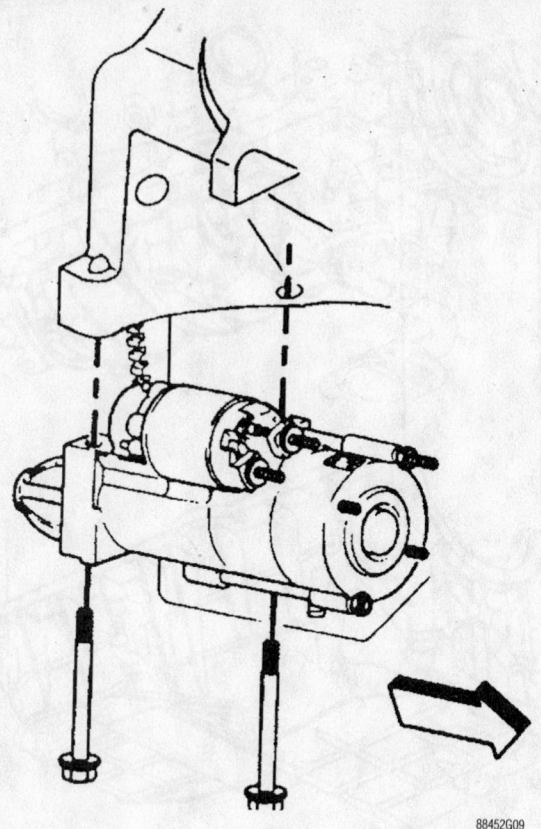

Starter mounting—4.3L

88452G09

2. Remove or disconnect the following:
- Negative battery cable
- Brush end mounting bracket, if removed
- Wiring from the starter solenoid
- Transfer case shield, if equipped
- Bolts that attach the brake pipe-to-transmission bracket to the transmission crossmember and the brackets
- Transmission crossmember bolts, (usually three on each side)
- Transmission mount bolts. Support the transmission assembly with a transmission jack and slide the transmission crossmember out of the way.
- Bracket that attaches the transmission cooler lines to the flywheel housing, brace rod to the flywheel housing, and/or the lower flywheel housing as necessary
- Starter motor mounting bolts
- Starter and if equipped, the starter shims

To install:
3. Install or connect the following:
- Shims in their original locations (if equipped), then place the starter motor into position

- Starter motor bolts and tighten them to 33 ft. lbs. (45 Nm)
- Lower flywheel cover, if removed
- Transmission line bracket to the housing and the brace rod to the housing, if equipped
- Crossmember and tighten the retaining bolts
- Transfer case shield, if equipped
- Solenoid wiring
- Brush end bracket and tighten the nuts to 97 inch lbs. (11 Nm), if equipped
- Negative battery cable

4. Start the vehicle to check for proper operation.

Oil Pan

REMOVAL & INSTALLATION

4.2L Engine

1. Before servicing the vehicle, refer to the precautions in the beginning of this section.
2. Disconnect the negative battery cable.

3. Remove or disconnect the following:
- A/C compressor bottom bolts and loosen the top bolts
- Oil dipstick and tube

4. Raise and safely support the vehicle.
5. Drain the engine crankcase oil.
6. Remove or disconnect the following:
- Left and right front tire and wheel assemblies
- Engine protection shield mounting bolts and shield
- Front steering gear crossmember
- Left and right driveshafts
- Front drive axle clutch fork assembly
- Prop shaft from the front axle pinion yoke
- Unclip the transmission cooler lines from the engine block
- Front differential bolts and position the differential aside
- 4 transmission bell housing-to-oil pan bolts
- Remaining oil pan bolts
- Oil pan, by placing 2 oil pan bolts in the jack screws on the oil pan and tighten evenly to release the oil pan from the engine

To install:
7. Clean the gasket mounting surfaces.

➡The alignment between the rear of the oil pan and the rear of the block is critical. When the oil pan is installed it could be inadvertently shifted front or back a small amount which could cause a transmission alignment problem. The back to the oil pan needs to be flush with the engine block.

8. Apply a 0.12 in. (3mm) bead of sealant to engine block, rather than the oil pan.

➡The oil pan MUST be installed within 10 minutes of applying the sealant to the engine block.

9. Install or connect the following:
- Oil pan, maneuvering it to clear the oil pump and screen assembly

➡After the bolts are installed, before tightening them to specifications, check the oil pan alignment. Use a straight edge on the back to the block and the oil pan transmission mounting surface.

- Oil pan bolts; tighten the side bolts to 18 ft. lbs. (25 Nm) and the end bolts to 89 inch lbs. (10 Nm)
- Transmission bell housing-to-oil

For Accessory Drive Belt illustrations, see Section 1 of this manual

pan bolts and tighten to 35 ft. lbs.
(47 Nm)

- A/C compressor bottom bolts.
  Tighten to 37 ft. lbs. (50 Nm)
- Front differential bolts and tighten
  to 63 ft. lbs. (85 Nm)
- Front drive axle and clutch fork
  assembly
- Transmission cooler lines to block
- Prop shaft to front differential
- Steering gear crossmember
- Left and right driveshaft
- Oil pan drain plug. Tighten to 19 ft.
  lbs. (26 Nm)
- Engine protection shield. Tighten
  the bolts to 18 ft. lbs. (25 Nm)
- Left and right front wheel and tire
  assemblies
10. Carefully lower the vehicle.
11. Refill the crankcase with fresh oil.
Start the engine, establish normal operating
temperatures and check for leaks.

4.3L Engine

2WD MODELS

1. Before servicing the vehicle, refer to
the precautions in the beginning of this sec-
tion.
2. Remove or disconnect the following:
  - Engine
  - Oil pan retainers (nuts, studs
    and/or bolts) and rail reinforce-
    ments, if equipped
  - Oil pan
  - Rubber bell housing plugs and
    gasket

To install:
3. Clean the gasket mounting surfaces.

➡The alignment between the rear of
the oil pan and the rear of the block is
critical. The oil pan must be flush or
slightly forward of the rear of the block
to allow for proper alignment with the
transmission housing. Use a feeler
gauge to measure the clearance
between the 3 oil pan-to-transmission
contact points. If the clearance exceeds
0.011 in. (0.3mm) at any of the 3
points, realign the oil pan.

4. Apply sealant to the oil pan rail where
it contacts the timing cover-to-block joint
(front) and the crankshaft rear seal retainer-
to-block joint (rear). Continue the bead of
sealant about 1 in. (25mm) in both directions from each of the 4 corners.
5. Install or connect the following:
  - Rubber bell housing plugs, if
    equipped
  - Oil pan using a new gasket

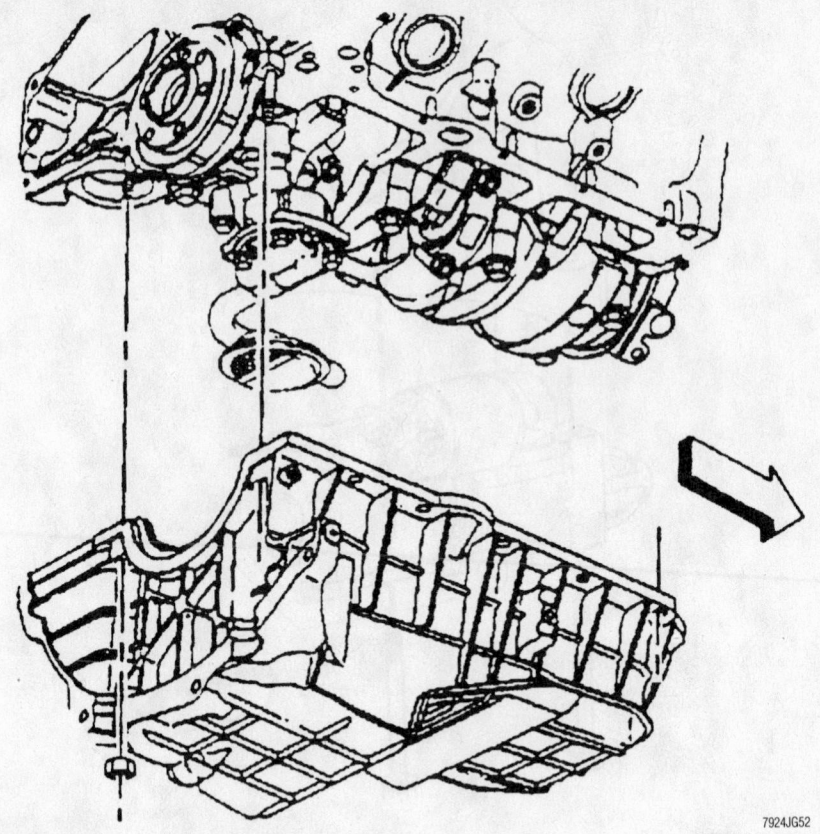

Oil pan mounting—4.3L engines

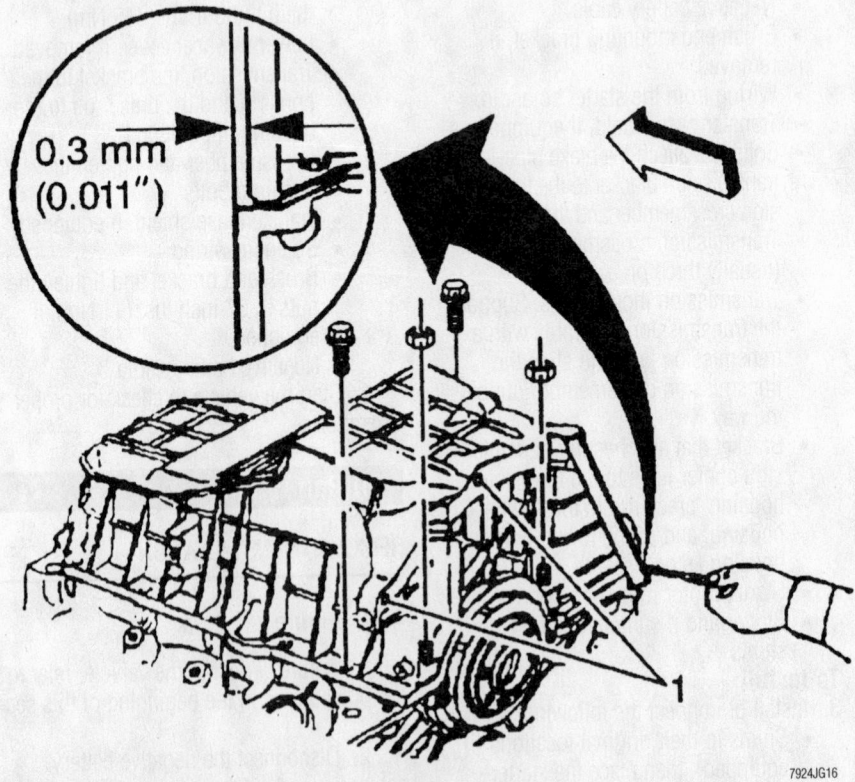

If the clearance between the 3 oil pan-to-transmission contact points exceeds 0.011 in.
(0.3mm) at any of the 3 points, realign the oil pan

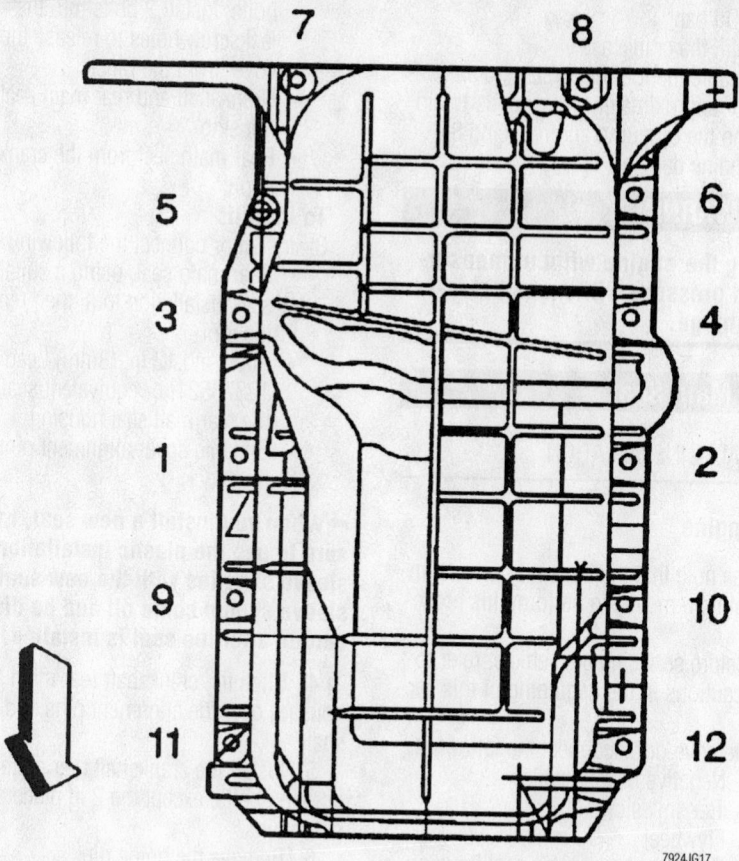

Tighten the bolts in sequence to prevent warping the sealing surface of the oil pan

➡ **The alignment between the rear of the pan and rear of the block is critical. The two surfaces must be flush to allow for proper alignment with the transmission housing.**

6. Use a feeler gauge to check the clearance between the oil pan-to-transmission contacts. If clearance exceeds 0.011 inch (0.3mm) at any of the three contact points, readjust the pan until the clearance is within specification.

7. Once the pan is in its correct position tighten the retainers to 18 ft. lbs. (25 Nm) using the proper sequence.

8. Install the engine into the vehicle. Refill the crankcase with fresh oil. Start the engine, establish normal operating temperatures and check for leaks.

4WD MODELS

1. Before servicing the vehicle, refer to the precautions in the beginning of this section.

2. Disconnect the negative battery cable.

3. Drain the engine crankcase oil.

4. Remove or disconnect the following:

- Dipstick
- Drivebelt splash shield, the front axle shield and the transfer case shield
- Front skid plate and the flywheel cover
- Left and right engine mount through-bolts

5. Raise the engine using a lifting device and block in position. This may be accomplished using large wooden blocks between the motor mounts and brackets.

➡ **Use extreme caution when blocking the engine in position. Get out from under the vehicle and rock the engine slightly once the blocks are in place to be sure the engine is properly supported.**

6. Remove or disconnect the following:

- Oil cooler line
- Pitman arm bolt and pitman arm
- Idler arm bolts and idler arm
- Front differential through-bolts
- Front driveshaft, if necessary
- Differential assembly by rolling it forward for clearance

- Starter motor
- Oil pan bolts, nuts and reinforcements
- Oil pan and discard the gasket

To install:

7. Clean the gasket mounting surfaces.

➡ **The alignment between the rear of the oil pan and the rear of the block is critical. The oil pan must be flush or slightly forward of the rear of the block to allow for proper alignment with the transmission housing. Use a feeler gauge to measure the clearance between the 3 oil pan-to-transmission contact points. If the clearance exceeds 0.011 in. (0.3mm) at any of the 3 points, realign the oil pan.**

8. Apply sealant to the oil pan rail where it contacts the timing cover-to-block joint (front) and the crankshaft rear seal retainer-to-block joint (rear). Continue the bead of sealant about 1 in. (25mm) in both directions from each of the 4 corners.

9. Install or connect the following:

- Oil pan, using a new gasket. Tighten the retainers, in sequence, to 18 ft. lbs. (25 Nm).
- Starter motor
- Differential by rolling it back into position
- Front driveshaft
- Front differential through-bolts
- Idler arm and secure using the retaining bolts
- Pitman arm and secure using the bolts
- Transfer case shield
- Flywheel cover
- Front skid plate
- Front axle shield
- Drive belt splash shield
- Dipstick
- Negative battery cable

10. Refill the engine crankcase.

11. Start the engine and check for leaks.

Oil Pump

REMOVAL & INSTALLATION

4.2L Engine

1. Before servicing the vehicle, refer to the precautions in the beginning of this section.

2. Remove or disconnect the following:

- Engine front cover
- Oil pump cover bolts

- Oil pump cover. Mark the inner and outer gears in relation to the pump housing.
- Inner and outer pump gears
- Oil pump pressure relief valve plug
- Oil pump pressure relief valve and spring

To install:

3. Install or connect the following:
- Oil pump pressure relief valve and spring
- Oil pump pressure relief valve plug. Tighten to 10 ft. lbs. (14 Nm).
- Oil pump outer and inner gears, as marked during removal
- Oil pump cover and bolts. Tighten the bolts to 89 inch lbs. (10 Nm).
- Front cover

4.3L Engine

1. Before servicing the vehicle, refer to the precautions in the beginning of this section.
2. Remove or disconnect the following:
- Oil pan
- Oil pump and the pickup tube/shaft, if equipped

➡ **Be careful not to crack the retainer.**

To install:

3. Ensure that the pump pickup tube is tight in the pump body. If the tube should come loose, oil pressure will be lost and oil starvation will occur. If the pickup tube is loose it should be replaced.

4. If the pump has been disassembled and is being replaced or for any reason oil has been removed, it must be primed. It can either be filled with oil before installing the cover plate and oil kept within the pump during handling or the entire pump cavity can be filled with petroleum jelly.

✳✳ WARNING

If the pump is not primed, the engine could be damaged upon start up.

5. Install or connect the following:
- Oil pump by aligning the pump shaft with the distributor drive gear as necessary. Tighten oil pump/pickup tube retainer(s) to 65 ft. lbs. (90 Nm).

➡ **If the oil pump does not build up oil pressure almost immediately, remove the pan and check for a loose oil pump-to-pickup tube attachment. If necessary dismantle the pump and pack the pump cavity with petroleum jelly.**

- Oil pan
6. Refill the crankcase.
7. Disable the ignition system; crank engine for approximately 10 seconds to aid in priming the oil pump and reducing the risk of engine damage.

✳✳ WARNING

Running the engine without measurable oil pressure will cause extensive damage.

Rear Main Seal

REMOVAL & INSTALLATION

4.2L Engine

Please note that the transmission assembly must be removed to perform this procedure.

1. Before servicing the vehicle, refer to the precautions in the beginning of this section.
2. Remove or disconnect the following:
- Negative battery cable
- Transmission
- Flywheel
- Crankshaft rear main seal housing

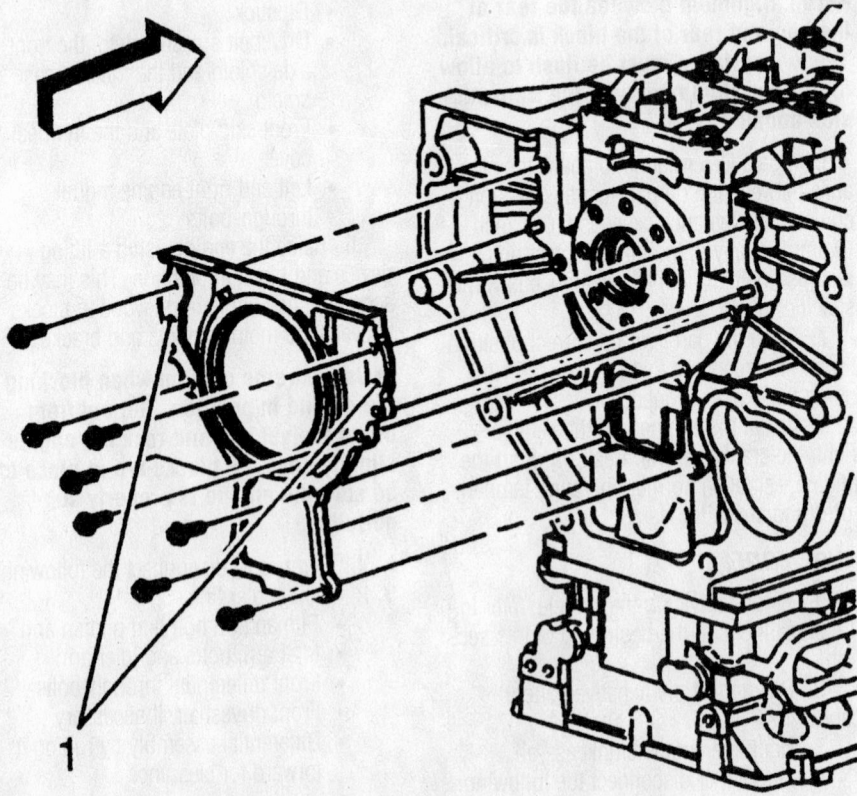

Install 2 bolts into the jackscrew holes (1) to push the cover off of the block

bolts. Install 2 bolts into the jackscrew holes to release the cover from the block
- Crankshaft and rear main seal housing
- Rear main seal from the crankshaft snout

To install:

3. Install or connect the following:
- Rear main seal, using a suitable seal installation tool, then remove the tool
- Apply a 0.12 in. (3mm) bead of 12378521, or equivalent sealant to the rear mail seal housing
- Suitable cover alignment pins into the block

➡ **When you install a new seal, make sure to use the plastic installation sleeve supplies with the new seal. The sleeve should come off and be discarded after the seal is installed.**

4. Slide the crankshaft rear main seal housing over the alignment pins and crankshaft.

5. Install the crankshaft rear main seal housing bolts, except the 2 in place of the guide pins.

6. Remove the guide pins.

9355MG04

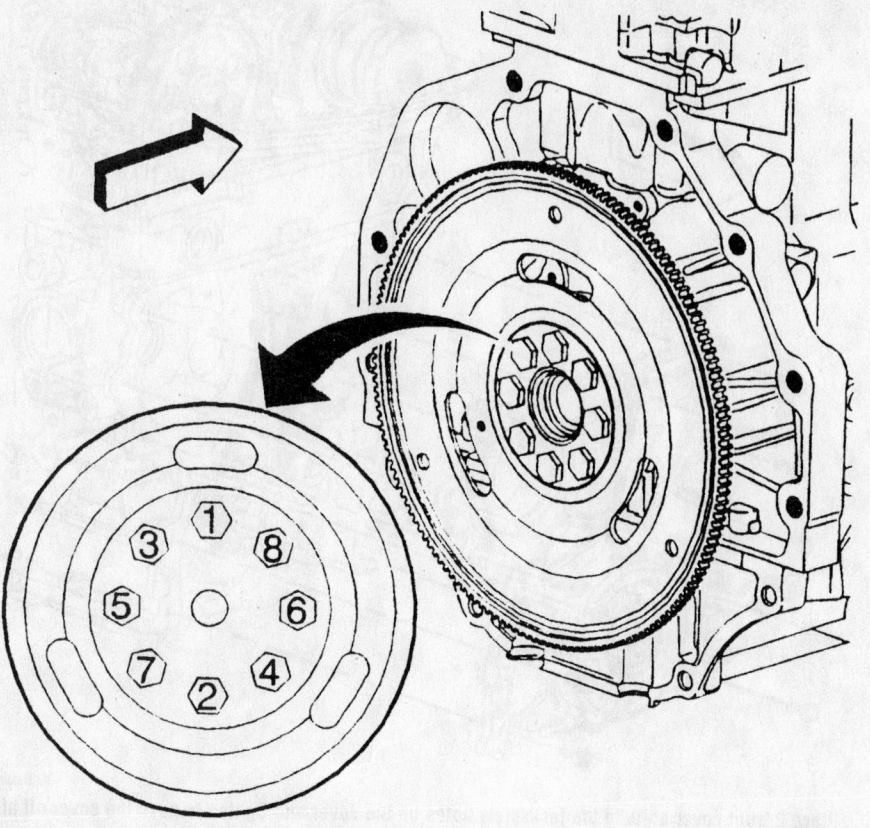

Flywheel bolt tightening sequence—4.2L engine

7. Install or connect the following:
   - Remaining 2 crankshaft rear main seal housing bolts and tighten to 89 inch lbs. (10 Nm). Wipe off any excess sealant.
   - Flywheel and secure with the mounting bolts. Tighten, in sequence, to 18 ft. lbs. (25 Nm), plus an additional 50 degrees.
   - Transmission

4.3L Engine

Please note that the transmission assembly and transfer case, if equipped, must be removed to perform this procedure.

1. Before servicing the vehicle, refer to the precautions in the beginning of this section.
2. Remove or disconnect the following:
   - Negative battery cable
   - Transfer case, if equipped
   - Transmission
   - Clutch assembly/flywheel or flexplate
3. Remove the crankshaft rear oil seal by inserting a suitable prying tool into the notches provided in the seal retainer and

prying the seal out. Take care not to damage the crankshaft sealing surface.

To install:

4. Inspect the crankshaft for grit, rust or burrs and correct as necessary.
5. Clean the running surface of the crankshaft with a non-abrasive cleaner.
6. Install or connect the following:
   - New rear seal lubricated with engine oil and a seal installer
   - Flywheel and clutch or flexplate
   - Transmission
   - Transfer case, if equipped
   - Negative battery cable
7. Start the engine and verify no oil leaks.

Timing Chain, Sprockets, Front Cover and Seal

REMOVAL & INSTALLATION

Front Cover and Seal

4.2L ENGINE

1. Before servicing the vehicle, refer to the precautions in the beginning of this section.
2. Remove or disconnect the following:
   - Negative battery cable
   - Drain the engine cooling system.

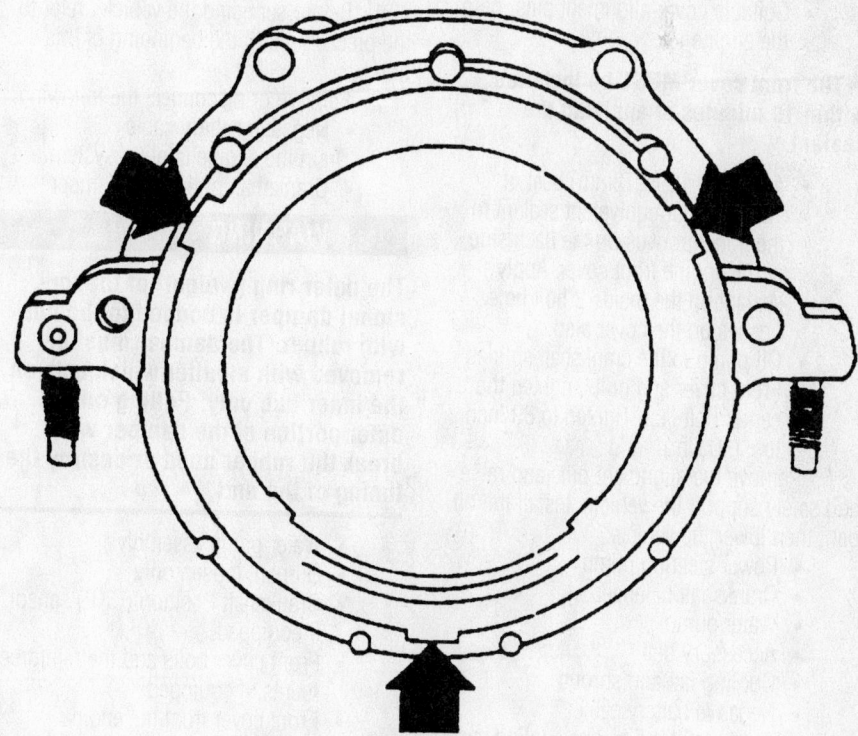

Carefully pry the rear main seal out of the retainer

- Cooling fan and shroud
- Accessory belt
- Water pump
- Crankshaft balancer

✳✳ WARNING

When removing the seal, be careful not to damage the front cover or crankshaft.

- Seal from the front cover, using a suitable prytool in the slots provided
- Power steering pump

3. Raise and safely support the vehicle.
- Oil pan, then carefully lower the vehicle
- 7mm center bolt
- Remaining front cover bolts. Place two of the front cover bolts in the jackscrew holes on the front cover and tighten the bolts evenly to release the front cover from the engine.
- 2 bolts from the front cover
- Oil pump

To install:

4. Clean the gasket mating surfaces of the engine and cover of all remaining gasket or sealer material. Be careful not to score or damage the surfaces.

5. Install or connect the following:
- Suitable cover alignment pins, onto the engine

➡**The front cover MUST be installed within 10 minutes of applying the sealant.**

- Apply a 0.12 in. (3mm) beat of 12378521 or equivalent sealant to the trace grooves on the back side of the engine front cover. Apply sealant on the inside 3 bolt hole bosses on the cover also.
- Oil pump to the crankshaft splines
- Front cover and bolts, tighten the center bolt last. Tighten to 89 inch lbs. (10 Nm).

6. Remove the alignment pins and raise and safely support the vehicle. Install the oil pan, then lower the vehicle.
- Power steering pump
- Crankshaft balancer
- Water pump
- Accessory belt
- Cooling fan and shroud
- Negative battery cable

7. Properly refill the engine cooling system.

8. Run the engine until normal operating temperature has been reached, then check for leaks.

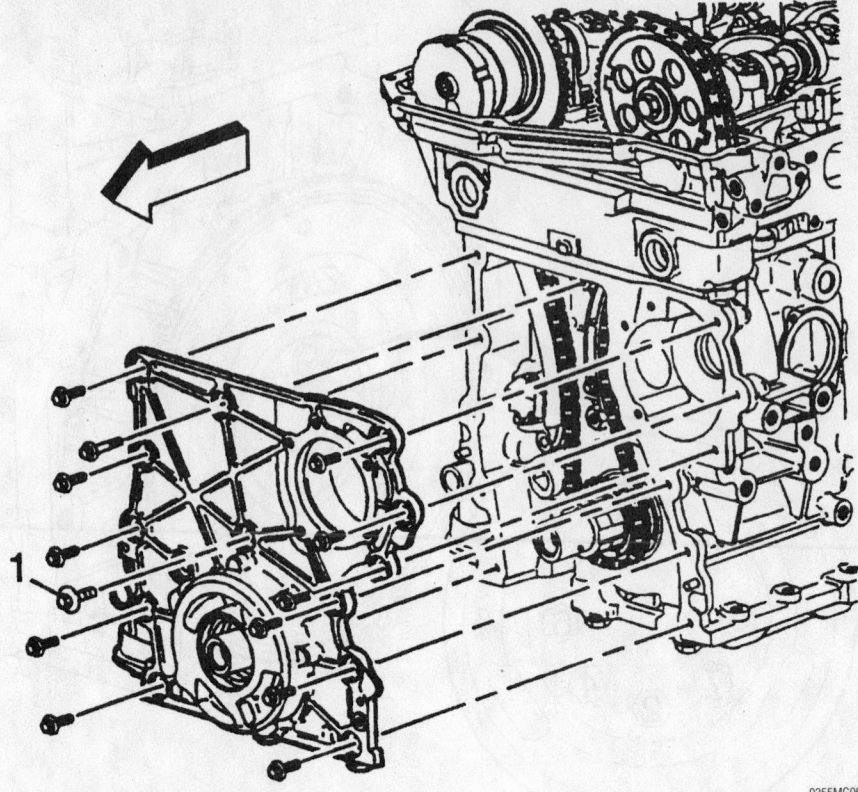

Place 2 front cover bolts in the jackscrew holes on the cover and tighten to push the cover off of the engine

9355MG06

4.3L ENGINE

1. Before servicing the vehicle, refer to the precautions in the beginning of this section.

2. Remove or disconnect the following:
- Negative battery cable

3. Drain the engine cooling system.
- Crankshaft pulley and damper

✳✳ WARNING

The outer ring (weight) of the torsional damper is bonded to the hub with rubber. The damper must be removed with a puller which acts on the inner hub only. Pulling on the outer portion of the damper will break the rubber bond or destroy the tuning of the unit.

- Water pump assembly
- Oil pan, loosen only
- Crankshaft Position (CKP) sensor, if equipped
- Front cover bolts and the reinforcements, if equipped
- Front cover from the engine

4. Pry the seal out of the front cover using a small prytool. Be very careful not to distort the front cover or to score the end of the crankshaft.

To install:

➡**Anytime the front cover is removed, the cover must be replaced upon reassembly. If you reuse the old cover, oil leaks may develop.**

5. Clean the gasket mating surfaces of the engine and cover of all remaining gasket or sealer material. Be careful not to score or damage the surfaces.

➡**The manufacturer suggests you wait until the front cover is mounted to the engine before you install the replacement crankshaft oil seal. This assures the cover is properly supported.**

6. Install or connect the following:
- New front cover gasket to the engine or cover using gasket cement to hold it in position. Lubricate the front of the oil pan seal with engine oil to aid in reassembly.
- Front cover to the engine. Take care while engaging the front of the oil pan seal with the bottom of the cover. On 1999–01 models, apply sealer 12346141 to the oil pan rail where it contacts the timing cover-to-block joint (front) and the crankshaft rear seal retainer-to-block

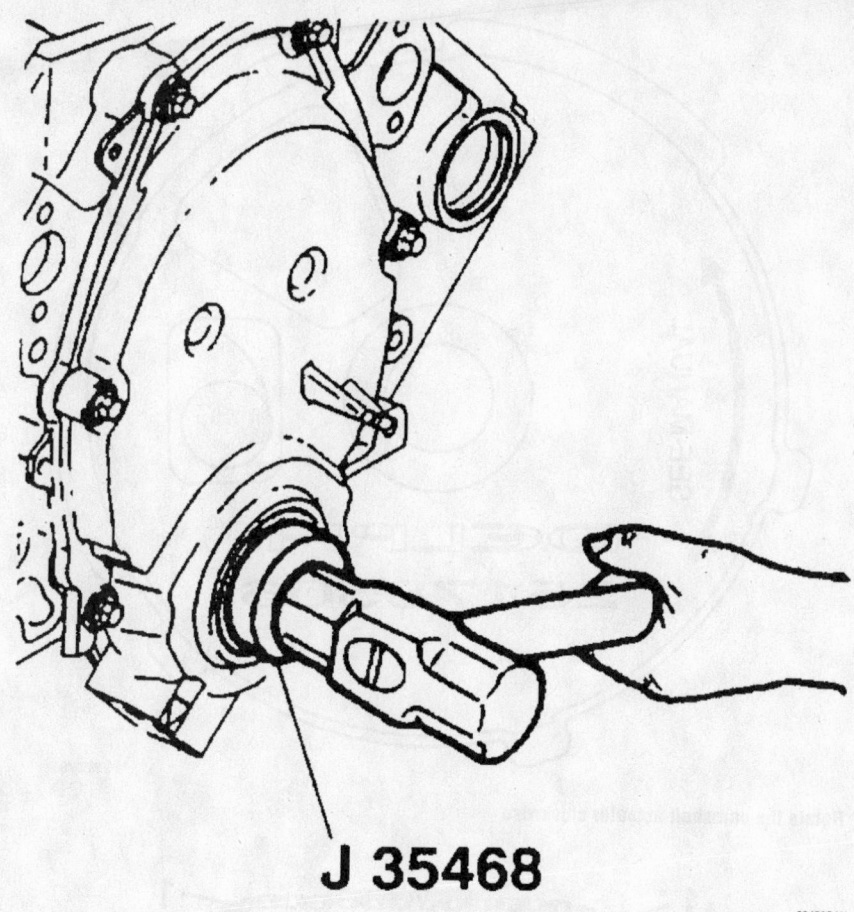

J 35468

88453GAU

Installing the crankshaft front oil seal

joint (rear). Continue the bead of sealant about 1 in. (25mm) in both directions from each of the four corners.

- Front cover retaining bolts and tighten to 106 inch. lbs. (12 Nm)

7. Lightly coat the lips of the replacement crankshaft seal with clean engine oil, then position the seal with the open end facing inward the engine. Use a suitable seal installation driver to position the seal in the front cover.

- CKP sensor O-ring and the sensor, if equipped
- Tighten the Oil pan bolts
- Water pump
- Crankshaft damper and pulley
- Negative battery cable

8. Properly refill the engine cooling system.

9. Run the engine until normal operating temperature has been reached, then check for leaks.

Timing Chain and Sprockets

4.2L ENGINE

➡ The following procedure requires the use of the Crankshaft Holding tool No. J-44221 and a suitable torque angle meter.

1. Before servicing the vehicle, refer to the precautions in the beginning of this section.

2. Remove or disconnect the following:

- Camshaft cover
- Timing chain (front) cover
- Tension on the timing chain by moving the tensioner shoe in. Place a tee into the tension to hold the shoe in place.
- Top chain guide bolts and guide
- Exhaust camshaft position actuator bolt and actuator
- Intake camshaft sprocket bolt and sprocket
- Timing chain
- Crankshaft sprocket
- Cylinder head access hole plugs
- Timing chain tensioner shoe bolt and shoe
- Timing chain tensioner guide bolts and guide
- Timing chain tensioner bolts and tensioner

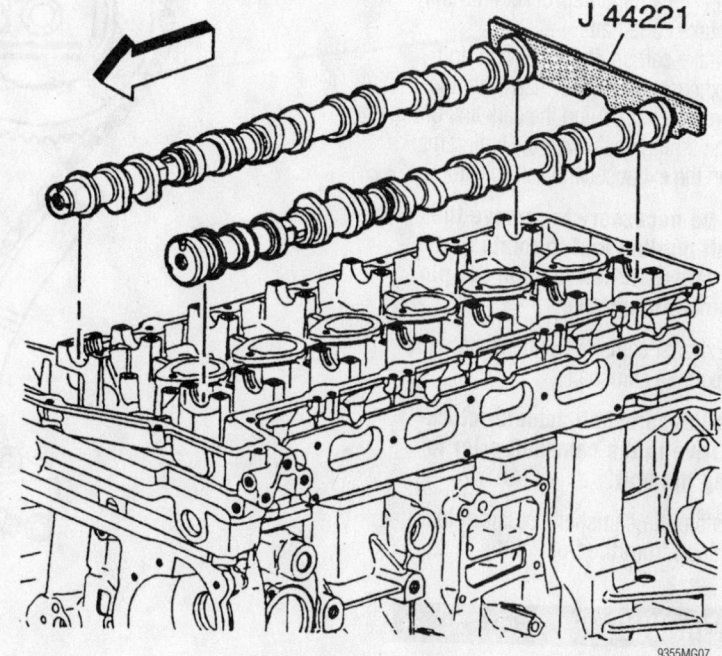

J 44221

9355MG07

Proper installation of the crankshaft holding tool with the No. 1 cylinder at TDC

To install:

➡ **Every seventh link of the timing chain is darkened to help in aligning the timing marks.**

3. Install or connect the following:
   - Timing chain tensioner and bolts. Tighten to 18 ft. lbs. (25 Nm).
   - Timing chain guide and bolts. Tighten to 89 inch lbs. (10 Nm).
   - Timing chain tensioner shoe and bolt. Tighten to 19 ft. lbs. (26 Nm).
   - Cylinder head access hole plugs and tighten to 44 inch lbs. (5 Nm)
   - Crankshaft Holding tool No. J-44221, or equivalent with the camshaft flats up and the No. 1 cylinder at Top Dead Center (TDC)
   - Crankshaft sprocket
   - Intake camshaft sprocket into the timing chain

4. Align the dark link of the timing chain with the timing mark on the intake camshaft sprocket. Feed the timing chain down through the opening in the head.
   - Timing chain onto the crankshaft sprocket. Align the dark link of the timing chain with the timing mark on the crankshaft sprocket.

➡ **It may be necessary to remove the crankshaft holding tool to rotate and hold the camshaft hex to align the pin to the camshaft sprocket**

   - Intake camshaft sprocket onto the intake camshaft
   - Intake camshaft washer and bolt
   - Exhaust camshaft actuator into the timing chain. Align the dark link of the timing chain with the timing mark on the exhaust camshaft actuator.

➡ **It may be necessary to remove the crankshaft holding tool to rotate and hold the camshaft hex to align the pin to the camshaft sprocket**

   - Exhaust camshaft actuator onto the exhaust camshaft

➡ **Rotate the camshaft actuator clockwise relative to the camshaft prior to tightening the bolt.**

5. Rotate the camshaft actuator clockwise (as seen from the front of the vehicle).

❋❋ WARNING

The camshaft actuator must be fully advanced during installation. Engine damage may occur if the camshaft actuator is not fully advanced.

9355MG08

Rotate the camshaft actuator clockwise

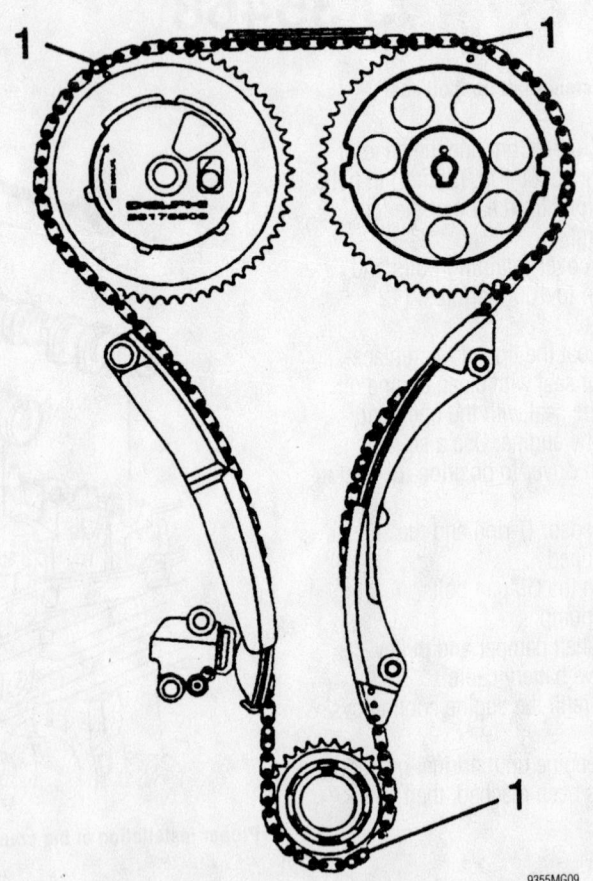

9355MG09

The dark lines on the timing chain should be aligned with the marks on the sprockets

6. Install the exhaust camshaft actuator bolt and tighten to 18 ft. lbs. (25 Nm), plus an additional 135 degrees, using a torque angle meter.

7. Tighten the intake camshaft sprocket bolt to 22 ft. lbs. (30 Nm), plus an additional 135 degrees, using a torque angle meter.

8. Remove the tee from the timing chain tensioner to regain tension on the timing chain.

9. Remove the crankshaft holding tool. The dark lines on the timing chain should be aligned with the marks on the sprockets.

10. Install or connect the following:
* Top chain guide
* Suitable threadlock to the top chain guide bolt threads, then install and tighten to 89 inch lbs. (10 Nm)
* Engine front cover
* Camshaft cover

4.3L ENGINE

➡The following procedure requires the use of the Crankshaft Sprocket Removal tool No. J-5825-A and the Crankshaft Sprocket Installation tool No. J-5590.

1. Before servicing the vehicle, refer to the precautions in the beginning of this section.

2. Remove the timing cover from the engine.

3. Rotate the crankshaft until the No. 4 cylinder is on the Top Dead Center (TDC) of its compression stroke and the camshaft sprocket mark aligns with the mark on the crankshaft sprocket (facing each other at a point closest together in their travel) and in line with the shaft centers.

4. Remove or disconnect the following:
* Crankshaft Position (CKP) sensor reluctor ring, if equipped
* Camshaft sprocket-to-camshaft nut and/or bolts
* Camshaft sprocket (along with the timing chain). If the sprocket is difficult to remove, use a plastic mallet to bump the sprocket from the camshaft.

➡The camshaft sprocket (located by a dowel) is lightly pressed onto the camshaft and should come off easily. The chain comes off with the camshaft sprocket.

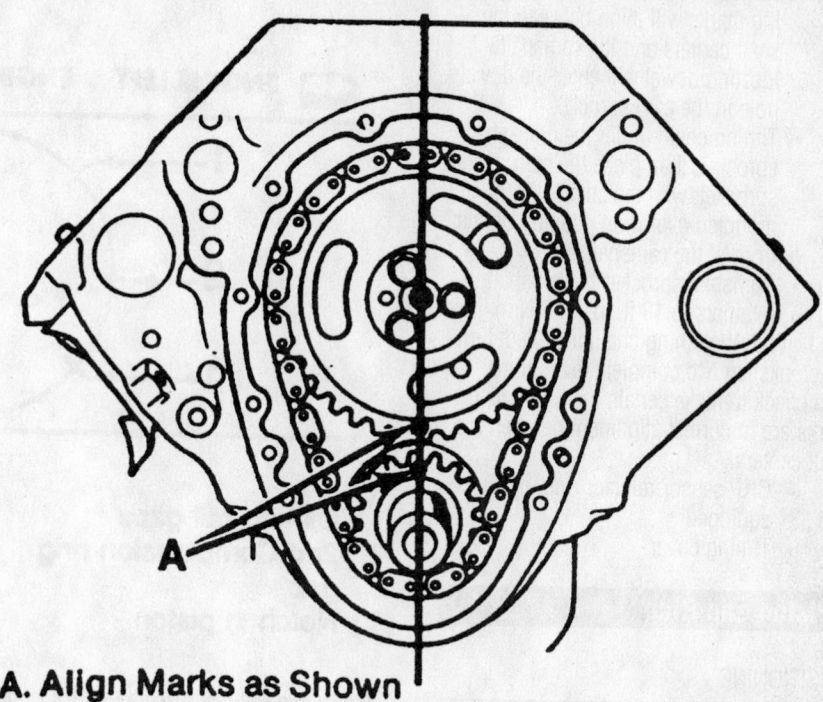

A. Align Marks as Shown

85383292

Timing mark alignment

5. If necessary use J-5825-A crankshaft sprocket removal tool to free the timing sprocket from the crankshaft.

6. If necessary, remove the crankshaft sprocket key.

To install:

7. Inspect the timing chain and the timing sprockets for wear or damage, replace the damaged parts as necessary.

8. Using a putty knife, clean the gasket mounting surfaces. Using solvent, clean the oil and grease from the gasket mounting surfaces.

9. Install or connect the following:
* Crankshaft sprocket key, if removed
* Crankshaft sprocket onto the crankshaft using J-5590 crankshaft sprocket installation tool and a hammer without disturbing the position of the engine

➡During installation, coat the thrust surfaces lightly with Molykote® or an equivalent pre-lube.

* Timing chain over the camshaft sprocket. Arrange the camshaft

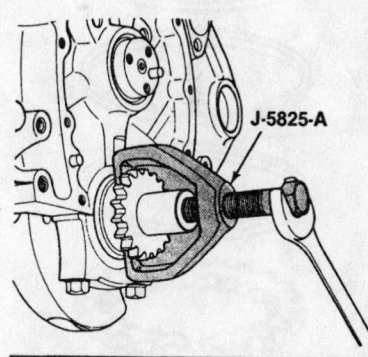

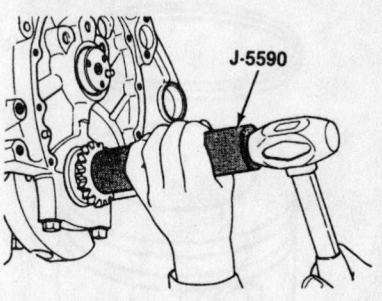

85383293

Removal (top) and installation of the crankshaft timing gear

sprocket in such a way that the timing marks will align between the shaft centers and the camshaft locating dowel will enter the dowel hole in the cam sprocket.

- Timing chain under the crankshaft sprocket, then place the cam sprocket, with the chain still mounted over it, in position on the front of the camshaft
- Camshaft sprocket-to-camshaft retainers to 18 ft. lbs. (25 Nm)

11. With the timing chain installed, turn the crankshaft two complete revolutions, then check to make certain that the timing marks are in correct alignment between the shaft centers.

- CKP sensor reluctor ring, if equipped
- Timing cover

Piston and Ring

POSITIONING

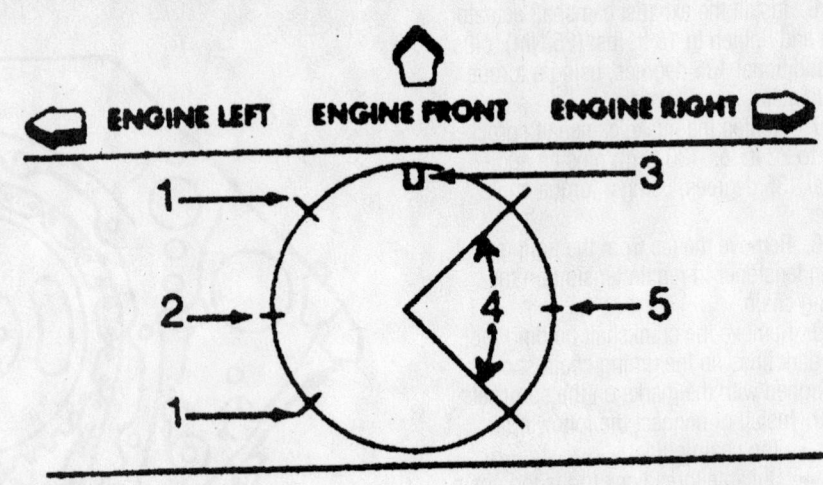

1. Oil ring rail gaps
2. 2nd Compression ring gap
3. Notch in piston
4. Oil ring spacer gap (tang in hole or slot with arc)
5. Top compression ring gap

Piston ring end-gap spacing—4.3L engine

Piston ring positioning—4.2L engine

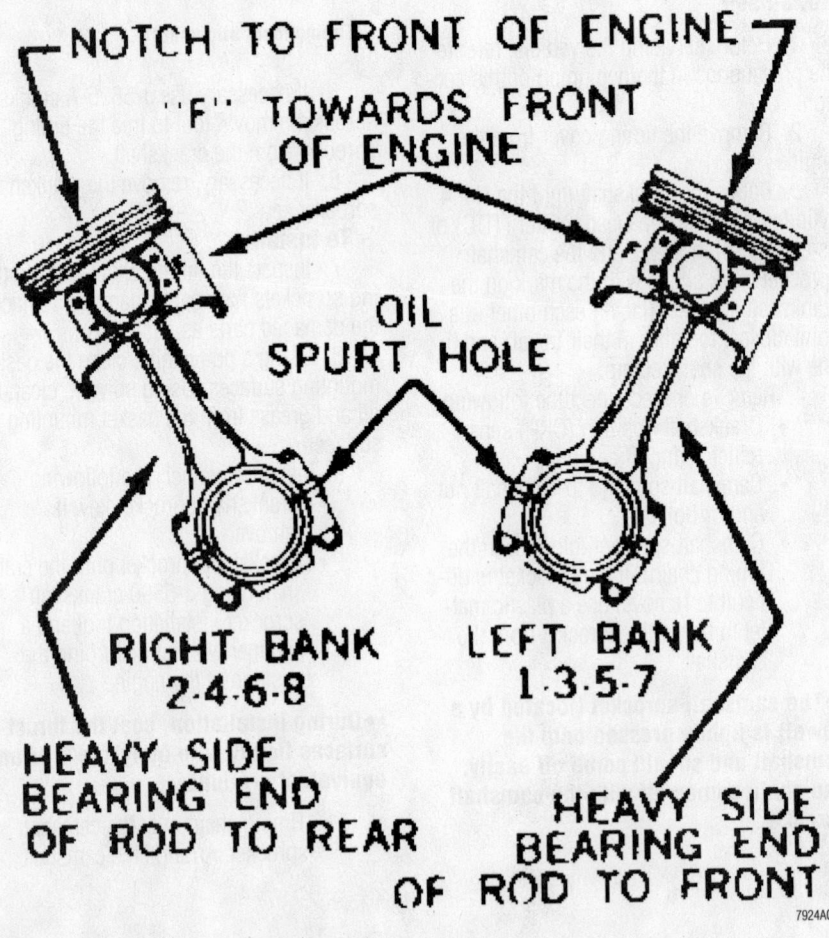

NOTCH TO FRONT OF ENGINE

"F" TOWARDS FRONT OF ENGINE

OIL SPURT HOLE

RIGHT BANK 2-4-6-8

LEFT BANK 1-3-5-7

HEAVY SIDE BEARING END OF ROD TO REAR

HEAVY SIDE BEARING END OF ROD TO FRONT

Piston and connecting rod assembly positioning—4.3L engine

FUEL SYSTEM

Fuel System Service Precautions

Safety is the most important factor when performing not only fuel system maintenance but also any type of maintenance. Failure to conduct maintenance and repairs in a safe manner may result in serious personal injury or death. Maintenance and testing of the vehicle's fuel system components can be accomplished safely and effectively by adhering to the following rules and guidelines.

• To avoid the possibility of fire and personal injury, always disconnect the negative battery cable unless the repair or test procedure requires that battery voltage be applied.

• Always relieve the fuel system pressure prior to disconnecting any fuel system component (injector, fuel rail, pressure regulator, etc.), fitting or fuel line connection. Exercise extreme caution whenever relieving fuel system pressure, to avoid exposing skin, face and eyes to fuel spray. Please be advised that fuel under pressure may penetrate the skin or any part of the body that it contacts.

• Always place a shop towel or cloth around the fitting or connection prior to loosening to absorb any excess fuel due to spillage. Ensure that all fuel spillage (should it occur) is quickly removed from engine surfaces. Ensure that all fuel soaked cloths or towels are deposited into a suitable waste container.

• Always keep a dry chemical (Class B) fire extinguisher near the work area.

• Do not allow fuel spray or fuel vapors to come into contact with a spark or open flame.

• Always use a back-up wrench when loosening and tightening fuel line connection fittings. This will prevent unnecessary stress and torsion to fuel line piping. Always follow the proper torque specifications.

• Always replace worn fuel fitting O-rings with new. Do not substitute fuel hose or equivalent where fuel pipe is installed.

Fuel System Pressure

RELIEVING

The fuel systems operate under high fuel pressures. It is very important that the pressure be properly relieved prior to servicing the system or any of its components.

1999–01 Vehicles

A Schrader valve is provided on these fuel systems to conveniently test or release the system pressure. A fuel pressure gauge and adapter will be necessary to connect the gauge to the fitting. Most of the MFI systems utilize a service valve on one end of the fuel rail assembly.

1. Before servicing the vehicle, refer to the precautions in the beginning of this section.

2. Disconnect the negative battery cable to assure the prevention of fuel spillage if the ignition switch is accidentally turned **ON** while a fitting is still detached.

3. Loosen the fuel filler cap to release the fuel tank pressure.

4. Be sure the release valve on the fuel gauge is closed, then connect the fuel gauge to the pressure fitting located on the inlet fuel pipe fitting.

> ❋❋ **CAUTION**
>
> **When connecting the gauge to the fitting, be sure to wrap a rag around the fitting to avoid spillage. After repairs, place the rag in an approved container.**

5. Install the bleed hose portion of the fuel gauge assembly into an approved container, then open the gauge release valve and bleed the fuel pressure from the system.

6. When the gauge is removed, be sure to open the bleed valve and drain all fuel from the gauge assembly.

7. When fuel service is finished, tighten the fuel filler cap and connect the negative battery cable.

2002–03 Vehicles

1. Before servicing the vehicle, refer to the precautions in the beginning of this section.

> ❋❋ **WARNING**
>
> **Do not perform this procedure for more than 2 minutes to avoid damaging the catalytic converter.**

2. Loosen the fuel filler cap to release the fuel tank pressure.

3. Remove the fuel pump relay from the junction block.

4. Crank the engine, allowing it to start and stall.

5. Crank the engine for an additional 3 seconds to relieve any remaining fuel pressure.

6. Disconnect the negative battery cable to avoid repressurizing the fuel system.

7. Install the fuel pump relay in the junction block.

8. Tighten the fuel filler cap.

9. After you are finished working on the fuel system, connect the negative battery cable.

Fuel Filter

REMOVAL & INSTALLATION

1. Before servicing the vehicle, refer to the precautions in the beginning of this section.

2. Properly relieve the fuel system pressure.

3. Remove or disconnect the following:
   • Negative battery cable and fuel filler cap, if not already done
4. Raise and support the vehicle.
   • Fuel tank shield, if equipped
   • Quick connect fittings from the filter
   • Filter feed nut and the clamp bolt
   • Filter and the clamp from the vehicle

To install:
5. Install or connect the following:
   • Filter and clamp with the directional arrow facing away from the fuel tank, towards the throttle body

➡ **The filter has an arrow (fuel flow direction) on the side of the case, be sure to install it correctly in the system, the with arrow facing away from the fuel tank.**

   • Tighten the fuel feed nut
   • Tighten the filter clamp assembly bolt
   • Fuel quick disconnect fittings to the filter
   • Fuel tank shield, if equipped
   • Fuel filler cap
   • Negative battery cable
6. Start the engine and check for leaks.

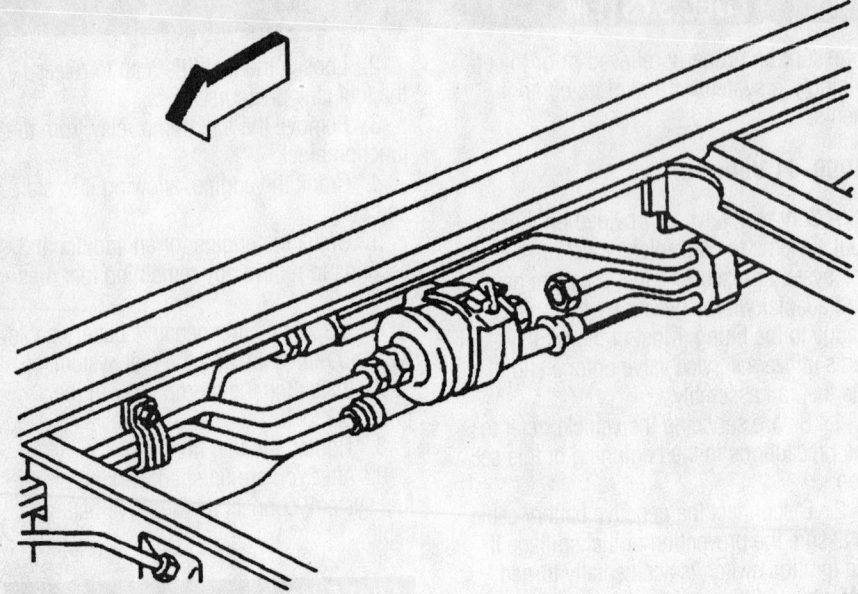

Typical fuel filter location along frame rail

Fuel Pump

REMOVAL & INSTALLATION

1. Before servicing the vehicle, refer to the precautions in the beginning of this section.
2. Properly relieve the fuel system pressure.
3. Drain the fuel tank.
4. Support the fuel tank.
5. Remove or disconnect the following:
   - Negative battery cable
   - Filler neck from the tank
   - Shield from tank and tank straps
   - Fuel lines and vapor hose from pump
   - Electrical connection from fuel pump
   - Fuel tank
   - Fuel pump/sending unit assembly by turning the locking ring (located on top of the fuel tank) counter-clockwise using a spanner wrench
   - Fuel pump from the fuel lever sending device

To install:

6. Install or connect the following:
   - Fuel pump in tank with new seal around opening

➡️**The fuel pump strainer must be in a horizontal position when the fuel sender is installed in the tank. When installing the sender assembly, make sure that the fuel pump strainer does not block full travel of the float arm.**

   - Tank and connect fuel lines and vapor hose
   - Tank to the frame. Torque the fasteners to 33 ft. lbs. (45 nm).
   - Shield
   - Fuel filler neck and clamp
   - Negative battery cable
7. Refill the tank.
8. Run the engine and check for leaks.

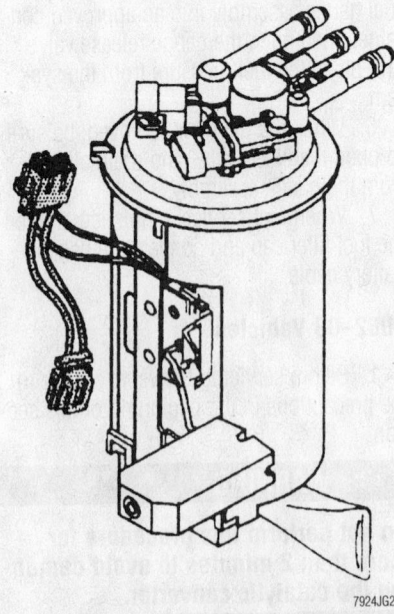

View of the in-tank fuel pump assembly

Fuel Injector

REMOVAL & INSTALLATION

4.2L Engine

1. Before servicing the vehicle, refer to the precautions in the beginning of this section.
2. Relieve the fuel system pressure. Refer to the fuel system relief procedure in this section.
3. Remove or disconnect the following:
   - Negative battery cable, if not done already
   - Intake manifold

➡️**Clean the fuel rail assembly with a suitable spray cleaner before proceeding. Never soak the fuel rail in a cleaning solvent.**

   - Fuel pressure regulator vacuum line
   - Fuel feed and return pipes
   - Fuel injector in-line electrical connector
   - Fuel rail attaching bolts and fuel rail
   - Fuel injector harness connector from the fuel injectors
   - Injector retaining clip
   - Injector from the fuel rail
   - Retainer clip and O-ring seals from each end of the injector and discard

To install:

➡️**Each injector is calibrated. When replacing the fuel injectors, be sure to replace it with the correct injector.**

4. Lubricate the new injector O-ring seats with engine oil.
5. Install or connect the following:
   - O-rings on the injector
   - New retainer clip on the injector
6. Push the fuel injector into the fuel rail socket, making sure the connector faces outward. The retainer clip locks to a flange on the fuel rail injector socket.
   - Fuel rail assembly. Tighten the bolts to 89 inch lbs. (10 Nm).
   - Fuel feed and return lines to the rail
   - Fuel injector electrical connectors
   - Fuel pressure regulator vacuum line
   - Intake manifold
   - Negative battery cable
7. Turn the ignition **ON** for 2 seconds and then turn it **OFF** for 10 seconds. Again turn the ignition **ON** and check for leaks.

4.3L Engine

1. Before servicing the vehicle, refer to the precautions in the beginning of this section.

2. Relieve the fuel system pressure. Refer to the fuel system relief procedure in this section.

3. Remove or disconnect the following:
- Negative battery cable
- Fuel meter body electrical connection and the fuel feed and return hoses from the engine fuel pipes
- Upper manifold assembly
- Poppet nozzle out of the casting socket
- Fuel meter body by releasing the locktabs

➡**Each injector is calibrated. When replacing the fuel injectors, be sure to replace it with the correct injector.**

- Lower hold-down plate and nuts

4. While pulling the poppet nozzle tube downward, push with a small prytool down between the injector terminals and remove the injectors.

To install:

5. Lubricate the new injector O-ring seats with engine oil.

6. Install or connect the following:
- O-rings on the injector
- Fuel injector into the fuel meter body injector socket.
- Lower hold-down plate and nuts. Torque the nuts to 27 inch lbs. (3 Nm).
- Fuel meter body assembly into the intake manifold. Torque the fuel meter bracket retainer bolts to 88 inch. lbs. (10 Nm).

✲✲ CAUTION

To reduce the risk of fire or injury ensure that the poppet nozzles are properly seated and locked in their casting sockets

- Fuel meter body into the bracket and lock all the tabs in place
- Poppet nozzles into the casting sockets
- Electrical connections
- New O-ring seals on the fuel return and feed hoses.
- Fuel feed and return hoses and tighten the fuel pipe nuts to 22 ft. lbs. (30 Nm)
- Negative battery cable

7. Turn the ignition **ON** for 2 seconds and then turn it **OFF** for 10 seconds. Again turn the ignition **ON** and check for leaks.

8. Install the manifold plenum.

DRIVE TRAIN

Manual Transmission Assembly

REMOVAL & INSTALLATION

1. Before servicing the vehicle, refer to the precautions in the beginning of this section.

2. Shift the transmission into 3rd or 4th gear position.

3. Remove or disconnect the following:
- Negative battery cable
- Shift lever and the if necessary, the shift housing
- Parking brake cable for clearance
- Propeller shaft
- Sid plate, if equipped
- Transfer case and shift lever, on 4WD models
- All wiring harness that would interfere with transmission removal
- Fuel line retainers from the rear crossmember
- Muffler from the catalytic converter
- Exhaust pipes from the exhaust manifold
- Catalytic converter hanger, if necessary
- Exhaust section
- Bolts and nuts attaching any transmission braces to the engine and transmission
- Hydraulic clutch quick-connect from the concentric slave cylinder following 1 of the 2 steps:

a. Use 2 small prytools at 180 degrees from each other to depress the white plastic sleeve on the quick connect to separate the clutch line from the concentric slave cylinder quick connect.

b. Use special tool J–36221 to depress the white plastic sleeve on the quick connect to separate the clutch line end from the concentric slave cylinder quick connect.

4. Remove or disconnect the following:
- Bolts securing the clutch housing cover to the transmission, if equipped
- Clutch plate and clutch cover, if necessary

5. Support the transmission with a suitable jack.
- Rear crossmember from the frame rail
- Wiring harness from the front crossmember, if equipped. Move the wiring harness away from the transmission oil pan. Lower the transmission enough to gain access to the top of the transmission.
- Fuel line retainers or wiring harness's from the top of the transmission
- Bolt, washer, and nut securing the wiring harness ground wires to the engine block

- Bolts retaining the transmission to the engine. Pull the transmission straight back on the clutch hub splines.

6. Lower the transmission using the transmission jack.

To install:

Installation is the reverse of removal, but please note the following important steps.

7. Place a THIN coat of high-temperature grease on the main drive gear (input shaft) splines.

8. Secure the transmission to the floor jack and raise the transmission into position.

➡**On some models, it may be necessary to rotate the transmission clockwise while inserting it into the clutch hub.**

9. Slowly insert the input shaft through the clutch. Rotate the output shaft slowly to engage the splines of the input shaft into the clutch while pushing the transmission forward into place. Do not force the transmission into position, the transmission should easily fall into place once everything is properly aligned.

10. Tighten the transmission mounting bolts to 35 ft. lbs. (47 Nm).

11. Do not remove the transmission jack until the crossmembers have been installed.

12. Check the transmission fluid level and replenish as necessary.

Automatic Transmission Assembly

REMOVAL & INSTALLATION

1999–01 Vehicles

1. Before servicing the vehicle, refer to the precautions in the beginning of this section.

2. Remove or disconnect the following:
   - Negative battery cable
3. Drain the transmission fluid.
   - Driveshaft from the transmission (2WD) and transfer case, if equipped (4WD)
4. Support the transmission with a suitable transmission jack.
   - Shift cable from the transmission control lever and bracket
   - Nut and washer securing the transmission mount to the crossmember
   - Bolts and washers securing the mount to the transmission
   - Exhaust pipe from the exhaust manifold(s)
   - Bolts securing the converter pan cover to the transmission, if equipped
   - 3 bolts securing the torque converter to the flywheel
   - Bolt, clip, and strap securing the three fuel lines and transmission vent hose to the transmission case
   - Bolts and nut securing the transmission to the engine
   - Oil filler tube and seal from the transmission
   - Transmission cooler lines from the transmission. Plug the lines and the ports in the transmission.
   - Wiring harness connectors from the transmission
5. Inspect for any other wiring, brackets etc. which may interfere with the removal of the transmission.

6. Since the transmission acts as a rear engine mount, properly support the rear of the engine with an underbody support or other suitable support before attempting to remove the transmission. Otherwise the rear of the engine may pitch downward and components on the rear of the engine and on the firewall may be damaged.

7. Remove the transmission from the engine by pulling the transmission rearward to disengage it from the locator dowel pins on the back of the block. Carefully lower the transmission from the vehicle. Use care that the torque converter does not fall out of the front of the transmission.

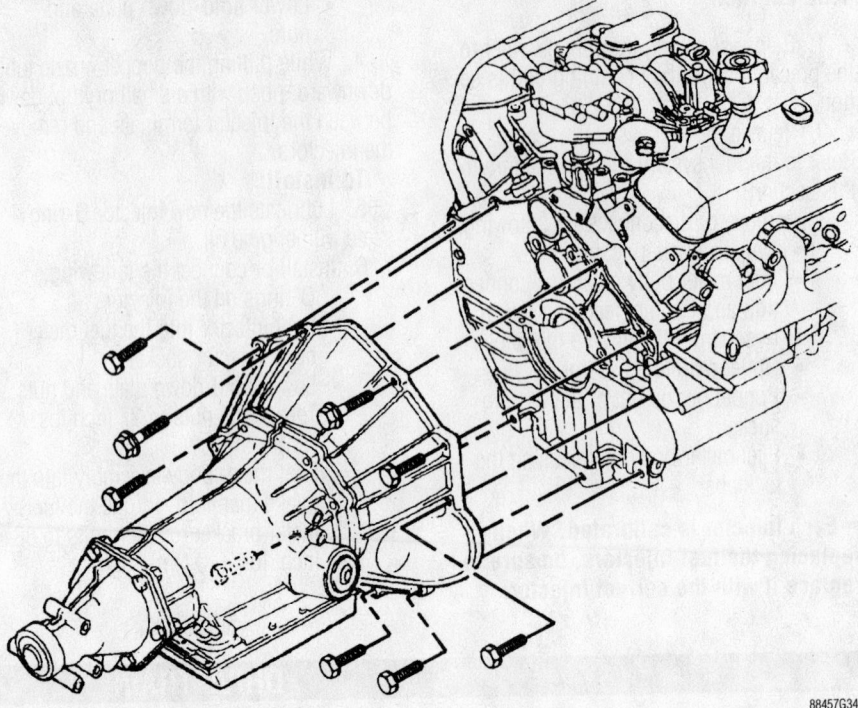

Transmission mounting on 4.3L engines

88457G34

➡**Use converter holding strap tool No. J-21366, to secure the torque converter to the transmission during removal and installation procedures.**

To install:

Installation is the reverse of removal, but please note the following important steps.

8. Make sure the torque converter is fully seated in the pump drive. If not, the transmission will not fit tightly to the rear of the engine block.

9. Raise the transmission into position and remove the torque converter holding strap. Carefully slide the transmission forward until the dowel pins are engaged.

10. The torque converter should be flush with the flywheel and turn freely by hand.

11. Install the transmission–to–engine bolts. Tighten the bolts to 34 ft. lbs. (47 Nm).

12. Tighten the torque converter-to-flywheel bolts to 46 ft. lbs. (63 Nm).

13. If equipped, tighten the converter pan cover to the transmission bolts to 37 ft. lbs. (50 Nm)

14. Tighten the bolts and washers securing the transmission mount to 35 ft. lbs. (47 Nm).

15. Tighten the nut and washer securing the transmission mount to the crossmember to 38 ft. lbs. (52 Nm).

16. Refill the transmission with the proper amount and type of fluid.

17. Connect the negative battery cable.

Start the vehicle and allow to warm while checking for leaks. Road test the vehicle to check for shift quality.

2002–03 Vehicles

➡**This procedure requires the use of a Converter Holding Strap tool No. J 21366 to secure the torque converter to the transmission during removal and installation.**

1. Before servicing the vehicle, refer to the precautions in the beginning of this section.

2. Remove or disconnect the following:
   - Negative, then the positive battery cables
   - Fill tube nut, located on the right side of the engine
3. Drain the transmission fluid.
   - Rear propeller shaft
4. Support the transmission with a transmission jack.
   - Nuts securing the transmission mount to the transmission support
   - Evaporative emission (EVAP) canister from its mounting bracket on the left inside of the frame to get access to the transmission support bolts. Do not disconnect the canister lines.
   - Fuel tank shield
   - Transmission support
   - Transmission mount bolts and mount
   - Front exhaust pipe assembly

5. Lower the transmission for access to the top and sides of the transmission.
- Transfer case, if equipped
- Range selector cable end from the transmission range selector lever ball stud and bracket
- Transmission heat shield, transmission vent hose park/neutral position switch connector, and main connector from the transmission
- Bolt that secures the fuel line bracket to the left side of the transmission
- Torque converter access plug

6. Matchmark the flywheel and torque converter orientation for reassembly.
- Flywheel-to-torque converter bolts. Be careful not to drop the bolts into the bell housing!
- Disconnect the transmission oil cooler lines from the transmission. Plug the transmission oil cooler lines connectors in the transmission case.

7. Install a safety chain around the transmission.
- Bolt that secures the fuel line bracket to the bell housing
- Bolts that secure the coolant pipe to the bell housing
- Remaining nuts, studs and/or bolts that secure the transmission to the engine

8. Install Converter Holding Strap tool No. J 21366 onto the transmission bell housing to hold the torque converter.

9. Pull the transmission straight back and remove it from the vehicle.

To install:

Installation is the reverse of removal, but please note the following important steps.

10. Make sure the torque converter is fully seated in the pump drive. If not, the transmission will not fit tightly to the rear of the engine block.

11. Raise the transmission into position and remove the torque converter holding strap. Carefully slide the transmission forward until the dowel pins are engaged while lining up the marks on the flywheel made during removal.

12. The torque converter should be flush with the flywheel and turn freely by hand.

13. Install the transmission-to-engine nuts, studs and or bolts. Tighten the studs and/or bolts to 37 ft. lbs. (50 Nm).

14. Tighten the bolts securing the heat shield to the transmission to 13 ft. lbs. (17 Nm).

15. Tighten the bolts and washers securing the transmission mount to 18 ft. lbs. (25 Nm).

16. Tighten the nut and washer securing the transmission mount to the transmission support to 35 ft. lbs. (46 Nm).

17. Refill the transmission with the proper amount and type of fluid.

18. Connect the negative battery cable. Start the vehicle and allow to warm while checking for leaks. Road test the vehicle to check for shift quality.

Clutch

REMOVAL & INSTALLATION

1. Before servicing the vehicle, refer to the precautions in the beginning of this section.

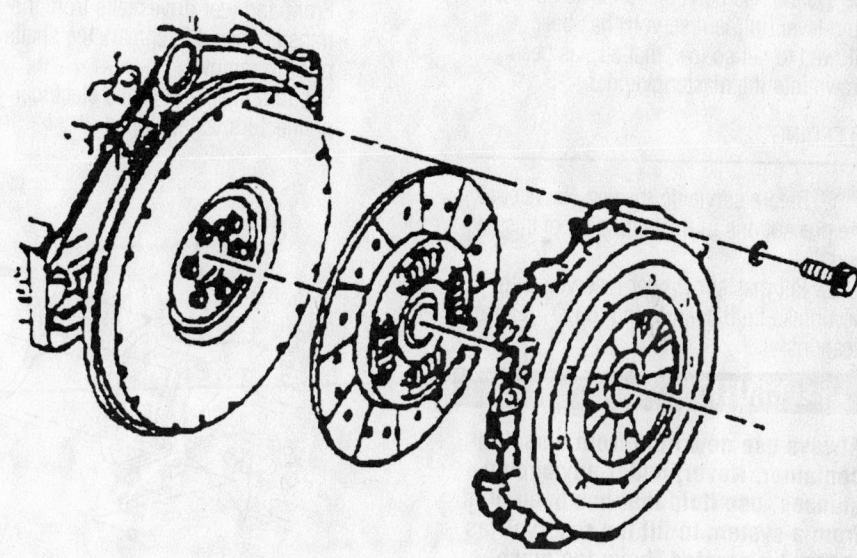

Exploded view of the clutch disc and related components

7924JG53

2. Remove or disconnect the following:
- Negative battery cable
- Transmission

3. Install a clutch alignment tool or a used transmission input shaft to support the clutch.

4. If the clutch assembly is going to be reused, mark the flywheel, clutch cover and a pressure plate lug for alignment when installing.

5. Remove or disconnect the following:
- Clutch cover bolts and washers
- Clutch cover assembly and the clutch plate
- Clutch alignment tool

6. Clean all parts and inspect for damage.

To install:

7. Install or connect the following:
- Clutch alignment tool, to support the clutch

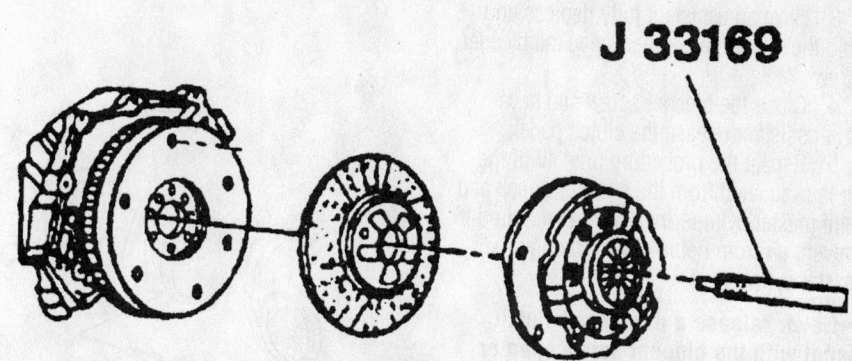

J 33169

Use the clutch alignment tool to center and support the clutch disc during installation

7924JG22

Timing belt service is covered in Section 3 of this manual

- Clutch cover by aligning the matchmarks or, if new, align the lightest part of the cover, identified by a yellow dot, with the heaviest part identified by an **X**.
- Clutch plate/clutch cover assembly to the flywheel. Tighten the bolts to 33 ft. lbs. (45 Nm) for 2.2L engines or to 29 ft. lbs. (40 Nm) for 4.3L engines.

➡ **Tighten each screw 1 turn at a time to avoid warping the clutch cover.**

8. Remove the clutch alignment tool.
9. Install or connect the following:
   - Transmission
   - Negative battery cable

Hydraulic Clutch System

Bleeding air from the hydraulic clutch system is necessary whenever any part of the system has been disconnected or the fluid level (in the reservoir) has been allowed to fall so low, that air has been drawn into the master cylinder.

BLEEDING

1. Before servicing the vehicle, refer to the precautions in the beginning of this section.
2. Fill master cylinder reservoir with new brake fluid conforming to DOT 3 specifications.

❄ CAUTION

Always use new fluid from a sealed container. Never, under any circumstances, use fluid that has been bled from a system to fill the reservoir as it may be aerated, have too much moisture content and possibly be contaminated.

3. Have an assistant fully depress and hold the clutch pedal, then open the bleeder screw.
4. Close the bleeder screw and have your assistant release the clutch pedal.
5. Repeat the procedure until all of the air is evacuated from the system. Check and refill master cylinder reservoir as required to prevent air from being drawn through the master cylinder.

➡ **Never release a depressed clutch pedal with the bleeder screw open or air will be drawn into the system.**

6. If the previous steps do not result in satisfactory pedal feel, remove the reservoir cap and pump the clutch pedal very fast for 30 seconds. Stop to let the air escape, then repeat the procedure as necessary to purge all remaining air.

7. Test the clutch for proper operation.

Transfer Case Assembly

REMOVAL & INSTALLATION

1999–01 Vehicles

1. Before servicing the vehicle, refer to the precautions in the beginning of this section.
2. Disconnect the negative battery cable.
3. Shift the transfer case into the **4HI** range.
4. Drain the transfer case fluid.
5. Support the transfer case.
6. Remove or disconnect the following:
   - Skid plate
   - Front and rear driveshafts from the transfer case. Matchmark the shafts prior to removal.
   - Vacuum lines and/or the electrical connectors, as equipped
   - Transfer case shift rod/cable from the case, if applicable
   - Support brace-to-transfer case bolts, if applicable
   - Transfer case
7. Remove all traces of old gasket material from the mating surfaces.

To install:
8. Install or connect the following:
   - New gasket using sealer to hold it in position
   - Transfer case. Torque the bolts to 33–35 ft. lbs. (45–47 Nm) on 1999–01 models.
   - Support brace bolts. Torque the bolts to 35–37 ft. lbs. (47–50 Nm), if equipped.
   - Shift rod to the case, if equipped
   - Vacuum lines and/or electrical connections, as necessary
   - Front and rear driveshafts by aligning the matchmarks
9. Refill the transfer case.
10. Install or connect the following:
    - Skid plate, if equipped
    - Negative battery cable

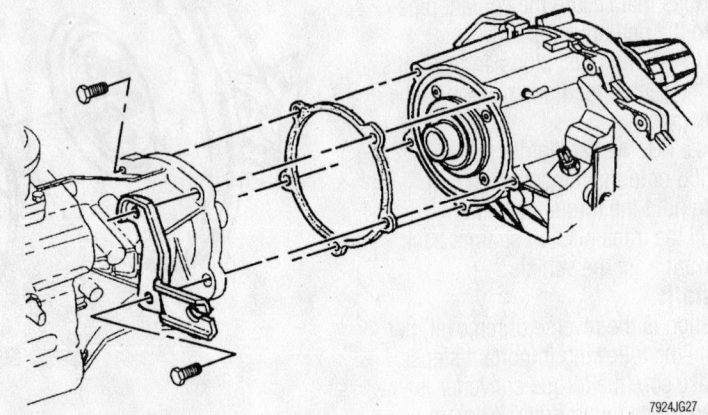

Transfer case-to-manual transmission mounting

7924JG27

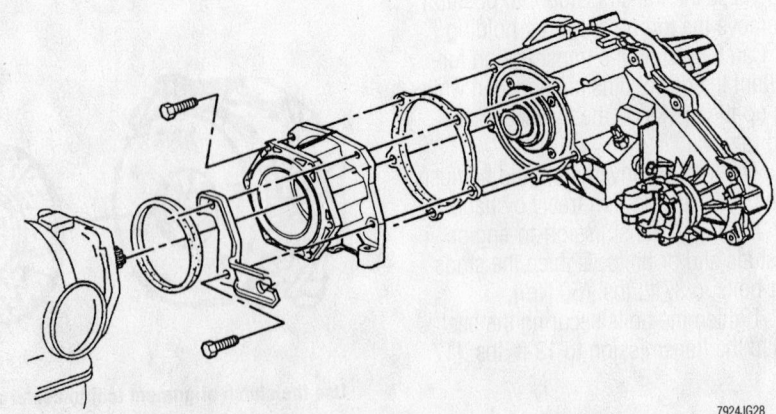

Transfer case-to-automatic transmission mounting

7924JG28

2002–03 Vehicles

1. Before servicing the vehicle, refer to the precautions in the beginning of this section.
2. Disconnect the negative battery cable.
3. Raise and support the vehicle. Drain the transfer case.
4. Remove or disconnect the following:
   - Fuel tank shield mounting bolts and shield
   - Front and rear propeller shaft. Matchmark the shafts prior to removal.
   - Fuel lines from the retainer
   - Electrical harness from the retainers on the right and left sides
   - Speed sensor electrical connectors
   - Motor/encoder electrical connector
   - Transfer case wiring harness
   - Vent hose
5. Install a transmission jack to support the transfer case.
   - Transfer case mounting bolts
   - Transfer case from the vehicle
   - Transfer case gasket and discard if damaged

To install:

6. Install or connect the following:

➡ **You must replace the transfer case gasket if it is damaged. Never use silicone sealant in place of, or with the transfer case gasket.**

   - Transfer case, using a new gasket if necessary
   - Transfer case mounting bolts and tighten to 35 ft. lbs. (47 Nm)
7. Remove the transmission jack.
8. The remainder of installation is the reverse of removal.
9. Refill the transfer case.

Halfshaft

REMOVAL & INSTALLATION

1999–01 Vehicles

1. Before servicing the vehicle, refer to the precautions in the beginning of this section.
2. Unlock the steering column so the steering linkage is free to move.
3. Remove or disconnect the following:
   - Negative battery cable
   - Front wheels

➡ **Place a drift through the caliper into the edge of the rotor to keep the rotor from turning when the nut is removed.**

   - Cotter pin, retainer, nut and washer.
   - Brake caliper and support it with a piece of wire to avoid damaging the brake hose
   - Brake rotor
   - Brake line support bracket and Anti-lock Brake System (ABS) wire bracket from the upper control arm
4. Place a jackstand or jack under the lower control arm.
   - Axle shaft from the hub by placing

a block of wood against the outer edge of the axle (to protect the threads), then strike the block of wood sharply with a hammer. Do not remove the axle at this time.
   - Tie rods from the steering knuckles
   - Lower shock absorber bolts
   - Upper ball joint from the steering knuckle and suspend the steering knuckle on a wire
   - Skid plate, if equipped

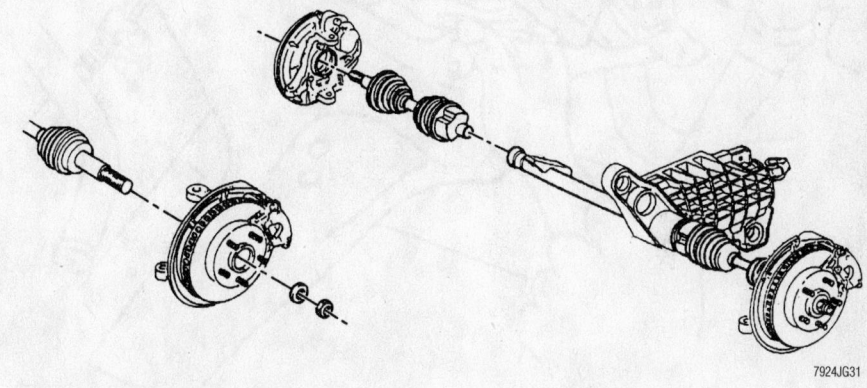

Halfshafts and related components

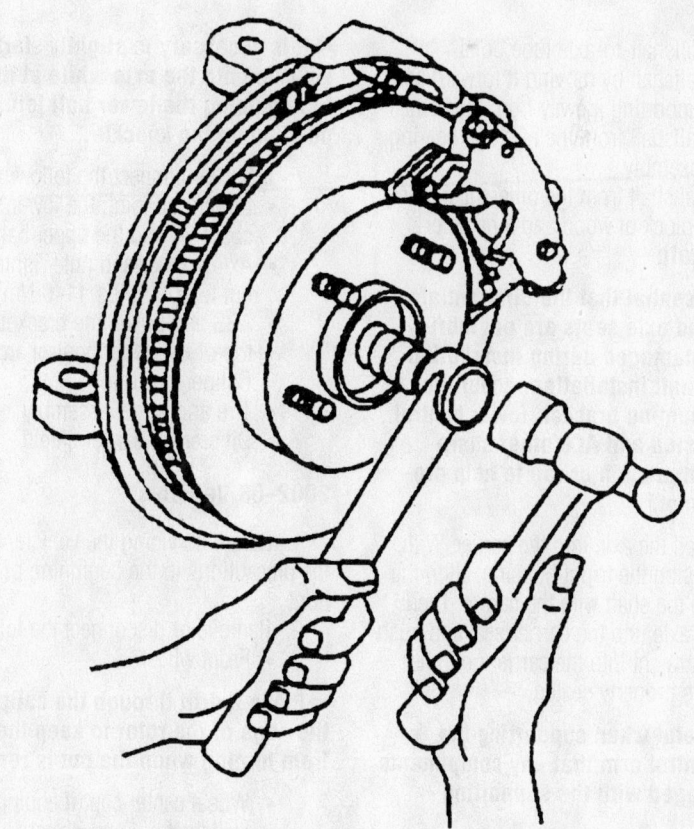

Tap the halfshaft out of the hub without damaging the threads

Heater Core replacement is covered in Section 2 of this manual

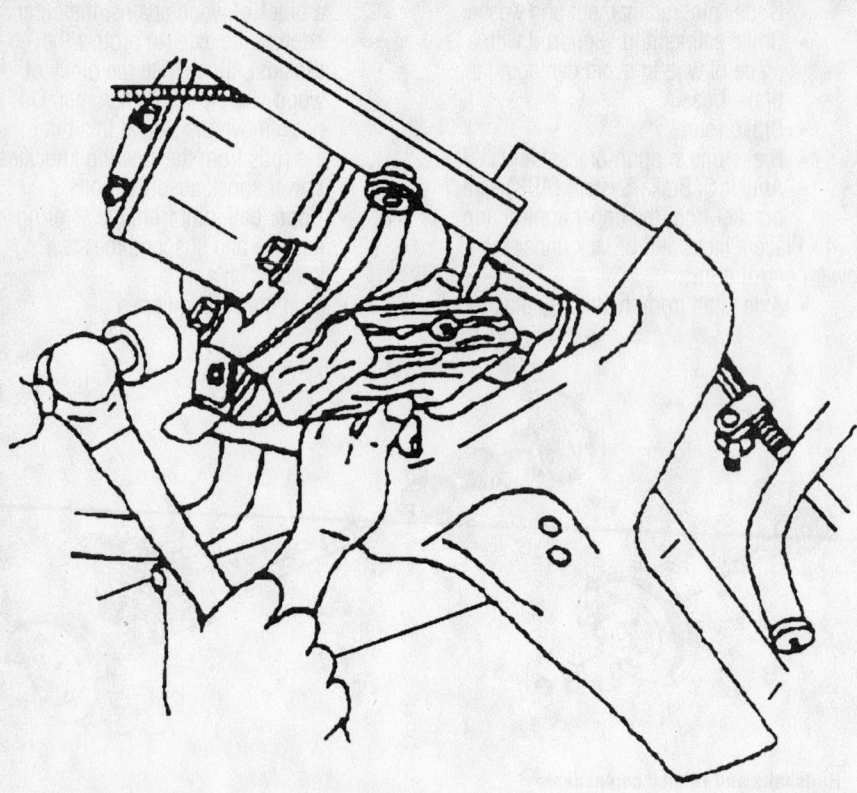

7924JG30

Using a block of wood and a mallet, disengage the halfshaft from the differential assembly

- Halfshaft-to-axle tube bolts
- Halfshaft by moving it forward and supporting it away from the frame
- Halfshaft from the hub and bearing assembly
- Halfshaft from the differential using a block of wood and a hammer

To install:

➡️**It is essential that the differential carrier and axle seals are not lubricated or damaged during installation. Prior to shaft installation, cover the shock mounting bracket, lower control arm ball stud and ALL other sharp edges with a cloth or rag to help protect the boot.**

5. Install the axle into the carrier. With both hands on the tripot housing, align the splines on the shaft with the carrier. Then center the axle into the carrier seal and push the shaft straight into the carrier until the snapring is properly seated.

➡️**Be careful when supporting the lower control arm that any components are damaged with the supporting device.**

6. Raise the lower control arm using a jackstand or jack until the full weight of the arm is supported.

➡️**It is necessary to slightly start the knuckle onto the axle while at the same time guiding the lower ball joint into position on the knuckle.**

7. Install or connect the following:
- Lower ball joint, the lower shock absorber and the upper ball joint
- Axle washer and nut. Tighten the nut to 103 ft. lbs. (140 Nm).
- ABS and brake line brackets to the top of the upper control arm
- Caliper and rotor
- Tire and wheel assembly
- Differential carrier shield

2002–03 Vehicles

1. Before servicing the vehicle, refer to the precautions in the beginning of this section.

2. Remove or disconnect the following:
- Front wheel

➡️**Place a drift through the caliper into the edge of the rotor to keep the rotor from turning when the nut is removed**

- Wheel center cap, if equipped
- Halfshaft nut and discard. A new nut must be used for installation.
- Drift from the rotor
- Brake caliper and support it with a

piece of wire to avoid damaging the brake hose
- Brake rotor

3. To remove the steering knuckle, remove or disconnect the following:
- Wheel hub and bearing
- Outer tie rod retaining nut
- Outer tie rod end from the steering knuckle using a puller
- Brake hose bracket retaining bolts
- Brake hose bracket
- Anti-lock Brake System (ABS) wheel speed sensor wiring harness bracket, if necessary
- Upper control arm-to-steering knuckle pinch bolt and nut
- Upper control arm from the steering knuckle
- Lower ball joint retaining nut
- Steering knuckle from the control arm using a puller
- Steering knuckle

4. Remove the left side halfshaft from differential carrier, or right halfshaft from the clutch fork housing as follows:
a. Place a brass drift against the tripot housing.
b. Use a hammer to strike the drift outward from the case, striking hard enough to overcome the snapring tension holding the halfshaft.

5. Pull the halfshaft straight out of the differential carrier or clutch fork housing.

To install:

6. Install the halfshaft as follows:
a. With both hands on the tripot housing, align the splines on the shaft with the differential carrier assembly (left) or clutch fork housing (right).
b. Center the halfshaft into the differential carrier or clutch fork housing assembly seal.
c. Firmly push the shaft straight into the differential carrier or clutch fork housing assembly until the snapring is properly seated.

7. To install the steering knuckle, install or connect the following:
- Steering knuckle to the lower control arm
- Lower ball joint retaining nut and tighten to 81 ft. lbs. (110 Nm)
- Upper control arm to the steering knuckle
- Upper control arm pinch bolt and nut and tighten to 30 ft. lbs. (40 Nm)
- ABS wheel speed sensor harness bracket
- Brake hose bracket. Tighten the bolts to 7 ft. lbs. (10 Nm).

- Outer tie rod to the steering knuckle and tighten the nut to 33 ft. lbs. (45 Nm)
- Hub and bearing

8. Install or connect the following:
- New halfshaft nut and tighten to 103 ft. lbs. (140 Nm)
- Wheel

9. Lower the vehicle. Adjust the front toe.

CV-Joints

OVERHAUL

Outer CV-Joint

1. Before servicing the vehicle, refer to the precautions in the beginning of this section.
2. Remove or disconnect the following:
- Front wheel
- Halfshaft and position it in a vise
- Large CV-joint boot clamp and discard it
- Small CV-joint boot clamp and discard it
- CV-joint boot and slide it back on the shaft
- Outer race from the halfshaft, by spreading the outer race-to-halfshaft retaining ring, using Snapring Pliers J-8059
- Retaining ring from the halfshaft and discard it
- CV-joint boot from the halfshaft and discard it, if damaged

3. Disassemble the chrome alloy balls from the CV-joint cage as follows:
   a. Position a brass drift against the CV-joint cage and tap it with a hammer to tilt the cage.
   b. Remove the 1st chrome alloy ball from the cage.
   c. Tilt the cage in the opposite direction.
   d. Remove the opposite chrome alloy ball.
   e. Repeat the procedure until all 6 balls are removed.

4. Disassemble the CV-joint cage and inner race as follows:
   a. Pivot the cage and race 90 degrees to the center line of the outer race.
   b. Align the cage windows with outer race lands.
   c. Remove the cage from the outer race.

   d. Rotate the inner race upward and remove it from the cage.

5. Thoroughly clean and inspect all parts.

To install:

6. Lubricate the parts with a light coat of grease.
7. Assemble the CV-joint cage and inner race, as follows:
   a. Rotate the inner race 90 degrees to the cage centerline.
   b. Align the cage windows with inner race lands.
   c. Insert the inner race into the cage by rotating the inner race downward.
   d. Insert the cage/inner race into the outer race.

8. Assemble the chrome alloy balls into the CV-joint cage, as follows:
   a. Position a brass drift against the CV-joint cage and tap it with a hammer to tilt the cage.
   b. Insert the 1st chrome alloy ball into the cage.
   c. Tilt the cage in the opposite direction.
   d. Insert the opposite chrome alloy ball.
   e. Repeat the procedure until all 6 balls are inserted.

9. Install ½ kit grease into the CV-joint.
10. Install or connect the following:
- Small ring clamp on the CV boot
- New retaining ring on the halfshaft
- Large ring clamp on the CV boot
- Outer race assembly onto the halfshaft until the ring engages the halfshaft groove

11. Slide the small end of the CV-joint boot/clamp into place, with the seal lip in the halfshaft groove

➡ **Make sure the boot lies flat against the halfshaft.**

12. Using the Crimp tool J-35910, a torque wrench and a breaker bar, crimp the small CV-joint boot clamp to 100 ft. lbs. (136 Nm).
13. Check the clamp gap dimension; if it is not 0.085 in. (2.15mm), continue tightening the clamp until it is.
14. Install ½ kit grease into the CV-joint boot.
15. Measure approximately 0.687 in. (17.5mm) up from the bottom edge of the outer CV-joint assembly.
16. Slide the large end of the CV boot/clamp into place, with the seal lip in place over the outer race.

➡ **Make sure the boot lies flat against the outer race.**

17. Using the Crimp tool J-35910, a torque wrench and a breaker bar, crimp the large CV-joint boot clamp to 130 ft. lbs. (176 Nm).
18. Check the clamp gap dimension; if it is not 0.102 in. (2.60mm), continue tightening the clamp until it is.
19. Install the halfshaft and the front wheel.

Inner (Tri-Pot) Joint

1. Before servicing the vehicle, refer to the precautions in the beginning of this section.
2. Remove or disconnect the following:
- Front wheel
- Halfshaft and place it in a vise
- Snapring from the stub shaft and discard it
- Small CV-joint boot clamp, cut and discard it
- Large CV-joint boot clamp, cut and discard it
- CV-joint boot by sliding it away from the tri-pot joint

3. Install a Stub Shaft Removal tool J-38868-A to the stub shaft snapring groove.
4. Using a slide hammer puller, press the stub shaft from the tri-pot housing.
5. Remove or disconnect the following:
- Tri-pot housing from the tri-pot spider
- Inboard spacer ring slide it rearward on the shaft using Snapring Pliers tool J-8059
- Outboard retaining ring using Snapring Pliers tool J-8059 and discard it
- Tri-pot joint spider assembly
- Inboard spacer ring and discard it
- CV-joint boot
- Trilobal tri-pot bushing from the housing

6. Thoroughly clean and inspect all parts.

To install:

7. Install or connect the following:
- New snapring onto the stub shaft
- Small boot clamp
- CV-joint boot

8. Using the Crimp tool J-35910, a torque wrench and a breaker bar, crimp the small CV-joint boot clamp to 100 ft. lbs. (136 Nm).
9. Install or connect the following:
- Inboard spacer ring slide it rearward on the shaft using Snapring

---

Brake service is covered in Section 4 of this manual

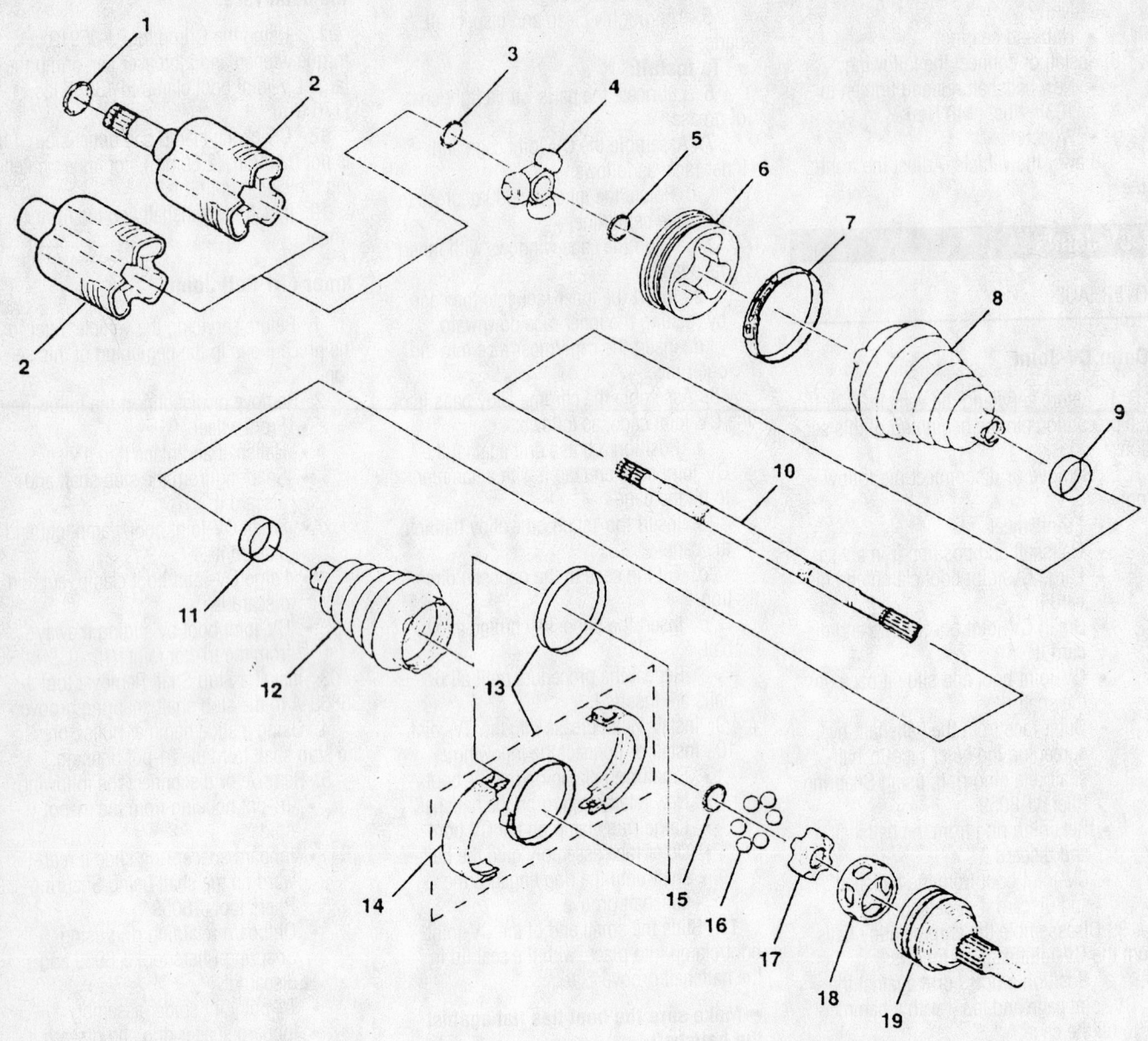

(1) Differential Shaft Ring
(2) Tripot Housing Assembly
(3) Spacer Ring
(4) Tripot Joint Spider Assembly
(5) Spacer Ring
(6) Tripot Bushing
(7) Boot Retaining Clamp
(8) Tripot Joint Boot
(9) Halfshaft Swage Ring
(10) Halfshaft Bar

(11) Halfshaft Swage Ring
(12) CV Joint Boot
(13) Swage Ring
(14) Clamp Protector
(15) Race Retaining Ring
(16) Ball
(17) CV Joint Inner Race
(18) CV Joint Cage
(19) CV Joint Outer Race

9308JG09

Exploded view of the CV-Joint Assembly

Pliers tool J-8059, past the 2nd groove
- Tri-pot joint spider assembly onto the shaft until it passes the 2nd groove
- Outboard retaining ring into the axle shaft groove using Snapring Pliers tool J-8059
- Tri-pot joint spider assembly, slide it against the outboard retaining ring
- Inboard spacer ring, seat it in the groove
- ½ kit grease into the boot
- ½ kit grease into the tri-pot housing
- Trilobal tip-pot bushing flush with the tri-pot housing face
- New large seal clamp onto the CV-joint boot
- Tri-pot housing, slide it over the tri-pot joint spider assembly
- CV-joint boot/clamp, slide it into place, over the trilobal tri-pot bushing with the seal lip in the groove

➡ **Make sure the boot lies flat against the trilobal bushing.**

10. Position the CV-joint boot so it measures 4.9 in. (125mm).

11. Using the Crimp tool J-35566, latch the large CV-joint boot clamp.

12. Install the halfshaft and the front wheel.

Axle Shaft, Bearing and Seal

REMOVAL & INSTALLATION

For the Axle Shaft, Bearing and Seal, Removal and Installation, please refer to Wheel Bearing procedure located in the section.

Pinion Seal

REMOVAL & INSTALLATION

1. Before servicing the vehicle, refer to the precautions in the beginning of this section.

➡ **The following procedure requires the use of the Pinion Holding tool J-8614-10, the Pinion Flange Removal tool J-8614-1, J-8614-2, J-8614-3 and the Pinion Seal Installation tool J-23911 or J-33782.**

2. Remove or disconnect the following:
- Driveshaft from the pinion flange. Matchmark the driveshaft prior to removal.
- Driveshaft from the rear axle pinion flange and support the shaft up in body tunnel by wiring it to the exhaust pipe.

➡ **If the U-joint bearings are not retained by a retainer strap, use a piece of tape to hold bearings on their journals.**

3. Mark the position of the pinion stem, flange and nut for reference.

4. Use an inch lbs. torque wrench to measure the amount of torque necessary to turn the pinion, then note this measurement as it is the combined pinion bearing, seal, carrier bearing, axle bearing and seal preload.

5. Remove or disconnect the following:
- Pinion flange nut and washer, using a Pinion Holding tool J-8614-10 and a Pinion Flange Removal tool J-8614-1, J-8614-2, J-8614-3, as applicable
- Pinion flange
- Pinion oil seal by driving it out of the differential with a blunt chisel; DO NOT damage the carrier

To install:

6. Examine the seal surface of pinion flange for tool marks, nicks or damage,

such as a groove worn by the seal. If damaged, replace flange.

7. Examine the carrier bore and remove any burrs that might cause leaks around the O.D. of the seal.

8. Apply GM seal lubricant 1050169 to the outside diameter of the pinion flange and sealing lip of new seal.

9. Install or connect the following:
- New pinion oil seal using a seal installer tool
- Pinion flange and tighten nut to the same position as marked earlier. Tighten the nut a little at a time and turn the pinion flange several times after each tightening in order to set the rollers.

10. Measure the torque necessary to turn the pinion and compare this to the reading taken during removal. Tighten the nut additionally, as necessary to achieve the same preload as measured earlier.

➡ **If fluid was lost from the differential housing during this procedure, be sure to check and add additional fluid, as necessary.**

11. Remove the support then align and secure the driveshaft assembly to the pinion flange.

➡ **The original matchmarks MUST be aligned to assure proper shaft balance and prevent vibration.**

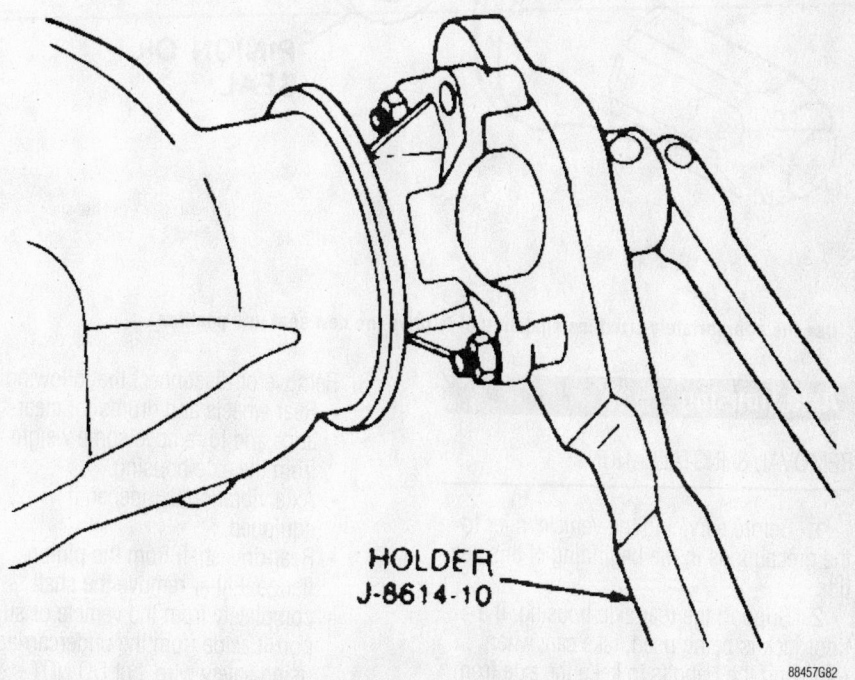

HOLDER
J-8614-10

Removing the pinion nut using a pinion holding fixture tool

88457G82

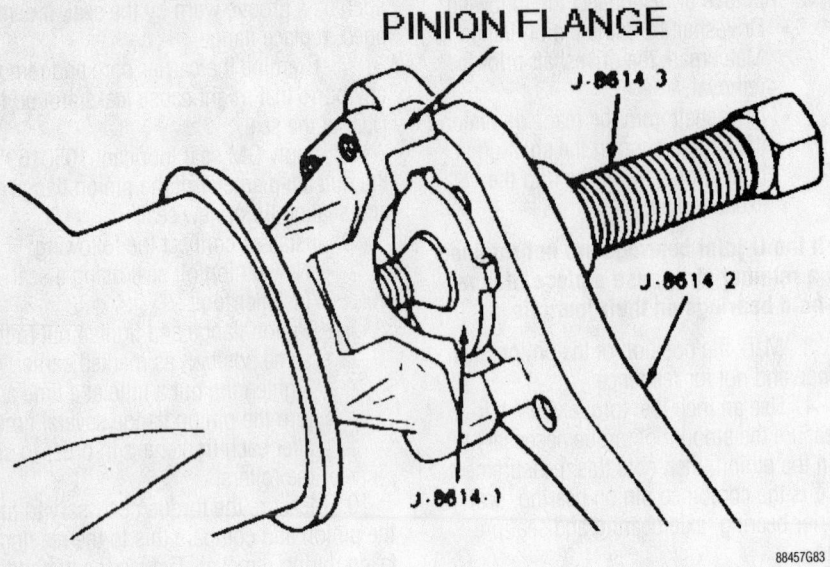

PINION FLANGE

J-8614 3
J-8614 2
J-8614-1

88457G83

A puller and adapter should be used to withdraw the pinion from the housing

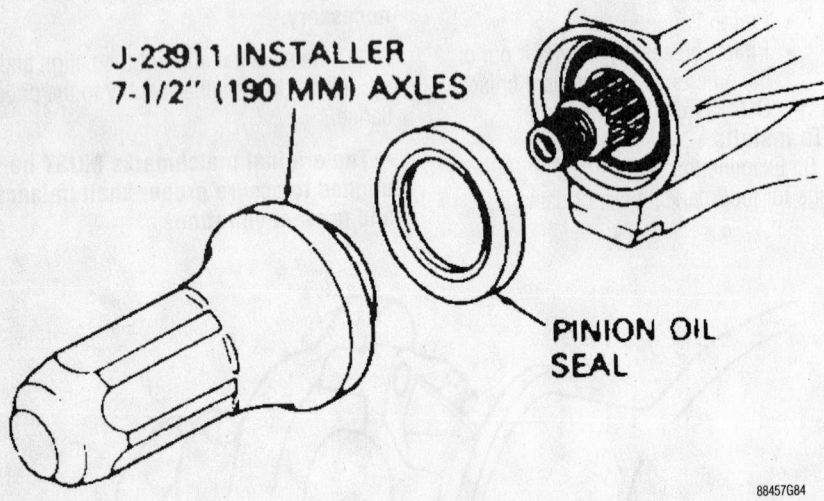

J-23911 INSTALLER
7-1/2" (190 MM) AXLES

PINION OIL
SEAL

88457G84

Use the appropriately sized installation tool to drive the new seal into position.

Axle Housing

REMOVAL & INSTALLATION

1. Before servicing the vehicle, refer to the precautions in the beginning of this section.

2. Support the rear axle housing. If a floor jack is being used, take care when removing the U-bolts to keep the axle from suddenly dislodging.

3. Remove or disconnect the following:
- Rear wheels and drums for clearance and to remove some weight from the axle housing
- Axle vibration dampener, if equipped
- Rear driveshaft from the pinion flange. Either remove the shaft completely from the vehicle or support it aside from the undercarriage using safety wire, but DO NOT allow the shaft to hang from the slip joint.
- Shock absorber-to-axle housing retainers, then swing the shock absorbers away from the axle housing
- Brake lines from the axle housing clips and the backing plates (wheel cylinders)

➡ **When disconnecting the brake lines from the wheel cylinders, immediately plug or cap the lines to prevent system contamination or excessive fluid loss.**

- Speed sensor connectors at the junction block, if applicable
- Parking brake cable(s)
- Axle housing-to-spring U-bolt nuts, washers, U-bolts and the anchor plates
- Vent hose from the top of the axle housing
- Axle with the help of an assistant by moving it to clear the leaf spring

To install:

4. With the help of an assistant, carefully position the rear axle into the vehicle.

5. Install or connect the following:
- Vent hose to the axle housing

6. Be sure the housing is properly positioned on the leaf spring, then loosely install the U-bolts, anchor plates, washers and nuts.

- Tighten the U-bolt nuts in a cross pattern to 18 ft. lbs. (25 Nm) to made sure everything is evenly seated. Then tighten the nuts in steps to 74 ft. lbs. (100 Nm).
- Brakes lines secure them to the axle housing
- Parking brake cable(s), if removed
- Speed sensor connectors to the junction block, if equipped
- Driveshaft assembly
- Shock absorbers to the lower mounts, then tighten the mount nuts
- Axle vibration dampener, if equipped
- Brake drums and the tire/wheel assemblies

7. Bleed the hydraulic brake system.

8. Check the fluid level in the rear axle assembly and add, as necessary. Make sure the vehicle is level when checking and adding fluid.

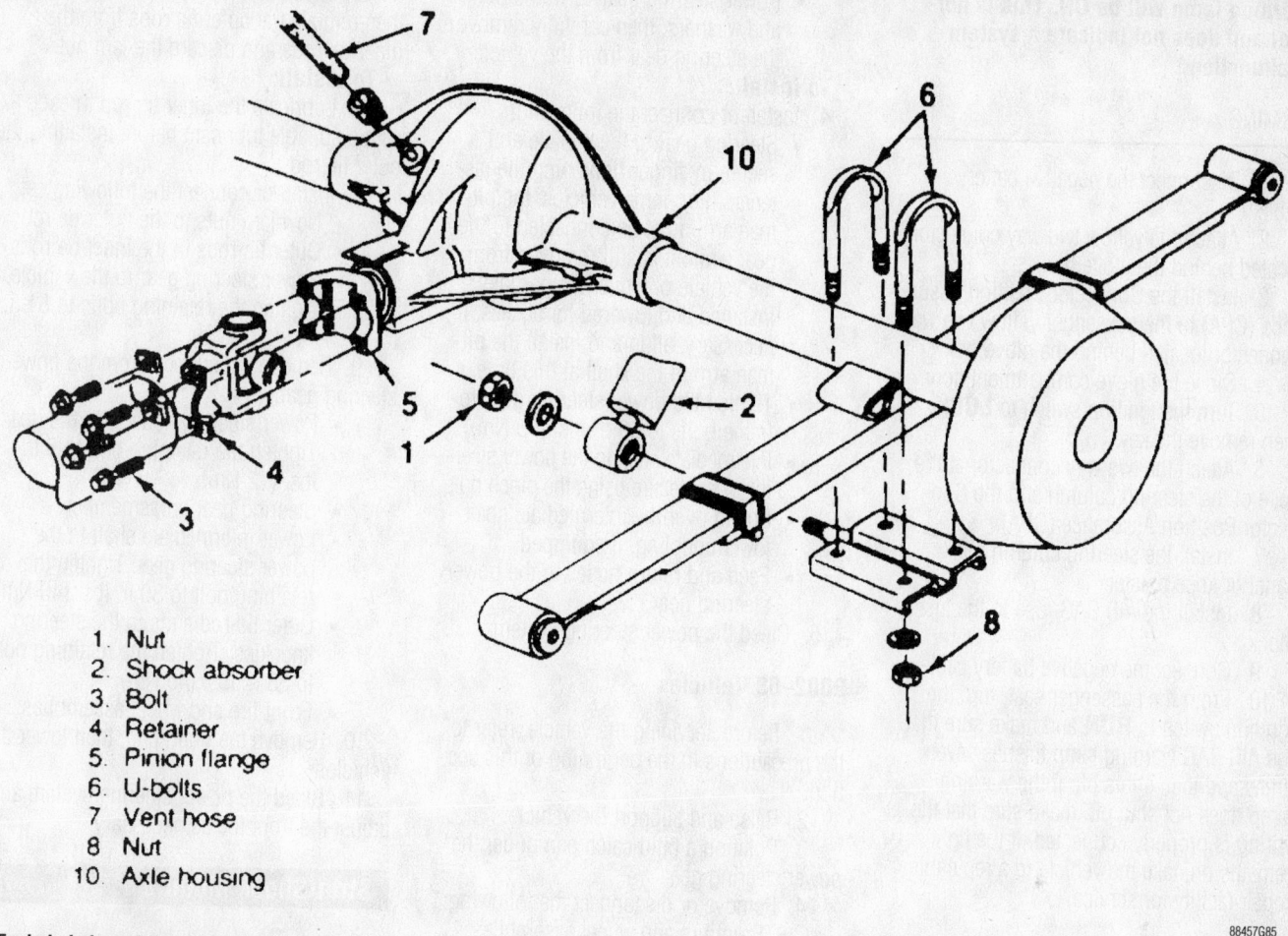

1. Nut
2. Shock absorber
3. Bolt
4. Retainer
5. Pinion flange
6. U-bolts
7. Vent hose
8. Nut
10. Axle housing

88457G85

Exploded view of the rear axle mounting

STEERING AND SUSPENSION

Air Bag

✳✳ CAUTION

Some vehicles are equipped with an air bag system, also known as the Supplemental Inflatable Restraint (SIR) system. The system must be disabled before performing service on or around system components, steering column, instrument panel components, wiring and sensors. Failure to follow safety and disabling procedures could result in accidental air bag deployment, possible personal injury and unnecessary system repairs.

PRECAUTIONS

Several precautions must be observed when handling the inflator module to avoid accidental deployment and possible personal injury.

• Never carry the inflator module by the wires or connector on the underside of the module.

• When carrying a live inflator module, hold securely with both hands, and ensure that the bag and trim cover are pointed away.

• Place the inflator module on a bench or other surface with the bag and trim cover facing up.

• With the inflator module on the bench, never place anything on or close to the module, that may be thrown in the event of an accidental deployment.

DISARMING

1. Turn the steering wheel so that the vehicle's wheels are pointing straight ahead.
2. Turn the ignition switch to **LOCK**, remove the key, then disconnect the negative battery cable.
3. Remove the AIR BAG or SIR fuse from the fuse block.
4. Remove the steering column filler panel or knee bolster.
5. Unplug the Connector Position Assurance (CPA) and yellow two way connector at the base of the steering column.
6. Open the glove compartment door, lift the stop and let the door fully open.
7. Remove the Connector Position Assurance (CPA) from the passenger yellow two way connector located behind the glove box.
8. Unplug the yellow two way connector located behind the glove box.
9. Connect the negative battery cable.

➡**With the AIR BAG fuse removed, the battery cable connected and the ignition in the ON position, the AIR BAG**

For Accessory Drive Belt illustrations, see Section 1 of this manual

warning lamp will be ON. This is normal and does not indicate a system malfunction.

ARMING

1. Disconnect the negative battery cable.

2. Attach the yellow two way connector located behind the glove box.

3. Install the Connector Position Assurance (CPA) to the passenger yellow two way connector located behind the glove box.

4. Close the glove compartment door.

5. Turn the ignition switch to **LOCK**, then remove the key.

6. Attach the two way connector at the base of the steering column and the Connector Position Assurance (CPA).

7. Install the steering column filler panel or knee bolster.

8. Install the AIR BAG fuse to the fuse block.

9. Connect the negative battery cable.

10. From the passenger seat, turn the ignition switch to **RUN** and make sure that the AIR BAG warning lamp flashes seven times and then shuts off. If the warning lamp does not shut off, make sure that the wiring is properly connected. If the light remains on, take the vehicle to a reputable repair facility for service.

Power Steering Gear

REMOVAL & INSTALLATION

1999–2001 Vehicles

1. Before servicing the vehicle, refer to the precautions in the beginning of this section.

2. Position a fluid catch pan under the power steering gear.

3. Remove or disconnect the following:
- Feed and return fluid hoses from the steering gear. Immediately cap or plug all openings to prevent system contamination or excessive fluid loss.
- Intermediate shaft lower coupling shield, if equipped
- Intermediate shaft-to-steering gear bolt. Matchmark the intermediate shaft-to-power steering gear and separate the shaft from the gear.
- Pitman arm from the gear pitman shaft

- Power steering gear-to-frame bolts and washers, then carefully remove the steering gear from the vehicle

To install:

4. Install or connect the following:
- Steering gear to the vehicle and secure by finger-tightening the fasteners. For some vehicles, the pitman arm must be connected to the gear while it is still removed from the vehicle or while it is partially installed and lowered for access. If necessary, align and install the pitman arm to the shaft at this time.
- Tighten the power steering gear-to-frame bolts to 55 ft. lbs. (75 Nm)
- Intermediate shaft to the power steering, then secure using the pinch bolt
- Shield over the intermediate shaft lower coupling, if equipped
- Feed and return hoses to the power steering gear

5. Bleed the power steering system.

2002–03 Vehicles

1. Before servicing the vehicle, refer to the precautions in the beginning of this section.

2. Raise and support the vehicle.

3. Position a fluid catch pan under the power steering gear.

4. Remove or disconnect the following:
- Front tire and wheel assemblies
- Outer tie rod retaining nuts

✵ WARNING

Do not try to separate a steering linkage joint by driving a wedge between the joint and the attached part. Doing this can cause seal damage and premature failure of the part.

- Outer tie rods from the steering knuckles using a suitable steering linkage and tie rod puller
- Lower intermediate shaft retaining bolt and shaft from the power steering gear
- Steering gear crossmember
- Feed and return fluid hoses from the steering gear. Immediately cap or plug all openings to prevent system contamination or excessive fluid loss.

5. Support the power steering gear.
- Power steering gear mounting bolts, then remove the gear from the vehicle

6. Loosen the outer tie rod jam nuts, then remove the outer tie rods from the inner tie rods and discard the jam nut.

To install:

7. Lubricate the inner tie rod threads with a suitable lubricant before installing the outer tie rod.

8. Install or connect the following:
- New jam nuts to the outer tie rods
- Outer tie rods to the inner tie rods
- Power steering gear to the vehicle. Tighten the retaining bolts to 81 ft. lbs. (110 Nm).

9. Remove the support from the power steering gear.
- Power steering hose(s) to the gear. Tighten the retaining bolt to 9 ft. lbs. (12 Nm).
- Steering gear crossmember
- Lower intermediate shaft to the power steering gear. Tighten the retaining bolt to 30 ft. lbs. (40 Nm).
- Outer tie rod ends to the steering knuckles. Tighten the retaining nuts to 33 ft. lbs. (45 Nm).
- Front tire and wheel assemblies

10. Remove the drain pan, then lower the vehicle.

11. Bleed the power steering system and adjust the front toe as necessary.

Strut/Shock Module

REMOVAL & INSTALLATION

Front—2002–03 Vehicles

➡In 2002–03, a "shock module", similar to a strut was used on these vehicles. This procedure requires the use of a suitable steering linkage and tie rod puller. ⬏

1. Before servicing the vehicle, refer to the precautions in the beginning of this section.

2. Remove or disconnect the following:
- Shock module upper retaining nuts
- Tire and wheel
- Shock module-to-lower control arm retaining nut
- Shock module yoke from the lower control arm using a suitable puller
- Shock module from the shock tower and lower control arm

To install:

3. Install or connect the following:
- Shock module to the shock tower and lower control arm

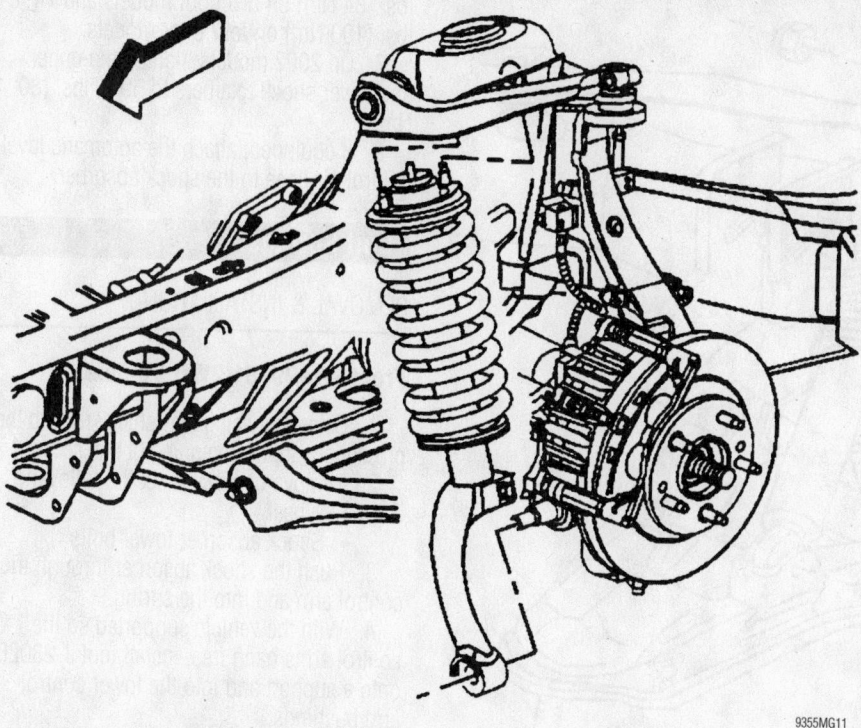

View of the shock module used on the front suspension of 2002 vehicles

9355MG11

- Shock module yoke to the lower control arm
- Shock module upper retaining nuts and tighten to 33 ft. lbs. (45 Nm)
- Shock module-to-lower control arm retaining nut and tighten to 81 ft. lbs. (110 Nm)
- Tire and wheel

Shock Absorbers

REMOVAL & INSTALLATION

Front—1999–01 Vehicles

2WD MODELS

1. Before servicing the vehicle, refer to the precautions in the beginning of this section.
2. Remove or disconnect the following:
   - Wheel
   - Mounting nut

➡ Hold the shock absorber stem with a wrench while backing the nut off.

   - Retaining nut and grommet
   - Shock absorber-to-lower control arm bolts
   - Shock absorber
   - Replace the parts, as necessary.

To install:

3. Fully extend the shock absorber stem, then push it up through the lower control arm and spring so that the upper stem passes through the mounting hole in the upper control arm frame bracket.
4. Install or connect the following:
   - Retaining nut and grommet on the stem. Tighten the nut to 106 inch lbs. (12 Nm).
   - Shock absorber-to-lower control arm bolts and tighten to 22 ft. lbs. (30 Nm)
   - Wheel

4WD MODELS

1. Before servicing the vehicle, refer to the precautions in the beginning of this section.
2. Remove or disconnect the following:
   - Wheel
   - Lower nut/bolt and collapse the shock absorber
   - Shock absorber upper nut and bolt
   - Shock absorber

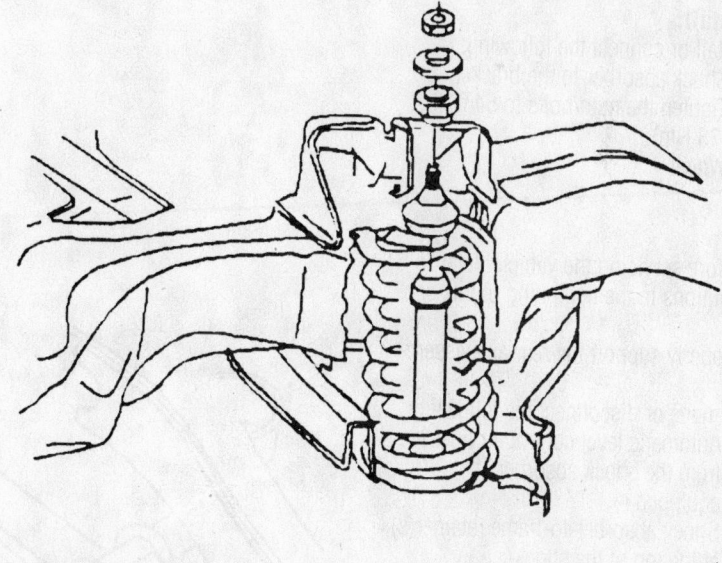

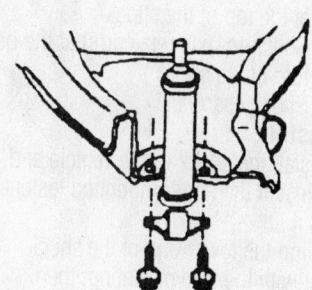

Front shock absorber mounting—2WD

7924JG33

For Tire, Wheel and Ball Joint specifications, see Section 1 of this manual

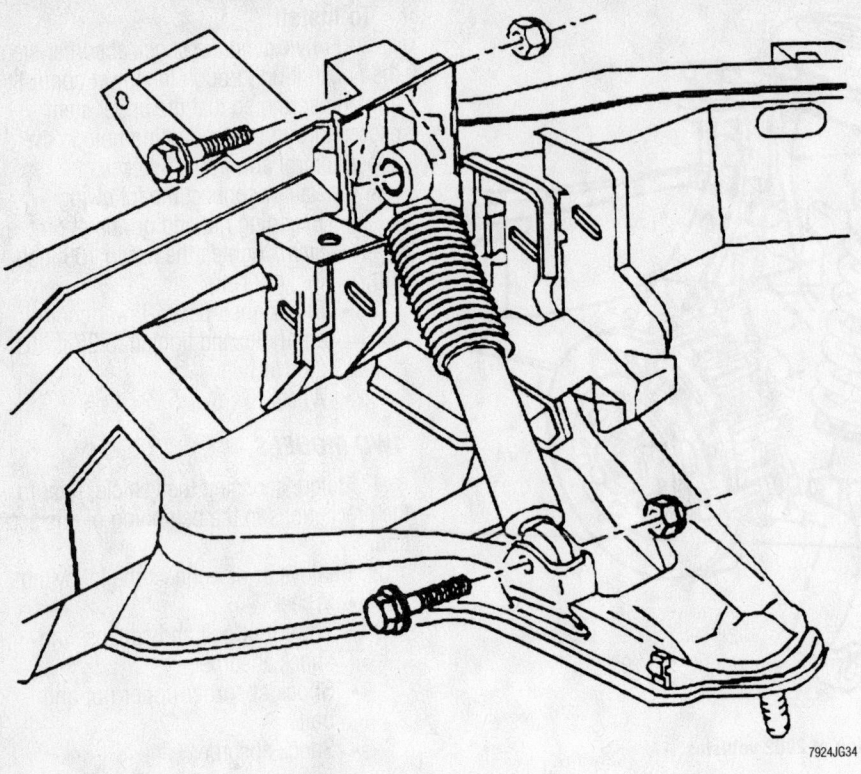

Front shock absorber mounting—4WD

To install:

3. Install or connect the following:
- Shock absorber to the bracket.
  Tighten the nuts/bolts to 54 ft. lbs.
  (73 Nm).
- Wheel

Rear

1. Before servicing the vehicle, refer to the precautions in the beginning of this section.

2. Properly support the rear axle assembly.

3. Remove or disconnect the following:
- Automatic level control air lines from the shock absorber, if equipped
- Shock absorber-to-frame retainer(s) at the top of the shock
- Shock-to-axle retainer(s) at the bottom of the shock
- Shock absorber

To install:

4. Install the shock in the vehicle and loosely install the upper mounting fasteners to retain it

5. Align the lower-end of the shock absorber with the axle mounting, then loosely install the retainers.

6. On 1999–01 models, tighten the upper shock retainers to 18 ft. lbs. (25 Nm). Tighten the lower shock retainers to 62 ft.

lbs. (84 Nm) on two door models and 74 ft. lbs. (100 Nm) on four door models.

7. On 2002 models, tighten the upper and lower shock retainers to 59 ft. lbs. (80 Nm).

8. If equipped, attach the automatic level control air lines to the shock absorber.

Coil Springs

REMOVAL & INSTALLATION

Front—1999–01 2WD Vehicles

1. Before servicing the vehicle, refer to the precautions in the beginning of this section.

2. Remove or disconnect the following:
- Wheel
- Shock absorber lower bolts

3. Push the shock absorber through the control arm and into the spring.

4. With the vehicle supported so the control arms hang free, install tool J-23028, onto a support and into the lower control arm bushings.
- Stabilizer bar from the control arm
- Stabilizer from the lower control arm

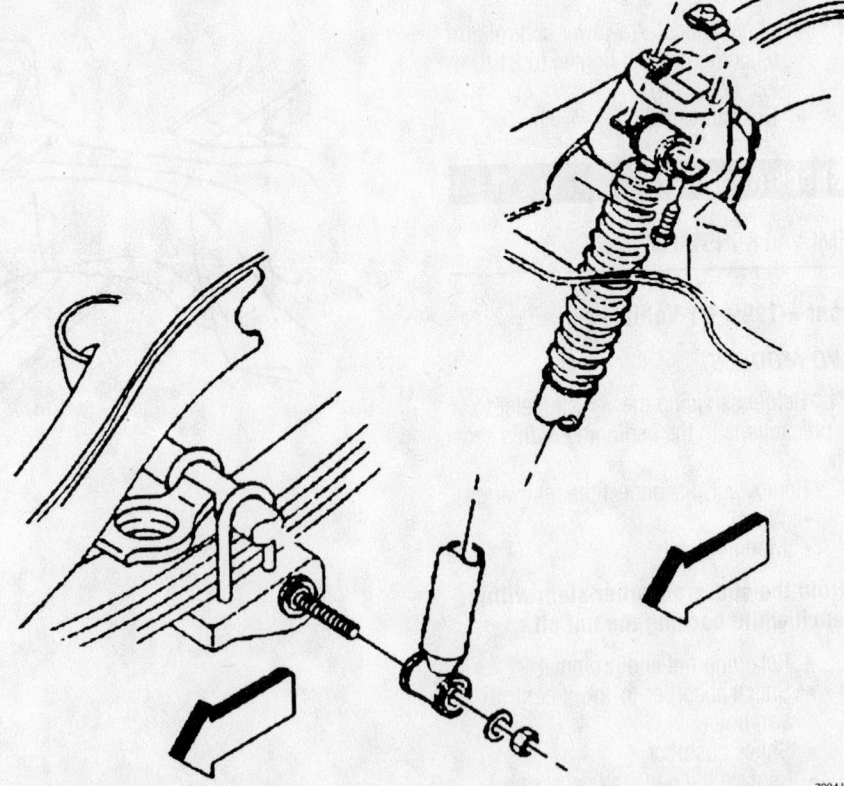

Rear shock absorber mounting

5. Raise and remove the tension on the lower control arm bolts.

6. Install a safety chain around the spring and through the lower control arm.
- Lower control arm pivot bolts, the rear first
- Lower control arm and allow it to hang free
- Spring assembly

To install:

➡ **When positioning the spring in the lower control arm, be sure the spring insulator is in the proper position before lifting the control arm in place.**

7. Install or connect the following:
- Spring assembly
- Lower control arm
- Lower control arm pivot bolts
- Stabilizer to the lower control arm

Front—2002–03 Vehicles

➡ **This procedure requires the use of a suitable spring compressor.**

1. Before servicing the vehicle, refer to the precautions in the beginning of this section.

2. Remove or disconnect the following:
- Wheel
- Shock module
- Shock module yoke-to-shock absorber pinch bolt and nut

3. Spread the shock module yoke at the pinch bolt using a suitable flat-bladed tool.
- Shock module yoke from the shock absorber

4. Install pieces of heater hose or equivalent material to the shock module spring where the spring compressor contacts the lower part of the spring.

5. Install the shock module into the spring compressor.

➡ **The spring is compressed when the shock absorber moves freely.**

6. Turn the spring compressor forcing screw until the coil spring is compressed.

7. Remove or disconnect the following:
- Shock absorber upper retaining nut
- Shock absorber from the shock module

8. Loosen the compressor forcing screw until the upper mounting plate and coil spring can be removed.
- Upper mounting plate and coil spring from the spring compressor

To install:

9. Install or connect the following:
- Coil spring and upper mounting plate to the spring compressor

10. Turn the compressor forcing screw until the coil spring is compressed.
- Shock absorber to the shock module. Tighten the retaining nut to 33 ft. lbs. (45 Nm).

11. Remove the shock module from the spring compressor. Remove the pieces of heater hose from the spring.
- Shock module yoke to the shock absorber
- Shock module yoke-to-shock pinch bolt and nut and tighten to 52 ft. lbs. (70 Nm)
- Shock module to the vehicle
- Tire and wheel

12. Lower the vehicle

Rear—2002–03 Vehicles

1. Before servicing the vehicle, refer to the precautions in the beginning of this section.

2. Raise and support the vehicle.

3. Support the rear axle.

4. Remove the shock absorber lower mounting bolts.

5. Lower the rear axle, then remove the coil springs.

To install:

6. Install the coil springs, then raise the rear axle.

7. Install the shock absorber lower mounting bolts and tighten to 59 ft. lbs. (80 Nm).

8. Remove the rear axle support.

9. Lower the vehicle.

Leaf Springs

REMOVAL & INSTALLATION

Rear—1999–01 Vehicles

1. Before servicing the vehicle, refer to the precautions in the beginning of this section.

➡ **The following procedure requires the use of two sets of jackstands.**

2. Support the rear axle with jackstands, support the axle and the body separately in order to relieve the load on the rear spring.

3. Remove or disconnect the following:
- Wheel
- Shock absorber
- U-bolt nuts, washers, anchor plate and bolts
- Spare tire, if equipped

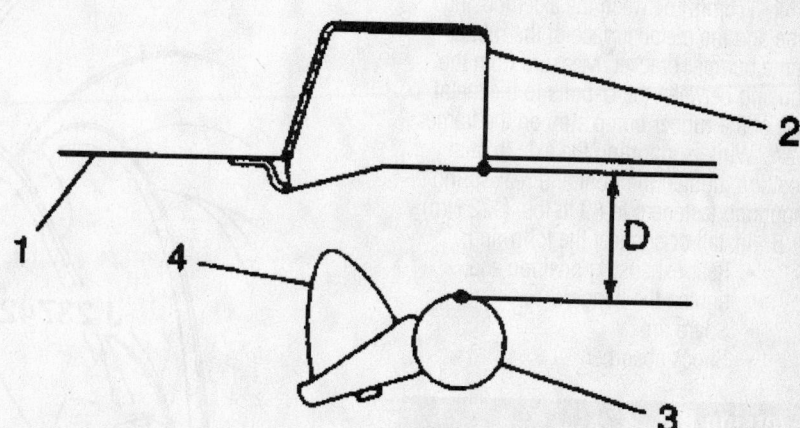

1. FRAME
2. AXLE STOP BRACKET
3. REAR AXLE (END VIEW)
4. BUMPER
5. TRIM HEIGHT 135 – 145MM (5.3 – 5.7 INCHES)

88268GB5

Typical rear suspension trim height measurement

For Wheel Alignment specifications, see Section 1 of this manual

- Rear exhaust hangers and lower the rear exhaust, if necessary
- Shackle-to-frame bolt, washers and nut
- Fuel tank, if necessary
- Front bracket nut, washers and bolt
- Spring
- Shackle from the spring, if necessary

To install:

4. Install or connect the following:
- Shackle to the rearward spring eye using the bolt, washers and nut, but do not fully tighten at this time
- Spring assembly
- Spring to the front bracket using the bolt, washers and nut, but do not fully tighten at this time
- Fuel tank, if removed
- Shackle-to-frame bolt, washers and nut, but do not fully tighten at this time. If used, remove the spring support.

5. Install the U-bolts, anchor plate, washers and U-bolt nuts. Torque the nuts using 2 passes of a diagonal sequence:
- a. Step 1: Torque to 18 ft. lbs. (25 Nm).
- b. Step 2: Torque to 73 ft. lbs. (100 Nm) in the sequence.

6. Position the axle to achieve an approximate gap of 6.46–6.94 in. (164–176mm) between the axle housing tube and the metal surface of the rubber frame bumper bracket. Measure from the housing between the U-bolts to the metal part of the rubber bump stop on the frame.

7. While supporting the axle in this position, tighten the front and rear spring mounting fasteners to 89 ft. lbs. (122 Nm).

8. Install or connect the following:
- Rear exhaust in position and tighten the hangers
- Spare tire
- Shock absorber

Torsion Bar

Instead of the coil spring used on the front suspension of 1999–01 2WD vehicles, the 4WD vehicles are equipped with a torsion bar.

REMOVAL & INSTALLATION

1. Before servicing the vehicle, refer to the precautions in the beginning of this section.

➡**The following procedure requires the use of the Torsion Bar Unloader tool J-36202.**

2. Remove or disconnect the following:
- Transmission shield, if equipped
- Torsion bar unloader tool to relax the tension on the torsion bar adjusting arm screw; record the number of turns necessary to properly install the tool. Remove the adjusting screw and the unloader tool.
- Lower link mount nut from one side
- Torsion bars by disengaging them

➡**Note the direction of the forward end and side of the torsion bar being removed**

- Lower link nut from the opposite side
- Lower link mount, upper link mount nut
- Upper link mount
- Torsion bar from the frame

To install:

3. Install or connect the following:
- Torsion bar and support
- Upper link mount. Torque the nut to 48 ft. lbs. (68 Nm).

4. Place a jack under the torsion bar to release tension.

5. Install or connect the following:
- Lower link mount bushing and nut. Torque the nut to 37 ft. lbs. (50 Nm).

- Torsion bar unloader tool. Tighten the tool against the adjusting arm the same number turns recorded earlier and remove the tool. This loads the torsion bars.
- Transmission shield, if removed

Upper Ball Joints

REMOVAL & INSTALLATION

1999–2001 2WD Vehicles

1. Before servicing the vehicle, refer to the precautions in the beginning of this section.

➡**The following procedure requires the use of a ball joint separator tool such as J-23742 and J-9519-E ball joint remover and installer set.**

2. Raise and support the front of the vehicle safely by placing stands securely under the lower control arms. Because the vehicle's weight is used to relieve spring tension on the upper control arm, the stands must be positioned between the spring seats and the lower control arm ball joints for maximum leverage.

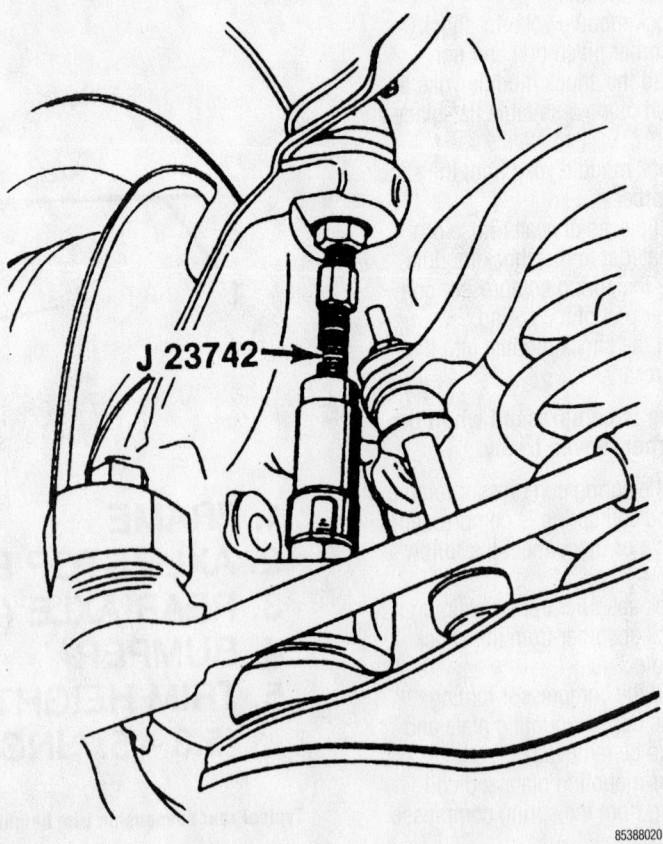

85388020

Use a ball joint separator tool to drive the upper ball joint from the steering knuckle

✽✽ CAUTION

With components unbolted, the stand is holding the lower control arm in place against the coil spring. Make sure the stand is firmly positioned and cannot move, or personal injury could result.

3. Remove or disconnect the following:
   • Tire and wheel assembly
   • Brake caliper and support it from the vehicle using a coat hanger or wire. Make sure the brake line is not stretched or damaged and that the caliper's weight is not supported by the line.
   • Cotter pin and retaining nut from the upper ball joint
   • Anti-lock brake sensor wire bracket, if equipped
   • Upper ball joint from the steering knuckle using tool J-23742 and pull the steering knuckle free of the ball joint

➡ After separating the steering knuckle from the upper ball joint, be sure to support the steering knuckle/hub assembly to prevent damaging the brake hose.

4. Remove the riveted upper ball joint from the upper control arm as follows:
   a. Drill a ⅛ in. (3mm) hole, about ¼ in. (6mm) deep into each rivet.
   b. Then use a ½ in. (13mm) drill bit, to drill off the rivet heads.
   c. Using a pin punch and the hammer, drive out the rivets in order to free the upper ball joint from the upper control arm assembly, then remove the upper ball joint.

5. Clean and inspect the steering knuckle hole. Replace the steering knuckle if the hole is out of round.

To install:

6. Install or connect the following:
   • Ball joint in the upper control arm
   • Ball joint retaining nuts and bolts. Position the bolts threaded upward from under the control arm. Tighten the ball joint retainers to 17 ft. lbs. (23 Nm).
   • Anti-lock brake sensor wire bracket, if removed
   • Ball joint to the knuckle. Make sure the joint is seated, then install the stud nut and tighten to 61 ft. lbs. (83 Nm). Insert a new cotter pin.

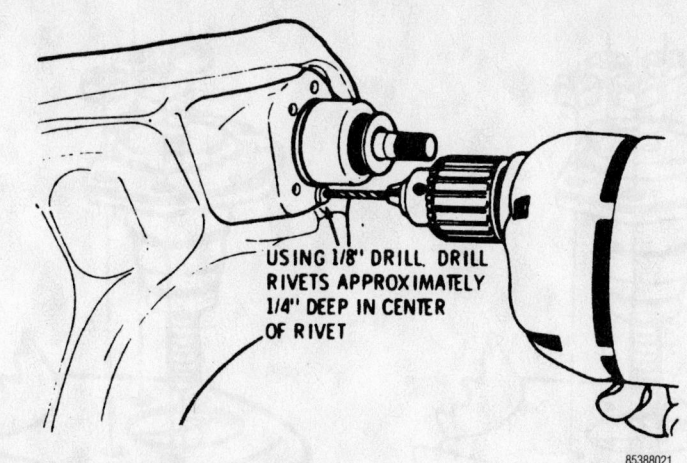

USING 1/8" DRILL. DRILL RIVETS APPROXIMATELY 1/4" DEEP IN CENTER OF RIVET

85388021

Drill a small guide hole into each ball joint rivet

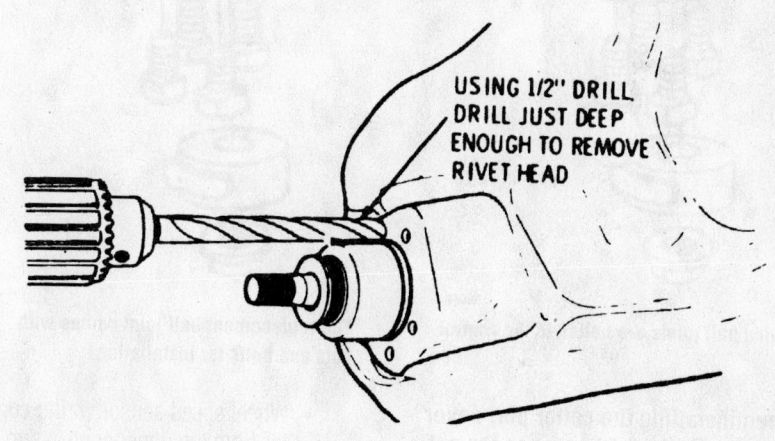

USING 1/2" DRILL DRILL JUST DEEP ENOUGH TO REMOVE RIVET HEAD

85388022

Then drill off the rivet heads

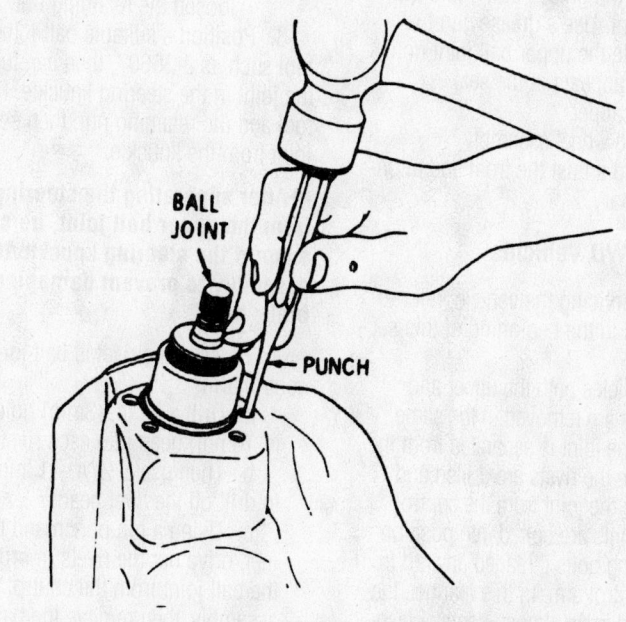

BALL JOINT

PUNCH

85388023

Punch the rivets out and remove the ball joint

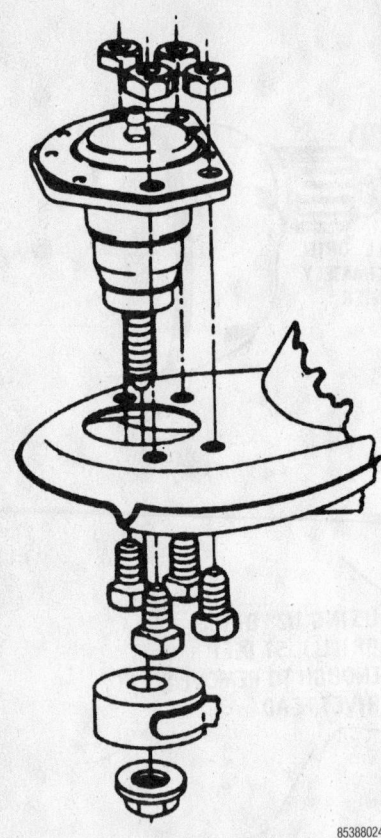

Service ball joints are bolted to the control arm

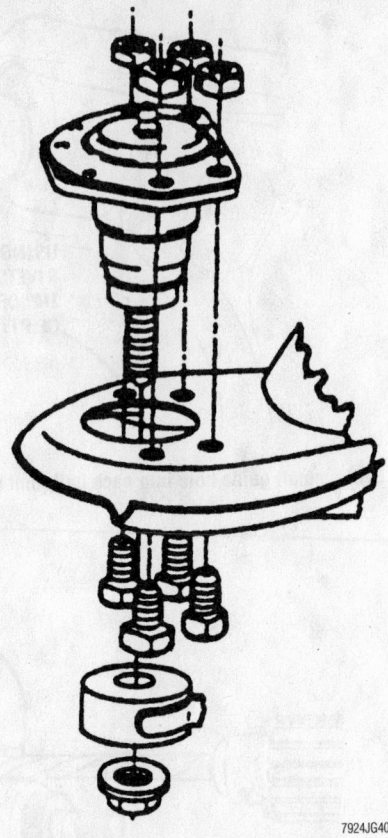

The replacement ball joint comes with nuts and bolts for installation

➡**When installing the cotter pin, never loosen the castle nut to expose the cotter pin hole.**

- Thread the grease fitting into the ball joint. Use a grease gun to lubricate the upper ball joint until grease appears at the seal.
- Brake caliper
- Tire and wheel assembly

7. Check and adjust the front end alignment, as necessary.

1999–2001 4WD Vehicles

1. Before servicing the vehicle, refer to the precautions in the beginning of this section.

On 4WD vehicles both the upper and lower ball joints are removed in the same manner. Once the joint is separated from the steering knuckle the rivets are drilled and punched to free the joint from the control arm. Service joints are bolted into position with the retaining bolts threaded upward from beneath the control arm. In this manner, the joint is replaced in an almost identical fashion to the upper joints on 2WD vehicles.

2. Remove or disconnect the following:
- Tire and wheel assembly

- Wheel speed sensor wiring connector from the upper control arm, if removing the upper ball joint
- Cotter pin from the ball joint, then loosen the retaining nut

3. Position a suitable ball joint separator tool such as J-36607, then carefully loosen the joint in the steering knuckle. Remove the tool and the retaining nut, then separate the joint from the knuckle.

➡**After separating the steering knuckle from the upper ball joint, be sure to support the steering knuckle/hub assembly to prevent damaging the brake hose.**

4. Remove the riveted ball joint from the control arm:

   a. Drill a ⅛ in. (3mm) hole, about ¼ in. (6mm) deep into each rivet.

   b. Then use a ½ in. (13mm) drill bit, to drill off the rivet heads.

   c. Using a pin punch and the hammer, drive out the rivets in order to free the ball joint from the control arm assembly, then remove the ball joint.

To install:

5. Install or connect the following:
- Ball joint in the control arm

- Ball joint retaining nuts and bolts. Position the bolts threaded upward from under the control arm. Tighten the ball joint retainers to 17 ft. lbs. (23 Nm).
- Ball joint to the knuckle. Make sure the joint is seated, tighten the lower nut to 79 ft. lbs. (108 Nm) and the upper nut to 61 ft. lbs. (83 Nm). Install a new cotter pin.

➡**When installing the cotter pin, never loosen the castle nut to expose the cotter pin hole, but DO NOT tighten more than an additional ⅙ turn.**

6. Use a grease gun to lubricate the upper ball joint.
- Wheel speed sensor wiring connector to the upper control arm, if the upper ball joint was removed
- Tire and wheel assembly

7. Check and adjust the front end alignment, as necessary.

2002–03 Vehicles

➡**This procedure requires the use of the following special tools: J 9519-E Lower Ball Joint Remover and Installer, J 21474-01 Control Arm Bushing Set and J 45117 Ball Joint Installation Spacer.**

1. On 4WD vehicles, remove the wheel center cap and drive axle nut.
2. Raise and support the vehicle.
3. Remove or disconnect the following:
- Tire and wheel
- Wheel hub and bearing, if necessary
- Outer tie rod retaining nut
- Out tie rod from the steering knuckle using a suitable puller
- Brake hose bracket retaining bolts and bracket
- Upper control arm-to-steering knuckle pinch bolt and nut
- Upper control arm from the steering knuckle
- Lower ball joint retaining nut
- Steering knuckle from the lower control arm using a suitable ball joint removal tool
- Steering knuckle from the vehicle
- Upper ball joint retaining clip
- Upper ball joint from the steering knuckle using Lower Ball Joint Removal and Installer tool No. J 9519-E

To install:

4. Install or connect the following:
- Upper ball joint to the steering knuckle using J 9519-E, J 21474-01 and J 45117

- Upper ball joint retaining clip
- Steering knuckle to the lower control arm
- Lower ball joint retaining nut and tighten to 81 ft. lbs. (110 Nm)
- Upper control arm to the steering knuckle
- Upper control arm pinch bolt and nut and tighten to 30 ft. lbs. (41 Nm)
- Brake hose bracket to the steering knuckle
- Brake hose bracket retaining nuts and tighten to 7 ft. lbs. (10 Nm)
- Outer tie rod to the steering knuckle
- Outer tie rod retaining nut and tighten to 33 ft. lbs. (45 Nm)
- Wheel hub and bearing, if removed
- Tire and wheel

5. Lower the vehicle
- Drive axle nut, if 4WD, and tighten to 103 ft. lbs. (140 Nm)
- Wheel enter cap, if removed

6. Check the front wheel alignment.

Lower Ball Joints

REMOVAL & INSTALLATION

1999–01 2WD Vehicles

1. Before servicing the vehicle, refer to the precautions in the beginning of this section.

➡The following procedure requires the use of a ball joint remover/installer set (the particular set may vary upon application but must include a clamping-type tool with the appropriately sized adapters) and a ball joint separator tool, such as J-23742.

2. Remove the tire and wheel assembly.
3. Position a jack under the spring seat of the lower control arm, then raise the jack to support the arm.

✳✳ CAUTION

The jack MUST remain under the lower control arm, during the removal and installation procedures, to retain the arm and spring positions. Make sure the jack is securely positioned and will not slip or release during the procedure or personal injury may result.

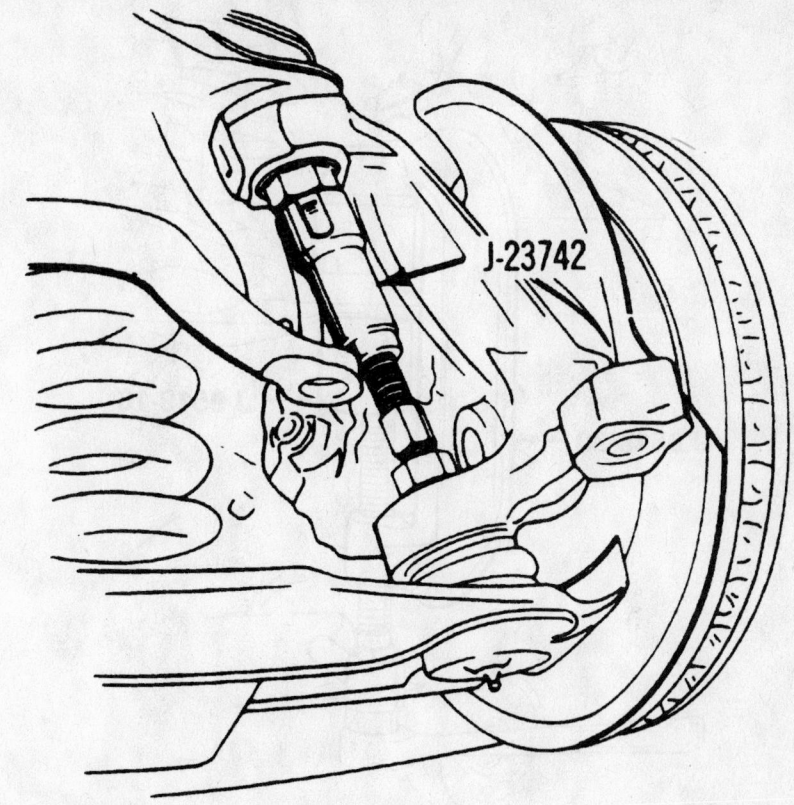

85388025

Use a ball joint separator to drive the lower joint from the knuckle

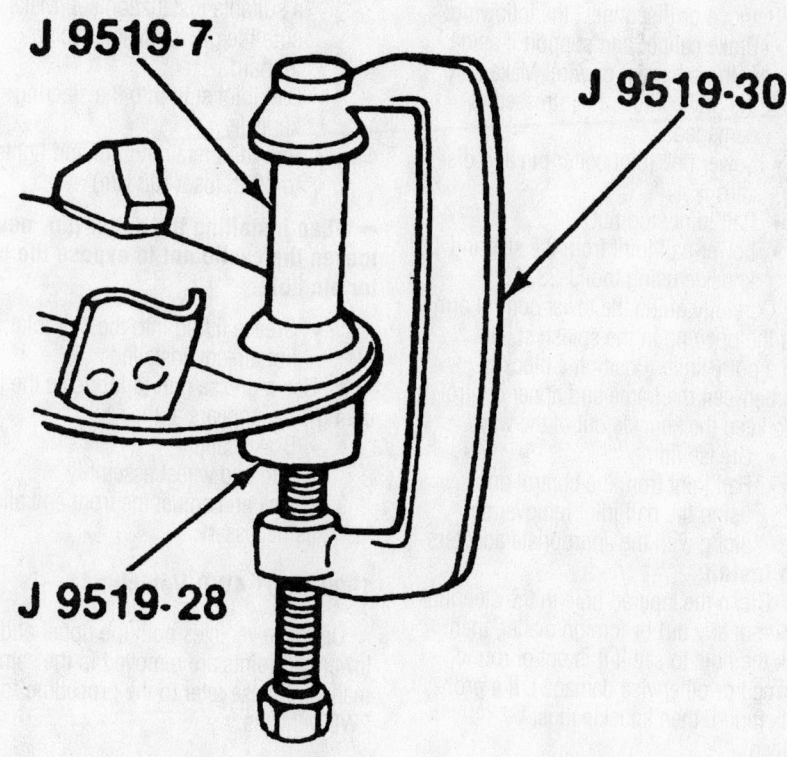

85388027

Driving the lower joint from the control arm

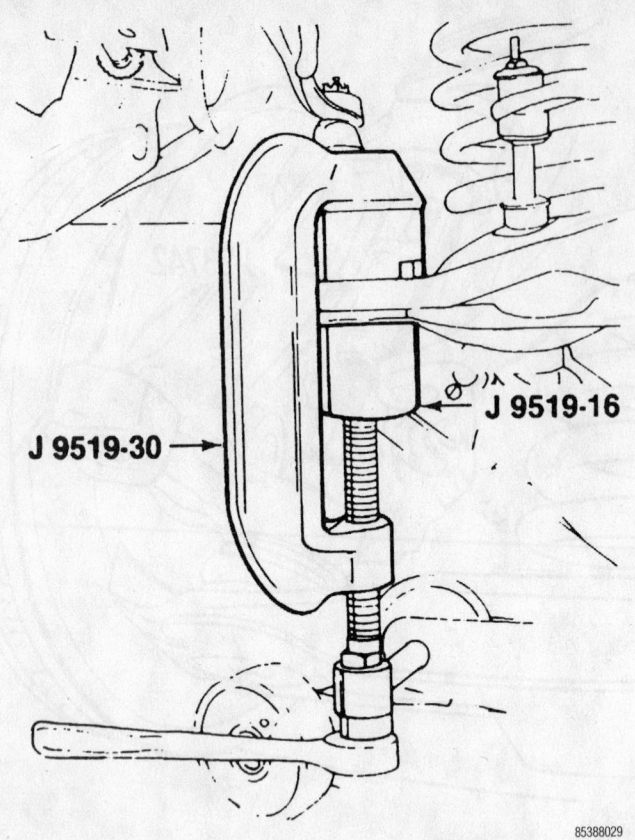

J 9519-30 →

← J 9519-16

85388029

Installing a new ball joint

4. Remove or disconnect the following:
   • Brake caliper and support it aside using a hanger or wire. Make sure the brake line is not stressed or damaged.
   • Lower ball joint cotter pin and discard
   • Ball joint stud nut
   • Lower ball joint from the steering knuckle using tool J-23742

5. Carefully guide the lower control arm out of the opening in the splash shield using a putty knife. Position a block of wood between the frame and upper control arm to keep the knuckle out of the way.
   • Grease fitting
   • Ball joint from the control arm using the ball joint remover set along with the appropriate adapters

To install:

6. Clean the tapered hole in the steering knuckle of any dirt or foreign matter, then check the hole to see if it is out of round, deformed or otherwise damaged. If a problem is found, then knuckle must be replaced.

7. Install or connect the following:
   • Press the new ball joint (with grease fitting pointing inward) until it bottoms in the control arm using

a suitable installation set. Make sure the grease seal is facing inboard.
   • Ball joint stud into the steering knuckle
   • Ball joint retaining nut and tighten to 79 ft. lbs. (108 Nm)

➡ **When installing the cotter pin, never loosen the castle nut to expose the cotter pin hole.**

   • Grease fitting into the ball joint, if not already installed

8. Use a grease gun to lubricate the joint until grease appears at the seal.
   • Brake caliper
   • Tire and wheel assembly

9. Check and adjust the front end alignment, as necessary.

1999–2001 4WD Vehicles

On these vehicles both the upper and lower ball joints are removed in the same manner. Please refer to the procedure for the 2WD vehicles.

2002–03 Vehicles

➡ **This procedure requires the use of the following special tools: J 9519-E Lower Ball Joint Remover and Installer,**

J 34874 Booster Seal Remover/Installer, J 41435 Ball Joint Installer, J 45105-1 Ball Joint Flaring Adapter and J 45105-2 Receiver.

1. On 4WD vehicles, remove the wheel center cap and drive axle nut.
2. Raise and support the vehicle.
3. Remove or disconnect the following:
   • Tire and wheel
   • Wheel hub and bearing, if necessary
   • Outer tie rod retaining nut
   • Out tie rod from the steering knuckle using a suitable puller
   • Brake hose bracket retaining bolts and bracket
   • Upper control arm-to-steering knuckle pinch bolt and nut
   • Upper control arm from the steering knuckle
   • Lower ball joint retaining nut
   • Steering knuckle from the lower control arm using a suitable ball joint removal tool
   • Steering knuckle from the vehicle
   • Lower ball joint flange with a chisel

4. Install tools J 9519-E and J 34874 to the lower ball joint, then use those tools to remove the lower ball joint from the lower control arm.

To install:

5. Install or connect the following:
   • Lower ball joint to the lower control arm, using tools J 9519-E, J 41435 and J 45105-2

6. Remove the tools from the lower control arm.
   • Tools J 9519-E and J 45105-1 to the lower ball joint

7. Flare the lower ball joint flange with J 9519-E and J 45105-1, then remove the tools from the lower ball joint.
   • Steering knuckle to the lower control arm
   • Lower ball joint retaining nut and tighten to 81 ft. lbs. (110 Nm)
   • Upper control arm to the steering knuckle
   • Upper control arm pinch bolt and nut and tighten to 30 ft. lbs. (41 Nm)
   • Brake hose bracket to the steering knuckle
   • Brake hose bracket retaining nuts and tighten to 7 ft. lbs. (10 Nm)
   • Outer tie rod to the steering knuckle
   • Outer tie rod retaining nut and tighten to 33 ft. lbs. (45 Nm)
   • Wheel hub and bearing, if removed
   • Tire and wheel

8. Lower the vehicle
   • Drive axle nut, if 4WD, and tighten to 103 ft. lbs. (140 Nm)
   • Wheel enter cap, if removed
9. Check the front wheel alignment.

Upper Control Arm

REMOVAL & INSTALLATION

1999–01 Vehicles

2WD MODELS

1. Before servicing the vehicle, refer to the precautions in the beginning of this section.
2. Remove or disconnect the following:
   • Negative battery cable
   • Wheel
   • Wheel speed sensor harness bracket retaining bolt and nut, if equipped
   • Steering knuckle from upper control arm ball joint
   • Mounting nuts/bolts and shims

➡**Make sure to note the location of the control arm shims prior to removal so that they may be installed in their original positions.**

   • Upper control arm
To install:
3. Install or connect the following:
   • Upper control arm

➡**Always tighten nut on the thinner shim pack first.**

   • Mounting nuts/bolts and shims. Torque the nuts to 81–85 ft. lbs. (110–115 Nm).
   • Steering knuckle to upper control arm ball joint
   • New cotter pin

➡**Tighten the nut to align the hole never loosen.**

   • Wheel speed sensor harness bracket retaining bolt and nut, if equipped
   • Wheel

4WD MODELS

1. Before servicing the vehicle, refer to the precautions in the beginning of this section.
2. Remove or disconnect the following:
   • Tire and wheel assembly
   • Cotter pin from the ball joint, then loosen the retaining nut

   • Steering knuckle from the upper ball joint. Be sure to support the steering knuckle/hub assembly to prevent damaging the brake hose.

➡**The 4WD vehicles do not use shims to adjust the front wheel alignment. Instead, the upper control arm bolts are equipped with cams, which are rotated to achieve caster and camber adjustments. In order to preserve adjustment and ease installation, matchmark the cams to the control arm before removal. If the control arm is being replaced, transfer the alignment marks to the new component before installation.**

   • Front and rear nuts retaining the control arm retaining bolts to the frame
   • Outer cams from the bolts
   • Bolts and inner cams
   • Control arm from the vehicle
   • Retaining nut and the bumper from the control arm, if necessary
3. If the bushings are being replaced, use a suitable bushing service set to remove the bushings from the arm.
To install:
4. Install or connect the following:
   • Bushing service set to drive the new bushings into the control arm, if removed
   • Bumper and retaining nut to the control arm, if removed. Tighten the bumper retaining nut to 20 ft. lbs. (27 Nm).
   • Control arm, retaining bolts (from the inside of the frame brackets facing outward) and the inner cams. The inner cams must be positioned on the bolts before they are inserted through the control arm and frame brackets.
   • Outer cams over the retaining bolts, then the nuts to the ends of the bolts at the front and rear of the control arm
5. Align the cams to the reference marks made earlier, then tighten the end nuts to 85 ft. lbs. (115 Nm).
   • Ball joint to the knuckle
   • Tire and wheel assembly
6. Check and adjust the front end alignment, as necessary.

2002–03 Vehicles

1. Before servicing the vehicle, refer to the precautions in the beginning of this section.

2. Remove or disconnect the following:
   • Tire and wheel assembly
   • Upper ball joint-to-upper control arm pinch bolt and nut
   • Upper control arm from the knuckle
   • Anti-lock Brake System (ABS) wheel speed sensor wiring harness
   • Upper control arm mounting bolts
   • Upper control arm
To install:
3. Install or connect the following:
   • Upper control arm and tighten the bolts to 111 ft. lbs. (150 Nm)
   • ABS wheel speed sensor wiring harness
   • Upper control arm to the steering knuckle
   • Upper ball joint-to-upper control arm pinch bolt and nut and tighten to 30 ft. lbs. (40 Nm)
   • Tire and wheel
4. Check the front wheel alignment.

CONTROL ARM BUSHING REPLACEMENT

2WD Vehicles

1. Before servicing the vehicle, refer to the precautions in the beginning of this section.

2. Remove or disconnect the following:
   • Upper control arm and place it in a vice
   • Upper control arm shaft nuts and retainers
   • Upper control arm bushings using tool J 22269-1, a slotted washer and a short piece if pipe that is slightly larger than the bushing
   • Upper control arm shaft
To install:
   • Upper control arm shaft
   • Upper control arm bushings using tool J 22269-1, a slotted washer and a short piece if pipe that is slightly larger than the bushing
3. Tighten J 22269-1 until the bushing is positioned on the shaft and the control arm as shown in the accompanying illustration. The measurement should be 0.48–0.52 inch (12.8–13.8mm) at both sides when the properly installed.
   • Upper control arm shaft nuts and retainers. Tighten to 85 ft. lbs. (115 Nm).
   • Upper control arm

4WD Vehicles

If the bushings require replacement, refer to the control arm removal and installation procedure for bushing replacement.

Lower Control Arm

REMOVAL & INSTALLATION

1999–01 Vehicles

2WD MODELS

1. Before servicing the vehicle, refer to the precautions in the beginning of this section.
2. Remove or disconnect the following:
   - Coil Spring
   - Lower ball joint from the steering knuckle
   - Lower control arm from the vehicle

To install:

3. Install or connect the following:
   - Lower control arm
   - Lower ball joint stud into the steering knuckle
   - Ball joint-to-steering knuckle nut and tighten to specification
   - New cotter pin to the lower ball joint stud
   - Coil spring
4. Align the vehicle.

4WD MODELS

1. Before servicing the vehicle, refer to the precautions in the beginning of this section.

➡Tools Needed: universal tie rod separator J–24319–01, torsion bar unloader J–36202, lower control arm bushing service kit J–36618 (if the control arm bushing are being replaced) and ball

joint C-clamp J–9519–23. Parts Needed: whether or not the control arm or bushing are being replaced, NEW control arm retaining nut should be used once the old ones have been loosened and removed.

2. Remove or disconnect the following:
   - Front wheels
   - 2 bolts from the front splash shield and pivot it in order to gain access to the tie rod
   - Stabilizer bar from the control arm (keeping all of the link hardware sorted for proper installation). If necessary, completely remove the bar from the vehicle for access.
   - Shock absorber
   - Inner tie rod from the relay rod using a tie rod separator
   - Outer halfshaft nut and washer
   - Bolts from the hub and bearing kit

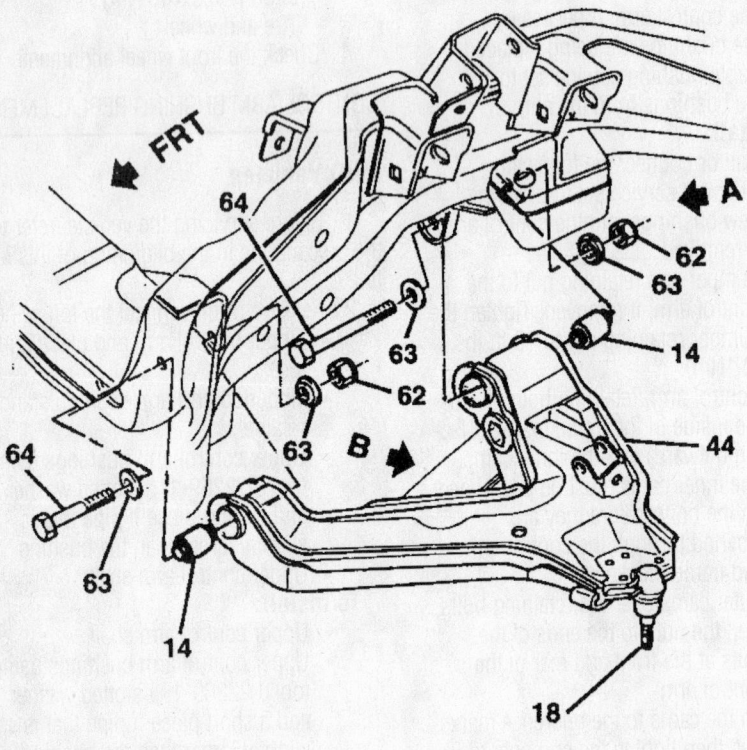

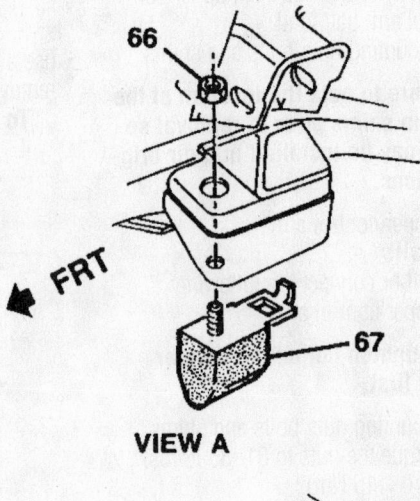

VIEW A

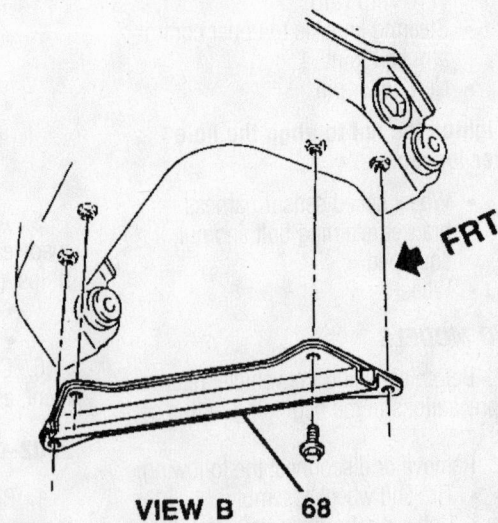

VIEW B

14. BUSHING
18. BALL JOINT, LOWER
44. ARM KIT, LOWER
62. NUT
63. WASHER
64. BOLT
66. NUT
67. BUMPER
68. BRACE

Exploded view of the lower control arm assembly mounting

88268GB1

3. Unload the torsion bar using the unloading tool J–36202. First, mark the adjuster for installation.

- Adjustment arm. Slide the bar forward and the adapter out of the rear to remove the adjusting arm.
- Lower ball joint cotter pin, nut and ball joint from the control arm using a ball joint separator
- Nuts and bolts and lower control arm with the torsion bar assembly. Note the direction which the control arm retaining bolts are facing for installation purposes.

To install:

4. Install or connect the following:

- Torsion bar to the lower control arm and place the assembly into the vehicle. Position the front leg of the lower control arm into the crossmember before installing the rear leg into the frame bracket.
- Control arm bolts (facing in the direction as noted during removal or shown in the accompanying illustration) with NEW nuts.

➡**The control arm retainers MUST be tightened with the vehicle suspension at normal ride height. This can either be accomplished by starting the nuts now, then installing the remaining components along with the wheels and lowering the vehicle, or by moving jackstands under the ends of the lower control arms and resting the vehicle on them. If the latter solution is tried, make sure front suspension is at actual ride height compression. If you are unsure, it is best to start the nuts now and tighten them to specification once the vehicle is lowered.**

- Ball joint stud in the knuckle

5. With the suspension at the correct height, tighten the control arm retaining nuts to 98 ft. lbs. (133 Nm).

➡**The lower ball joint retaining nut MUST be tightened with the vehicle suspension at normal ride height. This can either be accomplished by starting the nut now, then installing the remaining components along with the wheels and lowering the vehicle, or by moving jackstands under the ends of the lower control arms and resting the vehicle on them. If the latter solution is tried, make sure the FULL WEIGHT of the vehicle front end is on the suspension.**

6. Install or connect the following:

- Joint-to-control arm nut, then tighten the nut to 92 ft. lbs. (125 Nm) with the suspension at normal ride height and compression.
- New cotter pin to the castellated nut. Tighten the nut (but no more than an additional ⅛ turn) in order to align the cotter pin. DO NOT loosen the nut from the specified torque.
- Adjuster arm by sliding the adapter forward, over the torsion bar to install the sides of the nut. Load the torsion bar and install the adjuster bolt aligning the installation mark.
- Drive axle through the hub and bearing assembly
- Tighten the hub and bearing assembly retaining bolts
- Drive axle shaft nut and washer
- Inner tie rod end to the relay rod
- Shock absorber
- Stabilizer bar, if removed
- Stabilizer link(s) to the control arm(s)
- Splash shield
- Front wheels

7. Recheck all fasteners for proper torque and installation before road testing.

8. Refill the differential if any fluid was lost.

9. Check and adjust the front end alignment, as necessary.

2002–03 Vehicles

1. Before servicing the vehicle, refer to the precautions in the beginning of this section.

2. Raise the vehicle.

3. Remove or disconnect the following:

- Tire and wheel
- Upper ball joint-to-upper control arm pinch bolt and nut
- Upper control arm from the steering knuckle
- Anti-lock Brake System (ABS) wheel speed sensor harness
- Upper control arm bolt and control arm

To install:

4. Install or connect the following:

- Upper control arm and bolts. Tighten to 111 ft. lbs. (150 Nm).
- Connect the ABS wheel speed sensor wiring harness
- Upper control arm to steering knuckle. Tighten the pinch bolt to 30 ft. lbs. (40 Nm).
- Tire and wheel

5. Lower the vehicle and check the front end alignment.

CONTROL ARM BUSHING REPLACEMENT

1999–01 Vehicles

2WD MODELS

1. Before servicing the vehicle, refer to the precautions in the beginning of this section.

2. Remove lower control arm and place it in vise.

3. Install tools J 22269-01, 21474-8, 12 and 13 on the rear bushing and tighten until the bushing is removed.

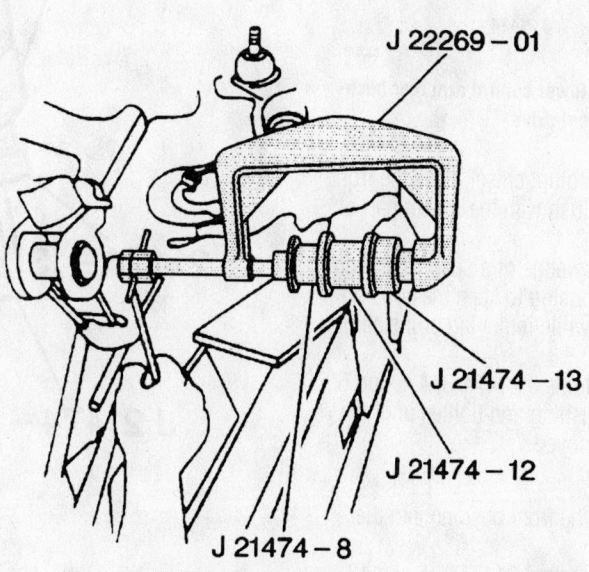

Removing the lower control arm rear bushing—all 2 wheel drive

9308JG06

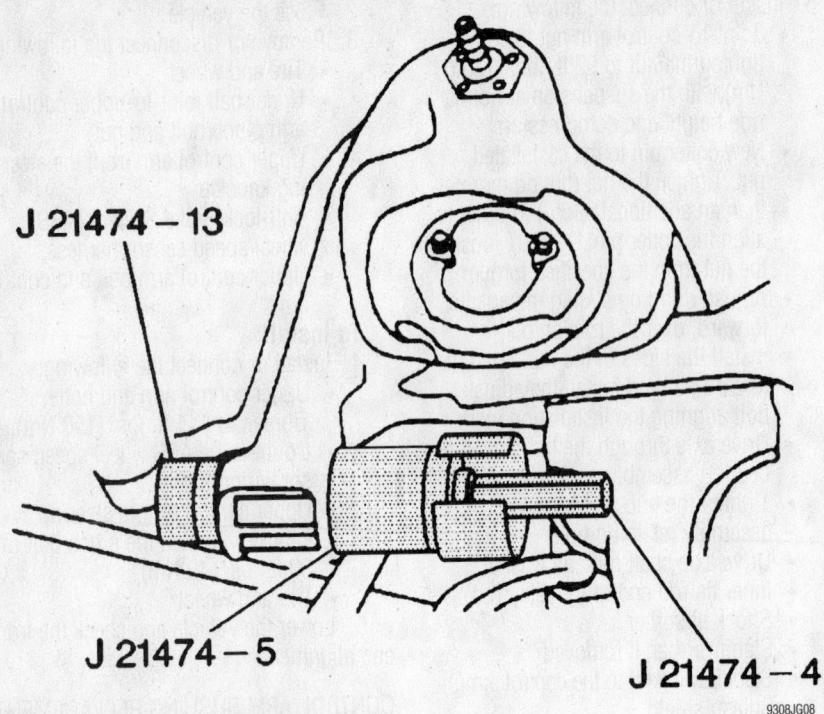

Installing the lower control arm front bushing—all 2 wheel drive

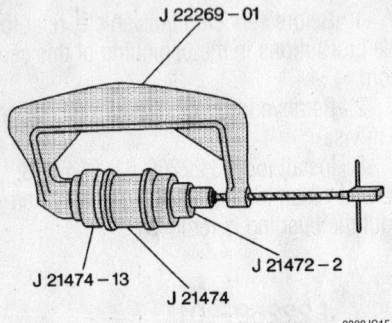

Installing the lower control arm rear bushing—all 2 wheel drive

4. Using a blunt chisel, drive the front bushing flare flush with the rubber part of the bushing.

5. Place a wedge or a spacer between the bushing housing to keep the housing from bending while removing or installing the bushing.

6. Install tools J 21474-3, 4, 5 and 6 on the front bushing and tighten until the bushing is removed.

To install:

7. Install the front bushing into the control arm.

8. Install tools J 21474-4, 5 and 13. Tighten until the bushing is fully seated.

9. Install the rear bushing into the control arm

10. Install tools J 22269-01, J 21474-2 and 13. Tighten until the bushing is fully seated.

11. Install the lower control arm.

4WD MODELS

1. Before servicing the vehicle, refer to the precautions in the beginning of this section.

2. Remove lower control arm and place it in vise.

3. Using bushing service set J 21474, remove the front and rear bushings.

To install:

4. Using bushing service set J 21474, install the front and rear bushings.

5. Install the lower control arm.

2002–03 Vehicles

➡**This procedure requires the use of Steering Linkage and Tie Rod Puller tool No. J 24319-B and Ball Joint Remover tool No. J 43631.**

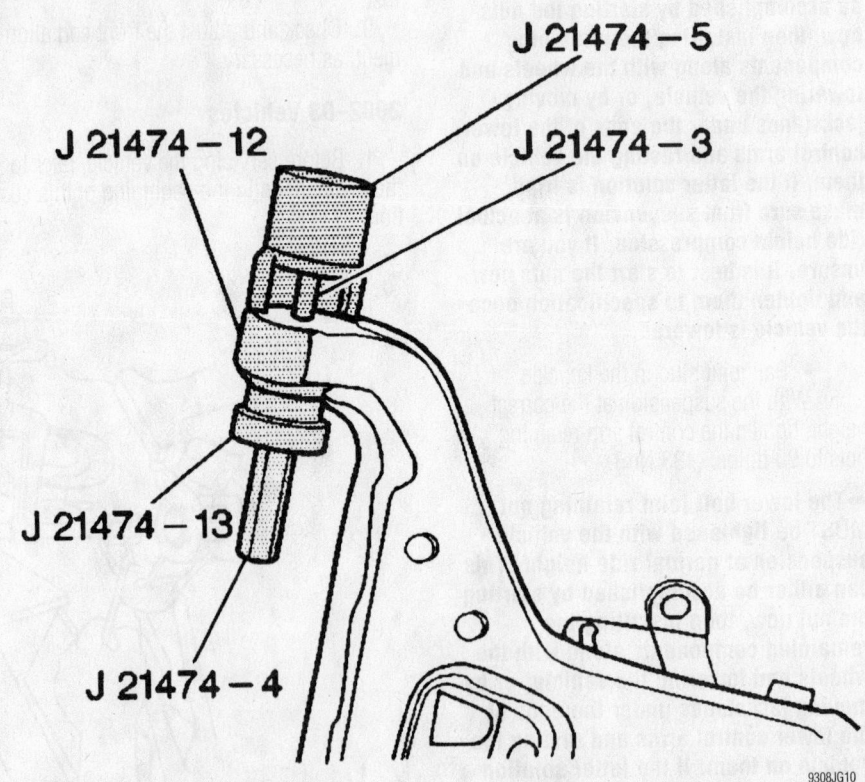

Removing the lower control arm front bushing—all 4 wheel drive

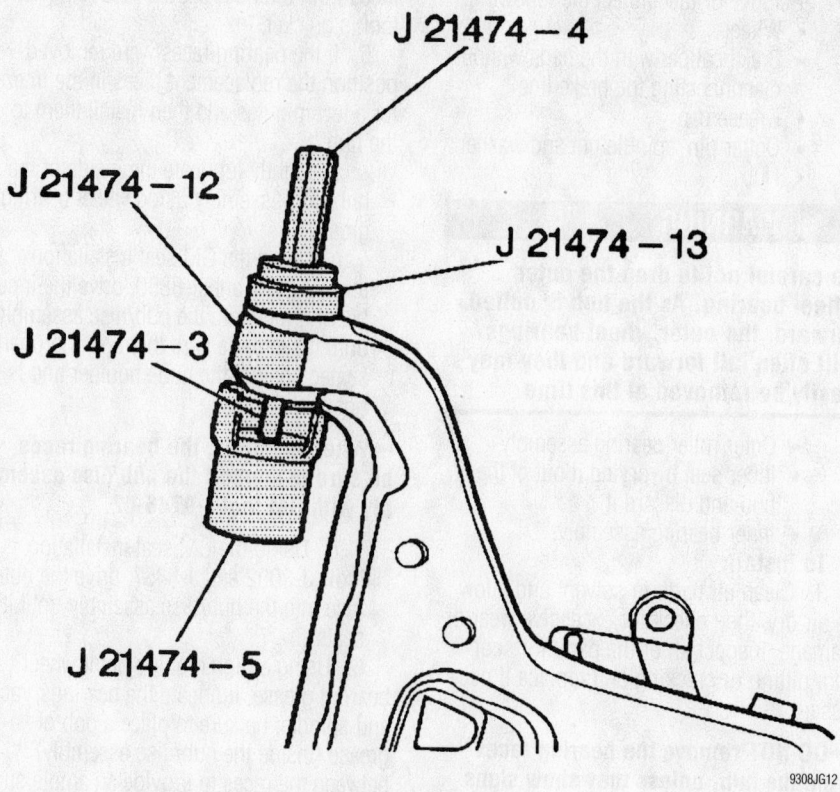

Installing the lower control arm front bushing—all 4 wheel drive

J 21474 – 4
J 21474 – 12
J 21474 – 13
J 21474 – 3
J 21474 – 5

9308JG12

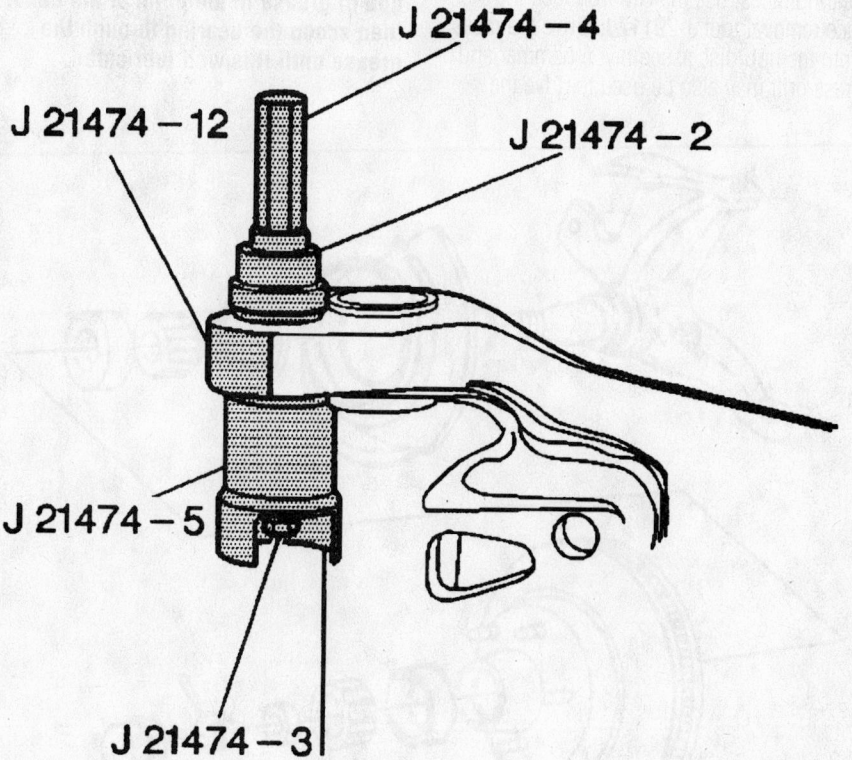

Installing the lower control arm rear bushing—all 4 wheel drive

J 21474 – 4
J 21474 – 12
J 21474 – 2
J 21474 – 5
J 21474 – 3

9308JG13

1. Before servicing the vehicle, refer to the precautions in the beginning of this section.
2. Raise the vehicle.
3. Remove or disconnect the following:
   - Tire and wheel
   - Outer tie rod retaining nut
   - Outer tie rod from the steering knuckle using Ball Joint Removal tool No. J 43631
   - Stabilizer shaft link lower nut, link and washer
   - Shock module yoke lower nut and shock module using Steering Linkage and Tie Rod Puller tool No. J 24319-B
   - Lower control arm-to-lower control arm bracket mounting bolts

➡Make sure to note the direction that the bolts are removed for installation.

   - Lower control arm-to-lower control arm bracket mounting bolts
   - Lower ball joint retaining nut
   - Lower ball joint from the steering knuckle using Ball Joint Removal tool No. J 43631

➡On 4WD vehicles, make sure not to disengage the axle shaft from the transmission.

4. Pivot the lower control arm out and down to disengage the lower control arm from the bracket, then remove the lower control arm from the knuckle.

To install:
5. Install or connect the following:
   - Lower control arm to the steering knuckle
   - Lower control to the bracket by pivoting it out and up

➡During installation and tightening of the bolts and nuts, make sure that the lower control arm is parallel to the bracket. This is to maintain proper alignment of the lower control arm bushings.

   - Lower control arm-to-bracket mounting bolts and tighten to 81 ft. lbs. (111 Nm)
   - Shock module yoke to the lower control arm
   - Shock module yoke lower mounting nut

➡If it becomes necessary to replace the washer, use only an identical hardened steel, felt lined washer. Standard washers must not be used.

- Stabilizer shaft link and washer to the lower control arm
- Stabilizer shaft link retaining bolt and tighten to 74 ft. lbs. (100 Nm)
- Outer tie rod to the steering knuckle. Tighten the nuts to 33 ft. lbs. (45 Nm).
- Tire and wheel

6. Lower the vehicle and check the front end alignment.

Wheel Bearings

ADJUSTMENT

1999–2000 Vehicles

2WD MODELS

1. Before servicing the vehicle, refer to the precautions in the beginning of this section.

2. If equipped, remove the wheel/hub cover for access, then remove the dust cap from the hub.

3. Remove the cotter pin and loosen the spindle nut.

4. Spin the wheel forward by hand and torque the nut to 12 ft. lbs. (16 Nm) in order to fully seat the bearings and remove any burrs from the threads.

5. Back off the nut until it is just loose, then finger-tighten the nut.

6. Loosen the nut ¼–½ turn until either hole in the spindle lines up with a slot in the nut, then install a new cotter pin. This may appear to be too loose, but it is the proper adjustment.

7. Proper adjustment creates 0.001–0.005 in. (0.025–0.127mm) end-play.

4WD MODELS

The front wheel bearings on the 4-wheel drive vehicles are not adjustable. If the bearings become loose or make noise, they must be replaced.

2001–03 Vehicles

The wheel bearings on these vehicles are not adjustable. If the bearings become loose or make noise, they must be replaced.

REMOVAL & INSTALLATION

Front

1999–00 2WD MODELS

1. Before servicing the vehicle, refer to the precautions in the beginning of this section.

2. Remove or disconnect the following:
- Wheel
- Brake caliper with the pads without disconnecting the brake line
- Grease cap
- Cotter pin, spindle nut and washer
- Hub

✳✳ WARNING

Be careful not to drop the outer wheel bearing. As the hub is pulled forward, the outer wheel bearings will often fall forward and they may easily be removed at this time.

- Outer roller bearing assembly
- Inner seal by prying it out of the hub and discard it
- Inner bearing assembly

To install:

3. Clean all parts in solvent and allow to air dry, then check for excessive wear or damage. Inspect all of the parts for scoring, pitting or cracking and replace if necessary.

➡**DO NOT remove the bearing races from the hub, unless they show signs of damage.**

4. If it is necessary to remove the wheel bearing races, use the GM front bearing race removal tool J-29117 to drive the races from the hub/disc assembly. A hammer and brass drift may also be used to drive the races from the hub, but the race removal tool is quicker.

5. If the bearing races were removed, position the replacement races in the freezer for a few minutes and then install them to the hub:

    a. Lightly lubricate the inside of the hub/disc assembly using wheel bearing grease.

    b. Using the GM seal installation tools J-8092 and J-8850, drive the inner bearing race into the hub/disc assembly until it seats. Be sure the race is properly seated against the hub shoulder and is not cocked.

➡**When installing the bearing races, be sure to support the hub/disc assembly with GM tool J-9746-02.**

    c. Using the GM seal installation tools J-8092 and J-8457, drive the outer race into the hub/disc assembly until it seats.

6. Using a high melting point wheel bearing grease, lubricate the bearings, races and spindle; be sure to place a gob of grease (inside the hub/disc assembly) between the races to provide an ample supply of lubricant.

➡**To lubricate each bearing, place a gob of grease in the palm of the hand, then scoop the bearing through the grease until it is well lubricated.**

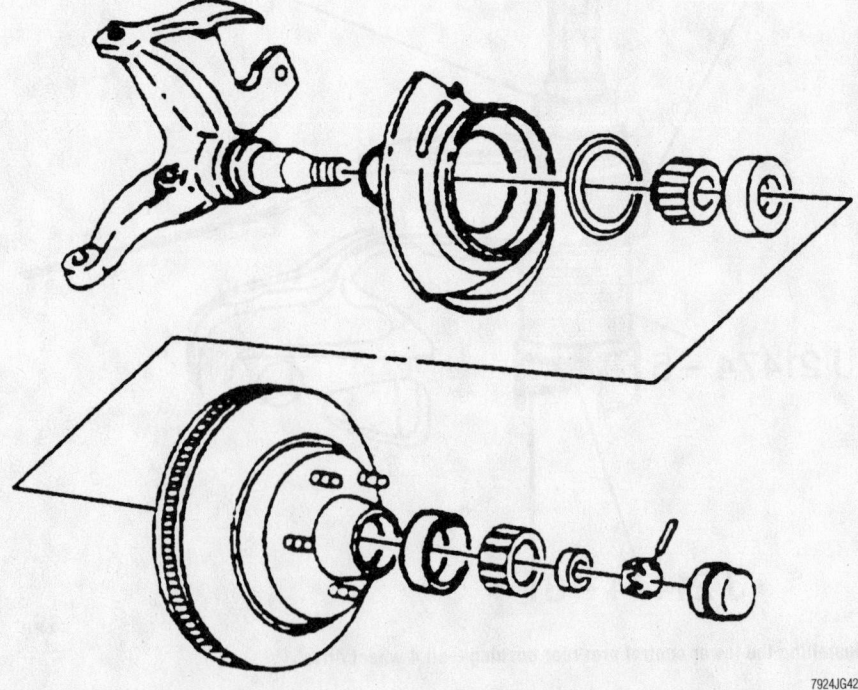

Wheel bearings, races and related components—2WD vehicles

7924JG42

7. Place the inner bearing in the hub, then apply a thin coating of grease to the sealing lip and install a new inner seal, making sure the seal flange faces the bearing cup.

➡ **Although a seal installation tool is preferable, a section of pipe with a smooth edge or a suitably sized socket may be used to drive the seal into position. Be sure the seal is flush with the outer surface of the hub assembly.**

8. Install or connect the following:
   • Wheel hub over the spindle
   • Outer bearing into the hub by hand
   • Spindle washer and nut
   • Brake caliper
   • Wheel
9. Properly adjust the wheel bearings
   • New cotter pin
   • Dust cap
   • Wheel cover

1999–00 4WD MODELS

1. Before servicing the vehicle, refer to the precautions in the beginning of this section.
2. Install Torsion Bar Unloading tool J 36202 on the torsion bar adjusting bolt and remove the bolt. To aid during installation, count the number of turns required to remove the bolt.
3. Remove the wheel.
4. Install an axle shaft boot seal protector to the Tri-pot axle joint.
5. Remove or disconnect the following:
   • Cotter pin and retainer
   • Castle nut and the thrust washer
   • Brake caliper and support it aside using wire or a coat hanger

➡ **Be sure the brake line is not stretched or damaged.**

   • Brake disc from the wheel hub
   • Halfshaft from the hub/bearing assembly, using a Spindle Remover tool J-28733-A to prevent damage to the shaft or hub/bearing assembly
   • Hub/bearing assembly from the knuckle
6. Clean and inspect the parts for nicks, scores and/or damage, then replace them as necessary.

To install:
7. Install or connect the following:
   • Hub and bearing assembly by aligning the threaded holes. Torque the bolts to 77 ft. lbs. (105 Nm).

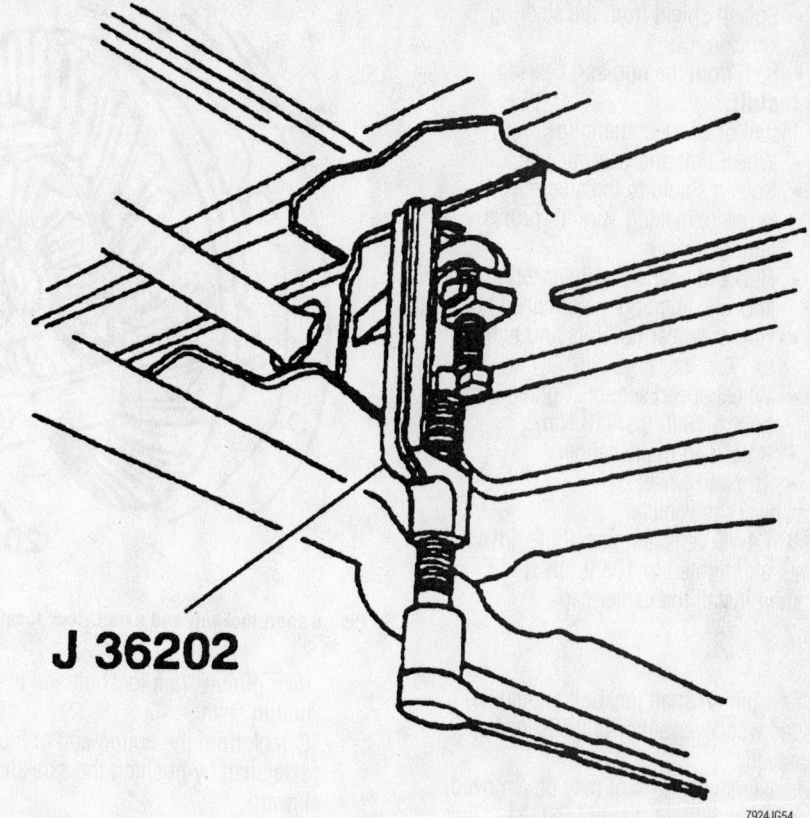

J 36202

7924JG54

Use Torsion Bar Unloading tool J 36202 to remove the adjusting bolt and unload the torsion bar

   • Tie rod end to the steering knuckle using the retaining nut
   • New cotter pin
   • Brake assembly
   • Halfshaft nut. Tighten the nut to 180 ft. lbs. (245 Nm).
   • Retainer and a new cotter pin but DO NOT back off specification in order to insert the cotter pin.
8. Remove the torsion bar unloader tool and the drive axle boot protector.
9. Install the wheel.
10. Check and/or adjust the vehicle trim height, as necessary.

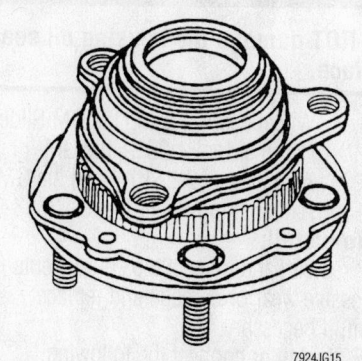

7924JG15

Hub and bearing assembly—4WD vehicles

2001–03 MODELS

1. Before servicing the vehicle, refer to the precautions in the beginning of this section.
2. On 4WD vehicles, remove wheel center cap, if equipped, and the drive axle nut and washer
3. Raise and support the vehicle.
4. Remove or disconnect the following:
   • Tire and wheel
   • Caliper, leaving the fluid lines connected
   • Brake rotor
   • Halfshaft from the hub and bearing on 4WD vehicles. Place a brass drift against the outer edge of the halfshaft to protect the shaft threads. Use a hammer to sharply strike the brass drift, but to do not remove the halfshaft at this time.
   • Wheel speed sensor
   • Wheel hub and bearing-to-steering knuckle bolts and hub and bearing

➡ **Lay the hub and bearing on the wheel studs on the outboard side. This will avoid damaging the bearing seal.**

Heater Core replacement is covered in Section 2 of this manual

- Splash shield from the steering knuckle
- Seal from the hub and bearing

To install:

5. Install or connect the following:
- Wheel hub and bearing seal
- Splash shield to the steering knuckle, making sure it's properly aligned
- Hub and bearing to the steering knuckle, aligning the threaded holes
- Hub and bearing bolts and tighten to 77 ft. lbs. (105 Nm)
- Wheel speed sensor. Tighten the bolt to 13 ft. lbs. (18 Nm).
- Rotor and brake caliper
- Tire and wheel

6. Lower the vehicle
7. On 4WD vehicles, install the drive axle nut and tighten to 103 ft. lbs. (140 Nm), then install the center cap.

Rear

A new pinion shaft lockbolt should be installed whenever either of the axle shafts is removed.

The axle shaft and seal may be removed and replaced without disturbing the bearing or seal but it is highly recommended to replace the seals when removing the axle shaft.

1. Before servicing the vehicle, refer to the precautions in the beginning of this section.
2. Remove or disconnect the following:
- Rear wheels
- Brake drums
3. Using a wire brush, clean the dirt/rust from around the rear axle cover.
4. Drain the fluid.
5. Remove or disconnect the following:

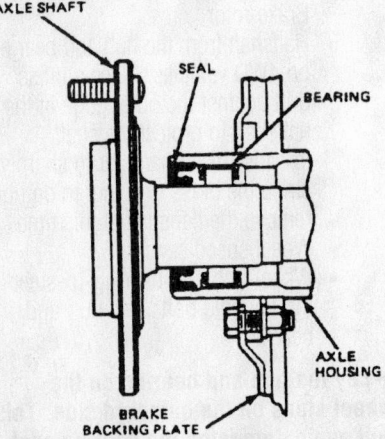

AXLE SHAFT
SEAL
BEARING
AXLE HOUSING
BRAKE BACKING PLATE

7924JG55

Cross-sectional view of the rear axle, bearing and seal assembly

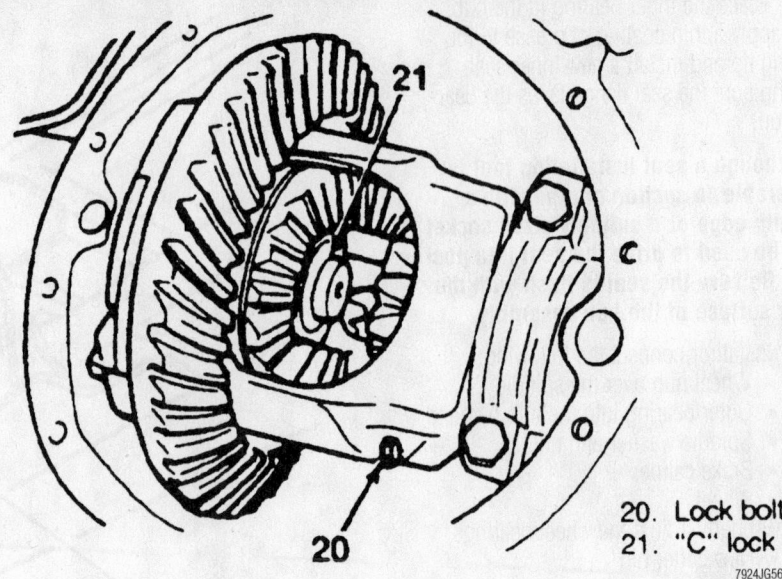

20. Lock bolt
21. "C" lock

7924JG56

Pinion shaft lockbolt and axle C-lock locations, inside the differential

- Rear pinion shaft lockbolt and the pinion shaft
- C-lock from the button end of the axle shaft by pushing the axle shaft inward
- Axle shaft from the axle housing

➡ **Be careful not to damage the oil seal.**

✲✲ WARNING

If equipped with an Anti-Lock Brake System (ABS), be careful not to damage the reflector ring on the axle shaft or the speed sensor bolted to the backing plate, immediately adjacent to the shaft.

6. Remove or disconnect the following:
- Oil seal by prying the it from the end of the rear axle housing

✲✲ WARNING

DO NOT damage the housing oil seal surface.

- Wheel bearing using the GM Slide Hammer tool J-2619, the GM Adapter tool J-2619-4 and the GM Axle Bearing Puller tool J-22813-01

To install:

7. Clean and inspect the components for excessive wear or damage and replace them, if necessary.
8. Install or connect the following:
- New or reused bearing, coated with gear lubricant, using the Axle Shaft Bearing Installer tool J-34974 to

drive the bearing in until it bottoms against the seat

✲✲ WARNING

Be sure the bearing installer does not contact and damage the speed sensor on ABS equipped vehicles.

- New seal lubricated with gear oil using the GM Axle Shaft Seal Installer tool J-33782 to seat it in the housing until it is flush with the axle tube

➡ **Be sure the seal installer does not contact and damage the speed sensor on ABS equipped vehicles.**

- Axle shaft into the housing by engaging the splines
- C-lock retainer on the axle shaft button end

✲✲ WARNING

BE CAREFUL not to damage the wheel bearing seal.

- Axle shaft by pulling it outward to seat the C-lock retainer in the counterbore of the side gears
- Pinion shaft through the case and the pinions. Tighten the new lockbolt to 27 ft. lbs. (36 Nm).
- New rear axle cover gasket
- Housing cover
- Brake drums
- Wheels

9. Refill the housing.

GENERAL MOTORS CORP.

14

Escalade • Denali • Denali XL • Suburban • Yukon • Yukon XL

PRECAUTIONS

Before servicing any vehicle, please be sure to read all of the following precautions, which deal with personal safety, prevention of component damage, and important points to take into consideration when servicing a motor vehicle:

• Never open, service or drain the radiator or cooling system when the engine is hot; serious burns can occur from the steam and hot coolant.

• Observe all applicable safety precautions when working around fuel. Whenever servicing the fuel system, always work in a well-ventilated area. Do not allow fuel spray or vapors to come in contact with a spark, open flame, or excessive heat (a hot drop light, for example) Keep a dry chemical fire extinguisher near the work area. Always keep fuel in a container specifically designed for fuel storage; also, always properly seal fuel containers to avoid the possibility of fire or explosion. Refer to the additional fuel system precautions later in this section.

• Fuel injection systems often remain pressurized, even after the engine has been turned **OFF**. The fuel system pressure must be relieved before disconnecting any fuel lines. Failure to do so may result in fire and/or personal injury.

• Brake fluid often contains polyglycol ethers and polyglycols. Avoid contact with the eyes and wash your hands thoroughly after handling brake fluid. If you do get brake fluid in your eyes, flush your eyes with clean, running water for 15 minutes. If eye irritation persists, or if you have taken brake fluid internally, IMMEDIATELY seek medical assistance.

• The EPA warns that prolonged contact with used engine oil may cause a number of skin disorders, including cancer! You should make every effort to minimize your exposure to used engine oil. Protective gloves should be worn when changing oil. Wash your hands and any other exposed skin areas as soon as possible after exposure to used engine oil. Soap and water, or waterless hand cleaner should be used.

• All new vehicles are now equipped with an air bag system. The system must be disabled before performing service on or around system components, steering column, instrument panel components, wiring and sensors. Failure to follow safety and disabling procedures could result in accidental air bag deployment, possible personal injury and unnecessary system repairs.

• Always wear safety goggles when working with, or around, the air bag system. When carrying a non-deployed air bag, be sure the bag and trim cover are pointed away from your body. When placing a non-deployed air bag on a work surface, always face the bag and trim cover upward, away from the surface. This will reduce the motion of the module if it is accidentally deployed. Refer to the additional air bag system precautions later in this section.

• Clean, high quality brake fluid from a sealed container is essential to the safe and proper operation of the brake system. You should always buy the correct type of brake fluid for your vehicle. If the brake fluid becomes contaminated, completely flush the system with new fluid. Never reuse any brake fluid. Any brake fluid that is removed from the system should be discarded. Also, do not allow any brake fluid to come in contact with a painted surface; it will damage the paint.

• Never operate the engine without the proper amount and type of engine oil; doing so WILL result in severe engine damage.

• Timing belt maintenance is extremely important! Many models utilize an interference-type, non-freewheeling engine. If the timing belt breaks, the valves in the cylinder head may strike the pistons, causing potentially serious (also time-consuming and expensive) engine damage. Refer to the maintenance interval charts in the front of this manual for the recommended replacement interval for the timing belt, and to the timing belt section for belt replacement and inspection.

• Disconnecting the negative battery cable on some vehicles may interfere with the functions of the on-board computer system(s) and may require the computer to undergo a relearning process once the negative battery cable is reconnected.

• When servicing drum brakes, only disassemble and assemble one side at a time, leaving the remaining side intact for reference.

GASOLINE ENGINE REPAIR

Distributor

REMOVAL

4.3L, 5.0L, 5.7L and 7.4L Engines

1. Before servicing the vehicle, refer to the precautions in the beginning of this section.
2. Remove or disconnect the following:
   - Negative battery cable
   - Spark plug wires and the coil leads from the distributor
   - Electrical connector at the base of the distributor
   - Distributor cap
3. Matchmark the rotor-to-housing and housing-to-engine block positions so that they can be matched during installation.
   - Distributor hold-down bolt
   - Distributor from the engine

4.8L, 5.3L and 6.0L Engines

➡ **If the Malfunction Indicator Lamp turns on, and a DTC code P1345 sets after installing the distributor, this indicates an incorrectly installed distributor. Engine damage or distributor damage may occur.**

1. Before servicing the vehicle, refer to the precautions in the beginning of this section.
2. Turn OFF the ignition switch.
3. Remove or disconnect the following:
   - Spark plug wires from the distributor cap
   - Electrical connector from the base of the distributor
   - Two screws that hold the distributor cap to the housing. Discard the screws.
   - Distributor cap from the housing
4. Use a grease pencil in order to note the position of the rotor in relation to the distributor housing.
5. Mark the distributor housing and the intake manifold with a grease pencil.
6. Remove or disconnect the following:
   - Mounting clamp hold-down bolt
   - Distributor
7. As the distributor is being removed from the engine, watch the rotor move in a counterclockwise direction about 42 degrees. This will appear as slightly more than the 1 o'clock position. Note the position of the rotor segment. Place a second mark on the base of the distributor. This will aid in achieving proper rotor alignment during the distributor installation.

INSTALLATION

4.3L, 5.0L, 5.7L and 7.4L Engines

TIMING NOT DISTURBED

1. Install or connect the following:
   - Distributor, aligning the match-marks properly
   - Distributor hold-down bolt
   - Distributor cap
   - Electrical connector at the base of the distributor
   - Spark plug wires and coil leads
   - Negative battery cable

TIMING DISTURBED

1. Remove the No. 1 cylinder spark plug. Turn the engine using a socket wrench on the large bolt on the front of the crank-

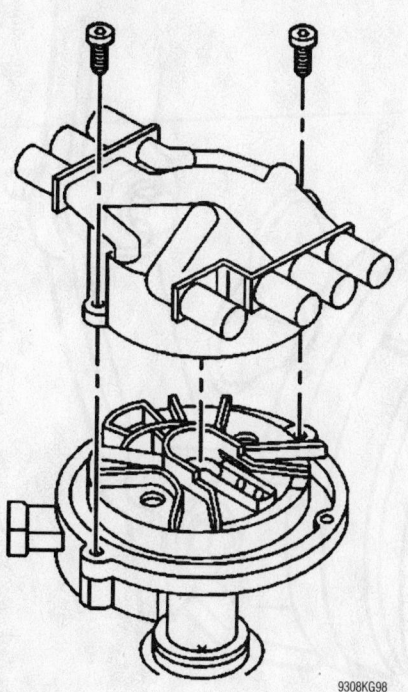

Distributor cap—4.8L, 5.3L and 6.0L engines

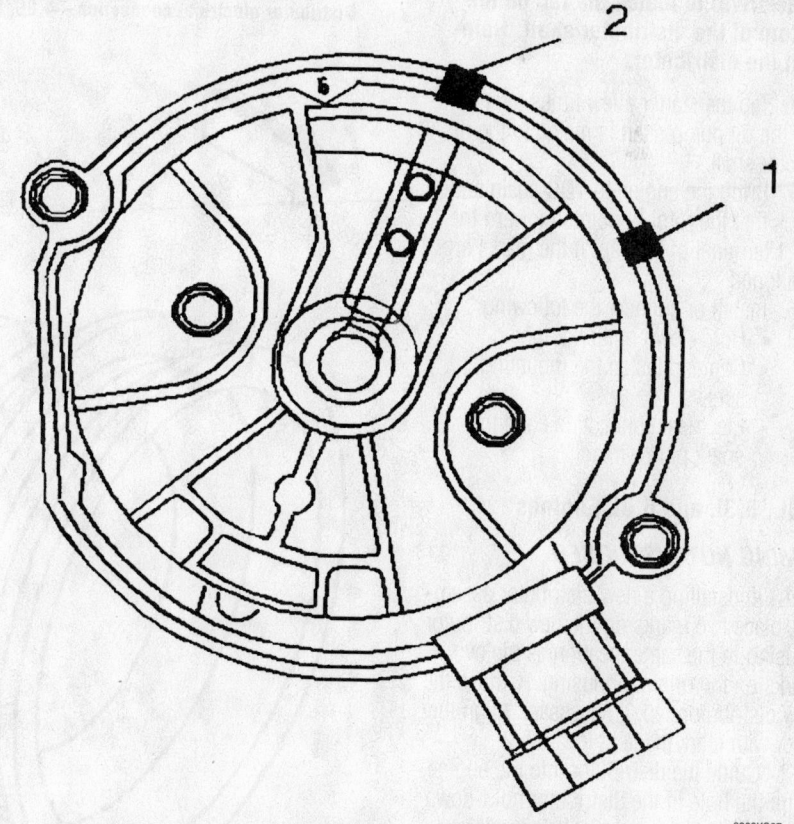

Distributor rotor starting point (1) and 42 degrees counterclockwise (2)—4.8L, 5.3L and 6.0L engines

For complete service labor times, order Nichols' Chilton Labor Guide

shaft pulley. Place a finger near the No. 1 spark plug hole and turn the crankshaft until the piston reaches TDC. As the engine approaches TDC, you will feel air being expelled through the No. 1 cylinder spark plug hole. The timing mark on the crankshaft pulley should now be aligned with the **0** mark on the timing scale. If the position is not being met, turn the engine another full turn (360 degrees) Once the engines position is correct, install the spark plug.

➡ **Before installation, position the rotor so it points to the No. 2 terminal on the cap. As the distributor is lowered into the engine, the rotor will rotate clockwise and stop at the No. 1 terminal. This is the desired position.**

2. Turn the rotor so that it will point to the No. 1 terminal of the distributor cap when it is fully seated in the engine.

3. Install the distributor. It may be necessary to turn the rotor a little in either direction, in order to engage the gears.

➡ **If the distributor will not seat completely in the engine, remove the distributor and align the groove on the top of the oil pump drive shaft with a long screwdriver to match the tab on the bottom of the distributor shaft. Reinstall the distributor.**

4. Tap the starter a few times to ensure that the oil pump shaft is mated to the distributor shaft.

5. Bring the engine to TDC again and check that the rotor is pointed toward the No. 1 terminal of the cap. If the marks are all aligned.

6. Install or connect the following:
- Hold-down bolt and tighten
- Cap and fasten the mounting screws
- Electrical connections and the spark plug wires

4.8L, 5.3L and 6.0L Engines

TIMING NOT DISTURBED

1. If installing a new distributor assembly, place two marks on the new distributor housing in the same location as the two marks on the original housing. Remove the new distributor cap, if necessary. Align the rotor with mark made at location 2.

2. Guide the distributor into the engine. Align the hole in the distributor hold-down base over the mounting hole in the intake manifold.

3. As the distributor is being installed, observe the rotor moving in a clockwise direction about 42 degrees. Once the dis-

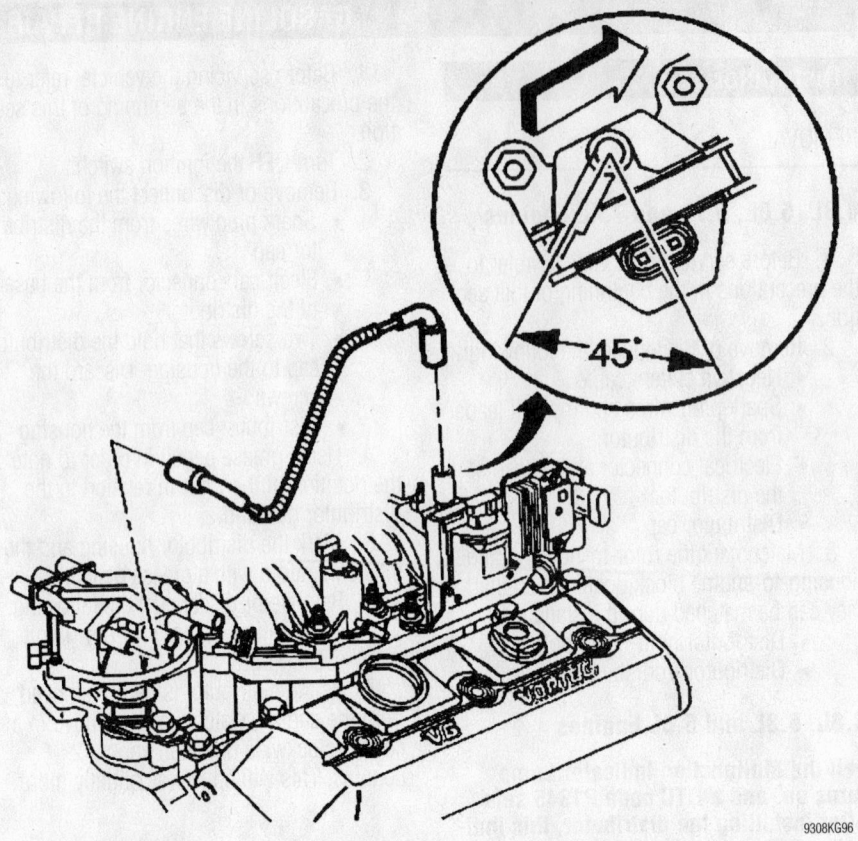

Distributor electrical connection—4.8L, 5.3L and 6.0L engines

9308KG96

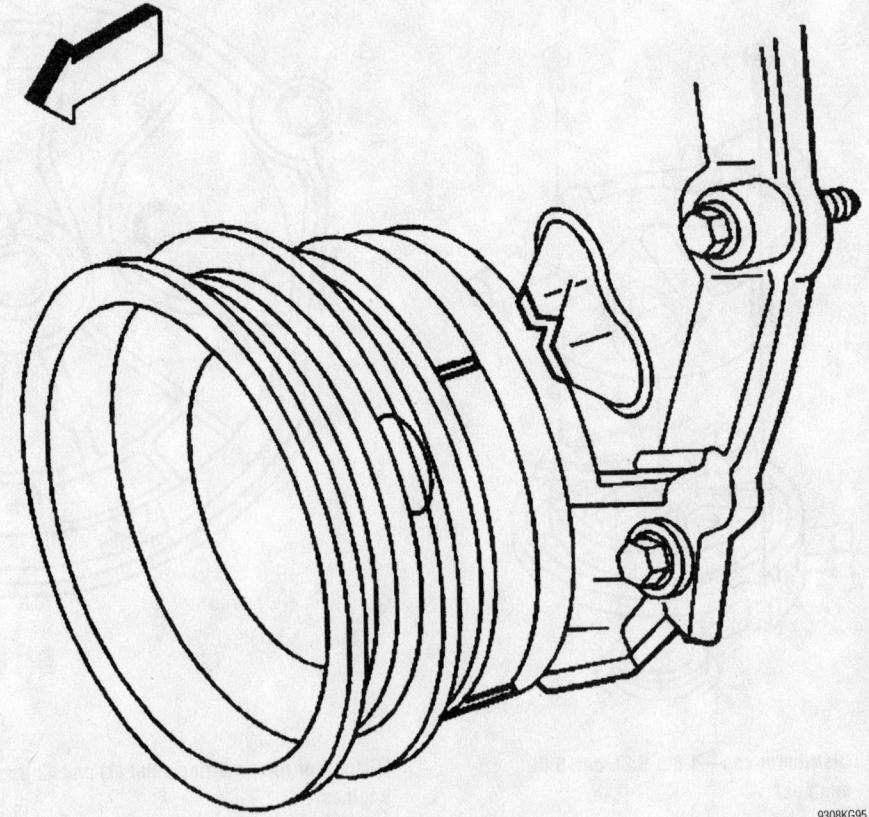

Engine at TDC compression—4.8L, 5.3L and 6.0L engines

9308KG95

tributor is completely seated, the rotor segment should be aligned with the mark on the distributor base in location number 1. If the rotor segment is not aligned with the number 1 mark, the driven gear teeth and the camshaft have meshed one or more teeth out of alignment.

4. Install or connect the following:
- Distributor mounting clamp bolt and tighten to 18 ft. lbs. (25 Nm)
- Distributor cap. Install two NEW distributor cap screws and tighten to 21 inch lbs. (2.4 Nm).
- Electrical connector to the distributor
- Spark plug wires to the distributor cap
- Ignition coil wire

➡**If the Malfunction Indicator lamp is turned on after installing the distributor, and a DTC P1345 is found, the distributor has been installed incorrectly.**

TIMING DISTURBED

1. Rotate the number 1 cylinder to TDC of the compression stroke. The engine front cover has 2 alignment tabs and the crankshaft balancer has 2 alignment marks (spaced 90 degrees apart) which are used for positioning number 1 piston at top dead center (TDC). With the piston on the compression stroke and at top dead center, the crankshaft balancer alignment mark must align with the engine front cover tab and the crankshaft balancer alignment mark must align with the engine front cover tab.

2. Align the white paint mark on the bottom stem of the distributor with the predrilled indent hole in the bottom of the gear. If the driven gear is installed incorrectly, the dimple will be approximately 180 degrees opposite of the rotor segment when it is installed in the distributor.

The OBD II ignition system distributor driven gear and rotor may be installed in multiple positions. In order to avoid mistakes, mark the distributor on the following components in order to ensure the same mounting position upon reassembly:
- The distributor driven gear
- The distributor shaft
- The rotor holes

Installing the driven gear 180 degrees out of alignment, or locating the rotor in the wrong holes, will cause a no-start condition. Premature engine wear or damage may result.

3. Using a long screwdriver, align the oil

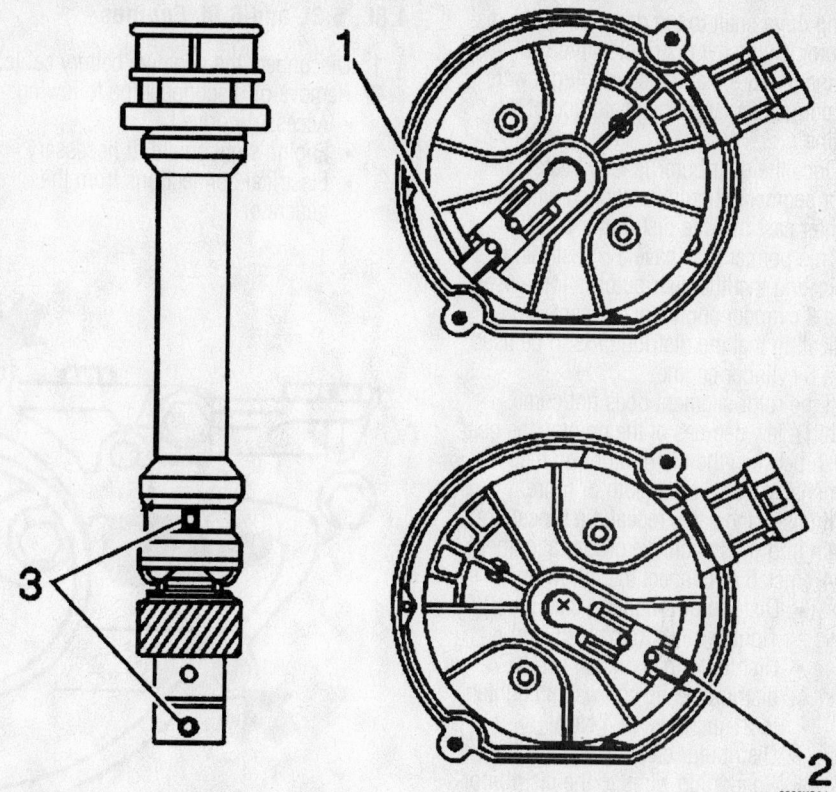

Distributor alignment. 1 is the starting point; 2 is installed; 3 are the shaft alignment marks—4.8L, 5.3L and 6.0L engines

9308KG94

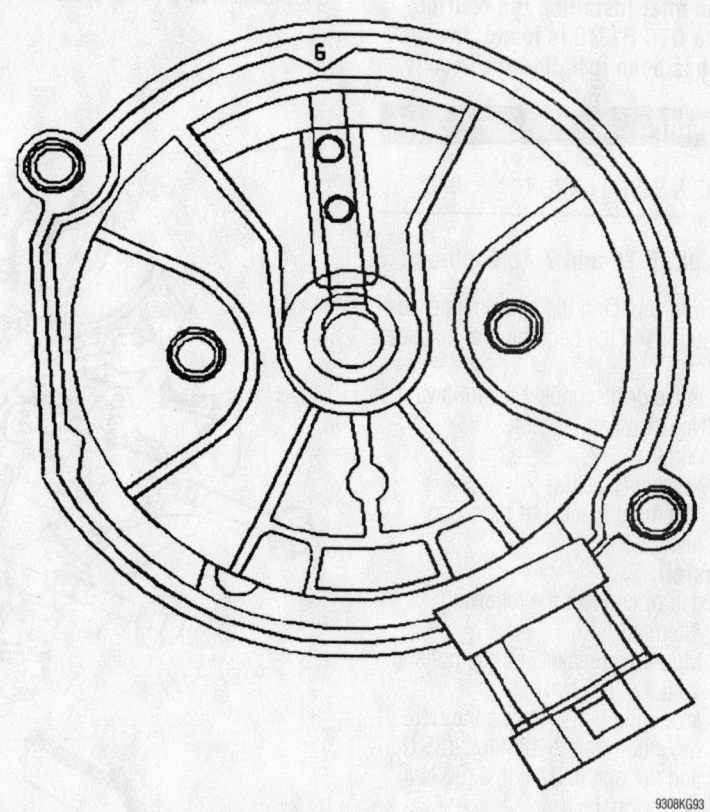

Distributor fully seated—4.8L, 5.3L and 6.0L engines

9308KG93

pump drive shaft to the drive tab of the distributor. Guide the distributor into the engine. Ensure that the spark plug towers are perpendicular to the centerline of the engine.

Once the distributor is fully seated, the rotor segment should be aligned with the pointer cast into the distributor base.

This pointer may have a 6 cast into it, indicating that the distributor is to be used on a 6 cylinder engine or a 8 cast into it, indicating that the distributor is to be used on a 8 cylinder engine.

If the rotor segment does not come within a few degrees of the pointer, the gear mesh between the distributor and the camshaft may be off a tooth or more.

If this is the case, repeat the procedure again in order to achieve proper alignment.

4. Install or connect the following:
   • Distributor mounting clamp bolt. Tighten to 18 ft. lbs. (25 Nm).
   • Distributor cap. Install two NEW distributor cap screws and tighten to 21 inch lbs. (2.4 Nm).
   • Distributor electrical connector
   • Spark plug wires to the distributor cap
   • Ignition coil wire

➥If the Malfunction Indicator lamp is turned on after installing the distributor, and a DTC P1345 is found, the distributor has been installed incorrectly.

Alternator

REMOVAL & INSTALLATION

4.3L, 5.0L, 5.7L and 7.4L Engines

1. Before servicing the vehicle, refer to the precautions in the beginning of this section.
2. Remove or disconnect the following:
   • Negative battery cable
   • Wires
   • Accessory belt(s)
   • Mounting bracket, if necessary
   • Alternator

To install:

3. Install or connect the following:
   • Alternator
   • Mounting bracket. Torque bolts to 18 ft lbs. (25 Nm).
   • Mounting bolts. Torque the right mounting bolt to 18 ft lbs. (25 Nm) and left bolt to 37 ft lbs. (50 Nm).
   • Accessory belt(s)
   • Wires. Torque the battery feed wire to 71 inch lbs. (8 Nm).
   • Negative battery cable

4.8L, 5.3L and 6.0L Engines

1. Disconnect the negative battery cable.
2. Remove or disconnect the following:
   • Accessory drive belt
   • Engine sight shield, if necessary
   • Electrical connections from the generator
   • Mounting bolts
   • Generator

To install:

3. Install or connect the following:
   • Generator
   • Generator mounting bolts and tighten to 37 ft. lbs. (50 Nm).

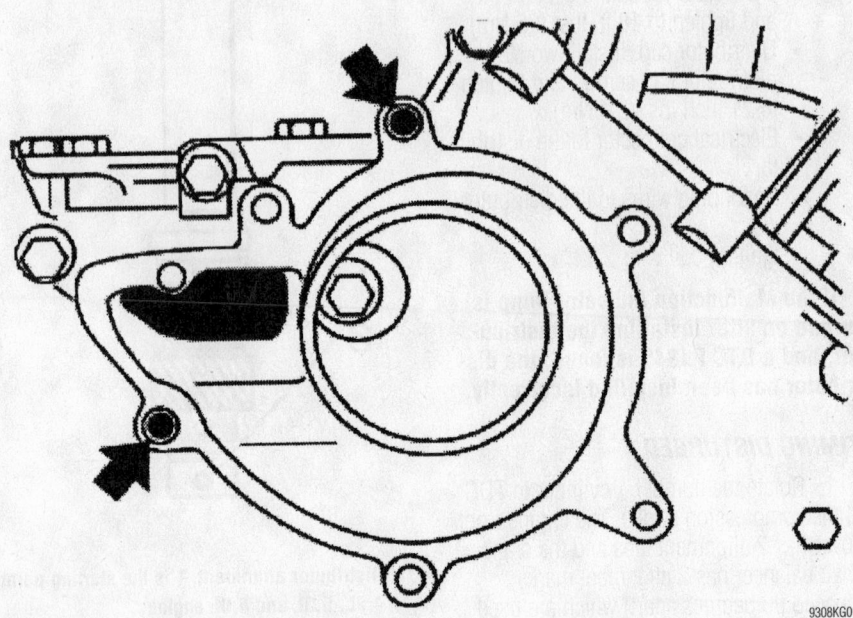

Exploded view of the alternator mounting

9308KG01

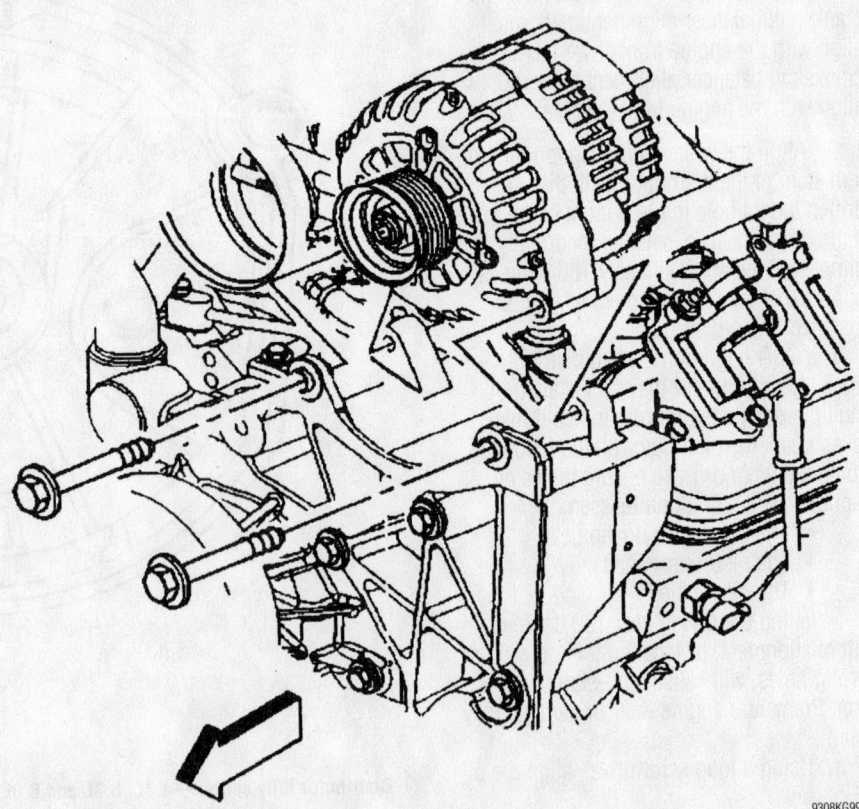

Alternator mounting—4.8L, 5.3L and 6.0L engines

9308KG99

- Electrical connections to the generator. Tighten the B+ nut to 13 ft. lbs. (18 Nm).
- Engine sight shield, if removed
- Accessory drive belt
- Negative battery cable

8.1L Engine

1. Disconnect the negative battery cable.
2. Remove or disconnect the following:
   - Electrical connections from the generator
3. Remove the cable from the generator as follows:
   a. Slide the boot down, to reveal the terminal stud.
   b. Unfasten the cable nut from the stud, then remove the generator cable.
4. Remove or disconnect the following:
   - Accessory drive belt
   - Mounting bolts
   - Generator
   - Mounting bolts securing the generator to the brace and bracket
   - Generator

To install:
5. Install or connect the following:
   - Generator
   - Generator mounting bolts and tighten to 37 ft. lbs. (50 Nm)
   - Accessory drive belt
   - Generator cable, secure with the nut and tighten to 80 inch lbs. (9 Nm)
   - Boot back over the terminal stud.
   - Electrical connections to the generator
   - Negative battery cable

Ignition Timing

ADJUSTMENT

Always refer to the Vehicle Emissions Control Information label in the engine compartment for base ignition timing specification and adjustment procedures.

Engine Assembly

REMOVAL & INSTALLATION

4.3L, 5.0L and 5.7L Engines

1. Before servicing the vehicle, refer to the precautions in the beginning of this section.
2. Drain the cooling system.

3. Drain the engine oil.
4. Remove or disconnect the following:
   - Negative battery cable
   - Hood
   - Air cleaner
   - Accessory drive belt
   - Fan
   - Water pump pulley
   - Radiator and shroud
   - Heater hoses at the engine
   - Accelerator, cruise control and detent linkage if used
   - Air conditioning compressor, if used, and lay aside
   - Power steering pump, if used, and lay aside
   - Wiring from the engine
   - Fuel line
   - Vacuum lines from the intake manifold
   - Exhaust pipes from the manifold
   - Strut rods at the engine mountings, if used
   - Flywheel or torque converter cover
   - Wiring along the oil pan rail
   - Starter
   - Wire for the fuel gauge
   - Converter-to-flex plate bolts, if equipped with automatic transmission
5. Support the transmission
   - Bell housing to engine bolts
   - Rear engine mounting to frame bolts and the front through bolts and the engine

To install:
6. Lower the engine.
7. Install or connect the following:
   - Engine mounting bolts. Torque the rear engine mounting to frame bolts or nuts to 45 ft. lbs. (54 Nm), the front through-bolts to 70 ft. lbs. (97 Nm) and the front nuts to 50 ft. lbs. (67 Nm)
   - Bell housing to engine bolts and torque to 35 ft. lbs. (47 Nm)
8. Remove the transmission support.
   - Converter-to-flex bolts and torque to 35 ft. lbs. (47 Nm)
   - Fuel gauge wiring
   - Starter
   - Flywheel or torque converter cover
   - Strut rods at the engine mountings, if used
   - Exhaust pipes at the manifold
   - Vacuum lines to the intake manifold
   - Fuel line
   - Engine wiring harness
   - Power steering pump, if used

- Air conditioning compressor, if used
- Accelerator, cruise control and detent linkage
- Heater hoses
- Radiator and shroud
- Accessory drive belts
- Hood
- Negative battery cable
9. Refill coolant and engine oil.

7.4L Engine

1. Before servicing the vehicle, refer to the precautions in the beginning of this section.
2. Drain the cooling system.
3. Remove or disconnect the following:
   - Hood
   - Negative battery cable
   - Air cleaner
   - Radiator and fan shroud
   - Engine wiring
   - Accelerator, cruise control linkage
   - Fuel supply lines
   - Vacuum wires
   - Air conditioning compressor, if used, and lay aside
   - Power steering pump and position it out of the way. It's not necessary to disconnect the fluid lines.
   - Exhaust pipes from the manifold
   - Starter
   - Torque converter cover
   - Converter-to-flexplate bolts
4. Support the transmission
   - Bellhousing-to-engine bolts
   - Rear engine mounting-to-frame bolts and the front through bolts
   - Engine

To install:
5. Lower the engine.
6. Install or connect the following:
   - Engine mounting bolts. Torque the rear engine mounting-to-frame bolts or nuts to 45 ft. lbs. (54 Nm), the front through bolts to 70 ft. lbs. (97 Nm) and the front nuts to 50 ft. lbs. (67 Nm)
   - Bellhousing-to-engine bolts. Torque the bolts to 35 ft. lbs. (47 Nm)
7. Remove transmission support.
   - Converter-to-flexplate bolts and torque them to 35 ft. lbs. (47 Nm)
   - Fuel gauge wiring
   - Starter
   - Torque converter cover
   - Exhaust pipes at the manifold
   - Power steering pump

Timing belt service is covered in Section 3 of this manual

- Air conditioning compressor
- Vacuum hoses
- Fuel supply line
- Accelerator, cruise control linkage
- Engine wiring
- Radiator and fan shroud
- Air cleaner
- Hood
- Negative battery cable

8. Refill the coolant.

4.8L, 5.3L and 6.0L Engines

❊❊ CAUTION

Before servicing any electrical component, the ignition key must be in the OFF or LOCK position and all electrical loads must be OFF, unless instructed otherwise in these procedures.

1. Before servicing the vehicle, refer to the precautions in the beginning of this section.

2. Remove or disconnect the following:
- Negative battery cable
- Coolant
- A/C refrigerant

3. Raise the hood to the servicing position. Move the hood hinge bolt to hold the hood in the servicing position.
- Upper and the lower radiator hoses from the engine
- Air cleaner duct from the engine
- A/C condenser mounting bolts
- Radiator support from the vehicle
- A/C compressor
- Coolant hose from the throttle body
- Heater hoses from the engine and the cowl
- Engine sight shield from the intake manifold
- Accelerator control cable mounting bracket from the intake manifold

❊❊ CAUTION

In order to avoid possible injury or vehicle damage, always replace the accelerator control cable with a NEW cable whenever you remove the engine from the vehicle. In order to avoid cruise control cable damage, position the cable out of the way while you remove or install the engine.

- Accelerator control cable and the cruise control cable, if equipped, from the throttle shaft

4. Open the large electrical harness retainer. Remove one 10mm nut in order to release the engine harness from the intake manifold.

5. Disconnect the electrical connectors from the following:
- Eight injector connectors
- Idle Air Control (IAC) motor
- Throttle Position (TP) sensor
- Evaporative Emissions (EVAP) canister purge solenoid
- Manifold Absolute Pressure (MAP) sensor
- Camshaft Position (CMP) sensor
- Ground splice at the rear of the right side of the block
- Ground splice and the ground strap at the rear of the left side of the block
- Coolant Temperature (CTS) sensor
- Oil pressure sensor/switch
- Intake electrical and disconnect from harness
- Junction block bracket from alternator bracket

6. Set the electrical harness aside.

7. Remove or disconnect the following:
- EVAP canister purge solenoid vent tube from the solenoid by squeezing the retainer, then release the tube from the solenoid
- Battery negative cable from the engine block

- Drive belt
- Bolts holding the alternator mounting bracket to the cylinder head and block
- Bolt behind the power steering pump to engine block
- Alternator mounting bracket. Position the bracket aside.
- Fuel pipes from the engine

8. Raise the vehicle.
- Steering linkage under body shield, if equipped
- Engine oil pan under body shield, if equipped
- Engine oil
- Starter motor

9. Disconnect the engine wiring harness from the following components:
- Crankshaft Position (CKP) sensor
- Engine oil level sensor
- Block heater, if equipped
- Wiring harness from the oil pan
- Reposition wiring from lower engine area

10. Remove or disconnect the following:
- Exhaust pipes from the exhaust manifolds
- Transmission cooler pipe retainer from the right side of the engine block, if equipped

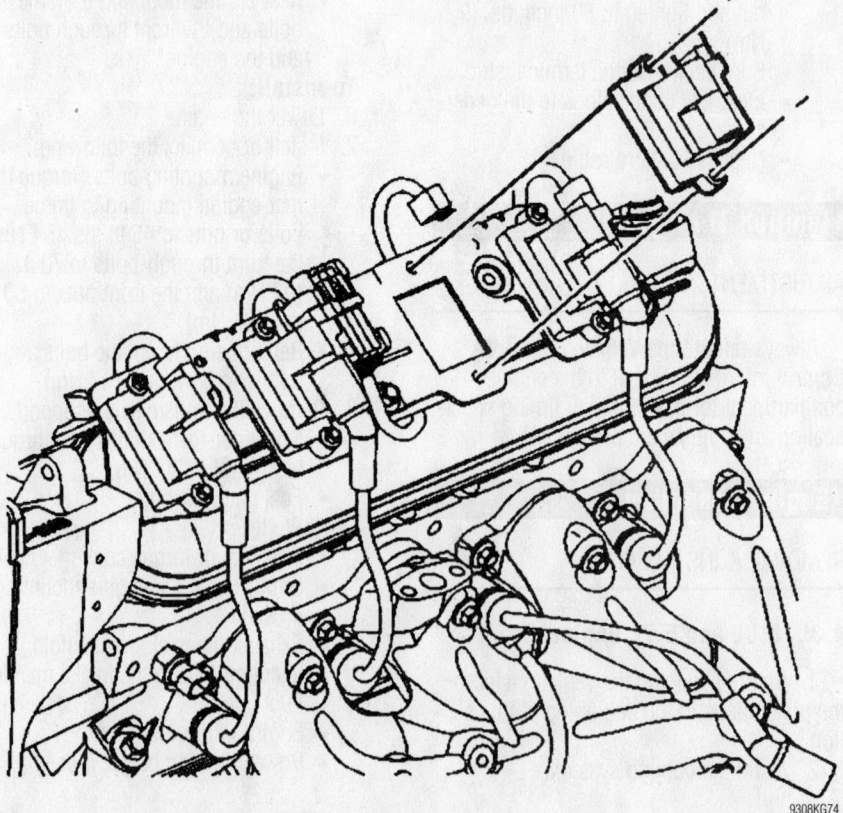

9308KG74

Ignition coil removal—4.8L, 5.3L and 6.0L engines

- Torque converter shield from the engine
- Torque converter bolts
- Nut and the transmission oil level indicator tube from the bellhousing stud
- Lower bellhousing studs from the engine

11. Lower the vehicle.
    - Remaining bellhousing bolts
    - Engine electrical harness; position aside
    - Ignition coil(s)

12. Install an engine crane.

13. Install a floor jack or stands to transmission for support.

14. Remove the engine mount bolts.

➡ **Use care while moving the engine assembly in order to avoid breaking the MAP sensor locating tabs. Broken MAP sensor tabs may result in decreased engine performance.**

15. Remove the engine from the vehicle.

To install:

16. Install or connect the following:
    - Engine to the vehicle
    - Engine mount bolts
    - Upper bellhousing bolts

17. Remove transmission support apparatus.

18. Remove the lifting device.

19. Remove the lift brackets from both cylinder heads.

20. Install the ignition coil(s) and the spark plug wire(s).

21. Route the engine wiring harness to the lower right hand side of the engine.

22. Raise the vehicle.

23. Install or connect the following:
    - Remaining bellhousing bolts
    - Torque converter bolts
    - Torque converter shield
    - Transmission oil level indicator tube and nut to bellhousing stud
    - A/C compressor
    - Transmission cooler pipe retainer to right side of engine block
    - Engine exhaust pipes to the exhaust manifolds

24. Reroute wiring to lower engine area and install bolt to oil pan.

25. Install or connect the following:
    - CKP sensor electrical connector
    - Engine oil level sensor and the block heater electrical connectors, if equipped.
    - Starter motor

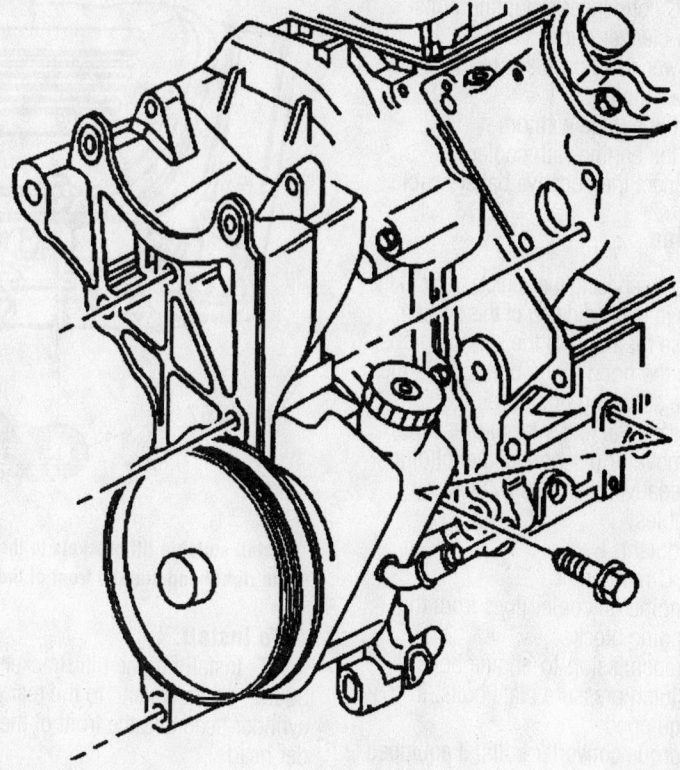

9308KG73

Power steering pump removal—4.8L, 5.3L and 6.0L engines

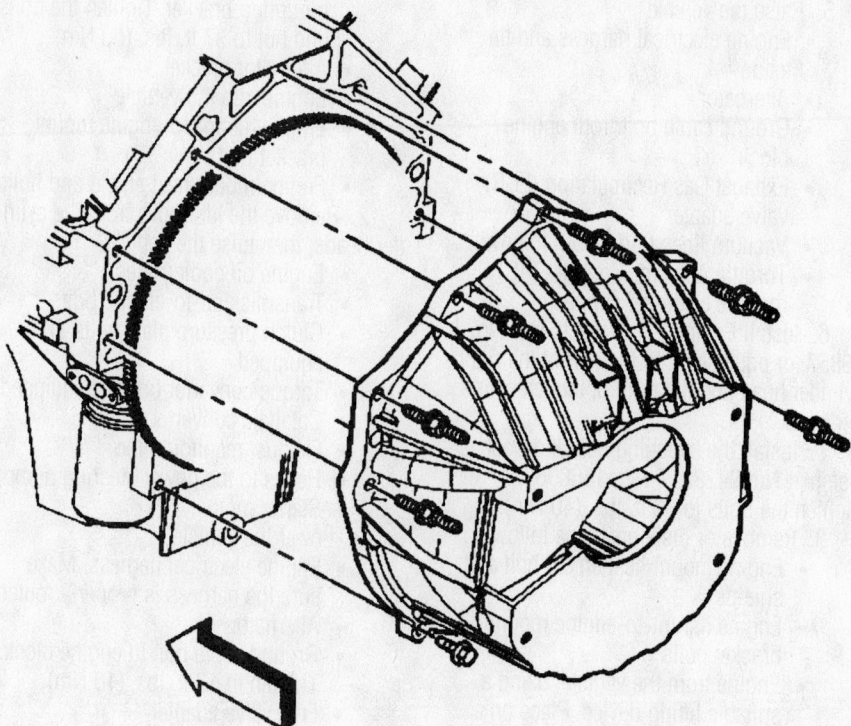

9308KG75

Bellhousing bolt removal—4.8L, 5.3L and 6.0L engines

Heater Core replacement is covered in Section 2 of this manual

- Engine oil pan under body shield, if equipped
- Steering linkage under body shield

26. Lower the vehicle.

- Fuel pipes to the engine
- Alternator mounting bracket to the cylinder head using the nuts and the bolts. Tighten the bolts to 37 ft. lbs. (50 Nm).
- Bolt at the rear of the power steering pump to the engine block and tighten to 37 ft. lbs. (50 Nm).
- Alternator
- Drive belt
- Negative battery cable to the engine block
- EVAP canister purge solenoid to the intake manifold

27. Route the engine harness over the top of the engine. Attach the connectors to the following components:

- Eight injector connectors
- IAC motor
- TP sensor
- EVAP canister purge solenoid
- MAP sensor
- CMP sensor
- Ground splice at the rear of the right side of engine block
- Ground splice and the ground strap at the rear of the left side of engine block
- CTS sensor
- Oil pressure sensor/switch

28. Install or connect the following:

- Nut to the engine wiring harness bracket and tighten to 89 inch lbs. (10 Nm)

✳✳ CAUTION

In order to avoid possible injury or vehicle damage, always replace the accelerator control cable with a NEW cable whenever you remove the engine from the vehicle. In order to avoid cruise control cable damage, position the cable out of the way while you remove or install the engine.

- NEW accelerator control cable
- Cruise control cable, if equipped, to the throttle shaft
- Bolts for the accelerator control cable mounting bracket and tighten to 89 inch lbs. (10 Nm)
- Engine sight shield to the intake manifold
- Heater hoses to the cowl and the engine
- Coolant hose to the throttle body
- Radiator support in the vehicle

- A/C condenser mounting bolts
- Air cleaner duct
- Lower radiator hoses to the engine

29. Lower the hood.
30. Fill the engine with oil.
31. Fill the engine with coolant.
32. Connect the negative battery cable.

8.1L Engine

1. Before servicing the vehicle, refer to the precautions in the beginning of this section.

2. Raise the hood to the servicing position. Move the hood hinge bolt to hold the hood in the servicing position.

3. Release the fuel system pressure.

4. Remove or disconnect the folloing:

- Negative, then positive battery cables
- Coolant
- A/C refrigerant
- Engine oil cooler lines from the engine block
- Transmission-to-engine bolts
- Clutch pressure plate bolts, if equipped
- Torque converter bolts, if equipped
- Catalytic converter
- Exhaust manifold pipe
- Hoses from power steering pump, then plug the lines and ports
- Starter motor

5. Raise the vehicle.

- Engine electrical harness and tie aside
- Alternator
- Ground cable bolt from engine block
- Exhaust Gas Recirculation (EGR) valve adapter
- Vacuum lines (tag before removal)
- Throttle Actuator Control (TAC) module electrical connector

6. Install Engine Lift Brackets part No. J 36857, or equivalent, to the rear of the right cylinder head and the front of the left cylinder head.

7. Install the attaching bolt and washer. Use part No. 9428217 with 1560963. Tighten the bolts to 30 ft. lbs. (40 Nm).

8. Remove or disconnect the following:

- Engine mount heat shield bolt and shields
- Engine mount-to-engine mount bracket bolts
- Engine from the vehicle, using a suitable lifting device. Place on a suitable stand.
- A/C compressor/power steering pump bracket from the cylinder head
- Lift brackets from the cylinder head

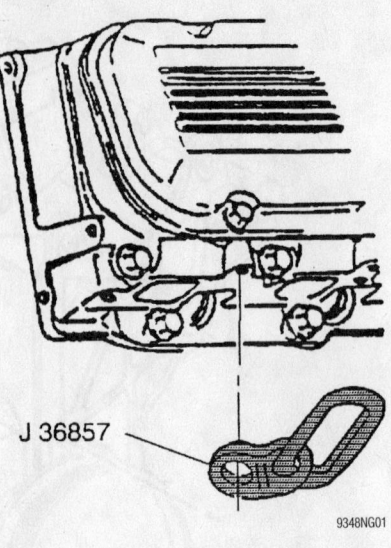

J 36857

9348NG01

Install suitable lift brackets to the rear of the right head and the front of the left head

To install:

9. Install Engine Lift Brackets part No. J 36857, or equivalent, to the rear of the right cylinder head and the front of the left cylinder head.

10. Install the attaching bolt and washer. Use part No. 9428217 with 1560963. Tighten the bolts to 30 ft. lbs. (40 Nm).

11. Install or connect the following:

- A/C compressor/power steering mounting bracket. Tighten the bolts and nut to 37 ft. lbs. (50 Nm)
- Alternator bracket
- Engine into the vehicle
- Engine mount-to-engine mount bracket bolts
- Engine mount heat shield and bolts

12. Remove the lift hooks from the cylinder heads, then raise the vehicle.

- Engine oil cooler lines
- Transmission-to-engine bolts
- Clutch pressure plate bolts, if equipped
- Torque converter bolts, if equipped
- Catalytic converter
- Exhaust manifold pipe
- Hoses to the power steering pump
- Starter motor

13. Lower the vehicle.

- Engine electrical harness. Make sure the harness is properly routed.
- Alternator
- Ground cable bolt to engine block. Tighten to 12 ft. lbs. (16 Nm).
- EGR valve adapter
- Vacuum lines, as tagged during removal
- TAC module electrical connector
- Radiator
- A/C compressor

- Fuel feed and return lines
- Ignition coils
- Positive, then negative battery cables
- Air cleaner outlet duct and secure with the clamp

14. Lower the hood from the service position.

15. Properly recharge the A/C system.

16. Fill the engine with oil.

17. Fill the engine with coolant.

18. Perform the Crankshaft Position (CKP) sensor variation learn procedure:

    a. Install a suitable scan tool and check for Diagnostic Trouble Codes (DTCs). If any DTCs, other than P1336 are set, resolve those codes first, before proceeding with this procedure.

    b. With the scan tool, select the crankshaft position variation learn procedure.

    c. Observe the fuel cut-off for the 8.1L engine.

    d. The scan tool will instruct you to perform certain steps, make sure you follow all directions given by the scan tool exactly.

    e. Enable the crankshaft position system variation learn procedure.

➡ **While the learn procedure is in progress, release the throttle immediately when the engine started to decelerate. The engine control is returned to the operator and the engine responds to throttle position after the learn procedure is complete.**

    f. Slowly increase the engine speed to the RPM that you observed.

    g. Immediately release the throttle when fuel cut-out is reached.

    h. The scan tool displays: Learn Status: Learned this ignition. If the scan tool does NOT display this message and not other DTCs set, you must perform further troubleshooting.

    i. Turn the ignition **OFF** for 30 seconds after the learn procedure has been completed successfully.

19. Start and run the engine, then check for leaks.

Water Pump

REMOVAL & INSTALLATION

4.3L, 5.0L, 5.7L and 7.4L Engines

1. Before servicing the vehicle, refer to the precautions in the beginning of this section.

2. Drain the radiator.

3. Remove or disconnect the following:
- Fan shroud
- Negative battery cable
- Drive belt(s)
- Alternator and other accessories, if necessary
- Fan, fan clutch and pulley
- Accessory brackets that might interfere with water pump removal
- Lower radiator hose from the water pump inlet
- Heater hose from the nipple on the pump
- Bypass hose, 7.4L engine only
- Water pump assembly away from the timing cover

To install:

4. Clean all old gasket material from the timing chain cover and water pump.

5. Install or connect the following:
- Pump assembly with a new gasket. Torque the bolts to 30 ft. lbs. (41 Nm)
- Hose between the water pump inlet and the pump
- Heater hose and the bypass hose (7.4L only)
- Fan, fan clutch and pulley
- Alternator and other accessories, if necessary
- Drive belt(s)
- Upper radiator shroud

6. Refill the cooling system.

7. Connect the battery.

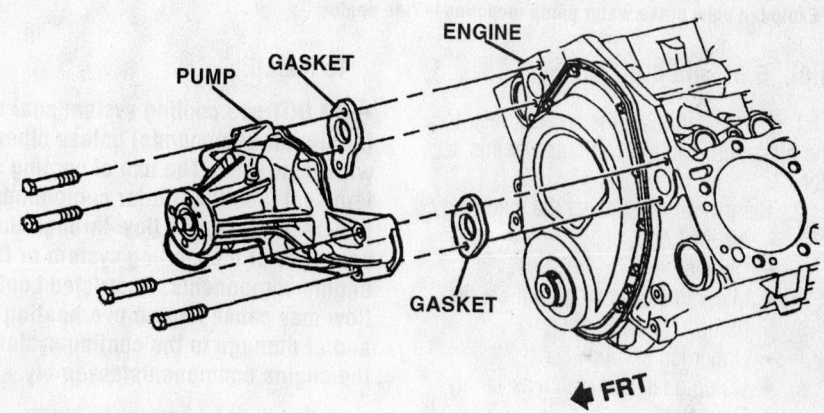

Exploded view of the water pump mounting—4.3L engine

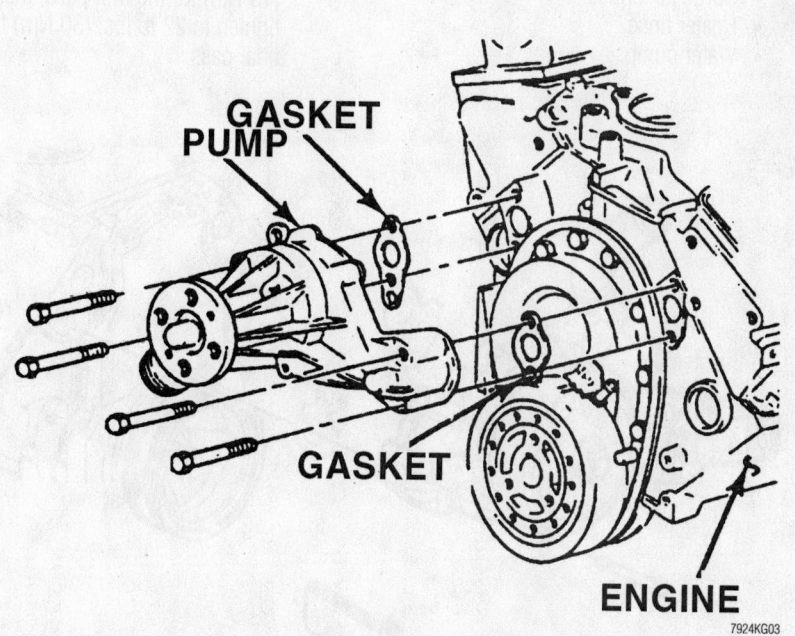

Exploded view of the water pump mounting—5.0L and 5.7L engines

Brake service is covered in Section 4 of this manual

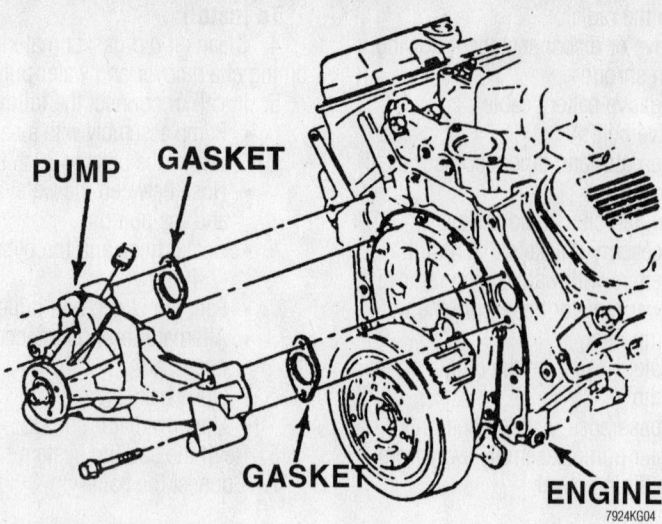

Exploded view of the water pump mounting—7.4L engine

4.8L, 5.3L and 6.0L Engines

1. Before servicing the vehicle, refer to the precautions in the beginning of this section.
2. Remove or disconnect the following:
   - Air inlet duct
   - Coolant
   - Inlet radiator hose from the water pump
   - Upper fan shroud
   - Cooling fan and clutch assembly
   - Drive belt
   - Radiator outlet hose from the coolant pump
   - Surge tank hose
   - Heater hose
   - Water pump

To install:

→ DO NOT use cooling system seal tabs (or similar compounds) unless otherwise instructed. The use of cooling system seal tabs (or similar compounds) may restrict coolant flow through the passages of the cooling system or the engine components. Restricted coolant flow may cause engine overheating and/or damage to the cooling system or the engine components/assembly.

3. Install or connect the following:
   - Water pump. Install the water pump bolts. Tighten the bolts to 11 ft. lbs. (15 Nm) for the first pass; then tighten to 22 ft. lbs. (30 Nm) for the final pass.

   - Water pump drive belt pulley and bolts (if applicable). Tighten the bolts to 89 inch lbs. (10 Nm) for the first pass; then tighten to 18 ft. lbs. (25 Nm) for the final pass.
   - Surge tank hose
   - Heater hose
   - Outlet radiator hose to the coolant pump
   - Drive belt
   - Cooling fan and clutch assembly
   - Upper fan shroud
   - Inlet radiator hose to the water pump
   - Air inlet duct
   - Coolant

8.1L Engines

1. Before servicing the vehicle, refer to the precautions in the beginning of this section.
2. Remove or disconnect the following:
   - Coolant
   - Drive belt
   - Fan clutch
   - Outlet hose clamp and hose
3. Reposition the bypass hose clamps at the water pump and water crossover.
   - Bypass hose
   - Water pump bolt and pump. Discard the water pump gaskets.

To install:

4. Install or connect the following:
   - New water pump gaskets
   - Water pump and bolts. Tighten the water pump bolts 37 ft. lbs. (50 Nm).
   - Bypass hose and clamps
   - Outlet hose and clamp
   - Fan clutch
   - Drive belt
   - Surge tank hose
   - Heater hose
   - Outlet radiator hose to the coolant pump
   - Drive belt
   - Cooling fan and clutch assembly
   - Upper fan shroud
   - Inlet radiator hose to the water pump
   - Air inlet duct
   - Coolant

Cylinder Head

REMOVAL & INSTALLATION

4.3L Engine

1. Before servicing the vehicle, refer to the precautions in the beginning of this section.

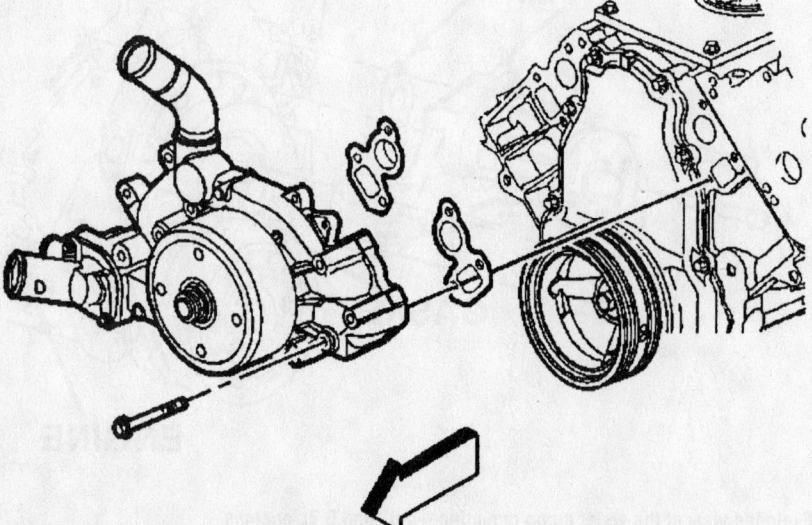

Exploded view of the water pump assembly—4.8L, 5.3L and 6.0L engines

2. Drain the coolant.
3. Remove or disconnect the following:
   - Negative battery cable
   - Engine cover, if equipped
   - Intake manifold
   - Exhaust manifold
   - Air pipe at the rear of the right cylinder head, if applicable
   - Alternator mounting bolt at the right cylinder head
   - Alternator, if necessary
   - Power steering pump and brackets from the left cylinder head and lay aside
   - Air conditioner compressor, and lay aside
   - Spark plug wires at their brackets
   - Ground strap from the right side and the coolant sensor wire from the left head
   - Cylinder cover
   - Spark plugs
   - Pushrods. Identify the pushrods so that they can be installed in their original positions.
   - Cylinder head bolts in the reverse order of the tightening sequence
   - Cylinder head and gasket

To install:

4. Clean all gasket mating surfaces.
5. Install or connect the following:
   - New gasket. Be sure the gasket has the word **HEAD** up.
   - Cylinder head

➡Coat a steel gasket on both sides with sealer. If a composition gasket is used, do not use sealer.

6. Clean the cylinder head bolts, apply sealer to the threads, and hand-tighten.

7. Install the cylinder head bolts in sequence to 22 ft. lbs. (30 Nm) The bolts must, then be tightened again in sequence in the following order:
   a. Step 1: Short bolts (11, 7, 3, 2, 6, 10), plus 55 degrees.
   b. Step 2: Medium bolts (12, 13), plus 65 degrees.
   c. Step 3: Long bolts (1, 4, 8, 5, 9), plus 75 degrees.
8. Install or connect the following:
   - Pushrods, in their original locations. Adjust the rocker arms, if necessary
   - Spark plugs
   - Rocker arm cover
   - Air conditioner compressor
   - Power steering pump and brackets
   - Alternator or the alternator mounting bolt at the cylinder head
   - Air pipe at the rear of the head if removed
   - Exhaust manifold
   - Intake manifold
   - Engine cover, if removed
9. Refill the engine with coolant.
10. Connect the negative battery cable.

4.8L, 5.3L and 6.0L Engines

RIGHT SIDE

> ※※ **CAUTION**
>
> **Before servicing any electrical component, the ignition key must be in the OFF or LOCK position and all electrical loads must be OFF, unless instructed otherwise in these procedures.**

1. Before servicing the vehicle, refer to the precautions in the beginning of this section.
2. Remove or disconnect the following:
   - Negative battery cable
   - Intake manifold
   - Push rods
   - Exhaust manifold(s)
   - Alternator
   - Three bolts holding the alternator mounting bracket to the cylinder head
   - Bolt behind the power steering pump
   - Alternator mounting bracket and set it aside
   - Bolt holding the oil level indicator tube to the right side cylinder head
   - Oil level indicator tube
   - Spark plugs from the cylinder head

➡**The M11 cylinder head bolts are NOT reusable. Install NEW M11 cylinder head bolts during assembly.**

   - Cylinder head bolts
   - Cylinder head(s) from the engine

➡**After removal, place the cylinder head on two wood blocks to prevent damage.**

3. Remove and discard the gasket. Discard the M11 cylinder head bolts.

To install:

➡**Do not use any type of sealant on the cylinder head gasket (unless specified). The cylinder head gaskets must be installed in the proper direction and position.**

4. Clean the engine block cylinder head bolt holes (if required). Thread repair tool J 42385-107 may be used to clean the threads of old threadlocking material.
5. Spray cleaner GM P/N 12346139, P/N 12377981, or equivalent into the hole.
6. Clean the cylinder head bolt holes with compressed air.
7. Check the cylinder head locating pins for proper installation.

➡**When properly installed, the tab on the right cylinder head gasket will be located right of center or closer to the front of the engine.**

8. Install or connect the following:
   - NEW right cylinder head gasket onto the locating pins
   - Cylinder head onto the locating pins and the gasket
   - NEW M11 cylinder head bolts.

Cylinder head bolt tightening sequence—4.3L engine

For complete Engine Mechanical specifications, see Section 1 of this manual

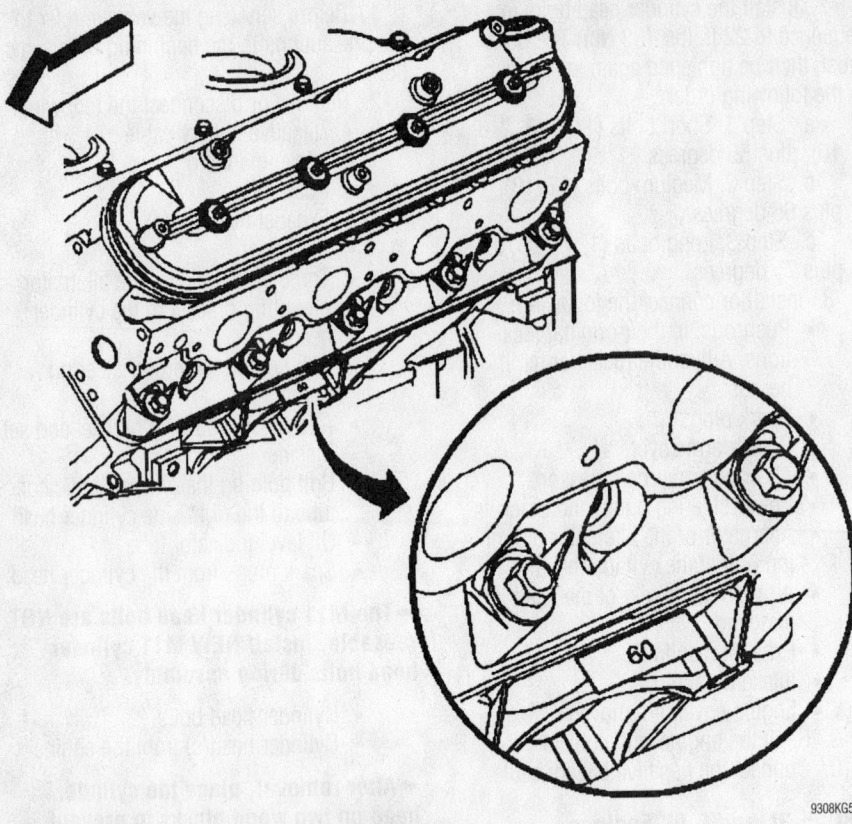

Locating tab—4.8L, 5.3L and 6.0L engines

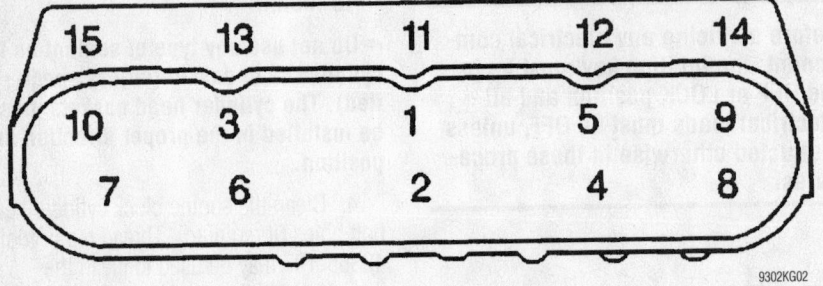

Cylinder head bolt tightening sequence—4.8L, 5.3L and 6.0L engines

Apply a 0.20 in. (5mm) band of threadlock GM P/N 12345382 or equivalent to the threads of the M8 cylinder head bolts.

- M8 cylinder head bolts.

9. Tighten the cylinder head bolts, as follows:

a. M11 bolts, first pass: in sequence to 22 ft. lbs. (30 Nm).

b. M11 bolts, second pass: in sequence + 90 degrees.

c. M11 bolts (1,2,3,4,5,6,7,8): + 90 degrees.

d. M11 bolts (9 and 10): + 50 degrees.

e. M8 bolts (11,12,13,14,15): to 22 ft.

lbs. (30 Nm). Begin with the center bolt (11) and alternating side-to-side, work outward tightening all of the bolts.

➡The cylinder head gasket displacement can be verified by markings visible on the underside of the right gasket locating tab. Some 4.8 and 5.3L head gaskets may have 53 stamped onto the locating tab. Some 6.0L head gaskets may have 60 stamped onto the locating tab.

- Install or connect the following:
- Alternator
- Exhaust manifold(s)
- Pushrods

- Intake manifold
- Negative battery cable

LEFT SIDE

✳✳ CAUTION

Before servicing any electrical component, the ignition key must be in the OFF or LOCK position and all electrical loads must be OFF, unless instructed otherwise in these procedures.

1. Before servicing the vehicle, refer to the precautions in the beginning of this section.

2. Remove or disconnect the following:
- Negative battery cable
- Intake manifold
- Pushrods
- Exhaust manifold(s)
- Alternator
- Three bolts holding the alternator mounting bracket to the cylinder head
- The bolt behind the power steering pump
- Alternator mounting bracket and set it aside
- Bolt holding the oil level indicator tube to the right side cylinder head
- Oil level indicator tube
- Spark plugs from the cylinder head

➡The M11 cylinder head bolts are NOT reusable. Install NEW M11 cylinder head bolts during assembly.

- Cylinder head bolts
- Cylinder head(s) from the engine

➡After removal, place the cylinder head on two wood blocks to prevent damage.

3. Remove and discard the gasket. Discard the M11 cylinder head bolts.

To install:

➡Do not use any type of sealant on the cylinder head gasket (unless specified). The cylinder head gaskets must be installed in the proper direction and position.

4. Clean the engine block cylinder head bolt holes (if required). Thread repair tool J 42385-107 may be used to clean the threads of old threadlocking material.

5. Spray cleaner GM P/N 12346139, P/N 12377981, or equivalent into the hole.

6. Clean the cylinder head bolt holes with compressed air.

7. Check the cylinder head locating pins for proper installation.

➡**When properly installed, the tab on the left cylinder head gasket will be located left of center or closer to the front of the engine.**

8. Install or connect the following:
- NEW left cylinder head gasket onto the locating pins
- Cylinder head onto the locating pins and the gasket
- NEW M11 cylinder head bolts. Apply a 0.20 in. (5mm) band of threadlock GM P/N 12345382 or equivalent to the threads of the M8 cylinder head bolts.
- M8 cylinder head bolts

9. Tighten the cylinder head bolts, as follows:
   a. M11 bolts, first pass: in sequence to 22 ft. lbs. (30 Nm).
   b. M11 bolts, second pass: in sequence + 90 degrees.
   c. M11 bolts (1,2,3,4,5,6,7,8): + 90 degrees.
   d. M11 bolts (9 and 10): + 50 degrees.
   e. M8 bolts (11,12,13,14,15): to 22 ft. lbs. (30 Nm). Begin with the center bolt (11) and alternating side-to-side, work outward tightening all of the bolts.

➡**The cylinder head gasket displacement can be verified by markings visible on the top side of the left gasket locating tab. Some 4.8 and 5.3L head gaskets may have 53 stamped onto the locating tab. Some 6.0L head gaskets may have 60 stamped onto the locating tab.**

10. Install or connect the following:
- Alternator mounting bracket. Tighten the four bolts to 37 ft. lbs. (50 Nm).
- Bolt at the rear of the power steering pump and tighten to 37 ft. lbs. (50 Nm).
- Exhaust manifold(s)
- Pushrods
- Intake manifold
- Negative battery cable

5.0L and 5.7L Engines

1. Before servicing the vehicle, refer to the precautions in the beginning of this section.
2. Drain the coolant.
3. Remove or disconnect the following:
- Negative battery cable
- Engine cover
- Coolant recovery reservoir, if applicable

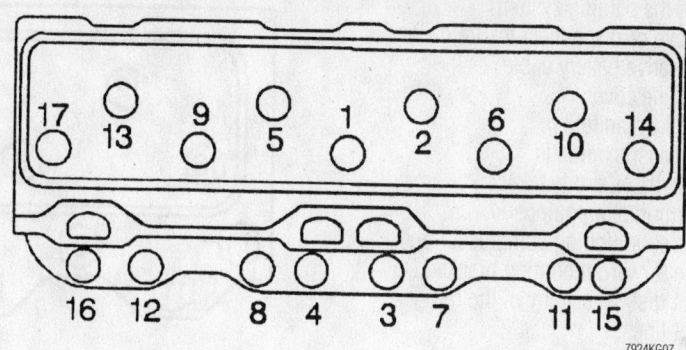

Cylinder head bolt tightening sequence—5.0L and 5.7L engines

- Intake manifold
- Exhaust manifolds and position them out of the way
- Ground strap at the rear of the right AIR pipe, if equipped
- Air conditioning compressor and the forward mounting bracket and lay the compressor aside, if equipped. Do not disconnect any of the refrigerant lines.
- Exhaust Gas Recirculation (EGR) inlet tube

4. On the right side cylinder head, disconnect the fuel pipe, spark plug wires and wiring harness bracket.
- Nut and stud attaching the main accessory bracket to the cylinder head.

➡**You may have to loosen the remaining bolts and studs in order to remove the head.**

- Coolant sensor wire
- Spark plug wire bracket
- Cylinder head covers
- Spark plugs
- Pushrods, Identify the pushrods so that they can be installed in their original positions.
- Cylinder head bolts in the reverse order of the tightening sequence
- Cylinder head(s)

To install:
5. Inspect the cylinder head and block mating surfaces. Clean all old gasket material.
6. Install or connect the following:
- Cylinder heads using new gaskets. Install the gaskets with the word **HEAD** up.

➡**Coat a steel gasket on both sides with sealer. If a composition gasket is used, do not use sealer.**

7. Clean the bolts, apply sealer to the threads, and hand-tighten.
8. Install the cylinder head bolts and tighten, in sequence, to 22 ft. lbs. (30 Nm). The bolts must be tightened once, then be tightened again in sequence in the following order:
   a. Step 1: Short bolts (3, 4, 7, 8, 11, 12, 15, 16), plus 55 degrees.
   b. Step 2: Medium bolts (14, 17), plus 65 degrees.
   c. Step 3: Long bolts (1, 2, 5, 6, 9, 10, 13), plus 75 degrees.
9. Install or connect the following:
- Pushrods, in their original positions
- Cylinder head covers
- Spark plugs
- Coolant sensor wire
- Spark plug wire bracket
- Main accessory bracket to the cylinder head
- EGR vent tube
- Fuel pipe
- Spark plug wires
- Wiring harness bracket
- Air conditioning compressor and forward mounting bracket, if equipped
- Ground strap to the rear of the right AIR pipe
- Exhaust manifolds
- Intake manifold
- Coolant recovery reservoir, if removed
- Engine cover.
- Negative battery cable
10. Refill the engine with coolant.

7.4L Engines

1. Before servicing the vehicle, refer to the precautions in the beginning of this section.

For Accessory Drive Belt illustrations, see Section 1 of this manual

2. Drain the cooling system.
3. Remove or disconnect the following:
- Negative battery cable
- Engine cover
- Intake manifold
- Exhaust manifolds
- Alternator and bracket
- Air pump, if equipped
- Air conditioning compressor and the forward mounting bracket. Do not disconnect any of the refrigerant lines.
- Rocker arm cover
- Spark plugs
- Air pipes at the rear of the head, if equipped
- Ground strap at the rear of the head
- Temperature sensor wire
- Pushrods. Identify the pushrods so that they can be installed in their original positions.
- Cylinder head bolts and the heads

To install:

➡**The cylinder head should be cleaned and inspected for warpage or damage before installation.**

4. Thoroughly clean the mating surfaces of the head and block. Clean the bolt holes thoroughly.

➡**Coat a steel gasket on both sides with sealer. If a composition gasket is used, do not use sealer.**

5. Install or connect the following:
- New gaskets, with the word **HEAD** up
- Cylinder heads
- Cylinder head bolts, apply sealer to the threads, and hand-tighten.
6. Tighten the head bolts a little at a time, in the proper sequence, in 3 stages:
  a. Step 1: Torque to 30 ft. lbs. (40 Nm).
  b. Step 2: Torque to 60 ft. lbs. (80 Nm).
  c. Step 3: Torque to 85 ft. lbs. (115 Nm).
- Intake and exhaust manifolds
- Pushrods
- Rocker arms
- Temperature sensor wire
- Ground strap at the rear of the head
- AIR pipes at the rear of the head
- Spark plugs
- Rocker arm cover
- Air conditioning compressor and the forward mounting bracket
- AIR pump
- Alternator
- Engine cover
7. Connect the battery cable and refill the cooling system.

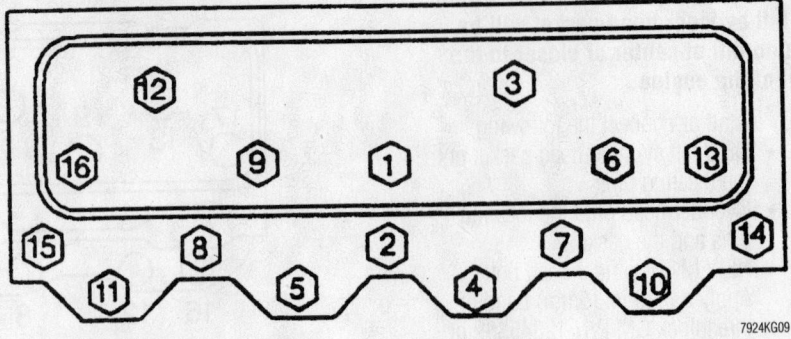

Cylinder head bolt tightening sequence—7.4L engines

8.1L Engine

LEFT SIDE

1. Before servicing the vehicle, refer to the precautions in the beginning of this section.
2. Drain the cooling system.
3. Remove or disconnect the following:
- Negative battery cable
- Intake manifold
- Valve cover
- Rocker arms and pushrods, keeping them in order for installation
- Engine harness ground bolts
4. Reposition the engine harness grounds and ground straps from the cylinder head.
- Water crossover
- Exhaust manifold
- Cylinder head bolts, then discard

➡**The cylinder head bolts must be replaced for installation.**

- Cylinder head. Place the head on 2 wood blocks to protect the sealing surfaces while it is removed.

To install:

➡**The cylinder head should be cleaned and inspected for warpage or damage before installation.**

5. Thoroughly clean the mating surfaces of the head and block. Clean the bolt holes thoroughly.

➡**If a composition gasket is used, do not use sealer.**

6. Align the cylinder head gasket locating marks to face up. Make sure that the gasket tabs are located of the no. 1 and 2 cylinder for proper installation.
7. Install or connect the following:
- New cylinder head gasket
- Cylinder head
- Sealer to the threads of new cylinder head bolts, if not pre-applied
- Cylinder head bolts and hand-tighten
8. Tighten the head bolts a little at a time, in the proper sequence, in 3 stages:
  a. Step 1: Torque to 30 ft. lbs. (40 Nm).
  b. Step 2: Tighten an additional 120 degrees using a torque angle meter.
  c. Step 3: Torque bolt numbers. 1, 2, 3, 6, 7, 8, 9, 10, 11, 14, 16 and 17 an additional 60 degrees. Tighten bolts 15 and 18 an additional 45 degrees, and bolt numbers 4, 5, 12 and 13 an additional 30 degrees.
9. Install or connect the following:
- Exhaust manifold

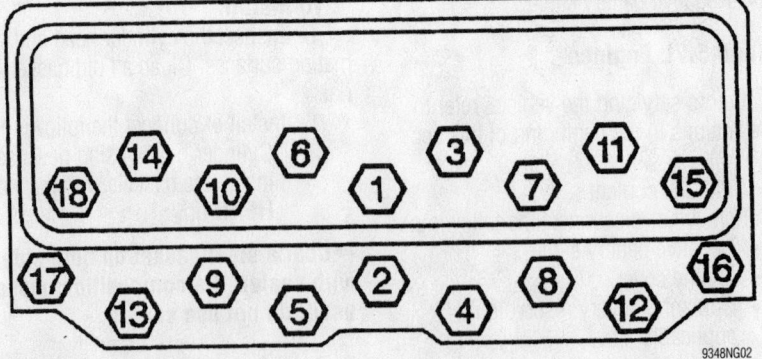

Cylinder head bolt tightening sequence—8.1L engine

- Water crossover
- Engine harness grounds and ground strap
- Rocker arms and pushrods
- Valve cover
- Intake manifold

10. Connect the battery cable and refill the cooling system.

RIGHT SIDE

1. Before servicing the vehicle, refer to the precautions in the beginning of this section.
2. Drain the cooling system.
3. Remove or disconnect the following:
   - Negative battery cable
   - Intake manifold
   - Valve cover
   - Rocker arms and pushrods, keeping them in order for installation
   - Engine Coolant Temperature (ECT) sensor clip from the bracket
   - ECT sensor
   - ECT sensor bracket bolt and bracket
   - Heater inlet and outlet hoses from the hose bracket
   - Water crossover
   - Exhaust manifold
   - Cylinder head bolts, then discard

➡ **The cylinder head bolts must be replaced for installation.**

- Cylinder head. Place the head on 2 wood blocks to protect the sealing surfaces while it is removed.

To install:

➡ **The cylinder head should be cleaned and inspected for warpage or damage before installation.**

4. Thoroughly clean the mating surfaces of the head and block. Clean the bolt holes thoroughly.

➡ **If a composition gasket is used, do not use sealer.**

5. Align the cylinder head gasket locating marks to face up. Make sure that the gasket tabs are located of the no. 1 and 2 cylinder for proper installation.
6. Install or connect the following:
   - New cylinder head gasket
   - Cylinder head
   - Sealer to the threads of new cylinder head bolts, if not pre-applied
   - Cylinder head bolts and hand-tighten
7. Tighten the head bolts a little at a time, in the proper sequence, in 3 stages:
   a. Step 1: Torque to 30 ft. lbs. (40 Nm).

b. Step 2: Tighten an additional 120 degrees using a torque angle meter.
c. Step 3: Torque bolt numbers. 1, 2, 3, 6, 7, 8, 9, 10, 11, 14, 16 and 17 an additional 60 degrees. Tighten bolts 15 and 18 an additional 45 degrees, and bolt numbers 4, 5, 12 and 13 an additional 30 degrees.
8. Install or connect the following:
   - Exhaust manifold
   - Water crossover
   - Heater hose bracket and bolts. Tighten the bolts to 37 ft. lbs. (50 Nm).
   - ECT sensor bracket and bolt. Tighten to 37 ft. lbs. (50 Nm).
   - ECT sensor
   - ECT sensor clip
   - Rocker arms and pushrods
   - Valve cover
   - Intake manifold
9. Connect the battery cable and refill the cooling system.

Rocker Arms

REMOVAL & INSTALLATION

4.3L, 5.0L and 5.7L Engines

1. Before servicing the vehicle, refer to the precautions in the beginning of this section.

2. Remove or disconnect the following:
   - Engine cover
   - Cylinder head cover.
   - Rocker arm nut. If you are only replacing the pushrod, back the nut off until you can swing the rocker out of the way.
   - Rocker arms and balls as a unit

➡ **Always remove each set of rocker arms (1 set per cylinder) as a unit.**

- Pushrods and pushrod guides

To install:

3. Install or connect the following:
   - Pushrods and their guides. Be sure that they seat properly in each lifter.
4. Position a set of rocker arms (for 1 cylinder) in the proper location.

➡ **Install the rocker arms for each cylinder only when the lifters are off the cam lobe and both valves are closed.**

5. Coat the replacement rocker arm with Molykote® or its equivalent, and the rocker arm and pivot with SAE 90 gear oil, and install the pivots.
6. Install or connect the following:
   - Nuts and tighten alternately
   - Engine cover

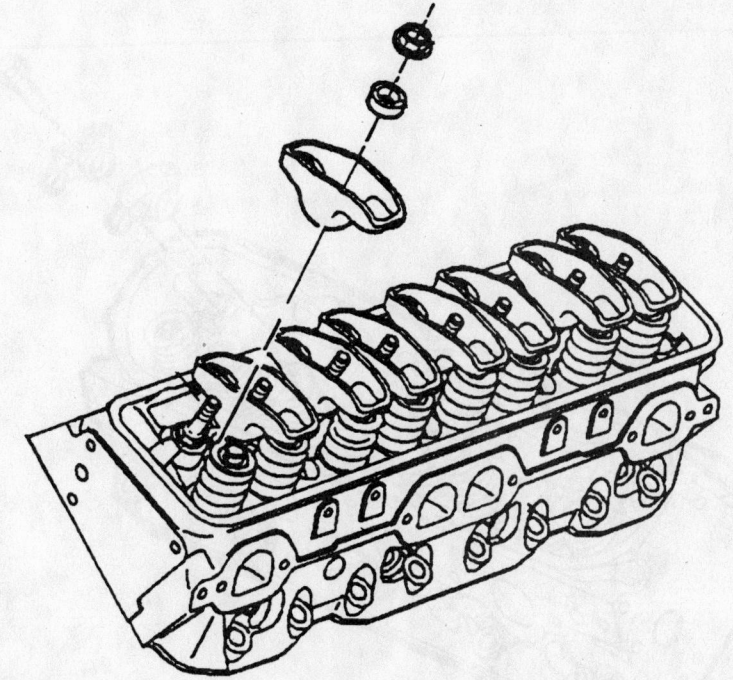

7924KG10

Exploded view to the rocker arm and related components—4.3L, 5.0L and 5.7L engines

4.8L, 5.3L and 6.0L Engines

➡ Do not remove the ignition coils from the valve rocker arm cover unless required. Do not remove the oil fill tube from the cover unless service is required. If the oil fill tube has been removed from the cover, install a NEW tube during assembly.

On the right side:

1. Remove or disconnect the folloing:
   - Ignition coil bracket bolts from the rocker arm cover (if required)
   - Ignition coil and bracket assembly from the cover
   - Valve rocker arm cover bolts
   - Valve rocker arm cover
   - Gasket from the cover. Discard the gasket. The bolt grommets may be reused if not damaged.
   - Oil fill cap from the oil fill tube
   - Oil fill tube (if required). Discard the oil fill tube.

On the left side:

➡ Do not remove the Positive Crankcase Ventilation (PCV) valve grommet from the cover unless service is required.

2. Remove or disconnect the following:
   - Ignition coil bracket bolts from the rocker arm cover (if required)
   - Ignition coil and bracket assembly from the cover
   - Valve rocker arm cover bolts
   - Valve rocker arm cover
   - Gasket from the cover. Discard the gasket. The bolt grommets may be reused if not damaged.
   - Valve rocker arm bolts
   - Valve rocker arms
   - Valve rocker arm pivot support
   - Pushrods

To install:

➡ Valve lash is built in. No valve adjustment is required.

3. Lubricate the valve rocker arms and pushrods with clean engine oil.
4. Lubricate the flange of the valve rocker arm bolts with clean engine oil.
5. Lubricate the flange or washer surface of the bolt that will contact the valve rocker arm.
6. Install or connect the following:
   - Valve rocker arm pivot support

➡ Make sure that the pushrods seat properly to the valve lifter sockets.

   - Pushrods

➡ Make sure that the pushrods seat properly to the ends of the rocker arms.

   - Rocker arms and bolts. DO NOT tighten the rocker arm bolts at this time

7. Rotate the crankshaft until number one piston is at top dead center of compression stroke. In this position, cylinder number one rocker arms will be off lobe lift, and the crankshaft sprocket key will be at the 1:30 position. If viewing from the rear of the engine, the additional crankshaft pilot hole (non-threaded) will be in the 10:30 position. The engine firing order is 1, 8, 7, 2, 6, 5, 4, 3. Cylinders 1, 3, 5 and 7 are left bank. Cylinders 2, 4, 6, and 8 are right bank.

8. With the engine in the number one firing position, tighten the following valve rocker arm bolts:
   a. Tighten exhaust valve rocker arm bolts 1, 2, 7, and 8 to 22 ft. lbs. (30 Nm)
   b. Tighten intake valve rocker arm bolts 1, 3, 4, and 5 to 22 ft. lbs. (30 Nm)
9. Rotate the crankshaft 360 degrees. Tighten the following valve rocker arm bolts:
   a. Tighten exhaust valve rocker arm bolts 3, 4, 5, and 6 to 22 ft. lbs. (30 Nm)
   b. Tighten intake valve rocker arm bolts 2, 6, 7, and 8 to 22 ft. lbs. (30 Nm)

On the right side:

➡ The valve rocker arm cover bolt grommets may be reused. If the oil fill tube has been removed from the valve rocker arm cover, install a NEW oil fill tube during assembly.

10. Lubricate the O-ring seal of the NEW oil fill tube with clean engine oil.
11. Install or connect the following:
    - NEW oil fill tube into the rocker arm cover and rotate the tube clockwise until locked in the proper position
    - Oil fill cap into the tube and rotate clockwise until locked in the proper position
    - NEW cover gasket into the valve rocker arm cover
    - Valve rocker arm cover onto the cylinder head
    - Cover bolts with grommets and tighten to 106 inch lbs. (12 Nm)
12. Apply threadlock GM P/N 12345382 or equivalent to the threads of the bracket bolts.
    - Ignition coil and bracket assembly and bolts. Tighten to 106 inch lbs. (12 Nm).

On the left side:

➡ DO NOT reuse the valve rocker arm cover gasket. The valve rocker arm cover bolt grommets may be reused. If the vapor vent grommet has been removed from the valve rocker arm cover, install a NEW vapor vent gourmet during assembly.

13. Install or connect the following:

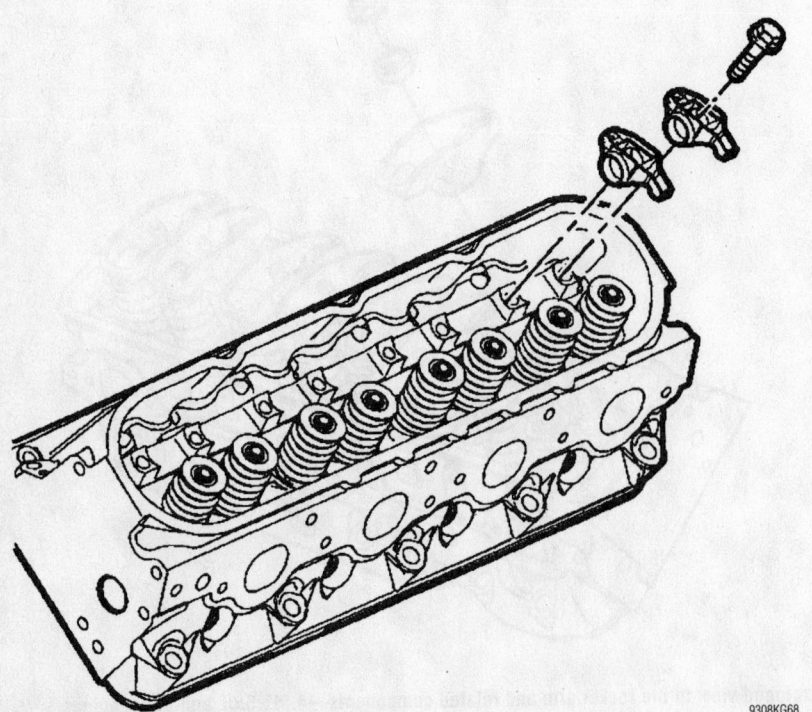

Rocker arm removal—4.8L, 5.3L and 6.0L engines

9308KG68

- NEW cover gasket (1) into the valve rocker arm cover
- Valve rocker arm cover onto the cylinder head
- Cover bolts with grommets and tighten to 106 inch lbs. (12 Nm)

14. Apply threadlock GM P/N 12345382 or equivalent to the threads of the bracket bolts.

- Ignition coils and bracket assembly and bolts. Tighten the bolts to 106 inch lbs. (12 Nm).

7.4L Engines

1. Before servicing the vehicle, refer to the precautions in the beginning of this section.

2. Remove or disconnect the following:
- Engine cover
- Cylinder head cover
- Rocker arm bolt. If you are only replacing the pushrod, back the nut off until you can swing the rocker out of the way.
- Rocker arms and balls as a unit

➡**Always remove each set of rocker arms (1 set per cylinder) as a unit.**

- Pushrods and pushrod guides

To install:

3. Install or connect the following:
- Pushrods and their guides, Be sure that they seat properly in each lifter.

4. Position a set of rocker arms (for 1 cylinder) in the proper location.

➡**Install the rocker arms for each cylinder only when the lifters are off the cam lobe and both valves are closed.**

5. Coat the replacement rocker arm with Molykote® or its equivalent, and the rocker arm and pivot with SAE 90 gear oil, and install the pivots.
- Rocker arm bolts and tighten to 45 ft. lbs. (61 Nm)
- Engine cover

8.1L Engine

➡**Always make sure to keep all removed valve train components in order for reassembly. They must be installed in the same position from which they were removed.**

1. Before servicing the vehicle, refer to the precautions in the beginning of this section.

2. Remove or disconnect the following:

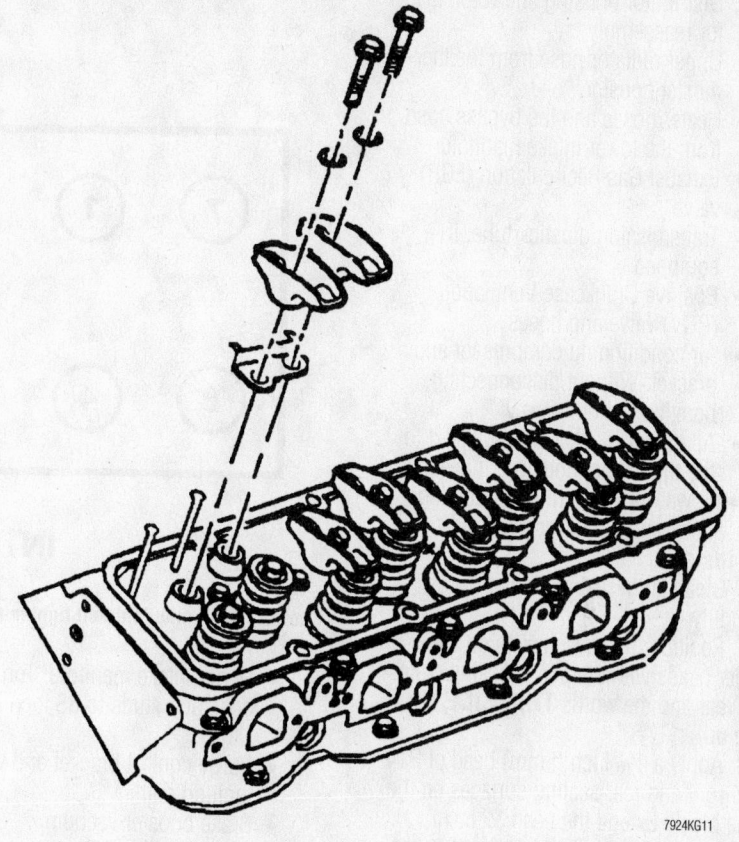

Exploded view of the rocker arms and related components—7.4L engines

- Valve (rocker arm) cover
- Rocker arm nuts, balls and rocker arms

➡**The intake pushrods are shorter than the exhaust pushrods.**

- Pushrods
- Rocker arm guides and pushrod guides

3. Clean and inspect all components for damage.

To install:

4. Apply a suitable sealer to the rocker arm stud-to-cylinder head threads.

5. Install or connect the following:
- Pushrod guides and rocker arm studs. Tighten to 37 ft. lbs. (50 Nm).
- Pushrods

6. Coat the rocker arm and ball bearing surfaces with a suitable prelube.
- Rocker arms, balls and nuts. Tighten the nuts slowly to 18 ft. lbs. (25 Nm) while guiding the tips of the rocker arms over the tips of the valves.
- Valve (rocker arm) cover

Intake Manifold

REMOVAL & INSTALLATION

4.3L Engine

1. Before servicing the vehicle, refer to the precautions in the beginning of this section.

2. Relieve the fuel system pressure

3. Remove or disconnect the following:
- Negative battery cable
- Air intake duct
- Wiring harness connectors and brackets from the manifold
- Throttle linkage and bracket from the upper manifold
- Cruise control cable, if equipped
- Fuel lines at the rear of the lower intake manifold
- Brake booster vacuum hose from the upper intake manifold
- Ignition coil and bracket
- Purge solenoid and bracket
- Studs and intake manifold attaching bolts, mark for reassembly
- Upper intake manifold

- Distributor housing and rotor, mark for reassembly
- Upper radiator hose from the thermostat housing
- Heater hoses and the bypass hose from the lower intake manifold
- Exhaust Gas Recirculation (EGR) valve
- Transmission dipstick tube, if equipped
- Positive Crankcase Ventilation (PCV) valve and hoses
- Air conditioning compressor and bracket. Without disconnecting, position out of the way
- Alternator bracket and bolt next to the thermostat housing, if needed
- Lower intake manifold mounting bolts and the lower manifold

To install:

4. Clean all gasket mating surfaces thoroughly.

5. Position the new gaskets on the cylinder heads with the port blocking plates at the rear and the words **THIS SIDE UP** facing up.

6. Apply a ³⁄₁₆ inch (5mm) bead of RTV to the front and rear sealing surfaces on the engine block. Extend the bead ½ inch (13mm) up each cylinder head to retain the gasket.

7. Carefully position the lower intake manifold onto the engine.

8. Apply GM 1052080 or equivalent sealer to the lower intake manifold bolts

9. Torque the bolts using 3 steps in the sequence shown:

   a. Step 1: Torque to 24 inch lbs. (3 Nm).

   b. Step 2: Torque to 108 inch lbs. (12 Nm).

   c. Step 3: Torque to 11 ft. lbs. (15 Nm).

10. Install or connect the following:
- Alternator bracket and bolts near the thermostat housing, if removed
- Air conditioning compressor
- PCV valve and hose
- Transmission dipstick tube, if equipped
- EGR valve
- Upper radiator and bypass hose to the thermostat housing
- Distributor

11. Position the upper intake manifold gasket on the lower manifold.

⁂ WARNING

Be careful not to pinch the injector tubes between the upper and lower manifolds.

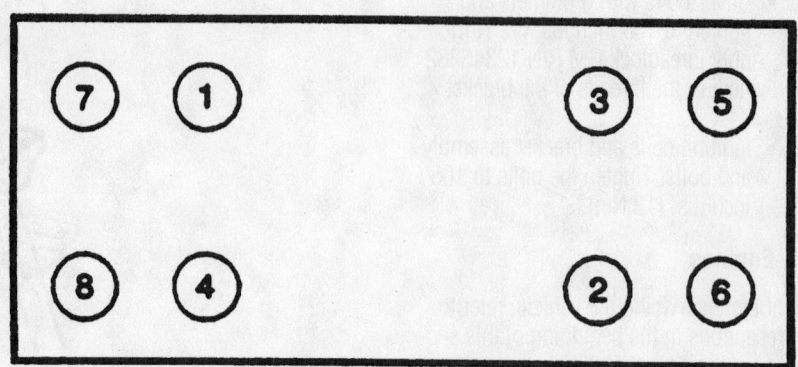

INTAKE SEQUENCE

7924KG14

Lower intake manifold bolt tightening sequence—4.3L engines

- Upper intake manifold. Torque the bolts and studs to 88 inch lbs. (10 Nm).
- Purge control bracket and valve
- Ignition coil
- Brake booster vacuum
- Fuel lines
- Accelerator cable
- Cruise control cable, if equipped
- Wiring harness brackets and connections
- Air intake duct
- Negative battery cable

12. Refill and bleed the cooling system.

13. Pressurize the fuel system and check for leaks.

4.8L, 5.3L and 6.0L Engines

➡ **The intake manifold, throttle body, fuel injection rail, and fuel injectors may be removed as an assembly. If not servicing the individual components,**

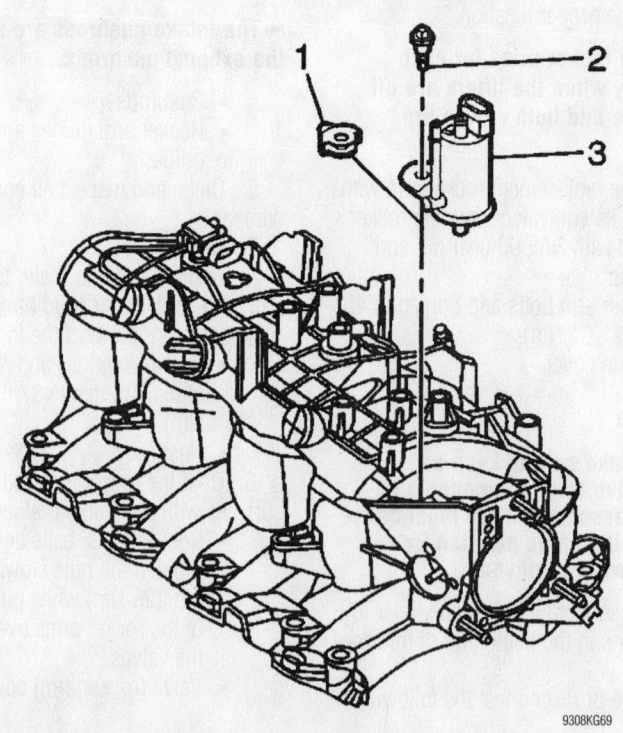

EVAP solenoid removal—4.8L, 5.3L and 6.0L engines

9308KG69

remove the manifold as a complete assembly.

1. Before servicing the vehicle, refer to the precautions in the beginning of this section.

2. Remove or disconnect the following:
- Positive Crankcase Ventilation (PCV) hose and valve
- Manifold Absolute Pressure (MAP) sensor, if required
- Engine coolant air bleed clamp and hose from the throttle body
- Accelerator control cable bracket and bolts, if required
- EVAP solenoid, bolt, and isolator
- Intake manifold bolts
- Intake manifold with gaskets
- Intake manifold-to-cylinder head gaskets from the manifold. Discard the intake manifold gaskets.
- Fuel rail with injectors
- Throttle body and gasket

3. Clean the intake manifold in solvent.

4. Dry the intake manifold with compressed air.

5. Inspect all components for damage, and replace the necessary parts.

6. Locate a straight edge across the intake manifold cylinder head deck surface. Position the straight edge across a minimum of two runner port openings.

7. Insert a feeler gauge between the intake manifold and the straight edge. A intake manifold with warpage in excess of 0.118 in. (3mm) over a 7.87 in. (200mm) area is warped and should be replaced.

To install:

8. Install or connect the following:
- MAP sensor
- EVAP solenoid, bolt, and isolator. Tighten the bolt to 89 inch lbs. (10 Nm).
- NEW intake manifold-to-cylinder head gaskets
- Intake manifold. Apply a 0.20 in. (5mm) band of threadlock GM P/N 12345382 or equivalent to the threads of the intake manifold bolts.
- Intake manifold bolts. Tighten intake manifold bolts, in sequence, to 44 inch lbs. (5 Nm). Then, tighten the bolts, in sequence, to 89 inch lbs. (10 Nm).
- PCV valve and hose
- Engine coolant air bleed hose and clamp onto the throttle body
- Accelerator control cable bracket and bolts, if applicable. Tighten the bolts to 89 inch lbs. (10 Nm).

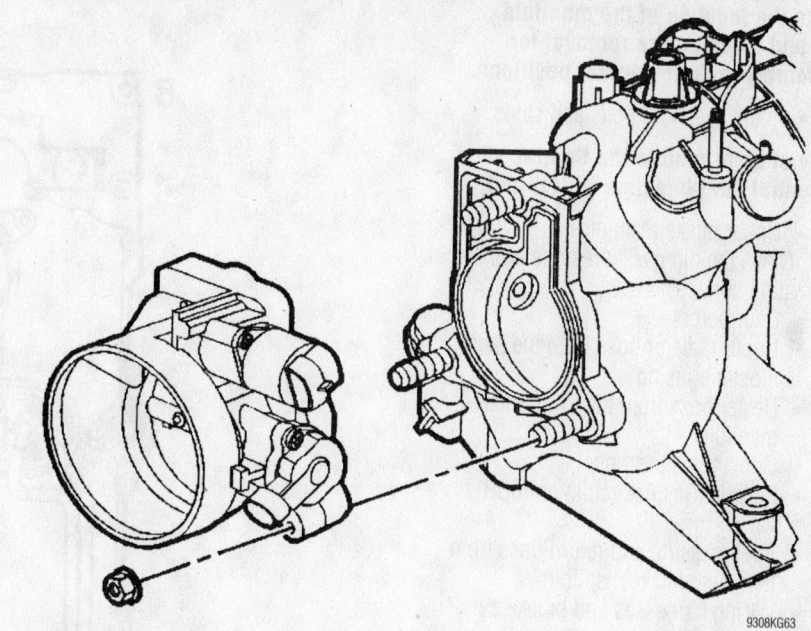

Throttle body removal—4.8L, 5.3L and 6.0L engines

9308KG63

5.0L, 5.7L and 7.4L Engines

1. Before servicing the vehicle, refer to the precautions in the beginning of this section.

2. Remove or disconnect the following:
- Negative battery cable
- Engine cover
- Air cleaner intake duct
- Coolant reservoir
- Wiring harness connectors and brackets
- Throttle linkage and bracket from the upper intake manifold
- Cruise control cable, if equipped
- Fuel lines and the bracket from the rear of the intake manifold
- Positive Crankcase Ventilation (PCV) valve and hoses
- Ignition coil and bracket
- Purge solenoid and bracket

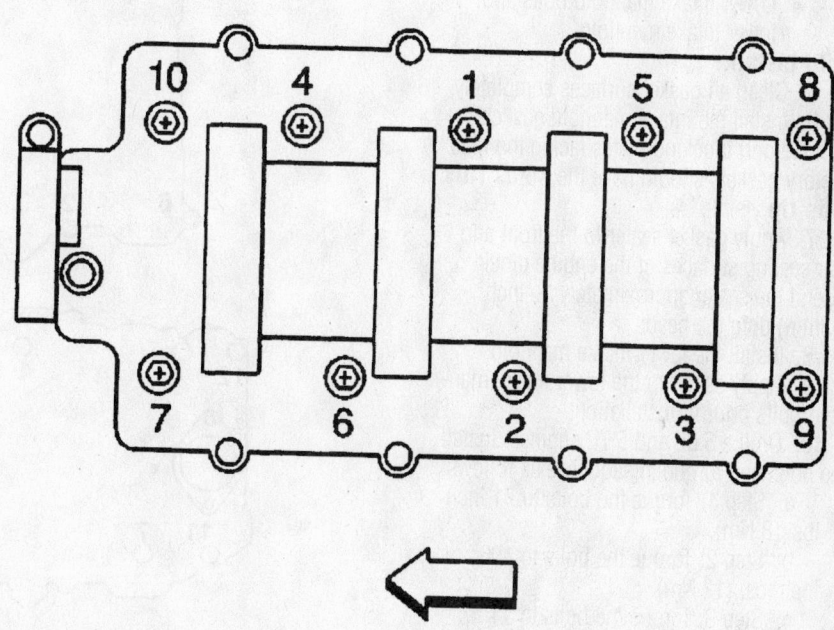

Lower intake manifold bolt tightening sequence—4.8L, 5.3L and 6.0L engines

9302KG03

➡Note the location of the manifold bolts and studs before removal for reassembly in their original positions.

- Intake manifold bolts and studs

➡**Do not disassemble the Central Sequential Fuel Injection (CSFI) unit.**

- Upper intake manifold

3. Clean the old gasket residue from both mating surfaces.

- Distributor
- Upper radiator hose from the thermostat housing
- Heater hose from the lower intake manifold
- Coolant bypass hose
- Exhaust Gas Recirculation (EGR) valve
- Fuel pressure and return lines from the lower intake manifold
- Wiring harnesses and brackets from the lower manifold
- Left side valve cover
- Transmission oil level indicator and tube, if equipped
- EGR tube, clamp and bolt
- Positive Crankcase Ventilation (PCV) valve and hoses
- Air conditioning compressor and bracket, but do not disconnect the lines

4. Loosen the compressor mounting bracket and slide it forward, but do not remove it.

- Power brake vacuum tube
- Lower intake manifold bolts and lower intake manifold

To install:

5. Clean all gasket surfaces completely.

6. Install the intake manifold gaskets with the port blocking plates facing the rear. Factory gaskets should have the words **This Side Up** visible.

7. Apply gasket sealer to the front and rear sealing surfaces of the engine block. Extend the sealer approximately ½ inch (13mm) onto the heads.

8. Install the lower intake manifold.

9. Apply sealer to the lower intake manifold bolts prior to installation.

10. On the 5.0L and 5.7L engines, install the bolts and torque in sequence as follows:

a. Step 1: Torque the bolts to 71 inch lbs. (8 Nm).

b. Step 2: Torque the bolts to 106 inch lbs. (12 Nm).

c. Step 3: Torque the bolts to 11 ft. lbs. (15 Nm).

11. On the 7.4L engine, torque the bolts to 30 ft. lbs. (40 Nm) in the sequence shown.

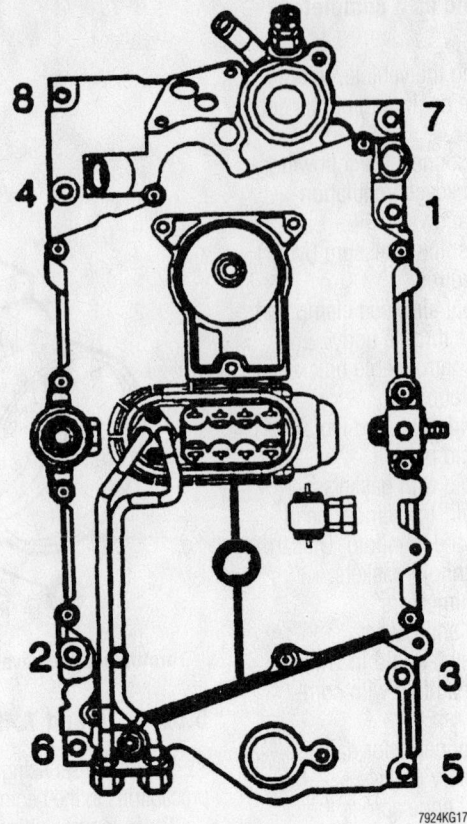

7924KG17

Lower intake manifold bolt tightening sequence—5.0L and 5.7L engines

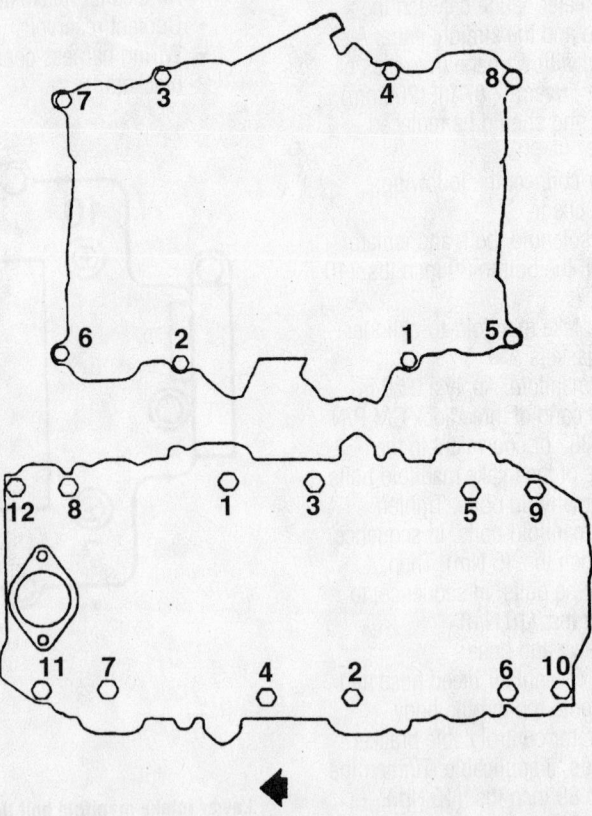

7924KG18

Upper and lower intake manifold bolt tightening sequence—7.4L engines

12. Install or connect the following:
- Power brake vacuum tube
- PCV valve and hose
- EGR tube, clamp and bolt
- Transmission oil level indicator and tube, if equipped
- Left side valve cover
- Wiring harnesses and brackets to the lower manifold
- Fuel pressure and return lines to the lower intake manifold
- EGR valve
- Coolant bypass hose
- Heater hose to the lower intake manifold
- Upper radiator hose to the thermostat housing
- Air conditioning compressor and bracket
- Distributor
- Upper intake manifold gasket
- Upper intake manifold

❋❋ WARNING

When installing the upper intake manifold be careful not to pinch the injector wires between the upper and lower intake manifolds.

- Upper intake manifold mounting bolts/studs, torque in a crisscross pattern as follows:
  a. Step 1: Torque to 44 inch lbs. (5 Nm).
  b. Step 2: Torque to 83 inch lbs. (10 Nm).
- Purge solenoid and bracket
- PCV hose
- Fuel lines and the bracket at the rear of the intake manifold
- Ignition coil and bracket
- Throttle linkage and bracket to the upper intake manifold
- Throttle linkage cable
- Cruise control cable, if equipped
- Wiring harness connectors and brackets
- Air cleaner intake duct
- Coolant recovery reservoir and the engine cover
- Negative battery cable

13. Start the vehicle and verify that there are no leaks.

8.1L Engine

➡ **The intake manifold, throttle body, fuel rail and injectors can be removed as an assembly. If you do not need to service these components individually,** **remove the manifold as a complete assembly.**

1. Before servicing the vehicle, refer to the precautions in the beginning of this section.

2. Relieve the fuel system pressure and drain the cooling system.

3. Remove or disconnect the following:
- Air cleaner outlet duct
- Intake manifold sight shield
- Fuel feed and return pipes
- Engine harness clips from the studs on the front of the dash
- Engine harness clip from the wheelhouse splash shield
- Pressure cycling switch, surge tank switch and Mass Air Flow (MAF) electrical connectors

4. Reposition the engine harness to the top of the engine
- Connector Position Assurance (CPA) retainer from the ignition coil harness
- Manifold Absolute Pressure (MAP) sensor electrical connector
- Ignition coil electrical connector
- Engine Coolant Temperature (ECT) sensor electrical connector
- Engine harness bolt and studs
- CPA retainer from the ignition coil harness
- Alternator, injector harness and ignition coil harness connectors
- Throttle Position (TP) sensor connector
- Electronic Throttle Control (ETC) and purge valve solenoid connectors

5. Reposition the engine harness to the drivers side of the engine compartment.
- Bypass valve vacuum hose from the intake manifold
- Exhaust Gas Recirculation (EGR) valve electrical connector
- EGR pipe bolts from the EGR adapter. Reposition the EGR pipe
- EGR valve pipe gasket and discard
- Secondary Air Injection (AIR) pipe nut from the fuel rail stud, if equipped
- AIR pipe bolts from the exhaust manifold
- AIR pipe from the AIR pump pipe
- AIR pipe gasket and discard
- AIR pipe nut from the fuel rail stud
- AIR pipe bolts from the exhaust manifold
- AIR pipe from the AIR pump pipe
- AIR pipe gasket

6. Reposition the AIR pump hose clamp, then remove the air pump hose from the pump pipe.
- AIR pump pipe bolt from the cylinder head
- AIR pump pipe
- Fuel pressure regulator vacuum hose
- Fuel rail studs and fuel rail, ONLY if replacing the manifold
- Intake manifold bolts

❋❋ WARNING

Do NOT try to remove the intake manifold by prying under the sealing surfaces.

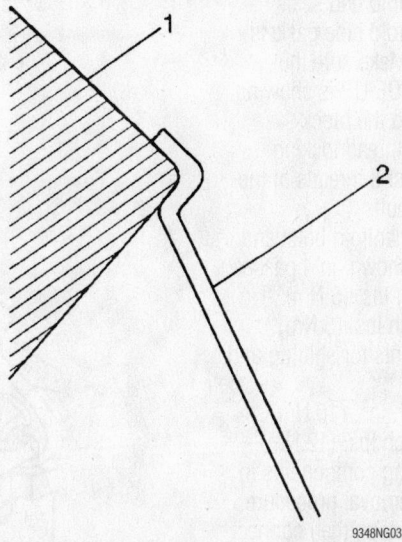

9348NG03

Make sure that the splash shield snap fits between the cylinder heads

---

For Tune-up, Capacities and Firing orders, see Section 1 of this manual

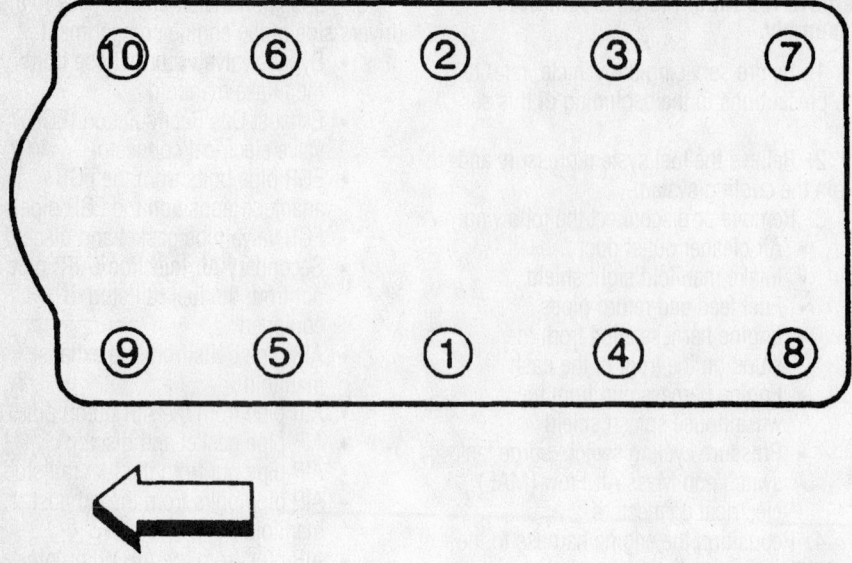

Intake manifold bolt tightening sequence—8.1L engine

9348NG04

- Intake manifold
- Intake manifold side gaskets and end seals and discard

➡**The splash shield is reusable and secured using a snap-in fit. Do not distort the shield during removal.**

- Splash shield

To install:

7. Clean all gasket surfaces completely.
8. Install or connect the following:
- Splash shield. Make sure the shield fits properly between the cylinder head.

➡**Make sure the manifold gasket tabs align with the hole in the head gasket.**

- New intake manifold end seals
- New intake manifold side gaskets onto the heads. Make sure the stamped THIS SIDE UP is showing.
- Intake manifold to the block
- Apply a suitable thread locking material to at least 8 threads of the intake manifold bolts

9. Install the intake manifold bolts and tighten, in the sequence shown, in 4 passes:
- a. 1st pass: 44 inch lbs. (5 Nm).
- b. 2nd pass: 44 inch lbs. (5 Nm). Check the manifold joints for shifting and fix as necessary.
- c. 3rd pass: 89 inch lbs. (10 Nm).
- d. 4th pass: 106 inch lbs. (12 Nm).

10. Install the remaining components in the reverse order of the removal procedure.
11. Fill the cooling system, then connect the negative battery cable
12. Start the vehicle and verify that there are no leaks.

Exhaust Manifold

REMOVAL & INSTALLATION

4.3L Engines

1. Before servicing the vehicle, refer to the precautions in the beginning of this section.

2. Remove or disconnect the following:
- Negative battery cable
- Engine cover, if equipped
- Exhaust Gas Recirculation (EGR) valve, inlet pipe (left side manifold), if necessary
- Exhaust pipe from the exhaust manifold
- Spark plug wires from the plugs and the retaining clips
- Heat shields
3. If removing the left side manifold:
a. Step 1: Remove the power steering/alternator rear bracket, if needed.
b. Step 2: Check for sufficient clearance between the manifold and the intermediate steering shaft. On some models it will be necessary to disconnect the intermediate shaft from the steering gear in order to reposition the shaft for clearance.
4. If removing the right side manifold:
a. Step 1: Remove air conditioning compressor and bracket, then position the assembly aside, if necessary. Do not disconnect the lines or allow them to become kinked or otherwise damaged.
b. Step 2: Remove the spark plugs, dipstick tube and wiring, if necessary.
5. Unbend the lock tangs.
6. Remove or disconnect the following:
- Exhaust manifold retaining bolts, washers and tab washers

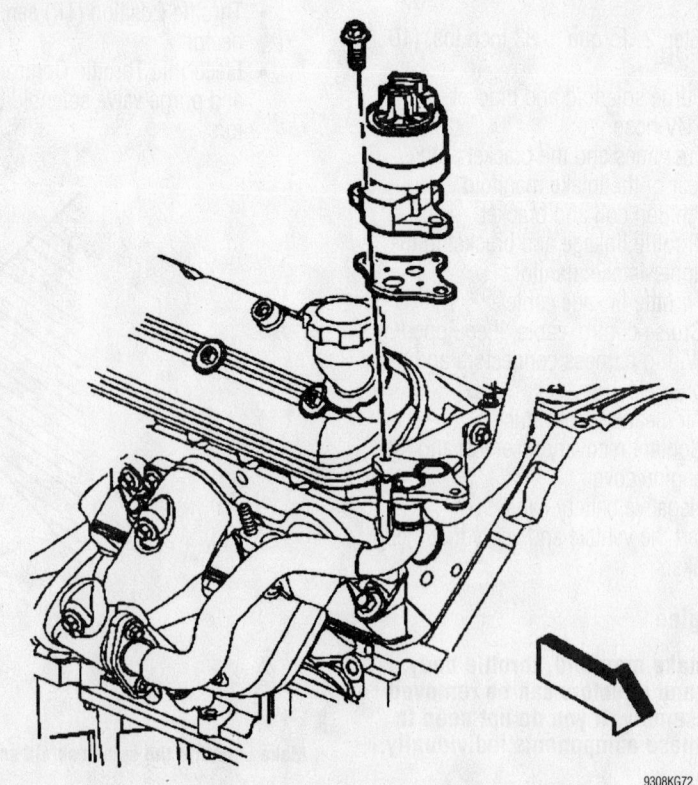

EGR valve removal—4.8L, 5.3L and 6.0L engines

9308KG72

- Exhaust manifold
- Old gaskets and discard

To install:

7. Clean the gasket mounting surfaces.

8. Inspect the exhaust manifold for distortion, cracks or damage; replace if necessary.

9. Install or connect the following:

- Exhaust manifold to the cylinder using a new gasket. Torque the exhaust manifold-to-cylinder head bolts to 26 ft. lbs. (36 Nm) on the center exhaust tube and to 20 ft. lbs. (28 Nm) on the front and rear exhaust tubes.

➡ **Once the bolts are tightened, bend the tabs on the washers back over the heads of all bolts in order to lock them in position.**

10. If the right manifold was removed, install the following:

- Spark plugs
- Dipstick tube
- Wiring
- Air conditioning compressor and bracket assembly, if unbolted

11. If the left manifold was removed, install the following:

- Intermediate shaft to the steering gear, if unbolted
- Power steering/alternator rear bracket
- Air cleaner along with the heat stove pipe and cold air intake pipe
- Spark plug wires to the retainer clips and plugs
- Exhaust pipe to the manifold
- Engine cover, if equipped
- Negative battery cable

4.8L, 5.3L and 6.0L Engines

RIGHT SIDE

➡ **Do not remove the Exhaust Gas Recirculation (EGR) valve from the pipe assembly unless service is required.**

1. Before servicing the vehicle, refer to the precautions in the beginning of this section.

2. Remove or disconnect the following:

- EGR valve, gasket, and bolts
- EGR pipe bolt from the intake manifold
- EGR pipe bolts and gasket from the exhaust manifold
- EGR pipe bolts from the cylinder head
- EGR pipe assembly. With mild

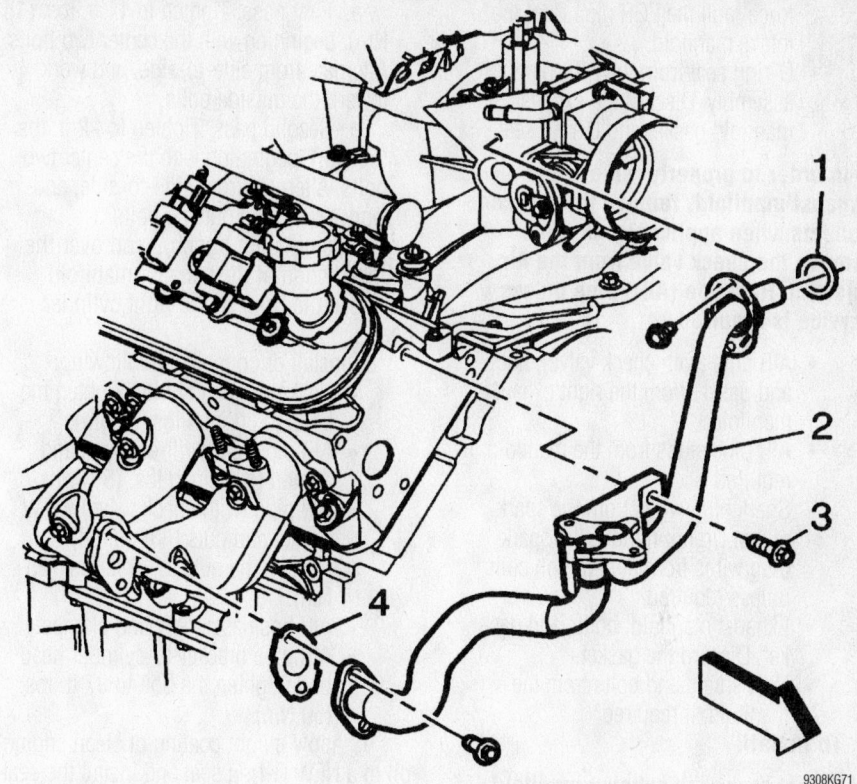

EGR pipe removal—4.8L, 5.3L and 6.0L engines

9308KG71

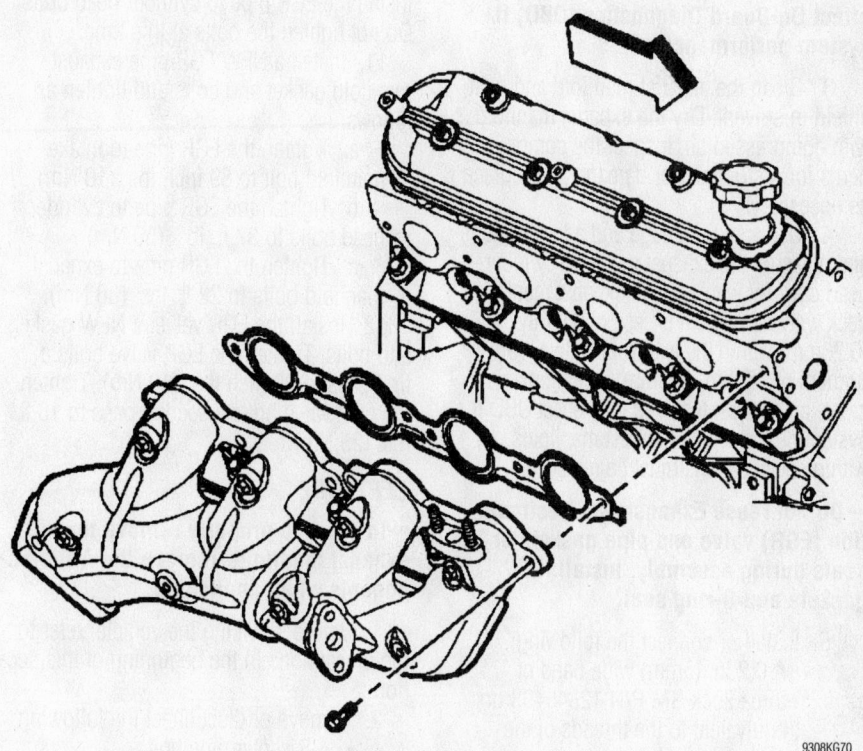

Right exhaust manifold removal—4.8L, 5.3L and 6.0L engines

9308KG70

For complete service labor times, order Nichols' Chilton Labor Guide

force, pull the EGR pipe from the intake manifold.

- O-ring seal from the EGR pipe assembly. Discard the exhaust manifold gasket and O-ring seal.

➡ **In order to properly remove the exhaust manifold, remove the AIR components when applicable. Do not remove the check valve from the Air Injection Reaction (AIR) pipe unless service is required.**

- AIR pipe (with check valve), nuts and gasket from the right exhaust manifold.
- AIR pipe studs from the manifold, if required
- Spark plug wires from the spark plugs. Do not remove the spark plug wires from the ignition coils unless required.
- Exhaust manifold, bolts, and gasket. Discard the gasket.
- Heat shield and bolts from the manifold, if required

To install:

➡ **Do not reuse the exhaust manifold-to-cylinder head gaskets. Upon installation of the exhaust manifold, install a NEW gasket. A improperly installed gasket or leaking exhaust system may effect On-Board Diagnostics (OBD) II system performance.**

3. Clean the exhaust manifold and heat shield in solvent. Dry the exhaust manifold with compressed air. Inspect the components for restrictions or damage and replace as necessary.

4. Use a straight edge and a feeler gauge and measure the exhaust manifold cylinder head deck for warpage. An exhaust manifold deck with warpage in excess of 0.01 in. (0.25mm) within the two front or two rear runners or 0.02 in. (0.5mm) overall, may cause an exhaust leak and may effect OBD II system performance. Exhaust manifolds not within specifications must be replaced.

➡ **Do not reuse Exhaust Gas Recirculation (EGR) valve and pipe gaskets or seals during assembly. Install NEW gaskets and O-ring seal.**

5. Install or connect the following:
- A 0.2 in. (5mm) wide band of threadlock GM P/N 12345493 or equivalent to the threads of the exhaust manifold bolts.
- Exhaust manifold gasket and exhaust manifold

6. Install the exhaust manifold bolts and tighten as follows:

a. First pass: Tighten to 11 ft. lbs. (15 Nm), beginning with the center two bolts. Alternate from side-to-side, and work toward the outside bolts.

b. Second pass: Tighten to 22 ft. lbs. (30 Nm), beginning with the center two bolts. Alternate from side-to-side, and work toward the outside bolts.

7. Using a flat punch, bend over the exposed edge of the exhaust manifold gasket at the front of the right cylinder head.

8. Install or connect the following:
- Heat shield and bolts. Tighten the bolts to 80 inch lbs. (9 Nm).
- AIR pipe studs (if required) and tighten to 45 inch lbs. (5 Nm).
- AIR pipe (with check valve), NEW gasket and nuts (if required). Tighten the nuts to 18 ft. lbs. (25 Nm).
- AIR hose assembly and clamps.
- AIR pipe bracket-to-cylinder head bolt. Tighten the bolt to 37 ft. lbs. (50 Nm).

9. Apply a light coating of clean engine oil to a NEW O-ring seal and install the seal onto the EGR pipe. Insert the EGR pipe into the intake manifold.

10. Start the EGR pipe to intake manifold bolt (1). Do not tighten the bolt at this time. Install the EGR pipe to cylinder head bolts. Do not tighten the bolts at this time.

11. Install a NEW EGR pipe exhaust manifold gasket and bolts and tighten as follows:

a. Tighten the EGR pipe to intake manifold bolt to 89 inch lbs. (10 Nm).

b. Tighten the EGR pipe to cylinder head bolts to 37 ft. lbs. (50 Nm).

c. Tighten the EGR pipe to exhaust manifold bolts to 22 ft. lbs. (30 Nm).

12. Install the EGR valve, a NEW gasket, and bolts. Tighten the EGR valve bolts a first pass to 89 inch lbs. (10 Nm). Tighten the EGR valve bolts a second pass to 18 ft. lbs. (25 Nm).

LEFT SIDE

➡ **In order to properly remove the exhaust manifold, remove the AIR components when applicable.**

1. Before servicing the vehicle, refer to the precautions in the beginning of this section.

2. Remove or disconnect the following:
- AIR center pipe bolt
- AIR hose clamps and remove the hose assembly

➡ **Do not remove the check valve from the AIR pipe unless service is required.**

- AIR pipe (with check valve), nuts and gasket from the left exhaust manifold
- AIR pipe studs from the manifold (if required)
- Spark plug wires from the spark plugs. Do not remove the spark plug wires from the ignition coils unless required.
- Exhaust manifold, bolts, and gasket. Discard the gasket.
- Heat shield and bolts from the manifold, if required

➡ **Do not reuse the exhaust manifold-to-cylinder head gaskets. Upon installation of the exhaust manifold, install a NEW gasket. An improperly installed gasket or leaking exhaust system may effect On-Board Diagnostics (OBD) II system performance.**

3. Clean the exhaust manifold and heat shield in solvent. Dry the exhaust manifold with compressed air. Inspect the components for restrictions or damage and replace as necessary.

4. Use a straight edge and a feeler gauge and measure the exhaust manifold cylinder head deck for warpage. An exhaust manifold deck with warpage in excess of 0.01 in. (0.25mm) within the two front or two rear runners or 0.02 in. (0.5mm) overall, may cause an exhaust leak and may effect OBD II system performance. Exhaust manifolds not within specifications must be replaced.

To install:

➡ **Do not apply sealant to the first three threads of the bolt.**

5. Apply a 0.2 in. (5mm) wide band of threadlock GM P/N 12345493 or equivalent to the threads of the exhaust manifold bolts. Install the exhaust manifold and NEW exhaust manifold gasket.

6. Install the exhaust manifold bolts and tighten as follows:

a. First pass: Tighten to 11 ft. lbs. (15 Nm), beginning with the center two bolts. Alternate from side-to-side, and work toward the outside bolts.

b. Second pass: Tighten to 22 ft. lbs. (30 Nm), beginning with the center two bolts. Alternate from side-to-side, and work toward the outside bolts.

7. Using a flat punch, bend over the exposed edge of the exhaust manifold gasket at the front of the right cylinder head.

8. Install or connect the following:
- Heat shield and bolts. Tighten the bolts to 80 inch lbs. (9 Nm).
- AIR pipe studs (if required) and tighten to 45 inch lbs. (5 Nm)

- AIR pipe (with check valve), NEW gasket and nuts (if required). Tighten the nuts to 18 ft. lbs. (25 Nm).

5.0L and 5.7L Engines

1. Before servicing the vehicle, refer to the precautions in the beginning of this section.

2. Remove or disconnect the following:
- Negative battery cable
- Engine cover
- Air cleaner, if needed
- Exhaust pipe at the manifold
- Oxygen Sensor (O_2S) wiring, if equipped
- AIR hose at the check valve
- Exhaust Gas Recirculation (EGR) valve, inlet pipe
- Heat stove pipe and the dipstick tube bracket, if working on the right side of the engine
- Power steering pump rear bracket at the manifold, if removing the left side manifold
- Loosen the alternator and remove the lower bracket, if necessary
- Air conditioner compressor rear bracket and the diverter valve and bracket. if needed

➡**On models with air conditioning, it may be necessary to remove the compressor, do not disconnect the compressor lines.**

- Manifold bolts and the manifold(s) Some models have lock tabs on the front and rear manifold bolts which must be removed before removing the bolts.

To install:

3. Clean gasket surfaces, and inspect manifold for cracks replace as necessary.

4. Install the manifold, then torque the bolts in 2 steps, as follows:
  a. Step 1: Torque to 15 ft. lbs. (20 Nm).
  b. Step 2: Torque to 22 ft. lbs. (30 Nm).

5. Install or connect the following:
- Alternator, if removed
- Air conditioning compressor, if removed
- Diverter, if removed
- Power steering brackets, if removed
- Dipstick tube on right side
- EGR inlet pipe
- Oxygen sensor connector, if equipped

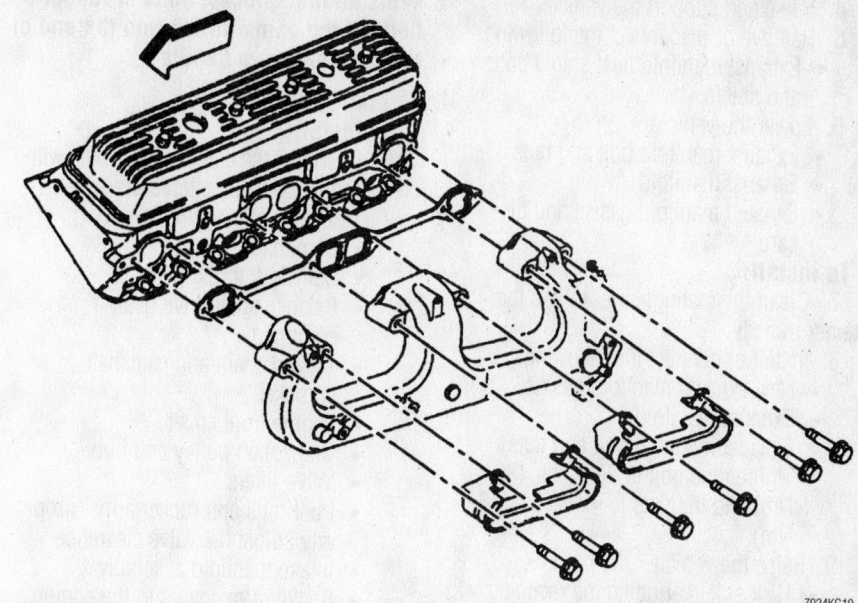

7924KG19

Exploded view of the left exhaust manifold, the right side is similar—5.0L and 5.7L engines

- Exhaust pipes
- Negative battery cable

7.4L Engines

1. Before servicing the vehicle, refer to the precautions in the beginning of this section.

2. Remove or disconnect the following:
- Negative battery cable
- Heat stove pipe and the dipstick tube, if removing the right side manifold
- Exhaust Gas Recirculation (EGR) valve inlet pipe
- Oxygen Sensor (O_2S) wiring, if equipped
- AIR hose at the check valve, if applicable
- Spark plugs and wires
- Exhaust manifold bolts and the spark plug heat shields

➡**Leave the front nut (left manifold) or rear nut (right manifold) in place for support.**

- Heat shield bolts from the engine mount and bell housing
- Heat shield, if equipped
- Exhaust pipe at the manifold
- Exhaust manifold

To install:

3. Clean the mating surfaces and the retainer threads.

4. Install or connect the following:
- Manifold, spark plug heat shields

and nuts. Torque the nuts to 22 ft. lbs. (30 Nm), starting from the center bolts and working outside.
- Exhaust pipe
- Spark plugs and wires
- AIR hose at check valve, if removed
- O_2S connector, if equipped
- EGR pipe and dipstick tube
- Negative battery cable

5. Run engine and check for leaks.

8.1L Engines

1. Before servicing the vehicle, refer to the precautions in the beginning of this section.

2. Remove or disconnect the following:
- Wheelhouse panel
- Oil dipstick tube, right side only
- Secondary Air Injection (AIR) pipe nut from the fuel rail stud, if equipped
- AIR pipe bolts from the exhaust manifold
- AIR pipe from the AIR pump pipe
- AIR pipe gasket and discard
- Spark plug wires
- Spark plugs

3. If removing the right exhaust manifold, perform the following:
  a. Remove the Exhaust Gas Recirculation (EGR) pipe nuts from the manifold.
  b. Remove the EGR pipe bracket bolt.
  c. Remove and discard the EGR pipe gasket.
  d. Reposition the EGR pipe.

4. Raise and support the vehicle.
5. Remove or disconnect the following:
   - Exhaust manifold heat shield bolts and shield
6. Lower the vehicle.
   - Exhaust manifold bolt and nuts
   - Exhaust manifold
   - Exhaust manifold gasket and discard

To install:

7. Clean the mating surfaces and the retainer threads.
8. Install or connect the following:
   - New exhaust manifold gasket
   - Exhaust manifold
   - Exhaust manifold bolt and nuts. Tighten the bolt to 26 ft. lbs. (35 Nm) and the nuts to 12 ft. lbs. (16 Nm).
9. Raise the vehicle.
   - Heat shield. Tighten the retaining bolts and nuts to 18 ft. lbs. (25 Nm).
10. Lower the vehicle.
    - EGR pipe, right side only. Tighten the pipe bracket bolt to 37 ft. lbs. (50 Nm) and the nuts to 22 ft. lbs. (30 Nm).
    - Spark plugs and plug wires
    - Air pipe, using a new gasket and reversing the removal procedure
    - Wheel house panel

Camshaft and Valve Lifters

REMOVAL & INSTALLATION

4.3L Engines

1. Before servicing the vehicle, refer to the precautions in the beginning of this section.
2. Properly relieve the fuel system pressure.
3. Drain the engine cooling system.
4. Remove or disconnect the following:
   - Negative battery cable
   - Radiator
   - Cooling fan
   - Water pump
   - Rocker arm covers from the engine
   - Intake manifold assembly
   - Rocker arms, pushrods and lifters
   - Crankshaft pulley and hub
   - Engine front cover
5. Align the timing marks on the crankshaft and camshaft sprockets.
   - Camshaft sprocket and timing chain
   - Balance shaft drive gear, if equipped
   - Camshaft thrust plate

➡️**Install the sprocket bolts or longer bolts of the same thread into the end of the camshaft as a handle.**

   - Camshaft

To install:

6. Lubricate the camshaft journals with clean engine oil or a suitable pre-lube.
7. Install or connect the following:
   - Camshaft
   - Camshaft thrust plate
   - Balance shaft drive gear, if equipped
   - Timing chain and camshaft sprocket
   - Engine front cover
   - Crankshaft pulley and hub
   - Valve lifters
   - Pushrods and rocker arms, properly adjust the valve clearance
   - Intake manifold assembly
   - Rocker arm covers to the engine
   - Radiator to the vehicle
   - Negative battery cable
8. Refill the engine cooling system.

4.8L, 5.3L and 6.0L Engines

1. Before servicing the vehicle, refer to the precautions in the beginning of this section.

2. Raise the hood to the servicing position and secure it. Move the hood hinge bolt to hold the hood in the servicing position.
3. Remove or disconnect the following:
   - Battery negative cable
   - Coolant
   - Upper and lower radiator hoses from the engine
   - Air cleaner duct from the engine
   - A/C condenser mounting bolts, if equipped
   - Radiator support and radiator form vehicle
   - Engine cooling fan
   - Drive belt
   - A/C drive belt, if equipped
   - Engine sight shield
   - Electrical wiring harness from the thermostat housing
   - Water pump
4. Raise the vehicle.
   - Starter motor
   - Right side closeout cover and bolt
   - Crankshaft balancer
   - Engine oil pan
   - Engine front cover
   - Cylinder heads from the engine
   - Valve lifters from the engine
5. Align the timing marks on the

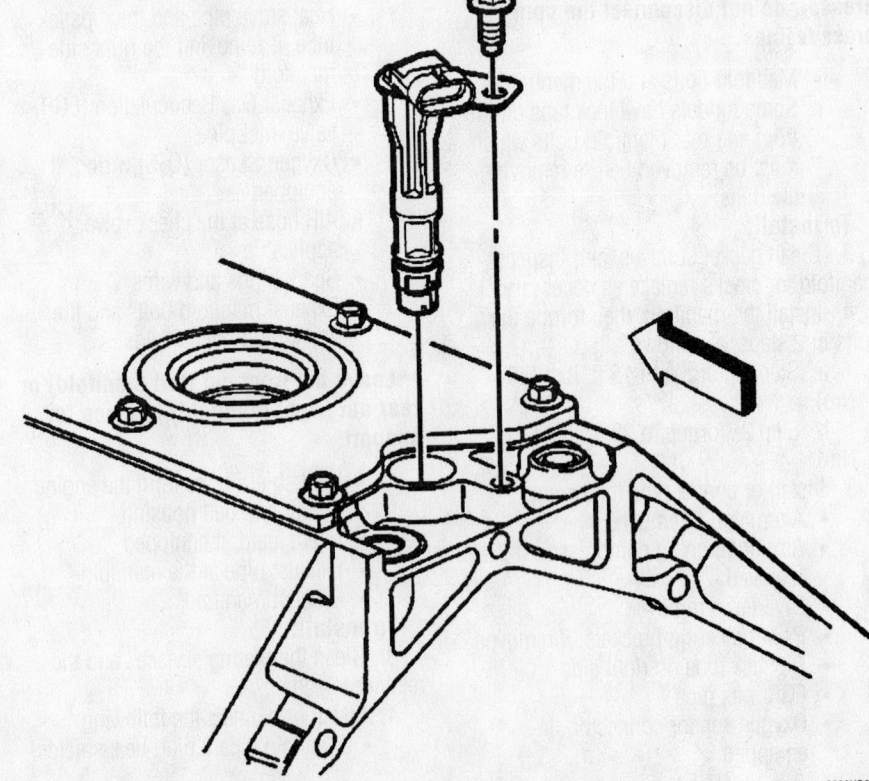

Camshaft sensor removal—4.8L, 5.3L and 6.0L engines

9308KG66

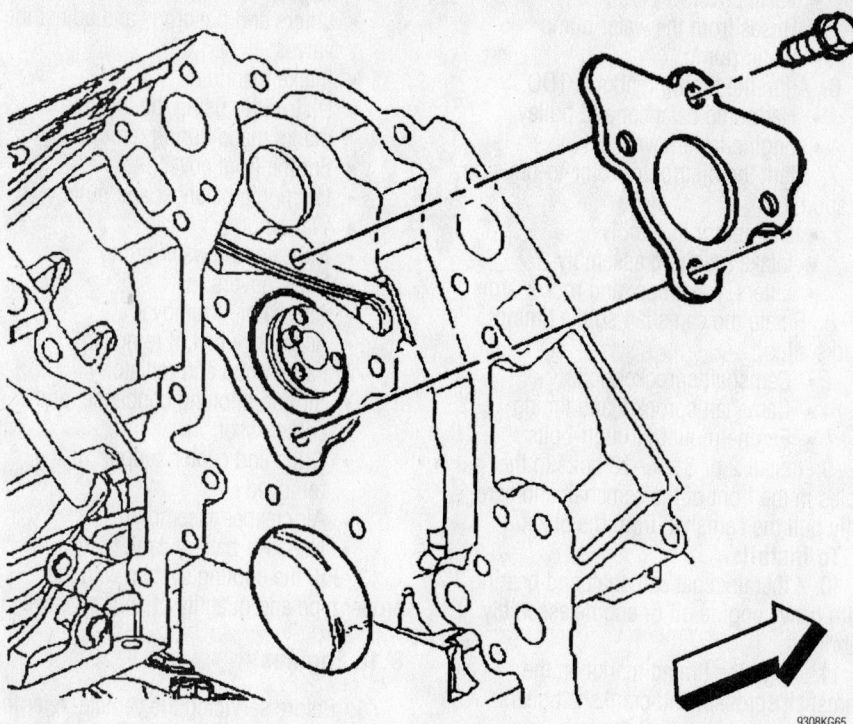

Camshaft retainer removal—4.8L, 5.3L and 6.0L engines

camshaft and crankshaft sprockets. Make sure that the number 1 piston is in the firing position.

- Camshaft sprocket
- Camshaft sensor bolt and the sensor
- Camshaft retainer bolts and the retainer

➡**All camshaft journals are the same diameter, so care must be used in removing or installing the camshaft to avoid damage to the camshaft bearings.**

6. Install the three M8-1.25 x 100 mm bolts in the camshaft front bolt holes. Using the bolts as a handle, carefully rotate and pull the camshaft out of the engine block. Remove the bolts from the front of the camshaft.

7. Clean and inspect all sealing surfaces.

To install:

➡**If camshaft replacement is required, the valve lifters must also be replaced.**

8. Lubricate the camshaft journals and the bearings with clean engine oil. Install three M8-1.25 x 100 mm (M8-1.25 x 4.0 in) bolts into the camshaft front bolt holes.

➡**All camshaft journals are the same diameter, so care must be used in removing or installing the camshaft to avoid damage to the camshaft bearings.**

9. Using the bolts as a handle, carefully install the camshaft into the engine block. Remove the three bolts from the front of the camshaft.

➡**Install the retainer plate with the sealing gasket facing the engine block. The gasket surface on the engine block should be clean and free of dirt or debris.**

10. Install or connect the following:
- Camshaft retainer and the bolts. Tighten the bolts to 18 ft. lbs. (25 Nm).

11. Inspect the camshaft sensor O-ring seal. If the O-ring seal is not cut or damaged, it may be reused. Lubricate the O-ring seal with clean engine oil.
- Camshaft sensor and bolt and tighten to 18 ft. lbs. (25 Nm)
- Camshaft sprocket and timing chain
- Valve lifters
- Cylinder heads
- Engine front cover to the engine
- Oil pan

- Right side closeout cover
- Starter motor
- Crankshaft balancer to the crankshaft
- Water pump
- Electrical wiring harness to the thermostat housing
- A/C drive belt, if equipped
- Drive belt
- Engine sight shield
- Radiator support and radiator
- A/C condenser mounting bolts
- Engine cooling fan
- Air cleaner duct
- Battery negative cable to the battery

5.0L and 5.7L Engines

1. Before servicing the vehicle, refer to the precautions in the beginning of this section.

2. Drain the cooling system.

3. Properly relieve the fuel system pressure.

4. Remove or disconnect the following:
- Air cleaner
- Air conditioning condenser and swing the condenser forward from its mounting, if equipped.
- Fan, the shroud and the radiator
- Valve covers
- Water pump assembly

5. Align the timing marks and remove the torsional damper.
- Timing chain cover
- Electrical and vacuum connections at the intake manifold
- Distributor assembly, mark the distributor rotor-to-housing location
- Intake manifold, pushrods and hydraulic lifters
- Camshaft sprocket bolts
- Camshaft sprocket and timing chain
- Crankshaft sprocket, as required
- Front engine mount through-bolts and raise the engine to gain sufficient clearance for camshaft removal, as required

6. Install 2 or 3 5⁄16–18 bolts 4–5 in. (102–127mm) long into the camshaft threaded holes.
- Camshaft

➡**Inspect the camshaft for signs of excessive wear or damage.**

To install:

➡**Liberally coat camshaft and bearing with heavy engine oil or engine assembly lubricant.**

Timing belt service is covered in Section 3 of this manual

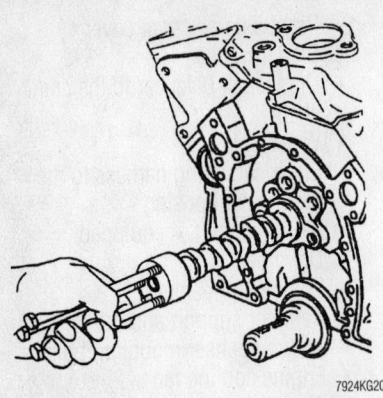

7924KG20

Install 2 or 3 long bolts into the camshaft to use as a handle for easy removal or installation—7.4L engine shown, other engines similar

7. Install or connect the following:
- Camshaft, align the timing marks on the camshaft and crankshaft gears
- Engine mount through-bolts
- Camshaft sprocket and chain. Torque the bolts to 18 ft. lbs. (25 Nm).
- Hydraulic lifters and pushrods
- Distributor assembly
- Timing chain cover
- Torsional damper
- Water pump
- Valve covers
- Fan, the shroud and radiator
- Air conditioning condenser, if equipped
- Air cleaner
- Negative battery cable
8. Refill the cooling system.

7.4L Engines

1. Before servicing the vehicle, refer to the precautions in the beginning of this section.
2. Drain the cooling system.
3. Properly relieve the fuel system pressure.
4. Properly discharge the air conditioning system.
5. Remove or disconnect the following:
- Negative battery cable
- Air cleaner assembly
- Grille and center support section, as required
- Air conditioning compressor, condenser and auxiliary fan, if equipped
- Fan, the shroud
- Radiator
- Accessory belt, as required
- Alternator, as required

- Valve covers
- Hoses from the water pump
- Water pump
6. Align the timing marks at TDC.
- Harmonic balancer and pulley
- Engine front cover
7. Mark the distributor rotor-to-housing location.
- Distributor assembly
- Intake manifold assembly
- Lifters, pushrods, and rocker arms
8. Rotate the camshaft so the timing marks align.
- Camshaft sprocket bolts
- Camshaft sprocket and timing
- Engine mount through-bolts
9. Install 2 or 3 $5/16$–18 bolts in the holes in the front of the camshaft and carefully pull the camshaft from the block.

To install:

10. Liberally coat camshaft and bearing with heavy engine oil or engine assembly lubricant
11. Align the timing marks on the camshaft sprocket and crankshaft gears.
12. Install or connect the following:
- Camshaft
- Camshaft sprocket and chain. Torque the bolts to 25 ft. lbs. (34 Nm).

- Engine mount bolts
- Lifters and pushrods and adjust the valves
- Intake manifold
- Distributor using the locating marks made during removal
- Engine front cover
- Harmonic balancer and pulley
- Water pump
- Hoses at the water pump
- Valve covers
- Alternator, if removed
- Accessory belt, if removed
- Fan shroud and radiator
- Air conditioning condenser and compressor
- Grille and center support, if removed
- Air cleaner assembly
- Negative battery cable
13. Fill the cooling system with the proper type and quantity of antifreeze.

8.1L Engines

1. Before servicing the vehicle, refer to the precautions in the beginning of this section.
2. Properly discharge the air conditioning system.

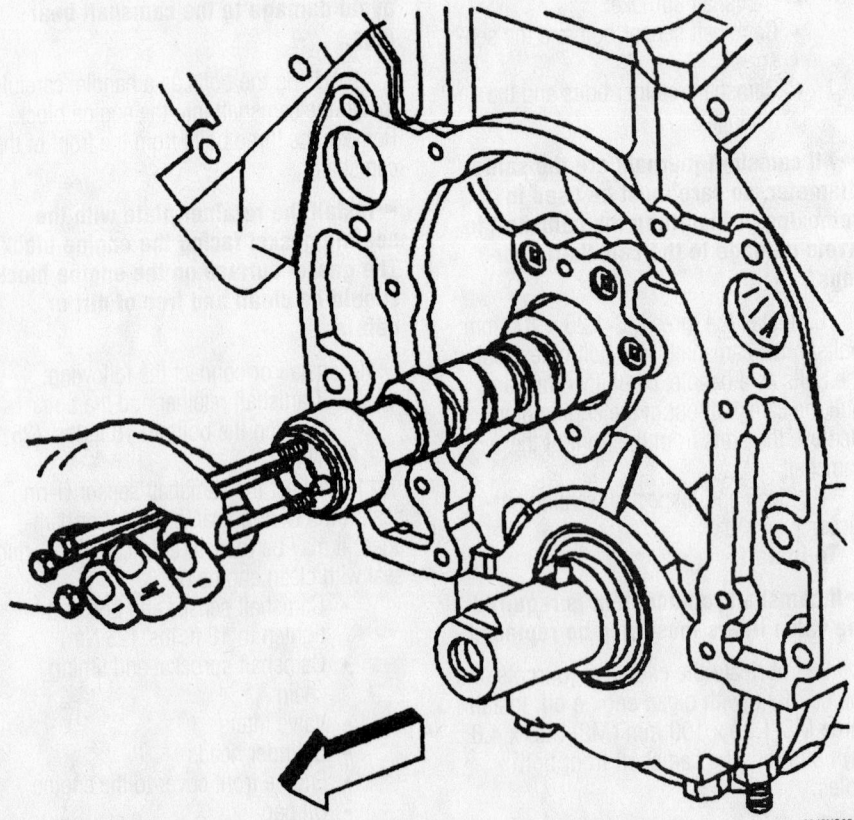

Use the 3 bolts as a handle to carefully remove and install the camshaft

9348NG05

3. Remove or disconnect the following:

- Grille
- A/C condenser
- Intake manifold
- Rocker arms and pushrods
- Valve lifter guide retainer bolts and retainer
- Valve lifter guides, keeping them in proper order for reassembly
- Valve lifters

➡ **If any lifters are stuck in their bores, use a suitable valve lifter to remove them.**

- Timing chain and sprocket
- Camshaft retaining bolts
- Camshaft retainer

❋❋ WARNING

All of the cam journals are the same size so be very careful when removing and installing the camshaft that you do not damage the bearings.

4. Install three 8-1.25 x 100mm bolts in the holes in the front of the camshaft and carefully pull the camshaft from the block.

5. Remove the bolts from the front of the camshaft.

6. Clean and inspect the camshaft for damage.

To install:

7. Liberally coat camshaft and bearings with heavy engine oil or engine assembly lubricant.

8. Install the camshaft, using the 3 bolts threaded into the camshaft bolt holes as a handle, then remove the bolts.

9. Install or connect the following:

- Camshaft retainer and bolts. Tighten to 106 inch lbs. (12 Nm).

➡ **If a new camshaft is installed, you MUST install new valve lifters.**

- Timing chain and sprocket
- Valve lifters
- Valve lifter guides over the flats on the lifters. Make sure the rollers of the lifters are properly aligned with the cam lobes.
- Valve lifter guide retainer and tighten the bolts to 18 ft. lbs. (25 Nm)
- Rocker arms and pushrods
- Intake manifold
- A/C condenser
- Grille

10. Recharge the A/C system.

Valve Lash

ADJUSTMENT

All engines use hydraulic lifters, which require no periodic adjustment.

Starter Motor

REMOVAL & INSTALLATION

4.3L and 5.0L Engines

1. Before servicing the vehicle, refer to the precautions in the beginning of this section.

2. Remove or disconnect the following:

- Negative battery cable
- Bracket and shield
- Wires
- Mounting bolts and shims
- Starter

To install:

3. Install or connect the following:

- Starter
- Mounting bolts and shim. Torque the bolts to 33 ft lbs. (45 Nm).
- Wires. Torque battery wire nut to 89 inch lbs. (10 Nm) and ignition nut to 18 inch lbs. (2 Nm).
- Bracket and shield. Torque the nuts to 53 inch lbs. (6 Nm).
- Negative battery cable

5.7L and 7.4L Engines

1. Before servicing the vehicle, refer to the precautions in the beginning of this section.

2. Remove or disconnect the following:
- Negative battery cable
- Mounting bolts and shims
- Wires
- Heat shield
- Starter

To install:

3. Install or connect the following:
- Starter
- Wires. Torque battery wire nut to 89 inch lbs. (10 Nm), and ignition nut to 18 inch lbs. (2 Nm)
- Heat shield. Torque the bolts to 53 inch lbs. (6 Nm) and the nuts to 35 inch lbs. (3 Nm)
- Mounting bolts and shim. Torque the bolts to 33 ft lbs. (45 Nm)
- Negative battery cable

4.8L, 5.3L and 6.0L Engines

❋❋ CAUTION

Before servicing any electrical component, the ignition key must be in the OFF or LOCK position and all electrical loads must be OFF, unless instructed otherwise in these procedures.

9308KG03

Exploded view of the starter motor.

Heater Core replacement is covered in Section 2 of this manual

1. Before servicing the vehicle, refer to the precautions in the beginning of this section.

2. Disconnect the negative battery cable.

3. Raise and support the vehicle.

4. Remove or disconnect the following:
- Protective shields, as necessary
- Starter solenoid shield
- Starter-to-transmission close out cover bolt
- Engine oil level sensor connection
- Front axle mounting bracket through bolt nut

5. Reposition the front axle mounting bracket through bolt until the bolt tip is flush with the support bushing. Do not remove the bolt.
- Mounting bolts from the engine block. Slide the starter forward until the starter clears the transmission.
- Starter transmission close out cover
- Positive battery cable and wiring harness from the starter
- Starter from the vehicle

To install:

6. Install or connect the following:
- Starter
- Positive battery cable to the starter. Tighten the nut to 12 ft. lbs. (16 Nm).
- Starter transmission close out cover
- Mounting bolts to the engine block and tighten to 37 ft. lbs. (50 Nm)

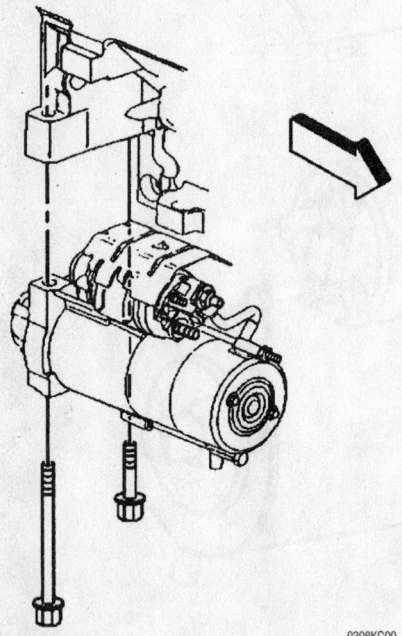

9308KG00

Starter removal—4.8L, 5.3L and 6.0L

7. Reposition the front axle mounting bracket through bolt until the bolt is fully seated.
- Front axle mounting bracket through bolt nut and tighten to 67 ft. lbs. (90 Nm)
- Engine oil level sensor connection
- Starter-to-transmission close out cover bolt
- Starter solenoid shield
- Protective shields as necessary

8. Remove the safety stands.

9. Lower the vehicle.

10. Connect the negative battery cable.

8.1L Engine

1. Before servicing the vehicle, refer to the precautions in the beginning of this section.

2. Remove or disconnect the following:
- Negative battery cable
- Positive battery cable nut
- Positive cable from the solenoid
- Engine harness ground nut and ground from the solenoid
- Mounting bolts and starter
- Heat shield bolts, nut and shield, if necessary

To install:

3. Install or connect the following:
- Heat shield, bolts and nut if removed. Tighten the bolts to 35 inch lbs. (3 Nm) and the nut to 44 inch lbs. (5 Nm).
- Starter and bolts. Tighten to 37 ft. lbs. (50 Nm).
- Ground wire and nut. Tighten to 30 inch lbs. (3.4 Nm).
- Positive cable and nut. Tighten to 80 inch lbs. (9 Nm).
- Negative battery cable

Oil Pan

REMOVAL & INSTALLATION

4.3L Engines

1. Before servicing the vehicle, refer to the precautions in the beginning of this section.

2. Drain the engine oil.

3. Remove or disconnect the following:
- Negative battery cable
- Exhaust crossover pipe
- Torque converter cover, if equipped with automatic transmission
- Cooler lines from guides and the oil filter adapter
- Strut rods at the flywheel/flexplate cover, if equipped

- Strut rod at the front engine mounts, if equipped
- Starter assembly
- Front drive axle tube nuts and lower axle bushing bolts
- Oil pan bolts/nuts and reinforcements
- Oil pan and gaskets

To install:

4. Thoroughly clean all gasket surfaces

5. Install or connect the following:
- New gasket
- Oil pan and new gaskets
- Oil pan bolts, nuts and reinforcements. Torque the bolts to 18 ft. lbs. (25 Nm).
- Front drive axle tube nuts and lower axle bushing bolts
- Starter
- Strut rod brackets at the front engine mounts
- Strut rods at the flywheel/flexplate cover
- Cooler lines into guides and oil filter adapter with new filter
- Torque converter cover, if equipped with automatic transmission
- Exhaust crossover pipe
- Negative battery cable

6. Refill the engine with oil.

4.8L, 5.3L and 6.0L Engines

➡The original oil pan gasket is retained and aligned to the oil pan by rivets. When installing a new gasket, it is not necessary to install new rivets. DO NOT reuse the oil pan gasket. When installing the oil pan, install a NEW oil pan gasket.

1. Before servicing the vehicle, refer to the precautions in the beginning of this section.

2. Remove or disconnect the following:
- Negative battery cable
- Front differential, if equipped with four wheel drive
- Underbody shield from the vehicle
- Oil pan shield
- Cross brace if equipped
- Engine oil and filter
- Transmission to oil pan bolts
- Oil level sensor electrical connector
- Two front wiring harness retainer bolts
- Engine wiring harness retainer bolts from the engine oil pan
- Engine oil cooler pipe to oil pan bolt
- Transmission oil cooler pipe retainer and the bolt from the oil pan

- Closeout covers and bolts (one each side of engine)
- Engine mount bolts each side
- Oil pan

To install:

➡The alignment of the structural oil pan is critical. The rear bolt hole locations of the oil pan provide mounting points for the transmission bellhousing. To ensure the rigidity of the powertrain and correct transmission alignment, it is important that the rear of the block and the rear of the oil pan must NEVER protrude beyond the engine block and transmission bellhousing plane.

3. Apply a 0.20 in. (5mm) bead of sealant GM P/N 12378190 or equivalent 0.8 in. (20mm) long to the engine block. Apply the sealant directly onto the tabs of the front cover gasket that protrudes into the oil pan surface.

4. Apply a 0.20 in. (5mm) bead of sealant GM P/N 12378190 or equivalent 0.8 in. (20mm) long to the engine block. Apply the sealant directly onto the tabs of the rear cover gasket that protrudes into the oil pan surface.

➡Be sure to align the oil gallery passages in the oil pan and engine block properly with the oil pan gasket.

5. Pre-assemble the oil pan gasket to the pan. Install the oil pan bolts to the pan through the gasket.

6. Install or connect the following:

- Oil pan, gasket and bolts to the engine block. Snug the oil pan bolts finger-tight. Do not over-tighten.
- Two lower bellhousing bolts to position the oil pan correctly.

7. Snug the lower bellhousing bolt finger-tight. Do not overtighten. Tighten the oil pan-to-block and oil pan-to-oil pan front cover bolts to 18 ft. lbs. (25 Nm). Tighten the oil pan-to-rear cover bolts to 106 inch lbs. (12 Nm). Tighten the bellhousing bolts to 37 ft. lbs. (50 Nm).

- Transmission oil cooler pipe retainer and the bolt to the oil pan
- Engine oil cooler pipe to oil pan bolt. Tighten the nut to 89 inch lbs. (10 Nm).
- Engine wiring harness retainer bolts to the oil pan
- Oil level sensor connector
- Transmission-to-oil pan bolts. Tighten the bolts to 41 ft. lbs. (55 Nm).

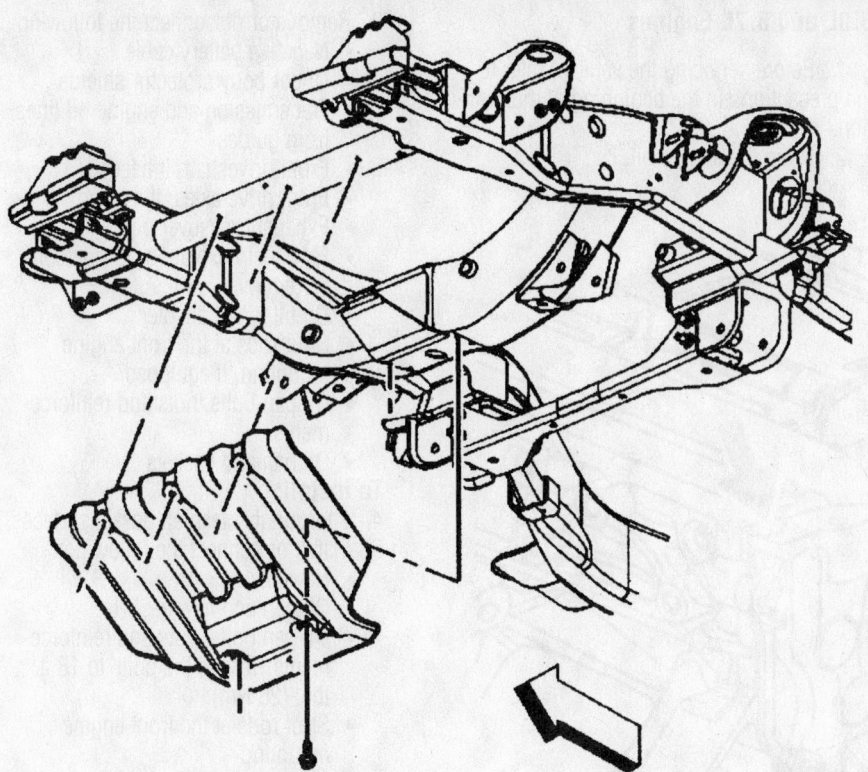

Oil pan shield—4.8L, 5.3L and 6.0L engines

9308KG81

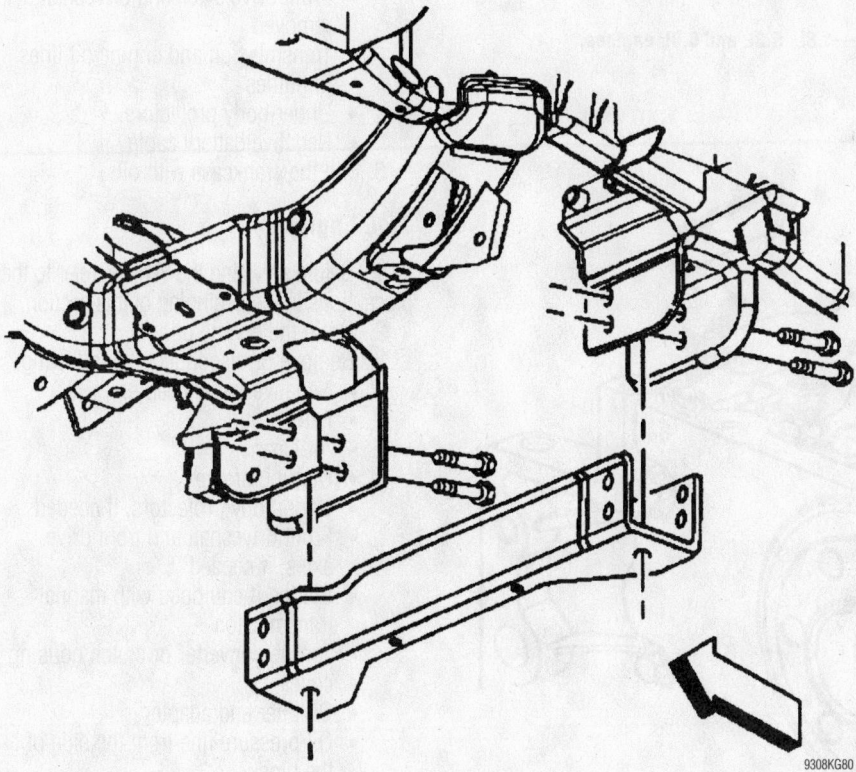

Cross brace—4.8L, 5.3L and 6.0L engines

9308KG80

Brake service is covered in Section 4 of this manual

- Front differential if equipped with four wheel drive
- Underbody shield

8. Lower the vehicle. Fill the engine with oil and install the engine oil filter.

9. Connect the negative battery cable.

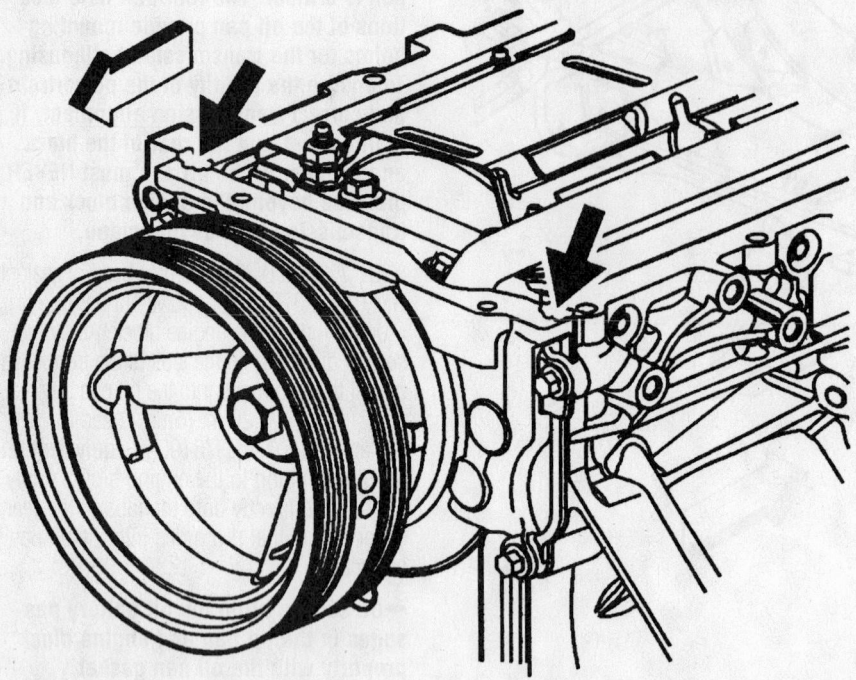

Apply sealant at these points at the front of the block—4.8L, 5.3L and 6.0L engines

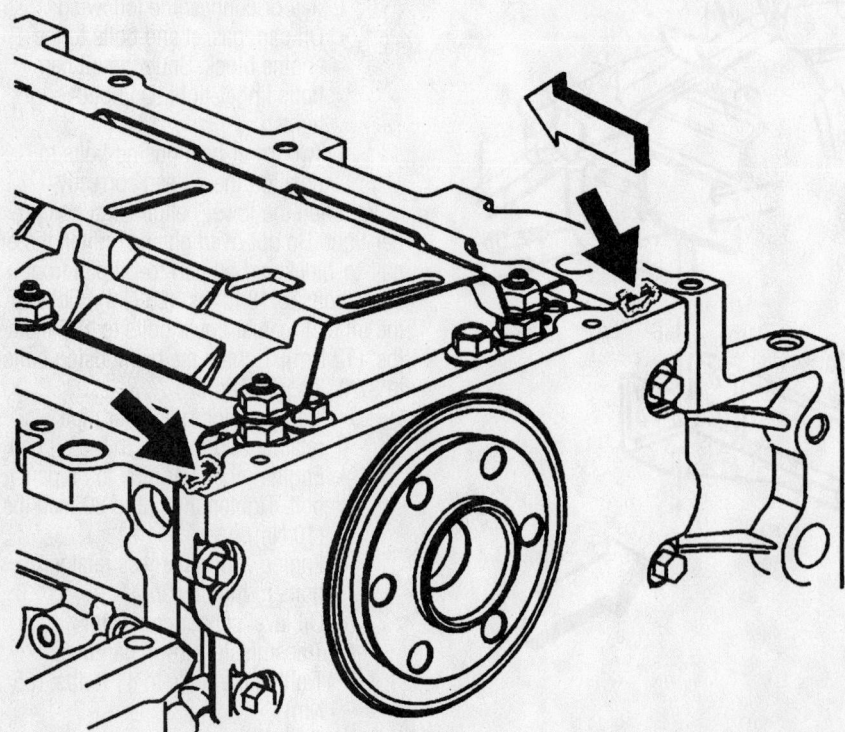

Apply sealant at these points at the rear of the block—4.8L, 5.3L and 6.0L engines

5.0L and 5.7L Engines

1. Before servicing the vehicle, refer to the precautions in the beginning of this section.

2. Drain the engine oil.

3. Remove or disconnect the following:
- Negative battery cable
- Under body protector shields
- Transmission and engine oil lines from guides
- Front driveshaft, if needed
- Front drive axles, if needed
- Exhaust crossover pipe
- Flywheel/flexplate or torque converter cover
- Oil filter and adapter
- Strut rods at the front engine mounting, if equipped
- Oil pan bolts, nuts and reinforcements
- Oil pan and gaskets

To install:

4. Thoroughly clean all gasket surfaces.

5. Install or connect the following:
- New gasket
- Oil pan and new gaskets
- Oil pan bolts, nuts and reinforcements. Torque the bolts to 18 ft. lbs. (25 Nm).
- Strut rods at the front engine mounting
- Oil filter and adapter
- Torque converter or flywheel/flexplate cover
- Exhaust crossover pipe
- Front drive axles and driveshaft, if removed
- Transmission and engine oil lines to guides
- Under body protectors
- Negative battery cable

6. Fill the crankcase with oil.

7.4L Engines

1. Before servicing the vehicle, refer to the precautions in the beginning of this section.

2. Drain the engine oil.

3. Remove or disconnect the following:
- Negative battery cable
- Fan shroud
- Air cleaner
- Distributor cap
- Underbody protectors, if needed
- Front driveshaft and front drive axles, if needed
- Starter, if equipped with manual transmission
- Torque converter or clutch housing cover
- Oil filter and adapter
- Oil pressure line from the side of the block

4. Support the engine
- Engine mount through-bolts
- Oil pan bolts
- Oil pan and discard the gaskets

To install:

5. Clean all sealing surfaces.

6. Apply RTV gasket material to the front and rear corners of the gaskets.

7. Install or connect the following:
- New gaskets, coat the gaskets with adhesive sealer and position them on the block
- Rear pan seal in the pan with the seal ends mating with the gaskets
- Front seal on the bottom of the front cover, pressing the locating tabs into the holes in the cover
- Oil pan
- Pan bolts, clips and reinforcements. Torque the bolts to 18 ft. lbs. (25 Nm).

8. Lower the engine onto the mounts.
- Engine mount through-bolts
- Oil pressure line
- Oil filter
- Starter, if removed
- Torque converter or clutch housing cover
- Front drive axles and driveshaft, if removed
- Underbody protectors, if removed
- Distributor cap
- Air cleaner
- Fan shroud
- Negative battery cable

9. Fill the crankcase with oil.

8.1L Engine

1. Before servicing the vehicle, refer to the precautions in the beginning of this section.

2. Disconnect the negative battery cable and drain the engine oil.

3. Remove or disconnect the following:
- Front differential, if equipped with 4WD
- Starter motor

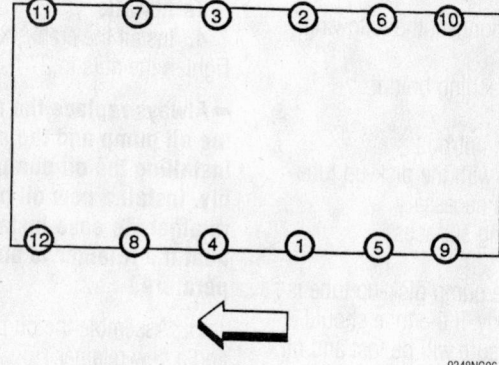

Oil pan bolt tightening sequence—8.1L engine

- Oil pan skid plate bolts and plate
- Crossbar bolt(s) and crossbar
- Oil level dipstick
- Oil level sensor electrical connector
- Engine harness clip from the oil pan
- Battery cable channel bolt
- Battery cable channel and reposition
- Oil pan bolts, oil pan and gasket

➡**You can reuse the oil pan gasket, if it is not damaged**

To install:

➡**You must install the oil pan within 5 minutes of applying the sealer.**

4. Apply sealant to the sides of the front and rear crankshaft bearing caps on the left and right sides.

5. Install or connect the following:
- Oil pan gasket into the oil pan groove
- Oil pan and bolts

6. Tighten the oil pan bolts, in sequence, as follows:
   a. 1st pass: 89 inch lbs. (10 Nm).
   b. 2nd pass: 18 ft. lbs. (25 Nm).

7. Install or connect the following:
- Battery cable channel and bolt. Tighten to 80 inch lbs. (9 Nm).
- Oil level sensor and tighten to 15 ft. lbs. (20 Nm)
- Engine harness clip
- Oil level sensor connector
- Oil level dipstick
- Crossbar and bolt(s). Tighten to 74 ft. lbs. (100 Nm).
- Skid plate. Tighten the bolts to 15 ft. lbs. (20 Nm).
- Starter motor
- Front differential
- Negative battery cable

8. Fill the crankcase with oil.

Oil Pump

REMOVAL & INSTALLATION

4.8L, 5.3L and 6.0L Engines

1. Before servicing the vehicle, refer to the precautions in the beginning of this section.

2. Remove or disconnect the following:
- Engine front cover
- Oil pan
- Oil pump screen bolt and nuts
- Oil pump screen with O-ring seal.
- O-ring seal from the pump screen. Discard the O-ring seal.
- Remaining crankshaft oil deflector nuts
- Crankshaft oil deflector
- Oil pump bolts

➡**Do not allow dirt or debris to enter the oil pump assembly, cap ends as necessary.**
- Oil pump

➡**The internal parts of the oil pump assembly are not serviced separately (excluding the spring). If the oil pump components are worn or damaged, replace the oil pump as an assembly. Do not attempt to repair the wire mesh portion of the pump and screen assembly.**

To install:

➡**Inspect the oil pump and engine block oil gallery passages. These surfaces must be clear and free of debris or restrictions.**

3. Align the splined surfaces of the crankshaft sprocket and the oil pump drive gear and install the oil pump. Install the oil pump onto the crankshaft sprocket until the pump housing contacts the face of the engine block.

4. Install or connect the following:
- Oil pump bolts. Tighten the oil pump bolts to 18 ft. lbs. (25 Nm).
- Crankshaft oil deflector.

➡**Lubricate a NEW oil pump screen O-ring seal with clean engine oil.**
- NEW O-ring seal onto the oil pump screen

➡**Push the oil pump screen tube completely into the oil pump prior to tightening the bolt. Do not allow the bolt to pull the tube into the pump.**

9348NG06

For complete Engine Mechanical specifications, see Section 1 of this manual

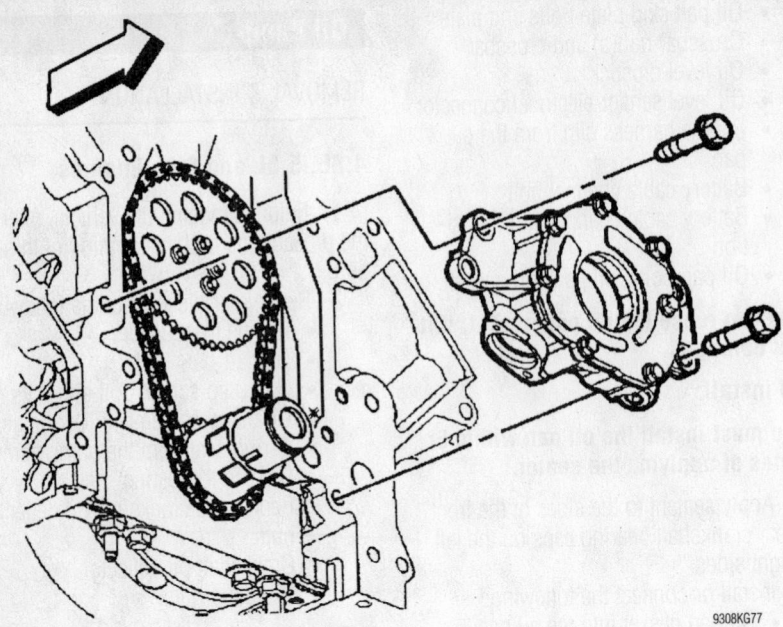

Oil pump removal—4.8L, 5.3L and 6.0L engines

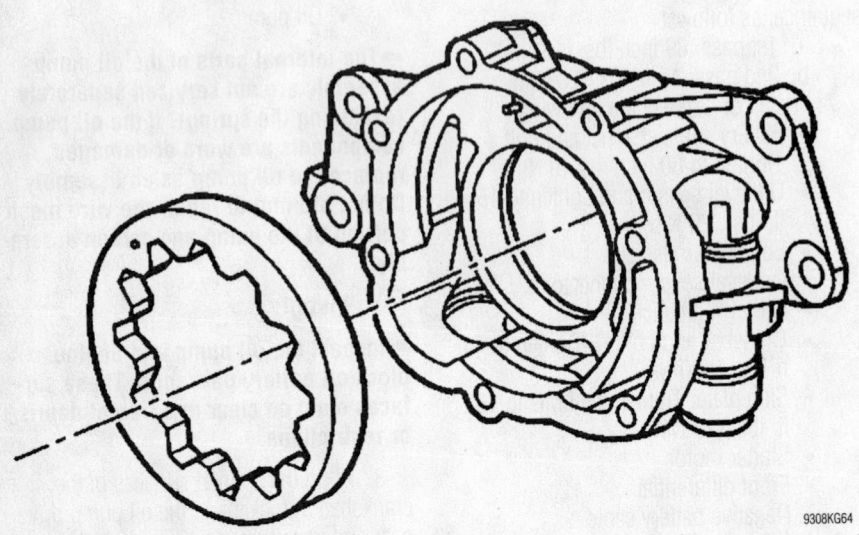

Oil pump disassembly—4.8L, 5.3L and 6.0L engines

5. Align the oil pump screen mounting brackets with the correct crankshaft bearing cap studs.
  - Oil pump screen, screen bolt and reflector nuts. Tighten the bolt (4) to 106 inch lbs. (12 Nm) and the oil deflector nuts (2) to 18 ft. lbs. (25 Nm).
  - Oil pan
  - Engine front cover

4.3L, 5.0L, 5.7L and 7.4L Engines

1. Before servicing the vehicle, refer to the precautions in the beginning of this section.

2. Remove or disconnect the following:
  - Oil pan
  - Oil pump attaching bolt, if equipped
  - Pick-up tube nut/bolt
  - Pump along with the pick-up tube and shaft, as necessary
3. Clean all sealing surfaces
To install:
4. Ensure that the pump pick-up tube is tight in the pump body. If the tube should come loose, oil pressure will be lost and oil starvation will occur. If the pick-up tube is loose it should be replaced.
5. If the pump has been disassembled

and is being replaced or for any reason oil has been removed, it must be primed. It can either be filled with oil before installing the cover plate and oil kept within the pump during handling or the entire pump cavity can be filled with petroleum jelly.

➡**If the pump is not primed, the engine could be damaged before it receives adequate lubrication when the engine is started.**

6. Install or connect the following:
  - Pump, aligning the pump shaft with the oil pump drive gear as necessary. Torque oil pump/pick-up tube retainer(s) to 65 ft. lbs. (90 Nm).
  - Oil pan
7. Refill the engine crankcase
8. Disable the ignition system; crank engine for approximately 10 seconds to aid in priming the oil pump and reducing the risk of engine damage.

➡**If the oil pump does not build up oil pressure almost immediately, remove the pan and check for a loose oil pump-to-pick-up tube attachment. If necessary dismantle the pump and pack the pump cavity with petroleum jelly. Running the engine without measurable oil pressure will cause extensive damage.**

8.1L Engine

1. Before servicing the vehicle, refer to the precautions in the beginning of this section.
2. Remove or disconnect the following:
  - Oil pan
  - Oil pump screen bolt
  - Oil pump, retainer and driveshaft. Discard the driveshaft retainer.
  - Crankshaft oil deflector nuts
  - Crankshaft oil deflector
  - Oil pump bolts
  - Oil pump
3. Clean and inspect the oil pump
To install:
4. Install the crankshaft oil deflector. Tighten the nuts to 37 ft. lbs. (50 Nm).

➡**Always replace the retainer between the oil pump and the shaft, when installing the oil pump. During assembly, install a new oil pump driveshaft retainer. To ease installation, slightly heat the retainer to above room temperature.**

5. Assemble the oil pump, driveshaft and a new retainer.
6. Install or connect the following:
  - Oil pump, positioning it on the locating pins

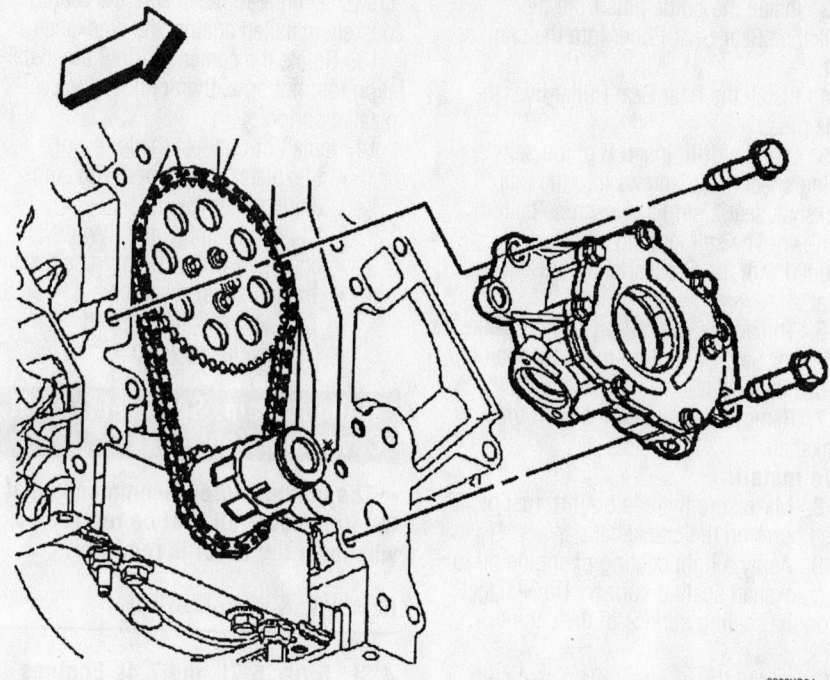

Exploded view of the oil pump mounting—4.8L, 5.3L and 6.0L engines

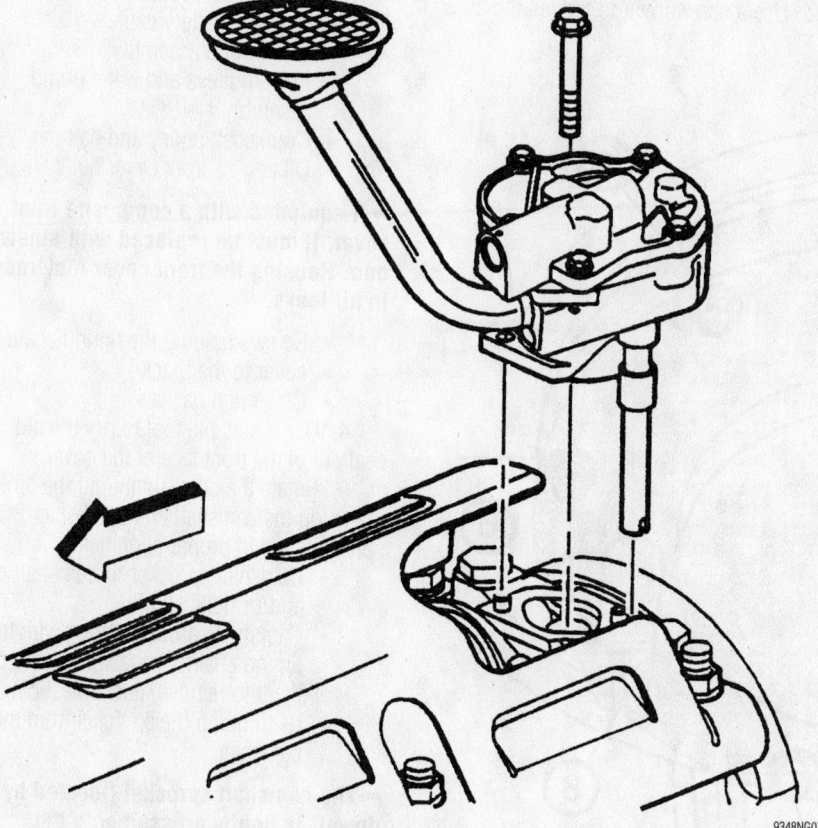

Oil pump removal—8.1L engine

- Oil pump bolt and tighten to 56 ft. lbs. (75 Nm)
- Oil pan

7. Refill the engine crankcase

8. Disable the ignition system; crank engine for approximately 10 seconds to aid in priming the oil pump and reducing the risk of engine damage.

➡ If the oil pump does not build up oil pressure almost immediately, remove the pan and check for a loose oil pump-to-pick-up tube attachment. If necessary dismantle the pump and pack the pump cavity with petroleum jelly. Running the engine without measurable oil pressure will cause extensive damage.

Rear Main Seal

REMOVAL & INSTALLATION

Except 8.1L Engine

➡ Please note that the entire transmission assembly and flywheel/flexplate must be removed to perform this procedure.

1. Before servicing the vehicle, refer to the precautions in the beginning of this section.

2. Remove or disconnect the following:
- Negative battery cable
- Transfer case, if equipped
- Transmission assembly
- Clutch assembly and flywheel, if equipped with manual transmission
- Flexplate, if equipped with automatic transmission
- Crankshaft rear main oil seal by inserting a suitable prying tool and prying the seal out. Take care not to damage the crankshaft sealing surface.

To install:

3. Clean the oil seal bore in the block thoroughly before installation of the new seal.

4. Inspect the crankshaft for grit, rust or burrs and correct as necessary. Also inspect the portion of the crankshaft where the oil seal makes contact, for wear due to the rubbing action of the oil seal.

5. Clean the seal running surface of the crankshaft with a non-abrasive cleaner.

6. Lubricate the inner diameter of the new seal and the outer diameter of the crankshaft with engine oil.

7. Install or connect the following:

For Accessory Drive Belt illustrations, see Section 1 of this manual

- Rear main oil seal, using installation tool J 38841, until the tool bottoms against the block and crankshaft rear main bearing cap.
- Flywheel and clutch
- Flexplate, as required
- Transmission assembly
- Transfer case, if equipped
- Negative battery cable

8. Start the engine and verify no oil leaks.

8.1L Engine

Please note that the entire transmission assembly and flywheel/flexplate must be removed to perform this procedure. This procedure requires the use of the following tools: Crankshaft Rear Seal Puller tool No. J 43320 and Crankshaft Rear Seal Installer tool No. J 42849.

1. Before servicing the vehicle, refer to the precautions in the beginning of this section.
2. Remove or disconnect the following:
- Negative battery cable
- Transfer case, if equipped
- Transmission assembly
- Clutch assembly and flywheel, if equipped with manual transmission
- Flexplate, if equipped with automatic transmission

3. Install the guide pins from the Crankshaft Rear Sear Puller into the crankshaft.
4. Install the Rear Seal Puller over the guide pins.
5. Using a drill, insert 8 of the self-drilling sheet metal screws into the rear crankshaft seal, using a crisscross pattern as shown. The self tapping screws are included with the Crankshaft Rear Seal Puller.
6. Thread the center bolt of the Crankshaft Rear Seal Puller into the crankshaft to remove the seal.
7. Remove the guide pins from the crankshaft.

To install:

8. Make sure there is no dirt, rust or loose burrs on the crankshaft.
9. Apply a light coating of engine oil to the crankshaft sealing surface. Do NOT get oil on the sealing surface of the engine block.
10. Install the new rear main seal onto the Crankshaft Rear Seal Installation Tool.
11. Position the Rear Seal Installation Tool against the crankshaft. Thread the attaching screws into the tapped holes in the crankshaft.
12. Use a screwdriver to tighten the

screws securely to make sure the seal is squarely installed against the crankshaft.

13. Rotate the center nut until the installation tool bottoms, then remove the seal installation tool.
14. Install or connect the following:
- Flexplate, if equipped with automatic transmission
- Clutch assembly and flywheel, if equipped with manual transmission
- Transmission assembly
- Transfer case, if equipped
- Negative battery cable

Timing Chain, Sprockets, Front Cover and Seal

➡ The manufacturer recommends that the front cover oil seal be replaced whenever the cover is removed.

REMOVAL & INSTALLATION

4.3L, 5.0L, 5.7L and 7.4L Engines

1. Before servicing the vehicle, refer to the precautions in the beginning of this section.
2. Drain the cooling system.
3. Remove or disconnect the following:
- Negative battery cable
- Fan shroud assembly
- Belts, pulleys and water pump assembly
- Crankshaft pulley and damper
- Oil pan-to-front cover bolts

➡ If equipped with a composite front cover, it must be replaced with a new one. Reusing the front cover may result in oil leaks.

- Screws holding the timing chain cover to the block
- Cover and gaskets

4. Use a suitable tool to pry the old seal out of the front face of the cover.
5. Rotate the crankshaft until the timing marks on the camshaft and crankshaft sprockets are in proper alignment.
- Camshaft sprocket-to-camshaft nut and/or bolts
- Camshaft sprocket (along with the timing chain). If the sprocket is difficult to remove, use a plastic mallet to bump the sprocket from the camshaft.

➡ The camshaft sprocket (located by a dowel) is lightly pressed onto the camshaft and should come off easily. The chain comes off with the camshaft sprocket.

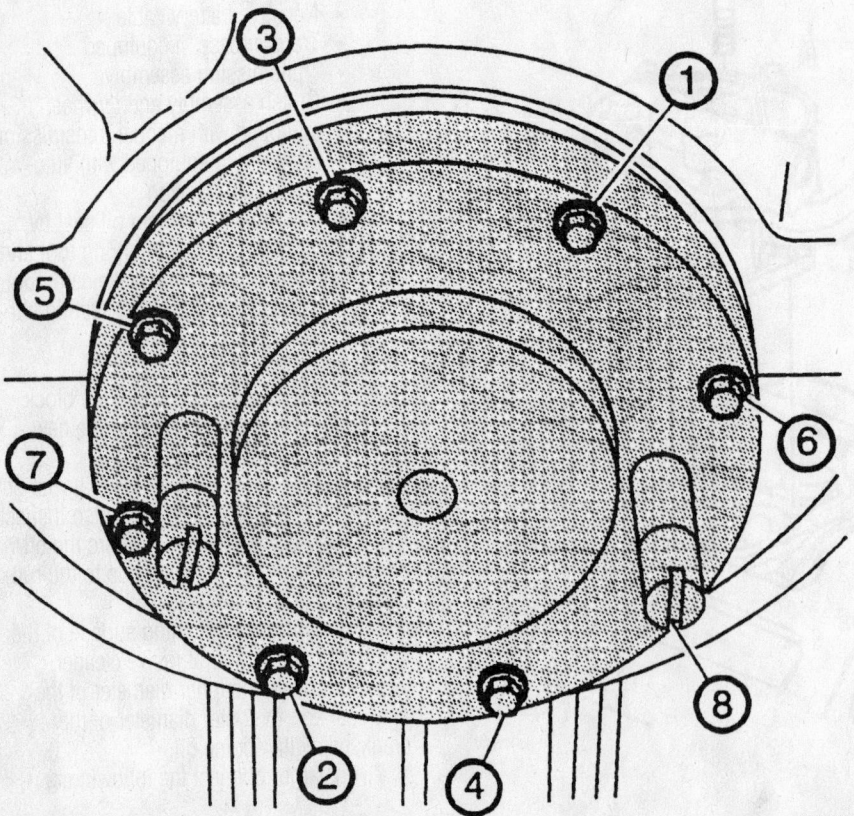

You must drill the screws into the rear main seal using a crisscross pattern—8.1L engine

9348NG08

front of the camshaft. Torque the camshaft sprocket-to-camshaft retainer bolts to 106 inch lbs. (12 Nm).

10. With the timing chain installed, turn the crankshaft 2 complete revolutions, then check to make certain that the timing marks are in correct alignment between the shaft centers.

➡ **Coat the lip of the new seal with oil prior to installation.**

- New seal so that the open end is toward the inside of the cover, Using seal driver J-22102, or equivalent.
- New front pan seal, cutting the tabs off

11. Coat a new cover gasket with adhesive sealer and position it on the block.

12. Apply a ⅛ in. (3mm) bead of RTV gasket material to the front cover.

- Cover carefully onto the locating dowels and tighten the attaching screws
- Oil pan, if removed
- Cover-to-pan bolts, Torque to 106 inch lbs. (12 Nm).
- Torsional damper
- Water pump assembly
- Negative battery cable

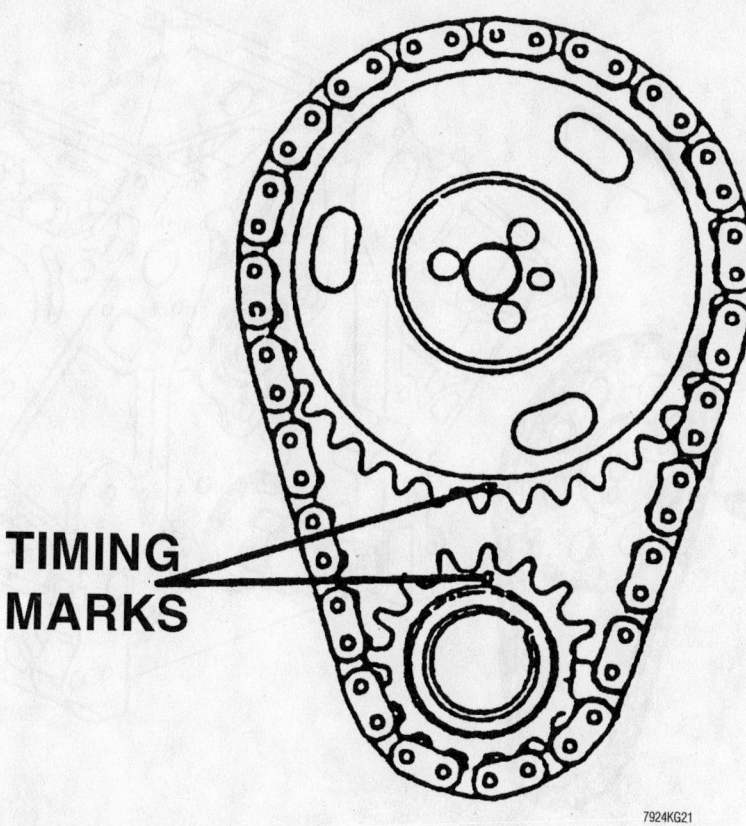

TIMING MARKS

7924KG21

Timing mark alignment for timing chain removal and installation—gasoline engines

6. If necessary use J-5825-A, or equivalent, crankshaft sprocket removal tool to free the timing sprocket from the crankshaft.

To install:

7. Inspect the timing chain and the timing sprockets for wear or damage, replace the damaged parts as necessary.

8. Clean the gasket mounting surfaces of all remaining traces of old gasket.

➡ **During installation, coat the thrust surfaces lightly with Molykote® or equivalent pre-lube.**

9. Install or connect the following:

- Crankshaft sprocket onto the crankshaft, use tool J-5590, or equivalent, crankshaft sprocket installation tool, and a hammer, without disturbing the position of the engine.
- Timing chain, arrange the camshaft sprocket in such a way that the timing marks will align between the shaft centers and the camshaft locating dowel will enter the dowel hole in the cam sprocket.
- Cam sprocket, with the chain mounted under it in position on the

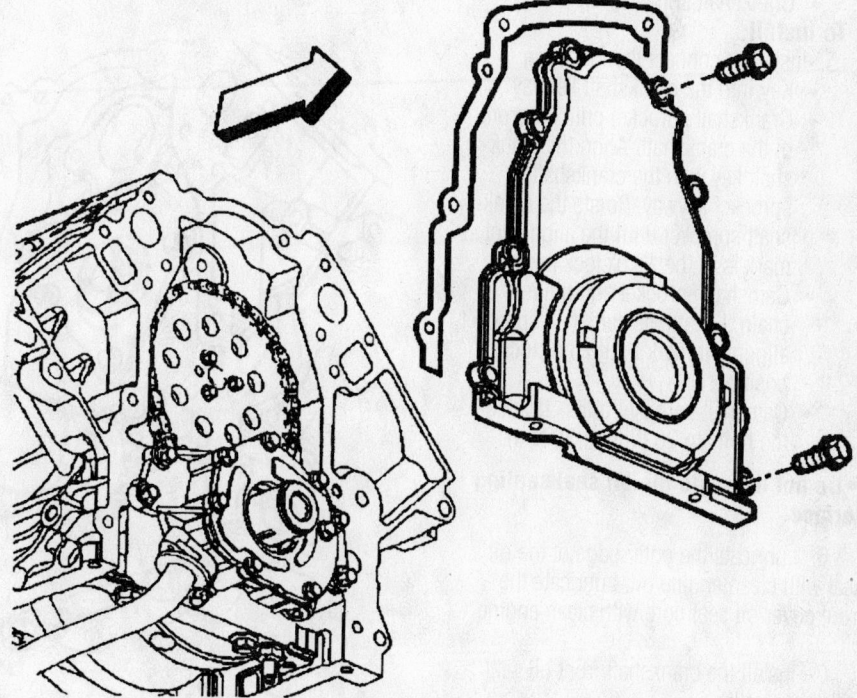

9308KG85

Front cover and gasket—4.8L, 5.3L and 6.0L engines

For Tire, Wheel and Ball Joint specifications, see Section 1 of this manual

13. Fill the cooling system with the proper type and quantity of antifreeze.

4.8L, 5.3L and 6.0L Engine

1. Before servicing the vehicle, refer to the precautions in the beginning of this section.

2. Drain the cooling system.

3. Remove or disconnect the following:

- Negative battery cable
- Water pump
- Crankshaft balancer from the crankshaft
- Front cover bolts
- Front cover and gasket. Discard the front cover gasket.
- Crankshaft front oil seal from the cover
- Oil pump

4. Rotate the crankshaft until the timing marks on the crankshaft and the camshaft sprockets are aligned.

➡ **Do not turn the crankshaft assembly after the timing chain has been removed in order to prevent damage to the piston assemblies or the valves.**

- Camshaft sprocket bolts
- Camshaft sprocket and timing chain
- Crankshaft sprocket
- Crankshaft sprocket key

To install:

5. Install or connect the following:
- Key into the crankshaft keyway
- Crankshaft sprocket onto the front of the crankshaft. Align the crankshaft key with the crankshaft sprocket keyway. Rotate the crankshaft sprocket until the alignment mark is in the 12 o'clock position.
- Camshaft sprocket and timing chain. Locate the camshaft sprocket alignment mark in the 6 o'clock position.
- Camshaft sprocket bolts. Tighten the bolts to 26 ft. lbs. (35 Nm).

➡ **Do not lubricate the oil seal sealing surface.**

6. Lubricate the outer edge of the oil seal with clean engine oil. Lubricate the front cover oil seal bore with clean engine oil.

7. Install the crankshaft front oil seal with an installer.

➡ **Do not apply any type sealant to the front cover gasket (unless specified). Special tools are used to properly align the engine front cover at the oil pan**

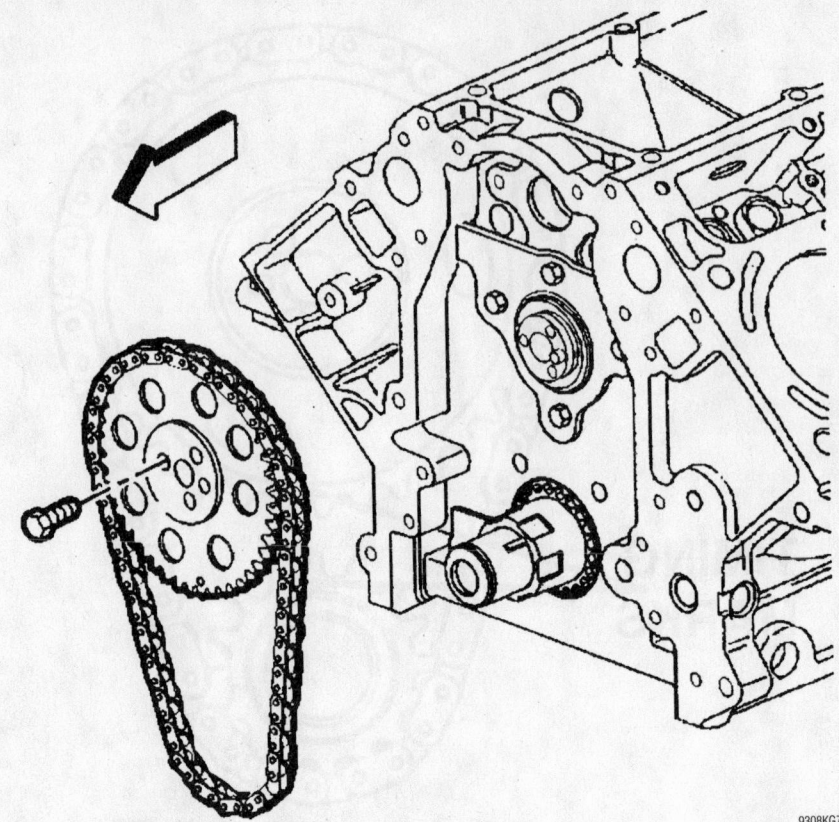

9308KG76

Sprocket and chain removal—4.8L, 5.3L and 6.0L engines

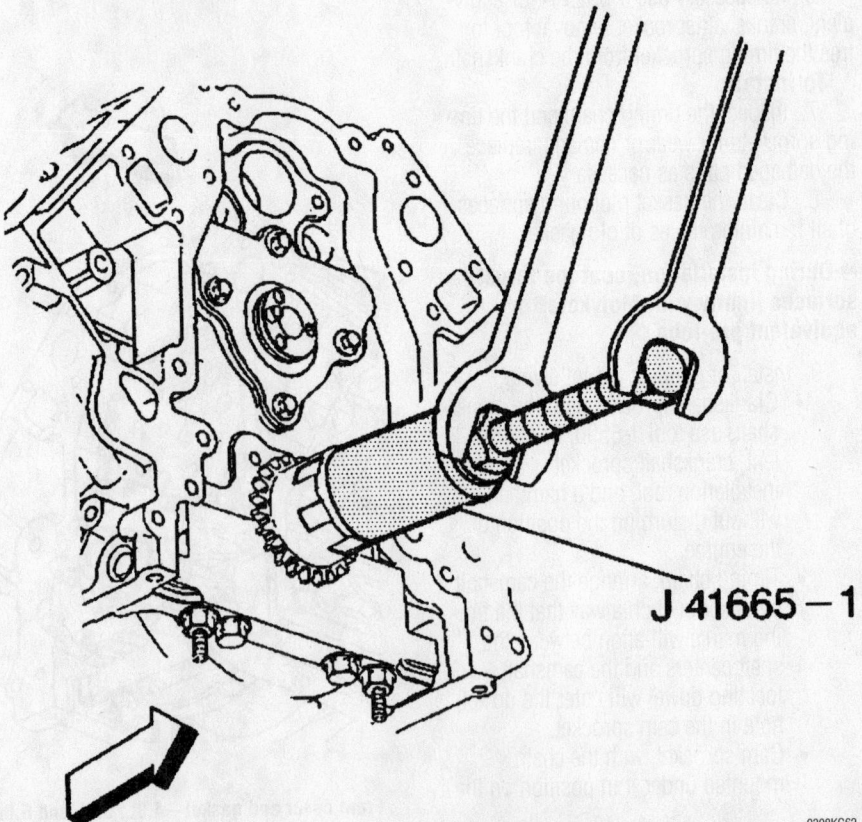

J 41665-1

9308KG62

Crankshaft sprocket installation—4.8L, 5.3L and 6.0L engines

surface and to center the crankshaft front oil seal.

8. Install the front cover gasket, cover, and bolts onto the engine. Tighten the cover bolts finger-tight. Do not overtighten.

9. Start the J41480 tool-to-front cover bolts. Don't tighten the bolts yet.

➡**Align the tapered legs of the tool with the machined alignment surfaces on the front cover.**

10. Install tool J41476 . Install the crankshaft balancer bolt. Tighten the crankshaft balancer bolt by hand until snug. Do not overtighten. Tighten the J41480 bolts and front cover bolts to 18 ft. lbs. (25 Nm).

11. Remove the tools.

12. Place a straight edge across the engine block and front cover oil pan sealing surfaces. Avoid contact with the portion of the gasket that protrudes into the oil pan surface. Insert a feeler gauge between the front cover and the straight edge tool. The cover must be flush with the oil pan surface or no more than 0.02 in. (0.5mm) below flush. If the front cover-to-engine block oil pan surface alignment is not within specifications, repeat the cover alignment procedure. If the correct front cover-to-engine block alignment cannot be obtained, replace the front cover.

13. Install the crankshaft balancer bolt. Tighten the crankshaft balancer bolt by hand until snug. Do not overtighten the bolt.

14. Snug the oil pan-to-cover bolts in order to position the cover at the pan rail.

15. Tighten the oil pan-to-front cover bolts to 18 ft. lbs. (25 Nm).

16. Tighten the front cover bolts to 18 ft. lbs. (25 Nm).

17. Install the water pump.

8.1L Engine

➡**This procedure requires the use of Crankshaft Sprocket Installer tool No. J 22102 and Crankshaft Protector Button tool No. J 42846.**

1. Before servicing the vehicle, refer to the precautions in the beginning of this section.

2. Drain the cooling system.

3. Remove or disconnect the following:
- Negative battery cable
- Water pump
- Crankshaft balancer from the crankshaft
- Camshaft Position (CMP) sensor connector

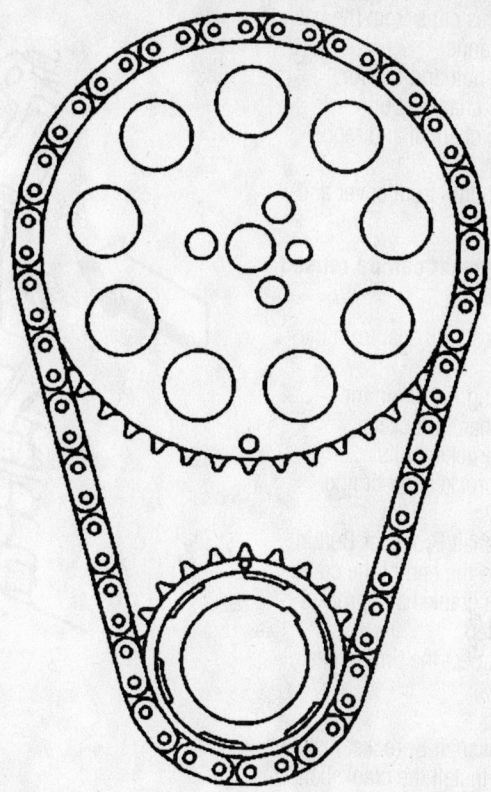

Timing mark alignment—4.8L, 5.3L and 6.0L engines

9308KG61

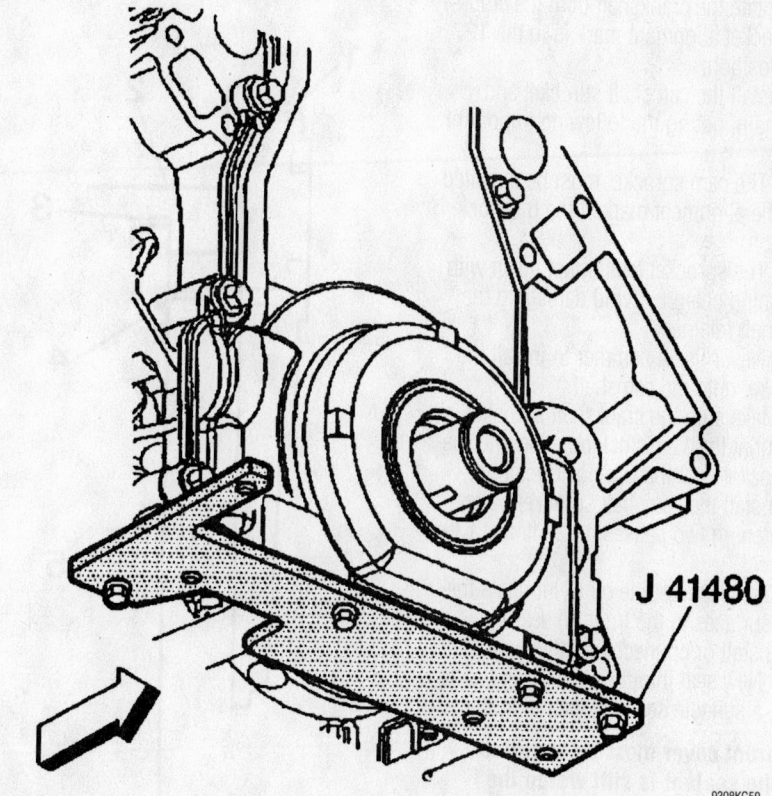

J 41480

J41480 installation—4.8L, 5.3L and 6.0L engines

9308KG59

For Wheel Alignment specifications, see Section 1 of this manual

- Engine harness clips from the battery cable channel
- CMP sensor bolt and sensor
- Battery cable channel bolt
- Battery cable channel and reposition
- Front cover bolts, front cover and gasket

➡ **The front cover gasket can be reused if it is not damaged.**

- Crankshaft front oil seal from the front cover

4. Align the timing marks on the camshaft and crankshaft sprockets.

- Camshaft sprocket bolts
- Camshaft sprocket and timing chain

5. Install Crankshaft Protector Button tool No. J 42846 into the end of the crankshaft and remove the crankshaft sprocket using a 3-jawed puller.

6. Clean and inspect the timing chain and sprockets.

To install:

7. Use the Crankshaft Sprocket Installer tool No. J 22102 to install the crankshaft sprocket. Align the keyway of the sprocket with the crankshaft pin.

8. Remove the installation tool.

9. Rotate the crankshaft until the crankshaft sprocket alignment mark is in the 12 o'clock position.

10. Install the camshaft sprocket and timing chain, noting the following important points:

   a. The cam sprocket must be installed with the alignment mark at the 6 o'clock position.

   b. The sprocket teeth must mesh with the timing chain to avoid damaging the camshaft retainer.

   c. Never use a hammer to install the sprocket onto the camshaft.

11. Make sure the crankshaft sprocket is alignment at the 12 o'clock position and the cam sprocket is at the 6 o'clock position.

12. Install the camshaft sprocket bolts and tighten, in two passes, to 22 ft. lbs. (30 Nm)

13. Use clean engine oil to lubricate the sealing surfaces of the front oil seal.

14. Install or connect the following:

- New seal into the front cover, using a suitable seal installation tool

➡ **The front cover must be installed while the sealant is still wet to the touch.**

- Sealant to the 2 places on the engine block where the front cover meets the oil pan

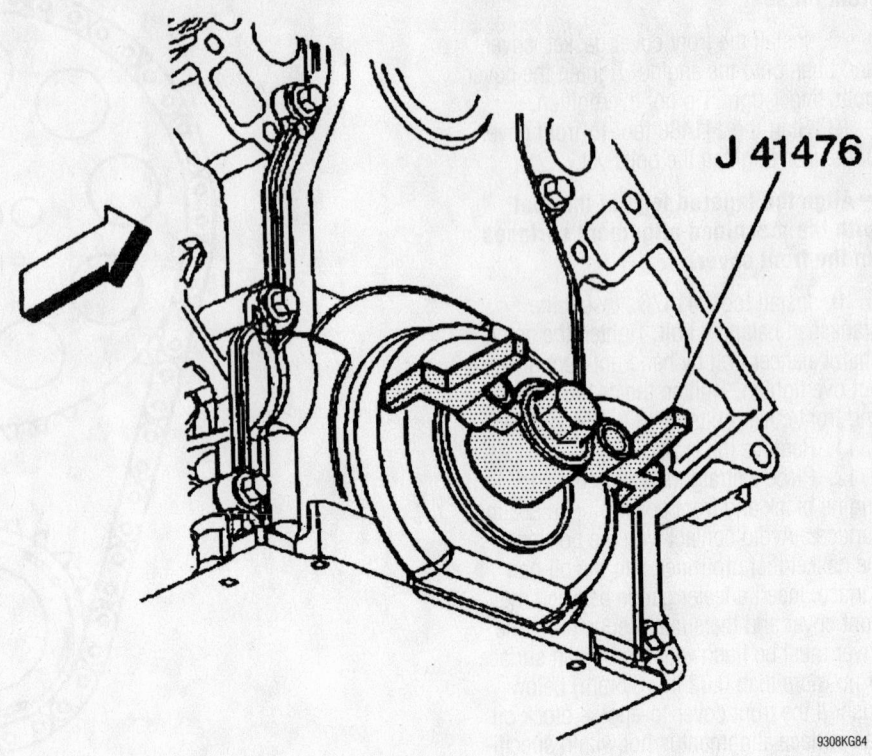

J 41476

9308KG84

Seal alignment tool installation—4.8L, 5.3L and 6.0L engines

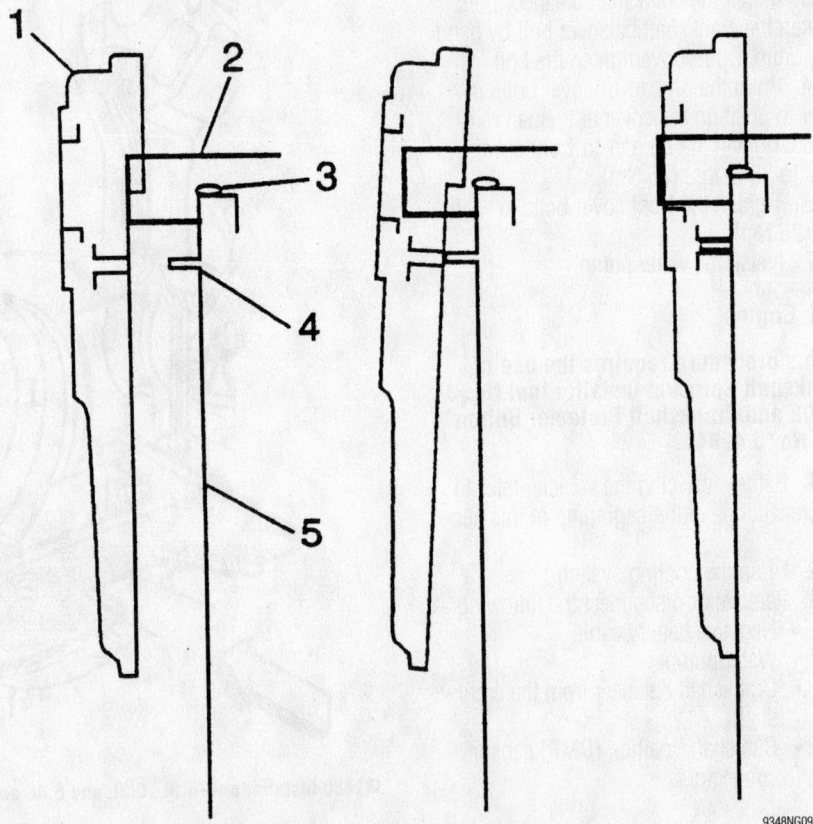

9348NG09

Proper front cover installation sequence—8.1L engine

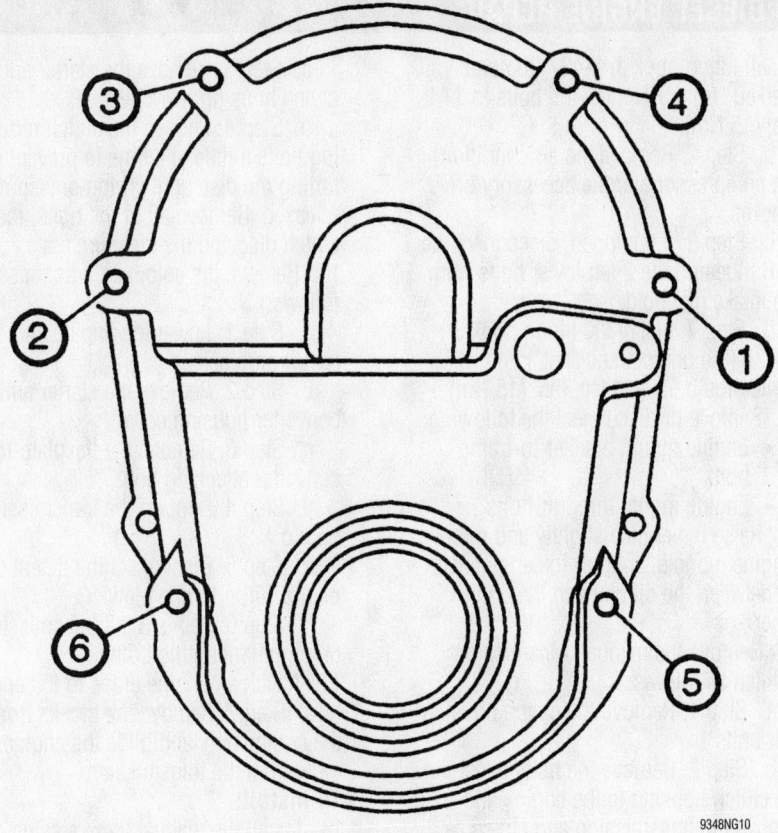

Engine front cover bolt tightening sequence—8.1L engine

9348NG10

• Front cover gasket into the cover
15. Install the front cover, referring to the accompanying figure and using the following steps only:

a. Hold the front cover (1) up to the crankshaft (2).

b. Lift the cover (1) while sliding the cover over the crankshaft (2).

c. Slide the front cover toward the engine block (5) while keeping the cover raised.

d. Lower the cover down over the dowel pin (4), allowing the front cover to rest on the sealant (3).

16. Install the front cover bolts and tighten, in sequence, as follows:

a. 1st pass: 53 inch lbs. (6 Nm).

b. 2nd pass: 106 inch lbs. (12 Nm).

17. Install or connect the following:

• Battery cable channel and bolt. Tighten to 80 inch lbs. (9 Nm).

• CMP sensor. Inspect the O-ring first, replace if necessary and coat with oil before installation.

• CMP sensor bolt to 106 inch lbs. (12 Nm).

• Engine harness clips to the battery cable channel

• CMP sensor electrical connector
• Crankshaft balancer
• Water pump
• Negative battery cable.

18. Fill the cooling system with the proper type and quantity of antifreeze.

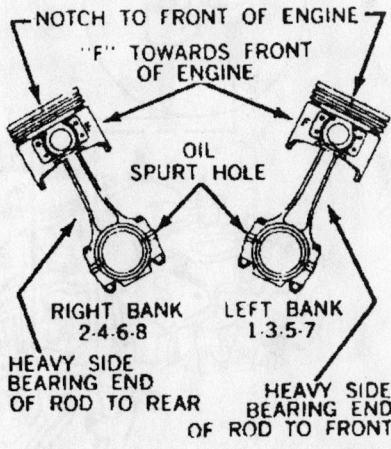

7924AG09

Piston and connecting rod assembly positioning —General Motors 4.3L, 4.8L, 5.0L, 5.3L, 5.7L and 6.0L engines

Piston and Ring

POSITIONING

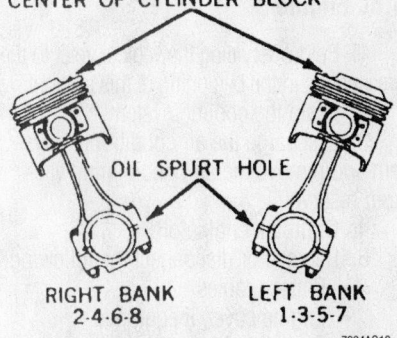

7924AG10

Piston and connecting rod assembly positioning —General Motors 7.4L engine

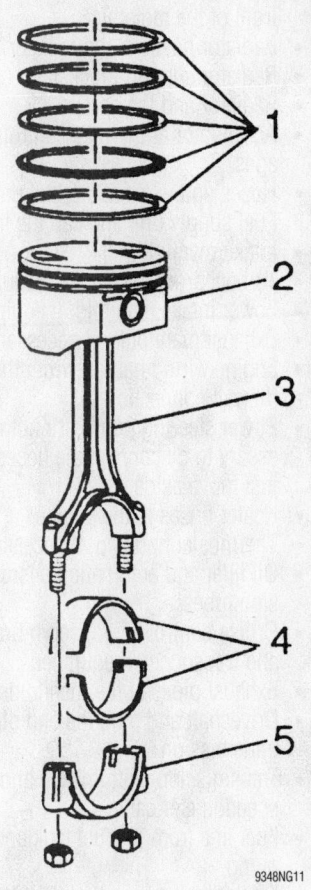

9348NG11

Piston rings (1), piston (2), connecting rod (3) and related components—General Motors 8.1L engine

DIESEL ENGINE REPAIR

Engine Assembly

REMOVAL & INSTALLATION

6.5L Engine

1. Before servicing the vehicle, refer to the precautions in the beginning of this section.
2. Drain the cooling system.
3. Discharge the air conditioning system and remove the air conditioning vacuum reservoir.
4. Drain the engine oil.
5. Remove or disconnect the following:
   - Battery cables
   - Engine cover, if equipped
   - Air cleaner
   - Radiator coolant reservoir bottle
   - Upper radiator support
   - Grille and the lower grille valance
   - Front bumper, if necessary
   - Air conditioning condenser from in front of the radiator
   - Radiator hoses at the radiator
   - Radiator support bracket
   - Radiator and the shroud
   - Accelerator and cruise control linkages
   - Hoses and wires at the fuel unit
   - Fuel supply unit and cap the lines
   - Intake manifold
   - Turbocharger assembly, if equipped
   - Lower intake manifold, if equipped
   - Exhaust manifolds, if necessary
   - Engine wiring harness from the firewall connection
   - Power steering pump, it's not necessary to disconnect the hoses; just move aside
   - Heater hoses at the engine
   - Thermostat housing, if necessary
   - Oil filler and automatic transmission tubes
   - Cruise control servo, servo bracket and transducer, if equipped
   - Exhaust pipes at the manifolds
   - Driveshaft and plug the end of the transmission
   - Transmission shift linkage and the speedometer cable
   - Fuel line from the fuel tank and pump
   - Transmission mounting bolts
6. Support the transmission and engine.
7. Install lifting hooks J-41427 as follows:
   a. Step 1: Remove the 2 right rear lower intake manifold retainers and install lifting hook J-41427 (the one marked "right") Tighten the bolts to 11 ft. lbs. (15 Nm).
   b. Step 2: Remove the air conditioning compressor and the accessory drive bracket.
   c. Step 3: If equipped, disconnect the EGR tube and the 2 left lower bolts from the intake manifold.
   d. Step 4: Install the lifting hook J-41427 (the one marked "left") and tighten the bolts to 11 ft. lbs. (15 Nm).
8. Remove or disconnect the following:
   - Engine mount bracket-to-frame bolts
   - Engine mount through-bolts
9. Raise the engine slightly and remove the engine mounts, support the engine with wood between the oil pan and the crossmember.
10. Remove the manual transmission and clutch as follows:
    a. Step 1: Remove the clutch housing rear bolts.
    b. Step 2: Remove the bolts attaching the clutch housing to the engine and remove the transmission and clutch as a unit.

➡Support the transmission as the last bolt is being removed to prevent damaging the clutch.

c. Step 3: Remove the starter and clutch housing rear cover.
   d. Step 4: Loosen the clutch mounting bolts a little at a time to prevent distorting the disc until spring pressure is released. Remove all of the bolts, the clutch disc and the pressure plate.
11. Remove the automatic transmission as follows:
    a. Step 1: Lower the engine and support it on blocks.
    b. Step 2: Remove the starter and converter housing cover.
    c. Step 3: Remove the flexplate-to-converter attaching bolts.
    d. Step 4: Support the transmission on blocks.
    e. Step 5: Disconnect the detent cable on the Turbo Hydra-Matic.
    f. Step 6: Remove the transmission-to-engine mounting bolts.
12. Attach an engine crane to the engine.
    a. Step 7: Remove the blocks from the engine only and glide the engine away from the transmission.

To install:

13. Install the manual transmission and clutch as follows:
    a. Step 1: Install the clutch disc and the pressure plate. Tighten the clutch mounting bolts a little at a time to prevent distorting the disc.

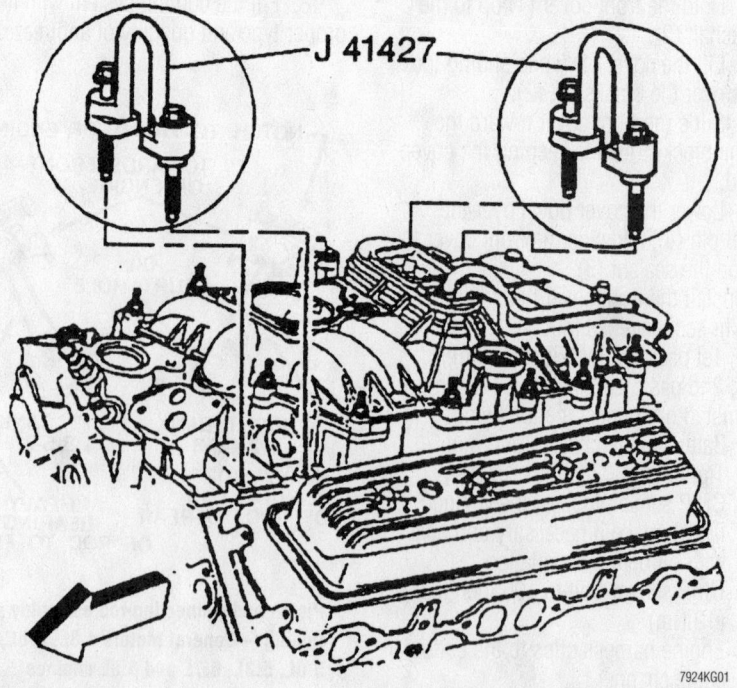

For engine removal and installation, universal lift brackets should be installed in place of the proper intake manifold bolts

7924KG01

b. Step 2: Install the starter and clutch housing rear cover.

c. Step 3: Install the bolts attaching the clutch housing to the engine and install the transmission and clutch as a unit. Tighten the bolts to specification.

d. Step 4: Install the clutch housing rear bolts.

14. Install the automatic transmission as follows:

a. Step 1: Position the transmission.

b. Step 2: Install the transmission-to-engine mounting bolts.

c. Step 3: Connect the throttle linkage and detent cable.

d. Step 4: Install the flexplate-to-converter attaching bolts. Torque the bolts to 65 ft lbs. (90 Nm).

e. Step 5: Install the starter and converter housing cover.

15. Install or connect the following:
- Engine mount through-bolts. Torque the bolts to 50 ft lbs. (68 Nm).
- Engine mount bracket-to-frame bolts. Torque the bolts to 44 ft lbs. (59 Nm).
- Clutch cross-shaft
- Transmission mounting bolts. Torque the bolts to 75 ft lbs. (100 Nm).

- Transmission shift linkage and the speedometer cable
- Driveshaft

16. Remove the lifting hooks.
- Compressor
- EGR valve tube
- Intake manifold retaining bolts
- Condenser
- Hood latch support
- Lower fan shroud and filler panel
- Transmission dipstick tube and the accelerator cable at the tube
- Coolant hose
- Cruise control servo, bracket and transducer
- Oil filler pipe and automatic transmission filler pipe
- Engine dipstick tube
- Thermostat housing, if removed
- Heater hoses at the engine
- Engine wiring harness to the firewall connection
- Radiator and the shroud
- Radiator support bracket
- Exhaust manifolds if removed
- Lower intake manifold if removed
- Intake manifold
- Fuel supply unit
- Lines to the fuel supply unit

- Accelerator and cruise control linkages
- Windshield wiper jar and bracket
- Air conditioning condenser
- Air conditioning vacuum reservoir, if equipped
- Fluid cooler lines at the radiator, if equipped with an automatic transmission
- Radiator coolant reservoir bottle
- Hoses at the radiator
- Upper radiator support
- Grille and the lower grille valance
- Air cleaner
- Air stove pipe
- Engine cover
- Battery cables

17. Refill the cooling system.

18. Recharge the air conditioning system.

Water Pump

REMOVAL & INSTALLATION

6.5L Engine

1. Before servicing the vehicle, refer to the precautions in the beginning of this section.

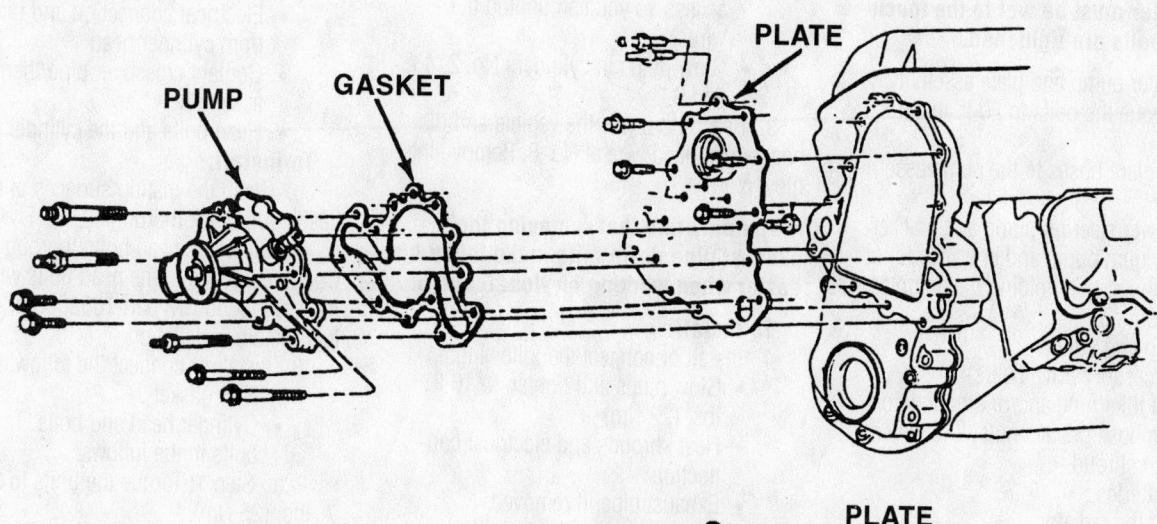

Exploded view of the water pump assembly and related components—6.5L diesel engines

7924KG05

2. Drain the engine coolant.
3. Remove or disconnect the following:
   - Negative battery cables
   - Fan and fan shroud
   - Air conditioning hose bracket and/or the oil filler tube, as required
   - Accessory drive belt(s)
   - Vacuum pump mounting bracket nuts/bolt
   - Vacuum pump and bracket
   - Power steering pump and bracket
   - Coolant hoses from the pump
   - Water pump plate retaining bolts
   - Pump and plate assembly from the engine

➡**Remove the bolt on the rear of the water pump plate.**

   - Separate the pump and gasket from the plate

To install:

4. Install or connect the following:
   - Water pump and a new gasket to the plate. Torque the retaining bolt (at the rear of the plate) to 20 ft. lbs. (28 Nm).

5. Be sure the block mating surface and the plate flanges are free of oil. Apply an anaerobic sealer GM part 1052357 or equivalent.

➡**The sealer must be wet to the touch when the bolts are tightened.**

   - Water pump and plate assembly. Torque the bolts to 20 ft. lbs. (28 Nm).
   - Coolant hoses to the pump assembly
   - Power steering pump and bracket
   - Vacuum pump and bracket, along with the bolt holding the pump and alternator
   - Fan and pulley
   - Accessory drive belt(s)
   - Oil filler tube and/or air conditioning hose bracket nuts. If removed
   - Fan shroud
   - Batteries

6. Refill the radiator.

Glow Plugs

REMOVAL & INSTALLATION

6.5L Engine

1. Before servicing the vehicle, refer to the precautions in the beginning of this section.
2. Remove or disconnect the following:

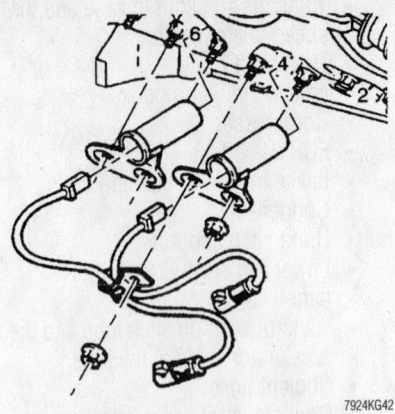

Exploded view of the heat shrouds and glow plug wiring—6.5L diesel engine

7924KG42

   - Negative battery cables
   - Glow plug lead wires
   - Plugs
   - Right front tire
   - Inner splash shield from the fender well
   - Lead wire from the plug at the No. 2 cylinder and the lead wires from plugs in the Nos. 4 and 6 cylinders at the harness connectors
   - Heat shroud for the plug in the No. 4 and 6 cylinder. Slide the shrouds back just far enough to allow access so you can unplug the wires.
   - Glow plugs in cylinders No. 2, 4 and 6.

3. Reach up under the vehicle and disconnect the lead wire at No. 8. Remove the glow plug.

➡**You may find that removing the exhaust pipe down mite make this a bit easier when working on Nos. 6 and 8.**

To install:

4. Install or connect the following:
   - Glow plugs and tighten to 16 ft. lbs. (22 Nm)
   - Heat shrouds and electrical connection
   - Exhaust pipe, if removed
   - Splash shields, if removed
   - Negative battery cable

Cylinder Head

REMOVAL & INSTALLATION

6.5L Engine

1. Before servicing the vehicle, refer to the precautions in the beginning of this section.

2. Relieve the fuel system pressure.
3. Drain the coolant system.
4. Discharge the air conditioning system.
5. Remove or disconnect the following:
   - Negative battery cables
   - Intake manifold
   - Fan upper shroud
   - Compressor assembly, if equipped
   - Turbocharger, if equipped
   - Exhaust manifold
   - Valve cover
   - Rocker arm assemblies and pushrods

➡**Mark all components so they may be returned to their original location.**

   - Air cleaner resonator and bracket
   - Transmission and oil dipstick tube; remove the oil fill tube from the coolant crossover pipe
   - Heater, radiator and bypass hoses
   - Alternator upper bracket
   - Alternator
   - Power steering pump
   - Vacuum pump
   - Fuel bleeder valve at the coolant crossover pipe
   - Fuel return crossover line clamp bolts from both cylinder heads
   - Wire connector from the sensor in the coolant crossover pipe
   - Electrical connection and brackets from cylinder head
   - Coolant crossover pipe/thermostat assembly
   - Head bolts and the cylinder heads

To install:

6. Clean the mating surfaces of the heads and block thoroughly.

7. Clean the head bolts thoroughly. Coat the threads of the head bolts with sealing compound GM part 1052080 or equivalent, before installation.

8. Install or connect the following:
   - New gasket
   - Cylinder head and bolts. Torque the bolts in the follows:

   a. Step 1: Torque the bolts to 20 ft. lbs. (25 Nm).

   b. Step 2: Torque the bolts to 50 ft. lbs. (65 Nm).

   c. Step 3: Then an additional 90 degrees (¼ turn).

   - Coolant crossover pipe and thermostat
   - Fuel valve
   - Bypass hose
   - Upper radiator hose
   - Heater hoses at the head
   - Transmission and oil dipstick tube
   - Air cleaner resonator and bracket

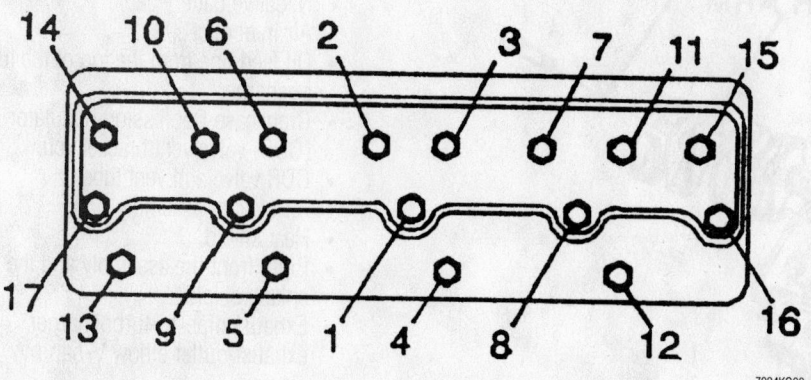

Tighten the cylinder head bolts according to the sequence shown for proper cylinder sealing—6.5L diesel engines

- Pushrods, hardened ends facing up
- Rocker arm assemblies
9. Adjust the valves.
- Valve cover
- Alternator and upper bracket
- Exhaust manifolds. Torque bolts to 22 ft. lbs. (30 Nm).
- Upper fan shroud
- Intake manifold
- Turbocharger, if equipped
- Vacuum pump, if equipped
- Air conditioning compressor, if equipped
- Engine electrical connection
- Negative battery cables
10. Refill the cooling system with the proper type and quantity of antifreeze.
11. Evacuate and recharge the air conditioning system.

Starter Motor

REMOVAL & INSTALLATION

1. Before servicing the vehicle, refer to the precautions in the beginning of this section.
2. Remove or disconnect the following:
- Negative battery cables
- Mounting bolts/nuts and shim, if used
- Starter
- Wires
- Heat shield and bracket
To install:
3. Install or connect the following:
- Heat shield and bracket. Torque the bolts to 13 ft lbs. (17 Nm).

- Wires. Torque battery wire nut to 89 inch lbs. (10 Nm), and ignition nut to 18 inch lbs. (2 Nm).
- Starter
- Mounting bolts/nuts and shim, if used. Torque the bolts to 33 ft lbs. (45 Nm) and the nut to 75 inch lbs. (8.5 Nm).
- Negative battery cables

Rocker Arms/Shaft

REMOVAL & INSTALLATION

6.5L Engine

1. Before servicing the vehicle, refer to the precautions in the beginning of this section.
2. Remove or disconnect the following:

➡Rotate the engine until the mark on the crankshaft balancer is at the 2 o'clock position. Rotate the crankshaft counterclockwise 3½ in. (88mm) aligning the crankshaft balancer mark with the first lower water pump bolt, at about the 12:30 position. This will ensure that no valves are close to a piston crown.

- Cylinder head cover
- Rocker shaft assembly

➡The rocker assemblies are mounted on 2 short rocker shafts per cylinder head, with each shaft operating 4 rockers. 2 bolts secure each rocker shaft assembly, Mark the shafts so they can be installed in their original locations.

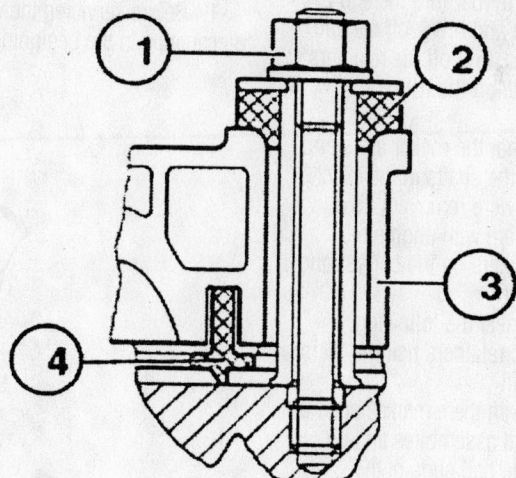

(1)   Nut

(2)   Decoupling element

(3)   Intake air manifold

(4)   Seal

Exploded view of the starter motor—6.5L engine shown

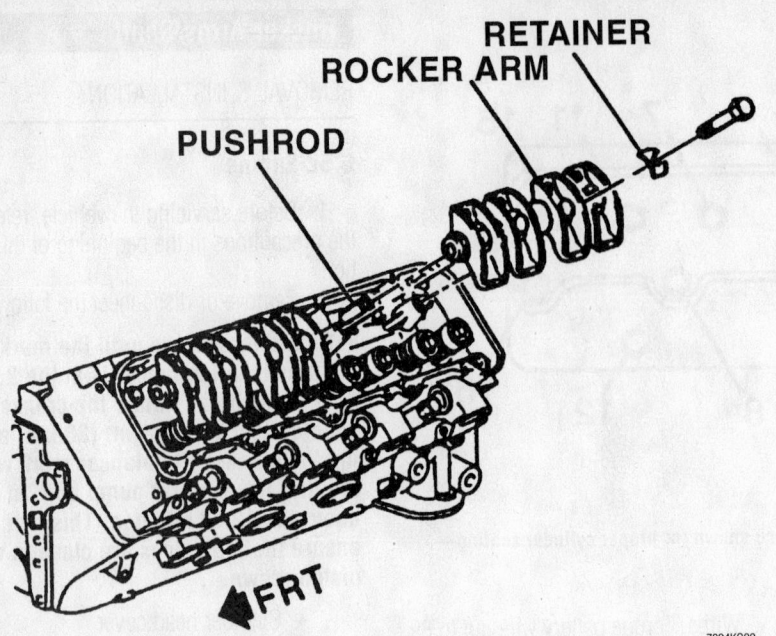

RETAINER
ROCKER ARM
PUSHROD

FRT

7924KG22

Rocker shaft assembly and related components—diesel engines

- Pushrods. The pushrods MUST be installed in the original direction! A paint stripe usually identifies the upper end of each rod, but if you can't see it, be sure to mark each rod yourself.

3. Insert a small prybar into the end of the rocker shaft bore and break off the end of the nylon retainers. Pull off the retainers with pliers, then slide off the rockers.

To install:

4. Be sure first that the rocker arms and springs go back on the shafts in the exact order in which they were removed. It's a good idea to coat them with engine oil.

5. Center the rockers on the corresponding holes in the shaft

6. Install or connect the following:
- New plastic retainers using a ½ in. (13mm) drift
- Pushrods with there marked ends up
- Rocker shaft assemblies and be sure that the ball ends of the pushrods seat themselves in the rockers

7. Rotate the engine clockwise until the mark on the torsional damper aligns with the **0** on the timing tab. Rotate the engine counterclockwise 3 ½ in. (88mm) measured at the damper. You can estimate this by checking that the mark on the damper is now aligned with the FIRST lower water pump bolt. BE CAREFUL! This ensures that the piston is away from the valves.
- Rocker shaft bolts. Torque them to 40 ft. lbs. (55 Nm).
- Cylinder head cover

Turbocharger

REMOVAL & INSTALLATION

6.5L Engine

1. Before servicing the vehicle, refer to the precautions in the beginning of this section.

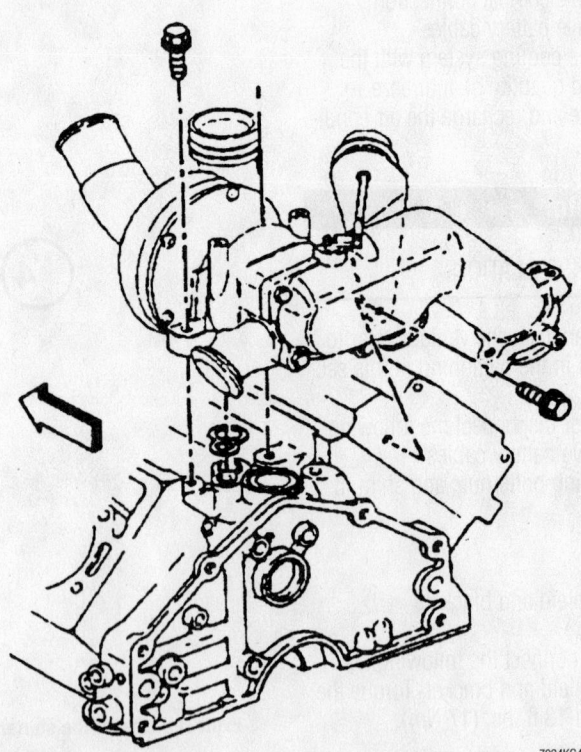

Turbocharger mounting—diesel engines

7924KG43

2. Remove or disconnect the following:

- Negative battery cable
- Air inlet duct
- Oil feed line from the top of the turbocharger
- Crankcase Depression Regulator (CDR) valve vent bracket screw
- CDR valve and vent tube
- Air cleaner assembly
- Heat shield
- Right front tire assembly and the splash shield
- Exhaust pipe-to-turbocharger exhaust outlet elbow V-band clamp
- Oil drain tube-to-turbocharger center bearing bolts
- Exhaust manifold-to-turbocharger nuts
- Turbocharger

To install:

➡**Use anti-seize compound on all threaded fasteners connected to the turbocharger**

3. Install or connect the following:
- Turbocharger to the exhaust manifold. Torque the nuts to 37 ft. lbs. (50 Nm).
- New oil drain tube flange gasket and the oil drain tube. Torque the bolts to 19 ft. lbs. (26 Nm).

➡**Use 0.03–0.07 fl. oz. (0.88–2ml) of engine oil to feed the oil feed hole at the top of the turbocharger and hand rotate the compressor wheel/shaft. This will pre-lube the shaft bearings**

- Oil feed line. Torque the connection to 13 ft. lbs. (17 Nm).
- Exhaust pipe to the turbocharger exhaust elbow V-band clamp. Torque the clamp to 71 inch. lbs. (8 Nm).

4. Disengage the injection pump fuel shutdown solenoid connector and crank the engine for no more 15 seconds to prime the oil system. Do not let the engine start.

- Right front wheel and splash shield
- Heat shield. Apply Loctite® or equivalent to the bolts. Torque to 56 inch. lbs. (6 Nm).
- Air cleaner assembly
- Turbocharger compressor outlet
- CDR valve, tube and bracket
- Air intake duct

➡**Operate the engine at idle for at least 3 minutes after installing the turbocharger**

Intake Manifold

REMOVAL & INSTALLATION

6.5L Engine

1. Before servicing the vehicle, refer to the precautions in the beginning of this section.
2. Recover air conditioning system, and reposition air conditioning lines.

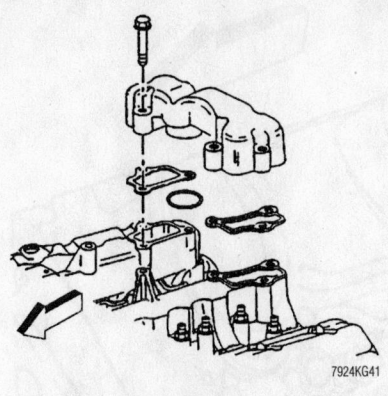

Center intake manifold mounting— 1999–00 6.5L engines

3. Remove or disconnect the following:
- Negative battery cable
- Air cleaner assembly
- Fuel lines, and electrical connections
- Engine and transmission oil level tubes

✳✳ WARNING

Do not remove the center intake and side intakes as an assembly.

Damage to the center intake and turbocharger may occur.

- Center intake assembly, and glow plug relay
- Side intake bolts and fuel retaining clips
- Side intakes

4. Clean all gaskets surface.

To install:

5. Install or connect the following:
- Side intakes and new gaskets. Torque the bolts to 31 ft. lbs. (42 Nm).
- Fuel lines retaining clips and electrical connection
- Center intake with new gaskets. Torque the bolts to 17 ft. lbs. (23 Nm).
- Engine oil and transmission oil level tubes
- Glow plug relay
- Air conditioning lines
- Air cleaner assembly
- Negative battery cable

6. Recharge air conditioning system.

Exhaust Manifold

REMOVAL & INSTALLATION

6.5L Engine

1. Before servicing the vehicle, refer to the precautions in the beginning of this section.

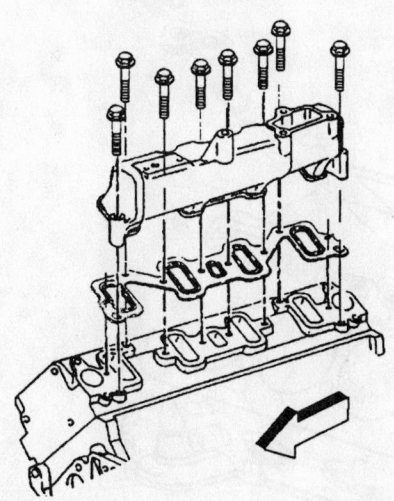

Exploded view of the side intake manifold mounting—1999–00 6.5L engines

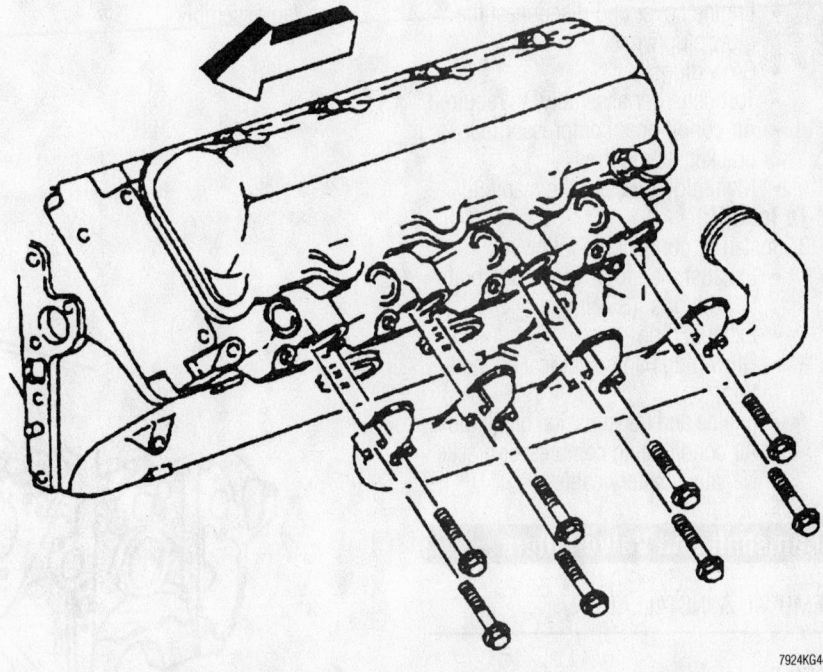

Exploded view of the left exhaust manifold mounting—6.5L diesel engines

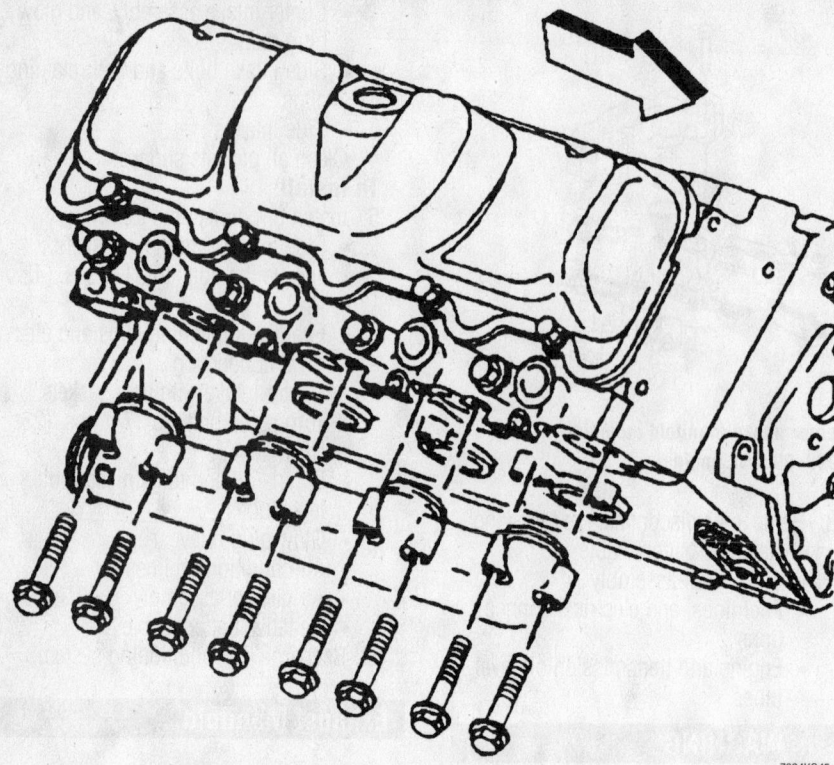

Exploded view of the right exhaust manifold mounting—6.5L diesel engines

7924KG45

- Grille and parking light assembly
- Hood latch and brace assembly
- Oil pump drive
- Power steering pump and position aside
- Alternator, air conditioner compressor and position aside
- Rocker arm covers
- Rocker arm assemblies and pushrods. Mark them so they can be returned to their original position.
- Cylinder heads
- Hydraulic lifters and keep them in order so they can be returned to their original bore
- Front cover
- Timing chain and camshaft sprocket
- Injector pump
- Front engine mounting through-bolts
- Air conditioner condenser mounting bolts and lift the condenser out
- Thrust plate bolts and thrust plate
- Camshaft from the block
- Thrust plate spacer, if necessary

To install:

6. Install the spacer with the ID chamfer toward the camshaft.

➡**It is recommended that the engine oil, oil filter and hydraulic lifters be replaced when installing a new camshaft.**

7. Coat the camshaft lobes with Molykote® or equivalent.

2. Remove or disconnect the following:
- Batteries cables
- Exhaust pipe from the manifold flange
- Engine oil and transmission oil fill tubes
- Engine cover and disconnect the glow plug wires
- Glow plugs
- Turbocharger assembly, as required
- Air conditioner compressor rear bracket, as required
- Manifold bolts and the manifold

To install:

3. Install or connect the following:
- Exhaust manifold. Torque the bolts to 26 ft. lbs. (35 Nm).
- Exhaust pipe
- Glow plugs and electrical connection
- Engine and transmission oil fill tubes
- Air conditioning compressor bracket
- Negative battery cable

Camshaft and Valve Lifters

REMOVAL & INSTALLATION

6.5L Engine

1. Before servicing the vehicle, refer to the precautions in the beginning of this section.

2. Drain the cooling system.
3. Discharge the air conditioning system.
4. Relieve the fuel system pressure.
5. Remove or disconnect the following:
- Battery cables
- Radiator, condenser, shroud and fan assembly

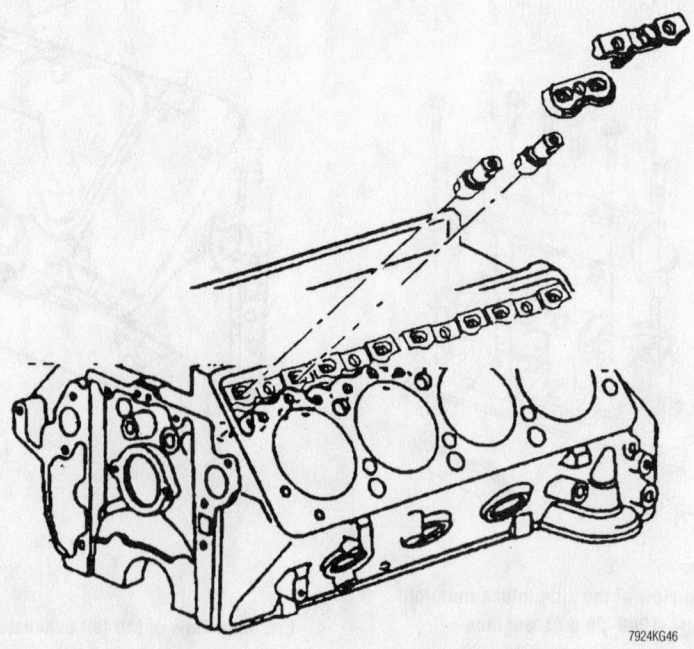

Exploded view of the lifter, guide plate and clamp—6.5L diesel engines

7924KG46

8. Lubricate the camshaft journals with engine oil.

9. Install or connect the following:
- Camshaft carefully into the block
- Thrust plate and bolts. Tighten to 17 ft. lbs. (23 Nm).
- Engine mount through-bolts
- Timing chain and sprockets, align the timing marks
- Air conditioner condenser, if equipped
- Injector pump
- Front cover
- Cylinder head
- Hydraulic lifters in the same bore as they were removed
- Rocker arm assemblies and pushrods in their original locations
- Rocker arm covers
- Power steering pump, alternator and air conditioner compressor
- Oil pump drive
- Hood latch and brace
- Grille and parking light assembly
- Radiator, the shroud and fan assembly
- Negative battery cables

10. Refill the cooling system with the proper type and quantity of antifreeze.

Valve Lash

ADJUSTMENT

All engines use hydraulic lifters, which require no periodic adjustment.

Oil Pan

REMOVAL & INSTALLATION

6.5L Engine

1. Before servicing the vehicle, refer to the precautions in the beginning of this section.

2. Drain the engine oil.

3. Remove or disconnect the following:
- Battery cables
- Oil dipstick
- Flywheel/flexplate cover
- Oil cooler line guides
- Front driveshaft
- Front axle, if needed
- Exhaust pipes from the manifolds
- Front engine mount through-bolts
- Oil pan bolts and the oil pan
- Oil pan rear seal

To install:

4. Clean all sealing surfaces.

5. Apply a ³⁄₁₆ in. (5mm) bead of RTV sealant to the oil pan sealing surface, inboard of the bolt holes. The sealant must be wet to the touch when the oil pan is to be installed.

6. Install or connect the following:
- Oil pan rear seal
- Oil pan to the engine. Torque all bolts except the rear 2 bolts to 84 inch lbs. (9.4 Nm) Tighten the rear bolts to 17 ft. lbs. (23 Nm).
- Engine mounting through-bolt and nut
- Front axles and front driveshaft, if removed
- Oil cooler lines in guides
- Oil dipstick
- Exhaust pipes to the manifolds
- Flywheel/flexplate cover
- Battery cables

7. Refill with the proper grade and quantity of oil.

Oil Pump

REMOVAL & INSTALLATION

6.5L Engine

1. Before servicing the vehicle, refer to the precautions in the beginning of this section.

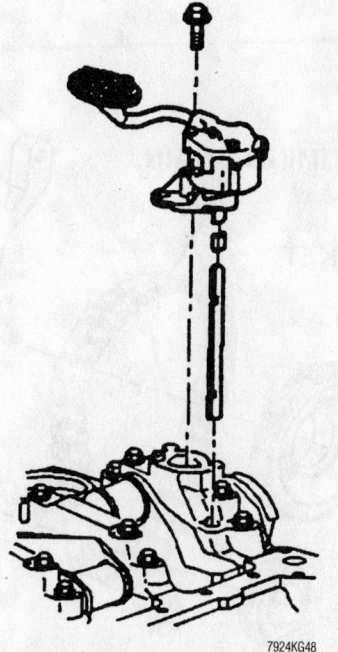

7924KG48

Exploded view of the oil pump mounting—6.5L diesel engines

2. Drain the engine oil

3. Remove or disconnect the following:
- Oil pan
- Oil pump to crankshaft rear main bearing attaching bolt
- Oil pump and hex drive

To install:

4. Inspect the oil pan pick up tube and screen for damage and the hex drive for cracks.

5. Install or connect the following:
- Oil pump and extension shaft to the engine. Align the extension shaft hex with the drive hex, the oil pump should push easily into place.
- Oil pump bolt. Torque the bolt to 65 ft. lbs. (90 Nm).
- Oil pan

6. Refill the crankcase with oil.

Rear Main Seal

REMOVAL & INSTALLATION

Please note that the entire transmission assembly must be removed before performing this procedure. Before a new seal is installed, the Crankcase Depression Regulator (CDR) and crankcase ventilation system should be cleaned and inspected. In addition, use care removing the flywheel. Some models use a heavy, dual mass flywheel that must be handled with care.

1. Before servicing the vehicle, refer to the precautions in the beginning of this section.

2. Remove or disconnect the following:
- Negative battery cables
- Transfer case, if equipped
- Transmission assembly
- Clutch assembly and flywheel, if equipped with manual transmission
- Flexplate, if equipped with automatic transmission
- Crankshaft rear main oil seal by inserting a suitable crankshaft seal removal tool and prying the seal out

To install:

3. Clean the oil seal bore in the block thoroughly before installation of the new seal.

4. Inspect the crankshaft for grit, rust or burrs and correct as necessary. Also inspect the portion of the crankshaft where the oil seal makes contact, for wear due to the rubbing action of the oil seal.

Timing belt service is covered in Section 3 of this manual

➡ **Because of rear crankshaft wear or grooving, the new oil seal should be seated in a new location. The J 39084 installation tool will control the seal positioning. This will provide a new surface on the crankshaft for the seal to ride on.**

5. Clean the running surface of the crankshaft with a non-abrasive cleaner.

6. Lubricate the inner diameter of the new seal and the outer diameter of the crankshaft with engine oil.

7. Install or connect the following:
- Rear main oil seal using a crankshaft rear oil seal installation tool
- Flywheel.
- Transmission assembly
- Transfer case, if equipped
- Negative battery cables

8. Start the engine and verify no oil leaks.

Timing Chain, Sprockets, Front Cover and Seal

REMOVAL & INSTALLATION

6.5L Engine

1. Before servicing the vehicle, refer to the precautions in the beginning of this section.

2. Drain the cooling system.

3. Remove or disconnect the following:
- Negative battery cables
- Water pump and pulleys

4. Rotate the crankshaft to align the marks on the torsional damper with the **0** mark on the timing tab.

5. Scribe a mark aligning the injection pump flange and the front cover, if not already marked.

➡ **The outer ring (weight) of the torsional damper is bonded to the hub with rubber. The damper must be removed with a puller that acts on the inner hub only. Pulling on the outer portion of the damper will break the rubber bond or destroy the tuning of the unit.**

6. Remove or disconnect the following:
- Crankshaft pulley and torsional damper
- Front cover-to-oil pan bolts (4)
- 2 fuel return line clips
- Injection pump gear
- Injection pump retaining nuts from the front cover
- Crankshaft sensor
- Baffle
- Cover bolts remaining and the front cover
- Injection pump gear

7. Align the camshaft timing gear marks

8. Remove the bolt and washer attaching the camshaft gear.
- Camshaft sprocket with the timing chain
- Crankshaft sprocket

To install:

9. Install or connect the following:
- Cam sprocket, timing chain and crankshaft sprocket as a unit, aligning the timing marks on the sprockets

10. Rotate the crankshaft to align the injection pump and camshaft gears.
- Injection pump gear

11. If the front cover oil seal is to be replaced, it can now be pried out of the cover with a suitable prying tool. Press the new seal into the cover evenly.

12. Clean both sealing surfaces until all traces of old sealer are gone. Apply a ³⁄₃₂ in. (2mm) bead of GM sealant 1052357 or equivalent to the sealing surface. Apply a ³⁄₁₆ in. (5mm) bead of RTV type sealer to the bottom portion of the front cover which attaches to the oil pan. Install the front cover.

13. Install or connect the following:
- Baffle
- Injection pump. Torque the nuts to 31 ft. lbs. (42 Nm), making sure

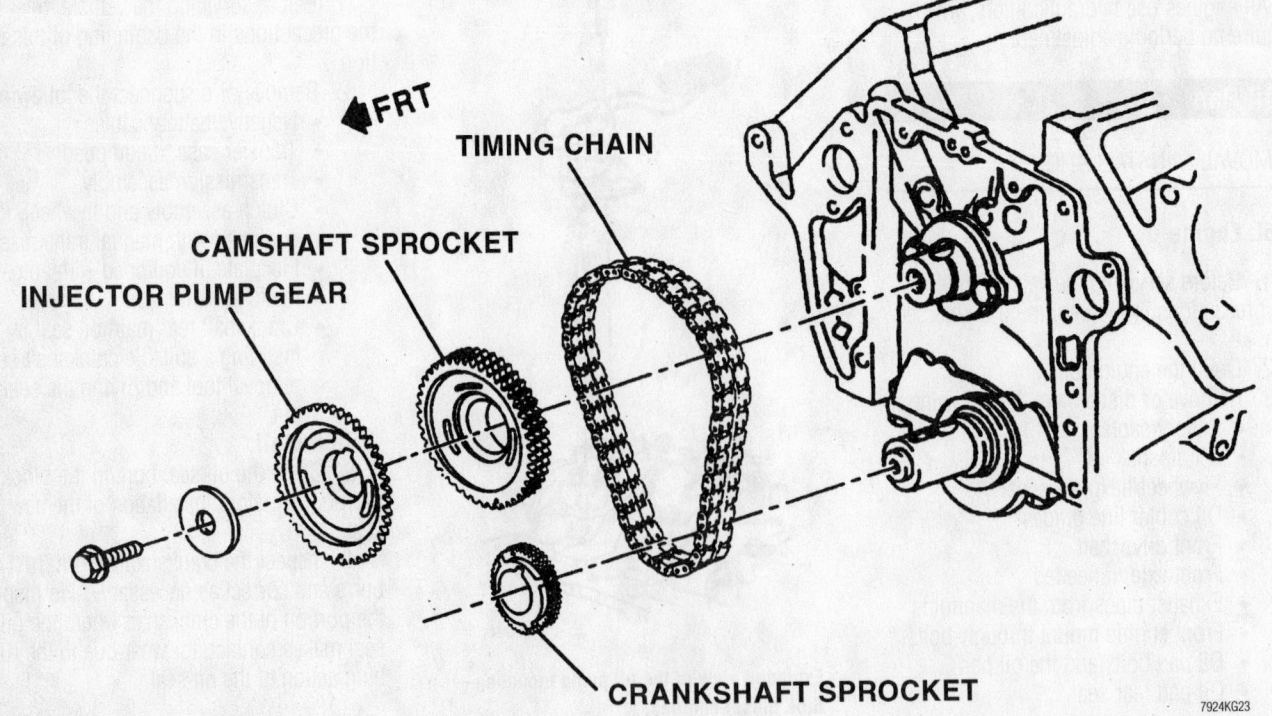

FRT

TIMING CHAIN

CAMSHAFT SPROCKET

INJECTOR PUMP GEAR

CRANKSHAFT SPROCKET

7924KG23

Timing chain and related components—6.5L diesel engines

the scribe marks on the pump and front cover are aligned.

- Injection pump driven gear. Torque the injection pump gear bolts to 17 ft. lbs. (23 Nm), making sure the marks on the cam gear and pump are aligned.

➡**Verify that there is a minimum clearance of 0.040 in. (1.0mm) between the injection pump gear and baffle or noise may be result.**

- Fuel line clips
- Front cover-to-oil bolts
- Torsional damper
- Crankshaft pulley. Torque the bolts to 80 inch lbs. (9 Nm).
- Oil pan bolts. Torque the bolts to 106 inch lbs. (12 Nm).
- Water pump
- Pulley assembly
- Negative battery cables

14. Refill the cooling system with the proper type and quantity of antifreeze.

15. Inspect the engine for leaks.

Piston and Ring

POSITIONING

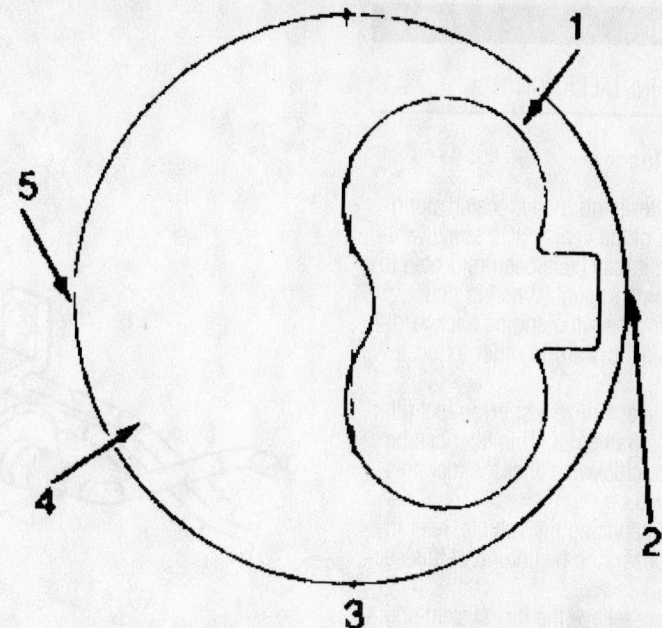

1. Oil control ring expander gap
2. Second compression ring gap
3. Centerline of piston pin
4. Oil control ring gap
5. Top compression ring gap

7924AG11

Piston ring end-gap spacing—6.5L diesel engines

GASOLINE FUEL SYSTEM

Fuel System Service Precautions

Safety is the most important factor when performing not only fuel system maintenance but any type of maintenance. Failure to conduct maintenance and repairs in a safe manner may result in serious personal injury or death. Maintenance and testing of the vehicle's fuel system components can be accomplished safely and effectively by adhering to the following rules and guidelines.

- To avoid the possibility of fire and personal injury, always disconnect the negative battery cable unless the repair or test procedure requires that battery voltage be applied.

- Always relieve the fuel system pressure prior to disconnecting any fuel system component (injector, fuel rail, pressure regulator, etc.), fitting or fuel line connection. Exercise extreme caution whenever relieving fuel system pressure, to avoid exposing skin, face and eyes to fuel spray. Please be advised that fuel under pressure may penetrate the skin or any part of the body that it contacts.

- Always place a shop towel or cloth around the fitting or connection prior to loosening to absorb any excess fuel due to spillage. Ensure that all fuel spillage (should it occur) is quickly removed from engine surfaces. Ensure that all fuel soaked cloths or towels are deposited into a suitable waste container.

- Always keep a dry chemical (Class B) fire extinguisher near the work area.

- Do not allow fuel spray or fuel vapors to come into contact with a spark or open flame.

- Always use a back-up wrench when loosening and tightening fuel line connection fittings. This will prevent unnecessary stress and torsion to fuel line piping. Always follow the proper torque specifications.

- Always replace worn fuel fitting O-rings with new. Do not substitute fuel hose or equivalent where fuel pipe is installed.

Fuel System Pressure

RELIEVING

A Schrader valve is provided on these fuel systems, in order to conveniently test or release the system pressure. A fuel pressure gauge and adapter will be necessary to connect the gauge to the fitting. Most of the MFI systems utilize a service valve on one end of the fuel rail assembly. The CMFI system covered here uses a valve located on the inlet pipe fitting, immediately before it enters the CMFI assembly (towards the rear of the engine)

1. Before servicing the vehicle, refer to the precautions in the beginning of this section.

2. Turn the ignition **OFF**.

3. Disconnect the negative battery cable.

4. Loosen the fuel filler cap in order to relieve the fuel tank vapor pressure.

5. Connect a fuel pressure gauge to the fuel pressure valve/fitting.

6. Wrap a shop towel around the fitting while connecting the gauge in order to avoid spillage.

7. Install the bleed hose of the gauge into an approved container.

8. Open the valve on the gauge to bleed the system pressure.

The fuel connections are now safe for servicing. Drain any fuel remaining in the gauge into an approved container.

Heater Core replacement is covered in Section 2 of this manual

Fuel Filter

REMOVAL & INSTALLATION

1999 Vehicles

The fuel filter is normally located along the frame rail of the vehicle. On some vehicles however, it may have been relocated to the engine compartment. When in doubt, trace a fuel line from the engine backwards or from the tank forward in order to locate the filter.

Some vehicles utilize a spin-on fuel filter located on the frame rail. This filter can be turned counterclockwise after the fuel pressure is relieved.

1. Before servicing the vehicle, refer to the precautions in the beginning of this section.

2. Properly relieve the fuel system pressure.

3. Remove or disconnect the following:
   - Negative battery cable
   - Fuel line connections from the filter or unscrew the filter in the case of the spin-on type
   - In line filters, remove the bolt from the filter mounting clamp, then remove the clamp and filter assembly. Separate the filter from the clamp.

To install:

➡ **The inline filter has an arrow (fuel flow direction) on the side of the case, be sure to install it correctly in the system, with the arrow facing away from the fuel tank.**

4. Install or connect the following:
   - In line filters place filter in clamp
   - Install filter and clamp
   - Spin-on filters, lubricate the gasket before installation. Then tighten the filter an additional ¾ of a turn from the point when the gasket touches

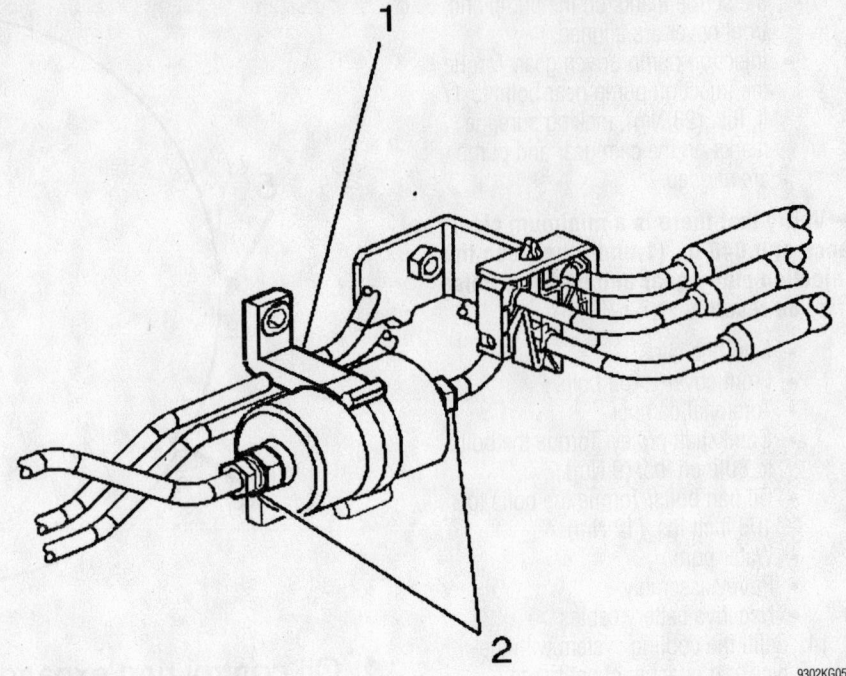

Typical in-line fuel filter mounting location

the filter adapter. Always check for leaks after a new filter is installed.

2001–03 Vehicles

1. Before servicing the vehicle, refer to the precautions in the beginning of this section.

2. Disconnect the negative battery cable.

3. Relieve the fuel system pressure.

4. Raise the vehicle.

5. Clean all the fuel filter connections and the surrounding areas before disconnecting the fuel pipes in order to avoid possible contamination of the fuel system.

6. Disconnect the threaded fittings from the fuel filter.

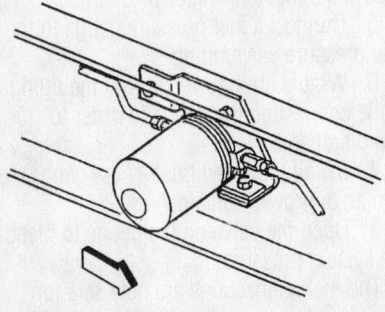

The spin-on fuel filter is serviced in the same manner as a spin-on oil filter

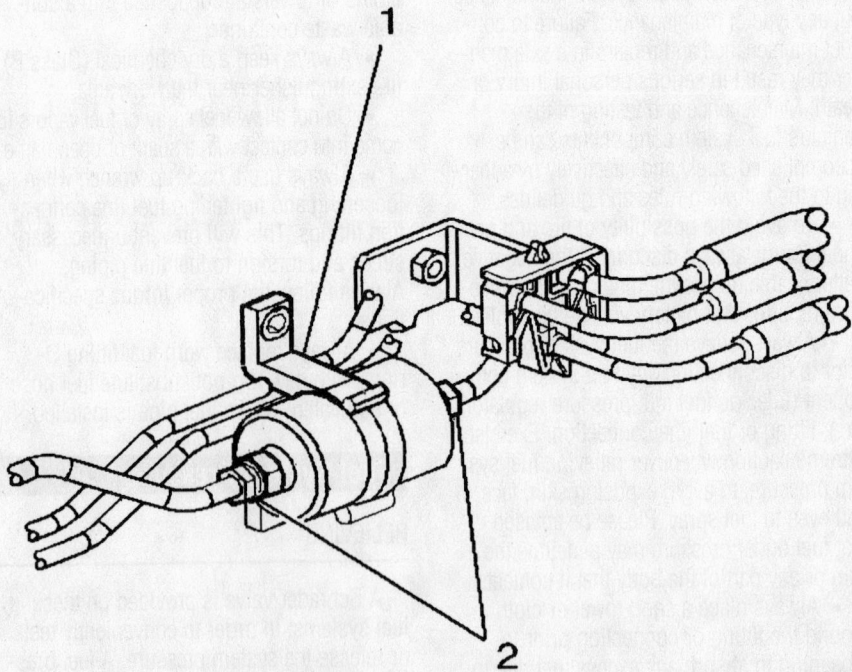

Fuel filter—4.8L, 5.3L and 6.0L engines

7. Cap the fuel pipes in order to prevent possible fuel system contamination.

8. Slide the fuel filter from the bracket.

9. Inspect the fuel pipe O-rings for cuts, nicks, swelling, or distortion. Replace the O-rings if necessary.

To install:

10. Slide the fuel filter into the bracket. Remove the caps from the fuel pipes.

11. Connect the threaded fittings to the fuel filter. Tighten the fittings to 18 ft. lbs. (25 Nm).

12. Lower the vehicle.

13. Tighten the fuel filler cap.

14. Connect the negative battery cable.

15. Turn the ignition **ON** for 2 seconds.

16. Turn the ignition **OFF** for 10 seconds.

17. Turn the ignition **ON**.

18. Inspect for fuel leaks.

Fuel Pump

REMOVAL & INSTALLATION

1999 Vehicles

1. Before servicing the vehicle, refer to the precautions in the beginning of this section.

2. Properly relieve the fuel system pressure.

3. Drain the fuel from the vehicle.

4. Remove or disconnect the following:

- Negative battery cable
- Fuel tank from the vehicle

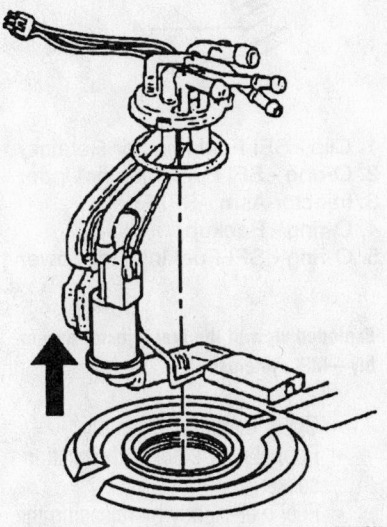

7924KG25

Lift the fuel pump assembly out of the tank after removing the locking ring

- Locking ring (located on top of the fuel tank) counterclockwise
- Lift the fuel pump assembly out of the tank

To install:

5. Install or connect the following:

- Fuel pump and secure with locking ring
- Fuel tank to vehicle and refill
- Negative battery cable

6. Run engine and check for fuel leaks.

2001–03 Vehicles

1. Before servicing the vehicle, refer to the precautions in the beginning of this section.

2. Remove or disconnect the following:

- Negative battery cable

3. Relieve the fuel system pressure.

4. Drain the fuel tank.

5. Remove or disconnect the following:

- Fuel tank

✷✷ WARNING

Do not handle the fuel sender assembly by the fuel pipes. The amount of leverage generated by handling the fuel pipes could damage the joints.

- Fuel sender assembly retaining ring using a fuel tank sending unit wrench. Remove the fuel sender assembly and the seal. Discard the seal.
- Note the position of the fuel strainer on the fuel sender. Support the fuel sender assembly with one hand and grasp the strainer with the other hand. Pull the strainer off the fuel sender. Discard the strainer after inspection. Inspect the strainer. Replace a contaminated strainer and clean the fuel tank.
- Fuel pump electrical connector
- Electrical connector retaining clip from the fuel level sensor
- Sensor electrical connector from under the fuel sender cover
- Fuel level sensor retaining clip

6. Squeeze the locking tangs and remove the fuel level sensor.

7. Remove the fuel pressure sensor.

To install:

8. Install or connect the following:

- Fuel pressure sensor
- Fuel level sensor
- Sensor retaining clip

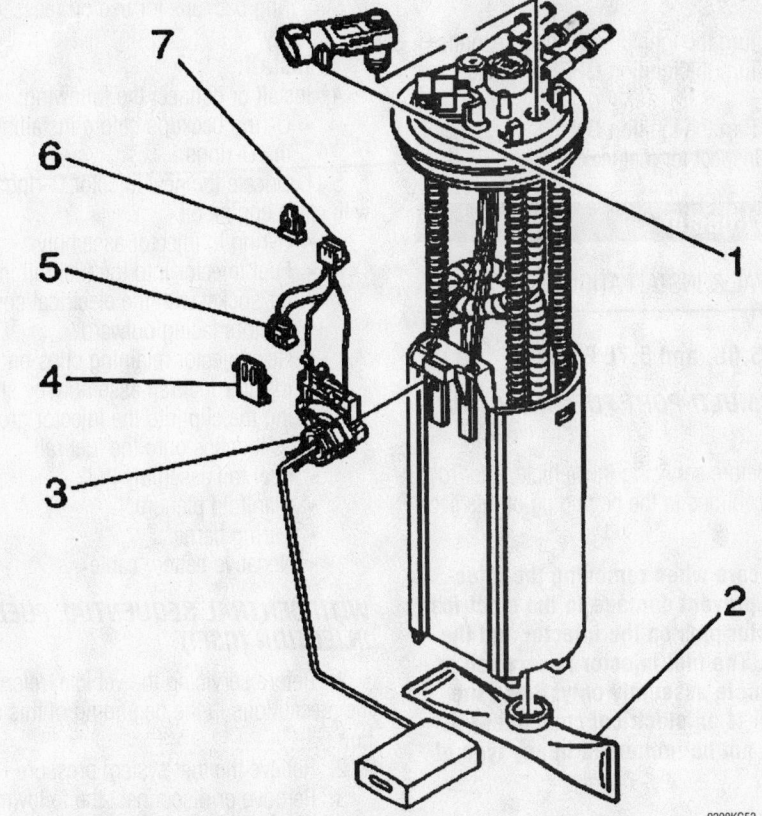

9308KG53

Fuel pump/sender assembly—4.8L, 5.3L and 6.0L engines

- Electrical connector to the fuel level sensor
- Electrical connector retaining clip to the fuel level sensor
- Fuel pump electrical connector

➡**Always install a new fuel strainer when replacing the fuel tank fuel pump module.**

- New fuel strainer in the same position as noted during disassembly. Push the strainer on the bottom of the fuel sender until the strainer is fully seated.
- New seal on the fuel tank

➡**The fuel pump strainer must be in a horizontal position when the fuel sender is installed in the tank. When installing the fuel sender assembly, assure that the fuel pump strainer does not block full travel of the float arm.**

- Fuel sender assembly into the fuel tank
- Fuel sender assembly retaining ring
- Fuel tank. Install the fuel tank strap attaching bolts. Tighten the bolts to 40 Nm (30 ft. lbs.).

9. Refill the fuel tank. Install the fuel filler cap. Connect the negative battery cable.
10. Turn the ignition **ON** for 2 seconds.
11. Turn the ignition **OFF** for 10 seconds.
12. Turn the ignition **ON**.
13. Inspect for fuel leaks.

Fuel Injector

REMOVAL & INSTALLATION

4.3L, 5.0L, and 5.7L Engines

WITH MULTI-PORT FUEL INJECTION (MFI)

1. Before servicing the vehicle, refer to the precautions in the beginning of this section.

➡**Use care when removing the injectors to prevent damage to the electrical connector pins on the injector and the nozzle. The fuel injector is serviced as a complete assembly only. Since the injector is an electrical component, it should not be immersed in any type of cleaner.**

2. Relieve the fuel system pressure.
3. Remove or disconnect the following:
   - Negative battery cable
   - Intake manifold plenum

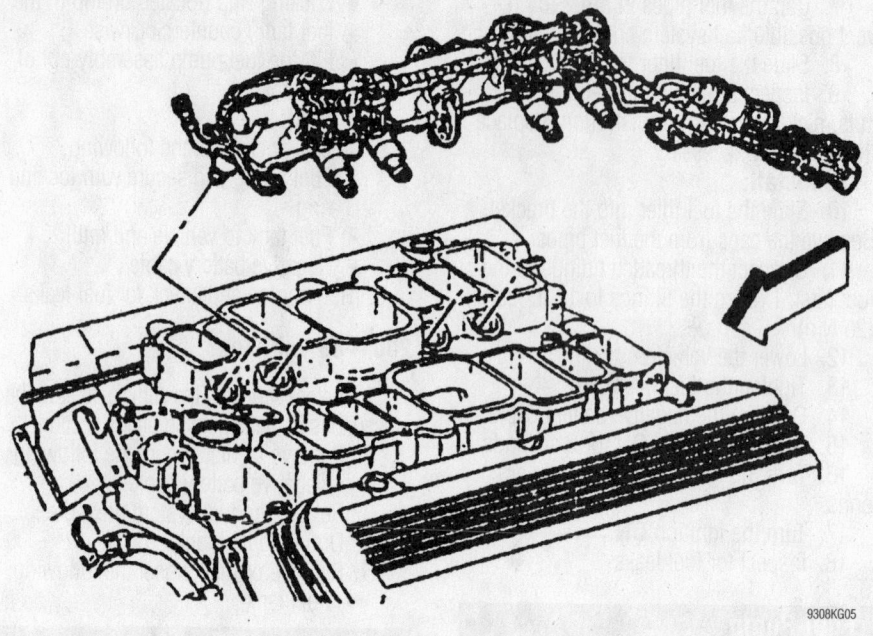

9308KG05

Exploded view of the fuel rail assembly—MFI systems

- Fuel rail assembly
- Wiring harness
- Injector clip and discard it
- Injector O-ring seals from both ends of the injector. Save the O-ring backups for use on reassembly.

To install:

4. Install or connect the following:
   - O-ring backups before installing the O-rings

5. Lubricate the new injector O-rings with clean engine oil
   - O-ring to injector assembly
   - Fuel injector into the fuel rail injector socket with the electrical connectors facing outward
   - New injector retaining clips on the injector fuel rail assembly by sliding the clip into the injector groove as it snaps onto the fuel rail
   - Fuel rail assembly
   - Manifold plenum
   - Wiring harness
   - Negative battery cable

WITH CENTRAL SEQUENTIAL FUEL INJECTION (CSFI)

1. Before servicing the vehicle, refer to the precautions in the beginning of this section.
2. Relieve the fuel system pressure.
3. Remove or disconnect the following:
   - Negative battery cable
   - Electrical connection
   - Fuel feed and return hoses from the engine fuel pipes

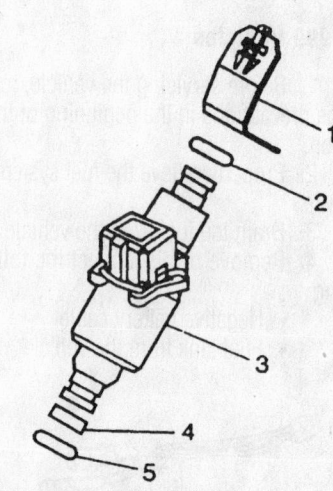

1. Clip - SFI Fuel Injector Retainer
2. O-ring - SFI Fuel Injector Upper
3. Injector Asm - SFI Fuel
4. O-ring - Backup
5. O-ring - SFI Fuel Injector Lower

9308KG06

Exploded view of the fuel injector assembly—MFI systems

- Upper manifold assembly
- Poppet nozzle out of the casting socket
- Fuel meter body by releasing the locktabs

➡**Each injector is calibrated. When replacing the fuel injectors, be sure to replace it with the correct injector.**

- Lower hold-down plate and nuts

4. While pulling the poppet nozzle tube downward, push with a small prytool down between the injector terminals and remove the injectors.

To install:

5. Lubricate the new injector O-ring seats with engine oil.

6. Install or connect the following:
- O-rings on the injector
- Fuel injector into the fuel meter body injector socket
- Lower hold-down plate and nuts, Torque the nuts to 27 inch lbs. (3 Nm).
- Fuel meter body assembly into the intake manifold. Torque the fuel meter bracket retainer bolts to 88 inch. lbs. (10 Nm).

✳✳ CAUTION

To reduce the risk of fire or injury ensure that the poppet nozzles are properly seated and locked in their casting sockets

- Fuel meter body into the bracket and lock all the tabs in place
- Poppet nozzles into the casting sockets
- Electrical connections
- New O-ring seals on the fuel return and feed hoses
- Fuel feed and return hoses. Torque the fuel pipe nuts to 22 ft. lbs. (30 Nm).
- Negative battery cable

7. Turn the ignition **ON** for 2 seconds and then turn it **OFF** for 10 seconds. Again turn the ignition **ON** and check for leaks.
- Manifold plenum

4.8L, 5.3L and 6.0L Engines

1. Before servicing the vehicle, refer to the precautions in the beginning of this section.

2. Relieve the fuel system pressure.

3. Remove or disconnect the following:
- Negative battery cable
- Engine sight shield bolts and bracket
- Accelerator control and cruise control cables from the cable bracket and throttle body
- Upper engine wire harness retainer nut
- Evaporative Emission (EVAP) purge valve harness connector

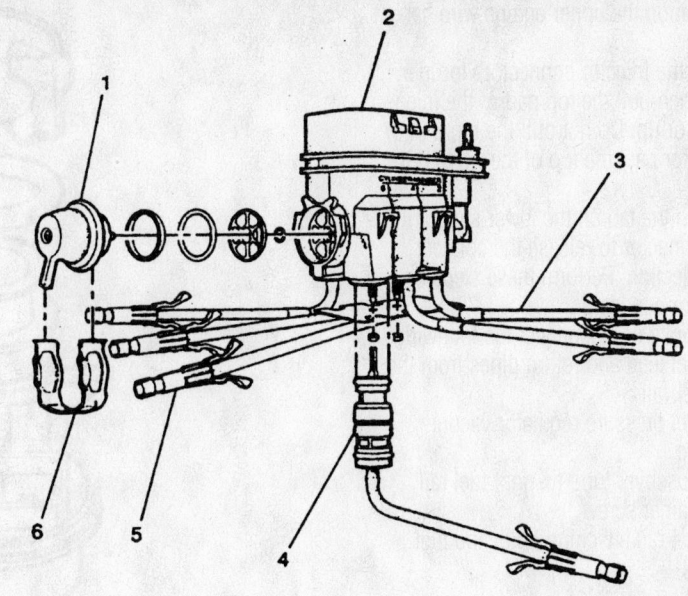

1. Regulator Assembly
2. Fuel Meter Body
3. Flexible Fuel Line
4. Injector Assembly
5. Poppet Nozzle
6. Regulator Retainer

9308KG07

Exploded view of the CSFI fuel meter body assembly

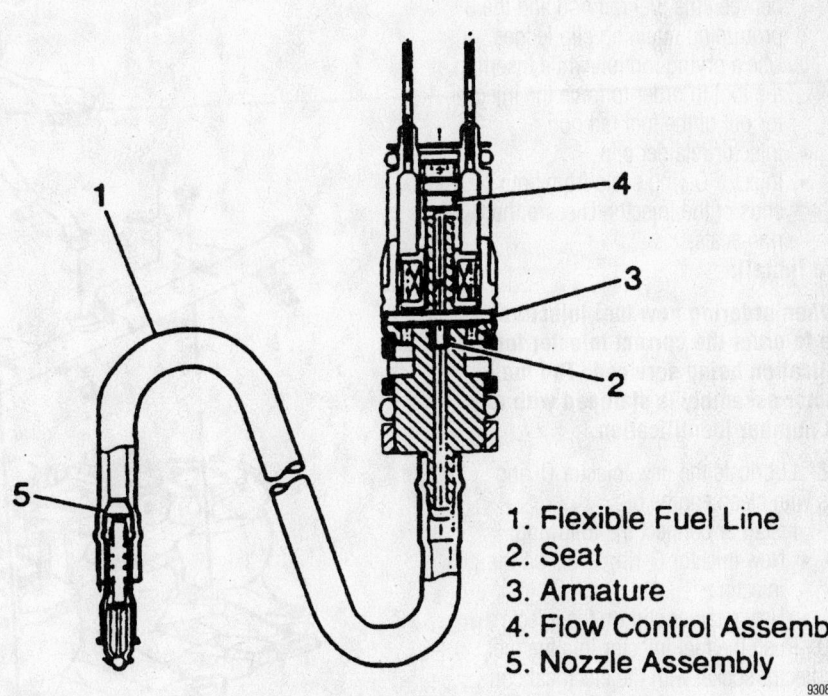

1. Flexible Fuel Line
2. Seat
3. Armature
4. Flow Control Assembly
5. Nozzle Assembly

9308KG08

Exploded view of the fuel injector assembly–CSFI systems

For complete Engine Mechanical specifications, see Section 1 of this manual

4. Position the upper engine wire harness aside

5. Tag the injector connectors for identification, then pull the top part of the injector connector up. Do not pull the top part of the connector past the top of the white portion.

6. Push the tab on the lower side of the injector connector to release the connect from the injection. Perform these steps on each injector connector.

7. Remove or disconnect the following:
- Fuel feed and return pipes from the fuel rail
- Fuel pressure regulator vacuum line
- Crossover tube-to-right fuel rail retainer screw
- Fuel rail attaching bolts and fuel rail

➡Use care in removing the fuel injectors in order to prevent damage to the electrical connector pins on the injector and to prevent damage to the nozzle. Service the fuel injector as a complete assembly only. The fuel injector is an electrical component. DO NOT immerse the fuel injector in any type of cleaner.

- Injector retainer clip. Insert the fork of a fuel injector assembly removal tool behind the injector connector between the fuel rail pod and the 3 protruding retaining clip ledges. Use a prying motion while inserting the tool in order to force the injector out of the fuel rail pod.
- Injector retainer clip
- Injector O-ring seals from both ends of the injector. Discard the O-ring seals.

To install:

➡When ordering new fuel injectors, be sure to order the correct injector for the application being serviced. The fuel injector assembly is stamped with a part number identification.

8. Lubricate the new injector O-ring seals with clean engine oil.

9. Install or connect the following:
- New injector O-ring seals on the injector
- New retainer clip on the injector

10. Push the fuel injector into the fuel rail injector socket with the electrical connector facing outward. The retainer clip locks on to a flange on the fuel rail injector socket.

11. Remove the crossover tube-to-right fuel rail retainer, then remove the crossover tube.

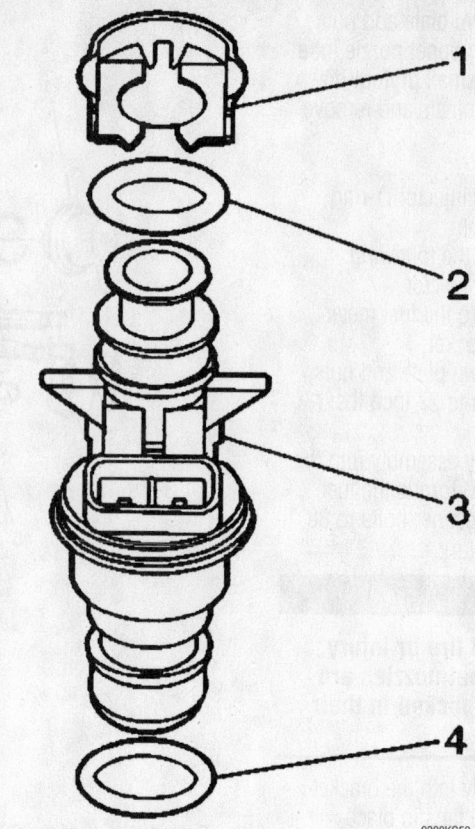

Fuel injector—4.8L, 5.3L and 6.0L engines

9308KG52

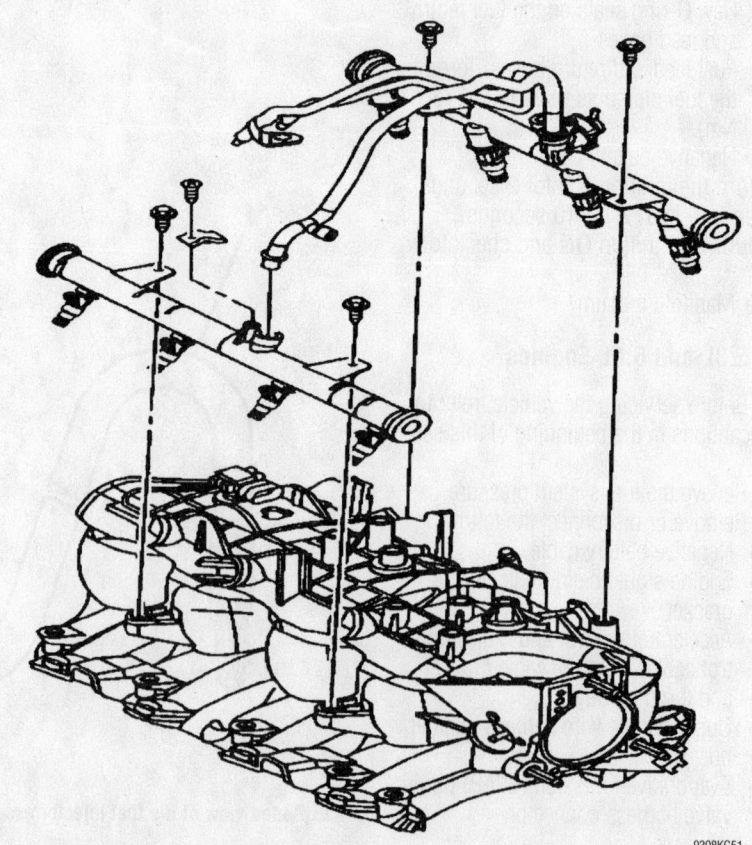

Fuel rail assembly—4.8L, 5.3L and 6.0L engines

9308KG51

12. Replace the crossover tube O-ring with a new, lubricated one.

13. Install or connect the following:
- Crossover tube and loosely install the retainer
- Fuel rail to the intake manifold
- Apply a 0.020 (5mm) band of threadlock to the fuel rail retaining bolts, then install and tighten to 89 inch lbs. (10 Nm). Tighten the crossover pipe retainer to 34 inch lbs. (3.8 Nm).
- Fuel pressure regulator vacuum line
- Fuel feed and return pipes
- Fuel injector electrical connectors, as tagged during removal. Rotate the injectors as necessary to avoid stretching the wire harness.
- Upper engine wire harness
- EVAP purge solenoid electrical connector
- Upper engine wire harness retainer nut and tighten to 49 inch lbs. (5.5 Nm)
- Accelerator control and cruise control cables to the bracket and throttle body
- Engine sight shield mounting bracket and bolts

14. Tighten the fuel cap.

15. Connect the negative battery cable.

16. Turn the ignition **ON** for 2 seconds.

17. Turn the ignition **OFF** for 10 seconds.

18. Turn the ignition **ON**.

19. Inspect for fuel leaks.

20. Install the engine sight shield. Tighten the engine sight shield bolts to 89 inch lbs. (10 Nm)

8.1L Engine

1. Before servicing the vehicle, refer to the precautions in the beginning of this section.

2. Relieve the fuel system pressure.

3. Remove or disconnect the following:
- Negative battery cable
- Engine sight shield nuts and bracket
- Alternator harness connector
- Evaporative Emission (EVAP) purge valve harness connector
- Throttle Position (TP) sensor electrical connector
- Electronic Throttle Control (ETC) electrical connector

4. Upper engine wire harness bracket studs, and position the harness aside.

5. Secondary Air Injection (AIR) crossover pipe, if equipped.

6. Tag the injector connectors for identification, then pull the top part of the injector connector up. Do not pull the top part of the connector past the top of the white portion.

7. Push the tab on the lower side of the injector connector to release the connect from the injection. Perform these steps on each injector connector.

8. Remove or disconnect the following:
- Fuel feed and return pipes from the fuel rail
- Fuel pressure regulator vacuum line
- Fuel rail attaching bolts and fuel rail

➡ **Use care in removing the fuel injectors in order to prevent damage to the electrical connector pins on the injector and to prevent damage to the nozzle. Service the fuel injector as a complete assembly only. The fuel injector is an electrical component. DO NOT immerse the fuel injector in any type of cleaner.**

- Injector retainer clip. Insert the fork of a fuel injector assembly removal tool behind the injector connector between the fuel rail pod and the 3 protruding retaining clip ledges. Use a prying motion while inserting the tool in order to force the injector out of the fuel rail pod.
- Injector retainer clip
- Injector from the fuel rail pod
- Injector O-ring seals from both ends of the injector. Discard the O-ring seals.

To install:

➡ **When ordering new fuel injectors, be sure to order the correct injector for the application being serviced. The fuel injector assembly is stamped with a part number identification.**

9. Lubricate the new injector O-ring seals with clean engine oil.

10. Install or connect the following:
- New injector O-ring seals on the injector
- New retainer clip on the injector

11. Push the fuel injector into the fuel rail injector socket with the electrical connector facing outward. The retainer clip locks on to a flange on the fuel rail injector socket.

- Fuel rail to the intake manifold
- Apply a 0.020 (5mm) band of threadlock to the fuel rail retaining bolts, then install and tighten to 106 inch lbs. (12 Nm)
- Fuel pressure regulator vacuum line
- Fuel feed and return pipes
- Fuel injector electrical connectors, as tagged during removal. Rotate the injectors as necessary to avoid stretching the wire harness.
- AIR crossover pipe, if equipped
- Upper engine wire harness bracket
- Retainer studs to the upper engine wire harness and tighten the nut to 89 inch lbs. (10 Nm)
- Alternator electrical connector
- EVAP purge solenoid electrical connector
- TP and ETC sensor connectors
- Engine sight shield mounting bracket and bolts

12. Tighten the fuel cap.

13. Connect the negative battery cable.

14. Turn the ignition **ON** for 2 seconds.

15. Turn the ignition **OFF** for 10 seconds.

16. Turn the ignition **ON**.

17. Inspect for fuel leaks.

18. Install the engine sight shield. Tighten the engine sight shield bolts to 89 inch lbs. (10 Nm)

For Accessory Drive Belt illustrations, see Section 1 of this manual

DIESEL FUEL SYSTEM

Fuel System Pressure

RELIEVING

Fuel system pressure can be released by wrapping a fuel fitting in a heavy shop towel and slightly loosening the fitting. NEVER perform this with any source of ignition nearby!

Fuel System Air

BLEEDING

1. Before servicing the vehicle, refer to the precautions in the beginning of this section.
2. Open the air bleed valve on the fuel manager/filter.
3. Connect a hose to the air bleed valve and place the other of the hose in a suitable container.

❊❊ CAUTION

The diesel/water mixture is flammable and may be hot. To avoid personal injury or property damage, do not allow the diesel/water mixture to come in contact with skin, open flame or a hot engine. Do not overfill the container holding the fuel mixture as heat from a warm engine or any another heat source may cause the fuel to expand and leak from the container that may lead to a fire.

4. Remove the F/SOL fuse from the fuse panel.
5. Crank the engine in short intervals of 10-to-15 seconds until clear fuel is observed at the air bleed hose (wait for 1 minute between cranking intervals)
6. Remove the hose and close the air bleed valve.
7. Install the F/SOL fuse and start the vehicle. Allow the vehicle to run at idle for 5 minutes.
8. Check for fuel leaks, and clear any Diagnostic Trouble Code's (DTC's)

Idle Speed

ADJUSTMENT

Idle speed and injection timing is controlled by the Powertrain Control Module (PCM). There is no provision for adjustment.

Fuel Filter

REMOVAL & INSTALLATION

6.5L Engine

1. Before servicing the vehicle, refer to the precautions in the beginning of this section.
2. Turn the ignition **OFF**. Remove the fuel tank cap to release any pressure or vacuum in the tank.

➡ **It is not necessary to drain all the fuel from the header in order to change the element since the fuel will remain in the header's cavity.**

3. Open the air bleed valve to relieve residual pressure.
4. Remove or disconnect the following:
   • Element nut, turning it by hand to the left. If necessary, a strap wrench may be used to loosen the nut.
   • Element by lifting straight up and out of the header assembly

To install:

5. Be sure the mating surface between the element assembly and the header assembly is clean.
6. Install or connect the following:

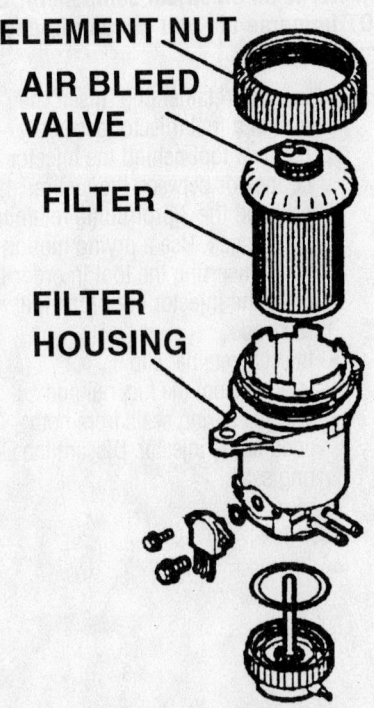

```
ELEMENT NUT
AIR BLEED
VALVE
FILTER
FILTER
HOUSING
```
7924KG26

Exploded view of the fuel filter assembly—6.5L diesel engines

• New element by aligning the widest key slot located under the element assembly cap with the widest key in the header assembly.
7. Carefully push the element downward until the mating surfaces make contact.
   • Element nut and tighten securely by hand
8. If not already done, open the air bleed valve on top of the fuel manager/filter assembly, then connect a length of hose placing the other end in a suitable container.

➡ **Be extremely cautious when handling diesel fuel. Do not expose the fuel to sparks or open flames. Also, be cautious as the fuel coming out of the drain hose could be hot.**

9. Disconnect the fuel injection pump shutdown solenoid wire.
10. Crank the engine for 10–15 seconds, then wait 1 minute for the starter motor to cool. Repeat until clear fuel is observed coming from the air bleed.
11. Close the air bleed valve, reconnect the injection pump solenoid wire and replace the fuel tank cap.
12. Start the engine, allow it to idle for 5 minutes and check the fuel manager/filter assembly for leaks.

Diesel Injection Pump

All vehicles are equipped with an electronically controlled pump. The electronic pump is driven by gears and rotates at the same speed as the camshaft. An electronic stepper motor used to control injection timing and a fuel solenoid driver used to control the fuel injection solenoid on the electronic model.

REMOVAL & INSTALLATION

6.5L Engine

1. Before servicing the vehicle, refer to the precautions in the beginning of this section.
2. Relieve the fuel system pressure.
3. Remove or disconnect the following:
   • Negative battery cables
   • Intake manifold
   • Fuel injection and inlet lines
   • Cables wires, and hoses at the injection pump
   • Fuel return line at the top of the injection pump

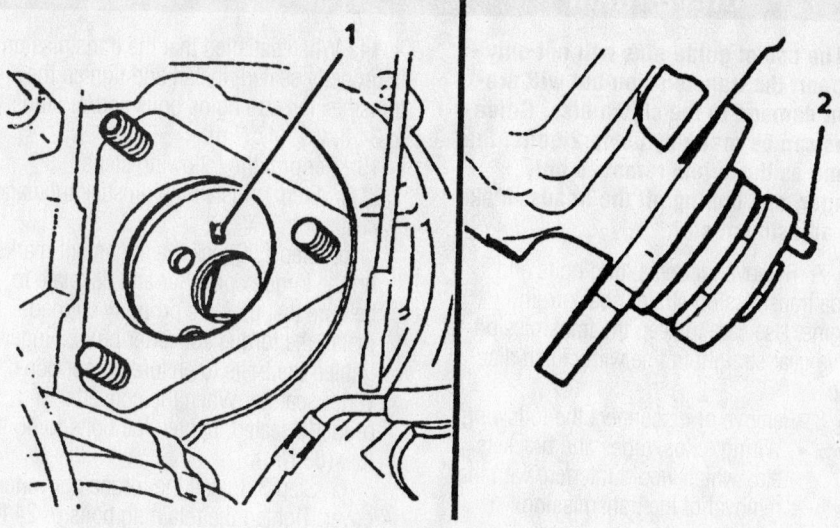

1 SLOT IN DRIVEN GEAR

2 PUMP HUB

7924KG27

Align the pin on the pump hub with the slot in the driven gear, NOT into the hole in the gear—diesel engines

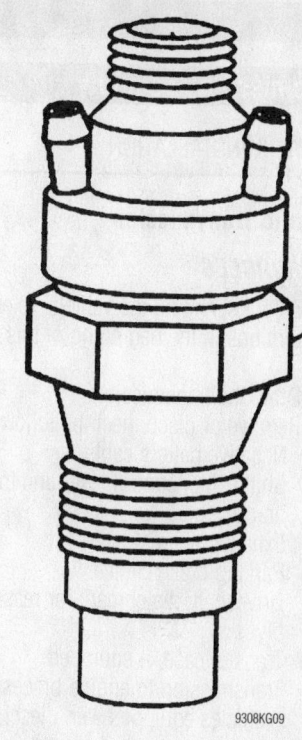

9308KG09

Exploded view of the fuel injector nozzle.

- Fuel feed line at the injection pump, if necessary
- Oil filler tube grommet

➡**Do not engage the starter in order to rotate the engine with the injection pump removed. The pump driven gear could jam in the front housing resulting in a sheared crankshaft or camshaft gear key and possible valvetrain damage.**

4. Scribe or paint a matchmark on the front cover and the injection pump flange.

5. Rotate the crankshaft by hand and remove the injection pump driven gear bolts, accessing the bolts through the oil filler neck hole.

6. Remove the injection pump-to-front cover attaching nuts. Remove the pump. Be sure to cap all open lines and nozzles in order to prevent system contamination and damage.

To install:

7. Align the locating pin on the pump hub with the slot in the injection pump driven gear (the SLOT not the hole in the gear) At the same time, align the timing marks.

8. Attach the injection pump to the front cover. Torque the nuts to 30 ft. lbs. (40 Nm)

checking the timing marks before tightening.

9. Install or connect the following:
   - Driven gear-to-injection pump bolts. Torque the bolts to 20 ft. lbs. (25 Nm).
   - Grommet and oil fill tube
   - Air conditioning bracket. If applicable
   - Fuel feed line. Torque to 20 ft. lbs. (25 Nm).
   - Fuel return line to the pump, if removed
   - Cables, wires and hoses previously removed
   - Injector lines
   - Intake manifold
   - Negative battery cables

10. Start the engine and check for leaks.

Fuel Injectors

REMOVAL & INSTALLATION

6.5L Engine

1. Before servicing the vehicle, refer to the precautions in the beginning of this section.

➡**Special tool J–29873, or its equivalent, an injection nozzle socket, will be necessary for this procedure.**

2. Remove or disconnect the following:
   - Batteries.
   - Fuel line clip
   - Fuel return hose
   - Fuel injection lines

3. Using GM special tool J–29873, remove the injector. Always remove the injector by turning the 30mm hex portion of the injector; turning the round portion will damage the injector. Always cap the injector and fuel lines when disconnected, to prevent contamination.

To install:

4. Always install the injector by turning the 30mm hex portion of the injector; turning the round portion will damage the injector.

5. Install or connect the following:
   - Injector with a new gasket. Torque to 50 ft. lbs. (70 Nm).
   - Injection line. Torque the nut to 20 ft. lbs. (25 Nm).
   - Fuel return hose
   - Fuel line clips
   - Batteries

DRIVE TRAIN

Transmission Assembly

REMOVAL & INSTALLATION

Automatic Transmission

1999 VEHICLES

1. Before servicing the vehicle, refer to the precautions in the beginning of this section.
2. Drain the transmission.
3. Remove or disconnect the following:
   - Negative battery cable
   - Shift cable, control lever and the bracket
   - Exhaust pipes
   - Parking brake cables
   - Driveshaft, matchmark for reassembly
   - Transfer case, if equipped
   - Transmission to engine braces (vehicles equipped with diesel engines have only 1 brace)
   - Wiring harness at the transmission
4. Support the transmission with a transmission jack.
   - Nut securing the transmission mount to the cross member
5. Position a transmission jack or equivalent, under the transmission for support.
   - Crossmember. Visually inspect to see if other equipment, brackets or lines, must be removed to permit removal of transmission.

➡**Mark position of crossmember when removing to prevent incorrect installation. The tapered surface should face the rear.**

6. Perform the following:
   a. Step 1: Remove the torque converter inspection cover.
   b. Step 2: Mark the alignment of the torque converter to the flexplate.
   c. Step 3: Remove the torque converter to flexplate bolts. Remove the dipstick tube and seal from the transmission. Plug the opening to avoid contamination.
   d. Step 4: Disconnect both transmission lines at the transmission and plug them to avoid contamination and leakage.
   e. Step 5: Position a J 21366 converter holding strap onto the transmission/torque converter to keep the torque converter from sliding off of the transmission turbine shaft.

➡**The use of guide pins will not only support the transmission but will prevent damage to the clutch disc. Guide pins can be made by using 2 bolts, the same as those just removed only longer, and cutting off the heads. Make an adjustment slot.**

7. Remove the remaining bolts and slide transmission straight back from engine. Use care to keep the transmission drive gear straight in line with clutch disc hub.
8. Remove or disconnect the following:
   - Wiring, clips, tubes and brackets etc., which would interfere with the removal of the transmission

➡**Ensure that the engine is supported with a jack stand before detaching the transmission from the engine.**

   - Transmission from the engine
9. Carefully lower the transmission using the transmission jack.
   To install:
10. Check the area behind the torque converter for leaks. Replace the front seal, if required.
11. It is good practice to examine the area around the rear crankshaft seal, checking for leaks. If necessary, remove the flexplate and replace the seal.
12. Inspect the flexplate ring gear teeth. If damaged, replace the flexplate.
13. Perform the following steps:
   a. Step 1: With tool J 21366 or equivalent, torque converter holding strap in place, raise the transmission into position with a transmission jack.
   b. Step 2: Remove the torque converter holding strap and slide the transmission into place. Slide the transmission straight onto the locating pins while lining up the marks on the flexplate and the torque converter. Be sure the transmission is fully seated against the rear of the engine block and the locating pins are completely engaged.

✲✲ **WARNING**

DO NOT attempt to draw the transmission to the block with the mounting bolts. If the transmission is not properly seated, the bolts will break the transmission case.

➡**The torque converter must be flush with the flexplate and rotate freely by hand.**

14. When satisfied that the transmission is properly seated, install and tighten the transmission-to-engine bolts and/or studs to 34 ft. lbs. (47 Nm).
15. Perform the following steps:
   a. Step 1: Install the dipstick tube and seal.
   b. Step 2: Check the alignment marks on the torque converter and flexplate to be sure that they are properly aligned. Install the torque converter bolts. Finger-tighten the bolts to ensure proper converter seating. When the converter is properly seated, tighten the bolts to 46 ft. lbs. (63 Nm).
   c. Step 3: Install the torque converter cover. Tighten the retaining bolts to 24 ft. lbs. (33 Nm) on the 4.3L engines or 89 inch lbs. (10 Nm) on the V8 engines.
16. Install or connect the following:
   - Wiring, clips, tubes and brackets etc
   - Transmission cooling lines
   - Shifter cable
   - Starter
   - Exhaust pipes
   - Transmission crossmember. Torque the bolts to 56 ft. lbs. (77 Nm).
   - Transmission mount on the transmission. Torque the bolts to 35 ft. lbs. (47 Nm).
   - Nut and washer that secure the transmission mount to the crossmember. Torque the nut to 38 ft. lbs. (52 Nm).
   - Transmission to engine brace(s) Torque the bolts to 41 ft. lbs. (55 Nm) for gasoline engines and to 51 ft. lbs. (70 Nm) for diesel engines.
   - Transfer case, if equipped
17. Remove the transmission jack and engine support stands.
   - Driveshaft
   - Shifter lever and boot, on manual transmissions
   - Negative battery cable
18. Refill the transmission with fluid.
19. Road test the vehicle and test for proper operation. Check for leaks.

2000–03 VEHICLES—4L60E

1. Before servicing the vehicle, refer to the precautions in the beginning of this section.
2. Remove or disconnect the following:
   - Transmission fluid
   - Transmission oil level indicator tube and seal from the transmission

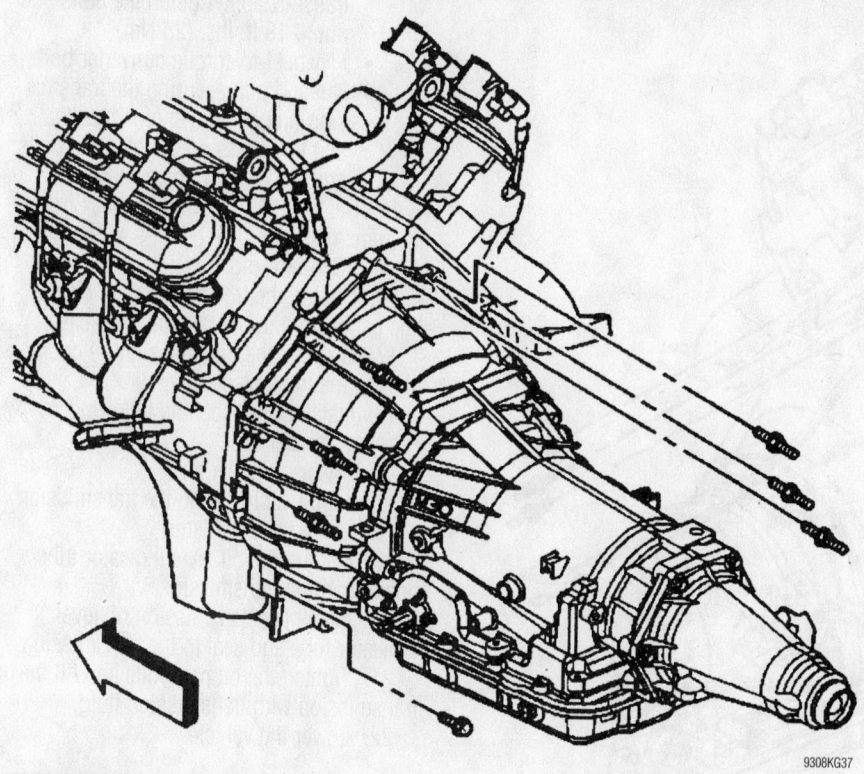

4L60E removal

9308KG37

- Plug the oil level indicator tube opening in the transmission.
3. Remove or disconnect the following:
  - Shift cable end from the transmission shift lever ball stud
  - Front propeller shaft, if equipped with a transfer case
  - Rear propeller shaft.
4. Plug the transmission oil cooler line connectors in the transmission case.
  - Starter motor
5. Support the transmission with a transmission jack.
  - Torque converter access plug
  - Flywheel-to-torque converter bolts
  - Two bolts and nut securing the transmission rear mount to the transmission
  - Two bolts securing the heat shield to the transmission
  - Transmission vent hose, fuel lines, and the wiring harness from the transmission
  - Stud and the bolt securing the transmission to the engine
  - Six studs and bolt securing the transmission to the engine. Install tool J21366 onto the transmission bell housing to retain the torque converter. Pull the transmission straight back.

- Transmission from the vehicle
6. Flush the transmission oil cooler and cooling lines when you remove the transmission.

To install:

7. Install Tool J21366 onto the transmission bell housing to retain the torque converter.
8. Support the transmission with a transmission jack.
9. Raise the transmission into place and remove the tool from the transmission.
10. Slide the transmission straight onto the locating pins while lining up the marks on the flywheel and the torque converter. The torque converter must be flush onto the flywheel and rotate freely by hand.
11. Install or connect the following:
  - Six studs and one bolt securing the transmission to the engine. Tighten the studs and the bolt to 37 ft. lbs. (50 Nm).
  - Stud and bolt securing the transmission to the engine. Tighten the stud and the bolt to 37 ft. lbs. (50 Nm).
  - Flywheel-to-torque converter bolts
  - Torque converter access plug
  - Transmission vent hose, fuel lines, and the wiring harness to the transmission

- Two bolts securing the heat shield to the transmission. Tighten the bolts to 13 ft. lbs. (17 Nm).
- Two bolts and nut securing the transmission rear mount to the transmission. Tighten the bolts and nut to 18 ft. lbs. (25 Nm).
12. Remove the transmission jack from the transmission.
13. Unplug the transmission oil cooler line connectors in the transmission case.
14. Install or connect the following:
  - Transmission oil cooler lines to the transmission
  - Front propeller shaft, if equipped with a transfer case
  - Rear propeller shaft
  - Shift cable end to the transmission shift lever ball stud
15. Unplug the oil level indicator tube opening in the transmission.
16. Install the transmission oil level indicator tube and seal to the transmission.
17. Tighten the oil pan bolts and fill the transmission with transmission fluid.
18. Lower the vehicle.

2000–03 VEHICLES—4L80E

1. Before servicing the vehicle, refer to the precautions in the beginning of this section.
2. Remove or disconnect the following:
  - Transmission fluid
  - Transmission oil level indicator tube and seal from the transmission
3. Plug the oil level indicator tube opening in the transmission.
4. Remove or disconnect the following:
  - Shift cable from the transmission shift lever ball stud
  - Propeller shaft(s), as necessary
  - Transmission oil cooler lines from the transmission
5. Plug the transmission oil cooler line connectors in the transmission case.
  - Starter motor
6. Support the transmission with a transmission jack.
  - Two bolts securing the heat shield to the transmission
  - Transmission vent hose, fuel lines, and the wiring harness from the transmission
  - Nut and bolt securing the transmission brace to the engine bracket and transmission
  - Two bolts securing the torque converter cover to the engine

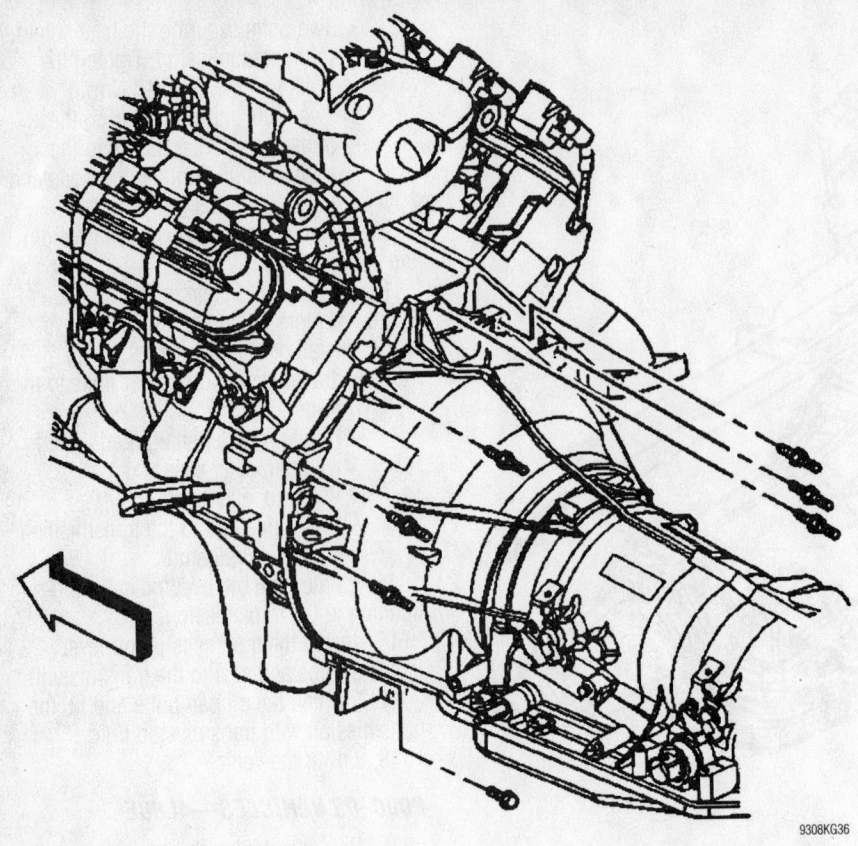

9308KG36

4L80E removal

- Four bolts securing the torque converter cover to the transmission
- Six flywheel-to-torque converter bolts
- Two bolts and nut securing the transmission rear mount to the transmission
- Stud and the bolt on the right side securing the transmission to the engine
- Remaining six studs and the one bolt securing the transmission to the engine

7. Install Tool J21366 onto the transmission bell housing to retain the torque converter

8. Pull the transmission straight back. Remove the transmission from the vehicle.

9. Flush the transmission oil cooler and cooling lines when you remove the transmission.

To install:

10. Install Tool J21366 onto the transmission bell housing to retain the torque converter

11. Support the transmission with a transmission jack.

12. Raise the transmission into place and remove the tool from the transmission.

13. Slide the transmission straight onto the locating pins while lining up the marks on the flywheel and the torque converter. The torque converter must be flush onto the flywheel and rotate freely by hand.

14. Install or connect the following:
- Six studs and one bolt securing the transmission to the engine. Tighten the studs and the bolt to 37 ft. lbs. (50 Nm).
- Stud and bolt on the right side securing the transmission to the engine. Tighten the stud and the bolt to 37 ft. lbs. (50 Nm).
- Six flywheel-to-torque converter bolts. Tighten the bolts to 44 ft. lbs. (60 Nm).
- Two bolts securing the torque converter cover to the engine. Tighten the bolt to 37 ft. lbs. (50 Nm).
- Four bolts securing the torque converter cover to the transmission. Tighten the stud and the bolt to 33 Nm (24 ft. lbs.).
- Transmission vent hose, fuel lines, and the wiring harness to the transmission
- Two bolts securing the heat shield to the transmission. Tighten the bolt to 13 ft. lbs. (17 Nm).
- Two bolts and nut securing the transmission rear mount to the transmission. Tighten the bolts and nut to 18 ft. lbs. (25 Nm).
- Flywheel-to-torque converter bolts
- Nut and bolt securing the transmission brace to the engine bracket and transmission. Tighten the bolts and nut to 37 ft. lbs. (50 Nm).

15. Remove the transmission jack from the transmission.
- Starter motor

16. Unplug the transmission oil cooler line connectors in the transmission case.

17. Connect the transmission oil cooler lines to the transmission.

18. Install or connect the following:
- Transfer case
- Rear propeller shaft
- Shift cable end to the transmission shift lever ball stud

19. Unplug the oil level indicator tube opening in the transmission.

20. Install the transmission oil level indicator tube and seal to the transmission.

21. Tighten the oil pan bolts and fill the transmission with transmission fluid.

22. Lower the vehicle.

Transfer Case Assembly

REMOVAL & INSTALLATION

1999 Vehicles

1. Before servicing the vehicle, refer to the precautions in the beginning of this section.

2. Drain transfer case of lubricant.

3. Remove or disconnect the following:
- Negative battery cable
- Skid plate, if equipped
- Vent hose clamp at the transfer case
- Front driveshaft at the transfer case and support it aside
- Rear driveshaft and support it aside
- Electrical connections at the transfer case
- Transfer case shift linkage

4. Support the transfer case with a transmission jack.
- Transmission to transfer case bolts and spring washers
- Transfer case assembly and gasket

5. Carefully lower the transfer case.

To install:

6. Carefully raise the transfer case into position.

7. Install or connect the following:
- New gasket to the transmission using gasket sealer to hold it in place

- Transfer case onto the transmission or transmission adapter. Torque the bolts to 33 ft. lbs. (45 Nm).
- Electrical harness connectors to the transfer case connections
- Transfer case shift linkage and make the proper adjustments
- Front and rear driveshafts

8. Refill the transfer case with DEXRON®IIE automatic transmission fluid.
- Skid plate, if equipped
- Negative battery cable

9. Test drive for proper operation.

2000–03 Vehicles

NVG261-NP2 2-SPEED MANUAL TRANSFER CASE

1. Before servicing the vehicle, refer to the precautions in the beginning of this section.

2. Remove or disconnect the following:
- Transfer case shields
- Front propeller shaft
- Rear propeller shaft
- Shift rod from the transfer case
- Vent hose from the transfer case
- Vehicle speed sensor electrical connectors
- Any wiring harness from the transfer case

3. Support the transfer case with a transmission jack.

4. If equipped with a NV3500 manual transmission, remove the bolt securing the left side support brace to the transmission.

5. If equipped with a NV3500 manual transmission, remove the bolt and stud securing the left side support brace to the transfer case.

6. If equipped with a NV3500 manual transmission, remove the 2 bolts securing the right side support brace to the transmission and transfer case.

7. For vehicles equipped with a manual transmission, remove the 6 nuts securing the transfer case and bracket to the transmission. Remove the transfer case.

8. For vehicles equipped with a automatic transmission, remove the 6 nuts securing the transfer case and bracket to the transmission adapter. Remove the transfer case.

9. Remove and discard the gasket.

To install:

10. Install a new gasket to the transmission. Use Teflon pipe sealant GM P/N 12346004 in order to hold the gasket in place.

11. Raise and position the transfer case to the vehicle

12. For vehicles equipped with a automatic transmission, install the 6 nuts securing the transfer case and bracket to the transmission adapter. Tighten the nuts to 37 ft. lbs. (50 Nm).

13. For vehicles equipped with a manual transmission, install the 6 nuts securing the transfer case and bracket to the transmission. Tighten the nuts to 37 ft. lbs. (50 Nm).

14. If equipped with a NVG 261 manual transmission, install the bolt securing the left side support brace to the transmission. Tighten the bolt to 37 ft. lbs. (50 Nm).

15. If equipped with a NVG 261 manual transmission, Install the bolt and stud securing the left side support brace to the transfer case. Tighten the bolt and stud to 37 ft. lbs. (50 Nm).

16. If equipped with a NVG 261 manual transmission, install the 2 bolts securing the right side support brace to the transmission and transfer case. Tighten the bolts to 37 ft. lbs. (50 Nm).

17. Install or connect the following:
- Vent hose to the transfer case

18. Check the transfer case oil level.
- Speed sensor electrical connectors
- Wiring harness to the transfer case
- Shift rod to the transfer case
- Rear propeller shaft
- Front propeller shaft
- Transfer case shields

19. Lower the vehicle.

NVG246-NP8 2-SPEED AUTOMATIC TRANSFER CASE

1. Before servicing the vehicle, refer to the precautions in the beginning of this section.

2. Remove or disconnect the following:
- Transfer case shields
- Front propeller shaft
- Rear propeller shaft
- Vent hose from the transfer case
- Vehicle speed sensor electrical connectors
- Electrical connectors from the transfer case motor/encoder
- Any wiring harness from the transfer case

3. Support the transfer case with a transmission jack.

4. If equipped with a NV3500 manual transmission, remove the bolt securing the left side support brace to the transmission.

5. If equipped with a NV3500 manual

transmission, remove the bolt and stud securing the left side support brace to the transfer case.

6. If equipped with a NV3500 manual transmission, remove the two bolts securing the right side support brace to the transmission and transfer case.

7. For vehicles equipped with a manual transmission, remove the six nuts securing the transfer case and bracket to the transmission. Remove the transfer case.

8. For vehicles equipped with a automatic transmission, remove the six nuts securing the transfer case and bracket to the transmission adapter. Remove the transfer case.

9. Remove and discard the gasket.

To install:

10. Install a new gasket to the transmission. Use Teflon Pipe Sealant GM P/N 12346004 in order to hold the gasket in place.

11. Raise and position the transfer case to the vehicle

12. For vehicles equipped with a automatic transmission, install the six nuts securing the transfer case and bracket to the transmission adapter. Tighten the nuts to 37 ft. lbs. (50 Nm).

13. For vehicles equipped with a manual transmission, install the six nuts securing the transfer case and bracket to the transmission. Tighten the nuts to 37 ft. lbs. (50 Nm).

14. If equipped with a NV3500 manual transmission, install the bolt securing the left side support brace to the transmission. Tighten the bolt to 37 ft. lbs. (50 Nm).

15. If equipped with a NV3500 manual transmission, Install the bolt and stud securing the left side support brace to the transfer case. Tighten the bolt and stud to 37 ft. lbs. (50 Nm).

16. If equipped with a NV3500 manual transmission, install the two bolts securing the right side support brace to the transmission and transfer case. Tighten the bolts to 37 ft. lbs. (50 Nm).

17. Install or connect the following:
- Vent hose to the transfer case
- Oil
- Speed sensor electrical connectors
- Electrical connectors to the transfer case motor/encoder
- Any wiring harness to the transfer case
- Rear propeller shaft
- Front propeller shaft
- Transfer case shields

Halfshaft

REMOVAL & INSTALLATION

1999 Vehicles

1. Before servicing the vehicle, refer to the precautions in the beginning of this section.
2. Remove or disconnect the following:
   - Front wheel and tire assembly
   - Skid plate, if equipped
   - Drive axle hub nut and washer
   - Brake line and wheel speed sensor support bracket from the upper control arm to allow extra travel of the control arm.
   - Left outer tie rod attaching nut and cotter pin. Separate the tie rod from the steering knuckle
3. Position the tie rod aside and push steering linkage to the opposite side of the vehicle.
   - Lower shock attaching nut and bolt; position the shock aside
   - Left stabilizer bar bracket and bushing at the frame
   - Stabilizer bar bolt, spacer and bushings at the lower control arm
4. Taking pressure off the upper control arm by placing a support below the lower control arm between the spring seat and the ball joint.
   - Upper ball joint cotter pin and loosen (do not remove) the upper ball joint attaching nut. Separate the ball joint stud from the steering knuckle. Remove the attaching nut.

➡ **Cover the shock mounting bracket and lower ball joint stud with a towel to prevent the axle boot from tearing during removal and installation.**

5. Separate the axle shaft from the hub and rotor using tool J-28733 or equivalent.
   - Axle shaft inner flange bolts and shaft

To install:

6. Lubricate the axle and hub splines with an approved high temperature wheel bearing grease.
7. Install or connect the following:
   - Axle shaft in the hub
   - Inboard CV-joint-to-flange bolts. Torque the bolts to 60 ft. lbs. (80 Nm).
   - Upper ball joint to steering knuckle. Torque the stud nut to 61 ft. lbs. (83 Nm).
   - New cotter pin through the upper ball joint stud and nut, lubricate the ball joint as required

   - Left stabilizer bar bracket and bushing at the frame
   - Stabilizer bar bolt, spacer and bushings at the lower control arm
   - Lower shock in the mount bracket and the attaching nut and bolt
   - Left tie rod end at the steering knuckle. Torque the nut to 35 ft. lbs. (47 Nm).
   - New cotter pin through the tie rod stud and nut
   - Brake line bracket to the control arm, ensuring the line and/or hose is not twisted or kinked
   - Skid plate, as required
   - Axle hub washer and nut. Insert a drift through the rotor vanes to keep the axle from turning. Toque the hub nut to 180 ft. lbs. (245 Nm).
   - Wheel and tire assembly

2000–03 Vehicles

1. Before servicing the vehicle, refer to the precautions in the beginning of this section.
2. Remove or disconnect the following:
   - Wheels
3. Insert a drift or a large screwdriver through the brake caliper into one of the brake rotor vanes in order to prevent the drive axle wheel drive shaft from turning.

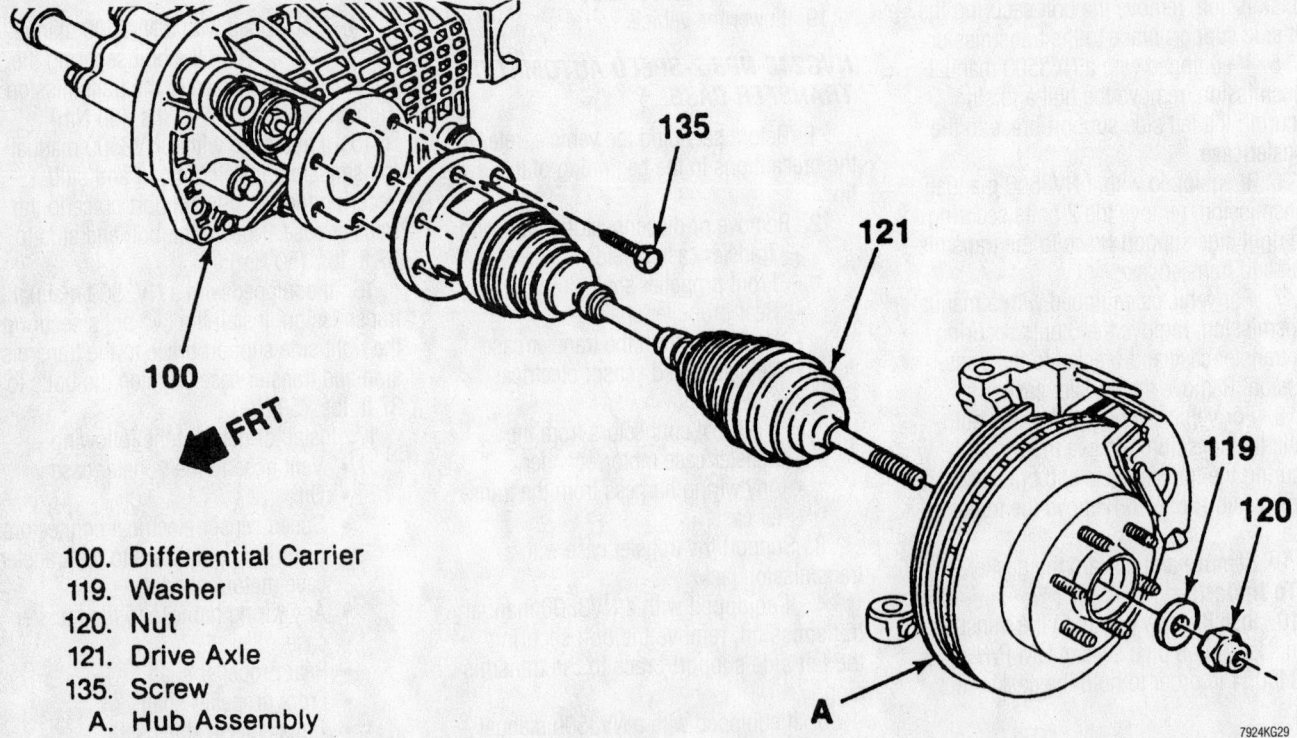

100. Differential Carrier
119. Washer
120. Nut
121. Drive Axle
135. Screw
A. Hub Assembly

7924KG29

The halfshaft is mounted to the flange on the differential and through the hub assembly—4-wheel drive models

- Nut and the washer from the hub. Do not reuse the nut. A new nut must be used when installing the wheel drive shaft.
- The 6 bolts securing the wheel drive shaft inboard flange to the output shaft flange
- The drift from the rotor
- The stabilizer shaft link from the lower control arm

4. Wrap shop towels around both the inner and the outer wheel drive shaft boots in order to avoid damage to the boots during removal and installation.

5. Pull the wheel drive shaft through the lower control arm opening.

To install:

6. Wrap shop towels around both the inner and the outer wheel drive shaft boots in order to avoid damage to the boots during removal and installation.

➡**Clean the steering knuckle and the wheel drive shaft splines and threads. These areas must be dry and free of grease, dirt, and contamination.**

7. Insert the wheel drive shaft splined shank into the knuckle hub.

➡**Use only a genuine GM front wheel drive shaft nut. Installation of anything but an OEM front wheel drive shaft nut could cause damage to the vehicle.**

8. Install or connect the following:

9. Install the washer and new hub nut to the wheel drive shaft. Do not tighten.

10. Install the wheel drive shaft inboard flange to the output shaft flange using the inboard flange bolts

11. Insert a drift or a large screwdriver through the brake caliper into 1 of the brake rotor vanes in order to prevent the wheel drive shaft from turning. Tighten the inboard flange bolts to 78 Nm (58 ft. lbs.). Tighten the hub nut to 210 Nm (155 ft. lbs.).

12. Remove the drift from the rotor.

13. Install the stabilizer shaft link.

14. Install the wheel and tire assembly.

CV-Joints

OVERHAUL

1999 Vehicles

OUTER CV-JOINT

1. Before servicing the vehicle, refer to the precautions in the beginning of this section.

2. Remove or disconnect the following:

- Front wheel
- Halfshaft and position it in a vise
- Large CV-joint boot clamp and discard it
- Small CV-joint boot clamp and discard it
- CV-joint boot and slide it back on the shaft
- Outer race from the halfshaft by spreading the outer race-to-halfshaft retaining ring using Snapring Pliers J-8059

- Retaining ring from the halfshaft and discard it
- CV-joint boot from the halfshaft and discard it if damaged

3. Disassemble the chrome alloy balls from the CV-joint cage as follows:

   a. Position a brass drift against the CV-joint cage and tap it with a hammer to tilt the cage.

   b. Remove the 1st chrome alloy ball from the cage.

   c. Tilt the cage in the opposite direction.

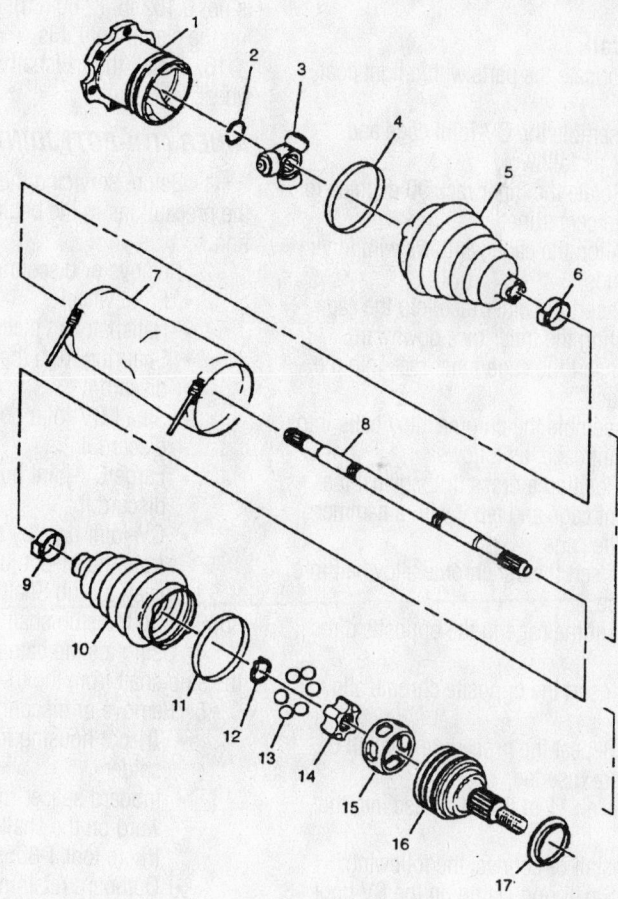

Legend
(1) Tripot Housing Assembly
(2) Spacer Ring
(3) Tripot Joint Spider Assembly
(4) Swage Ring
(5) Tripot Joint Seal
(6) Small Seal Retaining Clamp
(7) Drive Axle Seal Cover (Optional)
(8) Drive Axle Shaft
(9) CV Joint Seal
(10) Race Retaining Ring
(11) Ball
(12) CV Joint Inner Race
(13) CV Joint Cage
(14) CV Joint Outer Race
(15) Deflector Ring

9308GK10

Exploded view of the CV-Joint assemble

d. Remove the opposite chrome alloy ball.

e. Repeat the procedure until all 6 balls are removed.

4. Disassemble the CV-joint cage and inner race as follows:

a. Pivot the cage and race 90 degrees to the center line of the outer race.

b. Align the cage windows with outer race lands.

c. Remove the cage from the outer race.

d. Rotate the inner race upward and remove it from the cage.

5. Thoroughly clean and inspect all parts.

To install:

6. Lubricate the parts with a light coat of grease.

7. Assemble the CV-joint cage and inner race, as follows:

a. Rotate the inner race 90 degrees to the cage centerline.

b. Align the cage windows with inner race lands.

c. Insert the inner race into the cage by rotating the inner race downward.

d. Insert the cage/inner race into the outer race.

8. Assemble the chrome alloy balls into the CV-joint cage, as follows:

a. Position a brass drift against the CV-joint cage and tap it with a hammer to tilt the cage.

b. Insert the 1st chrome alloy ball into the cage.

c. Tilt the cage in the opposite direction.

d. Insert the opposite chrome alloy ball.

e. Repeat the procedure until all 6 balls are inserted.

9. Install ½ of the kit grease into the CV-joint.

10. Install or connect the following:
- Small ring clamp on the CV boot
- New retaining ring on the halfshaft
- Large ring clamp on the CV boot
- Outer race assembly onto the halfshaft until the ring engages the halfshaft groove

11. Slide the small end of the CV-joint boot/clamp into place, with the seal lip in the halfshaft groove

➡ **Make sure the boot lies flat against the halfshaft.**

12. Using the Crimp tool J-35910, a torque wrench and a breaker bar, crimp the small CV-joint boot clamp to 100 ft. lbs. (136 Nm).

13. Check the clamp gap dimension; if it is not 0.085 in. (2.15mm), continue tightening the clamp until it is.

14. Install ½ of the kit grease into the CV-joint boot.

15. Slide the large end of the CV boot/clamp into place, with the seal lip in place over the outer race.

➡ **Make sure the boot lies flat against the outer race.**

16. Using the Crimp tool J-35910, a torque wrench and a breaker bar, crimp the large CV-joint boot clamp to 130 ft. lbs. (176 Nm).

17. Check the clamp gap dimension; if it is not 0.102 in. (2.60mm), continue tightening the clamp until it is.

18. Install the halfshaft and the front wheel.

INNER (TRI-POT) JOINT

1. Before servicing the vehicle, refer to the precautions in the beginning of this section.

2. Remove or disconnect the following:
- Front wheel
- Halfshaft and place it in a vise
- Snapring from the stub shaft and discard it
- Small CV-joint boot clamp, cut and discard it
- Large CV-joint boot clamp, cut and discard it
- CV-joint boot by sliding it away from the tri-pot joint

3. Install a Stub Shaft Removal tool J-38868-A to the stub shaft snapring groove.

4. Using a slide hammer puller, press the stub shaft from the tri-pot housing.

5. Remove or disconnect the following:
- Tri-pot housing from the tri-pot spider
- Inboard spacer ring slide it rearward on the shaft using Snapring Pliers tool J-8059
- Outboard retaining ring using Snapring Pliers tool J-8059 and discard it
- Tri-pot joint spider assembly
- Inboard spacer ring and discard it
- CV-joint boot
- Trilobal tri-pot bushing from the housing

6. Thoroughly clean and inspect all parts.

To install:

7. Install or connect the following:
- New snapring onto the stub shaft
- Small boot clamp
- CV-joint boot

8. Using the Crimp tool J-35910, a torque wrench and a breaker bar, crimp the small CV-joint boot clamp to 100 ft. lbs. (136 Nm).
- Inboard spacer ring slide it rearward on the shaft using Snapring Pliers tool J-8059, past the 2nd groove
- Tri-pot joint spider assembly onto the shaft until it passes the 2nd groove
- Outboard retaining ring into the axle shaft groove using Snapring Pliers tool J-8059
- Tri-pot joint spider assembly, slide it against the outboard retaining ring
- Inboard spacer ring, seat it in the groove
- ½ of the kit grease into the boot
- ½ of the kit grease into the tri-pot housing
- Trilobal tip-pot bushing flush with the tri-pot housing face
- New large seal clamp onto the CV-joint boot
- Tri-pot housing, slide it over the tri-pot joint spider assembly
- CV-joint boot/clamp, slide it into place, over the trilobal tri-pot bushing with the seal lip in the groove

➡ **Make sure the boot lies flat against the trilobal bushing.**

9. Using the Crimp tool J-35910, a torque wrench and a breaker bar, crimp the large CV-joint boot clamp to 130 ft. lbs. (176 Nm)

10. Check the clamp gap dimension; if it is not 0.102 in. (2.60mm), continue tightening the clamp until it is.

11. Install the halfshaft and the front wheel.

2000–03 Vehicles

INNER JOINT

➡ **With removal of the halfshaft for any reason, the transmission sealing surface (the tripot male/female shank of the halfshaft) should be inspected for corrosion. If corrosion is evident, the surface should be cleaned with 320 grit cloth or equivalent. Transmission fluid may be used to clean off any remaining debris. The surface should be wiped dry and the halfshaft reinstalled free of any buildup.**

1. Before servicing the vehicle, refer to the precautions in the beginning of this section.

2. Use a hand grinder in order to cut through the swage ring.

3. Remove the tripot housing from the halfshaft. Wipe the grease off of the tripot assembly roller bearings and the tripot housing. Thoroughly degrease the tripot housing. Allow the tripot housing to dry prior to assembly.

➡ **Handle the tripot spider assembly with care. Tripot balls and needle rollers may separate from the spider trunnion if the tripot balls and needle rollers are not handled carefully.**

4. Before servicing the vehicle, refer to the precautions in the beginning of this section.

5. Use side cutters to cut away the small boot clamp.

6. Compress the tripot boot up the halfshaft away from the tripot spider assembly toward the outboard (CV joint assembly) end of the halfshaft.

7. Spread the spider spacer ring with tool J8059, or equivalent.

8. Remove the following items from the halfshaft bar:
   a. The spacer ring.
   b. The spider assembly.
   c. The tripot boot.

9. Clean the halfshaft bar. Use a wire brush in order to remove any rust in the boot mounting area (grooves).

10. Inspect the needle rollers, needle bearings, and trunnion. Check the tripot

housing for unusual wear, cracks, or other damage. Replace any damaged parts.

To assemble:

11. Place the new small boot clamp onto the small end of the joint boot.

12. Compress the joint boot and small boot clamp onto the halfshaft bar.

13. Position the small end of the joint boot into the joint boot groove on the halfshaft bar.

14. Secure the small boot clamp with tool J35910, or equivalent, a breaker bar, and a torque wrench. Tighten the small boot clamp (1) to 136 Nm (100 ft. lbs.).

15. Check the gap dimension on the clamp ear. Continue tightening until the gap dimension is reached.

➡ **Assemble the CV joint with the convolute retainer in the correct position, as illustrated.**

16. Install the convolute retainer over the inboard joint boot, being sure to capture three convolutions.

17. Install the tripot spider assembly onto the halfshaft bar with the counterbore towards the end of the halfshaft bar.

18. Install the spacer ring in the groove at the end of the halfshaft bar.

19. Push the spider assembly back toward the end of the halfshaft bar until the spacer ring is covered by the spider assembly counterbore.

20. Pack the tripot boot and the tripot housing with the grease supplied in the kit. The amount of grease supplied in this kit has been pre-measured for this application.

21. Reassemble the tripot housing and the tripot boot using the following procedure:
   a. Pinch the swage ring slightly by hand in order to distort it into an oval shape.
   b. Slide the distorted swage ring over the large diameter of the boot.
   c. Place the tripot housing over the spider assembly.
   d. Install the boot onto the tripot housing.
   e. Align the tripot boot with the swage ring in place, over the flat area on the tripot housing.

22. Mount tool J36652 in a vise. Install the bottom half of the split-plate swage clamp. For K15 models, use tool J36652-98. For K25 models, use tool J36652-1.

23. Check the inboard stroke position. Use measurement A for the K15 models. Use measurement B for the K25 models.

24. Position the inboard end (tripot end) of the halfshaft assembly in tool J36652. Install the top half of the proper size tool on the lower half of the tool. For K15 models, use tool J36652-98. For K25 models, use tool J36652-1.

25. Align the swage ring and the swage ring clamp. Insert the bolts. Hand tighten the bolts in tool J36652 until the bolts are snug.

26. Align the following during this procedure:
   a. The tripot boot.
   b. The housing.
   c. The swage ring. Tighten each bolt 180 degrees at a time. Alternate between the bolts until both sides of the top half of J36652 touch the bottom half of the tool.

27. Loosen the bolts and remove the halfshaft assembly from J36652.

28. Remove the convolute retainer from the boot.

OUTER JOINT

1. Before servicing the vehicle, refer to the precautions in the beginning of this section.

2. Place protective covers over the vise jaws. Place the halfshaft in the vise.

3. Use a hand grinder to cut through the swage ring. Use side cutters to cut off the small boot clamp.

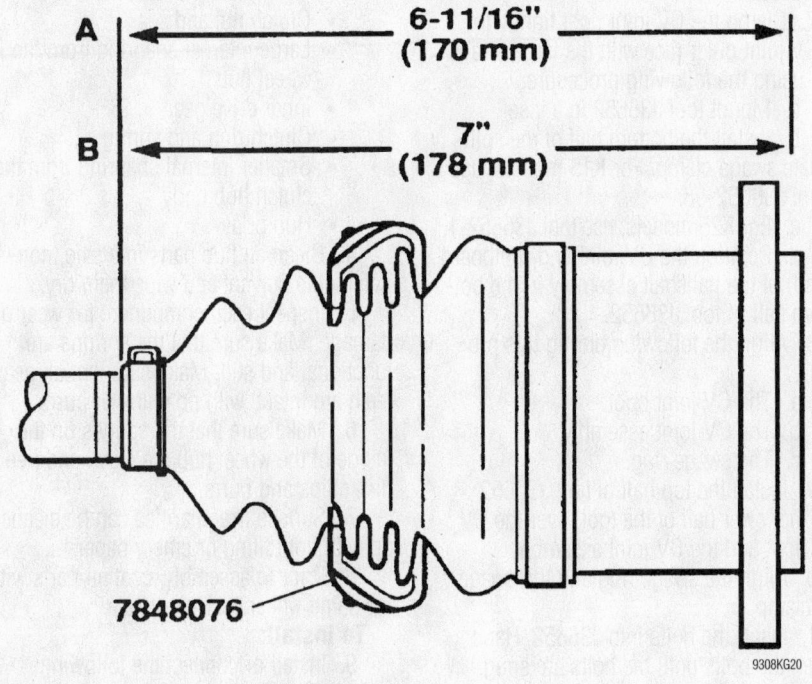

A 6-11/16" (170 mm)
B 7" (178 mm)
7848076
9308KG20
Assembled joint measurement—15 Series

4. Slide the boot down the halfshaft bar and away from the CV-joint outer race. Wipe all grease away from the face of the CV joint.

5. Find the halfshaft bar retaining snap ring, which is located in the inner race.

6. Spread the snapring ears apart.

7. Pull the CV joint and the CV joint boot from the halfshaft bar. Discard the old CV joint boot.

8. Place a brass drift against the CV joint cage. Tap gently on the brass drift with a hammer in order to tilt the cage.

9. Remove the first chrome alloy ball when the CV joint cage tilts. Tilt the CV joint cage (1) in the opposite direction to remove the opposing chrome alloy ball. Repeat this process to remove all six of the balls.

10. Pivot the CV joint cage and the inner race 90 degrees to the center line of the outer race. At the same time, align the cage windows with the lands of the outer race. Lift out the cage and the inner race.

11. Remove the inner race from the cage by rotating the inner race upward. Clean the following items thoroughly with cleaning solvent. Remove all traces of old grease and any contaminates.

   a. The inner and outer race assemblies.

   b. The CV joint cage.

   c. The chrome alloy balls.

12. Dry all the parts. Check the CV joint assembly for unusual wear, cracks, or other damage. Replace any damaged parts. Clean the halfshaft bar. Use a wire brush to remove any rust in the boot mounting area (grooves).

To assemble:

13. Inspect all of the parts for unusual wear, cracks, or other damage. Replace the CV joint assembly if necessary. Put a light coat of the recommended grease on the inner and the outer race grooves.

14. Hold the inner race at 90 degrees to the centerline of the cage. Align the lands of the inner race with the windows of the cage. Insert the inner race into the cage by rotating the inner race downward.

15. Insert the cage and inner race into the outer race.

16. Place a brass drift against the CV joint cage. Tap gently on the brass drift with a hammer in order to tilt the cage. Install the first chrome alloy ball when the CV joint cage tilts. Tilt the CV joint cage in the opposite direction to install the opposing chrome alloy ball. Repeat this process in order to install all six of the balls.

17. Pack the CV joint boot and the CV joint assembly with the grease supplied in the kit. The amount of grease supplied in this kit has been pre-measured for this application.

18. Place the new small boot clamp onto the CV joint boot.

19. Slide the CV joint boot onto the halfshaft bar.

20. Position the small end of the CV joint boot into the joint boot groove on the halfshaft bar.

21. Secure the small boot clamp, a breaker bar, and a torque wrench. Tighten the small clamp (1) to 136 Nm (100 ft. lbs.).

22. Check the gap dimension on the clamp ear. Continue tightening until the gap dimension is reached.

23. Pinch the new swage ring slightly by hand to distort it into an oval shape. Slide the distorted swage ring over the large diameter of the boot.

➡**Be sure that the retaining ring side of the CV joint inner race faces the halfshaft bar (3) before installation.**

24. Slide the CV joint onto the halfshaft bar. The retaining snap ring inside of the inner race engages in the halfshaft bar groove with a click when the CV joint is in the proper position.

25. Pull on the CV joint to verify engagement.

26. Slide the large diameter of the CV joint boot with the large swage ring in place, over the outside edge of the CV joint outer race.

27. Clamp the CV joint boot tightly to the CV joint outer race with the large swage ring, using the following procedure:

   a. Mount tool J36652 in a vise.

   b. Install the bottom half of the split-plate swage clamp. For K15 models, use tool J36652-98.

   c. For K25 models, use tool J36652-1.

   d. Position the CV joint end (outboard end) of the halfshaft assembly in the bottom half of tool J36652.

28. Align the following during this procedure:

   a. The CV joint boot.

   b. The CV joint assembly.

   c. The swage ring.

29. Install the top half of tool J36652 onto the lower half of the tool, over the CV joint boot and the CV joint assembly.

30. Align the swage ring and the swage ring clamp.

31. Insert the bolts into J36652. Hand tighten the bolts until the bolts are snug. Tighten each bolt 180 degrees at a time. Alternate between the bolts until both sides of the top half of the tool touch the bottom half of the tool.

32. Loosen the bolts and remove the halfshaft assembly from the tool.

Manual Locking Hubs

The engagement and disengagement of the hubs is a manual operation which must be performed at each hub assembly. The hubs should be placed FULLY in either Lock or Free position or damage will result.

✴✴ WARNING

Do not place the transfer case in either 4-wheel mode unless the hubs are in the Lock position!

Locking hubs should be run in the Lock position periodically for a few miles to assure proper differential lubrication.

REMOVAL & INSTALLATION

1. Before servicing the vehicle, refer to the precautions in the beginning of this section.

2. Remove or disconnect the following:
   • Wheels

3. Lock the hubs. Remove the outer retaining plate, Allen head bolts and take off the plate, O-ring, and knob assembly.
   • External snapring from the axle shaft
   • Compression spring
   • Clutch cup
   • O-ring and dial screw
   • Clutch nut and seal
   • Large internal snapring from the wheel hub
   • Inner drive gear
   • Clutch ring and spring
   • Smaller internal snapring from the clutch hub body
   • Hub body

4. Clean all hub parts in a safe, non-flammable solvent and wipe them dry.

5. Inspect each component for wear or damage. Make sure that the springs are functional and stiff. Make sure that all gear teeth are intact, with no chips or burrs.

6. Make sure that the splines on the inside of the wheel hub are clean and free of dirt, chips and burrs.

7. Surface irregularities can be cleaned up with light filing or emery paper.

8. Prior to assembly, coat all parts with the same wheel bearing grease.

To install:

9. Install or connect the following:
   • Hub body
   • Smaller internal snapring in the clutch hub body
   • Clutch ring and spring

- Inner drive gear
- Large internal snapring in the wheel hub
- External snapring on the axle shaft. If the snapring groove is not completely visible, reach around, inside the knuckle and push the axle shaft outwards.
- Clutch nut and seal
- O-ring and dial screw
- Clutch cup
- Compression spring

10. Place the hub dial in the Lock position.

11. Coat the hub dial assembly O-ring with wheel bearing grease and position the hub dial and retainer on the hub.

- Allen head bolts. Make sure that you used any washers that were there originally. Torque these bolts to 45 inch lbs. (5 Nm)

12. Rotate the hub dial to the free position and turn the wheel hubs to make sure that the axle is free.

- Wheels

Automatic Locking Hubs

REMOVAL & INSTALLATION

The following procedure covers removal & installation only, for the hub assembly. The hub should be disassembled ONLY if overhaul is necessary. In that event, an overhaul kit will be required. Follow the instructions in the overhaul kit to rebuild the hub.

1. Before servicing the vehicle, refer to the precautions in the beginning of this section.

2. Remove or disconnect the following:

- Capscrews and washer from the hub cap
- Hub cap and spring
- Bearing race, bearing and retainer
- Keeper from the outer clutch housing
- Large snapring to release the locking unit
- Locking unit from the hub. You can make this job easier by threading 2 hub cap screws into the outer clutch housing and hold these to pull out the unit.

To install:

3. Wipe clean all parts and check for wear or damage.

4. Coat all parts with the same wheel bearing grease you've used on the bearings.

5. Install or connect the following:

- Locking unit in the hub
- Large snapring, pull outward on the unit to make sure the snapring is fully seated in its groove
- Keepers
- Bearing retainer, bearing and race. Make sure that the bearing is fully pack with grease.

6. Coat the hub cap O-ring with wheel bearing grease and install the hub cap.

- Capscrews and washers. Tighten the screws to 45 inch lbs. (5 Nm).

Front Axle shaft, Bearing and seal

REMOVAL & INSTALLATION

1999 Vehicles

1. Before servicing the vehicle, refer to the precautions in the beginning of this section.

2. Drain the front axle.

3. Remove or disconnect the following:

- Electrical connectors, if equipped
- Drive axle (halfshaft)
- Axle shaft (output shaft)
- Axle shaft from case

- Deflector and seal
- Bearing

To install:

4. Install or connect the following:

- New Bearing, square shoulder in

➡ **Lubricate the new seal with grease.**

- New seal
- Deflector
- Axle shaft (output shaft)
- Drive axle (halfshaft)
- Electrical connectors, if equipped

5. Refill the front axle.

2000–03 Vehicles

1. Before servicing the vehicle, refer to the precautions in the beginning of this section.

2. Remove or disconnect the following:

- Halfshaft assembly
- Front axle fluid
- Electrical connectors
- Axle shaft (output shaft) tube nuts from the bracket
- Bracket bolts from the frame. Do not remove the bracket. The bolts are removed in order to provide clearance.
- Axle shaft (output shaft) bolts from the carrier

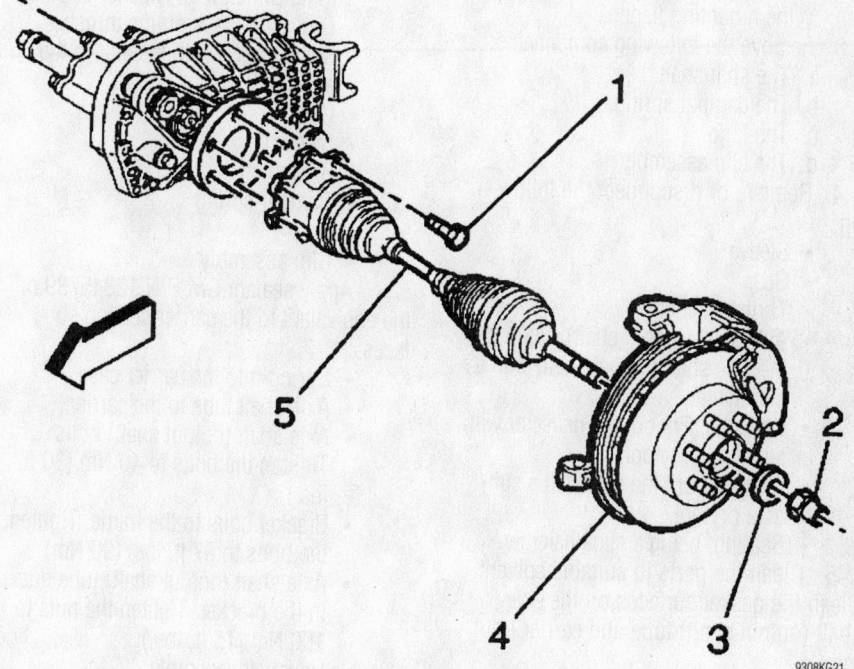

Front axle shaft removal

9308KG21

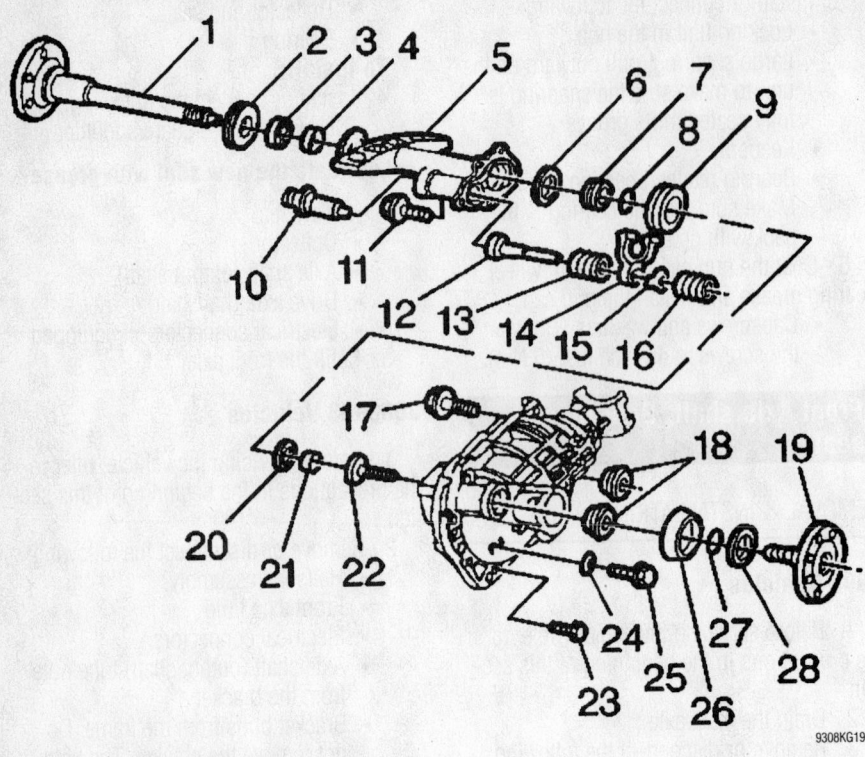

9308KG19

Front axle shaft exploded view

➡️**Keep the open end of the tube up.**

- Axle shaft (output shaft) tube from the carrier. Ensure the spring is not lost during removal. In a vise, hold the axle shaft (output shaft) tube by the mounting flange.

3. Remove the following components:
   a. The shift shaft.
   b. The damper spring.
   c. The fork.
   d. The clip assembly.

4. Remove or disconnect the following:
- Sleeve
- Gear
- Thrust washer
- Axle shaft (output shaft). Tap out the axle shaft (output shaft) with a soft mallet.
- Deflector. Pry out the deflector with a suitable prytool.
- Seal. Pry out the seal with a suitable prytool.
- Bearing, using a slide hammer.

5. Clean the parts in suitable solvent. Clean the gasket surfaces on the axle shaft (output shaft) tube and carrier housing.

To install:

6. Install or connect the following:
- New bearing into the axle shaft tube using a driver. Apply axle lubricant to the bearing.

- New seal using a driver. Coat the seal lips with axle lubricant.
- Deflector
- Axle shaft
- Thrust washer. Use grease in order to hold the thrush washer in place. Ensure the tabs on the thrush washer align with the slot in the axle shaft tube.
- Gear
- Sleeve
- Shift shaft
- Damper spring
- Fork
- Clip assembly

7. Apply sealant GM P/N 12345739 or the equivalent to the carrier sealing surfaces.

- Spring into the carrier case
- Axle shaft tube to the carrier
- Axle shaft (output shaft) bolts. Tighten the bolts to 40 Nm (30 ft. lbs.).
- Bracket bolts to the frame. Tighten the bolts to 67 ft. lbs. (90 Nm).
- Axle shaft (output shaft) tube nuts to the bracket. Tighten the nuts to 100 Nm (75 ft. lbs.).
- Halfshaft assembly
- Electrical connectors

8. Fill the front differential with lubricant until the level is 12mm (0.5 in) below the fill plug.

Rear Axle Shaft, Bearing and Seal

REMOVAL & INSTALLATION

1999 Vehicles

SEMI-FLOATING NON-LOCKING DIFFERENTIALS

1. Before servicing the vehicle, refer to the precautions in the beginning of this section.

2. Remove the wheels and brake drums.

3. Remove the differential cover

4. Turn the differential until you can reach the differential pinion shaft lockscrew. Remove the lockscrew and the pinion shaft.

5. Push in on the axle end. Remove the C-lock from the inner (button) end of the shaft.

6. Remove the shaft, being careful of the oil seal.

7. You can pry the oil seal out of the housing by placing the inner end of the axle shaft behind the steel case of the seal, then prying it out carefully.

8. A puller or a slide hammer is required to remove the bearing from the housing.

To install:

9. Pack the new or reused bearing with wheel bearing grease and lubricate the cavity between the seal lips with the same grease.

10. The bearing has to be driven into the housing. Don't use a drift, you might cock the bearing in its bore. Use a piece of pipe or a large socket instead. Drive only on the outer bearing race. In a similar manner, drive the seal in flush with the end of the tube.

11. Slide the shaft into place, turning it slowly until the splines are engaged with the differential. Be careful of the oil seal.

12. Install the C-lock on the inner axle end. Pull the shaft out so that the C-lock seats in the counterbore of the differential side gear.

13. Position the differential pinion shaft through the case and the pinion gears, aligning the lockscrew hole. Install the lockscrew.

14. Install the cover with a new gasket and tighten the bolts evenly in a criss-cross pattern.

15. Fill the axle with lubricant.

16. Replace the brake drums and wheels.

Remove the differential pinion shaft lockscrew

87987P14

Remove the axle shaft from the vehicle

87987P17

Remove the pinion shaft

87987P15

Use a puller to remove the oil seal

87987P18

Remove the C-lock from the inner (button) end of the shaft

87987P16

Install the oil seal using a seal installer

87987P19

Timing belt service is covered in Section 3 of this manual

SEMI-FLOATING LOCKING DIFFERENTIALS

This axle uses a thrust block on the differential pinion shaft.

1. Before servicing the vehicle, refer to the precautions in the beginning of this section.

2. Remove the wheels and brake drums.

3. Clean off the differential cover area, loosen the cover to drain the lubricant, and remove the cover.

4. Rotate the differential case so that you can remove the lockscrew and support the pinion shaft so it can't fall into the housing. Remove the differential pinion shaft lockscrew.

5. Carefully pull the pinion shaft partway out and rotate the differential case until the shaft touches the housing at the top.

6. Use a screwdriver to position the C-lock with its open end directly inward. You can't push in the axle shaft till you do this.

7. Push the axle shaft in and remove the C-lock.

8. Remove the shaft, being careful of the oil seal.

9. You can pry the oil seal out of the housing by placing the inner end of the axle shaft behind the steel case of the seal, then prying it out carefully.

10. A puller or a slide hammer is required to remove the bearing from the housing.

To install:

11. Pack the new or reused bearing with wheel bearing grease and lubricate the cavity between the seal lips with the same grease.

12. The bearing has to be driven into the housing. Don't use a drift, you might cock the bearing in its bore. Use a piece of pipe or a large socket instead. Drive only on the outer bearing race. In a similar manner, drive the seal in flush with the end of the tube.

13. Slide the shaft into place, turning it slowly until the splines are engaged with the differential. Be careful of the oil seal.

14. Keep the pinion shaft partway out of the differential case while installing the C-lock on the axle shaft. Put the C-lock on the axle shaft and carefully pull out on the axle shaft until the C-lock is clear of the thrust block.

15. Position the differential pinion shaft through the case and the pinion gears, aligning the lockscrew hole. Install the lockscrew.

16. Install the cover with a new gasket and tighten the bolts evenly in a criss-cross pattern.

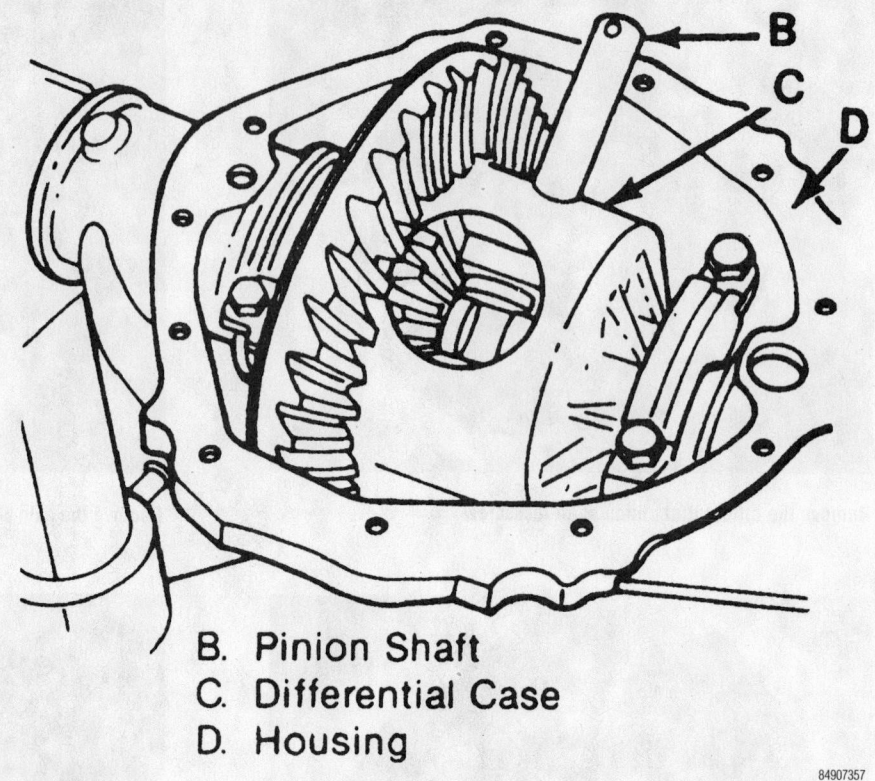

B. Pinion Shaft
C. Differential Case
D. Housing

84907357

Positioning the case for the best clearance—semi-floating axle w/locking differential

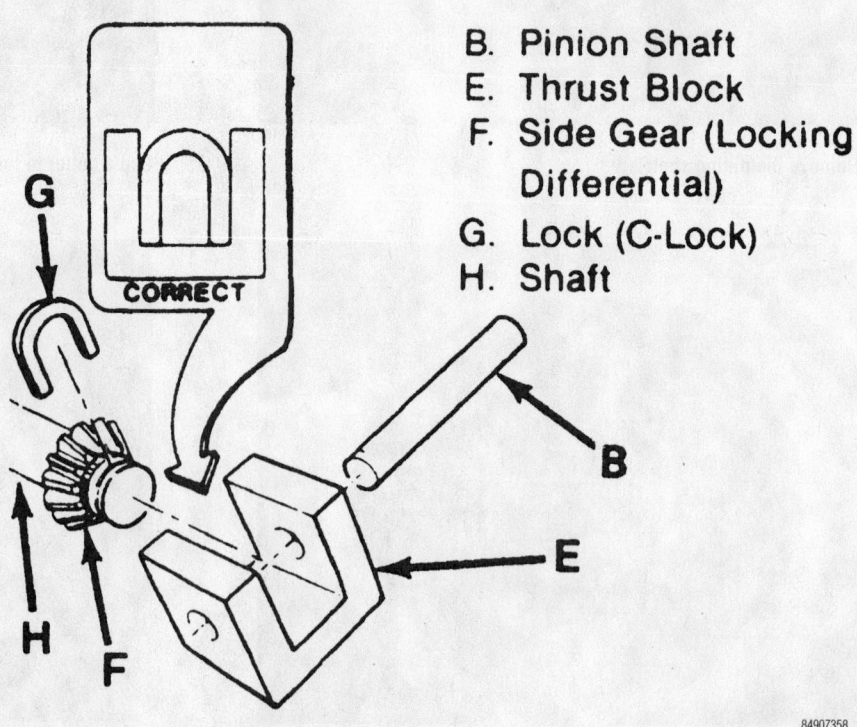

B. Pinion Shaft
E. Thrust Block
F. Side Gear (Locking Differential)
G. Lock (C-Lock)
H. Shaft

84907358

Aligning the lock—semi-floating axle w/locking differential

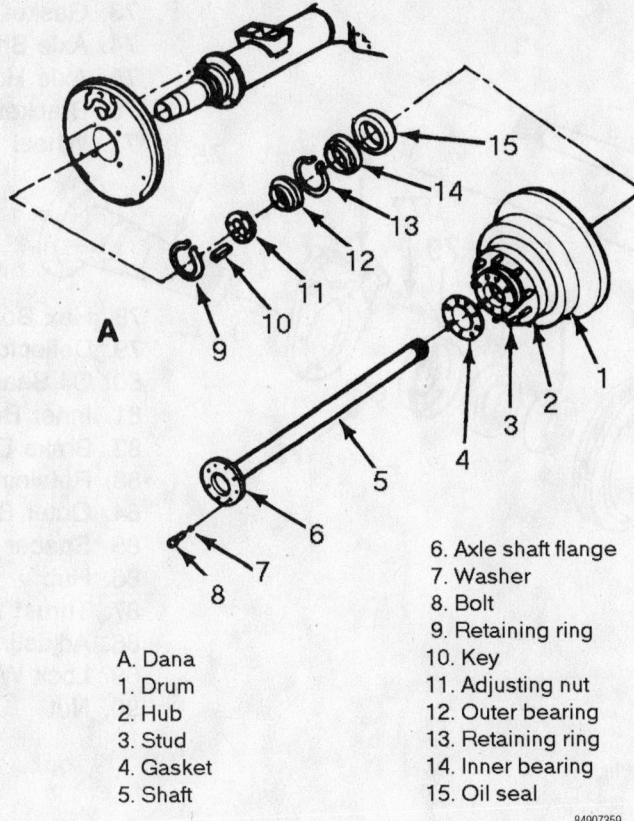

6. Axle shaft flange
7. Washer
8. Bolt
9. Retaining ring
10. Key
11. Adjusting nut
12. Outer bearing
13. Retaining ring
14. Inner bearing
15. Oil seal

A. Dana
1. Drum
2. Hub
3. Stud
4. Gasket
5. Shaft

84907359

Exploded view of the axle, hub and drum assembly—full-floating axle, 9¾ and 10½ in.

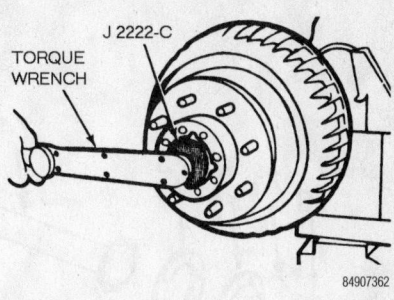

Tightening the adjusting nut—full-floating axle, 9¾ and 10½ in.

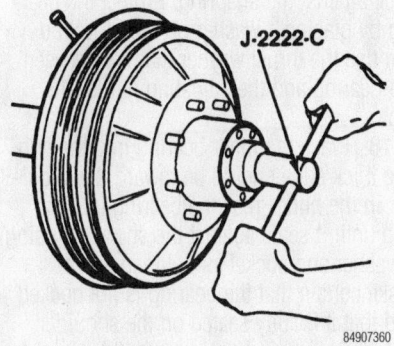

84907360

Removing the bearing adjusting nut—full-floating axle, 9¾ and 10½ in.

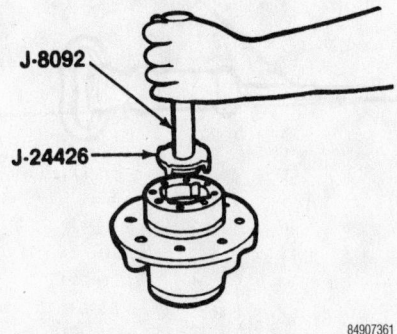

84907361

Removing the bearing outer cup—full-floating axle, 9¾ and 10½ in.

17. Fill the axle with lubricant.
18. Replace the brake drums and wheels.

FULL-FLOATING AXLES

The procedures are the same for locking and non-locking axles.

The best way to remove the bearings from the wheel hub is with an arbor press. Use of a press reduces the chances of damaging the bearing races, cocking the bearing in its bore, or scoring the hub walls. A local machine shop is probably equipped with the tools to remove and install bearings and seals. However, if one is not available, the hammer and drift method outlined can be used.

1. Before servicing the vehicle, refer to the precautions in the beginning of this section.
2. Support the axles on jackstands.
3. Remove the wheels.
4. Remove the bolts and lock washers that attach the axle shaft flange to the hub.
5. Rap on the flange with a soft faced

hammer to loosen the shaft. Grip the rib on the end of the flange with a pair of locking pliers and twist to start shaft removal. Remove the shaft from the axle tube.

6. The hub and drum assembly must be removed to remove the bearings and oil seals. You will need a large socket to remove and later adjust the bearing adjustment nut. There are also special tools available.

7. Disengage the tang of the locknut retainer from the slot or slat of the locknut, then remove the locknut from the housing tube.

8. Disengage the tang of the retainer from the slot or flat of the adjusting nut and remove the retainer from the housing tube.

9. Remove the adjusting nut from the housing tube.

10. Remove the thrust washer from the housing tube.

11. Pull the hub and drum straight off the axle housing.

12. Remove the oil seal and discard.

13. Use a hammer and a long drift to knock the inner bearing, cup, and oil seal from the hub assembly.

14. Remove the outer bearing snapring with a pair of pliers. It may be necessary to tap the bearing outer race away from the retaining ring slightly by tapping on the ring to remove the ring.

15. Drive the outer bearing from the hub with a hammer and drift.

To install:

16. Place the outer bearing into the hub. The larger outside diameter of the bearing should face the outer end of the hub. Drive the bearing into the hub using a washer that will cover both the inner and outer races of the bearing. Place a socket on top of this washer, then drive the bearing into place with a series of light taps. If available, an arbor press should be used for this job.

Heater Core replacement is covered in Section 2 of this manual

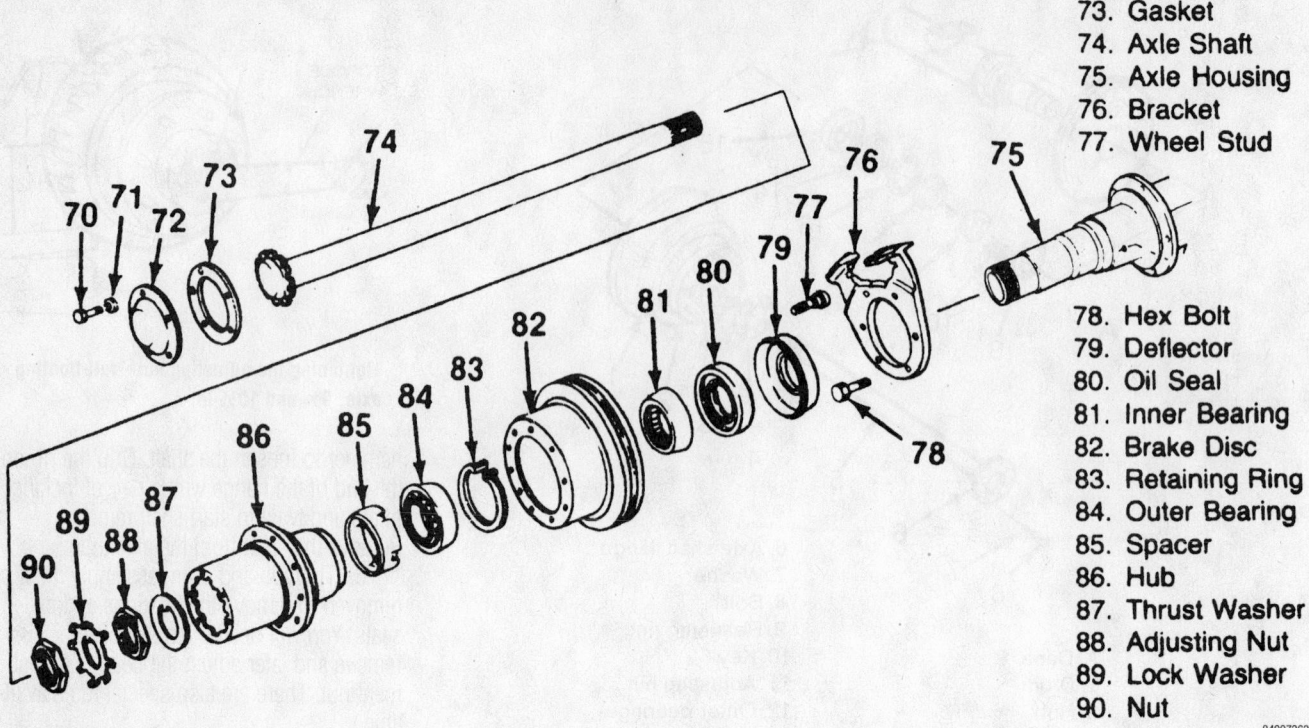

73. Gasket
74. Axle Shaft
75. Axle Housing
76. Bracket
77. Wheel Stud

78. Hex Bolt
79. Deflector
80. Oil Seal
81. Inner Bearing
82. Brake Disc
83. Retaining Ring
84. Outer Bearing
85. Spacer
86. Hub
87. Thrust Washer
88. Adjusting Nut
89. Lock Washer
90. Nut

84907363

Exploded view of the axle and hub assembly—full-floating axle, 12 in.

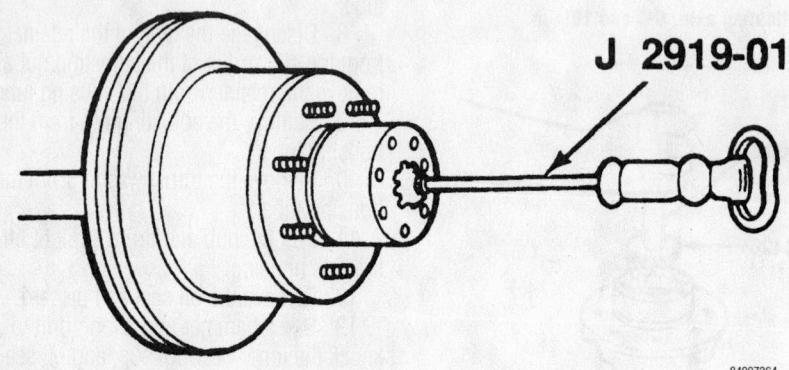

J 2919-01

84907364

Removing the axle shaft—full-floating axle, 12 in.

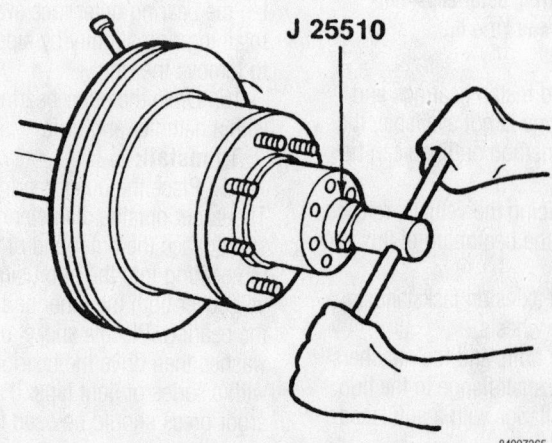

J 25510

84907365

Removing the wheel bearing nut—full-floating axle, 12 in.

17. Drive the bearing past the snapring groove, and install the snapring. Then, turning the hub assembly over, drive the bearing back against the snapring. Protect the bearing by placing a washer on top of it. You can use the thrust washer that fits between the bearing and the adjusting nut for the job.

18. Place the inner bearing into the hub. The thick edge should be toward the shoulder in the hub. Press the bearing into the hub until it seats against the shoulder, using a washer and socket as outlined earlier. Make certain that the bearing is not cocked and that it is fully seated on the shoulder.

19. Pack the cavity between the oil seal lips with wheel bearing grease, and position it in the hub bore. Carefully press it into place on top of the inner bearing.

20. Pack the wheel bearings with grease, and lightly coat the inside diameter of the hub bearing contact surface and the outside diameter of the axle housing tube.

21. Make sure that the inner bearing, oil seal, axle housing oil deflector, and outer bearing are properly positioned. Install the hub and drum assembly on the axle housing, being careful so as not to damage the oil seal or dislocate other internal components.

22. Install the thrust washer so that the tang on the inside diameter of the washer is in the keyway on the axle housing.

23. Install the adjusting nut. Tighten to 50 ft. lbs. (68 Nm) while rotating the hub. Back off the nut ¼ turn and retighten to 35 ft. lbs. (47 Nm) on models with the 11 inch ring gear and 13 ft. lbs. (17 Nm) on models with the 10 ½ inch ring gear.

24. Install the tanged retainer against the inner adjusting nut. Align the adjusting nut so that the short tang of the retainer will engage the nearest slot on the adjusting nut.

25. Install the outer locknut and tighten to 65 ft. lbs. (88 Nm). Bend the long tang of the retainer into the slot of the outer nut. This method of adjustment should provide 0.001–0.010 in. (0.0254–0.254mm) endplay.

26. Place a new gasket over the axle shaft and position the axle shaft in the housing so that the shaft splines enter the differential side gear. Position the gasket so that the holes are in alignment, and install the flange-to-hub attaching bolts. Tighten to 115 ft. lbs. (156 Nm) on models with the 10 ½ inch ring gear and tighten the axle cap bolts on models with the 11 inch ring gear to 15 ft. lbs. (20 Nm).

➡ **To prevent lubricant from leaking through the flange holes, apply a non-hardening sealer to the bolt threads. Use the sealer sparingly.**

27. Replace the wheels.

2000–03 Vehicles

8.5 INCH AND 9.5 INCH REAR AXLE

1. Before servicing the vehicle, refer to the precautions in the beginning of this section.

2. Raise and support the vehicle on a hoist.

3. Remove or disconnect the following:
- Tire and wheel assembly
- Brake caliper
- Rear cover and the gasket
- Pinion shaft locking screw
- Pinion shaft, on vehicles without a locking differential

4. On axles with a locking differential, remove the shaft part way. Rotate the case until the pinion shaft touches the housing.

5. On axles with a locking differential, use a screwdriver, or a similar tool, in order to enter the differential case and rotate the lock until the lock aligns with the thrust block.

6. Push the flange of the axle shaft toward the differential. Remove the lock from the button end of the axle shaft.

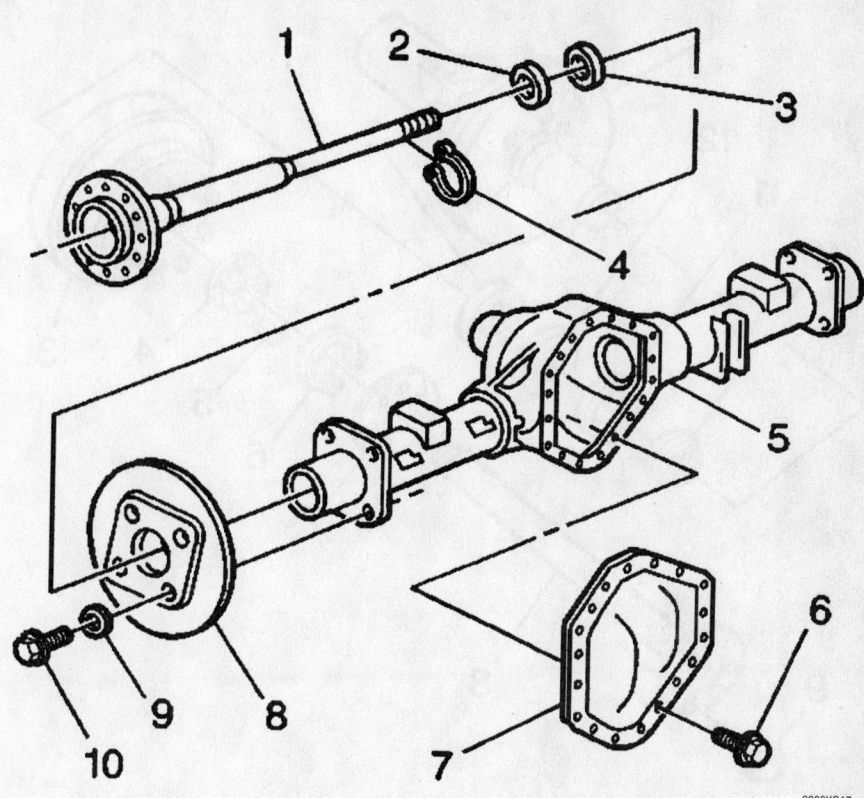

Rear axle shaft removal—8.5/9.5 inch 15 Series

➡ **When removing the axle shaft, do not rotate the shaft. Rotating the shaft will misalign the gears. Misaligning the gears will make the assembly difficult.**

7. Remove the axle shaft from the housing.

8. Inspect all the parts for damage. Replace the parts as necessary.

To install:

➡ **Carefully insert the axle shaft in order to not damage the seal.**

9. Install the axle shaft into the housing. Slide the axle shaft into place allowing the splines to engage the differential side gear.

10. On axles without a locking differential, place the lock on the button end of the axle shaft.

11. On axles with a locking differential, keep the pinion shaft partially withdrawn.

12. On axles with a locking differential, place the lock on the axle shaft so that the ends are flush with the thrust block. Pull the shaft flange outward in order to seat the lock in the differential gear.

➡ **Anytime you remove a differential pinion shaft locking screw, coat the screw threads with LOCTITE® 242 before reinstalling the screws. The screw has an adhesive coating in order to prevent the screw from loosening in the case. Removing the screw removes the adhesive on the screw.**

13. Align the hole in the pinion shaft with the screw hole in the differential case.

14. Install or connect the following:
- Pinion flange locking screw. Tighten the screw to 34 Nm (25 ft. lbs.).
- Rear cover and the gasket
- Brake caliper
- Tire and wheel assembly

15. Fill the rear axle.

16. Remove the supports and lower the vehicle.

10.5 INCH REAR AXLE

1. Before servicing the vehicle, refer to the precautions in the beginning of this section.

2. Remove or disconnect the following:

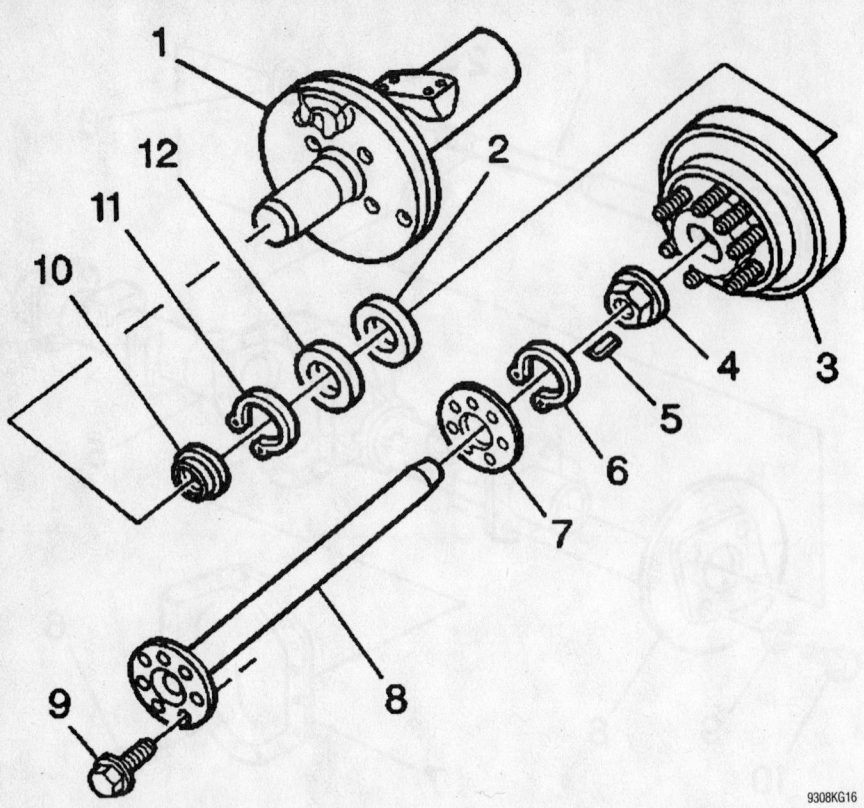

Rear axle shaft removal—10.5 inch 25 Series

- Tire and wheel
- Brake caliper
- Brake rotor
- Flange bolts

3. Lightly rap the axle shaft with a soft-faced hammer in order to loosen the shaft. Grip the rib on the axle shaft flange with a locking pliers. Twist the axle shaft flange in order to start the axle shaft removal. Remove the axle shaft from the tube.

4. Remove the gasket.

5. Clean the axle shaft flange and the outside face of the hub assembly. Inspect all the parts. Replace the parts as necessary.

To install:

6. Install the gasket onto the axle shaft.

7. Install the gasket and the axle shaft into the tube. Ensure the shaft splines mesh into the differential side gear. Align the holes in the axle flange and the gasket with the holes in the hub.

8. Install or connect the following:
- Axle flange bolts and tighten to 110 ft. lbs. (150 Nm)
- Brake rotor
- Brake caliper
- Wheel and tire

Front Drive Axle Pinion Seal

REMOVAL & INSTALLATION

1999 Vehicles

1. Before servicing the vehicle, refer to the precautions in the beginning of this section.

2. Raise and support the front end on jackstands.

3. Matchmark and disconnect the front driveshaft at the carrier.

4. Remove the wheels.

5. Dismount the calipers and wire them up, out of the way.

6. Position an inch pound torque wrench on the pinion nut. Measure the torque needed to rotate the pinion one full revolution. Record the figure.

7. Matchmark the pinion flange, shaft and nut. Count and record the number of exposed threads on the pinion shaft.

8. Hold the flange and remove the nut and washer.

9. Using a puller, remove the flange.

10. Carefully pry the seal from its bore. Be careful to avoid scratching the seal bore.

11. Remove the deflector from the flange.

To install:

12. Clean the seal bore thoroughly.

13. Remove any burrs from the deflector staking on the flange.

14. Tap the deflector onto the flange and stake it in three places.

15. Position the new seal in the carrier bore and drive it into place until flush. Coat the seal lips with wheel bearing grease.

16. Coat the outer edge of the flange neck with wheel bearing grease and slide it onto the pinion shaft.

17. Place a new nut and washer onto the pinion shaft and tighten it to the position originally recorded. That is, the alignment marks are aligned, and the recorded number of threads are exposed on the pinion shaft.

✳ WARNING

Never hammer the flange onto the pinion!

18. Measure the rotating torque of the

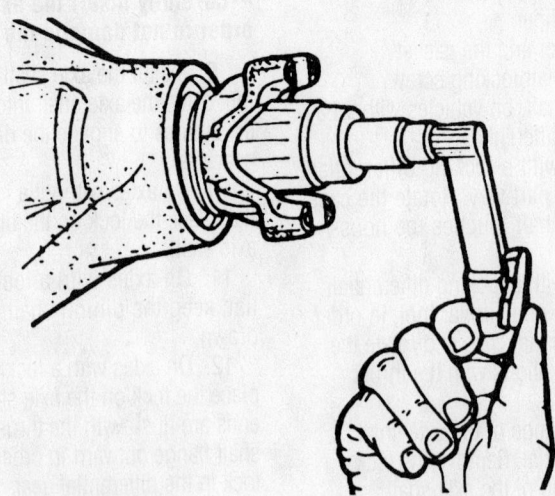

Measuring the pinion rotating torque

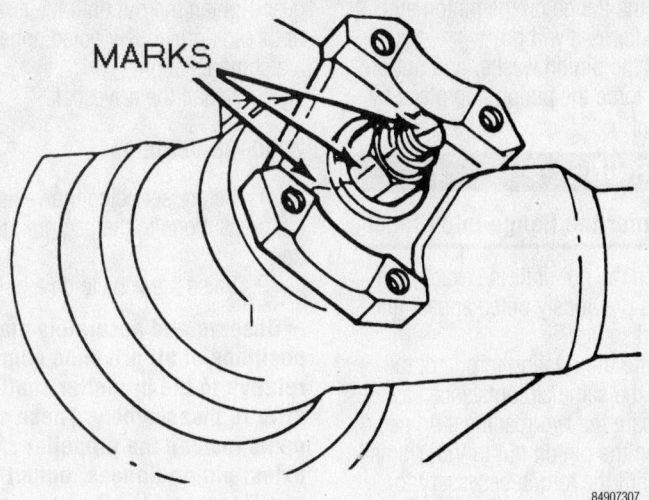

MARKS

Scribed marks

84907307

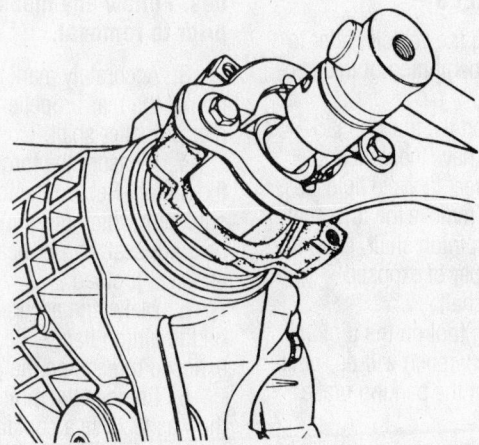

Removing the pinion nut

84907308

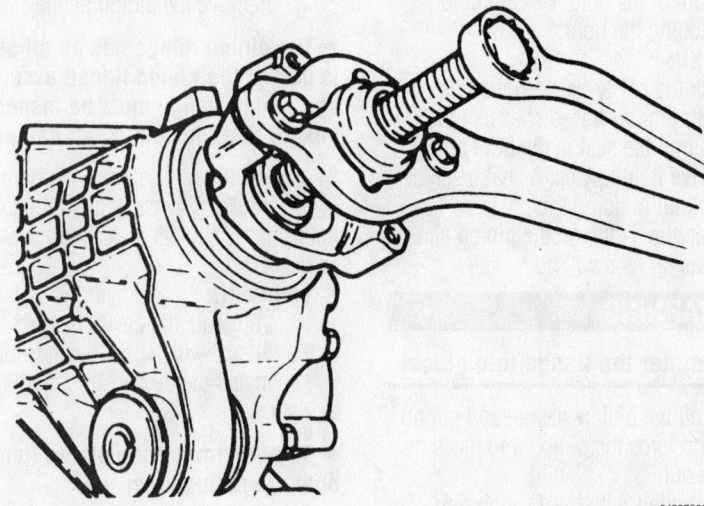

Removing the pinion flange

84907309

pinion. Compare this to the original torque. Tighten the pinion nut, in small increments, until the rotating torque is 3 inch lbs. (0.35 Nm) GREATER than the original torque.

19. Install the driveshaft.

20. Install the calipers and install the wheels.

2000–03 Vehicles

1. Before servicing the vehicle, refer to the precautions in the beginning of this section.

2. Raise the vehicle on a hoist.

3. Remove the propeller shaft from the axle.

4. Tie the propeller shaft to a frame rail or the crossmember.

5. Measure the torque required in order to rotate the pinion. Record the torque value for reassembly.

6. Scribe a line on the pinion stem, the pinion nut and the companion flange. Record the number of exposed threads on the pinion stem.

7. Remove the nut.

8. Position tool J8614-1 on the flange so that the 4 notches on the tool face the flange.

9. Remove the flange. Use the special nut and the forcing screw.

➡**Carefully pry the seal from the bore. Do not distort or scratch the aluminum case.**

10. Remove the oil seal.

11. Inspect the pinion flange for a smooth oil seal surface. Inspect the pinion flange for worn drive splines. Replace the pinion flange if necessary.

12. Remove the dust deflector.

To install:

➡**Stake the new deflector at 3 new equally spaced positions. You must stake the new deflector in such a way that you do not damage the seal operating surface.**

13. Install and stake the dust deflector on the flange.

14. Position the oil seal in the bore. Then place the a driver over the oil seal. Strike the driver with a hammer until the seal flange seats on the axle housing surface. Drive the seal in straight, not at an angle, as this will damage the aluminum housing.

➡**Do not hammer the pinion flange/yoke onto the pinion shaft. Pinion components may be damaged if the pinion flange/yoke is hammered onto the pinion shaft.**

For complete Engine Mechanical specifications, see Section 1 of this manual

15. Install the flange onto the pinion using tool J8614-01. Place the washer and a new nut on the pinion threads. Tighten the nut to the original scribed position using the scribe marks and the exposed threads as reference.

16. Measure the rotating torque of the pinion. Compare the measurement with the rotating torque recorded earlier. Tighten the pinion nut by small increments until the torque required in order to rotate the pinion is 0.35 Nm (3 inch lbs.) greater than the original torque.

17. Install the propeller shaft.

18. Lower the vehicle.

Rear Drive Axle Pinion Seal

REMOVAL & INSTALLATION

1999 Vehicles

SEMI-FLOATING AXLES

1. Before servicing the vehicle, refer to the precautions in the beginning of this section.

2. Raise and support the truck on jackstands. It would help to have the front end slightly higher than the rear to avoid fluid loss.

3. Matchmark and remove the driveshaft.

4. Release the parking brake.

5. Remove the rear wheels. Rotate the rear wheels by hand to make sure that there is absolutely no brake drag. If there is brake drag, remove the drums.

6. Using a torque wrench on the pinion nut, record the force needed to rotate the pinion.

7. Matchmark the pinion shaft, nut and flange. Count the number of exposed threads on the pinion shaft.

8. Install a holding tool on the pinion. A very large adjustable wrench will do, or, if one is not available, put the drums back on and set the parking brake as tightly as possible.

9. Remove the pinion nut.

10. Slide the flange off of the pinion. A puller may be necessary.

11. Centerpunch the oil seal to distort it and pry it out of the bore. Be careful to avoid scratching the bore.

To install:

12. Pack the cavity between the lips of the seal with lithium-based chassis lube.

13. Position the seal in the bore and carefully drive it into place. A seal installer is VERY helpful in doing this.

14. Pack the cavity between the end of the pinion splines and the pinion flange with Permatex No.2® sealer, or equivalent non-hardening sealer.

15. Place the flange on the pinion and push it on as far as it will go.

16. Install the pinion washer and nut on the shaft and force the pinion into place by turning the nut.

※ WARNING

Never hammer the flange into place!

17. Tighten the nut until the exact number of threads previously noted appear and the matchmarks align.

18. Measure the rotating torque of the pinion under the same circumstances as before. Compare the two readings. As necessary, tighten the pinion nut in VERY small increments until the torque necessary to rotate the pinion is 3 inch lbs. (0.35 Nm) higher than the originally recorded torque.

19. Install the driveshaft.

FULL-FLOATING AXLES

1. Before servicing the vehicle, refer to the precautions in the beginning of this section.

2. Raise and support the truck on jackstands. It would help to have the front end slightly higher than the rear to avoid fluid loss.

3. Matchmark and remove the driveshaft.

4. Matchmark the pinion shaft, nut and flange. Count the number of exposed threads on the pinion shaft.

5. Install a holding tool on the pinion. A very large adjustable wrench will do, or, if one is not available, set the parking brake as tightly as possible.

6. Remove the pinion nut.

7. Slide the flange off of the pinion. A puller may be necessary.

8. Centerpunch the oil seal to distort it and pry it out of the bore. Be careful to avoid scratching the bore.

To install:

9. Pack the cavity between the lips of the seal with lithium-based chassis lube.

10. Position the seal in the bore and carefully drive it into place. A seal installer is VERY helpful in doing this.

11. Place the flange on the pinion and push it on as far as it will go.

※ WARNING

Never hammer the flange into place!

12. Install the pinion washer and nut on the shaft and force the pinion into place by turning the nut.

13. On models with the 11 inch ring gear, tighten the nut to 440–500 ft. lbs. (596–678 Nm)

14. On models with the 10 ½ inch ring gear Tighten the nut until the exact number of threads previously noted appear and the matchmarks align.

15. Install the driveshaft.

2000–03 Vehicles

1. Before servicing the vehicle, refer to the precautions in the beginning of this section.

2. Raise the vehicle on a hoist.

➡Observe and accurately mark the positions of all driveline components relative to the propeller shaft and axles prior to disassembly. These components include the propeller shaft, drive axles, pinion flanges, output shafts, etc. Reassemble all components in the exact relationship the components had to each other during removal. Follow the specifications and the torque values. Follow any measurements made prior to removal.

3. Accurately mark the installed position of the rear propeller shaft. Remove the rear propeller shaft.

4. Measure the torque required to turn the pinion. Record the torque number measurement which gives the combined pinion bearing, seal, carrier bearing, axle bearing and seal preload.

5. Make and accurate alignment mark on the pinion flange. Record the number of exposed threads on the pinion stem.

6. Remove the pinion flange nut and the washer. Use a container in order to catch any lubricant.

➡Use care not to damage any of the machined surfaces.

7. Remove the pinion flange.

➡The pinion flange has an oil seal that is part of the pinion flange assembly. The pinion flange must be inspected to ensure that the seal is not damaged.

8. Pry the oil seal from the bore.

9. Thoroughly clean any foreign material from the contact area. Replace any parts as necessary.

To install:

10. Lubricate the cavity between the lips of the oil seal with wheel bearing lubricant.

11. Install the oil seal into the bore using a driver.

➡Do not hammer the pinion flange onto the pinion stem.

12. Install the pinion flange. Use the alignment marks in the installation of the pinion flange.

13. Install the washer and a new nut.

Tighten the nut on the pinion stem as close as possible to the alignment marks without going past the marks. Use the alignment marks and the thread count as a reference. Tighten the nut a little at a time. Turn the pinion flange several times after each tightening in order to seat the rollers.

➡**If the recorded preload torque value was less than 4 Nm (3 ft. lbs.), reset the torque specification to 3-5 Nm (4-7 ft. lbs.).**

14. Measure the torque required to rotate the pinion flange. Compare the measured torque with the recorded value. Continue tightening the pinion nut and measuring the torque until you achieve the recorded value.

15. Align the propeller shaft with the alignment marks. Connect the propeller shaft.

16. Install the retainers and the bolts. Tighten the bolts to 20 Nm (15 ft. lbs.).

17. Fill the rear axle.

18. Lower the vehicle.

Front Drive Axle Differential Carrier

REMOVAL & INSTALLATION

1999 Vehicles

1. Before servicing the vehicle, refer to the precautions in the beginning of this section.

2. Remove or disconnect the following:

- Wheels
- Skid plate
- Fluid

3. Matchmark and remove the front driveshaft.

4. Remove or disconnect the following:

- Right axle shaft at the tube flange
- Left axle shaft at the carrier flange

➡**Wire both axle shafts out of the way.**

- Connectors from the indicator switch and actuator
- Carrier vent hose
- Axle tube-to-frame bolts, washers and nuts
- Lower carrier mounting bolt
- Right side inner tie rod end at the relay rod

➡**Depending on the model, it may be necessary to remove the engine oil filter.**

5. Support the carrier on a floor jack

- Upper carrier mounting bolt

6. Lower the carrier assembly from the truck.

To install:

7. Raise the carrier into position.

8. Install or connect the following:

- Upper carrier mounting bolt, washers and nut. Then, install the lower carrier mounting bolt, washers and nut. Tighten the bolts to 80 ft. lbs. (110 Nm).
- Oil filter

- Tie rod end. Tighten the nut to 35 ft. lbs. (47 Nm).
- Axle tube-to-frame bolts, washers and nuts. Tighten the nuts to 75 ft. lbs. (100 Nm) for 15 and 25 series; 107 ft. lbs. (145 Nm) for 35 series.
- Vent hose
- Wiring
- Axle shafts at the flanges. Tighten the bolts to 59 ft. lbs. (80 Nm).
- Driveshaft. Tighten the bolts to 15 ft. lbs. (20 Nm).
- Gear oil
- Wheels

9. Add any engine oil lost when the filter was removed.

2000–03 Vehicles

1. Before servicing the vehicle, refer to the precautions in the beginning of this section.

2. Remove or disconnect the following:

- Front axle fluid
- Front propeller shaft
- Left and the right drive axle wheel drive shaft
- Axle tube nuts from the bracket
- Wiring at the axle
- Vent hose at the axle
- Carrier assembly lower mounting bolt and the nut
- Idler arm from the relay rod
- Pitman arm from the relay rod

3. Attach a transmission jack to the carrier assembly.

- Remove the upper carrier assembly mounting bolt and the nut.
- Remove the carrier assembly from the vehicle.

To install:

4. Install or connect the following:

- Carrier assembly in the vehicle
- Carrier assembly upper mounting bolt and the nut
- Lower carrier assembly mounting bolt and the nut. Tighten the bolts to 100 Nm (75 ft. lbs.).
- Pitman arm to the relay rod
- Idler arm to the relay rod
- Axle tube nuts to the bracket. Tighten the bolts to 100 Nm (75 ft. lbs.).
- Vent hose
- Wiring to the axle
- Left and the right drive axle wheel drive shaft
- Front propeller shaft to the pinion flange

5. Fill the front differential with lubricant.

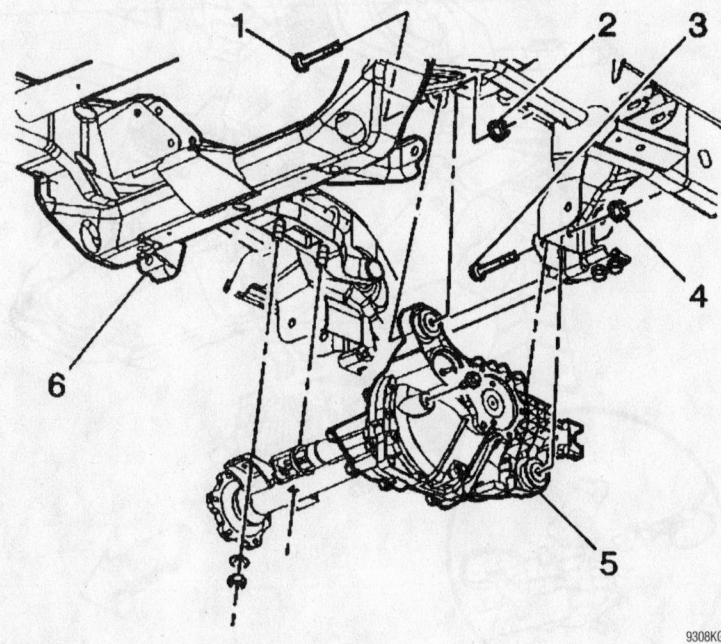

9308KG18

Front differential carrier removal—15 Series

Rear Drive Axle Housing

REMOVAL & INSTALLATION

1999 Vehicles

1. Before servicing the vehicle, refer to the precautions in the beginning of this section.
2. Drain the lubricant from the axle housing
3. Remove or disconnect the following:
   - Driveshaft
   - Wheel, the brake drum or hub and the drum assembly
   - Parking brake cable from the lever and at the brake flange plate
   - Hydraulic brake lines from the connectors
   - Shock absorbers from the axle brackets
   - Vent hose from the axle vent fitting, if equipped
   - Height sensing and brake proportional valve linkage, if equipped
   - Stabilizer shaft, if equipped
4. Support the axle assembly with a jack.
   - U-bolts
   - Spring plates and spacers
   - Axle assembly

To install:

5. Raise the axle assembly into position.

6. Install or connect the following:
   - U-bolts
   - Spring plates and spacers
   - Nuts and washers on the U-bolts. Torque the nuts to 81 ft. lbs. (110 Nm).
   - Stabilizer shaft, if equipped
   - Height sensing and brake proportional valve linkage, if equipped
   - Vent hose at the axle vent fitting, if equipped
   - Shock absorbers at the axle brackets
   - Hydraulic brake lines
   - Parking brake cable
   - Wheels
   - Driveshaft
7. Fill the axle housing.

2000–03 Vehicles

1. Before servicing the vehicle, refer to the precautions in the beginning of this section.
2. Remove or disconnect the following:

   - Axle lubricant
   - Propeller shaft
   - Wheel assemblies
   - Parking brake cable
   - Brake calipers

   - Shock absorbers from the axle brackets
   - Vent hose from the rear axle vent fitting
   - Nuts and the washers from the U-bolts
   - U-bolts, the spring plates and the spacers form the axle assembly
3. Lower the axle assembly.

To install:

4. Place the rear axle assembly under the vehicle. Align the rear axle assembly with the springs. Connect the spacers, the spring plates and the U-bolts to the rear axle. Raise the rear axle assembly into position.
5. Install or connect the following:
   - Washers and nuts to the U-bolts. Tighten the nuts to 80 Nm (59 ft. lbs.). first, then to 120 Nm (89 ft. lbs.).
   - Vent hose to the rear axle vent fitting
   - Shock absorbers to the rear axle
   - Brake calipers
   - Parking brake cable
   - Wheel assemblies
   - Propeller shaft
6. Fill the rear axle.
7. Bleed the brake system.
8. Remove the supports and lower the vehicle.

STEERING AND SUSPENSION

Air Bag

❊❊ CAUTION

Some vehicles are equipped with an air bag system. The system must be disabled before performing service on or around system components, steering column, instrument panel components, wiring and sensors. Failure to follow safety and disabling procedures could result in accidental air bag deployment, possible personal injury and unnecessary system repairs.

PRECAUTIONS

Several precautions must be observed when handling the inflator module to avoid accidental deployment and possible personal injury.
   - Never carry the inflator module by the wires or connector on the underside of the module
   - When carrying a live inflator module,

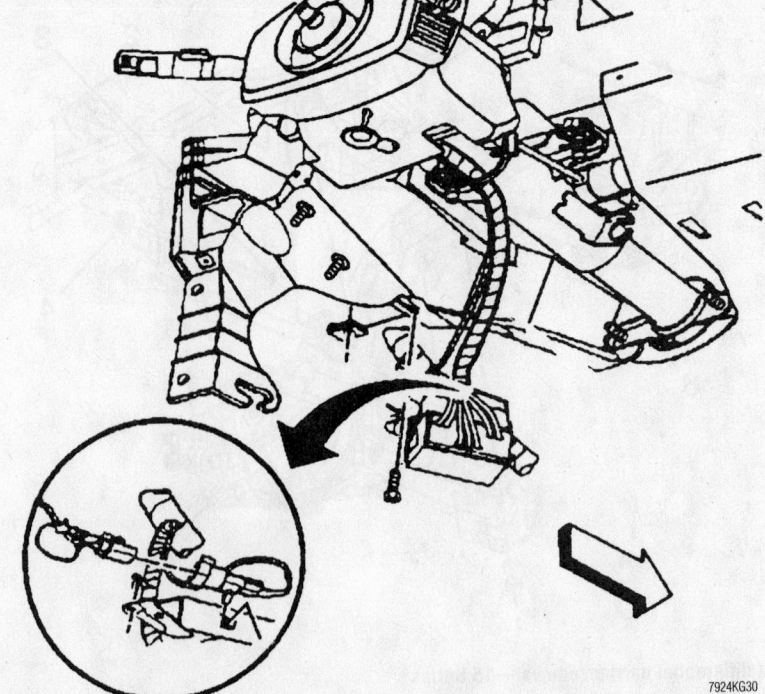

Typical air bag connector location—driver's side

7924KG30

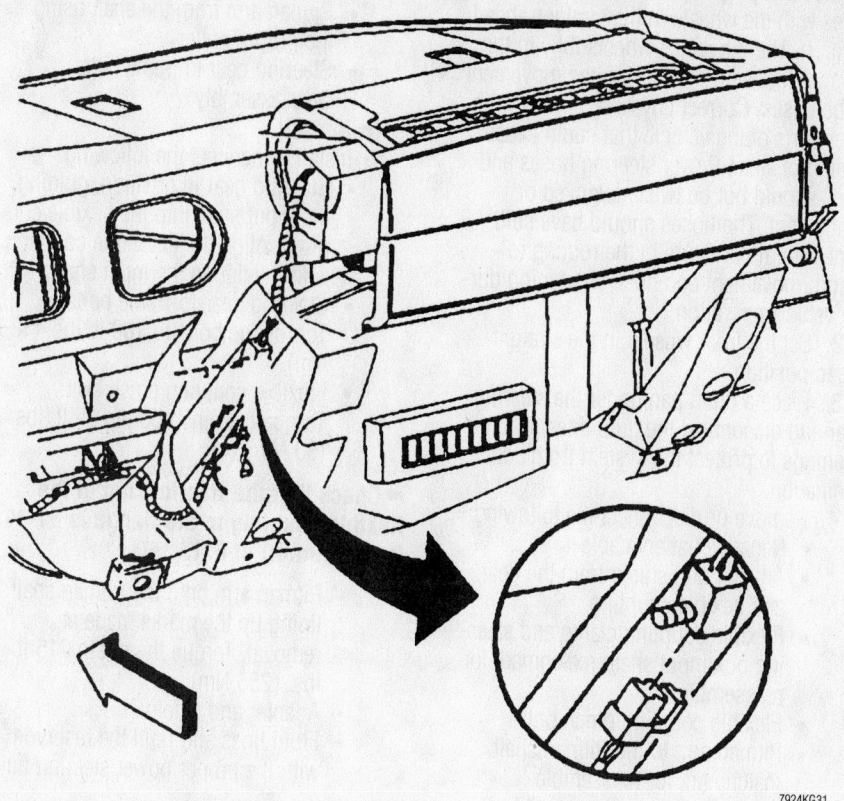

Typical air bag connector location—passenger's side

hold securely with both hands, and ensure that the bag and trim cover are pointed away

• Place the inflator module on a bench or other surface with the bag and trim cover facing up

• With the inflator module on the bench, never place anything on or close to the module that may be thrown in the event of an accidental deployment

DISARMING

1. Turn the front wheels to the straight-ahead position.

2. Turn the ignition switch to the **LOCK** position and remove the key.

➡ **If the key is in the RUN position when the Air Bag fuse is removed or open (blown), the Air Bag warning lamp in the dash will light up. This is normal operation, not a sign of a malfunction.**

3. Remove the Air Bag fuse from the fuse panel.

4. Remove the drivers side knee bolster and unplug the yellow 2-pin connector at the base of the steering column to disarm the driver's side Air Bag. Remove the pas- senger side knee bolster and unplug the yellow 2-pin connector to disable the pas- senger's side Air Bag.

5. Reverse the procedure to arm the Air Bag restraint system.

REMOVAL & INSTALLATION

4.3L, 5.0L, 5.7L, 6.5L and 7.4L Engines

1. Before servicing the vehicle, refer to the precautions in the beginning of this sec- tion.

2. Disconnect the hoses at the pump. When the hoses are disconnected, secure the ends in a raised position to prevent leakage. Cap the ends of the hoses to pre- vent the entrance of dirt.

3. Cap the pump fittings.

4. Loosen the belt tensioner.

5. Remove or disconnect the follow- ing:

• Pump drive belt
• Pulley with a puller such as J–29785–A
• Front mounting bolts, on the 4.3L, 5.0L and 5.7L engines
• Rear brace, on the 7.4L engine
• Front brace and rear mounting nuts, on the 6.5L diesel engine

6. Lift out the pump.

To install:

7. Observe the following torques:

• 4.3L, 5.0L and 5.7L engines: front mounting bolts to 37 ft. lbs. (50 Nm)
• 7.4L engine: rear brace nut: 61 ft. lbs. (82 Nm); rear brace bolt: 24 ft.

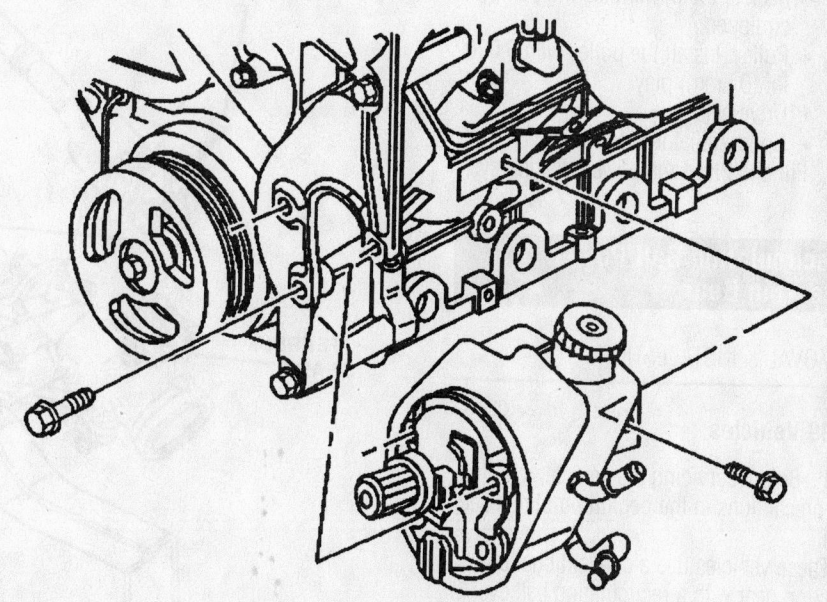

Power steering pump—4.8L, 5.3L and 6.0L engines shown

For Tire, Wheel and Ball Joint specifications, see Section 1 of this manual

lbs. (32 Nm); mounting bolts: 37 ft. lbs. (50 Nm)

- 6.5L diesel: front brace to 30 ft. lbs. (40 Nm); rear mounting nuts to 17 ft. lbs. (23 Nm).
- Pulley with J–25033–B
- Drive belt
- Hoses

8. Fill and bleed the system.

4.8L, 5.3L, 6.0L, and 8.1L Engines

1. Before servicing the vehicle, refer to the precautions in the beginning of this section.

2. Remove or disconnect the following:
- Upper radiator fan shroud, if necessary
- Drive belt
- Pulley
- Nut and clamp retaining the filler neck to the power steering pump, if equipped

3. Place a drain pan under the pump. Remove the hoses from the pump.
- Bolts from the rear of the pump
- Bolts from the front of the pump
- Pump from the vehicle

To install:

4. Install or connect the following:
- Power steering pump
- Bolts to the front and the rear of the pump. Tighten the bolts to 37 ft. lbs. (50 Nm).
- Hoses to the pump. Tighten the nut to 20 ft. lbs. (28 Nm).
- Nut and clamp retaining the filler neck to the power steering pump, if equipped
- Pulley. Install the pulley with 0.020 in. (0.5mm) play.
- Drive belt
- Upper radiator shroud

5. Fill and bleed the power steering system.

Recirculating Ball Power Steering Gear

REMOVAL & INSTALLATION

1999 Vehicles

1. Before servicing the vehicle, refer to the precautions in the beginning of this section.

These vehicles use a conventional power steering gear with a recirculating ball system. All tubes, hoses and fittings should be inspected for leakage at regular intervals. Fittings must be tight. Be sure the clips, clamps and supporting tubes and hoses are

in place and properly secured. Inspect the hoses with the wheels in the straight-ahead position. Then, turn the wheels fully to the left and right while observing the movement of the hoses. Correct any hose contact with other parts of the vehicle that could cause chafing or wear. Power steering hoses and pipes should not be twisted, kinked or tightly bent. The hoses should have sufficient natural curvature in the routing to absorb movement and hose shortening during vehicle operation.

2. Set the front wheels in the straight-ahead position.

3. Place a drain pan under the steering gear and disconnect the fluid lines. Cap the openings to protect the system from contamination.

4. Remove or disconnect the following:
- Negative battery cable
- Adapter and shield from the gear and flexible coupling
- Flexible coupling clamp and steering box input shaft, matchmark for reassemble
- Flexible coupling pinch bolt
- Pitman arm to the Pitman shaft, matchmark for reassemble

- Pitman shaft nut and lockwasher
- Pitman arm from the shaft using the proper puller
- Steering gear to frame bolts
- Gear assembly

To install:

5. Install or connect the following:
- Steering gear in position, guiding the input shaft into the flexible coupling. Align the flat in the coupling with the flat on the input shaft.
- Steering gear-to-frame bolts. Torque the bolts to 100 ft. lbs. (135 Nm).
- Flexible coupling pinch bolt. Torque the pinch bolt to 22 ft. lbs. (30 Nm).

➡**Check that the relationship of the flexible coupling to the flange is ¼–¾ in. (6–19mm) of flat.**

- Pitman arm onto the Pitman shaft, lining up the marks made at removal. Torque the nut to 215 ft. lbs. (285 Nm).
- Adapter and shield
- Fluid lines and refill the reservoir with the proper power steering fluid

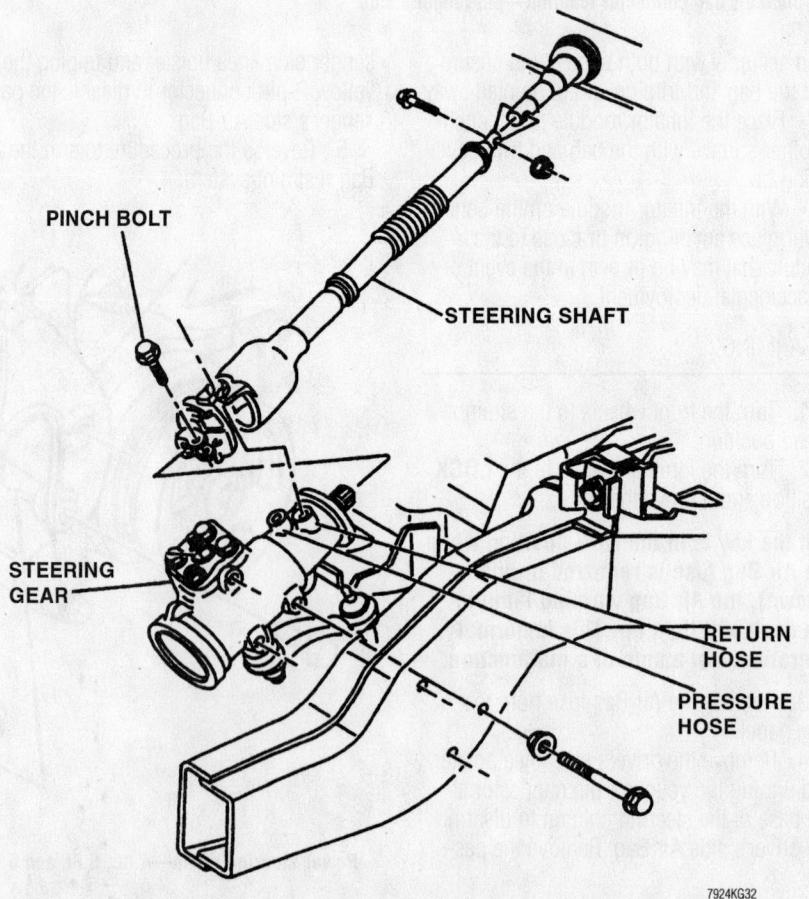

PINCH BOLT

STEERING SHAFT

STEERING GEAR

RETURN HOSE

PRESSURE HOSE

7924KG32

3 long bolts attach the power steering gear to the driver's side frame rail

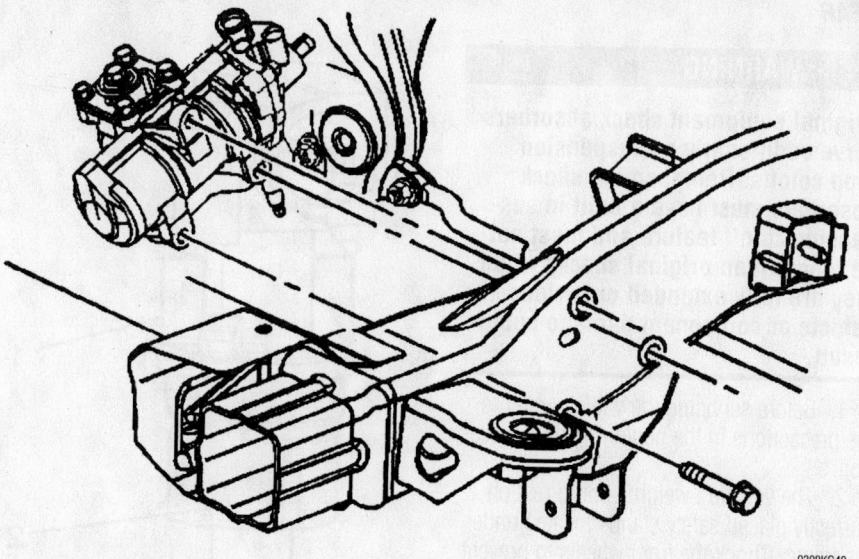

Recirculating ball gear

6. Properly bleed the system and verify no leaks.

7. Road test the vehicle for proper steering system operation.

2000–03 Vehicles

1. Before servicing the vehicle, refer to the precautions in the beginning of this section.

2. Raise the vehicle.

3. Remove the shield.

4. Place a drain pan below the steering gear.

5. Remove or disconnect the following:
   - Hoses from the steering gear
   - Intermediate shaft from the steering gear
   - Pitman arm from the relay rod
   - Steering gear frame bolts and the steering gear

To install:

6. Place the steering gear in position.

7. Install or connect the following:
   - Steering gear to the frame bolts. Tighten the bolts to 100 ft. lbs. (135 Nm).
   - Pitman arm
   - Intermediate shaft

8. Remove the plugs and the caps from the steering gear and the hoses. Connect the hoses to the steering gear. Tighten the hose connection to 20 ft. lbs. (28 Nm).

9. Install the shield.

10. Fill and bleed the system.

11. Lower the vehicle.

Rack & Pinion Steering Gear

REMOVAL & INSTALLATION

2000–03 Vehicles

1. Before servicing the vehicle, refer to the precautions in the beginning of this section.

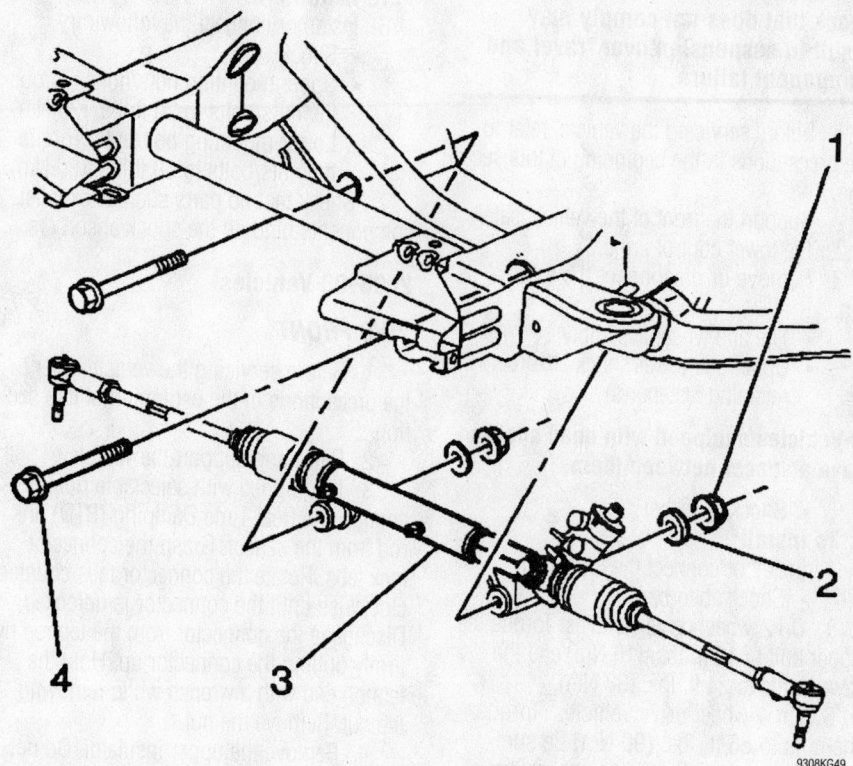

Rack and pinion gear

2. Remove or disconnect the following:
   - Wheel assemblies
   - Engine shield, if equipped
   - Stabilizer shaft
   - Power steering high pressure line from the rack and pinion assembly
   - Power steering low pressure line from the rack and pinion assembly
   - Coupler clamp bolt from the intermediate shaft
   - Intermediate shaft from the rack and pinion assembly
   - Rack and pinion assembly mounting nuts, the washers and the bolts
   - Rack and pinion assembly from the vehicle

To install:

3. Install or connect the following:
   - Rack and pinion assembly into the vehicle
   - Rack and pinion assembly mounting bolts, the washers and the nuts. Tighten the nuts to 185 Nm (136 ft. lbs.).
   - Intermediate shaft to the rack and pinion assembly. Install the coupler clamp bolt to the intermediate shaft. Tighten the coupler clamp bolt to 45 Nm (33 ft. lbs.).

For Wheel Alignment specifications, see Section 1 of this manual

- Power steering low pressure hose to the rack and pinion assembly
- Power steering high pressure hose to the rack and pinion assembly. Tighten the hoses to 27 Nm (28 ft. lbs.).
- Engine protection shield, if equipped
- Stabilizer shaft
- Wheels

4. Lower the vehicle.
5. Fill and bleed the power steering system.

Shock Absorber

REMOVAL & INSTALLATION

1999 Vehicles

FRONT

✹✹ WARNING

The front shock absorbers are multi-functional. They not only aid in a smooth ride, they serve as the suspension stop when the suspension is fully extended. When replacing front shocks, a shock of equivalent length and strength must be used. Use of a shock that does not comply may result in suspension over travel and component failure.

1. Before servicing the vehicle, refer to the precautions in the beginning of this section.
2. Support the front of the vehicle safely under the lower control arms.
3. Remove or disconnect the following:
   - Tire and wheel assembly
   - Upper and lower shock absorber retaining fastener(s)

➡Vehicles equipped with quad shocks have a spacer between them.

   - Shock absorber

To install:
4. Install or connect the following:
   - Shock absorber
5. On 2-wheel drive vehicles. Torque the upper bolt to 12 ft. lbs. (16 Nm) and the lower bolts to 24 ft. lbs. (33 Nm).
6. On 4-wheel drive vehicles. Torque the nuts to 66 ft. lbs. (90 Nm) Be sure the bolts are inserted in the proper direction. The upper bolt head should be forward; the bottom bolt head should be rearward.

REAR

✹✹ WARNING

Original equipment shock absorbers serve additionally as suspension drop cutoffs. Replacement shock absorbers must have a built in suspension cutoff feature and must not be longer than original shocks when they are fully extended or serious vehicle or component damage could result.

1. Before servicing the vehicle, refer to the precautions in the beginning of this section.
2. The vehicle's weight should rest on correctly placed safety stands located under the frame. Chock the front wheels to prevent vehicle movement.
3. Support the rear axle with a floor jack.
4. If the vehicle is equipped with air lift type shocks, bleed the air from the lines and disconnect the line from the shock absorber.
5. Remove or disconnect the following:
   - Shock absorber at the top by removing the 2 mounting bolts/nuts from the frame bracket
   - Nut, washers and bolt from the bottom mount
   - Shock

To install:
6. Install or connect the following:
   - Shock
   - Upper mounting nuts/bolts. Torque the nuts/bolts to 20 ft lbs. (25 Nm).
   - Lower mounting bolt/nuts. Torque the nuts/bolts to 60 ft lbs. (80 Nm).
7. Check that no parts such as exhaust components bind on the shock absorbers.

2000–03 Vehicles

2WD FRONT

1. Before servicing the vehicle, refer to the precautions in the beginning of this section.
2. Raise and support the vehicle.
3. If equipped with selectable ride, disconnect the Real Time Damping (RTD) link rod from the sensor. Grasp the connector lock tabs. Rotate the connector tabs counter clockwise until the connector is unlocked. Disengage the connector from the tennon by firmly pulling the connector up. Hold the tennon end with a wrench while removing the nut. Remove the nut.
4. Remove the upper insulator. Do not discard the plastic pilot ring.
5. Remove the shock absorber mounting bolts at the lower control arm. Remove

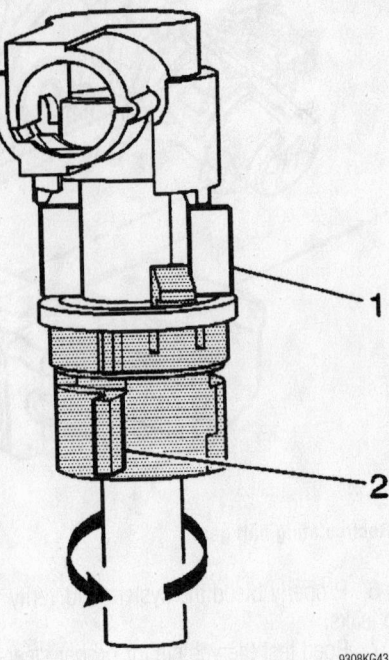

RTD connector

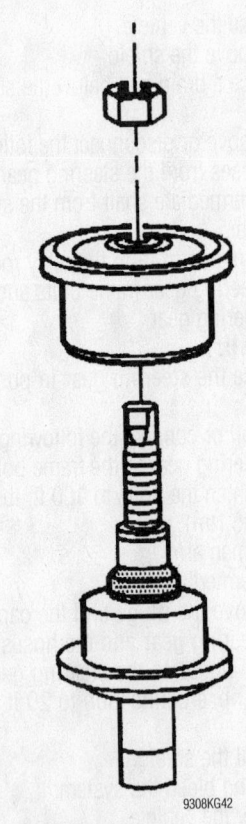

Upper shock insulator

the shock absorber through the lower control arm from below.

To install:

6. Support the lower control arm with a suitable jack in order to align the tennon with the mounting hole if equipped with selectable ride.

7. Install the shock absorber through the lower control arm from below. Insert the tennon through the mounting hole in the upper spring pocket. Align the shock absorber with the mounting holes in the lower control arm.

8. Install the shock absorber mounting bolts to the lower control arm. Tighten the bolts to 18 ft. lbs. (25 Nm).

➡**The upper insulators are substantially larger that the lower insulators. The upper insulator must be installed above the shock mounting bracket on the frame. The plastic pilot ring will assist the alignment of the isolators.**

9. Install the upper insulator to the shock absorber. Install the nut to the tennon end. Do not tighten the nut.

10. Connect the RTD link rod to the sensor (if equipped).

11. Remove the safety stands.

12. Lower the vehicle. Hold the tennon end with a wrench while torquing the nut. Tighten the nut to 20 Nm (15 ft. lbs.).

13. Connect the electrical connector using the following procedure:

   a. Verify that the connector is unlocked.

   b. Align the connector so that the tabs are perpendicular to the wrench flats on the tennon end.

   c. Engage the connector to the tennon by firmly pushing the connector down.

   d. Grasp the connector lock tabs. Rotate the connector counter clockwise.

14. The connector is locked into place when you hear an audible snap and the tabs are aligned.

4WD FRONT

1. Before servicing the vehicle, refer to the precautions in the beginning of this section.

2. Raise and support the vehicle.

3. Disconnect the (RTD) link rod from the sensor (if equipped).

4. Disconnect the electrical connector if equipped with selectable ride. Grasp the connector lock tabs. Rotate the connector tabs counter clockwise until the connector is unlocked. Disengage the connector from

the tennon by firmly pulling the connector up. Hold the tennon end with a wrench while removing the nut. Remove the nut.

5. Remove the upper insulator. Do not discard the plastic pilot ring.

6. Remove the shock absorber mounting bolt at the lower control arm (15 Series). The lower shock mounting bushing is serviceable by driving the bushing out with the appropriate tool.

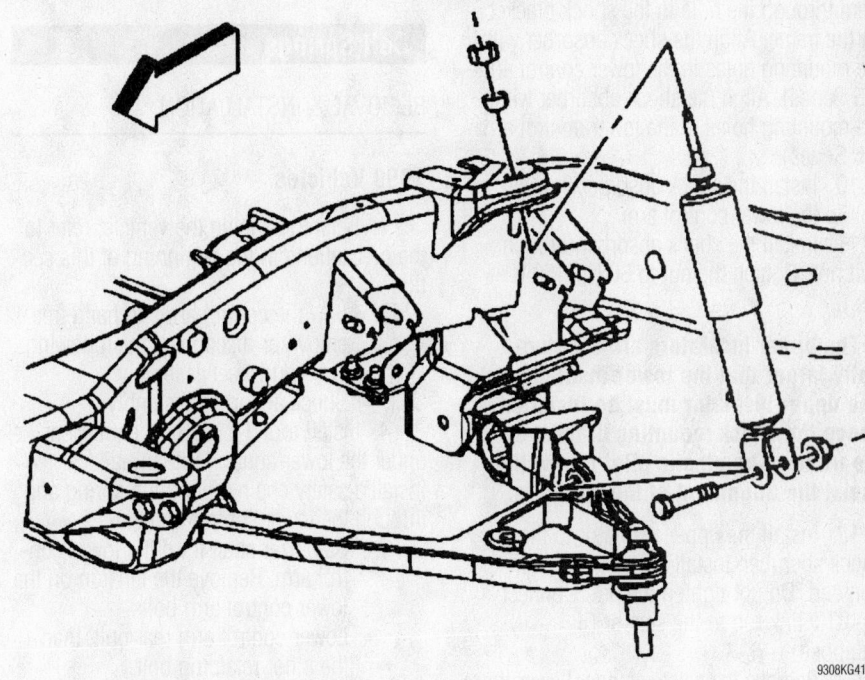

Shock absorber removal—4WD vehicles

9308KG41

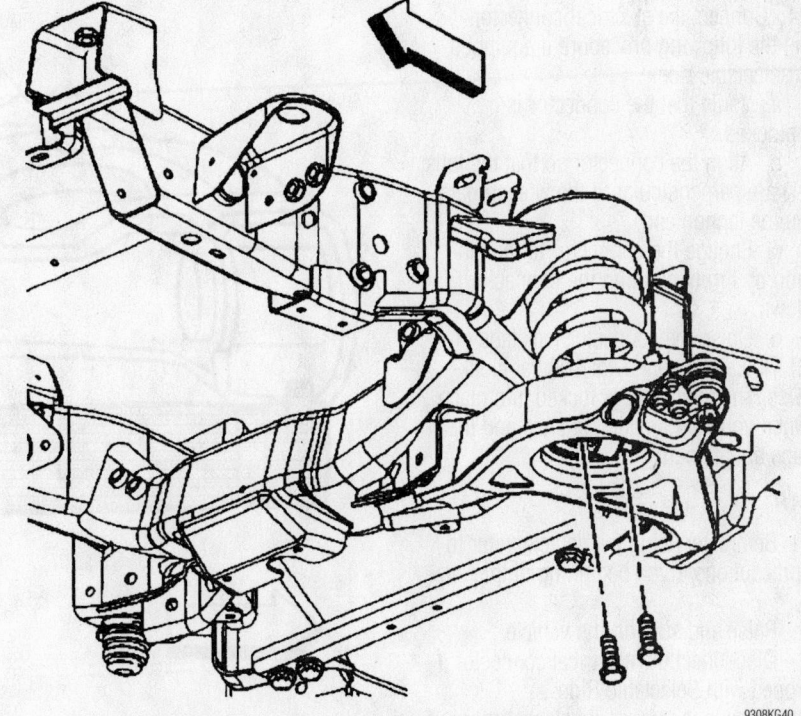

Shock absorber removal—2WD vehicles

9308KG40

7. Remove the shock absorber mounting bolt at the lower control arm (25 Series).

8. Remove the shock absorber.

To install:

9. Install the shock absorber. Insert the stem through the hole in the shock bracket on the frame. Align the shock absorber with the mounting holes in the lower control arm (15 Series). Align the shock absorber with the mounting holes in the lower control arm (25 Series).

10. Install the shock absorber through bolt to the lower control arm.

11. Install the shock absorber through bolt nut. Tighten the nut to 80 Nm (59 ft. lbs.).

➡**The upper insulators are substantially larger that the lower insulators. The upper insulator must be installed above the shock mounting bracket on the frame. The plastic pilot ring will assist the alignment of the isolators.**

12. Install the upper insulator to the shock absorber. Install the nut to the tennon end. Do not tighten the nut. Connect the RTD link rod to the sensor (if equipped).

13. Remove the safety stands. Lower the vehicle. Hold the tennon end with a wrench while torquing the nut. Tighten the nut to 20 Nm (15 ft. lbs.).

14. Connect the electrical connector using the following procedure if equipped with selectable ride.

   a. Verify that the connector is unlocked.

   b. Align the connector so that the tabs (1) are perpendicular to the wrench flats on the tennon end.

   c. Engage the connector to the tennon by firmly pushing the connector down.

   d. Grasp the connector lock tabs (1, 2). Rotate the connector counter clockwise. The connector is locked into place when you hear an audible snap and the tabs are aligned.

REAR

1. Before servicing the vehicle, refer to the precautions in the beginning of this section.

2. Raise and support the vehicle.

3. Disconnect the electrical connector if equipped with Selectable Ride.

4. Remove the upper shock absorber nut and the bolt.

5. Remove the lower shock absorber nut and the bolt.

6. Remove the shock absorber.

To install:

7. Installation is the reverse of removal. Tighten the nuts to 95 Nm (70 ft. lbs.).

8. Connect the electrical connector if equipped with Selectable Ride. Remove the safety stands. Lower the vehicle.

Coil Springs

REMOVAL & INSTALLATION

1999 Vehicles

1. Before servicing the vehicle, refer to the precautions in the beginning of this section.

2. Allow the control arms to hang free.

3. Remove or disconnect the following:
   • Tire and wheel assembly
   • Shock absorber assembly

4. Install tool J-23028, or equivalent, under the lower control arm and a jack. Install a safety chain around the spring and through the lower control arm.
   • Stabilizer shaft from the lower control arm. Remove the tension on the lower control arm bolts.
   • Lower control arm rear bolt, than the other retaining bolt
   • Spring assembly

To install:

5. Install or connect the following:

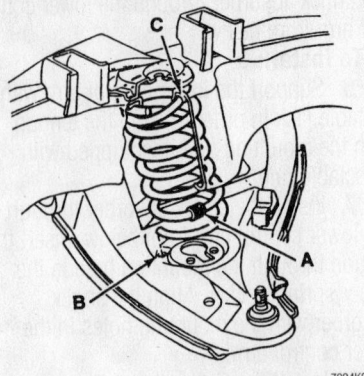

7924KG33

Position the coil spring so the bottom end of the spring covers only one drain hole— the other hole must remain open

   • Chain and spring. If you used spring compressors, install the spring and compressors.

6. Be sure the insulator is in place and the tape is towards the bottom of the spring. Position the gripper notch on the top coil in the frame bracket.

7. Be sure one drain hole in the lower arm is covered by the bottom coil and the other is open.

8. Slowly raise the lower control arm. Guide the control arm into place with a prybar.

9. Install or connect the following:
   • Pivot shaft bolts, front one first. The bolts must be installed with the

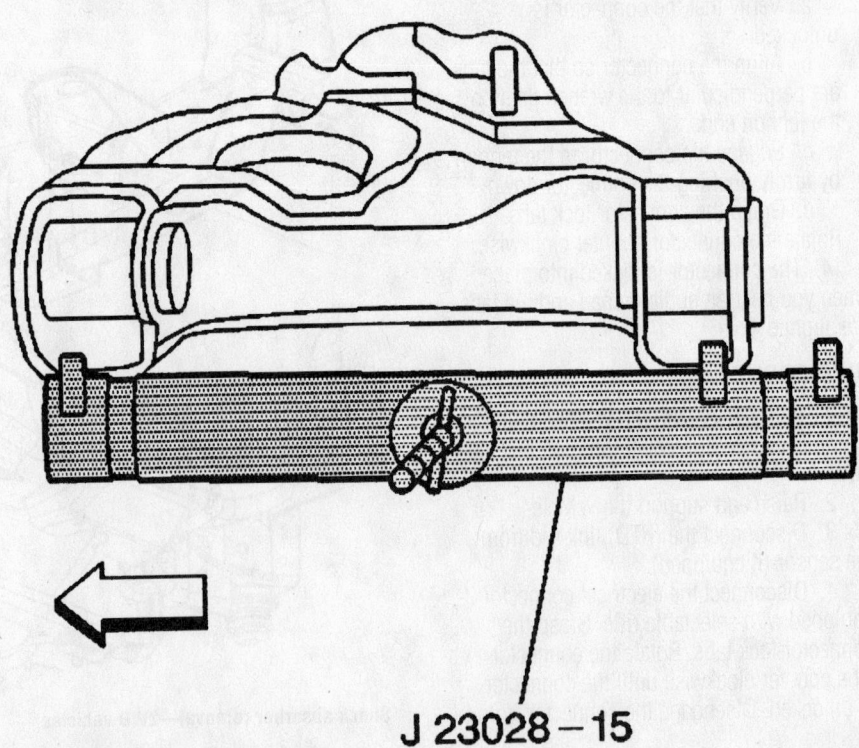

J 23028-15

Installing J23028-15 on the 25 Series

heads towards the front of the vehicle. Remove the safety chain or spring compressors.

➡**Do not tighten the bolts yet. The bolts must be tightened with the vehicle at its proper ride height.**

10. Remove the jack.
  • Stabilizer bar to the lower control arm. Torque the nuts to 24 ft lbs. (33 Nm).
  • Shock absorber
  • Wheel

11. Once the weight of the vehicle is on the wheels, bounce the vehicle 2 or 3 times by pushing down on the front bumper a couple of inches. When the vehicle settles, tighten the front nut first, then the rear nut to 101 ft. lbs. (137 Nm).

2000–03 Vehicles

FRONT

1. Before servicing the vehicle, refer to the precautions in the beginning of this section.

2. Raise and support the vehicle.

3. Remove or disconnect the following:
  • Engine protection shield
  • Frame cross bar (25 series only)
  • Tire and wheel assembly
  • Shock absorber
  • Front stabilizer shaft link

4. Install tool J23028-15 using the outboard locating tab (15 Series), or, the inboard locating tab (25 Series).

5. Attach the retaining hook to the control arm. Tighten the wing nut until you eliminate any free play.

6. Securely attach tool J23028-01 to a suitable transmission jack. Raise the jack until the yokes of tool J23028-01 line up with the notches in J23028-15.

7. Using the tools and the transmission jack, relieve the spring tension from the lower control arm pivot bolts.

8. Remove or disconnect the following:
  • Lower control arm pivot bolt nuts (15 Series)
  • Rear pivot bolt
  • Front pivot bolt
  • Lower control arm pivot bolt nuts (25 Series)
  • Rear pivot bolt
  • Front pivot bolt

9. Slowly lower the transmission jack in order to unload the front coil spring. It may be necessary to use a pry bar in order to guide the lower control arm out of position.
  • Coil spring and the insulator

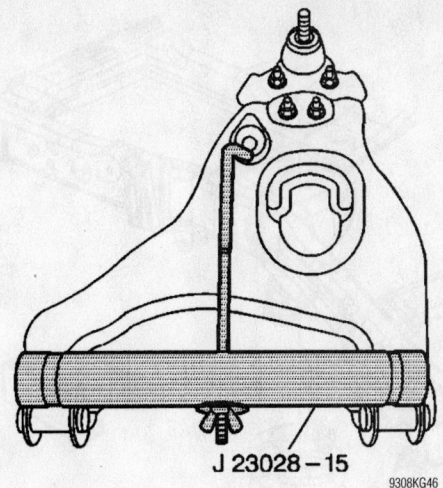

J 23028 – 15

9308KG46

Retaining hook installation

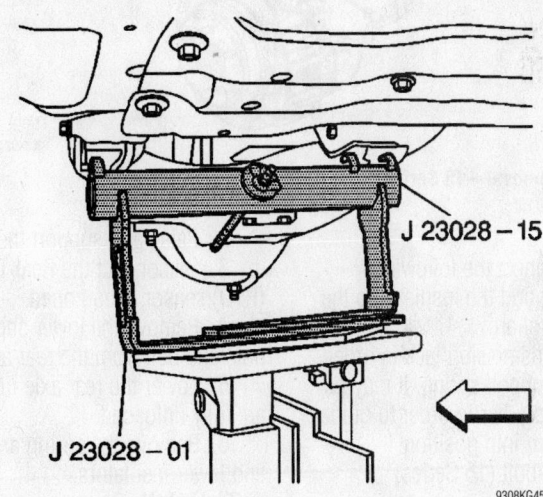

J 23028 – 15

J 23028 – 01

9308KG45

Tool attached to a jack

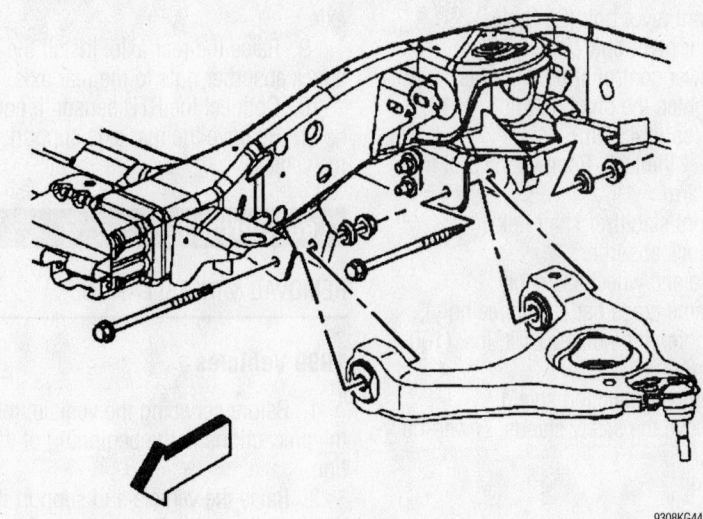

9308KG44

Lower control arm removal

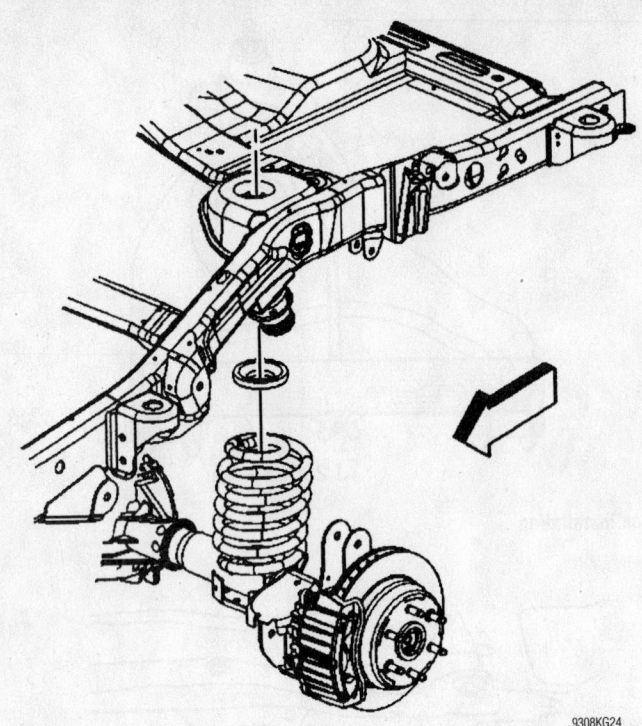

9308KG24

Rear coil spring removal—15 Series

To install:

10. Install or connect the following:
- Coil spring and the insulator to the lower control arm

11. Raise the transmission jack in order to compress the front coil spring. It may be necessary to use a pry bar in order to guide the lower control arm into position.
- Front pivot bolt (15 Series)
- Rear pivot bolt
- Lower control arm pivot nuts. Tighten the pivot bolt nuts to 107 ft. lbs. (145 Nm).
- Front pivot bolt (25 Series)
- Rear pivot bolt
- Lower control arm pivot nuts. Tighten the pivot bolt nuts to 107 ft. lbs. (145 Nm).

12. Lower the jack. Remove the tool from the control arm.
- Front stabilizer shaft link
- Shock absorber
- Tire and wheel assembly
- Frame cross bar (25 series only). Tighten the nuts to 74 ft. lbs. (100 Nm).
- Engine protection shield

13. Remove the safety stands. Lower the vehicle.

REAR

1. Before servicing the vehicle, refer to the precautions in the beginning of this section.

2. Raise and support the vehicle.

3. Disconnect the Real Time Damping (RTD) sensor, if equipped.

4. Remove the lower shock absorber nuts and bolt from the rear axle.

5. Lower the rear axle until the springs are fully unloaded.

6. Remove the spring and the upper and lower insulators.

To install:

7. Position the spring and the upper and lower insulators.

8. Install the rear spring to the rear axle.

9. Raise the rear axle. Install the lower shock absorber nuts to the rear axle.

10. Connect the RTD sensor, if equipped.

11. Remove the rear axle support. Lower the vehicle.

Leaf Springs

REMOVAL & INSTALLATION

1999 Vehicles

1. Before servicing the vehicle, refer to the precautions in the beginning of this section.

2. Raise the vehicle and support it so that there is no tension on the leaf spring assembly.

3. Remove or disconnect the following:
- U-bolt nuts, plates, and spacer(s)

- Anchor plate
- Spring-to-shackle retaining bolts, do not remove these bolts
- Bolts which attach the shackle to the rear bracket
- Bolt which attaches the spring to the front bracket
- Spring from the vehicle

4. Inspect the spring and replace any damaged components.

To install:

➡ **If the spring bushings are defective, use the following procedures for replacement. On bushings that are staked in place, the stakes must first be straightened. Using a press or vise, remove the bushing and install the new one. When a new, previously staked bushing is installed, stake it in 3 equally spaced locations.**

5. Place the spring assembly onto the axle housing. Position the front and rear of the spring at the brackets. Raise the axle with a floor jack as necessary to make the alignments.

6. Install or connect the following:
- Front and rear brackets bolts loosely
- Spacers and spring plate
- New u-bolts, washers and nuts
- Anchor plate
- U-bolt nuts. Torque them in a diagonal sequence, to 17 ft. lbs. (23 Nm) When the spring is evenly seated, tighten the nuts to 81 ft. lbs. (110 Nm).
- Hanger and shackle bolts are properly installed. All bolt heads should be inboard. Don't tighten them yet.

7. Using the floor jack, raise the axle until the distance between the bottom of the rebound bumper and its contact point on the axle is 182mm plus or minus 6mm.

8. When the spring is properly positioned, tighten all the hanger and shackle nuts to 70 ft. lbs. (95 Nm)
- Leaf spring-to-shackle nuts 15/25/35 series: 70 ft. lbs. (95 Nm)
- Leaf spring-to-shackle nuts C3HD series: 157 ft. lbs. (213 Nm)
- Shackle-to-bracket nuts C3HD series: 157 ft. lbs. (213 Nm)

2000–03 Vehicles

1. Before servicing the vehicle, refer to the precautions in the beginning of this section.

2. Raise and support the vehicle.

3. Support the rear axle independently in order to relieve the tension on the leaf springs.

4. Remove or disconnect the following:
- Real Time Damping (RTD) sensors, if equipped
- Trailer hitch if equipped
- Fuel tank for left side applications

- U-bolt nuts and U-bolts
- Spring spacer and anchor plate
- Shackle to the frame bracket nut and the bolt
- Front spring bracket bolt

- Leaf spring assembly from the vehicle
- Shackle from the spring

To install:

5. Loosely assemble the spring shackle bracket to the frame. Install the shackle bolt. Install the shackle nut.

6. Install the leaf spring assembly to the vehicle.

7. Loosely assemble the spring to the front hanger bracket.

8. Install or connect the following:
- Front spring hanger bracket bolt
- Front spring hanger bracket nut
- Shackle to the spring bolt
- Shackle to the spring nut

➡**Do not reuse the U-bolts.**

- Spring spacer
- U-bolts
- Anchor plate
- U-bolt nuts

9. Observe the following torques:
- 14mm U-bolt nuts to 80 Nm (59 ft. lbs.)
- 16mm U-bolt nuts to 120 Nm (89 ft. lbs.)
- Front hanger bracket nut to 125 Nm (92 ft. lbs.)
- Shackle to the frame nut to 95 Nm (70 ft. lbs.)
- Shackle to the spring nut to 95 Nm (70 ft. lbs.)

10. Install or connect the following:
- Fuel tank for left side applications
- Trailer hitch if equipped
- RTD sensors (25 series utilities if equipped)

11. Remove the rear axle support.

12. Remove the safety stands. Lower the vehicle.

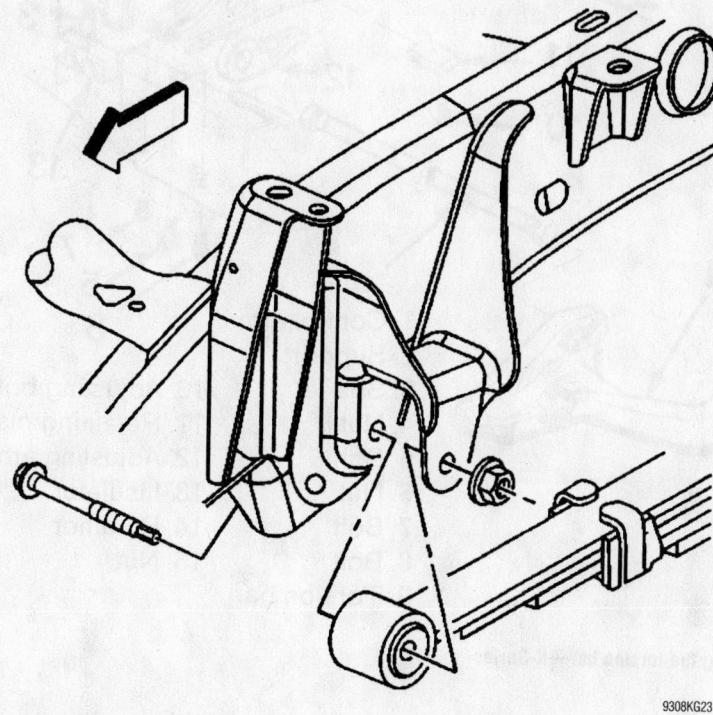

Rear leaf spring front shackle

9308KG23

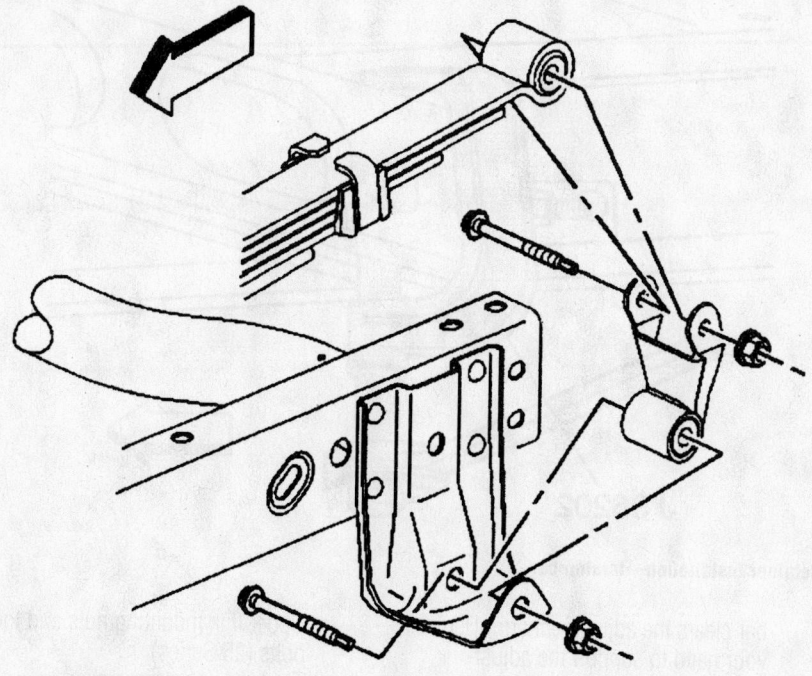

Rear leaf spring rear shackle

9308KG22

Torsion Bars

REMOVAL & INSTALLATION

1999 Vehicles

1. Before servicing the vehicle, refer to the precautions in the beginning of this section.

➡**Special tool J–36202, or its equivalent, is necessary for this procedure.**

2. Remove or disconnect the following:
- Wheels

3. Support the lower control arm with a floor jack.

4. Matchmark both torsion bar adjustment bolt positions.

5. Using tool J–36202, increase the tension on the adjusting arm.

6. Remove or disconnect the following:
- Adjustment bolt and retaining plate

7. Move the tool aside, and slide the torsion bars forward.
- Adjusting arms
- Torsion bar support crossmember and slide the support crossmember rearwards
- Torsion bars, matchmark the position. They are not interchangeable
- Support crossmember
- Retainer, spacer and bushing from the support crossmember

To install:

8. Install or connect the following:
- Retainer, spacer and bushing to the support crossmember
- Support assembly on the frame, out of the way
- Torsion bars, sliding them forward until they are supported. Align the marks made when removed.
- Support crossmember into position. Torque the center nut to 18 ft. lbs. (24 Nm), the edge nuts to 46 ft. lbs. (62 Nm).
- Adjuster retaining plate and bolt on each torsion bar

9. Using tool J–36202, increase tension on both torsion bars.

10. Set the adjustment bolt to the marked position.

11. Release the tension on the torsion bar until the load is taken up by the adjustment bolt.

12. Install both wheels

2000–03 Vehicles

1. Before servicing the vehicle, refer to the precautions in the beginning of this section.

➡**This procedure requires the removal of both torsion bars.**

2. Raise and support the vehicle.

3. Mark the adjustment bolt setting. Install tool J36202 to the adjustment arm and the crossmember.

4. Increase the tension on the adjustment arm until the load is removed from the adjustment bolt and the adjuster nut.

5. Remove or disconnect the following:
- Adjustment bolt and the adjuster nut
- Tool, allowing the torsion bar to unload
- Adjustment arm by sliding the torsion bar forward until the torsion

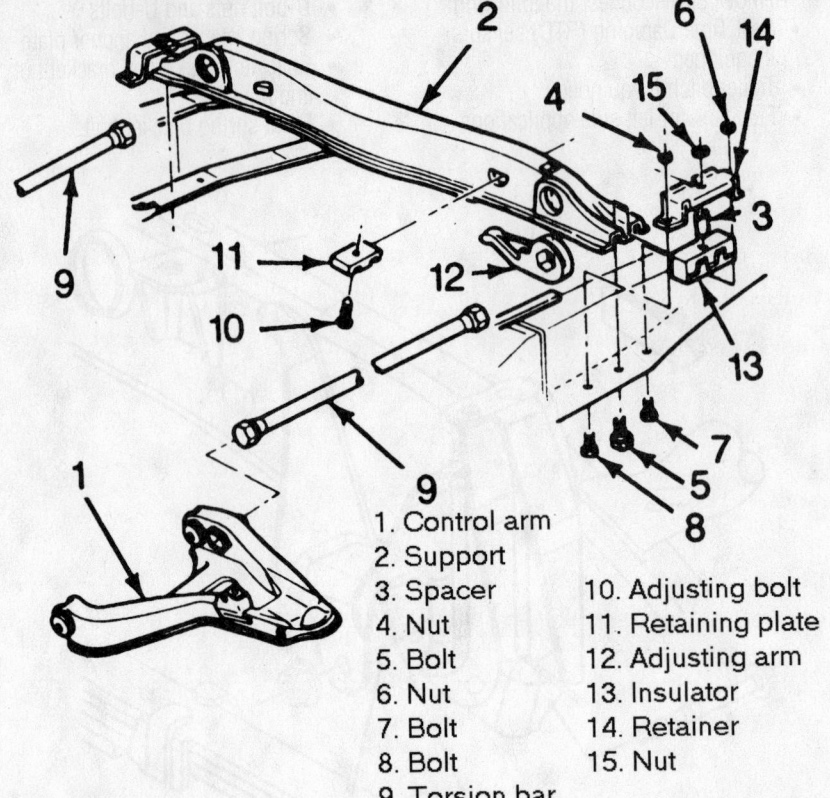

1. Control arm
2. Support
3. Spacer
4. Nut
5. Bolt
6. Nut
7. Bolt
8. Bolt
9. Torsion bar
10. Adjusting bolt
11. Retaining plate
12. Adjusting arm
13. Insulator
14. Retainer
15. Nut

84908058

Installing the torsion bar—K-Series

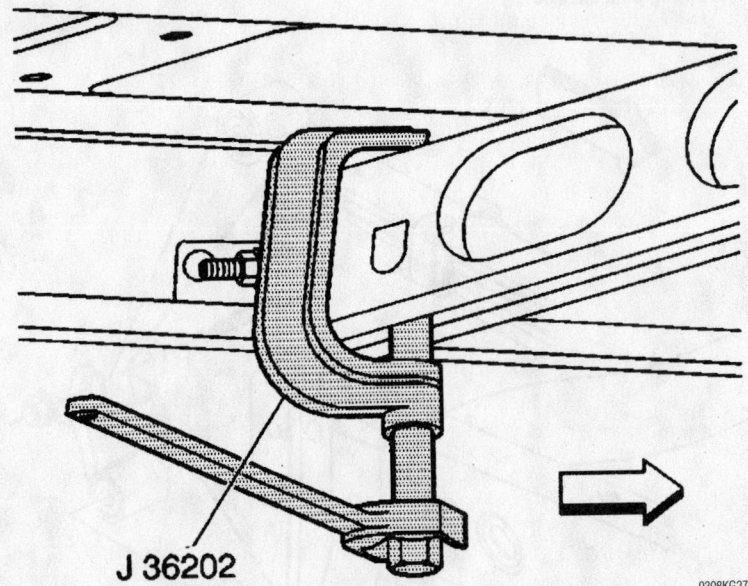

J 36202

9308KG27

Retainer installation—torsion bar

bar clears the adjustment arm. Use your hand to support the adjustment arm as the adjustment arm releases from the torsion bar.
- Torsion bar crossmember bolts from the weld nuts (15 Series)

- Upper link mounting nuts and the bolts (25 Series)
- Torsion bar crossmember

➡**Note the position of the torsion bars as the left and right bars are different.**

- Torsion bars

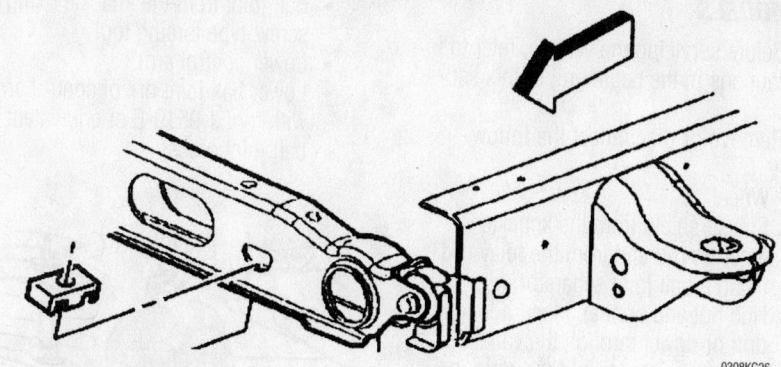

Adjuster nut removal—15 Series

9308KG26

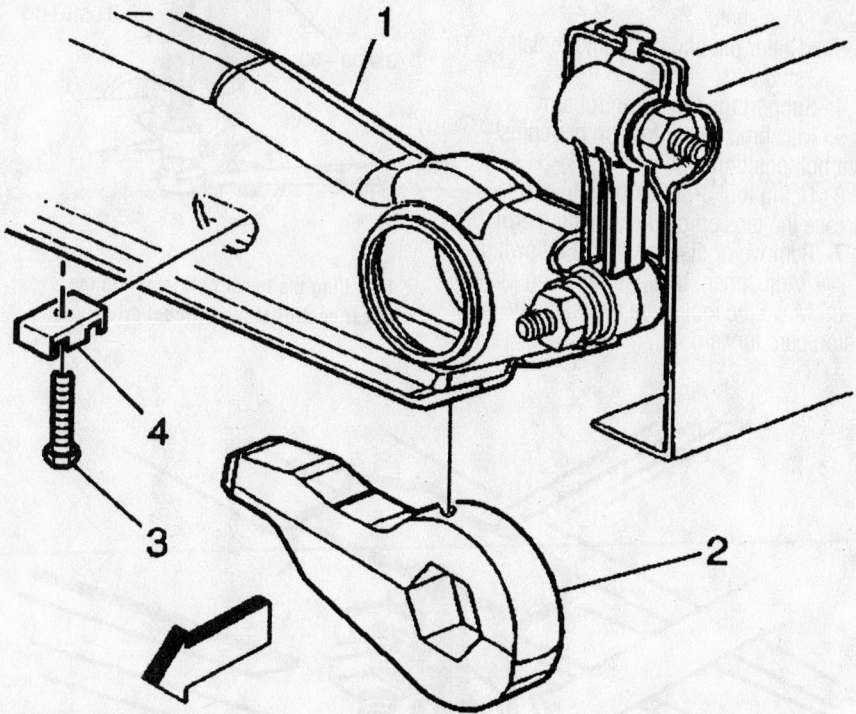

Adjuster bolt removal—15 Series

9308KG25

To install:

6. Install or connect the following:
- Torsion bars
- Torsion bar crossmember
- Torsion bar crossmember bolts to the weld nuts (15 Series). Tighten the bolt to 95 Nm (70 ft. lbs.)
- Upper link mounting nuts and the bolts (25 Series). Tighten the nut to 95 Nm (70 ft. lbs.)

7. While supporting the adjustment arm, slide the torsion bar rearward until the torsion bar fully engages the adjustment arm. Install tool J36202 to the adjustment arm and the crossmember. Increase the tension on the adjustment arm in order to load the torsion bar.
- Adjustment bolt and the adjuster nut

8. Remove the tool, releasing the tension on the torsion bar until the load is taken up by the adjustment bolt.

9. Remove the safety stands.

10. Lower the vehicle.

11. Measure the ride height.

12. Turn the adjustment bolt clockwise to increase the ride height and counterclockwise to decrease it.

Upper Ball Joint

REMOVAL & INSTALLATION

1999 Vehicles

1. Before servicing the vehicle, refer to the precautions in the beginning of this section.

2. Remove or disconnect the following:
- Wheel
- Brake hose bracket from the control arm

3. Using a ⅛ in. drill bit, drill a pilot hole through each ball joint rivet.

4. Drill out the rivets with a ½ in. drill bit. Punch out any remaining rivet material.
- Cotter pin and nut from the ball stud

5. Support the lower control arm.
- Stud from the knuckle using a ball joint separator

To install:

6. Install or connect the following:
- New ball joint on the control arm

➡ **Service replacement ball joints come with nuts and bolts to replace the rivets.**

- Bolts and nuts. Torque the nuts to 17 ft. lbs. (23 Nm) for 15- and 25-Series, 52 ft. lbs. (70 Nm) for 35-Series.

➡ **The bolts are inserted from the bottom.**

7. Start the ball stud into the knuckle. Ensure it is squarely seated. Install the ball stud nut and pull the ball stud into the knuckle with the nut. Tighten the nut after the vehicle wheels are on the ground and the suspension is loaded.
- Wheel

8. Once the weight of the vehicle is on the wheels tighten the nut to 84 ft. lbs. (115 Nm).

2000–03 Vehicles

1. Before servicing the vehicle, refer to the precautions in the beginning of this section.

2. Raise and support the vehicle.

3. Remove or disconnect the following:
- Tire and wheel assembly
- Upper control arm
- Upper ball joint, using a press

To install:

➡ **The ball joint must be installed with**

the flat edges or notches in the same position as the replaced ball joint. The ball joint is directional and damage will occur if this procedure is not followed.

4. Install or connect the following:
- Upper ball joint, using a press
- Upper control arm
- Tire and wheel assembly

5. Remove the safety stands.
6. Lower the vehicle.
7. Verify the wheel alignment.

Lower Ball Joint

REMOVAL & INSTALLATION

1999 Vehicles

2WD MODELS

1. Before servicing the vehicle, refer to the precautions in the beginning of this section.
2. Place jack under lower control arm, then raise the jack slightly.
3. Remove or disconnect the following:
- Tire and wheel assembly
- Brake caliper and position it to the side
- Cotter pin and the lower ball joint retaining nut. Using the proper tool separate the ball joint from its mounting. Support the knuckle assembly so its weight will not damage the brake hose.
- Ball joint out of the lower control arm, using tool J-9519-30-D or equivalent.

To install:
4. Start the new ball joint into the control arm. Position the bleed vent in the rubber boot facing inward.
5. Install or connect the following:
- Ball joint into the control arm until fully seated
- Lower ball joint stud into the steering knuckle
- Brake caliper, if removed
- Ball stud nut. Torque the nut to 90 ft. lbs. (122 Nm) plus the additional tighten necessary to align the cotter pin hole. Do not exceed 130 ft. lbs. (175 Nm) or back the nut off to align the holes with the pin.
- New lube fitting and lubricate the new joint
- Tire and wheel

4WD MODELS

1. Before servicing the vehicle, refer to the precautions in the beginning of this section.
2. Remove or disconnect the following:
- Wheel
- Splash shield from the knuckle
- Inner tie rod end from the relay rod using a ball joint separator
- Hub nut and washer. Insert a long drift or dowel through the vanes in the brake rotor to hold the rotor in place.
- Axle shaft inner flange bolts

3. Using a puller, force the outer end of the axle shaft out of the hub.
- Axle shaft
- Cotter pin and nut from the ball stud

4. Support the lower control arm.
5. Matchmark both torsion bar adjustment bolt positions.
6. Using tool J-36202 or equivalent, increase the tension on the adjusting arm.
7. Remove or disconnect the following:
- Adjustment bolt and retaining plate

8. Move the tool aside and slide the torsion bars forward.

- Ball joint from the knuckle using a screw-type forcing tool
- Lower control arm
- Lower ball joint out of control arm with tool J-9519-E or equivalent ball joint press

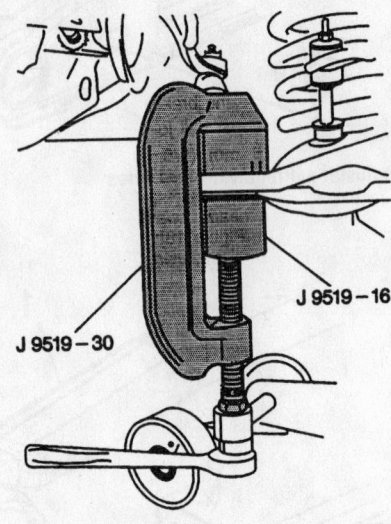

Installing the lower ball joint into the lower control arm—2-wheel drive

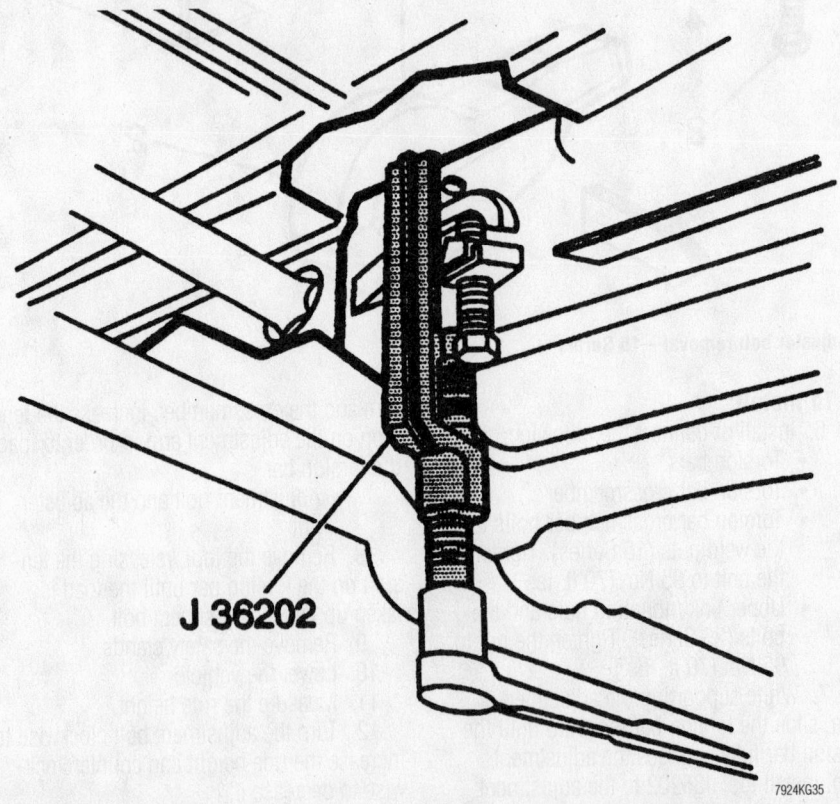

A special tool is available for removing or installing the torsion bar adjusting bolt—4-wheel drive

To install:

9. Install or connect the following:
 - New ball joint into the control arm with tool J-9519-E or equivalent
 - Lower control arm

10. Using tool J-36202 or equivalent, increase tension on both torsion bars.
 - Adjustment retainer plate and bolt on both torsion bars

11. Set the adjustment bolt to the marked position.

12. Release the tension on the torsion bar until the load is take up by the adjustment bolt and remove the tool.
 - Shaft in the hub and the washer and hub nut. Leave the drift in the rotor vanes and torque the hub nut to 175 ft. lbs. (238 Nm).
 - Flange bolts. Tighten them to 59 ft. lbs. (80 Nm), remove the drift.
 - Inner tie rod end at the steering relay rod. Torque the nut to 35 ft. lbs. (48 Nm).
 - Splash shield

 - Wheel

13. Once the weight of the vehicle is on the wheels follow these steps:
 a. Step 1: Lift the front bumper about 1 ½ in. (38mm) and let it drop.
 b. Step 2: Repeat this procedure 2–3 more times.
 c. Step 3: Draw a line on the side of the lower control arm from the centerline of the control arm pivot shaft, dead level to the outer end of the control arm.
 d. Step 4: Measure the distance

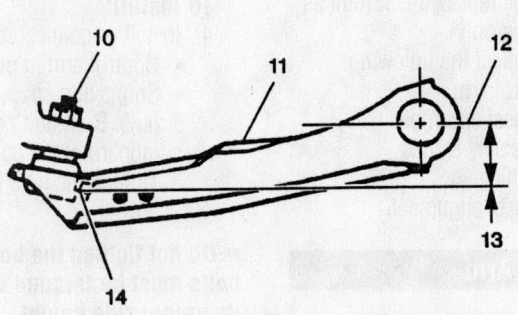

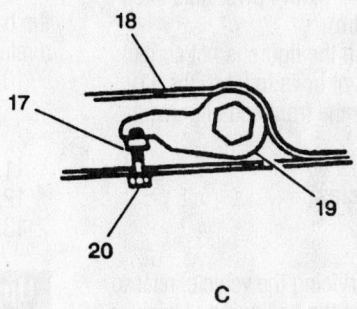

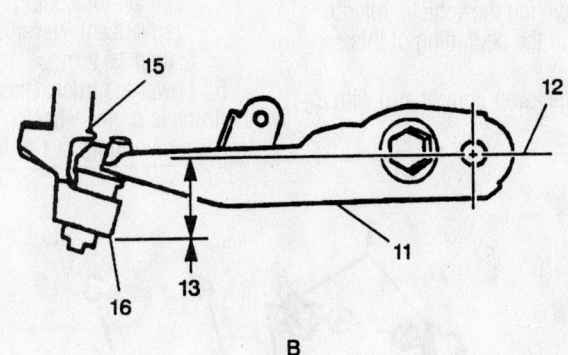

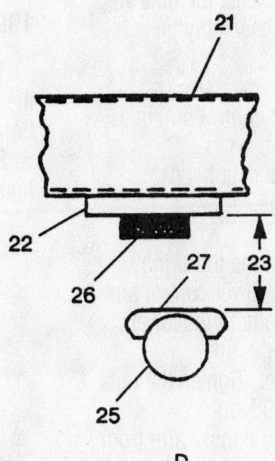

A. "C" MODEL	15. STEERING KNUCKLE
B. "K" MODEL	16. STEERING KNUCKLE LOWER CORNER
C. "K" MODEL TORSION BAR ADJUSTER	17. NUT
D. "CK" MODEL REAR SUSPENSION	18. TORSION BAR SUPPORT ASM.
10. LOWER BALL JOINT	19. TORSION BAR ADJUSTMENT ARM
11. LOWER CONTROL ARM	20. BOLT – ONE TURN EQUALS 6mm HEIGHT CHANGE
12. PIVOT BOLT CENTER LINE	21. FRAME
13. "Z" HEIGHT	22. BOTTOM SURFACE OF JOUNCE BRACKET
C 1,2,3 95.0 ± 6.0mm	23. "D" HEIGHT
K 1,2 157.0 ± 6.0mm	25. REAR AXLE
K 3 145.0 ± 6.0mm	26. JOUNCE BUMPER
14. LOWER BALL JOINT EXTRUSION	27. AXLE JOUNCE PAD

7924KG36

Use these specifications and diagrams to determine if the vehicle ride height is correct

Timing belt service is covered in Section 3 of this manual

between the lowest corner of the steering knuckle and the line on the control arm, record the figure.

e. Step 5: Push down about 1½ in. (38mm) on the front bumper and let it return. Repeat the procedure 2–3 more times.

f. Step 6: Re-measure the distance at the control arm.

g. Step 7: Determine the average of the 2 measurements. This is the "Z" height measurement. The "Z" height should be as specified in the chart.

h. Step 8: If the figure is correct, tighten the control arm pivot nuts to 94 ft. lbs. (128 Nm).

i. Step 9: If the figure is not correct, tighten the pivot bolts to 94 ft. lbs. (128 Nm) and have the front end alignment corrected.

2000–03 Vehicles

2WD MODELS

1. Before servicing the vehicle, refer to the precautions in the beginning of this section.
2. Raise and support the vehicle.
3. Remove or disconnect the following:
   • Tire and wheel assembly
   • Front coil spring
   • Lower control arm
4. Secure the lower control arm in a bench vice or equivalent.
5. Center punch the rivet heads.
6. Drill out the rivets.

To install:

7. Install or connect the following:
   • Ball joint to the lower control arm
   • Replacement bolts to the lower control arm
   • Nuts to the bolts. Tighten the nuts to 52 ft. lbs. (70 Nm).
8. Remove the lower control arm from the bench vice.
   • Lower control arm
   • Coil spring
   • Tire and wheel tire assembly
9. Remove the safety stands.
10. Lower the vehicle.
11. Verify the wheel alignment.

4WD MODELS

1. Before servicing the vehicle, refer to the precautions in the beginning of this section.
2. Raise and support the vehicle.
3. Remove or disconnect the following:
   • Tire and wheel assembly
   • Lower control arm
4. Place the lower control arm in a bench vice.

5. Using a chisel, remove the 4 securing crimps from the ball joint body (15 series only).
6. Using a press, remove the ball joint from the lower control arm.

To install:

➡ **Use the outer flange of the ball joint in order to press the ball joint into place.**

7. Install the new ball joint using a press.
8. Place the lower control arm in a bench vice.
9. Using a punch, install 4 crimps to the ball joint. Use the replaced ball joint as a reference (15 series only).
10. Install or connect the following:
    • Lower control arm
    • Tire and wheel assembly
11. Remove the safety stands.
12. Lower the vehicle.
13. Verify the wheel alignment.

Upper Control Arm

REMOVAL & INSTALLATION

1999 Vehicles

1. Before servicing the vehicle, refer to the precautions in the beginning of this section.
2. Support the lower control arm with a floor jack.

3. Remove or disconnect the following:
   • Wheel
   • Brake hose bracket from the control arm
   • Air cleaner extension (if necessary)
   • Brake hose bracket retainer and wire the hose aside
   • Cotter pin from the upper control arm ball stud and loosen the stud nut until the bottom surface of the nut is slightly below the end of the stud
   • Ball joint from the knuckle
   • Control arm to the frame brackets
   • Shims and spacers

To install:

4. Install or connect the following:
   • Control arm in position
   • Shims and spacers, bolts and new nuts. Both bolt heads **must** be inboard of the control arm brackets. Tighten the nuts finger-tight for now.

➡ **Do not tighten the bolts yet. The bolts must be torqued with the truck at its proper ride height.**

   • Ball joint to the knuckle. Torque the nut to 84 ft. lbs. (115 Nm).
   • Cotter pin. Never back off the nut to install the cotter pin. Always advance it. Never advance it more than ⅙ turn.
5. Lower the truck. Once the weight of the truck is on the wheels. Torque the control arm pivot nuts to 140 ft. lbs. (190 Nm)

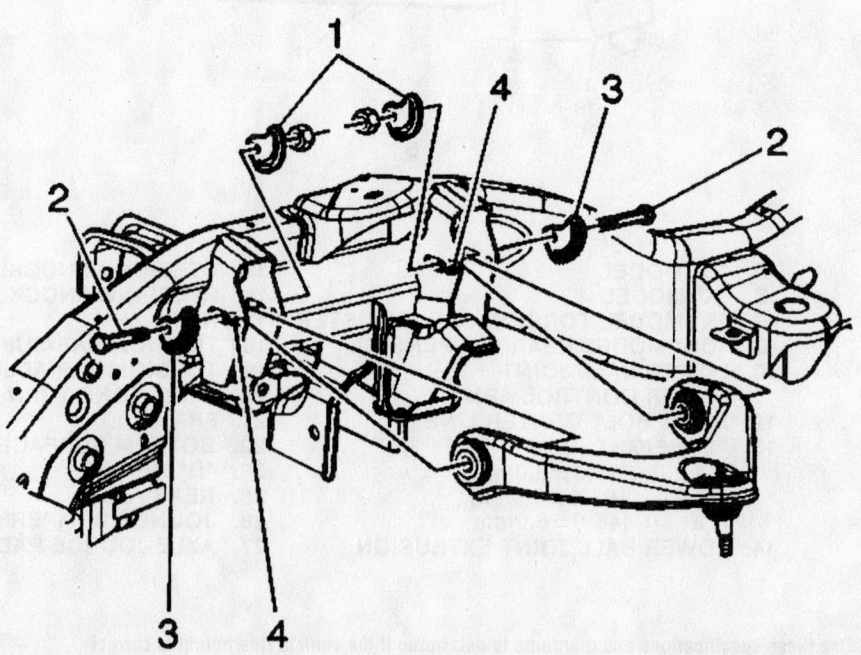

Upper control arm

9308KG31

2000–03 Vehicles

1. Before servicing the vehicle, refer to the precautions in the beginning of this section.
2. Raise and support the vehicle.
3. Remove or disconnect the following:
   - Tire and wheel assembly
   - Real Time Damping (RTD) link rod from the sensor, if equipped
   - Retaining bolt for the brake hose and the wheel speed sensor brackets
   - Halfshaft
   - Nut at the upper ball joint. Discard the nut.
   - Upper control arm from the steering knuckle
   - Upper control arm nuts and the adjustment cams (15 Series RWD, 4WD, and 25 Series RWD)
   - Upper control arm bolts (15 Series RWD, 4WD, and 25 Series RWD)
   - Upper control arm nuts and the adjustment cams (25 Series 4WD)
   - Upper control arm bolts (25 Series 4WD)
   - Upper control arm

To install:

4. Install or connect the following:
   - Upper control arm
   - Upper control arm bolts (25 Series 4WD)
   - Upper control arm nuts and the adjustment cams (25 Series 4WD). Tighten the nuts to 140 ft. lbs. (190 Nm).
   - Upper control arm bolts (15 Series RWD, 4WD, and 25 Series RWD)
   - Upper control arm nuts and the adjustment cams (2) (15 Series RWD, 4WD, and 25 Series RWD). Tighten the nuts to 140 ft. lbs. (190 Nm).
   - Upper control arm to the steering knuckle
   - Halfshaft
   - New nut to the upper ball joint stud. Tighten the nut to 37 ft. lbs. (50 Nm).
   - Retaining bolts for the brake hose and wheel speed sensor brackets. Tighten the bolts to 80 inch lbs. (9 Nm).
   - RTD link rod to the sensor, if equipped
   - Tire and wheel assembly
5. Remove the safety stands.
6. Lower the vehicle. Verify the wheel alignment.

CONTROL ARM BUSHING REPLACEMENT

1. The control arm bushings are removed and installed using a press.

Lower Control Arm and Bushing

REMOVAL & INSTALLATION

1999 Vehicles

1. Before servicing the vehicle, refer to the precautions in the beginning of this section.

➡**Special tools J–36202, J–36618–1, J–36618–2, J–36618–3, J–36618–4, J–36618–5, and J–9519–23, or their equivalents, are necessary for this procedure.**

2. Remove or disconnect the following:
   - Wheel
3. Matchmark the both torsion bar adjustment bolt positions.
4. Using tool J–36202, increase the tension on the adjusting arm.
   - Adjustment bolt and retaining plate, and move the tool aside
5. Slide the torsion bars forward.
   - Adjusting arm
   - Splash shield from the knuckle, if equipped
   - Hub nut and washer. Insert a long drift or dowel through the vanes in the brake rotor to hold the rotor in place.
   - Axle shaft out of the hub
   - Brake caliper and wire it aside
   - Rotor
   - Shock absorber from control arm
   - Inner tie rod end from the relay rod
6. Support the lower control arm with a floor jack.
   - Stabilizer bar from the control arm
   - Cotter pin from the lower ball stud and loosen the nut
   - Ball joint from control arm
   - Control arm-to-frame bracket bolts, nuts and washers
   - Lower control arm and torsion bar as a unit
7. Separate the control arm and torsion bar.

To install:

8. Install or connect the following:
   - Control arm assembly into position. Insert the front leg of the control arm into the crossmember first, then the rear leg into the frame bracket.
   - Mounting bolts, front one first. The bolts **must** be installed with the front bolt head heads towards the front of the truck and the rear bolt head towards the rear of the truck!

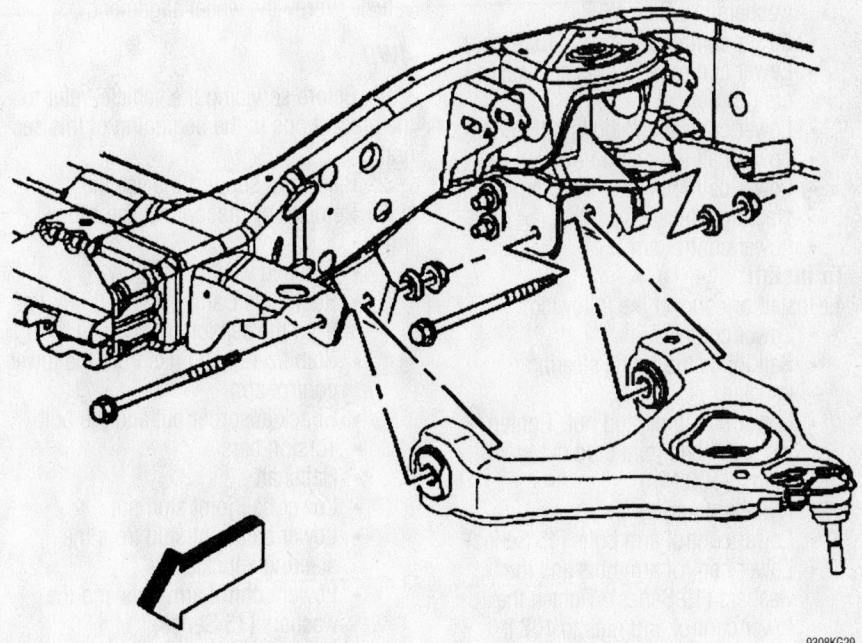

2WD lower control arm—15 Series

9308KG29

Heater Core replacement is covered in Section 2 of this manual

➡ **Do not tighten the bolts yet. The bolts must be torqued with the truck at its proper ride height.**

- Ball joint into the knuckle. Torque the nut to 94 ft lbs. (128 Nm).
- Adjuster arm

9. Using tool J–36202, increase tension on both torsion bars.

- Adjustment retainer plate and bolt on both torsion bars and set to marked positions. Release the tension on the torsion bar until the load is taken up by the adjustment bolt.
- Wheel

10. Lower the truck. Once the weight of the truck is on the wheels. Torque the bolts to 121 ft. lbs. (165 Nm).

2000–01 Vehicles

2WD

1. Before servicing the vehicle, refer to the precautions in the beginning of this section.

2. Raise and support the vehicle.

3. Remove or disconnect the following:

- Tire and wheel assembly
- Real Time Damping (RTD) link rod from the sensor, if equipped
- Shock absorber
- Front stabilizer shaft link
- Front coil spring
- Lower control arm nuts and the washers (15 Series)
- Lower control arm bolts (15 Series)
- Lower control arm nuts and washers (25 Series)
- Lower control arm bolts (25 Series)
- Lower ball joint stud nut
- Lower ball joint stud from the steering knuckle
- Lower control arm

To install:

4. Install or connect the following:

- Lower control arm
- Ball joint stud to the steering knuckle
- Lower ball joint stud nut. Tighten the lower ball joint stud nut to 74 ft. lbs. (100 Nm)
- Front coil spring
- Lower control arm bolts (15 Series)
- Lower control arm nuts and the washers (15 Series). Tighten the lower control arm nuts to 107 ft. lbs. (145 Nm).
- Lower control arm bolt (25 Series)
- Lower control arm nuts and the washers (25 Series). Tighten the

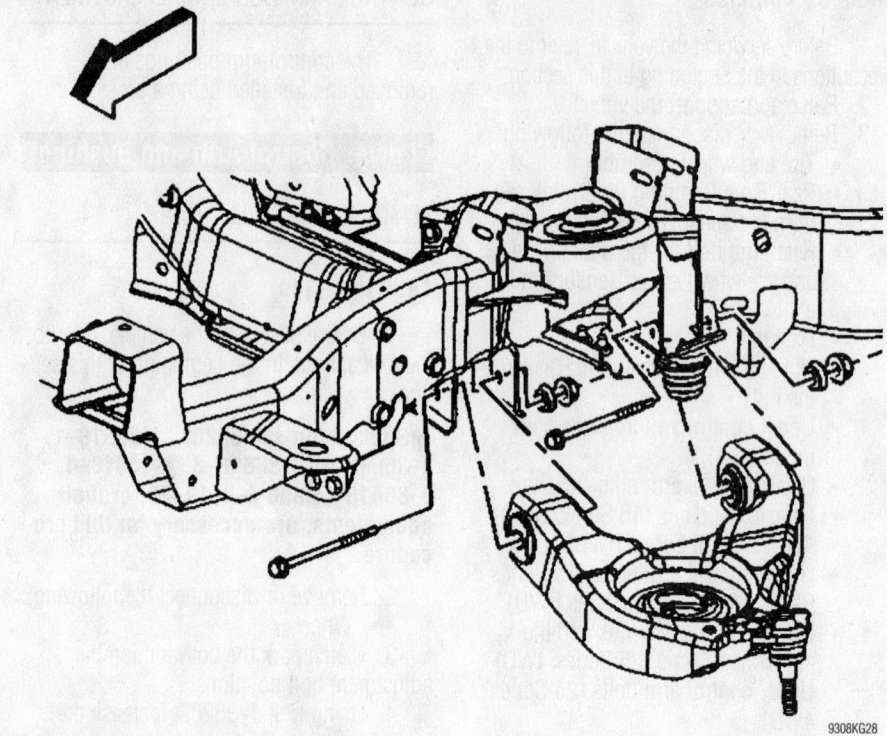

2WD lower control arm—25 Series

lower control arm nuts to 107 ft. lbs. (145 Nm).

- Front stabilizer shaft link.
- Shock absorber
- RTD sensor, if equipped
- Tire and wheel assembly

5. Remove the safety stands. Lower the vehicle. Verify the wheel alignment.

4WD

1. Before servicing the vehicle, refer to the precautions in the beginning of this section.

2. Raise and support the vehicle.

3. Remove or disconnect the following:

- Tire and wheel assembly
- Real Time Damping (RTD) link rod from the sensor, if equipped
- Stabilizer shaft links from the lower control arm
- Shock absorber nut and the bolt
- Torsion bars
- Halfshaft
- Lower ball joint stud nut
- Lower ball joint stud from the steering knuckle
- Lower control arm nuts and the washers (15 Series)
- Lower control arm bolts
- Lower control arm nuts and the washers (25 Series)
- Lower control arm bolts
- Lower control arm

To install:

- Lower control arm
- Lower control arm bolts (15 Series)
- Washers with the shoulder facing the arm
- Nuts. Tighten the nuts to 107 ft. lbs. (145 Nm).
- Halfshaft
- Lower ball joint stud to the steering knuckle. Install the nut to the ball joint stud. Tighten the nut to 74 ft. lbs. (100 Nm).
- Torsion bars
- Shock absorber through nut and bolt
- Stabilizer shaft links to the lower control arm
- RTD link rod to the sensor, if equipped
- Tire and wheel assembly

4. Remove the safety stands. Lower the vehicle. Verify the wheel alignment.

CONTROL ARM BUSHING REPLALEMENT

Front Bushing

1. Before servicing the vehicle, refer to the precautions in the beginning of this section.

2. On 15 and 25 Series, the bushings are not replaceable. If they are damaged, the control arm will have to be replaced.

3. Remove or disconnect the following:

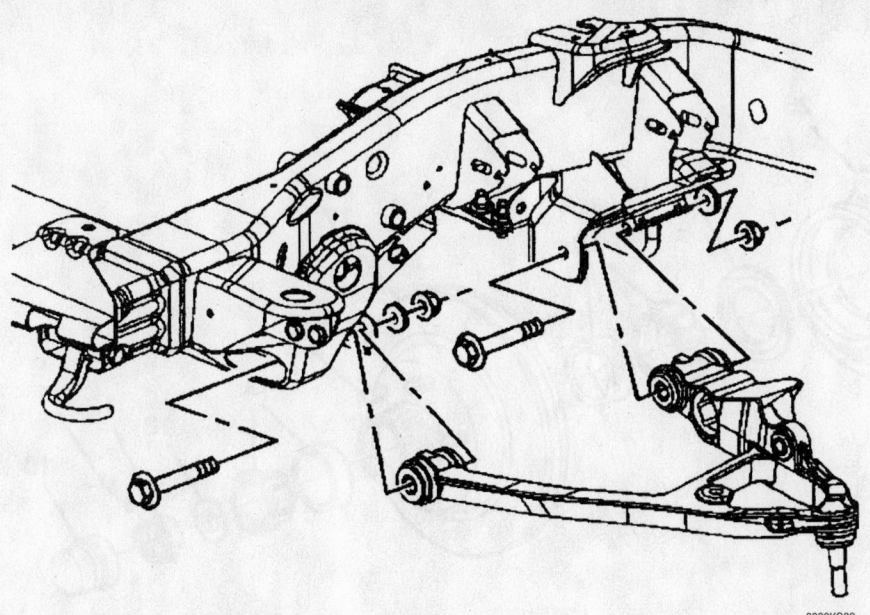

4WD lower control arm—15 Series

- Wheel
- Lower control arm
- Unbend the crimps with a punch on the front bushing

4. Press out the bushings with tools J–36618–2, J–9519–23, J–36618–4 and 36618–1.

To install:

5. Lubricate the outer case of the bushing.

6. Install or connect the following:
- Bushing into control arm

7. Press in the bushings with tools J–36618–2, J–9519–23, J–36618–4 and 36618–1 until the bushing is seated in.

8. After bushing is installed crimp it in place.

- Control arm and mounting bolts. Torque the front nut first then the rear to 140 ft lbs. (190 Nm)
- Wheel

Rear Bushing

➡**On 15 and 25 Series, the bushings are not replaceable. If they are damaged, the control arm will have to be replaced.**

1. Before servicing the vehicle, refer to the precautions in the beginning of this section.

2. Remove or disconnect the following:

- Wheel
- Lower control arm

3. Press out the bushings with tools.

J–36618–5, J–9519–23, J–36618–3 and J–36618–2. There are no crimps.

To install:

4. Lubricate the outer case of the bushing.

5. Install or connect the following:
- Bushing into control arm

6. Press in the bushings with tools J–36618–5, J–9519–23, J–36618–3 and J–36618–2. There are no crimps.
- Control arm and mounting bolts. Torque the front nut first then the rear to 140 ft lbs. (190 Nm).
- Wheel

Wheel Bearings

ADJUSTMENT

➡**The front wheel bearings on 2-wheel drive vehicles (exc. 2000–01 vehicles) are adjustable.**

1. Before servicing the vehicle, refer to the precautions in the beginning of this section.

2. Remove the dust cap, cotter pin.

3. Loosen the spindle nut.

4. Spin the wheel hub by hand and tighten the nut until it is just snug—12 ft. lbs. (16 Nm) Back off the nut until it is loose, then tighten it finger-tight. Loosen the nut until either hole in the spindle lines up with a slot in the nut and insert a new cotter pin. There should be 0.001–0.008 in. (0.025–0.200mm) end-play. This can be measured with a dial indicator, if you wish.

5. Replace the dust cap, wheel and tire.

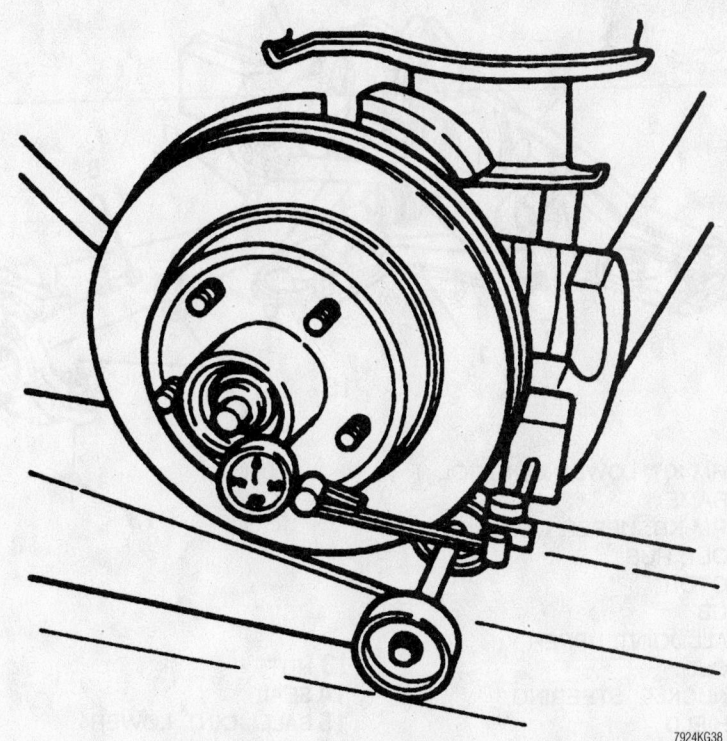

Use a dial indicator to measure the wheel bearing end-play—2-wheel drive vehicles

Brake service is covered in Section 4 of this manual

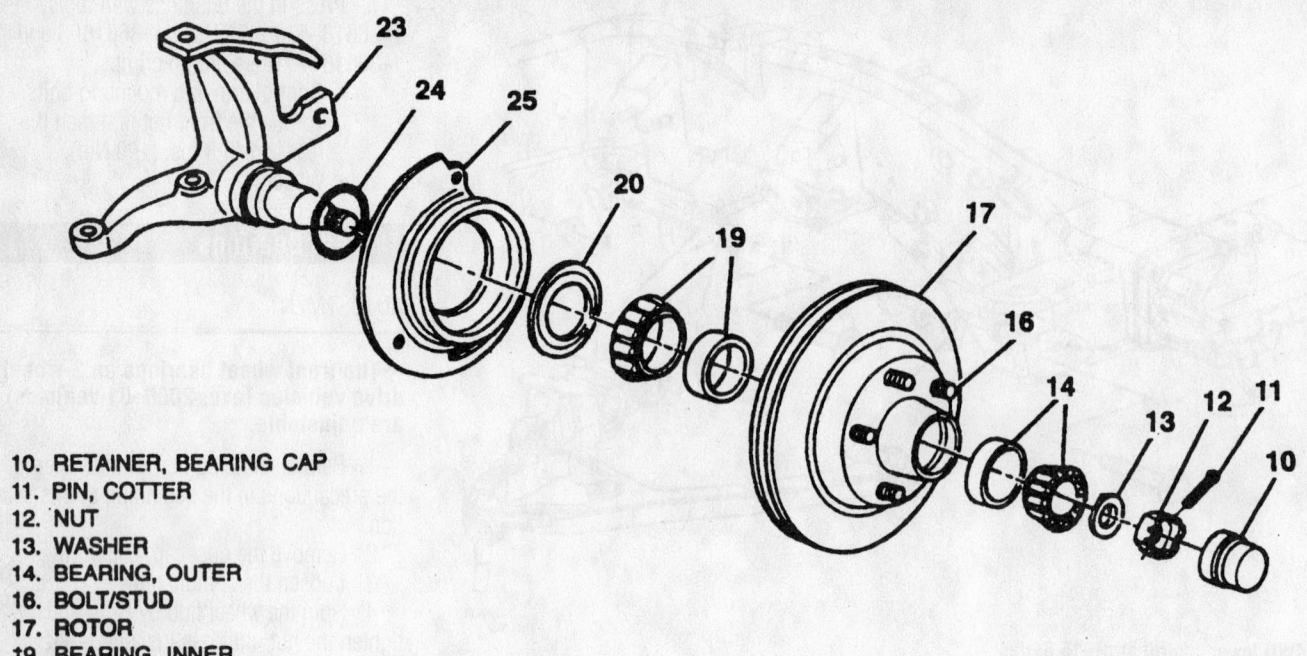

10. RETAINER, BEARING CAP
11. PIN, COTTER
12. NUT
13. WASHER
14. BEARING, OUTER
16. BOLT/STUD
17. ROTOR
19. BEARING, INNER
20. SEAL
23. KNUCKLE
24. GASKET
25. SHIELD

7924KG53

Exploded view of the front wheel bearing and related components—2-wheel drive models

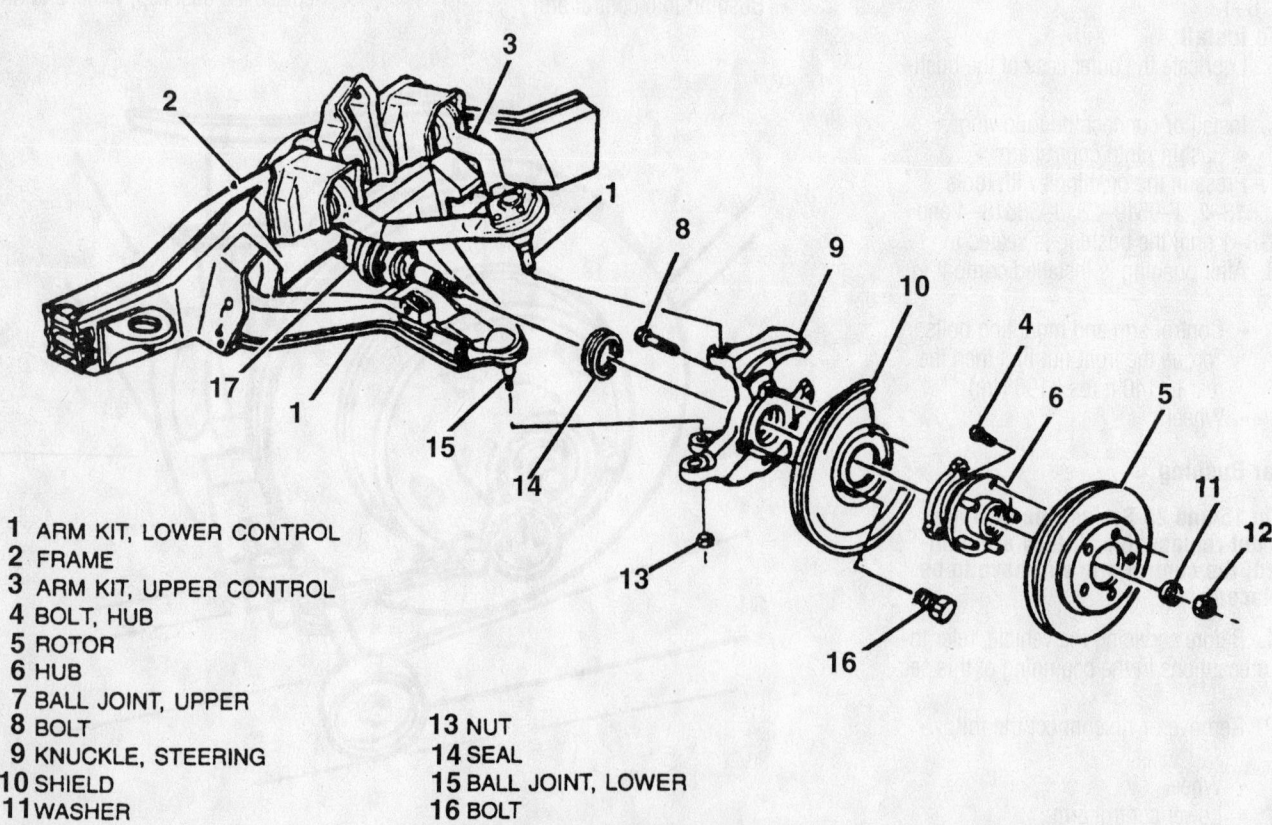

1 ARM KIT, LOWER CONTROL
2 FRAME
3 ARM KIT, UPPER CONTROL
4 BOLT, HUB
5 ROTOR
6 HUB
7 BALL JOINT, UPPER
8 BOLT
9 KNUCKLE, STEERING
10 SHIELD
11 WASHER
12 NUT
13 NUT
14 SEAL
15 BALL JOINT, LOWER
16 BOLT
17 JOINT KIT, FRONT AXLE

7924KG37

Exploded view of the front hub and knuckle assembly—4-wheel drive

REMOVAL & INSTALLATION

1999 Vehicles

FRONT—2WD

1. Before servicing the vehicle, refer to the precautions in the beginning of this section.

2. Remove or disconnect the following:
   - Wheel.
   - Caliper and wire it out of the way
   - Grease cap
   - Cotter pin, spindle nut, and washer
   - Hub.

✷✷ CAUTION

Do not drop the wheel bearings.

- Outer roller bearing assembly from the hub

3. The inner bearing assembly will remain in the hub and may be removed after prying out the inner seal. Discard the seal.

4. Clean all parts in a non-flammable solvent and let them air dry. Never spin-dry a bearing with compressed air! Check for excessive wear and damage.

5. If necessary for replacement, remove the bearing races from the hub using a hammer and drift. They are driven out from the inside out.

To install:

6. Install or connect the following:
   - New bearing races, if required. When installing new races, ensure that they are not cocked and that they are fully seated against the hub shoulder.

7. Pack both wheel bearings using high melting point wheel bearing grease for disc brakes.
   - Inner bearing in the hub and a new inner seal, making sure that the seal flange faces the bearing race.
   - Wheel hub over the spindle
   - Outer bearing into the hub
   - Spindle washer and nut

8. Spin the wheel hub by hand and tighten the nut until it is just snug—12 ft. lbs. (16 Nm) Back off the nut until it is loose, then tighten it finger-tight. Loosen the nut until either hole in the spindle lines up with a slot in the nut and insert a new cotter pin. There should be 0.001–0.008 in. (0.025–0.200mm) end-play. This can be measured with a dial indicator, if you wish.

9. Replace the dust cap, wheel and tire.

FRONT—4WD

1. Before servicing the vehicle, refer to the precautions in the beginning of this section.

"K" Series (4WD) vehicles have sealed front wheel bearings that are pre-adjusted and require no lubrication maintenance. Darkened areas on the bearing assembly are caused by a heat treatment process and do not require bearing replacement.

2. Remove or disconnect the following:
   - Wheel

➡**Wrap shop towels around the CV-Joint boots to protect them from damage during this procedure.**

   - Brake caliper and support it aside
   - Brake rotor
   - Cotter pin, retainer, castle nut, and thrust washer from the axle shaft
   - Tie rod end from the knuckle
   - Hub/bearing assembly retaining bolts

3. Using a J-28733-B hub/bearing puller tool or equivalent, press the hub/bearing assembly from the splined shaft.

✷✷ WARNING

After removal, lay the hub and bearing assembly on the outboard side. This will prevent damage and/or contamination of the bearing seal.

   - Splash shield
4. Support the lower control arm with a jack stand.
   - Upper and lower ball joints from the steering knuckle

   - Steering knuckle
   - Seal from the steering knuckle

To install:

5. Install or connect the following:
   - New seal in the steering knuckle, using a J 36605 seal installer
   - Steering knuckle on the ball joints and the retaining nuts. Torque the upper ball joint nut to 74 ft. lbs. (100 Nm) and the lower ball joint nut to 94 ft. lbs. (128 Nm) Tighten the nuts to align the holes for cotter pin insertion, but do NOT tighten more than an additional ⅙ turn.
   - Splash shield
   - Hub/bearing assembly over the splined shaft, making sure the splines line up correctly. Torque the bolts to 133 ft. lbs. (180 Nm)
   - Tie rod end at the steering knuckle

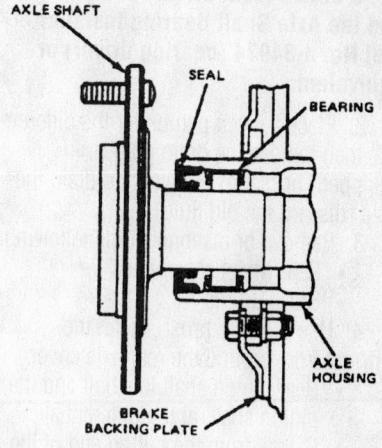

7924KG49

Cutaway view of the rear axle shaft and bearing assembly

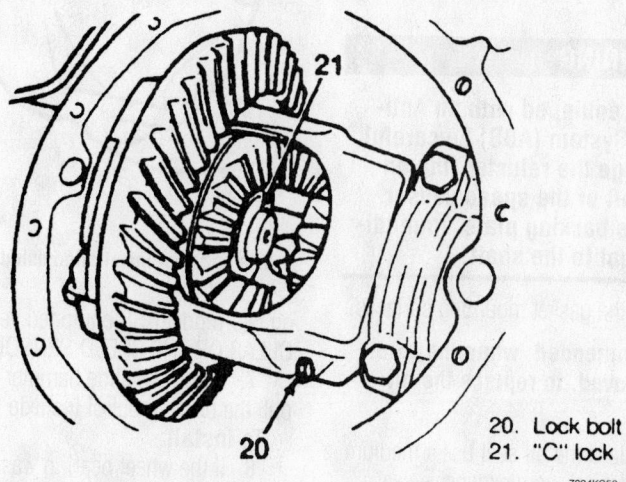

20. Lock bolt
21. "C" lock

7924KG50

Remove the lockbolt and pinion shaft, then push in the axle shaft and remove the C-lock

For complete Engine Mechanical specifications, see Section 1 of this manual

- Thrust washer and axle nut, Torque the nut to 165 ft. lbs. (225 Nm).
- Retainer and cotter pin
- Rotor and caliper

6. Remove the shop towels from the CV-Joint boot.

- Tire and wheel assemblies

7. Check and adjust the front end alignment and road test the vehicle.

REAR

1. Before servicing the vehicle, refer to the precautions in the beginning of this section.

A new pinion shaft lockbolt should be installed whenever either of the axle shafts is removed.

➡Axle shaft seal removal and installation uses the following special tools: the GM Axle Shaft Seal Installer tool No. J-33782 (seal driver) or equivalent and the Axle Shaft Bearing Installer tool No. J-34974 (bearing driver) or equivalent.

2. Place a catch pan under the differential, then remove the drain plug (if equipped) or rear axle cover and drain the fluid (discard the old fluid)

3. Remove or disconnect the following:

- Rear wheel assemblies
- Brake drums

4. Using a wire brush, clean the dirt/rust from around the rear axle cover.

- Rear pinion shaft lockbolt and the pinion shaft, at the differential
- C-lock from the button end of the axle shaft, push the axle shafts inward
- Axle shaft from the axle housing. Be careful not to damage the oil seal.

✳✳ WARNING

On vehicles equipped with an Anti-Lock Brake System (ABS) be careful not to damage the reluctor ring on the axle shaft or the speed sensor bolted to the backing plate, immediately adjacent to the shaft.

5. Clean the gasket mounting surfaces.

➡It is recommended, when the axle shaft is removed, to replace the oil seal.

6. To replace the oil seal use a medium prybar or, better yet, an inexpensive seal removal tool, to pry the oil seal from the end of the rear axle housing. DO NOT damage the housing oil seal surface. And again,

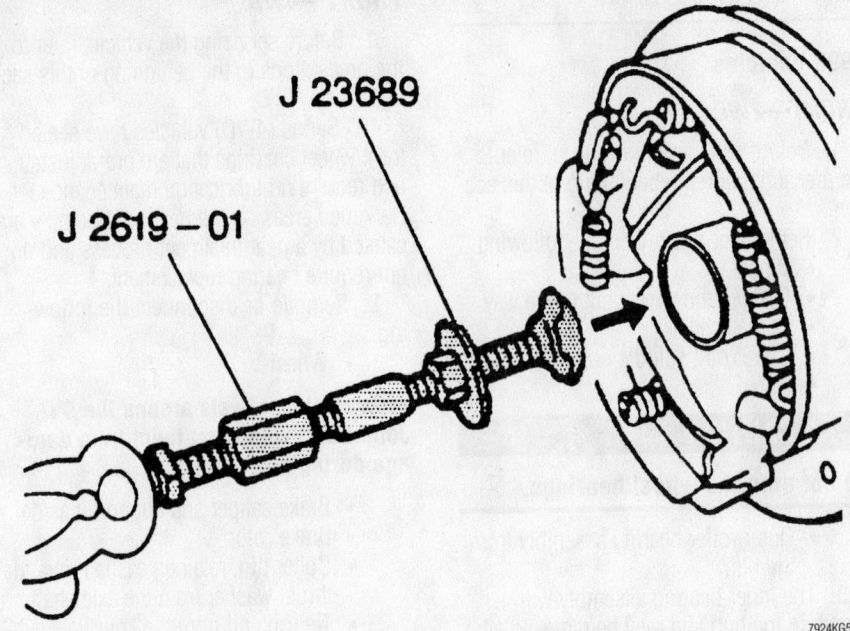

Using a slide hammer and adapters, remove the axle bearing and seal

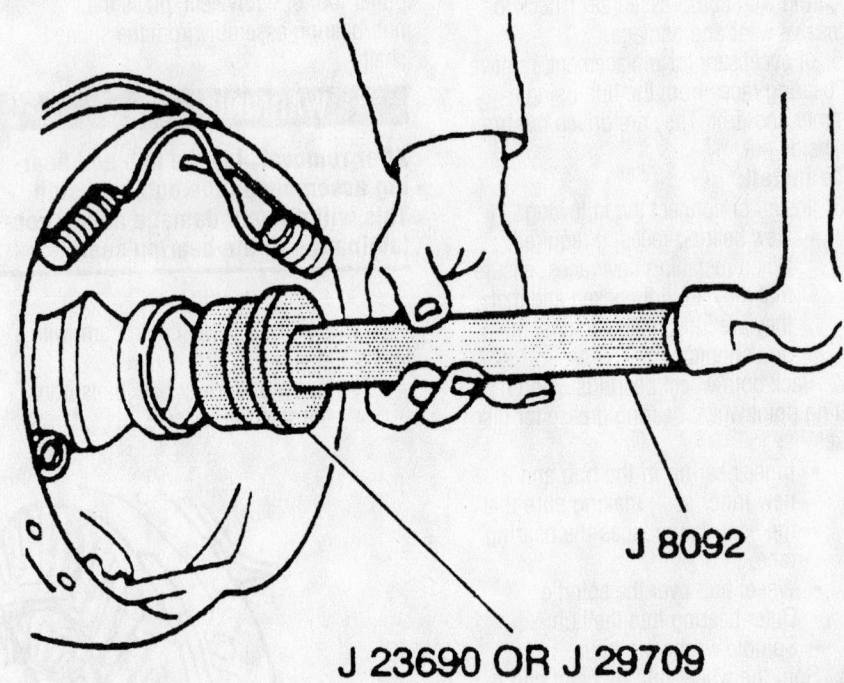

Axle and seal installation using a bearing driver

on late-model ABS equipped vehicles, STAY CLEAR OF THE SPEED SENSOR.

7. Using the slide hammer and adapter, pull the bearing out of the axle tube.

To install:

8. If the wheel bearing was removed:

a. Step 1: Using solvent, thoroughly clean the wheel bearing, then blow dry with compressed air. Inspect the wheel bearing for excessive wear or damage, then replace it, if necessary.

b. Step 2: With a new or the reused bearing, thoroughly coat the bearing with gear lubricant.

c. Step 3: Using the Axle Shaft Bearing Installer tool No. J-34974, or equivalent, drive the bearing into the axle housing until it bottoms against the seat.

Be sure the bearing installer does not contact and damage the speed sensor on ABS equipped vehicles.

9. If the axle shaft seal was removed:

a. Step 1: Clean and inspect the axle tube housing.

b. Step 2: Using the GM Axle Shaft Seal Installer tool No. J-33782, or an equivalent driver, seat the new seal into the housing until it is flush with the axle tube. Be sure the seal installer does not contact and damage the speed sensor on ABS equipped vehicles.

c. Step 3: Using gear oil, lubricate the new seal lips.

10. Slide the axle shaft into the rear axle housing and engage the splines of the axle shaft with the splines of the rear axle side gear, then install the C-lock retainer on the axle shaft button end.

✳✳ WARNING

BE CAREFUL not to damage the wheel bearing seal with the splines on the axle shaft.

11. After the C-lock is installed, pull the axle shaft outward to seat the C-lock retainer in the counterbore of the side gears.

12. Install or connect the following:
- Pinion shaft through the case and the pinions
- New pinion shaft lockbolt. Torque the new lockbolt to 25 ft. lbs. (34 Nm).
- Housing cover use a new rear axle cover gasket
- Brake drums
- Tire and wheel assemblies

13. Refill the housing. REMEMBER that the vehicle must be completely level, meaning that if the rear is still raised and supported, the front should also be raised.

2000–03 Vehicles

FRONT—2WD

1. Before servicing the vehicle, refer to the precautions in the beginning of this section.

2. Raise and support the vehicle.

3. Remove or disconnect the following:
- Tire and wheel assembly
- Rotor
- Wheel speed sensor and brake hose mounting bracket bolt from the steering knuckle
- Electrical connection for the wheel speed sensor
- Hub and bearing assembly mounting bolts
- Hub and bearing assembly

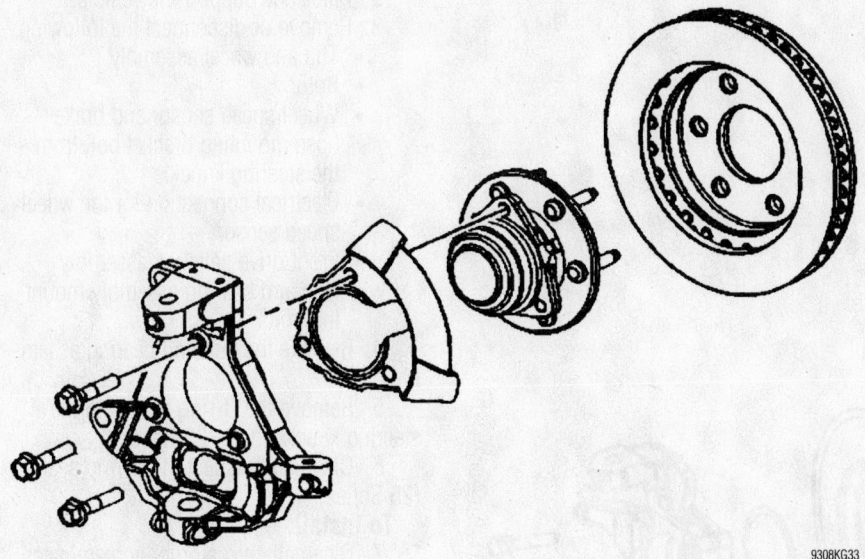

9308KG33

2WD front hub—15 Series

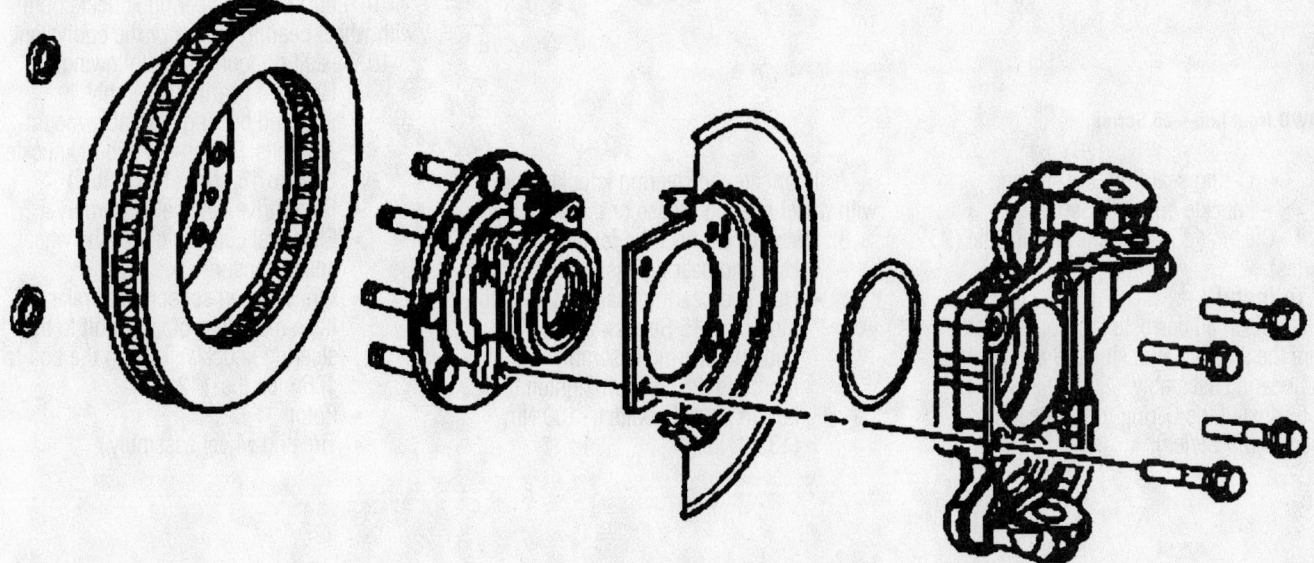

9308KG32

2WD front hub—25 Series

For Accessory Drive Belt illustrations, see Section 1 of this manual

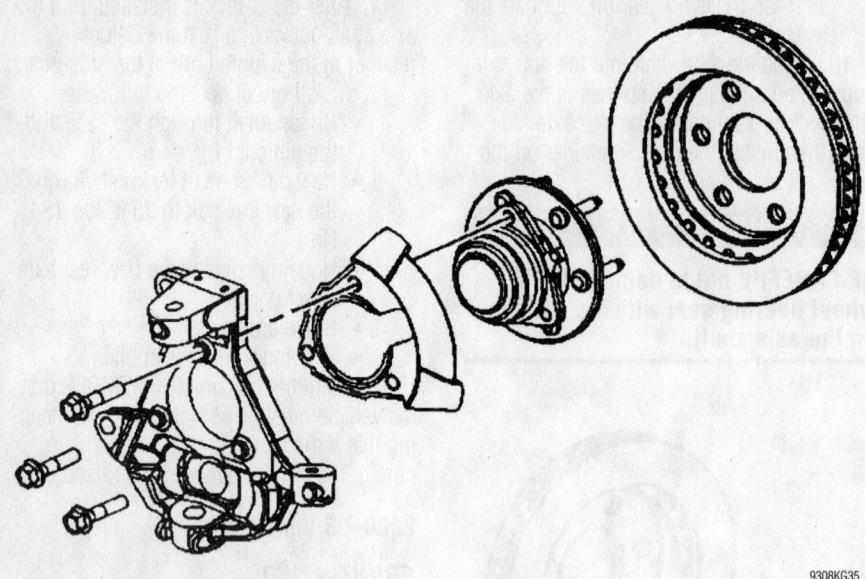

4WD front hub—15 Series

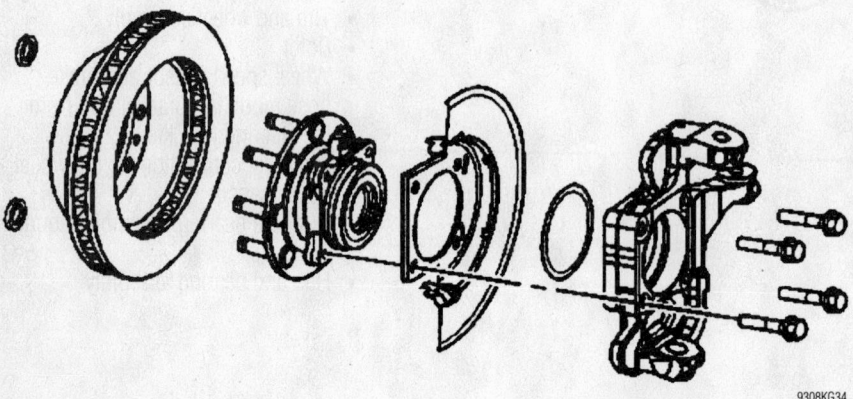

4WD front hub—25 Series

9308KG34

- O-ring seal from the steering knuckle bore (25 Series)

4. Clean and inspect the O-ring seal (25 Series).

To install:

5. Clean all corrosion or contaminates from the steering knuckle bore and the hub and bearing assembly.

6. Install the O-ring to the steering knuckle (25 Series).

7. Lubricate the steering knuckle bore with wheel bearing grease or the equivalent.

8. Install or connect the following:
- Hub and bearing assembly
- Hub and bearing assembly mounting bolts (15 Series). Install the hub and bearing assembly mounting bolts (25 Series). Tighten the hub to knuckle bolts to 180 Nm (133 ft. lbs.).

- Electrical connection for the wheel speed sensor
- Wheel speed sensor and brake hose mounting bracket bolt to the steering knuckle. Tighten the bolt to 106 inch lbs. (12 Nm).
- Rotor
- Tire and wheel assembly

9. Remove the safety stands. Lower the vehicle.

FRONT—4WD

1. Before servicing the vehicle, refer to the precautions in the beginning of this section.

2. Raise and support the vehicle.

3. Remove or disconnect the following:
- Tire and wheel assembly
- Rotor
- Wheel speed sensor and brake hose mounting bracket bolt from the steering knuckle
- Electrical connection for the wheel speed sensor
- Front drive halfshaft assembly
- Hub and bearing assembly mounting bolts

4. Remove the hub and bearing assembly.

5. Remove the O-ring seal from the steering knuckle bore (25 Series).

6. Clean and inspect the O-ring seal (25 Series).

To install:

7. Clean all corrosion or contaminates from the steering knuckle bore and the hub and bearing assembly.

8. Install the O-ring to the steering knuckle (25 Series).

9. Lubricate the steering knuckle bore with wheel bearing grease or the equivalent.

10. Install or connect the following:
- Hub and bearing assembly
- Hub and bearing assembly mounting bolts Tighten the hub to knuckle bolts to 180 Nm (133 ft. lbs.)
- Front drive halfshaft assembly
- Electrical connection for the wheel speed sensor
- Wheel speed sensor and brake hose mounting bracket bolt to the steering knuckle. Tighten the bolt to 106 inch lbs. (12 Nm).
- Rotor
- Tire and wheel assembly

HONDA

Honda-CR-V

15

PRECAUTIONS

Before servicing any vehicle, please be sure to read all of the following precautions, which deal with personal safety, prevention of component damage, and important points to take into consideration when servicing a motor vehicle:

• Never open, service or drain the radiator or cooling system when the engine is hot; serious burns can occur from the steam and hot coolant.

• Observe all applicable safety precautions when working around fuel. Whenever servicing the fuel system, always work in a well-ventilated area. Do not allow fuel spray or vapors to come in contact with a spark, open flame, or excessive heat (a hot drop light, for example). Keep a dry chemical fire extinguisher near the work area. Always keep fuel in a container specifically designed for fuel storage; also, always properly seal fuel containers to avoid the possibility of fire or explosion. Refer to the additional fuel system precautions later in this section.

• Fuel injection systems often remain pressurized, even after the engine has been turned **OFF**. The fuel system pressure must be relieved before disconnecting any fuel lines. Failure to do so may result in fire and/or personal injury.

• Brake fluid often contains polyglycol ethers and polyglycols. Avoid contact with the eyes and wash your hands thoroughly after handling brake fluid. If you do get brake fluid in your eyes, flush your eyes with clean, running water for 15 minutes. If eye irritation persists, or if you have taken brake fluid internally, seek medical assistance IMMEDIATELY.

• The EPA warns that prolonged contact with used engine oil may cause a number of skin disorders, including cancer. You should make every effort to minimize your exposure to used engine oil. Protective gloves should be worn when changing oil. Wash your hands and any other exposed skin areas as soon as possible after exposure to used engine oil. Soap and water, or waterless hand cleaner should be used.

• All new vehicles are now equipped with an air bag system. The system must be disabled before performing service on or around system components, steering column, instrument panel components, wiring and sensors. Failure to follow safety and disabling procedures could result in accidental air bag deployment, possible personal injury and unnecessary system repairs.

• Always wear safety goggles when working with, or around, the air bag system. When carrying a non-deployed air bag, be sure the bag and trim cover are pointed away from your body. When placing a non-deployed air bag on a work surface, always face the bag and trim cover upward, away from the surface. This will reduce the motion of the module if it is accidentally deployed. Refer to the additional air bag system precautions later in this section.

• Clean, high quality brake fluid from a sealed container is essential to the safe and proper operation of the brake system. You should always buy the correct type of brake fluid for your vehicle. If the brake fluid becomes contaminated, completely flush the system with new fluid. Never reuse any brake fluid. Any brake fluid that is removed from the system should be discarded. Also, do not allow any brake fluid to come in contact with a painted surface; it will damage the paint.

• Never operate the engine without the proper amount and type of engine oil; doing so WILL result in severe engine damage.

• Timing belt maintenance is extremely important. Many models utilize an interference-type, non-freewheeling engine. If the timing belt breaks, the valves in the cylinder head may strike the pistons, causing potentially serious (also time-consuming and expensive) engine damage. Refer to the maintenance interval charts in the front of this manual for the recommended replacement interval for the timing belt, and to the timing belt section for belt replacement and inspection.

• Disconnecting the negative battery cable on some vehicles may interfere with the functions of the on-board computer system(s) and may require the computer to undergo a relearning process once the negative battery cable is reconnected.

• When servicing drum brakes, only disassemble and assemble one side at a time, leaving the remaining side intact for reference.

• Only an MVAC-trained, EPA-certified automotive technician should service the air conditioning system or its components.

ENGINE REPAIR

➡ **Disconnecting the negative battery cable on some vehicles may interfere with the functions of the on board computer system. The computer may undergo a relearning process once the negative battery cable is reconnected.**

Distributor

The 2.4L engine does not have a distributor.

REMOVAL & INSTALLATION

1. Before servicing the vehicle, refer to the precautions in the beginning of this section.
2. Remove or disconnect the following:

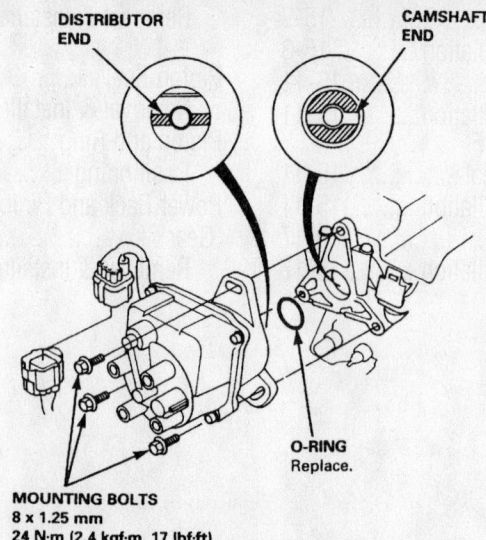

DISTRIBUTOR END

CAMSHAFT END

O-RING
Replace.

MOUNTING BOLTS
8 x 1.25 mm
24 N·m (2.4 kgf·m, 17 lbf·ft)

7924MG02

Exploded view of the distributor mounting—CR-V

- Negative battery cable
- Cruise control cable
- Air intake duct
- Distributor harness connector
- Spark plug wires
- Distributor

To install:

3. Install or connect the following:
- Distributor. Use a new O-ring seal.
- Spark plug wires
- Distributor harness connector
- Air intake duct
- Cruise control cable
- Negative battery cable

4. Set the ignition timing and tighten the mounting bolts to 13 ft. lbs. (18 Nm).

Alternator

REMOVAL

1999–01

1. Before servicing the vehicle, refer to the precautions in the beginning of this section.
2. Remove or disconnect the following:
- Negative battery cable
- Accessory drive belts
- Alternator wiring harness connectors
- Alternator

2002

1. Before servicing the vehicle, refer to the precautions in the beginning of this section.
2. Remove or disconnect the following:
- Negative battery cable
- Front cover
- Accessory drive belt
- The 3 bolts holding the alternator
- Alternator wiring harness connectors
- Alternator

INSTALLATION

1999–01

1. Install or connect the following:
- Alternator. Tighten the mounting nut to 33 ft. lbs. (44 Nm) and the adjustment locknut to 17 ft. lbs. (24 Nm).
- Alternator wiring harness connectors. Tighten the battery terminal nut to 70 inch lbs. (8 Nm).
- Accessory drive belts
- Negative battery cable

2002

1. Install or connect the following:
- Alternator. Tighten the bolts to 16 ft. lbs. (22 Nm).
- Alternator wiring harness connectors. Tighten the battery terminal nut to 70 inch lbs. (8 Nm).
- Accessory drive belts
- Front cover
- Negative battery cable

Ignition Timing

ADJUSTMENT

Adjustment is not possible on the 2.4L engine.

➡**Timing adjustments are made with the engine at operating temperature.**

1. Before servicing the vehicle, refer to the precautions in the beginning of this section.
2. Short the **2P** Service Check connector.
3. Connect a timing light to the No. 1 ignition wire.
4. The timing should be 13–17 degrees Before Top Dead Center (BTDC) (red timing mark on crankshaft pulley) at 650–750 rpm.
5. Adjust the timing as necessary and tighten the distributor bolts to 13 ft. lbs. (18 Nm).
6. Remove the **2P** connector jumper.

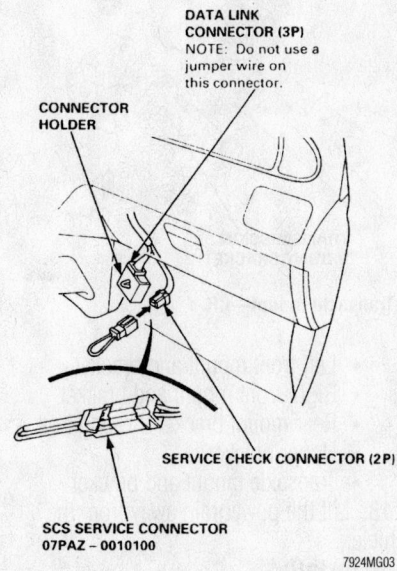

DATA LINK
CONNECTOR (3P)
NOTE: Do not use a
jumper wire on
this connector.

CONNECTOR
HOLDER

SERVICE CHECK CONNECTOR (2P)

SCS SERVICE CONNECTOR
07PAZ – 0010100

7924MG03

Service Check connector and shorting jumper

Engine Assembly

REMOVAL & INSTALLATION

1999–01

➡**The engine and transaxle are removed from the vehicle as a unit.**

1. Before servicing the vehicle, refer to the precautions in the beginning of this section.
2. Drain the cooling system.
3. Drain the transaxle fluid.
4. Drain the engine oil.
5. Relieve fuel system pressure.
6. Remove or disconnect the following:
- Negative battery cable
- Fuse/Relay box battery cables
- Battery and tray
- Air intake assembly
- Powertrain Control Module (PCM) connectors and grommet. Pull the PCM harness through the firewall.
- Left engine wire harness connectors
- Cruise control actuator
- Power steering pump belt and pump
- A/C compressor drive belt
- Fuel lines
- Brake booster vacuum line
- Accelerator cable
- Power Steering Pressure (PSP) switch
- Splash shield
- Radiator hoses
- Heater hoses

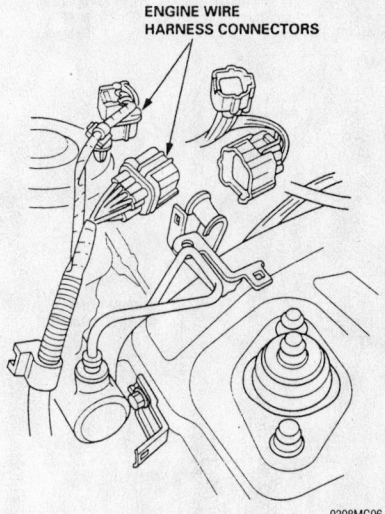

ENGINE WIRE
HARNESS CONNECTORS

9308MG06

Left engine wire harness connectors—CR-V

- Heated Oxygen Sensor (HO2S) connector
- Exhaust front pipe
- Right damper fork
- Lower ball joints

7. Separate the inner CV-joints from the transaxle and support the axle halfshafts out of the work area with safety wire.

8. If equipped with a manual transaxle, remove or disconnect the following:
- Clutch slave cylinder
- Clutch hose bracket
- Transaxle ground cable
- Shift cables

9. If equipped with an automatic transaxle, remove or disconnect the following:
- Shift cable cover
- Shift cable
- Transaxle ground cable and hose clamp
- Transaxle fluid cooler lines

10. For all vehicles, remove or disconnect the following:
- A/C hose clamp
- Radiator
- A/C compressor
- Rear driveshaft, if equipped

11. Attach a hoist to the engine lifting eyes and support the powertrain weight.

12. Remove or disconnect the following:

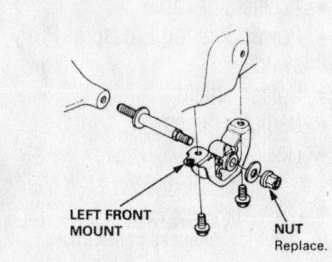

M/T:

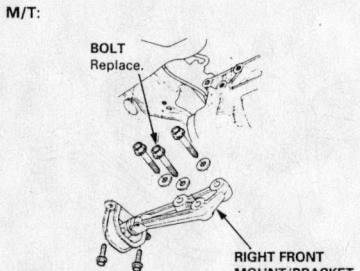

A/T:

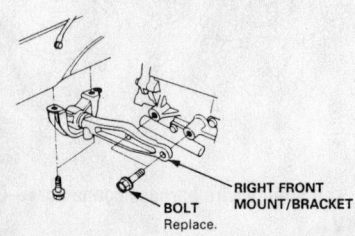

9308MG07

Front mounts—CR-V

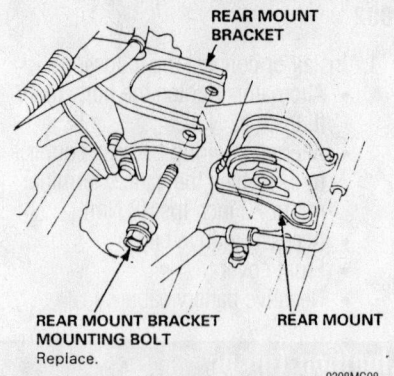

REAR MOUNT BRACKET

REAR MOUNT BRACKET MOUNTING BOLT
Replace.

REAR MOUNT

9308MG08

Rear mount—CR-V

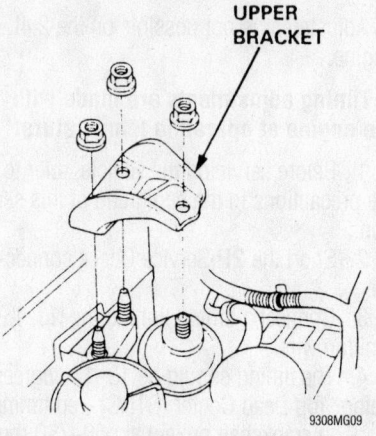

UPPER BRACKET

9308MG09

Upper bracket—CR-V

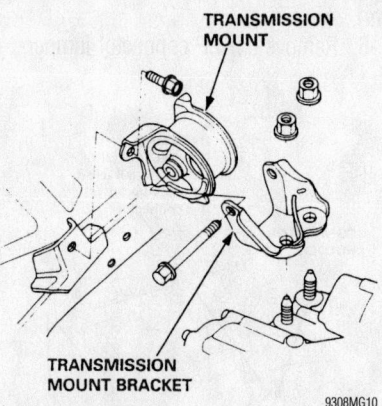

TRANSMISSION MOUNT

TRANSMISSION MOUNT BRACKET

9308MG10

Transaxle mount—CR-V

- Left front mount and bracket
- Right front mount and bracket
- Rear mount bracket through bolt
- Upper bracket
- Transaxle mount and bracket

13. Lift the powertrain away from the vehicle.

To install:

➡ Use new self-locking nuts and color-coded self-locking bolts when installing the engine mounts and suspension components.

➡ Do not tighten the engine or transaxle mount fasteners until instructed to do so.

14. Lower the powertrain into position.
15. Install or connect the following:
- Transaxle mount and bracket. Tighten the frame mounting bolts to 47 ft. lbs. (64 Nm).
- Upper bracket. Tighten the nuts in sequence to 54 ft. lbs. (74 Nm).
- Rear mount bracket through bolt
- Right front mount and bracket
- Left front mount and bracket

16. Tighten the remaining mount fasteners as follows:

a. Transaxle mount fasteners to 47 ft. lbs. (64 Nm) and the through bolt to 54 ft. lbs. (74 Nm).

b. Rear mount bracket through bolt to 43 ft. lbs. (59 Nm).

c. Right front mount 12mm bolts to 47 ft. lbs. (64 Nm) and the 10mm bolts to 33 ft. lbs. (44 Nm).

d. Left front mount 12mm stud bolt to 61 ft. lbs. (83 Nm), 10mm bolts to 33 ft. lbs. (44 Nm), and 12mm nut to 43 ft. lbs. (59 Nm).

e. Right front mount 12mm nut to 43 ft. lbs. (59 Nm).

17. Install or connect the following:
- Rear driveshaft, if equipped
- A/C compressor
- Radiator
- A/C hose clamp

18. If equipped with a manual transaxle, install or connect the following:
- Shift cables
- Transaxle ground cable
- Clutch hose bracket
- Clutch slave cylinder

19. If equipped with an automatic transaxle, install or connect the following:
- Transaxle fluid cooler lines

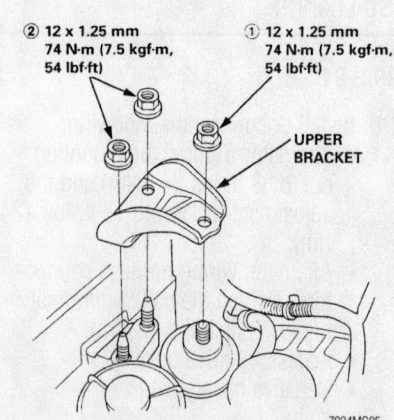

② 12 x 1.25 mm
74 N·m (7.5 kgf·m, 54 lbf·ft)

① 12 x 1.25 mm
74 N·m (7.5 kgf·m, 54 lbf·ft)

UPPER BRACKET

7924MG05

Upper bracket tightening sequence—CR-V

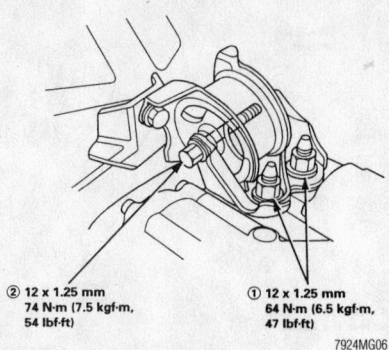

Transaxle mount fastener tightening sequence—CR-V

② 12 x 1.25 mm
74 N·m (7.5 kgf·m, 54 lbf·ft)

① 12 x 1.25 mm
64 N·m (6.5 kgf·m, 47 lbf·ft)

7924MG06

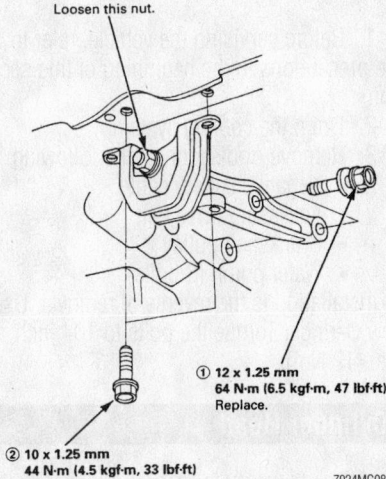

Loosen this nut.

① 12 x 1.25 mm
64 N·m (6.5 kgf·m, 47 lbf·ft)
Replace.

② 10 x 1.25 mm
44 N·m (4.5 kgf·m, 33 lbf·ft)

7924MG08

Right front mount tightening sequence—CR-V

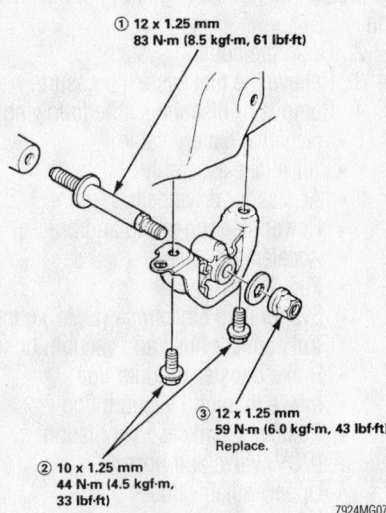

① 12 x 1.25 mm
83 N·m (8.5 kgf·m, 61 lbf·ft)

③ 12 x 1.25 mm
59 N·m (6.0 kgf·m, 43 lbf·ft)
Replace.

② 10 x 1.25 mm
44 N·m (4.5 kgf·m, 33 lbf·ft)

7924MG07

Left front mount tightening sequence—CR-V

- Transaxle ground cable and hose clamp
- Shift cable
- Shift cable cover

20. For all vehicles, install or connect the following:
- Axle halfshafts
- Lower ball joints
- Right damper fork
- Exhaust front pipe
- HO2S connector
- Heater hoses
- Radiator hoses
- Splash shield
- PSP switch
- Accelerator cable
- Brake booster vacuum line
- Fuel lines
- A/C compressor drive belt
- Power steering pump and belt
- Cruise control actuator
- Left engine wire harness connectors
- PCM connectors and grommet
- Air intake assembly
- Battery and tray
- Fuse/Relay box battery cables
- Negative battery cable

21. Fill the engine crankcase to the correct level.
22. Fill the transaxle to the correct level.
23. Fill the cooling system.
24. Start the engine and check for leaks.
25. Check the wheel alignment and adjust as necessary.

2002

2002

➡ **The engine and transaxle are removed from the vehicle as a unit.**

1. Before servicing the vehicle, refer to the precautions in the beginning of this section.
2. Drain the cooling system.
3. Drain the transaxle fluid.
4. Drain the engine oil.
5. Relieve fuel system pressure.
6. Remove or disconnect the following:
- Negative battery cable
- Fuse/Relay box battery cables
- Battery and tray
- Intake manifold cover
- IAT sensor connector
- Breather hose
- Intake duct
- Cables from the power distribution center

- Throttle and cruise cables
- Powertrain Control Module (PCM) connectors and grommet. Pull the PCM harness through the firewall.
- Fuel lines
- EVAP canister
- Brake booster vacuum line
- Clutch slave cylinder
- Clutch hose bracket
- Shift cables
- Drive belt
- Power steering pump, leaving the hoses connected

7. Attach a hoist to the engine lifting eyes and support the powertrain weight.
8. Remove or disconnect the following:
- Splash shield
- Wheels
- Catalytic converter
- Rear driveshaft
- Stabilizer links
- Right damper fork
- Halfshafts
- Shift cable
- Radiator
- Upper bracket
- Transaxle mount and bracket
- Front mount bolt
- Rear mount bracket bolts. Match-mark the sub-frame mounting bolt centers.

There is a special tool necessary for sub-frame removal. The Honda tool number is EQS02C000011. Attach the tool as explained in the tool instructions, attach a floor jack with adapter, remove the 4 sub-frame bolts and lower the sub-frame.

9. Remove or disconnect the following:
- A/C compressor without disconnecting the hoses

10. Check that all hoses and wires are disconnected.
11. Lower the engine about 6 inches and recheck all clearances.
12. Lower the engine all the way.
13. Remove the chain hoist.

To install:
14. Installation is the reverse of removal. Observe the following torques:
- Front engine mount bracket bolts: 33 ft. lbs. (44Nm)
- A/C compressor bracket: 33 ft. lbs. (44Nm)
- Stiffener 10mm bolts: 33 ft. lbs. (44Nm); 6mm bolts 9 ft. lbs. (12 Nm)
- A/C compressor bolts: 33 ft. lbs. (44 Nm)

- Subframe front bolt: 47 ft. lbs. (64 Nm)
- Subframe rear bolts: 43 ft. lbs. (59 Nm)
- Upper bracket bolt and nut: 40 ft. lbs. (54 Nm)
- Transmission mount bracket support bolts/nuts: 40 ft. lbs. (54 Nm)
- PS pump bolts: 16 ft. lbs. (22 Nm)
- Intake manifold cover: 9 ft. lbs. (12 Nm)

➡**Use new self-locking nuts and color-coded self-locking bolts when installing the engine mounts and suspension components.**

➡**Do not tighten the engine or transaxle mount fasteners until instructed to do so.**

15. Lower the powertrain into position.
16. Install or connect the following:
- Transaxle mount and bracket. Tighten the frame mounting bolts to 47 ft. lbs. (64 Nm).
- Upper bracket. Tighten the nuts in sequence to 54 ft. lbs. (74 Nm).
- Rear mount bracket through bolt
- Right front mount and bracket
- Left front mount and bracket

17. Tighten the remaining mount fasteners as follows:
a. Transaxle mount fasteners to 47 ft. lbs. (64 Nm) and the through bolt to 54 ft. lbs. (74 Nm).
b. Rear mount bracket through bolt to 43 ft. lbs. (59 Nm).
c. Right front mount 12mm bolts to 47 ft. lbs. (64 Nm) and the 10mm bolts to 33 ft. lbs. (44 Nm).
d. Left front mount 12mm stud bolt to 61 ft. lbs. (83 Nm), 10mm bolts to 33 ft. lbs. (44 Nm), and 12mm nut to 43 ft. lbs. (59 Nm).
e. Right front mount 12mm nut to 43 ft. lbs. (59 Nm).

18. Install or connect the following:
- Rear driveshaft, if equipped
- A/C compressor
- Radiator
- A/C hose clamp

19. If equipped with a manual transaxle, install or connect the following:
- Shift cables
- Transaxle ground cable
- Clutch hose bracket
- Clutch slave cylinder

20. If equipped with an automatic transaxle, install or connect the following:
- Transaxle fluid cooler lines
- Transaxle ground cable and hose clamp

- Shift cable
- Shift cable cover

21. For all vehicles, install or connect the following:
- Axle halfshafts
- Lower ball joints
- Right damper fork
- Exhaust front pipe
- HO2S connector
- Heater hoses
- Radiator hoses
- Splash shield
- PSP switch
- Accelerator cable
- Brake booster vacuum line
- Fuel lines
- A/C compressor drive belt
- Power steering pump and belt
- Cruise control actuator
- Left engine wire harness connectors
- PCM connectors and grommet
- Air intake assembly
- Battery and tray
- Fuse/Relay box battery cables
- Negative battery cable

22. Fill the engine crankcase to the correct level.
23. Fill the transaxle to the correct level.
24. Fill the cooling system.
25. Start the engine and check for leaks.
26. Check the wheel alignment and adjust as necessary.

Water Pump

REMOVAL & INSTALLATION

1999–01

1. Before servicing the vehicle, refer to the precautions in the beginning of this section.
2. Drain the cooling system.
3. Remove or disconnect the following:
- Negative battery cable
- Accessory drive belts
- Front cover
- Timing belt
- Water pump

To install:
4. Install or connect the following:
- Water pump. Use a new O-ring seal and tighten the bolts to 105 inch lbs. (12 Nm).
- Timing belt
- Front cover
- Accessory drive belts
- Negative battery cable

5. Fill the cooling system.
6. Start the engine and check for leaks.

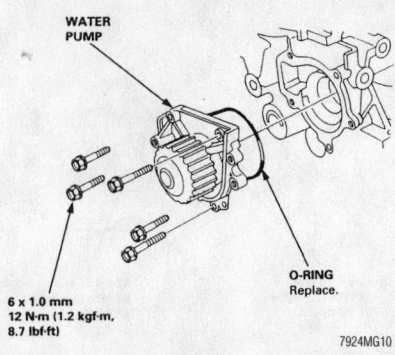

6 x 1.0 mm
12 N·m (1.2 kgf·m, 8.7 lbf·ft)

7924MG10

Exploded view of the water pump mounting

2002

1. Before servicing the vehicle, refer to the precautions in the beginning of this section.
2. Drain the cooling system.
3. Remove or disconnect the following:
- Negative battery cable
- Accessory drive belt
- Crankshaft pulley
- Water pump (6 bolts)

Installation is the reverse of removal. Use new O-rings. Torque the bolts to 104 inch lbs. (12 Nm).

Cylinder Head

REMOVAL & INSTALLATION

1999–01

1. Before servicing the vehicle, refer to the precautions in the beginning of this section.
2. Drain the cooling system.
3. Relieve the fuel system pressure.
4. Remove or disconnect the following:
- Negative battery cable
- Air intake assembly
- Accessory drive belts
- Power steering pump and bracket
- Accelerator cable
- Fuel lines
- Evaporative Emissions (EVAP) control canister hose and vacuum hose
- Brake booster vacuum line
- Intake manifold vacuum line
- Positive Crankcase Ventilation (PCV) valve and hose
- Upper radiator hose
- Heater hose
- Bypass hoses
- Fuel injector connectors
- Engine Coolant Temperature (ECT) sensor connector

- Radiator fan switch connector
- ECT gauge sending unit connector
- Throttle Position (TP) sensor connector
- Manifold Absolute Pressure (MAP) sensor connector
- Heated Oxygen Sensor (HO2S) connector
- Idle Air Control (IAC) valve connector
- Spark plug wires
- Distributor
- Cruise control actuator
- Engine side mount bracket
- Front cover
- Timing belt. Refer to the Timing Belt unit repair section.
- Camshaft sprockets
- Rear timing belt cover
- Valve cover

5. Loosen the valve adjuster locknuts and screws so that all valves are closed.

➡**Keep all valvetrain components in order for assembly.**

6. Remove or disconnect the following:
- Camshafts
- Rocker arms
- Exhaust front pipe
- Exhaust manifold bracket
- Intake manifold

7. Loosen the cylinder head bolts in sequence and 1/3 turns until all bolts are loose.

8. Remove the cylinder head.

To install:

9. Install the cylinder head with a new gasket.

10. Apply clean engine oil to the bolt threads and under the bolt heads.

11. Tighten the cylinder head bolts in sequence as follows:
   a. Step 1: 22 ft. lbs. (29 Nm)
   b. Step 2: 63 ft. lbs. (85 Nm)

12. Install or connect the following:
- Intake manifold. Tighten the bolts to 17 ft. lbs. (24 Nm).
- Exhaust manifold bracket
- Exhaust front pipe
- Rocker arms in their original positions
- Camshafts
- Rear timing belt cover
- Camshaft sprockets. Tighten the bolts to 27 ft. lbs. (37 Nm).
- Timing belt

13. Adjust the valve clearance.

14. Install or connect the following:
- Valve cover

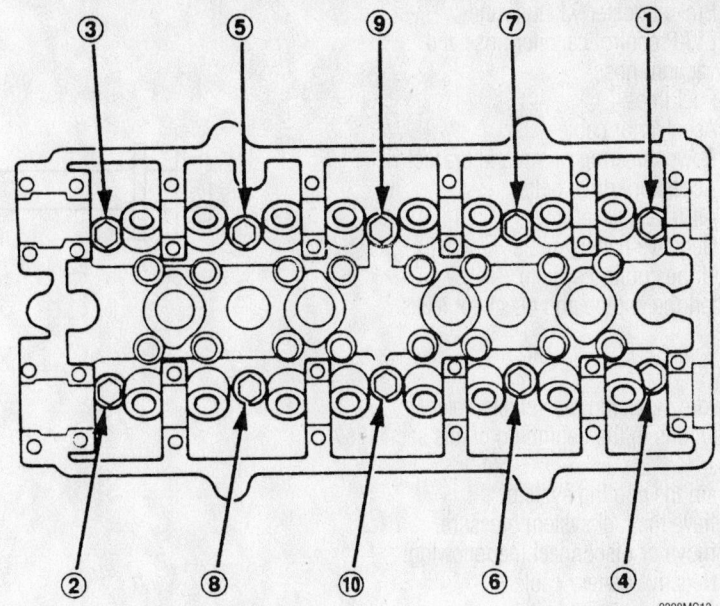

Cylinder head bolt loosening sequence—2.0L

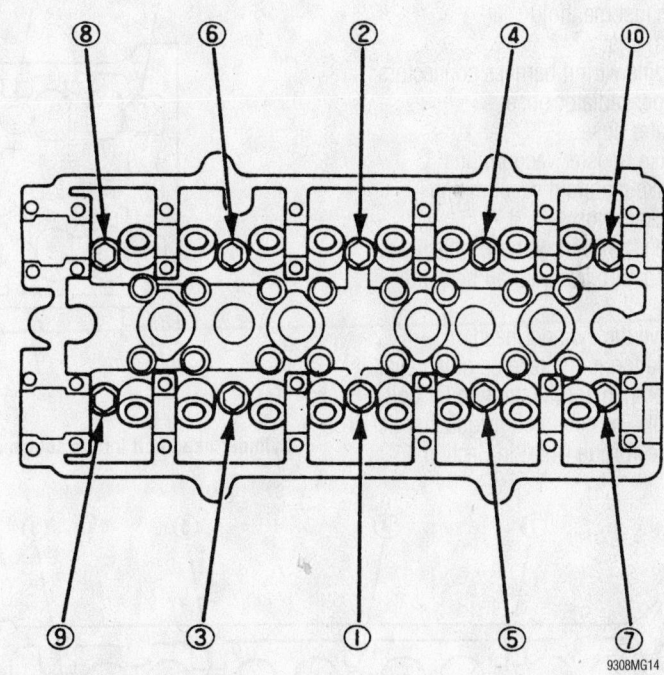

Cylinder head bolt tightening sequence—2.0L

- Front cover
- Engine side mount bracket
- Cruise control actuator
- Distributor
- Spark plug wires
- IAC valve connector
- HO2S connector
- MAP sensor connector
- TP sensor connector

- ECT gauge sending unit connector
- Radiator fan switch connector
- ECT sensor connector
- Fuel injector connectors
- Bypass hoses
- Heater hose
- Upper radiator hose
- PCV valve and hose
- Intake manifold vacuum line

Timing belt service is covered in Section 3 of this manual

- Brake booster vacuum line
- EVAP control canister hose and vacuum hose
- Fuel lines
- Accelerator cable
- Power steering pump and bracket
- Accessory drive belts
- Air intake assembly
- Negative battery cable

15. Fill the cooling system.
16. Start the engine and check for leaks.

2002

1. Before servicing the vehicle, refer to the precautions in the beginning of this section.
2. Drain the cooling system.
3. Relieve the fuel system pressure.
4. Remove or disconnect the following:
- Negative battery cable
- Accessory drive belt
- Fuel lines
- Intake manifold
- Bypass hoses
- Exhaust manifold
- Cam chain
- Engine wiring harness connectors
- Upper radiator hose
- Heater hose
- Brake booster vacuum line
- Intake manifold vacuum line
- Rocker arms

5. Loosen the cylinder head bolts in sequence and 1/3 turns until all bolts are loose.
6. Remove the cylinder head.
7. Installation is the reverse of removal. See the accompanying illustration for bolt measurement. For head bolt torque instructions, see the Torque Chart in Section 1.

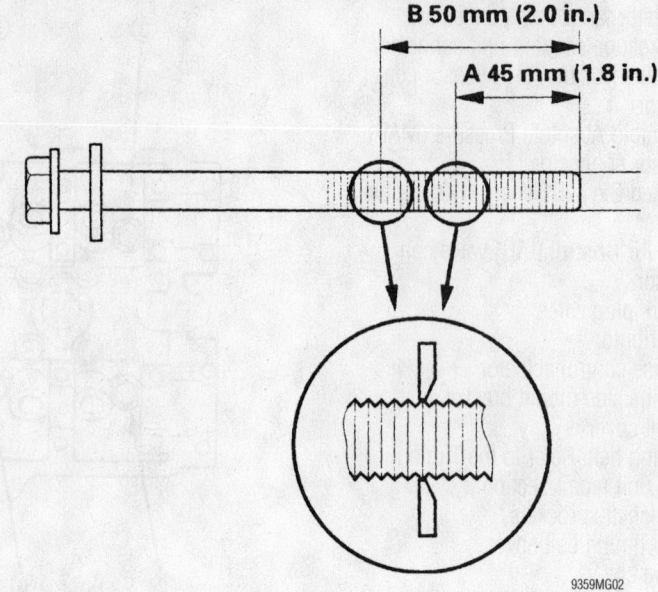

Cylinder head bolt inspection—2.4L

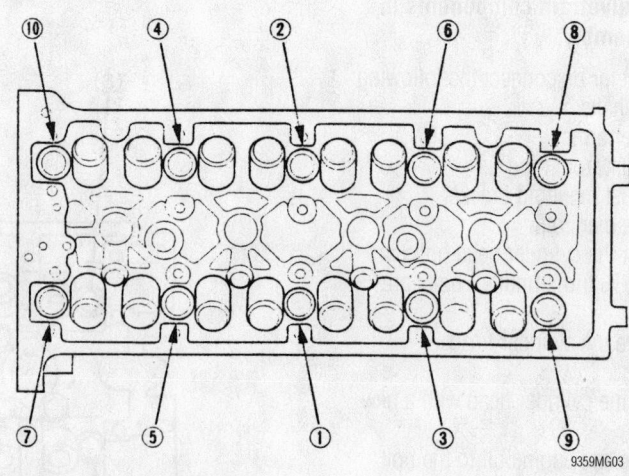

Cylinder head bolt torque sequence—2.4L

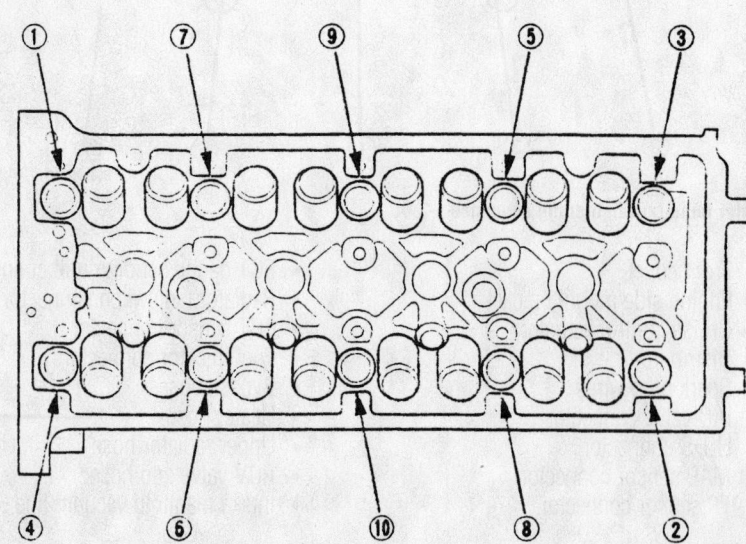

Cylinder head bolt loosening sequence—2.4L

Rocker Arms/Shafts

REMOVAL & INSTALLATION

2.0L

1. Before servicing the vehicle, refer to the precautions in the beginning of this section.
2. Remove or disconnect the following:
- Negative battery cable
- Spark plug wires
- Distributor
- Valve cover
- Accessory drive belts
- Front cover
- Timing belt. Refer to the Timing Belt unit repair section.

3. Loosen the valve adjuster locknuts and screws so that all valves are closed.

➡**Keep all valvetrain components in order for assembly.**

4. Remove or disconnect the following:
   • Camshaft sprockets
   • Rear timing belt cover
   • Camshafts
   • Rocker arms

To install:

5. Install or connect the following:
   • Rocker arms in their original positions
   • Camshafts
   • Rear timing belt cover
   • Camshaft sprockets. Tighten the bolts to 27 ft. lbs. (37 Nm).
   • Timing belt
6. Adjust the valve clearance.
7. Install or connect the following:
   • Front cover
   • Accessory drive belts
   • Valve cover
   • Distributor
   • Spark plug wires
   • Negative battery cable
8. Start the engine and check for leaks.

2.4L

1. Remove the camshaft chain
2. Loosen the rocker adjusting screws
3. Remove the camshaft holder bolts 2 turns at a time in the sequence shown.
4. Remove the camshaft chain guide, camshaft holders and camshafts.

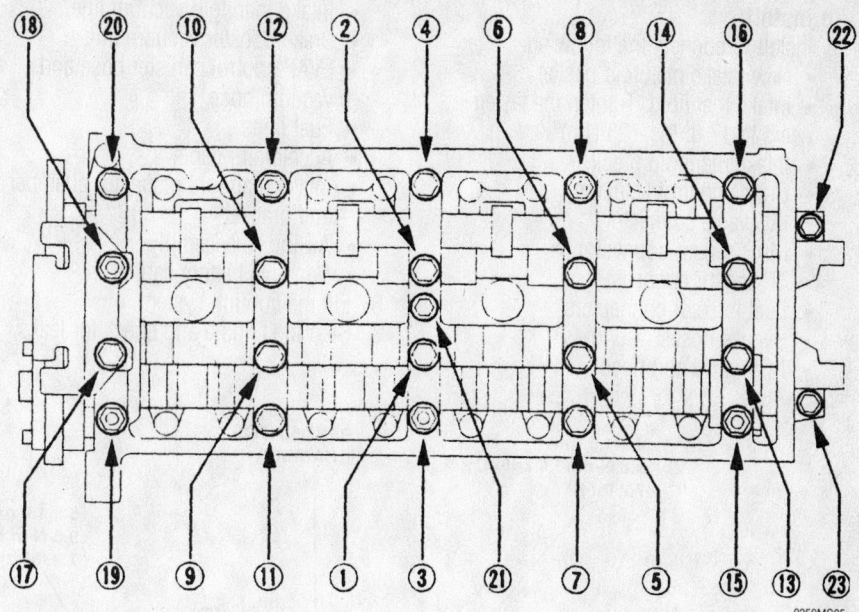

Camshaft holder bolt torque—2.4L

5. Remove the rocker arm assembly.
6. Installation is the reverse of removal. Prior to installation, clean the No.5 rocker shaft holder mating surface and apply RTV gasket sealer to the mounting point on the head. See the illustration for the torque sequence. Torque the 8mm bolts to 16 ft. lbs. (22 Nm) and the 6mm bolts to 9 ft. lbs. (12 Nm).

Intake Manifold

REMOVAL & INSTALLATION

2.0L

1. Before servicing the vehicle, refer to the precautions in the beginning of this section.
2. Drain the cooling system.
3. Relieve the fuel system pressure.
4. Remove or disconnect the following:
   • Negative battery cable
   • Air intake assembly
   • Intake manifold resonator chamber and bracket
   • Accelerator cable
   • Fuel lines
   • Evaporative Emissions (EVAP) control canister hose and vacuum hose
   • Brake booster vacuum line
   • Intake manifold vacuum line
   • Positive Crankcase Ventilation (PCV) valve and hose
   • Bypass hoses
   • Fuel injector connectors
   • Throttle Position (TP) sensor connector
   • Manifold Absolute Pressure (MAP) sensor connector
   • Idle Air Control (IAC) valve connector
   • Cruise control actuator
   • Intake manifold brackets
   • Intake manifold

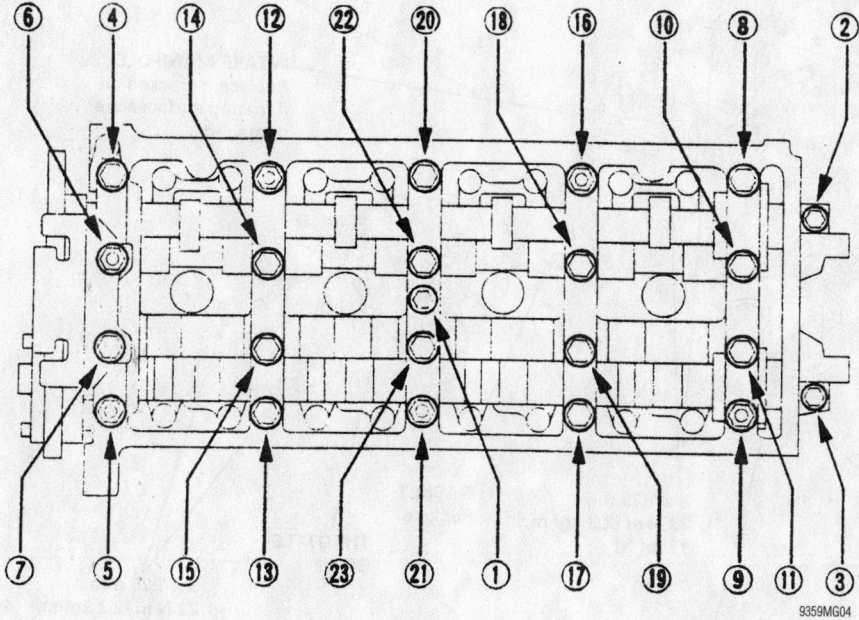

Camshaft holder bolt loosening sequence—2.4L

Heater Core replacement is covered in Section 2 of this manual

To install:

5. Install or connect the following:
- New intake manifold gasket
- Intake manifold. Tighten the fasteners to 17 ft. lbs. (23 Nm).
- Intake manifold brackets
- Cruise control actuator
- IAC valve connector
- MAP sensor connector
- TP sensor connector
- Fuel injector connectors
- Bypass hoses
- PCV valve and hose

- Intake manifold vacuum line
- Brake booster vacuum line
- EVAP control canister hose and vacuum hose
- Fuel lines
- Accelerator cable
- Intake manifold resonator chamber and bracket
- Air intake assembly
- Negative battery cable

6. Fill the cooling system.
7. Start the engine and check for leaks.

2.4L

1. Before servicing the vehicle, refer to the precautions in the beginning of this section.
2. Drain the cooling system.
3. Relieve the fuel system pressure.
4. Remove or disconnect the following:
- Negative battery cable
- Air intake assembly
- Accelerator cable
- Cruise control cable
- Evaporative Emissions (EVAP) control canister hose and vacuum hose

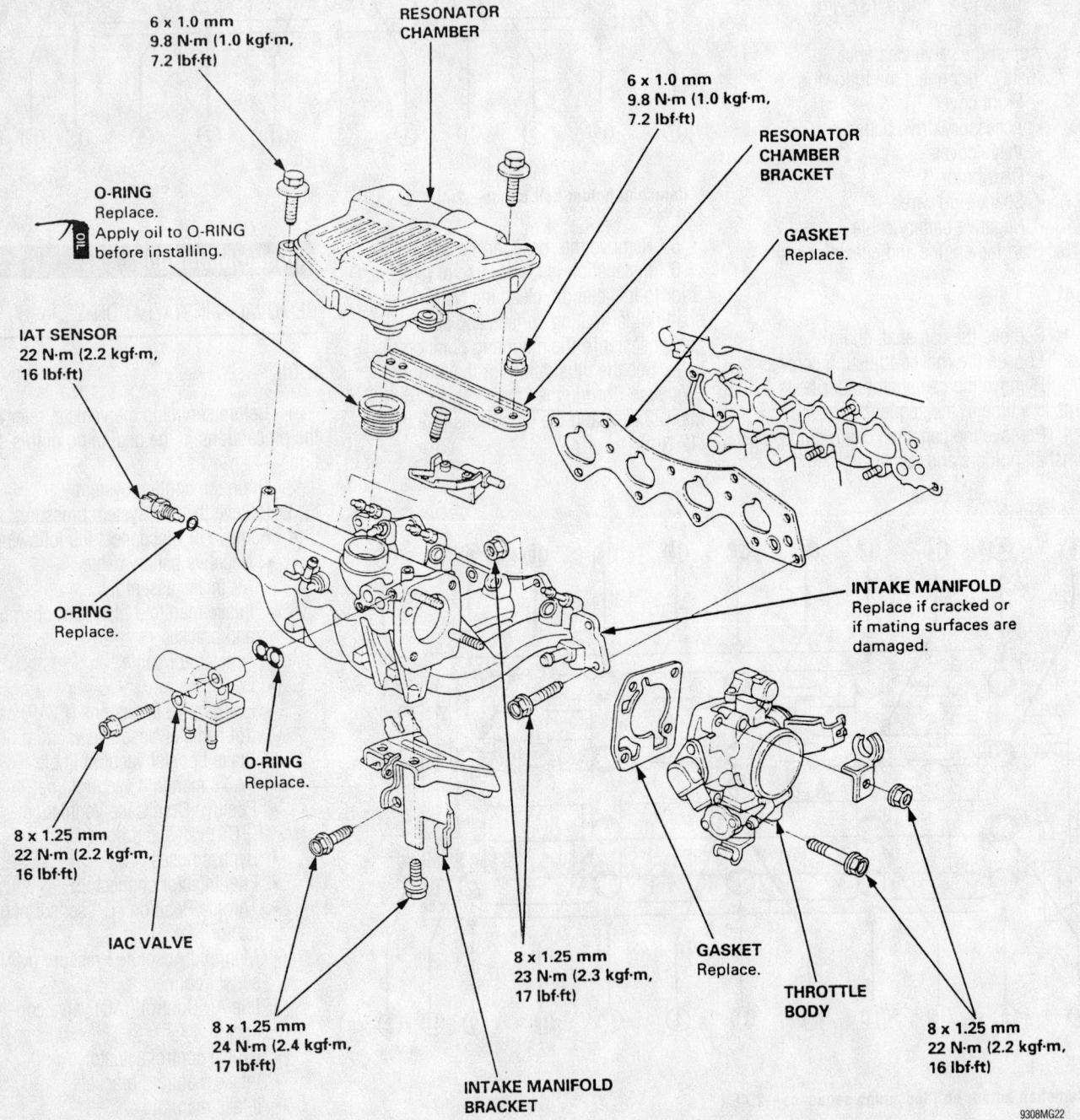

Intake manifold exploded view—1999–01 CR-V

9308MG22

- Brake booster vacuum line
- Bypass hoses
- Intake manifold resonator chamber and bracket, for 1999–01 models
- Fuel lines
- Intake manifold vacuum line
- Positive Crankcase Ventilation (PCV) valve and hose
- Fuel injector connectors
- Throttle Position (TP) sensor connector
- Manifold Absolute Pressure (MAP) sensor connector
- Idle Air Control (IAC) valve connector
- Intake manifold brackets
- Intake manifold

To install:

5. Install or connect the following:
- New intake manifold gasket
- Intake manifold. Tighten the fasteners to 16 ft. lbs. (22Nm).
- Intake manifold brackets
- Cruise control actuator
- IAC valve connector
- MAP sensor connector
- TP sensor connector
- Fuel injector connectors
- Bypass hoses
- PCV valve and hose
- Intake manifold vacuum line
- Brake booster vacuum line
- EVAP control canister hose and vacuum hose
- Fuel lines
- Accelerator cable
- Air intake assembly
- Negative battery cable

6. Fill the cooling system.
7. Start the engine and check for leaks.

Exhaust Manifold

REMOVAL & INSTALLATION

2.0L

1. Before servicing the vehicle, refer to the precautions in the beginning of this section.
2. Remove or disconnect the following:
- Negative battery cable
- Exhaust manifold heat shield
- Heated Oxygen Sensor (HO2S) connector
- Exhaust front pipe
- Exhaust manifold bracket, if equipped
- Exhaust manifold

To install:

3. Install or connect the following:
- Exhaust manifold. Tighten the fasteners to 23 ft. lbs. (31 Nm).
- Exhaust manifold bracket, if equipped. Tighten the bolts to 33 ft. lbs. (44 Nm).
- Exhaust front pipe. Tighten the nuts to 40 ft. lbs. (55 Nm).
- Heated Oxygen Sensor (HO2S) connector
- Exhaust manifold heat shield
- Negative battery cable

2.4L

1. Before servicing the vehicle, refer to the precautions in the beginning of this section.
2. Remove or disconnect the following:
- Negative battery cable
- VTEC solenoid valve
- Driveshaft heat shield
- Exhaust manifold heat shield
- Exhaust front pipe
- Exhaust manifold bracket, if equipped
- Exhaust manifold

To install:

3. Install or connect the following:
- Exhaust manifold. Tighten the fasteners to 3 ft. lbs. (44m).
- Exhaust manifold bracket, if equipped. Tighten the bolts to 33 ft. lbs. (44 Nm).
- Exhaust front pipe. Tighten the nuts to 16t. lbs. (22m).
- Exhaust manifold heat shield
- Driveshaft heat shield
- VTEC solenoid valve
- Negative battery cable

Front Crankshaft Seal

REMOVAL & INSTALLATION

For the 2.4L engine, see the Timing Chain Removal & Installation procedure.

2.0L

1. Before servicing the vehicle, refer to the precautions in the beginning of this section.
2. Remove or disconnect the following:
- Negative battery cable
- Accessory drive belts
- Side engine mount
- Valve cover
- Crankshaft pulley

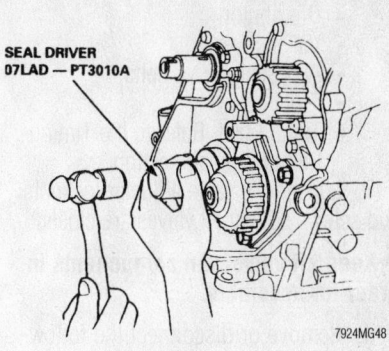

SEAL DRIVER
07LAD — PT3010A

7924MG48

Front crankshaft seal installation–2.0L

- Front cover
- Balance shaft belt, if equipped
- Timing belt. Refer to the Timing Belt unit repair section.
- Top Dead Center (TDC) sensor, if equipped
- Crankshaft timing sprocket
- Front crankshaft seal

To install:

3. Lubricate the crankshaft seal lip with grease prior to installation.
4. Install the front crankshaft seal so that it is flush with the surface of the oil pump housing.
5. Install or connect the following:
- Crankshaft timing sprocket
- Top Dead Center (TDC) sensor, if equipped
- Timing belt. Refer to the Timing Belt unit repair section.
- Balance shaft belt, if equipped
- Front cover
- Crankshaft pulley. Tighten the bolt to 130 ft. lbs. (177 Nm).
- Valve cover
- Side engine mount
- Accessory drive belts
- Negative battery cable

6. Check the engine oil level and add if necessary.
7. Start the engine and check for leaks.

Camshaft

REMOVAL & INSTALLATION

2.0L

1. Before servicing the vehicle, refer to the precautions in the beginning of this section.
2. Remove or disconnect the following:
- Negative battery cable
- Spark plug wires

- Distributor
- Valve cover
- Accessory drive belts
- Front cover
- Timing belt. Refer to the Timing Belt unit repair section.

3. Loosen the valve adjuster locknuts and screws so that all valves are closed.

➡**Keep all valvetrain components in order for assembly.**

4. Remove or disconnect the following:
- Camshaft sprockets
- Rear timing belt cover
- Camshafts

To install:

➡**Use new O-rings, seals and gaskets when installing the camshaft.**

5. Install or connect the following:
- Camshafts. Tighten the bearing cap bolts in sequence to 86 inch lbs. (10 Nm).
- Rear timing belt cover

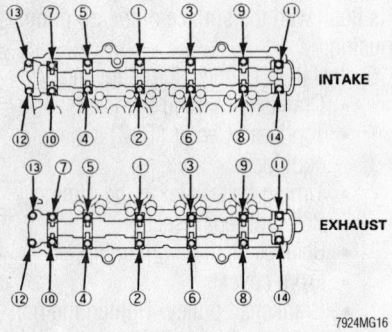

Camshaft bearing tightening sequence—2.0L

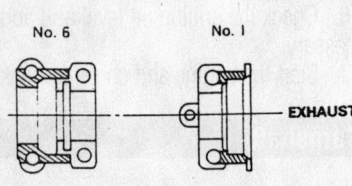

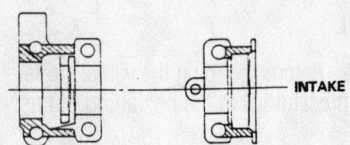

Apply liquid gasket to the shaded areas of the camshaft journals—2.0L

- Camshaft sprockets. Tighten the bolts to 27 ft. lbs. (37 Nm).
- Timing belt
6. Adjust the valve clearance.
7. Install or connect the following:
- Front cover
- Accessory drive belts
- Valve cover
- Distributor
- Spark plug wires
- Negative battery cable
8. Start the engine and check for leaks.

2.4L

See the Rocker Arm Shaft Removal & Installation procedure.

Valve Lash

ADJUSTMENT

Adjust the valves only when the cylinder head temperature is less than 100°F (38°C).

2.0L

1. Before servicing the vehicle, refer to the precautions in the beginning of this section.
2. Remove or disconnect the following:
- Negative battery cable
- Air intake tube
- Valve cover
3. Rotate the crankshaft so that the valves to be adjusted are closed and the rocker arm is contacting the camshaft lobe base circle.

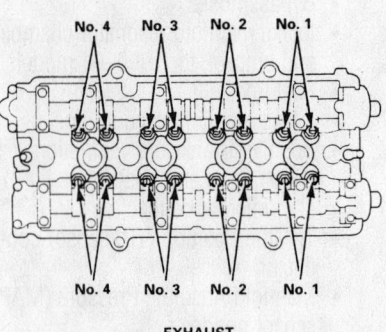

Intake and exhaust valve identification—2.0L

4. Measure the valve clearance. If adjustment is necessary, loosen the locknut and turn the adjusting screw as necessary to achieve the correct valve clearance.
5. The correct valve clearance is:
- Intake valves: 0.003–0.005 inches (0.08–0.12mm)
- Exhaust valves: 0.006–0.008 inches (0.16–0.20mm)
6. After adjustment, tighten the locknuts to 18 ft. lbs. (24 Nm).
7. Install or connect the following:
- Valve cover
- Air intake tube
- Negative battery cable
8. Start the engine and check for proper operation.

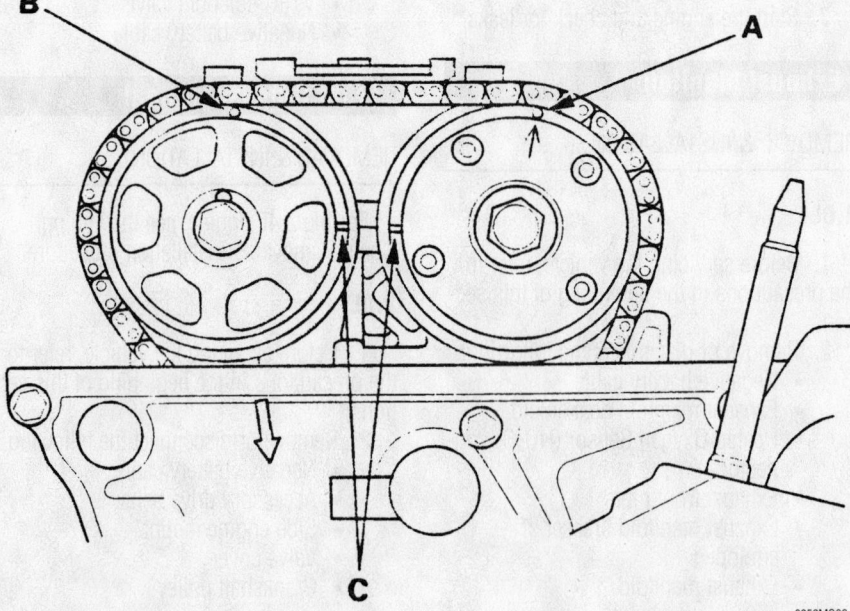

Align the timing marks—2.4L

2.4L

1. Before servicing the vehicle, refer to the precautions in the beginning of this section.
2. Remove or disconnect the following:
   - Negative battery cable
   - cylinder head cover
3. Set the timing marks as shown in the illustration with N0.1 at TDC. Check all clearances. Intake should be 0.008–0.010 in.; exhaust should be 0.011–0.013 in. Intake locknut torque is 14 ft. lbs.; exhaust is 10 ft. lbs.
4. Rotate the crankshaft 180 degrees clockwise and recheck No.3.
5. Rotate the crankshaft 180 degrees clockwise and recheck No.4.
6. Rotate the crankshaft 180 degrees clockwise and recheck No.2.

Starter Motor

REMOVAL & INSTALLATION

2.0L

1. Before servicing the vehicle, refer to the precautions in the beginning of this section.
2. Remove or disconnect the following:
   - Negative battery cable
   - Starter wiring harness connectors
   - Starter motor

To install:

3. Install or connect the following:
   - Starter motor. Tighten the bolts to 33 ft. lbs. (44 Nm).
   - Starter wiring harness connectors. Tighten the battery cable nut to 79 inch lbs. (9 Nm).
   - Negative battery cable

2.4L

1. Before servicing the vehicle, refer to the precautions in the beginning of this section.
2. Remove or disconnect the following:
   - Negative battery cable
   - Knock sensor connector
   - Starter wiring harness connectors
   - Starter motor

To install:

3. Install or connect the following:
   - Starter motor. Tighten the upper bolt to 33 ft. lbs. (44 Nm); the lower bolt to 47 ft. lbs. (64 Nm).
   - Starter wiring harness connectors. Tighten the battery cable nut to 84 inch lbs. (9 Nm).

   - Knock sensor connector
   - Negative battery cable

Timing Chain and Front Seal

REMOVAL & INSTALLATION

2.4L

1. Before servicing the vehicle, refer to the precautions in the beginning of this section.

2. Drain the engine oil.
3. Align the timing marks at TDC No.1.
4. Remove or disconnect the following:

   - Negative battery cable
   - Front splash shield
   - Drive belt
   - Cylinder head cover
   - Crankshaft pulley
   - CKP sensor
   - VTC oil control connector
   - VTC oil control solenoid valve

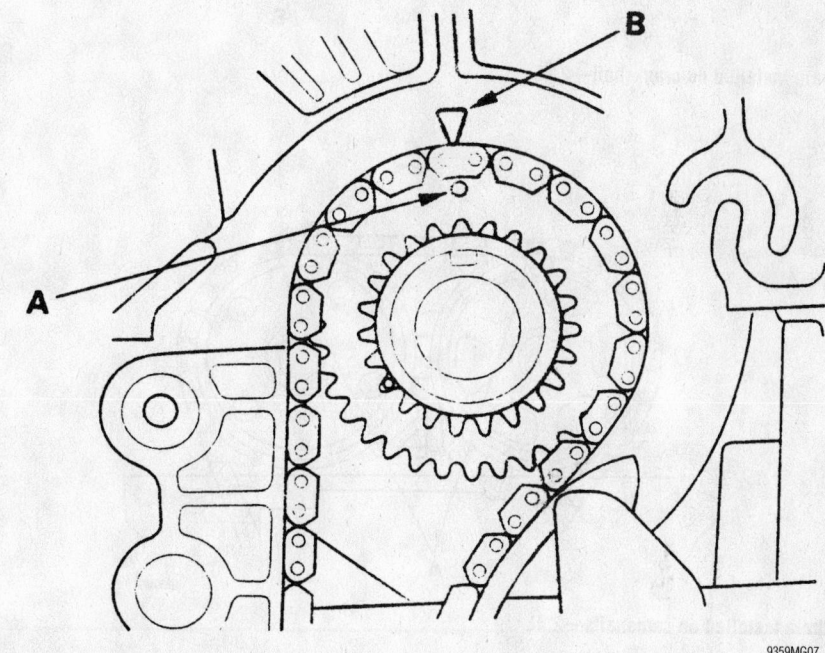

Align the crankshaft timing marks—2.4L

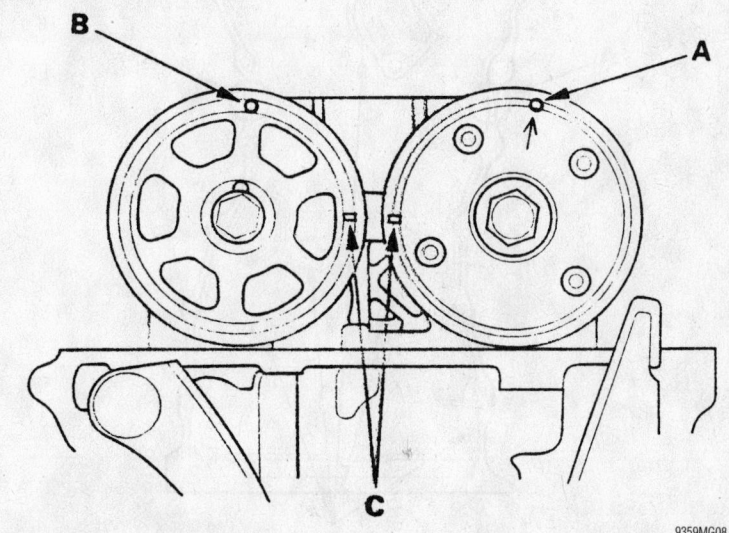

Align the camshaft timing marks—2.4L

For complete Engine Mechanical specifications, see Section 1 of this manual

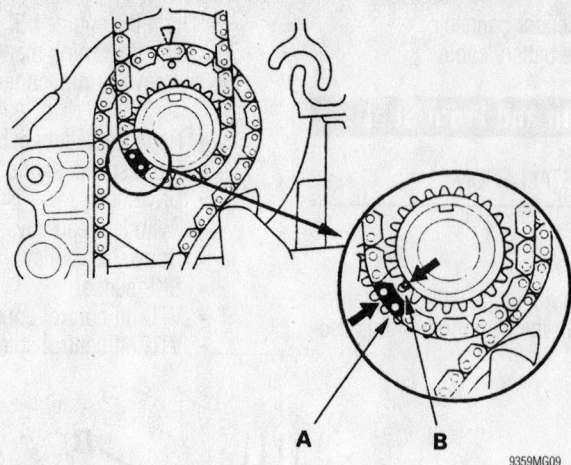

Chain installed on crankshaft—2.4L

9359MG09

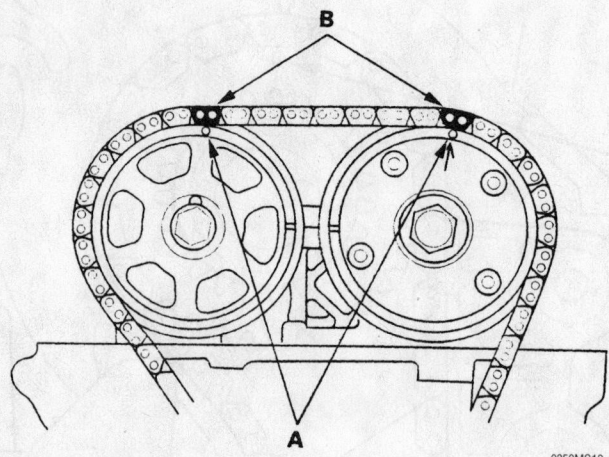

Chain installed on camshafts—2.4L

9359MG10

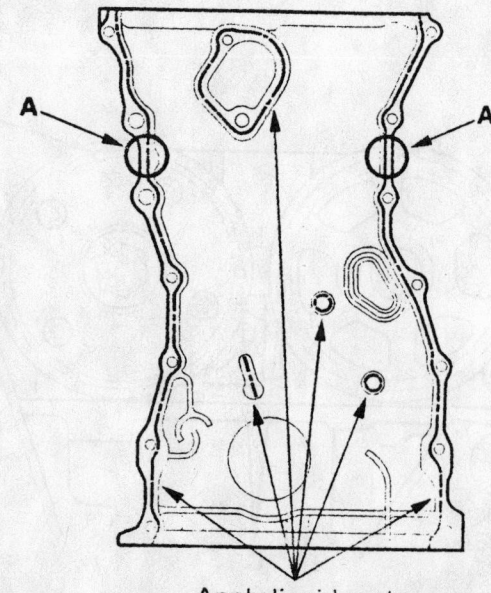

Apply liquid gasket along the broken line.

9359MG11

Chain case sealer application—2.4L

5. Support the engine with a block of wood and jack.

6. Remove or disconnect the following:
- Ground cable
- Upper support bracket
- Side engine mount
- Chain case

7. Loosely install the crank pulley. Turn the crankshaft counterclockwise to compress the auto-tensioner. Align the holes on the lock and auto-tensioner and insert a 1.5mm pin into the holes. Turn the crank clockwise to hold the pin.

8. Remove the auto-tensioner.

9. Remove the chain guides.

10. Remove the tensioner arm.

11. Remove the chain.

12. With the case removed, drive out the old seal and install a new one.

To install:

13. Set the crankshaft to TDC.

14. Set the camshafts to TDC.

15. Install the chain on the sprocket with the colored piece A aligned with the punch mark B.

16. The remainder of installation is the reverse of removal. See the accompanying illustration for sealer application. Observe the following torques:
- Chain guide: 9 ft. lbs. (12 Nm)
- Tensioner arm: 16 ft. lbs. (22 Nm)
- Auto-tensioner: 9 ft. lbs. (12 Nm)
- Upper chain guide: 16 ft. lbs. (22 Nm)
- Case: 9 ft. lbs. (12 Nm)
- Side mount: 33 ft. lbs. (44 Nm)
- Upper bracket: 40 ft. lbs. (54 Nm)

Oil Pan

REMOVAL & INSTALLATION

2.0L

1. Before servicing the vehicle, refer to the precautions in the beginning of this section.

2. Drain the engine oil.

3. Remove or disconnect the following:
- Negative battery cable
- Front splash shield
- Heated Oxygen Sensor (HO$_2$S) connector
- Exhaust front pipe
- Torque converter cover, if equipped with an automatic transaxle
- Oil pan

To install:

4. Install the oil pan.

5. Tighten the bolts in sequence to 105 inch lbs. (12 Nm).

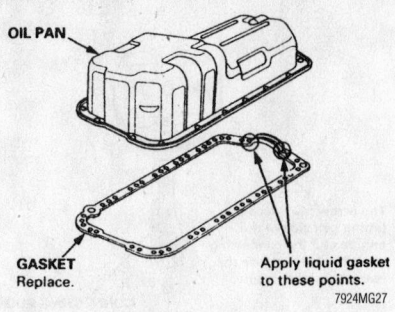

Oil pan gasket installation–2.0L

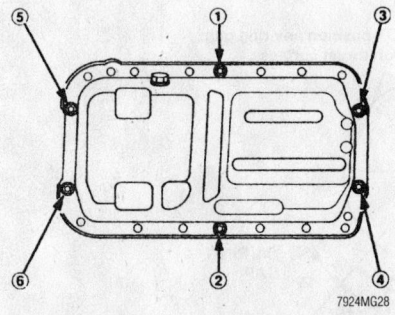

Oil pan fastener tightening sequence–2.0L

6. Install or connect the following:
- Torque converter cover, if removed
- Exhaust front pipe
- Subframe center beam, if removed
- HO$_2$S connector

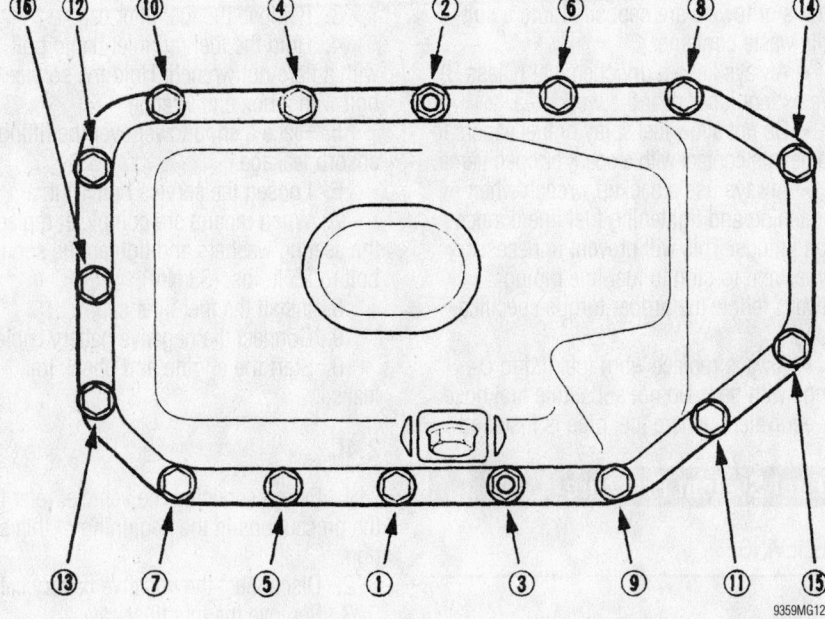

Oil pan fastener tightening sequence–2.4L

- Front splash shield
- Negative battery cable

2.4L

1. Before servicing the vehicle, refer to the precautions in the beginning of this section.
2. Drain the engine oil.
3. Remove or disconnect the following:

- Subframe. See engine Removal and Installation.
- With MT, the stiffener
- Oil pan bolts
- Oil pan. A gasket cutter will be needed.

4. Installation is the reverse of removal. Torque the bolts, in sequence, in 2 or 3 steps, to 9 ft. lbs. (12 Nm).

Oil Pump

REMOVAL & INSTALLATION

2.0L

1. Before servicing the vehicle, refer to the precautions in the beginning of this section.
2. Drain the engine oil.
3. Remove or disconnect the following:

- Negative battery cable
- Accessory drive belts
- Front cover
- Timing belt. Refer to the Timing Belt unit repair section.
- Crankshaft timing sprocket
- Oil pan
- Oil pump pickup tube
- Oil pump

To install:

➡**Use new gaskets and O-ring seals for assembly.**

4. Apply liquid gasket to the oil pump and to the bolt hole threads.
5. Install or connect the following:
- Oil pump. Tighten the 8mm bolts to 17 ft. lbs. (24 Nm) and the 6mm bolts to 86 inch lbs. (10 Nm).
- Oil pump pickup tube. Tighten the fasteners to 86 inch lbs. (10 Nm).
- Oil pan
- Crankshaft timing sprocket
- Timing belt. Refer to the Timing Belt unit repair section.
- Front cover
- Accessory drive belts
- Negative battery cable

6. Fill the crankcase to the correct level.
7. Start the engine and check for leaks.

2.4L

1. Before servicing the vehicle, refer to the precautions in the beginning of this section.
2. Drain the engine oil.
3. Remove or disconnect the following:

- Negative battery cable
- Oil pan
- Pump chain
- Pump sprocket
- Pump

To install:

4. Make sure that No.1 piston is at TDC.
5. Align the dowel pin on the rear balance shaft with the mark on the pump.
6. Install the pump and sprocket loosely.
7. Remove the balance shaft holding pin.
8. Torque the 10mm mounting bolts to 33 ft. lbs. (44 Nm); the 8mm bolts to 16 ft. lbs. (22 Nm).
9. Torque the pulley bolt to 33 ft. lbs. (44 Nm).
10. Torque the tensioner bolts to 9 ft. lbs. (12 nm).

For Accessory Drive Belt illustrations, see Section 1 of this manual

Rear Main Seal

REMOVAL & INSTALLATION

1. Before servicing the vehicle, refer to the precautions in the beginning of this section.
2. Remove or disconnect the following:
   - Transaxle
   - Clutch pressure plate and disc, if equipped
   - Flywheel
   - Oil seal

To install:

3. Install or connect the following:
   - Oil seal. Drive the seal square into the seal case.
   - Flywheel. Tighten the bolts in a crossing pattern to 54 ft. lbs. (73 Nm).
   - Clutch pressure plate and disc, if equipped
   - Transaxle
4. Check the fluid levels.
5. Start the engine and check for leaks.

Piston and Ring

POSITIONING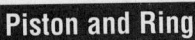

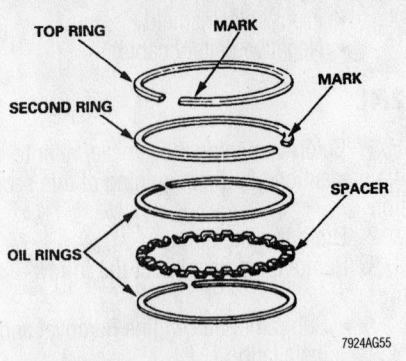

7924AG55

Piston ring positioning and top mark location

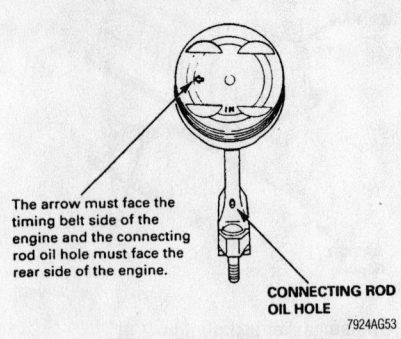

7924AG53

Piston and connecting rod assembly

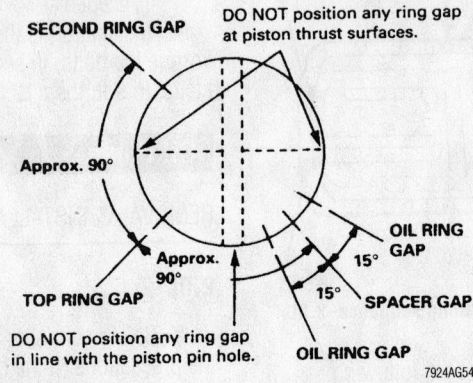

7924AG54

Piston ring end-gap spacing

FUEL SYSTEM

Fuel System Service Precautions

Safety is the most important factor when performing not only fuel system maintenance, but any type of maintenance. Failure to conduct maintenance and repairs in a safe manner may result in serious personal injury or death. Maintenance and testing of the vehicle's fuel system components can be accomplished safely and effectively by adhering to the following rules and guidelines:

- To avoid the possibility of fire and personal injury, always disconnect the negative battery cable unless the repair or test procedure requires that battery voltage be applied.
- Always relieve the fuel system pressure prior to disconnecting any fuel system component (injector, fuel rail, pressure regulator, etc.), fitting or fuel line connection. Exercise extreme caution whenever relieving fuel system pressure to avoid exposing skin, face and eyes to fuel spray. Please be advised that fuel under pressure may penetrate the skin or any part of the body that it contacts.
- Always place a shop towel or cloth around the fitting or connection prior to loosening to absorb any excess fuel due to spillage. Ensure that all fuel spillage

(should it occur) is quickly removed from engine surfaces. Ensure that all fuel soaked cloths or towels are deposited into a suitable waste container.

- Always keep a dry chemical (Class B) fire extinguisher near the work area.
- Do not allow fuel spray or fuel vapors to come into contact with a spark or open flame.
- Always use a backup wrench when loosening and tightening fuel line connection fittings. This will prevent unnecessary stress and torsion to fuel line piping. Always follow the proper torque specifications.
- Always replace worn fuel fitting O-rings with new. Do not substitute fuel hose or equivalent, where fuel pipe is installed.

Fuel System Pressure

RELIEVING

2.0L

1. Before servicing the vehicle, refer to the precautions in the beginning of this section.

2. Disconnect the negative battery cable.
3. Remove the fuel filler cap.
4. Hold the fuel rail inlet banjo bolt with a flare nut wrench. Hold the service bolt with a box end wrench.
5. Place a shop towel over the fitting to absorb leakage.
6. Loosen the service bolt 1 turn.
7. When repairs are complete, replace the sealing washers and tighten the service bolt to 25 ft. lbs. (33 Nm).
8. Install the fuel filler cap.
9. Connect the negative battery cable.
10. Start the engine and check for leaks.

2.4L

1. Before servicing the vehicle, refer to the precautions in the beginning of this section.
2. Disconnect the negative battery cable.
3. Remove the fuel filler cap.
4. Remove the engine cover.
5. Using a back up wrench and shop towel, turn the fuel pulsation damper one complete turn, slowly.

→**Replace all washers whenever the pulsation damper is loosened or removed.**

6. Tighten the damper to 16 ft. lbs. (22 Nm).

Fuel Filter

REMOVAL & INSTALLATION

2.0L

1. Before servicing the vehicle, refer to the precautions in the beginning of this section.
2. Relieve the fuel system pressure.
3. Remove or disconnect the following:
   - Negative battery cable
   - Wire harness bracket
   - Power steering hose bracket
   - Fuel lines
   - Fuel filter

To install:

4. Install or connect the following:
   - Fuel filter
   - Fuel lines. Use new sealing washers.
   - Power steering hose bracket
   - Wire harness bracket
   - Negative battery cable
5. Start the engine and check for leaks.

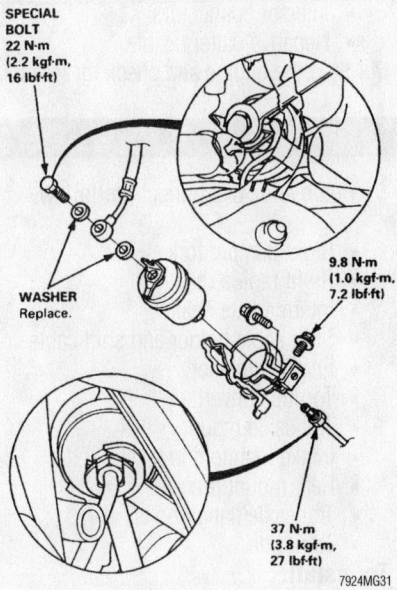

SPECIAL BOLT
22 N·m
(2.2 kgf·m,
16 lbf·ft)

WASHER
Replace.

9.8 N·m
(1.0 kgf·m,
7.2 lbf·ft)

37 N·m
(3.8 kgf·m,
27 lbf·ft)

7924MG31

Exploded view of the fuel filter mounting—2.0L

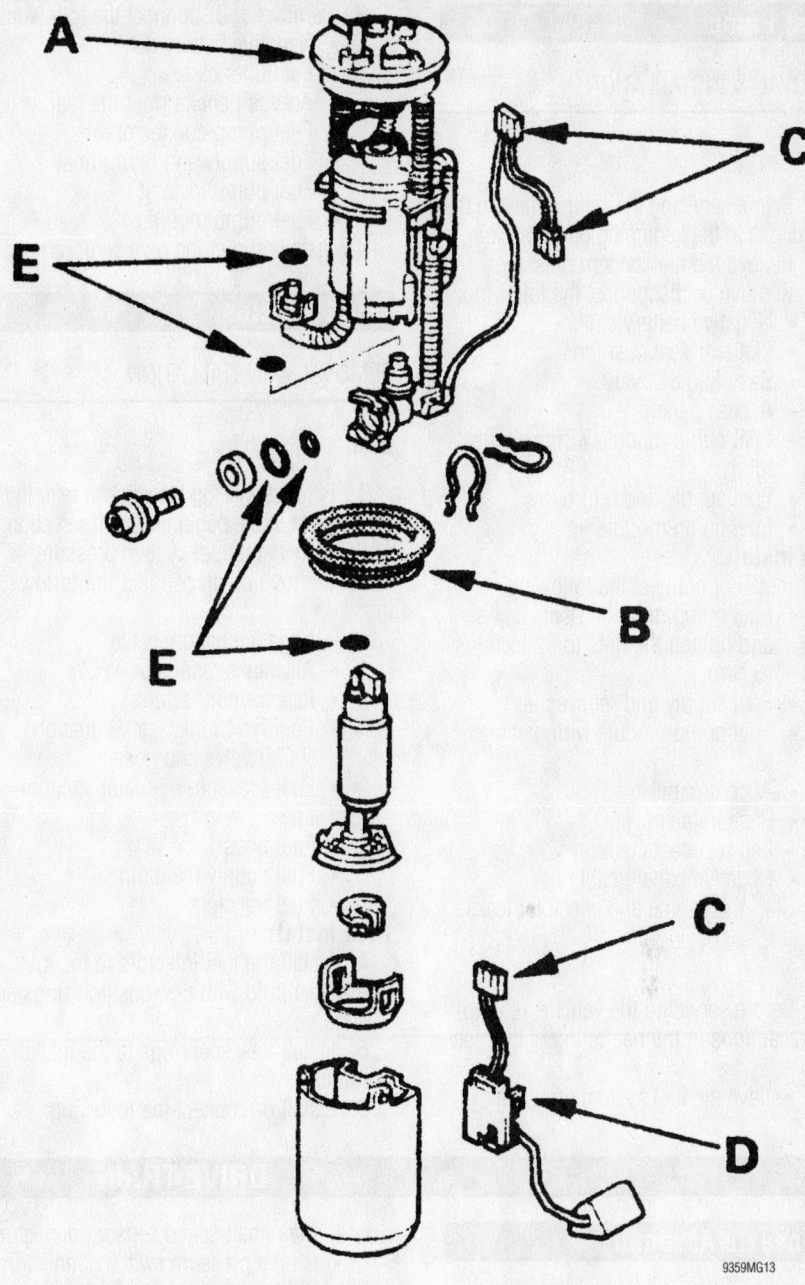

9359MG13

Exploded view of the fuel filter mounting—2.4L

2.4L

→**The fuel filter should be replaced whenever the fuel pressure drops below48 psi, after making sure that the fuel pump and fuel pressure regulator are okay.**

1. Before servicing the vehicle, refer to the precautions in the beginning of this section.
2. Relieve the fuel system pressure.
3. Remove or disconnect the following:
   - Negative battery cable
   - Fuel pump
   - Fuel filter carrier (A)
   - Fuel filter

To install:

4. Install or connect the following:
   - Fuel filter
   - Fuel lines
   - New gasket (B)
   - New o-rings (E)
   - Connectors (C)
   - Sending unit (D)
5. Start the engine and check for leaks.

For Tire, Wheel and Ball Joint specifications, see Section 1 of this manual

Fuel Pump

REMOVAL & INSTALLATION

2.4L

1. Before servicing the vehicle, refer to the precautions in the beginning of this section.
2. Relieve the fuel system pressure.
3. Remove or disconnect the following:
   - Negative battery cable
   - Left rear seat cushion
   - Base frame cover
   - Access panel
   - Fuel pump module wiring connector
   - Fuel supply and return lines
   - Fuel pump module

To install:

4. Install or connect the following:
   - Fuel pump module. Use a new seal and tighten the nuts to 52 inch lbs. (6 Nm).
   - Fuel supply and return lines
   - Fuel pump module wiring connector
   - Access panel
   - Base frame cover
   - Left rear seat cushion
   - Negative battery cable
5. Start the engine and check for leaks.

2.4L

1. Before servicing the vehicle, refer to the precautions in the beginning of this section.
2. Relieve the fuel system pressure.

3. Remove or disconnect the following:
   - Negative battery cable
   - Fuel filler cap
   - Access panel, under the rear seats
   - Fuel pump connector
   - Fuel supply and return lines
   - Fuel pump locknut
   - Fuel pump module
4. Installation is the reverse of removal.

Fuel Injector

REMOVAL & INSTALLATION

2.0L

1. Before servicing the vehicle, refer to the precautions in the beginning of this section.
2. Relieve the fuel system pressure.
3. Remove or disconnect the following:

   - Negative battery cable
   - Air intake resonator
   - Injector connectors
   - Positive Crankcase Ventilation (PCV) valve and hose
   - Fuel pressure regulator vacuum line
   - Fuel lines
   - Fuel supply manifold
   - Fuel injectors

To install:

4. Install the fuel injectors to the fuel supply manifold with new cushion rings and O-rings.
5. Install new seal rings to the intake manifold.
6. Install or connect the following:

   - Fuel supply manifold and injector assembly. Tighten the nuts to 105 inch lbs. (12 Nm).
   - Fuel lines
   - Fuel pressure regulator vacuum line
   - PCV valve and hose
   - Injector connectors
   - Air intake resonator
   - Negative battery cable
7. Start the engine and check for leaks.

2.4L

1. Before servicing the vehicle, refer to the precautions in the beginning of this section.
2. Relieve the fuel system pressure.
3. Remove or disconnect the following:
   - Negative battery cable
   - Injector connectors, ground cable and harness holder
   - Fuel line
   - Fuel supply manifold
   - Fuel injectors

To install:

4. Install the fuel injectors to the fuel supply manifold with new O-rings coated with clean engine oil.
5. Install new seal rings, coated with clean engine oil, in the intake manifold.
6. Install or connect the following:
   - Fuel supply manifold and injector assembly. Tighten the nuts to 16 ft. lbs. (22 Nm).
   - Fuel lines
   - Injector connectors
   - Negative battery cable
7. Start the engine and check for leaks.

DRIVE TRAIN

Transaxle Assembly

REMOVAL & INSTALLATION

Automatic Transaxle

1999–01

1. Before servicing the vehicle, refer to the precautions in the beginning of this section.
2. Drain the transaxle.
3. Remove or disconnect the following:
   - Battery
   - Battery tray
   - Air intake assembly
   - Starter motor
   - Transaxle ground cable
   - Clutch pressure control solenoid valve connector

- Mainshaft speed sensor connector
- Clutch pressure switch connectors
- Shift control solenoid valve connectors
- Lockup control solenoid connector
- Countershaft speed sensor connector
- Transaxle oil cooler lines
- Gear position switch connector
- Engine splash shield
- Front wheels
- Subframe center beam
- Rear driveshaft, if equipped
- Front motor mount bracket
- Lower ball joints
- Lower damper fork bolts

4. Separate the inner CV-joints from the transaxle and intermediate shaft and support the axle halfshafts out of the work area with safety wire.

5. Remove or disconnect the following:

- Right damper fork
- Right radius rod
- Intermediate shaft
- Shift cable holder and shift cable
- Engine stiffener
- Torque converter
- Transaxle mount
- Intake manifold bracket
- Rear mount bracket
- Transaxle flange bolts
- Transaxle

To install:

➡**Use new circlips, split pins and self-locking nuts for assembly.**

6. Install or connect the following:
   - Transaxle. Tighten the flange bolts to 47 ft. lbs. (64 Nm).

- Rear mount bracket. Tighten the bolts to 40 ft. lbs. (54 Nm).
- Intake manifold bracket. Tighten the bolts to 16 ft. lbs. (22 Nm).
- Transaxle mount. Tighten the stud bolt and nuts to 28 ft. lbs. (38 Nm) and the through bolt to 40 ft. lbs. (54 Nm).
- Front motor mount bracket. Tighten the bolts to 28 ft. lbs. (38 Nm).
- Torque converter. Tighten the drive-plate bolts to 105 inch lbs. (12 Nm).
- Engine stiffener. Tighten the bolts to 33 ft. lbs. (44 Nm).
- Shift cable holder and shift cable
- Intermediate shaft. Tighten the bolts to 29 ft. lbs. (39 Nm).
- Axle halfshafts
- Right damper fork. Tighten the pinch bolt to 32 ft. lbs. (43 Nm).
- Right radius rod. Tighten the bolts to 76 ft. lbs. (103 Nm) and the nut to 32 ft. lbs. (43 Nm).
- Lower damper fork bolts. Tighten the nut to 47 ft. lbs. (64 Nm).
- Lower ball joints. Tighten the nut to 36–43 ft. lbs. (49–59 Nm).
- Rear driveshaft, if equipped
- Subframe center beam. Tighten the bolts to 37 ft. lbs. (50 Nm).
- Front wheels
- Engine splash shield
- Gear position switch connector
- Transaxle oil cooler lines
- Countershaft speed sensor connector
- Lockup control solenoid connector
- Shift control solenoid valve connectors
- Clutch pressure switch connectors
- Mainshaft speed sensor connector
- Clutch pressure control solenoid valve connector
- Transaxle ground cable
- Starter motor. Tighten the bolts to 33 ft. lbs. (44 Nm).
- Air intake assembly
- Battery tray
- Battery

7. Fill the transaxle to the correct level.
8. Start the engine and check for leaks.
9. Check the wheel alignment and adjust as necessary.

2002

1. Before servicing the vehicle, refer to the precautions in the beginning of this section.

2. Drain the transaxle.
3. Remove or disconnect the following:
- Battery
- Battery tray
- Air intake assembly
- Splash shield
- Transaxle ground cable
- Starter motor
- Clutch pressure control solenoid valve connector
- Mainshaft speed sensor connector
- Clutch pressure switch connectors
- Shift control solenoid valve connectors
- Lockup control solenoid connector
- Countershaft speed sensor connector
- Transaxle oil cooler lines
- Engine wiring harness from the air cleaner bracket
- Water pipe mounting bolt
- Brake booster and EVAP line mounting bolts. Attach an engine support hanger to the head to support the weight of the engine.
- Stabilizer link from the lower arm
- Lower arms from the knuckles
- Torque converter nuts
- Shift cable
- Front mount bolt and nut
- Rear mount bolts
- Subframe (see the Engine Removal and Installation procedure)
- Rear driveshaft, if equipped

4. Separate the inner CV-joints from the transaxle and intermediate shaft and support the axle halfshafts out of the work area with safety wire.
- Intermediate shaft
- Engine stiffener
- Transaxle mount bolts and nuts
- Transaxle

5. Installation is the reverse of removal. Observe the following torques:
- Air cleaner housing bracket bolt: 16 ft. lbs. (22 Nm)
- Front mount bolts: 47 ft. lbs. (64 Nm)
- Rear mount bracket bolts: 40 ft. lb. (54 Nm)
- Transmission-to-engine bolts: 47 ft. lbs. (64 Nm)
- Upper transmission mount bolt: 40 ft. lbs. (54 Nm)
- Upper transmission mount nuts: 40 ft. lbs. (54 Nm)
- Rear driveshaft bolts: 24 ft. lbs. (32 Nm)
- Subframe bolts: 76 ft. lbs. (103 Nm)

Manual Transaxle

1999–01

1. Before servicing the vehicle, refer to the precautions in the beginning of this section.

2. Remove or disconnect the following:
- Negative battery cable
- Air intake assembly
- Clutch slave cylinder and hose bracket
- Starter motor
- Transaxle ground cable
- Reverse lamp switch connector
- Wire harness bracket
- Shift cables and bracket
- Vehicle Speed Sensor (VSS) connector
- Splash shield
- Heated Oxygen Sensor (HO2S) connector
- Exhaust front pipe
- Rear driveshaft
- Lower ball joints
- Right damper fork

3. Separate the inner CV-joints from the transaxle and intermediate shaft and support the axle halfshafts out of the work area with safety wire.

4. Remove or disconnect the following:
- Intermediate shaft
- Rear engine stiffener
- Clutch housing cover
- Right front mount and bracket
- Transaxle mount and bracket
- Rear engine mounting bolts
- Transaxle flange bolts
- Transaxle

To install:

→**Use new circlips, split pins and self-locking nuts for assembly.**

5. Install or connect the following:
- Transaxle. Tighten the flange bolts to 47 ft. lbs. (64 Nm) and the rear engine mounting bolts to 61 ft. lbs. (83 Nm).
- Transaxle mount and bracket. Tighten the bracket bolts to 47 ft. lbs. (64 Nm) and the through bolt to 54 ft. lbs. (74 Nm).
- Right front mount and bracket. Tighten the mount bolts to 33 ft. lbs. (44 Nm) and the bracket bolts to 47 ft. lbs. (64 Nm).
- Clutch housing cover. Tighten the 12mm bolts to 22 ft. lbs. (29 Nm) and the 6mm bolts to 105 inch lbs. (12 Nm).

- Rear engine stiffener. Tighten the 8mm bolts to 18 ft. lbs. (24 Nm) and the 10mm bolt to 33 ft. lbs. (44 Nm).
- Intermediate shaft. Tighten the bolts to 29 ft. lbs. (39 Nm).
- Axle halfshafts
- Right damper fork. Tighten the pinch bolt to 32 ft. lbs. (43 Nm) and the nut to 47 ft. lbs. (64 Nm).
- Lower ball joints. Tighten the nut to 36–43 ft. lbs. (49–59 Nm).
- Rear driveshaft. Tighten the bolts to 24 ft. lbs. (32 Nm).
- Exhaust front pipe
- HO$_2$S connector
- Splash shield
- VSS connector
- Shift cables and bracket. Tighten the bracket bolts to 20 ft. lbs. (27 Nm).
- Wire harness bracket
- Reverse lamp switch connector
- Transaxle ground cable
- Starter motor. Tighten the bolts to 32 ft. lbs. (44 Nm).
- Clutch slave cylinder and hose bracket
- Air intake assembly
- Negative battery cable

6. Fill the transaxle to the correct level.

7. Start the engine and check for leaks.

8. Check the wheel alignment and adjust as necessary.

2002

Manual Transaxle

1. Before servicing the vehicle, refer to the precautions in the beginning of this section.

2. Remove or disconnect the following:
- Negative battery cable
- Air intake assembly
- Transaxle ground cable
- Vehicle Speed Sensor (VSS) connector
- Splash shield
- Shift cables and bracket
- Clutch slave cylinder and hose bracket
- Wire harness bracket
- Water pipe mounting bolt
- Brake booster and EVAP line mounting bolts. Attach an engine support hanger to the head to support the weight of the engine.
- Upper transmission mounting bolts
- Transaxle mount and bracket

3. Separate the inner CV-joints from the transaxle and intermediate shaft and support the axle halfshafts out of the work area with safety wire.

4. Remove or disconnect the following:
- Intermediate shaft
- Right front mount and bracket
- Rear engine mounting bolts
- Rear driveshaft
- Subframe (see the Engine Removal and Installation procedure)
- Clutch housing cover
- Transaxle

5. Installation is the reverse of removal. Observe the following torques:
- Transaxle rear mount and bracket. Tighten the bracket bolts to 40 ft. lbs. (54 Nm) and the through bolt to 47 ft. lbs. (64 Nm)
- Transaxle. Tighten the flange bolts to 47 ft. lbs. (64 Nm)
- Front mount and bracket. Tighten the bolts to 47 ft. lbs. (64 Nm).
- Clutch housing cover. Tighten the bolts to 29 ft. lbs. (39 Nm)
- Subframe: 76 ft. lbs. (98 Nm)

Clutch

ADJUSTMENTS

The CR-V is equipped with a hydraulic clutch system. No adjustment is necessary.

REMOVAL & INSTALLATION

1. Before servicing the vehicle, refer to the precautions in the beginning of this section.

2. Remove or disconnect the following:
- Negative battery cable
- Transaxle
- Pressure plate. Loosen the bolts evenly in a crossing pattern.
- Clutch disc

To install:

3. Install the clutch disc and pressure plate. Tighten the pressure plate bolts in a crossing pattern and in several steps to 19 ft. lbs. (25 Nm).

4. Install or connect the following:
- Transaxle
- Negative battery cable

Hydraulic Clutch System

BLEEDING

1. Before servicing the vehicle, refer to the precautions in the beginning of this section.

2. Attach a hose to the bleeder screw and suspend the other end in a container of clean brake fluid.

3. Open the bleeder screw.

4. Slowly pump the clutch pedal until no more air bubbles appear at the bleeder hose.

5. Tighten the bleeder screw to 70 inch lbs. (8 Nm).

6. Refill the clutch master cylinder as necessary.

7. Check for leaks and proper clutch operation.

Transfer Assembly

REMOVAL & INSTALLATION

1999–01

1. Before servicing the vehicle, refer to the precautions in the beginning of this section.

2. Drain the transaxle fluid.

3. Remove or disconnect the following:
- Negative battery cable
- Heated Oxygen Sensor (HO$_2$S) connector on 1999–01 models
- Exhaust front pipe on 1999–01 models
- Rear driveshaft
- Transfer assembly and bracket

To install:

4. Install the transfer assembly and bracket with a new O-ring seal. Tighten the 10mm bolts to 33 ft. lbs. (44 Nm) and the 8mm bolts to 17 ft. lbs. (24 Nm).

5. Install or connect the following:
- Rear driveshaft. Tighten the bolts to 24 ft. lbs. (32 Nm).
- Exhaust front pipe
- HO$_2$S connector
- Negative battery cable

6. Fill the transaxle to the correct level and check for leaks.

Halfshaft

REMOVAL & INSTALLATION

Front

1999–01

1. Before servicing the vehicle, refer to the precautions in the beginning of this section.

2. Drain the transaxle.

3. Remove or disconnect the following:
- Negative battery cable
- Front wheels
- Damper fork

- Lower ball joint
- Spindle nut

4. Pry the inboard joint from the transaxle or intermediate shaft.

5. Remove the outer CV-joint stub shaft from the hub by tapping the stub shaft with a plastic hammer.

To install:

➡**Use new circlips, split pins and self-locking nuts for assembly.**

6. Install the outer CV-joint stub shaft into the hub.

7. Install the inner CV-joint to the transaxle or intermediate shaft until the circlip locks in the retaining groove.

8. Install or connect the following:
- Lower ball joint. Tighten the nut to 36–43 ft. lbs. (49–59 Nm).
- Damper fork. Tighten the pinch bolt to 32 ft. lbs. (43 Nm) and the nut to 47 ft. lbs. (64 Nm).
- Spindle nut. Tighten the nut to 181 ft. lbs. (245 Nm).
- Front wheels
- Negative battery cable

9. Fill the transaxle to the correct level and check for leaks.

2002

1. Before servicing the vehicle, refer to the precautions in the beginning of this section.

2. Drain the transaxle.

3. Remove or disconnect the following:
- Negative battery cable
- Front wheels
- Stabilizer bar
- Lower ball joint
- Spindle nut

4. On the left side, pry the inboard joint from the case with a prybar.

5. On the right side, drive the inboard shaft off the intermediate shaft with a drift and hammer.

6. Installation is the reverse of removal. Observe the following torques:
- Ball stud nuts: 40 ft. lbs. (54 Nm)
- Stabilizer link nuts: 29 ft. lbs. (39 Nm)
- Spindle nut: 181 ft. lbs. (245 Nm)

Rear

1999–02

1. Before servicing the vehicle, refer to the precautions in the beginning of this section.

2. Drain the differential.

3. Remove or disconnect the following:

- Negative battery cable
- Rear wheels
- Spindle nut

4. Pry the inboard joint from the differential.

5. Remove the outer CV-joint stub shaft from the hub by tapping the stub shaft with a plastic hammer.

To install:

➡**Use new circlips and self-locking nuts for assembly.**

6. Install the outer CV-joint stub shaft into the hub.

7. Install the inner CV-joint to the differential until the circlip locks in the retaining groove.

8. Install or connect the following:
- Spindle nut. Tighten the nut to 134 ft. lbs. (181 Nm).
- Rear wheels
- Negative battery cable

9. Fill the differential to the correct level and check for leaks.

CV-Joint

OVERHAUL

Front

OUTBOARD JOINT

1. Before servicing the vehicle, refer to the precautions in the beginning of this section.

2. Remove or disconnect the following:

- Axle halfshaft from the vehicle and place it in a vise
- Outboard joint boot clamps and push the boot back
- Outboard joint by driving it off the axle shaft with a brass drift and hammer
- Outboard joint boot

To install:

➡**Use new circlips and boot clamps for assembly.**

3. Install the outboard joint boot and clamps to the axle shaft.

4. Fill the outboard joint with grease. Install the outboard joint to the axle shaft. Tap the stub shaft with a brass hammer to seat the circlip.

5. Fill the outboard joint boot with grease and install the boot clamps.

6. Install the axle halfshaft to the vehicle.

INBOARD JOINT

1. Before servicing the vehicle, refer to the precautions in the beginning of this section.

2. Remove or disconnect the following:

- Axle halfshaft from the vehicle.
- Inboard joint boot clamps and push the boot back
- Inboard joint housing from the axle
- Rollers from the spider
- Snapring and the spider from the axle shaft
- Inboard joint boot

To install:

➡**Use new circlips and boot clamps for assembly.**

3. Install or connect the following:
- Inboard joint boot and clamps to the axle shaft
- Spider with a new snapring
- Rollers to the spider

4. Fill the joint housing with grease and install it.

5. Fill the inboard joint boot with grease and install the boot clamps.

6. Install the axle halfshaft to the vehicle.

Rear

1. Before servicing the vehicle, refer to the precautions in the beginning of this section.

2. Remove or disconnect the following:

- Axle halfshaft from the vehicle
- Joint boot clamps and push the boot back
- Joint housing from the axle
- Rollers from the spider
- Snapring and the spider from the axle shaft
- Joint boot

To install:

➡**Use new circlips and boot clamps for assembly.**

3. Install or connect the following:
- Joint boot and clamps to the axle shaft
- Spider with a new snapring
- Rollers to the spider

4. Fill the joint housing with grease and install it.

5. Fill the joint boot with grease and install the boot clamps.

6. Install the axle halfshaft to the vehicle.

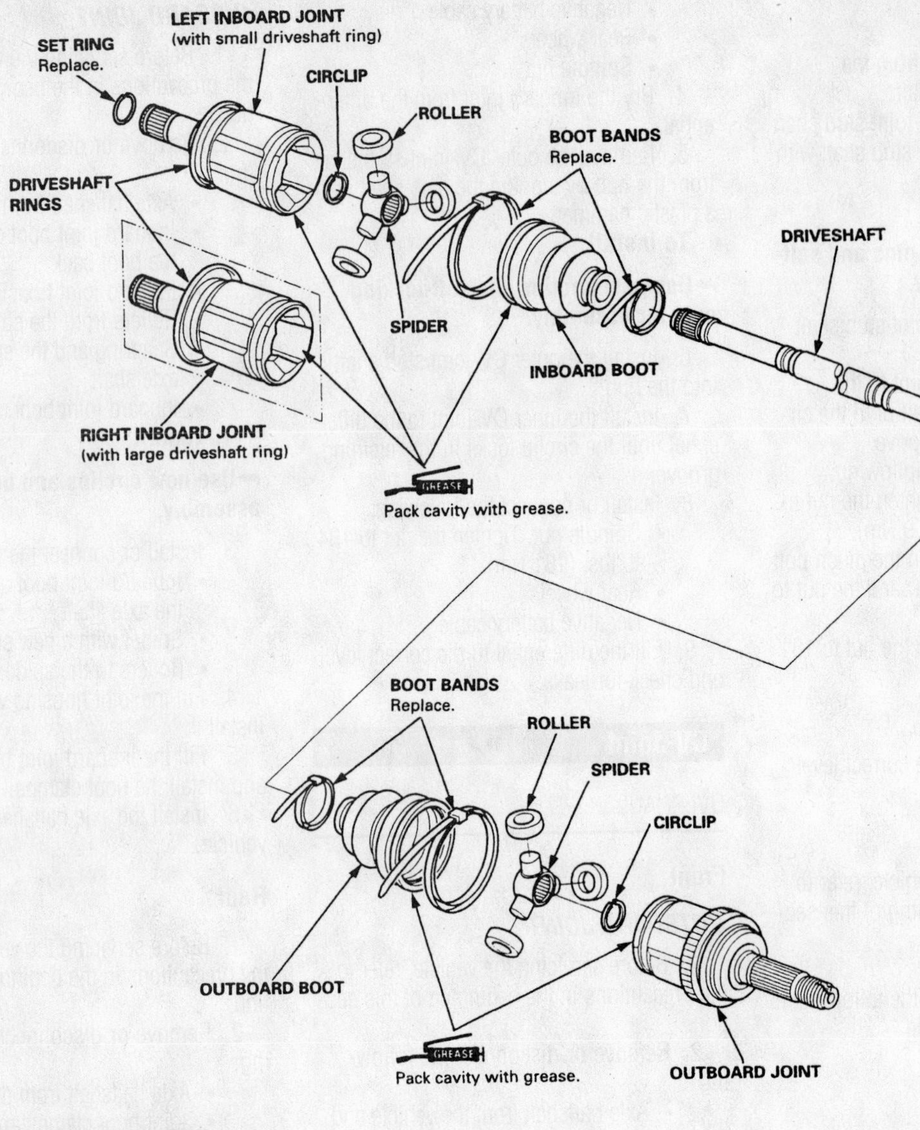

SET RING Replace.

LEFT INBOARD JOINT (with small driveshaft ring)

CIRCLIP

ROLLER

BOOT BANDS Replace.

DRIVESHAFT

DRIVESHAFT RINGS

SPIDER

INBOARD BOOT

RIGHT INBOARD JOINT (with large driveshaft ring)

GREASE
Pack cavity with grease.

BOOT BANDS Replace.

ROLLER

SPIDER

CIRCLIP

OUTBOARD BOOT

GREASE
Pack cavity with grease.

OUTBOARD JOINT

9308MG30

Exploded view of the rear axle—CR-V

Pinion Seal

REMOVAL & INSTALLATION

1. Before servicing the vehicle, refer to the precautions in the beginning of this section.
2. Remove or disconnect the following:
   - Driveshaft
   - Companion flange
   - Pinion seal

To install:

➡**Use a new locknut and O-ring for assembly.**

3. Install or connect the following:
   - Pinion seal. Drive the seal square into the bore.
   - Companion flange. Tighten the locknut to 87 ft. lbs. (118 Nm).
   - Driveshaft. Tighten the flange bolts to 24 ft. lbs. (32 Nm).

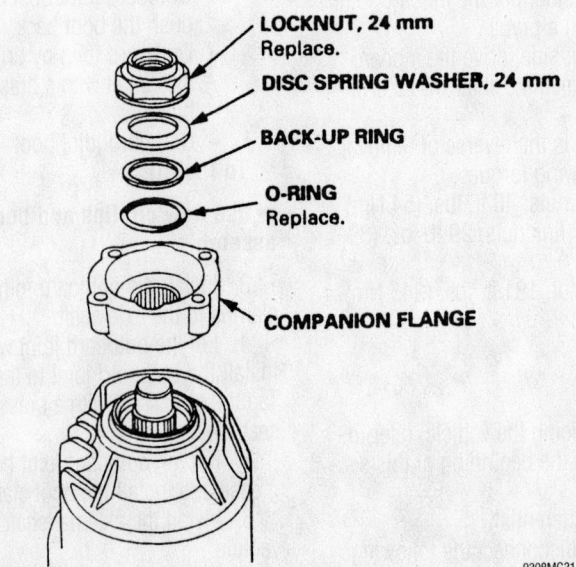

LOCKNUT, 24 mm Replace.

DISC SPRING WASHER, 24 mm

BACK-UP RING

O-RING Replace.

COMPANION FLANGE

9308MG31

Exploded view of the rear differential pinion components—CR-V

STEERING AND SUSPENSION

Air Bag

☀☀ CAUTION

Some vehicles are equipped with an air bag system. The system must be disarmed before performing service on, or around, system components, the steering column, instrument panel components, wiring and sensors. Failure to follow the safety precautions and the disarming procedure could result in accidental air bag deployment, possible injury and unnecessary system repairs.

PRECAUTIONS

Several precautions must be observed when handling the inflator module to avoid accidental deployment and possible personal injury.

• Never carry the inflator module by the wires or connector on the underside of the module.

• When carrying a live inflator module, hold securely with both hands, and ensure that the bag and trim cover are pointed away.

• Place the inflator module on a bench or other surface with the bag and trim cover facing up.

• With the inflator module on the bench, never place anything on or close to the module which may be thrown in the event of an accidental deployment.

Before servicing the vehicle, also make sure to refer to the precautions in the beginning of this section as well.

DISARMING

Disconnect and isolate the negative battery cable. Wait 3 minutes for the system capacitor to discharge before performing any service.

Power Rack and Pinion Steering Gear

REMOVAL & INSTALLATION

1999–01

☀☀ WARNING

Do not permit the steering wheel to turn whenever the steering gear is

disconnected from the steering column. Damage to the air bag wiring can result.

1. Before servicing the vehicle, refer to the precautions in the beginning of this section.

2. Center the steering wheel and lock it in position.

3. Remove or disconnect the following:
   • Negative battery cable
   • Air bag and steering wheel
   • Steering flexible joint
   • Front wheels
   • Outer tie rod ends
   • Heated Oxygen Sensor (HO2S) connector
   • Exhaust front pipe
   • Transmission shift cable
   • Power steering fluid lines

4. For 4 wheel drive vehicles, perform the following:

   a. Remove the rear driveshaft.

   b. Remove the right and left front mounts.

   c. Remove the rear mount and bracket.

   d. Tilt the engine back to lower the transfer assembly output flange about 1.57 inches (40mm).

5. For all vehicles, remove or disconnect the following:
   • Stiffener plate
   • Steering gear mounting brackets

6. Slide the steering gear to the right until the left end clears the subframe, then slide the gear to the left and out of the vehicle.

To install:

7. Position the steering gear in the vehicle, right end first.

8. Install or connect the following:

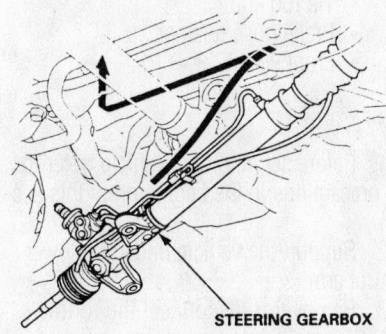

STEERING GEARBOX

9308MG27

Rack and pinion steering gear installation—1999–01

• Steering gear mounting brackets. Tighten the bolts to 29 ft. lbs. (39 Nm).
• Stiffener plate. Tighten the plate bolts to 28 ft. lbs. (38 Nm) and the steering bear bolt to 32 ft. lbs. (43 Nm).

➡ **Use new self-locking nuts and bolts for engine mount installation.**

9. For 4 wheel drive vehicles, install or connect the following:
   • Rear mount and bracket. Tighten the upper bracket bolt to 43 ft. lbs. (59 Nm), the lower bracket bolt to 61 ft. lbs. (83 Nm), the mount attaching bolts to 47 ft. lbs. (64 Nm), and the mount through bolt to 43 ft. lbs. (59 Nm).
   • Left and right front mounts. Tighten the bolts to 33 ft. lbs. (44 Nm) and the nuts to 43 ft. lbs. (59 Nm).
   • Rear driveshaft. Tighten the bolts to 24 ft. lbs. (32 Nm).

10. For all vehicles, install or connect the following:
   • Power steering fluid lines
   • Transmission shift cable
   • Exhaust front pipe
   • HO2S connector
   • Outer tie rod ends
   • Front wheels. Position the wheels straight ahead.
   • Steering flexible joint. Tighten the pinch bolts to 16 ft. lbs. (22 Nm).
   • Steering wheel and air bag
   • Negative battery cable

11. Fill the power steering system.

12. Check the wheel alignment and adjust as necessary.

2002

☀☀ WARNING

Do not permit the steering wheel to turn whenever the steering gear is disconnected from the steering column. Damage to the air bag wiring can result.

1. Before servicing the vehicle, refer to the precautions in the beginning of this section.

2. Center the steering wheel and lock it in position.

3. Remove or disconnect the following:
   • Negative battery cable

- Air bag and steering wheel
- Front wheels
- Driver's side dashboard lower cover and undercover
- Air cleaner housing
- Steering joint bolts
- Tie rod ends
- Steering hoses
- Left side flange bolts
- Mounting brackets

4. Lower the unit so the pinion shaft points outward. Remove the pinion shaft grommet. The steering gear is removed through the driver's side.

5. Installation is the reverse of removal. Observe the following torques:

- Mounting bracket and side flange bolts: 46 ft. lbs. (62 Nm)
- Supply line flare nut: 27 ft. lbs. (37 Nm)
- Tie rod ball stud nuts: 32 ft. lbs. (43 Nm)
- Steering joint bolts: 21 ft. lbs. (28 Nm)

Strut

REMOVAL & INSTALLATION

Front

1999–01

1. Before servicing the vehicle, refer to the precautions in the beginning of this section.

2. Remove or disconnect the following:

- Front wheel
- Brake hose retainer
- Damper fork
- Strut

To install:

➡**Use new self-locking fasteners for assembly.**

3. Install or connect the following:

- Strut. Tighten the mounting nuts to 43 ft. lbs. (59 Nm).
- Damper fork. Tighten the pinch bolt to 32 ft. lbs. (43 Nm) and the lower bolt to 47 ft. lbs. (64 Nm).
- Brake hose retainer
- Front wheel

2002

1. Before servicing the vehicle, refer to the precautions in the beginning of this section.

2. Remove or disconnect the following:

- Front wheel

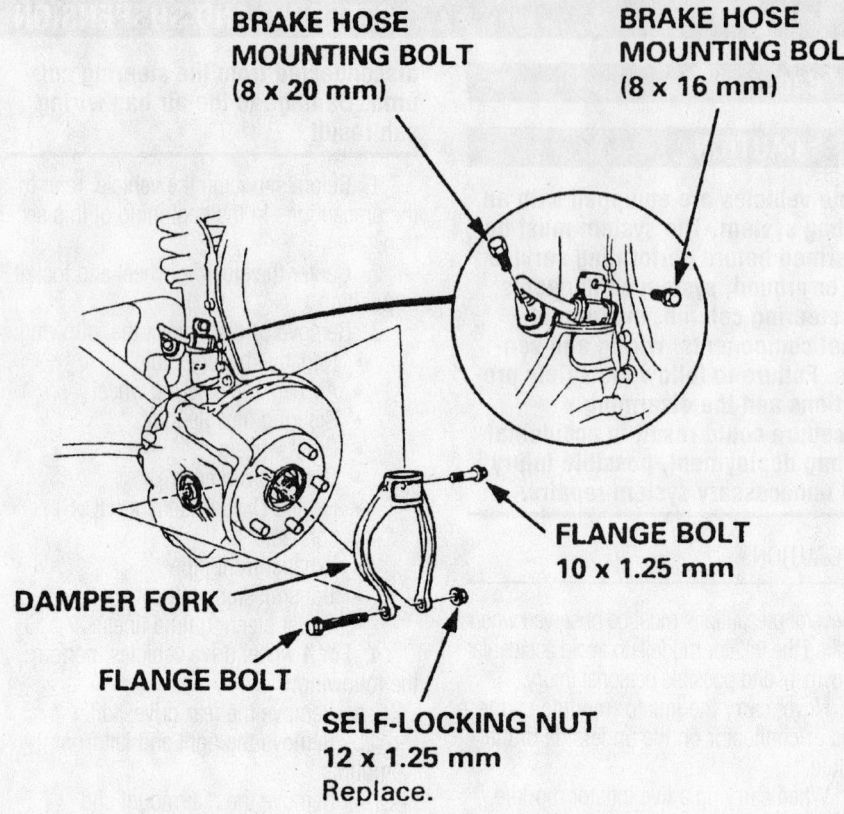

Identification of some of the front suspension components

- Tie rod end
- Brake hose retainer
- ABS sensor
- Strut

To install:

➡**Use new self-locking fasteners for assembly.**

3. Install or connect the following:

- Strut. Tighten the upper mounting nuts to 33 ft. lbs. (44 Nm).
- Tighten the pinch bolts to 116 ft. lbs. (157 Nm)
- ABS sensor
- Tie rod end
- Brake hose retainer
- Front wheel

Rear

1. Before servicing the vehicle, refer to the precautions in the beginning of this section.

2. Support the vehicle under the lower control arm.

3. Remove or disconnect the following:

- Rear wheel
- Interior access panel
- Damper cap
- Upper strut mount nuts

- Lower strut flange bolt
- Strut

To install:

4. Install or connect the following:

- Strut. Tighten the nuts to 36 ft. lbs. (49 Nm) for 1999–01; 54 ft. lbs. (74 Nm) for 2002 The bolt to 40 ft. lbs. (54 Nm) for 1999–01; 69 ft. lbs. (93 Nm).
- Damper cap
- Interior access panel
- Rear wheel

Coil Spring

REMOVAL & INSTALLATION

Front

1. Before servicing the vehicle, refer to the precautions in the beginning of this section.

2. Remove the strut from the vehicle and install in a strut spring compressor. Compress the spring until the end of the spring comes away from the spring seat.

3. Remove the upper strut mount, spring seat and related components.

4. Remove the coil spring from the strut spring compressor.

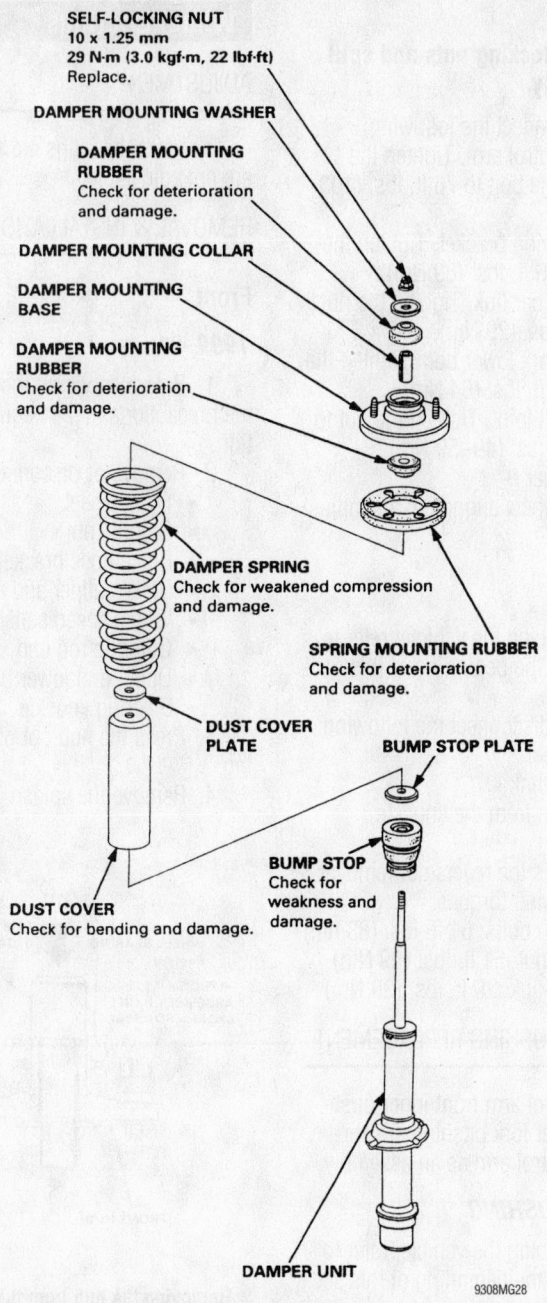

SELF-LOCKING NUT
10 x 1.25 mm
29 N·m (3.0 kgf·m, 22 lbf·ft)
Replace.

DAMPER MOUNTING WASHER

DAMPER MOUNTING
RUBBER
Check for deterioration
and damage.

DAMPER MOUNTING COLLAR

DAMPER MOUNTING
BASE

DAMPER MOUNTING
RUBBER
Check for deterioration
and damage.

DAMPER SPRING
Check for weakened compression
and damage.

SPRING MOUNTING RUBBER
Check for deterioration
and damage.

DUST COVER
PLATE

BUMP STOP PLATE

BUMP STOP
Check for
weakness and
damage.

DUST COVER
Check for bending and damage.

DAMPER UNIT

9308MG28

Front strut and spring exploded view—1999–01 shown

To install:

➡Use a new self-locking nut.

5. Compress the spring and position the strut so that the end of the spring aligns with the notch in the spring seat.

6. Install the upper strut mounting components and tighten the nut to 22 ft. lbs. (29 Nm) for 1999–01; 33 ft. lbs. (44 Nm) for 2002.

7. Install the strut to the vehicle.

8. Check the wheel alignment and adjust as necessary.

Rear

1. Before servicing the vehicle, refer to the precautions in the beginning of this section.

2. Remove the strut from the vehicle and install in a strut spring compressor. Compress the spring until the end of the spring comes away from the spring seat.

3. Remove or disconnect the following:

- Upper strut mount, spring seat and related components

- Coil spring from the strut spring compressor

To install:

➡Use a new self-locking nut.

4. Compress the spring and position the strut so that the end of the spring aligns with the notch in the spring seat.

5. Install or connect the following:

- Upper strut mounting components and tighten the nut to 22 ft. lbs. (29 Nm).
- Strut to the vehicle

6. Check the wheel alignment and adjust as necessary.

Upper Ball Joint

REMOVAL & INSTALLATION

The upper ball joints are replaced with the upper control arms as an assembly.

Lower Ball Joint

REMOVAL & INSTALLATION

1999–01

1. Before servicing the vehicle, refer to the precautions in the beginning of this section.

2. Remove or disconnect the following:

- Front wheel
- Spindle nut
- Brake hose bracket
- Brake caliper and rotor
- Wheel speed sensor, if equipped
- Outer tie rod end
- Upper and lower ball joints
- Steering knuckle
- Lower ball joint boot and set ring

3. Press the lower ball joint out of the steering knuckle.

To install:

➡Use new ball joint nuts, split pins, and a new spindle nut for assembly.

4. Press the lower ball joint into the steering knuckle.

5. Install or connect the following:

- Lower ball joint boot and set ring
- Steering knuckle. Tighten the upper ball joint nut to 29–35 ft. lbs. (39–47 Nm) and the lower ball joint nut to 36–43 ft. lbs. (49–59 Nm).
- Outer tie rod end. Tighten the nut to 32 ft. lbs. (43 Nm).

- Wheel speed sensor, if equipped
- Brake caliper and rotor. Tighten the caliper bracket bolts to 80 ft. lbs. (108 Nm).
- Brake hose bracket
- Spindle nut. Tighten the nut to 181 ft. lbs. (245 Nm).
- Front wheel

6. Check the wheel alignment and adjust as necessary.

2002

The ball joint is not replaceable.

Upper Control Arm

REMOVAL & INSTALLATION

1. Before servicing the vehicle, refer to the precautions in the beginning of this section.
2. Support the lower control arm assembly with a floor jack.
3. Remove or disconnect the following:
   - Upper ball joint
   - Inner control arm flange bolts.
   - Upper control arm

To install:

➡ **Use new self-locking nuts for assembly.**

4. Install the upper control arm. Tighten the ball joint nut to 29–35 ft. lbs. (39–47 Nm) and the inner flange bolts to 40 ft. lbs. (54 Nm).

CONTROL ARM BUSHING REPLACEMENT

The upper control arm bushings are serviced with the upper control arm as an assembly.

Lower Control Arm

REMOVAL & INSTALLATION

1999–01

1. Before servicing the vehicle, refer to the precautions in the beginning of this section.
2. Remove or disconnect the following:
   - Front wheel
   - Lower ball joint
   - Damper fork lower bolt
   - Stabilizer bar link
   - Front inner flange bolt
   - Rear bushing bracket bolts
   - Lower control arm

To install:

➡ **Use new self-locking nuts and split pins for assembly.**

3. Install or connect the following:
   - Lower control arm. Tighten the front flange bolt to 76 ft. lbs. (103 Nm).
   - Rear bushing bracket. Tighten the bolts to 66 ft. lbs. (89 Nm).
   - Stabilizer bar link. Tighten the nut to 22 ft. lbs. (29 Nm).
   - Damper fork lower bolt. Tighten the nut to 47 ft. lbs. (64 Nm).
   - Lower ball joint. Tighten the nut to 36–43 ft. lbs. (49–59 Nm).
   - Front wheel

4. Check the wheel alignment and adjust as necessary.

2002

1. Before servicing the vehicle, refer to the precautions in the beginning of this section.
2. Remove or disconnect the following:
   - Front wheel
   - Stabilizer link
   - Lower arm from the knuckle
   - Lower arm
3. Installation is the reverse of removal. Observe the following torques:
   - Lower arm bolts: 61 ft. lbs. (83 nm)
   - Ball stud nut: 51 ft. lbs. (69 Nm)
   - Stabilizer link: 29 ft. lbs. (39 Nm)

CONTROL ARM BUSHING REPLACEMENT

The lower control arm front inner bushing and the damper fork bushing are serviced with the control arm as an assembly.

REAR INNER BUSHING

1. Before servicing the vehicle, refer to the precautions in the beginning of this section.
2. Remove or disconnect the following:
   - Front wheel
   - Rear bushing bracket
   - Rear bushing

To install:

➡ **Use a new self-locking nut for assembly.**

3. Install or connect the following:
   - Rear bushing. Tighten the nut to 61 ft. lbs. (83 Nm).
   - Rear bushing bracket. Tighten the bolts to 66 ft. lbs. (89 Nm).
   - Front wheel

4. Check the wheel alignment and adjust as necessary.

Wheel Bearings

ADJUSTMENT

The wheel bearings are sealed units and are not adjustable.

REMOVAL & INSTALLATION

Front

1999–01

1. Before servicing the vehicle, refer to the precautions in the beginning of this section.
2. Remove or disconnect the following:
   - Front wheel
   - Spindle nut
   - Brake hose bracket
   - Brake caliper and rotor
   - Wheel speed sensor, if equipped
   - Outer tie rod end
   - Upper and lower ball joints
   - Steering knuckle
3. Press the hub out of the wheel bearing.
4. Remove the splash guard.

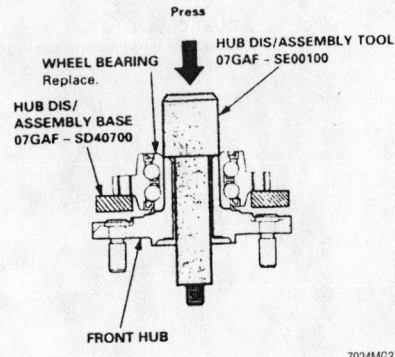

Removing the hub from the wheel bearing using the disassembly tools

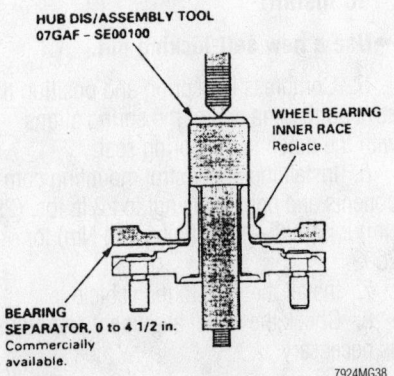

Pressing out the wheel bearing inner race

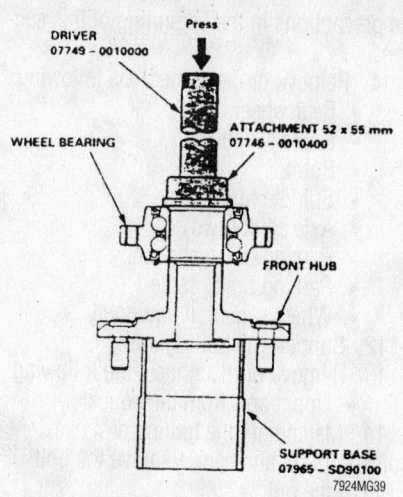

DRIVER
07749 – 0010000

Press

ATTACHMENT 52 x 55 mm
07746 – 0010400

WHEEL BEARING

FRONT HUB

SUPPORT BASE
07965 – SD90100

7924MG39

Utilizing the hub support base and driving attachment tools to install the new wheel bearing

5. Remove the snapring and press the wheel bearing out of the steering knuckle.

6. If necessary, press the inner bearing race off of the hub.

To install:

➡**Use new ball joint nuts, split pins, snapring and a new spindle nut for assembly.**

7. Press the bearing into the steering knuckle and install the snapring.

8. Install the splash guard.

9. Press the hub into the bearing.

10. Install or connect the following:
- Steering knuckle. Tighten the upper ball joint nut to 29–35 ft. lbs. (39–47 Nm) and the lower ball joint nut to 36–43 ft. lbs. (49–59 Nm).
- Outer tie rod end. Tighten the nut to 32 ft. lbs. (43 Nm).

- Wheel speed sensor, if equipped
- Brake caliper and rotor. Tighten the caliper bracket bolts to 80 ft. lbs. (108 Nm).
- Brake hose bracket
- Spindle nut. Tighten the nut to 181 ft. lbs. (245 Nm).
- Front wheel

11. Check the wheel alignment and adjust as necessary.

2002

1. Before servicing the vehicle, refer to the precautions in the beginning of this section.

2. Remove or disconnect the following:
- Front wheel
- Spindle nut
- Brake caliper and rotor. Forcing screws are needed to remove the rotor.

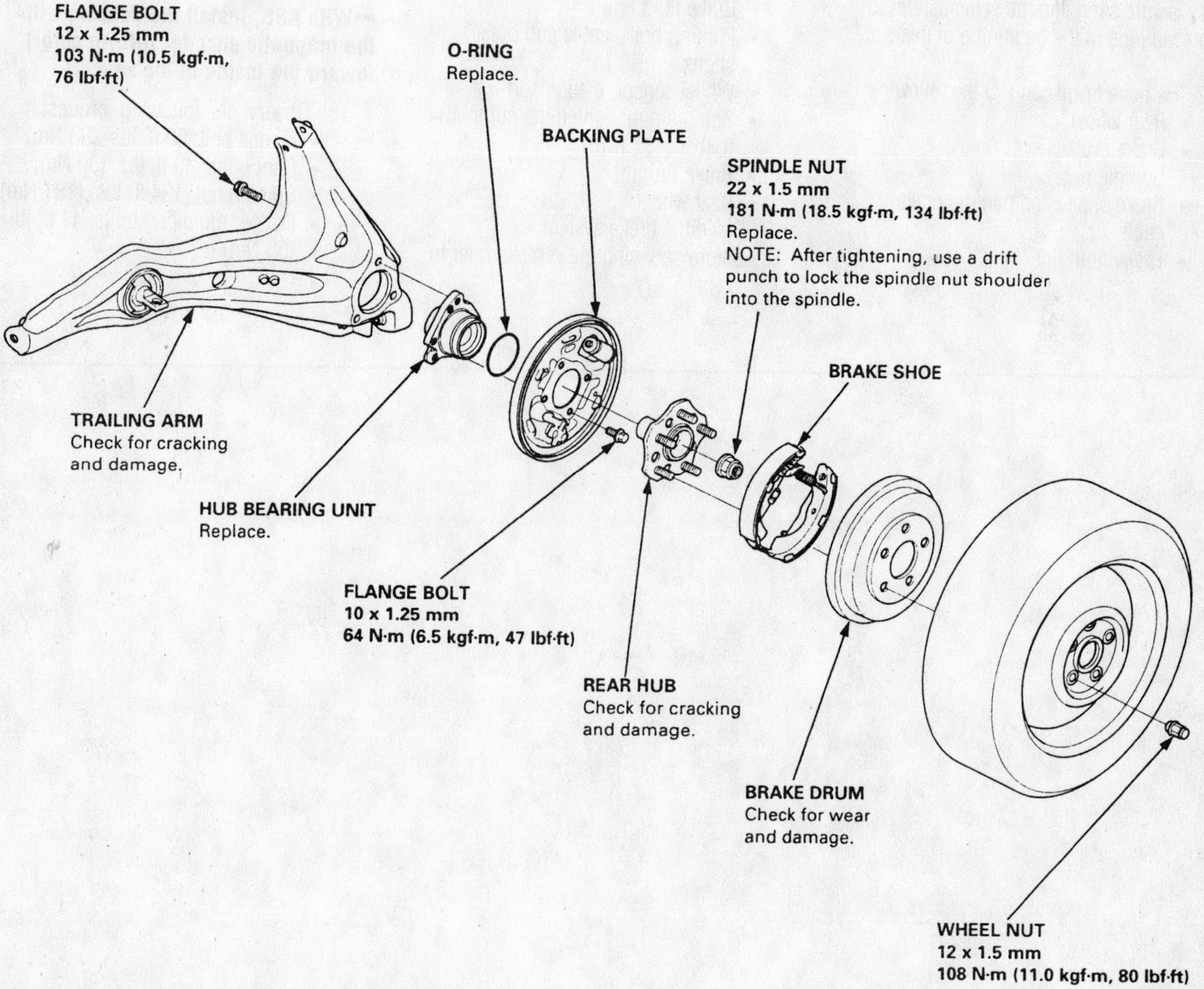

FLANGE BOLT
12 x 1.25 mm
103 N·m (10.5 kgf·m, 76 lbf·ft)

O-RING
Replace.

BACKING PLATE

SPINDLE NUT
22 x 1.5 mm
181 N·m (18.5 kgf·m, 134 lbf·ft)
Replace.
NOTE: After tightening, use a drift punch to lock the spindle nut shoulder into the spindle.

BRAKE SHOE

TRAILING ARM
Check for cracking and damage.

HUB BEARING UNIT
Replace.

FLANGE BOLT
10 x 1.25 mm
64 N·m (6.5 kgf·m, 47 lbf·ft)

REAR HUB
Check for cracking and damage.

BRAKE DRUM
Check for wear and damage.

WHEEL NUT
12 x 1.5 mm
108 N·m (11.0 kgf·m, 80 lbf·ft)

7924MG43

Exploded view of the rear hub and wheel bearing components—CR-V

- Brake hose bracket
- ABS sensor
- Stabilizer link
- Lower arm from the knuckle
- Strut-to-knuckle bolts
- Steering hub/knuckle assembly

3. Press the hub from the knuckle. The bearings and races can now be pressed out and replaced.

➡**With ABS, install the bearing with the magnetic encoder (brown color) toward the inside of the knuckle.**

4. Observe the following torques:
- Strut bolts: 116 ft. lbs. (157 Nm)
- Ball stud nuts: 51 ft. lbs. (69 Nm)
- Stabilizer bar link: 29 ft. lbs. (39 Nm)

Rear

1999–01

1. Before servicing the vehicle, refer to the precautions in the beginning of this section.

2. Remove or disconnect the following:
- Rear wheel
- Brake drum
- Spindle nut
- Brake shoes and parking brake cable
- Brake fluid line

- Wheel sensor, if equipped
- Wheel bearing flange bolts
- Hub, backing plate and bearing assembly

3. Press the hub out of the wheel bearing.

4. If necessary, press the inner bearing race off of the hub.

5. Remove the backing plate from the bearing assembly.

To install:

➡**Use a new spindle nut for assembly.**

6. Use a new O-ring and install the backing plate to the bearing assembly. Tighten the bolts to 47 ft. lbs. (64 Nm).

7. Press the hub into the bearing assembly.

8. Install or connect the following:
- Hub, backing plate and bearing assembly. Tighten the flange bolts to 76 ft. lbs. (103 Nm).
- Brake fluid line
- Parking brake cable and brake shoes
- Wheel sensor, if equipped
- Spindle nut. Tighten the nut to 134 ft. lbs. (181 Nm).
- Brake drum
- Rear wheel

9. Bleed the brake system.
10. Before servicing the vehicle, refer to

the precautions in the beginning of this section.

11. Remove or disconnect the following:
- Rear wheel
- Brake caliper
- Rotor
- Spindle nut
- Axle shaft (2wd)
- Parking brake shoes
- Parking brake cable
- Wheel sensor, if equipped

12. Support the trailing arm.

13. Remove or disconnect the following:
- Upper arm from the knuckle

14. Matchmark the trailing arm cam adjusting bolt and cam. Remove the bolt. Discard the nut.

15. Remove the flange bolt.
16. Remove the knuckle assembly.
17. Press the hub from the knuckle. The bearings and races can now be pressed out and replaced.

➡**With ABS, install the bearing with the magnetic encoder (brown color) toward the inside of the knuckle.**

18. Observe the following torques:
- Flange bolt: 69 ft. lbs. (93 Nm)
- Cam bolts: 43 ft. lbs. (59 Nm)
- Spindle nut: 134 ft. lbs. (181 Nm)
- Caliper mounting bolts: 41 ft. lbs. (55 Nm)

HYUNDAI

Santa Fe

16

PRECAUTIONS

Before servicing any vehicle, please be sure to read all of the following precautions, which deal with personal safety, prevention of component damage and important points to take into consideration when servicing a motor vehicle:

• Never open, service or drain the radiator or cooling system when the engine is hot; serious burns can occur from the steam and hot coolant.

• Observe all applicable safety precautions when working around fuel. Whenever servicing the fuel system, always work in a well-ventilated area. Do not allow fuel spray or vapors to come in contact with a spark, open flame, or excessive heat (a hot drop light, for example). Keep a dry chemical fire extinguisher near the work area. Always keep fuel in a container specifically designed for fuel storage; also, always properly seal fuel containers to avoid the possibility of fire or explosion. Refer to the additional fuel system precautions later in this section.

• Fuel injection systems often remain pressurized, even after the engine has been turned **OFF**. The fuel system pressure must be relieved before disconnecting any fuel lines. Failure to do so may result in fire and/or personal injury.

• Brake fluid often contains polyglycol ethers and polyglycols. Avoid contact with the eyes and wash your hands thoroughly after handling brake fluid. If you do get brake fluid in your eyes, flush your eyes with clean, running water for 15 minutes. If eye irritation persists, or if you have taken brake fluid internally, seek medical assistance IMMEDIATELY.

• The EPA warns that prolonged contact with used engine oil may cause a number of skin disorders, including cancer! You should make every effort to minimize your exposure to used engine oil. Protective gloves should be worn when changing oil. Wash your hands and any other exposed skin areas as soon as possible after exposure to used engine oil. Soap and water, or waterless hand cleaner should be used.

• All new vehicles are now equipped with an air bag system. The system must be disabled before performing service on or around system components, steering column, instrument panel components, wiring and sensors. Failure to follow safety and disabling procedures could result in accidental air bag deployment, possible personal injury and unnecessary system repairs.

• Always wear safety goggles when working with, or around, the air bag system. When carrying a non-deployed air bag, be sure the bag and trim cover are pointed away from your body. When placing a non-deployed air bag on a work surface, always face the bag and trim cover upward, away from the surface. This will reduce the motion of the module if it is accidentally deployed. Refer to the additional air bag system precautions later in this section.

• Clean, high quality brake fluid from a sealed container is essential to the safe and proper operation of the brake system. You should always buy the correct type of brake fluid for your vehicle. If the brake fluid becomes contaminated, completely flush the system with new fluid. Never reuse any brake fluid. Any brake fluid that is removed from the system should be discarded. Also, do not allow any brake fluid to come in contact with a painted surface; it will damage the paint.

• Never operate the engine without the proper amount and type of engine oil; doing so WILL result in severe engine damage.

• Timing belt maintenance is extremely important! Many models utilize an interference type, non-freewheeling engine. If the timing belt breaks, the valves in the cylinder head may strike the pistons, causing potentially serious (also time consuming and expensive) engine damage. Refer to the maintenance interval charts in the front of this manual for the recommended replacement interval for the timing belt and to the timing belt section for belt replacement and inspection.

• Disconnecting the negative battery cable on some vehicles may interfere with the functions of the on-board computer system(s) and may require the computer to undergo a relearning process once the negative battery cable is reconnected.

• When servicing drum brakes, only disassemble and assemble one side at a time, leaving the remaining side intact for reference.

• Only an MVAC-trained, EPA-certified automotive technician should service the air conditioning system or its components.

• The radio may contain a coded theft protection circuit. Always obtain the code number before disconnecting the battery.

ENGINE REPAIR

➡**Disconnecting the negative battery cable on some vehicles may interfere with the functions of the on board computer system. The computer may undergo a relearning process once the negative battery cable is reconnected.**

Distributor

The Santa Fe is equipped with a Distributorless Ignition System (DIS).

Alternator

REMOVAL

2.4L Engine

1. Before servicing the vehicle, refer to the precautions in the beginning of this section.

2. Remove or disconnect the following:

• Negative battery cable
• Drive belt
• Alternator electrical connections
• Alternator mounting bolts
• Alternator

2.7L Engine

1. Before servicing the vehicle, refer to the precautions in the beginning of this section.

2. Remove or disconnect the following:

• Negative battery cable
• Drive belt
• Alternator electrical connections
• Alternator mounting bolts
• Alternator

INSTALLATION

2.4L Engine

1. Install or connect the following:

• Alternator
• Alternator mounting bolts. Tighten the adjuster (upper) bolt to 15–18 ft. lbs. (20–25 Nm) and the lower bolt and nut to 26–41 ft. lbs. (34–54 Nm).
• Alternator electrical connections
• Drive belt
• Negative battery cable

2.7L Engine

1. Install or connect the following:

• Alternator
• Alternator mounting bolts. Tighten

[2.4 I4]

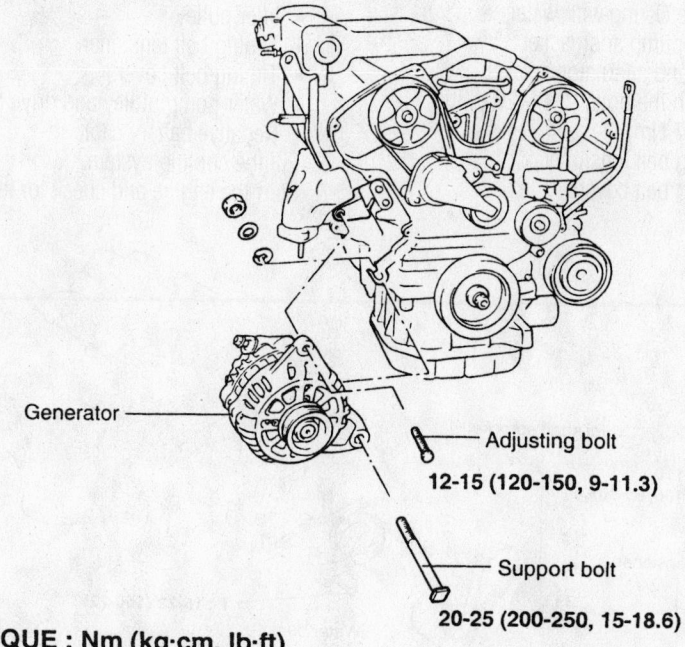

Adjusting bolt
20-25 (200-250, 15-18.6)

34-54 (340-540, 26-41)

Generator

Support bolt

[2.7 V6]

Generator

Adjusting bolt

12-15 (120-150, 9-11.3)

Support bolt

20-25 (200-250, 15-18.6)

TORQUE : Nm (kg·cm, lb·ft)

9355LG01

Exploded view of the alternator mounting for both engines used in the Santa Fe

the adjuster (upper) bolt to 9–11 ft. lbs. (12–15 Nm) and the lower bolt and nut to 15–18 ft. lbs. (20–25 Nm).

- Alternator electrical connections
- Drive belt
- Negative battery cable

Ignition Timing

ADJUSTMENT

The Santa Fe is equipped with a Distributorless Ignition System (DIS). The ignition timing is controlled by the Powertrain Con-

trol module (PCM). No adjustment is necessary.

Engine Assembly

REMOVAL & INSTALLATION

1. Before servicing the vehicle, refer to the precautions in the beginning of this section.
2. Remove the battery and air cleaner assembly.
3. Drain the cooling system.
4. Drain the engine oil.
5. Drain the transaxle fluid.
6. Relieve fuel system pressure.
7. Disconnect the following electrical connections:
- Starter
- Alternator
- Throttle Position Sensor (TPS)
- Power steering switch connector
- Oil pressure gauge connector
- Back-up lamp switch connector
- A/T solenoid inhibitor switch connector
- Coolant Temperature Sensor (CTS)
- Ignition coil
- Idle Speed Control (ISC) vale connector
- Manifold Absolute Pressure (MAP) sensor
- Oxygen (O$_2$S) sensor connector
8. If equipped with an automatic transmission, disconnect the oil cooler lines.
9. Remove or disconnect the following:
- Raditor hoses from the engine
- Radiator
- Engine ground
- Brake vacuum hose
- Heater hoses at the engine
- Throttle cable at the engine
- Cruise control cable at the engine, if equipped
- Main fuel line at the supply/return pipe
- Speedometer cable at the transaxle
- Clutch or control cable from the transaxle
- Power steering hoses from the pump
- Steering dust cover in the engine compartment
- Gear box universal joint bolt
- Front wheel
- Brake caliper and support with wire
- Strut lower bolt
- Front muffler bolts
- Transaxle control rod and extension

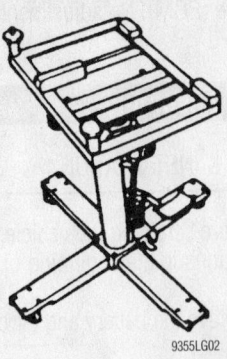

Attach the special tool to the transmission jack and support the transmission

rod, if equipped with a manual transmission

10. Support the transmission with a jack using the special attachment shown in the accompanying illustration.

11. Make sure all cable, harness connectors and hoses are disconnected from the engine and transmission.
   • Engine and transaxle mounting brackets
   • Sub frame bolts
   • Drive shaft

12. Lower the engine and transaxle assembly enough so the front and rear roll stoppers can be removed.

13. Remove the engine assembly

14. Installation is the reverse of removal but please note the following steps:
   • Tighten the roll stopper bolts to 36–47 ft. lbs. (50–65 Nm
   • Tighten the transaxle mounting bracket bolts to 65–80 ft. lbs. (90–110 Nm)
   • Tighten the engine mount bracket bolts to 43–58 ft. lbs. (60–80 Nm)

15. Fill the engine crankcase to the correct level.

16. Fill the transaxle to the correct level.

17. Fill the cooling system.

18. Fill the power steering system.

19. Start the engine and check for leaks.

20. Check the wheel alignment and adjust as necessary.

Water Pump

REMOVAL & INSTALLATION

2.4L Engine

1. Before servicing the vehicle, refer to the precautions in the beginning of this section.

2. Drain the cooling system.

3. Remove or disconnect the following:
   • Negative battery cable
   • Water pump inlet pipe

[DOHC ENGINE]

Engine coolant pump L= 65 (2.56)

Generator brace

L= 22 (0.86)

L= 22 (0.86)

L= 22 (0.86)

L=Length of bolt mm (in.)

9355LG05

Make sure the water pump bolts are positioned in their original positions as the bolts are different lengths—2.4L engine

   • Drive belt and water pump pulley
   • Timing belt covers
   • Timing belt tensioner
   • Water pump bolts
   • Alternator brace
   • Water pump and gasket

4. Clean the gasket mating surfaces.

To install:

5. Install or connect the following:
   • New O-ring onto the groove on the front end of the coolant pipe and wet the O-ring with water
   • Water pump and gasket
   • Bolts and alternator bracket. Tighten the bolts to 14–20 ft. lbs. (20–27 Nm).
   • Timing belt tensioner
   • Timing belt covers

   • Water pump pulley and drive belt
   • Water pump inlet pipe
   • Negative battery cable

6. Fill the cooling system.

7. Start the engine and check for leaks.

2.7L Engine

1. Before servicing the vehicle, refer to the precautions in the beginning of this section.

2. Drain the cooling system.

3. Remove or disconnect the following:
   • Negative battery cable
   • Drive belt and water pump pulley
   • Timing belt covers
   • Timing belt tensioner
   • Idler pulley
   • Water pump bolts
   • Water pump and gasket

4. Clean the gasket mating surfaces.

To install:

5. Install or connect the following:
   • Water pump and gasket
   • Bolts and alternator bracket. Tighten the bolts to 11–16 ft. lbs. (15–22 Nm).
   • Idler pulley
   • Timing belt tensioner
   • Timing belt covers
   • Water pump pulley and drive belt
   • Negative battery cable

6. Fill the cooling system.

7. Start the engine and check for leaks.

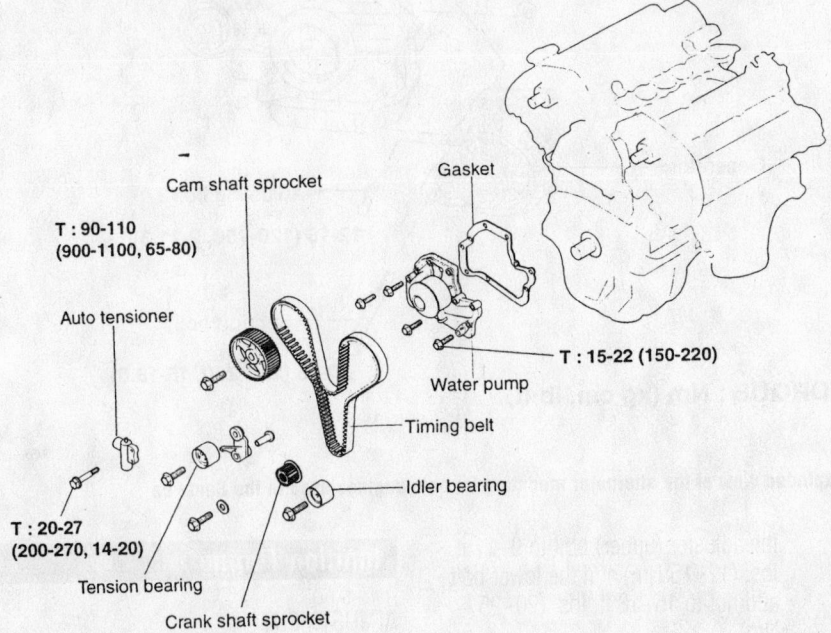

Cam shaft sprocket

T : 90-110 (900-1100, 65-80)

Auto tensioner

Gasket

T : 15-22 (150-220)

Water pump

Timing belt

Idler bearing

T : 20-27 (200-270, 14-20)

Tension bearing

Crank shaft sprocket

TORQUE : Nm (kg·cm, lb·ft)

9355LG06

Exploded view of the water pump mounting and related components—2.7L engine

Cylinder Head

REMOVAL & INSTALLATION

2.4L Engine

1. Before servicing the vehicle, refer to the precautions in the beginning of this section.
2. Drain the cooling system.
3. Relieve the fuel system pressure.
4. Remove or disconnect the following:
   - Negative battery cable
   - All necessary electrical connections, hoses and cables
   - Air cleaner
   - Intake manifold
   - Ignition coil
   - Timing belt
   - Exhaust manifold
   - Rocker cover
   - Camshafts
5. Loosen the cylinder head bolts in the sequence illustrated
6. Remove the cylinder head and gasket.

To install:
7. Clean the gasket mating surfaces
8. Install the cylinder head gasket so the surface with the identification mark faces towards the head.
9. Measure the head bolts. The bolt length should be 3.9 inch (99.4mm). If the

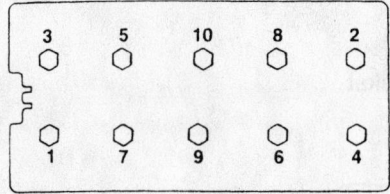

Cylinder head bolt loosening sequence—2.4L engine

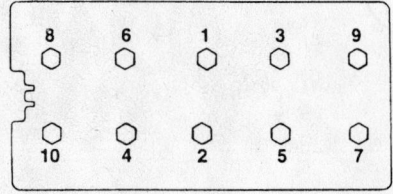

Cylinder head bolt tightening sequence—2.4L engine

bolts do not meet specification they must be replaced.
10. Install the cylinder head.
11. If using used parts (bolts, head or block), tighten the cylinder head bolts in sequence as follows:
   a. Step 1: 14 ft. lbs. (20 Nm).
   b. Step 2: plus an additional 90 degrees.
   c. Step 3: plus an additional 90 degrees.
12. If using new parts (even if only one thing is replaced), tighten the cylinder head bolts in sequence as follows:
   a. Step 1: 46 ft. lbs. (64 Nm).
   b. Step 2: release the bolts.
   c. Step 3: 14 ft. lbs. (20 Nm).
   d. Step 4: plus an additional 90 degrees.
   e. Step 5: plus an additional 90 degrees.
13. Install or connect the following:
   - Camshafts
   - Timing belt
   - Rocker cover
   - Ignition coil
   - Exhaust manifold
   - Intake manifold
   - Air cleaner
   - All necessary electrical connections, hoses and cables
   - Negative battery cable
14. Fill the cooling system.
15. Start the engine and check for leaks.

2.7L Engine

1. Before servicing the vehicle, refer to the precautions in the beginning of this section.
2. Drain the cooling system.
3. Relieve the fuel system pressure.
4. Remove or disconnect the following:
   - Negative battery cable
   - All necessary electrical connections, hoses and cables
   - Air cleaner
   - Intake manifold
   - Ignition coil
   - Timing belt
   - Exhaust manifold
   - Rocker cover
   - Camshafts
5. Loosen the cylinder head bolts in the reverse order of the tightening sequence.
6. Remove the cylinder head and gasket.

To install:
7. Clean the gasket mating surfaces
8. Install the cylinder head gasket so

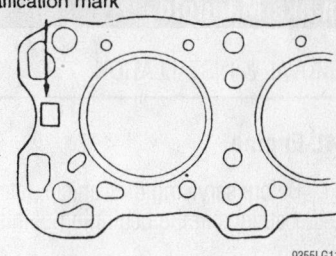

Location of the cylinder head gasket identification mark—2.7L engine

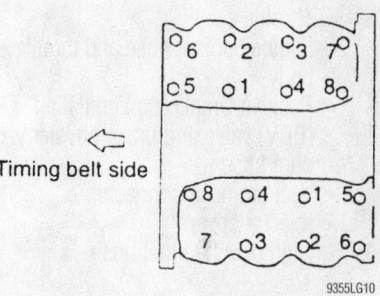

Cylinder head bolt tightening sequence—2.7L engine

the surface with the identification mark faces towards the head.
9. Install the cylinder head.
10. Tighten the cylinder head bolts in sequence as follows:
   a. Step 1: 18 ft. lbs. (25 Nm).
   b. Step 2: plus an additional 58–62 degrees.
   c. Step 3: plus an additional 43–47 degrees.
11. Install or connect the following:
   - Camshafts
   - Timing belt
   - Rocker cover
   - Ignition coil
   - Exhaust manifold
   - Intake manifold
   - Air cleaner
   - All necessary electrical connections, hoses and cables
   - Negative battery cable
12. Fill the cooling system.
13. Start the engine and check for leaks.

Rocker Arms/Shafts

REMOVAL & INSTALLATION

Refer to the camshaft removal and installation procedure.

Intake Manifold

REMOVAL & INSTALLATION

2.4L Engine

1. Before servicing the vehicle, refer to the precautions in the beginning of this section.
2. Remove or disconnect the following:
   - Negative battery cable
   - Air breather hose from the throttle body
   - Throttle cable
   - Engine coolant hose and throttle body
   - Positive Crankcase Ventilation (PCV) valve and brake booster vacuum hose
   - Vacuum hose connector
   - Injector cover
   - High pressure fuel hose
   - Fuel injector harness connector
   - Delivery pipe with the injectors and the pressure regulator as an assembly
   - Intake manifold stay
   - Intake manifold

To install:

3. Install or connect the following:
   - New intake manifold gasket
   - Intake manifold and bolts and nuts. Tighten the bolts to 11–14 ft. lbs. (15–20 Nm) and the nuts to 22–30 ft. lbs. (30–42 Nm).
   - Delivery pipe and injector assembly
   - Intake manifold stay and tighten the bolts to 13–18 ft. lbs. (18–25 Nm)
   - Fuel injector harness connector
   - High pressure fuel hose
   - Injector cover
   - Vacuum hose connector
   - PCV valve and brake booster vacuum hose
   - Throttle body and engine coolant hose
   - Throttle cable
   - Air breather hose from the throttle body
   - Negative battery cable
4. Start the engine and check for proper operation.

2.7L Engine

1. Before servicing the vehicle, refer to the precautions in the beginning of this section.
2. Remove or disconnect the following:
   - Negative battery cable
   - Air breather hose from the throttle body
   - Throttle and cruise control cables
   - Engine coolant hose and throttle body
   - Positive Crankcase Ventilation (PCV) valve and brake booster vacuum hose

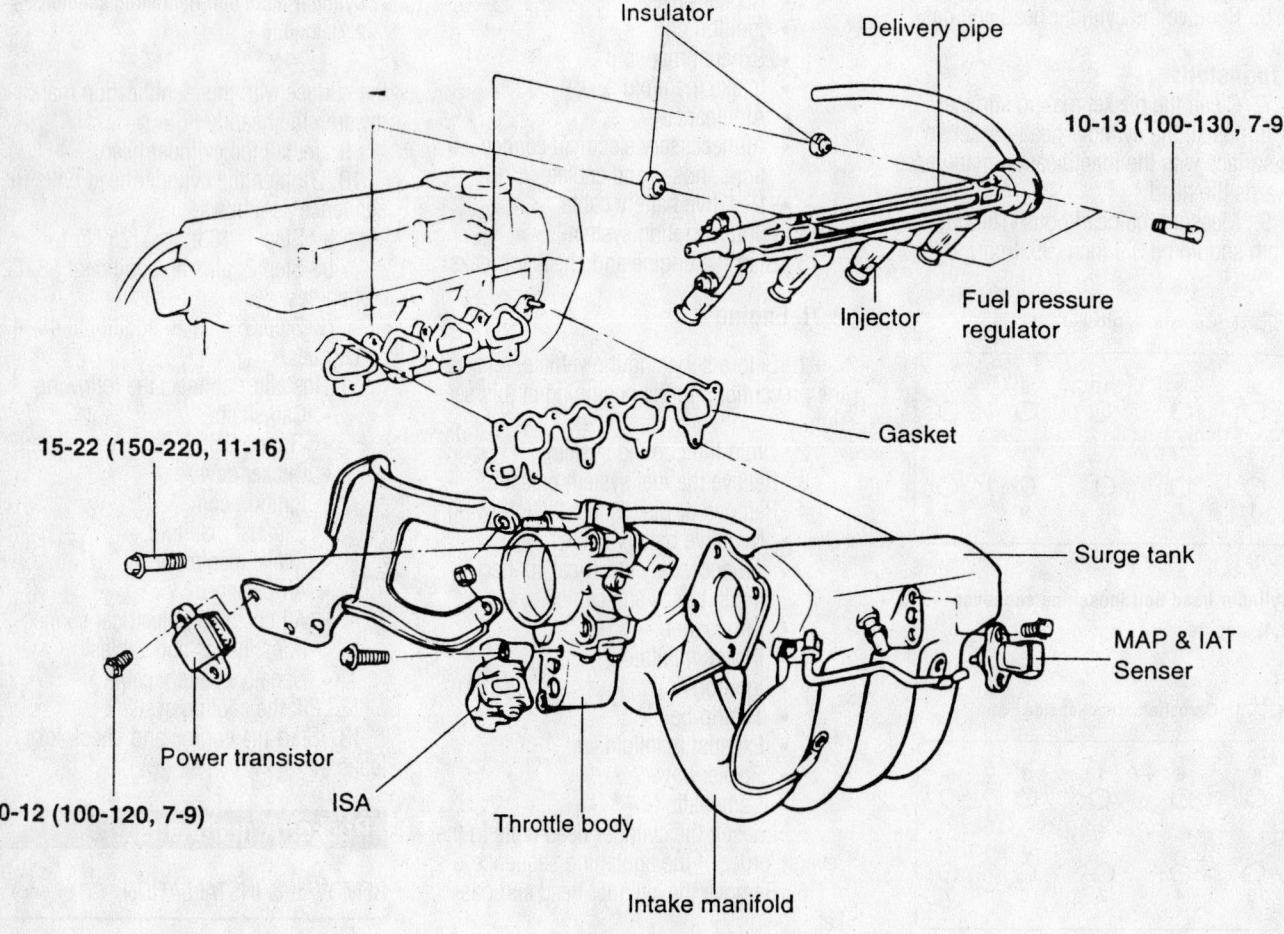

TORQUE : Nm (kg·cm, lb·ft)

Exploded view of the intake manifold—2.4L engine

9355LG12A

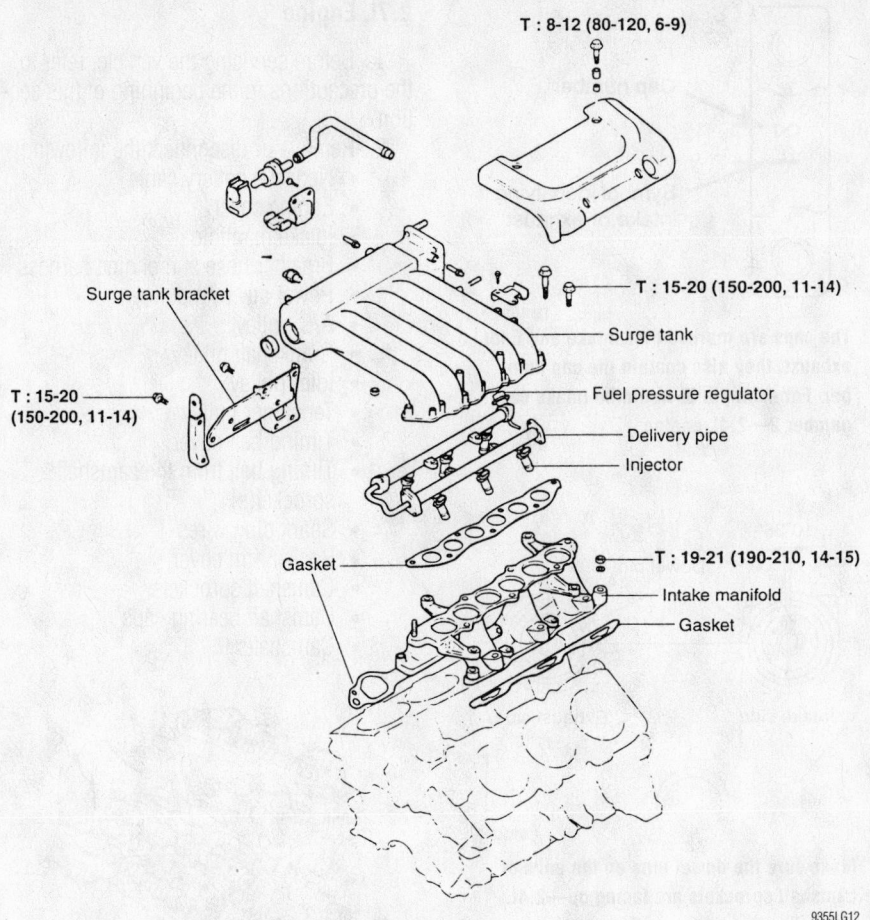

T : 8-12 (80-120, 6-9)

T : 15-20 (150-200, 11-14)

Surge tank bracket

T : 15-20 (150-200, 11-14)

Surge tank

Fuel pressure regulator

Delivery pipe

Injector

Gasket

T : 19-21 (190-210, 14-15)

Intake manifold

Gasket

9355LG12

Exploded view of the intake manifold—2.7L engine

- Vacuum hose connector
- Surge tank stay
- High pressure fuel hose
- Surge tank and gasket
- Fuel injector harness connector
- Delivery pipe with the injectors and the pressure regulator as an assembly
- Coolant Temperature Sensor (CTS) electrical connector
- Intake manifold

To install:

3. Install or connect the following:
- New intake manifold gasket
- Intake manifold and tighten the bolts to 14–15 ft. lbs. (19–21 Nm)
- CTS electrical connector
- Delivery pipe with the injectors and the pressure regulator as an assembly
- Fuel injector harness connector
- Surge tank and gasket
- High pressure fuel hose
- Surge tank stay
- Vacuum hose connector

- PCV valve and brake booster vacuum hose
- Engine coolant hose and throttle body
- Throttle and cruise control cables
- Air breather hose from the throttle body
- Negative battery cable

4. Start the engine and check for proper operation.

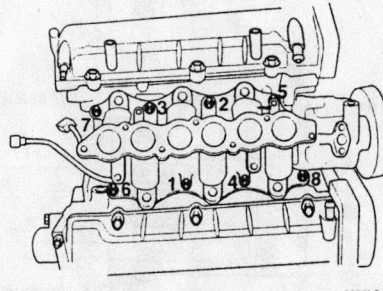

9355LG13

Intake manifold torque sequence—2.7L engine

Exhaust Manifold

REMOVAL & INSTALLATION

2.4L Engine

1. Before servicing the vehicle, refer to the precautions in the beginning of this section.
2. Remove or disconnect the following:
- Negative battery cable
- Heat shield
- Exhaust manifold retainers
- Manifold and gasket

To install:

3. Install or connect the following:
- New gasket and the manifold. Tighten the manifold M8 bolts to 18–22 ft. lbs. (25–30 Nm) and the M10 bolts to 25–40 ft. lbs. (35–55 Nm).
- Heat shield
- Negative battery cable

2.7L Engine

1. Before servicing the vehicle, refer to the precautions in the beginning of this section.
2. Remove or disconnect the following:
- Negative battery cable
- Heat shield
- Exhaust manifold retainers
- Manifold and gasket

To install:

3. Install or connect the following:
- New gasket and the manifold. Tighten the manifold bolts to 18–22 ft. lbs. (25–30 Nm).
- Heat shield
- Negative battery cable

Camshaft

REMOVAL & INSTALLATION

2.4L Engine

1. Before servicing the vehicle, refer to the precautions in the beginning of this section.
2. Drain the cooling system.
3. Remove or disconnect the following:
- Negative battery cable
- Breather hose from the air cleaner and rocker cover
- Air cleaner
- Timing belt cover
- Rocker cover and the Crankshaft Position (CKP) sensor

Timing belt service is covered in Section 3 of this manual

- Camshaft sprocket bolts and the sprockets
- Timing belt
- Camshaft bearing cap bolts using several passes
- Camshaft bearing caps, camshafts, rocker arms and lash adjusters

To install:

4. Inspect all parts for wear and damage.
5. Install or connect the following:
   - Camshafts and apply engine oil to the journals. Do not install the rocker arms yet.

➡ **The exhaust camshaft has a slit on the rear end for the CKP sensor.**

   - Bearing caps. The caps are marked **I** for intake and **E** for exhaust, they also contain the cap number. For example **I2** would be intake cap number 2.

6. Make sure the camshafts can turn freely, then remove the caps and the camshafts.
7. Make sure the dowel pins on the ends of camshaft sprockets are facing up.
   - Rocker arms
   - Camshafts and bearing caps. Tighten the bearing cap bolts uniformly to 14–15 ft. lbs. (19–21 Nm).
8. Using special tools, camshaft oil seal installer and guide 09221-21000, 09221-21100 install the oil seal. Coat the outside of the seal with oil prior to installation, then slide the seal along the front end of the camshaft and using the driver and a hammer install the seal until it is full seated.
   - Camshaft sprockets and bolts. Tighten the bolts to 58–72 ft. lbs. (80–100 Nm).
   - CKP sensor and rocker cover
   - Breather hose to the air cleaner and rocker cover
   - Air cleaner
   - Negative battery cable
9. Start the engine and check for leaks.

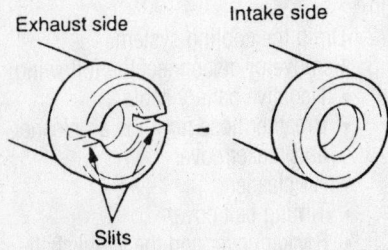

The exhaust camshaft has a slit on the rear end for the CKP sensor—2.4L engine

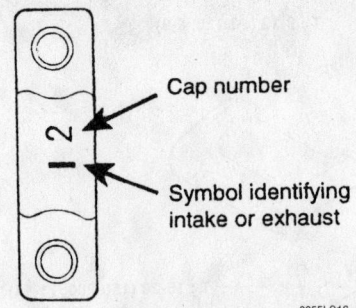

The caps are marked I for intake and E for exhaust, they also contain the cap number. For example I2 would be intake cap number 2—2.4L engine

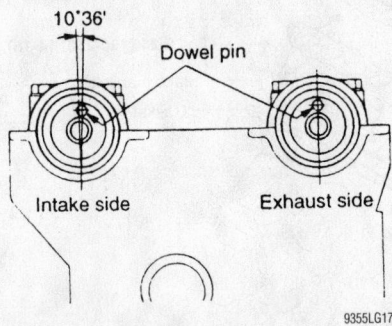

Make sure the dowel pins on the ends of camshaft sprockets are facing up—2.4L engine

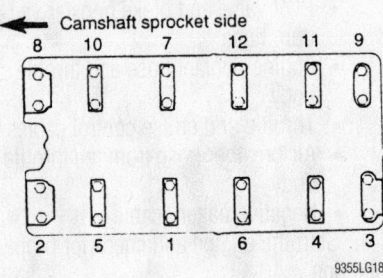

Camshaft bearing cap locations—2.4L engines

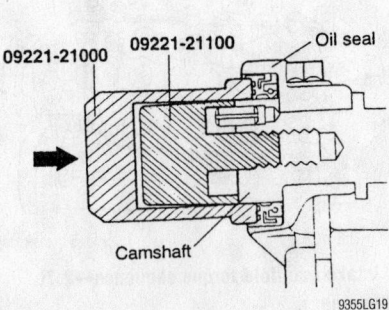

Installing the camshaft oil seal—2.4L engine

2.7L Engine

1. Before servicing the vehicle, refer to the precautions in the beginning of this section.
2. Remove or disconnect the following:
   - Negative battery cable
   - Engine cover
   - Intake manifold
   - Breather hose and engine harness
   - Power steering pulley
   - A/C pulley
   - Crankshaft pulley
   - Idler pulley
   - Tensioner pulley
   - Timing belt cover
   - Timing belt from the camshaft sprocket(s)
   - Spark plug wires
   - Rocker arm cover
   - Camshaft sprockets
   - Camshaft bearing caps
   - Camshafts

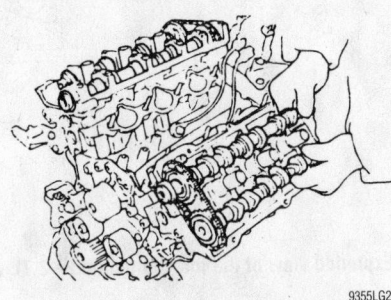

Remove the camshafts from the head—2.7L engine

To install:

3. Align the camshaft timing chain with the intake timing chain sprocket and exhaust sprocket as shown in the accompanying illustration.
4. Lubricate the camshaft journals with oil and install them.

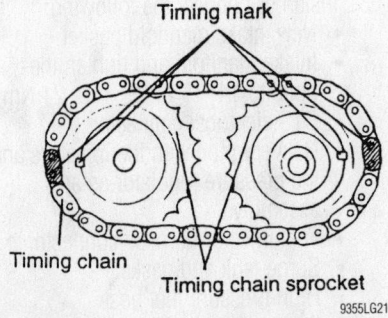

Align the camshaft timing chain with the intake timing chain sprocket and exhaust sprocket—2.7L engine

➥**To check the press fit, the camshaft (IN) and timing chain sprocket should be separable by a force greater than 1000kg minimum at room temperature.**

5. Install the bearing caps. The caps are marked **I** for intake and **E** for exhaust, they also contain the cap number. For example **I2** would be intake cap number 2.

6. Tighten the bearing caps M10 bolts to 10–12 ft. lbs. (14–16 Nm) and the M7 bolts to 7–9 ft. lbs. (10–12 Nm) using several passes.

7. Using special tool, camshaft oil seal installer 09221-21000 install the oil seal. Coat the outside of the seal with oil prior to installation, then insert the seal along the front end of the camshaft and using the driver and a hammer install the seal until it is full seated.

8. Install or connect the following:
- Camshaft sprocket and tighten the bolt to 65–80 ft. lbs. (90–110 Nm)
- Rocker cover
- Spark plug wires
- Timing belt
- Timing belt cover
- Power steering pulley
- A/C pulley
- Crankshaft pulley
- Idler pulley
- Tensioner pulley
- Breather hose and engine harness
- Intake manifold
- Engine cover
- Negative battery cable

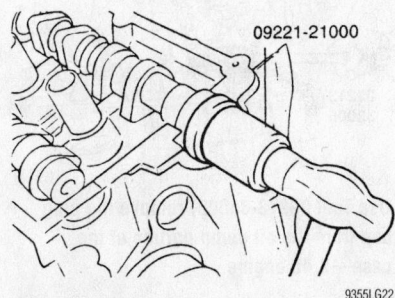

09221-21000

9355LG22

Install the camshaft oil seal—2.7L engine

Valve Lash

ADJUSTMENT

the valve lash is controlled by hydraulic adjusters and no adjustment is possible.

Starter Motor

REMOVAL & INSTALLATION

1. Before servicing the vehicle, refer to the precautions in the beginning of this section.

2. Remove or disconnect the following:
- Negative battery cable and wait at least 3 minutes
- Speedometer cable and shift cable from the transaxle
- Starter motor wiring
- Starter motor bolts and the starter

To install:

3. Installation is the reverse of removal. Tighten motor retainers to 20–25 ft. lbs. (27–34 Nm).

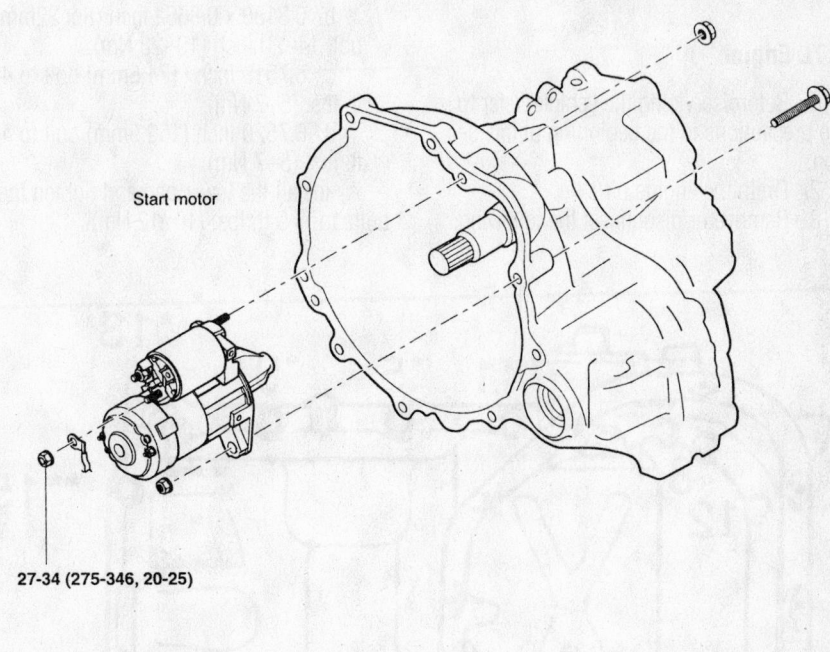

Start motor

27-34 (275-346, 20-25)

27-34 (275-346, 20-25)

TORQUE : Nm (kg.cm, lb.ft)

9355LG23

Starter motor mounting

Oil Pan

REMOVAL & INSTALLATION

2.4L Engine

1. Before servicing the vehicle, refer to the precautions in the beginning of this section.

2. Drain the engine oil.

3. Disconnect the negative battery cable.

4. Remove the oil pan bolts, note the bolts length and location.

5. Tap the oil pan with a rubber mallet and remove the upper and lower pan components.

6. Clean the gasket mating surfaces.

Heater Core replacement is covered in Section 2 of this manual

To install:

7. Apply 0.16 inch (4mm) of sealant to the oil pan groove as illustrated. Install the within 15 minutes of sealant installation.

8. Install the upper and lower oil pans and the bolts making sure the proper length is installed in its original position. Tighten the bolt in sequence to 7–9 ft. lbs. (10–12 Nm).

9. Connect the negative battery cable.

10. Refill the crankcase with oil.

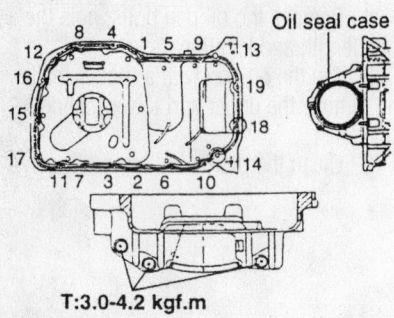

T:3.0-4.2 kgf.m

9355LG31

Oil pan torque sequence and sealant application points—2.4L engine

2.7L Engine

1. Before servicing the vehicle, refer to the precautions in the beginning of this section.

2. Drain the engine oil.

3. Remove or disconnect the following:

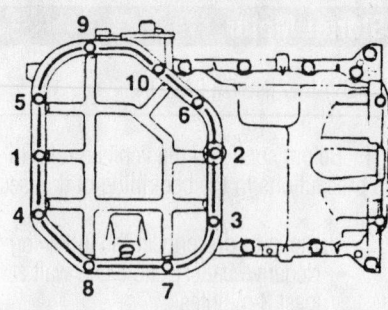

9355LG33

Lower oil pan bolt torque sequence—2.7L engine

- Negative battery cable
- Lower oil pan bolts and the pan
- Upper oil pan bolts and the pan

4. Clean the gasket mating surfaces

To install:

5. Apply 0.16 inch (4mm) of sealant to the lower oil pan groove. Install the within 15 minutes of sealant installation.

6. Install the upper oil pan and the tighten the bolts in sequence as follows:

    a. 0.937 x 1.4961 inch (10 x 38mm) bolt to 22–30 ft. lbs. (30–42 Nm).

    b. 0.3150 x 0.866a inch (8 x 22mm) bolt 14–20 inch (19–28 Nm).

    c. 6.7519 inch (171.5mm) bolt to 4–5 ft. lbs. (5–7 Nm).

    d. 6.7520 inch (152.5mm) bolt to 4–5 ft. lbs. (5–7 Nm).

7. Install the lower pan and tighten the bolts to 7–9 ft. lbs. (10–12 Nm).

8. Connect the negative battery cable.

9. Refill the crankcase with oil.

Oil Pump

REMOVAL & INSTALLATION

2.4L Engine

1. Before servicing the vehicle, refer to the precautions in the beginning of this section.

2. Drain the engine oil.

3. Remove or disconnect the following:
- Negative battery cable
- Timing belt
- Oil pan
- Oil screen and gasket
- Oil pressure switch
- Oil filter bracket and gasket

4. Using Tool 09213-33000, remove the plug cap from the oil pump portion of the case.

- Plug from the left side of the block and insert an 0.32 inch (8mm) screwdriver into the plug hole. The screwdriver must be inserted at least 2.4 inch (60mm).
- Pump driven gear the left counter balance shaft bolt
- Front case bolts (noting the bolt length and location), the case and gasket.

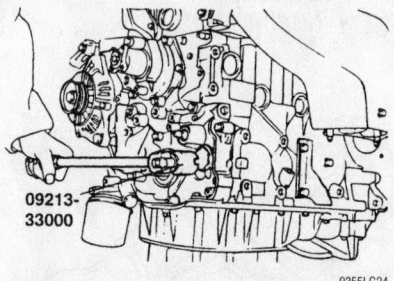

09213-33000

9355LG24

Use Tool 09213-33000, remove the plug cap from the oil pump portion of the case—2.4L engine

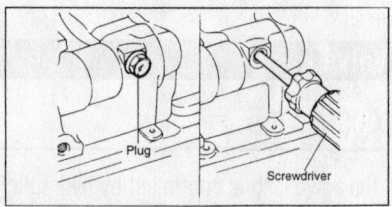

Plug

Screwdriver

9355LG25

Insert an 0.32 inch (8mm) screwdriver at least 2.4 inch (60mm) into the plug hole—2.4L engine

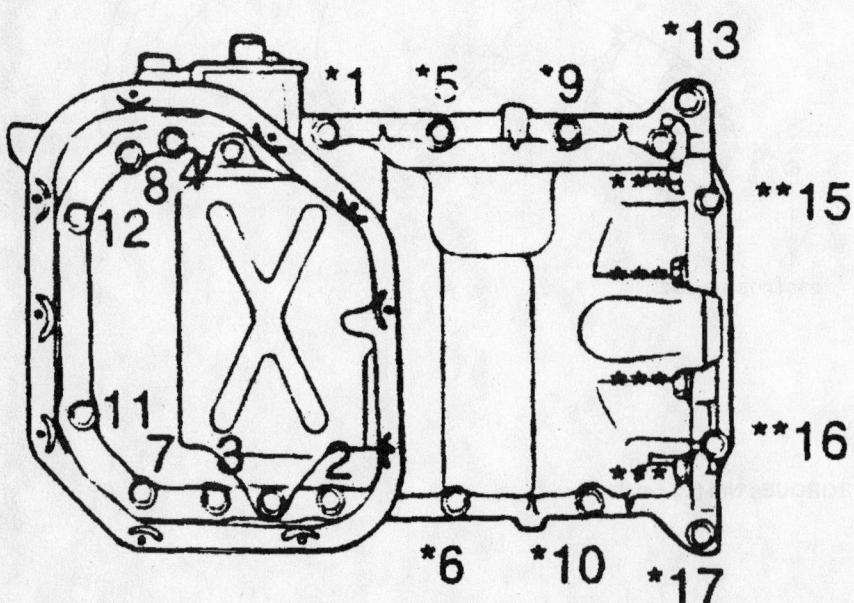

9355LG32

Upper oil pan torque sequence (tighten bolts indicated with * to 14–20 inch (19–28 Nm), bolts indicated with a ** to 4–5 ft. lbs. (5–7 Nm) and bolts indicated with a * to 22–30 ft. lbs. (30–42 Nm—2.7L engine**

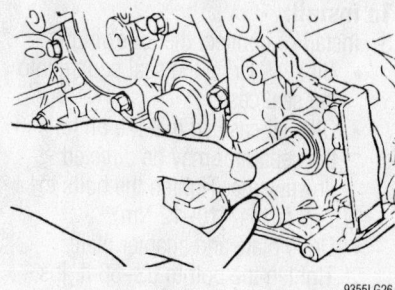

Remove the pump driven gear the left counter balance shaft bolt—2.4L engine

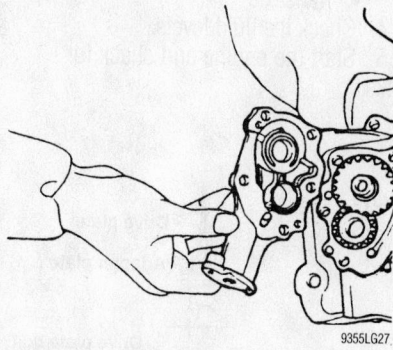

Remove the oil pump cover from the case—2.4L engine

- Two counter balance shafts from the block
- Oil pump cover from the case
- Oil pump gears from the case
- Screwdriver from the plug hole

To install:

5. Install the oil pump gears.

6. Inspect the tip clearance of the gears using a feeler gauge. The specifications are as follows:

a. Standard value drive gear: 0.0063–0.0083 inch (0.16–0.21mm).

b. Standard value driven gear: 0.0071–0.0083 inch (0.18–0.21mm).

c. Limit drive gear: 0.0098 inch (0.25mm).

d. Limit driven gear: 0.0098 inch (0.25mm).

7. Inspect the side clearance of the gears using a feeler gauge. The specifications are as follows:

a. Standard value drive gear: 0.0031–0.0055 inch (0.08–0.14mm).

b. Standard value driven gear: 0.0024–0.0047 inch (0.06–0.12mm).

c. Limit drive gear: 0.0098 inch (0.25mm).

d. Limit driven gear: 0.0098 inch (0.25mm).

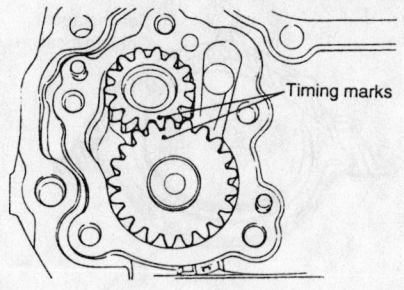

Apply engine oil to both gears and align the gear timing marks—2.4L engine

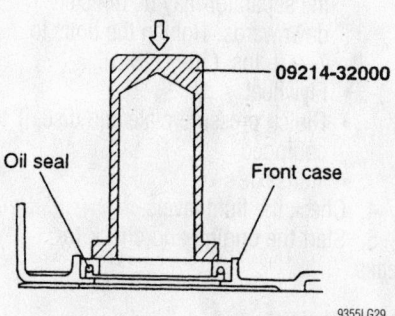

Using crankshaft front oil seal install Tool 09214-32000, install the oil seal into the case—2.4L engine

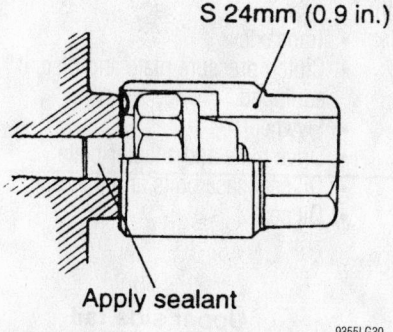

Front case bolt length and location—2.4L engine

8. Apply engine oil to both gears and align the gear timing marks.

9. Install the oil pump case.

10. Using crankshaft front oil seal install Tool 09214-32000, install the oil seal into the case.

11. Place special tool 09214-32100 on the front of the crankshaft and apply a coat of oil to the outside of the tool to aid in case installation.

12. Install or connect the following:
- New front case gasket and temporarily tighten the flange bolts

- Front case and tighten the bolts to 14–20 ft. lbs. (20–27 Nm), making sure the correct length bolt is installed in the correct location.

13. Insert an 0.32 inch (8mm) screwdriver into the plug hole. The screwdriver must be inserted at least 2.4 inch (60mm). Verify the shaft is in place and install the bolt.

- New O-ring on the groove on the front case
- Plug case and tighten to 14–20 ft. lbs. (20–27 Nm)
- Oil screen and gasket
- Oil pan
- Oil pressure switch using a 24mm deep socket. Apply Threebond 1104 sealant to the threads before installation and tighten to 6–9 ft. lbs. (8–12 Nm).
- Timing belt
- Negative battery cable

14. Fill the crankcase to the correct level.

15. Start the engine and check for leaks.

2.7L Engine

1. Before servicing the vehicle, refer to the precautions in the beginning of this section.

2. Drain the engine oil.

3. Remove or disconnect the following:
- Negative battery cable
- Oil pressure switch
- Oil filter and pans
- Oil screen and gasket
- Oil filter bracket and gasket
- Oil relief valve plug from the pump case
- Oil pump case

To install:

4. Install the oil pump gears.

5. Inspect the side clearance of the gears using a feeler gauge. The specifications are as follows:

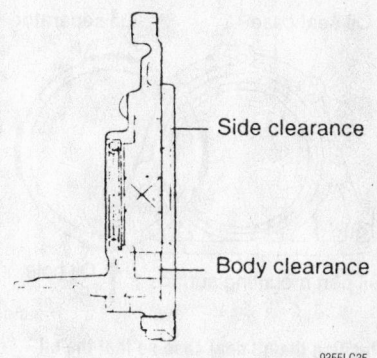

Check the oil pump gears side and body clearance—2.7L

Brake service is covered in Section 4 of this manual

a. Standard body clearance: 0.0039–0.0071 inch (0.100–0.181mm).

b. Standard side clearance: 0.0016–0.0037 inch (0.040–0.095mm).

6. Install the oil pump case with a new gasket. Tighten the bolt to 9–11 ft. lbs. (12–15 Nm) and the screw to 6–9 ft. lbs. (8–12 Nm).

7. Install a new oil seal into the pump as tightly as possible.

8. Using crankshaft front oil seal install Tool 09214-33000, install the oil seal into the case.

9. Install the relief plunger and spring and tighten the valve plug to 29–36 ft. lbs. (40–50 Nm).

10. Install the oil screen and a new gasket.

11. Oil pans and filter.
   • Negative battery cable

12. Fill the crankcase to the correct level.

13. Start the engine and check for leaks.

Rear Main Seal

REMOVAL & INSTALLATION

2.4L Engine

1. Before servicing the vehicle, refer to the precautions in the beginning of this section.

2. Remove or disconnect the following:
   • Transaxle
   • Clutch pressure plate and disc, if equipped
   • Flywheel
   • Oil seal case bolts and the case
   • Oil seal

To install:

3. Install or connect the following:
   • Oil seal. Drive the seal square into the seal case.

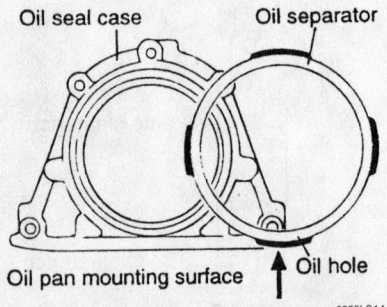

Position the oil seal case so that the oil hole in the separator may be directed downwards—2.4L engine

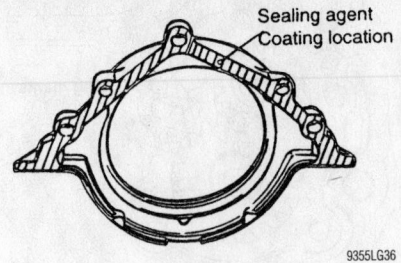

Apply sealant to the areas shown—2.7L engine

• Oil seal case so that the oil hole in the separator may be directed downwards. Tighten the bolts to 7–9 ft. lbs. (10–12 Nm).
• Flywheel
• Clutch pressure plate and disc, if equipped
• Transaxle

4. Check the fluid levels.

5. Start the engine and check for leaks.

2.7L Engine

1. Before servicing the vehicle, refer to the precautions in the beginning of this section.

2. Remove or disconnect the following:
   • Transaxle
   • Clutch pressure plate and disc, if equipped
   • Flywheel
   • Drive plate and adapter plate
   • Oil seal case bolts and the case
   • Oil seal

To install:

3. Install or connect the following:
   • Oil seal. Drive the seal square into the seal case.
   • Oil seal case so that the oil hole in the separator may be directed downwards. Tighten the bolts to 7–9 ft. lbs. (10–12 Nm).
   • Drive plate and adapter plate. Tighten the bolt to 53–56 ft. lbs. (73–77 Nm).
   • Flywheel
   • Clutch pressure plate and disc, if equipped
   • Transaxle

4. Check the fluid levels.

5. Start the engine and check for leaks.

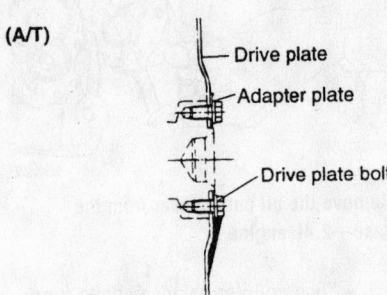

Install the drive plate and adapter plate—2.7L engine

Piston and Ring

POSITIONING

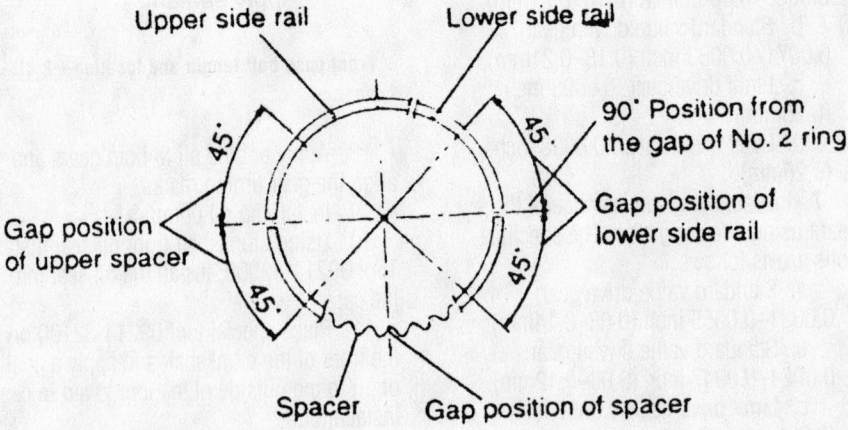

Ring end gap positioning

FUEL SYSTEM

Fuel System Service Precautions

Safety is the most important factor when performing not only fuel system maintenance, but any type of maintenance. Failure to conduct maintenance and repairs in a safe manner may result in serious personal injury or death. Maintenance and testing of the vehicle's fuel system components can be accomplished safely and effectively by adhering to the following rules and guidelines:

- To avoid the possibility of fire and personal injury, always disconnect the negative battery cable unless the repair or test procedure requires that battery voltage be applied.
- Always relieve the fuel system pressure prior to disconnecting any fuel system component (injector, fuel rail, pressure regulator, etc.), fitting or fuel line connection. Exercise extreme caution whenever relieving fuel system pressure to avoid exposing skin, face and eyes to fuel spray. Please be advised that fuel under pressure may penetrate the skin or any part of the body that it contacts.
- Always place a shop towel or cloth around the fitting or connection prior to loosening to absorb any excess fuel due to spillage. Ensure that all fuel spillage (should it occur) is quickly removed from engine surfaces. Ensure that all fuel soaked cloths or towels are deposited into a suitable waste container.
- Always keep a dry chemical (Class B) fire extinguisher near the work area.
- Do not allow fuel spray or fuel vapors to come into contact with a spark or open flame.
- Always use a backup wrench when loosening and tightening fuel line connection fittings. This will prevent unnecessary stress and torsion to fuel line piping. Always follow the proper torque specifications.
- Always replace worn fuel fitting O-rings with new. Do not substitute fuel hose or equivalent, where fuel pipe is installed.

Fuel System Pressure

RELIEVING

1. Before servicing the vehicle, refer to the precautions in the beginning of this section.

2. Remove the fuel filler cap.
3. Remove the fuel pump fuse and crank the engine until it stalls.
4. Disconnect the negative battery cable.
5. Replace the fuse.

Fuel Filter

The fuel filter is part of the fuel pump assembly located in the tank.

REMOVAL & INSTALLATION

1. Before servicing the vehicle, refer to the precautions in the beginning of this section.
2. Relieve the fuel system pressure.
3. Remove or disconnect the following:
   - Negative battery cable
   - Fuel pump connector
   - Fuel feed and return lines from the pump assembly
   - Pump cover screws
   - Pump assembly from the tank
   - Filter from the pump assembly

To install:

4. Install or connect the following:
   - Fuel filter
   - Pump into the tank
   - Pump cover screws
   - Fuel feed and return lines to the pump assembly
   - Fuel pump connector
   - Negative battery cable
5. Start the engine and check for leaks.

Fuel Pump

REMOVAL & INSTALLATION

1. Before servicing the vehicle, refer to the precautions in the beginning of this section.
2. Relieve the fuel system pressure.
3. Remove or disconnect the following:
   - Negative battery cable
   - Fuel pump connector
   - Fuel feed and return lines from the pump assembly
   - Pump cover screws
   - Pump assembly from the tank

To install:

4. Install or connect the following:
   - Pump into the tank
   - Pump cover screws
   - Fuel feed and return lines to the pump assembly

- Fuel pump connector
- Negative battery cable
5. Start the engine and check for leaks.

Fuel Injector

REMOVAL & INSTALLATION

1. Before servicing the vehicle, refer to the precautions in the beginning of this section.
2. Relieve the fuel system pressure.
3. Remove or disconnect the following:
   - Negative battery cable
   - Air breather hose from the throttle body
   - Throttle cable
   - Engine coolant hose and throttle body
   - Positive Crankcase Ventilation (PCV) valve and brake booster vacuum hose
   - Vacuum hose connector
   - Injector cover
   - High pressure fuel hose
   - Fuel injector harness connector
   - Delivery pipe with the injectors and the pressure regulator as an assembly
   - Injector from the delivery pipe
   - Injector O-ring, grommet and discard

To install:

4. Install a new grommet and O-ring
5. Apply a coating of spindle oil or gasoline to the injector O-ring
6. Install the injector into the delivery pipe while turning the injector left and right making sure the injector turns smoothly. If the injector does not turn smoothly check for a jammed O-ring, remove the injector and reinsert it again.
7. Install or connect the following:
   - Delivery pipe and injector assembly

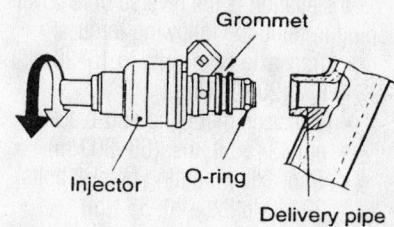

9355LG39

Install the injector using a twisting motion

For complete Engine Mechanical specifications, see Section 1 of this manual

- Intake manifold stay and tighten the bolts to 13–18 ft. lbs. (18–25 Nm)
- Fuel injector harness connector
- High pressure fuel hose
- Injector cover

- Vacuum hose connector
- PCV valve and brake booster vacuum hose
- Throttle body and engine coolant hose
- Throttle cable

- Air breather hose from the throttle body
- Negative battery cable

8. Start the engine and check for proper operation.

DRIVE TRAIN

Transaxle Assembly

REMOVAL & INSTALLATION

Manual

1. Before servicing the vehicle, refer to the precautions in the beginning of this section.
2. Drain the transaxle.
3. Remove or disconnect the following:
- Negative battery cable
- Air cleaner duct
- Air cleaner and air flow hose
- Back-up light connector
- Clutch line and clip
- Clutch release cylinder
- Speedometer cable
- Gear select or shift cables
- Starter motor mounting bolts
- Transaxle upper bolts
4. Attach an engine support fixture to the engine.
- Transaxle mounting bracket and insulator
- Front wheels
- Tie rod end, lower ball joint and drive shaft
- Gear box u-joint bolt and the return tube mounting bolts
- Front muffler
- Sub-frame mounting bolts and the frame
- Transaxle front and rear mounting brackets
- Transaxle side mounting bolts
5. Using a suitable jack, remove the transaxle from the vehicle.

To install:
6. Installation is the reverse of removal keeping in mind the following torques:
- Transaxle case bolts to 15–20 ft. lbs. (20–27 Nm)
- Transaxle mounting sub-bracket nut 43–58 ft. lbs. (60–80 Nm)
- Transaxle mounting bracket bolts 29–40 ft. lbs. (40–55 Nm)
- Transaxle mounting insulator bolt 65–80 ft. lbs. (90–110 Nm)
- Clutch release cylinder retainers to 11–16 ft. lbs. (15–22 Nm)

7. Fill the transaxle to the correct level.
8. Start the engine and check for leaks.
9. Check the wheel alignment and adjust as necessary.

Automatic

1. Before servicing the vehicle, refer to the precautions in the beginning of this section.
2. Drain the transaxle.
3. Remove or disconnect the following:
- Negative battery cable
- Air cleaner assembly
- Control cable
- Speedometer sensor connector
- Transaxle range switch, solenoid, and oil temperature sensor connectors
- Oil cooler hose
4. Attach an engine support fixture to the engine.
- Gear box, stabilizer bar, tie rod end, lower ball joint and drive shaft
- Gear box u-joint bolt and the return tube mounting bolts
- Sub-frame mounting bolts and the frame
- Starter motor
- Transaxle mounting bolts
5. Using a suitable jack, remove the transaxle from the vehicle.

To install:
6. Installation is the reverse of removal keeping in mind the following torques:
- Transaxle case bolts to 15–20 ft. lbs. (20–27 Nm)
- Transaxle mounting sub-bracket nut 43–58 ft. lbs. (60–80 Nm)
- Transaxle mounting bracket bolts 29–40 ft. lbs. (40–55 Nm)
- Transaxle mounting insulator bolt 65–80 ft. lbs. (90–110 Nm)
7. Adjust the control cable as follows:
  a. Move the shift lever and the transaxle range switch to the **N** position and install the cable.
  b. When attaching the control cable to the bracket, make sure the clip is installs so it contacts the cable.
  c. Adjust the nut to remove any free-

play in the cable and make sure the lever moves freely.
8. Fill the transaxle to the correct level.
9. Start the engine and check for leaks.
10. Check the wheel alignment and adjust as necessary.

Transfer Case

REMOVAL & INSTALLATION

1. Before servicing the vehicle, refer to the precautions in the beginning of this section.
2. Attach an engine support fixture to the engine.
3. Drain the transfer case fluid.
4. Remove or disconnect the following:
- Negative battery cable
- Transfer case and sub-frame mounting brackets
- Wheels
- Front muffler
- Steering tube from the sub-frame
- Sub-frame assembly
5. Support the transfer case with a suitable jack.
- Driveshafts
- Transfer case bolts and the case assembly
6. Installation is the reverse of removal

Clutch

ADJUSTMENTS

1. Before servicing the vehicle, refer to the precautions in the beginning of this section.
2. Measure the clutch pedal height from the face of the pedal pad to the floorboard. The proper measurement 218.9mm.
3. Measure the clutch pedal clevis pin play. The measurement should be 0.04–0.11 inch (1–3mm).
4. Adjust the height and clevis pin if necessary as follows:
  a. Turn and adjust the bolt, then secure it by tightening the lock nut.

➥**After adjustment, tighten the bolt until it reaches the pedal stopper and tighten the lock nut.**

    b. Turn the push rod until the proper specification is reached and tighten the lock nut.

➥**When adjusting the clevis pin play or the pedal height, make sure not to push the push rod towards the master cylinder.**

    c. After the adjustments are made, check that the clutch pedal free play is 0.2–0.5 inch (6–13mm). The free play is measured at the face of the pedal pad

    5. If the adjustments do not bring the pedal into specifications, check the hydraulic system for air or a faulty component.

REMOVAL & INSTALLATION

    1. Before servicing the vehicle, refer to the precautions in the beginning of this section.

    2. Disconnect the negative battery cable.

    3. Remove the transmission assembly.

    4. If the pressure plate is attached to the flywheel, remove the release bearing using snap-ring pliers as follows:

    a. Insert the snap-ring pliers under the wave washer and place it in the center of the snap-ring

    b. Spread the snap-ring by pushing down on the bearing assembly.

    c. Once the snap-ring is fully expanded, remove the release bearing

    5. Install a clutch disc alignment tool.

    6. Remove pressure plate retaining bolts using a star pattern in several passes.

    7. Remove the pressure plate with clutch disc.

To install:

    8. Apply multi-purpose grease to clutch splines.

    9. Install or connect the following:
- Clutch plate to the flywheel
- Pressure plate and cover. Do not torque the bolts at this time.

    10. Align the bearing to the release fork and install it until it is fully engaged.

    11. Torque the pressure plate bolts using a star pattern torque sequence to 11–16 ft. lbs. (15–22 Nm).

    12. Install transmission.

    13. Connect the negative battery cable.

Hydraulic Clutch System

BLEEDING

    Bleeding air from the hydraulic clutch system is necessary whenever any part of the system has been disconnected or the fluid level (in the reservoir) has been allowed to fall so low that air has been drawn into the master cylinder.

✳✳ WARNING

NEVER use fluid that has been bled from a clutch system to fill the master cylinder reservoir, as it may be aerated, contain excessive moisture and/or be contaminated in some other way.

    1. Before servicing the vehicle, refer to the precautions in the beginning of this section.

    2. Fill the clutch master cylinder reservoir with new hydraulic clutch fluid.

    3. Attach a hose to the bleeder on the clutch actuator and submerge the other end of the hose in a container of hydraulic clutch fluid.

    4. Have an assistant slowly depress and hold the clutch pedal.

    5. Loosen the bleeder to purge air.

    6. Tighten the bleeder.

    7. Repeat the above 3 steps until all air is completely purged from the system.

    8. Refill the clutch master cylinder reservoir.

Halfshaft

REMOVAL & INSTALLATION

Front

    1. Before servicing the vehicle, refer to the precautions in the beginning of this section.

    2. Drain the transaxle.

    3. Remove or disconnect the following:

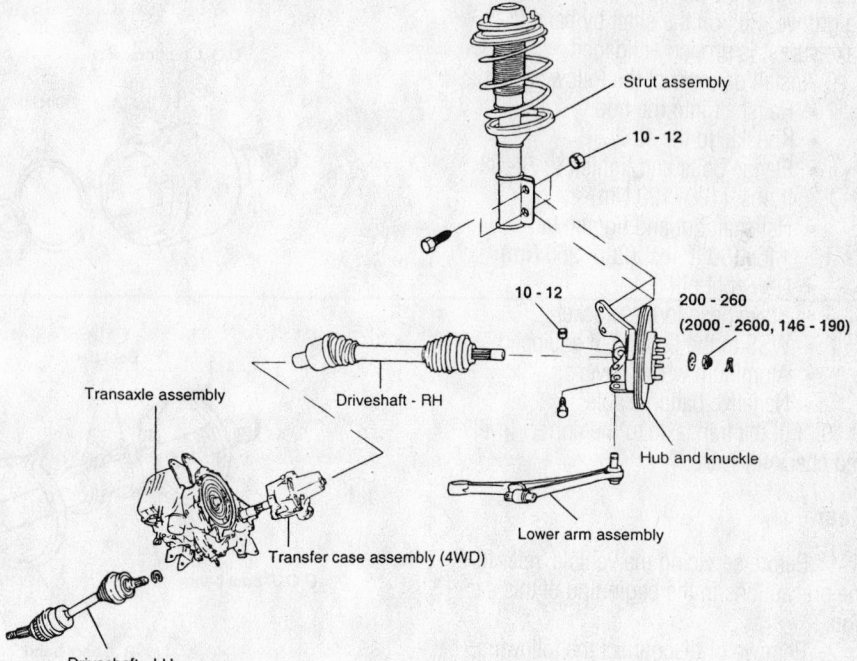

TORQUE : Nm (kg·cm, lb·ft)

Exploded view of the halfshaft mounting and related components

9355LG40

For Accessory Drive Belt illustrations, see Section 1 of this manual

- Negative battery cable
- Aluminum wheel cover
- Split pin and halfhaft nut
- Wheel Speed Sensor (WSS) from the bracket, if equipped
- Brake hose from the bracket
- Knuckle from the strut by removing the flange bolts
- Halfshaft from the hub by tapping the end with a plastic mallet
- Halfshaft from the transaxle using a pry bar.

4. Insert a plug in the transaxle opening.

To install:

➡**Use new circlips, split pins and self-locking nuts for assembly.**

5. Remove the plug from the transaxle opening.

6. Coat the halfshaft splines and sliding surfaces with gear oil.

7. Make sure the gap of the circlip is facing downwards.

8. Install the inner CV-joint to the transaxle until the circlip locks in the retaining groove. Pull on the shaft by hand to make sure it is properly engaged.

9. Install or connect the following:
- Halfshaft into the hub
- Knuckle to the strut
- Flange bolts and tighten to 74–88 ft. lbs. (100–120 Nm)
- Halfshaft nut and tighten to 146–190 ft. lbs. (200–260 Nm)
- New split pin
- Brake hose to the bracket
- WSS to the bracket, if equipped
- Aluminum wheel cover
- Negative battery cable

10. Fill the transaxle to the correct level and check for leaks.

Rear

1. Before servicing the vehicle, refer to the precautions in the beginning of this section.

2. Remove or disconnect the following:
- Rear wheel
- Split pin and nut
- Spare tire and support the hanger of the main muffler to avoid interfering with the carrier during right hand shaft removal

3. Matchmark the propeller rubber coupling and differential flange, then remove the bolts and nuts.

4. Support the differential carrier with a jack and remove the differential carrier mounting nuts and bolts.

- Shaft from the carrier by inserting a prybar between the carrier and the shaft
- Differential carrier to the rear after lowering the jack
- Shaft from the axle hub using a plastic mallet
- Shaft from the vehicle

To install:

5. Install or connect the following:
- Shaft into the axle hub
- Differential carrier to the rear using a jack
- Shaft to the carrier
- Differntial carrier mounting nuts and bolts. Tighten the carrier nuts and bolts to 51–58 ft. lbs. (70–80 Nm) and the carrier rear bracket bolt to 58–73 ft. lbs. (80–100 Nm).
- Propeller shaft
- Spare tire and remove the support from the main muffler hanger
- New shaft nut and tighten to 146–253 ft. lbs. (200–260 Nm)
- New split pin
- Rear wheel

CV-Joint

OVERHAUL

DOJ-BJ Type

The DOJ-BJ type joint is used on both the front and rear of the vehicle and the following overhaul procedure is used.

1. Before servicing the vehicle, refer to the precautions in the beginning of this section.

➡**Do not disassemble the BJ assembly.**

2. Remove or disconnect the following:
- Axle halfshaft from the vehicle and place it in a vise
- DOJ boot clamps and push the boot back from the outer race
- Circlip with a flat bladed prytool
- Driveshaft from the DOJ outer race
- Snap-ring and take out the inner race, cage and balls as an assembly

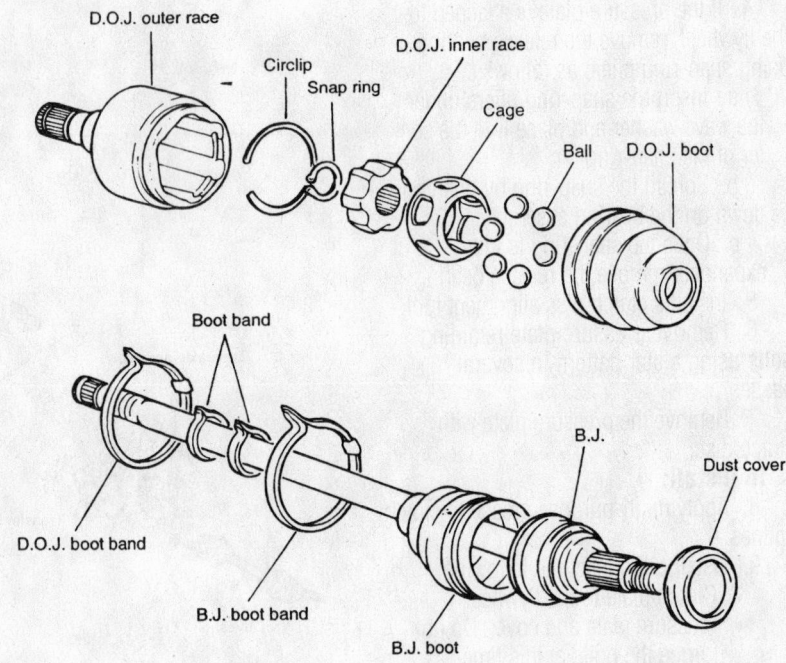

TORQUE : Nm (kg·cm, lb·ft)

Exploded view of the DOJ-BJ type CV-joint assembly

9355LG41

3. Clean the outer race, cage and balls without disassembling them.
- BJ boot clamps
- DOJ and BJ boots

To install:

➡**Use new circlips and boot clamps for assembly.**

4. Apply some of the grease supplied with the kit to the halfshaft..

5. Install the boots.

6. Apply the grease supplied with the kit to the inner race and cage.

7. Install the cage so that it is offset on the race.

8. Apply the grease supplied with the kit to the cage and install the balls

9. Install the chamfered side of the cage as shown in the accompanying illustration,

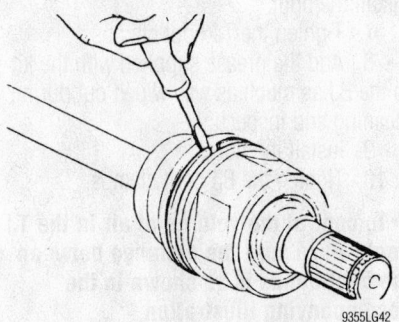

Remove the boot clamps from the DOJ

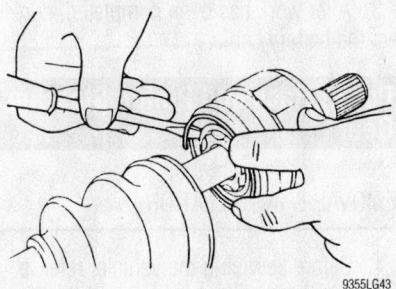

Remove the circlip with a suitable tool

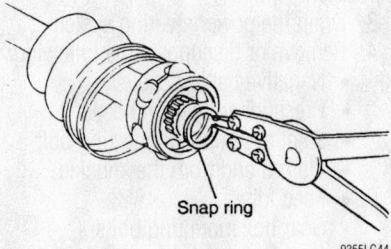

Snap ring

Remove snap-ring and take out the inner race, cage and balls as an assembly

then insert the inner race onto the shaft and install the snap-ring.

10. Apply the grease supplied with the kit to the BJ outer race (67–73 grams of grease in the joint and 62–68 grams of grease in the boot); and install the outer race onto the shaft.

11. Apply the grease supplied with the kit to the DOJ outer race (62–68 grams in the joint and 37–43 grams in the boot); and install the circlip.

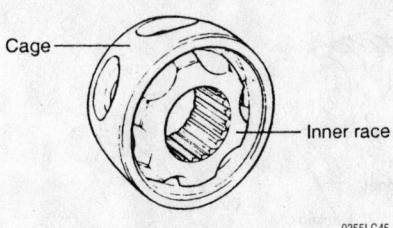

Cage

Inner race

9355LG45

Install the cage so that it is offset on the race

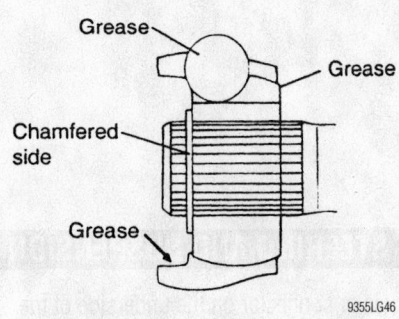

Grease

Grease

Chamfered side

Grease

9355LG46

Install the chamfered side of the cage as shown

12. Tighten the DOJ boot clamps.

13. Apply the grease supplied with the kit to the BJ.

14. Install the boots.

15. Tighten the BJ boot clamps

➡**To control the volume of air in the DOJ boot, make sure the distance between the boot bands is as shown in the accompanying illustration.**

16. Install the axle halfshaft to the vehicle.

TJ-BJ Type

1. Before servicing the vehicle, refer to the precautions in the beginning of this section.

2. Remove or disconnect the following:
- Axle halfshaft from the vehicle.
- TJ boot clamps
- TJ boot from the TJ case
- Snap-ring and spider assembly
- BJ boot clamps, then pull out the TJ boot and the BJ boot

3. Clean all the components properly

To install:

➡**Use new circlips and boot clamps for assembly.**

4. Apply the grease supplied with the kit to the shaft and install the boots.

5. If installing the dynamic damper, keep the BJ shaft in a straight line and attach the damper in the direction shown in the accompanying illustration, then install the boot clamp.

6. Apply the grease supplied with the kit to the TJ boot (97–103 grams in the

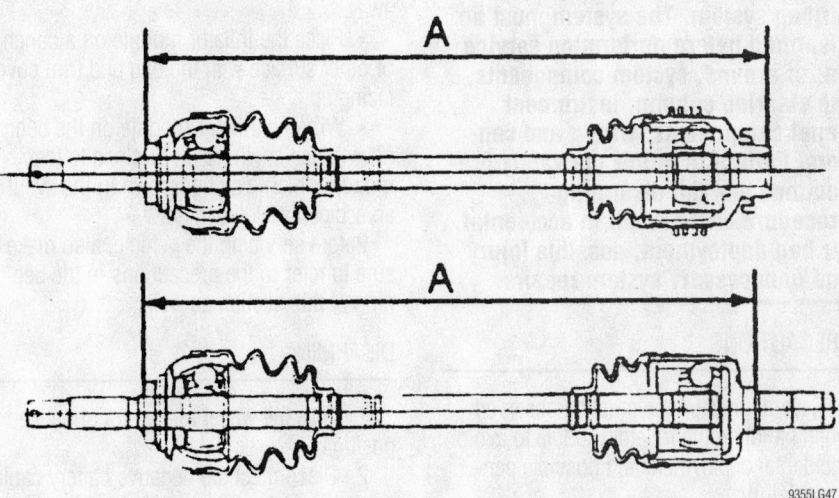

A

A

9355LG47

To control the volume of air in the DOJ boot, make sure the distance between the boot bands is as shown

For Tire, Wheel and Ball Joint specifications, see Section 1 of this manual

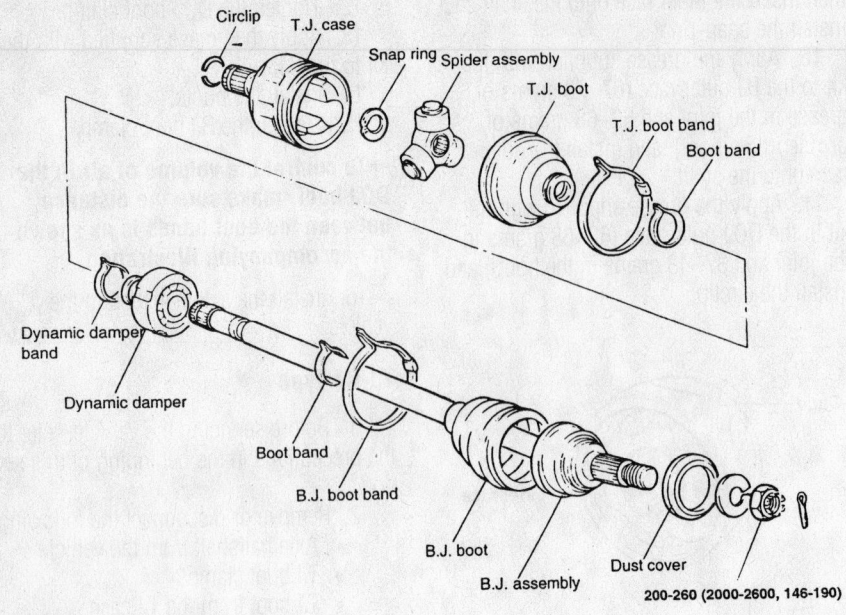

TORQUE : Nm (kg·cm, lb·ft)

9355LG48

Exploded view of the TJ-BJ type CV assembly

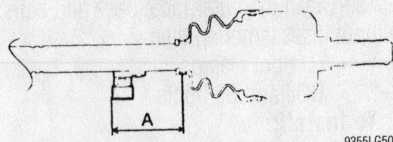

9355LG50

If installing the dynamic damper, keep the BJ shaft in a straight line and attach the damper in the direction shown

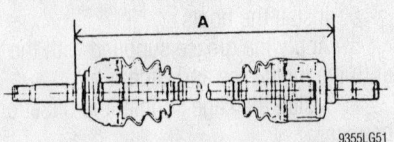

9355LG51

To control the volume of air in the TJ boot, make sure the distance between the boot bands is as shown

joint and 42–48 grams in the boot), then install the boot.

7. Tighten the TJ boot clamps.

8. Add the grease supplied with the kit to the BJ as much as was wiped out during cleaning and inspection.

9. Install the boots.

10. Tighten the BJ boot clamps.

➡To control the volume of air in the TJ boot, make sure the distance between the boot bands is as shown in the accompanying illustration

STEERING AND SUSPENSION

Air Bag

✳✳ CAUTION

Some vehicles are equipped with an air bag system. The system must be disarmed before performing service on, or around, system components, the steering column, instrument panel components, wiring and sensors. Failure to follow the safety precautions and the disarming procedure could result in accidental air bag deployment, possible injury and unnecessary system repairs.

PRECAUTIONS

Several precautions must be observed when handling the inflator module to avoid accidental deployment and possible personal injury.

• Never carry the inflator module by the wires or connector on the underside of the module.

• When carrying a live inflator module, hold securely with both hands, and ensure that the bag and trim cover are pointed away.

• Place the inflator module on a bench or other surface with the bag and trim cover facing up.

• With the inflator module on the bench, never place anything on or close to the module which may be thrown in the event of an accidental deployment.

Before servicing the vehicle, also make sure to refer to the precautions in the beginning of this section as well.

DISARMING

1. Turn the wheel to the straight ahead position.

2. Disconnect the negative battery cable and wait at least 30 seconds for the air bag energy to deplete.

3. After work has been completed, connect the battery cable.

Power Rack and Pinion Steering Gear

REMOVAL & INSTALLATION

1. Before servicing the vehicle, refer to the precautions in the beginning of this section.

2. Center the steering wheel and lock it in position.

3. Drain the power steering system.

4. Remove or disconnect the following:

• Negative battery cable
• Pressure and return hoses
• Joint assembly connecting bolt
• Tie rod end from the knuckle
• Feed tube
• Gear box mounting bolts
• Gear box assembly with the rubber mounts

To install:

5. Install or connect the following:
- Gear box assembly with the rubber mounts
- Gear box mounting bolts and tighten to 66–81 ft. lbs. (90–110 Nm)
- Feed tube and tighten to 7–11 ft. lbs. (10–16 Nm)
- Tie rod end from the knuckle and tighten to 18–25 ft. lbs. (24–34 Nm)
- Joint assembly connecting bolt and tighten to 11–14 ft. lbs. (15–20 Nm)
- Pressure and return hoses. Tighten the fittings to 9–13 ft. lbs. (12–18 Nm).
- Negative battery cable

6. Fill the power steering system.

7. Check the wheel alignment and adjust as necessary.

Strut

REMOVAL & INSTALLATION

Front

1. Before servicing the vehicle, refer to the precautions in the beginning of this section.

2. Remove or disconnect the following:
- Front wheel
- Brake hose clip from the strut mounting bracket
- Wheel Speed Sensor (WSS) from the knuckle
- Stabilizer bar link
- Strut-to-knuckle bolts
- 3 upper mounting nuts
- Strut

To install:

3. Install or connect the following:
- Strut. Tighten the upper mount nuts to 30–37 ft. lbs. (40–50 Nm).
- Strut-to-knuckle bolts and tighten to 74–88 ft. lbs. (100–120 Nm)
- Stabilizer bar link and tighten to 30–37 ft. lbs. (30–50 Nm)
- WSS from the knuckle
- Brake hose clip to the strut mounting bracket
- Front wheel

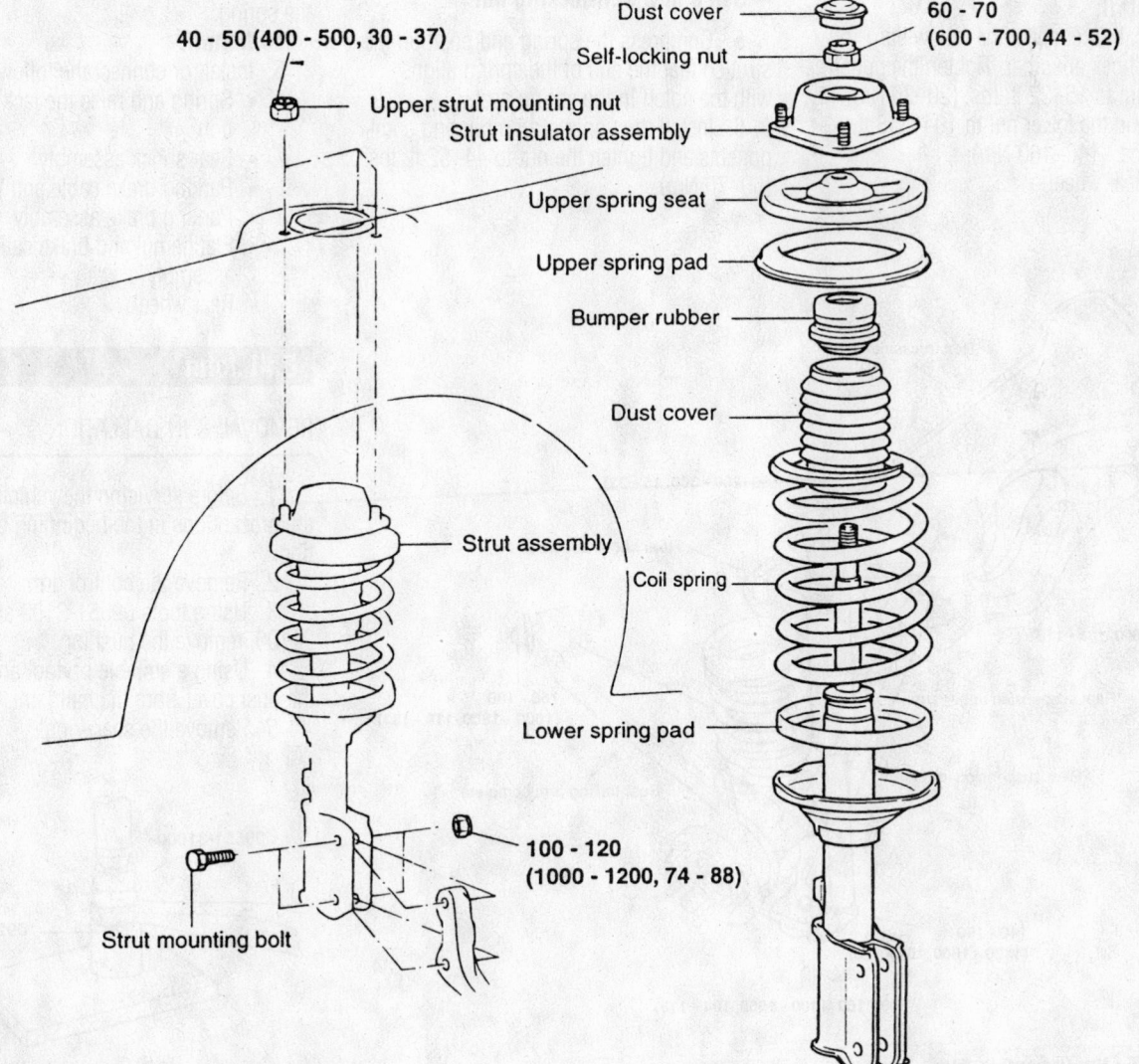

40 - 50 (400 - 500, 30 - 37)

Upper strut mounting nut

Dust cover

Self-locking nut

60 - 70
(600 - 700, 44 - 52)

Strut insulator assembly

Upper spring seat

Upper spring pad

Bumper rubber

Dust cover

Strut assembly

Coil spring

Lower spring pad

Strut mounting bolt

100 - 120
(1000 - 1200, 74 - 88)

TORQUE : Nm (kg·cm, lb·ft)

9355LG52

Exploded view of the front strut assembly and mounting

For Wheel Alignment specifications, see Section 1 of this manual

4. Check the wheel alignment and adjust as necessary.

Shock Absorber

REMOVAL & INSTALLATION

Rear

1. Before servicing the vehicle, refer to the precautions in the beginning of this section.

2. Support the vehicle under the lower control arm.

3. Remove or disconnect the following:
- Rear wheel
- Upper shock absorber upper nut
- Lower shock absorber nut
- Shock absorber

To install:

4. Install or connect the following:
- Shock absorber. Tighten the upper nut to 15–22 ft. lbs. (20–30 Nm) and the lower nut to 104–118 ft. lbs. (140–160 Nm).
- Rear wheel

Coil Spring

REMOVAL & INSTALLATION

Front

1. Before servicing the vehicle, refer to the precautions in the beginning of this section.

2. Remove the strut from the vehicle and install in a strut spring compressor. Compress the spring until the end of the spring comes away from the spring seat.

3. Remove the upper strut mount, spring seat and related components.

4. Remove the coil spring from the strut spring compressor.

To install:

➡**Use a new self-locking nut.**

5. Compress the spring and position the strut so that the end of the spring aligns with the notch in the spring seat.

6. Install the upper strut mounting components and tighten the nut to 44–52 ft. lbs. (60–70 Nm).

7. Install the strut to the vehicle.

8. Check the wheel alignment and adjust as necessary.

Rear

1. Before servicing the vehicle, refer to the precautions in the beginning of this section.

2. Support the vehicle under the lower control arm.

3. Remove or disconnect the following:
- Rear wheel
- Flange nut and brake caliper assembly
- Parking brake assembly
- Wheel Speed Sensor (WSS) and the parking brake cable
- Rear shock assembly

4. Lower the jack assembly and remove the spring.

To install:

5. Install or connect the following:
- Spring and raise the jack into position
- Rear shock assembly
- Parking brake cable and WSS
- Parking brake assembly
- Flange nut and brake caliper assembly
- Rear wheel

Ball Joint

REMOVAL & INSTALLATION

1. Before servicing the vehicle, refer to the precautions in the beginning of this section.

2. Remove the control arm.

3. Using tools 09551-3100 and 09216-21100, remove the bushing.

4. Using a suitable prytool and remove the dust cover from the ball joint.

5. Remove the snap-ring.

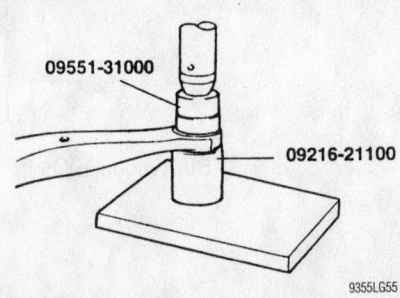

Removing the bushing from the control arm using the appropriate removal and installation tools

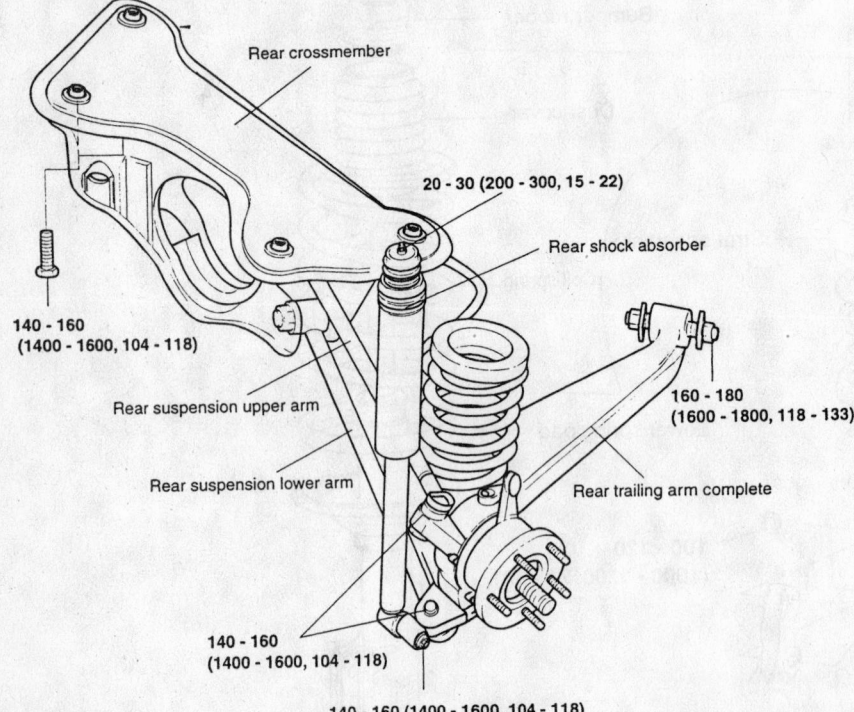

Rear crossmember

20 - 30 (200 - 300, 15 - 22)

Rear shock absorber

140 - 160 (1400 - 1600, 104 - 118)

Rear suspension upper arm

Rear suspension lower arm

160 - 180 (1600 - 1800, 118 - 133)

Rear trailing arm complete

140 - 160 (1400 - 1600, 104 - 118)

140 - 160 (1400 - 1600, 104 - 118)

TORQUE : Nm (kg·cm, lb·ft)

Exploded view of the rear suspension assembly

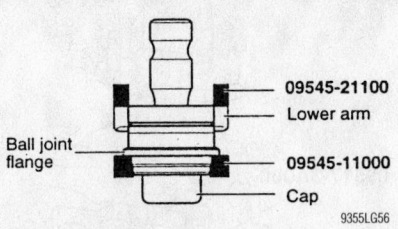

Installing the ball joint dust cover

09545-21100
Lower arm
Ball joint
flange
09545-11000
Cap
9355LG56

6. Remove the ball joint from the arm using a plastic hammer.

To install:

7. Install the ball joint.

8. Install the bushings into the arm using the appropriate tools. Make sure that the ball joint flange is supported while pressing down on the bushing until the flange touches the arm surface.

9. Install the ball joint snap-ring. Be careful to keep the snap-ring expansion as small as possible during installation.

10. Apply multi-purpose grease to the dust cover lip and inside of the cover.

11. Using tool 09545-11000, install the dust cover until it is completely seated on the snap ring.

12. Install the control arm.

Lower Control Arm

REMOVAL & INSTALLATION

Front

1. Before servicing the vehicle, refer to the precautions in the beginning of this section.

2. Remove or disconnect the following:
- Front wheel
- Ball joint-to-knuckle bolt
- Sub frame bolts and the frame
- Lower arm bolts and the arm

3. Installation is the reverse of removal. Tighten the fasteners as follows:

a. Tighten the arm bolt **A** to 74–88 ft. lbs. (100–120 Nm) and bolt **B** to 66–81 ft. lbs. (90–110 Nm).

b. Tighten the sub frame bolts to 118–148 ft. lbs. (160–200 Nm).

c. Tighten the ball joint-to-knuckle bolt to 74–88 ft. lbs. (100–120 Nm)

4. Check the wheel alignment and adjust as necessary.

Trailing Arm

REMOVAL & INSTALLATION

Rear

1. Before servicing the vehicle, refer to the precautions in the beginning of this section.

2. Support the vehicle under the lower control arm.

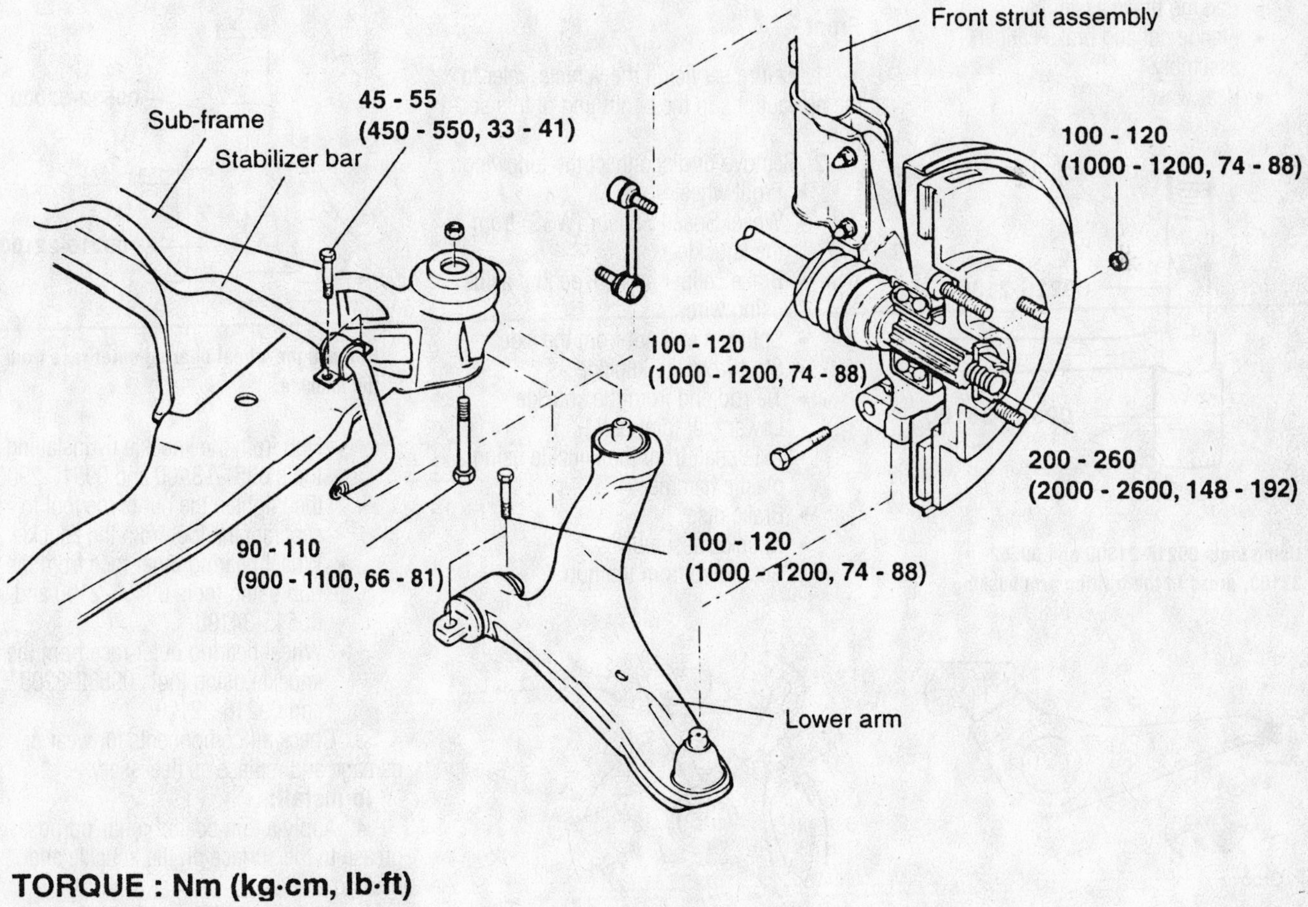

Sub-frame
Stabilizer bar
45 - 55
(450 - 550, 33 - 41)
Front strut assembly
100 - 120
(1000 - 1200, 74 - 88)
100 - 120
(1000 - 1200, 74 - 88)
200 - 260
(2000 - 2600, 148 - 192)
90 - 110
(900 - 1100, 66 - 81)
100 - 120
(1000 - 1200, 74 - 88)
Lower arm

TORQUE : Nm (kg·cm, lb·ft)

Lower control arm assembly

9355LG54

3. Remove or disconnect the following:
- Rear wheel
- Flange nut and brake caliper assembly
- Parking brake assembly
- Wheel Speed Sensor (WSS) and the parking brake cable
- Rear shock assembly
- Spring
- Rear driveshaft from the rear axle
- Upper and lower arm using tool 09517-43001.
- Trailing arm bolt and the arm

To install:

4. Install or connect the following:
- Trailing arm and tighten the trailing arm bolt to 118–133 (160–180 Nm)
- Upper and lower arm. Tighten the ball joint nut to 104–11 ft. lbs. (140–118 Nm).
- Spring and raise the jack into position
- Rear shock assembly
- Parking brake cable and WSS
- Parking brake assembly
- Flange nut and brake caliper assembly
- Rear wheel

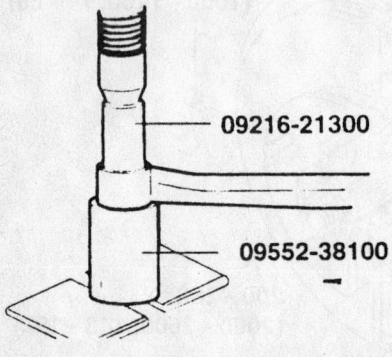

Using tools 09216-21300 and 09552-38100, press fit the trailing arm bushing

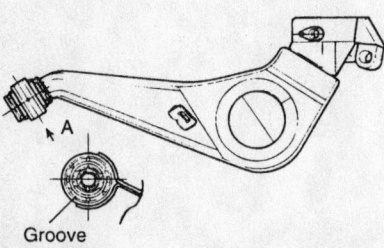

Position the groove in the arm bushing so that it is aligned as shown, before pressing the bushing into position

TRAILING ARM BUSHING REPLACEMENT

1. Before servicing the vehicle, refer to the precautions in the beginning of this section.
2. Remove the trailing arm.
3. Press the bushing from the trailing arm.

➡**Position the groove in the arm bushing so that it is aligned as shown in the accompanying illustration, then press fit the bushing.**

4. Using tools 09216-21300 and 09552-38100, press fit the bushing.

Wheel Bearings

ADJUSTMENT

The wheel bearings are sealed units and are not adjustable.

REMOVAL & INSTALLATION

Front

1. Before servicing the vehicle, refer to the precautions in the beginning of this section.
2. Remove or disconnect the following:
- Front wheel
- Wheel Speed Sensor (WSS) from the knuckle
- Brake caliper and suspend it aside using wire
- Split pin and nut from the axle
- Strut from the knuckle
- Tie rod end from the knuckle
- Lower ball joint bolt
- Axle shaft from the knuckle using a plastic hammer
- Brake disc
- Knuckle assembly
- Snap-ring from the hub

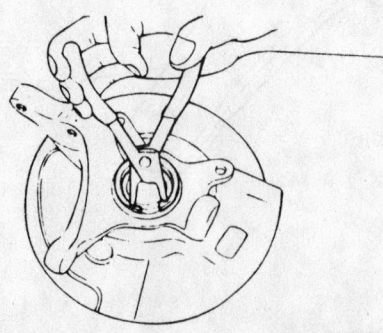

Remove the snap-ring from the hub

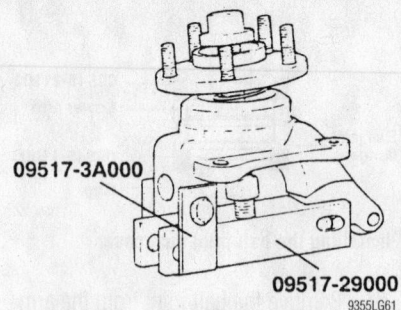

Remove the hub from the knuckle

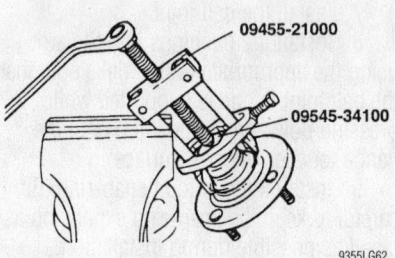

Remove the wheel bearing inner race from the hub

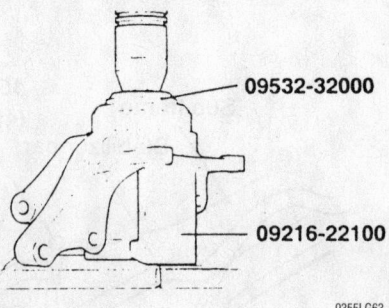

Remove the wheel bearing outer race from the knuckle

- Hub from the knuckle by installing tools 09517-3A00 and 09517-2900, then tighten the nut of the tool to separate the tool from the knuckle
- Wheel bearing inner race from the hub using tools 09455-2100 and 09545-34100
- Wheel bearing outer race from the knuckle using tools 09532-3200 and 09216-22100

3. Check all components for wear or damage and replace as necessary.

To install:

4. Apply a thin coat of multi-purpose grease to the surface on the knuckle and bearing.

➡**Do not press against the outer race of the bearing as this can cause bearing damage and always use a new bearing kit.**

5. Install or connect the following:
- Bearing onto the knuckle using tool 09216-21100.
- Snap-ring into the groove of the knuckle
- Backing plate onto the knuckle

➡ **Do not press against the outer race of the bearing as this can cause bearing damage and always use a new bearing kit.**

- Hub onto the knuckle by pressing it into position using tool 09431-3400

6. Rotate the bearing several times to seat the bearing.

7. Measure the wheel bearing torque using an inch lb. torque wrench. The measurement is 16.64 inch lbs. (1.88 Nm).

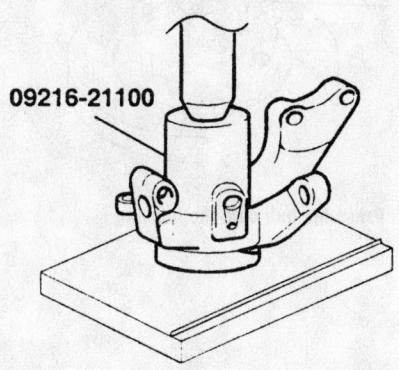

09216-21100

9355LG64

Install the bearing onto the knuckle

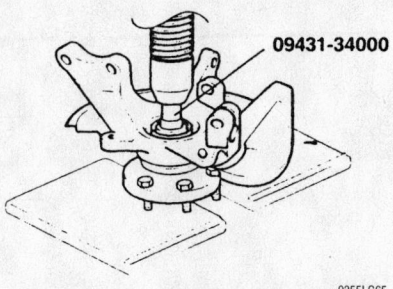

09431-34000

9355LG65

Press the hub onto the knuckle

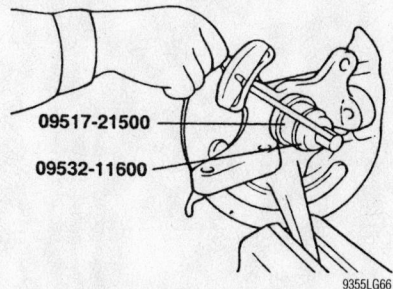

09517-21500
09532-11600

9355LG66

Check the wheel bearing starting torque

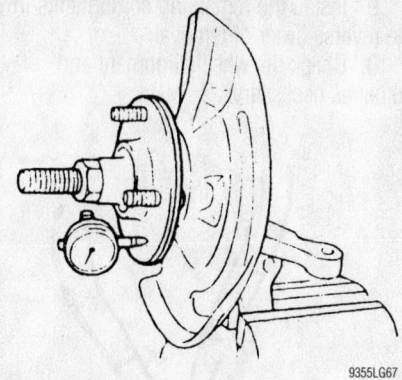

9355LG67

Check the hub end end-play

8. Measure the end-play of the hub using a dial gauge. The specification is 0.003-0.008mm.

9. Install the remaining components in the reverse order of removal.

10. Check the wheel alignment and adjust as necessary.

Rear

1. Before servicing the vehicle, refer to the precautions in the beginning of this section.

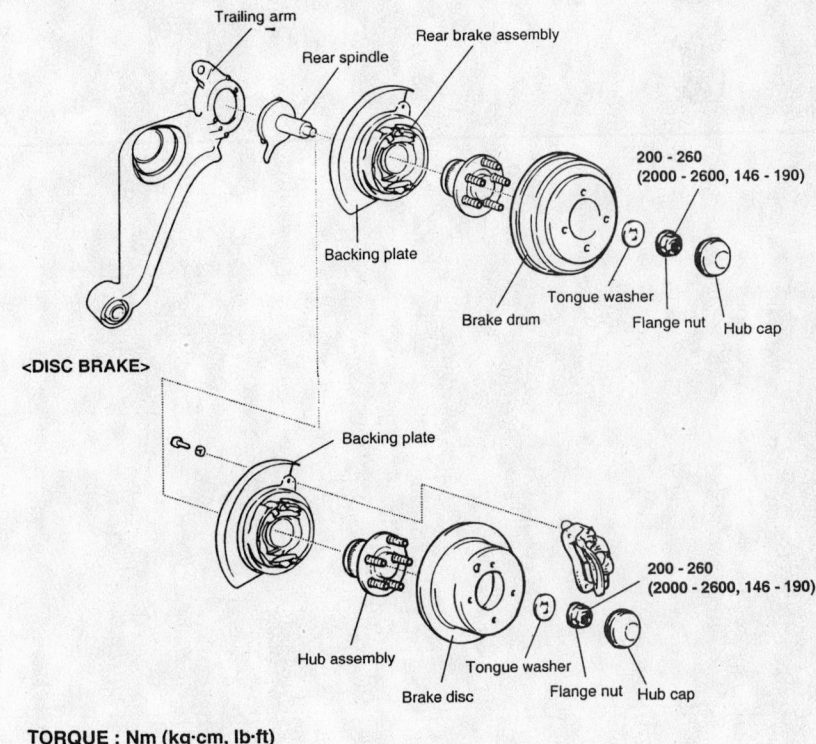

<DRUM BRAKE>

Trailing arm
Rear spindle
Rear brake assembly
Backing plate
Brake drum
Tongue washer
Flange nut
Hub cap
200 - 260
(2000 - 2600, 146 - 190)

<DISC BRAKE>

Backing plate
Hub assembly
Brake disc
Tongue washer
Flange nut
Hub cap
200 - 260
(2000 - 2600, 146 - 190)

TORQUE : Nm (kg·cm, lb·ft)

9355LG68

Exploded view of the rear hub assembly

2. Remove or disconnect the following:
- Rear wheel
- Flange nut and washer
- Drum or rotor
- Brake line
- Parking brake assembly
- Parking brake cable
- Spindle bolts and the spindle
- Rear hub from the housing using tool 09517-43001
- Wheel bearing snap-ring
- Wheel bearing inner race from the housing using tools 09500-2100, 09527-33000 and 09216-22100

3. Inspect the components for damage and replace as necessary.

To install:

4. Apply a thin coat of multi-purpose grease to the surface on the housing and bearing.

➡ **Do not press against the outer race of the bearing as this can cause bearing damage and always use a new bearing kit.**

5. Install or connect the following:
- Bearing onto the spindle using tools 09216-21100 and 09532-3200

- Snap-ring
- Backing plate, then press the hub onto the housing using tool 09517-21500

6. Rotate the bearing several times to seat the bearing.

7. Measure the wheel bearing torque using an inch lb. torque wrench. The measurement is 16.64 inch lbs. (1.88 Nm).

8. Measure the end-play of the hub using a dial gauge. The specification is 0.003-0.008mm.

9. Install the remaining components in the reverse order of removal.

10. Check the wheel alignment and adjust as necessary.

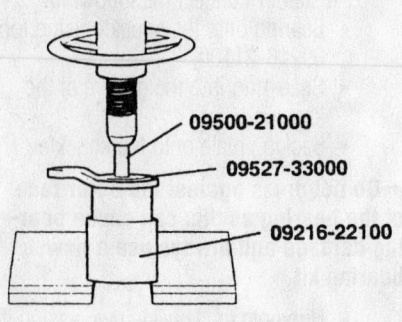

09500-21000
09527-33000
09216-22100

9355LG71

Removing the wheel bearing inner race from housing

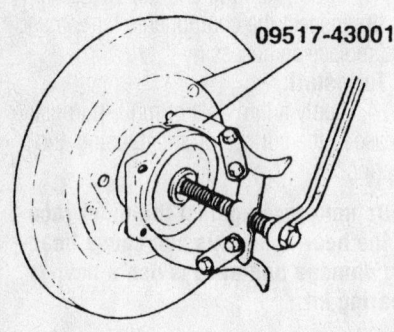

09517-43001

9355LG69

Removing the rear hub from housing

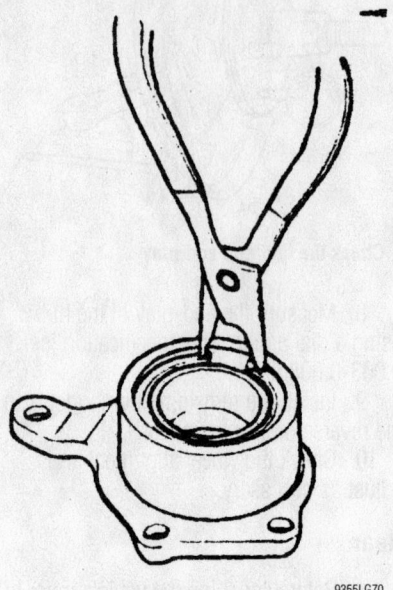

9355LG70

Removing the wheel bearing snap-ring

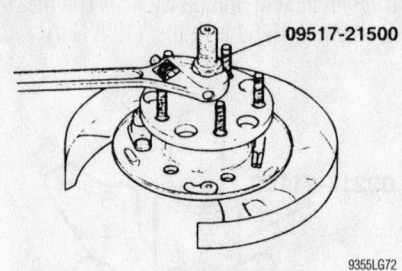

09517-21500

9355LG72

Press the hub onto the housing

PRECAUTIONS

Before servicing any vehicle, please be sure to read all of the following precautions, which deal with personal safety, prevention of component damage and important points to take into consideration when servicing a motor vehicle:

• Never open, service or drain the radiator or cooling system when the engine is hot; serious burns can occur from the steam and hot coolant.

• Observe all applicable safety precautions when working around fuel. Whenever servicing the fuel system, always work in a well-ventilated area. Do not allow fuel spray or vapors to come in contact with a spark, open flame, or excessive heat (a hot drop light, for example). Keep a dry chemical fire extinguisher near the work area. Always keep fuel in a container specifically designed for fuel storage; also, always properly seal fuel containers to avoid the possibility of fire or explosion. Refer to the additional fuel system precautions later in this section.

• Fuel injection systems often remain pressurized, even after the engine has been turned **OFF**. The fuel system pressure must be relieved before disconnecting any fuel lines. Failure to do so may result in fire and/or personal injury.

• Brake fluid often contains polyglycol ethers and polyglycols. Avoid contact with the eyes and wash your hands thoroughly after handling brake fluid. If you do get brake fluid in your eyes, flush your eyes with clean, running water for 15 minutes. If eye irritation persists, or if you have taken brake fluid internally, seek medical assistance IMMEDIATELY.

• The EPA warns that prolonged contact with used engine oil may cause a number of skin disorders, including cancer. You should make every effort to minimize your exposure to used engine oil. Protective gloves should be worn when changing oil. Wash your hands and any other exposed skin areas as soon as possible after exposure to used engine oil. Soap and water, or waterless hand cleaner should be used.

• All new vehicles are now equipped with an air bag system, often referred to as a Supplemental Restraint System (SRS) or Supplemental Inflatable Restraint (SIR) system. The system must be disabled before performing service on or around system components, steering column, instrument panel components, wiring and sensors. Failure to follow safety and disabling procedures could result in accidental air bag deployment, possible personal injury and unnecessary system repairs.

• Always wear safety goggles when working with, or around, the air bag system. When carrying a non-deployed air bag, be sure the bag and trim cover are pointed away from your body. When placing a non-deployed air bag on a work surface, always face the bag and trim cover upward, away from the surface. This will reduce the motion of the module if it is accidentally deployed. Refer to the additional air bag system precautions later in this section.

• Clean, high quality brake fluid from a sealed container is essential to the safe and proper operation of the brake system. You

should always buy the correct type of brake fluid for your vehicle. If the brake fluid becomes contaminated, completely flush the system with new fluid. Never reuse any brake fluid. Any brake fluid that is removed from the system should be discarded. Also, do not allow any brake fluid to come in contact with a painted surface; it will damage the paint.

• Never operate the engine without the proper amount and type of engine oil; doing so WILL result in severe engine damage.

• Timing belt maintenance is extremely important. Many models utilize an interference-type, non-freewheeling engine. If the timing belt breaks, the valves in the cylinder head may strike the pistons, causing potentially serious (also time-consuming and expensive) engine damage. Refer to the maintenance interval charts in the front of this manual for the recommended replacement interval for the timing belt and to the timing belt section for belt replacement and inspection.

• Disconnecting the negative battery cable on some vehicles may interfere with the functions of the on-board computer system(s) and may require the computer to undergo a relearning process once the negative battery cable is reconnected.

• When servicing drum brakes, only disassemble and assemble one side at a time, leaving the remaining side intact for reference.

• Only an MVAC-trained, EPA-certified automotive technician should service the A/C system or its components.

ENGINE REPAIR

➡ **Disconnecting the negative battery cable on some vehicles may interfere with the functions of the on board computer system. The computer may undergo a relearning process once the negative battery cable is reconnected.**

Distributor

REMOVAL

These engines are equipped with a Distributorless Ignition System (DIS).

Alternator

REMOVAL

2.2L Engine

1. Before servicing the vehicle, refer to the precautions in the beginning of this section.
2. Remove or disconnect the following:
   • Negative battery cable
   • Accessory drive belt
   • Alternator harness connectors
   • Alternator

3.2L Engine

1. Before servicing the vehicle, refer to the precautions in the beginning of this section.
2. Remove or disconnect the following:
   • Negative battery cable
   • Accessory drive belt
   • Alternator harness connectors
   • Alternator

3.5L Engine

1. Before servicing the vehicle, refer to the precautions in the beginning of this section.

2. Remove or disconnect the following:
- Negative battery cable
- Accessory drive belt
- Alternator wiring connectors
- Alternator

INSTALLATION

2.2L Engine

Install or connect the following:
- Alternator. Tighten the long bolt to 26 ft. lbs. (35 Nm) and the short bolt to 15 ft. lbs. (20 Nm).
- Alternator harness connectors
- Accessory drive belt
- Negative battery cable

3.2L Engine

Install or connect the following:
- Alternator. Tighten the 10mm bolt to 30 ft. lbs. (41 Nm) and the 8mm bolt to 15 ft. lbs. 21 Nm).
- Alternator harness connectors
- Accessory drive belt
- Negative battery cable

3.5L Engine

1. Before servicing the vehicle, refer to the precautions in the beginning of this section.
2. Install or connect the following:
- Alternator. Tighten the 10mm bolts to 30 ft. lbs. (41 Nm) and the 8mm bolts to 15 ft. lbs. (21 Nm).
- Alternator wiring connectors
- Accessory drive belt
- Negative battery cable

Ignition Timing

ADJUSTMENT

These engines are equipped with a Distributorless Ignition System (DIS). No adjustment is possible.

Engine Assembly

REMOVAL & INSTALLATION

2.2L engines

1. Before servicing the vehicle, refer to the precautions in the beginning of this section.
2. Drain the cooling system.
3. Relieve the fuel system pressure.

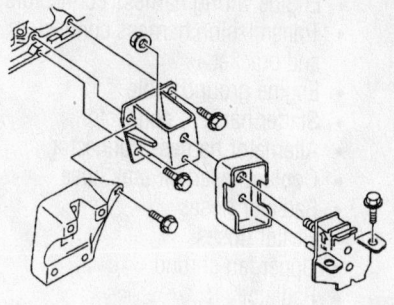

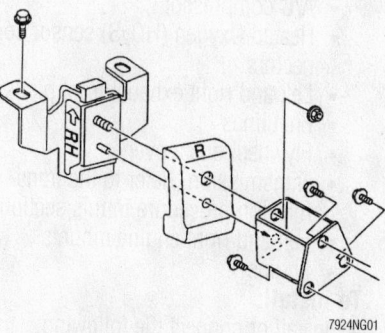

Left and right motor mounts—2.2L engine

7924NG01

4. Remove or disconnect the following:
- Battery
- Hood
- Accessory drive belt
- Accelerator cable
- Air intake assembly
- Engine wiring harness connectors at left rear of the engine compartment
- Brake booster vacuum line
- Engine ground cables
- Clutch fluid line bracket and slave cylinder
- Fuel lines and bracket
- Exhaust front pipe
- Transmission
- A/C compressor
- Power steering pump
- Chassis harness connectors at right rear of the engine compartment
- Frame ground cable
- Radiator hoses
- Heater hoses
- Cooling fan connector
- Cooling fan and shroud
- Radiator
- Left and right engine mounts
5. Lift the engine from the vehicle.

To install:

6. Position the engine in the engine compartment.

7. Install or connect the following:
- Left and right engine mounts. Tighten the fasteners to 30 ft. lbs. (41 Nm).
- Radiator
- Cooling fan and shroud
- Cooling fan connector
- Heater hoses
- Radiator hoses
- Frame ground cable
- Chassis harness connectors at right rear of the engine compartment
- Power steering pump
- A/C compressor
- Transmission
- Exhaust front pipe
- Fuel lines and bracket
- Clutch fluid line bracket and slave cylinder
- Engine ground cables
- Brake booster vacuum line
- Engine wiring harness connectors at left rear of the engine compartment
- Air intake assembly
- Accelerator cable
- Accessory drive belts
- Hood
- Battery
8. Fill the cooling system.
9. Start the engine and check for leaks.

3.2L Engines

1. Before servicing the vehicle, refer to the precautions in the beginning of this section.
2. Drain the cooling system.
3. Relieve the fuel system pressure.
4. Remove or disconnect the following:
- Battery
- Hood
- Accelerator cable
- Cruise control cable
- Air intake assembly
- Canister vacuum hose
- Brake booster vacuum hose
- Engine wiring harness connectors
- Front axle harness connector, if equipped
- Transmission harness connector and bracket
- Frame ground cable
- Firewall ground cable
- Starter harness connectors
- Alternator harness connectors
- Coolant overflow reservoir hose
- Radiator hoses
- Cooling fan and shroud
- Accessory drive belt
- Power steering pump

- A/C compressor
- Heated Oxygen (HO₂S) sensor connectors
- Exhaust front pipes
- Heater hoses
- Fuel lines
- Transmission
- Left and right engine mounts

5. Lift the engine from the vehicle.

To install:

6. Lower the engine into the vehicle.
7. Install or connect the following:
   - Left and right engine mounts. Tighten the bolts to 30 ft. lbs. (41 Nm) and the nuts to 37 ft. lbs. (50 Nm).
   - Transmission
   - Fuel lines
   - Heater hoses
   - Exhaust front pipes
   - Heated Oxygen (HO₂S) sensor connectors
   - A/C compressor
   - Power steering pump
   - Accessory drive belt
   - Cooling fan and shroud
   - Radiator hoses
   - Coolant overflow reservoir hose
   - Alternator harness connectors
   - Starter harness connectors
   - Firewall ground cable
   - Frame ground cable
   - Transmission harness connector and bracket
   - Front axle harness connector, if equipped
   - Engine wiring harness connectors
   - Brake booster vacuum hose
   - Canister vacuum hose
   - Air intake assembly
   - Cruise control cable
   - Accelerator cable
   - Hood
   - Battery

8. Fill the cooling system.
9. Start the engine and check for leaks.

3.5L Engine

EXCEPT AXIOM

1. Before servicing the vehicle, refer to the precautions in the beginning of this section.
2. Drain the cooling system.
3. Remove or disconnect the following:
   - Battery
   - Hood
   - Air cleaner assembly
   - Accelerator cable
   - Cruise control cable
   - Canister vacuum line
   - Brake booster vacuum line

- Engine wiring harness connectors
- Transmission harness connectors and bracket
- Engine ground cable
- Starter harness connector
- Alternator harness connector
- Coolant reservoir tank hose
- Radiator hoses
- Heater hoses
- Upper fan shroud
- Radiator
- Cooling fan
- Accessory drive belt
- Power steering pump
- A/C compressor
- Heated Oxygen (HO₂S) sensor connectors
- Left and right exhaust front pipes
- Fuel lines
- Flywheel dust cover
- Transmission. Refer to the transmission procedure in this section.
- Left and right engine mounts
- Engine

To install:

4. Install or connect the following:
   - Engine
   - Left and right engine mounts. Tighten the bolts to 30 ft. lbs. (41 Nm).
   - Transmission
   - Flywheel dust cover
   - Fuel lines
   - Left and right exhaust front pipes
   - HO₂S sensor connectors
   - A/C compressor
   - Power steering pump
   - Accessory drive belt
   - Cooling fan. Tighten the nuts to 16 ft. lbs. (22 Nm).
   - Radiator
   - Upper fan shroud
   - Heater hoses
   - Radiator hoses
   - Coolant reservoir tank hose
   - Alternator harness connector
   - Starter harness connector
   - Engine ground cable
   - Transmission harness connectors and bracket
   - Engine wiring harness connectors
   - Brake booster vacuum line
   - Canister vacuum line
   - Cruise control cable
   - Accelerator cable
   - Air cleaner assembly
   - Hood
   - Battery

5. Fill the cooling system. Check all fluid levels and adjust as necessary.
6. Start the engine and check for leaks.

AXIOM

1. Before servicing the vehicle, refer to the precautions in the beginning of this section.
2. Drain the cooling system.
3. Remove or disconnect the following:
   - Battery
   - Hood
   - Air cleaner assembly
   - Canister vacuum line
   - Brake booster vacuum line
   - Engine wiring harness connectors
   - Transmission harness connectors and bracket
   - Engine ground cable
   - Bonding cable connectors
   - Starter harness connector
   - Alternator harness connector
   - Coolant reservoir tank hose
   - Radiator hoses
   - Upper fan shroud
   - Cooling fan
   - Accessory drive belt
   - Power steering pump
   - A/C compressor
   - Heated Oxygen (HO₂S) sensor connectors
   - Left and right exhaust front pipes
   - Flywheel dust cover
   - Heater hoses
   - Fuel lines
   - Transmission. Refer to the transmission procedure in this section.
   - Accelerator cable
   - Cruise control cable
   - Left and right engine mounts
   - Engine

To install:

4. Install or connect the following:
   - Engine
   - Left and right engine mounts. Tighten the bolts to 30 ft. lbs. (41 Nm).
   - Transmission
   - Flywheel dust cover
   - Fuel lines
   - Left and right exhaust front pipes
   - HO₂S sensor connectors
   - A/C compressor
   - Power steering pump
   - Accessory drive belt
   - Cooling fan. Tighten the nuts to 16 ft. lbs. (22 Nm).
   - Radiator
   - Upper fan shroud
   - Heater hoses
   - Radiator hoses
   - Coolant reservoir tank hose
   - Alternator harness connector
   - Starter harness connector
   - Engine ground cable

- Transmission harness connectors and bracket
- Engine wiring harness connectors
- Brake booster vacuum line
- Canister vacuum line
- Cruise control cable
- Accelerator cable
- Air cleaner assembly
- Hood
- Battery

5. Fill the cooling system. Check all fluid levels and adjust as necessary.
6. Start the engine and check for leaks.

Water Pump

REMOVAL & INSTALLATION

2.2L Engine

1. Before servicing the vehicle, refer to the precautions in the beginning of this section.
2. Drain the cooling system.
3. Remove or disconnect the following:
   - Negative battery cable
   - Radiator hose
   - Accessory drive belt
   - Front cover
   - Timing belt. Refer to the Timing Belt unit repair section.
   - Water pump

To install:

4. Install a new O-ring and coat the water pump sealing surface with silicone grease.
5. Install or connect the following:
   - Water pump. Tighten the bolts to 18 ft. lbs. (25 Nm).

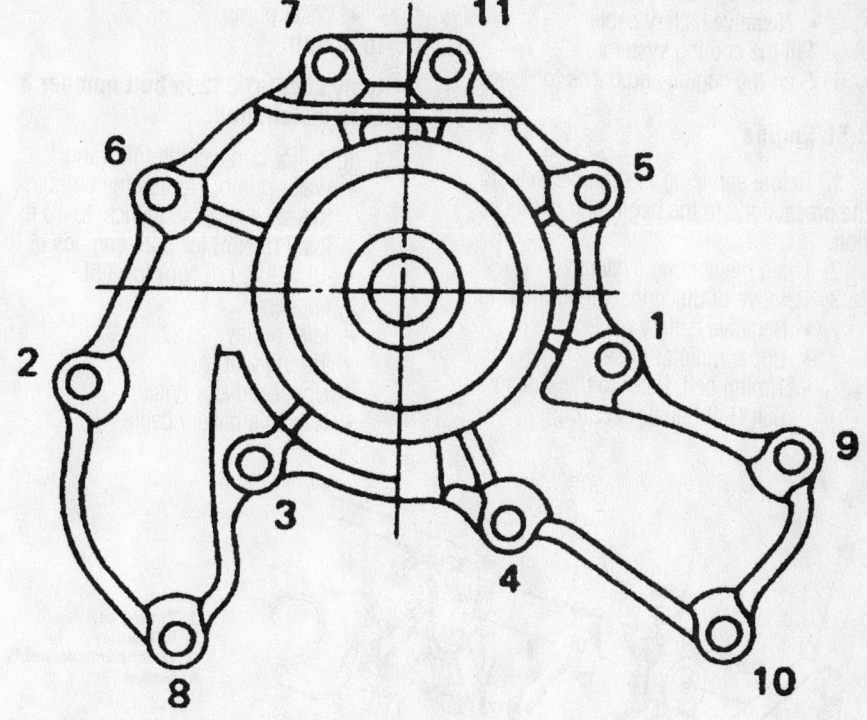

Water pump bolt tightening sequence—3.2L engines

- Timing belt
- Front cover
- Accessory
6. Install or connect the following:
   - Water pump. Tighten the bolts in two passes, in sequence, to 13 ft. lbs. (18 Nm) for 3.2L engines or to 18 ft. lbs. (25 Nm) for 3.5L engines.

- Idler pulley
- Timing belt
- Upper radiator hose
- Negative battery cable
7. Fill the cooling system.
8. Start the engine and check for leaks.

3.2L Engines

1. Before servicing the vehicle, refer to the precautions in the beginning of this section.
2. Drain the cooling system.
3. Remove or disconnect the following:
   - Negative battery cable
   - Radiator hose
   - Accessory drive belt
   - Front cover
   - Timing belt. Refer to the Timing Belt unit repair section.
   - Timing belt idler pulley
   - Water pump

To install:

4. Install or connect the following:
   - Water pump. Tighten the bolts in sequence to 18 ft. lbs. (25 Nm).
   - Timing belt idler pulley. Tighten the bolt to 38 ft. lbs. (52 Nm).
   - Timing belt
   - Front cover
   - Accessory drive belt

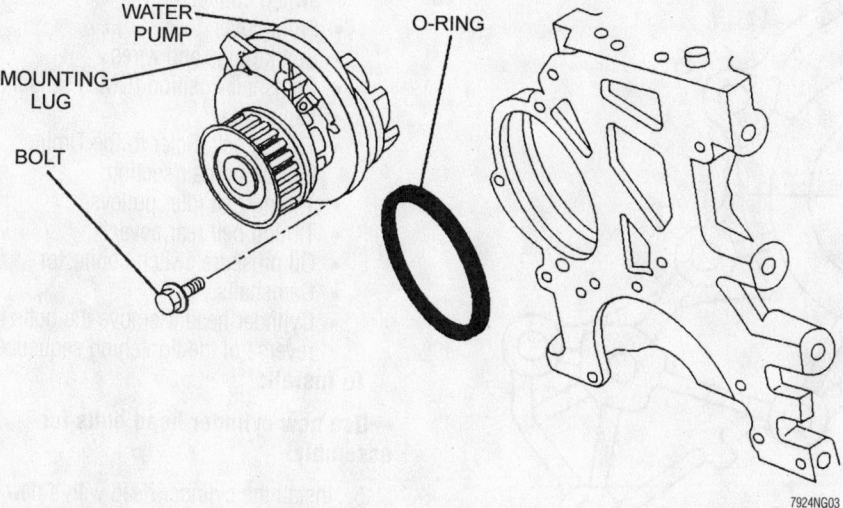

Exploded view of the water pump mounting, showing the location of the mounting lug—2.2L engine

- Radiator hose
- Negative battery cable

5. Fill the cooling system.
6. Start the engine and check for leaks.

3.5L Engine

1. Before servicing the vehicle, refer to the precautions in the beginning of this section.
2. Drain the cooling system.
3. Remove or disconnect the following:
   - Negative battery cable
   - Upper radiator hose
   - Timing belt. Refer to the Timing Belt Unit Repair Section.

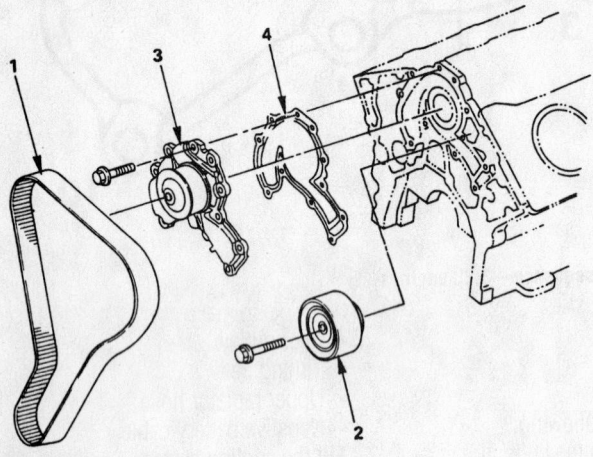

Exploded view of the water pump mounting

- Idler pulley
- Water pump

To install:

➡ **Apply Loctite® 262 to bolt number 3 prior to installation.**

4. Install or connect the following:
   - Water pump. Tighten the bolts in two passes, in sequence, to 13 ft. lbs. (18 Nm) for 3.2L engines or to 18 ft. lbs. (25 Nm) for 3.5L engines.
   - Idler pulley
   - Timing belt
   - Upper radiator hose
   - Negative battery cable

1. Timing belt
2. Idle pulley
3. Water pump assembly
4. Gasket

7924BG41

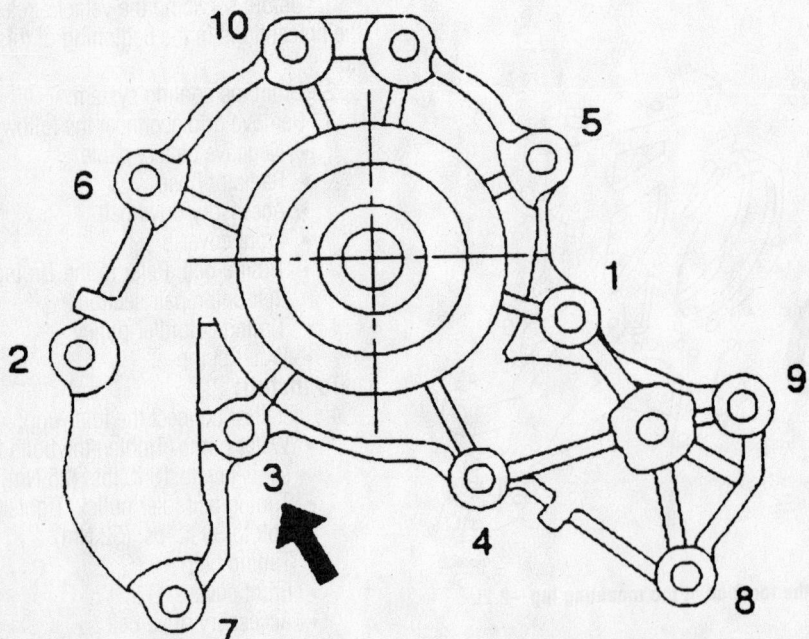

Water pump torque sequence. Apply LOCTITE® 262 to bolt number 3 (arrow)—3.5L engine

9302BG01

5. Fill the cooling system.
6. Start the engine and check for leaks.

Cylinder Head

REMOVAL & INSTALLATION

2.2L Engine

1. Before servicing the vehicle, refer to the precautions in the beginning of this section.
2. Drain the cooling system.
3. Relieve the fuel system pressure.
4. Remove or disconnect the following:
   - Negative battery cable
   - Intake Air Temperature (IAT) sensor connector
   - Positive Crankcase Ventilation (PCV) valve and hose
   - Air intake assembly
   - Upper radiator hose
   - Accessory drive belt
   - Exhaust front pipe
   - Alternator and brackets
   - Crankshaft Position (CKP) sensor connector
   - Knock sensor connector
   - Heater hoses
   - Water bypass hose
   - Fuel lines
   - Evaporative Emissions (EVAP) valve connector
   - Canister hose
   - Intake manifold
   - Engine wiring harness connectors at left rear of the engine compartment
   - Power steering pump pressure switch connector
   - Front cover
   - Spark plugs and wires
   - Camshaft Position (CMP) sensor
   - Valve cover
   - Timing belt. Refer to the Timing Belt unit repair section.
   - Timing belt idler pulleys
   - Timing belt rear cover
   - Oil pressure switch connector
   - Camshafts
   - Cylinder head. Remove the bolts in reverse of the tightening sequence.

To install:

➡ **Use new cylinder head bolts for assembly.**

5. Install the cylinder head with a new gasket. Tighten the bolts in sequence as follows:
   a. Step 1: 18 ft. lbs. (25 Nm)
   b. Step 2: Plus 90 degrees

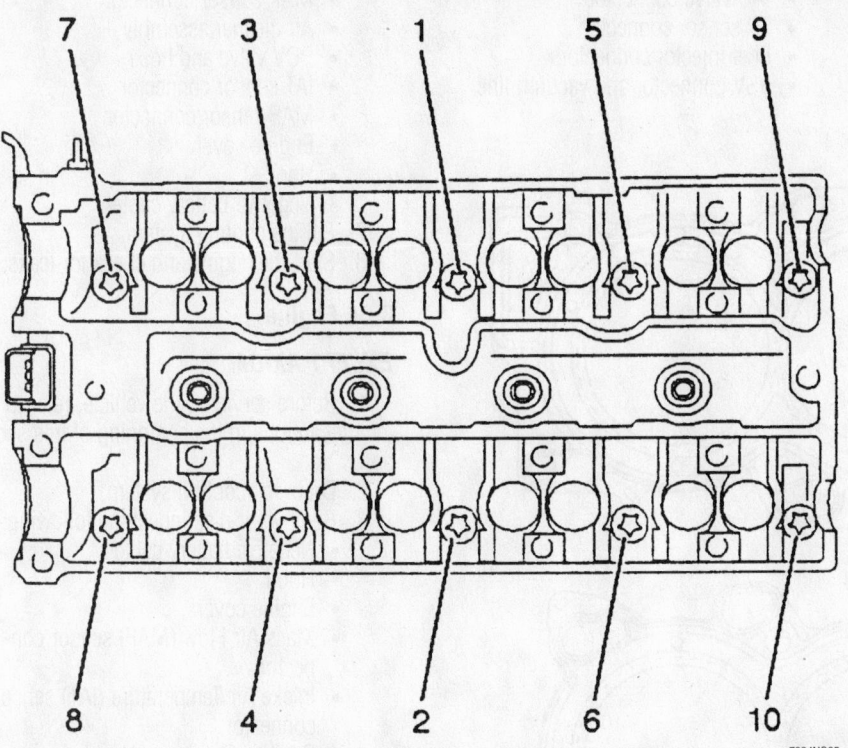

Cylinder head torque sequence—2.2L (VIN D) engine

7924NG05

c. Step 3: Plus 90 degrees
d. Step 4: Plus 90 degrees
6. Install or connect the following:
  • Camshafts
  • Oil pressure switch connector
  • Timing belt rear cover
  • Timing belt idler pulleys. Tighten the bolts to 18 ft. lbs. (25 Nm).
  • Timing belt
  • Valve cover
  • CMP sensor
  • Spark plugs and wires
  • Front cover
  • Power steering pump pressure switch connector
  • Engine wiring harness connectors at left rear of the engine compartment
  • Intake manifold
  • Canister hose
  • EVAP valve connector
  • Fuel lines
  • Water bypass hose
  • Heater hoses
  • Knock sensor connector
  • CKP sensor connector
  • Alternator and brackets
  • Exhaust front pipe
  • Accessory drive belt
  • Upper radiator hose

  • Air intake assembly
  • PCV valve and hose
  • IAT sensor connector
  • Negative battery cable
7. Fill the cooling system.
8. Start the engine and check for leaks.

3.2L Engines

1. Before servicing the vehicle, refer to the precautions in the beginning of this section.
2. Drain the cooling system.
3. Relieve the fuel system pressure.
4. Remove or disconnect the following:
  • Negative battery cable
  • Hood
  • Engine cover
  • Mass Air Flow (MAF) sensor connector
  • Intake Air Temperature (IAT) sensor connector
  • Positive Crankcase Ventilation (PCV) valve and hose
  • Air cleaner assembly
  • Manifold Absolute Pressure (MAP) sensor connector
  • Vacuum Switching Valve (VSV) connector and vacuum line
  • Fuel injector connectors
  • Throttle Position (TP) sensor connector

  • Idle Air Control (IAC) valve connector
  • Ignition coils
  • Brake booster vacuum line
  • Canister purge vacuum line
  • Duty solenoid valve
  • Fuel lines
  • Intake manifold
  • Radiator hoses
  • Engine coolant manifold
  • Upper fan shroud
  • Accessory drive belt and tensioner
  • Cooling fan and pulley
  • Alternator
  • Idler pulley
  • Power steering pump and bracket
  • A/C compressor
  • Crankshaft pulley
  • Oil cooler hoses
  • Timing belt cover
  • Valve covers
  • Timing belt. Refer to the Timing Belt Unit Repair Section.
  • Left and right exhaust front pipes
  • Oil dipstick tube
  • Cylinder heads

To install:

➡**Use new head bolts when installing the cylinder head.**

➡**The left and right cylinder head gaskets are not interchangeable.**

5. Install the cylinder heads with new gaskets. Tighten the bolts in sequence as follows:
  a. Step 1: 21 ft. lbs. (29 Nm)
  b. Step 2: 47 ft. lbs. (64 Nm)
6. Install or connect the following:
  • Oil dipstick tube
  • Left and right exhaust front pipes
  • Timing belt
  • Valve covers
  • Timing belt cover
  • Oil cooler hoses
  • Crankshaft pulley. Tighten the pulley bolt to 123 ft. lbs. (167 Nm).
  • A/C compressor
  • Power steering pump and bracket. Tighten the bolts to 34 ft. lbs. (46 Nm).
  • Idler pulley
  • Alternator
  • Cooling fan and pulley
  • Accessory drive belt and tensioner
  • Upper fan shroud
  • Engine coolant manifold
  • Radiator hoses
  • Intake manifold
  • Fuel lines

Timing belt service is covered in Section 3 of this manual

- Duty solenoid valve
- Canister purge vacuum line
- Brake booster vacuum line
- Ignition coils

- IAC valve connector
- TP sensor connector
- Fuel injector connectors
- VSV connector and vacuum line

- MAP sensor connector
- Air cleaner assembly
- PCV valve and hose
- IAT sensor connector
- MAF sensor connector
- Engine cover
- Hood
- Negative battery cable

7. Fill the cooling system.
8. Start the engine and check for leaks.

3.5L Engine

EXCEPT AXIOM

1. Before servicing the vehicle, refer to the precautions in the beginning of this section.

2. Drain the cooling system.

3. Remove or disconnect the following:

- Negative battery cable
- Hood
- Engine cover
- Mass Air Flow (MAF) sensor connector
- Intake Air Temperature (IAT) sensor connector
- Positive Crankcase Ventilation (PCV) valve and hose
- Air cleaner assembly
- Manifold Absolute Pressure (MAP) sensor connector
- Vacuum Switching Valve (VSV) connector and vacuum line
- Fuel injector connectors
- Throttle Position (TP) sensor connector
- Idle Air Control (IAC) valve connector
- Ignition coils
- Brake booster vacuum line
- Canister purge vacuum line
- Duty solenoid valve
- Fuel lines
- Intake manifold
- Radiator hoses
- Engine coolant manifold
- Upper fan shroud
- Accessory drive belt and tensioner
- Cooling fan and pulley
- Alternator
- Idler pulley
- Power steering pump and bracket
- A/C compressor
- Crankshaft pulley
- Oil cooler hoses
- Timing belt cover
- Valve covers
- Timing belt. Refer to the Timing Belt Unit Repair Section.
- Left and right exhaust front pipes
- Oil dipstick tube
- Cylinder heads

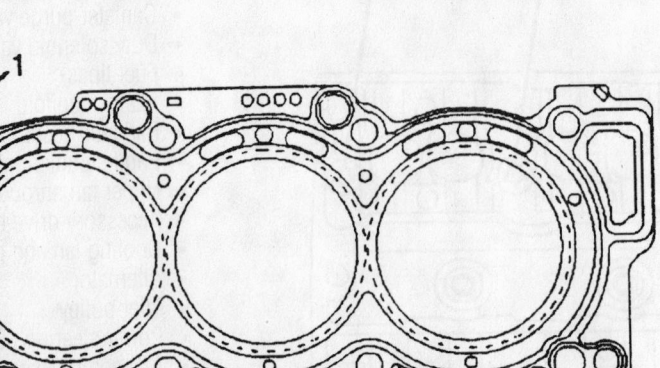

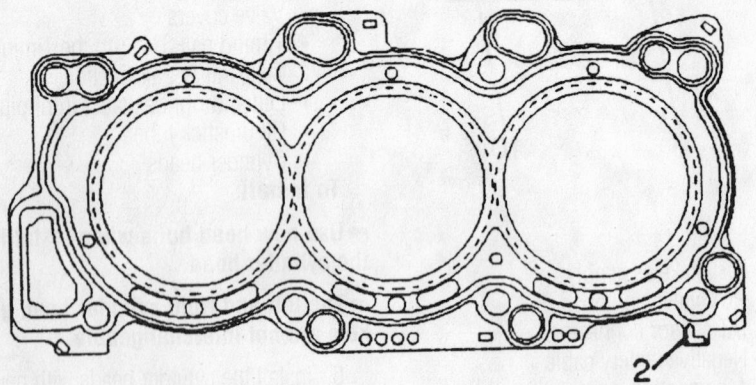

Right (1) and left (2) head gasket identification mark locations—3.2L DOHC engine

7924NG11

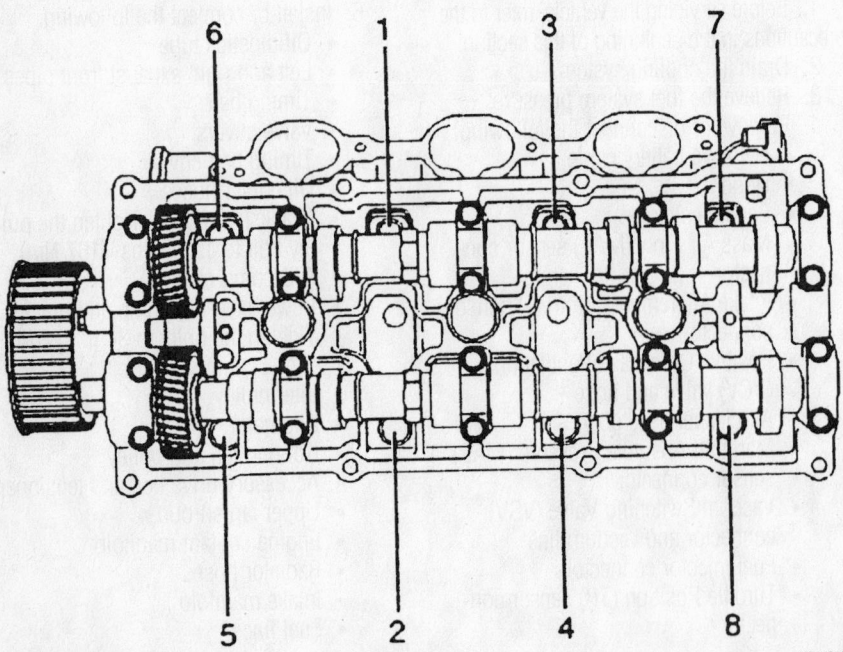

7924NG12

Cylinder head torque sequence—3.2L DOHC and 3.5L engines

To install:

➡**Use new head bolts when installing the cylinder head. Do not apply oil to the head bolt threads.**

➡**The left and right cylinder head gaskets are not interchangeable.**

4. Install the cylinder heads with new gaskets. Tighten the bolts to 47 ft. lbs. (64 Nm).

5. Install or connect the following:
- Oil dipstick tube
- Left and right exhaust front pipes
- Timing belt
- Valve covers
- Timing belt cover
- Oil cooler hoses
- Crankshaft pulley. Tighten the pulley bolt to 123 ft. lbs. (167 Nm).
- A/C compressor
- Power steering pump and bracket. Tighten the bolts to 34 ft. lbs. (46 Nm).
- Idler pulley
- Alternator

- Cooling fan and pulley
- Accessory drive belt and tensioner
- Upper fan shroud
- Engine coolant manifold
- Radiator hoses
- Intake manifold
- Fuel lines
- Duty solenoid valve
- Canister purge vacuum line
- Brake booster vacuum line
- Ignition coils
- IAC valve connector
- TP sensor connector
- Fuel injector connectors
- VSV connector and vacuum line
- MAP sensor connector
- Air cleaner assembly
- PCV valve and hose
- IAT sensor connector
- MAF sensor connector
- Engine cover
- Hood
- Negative battery cable

6. Fill the cooling system.
7. Start the engine. Check for leaks and proper operation.

AXIOM

1. Before servicing the vehicle, refer to the precautions in the beginning of this section.
2. Drain the cooling system.
3. Remove or disconnect the following:
- Negative battery cable
- Hood
- Engine cover
- Mass Air Flow (MAF) sensor connector
- Intake Air Temperature (IAT) sensor connector
- Positive Crankcase Ventilation (PCV) valve and hose
- Air cleaner assembly
- Manifold Absolute Pressure (MAP) sensor connector
- Vacuum Switching Valve (VSV) connector and vacuum line
- Fuel injector connectors
- Throttle Position (TP) sensor connector
- Idle Air Control (IAC) valve connector
- Ignition coils
- Brake booster vacuum line
- Canister purge vacuum line
- Duty solenoid valve
- Fuel lines
- Intake manifold
- Radiator hoses
- Engine coolant manifold
- Upper fan shroud
- Accessory drive belt and tensioner
- Cooling fan and pulley
- Alternator
- Idler pulley
- Power steering pump and bracket
- A/C compressor
- Crankshaft pulley
- Oil cooler hoses
- Timing belt cover
- Valve covers
- Timing belt. Refer to the Timing Belt Unit Repair Section.
- Left and right exhaust front pipes
- Oil dipstick tube
- Cylinder heads

To install:

➡**Use new head bolts when installing the cylinder head. Do not apply oil to the head bolt threads.**

➡**The left and right cylinder head gaskets are not interchangeable.**

4. Install the cylinder heads with new gaskets. Tighten the bolts to 22 ft. lbs. in sequence, then to 47 ft. lbs. (64 Nm) in sequence.

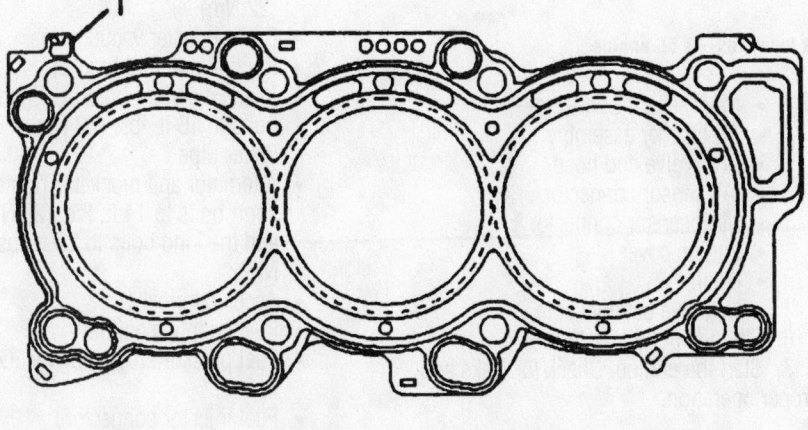

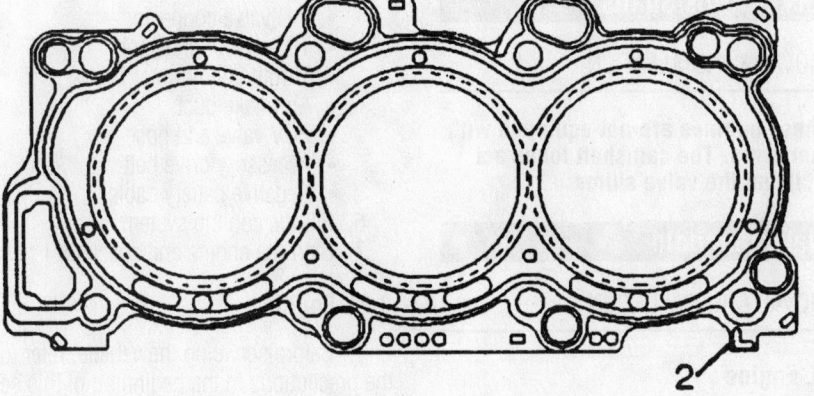

7924BG04

Right (1) and left (2) head gasket identification mark locations—3.5L engine

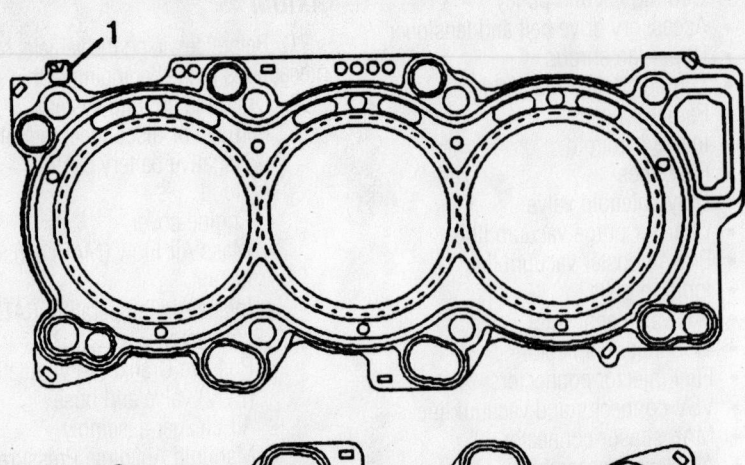

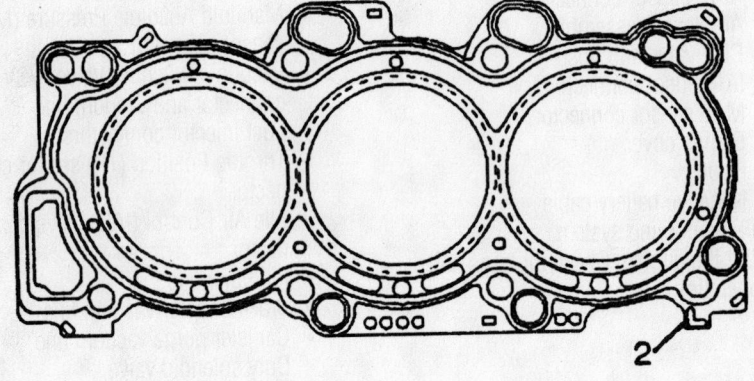

7924BG04

Right (1) and left (2) head gasket identification mark locations—3.5L engine

5. Install or connect the following:
- Oil dipstick tube
- Left and right exhaust front pipes
- Timing belt
- Valve covers
- Timing belt cover
- Oil cooler hoses
- Crankshaft pulley. Tighten the pulley bolt to 123 ft. lbs. (167 Nm).
- A/C compressor
- Power steering pump and bracket. Tighten the bolts to 34 ft. lbs. (46 Nm).
- Idler pulley
- Alternator
- Cooling fan and pulley
- Accessory drive belt and tensioner
- Upper fan shroud
- Engine coolant manifold
- Radiator hoses
- Intake manifold
- Fuel lines
- Duty solenoid valve
- Canister purge vacuum line
- Brake booster vacuum line
- Ignition coils
- IAC valve connector
- TP sensor connector
- Fuel injector connectors
- VSV connector and vacuum line

- MAP sensor connector
- Air cleaner assembly
- PCV valve and hose
- IAT sensor connector
- MAF sensor connector
- Engine cover
- Hood
- Negative battery cable

6. Fill the cooling system.
7. Start the engine. Check for leaks and proper operation.

Rocker Arms/Shafts

REMOVAL & INSTALLATION

➡**These engines are not equipped with rocker arms. The camshaft lobes act directly on the valve shims.**

Intake Manifold

REMOVAL & INSTALLATION

2.2L Engine

1. Before servicing the vehicle, refer to the precautions in the beginning of this section.
2. Drain the cooling system.

3. Relieve the fuel system pressure.
4. Remove or disconnect the following:
- Negative battery cable
- Accessory drive belt
- Positive Crankcase Ventilation (PCV) valve and hose
- Air intake duct
- Throttle body water hoses
- Throttle Position (TP) sensor connector
- Idle Air Control (IAC) valve connector
- Fuel lines
- Fuel injector connectors
- Fuel pressure regulator vacuum line
- Fuel supply manifold
- Accelerator cable
- Alternator and brackets
- Water pipe
- Intake manifold bracket
- Ignition coil and bracket
- Brake booster vacuum line
- Intake manifold

To install:
5. Install or connect the following:
- Intake manifold. Use a new gasket and tighten the bolts to 16 ft. lbs. (22 Nm).
- Brake booster vacuum line
- Ignition coil and bracket
- Intake manifold bracket. Tighten the bolts to 16 ft. lbs. (22 Nm).
- Water pipe
- Alternator and brackets. Tighten the short bolts to 14 ft. lbs. (20 Nm) and the long bolts to 25 ft. lbs. (35 Nm).
- Accelerator cable
- Fuel supply manifold
- Fuel pressure regulator vacuum line
- Fuel injector connectors
- Fuel lines
- IAC valve connector
- TP sensor connector
- Throttle body water hoses
- Air intake duct
- PCV valve and hose
- Accessory drive belt
- Negative battery cable

6. Fill the cooling system.
7. Start the engine and check for leaks.

3.2L Engine

1. Before servicing the vehicle, refer to the precautions in the beginning of this section.
2. Remove or disconnect the following:
- Negative battery cable
- Engine cover

- Air cleaner assembly
- Accelerator cable
- Cruise control cable
- Brake booster vacuum line
- Manifold Absolute Pressure (MAP) sensor connector
- Idle Air Control (IAC) valve connector
- Throttle Position (TP) sensor connector
- Canister purge solenoid connector
- Electronic Vacuum Sensing Valve (EVSV) connector and vacuum line
- Exhaust Gas Recirculation (EGR) valve
- Positive Crankcase Ventilation (PCV) valve and hose
- Pressure regulator vacuum line
- Ventilation hose
- Throttle body
- Fuel lines
- Fuel injector connectors
- Intake manifold

To install:

3. Install or connect the following:
   - Intake manifold. Tighten the fasteners to 18 ft. lbs. (25 Nm).
   - Fuel injector connectors
   - Fuel lines
   - Throttle body. Tighten the bolts to 88 inch lbs. (10 Nm).
   - Ventilation hose
   - Pressure regulator vacuum line
   - PCV valve and hose
   - EGR valve
   - EVSV connector and vacuum line
   - Canister purge solenoid connector
   - TP sensor connector
   - IAC valve connector
   - MAP sensor connector
   - Brake booster vacuum line
   - Cruise control cable
   - Accelerator cable
   - Air cleaner assembly
   - Engine cover
   - Negative battery cable

4. Start the engine and check for proper operation.

3.5L Engine

1. Before servicing the vehicle, refer to the precautions in the beginning of this section.

2. Remove or disconnect the following:
   - Negative battery cable
   - Engine cover
   - Air cleaner assembly
   - Accelerator cable
   - Cruise control cable

- Brake booster vacuum line
- Manifold Absolute Pressure (MAP) sensor connector
- Idle Air Control (IAC) valve connector
- Throttle Position (TP) sensor connector
- Canister purge solenoid connector
- Electronic Vacuum Sensing Valve (EVSV) connector and vacuum line
- Exhaust Gas Recirculation (EGR) valve
- Positive Crankcase Ventilation (PCV) valve and hose
- Pressure regulator vacuum line
- Ventilation hose
- Throttle body
- Fuel lines
- Fuel injector connectors
- Intake manifold

To install:

3. Install or connect the following:
   - Intake manifold. Tighten the fasteners to 18 ft. lbs. (25 Nm).
   - Fuel injector connectors
   - Fuel lines
   - Throttle body. Tighten the bolts to 88 inch lbs. (10 Nm).

- Ventilation hose
- Pressure regulator vacuum line
- PCV valve and hose
- EGR valve
- EVSV connector and vacuum line
- Canister purge solenoid connector
- TP sensor connector
- IAC valve connector
- MAP sensor connector
- Brake booster vacuum line
- Cruise control cable
- Accelerator cable
- Air cleaner assembly
- Engine cover
- Negative battery cable

4. Start the engine and check for proper operation.

Exhaust Manifold

REMOVAL & INSTALLATION

2.2L Engine

1. Before servicing the vehicle, refer to the precautions in the beginning of this section.

2. Remove or disconnect the following:

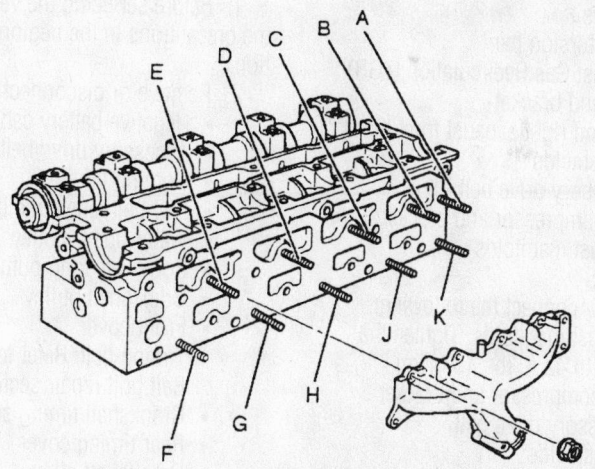

- **Tightening sequence:**
  Step1: J G H B D C J G B D
  Step2: A B C D E F G H J K
  Step3: A B C D E F G H J K
- **Tightening torque:**
  Step1: **14 N·m (10 lb ft)**
  Step2: **20 N·m (14 lb ft)**
  Step3: **20 N·m (14 lb ft)**

Exhaust manifold torque sequence—2.2L engine

7924NG17

Brake service is covered in Section 4 of this manual

- Negative battery cable
- Air intake duct
- Exhaust front pipe
- Exhaust manifold heat shield
- Exhaust manifold

To install:

3. Install the exhaust manifold. Tighten the nuts in sequence as follows:
   a. Step 1: 10 ft. lbs. (14 Nm)
   b. Step 2: 14 ft. lbs. (20 Nm)
   c. Step 3: 14 ft. lbs. (20 Nm)
4. Install or connect the following:
   - Exhaust manifold heat shield. Tighten the bolts to 71 inch lbs. (8 Nm).
   - Exhaust front pipe. Tighten the bolts to 18 ft. lbs. (25 Nm).
   - Air intake duct
   - Negative battery cable
5. Start the engine and check for leaks.

3.2L Engines

1. Before servicing the vehicle, refer to the precautions in the beginning of this section.
2. Remove or disconnect the following:
   - Negative battery cable
   - Air cleaner assembly
   - Heated Oxygen (HO2S) sensor connectors
   - Right torsion bar
   - Exhaust Gas Recirculation (EGR) pipe and bracket
   - Left and right exhaust front pipes
   - Heat shields
   - Accessory drive belt
   - A/C compressor and bracket
   - Exhaust manifolds

To install:

3. Install or connect the following:
   - Exhaust manifolds. Tighten the bolts to 42 ft. lbs. (57 Nm).
   - A/C compressor and bracket
   - Accessory drive belt
   - Heat shields
   - Left and right exhaust front pipes
   - EGR pipe and bracket
   - Right torsion bar
   - HO2S sensor connectors
   - Air cleaner assembly
   - Negative battery cable
4. Start the engine and check for leaks.

3.5L Engine

1. Before servicing the vehicle, refer to the precautions in the beginning of this section.
2. Remove or disconnect the following:
   - Negative battery cable
   - Air cleaner assembly
   - Heated Oxygen (HO2S) sensor connectors

- Right torsion bar
- Exhaust Gas Recirculation (EGR) pipe and bracket
- Left and right exhaust front pipes
- Heat shields
- Accessory drive belt
- A/C compressor and bracket
- Exhaust manifolds

To install:

3. Install or connect the following:
   - Exhaust manifolds. Tighten the bolts to 38 ft. lbs. (52 Nm).
   - A/C compressor and bracket
   - Accessory drive belt
   - Heat shields
   - Left and right exhaust front pipes
   - EGR pipe and bracket
   - Right torsion bar
   - HO2S sensor connectors
   - Air cleaner assembly
   - Negative battery cable
4. Start the engine and check for leaks.

Front Crankshaft Seal

REMOVAL & INSTALLATION

2.2L Engines

1. Before servicing the vehicle, refer to the precautions in the beginning of this section.
2. Remove or disconnect the following:
   - Negative battery cable
   - Accessory drive belts
   - Cooling fan
   - A/C belt tensioner, if equipped
   - Water pump pulley
   - Power steering pump
   - Crankshaft pulley
   - Front cover
   - Timing belt. Refer to the Timing Belt unit repair section.
   - Crankshaft timing sprocket
   - Rear timing cover
   - Crankshaft oil seal

To install:

3. Install or connect the following:
   - Crankshaft oil seal
   - Rear timing cover
   - Crankshaft timing sprocket. Tighten the bolt to 94 ft. lbs. (130 Nm) plus 45 degrees.
   - Timing belt. Refer to the Timing Belt unit repair section.
   - Front cover
   - Crankshaft pulley
   - Power steering pump
   - Water pump pulley
   - A/C belt tensioner, if equipped
   - Cooling fan

- Accessory drive belts
- Negative battery cable
4. Start the engine and check for leaks.

3.2L Engines

1. Before servicing the vehicle, refer to the precautions in the beginning of this section.
2. Remove or disconnect the following:
   - Negative battery cable
   - Air cleaner assembly
   - Upper fan shroud
   - Accessory drive belt and tensioner
   - Cooling fan and pulley
   - Idler pulley
   - Power steering pump
   - Crankshaft pulley
   - Timing belt cover
   - Timing belt. Refer to the Timing Belt Unit Repair Section.
   - Crankshaft timing sprocket
   - Oil seal

To install:

3. Install or connect the following:
   - Oil seal so that it is flush with the oil pump housing
   - Crankshaft timing sprocket
   - Timing belt
   - Timing belt cover
   - Crankshaft pulley. Tighten the bolt to 123 ft. lbs. (167 Nm).
   - Power steering pump
   - Idler pulley
   - Cooling fan and pulley
   - Accessory drive belt and tensioner
   - Upper fan shroud
   - Air cleaner assembly
   - Negative battery cable
4. Start the engine and check for leaks.

3.5L Engine

1. Before servicing the vehicle, refer to the precautions in the beginning of this section.
2. Remove or disconnect the following:
   - Negative battery cable
   - Air cleaner assembly
   - Upper fan shroud
   - Accessory drive belt and tensioner
   - Cooling fan and pulley
   - Idler pulley
   - Power steering pump and move it aside
   - Crankshaft pulley
   - Timing belt cover
   - Timing belt. Refer to the Timing Belt Unit Repair Section.
   - Crankshaft timing sprocket
   - Oil seal

To install:

3. Install or connect the following:

- Oil seal so that it is flush with the oil pump housing
- Crankshaft timing sprocket
- Timing belt
- Timing belt cover
- Crankshaft pulley. Tighten the bolt to 123 ft. lbs. (167 Nm).
- Power steering pump
- Idler pulley
- Cooling fan and pulley
- Accessory drive belt and tensioner
- Upper fan shroud
- Air cleaner assembly
- Negative battery cable

4. Start the engine and check for leaks.

Camshaft and Valve Lifters

REMOVAL & INSTALLATION

2.2L Engine

1. Before servicing the vehicle, refer to the precautions in the beginning of this section.

2. Remove or disconnect the following:
- Negative battery cable
- Positive Crankcase Ventilation (PCV) valve and hose
- Air intake duct and bracket
- Ground cables
- Engine wiring harness connectors

at left rear of the engine compartment
- Cooling fan harness connector
- Accessory drive belt
- Spark plug wire cover
- Spark plug wires
- Camshaft Position (CMP) sensor connector
- Crankshaft Position (CKP) sensor connector
- Crankshaft pulley
- Front cover
- Camshaft Position (CMP) sensor. Loosen the rear timing cover bolt for access.
- Valve cover

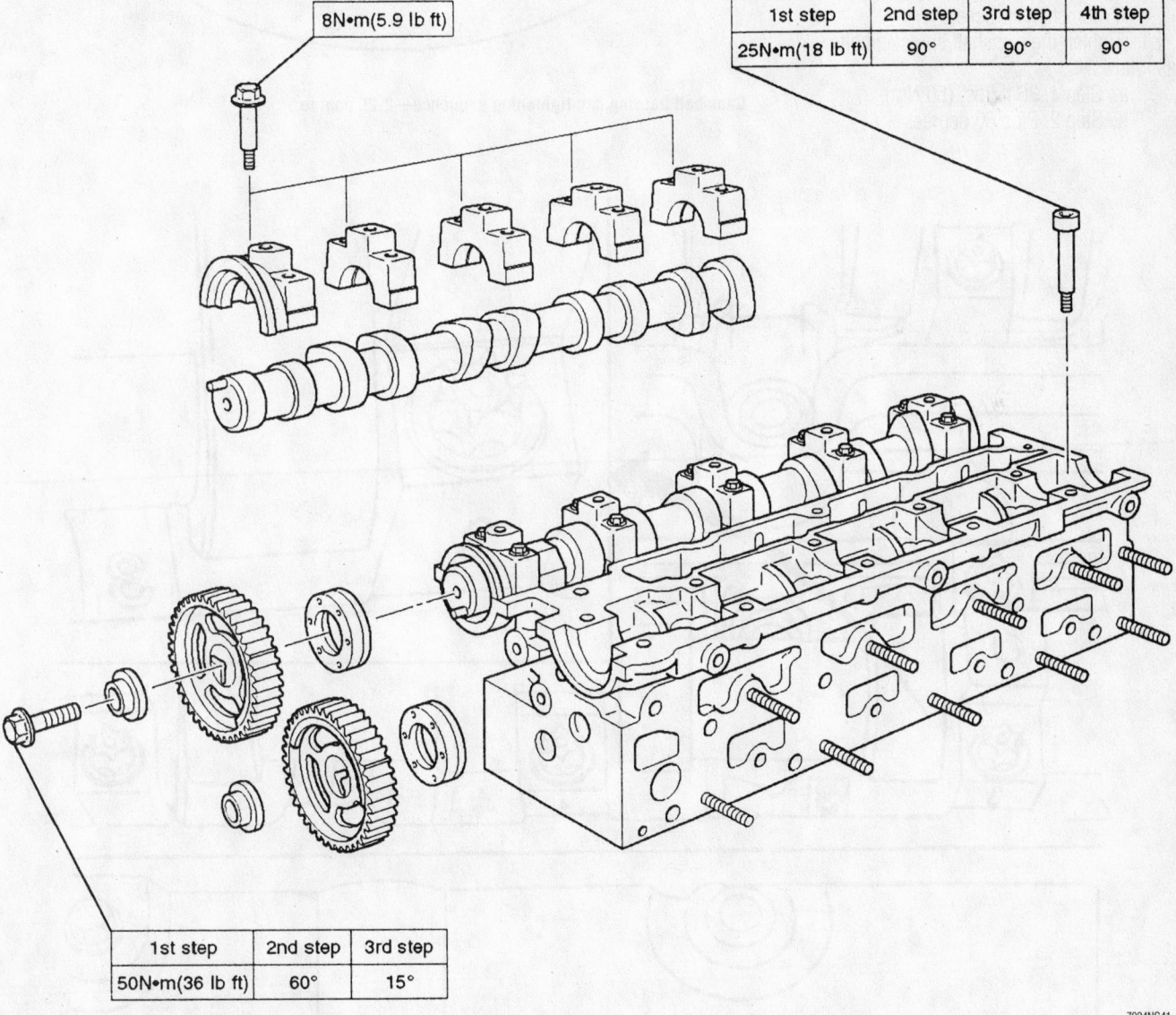

8N•m(5.9 lb ft)

1st step	2nd step	3rd step	4th step
25N•m(18 lb ft)	90°	90°	90°

1st step	2nd step	3rd step
50N•m(36 lb ft)	60°	15°

7924NG41

Exploded view of the cylinder head and camshaft components—2.2L engine

For complete Engine Mechanical specifications, see Section 1 of this manual

- Timing belt. Refer to the Timing Belt unit repair section.
- Camshaft sprockets
- Camshaft bearing caps
- Camshaft seals
- Camshafts
- Hydraulic tappets

➡**Keep all valvetrain components in order for assembly.**

To install:

3. Install or connect the following:
- Hydraulic tappets in their original locations
- Camshafts
- Camshaft bearing caps. Tighten the bolts in sequence to 71 inch lbs. (8 Nm).
- Camshaft seals
- Camshaft sprockets

4. Tighten the camshaft sprocket bolts as follows:
 a. Step 1: 36 ft. lbs. (50 Nm)
 b. Step 2: Plus 60 degrees

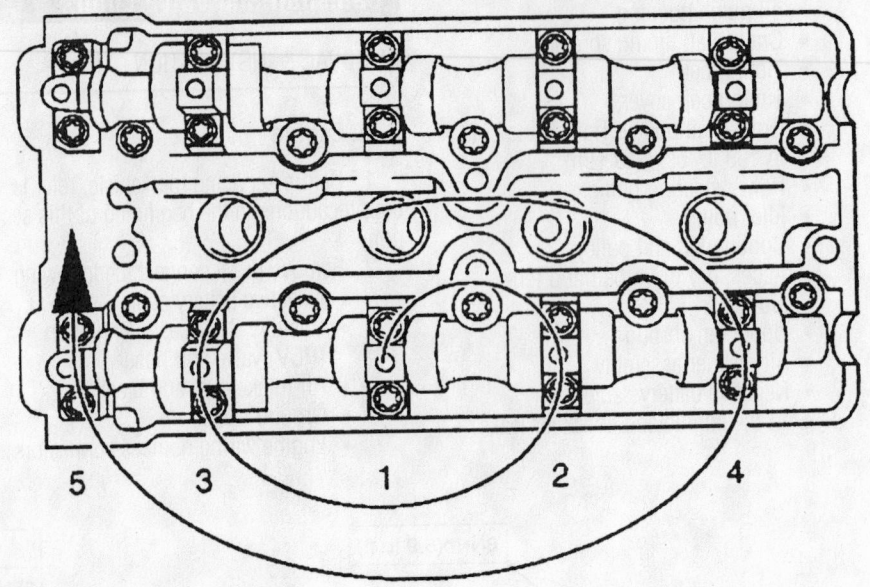

7924NG20

Camshaft bearing cap tightening sequence—2.2L engine

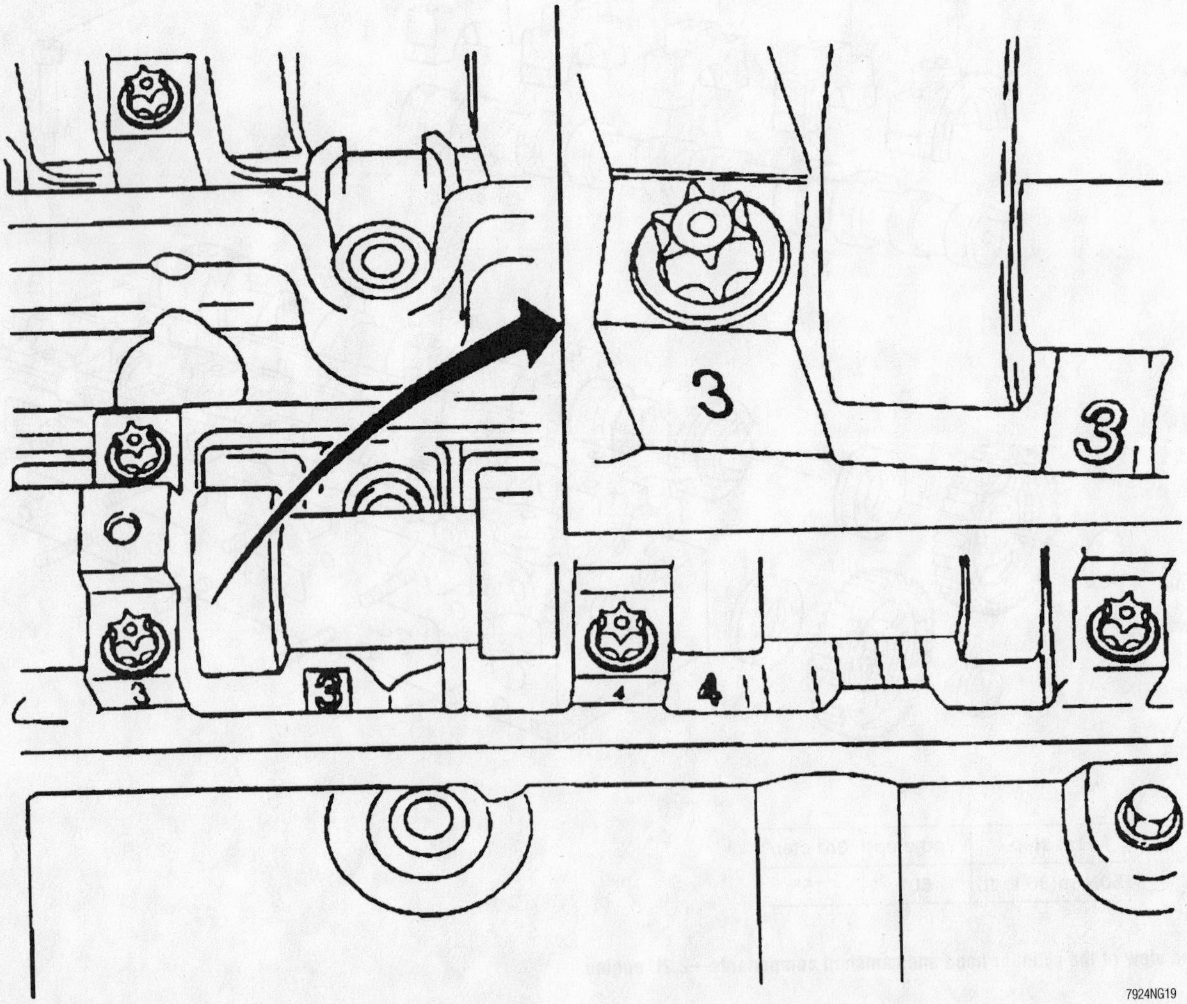

7924NG19

Camshaft bearing cap identification locations—2.2L engine

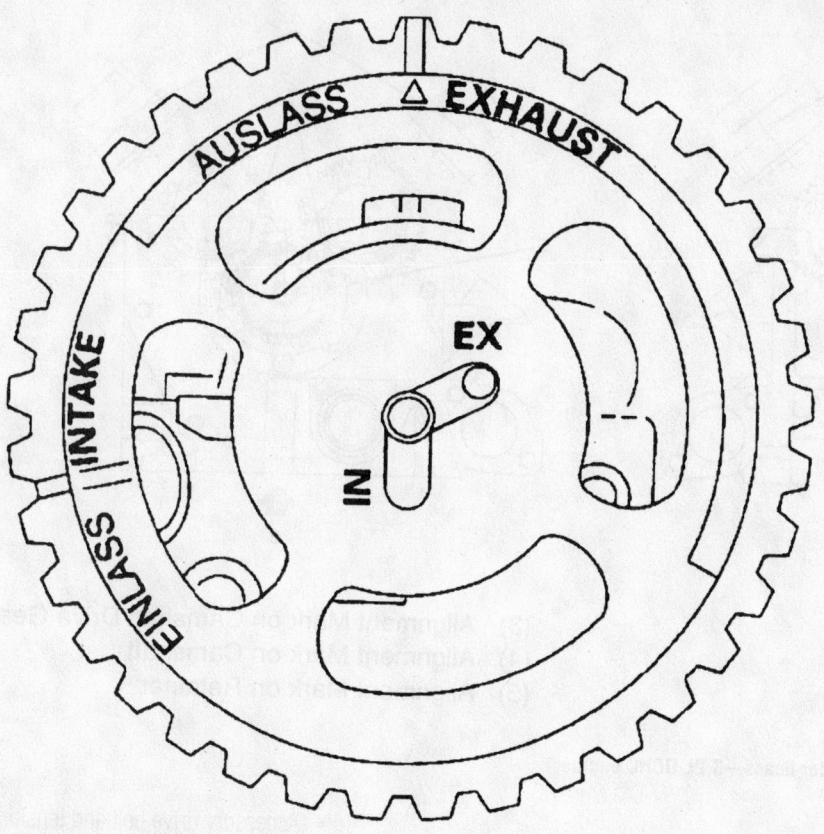

Guide pin location for the exhaust cam gear—2.2L engine

7924NG21

c. Step 3: Plus 15 degrees
5. Install or connect the following:
- Timing belt
- Valve cover
- CMP sensor. Loosen the rear timing cover bolt for access.
- Front cover
- Crankshaft pulley. Tighten the bolts to 14 ft. lbs. (20 Nm).
- CKP sensor connector
- CMP sensor connector
- Spark plug wires
- Spark plug wire cover
- Accessory drive belt
- Cooling fan harness connector
- Engine wiring harness connectors at left rear of the engine compartment
- Ground cables
- Air intake duct and bracket
- PCV valve and hose
- Negative battery cable

3.2L Engine

1. Before servicing the vehicle, refer to the precautions in the beginning of this section.

2. Remove or disconnect the following:
- Negative battery cable
- Air cleaner assembly
- Upper fan shroud
- Accessory drive belt and tensioner
- Cooling fan and pulley
- Idler pulley
- Power steering pump and move it aside
- Crankshaft pulley
- Timing belt cover
- Timing belt. Refer to the Timing Belt Unit Repair Section.
- Ignition coils
- Valve covers
- Camshafts
- Valve shims and tappets

➡**Keep the valve shims and tappets in order for installation.**

To install:

3. Install the valve tappets and shims in their original locations.

4. Using Gear Spring Lever J-42686, turn the sub gear clockwise to align the 5mm bolt holes in the sub gear and the camshaft driven gear. Tighten the 5mm bolt.

5. Install or connect the following:
- Camshafts by aligning the timing marks as shown. Tighten the bolts

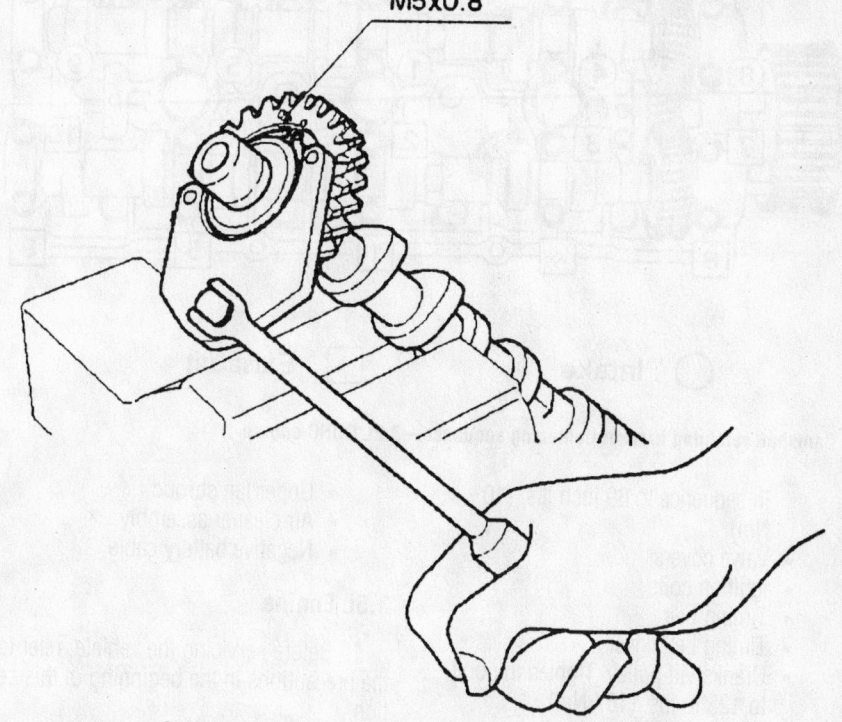

Aligning the sub gear with the Gear Spring Lever J-42686—3.2L DOHC engine

7924NG45

For Accessory Drive Belt illustrations, see Section 1 of this manual

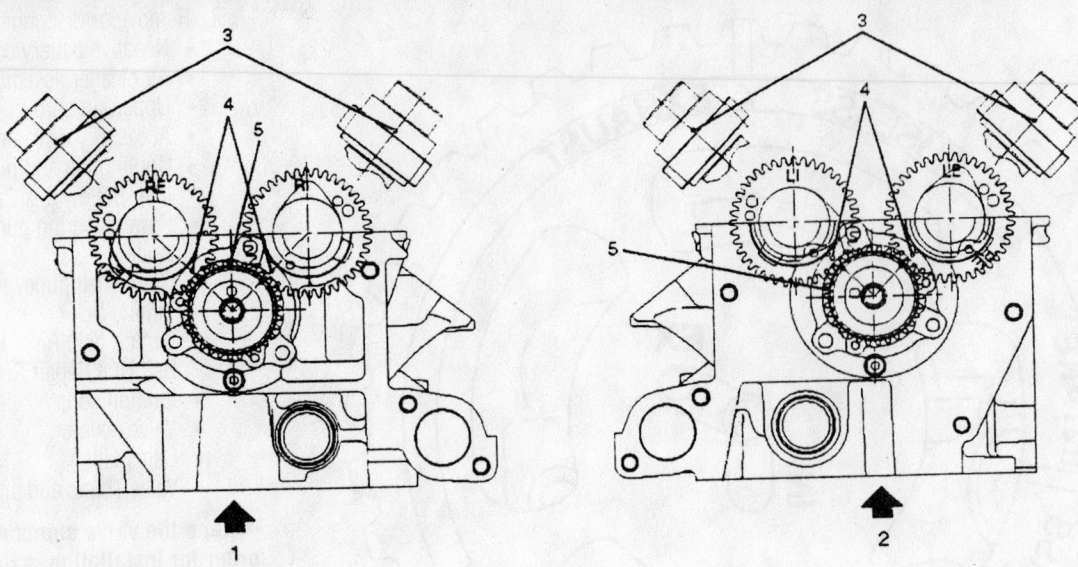

Legend
(1) Right Bank
(2) Left Bank

(3) Alignment Mark on Camshaft Drive Gear
(4) Alignment Mark on Camshaft
(5) Alignment Mark on Retainer

7924NG46

Camshaft alignment marks for the left and right cylinder heads—3.2L DOHC engine

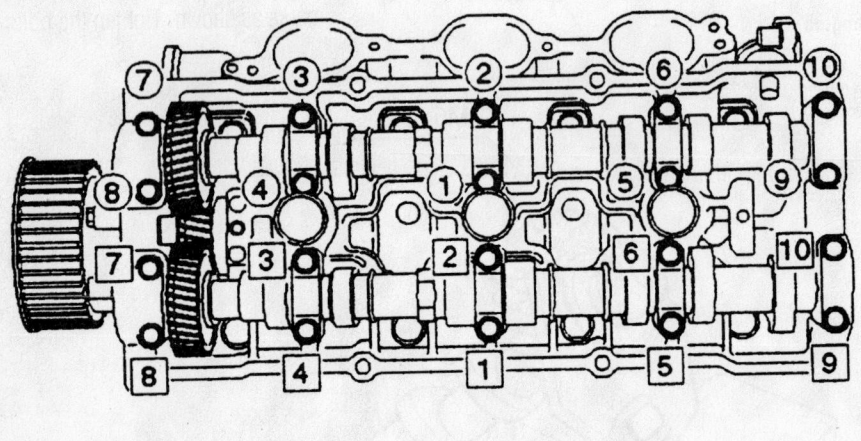

○ : Intake   □ : Exhaust

7924NG47

Camshaft retaining bracket tightening sequence—3.2L DOHC engine

in sequence to 89 inch lbs. (10 Nm).
- Valve covers
- Ignition coils
- Timing belt
- Timing belt cover
- Crankshaft pulley. Tighten the bolt to 123 ft. lbs. (167 Nm).
- Power steering pump
- Idler pulley
- Cooling fan and pulley
- Accessory drive belt and tensioner

- Upper fan shroud
- Air cleaner assembly
- Negative battery cable

3.5L Engine

1. Before servicing the vehicle, refer to the precautions in the beginning of this section.
2. Remove or disconnect the following:
- Negative battery cable
- Air cleaner assembly
- Upper fan shroud

- Accessory drive belt and tensioner
- Cooling fan and pulley
- Idler pulley
- Power steering pump and move it aside
- Crankshaft pulley
- Timing belt cover
- Timing belt. Refer to the Timing Belt Unit Repair Section.
- Ignition coils
- Valve covers
- Camshafts
- Valve shims and tappets

➡ Keep the valve shims and tappets in order for installation.

To install:
3. Install the valve tappets and shims in their original locations.
4. Using Gear Spring Lever J-42686, turn the sub gear clockwise to align the 5mm bolt holes in the sub gear and the camshaft driven gear. Tighten the 5mm bolt.
5. Install the camshafts. Align the timing marks as shown. Tighten the bolts in sequence to 89 inch lbs. (10 Nm).
6. Install or connect the following:
- Valve covers
- Ignition coils
- Timing belt. Refer to the Timing Belt Unit Repair Section.
- Timing belt cover
- Crankshaft pulley
- Power steering pump

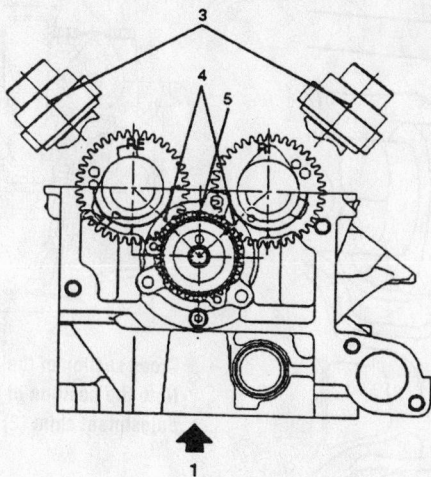

1 Right Bank	3 Alignment Mark on Camshaft Drive Gear
2 Left Bank	4 Alignment Mark on Camshaft
	5 Alignment Mark on Retainer

7924BG11

Camshaft alignment marks for the left and right cylinder heads—3.5L engine

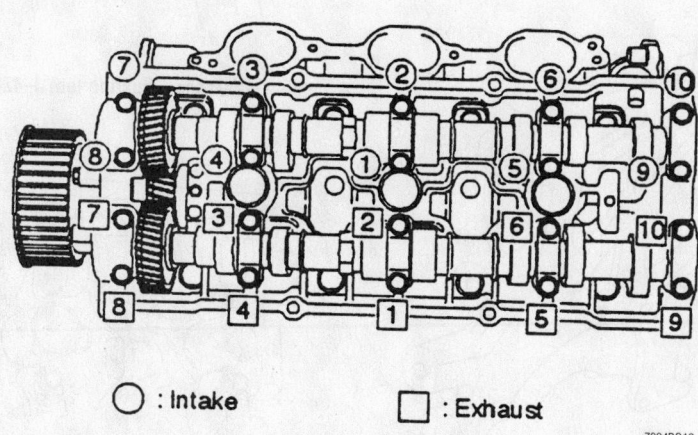

○ : Intake   □ : Exhaust

7924BG12

Camshaft retaining bracket tightening sequence—3.5L engine

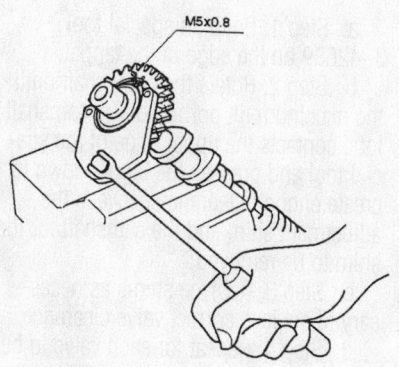

7924BG13

Aligning the sub gear with the Gear Spring Lever J-42686—3.5L engine

- Idler pulley
- Cooling fan and pulley
- Accessory drive belt and tensioner
- Upper fan shroud
- Air cleaner assembly
- Negative battery cable

Valve Lash

ADJUSTMENT

2.2L Engine

The 2.2L DOHC engine is equipped with hydraulic lash adjusters. No valve adjustment is necessary.

3.2L Engines

➡ **Measure valve clearance with the engine cold.**

1. Before servicing the vehicle, refer to the precautions in the beginning of this section.
2. Remove the valve covers.
3. Check the valve clearance with the camshafts positioned as shown. Intake valve clearance should be 0.0091–0.0130 in. (0.2311–0.3302mm). Exhaust valve clearance should be 0.0098–0.0138 in. (0.2489–0.3505mm).
4. If adjustment is required, replace the shims as follows:

a. Step 1: Position special tool J-42689 on the edge of the tappet.

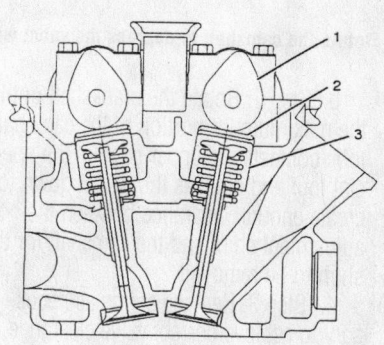

9302NG01

Cross section of the 3.2L DOHC cylinder head. Note the position of the camshaft lobe (1), adjustment shim (2) and the tappet (3)

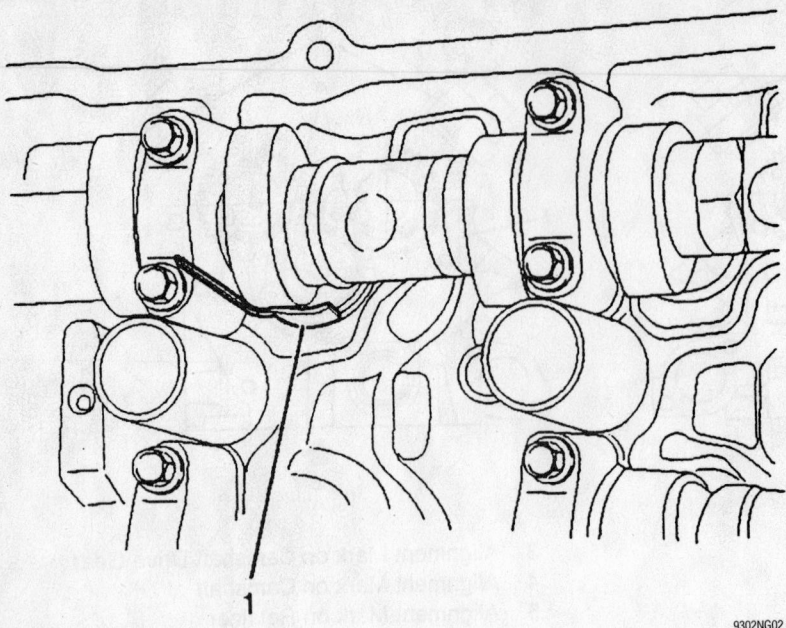

Insert special tool J-42689 (1) and use the camshaft to press the tappet down—3.2L DOHC engine

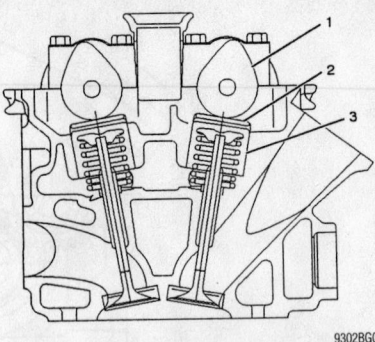

Cross section of the 3.5L cylinder head. Note the position of the camshaft lobe (1), adjustment shim (2) and the tappet (3)

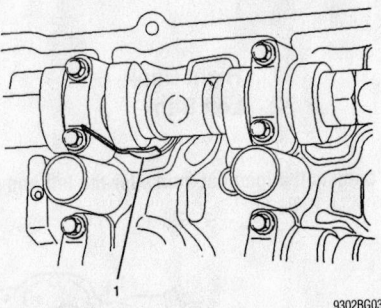

Valve clearance adjusting tool J–42689 (1)

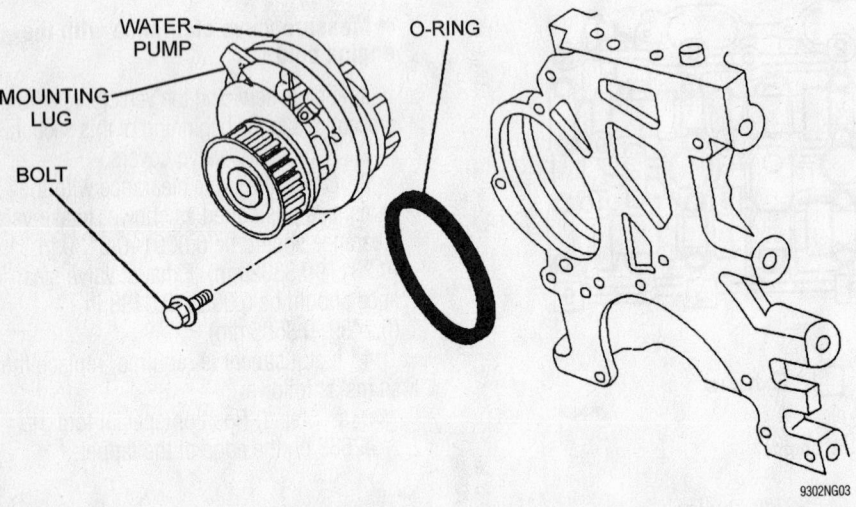

Rotate the camshaft to depress the valve with special tool J-42689—3.2L DOHC engine

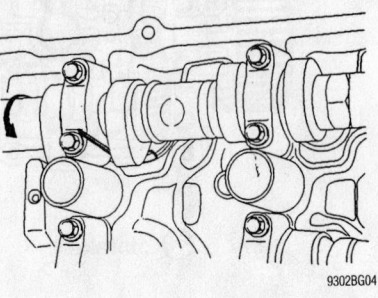

Using the valve clearance adjusting tool to hold the tappet for shim replacement

b. Step 2: Rotate the crankshaft until the maximum lift portion of the camshaft lobe contacts the upper edge of the special tool and presses the tappet down to create enough clearance between the adjustment shim and the camshaft for the shim to be removed.

c. Step 3: Replace shims as necessary to achieve correct valve clearance.

d. Step 4: Repeat for each valve to be adjusted.

5. Replace the valve covers. Tighten the bolts to 80 inch lbs. (9 Nm).

3.5L Engine

➡**Measure valve clearance with the engine cold.**

1. Before servicing the vehicle, refer to the precautions in the beginning of this section.

2. Remove the valve covers.

3. Check the valve clearance with the camshafts positioned as shown. Intake valve clearance should be 0.0091–0.0130 inches. Exhaust valve clearance should be 0.0098–0.0138 inches.

4. If adjustment is required, replace the shims as follows:

a. Step 1: Position special tool J–42689 on the edge of the tappet.

b. Step 2: Rotate the crankshaft until the maximum lift portion of the camshaft lobe contacts the upper edge of the special tool and presses the tappet down to create enough clearance between the adjustment shim and the camshaft for the shim to be removed.

c. Step 3: Replace shims as necessary to achieve correct valve clearance.

d. Step 4: Repeat for each valve to be adjusted.

5. Replace the valve covers. Tighten the bolts to 80 inch lbs. (9 Nm).

Starter Motor

REMOVAL & INSTALLATION

2.2L Engines

1. Before servicing the vehicle, refer to the precautions in the beginning of this section.
2. Remove or disconnect the following:
   - Negative battery cable
   - Starter harness connections
   - Starter motor

To install:

3. Install or connect the following:
   - Starter motor. Tighten the fasteners to 18 ft. lbs. (25 Nm) for the 2.2L engine or to 30 ft. lbs. (40 Nm) for the 2.6L engine.
   - Starter harness connections
   - Negative battery cable

3.2L Engines

1. Before servicing the vehicle, refer to the precautions in the beginning of this section.
2. Remove or disconnect the following:
   - Negative battery cable
   - Heated Oxygen (HO2S) sensor connectors
   - Exhaust front pipe
   - Heat shield
   - Starter wiring connectors
   - Starter motor

To install:

3. Install or connect the following:
   - Starter motor. Tighten the bolts to 30 ft. lbs. (40 Nm).
   - Starter wiring connectors
   - Heat shield
   - Exhaust front pipe
   - Heated Oxygen (HO2S) sensor connectors
   - Negative battery cable

3.5L Engine

1. Before servicing the vehicle, refer to the precautions in the beginning of this section.
2. Remove or disconnect the following:
   - Negative battery cable
   - Heated Oxygen (HO2S) sensor connectors
   - Exhaust front pipe
   - Heat shield
   - Starter wiring connectors
   - Starter motor

To install:

3. Install or connect the following:
   - Starter motor. Tighten the bolts to 30 ft. lbs. (40 Nm).
   - Starter wiring connectors
   - Heat shield
   - Exhaust front pipe
   - Heated Oxygen (HO2S) sensor connectors
   - Negative battery cable

Oil Pan

REMOVAL & INSTALLATION

2.2L Engines

1. Before servicing the vehicle, refer to the precautions in the beginning of this section.
2. Drain the engine oil.
3. Remove or disconnect the following:
   - Flywheel dust cover
   - Left and right engine mounts. Raise the engine for access.
   - Oil pan
   - Oil pan support, for 2.2L engine

To install:

4. Perform the following:
   a. Install the oil pan support and tighten the bolts to 14 ft. lbs. (20 Nm).
   b. Install the oil pan and tighten the bolts to 70 inch lbs. (8 Nm) plus 30 degrees.
5. Install or connect the following:
   - Left and right engine mounts. Tighten the nuts to 41 ft. lbs. (55 Nm).
   - Flywheel dust cover
6. Fill the crankcase to the correct level.
7. Start the engine and check for leaks.

3.2L Engines

1. Before servicing the vehicle, refer to the precautions in the beginning of this section.
2. Drain the engine oil.
3. Remove or disconnect the following:
   - Negative battery cable
   - Front wheels
   - Oil level dipstick
   - Stone guard
   - Radiator under fan shroud
   - Suspension crossmember
   - Flywheel dust cover
   - Pitman arm
   - Idler arm
4. If equipped with 4 wheel drive, unbolt and lower the front axle housing assembly for clearance.
5. Remove or disconnect the following:
   - Oil pan
   - Lower crankcase

To install:

6. Apply a bead of silicone sealant to the crankcase flange and install the crankcase. Tighten the fasteners in sequence to 89 inch lbs. (10 Nm).
7. Apply a bead of silicone sealant to the oil pan flange and install the oil pan.

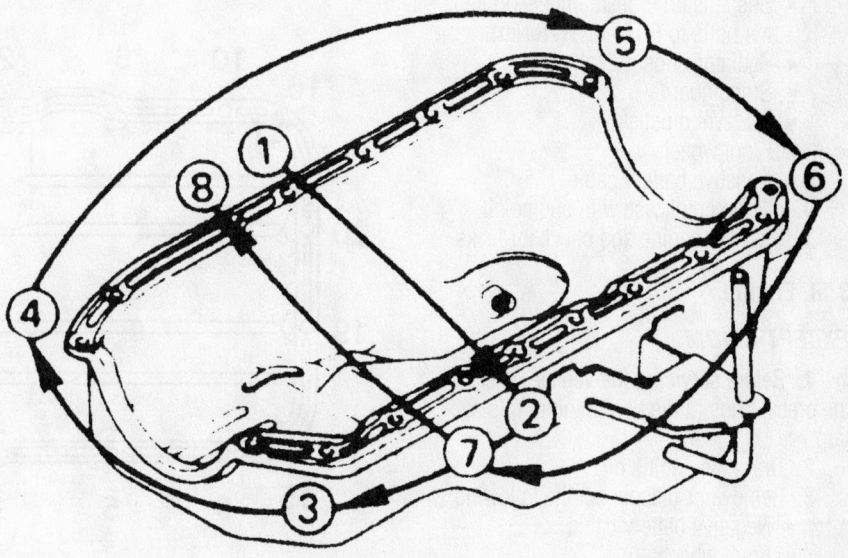

Oil pan bolt tightening sequence—2.2L engines

7924NG24

For Wheel Alignment specifications, see Section 1 of this manual

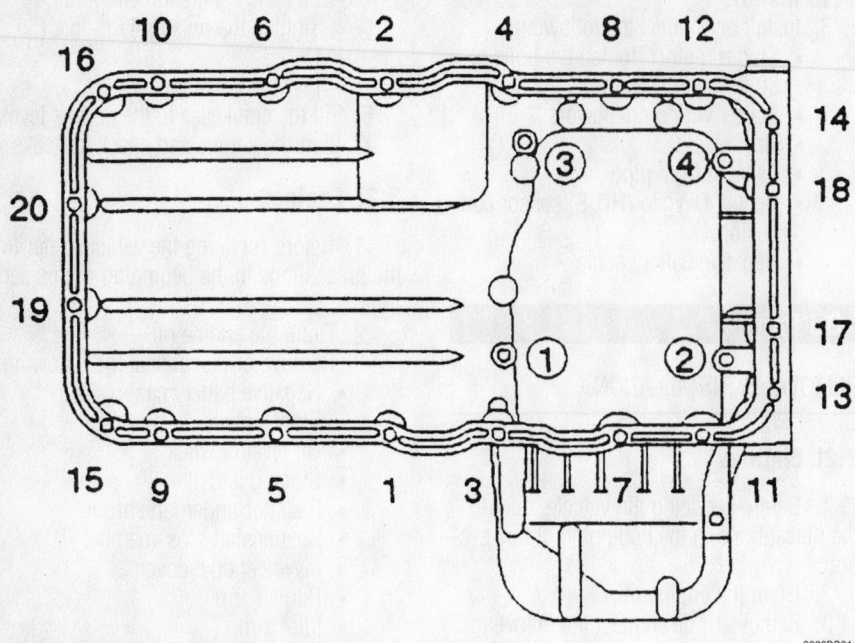

Lower crankcase torque sequence—3.2L DOHC engine

Tighten the fasteners to 89 inch lbs. (10 Nm).

8. If equipped, raise the axle housing assembly into position. Tighten the axle case bolts to 61 ft. lbs. (82 Nm) and the mounting bolts to 112 ft. lbs. (152 Nm).

9. Install or connect the following:
- Pitman arm. Tighten the nut to 159 ft. lbs. (216 Nm).
- Idler arm. Tighten the bolt to 33 ft. lbs. (44 Nm).
- Flywheel dust cover
- Suspension crossmember. Tighten the bolts to 58 ft. lbs. (78 Nm).
- Radiator under fan shroud
- Stone guard
- Oil level dipstick
- Front wheels
- Negative battery cable

10. Fill the crankcase with engine oil.
11. Start the engine and check for leaks.

3.5L Engine

EXCEPT AXIOM

1. Before servicing the vehicle, refer to the precautions in the beginning of this section.

2. Drain the engine oil.

3. Remove or disconnect the following:
- Negative battery cable
- Front wheels
- Oil level dipstick
- Stone guard
- Radiator under fan shroud
- Suspension crossmember

- Flywheel dust cover
- Pitman arm
- Idler arm

4. If equipped with 4 wheel drive, unbolt and lower the front axle housing assembly for clearance.

5. Remove or disconnect the following:
- Oil pan
- Lower crankcase

To install:

6. Apply a bead of silicone sealant to the crankcase flange and install the crankcase. Tighten the fasteners in sequence to 89 inch lbs. (10 Nm).

7. Apply a bead of silicone sealant to the oil pan flange and install the oil pan. Tighten the fasteners to 89 inch lbs. (10 Nm).

8. If equipped, raise the axle housing assembly into position. Tighten the axle case bolts to 61 ft. lbs. (82 Nm) and the mounting bolts to 112 ft. lbs. (152 Nm).

9. Install or connect the following:
- Pitman arm. Tighten the nut to 159 ft. lbs. (216 Nm).
- Idler arm. Tighten the bolt to 33 ft. lbs. (44 Nm).
- Flywheel dust cover
- Suspension crossmember. Tighten the bolts to 58 ft. lbs. (78 Nm).
- Radiator under fan shroud
- Stone guard
- Oil level dipstick
- Front wheels
- Negative battery cable

10. Fill the crankcase with engine oil.
11. Start the engine and check for leaks.

AXIOM

1. Before servicing the vehicle, refer to the precautions in the beginning of this section.

2. Drain the engine oil.

3. Remove or disconnect the following:
- Negative battery cable
- Front wheels
- Oil level dipstick
- Stone guard
- Radiator under fan shroud

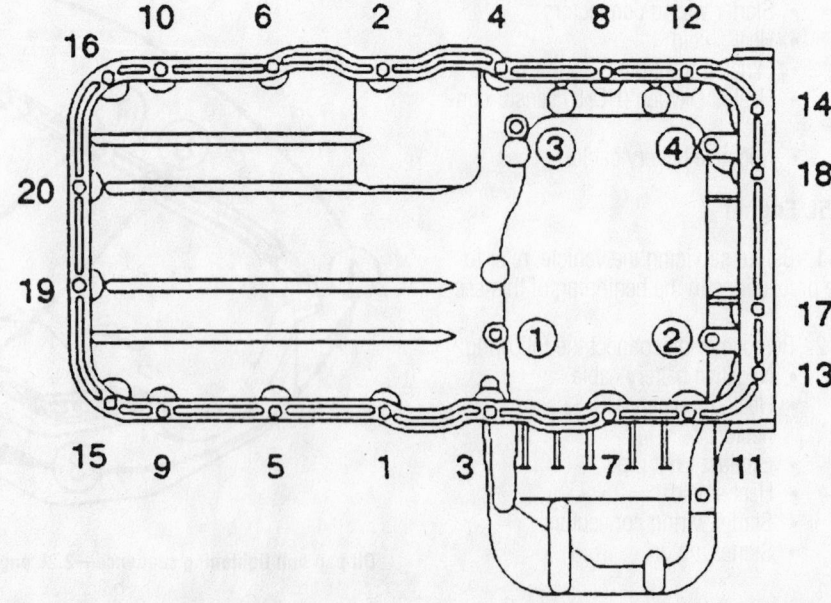

Lower crankcase torque sequence—3.5L engine

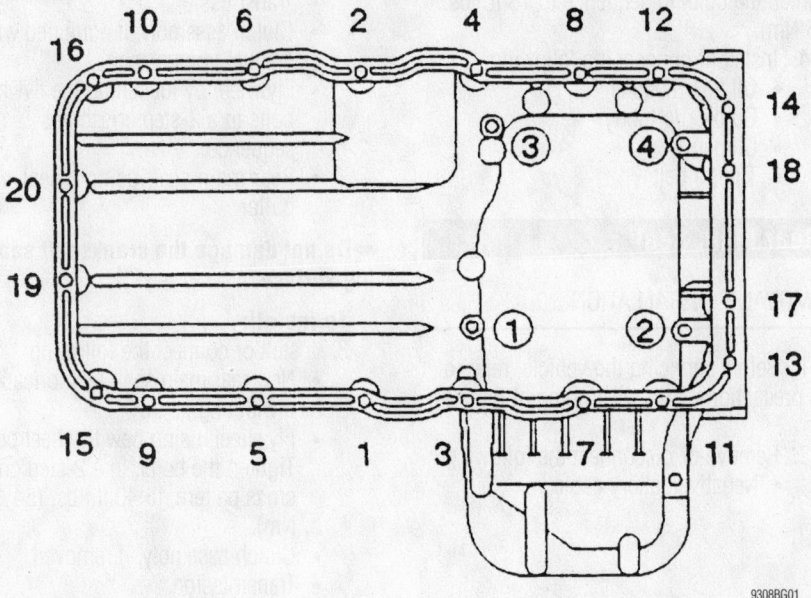

Lower crankcase torque sequence—3.5L engine

- Shift-on-the-fly from the axle
- Suspension crossmember

4. If equipped with 4 wheel drive, unbolt and lower the front axle housing assembly for clearance.

5. Remove or disconnect the following:
- Steering gear
- Starter
- Oil pan
- Lower crankcase

To install:

6. Apply a bead of silicone sealant to the crankcase flange and install the crankcase. Tighten the fasteners in sequence to 89 inch lbs. (10 Nm).

7. Apply a bead of silicone sealant to the oil pan flange and install the oil pan. Tighten the fasteners to 89 inch lbs. (10 Nm).

8. Install or connect the following:
- Starter. Torque to 30 ft. lbs. (40 Nm)

9. If equipped, raise the axle housing assembly into position. Tighten the axle case bolts to 61 ft. lbs. (82 Nm) and the mounting bolts to 112 ft. lbs. (152 Nm).

10. Install or connect the following:
- Steering gear
- Suspension crossmember. Tighten the bolts to 58 ft. lbs. (78 Nm).
- Radiator under fan shroud
- Stone guard
- Oil level dipstick
- Front wheels
- Negative battery cable

11. Fill the crankcase with engine oil.

12. Start the engine and check for leaks.

Oil Pump

REMOVAL & INSTALLATION

2.2L Engine

1. Before servicing the vehicle, refer to the precautions in the beginning of this section.

2. Drain the engine oil.

3. Remove or disconnect the following:
- Negative battery cable
- Accessory drive belts
- Cooling fan
- A/C belt tensioner, if equipped
- Water pump pulley
- Power steering pump
- Crankshaft pulley

- Front cover
- Timing belt. Refer to the Timing Belt unit repair section.
- Crankshaft timing sprocket
- Rear timing cover
- Crankshaft oil seal
- Oil pan
- Oil pump pickup tube
- Oil pump

To install:

4. Install or connect the following:
- Oil pump. Use a new gasket and tighten the bolts to 53 inch lbs. (6 Nm).
- Oil pump pickup tube. Tighten the bolts to 70 inch lbs. (8 Nm).
- Oil pan
- Crankshaft oil seal
- Rear timing cover
- Crankshaft timing sprocket. Tighten the bolt to 94 ft. lbs. (130 Nm) plus 45 degrees.
- Timing belt. Refer to the Timing Belt unit repair section.
- Front cover
- Crankshaft pulley. Tighten the bolts to 14 ft. lbs. (20 Nm).
- Power steering pump
- Water pump pulley
- A/C belt tensioner, if equipped
- Cooling fan
- Accessory drive belts
- Negative battery cable

5. Fill the crankcase to the correct level.

6. Start the engine and check for leaks.

3.2L Engine

1. Before servicing the vehicle, refer to the precautions in the beginning of this section.

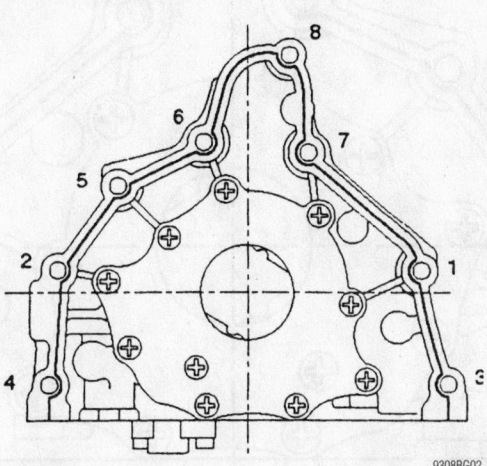

Oil pump torque sequence–3.2L engine

For Maintenance Interval recommendations, see Section 1 of this manual

2. Remove or disconnect the following:
- Timing belt
- Oil pan
- Oil pickup tube
- Oil filter adapter
- Oil pump

To install:

3. Apply silicone sealant to the oil pump mounting surface and install the oil pump. Tighten the bolts in sequence to 18 ft. lbs. (25 Nm).

4. Install or connect the following:
- Oil filter adapter
- Oil pickup tube
- Oil pan
- Timing belt

5. Fill the crankcase to the correct level.
6. Start the engine and check for leaks.

3.5L Engine

1. Before servicing the vehicle, refer to the precautions in the beginning of this section.

2. Remove or disconnect the following:
- Timing belt
- Oil pan
- Oil pick-up tube
- Oil filter adapter
- Oil pump

To install:

3. Apply silicone sealant to the oil pump mounting surface and install the oil pump.

Tighten the bolts in sequence to 18 ft. lbs. (25 Nm).

4. Install or connect the following:
- Oil filter adapter
- Oil pickup tube
- Oil pan
- Timing belt

Rear Main Seal

REMOVAL & INSTALLATION

1. Before servicing the vehicle, refer to the precautions in the beginning of this section.

2. Remove or disconnect the following:
- Negative battery cable

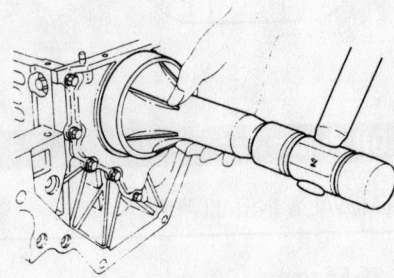

7924NG48

Installing a one-piece rear crankshaft oil seal

- Transmission
- Clutch assembly, if equipped with a manual transmission
- Flywheel by loosening the flywheel bolts in a 2-step crisscross sequence
- Rear main seal, using a seal puller

➡ **Do not damage the crankshaft sealing surface.**

To install:

3. Install or connect the following:
- New rear main seal, by lubricating it with engine oil
- Flywheel, using new flywheel bolts. Tighten the bolts, in a 2-step crisscross pattern, to 40 ft. lbs. (54 Nm).
- Clutch assembly, if removed
- Transmission
- Negative battery cable

4. Check the oil and refill as necessary.

Piston and Ring

POSITIONING

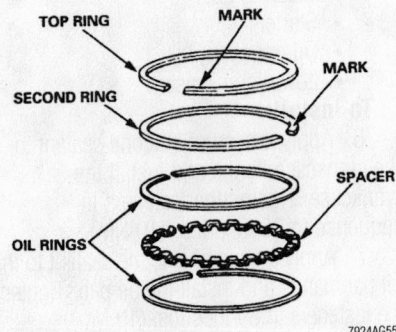

7924AG55

Piston ring positioning and top mark locations—all engines

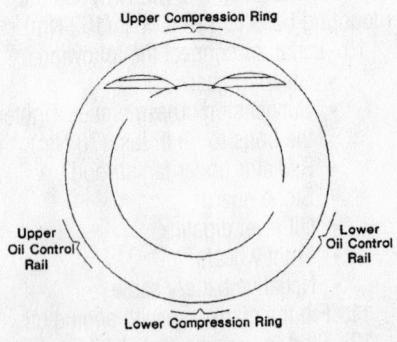

7924AG58

Piston ring end-gap spacing—2.2L engine

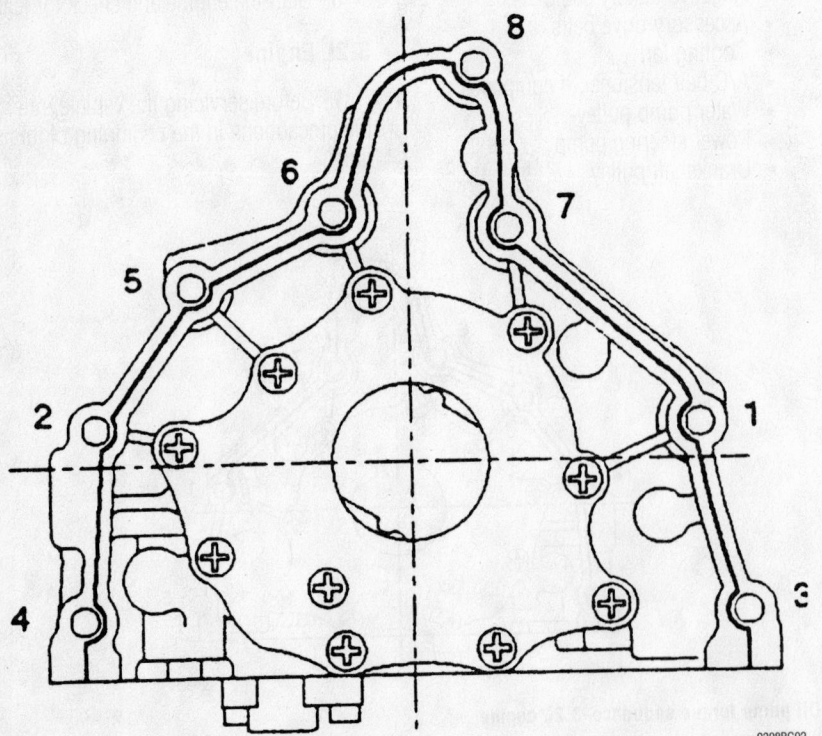

9308BG02

Oil pump torque sequence

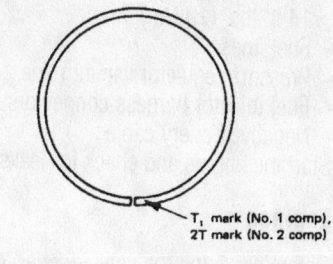

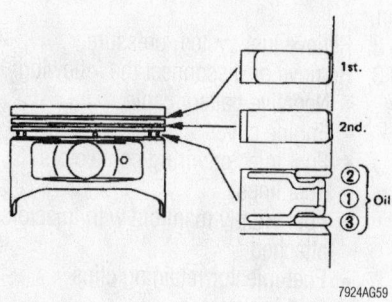

Piston ring positioning—3.2L and 3.5L engines

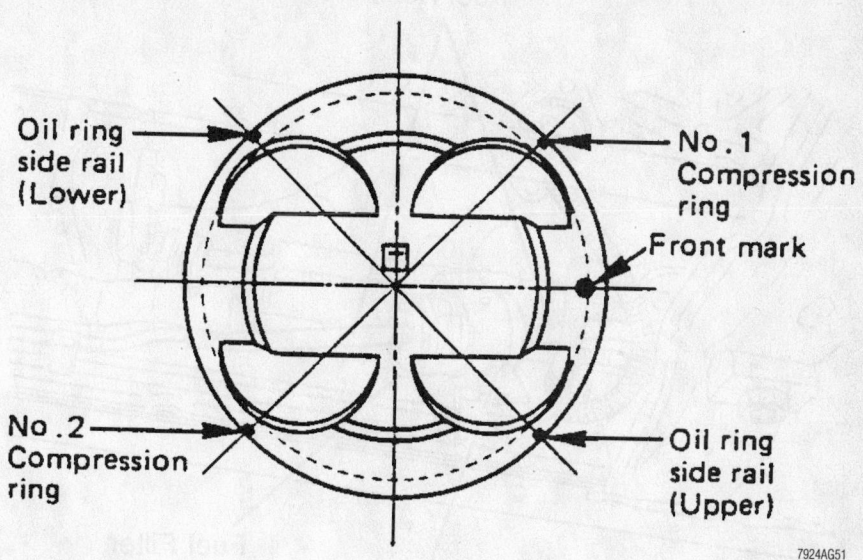

Piston ring end-gap spacing—3.2L and 3.5L engines

FUEL SYSTEM

Fuel System Service Precautions

Safety is the most important factor when performing not only fuel system maintenance but any type of maintenance. Failure to conduct maintenance and repairs in a safe manner may result in serious personal injury or death. Maintenance and testing of the vehicle's fuel system components can be accomplished safely and effectively by adhering to the following rules and guidelines:

• To avoid the possibility of fire and personal injury, always disconnect the negative battery cable unless the repair or test procedure requires that battery voltage be applied.

• Always relieve the fuel system pressure prior to disconnecting any fuel system component (injector, fuel rail, pressure regulator, etc.), fitting or fuel line connection. Exercise extreme caution whenever relieving fuel system pressure, to avoid exposing skin, face and eyes to fuel spray. Please be advised that fuel under pressure may penetrate the skin or any part of the body that it contacts.

• Always place a shop towel or cloth around the fitting or connection prior to loosening to absorb any excess fuel due to spillage. Ensure that all fuel spillage (should it occur) is quickly removed from engine surfaces. Ensure that all fuel soaked cloths or towels are deposited into a suitable waste container.

• Always keep a dry chemical (Class B) fire extinguisher near the work area.

• Do not allow fuel spray or fuel vapors to come into contact with a spark or open flame.

• Always use a backup wrench when loosening and tightening fuel line connection fittings. This will prevent unnecessary stress and torsion to fuel line piping. Always follow the proper tightening specifications.

• Always replace worn fuel fitting O-rings with new. Do not substitute fuel hose or equivalent, where fuel pipe is installed.

Fuel System Pressure

RELIEVING

1. Before servicing the vehicle, refer to the precautions in the beginning of this section.
2. Remove the fuel filler cap.
3. Remove the fuel pump relay from the underhood relay box.

4. Start the engine and let it run until it stalls, then crank the engine for an additional 30 seconds.
5. Turn the ignition switch to the **OFF** position and remove the key. Disconnect the negative battery cable.
6. When service is completed, install the fuel pump relay and connect the negative battery cable.

Fuel Filter

REMOVAL & INSTALLATION

1. Before servicing the vehicle, refer to the precautions in the beginning of this section.
2. Relieve the fuel system pressure.
3. Remove or disconnect the following:
   • Fuel lines from the fuel filter
   • Fuel filter

To install:
4. Install or connect the following:
   • Fuel filter and tighten the bracket bolt. Note the fuel flow directional arrow.
   • Fuel lines to the fuel filter
   • Negative battery cable
5. Start the engine and inspect the fuel filter connections for leaks.

For Tune-up, Capacities and Firing orders, see Section 1 of this manual

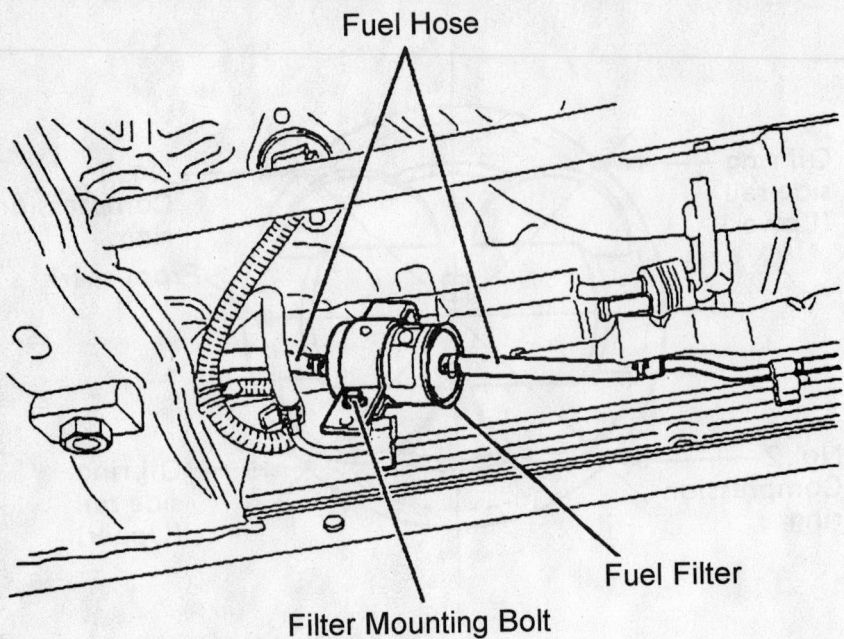

Fuel Hose

Fuel Filter

Filter Mounting Bolt

7924NG25

Fuel filter mounting location under the vehicle

Fuel Pump

REMOVAL & INSTALLATION

1. Before servicing the vehicle, refer to the precautions in the beginning of this section.
2. Relieve fuel system pressure.
3. Drain the fuel tank.
4. Remove or disconnect the following:
   - Negative battery cable
   - Fuel filler and vent hoses
   - Fuel tank skid plate
   - Fuel tank wiring connectors
   - Fuel supply and return lines
   - Fuel tank
   - Fuel pump assembly

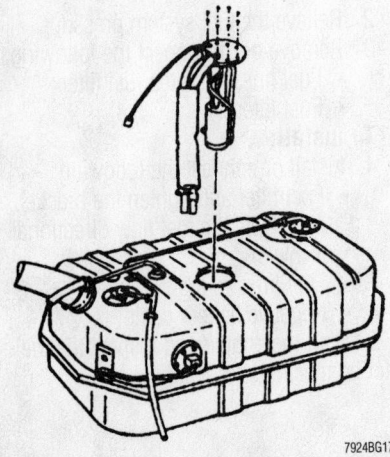

7924BG17

Fuel pump assembly mounting

To install:

5. Install or connect the following:
   - Fuel pump assembly
   - Fuel tank. Tighten the bolts to 27 ft. lbs. (36 Nm).
   - Fuel supply and return lines
   - Fuel tank wiring connectors
   - Fuel tank skid plate
   - Fuel filler and vent hoses
   - Negative battery cable
6. Start the engine and check for leaks.

Fuel Injector

REMOVAL & INSTALLATION

2.2L Engine

1. Before servicing the vehicle, refer to the precautions in the beginning of this section.
2. Relieve the fuel system pressure.
3. Remove or disconnect the following:
   - Negative battery cable
   - Fuel injector harness connectors
   - Pressure regulator vacuum line
   - Fuel lines
   - Fuel supply manifold with injectors attached
   - Fuel injector retaining clips
   - Fuel injectors

To install:

4. Install or connect the following:
   - Fuel injectors. Use new O-ring seals.
   - Fuel injector retaining clips
   - Fuel supply manifold with injectors

attached. Tighten the fasteners to 14 ft. lbs. (19 Nm).
   - Fuel lines
   - Pressure regulator vacuum line
   - Fuel injector harness connectors
   - Negative battery cable
5. Start the engine and check for leaks.

3.2L Engines

1. Before servicing the vehicle, refer to the precautions in the beginning of this section.
2. Relieve fuel system pressure.
3. Remove or disconnect the following:
   - Negative battery cable
   - Engine cover
   - Fuel injector wiring connectors
   - Fuel lines
   - Fuel supply manifold with injectors attached
   - Fuel injector retaining clips
   - Fuel injectors

To install:

4. Install or connect the following:
   - Fuel injectors. Use new O-ring seals.
   - Fuel injector retaining clips
   - Fuel supply manifold with injectors attached. Tighten the bolts to 60 inch lbs. (6.5 Nm).
   - Fuel lines
   - Fuel injector wiring connectors
   - Engine cover
   - Negative battery cable
5. Start the engine and check for leaks.

3.5L Engine

1. Before servicing the vehicle, refer to the precautions in the beginning of this section.
2. Relieve fuel system pressure.
3. Remove or disconnect the following:
   - Negative battery cable
   - Engine cover
   - Fuel injector wiring connectors
   - Fuel lines
   - Fuel supply manifold with injectors attached
   - Clips
   - Injectors from the supply manifold

To install:

4. Install or connect the following:
   - New O-rings on the fuel injectors
   - Fuel injectors
   - Fuel supply manifold with injectors attached. Tighten the bolts to 60 inch lbs. (6.5 Nm).
   - Fuel lines
   - Fuel injector wiring connectors
   - Engine cover
   - Negative battery cable
5. Start the engine and check for leaks.

DRIVE TRAIN

Transmission Assembly

REMOVAL & INSTALLATION

Manual Transmissions

2 WHEEL DRIVE

➡ **The transmission flange bolts vary in length. Note their locations for assembly.**

1. Before servicing the vehicle, refer to the precautions in the beginning of this section.
2. Remove the hood.

3. Install a support fixture to the engine lifting eyes.
4. Remove or disconnect the following:
   - Negative battery cable
   - Shift lever knob
   - Rear console assembly
   - Grommet assembly
   - Shift lever
   - Clutch slave cylinder and hose bracket
   - Driveshaft
   - Fuel line heat shield
   - Vehicle Speed (VSS) sensor connector
   - Reverse light switch connector

- Flywheel under cover
- Transmission mount and cross-member
- Transmission flange bolts
- Transmission

To install:

5. Install or connect the following:
   - Transmission. Tighten the large flange bolts to 52 ft. lbs. (71 Nm) and the small bolts to 30 ft. lbs. (41 Nm).
   - Crossmember. Tighten the bolts to 56 ft. lbs. (76 Nm).
   - Transmission mount. Tighten the fasteners to 30 ft. lbs. (41 Nm).

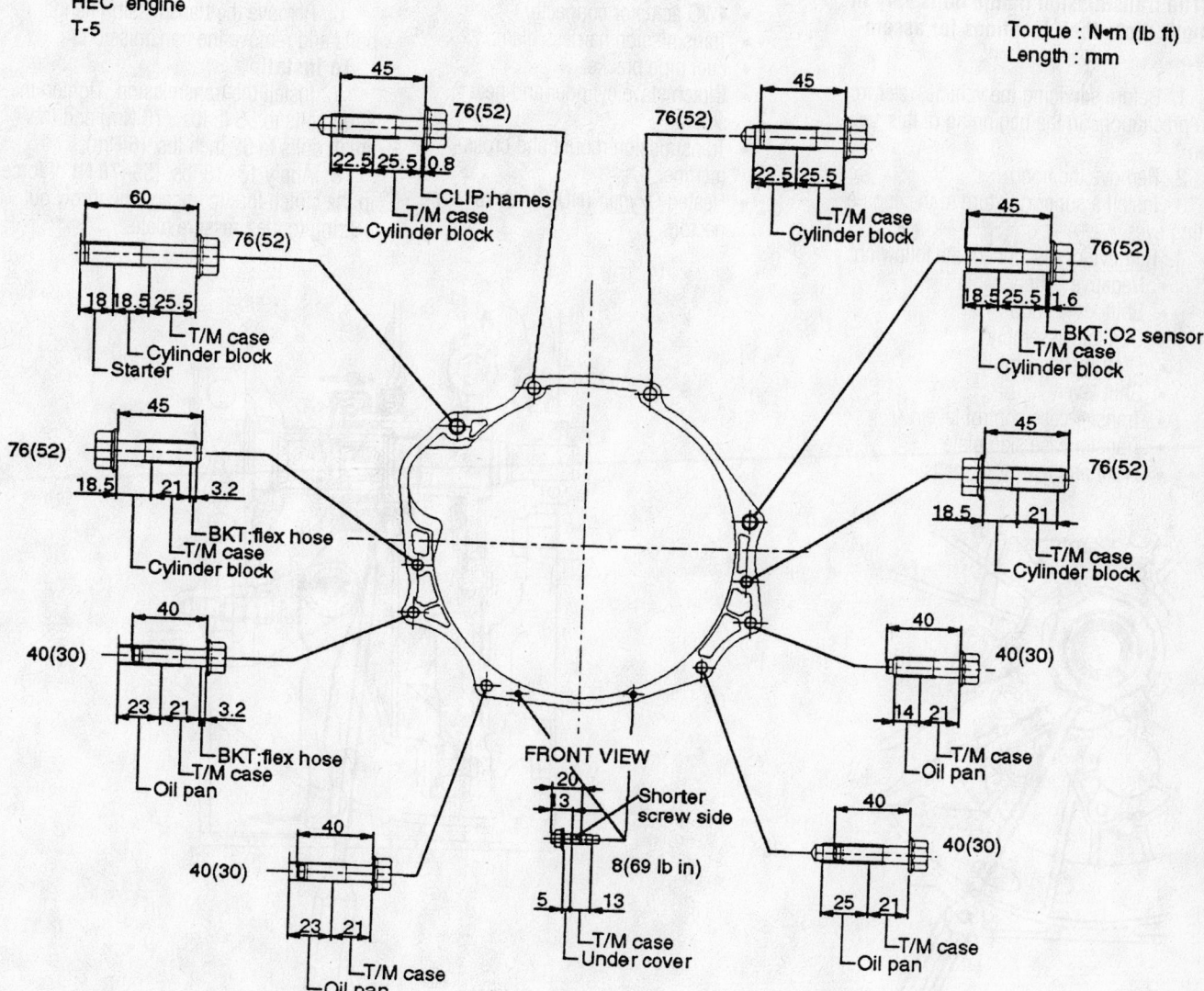

Transmission flange bolt identification and torque—2 wheel drive transmission

- Flywheel under cover. Tighten the bolts to 69 inch lbs. (8 Nm).
- Reverse light switch connector
- Vehicle Speed (VSS) sensor connector
- Fuel line heat shield
- Driveshaft. Tighten the bolts to 37 ft. lbs. (50 Nm).
- Clutch slave cylinder and hose bracket
- Shift lever
- Grommet assembly
- Rear console assembly
- Shift lever knob
- Negative battery cable

6. Remove the engine support fixture and install the hood.

4 WHEEL DRIVE

➡The transmission flange bolts vary in length. Note their locations for assembly.

1. Before servicing the vehicle, refer to the precautions in the beginning of this section.
2. Remove the hood.
3. Install a support fixture to the engine lifting eyes.
4. Remove or disconnect the following:
- Negative battery cable
- Shift lever knob
- Console assembly
- Grommet assembly
- Shift lever
- Transfer case control lever
- Transfer case skid plate
- Front and rear driveshafts

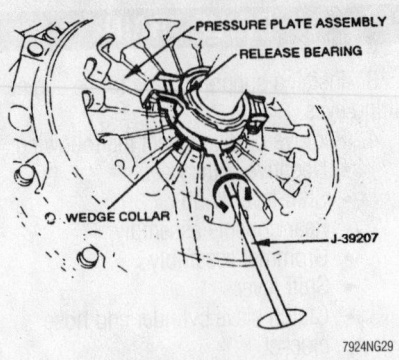

7924NG29

Turn the remover to separate the release bearing

- Reverse lamp switch connector
- Indicator switch connectors
- Vehicle Speed (VSS) sensor connector
- 4WD actuator connector
- Transmission harness clamps
- Fuel pipe bracket
- Clutch slave cylinder and heat shield
- Transmission mount and cross-member
- Heated Oxygen (HO2S) sensor connectors

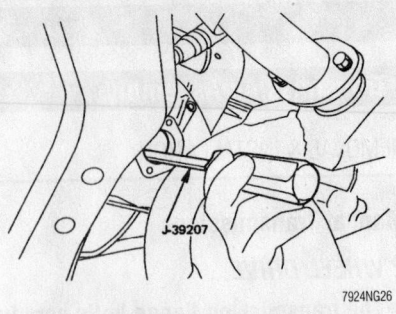

7924NG26

Insert the Release Bearing Remover tool J-39207 through the bell housing—4 wheel drive manual transmission

- Right exhaust front pipe
- Wiring harness heat shield
- Flywheel under cover

5. Release the throw out bearing from the pressure plate as shown.
6. Remove the transmission flange bolts and remove the transmission.

To install:

7. Install the transmission. Tighten the large bolts to 56 ft. lbs. (76 Nm) and the small bolts to 52 inch lbs. (6 Nm).
8. Apply 13–18 lbs. (59–78 N) of force to the clutch fork to engage the throw out bearing to the pressure plate.

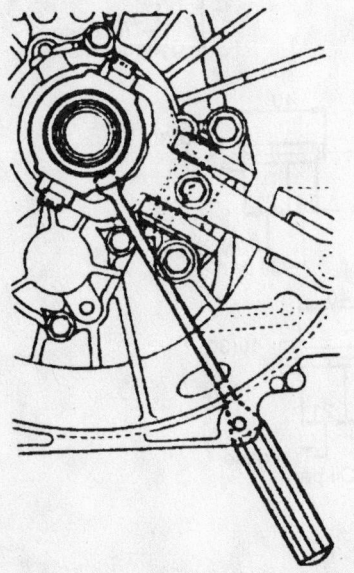

7924NG28

Insert the tool between the wedge collar and the release bearing

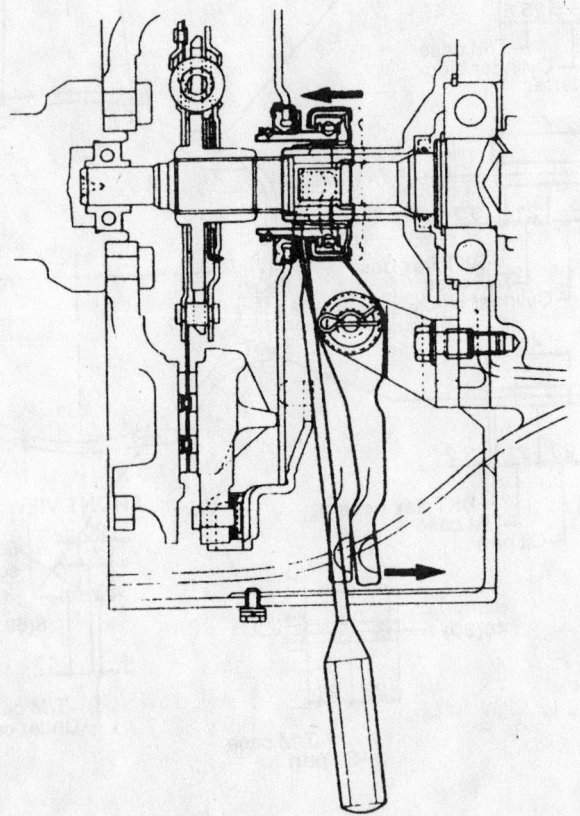

7924NG27

Push the release bearing fork toward the transmission to release the bearing from the pressure plate—4 wheel drive manual transmission

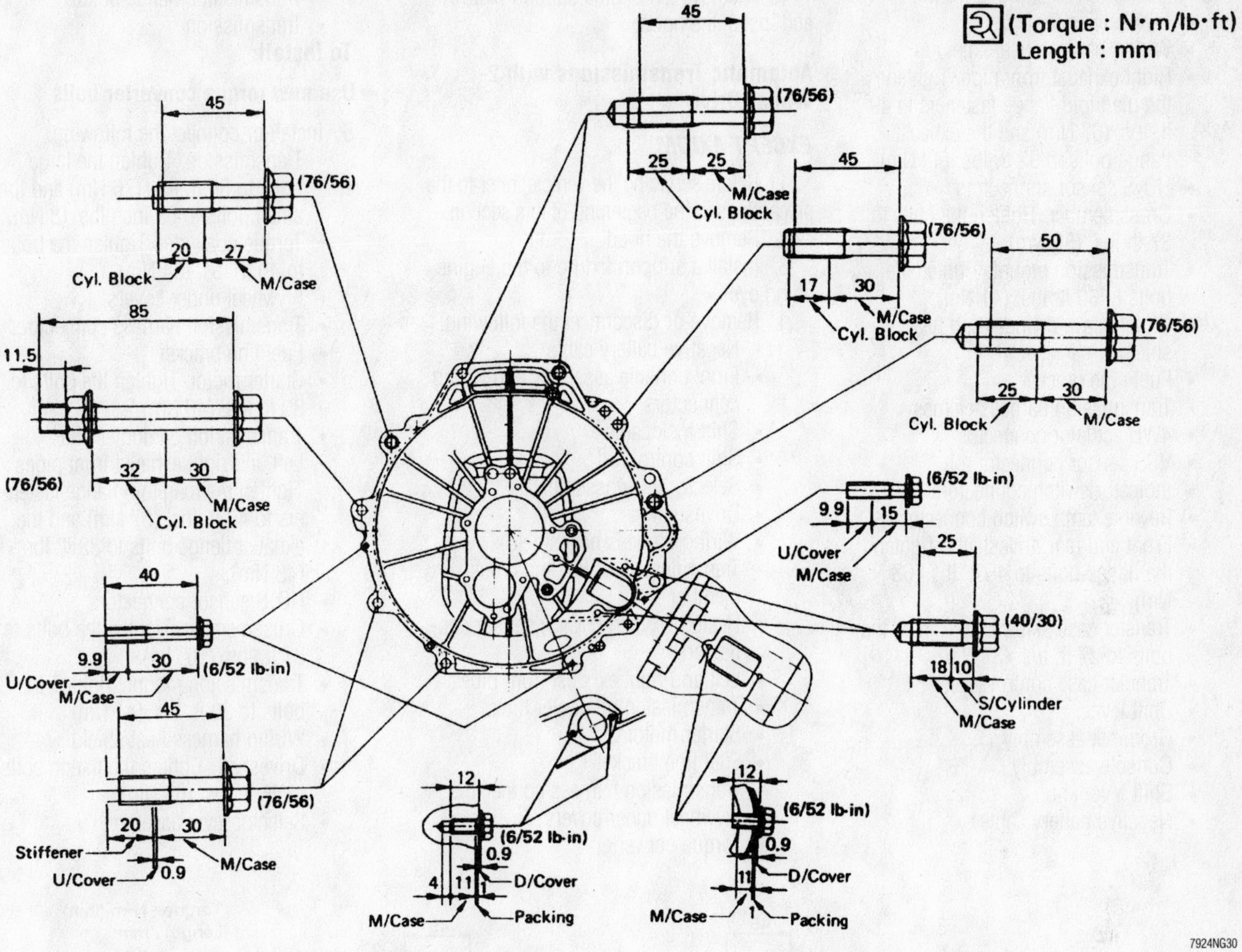

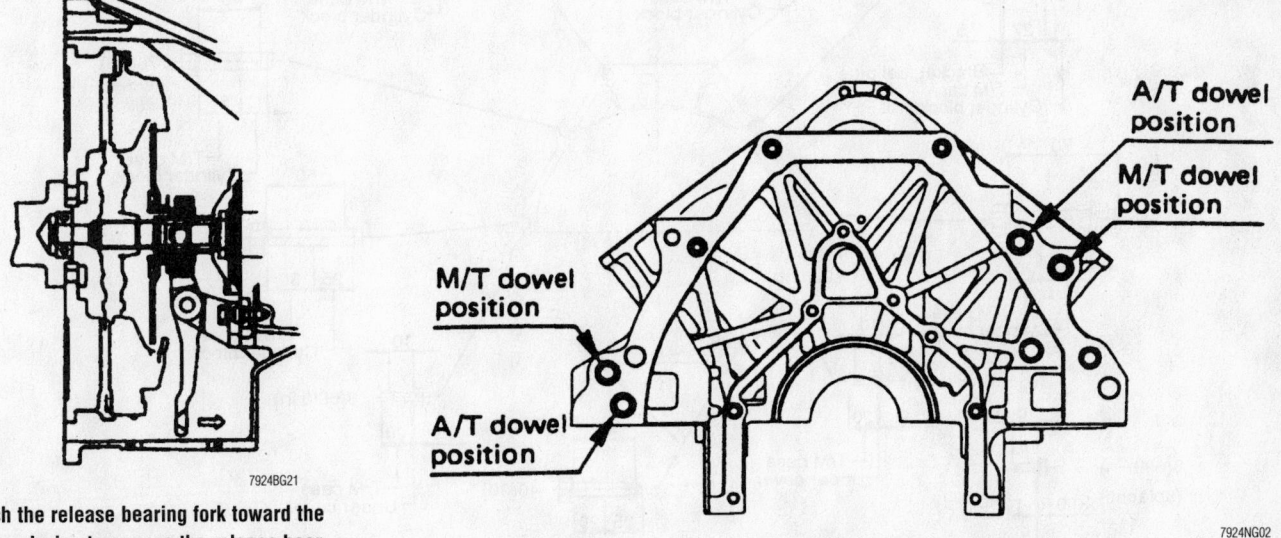

(Torque : N·m/lb·ft)
Length : mm

45
(76/56)
25 25
M/Case
Cyl. Block

45
(76/56)
17 30
M/Case
Cyl. Block

50
(76/56)
25 30
Cyl. Block M/Case

45
(76/56)
20 27
Cyl. Block M/Case

85
11.5
(76/56)
32 30
M/Case
Cyl. Block

(6/52 lb-in)
9.9 15
U/Cover
M/Case

25
(40/30)
18 10
S/Cylinder
M/Case

40
9.9 30
U/Cover (6/52 lb-in)
M/Case

45
(76/56)
20 30
Stiffener M/Case
U/Cover 0.9

12
(6/52 lb-in)
0.9
4 11 1 D/Cover
M/Case Packing

12
(6/52 lb-in)
0.9
11 D/Cover
M/Case 1 Packing

7924NG30

Transmission mounting bolt identification and torque specifications—V6 engine with 4 wheel drive transmission shown

7924BG21

Push the release bearing fork toward the transmission to engage the release bearing with the pressure plate

A/T dowel position
M/T dowel position
M/T dowel position
A/T dowel position

7924NG02

Dowel pin locations for automatic and manual transmissions—3.2L engine

9. Install or connect the following:
- Flywheel under cover
- Wiring harness heat shield
- Right exhaust front pipe. Tighten the manifold flange fasteners to 49 ft. lbs. (67 Nm) and the exhaust flange bolts to 32 ft. lbs. (43 Nm).
- HO2S sensor connectors
- Crossmember. Tighten the bolts to 37 ft. lbs. (50 Nm).
- Transmission mount. Tighten the bolts to 30 ft. lbs. (41 Nm).
- Clutch slave cylinder and heat shield
- Fuel pipe bracket
- Transmission harness clamps
- 4WD actuator connector
- VSS sensor connector
- Indicator switch connectors
- Reverse lamp switch connector
- Front and rear driveshafts. Tighten the flange bolts to 46 ft. lbs. (63 Nm).
- Transfer case skid plate. Tighten the bolts to 27 ft. lbs. (37 Nm).
- Transfer case control lever
- Shift lever
- Grommet assembly
- Console assembly
- Shift lever knob
- Negative battery cable

10. Remove the engine support fixture and install the hood.

Automatic Transmissions with 2-Wheel Drive

EXCEPT AXIOM

1. Before servicing the vehicle, refer to the precautions in the beginning of this section.
2. Remove the hood.
3. Install a support fixture to the engine lifting eyes.
4. Remove or disconnect the following:
- Negative battery cable
- Front console assembly and wiring connectors
- Shift lock cable
- Shift control rod
- Selector lever assembly
- Driveshaft
- Wiring harness heat shield
- Transmission mount and crossmember
- Heated Oxygen (HO2S) sensor connectors
- Left and right exhaust front pipes
- Transmission oil cooler lines
- Starter motor
- Fuel line bracket
- Transmission harness connectors
- Flywheel under covers
- Torque converter

- Transmission flange bolts
- Transmission

To install:

➡ **Use new torque converter bolts.**

5. Install or connect the following:
- Transmission. Tighten the large bolts to 56 ft. lbs. (76 Nm) and the small bolts to 69 inch lbs. (8 Nm).
- Torque converter. Tighten the bolts to 40 ft. lbs. (54 Nm).
- Flywheel under covers
- Transmission harness connectors
- Fuel line bracket
- Starter motor. Tighten the bolts to 30 ft. lbs. (40 Nm).
- Transmission oil cooler lines
- Left and right exhaust front pipes. Tighten the manifold flange fasteners to 49 ft. lbs. (67 Nm) and the exhaust flange bolts to 32 ft. lbs. (43 Nm).
- HO2S sensor connectors
- Crossmember. Tighten the bolts to 37 ft. lbs. (50 Nm).
- Transmission mount. Tighten the bolts to 30 ft. lbs. (41 Nm).
- Wiring harness heat shield
- Driveshaft. Tighten the flange bolts to 46 ft. lbs. (63 Nm).
- Selector lever assembly

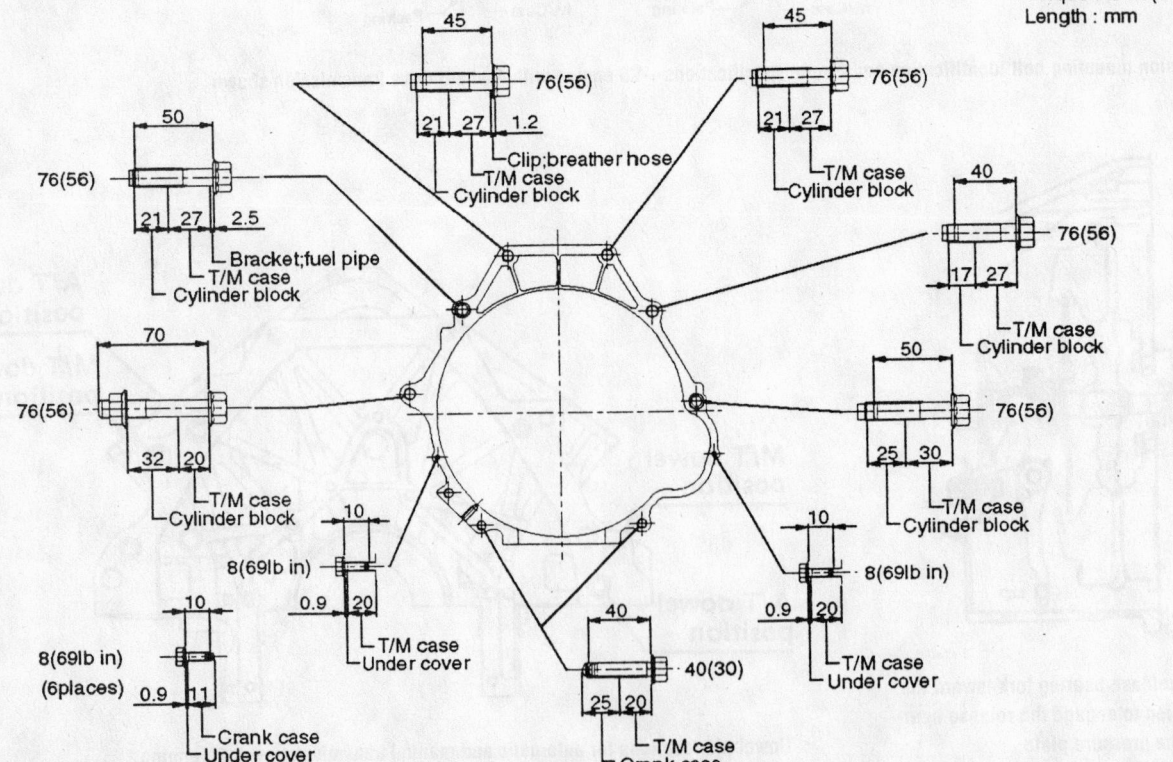

Torque : N•m (lb ft)
Length : mm

Automatic transmission mounting bolt locations and torque specifications

9308NG03

- Shift control rod
- Shift lock cable
- Front console assembly and wiring connectors
- Negative battery cable

6. Remove the engine support fixture and install the hood.

AXIOM

1. Before servicing the vehicle, refer to the precautions in the beginning of this section.
2. Remove or disconnect the following:
   - Negative battery cable
   - Driveshaft
   - Fuel line bracket from the transmission
   - Wiring harness heat shield
   - Transmission harness connectors
3. Support the transmission with a jack.
4. Remove or disconnect the following:
   - Transmission mount and crossmember
   - Transmission oil cooler lines
   - Selector cable
   - Starter motor
   - Undercovers
   - Torque converter bolts
   - Transmission flange bolts
   - Transmission

To install:

➡**Use new torque converter bolts.**

5. Install or connect the following:
   - Transmission. Tighten the large bolts to 56 ft. lbs. (76 Nm) and the small bolts to 69 inch lbs. (8 Nm).
   - Torque converter. Tighten the bolts to 40 ft. lbs. (54 Nm).
   - Under covers
   - Starter motor. Tighten the bolts to 30 ft. lbs. (40 Nm).
   - Selector cable
   - Transmission oil cooler lines
   - Crossmember. Tighten the bolts to 85 ft. lbs. (116 Nm).
   - Transmission mount. Tighten the bolts to 37 ft. lbs. (50 Nm).
   - Transmission harness connectors
   - Wiring harness heat shield
   - Fuel line bracket
   - Driveshaft. Tighten the flange bolts to 46 ft. lbs. (63 Nm).
   - Negative battery cable

Automatic Transmissions with 4-Wheel Drive

EXCEPT AXIOM

1. Before servicing the vehicle, refer to the precautions in the beginning of this section.

2. Remove the hood.
3. Install a support fixture to the engine lifting eyes.
4. Remove or disconnect the following:
   - Negative battery cable
   - Transfer case shift lever knob
   - Front console assembly and wiring connectors
   - Shift lock cable
   - Shift control rod
   - Selector lever assembly
   - Transfer case shift lever
   - Transfer case skid plate
   - Front and rear driveshafts
   - Wiring harness heat shield
   - Transmission mount and crossmember
   - Right torsion bar, if equipped with Torque On Demand (TOD) system
   - Heated Oxygen (HO$_2$S) sensor connectors
   - Left and right exhaust front pipes
   - Transmission oil cooler lines
   - Starter motor
   - Fuel line bracket
   - Transmission harness connectors
   - Flywheel under covers
   - Torque converter
   - Transmission flange bolts
   - Transmission

To install:

➡**Use new torque converter bolts.**

5. Install or connect the following:
   - Transmission. Tighten the large bolts to 56 ft. lbs. (76 Nm) and the small bolts to 30 ft. lbs. (40 Nm).
   - Torque converter. Tighten the bolts to 40 ft. lbs. (54 Nm).
   - Flywheel under covers
   - Transmission harness connectors
   - Fuel line bracket
   - Starter motor. Tighten the bolts to 30 ft. lbs. (40 Nm).
   - Transmission oil cooler lines
   - Left and right exhaust front pipes. Tighten the manifold flange fasteners to 49 ft. lbs. (67 Nm) and the exhaust flange bolts to 32 ft. lbs. (43 Nm).
   - HO$_2$S sensor connectors
   - Right torsion bar, if removed
   - Crossmember. Tighten the bolts to 37 ft. lbs. (50 Nm).
   - Transmission mount. Tighten the bolts to 30 ft. lbs. (41 Nm).
   - Wiring harness heat shield
   - Front and rear driveshafts. Tighten

the flange bolts to 46 ft. lbs. (63 Nm).
   - Transfer case skid plate. Tighten the bolts to 27 ft. lbs. (37 Nm).
   - Transfer case shift lever
   - Selector lever assembly
   - Shift control rod
   - Shift lock cable
   - Front console assembly and wiring connectors
   - Transfer case shift lever knob
   - Negative battery cable

6. Remove the engine support fixture and install the hood.

AXIOM

1. Before servicing the vehicle, refer to the precautions in the beginning of this section.
2. Remove or disconnect the following:
   - Negative battery cable
   - Skid plate
   - Driveshafts
   - Center exhaust pipe
   - Fuel line bracket from the transmission
   - Wiring harness heat shield
   - Transmission harness connectors
3. Support the transmission with a jack.
4. Remove or disconnect the following:
   - Transmission mount and crossmember
   - Transmission oil cooler lines
   - Selector cable
   - Starter motor
   - Undercovers
   - Torque converter bolts
   - Transmission flange bolts
   - Transmission

To install:

➡**Use new torque converter bolts.**

5. Install or connect the following:
   - Transmission. Tighten the large bolts to 56 ft. lbs. (76 Nm) and the small bolts to 69 inch lbs. (8 Nm).
   - Torque converter. Tighten the bolts to 40 ft. lbs. (54 Nm).
   - Under covers
   - Starter motor. Tighten the bolts to 30 ft. lbs. (40 Nm).
   - Selector cable
   - Transmission oil cooler lines
   - Crossmember. Tighten the bolts to 85 ft. lbs. (116 Nm).
   - Transmission mount. Tighten the bolts to 37 ft. lbs. (50 Nm).
   - Transmission harness connectors
   - Wiring harness heat shield

Timing belt service is covered in Section 3 of this manual

- Fuel line bracket
- Center exhaust pipe
- Driveshafts. Tighten the flange bolts to 46 ft. lbs. (63 Nm).
- Skid plate
- Negative battery cable

Clutch

ADJUSTMENTS

➡ **This vehicle is equipped with a hydraulic clutch linkage. No adjustment is necessary.**

REMOVAL & INSTALLATION

1. Before servicing the vehicle, refer to the precautions in the beginning of this section.
2. Remove the transmission.
3. Loosen the pressure plate mounting bolts in a 2-step crisscross sequence until the spring tension is relieved.
4. Remove the pressure plate and the clutch disc.
   To install:
5. Install a new wedge collar and wire snapring into the pressure plate.

6. Using a clutch alignment tool, assemble the clutch disc and pressure plate onto the flywheel.
7. Tighten the pressure plate bolts in sequence and in two passes to 13 ft. lbs. (8 Nm).
8. Install the transmission.
9. Road test the vehicle and check for proper clutch operation.

Hydraulic Clutch System

BLEEDING

1. Before servicing the vehicle, refer to the precautions in the beginning of this section.
2. Have an assistant pump the clutch pedal slowly several times and hold it depressed.
3. Open the slave cylinder bleeder screw and allow air to escape.
4. Close the bleeder screw before releasing the clutch pedal.
5. Repeat until all air is purged from the clutch hydraulic system.
6. Refill the reservoir to the full mark.

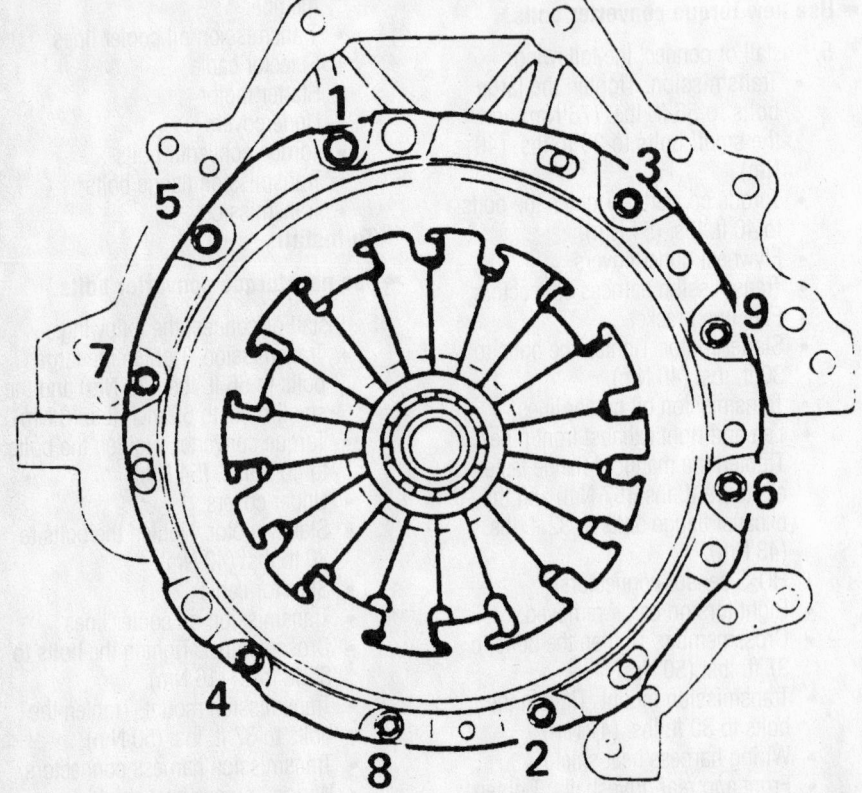

7924NG49

Pressure plate tightening sequence

Transfer Case Assembly

REMOVAL & INSTALLATION

Except Axiom

1. Before servicing the vehicle, refer to the precautions in the beginning of this section.
2. Remove or disconnect the following:
   - Negative battery cable
   - Transfer case skid plate
   - Front and rear driveshafts
   - Heated Oxygen (HO2S) sensor connectors
   - Left and right exhaust front pipes
   - Transfer case control lever knob
   - Selector lever assembly
   - Transfer case control lever
   - Vehicle Speed (VSS) sensor connector
   - 4 wheel drive switch connector
   - 4 wheel drive actuator connector
   - Transfer case flange fasteners
   - Transfer case

To install:
3. Install or connect the following:
   - Transfer case. Tighten the flange fasteners to 34 ft. lbs. (46 Nm).
   - 4 wheel drive actuator connector
   - 4 wheel drive switch connector
   - VSS sensor connector
   - Transfer case control lever
   - Selector lever assembly
   - Transfer case control lever knob
   - Left and right exhaust front pipes
   - HO2S sensor connectors
   - Front and rear driveshafts. Tighten the bolts to 46 ft. lbs. (63 Nm).
   - Transfer case skid plate. Tighten the bolts to 27 ft. lbs. (37 Nm).
   - Negative battery cable

Axiom

1. Before servicing the vehicle, refer to the precautions in the beginning of this section.
2. Remove or disconnect the following:
   - Negative battery cable
   - Transfer case skid plate
   - Front and rear driveshafts
   - Breather hose
   - Center exhaust pipe
   - Vehicle Speed (VSS) sensor connector
   - Harness connector
3. Support the transfer case
4. Remove or disconnect the following:
   - Transfer case flange fasteners
   - Transfer case

To install:

5. Install or connect the following:
- Transfer case. Tighten the flange fasteners to 34 ft. lbs. (46 Nm).
- Harness connector
- Breather hose
- VSS sensor connector
- Center exhaust pipe
- Front and rear driveshafts. Tighten the bolts to 46 ft. lbs. (63 Nm).

- Transfer case skid plate. Tighten the bolts to 27 ft. lbs. (37 Nm).
- Negative battery cable

Halfshaft

REMOVAL & INSTALLATION

1. Before servicing the vehicle, refer to the precautions in the beginning of this section.

2. Remove or disconnect the following:
- Negative battery cable
- Front wheel
- Radiator skid plate
- Transfer case skid plate
- Brake calipers and mounting bracket
- Brake rotor
- Wheel speed sensor
- Steering knuckle

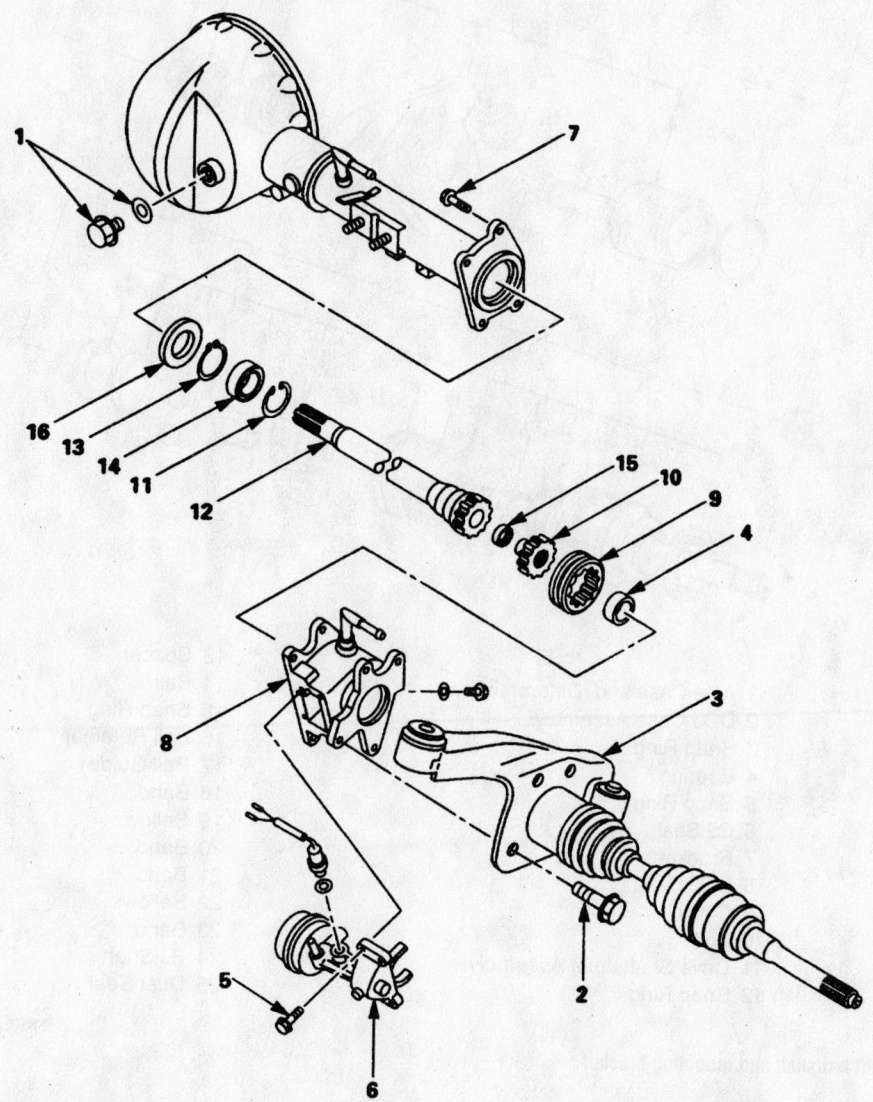

1. Filler plug
2. Bolt
3. Front axle drive shaft (LH side)
4. Spacer
5. Bolt
6. Actuator assembly
7. Bolt
8. Housing
9. Sleeve
10. Clutch gear
11. Snap ring
12. Inner shaft
13. Snap ring
14. Inner shaft bearing
15. Needle bearing
16. Oil seal

7924BG26

Exploded view of the left halfshaft, axle shaft and axle disconnect

Heater Core replacement is covered in Section 2 of this manual

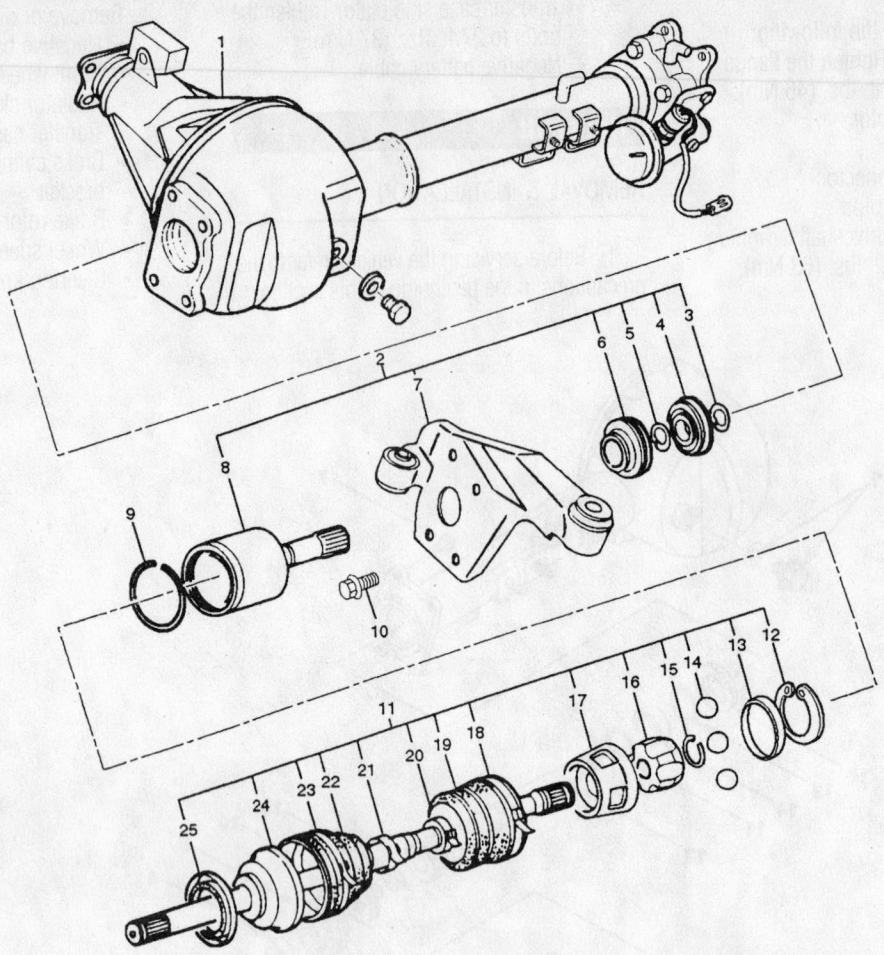

1 Axle Case and Differential
2 DOJ Case Assembly
3 Snap Ring
4 Bearing
5 Snap Ring
6 Oil Seal
7 Bracket
8 DOJ Case
9 Circlip
10 Bolt
11 Drive Shaft Joint Assembly
12 Snap Ring

13 Spacer
14 Ball
15 Snap Ring
16 Ball Retainer
17 Ball Guide
18 Band
19 Bellows
20 Band
21 Band
22 Bellows
23 Band
24 BJ Shaft
25 Dust Seal

9308BG03

Exploded view of the right halfshaft and mounting bracket

3. Support the axle housing with a jack. Unbolt the axle mounting bracket and remove the halfshaft/bracket assembly.

To install:

4. Install or connect the following:
- Axle/bracket assembly. Tighten the bracket flange bolts to 85 ft. lbs. (116 Nm) and the bracket mounting bolts to 112 ft. lbs. (152 Nm).
- Steering knuckle
- Wheel speed sensor
- Brake rotor

- Brake caliper and mounting bracket. Tighten the bracket bolts to 115 ft. lbs. (155 Nm).
- Transfer case skid plate. Tighten the bolts to 27 ft. lbs. (37 Nm).
- Radiator skid plate. Tighten the bolts to 58 ft. lbs. (78 Nm).
- Front wheel
- Negative battery cable

5. Check the wheel alignment and adjust as necessary.

CV-Joints

OVERHAUL

Outer CV-Joint

The outer CV-joint is serviced with the axle shaft as an assembly. The outer CV-joint boot can be serviced by removing the inner CV-joint.

Inner CV-Joint

1. Before servicing the vehicle, refer to the precautions in the beginning of this section.
2. Remove or disconnect the following:
   - Halfshaft from the vehicle
   - Snapring and bearing
   - Snapring and oil seal
   - Mounting bracket
   - CV-joint boot
   - Circlip and inner joint housing
   - Snapring and spacer
   - Inner joint balls
   - Snapring and inner CV-joint

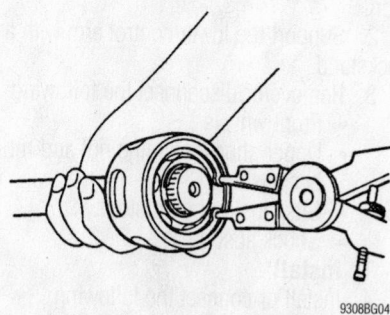

9308BG04

CV-joint spacer snapring—Inner CV-Joint

To install:

3. Install or connect the following:
   - Inner CV-joint and snapring
   - Inner joint balls
   - Spacer and snapring
   - Inner joint housing and circlip. Add 150 grams CV-joint grease.
   - CV-joint boot
   - Mounting bracket
   - Oil seal and snapring
   - Bearing and snapring
4. Install the halfshaft and mounting bracket to the vehicle.
5. Check the wheel alignment and adjust as necessary.

Rear Axle Shaft, Bearing and Seal

REMOVAL & INSTALLATION

1. Before servicing the vehicle, refer to the precautions in the beginning of this section.
2. Remove or disconnect the following:
   - Rear wheel
   - Disc brake caliper and bracket
   - Disc brake rotor
   - Wheel speed sensor bracket
   - Parking brake cable and bracket
   - Parking brake shoes
   - Axle shaft
   - Snapring and discard it
   - Bearing, press it off the axle shaft with the bearing holder and oil seal

To install:

3. Install or connect the following:
   - New oil seal into the bearing housing
   - Bearing housing onto the axle shaft
   - Bearing, press it onto the axle shaft
   - New snapring
   - Axle shaft. Use new lockwashers and tighten the bearing holder nuts to 54 ft. lbs. (74 Nm).
   - Parking brake shoes
   - Parking brake cable and bracket
   - Wheel speed sensor bracket
   - Disc brake rotor
   - Disc brake caliper and bracket. Tighten the bracket bolts to 76 ft. lbs. (103 Nm).
   - Rear wheel
4. Check the rear axle oil level and adjust as necessary.

Pinion Seal

REMOVAL & INSTALLATION

1. Before servicing the vehicle, refer to the precautions in the beginning of this section.

2. Remove or disconnect the following:
   - Driveshaft
   - Wheels
   - Brake calipers and pads

➡**The brake calipers and pads must be removed so that there is no additional drag when measuring pinion bearing preload.**

3. Use an inch lb. torque wrench and measure and record the amount of torque required to maintain pinion rotation through several revolutions.
4. Remove or disconnect the following:
   - Pinion flange
   - Pinion seal
   - Pinion bearing
   - Collapsible spacer

To install:

➡**Use a new collapsible spacer and flange nut for assembly.**

5. Install or connect the following:
   - Collapsible spacer
   - Pinion bearing
   - Pinion seal
   - Pinion flange
6. Rotate the pinion flange occasionally while tightening the flange nut to make sure the pinion bearings seat correctly.
7. Take frequent bearing preload torque readings. Tighten the flange nut to achieve the preload torque readings originally recorded.

✳✳ CAUTION

Never loosen the pinion nut to reduce bearing preload. If it is necessary to reduce bearing preload, install a new collapsible spacer and pinion nut.

8. Install or connect the following:
   - Driveshaft
   - Brake calipers and pads
   - Wheels
9. Fill the differential with gear lubricant and check for leaks.

STEERING AND SUSPENSION

Air Bag

✳✳ CAUTION

Some vehicles are equipped with an air bag system. The system must be disarmed before performing service on, or around, system components, the steering column, instrument panel components, wiring and sensors. Failure to follow the safety precautions and the disarming procedure could result in accidental air bag deployment, possible injury and unnecessary system repairs.

PRECAUTIONS

Several precautions must be observed when handling the inflator module to avoid accidental deployment and possible personal injury.

- Never carry the inflator module by the wires or connector on the underside of the module.

Brake service is covered in Section 4 of this manual

• When carrying a live inflator module, hold securely with both hands, and ensure that the bag and trim cover are pointed away from you.

• Place the inflator module on a bench or other surface with the bag and trim cover facing up.

• With the inflator module on the bench, never place anything on or close to the module which may be thrown in the event of an accidental deployment.

DISARMING

1. Before servicing the vehicle, refer to the precautions in the beginning of this section.

2. Turn the ignition switch to the **LOCK** position. Remove the key.

3. Disconnect the negative battery cable. Wait 1 minute before working around the air bags.

4. Disconnect the yellow 2-pin connector at the base of the steering column.

5. Disconnect the yellow 2-pin connector behind the glove box assembly.

6. When repairs are completed, connect the yellow 2-pin connectors.

7. Connect the negative battery cable.

8. Turn the ignition to the **ON** position, but don't start the engine. The AIR BAG warning light should turn on and flash on and off 7 times, and then turn off. This light sequence indicates that the SRS system is functioning normally. If the AIR BAG light doesn't come on, or stays on longer than 7 seconds, the system must be diagnosed.

Recirculating Ball Steering Gear

REMOVAL & INSTALLATION

1. Before servicing the vehicle, refer to the precautions in the beginning of this section.

2. Disable the air bag system.

3. Remove or disconnect the following:
   • Skid plates
   • Lower fan shroud
   • Stabilizer bar
   • Power steering pressure and return lines
   • Pitman arm
   • Steering column intermediate shaft
   • Steering gear

To install:

4. Install or connect the following:
   • Steering gear. Tighten the bolts to 33 ft. lbs. (44 Nm).
   • Steering column intermediate shaft. Tighten the pinch bolt to 18 ft. lbs. (25 Nm).

• Pitman arm. Tighten the nut to 159 ft. lbs. (216 Nm).
   • Power steering pressure and return lines. Tighten the fittings to 33 ft. lbs. (44 Nm).
   • Stabilizer bar
   • Lower fan shroud
   • Skid plates

5. Fill the power steering fluid reservoir.

6. Check the wheel alignment and adjust as necessary.

Rack and Pinion Steering Gear

REMOVAL & INSTALLATION

2-Wheel Drive

1. Before servicing the vehicle, refer to the precautions in the beginning of this section.

2. Disable the air bag system.

3. Remove or disconnect the following:
   • Stone guard
   • Tie rod from knuckle
   • Power steering pressure and return lines
   • Steering gear

To install:

4. Install or connect the following:
   • Steering gear. Torque the mounting bolts to 85 ft. lbs. (116 Nm)
   • Fluid lines. Torque the fitting to 18 ft. lbs. (25 Nm)
   • Tie rod ends. 87 ft. lbs. (118 Nm)
   • Stone guard

5. Fill and bleed the system.

4-Wheel Drive

1. Before servicing the vehicle, refer to the precautions in the beginning of this section.

2. Disable the air bag system.

3. Remove or disconnect the following:
   • Stone guard
   • Tie rod from knuckle
   • Power steering pressure and return lines
   • Torsion bar
   • Lower control arm frame side bolt
   • Crossmember bolts
   • Steering gear with crossmember
   • Steering gear

To install:

4. Install or connect the following:
   • Steering gear. Torque the mounting bolts to 85 ft. lbs. (116 Nm)
   • Crossmember Torque the bolts to 128 ft. lbs. (173 Nm)
   • Lower control arm bolt

• Torsion bar
   • Fluid lines. Torque the fitting to 18 ft. lbs. (25 Nm)
   • Tie rod ends. 87 ft. lbs. (118 Nm)
   • Stone guard

5. Fill and bleed the system.

Shock Absorber

REMOVAL & INSTALLATION

Front

EXCEPT AXIOM

1. Before servicing the vehicle, refer to the precautions in the beginning of this section.

2. Support the lower control arm with a jackstand.

3. Remove or disconnect the following:
   • Front wheels
   • Upper shock retaining nut and rubber bushing
   • Suspension bump stops
   • Shock absorber

To install:

4. Install or connect the following:
   • Shock absorber. Tighten the lower bolt to 60–61 ft. lbs. (82–84 Nm).
   • Bump stop. Tighten the bolts to 30 ft. lbs. (41 Nm).
   • Upper shock retaining nut and rubber bushing. Tighten the nut to 14–15 ft. lbs. (19–20 Nm).
   • Front wheels

AXIOM

1. Before servicing the vehicle, refer to the precautions in the beginning of this section.

2. Support the lower control arm with a jackstand.

3. Remove or disconnect the following:
   • Front wheels
   • Upper shock retaining nut and rubber bushing
   • ISC actuator and bracket
   • Suspension bump stops
   • Shock absorber

To install:

4. Install or connect the following:
   • Shock absorber. Tighten the lower bolt to 69 ft. lbs. (93 Nm).
   • ISC actuator and bracket
   • Bump stop. Tighten the bolts to 30 ft. lbs. (41 Nm).
   • Upper shock retaining nut and rubber bushing. Tighten the nut to 14 ft. lbs. (20 Nm).
   • Front wheels

Rear

EXCEPT AXIOM

1. Before servicing the vehicle, refer to the precautions in the beginning of this section.
2. Support the rear axle with jackstands.
3. Remove the rear shock absorbers.

To install:

4. Install the rear shock absorbers. Tighten the upper bolt to 70 ft. lbs. (95 Nm). Tighten the lower bolt to 58 ft. lbs. (78 Nm).
5. Remove the jackstands.

AXIOM

1. Before servicing the vehicle, refer to the precautions in the beginning of this section.
2. Support the rear axle with jackstands.
3. Remove the rear shock absorbers.

To install:

4. Install the rear shock absorbers. Tighten the upper bolt to 14 ft. lbs. (20 Nm). Tighten the lower bolt to 58 ft. lbs. (78 Nm).

➡ **With ISC, do not use any grease on or near the bushings.**

5. Remove the jackstands.

Coil Spring

REMOVAL & INSTALLATION

Except Axiom

1. Before servicing the vehicle, refer to the precautions in the beginning of this section.
2. Support the vehicle under the frame.
3. Support the rear axle with a jack.
4. Remove or disconnect the following:
   • Rear wheels
   • Stabilizer bar links
   • Parking brake cable brackets
   • Shock absorbers
5. Lower the rear axle with the jack to release the coil spring tension. Remove the coil springs and insulators.

To install:

6. Place the coil springs on the axle assembly and the insulators on top of the springs.
7. Raise the axle assembly into position.
8. Install or connect the following:
   • Shock absorbers
   • Parking brake cable brackets
   • Stabilizer bar links. Tighten the nuts to 37 ft. lbs. (50 Nm).
   • Rear wheels

Axiom

1. Before servicing the vehicle, refer to the precautions in the beginning of this section.
2. Support the vehicle under the frame.
3. Support the rear axle with a jack.
4. Remove or disconnect the following:
   • Rear wheels
   • Breather hose
   • Upper link fixing bolt, left side only
   • Stabilizer bar links
   • Shock absorbers
5. Lower the rear axle with the jack to release the coil spring tension. Remove the coil springs and insulators.

To install:

6. Place the coil springs on the axle assembly and the insulators on top of the springs.
7. Raise the axle assembly into position.
8. Install or connect the following:
   • Shock absorbers. Torque to 58 ft. lbs. (78 Nm)
   • Stabilizer bar links. Tighten the nuts to 23 ft. lbs. (31 Nm).
   • Upper link. Torque to 101 ft. lbs. (137 Nm)
   • Rear wheels

Torsion Bar

REMOVAL & INSTALLATION

1. Before servicing the vehicle, refer to the precautions in the beginning of this section.

2. Matchmark the adjusting bolt and end piece, then remove the bolt, end piece, and seat.
3. Matchmark the height control arm to the torsion bar, then remove the height control arm.
4. Matchmark the torsion bar to the lower control arm, then remove the torsion bar.

To install:

5. Apply grease to the torsion bar splines.
6. Apply grease to the contact points of the height control arm, adjusting bolt end piece and seat.
7. Align the matchmarks and install the torsion bar.
8. Align the matchmarks and install the height control arm.
9. Install the adjusting bolt, seat and end piece.
10. Tighten the adjusting bolt to align the matchmarks.

Upper Ball Joint

REMOVAL & INSTALLATION

1. Before servicing the vehicle, refer to the precautions in the beginning of this section.
2. Support the lower control arm with a floor jack.
3. Remove or disconnect the following:
   • Front wheel
   • Wheel speed sensor
   • Upper ball joint

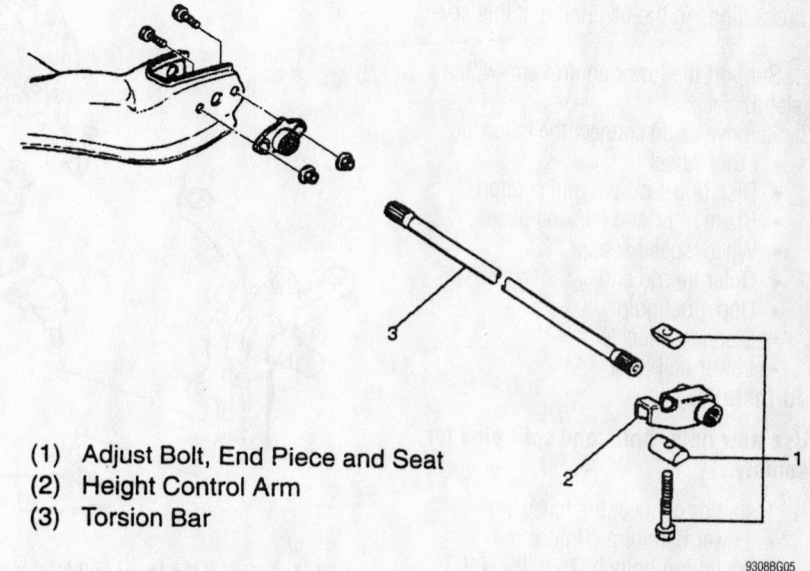

(1) Adjust Bolt, End Piece and Seat
(2) Height Control Arm
(3) Torsion Bar

9308BG05

Exploded view of the torsion bar assembly

For complete Engine Mechanical specifications, see Section 1 of this manual

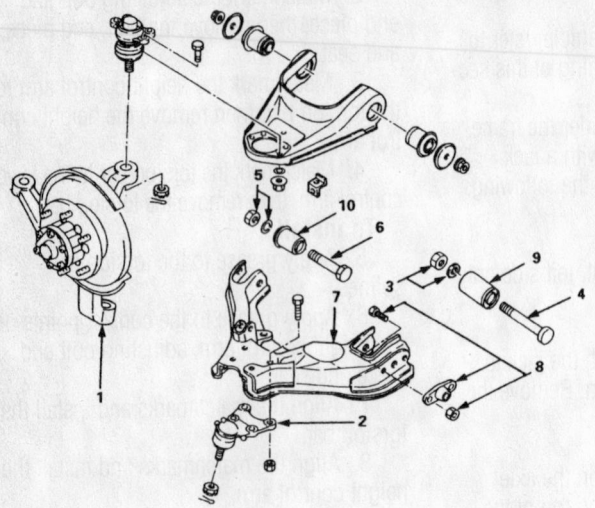

1. Knuckle
2. Lower end
3. Nut and washer, rear
4. Bolt, rear
5. Nut and washer, front
6. Bolt, front
7. Lower control arm assembly
8. Torsion bar arm bracket
9. Bushing, rear
10. Bushing, front

7924BG34

Exploded view of the control arm and ball joint components

To install:

➡ **Use new nuts, bolts and split pins for assembly.**

4. Install or connect the following:
   • Upper ball joint. Tighten the mounting bolts to 42 ft. lbs. (57 Nm) and the nut to 72 ft. lbs. (96 Nm).
   • Wheel speed sensor
   • Front wheel

Lower Ball Joint

REMOVAL & INSTALLATION

1. Before servicing the vehicle, refer to the precautions in the beginning of this section.
2. Support the lower control arm with a jackstand.
3. Remove or disconnect the following:
   • Front wheel
   • Disc brake caliper and support
   • Brake rotor and backing plate
   • Wheel speed sensor
   • Outer tie rod end
   • Upper ball joint
   • Steering knuckle
   • Lower ball joint

To install:

➡ **Use new nuts, bolts and split pins for assembly.**

4. Install or connect the following:
   • Lower ball joint. Tighten the mounting bolts to 76 ft. lbs. (103

Nm) on 1999–01 models; 85 ft. lbs. (116 Nm) on 2002 models.
   • Steering knuckle. Tighten the lower ball joint nut to 108 ft. lbs. (147 Nm).
   • Upper ball joint. Tighten the nut to 72 ft. lbs. (96 Nm).
   • Outer tie rod end. Tighten the nut to 72 ft. lbs. (98 Nm).
   • Wheel speed sensor
   • Brake rotor and backing plate
   • Disc brake caliper and support. Tighten the support bolts to 115 ft. lbs. (155 Nm).
   • Front wheel

5. Check the wheel alignment and adjust as necessary.

Upper Control Arm

REMOVAL & INSTALLATION

Except Axiom

1. Before servicing the vehicle, refer to the precautions in the beginning of this section.
2. Support the lower control arm with a jackstand.
3. Remove or disconnect the following:
   • Front wheel
   • Wheel speed sensor
   • Brake caliper
   • Upper ball joint
   • Upper control arm

➡ **Note the alignment shim location for assembly.**

To install:

4. Install or connect the following:
   • Upper control arm
   • Alignment shims in their original locations. Tighten the bolts to 112 ft. lbs. (152 Nm).
   • Upper ball joint. Tighten the nut to 72 ft. lbs. (98 Nm).
   • Brake caliper
   • Wheel speed sensor
   • Front wheel

5. Check the wheel alignment and adjust as necessary.

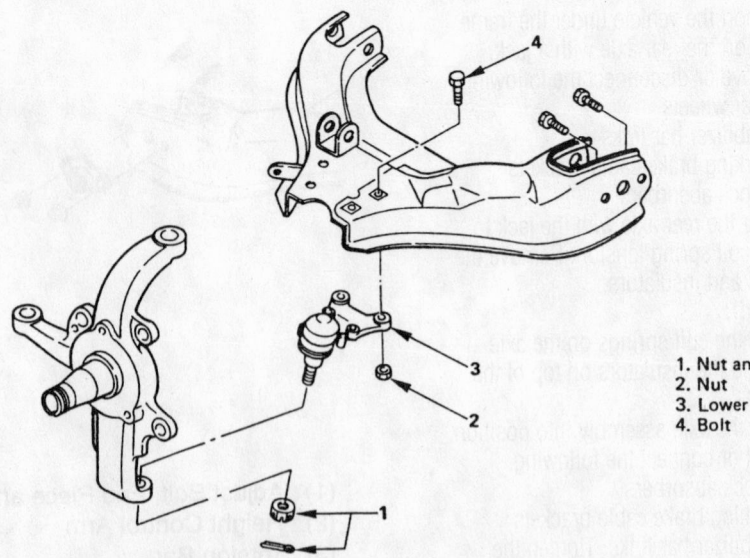

1. Nut and cotter pin
2. Nut
3. Lower ball joint
4. Bolt

7924BG35

Exploded view of the lower ball joint mounting and related components

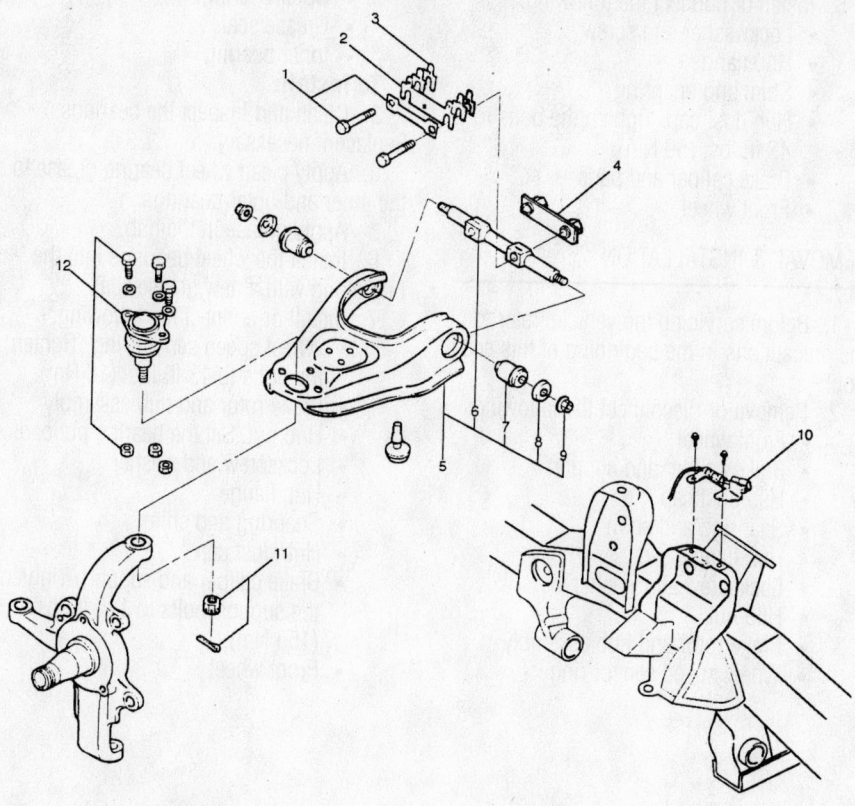

1	Bolt and Plate
2	Camber Shims
3	Caster Shims
4	Nut Assembly
5	Upper Control Arm Assembly
6	Fulcrum Pin

7	Bushing
8	Plate
9	Nut
10	Speed Sensor Cable
11	Nut and Cotter Pin
12	Upper Ball Joint

9308BG06

Upper control arm and related parts

Axiom

1. Before servicing the vehicle, refer to the precautions in the beginning of this section.

2. Support the lower control arm with a jackstand.

3. Remove or disconnect the following:
- Front wheel
- Wheel speed sensor
- Brake caliper
- Upper ball joint
- Upper control arm

➡**Note the alignment shim location for assembly.**

To install:

4. Install or connect the following:
- Upper control arm
- Alignment shims in their original locations. Tighten the bolts to 112 ft. lbs. (152 Nm).

- Upper ball joint. Tighten the nut to 72 ft. lbs. (98 Nm).
- Brake caliper
- Wheel speed sensor
- Front wheel

5. Check the wheel alignment and adjust as necessary.

CONTROL ARM BUSHING REPLACEMENT

1. Before servicing the vehicle, refer to the precautions in the beginning of this section.

2. Remove the upper control arm.

3. Remove the nuts and washers from the fulcrum pin.

4. Press the bushings out of the control arm.

To install:

5. Press the bushings into the control arm.

6. Install the fulcrum pin washers and nuts.

7. Install the upper control arm.

8. Raise the suspension so that there is 0.79 inches (20mm) between the bump stop and the lower control arm. Tighten the fulcrum pin nuts to 80 ft. lbs. (108 Nm).

9. Check the wheel alignment and adjust as necessary.

Lower Control Arm

REMOVAL & INSTALLATION

1. Before servicing the vehicle, refer to the precautions in the beginning of this section.

2. Remove or disconnect the following:
- Front wheel
- Torsion bar
- Lower ball joint
- Stabilizer bar link
- Shock absorber
- Lower control arm

To install:

3. Install or connect the following:
- Lower control arm
- Shock absorber
- Stabilizer bar link
- Lower ball joint
- Torsion bar
- Front wheel

4. Raise the suspension so that there is 0.79 inches (20mm) between the bump stop and the lower control arm. For 1999–01, tighten the front control arm bolt to 116 ft. lbs. (157 Nm) and tighten the rear control arm bolt to 145 ft. lbs. (196 Nm). For 2002, tighten the rear nut to 174 ft. lbs. (235 Nm); the rear nut to 137 ft. lbs. (186 Nm).

5. Check the wheel alignment and adjust as necessary.

CONTROL ARM BUSHING REPLACEMENT

1. Before servicing the vehicle, refer to the precautions in the beginning of this section.

2. Remove or disconnect the following:
- Lower control arm
- Bushings, press them from the control arm, using Remover/Installer J-36833 for the front bushing and Remover/Installer J-36834 for the rear bushing.

To install:

3. Install or connect the following:
- New bushings
- Lower control arm

For Accessory Drive Belt illustrations, see Section 1 of this manual

Wheel Bearings

ADJUSTMENT

2- and 4-Wheel Drive

1. Before servicing the vehicle, refer to the precautions in the beginning of this section.

2. Remove or disconnect the following:
- Front wheel
- Brake caliper and pads
- Hub dust cap
- Snapring and shim
- Hub flange
- Lockscrew and washer

3. Tighten the hub nut to 22 ft. lbs. (29 Nm) to seat the bearings and then fully loosen the nut.

4. Tighten the hub nut to achieve a bearing preload of 2.6–4.0 lbs. (1.2–1.8 Kg) for used bearings. If the bearings were replaced, set the preload to 4.4–5.5 lbs. (2.0–2.5 Kg).

5. Install or connect the following:
- Lockwasher and screw
- Hub flange
- Shim and snapring
- Hub dust cap. Tighten the bolts to 43 ft. lbs. (59 Nm).
- Brake caliper and pads
- Front wheel

REMOVAL & INSTALLATION

1. Before servicing the vehicle, refer to the precautions in the beginning of this section.

2. Remove or disconnect the following:
- Front wheel
- Brake caliper and support
- Hub dust cap
- Snapring and shim
- Hub flange
- Lockscrew and washer
- Hub nut
- Brake rotor and hub assembly
- Wheel speed sensor ring
- Outer bearing
- Grease seal
- Inner bearing

To install:

3. Clean and inspect the bearings. Replace if necessary.

4. Apply clean wheel bearing grease to the inner and outer bearings.

5. Apply grease in the hub.

6. Install the wheel bearings into the hub along with a new grease seal.

7. Install or connect the following:
- Wheel speed sensor ring. Tighten the bolts to 13 ft. lbs. (18 Nm).
- Brake rotor and hub assembly
- Hub nut. Set the bearing preload.
- Lockscrew and washer
- Hub flange
- Snapring and shim
- Hub dust cap
- Brake caliper and support. Tighten the support bolts to 115 ft. lbs. (155 Nm).
- Front wheel

JEEP

Cherokee • Grand Cherokee • Liberty • Wrangler

18

PRECAUTIONS

Before servicing any vehicle, please be sure to read all of the following precautions, which deal with personal safety, prevention of component damage, and important points to take into consideration when servicing a motor vehicle:

• Never open, service or drain the radiator or cooling system when the engine is hot; serious burns can occur from the steam and hot coolant.

• Observe all applicable safety precautions when working around fuel. Whenever servicing the fuel system, always work in a well-ventilated area. Do not allow fuel spray or vapors to come in contact with a spark, open flame, or excessive heat (a hot drop light, for example). Keep a dry chemical fire extinguisher near the work area. Always keep fuel in a container specifically designed for fuel storage; also, always properly seal fuel containers to avoid the possibility of fire or explosion. Refer to the additional fuel system precautions later in this section.

• Fuel injection systems often remain pressurized, even after the engine has been turned **OFF**. The fuel system pressure must be relieved before disconnecting any fuel lines. Failure to do so may result in fire and/or personal injury.

• Brake fluid often contains polyglycol ethers and polyglycols. Avoid contact with the eyes and wash your hands thoroughly after handling brake fluid. If you do get brake fluid in your eyes, flush your eyes with clean, running water for 15 minutes. If eye irritation persists, or if you have taken brake fluid internally, IMMEDIATELY seek medical assistance.

• The EPA warns that prolonged contact with used engine oil may cause a number of skin disorders, including cancer. You should make every effort to minimize your exposure to used engine oil. Protective gloves should be worn when changing oil. Wash your hands and any other exposed skin areas as soon as possible after exposure to used engine oil. Soap and water, or waterless hand cleaner should be used.

• All new vehicles are now equipped with an air bag system, often referred to as a Supplemental Restraint System (SRS) or Supplemental Inflatable Restraint (SIR) system. The system must be disabled before performing service on or around system components, steering column, instrument panel components, wiring and sensors. Failure to follow safety and disabling procedures could result in accidental air bag deployment, possible personal injury and unnecessary system repairs.

• Always wear safety goggles when working with, or around, the air bag system. When carrying a non-deployed air bag, be sure the bag and trim cover are pointed away from your body. When placing a non-deployed air bag on a work surface, always face the bag and trim cover upward, away from the surface. This will reduce the motion of the module if it is accidentally deployed. Refer to the additional air bag system precautions later in this section.

• Clean, high quality brake fluid from a sealed container is essential to the safe and proper operation of the brake system. You should always buy the correct type of brake fluid for your vehicle. If the brake fluid becomes contaminated, completely flush the system with new fluid. Never reuse any brake fluid. Any brake fluid that is removed from the system should be discarded. Also, do not allow any brake fluid to come in contact with a painted surface; it will damage the paint.

• Never operate the engine without the proper amount and type of engine oil; doing so WILL result in severe engine damage.

• Timing belt maintenance is extremely important. Many models utilize an interference-type, non-freewheeling engine. If the timing belt breaks, the valves in the cylinder head may strike the pistons, causing potentially serious (also time-consuming and expensive) engine damage. Refer to the maintenance interval charts in the front of this manual for the recommended replacement interval for the timing belt, and to the timing belt section for belt replacement and inspection.

• Disconnecting the negative battery cable on some vehicles may interfere with the functions of the on-board computer system(s) and may require the computer to undergo a relearning process once the negative battery cable is reconnected.

• When servicing drum brakes, only disassemble and assemble one side at a time, leaving the remaining side intact for reference.

• Only an MVAC-trained, EPA-certified automotive technician should service the air conditioning system or its components.

ENGINE REPAIR

➡Disconnecting the negative battery cable on some vehicles may interfere with the functions of the on-board computer system. The computer may undergo a relearning process once the negative battery cable is reconnected.

Distributor

➡The 3.7L and 4.7L engines do not use a distributor.

REMOVAL

2.5L and 4.0L Engines

1. Before servicing the vehicle, refer to the precautions in the beginning of this section.
2. Remove or disconnect the following:

• Negative battery cable
• Distributor cap
• Camshaft Position (CMP) sensor connector
• Distributor wiring harness from the main engine harness
• Cylinder No. 1 spark plug
3. Matchmark the distributor housing and the rotor.
4. Remove the distributor.

5.2L and 5.9L Engines

1. Before servicing the vehicle, refer to the precautions in the beginning of this section.
2. Remove or disconnect the following:

• Negative battery cable
• Air cleaner tube
• Distributor cap
• Camshaft Position (CMP) sensor connector
3. Matchmark the distributor housing and the rotor.
4. Matchmark the distributor housing and the intake manifold.
5. Remove the distributor.

INSTALLATION

Timing Not Disturbed

2.5L AND 4.0L ENGINES

1. Before servicing the vehicle, refer to the precautions in the beginning of this section.

➡The rotor will rotate clockwise as the gears engage.

2. Position the rotor slightly counterclockwise of the matchmark made during removal.

3. Install the distributor. Ensure that the rotor moves into alignment with the matchmark.

4. Align the locating fork with the clamp bolt hole. Install the clamp and bolt. Tighten the bolt to 17 ft. lbs. (23 Nm).

5. Install or connect the following:
- Cylinder No. 1 spark plug
- Distributor wiring harness to the main engine harness
- CMP sensor connector
- Distributor cap
- Air cleaner tube
- Negative battery cable

5.2L AND 5.9L ENGINES

1. Before servicing the vehicle, refer to the precautions in the beginning of this section.

➡ **The rotor will rotate clockwise as the gears engage.**

2. Position the rotor slightly counterclockwise of the matchmark made during removal.

3. Install the distributor.

4. Align the distributor housing and intake manifold matchmarks and check that the distributor housing matchmark and rotor are also aligned.

5. Install or connect the following:
- Distributor housing clamp and bolt. Tighten the bolt to 17 ft. lbs. (23 Nm).
- CMP sensor connector
- Distributor cap
- Air cleaner tube
- Negative battery cable

Timing Disturbed

2.5L AND 4.0L ENGINES

1. Before servicing the vehicle, refer to the precautions in the beginning of this section.

2. Set the engine at Top Dead Center (TDC) of the No. 1 cylinder compression stroke.

3. Position the slot in the oil pump drive gear as shown.

4. Locate the alignment holes in the plastic ring and align the correct hole with the mating hole in the distributor housing as shown. Install a locking pin.

➡ **The distributor will rotate clockwise as the gears engage.**

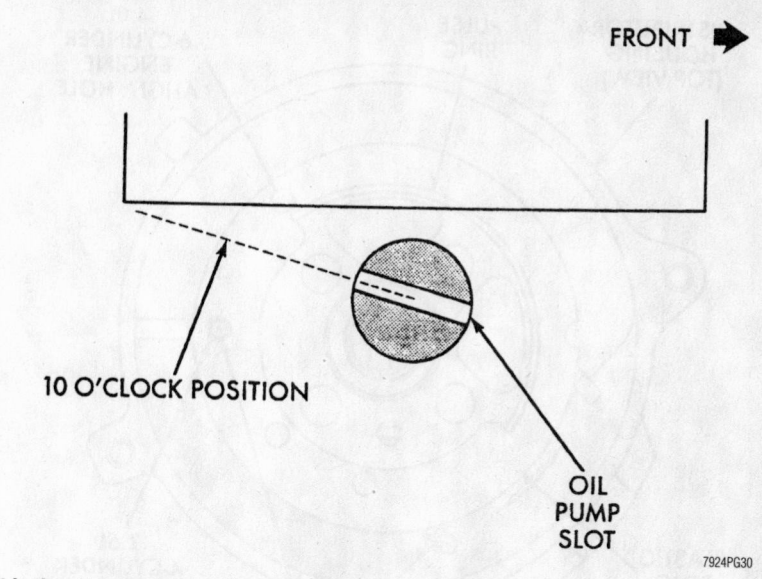

Slot in the oil pump gear at 10 o'clock position—2.5L engine

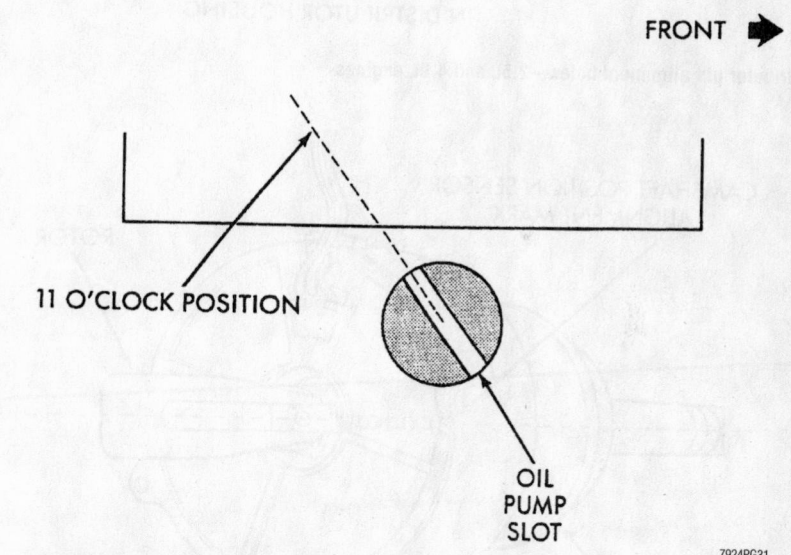

Slot in the oil pump gear at 11 o'clock position—4.0L engine

5. Position the base mounting slot at the 1 o'clock position and install the distributor.

6. Check that the centerline of the mounting slot aligns with the centerline of the clamp bolt hole.

7. Install the clamp and bolt. Tighten the bolt to 17 ft. lbs. (23 Nm).

8. Remove the locking pin.

9. Install or connect the following:
- CMP sensor connector
- Distributor cap
- Air cleaner tube
- Negative battery cable

5.2L AND 5.9L ENGINES

1. Before servicing the vehicle, refer to the precautions in the beginning of this section.

2. Set the engine at Top Dead Center (TDC) of the No. 1 cylinder compression stroke.

3. Install the distributor, clamp and bolt.

4. Rotate the distributor so that the rotor aligns with the No. 1 cylinder mark on the Camshaft Position (CMP) sensor. Tighten the clamp bolt to 17 ft. lbs. (23 Nm).

5. Install or connect the following:
- CMP sensor connector
- Distributor cap

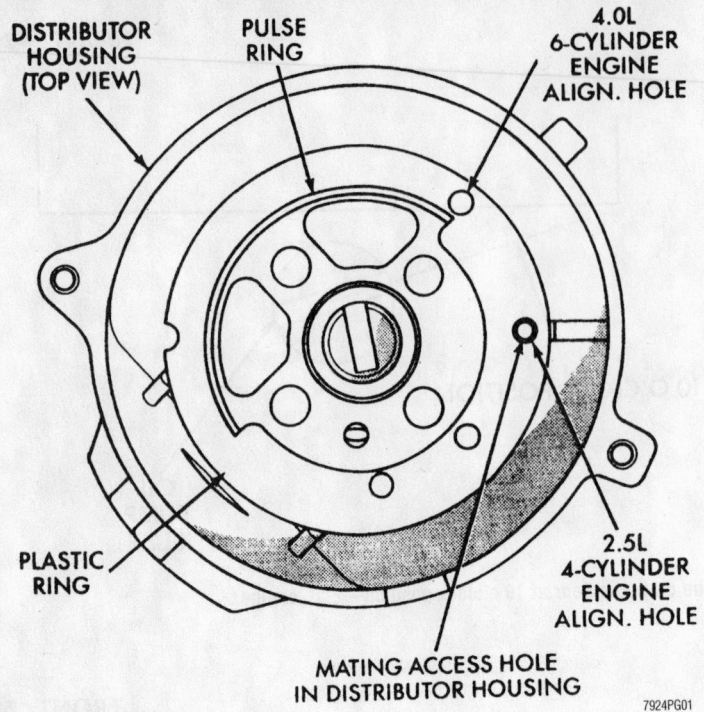

Distributor pin alignment holes—2.5L and 4.0L engines

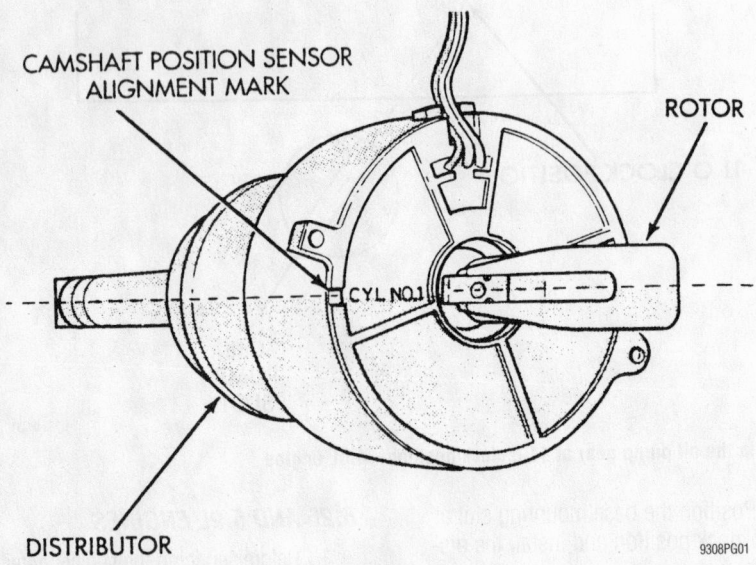

Distributor rotor alignment—5.2L and 5.9L engines

- Air cleaner tube
- Negative battery cable

6. The final distributor position must be set with a scan tool. Follow the instructions supplied by the scan tool manufacturer.

Alternator

REMOVAL

1. Before servicing the vehicle, refer to the precautions in the beginning of this section.

2. Remove or disconnect the following:
- Negative battery cable
- Accessory drive belt
- Alternator harness connectors
- Mounting bolts and alternator

➡ **The 3.7L engine has 1 vertical and 2 horizontal bolts.**

INSTALLATION

1. Before servicing the vehicle, refer to the precautions in the beginning of this section.

2. Install the alternator and tighten the bolts to the following specifications:
- 2.4L, 2.5L and 4.0L engines: 41 ft. lbs. (55 Nm).
- 3.7L engine: Tighten the horizontal bolts to 42 ft. lbs. (57 Nm), then the vertical bolt to 29 ft. lbs. (40 Nm)
- 4.7L engine: Vertical bolt and long horizontal bolt to 41 ft. lbs. (56 Nm), short horizontal bolt to 55 ft. lbs. (74 Nm)
- 5.2L and 5.9L engines: 30 ft. lbs. (41 Nm)

3. Install or connect the following:
- Alternator harness connectors
- Accessory drive belt
- Negative battery cable

Ignition Timing

ADJUSTMENT

Ignition timing is controlled by the Powertrain Control Module (PCM). No adjustment is possible.

Engine Assembly

REMOVAL & INSTALLATION

Liberty

3.7L ENGINE

1. Before servicing the vehicle, refer to the precautions in the beginning of this section.

2. Properly relieve the fuel system pressure.

3. Drain the cooling system.

4. Drain the engine oil.

5. Remove or disconnect the following:
- Negative battery cable
- Hood
- Air cleaner assembly
- Radiator
- Electric and mechanical fan assemblies
- A/C compressor, if equipped, and secure it out of the way with the lines attached. DO NOT DISCHARGE!
- Power steering pump, with the lines attached
- Alternator
- Coolant bottle
- Heater hoses
- Accelerator and speed control cables
- Lower and upper radiator hoses

- Engine ground straps
- Intake Air Temperature (IAT) sensor
- Fuel injection wiring connectors
- Throttle Position (TP) sensor
- Idle Air Control (IAC) motor
- Oil pressure sender connector
- Engine Coolant Temperature (ECT) sensor
- Manifold Absolute Pressure (MAP) sensor
- Camshaft position sensor
- Ignition coil wiring connector
- Crankshaft Position (CKP) sensor
- Coil pack
- Fuel rail
- PCV hose
- Vacuum hoses from the intake manifold
- Knock sensor connectors
- Oil dipstick tube
- Intake manifold
- Heated Oxygen (HO2S) sensor connector
- Block heater connector
- Front driveshaft at the differential
- Starter
- Structural cover
- With a manual transmission, remove the transmission
- Torque converter bolts and match-mark the converter
- Automatic transmission-to-engine bolts
- Exhaust front pipes
- Left and right engine mounts

6. Place a support stand under the transmission.

7. Install an engine lift plate

8. Lift the engine out of the vehicle.

To install:

9. If equipped with a manual transmission, install the transmission

10. Lower the engine and install the mounts. Don't tighten the bolts yet.

11. If equipped with an automatic transmission, perform the following steps:

a. Align the torque converter housing to the engine.

b. Torque the bolts to 30 ft. lbs. (41 Nm).

- Install the torque converter to flexplate bolts. Torque the bolts to 50 ft. lbs. (68 Nm).

12. Install or connect the following:

- Torque the through bolts to 45 ft. lbs. (61 Nm).
- Engine ground strap
- Starter motor
- CKP sensor

- Block heater cable
- Structural cover.

⁂ WARNING

The structural cover must be held tightly against the engine and bellhousing during tightening. The torque for all bolts is 40 ft. lbs. (54 Nm); the bolts must be tightened in the order shown.

- Exhaust pipes. New flange clamps MUST be used!
- HO2S sensor connectors
- KS sensors
- Intake Manifold
- Dipstick tube
- Vacuum hoses to the intake manifold
- PCV and breather hoses
- Fuel rail
- Ignition coil
- IAT sensor
- Fuel injector connectors
- TP sensor
- IAC motor
- Oil pressure sender
- ECT sensor electrical connector

- MAP sensor
- CMP sensor
- Radiator hoses
- Cruise control cable, if equipped
- Throttle cable
- Heater hoses
- Coolant bottle
- Power steering pump
- Alternator
- A/C compressor
- Radiator
- Fan assemblies
- Air cleaner assembly
- Negative battery cable

13. Fill and bleed the power steering system.

14. Fill the engine with clean oil.

15. Start the engine and check for leaks, repair if necessary.

Wrangler

1. Before servicing the vehicle, refer to the precautions in the beginning of this section.

2. Properly relieve the fuel system pressure.

3. Drain the cooling system.

4. Drain the engine oil.

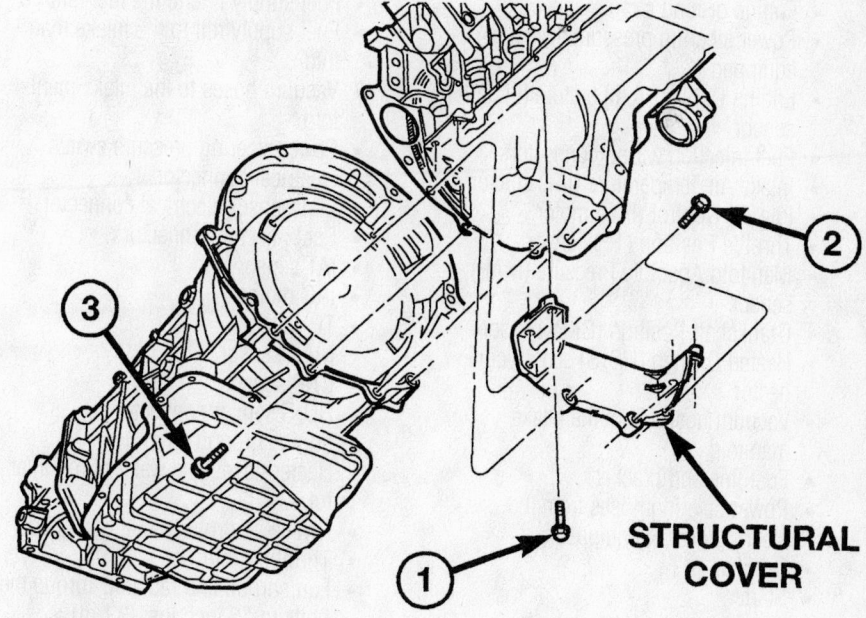

1 - BOLT
2 - BOLT
3 - BOLT

9355PG01

Tighten the structural cover bolts in this order—3.7L

5. Evacuate the A/C system.

6. Drain the power steering fluid, if equipped.

7. Remove or disconnect the following:

- Negative battery cable
- Air cleaner assembly
- Upper radiator hose
- Accessory drive belt and fan drive assembly
- Transmission cooler lines, if equipped
- Lower radiator hose
- Radiator and fan shroud
- Starter electrical connectors
- Alternator wiring connectors
- A/C compressor, if equipped
- Ignition coil wiring connector
- Oil pressure sender connector
- Wiring harness ground at the dipstick tube
- Heater hoses from the thermostat housing
- Water pump inlet tube
- Closed Crankcase Ventilation (CCV) hoses from the cylinder head and intake manifold
- Accelerator, transmission line pressure and speed control cables, if equipped
- Engine ground strap
- Power steering pressure switch, if equipped
- Engine Coolant Temperature (ECT) sensor
- Fuel injection wiring connectors
- Intake Air Temperature (IAT) sensor
- Idle Air Control (IAC) motor
- Throttle Position (TP) sensor
- Manifold Absolute Pressure (MAP) sensor
- Crankshaft Position (CKP) sensor
- Heated Oxygen (HO$_2$S) sensor connector
- Vacuum hoses from the intake manifold
- Fuel line and bracket
- Power steering hoses from the steering gear, if equipped
- Oil filter
- Starter
- Engine support through bolts
- Exhaust front pipe
- Flywheel cover
- Torque converter, if equipped with automatic transmission
- Transmission flange bolts
- Left and right motor mounts
- Engine shock damper bracket

8. Place a support stand under the transmission.

9. Lift the engine out of the vehicle.

To install:

10. If equipped with a manual transmission, perform the following steps:

   a. Insert the transmission shaft into the clutch spline.

   b. Align the flywheel housing to the engine.

   c. Install the flywheel housing bolts. Torque the bolts to 28 ft. lbs. (38 Nm).

11. If equipped with an automatic transmission, perform the following steps:

   a. Align the torque converter housing to the engine.

   b. Torque the bolts to 28 ft. lbs. (38 Nm).

- Install the torque converter to flexplate bolts. Torque the bolts to 50 ft. lbs. (68 Nm).

12. Install or connect the following:

- Engine onto the engine brackets. When properly aligned, torque the through bolts to 60 ft. lbs. (81 Nm).
- Inspection cover. Torque the bolts to 138 inch lbs. (16 Nm).
- Exhaust pipe. Torque the bolts to 23 ft. lbs. (31 Nm).
- Starter motor. Tighten the bolts to 33 ft. lbs. (45 Nm).
- Oil filter
- Power steering hoses
- Fuel supply line to the fuel rail
- Fuel supply rail to the intake manifold
- Vacuum hoses to the intake manifold
- Power steering pressure switch electrical connector
- ECT sensor electrical connector
- Fuel injector connectors
- IAT sensor
- IAC motor
- TP sensor
- MAP sensor
- CKP sensor
- HO$_2$S sensor connector
- Engine ground strap
- Heater hoses and water pump inlet tube
- Cruise control cable, if equipped
- Throttle cable
- Fan shroud and radiator. Torque the bolts to 75 inch lbs. (8 Nm).
- Transmission cooler lines, if equipped. Torque the bolts to 10 ft. lbs. (15 Nm).
- Fan drive assembly. Torque the bolts to 20 ft. lbs. (27 Nm).
- Drive belt
- Radiator hoses
- Ignition coil
- Distributor
- A/C compressor

- A/C high pressure switch
- Alternator
- Oil pressure sender
- Harness ground at the oil dipstick tube bracket
- A/C hoses
- Air cleaner assembly
- Negative battery cable

13. Fill and bleed the power steering system.

14. Fill the engine with clean oil.

15. Fill and bleed the cooling system.

16. Recharge the A/C system.

17. Start the engine and check for leaks, repair if necessary.

Cherokee

1. Before servicing the vehicle, refer to the precautions in the beginning of this section.

2. Drain the cooling system.

3. Recover the A/C refrigerant.

4. Drain the engine oil.

5. Drain the power steering system.

6. Properly relieve the fuel system pressure.

7. Remove or disconnect the following:

- Battery
- Hood
- Air cleaner assembly
- Radiator hoses
- Fan shroud
- Electric cooling fan, if equipped
- Radiator
- Cooling fan
- Heater hoses
- Throttle cable
- Cruise control cable
- Transmission cable
- Fuel injector harness
- Camshaft Position (CMP) sensor connector
- Ignition coil wiring connector
- Oil pressure sender connector
- Body ground cable
- Starter solenoid connectors
- Fuel injection wiring harness
- Fuel line and bracket
- A/C compressor suction/discharge hose assembly
- Brake booster vacuum line
- Power steering hoses
- Intake manifold vacuum lines
- Crankshaft Position (CKP) sensor connector
- Engine speed sensor
- Oil filter
- Starter motor
- Heated Oxygen (HO$_2$S) sensor connector
- Exhaust front pipe

- Flywheel housing access cover
- Torque converter, if equipped with automatic transmission
- Transmission flange bolts
- Left and right motor mounts

8. Support the transmission and lift the engine out of the vehicle.

To install:

9. Lower the engine in to the vehicle and position to the transmission.

10. Install or connect the following:
- Left and right motor mounts. Tighten the through bolts to 51 ft. lbs. (69 Nm).
- Transmission flange bolts. Tighten the bolts to 28 ft. lbs. (38 Nm).
- Torque converter, if equipped with automatic transmission. Tighten the bolts to 23 ft. lbs. (31 Nm).
- Flywheel housing access cover
- Exhaust front pipe
- Engine speed sensor electrical connector
- HO2S sensor connector
- Starter motor. Tighten the bolts to 33 ft. lbs. (45 Nm).
- Oil filter
- Vacuum hoses to the intake manifold
- CKP sensor connector
- Power steering hoses
- Brake booster vacuum line
- A/C compressor suction/discharge hose assembly
- Fuel line and bracket
- Fuel injection wiring harness
- Starter solenoid connectors
- Body ground cable
- Oil pressure sender connector
- Ignition coil wiring connector
- CMP sensor connector
- Fuel injector harness
- Transmission cable
- Cruise control cable
- Throttle cable
- Heater hoses
- Cooling fan
- Radiator
- Electric cooling fan, if equipped
- Fan shroud
- Radiator hoses
- Air cleaner assembly
- Hood
- Battery

11. Fill and bleed the cooling system.
12. Fill the engine with clean oil.
13. Fill and bleed the power steering system.
- Recharge the A/C system.

14. Start the engine, check for leaks and repair if necessary.

Grand Cherokee

4.0L ENGINE

1. Before servicing the vehicle, refer to the precautions in the beginning of this section.
2. Drain the cooling system.
3. Drain the engine oil.
4. Drain the power steering fluid.
5. Properly relieve the fuel system pressure.
6. Recover the A/C refrigerant.
7. Remove or disconnect the following:
- Negative battery cable
- Hood
- Radiator hoses
- Upper radiator support
- Cooling fan and shroud
- Radiator
- A/C condenser
- Heater hoses
- Accelerator cable
- Cruise control cable
- Transmission cable
- Body ground cable
- Power steering pressure switch connector
- Engine Coolant Temperature (ECT) sensor connector
- Intake Air Temperature (IAT) sensor
- Fuel injector connectors
- Throttle Position (TP) sensor connector
- Manifold Absolute Pressure (MAP) sensor connector
- Crankshaft Position (CKP) sensor connector
- Heated Oxygen (HO2S) sensor connector
- Camshaft Position (CMP) sensor connector
- Ignition coil wiring connector
- Oil pressure sender connector
- Fuel line and bracket
- Air cleaner assembly
- Brake booster vacuum line
- Power steering hoses
- Starter motor
- Exhaust front pipe
- Flywheel cover
- Torque converter
- Transmission flange bolts
- Left and right motor mounts

8. Place a support stand under the transmission.
9. Lift the engine out of the vehicle.

To install:

10. Lower the engine in to the vehicle and position to the transmission.
11. Install or connect the following:
- Left and right motor mounts. Tighten the through bolts to 51 ft. lbs. (69 Nm).
- Transmission flange bolts. Tighten the bolts to 30 ft. lbs. (41 Nm).
- Torque converter. Tighten the bolts to 23 ft. lbs. (31 Nm).
- Flywheel cover
- Exhaust front pipe
- Starter motor. Tighten the bolts to 33 ft. lbs. (45 Nm).
- Power steering hoses
- Brake booster vacuum line
- Air cleaner assembly
- Fuel line and bracket
- Oil pressure sender connector
- Ignition coil wiring connector
- CMP sensor connector
- HO2S sensor connector
- CKP sensor connector
- MAP sensor connector
- TP sensor connector
- Fuel injector connectors
- IAT sensor
- ECT sensor connector
- Power steering pressure switch connector
- Body ground cable
- Transmission cable
- Cruise control cable
- Accelerator cable
- Heater hoses
- A/C condenser
- Radiator
- Cooling fan and shroud
- Upper radiator support
- Radiator hoses
- Hood
- Negative battery cable

12. Fill and bleed the cooling system.
13. Fill and bleed the cooing system.
14. Fill the engine with clean oil.
15. Recharge the A/C system.
16. Start the engine, check for leaks and repair if necessary.

4.7L ENGINE

1. Before servicing the vehicle, refer to the precautions in the beginning of this section.
2. Drain the cooling system.
3. Drain the engine oil.
4. Drain the power steering system.
5. Properly relieve the fuel system pressure.

Timing belt service is covered in Section 3 of this manual

6. Recover the A/C refrigerant.
7. Remove or disconnect the following:
   - Negative battery cable
   - Front fascia
   - Exhaust crossover pipe
   - Engine ground straps
   - Crankshaft Position (CKP) sensor
   - Structural collar
   - Starter
   - Left and right inner fender liners
   - Headlamp mounting module
   - Air intake resonator
   - Accelerator cable
   - Cruise control cable
   - Crankcase breather tubes
   - Accessory drive belt
   - A/C compressor
   - Cooling fan assemblies
   - Radiator hoses
   - Transmission oil cooler lines
   - Radiator
   - A/C condenser
   - Alternator
   - Heater hoses
   - Throttle Position (TP) sensor connector
   - Intake Air Temperature (IAT) sensor connector
   - Fuel injector harness connectors

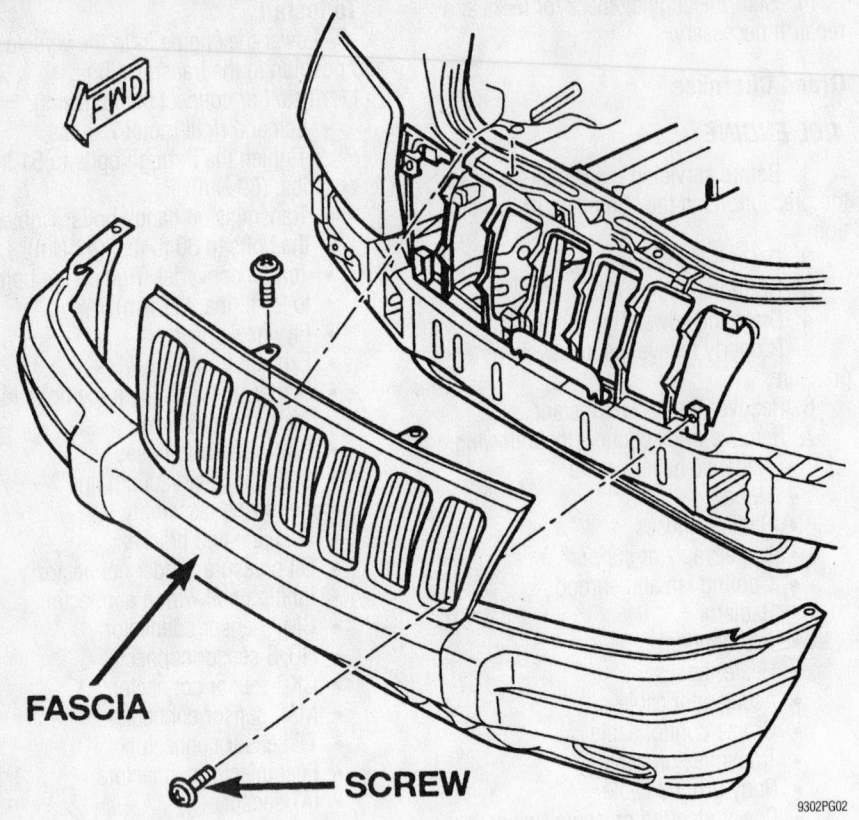

FWD

FASCIA

SCREW

9302PG02

Exploded view of the front fascia panel—1999–01 Grand Cherokee

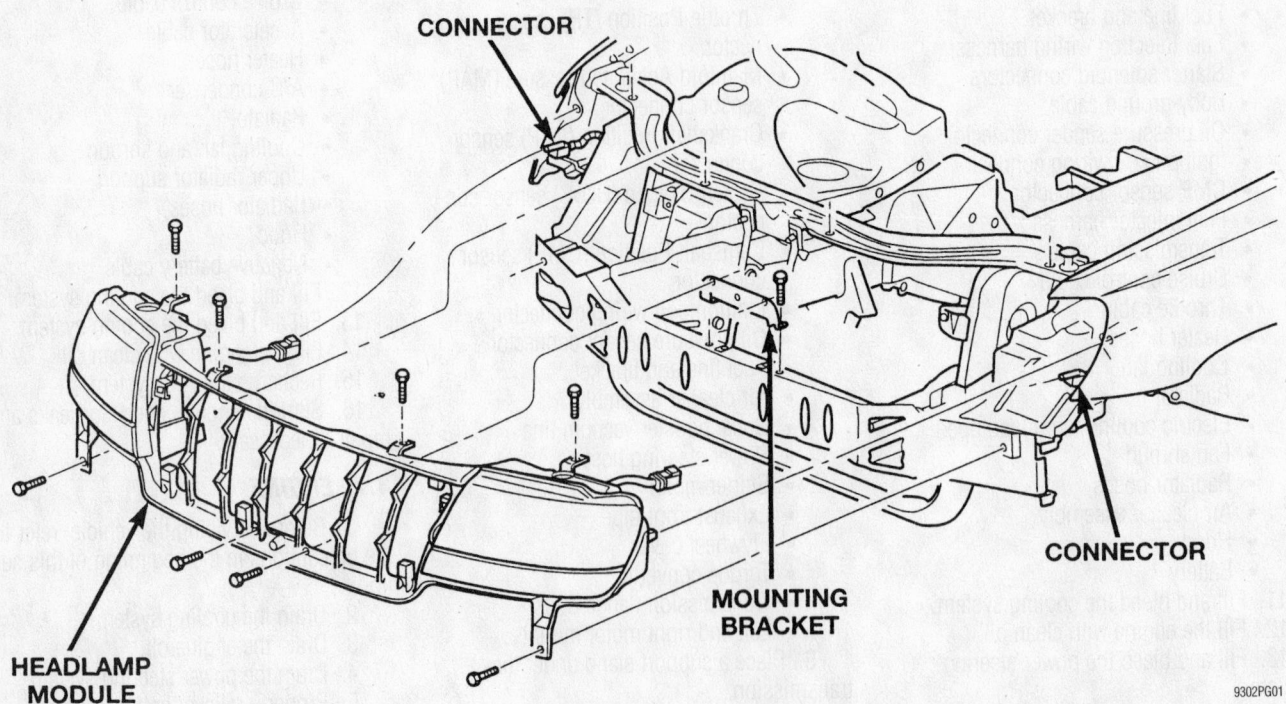

CONNECTOR

CONNECTOR

MOUNTING BRACKET

HEADLAMP MODULE

9302PG01

Exploded view of the headlamp module assembly—1999–01 Grand Cherokee

- Engine Coolant Temperature (ECT) sensor connector
- Idle Air Control (IAC) valve connector
- Manifold Absolute Pressure (MAP) sensor connector
- Ignition coils
- Fuel line
- Power steering pump
- Oil fill tube
- Oil dipstick tube
- Heated Oxygen (HO2S) sensor connectors
- Engine oil filter
- Exhaust crossover pipe
- Structural cover
- Rubber splash shield
- Starter motor
- Crankshaft Position (CKP) sensor connector
- Camshaft Position (CMP) sensor connector
- Torque converter
- Engine ground straps
- Left and right motor mounts
- Transmission flange bolts

8. Install Engine Lifting Fixture 8347 as shown.

9. Place a support stand under the transmission.

10. Lift the engine out of the vehicle.

To install:

11. Lower the engine in to the vehicle and position to the transmission.

12. Remove the engine lifting fixture.

13. Install or connect the following:
- Transmission flange bolts. Tighten the bolts to 50 ft. lbs. (68 Nm).
- Left and right motor mounts. Tighten the bolts to 45 ft. lbs. (61 Nm).
- Engine ground straps
- Torque converter. Tighten the bolts to 23 ft. lbs. (31 Nm).
- CMP sensor connector
- CKP sensor connector
- Starter motor
- Rubber splash shield

14. Install the structural cover as follows:

a. Install all the bolts finger-tight.

b. Hold the cover tightly against the transmission and the engine.

c. Tighten the bolts in sequence to 40 ft. lbs. (54 Nm).

15. Install or connect the following:
- Exhaust crossover pipe
- Engine oil filter
- HO2S sensor connectors

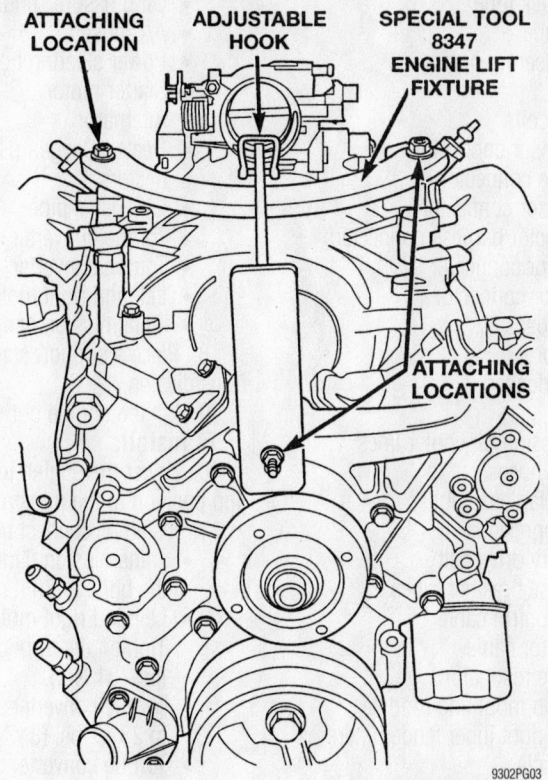

ATTACHING LOCATION   ADJUSTABLE HOOK   SPECIAL TOOL 8347 ENGINE LIFT FIXTURE

ATTACHING LOCATIONS

9302PG03

Engine Lifting Fixture—4.7L engine

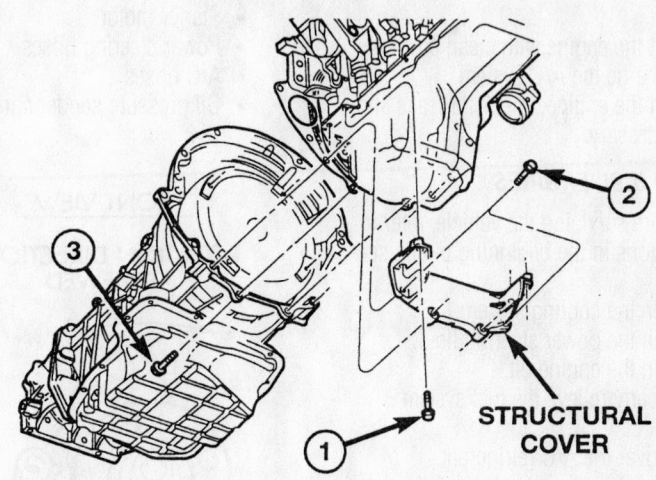

2

3

1

STRUCTURAL COVER

SEQUENCE	ITEM	TORQUE
1	BOLT (Qty 4)	54 N·m (40 ft. lbs.)
2	BOLT (Qty 2)	54 N·m (40 ft. lbs.)
3	BOLT (Qty 2)	54 N·m (40 ft. lbs.)

9308PG02

Structural cover torque sequence—4.7L engine

Heater Core replacement is covered in Section 2 of this manual

- Oil dipstick tube
- Oil fill tube
- Power steering pump
- Fuel line
- Ignition coils
- MAP sensor connector
- IAC valve connector
- ECT sensor connector
- Fuel injector harness connectors
- IAT sensor connector
- TP sensor connector
- Heater hoses
- Alternator
- A/C condenser
- Radiator
- Transmission oil cooler lines
- Radiator hoses
- Cooling fan assemblies
- A/C compressor
- Accessory drive belt
- Crankcase breather tubes
- Cruise control cable
- Accelerator cable
- Air intake resonator
- Headlamp mounting module
- Left and right inner fender liners
- Front fascia
- Negative battery cable

16. Fill and bleed the cooling system.
17. Fill and bleed the power steering system.
    - Fill the engine with clean oil.
18. Recharge the A/C system.
19. Start the engine, check for leaks and repair if necessary.

5.2L AND 5.9L ENGINES

1. Before servicing the vehicle, refer to the precautions in the beginning of this section.
2. Drain the cooling system.
3. Drain the power steering fluid.
4. Drain the engine oil.
5. Properly relieve the fuel system pressure.
6. Recover the A/C refrigerant.
7. Remove or disconnect the following:
   - Battery
   - Hood
   - Air cleaner assembly
   - Radiator hoses
   - Heater hoses
   - Radiator and fan shroud
   - Accessory drive belt
   - Distributor cap and spark plug wires
   - Vacuum lines
   - Engine control wiring harness connectors
   - Accelerator linkage
   - Fuel line
   - Throttle body

- Oil pressure sender connector
- A/C hoses
- Power steering hoses
- Starter motor
- Alternator
- Heated Oxygen (HO2S) sensor connectors
- Exhaust Y-pipe
- Torque converter cover
- Torque converter
- Left and right motor mounts
- Transmission flange bolts

8. Place a support stand under the transmission.
9. Lift the engine out of the vehicle.

To install:

10. Lower the engine in to the vehicle and position to the transmission.
11. Install or connect the following:
    - Transmission flange bolts. Tighten the bolts to 30 ft. lbs. (41 Nm).
    - Left and right motor mounts. Tighten the through bolts to 60 ft. lbs. (81 Nm).
    - Torque converter. Tighten the bolts to 23 ft. lbs. (31 Nm).
    - Torque converter cover
    - Exhaust Y-pipe
    - HO2S sensor connectors
    - Alternator
    - Starter motor
    - Power steering hoses
    - A/C hoses
    - Oil pressure sender connector

- Throttle body
- Fuel line
- Accelerator linkage
- Engine control wiring harness connectors
- Vacuum lines
- Distributor cap and spark plug wires
- Accessory drive belt
- Radiator and fan shroud
- Heater hoses
- Radiator hoses
- Air cleaner assembly
- Hood
- Battery

12. Fill and bleed the cooling system.
13. Fill and bleed the power steering system.
14. Fill the engine with clean oil.
15. Recharge the A/C system.
16. Start the engine, check for leaks and repair if necessary.

Water Pump

REMOVAL & INSTALLATION

2.5L and 4.0L Engines

➡The 2.5L and 4.0L engines covered use a reverse rotation water pump. The letter R is stamped on the impeller to identify. Engines from previous years may be equipped with forward rotation

FRONT VIEW

ROTATION DIRECTION AS VIEWED

BACK VIEW

ROTATION DIRECTION AS VIEWED

R STAMPED INTO IMPELLER

Reverse rotation water pump—2.5L and 4.0L engines

7924PG02

water pumps. Installation of the wrong water pump will cause engine over heating.

1. Before servicing the vehicle, refer to the precautions in the beginning of this section.
2. Drain the cooling system.
3. Remove or disconnect the following:
   • Negative battery cable
   • Electric cooling fan connector
   • Accessory drive belt
   • Engine cooling fan and pulley

➡Some 4.0L engines are equipped with a fan clutch that threads directly on to the water pump shaft. This fan clutch is equipped with right-hand threads.

➡Do not store the fan clutch assembly horizontally, silicone may leak into the bearing grease and cause contamination.

   • Water pump pulley
   • Power steering pump
   • Lower radiator hose from the water pump
   • Heater hose
   • Water pump and discard the gasket

➡One of the water pump bolts is longer than the others. Note the location for reassembly.

To install:
4. Clean the mating surfaces of all gasket material.
5. Install or connect the following:
   • Water pump using a new gasket. Torque the bolts to 17 ft. lbs. (23 Nm).
   • Heater hose
   • Lower radiator hose
   • Water pump pulley. Torque the bolts to 20 ft. lbs. (27 Nm).
   • Power steering pump
   • Engine cooling fan and shroud. Torque the bolts to 31 inch lbs. (3 Nm).
   • Accessory drive belt
   • Electric cooling fan connector
   • Negative battery cable
6. Fill and bleed the cooling system.
7. Start the engine, check for leaks and repair if necessary.

3.7L Engine

1. Before servicing the vehicle, refer to the precautions in the beginning of this section.

2. Drain the cooling system.
3. Remove or disconnect the following:
   • Negative battery cable
   • Fan and clutch assembly from the pump
   • Fan shroud and fan assembly. If you're reusing the fan clutch, keep it upright to avoid silicone fluid loss!
   • Lower hose
   • Water pump (8 bolts)
4. Installation is the reverse of removal. Tighten the bolts, in sequence, to 40 ft. lbs. (54 Nm).

4.7L Engine

   • Drive belt
   • Lower radiator hose from the water pump
   • Water pump
1. Clean the mating surfaces of all gasket material.

To install:
2. Install or connect the following:
   • Water pump using a new gasket. Torque the bolts to 40 ft. lbs. (54 Nm).
   • Lower radiator hose

   • Drive belt
   • Negative battery cable
3. Fill and bleed the cooling system.
4. Start the engine, check for leaks and repair if necessary.

5.2L and 5.9L Engines

1. Before servicing the vehicle, refer to the precautions in the beginning of this section.
2. Drain the cooling system.
3. Remove or disconnect the following:
   • Negative battery cable
   • Engine cooling fan and shroud

➡Do not store the fan clutch assembly horizontally, silicone may leak into the bearing grease and cause contamination.

   • Accessory drive belt
   • Water pump pulley
   • Lower radiator hose
   • Heater hose and tube
   • Bypass hose
   • Water pump

To install:
4. Install or connect the following:
   • Water pump, using a new gasket.

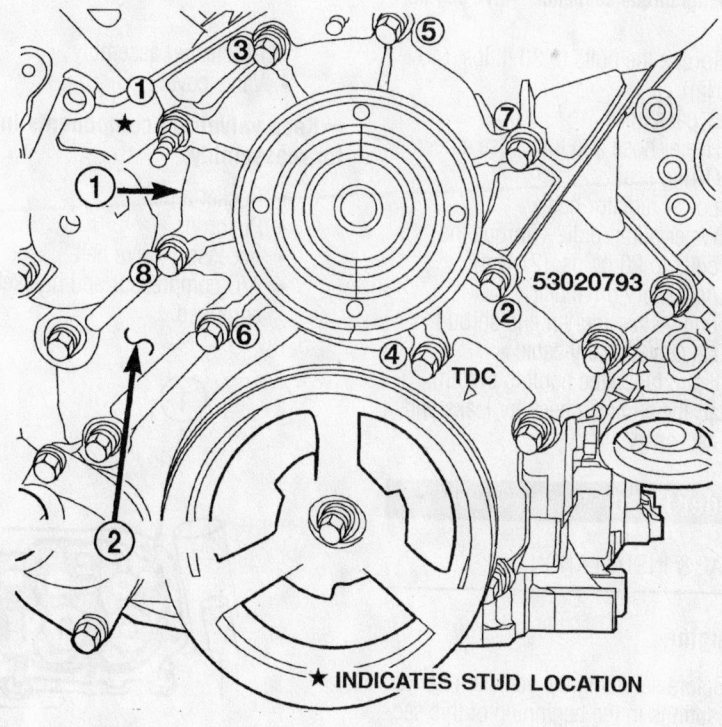

1 - WATER PUMP
2 - TIMING CHAIN COVER

★ INDICATES STUD LOCATION

53020793

TDC

Water pump tightening sequence—3.7L

9355PG02

Brake service is covered in Section 4 of this manual

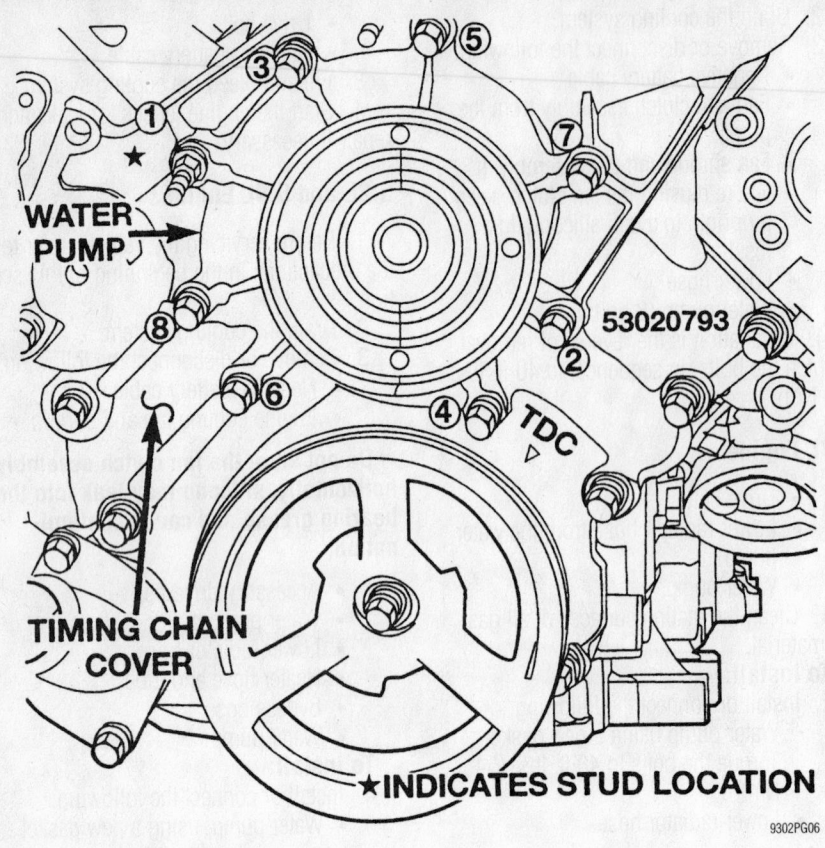

WATER PUMP

TIMING CHAIN COVER

53020793

TDC

★ INDICATES STUD LOCATION

9302PG06

Water pump torque sequence—4.7L engine

Torque the bolts to 30 ft. lbs. (40 Nm).
- Bypass hose
- Heater hose and tube. Use a new O-ring seal
- Lower radiator hose
- Water pump pulley. Torque the bolts to 20 ft. lbs. (27 Nm).
- Accessory drive belt
- Engine cooling fan and shroud
- Negative battery cable

5. Fill and bleed the cooling system.
6. Start the engine, check for leaks and repair if necessary.

Cylinder Head

REMOVAL & INSTALLATION

2.5L Engine

1. Before servicing the vehicle, refer to the precautions in the beginning of this section.
2. Properly relieve the fuel system pressure.
3. Drain the cooling system.
4. Remove or disconnect the following:
- Negative battery cable
- Crankcase Ventilation (CCV) hoses

- Air cleaner assembly
- Valve cover

➡ **Keep valvetrain components in order for reassembly.**

- Rocker arms
- Pushrods
- Accessory drive belt
- A/C compressor and bracket, if equipped

- Power steering pump and bracket, if equipped
- Fuel line
- Combination manifold
- Thermostat housing coolant hoses
- Spark plugs
- Engine Coolant Temperature (ECT) sensor connector
- Cylinder head

To install:

✳ WARNING

Cylinder head bolts may only be reused one time. If reusing a cylinder head bolt, place a paint mark on the bolt after installation. If a cylinder head bolt has a paint mark, discard it and use a new bolt.

5. Fabricate two alignment dowels from old cylinder head bolts. Cut the hex head off of the bolts, and cut a slot in each dowel to ease removal.
6. Install or connect the following:
- One dowel in bolt hole No. 8, and one dowel in bolt hole No. 10.
- Cylinder head and gasket.
- Cylinder head bolts except for No 8 and No 10. Coat the threads of bolt No. 7 with Loctite® 592 sealant.

7. Remove the alignment dowels and install the No. 8 and No. 10 head bolts.

✳ WARNING

During the final tightening sequence, bolt No. 7 will be tightened to a lower torque value than the rest of the bolts. Do not overtighten bolt No. 7.

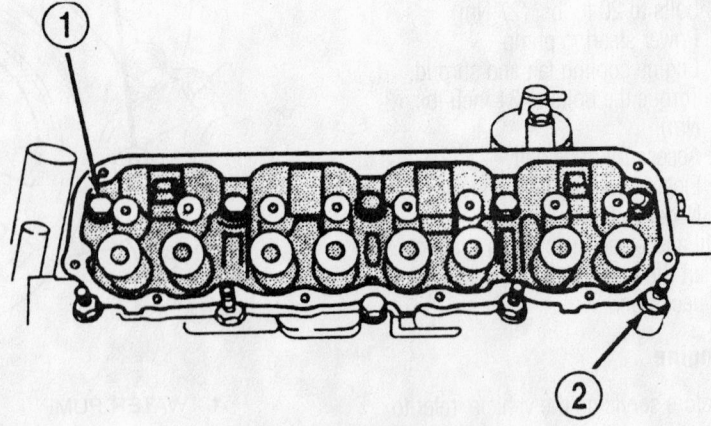

1 – ALIGNMENT DOWEL
2 – ALIGNMENT DOWEL

9308PG03

Alignment dowel locations—2.5L engine

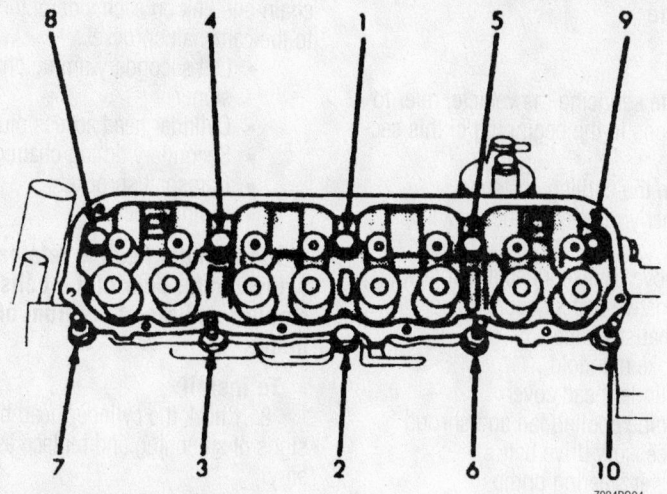

Cylinder head torque sequence—2.5L engine

8. Tighten the cylinder head bolts, in sequence, as follows:

a. Step 1: 22 ft. lbs. (30 Nm)

b. Step 2: 45 ft. lbs. (61 Nm)

c. Step 3: 45 ft. lbs. (61 Nm)

d. Step 4: Bolts 1–6 to 110 ft. lbs. (149 Nm)

e. Step 5: Bolt 7 to 100 ft. lbs. (136 Nm)

f. Step 6: Bolts 8–10 to 110 ft. lbs. (149 Nm)

g. Step 7: Repeat steps 4, 5 and 6

9. Install or connect the following:
- ECT sensor connector
- Spark plugs
- Thermostat housing coolant hoses
- Combination manifold
- Fuel line
- Power steering pump and bracket, if equipped
- A/C compressor and bracket, if equipped
- Accessory drive belt
- Pushrods and rocker arms in their original positions
- Valve cover
- Air cleaner assembly
- CCV hoses
- Negative battery cable

10. Fill and bleed the cooling system.

11. Start the engine, check for leaks and repair if necessary.

4.0L Engines

1. Before servicing the vehicle, refer to the precautions in the beginning of this section.

2. Drain the cooling system.

3. Properly relieve the fuel system pressure.

4. Remove or disconnect the following:

- Negative battery cable
- Crankcase Ventilation (CCV) hoses
- Air cleaner assembly
- Accelerator cable
- Cruise control cable, if equipped
- Transmission cable, if equipped
- Control cable bracket
- Valve cover

➡ **Keep valvetrain components in order for reassembly.**

- Rocker arms
- Pushrods
- Accessory drive belt
- A/C compressor and bracket, if equipped

- Power steering pump and bracket, if equipped
- Fuel line
- Combination manifold
- Thermostat housing coolant hoses
- Spark plugs
- Engine Coolant Temperature (ECT) sensor connector
- Cylinder head

To install:

✳✳ WARNING

Cylinder head bolts may only be reused one time. If reusing a cylinder head bolt, place a paint mark on the bolt after installation. If a cylinder head bolt has a paint mark, discard it and use a new bolt.

5. Install the cylinder head with a new gasket. Coat the threads of bolt No. 11 with Loctite® F 592 sealant.

✳✳ CAUTION

During the final tightening sequence, bolt No. 11 will be tightened to a lower torque value than the rest of the bolts. Do not overtighten bolt No. 11.

6. Tighten the cylinder head bolts, in sequence, as follows:

a. Step 1: 22 ft. lbs. (30 Nm)

b. Step 2: 45 ft. lbs. (61 Nm)

c. Step 3: 45 ft. lbs. (61 Nm)

d. Step 4: Bolts 1–10 to 110 ft. lbs. (149 Nm)

e. Step 5: Bolt 11 to 100 ft. lbs. (136 Nm)

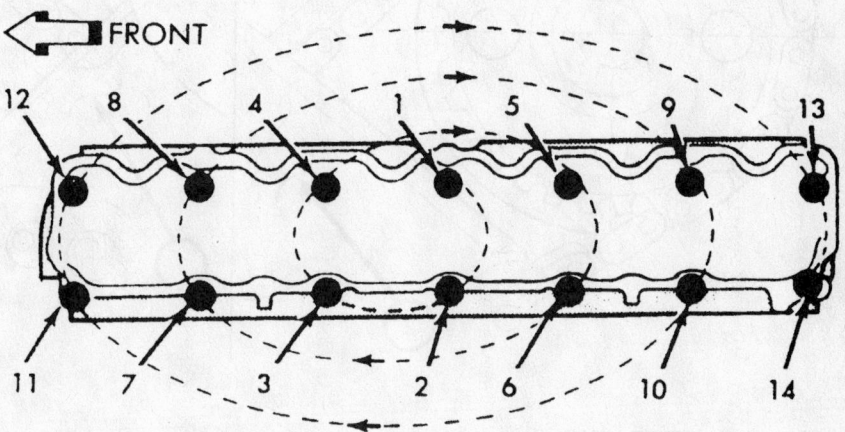

Cylinder head torque sequence—4.0L engine

For complete Engine Mechanical specifications, see Section 1 of this manual

f. Step 6: Bolts 12–14 to 110 ft. lbs. (149 Nm)

g. Step 7: Repeat steps 4, 5 and 6

7. Install or connect the following:
- ECT sensor connector
- Spark plugs
- Thermostat housing coolant hoses
- Combination manifold
- Fuel line
- Power steering pump and bracket, if equipped
- A/C compressor and bracket, if equipped. Torque the bolts to 30 ft. lbs. (40 Nm).
- Accessory drive belt
- Pushrods and rocker arms in their original positions
- Valve cover
- Control cable bracket
- Transmission cable, if equipped
- Cruise control cable, if equipped
- Accelerator cable
- Air cleaner assembly
- CCV hoses
- Negative battery cable

8. Fill and bleed the cooling system.

9. Start the engine, check for leaks and repair if necessary.

3.7L Engine

LEFT SIDE

1. Before servicing the vehicle, refer to the precautions in the beginning of this section.

2. Drain the cooling system.

3. Properly relieve the fuel system pressure.

4. Remove or disconnect the following:
- Negative battery cable
- Exhaust Y-pipe
- Intake manifold
- Cylinder head cover
- Engine cooling fan and shroud
- Accessory drive belt
- Power steering pump

5. Rotate the crankshaft so that the crankshaft timing mark aligns with the Top Dead Center (TDC) mark on the front cover, and the **V6** marks on the camshaft sprockets are at 12 o'clock as shown.
- Crankshaft damper
- Front cover

6. Lock the secondary timing chain to the idler sprocket with Timing Chain Locking tool 8429.

7. Matchmark the secondary timing chain one link on each side of the V6 mark to the camshaft sprocket.
- Left secondary timing chain tensioner
- Cylinder head access plug
- Secondary timing chain guide
- Camshaft sprocket
- Cylinder head

➡ **The cylinder head is retained by twelve bolts. Four of the bolts are smaller and are at the front of the head.**

To install:

8. Check the cylinder head bolts for signs of stretching and replace as necessary.

9. Lubricate the threads of the 11mm bolts with clean engine oil.

10. Coat the threads of the 8mm bolts with Mopar® Lock and Seal Adhesive.

11. Install the cylinder heads. Use new gaskets and tighten the bolts, in sequence, as follows:

a. Step 1: Bolts 1–8 to 20 ft. lbs. (27 Nm)

b. Step 2: Bolts 1–10 verify torque without loosening

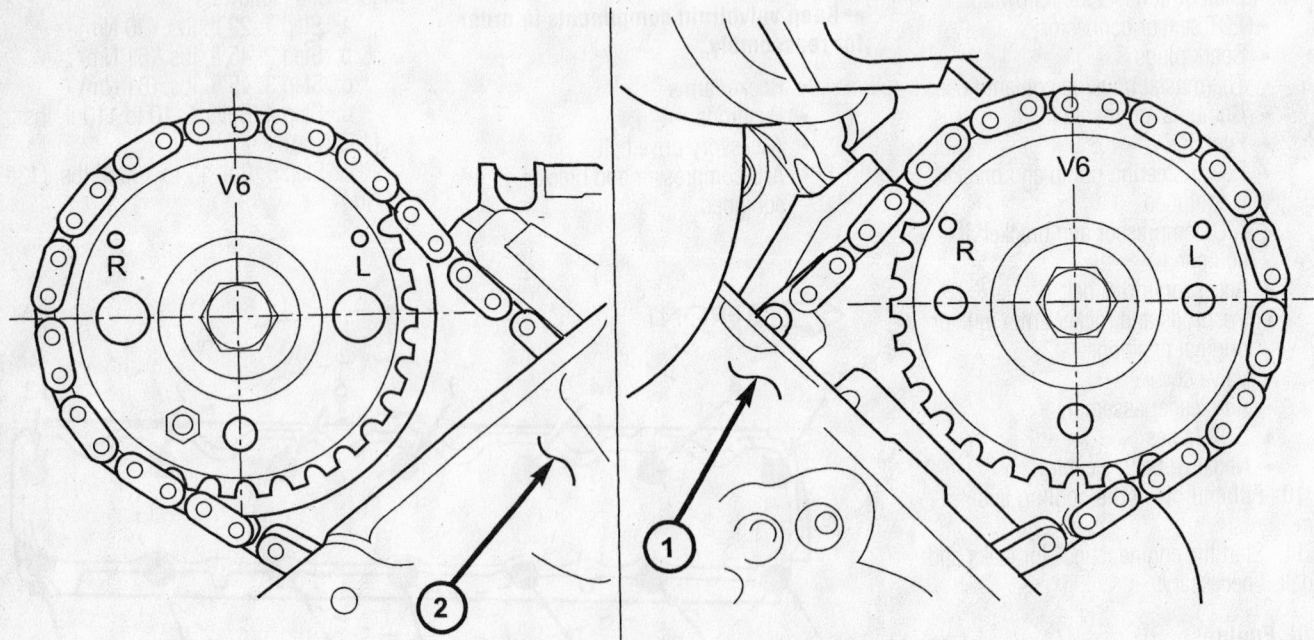

1 - LEFT CYLINDER HEAD
2 - RIGHT CYLINDER HEAD

Camshaft sprocket timing marks—3.7L

9355PG04

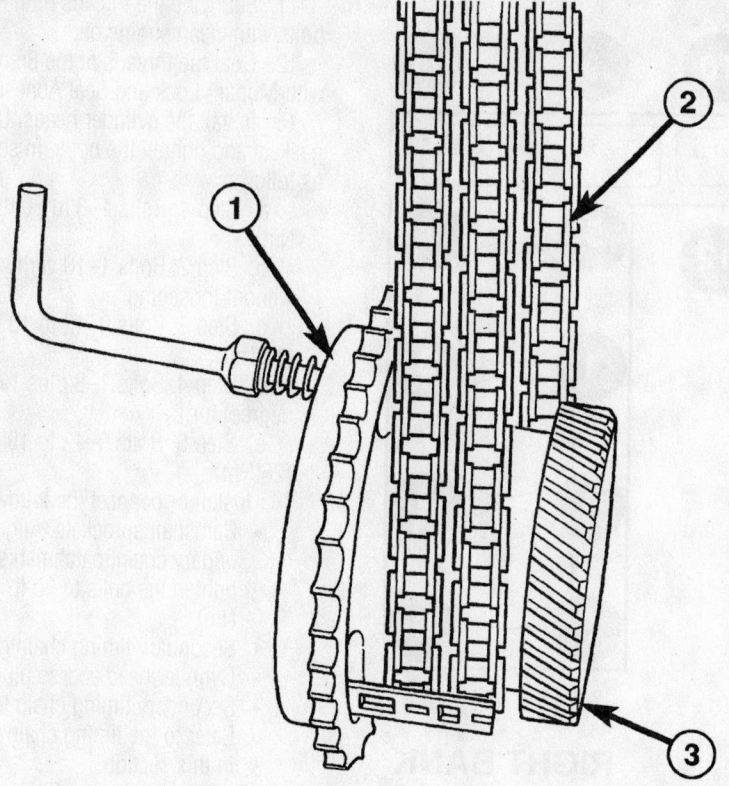

1 - SPECIAL TOOL 8429

2 - CAMSHAFT CHAIN

3 - CRANKSHAFT TIMING GEAR

9355PG05

Camshaft locking tool—3.7L

c. Step 3: Bolts 9–12 to 10 ft. lbs. (14 Nm)

d. Step 4: Bolts 1–8 plus ¼ (90 degree) turn

e. Step 5: Bolts 9–12 to 19 ft. lbs. (26 Nm)

12. Install or connect the following:
• Camshaft sprocket. Align the sec-

ondary chain matchmarks and tighten the bolt to 90 ft. lbs. (122 Nm).
• Secondary timing chain guide
• Cylinder head access plug
• Secondary timing chain tensioner. Refer to the timing chain procedure in this section.

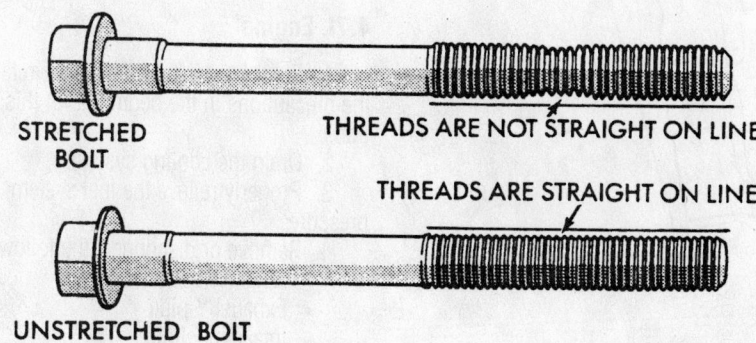

STRETCHED BOLT

THREADS ARE NOT STRAIGHT ON LINE

THREADS ARE STRAIGHT ON LINE

UNSTRETCHED BOLT

9302PG10

Examine the head bolts for signs of stretching—3.7L engine

13. Remove the Timing Chain Locking tool.

14. Install or connect the following:
• Front cover
• Crankshaft damper. Torque the bolt to 130 ft. lbs. (175 Nm).
• Power steering pump
• Accessory drive belt
• Engine cooling fan and shroud
• Cover
• Intake manifold
• Exhaust Y-pipe
• Negative battery cable

15. Fill and bleed the cooling system.

16. Start the engine, check for leaks and repair if necessary.

RIGHT SIDE

1. Before servicing the vehicle, refer to the precautions in the beginning of this section.

2. Drain the cooling system.

3. Properly relieve the fuel system pressure.

4. Remove or disconnect the following:
• Negative battery cable
• Exhaust Y-pipe
• Intake manifold
• Valve cover
• Engine cooling fan and shroud
• Accessory drive belt
• Oil fill housing
• Power steering pump

5. Rotate the crankshaft so that the crankshaft timing mark aligns with the Top Dead Center (TDC) mark on the front cover, and the **V6** marks on the camshaft sprockets are at 12 o'clock as shown.

6. Remove or disconnect the following:
• Crankshaft damper
• Front cover

7. Lock the secondary timing chains to the idler sprocket with Timing Chain Locking tool 8429.

8. Matchmark the secondary timing chains to the camshaft sprockets.

9. Remove or disconnect the following:
• Secondary timing chain tensioners
• Cylinder head access plugs
• Secondary timing chain guides
• Camshaft sprockets
• Cylinder heads

➡**Each cylinder head is retained by 8 11mm bolts and four 8mm bolts.**

To install:

10. Check the cylinder head bolts for signs of stretching and replace as necessary.

For Accessory Drive Belt illustrations, see Section 1 of this manual

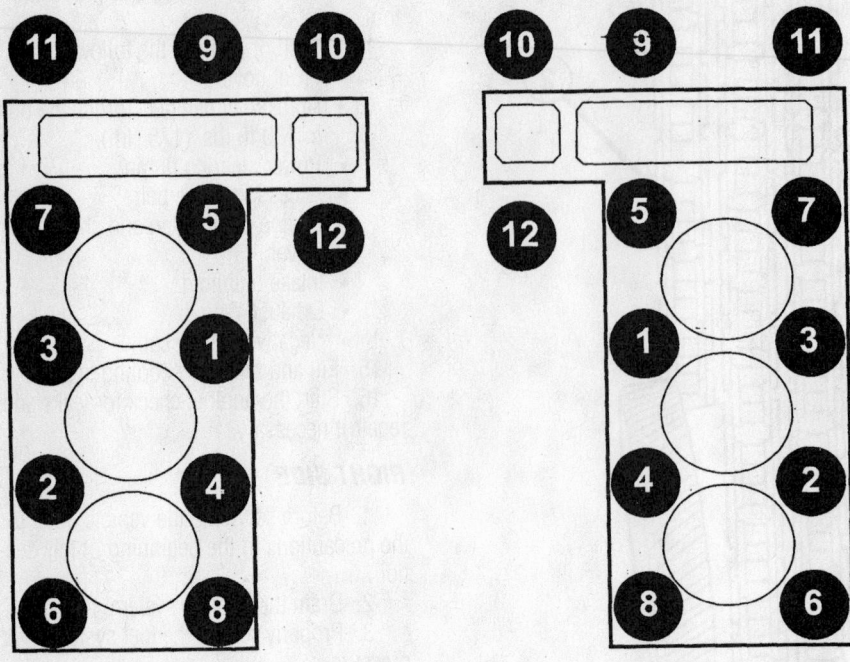

LEFT BANK

RIGHT BANK

9355PG03

Cylinder head bolt torque sequence—3.7L

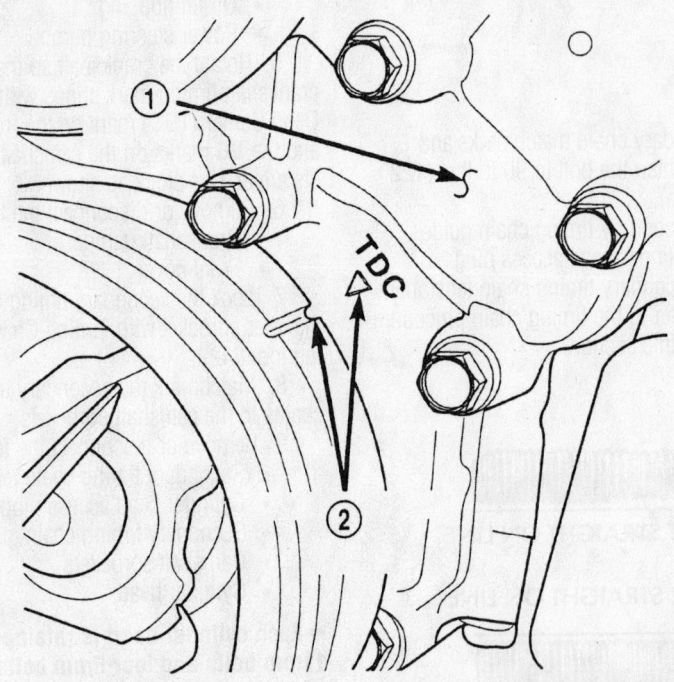

1 – TIMING CHAIN COVER
2 – CRANKSHAFT TIMING MARKS

9308PG04

Crankshaft timing marks—3.7L engine

11. Lubricate the threads of the 11mm bolts with clean engine oil.

12. Coat the threads of the 8mm bolts with Mopar® Lock and Seal Adhesive.

13. Install the cylinder heads. Use new gaskets and tighten the bolts, in sequence, as follows:

    a. Step 1: Bolts 1–8 to 20 ft. lbs. (27 Nm)

    b. Step 2: Bolts 1–10 verify torque without loosening

    c. Step 3: Bolts 9–12 to 10 ft. lbs. (14 Nm)

    d. Step 4: Bolts 1–8 plus ¼ (90 degree) turn

    e. Step 5: Bolts 9–12 to 19 ft. lbs. (26 Nm)

14. Install or connect the following:
- Camshaft sprockets. Align the secondary chain matchmarks and tighten the bolts to 90 ft. lbs. (122 Nm).
- Secondary timing chain guides
- Cylinder head access plugs
- Secondary timing chain tensioners. Refer to the timing chain procedure in this section.

15. Remove the Timing Chain Locking tool.

16. Install or connect the following:
- Front cover
- Crankshaft damper. Torque the bolt to 130 ft. lbs. (175 Nm).
- Rocker arms
- Power steering pump
- Oil fill housing
- Accessory drive belt
- Engine cooling fan and shroud
- Valve covers
- Intake manifold
- Exhaust Y-pipe
- Negative battery cable

17. Fill and bleed the cooling system.

18. Start the engine, check for leaks and repair if necessary.

4.7L Engine

1. Before servicing the vehicle, refer to the precautions in the beginning of this section.

2. Drain the cooling system.

3. Properly relieve the fuel system pressure.

4. Remove or disconnect the following:
- Negative battery cable
- Exhaust Y-pipe
- Intake manifold
- Valve covers
- Engine cooling fan and shroud
- Accessory drive belt
- Oil fill housing

- Camshaft sprockets
- Cylinder heads

➡**Each cylinder head is retained by ten 11mm bolts and four 8mm bolts.**

To install:

10. Check the cylinder head bolts for signs of stretching and replace as necessary.

11. Lubricate the threads of the 11mm bolts with clean engine oil.

12. Coat the threads of the 8mm bolts with Mopar® Lock and Seal Adhesive.

13. Install the cylinder heads. Use new gaskets and tighten the bolts, in sequence, as follows:

   a. Step 1: Bolts 1–10 to 15 ft. lbs. (20 Nm)

   b. Step 2: Bolts 1–10 to 35 ft. lbs. (47 Nm)

   c. Step 3: Bolts 11–14 to 18 ft. lbs. (25 Nm)

   d. Step 4: Bolts 1–10 plus ¼ (90 degree) turn

   e. Step 5: Bolts 11–14 to 19 ft. lbs. (26 Nm)

14. Install or connect the following:

   - Camshaft sprockets. Align the secondary chain matchmarks and tighten the bolts to 90 ft. lbs. (122 Nm).

   - Secondary timing chain guides

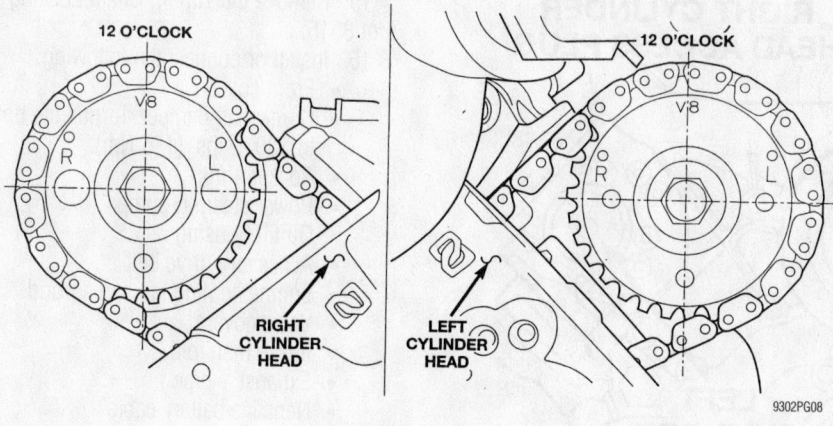

Camshaft positioning—3.7L engine

- Power steering pump
- Rocker arms

5. Rotate the crankshaft so that the crankshaft timing mark aligns with the Top Dead Center (TDC) mark on the front cover, and the **V8** marks on the camshaft sprockets are at 12 o'clock as shown.

6. Remove or disconnect the following:

   - Crankshaft damper
   - Front cover

7. Lock the secondary timing chains to the idler sprocket with Timing Chain Locking tool 8515.

8. Matchmark the secondary timing chains to the camshaft sprockets.

9. Remove or disconnect the following:

   - Secondary timing chain tensioners
   - Cylinder head access plugs
   - Secondary timing chain guides

1 – TIMING CHAIN COVER
2 – CRANKSHAFT TIMING MARKS

9308PG04

Crankshaft timing marks—4.7L engine

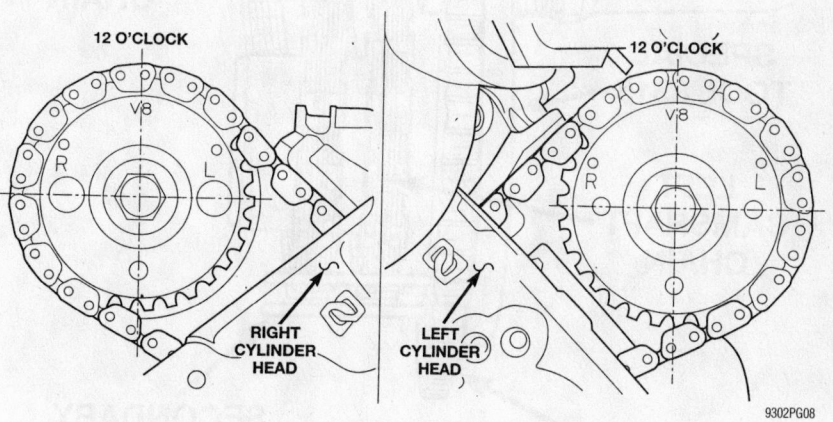

Camshaft positioning—4.7L engine

For Tire, Wheel and Ball Joint specifications, see Section 1 of this manual

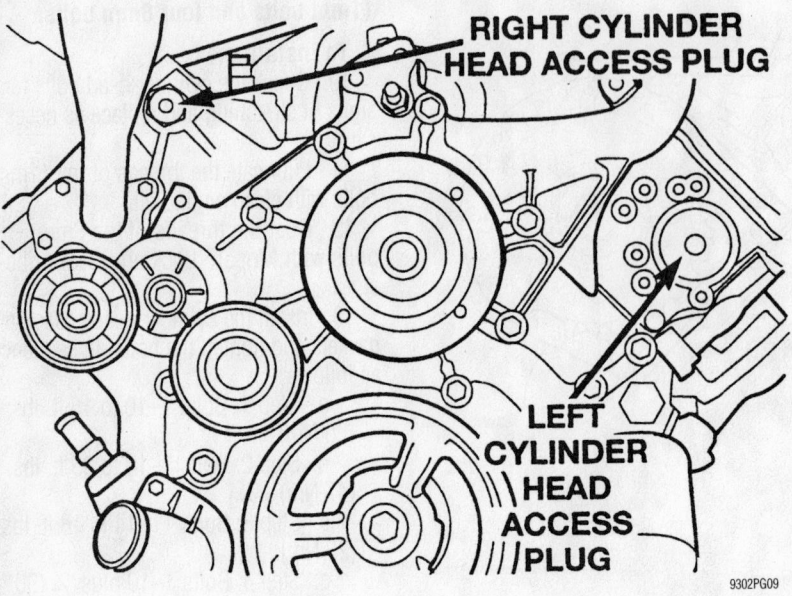

Cylinder head access plug locations—4.7L engine

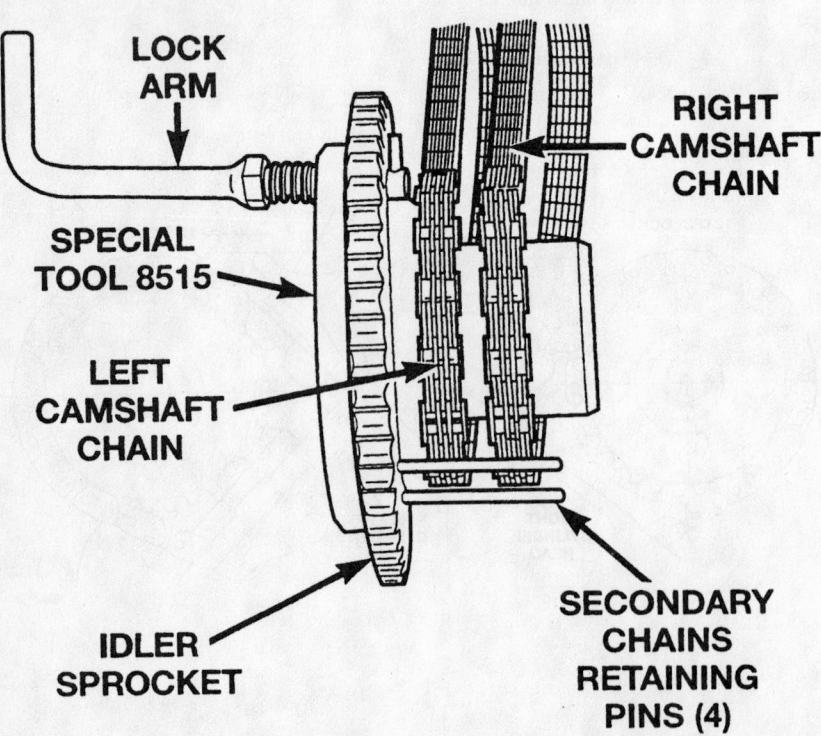

Use the special tool to lock the timing chains on the idler gear—4.7L engine

- Cylinder head access plugs
- Secondary timing chain tensioners. Refer to the timing chain procedure in this section.

15. Remove the Timing Chain Locking tool 8515.

16. Install or connect the following:
- Front cover
- Crankshaft damper. Torque the bolt to 130 ft. lbs. (175 Nm).
- Rocker arms
- Power steering pump
- Oil fill housing
- Accessory drive belt
- Engine cooling fan and shroud
- Valve covers
- Intake manifold
- Exhaust Y-pipe
- Negative battery cable

17. Fill and bleed the cooling system.

18. Start the engine, check for leaks and repair if necessary.

5.2L and 5.9L Engines

1. Before servicing the vehicle, refer to the precautions in the beginning of this section.

2. Drain the cooling system.

3. Properly relieve the fuel system pressure.

4. Remove or disconnect the following:
- Negative battery cable
- Accessory drive belt
- Alternator
- Air cleaner assembly
- Closed Crankcase Ventilation (CCV) system
- Evaporative emissions control system
- Fuel line
- Accelerator linkage
- Cruise control cable
- Transmission cable
- Distributor cap and wires
- Ignition coil wiring
- Engine Coolant Temperature (ECT) sensor connector
- Heater hoses
- Bypass hose
- Upper radiator hose
- Intake manifold
- Valve covers

➡Keep valvetrain components in order for reassembly.

- Rocker arms
- Pushrods
- Exhaust manifolds
- Spark plugs
- Cylinder heads

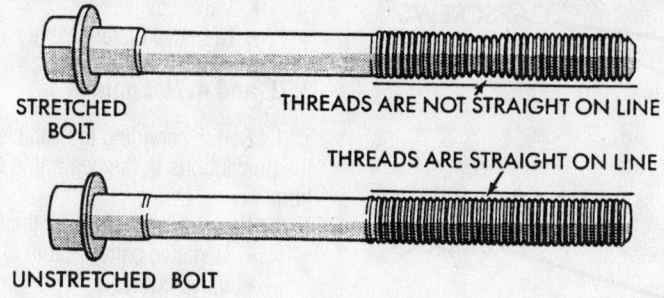

STRETCHED BOLT

THREADS ARE NOT STRAIGHT ON LINE

THREADS ARE STRAIGHT ON LINE

UNSTRETCHED BOLT

9302PG10

Examine the head bolts for signs of stretching—4.7L engine

◆ INDICATES SEALER APPLIED TO THREADS

Cylinder head torque sequence—4.7L engine

9302PG11

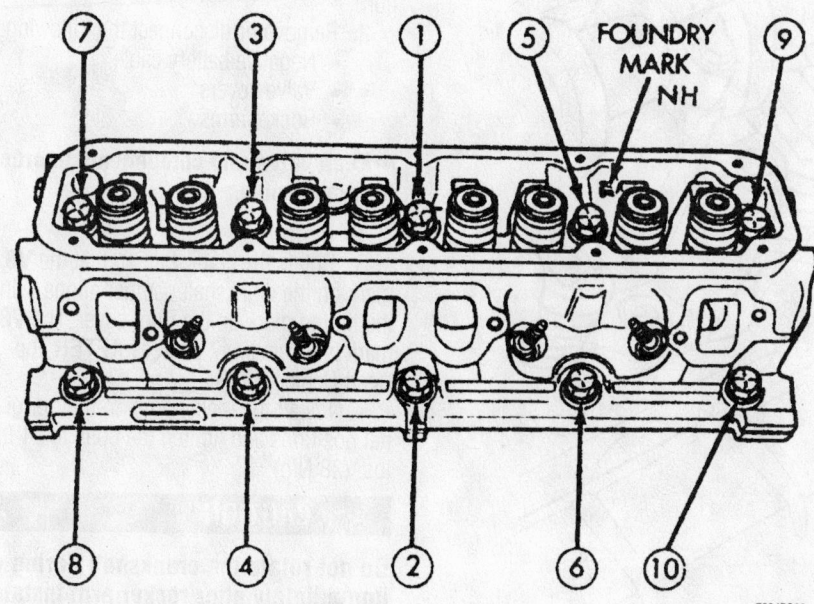

Cylinder head torque sequence—5.9L engine

7924PG06

To install:

5. Install the cylinder heads. Use new gaskets and tighten the bolts in sequence as follows:
   a. Step 1: 50 ft. lbs. (68 Nm)
   b. Step 2: 105 ft. lbs. (143 Nm)
   c. Step 3: 105 ft. lbs. (143 Nm)
6. Install or connect the following:
   • Spark plugs
   • Exhaust manifolds
   • Pushrods and rocker arms in their original positions
   • Valve covers
   • Intake manifold
   • Upper radiator hose
   • Bypass hose
   • Heater hoses
   • ECT sensor connector
   • Ignition coil wiring
   • Distributor cap and wires
   • Transmission cable
   • Cruise control cable
   • Accelerator linkage
   • Fuel line
   • Evaporative emissions control system
   • CCV system
   • Air cleaner assembly
   • Alternator
   • Accessory drive belt
   • Negative battery cable
7. Fill and bleed the cooling system.
8. Start the engine, check for leaks and repair if necessary.

Rocker Arms

REMOVAL & INSTALLATION

2.5L and 4.0L Engines

1. Before servicing the vehicle, refer to the precautions in the beginning of this section.
2. Remove or disconnect the following:
   • Negative battery cable
   • Valve cover
   • Rocker arm bolts, loosen them evenly to avoid damaging the alignment bridges
   • Rocker arms

➡**Keep valvetrain components in order for reassembly.**

To install:

3. Install or connect the following:
   • Rocker arms, pivots and bridges in their original positions. Torque the bolts for each bridged pair one turn at a time to 21 ft. lbs. (28 Nm).

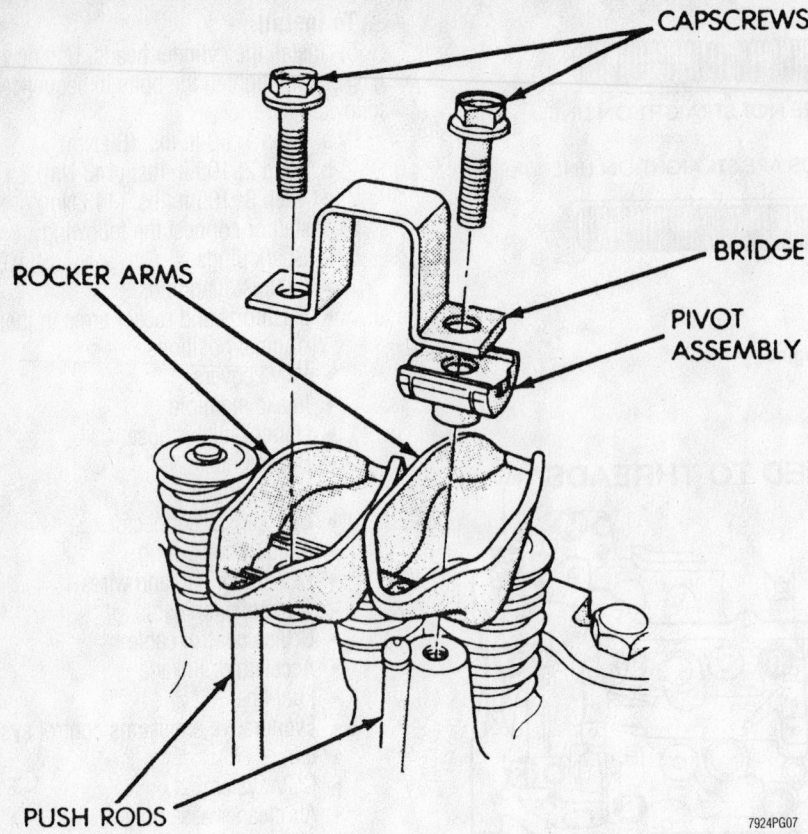

Exploded view of the rocker arm assembly—2.5L and 4.0L engines

CAPSCREWS

BRIDGE

PIVOT ASSEMBLY

ROCKER ARMS

PUSH RODS

7924PG07

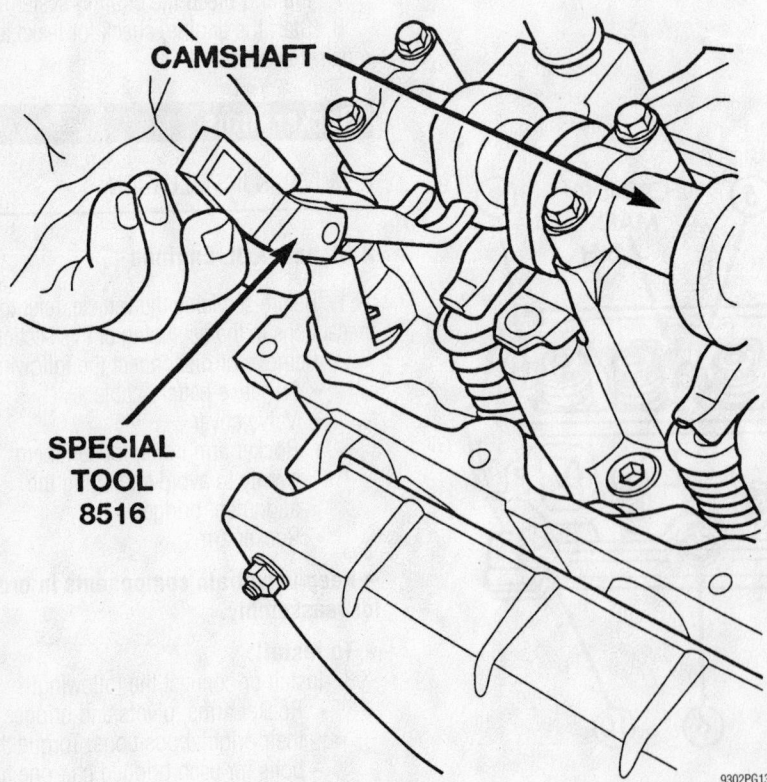

Rocker arm service—3.7L and 4.7L engines

CAMSHAFT

SPECIAL TOOL 8516

9302PG13

- Valve cover
- Negative battery cable

3.7L and 4.7L Engines

1. Before servicing the vehicle, refer to the precautions in the beginning of this section.

2. Remove or disconnect the following:
- Negative battery cable
- Valve covers

3. Rotate the crankshaft so that the piston of the cylinder to be serviced is at Top Dead Center (TDC) and both valves are closed.

4. Use special tool 8516 to depress the valve and remove the rocker arm.

5. Repeat for each rocker arm to be serviced.

➡**Keep valvetrain components in order for reassembly.**

To install:

6. Rotate the crankshaft so that the piston of the cylinder to be serviced is at BDC.

7. Compress the valve spring and install each rocker arm in its original position.

8. Repeat for each rocker arm to be installed.

9. Install or connect the following:
- Cylinder head cover
- Negative battery cable

5.2L and 5.9L Engines

1. Before servicing the vehicle, refer to the precautions in the beginning of this section.

2. Remove or disconnect the following:
- Negative battery cable
- Valve covers
- Rocker arms

➡**Keep valvetrain components in order for reassembly.**

To install:

3. Rotate the crankshaft so that the **V8** mark on the crankshaft damper aligns with the timing mark on the front cover. The **V8** mark is located 147 degrees **AFTER** Top Dead Center (TDC).

4. Install the rocker arms in their original positions and tighten the bolts to 21 ft. lbs. (28 Nm).

✳✳ CAUTION

Do not rotate the crankshaft during or immediately after rocker arm installation. Wait 5 minutes for the hydraulic lash adjusters to bleed down.

5. Install or connect the following:
- Valve covers
- Negative battery cable

Intake Manifold

REMOVAL & INSTALLATION

3.7L and 4.7L Engines

1. Before servicing the vehicle, refer to the precautions in the beginning of this section.
2. Drain the cooling system.
3. Properly relieve the fuel system pressure.

4. Remove or disconnect the following:
- Negative battery cable
- Air cleaner assembly
- Accelerator cable
- Cruise control cable
- Manifold Absolute Pressure (MAP) sensor connector
- Intake Air Temperature (IAT) sensor connector
- Throttle Position (TP) sensor connector
- Idle Air Control (IAC) valve connector
- Engine Coolant Temperature (ECT) sensor
- Positive Crankcase Ventilation (PCV) valve and hose
- Canister purge vacuum line
- Brake booster vacuum line
- Cruise control servo hose
- Accessory drive belt
- Alternator
- A/C compressor
- Engine ground straps
- Ignition coil towers
- Oil dipstick tube
- Fuel line
- Fuel supply manifold
- Throttle body and mounting bracket
- Cowl seal
- Right engine lifting stud
- Intake manifold. Remove the fasteners in reverse of the tightening sequence.

To install:

5. Install or connect the following:
- Intake manifold using new gaskets. Torque the bolts, in sequence, to 105 inch lbs. (12 Nm).
- Right engine lifting stud
- Cowl seal
- Throttle body and mounting bracket

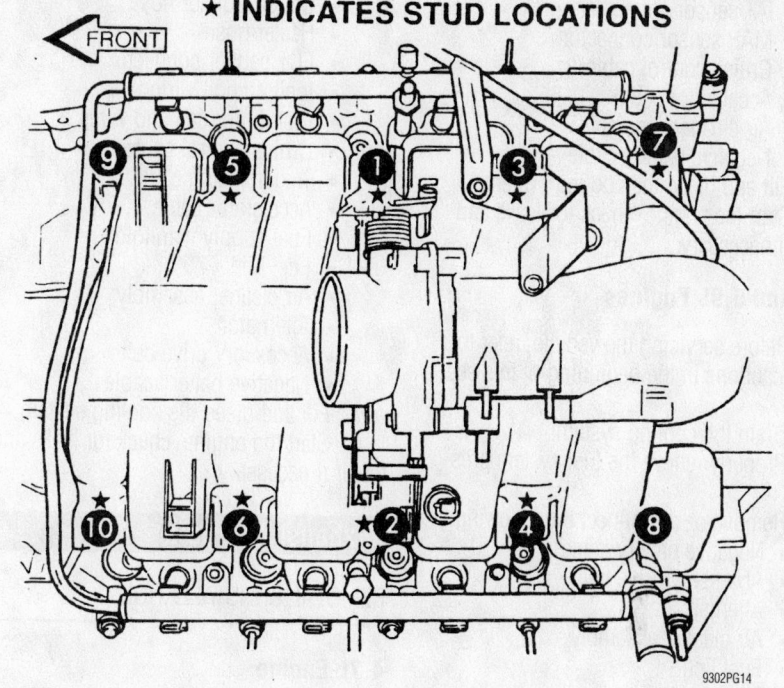

★ **INDICATES STUD LOCATIONS**

Intake manifold torque sequence—3.7 & 3.7L engines

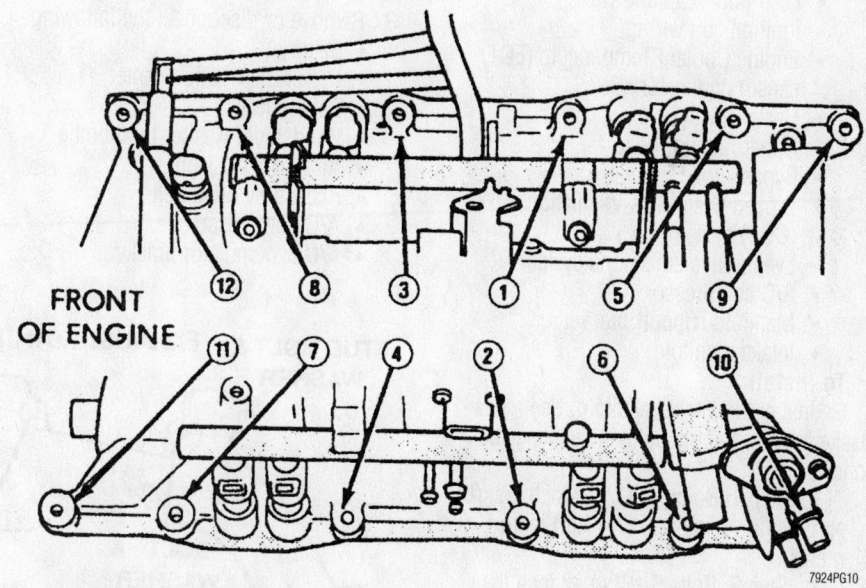

Intake manifold torque sequence—5.9L engine

- Fuel supply manifold
- Fuel line
- Oil dipstick tube
- Ignition coil towers
- Engine ground straps
- A/C compressor
- Alternator

- Accessory drive belt
- Cruise control servo hose
- Brake booster vacuum line
- Canister purge vacuum line
- PCV valve and hose
- ECT sensor
- IAC valve connector

For Maintenance Interval recommendations, see Section 1 of this manual

- TP sensor connector
- IAT sensor connector
- MAP sensor connector
- Cruise control cable
- Accelerator cable
- Air cleaner assembly
- Negative battery cable

6. Fill and bleed the cooling system.

7. Start the engine, check for leaks and repair if necessary.

5.2L and 5.9L Engines

1. Before servicing the vehicle, refer to the precautions in the beginning of this section.

2. Drain the cooling system.

3. Properly relieve the fuel system pressure.

4. Remove or disconnect the following:
- Negative battery cable
- Accessory drive belt
- Alternator
- Air cleaner assembly
- Fuel line
- Fuel supply manifold
- Accelerator cable
- Transmission cable
- Cruise control cable
- Distributor cap and wires
- Ignition coil wiring
- Engine Coolant Temperature (ECT) sensor connector
- Heater hose
- Upper radiator hose
- Bypass hose
- Closed Crankcase Ventilation (CCV) system
- Evaporative emissions system
- A/C compressor
- Manifold support bracket
- Intake manifold

To install:

5. Install the intake manifold. Use a new gasket and torque the bolts in sequence as follows:

   a. Step 1: Bolts 1–4 to 72 inch lbs. (8 Nm) using 12 inch lb. (1.4 Nm) increments

   b. Step 2: Bolts 5–12 to 72 inch lbs. (8 Nm)

   c. Step 3: Bolts 1–12 to 72 inch lbs. (8 Nm)

   d. Step 4: Bolts 1–12 to 12 ft. lbs. (16 Nm)

   e. Step 5: Bolts 1–12 to 12 ft. lbs. (16 Nm)

6. Install or connect the following:
- Manifold support bracket
- A/C compressor
- Evaporative emissions system
- CCV system

- Bypass hose
- Upper radiator hose
- Heater hose
- ECT sensor connector
- Ignition coil wiring
- Distributor cap and wires
- Cruise control cable
- Transmission cable
- Accelerator cable
- Fuel supply manifold
- Fuel line
- Air cleaner assembly
- Alternator
- Accessory drive belt
- Negative battery cable

7. Fill and bleed the cooling system.

8. Start the engine, check for leaks and repair if necessary.

Exhaust Manifold

REMOVAL & INSTALLATION

4.7L Engine

1. Before servicing the vehicle, refer to the precautions in the beginning of this section.

2. Drain the cooling system.

3. Remove or disconnect the following:
- Battery
- Power distribution center
- Battery tray
- Windshield washer fluid bottle
- Air cleaner assembly
- Accessory drive belt
- A/C compressor
- A/C accumulator bracket

- Heater hoses
- Exhaust manifold heat shields
- Exhaust Y-pipe
- Starter motor
- Exhaust manifolds

To install:

4. Install or connect the following:
- Exhaust manifolds, using new gaskets. Torque the bolts to 18 ft. lbs. (25 Nm), starting with the inner bolts and work out to the ends.
- Starter motor
- Exhaust Y-pipe
- Exhaust manifold heat shields
- Heater hoses
- A/C accumulator bracket
- A/C compressor
- Accessory drive belt
- Air cleaner assembly
- Windshield washer fluid bottle
- Battery tray
- Power distribution center
- Battery

5. Fill and bleed the cooling system.

6. Start the engine, check for leaks and repair if necessary.

3.7L, 5.2L and 5.9L Engines

1. Before servicing the vehicle, refer to the precautions in the beginning of this section.

2. Remove or disconnect the following:
- Negative battery cable
- Exhaust manifold heat shields
- Exhaust Gas Recirculation (EGR) tube
- Exhaust Y-pipe
- Exhaust manifolds

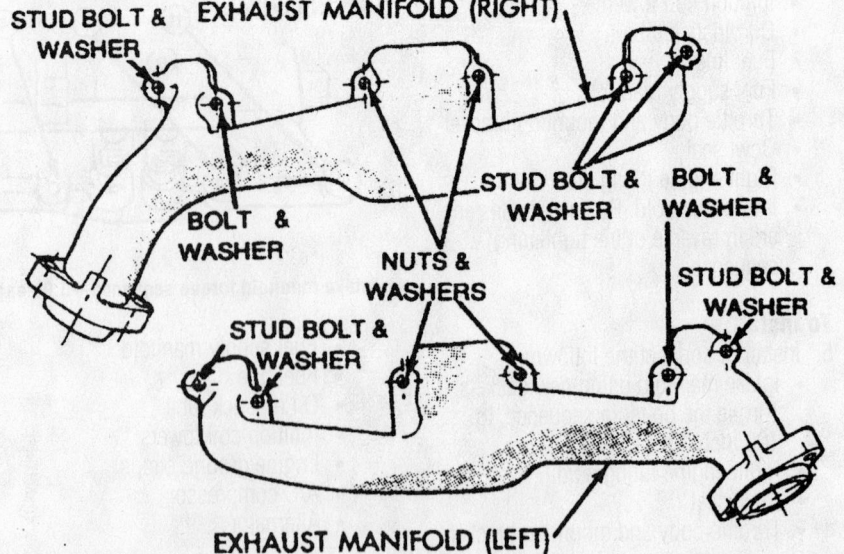

Exhaust manifold fastener locations—5.2L and 5.9L engines

7924PG11

To install:

➡️**If the exhaust manifold studs came out with the nuts when removing the exhaust manifolds, replace them with new studs.**

3. Install or connect the following:
- Exhaust manifolds. Torque the fasteners to 20 ft. lbs. (27 Nm), starting with the center nuts and work out to the ends.
- Exhaust Y-pipe
- EGR tube
- Exhaust manifold heat shields
- Negative battery cable

4. Start the engine, check for leaks and repair if necessary.

Combination Manifold

REMOVAL & INSTALLATION

2.5L Engine

1. Before servicing the vehicle, refer to the precautions in the beginning of this section.
2. Properly relieve the fuel system pressure.
3. Drain the cooling system.
4. Remove or disconnect the following:
- Negative battery cable
- Air intake hose
- Accessory drive belt
- Power steering pump and brackets, if equipped
- Fuel line
- Accelerator cable
- Cruise control cable, if equipped
- Transmission cable, if equipped
- Throttle Position (TP) sensor connector
- Idle Air Control (IAC) valve connector
- Engine Coolant Temperature (ECT) sensor connector
- Intake Air Temperature (IAT) sensor connector
- Heated Oxygen (HO2S) sensor connector
- Fuel injector connectors
- Closed Crankcase Ventilation (CCV) system
- Manifold Absolute Pressure (MAP) sensor vacuum line
- Brake booster vacuum line
- Exhaust front pipe
- Intake manifold and discard the gaskets

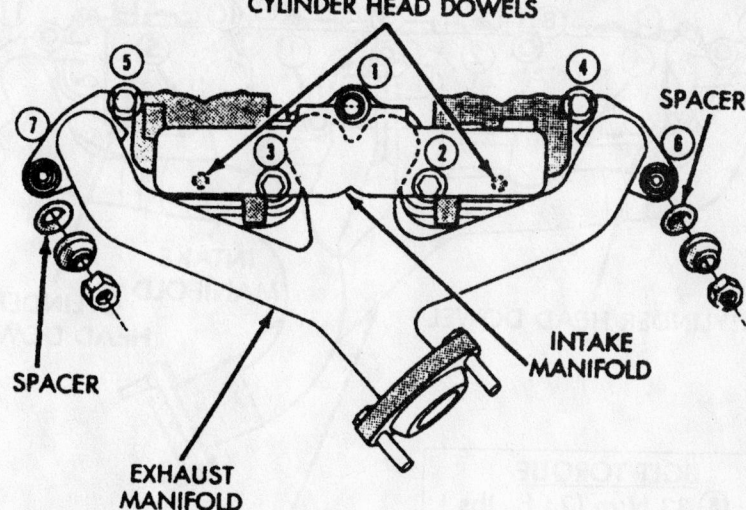

Combination manifold head torque sequence—2.5L engine

7924PG08

- Exhaust manifold and discard the gaskets

To install:
5. Install or connect the following:
- New gasket over the locating dowels
- Exhaust manifold to the studs and tighten the nuts finger-tight
6. Install the intake manifold. Torque the fasteners in sequence as follows:
   a. Step 1: Bolt 1 to 30 ft. lbs. (41 Nm).
   b. Step 2: Fasteners 2–7 to 23 ft. lbs. (31 Nm).
7. Install or connect the following:
- Exhaust front pipe
- Brake booster vacuum line
- MAP sensor vacuum line
- CCV system
- Fuel injector connectors
- HO2S sensor connector
- IAT sensor connector
- ECT sensor connector
- IAC valve connector
- TP sensor connector
- Transmission cable, if equipped
- Cruise control cable, if equipped
- Accelerator cable
- Fuel line
- Power steering pump and brackets, if equipped
- Accessory drive belt
- Air intake hose
- Negative battery cable
8. Fill and bleed the cooling system.
9. Start the engine, check for leaks and repair if necessary..

4.0L Engine

1. Before servicing the vehicle, refer to the precautions in the beginning of this section.
2. Drain the cooling system.
3. Properly relieve the fuel system pressure.
4. Remove or disconnect the following:
- Negative battery cable
- Air cleaner assembly
- Accessory drive belt
- Power steering pump and brackets, if equipped
- Fuel line
- Fuel injector connectors
- Fuel supply manifold and injectors
- Accelerator cable
- Cruise control cable, if equipped
- Transmission cable, if equipped
- Throttle Position (TP) sensor connector
- Idle Air Control (IAC) valve connector
- Engine Coolant Temperature (ECT) sensor connector
- Intake Air Temperature (IAT) sensor connector
- Heated Oxygen (HO2S) sensor connector
- Closed Crankcase Ventilation (CCV) system
- Manifold Absolute Pressure (MAP) sensor vacuum line
- Brake booster vacuum line
- Exhaust front pipe
- Intake and exhaust manifolds and discard the gaskets

For Tune-up, Capacities and Firing orders, see Section 1 of this manual

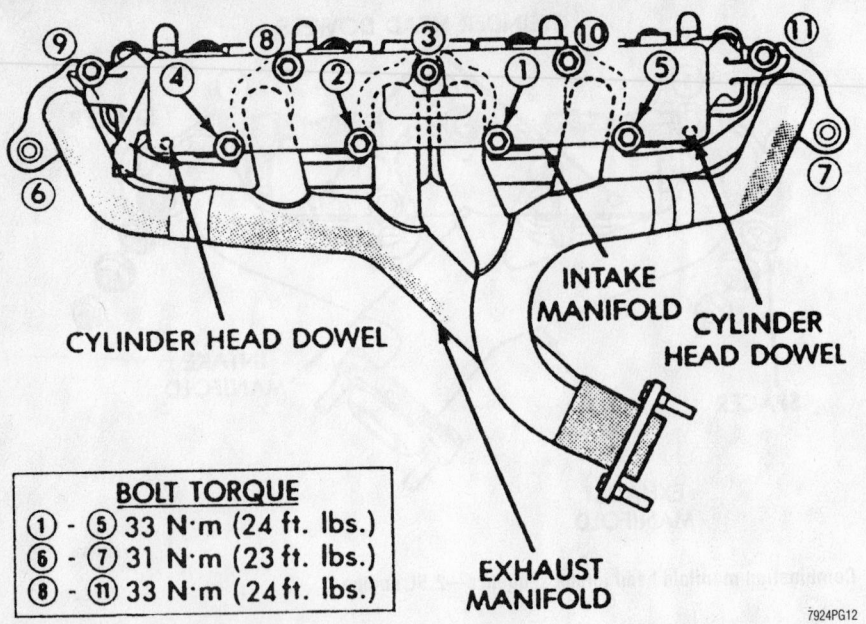

BOLT TORQUE
① - ⑤ 33 N·m (24 ft. lbs.)
⑥ - ⑦ 31 N·m (23 ft. lbs.)
⑧ - ⑪ 33 N·m (24 ft. lbs.)

Combination manifold torque sequence—4.0L engine

To install:

5. Install or connect the following:
   • New gasket over the locating dowels
   • Exhaust manifold to the studs and tighten the nuts finger-tight

6. Install the intake manifold. Torque the fasteners in sequence as follows:
   a. Step 1: Fasteners 1–5 to 30 ft. lbs. (41 Nm).
   b. Step 2: Fasteners 6 and 7 to 23 ft. lbs. (31 Nm).
   c. Step 3: Fasteners 8–11 to 24 ft. lbs. (33 Nm)

7. Install or connect the following:
   • Exhaust front pipe
   • Brake booster vacuum line
   • MAP sensor vacuum line
   • CCV system
   • HO$_2$S sensor connector
   • IAT sensor connector
   • ECT sensor connector
   • IAC valve connector
   • TP sensor connector
   • Transmission cable, if equipped
   • Cruise control cable, if equipped
   • Accelerator cable
   • Fuel supply manifold and injectors
   • Fuel injector connectors
   • Fuel line
   • Power steering pump and brackets, if equipped
   • Accessory drive belt
   • Air cleaner assembly
   • Negative battery cable

8. Fill and bleed the cooling system.

9. Start the engine, check for leaks and repair if necessary.

Camshaft and Valve Lifters

REMOVAL & INSTALLATION

2.5L Engine

1. Before servicing the vehicle, refer to the precautions in the beginning of this section.

2. Drain the cooling system.

3. Recover the A/C refrigerant, if equipped with air conditioning.

4. Remove or disconnect the following:
   • Negative battery cable
   • Grille, if necessary
   • Radiator

   • A/C condenser, if equipped
   • Distributor and ignition wires
   • Valve cover

➡**Keep all valvetrain components in order for assembly.**

   • Rocker arms and pushrods
   • Hydraulic valve tappets
   • Accessory drive belt
   • Crankshaft damper
   • Front cover
   • Timing chain and gears
   • Camshaft

To install:

➡**If the camshaft sprocket appears to have been rubbing against the cover, check the oil pressure relief holes in the rear cam journal for debris.**

5. Lubricate the camshaft with clean engine oil.

6. Install or connect the following:
   • Camshaft
   • Timing chain and gears
   • Front cover
   • Crankshaft damper
   • Accessory drive belt
   • Hydraulic valve tappets
   • Rocker arms and pushrods
   • Valve cover
   • Distributor
   • A/C condenser, if equipped
   • Radiator
   • Grille, if removed
   • Negative battery cable

7. Fill and bleed the cooling system.

8. Recharge the A/C system, if equipped.

9. Start the engine, check for leaks and repair if necessary.

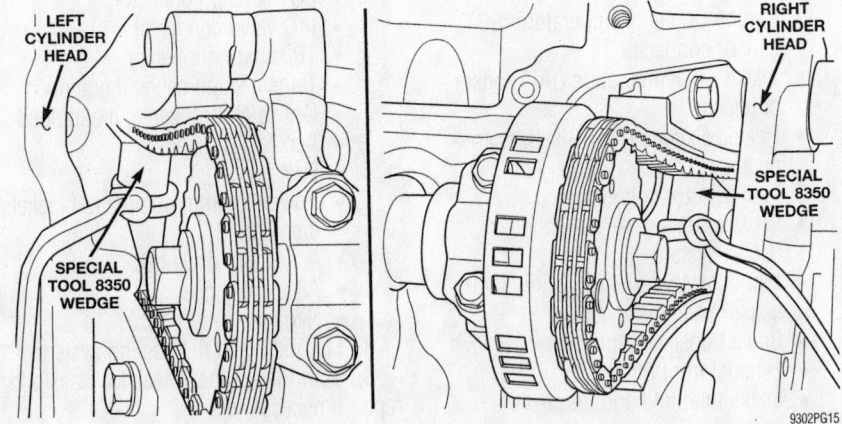

Chain Tensioner Retaining Wedges

4.0L Engine

1. Before servicing the vehicle, refer to the precautions in the beginning of this section.

2. Drain the cooling system.

3. Recover the A/C refrigerant, if equipped with air conditioning.

4. Remove or disconnect the following:
- Negative battery cable
- Grille, if necessary
- Radiator
- A/C condenser, if equipped
- Distributor or camshaft sensor housing
- Valve cover

➡**Keep all valvetrain components in order for assembly.**

- Rocker arms and pushrods
- Cylinder head
- Hydraulic valve tappets
- Accessory drive belt
- Crankshaft damper
- Front cover
- Timing chain and gears
- Thrust plate
- Camshaft

To install:

5. Lubricate the camshaft with clean engine oil.

6. Install or connect the following:
- Camshaft
- Thrust plate. Torque the bolts to 18 ft. lbs. (24 Nm).
- Timing chain and gears
- Front cover
- Crankshaft damper
- Accessory drive belt
- Hydraulic valve tappets
- Cylinder head
- Rocker arms and pushrods
- Valve cover
- Distributor or camshaft sensor housing
- A/C condenser, if equipped
- Radiator
- Grille, if removed
- Negative battery cable

7. Fill and bleed the cooling system.

8. Recharge the A/C system, if equipped.

9. Start the engine, check for leaks and repair if necessary.

3.7L and 4.7L Engines

1. Before servicing the vehicle, refer to the precautions in the beginning of this section.

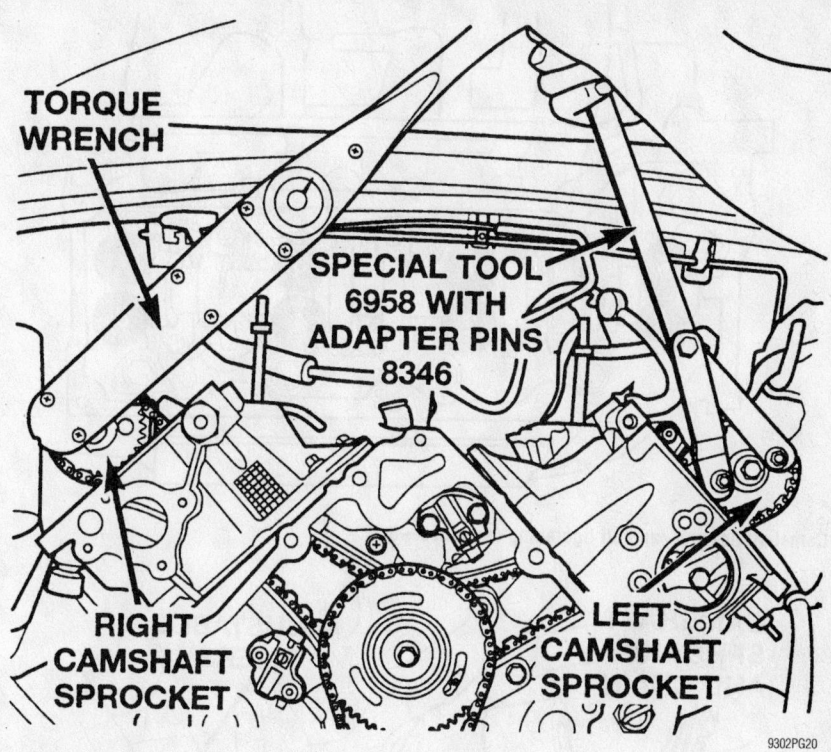

Hold the left camshaft sprocket with a spanner wrench while removing or installing the camshaft sprocket bolts—4.7L engine

2. Remove or disconnect the following:
- Negative battery cable
- Cylinder head covers
- Rocker arms
- Hydraulic lash adjusters

➡**Keep all valvetrain components in order for assembly.**

3. Set the engine at Top Dead Center (TDC) of the compression stroke for the No. 1 cylinder.

4. Install Timing Chain Wedge (8350 4.7L; 8379 3.7L) to retain the chain tensioners.

5. Matchmark the timing chains to the camshaft sprockets.

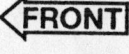
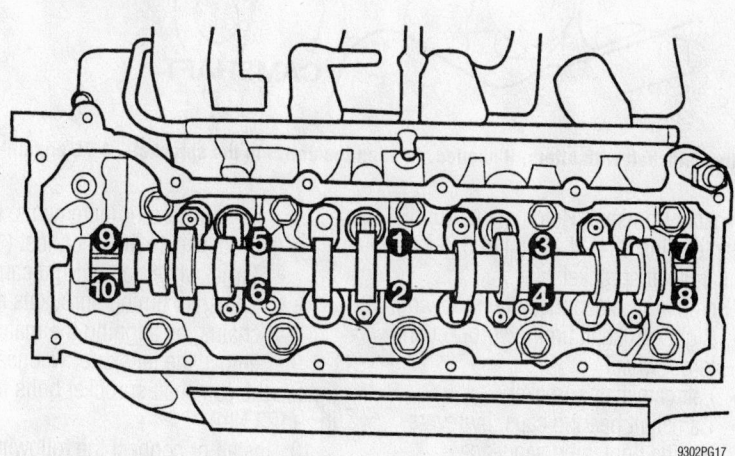

Camshaft bearing cap bolt tightening sequence—4.7L engine

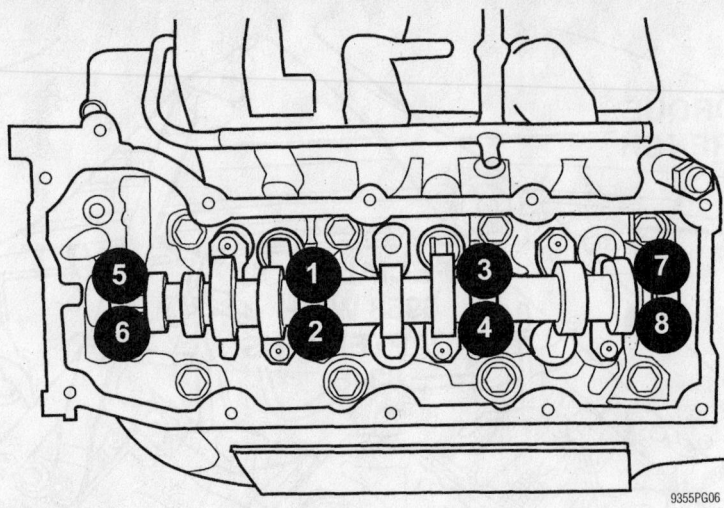

Camshaft bearing cap bolt tightening sequence—3.7L

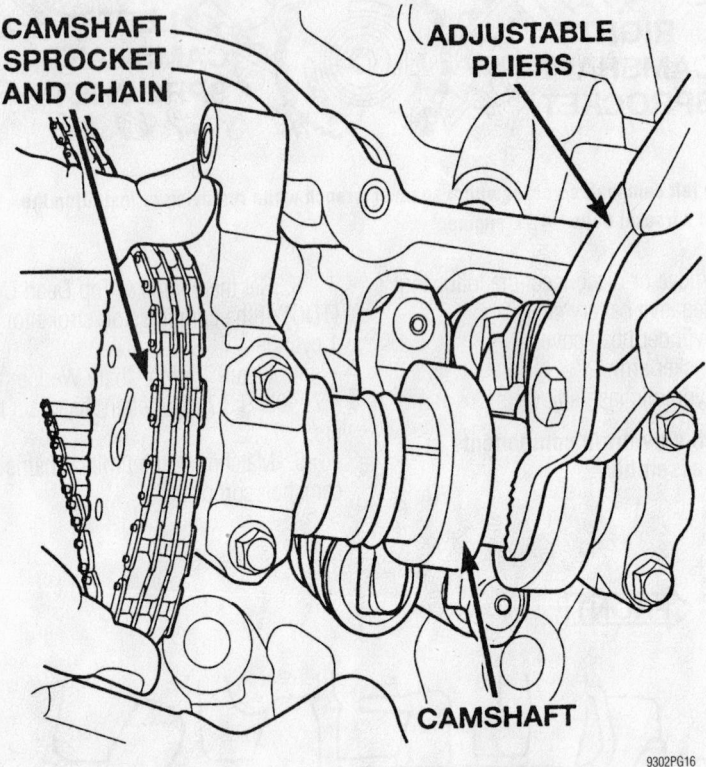

Turn the camshaft with pliers, if needed, to align the dowel in the sprocket—4.7L engine

6. Install Camshaft Holding Tool (6958 and Adapter Pins 8346 4.7L; 8428 3.7L) to the left camshaft sprocket.

7. Remove or disconnect the following:
- Right camshaft timing sprocket and target wheel
- Left camshaft sprocket
- Camshaft bearing caps, by reversing the tightening sequence
- Camshafts

To install:

8. Install or connect the following:
- Camshafts. Torque the bearing cap

bolts in ½ turn increments, in sequence, to 100 inch lbs. (11 Nm).
- Target wheel to the right camshaft
- Camshaft timing sprockets and chains, by aligning the matchmarks

9. Remove the tensioner wedges and tighten the camshaft sprocket bolts to 90 ft. lbs. (122 Nm).

10. Install or connect the following:
- Hydraulic lash adjusters in their original locations
- Rocker arms in their original locations

- Cylinder head covers
- Negative battery cable

5.2L and 5.9L Engines

1. Before servicing the vehicle, refer to the precautions in the beginning of this section.

2. Drain the cooling system.

3. Recover the A/C refrigerant.

4. Set the crankshaft to Top Dead Center (TDC) of the compression stroke for the No. 1 cylinder.

5. Remove or disconnect the following:

- Negative battery cable
- Accessory drive belt
- Power steering pump
- Water pump
- Radiator
- A/C condenser
- Grille
- Crankshaft damper
- Front cover
- Valve covers
- Distributor
- Intake manifold

➡**Keep all valvetrain components in order for assembly.**

- Rocker arms and pushrods
- Hydraulic lifters
- Timing chain and sprockets
- Camshaft thrust plate and chain oil tab
- Camshaft

To install:

6. Install or connect the following:
- Camshaft, secure it in position with the Camshaft Holding Tool C-3509
- Camshaft thrust plate and chain oil tab. Torque the bolts to 18 ft. lbs. (24 Nm).
- Timing chain and sprockets, by aligning the timing marks. Torque the camshaft sprocket bolt to 50 ft. lbs. (68 Nm).

7. Remove the camshaft holding tool.

8. Install or connect the following:
- Hydraulic lifters in their original positions
- Rocker arms and pushrods in their original positions
- Intake manifold
- Distributor
- Valve covers
- Front cover
- Crankshaft damper
- Grille
- A/C condenser
- Radiator

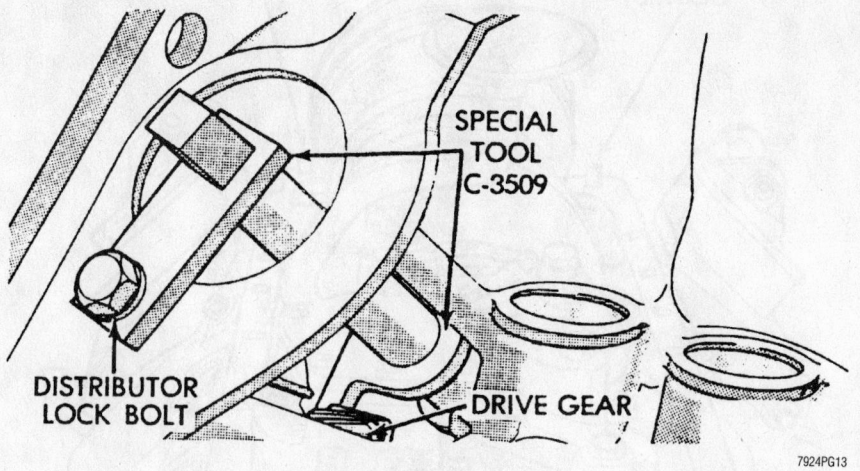

Camshaft holding tool C-3509—5.2L and 5.9L engines

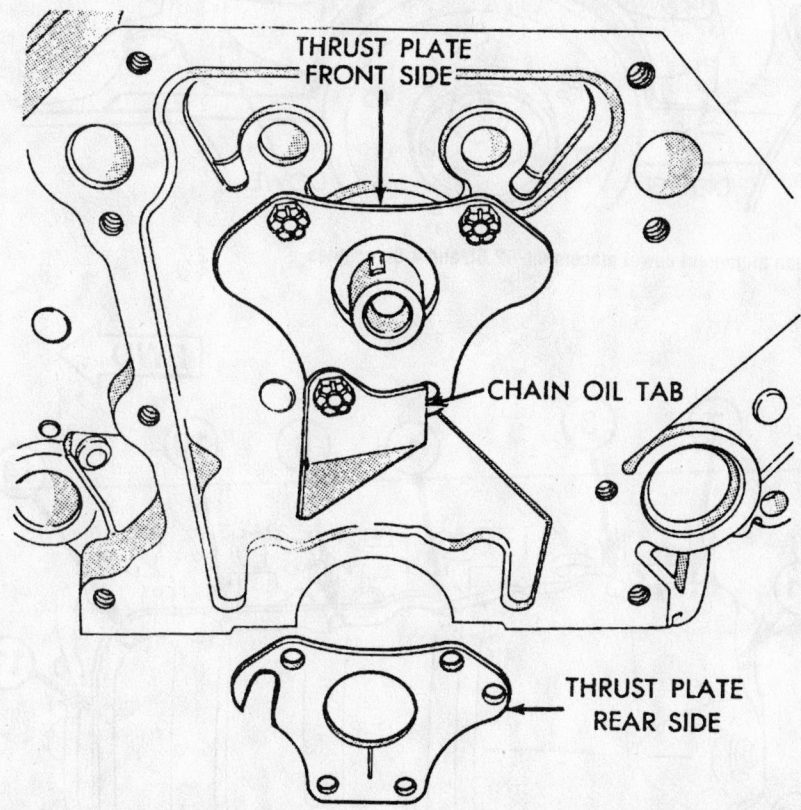

Thrust plate position—5.2L and 5.9L engines

- Water pump
- Power steering pump
- Accessory drive belt
- Negative battery cable
9. Fill and bleed the cooling system.
10. Recharge the A/C system.
11. Start the engine, check for leaks and repair if necessary.

Valve Lash

ADJUSTMENT

These engines are equipped with hydraulic valve lifters. No valve clearance adjustments are necessary.

Starter Motor

REMOVAL & INSTALLATION

1. Before servicing the vehicle, refer to the precautions in the beginning of this section.
2. Remove or disconnect the following:
- Negative battery cable
- Starter mounting bolts

➡On the 3.7L engine, the left side exhaust pipe and front driveshaft must be disconnected.

- Starter solenoid harness connections
- Starter

To install:
3. Connect the starter solenoid wiring connectors.
4. Install the starter and torque the bolts to the following specifications:
- 2.5L engine: 33 ft. lbs. (45 Nm).
- 4.0L engine: Upper bolt to 40 ft. lbs. (54 Nm) and lower bolt to 30 ft. lbs. (41 Nm).
- 2.4L, 3.7L and 4.7L engine: 40 ft. lbs. (54 Nm).
- 5.2L and 5.9L engines: 50 ft. lbs. (68 Nm).
- On the 3.7L engine, the left side exhaust pipe and front driveshaft
5. Install the negative battery cable and check for proper operation.

Oil Pan

REMOVAL & INSTALLATION

2.5L and 4.0L Engines

1. Before servicing the vehicle, refer to the precautions in the beginning of this section.
2. Drain the engine oil.
3. Remove or disconnect the following:
- Negative battery cable
- Exhaust front pipe
- Starter motor
- Bell housing access cover
- Oil level sensor connector, if equipped
- Left and right motor mounts
- Transmission oil cooler lines, if equipped
4. Place a jack under the crankshaft damper and raise the engine for clearance.
5. Remove the oil pan.

To install:

6. Fabricate 4 alignment dowels from 1½ in. x ¼ in. bolts. Cut the heads off the bolts and cut a slot into the top of the dowel to allow installation/removal with a screwdriver.

7. Install or connect the following:
   • Dowels
   • Oil pan, using a new gasket. Torque the ¼ inch bolts to 85 inch lbs. (9.5 Nm) and the ⁵⁄₁₆ inch bolts to 11 ft. lbs. (15 Nm).

8. Replace the alignment dowels with ¼ inch bolts and torque them to 85 inch lbs. (9.5 Nm).

9. Install or connect the following:
   • Left and right motor mounts
   • Oil level sensor connector, if equipped
   • Bell housing access cover
   • Starter motor
   • Exhaust front pipe
   • Negative battery cable

10. Fill the engine with clean oil.

11. Start the engine, check for leaks and repair if necessary.

3.7L Engine

1. Before servicing the vehicle, refer to the precautions in the beginning of this section.

2. Remove or disconnect the following:
   • Engine from the vehicle
   • Oil pan
   • Oil pump pickup tube
   • Oil pan gasket

3. Installation is the reverse of removal. Torque the bolts, in sequence, to 11 ft. lbs. (15 Nm).

4.7L Engine

1. Before servicing the vehicle, refer to the precautions in the beginning of this section.

2. Drain the engine oil.

3. Remove or disconnect the following:
   • Negative battery cable
   • Structural cover
   • Exhaust Y-pipe
   • Starter motor
   • Transmission oil cooler lines
   • Oil pan
   • Oil pump pickup tube
   • Oil pan gasket

To install:

4. Install or connect the following:
   • Oil pan gasket
   • Oil pump pickup tube, using a new O-ring. Torque the tube bolts to 20 ft. lbs. (28 Nm); torque the O-ring end bolt first.

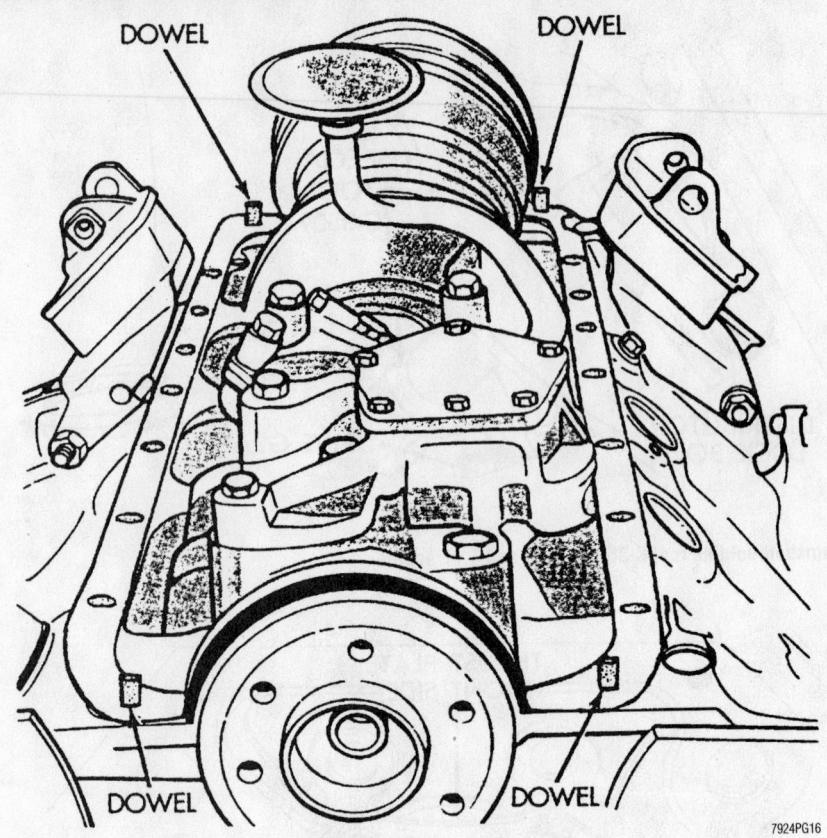

Oil pan alignment dowel placement—2.5L and 4.0L engines

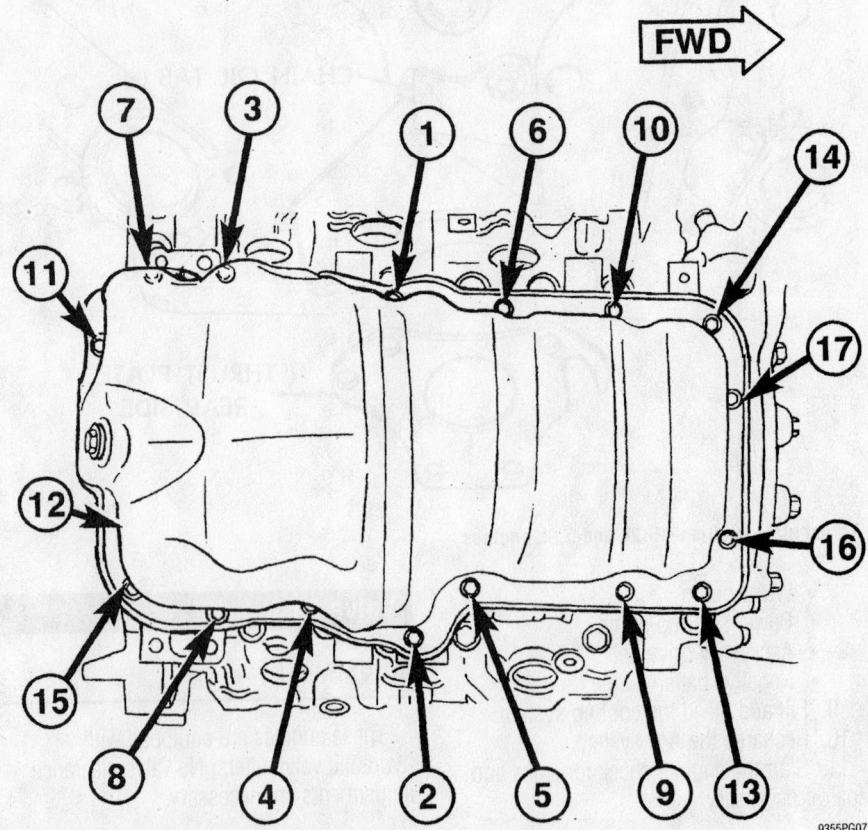

Oil pan bolt torque sequence—3.7L

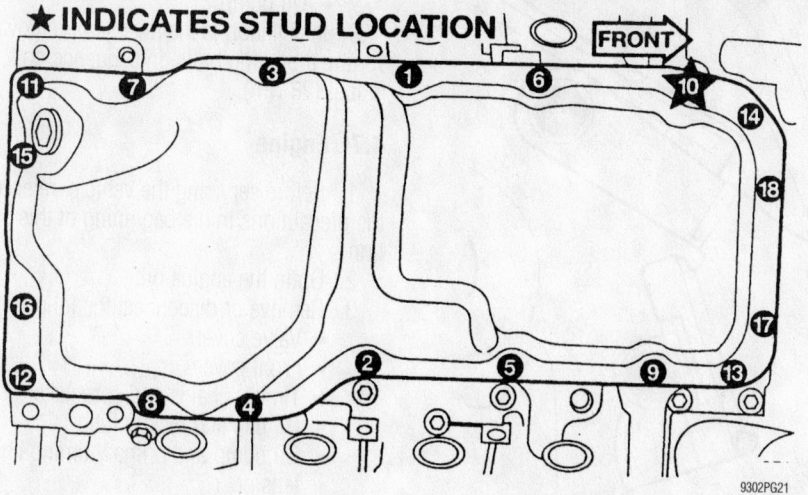

★ INDICATES STUD LOCATION

Oil pan mounting bolt tightening sequence—4.7L engine

- Oil pan. Torque the bolts, in sequence, to 11 ft. lbs. (15 Nm).
- Transmission oil cooler lines
- Starter motor
- Exhaust Y-pipe
- Structural cover
- Negative battery cable

5. Fill the engine to the proper level with clean oil.

6. Start the engine, check for leaks and repair if necessary.

5.2L and 5.9L Engines

1. Before servicing the vehicle, refer to the precautions in the beginning of this section.

2. Drain the engine oil.

3. Remove or disconnect the following:
- Oil filter
- Starter motor
- Cooler lines
- Oil level sensor connector
- Heated Oxygen (HO$_2$S) sensor connector
- Exhaust Y-pipe
- Oil pan

To install:

4. Install or connect the following:
- Oil pan, using a new gasket. Torque the bolts to 18 ft. lbs. (24 Nm).
- Exhaust Y-pipe
- Heated Oxygen (HO$_2$S) sensor connector
- Oil level sensor connector
- Cooler lines
- Starter motor
- Oil filter

5. Fill the engine with clean oil.

6. Start the engine, check for leaks and repair if necessary.

Oil Pump

REMOVAL & INSTALLATION

2.5L and 4.0L Engines

1. Before servicing the vehicle, refer to the precautions in the beginning of this section.

2. Drain the engine oil.

3. Remove or disconnect the following:
- Negative battery cable
- Oil pan
- Oil pump and pickup tube

➡ If the oil pump is not to be serviced, do not disturb the position of the oil inlet tube and strainer assembly in the pump body. If the tube is moved within the pump body, a replacement tube and strainer assembly must be installed to assure an airtight seal.

To install:

4. Install or connect the following:
- Oil pump. Torque the mounting bolts to 17 ft. lbs. (23 Nm).
- Oil pan
- Negative battery cable

5. Fill the engine with the proper type and quantity of oil.

6. Start the engine, check for leaks and repair if necessary.

3.7L Engine

1. Before servicing the vehicle, refer to the precautions in the beginning of this section.

2. Remove or disconnect the following:
- Oil Pan
- Timing chain cover
- Timing chains and tensioners

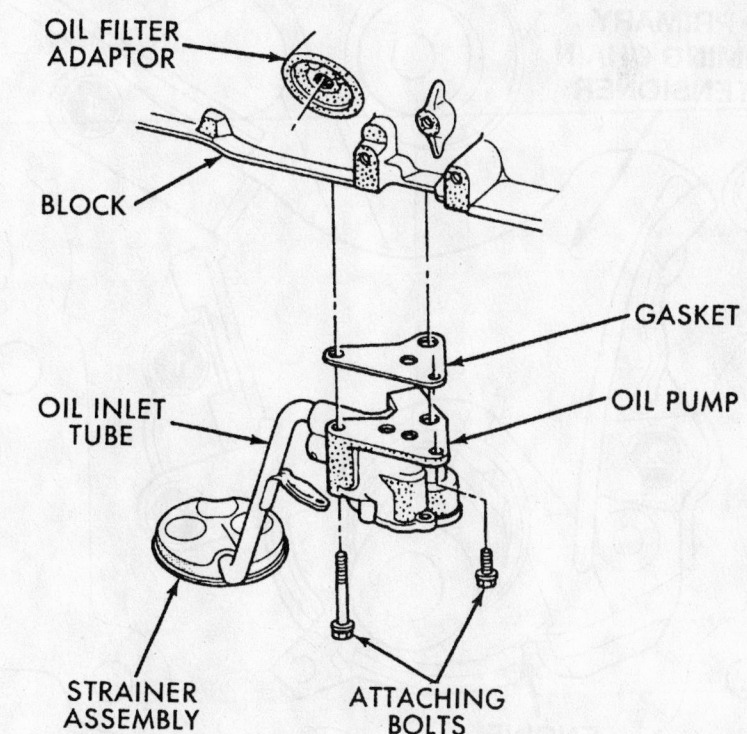

Exploded view of the oil pump assembly—2.5L and 4.0L engine

Timing belt service is covered in Section 3 of this manual

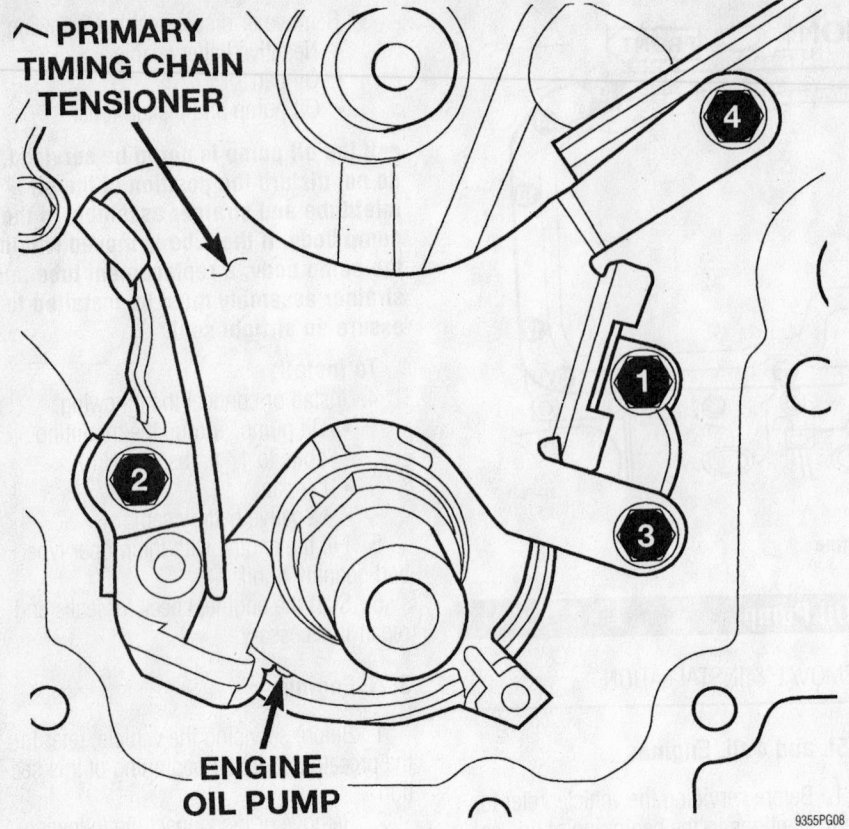

Oil pump bolt torque sequence—3.7L

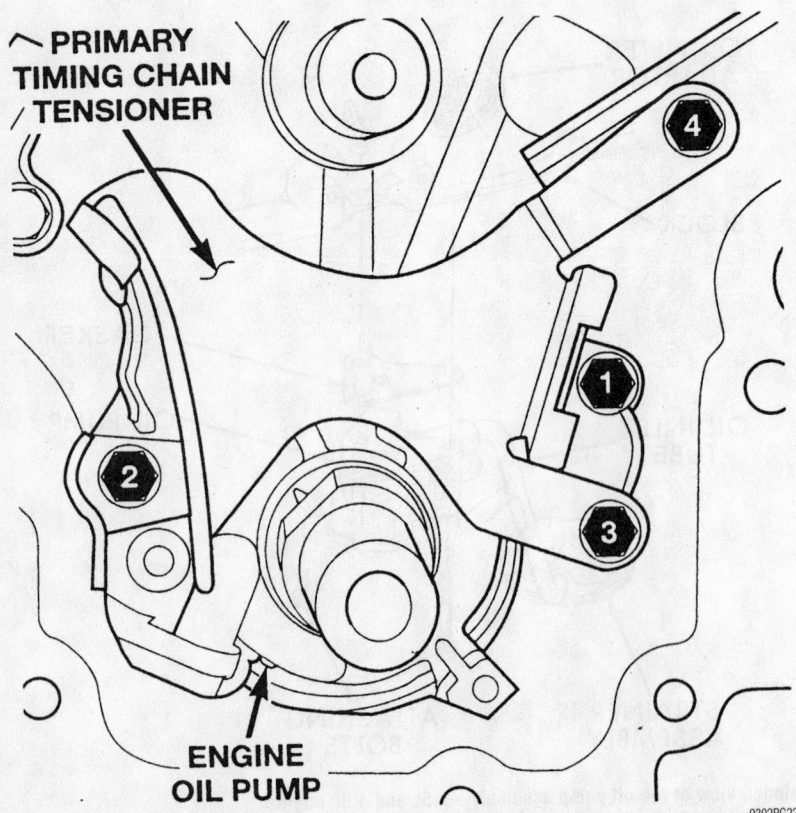

Oil pump and chain tensioner torque sequence—4.7L engine

- Oil pump
3. Installation is the reverse of removal. Torque the pump bolts, in sequence, to 21 ft. lbs. (28 Nm),

4.7L Engine

1. Before servicing the vehicle, refer to the precautions in the beginning of this section.
2. Drain the engine oil.
3. Remove or disconnect the following:
   - Valve covers
   - Front cover
   - Timing chains and sprockets
   - Oil pan and pickup tube
   - Oil pump and primary timing chain tensioner

To install:

4. Install or connect the following:
   - Oil pump and primary timing chain tensioner. Torque the bolts in sequence to 21 ft. lbs. (28 Nm).
   - Oil pan and pickup tube
   - Timing chains and sprockets
   - Front cover
   - Valve covers
5. Fill the engine with clean oil.
6. Start the engine, check for leaks and repair if necessary.

5.2L and 5.9L Engines

1. Before servicing the vehicle, refer to the precautions in the beginning of this section.
2. Drain the engine oil.
3. Remove or disconnect the following:
   - Negative battery cable
   - Oil pan
   - Oil pump

To install:

4. Install or connect the following:
   - Oil pump. Torque the bolts to 30 ft. lbs. (41 Nm).
   - Oil pan
   - Negative battery cable
5. Fill the engine with clean oil.
6. Start the engine, check for leaks and repair if necessary.

Rear Main Seal

REMOVAL & INSTALLATION

2.5L Engine

1. Before servicing the vehicle, refer to the precautions in the beginning of this section.
2. Remove or disconnect the following:
   - Negative battery cable

- Flywheel/converter drive plate and discard the bolt
- Rear main seal

To install:

3. Install or connect the following:
- Rear main seal so that it is flush with the cylinder block
- Flywheel/ converter drive plate. Torque the new to 50 ft. lbs. (68 Nm) plus a 60 degrees turn
- Negative battery cable

4. Start the engine, check for leaks and repair if necessary.

4.0L Engine

1. Before servicing the vehicle, refer to the precautions in the beginning of this section.
2. Drain the engine oil.
3. Remove or disconnect the following:
- Negative battery cable
- Transmission inspection cover
- Oil pan
- Main bearing cap brace
- No. 7 main bearing cap

4. Loosen the other main bearing cap bolts for clearance and remove the rear main seal halfs.

To install:

5. Install or connect the following:
- New upper seal half to the cylinder block
- New lower seal half to the bearing cap after applying sealant to the bearing cap
- No. 7 main bearing cap. Torque **all** main bearing cap bolts to 80 ft. lbs. (108 Nm).
- Main bearing cap brace. Tighten the nuts to 35 ft. lbs. (47 Nm).
- Oil pan
- Transmission inspection cover
- Negative battery cable

6. Fill the engine with clean oil.
7. Start the engine, check for leaks and repair if necessary.

3.7L and 4.7L Engines

1. Before servicing the vehicle, refer to the precautions in the beginning of this section.

2. Remove or disconnect the following:

- Transmission
- Flexplate

3. Thread Oil Seal Remover 8506 into the rear main seal as far as possible and remove the rear main seal.

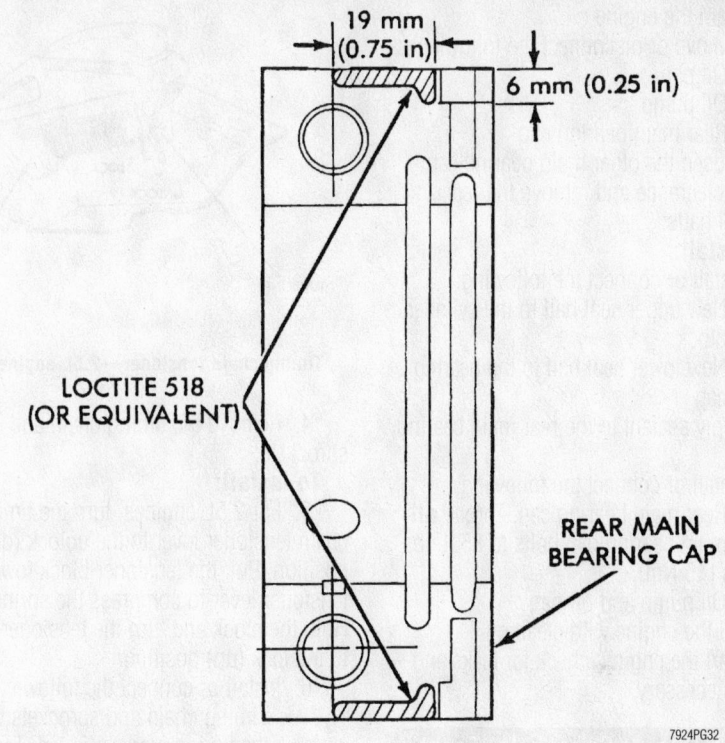

Sealant application locations—4.0L engine

To install:

4. Install or connect the following:
- Seal Guide 8349-2 onto the crankshaft
- Rear main seal on the seal guide
- Rear main seal, using the Crankshaft Rear Oil Seal Installer 8349 and Driver Handle C-4171; tap it into place until the installer is flush with the cylinder block

- Flexplate. Torque the bolts to 45 ft. lbs. (60 Nm).
- Transmission

5. Start the engine, check for leaks and repair if necessary.

5.2L and 5.9L Engines

1. Before servicing the vehicle, refer to the precautions in the beginning of this section.

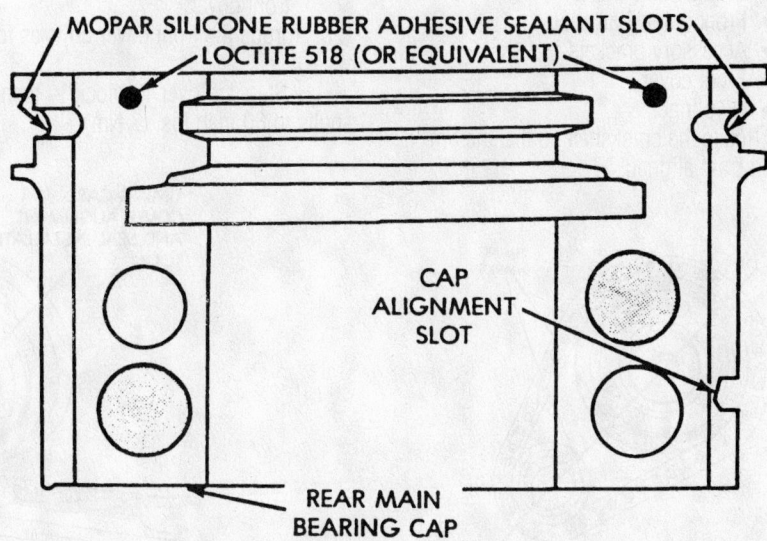

Sealant application locations —5.2L and 5.9L engines

Heater Core replacement is covered in Section 2 of this manual

2. Drain the engine oil.

3. Remove or disconnect the following:
- Oil pan
- Oil pump
- Rear main bearing cap

4. Loosen the other main bearing cap bolts for clearance and remove the rear main seal halfs.

To install:

5. Install or connect the following:
- New upper seal half to the cylinder block
- New lower seal half to the bearing cap

6. Apply sealant to the rear main bearing cap.

7. Install or connect the following:
- Rear main bearing cap. Torque **all** main bearing cap bolts to 85 ft. lbs. (115 Nm).
- Oil pump and oil pan

8. Fill the engine with clean oil.

9. Start the engine, check for leaks and repair if necessary.

Timing Chain, Sprockets, Front Cover and Seal

REMOVAL & INSTALLATION

2.5L and 4.0L Engines

1. Before servicing the vehicle, refer to the precautions in the beginning of this section.

2. Remove or disconnect the following:
- Negative battery cable
- Accessory drive belt
- Cooling fan and shroud
- Crankshaft damper
- Front crankshaft seal
- Accessory brackets
- Front cover
- Oil slinger

3. Rotate the crankshaft so that the timing marks are aligned.

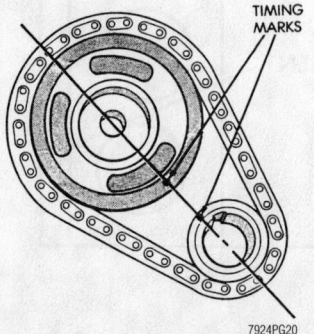

Timing chain alignment marks—2.5L and 4.0L engines

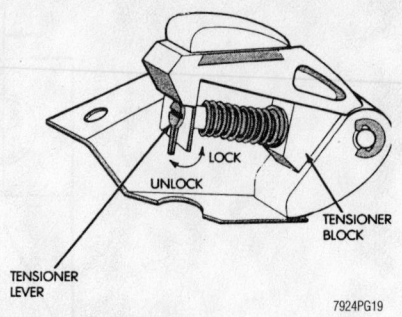

Timing chain tensioner—2.5L engines

4. Remove the timing chain and sprockets.

To install:

5. For 2.5L engines, turn the timing chain tensioner lever to the unlock (down) position. Pull the tensioner block toward the tensioner lever to compress the spring. Hold the block and turn the tensioner lever to the lock (up) position.

6. Install or connect the following:
- Timing chain and sprockets with the timing marks aligned. Torque the camshaft sprocket bolt to 80 ft. lbs. (108 Nm) for 2.5L engines or to 50 ft. lbs. (68 Nm) for 4.0L engines.

7. For 2.5L engines, release the timing chain tensioner.
- Oil slinger
- New front crankshaft seal to the front cover
- Front cover, using a new gasket
- Timing Case Cover Alignment and Seal Installation Tool 6139 in the crankshaft opening to center the front cover

8. Torque the front cover bolts as follows:

a. Step 1: Cover-to-block ¼ inch bolts to 60 inch lbs. (7 Nm).

b. Step 2: Cover-to-block ⁵⁄₁₆ inch bolts to 16 ft. lbs. (22 Nm).

c. Step 3: Oil pan-to-cover ¼ inch bolts to 85 inch lbs. (9.5 Nm).

d. Step 4: Oil pan-to-cover ⁵⁄₁₆ inch bolts to 11 ft. lbs. (15 Nm).

9. Install or connect the following:
- Accessory brackets
- Crankshaft damper. Torque the bolt to 80 ft. lbs. (108 Nm).
- Cooling fan and shroud
- Accessory drive belt
- Negative battery cable

10. Start the engine, check for leaks and repair if necessary.

3.7L and 4.7L Engines

1. Before servicing the vehicle, refer to the precautions in the beginning of this section.

2. Drain the cooling system.

3. Remove or disconnect the following:
- Negative battery cable
- Valve covers

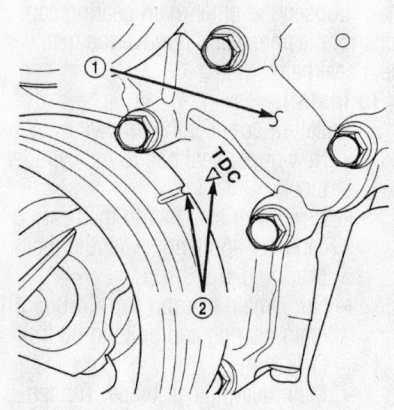

1 – TIMING CHAIN COVER
2 – CRANKSHAFT TIMING MARKS

Crankshaft timing marks—4.7L engine

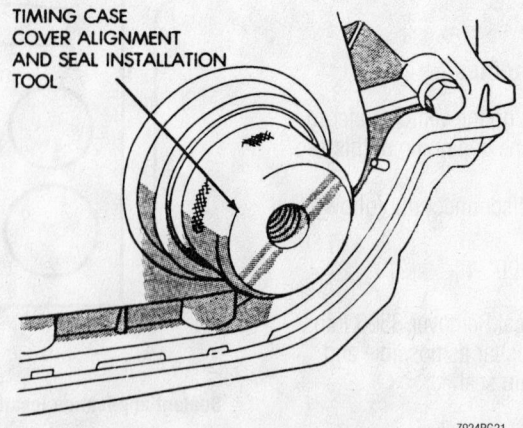

Timing Case Cover Alignment and Seal Installation Tool 6139—2.5L and 4.0L engines

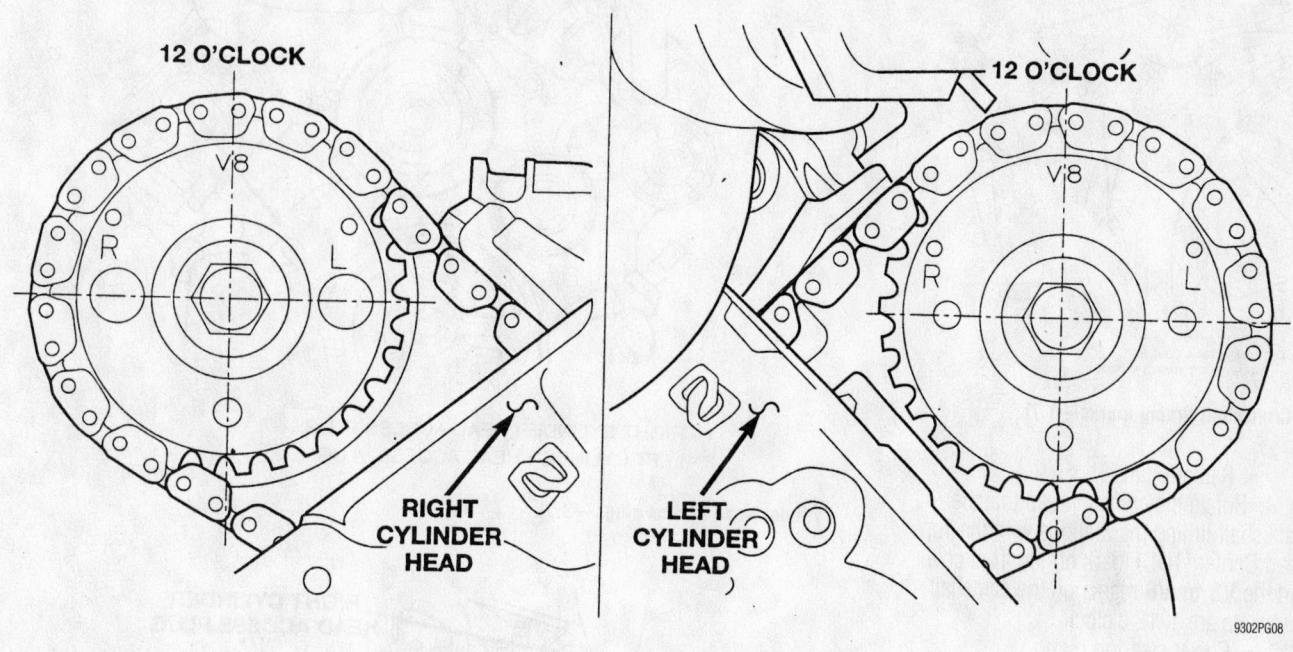

Camshaft positioning—4.7L engine

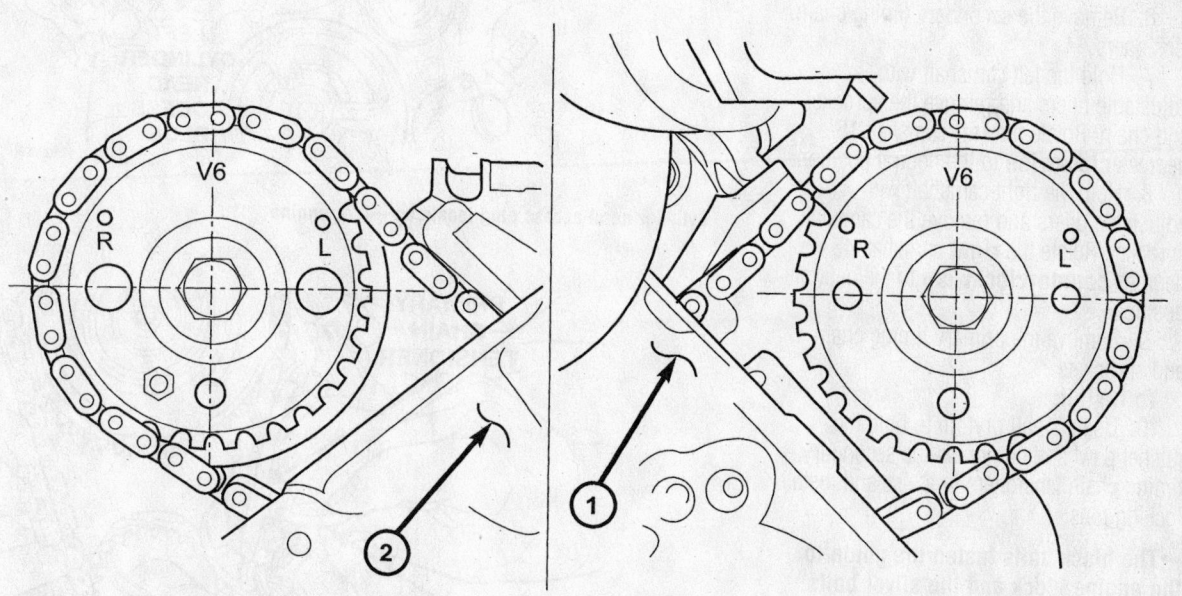

1 - LEFT CYLINDER HEAD
2 - RIGHT CYLINDER HEAD

Camshaft sprocket timing marks—3.7L

Brake service is covered in Section 4 of this manual

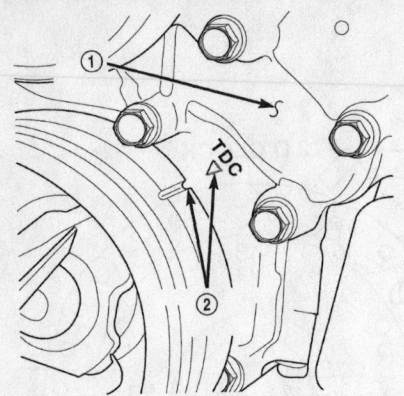

1 - TIMING CHAIN COVER
2 - CRANKSHAFT TIMING MARKS

9355PG10

Crankshaft timing marks—3.7L

- Radiator fan

4. Rotate the crankshaft so that the crankshaft timing mark aligns with the Top Dead Center (TDC) mark on the front cover, and the **V8 or V6** marks on the camshaft sprockets are at 12 o'clock.

- Power steering pump
- Access plugs from the cylinder heads
- Oil fill housing
- Crankshaft damper

5. Compress the primary timing chain tensioner and install a lockpin.

6. Remove the secondary timing chain tensioners.

7. Hold the left camshaft with adjustable pliers and remove the sprocket and chain. Rotate the **left** camshaft 15 degrees **clockwise** to the neutral position.

8. Hold the right camshaft with adjustable pliers and remove the camshaft sprocket. Rotate the **right** camshaft 45 degrees **counterclockwise** to the neutral position.

9. Remove the primary timing chain and sprockets.

To install:

10. Use a small prytool to hold the ratchet pawl and compress the secondary timing chain tensioners in a vise and install locking pins.

➡**The black bolts fasten the guide to the engine block and the silver bolts fasten the guide to the cylinder head.**

11. Install or connect the following:
- Secondary timing chain guides. Tighten the bolts to 21 ft. lbs. (28 Nm).
- Secondary timing chains to the idler sprocket so that the double plated links on each chain are visible through the slots in the primary idler sprocket

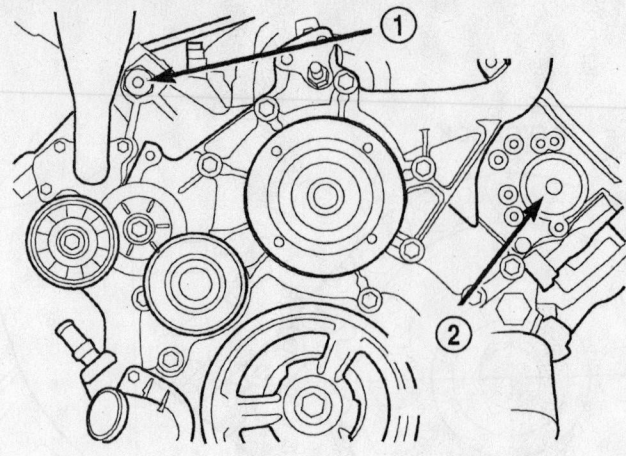

1 - RIGHT CYLINDER HEAD ACCESS PLUG
2 - LEFT CYLINDER HEAD ACCESS PLUG

9355PG11

Cylinder head access plugs—3.7L

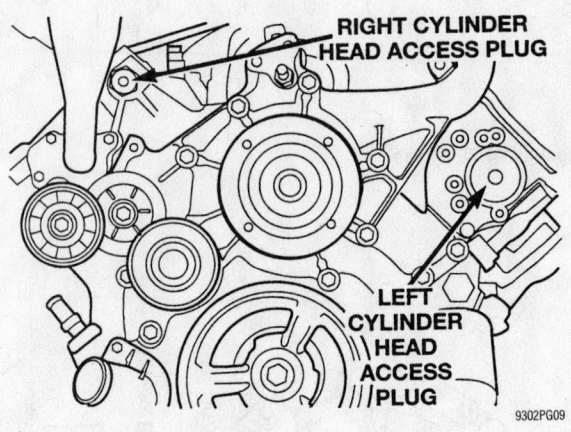

RIGHT CYLINDER HEAD ACCESS PLUG

LEFT CYLINDER HEAD ACCESS PLUG

9302PG09

Cylinder head access plug locations—4.7L engine

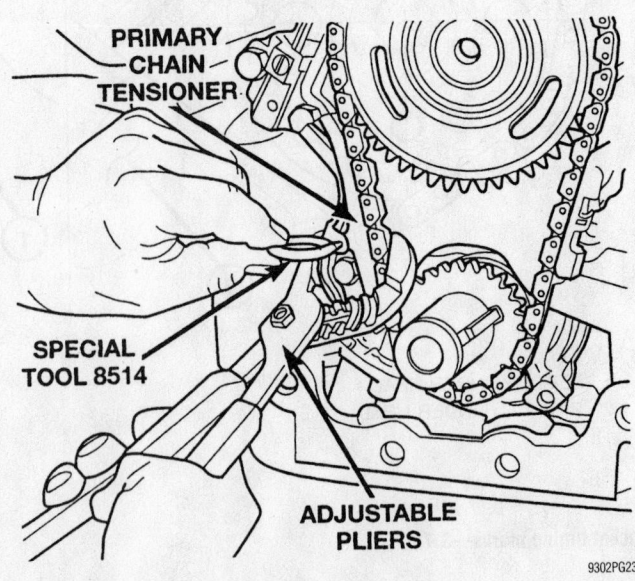

PRIMARY CHAIN TENSIONER

SPECIAL TOOL 8514

ADJUSTABLE PLIERS

9302PG23

Compress and lock the primary chain tensioner—4.7L engine

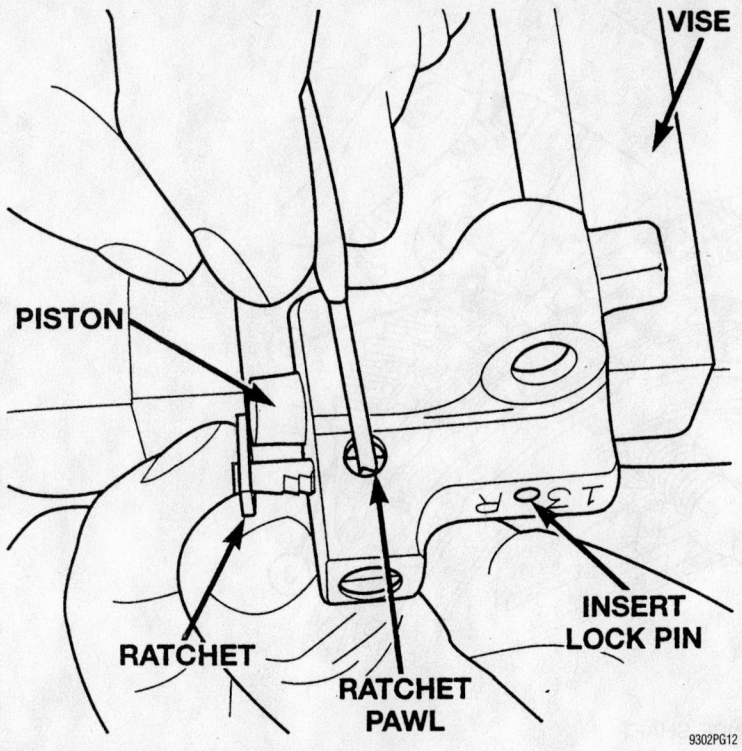

Secondary timing chain tensioner preparation—4.7L engine

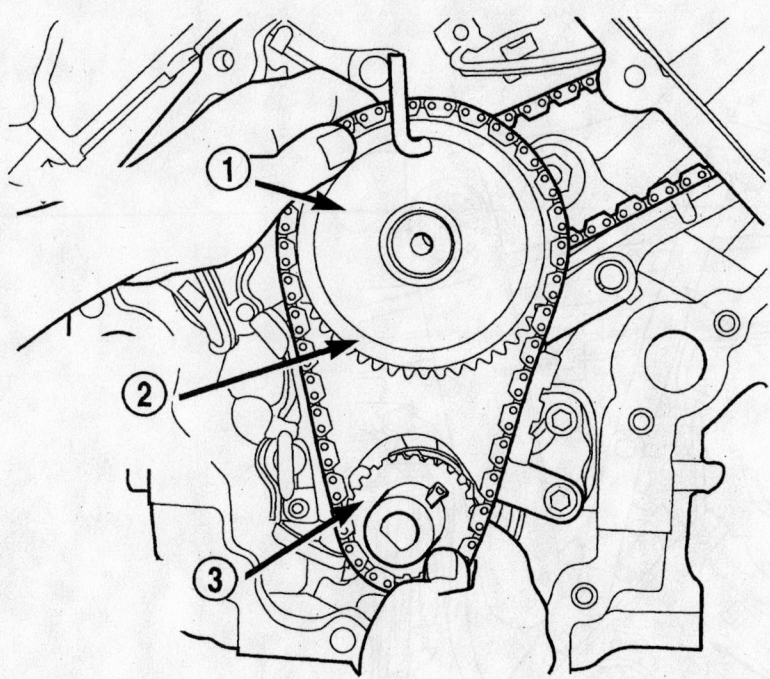

1 - SPECIAL TOOL 8429
2 - PRIMARY CHAIN IDLER SPROCKET
3 - CRANKSHAFT SPROCKET

Installing the idler gear and timing chains—3.7L

12. Lock the secondary timing chains to the idler sprocket with Timing Chain Locking tool as shown.

13. Align the primary chain double plated links with the idler sprocket timing mark and the single plated link with the crankshaft sprocket timing mark.

14. Install the primary chain and sprockets. Tighten the idler sprocket bolt to 25 ft. lbs. (34 Nm).

15. Align the secondary chain single plated links with the timing marks on the secondary sprockets. Align the dot at the **L** mark on the left sprocket with the plated link on the left chain and the dot at the **R** mark on the right sprocket with the plated link on the right chain.

16. Rotate the camshafts back from the neutral position and install the camshaft sprockets.

17. Remove the secondary chain locking tool.

18. Remove the primary and secondary timing chain tensioner locking pins.

19. Hold the camshaft sprockets with a spanner wrench and tighten the retaining bolts to 90 ft. lbs. (122 Nm).

20. Install or connect the following:
- Front cover. Tighten the bolts, in sequence, to 40 ft. lbs. (54 Nm).
- Front crankshaft seal
- Cylinder head access plugs
- A/C compressor
- Alternator
- Accessory drive belt tensioner. Tighten the bolt to 40 ft. lbs. (54 Nm).
- Oil fill housing
- Crankshaft damper. Tighten the bolt to 130 ft. lbs. (175 Nm).
- Power steering pump
- Lower radiator hose
- Heater hoses
- Accessory drive belt
- Engine cooling fan and shroud
- Camshaft Position (CMP) sensor
- Valve covers
- Negative battery cable

21. Fill and bleed the cooling system.

22. Start the engine, check for leaks and repair if necessary.

5.2L and 5.9L Engines

1. Before servicing the vehicle, refer to the precautions in the beginning of this section.

2. Drain the cooling system.

3. Remove or disconnect the following:

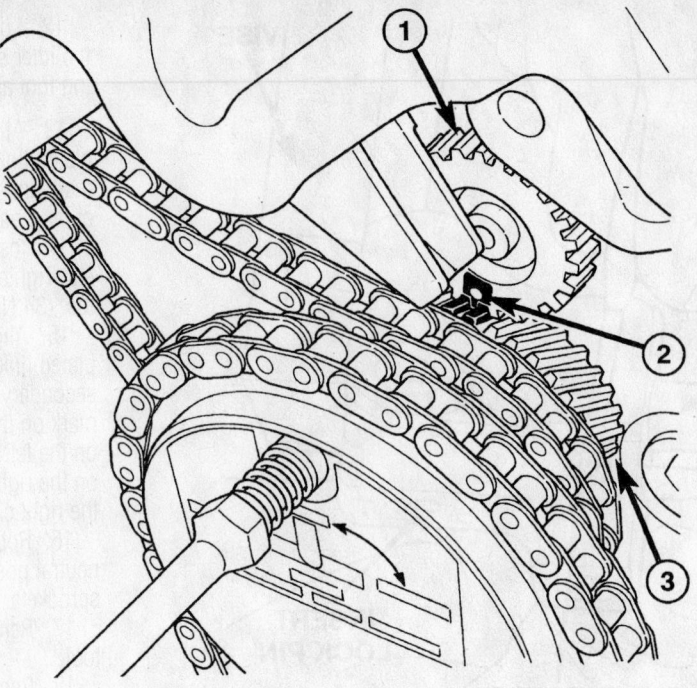

1 - COUNTERBALANCE SHAFT
2 - TIMING MARKS
3 - IDLER SPROCKET

9355PG13

Counterbalance shaft timing marks—3.7L

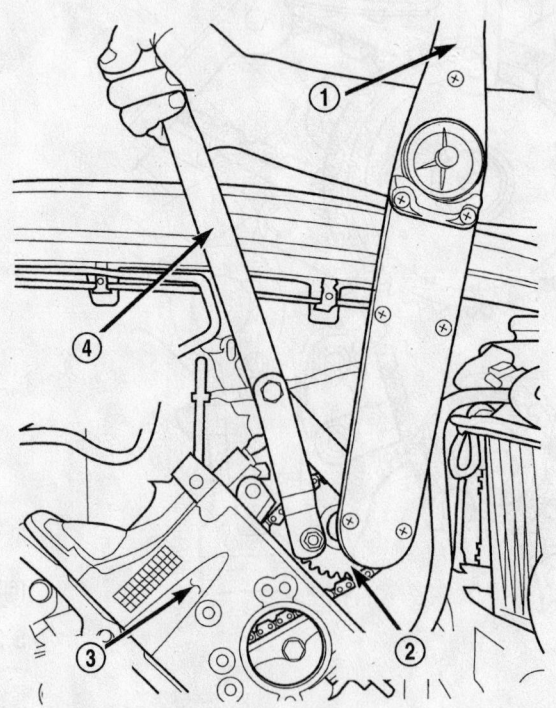

1 - TORQUE WRENCH
2 - CAMSHAFT SPROCKET
3 - LEFT CYLINDER HEAD
4 - SPECIAL TOOL 6958 SPANNER WITH ADAPTER PINS 8346
9355PG14

Tightening the left side camshaft sprocket—3.7L

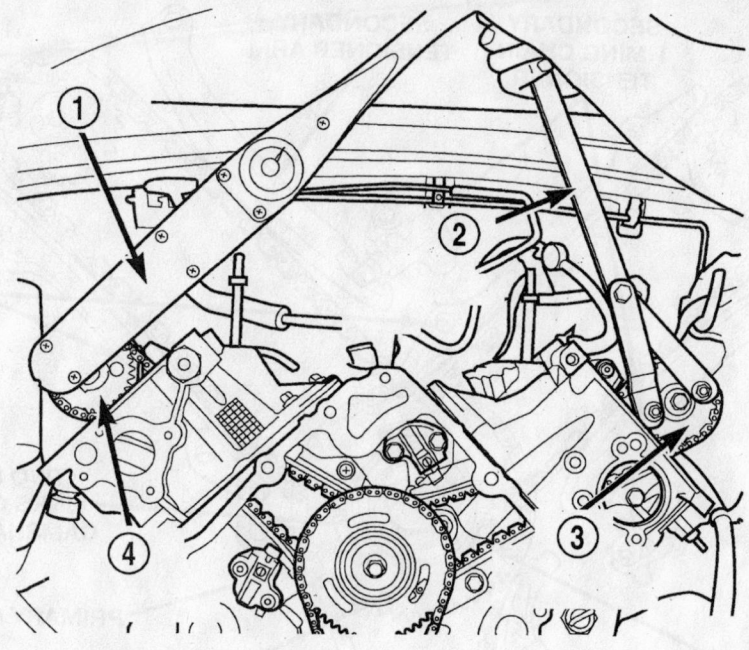

1 - TORQUE WRENCH
2 - SPECIAL TOOL 6958 WITH ADAPTER PINS 8346
3 - LEFT CAMSHAFT SPROCKET
4 - RIGHT CAMSHAFT SPROCKET

9355PG15

Tightening the right side camshaft sprocket—3.7L

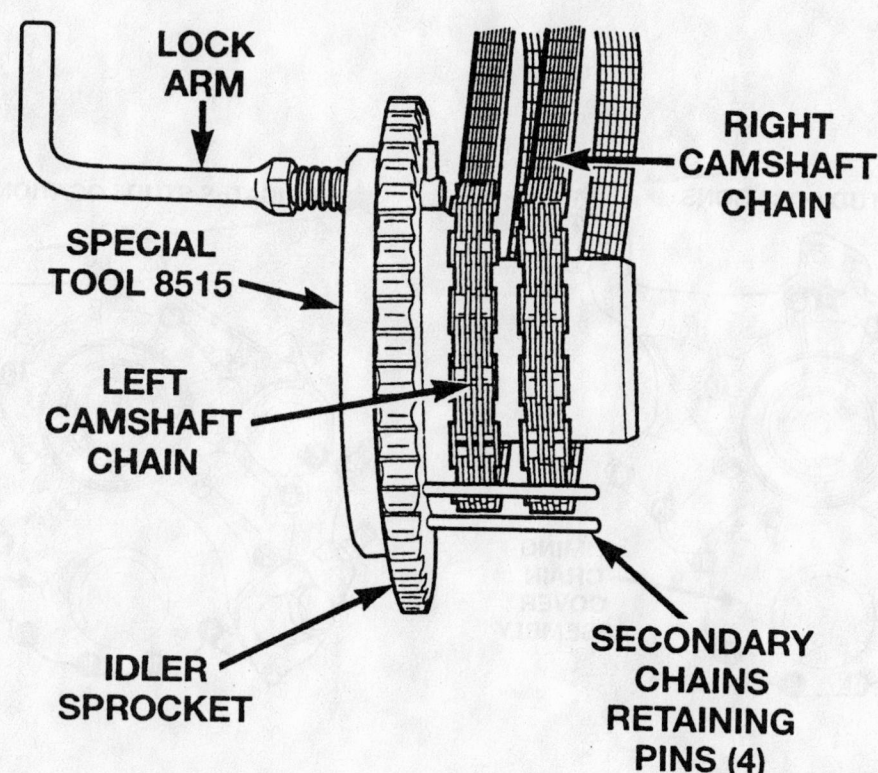

LOCK ARM

RIGHT CAMSHAFT CHAIN

SPECIAL TOOL 8515

LEFT CAMSHAFT CHAIN

IDLER SPROCKET

SECONDARY CHAINS RETAINING PINS (4)

9302PG07

Use the Timing Chain Locking tool to lock the timing chains on the idler gear—4.7L engine

For Accessory Drive Belt illustrations, see Section 1 of this manual

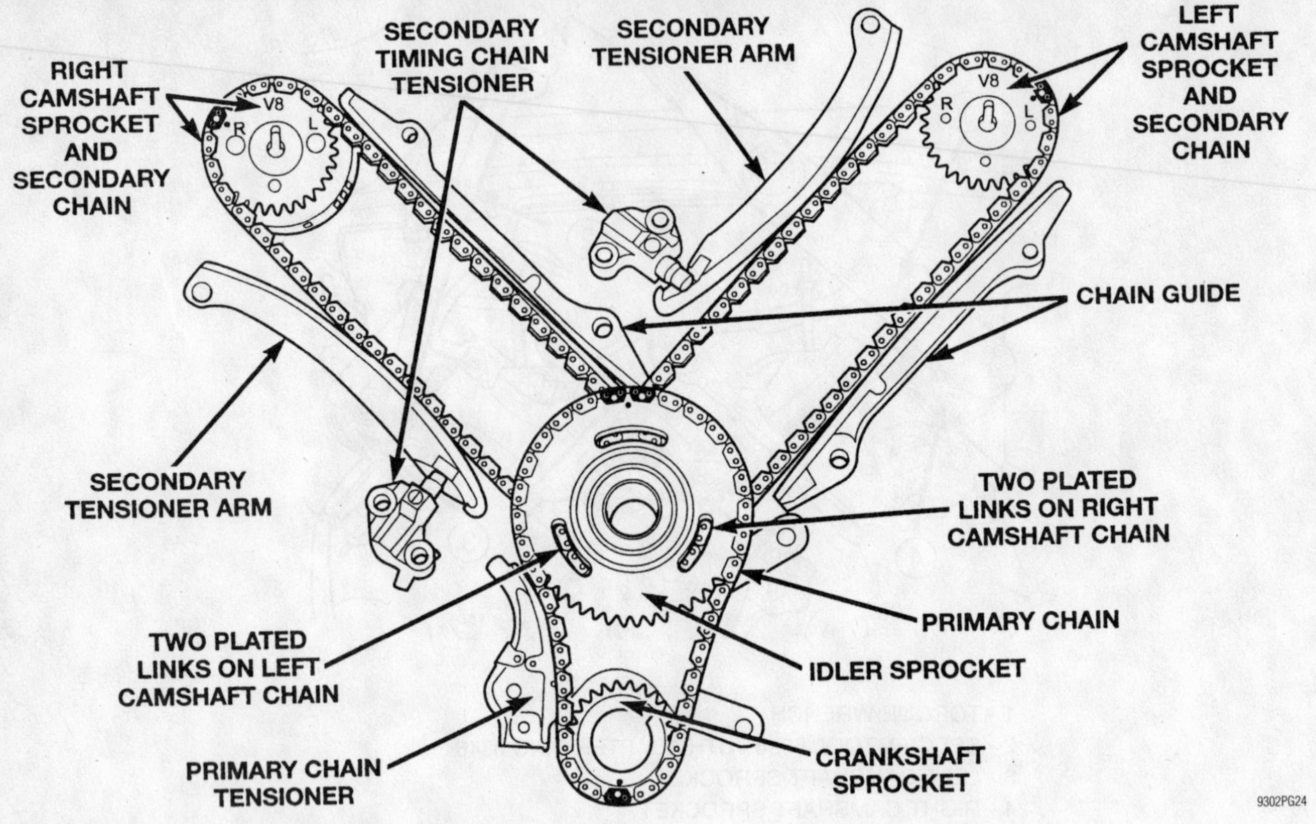

RIGHT CAMSHAFT SPROCKET AND SECONDARY CHAIN

SECONDARY TIMING CHAIN TENSIONER

SECONDARY TENSIONER ARM

LEFT CAMSHAFT SPROCKET AND SECONDARY CHAIN

CHAIN GUIDE

SECONDARY TENSIONER ARM

TWO PLATED LINKS ON RIGHT CAMSHAFT CHAIN

TWO PLATED LINKS ON LEFT CAMSHAFT CHAIN

PRIMARY CHAIN

IDLER SPROCKET

PRIMARY CHAIN TENSIONER

CRANKSHAFT SPROCKET

9302PG24

Timing chain system and alignment marks—4.7L engine

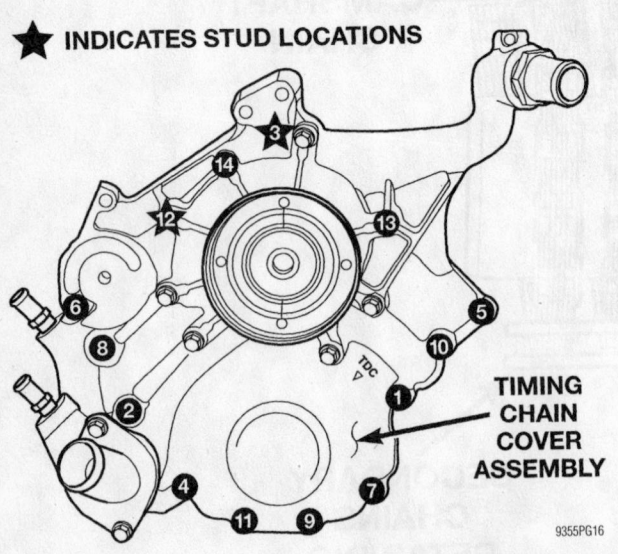

★ **INDICATES STUD LOCATIONS**

TIMING CHAIN COVER ASSEMBLY

9355PG16

Timing cover bolt torque sequence—3.7L

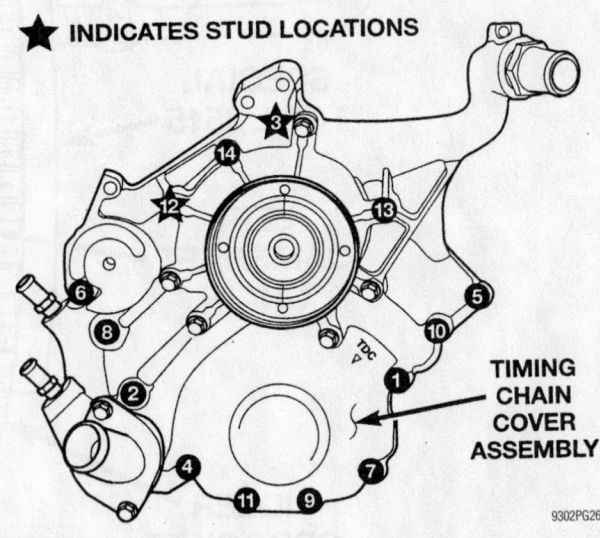

★ **INDICATES STUD LOCATIONS**

TIMING CHAIN COVER ASSEMBLY

9302PG26

Timing chain cover bolt torque sequence—4.7L engine

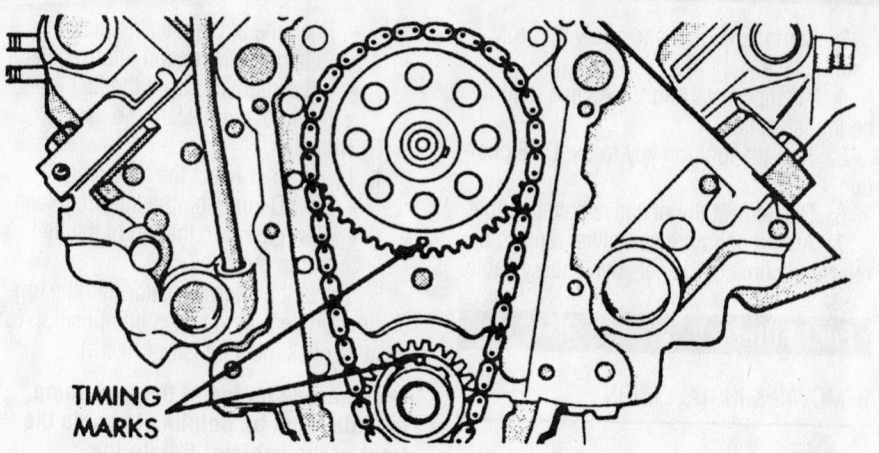

Timing chain alignment—5.2L and 5.9L engines

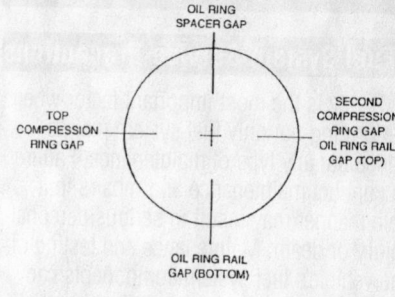

Piston ring end-gap spacing—2.5L, 5.2L and 5.9L engines

- Negative battery cable
- Accessory drive belt
- Cooling fan and shroud
- Water pump
- Power steering pump
- Crankshaft damper
- Front crankshaft seal
- Front cover

4. Rotate the crankshaft so that the camshaft sprocket and crankshaft sprocket timing marks are aligned.

5. Remove the timing chain and sprockets.

To install:

6. Install the timing chain and sprockets with the timing marks aligned. Tighten the camshaft sprocket bolt to 50 ft. lbs. (68 Nm).

7. Install or connect the following:
- Front cover. Tighten the cover bolts to 30 ft. lbs. (41 Nm) and the oil pan bolts to 18 ft. lbs. (24 Nm).
- Front crankshaft seal
- Crankshaft damper. Tighten the bolt to 135 ft. lbs. (183 Nm).
- Power steering pump
- Water pump
- Cooling fan and shroud
- Accessory drive belt
- Negative battery cable

8. Fill and bleed the cooling system.

9. Start the engine, check for leaks and repair if necessary.

Piston and Ring

POSITIONING

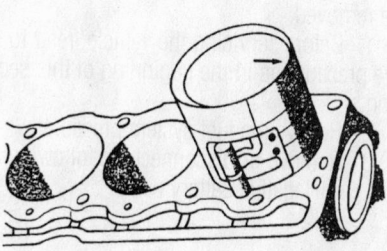

Piston to engine positioning—2.5L, 4.0L, 5.2L and 5.9L engines

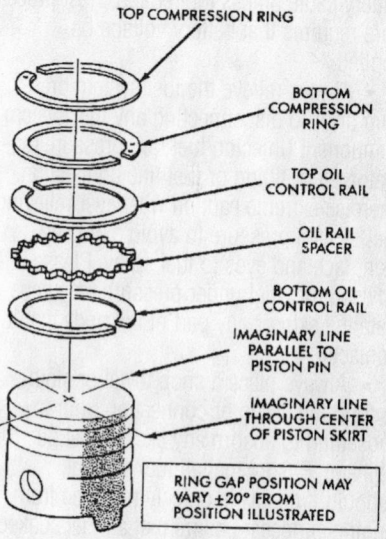

Piston ring end-gap spacing—4.0L engine

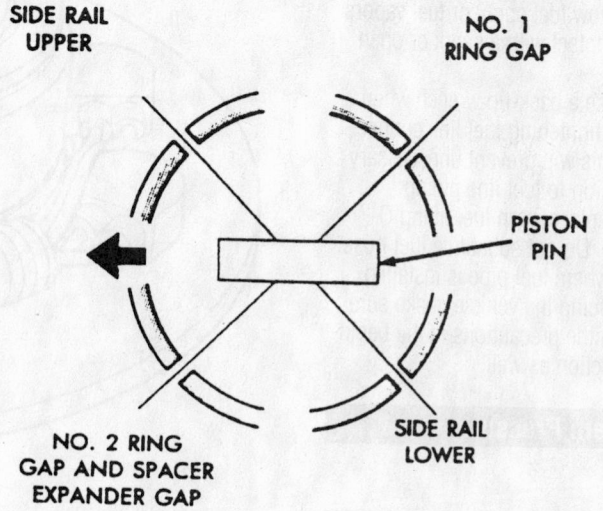

Piston ring end-gap spacing. Position raised "F" on piston towards front of engine—3.7L and 4.7L engines

For Tire, Wheel and Ball Joint specifications, see Section 1 of this manual

FUEL SYSTEM

Fuel System Service Precautions

Safety is the most important factor when performing not only fuel system maintenance but any type of maintenance. Failure to conduct maintenance and repairs in a safe manner may result in serious personal injury or death. Maintenance and testing of the vehicle's fuel system components can be accomplished safely and effectively by adhering to the following rules and guidelines.

• To avoid the possibility of fire and personal injury, always disconnect the negative battery cable unless the repair or test procedure requires that battery voltage be applied.

• Always relieve the fuel system pressure prior to disconnecting any fuel system component (injector, fuel rail, pressure regulator, etc.), fitting or fuel line connection. Exercise extreme caution whenever relieving fuel system pressure to avoid exposing skin, face and eyes to fuel spray. Please be advised that fuel under pressure may penetrate the skin or any part of the body that it contacts.

• Always place a shop towel or cloth around the fitting or connection prior to loosening to absorb any excess fuel due to spillage. Ensure that all fuel spillage (should it occur) is quickly removed from engine surfaces. Ensure that all fuel soaked cloths or towels are deposited into a suitable waste container.

• Always keep a dry chemical (Class B) fire extinguisher near the work area.

• Do not allow fuel spray or fuel vapors to come into contact with a spark or open flame.

• Always use a back-up wrench when loosening and tightening fuel line connection fittings. This will prevent unnecessary stress and torsion to fuel line piping.

• Always replace worn fuel fitting O-rings with new. Do not substitute fuel hose or equivalent where fuel pipe is installed.

Before servicing the vehicle, make sure to also refer to the precautions in the beginning of this section as well.

Fuel System Pressure

RELIEVING

1. Before servicing the vehicle, refer to the precautions in the beginning of this section.
2. Remove the fuel pump relay.

3. Start the engine and allow it to run until it stalls.
4. Attempt restarting the engine until it no longer runs.
5. Turn the ignition key to the **OFF** position.
6. Disconnect the negative battery cable.
7. After repairs are complete, replace the relay and connect the negative battery cable.

Fuel Filter

REMOVAL & INSTALLATION

1999 Models

The fuel filter is in combination with the fuel pressure regulator and is located on the fuel pump module. The module does not have to be removed, but the fuel tank must be removed.

1. Before servicing the vehicle, refer to the precautions in the beginning of this section.
2. Relieve the fuel system pressure.
3. Remove or disconnect the following:
   • Negative battery cable
   • Fuel tank

• Fuel line at the filter/regulator
• Retaining clamp and filter/regulator
• Filter/regulator from the fuel pump module and discard the gasket

To install:
4. Install or connect the following:
   • New O-rings on the filter/regulator
   • New gasket to the top of the fuel pump module
5. Press the filter/regulator into the top of the module until it snaps into position (a positive click must be felt or heard).

➡ **The arrow on top of the fuel pump module must be pointing towards the front of the vehicle. Rotate the filter/regulator until the fuel supply tube is pointed at the 10 o'clock position.**

6. Install or connect the following:
   • New retainer clamp to the top of the filter/regulator
   • Fuel line
   • Fuel tank
   • Negative battery cable
7. Start the engine, check for leaks and repair if necessary.

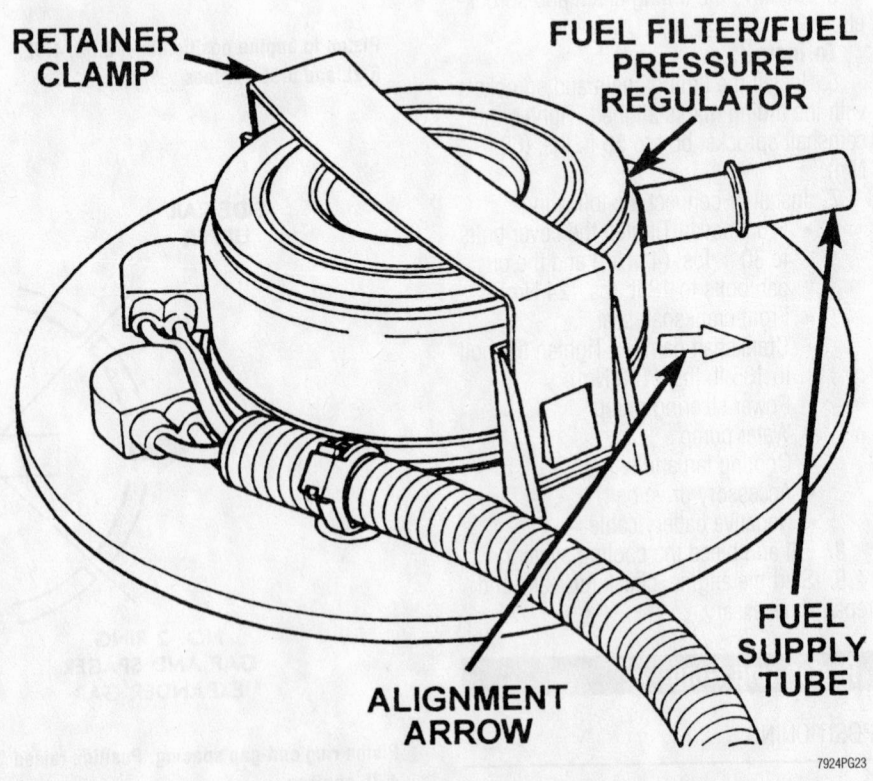

RETAINER CLAMP

FUEL FILTER/FUEL PRESSURE REGULATOR

FUEL SUPPLY TUBE

ALIGNMENT ARROW

7924PG23

The fuel filter/pressure regulator is located on top of the fuel pump module–1999 models

2000 Grand Cherokee and 2001 Cherokee

The fuel pump inlet filter is located on the bottom of the fuel pump module.

1. Before servicing the vehicle, refer to the precautions in the beginning of this section.
2. Relieve the fuel system pressure.
3. Remove or disconnect the following:
   - Negative battery cable
   - Fuel tank
   - Fuel pump module
   - Filter from the bottom of the fuel pump module

To install:

4. Install or connect the following:
   - New O-rings, lubricated with engine oil
   - Fuel filter
   - Fuel lines to the fuel filter
   - Negative battery cable
5. Start the engine and check for leaks.

2001 Grand Cherokee

The fuel filter/pressure regulator is mounted to the body above the rear axle and near the front of the fuel tank.

1. Before servicing the vehicle, refer to the precautions in the beginning of this section.
2. Relieve the fuel system pressure.
3. Remove or disconnect the following:
   - Negative battery cable
   - Fuel supply, return and pressure lines
   - Mounting bolts
   - Fuel filter/pressure regulator

To install:

4. Install or connect the following:
   - Fuel filter/pressure regulator to the body. Torque the bolts to 30 inch lbs. (3 Nm).
   - Fuel pressure, return and supply lines
   - Negative battery cable
5. Start the vehicle, check for leaks and repair if necessary.

Liberty

The fuel filter is attached to the front of the fuel tank.

1. Before servicing the vehicle, refer to the precautions in the beginning of this section.
2. Relieve the fuel system pressure.
3. Remove or disconnect the following:
   - Negative battery cable

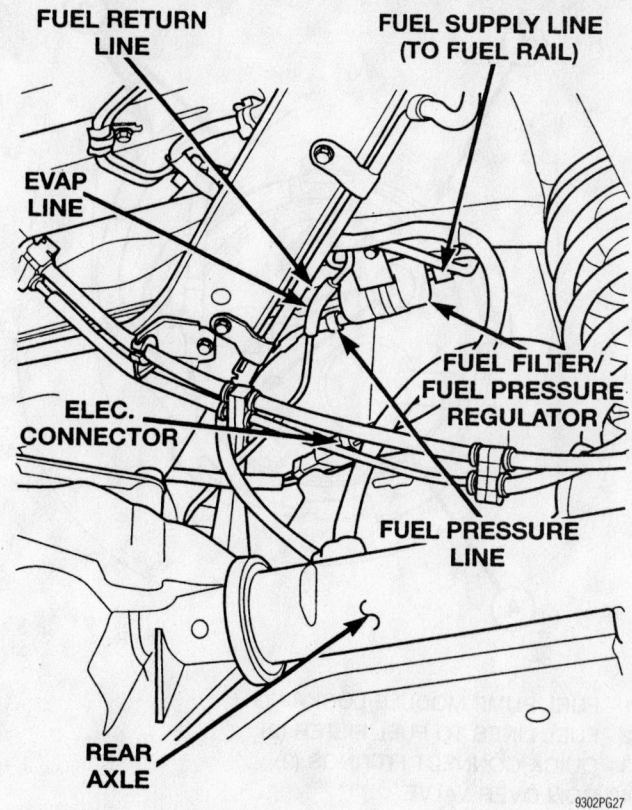

FUEL RETURN LINE
FUEL SUPPLY LINE (TO FUEL RAIL)
EVAP LINE
FUEL FILTER/ FUEL PRESSURE REGULATOR
ELEC. CONNECTOR
FUEL PRESSURE LINE
REAR AXLE

9302PG27

Fuel filter/pressure regulator location–1999–01 models

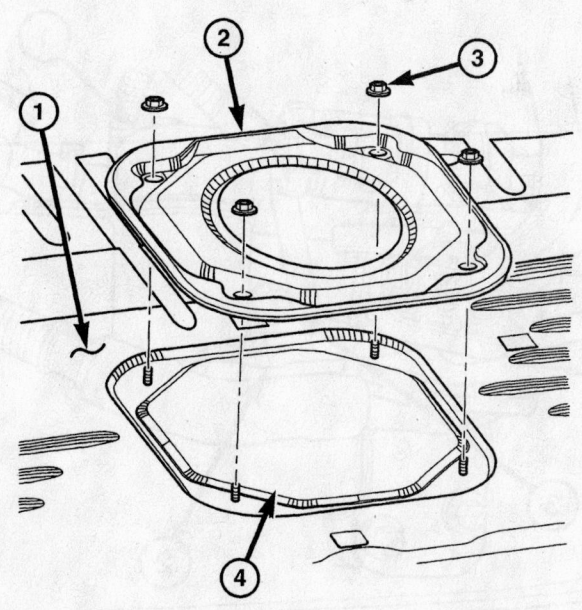

1 - FLOORPAN AT REAR
2 - FUEL PUMP MODULE ACCESS PLATE
3 - NUTS (4)
4 - OPENING TO PUMP MODULE

9355PG17

Access plate—Liberty

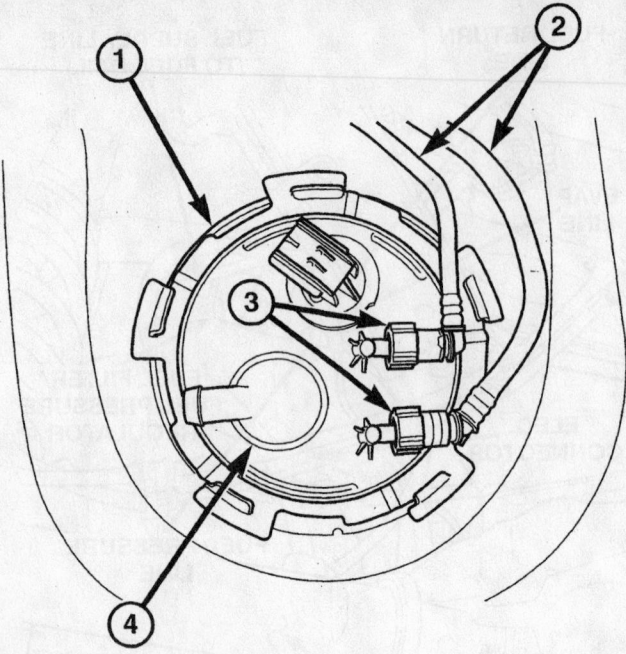

1 - FUEL PUMP MODULE LOCKRING
2 - FUEL LINES TO FUEL FILTER (2)
3 - QUICK-CONNECT FITTINGS (2)
4 - ROLLOVER VALVE

9355PG18

Fuel lines at the pump module—Liberty

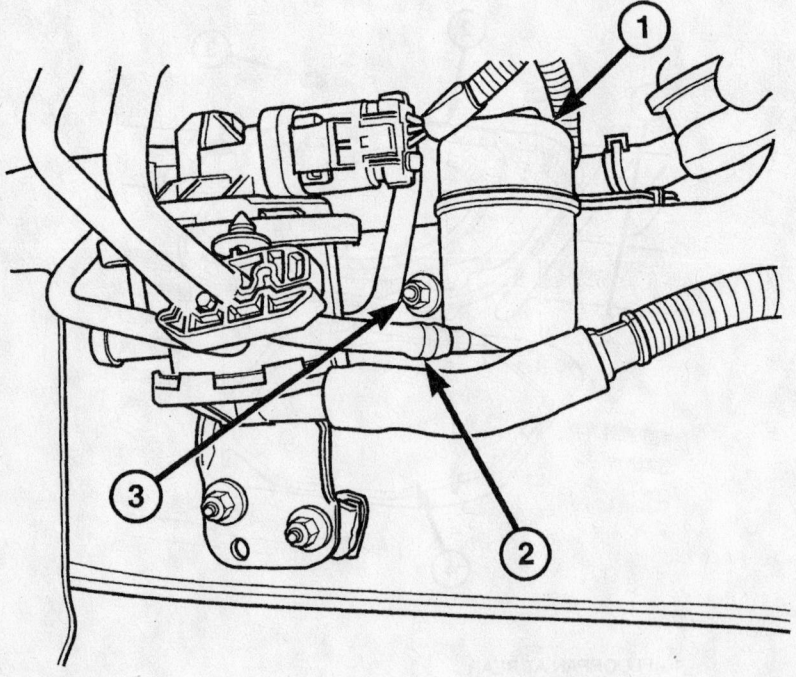

1 - FUEL FILTER
2 - 3RD FUEL LINE TO ENGINE
3 - FILTER MOUNTING NUT

9355PG19

Fuel filter location—Liberty

- 2 rear cargo hold-down clamps by drilling the rivets
- 4 fuel pump module access plate nuts

➡ **Once the nuts are removed, the plate must be removed by applying a heat gun to melt the adhesive. Take care to avoid bending the plate. Once removed, you can disconnect the 2 top hoses from the filter. The bottom hose must be removed from under the vehicle. The disconnect point on this hose is about a foot in front of the filter.**

- Ground strap
- Mounting nut and filter.
4. Installation is the reverse of removal.

Fuel Pump

REMOVAL & INSTALLATION

Except Liberty

1. Before servicing the vehicle, refer to the precautions in the beginning of this section.
2. Relieve the fuel system pressure.
3. Remove or disconnect the following:
- Negative battery cable
- Fuel tank
- Fuel lines
- Fuel pump module locknut
- Fuel pump module

To install:
4. Install or connect the following:
- Fuel pump module with a new gasket. Rotate the module until the arrow is pointed toward the front of the vehicle, on all models except Grand Cherokee. The arrow must pointed toward the rear of the vehicle.
- Locknut. Torque the locknut to 55 ft. lbs. (74 Nm).
- Fuel lines
- Fuel tank
- Negative battery cable
5. Start the engine, check for leaks and repair if necessary.

Liberty

The fuel filter is attached to the front of the fuel tank.

1. Before servicing the vehicle, refer to the precautions in the beginning of this section.
2. Relieve the fuel system pressure.
3. Remove or disconnect the following:
- Negative battery cable

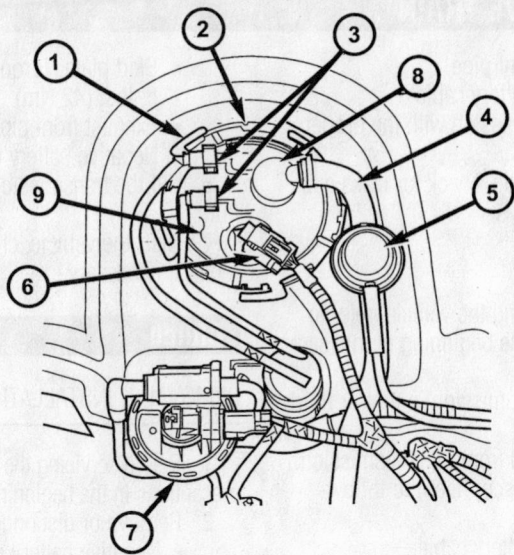

1 - LOCK RING
2 - ALIGNMENT NOTCH
3 - FUEL FILTER FITTINGS (2)
4 - ORVR SYSTEM HOSE AND CLAMP
5 - FLOW MANAGEMENT VALVE
6 - ELECTRICAL CONNECTOR
7 - LEAK DETECTION PUMP
8 - FUEL TANK CHECK (CONTROL) VALVE
9 - FUEL PUMP MODULE (UPPER SECTION)

9355PG20

Top view of the fuel pump module—Liberty

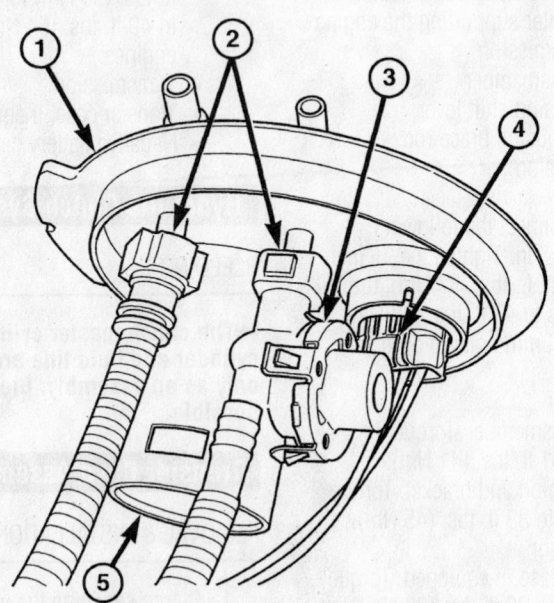

1 - UPPER SECTION OF PUMP MODULE
2 - QUICK-CONNECT FITTINGS
3 - FUEL PRESSURE REGULATOR
4 - 4-WIRE ELECTRICAL CONNECTOR
5 - FUEL TANK CHECK (CONTROL) VALVE

9355PG21

Fuel pressure regulator—Liberty

- 2 rear cargo hold-down clamps by drilling the rivets
- 4 fuel pump module access plate nuts

➡ Once the nuts are removed, the plate must be removed by applying a heat gun to melt the adhesive. Take care to avoid bending the plate.

- Fuel lines at the module
- Electrical connector by first sliding the red tab, then pushing the gray tab
- ORVR hose
- Module lockring
- Module from the tank

✳✳ WARNING

Lift the module out slowly and carefully until you can secure the rubber gasket. If you're not careful, the gasket will fall into the tank.

- Electrical connector from the upper module section
- Fule pressure regulator
- Fuel return line
- Upper module section

4. Drain the tank through the module opening.

5. Remove or disconnect the following:

- Lower module section by pushing gently on the release tab while sliding the lock tab upward

6. Installation is the reverse of removal.

Fuel Injector

REMOVAL & INSTALLATION

1. Before servicing the vehicle, refer to the precautions in the beginning of this section.

2. Relieve fuel system pressure.

3. Remove or disconnect the following:
- Negative battery cable
- Fuel rail
- Fuel injector retaining clips
- Fuel injectors

To install:

4. Install or connect the following:
- Fuel injectors, using new O-rings
- Fuel injector clips to the fuel rail
- Fuel rail
- Negative battery cable

5. Start the vehicle and check for leaks, repair if necessary.

For Maintenance Interval recommendations, see Section 1 of this manual

DRIVE TRAIN

Transmission Assembly

REMOVAL & INSTALLATION

Automatic

1. Before servicing the vehicle, refer to the precautions in the beginning of this section.
2. Drain the transmission fluid.
3. Remove or disconnect the following:
   - Negative battery cable
   - Exhaust front pipe
   - Transmission braces, if equipped
   - Transmission oil cooler lines
   - Starter motor
   - Crankshaft Position (CKP) sensor
   - Torque converter access cover
   - Transmission oil pan
   - Skid plate
   - Transmission oil dipstick tube
   - Propellar shafts after matchmarking them
   - Park/Neutral switch connector
   - Shift cable
   - Throttle valve cable
   - Shift rod from the transfer case shift lever
   - Crossmember
   - Transfer case vent hose
   - Transfer case
4. Slide the transmission/torque converter assembly rearward off the engine dowels
5. Lower and remove the transmission assembly

To install:
6. Install or connect the following:
   - Transmission. Torque the flange bolts to 55 ft. lbs. (75 Nm).
   - Front driveshaft and transfer case, if equipped
   - Transfer case vent hose
   - Transmission mount and cross-member
   - Throttle valve cable
   - Shift cable
   - Park/Neutral switch connector
   - Propellar shafts using the match-marks made during removal
   - Transmission oil dipstick tube
   - Skid plate
   - Transmission oil pan
   - Torque converter access cover
   - CKP sensor
   - Starter motor
   - Transmission oil cooler lines
   - Transmission braces, if equipped. Torque the bolts to 30 ft. lbs. (41 Nm).

   - Exhaust front pipe
   - Negative battery cable
7. Fill the transmission with the proper fluid.
8. Start the vehicle, check for leaks and repair if necessary.

Manual

1. Before servicing the vehicle, refer to the precautions in the beginning of this section.
2. Place the transmission in first or third gear.
3. Drain the fluid from the transmission.
4. Remove or disconnect the following:
   - Negative battery cable
   - Exhaust front pipe
   - Skid plate
   - Clutch slave cylinder
   - Rear propeller shaft
   - Front propeller shaft, if equipped
   - Wire harness from the transmission and transfer case, if equipped
   - Transfer case shift linkage, if equipped
   - Transfer case, if equipped
   - Crankcase Position (CKP) sensor
   - Rear cushion and transmission bracket after supporting the engine and transmission
   - Rear crossmember
   - Transmission shift lever
   - Clutch housing brace rod
   - Transmission

To install:
5. Install or connect the following:
   - Transmission. Tighten the ⅜ inch bolts to 27 ft. lbs. (37 Nm), the ⁷⁄₁₆ inch bolts to 43 ft. lbs. (58 Nm), and the 12mm bolts to 55 ft. lbs. (75 Nm).
   - Shift lever
   - Rear crossmember. Torque the bolts to 31 ft. lbs. (41 Nm).
   - Rear cushion and bracket. Torque the bolts to 33 ft. lbs. (45 Nm).
   - CKP sensor
   - Transfer case, if equipped. Torque the bolts to 26 ft. lbs. (35 Nm).
   - Transfer case shift linkage, if equipped
   - Wire harness to the transmission
   - Front driveshaft and transfer case, if equipped. Torque the bolts to 12 ft. lbs. (19 Nm).
   - Rear driveshaft. Torque the bolts to 12 ft. lbs. (19 Nm).
   - Clutch slave cylinder

   - Skid plate. Torque the bolts to 31 ft. lbs. (42 Nm).
   - Exhaust front pipe
   - Negative battery cable
6. Fill the transmission with the proper fluid.
7. Start the vehicle, check for leaks and repair if necessary.

Clutch

REMOVAL & INSTALLATION

1. Before servicing the vehicle, refer to the precautions in the beginning of this section.
2. Remove or disconnect the following:
   - Negative battery cable
   - Transfer case, if equipped
   - Transmission
   - Pressure plate. Loosen the bolts evenly in ½ turn steps.
   - Clutch disc

To install:
3. Install or connect the following:
   - Clutch disc and pressure plate. Tighten the pressure plate bolts evenly in ½ turns to 23 ft. lbs. (31 Nm) for 2.4L and 2.5L engines, 37 ft. lbs. (50 Nm) for 3.7L engines, or to 40 ft. lbs. (54 Nm) for 4.0L engines.
   - Transmission
   - Transfer case, if equipped
   - Negative battery cable

Hydraulic Clutch System

BLEEDING

➡**The clutch master cylinder, slave cylinder and fluid line are serviced only as an assembly. Bleeding is not possible.**

Transfer Case Assembly

REMOVAL & INSTALLATION

1. Before servicing the vehicle, refer to the precautions in the beginning of this section.
2. Shift the transfer case into **N**.
3. Drain the transfer case fluid.
4. Remove or disconnect the following:
   - Negative battery cable
   - Front and rear driveshafts
   - Transmission mount and cross-member. Support the transmission with a jackstand.

- Vehicle Speed (VSS) sensor connector
- Indicator switch connector
- Vacuum hose
- Shift linkage
- Vent hose
- Transfer case attaching nuts
- Transfer case

To install:

5. Install or connect the following:
   - Transfer case. Tighten the nuts to 26 ft. lbs. (35 Nm).
   - Indicator switch connector
   - Vacuum hose
   - Vent hose
   - Shift linkage
   - VSS sensor connector
   - Transmission mount and cross-member. Torque the bolts to 30 ft. lbs. (41 Nm).
   - Front and rear driveshafts
   - Negative battery cable
6. Fill the transfer case with the proper fluid.
7. Start the vehicle, check for leaks and repair if necessary.

CV-Joints

Driveshaft and CV-joints are serviced only as an assembly, as are halfshafts and CV-joints.

Halfshafts

REMOVAL & INSTALLATION

Liberty

1. Before servicing the vehicle, refer to the precautions in the beginning of this section.
2. Remove or disconnect the following:
   - Wheel
   - Hub nut
   - Stabilizer link
   - Lower clevis bolt
   - Ball joint from the lower arm
1. Pull on the hub and push the halfshaft from the knuckle

➡**The right side has a splined axle shaft that will stay in the axle.**

To install:

2. Apply a light coating of wheel bearing grease on the splines of the inner joint, and in the hub bearing bore.
3. Install or connect the following:
   - Halfshaft on the axle shaft splines

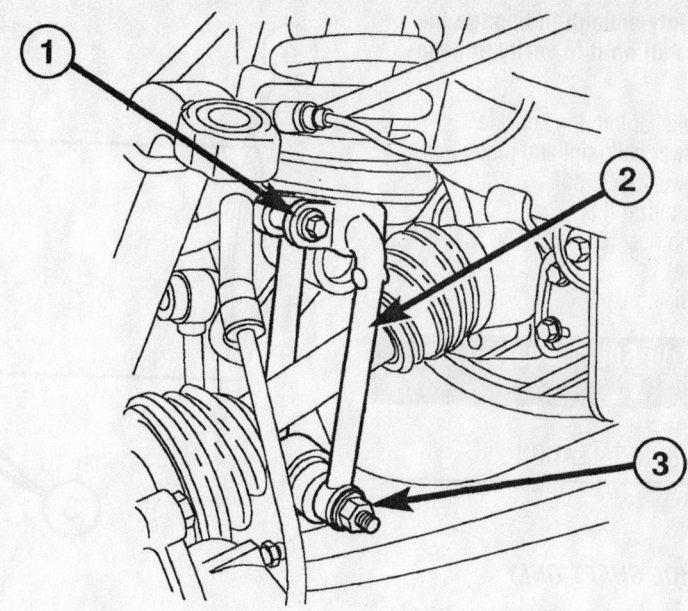

1 - UPPER BOLT
2 - CLEVIS BRACKET
3 - LOWER BOLT

Clevis bracket—Liberty

9355PG22

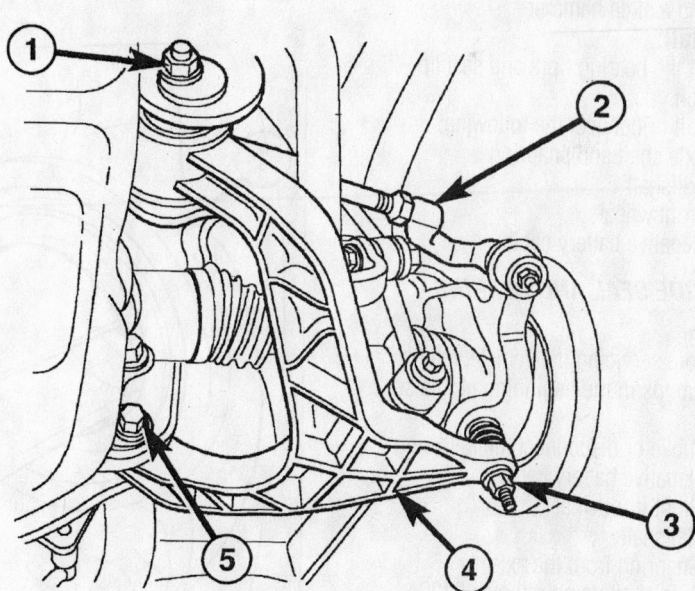

1 - FRONT CAM BOLT
2 - OUTER TIE ROD END
3 - LOWER BALL JOINT NUT
4 - LOWER CONTROL ARM
5 - REAR CAM BOLT

Lower control arm—Liberty

9355PG23

For Tune-up, Capacities and Firing orders, see Section 1 of this manual

➡ **Push firmly enough to engage the snapring. Pull on it to verify engagement.**

- Halfshat into the knuckle
- Lower ball joint and pinch bolt
- Lower clevis bolt
- Stabilizer link
- Hub nut. Torque to 100 ft. lbs. (136 Nm).
- Wheel

Front Axle Shaft, Bearing and Seal

REMOVAL & INSTALLATION

Liberty

RIGHT SIDE SHAFT ONLY

1. Before servicing the vehicle, refer to the precautions in the beginning of this section.
2. Remove or disconnect the following:
   - Negative battery cable
   - Right front wheel
   - Halfshaft
   - Snapring from the axle
   - Axle shaft using remover 8420A and a slide hammer

To install:

3. Coat the bearing bore and seal lip with gear oil.
4. Install or connect the following:
   - Axle shaft and snapring
   - Halfshaft
   - Front wheel
   - Negative battery cable

RIGHT SIDE SEAL AND BEARING ONLY

1. Before servicing the vehicle, refer to the precautions in the beginning of this section.
2. Remove or disconnect the following:
   - Negative battery cable
   - Right front wheel
   - Halfshaft
   - Snapring from the axle
   - Axle shaft using remover 8420A and a slide hammer
   - Seal, using Remover 7794A and a slide hammer
   - Bearing, using Remover 7794A and a slide hammer

To install:

3. Coat the bearing bore and seal lip with gear oil.
4. Install or connect the following:
   - New seal, using Installer 8806 and handle C4171, or their equivalents

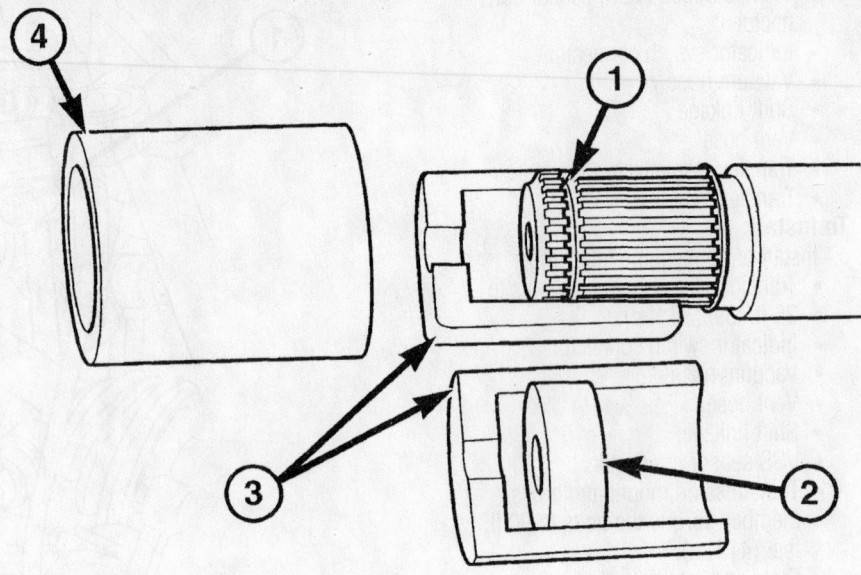

1 - SNAP RING GROVE
2 - SLID HAMMER THREADS
3 - REMOVER BLOCKS
4 - REMOVER COLLAR

9355PG24

Axle shaft remover tool—Liberty

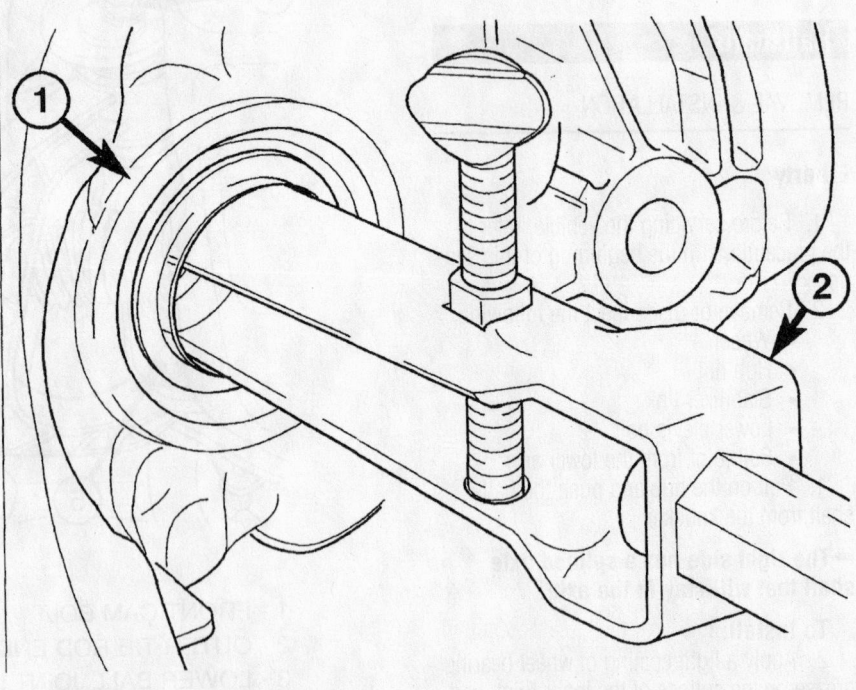

1 - SHAFT SEAL
2 - REMOVER

9355PG25

Seal removal—Liberty

- Axle shaft and snapring
- Halfshaft
- Front wheel
- Negative battery cable

Except Liberty

FRONT AXLE SHAFT

1. Before servicing the vehicle, refer to the precautions in the beginning of this section.
2. Remove or disconnect the following:
   - Negative battery cable
   - Front wheel
   - Brake caliper and rotor
   - Wheel speed sensor, if equipped
   - Axle hub nut
   - Wheel bearing and hub assembly
   - Axle shaft

To install:

3. Install or connect the following:
   - Axle shaft
   - Wheel bearing and hub assembly
   - Axle hub nut. Tighten the nut to 175 ft. lbs. (237 Nm).
   - Wheel speed sensor, if equipped

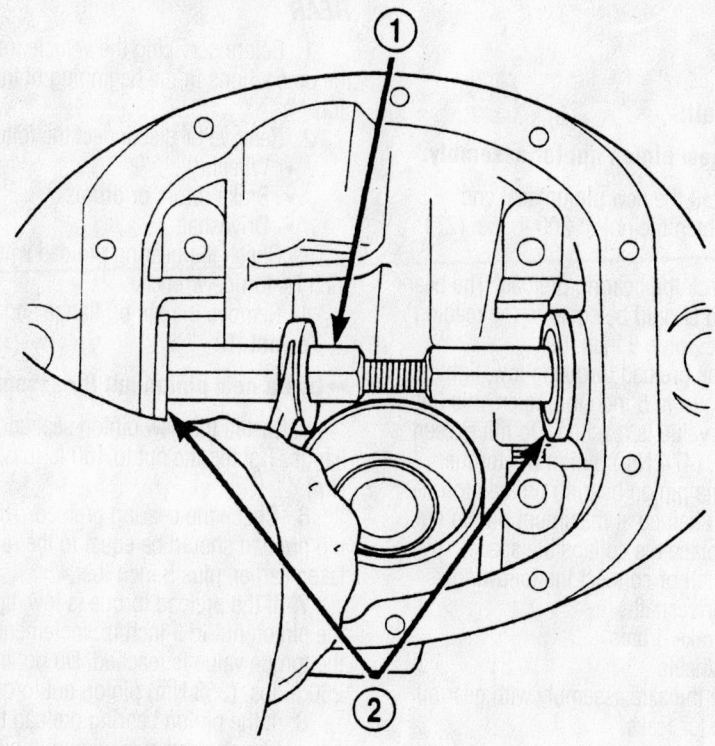

1 – TURNBUCKLE 6797
2 – DISCS 8110

9308PG05

Axle seal installation

- Brake caliper and rotor
- Front wheel
- Negative battery cable

FRONT SHAFT SEAL

1. Before servicing the vehicle, refer to the precautions in the beginning of this section.
2. Remove or disconnect the following:
   - Front axle shafts
   - Differential cover
   - Differential and ring gear assembly
   - Axle seals

To install:

3. Press the axle seals into the differential housing with Turnbuckle 6797 and Disc set 8110.
4. Install or connect the following:
   - Differential and ring gear assembly. Tighten the bearing cap bolts to 45 ft. lbs. (61 Nm).
   - Differential cover. Tighten the bolts to 30 ft. lbs. (41 Nm).
   - Front axle shafts
5. Fill the axle assembly with gear oil and check for leaks.

Rear Axle Shaft, Bearing and Seal

REMOVAL & INSTALLATION

C-Clip Type Rear Axle Shaft

1. Before servicing the vehicle, refer to the precautions in the beginning of this section.
2. Remove or disconnect the following:
   - Negative battery cable
   - Rear wheel
   - Brake drum
   - Differential cover
   - Differential gear shaft retainer
   - Differential gear shaft
   - C-clip
   - Axle shaft
   - Axle seal
   - Axle bearing

To install:

3. Install or connect the following:
   - Axle bearing
   - Axle seal
   - Axle shaft
   - C-clip
   - Differential gear shaft. Use Loctite® and tighten the retainer to 14 ft. lbs. (19 Nm).
   - Differential cover. Tighten the bolts to 30 ft. lbs. (41 Nm).
   - Brake drum
   - Rear wheel
4. Fill the axle assembly with gear oil and check for leaks.

Non C-Clip Type Rear Axle Shaft

1. Before servicing the vehicle, refer to the precautions in the beginning of this section.
2. Remove or disconnect the following:
   - Rear wheel
   - Brake caliper and rotor, if equipped
   - Brake drum, if equipped
   - Axle retainer nuts
   - Axle shaft, seal and bearing assembly
3. Split the bearing retainer with a chisel and remove the retainer ring.
4. Press the bearing off the axle shaft.
5. Remove the axle seal and retaining plate.

To install:

6. Install the retaining plate and axle seal onto the axle shaft.
7. Pack the wheel bearing with axle grease and press the bearing on to the axle shaft.

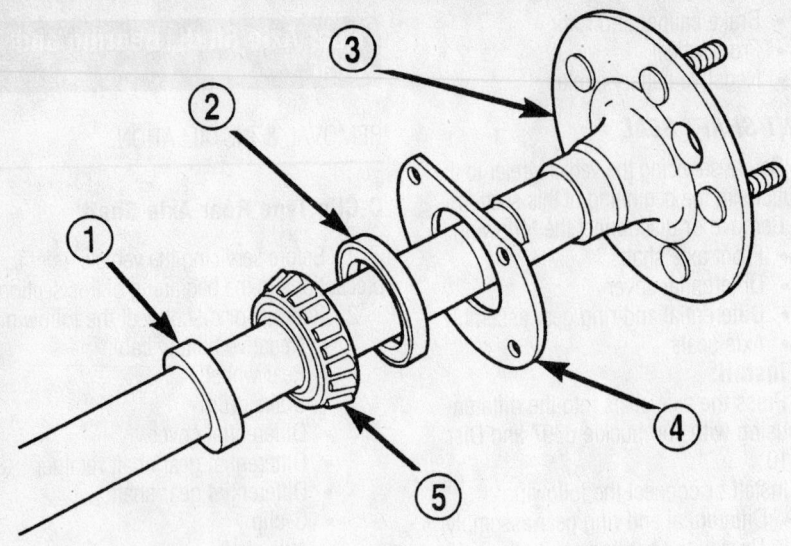

1 – RETAINING RING
2 – SEAL
3 – AXLE
4 – RETAINING PLATE
5 – AXLE BEARING

9308PG06

Rear axle seal and bearing components

8. Press the retaining ring onto the axle shaft.

9. Install or connect the following:
- Axle shaft, seal and bearing assembly. Tighten the nuts to 45 ft. lbs. (61 Nm).
- Brake caliper and rotor, if equipped
- Brake drum, if equipped
- Rear wheel

10. Fill the axle assembly with gear oil and check for leaks.

Pinion Seal

REMOVAL & INSTALLATION

C-Clip Type

1. Before servicing the vehicle, refer to the precautions in the beginning of this section.

2. Remove or disconnect the following:
- Wheels
- Brake drums
- Driveshaft

3. Check the bearing preload with an inch lb. torque wrench.

4. Remove the pinion flange and seal.

To install:

➡**Use a new pinion nut for assembly.**

5. Install the new pinion seal and flange. Tighten the nut to 200 ft. lbs. (271 Nm).

6. Check the bearing preload. The bearing preload should be equal to the reading taken earlier, plus 5 inch lbs.

7. If the preload torque is low, tighten the pinion nut in 5 inch lb. increments until the torque value is reached. Do not exceed 350 ft. lbs. (474 Nm) pinion nut torque.

8. If the pinion bearing preload torque cannot be attained at maximum pinion nut torque, replace the collapsible spacer.

9. Install or connect the following:
- Driveshaft
- Brake drums
- Wheels

10. Fill the axle assembly with gear oil and check for leaks.

Non C-Clip Type

FRONT

1. Before servicing the vehicle, refer to the precautions in the beginning of this section.

2. Remove or disconnect the following:
- Wheels
- Brake rotors
- Driveshaft

3. Check the bearing preload with an inch lb. torque wrench.

4. Remove the pinion flange and seal.

To install:

➡**Use a new pinion nut for assembly.**

5. Install the new pinion seal and flange. Tighten the nut to 160 ft. lbs. (217 Nm).

6. Check the bearing preload. The bearing preload should be equal to the reading taken earlier, plus 5 inch lbs.

7. If the preload torque is low, tighten the pinion nut in 5 inch lb. increments until the torque value is reached. Do not exceed 260 ft. lbs. (353 Nm) pinion nut torque.

8. If the pinion bearing preload torque can not be attained at maximum pinion nut torque, replace the collapsible spacer.

9. Install or connect the following:
- Driveshaft
- Brake rotors
- Wheels

10. Fill the axle assembly with gear oil and check for leaks.

REAR

1. Before servicing the vehicle, refer to the precautions in the beginning of this section.

2. Remove or disconnect the following:
- Wheels
- Brake rotors or drums
- Driveshaft

3. Check the bearing preload with an inch lb. torque wrench.

4. Remove the pinion flange and seal.

To install:

➡**Use a new pinion nut for assembly.**

5. Install the new pinion seal and flange. Tighten the nut to 160 ft. lbs. (217 Nm).

6. Check the bearing preload. The bearing preload should be equal to the reading taken earlier, plus 5 inch lbs.

7. If the preload torque is low, tighten the pinion nut in 5 inch lb. increments until the torque value is reached. Do not exceed 260 ft. lbs. (353 Nm) pinion nut torque.

8. If the pinion bearing preload torque can not be attained at maximum pinion nut torque, remove one or more pinion preload shims.

9. Install or connect the following:
- Driveshaft
- Brake rotors or drums
- Wheels

10. Fill the axle assembly with gear oil and check for leaks.

STEERING AND SUSPENSION

Air Bag

✳✳ CAUTION

Some vehicles are equipped with an air bag system. The system must be disarmed before performing service on, or around, system components, the steering column, instrument panel components, wiring and sensors. Failure to follow the safety precautions and the disarming procedure could result in accidental air bag deployment, possible injury and unnecessary system repairs.

PRECAUTIONS

Several precautions must be observed when handling the inflator module to avoid accidental deployment and possible personal injury.

• Never carry the inflator module by the wires or connector on the underside of the module.

• When carrying a live inflator module, hold securely with both hands, and ensure that the bag and trim cover are pointed away.

• Place the inflator module on a bench or other surface with the bag and trim cover facing up.

• With the inflator module on the bench, never place anything on or close to the module, which may be thrown in the event of an accidental deployment.

Before servicing the vehicle, also make sure to refer to the precautions in the beginning of this section as well.

DISARMING

Disconnect and isolate the negative battery cable. Wait 2 minutes for the system capacitor to discharge before performing any service.

ARMING

To arm the system, connect the negative battery cable.

Recirculating Ball Power Steering Gear

REMOVAL & INSTALLATION

1. Before servicing the vehicle, refer to the precautions in the beginning of this section.

2. Place the front wheels in the straight ahead position.
3. Drain the power steering system.
4. Remove or disconnect the following:
   • Negative battery cable
   • Air cleaner housing
   • Power steering pressure and return lines
   • Column coupler shaft bolt
   • Left front wheel assembly
   • Pitman arm
   • Windshield washer reservoir
   • Steering gear

To install:
5. Install or connect the following:
   • Steering gear. Tighten the bolts to 80 ft. lbs. (108 Nm).
   • Pitman arm. Tighten the nut to 185 ft. lbs. (251 Nm).
   • Windshield washer reservoir
   • Power steering pressure and return lines. Torque the fasteners to 14 ft. lbs. (20 Nm).
   • Intermediate shaft. Tighten the pinch bolt to 36 ft. lbs. (49 Nm).
   • Air cleaner assembly

• Left front wheel
• Negative battery cable
6. Fill and bleed the power steering fluid reservoir.
7. Start the engine, check for leaks and repair if necessary.

Rack and Pinion Power Steering Gear

REMOVAL & INSTALLATION

1. Before servicing the vehicle, refer to the precautions in the beginning of this section.
2. Remove or disconnect the following:
   • Negative battery cable
   • Fluid from the steering system

✳✳ CAUTION

Lock the steering wheel to prevent the clockspring from spinning.

• Skid plate
• Wheels

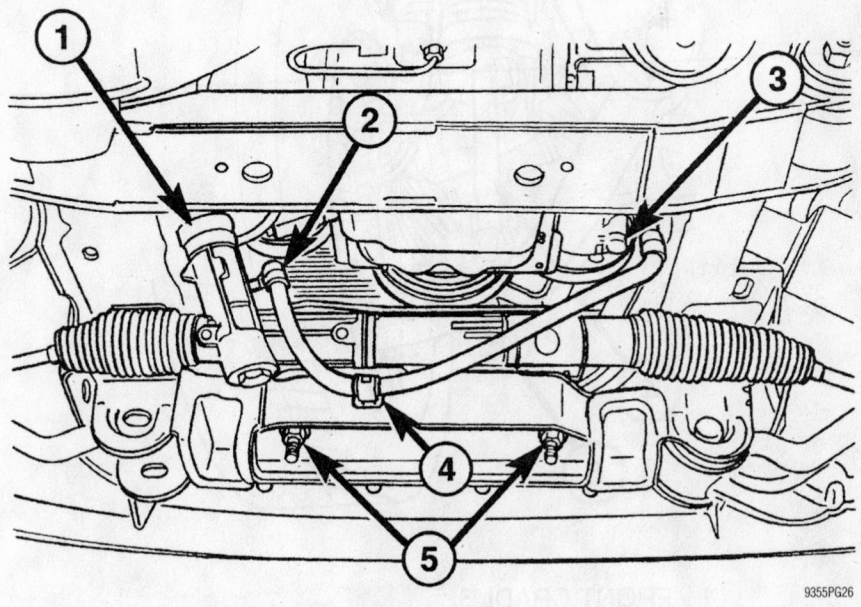

Rack mounting points—Liberty

9355PG26

➡ **Matchmark the alignment cams.**

- Lower control arms
- Front axle
- Tie rod end nuts
- Tie rod ends from the knuckles
- Intermediate shaft lower coupler from the gear
- High pressure hose bracket
- Hoses from the gear
- Gear mounting bolts and gear

3. Installation is the reverse of removal. Observe the following torques:

- Gear-to-frame bolts: 120 ft. lbs. (162 Nm)
- Intermediate shaft bolt: 36 ft. lbs. (49 Nm)
- Tie rod end-to-knuckle nut: 80 ft. lbs. (108 Nm)
- Tie rod end jam nut: 55 ft. lbs. (75 Nm)
- Pressure and return lines: 25 ft. lbs. (35 Nm)

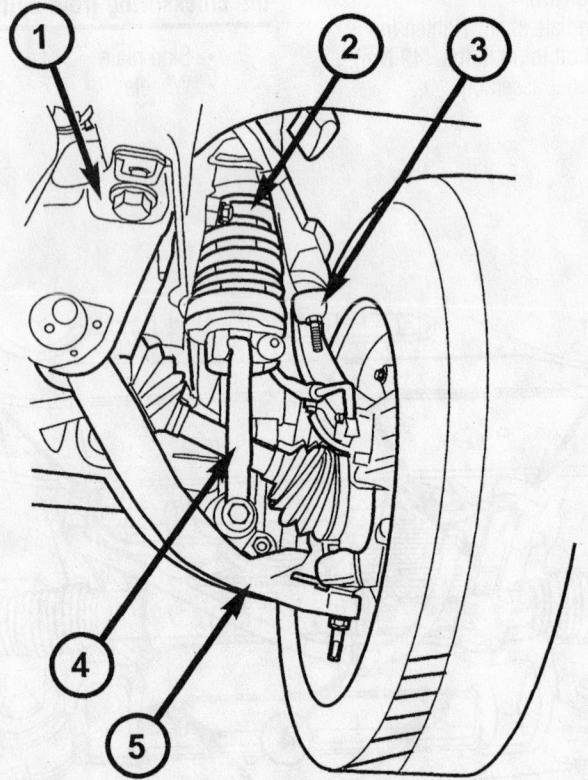

1 - FRONT CRADLE
2 - SPRING & SHOCK ASSEMBLY
3 - STEERING KNUCKLE
4 - CLEVIS BRACKET
5 - LOWER CONTROL ARM

9355PG27

Strut and clevis assembly—Liberty

Front Struts

REMOVAL & INSTALLATION

Liberty

LEFT SIDE

1. Before servicing the vehicle, refer to the precautions in the beginning of this section.

2. Remove or disconnect the following:
- Negative battery cable
- Battery
- Power center
- Battery tray
- Battery temperature sensor
- The 4 upper strut nuts
- Left wheel
- Lower clevis bracket bolt
- Stabilizer link
- Lower ball joint from the arm

- Clevis bracket
- Strut

3. Installation is the reverse of removal. Observe the following torques:
- Upper strut nuts: 80 ft. lbs. (108 Nm)
- Clevis bracket-to-strut bolt: 65 ft. lbs. (88 Nm)
- Ball joint nut: 60 ft. lbs. (81 Nm)
- Clevis bracket-to-arm bolt: 110 ft. lbs. (150 Nm)
- Stabilizer link-to-arm bolt: 100 ft. lbs. (136 Nm)

RIGHT SIDE

1. Before servicing the vehicle, refer to the precautions in the beginning of this section.

2. Remove or disconnect the following:
- Negative battery cable
- Air box
- Cruise control servo mounting nuts
- The 4 upper strut nuts
- Right wheel
- Lower clevis bracket bolt
- Stabilizer link
- Lower ball joint from the arm
- Clevis bracket
- Strut

3. Installation is the reverse of removal. Observe the following torques:
- Upper strut nuts: 80 ft. lbs. (108 Nm)
- Clevis bracket-to-strut bolt: 65 ft. lbs. (88 Nm)
- Ball joint nut: 60 ft. lbs. (81 Nm)
- Clevis bracket-to-arm bolt: 110 ft. lbs. (150 Nm)
- Stabilizer link-to-arm bolt: 100 ft. lbs. (136 Nm)

Shock Absorber

REMOVAL & INSTALLATION

Front

1. Before servicing the vehicle, refer to the precautions in the beginning of this section.

2. Remove or disconnect the following:
- Upper nut, washer and grommet from the upper stud
- Lower fasteners
- Shock absorber

To install:

3. Install or connect the following:
- Shock absorber. Torque the lower fasteners to 17 ft. lbs. (23 Nm).
- Upper grommet, washer, and nut to the stud. Torque it to 16 ft. lbs. (22 Nm).

Rear

1. Before servicing the vehicle, refer to the precautions in the beginning of this section.
2. Remove or disconnect the following:
   - Upper locknut and washer from the frame bracket stud, on the Grand Cherokee and Liberty
   - Upper mounting bolts, on the Wrangler and Cherokee
   - Lower bolt, nut and washers from the axle shaft tube bracket
   - Shock absorber

To install:

3. Place the shock absorber upper end in position and tighten the fasteners to the following specifications:
   - Wrangler: 23 ft. lbs. (31 Nm).
   - Cherokee: 17 ft. lbs. (23 Nm).
   - Liberty and Grand Cherokee: 80 ft. lbs. (108 Nm).
4. Place the shock absorber lower end in position and tighten the fasteners to the following specifications:
   - Wrangler: 74 ft. lbs. (100 Nm).
   - Cherokee: 46 ft. lbs. (62 Nm).
   - Liberty and Grand Cherokee: 85 ft. lbs. (115 Nm).

Coil Spring

REMOVAL & INSTALLATION

Front

EXCEPT LIBERTY

1. Before servicing the vehicle, refer to the precautions in the beginning of this section.
2. Remove or disconnect the following:
   - Front wheels
   - Front driveshaft, if equipped
   - Lower suspension arm
   - Stabilizer bar links
   - Track bar
   - Drag link
   - Brake hose brackets
   - Spring retainers
   - Coil spring

To install:

3. Install or connect the following:
   - Coil spring
   - Spring retainers. Torque the bolt to 16 ft. lbs. (21 Nm).
   - Stabilizer bar links. Torque the bolts to 70 ft. lbs. (95 Nm).

- Track bar. Torque the bolt to 35 ft. lbs. (47 Nm).
- Front propeller shaft to the axle
- Drag link
- Lower suspension arm. Torque the bolt to 133 ft. lbs. (180 Nm).
- Front driveshaft, if equipped
- Front wheels

LIBERTY

1. Before servicing the vehicle, refer to the precautions in the beginning of this section.
2. Remove the strut and place it in a Pentastar W7200 spring compressor, or equivalent.
3. Compress the spring, remove the strut mount nut and remove the strut from the compressor.
4. Installation is the reverse of removal. Torque the nut to 30 ft. lbs. (41 Nm).

Rear

EXCPET LIBERTY

1. Before servicing the vehicle, refer to the precautions in the beginning of this section.

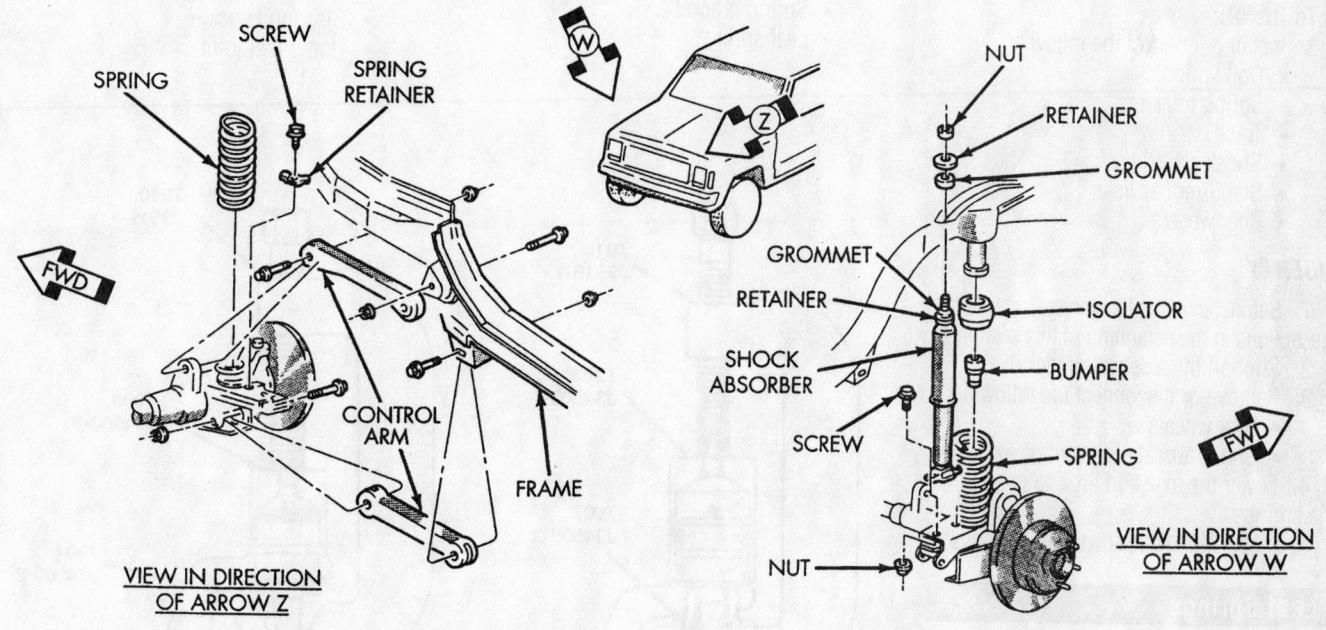

Exploded view of the front suspension—Cherokee

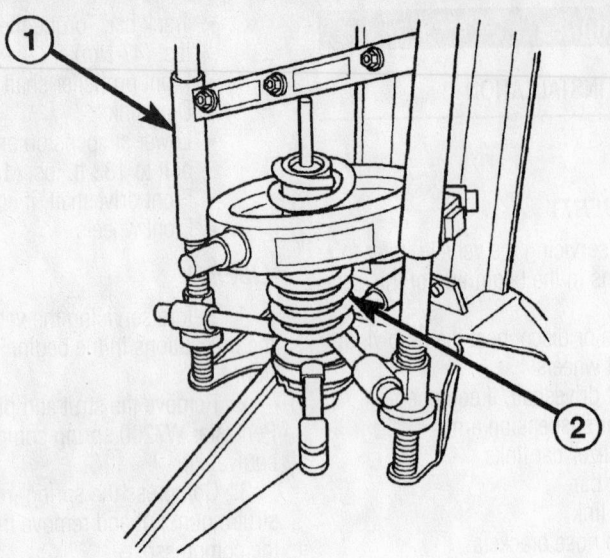

1 - SPRING COMPRESSOR
2 - SPRING

9355PG28

Front coil spring removal—Liberty

2. Remove or disconnect the following:
- Rear wheels
- Stabilizer bar links
- Shock absorbers
- Track bar
- Spring retainers
- Coil springs

To install:

3. Install or connect the following:
- Coil springs
- Spring retainers
- Track bar
- Shock absorbers
- Stabilizer bar links
- Rear wheels

LIBERTY

1. Before servicing the vehicle, refer to the precautions in the beginning of this section.
2. Support the axle with a jack.
3. Remove or disconnect the following:
- Rear wheels
- Shock absorbers from the axle
4. Lower the axle and tilt it to remove the springs.
5. Installation is the reverse of removal.

Leaf Springs

REMOVAL & INSTALLATION

Cherokee

1. Before servicing the vehicle, refer to the precautions in the beginning of this section.

2. Support the vehicle at the frame rails.
3. Support the rear axle with a jack.
4. Remove or disconnect the following:
- Rear wheel
- Stabilizer bar link
- Axle U-bolts
- Spring bracket
- Leaf spring

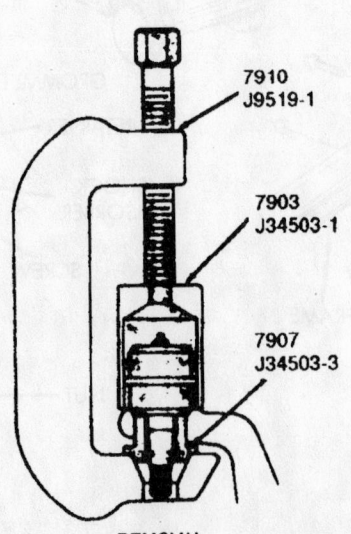

7910
J9519-1

7903
J34503-1

7907
J34503-3

REMOVAL

Upper ball joint removal and installation

To install:

➡ **The weight of the vehicle must be supported by the springs when the spring eye and stabilizer bar fasteners are tightened.**

5. Install or connect the following:
- Leaf spring
- Spring bracket
- Axle U-bolts. Torque the nuts to 52 ft. lbs. (70 Nm).
- Stabilizer bar link
- Rear wheel
6. Torque the front spring eye bolt and nut to 115 ft. lbs. (156 Nm).
7. Torque the rear spring eye bolt and nut to 80 ft. lbs. (108 Nm).
8. Torque the stabilizer bar nuts 55 ft. lbs. (74 Nm).

Upper Ball Joint

REMOVAL & INSTALLATION

Except Liberty

1. Before servicing the vehicle, refer to the precautions in the beginning of this section.
2. Remove or disconnect the following:
- Front wheel
- Disc brake caliper and rotor
- Wheel bearing
- Axle shaft
- Steering knuckle
- Upper ball joint

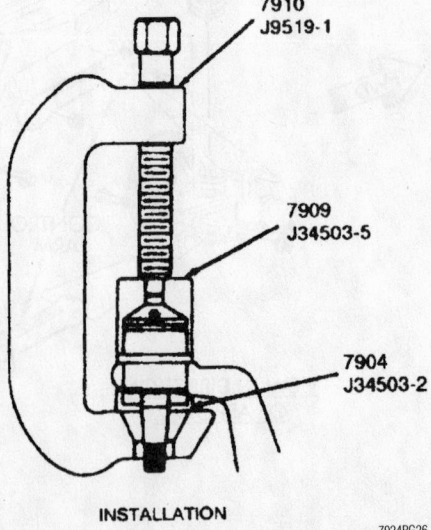

7910
J9519-1

7909
J34503-5

7904
J34503-2

INSTALLATION

7924PG26

To install:

3. Install or connect the following:
- Upper ball joint
- Steering knuckle. Torque the nuts to 100 ft. lbs. (135 Nm).
- Axle shaft
- Wheel bearing
- Disc brake caliper and rotor
- Front wheel

Liberty

The upper ball joint is serviced as an assembly with the control arm.

Lower Ball Joint

REMOVAL & INSTALLATION

Except Liberty

1. Before servicing the vehicle, refer to the precautions in the beginning of this section.
2. Remove or disconnect the following:
- Front wheel
- Disc brake caliper and rotor
- Wheel bearing
- Axle shaft
- Steering knuckle
- Lower ball joint

To install:

3. Install or connect the following:
- Lower ball joint
- Steering knuckle. Torque the nuts to 100 ft. lbs. (135 Nm).
- Axle shaft

- Wheel bearing
- Disc brake caliper and rotor
- Front wheel

Liberty

1. Before servicing the vehicle, refer to the precautions in the beginning of this section.
2. Remove or disconnect the following:
- Front wheel
- Lower clevis bracket bolt from the control arm
- Stabilizer link at the control arm
- Lower ball joint nut
- Control arm from lower ball joint with tool C4150A
3. Installation is the reverse of removal. Torque the ball joint nut to 60 ft. lbs. (81 Nm); the stabilizer link bolt to 100 ft. lbs. (136 Nm); the lower clevis bracket bolt to 110 ft. lbs. (150 Nm).

Upper Control Arm

REMOVAL & INSTALLATION

Front

EXCEPT LIBERTY

1. Before servicing the vehicle, refer to the precautions in the beginning of this section.
2. Support the axle with a jackstand.
3. Unbolt and remove the upper control arm.

To install:

➡**The weight of the vehicle must be supported by the springs before tightening the control arm fasteners.**

4. Install the control arms.
5. Torque the axle fastener to 55 ft. lbs. (75 Nm) and the frame fastener to 66 ft. lbs. (90 Nm).

LIBERTY-RIGHT SIDE

1. Before servicing the vehicle, refer to the precautions at the beginning of this section.
2. Remove or disconnect the following:
- Wheel
- Upper ball joint nut
- Upper Ball joint from the knuckle
- Air box
- Cruise control servo mounting nuts
- Upper arm rear bolt
- Upper arm front bolt
- Upper arm
3. Installation is the reverse of removal. Observe the following torques:
- Front and rear bolts: 90 ft. lbs. (122 Nm)
- Ball joint stud nut: 60 ft. lbs. (81 Nm)

LIBERTY-LEFT SIDE

1. Before servicing the vehicle, refer to the precautions at the beginning of this section.
2. Remove or disconnect the following:
- Wheel
- Ball joint nut
- Ball joint from the knuckle
- Battery
- Power center
- Battery tray
- Battery temperature sensor
- Control arm rear bolt, by using a ratchet and extension under the steering shaft, positioned by the power steering reservoir
- Control arm front bolt
- Control arm
3. Installation is the reverse of removal. Observe the following torques:
- Front and rear bolts: 90 ft. lbs. (122 Nm)
- Ball joint stud nut: 60 ft. lbs. (81 Nm)

Rear

EXCEPT LIBERTY AND GRAND CHEROKEE

1. Before servicing the vehicle, refer to the precautions in the beginning of this section.

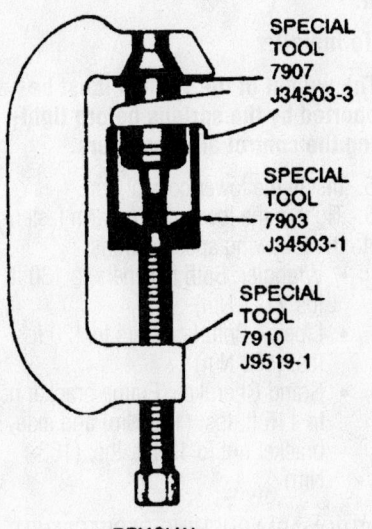

SPECIAL TOOL 7907 J34503-3

SPECIAL TOOL 7903 J34503-1

SPECIAL TOOL 7910 J9519-1

REMOVAL

Lower ball joint removal and installation

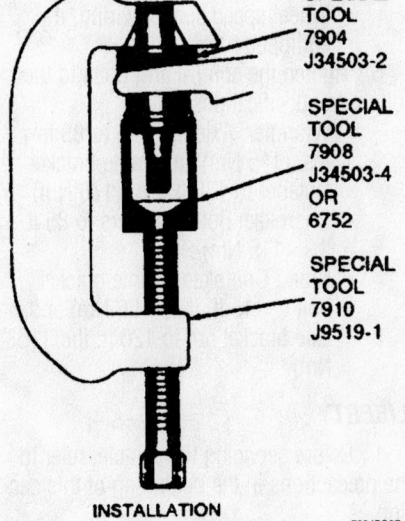

SPECIAL TOOL 7904 J34503-2

SPECIAL TOOL 7908 J34503-4 OR 6752

SPECIAL TOOL 7910 J9519-1

INSTALLATION

7924PG27

Heater Core replacement is covered in Section 2 of this manual

2. Support the axle with a jackstand.
3. Remove or disconnect the following:
- Parking brake cable and bracket
- Wheel speed sensor wiring bracket, if equipped
- Upper control arm

To install:

➡**The weight of the vehicle must be supported by the springs before tightening the control arm fasteners.**

4. Install or connect the following:
- Upper control arm
- Wheel speed sensor wiring bracket, if equipped
- Parking brake cable and bracket

5. Torque the control arm fasteners to 55 ft. lbs. (75 Nm).

GRAND CHEROKEE

1. Before servicing the vehicle, refer to the precautions in the beginning of this section.
2. Support the axle with a jackstand.
3. Remove or disconnect the following:
- Parking brake cable brackets
- Brake hose brackets
- Axle ball joint
- Upper control arm

To install:

➡**Use a new axle ball joint nut.**

4. Install or connect the following:
- Upper control arm. Torque the frame bracket bolts to 74 ft. lbs. (100 Nm) and the axle ball joint nut to 105 ft. lbs. (142 Nm).
- Brake hose brackets
- Parking brake cable brackets

LIBERTY

1. Before servicing the vehicle, refer to the precautions in the beginning of this section.
2. Support the axle with a jackstand.
3. Remove or disconnect the following:
- Ball joint pinch bolt from the top of the differential housing bracket.
- Heat shield nuts and lower the shield
- Upper arm-to-body bolts and arm

4. Installation is the reverse of removal. Observe the following torques:
- Upper arm mounting bolts: 74 ft. lbs. (100 Nm)
- Pinch bolt: 70 ft. lbs. (95 Nm)

CONTROL ARM BUSHING REPLACEMENT

The upper control arm bushings are serviced with the control arms as complete

assemblies, with the exception of the front upper axle bushing, which may be replaced after removing the upper control arm.

Front Upper Control Arm Axle Bushing

1. Before servicing the vehicle, refer to the precautions in the beginning of this section.
2. Remove the upper control arm.
3. Press the old bushing out of the axle housing.
4. Press the new bushing into the axle housing.
5. Install the upper control arm.

Lower Control Arms

REMOVAL & INSTALLATION

Front

EXCEPT LIBERTY

1. Before servicing the vehicle, refer to the precautions in the beginning of this section.
2. Support the axle with a jackstand.
3. Remove or disconnect the following:
- Wheel speed sensor wiring, if equipped
- Lower control arm

To install:

➡**The weight of the vehicle must be supported by the springs before tightening the control arm fasteners.**

4. Install or connect the following:
- Lower control arm
- Wheel speed sensor wiring, if equipped

5. Tighten the control arm bolts to the following specifications:
- Wrangler: Axle fastener to 85 ft. lbs. (115 Nm) and frame bracket fastener to 130 ft. lbs. (176 Nm).
- Cherokee: Both fasteners to 85 ft. lbs. (115 Nm).
- Grand Cherokee: Frame bracket bolt to 115 ft. lbs. (156 Nm) and axle bracket nut to 120 ft. lbs. (163 Nm).

LIBERTY

1. Before servicing the vehicle, refer to the precautions in the beginning of this section.
2. Remove or disconnect the following:
- Front wheel
- Lower clevis bracket bolt from the control arm

- Stabilizer link at the control arm
- Lower ball joint nut
- Control arm from lower ball joint with tool C4150A

➡**Matchmark the front and rear control arm pivot bolts.**

- Front pivot bolt
- Rear pivot bolt
- Control arm

To install:

3. Install or connect the following:
- Lower control arm
- Rear pivot bolt
- Front pivot bolt
- Ball joint nut. Torque the ball joint nut to 60 ft. lbs. (81 Nm)

4. Align the matchmarks and tighten the pivot bolts to 125 ft. lbs. (170 Nm).

5. The remainder of the installation is the reverse of removal. Torque the stabilizer link bolt to 100 ft. lbs. (136 Nm); the lower clevis bracket bolt to 110 ft. lbs. (150 Nm).

Rear

1. Before servicing the vehicle, refer to the precautions in the beginning of this section.
2. Support the axle with a jackstand.
3. On the Liberty, disconnect the stabilizer bar from the arm.
4. Unbolt and remove the lower control arm.

➡**On the Liberty's right arm, it will be necessary to pry the exhaust pipe slightly to get to the frame rail-to-arm bolt.**

To install:

➡**The weight of the vehicle must be supported by the springs before tightening the control arm fasteners.**

5. Install the lower control arm.
6. Tighten the lower control arm fasteners to the following specifications:
- Wrangler: Both fasteners to 130 ft. lbs. (177 Nm)
- Liberty: Both fasteners to 120 ft. lbs. (163 Nm)
- Grand Cherokee: Frame bracket nut to 115 ft. lbs. (156 Nm) and axle bracket nut to 120 ft. lbs. (163 Nm)

CONTROL ARM BUSHING REPLACEMENT

The lower control arm bushings are serviced with the control arms as complete assemblies.

Front Wheel Bearings

ADJUSTMENT

2-Wheel Drive

EXCEPT LIBERTY

1. Before servicing the vehicle, refer to the precautions in the beginning of this section.

2. Remove or disconnect the following:
   - Front wheel, grease cap, cotter pin and nut cap
   - Wheel bearing nut, loosen it

3. While turning the rotor, tighten the nut to 25 ft. lbs. (34 Nm) to seat the bearings.

4. Back the nut off ½ turn.

5. While turning the rotor, tighten the nut to 19 inch lbs. (2 Nm).

6. Install or connect the following:
   - Nut cap, new cotter pin and the grease cap
   - Wheel

4-WHEEL DRIVE AND ALL LIBERTY MODELS

The front wheel bearings are not adjustable.

REMOVAL & INSTALLATION

2WD Models

EXCEPT LIBERTY

1. Before servicing the vehicle, refer to the precautions in the beginning of this section.

2. Remove or disconnect the following:
   - Front wheel
   - Brake caliper
   - Grease cap
   - Split pin
   - Nut retainer
   - Nut and washer
   - Outer bearing
   - Brake rotor and hub assembly
   - Grease seal
   - Inner bearing

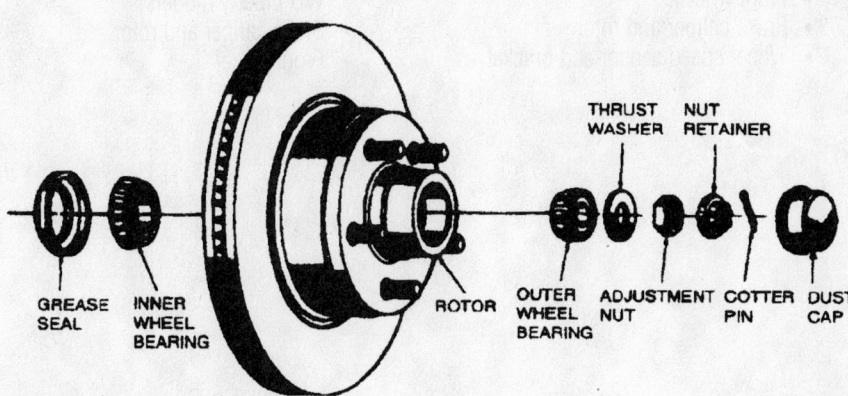

Exploded view of the front wheel bearings—2WD models

7924PG28

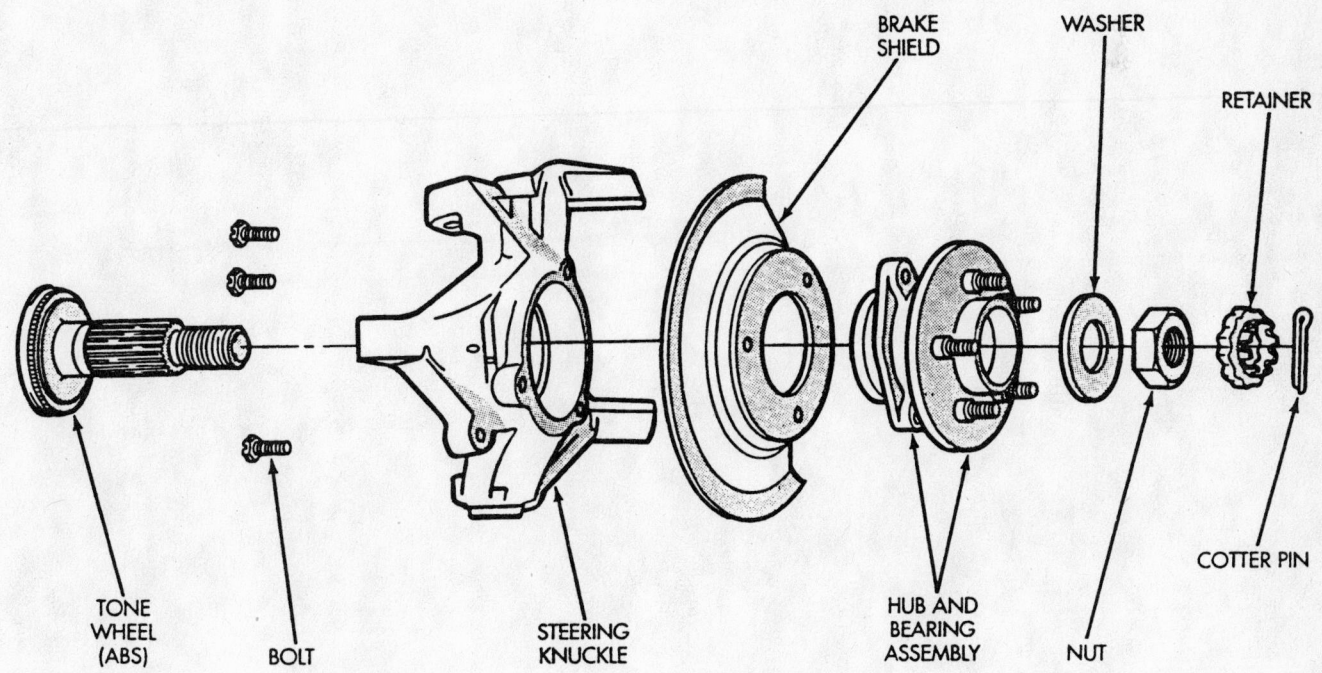

Exploded view of the hub assembly—4WD models

7924PG29

Brake service is covered in Section 4 of this manual

To install:

➡ **Use new grease seals and split pins.**

3. Pack the wheel bearings and the inside of the hub with high temperature wheel bearing grease. Add grease to the hub until it is flush with the inside diameter of the bearing cup.

4. Install or connect the following:
- Inner bearing
- Grease seal
- Brake rotor and hub assembly
- Outer bearing
- Nut and washer

- Nut retainer
- Split pin
- Grease cap
- Brake caliper
- Front wheel

4WD MODELS AND ALL LIBERTY MODELS

1. Before servicing the vehicle, refer to the precautions in the beginning of this section.

2. Remove or disconnect the following:
- Front wheel
- Brake caliper and rotor
- Wheel speed sensor and bracket

- Axle stub shaft nut (except 2-WD Liberty)
- Hub and wheel bearing assembly

To install:

3. Install or connect the following:
- Hub assembly over the axle stub shaft. Tighten the hub bolts to 75 ft. lbs. (102 Nm) except Liberty; 96 ft. lbs. (130 Nm) for Liberty models
- Axle stub shaft nut. Torque the nut to 175 ft. lbs. (237 Nm) except Liberty; 100 ft. lbs. (136 Nm) for 4-WD Liberty models
- Brake caliper and rotor
- Front wheel

KIA

Sportage

PRECAUTIONS

Before servicing any vehicle, please be sure to read all of the following precautions, which deal with personal safety, prevention of component damage, and important points to take into consideration when servicing a motor vehicle:

• Never open, service or drain the radiator or cooling system when the engine is hot; serious burns can occur from the steam and hot coolant.

• Observe all applicable safety precautions when working around fuel. Whenever servicing the fuel system, always work in a well-ventilated area. Do not allow fuel spray or vapors to come in contact with a spark, open flame, or excessive heat (a hot drop light, for example). Keep a dry chemical fire extinguisher near the work area. Always keep fuel in a container specifically designed for fuel storage; also, always properly seal fuel containers to avoid the possibility of fire or explosion. Refer to the additional fuel system precautions later in this section.

• Fuel injection systems often remain pressurized, even after the engine has been turned **OFF**. The fuel system pressure must be relieved before disconnecting any fuel lines. Failure to do so may result in fire and/or personal injury.

• Brake fluid often contains polyglycol ethers and polyglycols. Avoid contact with the eyes and wash your hands thoroughly after handling brake fluid. If you do get brake fluid in your eyes, flush your eyes with clean, running water for 15 minutes. If eye irritation persists, or if you have taken brake fluid internally, IMMEDIATELY seek medical assistance.

• The EPA warns that prolonged contact with used engine oil may cause a number of skin disorders, including cancer! You should make every effort to minimize your exposure to used engine oil. Protective gloves should be worn when changing oil. Wash your hands and any other exposed skin areas as soon as possible after exposure to used engine oil. Soap and water, or waterless hand cleaner should be used.

• All new vehicles are now equipped with an air bag system, often referred to as a Supplemental Restraint System (SRS) or Supplemental Inflatable Restraint (SIR) system. The system must be disabled before performing service on or around system components, steering column, instrument panel components, wiring and sensors. Failure to follow safety and disabling procedures could result in accidental air bag deployment, possible personal injury and unnecessary system repairs.

• Always wear safety goggles when working with, or around, the air bag system. When carrying a non-deployed air bag, be sure the bag and trim cover are pointed away from your body. When placing a non-deployed air bag on a work surface, always face the bag and trim cover upward, away from the surface. This will reduce the motion of the module if it is accidentally deployed. Refer to the additional air bag system precautions later in this section.

• Clean, high quality brake fluid from a sealed container is essential to the safe and proper operation of the brake system. You should always buy the correct type of brake fluid for your vehicle. If the brake fluid becomes contaminated, completely flush the system with new fluid. Never reuse any brake fluid. Any brake fluid that is removed from the system should be discarded. Also, do not allow any brake fluid to come in contact with a painted surface; it will damage the paint.

• Never operate the engine without the proper amount and type of engine oil; doing so WILL result in severe engine damage.

• Timing belt maintenance is extremely important! Many models utilize an interference-type, non-freewheeling engine. If the timing belt breaks, the valves in the cylinder head may strike the pistons, causing potentially serious (also time-consuming and expensive) engine damage. Refer to the maintenance interval charts in the front of this manual for the recommended replacement interval for the timing belt, and to the timing belt section for belt replacement and inspection.

• Disconnecting the negative battery cable on some vehicles may interfere with the functions of the on-board computer system(s) and may require the computer to undergo a relearning process once the negative battery cable is reconnected.

• When servicing drum brakes, only disassemble and assemble one side at a time, leaving the remaining side intact for reference.

• Only an MVAC-trained, EPA-certified automotive technician should service the A/C system or its components.

ENGINE REPAIR

Alternator

REMOVAL

1. Before servicing the vehicle, refer to the precautions in the beginning of this section.

2. Remove or disconnect the following:
• Negative battery cable
• Air cleaner inlet pipe front bolts
• Top hose from the resonance chamber
• Air cleaner inlet pipe
• Alternator electrical connectors
• Loosen the pivot and tensioner mounting bolts, do mot remove them
• Drive belt from the alternator pulley
• Drive belt tensioner
• Alternator pivot bolt
• Loosen the bolt at the base of the adjusting bracket and rotate the bracket up
• Alternator

To install:
3. Install or connect the following:
• Alternator
• Pivot bolt and hand tighten
• Rotate the bracket down on top of the alternator
• Belt tensioner on the adjustment bracket
• Tensioner mounting bolt and hand tighten it
• Drive belt
• Torque the tensioner bolt to 19 ft. lbs. (26 Nm).

• Torque the pivot bolt to 38 ft. lbs. (51 Nm).
• Alternator electrical connectors
• Air cleaner inlet pipe and tighten the clamp
• Hose to the resonance chamber
• Air inlet pipe bolts and tighten
• Negative battery cable

Ignition Timing

ADJUSTMENT

The 2.0L engine in the Sportage is equipped with a distributorless ignition system. The ignition timing is controlled by the Powertrain Control Module (PCM) through the input of engine control system sensors. The ignition timing is set at 4 degrees

BTDC for vehicles equipped with manual or automatic transmissions. The ignition timing cannot be adjusted.

Engine Assembly

REMOVAL & INSTALLATION

1. Before servicing the vehicle, refer to the precautions in the beginning of this section .
2. Properly relieve the fuel system pressure.
3. Drain the cooling system.
4. Drain the engine oil.
5. Drain the transmission fluid.
6. Remove or disconnect the following:
   - Both battery cables
   - Windshield washer hose from the hood
   - Hood
   - 2 air duct mounting bolts from the top of the radiator. Loosen the clamp at the air intake housing and remove the duct
   - Accelerator cable by pulling the throttle back and rotating the cable until it aligns with the slot in the pulley
   - Transmission control cable
   - Resonance chamber mounting bolt, chamber bolt and air silencer
   - Idle Air Control (IAC) air hose, breather hose and vacuum line from the air intake tube
   - Manifold Air Flow (MAF) sensor connector
   - Loosen the air inlet hose clamp from the MAF sensor
   - 3 bolts from the air intake tube to the throttle body
   - Air intake hose and tube as an assembly
   - Upper radiator hose
   - Clutch fan nuts
   - Cooling fan shroud bolts
   - Fan and shroud at the same time
   - Alternator drive belt
   - Fan pulley
   - Alternator electrical connectors
   - Exhaust Gas Recirculation (EGR) solenoid valve connector on the intake manifold in front of the dynamic chamber
   - Both heater hoses from the pipes
   - Engine-to-body ground wire from the intake manifold and the harness bracket
   - Brake booster vacuum hose from the dynamic chamber

- Fuel lines and fuel pressure regulator from the rear of the dynamic chamber
- Vacuum hose from the bottom of the EGR valve
- Purge solenoid valve vacuum hose from dynamic chamber
- Vacuum hoses from the top of the charcoal canister and slide the charcoal canister up and out of the bracket
- Lower radiator hose
- Cooling lines from the radiator, if equipped with an automatic transmission
- Radiator and raise and safely support the vehicle
- Lower splash panel
- Drive belt
- A/C pulley assembly
- A/C compressor mounting bolts and move the A/C compressor out of the way
- Power steering drive belt
- Intake manifold support bracket
- Starter wiring harness
- Starter
- Front exhaust pipe from the exhaust manifold
- Bracket bolt from the front exhaust pipe
- Exhaust-to-clutch (manual transmission) or converter (automatic transmission) housing bolts and the bracket
- Clutch housing (manual transmission) or converter (automatic transmission) housing bolts.
- Drive plate-to-torque converter bolts, if equipped

7. Support the transmission from underneath the vehicle.
8. Connect the engine hoist to the engine assembly.
   - Left and right side engine mounting bolts
9. Lift the engine up and forward slightly to provide access to three electrical connectors on the rear of the cylinder head.
   - Electrical connectors from the Camshaft Position (CMP) sensor, coil and condenser on the rear of the cylinder head
   - Engine from the vehicle

To install:

10. Lower the engine enough to connect the three electrical connectors to the CMP sensor, coil and condenser on the rear of the cylinder head.

11. Position the engine to the transmission. Install the transmission bolts and tighten the bolts. Torque according to bolt size:
   - 14mm bolts to 80 ft. lbs. (108 Nm)
   - 10mm bolts to 28 ft. lbs. (38 Nm)
   - 6mm bolts to 60 inch lbs. (7 Nm)
   - Right and left side engine mounting bolts. Torque the bolt s to 38 ft. lbs. (52 Nm).
12. Disconnect the engine hoist from the engine assembly.
13. Raise and safely support the vehicle.
   - Drive plate-to-torque converter bolts, if equipped.
   - Connect the front exhaust pipe to the exhaust manifold. Torque the flange bolts to 24 ft. lbs. (31 Nm).
   - Front exhaust pipe. Torque the bolts to 20 ft. lbs. (27 Nm).
   - Starter
   - Connect the starter wiring harness
   - Intake manifold support bracket bolts and the bracket.
14. Install the power steering pump lock and mounting bolts. Install the power steering drive belt. Torque the bolts to 30 ft. lbs. (42 Nm).
   - A/C compressor mounting bolts. Torque the bolts to 18 ft. lbs. (24 Nm).
   - A/C belt pulley assembly and drive belt. Install the two A/C idler pulley bracket bolts and torque to 24 ft. lbs. (32 Nm).
   - Lower splash panel
   - Radiator
   - Cooling lines to the radiator, if equipped
   - Lower radiator hose
   - Slide the charcoal canister in the bracket
   - Vacuum hoses to the top of the charcoal canister
   - Engine-to-body ground wire to the intake manifold and the harness bracket
   - Brake booster vacuum hose to the dynamic chamber
   - Fuel lines and fuel pressure regulator to the rear of the dynamic chamber
   - Vacuum hose to the bottom of the EGR valve
   - Purge solenoid valve vacuum hose to dynamic chamber
   - EGR solenoid valve connector on the intake manifold in front of the dynamic chamber

- Heater hoses
- Electrical terminal connectors to the alternator
- Fan pulley
- Alternator drive belt. Torque the adjusting bolt 16 ft. lbs. (22 Nm) and the mounting bolt to 32 ft. lbs. (45 Nm).
- Fan and shroud as an assembly. Torque the five cooling fan shroud bolts to 72 inch lbs. (8 Nm)
- Clutch fan nuts. Torque the nuts to 27 ft. lbs. (37 Nm).
- Upper radiator hose

- Air intake hose and tube as an assembly
- Air intake tube to the throttle body and tighten the air inlet hose clamp to the MAF sensor
- MAF sensor electrical connector
- Resonance chamber mounting bolt, chamber bolt and air silencer
- IAC air hose, breather hose and vacuum line to the air intake tube
- Accelerator cable by pulling the throttle back and rotating the cable until it aligns with the slot in the pulley

- Transmission control cable
- Air duct mounting bolts to the top of the radiator. Torque the clamp at the air intake housing
- Hood
- Windshield washer hose to the hood

15. Fill the engine with clean engine oil.
16. Connect the battery cables.
17. Fill the cooling system.
18. Fill the transmission fluid.
19. Recharge the A/C system.
20. Start the vehicle. Check for leaks, repair if necessary.

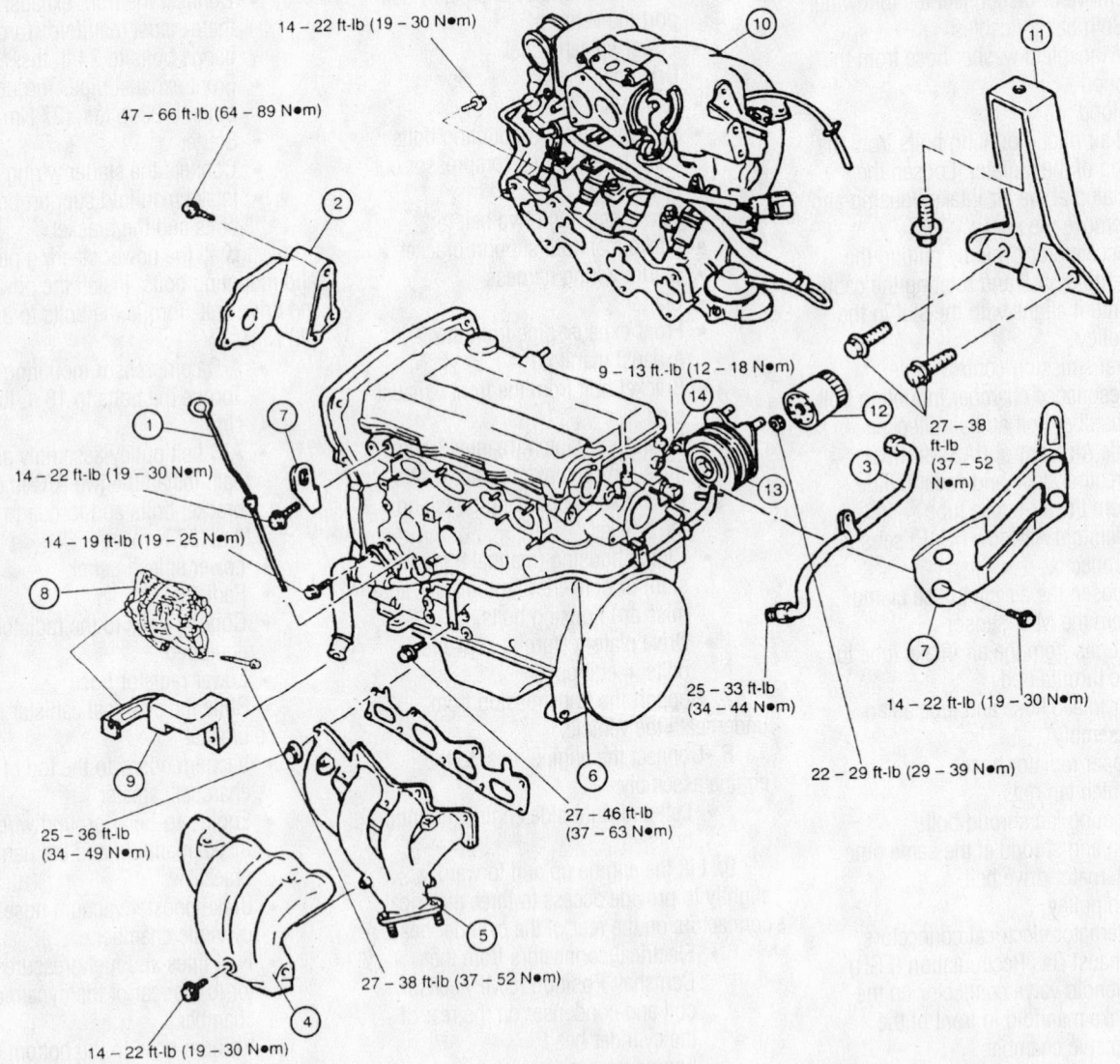

14 – 22 ft-lb (19 – 30 N•m)

47 – 66 ft-lb (64 – 89 N•m)

9 – 13 ft.-lb (12 – 18 N•m)

14 – 22 ft-lb (19 – 30 N•m)

14 – 19 ft-lb (19 – 25 N•m)

27 – 38 ft-lb (37 – 52 N•m)

25 – 33 ft-lb (34 – 44 N•m)

14 – 22 ft-lb (19 – 30 N•m)

22 – 29 ft-lb (29 – 39 N•m)

25 – 36 ft-lb (34 – 49 N•m)

27 – 46 ft-lb (37 – 63 N•m)

27 – 38 ft-lb (37 – 52 N•m)

14 – 22 ft-lb (19 – 30 N•m)

1. Oil Level Gauge
2. Thermo-Modulated Fan Bracket
3. EGR Pipe
4. Exhaust Manifold Heat Shield
5. Exhaust Manifold
6. Coolant Inlet Pipe and Bypass Pipe
7. Engine Hanger
8. Generator
9. Generator Strap and Bracket
10. Intake Manifold Assembly
11. Intake Manifold Support Bracket
12. Oil Filter
13. Oil Cooler
14. Oil Pressure Switch

7924QG36

Exploded view of some peripheral engine component mountings

21. Road test the vehicle to check engine performance.

Water Pump

REMOVAL & INSTALLATION

1. Before servicing the vehicle, refer to the precautions in the beginning of this section .
2. Drain the cooling system.
3. Remove or disconnect the following:

- Negative battery cable
- Lower splash shield
- Upper and lower radiator hoses
- Coolant reservoir tank hose
- Fresh air duct
- Fan and shroud
- Loosen the alternator mounting and adjusting bolts
- Alternator belt
- Fan pulley and bracket
- Upper and lower timing belt covers and turn the crankshaft until No. 1 cylinder is at Top Dead Center (TDC).
- Loosen the tensioner lockbolt and pry the tensioner away from the belt
- Timing belt
- Loosen the tensioner bolt

- Water pump
- Tensioners from the water pump

To install:
4. Clean the surface of any old gasket material.
5. Install or connect the following:

- Tensioners on the water pump
- Water pump and gasket. Torque the bolts to 19 ft. lbs. 25 Nm).
- Timing belt
- Loosen the tensioner lockbolt and allow the tensioner to rest against the belt. Torque the tensioner lockbolt to 32 ft. lbs. (43 Nm).
- Upper and lower timing belt covers
- Fan bracket assembly and fan pulley
- Drive belt
- Cooling fan and shroud
- Position the radiator and torque the bracket bolts to 89 inch lbs. (10 Nm).
- Torque the shroud bolts to 89 inch lbs. (10 Nm) and torque the alternator adjusting and mounting bolts
- Position the fresh air duct over the radiator and tighten the retaining bolt 89 inch lbs. (10 Nm)
- Radiator hoses and tighten the clamps

- Lower splash shield
- Negative battery cable

6. Fill the cooling system.
7. Start the vehicle and bring the engine to operating temperature. Check for leaks and repair if necessary.

Cylinder Head

REMOVAL & INSTALLATION

1. Before servicing the vehicle, refer to the precautions in the beginning of this section .
2. Properly relieve the fuel system pressure.
3. Drain the cooling system.
4. Remove or disconnect the following:

- Negative battery cable
- Brake booster vacuum hose from the dynamic chamber
- Fuel line from the pressure regulator and the return line from the rear of the dynamic chamber
- Ground wire from the intake manifold
- Purge solenoid valve vacuum hose
- Upper radiator hose
- Intake manifold support bracket
- Converter flange inlet pipe
- Timing belt
- Cylinder head cover
- Cylinder head with the intake and exhaust manifolds attached
- 3 wire harness connectors at the rear of the cylinder head

To install:
5. Place the new head gasket on the engine block.
6. Install or connect the following:

- Cylinder head with the intake and exhaust manifolds attached
- 3 wiring connectors at the rear of the cylinder head
- Cylinder head bolts in the proper sequence. Torque the bolts to 64 ft. lbs. (87 Nm).
- Cylinder head cover
- Timing belt
- Converter inlet pipe flange nuts. Torque the nuts to 24 ft. lbs. (33 Nm).
- Upper radiator hose
- Vacuum hose from the intake manifold to the charcoal canister
- Purge solenoid vacuum hose
- Ground wire and harness bracket to

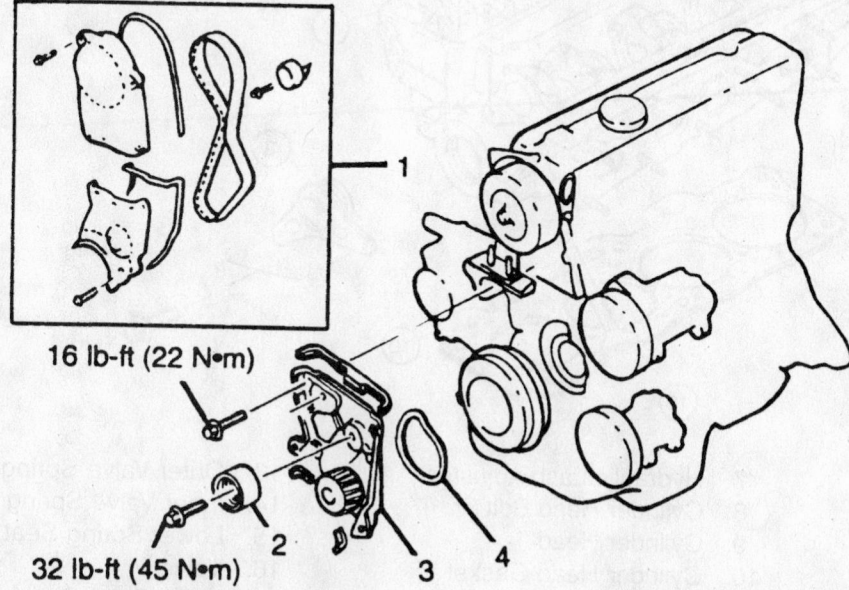

16 lb-ft (22 N•m)

32 lb-ft (45 N•m)

1 TIMING BELT COVERS, GASKETS AND TIMING BELT
2 IDLER PULLEY

3 COOLANT PUMP
4 GASKET

79240G01

Exploded view of the water pump mounting

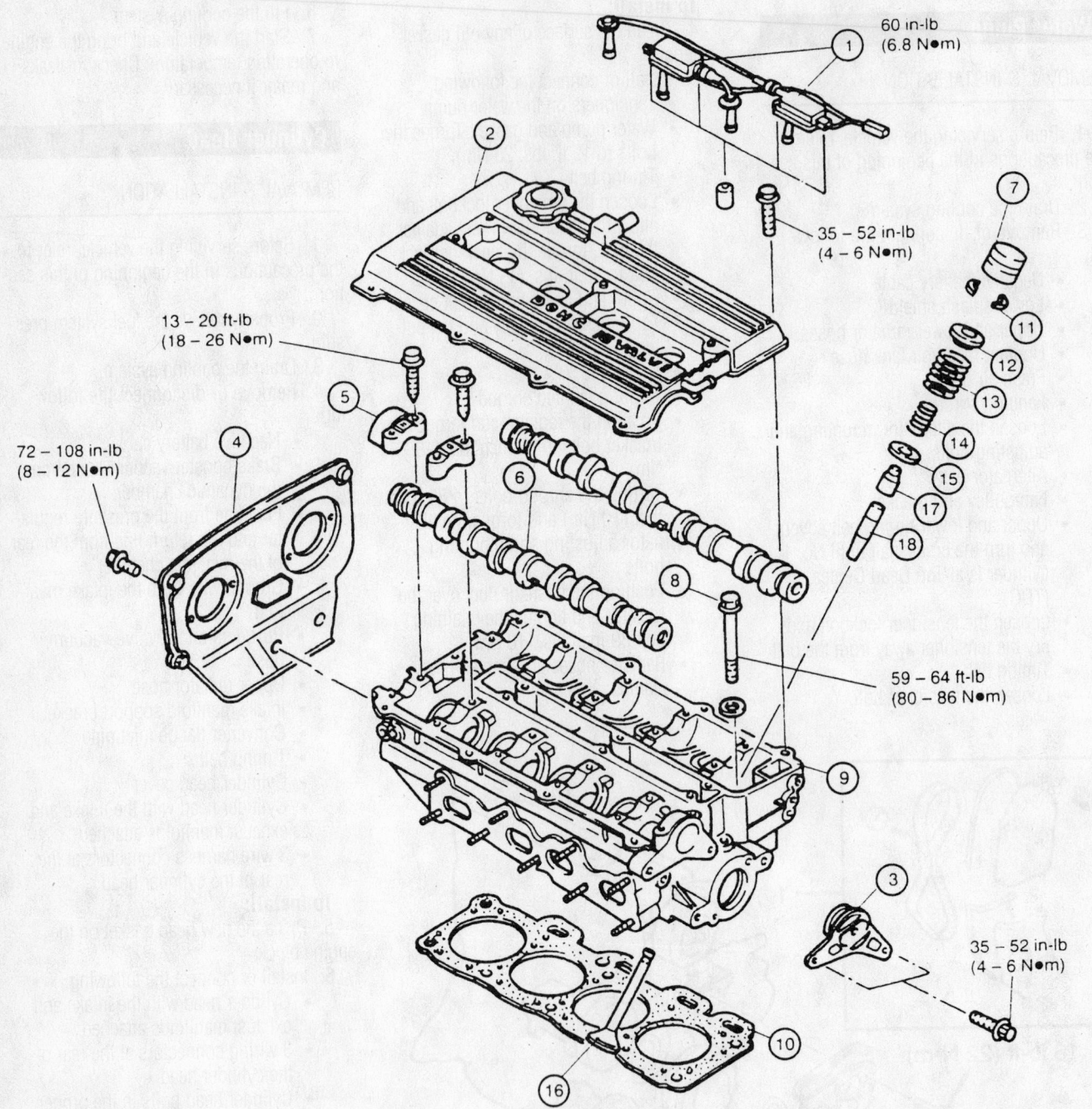

60 in-lb
(6.8 N•m)

35 – 52 in-lb
(4 – 6 N•m)

13 – 20 ft-lb
(18 – 26 N•m)

72 – 108 in-lb
(8 – 12 N•m)

59 – 64 ft-lb
(80 – 86 N•m)

35 – 52 in-lb
(4 – 6 N•m)

1. Ignition Coils and High Tension Leads
2. Cylinder Head Cover
3. Camshaft Position Sensor
4. Seal Plate
5. Camshaft Caps
6. Camshafts
7. Hydraulic Lash Adjuster
8. Cylinder Head Bolt
9. Cylinder Head
10. Cylinder Head Gasket
11. Valve Locks
12. Upper Spring Seat
13. Outer Valve Spring
14. Inner Valve Spring
15. Lower Spring Seat
16. Valve
17. Valve Stem Seal
18. Valve Guide

79240G37

Exploded view of the cylinder head assembly

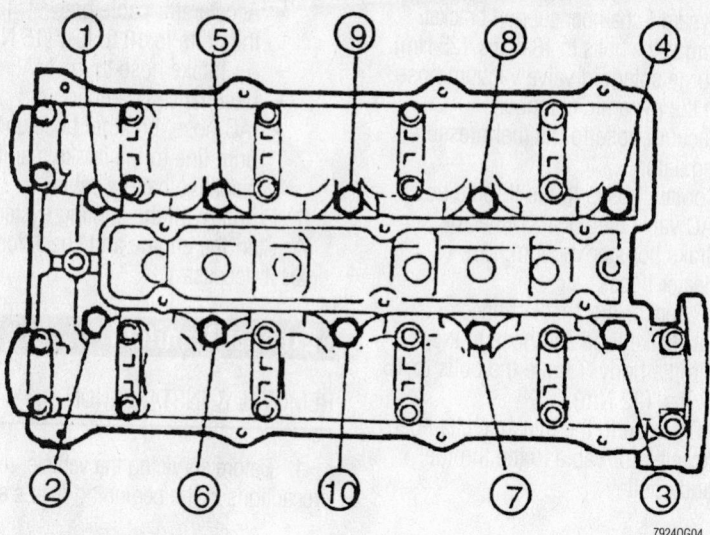

Cylinder head removal sequence

7924QG04

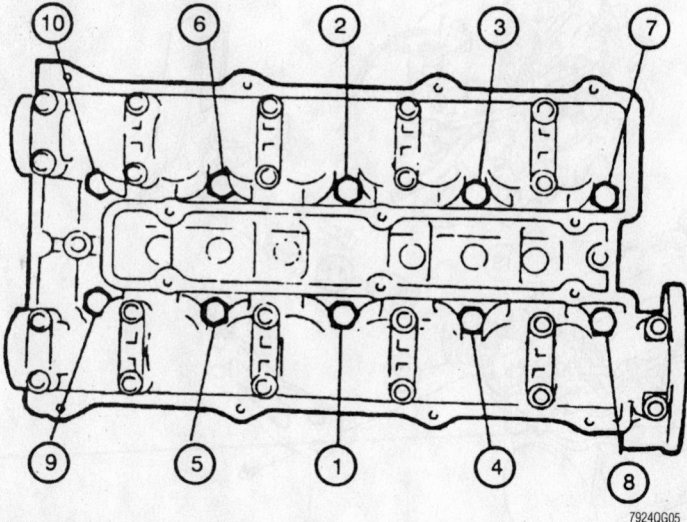

Cylinder head torque sequence—Kia Sportage

7924QG05

the intake manifold. Torque the bolts to 18 ft. lbs. (25 Nm).
- Fuel line to the pressure regulator and the return line to the fuel rail
- Brake booster vacuum hose
- Negative battery cable

7. Properly fill the cooling system.

8. Start the engine and check for leaks, repair if necessary..

Intake Manifold

REMOVAL & INSTALLATION

1. Before servicing the vehicle, refer to the precautions in the beginning of this section .

2. Properly relieve the fuel system pressure.

3. Drain the coolant.

4. Remove or disconnect the following:
- Negative battery cable
- Accelerator cable bracket bolts from the valve cover
- Air intake tube to cylinder head cover bolts
- Air intake tube to throttle body bolts
- Loosen the clamp attaching the air tube to the Mass Air Flow (MAF) sensor
- Idle Air Control (IAC) valve, breather hose and vacuum line from the air intake tube
- Air intake tube
- Positive Crankcase Ventilation hose (PCV) from the dynamic chamber
- Purge solenoid valve vacuum hose from the dynamic chamber
- Throttle Position (TP) sensor electrical connector
- IAC valve electrical connector
- Heater hoses
- Engine-to-body ground strap from the intake manifold and the harness bracket below it
- Brake booster vacuum line
- Vacuum hose from the fuel pressure regulator hose
- Dynamic chamber support bracket bolts
- Fuel injector electrical connectors
- Fuel pressure and return lines
- Intake manifold support bracket
- Oil filter
- Intake manifold bolts
- Bypass pipe from the heater hose
- Intake manifold and discard the gasket

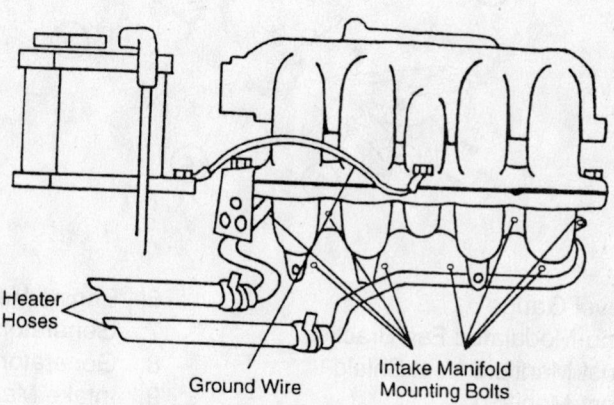

Heater Hoses

Ground Wire

Intake Manifold Mounting Bolts

7924QG12

Intake manifold mounting bolt locations. Be sure to connect the ground cable

Timing belt service is covered in Section 3 of this manual

To install:

5. Install or connect the following:
- Intake manifold with a new gasket to the cylinder head
- Heater hose to the bypass pipe
- Bolts and nuts attaching the intake manifold to the cylinder head. Torque the bolts to 14–22 ft. lbs. (19–30 Nm).
- New oil filter
- Intake manifold support bracket. Torque the bolts to 27–38 ft. lbs. (37–52 Nm)
- Fuel lines
- Fuel injector electrical connectors
- Engine to body ground wire. Torque the bolt to 18 ft. lbs. (25 Nm).

- Dynamic chamber support bracket. Torque the bolts to 18 ft. lbs. (25 Nm).
- Purge solenoid valve vacuum hose to the dynamic chamber
- Vacuum hose to the fuel pressure regulator
- Coolant hoses to the throttle body
- IAC valve electrical connector
- Brake booster vacuum hose
- Heater hoses
- TP sensor electrical connector
- Air intake tube and hose to the throttle body. Torque the bolts to 16 ft. lbs. (22 Nm).
- PCV hose to the dynamic chamber
- Accelerator cable to the throttle body pulley

- Accelerator cable bracket. Torque the bolts to 10 ft. lbs. (15 Nm).
- Air intake hose to the MAF sensor
- Resonance chamber
- IAC hose, breather hose and vacuum line to the intake manifold
- Negative battery cable

6. Properly fill the cooling system.
7. Start the engine and check for leaks, repair if necessary.

Exhaust Manifold

REMOVAL & INSTALLATION

1. Before servicing the vehicle, refer to the precautions in the beginning of this section .

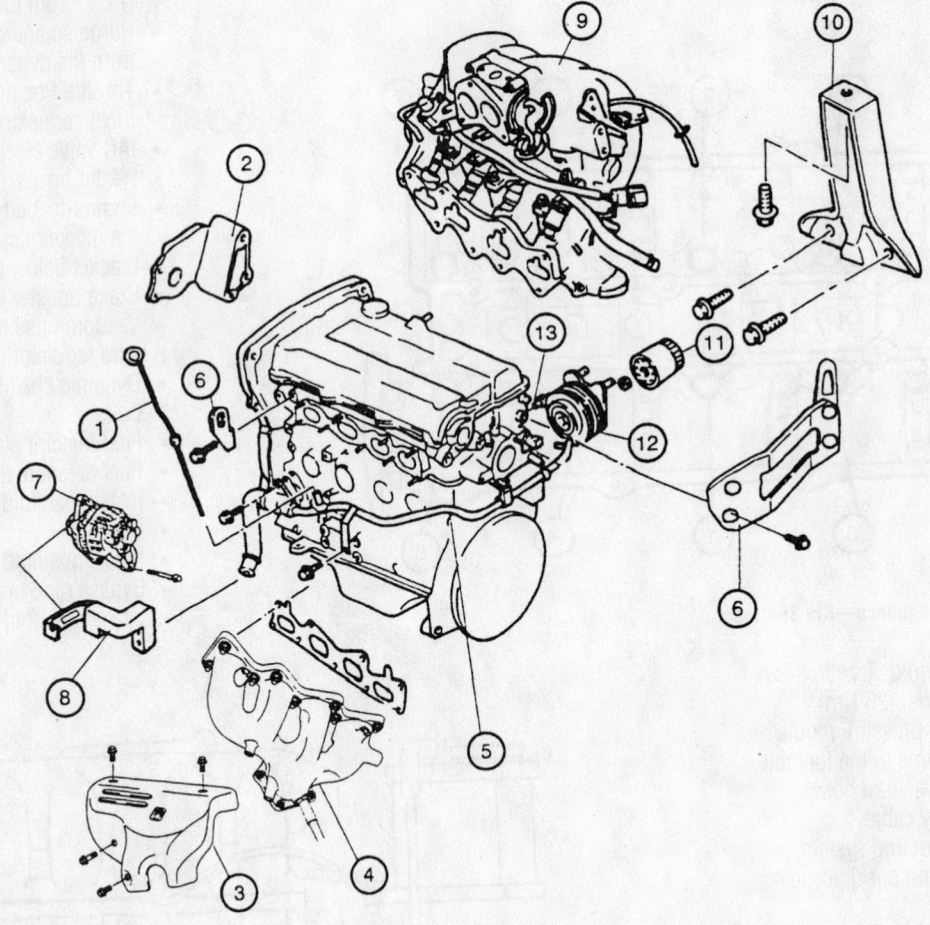

1. Oil Level Gauge
2. Thermo-Modulated Fan Bracket
3. Exhaust Manifold Heat Shield
4. Exhaust Manifold
5. Coolant Inlet Pipe and Bypass Pipe
6. Engine Hanger
7. Generator
8. Generator Strap and Bracket
9. Intake Manifold Assembly
10. Intake Support Bracket
11. Oil Filter
12. Oil Cooler
13. Oil Pressure Switch

9308QG01

Exploded view of the intake, exhaust manifold and related components

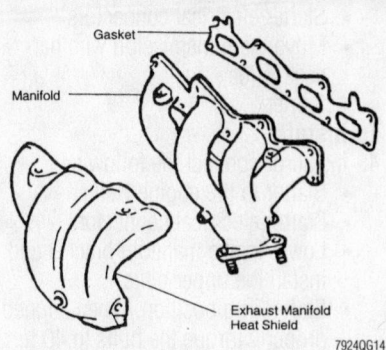

**Exploded view of the exhaust manifold
assembly**

2. Remove or disconnect the following:
- Negative battery cable
- Air intake hose
- Exhaust manifold heat shield
- Converter inlet pipe flange locknuts
- Exhaust manifold and discard the gasket
3. Clean the mating surfaces.

To install:

4. Install or connect the following:
- Exhaust manifold with a new gasket. Torque the bolts to 31 ft. lbs. (42 Nm).
- New flange gasket and install the converter inlet pipe. Torque the bolts to 24 ft. lbs. (33 Nm).
- Exhaust manifold heat shield. Torque the bolts to 18 ft. lbs. (25 Nm).
- Air intake hose
- Negative battery cable

5. Start the vehicle and check for leaks, repair if necessary.

Front Crankshaft Seal

REMOVAL & INSTALLATION

1. Before servicing the vehicle, refer to the precautions in the beginning of this section .

2. Remove or disconnect the following:
- Negative battery cable
- Engine under cover
- Timing belt
- Timing belt pulley lock bolt
- Timing belt pulley
- Pulley woodruff key
- Oil seal by carefully cutting it out of the oil pump housing

To install:

3. Lubricate the lip of the new seal with clean engine oil.

4. Install or connect the following:

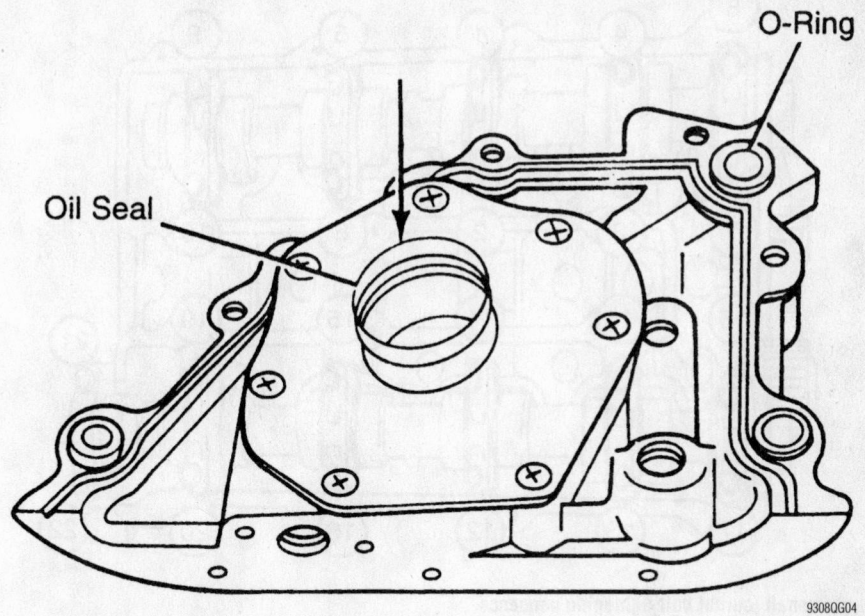

Install the oil seal into the oil pump housing

- New oil seal into the oil pump housing by hand
- Press the oil seal into pump until it is flush with the edge of the oil pump body
- Timing belt pulley
- Pulley woodruff key
- Pulley lock bolt. Torque the bolt to 18 ft. lbs. (25 Nm).
- Timing belt
- Engine under cover. Torque the bolts to 18 ft. lbs. (25 Nm).

- Negative battery cable

5. Start the engine and check for leaks, repair if necessary.

Camshaft

REMOVAL & INSTALLATION

1. Before servicing the vehicle, refer to the precautions in the beginning of this section .

2. Properly relieve the fuel system pressure.

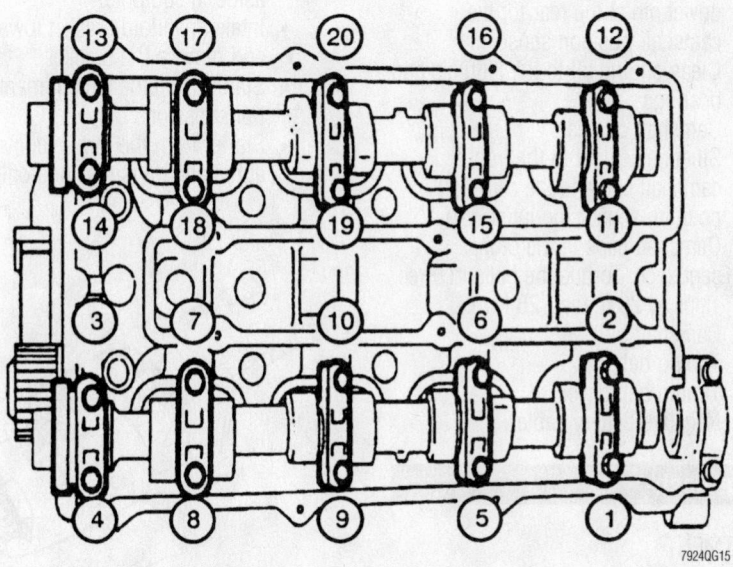

Camshaft cap bolt removal sequence

Heater Core replacement is covered in Section 2 of this manual

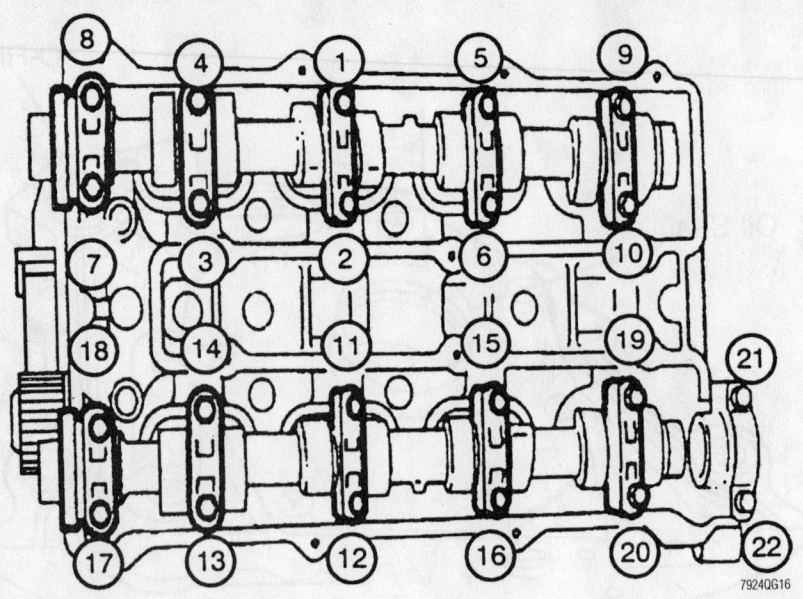

Camshaft journal bolt tightening sequence

3. Drain the coolant into a suitable container.

4. Remove or disconnect the following:
- Negative battery cable
- Upper timing belt cover
- Timing belt from the camshaft pulley
- Camshaft pulleys
- Camshaft cap bolts in the proper sequence
- Camshaft caps
- Camshafts

To install:

5. Install or connect the following:
- Camshafts into the cylinder head. The exhaust camshaft has a steel dowel pin at the rear for the camshaft position sensor
- Clean engine oil to the journals and bearings
- Camshaft oil seal
- Silicone sealant to the front camshaft cap and the camshaft position sensor mounting cap
- Camshaft caps in the proper sequence. Torque the bolts in three steps to 20 ft. lbs. (26 Nm).
- Camshaft pulleys
- Timing belt
- Timing belt cover
- Negative battery cable

Valve Lash

ADJUSTMENT

The DOHC engine uses Hydraulic Lash Adjusters (HLA's), which automatically maintain the proper amount of valve lash. Therefore, the DOHC engine does not need manual valve lash adjustment.

Starter

REMOVAL & INSTALLATION

1. Before servicing the vehicle, refer to the precautions in the beginning of this section .

2. Drain the engine oil.

3. Remove or disconnect the following:
- Negative battery cable
- Intake manifold bracket upper bolts
- Clutch release cylinder and move it aside, if equipped
- Intake manifold bracket lower bolts and remove the bracket
- Starter from the clutch, manual transmissions only
- Starter from the torque converter, automatic transmissions only

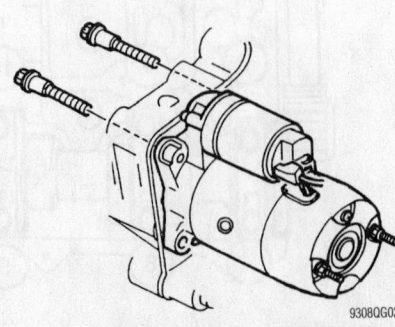

Exploded view of the starter

- Starter electrical connectors
- Move the transmission wire harness aside
- Starter

To install:

4. Install or connect the following:
- Starter to the engine well
- Starter electrical connectors
- Lower intake manifold bracket and install the upper bolts
- Starter into position. When aligned properly torque the bolts to 40 ft. lbs. (54 Nm).
- Torque the intake manifold bracket bolts to 40 ft. lbs. (54 Nm).
- Properly position the clutch release cylinder, if equipped. Torque the bolts to 40 ft. lbs. (54 Nm).
- Torque the intake manifold bracket upper bolts to 40 ft. lbs. (54 Nm).
- Negative battery cable

Oil Pan

REMOVAL & INSTALLATION

1. Before servicing the vehicle, refer to the precautions in the beginning of this section .

2. Drain the engine oil.

3. Remove or disconnect the following:
- Negative battery cable
- 2 Top intake manifold bracket bolts
- Front 3 axle housing mounting bolts, 4WD only
- Left front bushing from the axle housing mount and lower the front axle housing
- Both gusset plates from the engine
- Transmission under cover
- Engine under cover
- Oil pan mounting bolts and using a scrapper tool separate the oil pan
- Oil pan
- Oil strainer assembly
- Oil baffle

To install:

4. Clean the engine block, oil pan and baffle pan surfaces of any gasket material.

5. Apply a continuous bead of Loctite Ultra Blue 587® silicone sealant around the baffle pan.

6. Install or connect the following:
- Oil baffle. Torque the bolt to 84 inch lbs. (9.5 Nm).
- Oil strainer. Torque the bolts to 84 inch lbs. (9.5 Nm).

7. Apply a continuous bead of Loctite Ultra Blue 587® silicone sealant around the oil pan.
- Oil pan. Torque the bolts to 84 inch lbs. (9.5 Nm).

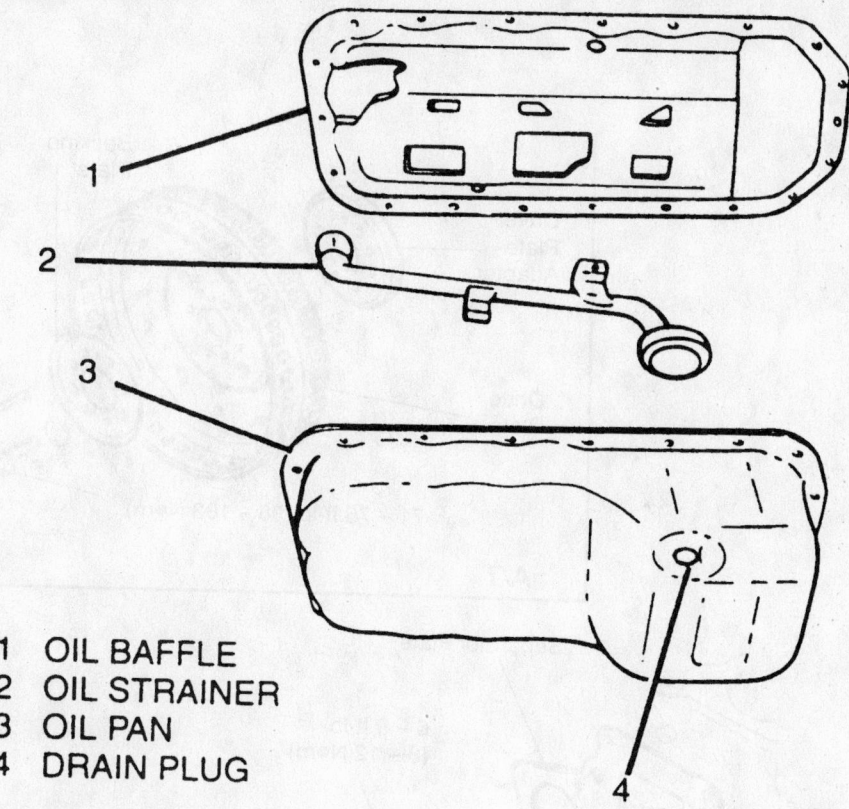

1 OIL BAFFLE
2 OIL STRAINER
3 OIL PAN
4 DRAIN PLUG

Exploded view of the oil pan assembly mounting

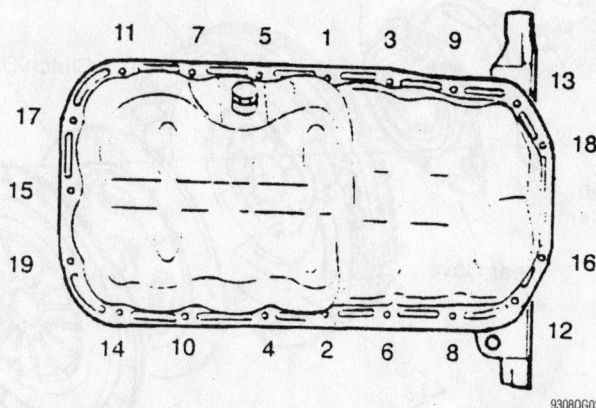

Tighten the oil pan bolts in sequence

- Transmission under cover. Torque the bolts to 84 inch lbs. (9.5 Nm).
- Gusset plates to the engine. Torque the bolts to 33 ft. lbs. (45 Nm).
- Engine under cover
- Front axle housing into position. When properly aligned, torque the bolts to 48 ft. lbs. (65 Nm).
- Intake manifold bracket bolts. Torque the bolts to 34 ft. lbs. (65 Nm).
- Negative battery cable

8. Fill the engine with clean oil.
9. Start the vehicle and check for leaks, repair if necessary.

Oil Pump

REMOVAL & INSTALLATION

➡**The oil pump is externally-mounted, but still requires the removal of the oil pan to disconnect the oil pump strainer.**

1. Before servicing the vehicle, refer to the precautions in the beginning of this section .
2. Properly relieve the fuel system pressure.
3. Drain the engine oil.
4. Drain the cooling system.
5. Remove or disconnect the following:
- Negative battery cable
- Alternator belt
- Fresh air duct from the radiator
- Upper radiator hose
- Clutch fan and shroud
- Splash guard
- Loosen the A/C drive belt
- Power steering belt
- Timing belt covers
- Lower timing belt pulley and lock bolt and place a support under the front axle
- Axle attaching bolts and lower the axle enough to gain access to the oil pan
- Transmission under cover
- Oil pan
- Oil pump

To install:

6. Clean the engine block, oil pan and baffle pan surfaces of any gasket material.
7. Apply a continuous bead of silicone sealant around the oil pump.

➡**Do not allow sealant to get in the oil passages when applying sealant to the contact surface.**

8. Install or connect the following:
- New O-ring and mount the oil pump to the engine. Torque the "A" bolts to 16 ft. lbs. (22 Nm) and the "B" bolts to 33 ft. lbs. (45 Nm).
9. Remove the upper and lower A/C compressor mounting bolts.
10. Loosen the A/C compressor bracket.
- Power steering pump bracket and hand tighten the bolts

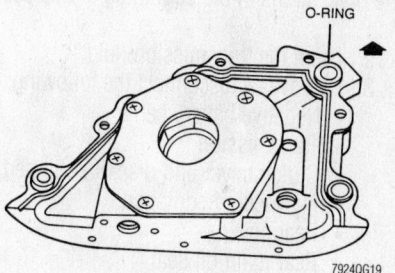

Be sure the oil pump O-ring is in the proper location prior to installation

Brake service is covered in Section 4 of this manual

- A/C compressor bracket
- A/C compressor. Torque the mounting bolts to 17 ft. lbs. (23 Nm).
- Torque the power steering pump bracket bolts to 24 ft. lbs. (33 Nm).
- Power steering pump. Torque the bolts to 43 ft. lbs. (58 Nm).
- Timing belt gear on the crankshaft. Torque the large crank bolt to 119 ft. lbs. (162 Nm).
- Oil baffle after applying sealant to the mating surface. Torque the bolt to 84 inch lbs. (9.5 Nm).
- Oil strainer. Torque the bolts to 84 inch lbs. (9.5 Nm).
- Oil pan
- Transmission under cover. Torque the bolts to 84 inch lbs. (9.5 Nm).
- Both gusset plates. Torque the bolts to 33 ft. lbs. (45 Nm).
- Raise the front axle into position. When aligned properly, torque the bolts to 123 ft. lbs. (167 Nm).
- Timing belt and cover
- Alternator belt
- A/C and power steering belt and adjust as needed
- Splash shield
- Upper radiator hose
- Clutch fan and shroud as an assembly
- Air duct to the top of the radiator
- Engine under cover. Torque the bolts to 18 ft. lbs. (25 Nm).
- Negative battery cable

11. Properly fill the cooling system.
12. Fill the engine with clean oil.
13. Start the engine, check for leaks, and repair if necessary.

Rear Main Seal

REMOVAL & INSTALLATION

1. Before servicing the vehicle, refer to the precautions in the beginning of this section.
2. Drain the transmission fluid.
3. Remove or disconnect the following:
   - Negative battery cable
   - Transmission
   - Clutch cover and disc, if equipped
   - Flywheel, if equipped
   - Rear cover
   - Rear main oil seal

To install:

4. Coat the new seal with clean oil and press the seal into the cover.
5. Install or connect the following:
   - Rear cover

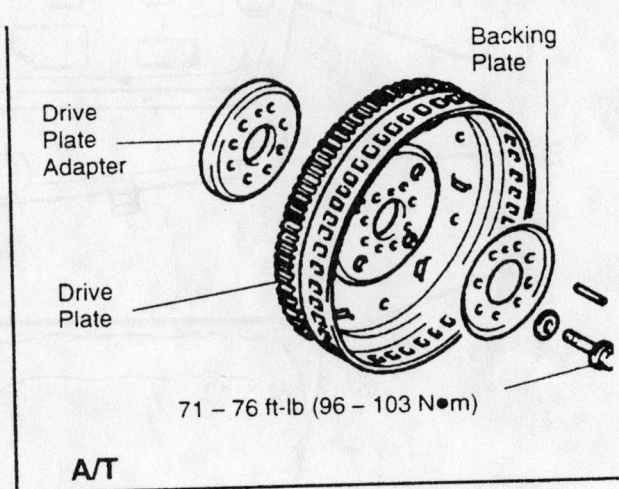

71 – 76 ft-lb (96 – 103 N•m)

A/T

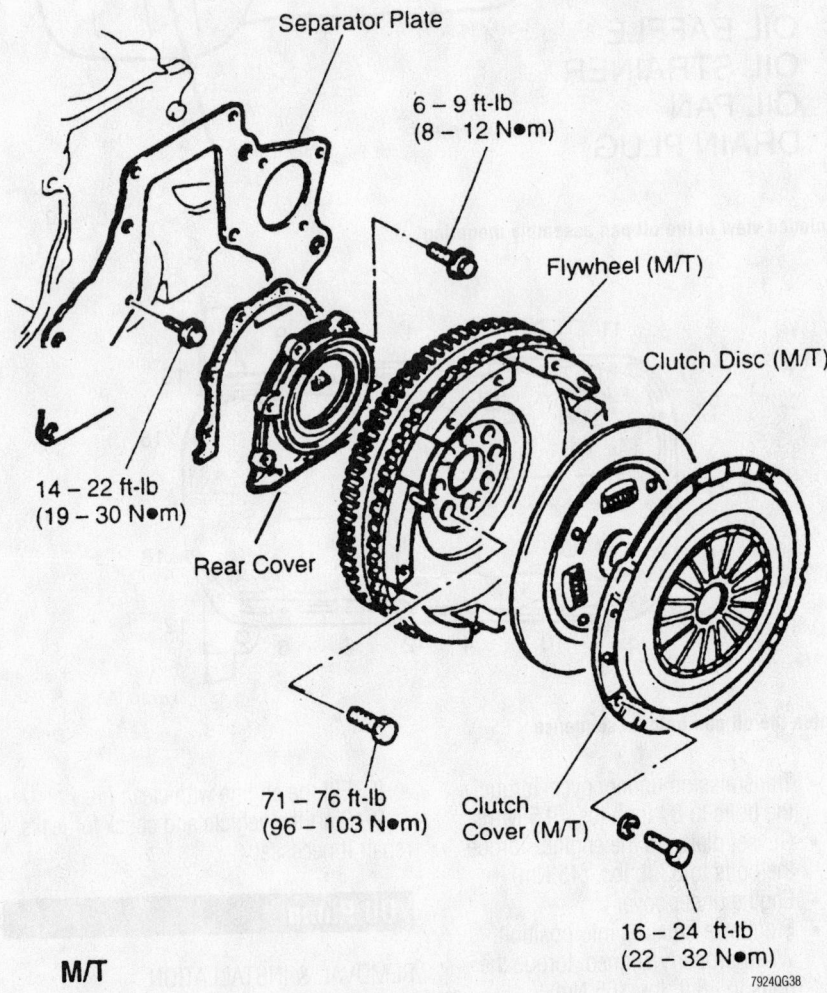

Separator Plate

6 – 9 ft-lb (8 – 12 N•m)

Flywheel (M/T)

Clutch Disc (M/T)

14 – 22 ft-lb (19 – 30 N•m)

Rear Cover

71 – 76 ft-lb (96 – 103 N•m)

Clutch Cover (M/T)

16 – 24 ft-lb (22 – 32 N•m)

M/T

79240G38

Exploded view of the rear main seal and related components

- Flywheel onto the crankshaft. While holding the flywheel torque the bolts in sequence:
  a. Step 1: 30 ft. lbs. (41 Nm).
  b. Step 2: 60 ft. lbs. (81 Nm).
  c. Step 3: 73 ft. lbs. (99 Nm).
- Clutch disc and cover. Torque the bolts to 16 ft. lbs. (22 Nm).
- Transmission
- Negative battery cable

6. Fill the transmission to the proper level.

7. Start the vehicle and check for leaks, repair if necessary.

Piston and Ring

POSITIONING

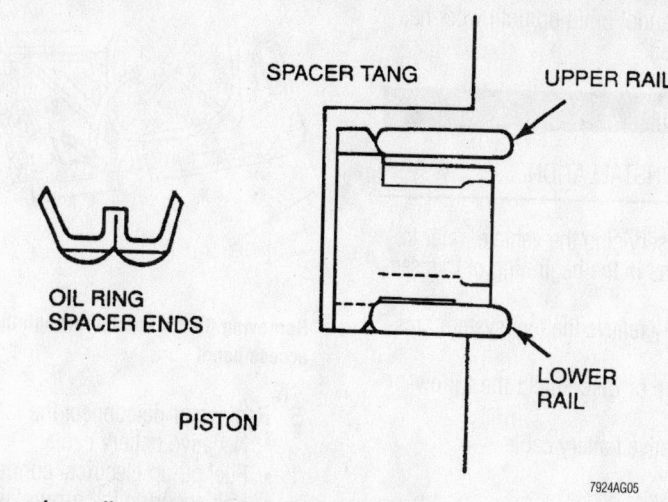

Kia 2.0L engine—oil control ring rail and spacer positioning

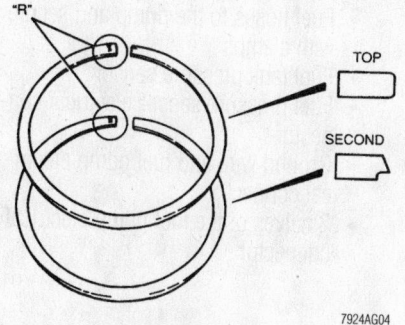

Kia 2.0L engine—compression ring positioning mark locations

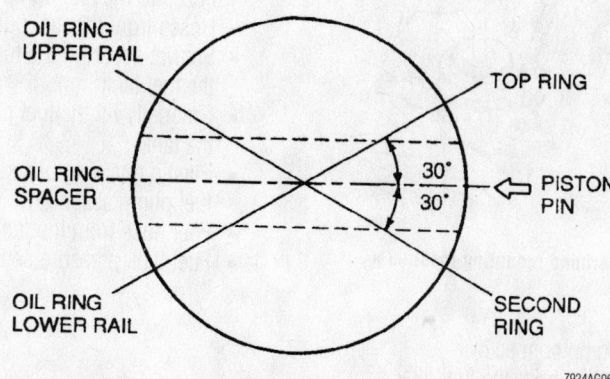

Kia 2.0L engine—piston ring end-gap spacing

FUEL SYSTEM

Fuel System Service Precautions

Safety is the most important factor when performing not only fuel system maintenance but any type of maintenance. Failure to conduct maintenance and repairs in a safe manner may result in serious personal injury or death. Maintenance and testing of the vehicle's fuel system components can be accomplished safely and effectively by adhering to the following rules and guidelines.

- To avoid the possibility of fire and personal injury, always disconnect the negative battery cable unless the repair or test procedure requires that battery voltage be applied.

- Always relieve the fuel system pressure prior to disconnecting any fuel system component (injector, fuel rail, pressure regulator, etc.), fitting or fuel line connection.

Exercise extreme caution whenever relieving fuel system pressure to avoid exposing skin, face and eyes to fuel spray. Please be advised that fuel under pressure may penetrate the skin or any part of the body that it contacts.

- Always place a shop towel or cloth around the fitting or connection prior to loosening to absorb any excess fuel due to spillage. Ensure that all fuel spillage (should it occur) is quickly removed from engine surfaces. Ensure that all fuel soaked cloths or towels are deposited into a suitable waste container.

- Always keep a dry chemical (Class B) fire extinguisher near the work area.

- Do not allow fuel spray or fuel vapors to come into contact with a spark or open flame.

- Always use a back-up wrench when loosening and tightening fuel line connection fittings. This will prevent unnecessary stress and torsion to fuel line piping.

- Always replace worn fuel fitting O-rings with new. Do not substitute fuel hose or equivalent where fuel pipe is installed.

Fuel System Pressure

RELIEVING

1. Before servicing the vehicle, refer to the precautions in the beginning of this section.

2. Disconnect the fuel pump harness connector located behind the rear seat.

3. Start the engine and allow the engine to run out of fuel.

4. Once the engine has stalled, turn the key to the **OFF** position and connect the electrical connector.

5. Disconnect the negative battery cable

For complete Engine Mechanical specifications, see Section 1 of this manual

so pressure cannot build up until work has been completed.

Fuel Filter

REMOVAL & INSTALLATION

1. Before servicing the vehicle, refer to the precautions in the beginning of this section.
2. Properly relieve the fuel system pressure.
3. Remove or disconnect the following:
   - Negative battery cable

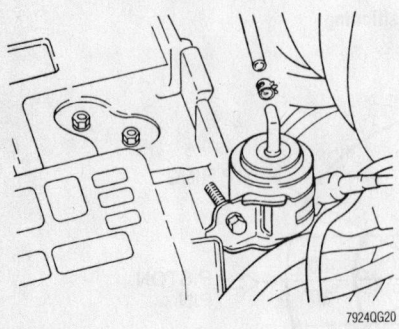

Fuel filter underhood mounting location

   - Fuel pump connector
   - Fuel hoses from the fuel filter
   - Fuel filter from the bracket

To install:

4. Install or connect the following:
   - Fuel filter in the bracket
   - Fuel filter. Torque the bolts to 95 ft. lbs. (129 Nm).
   - Fuel hoses on the filter and make certain that the hoses are seated properly
   - Fuel pump connector
   - Negative battery cable

5. Start the engine and check for fuel leaks, repair if necessary.

Fuel Pump

REMOVAL & INSTALLATION

1. Before servicing the vehicle, refer to the precautions in the beginning of this section .
2. Properly relieve the fuel system pressure.
3. Release the catch for the back seat and tilt the seat out of the way.
4. Move the carpet behind the seat that covers the fuel pump access panel.

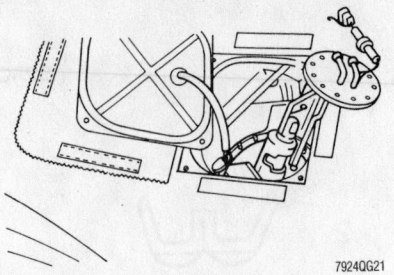

Removing the fuel pump through the access panel

5. Remove or disconnect the following:
   - Negative battery cable
   - Fuel pump electrical connectors
   - Bolt securing the ground wire
   - Fuel pump access panel
   - Hose clamps connecting the fuel hoses to the fuel pump
   - Hoses from the fuel pump
   - Screws securing the fuel pump to the fuel tank
   - Gradually lift the fuel pump from the tank
   - Plastic retaining bracket from the fuel pump assembly
   - Fuel hose from the fuel pump
   - Fuel tank pressure sensor

6. Wrap the fuel pump assembly in a rag before removing the assembly from the vehicle.
7. Cover or seal the fuel tank until installing the fuel pump assembly.

To install:

➡ **The fuel pump is part of the assembly and is replaced as a complete unit.**

8. Install or connect the following:
   - Plastic mounting bracket to the fuel pump and secure the bracket to the pump with 4 screws
   - Fuel hose to the pump and secure with a new clamp
   - Fuel pump into the access port on top of the fuel tank
   - Twist the fuel pump as necessary to properly position it in the fuel tank
   - 8 screws around the top surface of the fuel pump
   - Fuel hoses to the pump and secure with clamps
   - Fuel tank pressure sensor
   - Fuel pressure sensor electrical connector
   - Ground wire and fuel pump electrical connector
   - 2 halves of the fuel pump electrical connector

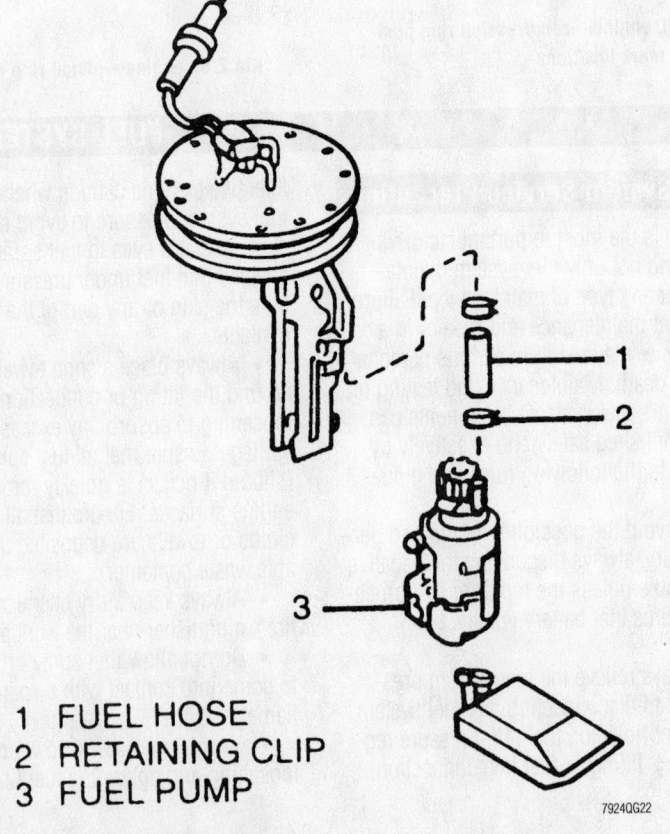

1 FUEL HOSE
2 RETAINING CLIP
3 FUEL PUMP

Exploded view of the fuel pump assembly

- Access plate and reposition the carpet at seat
- Negative battery cable

Fuel Injector

REMOVAL & INSTALLATION

1. Before servicing the vehicle, refer to the precautions in the beginning of this section .
2. Properly relieve the fuel system pressure.
3. Drain the cooling system.
4. Remove or disconnect the following:
   - Negative battery cable
   - Air intake hose assembly
   - Breather hoses from the air intake duct
   - Mass Air Flow (MAF) sensor electrical connector
   - Air intake hose bracket
   - Accelerator cable
   - Vacuum hose from the intake manifold to the vacuum pipe
   - Throttle Position (TP) sensor and

Idle Air Control (IAC) valve electrical connectors
   - Coolant hoses from the throttle body
   - Clamp and hoses from the IAC valve
   - Brake booster vacuum hose from the dynamic chamber
   - Cruise control vacuum hose, if equipped
   - Bracket from the dynamic chamber
   - Manifold bracket
   - heater inlet hose
   - Dynamic chamber
   - Fuel hose from the pressure regulator
   - Fuel injector rail clips
   - Fuel rail
   - Fuel injector insulators
   - Pressure regulator
   - Fuel injectors

To install:

5. Install or connect the following:
   - Fuel injectors to the fuel rail
   - New insulators
   - New injector clips

- Fuel rail
- Clamps and air hose to the fuel rail
- Fuel hose to the pressure regulator
- Dynamic chamber with a new gasket. Torque the bolts to 16 ft. lbs. (22 Nm).
- IAC valve bracket bottom bolt
- IAC valve and TP sensor electrical connectors
- Heater inlet hose
- Cruise control hose, if equipped
- Vacuum hose to the pressure regulator
- Air hose and clamp to the air rail
- IAC valve vacuum hose
- Manifold bracket bolts
- Coolant hoses to the throttle body
- Vacuum hose to the vacuum pipe
- MAF sensor bracket
- MAF sensor electrical connector
- Accelerator cable
- Breather hoses
- Negative battery cable

6. Fill the cooling system.
7. Start the vehicle and check for leaks, repair if necessary.

DRIVE TRAIN

Transmission Assembly

REMOVAL & INSTALLATION

Manual

➡ **The removal of the manual transmission is virtually the same for 4WD and 2WD vehicles.**

1. Before servicing the vehicle, refer to the precautions in the beginning of this section .
2. Drain the transmission fluid.
3. Drain the transfer case.
4. Remove or disconnect the following:
   - Negative battery cable
   - Rear portion of the center console
   - Shift lever and transfer lever knobs
   - Slide the boot cover over the shifters and remove the center console
   - Shift lever
   - Transfer lever
   - Front driveshaft by removing the bolts at the front differential and the bolts at the transfer case, if equipped
   - Bolts from the rear differential flange and the center support. Pull

the driveshaft out of the tail shaft housing, if equipped with a 4 x 4
   - Bolts from the rear differential flange and center support. Pull the driveshaft out of the tail shaft housing, if equipped with a 4 x 2
   - Back-up light electrical connector and the Vehicle Speed Sensor (VSS) electrical connectors and move the wire harness aside
   - Crankshaft Position (CKP) sensor from the transmission housing
   - Clutch release cylinder and move it aside
   - Front exhaust pipe bracket
   - Front lower transmission housing bolts
   - Transfer case side mount, if equipped and properly support the transmission
   - Transmission crossmember
   - transmission mount bolts
   - Starter from the front housing
   - Transmission and transfer case, if equipped

To install:

5. Install or connect the following:
   - Transmission into position at the rear of the engine
   - Wire harness along the right side

of the transmission and route the VSS wire over the transmission to the rear of the control rod extension
   - Transmission to engine 14mm mounting bolts. Torque the bolts to 80 ft. lbs. (108 Nm).
   - Transmission to engine 10mm mounting bolts. Torque the bolts to 29 ft. lbs. (39 Nm).
   - Transmission to engine 6mm mounting bolts. Torque the bolts to 5 ft. lbs. (7 Nm).
   - Exhaust pipe to the bracket. Torque the bolts to 20 ft. lbs. (27 Nm).
   - Starter and ground wire. Torque the bolts to 29 ft. lbs. (39 Nm).
   - Transmission crossmember mount. Torque the bolts to 80 ft. lbs. (108 Nm).
   - Crossmember to the chassis. Torque the bolts to 32 ft. lbs. (44 Nm).
   - CKP sensor. Torque the bolt to 5 ft. lbs. (7 Nm).
   - 4WD indicator switch connector
   - Back-up light electrical connector
   - VSS electrical connector
   - Clutch release cylinder. Torque the bolts to 29 ft. lbs. (39 Nm).

For Accessory Drive Belt illustrations, see Section 1 of this manual

- Driveshaft to the rear differential flange, 4 x 4 only
- Forward end of the driveshaft into the extension housing and attach the center support to the chassis
- Shaft to the rear differential flange. Torque the bolts to 27 ft. lbs. (36 Nm).
- Front driveshaft at the transfer case and install the bolts to the front differential, 4 x 4 only
- Transfer case side mount, if equipped. Torque the bolts to 38 ft. lbs. (52 Nm).
- Transfer case side mount to the chassis. Torque the bolts to 38 ft. lbs. (52 Nm).
- Shift lever assembly. Torque the shift lever bracket bolts to 18 ft. lbs. (25 Nm).
- Transfer lever assembly, if equipped. Torque the bolts to 18 ft. lbs. (25 Nm).
- Dust cover plate over the shifter lever handles. Torque the bolts to 15 ft. lbs. (20 Nm).
- Front console
- Shifter lever knobs
- Negative battery cable
- Fill the transfer case.

6. Fill the transmission assembly.
7. Start the vehicle and check for leaks, repair if necessary.

Automatic

1. Before servicing the vehicle, refer to the precautions in the beginning of this section .
2. Drain the transmission fluid.
3. Drain the cooling system.
4. Remove or disconnect the following:
   - Negative battery cable
   - Automatic transmission control cable from the throttle body
5. Slide the front seats forward and remove the 2 rear shift console mounting screws and set the parking brake.
   - Rear console and slide the front seats rearward
   - Front console mounting screws and untie the shifter boot draw strings
   - Loosen the transfer case lock nut and remove the transfer case shifter lever knob
   - Power/economy switch electrical connector
   - Front console and shift the transfer lever to the 4L position
   - transfer shift lever cover plate
   - transfer case shift lever assembly and place the transmission in the park position

- Selector lever nuts
- Split pin from the shift selector lever
- Shift selector rod and spring washers
- Electrical connectors from the base of the selector lever

➡ **There is a fifth wire connection (Park Position). This wire is hard-wired, do not disconnect it.**

- Selector lever assembly
- Control cable from the throttle linkage and slide the cable from the bellcrank
- Upper dipstick tube from the lower dipstick tube
- Input/Turbine speed sensor from the top rear of the transmission
- Shift solenoids from the lower left side of the transmission
- Vehicle Speed Sensor (VSS) from the center of the transfer case
- Input speed sensor electrical connectors
- Matchmark the front and rear driveshafts. Remove the attaching bolts from both flanges and remove the driveshafts
- Oil cooler pipes at the transmission
- Starter
- Transfer case mounting bolts
- Transfer case nuts from the crossmember and support the transmission
- Crossmember
- Transmission to engine mounting bolts
- Front lower splash shield
- Left side lower gusset
- Torque converter inspection cover
- Torque converter to drive plate bolts

- Slide the transmission away from the engine and lower the transmission slightly
- Crankshaft Position (CKP) sensor attaching bolt and sensor
- 4WD, 4WD LOW indicator switches and the VSS from the transfer case

6. Lower the transmission. Be sure all wiring is clear and disconnected. Be sure the throttle cables come out without binding or attaching to anything.

To install:

7. Install or connect the following:
   - Transmission into position to attach the sensor and indicator switch wiring. Be sure the throttle cables are guided into the engine compartment without binding or attaching to anything.
   - 4WD, 4WD LOW indicator switches and the VSS to the transfer case
   - Input/turbine speed sensor and CKP sensor. Torque the bolt to 12 ft. lbs. (16 Nm).
   - Raise the transmission and position it to the engine. Install the upper housing bolts. Torque the bolts in the following sequence:

8. 10mm bolts to 38–60 ft. lbs. (57–81 Nm).
9. 12mm bolts to 51–65 ft. lbs. (69–88 Nm).

   - Exhaust hanger and bracket. Torque the bolts to 38–60 ft. lbs. (57–81 Nm).
   - Oil cooler pipe to the lower engine mount. Torque the bolt to 60 ft. lbs. (81 Nm).
   - Crossmember. Torque the bolts to 23–34 ft. lbs. (31–46 Nm).
   - Two transfer case mounting bolts located at the right center of the

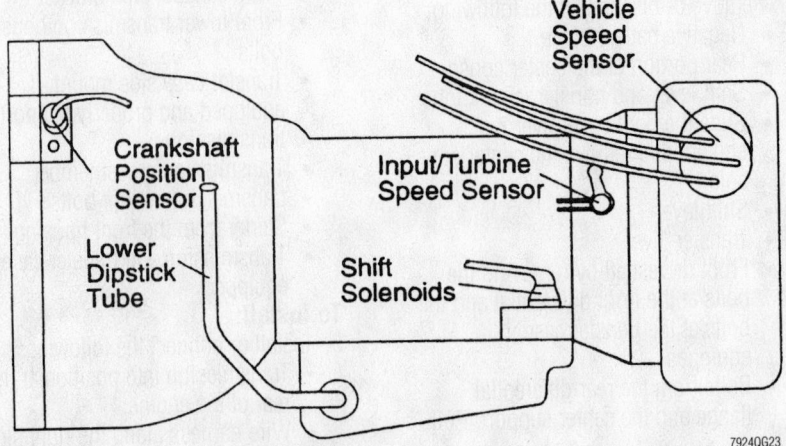

Automatic transmission wiring connections

Radiator Side

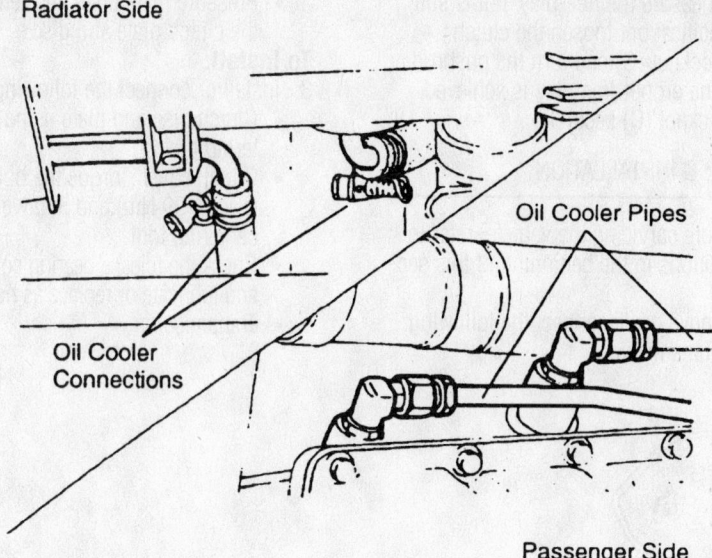

Oil Cooler Connections

Oil Cooler Pipes

Passenger Side

7924QG24

Exploded view of the oil cooler pipe connections

transfer case. Torque the bolts to 23–34 ft. lbs. (31–46 Nm).
- Four transfer case nuts to the crossmember. Torque the nuts to 23–34 ft. lbs. (31–46 Nm).
- Torque converter-to-drive plate bolts. Torque the bolts to 12–20 ft. lbs. (16–27 Nm).
- Torque converter inspection cover. Torque the bolts to 41–62 inch lbs. (5–7 Nm).
- Front splash guard. Torque the bolts to 41–62 inch lbs. (5–7 Nm).
- Starter. Torque bolts to 27–40 ft. lbs. (37–54 Nm).
- Left side lower gusset. Torque the bolts to 38–60 ft. lbs. (57–81 Nm).
- Right side lower gusset in 3 steps:

a. Install the bottom mounting bolts, but do not tighten.

b. Install the top bolt to the intake manifold support bracket and manifold. Tighten to 27–40 ft. lbs. (37–54 Nm).

c. Secure the manifold intake bracket by tightening the two attaching bolts to 38–60 ft. lbs. (57–81 Nm).
- Oil cooler pipes at the transmission. Torque the lines to 42–62 inch lbs. (5–7 Nm).
- Two oil cooling tube clamps to the lines
- Driveshafts with the matchmarks aligned. Torque the bolts to the differential flanges to 20–22 ft. lbs. (27–30 Nm) and the transfer case flange bolts to 36–43 ft. lbs. (49–59 Nm).

- Undercover splash shield. Torque the bolts to 42–62 inch lbs. (5–7 Nm).
- Upper dipstick tube to the lower tube and lower the vehicle

10. Provide automatic transmission control cable slack by gently pulling the cable to the left until the cable pin has rotated sufficiently to line the automatic transmission control cable up with the slot in the rear of the throttle body bellcrank.

11. Slide the automatic transmission control cable and cable pin into the bellcrank.

12. Tighten the locknut.
- Throttle kickdown cable to the mounting bracket
- Automatic transmission control cable to the throttle body
- Air silencer
- Shifter lever assembly to the transfer case. Torque the bolts to 72–102 inch lbs. (22–28 Nm).
- Four wiring connectors under the shift selector lever
- Shift selector lever, do not exert any force when installing the shift selector lever
- Four shift selector lever nuts. Tighten to 72–102 inch lbs. (22–28 Nm).
- Shift rod and washers to the selector lever and install the split pin to the shift selector lever
- Power/Economy switch wiring connector
- Rear console

- Front console and tie the shift boot draw strings
- Negative battery cable

13. Fill the transmission to the proper level.

14. Fill the cooling system.

15. Start the vehicle, check for leaks, and repair if necessary.

Clutch

ADJUSTMENT

Clutch Pedal Height

1. Before servicing the vehicle, refer to the precautions in the beginning of this section .

2. Pull back the carpet to measure the distance from the firewall to the top of the pedal. The standard height is 9.84 in. (250 mm).

3. If adjustment is required, loosen the locknut and turn the stopper bolt.

4. After adjustment is made tighten the locknut to 12 ft. lbs. (16 Nm).

Clutch Pedal Free-Play

1. Before servicing the vehicle, refer to the precautions in the beginning of this section .

2. Depress the clutch pedal gently by hand and measure the amount of free-play

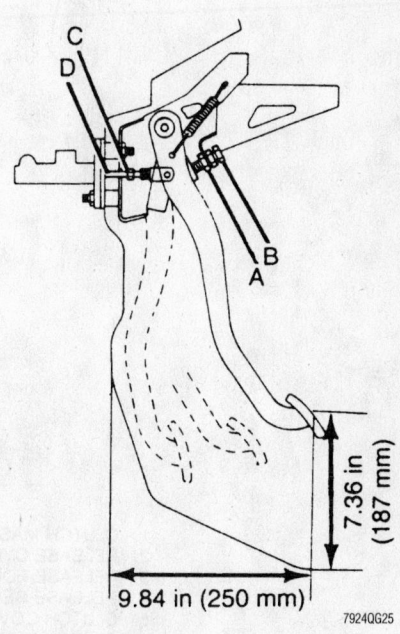

7.36 in (187 mm)

9.84 in (250 mm)

7924QG25

Clutch pedal height and free-play adjustment points

For Tire, Wheel and Ball Joint specifications, see Section 1 of this manual

(distance the pedal travels before resistance is felt). The proper amount of free-play is 0.5 in. (12.7mm). If the free-play is not within the proper specifications, continue with the procedure.

3. Measure from the floor pan to the middle point of the clutch pedal when the pedal is in the fully released position. The proper clutch pedal height is 7.25 in. (184mm).

4. If the pedal height is incorrect, loosen locknut (A) and turn the pedal adjusting bolt (B) until the proper height is achieved, then retighten the locknut.

5. Remeasure the free-play. If it is still out of specification, loosen the clutch pushrod locknut (C) and turn the pushrod (D) until the proper free-play is achieved. Tighten locknut (C) securely.

REMOVAL & INSTALLATION

1. Before servicing the vehicle, refer to the precautions in the beginning of this section .

2. Remove or disconnect the following:
- Transmission

- Pressure plate bolts and remove the clutch plate and disc

To install:

3. Install or connect the following:
- Clutch disc and plate using a centering tool
- Clutch cover. Torque the bolts to 73 ft. lbs. (99 Nm) and remove the centering tool
- Check the release bearing condition and lubricate or replace as necessary
- Transmission

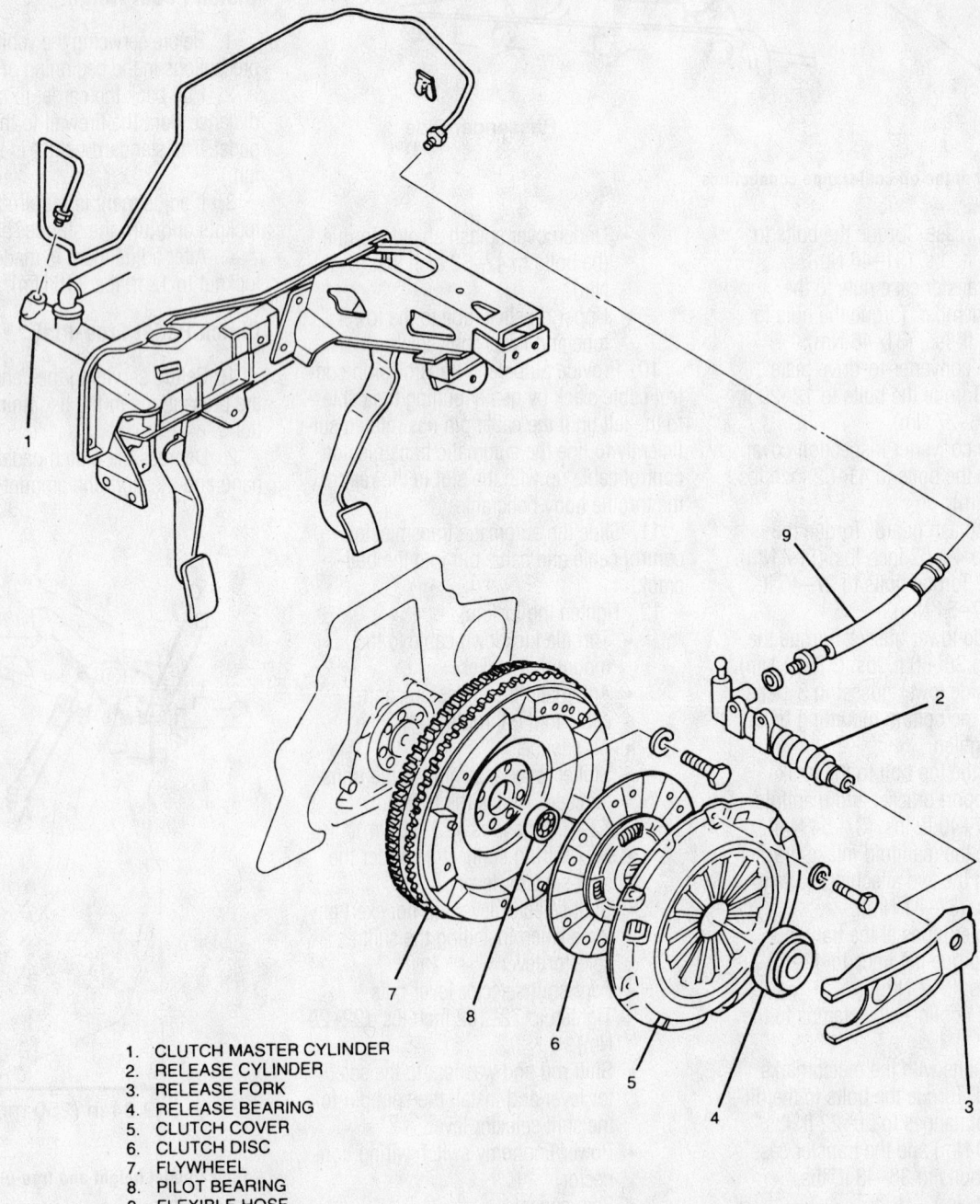

1. CLUTCH MASTER CYLINDER
2. RELEASE CYLINDER
3. RELEASE FORK
4. RELEASE BEARING
5. CLUTCH COVER
6. CLUTCH DISC
7. FLYWHEEL
8. PILOT BEARING
9. FLEXIBLE HOSE

79240G26

Exploded view of the clutch assembly

Hydraulic Clutch System

BLEEDING

1. Before servicing the vehicle, refer to the precautions in the beginning of this section .

2. With an assistant in the vehicle, raise and safely support the vehicle.

3. Have your assistant pump the clutch pedal three times and hold the pedal to the floor.

4. Open the bleeder valve on the clutch slave cylinder until the air is purged from the cylinder.

5. Tighten the bleeder valve.

6. Have your assistant release the clutch pedal.

7. Fill the clutch master cylinder if below minimum.

8. Repeat Steps 2 through 6 until no air exits from the bleeder valve.

9. Lower the vehicle.

10. Fill the clutch master cylinder fluid reservoir.

Transfer Case Assembly

REMOVAL & INSTALLATION

1. Before servicing the vehicle, refer to the precautions in the beginning of this section .

2. Drain the transfer case.

3. Remove or disconnect the following:
   • Negative battery cable
   • Two rear console mounting screws. Slide the console forward to clear

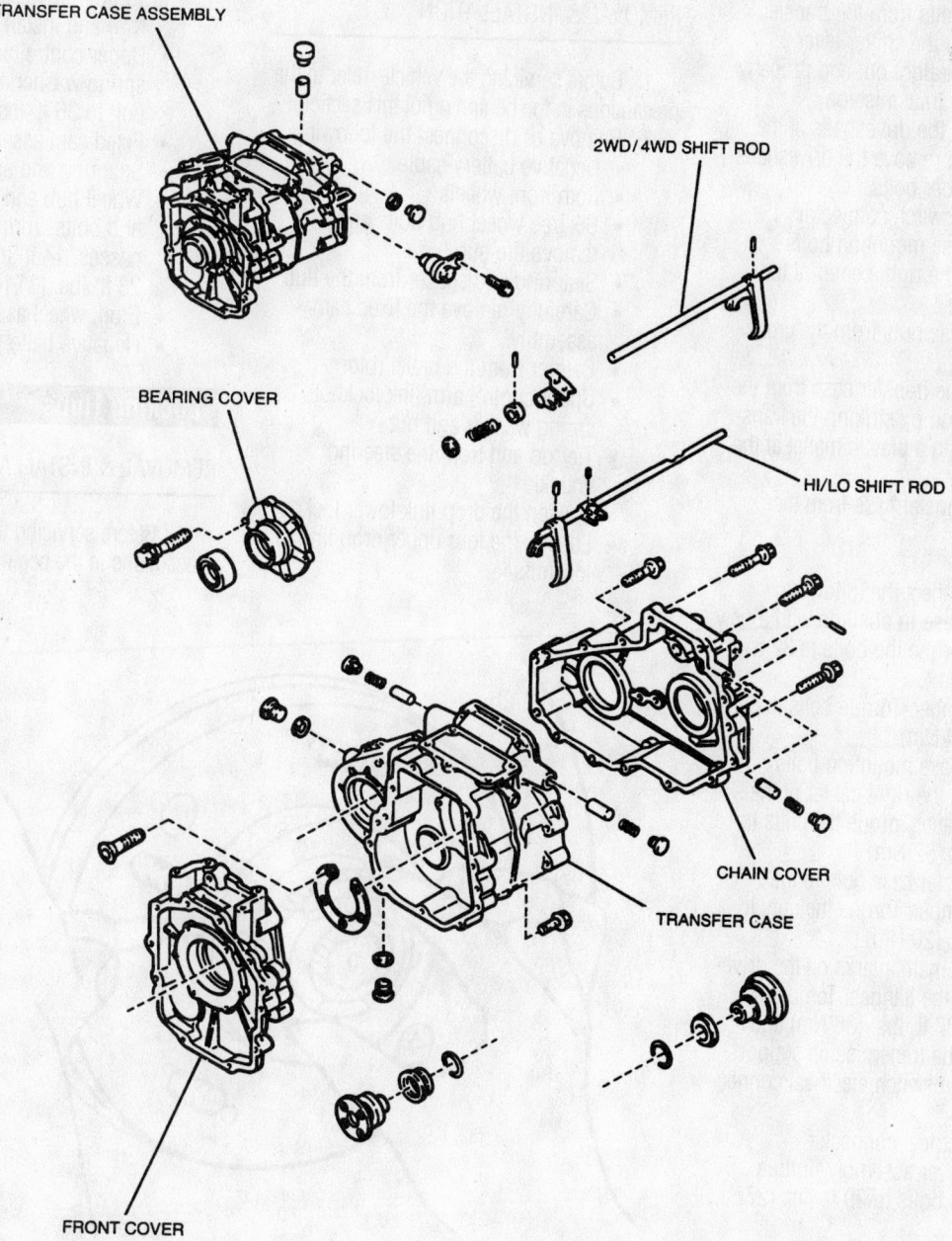

TRANSFER CASE ASSEMBLY

2WD/4WD SHIFT ROD

BEARING COVER

HI/LO SHIFT ROD

CHAIN COVER

TRANSFER CASE

FRONT COVER

9308QG05

Exploded view of the transfer case assembly

For Wheel Alignment specifications, see Section 1 of this manual

the parking brake handle and set aside

- Three mounting screws from the front console. Untie the shift boot draw strings and open the boot
- Loosen the transfer case shift lever locknut and remove the lever knob
- Pull the console up to access the Power/Economy switch wiring connector. Unplug the connector and remove the console

4. Shift the transfer lever to the 4L position.

- Cover plate
- Retaining bolts from the transfer case and lift the shifter lever assembly straight out and properly support the transmission
- Matchmark the driveshafts at the flanges and remove the driveshafts
- Crossmember bolts
- 4WD light switch connector
- Transfer case mounting bolts located at the right center of the transfer case
- Transfer case nuts from the crossmember
- Separate the transfer case from the transmission by striking the transfer case with a plastic mallet at the seal area

5. Lower the transfer case from the vehicle.

To install:

6. Install or connect the following:

- Transfer case in position with a new gasket. Torque the bolts to 32 ft. lbs. (44 Nm).
- Crossmember. Torque bolts to 32 ft. lbs. (44 Nm).
- Transfer case mounting bolts located at the right center of the transfer case. Torque the bolts to 38 ft. lbs. (52 Nm).
- Four transfer case nuts to the crossmember. Torque the nuts to 15 ft. lbs. (20 Nm).
- Align the matchmarks on the driveshafts to the flanges. Torque the bolts to 27 ft. lbs. (36 Nm) and remove the transmission support
- 4WD light switch electrical connector
- VSS electrical connector
- Shifter lever assembly. Torque retaining bolts to 20 ft. lbs. (27 Nm).
- Cover plate. Torque the bolts to 15 ft. lbs. (20 Nm).
- Power/Economy switch wiring connector
- Front console

- Lever knobs
- Front console and tie the shift boot draw strings
- Slide the console over the parking brake handle
- Rear console
- Negative battery cable

7. Fill the transfer case to the proper level

8. Start the vehicle and check for leaks, repair if necessary.

Halfshaft

REMOVAL & INSTALLATION

1. Before servicing the vehicle, refer to the precautions in the beginning of this section .

2. Remove or disconnect the following:

- Negative battery cable
- Both front wheels
- Six free wheel hub bolts and remove the hub
- Snapring and spacer from the hub
- Carefully remove the fixed cam assembly
- Caliper from the brake rotor
- Upper control arm link lockbolt, spring washer and nut
- Tie rod end from the steering knuckle
- Loosen the drop link lower locknut
- Loosen the four upper drop link locknuts

- Spread open the drop link fork with a rubber mallet
- Matchmark the halfshaft and differential.
- Carefully pry the halfshaft from the differential
- Halfshaft

To install:

3. Install or connect the following:

- Halfshaft with the matchmarks aligned with the differential
- Torque the upper and lower drop link nuts to 36 ft. lbs. (49 Nm).
- Tie rod end to the steering knuckle. Torque the locknut to 27 ft. lbs. (36 Nm) and install a new cotter pin
- Upper control arm link lockbolt, spring washer and nut. Torque the bolt to 36 ft. lbs. (49 Nm).
- Fixed cam assembly
- Snapring and spacer in the hub
- Wheel hub and the six free wheel hub bolts. Torque the bolts in two passes, 14 ft. lbs. (17 Nm), then to 23 ft. lbs. (31 Nm).
- Front wheel assemblies
- Negative battery cable

Locking Hubs

REMOVAL & INSTALLATION

1. Before servicing the vehicle, refer to the precautions in the beginning of this section.

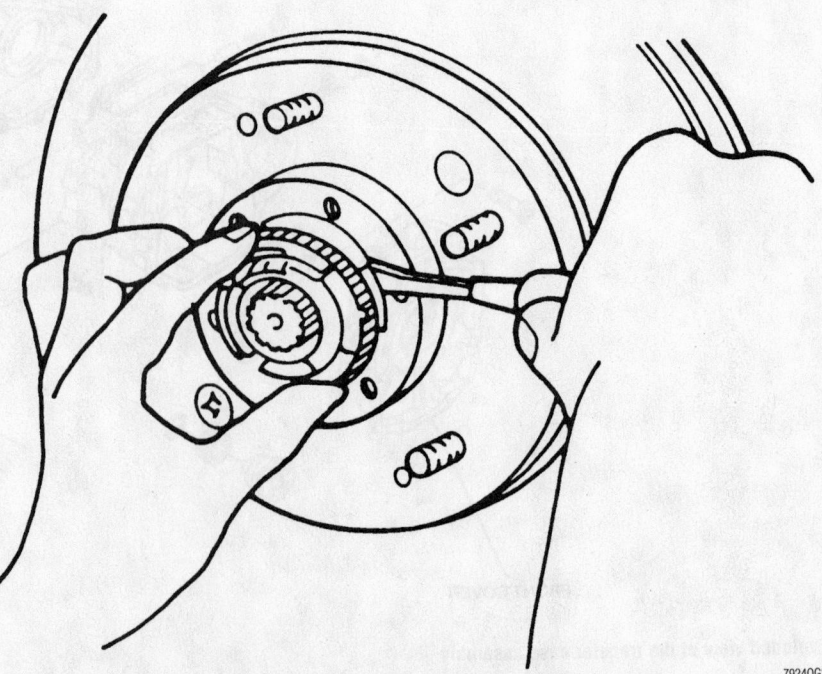

Removing the 4WD fixed cam assembly

79240G27

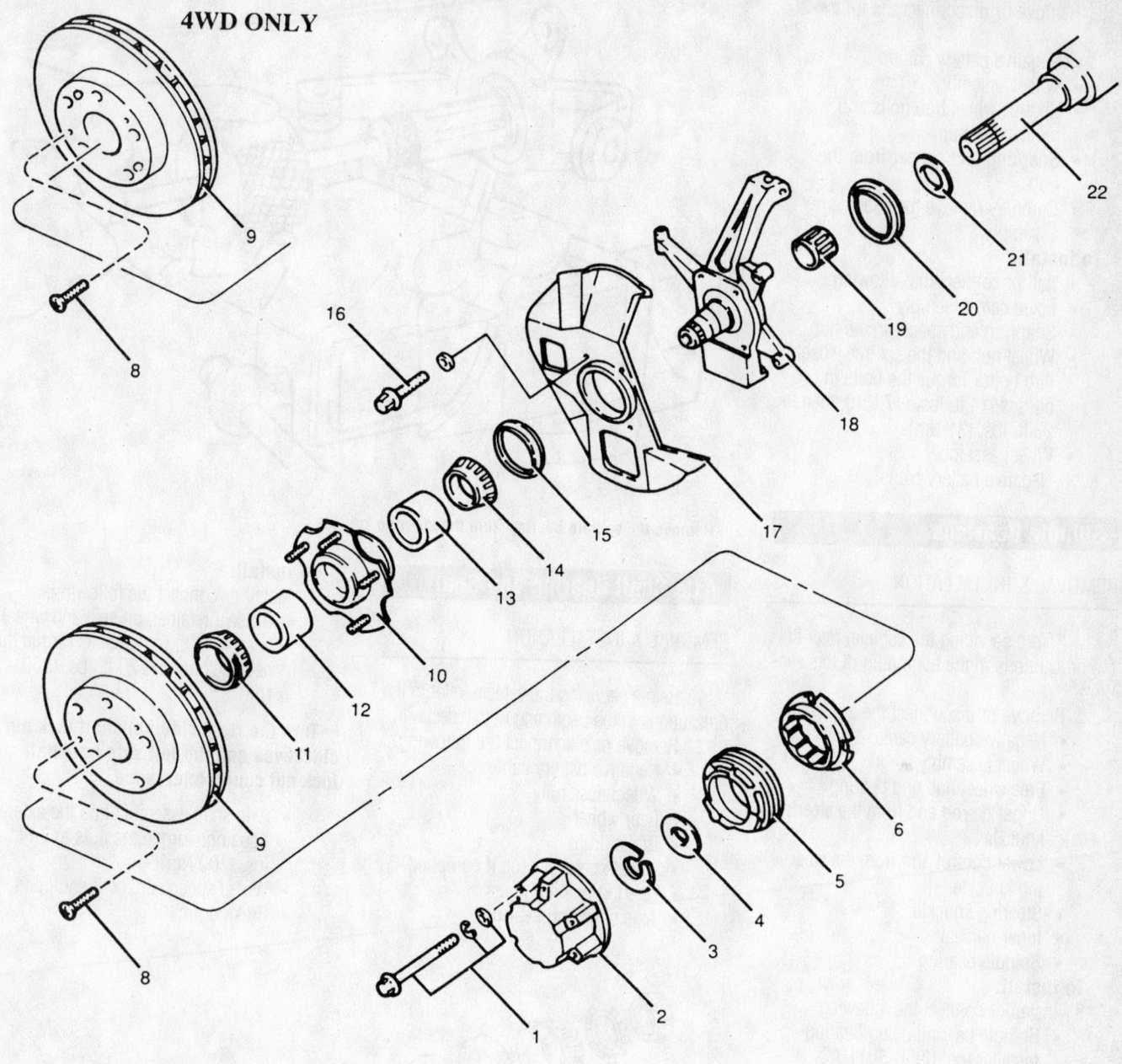

4WD ONLY

1	BOLT/WASHER	9	ROTOR	16	BOLT & SPRING WASHER
2	FREE WHEEL HUB BODY	10	WHEEL HUB	17	DUST COVER
3	SNAP RING	11	INNER BEARING INNER RACE	18	KNUCKLE
4	SPACER	12	INNER BEARING OUTER RACE	19	NEEDLE BEARING
5	FIXED CAM ASSEMBLY	13	OUTER BEARING OUTER RACE	20	OIL SEAL
6	LOCK NUT	14	OUTER BEARING INNER RACE	21	SPACER
8	SCREW	15	OIL SEAL	22	DRIVE SHAFT (LH)

7924QG28

Exploded view of the 4WD locking hub assembly

For Maintenance Interval recommendations, see Section 1 of this manual

2. Remove or disconnect the following:

- Negative battery cable
- Wheel assembly
- Six free wheel hub bolts and remove the hub
- Snapring and spacer from the hub
- Carefully remove the fixed cam assembly

To install:

3. Install or connect the following:

- Fixed cam assembly
- Snapring and spacer in the hub
- Wheel hub and the six free wheel hub bolts. Torque the bolts in two passes, 14 ft. lbs. (17 Nm), then to 23 ft. lbs. (31 Nm).
- Wheel assembly
- Negative battery cable

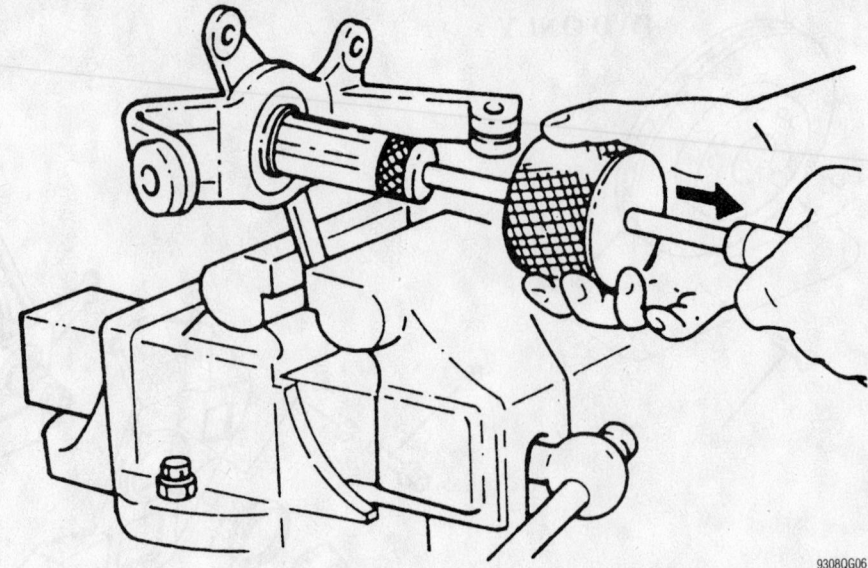

Remove the spindle bearing from the steering knuckle

Spindle Bearings

REMOVAL & INSTALLATION

1. Before servicing the vehicle, refer to the precautions in the beginning of this section.
2. Remove or disconnect the following:

- Negative battery cable
- Wheel assembly
- Free wheel hub and bearing
- Upper tie rod end from the steering knuckle
- Lower control arm from the steering knuckle
- Steering knuckle
- Inner oil seal
- Spindle bearing

To install:

3. Install or connect the following:

- Spindle bearing using Bearing Installer tool K95B-5011-A
- New oil seal and apply grease to the bearing and seal lip
- Halfshaft end
- Steering knuckle to the halfshaft with the upper and lower ball joints in the mounting holes
- Lower control arm. Torque the lock nut to 110 ft. lbs. (148 Nm) and install a cotter pin
- Tie rod end to the steering knuckle. Torque the nut to 27 ft. lbs. (36 Nm) and install a cotter pin
- Upper control arm. Torque the lock bolt to 36 ft. lbs. (49 Nm).
- Free wheel hub and bearing assembly
- Front wheel
- Negative battery cable

Axle Shaft Bearing and Seal

REMOVAL & INSTALLATION

1. Before servicing the vehicle, refer to the precautions in the beginning of this section.
2. Remove or disconnect the following:

- Negative battery cable
- Wheel assembly
- Rear wheel
- Brake drum
- wheel speed sensor, if equipped
- Bearing retaining nuts
- Axle shaft and bearing

To install:

3. Install or connect the following:

- Oil seal retainer, oil seal and wheel bearing to the axle shaft. Torque the new lock nut to 220 ft. lbs. (300 Nm).

➡**Turn the right side halfshaft lock nut clockwise and the left side halfshaft lock nut counterclockwise.**

- Axle shaft assembly into the axle housing. Torque the nuts to 74 ft. lbs. (100 Nm).
- Wheel speed sensor, if equipped
- Brake drum

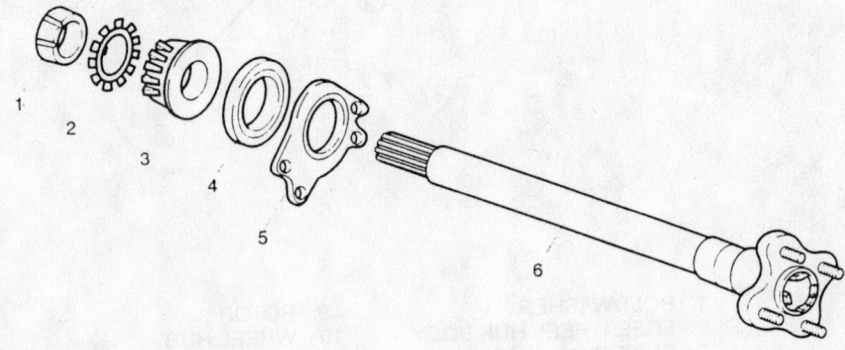

1. Lock Nut
2. Lock Washer
3. Bearing
4. Oil Seal
5. Oil Seal Retainer
6. Axle Shaft

Exploded view of the axle shaft, bearing and seal

- Rear wheel assembly
- Negative battery cable

4. Check the fluid level and top off if necessary.

Pinion Seal

REMOVAL & INSTALLATION

1. Before servicing the vehicle, refer to the precautions in the beginning of this section .

2. Drain the gear oil.

3. Remove or disconnect the following:
   - Negative battery cable
   - Wheel assemblies

- Brake drums
- driveshaft

➡**Use an inch lb. (Nm) torque wrench , measure and record the amount of torque required to maintain rotation of the pinion.**

- Pinion flange
- Pinion seal

To install:

4. Install or connect the following:
   - New pinion seal lightly coated with clean gear oil
   - Pinion flange

5. Rotate the pinion flange occasionally while tightening the flange nut and make certain that the pinion bearings are seated properly.

6. Take several bearing preload torque readings. Tighten the flange nut to achieve the preload torque reading. The maximum torque reading should not exceed 14 inch lbs. (1.6 Nm).
   - Driveshaft after aligning the match-marks
   - Brake drums
   - Wheel assemblies
   - Negative battery cable

7. Fill the gear oil to the proper level.

8. Start the vehicle and check for leaks; repair if necessary.

STEERING AND SUSPENSION

Air Bag

PRECAUTIONS

Several precautions must be observed when handling the inflator module to avoid accidental deployment and possible personal injury.

1. Never carry the inflator module by the wires or connector on the underside of the module.

2. When carrying a live inflator module, hold securely with both hands, and ensure that the bag and trim cover are pointed away from you.

3. Place the inflator module on a bench or other surface with the bag and trim cover facing up.

4. With the inflator module on the bench, never place anything on or close to the module which may be thrown in the event of an accidental deployment.

DISARMING

1. Before servicing the vehicle, refer to the precautions in the beginning of this section .

2. Turn the ignition switch to the **LOCK** position.

3. Disconnect the negative battery cable.

4. Wait 10 minutes for the back-up power to discharge.

ARMING

Assuming the system components (air bag control module, sensors, air bag, etc.) are installed correctly and are in good working order, the system is armed whenever the battery positive and negative battery cables are connected.

If you have disarmed the air bag system for any reason, to rearm, be sure no one is in the vehicle (as an added safety measure), then connect the negative battery cable.

Power Steering Gear

ADJUSTMENT

1. Before servicing the vehicle, refer to the precautions in the beginning of this section .

2. Place the steering gear in a vise with protective jaws.

3. Place a torque wrench on the Pitman arm end of the shaft.

4. Loosen the locknut on the adjusting bolt.

5. Slowly turn the adjusting bolt to until the breakaway torque is 65 ft. lbs. (88 Nm).

6. Hold the adjusting bolt in position and tighten the locknut to 25 ft. lbs. (34 Nm).

REMOVAL & INSTALLATION

1. Before servicing the vehicle, refer to the precautions in the beginning of this section .

2. Center the steering wheel.

3. Drain the power steering fluid.

4. Remove or disconnect the following:
   - Negative battery cable
   - Left front wheel
   - Pitman arm-to-centerlink attaching nut
   - Separate the Pitman arm from the centerlink with a ball joint puller
   - Power steering hoses

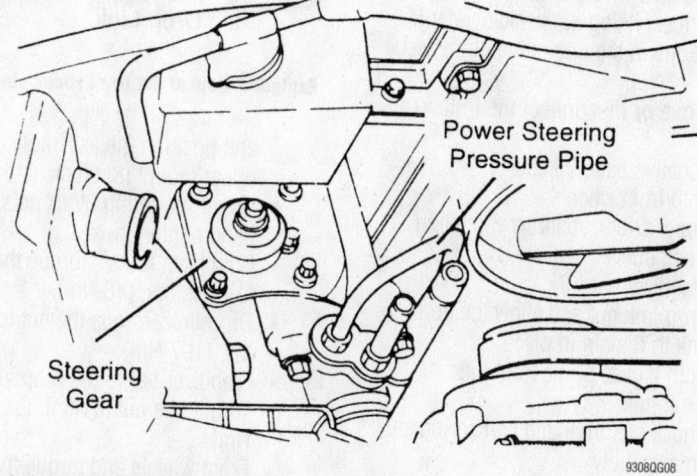

Power Steering Pressure Pipe

Steering Gear

Exploded view of the steering gear

93080G08

- Coolant recovery tank and the power steering reserve tank
- Set bolt from the intermediate shaft
- Intermediate shaft from the steering gear
- Steering gear-to-frame bolts
- Steering gear

To install:

5. Install or connect the following:
- Position the steering gear to the frame. Torque the bolts to 159 ft. lbs. (215 Nm).
- Pitman arm to the centerlink. Torque the bolt to 36 ft. lbs. (49 Nm) and install a new cotter pin
- Intermediate shaft to the steering gear shaft. Torque the set bolt o 25 ft. lbs. (34 Nm).
- Power steering hoses/lines. Torque the fasteners to 29 ft. lbs. (34 Nm).
- Power steering reserve tank over the bracket and press until full engagement is reached
- Coolant recovery tank
- Left front wheel
- Negative battery cable

6. Fill the power steering fluid to the proper level and bleed the system.

7. Start the vehicle and check for leaks, repair if necessary.

8. Road test the vehicle to check that the steering wheel is straight.

Shock Absorber

REMOVAL & INSTALLATION

Front

The front shock absorber and coil spring are removed as a single unit.

1. Before servicing the vehicle, refer to the precautions in the beginning of this section .

2. Remove or disconnect the following:
- Negative battery cable
- Both front wheels
- Upper shock absorber mounting block nuts
- Stabilizer bar
- Drop link nut and allow the drop link to remain in place
- Both halves of the front fork
- Drop link
- Shock absorber and coil spring as an assembly

To install:

3. Install or connect the following:
- Coil spring to the shock absorber

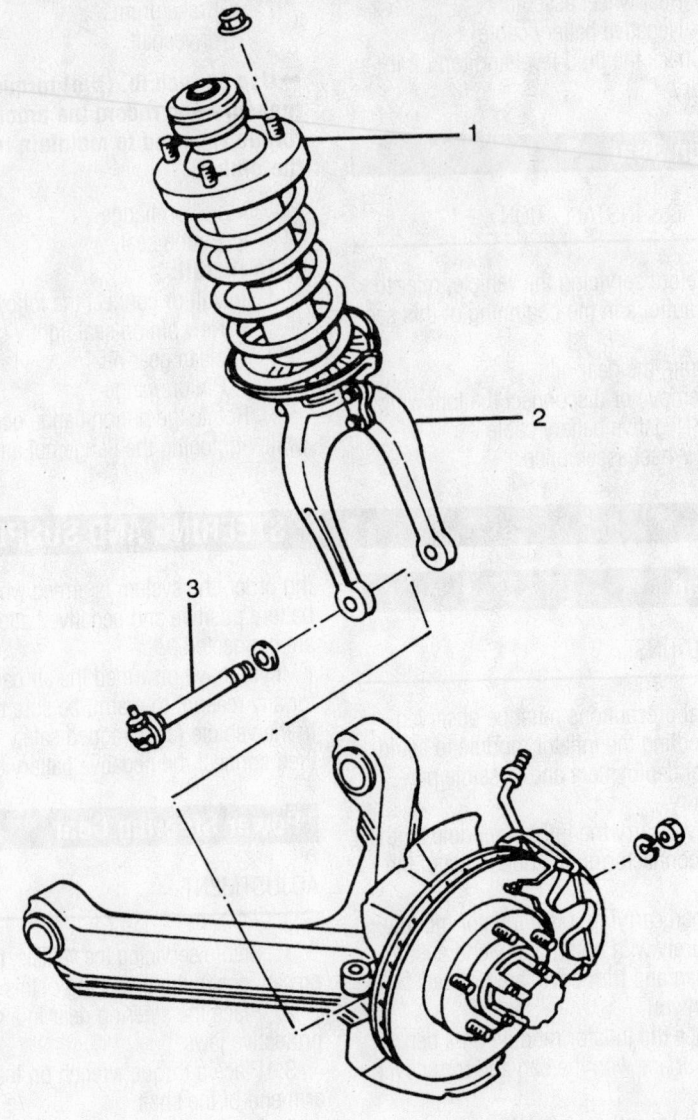

1 Front Shock Absorber & Coil Spring Assembly
2 Front Fork
3 Drop Link

9308QG09

Exploded view of the front shock absorber assembly

and position the assembly to the upper mounting block
- Upper mounting block nuts and hand tighten them
- Both front forks . Torque the bolts to 36 ft. lbs. (48 Nm).
- Drop link. Torque the nut to 145 ft. lbs. (197 Nm).
- Stabilizer bar to the drop link. Torque the nut to 36 ft. lbs. (48 Nm).
- Front wheels and torque the mounting block nuts to 18 ft. lbs. (25 Nm).
- Negative battery cable

Rear

1. Before servicing the vehicle, refer to the precautions in the beginning of this section .

2. Remove or disconnect the following:
- Negative battery cable
- Rear wheels
- Raise the rear axle with a floor jack to relax the shock absorbers and support the rear axle when the shock absorber is removed.
- Rear safety nut, upper nut and washer
- Upper rubber plate

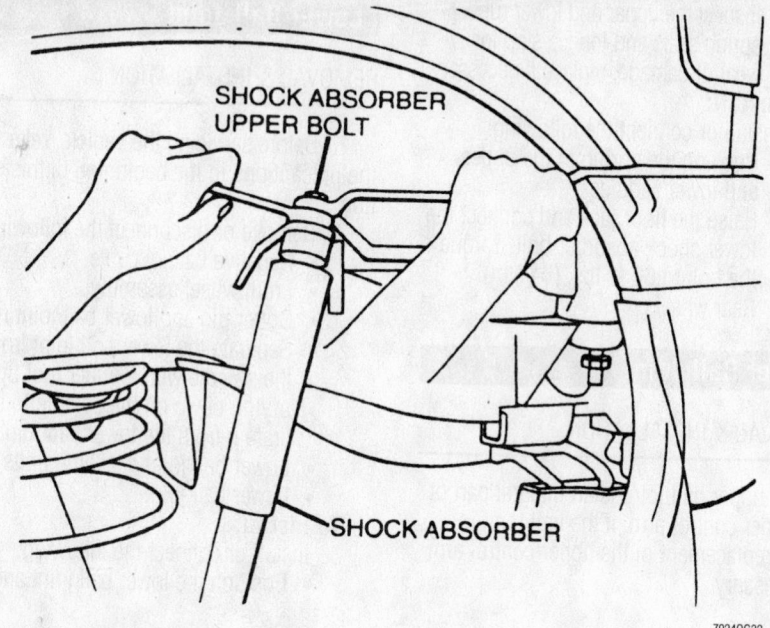

SHOCK ABSORBER
UPPER BOLT

SHOCK ABSORBER

7924QG32

Upper shock absorber mounting nut

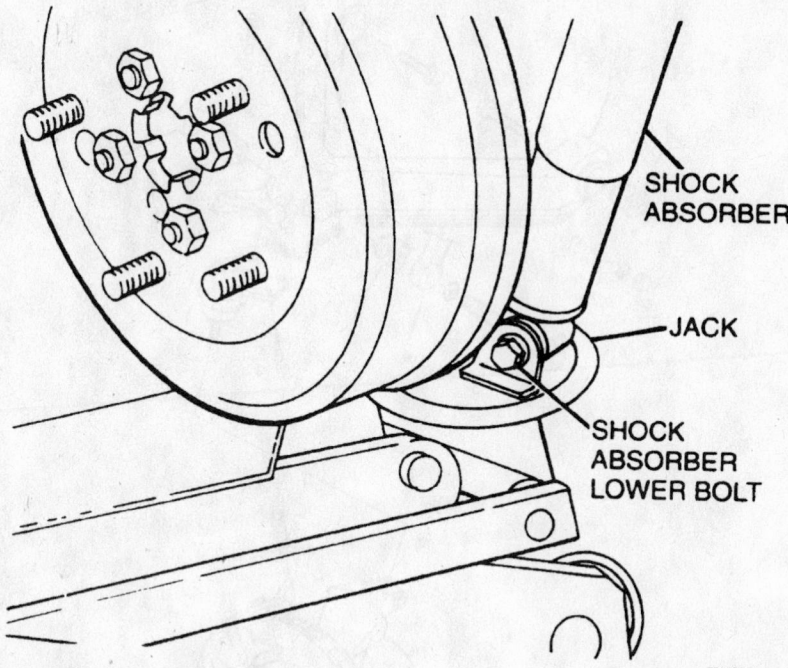

SHOCK
ABSORBER

JACK

SHOCK
ABSORBER
LOWER BOLT

7924QG33

Lower shock absorber mounting bolt

- Lower bolt from the shock absorber
- Lower the rear axle housing
- Shock absorber

3. Remove the lower mounting bolt and remove the shock absorber.

To install:

4. Install or connect the following:
- Bottom washer and rubber cushion on the top of the shock absorber.

Position the shock absorber on the vehicle
- Lower bolt
- Rubber cushion, washer and nut. Tighten to 53 ft. lbs. (72 Nm)
- Safety nut. Torque the lower bolt to 62 ft. lbs. (84 Nm).
- Rear wheels
- Negative battery cable

Coil Spring

REMOVAL & INSTALLATION

Front

1. Before servicing the vehicle, refer to the precautions in the beginning of this section .

2. Remove or disconnect the following:
- Negative battery cable
- Shock absorber and coil spring as an assembly
- Place the shock absorber in a vice
- Loosen the pivot rod nut several turns
- While still secured in a vise compress the coil spring
- Piston rod nut and disassemble the coil spring as needed

To install:

3. Install or connect the following:
- Bottom portion of the shock absorber in a vice and compress the coil spring
- End of the coil spring to the rubber seat and install the spring
- Assemble the dust boot and lower retainer, lower insulator, spring seat, boss, center washer, upper insulator and install the coil spring
- Hand tighten the piston rod nut.
- Carefully loosen the spring compressor tool and remove the tool
- Torque the piston rod nut to 31 ft. lbs. (42 Nm).
- Coil spring and shock absorber as an assembly

Rear

1. Before servicing the vehicle, refer to the precautions in the beginning of this section .

2. Remove or disconnect the following:
- Both rear wheels
- Raise the rear axle with a floor jack to relax the shock absorbers and support the rear axle when the shock absorber is removed.

➡**For easier installation, complete one side at a time.**

- Lower mounting bolt, then the shock absorber
- Lower the floor jack until the coil spring is fully expanded
- Coil spring

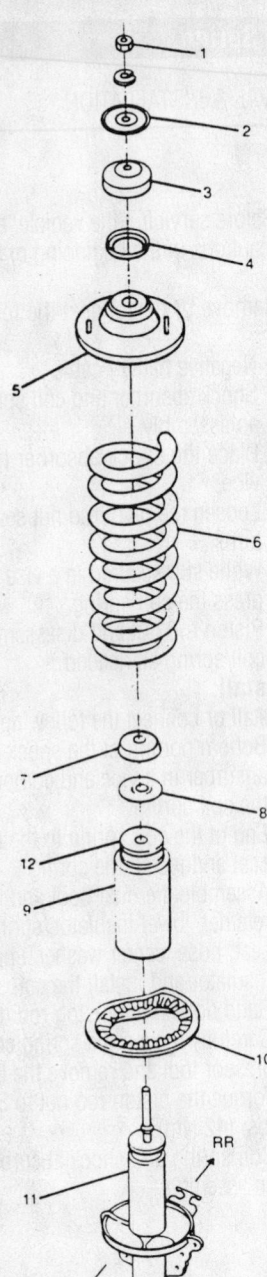

1. Nut
2. Upper retainer
3. Upper insulator
4. Centering washer
5. Spring seat
6. Coil spring
7. Lower insulator
8. Lower retainer
9. Dust boot
10. Rubber seat
11. Shock absorber
12. Front jounce stop

9308QG10

**Exploded view of the front shock absorber
and coil spring assembly**

• Inspect the upper and lower rubber
spring seats and jounce stop for
wear or damage, replace if necessary

To install:
3. Install or connect the following:
• Position the spring in the upper
and lower saddles
• Raise the floor jack and connect the
lower shock absorber bolt. Torque
the bolt to 62 ft. lbs. (84 Nm).
• Rear wheels

Upper Ball Joint

REMOVAL & INSTALLATION

The upper ball joint is an integral part of
the upper control arm. If the ball joint is
worn, replacement of the upper control arm
is necessary.

Lower Ball Joint

REMOVAL & INSTALLATION

1. Before servicing the vehicle, refer to
the precautions in the beginning of this sec-
tion .
2. Remove or disconnect the following:
• Negative battery cable
• Front wheel assembly
• Cotter pin and lower ball joint nut
• Separate the lower ball joint from
the spindle with a puller tool by
prying down on the spindle to sep-
arate it from the lower ball joint
• Lower ball joint attaching bolts
• Lower ball joint
To install:
3. Install or connect the following:
• Position the lower ball joint and

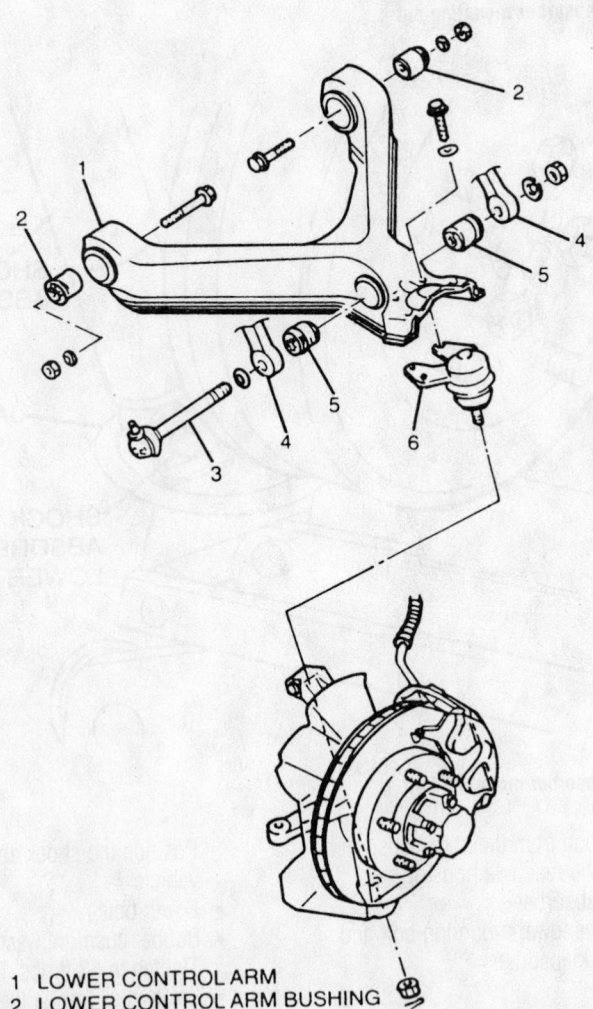

1 LOWER CONTROL ARM
2 LOWER CONTROL ARM BUSHING
3 DROP LINK
4 FRONT FORK
5 DROP LINK BUSHING
6 LOWER CONTROL ARM BALL JOINT

79240G34

Exploded view of the lower control arm and ball joint assembly

install the attaching nuts and bolts. Torque the fasteners to 36 ft. lbs. (48 Nm).

- Pry down and guide the spindle onto the lower ball joint. Torque the nut 87 ft. lbs. (118 Nm) and install a new cotter pin
- Front wheel assembly.
- Negative battery cable

Upper Control Arm

REMOVAL & INSTALLATION

1. Before servicing the vehicle, refer to the precautions in the beginning of this section .
2. Remove or disconnect the following:

- Negative battery cable
- Front wheel assembly
- Bolt securing the upper ball joint to the steering knuckle

➡**Note the matchmark setting on the upper control arm mounting bolts before removal.**

- Upper control arm mounting bolts
- Upper control arm from the vehicle.

To install:

3. Install or connect the following:
- Position the upper control arms in the frame mounting. Hand tighten the bolts
- Position the ball joint in the spindle. Torque the through bolt to 36 ft. lbs. (48 Nm).

➡**Be sure the slot in the ball joint aligns with the through-bolt during installation.**

- Align the upper control arm bolts to the previous settings. Torque bolts to 62 ft. lbs. (108 Nm).
- Front wheel assembly

4. Check and adjust the alignment, if necessary.

UPPER CONTROL ARM BUSHING REPLACEMENT

1. Before servicing the vehicle, refer to the precautions in the beginning of this section .
2. Remove or disconnect the following:
- Upper control arm assembly
- Secure the control arm in a vise
- Using a standard press, remove the bushing

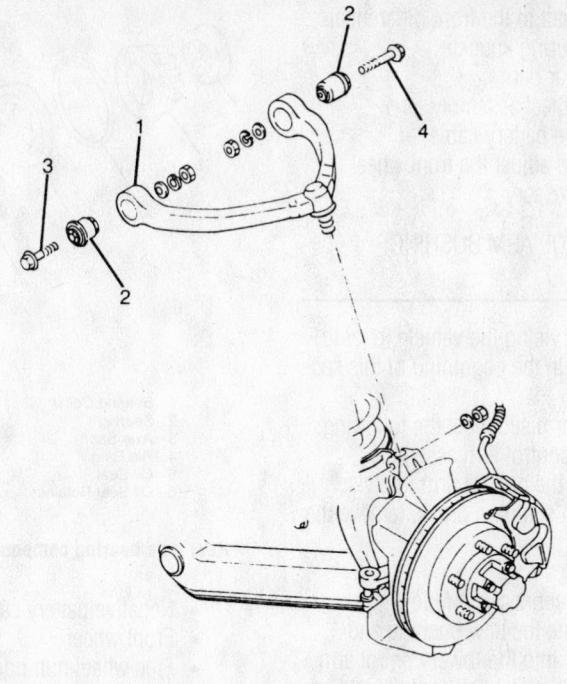

1 UPPER CONTROL ARM
2 UPPER CONTROL ARM BUSHING
3 FRONT SPINDLE
4 REAR SPINDLE

7924QG31

Exploded view of the upper control arm assembly

To install:

3. Install or connect the following:
- Lubricate the new bushing and press it into the upper control arm
- Upper control arm to the vehicle

4. Check the wheel alignment and adjust if necessary.

Lower Control Arm

REMOVAL & INSTALLATION

1. Before servicing the vehicle, refer to the precautions in the beginning of this section .
2. Remove or disconnect the following:
- Negative battery cable
- Front wheel assembly
- Stabilizer bar
- Driveshaft from the differential and steering knuckle
- Shock absorber and coil spring from the front half of the fork
- Cotter pin and castle nut from the lower control arm ball joint

- Lower control arm ball joint from the steering knuckle
- Lower control arm bushing bolts from the front frame crossmember brackets
- Lower control arm

To install:

3. Install or connect the following:
- Lower control arm to the front frame crossmember brackets and hand tighten the bushing bolts
- Lower control arm ball joint and bolt to the steering knuckle. Torque the bolt to 87 ft. lbs. (118 Nm).
- New cotter pin and castle nut
- Torque the lower control arm bushing bolts to 206 ft. lbs. (280 Nm).
- Front fork halves to the shock absorber and coil spring and position the lower portion over the lower control arm drop link holes
- Drop link. Torque the nut to 145 ft. lbs. (197 Nm).

- Driveshaft to the front differential and steering knuckle
- Stabilizer bar
- Front wheel assembly
- Negative battery cable

4. Check and adjust the front wheel alignment if necessary.

LOWER CONTROL ARM BUSHING REPLACEMENT

1. Before servicing the vehicle, refer to the precautions in the beginning of this section .
2. Remove or disconnect the following:
   - Lower control arm assembly
   - Secure the control arm in a vise
   - Using a standard press, remove the bushing

To install:

3. Install or connect the following:
   - Lubricate the new bushing and press it into the lower control arm
   - lower control arm to the vehicle
4. Check the wheel alignment and adjust if necessary.

Wheel Bearings

ADJUSTMENT

Front

1. Remove or disconnect the following:
2. Before servicing the vehicle, refer to the precautions in the beginning of this section .
   - Negative battery cable
   - Front wheel
   - Brake caliper from the rotor and hang it out of the way
3. Attach a dial indicator to the axle hub and measure the bearing play
4. If the play exceeds.004 inch (.10mm), check and adjust locknut torque. The bolts should be tightened to 23 ft. lbs. (31 Nm).

Rear

The rear wheel bearings are not adjustable.

REMOVAL & INSTALLATION

Front

1. Before servicing the vehicle, refer to the precautions in the beginning of this section .
2. Remove or disconnect the following:

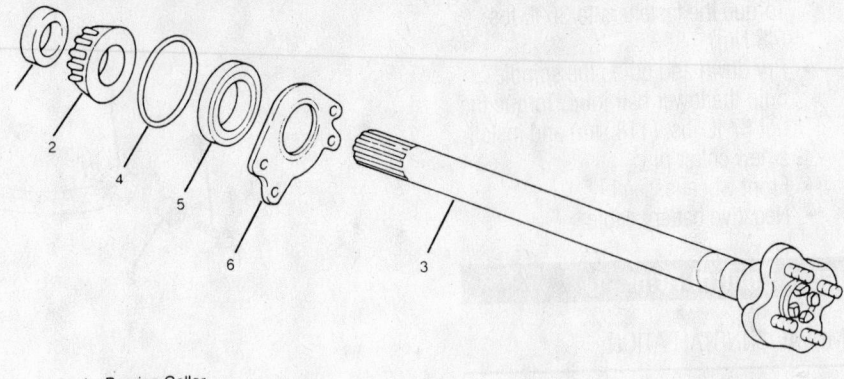

1 Bearing Collar
2 Bearing
3 Axle Shaft
4 Rib Ring
5 Oil Seal
6 Oil Seal Retainer

79240G35

Rear axle bearing component identification

- Negative battery cable
- Front wheel
- Free wheel hub body
- Brake caliper
- Brake rotor
- Wheel bearing and using a screw driver, pry out the oil seal
- Inner and outer bearings
- Using a drift punch, remove the inner and outer bearing race

To install:

3. Pack the new bearings with grease.
4. Install or connect the following:
   - Inner bearing and race and a new oil seal
   - Outer bearing and race and secure the dust cover with four screws
   - Apply grease to the new bearings and the lip of the oil seal
   - Hub assembly in the steering knuckle
   - Screw the locknut against the hub assembly until there is 10 inch lbs. (1.3 Nm) of preload on the hub
   - Brake rotor and retaining screws
   - Attach a run-out gauge to check the rotor run-out. The run-out should not exceed 0.004 inch (0.10mm)
   - Brake caliper
   - Free wheel hub fixed cam key with the locknut groove and push the on the fixed cam assembly
   - Axle retainer snap ring
   - Apply a light coat of sealant on the free wheel hub body. Install the body on the hub. Torque the bolts in 2 passes. Tighten the bolts on the first pass to 18 ft. lbs. (25 Nm).

Tighten the bolts on the second pass to 23 ft. lbs. (31 Nm).
- Negative battery cable

Rear

1. Before servicing the vehicle, refer to the precautions in the beginning of this section .
2. Remove or disconnect the following:
   - Negative battery cable
   - Rear wheels
   - Brake drum
   - Oil seal retainer flange

➡**The axle shafts are different from side-to-side, mark the to ensure they are returned to the proper side.**

- Using a slide hammer, remove the axle shaft assembly
- Using a hydraulic press, remove the bearing collar and bearing from the axle
- oil seal from the differential

To install:

3. Install or connect the following:
   - Using the appropriate seal driver, install the new axle seal into the differential

➡**The left-hand axle is 25.5 inches (647mm) long, and the right-hand axle is 27.4 inches (697mm) long.**

- Using a hydraulic press, install the new wheel bearing and retainer collar to the axle shaft
- Axle shaft into the carrier. Torque the nuts to 75 ft. lbs. (100 Nm).
- Brake drum and rear wheels
- Negative battery cable

LAND ROVER

20

Discovery • Discovery Series II • Freelander • Range Rover

PRECAUTIONS

Before servicing any vehicle, please be sure to read all of the following precautions, which deal with personal safety, prevention of component damage, and important points to take into consideration when servicing a motor vehicle:

• Never open, service or drain the radiator or cooling system when the engine is hot; serious burns can occur from the steam and hot coolant.

• Observe all applicable safety precautions when working around fuel. Whenever servicing the fuel system, always work in a well-ventilated area. Do not allow fuel spray or vapors to come in contact with a spark, open flame, or excessive heat (a hot drop light, for example). Keep a dry chemical fire extinguisher near the work area. Always keep fuel in a container specifically designed for fuel storage; also, always properly seal fuel containers to avoid the possibility of fire or explosion. Refer to the additional fuel system precautions later in this section.

• Fuel injection systems often remain pressurized, even after the engine has been turned **OFF**. The fuel system pressure must be relieved before disconnecting any fuel lines. Failure to do so may result in fire and/or personal injury.

• Brake fluid often contains polyglycol ethers and polyglycols. Avoid contact with the eyes and wash your hands thoroughly after handling brake fluid. If you do get brake fluid in your eyes, flush your eyes with clean, running water for 15 minutes. If eye irritation persists, or if you have taken brake fluid internally, IMMEDIATELY seek medical assistance.

• The EPA warns that prolonged contact with used engine oil may cause a number of skin disorders, including cancer! You should make every effort to minimize your exposure to used engine oil. Protective gloves should be worn when changing oil. Wash your hands and any other exposed skin areas as soon as possible after exposure to used engine oil. Soap and water, or waterless hand cleaner should be used.

• All new vehicles are now equipped with an air bag system. The system must be disabled before performing service on or around system components, steering column, instrument panel components, wiring and sensors. Failure to follow safety and disabling procedures could result in accidental air bag deployment, possible personal injury and unnecessary system repairs.

• Always wear safety goggles when working with, or around, the air bag system. When carrying a non-deployed air bag, be sure the bag and trim cover are pointed away from your body. When placing a non-deployed air bag on a work surface, always face the bag and trim cover upward, away from the surface. This will reduce the motion of the module if it is accidentally deployed. Refer to the additional air bag system precautions later in this section.

• Clean, high quality brake fluid from a sealed container is essential to the safe and proper operation of the brake system. You should always buy the correct type of brake fluid for your vehicle. If the brake fluid becomes contaminated, completely flush the system with new fluid. Never reuse any brake fluid. Any brake fluid that is removed from the system should be discarded. Also, do not allow any brake fluid to come in contact with a painted surface; it will damage the paint.

• Never operate the engine without the proper amount and type of engine oil; doing so WILL result in severe engine damage.

• Timing belt maintenance is extremely important! Many models utilize an interference-type, non-freewheeling engine. If the timing belt breaks, the valves in the cylinder head may strike the pistons, causing potentially serious (also time-consuming and expensive) engine damage. Refer to the maintenance interval charts in the front of this manual for the recommended replacement interval for the timing belt, and to the timing belt section for belt replacement and inspection.

• Disconnecting the negative battery cable on some vehicles may interfere with the functions of the on-board computer system(s) and may require the computer to undergo a relearning process once the negative battery cable is reconnected.

• When servicing drum brakes, only disassemble and assemble one side at a time, leaving the remaining side intact for reference.

• Only an MVAC-trained, EPA-certified automotive technician should service the air conditioning system or its components.

ENGINE REPAIR

Distributor

The vehicles covered in this section are equipped with DIS.

Alternator

REMOVAL & INSTALLATION

Freelander

1. Before servicing the vehicle, refer to the precautions in the beginning of this section.
2. Disconnect the battery ground cable.
3. Remove the top arm.
4. Remove the drive belt.
5. Disconnect the wiring from the alternator.

6. Remove the lower bolt, upper nut and front plate bolt
7. Remove the alternator.
8. Installation is the reverse of removal. Tighten all mounting bolts/nut to 33 ft. lbs. (45Nm).

Discovery/Discovery Series II

1. Before servicing the vehicle, refer to the precautions in the beginning of this section.
2. Disconnect the battery ground cable.
3. Remove the drive belt.
4. Remove the 2 mounting bolts.
5. Lift the alternator from the bracket and disconnect the wires.
6. Installation is the reverse of removal. If the pulley was removed, tighten the pulley bolt to 30 ft. lbs. (40 Nm). Torque the mounting bolts to 18 ft. lbs. (25 Nm).

Ignition Timing

➡**The ignition timing is not adjustable. It is controlled by the PCM.**

Engine Assembly

REMOVAL & INSTALLATION

Range Rover

The engine and transmission are removed and installed, as an assembly, from the vehicle.

1. Before servicing the vehicle, refer to the precautions in the beginning of this section.
2. Remove or disconnect the following:
   • Battery

- Fuel system pressure
- Coolant
- ECM located next to the battery
- Wiring harness from the starter and alternator
- Fuel supply and return lines from the fuel rail, then plug the openings to prevent contaminants from entering.
- Purge valve
- Intake hose/air flow meter assembly
- Throttle and cruise control cables from the throttle linkage
- All coolant hoses related to engine/transmission service
- Battery tray
- The 2 fuse box mounting bolts and pivot the fuse box aside.
- Engine wiring harness connector from the base of the fuse box
- Ground cable from the valance stud
- The 2 engine wiring harness connectors from the main harness
- Hood struts from the body locations. Raise the hood in the vertical position and stabilize.

3. Properly discharge the air conditioning system.

4. Remove or disconnect the following:
- Cooling fan and viscous coupling
- Grille
- Hood release cable strap from the upper radiator support
- Radiator upper support
- Left and right radiator air deflectors
- Washer bottle filler neck
- Engine and transmission oil coolers
- Coolant hoses from the radiator and thermostat housing
- The 2 fog lamp breather hoses from the clips on either side of the radiator
- Power steering fluid
- Transmission oil temperature sensor connector
- Refrigerant lines from the air conditioning condenser
- The 2 nuts and bolts securing the radiator mountings to chassis

5. With the aid of an assistant, raise the radiator assembly for access to the condenser cooling fan connectors.

6. Remove or disconnect the following:
- The 2 condenser cooling fan connectors
- Radiator/condenser/oil cooler assembly
- Window switch pack
- Handbrake and cable clevis pin

- Transmission fluid
- Transfer case fluid
- Engine oil
- Exhaust pipes from the manifolds
- Hand brake cable from the grommet in tunnel
- Rear driveshaft shaft guard
- Driveshafts
- Gear selector cable from the transmission lever
- The 2 bolts securing selector cable bracket to the transmission
- Transfer case fluid temperature sensor
- High/Low motor and output shaft speed sensor connectors
- Gear selection position switch and transmission speed sensor connectors
- Engine wiring harness-to-transmission harness connector
- The 4 nuts securing the engine mounts to the chassis and engine brackets

➡**When attaching the engine lifting chain and hoist, it may be necessary to remove the oil filler cap to prevent it from being damaged. Place cloth over plenum chamber to protect from damage during lifting.**

7. Install a suitable engine lifting chain and hoist.

8. Raise the engine slightly. Ensure that lifting bracket does not foul the bulkhead. Remove both engine mountings.

➡**It may be necessary to lower the transmission support slightly during the above operation.**

9. Raise the engine/transmission and pull forward, while lowering the transmission support.

➡**The engine/transmission must be tilted at an angle of approximately 45 degrees before it can be removed from the engine compartment.**

10. Remove the engine/transmission assembly.

To install:

11. Guide the engine into position.

12. With the aid of an assistant, raise the transmission and lower the engine until the engine mountings can be installed.

13. Attach the engine mounts to the chassis with new flange nuts, but do not tighten at this stage.

14. Lower and guide the engine onto the mounting studs.

15. Remove the engine lifting chain and hoist.

16. Route the transmission wiring harness.

17. The remaining step of the installation is the reverse of the removal procedure while keeping the following items in mind:
- Attach all wiring harness and electrical connectors.
- Align the driveshaft matchmarks and tighten the mounting nuts and bolts to 35 ft. lbs. (48 Nm).
- Attach all cables to the transmission.
- Tighten the engine mounting nuts to 33 ft. lbs. (45 Nm).
- Connect the handbrake cable and associated components.
- Install and connect the radiator/condenser/oil cooler assembly, using new sealing rings.
- Tighten the fittings on the air conditioning compressor to 17 ft. lbs. (23 Nm) and condenser to 11 ft. lbs. (15 Nm).
- Tighten the fittings for the power steering lines 12 ft. lbs. (16 Nm).
- Tighten the oil cooler lines to 22 ft. lbs. (30 Nm).

18. Evacuate and recharge the air conditioning system.

19. Connect the fuel supply and return lines to the fuel rail and tighten to 12 ft. lbs. (16 Nm).

20. Check and correct the fluid levels in the engine, transmission and transfer case.

21. Install and connect the battery.

22. Start the engine. Check for fuel, coolant and oil leaks.

Discovery/Discovery Series II

1. Before servicing the vehicle, refer to the precautions in the beginning of this section.

2. Remove or disconnect the following:
- Negative battery cable
- Coolant
- Fuel system pressure
- Engine oil and filter
- Radiator hoses
- Radiator
- Upper intake manifold and ignition coil assemblies
- Fuel supply and return lines from the fuel rail
- Auxiliary drive belt
- The 3 bolts securing the Active Cornering Enhancement (ACE)

pump, then position it aside leaving the hoses attached.

- The 4 bolts securing the air conditioning compressor, and position it aside leaving the hoses attached.
- Power steering hoses from the pump
- Coolant hoses at the water pump and rail
- Bolt securing the coolant rail and position the rail aside
- Engine ground and power supply cables
- Starter wiring harness
- The 2 engine wiring harness connectors from the fuse box
- EVAP solenoid connector
- Nut mounting the engine harness ground-to-body and detach the engine harness-to-main harness connector.
- Right side interior kick panel
- The 5 connectors attaching the engine harness to the ECM

3. Release the engine wiring harness, pull it into the engine bay and coil on top of engine.

4. Raise and safely support the vehicle.

5. Remove the 3 bolts securing the oil cooler lines to the engine block.

6. Detach the engine oil cooler lines and tie the lines aside.

7. Disconnect the front exhaust pipes from the manifolds.

8. Remove the torque converter access plug, matchmark the torque converter-to-driveplate relationship, then remove the 4 bolts securing the torque converter to the driveplate.

9. Remove the 12 engine-to-transmission mounting bolts.

10. Attach suitable lifting equipment to the engine.

11. Remove the 4 nuts securing the engine mounts, raise the engine and remove the engine mounts.

12. Support the transmission on a jack.

13. Separate the engine from the transmission dowels.

14. With the aid of an assistant, remove the engine from the engine bay.

To install:

15. Clean the mating faces of the engine and transmission, dowel and dowel holes.

16. With the aid of an assistant, position the engine in the engine bay, align to gearbox and locate on dowels.

17. Install the engine-to-transmission bolts and tighten to 37 ft. lbs. (50 Nm).

18. Install the engine mounts and tighten the fasteners to 63 ft. lbs. (85 Nm).

19. Remove the engine lifting equipment.

20. Align the torque converter to the driveplate, install the bolts and tighten to 37 ft. lbs. (50 Nm).

21. Install or connect the following:

- Torque converter access plug
- Front exhaust pipes to the manifolds
- All electrical connectors and replace any ties.
- Power steering hoses and check the fluid level
- Coolant hoses
- Radiator
- Coolant
- Air conditioning compressor mounting bolts to 16 ft. lbs. (22 Nm)
- ACE pump mounting bolts to 16 ft. lbs. (22 Nm)
- Accessory drive belt
- Fuel supply and return lines to the fuel rail
- Ignition coils
- Spark plug wires
- Upper intake manifold
- Oil filter
- Engine oil
- Transmission fluid level
- Negative battery cable

Freelander

➡ **The engine and transaxle are removed as an assembly.**

1. Before servicing the vehicle, refer to the precautions in the beginning of this section.

2. Remove the hood.

3. Remove the battery and battery box.

4. Drain the cooling system and engine oil.

5. Drain the transaxle fluid.

6. Release the fuel system pressure.

7. Remove or disconnect the following:

- Acoustic cover
- ECM
- Engine wiring harness connectors
- Air box
- Fuse box cover
- Battery and starter leads from the fuse box
- All wiring from the fuse box
- Engine ground strap
- Wiring from the starter
- Wiring from the transaxle
- Fuel line from the fuel rail
- Hose from the purge valve
- Throttle cable
- Heater hoses
- Radiator hoses
- All remaining wiring and hoses
- Front wheels

- Halfshaft hub nuts
- Front suspension rear beam
- Exhaust pipe from the manifold

8. Matchmark the driveshaft-to-IRD unit. Remove the 6 nuts and bolts and disconnect the driveshaft from the IRD. Tie it aside.

✳✳ WARNING

Do not allow the driveshaft to hang. Damage to the Tripod joint will occur.

9. Drain the IRD fluid.

10. Remove the right and left splash shields.

11. Remove or disconnect the following:

- Clips securing the brake hoses to the struts
- ABS sensor wiring
- Track rod ball studs from the steering arms
- Halfshafts
- Selector lever from the gearbox
- Accessory drive belts
- A/C compressor and tie it out of the way

12. Attach an engine crane to the engine using an equalizer.

13. Raise the engine just far enough to take the weight off the mounts.

14. Remove the left side engine mount and left transaxle mount.

15. Secure a lifting bracket to the transaxle using the mount bolts.

16. Place a wood block on a jack and raise the transaxle just enough to take its weight. Unhook the chain from the rear engine lift bracket and hook it to the transaxle bracket.

17. Remove the right side engine brackets.

18. Disconnect the power steering pressure hose from the front plate and top arm.

19. Remove the top arm.

20. Remove the power steering pulley.

21. Remove the power steering pump and position it out of the way.

22. Lift the engine from the vehicle.

To install:

23. With assistance, raise engine and gearbox into engine compartment.

24. Position PAS pump to front mounting plate, fit and tighten bolts to 25 Nm (18 ft. lbs.).

25. Position PAS pump pulley, fit and tighten Torx screws to 9 Nm (7 ft. lbs.).

26. Position top arm to engine mounting bracket and right side hydramount, fit and tighten bolts to 100 Nm (74 ft. lbs.).

27. Position PAS pipe support bracket to

RIGHT SIDE hydramount, fit and tighten nut to 85 Nm (63ft. lbs.).

28. Place a wooden block on jack, position jack under gearbox and raise jack sufficient to support weight of gearbox. Release lifting hook from lifting bracket on gearbox and connect to rear engine lifting bracket.

29. Lower and remove jack supporting gearbox.

30. Remove bolts securing lifting bracket to gearbox and remove bracket.

31. Position left side mounting bracket to gearbox, fit and tighten bolts to 85 Nm (63 ft. lbs.).

32. Position LEFT SIDE mounting to body, fit and tighten bolts to 48 Nm.

33. Align gearbox bracket to LEFT SIDE body mounting, fit and tighten through bolt to 100 Nm (74 ft. lbs.).

34. Position upper RIGHT SIDE engine steady to top arm, fit and tighten bolt to 100 Nm (74 ft. lbs.).

35. Tighten bolt securing upper RIGHT SIDE engine steady to body to 100 Nm (74 ft. lbs.).

36. Position PAS pipe to engine front mounting plate, fit and tighten bolt to 25 Nm (18 ft. lbs.).

37. Lower hoist, disconnect and remove lifting bracket.

38. Raise vehicle on ramp.

39. Position A/C compressor to front mounting plate and cylinder block, align heat shield, fit and tighten bolts to 25 Nm (18 ft. lbs.).

40. Connect multiplug to A/C compressor.

41. Using a ⅜ square drive socket bar, raise ancillary drive belt tensioner and fit drive belt to pulleys.

42. Position selector cable to gearbox bracket and secure with clip.

43. Position selector lever to selector shaft, fit and tighten nut to 25 Nm (18 ft. lbs.).

44. Clean splines and seal areas on each driveshaft and mating faces in front hubs.

45. Install new circlips to RIGHT SIDE and LEFT SIDE driveshaft inner joint splines.

46. Install driveshafts to IRD and gearbox, ensuring that the circlip on each driveshaft is fully engaged.

47. Engage LEFT SIDE and RIGHT SIDE driveshafts into front hubs.

48. Install new driveshaft flange nuts but do not tighten at this stage.

49. Clean ball joint tapers and taper seats.

50. Position LEFT SIDE and RIGHT SIDE track rod ends to steering arms, fit new nuts and tighten to 55 Nm (40 ft. lbs.).

51. Clean ABS sensors and mating faces.

52. Apply anti-seize grease to both ABS sensors and position sensors in front hubs.

➡**Ensure ABS sensor is fully located into hub, so that sensor touches pole wheel teeth.**

53. Position LEFT SIDE and RIGHT SIDE brake hoses to front damper brackets and secure with clips.

54. Position LEFT SIDE and RIGHT SIDE splash shields, fit and tighten bolts.

55. Ensure mating face of propeller shaft and IRD drive flange are clean.

56. Install propeller shaft to IRD flange and align marks. Tighten nuts and bolts to 42 Nm (31 ft. lbs.).

57. Install rear beam.

58. Fill IRD to correct level with fluid.

59. Install exhaust front pipe.

60. With assistant depressing the brake pedal, tighten front hub nuts to 400 Nm (295 ft. lbs.).

61. Stake nut to shaft.

62. Install front road wheels, fit and tighten nuts to 115 Nm (85 ft. lbs.).

63. Lower vehicle on ramp.

64. Connect brake servo vacuum hose to inlet manifold chamber.

65. Connect coolant hose to underside of expansion tank and secure with clip.

66. Connect expansion tank hose to inlet manifold and secure clip.

67. Connect top hose to radiator and secure with clip. Position hose in bracket.

68. Connect heater feed and return hoses and secure with clips.

69. Connect throttle inner cable to throttle cam and secure outer cable in abutment bracket, if fitted.

70. Secure throttle cable in clips on harness brackets, if fitted.

71. Adjust throttle cable, if fitted.

72. Connect hose to purge control valve.

73. Connect fuel hose to fuel rail pipe, fit rubber sleeve over hose connector.

74. Connect gearbox harness multiplugs and secure multiplugs in mounting bracket clips.

75. Connect Lucar connector to starter solenoid.

76. Position ground lead to gearbox housing, fit and tighten bolt to 25 Nm (18 ft. lbs.).

77. Position engine harness to air box mounting bracket and secure with clips.

78. Connect ground header multiplug.

79. Connect multiplug to underhood fuse box.

80. Position battery and starter motor lead to underhood fuse box, fit and tighten bolts to 8 Nm (6 ft. lbs.).

81. Install underhood fuse box cover.

82. Position air box, secure in retaining clip, fit and tighten nut to 9 Nm (7 ft. lbs.).

83. Position carrier in air box and secure with clips.

84. Position and secure air duct and harness rubber sleeve in air box.

85. Connect multiplugs securing main harness to engine harness.

86. Position ECM harness and multiplug to air box, align harness clamp and secure screws to air box.

87. Install engine ECM.

88. Install battery carrier.

89. Fill cooling system.

90. Connect battery ground lead.

91. Fill gearbox with fluid.

92. Install engine acoustic cover.

Water Pump

REMOVAL & INSTALLATION

Freelander

1. Before servicing the vehicle, refer to the precautions in the beginning of this section.

2. Disconnect battery ground lead.

3. Drain cooling system.

4. Remove camshaft timing belt.

5. Remove and discard 7 bolts securing coolant pump to cylinder block.

6. Release coolant pump from cylinder block and remove.

7. Remove and discard O-ring from coolant pump.

To install:

8. Clean coolant pump and mating face of cylinder block.

9. Lubricate new O-ring with rubber grease and fit to coolant pump.

10. Fit coolant pump to cylinder block. Fit new bolts and tighten progressively in the sequence illustrated to 9 Nm (7 ft. lbs.).

11. Fit camshaft timing belt.

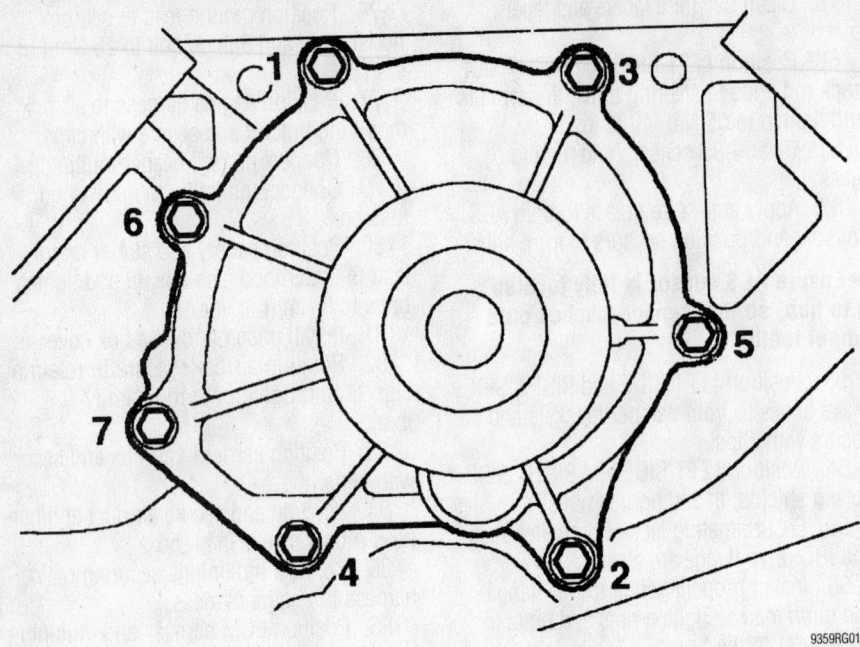

Water pump mounting bolt location torque sequence—Freelander

9359RG01

12. Fill cooling system.
13. Connect battery ground lead.

Discovery/Discovery Series II and Range Rover

1. Before servicing the vehicle, refer to the precautions in the beginning of this section.
2. Disconnect the negative battery cable.
3. Drain and recycle the engine coolant.
4. Remove the accessory drive belt.

5. Remove the 3 bolts securing the pulley to the coolant pump, and remove the pulley.
6. Release the clip and disconnect the feed hose from the coolant pump.
7. Remove the 9 bolts securing the coolant pump, remove the pump and discard the gasket.
To install:
8. Clean the coolant pump and mating face.

9. Install new gasket and coolant pump to the cylinder block.
10. Install the mounting bolts and tighten to 16 ft. lbs. (22 Nm).
11. Connect the feed hose to the coolant pump, and secure with clip.
12. Ensure the mating faces of coolant pump pulley and flange are clean.
13. Install the pulley, and tighten the bolts to 16 ft. lbs. (22 Nm).
14. Install the auxiliary drive belt.
15. Connect the negative battery cable.

Cylinder Head

REMOVAL & INSTALLATION

Freelander

LEFT SIDE

1. Before servicing the vehicle, refer to the precautions in the beginning of this section.
2. Disconnect battery ground lead.
3. Drain cooling system.
4. Remove inlet manifold chamber.
5. Remove left side exhaust manifold gasket.
6. Remove camshaft timing belt.
7. Remove 4 bolts securing left side front camshaft timing belt cover backplate to cylinder head and remove backplate.
8. Disconnect 3 high tension leads from spark plugs.
9. Release high tension leads from guide bracket.
10. Remove special bolt securing high tension lead guide bracket to left side camshaft cover and remove bracket.
11. Disconnect CMP sensor multiplug and release male multiplug from mounting bracket.
12. Remove bolt securing multiplug bracket, remove bracket.
13. Remove 3 nuts and 3 bolts securing ignition coils to left side inlet manifold and release coil ground lead.
14. Remove ignition coils and position aside.
15. Release clip and disconnect coolant bleed hose from inlet manifold and move hose aside.
16. Depress locking collar and release breather hose from left side inlet manifold.
17. Disconnect clips securing right side injector harness to injector protection cover.

➡**Protection covers are NOT fitted to vehicles.**

Water pump mounting bolt location—Except Freelander

9302RG03

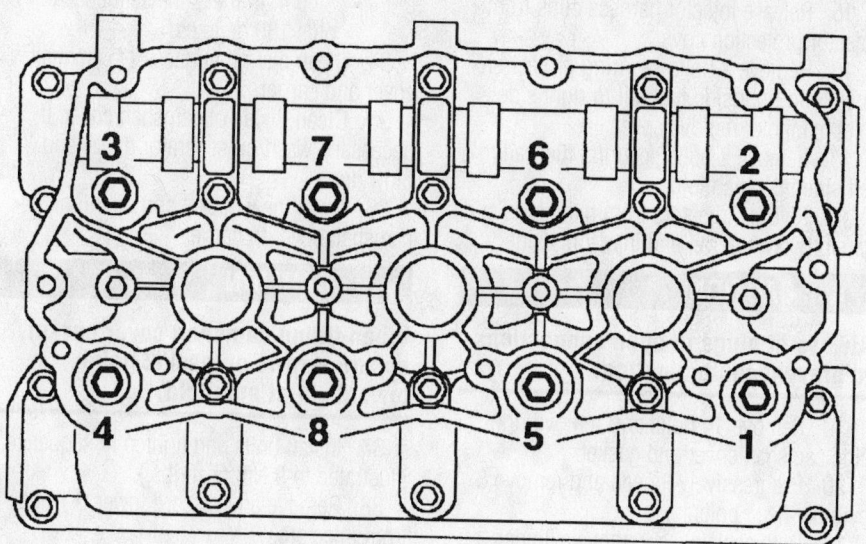

Cylinder head bolt loosening sequence—Freelander

9359RG06

18. Remove 2 bolts securing protection cover and fuel rail to right side inlet manifold, remove cover.

19. Remove 2 bolts securing fuel rail to left side inlet manifold.

20. Release injectors from manifolds and carefully lay fuel rail and injectors aside.

✳✳ CAUTION

Always fit plugs to open connections to prevent contamination.

21. Remove 14 bolts and remove left side camshaft cover and gasket.

22. Progressively loosen and remove 8 cylinder head bolts.

23. With assistance, remove cylinder head assembly.

✳✳ CAUTION

Support both ends of cylinder head on blocks of wood.

24. Remove and discard cylinder head gasket.

25. Install cylinder liner clamps LRT-12-144 and secure with bolts. Ensure that feet of clamps do not protrude over bores.

To install:

26. Clean cylinder head face.

27. Remove bolts securing cylinder liner clamps LRT-12-144 to cylinder block and remove clamps.

28. Clean cylinder block face, dowels and dowel holes.

29. Clean cylinder head bolts and wipe dry.

30. Lightly lubricate threads and beneath heads of cylinder head bolts with clean engine oil.

31. Ensure locating dowels are in cylinder block.

32. Install new cylinder head gasket onto cylinder block with the word 'TOP' uppermost.

33. With assistance, fit cylinder head and carefully position left side inlet to right side inlet manifold, cylinder head to dowels.

34. Carefully enter cylinder head bolts, DO NOT DROP. Screw bolts into place by hand.

35. Tighten cylinder head bolts progressively in the sequence illustrated to:
- Step 1: 25 Nm (18 ft. lbs.)
- Step 2: Tighten in the same sequence to 25 Nm (18 ft. lbs.)
- Step 3: Tighten in the same sequence to 25 Nm (18 ft. lbs.)
- Step 4: In the same sequence to plus 180 degrees.

36. Clean mating surfaces of camshaft cover and carrier.

37. Clean inside of camshaft cover. If necessary, wash oil separator gauze and blow dry.

38. Install gasket and position camshaft cover to carrier.

✳✳ CAUTION

When fitting camshaft cover gasket, ensure arrows on gasket point towards inlet manifold.

39. Install bolts and tighten in sequence illustrated to 9 Nm (7 ft. lbs.).

40. Remove and discard O-rings from injectors.

41. Clean injectors and injector locations in fuel rail.

42. Lubricate new O-rings with castor oil and fit to injectors.

43. Position fuel rail assembly and secure injectors to inlet manifolds.

44. Position injector protection cover to right side fuel rail and secure injector harness to protection cover with clips.

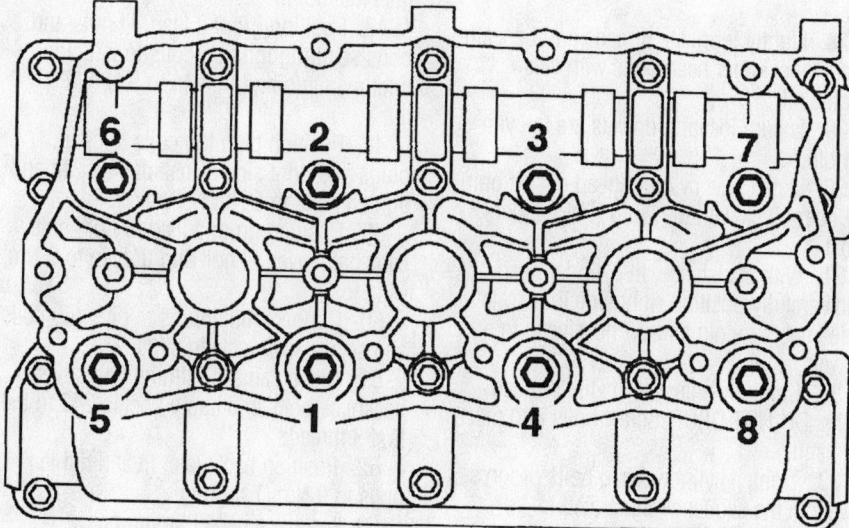

Cylinder head bolt torque sequence—Freelander

9359RG05

Timing belt service is covered in Section 3 of this manual

45. Install bolts securing fuel rail to inlet manifolds and tighten to 9 Nm (7 ft. lbs.).

46. Position CMP sensor bracket, fit and tighten bolt to 9 Nm (7 ft. lbs.).

47. Connect CMP sensor multiplug and secure to bracket.

48. Connect breather hose to left side inlet manifold.

49. Connect coolant bleed hose to inlet manifold and secure with clip.

50. Position ignition coils to left side inlet manifold.

51. Position ground lead, fit nuts and bolts securing ignition coils to left side inlet manifold and tighten to 9 Nm (7 ft. lbs.).

52. Position high tension lead guide bracket, fit and tighten special bolt.

53. Connect high tension leads to spark plugs and secure high tension leads in guide clips.

54. Clean camshaft timing belt cover backplate bolts and apply Loctite 242 to the first 3 threads.

55. Position backplate, fit and tighten bolts to 9 Nm.

56. Install camshaft timing belt.

57. Install left side exhaust manifold gasket.

58. Install inlet manifold chamber.

59. Fill cooling system.

RIGHT SIDE

1. Before servicing the vehicle, refer to the precautions in the beginning of this section.

2. Disconnect battery ground lead.

3. Drain cooling system.

4. Remove inlet manifold chamber.

5. Remove right side exhaust manifold gasket.

6. Remove camshaft timing belt.

7. Remove 4 bolts securing right side front timing belt cover backplate to cylinder head and remove backplate.

8. Depress locking collars and release breather hoses from right side camshaft cover and right side inlet manifold.

9. Release 3 locking clips and disconnect 3 multiplugs from plug top coils.

10. Remove bolt securing ground lead to right side camshaft cover and release ground lead.

11. Remove 6 bolts securing plug top coils to camshaft cover and remove coils.

12. Remove 3 nuts and 3 bolts securing ignition coils to left side inlet manifold and release coil ground lead.

13. Remove ignition coils and position aside.

14. Release clip and disconnect coolant bleed hose from right side inlet manifold, position hose aside.

15. Release injector harness clips from injector protection cover.

16. Remove 2 bolts securing protection cover and right side fuel rail to right side inlet manifold, remove cover.

17. Remove 2 bolts securing fuel rail to left side inlet manifold.

18. Release injectors from manifolds and carefully lay fuel rail and injectors aside.

✲✲ CAUTION

Always fit plugs to open connections to prevent contamination.

19. Remove 14 bolts and remove right side camshaft cover and gasket.

20. Progressively loosen and remove 8 cylinder head bolts.

21. With assistance, remove cylinder head assembly.

✲✲ CAUTION

Support both ends of cylinder head on blocks of wood.

22. Remove and discard cylinder head gasket.

23. Install cylinder liner clamps LRT-12-144 and secure with bolts. Ensure that feet of clamps do not protrude over bores.

To install:

24. Clean cylinder head face.

25. Remove bolts securing cylinder liner clamps LRT-12-144 to cylinder block and remove clamps.

26. Clean cylinder block face, dowels and dowel holes.

27. Clean cylinder head bolts and wipe dry.

28. Lightly lubricate threads and beneath heads of cylinder head bolts with clean engine oil.

29. Ensure locating dowels are in cylinder block.

30. Install new cylinder head gasket onto cylinder block with the word 'TOP' uppermost.

31. With assistance, fit cylinder head and carefully position right side inlet to left side inlet manifold and cylinder head to dowels.

32. Carefully enter new cylinder head bolts, DO NOT DROP. Screw bolts into place by hand.

33. Tighten cylinder head bolts progressively in the sequence illustrated to:
- Step 1: 25 Nm (18 ft. lbs.)
- Step 2: Tighten in the same sequence to 25 Nm (18 ft. lbs.)
- Step 3: Tighten in the same sequence to 25 Nm (18 ft. lbs.)
- Step 4: In the same sequence to plus 180 degrees.

34. Clean mating surfaces of camshaft cover and carrier.

35. Clean inside of camshaft cover. If necessary, wash oil separator gauze and blow dry.

36. Install new gasket and position camshaft cover to carrier.

✲✲ CAUTION

When fitting camshaft cover gasket, ensure arrows on gasket point towards inlet manifold.

37. Install bolts and tighten in sequence illustrated to 9 Nm (7 ft. lbs.).

38. Remove and discard lower O-rings from injectors.

39. Clean injectors and injector locations in fuel rail.

40. Lubricate new O-rings with castor oil and fit to injectors.

41. Position fuel rail assembly and secure injectors to inlet manifolds.

42. Position injector protection cover to right side fuel rail and secure injector harness to protection cover with clips.

43. Install bolts securing fuel rail to inlet manifolds and tighten to 9 Nm (7 ft. lbs.).

44. Connect coolant bleed hose to inlet manifold and secure with clip.

45. Connect breather hoses to right side camshaft cover and inlet manifold.

46. Position ignition coils to left side inlet manifold.

47. Position ground lead, fit nuts and bolts securing ignition coils to left side inlet manifold and tighten to 9 Nm (7 ft. lbs.).

48. Position plug top coils to spark plugs and right side camshaft cover, fit and tighten bolts to 9 Nm (7 ft. lbs.).

49. Position ground lead to right side camshaft cover, fit bolt and tighten to 9 Nm (7 ft. lbs.).

50. Connect multiplugs to plug top coils and secure with locking clips.

51. Clean camshaft timing belt cover backplate bolts and apply Loctite 242 to the first 3 threads.

52. Position backplate, fit and tighten bolts to 9 Nm (7 ft. lbs.).

53. Install camshaft timing belt.

54. Install right side exhaust manifold gasket.

55. Install inlet manifold chamber.

56. Fill cooling system.

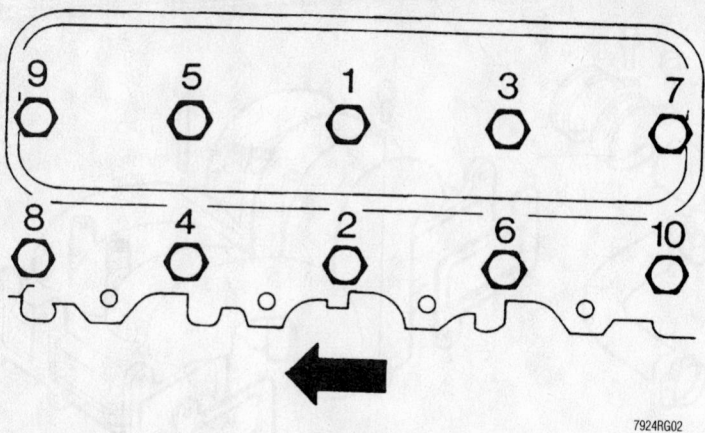

7924RG02

Cylinder head bolt torque sequence—4.0L and 4.6L engines

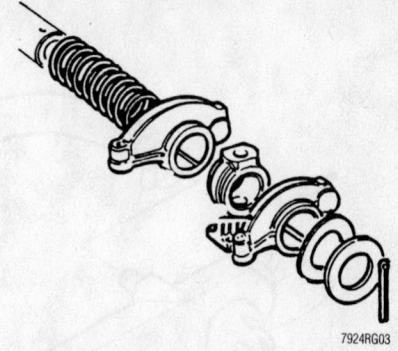

7924RG03

Exploded view of the rocker arm shaft components—Except Freelander

Discovery/Discovery Series II and Range Rover

1. Before servicing the vehicle, refer to the precautions in the beginning of this section.

2. Remove or disconnect the following:
- Negative battery cable
- Coolant
- Intake manifold
- Alternator
- Air conditioning compressor and position it aside leaving the hoses attached.
- Rocker arm cover
- Rocker shafts and pushrods
- Front exhaust pipes from the manifolds
- Exhaust manifolds
- Air cleaner and intake air duct assembly

3. On the left cylinder head, remove the ground cable attached to the left-hand cylinder head.

4. On the right cylinder head, remove the breather pipe from the engine lifting bracket.

5. Loosen the cylinder head bolts, reversing the tightening sequence.

6. Lift the cylinder head off the engine block.

To install:

7. Remove the cylinder head gaskets, then clean the gasket mating surfaces.

8. Clean the exhaust gasket mating faces.

9. Ensure that the bolt holes in cylinder block are clean and dry.

10. Install new cylinder head gaskets with the word TOP facing out.

11. Clean, then lightly oil the threads of the head bolts.

12. Install the cylinder heads on the block.

13. Install the cylinder head bolts in the positions as illustrated.
- 96mm long bolts: 2, 4, 6, 7, 8, 9, 10
- 66mm long bolts: 1, 3, 5

➡**There are no bolts fitted in the 4 lower holes in each cylinder head.**

14. Tighten the cylinder head bolts progressively in sequence, as shown, as follows:
- 15 ft. lbs. (20 Nm)
- + ¼ turn (90°)
- + ¼ (90°)

15. The remaining steps are the reverse of the removal procedure.

16. Connect the negative battery cable.

17. Fill and bleed the cooling system. Run the engine and check for leaks.

Rocker Arms/Shafts

REMOVAL & INSTALLATION

Discovery/Discovery Series II and Range Rover

1. Before servicing the vehicle, refer to the precautions in the beginning of this section.

2. Disconnect the negative battery cable.

3. Relieve fuel system pressure.

4. Drain the cooling system.

5. Remove the rocker arm covers.

6. Remove the 4 rocker arm shaft bolts and lift off the rocker arm shaft assemblies.

7. Remove the cotter pin from either end of the shaft and slide the components from the shaft, keeping them in order for reassembly.

8. Inspect the parts for wear or damage,

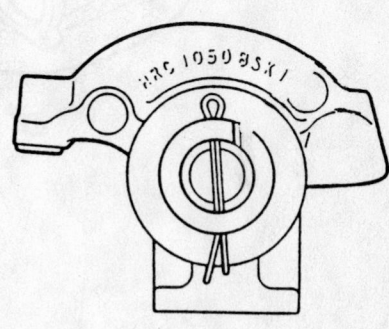

7924RG04

The end notches of the rocker shaft must face up—Except Freelander

and replace any suspect parts. Discard any weak springs.

To install:

9. Reassemble the shaft and components. Note the position of the oil feed holes. Use new cotter pin(s).

10. Position the rocker shaft assembly on the head. Be sure that the shaft is installed with the notches at each end on the upper side. Be sure that each rocker ball stud engages its respective pushrod.

11. Install the bolts and tighten them, gradually, to 28 ft. lbs. (38 Nm), starting with the 2 inner, then the 2 outer bolts.

Intake Manifold

REMOVAL & INSTALLATION

Freelander

LEFT SIDE

1. Before servicing the vehicle, refer to the precautions in the beginning of this section.

2. Disconnect battery ground lead.

3. Drain cooling system.

Heater Core replacement is covered in Section 2 of this manual

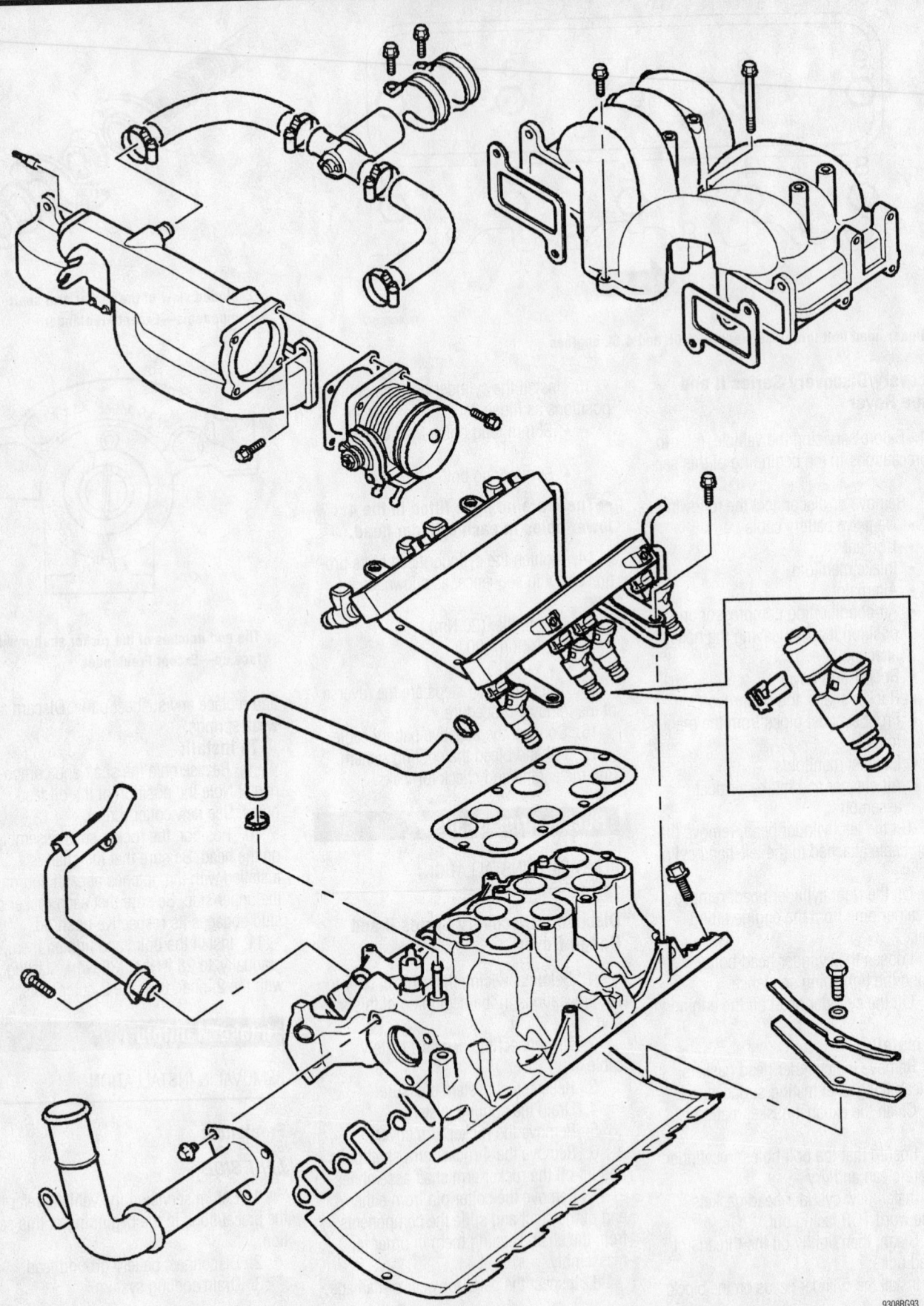

Intake manifold components—Except Freelander

9308RG93

4. Remove fuel rail.

5. Depress locking collar and release breather hose from left side inlet manifold.

6. Release clip and disconnect coolant hose from left side inlet manifold.

7. Progressively loosen and remove 7 bolts securing left side inlet manifold to cylinder head.

8. Remove inlet manifold and discard gasket.

To install:

9. Clean inlet manifold and cylinder head mating faces.

10. Install new inlet manifold gasket to cylinder head, position inlet manifold, fit and tighten bolts to 25 Nm (18 ft. lbs.).

11. Connect breather hose to left side inlet manifold.

12. Connect coolant hose to inlet manifold and secure clip.

13. Install fuel rail.

14. Fill cooling system.

15. Connect battery ground lead.

RIGHT SIDE

1. Before servicing the vehicle, refer to the precautions in the beginning of this section.

2. Disconnect battery ground lead.

3. Drain cooling system.

4. Remove fuel rail.

5. Release clip and disconnect coolant hose from right side inlet manifold.

6. Depress locking collar and release breather hose from right side inlet manifold.

7. Progressively loosen and remove 7 bolts securing right side inlet manifold to cylinder head.

8. Remove inlet manifold and gasket.

9. Remove bolt securing coolant breather hose mounting bracket to inlet manifold and remove bracket.

To install:

10. Clean inlet manifold and cylinder head face.

11. Position coolant/breather hose bracket to inlet manifold, fit and tighten bolt to 9 Nm (7 ft. lbs.).

12. Install new inlet manifold gasket to cylinder head, position inlet manifold, fit and tighten bolts to 25 Nm (18 ft. lbs.).

13. Connect coolant hose to inlet manifold and secure clip.

14. Connect breather hose to manifold and secure into collar.

15. Install fuel rail.

16. Fill cooling system.

17. Connect battery ground lead.

Discovery/Discovery Series II and Range Rover

1. Before servicing the vehicle, refer to the precautions in the beginning of this section.

2. Remove or disconnect the following:
- Negative battery cable
- Coolant
- Fuel system pressure
- Alternator
- Air intake hose from the plenum chamber
- Throttle and cruise control cables from the throttle linkage and mounting brackets
- Breather hose from the plenum chamber
- Purge hose from plenum chamber
- TP sensor
- Stepper motor electrical connector
- The 6 bolts securing the plenum chamber to the ram housing
- Coolant hoses attached to the plenum chamber and intake manifold
- Plenum chamber assembly
- Breather hoses, as necessary
- ECT and temperature gauge sensors
- The 8 fuel injector connectors
- Fuel temperature sensor connector
- Fuel supply and return lines
- The 6 nuts securing fuel rail and ignition coil bracket to the inlet manifold

3. Lift the fuel rail slightly to remove the ignition coil bracket from the inlet manifold studs and place aside.

4. Using the sequence shown, remove the 12 bolts securing the intake manifold to the cylinder heads, then remove the manifold.

5. Remove the bolts and clamps securing the manifold gasket to the cylinder block.

6. Remove the inlet manifold gasket and discard.

7. Remove the gasket seals and discard.

To install:

8. Clean all gasket mating faces.

9. Apply a thin bead of Loctite® Superflex (black) sealant to the 4 notches between cylinder head and block.

10. Install a new gasket seals. Be sure the ends engage correctly in the notches.

11. Install a new inlet manifold gasket and tighten the manifold gasket clamps to 6 inch lbs. (0.7 Nm).

12. With the aid of an assistant, hold the harness and ignition coils aside, position the inlet manifold assembly.

➡**Tighten the manifold bolts in the reverse order of removal.**

13. Install the inlet manifold bolts and tighten as follows:
- 84 inch lbs. (10 Nm).
- 37 ft. lbs. (50 Nm).

14. Tighten the gasket clamp bolts to 13 ft. lbs. (17 Nm).

15. Install or connect the following:
- Ignition coil bracket on the inlet

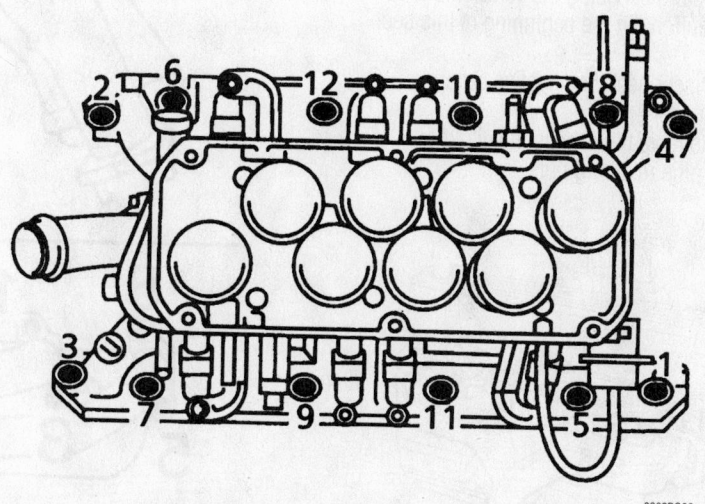

9302RG06

Intake manifold bolt removal sequence—Except Freelander

Brake service is covered in Section 4 of this manual

manifold studs, and tighten the mounting bolts to 72 inch lbs. (8 Nm).

- Fuel lines
- Connectors to the fuel injectors and fuel temperature sensor
- ECT sensor and temperature gauge sensor
- Breather hoses as removed

16. Clean the mating faces of the plenum chamber.

17. Connect any coolant hoses that were removed.

18. Apply a thin, uniform coating of Loctite® 577 sealant to the mating face of the plenum chamber.

19. Install or connect the following:

- Plenum chamber to the ram housing and tighten to 18 ft. lbs. (24 Nm)
- Stepper motor and TP sensor connectors
- Purge hose to the plenum chamber
- Breather hose to the plenum chamber
- Throttle and cruise control cables
- Air intake hose to the plenum chamber and secure with clip
- Coolant
- Alternator
- Negative battery cable

Exhaust Manifold

REMOVAL & INSTALLATION

Discovery/Discovery Series II

1. Before servicing the vehicle, refer to the precautions in the beginning of this section.

2. Disconnect the negative battery cable.

3. Remove the exhaust manifold-to-exhaust pipe retaining nuts.

4. Remove the 8 manifold bolts. Remove the exhaust manifold and discard the gasket.

To install:

5. Clean all mating surfaces.

6. Place the exhaust manifold in position on the cylinder head along with a new gasket. Install the exhaust manifold bolts and tighten to 15 ft. lbs. (20 Nm).

7. Install the exhaust pipe to the exhaust manifold, and tighten the exhaust pipe retaining nuts to 25–34 ft. lbs. (34–47 Nm).

8. Lower the vehicle.

9. Connect the negative battery cable.

10. Start the engine and check for exhaust leaks.

11. Road test the vehicle and check for proper engine operation.

Range Rover

1. Before servicing the vehicle, refer to the precautions in the beginning of this section.

2. Remove or disconnect the following:
- Negative battery cable
- Front exhaust pipes from the exhaust manifolds

3. Lower the vehicle.

4. For the left manifold, perform the following:

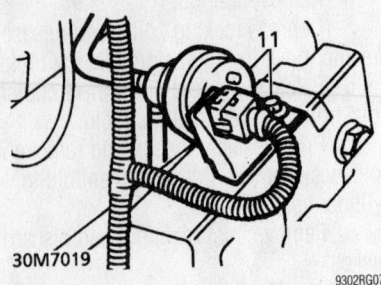

30M7019 9302RG07

Purge valve mounting bolt location— Range Rover

a. Release the intake hose from the plenum chamber.

b. Release the harness from the intake hose clip.

c. Remove the air flow meter.

d. Disconnect the purge valve (11) and position the valve aside.

5. For the right manifold, perform the following:

a. Disconnect the spark plug wires and position them out of the way.

b. Unscrew the right shock absorber top mounting bolt to provide additional clearance for the removal of the heat shield.

6. Remove or disconnect the following:

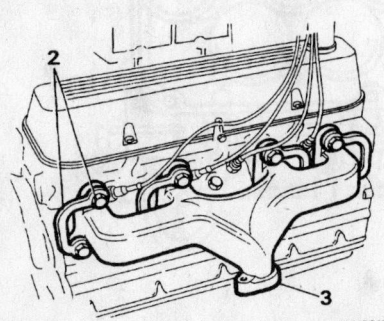

7924RG07

View of one of the exhaust manifolds (3), showing the mounting bolt locktabs (2)— Discovery/Discovery Series II models

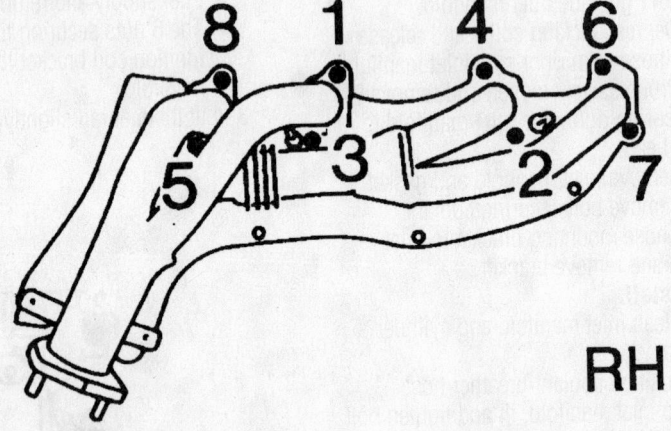

RH

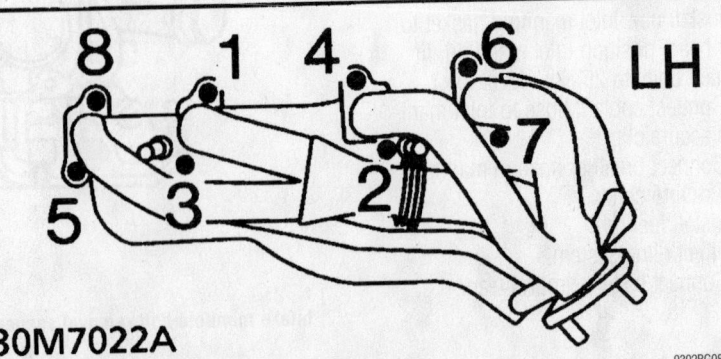

LH

30M7022A 9302RG08

Exhaust manifold torque sequence—Range Rover model

- The 8 bolts (for the right manifold) or 7 bolts (for the left manifold) securing the outer heat shield to the manifold.
- The 8 exhaust manifold-to-cylinder head mounting bolts, then remove manifolds.

To install:

7. Ensure all gasket mating surfaces are clean.

8. Install or connect the following:

- Manifolds using new gaskets. Tighten the bolts 40 ft. lbs. (55 Nm), in the sequence as shown.
- Heat shields and tighten the mounting bolts to 72 inch lbs. (8 Nm).
- Purge valve on the shock absorber turret
- Air flow meter
- Right shock absorber top mounting bolt to 63 ft. lbs. (85 Nm).
- Spark plug wires
- Front exhaust pipes to the manifolds and tighten the mounting nuts to 37 ft. lbs. (50 Nm).
- Negative battery cable

Freelander

RIGHT SIDE

1. Before servicing the vehicle, refer to the precautions in the beginning of this section.

2. Disconnect battery ground lead.

3. Remove right side catalytic converter.

4. Remove 4 nuts securing exhaust manifold to cylinder head, remove manifold and discard gasket.

To install:

5. Clean exhaust manifold and mating face.

6. Using new gasket, fit exhaust manifold and tighten nuts to 45 Nm (33 ft. lbs.).

7. Fit right side catalytic converter.

8. Connect battery ground lead.

LEFT SIDE

1. Before servicing the vehicle, refer to the precautions in the beginning of this section.

2. Position vehicle on 4 post ramp.

3. Disconnect battery ground lead.

4. Remove engine acoustic cover.

5. Remove underbelly panel.

6. Using a ⅜ in. socket bar raise ancillary drive belt tensioner and release drive belt from alternator and PAS pump pulleys.

7. Remove 3 bolts securing A/C compressor to fixing brackets and position compressor aside.

8. Remove compressor heat shield.

9. Release and disconnect HO2S multiplug.

10. Release HO2S harness from clip.

11. Release and disconnect post catalyst HO2S multiplug.

12. Release HO2S harness from clips.

13. Remove 2 nuts securing exhaust front pipe to exhaust manifold.

14. Remove 4 nuts securing manifold and remove exhaust manifold.

15. Remove and discard manifold and down pipe gaskets.

To install:

16. Clean exhaust manifold and mating faces.

17. Using new gaskets, fit exhaust manifold and tighten nuts to cylinder head to 45 Nm (33 ft. lbs.) and nuts to front pipe to 50 Nm (37 ft. lbs.).

18. Connect HO2S multiplugs, secure multiplugs to brackets and harness to clips.

19. Clean compressor and mounting bracket mating faces.

20. Fit compressor heat shield.

21. Position compressor to mounting and align heat shield. Fit and tighten bolts to 25 Nm (18 ft. lbs.).

22. Using a ⅜ in. square drive socket bar, raise ancillary drive belt tensioner and fit drive belt to pulleys.

23. Fit underbelly panel.

24. Fit engine acoustic cover.

25. Connect battery ground lead.

REMOVAL & INSTALLATION

Freelander

LEFT SIDE

1. Before servicing the vehicle, refer to the precautions in the beginning of this section.

2. Disconnect battery ground lead.

3. Remove camshaft timing belt.

4. Position LRT-12-175 to rear camshaft gears as illustrated, remove and discard bolts securing gears to camshafts.

5. If camshaft timing belt is to be refitted, mark direction of rotation on timing belt.

6. Remove rear camshaft gears and drive belt as an assembly.

7. Depress locking collars and release 2 breather hoses from left side camshaft cover.

8. Disconnect 3 high tension leads from spark plugs.

9. Release high tension leads from guide bracket.

10. Remove special bolt securing high tension lead guide bracket to left side camshaft cover and position ht leads aside.

11. Disconnect CMP sensor multiplug and release male multiplug from mounting bracket.

12. Remove bolt securing multiplug bracket which also retains front right side

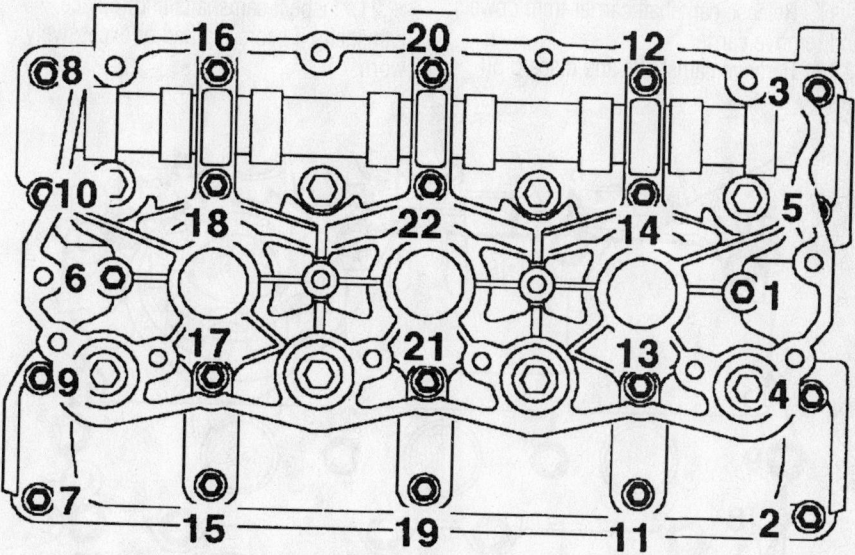

Camshaft carrier bolt loosening sequence—Freelander

9359RG07

For complete Engine Mechanical specifications, see Section 1 of this manual

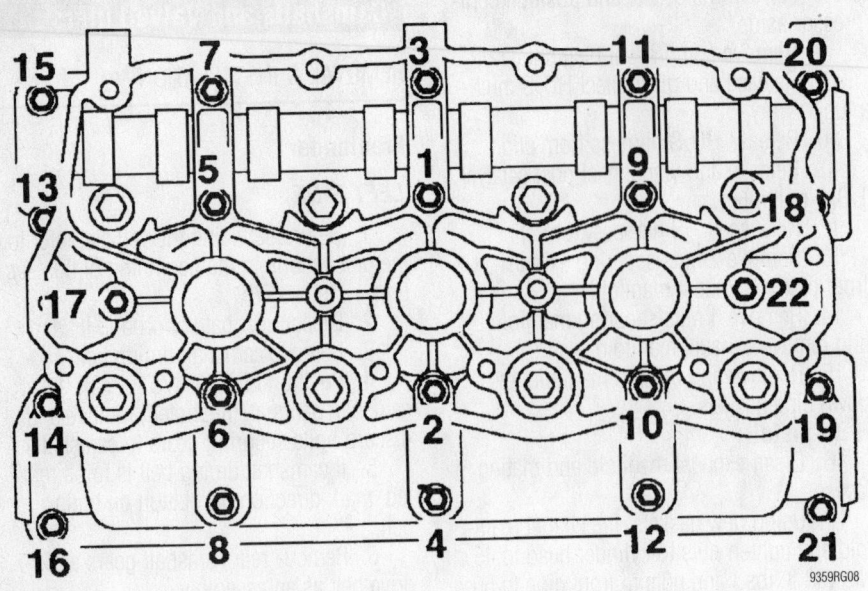

Camshaft carrier bolt torque sequence—Freelander

O$_2$S multiplug, move bracket and multiplug aside.

13. Remove 14 bolts and remove left side camshaft cover and gasket.

14. Remove 4 bolts and remove front camshaft drive belt cover backplate from cylinder head.

15. Remove 2 bolts securing left side rear camshaft drive belt cover backplate to cylinder head and remove backplate.

16. Using sequence shown, progressively loosen 22 bolts securing camshaft carrier to cylinder head until valve spring pressure is released, remove bolts.

17. Release camshaft carrier from dowels and remove carrier.

18. Remove camshafts and discard oil seals.

19. Using a stick magnet, remove 12 hydraulic tappets from cylinder head.

✳✳ CAUTION

Store hydraulic tappets in their fitted order and store upright. Maintain absolute cleanliness when handling hydraulic tappets. Failure to observe these precautions can result in engine failure.

20. Clean camshafts and bearing running surfaces in camshaft carrier and cylinder head.

21. Inspect camshafts and replace camshafts if scored, pitted or excessively worn.

22. Use a lint-free cloth and solvent, clean sealing surfaces on cylinder head and camshaft carrier.

23. Position camshafts in cylinder head and place Plastigage® across each journal.

24. Refit camshaft carrier and tighten bolts in sequence to 10 Nm (7.5 ft. lbs.). DO NOT rotate camshaft.

25. Using the sequence shown, progressively loosen and remove camshaft carrier bolts.

Release and remove camshaft carrier from cylinder head.

26. Measure widest part of Plastigage® on each journal. Camshaft bearing clearance = 0.025 to 0.059 mm. If clearance is excessive, fit new camshaft and repeat check. If still out of tolerance, renew cylinder head.

To install:

27. Remove all traces of Plastigage® from camshaft bearing journals.

28. Thoroughly clean and lubricate hydraulic tappets with clean engine oil. Fit hydraulic tappets to original bores in cylinder head.

29. Ensure that mating faces of camshaft carrier and cylinder head are clean and dry.

30. Lubricate camshafts and bearing journals with clean engine oil.

31. Position camshafts in cylinder head with rear timing gear drive slots in each camshaft facing towards the center as shown.

32. Apply continuous thin beads of sealant, Part No. STC 4600, to paths on camshaft carrier as shown. Spread sealant to an even film using a roller.

✳✳ CAUTION

To avoid contamination, assembly should be completed immediately after application of sealant.

33. Position camshaft carrier, fit and progressively tighten bolts in the sequence shown to 10 Nm (7.5 ft. lbs.).

34. Noting that the front camshaft oil seals are black in color and the rear oil seals are red, fit new camshaft oil seals using LRT-12-203 and LRT 12-148A

✳✳ CAUTION

Oil seal is waxed on outer diameter and must not be lubricated before fitting.

35. Clean camshaft timing belt cover backplate bolts and apply Loctite 242 to the first 3 threads.

36. Position camshaft timing belt rear

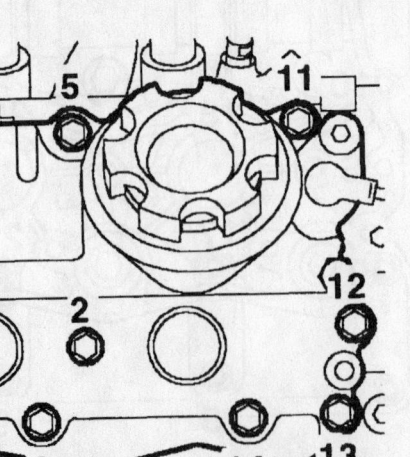

Camshaft cover bolt torque sequence—Freelander

cover backplates to cylinder head, fit and tighten bolts to 9 Nm (7 ft. lbs.).

37. Clean mating surfaces of camshaft cover and carrier.

38. Clean inside of camshaft cover. If necessary, wash oil separator elements in solvent and blow dry.

39. Position new gasket to camshaft carrier with arrows on gasket pointing towards inlet manifold.

40. Position camshaft cover, fit and tighten bolts progressively in the sequence shown to 9 Nm (7 ft. lbs.).

41. Position multiplug bracket to camshaft cover, fit and tighten bolt to 9 Nm (7 ft. lbs.).

42. Secure male end of CMP sensor multiplug to bracket and connect multiplug.

43. Position high tension lead guide bracket, fit and tighten special bolt.

44. Connect high tension leads to spark plugs and secure ht leads in guide clips.

45. Connect breather hoses to left side camshaft cover.

46. Fit camshaft timing belt.

47. Clean rear camshaft gears and mating faces on camshafts.

48. Place gears inverted on a flat surface, with the locating lugs on the gears positioned as illustrated.

49. Keeping the timing marks aligned, position timing belt onto gears.

50. Position LRT-12-195 between the gears, turn center nut sufficiently to spread drive belt and position LRT-12-175 to camshaft gears, remove LRT-12-195 from between camshaft gears.

51. Fit LRT-12-198 alignment pins into the end of each camshaft.

52. Position timing belt and gears over LRT-12-198 and locate gears onto camshafts.

53. Position LRT-12-197 into the front end of the left side exhaust camshaft.

54. With assistance, using a 30mm socket on LRT-12-197, turn the left side exhaust camshaft sufficiently to align camshaft gears to the drive slots in each camshaft.

55. Remove LRT-12-198 alignment pins and fit new camshaft gear retaining bolts.

56. Tighten camshaft gear bolts to 27 Nm (20 ft. lbs.) then a further 90 degrees. Remove LRT-12-175 from camshaft gears.

57. Remove LRT-12-197 from front end of exhaust camshaft.

58. Clean left side exhaust camshaft cap seal recess and fit new cap seal.

59. Clean left side rear timing belt cover.

60. Position left side rear timing belt cover, fit and tighten bolts to 4 Nm (3 ft. lbs.).

61. Connect battery ground lead.

RIGHT SIDE

1. Before servicing the vehicle, refer to the precautions in the beginning of this section.

2. Disconnect battery ground lead.

3. Remove right side rear camshaft timing belt.

4. Depress locking collar and release breather hose from right side camshaft cover.

5. Remove 6 bolts securing plug top coils to camshaft cover and remove coils.

6. Remove bolt securing ground lead to right side camshaft cover and release ground lead.

7. Progressively loosen and remove 14 bolts securing camshaft cover.

8. Remove camshaft cover and gasket.

9. Remove 4 bolts securing right side front timing belt cover backplate to cylinder head and remove backplate.

10. Remove 2 bolts securing right side rear timing belt cover backplate and remove backplate.

11. Using sequence shown, progressively loosen 22 bolts securing camshaft carrier to cylinder head until valve spring pressure is released and remove bolts.

12. Release camshaft carrier from dowels and remove carrier.

13. Remove both camshafts and discard oil seals.

14. Using a stick magnet, remove 12 hydraulic tappets from cylinder head.

✳✳ CAUTION

Store hydraulic tappets in their fitted order and store upright. Maintain absolute cleanliness when handling hydraulic tappets. Failure to observe these precautions can result in engine failure.

15. Clean camshafts and bearing running surfaces in camshaft carrier and cylinder head.

16. Inspect camshafts and replace camshafts if scored, pitted or excessively worn.

17. Use a lint-free cloth and solvent, clean sealing surfaces on cylinder head and camshaft carrier.

18. Position camshafts in cylinder head and place Plastigage® across each journal.

19. Refit camshaft carrier and tighten bolts in sequence to 10 Nm (7.5 ft. lbs.). DO NOT rotate camshaft.

20. Using the sequence shown, progressively loosen and remove camshaft carrier bolts.

Release and remove camshaft carrier from cylinder head.

21. Measure widest part of Plastigage® on each journal. Camshaft bearing clearance = 0.025 to 0.059 mm. If clearance is excessive, fit new camshaft and repeat check. If still out of tolerance, renew cylinder head.

To install:

22. Remove all traces of Plastigage® from camshaft bearing journals.

23. Thoroughly clean and lubricate hydraulic tappets with clean engine oil. Fit hydraulic tappets to original bores in cylinder head.

24. Ensure that mating faces of camshaft carrier and cylinder head are clean and dry.

25. Lubricate camshafts and bearing journals with clean engine oil.

26. Position camshafts in cylinder head with rear timing gear drive slots in each camshaft facing towards the center as shown.

27. Apply continuous thin beads of sealant, Part No. STC 4600, to paths on camshaft carrier as shown. Spread sealant to an even film using a roller.

✳✳ CAUTION

To avoid contamination, assembly should be completed immediately after application of sealant.

28. Position camshaft carrier, fit and progressively tighten bolts in the sequence shown to 10 Nm (7.5 ft. lbs.).

29. Noting that the front camshaft oil seals are black in color and the rear oil seals are red, fit new camshaft oil seals using LRT-12-203 and LRT -12-148A

✳✳ CAUTION

Oil seal is waxed on outer diameter and must not be lubricated before fitting.

30. Clean camshaft timing belt cover backplate bolts and apply Loctite 242 to the first 3 threads.

31. Position camshaft timing belt rear cover backplates to cylinder head, fit and tighten bolts to 9 Nm (7 ft. lbs.).

For Accessory Drive Belt illustrations, see Section 1 of this manual

32. Clean mating surfaces of camshaft cover and carrier.

33. Clean inside of camshaft cover. If necessary, wash oil separator elements in solvent and blow dry.

34. Position new gasket to camshaft carrier with arrows on gasket pointing towards inlet manifold.

35. Position camshaft cover, fit and tighten bolts progressively in the sequence shown to 9 Nm (7 ft. lbs.).

36. Position plug top coils to spark plugs and right side camshaft cover, fit and tighten bolts to 9 Nm (7 ft. lbs.).

37. Position ground lead to right side camshaft cover, fit bolt and tighten to 9 Nm (7 ft. lbs.).

38. Connect breather hose to right side camshaft cover.

39. Fit camshaft timing belt.

40. Fit right side rear camshaft timing belt.

41. Connect battery ground lead.

Discovery/Discovery Series II and Range Rover

1. Before servicing the vehicle, refer to the precautions in the beginning of this section.

2. Remove or disconnect the following:
- Engine from the vehicle
- Intake manifold and gasket
- Both rocker arm shaft assemblies

➡**Identify each rocker shaft assembly to ensure installation on original cylinder bank.**

- Pushrods and place them in order
- Valve lifters and place them in order
- Front cover and timing chain
- Camshaft thrust plate
- Camshaft

➡**The camshaft installed in the 4.0L engines is color-coded orange, while the camshaft in the 4.6L engines is color-coded red.**

To install:

3. Lubricate the camshaft journals with engine oil.

4. Install or connect the following:
- Camshaft into cylinder block. Tighten the thrust plate mounting bolts to 18 ft. lbs. (25 Nm).
- Timing chain and front cover

5. Soak the lifters in engine oil. Before installing each lifter, pump the inner sleeve of the lifter several times using a pushrod to prime the lifter; this will reduce lifter noise when the engine is first started.

6. Lubricate the lifter bores with engine oil, then install the lifters.

➡**Some lifter noise may still be heard on initial start-up. If necessary, run the engine at 2500 rpm for a few minutes until the noise clears.**

7. Install or connect the following:
- Pushrods
- Rocker arm shaft assemblies
- Intake manifold
- Engine

Starter Motor

REMOVAL & INSTALLATION

Freelander

1. Before servicing the vehicle, refer to the precautions in the beginning of this section.

2. Remove battery carrier.

3. Release starter motor solenoid terminal cover, remove nut securing battery lead to solenoid and disconnect battery lead.

4. Disconnect connector from starter solenoid.

5. Remove 3 bolts securing starter motor to gearbox noting that the left-hand bolt also secures the mounting bracket for the CKP sensor multiplug.

6. Maneuver and remove starter motor.

To install:

7. Clean starter motor and mating face on gearbox.

8. Position starter motor to gearbox, align CKP sensor multiplug bracket, install and tighten bolts to 45 Nm (33 ft. lbs.).

9. Connect battery lead to solenoid terminal, install nut and tighten to 13 Nm (9.5 ft. lbs.). Secure terminal cover.

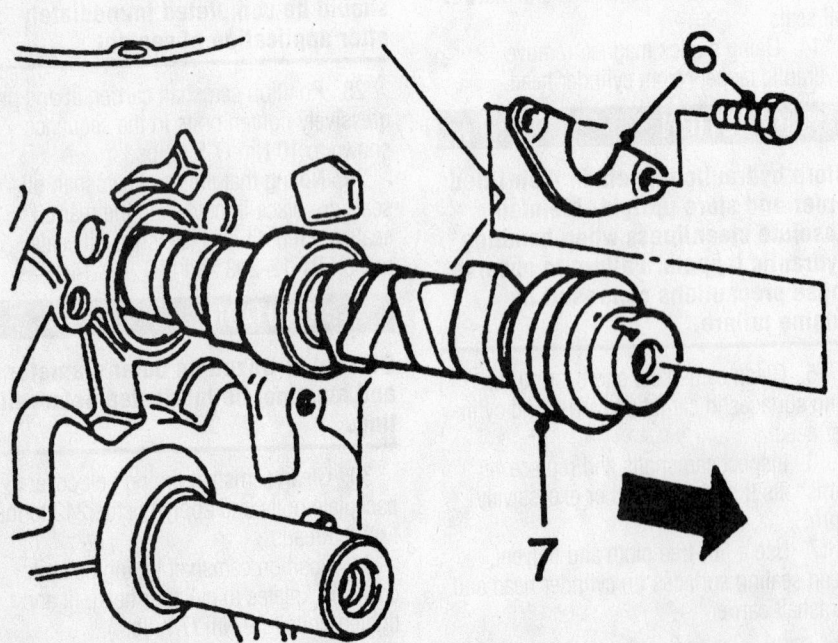

7924RG09

Exploded view of the camshaft (7) and thrust plate (6) mounting—Except Freelander

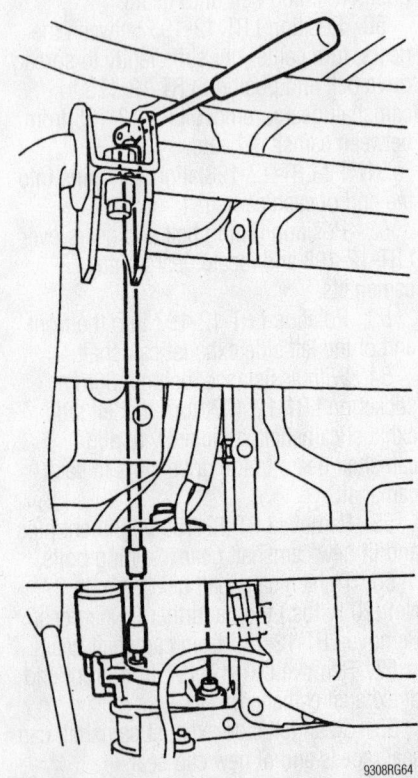

9308RG64

Here's how you reach the starter mounting bolts—Except Freelander

10. Connect connector to starter solenoid.

11. Install battery carrier.

Discovery/Discovery Series II and Range Rover

1. Before servicing the vehicle, refer to the precautions in the beginning of this section.

2. Disconnect the battery ground cable.

3. Remove the right side transmission sound shield.

4. Disconnect the wiring from the solenoid.

5. Remove the 2 starter mounting bolts.

6. Installation is the reverse of removal. Torque the mounting bolts to 33 ft. lbs. (45 Nm).

Oil Pan

REMOVAL & INSTALLATION

Range Rover

1. Before servicing the vehicle, refer to the precautions in the beginning of this section.

2. Lift the vehicle on a 4-post lift, or, on jackstands under the chassis with a jack under the axle.

3. Remove or disconnect the following:
- Battery ground cable
- Engine acoustic cover
- Transmission acoustic cover
- Engine oil dipstick
- Engine oil

4. Position a support under the front crossmember.

5. Lower the axle for clearance.

6. Remove or disconnect the following:
- O$_2$ sensor connectors
- The 3 nuts and 14 bolts securing the pan
- Oil Pan

To install:

7. Clean all mating surfaces.

8. Apply RTV Hylosil 101 or 106, or equivalent sealant, to the oil pan. Using the illustration, the bead width and length should be:
- Width at A, B, C and D: 12mm
- Width at remaining areas: 5mm
- Length at A and B: 32mm
- Length at remaining areas: 19mm

➡**Do not spread the bead. Install the pan immediately!**

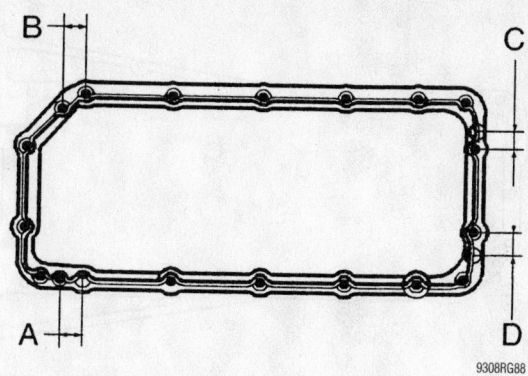

Sealant bead application—Range Rover

9308RG88

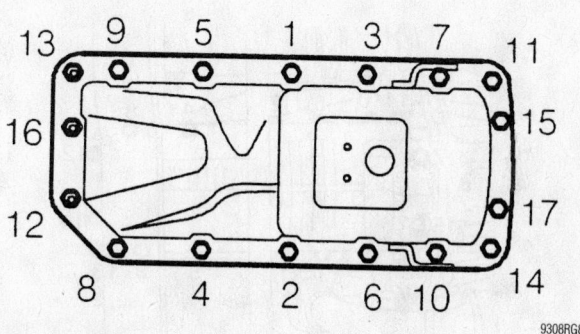

Oil pan fastener torque sequence—Range Rover

9308RG87

9. Position the pan, taking care to avoid disturbing the bead.

10. Install the oil pan and tighten the nuts bolts to 17 ft. lbs. (23 Nm) in sequence as shown.

11. Install the drain plug. Torque to 33 ft. lbs. (45 Nm).

12. Connect the sensors.

13. Install the dipstick tube to the rocker cover.

14. Install the covers.

15. Fill the engine with oil, start the engine and check for leaks.

Discovery/Discovery Series II

1. Before servicing the vehicle, refer to the precautions in the beginning of this section.

2. Lift the vehicle on a 4-post lift, or, on jackstands under the chassis with a jack under the axle.

3. Remove or disconnect the following:
- Battery ground cable
- Engine oil dipstick
- Engine oil
- Front crossmember

4. Lower the axle for clearance.

5. Remove or disconnect the following:
- Transmission cooler lines

- The 2 forward-facing and 4 rear-facing bolts securing the oil pan to the bell housing
- The 2 bolts in the sump oil pan recess
- The 3 nuts and 12 bolts securing the pan
- Oil Pan

To install:

6. Clean all mating surfaces.

7. Apply RTV silicone sealant, 5mm wide, across the front cover-to-block joint and at the rear main bearing joint. Also, apply a glob of RTV to cover the ends of the seal.

8. Position a new gasket on the pan, making sure that the tabs are correctly located.

9. Position the pan, taking care to avoid disturbing the bead.

10. Install the oil pan and tighten the nuts bolts to 16 ft. lbs. (22 Nm) in sequence as shown.

11. Install the drain plug. Torque to 33 ft. lbs. (45 Nm).

12. Connect all wiring.

13. Install the dipstick tube to the rocker cover.

14. Install the crossmember. Torque the bolts to 18 ft. lbs. (25 Nm).

For Tire, Wheel and Ball Joint specifications, see Section 1 of this manual

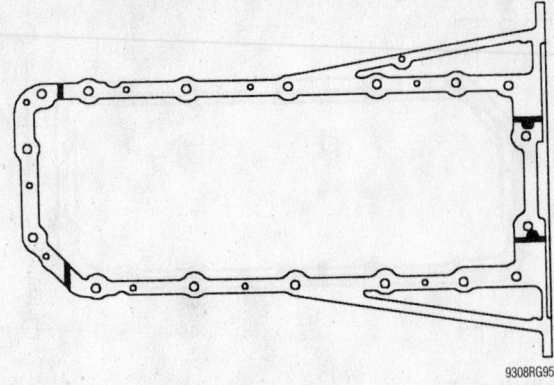

Oil pan sealant application—Discovery/Discovery Series II

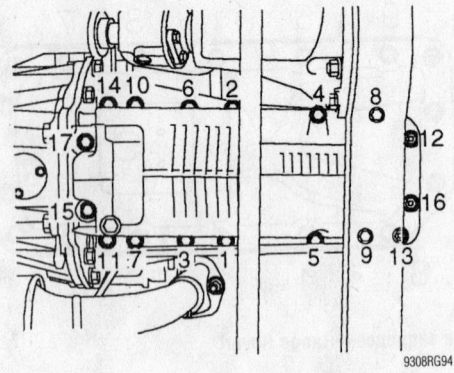

Oil pan fastener torque sequence—Discovery/Discovery Series II

15. Fill the engine with oil, start the engine and check for leaks.

Freelander

1. Before servicing the vehicle, refer to the precautions in the beginning of this section.
2. Disconnect battery ground.
3. Remove engine acoustic cover.
4. Drain engine oil.
5. Remove 3 bolts securing right side splash shield to body and remove shield.
6. Remove 3 nuts securing engine oil cooler to sump and position oil cooler aside.
7. Remove bolt securing dipstick tube to cylinder block.
8. Depress collar and remove dipstick tube from sump.
9. Remove 3 bolts securing IRD support bracket to sump.
10. Noting their installed position, remove 10 bolts securing sump to lower crankcase.
11. Release and remove sump from lower crankcase.
   To install:
12. Using and suitable cleaning solvent, clean sump and mating face on lower crankcase. DO NOT use a metal scraper on sealing surfaces.

13. Apply a 2 mm bead of sealant, Part No. STC 4600, or equivalent, along center of sump flange, then spread to an even film using a roller.

> ❊❊ **CAUTION**
>
> **To avoid contamination, assembly should be completed immediately after application of sealant.**

14. Position sump, install bolts and tighten progressively in the sequence shown to 35 Nm (26 ft. lbs.).
15. Fit bolts securing IRD support bracket to sump and tighten to 45 Nm (33 ft. lbs.).
16. Position engine oil cooler to sump mounting bracket, install and tighten nuts to 25 Nm (18 ft. lbs.).
17. Position dipstick tube to sump and cylinder block, install bolt and tighten to 9 Nm (7 ft. lbs.).
18. Install splash shield and secure with bolts.
19. Fill engine with correct quantity and grade of oil.
20. Install engine acoustic cover.
21. Connect battery ground.

Oil Pump

REMOVAL & INSTALLATION

Freelander

1. Before servicing the vehicle, refer to the precautions in the beginning of this section.
2. Disconnect battery ground lead.

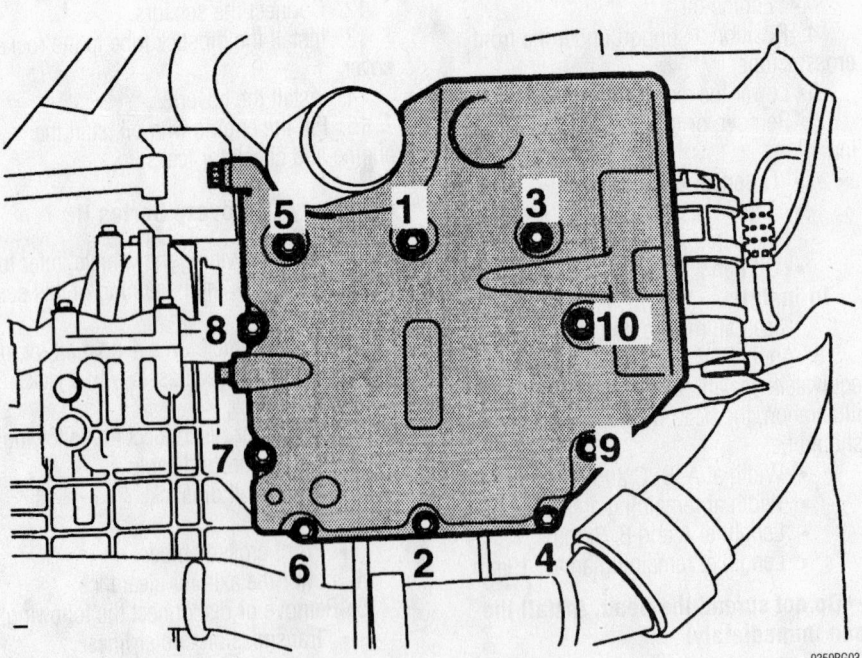

Oil pan mounting bolt location torque sequence—Freelander

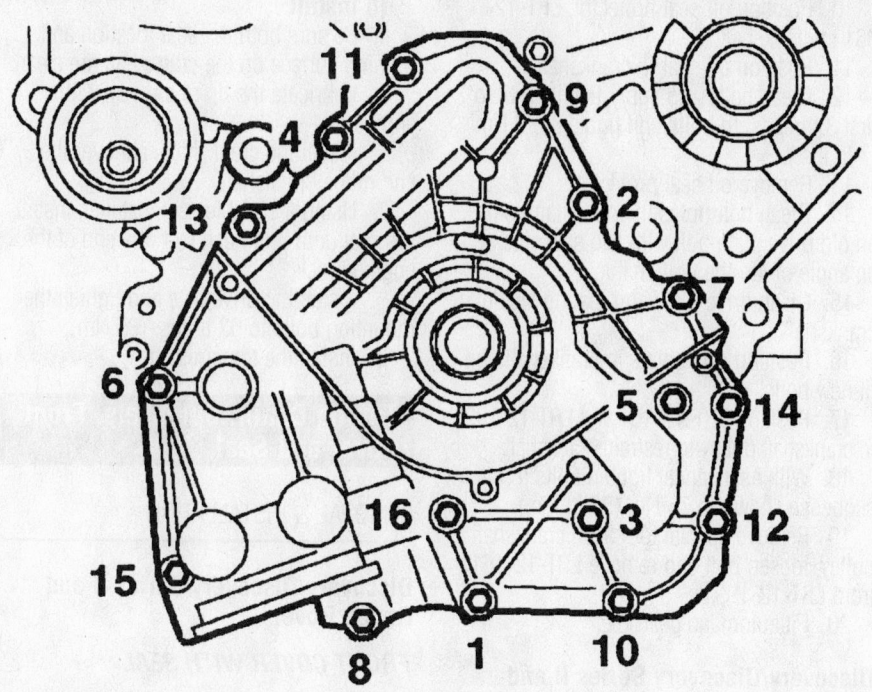

Oil pump mounting bolt location torque sequence—Freelander

3. Drain engine oil.

4. Remove camshaft timing belt.

5. Remove crankshaft gear.

6. Disconnect multiplug from oil pressure switch.

7. Loosen and remove oil cooler pipe unions from oil filter housing, remove and discard 2 O-rings.

8. Using strap wrench, remove and discard oil filter.

9. Remove and discard 16 bolts securing oil pump to cylinder block.

10. Release oil pump from locating dowels and remove oil pump.

11. Remove and discard oil pump gasket.

12. Remove and discard crankshaft front oil seal from oil pump housing.

To install:

13. Using a lint free cloth and a suitable cleaning solvent, clean oil pump and mating face on cylinder block.

14. Using a lint free cloth, thoroughly clean oil seal recess in oil pump and running surface on crankshaft.

15. Fit new oil pump gasket, dry, to cylinder block.

16. Fit oil seal guide, from seal kit, over end of crankshaft.

17. Position oil pump, aligning flats on oil pump drive to flats on crankshaft. Fit new Patchlok bolts and tighten progressively in sequence illustrated to 25 Nm (18 ft. lbs.).

18. Position new seal on crankshaft up against oil pump housing. Drift seal into place using tool LRT-12-202.

19. Remove oil seal guide from crankshaft.

20. Connect multiplug to oil pressure switch.

21. Lubricate oil filter sealing ring with clean engine oil.

22. Fit oil filter and tighten by hand until it seats then tighten a further half turn.

23. Lubricate new O-rings with clean engine oil and fit to oil cooler pipe unions.

24. Connect oil cooler pipes to oil filter housing and tighten unions to 26 Nm (19 ft. lbs.).

25. Clean crankshaft gear and wipe end of crankshaft.

26. Fit crankshaft gear.

27. Fit camshaft timing belt.

28. Fill engine with oil.

29. Connect battery ground lead.

Discovery/Discovery Series II and Range Rover

1. Before servicing the vehicle, refer to the precautions in the beginning of this section.

2. Remove or disconnect the following:
- Coolant
- Oil pan
- Oil filter
- Oil pump pick-up tube
- Crankshaft pulley
- Drive belt
- Front cover
- Remove the timing chain and gears.
- Remove the 7 bolts/screws mounting the oil pump cover plate.

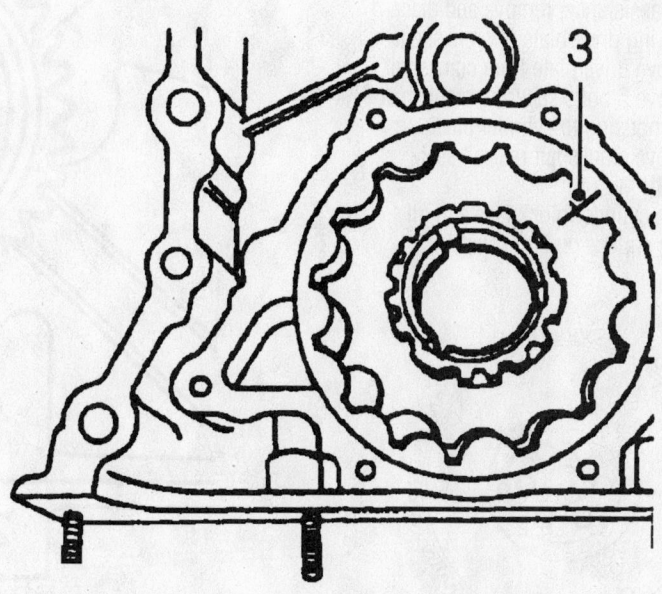

Place matchmarks on the rotors (3) before removing—Except Freelander

- Matchmark the inner and outer oil pump rotors, then remove the rotors and oil pump drive gear as an assembly.

To install:

3. Lubricate the rotors, oil pump drive gear, cover plate and housing with engine oil.

4. Align the rotor matchmarks, then install the oil pump drive gear and rotors as an assembly.

5. Apply Loctite® 222 to the mounting bolts/screws for the oil pump cover plate, then install the plate and tighten the bolts to 72 inch lbs. (8 Nm) and the screws to 36 inch lbs. (4 Nm).

6. Install the timing chain and gears.

7. Connect the oil pump pick-up tube and install the oil pan.

8. Install a new oil filter and fill the engine with oil.

9. Start the vehicle and check for leaks.

Rear Main Seal

REMOVAL & INSTALLATION

Freelander

1. Before servicing the vehicle, refer to the precautions in the beginning of this section.

2. Remove automatic gearbox.

3. Assemble LRT-12-161 to LRT-12-199 and secure with clamp bolt.

4. Position LRT-12-161 and LRT-12-199 to crankshaft pulley to restrain crankshaft.

5. With assistance, remove and discard 6 bolts securing drive plate to crankshaft.

6. Remove drive plate from crankshaft.

7. Remove 5 bolts securing crankshaft rear oil seal housing to cylinder block.

8. Remove crankshaft rear oil seal.

To install:

9. Clean cylinder block face and oil seal running surface on crankshaft.

10. Position oil seal protector, LRT-12-061 to crankshaft.

11. Position oil seal to crankshaft.

12. Clean bolts and apply Loctite 242 to first 3 threads, fit bolts and tighten to 9 Nm (7 ft. lbs.).

13. Remove oil seal protector.

14. Clean bolt holes in crankshaft using an old drive plate bolt with two saw cuts at an angle of 4to the bolt shank.

15. Clean drive plate and mating face of crankshaft.

16. Position drive plate to crankshaft and fit new bolts.

17. Position LRT-12-161 and LRT-12-199 to crankshaft pulley to restrain crankshaft.

18. With assistance, tighten bolts in the sequence shown to 75 Nm (55 ft. lbs.).

19. Remove special tool from crankshaft pulley, loosen bolt and remove LRT-12-161 from LRT-12-199.

20. Fit automatic gearbox.

Discovery/Discovery Series II and Range Rover

1. Before servicing the vehicle, refer to the precautions in the beginning of this section.

2. Remove the transmission and drive-plate.

3. Using a suitable seal removal tool, remove the oil seal from the engine block.

To install:

4. Be sure both the seal location and running surface on the crankshaft are clean.

5. Lubricate the lip of the seal with engine oil.

6. Install the crankshaft seal over the end of the crankshaft.

7. Using a suitable seal installer, install the seal until it is flush with the end of the engine block.

8. Install the driveplate and tighten the mounting bolts to 63 ft. lbs. (85 Nm).

9. Install the transmission.

Timing Chain, Sprockets, Front Cover and Seal

REMOVAL & INSTALLATION

Discovery/Discovery Series II and Range Rover

FRONT COVER WITH SEAL

➡**For seal replacement, see the next procedure.**

1. Before servicing the vehicle, refer to the precautions in the beginning of this section.

2. Remove or disconnect the following:

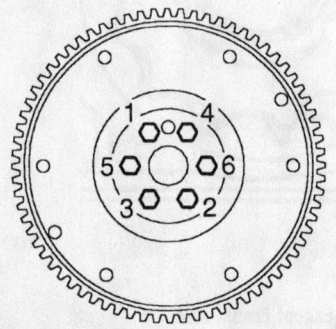

Drive plate bolt torque sequence—Freelander

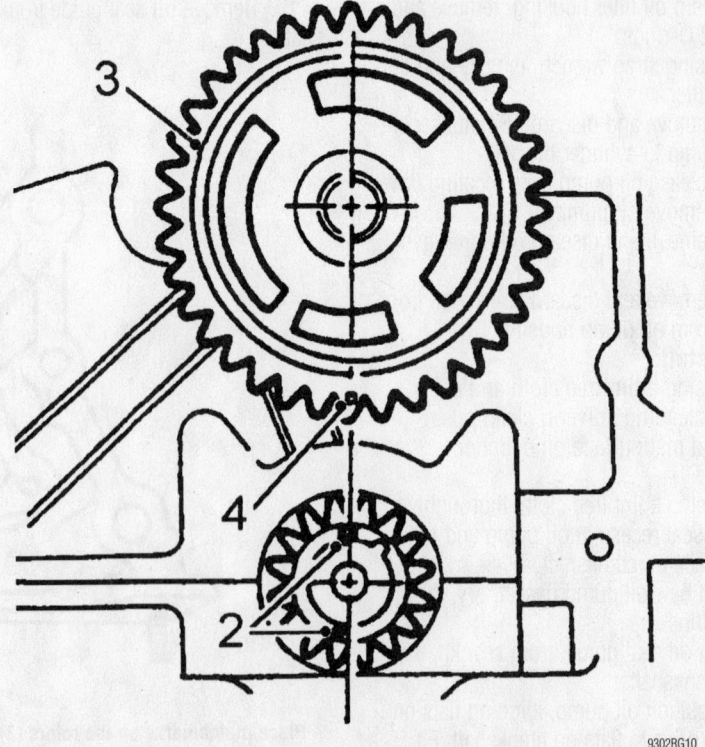

Be sure to align the cam gear (3) timing mark (4) with the crankshaft gear (2), as shown—Except Freelander

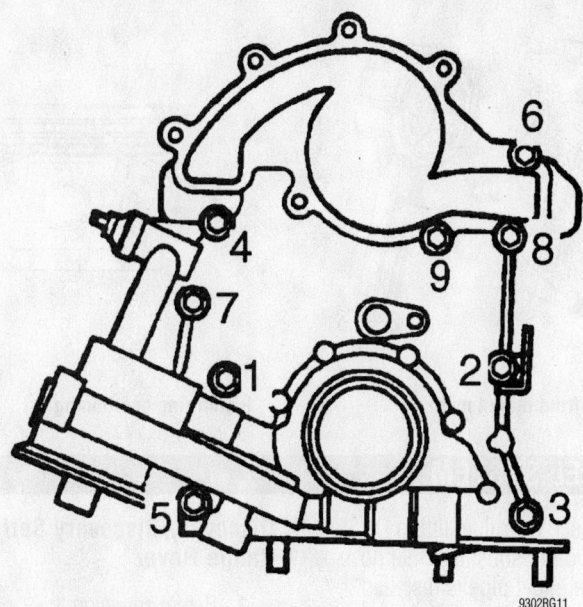

Front timing chain cover bolt torque sequence

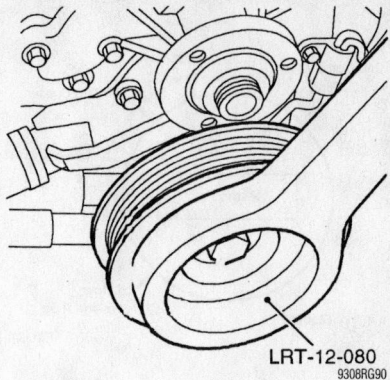

LRT-12-080
9308RG90

Crankshaft pulley holding tool

- Negative battery cable
- Coolant
- Oil pan
- Oil pick up strainer
- Crankshaft pulley and drive belt tensioner
- Hose from the water pump
- Oil cooler hoses from the front cover, then plug the hoses and connections to prevent dirt from entering.
- Oil pressure switch and CMP sensor electrical connectors
- Timing cover mounting bolts, then remove the cover.

3. Clean the gasket mating surfaces and drive out the crankshaft seal.

4. Clean the timing gears and turn the crankshaft until the timing mark on the crankshaft and camshaft face each other (camshaft at 6 o'clock and crankshaft at twelve o'clock).

5. Remove the camshaft timing gear mounting bolt.

6. Remove the timing gears and chain as an assembly.

To install:

7. Assemble the timing chain and gears on a work bench, with the timing marks aligned.

8. Install the timing chain and gear assembly onto the engine with the timing marks facing outwards.

9. Install the camshaft timing gear mounting bolt and tighten to 37 ft. lbs. (50 Nm).

10. Lubricate the new timing cover oil seal with Shell Retinax LX, or equivalent grease, ensuring that the space between seal lips is filled with grease.

11. Install or connect the following:

- Seal with a suitable seal driver
- Front cover with a new gasket and tighten the cover bolts, in the sequence shown, to 16 ft. lbs. (22 Nm).
- Oil pressure switch and CMP sensor electrical connectors
- Plugs from the oil cooler hoses
- Oil cooler hoses using new O-ring seals and tighten to 11 ft. lbs. (15 Nm)
- Drive belt tensioner and tighten the bolt to 37 ft. lbs. (50 Nm)
- Hose to the water pump
- Oil filter
- Crankshaft pulley and tighten the bolt 200 ft. lbs. (270 Nm).
- Oil pump with a new O-ring and tighten the mounting bolt to 72 inch lbs. (8 Nm).
- Oil pan
- Engine oil
- Coolant
- Negative battery cable

SEAL REPLACEMENT ONLY

1. Before servicing the vehicle, refer to the precautions in the beginning of this section.

2. Remove or disconnect the following:

- Battery ground cable
- Cooling fan
- Water pump pulley bolts
- Drive belt
- Engine under-cover

3. Install a holding tool, such as LRT-12-080, on the pulley

4. Remove the pulley bolt and pulley.

5. Using a seal remover, such as LRT-12-088, remove the seal.

To install:

6. Clean all sealing surfaces.

7. Coat the outer edge of the seal with engine oil.

8. Using an installer, such as LRT-12-089, install the seal.

9. Coat the seal lip with engine oil.

10. Install the holding tool, position the pulley on the crankshaft and install the bolt. Torque the bolt to 200 ft. lbs. (270 Nm).

11. The remainder of installation is the reverse of removal.

Timing Belt, Covers, and Front Seal

REMOVAL & INSTALLATION

For the Freelander timing belt and front cover procedures, see the Timing Belt Unit Repair Section in the front of this book.

Piston and Ring Positioning

Before servicing the vehicle, refer to the precautions in the beginning of this section.

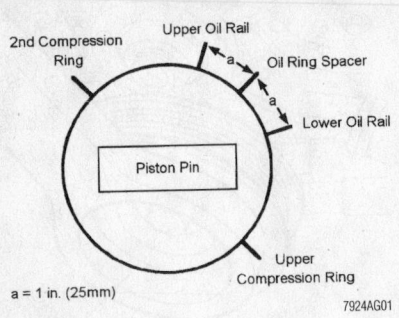

Piston ring end-gap spacing

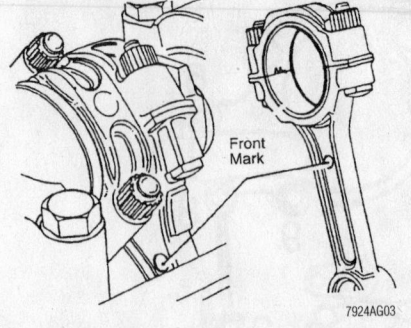

Connecting rod front mark location

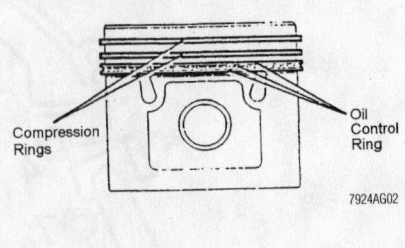

Piston ring positioning

FUEL SYSTEM

Fuel System Service Precautions

Safety is the most important factor when performing not only fuel system maintenance but any type of maintenance. Failure to conduct maintenance and repairs in a safe manner may result in serious personal injury or death. Maintenance and testing of the vehicle's fuel system components can be accomplished safely and effectively by adhering to the following rules and guidelines.

- To avoid the possibility of fire and personal injury, always disconnect the negative battery cable unless the repair or test procedure requires that battery voltage be applied.
- Always relieve the fuel system pressure prior to disconnecting any fuel system component (injector, fuel rail, pressure regulator, etc.), fitting or fuel line connection. Exercise extreme caution whenever relieving fuel system pressure, to avoid exposing skin, face and eyes to fuel spray. Please be advised that fuel under pressure may penetrate the skin or any part of the body that it contacts.
- Always place a shop towel or cloth around the fitting or connection prior to loosening to absorb any excess fuel due to spillage. Ensure that all fuel spillage (should it occur) is quickly removed from engine surfaces. Ensure that all fuel soaked cloths or towels are deposited into a suitable waste container.
- Always keep a dry chemical (Class B) fire extinguisher near the work area.
- Do not allow fuel spray or fuel vapors to come into contact with a spark or open flame.
- Always use a back-up wrench when loosening and tightening fuel line connection fittings. This will prevent unnecessary stress and torsion to fuel line piping.

- Always replace worn fuel fitting O-rings with new. Do not substitute fuel hose or equivalent where fuel pipe is installed.

Before servicing the vehicle, be sure to also refer to the precautions in the beginning of this section as well.

Fuel System Pressure

RELIEVING

The fuel injection system operates under high pressure. This makes it necessary to first relieve the system of pressure before servicing. The pressurized fuel, when released, may ignite or cause personal injury.

Freelander

1. Before servicing the vehicle, refer to the precautions in the beginning of this section.
2. Remove acoustic cover.
3. Remove Schrader valve cap.
4. Position absorbent cloth around fuel feed pipe connection to collect spillage.

✳✳ WARNING

The spilling of fuel is unavoidable during this operation. Ensure that all necessary precautions are taken to prevent fire and explosion.

5. Connect adapter LRT-19-006 to Schrader valve.
6. Position opposite end of adapter LRT-19-006 into container, turn tap to release pressure.
7. Remove adapter LRT-19-006 from Schrader valve.
8. Fit cap to Schrader valve.
9. Fit acoustic cover.

Discovery/Discovery Series II and Range Rover

1. Before servicing the vehicle, refer to the precautions in the beginning of this section.
2. Disconnect the power to the fuel pump by removing the relay or the fuel pump fuse. Check the list on the fuse box lid to be sure. The fuse can be removed to stop the fuel pump from running. With the engine operating at idle, wait until the engine stalls from fuel starvation.
3. Switch the ignition **OFF** and remove the negative battery cable.
4. Carefully loosen the fuel line on the control pressure regulator or component to be serviced.
5. Wrap a clean rag around the connection, while loosening, to catch any fuel.
6. After service is complete, discard the fuel soaked rag in the proper manner and reconnect negative battery cable, relay or fuses.

Fuel Filter

REMOVAL & INSTALLATION

Freelander

1. Before servicing the vehicle, refer to the precautions in the beginning of this section.
2. Disconnect battery ground lead.
3. Remove sender unit.
4. Disconnect 2 Lucars from top of pump unit assembly.
5. Release 3 slots in top of pump unit assembly from lugs in base.
6. Carefully maneuver top of pump unit assembly away from base, ensuring that fuel feed hose does not become strained.
7. Collect compression spring from fuel filter.

8. Carefully release 3 sprag clips securing fuel filter.

9. Release fuel filter from inlet and outlet connections.

10. Remove fuel filter.

11. Collect O-rings.

To install:

12. Lubricate NEW O-rings with silicone grease and fit to ports.

13. Carefully fit fuel filter to ports and push fully home, ensuring that sprag clips engage fully.

✳✳ WARNING

Ensure ground spring fitted to fuel pressure regulator is correctly located.

14. Position spring to fuel filter recess and engage in top location.

✳✳ WARNING

Ensure that filter ground tag is correctly located to contact the base of fuel filter.

15. Engage pump top to base, ensuring that slots engage correctly with lugs.

✳✳ WARNING

During refit, ensure that all electrical connections are made correctly. Earth tag on fuel pump negative terminal must not become distorted.

16. Fit sender unit.

17. Connect battery ground lead.

Discovery/Discovery Series II and Range Rover

1. Before servicing the vehicle, refer to the precautions in the beginning of this section.

2. Disconnect the negative battery cable and relieve the fuel system pressure.

3. Raise and support the vehicle safely.

4. Remove the bracket cover over the filter, if equipped.

5. Place a pan under the filter. Using 2 wrenches, disconnect the fuel lines from the filter.

6. Remove the fuel filter from the bracket and retainer, if equipped. Note the direction of the flow arrow so the replacement filter can be installed correctly.

To install:

7. Install the fuel filter into the bracket making sure the flow direction is correct.

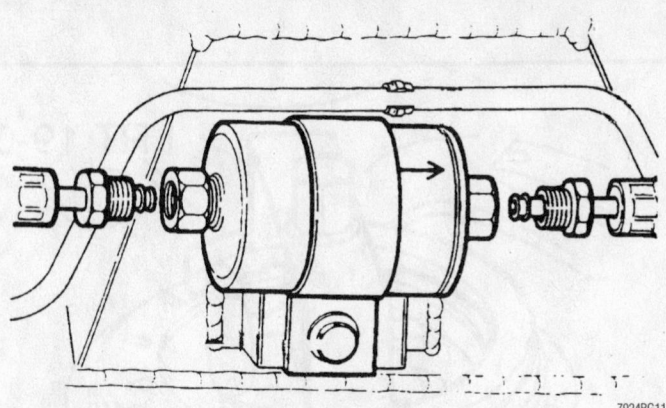

Exploded view of the fuel filter mounting—Except Freelander

Tighten the clamp to 15–25 inch lbs. (2–3 Nm).

8. Connect the fuel lines, using new O-rings. Tighten the fittings to 15 ft. lbs. (20 Nm).

9. Lower the vehicle.

10. Start the engine and check for leaks.

Fuel Pump

REMOVAL & INSTALLATION

Freelander

1. Before servicing the vehicle, refer to the precautions in the beginning of this section.

2. Disconnect battery ground lead.

3. Open right side rear and tail doors.

4. Fold right side rear seat forward.

5. Remove 2 fasteners securing front and rear carpets.

6. Pull back load space carpet from fuel pump access panel.

7. Remove 6 screws securing access panel.

✳✳ WARNING

Depressurize the system before disconnecting any components. Fuel pressure will be present in the system even if the ignition has been switched off for some time.

8. Remove access panel.

✳✳ WARNING

The spilling of fuel is unavoidable during this operation. Ensure that all necessary precautions are taken to prevent fire and explosion.

9. Disconnect multiplug and fuel hose from fuel pump housing.

✳✳ CAUTION

Always fit plugs to open connections to prevent contamination.

10. Use LRT-19-009 to remove locking ring from fuel pump housing.

11. Remove fuel pump housing and discard sealing ring.

To install:

12. Clean fuel pump housing and mating face on fuel tank.

13. Fit new seal to fuel pump housing.

14. Fit fuel pump housing to fuel tank, fit and tighten locking ring to 35 Nm (26 ft. lbs.) using LRT-19-009.

15. Connect multiplug and fuel hose to fuel pump housing.

16. Fit access panel and secure with screws.

17. Position carpet and secure with fasteners.

18. Reposition rear seat.

19. Close rear and tail doors.

20. Connect battery ground lead.

Discovery/Discovery Series II and Range Rover

1. Before servicing the vehicle, refer to the precautions in the beginning of this section.

2. Relieve the fuel system pressure.

3. Remove or disconnect the following:

- Negative battery cable
- Fuel from the tank
- Fuel filler tube from the tank
- Feed pipe at the rear of the filter
- Return line

4. Support the fuel tank with a jack.

5. Remove the 3 nuts and 2 bolts

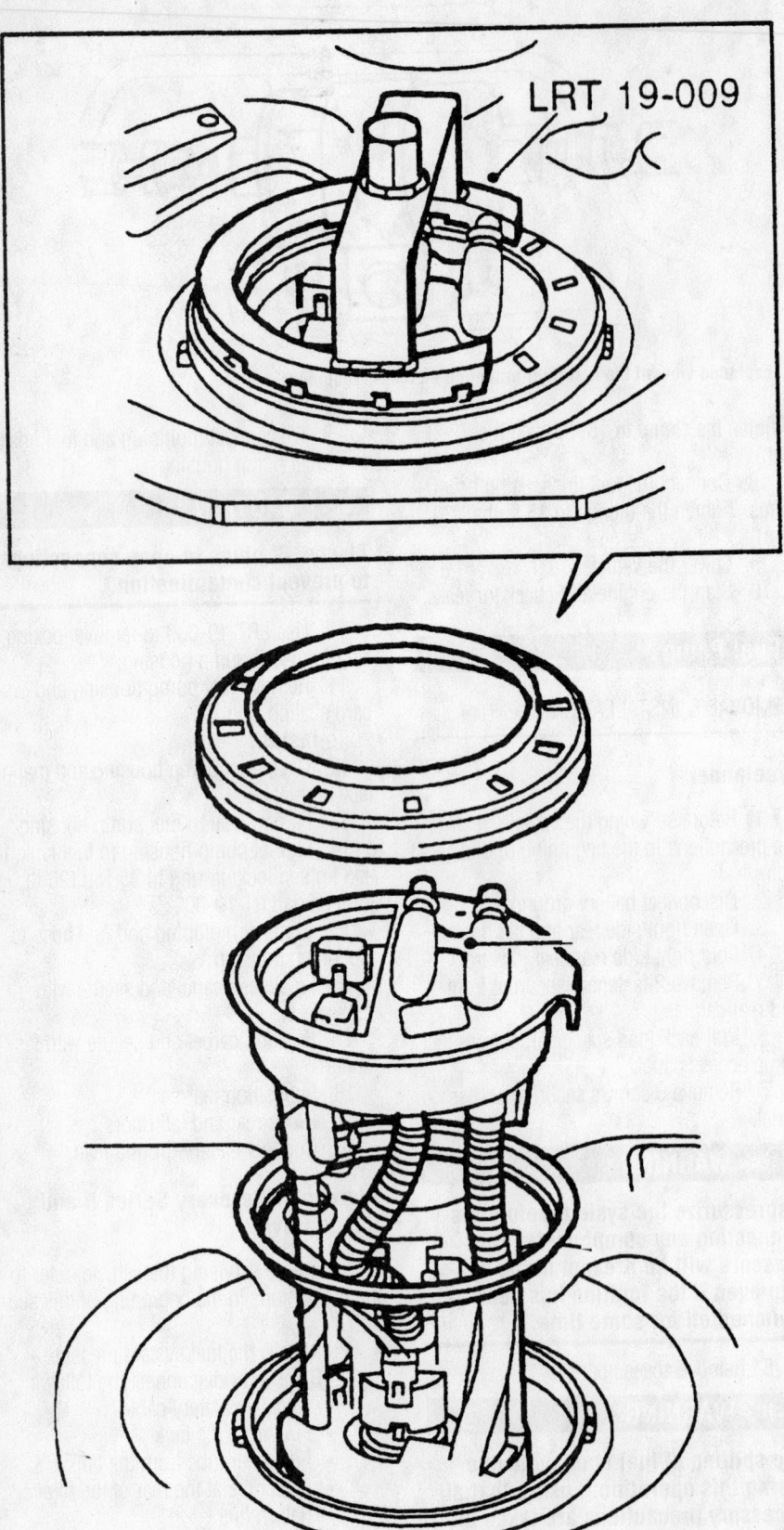

LRT 19-009

Fuel pump mounting—Except Freelander

9308RG86

securing the fuel tank cradle-to-the floor pan.

6. Lower the tank about 6 inches (150mm) and unplug the wiring connectors.

7. Remove the tank/cradle assembly.

8. Remove the tank from the cradle.

9. Remove any dirt that has accumulated around the fuel pump flange so it will not enter the fuel tank during removal and installation.

10. Disconnect the breather hose from the pressure sensor.

11. Disconnect the hoses from the pump.

12. Using a locking ring tool, such as LRT 19-009, unscrew the retaining ring from the pump flange.

➥**A lifting ring is provided to pull the unit out. Don't pull on any other parts!**

13. Remove the fuel pump and discard the seal ring. Separate the fuel pump from the sending unit, if required.

To install:

14. Clean the fuel pump mounting flange and tank mounting surface.

15. Apply a light coating of sealer on a new seal ring.

16. Install the fuel pump on the sending unit, if removed. Install the fuel pump assembly in the tank and tighten it to 26 ft. lbs. (35 Nm).

17. Install the fuel tank in the vehicle, tighten the mounting bolts to 33 ft. lbs. (45 Nm), and be sure to connect the fuel tank vent hose and filler neck.

18. Lower the vehicle and fill the fuel tank with at least 10 gallons of fuel. Connect the negative battery cable. Turn the ignition key to **RUN** for 3 seconds repeatedly, 5–10 times, to pressurize the system. Check for leaks.

19. Start the engine and check for leaks.

Fuel Rail and Injectors

REMOVAL & INSTALLATION

Freelander

1. Before servicing the vehicle, refer to the precautions in the beginning of this section.

2. Disconnect battery ground lead.

3. Remove inlet manifold chamber.

4. Remove underbelly panel.

5. Using a ⅜ in. socket bar raise ancillary drive belt tensioner and release drive belt from alternator and PAS pump pulleys.

6. Remove nuts and bolts securing alternator.

7. Release alternator and place aside.

8. Note position of coil ground lead, remove 3 nuts and 3 bolts securing ignition coils to left side inlet manifold.

9. Remove ignition coils and position aside.

10. Disconnect clips securing right side injector harness to injector protection cover.

11. Remove 2 bolts securing right side fuel rail to right side inlet manifold.

12. Release clips and disconnect injector multiplugs.

13. Remove 2 bolts securing fuel rail to left side inlet manifold.

14. Release clips and disconnect injector multiplugs.

15. Release injectors from manifolds, raise fuel rail assembly, disconnect 4 clips securing injector harness to fuel rails and move harness aside.

16. Remove fuel rail and discard O-rings.

✳✳ CAUTION

Always fit plugs to open connections to prevent contamination.

17. Remove clip securing injector to fuel rail.

18. Remove injector from fuel rail, remove and discard O-ring from injector.

✳✳ CAUTION

Always fit plugs to open connections to prevent contamination.

To install:

19. Clean injector and mating face in fuel rail.

20. Remove cap from new injector and fit to old injector.

21. Lubricate new O-ring with castor oil and fit to injector.

22. Install injector to fuel rail and secure with clip.

23. Lubricate new O-rings with castor oil and fit to injectors.

24. Remove plugs from inlet manifolds.

25. Position injector harness to fuel rail and secure with clips.

26. Position injectors and fuel rail to inlet manifolds.

27. Connect multiplugs to injectors.

28. Install and tighten 4 bolts securing fuel rail to inlet manifold to 9 Nm (7 ft. lbs.).

29. Position ignition coils to left side inlet manifold.

30. Position ground lead, fit nuts and bolts securing ignition coils to left side inlet manifold and tighten to 9 Nm (7 ft. lbs.).

31. Position alternator and secure with nuts and bolts, tighten to 45 Nm (33 ft. lbs.).

32. Using a ⅜ square drive socket bar, raise ancillary drive belt tensioner and fit drive belt to pulleys.

33. Install underbelly panel.

34. Install inlet manifold chamber.

35. Connect battery ground lead.

Discovery/Discovery Series II and Range Rover

1. Before servicing the vehicle, refer to the precautions in the beginning of this section.

2. Remove or disconnect the following:
- Battery ground cable
- Fuel system pressure
- Throttle cable
- Cruise control cable
- Harness clip from the throttle linkage bracket
- Breather hose form the plenum
- IAC connector
- TPS connector
- Plenum chamber
- Purge hose
- Crankcase breather hose
- Pressure regulator vacuum hose

- Ram housing from the intake manifold

➡ **It may be necessary to pry on the housing to break the seal. If so, use a small block of wood as a pry point between the manifold and ram housing. NEVER pry on the fuel rail!**

- The 8 injector plugs
- Fuel temperature sensor connection
- Fuel feed hose form the fuel rail
- Fuel return hose from the pressure regulator

➡ **Advanced EVAPS vehicles have a threaded connector.**

- Fuel rail and ignition coil bracket from the manifold
- Coil bracket
- Fuel rail and injectors
- Injector retaining clips and injectors
- Two O-rings from each injector

To install:

3. Install or connect the following:
- New O-rings coated with silicone grease
- Injectors and clips on the rail
- Fuel rail and injectors on the manifold
- Ignition coil and bracket
- Return hose
- Fuel feed line. Torque the union to 12 ft. lbs. (16 Nm).
- All injector wiring
- Ram housing. Use Loctite 577, or equivalent, as a sealant. Torque the bolts to 18 ft. lbs. (24 Nm).
- All vacuum hoses
- All remaining wiring
- Plenum chamber. Use Loctite 577, or equivalent, as a sealant. Torque the bolts to 18 ft. lbs. (24 Nm).
- Coolant hoses
- All remaining hoses and wires
- Battery ground

DRIVE TRAIN

Transmission Assembly

REMOVAL & INSTALLATION

Discovery/Discovery Series II

1. Before servicing the vehicle, refer to the precautions in the beginning of this section.
2. Remove or disconnect the following:
   - Negative battery cable
   - Fan shroud from the radiator
   - Transmission breather pipes from the right cylinder head at the rear and the dipstick
   - Kickdown cable from the throttle linkage
   - Shift boot knob and boot from the center console
   - Fluid from the transmission and the transfer case
   - Exhaust system, from the manifolds back
   - Speedometer from the transfer case
   - Driveshafts at the transmission and transfer case and support them out of the way.
   - Transmission oil cooler lines and secure them out of the way.
   - Transmission shift cable and the wiring harness from the transmission
3. Secure a transmission jack to the transmission.
4. Remove or disconnect the following:
   - Transmission crossmember
   - Transfer case side mounts and mounting brackets
   - Parking brake cable from the lever
   - Driveplate inspection cover and matchmark the torque converter to the driveplate.
   - Torque converter bolts and the fill tube from the transmission
   - Bell housing to engine bolts
5. Pull the transmission rearward, slightly, secure the torque converter to the transmission.
6. Remove the transmission from the vehicle.

To install:

7. Install or connect the following:
   - Transmission to the vehicle
   - Bell housing-to-engine bolts and tighten to 31 ft. lbs. (42 Nm)
8. Align the matchmarks for the torque converter to the driveplate.
9. Apply Loctite® to the torque converter bolts and tighten to 29 ft. lbs. (39 Nm).

10. Install or connect the following:
    - Fill tube to the transmission
    - Driveplate inspection cover
    - Parking brake cable to the lever
    - Transfer case side mounts and tighten the bolts to 33 ft. lbs. (45 Nm).
    - Transmission crossmember
    - Transmission shift cable and the wiring harness to the transmission
    - Transmission oil cooler lines
    - Driveshafts at the transmission and transfer case
    - Speedometer to the transfer case
    - Exhaust system
    - Transmission and transfer case with Dexron®II ATF
    - Shift boot and knob
    - Kickdown cable to the throttle linkage
    - Transmission breather pipes to the right cylinder head at the rear
    - Fan shroud to the radiator
    - Negative battery cable

Range Rover

1. Before servicing the vehicle, refer to the precautions in the beginning of this section.
2. Disconnect the negative battery cable.
3. Remove or disconnect the following:
   - Fan shroud
   - Transmission filler tube from the engine
   - Window switch pack
   - Handbrake cable clevis pin
   - Handbrake cable from the tunnel grommet
   - Transmission and transfer case fluid
   - Exhaust front pipe

   - Transmission mount
4. Secure a transmission jack to the transmission.
5. Remove or disconnect the following:
   - Front and rear driveshafts
6. Place a block of wood between the axle housing and the oil pan for support, and slightly lower the assembly.
7. Remove or disconnect the following:
   - Gear selector cable from the transmission lever
   - Selector cable
   - Transmission temperature sensor wiring
   - High/low motor wiring
   - Speed sensor wiring
   - Position sensor wiring
   - Wiring harness from the clips
   - Cooler pipe clamp bolt
   - Cooler pipes
   - Filler pipe from the transmission
   - Breather pipes
   - Converter housing cover
8. Matchmark the drive plate and converter.
9. Remove or disconnect the following:
   - Drive plate-to-converter bolts
   - Converter housing-to-engine bolts
10. Pull the transmission back, making sure that you don't drop the converter.
11. Lower and remove the transmission/transfer case assembly.
12. Support the assembly, nose down, on a bench.
13. Remove the 6 transfer case-to-transmission bolts.
14. Release the 2 ring dowels and separate the transfer case from the transmission.

To install:

15. Clean all mating surfaces.
16. Support the transmission, nose down, on a bench.

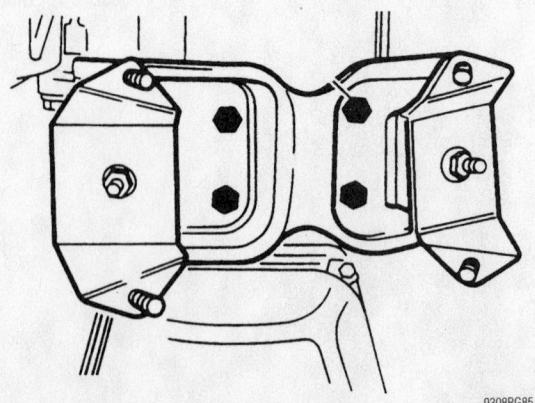

9308RG85

Transmission rear mount—Range Rover

17. Position the transfer case and ring dowels. Install the bolts and tighten them to 33 ft. lbs. (45 Nm).

18. Raise the assembly into position, aligning the torque converter and drive plate. Tighten the transmission-to-engine bolts to 31 ft. lbs. (42 Nm).

19. Install the converter-to-flexplate bolts and torque to 33 ft. lbs. (45 Nm).

20. Install or connect the following:
   • Converter cover plate, with a new gasket
   • Breather pipes
   • Filler pipe
   • Cooler lines
   • Cooler line clamp
   • Wiring
   • Selector cable
   • Transmission mount. Torque the bolts to 33 ft. lbs. (45 Nm).

Freelander

1. Before servicing the vehicle, refer to the precautions in the beginning of this section.

2. Tie hood back in upright position.

3. Disconnect battery ground lead.

4. Remove engine acoustic cover.

5. Remove IRD.

6. Loosen selector cable trunnion nut.

7. Release clip securing selector cable to gearbox bracket, remove selector cable and collect trunnion from selector lever.

8. Remove 3 screws securing left side splash shield and remove shield.

9. Secure LRT-54-026 to drive shaft inboard joint. Using a suitable lever, release inboard joint from gearbox.

10. With assistance pull hub outwards and release drive shaft from gearbox.

❊❊ CAUTION

Care must be taken not to damage oil seal when removing drive shaft from gearbox

11. Remove and discard circlip from drive shaft.

12. Remove 2 bolts securing torque converter access plate.

13. Remove access plate.

14. Mark drive plate to torque converter, for refit purposes.

15. Remove 4 bolts securing drive plate to converter.

16. Remove bolt securing IRD cooling hose retainer. Remove retainer.

17. Remove 2 bolts securing gearbox to engine.

18. Release HOS multiplug from support bracket on left side camshaft cover, disconnect multiplug.

19. Remove 4 nuts securing left side exhaust manifold to cylinder head.

20. Remove exhaust manifold, remove and discard gasket.

21. Position container to collect fluid spillage.

22. Disconnect 2 fluid cooler hose unions and discard O-rings.

❊❊ CAUTION

Always fit plugs to open connections to prevent contamination.

23. Remove 3 bolts securing fluid cooler bracket.

24. Move fluid cooler aside.

25. Remove bolt securing CKP sensor to gearbox, release sensor and position aside.

26. Remove nut and bolt, adjacent to CKP sensor, securing gearbox to engine.

27. Release throttle cable from abutment bracket and disconnect cable from throttle body cam, if fitted.

28. Depress collars and release 2 breather pipes from throttle housing, if fitted.

29. Vehicles with cruise control: Disconnect vacuum hose from cruise control actuator.

30. Remove 4 Torx screws securing throttle housing, release throttle housing and position aside.

31. Remove and discard O-ring from throttle housing.

32. Remove starter motor.

33. Remove bolt securing engine ground lead.

34. Release multiplugs from clips attached to fluid pan.

35. Disconnect 2 gearbox harness to main harness multiplugs.

36. Using a hoist, connect adjustable lifting bracket, LRT-12-138 to engine.

37. Raise hoist to take weight without exerting any load on the engine mountings.

38. Install suitable lifting brackets to gearbox and secure with nuts and bolts.

39. Connect lifting equipment to brackets.

40. Remove through bolt securing left side engine mounting to gearbox bracket.

41. Remove 4 bolts securing left side mounting to body and remove mounting.

42. Remove 4 bolts securing left side mounting bracket to gearbox and remove bracket.

43. Remove 2 top bolts securing gearbox to engine.

44. Release gearbox from 2 dowels.

45. Maneuver and lower gearbox to floor.

46. Install converter retaining plate and secure with bolts.

To install:

47. Remove torque converter retaining plate.

48. Ensure converter is fully located in oil pump drive by checking depth 'A' as illustrated. Depth A = 4 mm.

49. Clean gearbox to engine mating faces, dowels and dowel holes.

50. Install gearbox assembly.

51. Install bolts securing gearbox and tighten to 85 Nm (63 ft. lbs.).

52. Position left side mounting bracket to gearbox, fit and tighten bolts to 85 Nm (63 ft. lbs.).

53. Position left side mounting to body, fit and tighten bolts to 48 Nm (35 ft. lbs.).

54. Align gearbox bracket to left side body mounting, fit and tighten through bolt to 100 Nm (74 ft. lbs.).

55. Disconnect lifting equipment.

56. Remove nuts and bolts securing lifting brackets to gearbox and remove brackets.

57. Connect engine and gearbox harness multiplugs to main harness.

58. Secure multiplugs to clips.

59. Position engine ground lead and secure with bolt.

60. Install starter motor.

61. Clean throttle housing and manifold chamber mating faces.

62. Install new seal to inlet manifold chamber.

63. Position throttle housing to manifold chamber, fit Torx screws and tighten to 7 Nm (7 ft. lbs.).

64. Connect throttle inner cable to throttle cam and secure outer cable in abutment bracket, if fitted.

65. Connect hose to cruise control actuator.

66. Secure breather hoses to throttle housing, if fitted.

67. Adjust throttle cable, if fitted.

68. Clean CKP sensor and mating face.

69. Install CKP sensor, fit bolt and tighten to 9 Nm (7 ft. lbs.).

70. Position fluid cooler, tighten M12 bolts to 85 Nm (63 ft. lbs.) and M8 bolt to 25 Nm (18 ft. lbs.).

71. Clean fluid cooler unions.

72. Lubricate new O-rings with clean

transmission fluid and fit O-rings to fluid cooler hoses.

73. Connect fluid cooler hoses to gearbox and tighten unions to 18 Nm (13 ft. lbs.).

74. Clean exhaust manifold and mating face on cylinder head.

75. Install exhaust manifold gasket.

76. Position exhaust manifold, fit nuts and progressively tighten, from center outwards to 45 Nm (33 ft. lbs.).

77. Connect HOS multiplug and secure to support bracket.

78. Position IRD cooling hose retainer, fit bolt and tighten to 25 Nm (18 ft. lbs.).

79. Align marks on drive plate to torque converter.

80. Install bolts securing drive plate to torque converter and tighten bolts to 45 Nm (33 ft. lbs.).

81. Clean torque converter access plate.

82. Position access plate, fit bolts and tighten to 9 Nm (7 ft. lbs.).

83. Clean end of drive shaft and mating splines in gearbox.

84. Install new circlip to left side drive shaft.

85. With assistance pull hub outwards, align drive shaft and fit to gearbox, taking care not to damage drive shaft oil seal.

✳✳ CAUTION

Pull the drive shaft to ensure the circlip is fully engaged and retains the shaft.

86. Install splash shield and secure with bolts.

87. Position trunnion to selector lever, locate inner cable through trunnion, do not tighten nut at this stage.

88. Position selector cable to gearbox bracket and secure with clip.

89. Adjust selector cable.

90. Install IRD.

91. Connect battery ground lead.

92. Install engine acoustic cover.

93. Untie and close hood.

Transfer Case Assembly

REMOVAL & INSTALLATION

Discovery/Discovery Series II

1. Before servicing the vehicle, refer to the precautions in the beginning of this section.

2. Remove or disconnect the following:
- Negative battery cable
- Radiator fan shroud
- Transfer case shift knob and boot
- Transfer case fluid
- Heat shield from the front exhaust pipe
- Catalytic converter assembly
- Crossmember from under the transfer case
- Speedometer cable and tie it out of the way

3. Matchmark the front and rear drive-shafts-to-flanges relation.

4. Disconnect the front and rear drive-shafts and tie them out of the way.

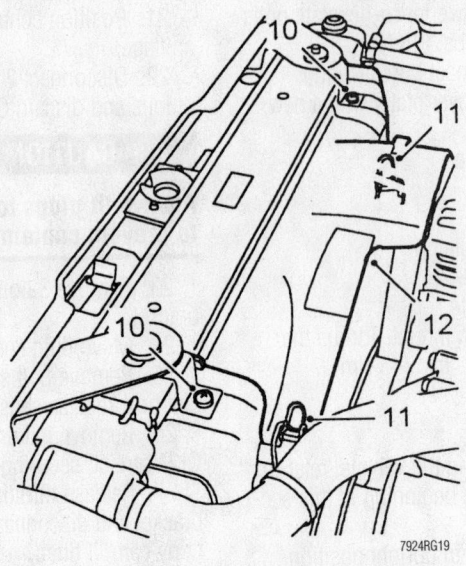

Remove the fan shroud to prevent the fan from damaging it—Discovery/Discovery Series II models

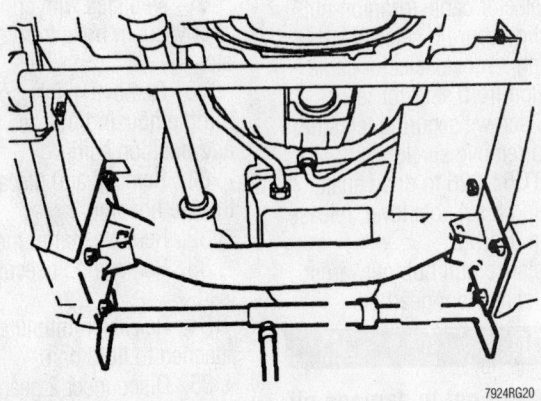

Be sure to support the transmission before removing the crossmember bolts—Discovery/Discovery Series II models

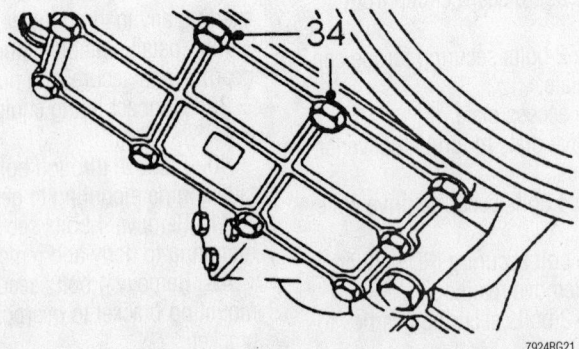

Location of the 4 central bolts—mount the adapter plate after removing these bolts—Discovery/Discovery Series II

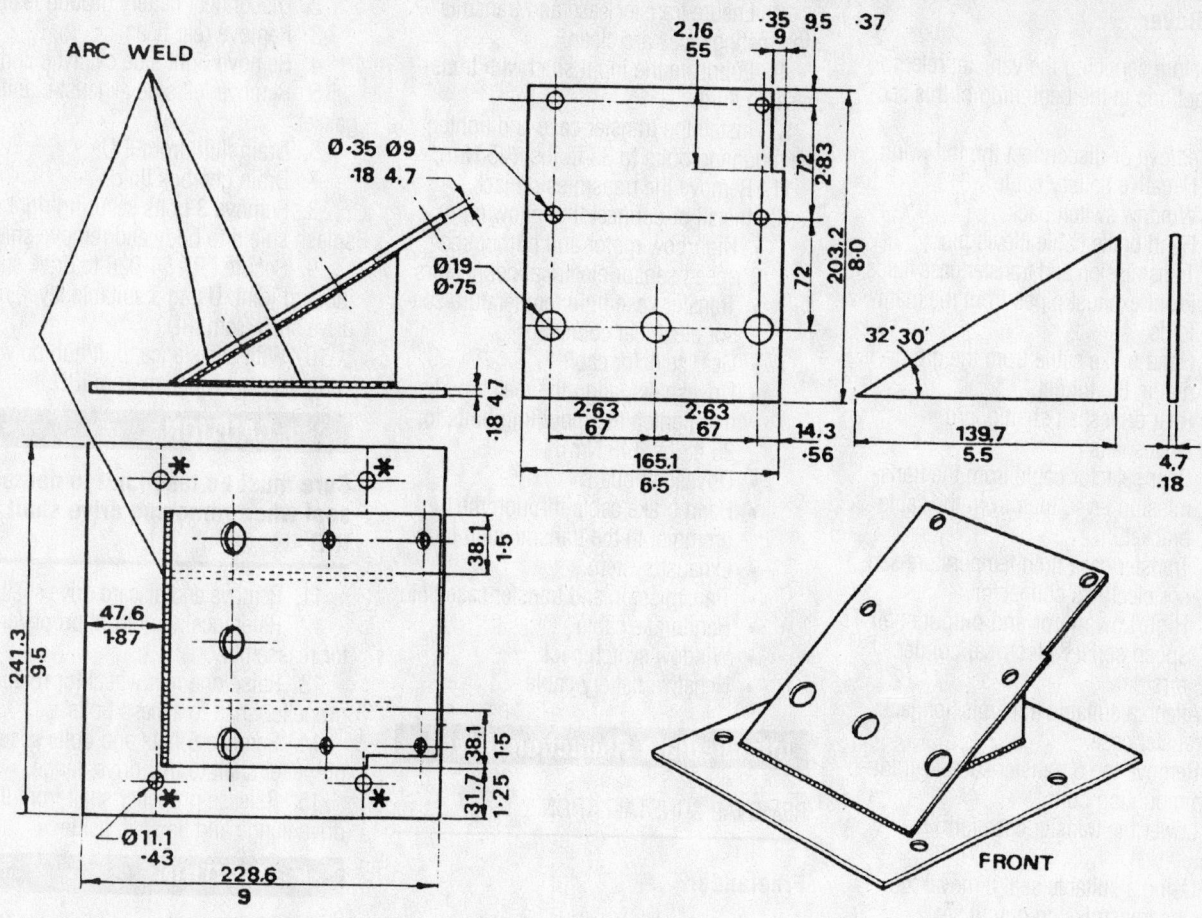

ARC WELD

MATERIAL: STEEL PLATE

✱ = TO BE DRILLED TO FIT TRANSMISSION JACK BEING USED

Dimensions for the adapter plate—Discovery/Discovery Series II

7924RG22

5. Construct an adapter plate for removing the transfer case, as indicated in the accompanying illustration.

6. Place 4 1³⁄₁₆ in. (30mm) spacers between the top of the hoist and the adapter plate, then secure the plate to the hoist.

7. Remove the 4 central bolts from the transfer case and secure the hoist with the adapter plate to the unit.

8. Raise the hoist to take the weight off the transfer case.

9. Remove the left and right transfer case rubber mounts.

10. Slowly lower the hoist until the park brake drum clears the passenger footwell. Be sure the engine does not crush any components.

11. Loosen the park brake adjustment nut and remove the park brake drum assembly from the rear output flange.

12. Label and unplug all sensors and switches from the transfer case.

13. Remove the transfer case breather banjo bolt and position it aside.

14. Disconnect the differential lock engaging rod.

15. Place the transfer case in low range.

16. Remove the range selector rod lower nut, then the rod from the yoke.

17. Support the transmission with a wooden block.

18. Remove the upper and lower transfer case mounting bolts.

19. Fit guide studs 18G 1425 to the transmission and move the transfer case rearward to remove.

To install:
20. Be sure the mating surface of the transmission and transfer case is free from dirt and debris.

21. Raise the transfer case until it is located over the guide studs 18G 1425, then slide it forward onto the transmission.

22. Remove the guide studs and bolt the transfer case to the transmission.

23. Complete the installation procedure in the reverse order of the removal procedure, noting the following items:
- After removing the adapter plate, clean the 4 bottom cover bolts and coat them with Loctite ® 290, then tighten to 19 ft. lbs. (25 Nm).
- Fill the transfer case with 90W oil.
- Check and adjust the park brake cable.

Timing belt service is covered in Section 3 of this manual

Range Rover

1. Before servicing the vehicle, refer to the precautions in the beginning of this section.

2. Remove or disconnect the following:
- Negative battery cable
- Window switch pack
- Hand brake cable clevis pin
- Transmission and transfer case fluids
- Front exhaust pipes from the manifolds
- Hand brake cable from the grommet in the tunnel
- Rear driveshaft shaft guard
- Driveshafts
- Gear selector cable from the transmission lever, then from the cable bracket.
- Transfer case fluid temperature sensor electrical connector
- High/Low motor and output shaft speed sensor electrical connectors

3. Attach a suitable transmission jack to the transfer case.

4. Remove the 6 transfer case-to-transmission mounting bolts.

5. Lower the transfer case form the vehicle.

6. Using a suitable seal removal tool, remove the transmission output seal.

To install:

7. Install a new transmission output seal, using a suitable seal installer.

8. Ensure transfer case and transmission mating faces are clean.

9. Lubricate the input shaft with transmission fluid.

10. Install the transfer case and tighten the mounting bolts to 33 ft. lbs. (45 Nm).

11. Remove the transmission jack.

12. Install or connect the following:
- High/Low motor and output shaft speed sensor electrical connectors
- Transfer case fluid temperature sensor electrical connector
- Gear selector cable
- Driveshafts, align the matchmarks and tighten the mounting bolts to 35 ft. lbs. (48 Nm).
- Driveshaft guard
- Hand brake cable through the grommet in the transmission tunnel
- Exhaust system
- Transmission and transfer case fluid
- Handbrake cable
- Window switch pack
- Negative battery cable

Intermediate Reduction Drive

REMOVAL & INSTALLATION

Freelander

1. Before servicing the vehicle, refer to the precautions in the beginning of this section.

2. Disconnect battery ground lead.

3. Remove rear beam.

4. Remove right side catalytic converter.

5. Remove left side exhaust manifold gasket.

6. Drain fluid from IRD.

7. Drain gearbox fluid.

8. Remove 3 bolts securing right side splash shield to body and remove shield.

9. Secure LRT-54-026 to drive shaft inboard joint. Using a suitable lever, release drive shaft from IRD.

10. With assistance, pull hub outwards and release drive shaft from IRD.

✳✳ CAUTION

Care must be taken not to damage oil seal when removing drive shaft from IRD.

11. Remove and discard drive shaft circlip.

12. Reference mark front propeller shaft for reassembly.

13. Raise one rear wheel for rotation of propeller shaft to access bolts.

14. Remove 6 nuts and bolts securing propeller shaft to IRD drive flange.

15. Release propeller shaft from IRD drive flange and tie shaft aside.

✳✳ CAUTION

Care must be taken to support the Tripod joint when removed from the IRD unit. To avoid damage to gaiter or steel can, the joint should not be allowed to fully extend or be dropped.

16. Remove nut securing manifold heat shield to IRD unit.

17. Remove nut securing heat shield to IRD pinion housing.

18. Remove 2 bolts securing heat shield and remove heat shield.

19. Disconnect breather hose from IRD housing.

20. Position container to collect coolant spillage.

21. Release clips and disconnect coolant hoses from IRD.

22. Remove bolt securing engine lower steady to IRD support bracket.

23. Remove lower engine steady noting that 'TOP' mark on engine steady faces uppermost.

24. Remove 3 bolts securing IRD support bracket to sump.

25. Remove 2 bolts securing IRD support bracket to engine front mounting plate.

26. Remove 5 bolts securing support bracket to IRD.

27. Remove support bracket.

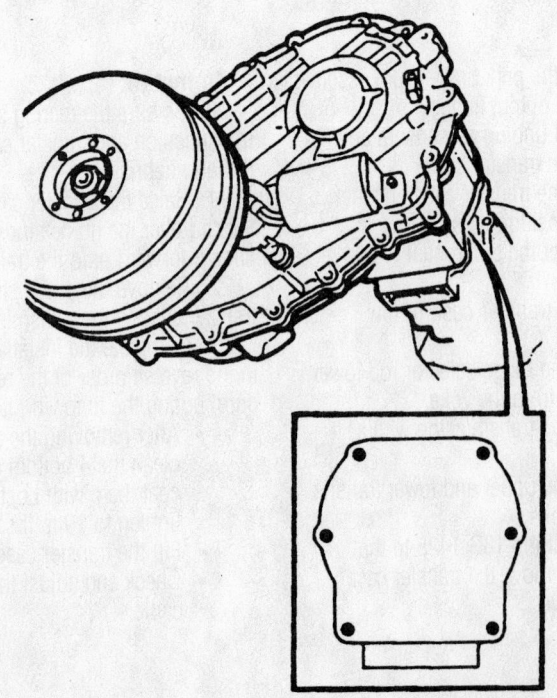

9308RG84

Transfer case bolt locations—Range Rover

28. Remove 4 bolts securing IRD.

29. With assistance, release IRD from gearbox and remove.

30. Remove and discard O-ring from IRD.

To install:

31. Clean mating faces of IRD and gearbox.

32. Lubricate and fit new O-ring

33. With assistance, fit IRD.

34. Install bolts securing IRD to gearbox and tighten sufficiently only to pull mating faces of IRD and gearbox together at this stage.

35. Install IRD support bracket and tighten bolts sufficiently only to pull mating faces together.

36. Final tighten bolts securing IRD to gearbox to 80 Nm (59 ft. lbs.).

37. Final tighten bolts securing IRD support bracket in following sequence:

- 5 bolts securing support bracket to IRD 50 Nm (37 ft. lbs.)
- 2 bolts securing support bracket to engine front mounting bracket 50 Nm (37 ft. lbs.)
- 3 bolts securing support bracket to sump 45 Nm (33 ft. lbs.).

38. Position lower engine steady, 'TOP' mark uppermost. Install bolt but do not tighten at this stage.

39. Connect coolant hoses and secure with clips.

40. Connect breather hose to IRD housing.

41. Install manifold heat shield and fit nut securing heat shield to pinion housing finger tight.

42. Install bolts securing manifold heat shield to IRD support bracket and tighten to 9 Nm (7 ft. lbs.).

43. Install nut securing heat shield to IRD and tighten to 45 Nm (33 ft. lbs.).

44. Final tighten nut securing manifold heat shield to IRD pinion housing to 25 Nm (18 ft. lbs.).

45. Clean propeller shaft flange and mating face.

46. Install propeller shaft to IRD flange and align marks. Tighten nuts and bolts to 40 Nm (30 ft. lbs.).

47. Inspect drive shaft oil seal, renew if worn or damaged.

48. Clean drive shaft and flange splines.

49. Install new circlip to driveshaft.

50. With assistance, pull hub outwards, align drive shaft and fit to IRD taking care not to damage oil seal.

> ※※ **CAUTION**
>
> **Pull the drive shaft to ensure the circlip is fully engaged and retains the shaft.**

51. Install splash shield and secure with bolts.

52. Install exhaust front pipe.

53. Install rear beam.

54. Final tighten bolt securing lower engine steady to IRD support bracket to 100 Nm (74 ft. lbs.).

55. Fill IRD to correct level with fluid.

56. Install right side catalytic converter.

57. Install left side exhaust manifold gasket.

58. Fill gearbox with fluid.

59. Connect battery ground lead.

60. Refill cooling system.

Front Axle Swivel Hubs

REMOVAL & INSTALLATION

Range Rover

> ※※ **CAUTION**
>
> **Before beginning, depressurize the air suspension.**

1. Before servicing the vehicle, refer to the precautions in the beginning of this section.

2. Remove or disconnect the following:

- Drive hub
- Axle shaft
- Track rod from the swivel hub
- Drag link from the swivel hub
- Swivel hub. Use a forcing tool, such as the one illustrated. If the joint pin turns in its socket, use a 6mm Allen wrench to hold it.
- Adjusting collar from the hub

To install:

3. Install or connect the following:

- New adjusting collar into the hub. Install the collar until a 4mm gap exists between the shoulder of the collar and the top of the hub.
- Hub onto the axle. Install the upper nut, only. Torque the nut to 81 ft. lbs. (110 Nm).

4. Clean the seal surface in the axle.

5. Turn the clamp screw of tool LRT-54-006/1 fully counterclockwise. Make sure that the clamp toggle rotates freely. Position the tool in the axle with the **TOP** mark upwards.

6. Make sure that the tool is located

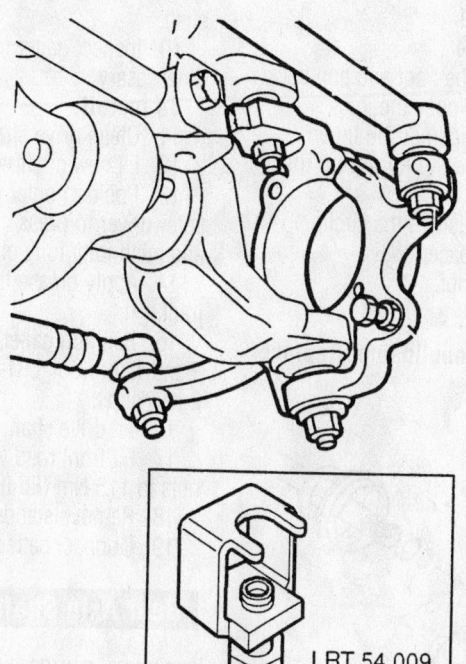

LRT 54 009

9308RG74

Front axle swivel hub. The inset shows the ball stud removal tool—Except Freelander

---

Heater Core replacement is covered in Section 2 of this manual

squarely on the sealing surface. Use a plastic or brass mallet to tap the end of the screw. This will ensure that the tool is properly positioned.

7. Install and tighten the lower nut, until the ball stud is squarely seated in the joint but the collar can still turn. Adjust the height of the collar until the tool is a slide-fit in the hub.

8. Remove the tool. Tighten the collar 1¼ turn while tightening the lower nut to 100 ft. lbs. (135 Nm). The hub should turn smoothly and evenly. If not, repeat the assembly procedure.

9. Install or connect the following:
- Drag link. Torque the nut to 59 ft. lbs. (80 Nm).
- Track rod. Torque the nut to 59 ft. lbs. (80 Nm).
- Axle shaft
- Drive hub

Front Axle Shaft CV-Joints

REMOVAL & INSTALLATION

Discovery/Discovery Series II

1. Before servicing the vehicle, refer to the precautions in the beginning of this section.

2. Remove or disconnect the following:
- Axle shaft
- Bands from the boot and pull back the boot to expose the joint.

3. Using a drift against the inner part of the joint, drive the joint from the shaft.

4. Remove and discard the circlip.

5. Remove the spacer.

6. Remove the boot.

To install:

➡ **The joint is not rebuildable. Replace it and the boot.**

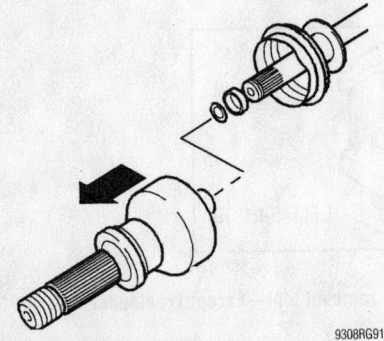

Front axle shaft CV-joint—Except Freelander

9308RG91

7. Install a new boot and bands.

8. Install the circlip and spacer.

9. Place the new joint into position and press it into place.

➡ **A new joint will come with a grease packet. Apply the grease as directed in the kit.**

10. Position the boot and secure the bands.

11. Install the axle shaft.

Freelander

1. Before servicing the vehicle, refer to the precautions in the beginning of this section.

2. Disconnect battery ground lead.

3. Raise front of vehicle.

✳✳ WARNING

Do not work on or under a vehicle supported only by a jack. Always support the vehicle on safety stands.

4. Remove road wheel.

5. Remove drive shaft.

6. Place drive shaft in vice.

7. Release both gaiter clips and discard.

8. Slide gaiter along shaft to gain access to outer joint.

9. Using a suitable drift against the inner part of the joint, remove joint from shaft.

10. Inspect gaiter for damage and renew if necessary.

To install:

11. Clean drive shaft and gaiter.

12. Fit new circlip to drive shaft.

13. Position outer joint to shaft, use a screwdriver to press circlip into its groove and push joint fully onto shaft.

14. Apply grease from the sachet to the joint.

15. Position gaiter to joint and use a 'Band-it thriftool' LRT-99-019 to secure the 2 new clips.

16. Fit drive shaft.

17. Fit front road wheels, fit and tighten nuts to 115 Nm (85 ft. lbs.).

18. Remove stands and lower vehicle.

19. Connect battery ground lead.

Rear Axle Shaft CV-Joints

REMOVAL & INSTALLATION

Freelander

1. Before servicing the vehicle, refer to the precautions in the beginning of this section.

2. Disconnect battery ground lead.

3. Raise rear of vehicle.

✳✳ WARNING

Do not work on or under a vehicle supported only by a jack. Always support the vehicle on safety stands.

4. Remove road wheel.

5. Remove drive shaft.

6. Place drive shaft in vice.

7. Release both gaiter clips and discard.

8. Slide gaiter along shaft to gain access to outer joint.

9. Using a suitable drift against the inner part of the joint, remove joint from shaft.

10. Inspect gaiter for damage and renew if necessary.

11. Remove and discard circlip from drive shaft.

To install:

12. Clean drive shaft and gaiter.

13. Fit new circlip to drive shaft.

14. Position outer joint to shaft, use a screwdriver to press circlip into its groove and push joint fully onto shaft.

15. Apply grease from the sachet to the joint.

16. Position gaiter to joint and use a 'Band-it thriftool' LRT-99-019 to secure the 2 new clips.

17. Fit drive shaft.

18. Fit road wheel(s) and tighten nuts to 115 Nm (85 ft. lbs.).

19. Remove stands and lower vehicle.

20. Connect battery ground lead.

Front Axle Drive Hub

REMOVAL & INSTALLATION

Range Rover

✳✳ CAUTION

If so equipped, before beginning, depressurize the air suspension.

1. Before servicing the vehicle, refer to the precautions in the beginning of this section.

2. Remove the center cap from the wheel.

3. Loosen the axle shaft nut.

4. Remove or disconnect the following:
- Wheel
- Brake rotor shield
- ABS sensor harness from the brackets

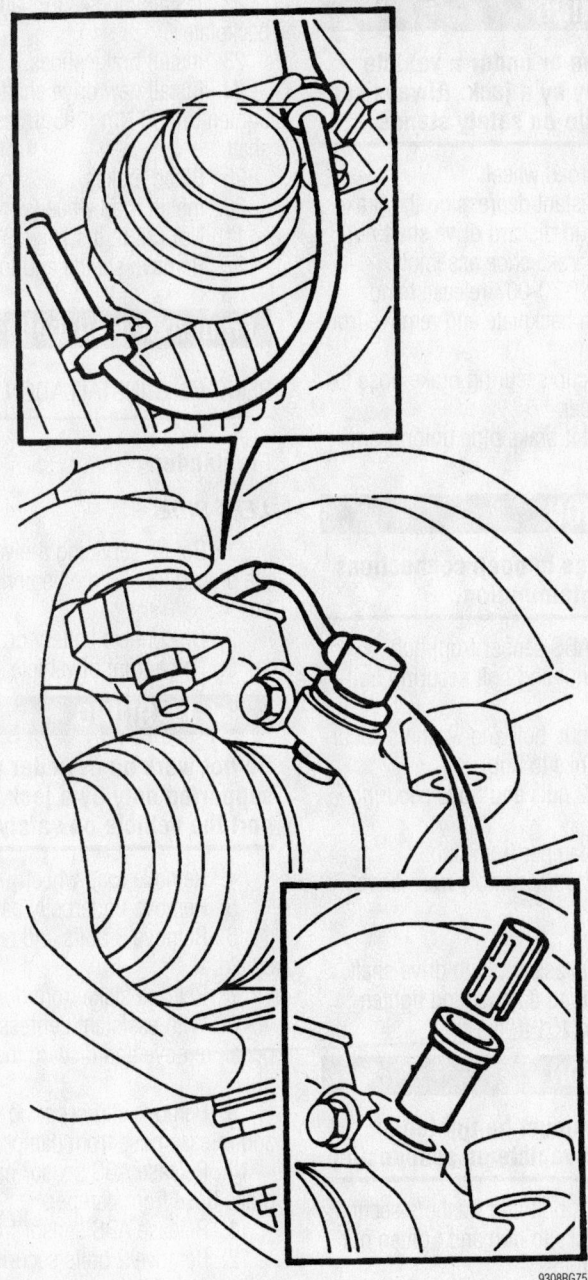

Front axle ABS sensor installation—Range Rover

- Sensor from the hub
- Sensor bushing
- 4 bolts securing the hub to the carrier
- Axle shaft nut
- Drive hub

To install:

5. Install or connect the following:
- Hub onto the shaft. Install, but don't tighten, the nut.
- Hub-to-carrier bolts. Torque to 100 ft. lbs. (135 Nm).

- ABS sensor bushing and sensor, coated with silicone grease
- Wiring harness to brackets
- Rotor shield
- Wheel

6. Lower the vehicle and torque the axle shaft nut to 192 ft. lbs. (260 Nm). Stake the nut.

Discovery/Discovery Series II

1. Before servicing the vehicle, refer to the precautions in the beginning of this section.

✳✳ CAUTION

If so equipped, before beginning, depressurize the air suspension.

2. Remove the center cap from the wheel.

3. Loosen the axle shaft nut.

4. Remove or disconnect the following:
- Wheel
- Brake rotor shield
- Brake rotor
- ABS sensor harness from the brackets
- Sensor from the hub
- Sensor bushing
- 4 bolts securing the hub to the carrier
- Axle shaft nut
- Drive hub

To install:

5. Apply anti-seize compound to the mating surfaces of the hub and knuckle.

6. Install or connect the following:
- Hub onto the shaft. Install, but don't tighten, the nut.
- Hub-to-carrier bolts. Torque to 74 ft. lbs. (100 Nm).
- ABS sensor bushing and sensor, coated with silicone grease
- Wiring harness to brackets
- Rotor
- Rotor shield
- Wheel

7. Lower the vehicle and torque the axle shaft nut to 360 ft. lbs. (490 Nm). Stake the nut.

Freelander

1. Before servicing the vehicle, refer to the precautions in the beginning of this section.

2. Raise front of vehicle.

✳✳ WARNING

Do not work on or under a vehicle supported only by a jack. Always support the vehicle on safety stands.

3. Remove road wheel.

4. Release stake in drive shaft nut.

5. With an assistant applying brakes, remove and discard drive shaft hub nut.

6. Remove clip securing RIGHT SIDE brake hose to support bracket, release hose from bracket.

Release ABS sensor and pad wear sensor harnesses from bracket.

7. Remove 2 bolts securing brake

Brake service is covered in Section 4 of this manual

caliper to hub. Release caliper from hub and tie aside.

✳✳ CAUTION

Do not allow caliper to hang on brake hose.

8. Mark brake disc to hub relationship.
9. Remove 2 screws securing brake disc and remove brake disc.
10. Remove 2 nuts and bolts securing hub to damper.
11. Release hub from damper.
12. Release drive shaft from hub.
13. Restrain hub from rotating and remove nut from lower swivel joint.
14. Break taper joint using LRT-57-043.
15. Remove swivel hub.
To install:
16. Clean drive shaft.
17. Fit hub assembly to lower joint, fit new nut and tighten to 65 Nm (48 ft. lbs.).
18. Fit drive shaft to hub.
19. Fit hub to damper, fit nuts and bolts and tighten to 205 Nm (151 ft. lbs.).
20. Clean brake disc to drive flange mating faces.
21. Fit disc to drive flange, align reference marks, fit screws and tighten to 5 Nm (4 ft. lbs.).
22. Clean mating faces of caliper and hub.
23. Position caliper to brake disc fit bolts and tighten to 100 Nm (74 ft. lbs.).
24. Fit brake hose to abutment bracket and fit clip.
25. Clean ABS sensor, smear sensor with an anti-seize grease and fit sensor to hub.

✳✳ CAUTION

Ensure ABS sensor is fully located into hub, so that sensor touches pole wheel teeth.

26. Fit ABS sensor lead to bracket.
27. Fit new drive shaft nut and tighten to 400 Nm (295 ft. lbs.). Stake nut to shaft.
28. Stake drive shaft hub nut.
29. Fit road wheel(s) and tighten nuts to 115 Nm (85 ft. lbs.).
30. Remove stands and lower vehicle.

Rear Axle Drive Hub

REMOVAL & INSTALLATION

Freelander

1. Before servicing the vehicle, refer to the precautions in the beginning of this section.
2. Raise rear of vehicle.

✳✳ WARNING

Do not work on or under a vehicle supported only by a jack. Always support the vehicle on safety stands.

3. Remove road wheel.
4. With assistant depressing the brake pedal, remove and discard drive shaft nut.
5. Remove brake shoe assembly.
6. Using LRT-70-007 release handbrake cable from backplate and remove from backplate.
7. Remove clip securing brake hose to bracket on damper.
8. Disconnect brake pipe union from wheel cylinder.

✳✳ CAUTION

Always fit plugs to open connections to prevent contamination.

9. Release ABS sensor from hub.
10. Remove nut and bolt securing trailing link to hub.
11. Remove nut, bolt and washers securing transverse links to hub.
12. Remove 2 nuts and bolts securing hub to damper.
13. Release damper from hub.
14. Remove hub assembly from drive shaft.
To install:
15. Install hub assembly to drive shaft.
16. Install hub to damper and tighten bolts to 205 Nm (151 ft. lbs.).

✳✳ CAUTION

Nuts and bolts must be tightened with weight of vehicle on suspension.

17. Install nut, bolt and washers securing transverse links to hub and tighten nut to 120 Nm (89 ft. lbs.).
18. Install trailing link to hub and tighten nut and bolt to 120 Nm (89 ft. lbs.).

➡ **Ensure that washers are fitted to both ends of bolts**

19. Clean ABS sensor, smear sensor with an anti-seize grease and fit sensor to hub.

✳✳ CAUTION

Ensure ABS sensor is fully located into hub, so that sensor touches pole wheel teeth.

20. Install brake pipe to wheel cylinder and tighten union to 14 Nm (10 ft. lbs.).
21. Install clip securing brake pipe to bracket.

22. Install and secure handbrake cable to backplate.
23. Install brake shoes.
24. Install new drive shaft nut and tighten to 400 Nm (295 ft. lbs.). Stake nut to shaft.
25. Bleed brakes.
26. Install road wheel(s) and tighten nuts to 115 Nm (85 ft. lbs.).
27. Remove stands and lower vehicle.

Front Axle Halfshafts

REMOVAL & INSTALLATION

Freelander

LEFT SIDE

1. Before servicing the vehicle, refer to the precautions in the beginning of this section.
2. Disconnect battery ground lead.
3. Raise front of vehicle.

✳✳ WARNING

Do not work on or under a vehicle supported only by a jack. Always support the vehicle on safety stands.

4. Remove road wheel.
5. Remove underbelly panel.
6. Remove 3 bolts and remove splash shield.
7. Release stake from drive shaft nut.
8. With assistant depressing the brake pedal, remove and discard the drive shaft nut.
9. Remove clip securing brake hose and release hose from damper bracket.
10. Release ABS sensor harness and brake hose from damper.
11. Release ABS sensor from hub.
12. Remove 2 bolts securing brake caliper to hub.
Release caliper from hub and tie aside.

✳✳ CAUTION

Do not allow caliper to hang on brake hose.

13. Remove 2 bolts securing damper to hub.
14. Release hub from damper.
15. Release drive shaft from hub.
16. Position container to catch oil spillage
17. Secure LRT-54-026 to drive shaft inboard joint. Using a suitable lever, release drive shaft from gearbox.
18. Remove drive shaft.

19. Remove and discard circlip from drive shaft.

To install:

20. Inspect gearbox seal for signs of wear or damage.

21. Wipe drive shaft ends, gearbox oil seal and hub.

22. Lubricate oil seal running surfaces.

23. Install new circlip to drive shaft.

24. Install drive shaft ensuring circlip is fully engaged.

✳✳ CAUTION

Drive shaft must be fitted with care to prevent damage to gearbox oil seal.

25. Position drive shaft in hub.

26. Install new hub nut but do not tighten at this stage.

27. Install hub to damper, fit nuts and bolts and tighten to 205 Nm (150 ft. lbs.).

28. Position caliper to brake disc fit bolts and tighten to 100 Nm (74 ft. lbs.).

29. Clean ABS sensor and mating face.

30. Lubricate ABS sensor with anti-seize grease.

31. Install ABS sensor .

✳✳ CAUTION

Ensure ABS sensor is fully located into hub, so that sensor touches reluctor ring.

32. Position ABS harness and brake hose in bracket and secure with clip.

33. Install splash shield and secure with bolts.

34. Tighten front hub nut to 400 Nm (295 ft. lbs.).

35. Stake front hub nut.

36. Install road wheel(s) and tighten nuts to 115 Nm (85 ft. lbs.).

37. Check and top up oil level as required.

38. Install underbelly panel.

39. Remove stands and lower vehicle.

40. Connect battery ground lead.

RIGHT SIDE

1. Before servicing the vehicle, refer to the precautions in the beginning of this section.

2. Disconnect battery ground lead.

3. Raise front of vehicle.

✳✳ WARNING

Do not work on or under a vehicle supported only by a jack. Always support the vehicle on safety stands.

4. Remove road wheel.

5. Remove underbelly panel.

6. Remove 3 bolts securing right side splash shield to body and remove shield.

7. Release stake from drive shaft nut.

8. With assistant depressing the brake pedal, remove and discard the drive shaft nut.

9. Remove clip securing brake hose and release hose from damper bracket.

10. Release ABS sensor harness and brake hose from damper.

11. Release ABS sensor from hub.

12. Remove 2 bolts securing brake caliper to hub.

Release caliper from hub and tie aside.

✳✳ CAUTION

Do not allow caliper to hang on brake hose.

13. Remove 2 bolts securing damper to hub.

14. Release hub from damper.

15. Release drive shaft from hub.

16. Position container to catch oil spillage.

17. Secure LRT-54-026 to drive shaft inboard joint.

Using a suitable lever, release drive shaft from gearbox.

18. Remove drive shaft.

✳✳ CAUTION

Care must be taken not to damage oil seal when removing drive shaft from gearbox

19. Remove and discard circlip from drive shaft.

To install:

20. Inspect gearbox seal for signs of wear or damage.

21. Wipe drive shaft ends, gearbox oil seal and hub.

22. Lubricate oil seal running surfaces.

23. Install new circlip to drive shaft.

24. Install drive shaft ensuring circlip is fully engaged.

✳✳ CAUTION

Drive shaft must be fitted with care to prevent damage to gearbox oil seal.

25. Position drive shaft in hub.

26. Install new hub nut but do not tighten at this stage.

27. Install hub to damper, fit nuts and bolts and tighten to 205 Nm (150 ft. lbs.).

28. Position caliper to brake disc fit bolts and tighten to 100 Nm (74 ft. lbs.).

29. Clean ABS sensor and mating face.

30. Lubricate ABS sensor with anti-seize grease.

31. Install ABS sensor .

✳✳ CAUTION

Ensure ABS sensor is fully located into hub, so that sensor touches reluctor ring.

32. Position ABS harness and brake hose in bracket and secure with clip.

33. Install splash shield and secure with bolts.

34. Tighten front hub nut to 400 Nm (295 ft. lbs.).

35. Stake front hub nut.

36. Install road wheel(s) and tighten nuts to 115 Nm (85 ft. lbs.).

37. Check and top up oil level as required.

38. Install underbelly panel.

39. Remove stands and lower vehicle.

40. Connect battery ground lead.

Rear Axle Halfshafts

REMOVAL & INSTALLATION

Freelander

1. Before servicing the vehicle, refer to the precautions in the beginning of this section.

2. Disconnect battery ground lead.

3. Raise rear of vehicle.

✳✳ WARNING

Do not work on or under a vehicle supported only by a jack. Always support the vehicle on safety stands.

4. Remove road wheel.

5. Release stake from drive shaft nut.

6. With assistant depressing the brake pedal, remove and discard drive shaft nut.

7. Remove nut and bolt securing trailing link to rear hub, collect spacer from under bolt head.

8. Remove nut and bolt securing fixed transverse link to subframe. Collect dynamic damper.

9. Remove nut and bolt securing adjustable transverse link to subframe.

10. Position container to catch oil spillage.

11. With assistance pull hub assembly

outwards and release drive shaft outer joint from hub assembly.

12. Taking care not to damage 'Flinger', release drive shaft inner joint from differential using LRT-51-014 and remove drive shaft.

13. Remove and discard drive shaft circlip.

To install:

14. Inspect differential seal, renew if worn or damaged.

15. Clean ends of drive shaft and locations in hub and differential.

16. Lubricate oil seal running surface with transmission oil.

17. Install new circlip to drive shaft.

18. Install drive shaft to differential and push fully home.

✳✳ CAUTION

Pull the drive shaft to ensure the circlip is fully engaged and retains the shaft.

19. With assistance fit drive shaft to hub.

20. Install new drive shaft nut but do not tighten at this stage.

21. Install nut, bolt and dynamic damper to adjustable transverse link and tighten to 120 Nm (89 ft. lbs.).

✳✳ CAUTION

Nuts and bolts must be tightened with the weight of the vehicle on the

suspension.

22. Install nut and bolt to fixed transverse link and tighten to 120 Nm (89 ft. lbs.).

23. Install spacer, nut and bolt to trailing link and tighten to 120 Nm (89 ft. lbs.).

24. Install new drive shaft nut and tighten to 400 Nm (295 ft. lbs.). Stake nut to shaft.

25. Check and top up oil level.

26. Install road wheel(s) and tighten nuts to 115 Nm 85 ft. lbs.).

27. Remove stands and lower vehicle.

28. Connect battery ground lead.

Front Axle Shaft and Seal

REMOVAL & INSTALLATION

Range Rover

1. Before servicing the vehicle, refer to the precautions in the beginning of this section.

✳✳ CAUTION

Before beginning, depressurize the air suspension.

➡️**If you're not separating the hub and shaft, don't loosen the axle shaft nut.**

2. Remove the center cap from the wheel.

3. Loosen the axle shaft nut.

4. Remove or disconnect the following:
- Wheel
- Brake rotor shield
- ABS sensor harness from the brackets
- Sensor from the hub
- Sensor bushing
- 4 bolts securing the hub to the carrier
- Axle shaft nut, if loosened
- Drive hub and axle shaft
- Seal from the axle tube

To install:

5. Clean all surfaces.

6. Lubricate the new seal lip and running surface with silicone grease. Drive a new seal into place.

7. Install or connect the following:
- Axle shaft into the axle tube
- Hub onto the shaft. Install, but don't tighten, the nut.
- Hub-to-carrier bolts. Torque to 100 ft. lbs. (135 Nm).
- ABS sensor bushing and sensor, coated with silicone grease
- Wiring harness to brackets
- Rotor shield
- Wheel

8. Lower the vehicle and torque the axle shaft nut to 192 ft. lbs. (260 Nm). Stake the nut.

Discovery/Discovery Series II

1. Before servicing the vehicle, refer to the precautions in the beginning of this section.

✳✳ CAUTION

If so equipped, before beginning, depressurize the air suspension.

2. Remove the center cap from the wheel.

3. Loosen the axle shaft nut.

4. Remove or disconnect the following:
- Wheel
- Brake rotor shield
- Brake rotor
- ABS sensor harness from the brackets
- Sensor from the hub

- Sensor bushing
- 4 bolts securing the hub to the carrier
- Axle shaft nut
- Drive hub
- Axle shaft
- Oil Seal

To install:

5. Clean all mating surfaces thoroughly.

6. Apply gear oil to the new seal's lip and outer edge.

7. Drive the new seal into place.

8. Slide the axle shaft into place.

9. Apply anti-seize compound to the mating surfaces of the hub and knuckle.

10. Install or connect the following:
- Hub onto the shaft. Install, but don't tighten, the nut.
- Hub-to-carrier bolts. Torque to 74 ft. lbs. (100 Nm).
- ABS sensor bushing and sensor, coated with silicone grease
- Wiring harness to brackets
- Rotor
- Rotor shield
- Wheel

11. Lower the vehicle and torque the axle shaft nut to 360 ft. lbs. (490 Nm). Stake the nut.

Rear Axle Shaft and Bearing

REMOVAL & INSTALLATION

Range Rover

1. Before servicing the vehicle, refer to the precautions in the beginning of this section.

✳✳ CAUTION

Before beginning, depressurize the air suspension.

➡️**If you're not separating the hub and shaft, don't loosen the axle shaft nut.**

2. Remove the center cap from the wheel.

3. Loosen the axle shaft nut.

4. Remove or disconnect the following:
- Wheel
- Brake rotor shield
- 2 bolts and remove the backstrap from the hub.
- Sensor from the hub
- Sensor bushing
- 6 bolts securing the hub to the axle carrier
- Drive hub and axle shaft
- Axle shaft nut, if loosened
- Seal from the axle tube

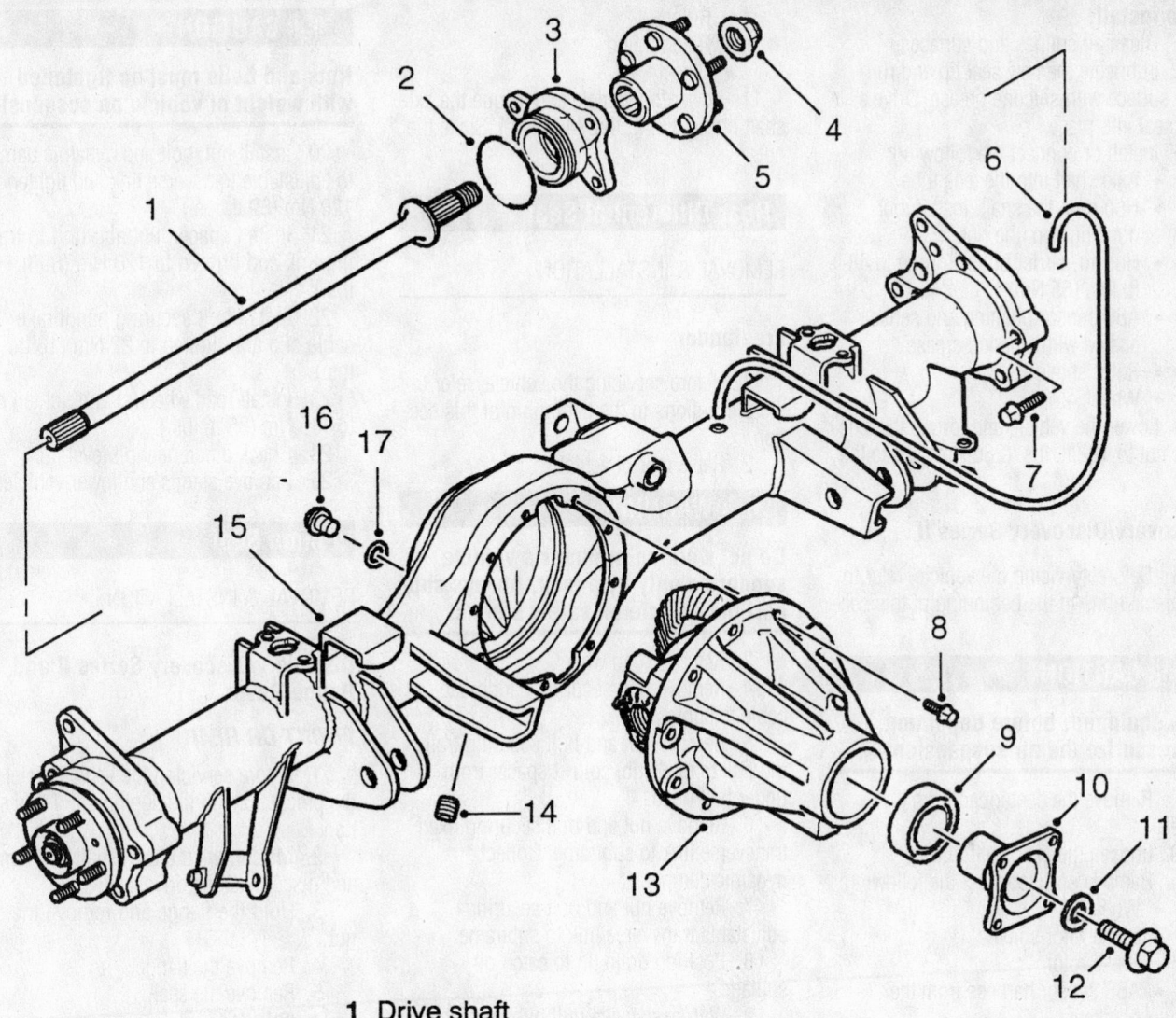

1 Drive shaft
2 'O' ring
3 Hub bearing
4 Stake nut
5 Hub flange
6 Breather tube
7 Bolt
8 Bolt
9 Oil seal
10 Drive flange
11 Washer
12 Bolt
13 Differential unit
14 Drain plug
15 Axle casing
16 Oil level plug
17 'O' ring

Rear axle components—Except Freelander

9308RG92

For Accessory Drive Belt illustrations, see Section 1 of this manual

To install:

5. Clean all splines and surfaces.

6. Lubricate the new seal lip and running surface with silicone grease. Drive a new seal into place.

7. Install or connect the following:
- Axle shaft into the axle tube
- Hub onto the shaft. Install, but don't tighten, the nut.
- Hub-to-carrier bolts. Torque to 48 ft. lbs. (65 Nm).
- ABS sensor bushing and sensor, coated with silicone grease
- Rotor shield
- Wheel

8. Lower the vehicle and torque the axle shaft nut to 192 ft. lbs. (260 Nm). Stake the nut.

Discovery/Discovery Series II

1. Before servicing the vehicle, refer to the precautions in the beginning of this section.

✳✳ CAUTION

If so equipped, before beginning, depressurize the air suspension.

2. Remove the center cap from the wheel.

3. Loosen the axle shaft nut.

4. Remove or disconnect the following:
- Wheel
- Brake rotor shield
- Brake rotor
- ABS sensor harness from the brackets
- Sensor from the hub
- Sensor bushing
- 4 bolts securing the hub to the carrier
- Axle shaft nut
- Hub/bearing assembly
- Axle shaft
- Oil Seal

To install:

5. Clean all mating surfaces thoroughly.

6. Apply gear oil to the new seal's lip and outer edge.

7. Drive the new seal into place.

8. Slide the axle shaft into place.

9. Apply anti-seize compound to the mating surfaces of the hub and knuckle.

10. Install or connect the following:
- Hub onto the shaft. Install, but don't tighten, the nut.
- Hub-to-axle carrier bolts. Torque to 74 ft. lbs. (100 Nm).
- ABS sensor bushing and sensor, coated with silicone grease
- Wiring harness to brackets

- Rotor
- Rotor shield
- Wheel

11. Lower the vehicle and torque the axle shaft nut to 360 ft. lbs. (490 Nm). Stake the nut.

Rear Differential Seal

REMOVAL & INSTALLATION

Freelander

1. Before servicing the vehicle, refer to the precautions in the beginning of this section.

2. Raise rear of vehicle.

✳✳ WARNING

Do not work on or under a vehicle supported only by a jack. Always support the vehicle on safety stands.

3. Remove road wheel.

4. Remove bolt securing handbrake cable to subframe.

5. Remove nut and bolt securing trailing link to rear hub, collect spacer from under bolt head.

6. Remove nut and bolt securing fixed transverse link to subframe. Collect dynamic damper.

7. Remove nut and bolt securing adjustable transverse link to subframe.

8. Position drain tin to catch oil spillage.

9. With assistance pull hub assembly outwards.

10. Taking care not damage oil seal 'flinger', release drive shaft from differential using LRT-51-014 and position shaft aside.

11. Remove and discard circlip from drive shaft.

12. Remove differential oil seal.

To install:

13. Clean drive shaft oil seal recess in axle casing.

14. Install new oil seal using LRT-51-012.

15. Clean end of drive shaft and location in differential.

16. Check condition of oil seal 'Flinger', renew if damaged.

17. Install new circlip to drive shaft.

18. With assistance fit drive shaft to differential, push drive shaft fully home to engage circlip.

19. Install nut and bolt to fixed transverse link and tighten to 120 Nm (89 ft. lbs.).

✳✳ CAUTION

Nuts and bolts must be tightened with weight of vehicle on suspension.

20. Install nut, bolt and dynamic damper to adjustable transverse link and tighten to 120 Nm (89 ft. lbs.).

21. Install spacer, nut and bolt to trailing link and tighten to 120 Nm (89 ft. lbs.).

22. Install bolt securing handbrake cable clip and tighten to 22 Nm (16 ft. lbs.).

23. Install road wheel(s) and tighten nuts to 115 Nm (85 ft. lbs.).

24. Check differential oil level.

25. Remove stands and lower vehicle.

Pinion Seal

REMOVAL & INSTALLATION

Discovery/Discovery Series II and Range Rover

FRONT OR REAR

1. Before servicing the vehicle, refer to the precautions in the beginning of this section.

2. Matchmark the driveshaft and flange and disconnect the driveshaft.

3. Hold the flange and remove the nut.

4. Remove the flange.

5. Remove the seal.

To install:

6. Clean all mating surfaces.

7. Lubricate the oil seal lips with gear oil.

8. Drive the seal into place.

9. Position the flange and install the holding tool.

10. Install the nut or bolt. Torque the nut to 100 ft. lbs. (135 Nm); the bolt to 74 ft. lbs. (100 Nm).

11. Connect the driveshaft. Torque the nuts to 35 ft. lbs. (48 Nm).

9308RG83

Pinion seal and flange—Except Freelander

Freelander

REAR

1. Before servicing the vehicle, refer to the precautions in the beginning of this section.

2. Release both drive shafts from differential assembly.

3. Reference mark propeller shaft and pinion flanges to aid reassembly.

4. Remove 4 nuts and bolts securing propeller shaft to differential. Release propeller shaft and tie aside.

5. Check and record the torque required to rotate the pinion and differential.

✳✳ CAUTION

Drive shafts must be removed to obtain correct torque to turn figure.

6. Using LRT-51-003 to restrain differential flange, remove nut and washer securing pinion flange. Discard nut.

7. Remove pinion flange.

8. Carefully remove and discard oil seal, take care not to damage oil seal recess.

9. Remove oil thrower.

10. Remove pinion bearing inner race.

11. Remove and discard collapsible spacer.

To install:

12. Fit new collapsible spacer.

13. Fit pinion bearing and oil thrower.

14. Clean pinion flange and oil seal recess.

15. Fit new oil seal using LRT-51-010.

16. Fit pinion flange and washer.

17. Restrain pinion flange using LRT-51-003.

18. Fit new pinion nut and tighten to 190 Nm (140 ft. lbs.).

19. Check for end float on pinion. If end float exists continue to tighten pinion nut until end float is removed.

20. Continue to tighten pinion nut until correct preload is obtained.

21. Pinion preload is 1.7—2.8 Nm (1.2—2.1 ft. lbs.), if higher replace collapsible spacer.

✳✳ CAUTION

Do not tighten pinion nut to more than 373 Nm (275 ft. lbs.), or the collapsible spacer will compress too far.

22. Clean propeller shaft flange and mating face.

23. Position propeller shaft to rear axle and align reference marks.

24. Tighten propeller shaft nuts and bolts to 65 Nm (48 ft. lbs.).

25. Fit drive shafts.

26. Check differential oil level.

Rear Axle Housing Assembly

REMOVAL & INSTALLATION

Discovery/Discovery Series II and Range Rover

1. Before servicing the vehicle, refer to the precautions in the beginning of this section.

✳✳ CAUTION

Before beginning, depressurize the air suspension.

2. Raise and support the chassis.

3. Support the axle assembly.

4. Matchmark the driveshaft and flange. Disconnect the driveshaft.

5. Remove or disconnect the following:
   - Wheels
   - Shock absorbers from the axle
   - Clips securing the air springs to the axle
   - Panhard rod from the axle

6. Disconnect the brake pipes from the body bracket. Plug the pipes and connections.

7. Remove or disconnect the following:
   - 2 clips securing the brake pipes from the body bracket
   - Banjo bolt and strap securing the breather hose to the axle
   - Height sensors from the trailing arms
   - Trailing arms from the body
   - Trailing arms from the axle

8. Lower the axle from the vehicle.

To install:

9. Install or connect the following:
   - Axle into position
   - Trailing arms to the axle. M16 8.8 grade bolts are torqued to 118 ft. lbs. (160 Nm); M16 10.9 grade bolts are torqued to 177 ft. lbs. (240 Nm); M12 bolts are torqued to 92 ft. lbs. (125 Nm).
   - Trailing arms to the chassis. Torque the bolts to 118 ft. lbs. (160 Nm).
   - Air spring clips
   - Height sensors
   - Shock absorbers to the axle. Torque the nuts to 33 ft. lbs. (45 Nm).
   - Breather hose

   - Brake pipes
   - ABS wiring
   - Panhard rod. Torque the bolt to 148 ft. lbs. (200 Nm).
   - Driveshaft. Torque the nuts to 35 ft. lbs. (48 Nm).

10. Refill the axle.

11. Bleed the brakes.

Front Drive Axle Housing Assembly

REMOVAL & INSTALLATION

Discovery/Discovery Series II and Range Rover

1. Before servicing the vehicle, refer to the precautions in the beginning of this section.

✳✳ CAUTION

Before beginning, depressurize the air suspension.

2. Remove or disconnect the following:
   - Brake pads
   - Caliper. Tie it out of the way.
   - ABS sensors
   - Brake hoses from the knuckles
   - Drag link from the knuckle
   - Panhard rod from the axle
   - Sway bar
   - Track rod
   - Axle oil
   - Driveshaft
   - Height sensors
   - Breather hose

3. Support the axle

4. Remove or disconnect the following:
   - Air spring retaining clips

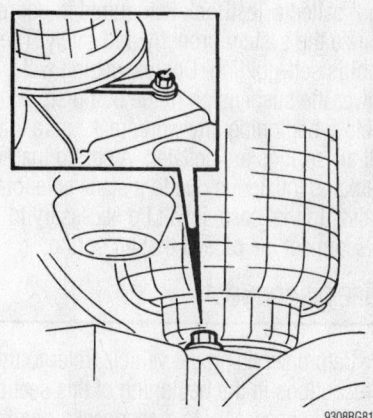

9308RG81

Lower air spring mounting bolt

For Tire, Wheel and Ball Joint specifications, see Section 1 of this manual

- Air springs from the axle
- Shock absorbers from the axle
- Radius arms-to-chassis nuts

5. Move the axle forward and release the radius arms from the chassis brackets.

6. Remove the axle/radius arms assembly.

7. Remove the radius arms from the axle.

To install:

8. Clean all mating surfaces.

9. Install or connect the following:
- Radius arms to the axle. Torque the nuts to 92 ft. lbs. (125 Nm).
- Axle into position. Locate the radius arms, with bushings, into the chassis brackets. Tighten the nuts to 118 ft. lbs. (160 Nm).
- Shock absorbers. Torque the nuts to 33 ft. lbs. (45 Nm).
- Air springs
- Air spring retaining pins and bolts. Torque the bolts to 15 ft. lbs. (20 Nm).
- Breather hose
- Height sensors
- Driveshaft. Torque the nuts to 35 ft. lbs. (48 Nm).
- Track rod. Torque the nuts to 59 ft. lbs. (80 Nm).
- Sway bar

- Panhard rod. Torque the bolt to 148 ft. lbs. (200 Nm).
- Drag link. Torque the nut to 59 ft. lbs. (80 Nm).
- ABS sensors, coated with silicone grease
- Brake hoses
- Calipers. Torque the bolts to 162 ft. lbs. (220 Nm).
- Brake pads
- Axle fluid

Rear Differential Assembly

REMOVAL & INSTALLATION

Freelander

1. Before servicing the vehicle, refer to the precautions in the beginning of this section.

2. Remove both drive shafts.

3. Reference mark rear propeller shaft for reassembly.

4. Position container to catch oil spillage.

5. Remove 4 nuts and bolts securing propeller shaft to differential. Release propeller shaft and tie aside.

6. Support weight of differential assembly on a jack.

7. Remove 2 bolts securing differential to front mounting.

8. Depress red locking collar and disconnect breather pipe from differential casing.

9. Remove 4 bolts securing differential assembly to rear mountings.

10. With assistance, rotate differential assembly through 90 degrees and remove from subframe.

To install:

11. With assistance position differential assembly to subframe and locate in mountings, fit bolts but do not tighten at this stage.

12. Position centralizing jig LRT-51-013 to align differential assembly.

13. Tighten forward bolts to 65 Nm (48 ft. lbs.).

14. Tighten rearward bolts to 65 Nm (48 ft. lbs.).

15. Remove LRT-51-013.

16. Connect breather pipe.

17. Position propeller shaft to rear axle and align reference marks.

18. Install and tighten nuts and bolt securing propeller shaft to rear axle to 65 Nm (48 ft. lbs.).

19. Install drive shafts.

20. Check differential oil level.

STEERING AND SUSPENSION

Air Suspension

DEPRESSURIZING

The air suspension is pressurized up to 150 psi (1034 kPa). Before beginning work on suspension or steering components, the system should be depressurized. A special tool called a TestBook is required to depressurize the system properly and safely. The tool is self-guiding. Depressurizing will lower the suspension to the bump stops. Before beginning any work, make sure that all air springs are deflated. A spring that remains inflated is due to a stuck solenoid valve. In that case, it will be necessary to disconnect the pipe at that air spring.

PIPE DISCONNECT

Before servicing the vehicle, refer to the precautions in the beginning of this section. If it is necessary to disconnect a pipe:

a. Wear hand, ear and eye protection. Wrap a cloth around the connection.

b. Clean the connection with a stiff wire brush.

c. Peel back the boot.

d. Apply equal downward pressure on the collet flange at "A", as shown.

e. Pull the pipe firmly through the center of the collet.

f. To connect the pipe, push it firmly

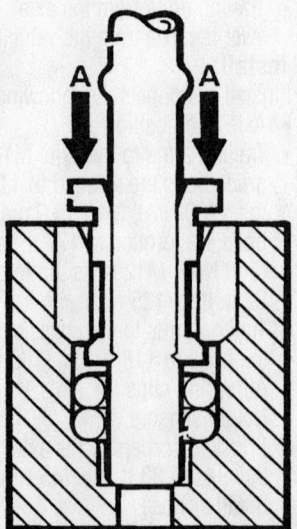

Air pipe disconnection

9308RG78

through the 2 O-rings until it contacts the base of the housing. Gently pull back on the pipe. Some slight movement will be noticed.

g. Reposition the boot.

When the job is done, run the engine to repressurize the system. The system must be recalibrated using the TestBook tool. Recalibration must be done on a floor surface that is level and smooth in all directions.

Air Bag

❋❋ CAUTION

Some vehicles may be equipped with an air bag system. The system must be disabled before performing service on or around system components, steering column, instrument panel components, wiring and sensors. Failure to follow safety and disabling procedures could result in accidental air bag deployment, possible personal injury and unnecessary system repairs.

PRECAUTIONS

Several precautions must be observed when handling the inflator module to avoid accidental deployment and possible personal injury.

• Never carry the inflator module by the wires or connector on the underside of the module.

• When carrying a live inflator module, hold securely with both hands, and ensure that the bag and trim cover are pointed away.

• Place the inflator module on a bench or other surface with the bag and trim cover facing up.

• With the inflator module on the bench, never place anything on or close to the module which may be thrown in the event of an accidental deployment.

Before servicing the vehicle, be sure to also refer to the precautions in the beginning of this section as well.

DISARMING

1. Before servicing the vehicle, refer to the precautions in the beginning of this section.
2. Remove the key from the ignition.
3. Disconnect the negative battery cable first, then the positive cable.
4. Wait 20 minutes for the back-up power to discharge.

ARMING

1. After performing the required service, rearm the SRS by reconnecting the battery.
2. Start the vehicle and the SRS service light should go OFF after 5 seconds.

Power Steering Pump

REMOVAL & INSTALLATION

Freelander

1. Before servicing the vehicle, refer to the precautions in the beginning of this section.
2. Disconnect battery ground lead.
3. Remove engine mounting top arm.
4. Loosen PAS pump pulley bolts.
5. Using a ⅜ in. socket bar, raise ancillary drive belt tensioner and release drive belt from alternator and PAS pump pulleys.
6. Position container to collect PAS fluid spillage.

7. Release clip and disconnect fluid inlet hose from PAS pump.
8. Remove banjo bolt securing fluid outlet hose to PAS pump, release hose, remove and discard sealing washers.

❊❊ CAUTION

To prevent damage to components, use two spanners when loosening or tightening unions.

9. Remove bolt securing PAS outlet pipe support bracket to cylinder head and move pipe aside.
10. Remove 3 bolts securing PAS pump pulley to PAS pump and remove pulley.

❊❊ CAUTION

Always fit plugs to open connections to prevent contamination.

11. Remove 3 bolts securing PAS pump to mounting bracket.
12. Remove PAS pump.
To install:
13. Position PAS pump and align to mounting bracket.
14. Fit bolts securing PAS pump to mounting bracket and tighten to 25 Nm (18 ft. lbs.).
15. Fit PAS pump pulley to PAS pump and tighten bolts to finger tight.
16. Remove plug from fluid outlet hose.
17. Clean banjo bolt and mating faces.
18. Using new sealing washers, fit PAS outlet hose to pump and tighten banjo bolt to 28 Nm.

❊❊ CAUTION

To prevent damage to components, use two spanners when loosening or tightening unions.

19. Align outlet pipe clip. Fit bolt and tighten to 25 Nm (18 ft. lbs.).
20. Clean elbow on PAS pump.
21. Remove plug from fluid inlet hose, fit new clip and connect hose to PAS pump.
22. Remove container.
23. Clean pulley 'V's and tensioner pulley running surface.
24. Using a ⅜ in. square drive socket bar, raise ancillary drive belt tensioner and fit drive belt to pulleys.
25. Tighten PAS pump pulley bolts to 10 Nm (7.5 ft. lbs.).
26. Fit engine mounting top arm.
27. Connect battery ground lead.
28. Bleed PAS system.

Discovery/Discovery Series II and Range Rover

1. Before servicing the vehicle, refer to the precautions in the beginning of this section.
2. Remove the drive belt.
3. Remove the 3 pulley bolts.
4. Disconnect the return hose.
5. Disconnect the high pressure hose.
6. Remove the 4 bolts securing the pump/compressor bracket to the engine.
7. Remove the 3 bolts securing the mounting plate to the pump.
8. Remove the pump.
9. Installation is the reverse of removal. Tighten as follows:

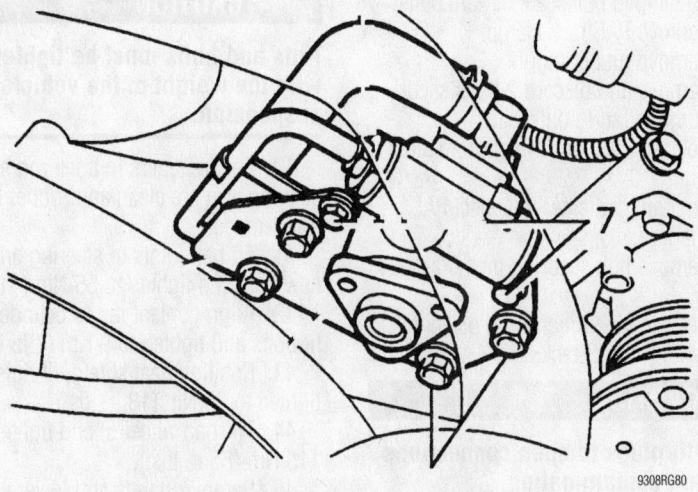

9308RG80

Power steering pump/compressor mounting bracket—Except Freelander

For Wheel Alignment specifications, see Section 1 of this manual

- Lift bracket to the pump: 13 ft. lbs. (18 Nm).
- Bracket to the engine: 30 ft. lbs. (40 Nm).
- Pump mounting bolts: 13 ft. lbs. (18 Nm).
- High pressure fitting: 12 ft. lbs. (16 Nm).

Power Steering Gear

REMOVAL & INSTALLATION

Freelander

1. Before servicing the vehicle, refer to the precautions in the beginning of this section.
2. Raise front of vehicle.

✳✳ WARNING

Do not work on or under a vehicle supported only by a jack. Always support the vehicle on safety stands.

3. Remove front road wheels.
4. Remove and discard nuts securing track rod ball joints to steering arms.
5. Fit an M12 nut to each ball pin, flush with end of each pin.
6. Using LRT-57-043, separate ball pins from right side and left side steering arms. Remove M12 nuts and release ball pins from steering arms.
7. Remove pinch bolt securing steering column to PAS rack pinion.
8. Remove 2 nuts securing steering rack heat shield and remove heat shield.
9. Remove 2 bolts securing coolant rail to cylinder block.
10. Remove 2 bolts and washers securing PAS rack clamp to bulkhead, discard bolts.
11. Remove PAS rack clamp.
12. Remove rubber mount.
13. Remove and discard 2 bolts securing PAS rack mounting to bulkhead.
14. Release PAS rack pinion from steering column.
15. Position container to collect PAS fluid spillage.
16. Remove bolt securing pipe bracket to PAS rack.
17. Release pipe unions and disconnect fluid pipes from PAS rack.

✳✳ CAUTION

Always fit plugs to open connections to prevent contamination.

18. Remove and discard O-rings.
19. Remove bolt securing PAS pipes to clamp and loosen clamp bolt.

20. With assistance remove PAS rack from passenger side of vehicle.
21. Remove dust seal from pinion housing.

To install:
22. Fit PAS rack to vehicle from passengers side.
23. Fit dust shield to pinion housing.
24. Ensure pipe unions are clean.
25. Fit new O-rings to fluid pipes.
26. Fit fluid pipes to PAS rack but do not tighten at this stage.
27. Align fluid pipe bracket to PAS rack, fit bolt but do not tighten at this stage.
28. With assistance fit PAS rack pinion to steering column, ensuring column coupling is aligned with gear input flag.
29. Fit washers and new bolts securing steering rack mounting to bulkhead, but do not tighten at this stage. Ensure large washer is fitted to lower bolt.
30. Fit rubber mount and clamp to PAS rack.
31. Fit bolts securing clamp to bulkhead but do not tighten at this stage.
32. Tighten PAS rack mounting bolts to 45 Nm (33 ft. lbs.).
33. Tighten PAS rack clamp bolts to 45 Nm (33 ft. lbs.).
34. Tighten PAS rack fluid feed pipe union to 18 Nm (13 ft. lbs.).
35. Tighten PAS rack fluid return pipe union to 22 Nm (16 ft. lbs.).
36. Tighten fluid pipe bracket to 10 Nm (7.5 ft. lbs.).
37. Align PAS rack clamp and tighten bolt.
38. Align pipes to clamp fit bolt and tighten to 10 Nm (7.5 ft. lbs.).
39. Fit pinch bolt to steering column and tighten to 32 Nm (24 ft. lbs.).

✳✳ CAUTION

Nuts and bolts must be tightened with the weight of the vehicle on the suspension.

40. Ensure tapers in track rod end and steering arm are clean and rubber boot is not damaged.
41. Fit ball joints to steering arms, fit new nuts and tighten to 55 Nm (41 ft. lbs.).
42. Align coolant rail to cylinder block, fit bolts and tighten to 9 Nm (7 ft. lbs.).
43. Position heat shield, fit nuts and tighten to 25 Nm (18 ft. lbs.)
44. Fit road wheel(s) and tighten nuts to 115 Nm (85 ft. lbs.).
45. Remove stands and lower vehicle.
46. Bleed PAS system.
47. Check and adjust front wheel alignment.

Discovery/Discovery Series II and Range Rover

1. Before servicing the vehicle, refer to the precautions in the beginning of this section.
2. Disconnect the negative battery cable.
3. Drain the fluid from the power steering fluid reservoir.
4. Clean the steering box to prevent dirt from entering when the hoses are removed.
5. Disconnect the feed and return lines from the steering box.
6. Raise and safely support the vehicle.
7. Remove the under tray
8. Disconnect the drag link from the drop arm.
9. Remove the universal joint connecting the steering column to the steering box.
10. Remove the power steering box mounting bolts, then remove the box from the vehicle.

To install:
11. Install the steering box and tighten the mounting bolts as follows:
- Discovery/Discovery Series II: 66 ft. lbs. (90 Nm)
- Range Rover: 92 ft. lbs. (125 Nm)
12. Connect the universal joint and tighten the pinch bolts 18 ft. lbs. (25 Nm).
13. Connect the drag link to the drop arm and tighten the retaining nut as follows:
- Discovery/Discovery Series II: 30 ft. lbs. (40 Nm)
- Range Rover: 59 ft. lbs. (80 Nm)
14. Lower the vehicle.
15. Connect the hydraulic hoses to the steering box and tighten the 16mm thread to 37 ft. lbs. (50 Nm) and the 14mm thread to 22 ft. lbs. (30 Nm).
16. Fill the power steering fluid reservoir.
17. With engine running, test steering system for leaks by holding steering in both full lock directions.
18. Ensure steering wheel is correctly aligned when wheels are positioned straight ahead.
19. If necessary, reposition the steering wheel.
20. Road test vehicle.

Shock Absorbers

REMOVAL & INSTALLATION

Front

RANGE ROVER

✳✳ CAUTION

Be sure to support the axle when the shock is removed, otherwise the

pressurized air spring could fail and cause component damage and possible personal injury. It is possible to remove the shock absorber without depressurizing air springs, BUT the distance between the axle and chassis must be held as if the shock absorber was still fitted. This is achieved by supporting the vehicle on supports, with a jack under the axle.

1. Before servicing the vehicle, refer to the precautions in the beginning of this section.
2. Raise and safely support the vehicle.
3. Support the front axle on a jack.
4. Remove front wheel.

❋❋ WARNING

Do not lower axle when shock absorber is removed. This may result in air spring damage.

5. Remove the lower shock absorber retaining nut.
6. Remove the upper shock absorber retaining bolt.
7. Remove the shock absorber.

To install:
8. Install the shock absorber.
9. Install the upper mounting bolt and tighten to 92 ft. lbs. (125 Nm).
10. Install the lower mounting bolt and tighten to 33 ft. lbs. (45 Nm).
11. Install the front wheel and tighten the lug nuts to 80 ft. lbs. (108 Nm).
12. Remove the jack supporting the axle.
13. Remove the supports and lower vehicle.

DISCOVERY

1. Before servicing the vehicle, refer to the precautions in the beginning of this section.
2. Raise and support the front end securely.
3. Remove the wheel.
4. On the right side, remove the coolant reservoir.
5. Support the weight of the axle with a jack.
6. Loosen the upper bolt securing the shock to the tower.
7. Remove the 4 nuts securing the shock tower to the chassis.
8. Remove the 2 lower shock absorber bolts.

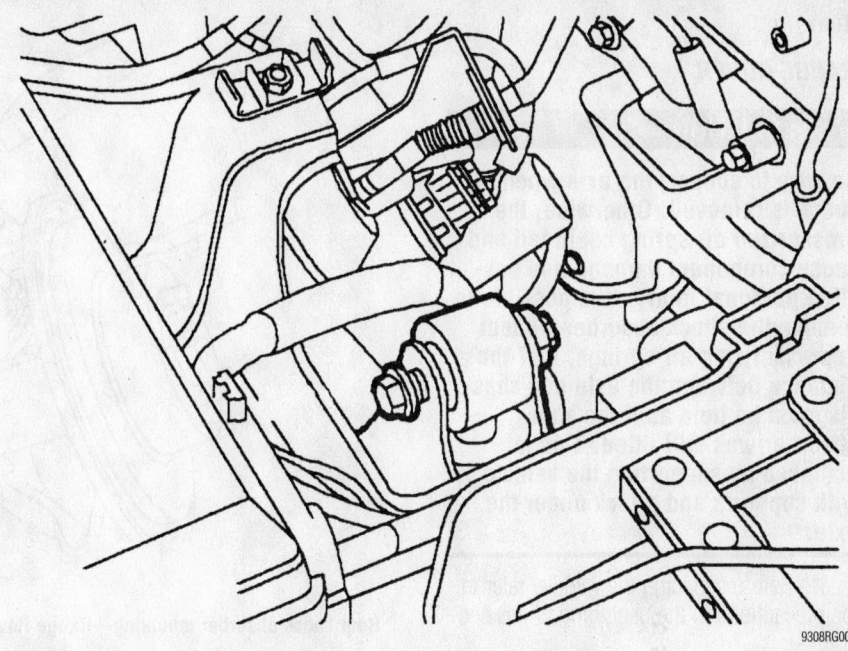

View of the upper front shock absorber bolt—Discovery/Discovery Series II

9308RG00

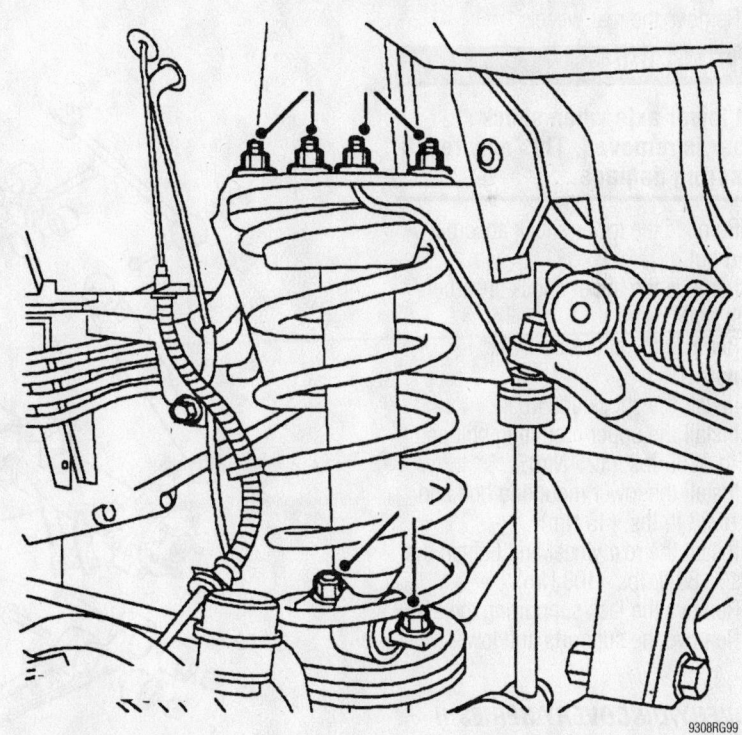

View of the lower front shock absorber attachments—Discovery/Discovery Series II

9308RG99

9. On ACE equipped models, remove the ACE pipe clamp bolt and clamp.
10. Raise the tower and remove the upper bolt.
11. Compress the shock and remove the assembly.

12. Installation is the reverse of removal. Observe the following torques:
- Lower shock bolts: 92 ft. lbs. (125 Nm).
- Tower to chassis nuts: 17 ft. lbs. (23 Nm).
- Upper bolt: 92 ft. lbs. (125 Nm).

For Maintenance Interval recommendations, see Section 1 of this manual

Rear

RANGE ROVER

❋❋ CAUTION

Be sure to support the axle when the shock is removed. Otherwise, the pressurized air spring could fail and cause component damage and possible personal injury. It is possible to remove the shock absorber without depressurizing air springs, BUT the distance between the axle and chassis must be held as if the shock absorber was still fitted. This is achieved by supporting the vehicle with supports and a jack under the axle.

1. Before servicing the vehicle, refer to the precautions in the beginning of this section.
2. Raise and safely support the vehicle.
3. Support the axle on a jack.
4. Remove the rear wheels.

❋❋ WARNING

Do not lower axle when shock absorber is removed. This may result in air spring damage.

5. Remove the lower shock absorber retaining nut.
6. Remove the upper shock absorber retaining bolt.
7. Remove the shock absorber.

To install:
8. Install the shock absorber.
9. Install the upper mounting bolt and tighten to 92 ft. lbs. (125 Nm).
10. Install the lower mounting bolt and tighten to 33 ft. lbs. (45 Nm).
11. Install the rear wheel and tighten the lug nuts to 80 ft. lbs. (108 Nm).
12. Remove the jack supporting the axle.
13. Remove the supports and lower vehicle.

DISCOVERY/DISCOVERY SERIES II

1. Before servicing the vehicle, refer to the precautions in the beginning of this section.
2. Raise and safely support the vehicle.
3. Place a jack under the rear axle and raise it slightly to take the load off the shock absorbers.
4. Remove the shock absorber lower attaching nut, then pull the lower end free of the mounting bracket on the axle housing.
5. Remove the upper attaching nut and remove the shock absorber.

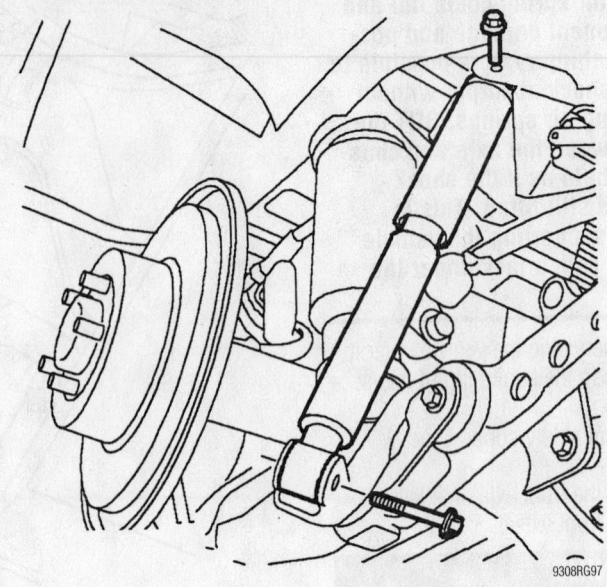

Rear shock absorber mounting—Range Rover

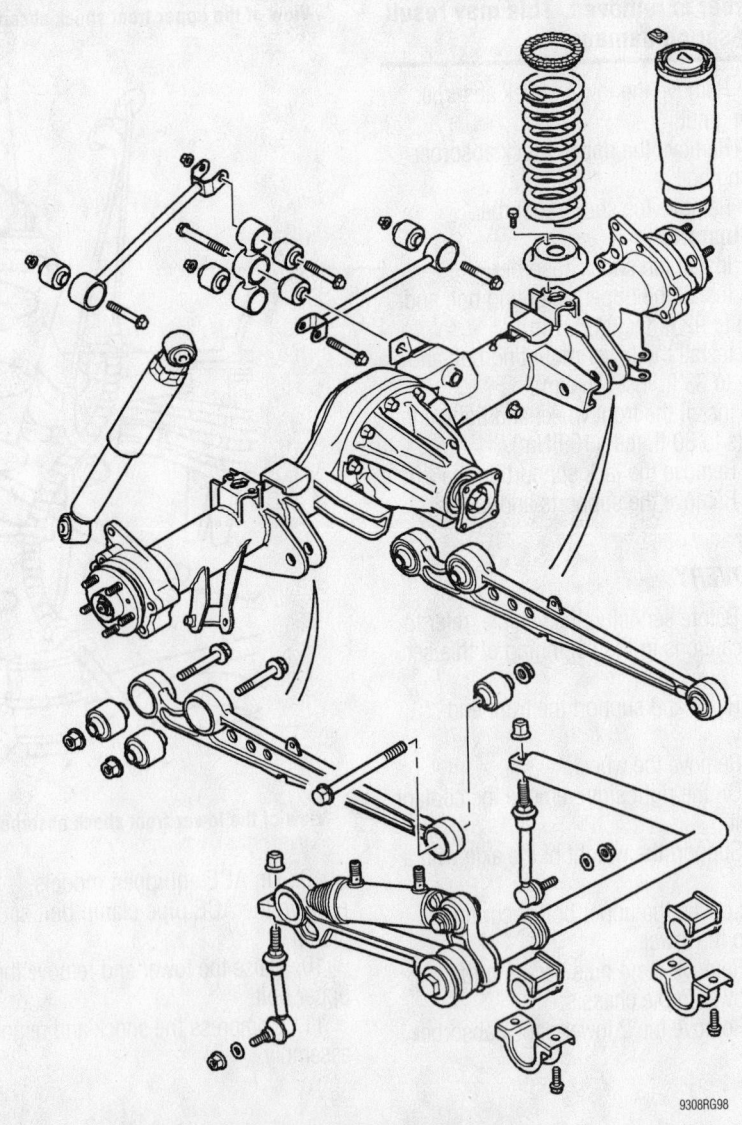

Rear suspension components—Discovery/Discovery Series II

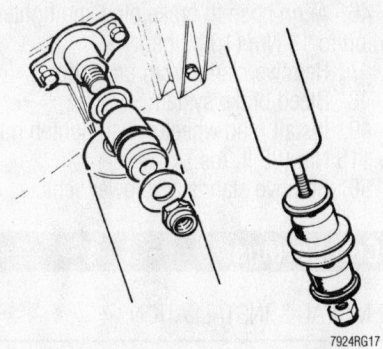

Exploded view of the upper and lower mountings for the rear shock absorber—Except Freelander

To install:

6. Install the shock absorber onto the upper bracket and screw on the attaching nut. Don't tighten it yet.

7. Swing the shock absorber down and position in the lower mounting bracket. Attach the lower mounting nut.

8. Tighten the upper and lower attaching bolts to 92 ft. lbs. (125 Nm).

MacPherson Struts and Springs

REMOVAL & INSTALLATION

Freelander

FRONT

1. Before servicing the vehicle, refer to the precautions in the beginning of this section.

2. Disconnect battery ground lead.

3. Raise front of vehicle.

✳✳ WARNING

Do not work on or under a vehicle supported only by a jack. Always support the vehicle on safety stands.

4. Remove road wheel.

5. Remove clip securing brake hose to bracket on damper.

6. Release ABS sensor harness and brake hose from damper.

7. Release ABS sensor from hub.

8. Remove 2 bolts securing brake caliper to hub. Release caliper from hub and tie aside.

✳✳ CAUTION

Do not allow caliper to hang on brake hose.

9. Remove nut securing track rod to steering arm and break taper joint using LRT-57-043.

10. Remove upper nut from anti-roll bar link, release link and position aside.

✳✳ CAUTION

Use a spanner and an Allen key to prevent ball joint rotating when undoing link.

11. Remove 2 nuts and bolts securing hub to damper.

12. Release damper from hub.

13. On the left side: Remove 2 bolts securing positive and negative leads to fusebox.

14. On the left side: Disconnect multi-plug from engine compartment fuse box.

15. On the left side: Release leads and position aside.

16. On the left side: Remove 3 nuts securing fusebox and position fusebox aside.

17. On the right side: Remove bolt securing coolant reservoir and position reservoir aside.

18. Reference mark top mounting in relationship to body.

19. Remove 3 nuts from damper top mounting and remove spring and damper assembly.

20. Position a suitable spring compressor in vice.

21. Position spring and damper assembly to spring compressor. Compress spring.

✳✳ CAUTION

Note alignment of top mounting, spring and damper dust cover.

22. Compress spring by 2 to 3 cm until loose, hold damper shaft with Allen key, remove and discard mounting plate nut.

23. Remove rebound washer and mounting plate.

24. Remove spring aid and bump plate.

25. Remove spring seat, dust cover and bump stop cup.

26. Remove damper from spring.

27. Release and remove spring from compressor.

To install:

28. Inspect damper, spring mounting rubbers and bearing for deterioration and damage.

29. Clean mating faces of spring, mounting and mounting plate.

30. Clean damper shaft and bump stop plate.

31. Position spring and damper assembly to spring compressor. Compress spring.

32. Install damper to spring, ensure spring locates in cut recess in damper plate.

33. Install bump cup, bump stop and dust cover to damper.

34. Install spring aid and bump plate.

35. Install mounting plate and rebound washer.

36. Using new nut, hold damper shaft with Allen key and tighten nut to 57 Nm (42 ft. lbs.).

✳✳ CAUTION

Note alignment of top mounting, spring and damper dust cover.

37. Release and remove spring from compressor.

38. Clean mating face of top mounting plate.

39. Position damper assembly and align top mounting to body, fit nuts and tighten to 45 Nm (33 ft. lbs.).

40. On the left side: Position fusebox, fit nuts and tighten to 8 Nm (6 ft. lbs.).

41. On the left side: Connect positive and negative leads, fit bolts and tighten to 8 Nm (6 ft. lbs.).

42. On the left side: Connect multiplug to fusebox.

43. On the right side: Position coolant reservoir and secure with bolt.

44. Install hub to damper, fit nuts and bolts and tighten to 205 Nm (151 ft. lbs.).

45. Clean anti-roll bar link taper and mating face.

46. Connect anti-roll bar link, fit new nut and tighten to 55 Nm (41 ft. lbs.).

✳✳ CAUTION

Use a spanner and an Allen key to prevent ball joint rotating when undoing link.

47. Clean track rod taper and mating face.

48. Connect track rod end to steering arm using new nut and tighten nut to 55 Nm (41 ft. lbs.).

49. Clean ABS sensor, smear sensor with an anti-seize grease and fit sensor to hub.

✳✳ CAUTION

Ensure ABS sensor is fully located into hub, so that sensor touches pole wheel teeth.

50. Position ABS harness and brake hose in bracket and secure with clip.

51. Install road wheel(s) and tighten nuts to 115 Nm (85 ft. lbs.).

52. Remove stand and lower vehicle.

53. Connect battery ground lead.

REAR

1. Before servicing the vehicle, refer to the precautions in the beginning of this section.

2. Raise rear of vehicle.

✳✳ WARNING

Do not work on or under a vehicle supported only by a jack. Always support the vehicle on safety stands.

3. Remove road wheel.

4. Clamp brake hose to prevent fluid loss.

5. Position absorbent cloth to catch spillage.

6. Loosen brake pipe union to hose and release union.

✳✳ CAUTION

Always fit plugs to open connections to prevent contamination.

7. Remove clip securing brake hose to bracket on damper. Release brake hose from bracket.

8. Release ABS sensor harness and brake hose from damper.

9. Release ABS sensor from hub.

10. Remove 2 nuts and bolts securing hub to damper.

11. Release damper from hub.

12. Remove rear quarter lower casing.

13. Remove 3 nuts from damper top mounting and remove spring and damper assembly.

14. Remove rubber seal from top mounting.

15. Position a suitable spring compressor in vice.

✳✳ CAUTION

Note alignment of top mounting, spring and damper dust cover.

16. Position spring and damper assembly to spring compressor. Compress spring.

17. Reference mark between top mounting and spring.

18. Remove cover from top mounting.

19. Compress spring by 2 to 3 cm until loose, hold damper shaft with Allen key, remove and discard mounting plate nut.

20. Remove top mounting plate.

21. Remove rebound washer and mounting plate.

22. Remove spring aid and bump plate.

23. Remove spring seat, dust cover and bump stop cup.

24. Remove damper from spring.

25. Release and remove spring from compressor.

To install:

26. Inspect damper, spring mounting rubbers and bearing for deterioration and damage.

27. Clean mating faces of spring, mounting and mounting plate.

28. Clean damper shaft and bump stop plate.

29. Position spring and damper assembly to spring compressor. Compress spring.

30. Install damper to spring, ensure spring locates in cut recess in damper plate.

31. Install bump stop, bump stop cup and dust cover to damper.

32. Install spring aid and bump plate.

33. Install mounting plate and rebound washer.

34. Using new nut, hold damper shaft with Allen key and tighten nut to 57 Nm (42 ft. lbs.).

✳✳ CAUTION

Note alignment of top mounting, spring and damper dust cover.

35. Install top mounting cover.

36. Release and remove spring from compressor.

37. Clean mating face of top mounting plate.

38. Install rubber seal to top mounting.

39. Position damper assembly and align top mounting to body, fit nuts and tighten to 45 Nm (33 ft. lbs.).

40. Install rear quarter lower trim casings.

41. Install hub to damper and tighten bolts to 205 Nm (151 ft. lbs.).

42. Clean ABS sensor, smear sensor with an anti-seize grease and fit sensor to hub.

✳✳ CAUTION

Ensure ABS sensor is fully located into hub, so that sensor touches pole wheel teeth.

43. Secure brake hose and ABS sensor harness to damper.

44. Secure brake hose with 'C' clip.

45. Remove plugs and clean brake pipe male end.

46. Align hose to brake pipe and tighten union to 14 Nm (10 ft. lbs.).

47. Remove clamp from brake hose.

48. Bleed brake system.

49. Install road wheel(s) and tighten nuts to 115 Nm (85 ft. lbs.).

50. Remove stands and lower vehicle.

Coil Springs

REMOVAL & INSTALLATION

The Discovery models are equipped with a coil spring suspension. The Range Rover and Discovery Series II models are equipped with an air spring suspension.

Discovery and Range Rover

FRONT

1. Before servicing the vehicle, refer to the precautions in the beginning of this section.

2. Raise and safely support the vehicle.

3. Disconnect the sway bar end links.

4. Remove the shock absorber-to-axle bolts.

5. Slowly lower the axle to relieve the spring tension. DO NOT STRETCH THE BRAKE HOSE!

6. Remove the spring.

To install:

7. Install the spring in the upper seat.

8. Raise the axle until the spring is seated in the lower spring seat.

9. Connect the end links.

10. Install the shock absorber.

REAR

1. Before servicing the vehicle, refer to the precautions in the beginning of this section.

2. Raise and safely support the vehicle.

3. Remove the wheels.

4. Support the axle on a jack.

5. Disconnect the shock absorber from the axle bracket.

6. Unclip the brake pipe from its bracket.

7. Disconnect the ABS sensor.

8. Lower the rear axle until the coil springs are no longer under compression. DO NOT STRETCH THE BRAKE LINE!

To install:

9. Install the coil spring, top end first, and position it in the lower perch using a twisting motion.

10. Raise the axle to the normal ride position and install the shock absorber.

11. Connect the ABS sensor and brake pipe.

7924RG18

View of the rear spring mounting—Except Freelander

12. Install the wheels. Tighten the lug nuts to 103 ft. lbs. (140 Nm).

13. Lower the vehicle.

Air Springs

REMOVAL & INSTALLATION

Front

1. Before servicing the vehicle, refer to the precautions in the beginning of this section.

2. Depressurize the system.

3. Remove the wheel well liner.

4. Disconnect the air tube from the spring. Seal the openings.

5. Place supports under the front crossmember.

6. Remove the clips securing the air spring to the chassis.

7. Remove the bolt securing the air spring to the axle.

8. Remove the pin.

9. Raise the chassis slightly with a jack to provide clearance.

10. Remove the spring.

❋❋ WARNING

Keep the chassis supported until the system is repressurized. Never allow the chassis to rest on a deflated spring.

To install:

11. Clean the mating surfaces of the chassis, axle and spring.

12. Place the air spring on the axle and install the pin and bolt.

13. Lower the chassis just enough to connect the air spring.

14. Connect the pipe.

15. Repressurize the system.

16. Install the liner.

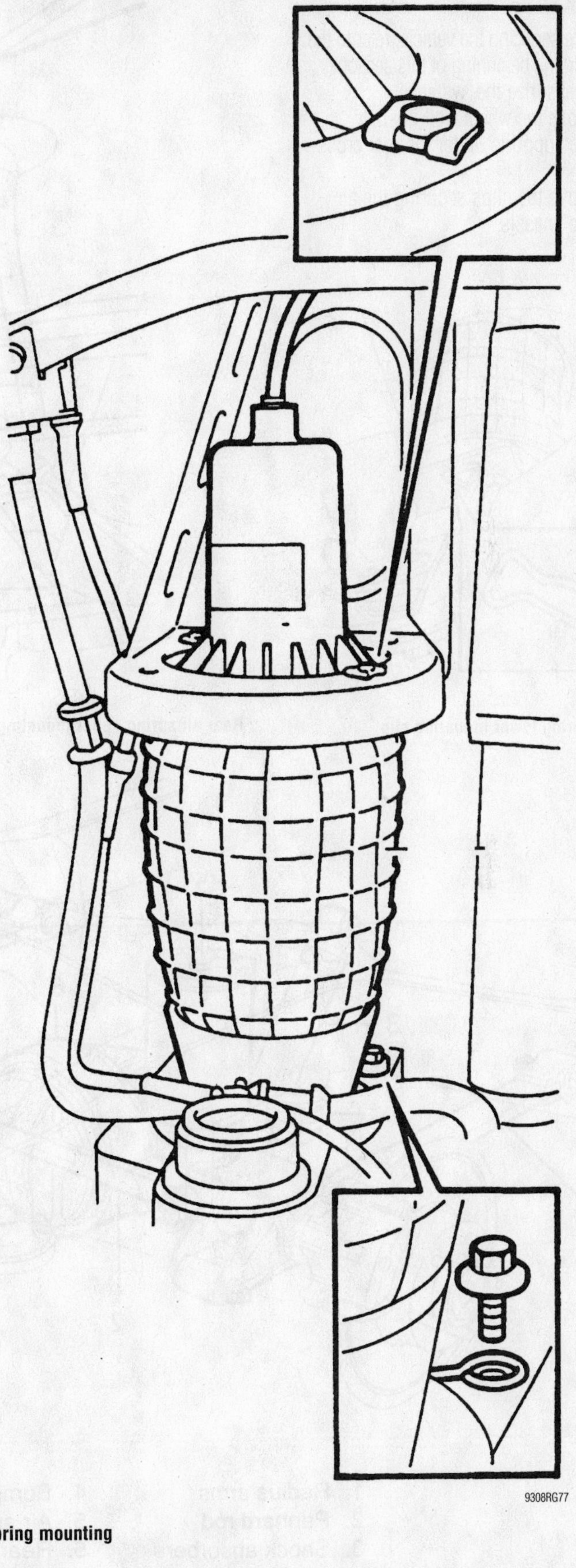

9308RG77

Front air spring mounting

Rear

1. Before servicing the vehicle, refer to the precautions in the beginning of this section.
2. Depressurize the system.
3. Remove the wheel well liner.
4. Place supports under the rear cross-member.
5. Remove the clips securing the air spring to the chassis.

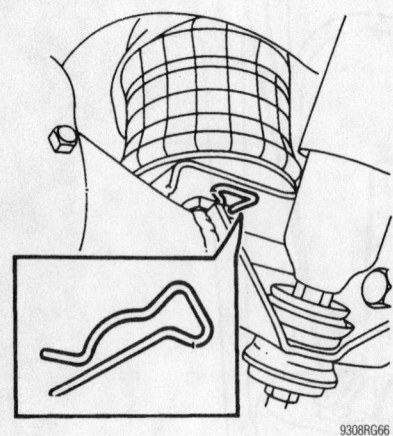

9308RG66

Rear air spring lower mounting clip

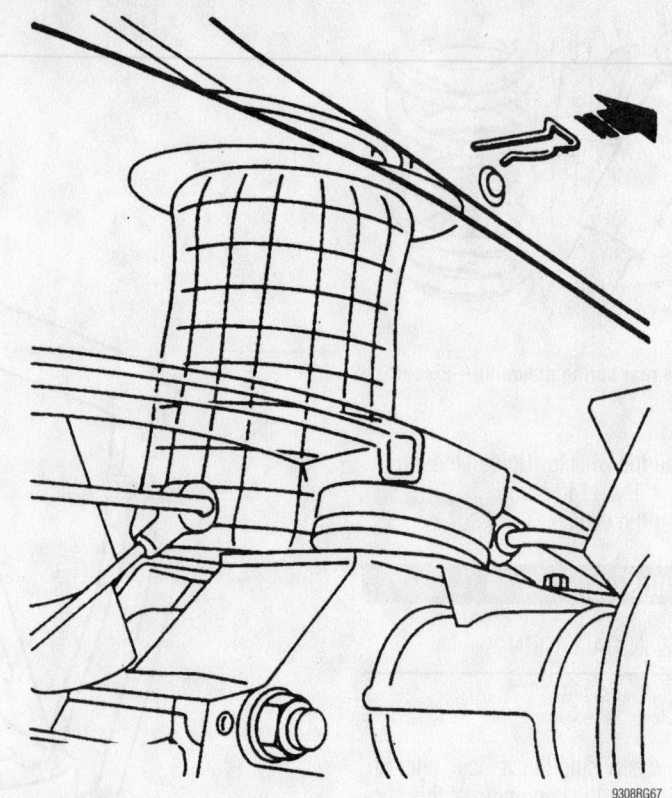

9308RG67

Rear air spring upper mounting clip

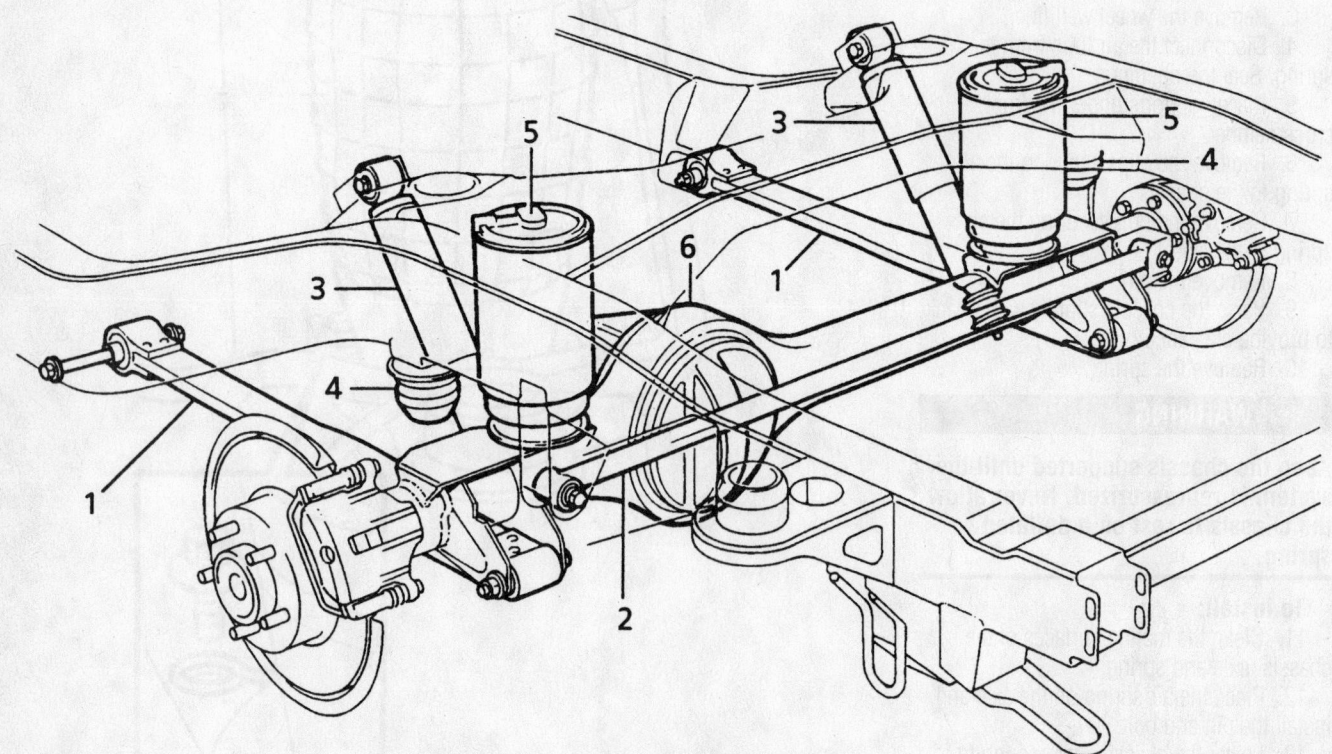

9308RG68

1. Radius arms
2. Panhard rod
3. Shock absorbers
4. Bump stops
5. Air springs
6. Rear axle

Rear air suspension

6. Remove the bolt securing the air spring to the axle.

7. Raise the chassis slightly with a jack to provide clearance.

8. Move the air spring away from the chassis.

9. Disconnect the air tube from the spring. Seal the openings.

10. Remove the spring.

✳✳ WARNING

Keep the chassis supported until the system is repressurized. Never allow the chassis to rest on a deflated spring.

To install:

11. Clean the mating surfaces of the chassis, axle and spring.

12. Connect the pipe.

13. Connect the spring to the chassis.

14. Lower the chassis just enough to connect the air spring to the axle.

15. Place the air spring on the axle and install the clip.

16. Repressurize the system.

17. Install the liner.

Upper Ball Joint

REMOVAL & INSTALLATION

➡Each ball joint can be replaced up to 3 times before the yoke bore becomes over-sized. At each replacement, the yoke should be marked. Factory-trained technicians mark the yoke with yellow paint at each replacement.

Range Rover

1. Before servicing the vehicle, refer to the precautions in the beginning of this section.

2. Remove the swivel hub.

3. Using a ball joint forcing tool, such as LRT-54-008/4 and /5, remove the ball joint.

To install:

4. Clean the yoke thoroughly.

5. Place a yellow paint stripe on the yoke adjacent to the bore.

6. Place a forcing tool, such as LRT-54-008-8, on the yoke.

7. Place base LRT-54-008/7 on the yoke. Position the ball joint and align the tool.

8. Press the ball joint into the yoke.

9. Install the swivel.

Discovery

1. Before servicing the vehicle, refer to

the precautions in the beginning of this section.

2. Remove or disconnect the following:
- Axle shaft
- Mud shield
- Track rod
- Drag link
- Ball joint nuts
- Break loose the knuckle

3. Using forcing tools, such as LRT-54-008 and 008/4 and /5, force the joint from the yoke.

To install:

4. Clean the yoke thoroughly.

5. Place a yellow paint stripe on the yoke adjacent to the bore.

6. Place a forcing tool, such as LRT-54-008-8, on the yoke.

7. Place base LRT-54-008/7 on the yoke. Position the ball joint and align the tool.

8. Press the ball joint into the yoke.

9. Position the knuckle on the ball studs. Torque the upper nut to 81 ft. lbs. (110 Nm); the lower nut to 100 ft. lbs. (135 Nm).

10. Connect the track rod and drag link. Torque the nuts to 59 ft. lbs. (81 Nm).

11. The remainder of assembly is the reverse of disassembly.

Lower Ball Joint

REMOVAL & INSTALLATION

Freelander

The ball joints are not replaceable. If a ball joint becomes defective, the arm must be replaced.

Range Rover

➡Each ball joint can be replaced up to 3 times before the yoke bore becomes over-sized. At each replacement, the yoke should be marked. Factory-trained technicians mark the yoke with yellow paint at each replacement.

✳✳ CAUTION

Each ball joint can be replaced up to 3 times before the axle yoke bore becomes oversize. Before commencing work, clean the surrounding area of the joint to be renewed and check for yellow paint marks. If more than 2 marks are found, the axle case must be renewed.

1. Before servicing the vehicle, refer to the precautions in the beginning of this section.

2. Remove the swivel hub.

3. Assemble the ball joint press tools LRT-54-008/10 and /11 as shown, then press the joint out of the axle yoke.

➡When the ram lead screw reaches the end of the stroke, retract the lead screw, screw the ram into the base tool and repeat the operation the until joint is free from the axle.

4. Remove the screw and collect the adapter from the base tool.

To install:

5. Clean the joint location and the surrounding area of the axle yoke, then make a 12mm wide yellow paint stripe on the axle yoke, adjacent to joint location.

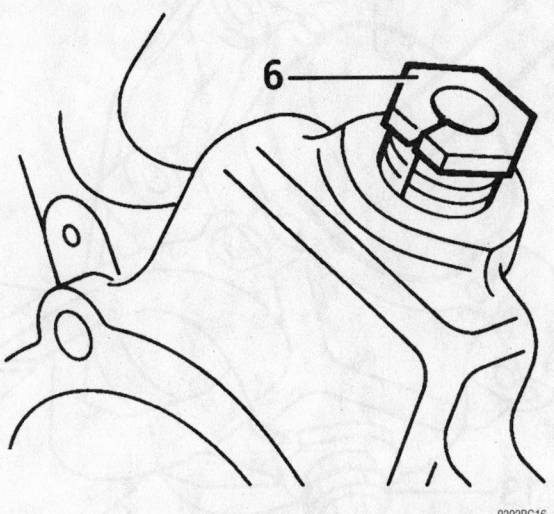

6

9302RG16

Be sure to install the adjustable collet (6) into the swivel hub—Range Rover

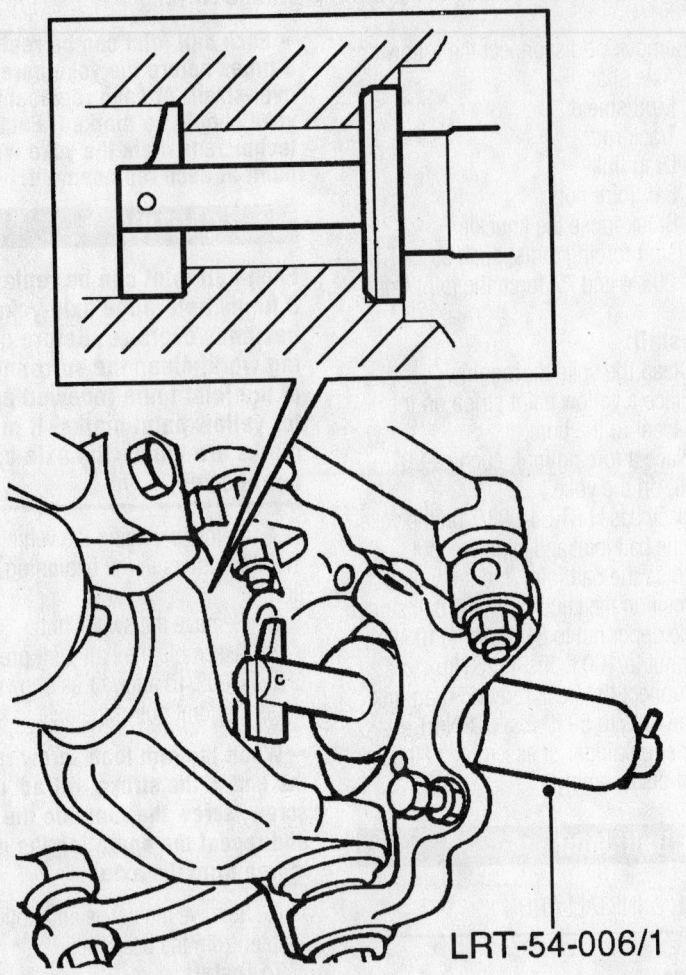

Install the tool into the axle casing with the TOP mark facing out upwards—Range Rover

LRT-54-006/1

9302RG17

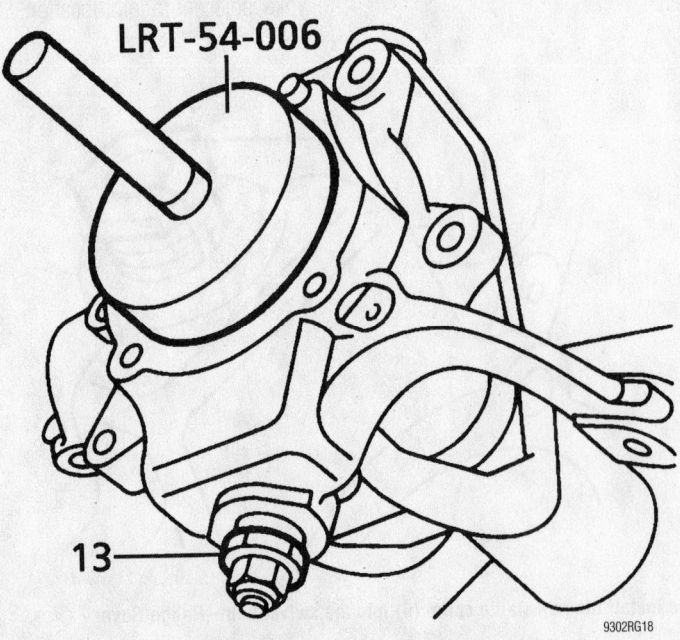

Using tool LTR-54-006/2 to adjust the height of the hub—Range Rover

9302RG18

6. Align the tool assembly and press the joint into the axle yoke.

7. Remove the tools from the axle yoke.

8. Install the swivel hub onto the axle.

Discovery

1. Before servicing the vehicle, refer to the precautions in the beginning of this section.

2. Remove or disconnect the following:

- Axle shaft
- Mud shield
- Track rod
- Drag link
- Ball joint nuts
- Break loose the knuckle

3. Using forcing tools, such as LRT-54-008 and 008/11 and /12, force the joint from the yoke.

To install:

4. Clean the yoke thoroughly.

5. Place a 12mm yellow paint stripe on the yoke adjacent to the bore.

6. Place a forcing tools, such as LRT-54-008, 022 and 008/14on the yoke.

7. Position the ball joint and align the tools.

8. Press the ball joint into the yoke.

9. Position the knuckle on the ball studs. Torque the upper nut to 81 ft. lbs. (110 Nm); the lower nut to 100 ft. lbs. (135 Nm).

10. Connect the track rod and drag link. Torque the nuts to 59 ft. lbs. (81 Nm).

11. The remainder of assembly is the reverse of disassembly.

Lower Control Arm

REMOVAL & INSTALLATION

Freelander

FRONT

1. Before servicing the vehicle, refer to the precautions in the beginning of this section.

2. Raise front of vehicle.

✸✸ WARNING

Do not work on or under a vehicle supported only by a jack. Always support the vehicle on safety stands.

3. Remove road wheel.

4. Remove underbelly panel.

5. Release stake in drive shaft nuts.

6. With an assistant applying brakes,

remove and discard drive shaft hub nut.

7. Remove clip securing brake hose to bracket on damper.

8. Release ABS sensor harness and brake hose from damper.

9. Release ABS sensor from hub.

10. Remove 2 bolts securing brake caliper to hub. Release caliper from hub and tie aside.

✳✳ CAUTION

Do not allow caliper to hang on brake hose.

11. Remove 2 nuts and bolts securing hub to damper.

12. Release drive shaft from hub.

13. Tie drive shaft aside.

14. Remove nut securing lower arm ball joint, discard nut.

15. Break taper joint using LRT-57-043.

16. Remove hub assembly.

17. Remove 2 bolts securing lower arm rear bush housing.

18. Remove bolt securing lower arm front mounting.

19. Remove lower arm.

20. Remove nut from rear mounting, remove snubber washer and remove mounting.

✳✳ CAUTION

Note orientation of snubber washer.

To install:

21. Clean rear mounting mating faces.

22. Install rear mounting and snubber rubber to lower arm, fit nut but do not tighten at this stage.

✳✳ CAUTION

Ensure correct orientation of snubber washer. Ensure that 'OUT' is visible on snubber washer when fitted.

23. Position lower arm and align to subframe fit bolt but do not tighten at this stage.

24. Clean hub to lower arm ball joint mating faces.

25. Install hub assembly to lower joint, fit new nut and tighten to 65 Nm (48 ft. lbs.).

26. Clean drive shaft and flange splines.

27. Install drive shaft to hub.

28. Install new hub nut but do not tighten at this stage.

29. Install hub to damper, fit nuts and bolts and tighten to 205 Nm (151 ft. lbs.).

30. Position caliper to brake disc fit bolts and tighten to 100 Nm (74 ft. lbs.).

31. Install brake hose to bracket on damper and secure with clip.

32. Clean ABS sensor, smear sensor with an anti-seize grease and fit sensor to hub.

✳✳ CAUTION

Ensure ABS sensor is fully located into hub, so that sensor touches pole wheel teeth.

33. Install ABS sensor lead to bracket.

34. Tighten lower arm front bush bolts to 190 Nm (140 ft. lbs.).

✳✳ CAUTION

Nuts and bolts must be tightened with weight of vehicle on suspension.

35. Align bush housings ensuring roll pin is correctly located. Install bolts and tighten to 105 Nm (77 ft. lbs.).

36. Tighten rear bush housing nut to 140 Nm (103 ft. lbs.).

37. Install new drive shaft nut and tighten to 400 Nm (295 ft. lbs.). Stake nut to shaft.

38. Install underbelly panel.

39. Install road wheel(s) and tighten nuts to 115 Nm (85 ft. lbs.).

40. Remove stands and lower vehicle.

REAR

1. Before servicing the vehicle, refer to the precautions in the beginning of this section.

2. Raise rear of vehicle.

✳✳ WARNING

Do not work on or under a vehicle supported only by a jack. Always support the vehicle on safety stands.

3. Remove road wheel.

4. Remove nut and bolt securing trailing link to hub.

5. Remove bolt securing trailing link to bracket.

6. Remove trailing link.

To install:

7. Fit trailing link.

8. Fit bolt to bracket but do not tighten at this stage.

9. Align trailing link to hub, fit nut and bolt but do not tighten at this stage.

10. Support weight of vehicle with jack under rear hub.

✳✳ CAUTION

Nuts and bolts must be tightened with weight of vehicle on suspension.

11. Tighten nuts and bolts to 120 Nm (89 ft. lbs.).

12. Fit road wheel(s) and tighten nuts to 115 Nm (85 ft. lbs.).

13. Remove stands and lower vehicle.

14. Check and if necessary adjust rear wheel alignment.

BUSHING REPLACEMENT

Freelander

LOWER ARM FRONT

1. Before servicing the vehicle, refer to the precautions in the beginning of this section.

2. Remove underbelly panel.

3. Raise front of vehicle.

✳✳ WARNING

Do not work on or under a vehicle supported only by a jack. Always support the vehicle on safety stands.

4. Remove 2 bolts securing rear bush housing to rear beam.

5. Release rear bush housing from body dowel.

6. Remove bolt securing lower arm front mounting.

7. Release lower arm front mounting from rear beam.

8. Using LRT-60-008 and adapters remove lower arm bush.

To install:

9. Ensure bush bore in hub is clean.

10. Using LRT-60-008 fit new bush into lower arm.

11. Align lower arm to rear beam, fit bolt but do not tighten at this stage.

12. Align rear bush housing to body dowel, fit bolts but do not tighten at this stage.

13. Tighten rear bush housing bolts to 105 Nm (77 ft. lbs.).

✳✳ CAUTION

Nuts and bolts must be tightened with the weight of the vehicle on the suspension.

14. Tighten lower arm front bush bolts to 190 Nm (140 ft. lbs.).

15. Fit underbelly panel.

Timing belt service is covered in Section 3 of this manual

16. Remove stands and lower vehicle.

17. Check and, if necessary, adjust wheel alignment.

LOWER ARM REAR

1. Before servicing the vehicle, refer to the precautions in the beginning of this section.

2. Raise front of vehicle.

❄❄ WARNING

Do not work on or under a vehicle supported only by a jack. Always support the vehicle on safety stands.

3. Remove nut securing lower arm rear bush housing and remove snubber rubber.

❄❄ CAUTION

Note orientation of snubber washer.

4. Remove 2 bolts securing lower arm rear bush housing.

5. Release bush locating pin from body and remove bush housing.

To install:

6. Clean lower arm and bush housing mating faces.

7. Fit bush housing to lower arm and locate dowel into body.

8. Fit bolts securing lower arm rear bush housings and tighten to 105 Nm (77 ft. lbs.).

9. Fit snubber rubber and nut but do not tighten at this stage.

❄❄ CAUTION

Ensure correct orientation of snubber washer. Ensure that 'OUT' is visible on snubber washer when fitted.

10. Tighten rear bush housing nut to 140 Nm (103 ft. lbs.).

❄❄ CAUTION

Nuts and bolts must be tightened with the weight of the vehicle on the suspension.

11. Remove stand and lower vehicle.

12. Check front wheel alignment.

Wheel Bearings

ADJUSTMENT

The front and rear wheel bearings are not adjustable.

REMOVAL & INSTALLATION

Front

DISCOVERY/DISCOVERY SERIES II AND RANGE ROVER

1. Before servicing the vehicle, refer to the precautions in the beginning of this section.

2. Raise and safely support the vehicle and remove the front wheels.

3. Unbolt the brake caliper and position it aside, leaving the hose attached.

4. Remove the dust cap.

5. Remove the circlip and shim from driveshaft.

6. Remove the 5 bolts, then remove the driving member and joint washer.

7. Bend back the lockwasher tabs, then remove the locknut and washer.

8. Remove the hub adjusting nut.

9. Remove the spacing washer.

10. Remove the hub and brake disc assembly with bearings.

11. Remove the outer bearing.

12. Matchmark the hub to the brake disc, then remove the 5 hub-to-brake disk attaching bolts and separate the hub from the brake disc.

13. Remove the grease seal and inner bearing from the hub and discard the seal.

14. Using a punch and hammer, drive out the inner and outer bearing races.

15. Clean the hub of any old grease, and dry it.

To install:

16. Using a suitable bearing driver, install the inner and outer bearing races.

17. Pack the hub inner bearing with grease and install.

18. Using a suitable seal driver, install the seal until it is flush with the rear face of the hub. Apply grease between the seal lips.

19. Align the brake disc-to-hub matchmarks, apply Loctite® 270 to the mounting bolts and tighten to 54 ft. lbs. (73 Nm).

20. Grease and install the outer bearing to the hub.

21. Clean the stub axle and driveshaft

22. Install the hub assembly onto the axle and place the spacing washer .

23. Install the hub adjusting nut and tighten to 45 ft. lbs. (61 Nm).

24. Loosen the adjusting nut 90˚, then tighten to 35 inch lbs. (4 Nm). This will give the required hub end-float of 0.0004 inch (0.010mm)

25. Install a new lockwasher and locknut.

26. Tighten the locknut to 45 ft. lbs. (61 Nm).

27. Fold the tab over the lockwasher to secure the adjusting nut and locknut.

28. Install driving member with a new joint washer and tighten the hub retaining bolts to 48 ft. lbs. (65 Nm).

29. Install the original driveshaft shim and secure the shim with a circlip.

30. Check the driveshaft end-play by mounting a dial gauge as shown.

31. Install a suitable bolt to the threaded end of the driveshaft. Move the driveshaft in and out noting the dial gauge reading. The end-play should be between 0.0032–0.0098 inch (0.08–0.25 mm).

32. If out of specification, remove the circlip, measure the shim thickness and install an appropriate

shim to give required end-play.

33. Remove the bolt from the driveshaft, install the circlip and dust cap.

34. Install the brake caliper and tighten the mounting bolts to 60 ft. lbs. (82 Nm).

35. Install the wheels, lower the vehicle and tighten the lug nuts to 93 ft. lbs. (126 Nm).

36. Operate the brake pedal to locate brake pads before taking vehicle for a road test.

FREELANDER

1. Before servicing the vehicle, refer to the precautions in the beginning of this section.

2. Raise front of vehicle.

❄❄ WARNING

Do not work on or under a vehicle supported only by a jack. Always support the vehicle on safety stands.

3. Remove road wheel.

4. Release stake in drive shaft nut.

5. With an assistant applying brakes, remove and discard drive shaft hub nut.

6. Remove clip securing RIGHT SIDE brake hose to support bracket, release hose from bracket.

Release ABS sensor and pad wear sensor harnesses from bracket.

7. Remove 2 bolts securing brake caliper to hub. Release caliper from hub and tie aside.

❄❄ CAUTION

Do not allow caliper to hang on brake hose.

8. Mark brake disc to hub relationship.

9. Remove 2 screws securing brake disc and remove brake disc.

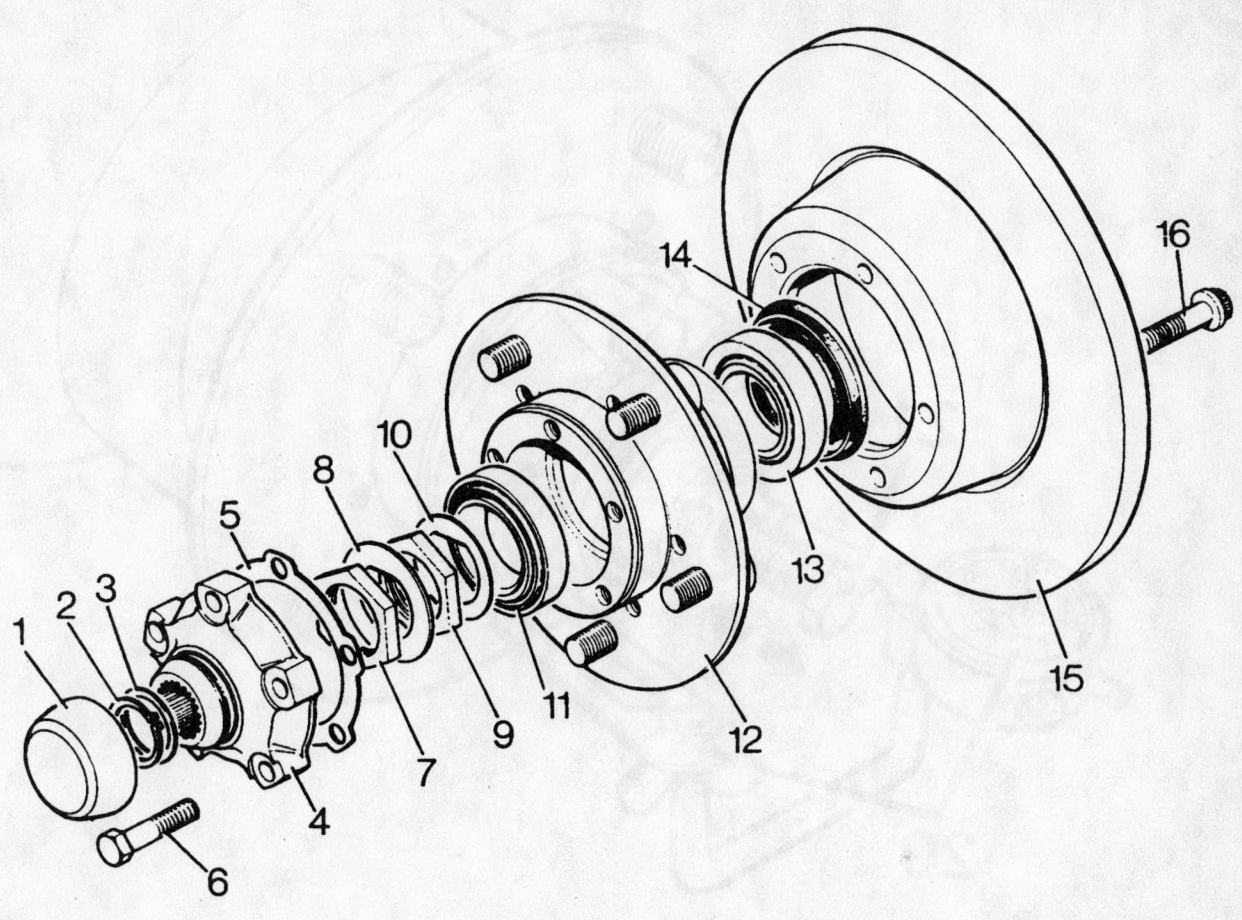

HUB COMPONENTS

1. Dust cap.
2. Drive shaft circlip.
3. Drive shaft shim.
4. Drive member.
5. Drive member joint washer.
6. Drive member retaining bolt.
7. Lock nut.
8. Lock washer.

9. Hub adjusting nut.
10. Spacing washer.
11. Outer bearing.
12. Hub.
13. Inner bearing.
14. Grease seal.
15. Brake disc.
16. Disc retaining bolt.

9302RG19

Exploded view of the front hub and related components—Discovery/Discovery Series II

10. Remove 2 nuts and bolts securing hub to damper.
11. Release hub from damper.
12. Release drive shaft from hub.
13. Restrain hub from rotating and remove nut from lower swivel joint.

14. Break taper joint using LRT-57-043.
15. Remove swivel hub.

➡**Do not carry out further dismantling if component is removed for access only.**

16. Remove 3 bolts securing brake disc shield.
17. Position hub assembly to press, support on tools LRT-54-017 and press out drive flange using tool LRT-54-014.
18. Remove brake disc shield.

Heater Core replacement is covered in Section 2 of this manual

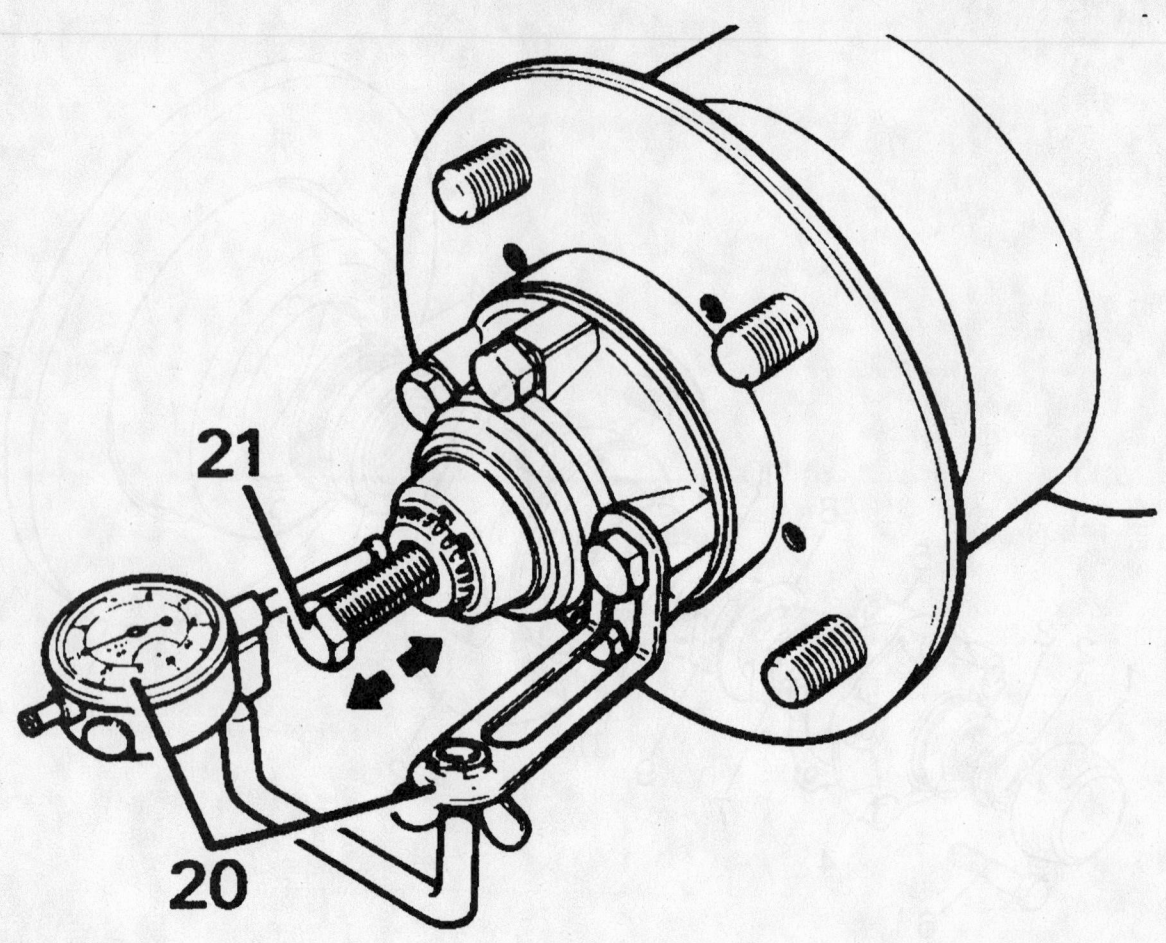

Install the dial indicator and bracket (20), then push and pull on bolt (21) to check the end-play—Discovery/Discovery Series II

9302RG20

→Outer bearing track will remain on drive flange.

19. Remove bearing sealing plate from inner track.

20. Position drive flange in a vice.

21. Clamp both halves of a suitable bearing separator around inner track ensuring that inner lip fits in groove on inner track.

→From 2002MY the groove in the inner track was deleted. To remove the inner track, clamp the separator around the bearing inner track surface.

22. Using tool LRT-99-500 and thrust pad LRT-54-014 withdraw inner track from drive flange.

23. Remove circlip from bearing.

24. Position hub to press and press out bearing using tool LRT-54-015 and LRT-54-017, discard bearing.

✳✳ CAUTION

Never re-use existing bearing.

To install:

25. Clean hub and drive flange.

26. Support hub on tool LRT-54-016 and press in new bearing using LRT-54-015.

27. Fit circlip to hub.

28. Fit brake disc shield, fit bolts and tighten to 8.5 Nm (6.5 ft. lbs.).

29. Support bearing on tool LRT-54-015 and press drive flange into bearing using LRT-54-014.

30. Clean drive shaft.

31. Fit hub assembly to lower joint, fit new nut and tighten to 65 Nm (48 ft. lbs.).

32. Fit drive shaft to hub.

33. Fit hub to damper, fit nuts and bolts and tighten to 205 Nm (151 ft. lbs.).

34. Clean brake disc to drive flange mating faces.

35. Fit disc to drive flange, align reference marks, fit screws and tighten to 5 Nm (4 ft. lbs.).

36. Clean mating faces of caliper and hub.

37. Position caliper to brake disc fit bolts and tighten to 100 Nm (74 ft. lbs.).

38. Fit brake hose to abutment bracket and fit clip.

39. Clean ABS sensor, smear sensor with an anti-seize grease and fit sensor to hub.

✳✳ CAUTION

Ensure ABS sensor is fully located into hub, so that sensor touches pole wheel teeth.

40. Fit ABS sensor lead to bracket.

41. Fit new drive shaft nut and tighten to 400 Nm (295 ft. lbs.). Stake nut to shaft.

42. Stake drive shaft hub nut.

43. Fit road wheel(s) and tighten nuts to 115 Nm (85 ft. lbs.).

44. Remove stands and lower vehicle.

Rear

FREELANDER

1. Before servicing the vehicle, refer to the precautions in the beginning of this section.

2. Raise rear of vehicle.

Do not work on or under a vehicle supported only by a jack. Always support the vehicle on safety stands.

3. Remove road wheel.
4. With assistant depressing the brake pedal, remove and discard drive shaft nut.
5. Remove brake shoe assembly.
6. Using LRT-70-007 release handbrake cable from backplate and remove from backplate.
7. Remove clip securing brake hose to bracket on damper.
8. Disconnect brake pipe union from wheel cylinder.

※※ CAUTION

Always fit plugs to open connections to prevent contamination.

9. Release ABS sensor from hub.
10. Remove nut and bolt securing trailing link to hub.
11. Remove nut, bolt and washers securing transverse links to hub.
12. Remove 2 nuts and bolts securing hub to damper.
13. Release damper from hub.
14. Remove hub assembly from drive shaft.

➡**Do not carry out further dismantling if component is removed for access only.**

15. Position hub assembly to press, support on tools LRT-54-017 and press out drive flange using tool LRT-54-014.

➡**Outer bearing track will remain on drive flange.**

16. Remove bearing sealing plate from inner track.
17. Position drive flange in a vice.
18. Clamp both halves of a suitable bearing separator around inner track ensuring that inner lip fits in groove on inner track.

➡**From 2002MY the groove in the inner track was deleted. To remove the inner track, clamp the separator around the bearing inner track surface.**

19. Using tool LRT-99-500 and thrust pad LRT-54-014 withdraw inner track from drive flange.

20. Install hub to vice and remove 4 bolts securing backplate to hub.
21. Remove backplate.
22. Remove circlip from bearing.
23. Position hub to press and press out bearing using tool LRT-54-015 and LRT-54-017, discard bearing.

※※ CAUTION

Never re-use existing bearing.

To install:

24. Clean hub and drive flange.
25. Support hub on tool LRT-54-016 and press in new bearing using LRT-54-015.
26. Install circlip to hub.
27. Install hub to vice, fit backplate and tighten bolts to 45 Nm (33 ft. lbs.).
28. Support bearing on tool LRT-54-015 and press drive flange into bearing using LRT-54-014.
29. Install hub assembly to drive shaft.
30. Install hub to damper and tighten bolts to 205 Nm (151 ft. lbs.).

※※ CAUTION

Nuts and bolts must be tightened with weight of vehicle on suspension.

31. Install nut, bolt and washers securing transverse links to hub and tighten nut to 120 Nm (89 ft. lbs.).
32. Install trailing link to hub and tighten nut and bolt to 120 Nm (89 ft. lbs.).

➡**Ensure that washers are fitted to both ends of bolts**

33. Clean ABS sensor, smear sensor with an anti-seize grease and fit sensor to hub.

※※ CAUTION

Ensure ABS sensor is fully located into hub, so that sensor touches pole wheel teeth.

34. Install brake pipe to wheel cylinder and tighten union to 14 Nm (10 ft. lbs.).
35. Install clip securing brake pipe to bracket.
36. Install and secure handbrake cable to backplate.
37. Install brake shoes.
38. Install new drive shaft nut and tighten to 400 Nm (295 ft. lbs.). Stake nut to shaft.
39. Bleed brakes.

40. Install road wheel(s) and tighten nuts to 115 Nm (85 ft. lbs.).
41. Remove stands and lower vehicle.

DISCOVERY AND RANGE ROVER

1. Before servicing the vehicle, refer to the precautions in the beginning of this section.
2. Raise and safely support the vehicle.
3. Remove the rear wheels.
4. Release the brake hose clip, then remove the caliper and position it aside taking care not to kink the brake hose.
5. Remove the 5 axle retaining bolts, then withdraw the axle shaft.
6. Remove the joint washer.
7. Straighten the lockwasher tabs, then remove the locknut and lockwasher.
8. Remove the hub adjusting nut and spacing washer.
9. Remove the hub and brake rotor as an assembly.
10. Remove the outer bearing.
11. remove the 5 Nyloc® nuts, then the ABS tone ring.
12. Matchmark the hub to the rotor for reassembly.
13. Remove the 5 bolts and separate the hub from the brake rotor.
14. Remove the grease seal with the appropriate seal puller, then the inner bearing.
15. Remove the inner and outer bearing races.
16. Clean the hub of any old grease and dry.

To install:

17. Install the inner and outer races.
18. Pack the inner bearing with grease and install it to the hub.
19. Using a seal driver, install the inner grease seal, and lubricate the seal lips to prevent premature wear.
20. Install the brake rotor to the hub, align the matchmarks, and apply Loctite® 270 or equivalent to the mounting bolts and tighten to 54 ft. lbs. (73 Nm).
21. install the ABS tone ring and tighten the new Nyloc® nuts to 80 inch lbs. (9 Nm).
22. Pack the outer bearing with grease and install it to the hub.
23. Retract the ABS sensor slightly.
24. Install the hub and brake rotor assembly.
25. Install the spacing washer and nut. tighten the nut to 45 ft. lbs. (61 Nm), loosen ½ turn (90˚), then tighten to 35 inch lbs. (4 Nm).

26. Install a new lockwasher.
27. Install the locknut and tighten to 45 ft. lbs. (61 Nm).
28. Fold the tab of the lockwasher over the locknut.
29. Using a new joint washer install the axle shaft to the hub, and tighten the 5 bolts to 48 ft. lbs. (65 Nm).
30. Install the brake caliper and tighten the mounting bolt to 60 ft. lbs. (82 Nm), and install the brake hose clip.
31. Reinstall the ABS sensor.
32. Install the wheels and tighten the nuts to 93 ft. lbs. (126 Nm), and operate the foot brake to seat the brake pads.
33. Check the fluid level in the axle.
34. Lower the vehicle and test drive.

PRECAUTIONS

Before servicing any vehicle, please be sure to read all of the following precautions, which deal with personal safety, prevention of component damage, and important points to take into consideration when servicing a motor vehicle:

• Never open, service or drain the radiator or cooling system when the engine is hot; serious burns can occur from the steam and hot coolant.

• Observe all applicable safety precautions when working around fuel. Whenever servicing the fuel system, always work in a well-ventilated area. Do not allow fuel spray or vapors to come in contact with a spark, open flame, or excessive heat (a hot drop light, for example). Keep a dry chemical fire extinguisher near the work area. Always keep fuel in a container specifically designed for fuel storage; also, always properly seal fuel containers to avoid the possibility of fire or explosion. Refer to the additional fuel system precautions later in this section.

• Fuel injection systems often remain pressurized, even after the engine has been turned **OFF**. The fuel system pressure must be relieved before disconnecting any fuel lines. Failure to do so may result in fire and/or personal injury.

• Brake fluid often contains polyglycol ethers and polyglycols. Avoid contact with the eyes and wash your hands thoroughly after handling brake fluid. If you do get brake fluid in your eyes, flush your eyes with clean, running water for 15 minutes. If eye irritation persists, or if you have taken brake fluid internally, IMMEDIATELY seek medical assistance.

• The EPA warns that prolonged contact with used engine oil may cause a number of skin disorders, including cancer! You should make every effort to minimize your exposure to used engine oil. Protective gloves should be worn when changing oil. Wash your hands and any other exposed skin areas as soon as possible after exposure to used engine oil. Soap and water, or waterless hand cleaner should be used.

• All new vehicles are now equipped with an air bag system. The system must be disabled before performing service on or around system components, steering column, instrument panel components, wiring and sensors. Failure to follow safety and disabling procedures could result in accidental air bag deployment, possible personal injury and unnecessary system repairs.

• Always wear safety goggles when working with, or around, the air bag system. When carrying a non-deployed air bag, be sure the bag and trim cover are pointed away from your body. When placing a non-deployed air bag on a work surface, always face the bag and trim cover upward, away from the surface. This will reduce the motion of the module if it is accidentally deployed. Refer to the additional air bag system precautions later in this section.

• Clean, high quality brake fluid from a sealed container is essential to the safe and proper operation of the brake system. You should always buy the correct type of brake fluid for your vehicle. If the brake fluid becomes contaminated, completely flush the system with new fluid. Never reuse any brake fluid. Any brake fluid that is removed from the system should be discarded. Also, do not allow any brake fluid to come in contact with a painted surface; it will damage the paint.

• Never operate the engine without the proper amount and type of engine oil; doing so WILL result in severe engine damage.

• Timing belt maintenance is extremely important! Many models utilize an interference-type, non-freewheeling engine. If the timing belt breaks, the valves in the cylinder head may strike the pistons, causing potentially serious (also time-consuming and expensive) engine damage. Refer to the maintenance interval charts in the front of this manual for the recommended replacement interval for the timing belt, and to the timing belt section for belt replacement and inspection.

• Disconnecting the negative battery cable on some vehicles may interfere with the functions of the on-board computer system(s) and may require the computer to undergo a relearning process once the negative battery cable is reconnected.

• When servicing drum brakes, only disassemble and assemble one side at a time, leaving the remaining side intact for reference.

ENGINE REPAIR

➡**Disconnecting the negative battery cable on some vehicles may interfere with the functions of the on board computer systems and may require the computer to undergo a relearning process, once the negative battery cable is reconnected.**

Distributor

All of the engines covered in this section are distributorless.

Alternator

REMOVAL

1. Before servicing the vehicle, refer to the precautions in the beginning of this section.
2. Remove or disconnect the following:

• Negative battery cable
• Under cover
• Air cleaner assembly, ducts and air intake hose
• Drive belt(s)
• Wires
• Mounting bracket, if equipped
• Alternator

To install:

3. Install or connect the following:

• Alternator. On the 2.4L torque the through-bolt to 14–18 ft. lbs. (20–25 Nm) and the bracket bolt to 106–133 inch lbs. (12–15 Nm). On the 3.0L and 3.5L torque the through-bolt to 38 ft. lbs. (52 Nm) and the mounting bolt to 16 ft. lbs. (22 Nm).
• Mounting bracket, if equipped. Torque the bolt to 14–18 ft. lbs. (20–25 Nm).

• Wires. Torque the nut to 124 inch lbs. (14 Nm).
• Drive belt(s)
• Air cleaner assembly, ducts and air intake hose
• Under cover
• Negative battery cable

Ignition Timing

ADJUSTMENT

The ignition timing is controlled by the ECM and is not adjustable. The ECM determines the timing based on input from the crankshaft position sensor.

TIMING CHECK

1. Before servicing the vehicle, refer to the precautions in the beginning of this section.

Before attempting to adjust the ignition timing, be sure of the following:

- The engine should be at normal operating temperature.
- The lights and all accessories should be OFF.
- If equipped with an automatic transmission, the transmission should be in **P** or **N**.
- Connect scan tool MB991502 to the data link connector
- Set up the timing light.

- Start the engine and run at idle.
- Verify that the idle speed is 600–800 rpm.
- Select scan tool MB991502 actuator test "item number 17."
- Check that basic timing is with in standard, it should be 3–7° BTDC.
- Press the clear key on the scan tool, select forced drive stop mode and cancel the actuator test.

✳✳ CAUTION

If the actuator test is not canceled, the forced drive will continue for 27 minutes. Driving in this state could lead to engine failure.

2. If the base timing is out of specification:

3. Check to see if the distributor is aligned properly

4. Check to see if the timing belt cover

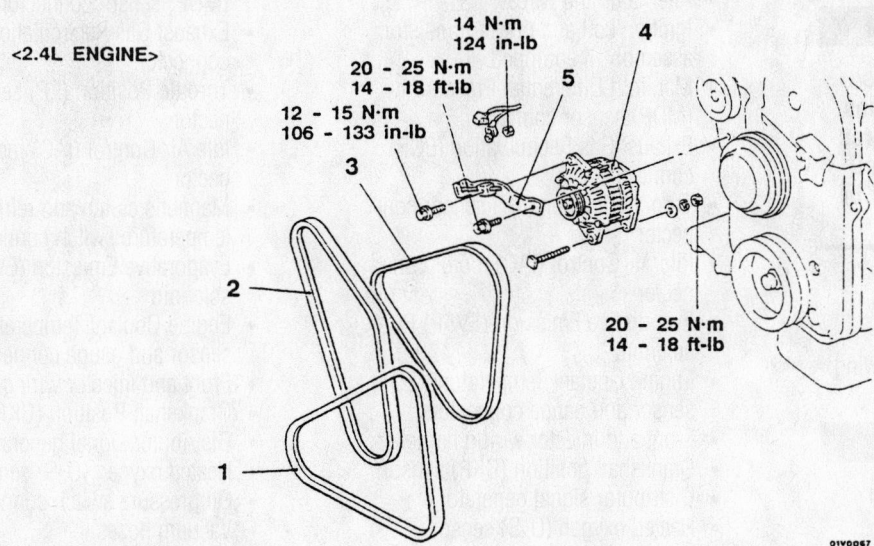

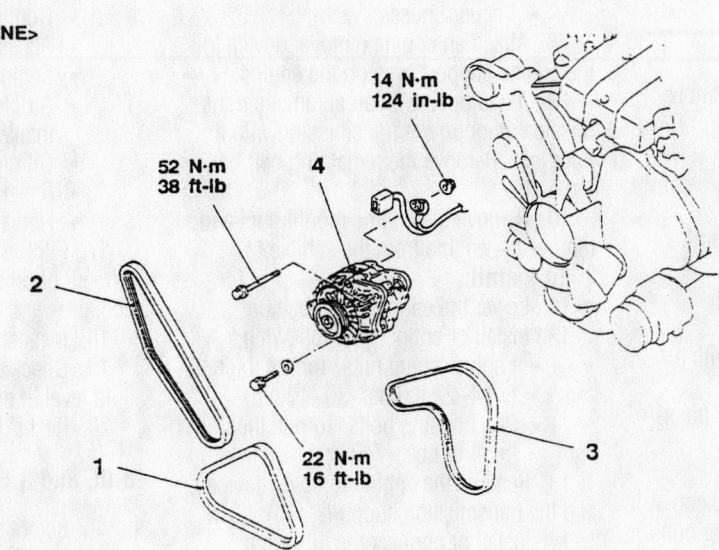

1. DRIVE BELT (FOR A/C)
2. DRIVE BELT (FOR POWER STEERING)
3. DRIVE BELT (FOR GENERATOR)

4. GENERATOR
5. GENERATOR BRACE ASSEMBLY <2.4L ENGINE>

9308UG02

Alternator mounting and related components

and Crankshaft Position (CKP) sensor installation is conditions.

5. Crankshaft sensing blade conditions.

Engine Assembly

REMOVAL & INSTALLATION

2.4L Engine

1. Before servicing the vehicle, refer to the precautions in the beginning of this section.

✳✳ **CAUTION**

The fuel injection system remains under pressure after the engine has been OFF. Properly relieve fuel pressure before disconnecting any fuel lines. Failure to do so may result in fire or personal injury.

2. Relieve the fuel system pressure.
3. Drain the engine oil and coolant.
4. Drain the cooling system.
5. Remove or disconnect the following:
   • Negative Battery Cable

✳✳ **CAUTION**

Wait at least 90 seconds after the negative battery cable is disconnected to prevent possible deployment of the air bag.

   • Battery
   • Hood, matchmark for reassembly
   • Oil dipstick
   • Engine undercover
   • Lower radiator hose
   • Starter
   • Exhaust pipe from the exhaust manifold
   • Transfer case from vehicle, if equipped with 4wd
   • Transmission, if equipped with a manual transmission

6. If equipped with an automatic transmission and 2wd:
   a. Remove the inspection plate.
   b. Matchmark the flexplate and converter. Remove the torque converter bolts and move the torque converter back as far as it will go.
   c. Remove the lower bell housing bolts.

7. Remove or disconnect the following:
   • Air cleaner assembly, ducts and air intake hose
   • Linkages and cables from the throttle body
   • Fuel lines and plug the lines
   • Air conditioning compressor, if equipped and position it aside. It is not necessary to remove the lines from the compressor.
   • Radiator and shroud
   • Cooling fan
   • Heater hoses
   • Accessory belts
   • Power steering pump and wires from its brackets and position it to the side. Do not remove the hoses from the pump.
   • Alternator and wires
   • Ignition coil and power transistor assembly, if equipped
   • Manifold Differential Pressure (MDP) sensor connector
   • Exhaust Gas Recirculation (EGR) connector
   • Throttle Position (TP) sensor connector
   • Idle Air Control (IAC) motor connector
   • Evaporative Emission (EVAP) Purge solenoid
   • Engine Coolant Temperature (ECT) sensor and gauge connectors
   • Front and injector wiring harness
   • Crankshaft position (CKP) sensor
   • Distributor signal generator
   • Heated oxygen (O2S) sensor
   • Oil pressure switch connector
   • Vacuum hoses

8. Attach an engine removal device to the engine support eyes on the engine.

9. If equipped with an automatic transmission, support the transmission with a floor jack. Remove the remaining bell housing bolts.

10. Remove the engine mount nuts and remove the engine from the vehicle.

To install:

11. Lower the engine into position

12. Install or connect the following:
   • Engine mount nuts. Torque the nuts to 14–22 ft. Lbs. (30–40 nm).
   • Bell housing bolts. Torque the bolts to 54 ft. Lbs. (74 nm).

13. Remove the engine removal device and the transmission support.

14. Install or connect the following:
   • Transfer case, if equipped
   • Manual transmission, if equipped

15. If equipped with an automatic transmission, align the torque converter and flexplate and install the bolts. Torque the bolts to 33–38 ft. Lbs. (45–52 nm).

16. Install or connect the following:
   • Inspection plate
   • Starter
   • Exhaust pipe to the exhaust manifold using new gaskets
   • Lower radiator hose
   • Heater hoses
   • Alternator
   • Power steering pump and all brackets
   • Air conditioning compressor
   • Linkages and cables to the throttle body
   • Ignition coil and power transistor assembly
   • Manifold Differential Pressure (MDP) sensor connector
   • Exhaust Gas Recirculation (EGR) connector
   • Throttle Position (TP) sensor connector
   • Idle Air Control (IAC) motor connector
   • Magnetic clutch and refrigerant temperature switch connector
   • Evaporative Emission (EVAP) Purge solenoid
   • Engine Coolant Temperature (ECT) sensor and gauge connectors
   • Front and injector wiring harness
   • Crankshaft Position (CKP) sensor
   • Distributor signal generator
   • Heated oxygen (O2S) sensor
   • Oil pressure switch connector
   • Vacuum hoses
   • Radiator and shroud
   • Cooling fan
   • Accessory belts
   • Engine undercover
   • Air cleaner assembly, ducts and air intake hose
   • Oil dipstick
   • Battery and cables
   • Hood

17. Fill the engine with the specified amount of oil and fill the radiator with coolant.

18. Inspect the fuel system for leaks.

19. Check the automatic transmission fluid level, if equipped.

20. Recheck all engine adjustments.

3.0L and 3.5L Engines

1. Before servicing the vehicle, refer to the precautions in the beginning of this section.

2. Relieve the fuel system pressure.

✳✳ **CAUTION**

The fuel injection system remains under pressure after the engine has been OFF. Properly relieve fuel pressure before disconnecting any fuel

lines. Failure to do so may result in fire or personal injury.

3. Drain the engine oil.
4. Drain the cooling system.
5. Remove or disconnect the following:
- Battery
- Hood, matchmark for reassembly
- Oil dipstick
- Engine undercover
- Starter
- Exhaust pipe from the exhaust manifolds
- Transfer case, if equipped with 4WD
- Transmission, if equipped with a manual transmission

6. If equipped with an automatic transmission and 2WD:
   a. Remove the inspection plate.
   b. Matchmark the flexplate to the converter; remove the torque converter bolts and move the torque converter back as far as it will go.
   c. Remove the lower bell housing bolts.

7. Remove or disconnect the following:
- Air cleaner assembly, ducts and air intake hose
- Linkages and cables from the throttle body
- Fuel lines and plug the lines
- Air conditioning compressor, if equipped and position it aside. It is not necessary to remove the lines from the compressor.
- Radiator, shroud
- Cooling fan
- Heater hoses
- Accessory belts
- Power steering pump and wires from its brackets and position it to the side. Do not remove the hoses from the pump.
- Alternator and wires
- Ignition coil and power transistor assembly, if equipped
- MDP sensor connector
- EGR connector
- TP sensor connector
- IAC motor connector
- Magnetic clutch and refrigerant temperature switch connector
- EVAP Purge Solenoid
- ECT sensor and gauge connectors
- Front and injector wiring harness
- CMP sensor
- CKP sensor
- Distributor signal Generator

- Compactor connector
- Left and right heated O2S sensor
- Oil pressure switch connector
- Vacuum hoses

8. Attach an engine removal device to the engine support eyes on the engine.

9. If equipped with an automatic transmission, support the transmission with a floor jack. Remove the remaining bell housing bolts.

10. Remove the engine mount nuts and remove the engine from the vehicle.

To install:

11. Lower the engine into position and install the engine mount nuts. Tighten the nuts to 20 ft. lbs. (27 Nm), on the Montero, tighten to 33 ft. lbs. (44 Nm).

12. Install or connect the following:
- Bell housing bolts

13. Remove the engine removal device and the transmission support.

14. Install or connect the following:
- Transfer case, if equipped
- Manual transmission, if equipped
- Automatic transmission, if equipped align the torque converter and flexplate and the bolts.
- Inspection plate
- Starter motor
- Exhaust pipe to the exhaust manifolds using new gaskets
- Lower radiator hose
- Heater hoses
- Alternator and wires
- Power steering pump and brackets
- Air conditioning compressor
- Linkages and cables to the carburetor or throttle body
- Ignition coil and power transistor assembly, if equipped
- MDP sensor connector
- EGR connector
- TP sensor connector
- IAC motor connector
- Magnetic clutch and refrigerant temperature switch connector
- EVAP Purge Solenoid
- ECT sensor and gauge connectors
- Front and injector wiring harness
- CMP sensor
- CKP sensor
- Distributor signal Generator
- Compactor connector
- Left and right heated O2S sensor
- Oil pressure switch connector
- Vacuum hoses
- Air cleaner assembly, ducts and air intake hose
- Accessory belts

- Radiator, shroud and upper hose
- Cooling fan
- Battery
- Oil dipstick
- Hood

15. Refill the engine with the specified amount of oil.
16. Refill the radiator with coolant.
17. Check fuel system for leaks.
18. Check the automatic transmission fluid level, if equipped.
19. Recheck all engine adjustments.

Water Pump

REMOVAL & INSTALLATION

1. Before servicing the vehicle, refer to the precautions in the beginning of this section.
2. If necessary, properly release the fuel pressure.
3. Drain the cooling system.
4. Remove or disconnect the following:
- Negative battery cable

✳✳ CAUTION

Wait at least 90 seconds after the negative battery cable is disconnected to prevent possible deployment of the air bag.

- Upper radiator shroud
- Accessory belts
- Air conditioning compressor tensioner pulley, if equipped
- Cooling fan and clutch assembly and the water pump pulley
- Thermostat and housing on 3.0L, 3.5L engines
- Water outlet, gasket and houses
- Radiator hoses from the water pump
- Crankshaft pulley(s)
- Timing belt covers. If the same timing belt will be reused, mark the direction of the timing belt's rotation, for installation in the same direction. Be sure the engine is positioned so the No. 1 cylinder is at the TDC of its compression stroke and the sprockets timing marks are aligned with the engine's timing mark indicators.
- Timing belt
- Water pump bolts are different lengths, note their positions before removing.
- Water pump from the block
- Water pipe connection and O-ring

1. Alternator brace
2. Water pump
3. Gasket
4. O-ring

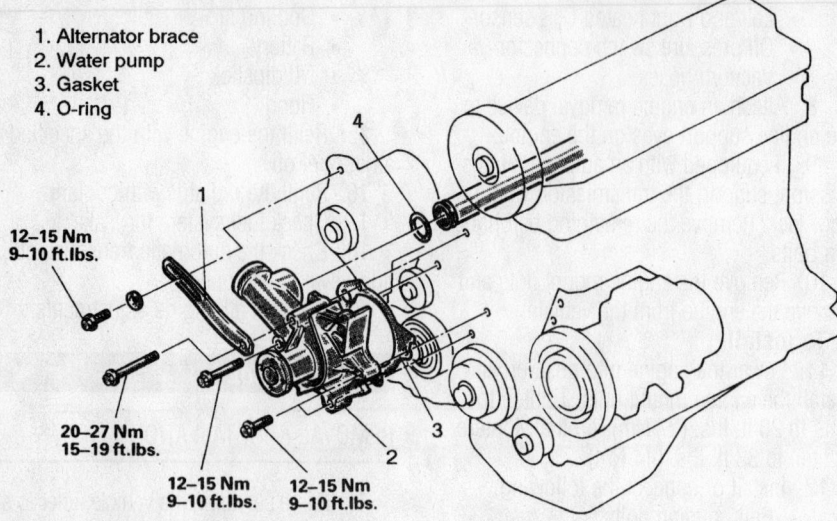

12–15 Nm
9–10 ft.lbs.

20–27 Nm
15–19 ft.lbs.

12–15 Nm
9–10 ft.lbs.

12–15 Nm
9–10 ft.lbs.

7924UG06

Water pump and related components—2.4L engine

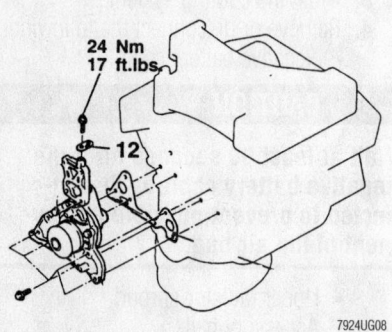

24 Nm
17 ft.lbs.

12

7924UG08

Water pump mounting—3.0L engine

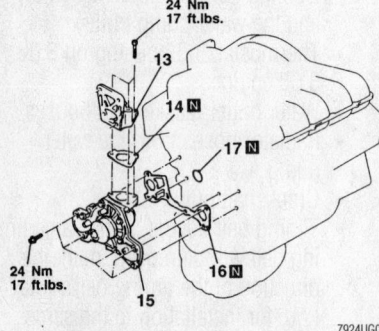

24 Nm
17 ft.lbs.

13

14 N

17 N

24 Nm
17 ft.lbs.

16 N

15

7924UG09

Water pump and related components— 3.5L engine

To install:

5. Clean and dry the mating surfaces of the block and water pump

6. Install or connect the following:
   - New O-ring on the water pipe connection, wet the new O-ring with water to aid in installation
   - Water pump, with a new gasket, Torque the bolts to 106–133 inch lbs. (12–15 Nm) on 2.4L engine,

17 ft. lbs. (23 Nm) on 3.0L and 3.5L engines
   - Alternator bracket bolt to 17 ft. lbs. (23 Nm)
   - Timing belt(s) and covers
   - Crankshaft pulley(s)
   - Thermostat and housing on 3.0L, 3.5L engines. Torque the bolts to 12–14 ft. lbs. (17–20 Nm).
   - Radiator hose to the water pump
   - Water outlet, new gasket and houses. Torque the bolts to 12–14 ft. lbs. (17–20 Nm).
   - Water pump pulley
   - Cooling fan and clutch assembly
   - Air conditioning compressor tensioner pulley, if equipped
   - Accessory belts
   - Upper radiator shroud
   - Thermostat and housing on 3.0L, 3.5L engines
   - Negative battery cable

7. Refill the radiator with coolant. This cooling system has a self-bleeding thermostat, so system bleeding is not required.

8. Run the vehicle until the thermostat opens and fill the overflow tank. Check for leaks.

9. Once the vehicle has cooled, recheck the coolant level.

Cylinder Head

REMOVAL & INSTALLATION

2.4L Engine

1. Before servicing the vehicle, refer to the precautions in the beginning of this section.

2. Drain the cooling system.

✳✳ CAUTION

Some models covered by this manual may be equipped with an air bag. Whenever working near any of the Supplement Restraint System (SRS) components, such as the impact sensors, the air bag module, steering column and instrument panel, disable the SRS.

3. Properly relieve the fuel system pressure.

4. Remove or disconnect the following:
   - Negative battery cable
   - Air cleaner assembly, ducts and air intake hose
   - Accelerator, and if equipped, kickdown cable
   - Upper radiator hose
   - Heater hoses
   - Fuel lines and plug
   - Power steering, if equipped unbolt the power steering pump from its brackets and position it to the side. Do not disconnect the power steering lines.
   - Timing belt upper cover and valve cover
   - Manifold Differential Pressure (MDP) sensor connector
   - Exhaust Gas Recirculation (EGR) connector
   - Throttle Position (TP) sensor connector
   - Idle Air Control (IAC) motor connector
   - Magnetic clutch and refrigerant temperature switch connector
   - Evaporative Emission (EVAP) Purge Solenoid
   - Engine Coolant Temperature (ECT) sensor and gauge connectors
   - Front and injector wiring harness
   - Camshaft Position (CMP) sensor
   - Crankshaft Position (CKP) sensor
   - Compactor connector
   - Oil pressure switch connector
   - Vacuum hoses
   - Oil dipstick

5. Rotate the crankshaft clockwise and align the timing mark on the camshaft sprocket with the timing mark on the cylinder head.

6. Remove or disconnect the following:
   - Camshaft bolt
   - Sprocket from the camshaft (with the timing belt attached) and allow it to rest on the lower cover. Secure the belt to the sprocket so that they do not become disengaged.

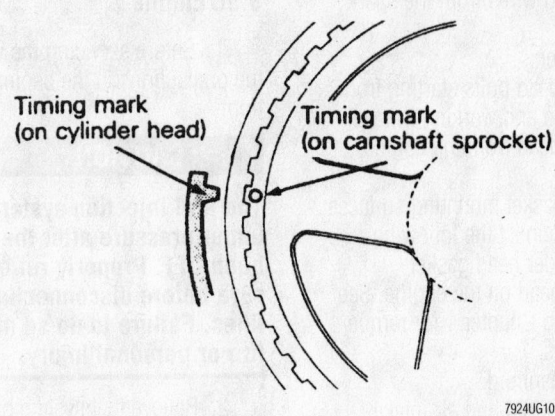

Align the camshaft sprocket timing mark with the timing mark on the cylinder head—2.4L engines

7924UG10

> ※※ **WARNING**
>
> **Do not rotate the crankshaft after the camshaft sprocket is removed from the camshaft. Secure the sprocket and timing belt so there is no slack in the belt. Be sure the sprocket does not become disengaged from the timing belt. If the engine is disturbed or the timing belt moved, the camshaft timing will have to be reset.**

- Spark plug wires
- Distributor cap and wires

7. Mark the position of the rotor and distributor housing in relation to the cylinder head and remove the distributor.

8. Remove or disconnect the following:

- Exhaust pipe from the exhaust manifold
- Cylinder head bolts, starting from the outside and working inward
- Cylinder head from the engine
- Intake and exhaust manifolds from the cylinder head, if necessary

To install:

9. Clean the cylinder head gasket mating surfaces.

10. Install or connect the following:

- Intake and exhaust manifolds to the cylinder head, if removed
- New head gasket to the block and position the cylinder head assembly with all head bolts and washers.

11. Tighten the bolts in the following sequence:

a. Step 1: Torque the bolts in sequence to 58 ft. lbs. (78 Nm).
b. Step 2: Loosen all bolts in sequence to 0 ft. lbs.
c. Step 3: Torque the bolts in sequence to 14 ft. lbs. (20 Nm).
d. Step 4: Turn the bolts an additional 90 degrees.
e. Step 5: Turn an additional 90 degrees.

12. Install or connect the following:

- Camshaft sprocket to the camshaft. Torque bolt to 65 ft. lbs. (88 Nm).
- Distributor, aligning the marks made during removal
- Distributor cap and spark plug wires
- Power steering pump and adjust the belt tension
- Heater hoses and upper radiator hose
- Valve cover. Torque the bolts to 30 ft. lbs. (34 Nm).
- Upper timing belt cover. Torque the bolts to 95 inch lbs. (11 Nm).
- Fuel lines
- Accelerator and kickdown cable, if equipped
- MDP sensor connector
- EGR connector
- TP sensor connector
- IAC motor connector
- Magnetic clutch and refrigerant temperature switch connector
- EVAP Purge Solenoid
- ECT sensor and gauge connectors
- Front and injector wiring harness
- CMP sensor
- CKP sensor
- Compactor connector
- Oil pressure switch connector
- Oil dipstick
- Air cleaner assembly, ducts and air intake hose
- Negative battery cable

13. Refill the cooling system.

14. Start the engine and check for leaks. Check the ignition timing.

3.0L Engines

1. Before servicing the vehicle, refer to the precautions in the beginning of this section.

2. Relieve the fuel system pressure.

3. Drain the cooling system.

4. Remove or disconnect the following:

- Negative battery cable
- Air cleaner assembly, ducts and air intake hose

Cylinder head bolt tightening sequence—2.4L engines

7924UG11

Timing belt service is covered in Section 3 of this manual

- Upper radiator hose
- Accessory drive belts
- Cooling fan and pulleys
- Air conditioning compressor, if equipped
- Power steering pump and mounting brackets and position them to the side, without disconnecting the lines.
- Timing belt covers

5. Remove the timing belt as follows:

   a. Rotate the crankshaft and bring the No. 1 piston to Top Dead Center (TDC) on the compression stroke. Align the camshaft and crankshaft sprocket timing marks.

   b. Mark the timing belt in the direction of rotation for reinstallation purposes.

   c. Loosen the timing belt tensioner bolt and turn the tensioner counterclockwise.

✳✳ WARNING

Do not rotate the crankshaft or camshaft sprockets after the timing belt has been removed.

6. Remove or disconnect the following:
   - Timing belt
   - Fuel lines and plug
   - Wiring connectors, vacuum lines and hoses from the air intake plenum, intake manifold and cylinder head.
   - Air intake plenum
   - Intake manifold
   - Exhaust manifold
   - Camshaft sprocket bolt and camshaft sprocket, if necessary
   - Alternator bracket and/or timing belt rear cover
   - Oil dipstick, on left side only
   - Crankshaft position (CKP) sensor on left side only

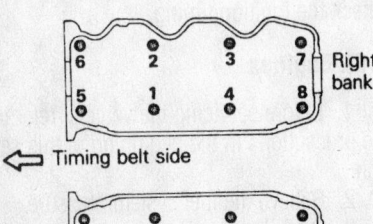

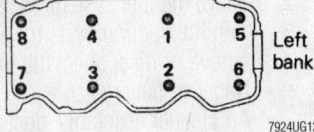

Cylinder head bolt tightening sequence— 3.0L engines

- Spark plug wires from the spark plugs
- Valve cover
- Cylinder head bolts starting from the outside and working inward
- Cylinder head from the engine

To install:

7. Clean the gasket mounting surfaces.
8. Install or connect the following:
   - New cylinder head gasket
   - Cylinder head on the engine. See the chart in Chapter 1 for torque procedures.
   - Exhaust manifold
   - Intake manifold and air intake plenum
   - Fuel lines
   - Wiring connectors, vacuum lines and hoses to the air intake plenum, intake manifold and cylinder head
   - Valve cover
   - Spark plug wires
   - CKP sensor, if removed
   - Oil dipstick, if removed
   - Alternator bracket and/or timing belt rear cover
   - Camshaft sprocket bolt and camshaft sprocket, if necessary

9. Be sure the camshaft and crankshaft sprocket timing marks are aligned.

10. Turn the timing belt tensioner to the extreme counter-clockwise position and temporarily tighten the bolt.

11. Install the timing belt in the original rotation direction. Loosen the timing belt tensioner bolt and allow the spring force of the tensioner to tension the belt.

12. Turn the crankshaft 2 turns in the normal direction of rotation and check the timing mark alignment.

13. If the timing is correct, tighten the tensioner bolt to 21 ft. lbs. (30 Nm). If the timing is incorrect, repeat the belt installation procedure.

14. Install or connect the following:
    - Timing belt covers
    - Alternator, alternator cover and alternator stay, if removed
    - Air conditioning compressor
    - Power steering pump with the brackets
    - Pulleys. Torque the crankshaft pulley bolt to 134 ft. lbs. (181 Nm).
    - Cooling fan
    - Accessory drive belts
    - Air cleaner assembly, ducts and air intake hose
    - Upper radiator hose
    - Negative battery

15. Refill the cooling system.
16. Start the engine and check for leaks. Check the ignition timing.

3.5L Engine

1. Before servicing the vehicle, refer to the precautions in the beginning of this section.

✳✳ CAUTION

The fuel injection system remains under pressure after the engine has been OFF. Properly relieve fuel pressure before disconnecting any fuel lines. Failure to do so may result in fire or personal injury.

2. Relieve fuel system pressure.
3. Drain the cooling system.
4. Remove or disconnect the following:
   - Negative battery cable

✳✳ CAUTION

Work must be started after 90 seconds from the time the ignition switch is turned to the LOCK position and the negative battery cable is disconnected.

- Air intake hoses
- Air intake plenum and intake manifold
- Exhaust manifold
- Engine under cover
- Radiator and shroud
- Alternator
- Cooling fan
- Timing belt
- Breather hose
- Oil dipstick
- Camshaft Position (CMP) sensor
- Spark plug cable center cover and remove the spark plug cables
- Valve cover
- Intake camshaft sprocket
- Rear timing belt cover
- Ignition coil
- Water hoses from the thermostat housing and the housing

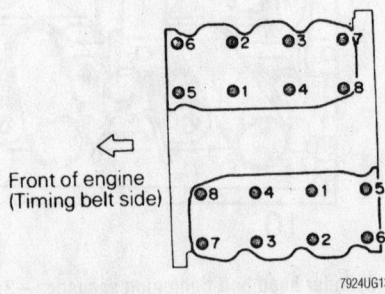

Cylinder head bolt tightening sequence— 3.5L engines

- Water inlet from the front head and discard O-ring
- Water passage

5. Loosen the cylinder head mounting bolts in 3 steps, starting from the outside and working inward. Lift off the cylinder head assembly and remove the head gasket.

To install:

6. Thoroughly clean and dry the mating surfaces of the head and block. Check the cylinder head for cracks, damage or engine coolant leakage. Remove scale, sealing compound and carbon. Clean oil passages thoroughly. Check the head for flatness. End to end, the head should be within 0.0012 in. (0.030mm), normally with 0.008 in. (0.203mm) the maximum allowed out of true. The total thickness allowed to be removed from the head and block is 0.008 in. (0.203mm) maximum.

7. Place a new head gasket on the cylinder block with the identification marks in the front top (upward) position. Do not use sealer on the gasket.

8. Install or connect the following:
- Cylinder head on the block. Be sure the head bolt washers are installed with the chamfered edge upward. See the chart in Chapter 1 for torque procedures.
- New O-ring and the water inlet to the front head
- New gaskets, thermostat housing and connect the hoses
- Water passage and new gaskets
- Ignition coil and center rear timing belt cover
- Intake camshaft sprocket. Use hex flange on camshaft to secure and tighten the retaining bolt to 65 ft. lbs. (90 Nm).
- New gasket and the valve cover. Torque the bolts to 84 inch lbs. (10 Nm).
- Spark plug cables and the center cover
- Oil dipstick
- CMP sensor
- Breather hose
- Radiator and shroud
- Timing belt
- Cooling fan
- Alternator
- Engine under cover
- Intake manifold and new gasket. Torque the nuts to 16 ft. lbs. (21 Nm).
- Air intake plenum and new gaskets.

Torque the bolts to 13 ft. lbs. (18 Nm).
- Exhaust manifold and new gaskets. Torque the nuts to 22 ft. lbs. (29 Nm).
- Air intake hoses
- Negative battery cable

9. Change the engine oil and oil filter.
10. Refill the system with coolant.
11. Run the vehicle until the thermostat opens.
12. Once the vehicle has cooled, recheck the coolant level.

Rocker Arms/Shafts

REMOVAL & INSTALLATION

2.4L Engines

1. Before servicing the vehicle, refer to the precautions in the beginning of this section.

2. Remove or disconnect the following:
- Breather hose
- Positive Crankcase Ventilation (PCV) hose
- Valve cover

➡**Install the special clips MB 998443-01 to hold the auto-adjusters in place.**

- Rocker arms and rocker arm shafts
- Rocker shaft springs

➡**Take care during removal that no oil, grease or dirt comes into contact with the timing belt.**

To install:

3. Disassemble the rocker assembly, checking each component for wear, scoring, or plugged oil passages. Check the roller for correct and smooth rotation. Inspect the inner diameter of each rocker for any scoring or enlargement. If wear is found inside the rocker, replace it and inspect the shaft for damage.

4. Before reassembly, coat the contact faces liberally with clean motor oil. Observe the numbers on the bearing caps so that they are replaced in the correct location. Reassemble the rocker shafts into the front bearing cap so that the notches face outward. As you continue to assemble the springs, rockers and bearing caps, remember that the arrows on the bearing caps must point in the same direction as the arrow on the head. This is particularly important if the rockers have been removed from both heads.

5. Install or connect the following:
- Rocker arms and rocker arm shafts. Torque the bolts to 23 ft. lbs. (31 Nm).
- Valve cover Torque the bolts to 30 inch lbs. (3.4 Nm).
- PCV hose
- Breather hose

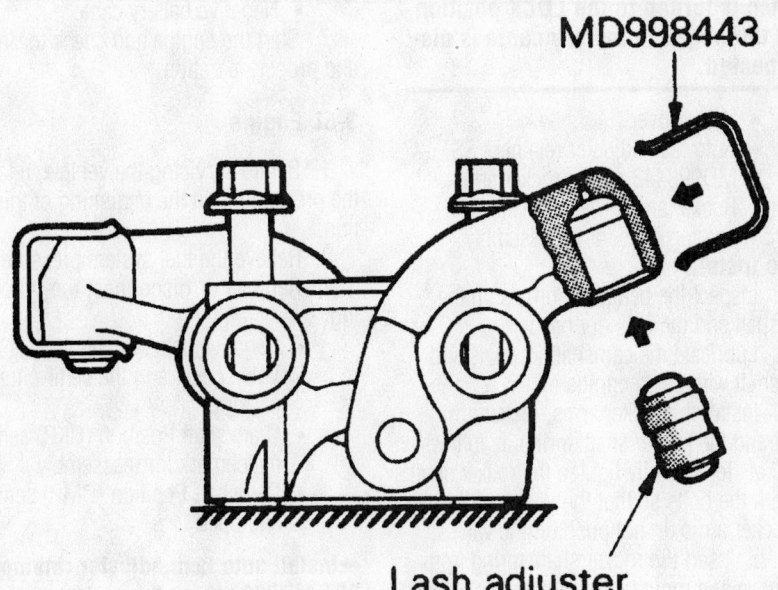

MD998443

Lash adjuster

7924UG14

Insert the Lash Adjuster Holder tool MD998443 to prevent the lash adjuster from falling out—2.4L engines

Heater Core replacement is covered in Section 2 of this manual

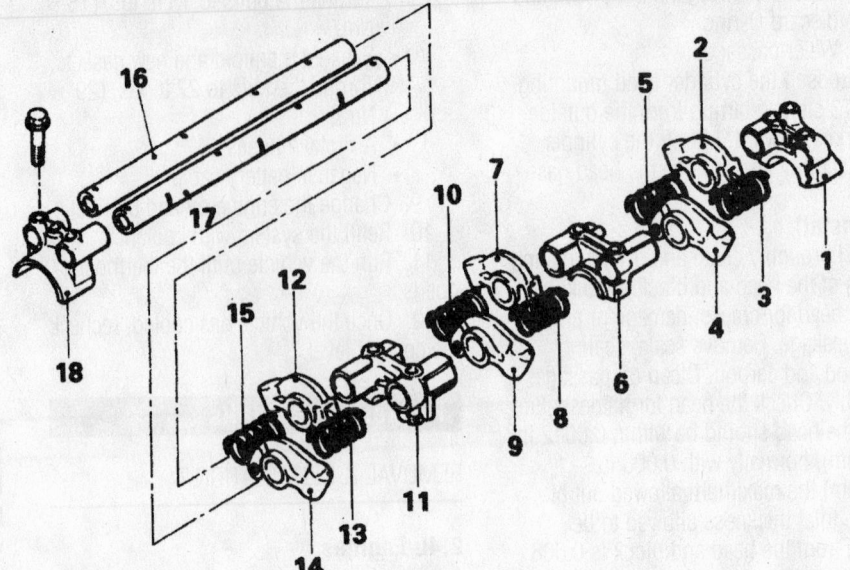

1. Bearing cap No. 4
2. Rocker arm (B)
3. Spring
4. Rocker arm (A)
5. Spring
6. Bearing cap No. 3
7. Rocker arm (B)
8. Spring
9. Rocker arm (A)
10. Spring
11. Bearing cap No. 2
12. Rocker arm (B)
13. Spring
14. Rocker arm (A)
15. Spring
16. Rocker arm shaft (B)
17. Rocker arm shaft (A)
18. Bearing cap No. 1

7924UG15

Exploded view of the rocker arms and shafts—3.0L engine

3.0L Engines

1. Before servicing the vehicle, refer to the precautions in the beginning of this section.
2. Remove or disconnect the following:
   • Negative battery cable

> ※※ **CAUTION**
>
> **Work must be started after 90 seconds from the time the ignition switch is turned to the LOCK position and the negative battery cable is disconnected.**

   • Valve cover
   • Auto lash adjuster retainers SST MD998443 on the rocker arms
   • Rocker arms, rocker shafts and bearing caps, as an assembly

To install:

3. Inspect the bearing journals on the camshaft and the cylinder head.
4. Lubricate the camshaft journals and camshaft with clean engine oil.
5. Install the rocker arms, rocker arm shaft and the rocker shaft spring as follows:
   a. Temporarily tighten the rocker shaft with the bolts so that the intake valve rocker arms do not push on the valves.
   b. Insert the rocker shaft spring from above and mount it at right angles to the plug guide.
   c. Before installing the exhaust rocker arms and the rocker arm shaft, mount the rocker shaft spring.
   d. Remove tool SST MD998443 used to hold the lash adjuster in position.

   e. Check to ensure that the flat side of the rocker shaft is perpendicular to the cylinder head, and facing the valves.
   f. Gradually tighten the bearing caps in 2 or 3 steps. In the final step, tighten to 23 ft. lbs. (31 Nm).
6. Install or connect the following:
   • Valve cover and new gasket. Torque the bolt to 2–3 ft. lbs. (3–4 Nm).
   • Negative battery cable
7. Start the engine and check for leaks and proper operation.

3.5L Engine

1. Before servicing the vehicle, refer to the precautions in the beginning of this section.
2. Relieve the fuel system pressure.
3. Remove or disconnect the following:

   • Negative battery cable
   • Valve cover and the semi-circular packing.
   • Crankshaft Position (CKP) sensor, matchmark for reassembly
   • Camshaft Position (CMP) sensor, if equipped

➡**Install auto lash adjuster retainers SST MD998443 on the rocker arms**

   • Rocker arms and shafts
   • Lash adjusters
4. Check the camshaft journals for wear or damage. Check the cam lobes for damage. Also, check the cylinder head oil holes for clogging.

To install:

➡**Lubricate the valve train components with clean engine oil.**

5. Bleed and install the lash adjusters to the to the original bores in the cylinder head.
6. Install or connect the following:
   • Rocker arms and shafts. Torque the bolts to 23 ft. lbs. (31 Nm).
   • Camshaft position sensor, if removed. Torque the mounting bolts to 78 inch lbs. (9 Nm).
   • Camshaft Position (CMP) sensor, if equipped
   • Valve cover and the semi-circular packing. Torque the bolts to 2.5 ft. lbs. (3.5 Nm).
   • Negative battery cable
7. Run vehicle and check for leaks.

Intake Manifold

REMOVAL & INSTALLATION

2.4L Engine

1. Before servicing the vehicle, refer to the precautions in the beginning of this section.
2. Drain the engine coolant.
3. Relieve the fuel pressure.
4. Remove or disconnect the following:
   • Negative battery cable

> ※※ **CAUTION**
>
> **Wait at least 90 seconds after the negative battery cable is discon-**

nected to prevent possible deployment of the air bag.

- Upper radiator hose from the thermostat housing
- Air intake hoses, breather hose and the air intake pipe
- Wires, hoses and linkages from the throttle body
- Ignition coil
- Brake booster hose and vacuum

hose cluster from the air intake plenum
- Air intake plenum from the intake manifold
- Fuel lines, keep the line covered or plugged
- Fuel rail assembly with injectors intact
- Heater hose from the manifold
- Engine Coolant Temperature (ECT) sensor

- Water outlet fitting
- Distributor, matchmark for reassembly
- Intake manifold from the cylinder head and engine
- Manifold Differential Pressure (MDP) sensor, if equipped

5. Clean and dry the mating surfaces of the manifold and cylinder head.

To install:

6. Install or connect the following:

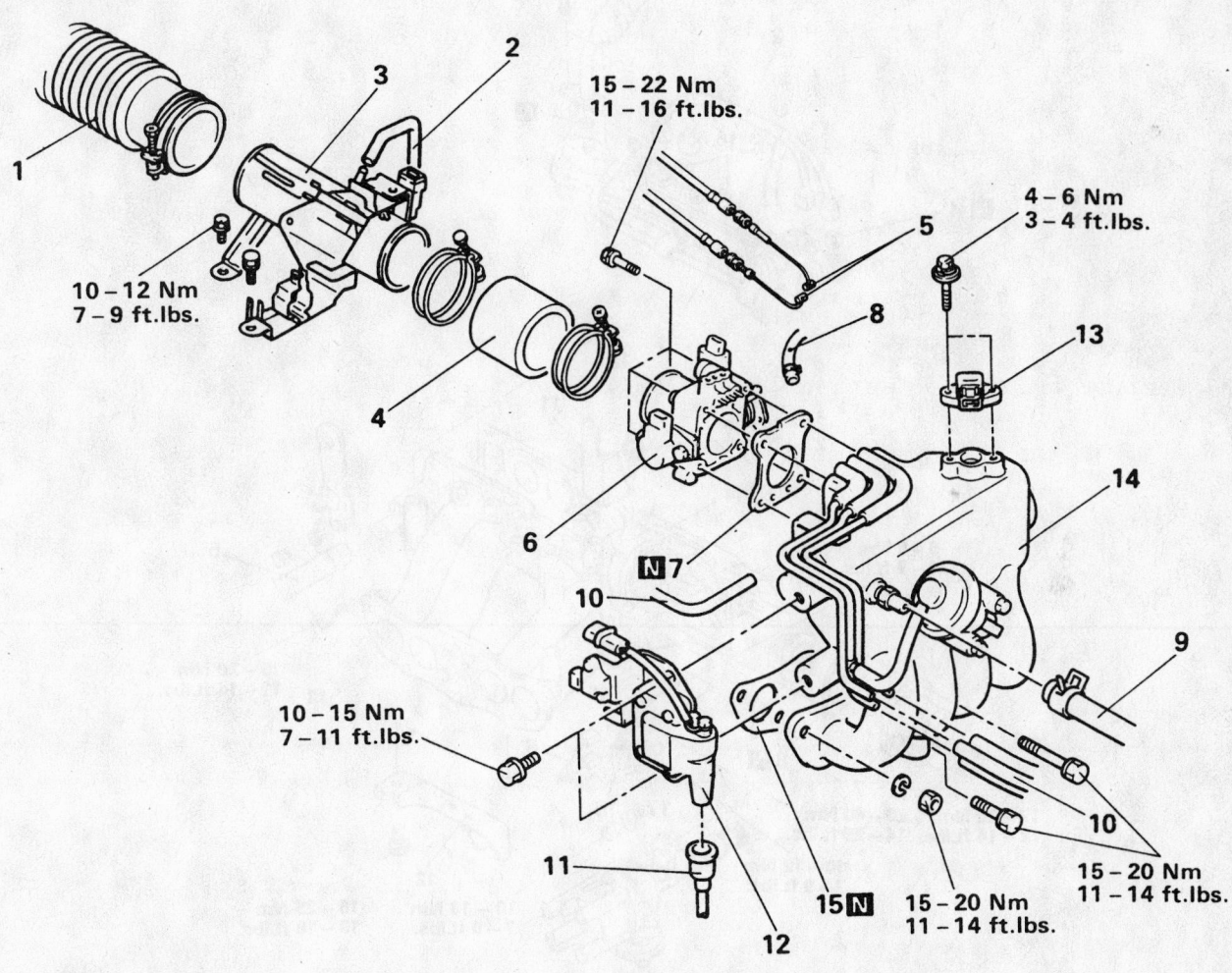

1. Air intake hose
2. Breather hose
3. Air intake pipe
4. Air hose
5. Accelerator cable and kick down plate
6. Throttle body
7. Gasket
8. Water hose

9. Brake booster vacuum hose
10. Vacuum hose connection
11. High tension cable
12. Ignition coil
13. Manifold difference pressure sensor
14. Intake manifold plenum assembly
15. Intake manifold plenum gasket

7924UG39

Exploded view of the intake plenum chamber and related components—2.4L engine shown

Brake service is covered in Section 4 of this manual

- New gasket and the intake manifold to the head. Starting from the middle and working outward, tighten the retaining nuts to 14 ft. lbs. (19 Nm).
- ECT sensor
- Water outlet fitting. Torque the bolts to 14 ft. lbs. (19 Nm).

- Distributor with matchmarks aligned
- Fuel rail assembly to the manifold using a new O-ring
- Fuel lines
- Heater hose to the manifold
- Air intake plenum with a new gasket. Tighten the retaining bolts to 12 ft. lbs. (16 Nm).

- Vacuum hoses cluster, brake booster hose and all wires, hoses and linkages to the throttle body
- Ignition coil
- Air intake hoses, breather hose and the air intake pipe
- Manifold Differential Pressure (MDP) sensor, if equipped

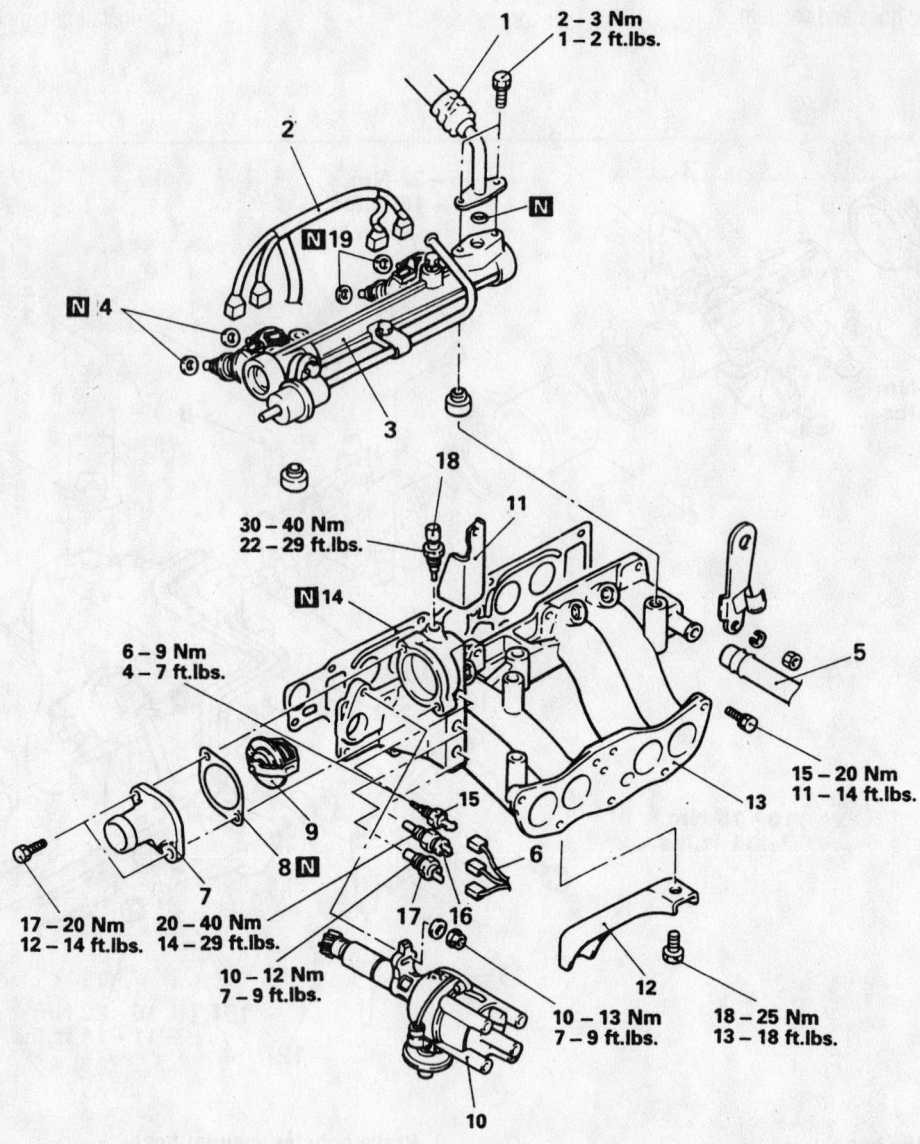

1. High pressure fuel hose connection
2. Fuel injector harness connector
3. Fuel rail
4. Insulator
5. Heater hose
6. Wiring harness connector
7. Water outlet fitting
8. Water outlet fitting gasket
9. Thermostat
10. Distributor
11. Intake manifold plenum stay
12. Intake manifold stay
13. Intake manifold
14. Intake manifold gasket
15. Thermal switch <A/T>
16. Engine coolant temperature sensor
17. Engine coolant temperature gauge unit
18. Air conditioning engine coolant temperature switch <A/C>

7924UG40

Exploded view of the intake manifold and related components—2.4L engine shown

- Upper radiator hose
- Negative battery cable

7. Refill the radiator with coolant.
8. Check the system for leaks.

3.0L Engine

1. Before servicing the vehicle, refer to the precautions in the beginning of this section.
2. Relieve the fuel pressure.

❋❋ CAUTION

The fuel injection system remains under pressure after the engine has been OFF. Properly relieve fuel pressure before disconnecting any fuel lines. Failure to do so may result in fire or personal injury.

3. Drain the engine coolant.
4. Remove or disconnect the following:
   - Negative battery cable

❋❋ CAUTION

Work must be started after 90 seconds from the time the ignition switch is turned to the LOCK position and the negative battery cable is disconnected.

- Air intake hose from the throttle body
- Positive Crankcase Ventilation (PCV) hose

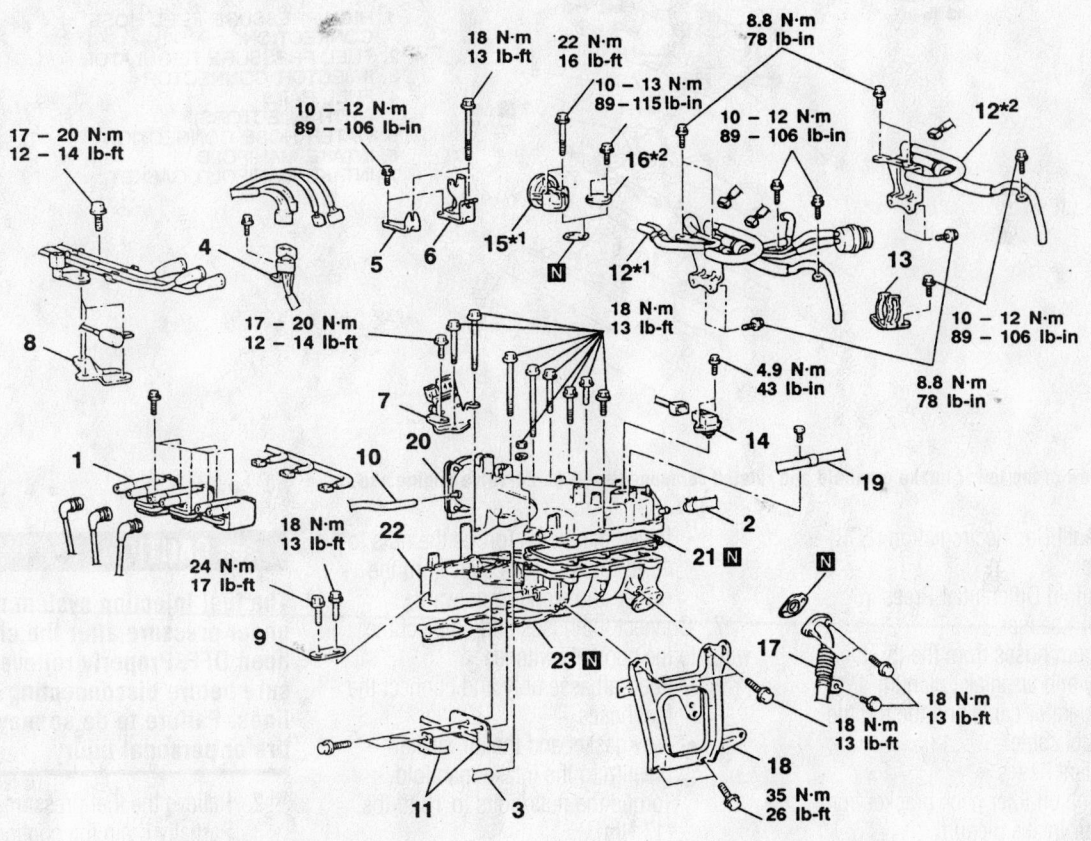

1. IGNITION COILS
2. BRAKE BOOSTER VACUUM HOSE CONNECTION
3. PCV HOSE CONNECTION
4. CRANKSHAFT POSITION SENSOR AND CAM POSITION SENSOR CONNECTOR
5. ACCELERATOR CABLE BRACKET <M/T>
6. THROTTLE CABLE BRACKET <A/T>
7. IGNITION POWER TRANSISTOR
8. WATER OUTLET FITTING BRACKET
9. WATER PUMP STAY
10. VACUUM HOSE CONNECTION
11. FUEL PIPE CONNECTION
12. SOLENOID VALVE AND VACUUM HOSE ASSEMBLY

13. VCV BRACKET
14. MDP SENSOR
15. EGR VALVE
16. COVER
17. EGR PIPE CONNECTION
18. INTAKE MANIFOLD PLENUM STAY
19. THROTTLE CABLE CONNECTION
20. AIR INTAKE FITTING
21. AIR INTAKE FITTING GASKET
22. UPPER INTAKE MANIFOLD
23. INTAKE MANIFOLD PLENUM GASKET

NOTE
*1: Vehicles for Federal
*2: Vehicles for California

7924UG41

Exploded view of the upper intake manifold and related components—3.0L 24-valve engine shown

For complete Engine Mechanical specifications, see Section 1 of this manual

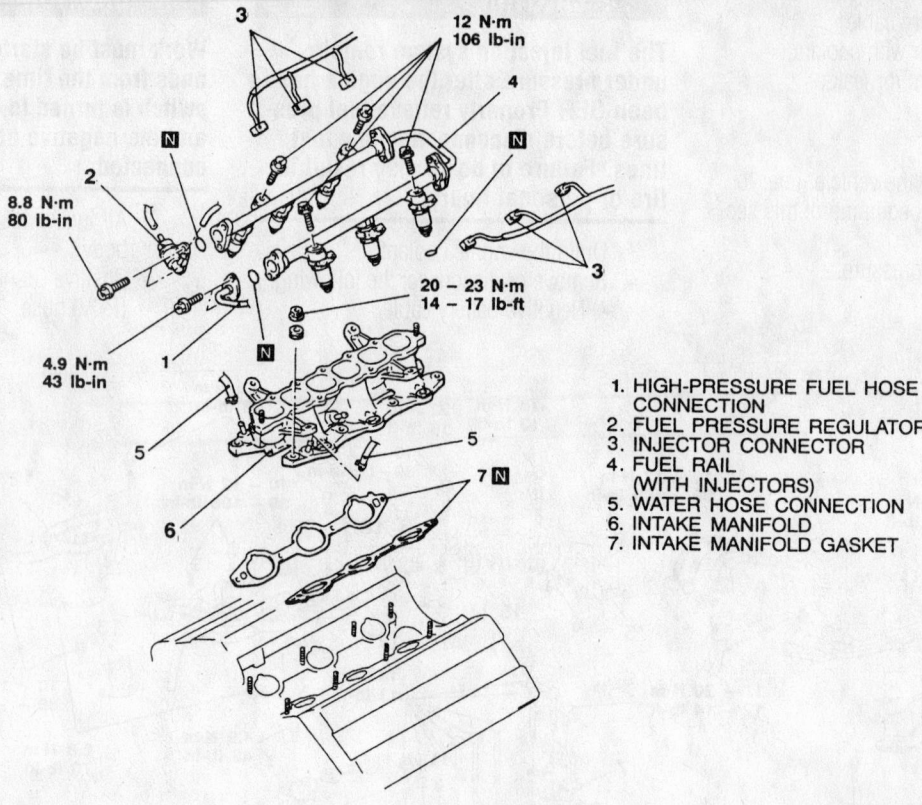

1. HIGH-PRESSURE FUEL HOSE CONNECTION
2. FUEL PRESSURE REGULATOR
3. INJECTOR CONNECTOR
4. FUEL RAIL (WITH INJECTORS)
5. WATER HOSE CONNECTION
6. INTAKE MANIFOLD
7. INTAKE MANIFOLD GASKET

Exploded view of the lower intake manifold and related components—3.0L 24-valve engine shown

7924UG42

- Exhaust Gas Recirculation (EGR) valve
- Manifold Differential Pressure (MDP) sensor
- Vacuum hoses from the throttle body and air intake plenum
- Accelerator cable and the throttle control cable
- Coolant hoses
- Engine oil filler neck bracket from the air intake plenum
- EGR tube from the air intake plenum
- Plenum brackets
- Air intake plenum assembly from the intake manifold and remove. Note the position of the mounting bolts as they are removed.
- Fuel hose from the fuel rail
- Fuel return line and vacuum hose from the fuel pressure regulator
- Electrical connectors from the injectors
- Fuel rail and injectors
- Intake manifold

5. Remove the gaskets and thoroughly clean and dry the mating surfaces of the manifold and heads.

To install:

6. Install or connect the following:

- Intake manifold. Torque the nuts to 16 ft. lbs. (21 Nm) start from the center and working outward.

7. Connect the hoses and connect the wires to the coolant switches.

- Fuel rail assembly and connect the fuel hoses
- New gasket and the air intake plenum to the intake manifold. Torque the nuts/bolts to 13 ft. lbs. (17 Nm).
- Plenum brackets
- PCV hose and vacuum hose cluster to the plenum
- EGR tube
- EGR temperature sensor wire
- Wires, hoses and linkages to the throttle body
- Air intake hose to the throttle body
- Upper radiator hose to the thermostat housing
- Negative battery cable

8. Refill the radiator with coolant.
9. Check fuel system for leaks.

3.5L Engines

1. Before servicing the vehicle, refer to the precautions in the beginning of this section.

> **⁜⁜ CAUTION**
>
> **The fuel injection system remains under pressure after the engine has been OFF. Properly relieve fuel pressure before disconnecting any fuel lines. Failure to do so may result in fire or personal injury.**

2. Relieve the fuel pressure.
3. Partially drain the cooling system.
4. Remove or disconnect the following:
   - Negative battery cable

> **⁜⁜ CAUTION**
>
> **Wait at least 90 seconds after the negative battery cable is disconnected to prevent possible deployment of the air bag.**

- Air intake hose from the throttle body
- Electrical connectors and vacuum hoses from the throttle body and air intake plenum
- Accelerator cable and the throttle control cable
- Coolant hoses
- Positive Crankcase Ventilation (PCV) hose

- Exhaust Gas Recirculation (EGR) temperature sensor connector
- EGR tube from the air intake plenum
- Intake manifold plenum cover
- Intake manifold plenum stay brackets
- Air intake plenum assembly from the intake manifold and remove. Note the position of the mounting bolts as they are removed
- Induction control valve assembly
- Fuel hose from the fuel rail
- Fuel return line and vacuum hose from the fuel pressure regulator
- Electrical connectors from the injectors
- Fuel rail and injectors
- Intake manifold

5. Remove the gaskets and thoroughly clean and dry the mating surfaces of the manifold and heads.

To install:

6. Install or connect the following:
- Intake manifold. Tighten the nuts to 16 ft. lbs. (21 Nm). Start from the center and work outward.
- Fuel rail assembly and connect the fuel hoses
- Induction control valve assembly and tighten to 72 inch lbs. (9 Nm).
- New gasket and air intake plenum to the intake manifold. Torque the nuts/bolts to 13 ft. lbs. (18 Nm).
- Plenum to engine brackets
- Hoses and wires to the coolant switches
- PCV hose and vacuum hose cluster to the plenum
- EGR tube. Torque the bolts to 13 ft. lbs. (18 Nm).
- EGR temperature sensor wire
- Wires, hoses and linkages to the throttle body
- Air intake hose to the throttle body
- Upper radiator hose to the thermostat housing
- Negative battery cable

7. Refill the radiator with coolant.
8. Check the system for fuel leaks.
9. Set all adjustments to specifications.

Exhaust Manifold

REMOVAL & INSTALLATION

2.4L Engine

1. Remove or disconnect the following:
- Negative battery cable

✱✱ CAUTION

Wait at least 90 seconds after the negative battery cable is disconnected to prevent possible deployment of the air bag.

- Heat cowl from the exhaust manifold
- Aspirator valve assembly, if equipped
- Oil dipstick and guide
- Oxygen (O_2S) sensor connector and ground cable, if equipped
- Exhaust pipe from the manifold
- Exhaust manifold and gasket from the engine

To install:

2. Install or connect the following:
- Exhaust manifold. Torque the M8 mounting nuts to 22 ft. lbs. (29 Nm) and the M10 mounting bolts 36 ft. lbs. (49 Nm). starting from the middle and working outward.
- O_2S sensor connector and ground cable, if equipped
- Exhaust pipe to the manifold. Torque the nuts to 35ft. (49 Nm).
- Aspirator valve assembly, if removed
- Heat cowl to the exhaust manifold. Torque the bolts to 117 inch lbs. (13 Nm).
- Negative battery cable

3. Start the engine and check for exhaust leaks.

3.0L and 3.5L Engines

1. Remove or disconnect the following:
- Negative battery cable
- Exhaust pipe from the exhaust manifolds
- Oil dipstick, guide and O-ring
- Heat shields
- Exhaust manifolds

2. Clean the gasket mounting surfaces. Inspect the manifolds for cracks, flatness and/or damage.

To install:

3. Install or connect the following:
- New gasket and exhaust manifold. Torque the nuts to 33 ft. lbs. (44 Nm) on 3.0L engines and 22 ft. lbs. (29 Nm) on 3.5L engines.
- Heat shield, Torque the bolts to 10 ft. lbs. (14 Nm).
- Exhaust pipe to the exhaust manifolds. Torque the nuts to 35ft. (49 Nm).

- Oil dipstick, guide and new O-ring
- Negative battery cable

4. Start the engine and check for exhaust leaks.

Front Crankshaft Seal

REMOVAL & INSTALLATION

2.4L Engines

1. Before servicing the vehicle, refer to the precautions in the beginning of this section.
2. Drain the crankcase.
3. Drain and recycle the engine coolant.
4. Remove or disconnect the following:
- Negative battery cable

✱✱ CAUTION

Wait at least 90 seconds after the negative battery cable is disconnected to prevent possible deployment of the air bag.

- Fan shroud
- Radiator and cooling fan
- Accessory drive belts
- Air conditioner tension pulley, if equipped
- Ignition coil, if equipped
- Crankshaft pulley and the timing belt front covers

5. Reinstall the crankshaft pulley bolt and use it to rotate the engine clockwise until the timing marks are aligned.

6. Remove or disconnect the following:
- Timing belt and belt tension
- Crankshaft sprocket
- Crankshaft sensor blade
- Inner timing belt and inner crankshaft sprocket
- Crankshaft seal without scratching the crankshaft

To install:

7. Place the seal guide over the crankshaft and coat it with engine oil.

8. Slide the seal over the guide until it touches the front case assembly, then use the appropriate seal driver to install the seal. Remove the seal guide.

9. Install or connect the following:
- Inner crankshaft sprocket
- Inner timing belt and belt tension
- Crankshaft sensor blade

10. Install the flange and the timing belt sprocket on the crankshaft.

11. Install or connect the following:
- Timing belt. Adjust the belt tension

For Accessory Drive Belt illustrations, see Section 1 of this manual

- Timing belt front covers and the crankshaft pulleys
- Ignition coil, if equipped
- Air conditioner belt tension pulley if equipped
- Accessory drive belts
- Radiator, cooling fan and fan shroud
- Negative battery cable

12. Refill the crankcase.
13. Refill the cooling system.
14. Start the engine and check for leaks.

3.0L and 3.5L Engines

1. Before servicing the vehicle, refer to the precautions in the beginning of this section.
2. Drain the crankcase.
3. Drain and recycle the engine coolant.
4. Remove or disconnect the following:
   - Negative battery cable

✳✳ CAUTION

Wait at least 90 seconds after the negative battery cable is disconnected to prevent possible deployment of the air bag.

- Cooling fan
- Accessory drive belts
- Alternator
- Engine undercover, if equipped
- Power steering oil pump assembly
- Air conditioner compressor and bracket, if equipped
- Timing indicator bracket
- Accessory mount Assembly
- Crankshaft pulley
- Timing belt covers and the timing belt
- Crankshaft sprocket

5. Cut out a portion in the crankshaft oil seal lip and pry out the oil seal with a flat prying tool, being careful not to damage the crankshaft.

To install:

6. Coat the lip of the new seal with oil and install the seal using the proper seal driver.
7. Install or connect the following:
   - Crankshaft sprocket and the timing belt
   - Timing belt covers
   - Crankshaft pulley. Torque the bolt to 134 ft. lbs. (181Nm).
   - Accessory mount Assembly. Torque the bolts to 33 ft. lbs. (44 Nm).
   - Timing indicator bracket. Torque the bolts to 97 inch lbs. (11 Nm).
   - Air conditioner compressor and bracket, if equipped

- Power steering oil pump assembly
- Engine undercover, if equipped
- Alternator
- Accessory drive belts
- Cooling fan
- Negative battery cable

8. Refill the crankcase.
9. Refill the cooling system.
10. Start the engine and check for proper operation.

Camshaft and Valve Lifters

REMOVAL & INSTALLATION

2.4L Engine

1. Before servicing the vehicle, refer to the precautions in the beginning of this section.
2. Drain and recycle the engine coolant.
3. Drain and recycle the engine crankcase.
4. Remove or disconnect the following:
   - Negative battery cable

✳✳ CAUTION

Wait at least 90 seconds after the negative battery cable is disconnected to prevent possible deployment of the air bag.

- Positive Crankcase Valve (PCV) hose
- Valve cover and the upper timing belt cover

5. Rotate the crankshaft clockwise and align the camshaft sprocket timing mark with the timing mark on the cylinder head.
6. Remove or disconnect the following:
   - Distributor, matchmark for reassembly

7. Install auto lash adjuster retainer MD998443 to each rocker arm to hold the auto lash adjusters in place
8. Remove or disconnect the following:
   - Rocker arm and shafts

✳✳ WARNING

Do not rotate the crankshaft after the sprocket is removed from the camshaft. Be sure there is no slack in the timing belt. Be sure the timing belt does not disengage from the sprocket. If the crankshaft is rotated or the timing belt position is disturbed, timing belt and sprocket alignment will have to be set.

- Thrust cage and o-ring
- Camshaft and the front seal from the engine

To install:

9. Install or connect the following:
   - Front seal
   - Camshaft, lubricate with engine oil
   - Thrust cage and o-ring. Torque the bolts to 13 ft. lbs. (18 Nm).
   - Rocker arm and shafts. Torque the bolts to 23 ft. lbs. (31 Nm).

10. Remove the auto lash adjuster retainer MD998443 from each rocker arm.
11. Install or connect the following:
   - Valve cover and upper timing belt cover
   - Positive Crankcase Valve (PCV) hose
   - Distributor, aligning the mark made during removal
   - Negative battery cable

12. Start the engine and check for leaks.

3.0L, 3.5L Engines

1. Before servicing the vehicle, refer to the precautions in the beginning of this section.
2. Relieve the fuel system pressure.
3. Drain and recycle the engine coolant.
4. Drain and recycle the engine crankcase.
5. Remove or disconnect the following:
   - Negative battery cable

✳✳ CAUTION

Work must be started after 90 seconds from the time the ignition switch is turned to the LOCK position and the negative battery cable is disconnected.

- Intake manifold plenum
- Valve cover
- Timing belt
- Sprocket from the camshaft

6. Install auto lash adjuster retainers SST MD998443 on the rocker arms.
7. Remove or disconnect the following:
   - Distributor and the distributor extension, if equipped
   - Rocker arms, rocker shafts and bearing caps, as an assembly
   - Thrust cage and o-ring
   - Camshaft from the cylinder head

8. Inspect the bearing journals on the camshaft and the cylinder head.

To install:

9. Lubricate the camshaft journals and camshaft with clean engine oil
10. Install or connect the following:
   - Camshaft in the cylinder head.

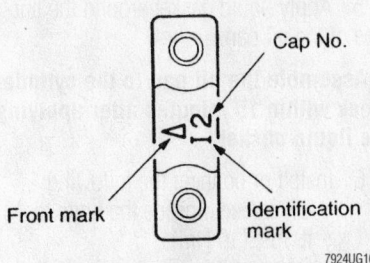

The camshaft bearing caps have identification marks on them—3.5L engine

- Thrust cage and o-ring. Torque the bolts to 109 inch lbs. (12 Nm).
- Rocker arms, rocker arm shaft and the rocker shaft spring.

11. Temporarily tighten the rocker shaft with the bolts positioned so that the intake valve rocker arms do not push the valves.

12. Install or connect the following:
- Rocker shaft spring from above and mount it at right angles to the plug guide.

13. Before installing the exhaust rocker arms and the rocker arm shaft, mount the rocker shaft spring.

14. Remove the SST used to hold the lash adjuster in position.

15. Check to ensure that the flat side of the rocker shaft is perpendicular to the cylinder head, and facing the valves.

16. Gradually tighten the bearing caps in 2 or 3 steps. In the final step tighten to 23 ft. lbs. (31 Nm).

17. Install or connect the following:
- Distributor, if removed
- Sprockets. Torque the bolts to 65 ft. lbs. (88 Nm).
- Timing belt and timing belt cover
- Valve cover. Torque the bolts to 26 inch lbs. (3.4 Nm).
- Intake manifold plenum
- Negative battery cable

18. Start the engine and check for leaks and proper operation.

19. Refill the coolant and crankcase.

Starter Motor

REMOVAL & INSTALLATION

1. Before servicing the vehicle, refer to the precautions in the beginning of this section.

2. Remove or disconnect the following:
- Negative battery cable
- Engine under cover, if equipped
- Front engine mount heat protector
- Starter cover
- Wires
- Starter motor

To install:

3. Install or connect the following:
- Starter motor and cover. Torque the bolts to 20–25 ft. lbs. (26–33 Nm).
- Wires
- Front engine mount heat protector. Torque the nut to 89 inch lbs. (10 Nm).
- Engine under cover, if equipped
- Negative battery cable

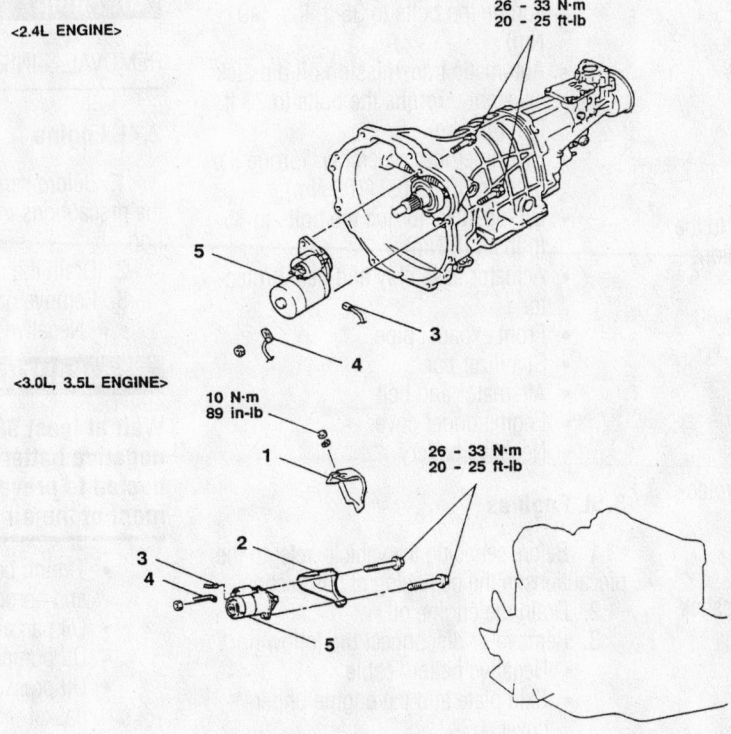

1. FRONT ENGINE MOUNT HEAT PROTECTOR <RH>
2. STARTER COVER
3. STARTER CONNECTOR
4. BATTERY CABLE
5. STARTER ASSEMBLY

Starter motor mounting

For Tire, Wheel and Ball Joint specifications, see Section 1 of this manual

Oil Pan

REMOVAL & INSTALLATION

2.4L Engine

1. Before servicing the vehicle, refer to the precautions in the beginning of this section.
2. Drain the engine oil.
3. Remove or disconnect the following:
   - Negative battery
   - Engine under cover
   - Bell housing cover
   - Oil pan

To install:

4. Before installing, thoroughly clean the oil pan and cylinder block mating surfaces.
5. Apply liquid gasket around the surface of the oil pan.

➡**Assemble the oil pan to the cylinder block within 15 minutes after applying the liquid gasket.**

6. Install or connect the following:
   - Oil pan. Torque the bolts to 61 inch lbs. (6.9 Nm).
   - Bell housing cover. Torque the bolts to 78 inch lbs. (8.8 Nm).
   - Engine under cover
   - Negative battery cable

3.0L Engines

1. Before servicing the vehicle, refer to the precautions in the beginning of this section.
2. Drain the engine oil.
3. Remove or disconnect the following:
   - Negative battery
   - Engine under cover
   - Alternator and belt
   - Stabilizer bar
   - Front exhaust pipe
   - Actuator assembly and heat protector
   - Oil dipstick
   - Crossmember assembly
   - Automatic transmission oil dipstick assembly
   - Exhaust pipe support bracket
   - Transmission stay
   - Oil pan, lower
   - Oil screen and baffle plate
   - Oil pan upper

To install:

4. Before installing, thoroughly clean the oil pan and cylinder block mating surfaces.
5. Apply liquid gasket around the surface of the oil pan.

➡**Assemble the oil pan to the cylinder block within 15 minutes after applying the liquid gasket.**

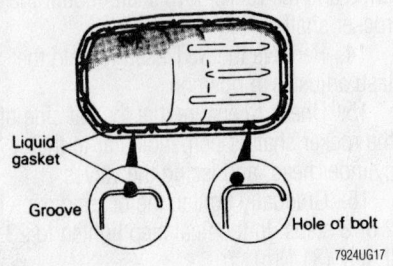

Apply a bead of sealant around the oil pan flange as shown—all engines are similar

6. Install or connect the following:
   - Oil pan upper. Torque the bolts to 53 inch lbs. (6.0 Nm).
   - Oil screen and baffle plate. Torque the bolts to 14 ft. lbs. (19 Nm).
   - Oil pan, lower. Torque the bolts to 53 inch lbs. (6.0 Nm).
   - Transmission stay. Torque the bolts to 26 ft. lbs. (35 Nm).
   - Exhaust pipe support bracket. Torque the bolts to 35 ft. lbs. (49 Nm).
   - Automatic transmission oil dipstick assembly. Torque the bolts to 33 ft. lbs. (44 Nm).
   - Crossmember assembly. Torque the bolts to 80 ft. lbs. (108 Nm).
   - Oil dipstick. Torque the bolts to 35 ft. lbs. (48 Nm).
   - Actuator assembly and heat protector
   - Front exhaust pipe
   - Stabilizer bar
   - Alternator and belt
   - Engine under cover
   - Negative battery

3.5L Engines

1. Before servicing the vehicle, refer to the precautions in the beginning of this section.
2. Drain the engine oil.
3. Remove or disconnect the following:
   - Negative battery cable
   - Skid plate and the engine undercover
   - Front exhaust pipe, if necessary
   - Catalytic converter
   - Lower oil pan
   - Front differential carrier
   - Cover
   - Oil dipstick
   - Oil pan upper
   - Oil screen

To install:

4. Before installing, thoroughly clean the oil pan and cylinder block mating surfaces.

5. Apply liquid gasket around the surface of the oil pan.

➡**Assemble the oil pan to the cylinder block within 15 minutes after applying the liquid gasket.**

6. Install or connect the following:
   - Oil screen. Torque the bolts to 13 ft. lbs. (19 Nm).
   - Oil pan upper. Torque the bolts to 48 inch lbs. (6 Nm).
   - Oil dipstick. Torque the bolts to 39 inch lbs. (4.8 Nm).
   - Cover. Torque the bolts to 84–108 inch lbs. (10–12 Nm).
   - Front differential carrier
   - Lower oil pan. Torque the bolts to 84–108 inch lbs. (10–12 Nm).
   - Catalytic converter
   - Front exhaust pipe, if necessary. Torque the bolts to 35 ft. lbs. (49 Nm).
   - Skid plate and the engine undercover
   - Negative battery cable

Oil Pump

REMOVAL & INSTALLATION

2.4L Engine

1. Before servicing the vehicle, refer to the precautions in the beginning of this section.
2. Drain the oil and remove the oil filter.
3. Remove or disconnect the following:
   - Negative battery cable

❊ CAUTION

Wait at least 90 seconds after the negative battery cable is disconnected to prevent possible deployment of the air bag.

   - Timing belt covers, timing belts and sprockets
   - Oil pan and gasket
   - Oil pump pick-up and gasket
   - Oil pressure relief plunger plug and gasket
   - Spring and plunger from the oil filter bracket
   - Bracket mounting bolts and the oil filter mount and gasket
   - Plug cap and gasket that covers the oil pump driven gear shaft. Using special tool MD998162 this is located on the right side of the front case, just above the protruding drive gear shaft.
   - Retaining bolt from the oil pump

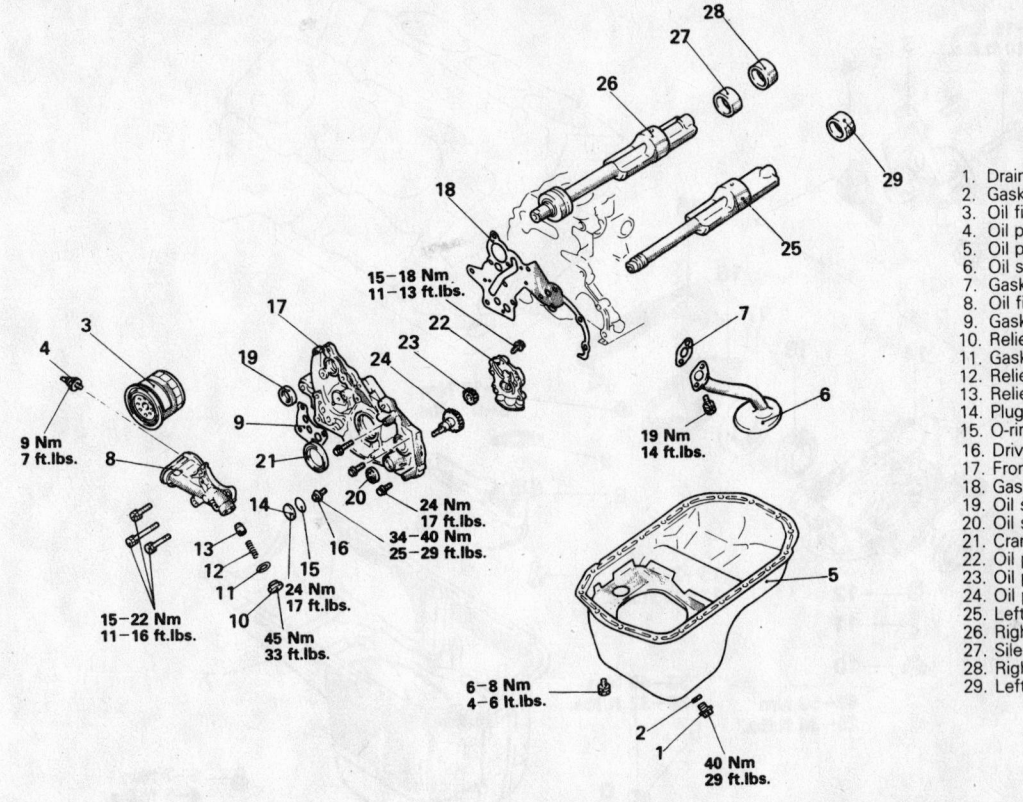

1. Drain plug
2. Gasket
3. Oil filter
4. Oil pressure switch
5. Oil pan
6. Oil screen
7. Gasket
8. Oil filter bracket
9. Gasket
10. Relief plug
11. Gasket
12. Relief spring
13. Relief plunger
14. Plug cap
15. O-ring
16. Driven gear bolt
17. Front case
18. Gasket
19. Oil seal
20. Oil seal
21. Crankshaft front oil seal
22. Oil pump cover
23. Oil pump driven gear
24. Oil pump drive gear
25. Left silent shaft
26. Right silent shaft
27. Silent shaft front bearing
28. Right silent shaft rear bearing
29. Left silent shaft rear bearing

7924UG18

Exploded view of the oil pump, oil pan and related components—2.4L engine shown; other engines are similar

driven gear located behind the plug removed earlier

- Front case mounting bolts and the case from the block
- Case gasket from the block

To install:

4. Prime the pump by pouring fresh oil into the pump intake and turning the driveshaft until oil comes out of the pressure port. Repeat this a few times until no air bubbles are present. Replace all seals on the case assembly.

5. Install a special seal guide to the crankshaft, MD998285, so the smaller diameter faces outward. Coat the outer diameter of the seal with clean engine oil.

6. Install or connect the following:
- New front case gasket and the front case by carefully positioning the crankshaft seal over the seal guide and lining up all bolt holes. Torque the bolts to 17 ft. lbs. (23 Nm).

7. Remove the plug from the left side of the block. Hold the left side silent shaft by inserting a tool in the plug hole and tighten the driven gear bolt to 26 ft. lbs. (35 Nm)
- New O-ring and the plug cover
- Oil filter mounting bracket gasket

- Mounting bracket and bolts tightening the oil filter mounting bracket bolts to 12 ft. lbs. (16 Nm).

8. Clean or replace the oil pick-up screen and install with a new gasket.

9. Install or connect the following:
- Oil pan using a new gasket
- Timing sprockets, belts and covers
- Negative battery cable

10. Refill the engine with the proper amount of engine oil.

11. Start the engine and check for proper oil pressure. Check for leaks.

3.0L and 3.5L Engines

1. Before servicing the vehicle, refer to the precautions in the beginning of this section.

2. Drain the engine oil.

3. Remove or disconnect the following:
- Negative battery cable
- Timing belt
- Oil pressure switch
- Oil dipstick
- Oil pans from the engine
- Oil baffle and screen
- Oil pump mounting bolts and the pump from the front of the engine

➡**Note the position of each oil pump case retaining bolts to facilitate installation. The bolts are of different length.**

To install:

4. Clean the gasket mounting surfaces of the pump and engine block.

5. Prime the pump by pouring fresh oil into the inlet and turning the rotors or by packing pump with petroleum jelly. Using a new gasket, install the oil pump on the engine and tighten all bolts to 10 ft. lbs. (14 Nm).

6. Clean out the oil pick-up or replace as required. Replace the oil pick-up gasket ring and install the pick-up to the pump.

7. Install or connect the following:
- Oil filter and the bracket. Torque the bolts to 17 ft. lbs. (23 Nm).
- Oil baffle and screen. Torque the bolts to 13 ft. lbs. (18 Nm).
- Oil pans. Torque the engines to 52 inch lbs. (5.9 Nm).
- Oil pressure switch. Torque the switch to 87 inch lbs. (9.8 Nm).
- Timing belt
- Dipstick
- Negative battery cable

8. Refill the engine with the proper amount of oil.

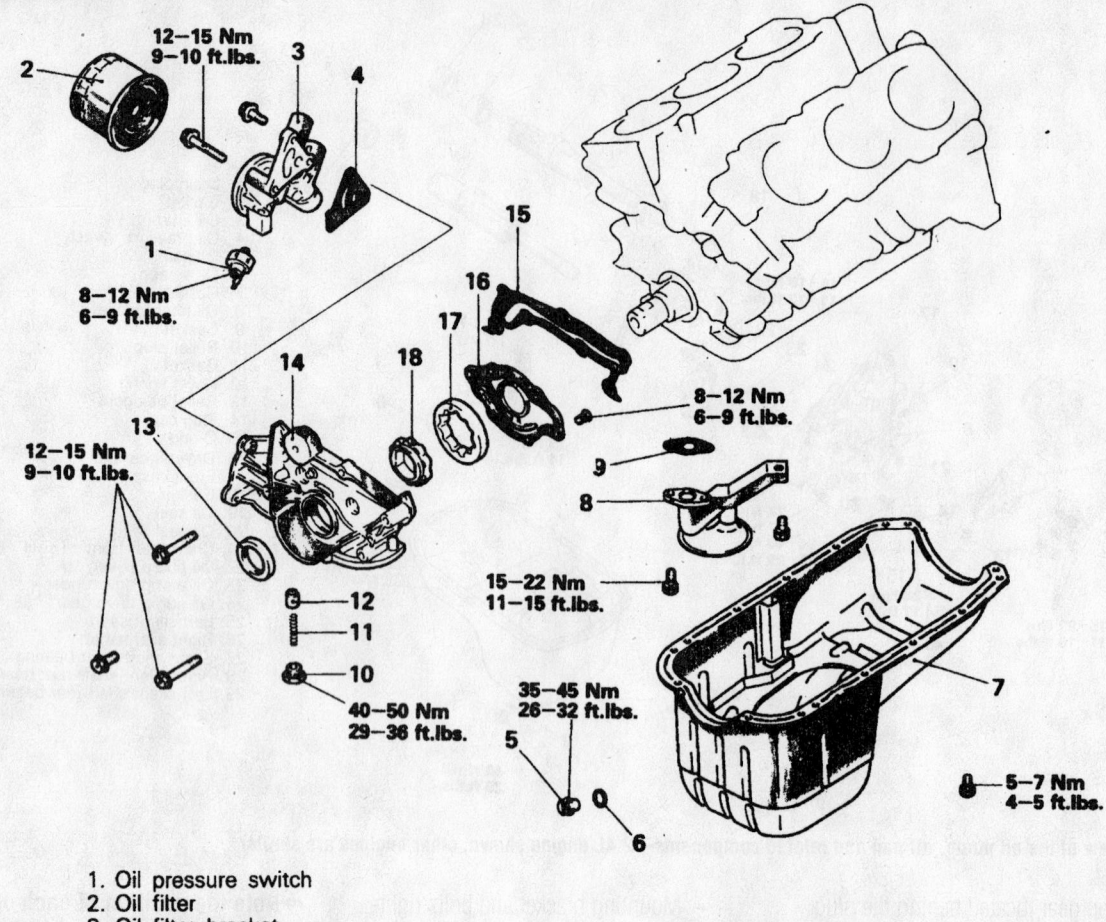

1. Oil pressure switch
2. Oil filter
3. Oil filter bracket
4. Oil filter bracket gasket
5. Drain plug
6. Drain plug gasket
7. Oil pan
8. Oil screen
9. Oil screen gasket
10. Plug
11. Relief spring
12. Relief plunger
13. Crankshaft front oil seal

14. Oil pump case
15. Oil pump gasket
16. Oil pump cover
17. Oil pump outer rotor
18. Oil pump inner rotor

7924UG19

Exploded view of the oil pump, oil pan and related components—2.4L engine shown; other engines are similar

9. Start the engine and check for proper oil pressure. Check for leaks.

Rear Main Seal

REMOVAL & INSTALLATION

2.4L Engine

1. Before servicing the vehicle, refer to the precautions in the beginning of this section.

2. Remove or disconnect the following:
- Transmission and clutch assembly, if equipped.
- Flywheel or the driveplate and the adapter plate, if equipped. Match-mark for reassembly.
- Oil seal

➡Take care not to gouge or damage the metal surrounding the seal. Inspect the sealing surface at the rear of the crankshaft. If a deep groove is worn into the surface, the crankshaft will have to be replaced. Lubricate the sealing surface with clean engine oil.

To install:

3. Using a seal installer of the correct size, install the new seal into the bore of rear oil seal case. Make certain that the flat side of the seal will face outward when the case is installed on the engine. The inside of the seal must be flush with the inside surface of the seal case.

4. Install or connect the following:
- Rear plate and bell housing cover
- Flywheel or drive plate, observing the matchmarks made earlier
- Transmission and related components as necessary

3.0L and 3.5L Engines

1. Before servicing the vehicle, refer to the precautions in the beginning of this section.

2. Remove or disconnect the following:
- Transmission and clutch assembly, if so equipped.

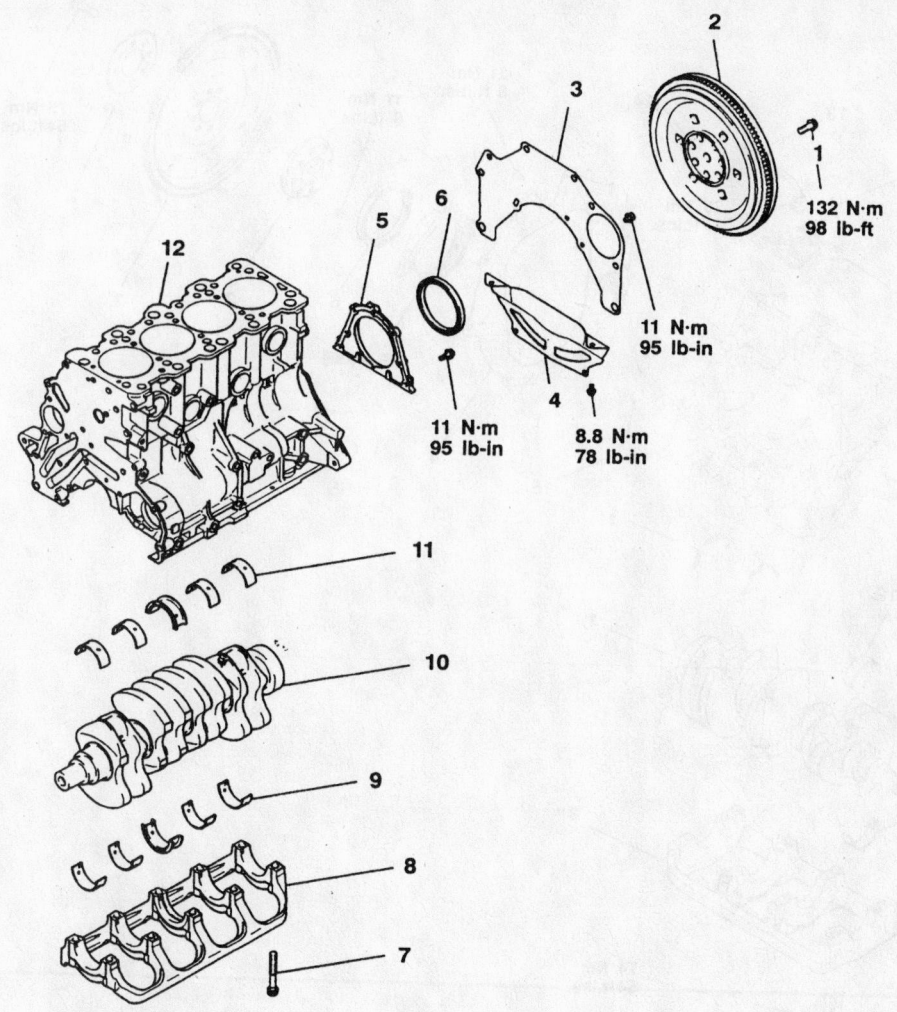

1. FLYWHEEL BOLT
2. FLYWHEEL
3. REAR PLATE
4. BELL HOUSING COVER
5. OIL SEAL CASE
6. OIL SEAL

7. BEARING CAP BOLT
8. BEARING CAP
9. CRANKSHAFT BEARING (LOWER)
10. CRANKSHAFT
11. CRANKSHAFT BEARING (UPPER)
12. CYLINDER BLOCK

7924UG20

Exploded view of the crankshaft, rear main seal, and related components—2.4L engines

- Flywheel or driveplate and adapter plate, matchmark for reassembly. For the 3.0L engines, use the Mitsubishi tools (MB990767-01 and MIT308239) to hold the crankshaft and flywheel stationary while loosening the flywheel bolts. For the 3.5L engine, use Mitsubishi tool (MD998781) to hold the flywheel in position.
3. Remove the rear oil seal as follows:
   a. Cut out a portion in the crankshaft oil seal lip.

b. Cover the tip of a small prytool with a cloth and apply it to the cutout in the oil seal to pry the oil seal out.

⁕⁕ CAUTION

Take care not to damage the crankshaft and oil seal case.

To install:
4. Inspect the sealing surface at the rear of the crankshaft. If a deep groove is worn into the surface, the crankshaft will have to be replaced. Coat the sealing lip of the seal

with fresh, clean engine oil. Press the new seal into the case with a seal installing tool. The seal must be pressed in squarely until it bottoms in the case. It is necessary to use the proper tool (MD998718-01) to fit the seal into place.
5. Install or connect the following:
   - Rear plate
   - Transmission mounting plate
   - Flywheel or drive plate and adapter
   - Transmission and related components as necessary

For Maintenance Interval recommendations, see Section 1 of this manual

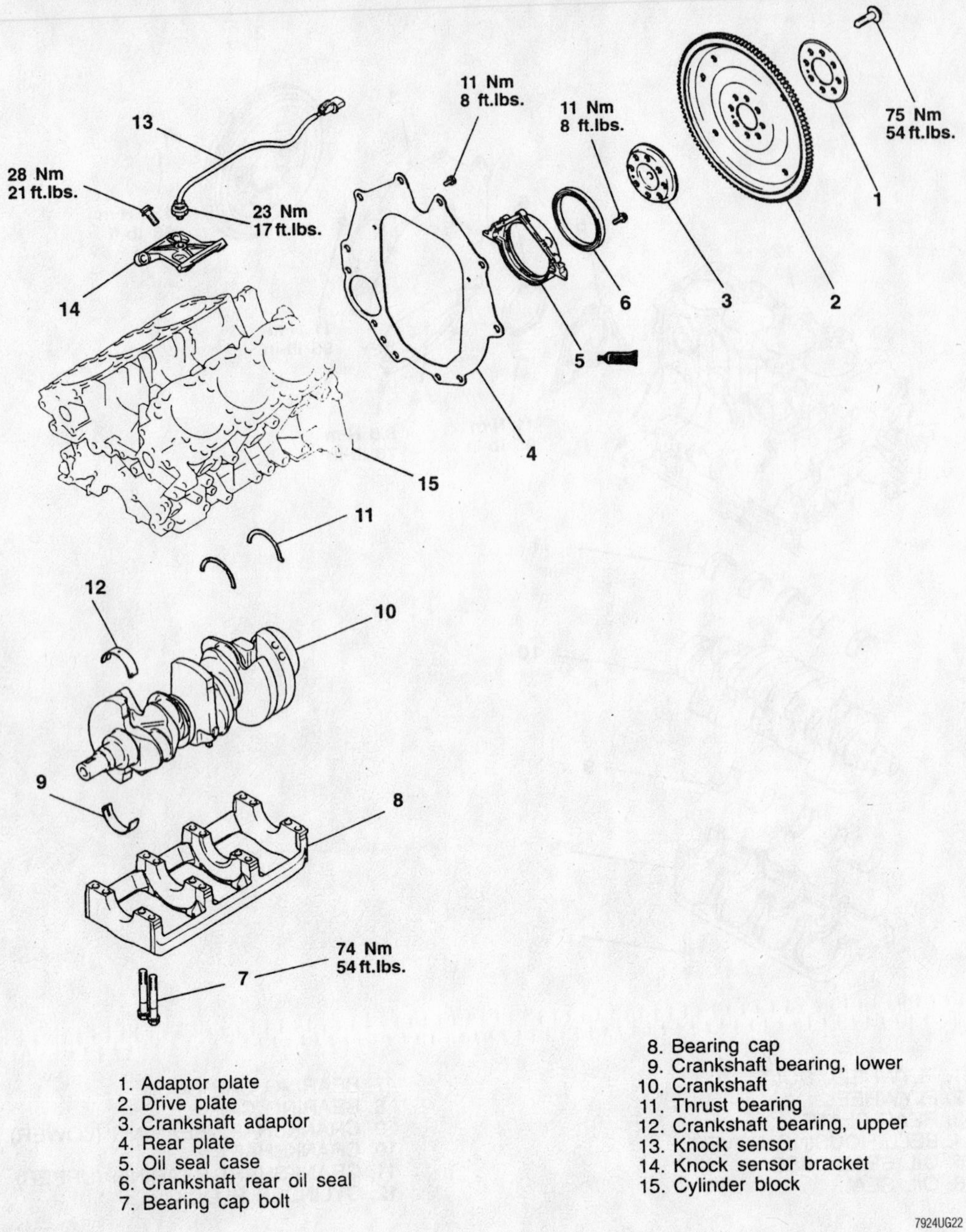

11 Nm
8 ft.lbs.
11 Nm
8 ft.lbs.
75 Nm
54 ft.lbs.
13
28 Nm
21 ft.lbs.
23 Nm
17 ft.lbs.
14
1
2
3
4
5
6
15
11
12
10
9
8
74 Nm
54 ft.lbs.
7
1. Adaptor plate
2. Drive plate
3. Crankshaft adaptor
4. Rear plate
5. Oil seal case
6. Crankshaft rear oil seal
7. Bearing cap bolt
8. Bearing cap
9. Crankshaft bearing, lower
10. Crankshaft
11. Thrust bearing
12. Crankshaft bearing, upper
13. Knock sensor
14. Knock sensor bracket
15. Cylinder block
7924UG22

Exploded view of the crankshaft, rear main seal and related components—3.5L engine shown; 3.0L engine is similar

Piston and Ring

POSITIONING

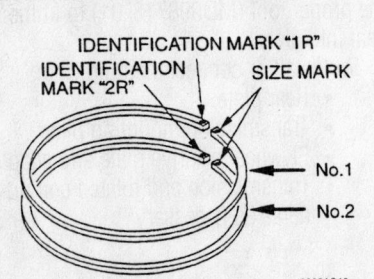

IDENTIFICATION MARK "1R"
IDENTIFICATION MARK "2R"
SIZE MARK
No.1
No.2
9302AG13

Piston ring identification

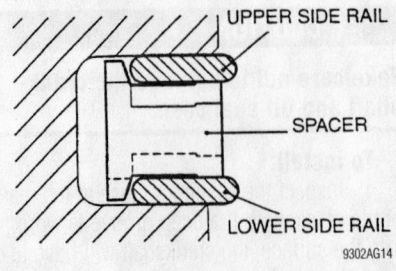

UPPER SIDE RAIL
SPACER
LOWER SIDE RAIL
9302AG14

Oil ring identification

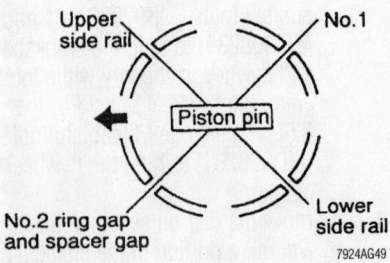

Upper side rail
No.1
Piston pin
Lower side rail
No.2 ring gap and spacer gap
7924AG49

Piston ring end-gap spacing

FUEL SYSTEM

Fuel System Service Precautions

Safety is the most important factor when performing not only fuel system maintenance but any type of maintenance. Failure to conduct maintenance and repairs in a safe manner may result in serious personal injury or death. Maintenance and testing of the vehicle's fuel system components can be accomplished safely and effectively by adhering to the following rules and guidelines.

• To avoid the possibility of fire and personal injury, always disconnect the negative battery cable unless the repair or test procedure requires that battery voltage be applied.

• Always relieve the fuel system pressure prior to disconnecting any fuel system component (injector, fuel rail, pressure regulator, etc.), fitting or fuel line connection. Exercise extreme caution when relieving fuel system pressure, to avoid exposing your skin, face and eyes to fuel spray. Please be advised that fuel under pressure may penetrate the skin or any part of the body that it contacts.

• Always place a shop towel or cloth around the fitting or connection prior to loosening to absorb any excess fuel due to spillage. Ensure that all fuel spillage (should it occur) is quickly removed from engine surfaces. Ensure that all fuel soaked cloths or towels are deposited into a suitable waste container.

• Always keep a dry chemical (Class B) fire extinguisher near the work area.

• Do not allow fuel spray or fuel vapors to come into contact with a spark or open flame.

• Always use a back-up wrench when loosening and tightening fuel line connection fittings. This will prevent unnecessary stress and torsion to fuel line piping. Always follow the proper torque specifications.

• Always replace worn fuel fitting O-rings with new. Do not substitute fuel hose where fuel pipe is installed.

Fuel System Pressure

RELIEVING

❊❊ CAUTION

The fuel system is under constant pressure, even with the engine off. This pressure must be relieved

before disconnecting any fuel system component, fitting or fuel line connection. Failure to do so may result in personal injury.

1999–02 Montero Sport and 1999–00 Montero

1. Disconnect the fuel pump electrical connector, located at the rear side of the fuel tank.
2. Start the engine.
3. After the engine stalls, turn the ignition switch **OFF** and reconnect the fuel pump connector.
4. Disconnect the negative battery cable, then continue with the service procedure.

2001–02 Montero

1. Turn the ignition switch to **LOCK**.
2. Fold down the second seat.
3. Remove the upper and lower service hole cover.
4. Disconnect the fuel pump module connector.
5. Start the engine and let it run out of fuel.

Fuel Filter

REMOVAL & INSTALLATION

❊❊ CAUTION

The fuel injection system remains under pressure after the engine has been OFF. Properly relieve fuel pressure before disconnecting any fuel lines. Failure to do so may result in fire or personal injury.

❊❊ CAUTION

Do not allow fuel spray or fuel vapors to come in contact with a spark or open flame. Keep a dry chemical fire extinguisher nearby. Never store fuel in an open container due to risk of fire or explosion.

1. Relieve the fuel system pressure.
2. Before servicing the vehicle, refer to the precautions in the beginning of this section.
3. Disconnet the negative battery cable.
4. Remove the fuel filter protector if equipped.

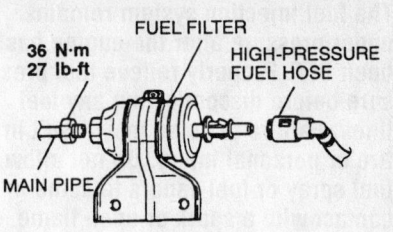

Fuel filter removal—Montero Sport shown

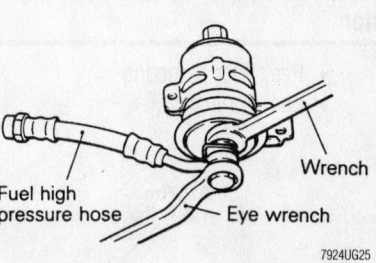

Always use a back-up wrench when removing or installing fuel lines to the filter

5. Using a back-up wrench disconnect the fuel line(s) from the filter. If the filter uses a push-on type connector, press the retainer to release the connection.
6. Remove the filter from the mounting bracket.

To install:

7. Position the filter to the mounting bracket in the proper direction.
8. Connect the fuel lines to the filter. Use a back-up wrench to hold the fuel filter. Torque the banjo bolt(s) to 18–25 ft. lbs. (25–35 Nm) or the line fitting to 27 ft. lbs. (36 Nm).
9. Install the fuel filter protector if equipped.
10. Connect the negative battery cable.
11. Start the engine and check for leaks.

Fuel Pump

REMOVAL & INSTALLATION

Montero

➡**The manufacturer recommends draining of the fuel tank.**

1. Before servicing the vehicle, refer to the precautions in the beginning of this section.
2. Relieve the fuel system pressure.
3. Remove or disconnect the following:
   • Negative battery cable

✳✳ CAUTION

The fuel injection system remains under pressure after the engine has been OFF. Properly relieve fuel pressure before disconnecting any fuel lines. Failure to do so may result in fire or personal injury. Do not allow fuel spray or fuel vapors to come in contact with a spark or open flame. Keep a dry chemical fire extinguisher nearby. Never store fuel in an open container due to risk of fire or explosion.

- Rear floor carpeting
- Fuel pump cover

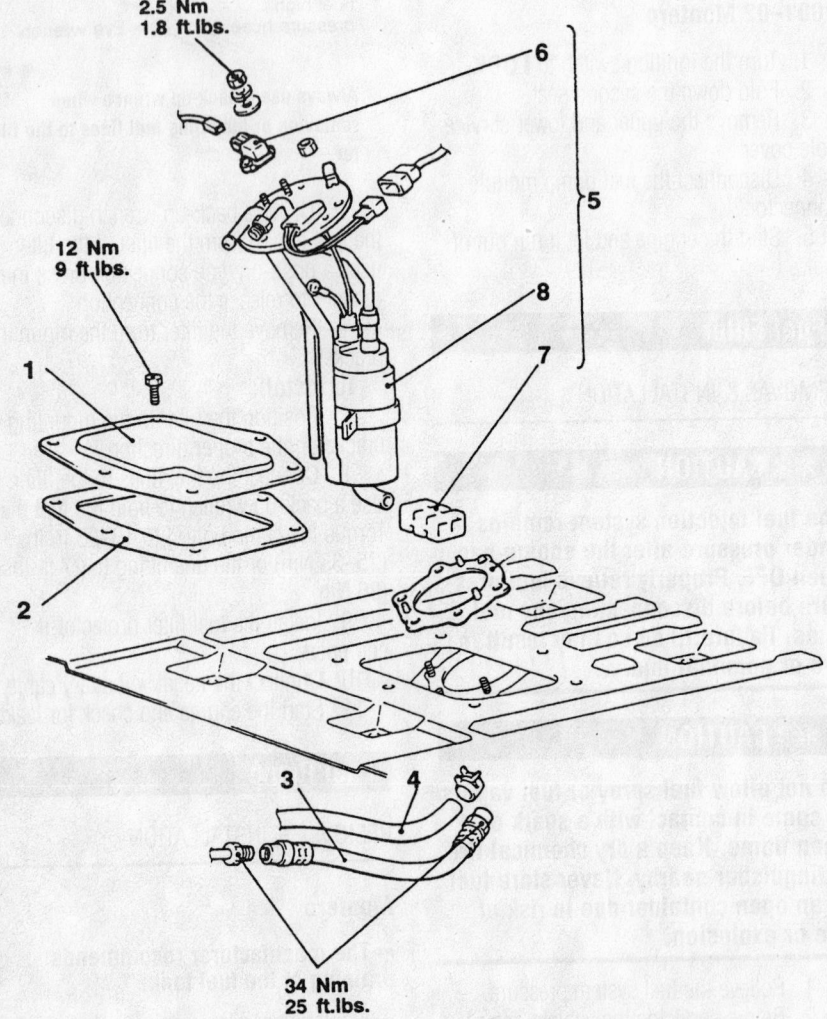

2.5 Nm
1.8 ft.lbs.

12 Nm
9 ft.lbs.

34 Nm
25 ft.lbs.

1. Floor cover
2. Packing
3. High-pressure fuel hose
4. Fuel return hose connection
5. Fuel pump and filter assembly
6. Fuel tank differential pressure sensor
7. Filter
8. Fuel pump assembly

7924UG26

The fuel pump on the Montero is removed through the rear floor pan

- Fuel pump connector and the fuel hoses
- Fuel pump assembly

To install:
4. Install or connect the following:
- Fuel pump assembly into the fuel tank. Torque the nuts to 24 inch lbs. (2.5 Nm).
- Fuel lines and the fuel pump connector.
- Fuel pump cover. Torque the bolts to 108 inch lbs. (12 Nm).
- Rear floor carpeting.
- Negative battery cable
5. Refill the fuel tank, if drained
6. Start the vehicle; check for leaks and proper operation.

Montero Sport

1. Before servicing the vehicle, refer to the precautions in the beginning of this section.
2. Properly relieve the fuel system pressure.
3. Remove the fuel tank drain plug and drain the fuel from the tank.

✳✳ CAUTION

The fuel injection system remains under pressure after the engine has been OFF. Properly relieve fuel pressure before disconnecting any fuel lines. Failure to do so may result in fire or personal injury.

4. Remove or disconnect the following:
- Negative battery cable

✳✳ CAUTION

Wait at least 90 seconds after the negative battery cable is disconnected to prevent possible deployment of the air bag.

- Fuel tank protector, if equipped
- Fuel tank from the vehicle
- Fuel pump retaining screws and the pump from the tank

To install:
5. Clean the seal area of the tank.
6. Install or connect the following:
- New gasket
- Fuel pump in the same position as originally installed.
- Fuel pump retaining screws, Torque the nuts to 22 inch lbs. (2.5 Nm).
- Fuel tank. Torque the bolts to 20 ft. lbs. (27 Nm).
- Fuel tank drain plug and the fuel tank protector, if equipped
- Negative battery cable
7. Refill the fuel tank and install the cap.
8. Check fuel system for leaks.

Fuel Injector

REMOVAL & INSTALLATION

2.4L Engines

1. Before servicing the vehicle, refer to the precautions in the beginning of this section.
2. Properly relieve the fuel system pressure.
3. Remove or disconnect the following:
- Accelerator cable
- Air intake tube and hose

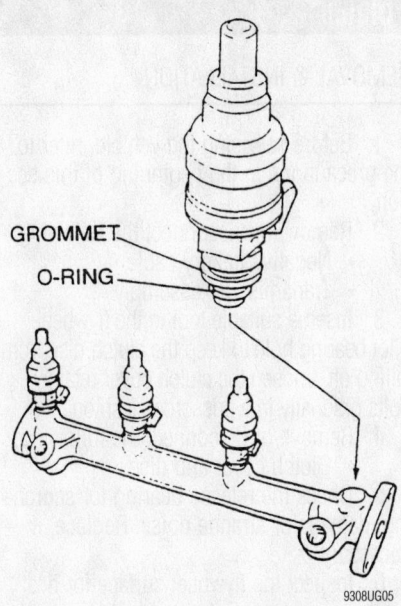

GROMMET

O-RING

9308UG05

Exploded view of fuel injector and rail on 2.4L engines other engines are similar

- High pressure fuel hose and o-ring
- Fuel return line
- Pressure regulator, vacuum line and o-ring

- Injector connector
- Fuel rail and insulators
- Fuel injectors and insulators
- O-rings and grommets

To install:

4. Install or connect the following:
- O-rings and grommets
- Fuel injectors and insulators
- Fuel rail and insulators. Torque the bolts to 106 inch lbs. (12 Nm).
- Injector connector
- Pressure regulator, vacuum line and o-ring. Torque the bolts to 78 inch lbs. (8.8 Nm).
- Fuel return line
- High pressure fuel hose and o-ring. Torque the bolts to 43 inch lbs. (4.9 Nm).
- Air intake tube and hose
- Accelerator cable

3.0L and 3.5L engines

1. Before servicing the vehicle, refer to the precautions in the beginning of this section.

2. Properly relieve the fuel system pressure.

3. Remove or disconnect the following:

- Air cleaner assembly, ducts and air intake hose
- Intake manifold plenum
- Fuel return line
- Pressure regulator, vacuum line and o-ring
- High pressure fuel hose and o-ring
- Injector connector
- Fuel pipe and o-rings
- Fuel rails and insulators
- Fuel injectors and insulators
- O-rings and grommets

To install:

4. Install or connect the following:
- O-rings and grommets
- Fuel injectors and insulators
- Fuel rail and insulators. Torque the bolts to 106 inch lbs. (12 Nm).
- Injector connector
- High pressure fuel hose and o-ring. Torque the bolts to 43 inch lbs. (4.9 Nm).
- Pressure regulator, vacuum line and o-ring. Torque the bolts to 78 inch lbs. (8.8 Nm).
- Fuel return line
- Intake manifold plenum
- Air cleaner assembly, ducts and air intake hose

DRIVE TRAIN

Transmission Assembly

REMOVAL & INSTALLATION

1. Before servicing the vehicle, refer to the precautions in the beginning of this section.

2. Drain the transmission fluid.

3. Remove or disconnect the following:
- Negative battery cable
- Transmission and transfer case shift lever assembly. On manual transmissions
- Transfer case protector, if equipped
- Front exhaust pipe from the 2 exhaust manifolds, then disconnect it from the intermediate pipe/catalytic converter (make certain to retain the bolts and nuts for reassembly).
- Rear driveshaft at both the rear axle and the transfer case flanges. Matchmark for reassembly.
- Front driveshaft from the front axle, by sliding it forward, plug the transfer case to prevent residual

fluid leakage. On 4-wheel drive vehicles
- Dust seal from the rear extension housing.
- Ground cables
- 4WD indicator light switch connector
- Pulse generator connector
- Speed Sensor Connector
- Oxygen (O_2S) sensor connector
- Back-up light switch connector
- HI/LO detection switch connector
- Center differential lock detection switch connection.

There may be others depending on the particular year, model and engine with which the vehicle came equipped.

4. Remove or disconnect the following:

- Speedometer cable out of the transmission

5. On manual transmissions:

6. Remove or disconnect the following:
- Clutch cylinder heat protector
- Clutch release cylinder (with the clutch hose connected to it) from the transmission. Suspend it from

the body by using a piece of wire or a similarly safe method.
- Starter motor and the heat shield

7. On automatic transmissions:

8. Remove or disconnect the following:
- Bolts attaching the torque converter to the flexplate
- Dipstick
- Fluid cooling lines
- Shift linkage at the transmission

9. Place a floor jack and a block of wood below the engine oil pan.

10. Lift the floor jack under the engine just until the weight of the engine is taken onto the jack—the engine should only barely be lifted by the jack.

11. Use a transmission jack or second floor jack to place under the transmission. Don't support the transmission yet, only lift the jack until it is slightly below the transmission.

12. Remove or disconnect the following:

- Left-hand and right-hand side transmission stays from the front of the transmission, if equipped
- Bell housing lower cover

- Transfer case mounting bracket, if equipped

13. Lift the floor jack up until the transmission is being slightly supported by it.

14. Remove or disconnect the following:
- Transmission-to-crossmember bolts. Lift the jack about ¼ in. (6mm) off of the crossmember support.
- Crossmember from the vehicle
- Transmission mounting blots. Pull the transmission away from the engine and lower it from the vehicle.

To install:

15. Lift the transmission and transfer assembly into position with the floor jack.

16. On the engine side, there are 2 centering locations. Be sure that the transmission mounting bolt holes are aligned with them before mounting the transmission and transfer assembly to the engine. Lowering the rear of the engine SLIGHTLY may help align the 2 assemblies.

17. Install or connect the following:
- Transmission assembly onto the engine making sure the aligning areas stay aligned. Torque the bolts to 54 ft. lbs. (75 Nm).

18. Lift the transmission/transfer assembly with the floor jack. Since the engine is now attached to the transmission, it also will rise slightly. Adjust its jack to keep only slight support.

19. Install or connect the following:
- Crossmember in place and secure with the mounting bolts. Torque the bolts to 47 ft. lbs. (65 Nm).
- Transmission and transfer case assembly onto the crossmember
- Crossmember-to-transmission bolts. Torque the bolts to 15–18 ft. lbs. (20–24 Nm) on Montero sport and to 36 ft. lbs. (49 Nm) on Montero.
- Mounting bracket back onto the transfer case, if equipped
- Bell housing lower cover
- Left-hand and right-hand side transmission stays
- Starter motor and heat shield
- Clutch release cylinder
- Speedometer cable into the transmission and secure it there with the retaining ring.
- Center differential lock detection switch connection
- HI/LO detection switch connector
- Back-up light switch connector
- O2S sensor connector
- Speed Sensor Connector
- Pulse generator connector
- 4WD indicator light switch connector

- Ground cables
- Flexplate-to-torque converter bolts. Torque the bolts to 25–30 ft. lbs. (35–42 Nm). On automatic transmissions
- Dipstick tube
- Shift linkage
- Fluid cooler lines. Torque the line fittings to 32 ft. lbs. (44 Nm). On automatic transmissions

20. Tap the dust seal guard back onto the rear extension housing with a rubber or plastic mallet.

21. Install or connect the following:
- Front driveshaft into the transfer case, then attach it to the front differential
- Rear driveshaft, make certain that the matchmarks line up.
- Front exhaust pipe to the catalytic converter and the exhaust manifolds
- Transfer case protector, if equipped
- Transmission and transfer case shift lever assembly. On manual transmissions
- Negative battery cable to the battery

22. Refill the transmission and transfer case with oil.

23. Start the vehicle and check for any leaks.

Clutch

REMOVAL & INSTALLATION

1. Before servicing the vehicle, refer to the precautions in the beginning of this section.

2. Remove or disconnect the following:
- Negative battery cable
- Transmission assembly

3. Insert a suitable tool in the flywheel pilot bearing hole to keep the clutch disc from falling off. Loosen the clutch cover retainer bolts gradually in a crisscross fashion.

4. Remove or disconnect the following:
- Clutch cover and disc

5. Check the release bearing for scorching, damage or strange noise. Replace, if necessary.

6. Inspect the flywheel surface for heat cracks or scoring. Reface or replace the flywheel as required.

To install:

7. Apply high temperature grease to the clutch disc splines, input shaft, contact points of the release fork and inside diameter of the release bearing.

➡**Do not allow oil or grease to contact the clutch facing and pressure plate.**

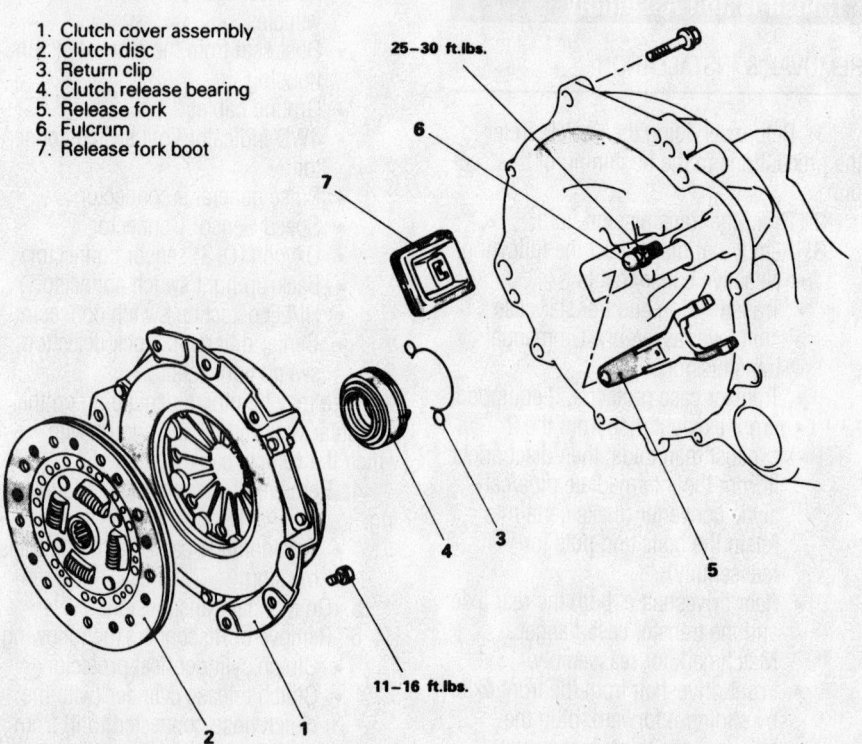

1. Clutch cover assembly
2. Clutch disc
3. Return clip
4. Clutch release bearing
5. Release fork
6. Fulcrum
7. Release fork boot

25–30 ft.lbs.

11–16 ft.lbs.

Exploded view of the typical clutch assembly components

8. Install or connect the following:
- Flywheel, align using a suitable tool

➡ **When installing the clutch disc, be sure that the surface having the manufacturer's stamped mark is on the pressure plate side.**

- Clutch cover with the dowel pin holes in alignment with the dowel pins in the flywheel and tighten the bolt gradually in a crisscross fashion. Torque the bolts to 14 ft. lbs. (19 Nm).
- Transmission assembly
- Negative battery cable

9. Road test the vehicle for proper operation.

Hydraulic Clutch System

BLEEDING

✳✳ WARNING

When bleeding, keep the facial area well away from the slave cylinder and protect all painted surfaces from fluid contact. Brake fluid will damage painted surfaces and could cause physical injury.

1. Fill the clutch master cylinder with fresh DOT 3 brake fluid.
2. Have a helper sit in the vehicle.
3. Remove the bleeder screw cap.
4. If the system is empty, the most efficient way to get fluid down to the cylinder is:

   a. Loosen the bleeder about ½ – ¾ turn

   b. Place a finger firmly over the bleeder

   c. Have a helper pump the brakes slowly until fluid pressure is felt at the bleeder

   d. Once fluid is at the bleeder, close it before the pedal is released.

➡ **If the pedal is pumped rapidly, the fluid will churn and create small air bubbles, which are difficult and time consuming to remove from the system. These air bubbles will eventually congregate and will result in a spongy pedal.**

5. Once fluid has been pumped to the slave cylinder:
- Open the bleeder screw

- Have a helper depress the clutch pedal
- Lock the bleeder and have the helper release the pedal
- Wait 15 seconds and repeat the procedure (including the 15 second wait) until no air bubbles flow from the bleeder.

Remember to close the bleeder before the pedal is released. If the bleeder is left open when the pedal is released, air will be induced into the system.

6. If a helper is not available, connect a small hose to the bleeder, submerge the other end in a clean container of fresh brake fluid placed in a position that is visible from the driver's seat. Pump the pedal until no air comes out of the tube.

Transfer Case Assembly

REMOVAL & INSTALLATION

The transfer case is removed from the vehicle along with the transmission. Refer to the Transmission Removal and Installation procedure for information.

Front Halfshaft

REMOVAL & INSTALLATION

Outer Axle Shafts

1. Before servicing the vehicle, refer to the precautions in the beginning of this section.

2. Remove or disconnect the following:
- Negative battery cable
- Undercover
- Wheels
- Hub cover dust cap
- Snapring from the inside of the hub and the shim
- Front brake caliper assembly and support with mechanics wire
- Speed sensor, if equipped with ABS
- Tie rod from the steering knuckle assembly
- Upper and lower ball joints from the steering knuckle assembly
- Front hub/knuckle assembly with the inner and outer bearings intact

3. On the left side, pull the halfshaft from the differential carrier.

4. For the right side, remove the fasteners and the halfshaft from the vehicle.

To install:

5. Install or connect the following:
- New circlip, on the left side halfshaft
- Inner shaft. Torque the nuts to 36–43 ft. lbs. (49–59 Nm), on the right side halfshaft
- Front hub/knuckle and bearing assembly
- Upper ball joint to the knuckle. Torque the nut to 54 ft. lbs. (74 Nm).
- Lower ball joint to knuckle. Torque the nut to 108 ft. lbs. (147 Nm).
- New cotter pins
- Tie rod end to the steering knuckle. Torque the nut to 33 ft. lbs. (44 Nm) and a new cotter pin.
- Speed sensor, if removed
- Front brake assembly
- Shim and snapring to the axle shaft. Install the front hub dust cover
- Wheels and the undercover
- Negative battery cable

Inner Axle Shafts

1. Before servicing the vehicle, refer to the precautions in the beginning of this section.

2. Remove or disconnect the following:
- Negative battery cable
- Undercover
- Right side wheel
- Right outer halfshaft
- Lower shock absorber mounting bolts
- Inner shaft from housing, using a slide hammer with tool MB990241

To install:

3. Install or connect the following:
- New circlip on the inner halfshaft
- Inner shaft into the housing, drive the axle into position.
- Lower shock absorber mounting bolts. Torque the bolts to 65–76 ft. lbs. (88–103 Nm).
- Right halfshaft assembly
- Undercover
- Wheel
- Negative battery cable

Rear Halfshaft

REMOVAL & INSTALLATION

1. Before servicing the vehicle, refer to the precautions in the beginning of this section.

2. Remove or disconnect the following:
- Negative battery cable
- Wheels
- Hub cover dust cap and nut
- Caliper assembly and support with mechanics wire
- Speed sensor, if equipped with ABS
- Upper and lower arms from the knuckle assembly
- Knuckle assembly

3. On the left side, pull the halfshaft from the differential carrier.

4. For the right side, remove the fasteners and the halfshaft from the vehicle.

To install:

5. Install or connect the following:
- Hub/knuckle and bearing assembly
- Upper arm to the knuckle. Torque the nut to 111 ft. lbs. (150 Nm).
- Lower arm to knuckle. Torque the nut to 170 ft. lbs. (231 Nm).
- New cotter pins
- Toe control arm. Torque the nut to 50 ft. lbs. (67 Nm)
- Shaft nut and a new cotter pin. Torque to 188 ft. lbs. (255 Nm)
- Speed sensor, if removed

- Front brake assembly
- Install the front hub dust cover
- Wheels
- Negative battery cable

CV-Joints

OVERHAUL

1. Before servicing the vehicle, refer to the precautions in the beginning of this section.

2. Remove or disconnect the following:

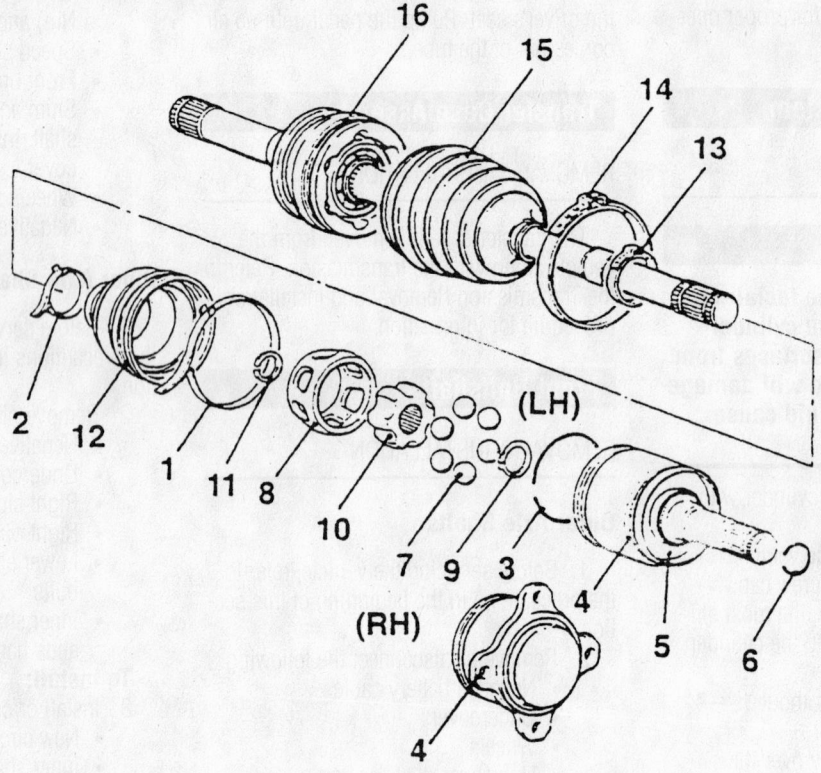

1. D.O.J. boot band (large)
2. D.O.J. boot band (small)
3. Circlip
4. D.O.J. outer race
5. Dust cover
6. Circlip
7. Ball
8. D.O.J. cage
9. Snap ring
10. D.O.J. inner race
11. Circlip
12. D.O.J. boot
13. B.J. boot band (small)
14. B.J. boot band (large)
15. B.J. boot
16. B.J. assembly

CV-Joint, exploded view

- Front wheel
- Driveshaft from the car
- Small and larger band
- Circlip
- Double Offset Joint (DOJ) outer race
- Dust cover
- Circlip
- Balls from the cage
- Cage from the inner race. Turn the cage so that the projections of the inner race align with the recesses of the cage.
- Snapring from the shaft
- DOJ inner race
- Slide the boot off
- Birfield Joint (BJ) small and larger bands
- BJ boot
- BJ assembly

To install:

3. Check the shaft and splines for damage or wear. Inspect the cage, race and balls for any sign of corrosion, wear, cracking or damage. Clean all the parts thoroughly and air dry them completely before installation. Any remaining cleaning solvent can dissolve the lubricating grease.

4. Tool MB991561 can be used to crimp the bands in place.

5. Install or connect the following:
- BJ assembly
- BJ boot, slid the small end of the boot until only one shaft groove cone be seen.
- BJ small band, crimp the band. Fill the BJ boot with 4.6 oz (130 g) of grease.
- BJ larger band, crimp the band
- DOJ small band and boot, fill the boot with grease
- DOJ cage onto the driveshaft so that the smaller diameter side is installed first.
- Circlip
- DOJ inner race and new snap ring, apply grease to the inner race
- Balls into the cage, grease to the ball areas of the cage and race

- Outer race, fill the outer race about ⅓ full of grease.
- Dust cover
- Circlip
- Large boot band, release the air from the boot then crimp
- Driveshaft into the car

Automatic Locking Hubs

REMOVAL & INSTALLATION

1. Place the locking hub in the free position. To do this, shift the transfer shift lever to the 2H position, then move the vehicle 4–7 ft. (1–2 m) backwards.

2. Before servicing the vehicle, refer to the precautions in the beginning of this section.

3. Remove or disconnect the following:
- Front wheels of the vehicle
- Hub cover
- Snapring from the axle shaft
- Shim

1. Cover
   Adjustment of drive shaft end play
2. Snap ring
3. Shim
4. Front brake assembly

5. Bolts
6. Automatic free-wheeling hub assembly
7. Shim
8. Lock washer
9. Lock nut
10. Front hub assembly

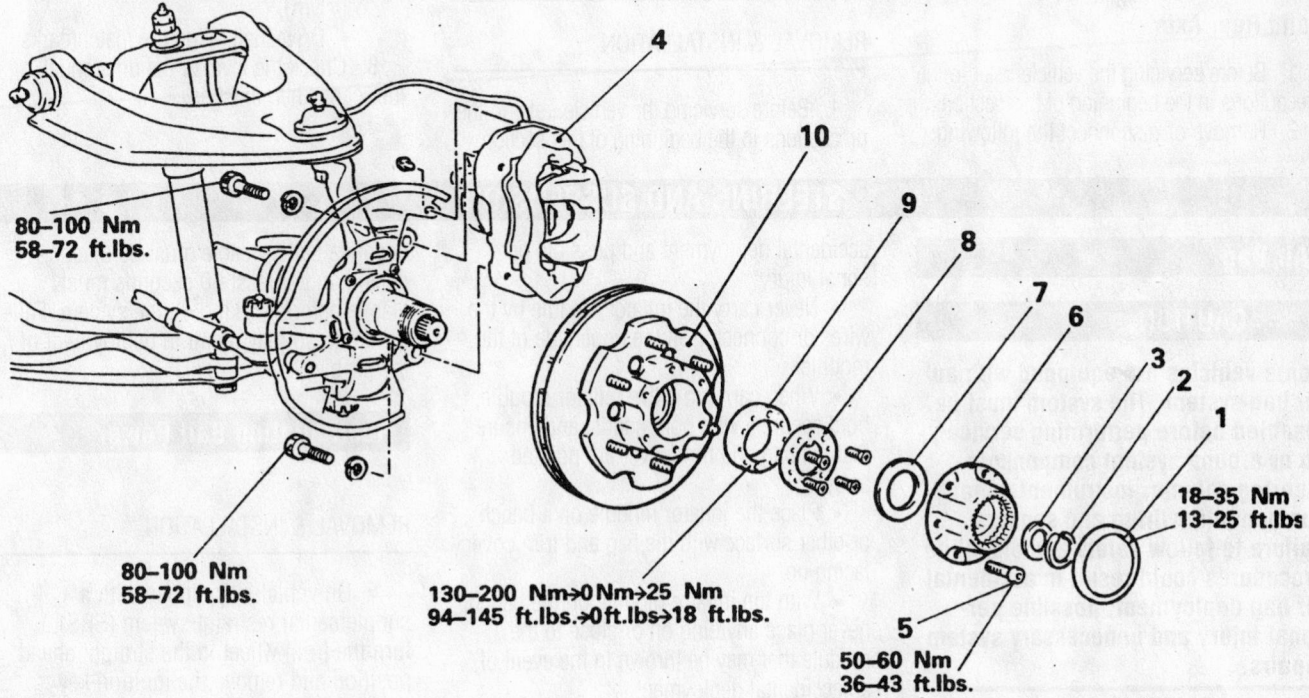

80–100 Nm
58–72 ft.lbs.

80–100 Nm
58–72 ft.lbs.

130–200 Nm→0 Nm→25 Nm
94–145 ft.lbs.→0 ft.lbs.→18 ft.lbs.

50–60 Nm
36–43 ft.lbs.

18–35 Nm
13–25 ft.lbs

7924UG45B

Axle hub and locking hub removal and installation—automatic hubs

Timing belt service is covered in Section 3 of this manual

- Drive flange
- Front brake assembly
- Speed sensor, if equipped
- Lock washer
- Lock nut
- Front hub assembly

To install:

4. Install or connect the following:
- Front hub assembly
- Lock nut. Torque the nut as follows:
  a. Step 1: Torque the nut to 119 ft. lbs. (162 Nm).
  b. Step 2: Loosen to 0 ft. lbs. (0 Nm).
  c. Step 3: Torque the nut to 18 ft lbs. (25 Nm).
  d. Step 4: Loosen the nut 30–40 degrees.
5. Install or connect the following:
- Lock washer
- Speed sensor, if equipped
- Front brake assembly. Torque the bolts to 65 ft. lbs. (88 Nm).
- Drive flange. Torque the bolts to 36–43 ft. lbs. (49–59 Nm).
- Shim
- Snapring to the axle shaft
- Hub cover
- Front wheels of the vehicle

Axle shaft, Bearing and Seal

REMOVAL & INSTALLATION

Solid Rear Axle

1. Before servicing the vehicle, refer to the precautions in the beginning of this section.
2. Remove or disconnect the following:

- Wheel assembly
- Brake line
- Rear brake assembly
- Parking brake cable and assembly
- Axle shaft
- Snapring
- Retainer
- Bearing inner race, inner and outer
- Oil seal
- Bearing case
- O-ring
- Oil seal

To install:

3. Install or connect the following:
- Bearing case
- Bearing inner race, outer. Press the bearing into the bearing case.
- Oil seal. Press the seal using tools MB990932 and MB990938.
- Bearing inner race, inner. Press the bearing into the bearing case.
- Axle shaft, Place into bearing case
- Retainer, press onto shaft
- Snapring
- Oil seal and O-ring into axle shaft
4. Axle shaft assembly into axle.
- Parking brake cable end
- Parking brake cable attaching bolt
- Rear brake assembly
- Brake line
- Wheel assembly

Pinion Seal

REMOVAL & INSTALLATION

1. Before servicing the vehicle, refer to the precautions in the beginning of this section.

2. Remove or disconnect the following:

- Driveshaft, matchmark for reassembly
3. Check the turning torque of the pinion before proceeding. It should be 2.6–4.5 inch lbs. (0.4–0.5 Nm). This is the torque that must be reached during installation of the pinion nut.
- Pinion nut and washer using a suitable pinion flange holding tool
- Companion flange from the drive pinion
4. Pry the pinion seal out of the differential carrier.

To install:

5. Clean and inspect the sealing surface of the housing.
6. Install or connect the following:
- New seal into the housing until the flange on the seal is flush with the carrier. Using a seal driver.
7. With the seal installed, the pinion bearing preload must be set.
- Pinion nut (a new self-locking pinion nut must be used) while holding the flange, until the turning torque is the same as before removal. The final pinion nut torque must be between 137–181 ft. lbs. (190–250 Nm).
- Driveshaft, align the matchmarks
8. Check the level of the differential lubricant when finished.

STEERING AND SUSPENSION

Air Bag

❊❊ CAUTION

Some vehicles are equipped with an air bag system. The system must be disabled before performing service on or around system components, steering column, instrument panel components, wiring and sensors. Failure to follow safety and disabling procedures could result in accidental air bag deployment, possible personal injury and unnecessary system repairs.

PRECAUTIONS

Several precautions must be observed when handling the inflator module to avoid accidental deployment and possible personal injury.

- Never carry the inflator module by the wires or connector on the underside of the module.
- When carrying a live inflator module, hold securely with both hands, and ensure that the bag and trim cover are pointed away.
- Place the inflator module on a bench or other surface with the bag and trim cover facing up.
- With the inflator module on the bench, never place anything on or close to the module that may be thrown in the event of an accidental deployment.

DISARMING

To avoid personal injury when working on vehicles equipped with an air bag, the negative battery cable must be disconnected and at least 60 seconds must elapse before working on the system. Failure to do so may result in deployment of the air bag.

Recirculating Ball Power

REMOVAL & INSTALLATION

1. On vehicles equipped with a supplemental restraint system (SRS), turn the front wheel to the straight ahead position and remove the ignition key to prevent the steering wheel from turning.
2. Drain the power steering fluid.
3. Remove or disconnect the following:

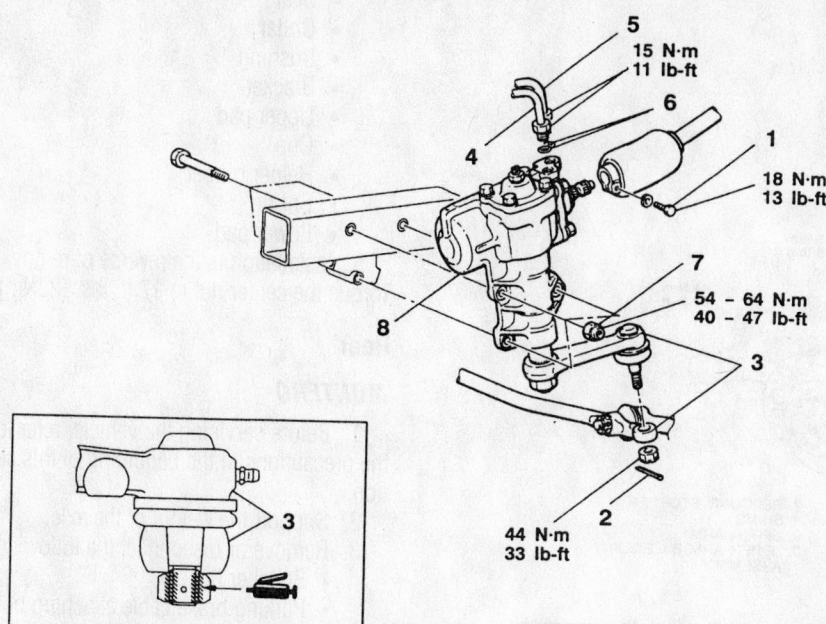

5
15 N·m
11 lb-ft

6

4

1

18 N·m
13 lb-ft

7

54 – 64 N·m
40 – 47 lb-ft

8

3

2

44 N·m
33 lb-ft

1. **CONNECTING BOLT FOR STEERING GEAR BOX AND STEERING SHAFT**
2. **COTTER PIN**
3. **CONNECTION FOR PITMAN ARM AND RELAY ROD**

4. **PRESSURE TUBE**
5. **RETURN TUBE**
6. **O-RING**
7. **SELF-LOCKING NUT**
8. **POWER STEERING GEAR BOX**

7924UG33

Exploded view of a typical power steering gear mounting

- Negative battery cable
- Pinch bolt securing the steering shaft to the steering gear
- Pitman arm from the relay rod
- Fluid lines from the steering gear
- Mounting bolts securing the gear to the frame rail and steering gear

To install:
4. Install or connect the following:
- Steering gear on the frame rail. Torque the nuts to 40–47 ft. lbs. (54–64 Nm).
- Fluid lines to the steering gear use a new O-rings. Torque the fittings to 11 ft. lbs. (15 Nm).
- Relay rod on the Pitman arm. Torque the nut to 33 ft. lbs. (44 Nm).
- Steering shaft on the steering gear. Torque the bolt to 13 ft. lbs. (18 Nm).
- Negative battery cable
5. Refill and bleed the power steering system.

Rack and Pinion Steering Gear

REMOVAL & INSTALLATION

1. On vehicles equipped with a supplemental restraint system (SRS), turn the front wheel to the straight ahead position and remove the ignition key to prevent the steering wheel from turning.
2. Drain the power steering fluid.
3. Remove or disconnect the following:
- Negative battery cable
- Engine under cover
- Tie rod ends
- Steering hoses
- Left side differential mount bracket
- Intermediate shaft-to-gear box bolt
- Gear mounting clamps and gear
4. Installation is the reverse of removal. Observe the following torques:
- Mounting clamp bolts: 51 ft. lbs. (69Nm)
- Tie rod end ball stud nuts: 29 ft. lbs. (39Nm)
- Intermediate shaft pinch bolt: 13 ft. lbs. (18Nm)

Shock Absorber

REMOVAL & INSTALLATION

Front

1. Before servicing the vehicle, refer to the precautions in the beginning of this section.
2. Remove the upper shock mounting nut, washer and bushing.
3. Remove the lower mounting bolts.
4. Remove the shock absorber.
To install:

➡**If the shock absorber has a white paint mark on the lower end, be sure the mark faces the outside of the vehicle when installed.**

5. Install the shock absorber. Torque the lower nut to 65–76 ft. lbs. (88–103 Nm) and the upper nut to 11 ft. lbs. (15 Nm).
6. Test drive the vehicle and check the alignment.

Rear

1. Before servicing the vehicle, refer to the precautions in the beginning of this section.
2. Support the rear axle assembly with a hydraulic floor jack, so that the shock absorber may be removed.
3. Remove the upper and lower mounting nuts and bolts that attach the shock to the frame and bracket.
4. Remove the shock absorber from the vehicle.
To install:
5. Install the shock absorber
- 1999–00 lower bolt: 159–181 ft. lbs. (216–245 Nm) on the Montero and 16 ft. lbs. (22 Nm) on the Montero Sport
- 2001–02 lower bolt: 159–181 ft. lbs. (216–245 Nm) on the Montero Sport and 113 ft. lbs. (152Nm) on the Montero
- 1999–00 upper mounting nut: 11 ft. lbs. (15 Nm) on the Montero and 16 ft. lbs. (22 Nm) on the Montero Sport
- 2001–02 upper mounting nut: 16 ft. lbs. (22Nm) on the Montero Sport and 33 ft. lbs. (44Nm) on the Montero
6. Remove the floor jack from under the axle assembly.

Heater Core replacement is covered in Section 2 of this manual

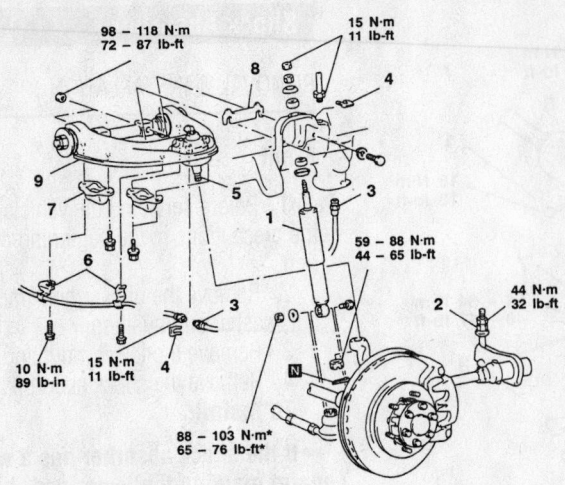

98 – 118 N·m
72 – 87 lb-ft

15 N·m
11 lb-ft

59 – 88 N·m
44 – 65 lb-ft

44 N·m
32 lb-ft

10 N·m
89 lb-in

15 N·m
11 lb-ft

88 – 103 N·m*
65 – 76 lb-ft*

1. SHOCK ABSORBER
● BUMP STOPPER AND BUMP
   STOPPER BRACKET CLEARANCE
   ADJUSTMENT
2. REAR ANCHOR ARM ADJUSTING
   NUT
3. BRAKE HOSE CONNECTION
4. HOSE CLIP
5. UPPER ARM BALL JOINT
   CONNECTION
6. SPEED SENSOR BRACKET
   <VEHICLES WITH ABS>

7. REBOUND STOPPER
8. SHIMS
9. UPPER ARM
10. UPPER ARM BALL JOINT
    ASSEMBLY

Caution
*: Indicates parts which should be temporarily
tightened, and then fully tightened with the vehicle
on the ground in an unladen condition.

7924UG34

Common shock absorber and upper control arm components

Strut

REMOVAL & INSTALLATION

1. On vehicles equipped with a supplemental restraint system (SRS), turn the front wheel to the straight ahead position and remove the ignition key to prevent the steering wheel from turning.

2. Remove or disconnect the following:
   • Negative battery cable
   • Upper control arm
   • Battery and battery tray
   • A/C condenser
   • Air cleaner
   • Wheel
   • Upper mounting nuts
   • Lower mounting bolt/nut

3. Installation is the reverse of removal. Torque the upper mounting nuts to 32 ft. lbs. (44Nm); the lower mount nut to 119 ft. lbs. (162Nm)

Coil Spring

REMOVAL & INSTALLATION

Front

MONTERO WITH MACPHERSON STRUTS

1. Before servicing the vehicle, refer to the precautions in the beginning of this section.

2. Remove or disconnect the following:
   • Strut

3. Compress the coil spring until there is a clearance on both ends

4. Remove or disconnect the following:
   • Strut center nut

   • Seat
   • Collar
   • Bushing
   • Bracket
   • Upper pad
   • Cup
   • Helper rubber
   • Spring
   • Lower pad

5. Installation is the reverse of removal. Torque the center nut to 17 ft. lbs. (22Nm).

Rear

MONTERO

1. Before servicing the vehicle, refer to the precautions in the beginning of this section.

2. Support the weight of the axle.

3. Remove or disconnect the following:
   • Breather hose
   • Parking brake cable attaching bolt
   • ABS speed sensor attaching bolt
   • Brake hose connection
   • Lower shock mounting bolt
   • Bolt that attaches the lateral rod to the body
   • Stabilizer bar

4. Lower the axle and remove the coil spring and seat

5. Installation is the reverse of removal. Torque the lateral rod-to-body bolt to 159–181 ft. lbs. (216–245Nm).

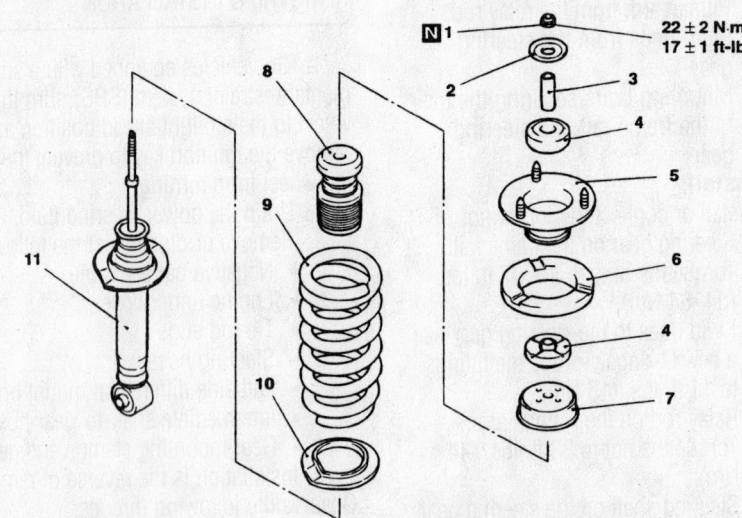

22 ± 2 N·m
17 ± 1 ft-lb

1. SELF-LOCKING NUT
2. SEAT
3. COLLAR
4. UPPER BUSHING
5. SPRING BRACKET ASSEMBLY
6. SPRING UPPER PAD
7. CUP ASSEMBLY
8. HELPER RUBBER

9. COIL SPRING
10. SPRING LOWER PAD
11. SHOCK ABSORBER ASSEMBLY

9355UG01

Strut exploded view

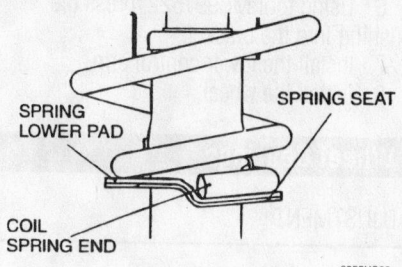

Coil spring installation on strut

MONTERO SPORT

1. Before servicing the vehicle, refer to the precautions in the beginning of this section.
2. Remove or disconnect the following:
• Shock absorber lower bolt
• Lower arm mounting bolt
• Coil spring
3. Installation is the reverse of removal. Torque the lower arm bolt to 113 ft. lbs. (152Nm).

Torsion Bars

REMOVAL & INSTALLATION

1. Before servicing the vehicle, refer to the precautions in the beginning of this section.
2. Support the lower arm with a jack.
3. Remove or disconnect the following:
• Heat protector, right side only
• Bump stopper
• Anchor adjustment nut and arm assembly
• Anchor collar
• Torsion bar
• Dust covers
• Heat covers, right side only

To install:
4. Install or connect the following:
• Heat covers, right side only
• Dust covers
• Torsion bar
• Anchor collar
• Anchor adjustment nut and arm assembly. Torque the nut to 32 ft. lbs. (44 Nm).
• Heat protector, right side only

Upper Ball Joint

REMOVAL & INSTALLATION

1. Before servicing the vehicle, refer to the precautions in the beginning of this section.
2. Remove the front wheel.

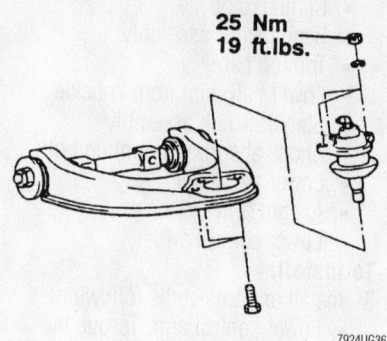

25 Nm
19 ft.lbs.

Exploded view of the upper ball joint and related components—1999–00 models

3. Support the lower control arm.
4. Remove the upper ball joint from the steering knuckle.
5. Remove the ball joint from the upper control arm.

To install:
6. Install the ball joint in the upper control arm. Tighten the bolts to 18 ft. lbs. (25 Nm) for 1999–00 models; 22 ft. lbs. (30Nm) for 2001–02 models.
7. Install the ball joint stud to the steer-

⚠ **CAUTION**
*: Indicates parts which should be temporarily tightened, and then fully tightened with the vehicle on the ground in an unladen condition.

ing knuckle. Torque the nut to 54 ft. lbs. (74 Nm).
8. Install a new cotter pin.
9. Install the front wheel.
10. Grease the upper ball joint and all other suspension components with a grease fitting.

Lower Ball Joint

REMOVAL & INSTALLATION

1. Before servicing the vehicle, refer to the precautions in the beginning of this section.
2. Apply upward pressure to the lower control arm with a jack or an adjustable stand.

✳✳ CAUTION

Do not disconnect the lower ball joint stud from the steering knuckle unless the lower control arm has a stand or a jack under it.

3. Remove the ball joint stud nut/stud from the steering knuckle.

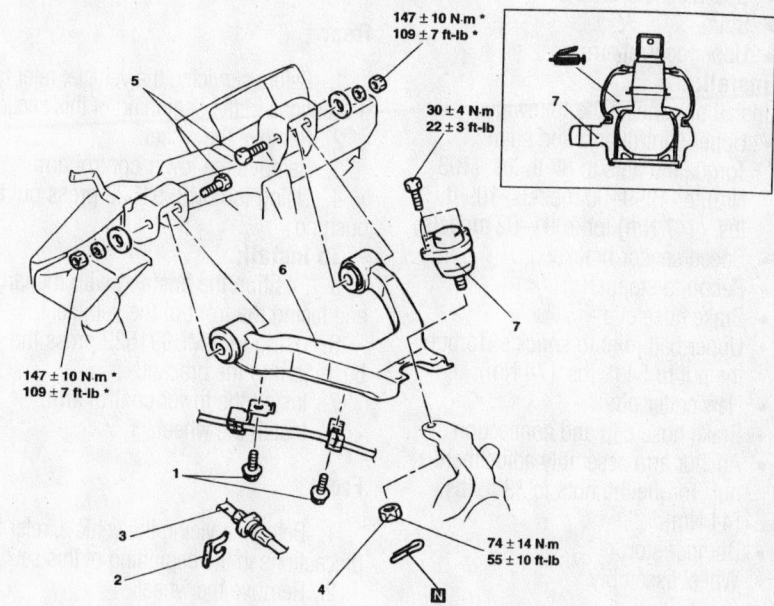

147 ± 10 N·m
109 ± 7 ft-lb *

30 ± 4 N·m
22 ± 3 ft-lb

147 ± 10 N·m *
109 ± 7 ft-lb *

74 ± 14 N·m
55 ± 10 ft-lb

1. FRONT WHEEL SPEED SENSOR BRACKET MOUNTING BOLT
2. CLIP
3. BRAKE HOSE
4. UPPER ARM ASSEMBLY AND KNUCKLE CONNECTION
5. UPPER ARM ASSEMBLY AND FRONT FRAME CONNECTION
6. UPPER ARM ASSEMBLY
7. UPPER ARM BALL JOINT ASSEMBLY

Exploded view of the upper ball joint and related components—2001–02 models

Brake service is covered in Section 4 of this manual

4. Remove the ball joint retaining nuts/bolts and the ball joint from the arm

To install:

5. Install the lower ball joint on the control arm. Torque the ball joint retaining nuts/bolts to 60 ft. lbs. (81 Nm) on 1999–00 models and 2001–02 Montero Sport; 70 ft. lbs. (95Nm) on 2001–02 Montero models.

6. Install the ball stud to the knuckle. Torque the nut to 108 ft. lbs. (147 Nm) and a new cotter pin.

7. Lubricate the ball joint with a grease gun.

8. Check and adjust the alignment if necessary.

Upper Control Arm

REMOVAL & INSTALLATION

1. Before servicing the vehicle, refer to the precautions in the beginning of this section.

2. Remove or disconnect the following:
- Wheel assembly
- Bumper stop
- Anchor arm assembly adjustment nut
- Brake hose clip and connection
- Upper ball joint from knuckle
- Brake hose clip
- Rebound stopper(s)
- Speed sensor bracket
- Shim
- Upper control arm

To install:

3. Install or connect the following:
- Upper control arm and shim. Torque the nuts to 80 ft. lbs. (108 Nm) for 1999–00 models; 109 ft. lbs. (147 Nm) for 2001–02 models
- Speed sensor bracket
- Rebound stopper(s)
- Brake hose clip
- Upper ball joint to knuckle. Torque the nut to 54 ft. lbs. (74 Nm).
- New cotter pin
- Brake hose clip and connection
- Anchor arm assembly adjustment nut. Torque the nuts to 33 ft. lbs. (44 Nm).
- Bumper stop
- Wheel assembly

Lower Control Arm

REMOVAL & INSTALLATION

1. Before servicing the vehicle, refer to the precautions in the beginning of this section.

2. Remove or disconnect the following:
- Skid plate and undercover

- Bumper stop
- Rear anchor assembly
- Torsion bar
- Lower ball joint from knuckle
- Stabilizer link assembly
- Shock absorber mounting bolts
- Lower arm shaft
- Anchor arm
- Lower control arm

To install:

3. Install or connect the following:
- Lower control arm. Torque the mounting bolts to 108 ft. lbs. (147 Nm) for 1999–00 models; 91 ft. lbs. (123 Nm) for 2001–02 models
- Anchor arm. Torque the bolt to 33 ft. Lbs. (44 Nm).
- Lower arm shaft. Torque the bolt to 108 ft. lbs. (147 Nm).
- Shock absorber mounting bolts
- Stabilizer link assembly
- Lower ball joint to knuckle. Torque the nut to 108 ft. lbs. (147 Nm).
- Torsion bar
- Rear anchor assembly. Torque the nuts to 33ft. lbs. (44 Nm).
- Bumper stop. Torque the nut to 18 ft. lbs. (25 Nm).
- Skid plate and undercover

CONTROL ARM BUSHING REPLACEMENT

Rear

1. Before servicing the vehicle, refer to the precautions in the beginning of this section.

2. Remove the wheel.

3. Remove the lower control arm.

4. Using tool MB991522 press out the bushing

To install:

5. Position the bushing with the larger end facing the front of the vehicle.

6. Using tool MB991522 press the bushing into the bracket.

7. Install the lower control arm.

8. Install the wheel.

Front

1. Before servicing the vehicle, refer to the precautions in the beginning of this section.

2. Remove the wheel.

3. Remove the lower control arm and place in a vise.

4. Using tool MB990883 remove the bushing

To install:

5. Position the bushing with the larger end facing the front of the vehicle.

➡ **Coat the bushing with a soap solution and take care not to twist.**

6. Using tool MB991522, press the bushing into the bracket.

7. Install the lower control arm.

8. Install the wheel.

Wheel Bearings

ADJUSTMENT

Front

1999–00

1. Tighten the wheel bearing nut to 119 ft. lbs. (162 Nm) while turning the rotor.

2. Loosen the wheel bearing adjusting nut completely.

3. Tighten the nut to 18 ft. lbs. (25 Nm), then loosen the nut approximately 30°.

4. Using a dial indicator, check the wheel bearing end-play. The specification is 0.002 in. (0.05mm).

5. Install the locknut.

2001–02 MONTERO

The bearings are integral with the hub. No adjustment is possible.

2001–02 MONTERO SPORT

1. With the caliper removed, check the rotational starting torque. Rotational torque should be 2.7–11.5 inch lbs. (4–19Nm). Rotational torque can be adjusted by tightening or loosening the adjusting nut.

2. Check the hub axial. Endplay should not exceed 0.002 inch (0.05mm). If adjusting nut tightening does not bring the axial play within specifications, the bearings must be replaced.

3. Check hub endplay. Endplay should be 0.02–0.03 inch (0.4–0.7mm). Shims are available to adjust endplay.

4. Install the hub assembly. Tighten the nut to 94–145 ft. lbs. (127–196Nm). Loosen it completely. Tighten the nut to 18 ft. lbs. (25Nm), then back it off 30 degrees.

Rear

The rear wheel bearings are not adjustable. If the bearings are noisy or become loose, they must be replaced.

REMOVAL & INSTALLATION

Front

1999–00 MONTERO WITH 2WD AND 1999–02 MONTERO SPORT WITH 2WD

1. Before servicing the vehicle, refer to the precautions in the beginning of this section.

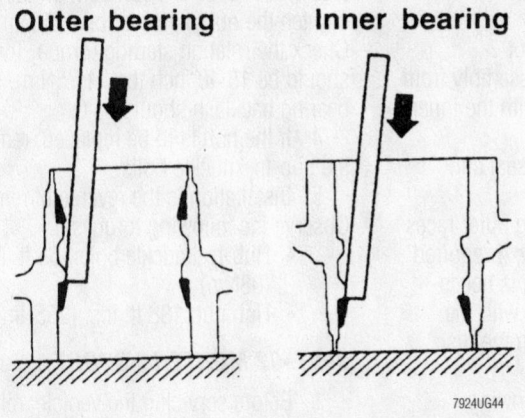

Outer bearing   **Inner bearing**

7924UG44

The bearing races can be removed from the hub using a drift and hammer

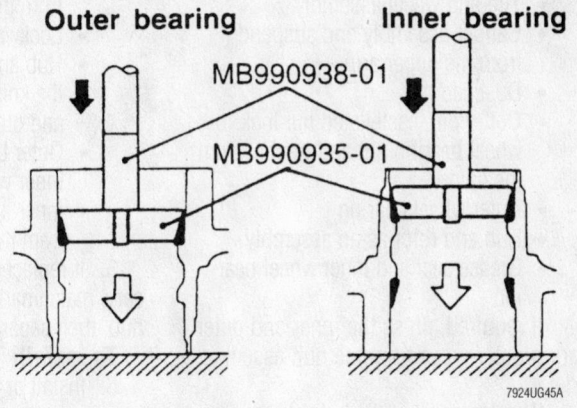

Outer bearing   **Inner bearing**

MB990938-01

MB990935-01

7924UG45A

Install the new races into the hub using the proper size driver

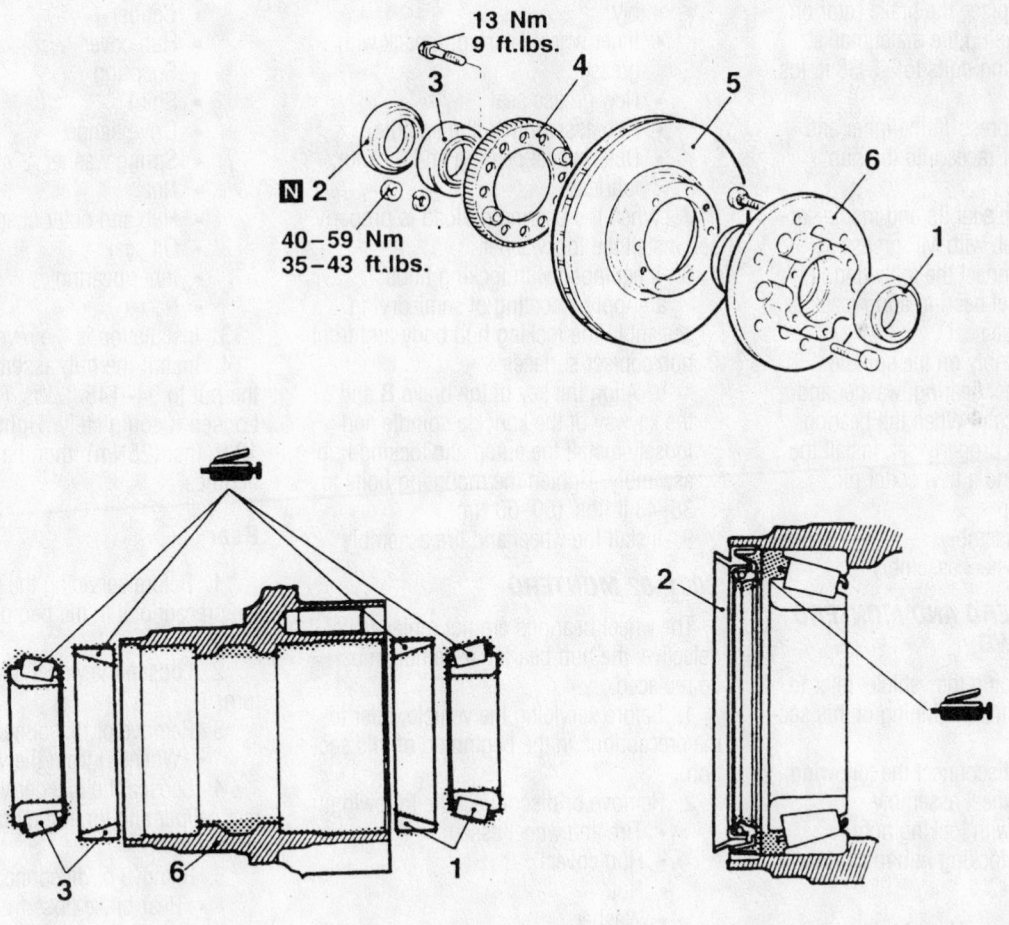

13 Nm
9 ft.lbs.

N 2

40–59 Nm
35–43 ft.lbs.

1. Outer bearing
2. Oil seal
3. Inner bearing

4. Rotor
5. Brake disc
6. Front hub

7924UG43

Exploded view of the hub and wheel bearing assembly—Montero

For complete Engine Mechanical specifications, see Section 1 of this manual

2. Remove or disconnect the following:
- Tire and wheel assembly
- Caliper assembly and suspend it from the upper arm
- Dust cap
- Cotter pin, castellated nut lock, wheel bearing nut and washer from the spindle
- Outer wheel bearing
- Hub and rotor as an assembly
- Grease seal and inner wheel bearing

3. If required, press the inner and outer bearing outer races from the hub assembly.

4. If replacement of the hub is necessary, matchmark the brake disc with the hub, then separate the hub from the disc.

To install:

5. If removed, place the brake rotor on the hub, while aligning the matchmarks. Tighten the mounting bolts to 34–38 ft. lbs. (47–52 Nm).

6. If removed, press-fit the inner and outer bearing outer races into the hub assembly.

7. Lubricate the seal lip and inside surface of the front hub with MP grease.

8. Install or connect the following:
- Inner wheel bearing and repack
- New grease seal
- Hub assembly on the spindle
- Outer wheel bearing, washer and nut, lubricate. When the bearing preload is properly set, install the nut lock and a new cotter pin.
- Grease cap
- Caliper assembly
- Tire and wheel assembly

1999–00 MONTERO AND MONTERO SPORT WITH 4WD

1. Before servicing the vehicle, refer to the precautions in the beginning of this section.

2. Remove or disconnect the following:
- Tire and wheel assembly

3. If equipped with locking hub:

a. Place the locking hub in the free position.

➡ **A free position can be obtained by shifting the transfer shift lever to the 2H position, then moving the vehicle in reverse for approximately 3–6 ft. (1–2 m).**

b. Remove the hub cover.

c. Remove the snapring from the axle shaft.

d. Remove the bolts and remove the automatic locking hub.

4. Remove or disconnect the following:

- Caliper assembly and suspend it from the upper arm.
- Lockwasher and locknut
- Hub and rotor as an assembly from the knuckle together with the inner and outer bearings
- Outer bearing, grease seal and inner wheel bearing
- Inner and outer bearing outer races from the hub assembly, if required

5. If replacement of the hub is necessary, matchmark the brake disc with the hub, then separate the hub from the disc.

To install:

6. Install or connect the following:
- Brake rotor on the hub, while aligning the matchmarks, if removed
- Press-fit the inner and outer bearing outer races into the hub assembly
- Inner wheel bearing, repack with grease
- New grease seal
- Hub assembly to the spindle
- Outer wheel bearing and locknut, lubricate

7. When the bearing preload is properly set, install the lockwasher.

8. If equipped with locking hubs:

a. Apply a coating of semi-drying sealant to the locking hub body and front hub contact surfaces.

b. Align the key of the brake **B** and the keyway of the knuckle spindle and loosely install the automatic locking hub assembly. Tighten the mounting bolts to 36–43 ft. lbs. (50–60 Nm).

9. Install the wheel and tire assembly

2001–02 MONTERO

The wheel bearings are not replaceable. If defective, the hub/bearing assembly must be replaced.

1. Before servicing the vehicle, refer to the precautions in the beginning of this section.

2. Remove or disconnect the following:
- Tire and wheel assembly
- Hub cover
- Nut
- Washer
- Brake hose
- Speed sensor
- Caliper
- Rotor
- Dust cover
- Tie rod end
- Upper and lower arms from the knuckle
- Rotor shield
- Hub/knuckle assembly

3. Mount the assembly in a vise. Install

tool MB990998, or equivalent on the hub. Tighten the nut to 188 ft. lbs. (255Nm). Check the rotation starting torque. Torque should be 15.48 inch lbs. (1.75Nm). Wheel bearing backlash should be 0.

4. If the hub is to be replaced, remove the hub-to-knuckle bolts.

5. Installation is the reverse of removal. Observe the following torques:
- Hub-to-knuckle bolts: 65 ft. lbs. (88Nm)
- Hub nut: 188 ft. lbs. (255Nm)

2001–02 MONTERO SPORT

1. Before servicing the vehicle, refer to the precautions in the beginning of this section.

2. Remove or disconnect the following:
- Wheel
- Caliper
- Hub cover
- Snapring
- Shim
- Drive flange
- Spring washer
- Nut
- Hub and outer bearing
- Oil seal
- Inner bearing
- Races

3. Installation is the reverse of removal.

4. Install the hub assembly. Tighten the nut to 94–145 ft. lbs. (127–196Nm). Loosen it completely. Tighten the nut to 18 ft. lbs. (25Nm), then back it off 30 degrees.

Rear

1. Before servicing the vehicle, refer to the precautions in the beginning of this section.

2. Loosen the wheel lug nuts only ½ a turn.

3. Remove or disconnect the following:
- Wheel(s) from the vehicle

4. Loosen the bleeder valve on the right rear caliper and drain the brake fluid into a container.

5. Remove or disconnect the following:
- Rear brake hose from the hard line on the frame
- Rear brake caliper
- Rear disc off of the rear axle
- Parking cable attaching bolt and cable end from the brake assembly
- Parking brake assembly from the end of the axle
- Speed sensor, on vehicles with Anti-lock Brakes (ABS)
- Rear axle shaft out of the axle housing. If the rear axle shaft is dif-

ficult to remove, use a slide hammer (impact puller) to remove it.

✳✳ WARNING

Do not damage the oil seal during removal.

- Snapring from the inside end of the axle shaft. Remove 1 retainer bolt from the backing plate with a plastic mallet. Apply cloth tape around the edge of the bearing case for

protection. Position the axle shaft in a vise or with a similar method. Using a grinder, grind down the retainer flat, on one side, until the thickness of the retainer is only 0.04–0.08 in. (1–2mm). That is that the retainer is ground down toward the axle shaft, not toward the flange. Cut, with a chisel, the place where the retainer ring has been shaven down and remove the retainer.

✳✳ CAUTION

Be careful not to damage the bearing case and the axle shaft.

➡ Only the retainer ring is to be ground down, NOT the axle shaft, the axle flange, the bearing or any other component.

6. Grind the plate of special tool MB990861 with a grinder (see illustration)

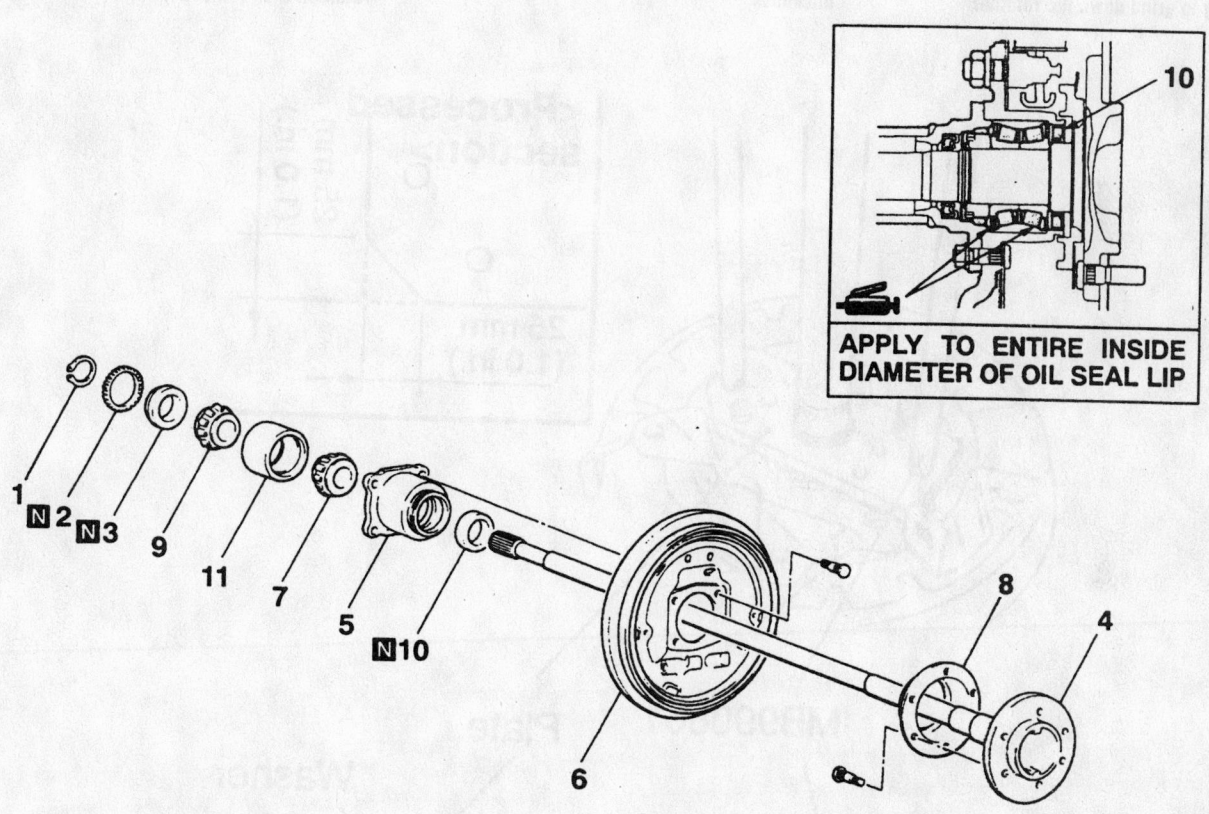

APPLY TO ENTIRE INSIDE DIAMETER OF OIL SEAL LIP

DISASSEMBLY STEPS
1. SNAP RING
2. ABS ROTOR
3. RETAINER
4. AXLE SHAFT
5. BEARING CASE
6. BACKING PLATE
7. OUTER BEARING INNER RACE
8. DUST COVER
9. INNER BEARING INNER RACE
10. OIL SEAL
11. BEARING OUTER RACE

ASSEMBLY STEPS
11. BEARING OUTER RACE
9. INNER BEARING INNER RACE
7. OUTER BEARING INNER RACE
10. OIL SEAL
8. DUST COVER
6. BACKING PLATE
5. BEARING CASE
4. AXLE SHAFT
3. RETAINER
2. ABS ROTOR
1. SNAP RING

Exploded view of the typical rear axle shaft, bearings and races

7924UG48

For Accessory Drive Belt illustrations, see Section 1 of this manual

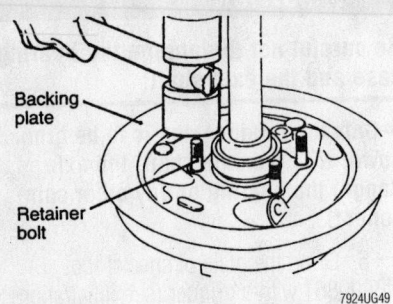

7924UG49

Remove one of the rear axle studs before attempting to grind down the retainer

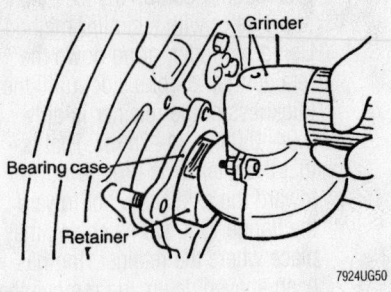

7924UG50

Using a grinder, grind the retainer, on one side, down to 1–2mm (0.04–0.08 in.) thickness

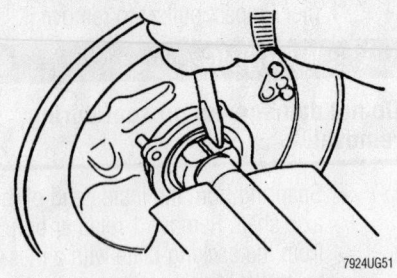

7924UG51

Use a chisel on the ground-down spot on the rear axle bearing retainer to split the retainer, then remove it

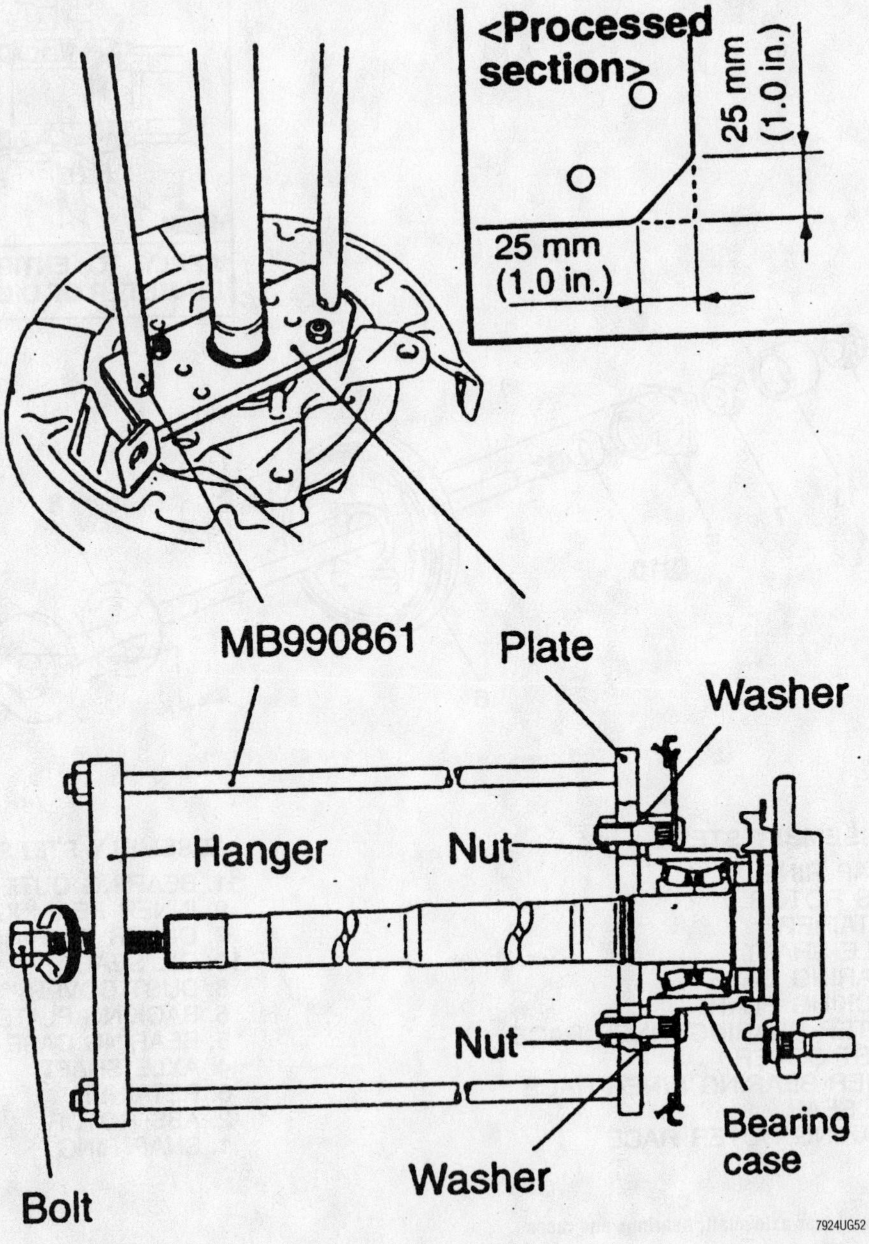

Use the special tool MB990861 to remove the rear axle shaft from the bearing case

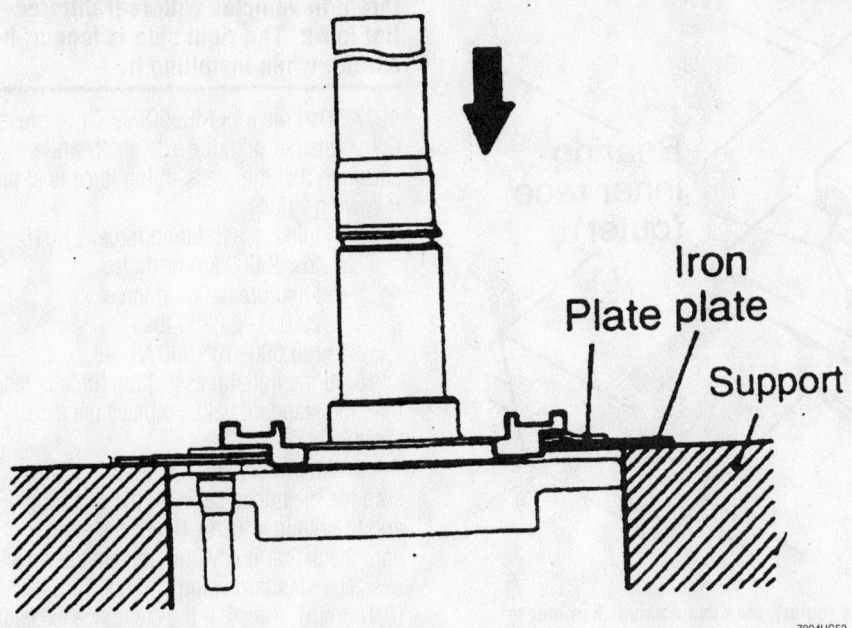

Iron
Plate plate

Support

Use an iron plate and supports to remove the rotor assembly

so that there will be no interference between the plate and the bearing case. While adjusting the height of the hanger, secure the washers, plate and nuts in order so that the processed plate is as shown in the illustration.

➡**The washers are used to eliminate the difference in height of the bearing case so that the plate and the bearing case are parallel.**

Place the end of the bolt against the center of the axle shaft, then tighten the nuts to remove the axle shaft from the bearing case assembly.

➡**The hanger and plate must be placed so that they are parallel.**

7. Remove the bearing inner race and the bearing outer race. To remove the races, install the tool MB990560 and use a press to remove the bearing race from the axle shaft.

8. Remove the oil seal and the dust cover on vehicles without ABS.

9. On vehicles without ABS, insert an iron plate of approximately 0.04 in. (1mm) thickness between the rotor assembly and the axle shaft, then use a press to remove the rotor assembly.

✳✳ WARNING

In order not to bend the rotor assembly plate, place the support in con-tact with the axle shaft when using the press.

10. Remove or disconnect the following:
• Axle shaft from the remaining bearings and components
• Backing plate

11. Reinstall the bearing inner race that was removed previously, then use the tool MB990799-01 and press to remove the bearing outer race.

12. Remove or disconnect the following:
• Bearing case
• O-ring from the end of the axle housing tube
• Oil seal from the end of the rear axle housing using the tool MB990211-01 (slide hammer with a hooked end), if necessary

13. Check the dust cover for deformation or damage. Check the oil seal for damage. Check the inner and outer bearings for seizure, discoloration and rough raceway surface. Check the axle shaft for cracks, wear and damage. For there are any of these indications, replace the part with a new one. The retainer, the bearing inner (inner and outer) and outer races and the oil seal need to be replaced with new components upon reassembly. After all of this work, it is probably a good idea to replace the bearings and the axle housing tube oil seals.

To install:
14. Install or connect the following:
• New oil seal into the end of the rear axle housing using the tools MB990932-01 and MB990938-01, if necessary.
• New O-ring into the axle tube

15. Apply multi-purpose grease to the external surface of the bearing out race. Press-fit the bearing outer race into the bearing case by using the tool MB990890-01.

16. Install or connect the following:
• Speed sensor bracket to the back of the backing plate

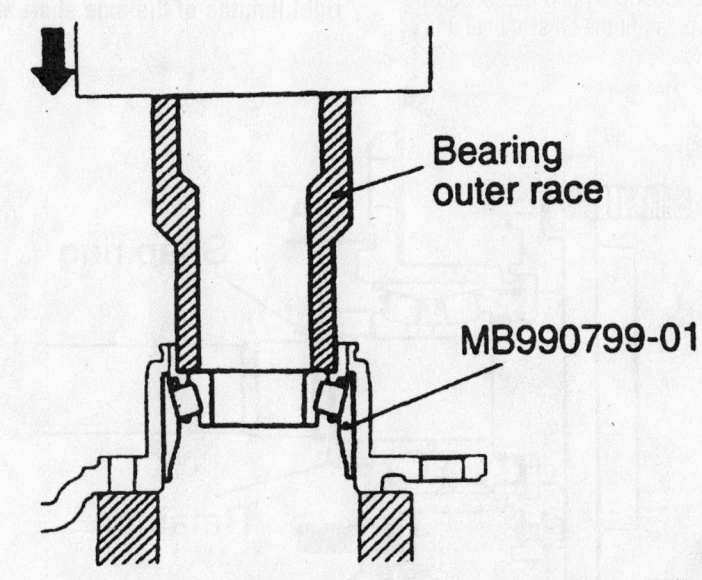

Bearing
outer race

MB990799-01

Use the tool MB990799-01 to install and remove the rear axle bearing races

For Tire, Wheel and Ball Joint specifications, see Section 1 of this manual

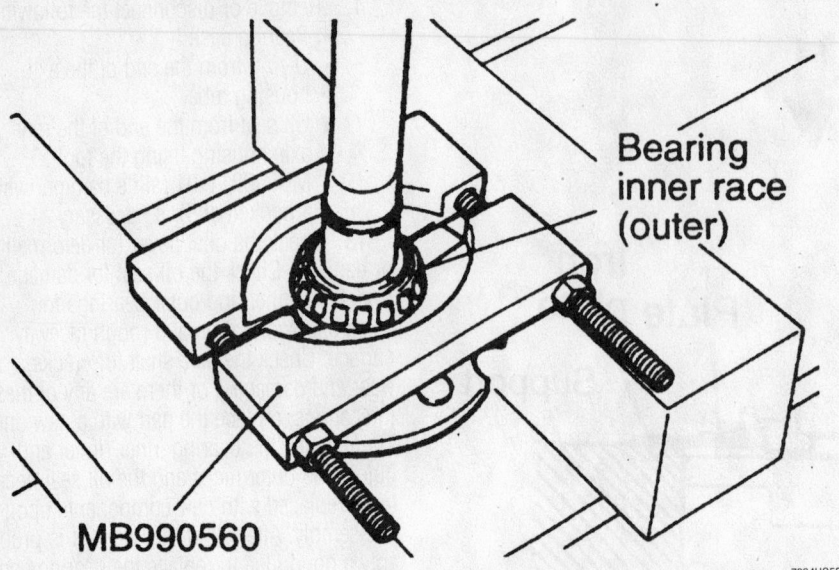

Use the MB990560 tool to hold the bearing inner race (outer), then use a plastic hammer to drive the axle out of the race—do not let the axle fall onto a hard floor

- Rotor assembly to the axle shaft by press-fitting (plastic mallet will also work) it on using the special tool MB991388
- Backing plate onto the axle shaft
- Dust cover to the backing plate if the vehicle is equipped with ABS.
- Bearing inner race (outer) to the bearing case
- Oil seal to the front end of the bearing case. To do this, apply multi-purpose grease to the outside of the oil seal. Use the special tools MB990936-01 and MB990938-01 to press-fit the oil seal until it is

flush with the end of the bearing case. Apply multi-purpose grease to the lip of the oil seal.
- Axle shaft through the bearing inner race, the bearing case and the second bearing inner race in that order. Use the special tool MB990799 to press-fit the bearing inner race to the axle shaft.

✳✳ WARNING

Both bearing inner race sets should be press-fitted together. The left and right lengths of the axle shaft are dif-

ferent in vehicles with rear differential locks. The right side is longer; be careful when installing it.

17. Use the tool MB990799-01 to press-fit the retainer onto the axle shaft, while checking that the press-fitting force is at the following values:
- Initial press-fitting force: 11,016 lbs. (5000 kg) or more.
- Final press-fitting force: 22,031–24,280 lbs. (98,000–108,000 N).

18. If the initial press-fitting force is less than the standard value, replace the axle shaft.

19. After installing the snapring, measure the clearance between the snapring and the retainer with a thickness gauge, and check that it is within the standard values. The standard value is 0.0065 in. (0.166mm) or less. If the clearance exceeds the standard value, change the snapring so that the clearance is at the standard value. Use the following adjusting snapring thicknesses:
- 0.0854 in. (2.17mm): no color.
- 0.0791 in. (2.01mm): yellow.
- 0.0728 in. (1.85mm): blue.
- 0.0665 in. (1.69mm): purple.
- 0.0602 in. (1.53mm): red.

20. Install or connect the following:
- Axle assembly into the axle housing. Be sure that the grooves on the end of the axle shaft line up in the differential. Use a plastic or rubber mallet to help drive the axle shaft into the differential unit. Tighten the 4 retaining bolts for the axle shafts to 36–43 ft. lbs. (49–59 Nm).
- Speed sensor
- Parking brake assembly components to the axle flange.
- Parking brake cable to the parking brake assembly, then secure it in place with the cable bracket.
- Brake rotor onto the axle shaft, and the brake caliper. Torque the caliper bolts to 65 ft. lbs. (88 Nm).
- Brake hose to the frame brake line. Torque the flare nut to 11 ft. lbs. (15 Nm).
- Wheels. Torque the lug nuts as tight as possible with the vehicle not on the ground.

21. Bleed the brake system.

22. Lower the vehicle until the wheels are touching the ground, then finish tightening the lug nuts. Lower the vehicle the rest of the way to the ground.

23. Road test the vehicle and check for leaks.

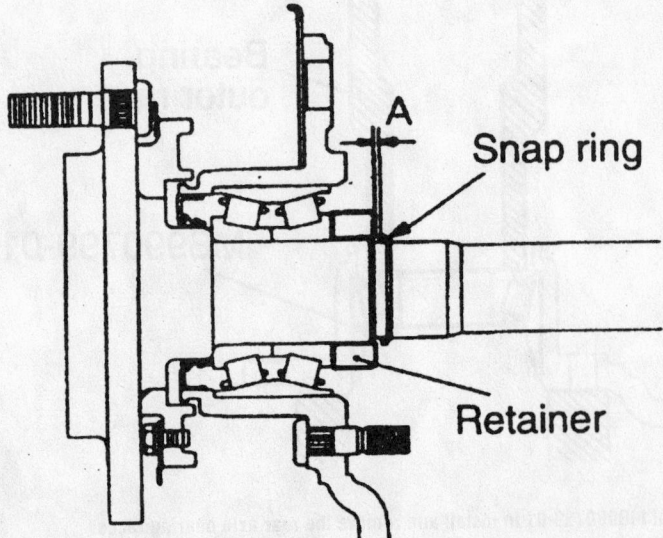

Measure the clearance (A) between the snapring and the retainer edge

NISSAN AND INFINITI

Nissan-Pathfinder • Xterra • Infiniti-QX4

PRECAUTIONS

Before servicing any vehicle, please be sure to read all of the following precautions, which deal with personal safety, prevention of component damage, and important points to take into consideration when servicing a motor vehicle:

• Never open, service or drain the radiator or cooling system when the engine is hot; serious burns can occur from the steam and hot coolant.

• Observe all applicable safety precautions when working around fuel. Whenever servicing the fuel system, always work in a well-ventilated area. Do not allow fuel spray or vapors to come in contact with a spark, open flame, or excessive heat (a hot drop light, for example). Keep a dry chemical fire extinguisher near the work area. Always keep fuel in a container specifically designed for fuel storage; also, always properly seal fuel containers to avoid the possibility of fire or explosion. Refer to the additional fuel system precautions later in this section.

• Fuel injection systems often remain pressurized, even after the engine has been turned **OFF**. The fuel system pressure must be relieved before disconnecting any fuel lines. Failure to do so may result in fire and/or personal injury.

• Brake fluid often contains polyglycol ethers and polyglycols. Avoid contact with the eyes and wash your hands thoroughly after handling brake fluid. If you do get brake fluid in your eyes, flush your eyes with clean, running water for 15 minutes. If eye irritation persists, or if you have taken brake fluid internally, IMMEDIATELY seek medical assistance.

• The EPA warns that prolonged contact with used engine oil may cause a number of skin disorders, including cancer! You should make every effort to minimize your exposure to used engine oil. Protective gloves should be worn when changing oil. Wash your hands and any other exposed skin areas as soon as possible after exposure to used engine oil. Soap and water, or waterless hand cleaner should be used.

• All new vehicles are now equipped with an air bag system. The system must be disabled before performing service on or around system components, steering column, instrument panel components, wiring and sensors. Failure to follow safety and disabling procedures could result in accidental air bag deployment, possible personal injury and unnecessary system repairs.

• Always wear safety goggles when working with, or around, the air bag system. When carrying a non-deployed air bag, be sure the bag and trim cover are pointed away from your body. When placing a non-deployed air bag on a work surface, always face the bag and trim cover upward, away from the surface. This will reduce the motion of the module if it is accidentally deployed. Refer to the additional air bag system precautions later in this section.

• Clean, high quality brake fluid from a sealed container is essential to the safe and proper operation of the brake system. You should always buy the correct type of brake fluid for your vehicle. If the brake fluid becomes contaminated, completely flush the system with new fluid. Never reuse any brake fluid. Any brake fluid that is removed from the system should be discarded. Also, do not allow any brake fluid to come in contact with a painted surface; it will damage the paint.

• Never operate the engine without the proper amount and type of engine oil; doing so WILL result in severe engine damage.

• Timing belt maintenance is extremely important! Many models utilize an interference-type, non-freewheeling engine. If the timing belt breaks, the valves in the cylinder head may strike the pistons, causing potentially serious (also time-consuming and expensive) engine damage. Refer to the maintenance interval charts in the front of this manual for the recommended replacement interval for the timing belt, and to the timing belt section for belt replacement and inspection.

• Disconnecting the negative battery cable on some vehicles may interfere with the functions of the on-board computer system(s) and may require the computer to undergo a relearning process once the negative battery cable is reconnected.

• When servicing drum brakes, only disassemble and assemble one side at a time, leaving the remaining side intact for reference.

ENGINE REPAIR

➡**Disconnecting the negative battery cable on some vehicles may interfere with the functions of the on board computer system. The computer may undergo a relearning process once the negative battery cable is reconnected.**

Distributor

REMOVAL

1. Before servicing the vehicle, refer to the precautions in the beginning of this section.
2. Remove or disconnect the following:
   • Negative battery cable
   • Distributor cap
   • Distributor wiring harness connector
3. Matchmark the rotor to the distributor housing and the distributor housing to the cylinder head.
4. Remove the distributor.

INSTALLATION

Timing Not Disturbed

1. Install or connect the following:
   • Distributor by aligning the matchmarks made during removal
   • Distributor wiring harness connector
   • Distributor cap
   • Negative battery cable
2. Check the ignition timing and adjust, as necessary.

Timing Disturbed

2.4L ENGINE

1. Set the engine to Top Dead Center (TDC) of the compression stroke for the No. 1 cylinder.
2. Install the distributor so that the distributor shaft engages the oil pump driveshaft.
3. Check that the distributor rotor is aligned, as shown.

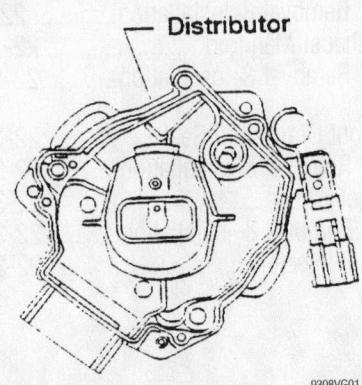

— Distributor

9308VG01

Distributor rotor alignment with the engine at Top Dead Center (TDC)—2.4L engine

4. Install or connect the following:
 - Distributor cap
 - Distributor harness connector
5. Check the ignition timing and adjust, as necessary.

3.3L ENGINE

1. Set the engine to Top Dead Center (TDC) of the compression stroke for the No. 1 cylinder.
2. Align the index mark on the distributor shaft with the protrusion on the distributor housing.
3. Install the distributor and check that the distributor rotor is aligned.
4. Install or connect the following:
 - Distributor cap
 - Distributor harness connector
5. Check the ignition timing and adjust, as necessary.

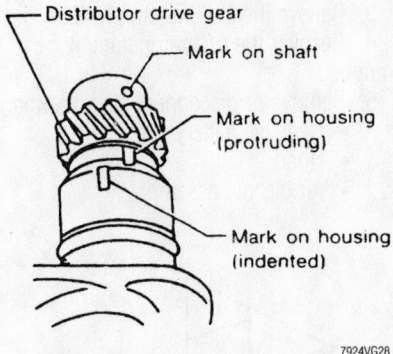

Distributor shaft alignment—3.3L engine

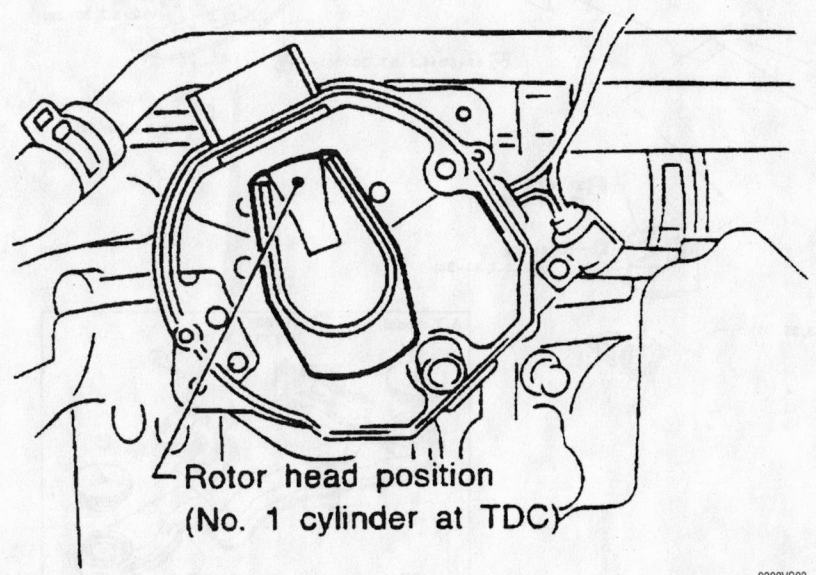

Distributor rotor alignment—3.3L engine

Alternator

REMOVAL & INSTALLATOIN

2.4L Engine

1. Before servicing the vehicle, refer to the precautions in the beginning of this section.
2. Remove or disconnect the following:
 - Negative battery cable
 - Engine under cover
 - Right side splash shield
 - Electrical connectors
 - Belt, by loosening the adjusting bolt
 - Alternator mounting bolts
 - Alternator

To install:

3. Install or connect the following:
 - Alternator
 - Alternator mounting bolts. Tighten the short bolt to 12–14 ft. lbs. (16–19 Nm) and the long bolt to 32–38 ft. lbs. (44–52 Nm)
 - Belt
 - Electrical connectors
 - Right side splash shield
 - Engine under cover
 - Negative battery cable

3.3L and 3.5L Engines

1. Before servicing the vehicle, refer to the precautions in the beginning of this section.

2. Remove or disconnect the following:
 - Negative battery cable
 - Alternator harness connectors
 - Engine under cover
 - Alternator belt
 - Alternator

To install:

3. Install or connect the following:
 - Alternator
 - Alternator belt. Tighten the adjustment bolt to 12–14 ft. lbs. (16–19 Nm) and the pivot bolts to 16–22 ft. lbs. (22–30 Nm).
 - Engine under cover
 - Alternator harness connectors
 - Negative battery cable

Ignition Timing

ADJUSTMENT

➡ **Ignition timing is set with the engine at operating temperature, transmission in Neutral and all electrical accessories OFF.**

1. Before servicing the vehicle, refer to the precautions in the beginning of this section.
2. Attach a timing light to the No. 1 spark plug wire.
3. Start the engine and allow it to reach normal operating temperature.
4. Check that the idle speed is less than 1000 rpm.
5. Run the engine at 2000 rpm for 2 minutes.
6. Rev the engine to 3000 rpm 2–3 times and allow it to idle for 1 minute.
7. Check for the presence of Diagnostic Trouble Codes (DTC) and service as necessary.
8. Run the engine at 2000 rpm for 2 minutes.
9. Stop the engine and disconnect the Throttle Position (TP) sensor.

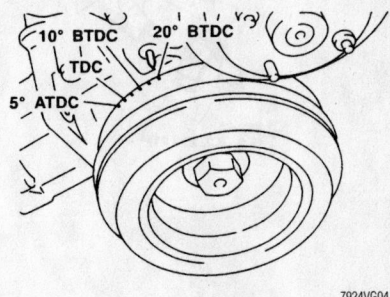

Timing indicator—2.4L and 3.3L engines

10. Start the engine and rev it to 3000 rpm 2–3 times and allow it to idle.

11. Set the base timing to 18–22 degrees Before Top Dead Center (BTDC) for 2.4L engine, or 8–12 degrees Before Top Dead Center (BTDC) for 3.3L engines.

12. Tighten the distributor lockbolt to 83–113 inch lbs. (9–13 Nm).

13. Set the base idle speed to 700–800 rpm.

14. Stop the engine and connect the TP sensor.

Engine Assembly

REMOVAL & INSTALLATION

2.4L Engine

1. Before servicing the vehicle, refer to the precautions in the beginning of this section.

2. Drain the cooling system.

3. Relieve the fuel system pressure.

4. Recover the A/C refrigerant, if equipped.

5. Remove or disconnect the following:
- Negative battery cable
- Hood
- Air cleaner
- Accessory drive belts
- Radiator
- Exhaust manifold heat shield
- Exhaust system at the rear of the manifold
- A/C compressor from the bracket
- Accelerator wire
- Vacuum hoses

➡ **Tag or label all lines and connections before removing them.**

- Electrical connectors
- Heater hoses
- Vacuum booster hose
- Power steering pump bolts and position pump aside
- Transmission

6. Install slingers and connect a suitable engine lift.
- Left and right side engine mounts
- Engine

To install:

7. Install or connect the following:
- Engine. Tighten the engine mount nuts to 32–42 ft. lbs. (43–58 Nm).

8. Remove the slingers
- Transmission
- Power steering pump
- Vacuum booster hose
- Heater hoses
- Electrical connectors
- Vacuum hoses
- Accelerator wire
- A/C compressor to the bracket
- Exhaust system to the rear of the manifold
- Exhaust manifold heat shield
- Radiator
- Accessory drive belts
- Air cleaner
- Negative battery cable
- Hood

9. Fill the cooling system.

10. Recharge the A/C system, if equipped.

11. Start the engine and check for leaks.

3.3L Engine

1. Before servicing the vehicle, refer to the precautions in the beginning of this section.

2. Drain the cooling system.

3. Relieve the fuel system pressure.

4. Recover the A/C refrigerant, if equipped.

5. Remove or disconnect the following:
- Negative battery cable
- Hood
- Air cleaner assembly

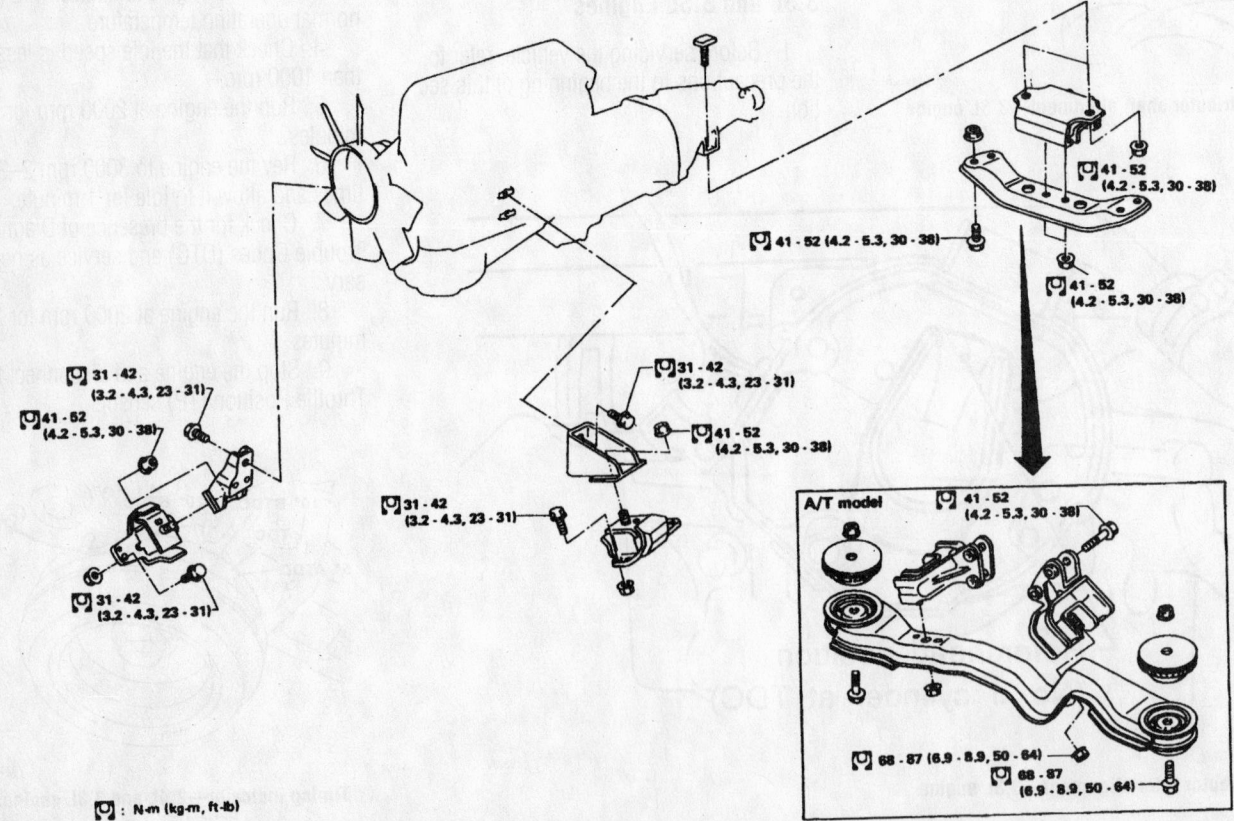

⬚ : N·m (kg·m, ft·lb)

Engine and transmission mounting—2.4L engines

7924VG05

- Idle Air Control (IAC) valve and solenoid connectors
- Throttle Position (TP) sensor and switch connectors
- Engine Coolant Temperature (ECT) sensor connector
- Manifold Absolute Pressure (MAP) sensor connector and vacuum line
- Evaporative Emissions (EVAP) canister purge valve connector and vacuum line
- Mass Air Flow (MAF) sensor connector
- Brake booster vacuum line
- Fuel lines
- Exhaust Gas Recirculation (EGR) temperature sensor connector
- Throttle cable
- Accessory drive belts
- Cooling fan and shroud
- Radiator and hoses
- Engine under cover
- A/C compressor manifold

- Power steering pump
- Heated Oxygen (HO2S) sensor connectors
- Exhaust front pipes
- Crankshaft Position (CKP) sensor
- Starter motor
- Transmission
- Left and right engine mounts
- Engine

➡ **When removing the engine mounts, do not loosen the 4 mount cover nuts. The mount is fluid filled and will not function if the fluid leaks out.**

To install:
6. Install or connect the following:
- Engine. Tighten the engine mount nuts to 43–58 ft. lbs. (59–78 Nm).
- Transmission
- Starter motor
- CKP sensor
- Exhaust front pipes
- HO2S sensor connectors
- Power steering pump

- A/C compressor manifold
- Engine under cover
- Radiator and hoses
- Cooling fan and shroud
- Accessory drive belts
- Throttle cable
- EGR temperature sensor connector
- Fuel lines
- Brake booster vacuum line
- MAF sensor connector
- EVAP canister purge valve connector and vacuum line
- MAP sensor connector and vacuum line
- ECT sensor connector
- TP sensor and switch connectors
- IAC valve and solenoid connectors
- Air cleaner assembly
- Hood
- Negative battery cable
7. Fill the cooling system.
8. Recharge the A/C system, if equipped.
9. Start the engine and check for leaks.

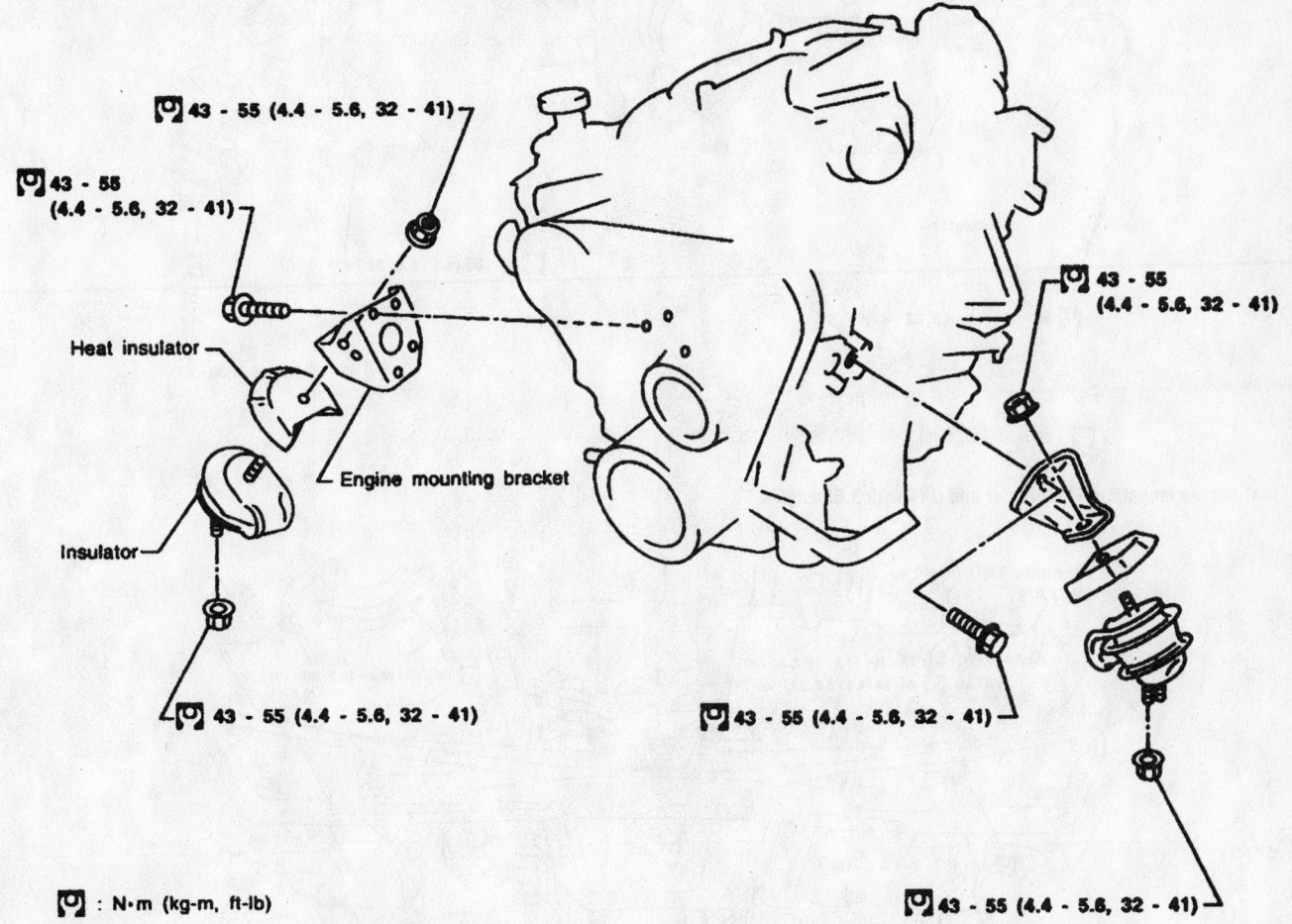

Engine mounts and related components—3.3L engine

3.5L Engine

PATHFINDER AND QX4

1. Release fuel pressure.
2. Remove engine hood and front RH and LH wheels.
3. Remove engine undercover and suspension member stay.
4. Drain coolant from radiator.

5. Remove the following parts.
   - Radiator shroud
   - Radiator
   - Cooling fan
   - Drive belts
   - Battery
   - Engine cover
   - Throttle wires
6. Air duct with air cleaner case.

7. Disconnect vacuum hoses, fuel hoses, heater hoses, EVAP canister hoses, harnesses, connectors and so on.
8. Remove air conditioner compressor from bracket, then put it aside holding with a suitable wire.
9. Remove power steering oil pump and reservoir tank with bracket, then put it aside holding with a suitable wire.

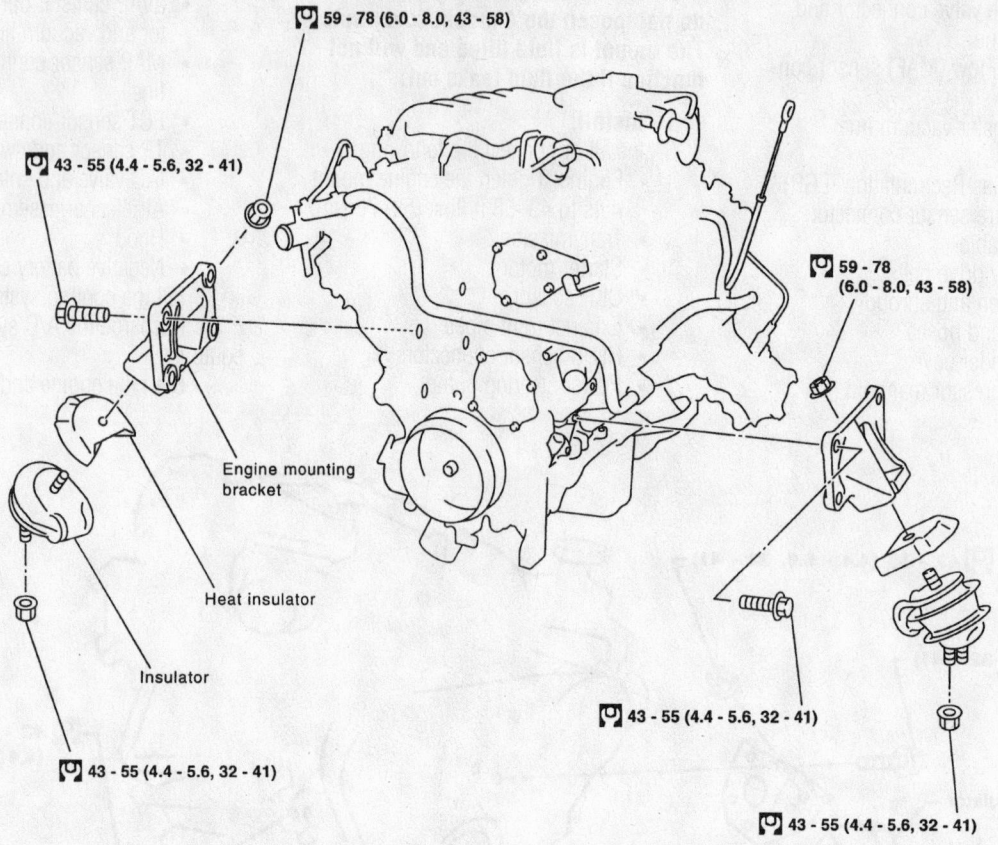

Front engine mounting—Pathfinder and QX4 with 3.5L engine

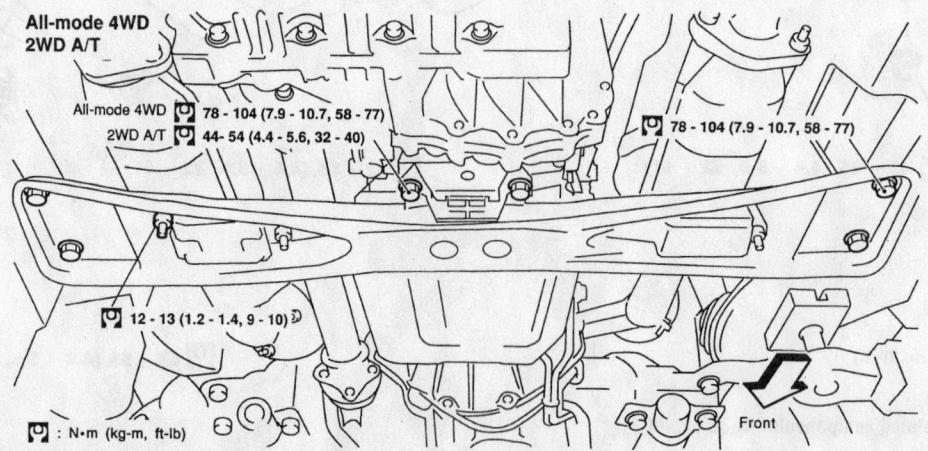

Rear engine mounting— Pathfinder and QX4 with 3.5L engine

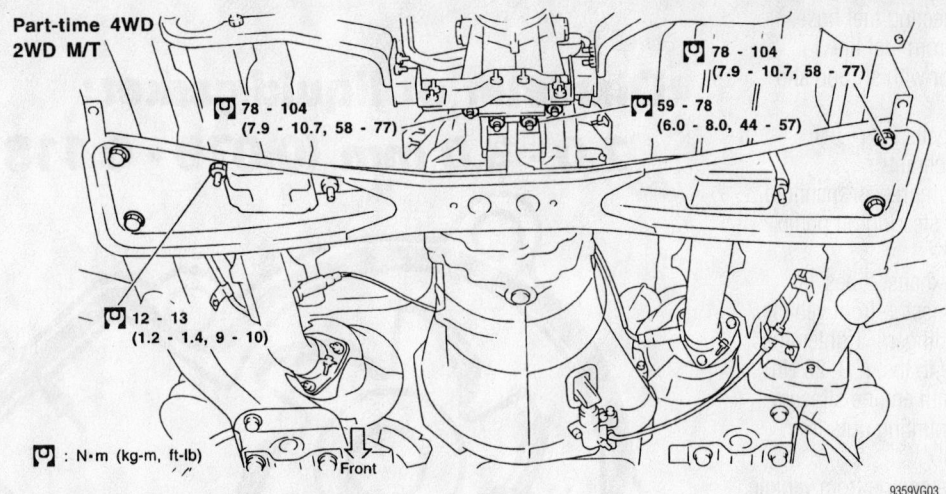

Rear engine mounting— Pathfinder and QX4 with 3.5L engine

10. Remove alternator.

11. Remove exhaust front tube heat insulators, then remove rear heated oxygen sensors.

12. Remove exhaust front and rear tubes.

13. Remove transmission.

14. Remove TWC (manifold) heat insulators, then remove TWC (manifold).

15. Install engine slingers.

16. Hoist engine with engine slingers and remove front engine mounting nuts.

17. Remove engine from vehicle.

To install:

Installation is in the reverse order of removal. Observe the following torques:

• Front engine mount-to-bracket: 43–58 ft. lbs.

• Front mount-to-frame: 32–41 ft. lbs.

• Front bracket-to-block: 32–41 ft. lbs.

• Rear engine mount-to-bracket: all exc. 2wd with AT: 58–77 ft. lbs.; 2wd with AT: 32–40 ft. lbs.

• Crossmember-to-frame: 58–77 ft. lbs.

XTERRA

➡ Do not loosen front engine mounting insulator cover securing bolts. When cover is removed, damper oil flows out and mounting insulator will not function.

1. Remove engine undercover and hood.

2. Drain coolant from cylinder block and radiator.

3. Remove vacuum hoses, fuel tubes, wires, harnesses and connectors.

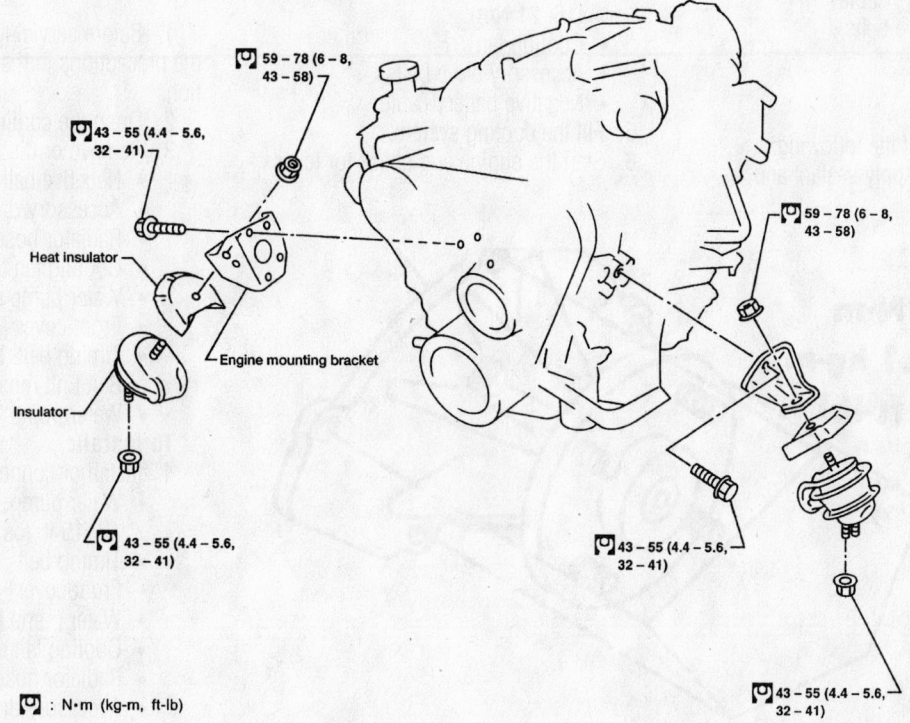

Engine mounting—Xterra with 3.5L engine

4. Before disconnecting fuel hose, release fuel pressure from fuel line.

5. Remove radiator with shroud and cooling fan.

6. Remove drive belts.

7. Discharge refrigerant.

8. Remove A/C compressor manifold.

9. Remove power steering oil pump from engine.

10. Remove front exhaust tubes.

11. Remove transmission from vehicle.

12. Install engine slingers. Tighten the slinger bolts to 15–20 ft. lbs. (20–26 Nm).

13. Hoist engine with engine slingers and remove engine mounting nuts from both sides.

14. Lift and remove engine from vehicle.

15. Installation is the reverse of removal. See the accompanying illustration for installation torques.

Water Pump

REMOVAL & INSTALLATION

2.4L Engine

1. Before servicing the vehicle, refer to the precautions in the beginning of this section.

2. Drain the cooling system.

3. Remove or disconnect the following:
   - Negative battery cable
   - Accessory drive belts
   - Cooling fan
   - Water pump

To install:

4. Install or connect the following:
   - Water pump. Apply sealant and

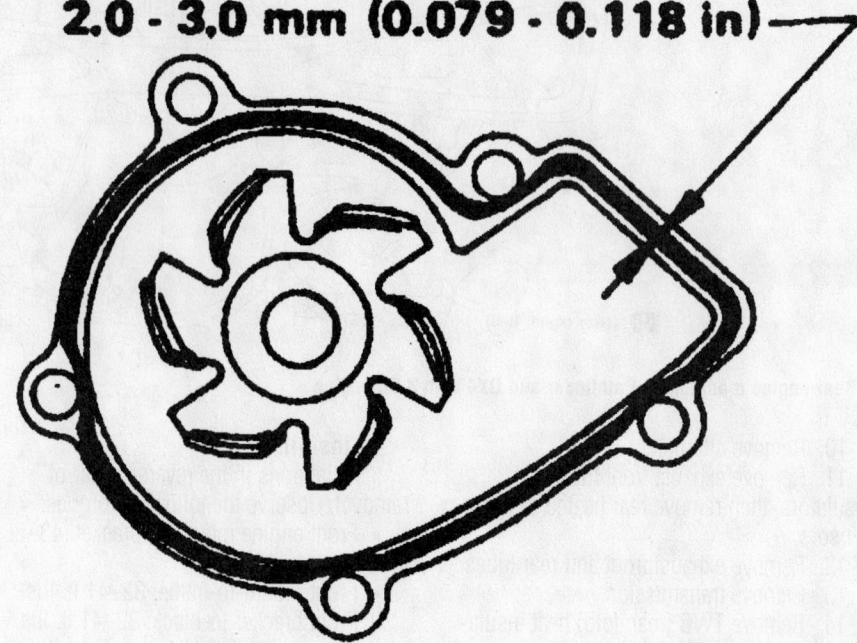

Diameter of liquid gasket: 2.0 - 3.0 mm (0.079 - 0.118 in)

7924VG17

Liquid gasket application—2.4L engine

tighten the bolts to 12–15 ft. lbs. (16–21 Nm).
   - Cooling fan
   - Accessory drive belts
   - Negative battery cable

5. Fill the cooling system.

6. Start the engine and check for leaks.

3.3L Engine

1. Before servicing the vehicle, refer to the precautions in the beginning of this section.

2. Drain the cooling system.

3. Remove or disconnect the following:
   - Negative battery cable
   - Accessory drive belts
   - Radiator hoses
   - Cooling fan and shroud
   - Water pump pulley
   - Front cover
   - Timing belt. Refer to the Timing Belt unit repair section.
   - Water pump

To install:

4. Install or connect the following:
   - Water pump. Tighten the bolts to 12–15 ft. lbs. (16–21 Nm).
   - Timing belt
   - Front cover
   - Water pump pulley
   - Cooling fan and shroud
   - Radiator hoses
   - Accessory drive belts
   - Negative battery cable

5. Fill the cooling system.

6. Start the engine and check for leaks.

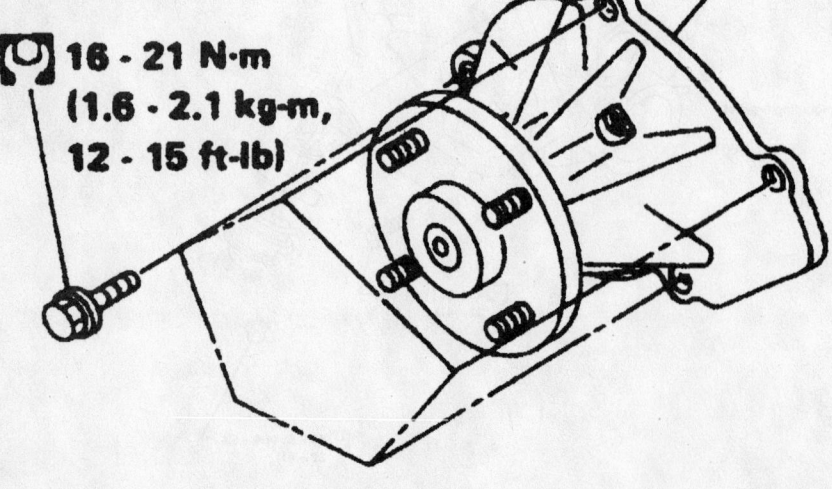

16 - 21 N·m (1.6 - 2.1 kg-m, 12 - 15 ft-lb)

7924VG16

Water pump assembly—2.4L engine

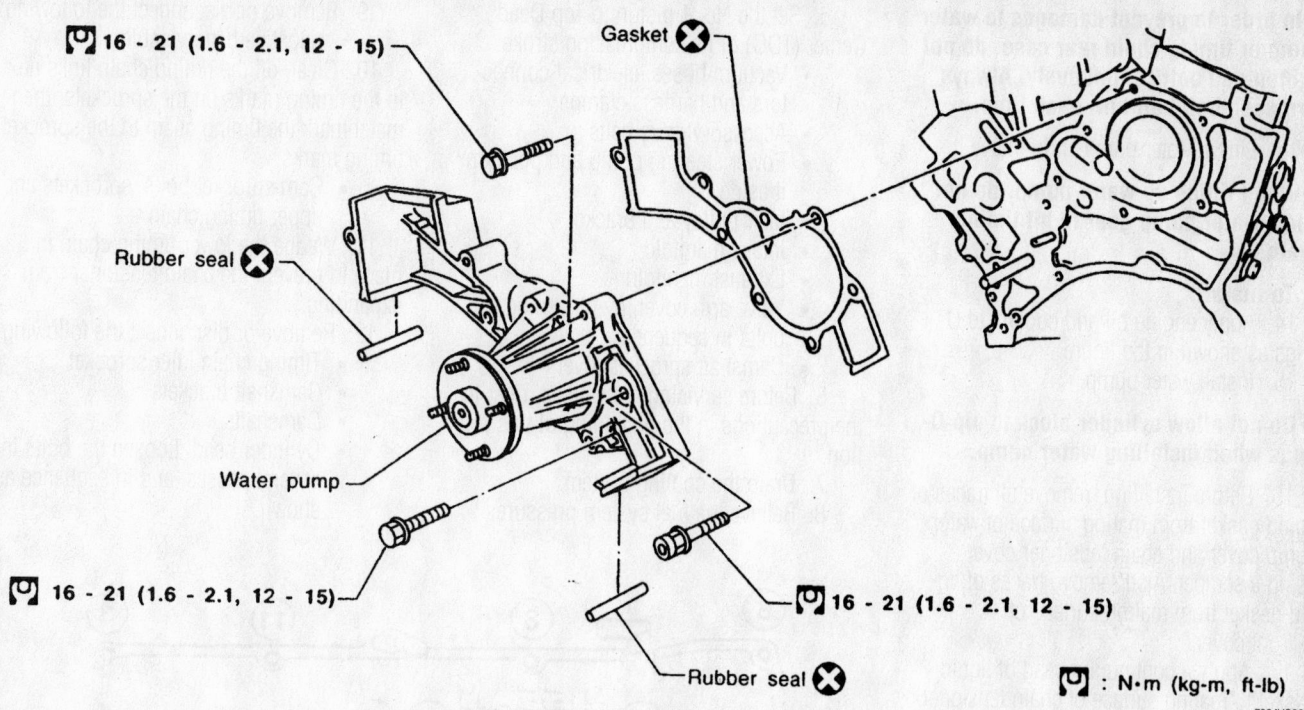

16 - 21 (1.6 - 2.1, 12 - 15)

Gasket ⊗

Rubber seal ⊗

Water pump

16 - 21 (1.6 - 2.1, 12 - 15)

16 - 21 (1.6 - 2.1, 12 - 15)

Rubber seal ⊗

: N•m (kg-m, ft-lb)

7924VG20

Exploded view of the water pump assembly—3.3L engine

3.5L Engine

1. Remove undercover.
2. Remove suspension member stay.
3. Drain coolant from radiator.
4. Remove radiator shrouds.
5. Remove drive belts.
6. Remove cooling fan.

7. Remove water drain plug on water pump side of cylinder block.
8. Remove chain tensioner cover and water pump cover.
9. Pushing timing chain tensioner sleeve, apply a stopper pin so it does not return. Then remove the chain tensioner assembly.
10. Remove the 3 water pump fixing

bolts. Secure a gap between water pump gear and timing chain, by turning crankshaft pulley 20° backwards.
11. Put M8 bolts to two water pump fixing bolt holes.
12. Tighten M8 bolts by turning half turn alternately until they reach timing chain rear case.

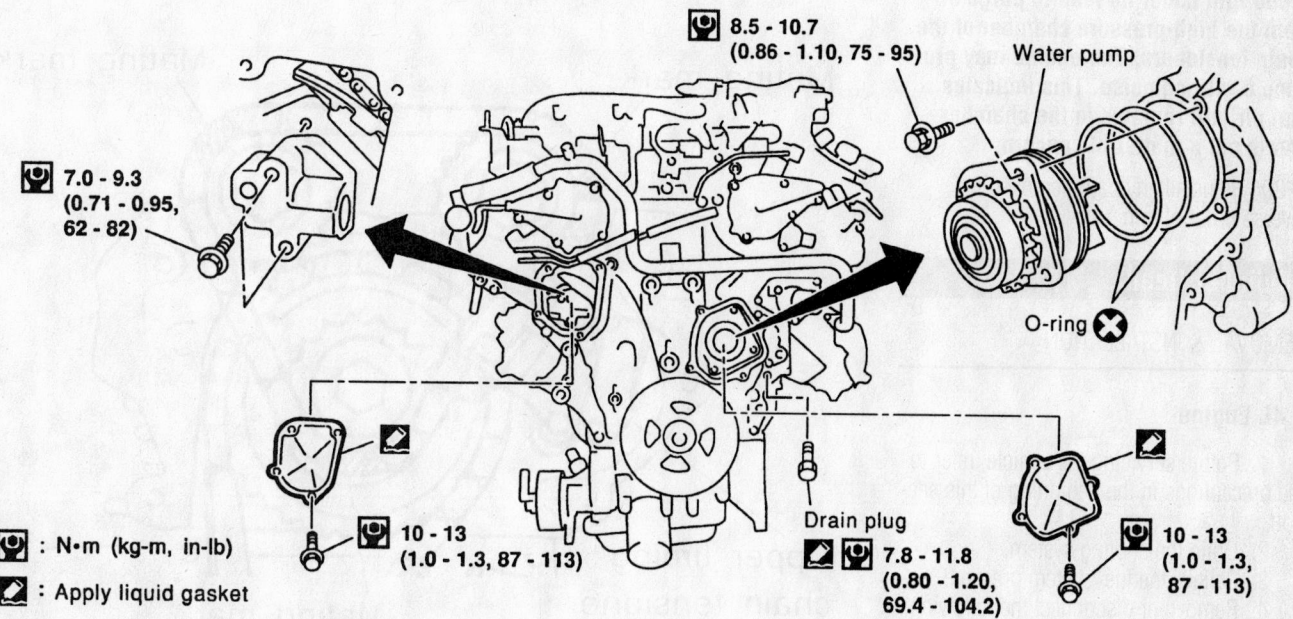

8.5 - 10.7
(0.86 - 1.10, 75 - 95)

Water pump

7.0 - 9.3
(0.71 - 0.95, 62 - 82)

O-ring ⊗

: N•m (kg-m, in-lb)

: Apply liquid gasket

10 - 13
(1.0 - 1.3, 87 - 113)

Drain plug

7.8 - 11.8
(0.80 - 1.20, 69.4 - 104.2)

10 - 13
(1.0 - 1.3, 87 - 113)

Exploded view of the water pump assembly—3.5L engine

9359VG04

Heater Core replacement is covered in Section 2 of this manual

➡ **In order to prevent damages to water pump or timing chain rear case, do not tighten one bolt continuously. Always turn each bolt half turn each time.**

13. Lift up water pump and remove it.

➡ **When lifting up water pump, do not allow water pump gear to hit timing chain.**

To install:

14. Apply engine oil and coolant to O-rings as shown in the figure.

15. Install water pump.

➡ **Do not allow cylinder block to nip O-rings when installing water pump.**

16. Before installing, remove all traces of liquid gasket from mating surface of water pump cover and chain tensioner cover using a scraper. Also remove traces of liquid gasket from mating surface of front cover.

17. Apply a continuous bead of liquid gasket to mating surface of chain tensioner cover and water pump cover. Use Genuine RTV silicone sealant or equivalent.

18. Return the crankshaft pulley to its original position by turning it 20° forward.

19. Install timing chain tensioner, then remove the stopper pin.

➡ **When installing the timing chain tensioner, engine oil should be applied to the oil hole and tensioner.**

➡ **After starting engine, let idle for three minutes, then rev engine up to 3,000 rpm under no load to purge air from the high-pressure chamber of the chain tensioners. The engine may produce a rattling noise. This indicates that air still remains in the chamber and is not a matter of concern.**

20. Reinstall any parts removed in reverse order of removal.

Cylinder Head

REMOVAL & INSTALLATION

2.4L Engine

1. Before servicing the vehicle, refer to the precautions in the beginning of this section.

2. Drain the cooling system.

3. Relieve the fuel system pressure.

4. Remove or disconnect the following:
   • Negative battery cable
   • Air cleaner assembly
   • Spark plug wires

5. Set the No. 1 piston to Top Dead Center (TDC) of its compression stroke.
   • Vacuum hoses, electrical connectors and harness clamps
   • Accessory drive belts
   • Power steering pump and position it aside
   • Idler pulley and bracket
   • Intake manifold
   • Exhaust manifold
   • Valve arm cover, by loosening the bolts in sequence
   • Camshaft sprocket cover

6. Before servicing the vehicle, refer to the precautions in the beginning of this section.

7. Drain the cooling system.

8. Relieve the fuel system pressure.

9. Remove or disconnect the following:
   • Negative battery cable

10. Clean off the timing chain links next to the timing marks on the sprockets, then matchmark the timing chain to the sprocket timing marks.
    • Cam sprocket bolts, sprockets and upper timing chain

11. Wedge the lower timing chain in place to prevent the chain tensioner from expanding.

12. Remove or disconnect the following
    • Timing chain idler sprocket
    • Camshaft brackets
    • Camshafts
    • Cylinder head. Loosen the bolts in several passes, and in sequence as shown.

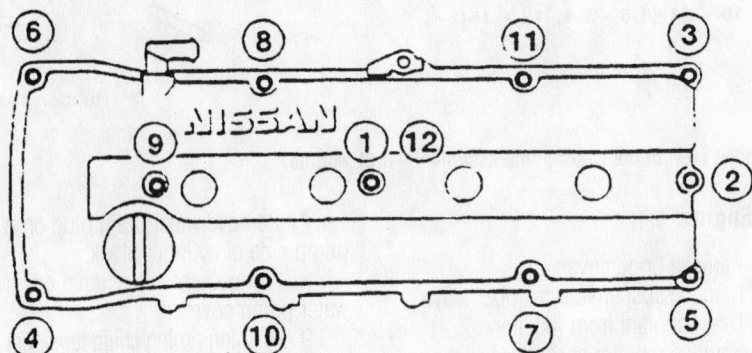

Loosen in numerical order.

9308VG06

Valve cover bolt loosening sequence—2.4L engine

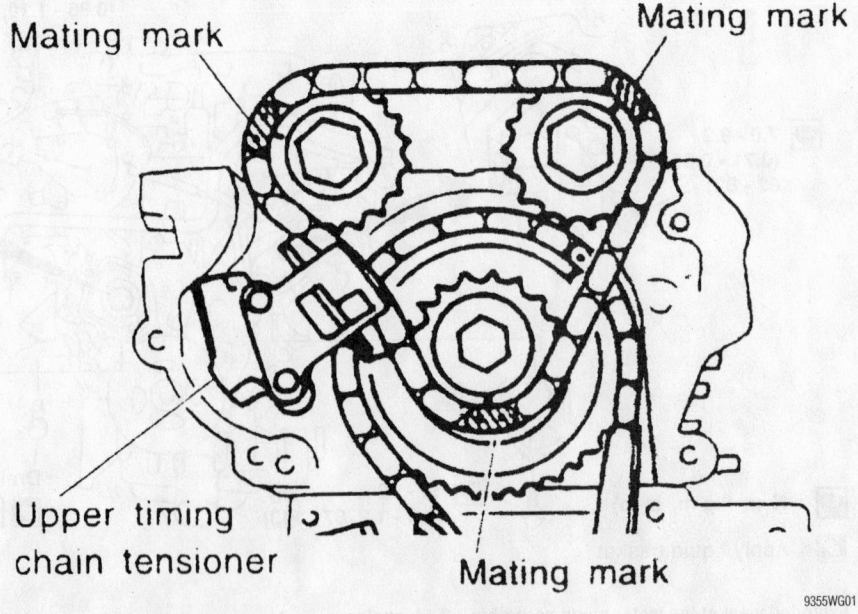

Matchmark the timing chain to the timing marks—2.4L engine

9355WG01

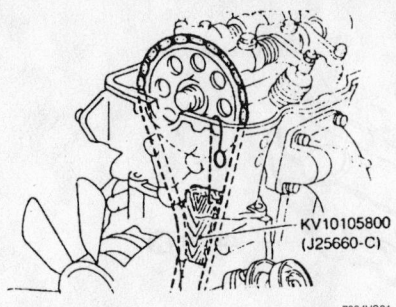

KV10105800
(J25660-C)

7924VG21

Wedge the chain in place so that it will not fall down into the front cover—2.4L engine

To install:

13. Install the cylinder head with a new gasket. Tighten the bolts in sequence, as follows:
    a. Step 1: 22 ft. lbs. (30 Nm)
    b. Step 2: 59 ft. lbs. (79 Nm)
    c. Step 3: Loosen all bolts completely
    d. Step 4: 18–25 ft. lbs. (25–34 Nm)
    e. Step 5: Plus 86–91 degrees
14. Install or connect the following:
    • Camshafts and camshaft brackets.
    • Timing chain idler sprocket. Remove the wedge and tighten the bolt to 48–61 ft. lbs. (66–83 Nm).

• Camshaft sprockets and upper timing chain. Tighten the bolts to 123–130 ft. lbs. (167–177 Nm).
• Camshaft sprocket cover

• Valve cover. Tighten the bolts in sequence to 69–95 inch lbs. (8–11 Nm)
• Exhaust manifold
• Intake manifold
• Idler pulley and bracket
• Power steering pump
• Accessory drive belts
• Vacuum hoses, electrical connectors and harness clamps
• Spark plug wires
• Air cleaner assembly
• Negative battery cable
15. Fill the cooling system.
16. Start the engine and check for leaks.

3.3L Engine

1. Before servicing the vehicle, refer to the precautions in the beginning of this section.
2. Drain the cooling system.
3. Relieve the fuel system pressure.
4. Remove or disconnect the following:
    • Negative battery cable
    • Accessory drive belts
    • Front cover
    • Timing belt. Refer to the Timing Belt unit repair section.
    • Upper intake manifold
    • Lower intake manifold
    • Camshaft sprockets
    • Rear timing cover
    • Distributor
    • Exhaust front pipes
    • A/C compressor
    • Alternator
    • Power steering pump
    • Accessory brackets
    • Valve covers. Loosen the bolts in several passes and in sequence.
    • Cylinder heads with the exhaust manifolds attached. Loosen the bolts in several passes and in sequence.

➡ **The cylinder head bolts vary in length. Note the bolt locations for assembly.**

To install:

5. Install the cylinder heads and the lower intake manifold at the same time. Tighten the bolts in sequence as follows:
    a. Step 1: Tighten the cylinder head bolts to 22 ft. lbs. (29 Nm)

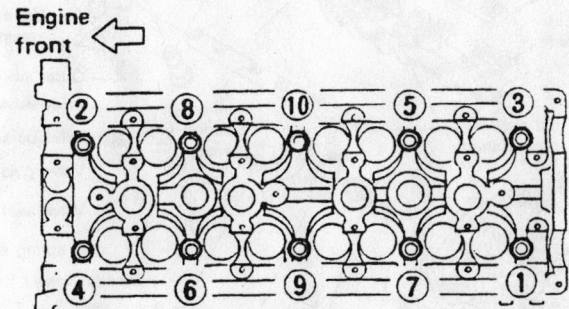

Loosen in numerical order.

9308VG04

Cylinder head bolt loosening sequence—2.4L engine

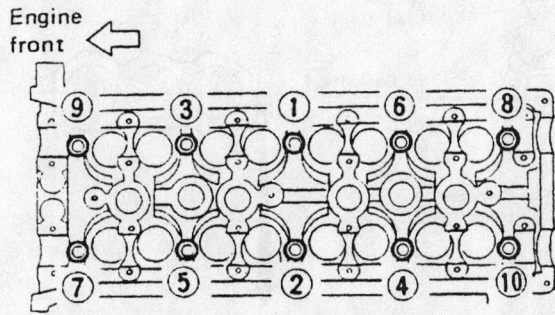

Tighten in numerical order.

9308VG05

Cylinder head bolt tightening sequence—2.4L engine

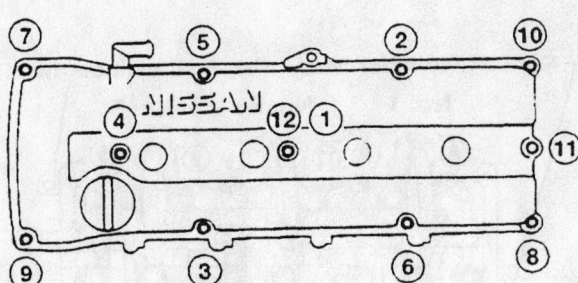

Tighten in numerical order.

9308VG07

Valve cover bolt tightening sequence—2.4L engine

Brake service is covered in Section 4 of this manual

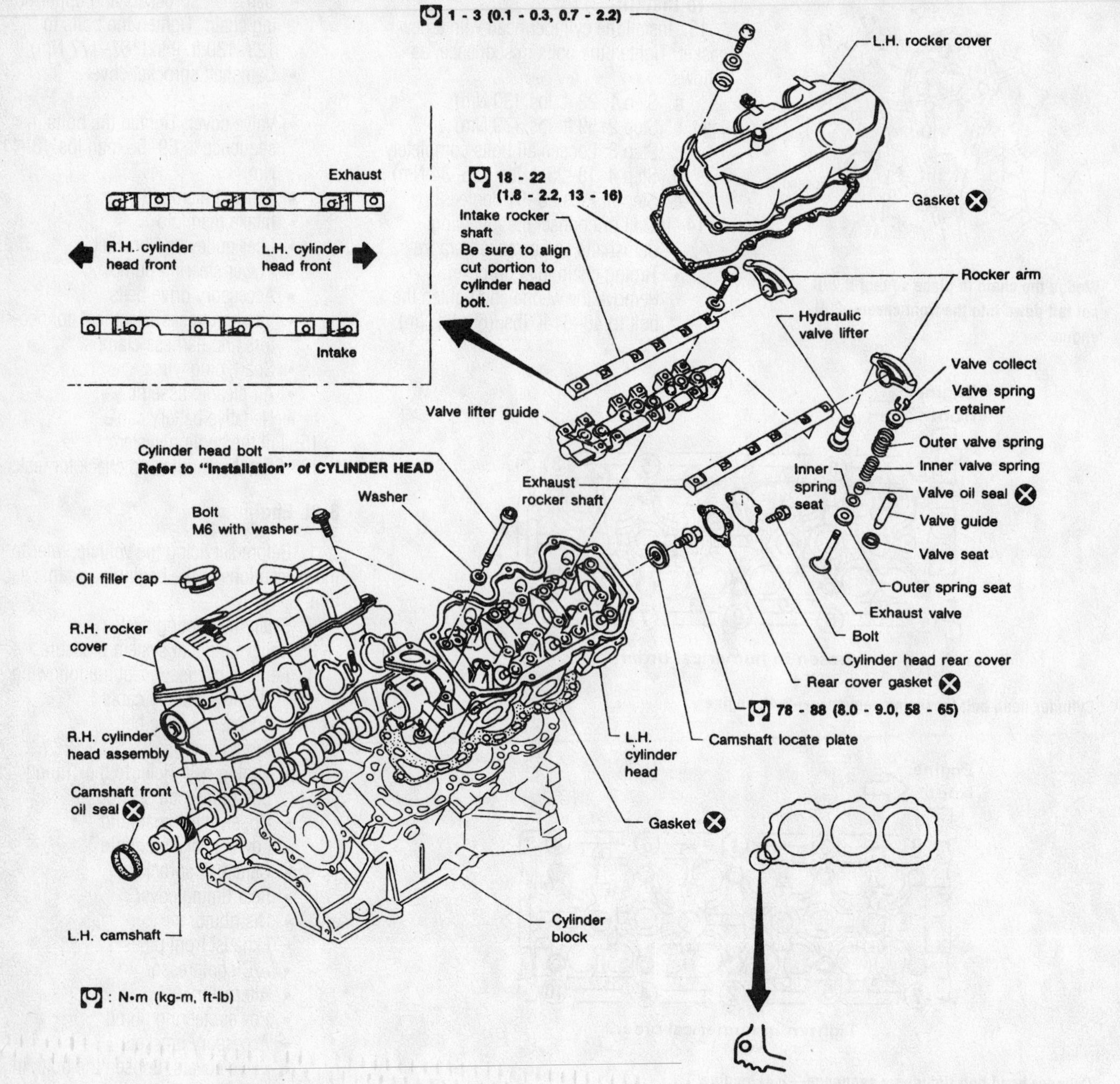

⬡ 1 - 3 (0.1 - 0.3, 0.7 - 2.2) ── L.H. rocker cover

Exhaust

R.H. cylinder head front ⟵ ⟶ L.H. cylinder head front

Intake

⬡ 18 - 22 (1.8 - 2.2, 13 - 16)
Intake rocker shaft
Be sure to align cut portion to cylinder head bolt.

Gasket ⊗

Rocker arm

Hydraulic valve lifter

Valve collect

Valve spring retainer

Outer valve spring

Inner valve spring

Valve oil seal ⊗

Valve guide

Valve seat

Outer spring seat

Exhaust valve

Bolt

Cylinder head rear cover

Rear cover gasket ⊗

⬡ 78 - 88 (8.0 - 9.0, 58 - 65)

Camshaft locate plate

Valve lifter guide

Exhaust rocker shaft

Inner spring seat

Cylinder head bolt
Refer to "Installation" of CYLINDER HEAD

Washer

Bolt M6 with washer

Oil filler cap

R.H. rocker cover

R.H. cylinder head assembly

Camshaft front oil seal ⊗

L.H. camshaft

L.H. cylinder head

Gasket ⊗

Cylinder block

⬡ : N•m (kg-m, ft-lb)

Exploded view of the cylinder head assembly—3.3L engine

7924VG25

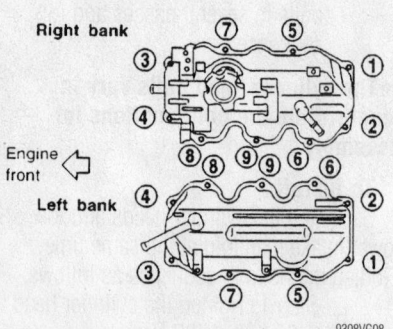

Right bank

Engine front

Left bank

9308VG08

Valve cover bolt loosening sequence—3.3L engine

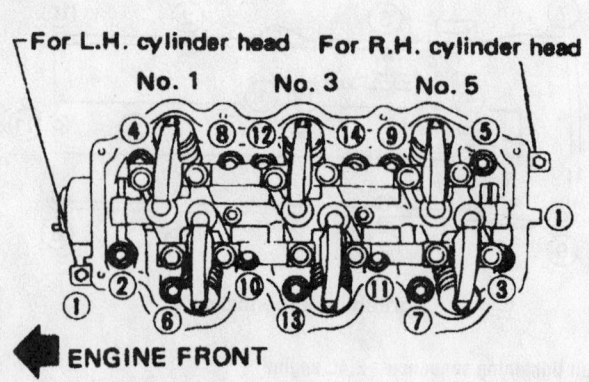

For L.H. cylinder head   For R.H. cylinder head

No. 1   No. 3   No. 5

⟵ ENGINE FRONT

7924VG26

Cylinder head loosening sequence—3.3L engine

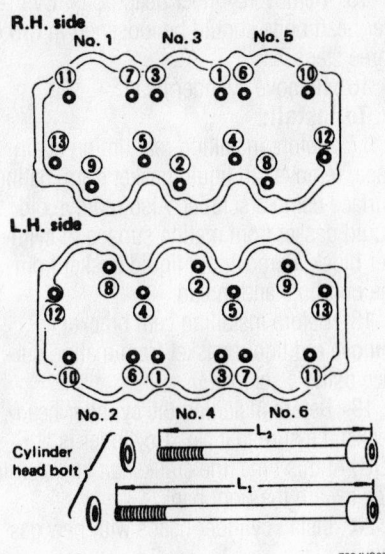

R.H. side

L.H. side

Cylinder head bolt

7924VG27

Cylinder head torque sequence—3.3L engine

- Distributor
- Rear timing cover
- Camshaft sprockets
- Upper intake manifold
- Timing belt
- Front cover
- Accessory drive belts
- Negative battery cable
7. Fill the cooling system.
8. Start the engine and check for leaks.

3.5L Engine

1. Remove engine from vehicle.
2. Remove exhaust manifolds in reverse order of installation.
3. Place engine on a work stand.
4. Remove aluminum oil pan
5. Remove timing chain.

6. Remove intake manifold in reverse order of installation.
7. Remove water outlet.
8. Remove rear timing chain case bolts. Loosen in numerical order as shown in the figure.
9. Remove rear timing chain case.
10. Remove O-rings to cylinder head.
11. Remove O-rings to cylinder block.
12. Remove intake valve timing control solenoid valves.
13. Remove intake and exhaust camshafts and camshaft brackets. Equally loosen camshaft bracket bolts in several steps in the numerical order shown in the figure. For reinstallation, be sure to put marks on camshaft bracket before removal.
14. Remove RH and LH camshaft chain tensioners from cylinder head.

b. Step 2: Tighten the cylinder head bolts to 43 ft. lbs. (59 Nm)

c. Step 3: Loosen all cylinder head bolts completely

d. Step 4: Tighten the cylinder head bolts to 84 inch lbs. (10 Nm)

e. Step 5: Tighten the intake manifold fasteners to 35 inch lbs. (4 Nm)

f. Step 6: Tighten the intake manifold fasteners to 13 ft. lbs. (18 Nm)

g. Step 7: Tighten the intake manifold fasteners to 12–14 ft. lbs. (16–20 Nm)

h. Step 8: Loosen all intake fasteners completely

i. Step 9: Tighten the cylinder head bolts to 22 ft. lbs. (29 Nm)

j. Step 10: Tighten the cylinder head bolts 60–65 degrees **OR** tighten to 40–47 ft. lbs. (54–64 Nm)

k. Step 11: Tighten the cylinder head sub-bolts to 80–105 inch lbs. (9–12 Nm)

l. Step 12: Tighten the intake manifold fasteners to 35 inch lbs. (4 Nm)

m. Step 13: Tighten the intake manifold fasteners to 78 inch lbs. (9 Nm)

n. Step 14: Tighten the intake manifold fasteners to 70–84 inch lbs. (6–7 Nm)

6. Install or connect the following:
- Valve covers
- Accessory brackets
- Power steering pump
- Alternator
- A/C compressor
- Exhaust front pipes

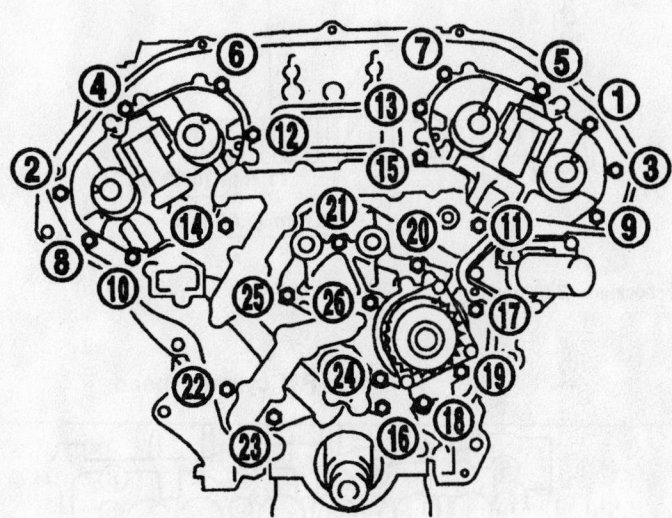

9359VG05

Rear timing case loosening sequence—3.5L engine

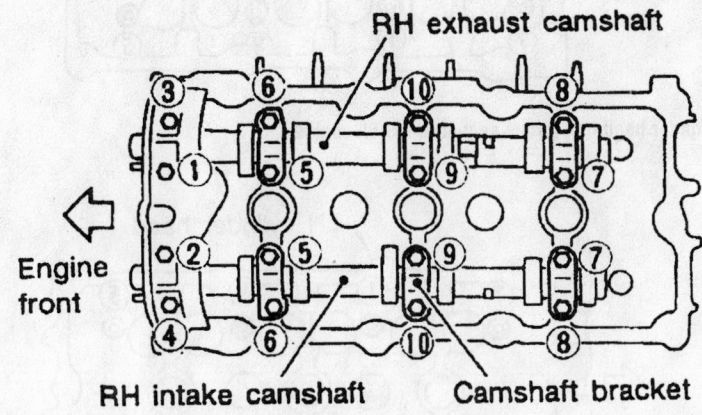

RH exhaust camshaft

Engine front

RH intake camshaft   Camshaft bracket

Loosen in numerical order.

9359VG06

Right camshaft loosening sequence—3.5L engine

For complete Engine Mechanical specifications, see Section 1 of this manual

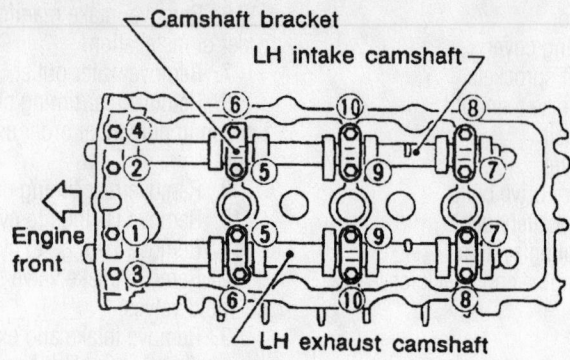

Left camshaft loosening sequence—3.5L engine

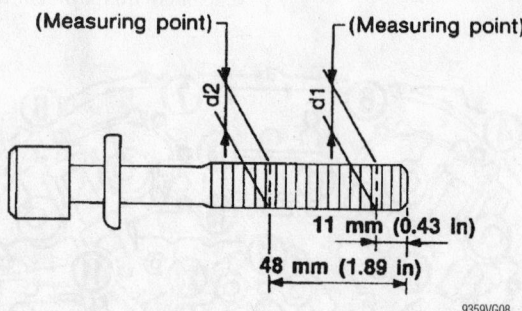

Head bolt checking—3.5L engine

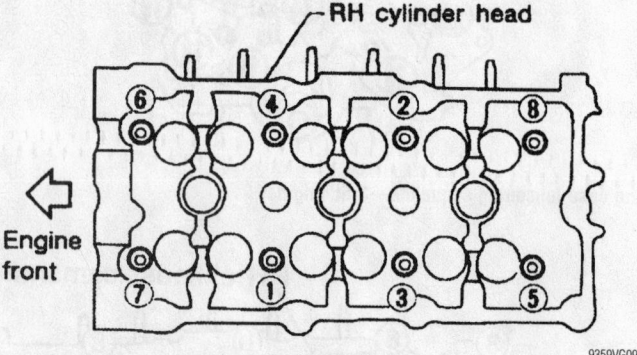

Right cylinder head bolt torque sequence—3.5L engine

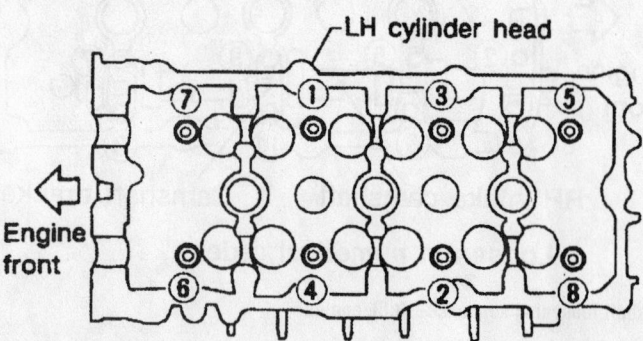

Left cylinder head bolt torque sequence—3.5L engine

15. Remove cylinder head bolts. Cylinder head bolts should be loosened in two or three steps.

16. Remove cylinder head.

To install:

17. Before installing rear timing chain case, remove old liquid gasket from mating surface using a scraper. Also remove old liquid gasket from mating surface of cylinder block. Remove old liquid gasket from the bolt hole and thread.

18. Before installing cam bracket, remove old liquid gasket from mating surface using a scraper.

19. Before installing the cylinder head gasket, be sure that No. 1 cylinder is at TDC. At this time, the crankshaft key should face toward the right bank.

20. Install cylinder heads with new gaskets.

➡**Do not rotate crankshaft and camshaft separately, or valves will strike piston heads.**

❊❊ CAUTION

Cylinder head bolts are tightened by plastic zone tightening method. Whenever the size difference between d1 and d2 exceeds the limit, replace them with new ones. Limit (d1—d2): 0.0043 in. Lubricate threads and seat surfaces of the bolts with new engine oil.

21. Install cylinder head outside bolts Tighten in numerical order shown in the figure. Tightening procedure:

   a. Tighten all bolts to 98 N·m (10 kg-m, 72 ft-lb).

   b. Completely loosen all bolts.

   c. Tighten all bolts to 34 to 44 N·m (3.5 to 4.5 kg-m, 25 to 33 ft-lb).

   d. Turn all bolts 90 to 95 degrees clockwise.

   e. Turn all bolts 90 to 95 degrees clockwise.

22. Install camshaft chain tensioners on both sides of cylinder head.

23. Install exhaust and intake camshafts and camshaft brackets.

➡**Intake camshaft has a drill mark on camshaft sprocket mounting flange. Install it on the intake side. Position camshaft. RH exhaust camshaft dowel pin at about 10 o'clock; LH exhaust camshaft dowel pin at about 2 o'clock**

24. Before installing camshaft brackets, apply sealant to mating surface of No. 1 journal head. Use Genuine RTV silicone sealant or equivalent. Install camshaft brack-

- Identification marks are present on camshafts.

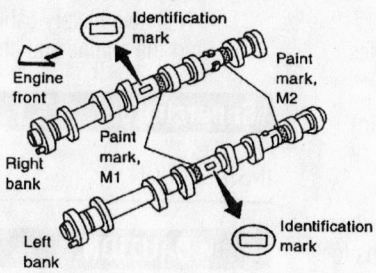

Camshaft identification—3.5L engine

Bank	INT/EXH	ID mark	Drill mark	Paint mark	
				M1	M2
RH	INT	R3	Yes	Yes	No
	EXH	R3	No	No	Yes
LH	INT	L3	Yes	Yes	No
	EXH	L3	No	No	Yes

9359VG11

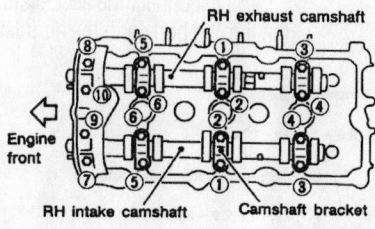

Right camshaft bolt torque sequence—3.5L engine

- Tighten the camshaft brackets in the following steps.

Step	Tightening torque	Tightening order
1	1.96 N·m (0.2 kg-m, 17 in-lb)	Tighten in the order of 7 to 10, then tighten 1 to 6.
2	5.88 N·m (0.6 kg-m, 52 in-lb)	Tighten in the numerical order.
3	9.02 - 11.8 N·m (0.92 - 1.20 kg-m, 79.9 - 104.2 in-lb)	Tighten in the order of 1 to 6.
	8.3 - 10.3 N·m (0.9 - 1.0 kg-m, 74 - 91 in-lb)	Tighten in the order of 7 to 10.

9359VG12

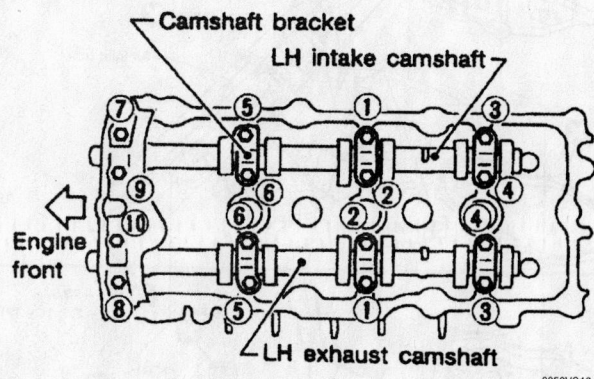

Left camshaft bolt torque sequence—3.5L engine

9359VG13

ets in their original positions. Align stamp mark as shown in the figure. If any part of valve assembly or camshaft is replaced, check valve clearance according to reference data. After completing assembly check valve clearance. Valve clearance (Cold):

- Intake 0.26—0.34 mm (0.010—0.013 in)
- Exhaust 0.29—0.37 mm (0.011—0.015 in)

➡ **Lubricate threads and seat surfaces of camshaft bracket bolts with new engine oil before installing them.**

25. Install intake valve timing control solenoid valves.

26. Install O-rings to cylinder block.

27. Install O-rings to cylinder head.

28. Apply sealant to the hatched portion of rear timing chain case. Apply continuous bead of liquid gasket to mating surface of rear timing chain case. Before installation, wipe off the protruding sealant.

29. Align rear timing chain case with dowel pins, then install on cylinder head and block.

30. Tighten rear chain case bolts.

a. Tighten bolts in numerical order shown in the figure.

b. Repeat above step a.

31. Reinstall all removed parts in reverse order of removal.

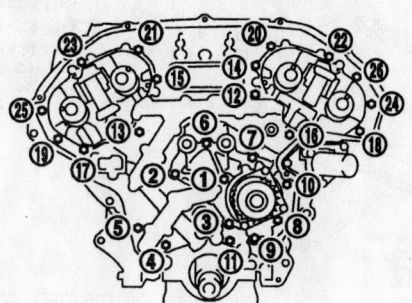

⌷ 12 - 13 N·m
(1.2 - 1.4 kg-m, 9 - 10 ft-lb)

9359VG14

Rear timing case bolt torque sequence—3.5L engine

For Accessory Drive Belt illustrations, see Section 1 of this manual

Rocker Arms/Shafts

REMOVAL & INSTALLATION

2.4L Engine

This engine is not equipped with rocker arms. The camshafts act directly on the valve lifters.

3.3L Engine

1. Before servicing the vehicle, refer to the precautions in the beginning of this section.
2. Remove or disconnect the following:
   • Negative battery cable
   • Upper intake manifold
   • Valve covers
   • Rocker arm and shaft assemblies
   • Rocker arms from the shafts

➡ Keep all valvetrain components in order for assembly.

To install:

3. Lubricate all contact points with clean engine oil and assemble the rocker arms to the shafts in their original positions.
4. Install or connect the following:
   • Rocker arm and shaft assemblies. Tighten the bolts to 13–16 ft. lbs. (18–22 Nm).
   • Valve covers
   • Upper intake manifold
   • Negative battery cable
5. Start the engine and check for leaks.

Supercharger

INSTALLATION

✳✳ CAUTION

Do not disassemble or adjust the supercharger.

1. Disconnect the negative battery cable.
2. Disconnect the accelerator cable from the throttle body and the air inlet tube bracket.

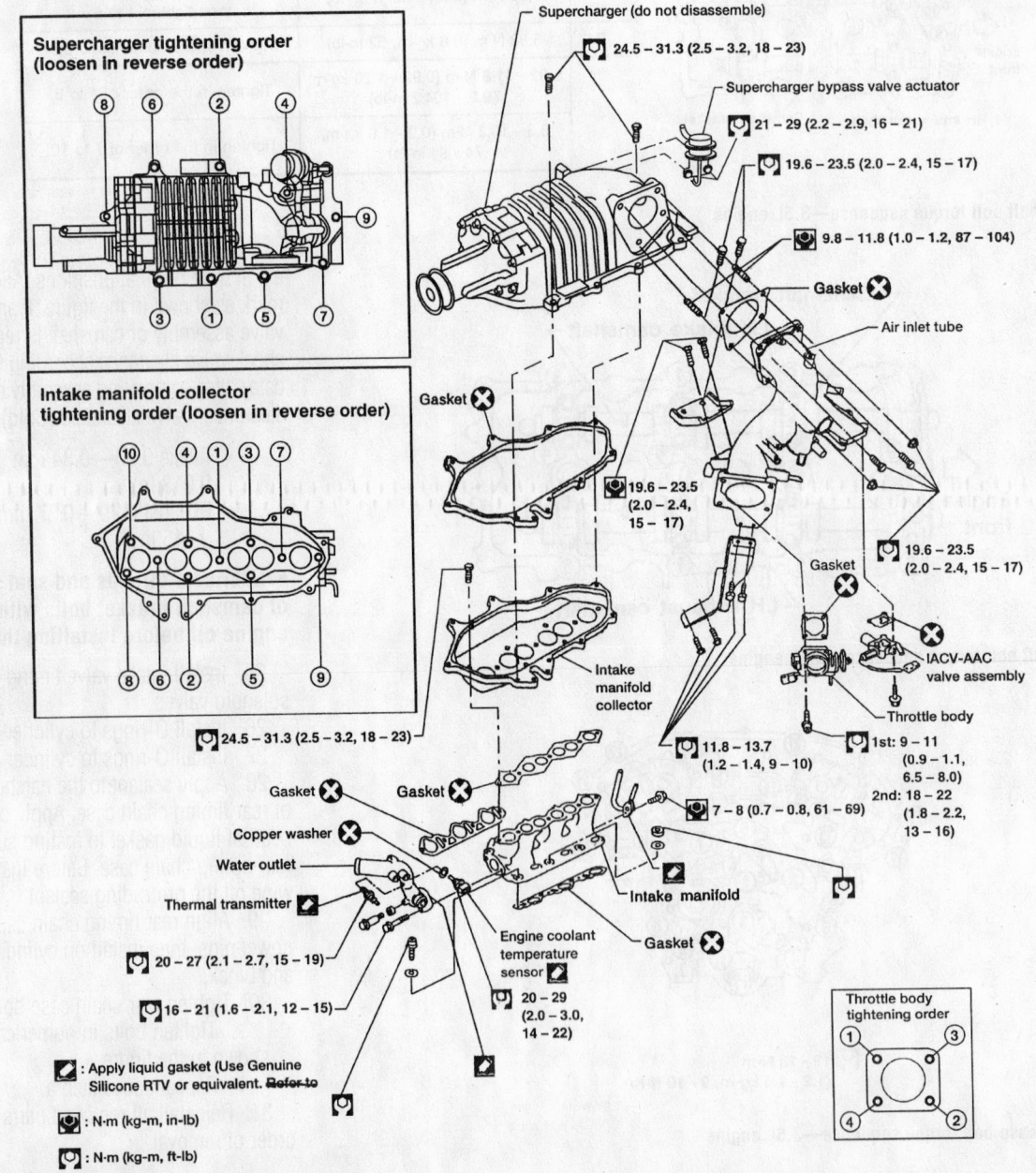

Supercharger tightening order (loosen in reverse order)

Intake manifold collector tightening order (loosen in reverse order)

Supercharger (do not disassemble)

24.5 – 31.3 (2.5 – 3.2, 18 – 23)

Supercharger bypass valve actuator

21 – 29 (2.2 – 2.9, 16 – 21)

19.6 – 23.5 (2.0 – 2.4, 15 – 17)

9.8 – 11.8 (1.0 – 1.2, 87 – 104)

Gasket

Air inlet tube

Gasket

19.6 – 23.5 (2.0 – 2.4, 15 – 17)

19.6 – 23.5 (2.0 – 2.4, 15 – 17)

Gasket

IACV-AAC valve assembly

Intake manifold collector

Throttle body

11.8 – 13.7 (1.2 – 1.4, 9 – 10)

1st: 9 – 11 (0.9 – 1.1, 6.5 – 8.0)
2nd: 18 – 22 (1.8 – 2.2, 13 – 16)

7 – 8 (0.7 – 0.8, 61 – 69)

24.5 – 31.3 (2.5 – 3.2, 18 – 23)

Gasket

Gasket

Copper washer

Water outlet

Thermal transmitter

Intake manifold

Gasket

Engine coolant temperature sensor

20 – 27 (2.1 – 2.7, 15 – 19)

16 – 21 (1.6 – 2.1, 12 – 15)

20 – 29 (2.0 – 3.0, 14 – 22)

: Apply liquid gasket (Use Genuine Silicone RTV or equivalent. Refer to

: N·m (kg-m, in-lb)

: N·m (kg-m, ft-lb)

Throttle body tightening order

Supercharger components

9359VG35

3. Disconnect the ASCD cable from the throttle body and the air inlet tube bracket, if equipped.

4. Remove the air inlet duct
   a. Disconnect the PCV hoses.
   b. Disconnect the resonator hose.

5. Partially drain the cooling system.

6. Remove the supercharger pulley cover and the supercharger/air conditioning drive belt.

7. Remove the air inlet tube upper and lower supports.

8. Remove the air inlet tube bolts, nuts, and studs. Position the air inlet tube aside.
   a. Disconnect the evaporative emission vacuum hose.
   b. Disconnect the brake booster vacuum hose.
   c. Disconnect the TPS sensor electrical connector.
   d. Disconnect the TPS switch electrical connector.

9. Remove the supercharger bolts and the supercharger assembly.
   a. Disconnect the boost control valve vacuum hose.
   b. Disconnect the PCV hose.

INSPECTION

Supercharger Flange

1. Clean the mating surface of the supercharger flange.

2. Check the flange surface for any deformation and flatness.

Use a reliable straightedge and feeler gauge, or attach the supercharger flange to the intake collector mating flange, and check that the flatness is within specification. Flange flatness limit: 0.12 mm (0.005 in).

Rotor System

1. Check that the supercharger pulley rotates smoothly when turning it by hand in a clockwise direction. Rotating torque must not exceed specification. Rotating torque: 0.5 N.m (0.05 kg-m, 4 in-lb).

2. Check that both the left and right rotors are free from any cracks or contamination.

Supercharger Bypass Valve Actuator

1. Apply air pressure of less than 12 kPa (90 mmHg, 3.54 inHg) to the supercharger bypass valve actuator's lower side hose port and check for any leakage.

2. Check the supercharger bypass valve

actuator rod for smooth movement while maintaining the pressure at the specified levels below:
- Rod starts to extend at approximately: 12 Kpa (90 mmHg, 3.54 inHg)
- Rod is fully extended at approximately: 33.3 kPa (250 mmHg, 9.84 inHg)
- Rod full extended length: 20.83–22.71 mm (0.82–0.89 in)

INSTALLATION

To install the supercharger, follow the removal steps in reverse order. Replace all gaskets; make sure that all gasket surfaces are clean and undamaged. Follow all torque sequences for tightening. Refill the cooling system.

Intake Manifold

REMOVAL & INSTALLATION

2.4L Engine

1. Before servicing the vehicle, refer to the precautions in the beginning of this section.

2. Drain the cooling system.

3. Relieve the fuel system pressure.

4. Remove or disconnect the following:
   - Negative battery cable
   - Air cleaner assembly
   - Coolant hoses
   - Fuel lines
   - Accelerator cable
   - Cruise control cable, if equipped
   - Positive Crankcase Ventilation (PCV) valve and hose
   - Exhaust Gas Recirculation (EGR) tube
   - EGR temperature sensor connector
   - Idle Air Control (IAC) valve and solenoid connectors
   - Throttle Position (TP) sensor and switch connectors
   - Engine Coolant Temperature (ECT) sensor connector
   - Manifold Absolute Pressure (MAP) sensor connector and vacuum lines
   - Evaporative Emission (EVAP) canister purge valve vacuum line
   - Brake booster vacuum line
   - Fuel injector connectors
   - Intake manifold bracket
   - Intake manifold. Loosen the fasteners in the reverse of the tightening sequence.

To install:

5. Install or connect the following:
   - Intake manifold. Tighten the bolts to 12–14 ft. lbs. (16–19 Nm).
   - Intake manifold bracket. Tighten the bolts to 24–28 ft. lbs. (32–38 Nm).
   - Fuel injector connectors
   - Brake booster vacuum line
   - EVAP canister purge valve vacuum line

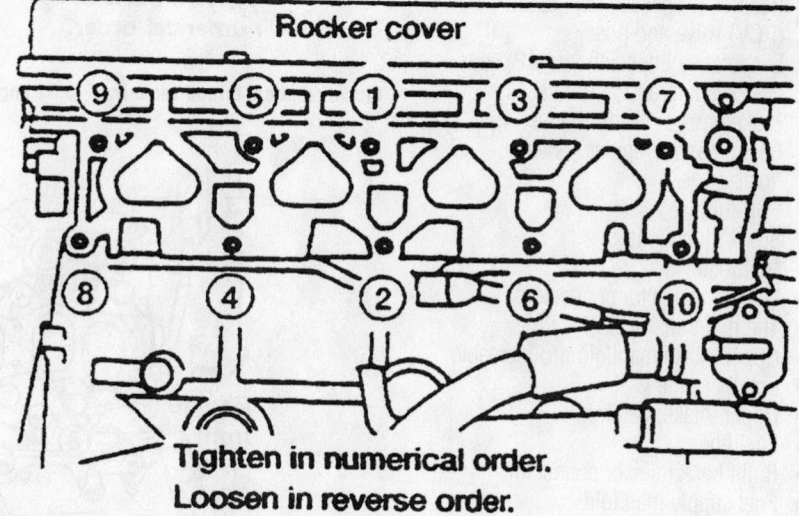

Intake manifold bolt tightening sequence—2.4L engine

9308VG02

For Tire, Wheel and Ball Joint specifications, see Section 1 of this manual

- MAP sensor connect and vacuum line
- ECT sensor connector
- TP sensor and switch connectors
- IAC valve and solenoid connectors
- EGR temperature sensor connector
- EGR tube
- PCV valve and hose
- Cruise control cable, if equipped
- Accelerator cable
- Fuel lines
- Coolant hoses
- Air cleaner assembly
- Negative battery cable

6. Fill the cooling system.
7. Start the engine and check for leaks.

3.3L and 3.5L Engines

1. Before servicing the vehicle, refer to the precautions in the beginning of this section.
2. Drain the cooling system.
3. Relieve the fuel system pressure.
4. Remove or disconnect the following:

- Negative battery cable
- Air intake duct
- Accelerator cable
- Cruise control cable
- Idle Air Control (IAC) valve connector
- Throttle Position (TP) sensor and switch connectors
- Ignition coil and power transistor connectors
- Exhaust Gas Recirculation (EGR) Solenoid valve connector
- EGR temperature sensor connector
- Radiator hoses
- Heater hoses
- Positive Crankcase Ventilation (PCV) valve and hose
- Evaporative Emissions (EVAP) canister vacuum and purge hoses
- Brake booster vacuum hose
- Fuel pressure regulator vacuum hose
- EGR tube
- Spark plug wires
- Distributor
- Left bank injector connectors
- Thermal transmitter
- Upper intake manifold ground cable
- Breather pipe
- Upper intake manifold
- Fuel lines
- Right bank injector connectors
- Fuel supply manifold
- Engine Coolant Temperature (ECT) sensor connector
- Lower intake manifold. Loosen the fasteners in the sequence shown.

To install:

5. Install the lower intake manifold with a new gasket.
6. For 3.3L engines, tighten the fasteners in sequence as follows:
   a. Step 1: 35 inch lbs. (4 Nm)
   b. Step 2: 78 inch lbs. (9 Nm)
   c. Step 3: 70–84 inch lbs. (8–10 Nm)

7. For 3.5L engines, tighten the fasteners in sequence as follows:
   a. Step 1: 86 inch lbs. (4 Nm)
   b. Step 2: 23 ft. lbs. (9 Nm)
8. Install or connect the following:
   - ECT sensor connector
   - Fuel supply manifold
   - Right bank injector connectors

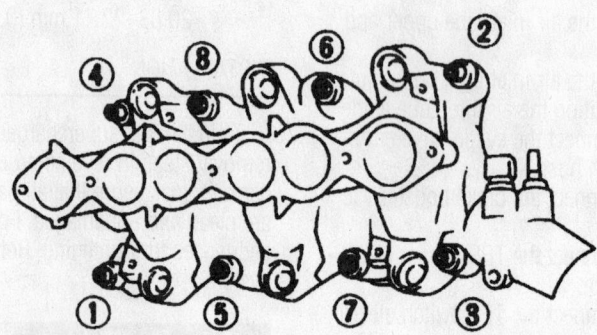

Loosen bolts in numerical order.

7924VG32

Intake manifold loosening sequence—3.3L engine

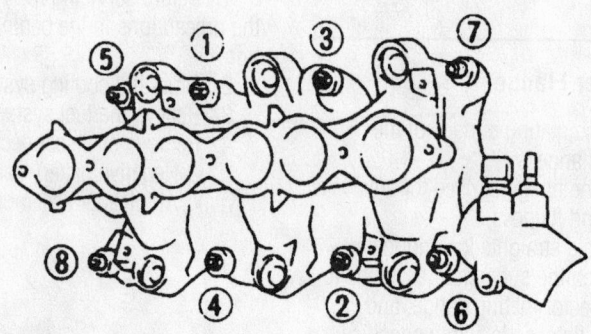

Tighten bolts in numerical order.

7924VG33

Intake manifold torque sequence—3.3L engine

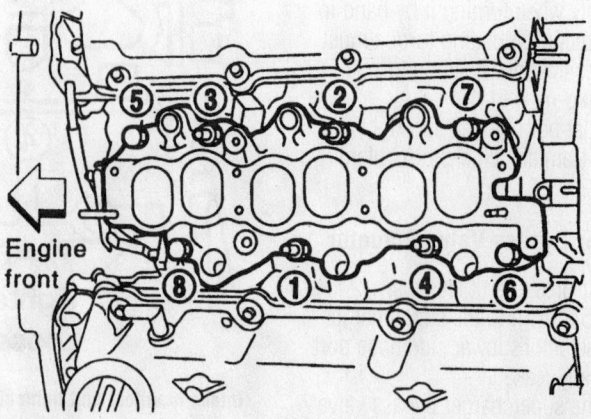

Engine front

9359VG15

Lower intake manifold torque sequence—3.5L engine

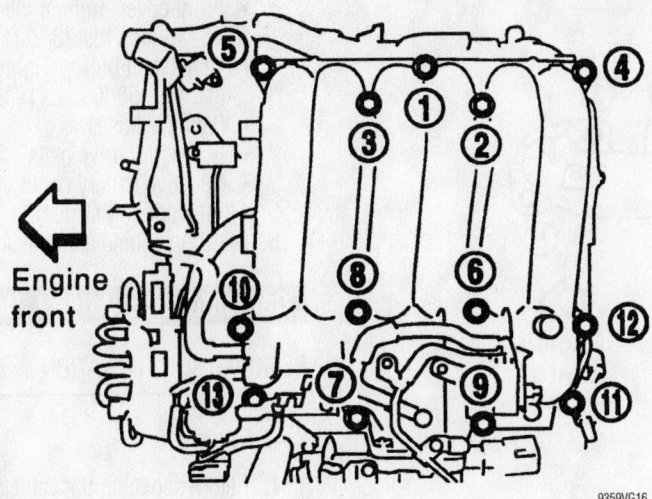

Upper intake manifold torque sequence—3.5L engine

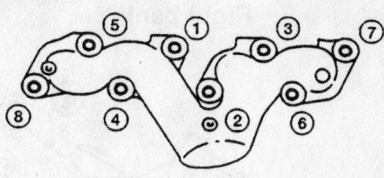

Exhaust manifold bolt torque sequence—2.4L engine

- Fuel lines
- Upper intake manifold
- Breather pipe
- Upper intake manifold ground cable
- Thermal transmitter
- Left bank injector connectors
- Distributor
- Spark plug wires
- EGR tube
- Fuel pressure regulator vacuum hose
- Brake booster vacuum hose
- EVAP canister vacuum and purge hoses
- PCV valve and hose
- Heater hoses
- Radiator hoses
- EGR temperature sensor connector
- EGR Solenoid valve connector
- Ignition coil and power transistor connectors
- TP sensor and switch connectors
- IAC valve connector
- Cruise control cable
- Accelerator cable
- Air intake duct
- Negative battery cable
9. Fill the cooling system.
10. Start the engine and check for leaks.

Exhaust Manifold

REMOVAL & INSTALLATION

2.4L Engine

1. Before servicing the vehicle, refer to the precautions in the beginning of this section.

2. Remove or disconnect the following:
- Negative battery cable
- Heated Oxygen (HO$_2$S) sensor connector
- Exhaust manifold heat shields
- Exhaust Gas Recirculation (EGR) tube
- Exhaust front pipe
- Exhaust manifold. Loosen the nuts in the reverse of the torque sequence.

To install:
3. Install or connect the following:
- Exhaust manifold. Tighten the nuts in sequence to 28–35 ft. lbs. (37–48 Nm)
- Exhaust front pipe. Tighten the fasteners to 32–37 ft. lbs. (43–50 Nm)
- EGR tube. Tighten the flange fittings to 29–36 ft. lbs. (39–49 Nm)
- Exhaust manifold heat shield. Tighten the bolts to 45–57 inch lbs. (5–7 Nm)

- HO$_2$S) sensor connector
- Negative battery cable
4. Start the engine and check for leaks.

3.3L and 3.5L Engines

1. Before servicing the vehicle, refer to the precautions in the beginning of this section.

2. Remove or disconnect the following:

- Negative battery cable
- Exhaust manifold heat shields
- Exhaust Gas Recirculation (EGR) tube
- Heated Oxygen (HO$_2$S) sensor connectors
- Exhaust front pipes
- Exhaust manifolds with catalytic converters attached. Loosen the nuts in the reverse of the torque sequence.

To install:
3. Install or connect the following:
- Exhaust manifolds with catalytic converters attached. Tighten the nuts in sequence to 21–25 ft. lbs. (28–33 Nm).
- Exhaust front pipes. Tighten the bolts to 21–25 ft. lbs. (28–33 Nm).

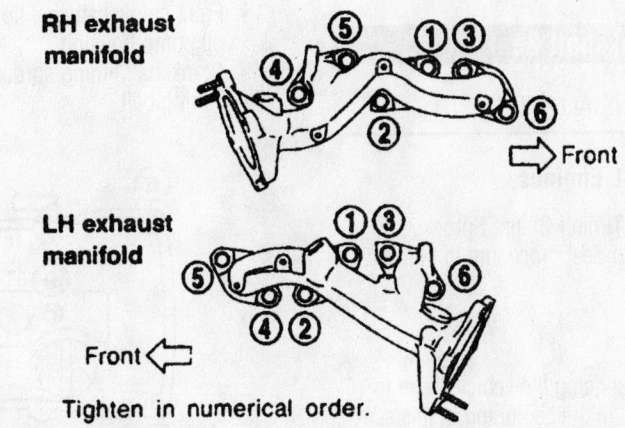

Tighten in numerical order.

Exhaust manifold torque sequence—3.3L engine

For Wheel Alignment specifications, see Section 1 of this manual

Right bank

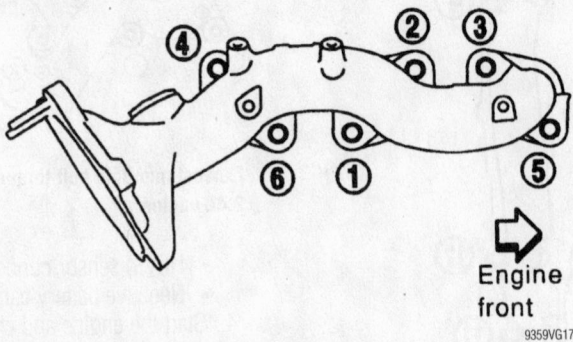

Right exhaust manifold torque sequence—3.5L engine

9359VG17

Left bank

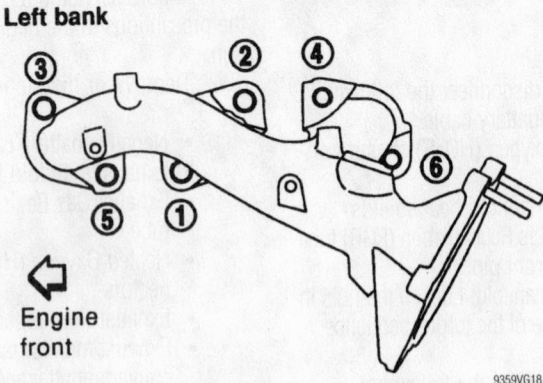

Exhaust manifold torque sequence—3.5L engine

9359VG18

- Heated Oxygen (HO2S) sensor connectors
- EGR tube. Tighten the flange fittings to 29–36 ft. lbs. (39–49 Nm).
- Exhaust manifold heat shields. Tighten the bolts to 84–96 inch lbs. (9–11 Nm)
- Negative battery cable
4. Start the engine and check for leaks.

Front Crankshaft Seal

REMOVAL & INSTALLATION

2.4L and 3.5L Engines

Refer to the Timing Chain, Sprockets, Front Cover and Seal procedure in this section.

3.3L Engine

1. Before servicing the vehicle, refer to the precautions in the beginning of this section.
2. Drain the cooling system.
3. Remove or disconnect the following:
   - Negative battery cable
   - Accessory drive belts

- Radiator hoses
- Crankshaft pulley
- Front cover
- Timing belt. Refer to the Timing Belt unit repair section.
- Crankshaft timing sprocket
- Front crankshaft seal

To install:
4. Install or connect the following:
   - Front crankshaft seal flush with the oil pump housing
   - Crankshaft timing sprocket
   - Timing belt

- Front cover. Tighten the bolts to 26–43 inch lbs. (3–5 Nm).
- Crankshaft pulley. Tighten the bolt to 141–156 ft. lbs. (191–211 Nm).
- Radiator hoses
- Accessory drive belts
- Negative battery cable
5. Fill the cooling system.
6. Start the engine and check for leaks.

Camshaft and Valve Lifters

REMOVAL & INSTALLATION

2.4L Engine

1. Before servicing the vehicle, refer to the precautions in the beginning of this section.
2. Remove or disconnect the following:
   - Negative battery cable
   - Air cleaner assembly
   - Spark plug wires
   - Valve cover. Remove the bolts in the sequence shown.
   - Camshaft sprocket cover
   - Camshaft sprocket bolts and upper timing chain

➡ **Keep all valvetrain components in order for assembly.**

- Camshaft bearing caps. Loosen the bolts, in several passes, in reverse of the torque sequence.
- Camshafts
- Valve lifters and shims

To install:
3. Install or connect the following:
   - Valve lifters and shims in their original positions
   - Camshafts
4. Install the bearing caps. Tighten the bolts in sequence in follows:
   a. Step 1: 17 inch lbs. (2 Nm)
   b. Step 2: 80–104 inch lbs. (9–12 Nm)

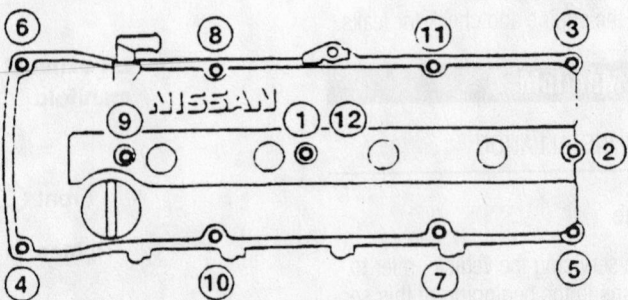

Loosen in numerical order.

Valve cover loosening sequence—2.4L engine

9308VG06

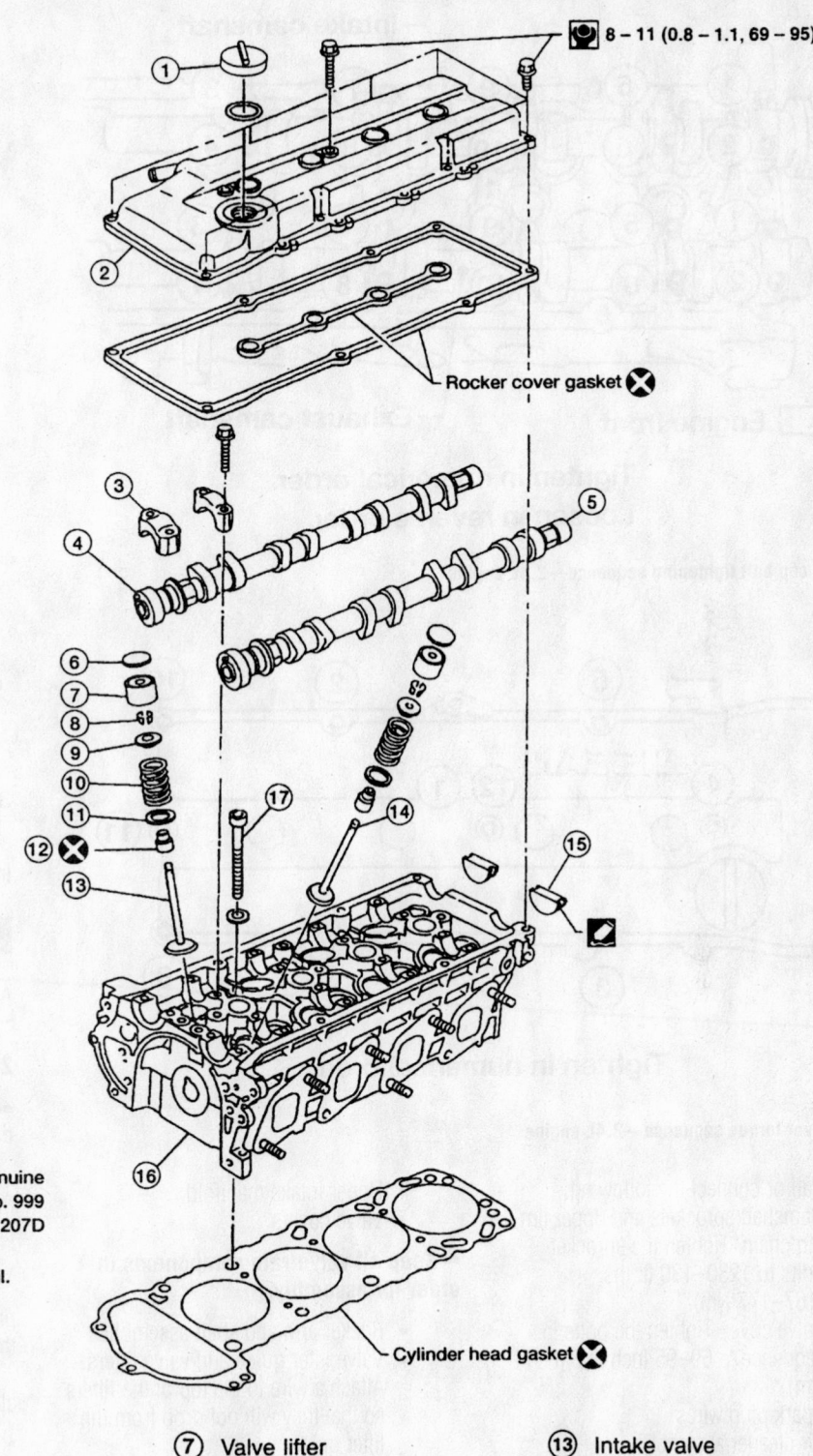

8 – 11 (0.8 – 1.1, 69 – 95)

Rocker cover gasket ⊗

Cylinder head gasket ⊗

: Apply liquid gasket. Use Genuine
RTV silicone sealant, Part No. 999
MP-A7007, Three Bond TB 1207D
or equivalent.

: Lubricate with new engine oil.

: N·m (kg-m, in-lb)

: N·m (kg-m, ft-lb)

① Oil filler cap
② Rocker cover
③ Camshaft bracket
④ Intake camshaft
⑤ Exhaust camshaft
⑥ Shim
⑦ Valve lifter
⑧ Valve cotter
⑨ Spring retainer
⑩ Valve spring
⑪ Spring seat
⑫ Valve oil seal
⑬ Intake valve
⑭ Exhaust valve
⑮ Rubber plug
⑯ Cylinder head
⑰ Cylinder head bolt

7924VG53

Exploded view of the camshafts and related components—2.4L engine

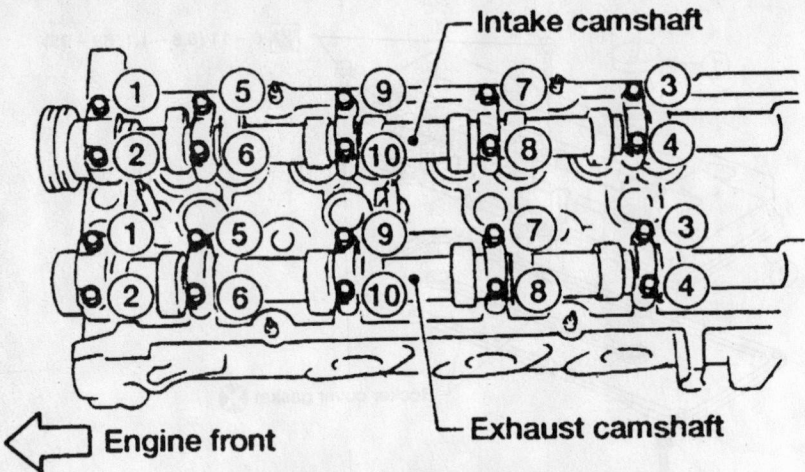

**Tighten in numerical order.
Loosen in reverse order.**

7924VG51

Bearing cap bolt tightening sequence—2.4L engine

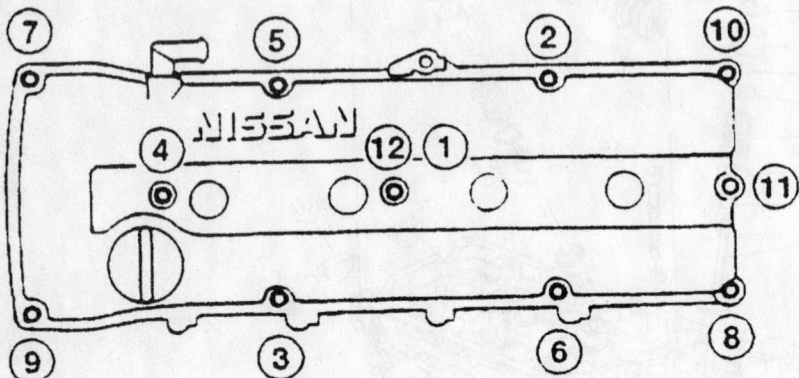

Tighten in numerical order.

9308VG07

Valve cover torque sequence—2.4L engine

5. Install or connect the following:
  - Camshaft sprockets and upper timing chain. Tighten the sprocket bolts to 1230–130 ft. lbs. (167–177 Nm).
  - Valve cover. Tighten the bolts in sequence to 69–95 inch lbs. (8–11 Nm)
  - Spark plug wires
  - Air cleaner assembly
  - Negative battery cable.

3.3L Engines

1. Before servicing the vehicle, refer to the precautions in the beginning of this section.
2. Drain the cooling system.
3. Remove or disconnect the following:
  - Negative battery cable

  - Upper intake manifold
  - Valve covers

➡**Keep all valvetrain components in order for assembly.**

  - Rocker arm and shaft assemblies
  - Valve lifter guide and valve lifters. Attach a wire to the top of the lifters so that they will not drop from the lifter guide.
  - Radiator
  - Accessory drive belts
  - Front cover
  - Timing belt. Refer to the Timing Belt unit repair section.
  - Camshaft sprockets
  - Camshaft seals
  - Rear timing cover
  - Distributor
  - Cylinder head rear covers

  - Camshaft locating plates
  - Camshafts

To install:

4. Install or connect the following:
  - Camshafts
  - Camshaft locating plates. Tighten the bolts to 58–65 ft. lbs. (78–88 Nm).
  - Cylinder head rear covers
  - Distributor
  - Rear timing cover
  - Camshaft seals
  - Camshaft sprockets. Tighten the bolts to 58–65 ft. lbs. (78–88 Nm).
  - Timing belt
  - Front cover
  - Accessory drive belts
  - Radiator
  - Valve lifter guide and valve lifters
  - Rocker arm and shaft assemblies. Tighten the bolts to 13–16 ft. lbs. (18–22 Nm).
  - Valve covers
  - Upper intake manifold
  - Negative battery cable
5. Fill the cooling system.
6. Start the engine and check for leaks.

3.5L Engines

See the Cylinder Head Removal and Installation procedure.

Valve Lash

ADJUSTMENT

2.4L Engines

➡**Measure the valve clearance with the engine warm.**

1. Before servicing the vehicle, refer to the precautions in the beginning of this section.
2. Remove the valve cover.
3. Set the engine to the top of the compression stroke with the valves closed for the cylinder to be measured.
4. Check the valve clearance. The valve clearance specifications are as follows:
  - Intake: 0.012–0.015 in. (0.31–0.39mm)
  - Exhaust: 0.013–0.016 in. (0.33–0.41mm)
5. If adjustment is necessary, compress the valve spring with Tool **A** and insert Tool **B** to hold the valve in the open position as shown.
6. Replace the shims as necessary to achieve the correct valve clearance.
7. Repeat for each valve to be adjusted.

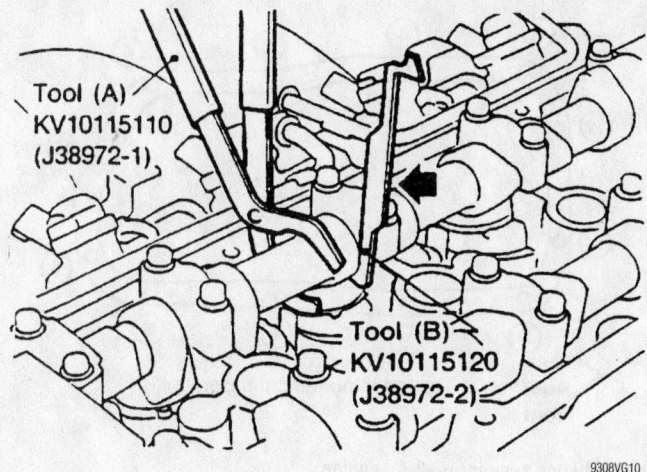

Valve adjustment tools (A) and (B)—2.4L engine

3.3L Engines

These engines are equipped with hydraulic valve lifters that do not require periodic adjustment.

3.5L Engines

➡**Adjust valve clearance while engine is cold.**

1. Turn crankshaft, to position cam lobe on camshaft of valve that must be adjusted upward.

2. Thoroughly wipe off engine oil around adjusting shim using a rag.

3. Using an extra-fine screwdriver, turn the round hole of the adjusting shim in the direction of the arrow.

4. Place Tool (A) around camshaft as shown in figure.

Before placing Tool (A), rotate notch toward center of cylinder head (See figure.), to simplify shim removal later.

※※ CAUTION

Be careful not to damage cam surface with Tool (A).

5. Rotate Tool (A) (See figure.) so that valve lifter is pushed down.

6. Place Tool (B) between camshaft and the edge of the valve lifter to retain valve lifter.

※※ CAUTION

Tool (B) must be placed as close to camshaft bracket as possible. Be careful not to damage cam surface with Tool (B).

7. Remove Tool (A).

8. Blow air into the hole to separate adjusting shim from valve lifter.

9. Remove adjusting shim using a small screwdriver and a magnetic finger.

10. Determine replacement adjusting shim size following formula. Using a micrometer determine thickness of removed shim. Calculate thickness of new adjusting shim so valve clearance comes within specified values.

- R = Thickness of removed shim
- N = Thickness of new shim
- M = Measured valve clearance
- Intake: $N = R + [M — 0.30 \text{ mm} (0.0118 \text{ in})]$
- Exhaust: $N = R + [M — 0.33 \text{ mm} (0.0130 \text{ in})]$

Shims are available in 64 sizes from 2.32 mm (0.0913 in) to 2.95 mm (0.1161 in), in steps of 0.01 mm (0.0004 in). Select new shim with thickness as close as possible to calculated value.

11. Install new shim using a suitable tool. Install with the surface on which the thickness is stamped facing down.

12. Place Tool (A) as mentioned in steps 2 and 3.

13. Remove Tool (B).

14. Remove Tool (A).

15. Recheck valve clearance.

Valve clearance (Cold)
- Intake: 0.010—0.013
- Exhaust: 0.011—0.015

Starter Motor

REMOVAL & INSTALLATION

1. Before servicing the vehicle, refer to the precautions in the beginning of this section.

2. Remove or disconnect the following:
- Negative battery cable
- Engine under cover
- Starter harness connectors
- Starter motor

To install:

3. Install or connect the following:
- Starter motor. Tighten the bolts to 22–27 ft. lbs. (30–36 Nm) on the 2.4L and 3.3L; 37–45 ft. lbs. (61-69NM) on the 3.5L.
- Starter harness connectors
- Engine under cover
- Negative battery cable

Oil Pan

REMOVAL & INSTALLATION

2.4L Engine

1. Before servicing the vehicle, refer to the precautions in the beginning of this section.

2. Drain the engine oil.

3. Remove or disconnect the following:

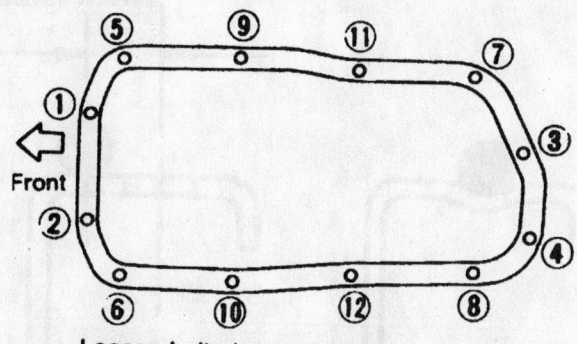

Loosen bolts in reverse order.

Oil pan bolt removal sequence—2.4L engine

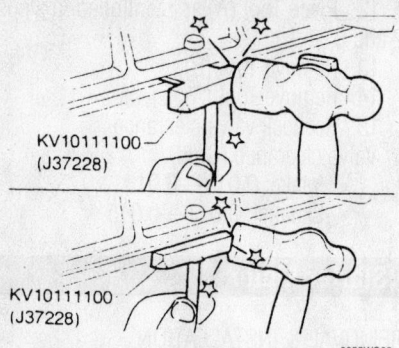

KV10111100
(J37228)

KV10111100
(J37228)

9355WG02

Insert the tool between the oil pan and the block and tap it with a hammer, while sliding it around the edge to loosen the oil pan

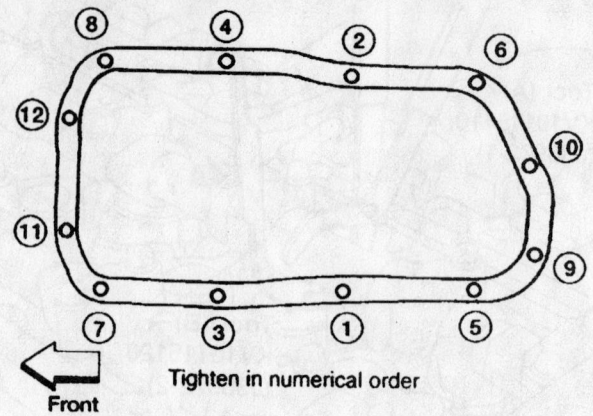

Tighten in numerical order

Front

7924VG40

Oil pan bolt installation sequence—2.4L engine

- Negative battery cable
- Engine under cover
- Stabilizer bar/Front suspension member
- Oil pan bolts, in the sequence shown

4. Remove the oil pan by inserting the tool shown between the cylinder block and oil pan. Do NOT damage the pan or block

mating surfaces! Slide the tool by tapping on the side of it with a hammer. Pull the oil pan front the front side.

To install:

5. Thoroughly clean all gasket mating surfaces.

6. Apply a continuous bead of sealant 0.0138–0.0177 in. (3.5–4.5mm) to the oil pan mating surface.

7. Install or connect the following:
- Oil pan. Tighten the bolts in sequence to 60–72 inch lbs. (7–8 Nm).
- Stabilizer bar. Tighten the bracket bolts to 38–45 ft. lbs. (51–61 Nm) and the link nuts to 12–16 ft. lbs. (16–22 Nm).
- Engine undercover
- Negative battery cable

➡ **Wait about 30 minutes after installation of the oil pan to allow the sealant to cure before adding oil.**

8. Fill the crankcase to the correct level.
9. Start the engine and check for leaks.

3.3L Engine

2WD MODELS

1. Before servicing the vehicle, refer to the precautions in the beginning of this section.
2. Drain the engine oil.
3. Remove or disconnect the following:
- Negative battery cable
- Engine under cover
- Stabilizer bar
- Front crossmember
- Starter motor
- Transmission mount
- Left and right motor mounts
- Power steering gear
4. Raise and support the engine for clearance.
5. Remove or disconnect the following:
- Oil pan bolts in the sequence
- Oil pan

To install:

6. Apply a continuous bead of sealant 0.138–0.177 in. (3.5–4.5mm) to the oil pan mating surface.
7. Install or connect the following:
- Oil pan. Tighten the bolts in reverse

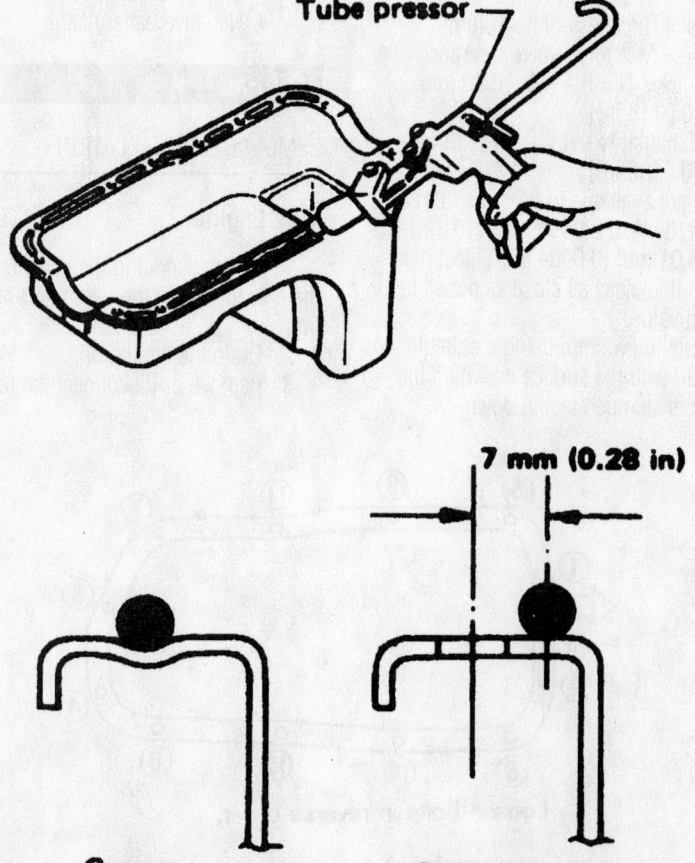

Tube pressor

7 mm (0.28 in)

Groove

Bolt hole

7924VG39

Oil pan sealant application—2.4L engine shown

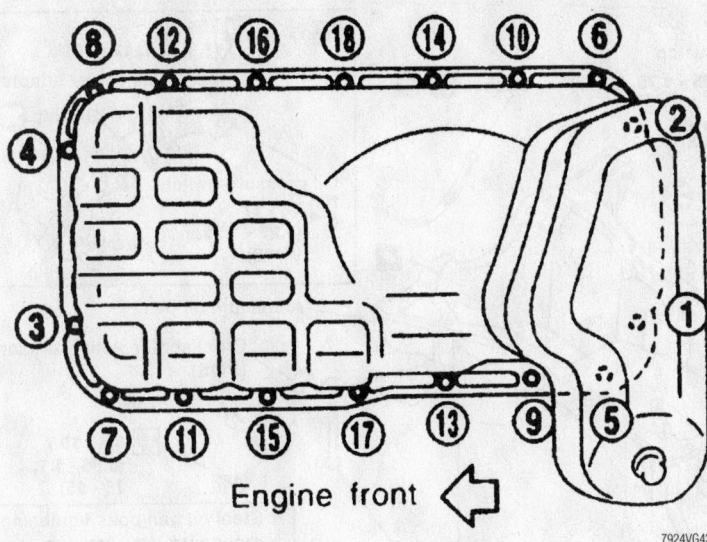

Oil pan bolt removal sequence—3.3L engine

of the removal sequence to 62 inch lbs. (7 Nm).
- Power steering gear
- Left and right motor mounts
- Transmission mount
- Starter motor
- Front crossmember
- Stabilizer bar
- Engine under cover
- Negative battery cable

➡ **Wait 30 minutes after installation of the oil pan to allow the sealant to cure before adding oil.**

8. Fill the crankcase to the correct level.
9. Start the engine and check for leaks.

4WD MODELS

1. Before servicing the vehicle, refer to the precautions in the beginning of this section.
2. Drain the engine oil.
3. Remove or disconnect the following:
- Negative battery cable
- Engine under cover
- Stabilizer bar brackets
- Front driveshaft
- Axle halfshafts
- Front suspension crossmember
- Front differential and mounting bracket
- Starter motor
- Transmission mount
- Left and right motor mounts
- Power steering gear
- Relay rod
4. Raise and support the engine for clearance.

5. Remove or disconnect the following:
- Oil pan bolts in the sequence
- Oil pan

To install:

6. Apply a continuous bead of sealant 0.138–0.177 in. (3.5–4.5mm) to the oil pan mating surface.
7. Install or connect the following:
- Oil pan. Tighten the bolts in reverse of the removal sequence to 62 inch lbs. (7 Nm).
- Relay rod
- Power steering gear
- Left and right motor mounts
- Transmission mount
- Starter motor
- Front differential and mounting bracket
- Front suspension crossmember
- Axle halfshafts
- Front driveshaft
- Stabilizer bar brackets
- Engine under cover
- Negative battery cable

➡ **Wait 30 minutes after installation of the oil pan to allow the sealant to cure before adding oil.**

8. Fill the crankcase to the correct level.
9. Start the engine and check for leaks.

3.5L Engines

1. Remove front RH and LH wheels.
2. Remove battery.
3. Remove oil level gauge.
4. Remove engine undercover.

5. Remove suspension member stay.
6. Drain engine coolant from radiator drain plug.
7. Disconnect A/T oil cooler hoses. (A/T)
8. Drain engine oil.
9. Remove the crankshaft position sensors (REF and POS).
10. Remove drive belts and idler pulley with bracket.
11. Remove power steering oil pump, then put it aside holding with a suitable wire.
12. Remove alternator.
13. Install engine slingers.
14. Remove front propeller shaft. (4WD)
15. Remove exhaust front tube heat insulators, then remove rear heat oxygen sensors.
16. Remove exhaust front tube from both sides.
17. Remove front final drive. (4WD)
18. Remove starter motor.
19. Disconnect oil pressure switch harness connector.
20. Loosen and disconnect the bolts fixing the steering column assembly lower joint and the power steering gear.
21. Set a suitable transmission jack under the front suspension member and hoist engine with engine slingers.
22. Remove front engine mounting nuts from both sides.
23. Remove front suspension member bolts.
24. Lower the transmission jack carefully to secure clearance between the oil pan and suspension member.
25. Remove A/T oil cooler tube. (A/T)
26. Remove water hose and tube. (A/T)
27. Remove the four engine-to-transmission bolts.
28. Remove aluminum oil pan bolts in numerical order.
29. Remove aluminum oil pan.
   a. Insert tool between aluminum oil pan and cylinder block.

➡ **Be careful not to damage aluminum mating surface. I Do not insert screwdriver, or oil pan flange will be deformed.**

   b. Slide tool by tapping its side with a hammer.
30. Remove O-rings from cylinder block and oil pump body.
31. Remove front cover gasket and rear oil seal retainer gasket.

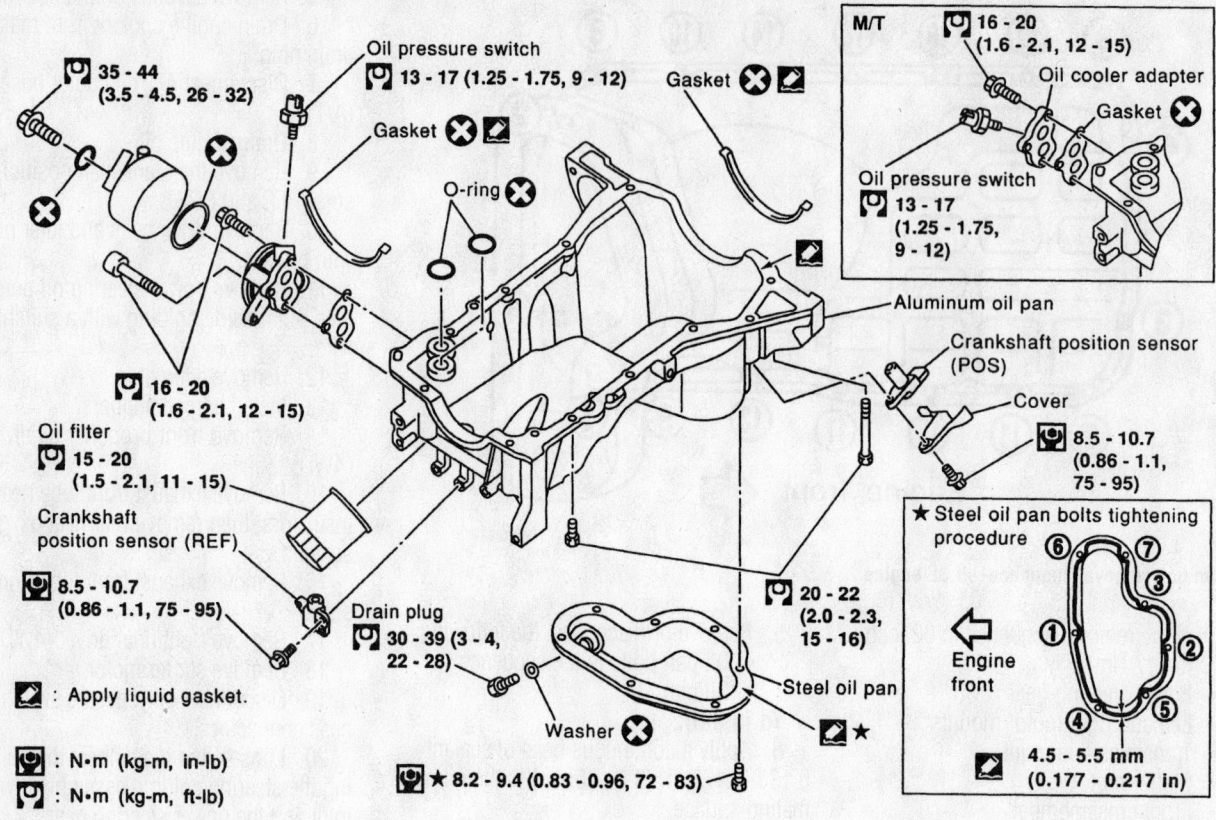

Oil pan exploded view—3.5L engine

To install:

32. Before installing oil pan, remove old liquid gasket from mating surface using a scraper. Also remove old liquid gasket from mating surface of cylinder block. Remove old liquid gasket from the bolt hole and thread.

33. Apply sealant to front cover gasket and rear oil seal retainer gasket.

34. Install front cover gasket and rear oil seal retainer gasket.

35. Apply a continuous bead of liquid gasket to mating surface of aluminum oil pan. Use RTV silicone sealant or equivalent.

36. Apply liquid gasket to inner sealing surface as shown in figure. Be sure liquid gasket is 4.0 to 5.0 mm (0.157 to 0.197 in) or 4.5 to 5.5 mm (0.177 to 0.217 in) wide. Attaching should be done within 5 minutes after coating.

37. Install O-rings, cylinder block and oil pump body.

38. Install aluminum oil pan. Tighten bolts in numerical order. Wait at least 30 minutes before refilling engine oil.

39. Install the four engine-to-transmission bolts.

40. Reinstall in the reverse order of removal.

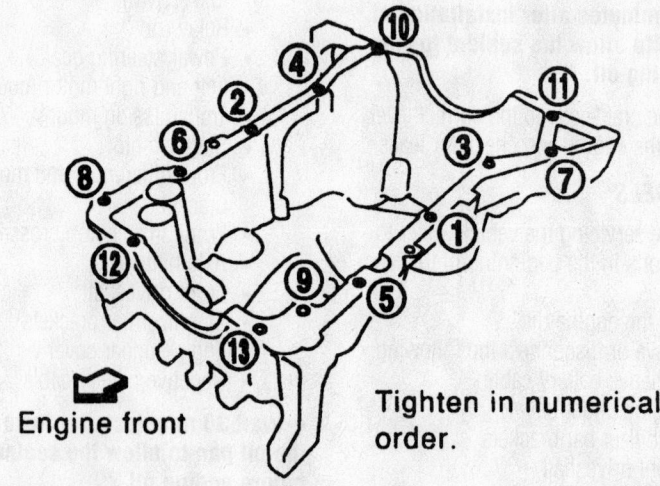

Oil pan bolt torque sequence—3.5L engine

Oil Pump

REMOVAL & INSTALLATION

2.4L Engine

1. Before servicing the vehicle, refer to the precautions in the beginning of this section.

2. Set the engine to Top Dead Center (TDC) of the compression stroke for the No. 1 cylinder.

3. Remove or disconnect the following:
- Negative battery cable
- Distributor cap
- Distributor
- Engine under cover
- Stabilizer bar
- Oil pump and drive spindle

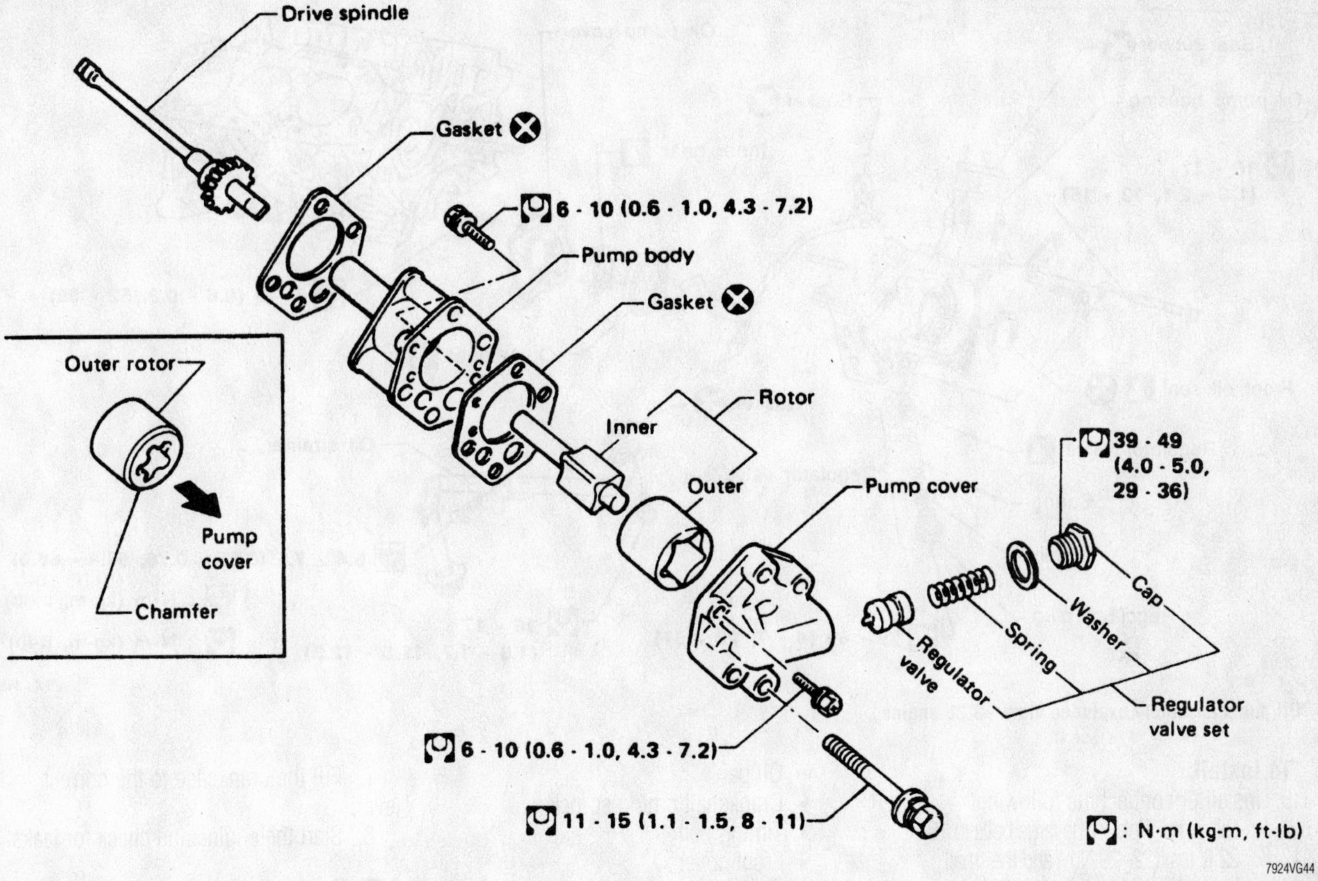

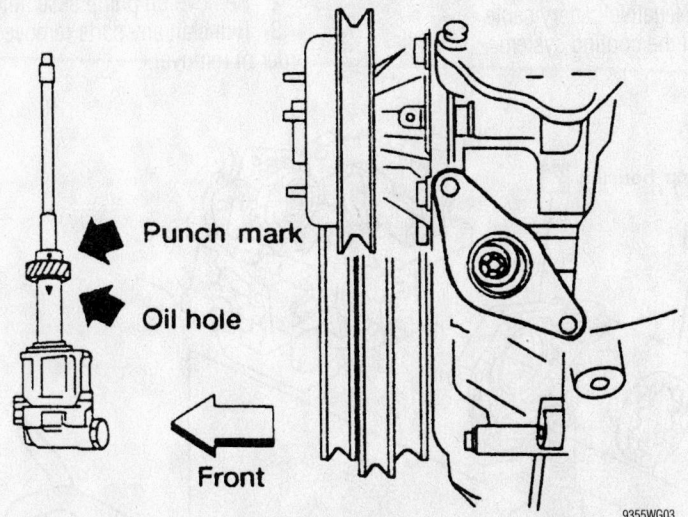

Drive spindle

Gasket ⊗

6 - 10 (0.6 - 1.0, 4.3 - 7.2)

Pump body

Gasket ⊗

Rotor

Inner

Outer

Pump cover

39 - 49
(4.0 - 5.0,
29 - 36)

Cap

Washer

Spring

Regulator
valve

Regulator
valve set

Outer rotor

Pump
cover

Chamfer

6 - 10 (0.6 - 1.0, 4.3 - 7.2)

11 - 15 (1.1 - 1.5, 8 - 11)

⟨⟩ : N·m (kg-m, ft-lb)

7924VG44

Exploded view of the oil pump assembly—2.4L engine

Punch mark

Oil hole

Front

9355WG03

Align the punch mark with the oil hole before oil pump installation—2.4L engine

To install:

4. Fill the pump housing with engine oil, then align the punch mark on the spindle and the hole in the oil pump as shown.

5. Install or connect the following:

- Oil pump and drive spindle. Tighten the mounting bolts to 96–132 inch lbs. (11–15 Nm)
- Stabilizer bar
- Engine under cover

- Distributor
- Distributor cap
- Negative battery cable

6. Start the engine and check for leaks.

7. Check the ignition timing and adjust as necessary.

3.3L Engine

1. Before servicing the vehicle, refer to the precautions in the beginning of this section.

2. Drain the engine oil.

3. Drain the cooling system.

4. Remove or disconnect the following:

- Negative battery cable
- Accessory drive belts
- Radiator hoses
- Crankshaft pulley
- Front cover
- Timing belt. Refer to the Timing Belt unit repair section.
- Crankshaft timing sprocket
- Oil pan
- Oil pump pickup tube
- Oil pump

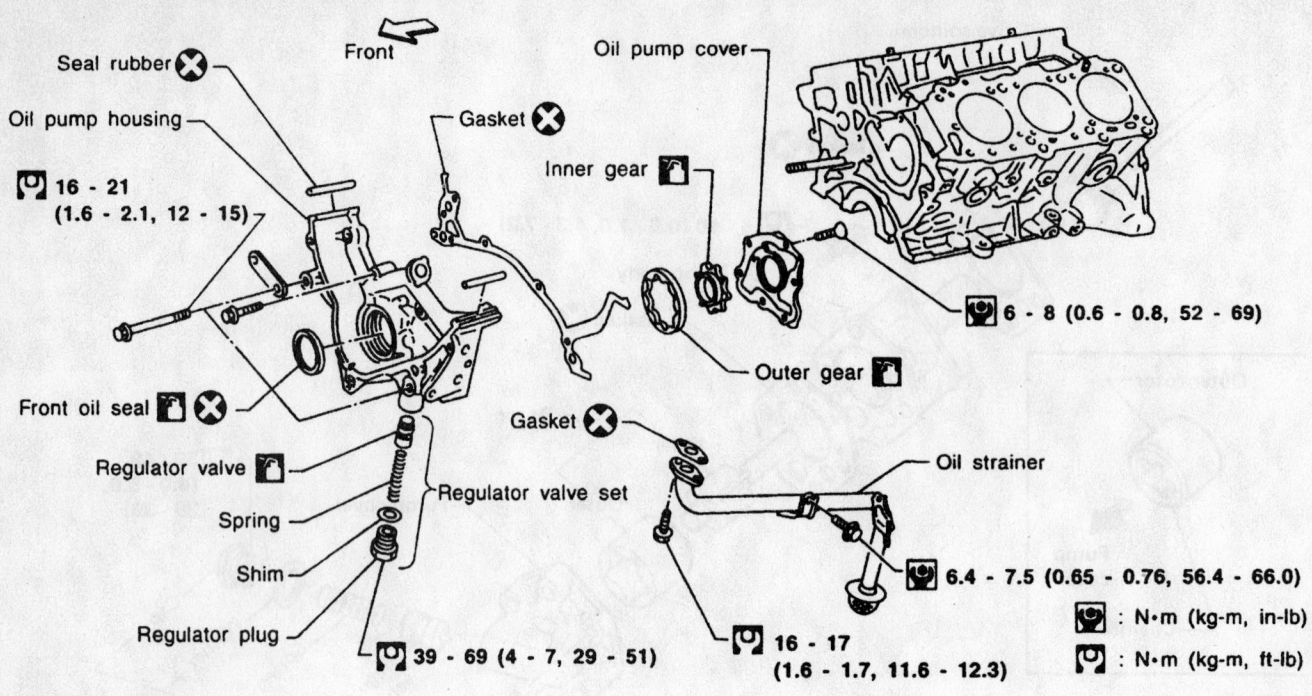

Oil pump assembly exploded view—3.3L engine

To install:

5. Install or connect the following:
  - Oil pump. Tighten the large bolts to 16–22 ft. lbs. (22–29 Nm) and the small bolts to 55–74 inch lbs. (6–8 Nm).
  - Oil pump pickup tube. Tighten the flange bolts to 12 ft. lbs. (16 Nm) and the bracket bolt to 55–74 inch lbs. (6–8 Nm).

  - Oil pan
  - Crankshaft timing sprocket
  - Timing belt
  - Front cover
  - Crankshaft pulley
  - Radiator hoses
  - Accessory drive belts
  - Negative battery cable
6. Fill the cooling system.

7. Fill the crankcase to the correct level.

8. Start the engine and check for leaks.

3.5L Engine

1. Remove timing chain.
2. Remove oil pump assembly.
3. Reinstall any parts removed in reverse order of removal.

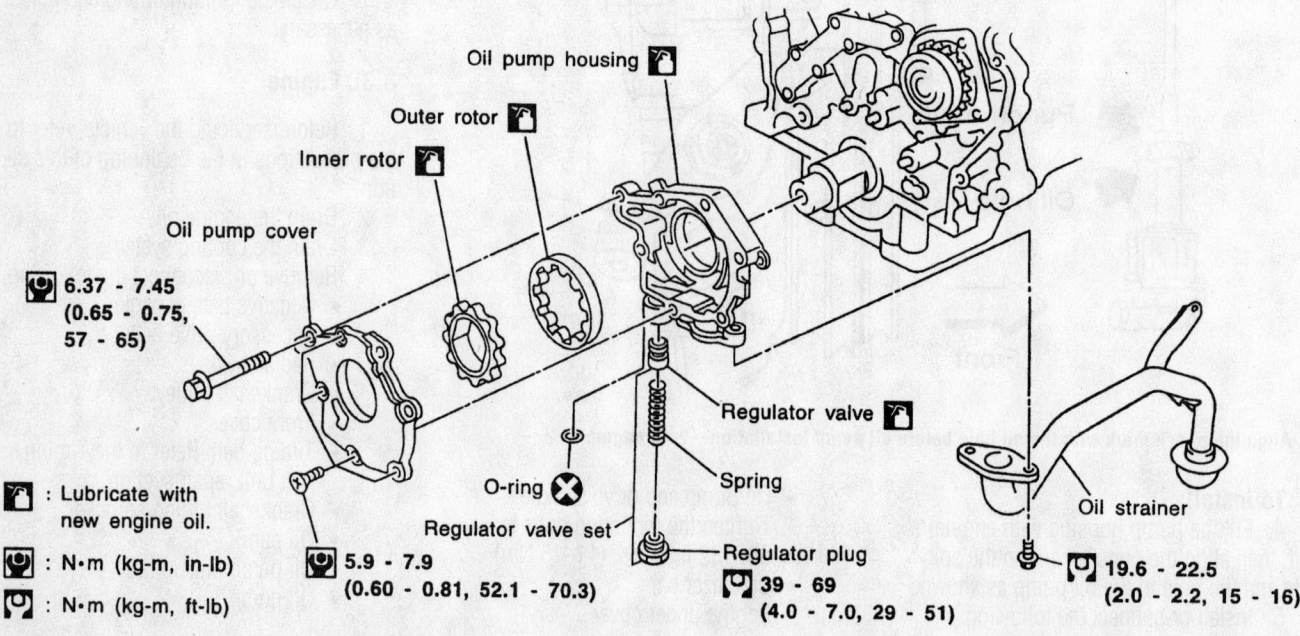

Oil pump assembly exploded view—3.5L engine

Rear Main Seal

REMOVAL & INSTALLATION

2.4L and 3.3L Engines

1. Before servicing the vehicle, refer to the precautions in the beginning of this section.
2. Remove or disconnect the following:
   - Transmission
   - Flywheel
   - Rear main seal

To install:

3. Install the seal so that it is flush with the retainer housing.
4. Install or connect the following:
   - Flywheel. Tighten the bolts to 61–69 ft. lbs. (83–93 Nm).
   - Transmission

3.5L Engine

1. Remove transmission.
2. Remove flywheel or drive plate.
3. Remove oil pan.
4. Remove rear oil seal retainer.
5. Remove old liquid gasket using scraper. Remove old liquid gasket from the bolt hole and thread.
6. Apply liquid gasket to rear oil seal retainer.

Timing Chain, Sprockets, Front Cover and Seal

REMOVAL & INSTALLATION

2.4L Engine

1. Before servicing the vehicle, refer to the precautions in the beginning of this section.
2. Drain the cooling system.
3. Drain the engine oil.
4. Set the engine to Top Dead Center (TDC) of the compression stroke for the No. 1 cylinder.
5. Remove or disconnect the following:
   - Negative battery cable
   - Air cleaner assembly
   - Spark plug wires
   - Cooling fan and shroud
   - Distributor
   - Valve cover
   - Accessory drive belts
   - Power steering pump and brackets
   - A/C compressor and bracket
   - Idler pulleys

- Water pump pulley
- Crankshaft pulley
- Front crankshaft seal
- Oil pump and drive spindle
- Oil pan
- Upper timing cover
- Lower timing cover
- Upper timing chain tensioner
- Upper timing chain and camshaft sprockets. Matchmark the timing chain to the sprockets.
- Lower timing chain tensioner.
- Lower timing chain and idler sprocket. Matchmark the timing chain to the sprockets.

To install:

6. Install or connect the following:
   - Lower timing chain and idler sprocket with the timing marks aligned as shown. tighten the idler sprocket bolt to 48–61 ft. lbs. (66–83 Nm).
   - Lower timing chain tensioner. Tighten the bolts to 56–66 inch lbs. (6.5–7.5 Nm).
   - Upper timing chain and camshaft sprockets with the timing marks aligned as shown. Tighten the camshaft sprocket bolts to 123–130 ft. lbs. (167–177 Nm).

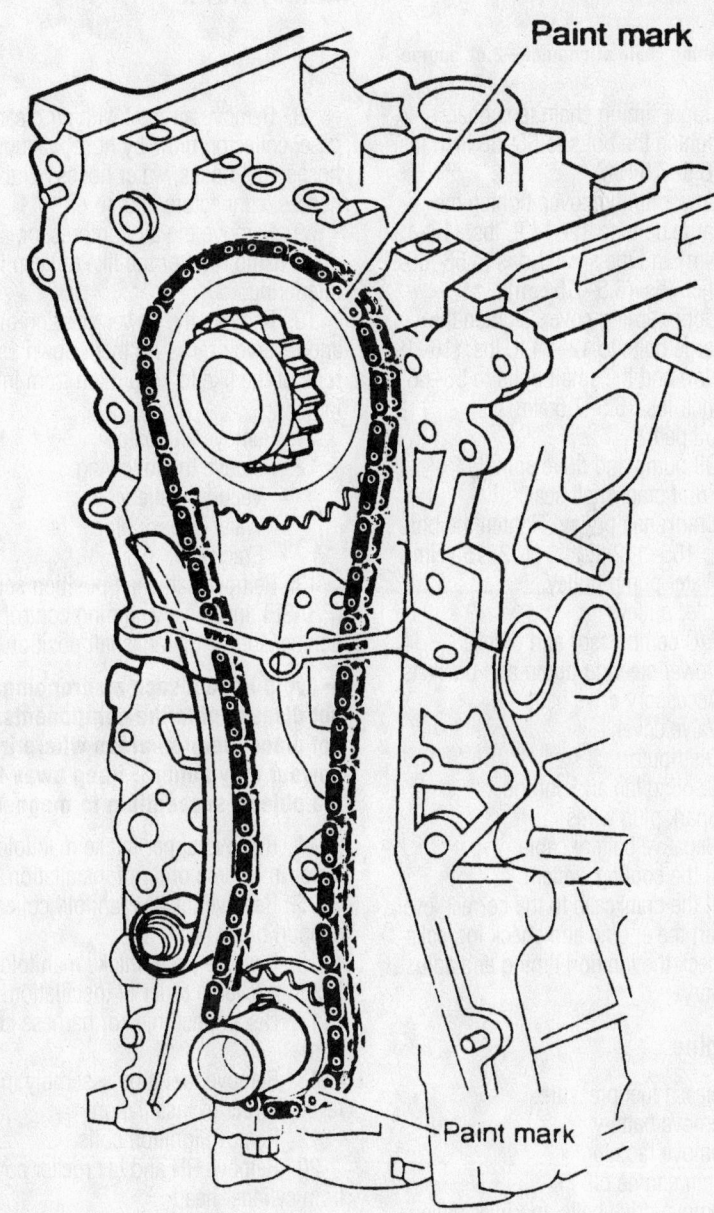

Lower timing chain alignment—2.4L engine

9355WG04

Timing belt service is covered in Section 3 of this manual

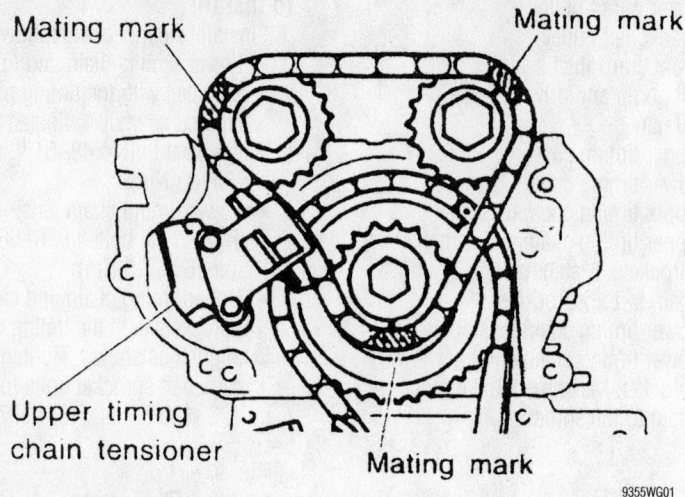

Upper timing chain alignment—2.4L engine

- Upper timing chain tensioner. tighten the bolts to 56–66 inch lbs. (6.5–7.5 Nm).
- Lower timing cover. tighten the large bolts to 12–14 ft. lbs. (16–19 Nm) and the small bolts to 56–66 inch lbs. (6.5–7.5 Nm).
- Upper timing cover. Tighten the large bolts to 12–14 ft. lbs. (16–19 Nm) and the small bolts to 56–66 inch lbs. (6.5–7.5 Nm).
- Oil pan
- Oil pump and drive spindle
- Front crankshaft seal
- Crankshaft pulley. Tighten the bolt to 105–112 ft. lbs. (142–152 Nm).
- Water pump pulley
- Idler pulleys
- A/C compressor and bracket
- Power steering pump and brackets
- Accessory drive belts
- Valve cover
- Distributor
- Cooling fan and shroud
- Spark plug wires
- Negative battery cable
7. Fill the cooling system.
8. Fill the crankcase to the correct level.
9. Start the engine and check for leaks.
10. Check the ignition timing and adjust, as necessary.

3.5L Engine

1. Release fuel pressure.
2. Remove battery.
3. Remove radiator.
4. Drain engine oil.
5. Remove drive belts and idler pulley with brackets.
6. Remove cooling fan with bracket.
7. Remove engine cover.

8. Remove air duct with air cleaner case, collector, blow-by hose, vacuum hoses, fuel hoses, water hoses, wires, harnesses, connectors and so on.
9. Remove the air compressor, and tie it down using rope or the like to keep it from interfering.
10. Remove the power steering oil pump and reservoir tank. Tie them down using rope or the like to keep them from interfering.
11. Remove alternator.
12. Remove the following.
    - Vacuum gallery
    - Water bypass pipe
    - Brackets
13. Remove camshaft position sensor (PHASE), intake valve timing control position sensors and crankshaft position sensor.

☞Avoid impact such as dropping. Do not disassemble the components. Do not place them on areas where iron powder may adhere. Keep away from the objects susceptible to magnetism.

14. Remove upper intake manifold collector in reverse order of installation.
15. Remove intake manifold collector support bolts.
16. Remove lower intake manifold collector in reverse order of installation.
17. Disconnect injector harness connectors.
18. Remove fuel tube assembly in reverse order of installation.
19. Remove ignition coils.
20. Remove RH and LH rocker covers from cylinder head.
21. Set No. 1 piston at TDC on the compression stroke by rotating crankshaft. Align pointer with TDC mark on crankshaft pulley. Check that intake and exhaust cam nose on

No. 1 cylinder are installed as shown left. If not, turn the crankshaft one revolution (360°) and align as above.
22. Remove starter motor, and set ring gear stopper using the mounting bolt hole. Be careful not to damage the signal plate teeth.
23. Loosen the crankshaft pulley bolt.
24. Remove crankshaft pulley with a suitable puller.
25. Remove aluminum oil pan.
26. Temporarily install the suspension member bolts and engine mounting nuts.
27. Remove intake valve timing control valve covers. Loosen bolts in numerical order as shown in the figure. In the cover, the shaft is engaged with the center hole of the intake cam sprocket. Remove it straight out until the engagement comes off.
28. Remove front timing chain case bolts. Loosen bolts in numerical order as shown in the figure.
29. Remove front timing chain case. Do not scratch sealing surfaces.
30. Remove internal chain guide.
31. Remove upper tension guide.
32. Remove timing chain tensioner and slack guide. Remove timing chain tensioner. (Push piston and insert a suitable pin into pinhole.)
33. Attach a suitable stopper pin to RH and LH camshaft chain tensioners.
34. Remove intake and exhaust camshaft sprocket bolts. I Apply paint to timing chain and camshaft sprockets for alignment during installation. Secure the hexagonal head of the camshaft using a spanner to loosen mounting bolts.
35. Remove primary and secondary timing chains along with the camshaft sprockets. Do not disassemble the intake camshaft sprocket. Avoid damaging the signal mark protrusion area at the front of the left bank intake camshaft sprocket. Keep it away from magnetized objects.
36. Remove lower chain guide.
37. Remove crankshaft sprocket.
38. Use a scraper to remove all traces of liquid gasket from front timing chain case. Remove old liquid gasket from the bolt hole and thread.
39. Use a scraper to remove all traces of liquid gasket from intake valve timing control valve cover.

To install:
40. Position crankshaft so that No. 1 piston is set at TDC on compression stroke.
41. Install crankshaft sprocket on crankshaft. Make sure that mating marks on crankshaft sprocket face front of engine.
42. Install lower chain guide on dowel

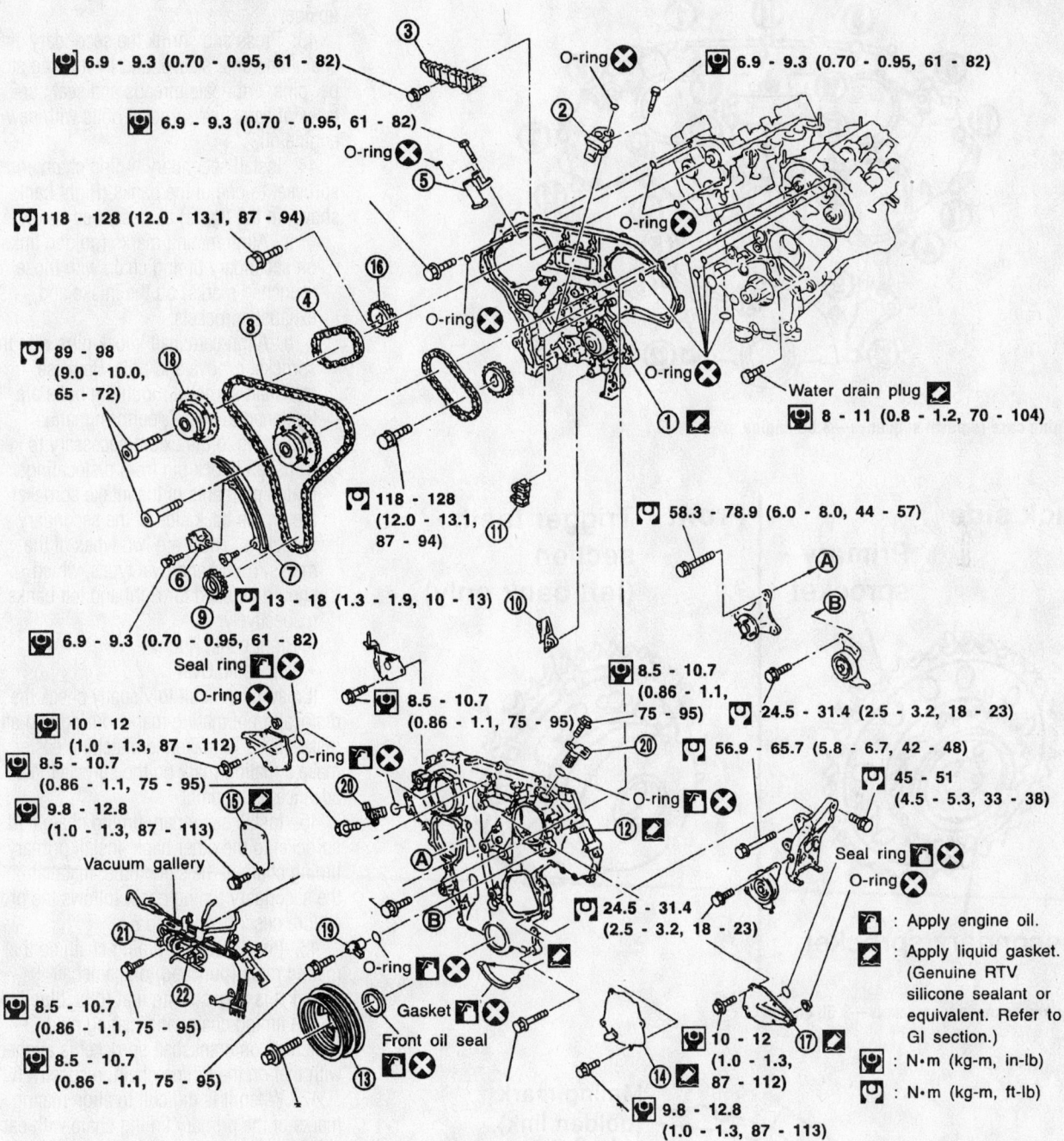

6.9 - 9.3 (0.70 - 0.95, 61 - 82)

O-ring ⊗

6.9 - 9.3 (0.70 - 0.95, 61 - 82)

6.9 - 9.3 (0.70 - 0.95, 61 - 82)

O-ring ⊗

O-ring ⊗

118 - 128 (12.0 - 13.1, 87 - 94)

O-ring ⊗

O-ring ⊗

Water drain plug

8 - 11 (0.8 - 1.2, 70 - 104)

89 - 98 (9.0 - 10.0, 65 - 72)

118 - 128 (12.0 - 13.1, 87 - 94)

13 - 18 (1.3 - 1.9, 10 - 13)

58.3 - 78.9 (6.0 - 8.0, 44 - 57)

6.9 - 9.3 (0.70 - 0.95, 61 - 82)

Seal ring ⊗

O-ring ⊗

10 - 12 (1.0 - 1.3, 87 - 112)

8.5 - 10.7 (0.86 - 1.1, 75 - 95)

9.8 - 12.8 (1.0 - 1.3, 87 - 113)

Vacuum gallery

O-ring ⊗

8.5 - 10.7 (0.86 - 1.1, 75 - 95)

8.5 - 10.7 (0.86 - 1.1, 75 - 95)

56.9 - 65.7 (5.8 - 6.7, 42 - 48)

O-ring ⊗

24.5 - 31.4 (2.5 - 3.2, 18 - 23)

45 - 51 (4.5 - 5.3, 33 - 38)

Seal ring ⊗

O-ring ⊗

8.5 - 10.7 (0.86 - 1.1, 75 - 95)

8.5 - 10.7 (0.86 - 1.1, 75 - 95)

O-ring ⊗

Gasket ⊗

Front oil seal

24.5 - 31.4 (2.5 - 3.2, 18 - 23)

10 - 12 (1.0 - 1.3, 87 - 112)

9.8 - 12.8 (1.0 - 1.3, 87 - 113)

: Apply engine oil.

: Apply liquid gasket. (Genuine RTV silicone sealant or equivalent. Refer to GI section.)

: N•m (kg-m, in-lb)

: N•m (kg-m, ft-lb)

1. Rear timing chain case
2. Left camshaft chain tensioner
3. Internal guide
4. Timing chain (Secondary)
5. Right camshaft chain tensioner
6. Timing chain tensioner
7. Slack guide
8. Timing chain (Primary)
9. Crankshaft sprocket

10. Lower tension guide
11. Upper tension guide
12. Front timing chain case
13. Crankshaft pulley
14. Water pump cover
15. Chain tensioner cover
16. Exhaust camshaft sprocket
17. Intake valve timing control valve cover

18. Intake camshaft sprocket
19. Camshaft position sensor (PHASE)
20. Intake valve timing control position sensor
21. Power valve actuator (A/T)
22. Swirl control valve control solenoid valve

9359VG30

Timing chain components—3.5L engine

Heater Core replacement is covered in Section 2 of this manual

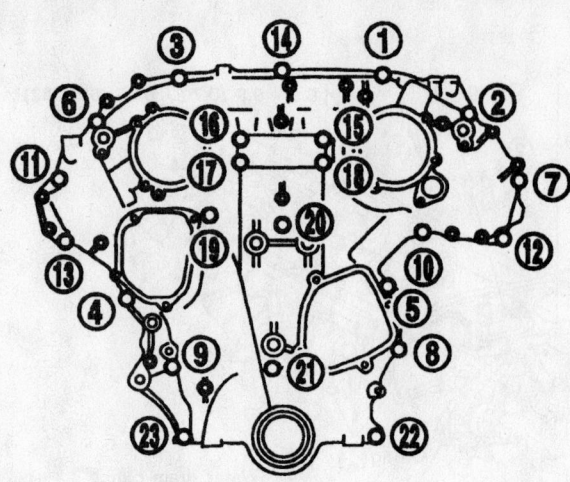

Rear timing case removal sequence—3.5L engine

9359VG22

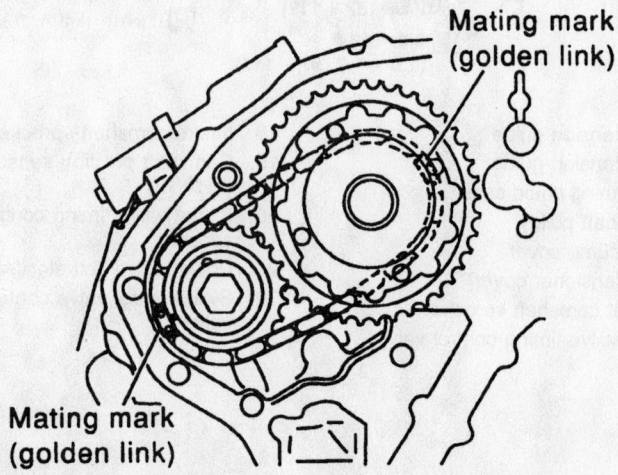

Primary and secondary sprockets—3.5L engine

9359VG23

Back side
Primary sprocket
Secondary sprocket

Front Trigger teeth section (left bank only)

Mating mark (golden link)

Mating mark (golden link)

Secondary timing chain installed—3.5L engine

9359VG24

pin, with front mark on the guide facing upside.

43. Press and shrink the secondary chain tensioner sleeve, and fix it using stopper pins. Lubricate threads and seat surfaces of camshaft sprocket bolts with new engine oil.

44. Install secondary timing chain and sprocket to one of the banks (Right bank shown in the figure) as described below.

a. Align mating marks (golden links) on secondary timing chain with those (punched marks) on the intake and exhaust sprockets.

b. Align camshaft knock pins with the sprocket groove and hole. Because camshaft sprocket mounting bolts are tightened in step 7, perform manual tightening to the extent necessary to keep camshaft knock pin from dislocating. Matching marks of the intake sprocket are on the back side of the secondary sprockets. There are two types of the marks; round and oval types, which should be used for right and left banks respectively.
• Right bank: Round
• Left bank: Oval

It may be difficult to visually check the dislocation of mating marks during and after installation. To make the matching easier, make a mating mark on the sprocket teeth in advance using paint.

45. Install secondary timing chain and sprocket to the other bank. Install primary timing chain at the same time. Installation of the secondary timing chain follows the procedure described in step 5.

46. Install primary timing chain so that mating mark (punched) on camshaft sprocket is aligned with that (dark blue link) on the timing chain, and mating mark (notched) on crankshaft sprocket is aligned with that on the timing chain, respectively.

47. When it is difficult to align mating marks of the primary timing chain with each sprocket, gradually turn the camshaft hexagonal head using a spanner so it is aligned with the mating mark.

48. During alignment, be careful to prevent dislocation of mating marks on the secondary timing chain.

49. After confirming the mating marks are aligned, tighten the camshaft sprocket mounting bolts. Secure the camshaft hexagonal head using a spanner to tighten mounting bolts.

50. Pull out the stopper pin from the secondary timing chain tensioner.

51. Install internal guide.

52. Install upper tension guide and slack guide.

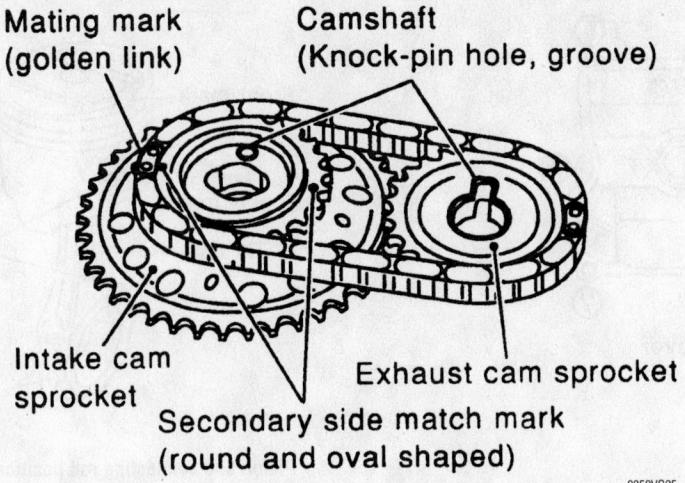

Intake sprocket mating marks—3.5L engine

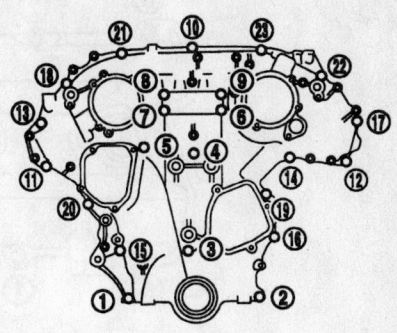

① - ② 8 mm dia. bolts
25.5 - 31.4 N·m
(2.6 - 3.2 kg-m, 18.8 - 23.1 ft-lb)
③ - ㉑ 6 mm dia. bolts
11.8 - 13.7 N·m
(1.2 - 1.4 kg-m, 8.7 - 10.1 ft-lb)

9359VG27

Rear timing case installation—3.5L engine

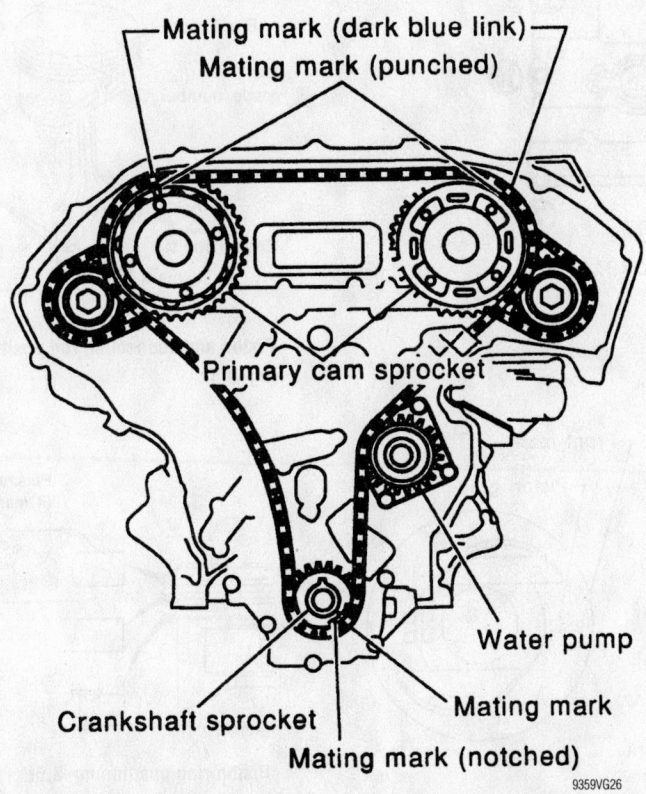

Primary timing chain installation—3.5L engine

53. Install timing chain tensioner, then remove the stopper pin. When installing the timing chain tensioner, engine oil should be applied to the oil hole and tensioner.

54. Install O-rings on rear timing chain case.

55. Apply liquid gasket to front timing chain case. Before installation, wipe off the protruding sealant.

56. Install rear case pin into dowel pin hole on front timing chain case.

57. Tighten bolts to the specified torque in order shown in the figure. Leave the bolts unattended for 30 minutes or more after tightening.

58. Install intake valve timing control valve cover.

a. Install O-rings at front timing chain case.

b. Install seal ring at intake valve timing control valve covers.

c. Apply liquid gasket to intake valve timing control valve covers. Use RTV silicone sealant or equivalent. I Being careful not to move the seal ring from the installation groove, align the dowel pins on the chain case with the holes to install the intake valve timing control valve cover. Tighten in numerical order as shown in the figure.

59. Install RH and LH rocker covers. Rocker cover tightening procedure:

• Tighten in numerical order as shown in the figure.
• Tighten bolts 1 to 10 in that order to 6.9 to 8.8 N·m (0.7 to 0.9 kg-m, 61 to 78 in-lb).
• Then tighten bolts 1 to 10 as indicated in figure to 6.9 to 8.8 N·m (0.7 to 0.9 kg-m, 61 to 78 in-lb).

60. Hang engine using the right and left side engine slingers with a suitable hoist.

61. Set a suitable transmission jack under the suspension member.

62. Remove right and left side engine mounting nuts.

63. Remove right and left side suspension member bolts.

64. Install aluminum oil pan.

65. Set ring gear stopper using the mounting bolt hole. Be careful not to damage the signal plate teeth.

66. Install crankshaft pulley to crankshaft. Align pointer with TDC mark on crankshaft pulley.

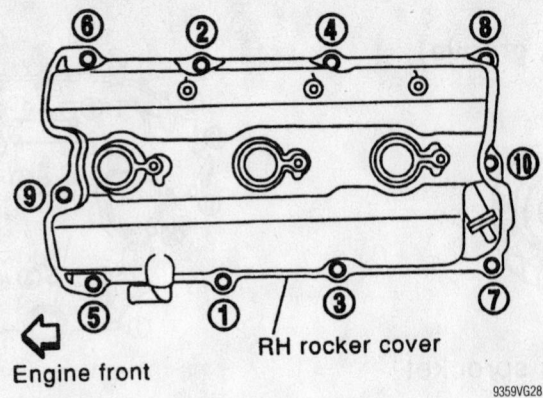

Right rocker cover installation—3.5L engine

9359VG28

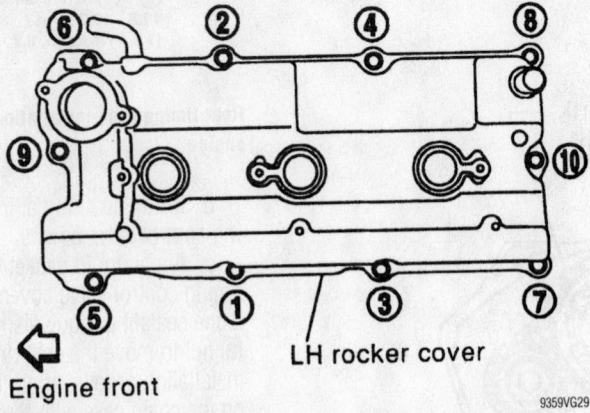

Left rocker cover installation—3.5L engine

9359VG29

67. Install crankshaft pulley bolt. Lubricate thread and seat surface of the bolt with new engine oil. Tighten to 39 to 49 N·m (4.0 to 5.0 kg-m, 29 to 36 ft-lb). Put a paint mark on the crankshaft pulley. Again tighten by turning 60° to 66°, about the angle from one hexagon bolt head corner to another.

68. Install camshaft position sensor (PHASE), crankshaft position sensors (REF)/(POS) and intake valve timing control position sensors.

69. Reinstall removed parts in the reverse order of removal. After starting engine, keep idling for three minutes. Then rev engine up to 3,000 rpm under no load to purge air from the high-pressure chamber of the chain tensioners. The engine may produce a rattling noise. This indicates that air still remains in the chamber and is not a matter of concern.

Piston and Ring

POSITIONING

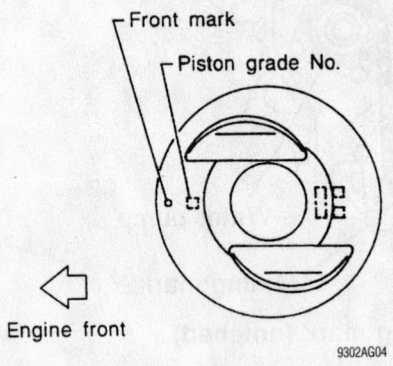

Piston ring positioning—3.3L engine

9302AG04

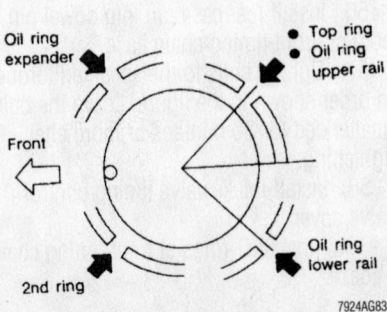

Piston ring end-gap spacing

7924AG83

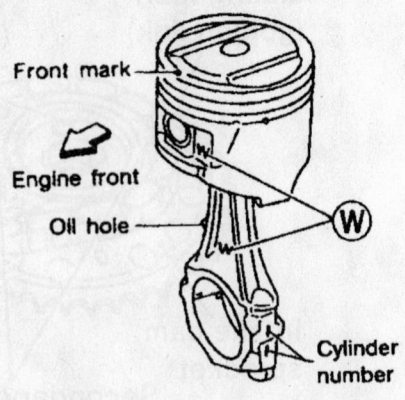

Piston and connecting rod positioning

7924AG84

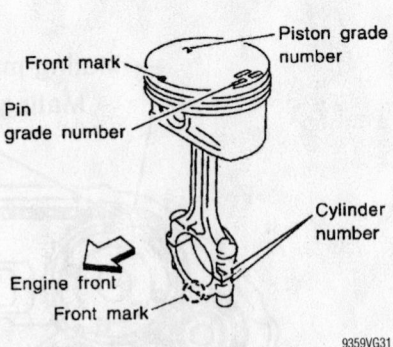

Piston and connecting rod positioning—3.5L

9359VG31

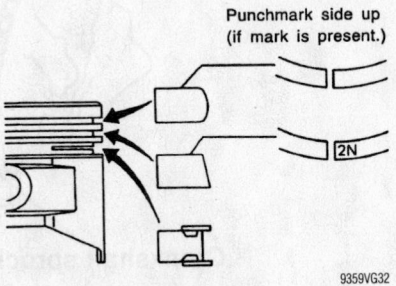

Piston ring positioning—3.5L

9359VG32

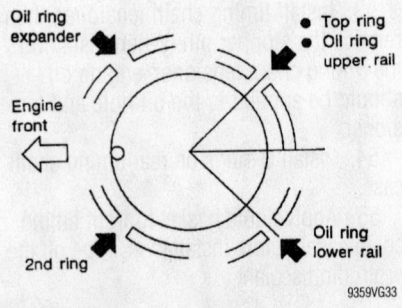

Piston ring positioning—3.5L

9359VG33

FUEL SYSTEM

Fuel System Service Precautions

Safety is the most important factor when performing not only fuel system maintenance but any type of maintenance. Failure to conduct maintenance and repairs in a safe manner may result in serious personal injury or death. Maintenance and testing of the vehicle's fuel system components can be accomplished safely and effectively by adhering to the following rules and guidelines.

• To avoid the possibility of fire and personal injury, always disconnect the negative battery cable unless the repair or test procedure requires that battery voltage be applied.

• Always relieve the fuel system pressure prior to disconnecting any fuel system component (injector, fuel rail, pressure regulator, etc.), fitting or fuel line connection. Exercise extreme caution whenever relieving fuel system pressure, to avoid exposing skin, face and eyes to fuel spray. Please be advised that fuel under pressure may penetrate the skin or any part of the body that it contacts.

• Always place a shop towel or cloth around the fitting or connection prior to loosening to absorb any excess fuel due to spillage. Ensure that all fuel spillage (should it occur) is quickly removed from engine surfaces. Ensure that all fuel soaked cloths or towels are deposited into a suitable waste container.

• Always keep a dry chemical (Class B) fire extinguisher near the work area.

• Do not allow fuel spray or fuel vapors to come into contact with a spark or open flame.

• Always use a back-up wrench when loosening and tightening fuel line connection fittings. This will prevent unnecessary stress and torsion to fuel line piping. Always follow the proper torque specifications.

• Always replace worn fuel fitting O-rings with new. Do not substitute fuel hose or equivalent, where fuel pipe is installed.

Fuel System Pressure

RELIEVING

1. Before servicing the vehicle, refer to the precautions in the beginning of this section.

2. Remove the fuel pump fuse from the panel.

3. Start the engine and allow it to run until it stalls. Crank the engine for a few seconds to relieve additional fuel pressure.

4. Disconnect the negative battery cable.

5. When repairs are complete, replace the fuel pump fuse and connect the negative battery cable.

Fuel Filter

REMOVAL & INSTALLATION

➡**The fuel filter is located under the vehicle near the fuel tank.**

1. Before servicing the vehicle, refer to the precautions in the beginning of this section.

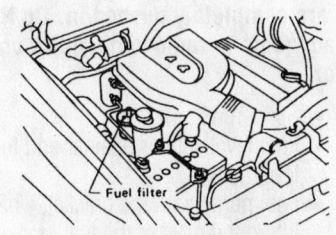

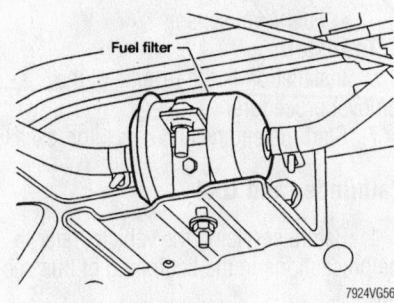

7924VG56

Typical fuel filter locations

2. Relieve the fuel system pressure.
3. Remove or disconnect the following:
• Fuel filter shield, if equipped
• Fuel lines
• Fuel filter from the bracket

To install:

4. Install or connect the following:
• Fuel filter to the bracket
• Fuel lines
• Fuel filter shield, if equipped

5. Start the engine and check for leaks.

Fuel Pump

REMOVAL & INSTALLATION

Xterra

2000 VEHICLES

1. Before servicing the vehicle, refer to the precautions in the beginning of this section.

2. Relieve the fuel system pressure.

3. Remove the rear seat and the access panel.

4. Drain the fuel tank.

5. Remove or disconnect the following:
• Negative battery cable
• Fuel pump module harness connectors
• Filler hose shield
• Fuel pressure and return lines
• Filler hose
• Vent hose
• Evaporative Emissions (EVAP) hose
• Fuel tank skid plate
• Fuel tank
• Fuel level sender
• Fuel pump

To install:

6. Install or connect the following:
• Fuel pump

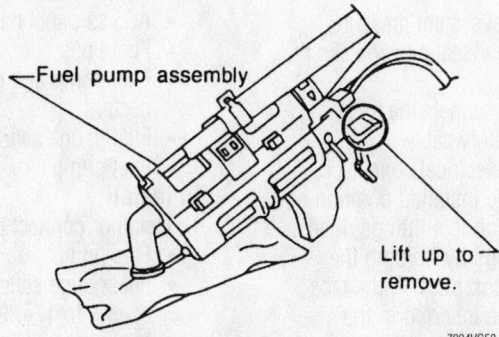

7924VG58

Remove the fuel pump with bracket while lifting the pawl of the pump bracket upward

For complete Engine Mechanical specifications, see Section 1 of this manual

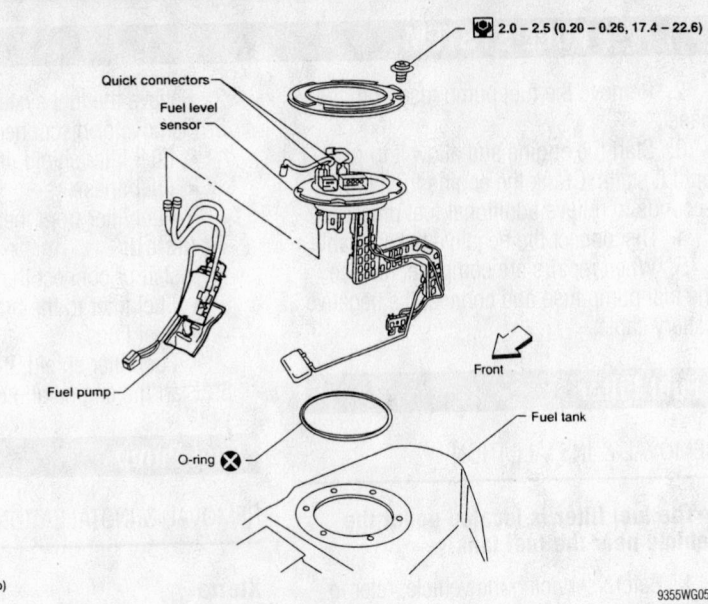

2.0 – 2.5 (0.20 – 0.26, 17.4 – 22.6)

Quick connectors

Fuel level sensor

Fuel pump

Front

Fuel tank

O-ring

: N·m (kg-m, in-lb)

9355WG05

Exploded view of the fuel pump—2001–02 Xterra shown

- Fuel level sender. Tighten the screws to 17–23 inch lbs. (2.0–2.5 Nm).
- Fuel tank. Tighten the bolts to 27–36 ft. lbs. (37–49 Nm).
- Fuel tank skid plate. Tighten the bolts to 27–36 ft. lbs. (37–49 Nm).
- EVAP hose
- Vent hose
- Filler hose
- Fuel pressure and return lines
- Filler hose shield
- Fuel pump module harness connectors
- Negative battery cable

7. Install the access panel and the rear seat.
8. Fill the fuel tank.
9. Start the engine and check for leaks.

2001–02 VEHICLES

1. Before servicing the vehicle, refer to the precautions in the beginning of this section.
2. Relieve the fuel system pressure.
3. Remove the rear seat cushion and the access panel.
4. Remove or disconnect the following:
   - Negative battery cable
   - Fuel pump electrical connectors
5. Matchmark the installed position of the fuel line quick connect fittings, then disconnect the fittings by holding the sides of the connector, push in the tabs and pull out the tube inserted in the retainer.

➡The tube can be removed when the tabs are completely pushed in. Do NOT use any tools to remove the quick connector.

- Six screws
- Fuel level sensor retainer and fuel level sensor
- Fuel pump with the bracket, while lifting the pawl of the fuel bracket upward
- Fuel level sensor

To install:
6. Installation is the reverse of the removal procedure.
7. Start the engine and check for leaks.

Pathfinder and QX4

1. Before servicing the vehicle, refer to the precautions in the beginning of this section.
2. Relieve the fuel system pressure.
3. Remove or disconnect the following:
   - Negative battery cable
   - Access panel behind the rear seat
   - Fuel lines
   - Fuel pump and gauge harness connectors
   - Fuel gauge sender
   - Fuel pump

To install:
4. Install or connect the following:
   - Fuel pump
   - Fuel gauge sender. Tighten the screws to 17–23 inch lbs. (2.0–2.5 Nm).

- Fuel pump and gauge harness connectors
- Fuel lines
- Access panel
- Negative battery cable

5. Start the engine and check for leaks.

Fuel Injectors

REMOVAL & INSTALLATION

2.4L Engine

1. Before servicing the vehicle, refer to the precautions in the beginning of this section.
2. Relieve the fuel system pressure.
3. Remove or disconnect the following:
   - Negative battery cable
   - Air cleaner assembly
   - Fuel lines
   - Fuel pressure regulator vacuum line
   - Fuel injector connectors
   - Fuel supply manifold, with the injectors attached
   - Fuel injector caps
   - Fuel injectors

To install:

➡Use new insulators and O-ring seals during reassembly.

4. Install or connect the following:
   - Fuel injectors
   - Fuel injector caps. Tighten the screws to 26–34 inch lbs. (3–4 Nm).
   - Fuel supply manifold with the injectors attached. Tighten the bolts to 96–132 inch lbs. (11–15 Nm).
   - Fuel injector connectors
   - Fuel pressure regulator vacuum line
   - Fuel lines
   - Air cleaner assembly
   - Negative battery cable

5. Start the engine and check for leaks.

3.3L Engine

1. Before servicing the vehicle, refer to the precautions in the beginning of this section.
2. Drain the cooling system.
3. Relieve the fuel system pressure.
4. Remove or disconnect the following:
   - Negative battery cable
   - Air intake duct

- Accelerator cable
- Cruise control cable
- Idle Air Control (IAC) valve connector
- Throttle Position (TP) sensor and switch connectors
- Ignition coil and power transistor connectors
- Exhaust Gas Recirculation (EGR) Solenoid valve connector
- EGR temperature sensor connector
- Radiator hoses
- Heater hoses
- Positive Crankcase Ventilation (PCV) valve and hose
- Evaporative Emissions (EVAP) canister vacuum and purge hoses
- Brake booster vacuum hose
- Fuel pressure regulator vacuum hose
- EGR tube
- Left bank injector connectors
- Thermal transmitter
- Upper intake manifold ground cable
- Breather pipe
- Upper intake manifold
- Fuel lines
- Right bank injector connectors
- Fuel supply manifold with the injectors attached
- Fuel injector caps
- Fuel injectors

To install:

➡**Use new insulators and O-ring seals for assembly.**

5. Install or connect the following:
- Fuel injectors
- Fuel injector caps. Tighten the

screws to 26–34 inch lbs. (3–4 Nm).
- Fuel supply manifold with the injectors attached. Tighten the bolts to 96–132 inch lbs. (11–15 Nm).
- Right bank injector connectors
- Fuel lines
- Upper intake manifold
- Breather pipe
- Upper intake manifold ground cable
- Thermal transmitter
- Left bank injector connectors
- EGR tube
- Fuel pressure regulator vacuum hose
- Brake booster vacuum hose
- EVAP canister vacuum and purge hoses
- PCV valve and hose
- Heater hoses
- Radiator hoses
- EGR temperature sensor connector
- EGR Solenoid valve connector
- Ignition coil and power transistor connectors
- TP sensor and switch connectors
- IAC valve connector
- Cruise control cable
- Accelerator cable
- Air intake duct
- Negative battery cable

6. Fill the cooling system.
7. Start the engine and check for leaks.

3.5L

1. Release fuel pressure to zero.
2. Remove intake manifold collector.
3. Remove fuel tube assemblies in

numerical sequence as shown in the figure at left.
4. Expand and remove clips securing fuel injectors.
5. Extract fuel injectors straight from fuel tubes.

➡**Be careful not to damage injector nozzles during removal. Do not bump or drop fuel injectors.**

6. Carefully install O-rings, including the one used with the pressure regulator. Lubricate O-rings with a smear of engine oil.

➡**Be careful not to damage O-rings with service tools, finger nails or clips. Do not expand or twist O-rings. Discard old clips; replace with new ones.**

7. Position clips in grooves on fuel injectors. Make sure that protrusions of fuel injectors are aligned with cutouts of clips after installation.
8. Align protrusions of fuel tubes with those of fuel injectors. Insert fuel injectors straight into fuel tubes.
9. After properly inserting fuel injectors, check to make sure that fuel tube protrusions are engaged with those of fuel injectors, and that flanges of fuel tubes are engaged with clips.
10. Tighten fuel tube assembly mounting nuts in numerical sequence (indicated in the figure at left) and in two stages. Tighten to:
   Step 1: 84–96 inch lbs.
   Step 2: 16–19 ft. lbs.
11. Install all parts removed in reverse order of removal.

DRIVE TRAIN

Manual Transmission

REMOVAL & INSTALLATION

2 Wheel Drive

1. Before servicing the vehicle, refer to the precautions in the beginning of this section.
2. Remove or disconnect the following:
- Negative battery cable
- Shift lever
- Crankshaft Position (CKP) sensor
- Clutch slave cylinder
- Vehicle Speed (VSS) sensor connector

- Back-up lamp switch connector
- Park/Neutral Position (PNP) switch connector
- Rear Heated Oxygen (HO$_2$S) sensor connector
- Starter motor
- Driveshaft
- Exhaust mounting bracket
- Transmission mount and crossmember. Support the transmission.
- Transmission flange bolts
- Transmission

➡**The transmission flange bolts vary in length. Note their positions for assembly.**

To install:

3. Apply sealant to the transmission flange, engine block and engine rear plate as shown.
4. Install or connect the following:
- Transmission. Tighten the large bolts to 29–36 ft. lbs. (39–49 Nm) and the small bolts to 12–16 ft. lbs. (16–22 Nm).
- Transmission mount and crossmember. Tighten the mount and crossmember fasteners to 30–38 ft. lbs. (41–52 Nm).
- Exhaust mounting bracket
- Driveshaft
- Starter motor

For Accessory Drive Belt illustrations, see Section 1 of this manual

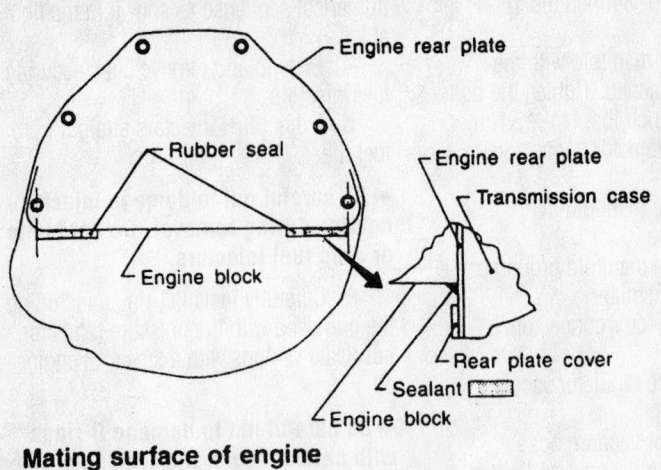

Mating surface of engine block and engine rear plate

Mating surface of engine rear plate and transmission case

45 (1.77)
45 (1.77)

Do not apply sealant in this range.

▨ : Apply recommended sealant (Nissan genuine part: KP510-00150) or equivalent.

▨ : Apply recommended sealant (Nissan genuine part: KP610-00250) or equivalent.

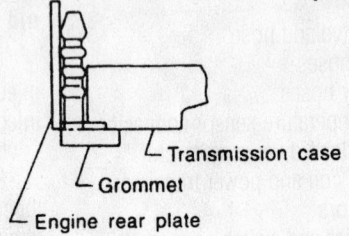

Unit: mm (in)

7924VG61

Apply sealant to the indicated areas between the engine block, transmission and engine rear plate—4 Wheel Drive shown

- HO2S sensor connector
- PNP switch connector
- Back-up lamp switch connector
- VSS sensor connector
- Clutch slave cylinder
- CKP sensor
- Shift lever
- Negative battery cable

4 Wheel Drive

XTERRA

1. Before servicing the vehicle, refer to the precautions in the beginning of this section.
2. Remove or disconnect the following:
   - Negative battery cable
   - Shift lever
   - Transfer case select lever
   - Crankshaft Position (CKP) sensor
   - Clutch slave cylinder
   - Vehicle Speed (VSS) sensor connector
   - Back-up lamp switch connector
   - Park/Neutral Position (PNP) switch connector
   - Rear Heated Oxygen (HO2S) sensor connector
   - Starter motor
   - Front and rear driveshafts
   - Exhaust front pipes
   - Exhaust center pipe
   - Torsion bars and mounts

- Rear torsion bar cross mount
- Transmission mount and crossmember. Support the transmission.
- Transmission flange bolts
- Transmission

➡The transmission flange bolts vary in length. Note their positions for assembly.

To install:

3. Apply sealant to the transmission flange, engine block, and engine rear plate as shown.
4. Install or connect the following:
   - Transmission. Tighten the large bolts to 29–36 ft. lbs. (39–49 Nm) and the small bolts to 22–29 ft. lbs. (29–39 Nm).
   - Transmission mount and crossmember. Tighten the mount and crossmember fasteners to 30–38 ft. lbs. (41–52 Nm).
   - Rear torsion bar cross mount
   - Torsion bars and mounts
   - Exhaust center pipe
   - Exhaust front pipes
   - Front and rear driveshafts
   - Starter motor
   - HO2S sensor connector
   - PNP switch connector
   - Back-up lamp switch connector
   - VSS sensor connector
   - Clutch slave cylinder

- CKP sensor
- Transfer case select lever
- Shift lever
- Negative battery cable

PATHFINDER

1. Before servicing the vehicle, refer to the precautions in the beginning of this section.
2. Remove or disconnect the following:
   - Negative battery cable
   - Shift lever
   - Transfer case select lever
   - Crankshaft Position (CKP) sensor
   - Clutch slave cylinder
   - Vehicle Speed (VSS) sensor connector
   - Back-up lamp switch connector
   - Park/Neutral Position (PNP) switch connector
   - Rear Heated Oxygen (HO2S) sensor connector
   - Starter motor
   - Front and rear driveshafts
   - Exhaust front pipes
   - Exhaust center pipe
   - Transmission mount and crossmember. Support the transmission.
   - Transmission flange bolts
   - Transmission

➡The transmission flange bolts vary in length. Note their positions for assembly.

To install:

3. Apply sealant to the transmission flange, engine block, and engine rear plate as shown.

4. Install or connect the following:
- Transmission. Tighten the large bolts to 29–36 ft. lbs. (39–49 Nm) and the small bolts to 22–29 ft. lbs. (29–39 Nm).

- Transmission mount and cross-member. Tighten the mount and crossmember fasteners to 30–38 ft. lbs. (41–52 Nm).
- Exhaust center pipe
- Exhaust front pipes
- Front and rear driveshafts
- Starter motor
- HO2S sensor connector

- PNP switch connector
- Back-up lamp switch connector
- VSS sensor connector
- Clutch slave cylinder
- CKP sensor
- Transfer case select lever
- Shift lever
- Negative battery cable

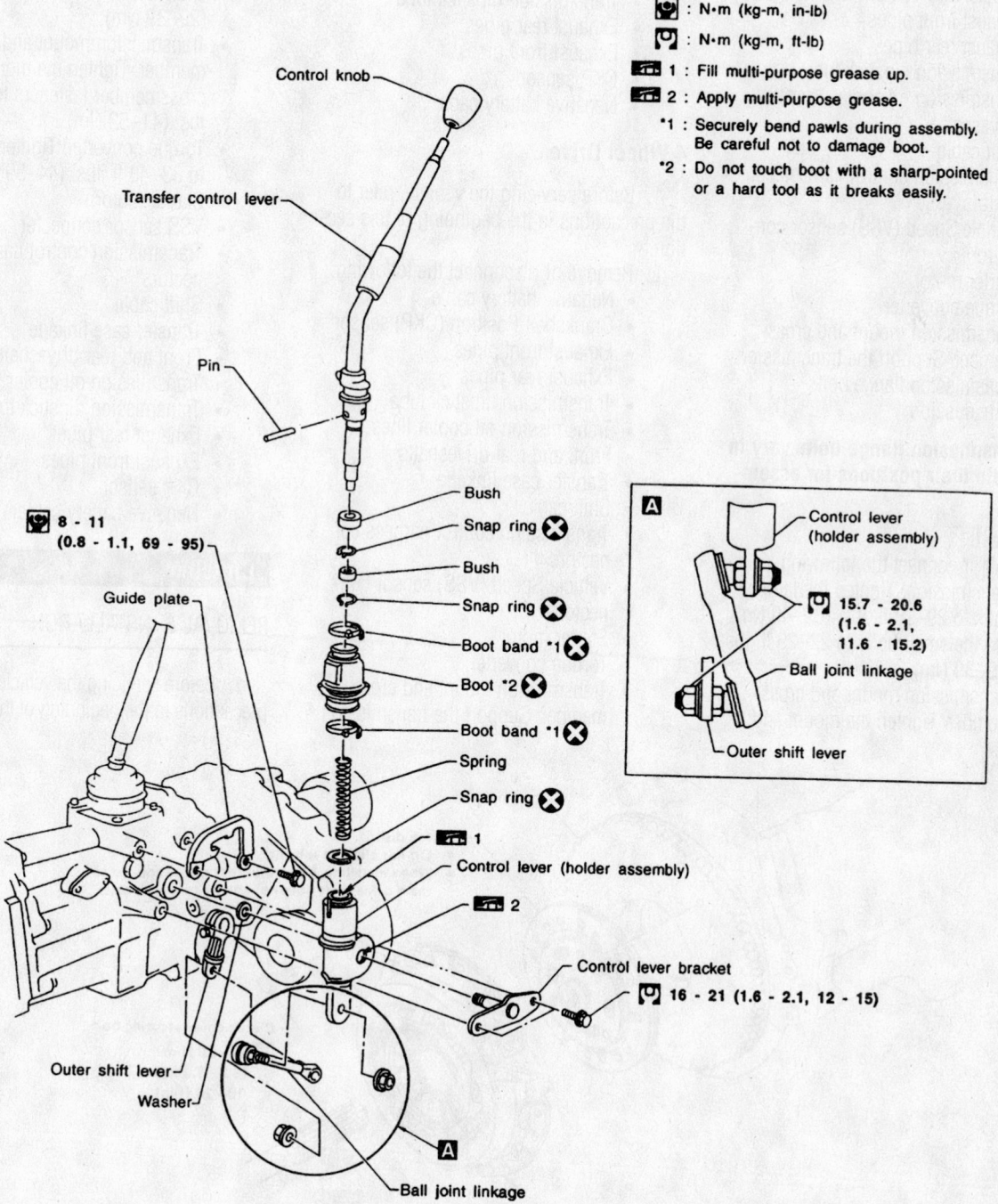

: N•m (kg-m, in-lb)

: N•m (kg-m, ft-lb)

1 : Fill multi-purpose grease up.

2 : Apply multi-purpose grease.

*1 : Securely bend pawls during assembly. Be careful not to damage boot.

*2 : Do not touch boot with a sharp-pointed or a hard tool as it breaks easily.

Control knob

Transfer control lever

Pin

8 - 11 (0.8 - 1.1, 69 - 95)

Guide plate

Bush

Snap ring ⊗

Bush

Snap ring ⊗

Boot band *1 ⊗

Boot *2 ⊗

Boot band *1 ⊗

Spring

Snap ring ⊗

1

Control lever (holder assembly)

2

Control lever bracket

16 - 21 (1.6 - 2.1, 12 - 15)

Outer shift lever

Washer

Ball joint linkage

A

Control lever (holder assembly)

15.7 - 20.6 (1.6 - 2.1, 11.6 - 15.2)

Ball joint linkage

Outer shift lever

7924VG60

Exploded view of the transfer case shifter lever and related components—Pathfinder 4WD

For Tire, Wheel and Ball Joint specifications, see Section 1 of this manual

Automatic Transmission

REMOVAL & INSTALLATION

2 Wheel Drive

1. Before servicing the vehicle, refer to the precautions in the beginning of this section.
2. Remove or disconnect the following:
   - Negative battery cable
   - Crankshaft Position (CKP) sensor
   - Exhaust front pipes
   - Exhaust rear pipes
   - Transmission dipstick tube
   - Transmission oil cooler lines
   - Driveshaft
   - Shift cable
   - Transmission control harness connectors
   - Vehicle Speed (VSS) sensor connector
   - Starter motor
   - Torque converter
   - Transmission mount and crossmember. Support the transmission.
   - Transmission flange bolts
   - Transmission

➡ **The transmission flange bolts vary in length. Note their positions for assembly.**

To install:

3. Install or connect the following:
   - Transmission. Tighten the large bolts to 29–36 ft. lbs. (39–49 Nm) and the small bolts to 22–29 ft. lbs. (29–39 Nm).
   - Transmission mount and crossmember. Tighten the mount and crossmember fasteners to 30–38 ft. lbs. (41–52 Nm).
   - Torque converter. Tighten the bolts to 33–43 ft. lbs. (44–59 Nm).
   - Starter motor
   - VSS sensor connector
   - Transmission control harness connectors
   - Shift cable
   - Driveshaft
   - Transmission oil cooler lines
   - Transmission dipstick tube
   - Exhaust rear pipes
   - Exhaust front pipes
   - CKP sensor
   - Negative battery cable

4 Wheel Drive

1. Before servicing the vehicle, refer to the precautions in the beginning of this section.
2. Remove or disconnect the following:
   - Negative battery cable
   - Crankshaft Position (CKP) sensor
   - Exhaust front pipes
   - Exhaust rear pipes
   - Transmission dipstick tube
   - Transmission oil cooler lines
   - Front and rear driveshafts
   - Transfer case linkage
   - Shift cable
   - Transmission control harness connectors
   - Vehicle Speed (VSS) sensor connector
   - Starter motor
   - Torque converter
   - Transmission mount and crossmember. Support the transmission.
   - Transmission flange bolts
   - Transmission

➡ **The transmission flange bolts vary in length. Note their positions for assembly.**

To install:

3. Install or connect the following:
4. Install or connect the following:
   - Transmission. Tighten the large bolts to 29–36 ft. lbs. (39–49 Nm) and the small bolts to 22–29 ft. lbs. (29–39 Nm).
   - Transmission mount and crossmember. Tighten the mount and crossmember fasteners to 30–38 ft. lbs. (41–52 Nm).
   - Torque converter. Tighten the bolts to 33–43 ft. lbs. (44–59 Nm).
   - Starter motor
   - VSS sensor connector
   - Transmission control harness connectors
   - Shift cable
   - Transfer case linkage
   - Front and rear driveshafts
   - Transmission oil cooler lines
   - Transmission dipstick tube
   - Exhaust rear pipes
   - Exhaust front pipes
   - CKP sensor
   - Negative battery cable

Clutch

REMOVAL & INSTALLATION

1. Before servicing the vehicle, refer to the precautions in the beginning of this section.

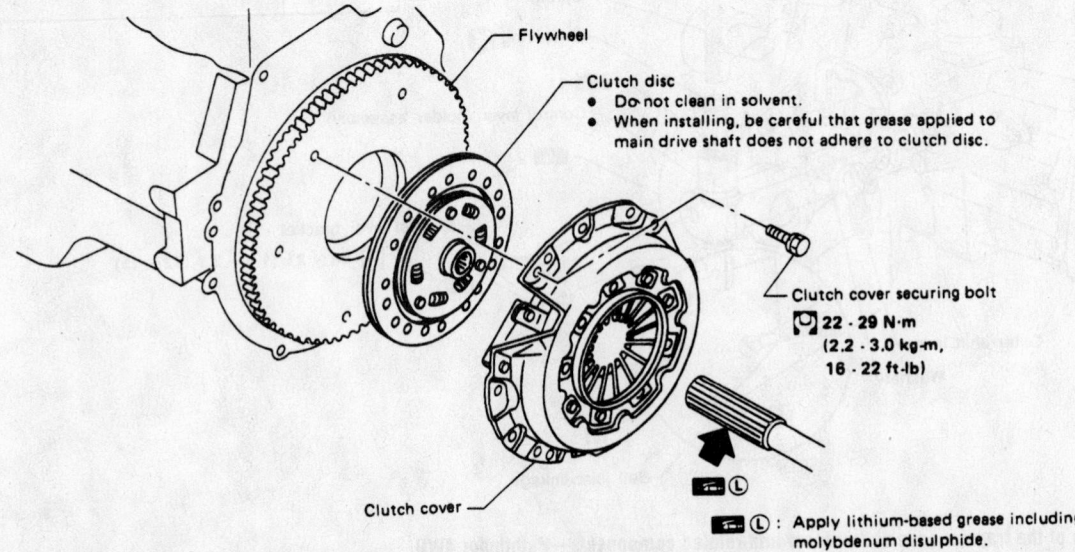

Flywheel

Clutch disc
- Do not clean in solvent.
- When installing, be careful that grease applied to main drive shaft does not adhere to clutch disc.

Clutch cover securing bolt
22 - 29 N·m
(2.2 - 3.0 kg-m,
16 - 22 ft-lb)

Clutch cover

⊡ Ⓛ : Apply lithium-based grease including molybdenum disulphide.

7924VG63

Exploded view of the pressure plate and clutch disc and related components—all models

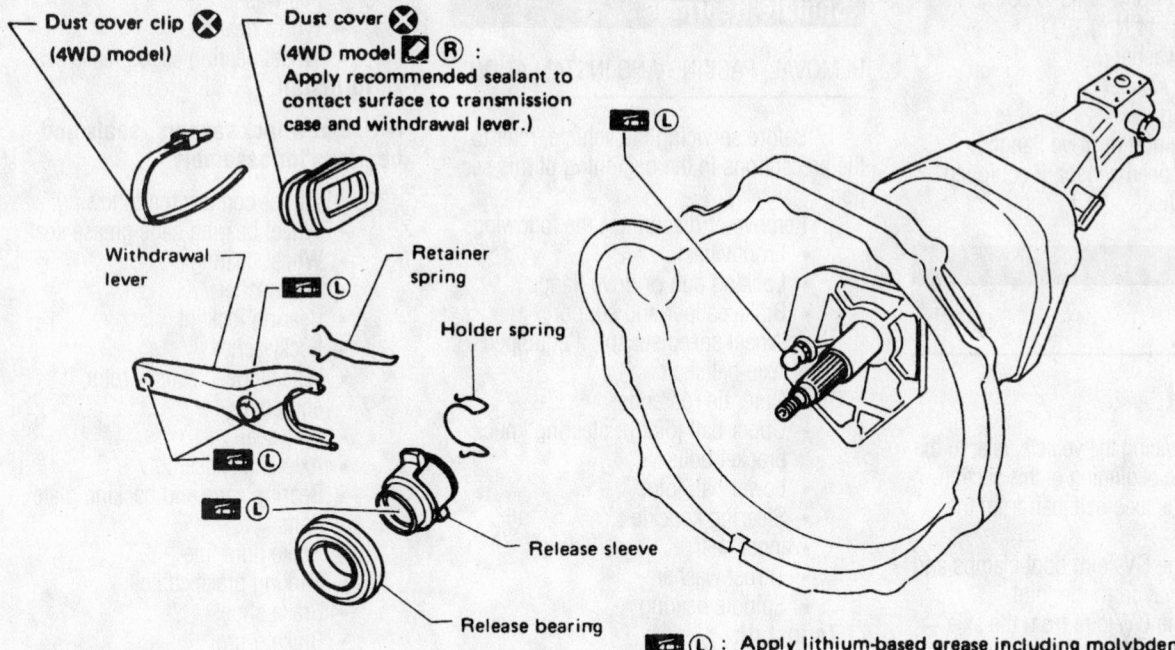

Dust cover clip ⊗ (4WD model)

Dust cover ⊗ (4WD model ◨ Ⓡ : Apply recommended sealant to contact surface to transmission case and withdrawal lever.)

Withdrawal lever

Retainer spring

Holder spring

Release sleeve

Release bearing

▣ Ⓛ : Apply lithium-based grease including molybdenum disulphide

7924VG64

Clutch release mechanism exploded view—all models

2. Remove or disconnect the following:
   • Negative battery cable
   • Transmission
   • Pressure plate. Loosen the bolts evenly in ½ turn steps.
   • Clutch disc

To install:

3. Install or connect the following:
   • Clutch disc and pressure plate. Tighten the pressure plate bolts evenly in ½ turns to 16–22 ft. lbs. (22–29 Nm).
   • Transmission
   • Negative battery cable

Hydraulic Clutch System

BLEEDING

1. Before servicing the vehicle, refer to the precautions in the beginning of this section.

2. Fill the clutch master cylinder reservoir with fresh clean brake fluid.

3. Connect a clear plastic hose to the air bleeder.

4. Have an assistant pump the clutch pedal slowly several times and hold it depressed.

5. Open the slave cylinder bleeder screw and allow air to escape.

6. Close the bleeder screw before releasing the clutch pedal.

7. Repeat until all air is purged from the clutch hydraulic system.

8. Refill the reservoir to the full mark.

Transfer Case Assembly

REMOVAL & INSTALLATION

1. Before servicing the vehicle, refer to the precautions in the beginning of this section.

2. Remove or disconnect the following:
   • Negative battery cable
   • Front and rear driveshafts
   • Torsion bars and mounts
   • Rear torsion bar crossmember
   • Exhaust front pipes
   • Exhaust rear pipes
   • Vehicle Speed (VSS) sensor connector
   • Transfer case shift linkage
   • Transfer case neutral switch connector
   • 4 wheel drive switch connector
   • Vent hose
   • Transfer case flange bolts
   • Transfer case

To install:

3. Install or connect the following:
   • Transfer case. Tighten the flange bolts to 23–30 ft. lbs. (31–41 Nm).
   • Vent hose
   • 4 wheel drive switch connector

• Transfer case neutral switch connector
• Transfer case shift linkage
• VSS sensor connector
• Exhaust rear pipes
• Exhaust front pipes
• Rear torsion bar crossmember
• Torsion bars and mounts
• Front and rear driveshafts
• Negative battery cable

Halfshaft

REMOVAL & INSTALLATION

1. Before servicing the vehicle, refer to the precautions in the beginning of this section.

2. Remove or disconnect the following:
   • Front wheel
   • Wheel speed sensor, if equipped
   • Locking hub or drive flange
   • Snapring
   • Spindle washer
   • Thrust washer
   • Inner CV-joint bolts
   • Axle halfshaft. Separate the stub shaft from the spindle by tapping with a plastic hammer.

To install:

3. Install or connect the following:
   • Axle halfshaft. Guide the stub shaft into the spindle and tighten the

For Wheel Alignment specifications, see Section 1 of this manual

inner CV-joint bolts to 25–33 ft. lbs. (34–44 Nm).
- Thrust washer
- Spindle washer
- Snapring
- Locking hub or drive flange
- Wheel speed sensor, if equipped
- Front wheel

CV-Joints

OVERHAUL

Outer CV-Joint

1. Before servicing the vehicle, refer to the precautions in the beginning of this section.
2. Remove the axle halfshaft from the vehicle.
3. Remove the CV-joint boot clamps and push the boot away from the joint.
4. Remove the CV-joint from the axle shaft by tapping it with a brass hammer.

To install:

➡**Use new circlips and boot clamps for assembly.**

5. Install the CV-joint to the axle shaft by tapping it with a brass hammer.
6. Pack the joint with grease.
7. Install the boot clamps.
8. Install the axle halfshaft to the vehicle.

Inner Tri-Pot Joint

1. Before servicing the vehicle, refer to the precautions in the beginning of this section.
2. Remove the axle halfshaft from the vehicle.
3. Remove the plug seal by tapping around the joint housing flange with a brass hammer.
4. Remove or disconnect the following:
- CV-joint boot clamps
- Snapring
- Spider assembly
- CV-joint housing
- CV-joint boot

To install:

➡**Use new snaprings and plug seals for assembly.**

5. Install or connect the following:
- CV-joint boot
- CV-joint housing
- Spider assembly
- Snapring. Pack the joint with grease.
- CV-joint boot clamps
- Plug seal
6. Install the axle halfshaft to the vehicle.

Spindle Bearings

REMOVAL, PACKING AND INSTALLATION

1. Before servicing the vehicle, refer to the precautions in the beginning of this section.
2. Remove or disconnect the following:
- Front wheel
- Locking hub or drive flange
- Brake caliper and support
- Wheel speed sensor, if equipped
- Axle halfshaft
- Outer tie rod ends
- Upper ball joint or steering knuckle bracket bolts
- Lower ball joint
- Steering knuckle
- Inner seal
- Thrust washer
- Spindle bearing

To install:

3. Install or connect the following:
- Spindle bearing. Coat the bearing with multi-purpose grease.
- Thrust washer
- Inner seal
- Steering knuckle
- Lower ball joint
- Upper ball joint or steering knuckle bracket bolts
- Outer tie rod ends
- Axle halfshaft
- Wheel speed sensor, if equipped
- Brake caliper and support
- Locking hub or drive flange
- Front wheel

Axle Shaft, Bearing and Seal

REMOVAL & INSTALLATION

1. Before servicing the vehicle, refer to the precautions in the beginning of this section.
2. Remove or disconnect the following:
- Rear wheel
- Wheel speed sensor, if equipped
- Brake drum
- Brake shoes
- Parking brake cable
- Brake fluid line
- Bearing cage and backing plate bolts
- Axle shaft assembly
- Axle seal
- Wheel speed sensor rotor, if equipped
- Lockwasher
- Bearing locknut
- Flat washer
- Wheel bearing
- Wheel bearing cage grease seal

To install:

➡**Use new lockwashers, seals and bearings for assembly.**

3. Install or connect the following:
- Wheel bearing cage grease seal
- Wheel bearing
- Flat washer
- Bearing locknut
- Lockwasher
- Wheel speed sensor rotor, if equipped
- Axle seal
- Axle shaft assembly
- Bearing cage and backing plate bolts
- Brake fluid line
- Parking brake cable
- Brake shoes
- Brake drum
- Wheel speed sensor, if equipped
- Rear wheel
4. Bleed the rear brakes and check the rear axle lubricant level.

Pinion Seal

Front

1. Before servicing the vehicle, refer to the precautions in the beginning of this section.
2. Remove or disconnect the following:
- Driveshaft
- Front wheels
- Front brake calipers

➡**The front brake calipers must be removed so that there is no additional drag when measuring pinion bearing preload.**

3. Use an inch lb. torque wrench and measure the amount of torque required to maintain pinion rotation through several revolutions.
4. Remove or disconnect the following:
- Pinion flange
- Oil seal

To install:

5. Install or connect the following:
- Pinion seal
- Pinion flange
6. Rotate the pinion flange occasionally while tightening the flange nut to make sure the pinion bearings seat correctly.
7. Take frequent bearing preload torque readings. Tighten the flange nut to achieve the preload torque readings originally recorded. Do not exceed 137–217 ft. lbs.

(186–294 Nm) torque when tightening the pinion flange nut.

❊❊ CAUTION

If the bearing preload can not be achieved at the specified torque, remove the pinion bearing and install a new adjustment spacer.

8. Install or connect the following:
   • Front brake calipers
   • Front wheels
   • Driveshaft. Tighten the fasteners to 29–33 ft. lbs. (39–44 Nm).
9. Fill the differential with gear lubricant and check for leaks.

Rear

2 WHEEL DRIVE

1. Before servicing the vehicle, refer to the precautions in the beginning of this section.
2. Remove or disconnect the following:
   • Driveshaft
   • Rear wheels
   • Brake drums

➡The rear brake drums must be removed so that there is no additional drag when measuring pinion bearing preload.

3. Use an inch lb. torque wrench and measure the amount of torque required to maintain pinion rotation through several revolutions.
4. Remove or disconnect the following:
   • Pinion flange
   • Wheel speed sensor and rotor, if equipped
   • Oil seal

   • Pinion bearing
   • Collapsible spacer

To install:

➡**Use a new collapsible spacer and wheel speed sensor rotor for assembly.**

5. Install or connect the following:
   • Collapsible spacer
   • Pinion bearing
   • Pinion seal
   • Pinion flange
6. Rotate the pinion flange occasionally while tightening the flange nut to make sure the pinion bearings seat correctly.
7. Take frequent bearing preload torque readings. Tighten the flange nut to achieve the preload torque readings originally recorded. Do not exceed 137–217 ft. lbs. (186–294 Nm) torque when tightening the pinion flange nut.

❊❊ CAUTION

Never loosen the pinion nut to reduce bearing preload. If it is necessary to reduce bearing preload, install a new collapsible spacer.

8. Install or connect the following:
   • Brake drums
   • Rear wheels
   • Driveshaft. Tighten the fasteners to 58–65 ft. lbs. (78–88 Nm).
9. Fill the differential with gear lubricant and check for leaks.

4 WHEEL DRIVE

1. Before servicing the vehicle, refer to the precautions in the beginning of this section.
2. Remove or disconnect the following:
   • Driveshaft

   • Rear wheels
   • Brake drums

➡The rear brake drums must be removed so that there is no additional drag when measuring pinion bearing preload.

3. Use an inch lb. torque wrench and measure the amount of torque required to maintain pinion rotation through several revolutions.
4. Remove or disconnect the following:
   • Pinion flange
   • Oil seal

To install:
5. Install or connect the following:
   • Pinion seal
   • Pinion flange
6. Rotate the pinion flange occasionally while tightening the flange nut to make sure the pinion bearings seat correctly.
7. Take frequent bearing preload torque readings. Tighten the flange nut to achieve the preload torque readings originally recorded. Do not exceed 137–217 ft. lbs. (186–294 Nm) torque when tightening the pinion flange nut.

❊❊ CAUTION

If the bearing preload can not be achieved at the specified torque, remove the pinion bearing and install a new adjustment spacer.

8. Install or connect the following:
   • Brake drums
   • Rear wheels
   • Driveshaft. Tighten the fasteners to 58–65 ft. lbs. (78–88 Nm).
9. Fill the differential with gear lubricant and check for leaks.

STEERING AND SUSPENSION

Air Bag

❊❊ CAUTION

Some vehicles are equipped with an air bag system. The system must be disarmed before performing service on, or around, system components, the steering column, instrument panel components, wiring and sensors. Failure to follow the safety precautions and the disarming procedure could result in accidental air bag deployment, possible injury and unnecessary system repairs.

PRECAUTIONS

Several precautions must be observed when handling the inflator module to avoid accidental deployment and possible personal injury.
• Never carry the inflator module by the wires or connector on the underside of the module.
• When carrying a live inflator module, hold securely with both hands, and ensure that the bag and trim cover are pointed away.
• Place the inflator module on a bench or other surface with the bag and trim cover facing up.

• With the inflator module on the bench, never place anything on or close to the module which may be thrown in the event of an accidental deployment.

DISARMING

To disarm the **SRS** system turn the ignition switch to the **OFF** position. Then, disconnect both battery cables starting with the negative cable first and wait at least 3 minutes after
the cables are disconnected.
To rearm the **SRS** system, turn the ignition switch to the **OFF** position. Connect

both battery cables starting with the positive cable first.

Recirculating Ball Power Steering Gear

REMOVAL & INSTALLATION

1. Before servicing the vehicle, refer to the precautions in the beginning of this section.
2. Remove or disconnect the following:
   - Pitman arm
   - Steering column intermediate shaft
   - Power steering hoses
   - Steering gear

To install:

3. Install or connect the following:
   - Steering gear. Tighten the bolts to 62–71 ft. lbs. (84–96 Nm).
   - Power steering hoses. Tighten the banjo fittings to 29–38 ft. lbs. (39–51 Nm).
   - Steering column intermediate shaft. Tighten the pinch bolt to 17–22 ft. lbs. (24–29 Nm).
   - Pitman arm. Tighten the nut to 174–195 ft. lbs. (235–265 Nm).
4. Check the wheel alignment and adjust, as necessary.

Rack and Pinion Steering Gear

REMOVAL & INSTALLATION

1. Before servicing the vehicle, refer to the precautions in the beginning of this section.
2. Remove or disconnect the following:
   - Front wheels
   - Outer tie rod ends
   - Steering shaft coupler
   - Power steering hoses
   - Steering gear

To install:

3. Install or connect the following:
   - Steering gear. Tighten the bolts to 101 ft. lbs. (137 Nm).
   - Power steering hoses. Tighten the fittings to 25 ft. lbs. (35 Nm).
   - Steering shaft coupler. Tighten the bolt to 22 ft. lbs. (29 Nm).
   - Outer tie rod ends. Tighten the nuts to 65 ft. lbs. (88 Nm).
   - Front wheels

Strut

REMOVAL & INSTALLATION

Front

PATHFINDER AND QX4

1. Before servicing the vehicle, refer to the precautions in the beginning of this section.

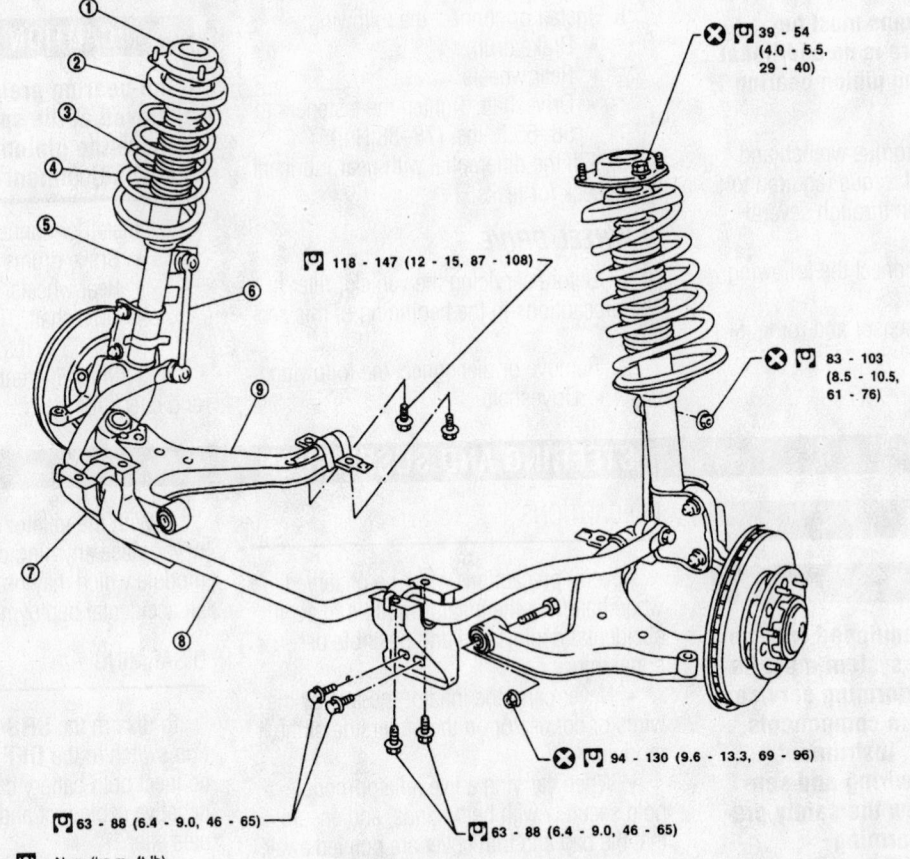

When installing rubber parts, final tightening must be carried out under unladen condition* with tires on ground.
Fuel, radiator coolant and engine oil full.
Spare tire, jack, hand tools and mats in designated positions.

118 - 147 (12 - 15, 87 - 108)

39 - 54 (4.0 - 5.5, 29 - 40)

83 - 103 (8.5 - 10.5, 61 - 76)

94 - 130 (9.6 - 13.3, 69 - 96)

63 - 88 (6.4 - 9.0, 46 - 65)

63 - 88 (6.4 - 9.0, 46 - 65)

: N·m (kg-m, ft-lb)

① Strut mounting insulator
② Spring upper seat
③ Bound bumper
④ Coil spring
⑤ Strut assembly
⑥ Stabilizer connecting rod
⑦ Bracket
⑧ Stabilizer bar
⑨ Transverse link

7924VG66

Exploded view of the front suspension—2WD Pathfinder shown

2. Remove or disconnect the following:
- Front wheel
- Stabilizer bar link
- Steering knuckle bracket bolts
- Upper strut mount nuts
- Strut

To install:

➡**Use new nuts and bolts for assembly.**

3. Install or connect the following:
- Strut. Tighten the upper strut mount nuts to 29–40 ft. lbs. (39–54 Nm) and the knuckle bracket bolts to 111–122 ft. lbs. (151–165 Nm).
- Stabilizer bar link. Tighten the nut to 61–76 ft. lbs. (83–103 Nm).
- Front wheel

4. Check the wheel alignment and adjust, as necessary.

Shock Absorber

REMOVAL & INSTALLATION

Front

XTERRA

1. Before servicing the vehicle, refer to the precautions in the beginning of this section.
2. Support the lower control arm.
3. Remove or disconnect the following:
- Front wheel
- Lower shock absorber mounting bolt
- Upper shock absorber mounting nut
- Shock absorber

To install:

4. Install or connect the following:
- Shock absorber
- Upper shock absorber mounting nut. Tighten the nut to 12–16 ft. lbs. (16–22 Nm).
- Lower shock absorber mounting bolt. Tighten the bolt to 87–108 ft. lbs. (118–147 Nm).
- Front wheel

Rear

PATHFINDER AND QX4

1. Before servicing the vehicle, refer to the precautions in the beginning of this section.
2. Support the rear axle.
3. Remove or disconnect the following:
- Lower shock absorber bolt
- Upper shock absorber bolt
- Shock absorber

To install:

➡**Use new fasteners for assembly.**

4. Install the shock absorber and tighten the bolts to 49–65 ft. lbs. (67–88 Nm).

XTERRA

1. Before servicing the vehicle, refer to the precautions in the beginning of this section.
2. Remove or disconnect the following:
- Upper and lower shock absorber nuts
- Shock absorber

To install:

➡**Use new nuts for assembly.**

3. Install the shock absorber and tighten the nuts to 30–37 ft. lbs. (40–50 Nm).

Coil Spring

REMOVAL & INSTALLATION

Pathfinder and QX4

FRONT

1. Before servicing the vehicle, refer to the precautions in the beginning of this section.
2. Remove the strut assembly.
3. Compress the coil spring and remove the piston rod nut.
4. Remove or disconnect the following:
- Upper strut mount
- Strut mount bracket
- Upper strut bearing
- Spring upper seat
- Coil spring

To install:

➡**Use new fasteners for assembly.**

5. Install or connect the following:
- Coil spring
- Spring upper seat
- Upper strut bearing
- Strut mount bracket
- Upper strut mount. Tighten the piston rod nut to 43–58 ft. lbs. (59–78 Nm).

6. Remove the spring compressor and install the strut assembly to the vehicle.
7. Check the wheel alignment and adjust, as necessary.

REAR

1. Before servicing the vehicle, refer to the precautions in the beginning of this section.
2. Support the vehicle at the frame.
3. Support the axle with a floor jack.
4. Remove or disconnect the following:
- Rear wheels
- Shock absorbers
- Stabilizer bar links
- Lateral control rod
- Coil springs

To install:

➡**Use new fasteners for assembly.**

5. Install or connect the following:
- Coil springs
- Lateral control rod. Tighten the nut to 80–94 ft. lbs. (108–127 Nm).
- Stabilizer bar links. Tighten the nuts to 30–35 ft. lbs. (41–47 Nm).
- Shock absorbers
- Rear wheels

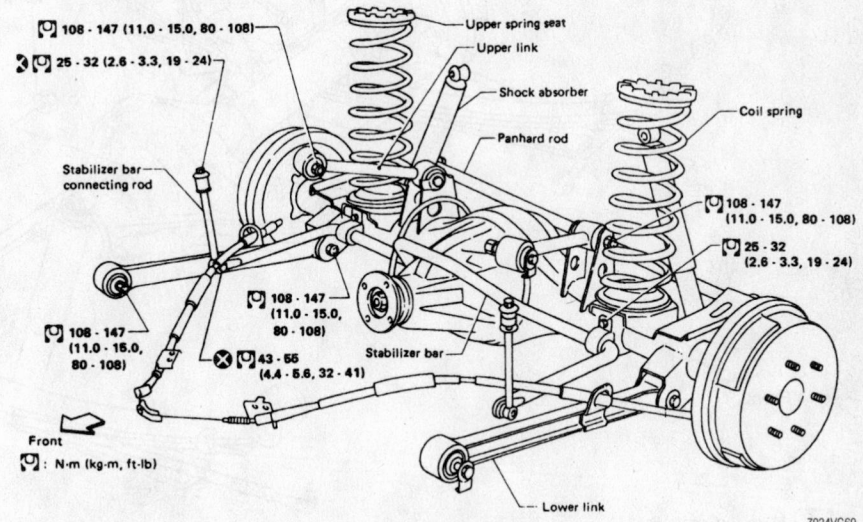

108 - 147 (11.0 - 15.0, 80 - 108)
25 - 32 (2.6 - 3.3, 19 - 24)
Upper spring seat
Upper link
Shock absorber
Coil spring
Panhard rod
Stabilizer bar connecting rod
108 - 147 (11.0 - 15.0, 80 - 108)
25 - 32 (2.6 - 3.3, 19 - 24)
108 - 147 (11.0 - 15.0, 80 - 108)
108 - 147 (11.0 - 15.0, 80 - 108)
43 - 55 (4.4 - 5.6, 32 - 41)
Stabilizer bar
Front
: N·m (kg-m, ft-lb)
Lower link

7924VG69

Rear suspension component identification—Pathfinder

For Tune-up, Capacities and Firing orders, see Section 1 of this manual

Leaf Springs

REMOVAL & INSTALLATION

1. Before servicing the vehicle, refer to the precautions in the beginning of this section.
2. Support the vehicle at the frame.
3. Support the axle with a floor jack.
4. Remove or disconnect the following:
   - Rear wheels
   - Shock absorbers
   - Axle U-bolts and spring pad
   - Spring shackle
   - Front mount bolt/pin
   - Leaf spring

To install:

➡**Use new fasteners for assembly.**

5. Install or connect the following:
   - Leaf spring. Tighten the front mount bolt to 86–108 ft. lbs. (117–147 Nm).
   - Spring shackle. Tighten the nuts to 58–72 ft. lbs. (78–98 Nm).
   - Axle U-bolts and spring pad. Tighten the nuts to 72–80 ft. lbs. (98–108 Nm).

- Shock absorbers
- Rear wheels

Torsion Bar

1. Before servicing the vehicle, refer to the precautions in the beginning of this section.
2. Move the dust cover, if equipped.
3. Matchmark the torsion bar to the control arm mount and the anchor arm.
4. Measure the adjustment bolt protrusion as shown and note the length (L) for assembly.
5. Loosen the adjustment bolt so that all tension is released.
6. Remove the torsion bar mount from the control arm and remove the torsion bar.

To install:

7. Align the matchmarks and install the torsion bar. Tighten the large mount nut to 66–87 ft. lbs. (89–118 Nm) and the small nut to 33–44 ft. lbs. (45–60 Nm).
8. Tighten the adjustment bolt to achieve the measurement (L) noted earlier. Tighten the locknut to 22–30 ft. lbs. (30–40 Nm).
9. If a new torsion bar is being installed, set length (L) as follows:

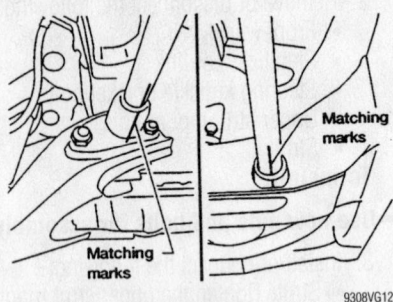

Torsion bar matchmarks

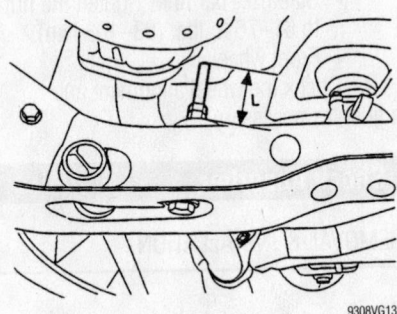

Adjustment bolt measurement (L)

SEC. 380 • 430 • 431

⬙⬚ 78 – 98 (8.0 – 10.0, 58 – 72)

When installing rubber parts, final tightening must be carried out under unladen condition* with tires on ground.

* Fuel, radiator coolant and engine oil full. Spare tire, jack, hand tools and mats in designated positions.

⊗⬚ 117 – 147 (12 – 15, 86 – 108)

⊗⬚ 40 – 50 (4.1 – 5.1, 30 – 37)

⬚ 43 – 55 (4.4 – 5.6, 32 – 41)

⬚ 41.2 – 47.1 (4.2 – 4.8, 30 – 35)

Front

⬚ : N·m (kg-m, ft-lb)

⬚ 118 – 147 (12 – 15, 87 – 108)

9355WG06

Leaf springs and rear suspension components—Xterra with 2.4L engine shown

- 2 wheel drive: 2.13 inches (54mm)
- 4 wheel drive: 2.76 inches (70mm)

Upper Ball Joint

REMOVAL & INSTALLATION

Xterra

The upper ball joint is serviced with the upper control arm as an assembly.

Lower Ball Joint

REMOVAL & INSTALLATION

Xterra

The lower ball joint is serviced with the lower control arm as an assembly.

Pathfinder and QX4

1. Before servicing the vehicle, refer to the precautions in the beginning of this section.
2. Support the lower control arm.
3. Remove or disconnect the following:
   - Front wheel
   - Lower ball joint

To install:

4. Install or connect the following:
   - Lower ball joint. Tighten the control arm bolts to 76–94 ft. lbs. (103–127 Nm) and the stud nut to 87–123 ft. lbs. (118–167 Nm).
   - Front wheel

Upper Control Arm

REMOVAL & INSTALLATION

1. Before servicing the vehicle, refer to the precautions in the beginning of this section.
2. Support the lower control arm.
3. Remove or disconnect the following:
   - Front wheel
   - Shock absorber
   - Upper ball joint
   - Control arm mounting bolts
   - Upper control arm

To install:

4. Install or connect the following:
   - Upper control arm. Tighten the mounting bolts to 72–87 ft. lbs. (98–118 Nm).

- Upper ball joint. Tighten the nut to 58–108 ft. lbs. (78–147 Nm).
- Shock absorber
- Front wheel

5. Check the wheel alignment and adjust, as necessary.

CONTROL ARM BUSHING REPLACEMENT

1. Before servicing the vehicle, refer to the precautions in the beginning of this section.
2. Remove the control arm from the vehicle.
3. Remove the control arm bushing with a press.

To install:

4. Lubricate the control arm bushings with liquid soap.
5. Install the bushings with a press.
6. Install the control arm to the vehicle.
7. Check the wheel alignment and adjust, as necessary.

Lower Control Arm

REMOVAL & INSTALLATION

1. Before servicing the vehicle, refer to the precautions in the beginning of this section.
2. Remove or disconnect the following:
   - Front wheel
   - Torsion bar
   - Shock absorber
   - Stabilizer bar link
   - Axle halfshaft, if equipped
   - Lower ball joint
   - Control arm mounting bolts
   - Lower control arm

To install:

3. Install or connect the following:
   - Lower control arm. Tighten the mount bolts to 80–105 ft. lbs. (108–142 Nm) for 1999–01; 69–96 ft. lbs. (94–130 Nm) for 2002 models..
   - Lower ball joint. Tighten the nut to 87–141 ft. lbs. (118–191 Nm) for 1999–01; 87–123 ft. lbs. (118–167 Nm) for 2002 models.
   - Axle halfshaft, if equipped
   - Stabilizer bar link
   - Shock absorber
   - Torsion bar
   - Front wheel

4. Check the wheel alignment and adjust, as necessary.

CONTROL ARM BUSHING REPLACEMENT

1. Before servicing the vehicle, refer to the precautions in the beginning of this section.
2. Remove the control arm from the vehicle.
3. Remove the control arm bushing with a press.

To install:

4. Lubricate the control arm bushings with liquid soap.
5. Install the bushings with a press.
6. Install the control arm to the vehicle.
7. Check the wheel alignment and adjust, as necessary.

Wheel Bearings

ADJUSTMENT

2 Wheel Drive

➡**Use a new split pin for assembly.**

1. Before servicing the vehicle, refer to the precautions in the beginning of this section.
2. Remove or disconnect the following:
   - Dust cap
   - Split pin
   - Spindle nut cap
3. Tighten the spindle nut to 25–29 ft. lbs. (34–39 Nm).
4. Spin the hub several times to fully seat the bearings.
5. Retighten the spindle nut to 25–29 ft. lbs. (34–39 Nm).
6. Loosen the spindle nut 45–60 degrees and install the spindle nut cap and split pin.
7. Install the dust cap.

4 Wheel Drive

1. Before servicing the vehicle, refer to the precautions in the beginning of this section.
2. Remove or disconnect the following:
   - Locking hub or driveplate
   - Snapring
   - Spindle washer
   - Thrust washer
   - Lockwasher
3. Tighten the wheel bearing locknut to 58–72 ft. lbs. (78–98 Nm).

4. Loosen the locknut fully.

5. Tighten the wheel bearing locknut to 4–13 inch lbs. (0.5–1.5 Nm).

6. Spin the hub several times to fully seat the bearings.

7. Retighten the wheel bearing locknut to 4–13 inch lbs. (0.5–1.5 Nm).

8. Install or connect the following:
   - Lockwasher. Tighten the retaining screw to 10–16 inch lbs. (1–2 Nm).
   - Thrust washer
   - Spindle washer
   - Snapring
   - Locking hub or driveplate

REMOVAL & INSTALLATION

2 Wheel Drive

1. Before servicing the vehicle, refer to the precautions in the beginning of this section.

2. Remove or disconnect the following:
   - Front wheel
   - Brake caliper and support
   - Dust cap
   - Split pin

- Spindle nut cap
- Spindle nut
- Bearing washer
- Outer bearing
- Hub and brake rotor assembly
- Inner grease seal
- Inner wheel bearing

To install:

3. Install or connect the following:
   - Inner wheel bearing
   - Inner grease seal
   - Hub and brake rotor assembly
   - Outer bearing
   - Bearing washer
   - Spindle nut. Adjust the wheel bearings.
   - Spindle nut cap
   - Split pin
   - Dust cap
   - Brake caliper and support
   - Front wheel

4 Wheel Drive

1. Before servicing the vehicle, refer to the precautions in the beginning of this section.

2. Remove or disconnect the following:

- Front wheel
- Brake caliper and support
- Locking hub or driveplate
- Snapring
- Spindle washer
- Thrust washer
- Lockwasher
- Wheel bearing locknut
- Outer bearing
- Hub and brake rotor assembly
- Inner grease seal
- Inner wheel bearing

To install:

3. Install or connect the following:
   - Inner wheel bearing
   - Inner wheel bearing
   - Inner grease seal
   - Hub and brake rotor assembly
   - Outer bearing
   - Wheel bearing locknut. Adjust the wheel bearings.
   - Lockwasher
   - Thrust washer
   - Spindle washer
   - Snapring
   - Locking hub or driveplate
   - Brake caliper and support
   - Front wheel

GENERAL MOTOR CORP.

23

Pontiac-Vibe

PRECAUTIONS

Before servicing any vehicle, please be sure to read all of the following precautions, which deal with personal safety, prevention of component damage, and important points to take into consideration when servicing a motor vehicle:

• Never open, service or drain the radiator or cooling system when the engine is hot; serious burns can occur from the steam and hot coolant.

• Observe all applicable safety precautions when working around fuel. Whenever servicing the fuel system, always work in a well-ventilated area. Do not allow fuel spray or vapors to come in contact with a spark, open flame, or excessive heat (a hot drop light, for example). Keep a dry chemical fire extinguisher near the work area. Always keep fuel in a container specifically designed for fuel storage; also, always properly seal fuel containers to avoid the possibility of fire or explosion. Refer to the additional fuel system precautions later in this section.

• Fuel injection systems often remain pressurized, even after the engine has been turned **OFF**. The fuel system pressure must be relieved before disconnecting any fuel lines. Failure to do so may result in fire and/or personal injury.

• Brake fluid often contains polyglycol ethers and polyglycols. Avoid contact with the eyes and wash your hands thoroughly after handling brake fluid. If you do get brake fluid in your eyes, flush your eyes with clean, running water for 15 minutes. If eye irritation persists, or if you have taken brake fluid internally, IMMEDIATELY seek medical assistance.

• The EPA warns that prolonged contact with used engine oil may cause a number of skin disorders, including cancer! You should make every effort to minimize your exposure to used engine oil. Protective gloves should be worn when changing oil. Wash your hands and any other exposed skin areas as soon as possible after exposure to used engine oil. Soap and water, or waterless hand cleaner should be used.

• All new vehicles are now equipped with an air bag system. The system must be disabled before performing service on or around system components, steering column, instrument panel components, wiring and sensors. Failure to follow safety and disabling procedures could result in accidental air bag deployment, possible personal injury and unnecessary system repairs.

• Always wear safety goggles when working with, or around, the air bag system. When carrying a non-deployed air bag, be sure the bag and trim cover are pointed away from your body. When placing a non-deployed air bag on a work surface, always face the bag and trim cover upward, away from the surface. This will reduce the motion of the module if it is accidentally deployed. Refer to the additional air bag system precautions later in this section.

• Clean, high quality brake fluid from a sealed container is essential to the safe and proper operation of the brake system. You should always buy the correct type of brake fluid for your vehicle. If the brake fluid becomes contaminated, completely flush the system with new fluid. Never reuse any brake fluid. Any brake fluid that is removed from the system should be discarded. Also, do not allow any brake fluid to come in contact with a painted surface; it will damage the paint.

• Never operate the engine without the proper amount and type of engine oil; doing so WILL result in severe engine damage.

• Timing belt maintenance is extremely important! Many models utilize an interference-type, non-freewheeling engine. If the timing belt breaks, the valves in the cylinder head may strike the pistons, causing potentially serious (also time-consuming and expensive) engine damage. Refer to the maintenance interval charts in the front of this manual for the recommended replacement interval for the timing belt, and to the timing belt section for belt replacement and inspection.

• Disconnecting the negative battery cable on some vehicles may interfere with the functions of the on-board computer system(s) and may require the computer to undergo a relearning process once the negative battery cable is reconnected.

• When servicing drum brakes, only disassemble and assemble one side at a time, leaving the remaining side intact for reference.

ENGINE REPAIR

Alternator

REMOVAL & INSTALLATION

1. Before servicing the vehicle, refer to the precautions in the beginning of this section.
2. Remove or disconnect the following:
   • Negative battery cable
   • Drive belt
   • Wire clamp from the clip on the rectifire end frame
   • Rubber clamp and nut
   • Alternator wiring and connector
   • Alternator

To install:

3. Install or connect the following:

   • Alternator. Torque the 12mm bolt to 18 ft. lbs. (25 Nm) and the 14mm bolt to 39 ft. lbs. (54 Nm).
   • Alternator connector and wiring
   • Rubber clamp and nut
   • Wire clamp
   • Drive belt
   • Negative battery cable

Ignition Timing

ADJUSTMENT

➡**The timing on engines equipped with Distributorless Ignition Systems (DIS) is not adjustable.**

Engine Assembly

REMOVAL & INSTALLATION

1. Before servicing the vehicle, refer to the precautions in the beginning of this section.
2. Relieve the fuel system pressure.
3. Drain the cooling system.
4. Drain the engine oil.
5. Drain the transaxle fluid and transfer fluid, if equipped.
6. Remove or disconnect the following:
   • Negative battery cable. Wait at least 90 seconds before proceeding.
   • Battery
   • Hood

- Undercovers
- Radiator inlet and outlet hoses
- Radiator hose outlet
- Oil cooler inlet and outlet tubes
- Upper radiator support and radiator, if equipped with A/C
- Battery
- Air cleaner assembly
- Fuel pipe clamp
- Fuel tube sub-assembly
- Accelerator control cable
- Cruise control actuator, if equipped
- Union-to-connector tube hose
- Heater inlet and outlet hoses
- Transmission shift cable(s)
- Clutch release cylinder, on manual transaxle
- Glove compartment door
- Engine relay block cover
- 3 connectors from the relay block
- 2 ground cables
- Engine wire from the Engine Control Module (ECM) and junction block
- Engine wire from the cabin
- Drive belt
- Compressor and magnetic clutch, if equipped with A/C. Unbolt and position aside, DO NOT disconnect the lines.
- Vane pump oil reservoir from the bracket
- Return tube
- Right side front door scuff plate
- Right side cowl side trim plate
- Right side rear door scuff plate, AWD
- Lower right side center pillar garnish, AWD
- Right front seat, AWD
- Column hole cover silencer sheet
- Steering intermediate shaft
- Front floor panel brace, FWD
- Center exhaust pipe, AWD
- Propeller shaft with center bearing shaft, AWD
- Front exhaust pipe
- Front hub nuts
- Tie rod ends from the steering knuckles

7. Separate the front stabilizer links and lower control arm ball joints
- Front halfshafts

8. Remove the engine from the vehicle, as follows:

a. Set the engine lifter.

b. Remove the bolts and nuts, then remove the engine mounting insulator.

c. Remove the through bolt and nut,

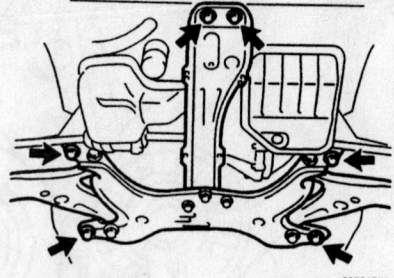

Remove the 6 bolts, as indicated by arrows

then detach the engine mounting insulator from the vehicle.

d. Remove the 6 bolts as shown.

e. Use a suitable tool to suspend the engine assembly, as shown in the figure.

f. No. 1 engine hanger: P/N 12281-15040 (1ZZ-FE), 12281-88600 (2ZZ-GE).

g. No. 2 engine hanger: P/N 12281-22021 (1ZZ-FE), 12281-88600 (2ZZ-GE).

h. Bolt: P/N 91512-B1016.

i. Torque the bolts to 28 ft. lbs. (38 Nm).

✲✲ CAUTION

Do not try to suspend the engine by hooking the chain to any other part.

j. Attach an engine chain hoist to the hangers.

k. Using the chain block and sling device, suspend the engine.

l. Remove the engine and transaxle assembly from the vehicle.

9. Remove or disconnect the following components, as necessary:
- Vane pump
- Steering gear
- Crossmember
- Manifold stay
- Oxygen (O_2) sensor
- Exhaust manifold
- Starter
- Transaxle

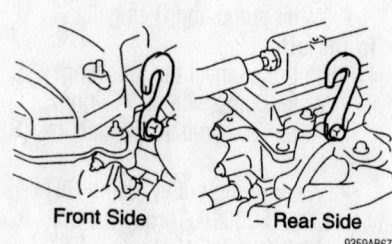

Front Side — **Rear Side**

Install the engine hangers—1ZZ-FE shown, 2ZZ-GE similar

- Transfer case
- Clutch
- Flywheel
- Alternator
- Ignition coil
- Fuel delivery pipe
- Intake manifold
- Oil level gauge
- Water inlet and bypass pipes
- Thermostat
- Oil pressure switch
- Crankshaft Position (CKP) sensor
- Knock Sensor (KS)
- Drive belt tensioner
- Engine mounts and brackets
- Coolant Temperature Sensor (CTS)

To install:

10. Install any removed components to the engine and transaxle assembly.

11. To install the engine:

a. Place the engine and transaxle on an engine lifter.

b. Install the engine with the transaxle to the vehicle.

c. Temporarily install the crossmember and 6 bolts.

d. Install the left engine mounting insulator. Tighten the bolts to 59 ft. lbs. (80 Nm).

e. Install the right engine mounting

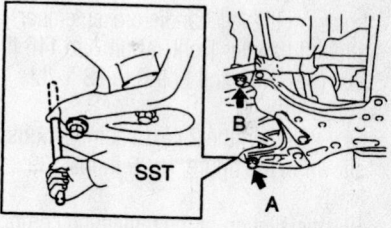

Insert the SST to the positioning holes of the right handle crossmember and on the right handle of the vehicle. Temporarily tighten bolt A, then bolt B

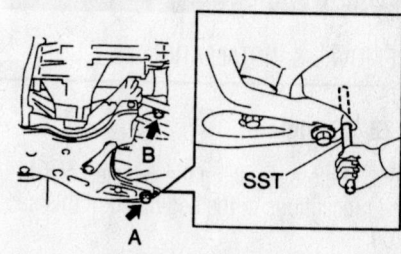

Insert the SST to the positioning holes of the left handle crossmember and on the left handle of the vehicle. Temporarily tighten bolt A, then bolt B

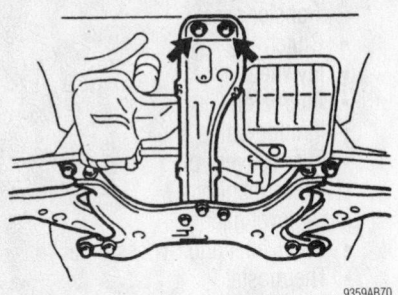

Tighten the 2 crossmember bolts, indicated by arrows

9359AB70

insulator. Tighten the bolts to 38 ft. lbs. (52 Nm).

f. Insert SST 09670-00010 to the positioning holes of the right handle crossmember and on the right handle of the vehicle. Temporarily tighten bolt A, then bolt B.

g. Insert SST 09670-00010 to the positioning holes of the left handle crossmember and on the left handle of the vehicle. Temporarily tighten bolt A, then bolt B.

h. Insert the SST to the positioning holes on the right-handle crossmember and right handle. Tighten bolt A to 116 ft. lbs. (157 Nm) and bolt B to 83 ft. lbs. (113 Nm).

i. Insert the SST to the positioning holes on the left-handle crossmember and left handle. Tighten bolt A to 116 ft. lbs. (157 Nm) and bolt B to 83 ft. lbs. (113 Nm).

j. Tighten the 2 crossmember bolts, shown in the figure, to 29 ft. lbs. (39 Nm).

12. Installation of the remaining components is the reverse of the removal procedure.

13. Make sure all fluid levels are accurate, then start the engine check for leaks.

Water Pump

REMOVAL & INSTALLATION

1ZZ-FE Engine

1. Before servicing the vehicle, refer to the precautions in the beginning of this section.
2. Drain the cooling system.
3. Remove or disconnect the following:
   - Negative battery cable
   - Right-hand engine under cover
   - Drive belt
   - Alternator
   - Water pump

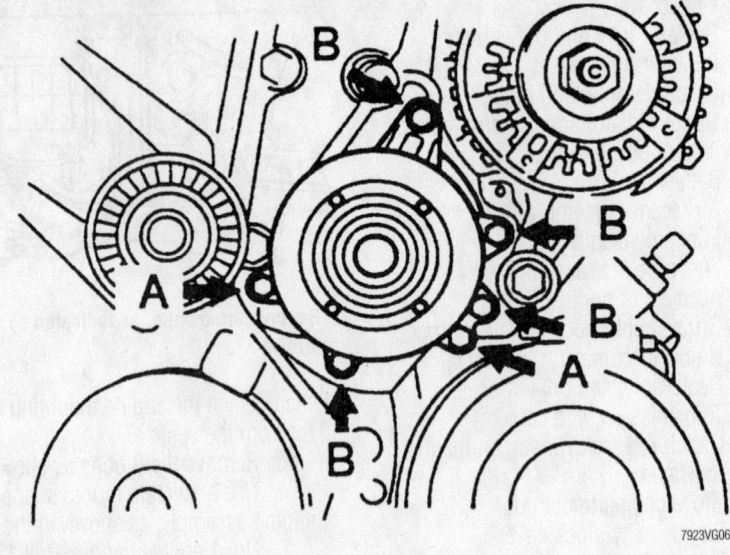

Water pump bolt identification—1.8L (1ZZ-FE) engine

7923VG06

To install:
4. Install or connect the following:
   - Water pump. Torque bolts marked **A** (short) to 80 inch lbs. (9 Nm) and bolts marked **B** (long) to 96 inch lbs. (11 Nm).
   - Alternator
   - Drive belt
   - Right engine under cover
   - Negative battery cable
5. Fill the cooling system to the proper level.
6. Start the vehicle, check for leaks and repair if necessary.

2ZZ-GE Engine

1. Before servicing the vehicle, refer to the precautions in the beginning of this section.
2. Drain the cooling system.
3. Remove or disconnect the following:
   - Negative battery cable
   - Right-hand engine under cover
   - Drive belt
   - Alternator
   - Water pump pulley, using SST 09960-10010
   - Water pump and O-ring

To install:
4. Install or connect the following:
   - Water pump with new O-ring. Torque the bolts to 80 inch lbs. (9 Nm).
   - Water pump pulley, using SST 09960-10010. Torque the bolts to 11 ft. lbs. (15 Nm).
   - Alternator
   - Drive belt
   - Right engine under cover
   - Negative battery cable

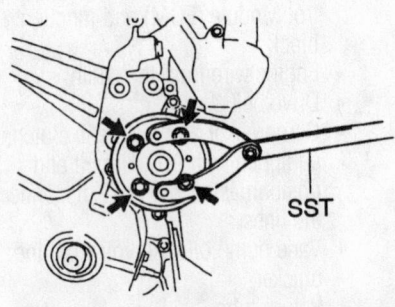

9359AB02

View of the special tool needed to remove and install the water pump pulley

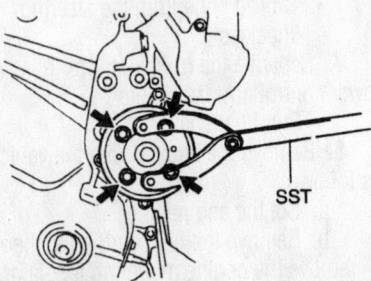

9359AB01

Water pump mounting and bolt locations—2ZZ-GE engine

5. Fill the cooling system to the proper level.

6. Start the vehicle, check for leaks and repair if necessary.

Cylinder Head

REMOVAL & INSTALLATION

1. Before servicing the vehicle, refer to the precautions in the beginning of this section.

2. Drain the cooling system.

3. Remove or disconnect the following:
   - Right side engine under cover
   - Right front wheel and tire
   - Cylinder head cover
   - Air cleaner assembly with hose
   - Accelerator control cable
   - Wire harness clamp and suction hose assembly, 2ZZ-GE engine only
   - Water bypass hoses
   - Fuel pipe clamp
   - Fuel tube sub-assembly
   - Union-to-connector tube hose
   - Radiator and heater inlet hoses
   - Drive belt

4. Separate the vane pipe assembly, but do not disconnect the hose, 1ZZ-FE engine.
   - Alternator bracket, 2ZZ-GE
   - Alternator

5. Separate the compressor and magnetic clutch, on 2ZZ-GE engines with air conditioning.
   - Front exhaust pipe assembly
   - Power steering pump reservoir and position it aside, 1ZZ-FE engine

6. Place a jack with a wooden block under the vehicle for support, then remove the 4 bolts and 2 nuts and remove the right side engine mount.

7. Remove the engine wire, on 1ZZ-FE engines as follows:
   a. Remove the 5 clamps from the brackets.

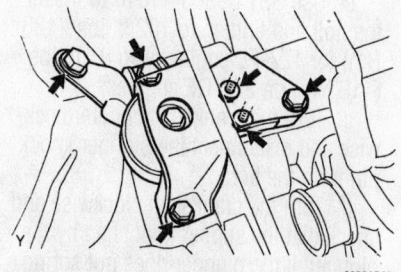

With the engine supported, remove the right side engine mount—1ZZ-FE engine shown, 2ZZ-GE similar

b. Detach the connectors.
   c. Remove the ignition coil connectors.
   d. Bolt and nut holding the engine wire.

8. Remove or disconnect the following:
   - Ignition coil assembly
   - Positive Crankcase Ventilation (PCV) hoses
   - Valve (cylinder head) cover sub-assembly

9. Set the No. 1 cylinder to Top Dead Center (TDC) of the compressor stroke as follows:
   a. Turn the crankshaft pulley, and align its groove with the "0" timing mark of the timing chain cover.
   b. Make sure the point marks of the camshaft timing sprockets and Variable Valve Timing (VVT) timing sprockets are in a straight line as shown. If not, turn the crankshaft 1 complete revolution (360°) and align the marks.

10. Remove or disconnect the following:
   - Crankshaft pulley, using SST 09960-10010
   - Belt tensioner
   - Exhaust manifold stay and head insulator, 2ZZ-GE engine
   - Water pump pulley and pump
   - Transverse engine mounting bracket
   - Crankshaft Position (CKP) sensor
   - No. 1 chain tensioner assembly, making sure not to revolve the crankshaft without the tensioner
   - Timing chain or belt cover

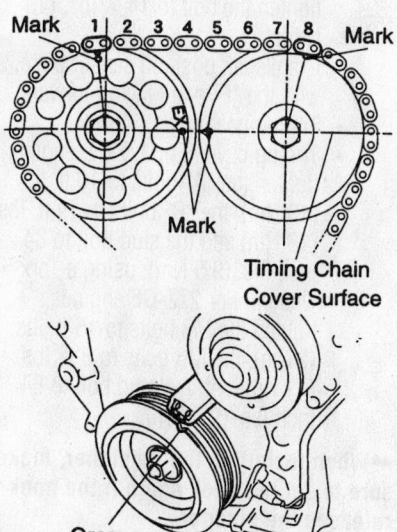

Proper timing mark alignment for TDC

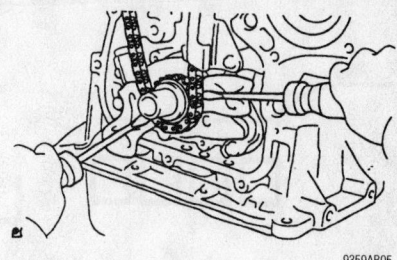

Remove the timing chain with the crankshaft gear

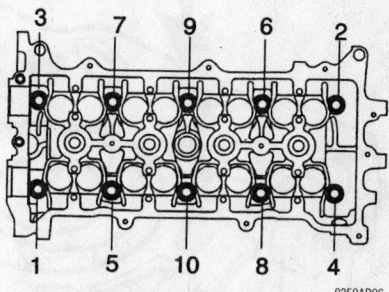

Cylinder head bolt loosening sequence

   - Timing gear cover oil seal
   - CKP sensor plate No. 1
   - Timing chain tensioner slipper
   - Timing chain vibration damper No. 1

➡ **In case you turn the camshafts with the timing chain removed, turn the crankshaft ¼ turn for the valve to avoid contact with the pistons.**

   - Timing chain sub-assembly. Remove the chain with the crankshaft gear, using screwdrivers as shown.
   - Surge tank stay, 2ZZ-GE engine
   - Intake manifold
   - Oil level gauge
   - Water bypass pipe bolts and pipe, 1ZZ-FE engine
   - Camshafts
   - Camshaft timing oil control valve, 1ZZ-FE engine
   - Manifold stay, 1ZZ-FE engine
   - Cylinder head bolts in sequence. To prevent damage to the cylinder head, loosen each bolt about ¼ of a turn during each pass until the bolts are loose.
   - Cylinder head

To install:

11. Clean and degrease the surface of the cylinder head and engine block.

12. Check the length of the cylinder head bolts. They should be 5.780–5.835 in. (146.8–148.2mm) long. If they are longer

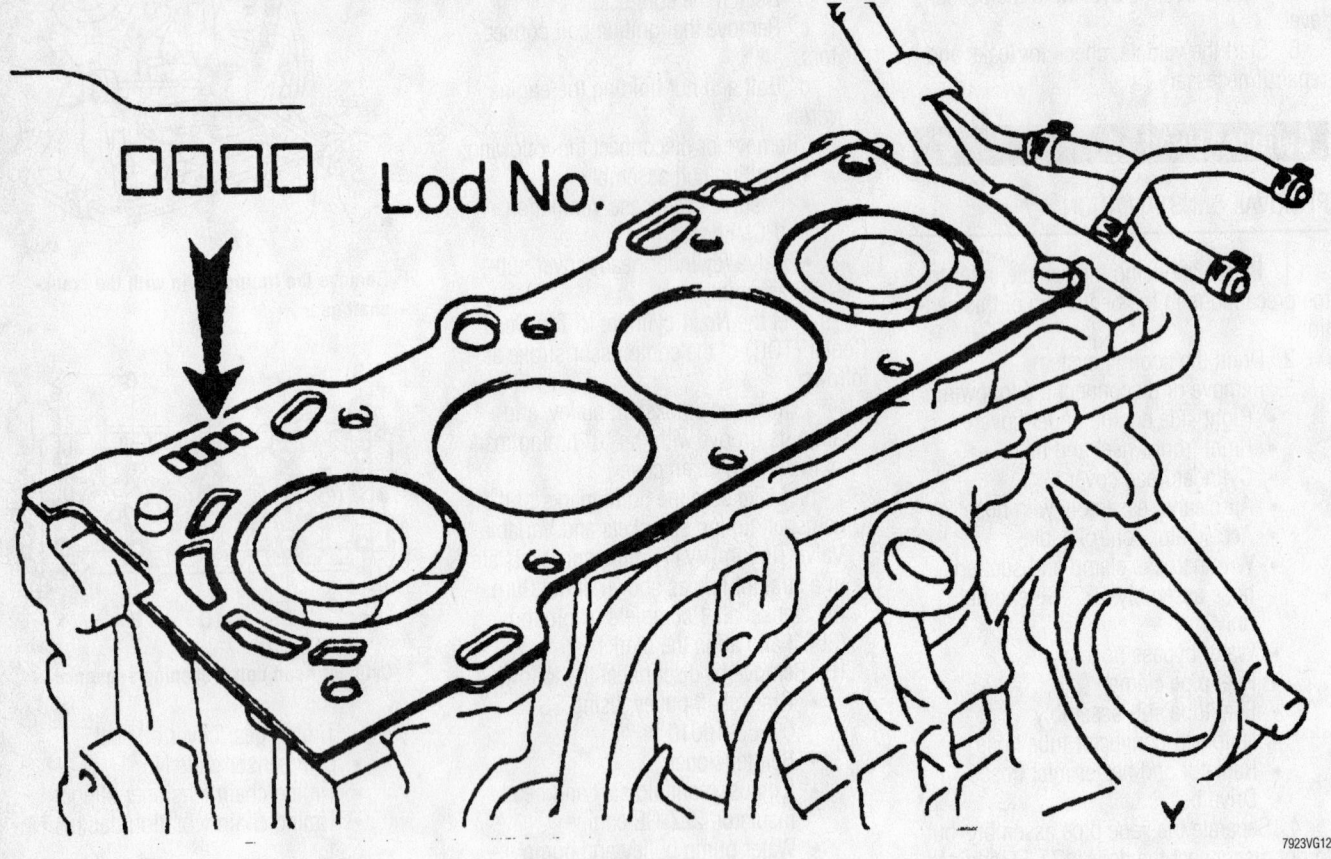

Lod No.

Position the head gasket correctly on the cylinder head—1.8L (1ZZ-FE) engine

7923VG12

than 5.846 in. (148.5mm), they must be replaced.

13. Install or connect the following:
- New gasket on the engine block with the Lot No. stamp facing up.
- Cylinder head
- Apply a light coat of oil to cylinder head bolt threads and tighten in sequence, in several passes, to 36 ft. lbs. (49 Nm) for 1ZZ-FE engines or to 26 ft. lbs. (35 Nm) for 2ZZ-GE engines.
- Tighten each head bolt, in sequence, an additional 90 degree turn for 1ZZ-FE engines, or 180 degree turn for 2ZZ-GE engines.
- Manifold stay, 1ZZ-FE engine. Tighten the bolts to 36 ft. lbs. (49 Nm).
- Camshaft timing oil control valve, on 1ZZ-FE engines, and tighten to 80 inch lbs. (9 Nm)
- Camshaft
- Water by-pass pipe, on 1ZZ-FE engines, and tighten to 80 inch lbs. (9 Nm)
- Oil level gauge
- Intake manifold
- Surge tank stay, 2ZZ-GE engine.

- Tighten to 18 ft. lbs. (24 Nm).
- Timing chain
- Timing chain vibration damper. Tighten the bolts to 80 inch lbs. (9 m).
- Timing chain tensioner slipper and tighten the bolt to 14 ft. lbs. (19 Nm).
- Crankshaft position sensor plate, with the "F" mark facing forward.
- Timing gear cover oil seal
- Timing cover. For 1ZZ-FE engine, tighten the "A" bolts to 10 ft. lbs. (13 Nm), the "B" bolts to 14 ft. lbs. (19 Nm) and the stud bolt to 84 inch lbs. (9.5 Nm), using a Torx® wrench. For 2ZZ-GE engines, tighten the M8 bolts to 15 ft. lbs. (21 m), the M6 bolts to 8 ft. lbs. (11 Nm) and the stud bolt to 84 inch lbs. (9.5 Nm).

➡ **When installing the tensioner, make sure to set the hook again if the hook releases the plunger.**

- Timing chain tensioner. Torque the nuts to 80 inch lbs. (9 Nm).
- CKP sensor and tighten the bolts to 80 inch lbs. (9 Nm)

- Transverse engine mounting bracket. Tighten the bolts to 35 ft. lbs. (47 Nm).
- Water pump and pulley
- Exhaust manifold stay and heat insulator, 2ZZ-GE engine
- Belt tensioner. Tighten the nut to 21 ft. lbs. (29 Nm) and the bolt to 51 ft. lbs. (69 Nm) on 1ZZ-FE engines or to 74 ft. lbs. (100 Nm) on 2ZZ-GE engines.

14. Install the crankshaft pulley, as folows:
   a. Align the pulley set key with the key groove of the pulley and slide on the pulley.
   b. Use SST 09960-11010 to install the bolt and tighten to 102 ft. lbs. (138 Nm) for 1ZZ-FE engine or to 87 ft. lbs. (118 Nm) on 2ZZ-GE engines.
   c. Turn the crankshaft counterclockwise and disconnect the plunger knock pin from the hook.
   d. Turn the crankshaft clockwise and check that the slipper is pushed by the plunger. If the plunger does not spring out, press the slipper into the chain tensioner with a screwdriver so that the hook is released from the knock pin and the plunder springs out.

15. Install or connect the following:

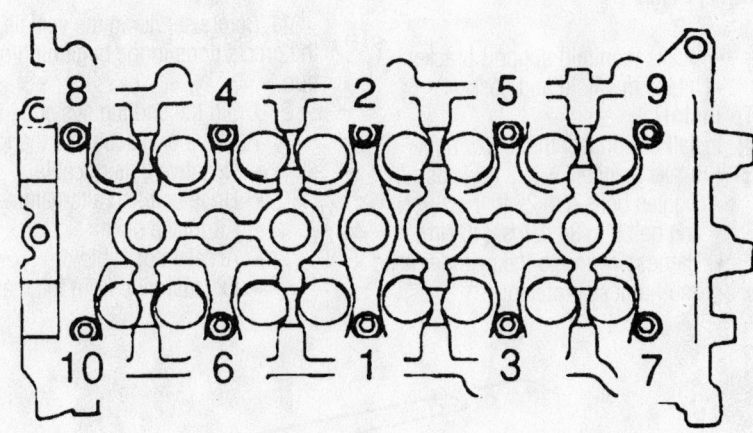

10 mm
Bi–Hexagon
Wrench

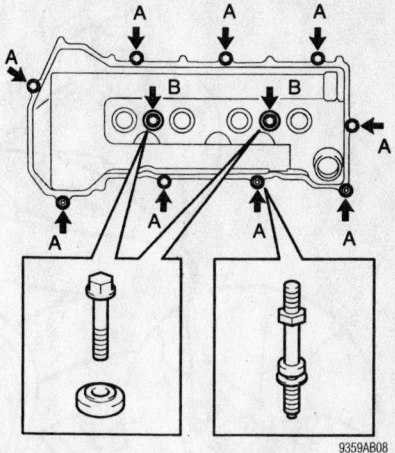

Cylinder head (valve) cover bolt locations—1ZZ-FE engine

16. Fill the cooling system to the proper level.

17. Start the vehicle, check for leaks and repair if necessary.

Cylinder head bolt tightening sequence—1ZZ-FE and 2ZZ-GE engines

- Cylinder head sub-assembly cover. Install seal packing into the locations shown and install within 3 minutes. Tighten the "A" bolts to 8 ft. lbs. (11 Nm) and the "B" bolts to 80 inch lbs. (9 Nm) for 1ZZ-FE engines. For 2ZZ-GE engines, tighten the bolts to 7 ft. lbs. (10 Nm).
- Ignition coil assembly. Torque the bolts to 80 inch lbs. (9 Nm).
- Engine wire and tighten to 80 inch lbs. (9 Nm), 1ZZ-FE
- Right side engine mount. Tighten to 38 ft. lbs. (52 Nm).
- Front exhaust pipe
- Vane pump, 1ZZ-FE
- Compressor and magnetic clutch, 2ZZ-GE

- Alternator bracket, 2ZZ-GE engine
- Alternator
- Suction hose and wire harness clamp, 2ZZ-GE engine
- Air cleaner and hose
- Main cylinder head cover and tighten to 62 inch lbs. (7 Nm)
- Right front wheel and tire. Tighten the lug nuts to 76 ft. lbs. (103 Nm).

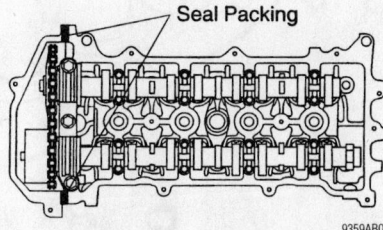

Seal packing installation locations

Intake Manifold

REMOVAL & INSTALLATION

1ZZ-FE Engine

1. Before servicing the vehicle, refer to the precautions in the beginning of this section.
2. Drain the cooling system.
3. Remove or disconnect the following:
   - Negative battery cable
   - Drive belt and alternator
   - Air intake duct
   - Accelerator cable
   - Exhaust pipe from the manifold.
   - Exhaust manifold support bracket
   - Spark plug wires, then ignition coils
   - Spark plugs
   - Positive Crankcase Ventilation (PCV) hoses
   - Throttle body assembly
   - 2 bolts securing the wiring harness protector
   - Wiring connectors and ground wires
   - Intake manifold support bracket
   - Intake manifold and gasket

To install:

4. Install or connect the following:
   - Intake manifold with a new gasket. Torque the bolts to 22 ft. lbs. (30 Nm).
   - Harness wiring to the cylinder head and harness protector

Timing belt service is covered in Section 3 of this manual

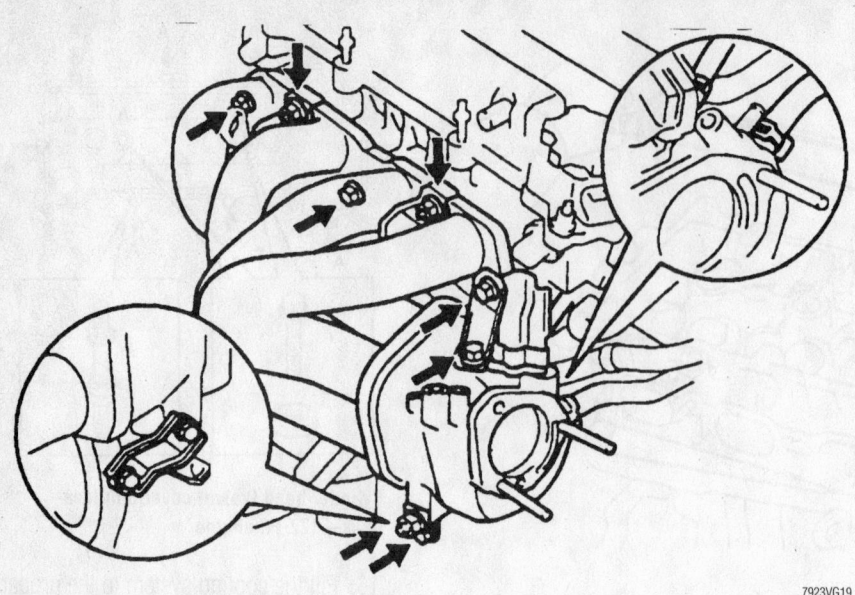

7923VG19

Intake manifold mounting fastener locations—1.8L (1ZZ-FE) engine

- Fuel injectors, throttle body and the PCV hoses
- Spark plugs and ignition coils. Tighten the bolts and nuts to 80 inch lbs. (9 Nm).
- Exhaust manifold and support bracket. Tighten the bolts to 37 ft. lbs. (49 Nm).
- Front exhaust pipe to the manifold. Tighten the bolts to 46 ft. lbs. (62 Nm).
- Oxygen Sensor ($O_{2>S}$S). Tighten the nuts to 14 ft. lbs. (20 Nm).
- Accelerator cable and air intake duct
- Alternator and drive belt
- Negative battery cable

5. Fill the cooling system.

6. Start the vehicle, check for leaks and repair if necessary.

2ZZ-GE Engine

1. Before servicing the vehicle, refer to the precautions in the beginning of this section.

2. Drain the cooling system.

3. Remove or disconnect the following:
- Negative battery cable
- Drive belt and alternator
- Air intake duct
- Accelerator cable
- Spark plug wires, then ignition coils
- Spark plugs
- Positive Crankcase Ventilation (PCV) hoses
- Throttle body assembly
- Wiring harness
- Hoses and tubes connected to the head

- Intake manifold support bracket
- Intake manifold and gasket

To install:

4. Install or connect the following:
- Intake manifold with a new gasket. Tighten bolts A to 25 ft. lbs. (34 Nm) and bolt B to 34 ft. lbs. (46 Nm).
- Harness wiring to the cylinder head and harness protector

- Fuel injectors, throttle body and the PCV hoses
- Spark plugs and ignition coils. Tighten the bolts and nuts to 80 inch lbs. (9 Nm).
- Oxygen Sensor (O_2S). Tighten the nuts to 14 ft. lbs. (20 Nm).
- Accelerator cable and air intake duct
- Alternator and drive belt
- Negative battery cable

5. Fill the cooling system.

6. Start the vehicle, check for leaks and repair if necessary.

Exhaust Manifold

REMOVAL & INSTALLATION

1ZZ-FE Engine

1. Before servicing the vehicle, refer to the precautions in the beginning of this section.

2. Drain the cooling system.

3. Remove or disconnect the following:
- Negative battery cable
- Drive belt and alternator
- Air intake duct
- Accelerator cable
- Exhaust pipe from the manifold

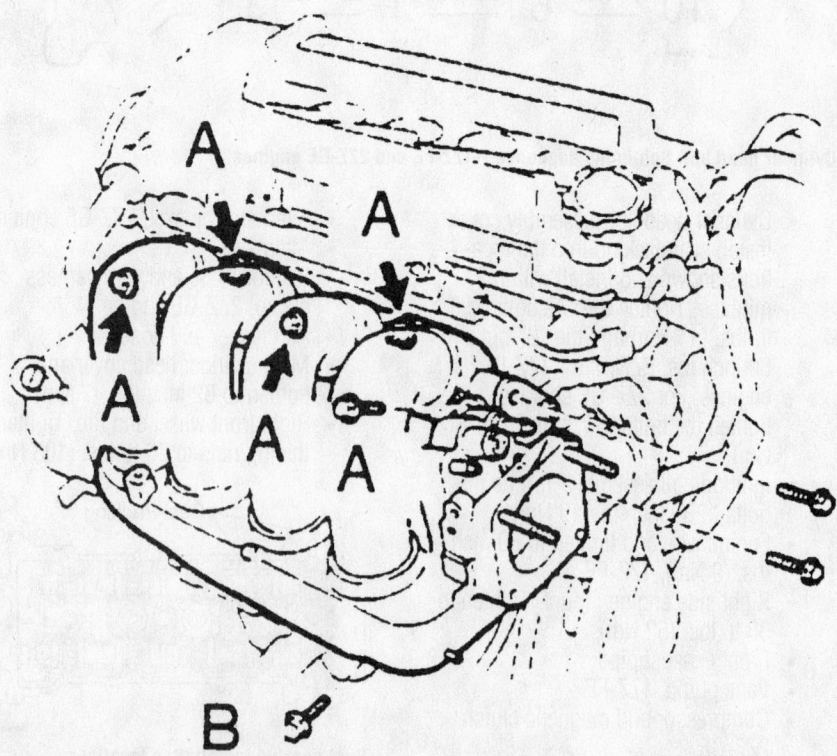

9307WG93

Intake manifold bolt installation—2ZZ-GE engine

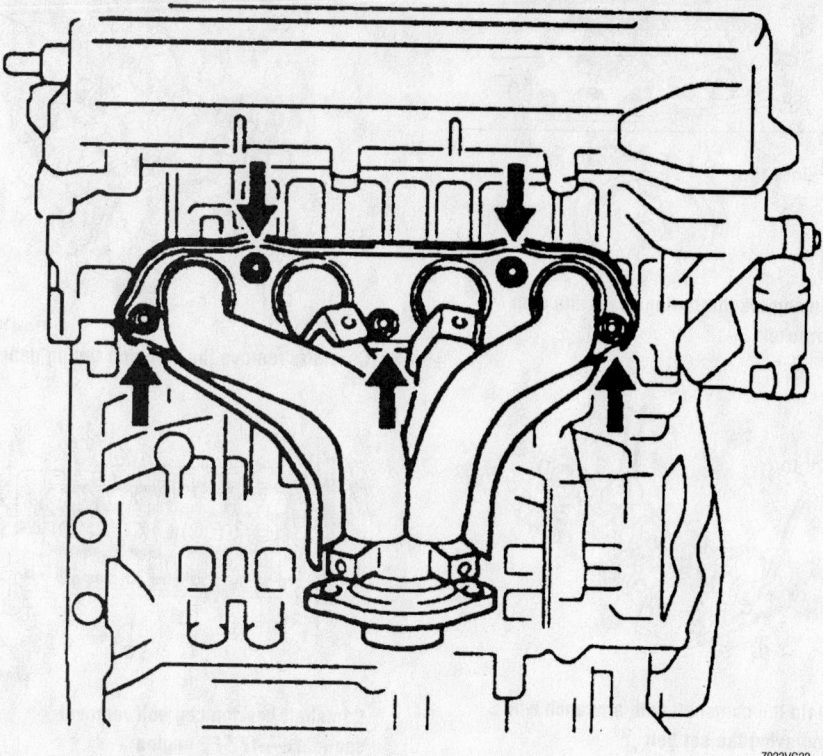

Exhaust manifold mounting nut locations—1.8L (1ZZ-FE) engine

- Exhaust manifold support bracket
- Heat insulator from the dash panel
- Upper heat insulator
- Exhaust manifold and gasket
- If necessary, the lower heat insulator from the exhaust manifold.

To install:

4. Install or connect the following:
   - Lower heat insulator on the exhaust manifold. Tighten the bolts to 108 inch lbs. (12 Nm).
   - Exhaust manifold using a new gasket. Tighten the nuts, in several passes, to 27 ft. lbs. (37 Nm).
   - Upper heat insulator. Tighten the bolts to 108 inch lbs. (12 Nm).
   - Heat insulator on the dash panel
   - Exhaust manifold support bracket. Tighten the bolts, in an alternating pattern, to 37 ft. lbs. (49 Nm).
   - Front exhaust pipe to the manifold. Tighten the bolts to 46 ft. lbs. (62 Nm).
   - Oxygen Sensor (O₂S). Tighten the nuts to 14 ft. lbs. (20 Nm).
   - Accelerator cable and air intake duct
   - Alternator and drive belt
   - Negative battery cable

5. Fill the cooling system.

6. Start the vehicle, check for leaks and repair if necessary.

2ZZ-GE Engine

1. Before servicing the vehicle, refer to the precautions in the beginning of this section.

2. Drain the cooling system.

3. Remove or disconnect the following:
   - Negative battery cable
   - Drive belt and alternator
   - Air intake duct
   - Accelerator cable
   - Exhaust pipe from the manifold
   - Exhaust manifold support bracket
   - Heat insulator from the dash panel
   - Upper heat insulator
   - Exhaust manifold and gasket
   - If necessary, the lower heat insulator from the exhaust manifold.

To install:

4. Install or connect the following:
   - Lower heat insulator on the exhaust manifold. Tighten the bolts to 15 ft. lbs. (20 Nm).
   - Exhaust manifold using a new gasket. Tighten the nuts, in several passes to 37 ft. lbs. (50 Nm).
   - Upper heat insulator. Tighten the bolts to 15 ft. lbs. (20 Nm).

- Heat insulator on the dash panel.
- Exhaust manifold support bracket. Tighten the bolts to 37 ft. lbs. (49 Nm).
- Front exhaust pipe to the manifold. Tighten the bolts to 46 ft. lbs. (62 Nm).
- Oxygen Sensor (O₂S). Tighten the nuts to 14 ft. lbs. (20 Nm).
- Accelerator cable and air intake duct.
- Alternator and drive belt.
- Negative battery cable

5. Fill the cooling system.

6. Start the vehicle, check for leaks and repair if necessary.

Camshaft(s)

REMOVAL & INSTALLATION

1ZZ-FE Engine

1. Before servicing the vehicle, refer to the precautions in the beginning of this section.

2. Remove or disconnect the following:
   - Negative battery cable
   - Right side engine under cover
   - Cylinder head cover
   - Suction hose sub-assembly, 2ZZ-GE engine
   - Drive belt
   - Power steering pump reservoir and position it aside, 1ZZ-FE engine

3. Place a jack with a wooden block under the vehicle for support, then remove the 4 bolts and 2 nuts and remove the right side engine mount.

4. Remove the engine wire, on 1ZZ-FE engines:
   a. Remove the 5 clamps from the brackets.
   b. Detach the connectors.
   c. Remove the ignition coil connectors.

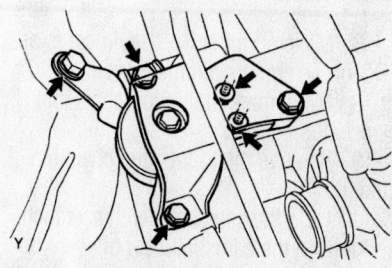

With the engine supported, remove the right side engine mount—1ZZ-FE engine shown, 2ZZ-GE similar

Heater Core replacement is covered in Section 2 of this manual

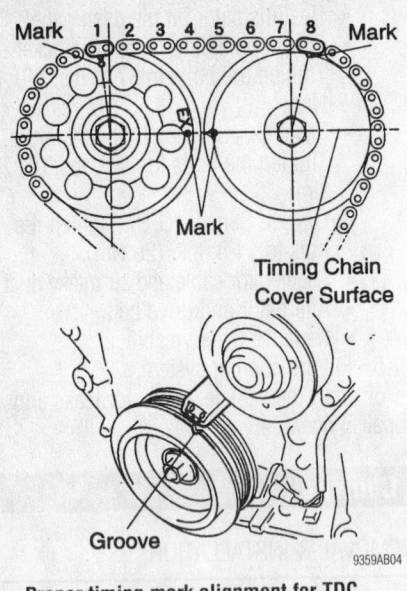

Proper timing mark alignment for TDC

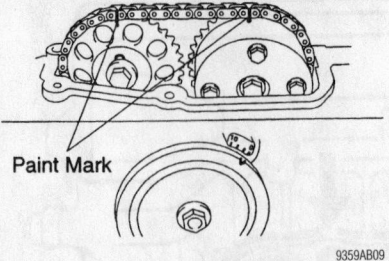

Matchmark the timing chain and cam sprockets

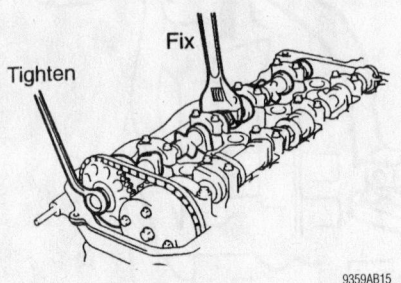

Hold the camshaft with a wrench while removing the set bolt

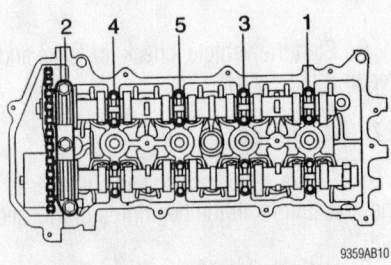

Camshaft bearing cap bolt removal sequence—1ZZ-FE engine

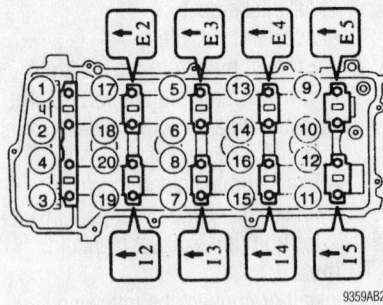

Camshaft bearing cap bolt removal sequence—2ZZ-GE engine

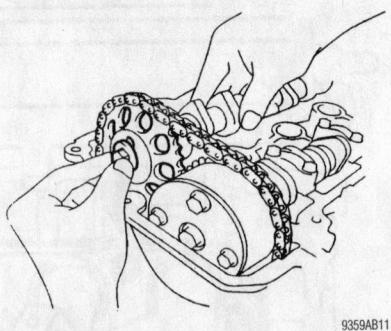

Carefully remove the cam and timing gear

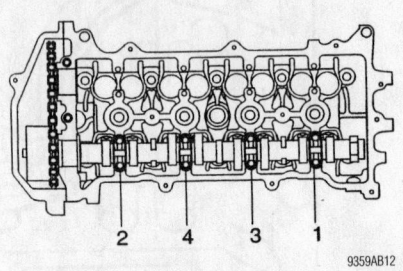

Camshaft bearing cap bolt removal sequence—1ZZ-FE engine

d. Bolt and nut holding the engine wire.

5. Remove or disconnect the following:
• Ignition coil assembly
• Positive Crankcase Ventilation (PCV) hoses from the valve cover
• Valve (cylinder head) cover sub-assembly

6. Set the No. 1 cylinder to Top Dead Center (TDC) of the compressor stroke as follows:

a. Turn the crankshaft pulley, and align its groove with the "0" timing mark of the timing chain cover.

b. Make sure the point marks of the camshaft timing sprockets and VVT timing sprockets are in a straight line as shown. If not, turn the crankshaft 1 complete revolution (360°) and align the marks.

7. Remove the drive belt tensioner.

✳✳ WARNING

Do not turn the crankshaft without the tensioner installed.

8. Make sure the No. 1 cylinder is at TDC of the compression stroke.

9. Matchmark the timing chain and camshaft sprockets

10. Remove the 2 nuts and chain tensioner.

11. Hold the camshafts with a wrench and loosen the camshaft set bolt.

12. Using several passes, gradually remove the bearing cap bolts from the No. 2 camshaft, in the proper sequence.

13. Remove the camshaft and timing gear as shown.

14. Using several passes, gradually remove the bearing cap bolts from the other camshaft, in the proper sequence.

15. Remove the camshaft while holding the timing chain.

✳✳ WARNING

Do not let anything drop down into the timing chain cover while the camshafts are removed.

16. Tie the timing chain with a string as shown, to prevent it from dropping down into the timing chain cover.

To install:

17. Position the camshaft on the cylinder head, then install the timing chain on the cam timing gear, with the painted links aligned with the marks on the timing gear.

18. Check the front marks and numbers and torque the camshaft cap bolts, in sequence, to 10 ft. lbs. (13 Nm) for 1ZZ-FE engine, or to 14 ft. lbs. (19 Nm) for 2ZZ-GE engines.

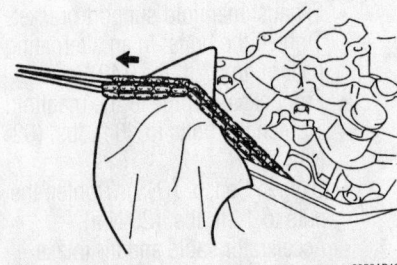

Secure the timing chain with string to prevent it from slipping down into the timing chain cover

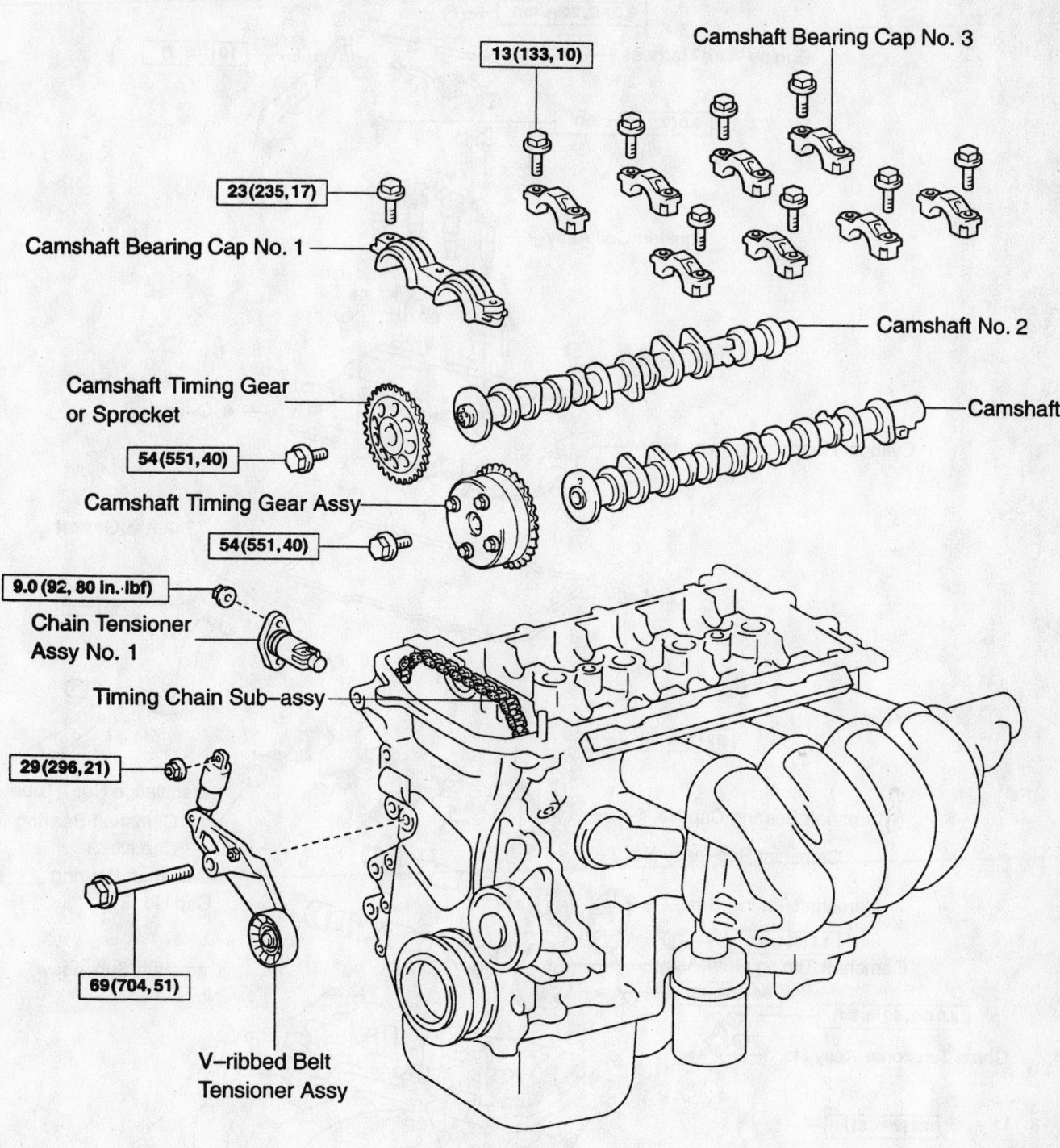

13(133,10)

Camshaft Bearing Cap No. 3

23(235,17)

Camshaft Bearing Cap No. 1

Camshaft No. 2

Camshaft Timing Gear
or Sprocket

Camshaft

54(551,40)

Camshaft Timing Gear Assy

54(551,40)

9.0 (92, 80 in.·lbf)

Chain Tensioner
Assy No. 1

Timing Chain Sub-assy

29(296,21)

69(704,51)

V-ribbed Belt
Tensioner Assy

N·m (kgf·cm, ft·lbf) : Specified torque

Exploded view of the camshafts and related components—1ZZ-FE engine

9359AB21

Brake service is covered in Section 4 of this manual

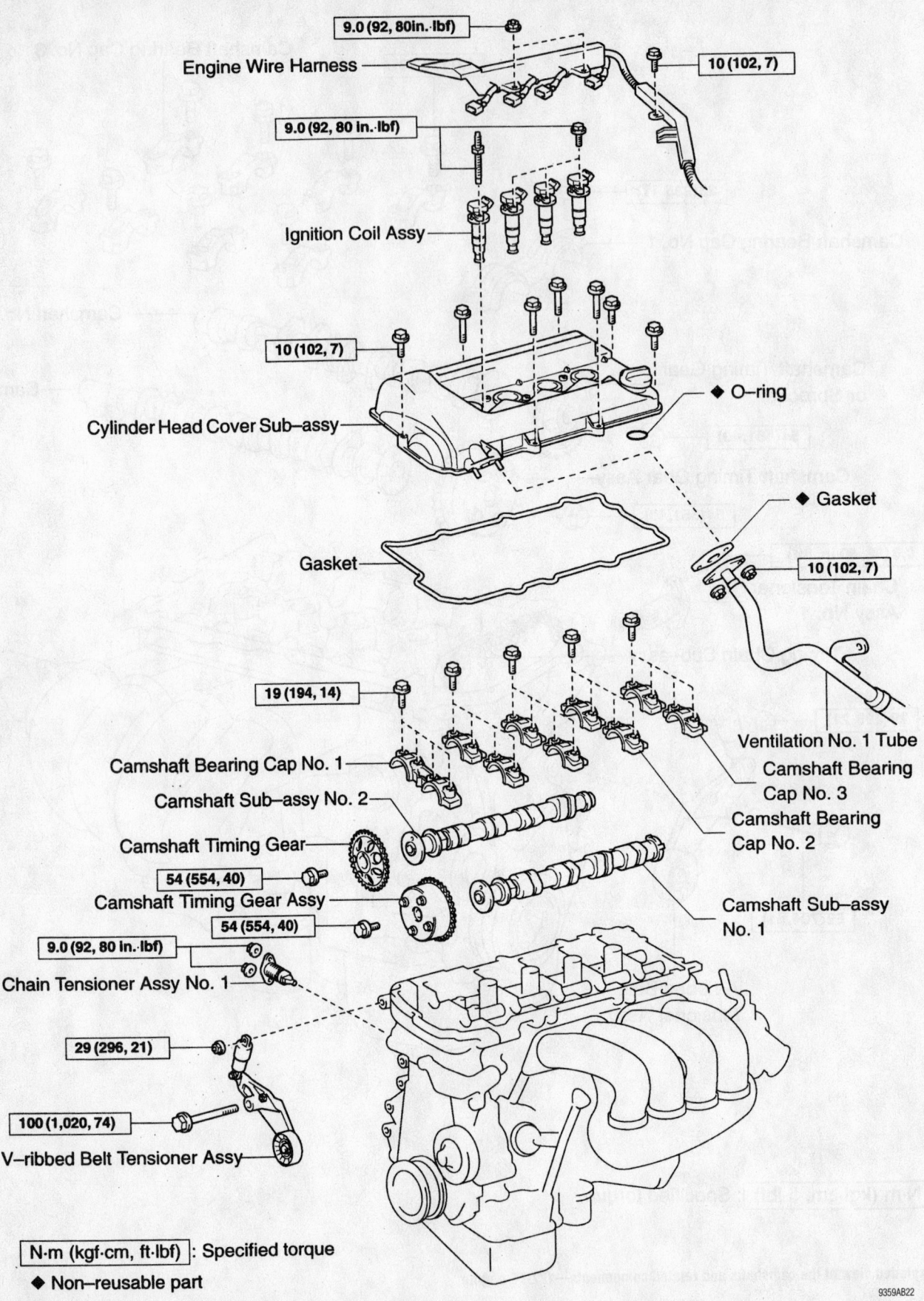

9.0 (92, 80 in.·lbf)

Engine Wire Harness

10 (102, 7)

9.0 (92, 80 in.·lbf)

Ignition Coil Assy

10 (102, 7)

Cylinder Head Cover Sub–assy

◆ O–ring

◆ Gasket

Gasket

10 (102, 7)

19 (194, 14)

Ventilation No. 1 Tube

Camshaft Bearing Cap No. 1

Camshaft Bearing Cap No. 3

Camshaft Sub–assy No. 2

Camshaft Bearing Cap No. 2

Camshaft Timing Gear

54 (554, 40)

Camshaft Timing Gear Assy

Camshaft Sub–assy No. 1

54 (554, 40)

9.0 (92, 80 in.·lbf)

Chain Tensioner Assy No. 1

29 (296, 21)

100 (1,020, 74)

V–ribbed Belt Tensioner Assy

N·m (kgf·cm, ft·lbf) : Specified torque

◆ Non–reusable part

9359AB22

Exploded view of the camshafts and related components—2ZZ-GE engine

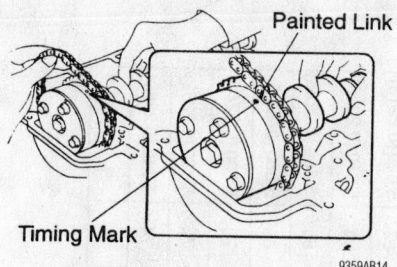

Make sure the alignment marks on the timing chain and camshaft gear match up

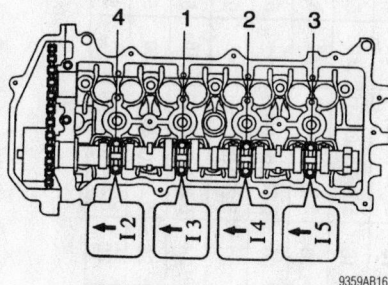

Camshaft cap bolt tightening sequence—1ZZ-FE engine

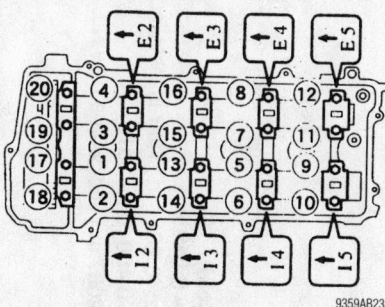

Camshaft cap bolt tightening sequence—2ZZ-GE engine

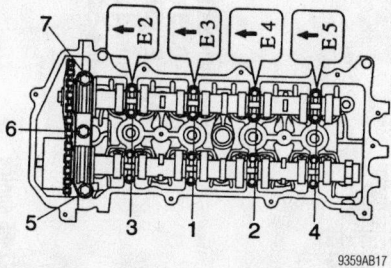

Camshaft cap bolt tightening sequence—1ZZ-FE

19. Put camshaft No. 2 on the cylinder head, with the painted links of the chain aligned with the mark on the timing gear.

20. Tighten the camshaft gear set bolt temporarily.

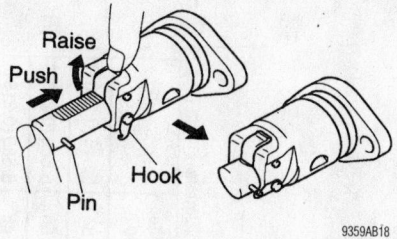

Set the timing chain tensioner hook properly

21. Check the front marks and numbers and torque the camshaft cap bolts, in sequence, to 10 ft. lbs. (13 Nm). Install the No. 1 bearing cap and tighten to 17 ft. lbs. (23 Nm).

22. Hold the camshaft secure with a wrench and tighten the set bolt to 40 ft. lbs. (54 Nm). Be careful not the damage the lifters.

23. Check to be sure the matchmarks on the timing chain and cam sprockets, and the alignment of the pulley groove with the timing mark on the cover are still aligned.

24. Install the chain tensioner:
   a. Make sure the O-ring is clean, then set the hook as shown.
   b. Oil the tensioner, then install and tighten to 80 inch lbs. (9 Nm).

➡ **When installing the tensioner, set the hook again if the hook releases the plunger.**

   c. Turn the crankshaft counterclockwise, and disconnect the plunger knock pin from the hook.
   d. Turn the crankshaft clockwise and check that the slipper is pushed by the plunger. If the plunger does not spring out, press the slipper into the chain tensioner with a screwdriver so that the hook is released from the knock pin and the plunger springs out.

25. Check the valve clearance and make adjustments as needed.

26. Install or connect the following:
   • Belt tensioner. Tighten the nut to 21 ft. lbs. (29 Nm) and the bolt to 51 ft. lbs. (69 Nm).
   • Cylinder head sub-assembly cover. Install seal packing into the locations shown and install within 3 minutes. Tighten the "A" bolts to 8 ft. lbs. (11 Nm) and the "B" bolts to 80 inch lbs. (9 Nm) for 1ZZ-FE engine and to 7 ft. lbs. (10 Nm) for 2ZZ-GE engines.

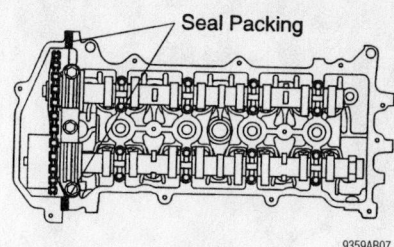

Seal packing installation locations

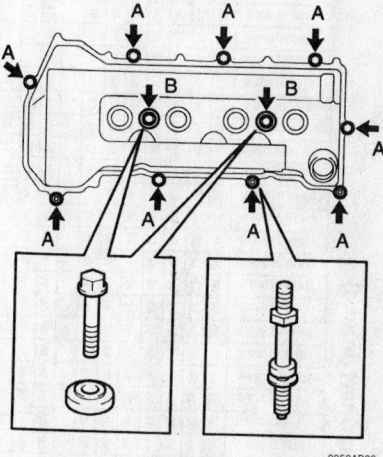

Cylinder head (valve) cover bolt locations—1ZZ-FE engine

   • Ignition coil assembly. Torque the bolts to 80 inch lbs. (9 Nm).
   • Engine wire and tighten to 80 inch lbs. (9 Nm).
   • Right side engine mount. Tighten to 38 ft. lbs. (52 Nm).
   • Cylinder head (valve) cover
   • Negative battery cable

Valve Lash

ADJUSTMENT

1ZZ-FE Engine

➡ **Adjust the valve clearance when the engine is cold.**

1. Before servicing the vehicle, refer to the precautions in the beginning of this section.
2. Remove or disconnect the following:
   • Negative battery cable.
   • Cylinder head covers
   • Engine wire
   • Ignition coil
   • Positive Crankcase Ventilation (PCV) hoses
   • Cylinder head cover sub-assembly

For complete Engine Mechanical specifications, see Section 1 of this manual

1ZZ-FE: Valve Lifter Selection Chart (Intake)

New lifter thickness

Lifter No.	Thickness mm (in.)	Lifter No.	Thickness mm (in.)	Lifter No.	Thickness mm (in.)
06	5.060 (0.1992)	30	5.300 (0.2087)	54	5.540 (0.2181)
08	5.080 (0.2000)	32	5.320 (0.2094)	56	5.560 (0.2189)
10	5.100 (0.2008)	34	5.340 (0.2102)	58	5.580 (0.2197)
12	5.120 (0.2016)	36	5.360 (0.2110)	60	5.600 (0.2205)
14	5.140 (0.2024)	38	5.380 (0.2118)	62	5.620 (0.2213)
16	5.160 (0.2031)	40	5.400 (0.2126)	64	5.640 (0.2220)
18	5.180 (0.2039)	42	5.420 (0.2134)	66	5.660 (0.2228)
20	5.200 (0.2047)	44	5.440 (0.2142)	68	5.680 (0.2236)
22	5.220 (0.2055)	46	5.460 (0.2150)	70	5.700 (0.2244)
24	5.240 (0.2063)	48	5.480 (0.2157)	72	5.720 (0.2252)
26	5.260 (0.2071)	50	5.500 (0.2165)	74	5.740 (0.2260)
28	5.280 (0.2079)	52	5.520 (0.2173)		

Intake valve clearance (Cold):
0.15 – 0.25 mm (0.006 – 0.010 in.)

EXAMPLE: The 5.250 mm (0.2067 in.) lifter is installed, and the measured clearance is 0.400 mm (0.0157 in.).
Replace the 5.250 mm (0.2067 in.) lifter with a new No. 48 lifter.

Adjusting shim chart (intake)—1ZZ-FE engine

The Valve Lifter Selection Chart (Intake) cross-references the **Measured clearance mm (in.)** rows against the **Installed lifter thickness mm (in.)** columns to determine the new lifter number.

Measured clearance mm (in.) rows:

Measured clearance mm	(in.)
0.000 – 0.030	(0.0000 – 0.0012)
0.031 – 0.050	(0.0012 – 0.0020)
0.051 – 0.070	(0.0020 – 0.0028)
0.071 – 0.090	(0.0028 – 0.0035)
0.091 – 0.110	(0.0036 – 0.0043)
0.111 – 0.130	(0.0044 – 0.0051)
0.131 – 0.149	(0.0052 – 0.0059)
0.150 – 0.250	(0.0059 – 0.0098)
0.251 – 0.270	(0.0099 – 0.0106)
0.271 – 0.290	(0.0107 – 0.0114)
0.291 – 0.310	(0.0115 – 0.0122)
0.311 – 0.330	(0.0122 – 0.0130)
0.331 – 0.350	(0.0130 – 0.0138)
0.351 – 0.370	(0.0138 – 0.0146)
0.371 – 0.390	(0.0146 – 0.0154)
0.391 – 0.410	(0.0154 – 0.0161)
0.411 – 0.430	(0.0162 – 0.0169)
0.431 – 0.450	(0.0170 – 0.0177)
0.451 – 0.470	(0.0178 – 0.0185)
0.471 – 0.490	(0.0185 – 0.0193)
0.491 – 0.510	(0.0193 – 0.0201)
0.511 – 0.530	(0.0201 – 0.0209)
0.531 – 0.550	(0.0209 – 0.0217)
0.551 – 0.570	(0.0217 – 0.0224)
0.571 – 0.590	(0.0225 – 0.0232)
0.591 – 0.610	(0.0233 – 0.0240)
0.611 – 0.630	(0.0241 – 0.0248)
0.631 – 0.650	(0.0248 – 0.0256)
0.651 – 0.670	(0.0256 – 0.0264)
0.671 – 0.690	(0.0264 – 0.0272)
0.691 – 0.710	(0.0272 – 0.0280)
0.711 – 0.730	(0.0280 – 0.0287)
0.731 – 0.750	(0.0288 – 0.0295)
0.751 – 0.770	(0.0296 – 0.0303)
0.771 – 0.790	(0.0304 – 0.0311)
0.791 – 0.810	(0.0311 – 0.0319)
0.811 – 0.830	(0.0319 – 0.0327)
0.831 – 0.850	(0.0327 – 0.0335)
0.851 – 0.870	(0.0335 – 0.0343)
0.871 – 0.890	(0.0343 – 0.0350)
0.891 – 0.910	(0.0351 – 0.0358)
0.911 – 0.930	(0.0359 – 0.0366)

9307WG70

1ZZ–FE: Valve Lifter Selection Chart (Exhaust)

New lifter thickness   mm (in.)

Lifter No.	Thickness	Lifter No.	Thickness	Lifter No.	Thickness
06	5.060 (0.1992)	30	5.300 (0.2087)	54	5.540 (0.2181)
08	5.080 (0.2000)	32	5.320 (0.2094)	56	5.560 (0.2189)
10	5.100 (0.2008)	34	5.340 (0.2102)	58	5.580 (0.2197)
12	5.120 (0.2016)	36	5.360 (0.2110)	60	5.600 (0.2205)
14	5.140 (0.2024)	38	5.380 (0.2118)	62	5.620 (0.2213)
16	5.160 (0.2031)	40	5.400 (0.2126)	64	5.640 (0.2220)
18	5.180 (0.2039)	42	5.420 (0.2134)	66	5.660 (0.2228)
20	5.200 (0.2047)	44	5.440 (0.2142)	68	5.680 (0.2236)
22	5.220 (0.2055)	46	5.460 (0.2150)	70	5.700 (0.2244)
24	5.240 (0.2063)	48	5.480 (0.2157)	72	5.720 (0.2252)
26	5.260 (0.2071)	50	5.500 (0.2165)	74	5.740 (0.2260)
28	5.280 (0.2079)	52	5.520 (0.2173)		

Exhaust valve clearance (Cold):
0.25 – 0.35 mm (0.010 – 0.014 in.)
EXAMPLE: The 5.340 mm (0.2102 in.) lifter is installed, and the measured clearance is 0.440 mm (0.0173 in.).
Replace the 5.340 mm (0.2102 in.) lifter with a new No. 48 lifter.

Adjusting shim chart (exhaust)—1ZZ-FE engine

9307WG71

Installed lifter thickness (columns), mm (in.): 5.060 (0.1992), 5.080 (0.2000), 5.100 (0.2008), 5.120 (0.2016), 5.160 (0.2031), 5.180 (0.2039), 5.200 (0.2047), 5.210 (0.2051), 5.220 (0.2055), 5.230 (0.2059), 5.240 (0.2063), 5.250 (0.2067), 5.260 (0.2071), 5.270 (0.2075), 5.280 (0.2079), 5.290 (0.2083), 5.300 (0.2087), 5.310 (0.2091), 5.320 (0.2094), 5.330 (0.2098), 5.340 (0.2102), 5.350 (0.2106), 5.360 (0.2110), 5.370 (0.2114), 5.380 (0.2118), 5.390 (0.2122), 5.400 (0.2126), 5.410 (0.2130), 5.420 (0.2134), 5.430 (0.2138), 5.440 (0.2142), 5.450 (0.2146), 5.460 (0.2150), 5.470 (0.2154), 5.480 (0.2157), 5.490 (0.2161), 5.500 (0.2165), 5.510 (0.2169), 5.520 (0.2173), 5.530 (0.2177), 5.540 (0.2181), 5.550 (0.2185), 5.560 (0.2189), 5.570 (0.2193), 5.580 (0.2197), 5.590 (0.2201), 5.600 (0.2205), 5.620 (0.2213), 5.640 (0.2220), 5.660 (0.2228), 5.680 (0.2236), 5.700 (0.2244), 5.720 (0.2252), 5.740 (0.2260)

Measured clearance (rows), mm (in.): 0.000 – 0.030 (0.0000 – 0.0012), 0.031 – 0.050 (0.0012 – 0.0020), 0.051 – 0.070 (0.0020 – 0.0028), 0.071 – 0.090 (0.0028 – 0.0035), 0.091 – 0.110 (0.0036 – 0.0043), 0.111 – 0.130 (0.0044 – 0.0051), 0.131 – 0.150 (0.0052 – 0.0059), 0.151 – 0.170 (0.0059 – 0.0067), 0.171 – 0.190 (0.0067 – 0.0075), 0.191 – 0.210 (0.0075 – 0.0083), 0.211 – 0.230 (0.0083 – 0.0091), 0.231 – 0.249 (0.0091 – 0.0098), 0.250 – 0.350 (0.0098 – 0.0138), 0.351 – 0.370 (0.0138 – 0.0146), 0.371 – 0.390 (0.0146 – 0.0154), 0.391 – 0.410 (0.0154 – 0.0161), 0.411 – 0.430 (0.0162 – 0.0169), 0.431 – 0.450 (0.0170 – 0.0177), 0.451 – 0.470 (0.0178 – 0.0185), 0.471 – 0.490 (0.0185 – 0.0193), 0.491 – 0.510 (0.0193 – 0.0201), 0.511 – 0.530 (0.0201 – 0.0209), 0.531 – 0.550 (0.0209 – 0.0217), 0.551 – 0.570 (0.0217 – 0.0224), 0.571 – 0.590 (0.0225 – 0.0232), 0.591 – 0.610 (0.0232 – 0.0240), 0.611 – 0.630 (0.0241 – 0.0248), 0.631 – 0.650 (0.0248 – 0.0256), 0.651 – 0.670 (0.0256 – 0.0264), 0.671 – 0.690 (0.0264 – 0.0272), 0.691 – 0.710 (0.0272 – 0.0280), 0.711 – 0.730 (0.0280 – 0.0287), 0.731 – 0.750 (0.0288 – 0.0295), 0.751 – 0.770 (0.0296 – 0.0303), 0.771 – 0.790 (0.0304 – 0.0311), 0.791 – 0.810 (0.0311 – 0.0319), 0.811 – 0.830 (0.0319 – 0.0327), 0.831 – 0.850 (0.0327 – 0.0335), 0.851 – 0.870 (0.0335 – 0.0343), 0.871 – 0.890 (0.0343 – 0.0350), 0.891 – 0.910 (0.0351 – 0.0358), 0.911 – 0.930 (0.0359 – 0.0366), 0.931 – 0.950 (0.0367 – 0.0374), 0.951 – 0.970 (0.0374 – 0.0382), 0.971 – 0.990 (0.0382 – 0.0390), 0.991 – 1.010 (0.0390 – 0.0398), 1.011 – 1.030 (0.0398 – 0.0406)

New Shim thickness   mm (in.)

Shim No.	Thickness	Shim No.	Thickness	Shim No.	Thickness
00	2.000 (0.0787)	28	2.280 (0.0898)	56	2.560 (0.1008)
02	2.020 (0.0795)	30	2.300 (0.0906)	58	2.580 (0.1016)
04	2.040 (0.0803)	32	2.320 (0.0913)	60	2.600 (0.1024)
06	2.060 (0.0811)	34	2.340 (0.0921)	62	2.620 (0.1031)
08	2.080 (0.0819)	36	2.360 (0.0929)	64	2.640 (0.1039)
10	2.100 (0.0827)	38	2.380 (0.0937)	66	2.660 (0.1047)
12	2.120 (0.0835)	40	2.400 (0.0945)	68	2.680 (0.1055)
14	2.140 (0.0843)	42	2.420 (0.0953)	70	2.700 (0.1063)
16	2.160 (0.0850)	44	2.440 (0.0961)	72	2.720 (0.1071)
18	2.180 (0.0858)	46	2.460 (0.0969)	74	2.740 (0.1079)
20	2.200 (0.0866)	48	2.480 (0.0976)	76	2.760 (0.1087)
22	2.220 (0.0874)	50	2.500 (0.0984)	78	2.780 (0.1094)
24	2.240 (0.0882)	52	2.520 (0.0992)	80	2.800 (0.1102)
26	2.260 (0.0890)	54	2.540 (0.1000)		

Intake valve clearance (Cold):
0.08 – 0.18 mm (0.0031 – 0.0071 in.)
EXAMPLE: The 2.200 mm (0.0826 in.) shim is installed, and
the measured clearance is 0.400 mm (0.0157 in.).
Replace the 2.600 mm (0.1024 in.) shim with a new No. 60 shim.

Adjusting shim chart (intake)—2ZZ-GE engine

Measure clearance mm(in.) \ Installed shim thickness mm(in.)	2.000 (0.0787)	2.020 (0.0795)	2.040 (0.0803)	2.060 (0.0811)	2.080 (0.0819)	2.100 (0.0827)	2.120 (0.0835)	2.140 (0.0843)	2.160 (0.0850)	2.180 (0.0858)	2.190 (0.0862)	2.200 (0.0866)	2.210 (0.0870)	2.220 (0.0874)	2.230 (0.0878)	2.240 (0.0882)	2.250 (0.0886)	2.260 (0.0890)	2.270 (0.0894)	2.280 (0.0898)	2.290 (0.0902)	2.300 (0.0906)	2.310 (0.0909)	2.320 (0.0913)	2.330 (0.0917)	2.340 (0.0921)	2.340 (0.0921)	2.350 (0.0925)	2.360 (0.0929)	2.370 (0.0933)
0.000 - 0.030 (0.0000 - 0.0012)																														
0.031 - 0.050 (0.0012 - 0.0020)																												00	04	04
0.051 - 0.070 (0.0020 - 0.0028)																								00	02	02	04	04	06	06
0.071 - 0.090 (0.0028 - 0.0035)																					00	00	02	02	04	04	06	06	08	08
0.091 - 0.110 (0.0036 - 0.0043)																		00	02	02	04	04	06	06	08	08	10	10	12	12
0.111 - 0.130 (0.0044 - 0.0051)															00	00	02	02	04	04	06	06	08	08	10	10	12	12	14	14
0.131 - 0.150 (0.0052 - 0.0059)												00	02	02	04	04	06	06	08	08	10	10	12	12	14	14	16	16	18	18
0.151 - 0.170 (0.0059 - 0.0067)										00	02	02	04	04	06	06	08	08	10	10	12	12	14	14	16	16	18	18	20	20
0.171 - 0.190 (0.0067 - 0.0075)								00	02	04	06	06	08	08	10	10	12	12	14	14	16	16	18	18	20	20	22	22	24	24
0.191 - 0.210 (0.0075 - 0.0083)							00	02	04	06	08	08	10	10	12	12	14	14	16	16	18	18	20	20	22	22	24	24	26	26
0.211 - 0.230 (0.0083 - 0.0091)					00	02	04	06	08	10	12	12	14	14	16	16	18	18	20	20	22	22	24	24	26	26	28	28	30	30
0.231 - 0.239 (0.0091 - 0.0094)				00	02	04	06	08	10	12	14	14	16	16	18	18	20	20	22	22	24	24	26	26	28	28	30	30	32	32
0.301 - 0.320 (0.0119 - 0.0126)	06	08	10	12	14	16	18	20	22	24	26	26	28	28	30	30	32	32	34	34	36	36	38	38	40	40	42	42	44	44
0.321 - 0.340 (0.0126 - 0.0134)	08	10	12	14	16	18	20	22	24	26	28	28	30	30	32	32	34	34	36	36	38	38	40	40	42	42	44	44	46	46
0.341 - 0.360 (0.0134 - 0.0142)	12	14	16	18	20	22	24	26	28	30	32	32	34	34	36	36	38	38	40	40	42	42	44	44	46	46	48	48	50	50
0.361 - 0.380 (0.0142 - 0.0150)	16	18	20	22	24	26	28	30	32	34	36	36	38	38	40	40	42	42	44	44	46	46	48	48	50	50	52	52	54	54
0.381 - 0.400 (0.0150 - 0.0157)	18	20	22	24	26	28	30	32	34	36	38	38	40	40	42	42	44	44	46	46	48	48	50	50	52	52	54	54	56	56
0.401 - 0.420 (0.0158 - 0.0165)	22	24	26	28	30	32	34	36	38	40	42	42	44	44	46	46	48	48	50	50	52	52	54	54	56	56	58	58	60	60
0.421 - 0.440 (0.0166 - 0.0173)	24	26	28	30	32	34	36	38	40	42	44	44	46	46	48	48	50	50	52	52	54	54	56	56	58	58	60	60	62	62
0.441 - 0.460 (0.0174 - 0.0181)	28	30	32	34	36	38	40	42	44	46	48	48	50	50	52	52	54	54	56	56	58	58	60	60	62	62	64	64	66	66
0.461 - 0.480 (0.0181 - 0.0189)	30	32	34	36	38	40	42	44	46	48	50	50	52	52	54	54	56	56	58	58	60	60	62	62	64	64	66	66	68	68
0.481 - 0.500 (0.0189 - 0.0197)	34	36	38	40	42	44	46	48	50	52	54	54	56	56	58	58	60	60	62	62	64	64	66	66	68	68	70	70	72	72
0.501 - 0.520 (0.0197 - 0.0205)	36	38	40	42	44	46	48	50	52	54	56	56	58	58	60	60	62	62	64	64	66	66	68	68	70	70	72	72	74	74
0.521 - 0.540 (0.0205 - 0.0213)	40	42	44	46	48	50	52	54	56	58	60	60	62	62	64	64	66	66	68	68	70	70	72	72	74	74	76	76	78	78
0.541 - 0.560 (0.0213 - 0.0220)	42	44	46	48	50	52	54	56	58	60	62	62	64	64	66	66	68	68	70	70	72	72	74	74	76	76	78	78	80	80
0.561 - 0.580 (0.0221 - 0.0228)	46	48	50	52	54	56	58	60	62	64	66	66	68	68	70	70	72	72	74	74	76	76	78	78	80	80				
0.581 - 0.600 (0.0229 - 0.0236)	48	50	52	54	56	58	60	62	64	66	68	68	70	70	72	72	74	74	76	76	78	78	80	80						
0.601 - 0.620 (0.0237 - 0.0244)	52	54	56	58	60	62	64	66	68	70	72	72	74	74	76	76	78	78	80	80										
0.621 - 0.640 (0.0244 - 0.0252)	54	56	58	60	62	64	66	68	70	72	74	74	76	76	78	78	80	80												
0.641 - 0.660 (0.0252 - 0.0260)	58	60	62	64	66	68	70	72	74	76	78	78	80	80																
0.661 - 0.680 (0.0260 - 0.0268)	60	62	64	66	68	70	72	74	76	78	80	80																		
0.681 - 0.700 (0.0268 - 0.0276)	64	66	68	70	72	74	76	78	80																					
0.701 - 0.720 (0.0276 - 0.0283)	66	68	70	72	74	76	78	80																						
0.721 - 0.740 (0.0284 - 0.0291)	70	72	74	76	78	80																								
0.741 - 0.760 (0.0292 - 0.0299)	72	74	76	78	80																									
0.761 - 0.780 (0.0300 - 0.0307)	76	78	80																											
0.781 - 0.800 (0.0307 - 0.0315)	80																													
0.801 - 0.820 (0.0315 - 0.0323)	80																													

Measure clearance mm(in.) \ Installed shim thickness mm(in.)	2.380 (0.0937)	2.390 (0.0941)	2.400 (0.0945)	2.410 (0.0949)	2.420 (0.0953)	2.430 (0.0957)	2.440 (0.0961)	2.450 (0.0965)	2.460 (0.0969)	2.470 (0.0972)	2.480 (0.0976)	2.490 (0.0980)	2.500 (0.0984)	2.510 (0.0988)	2.520 (0.0992)	2.530 (0.0996)	2.540 (0.1000)	2.550 (0.1004)	2.560 (0.1008)	2.580 (0.1016)	2.600 (0.1024)	2.620 (0.1031)	2.640 (0.1039)	2.660 (0.1047)	2.680 (0.1055)	2.700 (0.1063)	2.720 (0.1071)	2.740 (0.1079)	2.760 (0.1087)	2.780 (0.1094)	2.800 (0.1102)
0.000 - 0.030 (0.0000 - 0.0012)	00	00	02	02	04	04	06	06	08	08	10	10	12	12	14	14	16	16	18	20	22	24	26	28	30	32	34	36	38	40	42
0.031 - 0.050 (0.0012 - 0.0020)	06	06	08	08	10	10	12	12	14	14	16	16	18	18	20	20	22	22	24	26	28	30	32	34	36	38	40	42	44	46	48
0.051 - 0.070 (0.0020 - 0.0028)	08	08	10	10	12	12	14	14	16	16	18	18	20	20	22	22	24	24	26	28	30	32	34	36	38	40	42	44	46	48	50
0.071 - 0.090 (0.0028 - 0.0035)	10	10	12	12	14	14	16	16	18	18	20	20	22	22	24	24	26	26	28	30	32	34	36	38	40	42	44	46	48	50	52
0.091 - 0.110 (0.0036 - 0.0043)	14	14	16	16	18	18	20	20	22	22	24	24	26	26	28	28	30	30	32	34	36	38	40	42	44	46	48	50	52	54	56
0.111 - 0.130 (0.0044 - 0.0051)	16	16	18	18	20	20	22	22	24	24	26	26	28	28	30	30	32	32	34	36	38	40	42	44	46	48	50	52	54	56	58
0.131 - 0.150 (0.0052 - 0.0059)	20	20	22	22	24	24	26	26	28	28	30	30	32	32	34	34	36	36	38	40	42	44	46	48	50	52	54	56	58	60	62
0.151 - 0.170 (0.0059 - 0.0067)	22	22	24	24	26	26	28	28	30	30	32	32	34	34	36	36	38	38	40	42	44	46	48	50	52	54	56	58	60	62	64
0.171 - 0.190 (0.0067 - 0.0075)	26	26	28	28	30	30	32	32	34	34	36	36	38	38	40	40	42	42	44	46	48	50	52	54	56	58	60	62	64	66	68
0.191 - 0.210 (0.0075 - 0.0083)	28	28	30	30	32	32	34	34	36	36	38	38	40	40	42	42	44	44	46	48	50	52	54	56	58	60	62	64	66	68	70
0.211 - 0.230 (0.0083 - 0.0091)	32	32	34	34	36	36	38	38	40	40	42	42	44	44	46	46	48	48	50	52	54	56	58	60	62	64	66	68	70	72	74
0.231 - 0.239 (0.0091 - 0.0094)	34	34	36	36	38	38	40	40	42	42	44	44	46	46	48	48	50	50	52	54	56	58	60	62	64	66	68	70	72	74	76
0.301 - 0.320 (0.0119 - 0.0126)	46	46	48	48	50	50	52	52	54	54	56	56	58	58	60	60	62	62	64	66	68	70	72	74	76	78	80	80			
0.321 - 0.340 (0.0126 - 0.0134)	48	48	50	50	52	52	54	54	56	56	58	58	60	60	62	62	64	64	66	68	70	72	74	76	78	80					
0.341 - 0.360 (0.0134 - 0.0142)	52	52	54	54	56	56	58	58	60	60	62	62	64	64	66	66	68	68	70	72	74	76	78	80							
0.361 - 0.380 (0.0142 - 0.0150)	56	56	58	58	60	60	62	62	64	64	66	66	68	68	70	70	72	72	74	76	78	80									
0.381 - 0.400 (0.0150 - 0.0157)	58	58	60	60	62	62	64	64	66	66	68	68	70	70	72	72	74	74	76	78	80										
0.401 - 0.420 (0.0158 - 0.0165)	62	62	64	64	66	66	68	68	70	70	72	72	74	74	76	76	78	78	80												
0.421 - 0.440 (0.0166 - 0.0173)	64	64	66	66	68	68	70	70	72	72	74	74	76	76	78	78	80	80													
0.441 - 0.460 (0.0174 - 0.0181)	68	68	70	70	72	72	74	74	76	76	78	78	80	80																	
0.461 - 0.480 (0.0181 - 0.0189)	70	70	72	72	74	74	76	76	78	78	80	80																			
0.481 - 0.500 (0.0189 - 0.0197)	74	74	76	76	78	78	80	80																							
0.501 - 0.520 (0.0197 - 0.0205)	76	76	78	78	80	80																									
0.521 - 0.540 (0.0205 - 0.0213)	80	80																													

New Shim thickness mm (in.)

Shim No.	Thickness	Shim No.	Thickness	Shim No.	Thickness
00	2.000 (0.0787)	28	2.280 (0.0898)	56	2.560 (0.1008)
02	2.020 (0.0795)	30	2.300 (0.0906)	58	2.580 (0.1016)
04	2.040 (0.0803)	32	2.320 (0.0913)	60	2.600 (0.1024)
06	2.060 (0.0811)	34	2.340 (0.0921)	62	2.620 (0.1031)
08	2.080 (0.0819)	36	2.360 (0.0929)	64	2.640 (0.1039)
10	2.100 (0.0827)	38	2.380 (0.0937)	66	2.660 (0.1047)
12	2.120 (0.0835)	40	2.400 (0.0945)	68	2.680 (0.1055)
14	2.140 (0.0843)	42	2.420 (0.0953)	70	2.700 (0.1063)
16	2.160 (0.0850)	44	2.440 (0.0961)	72	2.720 (0.1071)
18	2.180 (0.0858)	46	2.460 (0.0969)	74	2.740 (0.1079)
20	2.200 (0.0866)	48	2.480 (0.0976)	76	2.760 (0.1087)
22	2.220 (0.0874)	50	2.500 (0.0984)	78	2.780 (0.1094)
24	2.240 (0.0882)	52	2.520 (0.0992)	80	2.800 (0.1102)
26	2.260 (0.0890)	54	2.540 (0.1000)		

Exhaust valve clearance (Cold):
0.22 – 0.32 mm (0.0087 – 0.0126 in.)

EXAMPLE: The 2.200 mm (0.0862 in.) shim is installed, and the measured clearance is 0.500 mm (0.0197 in.). Replace the 2.540 mm (0.1000 in.) shim with a new No. 54 shim.

9359AB27

Adjusting shim chart (exhaust)—2ZZ-GE engine

For Tire, Wheel and Ball Joint specifications, see Section 1 of this manual

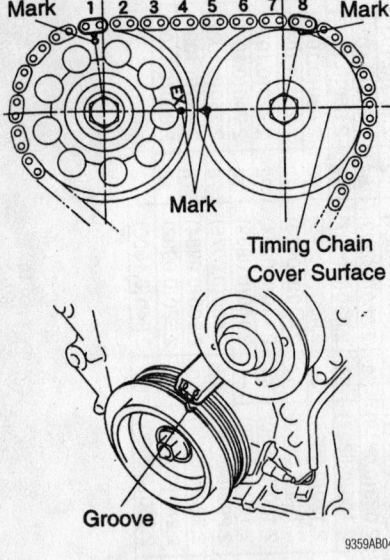

Timing Chain
Cover Surface

Mark

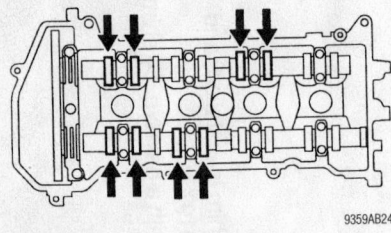

Groove

9359AB04

**Proper timing mark alignment for TDC—
1ZZ-FE and 2ZZ-GE engines**

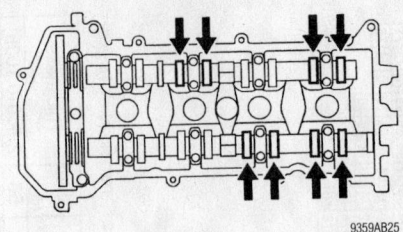

9359AB24

**Check the clearance of the 1st set of
valves—1ZZ-FE engine**

3. Set the No. 1 cylinder to Top Dead Center (TDC) of the compressor stroke as follows:

a. Turn the crankshaft pulley, and align its groove with the "0" timing mark of the timing chain cover.

b. Make sure the point marks of the camshaft timing sprockets and VVT timing sprockets are in a straight line as shown. If not, turn the crankshaft 1 complete revolution (360°) and align the marks.

4. Check the valve clearance of the first set of the valves shown:

a. Use a feeler gauge to measure the clearance between the valve lifter and camshaft. The clearance of the intake valves should be 0.0059–0.0098 in. (0.15–0.25mm). The clearance of the exhaust valves should be 0.0098–0.0138 in. (0.25–0.35mm).

b. Note the out-of-specification valve clearance measurements. You will need them later to determine the required replacement valve lifter.

c. Turn the crankshaft 1 revolution (360°) to set the No. 4 cylinder to TDC.

9359AB25

**Check the clearance of the 2nd set of
valves—1ZZ-FE engine**

5. Check the valve clearance of the second set of the valves shown:

a. Use a feeler gauge to measure the clearance between the valve lifter and camshaft. The clearance of the intake valves should be 0.0059–0.0098 in. (0.15–0.25mm). The clearance of the exhaust valves should be 0.0098–0.0138 in. (0.25–0.35mm).

b. Note the out-of-specification valve clearance measurements. You will need them later to determine the required replacement valve lifter.

6. Remove or disconnect the following:
- Drive belt
- Right side engine mount
- Drive belt tensioner

✱✱ WARNING

DO NOT turn the crankshaft while the tensioner is removed!

7. Set the No. 1 cylinder to TDC of the compression stroke.
- Camshafts
- Valve lifters.

8. Use a micrometer to measure the thickness of the used lifter. Calculate the thickness of a new lifter. so the valve clearance comes within the specified value:

a. A: Thickness of new lifter.

b. B: Thickness of used lifter.

c. C: Measured valve clearance.

d. Intake valve clearance: A = B + (C—0.0079 in. (0.20mm).

e. Exhaust valve clearance: A = B + (C—0.0118 in. (0.30mm).

f. Select a new lifter with a thickness as close as possible to the calculated values. Lifters come in 35 sizes in increments of 0.0008 in. (0.020mm) from 0.1992–0.2260 in (5.060–5.740mm).

9. Install or connect the following:
- Camshafts
- Drive belt tensioner
- Right hand engine mount
- Cylinder head (valve) cover sub-assembly
- Ignition coil
- Engine wire

- Cylinder head (valve) cover
- Negative battery cable

2ZZ-GE Engine

➡ **Adjust the valve clearance when the engine is cold.**

1. Before servicing the vehicle, refer to the precautions in the beginning of this section.

2. Remove or disconnect the following:
- Negative battery cable.
- Right side engine under cover
- Cylinder head cover
- Ignition coil assembly
- Wire harness clamp
- Suction hose sub-assembly
- Cylinder head cover sub-assembly
- Drive belt
- Right side engine mount

3. Set the No. 1 cylinder to Top Dead Center (TDC) of the compressor stroke as follows:

a. Turn the crankshaft pulley, and align its groove with the "0" timing mark of the timing chain cover.

b. Make sure the point marks of the camshaft timing sprockets and VVT timing sprockets are in a straight line as shown. If not, turn the crankshaft 1 complete revolution (360°) and align the marks.

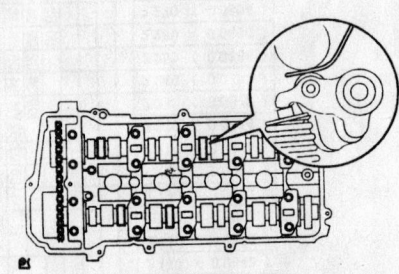

9359AB28

**Check the clearance of the 1st set of
valves—2ZZ-GE engine**

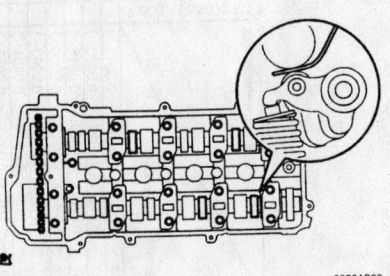

9359AB29

**Check the clearance of the 2nd set of
valves—2ZZ-GE engine**

4. Check the valve clearance of the first set of the valves shown:

a. Use a feeler gauge to measure the clearance between the valve lifter and camshaft. The clearance of the intake valves should be 0.0031–0.0071 in. (0.08–0.18mm). The clearance of the exhaust valves should be 0.0087–0.0126 in. (0.22–0.32mm).

b. Note the out-of-specification valve clearance measurements. You will need them later to determine the required replacement valve lifter.

c. Turn the crankshaft 1 revolution (360°) to set the No. 4 cylinder to TDC.

5. Check the valve clearance of the second set of the valves shown:

a. Use a feeler gauge to measure the clearance between the valve lifter and camshaft. The clearance of the intake valves should be 0.0031–0.0071 in. (0.08–0.18mm). The clearance of the exhaust valves should be 0.0087–0.0126 in. (0.22–0.32mm).

b. Note the out-of-specification valve clearance measurements. You will need them later to determine the required replacement valve lifter.

6. To adjust the intake valve clearance:

a. Set the SST. Turn the crankshaft so the related rocker arm, where the valve clearance is adjusted, is fully pushed down.

➡**Remove he spark plug and take off the compression.**

b. Insert SST 09248-77010 into the plug tube. The tool cannot be inserted unless the set screw is loosened.

c. Operate the lever so that the SST's seat surface comes to contact with the valve retainer and lock them with the set screw. Clearance between the valve retainer and SST's set surface is not allowed. Be careful not to make clearance when inserting the SST, since clearance may unlock the keeper.

d. Lock the set screw on the tube side of the SST.

e. Rotate the crankshaft so that the camshaft is position as shown. During rotation, pay attention to the direction, to prevent the nose of the camshaft from interfering with the SST's shaft. Do not rotate the crankshaft excessively.

f. Lift the rocker arm to make room and remove the adjusting shim using SST 09248-77010.

7. Determine the size of the replaced

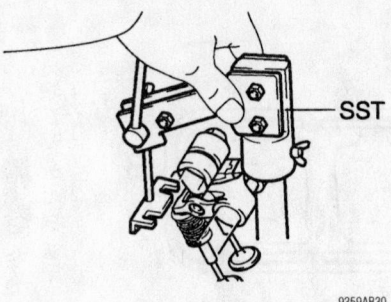

Insert the special tool into the plug tube—2ZZ-GE

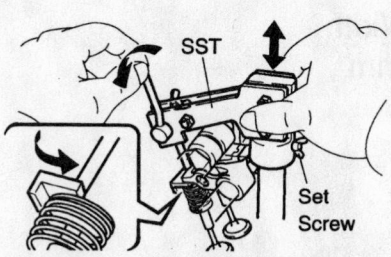

Operate the lever so that the SST's seat surface comes to contact with the valve retainer and lock them with the set screw

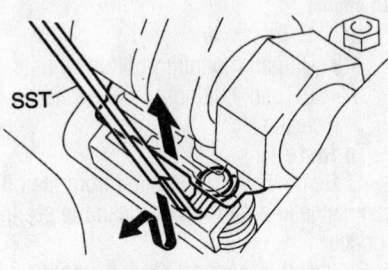

Setting the tool from the right side, makes shim removal easier—2ZZ-GE

shim according to the chart or the following formula:

a. Use a dial indicator to measure the thickness of the removed shim.

b. Calculate the thickness of a new shim so that the valve clearance comes within the specified value.

c. A: Thickness of new shim.

d. B: Thickness of used shim.

e. C: Measured valve clearance.

f. Intake: A = B + (C—0.005 in. [0.13mm])

g. Exhaust: A = B + (C—0.011 in. [0.27mm])

h. Select a new shim with a thickness as close as possible to the calculated

values. Shims come in 41 sizes in increments of 0.0008 in. (0.020mm) from 0.0787–0.1102 in (2.0–2.8mm).

8. Lift the rocker arm to make room, then install the adjusting shim using the SST. To remove the tool from the shim, push down on the rocker arm.

9. Turn the crankshaft so the related rocker arm, where the valve clearance is adjusted, is fully pushed down.

10. Loosen the 2 set-screws, then remove the SST.

11. Install all components in the reverse of the removal procedure.

Starter

REMOVAL & INSTALLATION

1. Before servicing the vehicle, refer to the precautions in the beginning of this section.

2. Remove or disconnect the following:
- Negative battery cable
- Right side engine undercover
- Starter wiring
- Starter

3. Installation is the reverse of removal. Torque the bolts to 27 ft. lbs. (37 Nm) and the nut to 7 ft. lbs. (10 Nm).

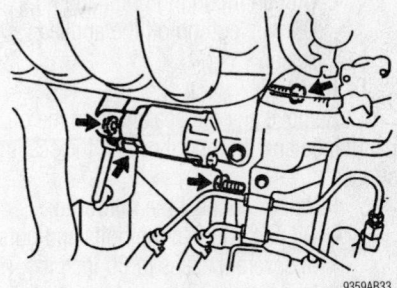

Starter mounting—Vibe

Oil Pan

REMOVAL & INSTALLATION

1ZZ-FE Engine

1. Before servicing the vehicle, refer to the precautions in the beginning of this section.

2. Drain the engine oil.

3. Remove or disconnect the following:
- Negative battery cable
- Undercovers
- Front exhaust pipe

For Wheel Alignment specifications, see Section 1 of this manual

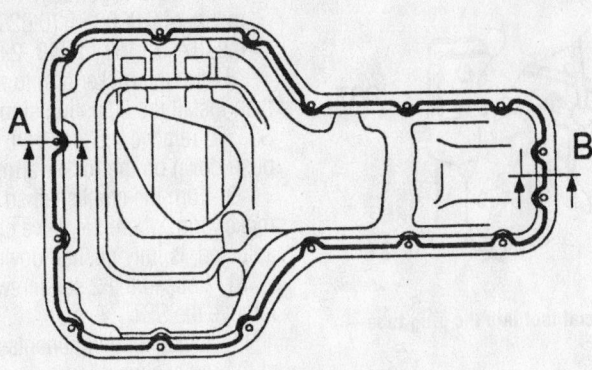

**Seal Width
4 – 5 mm**

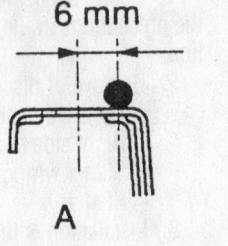

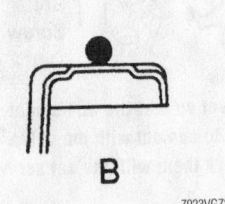

Apply sealant to the oil pan as shown—1.8L (1ZZ-FE) engine

- Oil pan mounting bolts and nuts
- Oil pan, cutting off the applied sealer.

To install:

4. Remove any old sealant from the oil pan flange and thoroughly clean the sealing surface.
5. Install or connect the following:
   - Oil pan. Tighten the bolts and nuts in several passes to 80 inch lbs. (9 Nm).
   - Front exhaust pipe
   - Negative battery cable
   - Undercovers
6. Fill the engine with clean oil.
7. Start the vehicle, check for leaks and repair if necessary.

2ZZ-GE Engine

1. Before servicing the vehicle, refer to the precautions in the beginning of this section.
2. Drain the engine oil.
3. Remove or disconnect the following:
   - Negative battery cable. On vehicles equipped with an air bag, wait at least 90 seconds before proceeding.
   - Undercovers
   - Front exhaust pipe

- Oil pan mounting bolts and nuts
- Oil pan, cutting off the applied sealer

To install:

4. Remove any old sealant from the oil pan flange and thoroughly clean the sealing surface.
5. Install or connect the following:
   - Oil pan. Tighten the bolts and nuts in several passes to 80 inch lbs. (9 Nm).
   - Front exhaust pipe
   - Negative battery cable
   - Undercovers
6. Fill the engine with clean oil.
7. Start the vehicle, check for leaks and repair if necessary.

Oil Pump

REMOVAL & INSTALLATION

1ZZ-FE Engine

1. Before servicing the vehicle, refer to the precautions in the beginning of this section.
2. Drain the engine oil.
3. Remove or disconnect the following:
   - Negative battery cable

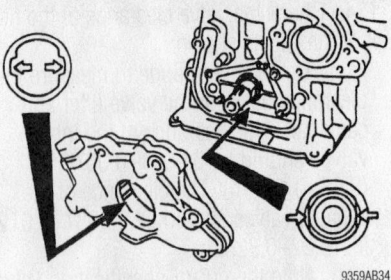

Oil pump mounting—1ZZ-FE and 2ZZ-GE engines

- Timing chain and crankshaft sprocket
- Timing chain vibration damper
- Oil pump bolts, pump and gasket

To install:

4. Clean the mounting surface.
5. Install or connect the following:
   - Oil pump, with new gasket. Engage the spline teeth of the oil pump drive rotor with the larger teeth of the crankshaft, and slide the pump on.
   - Oil pump bolts and tighten to 80 inch lbs. (9 Nm)
   - Crankshaft vibration damper and tighten to 80 inch lbs. (9 Nm)
   - Crankshaft sprocket and timing chain
   - Negative battery cable
6. Fill the engine with clean oil.
7. Start the vehicle, check for leaks and repair if necessary.

2ZZ-GE Engine

1. Before servicing the vehicle, refer to the precautions in the beginning of this section.
2. Drain the engine oil.
3. Remove or disconnect the following:
   - Negative battery cable
   - Timing chain and crankshaft sprocket
   - Oil pump and gasket

To install:

4. Clean the mounting surface.
5. Install or connect the following:
   - Oil pump, with new gasket. Engage the spline teeth of the oil pump drive rotor with the larger teeth of the crankshaft, and slide the pump on.
   - Oil pump bolts and tighten to 80 inch lbs. (9 Nm)
   - Crankshaft sprocket and timing chain
   - Negative battery cable
6. Fill the engine with clean oil.
7. Start the vehicle, check for leaks and repair if necessary.

Rear Main Seal

REMOVAL & INSTALLATION

1. Remove or disconnect the following:
   - Transaxle
   - Clutch assembly
   - Flywheel or flexplate
2. Use a small sharp knife to cut off the lip of the oil seal. Take great care not to score any metal with the knife.
3. Use a small prytool to pry the old seal from the retaining plate. Be careful not to damage the plate. Protect the tip of the tool with tape and pad the fulcrum point with cloth.
4. Inspect the crankshaft and seal lip contact surfaces for any sign of damage.

To install:

5. Apply a light coat of multi-purpose grease to the lip of a new oil seal. Loosely fit the seal into place by hand, making sure it is not crooked.
6. Use a seal driver of the correct size to install the seal. Tap it into place until the surface of the seal is flush with the edge of the housing.

Timing Chain, Sprockets, Front Cover and seal

REMOVAL & INSTALLATION

1. Before servicing the vehicle, refer to the precautions in the beginning of this section.
2. Drain the cooling system.
3. Remove or disconnect the following:
   - Right side engine under cover
   - Right front wheel and tire
   - Cylinder head cover
   - Wire harness clamp and suction hose assembly, 2ZZ-GE engine
   - Drive belt
4. Separate the vane pipe assembly, but do not disconnect the hose, 1ZZ-FE engine.
   - Alternator bracket, 2ZZ-GE
   - Alternator
   - Power steering pump reservoir and position it aside, 1ZZ-FE engine
5. Place a jack with a wooden block under the vehicle for support, then remove the 4 bolts and 2 nuts and remove the right side engine mount.
6. Remove the engine wire as follows, on 1ZZ-FE engines:
   a. Remove the 5 clamps from the brackets.

b. Detach the connectors.
   c. Remove the ignition coil connectors.
   d. Bolt and nut holding the engine wire.
7. Remove the engine wire as follows, on 2ZZ-GE engines:
   a. Detach the ignition coil, oil control valve and Crankshaft Position Sensor (CKP) sensor electrical connectors.
   b. Bolt and nut for the engine ground, then position the engine wire aside
8. Remove or disconnect the following:
   - Ignition coil assembly
   - Positive Crankcase Ventilation (PCV) hoses from the cylinder head cover, if necessary
   - Cylinder head (valve) cover sub-assembly
9. Set the No. 1 cylinder to Top Dead Center (TDC) of the compressor stroke as follows:
   a. Turn the crankshaft pulley, and

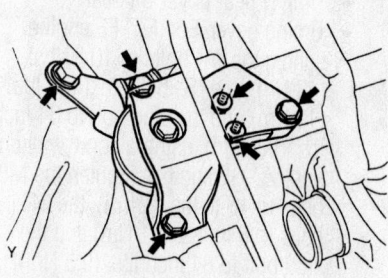

With the engine supported, remove the right side engine mount—1ZZ-FE engine shown, 2ZZ-GE similar

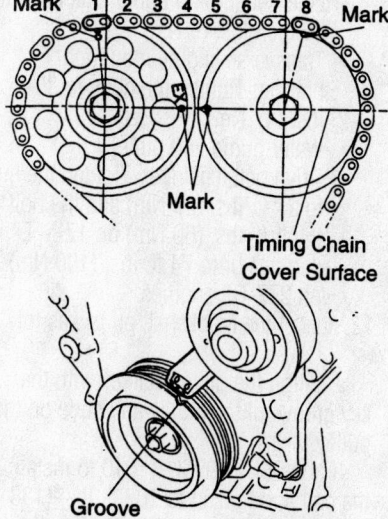

Proper timing mark alignment for TDC

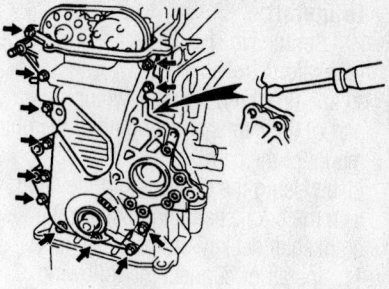

Timing chain cover mounting—1ZZ-FE engine shown, 2ZZ-GE similar

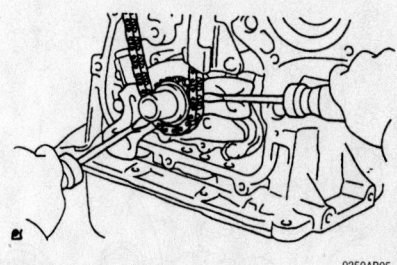

Remove the timing chain with the crankshaft gear

align its groove with the "0" timing mark of the timing chain cover.
   b. Make sure the point marks of the camshaft timing sprockets and VVT timing sprockets are in a straight line as shown. If not, turn the crankshaft 1 complete revolution (360°) and align the marks.
   - Crankshaft pulley, using SST 09960-10010
   - Belt tensioner
   - Water pump pulley, if equipped, and pump
   - Transverse engine mounting bracket
   - Crankshaft Position (CKP) sensor
   - No. 1 chain tensioner assembly, making sure not to revolve the crankshaft without the tensioner
   - Timing chain cover. The cover is retained with 11 bolts and nuts and a Torx® stud bolt. Pry the cover between the cylinder head and block to remove it.
   - Timing gear cover oil seal
   - CKP sensor plate No. 1
   - Timing chain tensioner slipper

➡ **In case you turn the camshafts with the timing chain removed, turn the crankshaft ¼ turn for the valve to avoid contact with the pistons.**

   - Timing chain sub-assembly. Remove the chain with the crankshaft gear, using screwdrivers as shown.

To install:

10. Set the No. 1 cylinder to TDC of the compression stroke:

a. Turn the hexagonal wrench head part of the camshafts, and align the point marks of the cam sprockets.

b. Using the crankshaft pulley bolt, turn the crankshaft and position the crankshaft set key upward.

11. Install or connect the following:

- Timing chain on the crank sprocket with the yellow link aligned with the mark on the crank sprocket. There are 3 yellow links on the timing chain.
- Crankshaft sprocket, using SST 09223-22010

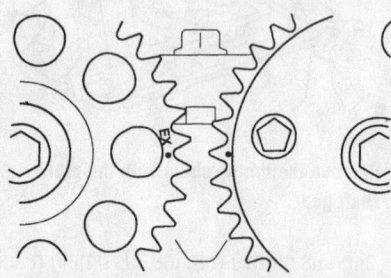

Proper alignment of the camshaft sprockets—1ZZ-FE engine

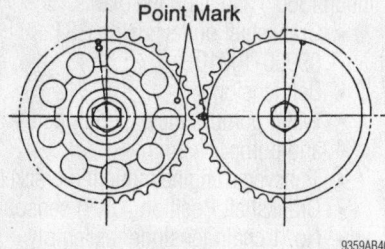

Proper alignment of the camshaft sprockets—2ZZ-GE engine

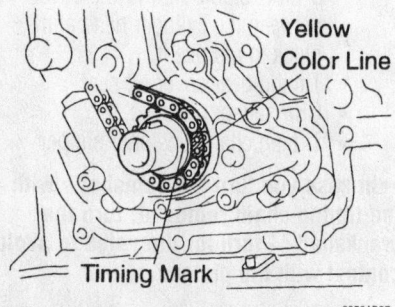

Make sure the yellow link is aligned with the crankshaft sprocket timing mark—1ZZ-FE and 2ZZ-GE engines

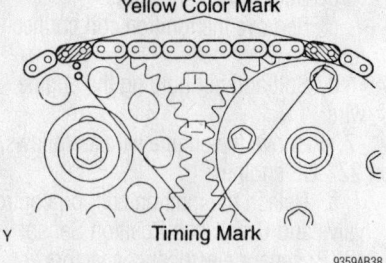

The yellow links of the timing chain must align with the camshaft sprocket timing marks—1ZZ-FE and 2ZZ-GE engines

- Timing chain on the camshaft sprockets with the yellow links aligned with the marks on the cam sprockets
- Timing chain tensioner slipper and tighten the bolt to 14 ft. lbs. (19 Nm)
- Crankshaft position sensor plate, with the "F" mark facing forward
- Timing gear cover oil seal
- Timing cover. For 1ZZ-FE engine, tighten the "A" bolts to 10 ft. lbs. (13 Nm), the "B" bolts to 14 ft. lbs. (19 Nm) and the stud bolt to 84 inch lbs. (9.5 Nm), using a Torx® wrench. For 2ZZ-GE engines, tighten the M8 bolts to 15 ft. lbs. (21 m), the M6 bolts to 8 ft. lbs. (11 Nm) and the stud bolt to 84 inch lbs. (9.5 Nm).

➡ **When installing the tensioner, make sure to set the hook again if the hook releases the plunger.**

- Timing chain tensioner. Torque the nuts to 80 inch lbs. (9 Nm).
- CKP sensor and tighten the bolts to 80 inch lbs. (9 Nm)
- Transverse engine mounting bracket. Tighten the bolts to 35 ft. lbs. (47 Nm).
- Water pump and pulley
- Drive belt tensioner. Tighten the nut to 21 ft. lbs. (29 Nm) and the bolt to 51 ft. lbs. (69 Nm) on 1ZZ-FE engines or to 74 ft. lbs. (100 Nm) on 2ZZ-GE engines.

12. Install the crankshaft pulley, as follows:

a. Align the pulley set key with the key groove of the pulley and slide on the pulley.

b. Use SST 09960-11010 to install the bolt and tighten to 102 ft. lbs. (138 Nm) for 1ZZ-FE engine or to 87 ft. lbs. (118 Nm) on 2ZZ-GE engines.

c. Turn the crankshaft counterclockwise and disconnect the plunger knock pin from the hook.

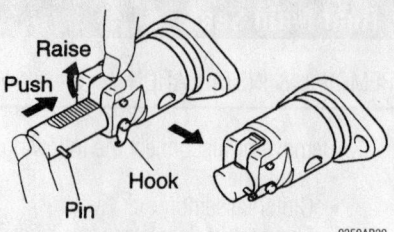

Timing chain tensioner—1ZZ-FE engine

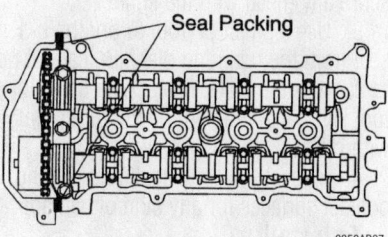

Seal packing installation locations

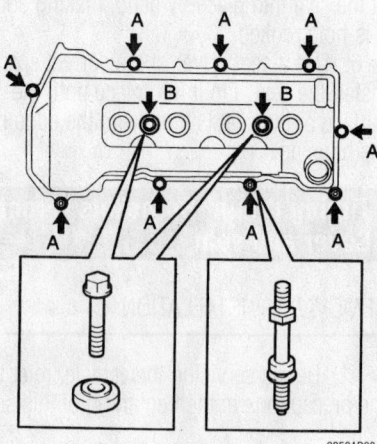

Cylinder head (valve) cover bolt locations—1ZZ-FE engine

d. Turn the crankshaft clockwise and check that the slipper is pushed by the plunger. If the plunger does not spring out, press the slipper into the chain tensioner with a screwdriver so that the hook is released from the knock pin and the plunger springs out.

- Cylinder head sub-assembly cover. Install seal packing into the locations shown and install within 3 minutes. Tighten the "A" bolts to 8 ft. lbs. (11 Nm) and the "B" bolts to 80 inch lbs. (9 Nm) for 1ZZ-FE engines. For 2ZZ-GE engines, tighten the bolts to 7 ft. lbs. (10 Nm).
- Ignition coil assembly. Torque the bolts to 80 inch lbs. (9 Nm).
- Engine wire and tighten to 80 inch lbs. (9 Nm)
- Right side engine mount. Tighten to 38 ft. lbs. (52 Nm).

- Alternator bracket, 2ZZ-GE engine
- Alternator
- Vane pump, 1ZZ-FE
- Main cylinder head cover and tighten to 62 inch lbs. (7 Nm)
- Right front wheel and tire. Tighten the lug nuts to 76 ft. lbs. (103 Nm).

13. Fill the cooling system to the proper level.

14. Start the vehicle, check for leaks and repair if necessary.

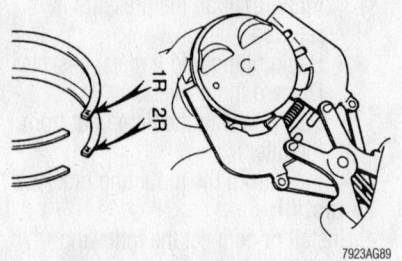

Piston ring identification mark locations—1ZZ-FE and 2ZZ-GE engines

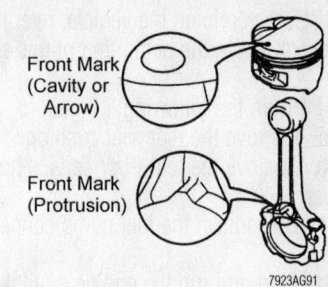

Piston-to-connecting rod assembly —1ZZ-FE and 2ZZ-GE engines

Piston and Ring Positioning

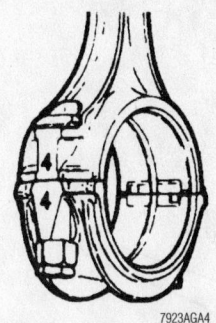

Before removing the caps from the connecting rods, be sure to matchmark them as shown

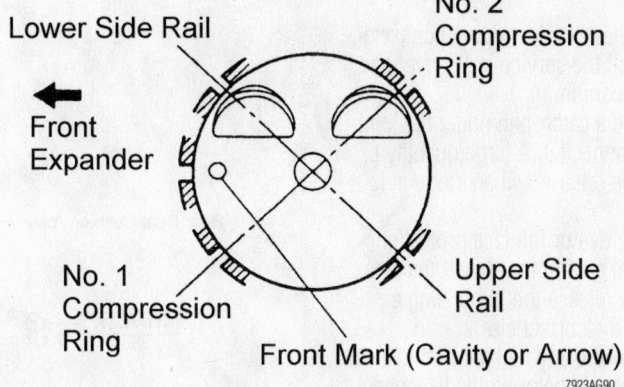

Piston ring end-gap spacing —1ZZ-FE and 2ZZ-GE engines

FUEL SYSTEM

Fuel System Service Precautions

Safety is the most important factor when performing not only fuel system maintenance, but any type of maintenance. Failure to conduct maintenance and repairs in a safe manner may result in serious personal injury or death. Work on a vehicle's fuel system components can be accomplished safely and effectively by adhering to the following rules and guidelines.

- To avoid the possibility of fire and personal injury, always disconnect the negative battery cable unless the repair or test procedure requires that battery voltage by applied.
- Always relieve the fuel system pressure prior to disconnecting any fuel system component (injector, fuel rail, pressure regulator, etc.) fitting or fuel line connection. Exercise extreme caution whenever relieving fuel system pressure, to avoid exposing skin, face and eyes to fuel spray. Please be advised that fuel under pressure may penetrate the skin or any part of the body that it contacts.

- Always place a shop towel or rag around the fitting or connection prior to loosening to absorb any excess fuel due to spillage. Ensure that all fuel spillage is quickly remove from engine surfaces. Ensure that all fuel-soaked cloths or towels are deposited into a flame-proof waste container with a lid.
- Always keep a dry chemical (Class B) fire extinguisher near the work area.
- Do not allow fuel spray or fuel vapors to come into contact with a light bulb, spark or open flame.
- Always use a second wrench when loosening or tightening fuel line connections fittings. This will prevent unnecessary stress and torsion to fuel piping. Always follow the proper torque specifications.
- Always replace worn fuel fitting O-rings with new ones. Do not substitute fuel hose where rigid pipe is installed.

Fuel System Pressure

RELIEVING

✳✳ CAUTION

Failure to relieve fuel pressure before repairs or disassembly can cause serious personal injury and/or property damage. Fuel pressure is maintained within the fuel lines, even if the engine is OFF or has not been run in a period of time. This pressure must be safely relieved before any fuel-bearing line or component is loosened or removed. On vehicles equipped with inflatable restraints or air bag systems, wait at least 90 seconds after disconnecting the battery cable before performing any other work. The back-up power will keep the restraint system energized for a period of time after the battery is disconnected.

1. Before servicing the vehicle, refer to the precautions in the beginning of this section.

2. Perform the following:

   a. Remove the rear seat cushion.

   b. Remove the rear floor service hole cover.

   c. Disconnect the fuel pump connector.

   d. Start and run the engine, until it stalls.

   e. Turn the ignition key to the **LOCK** position.

   f. Disconnect the negative battery cable.

   g. Connect the fuel pump connector.

   h. Install the service hole cover and rear seat cushion.

   i. Place a catch-pan under the joint to be disconnected. A large quantity of fuel may be released when the joint is opened.

   j. Wear eye or full face protection.

   k. Place a shop towel over the area and slowly release the joint using a wrench of the correct size.

   l. Allow the any fuel left in the line to bleed off slowly before fully disconnecting the joint.

   m. Plug the opened lines.

Fuel Filter

REMOVAL & INSTALLATION

1. Before servicing the vehicle, refer to the precautions in the beginning of this section.

2. Relieve the fuel system pressure.

3. Remove or disconnect the following:

   - Negative battery cable
   - Protective shield for the fuel filter
   - Air cleaner hose and cap, if necessary
   - Charcoal canister, if necessary
   - Slowly loosen the lower flare nut

fitting until all the pressure is relieved

   - Banjo fitting and 2 metal gaskets. Discard the gaskets.
   - Fuel line with the flared nut from the filter
   - Filter from the mounting bracket

To install:

4. Install or connect the following:

   - New fuel filter
   - Banjo fitting with a new metal gasket on each side and install the union bolt. Bolt: 22 ft. lbs. (30 Nm).
   - Flare nut to the lower connection. Nut: 22 ft. lbs. (30 Nm).
   - Charcoal canister
   - Air cleaner hose and cap
   - Protective shield
   - Negative battery cable

Fuel Pump

REMOVAL & INSTALLATION

1. Before servicing the vehicle, refer to the precautions in the beginning of this section.

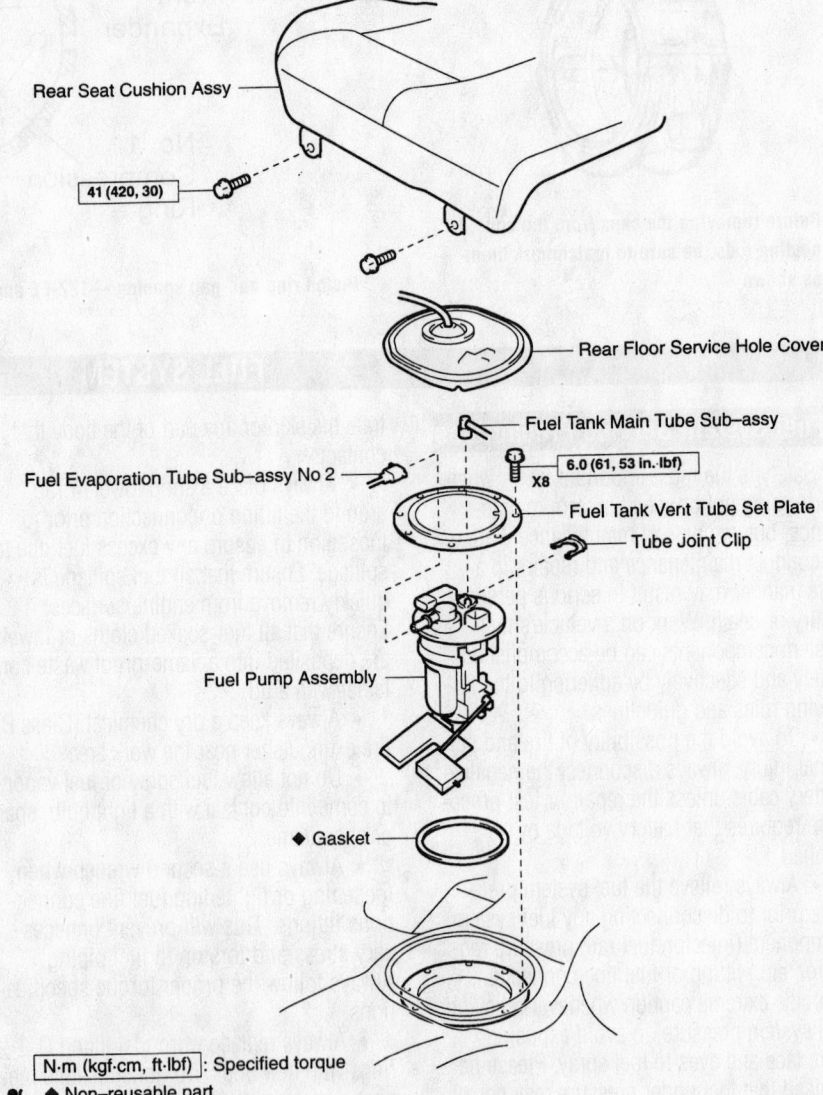

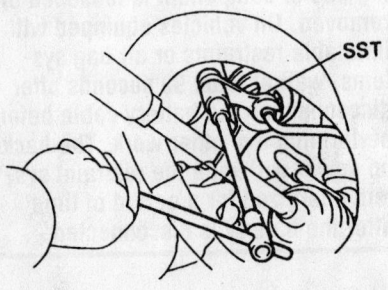

A line wrench with an extension may be needed to loosen the inlet line at the filter

7923VG85

Exploded view of the fuel pump mounting—FWD shown, AWD similar

9359AB41

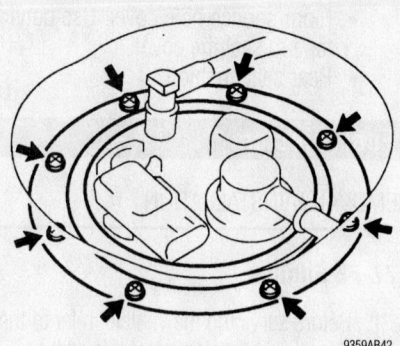

The fuel tank vent tube set plate is secured with 8 bolts on FWD vehicles

2. Remove or disconnect the following:
   - Negative battery cable
   - Rear seat cushion and floor service hole cover
   - Fuel pump and vapor pressure sensor connectors
   - Start and run the engine, until it stalls
3. Turn the ignition key to the **LOCK** position.
   - Negative battery cable
4. Connect the fuel pump connector.
   - Fuel tank protector, AWD vehicles

- Fuel tank main tube sub-assembly
- Fuel emission tube sub-assembly No. 1, FWD vehicles
- Fuel tank vent tube set plate. The plate is secured with 8 bolts on FWD vehicles, or 5 bolts on AWD vehicles.
- Fuel pump assembly, being careful not to damage the filter or bend the arm of the fuel sender gauge
- Fuel suction tube set gasket
- Fuel suction support No. 2

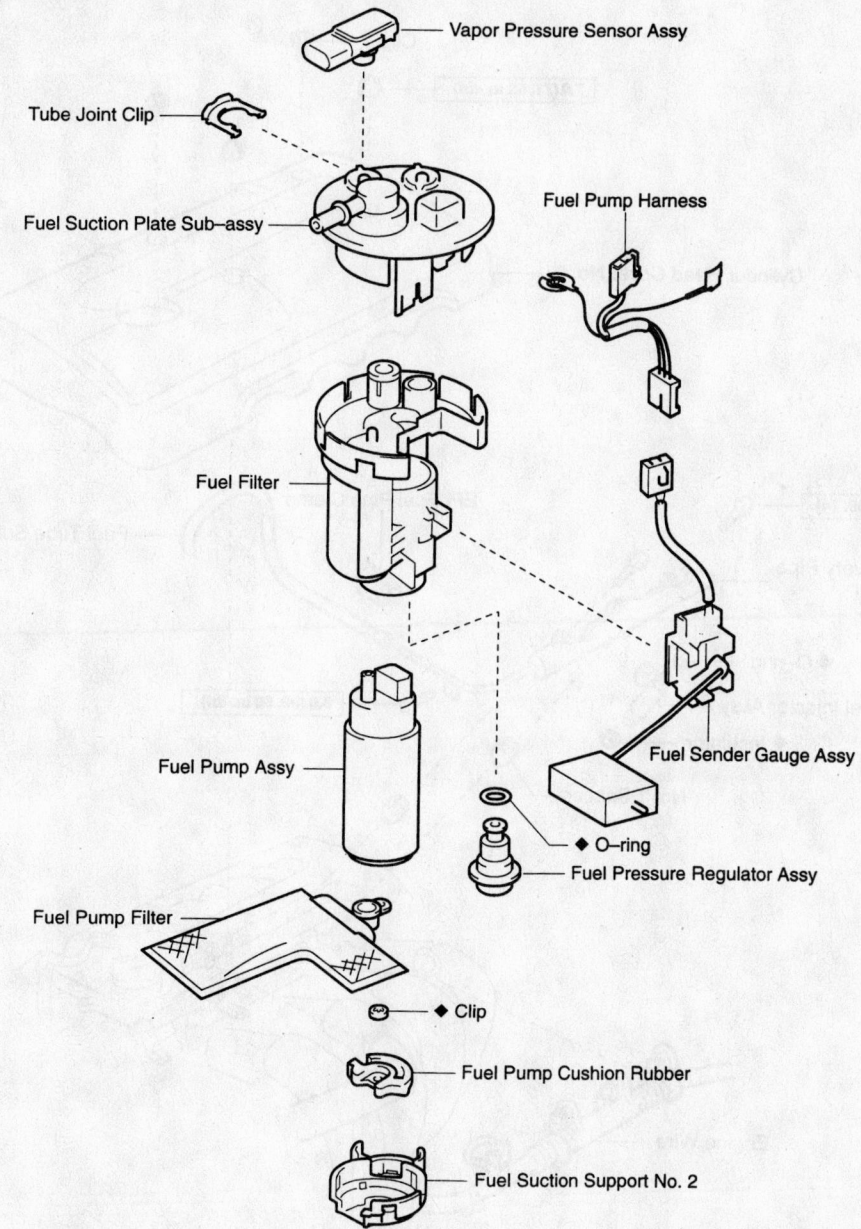

Fuel pump assembly components—FWD vehicles shown, AWD similar

- Fuel pump rubber cushion
- Fuel sender gauge assembly. Unplug the connector, then use a screwdriver to unlock the gauge and slide it to remove.
- Fuel section plate sub-assembly
- Vapor pressure sensor
- Fuel pump harness
- Fuel pump
- Fuel pump filter
- Fuel pressure regulator and O-ring

To install:

5. Install or connect the following:
   - New regulator O-ring and regulator

- Fuel pump filter
- Fuel pump
- Vapor pressure sensor
- Fuel suction tube set gasket
- Fuel pump assembly
- Fuel tank vent tube set plate. Tighten the bolts to 53 inch lbs. (6 Nm).
- Connect the fuel emission tube sub-assembly
- Fuel tank main tube sub-assembly
- Fuel tank protector No. 2, AWD vehicles
- Negative battery cable. Check for fuel leaks.

- Floor service hole cover. Use butyl tape to seal the cover.
- Rear seat cushion

Fuel Injectors

REMOVAL & INSTALLATION

1ZZ-FE Engine

1. Before servicing the vehicle, refer to the precautions in the beginning of this section.
2. Properly relieve the fuel system pressure.
3. Remove or disconnect the following:

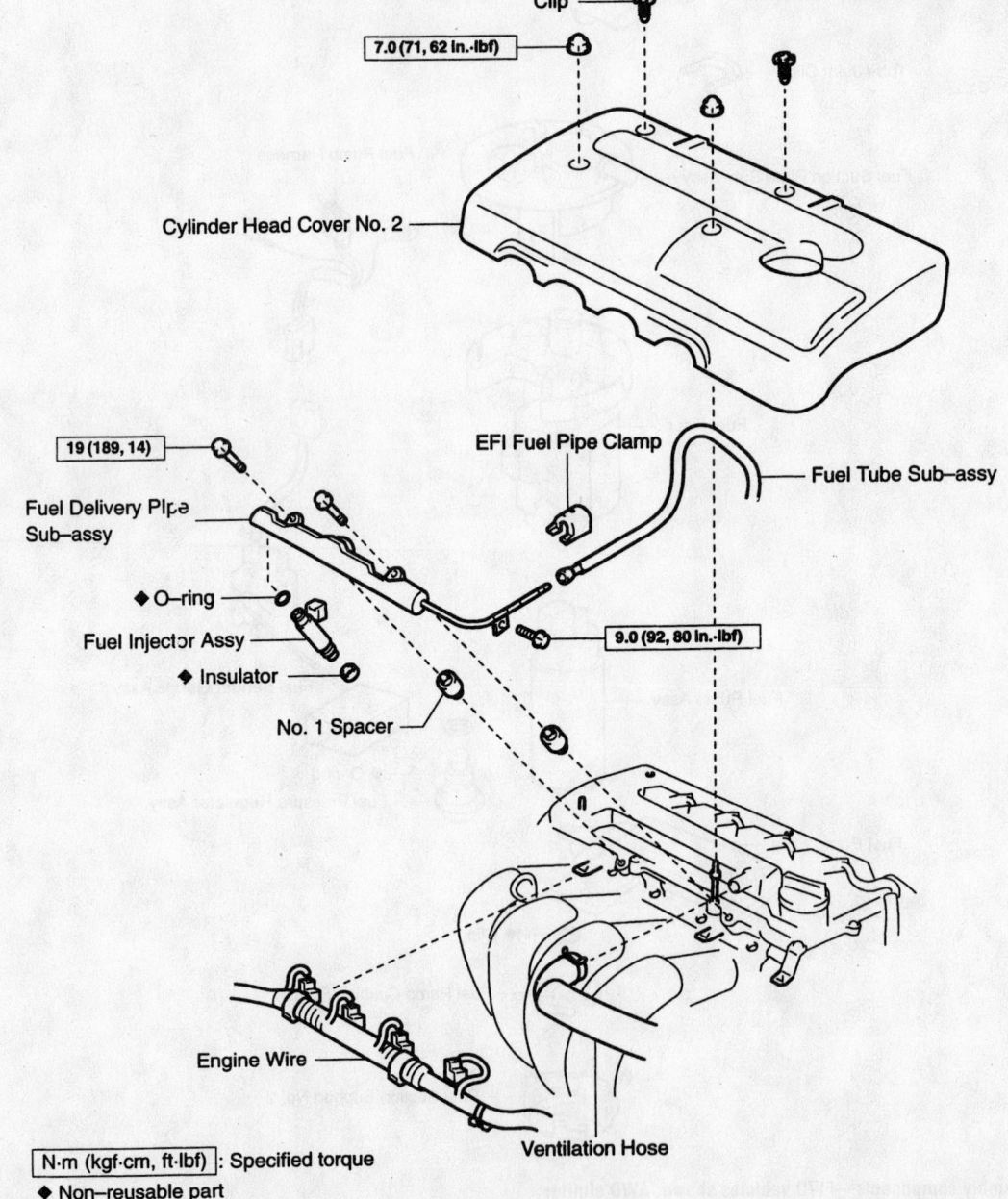

N·m (kgf·cm, ft·lbf): Specified torque
◆ Non-reusable part

Fuel injector removal and installation—1ZZ-FE engine

9359AB44

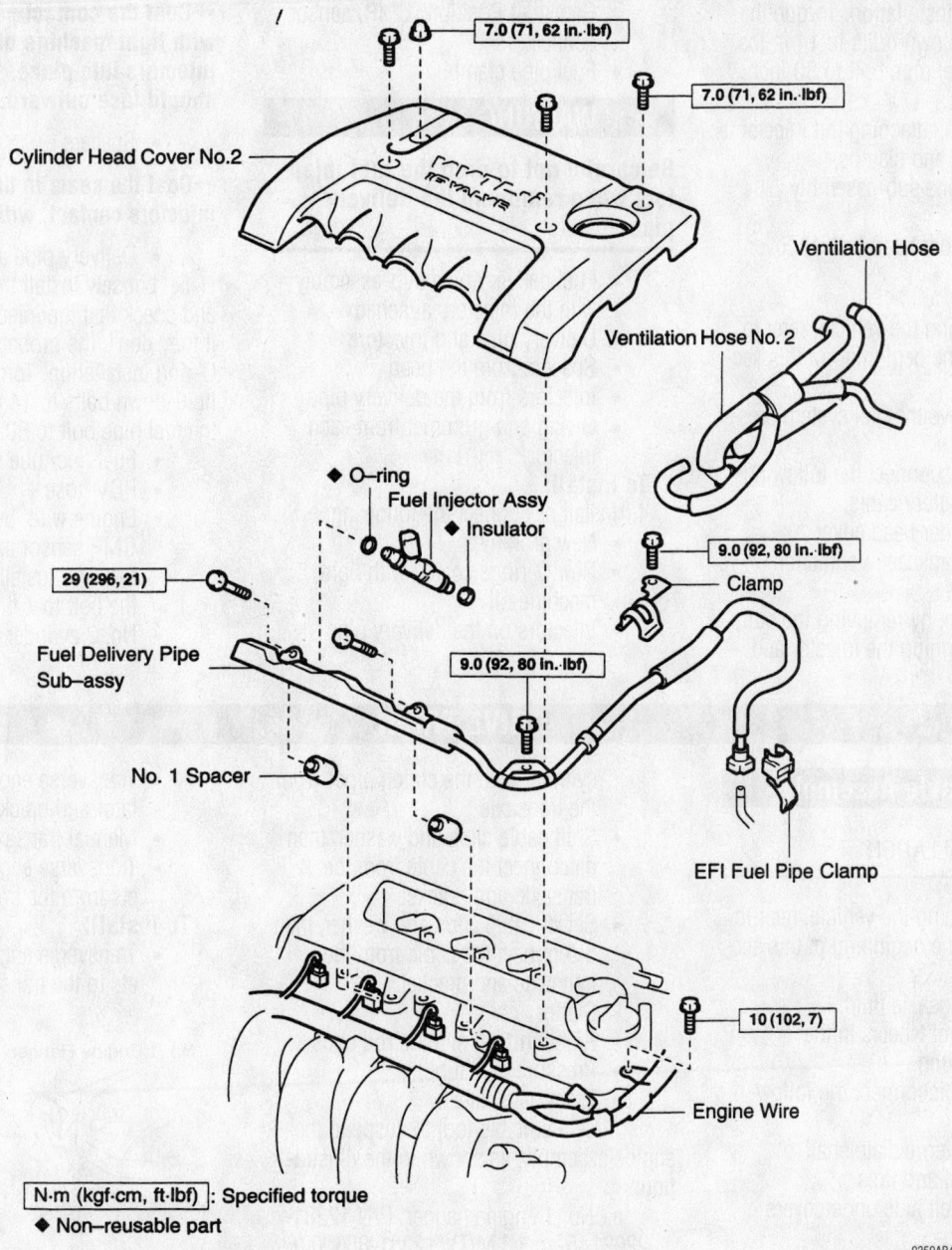

7.0 (71, 62 in.·lbf)

7.0 (71, 62 in.·lbf)

Cylinder Head Cover No.2

Ventilation Hose

Ventilation Hose No. 2

◆ O-ring
Fuel Injector Assy
◆ Insulator

9.0 (92, 80 in.·lbf)

Clamp

29 (296, 21)

9.0 (92, 80 in.·lbf)

Fuel Delivery Pipe
Sub-assy

No. 1 Spacer

EFI Fuel Pipe Clamp

10 (102, 7)

Engine Wire

N·m (kgf·cm, ft·lbf) : Specified torque
◆ Non–reusable part

9359AB45

Fuel injector removal and installation—2ZZ-GE engine

- Negative battery cable.
- No. 2 cylinder head cover
- Positive Crankcase Ventilation (PCV) hose
- Engine wire, unplugging the injector connectors and clamps
- Fuel pipe clamp
- Fuel line/tube sub-assembly

✴✴ WARNING

Be careful not to drop the fuel injectors when removing the delivery pipe.

- Fuel delivery pipe sub-assembly with the injectors attached
- Delivery pipe and injectors
- Spacers from the head
- Injectors from the delivery pipe
- O-ring and grommet from each injector

To install:
4. Install or connect the following:
- New grommets
- New O-rings coated with light machine oil
- Injectors on the delivery pipe

➡**Coat the contact point on the pipe with light machine oil and twist the injectors into place. The connector should face outward.**

- Spacers

➡**Coat the seats in the head where the injectors contact, with light machine oil.**

- Delivery pipe and injectors
5. Loosely install the hold-down bolts and check that the injectors rotate smoothly. If they don't, the probable cause

is incorrect O-ring installation. Torque the delivery pipe hold-down bolts to 14 ft. lbs. (19 Nm) and the fuel pipe bolt to 80 inch lbs. (9 Nm).

- Engine wire, attaching the injector connectors and clamps
- Fuel line/tube sub-assembly
- PCV hose
- No. 2 cylinder head (valve) cover

2ZZ-GE Engine

1. Before servicing the vehicle, refer to the precautions in the beginning of this section.

2. Properly relieve the fuel system pressure.

3. Remove or disconnect the following:
- Negative battery cable.
- No. 2 cylinder head cover
- Positive Crankcase Ventilation (PCV) hose
- Engine wire, by removing the bolt, then unplugging the injector and

Camshaft Position (CMP) sensor connectors
- Fuel pipe clamp

✷✷ WARNING

Be careful not to drop the fuel injectors when removing the delivery pipe.

- Fuel delivery pipe sub-assembly with the injectors attached
- Delivery pipe and injectors
- Spacers from the head
- Injectors from the delivery pipe
- O-ring and grommet from each injector

To install:

4. Install or connect the following:
- New grommets
- New O-rings coated with light machine oil
- Injectors on the delivery pipe

➡ Coat the contact point on the pipe with light machine oil and twist the injectors into place. The connector should face outward.

- Spacers

➡ Coat the seats in the head where the injectors contact, with light machine oil.

- Delivery pipe and injectors

5. Loosely install the hold-down bolts and check that the injectors rotate smoothly. If they don't, the probable cause is incorrect O-ring installation. Torque the delivery pipe hold-down bolts to 14 ft. lbs. (19 Nm) and the fuel pipe bolt to 80 inch lbs. (9 Nm).
- Fuel line/tube sub-assembly
- PCV hose
- Engine wire, by connecting the CMP sensor and injector connectors and installing the bolt. Tighten the bolt to 7 ft. lbs. (10 Nm).
- No. 2 cylinder head (valve) cover

DRIVE TRAIN

Manual Transaxle Assembly

REMOVAL & INSTALLATION

1. Before servicing the vehicle, refer to the precautions in the beginning of this section.

2. Drain the transaxle fluid.

3. Place the front wheels in the straight-ahead position.

4. Remove or disconnect the following:

- Steering intermediate shaft
- Front wheel and tires
- Right and left side undercovers
- Exhaust pipe
- Hood
- Cylinder head (valve) cover
- Air cleaner assembly
- Battery clamp, battery, battery tray and battery carrier
- Cruise control actuator assembly, if equipped

5. Remove the wire harness as follows:

a. Remove the wire harness clamp, 2 bolts and wire harness brackets.

b. Remove the 2 bolts and 2 ground cables.

6. Remove or disconnect the following:
- Back-up lamp switch connector, with ABS
- Speed sensor connector, without ABS
- 5 bolts, then separate the release

cylinder with the clutch pipes from the transaxle
- Shift cable clips and washer, then disconnect the cable from the transaxle and bracket
- Select cable clips and washer, then disconnect the cable from the transaxle and bracket
- Starter
- Right and left side tie rod ends
- Pressure feed tube
- Front halfshafts

7. Use a suitable tool to suspend the engine assembly, as shown in the illustration:

a. No. 1 engine hanger: P/N 12281-22021 (5-speed M/T), 12281-88600 (6-speed M/T).

b. No. 2 engine hanger: P/N 12281-15040 (5-speed M/T), 12281-88600 (6-speed M/T).

c. Bolt: P/N 91512-B1016.

d. Torque the bolts to 28 ft. lbs. (38 Nm).

✷✷ CAUTION

Do not try to suspend the engine by hooking the chain to any other part.

e. Attach an engine chain hoist to the hangers.

8. Remove or disconnect the following:
- Front suspension crossmember

9. Support the transaxle with a floor jack.

- Transverse engine mounting insulator and brackets
- Manual transaxle assembly
- Transverse engine mounting brackets from the transaxle, if necessary

To install:
- Transverse engine mounting brackets to the transaxle, if necessary

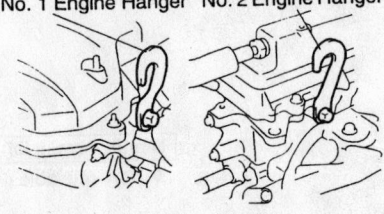

Secure the engine using the proper tools—5-speed manual transmission shown

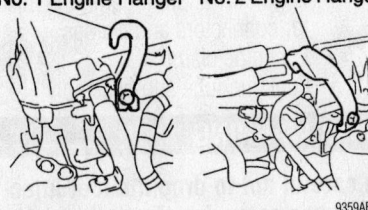

Secure the engine using the proper tools—6-speed manual transmission shown

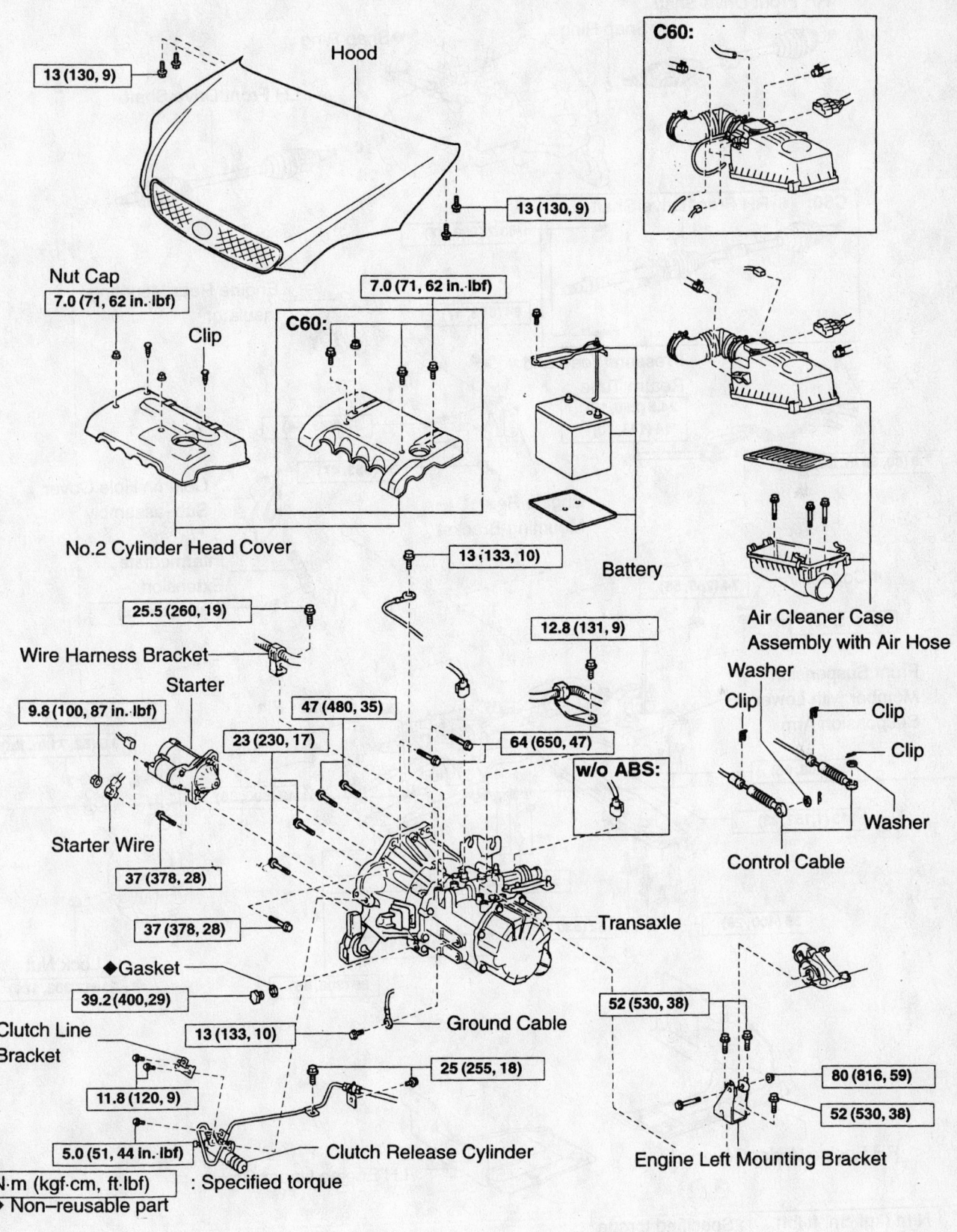

C60:

Hood

13 (130, 9)

13 (130, 9)

Nut Cap
7.0 (71, 62 in.·lbf)

7.0 (71, 62 in.·lbf)

Clip

C60:

No.2 Cylinder Head Cover

Battery

13 (133, 10)

Air Cleaner Case Assembly with Air Hose

25.5 (260, 19)

Wire Harness Bracket

12.8 (131, 9)

Washer

Clip

Clip

Starter

Clip

9.8 (100, 87 in.·lbf)

47 (480, 35)

Washer

23 (230, 17)

64 (650, 47)

Control Cable

w/o ABS:

Starter Wire

37 (378, 28)

Transaxle

37 (378, 28)

◆**Gasket**

39.2 (400, 29)

52 (530, 38)

Clutch Line Bracket

Ground Cable

11.8 (120, 9)

25 (255, 18)

80 (816, 59)

52 (530, 38)

5.0 (51, 44 in.·lbf)

Clutch Release Cylinder

Engine Left Mounting Bracket

N·m (kgf·cm, ft·lbf) : Specified torque
◆ Non–reusable part

9359AB47

Exploded view of the manual transaxle (1 of 2)

Timing belt service is covered in Section 3 of this manual

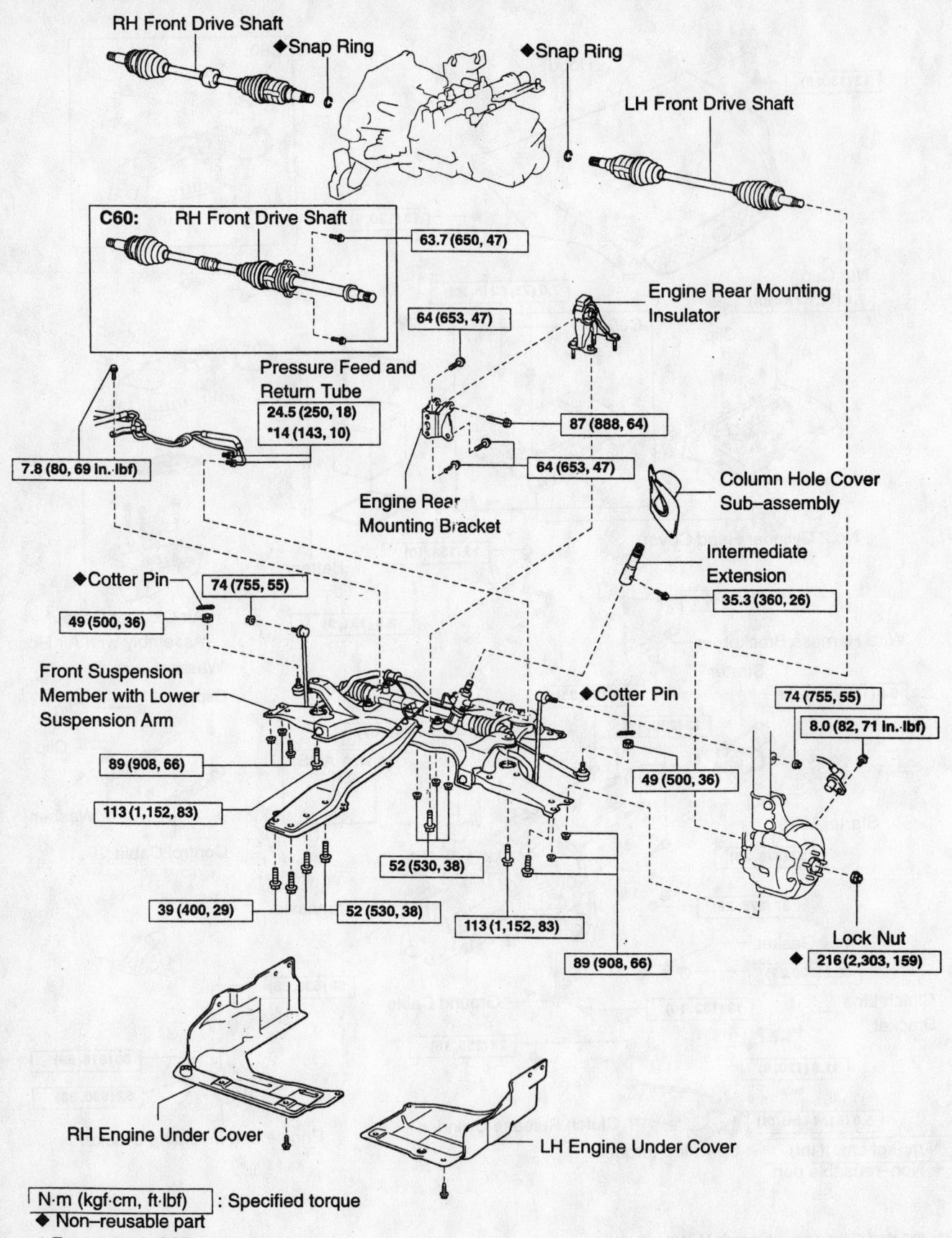

RH Front Drive Shaft

◆Snap Ring

◆Snap Ring

LH Front Drive Shaft

C60: RH Front Drive Shaft

63.7 (650, 47)

Engine Rear Mounting
Insulator

64 (653, 47)

Pressure Feed and
Return Tube

24.5 (250, 18)
*14 (143, 10)

87 (888, 64)

7.8 (80, 69 in. lbf)

64 (653, 47)

Column Hole Cover
Sub–assembly

Engine Rear
Mounting Bracket

Intermediate
Extension

35.3 (360, 26)

◆Cotter Pin

74 (755, 55)

49 (500, 36)

Front Suspension
Member with Lower
Suspension Arm

◆Cotter Pin

74 (755, 55)

8.0 (82, 71 in. lbf)

89 (908, 66)

113 (1,152, 83)

49 (500, 36)

52 (530, 38)

39 (400, 29)

52 (530, 38)

113 (1,152, 83)

89 (908, 66)

Lock Nut

◆ 216 (2,303, 159)

RH Engine Under Cover

LH Engine Under Cover

N·m (kgf·cm, ft·lbf) : Specified torque
◆ Non–reusable part
* For use with SST

9359AB48

Exploded view of the manual transaxle (2 of 2)

- Manual transaxle, by aligning the input shaft with the clutch disc. Torque the "A" bolts to 47 ft. lbs. (64 Nm), the "B" bolts to 35 ft. lbs. (47 Nm) and the "C" bolts to 17 ft. lbs. (23 Nm).
- Transverse engine mounting bracket. Tighten to 38 ft. lbs. (52 Nm).
- Transverse engine mounting insulator. Tighten the "A" bolts to 38 ft. lbs. (52 Nm) and the "B" bolts to 59 ft. lbs. (80 Nm).

10. The remainder of installation is the reverse of the removal procedure, noting the following specifications:

a. Starter mounting bolts: 27 ft. lbs. (37 Nm).

b. Clutch release cylinder bolts: "A" bolts 19 ft. lbs. (25 Nm), "B" bolts 9 ft. lbs. (12 Nm) and "C" bolts 44 inch lbs. (5 Nm).

c. Battery carrier bolts: 10 ft. lbs. (13 Nm).

d. Battery clamp bolt: 44 inch lbs. (5 Nm).

e. Battery clamp nut: 31 inch lbs. (3.5 Nm).

f. Cylinder head cover bolts: 62 inch lbs. (7 Nm).

g. Hood bolts: 10 ft. lbs. (13 Nm).

h. Wheel lug nuts: 76 ft. lbs. (103 Nm).

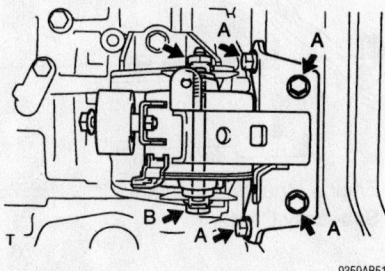

Transverse engine mounting insulator bolt locations

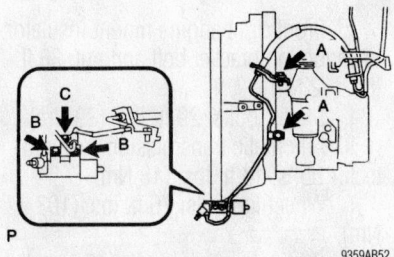

Clutch release cylinder bolt locations

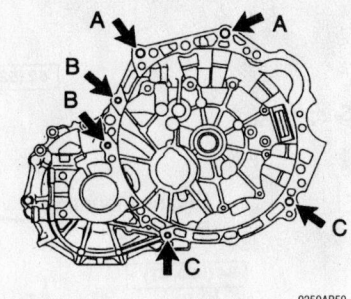

Manual transaxle bolt installation locations

11. Fill the transaxle fluid to the proper level.

12. Start the vehicle, check for leaks and repair if necessary.

Automatic Transaxle

REMOVAL & INSTALLATION

FWD—A246E & U240E Transaxles

1. Before servicing the vehicle, refer to the precautions in the beginning of this section.

2. Drain the transaxle fluid.

3. Remove or disconnect the following:
- Negative battery cable
- Hood
- No. 2 cylinder head cover
- Battery and battery carrier
- Air cleaner assembly with hose
- Floor shift cable transmission control shift
- Transmission control cable support
- No. 1 transmission control cable bracket
- Wiring harness and brackets
- Transmission wire connector
- Park/neutral position switch connector, with Anti-lock Brake System (ABS)
- Speedometer sensor connector, without ABS
- Transmission revolution sensor connectors, if equipped
- Transmission fluid filler tube
- No. 1 oil cooler inlet and outlet tubes
- Foot rest
- Floor carpet
- Oxygen (O$_2$) sensor connector

4. Suspend the engine as follows:

a. Disconnect the 2 Positive Crankcase Ventilation (PCV) hoses.

b. Install the No. 1 and No. 2 engine hangers in the correct direction.

c. No. 1 engine hanger: P/N 12281-22021 (A246E) or 12281-88600 (U240E).

d. No. 2 engine hanger: P/N 12281-15040 (A246E) or 12281-88600 (U240E)

e. Bolt: P/N 91512-B1016.

f. Torque the bolt to 28 ft. lbs. (38 Nm).

g. Attach an engine chain hoist to the engine hangers.
- Front wheels
- Right and left engine undercovers
- Front floor panel brace, U240E transaxle
- Front exhaust pipe
- Front halfshafts
- Automatic transmission case protector
- Starter

5. Support the transaxle with a floor jack
- Left side transverse engine mounting insulator and bracket
- Right side front and rear engine mount insulators
- 4 bolts, dynamic damper and member sub-assembly
- Front and rear right side transverse engine mounting brackets
- Flywheel housing undercover
- Automatic transaxle. Turn the crankshaft for access to the 6 bolts while holding the crankshaft pulley bolt with a wrench.
- Torque converter clutch

6. Installation is the reverse of the removal procedure, noting the following specifications:

a. Automatic transaxle: Bolt "A" to 47 ft. lbs. (64 Nm), bolt "B" to 34 ft. lbs. (47 Nm) and bolt "C" to 17 ft. lbs. (23 Nm).

b. Torque converter bolts: 20 ft. lbs. (28 Nm).

c. Front and rear right transverse engine mounting bracket bolts: 47 ft. lbs. (64 Nm).

d. Member sub-assembly center bolts: "A" bolts to 29 ft. lbs. (39 Nm) and "B" bolts to 38 ft. lbs. (52 Nm).

e. Right rear engine mounting insulator-to-engine mounting bracket bolt: 64 ft. lbs. (87 Nm).

f. Right rear engine mount insulator nuts and bolt: 38 ft. lbs. (52 Nm).

g. Left side engine mounting bracket-to-transaxle bolts: 38 ft. lbs. (52 Nm).

h. Left side engine mounting insulator bolts and nut: Bolt "A" to 38 ft. lbs. (52 Nm), Bolt "B" and Nut "B" to 59 ft. lbs. (80 Nm).

Heater Core replacement is covered in Section 2 of this manual

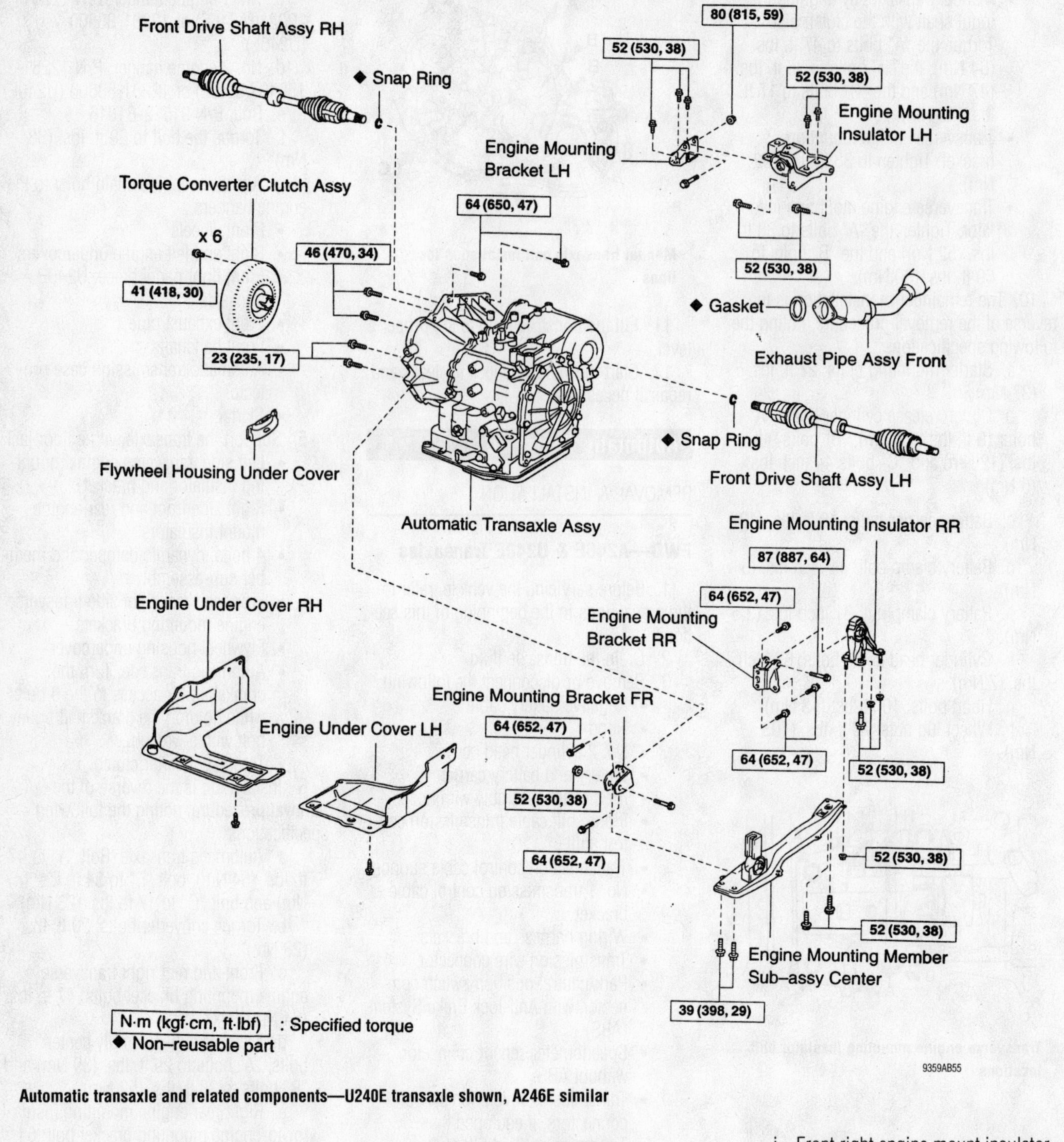

Front Drive Shaft Assy RH

◆ Snap Ring

80 (815, 59)

52 (530, 38)

52 (530, 38)

Engine Mounting Insulator LH

Engine Mounting Bracket LH

Torque Converter Clutch Assy

x 6

64 (650, 47)

52 (530, 38)

41 (418, 30)

46 (470, 34)

◆ Gasket

Exhaust Pipe Assy Front

23 (235, 17)

Flywheel Housing Under Cover

◆ Snap Ring

Front Drive Shaft Assy LH

Automatic Transaxle Assy

Engine Mounting Insulator RR

87 (887, 64)

Engine Mounting Bracket RR

64 (652, 47)

Engine Mounting Bracket RR

Engine Under Cover RH

Engine Mounting Bracket FR

Engine Under Cover LH

64 (652, 47)

64 (652, 47)

52 (530, 38)

52 (530, 38)

64 (652, 47)

52 (530, 38)

Engine Mounting Member Sub-assy Center

52 (530, 38)

39 (398, 29)

N·m (kgf·cm, ft·lbf) : Specified torque
◆ Non-reusable part

Automatic transaxle and related components—U240E transaxle shown, A246E similar

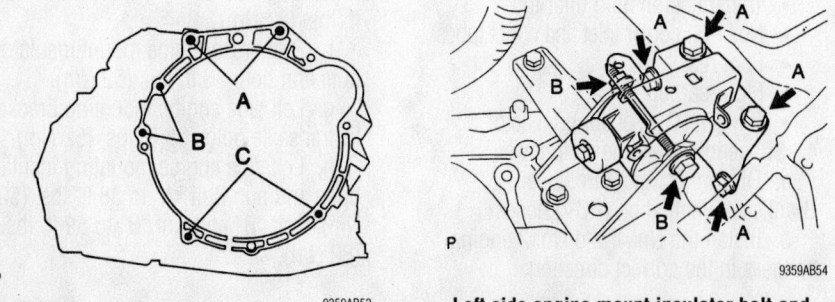

P

Automatic transaxle bolt locations

9359AB53

P

Left side engine mount insulator bolt and nut locations

9359AB54

9359AB55

i. Front right engine mount insulator-to-mounting bracket bolt and nut: 38 ft. lbs. (52 Nm).

j. Starter bolts: 29 ft. lbs. (39 Nm).

k. Automatic transmission case protector bolts: 14 ft. lbs. (18 Nm).

l. Wheel lug nuts: 76 ft. lbs. (103 Nm).

m. Oil cooler clamp bolts: 49 inch lbs. (5.5 Nm).

n. Oil cooler inlet and outlet tubes: 25 ft. lbs. (34 Nm).

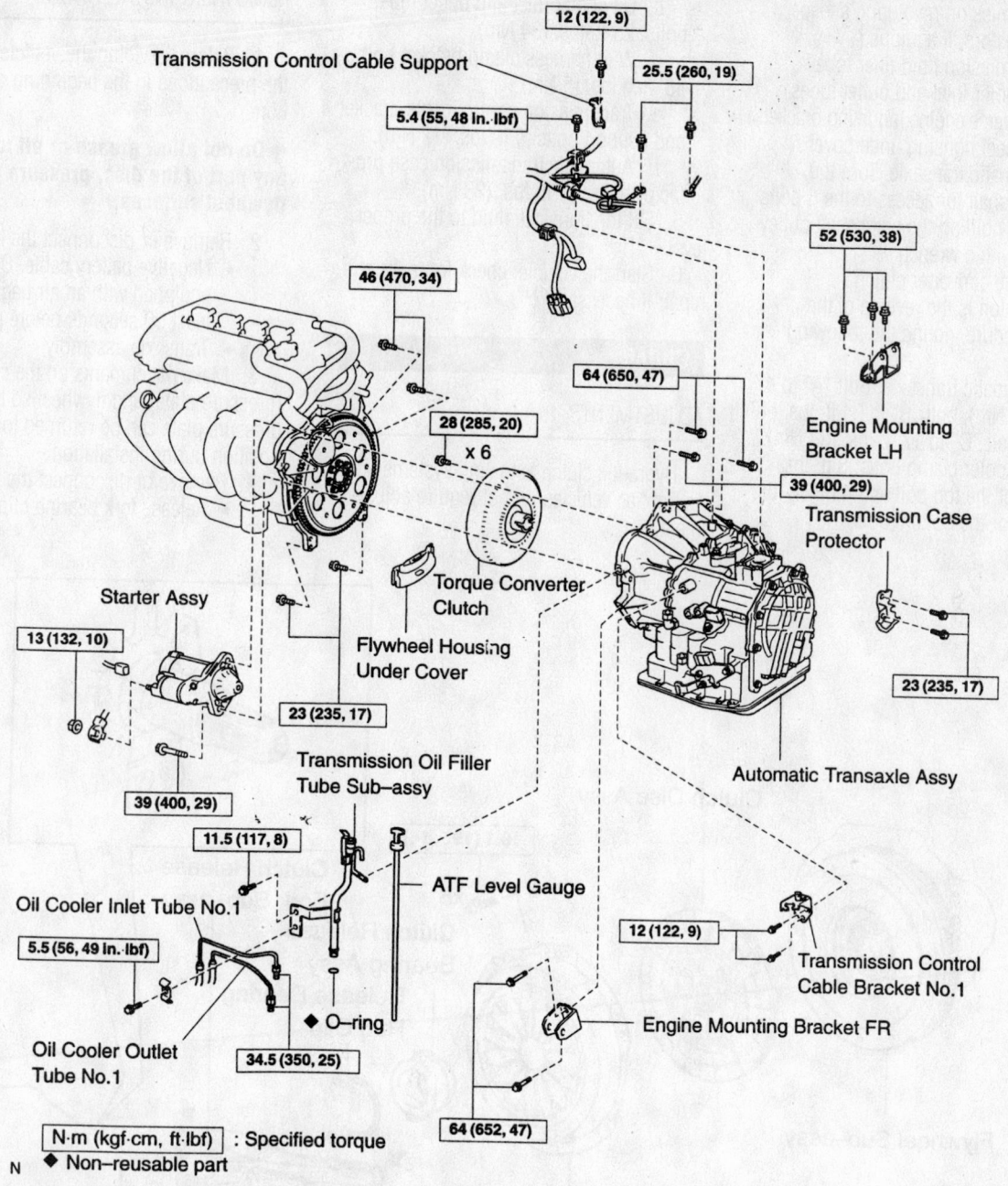

Transmission Control Cable Support

12 (122, 9)

25.5 (260, 19)

5.4 (55, 48 in.·lbf)

52 (530, 38)

46 (470, 34)

64 (650, 47)

Engine Mounting Bracket LH

28 (285, 20)
x 6

39 (400, 29)

Transmission Case Protector

Torque Converter Clutch

Starter Assy

13 (132, 10)

Flywheel Housing Under Cover

23 (235, 17)

23 (235, 17)

Automatic Transaxle Assy

39 (400, 29)

Transmission Oil Filler Tube Sub–assy

11.5 (117, 8)

ATF Level Gauge

Oil Cooler Inlet Tube No. 1

5.5 (56, 49 in.·lbf)

12 (122, 9)

Transmission Control Cable Bracket No. 1

◆ O–ring

Oil Cooler Outlet Tube No. 1

34.5 (350, 25)

Engine Mounting Bracket FR

64 (652, 47)

N·m (kgf·cm, ft·lbf) : Specified torque
N ◆ Non–reusable part

9359AB56

Automatic transaxle and related components—U341F transaxle

o. Wire harness bracket bolt: 9 ft. lbs. (13 Nm).

p. Transmission control cable bracket bolts: 9 ft. lbs. (12 Nm).

q. Transmission control cable support: 9 ft. lbs. (12 Nm).

r. Battery carrier: 10 ft. lbs. (13 Nm).

s. Air cleaner assembly: 62 inch lbs. (7 Nm).

t. Cylinder head cover bolts: 62 inch lbs. (7 Nm).

u. Hood bolts: 10 ft. lbs. (13 Nm).

7. Fill the transaxle fluid to the proper level.

8. Start the vehicle, check for leaks and repair if necessary.

AWD—U341F Transaxle

1. Before servicing the vehicle, refer to the precautions in the beginning of this section.

2. Drain the transaxle fluid.

3. Remove or disconnect the following:
- Negative battery cable

- Engine and transaxle assembly
- Transfer case
- Automatic transmission case protector
- Front left side halfshaft
- Transmission control cable support and bracket
- Wire harness clamp bracket, bolts and 2 wire harnesses
- Transmission wire connector
- Park/neutral position switch connector

Brake service is covered in Section 4 of this manual

- Transmission revolution sensor connectors, if equipped
- Transmission fluid filler tube
- Oil cooler inlet and outlet tubes
- Transverse engine mounting brackets
- Flywheel housing undercover
- Automatic transaxle. Turn the crankshaft for access to the 6 bolts while holding the crankshaft pulley bolt with a wrench.
- Torque converter clutch

4. Installation is the reverse of the removal procedure, noting the following specifications:

a. Automatic transaxle: Bolt "A" to 47 ft. lbs. (64 Nm), bolt "B" to 34 ft. lbs. (47 Nm) and bolt "C" to 17 ft. lbs. (23 Nm).

b. Oil cooler clamp bolts: 8 ft. lbs. (11 Nm) for the top bolt and 49 inch lbs. (5.5 Nm) for the bottom bolt

c. Oil cooler inlet and outlet tube bolts: 25 ft. lbs. (34 Nm).

d. Wire harness clamp bracket bolt: 48 inch lbs. (5 Nm).

e. Transmission control cable bracket and support bolts: 9 ft. lbs. (12 Nm).

f. Automatic transmission case protector bolts: 17 ft. lbs. (23 Nm).

5. Fill the transaxle fluid to the proper level.

6. Start the vehicle, check for leaks and repair if necessary.

Clutch

ADJUSTMENTS

Hydraulic clutch actuating systems used in Pontiac vehicles do not require adjustment.

REMOVAL & INSTALLATION

1. Before servicing the vehicle, refer to the precautions in the beginning of this section.

➡ **Do not allow grease or oil to get on any part of the disc, pressure plate, or flywheel surfaces.**

2. Remove or disconnect the following:
- Negative battery cable. On vehicles equipped with an air bag, wait at least 90 seconds before proceeding
- Transaxle assembly

3. Make matchmarks on the clutch cover (pressure plate) and flywheel so that the pressure plate can be returned to its original position during installation.

4. Remove or disconnect the following:
- Release fork bearing clips

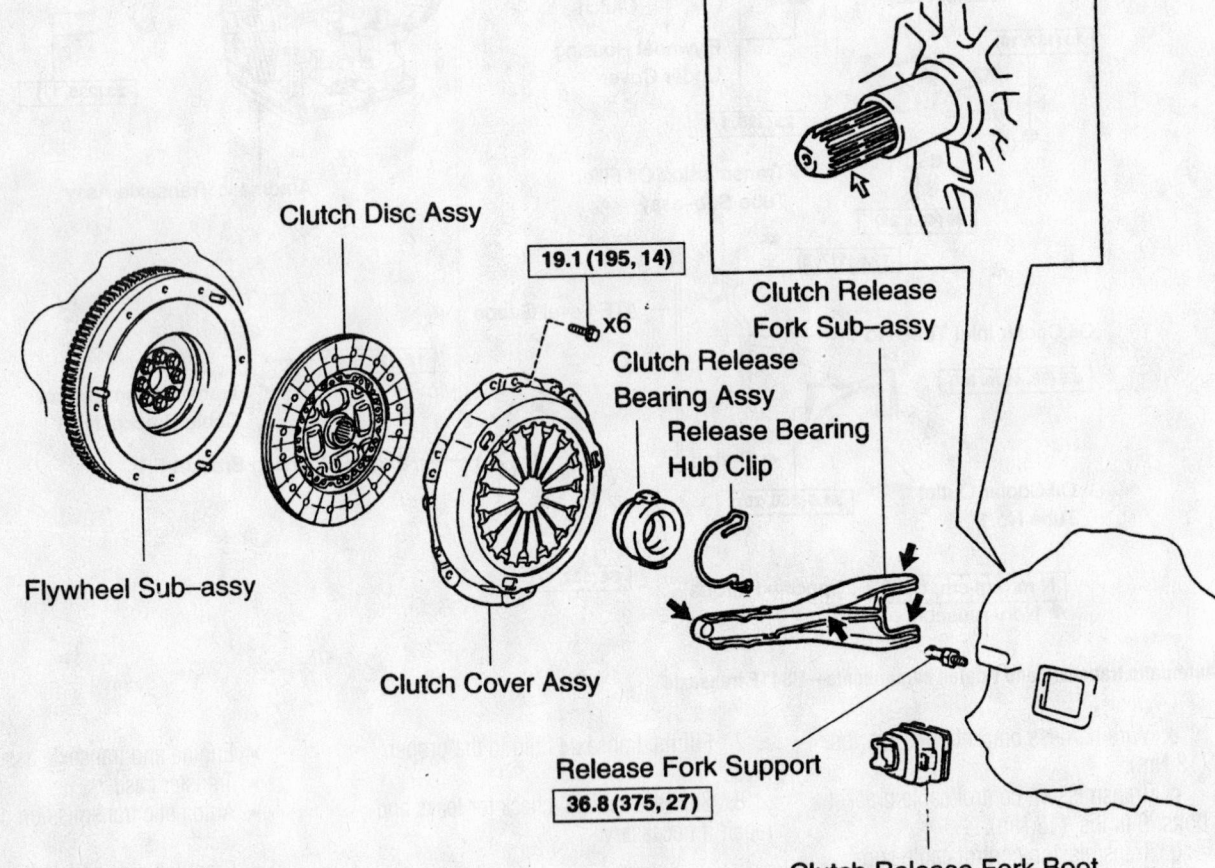

Flywheel Sub–assy

Clutch Disc Assy

19.1 (195, 14)

×6

Clutch Cover Assy

Clutch Release Bearing Assy

Release Bearing Hub Clip

Clutch Release Fork Sub–assy

Release Fork Support
36.8 (375, 27)

Clutch Release Fork Boot

N·m (kgf·cm, ft·lbf) : Specified torque
◆ Non–reusable part
⇐ Clutch spline grease
⬅ Release hub grease

Exploded view of the clutch components

- Release bearing hub, complete with the release bearing
- Release fork and support

✳✳ CAUTION

Slowly unfasten the bolts which attach the pressure plate. Loosen each bolt 1 turn at a time until the spring tension is released. If the bolts are released improperly the clutch assembly could fly apart, causing possible injury.

- Pressure plate from the clutch cover/spring assembly

5. Inspect the disc, pressure plate and flywheel for damage and wear using a caliper to measure depth and width and a dial indicator to measure runout.

   a. The minimum clutch disc rivet head depth is 0.012 in. (0.3mm).

   b. The maximum clutch disc runout is 0.031 in. (0.8mm).

   c. The maximum pressure plate spring depth is 0.024 in. (0.6mm).

   d. The maximum pressure plate spring width is 0.197 in. (5.0mm).

   e. The maximum flywheel runout is 0.004 in. (0.1mm).

6. Replace or machine parts as necessary.

To install:

7. When reassembling, apply a thin coating of multipurpose grease to the release bearing hub and release fork contact points. Also, pack the groove inside the clutch hub with multipurpose grease and lubricate the pivot points of the release fork.

8. Install or connect the following:
- Clutch disc and pressure plate. The bolts should be tightened in 2 or 3 steps, gradually and evenly. Final bolt torque is 14 ft. lbs. (19 Nm).
- Release bearing, fork and boot
- Transaxle assembly
- Negative battery cable

Hydraulic Clutch System

BLEEDING

➡**If any maintenance on the clutch system was performed or the system is suspected of containing air, bleed the system. Use care; brake fluid will remove the paint from any surface. If the brake fluid spills onto any painted surface, wash it off immediately with soap and water.**

1. Before servicing the vehicle, refer to the precautions in the beginning of this section.

2. Fill the clutch reservoir with brake fluid. Check the reservoir level frequently and add fluid as needed.

3. Connect one end of a vinyl tube to the bleeder plug on the slave cylinder and submerge the other end into a clear container half-filled with brake fluid.

4. Slowly pump the clutch pedal several times.

5. Have an assistant hold the clutch pedal down and loosen the bleeder plug until fluid and/or air starts to run out of the bleeder plug. Close the bleeder plug while the pedal is held to the floor.

➡**Do not allow the pedal to rise back-up while the bleeder is still open. If this happens, it will allow air to re-enter the slave cylinder and cause the clutch system not to work properly.**

6. Repeat Steps 2 and 3 until all the air bubbles are removed from the system.

7. Tighten the bleeder plug when all the air is gone.

8. Refill the master cylinder to the proper level as required.

9. Check the system for leaks.

Transfer Case

REMOVAL & INSTALLATION

1. Before servicing the vehicle, refer to the precautions in the beginning of this section.

2. Drain the transfer case fluid.

3. Remove or disconnect the following:

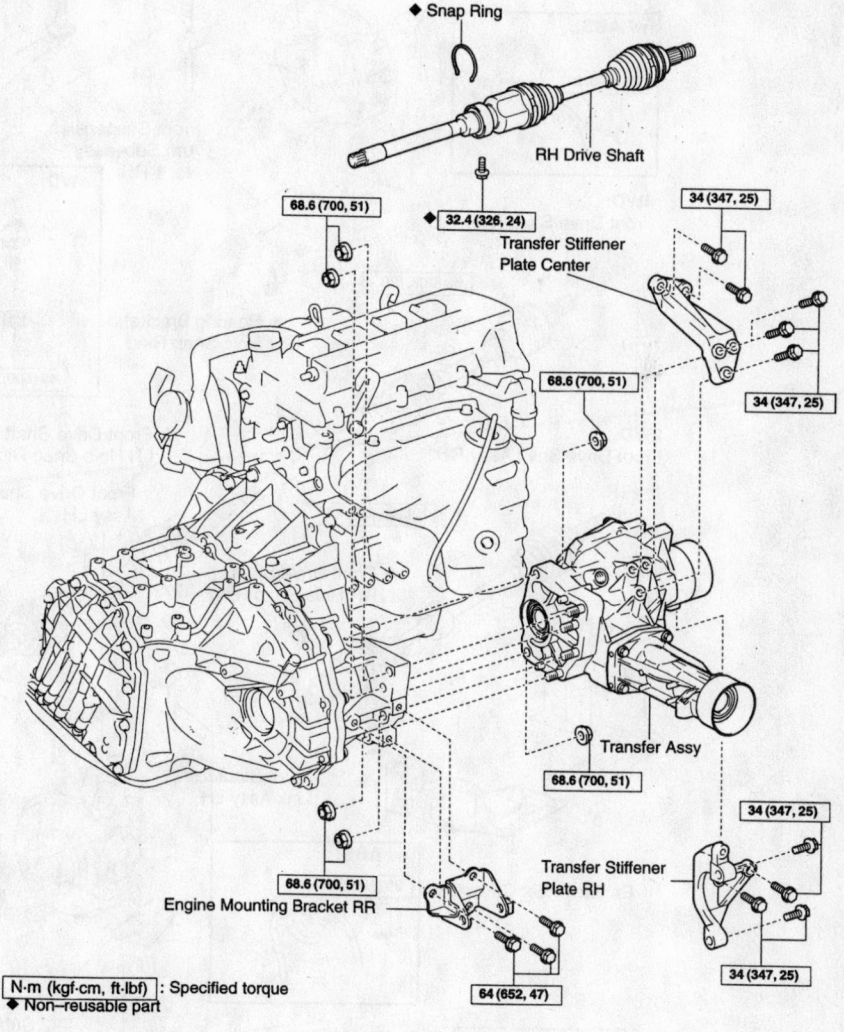

Exploded view of the transfer case mounting

9359AB65

For complete Engine Mechanical specifications, see Section 1 of this manual

- Negative battery cable. Due to the air bag system, wait at least 90 seconds before proceeding
- Engine and transaxle assembly
- Separate vane pump
- Steering gear
- Crossmember
- Manifold stay
- Oxygen (O_2) sensor
- Exhaust manifold heat shield
- Exhaust manifold
- Starter
- Right side halfshaft
- Transverse engine mounting bracket
- Center and right side transfer stiffener plates

✳✳ WARNING

When removing the transfer case, DO NOT touch the oil seal.

- Transfer case bolts, and transfer assembly, using a mallet to dislodge it from the transaxle

4. Installation is the reverse of the removal procedure, noting the following specifications:

   a. Transfer case stiffener case bolts: 25 ft. lbs. (34 Nm).

   b. Engine mounting bracket bolts: 47 ft. lbs. (64 Nm).

5. Add fluid to the transfer case, and check for leaks.

Halfshaft

REMOVAL & INSTALLATION

➡The hub bearing could be damaged if subjected to the full weight of the vehicle, such as if the vehicle is moved without the halfshafts. If it is absolutely necessary to place the full vehicle weight on the hub bearing, first support the bearing with SST No. 09608–16041.

1. Before servicing the vehicle, refer to the precautions in the beginning of this section.

2. Drain the transaxle fluid.

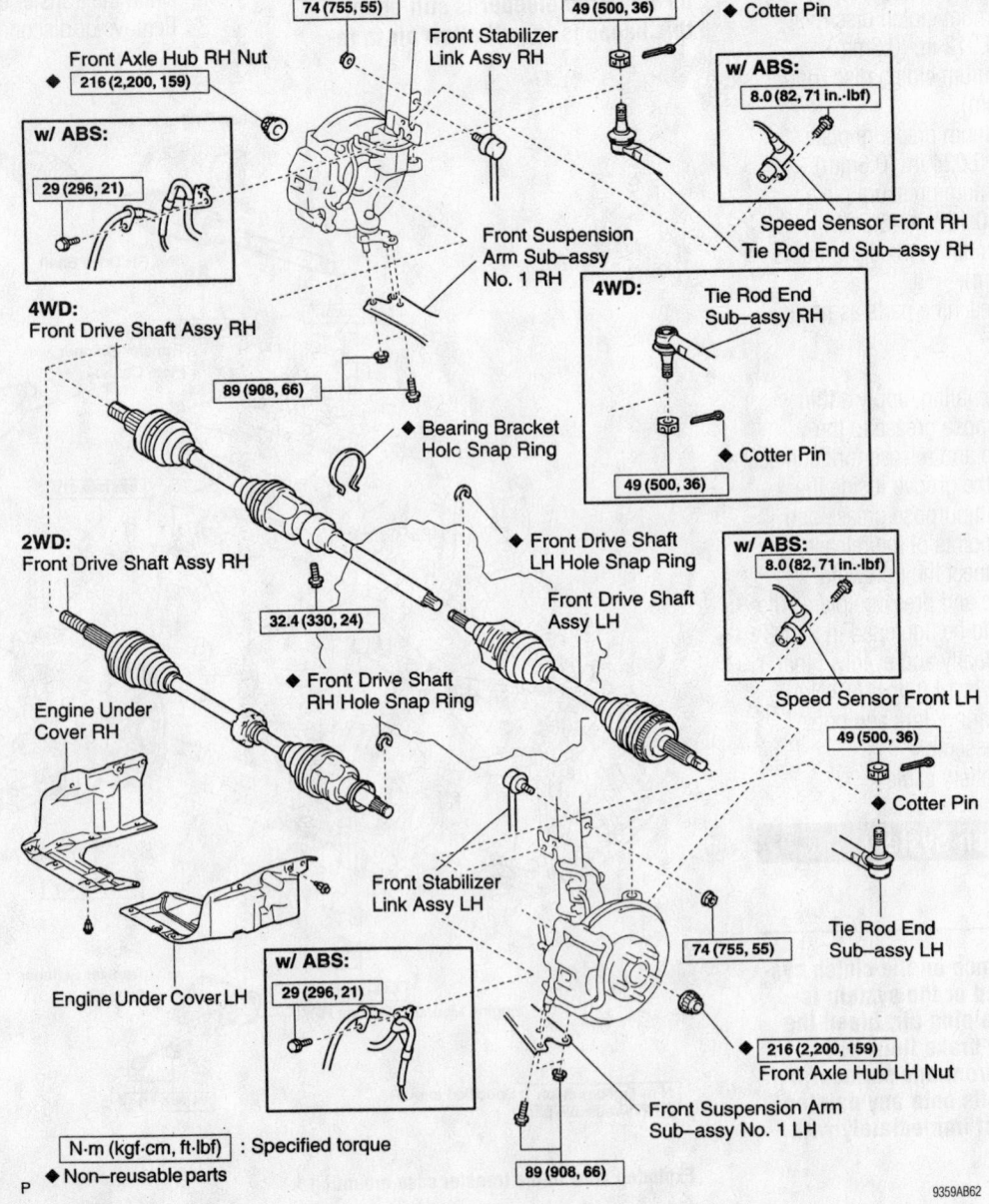

74 (755, 55)
49 (500, 36)
◆ Cotter Pin
Front Axle Hub RH Nut
◆ 216 (2,200, 159)
Front Stabilizer Link Assy RH
w/ ABS: 8.0 (82, 71 in.-lbf)
w/ ABS: 29 (296, 21)
Speed Sensor Front RH
Tie Rod End Sub–assy RH
Front Suspension Arm Sub–assy No. 1 RH
4WD: Front Drive Shaft Assy RH
4WD: Tie Rod End Sub–assy RH
89 (908, 66)
◆ Bearing Bracket Holc Snap Ring
◆ Cotter Pin
49 (500, 36)
2WD: Front Drive Shaft Assy RH
◆ Front Drive Shaft LH Hole Snap Ring
Front Drive Shaft Assy LH
w/ ABS: 8.0 (82, 71 in.-lbf)
32.4 (330, 24)
◆ Front Drive Shaft RH Hole Snap Ring
Speed Sensor Front LH
Engine Under Cover RH
49 (500, 36)
◆ Cotter Pin
Front Stabilizer Link Assy LH
w/ ABS: 29 (296, 21)
Engine Under Cover LH
74 (755, 55)
Tie Rod End Sub–assy LH
◆ 216 (2,200, 159)
Front Axle Hub LH Nut
Front Suspension Arm Sub–assy No. 1 LH
89 (908, 66)

N·m (kgf·cm, ft·lbf) : Specified torque
P
◆Non–reusable parts

9359AB62

Halfshafts and related components

3. Remove or disconnect the following:

- Negative battery cable. Due to the air bag system, wait at least 90 seconds before proceeding.
- Both front wheels
- Cotter pin, locknut cap, and the hub nut
- Undercovers
- Speed sensors
- Tie rod ball joint from the steering knuckle
- Stabilizer bar link from the lower suspension arm
- Lower ball joint from the lower suspension arm
- Halfshaft from the knuckle

➡**Be careful not to damage the inner oil seal or the ABS sensor rotor on the halfshaft.**

4. To remove the left side halfshaft, separate the halfshaft from the transaxle.

5. To remove the right side halfshaft perform the following steps:

- Remove the 2 bolts of the center bearing bracket
- Pull the halfshaft out together with the center bearing case and the center halfshaft
- Remove the center shaft with the right-hand halfshaft from the transaxle through the bearing bracket.

➡**Do not damage the oil seal lip.**

To install:

6. Install or connect the following:

- Snapring opening side facing downward, on the oiled inboard joint tulip
- Left side halfshaft into the transaxle
- Right side halfshaft, with the bearing case and center shaft, into the transaxle
- Center bearing case (right side).

7. After installing either halfshaft, check that there is 0.08–0.12 in. (2–3mm) of axial play. Check that the halfshaft is making contact with the pinion shaft and that the halfshaft cannot be pulled out.

8. Install or connect the following:

- Halfshaft into the knuckle
- Lower suspension arm to the lower ball joint. Torque the bolt and nuts to 66 ft. lbs. (89 Nm).
- Tie rod end to the steering knuckle. Tighten the nut to 36 ft. lbs. (49 Nm).

- Stabilizer bar link to the lower suspension arm. Torque the nuts to 55 ft. lbs. (74 Nm).
- Front wheels
- Hub nut and washer and tighten to 159 ft. lbs. (216 Nm)
- Negative battery cable
- Locknut cap and a new cotter pin.
- Speed sensors
- Undercover

9. Fill the transaxle fluid to the proper level

10. Start the vehicle, check for leaks and repair if necessary.

CV-Joints

OVERHAUL

1. Before servicing the vehicle, refer to the precautions in the beginning of this section.

2. Remove or disconnect the following:

- Inboard joint boot clips
- Inboard joint tulip from the driveshaft
- Snapring
- Using a brass rod and hammer, the tri-pot joint off the driveshaft without hitting the joint roller

2WD:

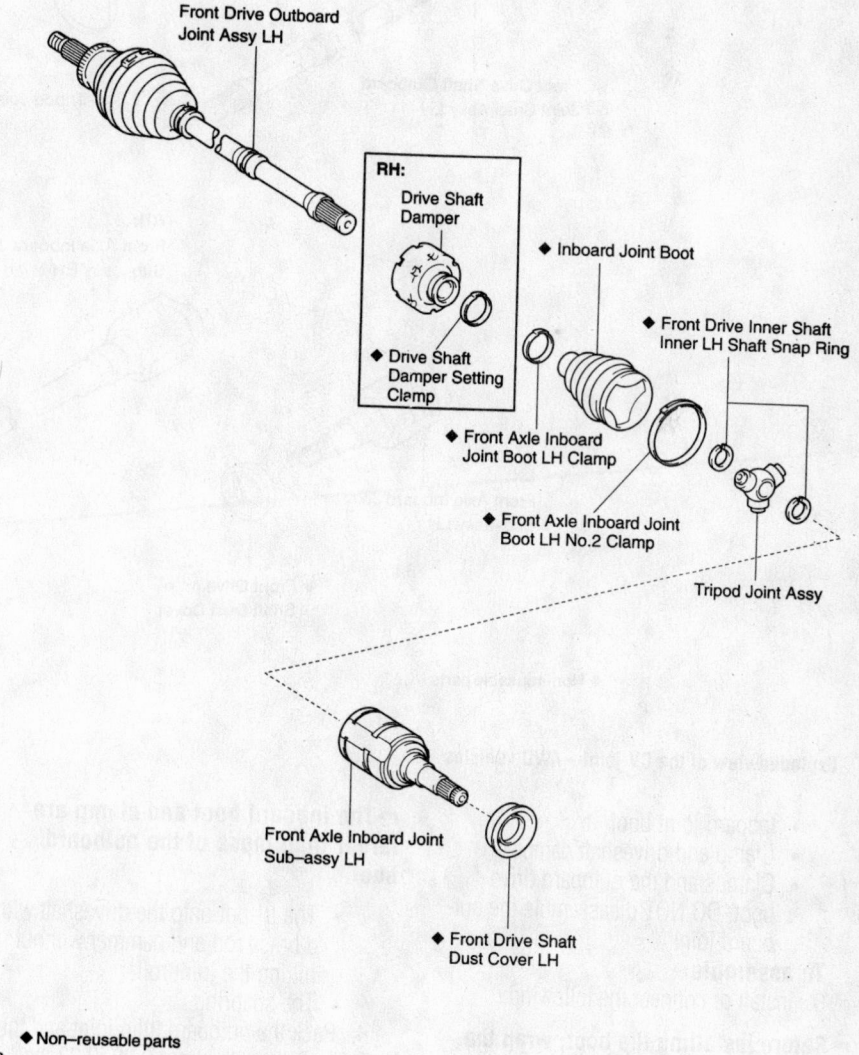

Front Drive Outboard Joint Assy LH

RH:

Drive Shaft Damper

◆ Drive Shaft Damper Setting Clamp

◆ Inboard Joint Boot

◆ Front Drive Inner Shaft Inner LH Shaft Snap Ring

◆ Front Axle Inboard Joint Boot LH Clamp

◆ Front Axle Inboard Joint Boot LH No.2 Clamp

Tripod Joint Assy

Front Axle Inboard Joint Sub-assy LH

◆ Front Drive Shaft Dust Cover LH

◆ Non-reusable parts

P

Exploded view of the CV-joint—FWD vehicles

9359AB63

For Accessory Drive Belt illustrations, see Section 1 of this manual

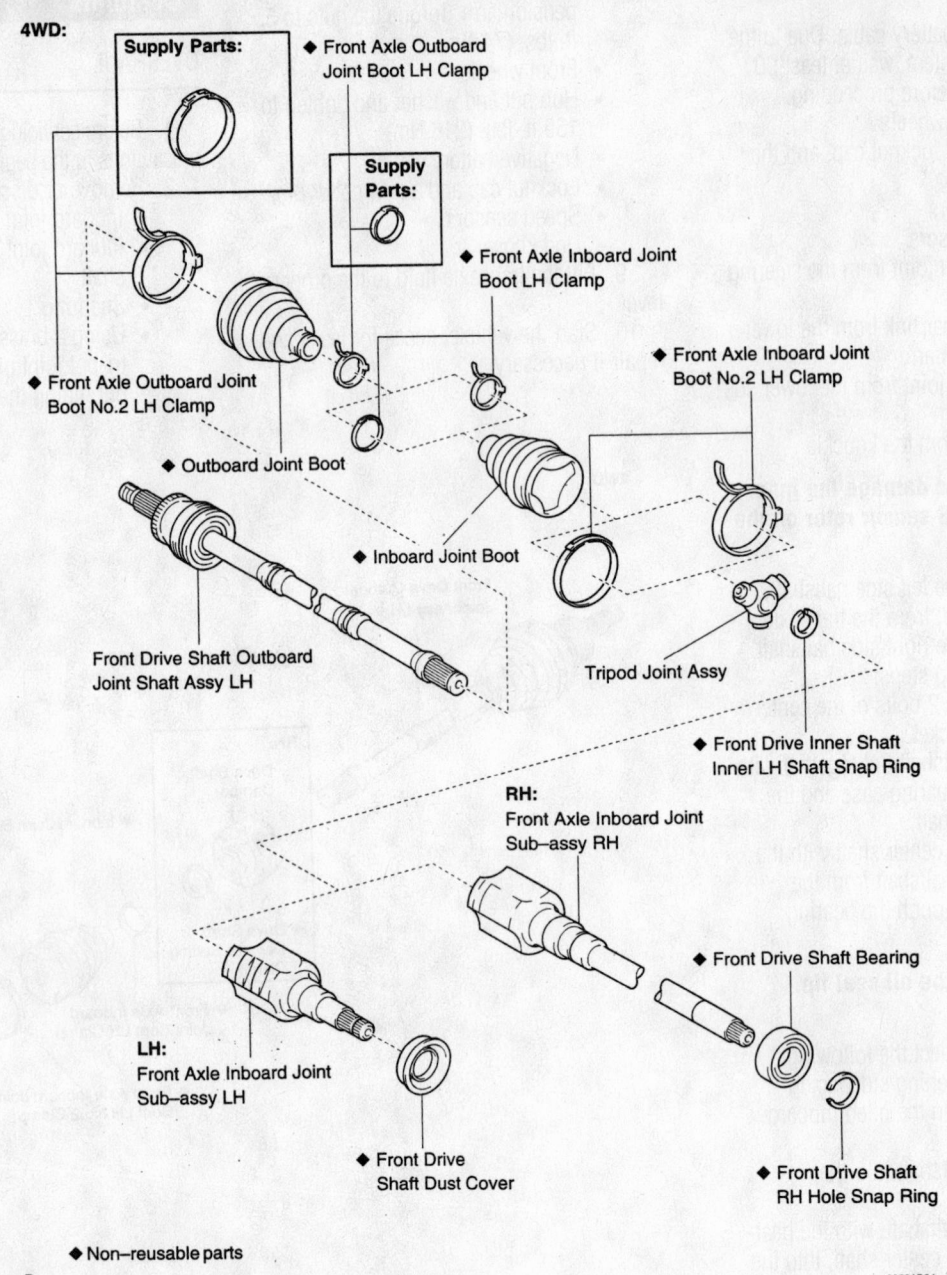

4WD:

Supply Parts:

◆ Front Axle Outboard
Joint Boot LH Clamp

Supply Parts:

◆ Front Axle Inboard Joint
Boot LH Clamp

◆ Front Axle Inboard Joint
Boot No.2 LH Clamp

◆ Front Axle Outboard Joint
Boot No.2 LH Clamp

◆ Outboard Joint Boot

◆ Inboard Joint Boot

Front Drive Shaft Outboard
Joint Shaft Assy LH

Tripod Joint Assy

◆ Front Drive Inner Shaft
Inner LH Shaft Snap Ring

RH:
Front Axle Inboard Joint
Sub–assy RH

◆ Front Drive Shaft Bearing

LH:
Front Axle Inboard Joint
Sub–assy LH

◆ Front Drive
Shaft Dust Cover

◆ Front Drive Shaft
RH Hole Snap Ring

◆ Non–reusable parts

P

9359AB64

Exploded view of the CV-joint—AWD vehicles

- Inboard joint boot
- Clamp and driveshaft damper
- Clamps and the outboard drive boot. DO NOT disassemble the outboard joint.

To assemble:

3. Install or connect the following:

➡**Before installing the boot, wrap the spline end of the shaft with masking tape to prevent damage to the boot.**

- Driveshaft damper with a new clamp
- Temporarily, the inboard boot with new clamp to the drive joint

➡**The inboard boot and clamp are larger than those of the outboard boot.**

- The tri-pot onto the driveshaft with a brass rod and hammer without hitting the joint roller
- The snapring

4. Pack the outboard tulip joint and the outboard boot with about 0.26–0.33 lbs. ounces of grease that was supplied with the boot kit.

5. Install or connect the following:
- Boot onto the outboard joint

6. Pack the inboard tulip joint and boot with ½ lb. of grease that was supplied with the boot kit.

- Inboard tulip joint onto the driveshaft
- Boot onto the driveshaft

7. Before checking the standard length, bend the band and lock it. Make sure that the boot is not stretched or squashed when the driveshaft is at standard length. Standard driveshaft length: LH: 540.2 mm (21.268 in.); RH: 857.4 mm (33.756 in.)

STEERING AND SUSPENSION

Air Bag

PRECAUTIONS

Several precautions must be observed when handling the inflator module to avoid accidental deployment and possible personal injury.

• Never carry the inflator module by the wires or connector on the underside of the module.

• When carrying a live inflator module, hold securely with both hands, and ensure that the bag and trim cover are pointed away.

• Place the inflator module on a bench or other surface with the bag and trim cover facing up.

• With the inflator module on the bench, never place anything on or close to the module that may be thrown in the event of an accidental deployment.

DISARMING

To avoid personal injury when working on vehicles equipped with an air bag, the negative battery cable must be disconnected and at least 90 seconds must elapse before working on the system. Failure to do so may result in deployment of the air bag.

REARMING

After vehicle service is completed, reattach the battery cables (positive cable first!) to rearm the air bag system.

Rack and Pinion Steering Gear

REMOVAL & INSTALLATION

1. Before servicing the vehicle, refer to the precautions in the beginning of this section.
2. Position the front wheels straight ahead.
3. Remove or disconnect the following:
   • Negative battery cable. Because these vehicles are equipped with air bags, wait at least 90 seconds before proceeding.
   • Horn button
   • Steering wheel
   • Front wheels
   • Left and right engine undercovers

• Left and right tie rod ends
• Column hose cover silencer sheet
• Steering intermediate shaft
• Pressure feed and return tubes
• Left and right side front stabilizer links
• Right and left front lower control arms from the ball joints
• Hood
• No. 2 cylinder head (valve) cover

4. Install an engine support and tension it to support the engine without raising it.
   a. No. 1 engine hanger: P/N 12281-22021 1ZZ-FE, 12281-88600 2ZZ-GE.
   b. No. 2 engine hanger: P/N 12281-15040 1ZZ-FE, 12281-88600 2ZZ-GE.
   c. Bolt: P/N 91512-B1016.
   d. Torque the bolts to 28 ft. lbs. (38 Nm).

✳✳ CAUTION

Do not try to suspend the engine by hooking the chain to any other part.

e. Attach an engine chain hoist to the hangers.

✳✳ CAUTION

The engine hoist is now in place and under tension. Use care when repositioning the vehicle and make necessary adjustments to the engine support.

5. Remove or disconnect the following:

• Bolt and nuts holding in the middle of the crossmember and support the crossmember with a jack
• Bolts from the outer side of the suspension crossmember
• Suspension crossmember with the steering gear assembly
• Steering intermediate shaft, after matchmarking it
• Rack and pinion steering gear from the crossmember

1ZZ-FE:

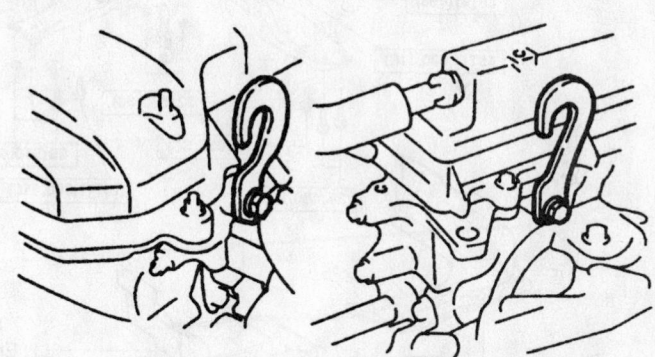

2ZZ-GE:

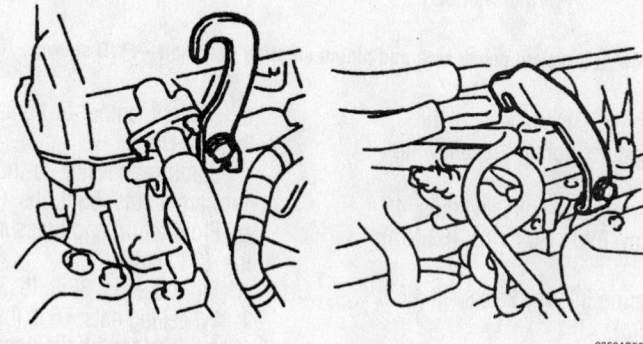

9359AB58

Proper installation of engine hangers

For Tire, Wheel and Ball Joint specifications, see Section 1 of this manual

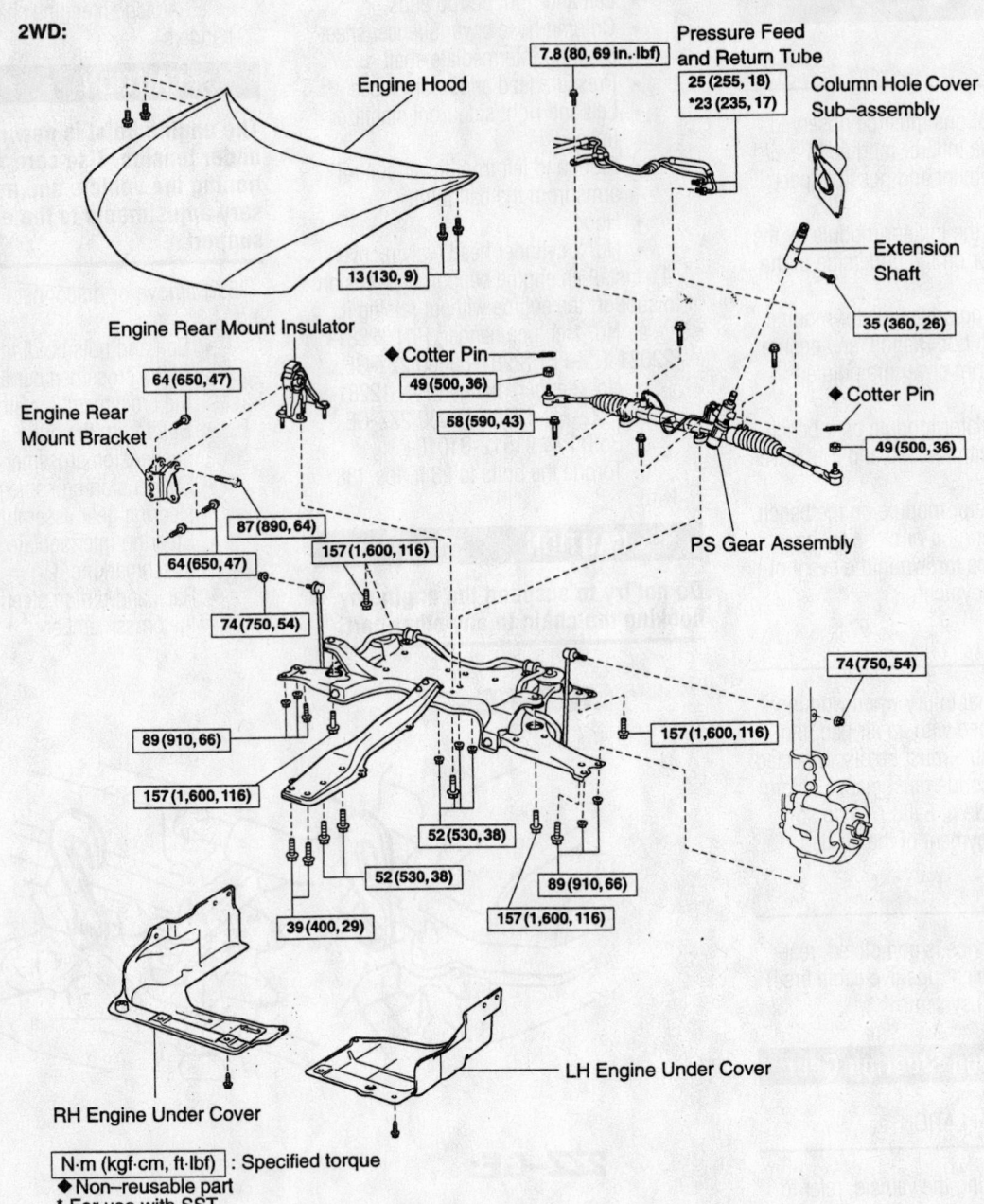

2WD:

Engine Hood

Pressure Feed and Return Tube

Column Hole Cover Sub–assembly

7.8 (80, 69 in.·lbf)

25 (255, 18)
*23 (235, 17)

Extension Shaft

35 (360, 26)

13 (130, 9)

Engine Rear Mount Insulator

◆ Cotter Pin

64 (650, 47)

49 (500, 36)

◆ Cotter Pin

Engine Rear Mount Bracket

58 (590, 43)

49 (500, 36)

87 (890, 64)

PS Gear Assembly

64 (650, 47)

157 (1,600, 116)

74 (750, 54)

74 (750, 54)

89 (910, 66)

157 (1,600, 116)

157 (1,600, 116)

52 (530, 38)

89 (910, 66)

52 (530, 38)

39 (400, 29)

157 (1,600, 116)

LH Engine Under Cover

RH Engine Under Cover

N·m (kgf·cm, ft·lbf) : Specified torque
◆Non–reusable part
* For use with SST

9359AB59

Exploded view of a typical power rack and pinion steering gear unit—FWD shown

6. Installation is the reverse of the removal procedure, noting the following specifications:

   a. Steering gear bolts and nuts: 43 ft. lbs. (58 Nm) FWD, 60 ft. lbs. (82 Nm) AWD.

   b. Steering intermediate shaft: 26 ft. lbs. (35 Nm).

   c. Suspension crossmember bolts: 116 ft. lbs. (157 Nm).

   d. Engine mount insulator bolts: 38 ft. lbs. (52 Nm).

   e. Center member-to-frame bolts: 29 ft. lbs. (39 Nm).

   f. Stabilizer bar link-to-the lower control arms nuts: 55 ft. lbs. (74 Nm).

   g. Fluid return and pressure tubes: 17 ft. lbs. (23 Nm).

   h. Tie rod ends: 36 ft. lbs. (49 Nm).

   i. Wheel lug nuts: 76 ft. lbs. (103 Nm).

7. Check and top off the power steering fluid.

8. Check and adjust the alignment, if needed.

Strut and Coil Spring

REMOVAL & INSTALLATION

Front

1. Before servicing the vehicle, refer to the precautions in the beginning of this section.

2. Remove or disconnect the following:

   • Negative battery cable. Because of the air bag system, wait at least 90 seconds before proceeding

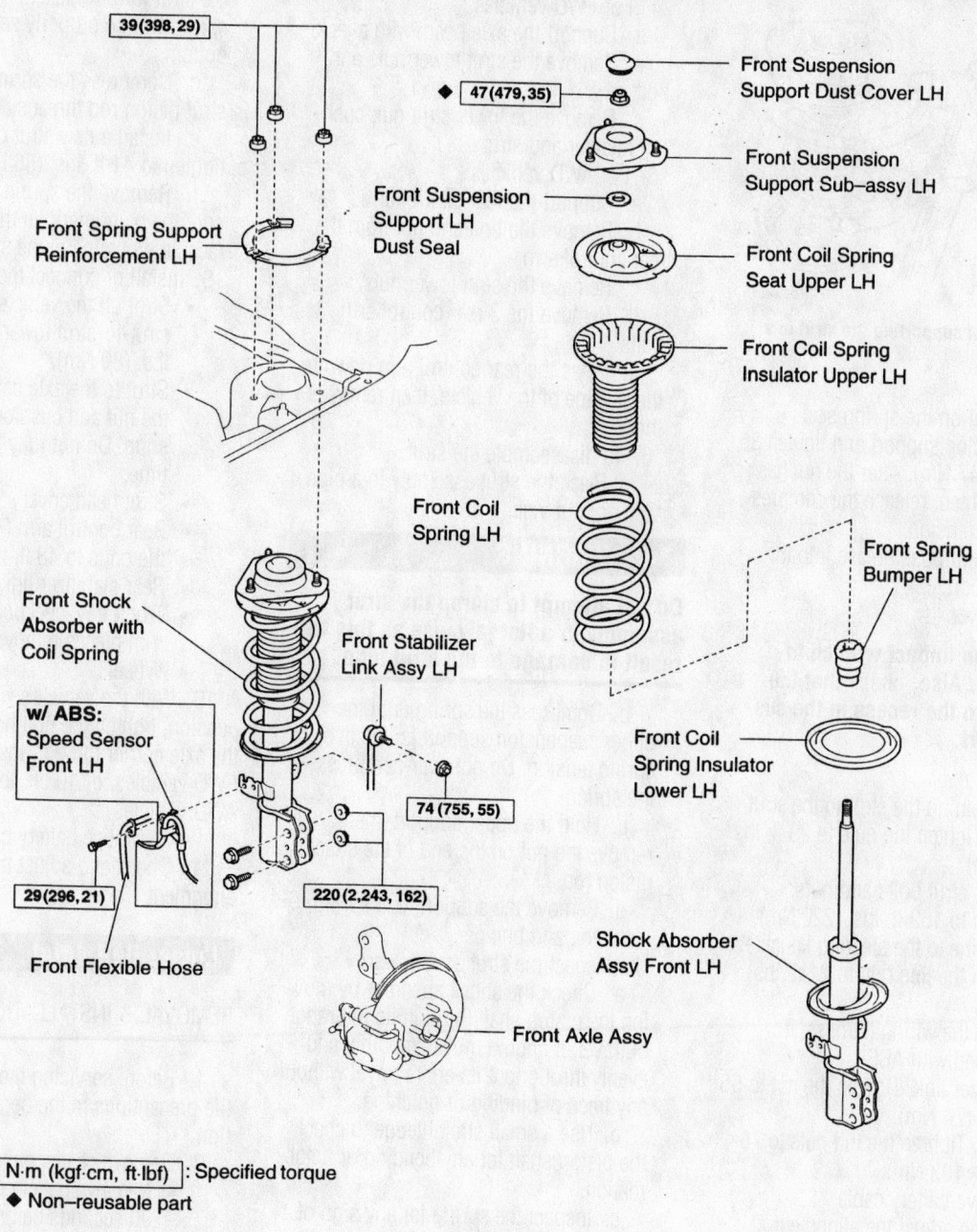

39(398, 29)

◆ 47(479, 35)

Front Suspension
Support Dust Cover LH

Front Suspension
Support Sub–assy LH

Front Suspension
Support LH
Dust Seal

Front Spring Support
Reinforcement LH

Front Coil Spring
Seat Upper LH

Front Coil Spring
Insulator Upper LH

Front Coil
Spring LH

Front Spring
Bumper LH

Front Shock
Absorber with
Coil Spring

Front Stabilizer
Link Assy LH

w/ ABS:
Speed Sensor
Front LH

Front Coil
Spring Insulator
Lower LH

74 (755, 55)

29 (296, 21)

220 (2,243, 162)

Front Flexible Hose

Shock Absorber
Assy Front LH

Front Axle Assy

N·m (kgf·cm, ft·lbf) : Specified torque
P ◆ Non–reusable part

9359AB60

Common coil spring and strut component assembly

✴✴ WARNING

**Do not support the weight of the
vehicle on the suspension arm; the
arm will deform under its weight.**

- Wheel
- Stabilizer link from the strut
- Bolt, and disconnect the brake hose
  from the strut
- With ABS brakes, speed sensor
  wiring harness from the strut

- Lower strut bolts and nuts
- Upper strut nuts
- Strut from the steering knuckle
- Strut
3. To disassemble the strut:
   - Install a bolt and 2 nuts to the
     bracket at the lower portion of the
     strut shell and secure it in a vise
   - Compress the coil spring
   - Dust cover and hold the spring seat
     so that it will not turn
   - Nut on the top of the strut

- Suspension support, bearing, dust
  seal, spring seat, spring, insulators
  and bumper
To install:
4. To assemble the strut:
   - Install the spring bumper to piston
5. Using a spring compressor, compress
   the spring.
   - Coil spring to the strut. Fit the
     lower end of the coil spring into the
     gap of the lower seat.
   - Spring seat with the insulator

For Wheel Alignment specifications, see Section 1 of this manual

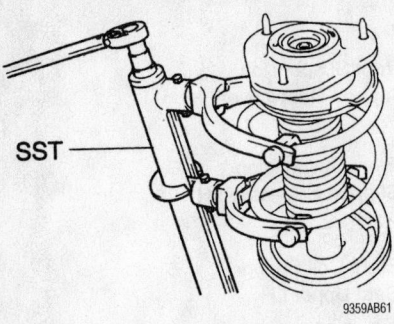

Proper method of supporting the strut in a vise

- Dust seal on the spring seat
- Suspension support and tighten 35 ft. lbs. (47 Nm). After the nut has been tighten, release the compressor tool tension.

6. Pack multipurpose grease into the suspension support.
- Dust cover.

→**Do not use an impact wrench to tighten the nut. Also, check that the bearing fits into the recess in the suspension support.**

- Strut
- Nuts holding the strut to the strut tower. Tighten the nuts to 29 ft. lbs. (39 Nm).
- 2 lower strut bolts and nuts. Tighten to 162 ft. lbs. (220 Nm).
- Brake line to the steering knuckle. Tighten the line bolt to 21 ft. lbs. (29 Nm).
- Secure the wiring harness, if equipped with ABS
- Stabilizer link. Tighten the nut to 55 ft. lbs. (74 Nm).
- Wheel. Tighten the lug nuts to 76 ft. lbs. (103 Nm).
- Negative battery cable

7. Check and adjust the alignment, if needed.

Rear

1. Before servicing the vehicle, refer to the precautions in the beginning of this section.
2. Remove or disconnect the following:
- Negative battery cable. Because of the air bag system, wait at least 90 seconds before proceeding.
- Rear wheel
- Rear deck board, luggage compartment tray and any trim necessary to access the strut towers
- Shock absorber head cover
3. On AWD vehicles, separate the rear stabilizer link.

4. For FWD vehicles:
   a. Support the axle beam with a jack.
   b. Remove the strut tower nuts and bolt.
   c. Remove the lower strut nut, cushion retainer and strut.
5. For AWD vehicles:
   a. Support the rear control arm.
   b. Remove the bolt and nut from the rear control arm.
   c. Remove the strut tower nuts.
   d. Remove the 3 rear control arm bolts.
   e. Press the rear control arm down to the outside of the vehicle, then remove the strut.
6. To disassemble the strut:
   a. Place the strut assembly in a pipe vise or strut vise.

✷✷ WARNING

Do not attempt to clamp the strut assembly in a flat jaw vise as this will result in damage to the strut tube.

   b. Compress the spring until the upper suspension support is free of any spring tension. Do not over-compress the spring.
   c. Hold the upper support, then remove the nut on the end of the shock piston rod.
   d. Remove the support, coil spring, insulator, and bumper.
7. Inspect the strut as follows:
   a. Check the shock absorber by moving the piston shaft through its full range of travel. It should move smoothly and evenly throughout its entire travel without any trace of binding or notching.
   b. Use a small straightedge to check the piston shaft for any bending or deformation.
   c. Inspect the spring for any sign of deterioration or cracking. The waterproof coating on the coils should be intact to prevent rusting.
To install:

→**Never reuse a self-locking nut. Always replace self-locking nuts and cotter pins as applicable.**

8. Assemble the strut as follows:
   a. Loosely assemble all components onto the strut assembly. Be sure the spring end aligns with the hollow in the lower seat.
   b. Align the upper suspension support with the piston rod and install the support.
   c. Align the suspension support with

the strut lower bracket. This assures the spring will be properly seated top and bottom.
   d. Compress the spring to expose the strut piston rod threads.
   e. Install a new strut piston nut and tighten to 41 ft. lbs. (56 Nm).
   f. Remove the spring compressor. Be sure the paint mark on the upper support faces the outside of the strut.
9. Install or connect the following:
- Strut on the vehicle. Tighten the strut-to-strut tower nuts to 59 ft. lbs. (80 Nm).
- Strut to the axle carrier and install the nut and cushion retainer/bolt snug. Do not fully tighten at this time.
- Strut head cover
- Rear control arm (AWD). Tighten the bolts to 48 ft. lbs. (65 Nm).
- Rear stabilizer link (AWD)
- Trunk tray, deckboard and any other trim pieces removed
- Wheel

10. With the vehicle's weight on the suspension, tighten the bolt holding the strut to the axle carrier to 59 ft. lbs. (80 Nm) for FWD vehicles, or 103 ft. lbs. (140 Nm) for AWD vehicles.
- Negative battery cable

11. Check and adjust the rear wheel alignment.

Lower Ball Joint

REMOVAL & INSTALLATION

1. Before servicing the vehicle, refer to the precautions in the beginning of this section.
2. Remove or disconnect the following:
- Negative battery cable. Wait at least 90 seconds before proceeding.
- Front wheel
3. Depress the brake pedal and loosen the hub nut

Removing the ball joint from the knuckle

- ABS speed sensor, if equipped
- Cotter pin and nut from the tie rod end. Using a tie rod end removal tool, separate the tie rod end from the steering knuckle.
- Lower control arm ball joint, using a suitable puller
- Separate the front halfshaft
- Lower ball joint cotter pin and castle nut
- Lower ball joint from the steering knuckle using a puller

To install:

4. Install or connect the following:
- Lower ball joint to the lower arm. Tighten the castle nut to 76 ft. lbs. (103 Nm).
- New cotter pin
- Front halfshaft
- Lower control arm
- Tie rod end to the knuckle
- ABS speed sensor
- Hub nut
- Wheel
- Negative battery cable

5. Check and adjust the alignment, if needed.

Upper Control Arm

REMOVAL & INSTALLATION

Rear—AWD Only

1. Before servicing the vehicle, refer to the precautions in the beginning of this section.
2. Remove or disconnect the following:
- Negative battery cable. Wait at least 90 seconds before proceeding
- Rear wheel
- Exhaust pipe
- Propeller shaft with center bearing shaft
- Rear stabilizer links
- Rear hub nuts
- Rear brake drum
- Speed sensor
- Front brake shoe
- Parking brake shoe strut set
- Rear brake shoe
- Parking brake cables
- Rear brake hoses
- Separate the rear suspension arms
- Separate the upper control arm
- Rear drive axle assembly
- Rear strut nut and bolt
- Rear strut
- Rear suspension arm
- Rear suspension member

- Upper control arm assembly. Matchmark the camber adjust cams and rear suspension member prior to removal.

3. Installation is the reverse of the removal procedure.

Lower Control Arm

REMOVAL & INSTALLATION

1. Before servicing the vehicle, refer to the precautions in the beginning of this section.
2. Remove or disconnect the following:
- Negative battery cable. Wait at least 90 seconds before proceeding..
- Front wheel
- Stabilizer link
- Bolt and nuts and separate the lower control arm from the lower ball joint
- Bolts and nuts, then separate the steering gear. Loosen the bolt, since the nut cannot be rotated, then suspend the steering gear.

3. Support the engine, using the engine lifting hooks and the procedure under Engine Removal & Installation.
- Crossmember
- Lower control arm from the crossmember

4. Installation is the reverse of the removal procedure.

Wheel Bearings

REMOVAL & INSTALLATION

Front

1. Before servicing the vehicle, refer to the precautions in the beginning of this section.
2. Remove or disconnect the following:
- Negative battery cable. On vehicles equipped with an air bag, wait at least 90 seconds before proceeding.
- Wheels
- Hub nut
- Front stabilizer link
- Anti-lock Brake System (ABS) speed sensor
- Brake caliper
- Rotor
- Tie rod end from the steering knuckle

- Lower control arm ball joint
- Front halfshaft from the hub, using a mallet to tap it out. Be careful not to damage the boot or speed sensor.

3. Loosen the nuts on the lower side of the strut assembly. Do not remove at this time.
- Lower ball joint using a puller
- Tie rod end from the steering knuckle
- Steering knuckle from the lower control arm
- Knuckle from the strut assembly
- Hub

➡**Cover the halfshaft boot with a shop rag to protect it from any damage.**

4. Clamp the steering knuckle in a vise and remove the dust deflector. Remove the nut holding the steering knuckle to the ball joint. Press the ball joint out of the steering knuckle.

5. Remove the inner axle seal.

6. Using a Torx® wrench, remove the bolts securing the dust cover.

7. Using hub puller, remove the hub and backing plate from the steering knuckle.

8. Using a proper sized driver and a press, remove the inner hub race from the axle hub.

9. Using seal removal tool, remove the outer axle seal.

10. Using snapring pliers, remove the snapring from the inner side of the steering knuckle.

11. Using a proper sized driver and a press, remove the bearing from the steering knuckle. The bearing is pressed from the front of the steering knuckle and is removed through the back of the steering knuckle.

To install:

12. Perform the following:

13. Using a proper sized driver and a press, install a new bearing to the steering knuckle.

14. Install the snapring to the steering knuckle using snapring pliers.

15. Using a seal driver and a hammer, install a new outer oil seal. Apply multipurpose grease to the oil seal lip.

16. Place the dust cover on the steering knuckle. Tighten the bolts: 78 inch lbs. (9 Nm).

17. Using a press and a proper sized driver, install the axle hub to the steering knuckle.

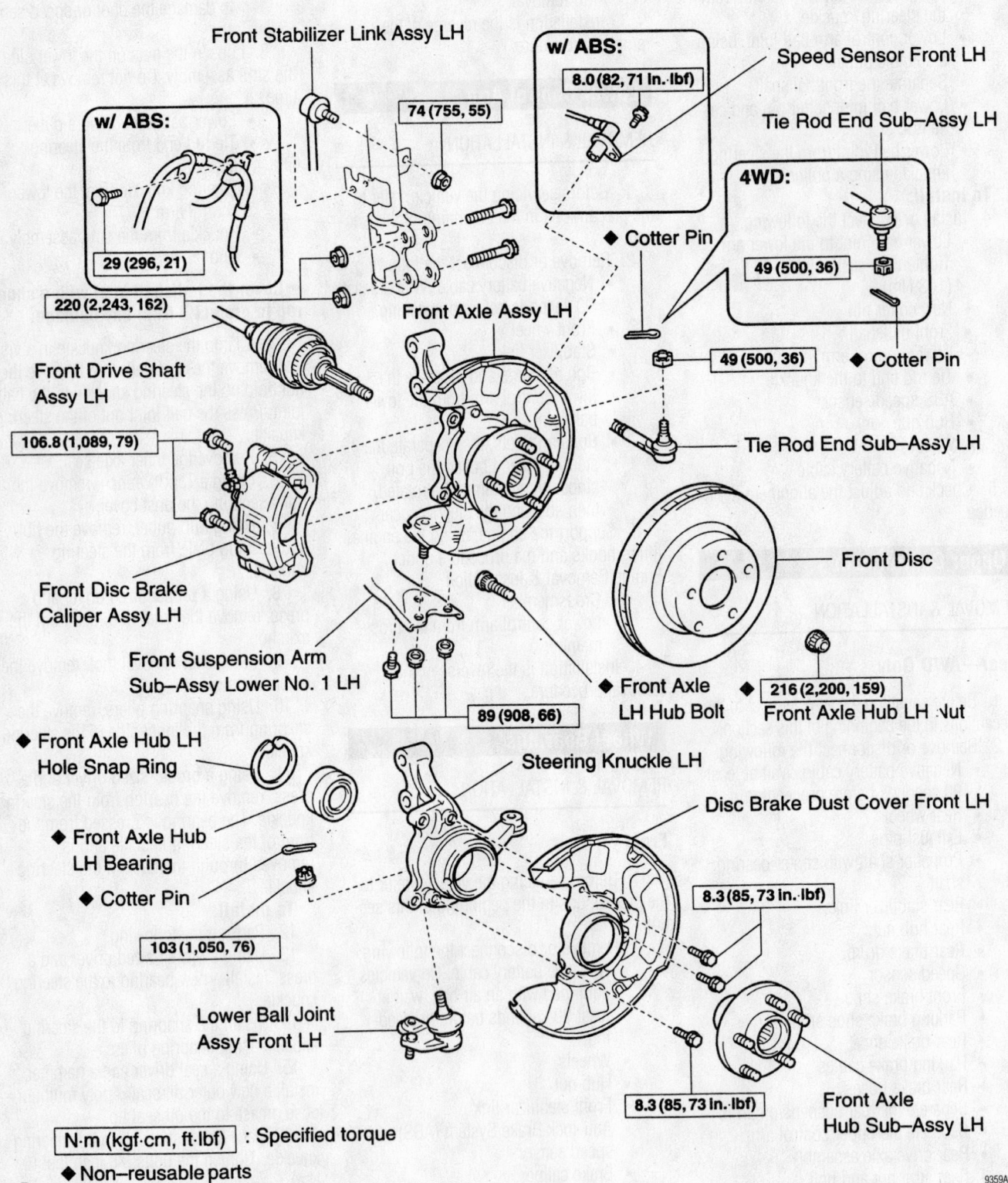

Front Stabilizer Link Assy LH

w/ ABS:
8.0 (82, 71 In.·lbf)

Speed Sensor Front LH

Tie Rod End Sub–Assy LH

w/ ABS:

74 (755, 55)

4WD:

29 (296, 21)

49 (500, 36)

220 (2,243, 162)

◆ Cotter Pin

Front Axle Assy LH

◆ Cotter Pin

Front Drive Shaft
Assy LH

49 (500, 36)

106.8 (1,089, 79)

Tie Rod End Sub–Assy LH

Front Disc

Front Disc Brake
Caliper Assy LH

Front Suspension Arm
Sub–Assy Lower No. 1 LH

◆ Front Axle
LH Hub Bolt

◆ Front Axle Hub LH Nut

89 (908, 66)

216 (2,200, 159)

◆ Front Axle Hub LH
Hole Snap Ring

Steering Knuckle LH

◆ Front Axle Hub
LH Bearing

Disc Brake Dust Cover Front LH

◆ Cotter Pin

8.3 (85, 73 In.·lbf)

103 (1,050, 76)

Lower Ball Joint
Assy Front LH

8.3 (85, 73 In.·lbf)

Front Axle
Hub Sub–Assy LH

N·m (kgf·cm, ft·lbf) : Specified torque

◆ Non–reusable parts

P

9359AB72

Exploded view of the front hub and bearing, and related components

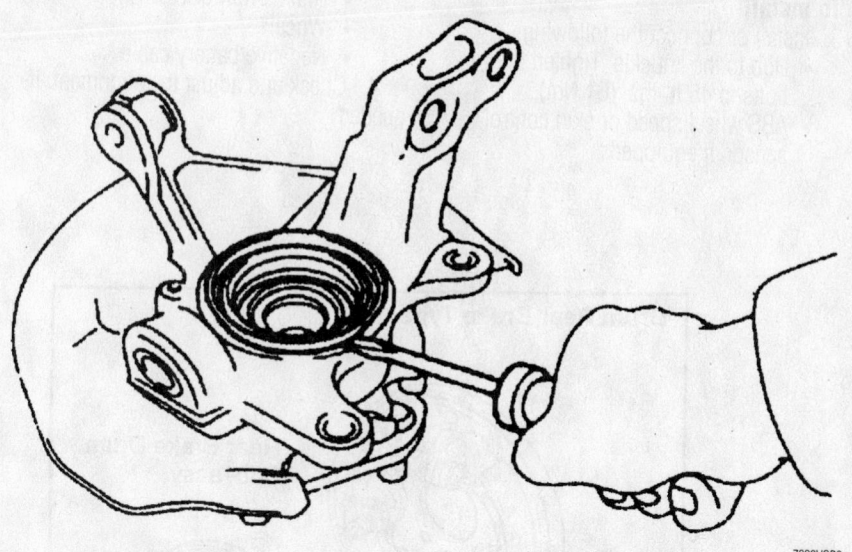

Removing the inner axle seal from the hub assembly

7923VGB3

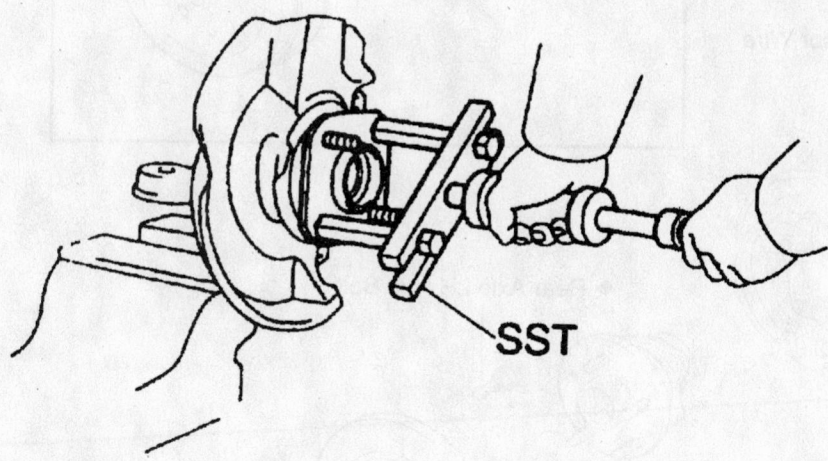

Removing the axle hub from the knuckle

SST

7923VGB4

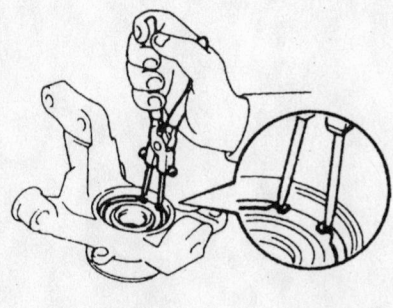

Removing the snapring from the knuckle
before pressing out the bearing

7923VGB5

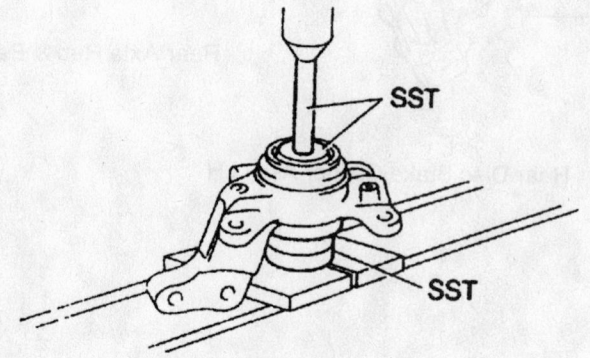

Removing the bearing from the steering knuckle using a press

SST

SST

7923VGB6

18. Attach the ball joint to the steering knuckle. Install a new cotter pin.

19. Using a seal driver and a hammer, install a new inner oil seal. Apply multipurpose grease to the oil seal lip.

20. Install the knuckle and hub assembly to the axle and temporarily tighten the axle nut.

21. Connect the knuckle assembly to the lower strut bracket. Temporarily insert the mounting bolts from the rear and install the nuts making sure the matchmarks made earlier are in alignment.

22. Connect the lower ball joint to lower arm.

23. Connect the tie rod end to the knuckle.

24. Tighten the bolts on the lower side of the strut assembly.

25. If equipped, install the ABS speed sensor.

26. Install the brake disc and the caliper.

27. Tighten the axle nut while someone depresses the brake pedal.

28. Install the wheels to the vehicle. Verify that the wheel turns freely.

29. Connect the negative battery cable to the battery.

30. Check alignment.

Rear

1. Before servicing the vehicle, refer to the precautions in the beginning of this section.

2. Remove or disconnect the following:
   - Negative battery cable. On vehicles equipped with an air bag, wait at least 90 seconds before proceeding.
   - Wheel

For Tune-up, Capacities and Firing orders, see Section 1 of this manual

- Brake drum or rotor
- With ABS brakes, ABS wheel speed sensor or skid control sensor, as applicable
- 4 hub retaining bolts
- Hub

To install:

3. Install or connect the following:
   - Hub to the knuckle. Tighten the bolts to 45 ft. lbs. (61 Nm).
   - ABS wheel speed or skid control sensor, if equipped

- Brake drum or rotor
- Wheel
- Negative battery cable

4. Check and adjust the alignment, if needed.

Disc Rear Brake Type:

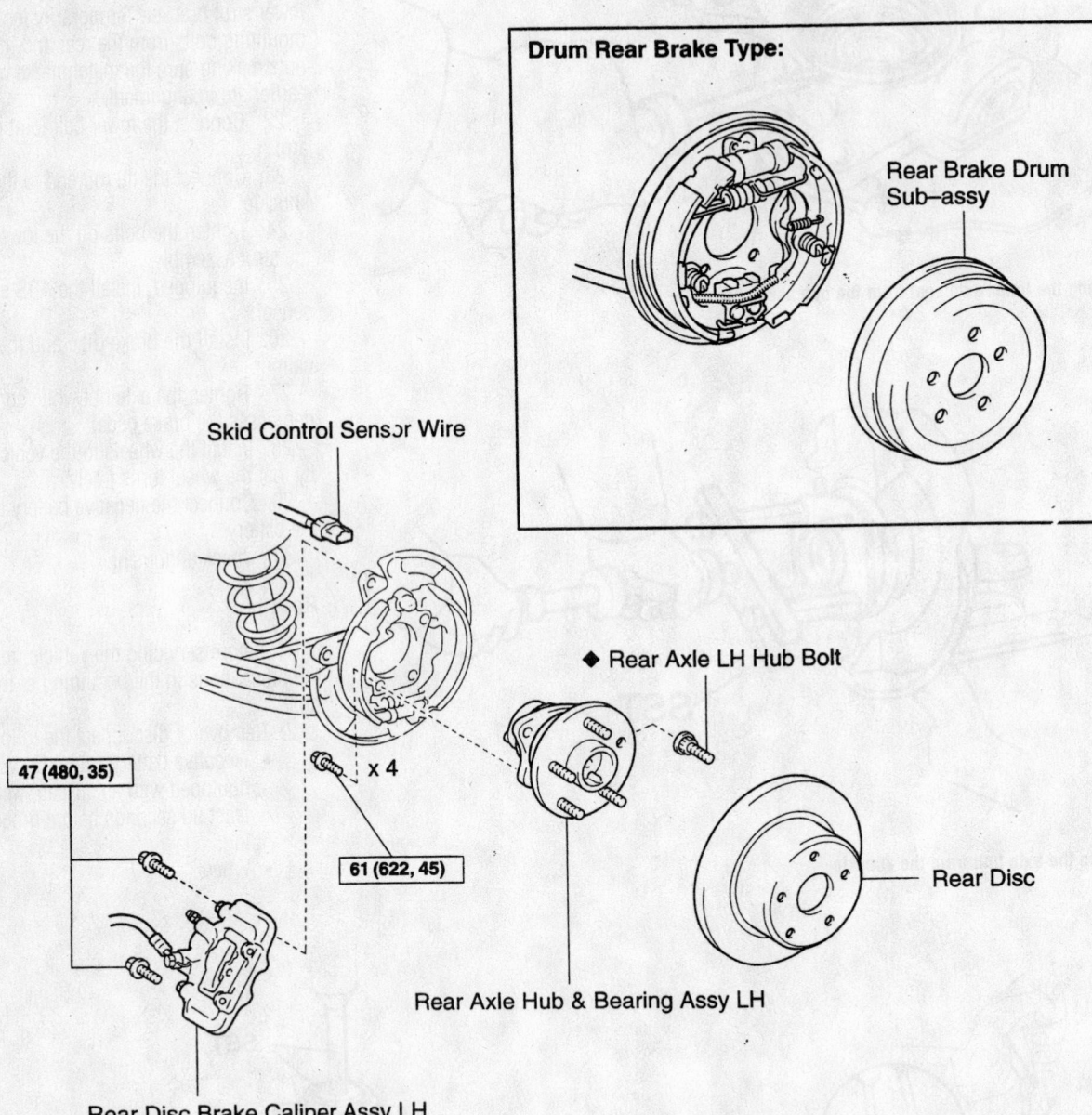

Drum Rear Brake Type:

Rear Brake Drum Sub–assy

Skid Control Sensor Wire

◆ Rear Axle LH Hub Bolt

47 (480, 35)

61 (622, 45)

x 4

Rear Disc

Rear Axle Hub & Bearing Assy LH

Rear Disc Brake Caliper Assy LH

N·m (kgf·cm, ft·lbf) : Specified torque

◆ Non–reusable part

Exploded view of the hub and wheel bearing assembly

9359AB73

PRECAUTIONS

Before servicing any vehicle, please be sure to read all of the following precautions, which deal with personal safety, prevention of component damage, and important points to take into consideration when servicing a motor vehicle:

• Never open, service or drain the radiator or cooling system when the engine is hot; serious burns can occur from the steam and hot coolant.

• Observe all applicable safety precautions when working around fuel. Whenever servicing the fuel system, always work in a well-ventilated area. Do not allow fuel spray or vapors to come in contact with a spark, open flame or excessive heat (a hot drop light, for example). Keep a dry chemical fire extinguisher near the work area. Always keep fuel in a container specifically designed for fuel storage; also, always properly seal fuel containers to avoid the possibility of fire or explosion. Refer to the additional fuel system precautions later in this section.

• Fuel injection systems often remain pressurized, even after the engine has been turned **OFF**. The fuel system pressure must be relieved before disconnecting any fuel lines. Failure to do so may result in fire and/or personal injury.

• Brake fluid often contains polyglycol ethers and polyglycols. Avoid contact with the eyes and wash your hands thoroughly after handling brake fluid. If you do get brake fluid in your eyes, flush your eyes with clean, running water for 15 minutes. If eye irritation persists, or if you have taken

brake fluid internally, IMMEDIATELY seek medical assistance.

• The EPA warns that prolonged contact with used engine oil may cause a number of skin disorders, including cancer! You should make every effort to minimize your exposure to used engine oil. Protective gloves should be worn when changing oil. Wash your hands and any other exposed skin areas as soon as possible after exposure to used engine oil. Soap and water, or waterless hand cleaner should be used.

• All new vehicles are now equipped with an air bag system. The system must be disabled before performing service on or around system components, steering column, instrument panel components, wiring and sensors. Failure to follow safety and disabling procedures could result in accidental air bag deployment, possible personal injury and unnecessary system repairs.

• Always wear safety goggles when working with, or around, the air bag system. When carrying a non-deployed air bag, be sure the bag and trim cover are pointed away from your body. When placing a non-deployed air bag on a work surface, always face the bag and trim cover upward, away from the surface. This will reduce the motion of the module if it is accidentally deployed. Refer to the additional air bag system precautions later in this section.

• Clean, high quality brake fluid from a sealed container is essential to the safe and proper operation of the brake system. You should always buy the correct type of brake

fluid for your vehicle. If the brake fluid becomes contaminated, completely flush the system with new fluid. Never reuse any brake fluid. Any brake fluid that is removed from the system should be discarded. Also, do not allow any brake fluid to come in contact with a painted surface; it will damage the paint.

• Never operate the engine without the proper amount and type of engine oil; doing so WILL result in severe engine damage.

• Timing belt maintenance is extremely important! Many models utilize an interference-type, non-freewheeling engine. If the timing belt breaks, the valves in the cylinder head may strike the pistons, causing potentially serious (also time-consuming and expensive) engine damage. Refer to the maintenance interval charts in the front of this manual for the recommended replacement interval for the timing belt, and to the timing belt section for belt replacement and inspection.

• Disconnecting the negative battery cable on some vehicles may interfere with the functions of the on-board computer system(s) and may require the computer to undergo a relearning process once the negative battery cable is reconnected.

• When servicing drum brakes, only disassemble and assemble one side at a time, leaving the remaining side intact for reference.

• Only an MVAC-trained, EPA-certified automotive technician should service the air conditioning system or its components.

ENGINE REPAIR

Ignition Timing

ADJUSTMENT

The engines covered in this section utilize a Distributorless Ignition System (DIS), no adjustment is possible.

Alternator

REMOVAL

2.2L Engine

1. Before servicing the vehicle, refer to the precautions in the beginning of this section.
2. Remove or disconnect the following:

• Negative battery cable
• Throttle body air duct
• Accessory drive belt
• Alternator electrical connectors
• Alternator bolts
• Alternator

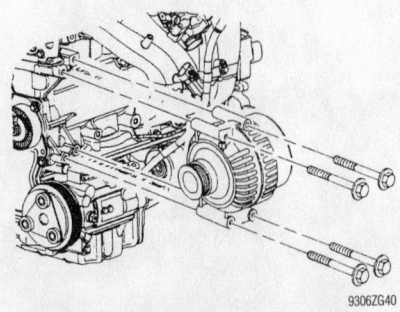

9306ZG40

Typical alternator mounting–2.2L engine

3.0L Engine

1. Before servicing the vehicle, refer to the precautions in the beginning of this section.
2. Remove or disconnect the following:

• Negative battery cable
• Accessory drive belt and tensioner
• Upper alternator bolts
• Alternator lower bolts
• Alternator electrical connectors
• Alternator

INSTALLATION

2.2L Engine

Install or connect the following:
• Alternator and torque the bolts to 26 ft. lbs. (35 Nm)

- Alternator electrical connectors
- Accessory drive belt
- Throttle body air duct
- Negative battery cable

3.0L Engine

Install or connect the following:
- Alternator electrical connectors
- Alternator and torque the bolts to 26 ft. lbs. (35 Nm)
- Drive belt tensioner and torque the bolts to 30 ft lbs. (40 Nm)
- Accessory drive belt
- Negative battery cable

Engine Assembly

REMOVAL & INSTALLATION

2.2L Engine

AUTOMATIC TRANSMISSION

1. Before servicing the vehicle, refer to the precautions in the beginning of this section.
2. Properly relieve the fuel system pressure.
3. Drain the engine coolant.
4. Drain the engine oil.
5. Drain the power steering fluid.
6. Remove or disconnect the following:
- Both battery cables
- Battery
- Air cleaner and intake duct assembly
- Intake Air Temperature (IAT) sensor electrical connector
- Underhood fuse panel cover
- 3 connector through bolts from the fuse block
- Battery and electronic power steering feed wire nut and the wires from the stud
- Fuse block connectors from the fuse block
- Fuse block and battery tray bolts, then the fuse block and the battery tray
- Electrical connectors from the Transmission Control Module (TCM), if equipped
- Rear Heated Oxygen (HO2S) sensor electrical connector
- 8-way electrical connection
- Main harness connector

➡**Do not remove the shifter cable from the bracket before removing the cable from the switch.**

- Shift cable from the transaxle range switch by slightly prying between the cable plastic retainer and the switch, if equipped
- Shift lever cables from the shift control housing using tool J36346, manual transmission only
- Shift lever cable from the bracket, manual transmission only
- Pressure line from the clutch actuator
- Back-up lamp switch
- Upper radiator hose from the cylinder head
- Lower radiator hose at the coolant pipe
- De-gas hose at the surge tank
- Surge hose at the surge tank
- Heater hoses from the core at the firewall
- Fuel lines
- Purge hoses from the solenoid
- Headlamp assemblies

7. Attach the radiator/condenser assembly to the radiator support using tie straps as this assembly stays in the vehicle.
- Left front wheel and splash shield

8. Install a block of wood 1 inch x 2 inch x 4 inch between the transfer case and the engine cradle.
- Right front wheel and splash shield

9. Install a block of wood 1 inch x 2 inch x 4 inch between the oil pane and the engine cradle. Do not place the wood under the oil pan and plug boss.
- Drive belt
- Electrical connections from the A/C compressor and pressure transducer
- A/C compressor bolts and position the compressor aside
- Push pins that attach the air deflector to the cradle

10. Drain the transaxle fluid.
- Transaxle lines from the transaxle and discard the seals. Replace with new seals during assembly.
- Rear HO2S sensor
- Exhaust pipe-to-manifold flange and intermediate fasteners
- Converter pipe and support the intermediate pipe
- Propshaft bolt to Power Take Off (PTO) on all wheel drive models
- Propshaft bolts, bracket and the shaft, on all wheel drive models
- Shifter cable from the bracket, on automatic transaxle models
- Power steering gear-to-intermediate shaft bolt

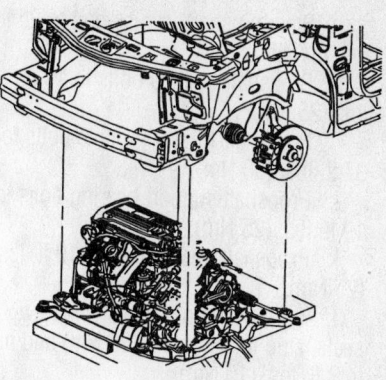

9359ZG01

Remove the engine with a support table–2.2L engine

- Tie rod from the knuckles
- Lower control arms from the knuckles
- Stabilizer link nuts
- Left hand axle shaft from the transaxle
- Right hand shaft from the intermediate shaft using tool J45341
- Right hand engine mount and the left hand transaxle mount and let the engine rest on the wood blocks

➡**Support the rear of the vehicle with a jackstand prior to engine removal.**

11. Place an engine support table under the engine.
12. Place blocks of wood to level the powertrain assembly, if necessary. The blocks can be placed between the oil pan and the cradle.
13. Raise the support table until it supports the powertrain assembly.
14. Remove the cradle-to-body bolts.
15. Check that all hoses, lines and wiring is free, then lower the engine table and raise the body on a hoist to remove the assembly

To install:

16. Installation is the reverse of removal, please note the following torque specifications.

   a. Frame-to-body bolts to 114 ft. lbs. (155 Nm).

   b. Right hand engine mount-to-bracket bolts to 81 ft. lbs. (110 Nm).

   c. Left hand transaxle mount-to-bracket bolts to 37 ft. lbs. (50 Nm).

   d. Lower control arm to knuckle bolts to 89 inch lbs. (10 Nm) plus 150 degrees.

   e. Stabilizer link nuts to 48 ft. lbs. (65 Nm).

   f. Tie rod-to-knuckle assembly to 37 ft. lbs. (50 Nm).

g. Steering shaft-to-rack bolt to 25 ft. lbs. (34 Nm).

h. Propshaft-to-PTU bolts to 19 ft. lbs. (25 Nm).

i. Propshaft-to-rear module bolts to 37 ft. lbs. (50 Nm).

j. Propshaft support bearing bolts to 19 ft. lbs. (25 Nm).

k. Propshaft guard strap to 19 ft. lbs. (25 Nm).

l. Transaxle cooler lines with new seals, stud to 15 ft. lbs. (20 Nm) and nut to 7 ft. lbs. (10 Nm).

m. Down pipe-to-manifold nuts to 22 ft. lbs. (30 Nm).

n. Down pipe-to-intermediate pipe nuts to 37 ft. lbs. (50 Nm).

o. A/C compressor bolts to 18 ft. lbs. (25 Nm).

p. Engine ground-to-body bolt to 15 ft. lbs. (20 Nm).

17. Fill the engine with coolant.

18. Fill the engine with new oil.

19. Prime the fuel system by cycling the ignition **ON** for 5 seconds and **OFF** for 10 seconds a few times without cranking the engine.

20. Start the engine, check for leaks, and repair if necessary.

3.0L Engine

1. Before servicing the vehicle, refer to the precautions in the beginning of this section.

2. Properly relieve the fuel system pressure.

3. Drain the engine coolant.

4. Drain the engine oil.

5. Drain the power steering fluid.

6. Remove or disconnect the following:

- Both battery cables
- Battery
- Air cleaner and intake duct assembly
- Intake Air Temperature (IAT) sensor electrical connector
- Underhood fuse panel cover
- 3 connector through bolts from the fuse block
- Battery and electronic power steering feed wire nut and the wires from the stud
- Fuse block connectors from the fuse block
- Electrical connectors from the Transmission Control Module (TCM), if equipped
- Rear Heated Oxygen (HO₂S) sensor electrical connector
- Vacuum hose with the check valve

from the brake booster and lay on the engine
- Main harness connector
- Fuse block and battery tray bolts, then the fuse block and the battery tray
- Shift cable from the PRNDL switch by slightly prying between the cable plastic retainer and the switch
- Upper radiator hose from the cylinder head
- Lower radiator hose at the coolant pipe
- De-gas hose at the surge tank
- Surge hose at the surge tank
- Heater hoses from the core at the firewall
- Fuel lines
- Purge hoses from the solenoid

7. Attach the radiator/condenser assembly to the radiator support using tie straps as this assembly stays in the vehicle.

8. Evacutae the A/C system using approved equipment and disconnect the transducer and connector.

- Left front wheel and splash shield

9. Install a block of wood 1 inch x 2 inch x 4 inch between the transfer case and the engine cradle.

- Right front wheel and splash shield

10. Install a block of wood 1 inch x 2 inch x 4 inch between the oil pane and the engine cradle. Do not place the wood under the oil pan and plug boss.

- Drive belt
- A/C compressor and position the A/C line aside
- Push pins that attach the air deflector to the cradle

11. Drain the transaxle fluid.

- Transaxle lines from the transaxle and discard the seals. Replace with new seals during assembly.
- Exhaust pipe-to-manifold flange fasteners
- Exhaust pipe-to-muffler nuts
- Converter pipe and support the intermediate pipe
- Propshaft bolt-to-Power Take Off (PTO) ,on all wheel drive models
- Propshaft bolts, bracket and the shaft, on all wheel drive models
- Shifter cable from the bracket
- Power steering gear-to-intermediate shaft bolt
- Tie rod from the knuckles
- Lower control arms from the knuckles
- Stabilizer link nuts
- Left hand axle shaft from the transaxle
- Right hand shaft from the intermediate shaft using tool J45341

- Right hand engine mount and the left hand transaxle mount and let the engine rest on the wood blocks

➡**Support the rear of the vehicle with a jackstand prior to engine removal.**

12. Place an engine support table under the engine.

13. Place blocks of wood to level the powertrain assembly, if necessary. The blocks can be placed between the oil pan and the cradle.

14. Raise the support table until it supports the powertrain assembly.

15. Remove the cradle-to-body bolts.

16. Check that all hoses, lines and wiring is free, then lower the engine table and raise the body on a hoist to remove the assembly

To install:

17. Installation is the reverse of removal, please note the following torque specifications.

a. Frame-to-body bolts to 114 ft. lbs. (155 Nm).

b. Right hand engine mount-to-bracket bolts to 37 ft. lbs. (50 Nm).

c. Left hand transaxle mount-to-bracket bolts to 37 ft. lbs. (50 Nm).

d. Lower control arm to knuckle bolts to 89 inch lbs. (10 Nm) plus 150 degrees.

e. Stabilizer link nuts to 48 ft. lbs. (65 Nm).

f. Tie rod-to-knuckle assembly to 37 ft. lbs. (50 Nm).

g. Steering shaft-to-rack bolt to 25 ft. lbs. (34 Nm).

h. Propshaft-to-PTU bolts to 19 ft. lbs. (25 Nm).

i. Propshaft-to-rear module bolts to 37 ft. lbs. (50 Nm).

j. Propshaft support bearing bolts to 19 ft. lbs. (25 Nm).

k. Propshaft guard strap to 19 ft. lbs. (25 Nm).

l. Transaxle cooler lines with new seals, stud to 15 ft. lbs. (20 Nm) and nut to 7 ft. lbs. (10 Nm).

m. Down pipe-to-manifold nuts to 22 ft. lbs. (30 Nm).

n. Down pipe-to-intermediate pipe nuts to 22 ft. lbs. (30 Nm).

o. A/C compressor bolts to 18 ft. lbs. (25 Nm).

p. Engine ground-to-body bolt to 15 ft. lbs. (20 Nm).

18. Fill the engine with coolant.

19. Fill the engine with new oil.

20. Prime the fuel system by cycling the ignition **ON** for 5 seconds and **OFF** for 10 seconds a few times without cranking the engine.

21. Start the engine, check for leaks, and repair if necessary.

Water Pump

REMOVAL & INSTALLATION

2.2L Engine

1. Before servicing the vehicle, refer to the precautions in the beginning of this section.

2. Drain the cooling system.

3. Remove or disconnect the following:
   - Negative battery cable
   - Air cleaner assembly
   - Thermostat housing pipe-to-cylinder head bolt (near the front of the engine)
   - Exhaust manifold heat shield
   - Water pump access plate
   - Right hand wheel and splash shield

4. Remove the drain plug from the bottom of the pump and drain the remaining coolant.
   - Engine Coolant Temperature (ECT) sensor connection
   - Thermostat housing bolts, then move the housing towards the left hand side of the vehicle while twisting the feed pipe from the rear of the pump. Leave the coolant hoses and the housing connected.
   - Water feed pipe and discard the seals

5. Install a Water Pump Holding Tool J43651. Tighten the bolts on the tool into threads on the pump sprocket, then install

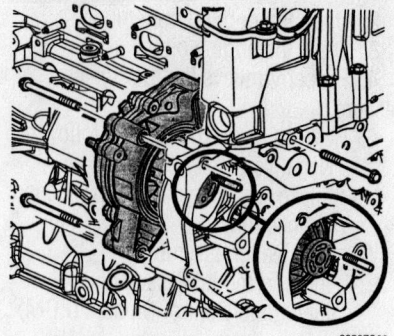

Exploded view of the water pump mounting—2.2L engine

some of the access plate bolts to attach the tool to the front cover.
   - Water pump retaining bolts
   - Water pump

To install:

6. Install or connect the following:
   - Water pump with a new seal and torque the bolts to 18 ft. lbs. (25 Nm)
   - Water pump sprocket and torque the bolts to 89 inch lbs. (10 Nm)
   - Water pump sprocket access plate and torque the bolts to 89 inch lbs. (10 Nm)

   - Water feed tube after lubricating the O-ring
   - Thermostat housing and torque the bolts to 89 inch lbs. (10 Nm)
   - Exhaust manifold heat shield
   - Air cleaner assembly
   - Negative battery cable

7. Fill the cooling system.

8. Start the vehicle and check for leaks, repair if necessary.

3.0L Engine

1. Before servicing the vehicle, refer to the precautions in the beginning of this section.

2. Drain the cooling system.

3. Remove or disconnect the following:
   - Negative battery cable
   - Air cleaner assembly
   - Left front wheel
   - Wheel well splash shield
   - Loosen the water pump pulley and power steering pulley bolts, but do not remove them

4. Install an engine support fixture.
   - Right front engine mount
   - Accessory drive belt

5. Release the retaining tabs on the wiring harness channel and remove the front cover.

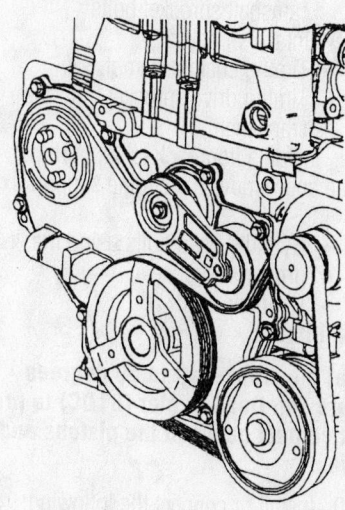

Water pump holding tool J43651—2.2L engine

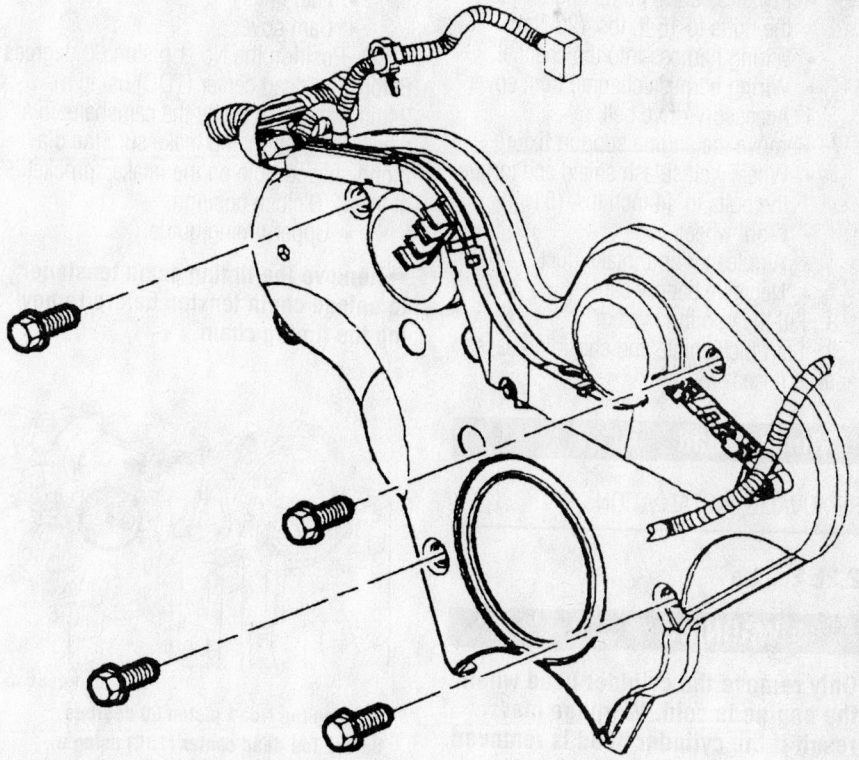

Timing belt cover bolt locations—3.0L engine

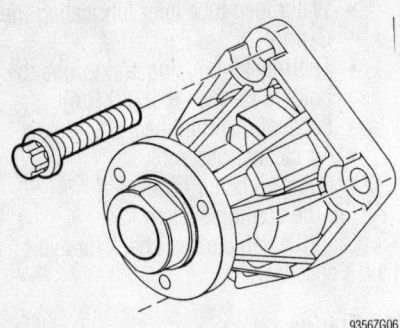

Water pump mounting–3.0L engine

- Wiring harness from the channel
- Water pump pulley
- Power steering pump pulley
- Drive belt tensioner
- Front timing belt cover
- Water pump

To install:

6. Install or connect the following:
   - Water pump with a new O-ring and torque the bolts to 18 ft. lbs. (25 Nm)
   - Front timing belt cover and torque the bolts to 71 inch lbs. (8 Nm)
   - Drive belt tensioner and torque the bolts to 30 ft. lbs. (40 Nm)
   - Water pump pulley's and torque the bolts to 71 inch lbs. (8 Nm)
   - Power steering pump and torque the bolts to 15 ft. lbs. (20 Nm)
   - Wiring harness into the channel
   - Wiring harness channel front cover
   - Accessory drive belt
7. Remove the engine support fixture.
   - Wheel well splash shield and torque the bolts to 44 inch lbs. (5 Nm)
   - Front wheel
   - Air cleaner and intake duct
   - Negative battery cable
8. Fill the cooling system.
9. Start the vehicle and check for leaks, repair if necessary.

Cylinder Head

REMOVAL & INSTALLATION

2.2L Engine

※※ WARNING

Only remove the cylinder head when the engine is cold. Warpage may result if the cylinder head is removed while the engine is hot.

1. Before servicing the vehicle, refer to the precautions in the beginning of this section.

2. Drain the cooling system.
3. Drain the engine oil.
4. Properly relieve the fuel system pressure.
5. Remove or disconnect the following:
   - Negative battery cable
   - Intake Air Temperature (IAT) sensor connection
   - Air cleaner assembly
   - Ignition module assembly
   - Electronic Control Module (ECM) connections
   - Oil dipstick tube bolt
   - Throttle body electrical connection
   - Electrical connector from the fuel injector harness and attachment at the bottom of the intake manifold
   - Electrical connector at the purge solenoid and Manifold Absolute Pressure (MAP) sensor
   - Vacumm hose at the brake booster
   - Coolant pipe bracket bolts from the cylinder head
   - De-gas hose clamp from the cylinder head and fuel rail and position aside
   - Ground strap from the rear cam cover
   - Fuel rail bracket from the cam cover
   - Fuel lines
   - Cam cover
6. Position the No. 1 piston 60 degrees Before Top dead center (TDC) using a 24mm wrench to rotate the camshafts in a clockwise motion and make sure the diamond shaped hole on the intake sprocket is at the 12 O'clock position.
   - Upper timing guide

➡**Remove the timing chain tensioner to unload chain tension before removing the timing chain.**

Position the No. 1 piston 60 degrees Before Top dead center (TDC) using a 24mm wrench to rotate the camshafts in a clockwise motion and make sure the diamond shaped hole on the intake sprocket is at the 12 O'clock position—2.2L engine

- Fixed timing chain guide access plug
- Upper fixed guide bolt using a magnetic socket
7. Install a three bar engine support fixture.
   - Right hand engine mount
   - Right hand mount bracket
   - Right wheel splash shield
   - Install a 1 x 2 x 4 inch block of wood between the oil pan and cradle
   - Drive belt
   - Drive belt tensioner assembly
8. Install crankshaft pullet holder J38122A..
   - Crankshaft balancer bolt and pulley
   - Front cover bolts
   - Lower water pump bolt
   - Front cover and gasket
   - Lower fixed guide
   - Upper radiator hose from the cylinder head
   - Exhaust manifold pipe nuts
   - Front and rear Oxygen (O_2S) sensor electrical connector
   - Down pipe-to-intermediate pipe nuts
   - Exhaust camshaft sprocket bolts while holding the camshaft with a 24mm wrench and discard the camshaft sprocket bolts
   - Exhaust sprocket
   - Adjustable guide through the top of the cylinder head
   - Intake camshaft sprocket bolts while holding the camshaft with a 24mm wrench and discard the camshaft sprocket bolts
   - Intake sprocket
   - Timing chain assembly
   - Timing drive sprocket from the crankshaft
9. Install a floor jack to support the engine and remove the engine support fixture.
   - Cylinder head bolts using the proper sequence
   - Cylinder head

To install:

➡**Set the crankshaft to 60 degrees Before Top Dead Center (BTDC) to prevent contact between the pistons and valves.**

10. Install or connect the following:
    - New cylinder head gasket with the side imprinted **OBEN**facing up
    - Cylinder head and align it on the dowels
    - New cylinder head bolts and torque

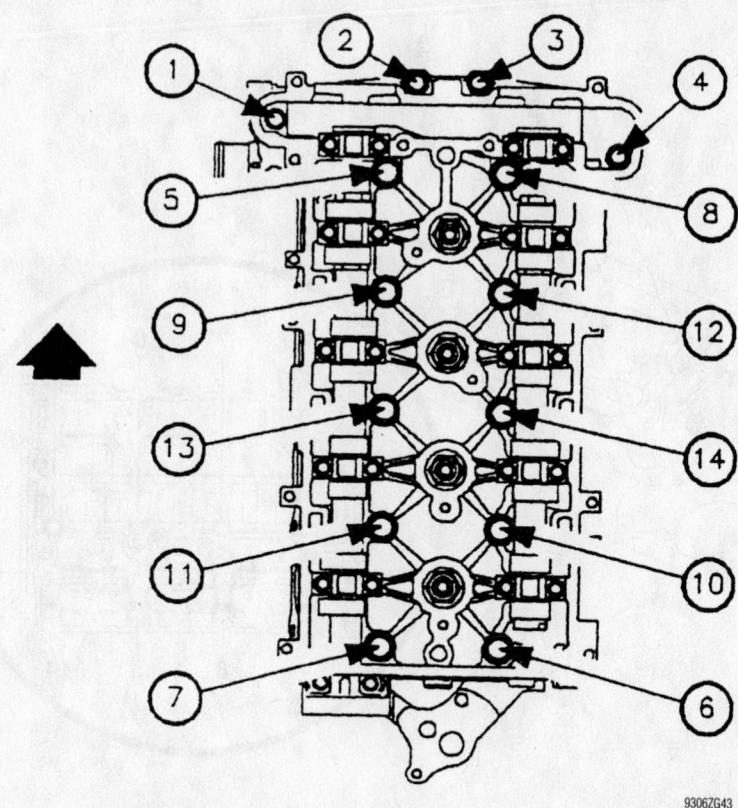

Cylinder head bolt loosening sequence—2.2L engine

9306ZG43

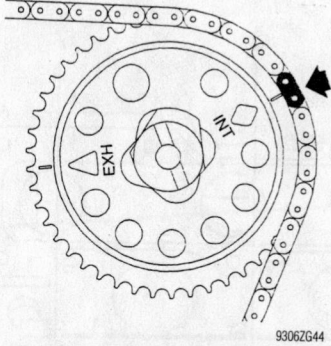

Align the copper link on the timing chain with the INT diamond timing mark

9306ZG44

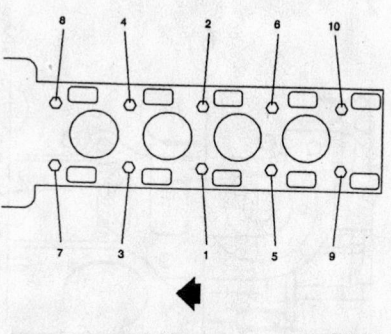

Cylinder head bolt tightening sequence—2.2L engine

9359ZG04

them in sequence to 22 ft. lbs. (30 Nm) plus 155 degrees
  • Front 4 cylinder head bolts coated with Loctite® and torque them to 26 ft. lbs. (35 Nm)
11. Position the exhaust camshaft with the offset slot in the 2 o'clock position and the intake camshaft with the offset slot in the 11 o'clock position.
  • Timing chain around the intake camshaft sprocket with the copper link aligned with the **INT** diamond timing mark

  • Sprocket to the camshaft and align it with the offset slot
  • New camshaft sprocket bolt but do not tighten
  • Timing chain around the crankshaft sprocket and align the silver link to the timing mark
  • Adjustable timing chain guide through the opening on top of the cylinder head and torque the bolt to 89 inch lbs. (10 Nm)
  • Timing chain around the exhaust camshaft sprocket with the silver link

aligned with the offset slot. Install but do not tighten a new sprocket bolt

> **✳✳ CAUTION**
>
> **Make certain that all timing marks and colored links are aligned properly before proceeding to the next step. If the timing chain is not aligned properly, severe engine damage may occur.**

12. Torque the intake and exhaust camshaft bolts to 63 ft. lbs. (85 Nm) plus a 30 degree turn.
13. Install or connect the following:
  • Fixed timing guide and torque the bolt to 89 inch lbs. (10 Nm)
  • Fixed timing guide bolt access plug after applying Loctite® to the threads and torque it to 30 ft. lbs. (40 Nm)
  • Timing chain tensioner and torque the bolts 55 ft. lbs. (75 Nm)
14. Tap the top of the timing chain between the camshaft sprockets to engage the tensioner.
  • Upper timing chain guide and torque the bolts to 89 inch lbs. (10 Nm)
  • Front cover with a new gasket and torque the bolts to 18 ft. lbs. (25 Nm)
  • Water pump bolt and torque it to 18 ft. lbs. (25 Nm)
  • Crankshaft damper and torque the bolt to 74 ft. lbs. (100 Nm) plus 75 degrees
  • Drive belt tensioner and torque the bolts to 30 ft. lbs. (40 Nm)
  • Exhaust manifold pipe to the manifold and tighten the nuts to 22 ft. lbs. (30 Nm).
  • Oil dipstick tube bolt to 89 inch lbs. (10 Nm).
  • Engine mount bracket and torque the bolts to 66 ft. lbs. (90 Nm)
  • Engine mount to body nuts to 81 ft. lbs. (110 Nm)
  • Engine mount to the bracket and torque the bolts to 41 ft. lbs. (55 Nm)
15. Remaining components in the reverse order of removal.
16. Fill the engine with clean oil.
17. Fill the cooling system.
18. Prime the fuel system by cycling the ignition **ON** for 5 seconds and **OFF** for 10 seconds a few times without cranking the engine.
19. Start the engine, check for leaks, and repair if necessary.

Timing belt service is covered in Section 3 of this manual

3.0L Engine

FRONT

❊❊ WARNING

Only remove the cylinder head when the engine is cold. Warpage may result if the cylinder head is removed while the engine is hot.

1. Before servicing the vehicle, refer to the precautions in the beginning of this section.
2. Drain the cooling system.
3. Drain the engine oil.
4. Properly relieve the fuel system pressure.
5. Remove or disconnect the following:
   - Negative battery cable
   - Air cleaner assembly
   - Intake plenum
   - Intake manifold
   - Intake manifold spacer
   - Coolant bridge
   - Upper radiator hose from the coolant extension housing
6. Properly support the powertrain assembly.
   - Front transmission mount through bolt
   - Extension housing over the coolant module
   - Oil level indicator tube
   - Coolant extension housing by twisting it off
   - Front camshaft cover
   - Electornic Control Module (ECM) from the bracket
   - Ground wires from the lift bracket
   - Oxygen (O$_2$S) sensor electrical connector
   - Rear lift bracket
   - Down pipe from the exhaust manifold
   - Front timing belt cover
   - Timing belt
   - Timing belt tensioner bracket
   - Rear timing belt cover
   - Camshaft position (CMP) sensor electrical connector
   - Exhaust camshaft
   - Loosen the cylinder head bolts in stages as shown
   - Cylinder head
   - Exhaust manifold from the cylinder head (if necessary)

To install:

7. Install or connect the following:
   - Exhaust manifold with a new gasket and torque the bolts to 15 ft. lbs. (20 Nm)

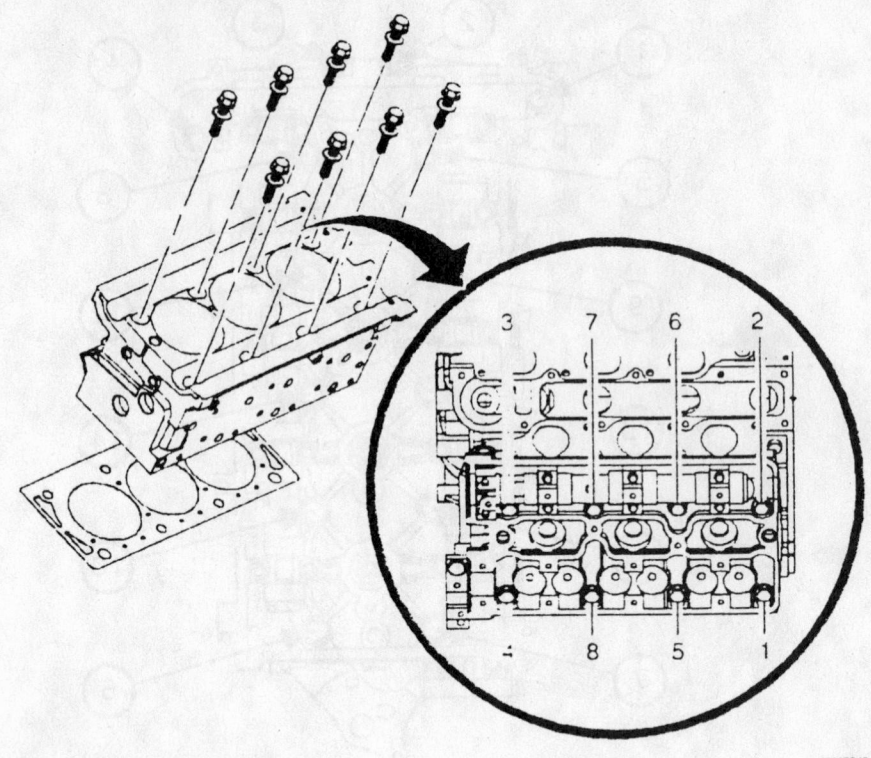

Cylinder head bolt removal for the front and rear cylinder heads—3.0L engine

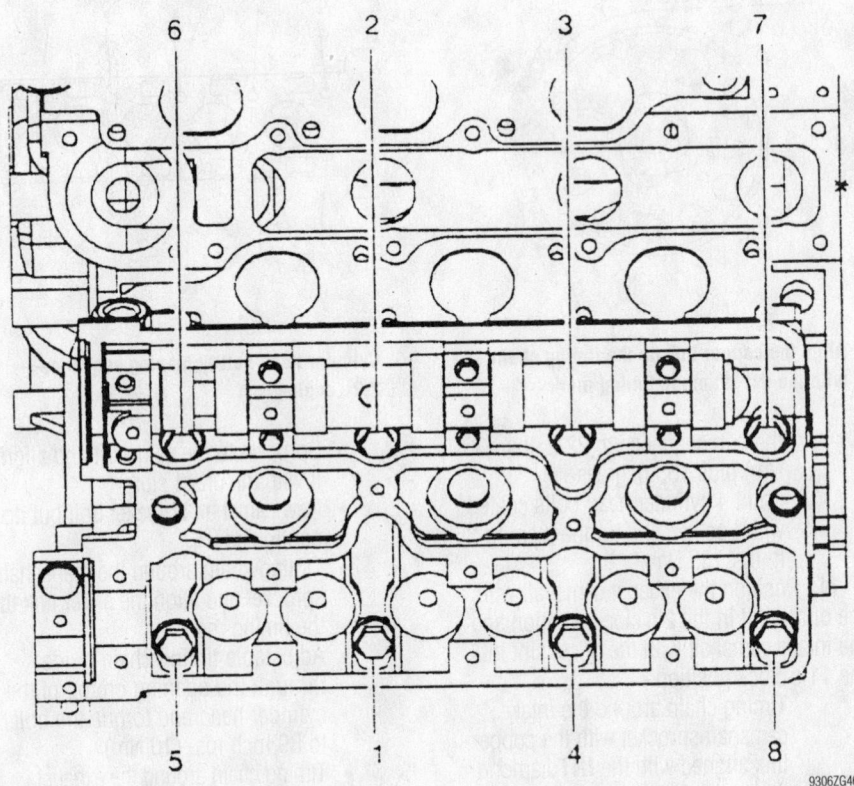

Front and rear cylinder head bolt tightening sequence—3.0L engine

- New cylinder head gasket with the part number imprint facing the top of the engine
- Cylinder head
8. Torque the new cylinder head bolts, in sequence, as follows:
   a. 18 ft. lbs. (25 Nm).
   b. 90 degree turn.
   c. 90 degree turn.
   d. 90 degree turn.
   e. 15 degree turn.
9. Install or connect the following:
   - Coolant pipe with new sealing rings
   - Engine lift bracket bolt and torque it to 15 ft. lbs. (20 Nm)
   - Upper radiator hose to the coolant pipe
   - Front transmission mount through bolt
   - Exhaust camshaft
   - CMP sensor electrical connector
   - Rear timing belt cover and torque the bolts to 71 inch lbs. (8 Nm)
   - Camshaft gears
   - Timing belt tensioner bracket
   - Timing belt
   - Front timing belt cover and torque the bolts to 71 inch lbs. (8 Nm)
   - Down pipe to the exhaust manifold
   - O₂S electrical connector
   - Front camshaft cover and torque the bolts to 71 inch lbs. (8 Nm)
   - Coolant bridge and torque the bolt to 22 ft. lbs. (33 Nm)
   - Intake manifold spacer and torque the bolts in a spiral direction from the inside and working out to 11 ft. lbs. (16 Nm)
   - Intake manifold and torque the bolts to 15 ft. lbs. (20 Nm)
   - Intake plenum and torque the bolts to 71 inch lbs. (8 Nm)
   - Air cleaner assembly
   - Negative battery cable
10. Fill the engine with clean oil.
11. Fill the cooling system.
12. Prime the fuel system by cycling the ignition **ON** for 5 seconds and **OFF** for 10 seconds a few times without cranking the engine.
13. Start the engine, check for leaks, and repair if necessary.

REAR

❊❊ WARNING

Only remove the cylinder head when the engine is cold. Warpage may

result if the cylinder head is removed while the engine is hot.

1. Before servicing the vehicle, refer to the precautions in the beginning of this section.
2. Drain the cooling system.
3. Drain the engine oil.
4. Properly relieve the fuel system pressure.
5. Remove or disconnect the following:
   - Negative battery cable
   - Air cleaner assembly
   - Intake plenum
   - Intake manifold
   - Intake manifold spacer
   - Coolant bridge
   - Rear camshaft cover
   - Oxygen (O₂S) sensor electrical connector
   - Front timing belt cover
   - Timing belt
   - Timing belt tensioner bracket
   - Camshaft gears
   - Rear timing belt cover
   - Exhaust manifold pipe heat shield
   - Front exhaust manifold pipe-to-rear exhaust manifold pipe fasteners
   - Rear exhaust manifold pipe nuts, pull the manifold pipe down and discard the gasket
   - Exhaust Gas Recirculation (EGR)-to-exhaust manifold pipe
   - Exhaust camshaft
   - Cylinder head bolts
   - Cylinder head
   - Exhaust manifold from the cylinder head

To install:
6. Install or connect the following:
   - Exhaust manifold with a new gasket and torque the bolts to 15 ft. lbs. (20 Nm), if removed
   - New cylinder head gasket with the part number imprint facing the top of the engine
   - Cylinder head
7. Torque the new cylinder head bolts, in sequence, as follows:
   a. 18 ft. lbs. (25 Nm).
   b. 90 degree turn.
   c. 90 degree turn.
   d. 90 degree turn.
   e. 15 degree turn.
8. Install or connect the following:
   - Exhaust camshaft
   - Rear timing belt cover and torque the bolts to 71 inch lbs. (8 Nm)

- Camshaft gears and torque the bolts to 37 ft. lbs. (50 Nm) plus a 60 degree turn plus another 15 degree turn
- Timing belt tensioner bracket and torque the bolts to 30 ft. lbs. 940 Nm)
- Timing belt
- Front timing belt cover and torque the bolts to 71 inch lbs. (8 Nm)
- Rear camshaft cover and torque the bolts to 71 inch lbs. (8 Nm)
- Exhaust manifold pipe to the manifold
- Exhaust manifold pipe heat shield
- EGR-to-exhaust manifold pipe and torque the nut to 18 ft. lbs. (25 Nm)
- Coolant bridge and torque the bolt to 22 ft. lbs. (33 Nm)
- Engine ventilation chamber and torque the bolts to 71 inch lbs. (8 Nm)
- Intake manifold spacer and torque the bolts to 25 ft. lbs. (20 Nm)
- Intake manifold and torque the bolts to 15 ft. lbs. (20 Nm)
- Intake plenum and torque the bolts to 71 inch lbs. (8 Nm)
- Air cleaner assembly
- Negative battery cable
9. Fill the engine with clean oil.
10. Fill the cooling system.
11. Prime the fuel system by cycling the ignition **ON** for 5 seconds and **OFF** for 10 seconds a few times without cranking the engine.
12. Start the engine, check for leaks, and repair if necessary.

Intake Manifold

REMOVAL & INSTALLATION

2.2L Engine

1. Before servicing the vehicle, refer to the precautions in the beginning of this section.
2. Remove or disconnect the following:
   - Negative battery cable
   - Intake Air Temperature (IAT) sensor electrical connector
   - Air cleaner assembly
   - Throttle body electrical connection
   - Throttle cable and automatic transmission downshift cable from the throttle body
   - Throttle body
   - Intake manifold

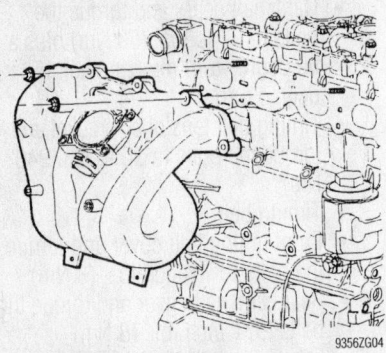

Intake manifold mounting—2.2L engines

To install:

3. Install or connect the following:
- Intake manifold with a new gasket and torque the nuts to 89 inch lbs. (10 Nm) starting from the center and working outward
- Throttle body to the intake manifold and torque the bolts to 89 inch lbs. (10 Nm)
- Throttle cable and automatic transmission downshift cable from the throttle body
- Throttle body electrical connection
- Air cleaner assembly
- Intake Air Temperature (IAT) sensor electrical connector
- Negative battery cable

4. Start the engine, check for leaks, and repair if necessary.

3.0L Engine

1. Before servicing the vehicle, refer to the precautions in the beginning of this section.
2. Properly relieve the fuel system pressure.
3. Remove or disconnect the following:
- Negative battery cable
- Air cleaner assembly
- Fuel line
- Manifold electrical connector
- Attachment of the purge solenoid and line from the rear of the manifold
- Brake booster line from the manifold
- Fuel injector and manifold control solenoid connections
- Resonator and bracket
- Positive Crankcase Valve (PCV) hose from the throttle body
- Throttle body bolts
- Throttle body cooling hose bracket bolts

4. Lay the throttle body over the master cylinder with the coolant hoses attached.

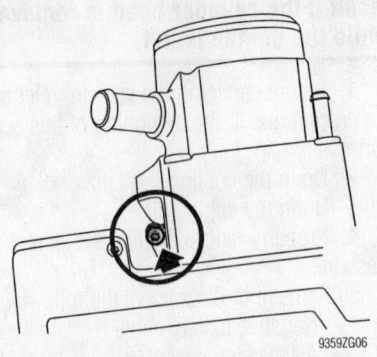

Remove the lower manifold bolt located near the throttle body—3.0L engines

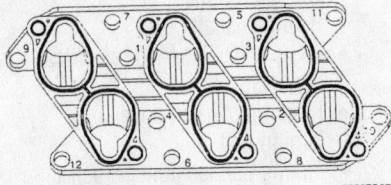

Tighten the intake manifold spacer bolts as shown—3.0L engines

- Intake manifold-to-spacer bolts using a T–30 Torx bit and a long ¼ inch extension
- Lower manifold bolt located near the throttle body
- Intake manifold
- Spacer plate bolts, spacer and seals

To install:

5. Install or connect the following:
- Intake manifold spacer with new seals. Apply Loctite &reg 242 to the bolts and torque the bolts to 12 ft. lbs. (16 Nm) in sequence.
- Intake manifold with a new gasket and torque the bolts to 66 inch lbs. (7.5 Nm)
- Throttle body and gasket, tighten the bolts to 66 inch lbs. (7.5 Nm)
- PCV hose to the throttle body
- Resonator and bracket
- Fuel injector and manifold control solenoid connections
- Brake booster line to the manifold
- Fuel line
- Attachment of the purge solenoid and line to the rear of the manifold
- Manifold electrical connector
- Air cleaner assembly
- Negative battery cable

6. Prime the fuel system by cycling the ignition **ON** for 5 seconds and **OFF** for 10 seconds a few times without cranking the engine.

7. Start the engine, check for leaks, and repair if necessary.

Exhaust Manifold

REMOVAL & INSTALLATION

2.2L Engine

1. Before servicing the vehicle, refer to the precautions in the beginning of this section.
2. Remove or disconnect the following:
- Negative battery cable
- Air cleaner assembly
- Exhaust manifold heat shield
- Oxygen (O2S) sensor from the manifold
- Exhaust pipe from the manifold
- Exhaust pipe-to-resonator pipe nuts from behind the converter
- Exhaust manifold pipe and resonator pipe
- Exhaust manifold

To install:

3. Install or connect the following:
- Exhaust manifold with a new gasket and torque the bolts, starting from the center and working outward, to 13 ft. lbs. (18 Nm)
- O2S sensor and torque it to 33 ft. lbs. (45 Nm)
- Exhaust pipe to the manifold with a new gasket and torque the nuts to 22 ft. lbs. (30 Nm)
- Exhaust pipe-to-resonator pipe nuts and tighten to 31 ft. lbs. (42 NM)
- Exhaust manifold heat shield and torque the bolts to 18 ft. lbs. (25 Nm)
- air cleaner assembly
- Negative battery cable

4. Start the vehicle and check for leaks, repair if necessary.

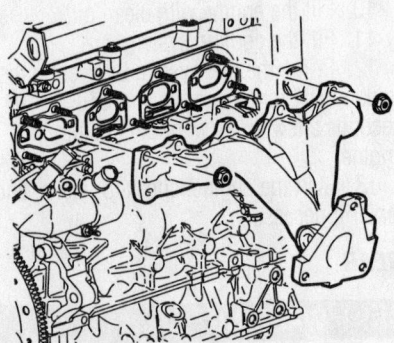

Remove the exhaust manifold and gasket—2.2L engine

3.0L Engine

FRONT MANIFOLD

1. Before servicing the vehicle, refer to the precautions in the beginning of this section.
2. Drain the cooling system.
3. Remove or disconnect the following:
   - Negative battery cable
   - Exhaust manifold Oxygen (O₂S) sensor
   - Coolant extension pipe and engine lift bracket bolts
   - Oil level indicator tube
   - Upper exhaust manifold nuts
   - Front down pipe assembly
   - Right hand splash shield
   - Accessory drive belt
   - A/C compressor and pressure transducer connections
   - A/C compressor-to-bracket bolts and position the compressor aside without disconnecting the lines
   - Lower exhaust manifold nuts
   - Exhaust manifold

To install:

4. Install or connect the following:
   - Exhaust manifold with a new gasket and torque the bolts to 15 ft. lbs. (20 Nm)
   - Front down pipe and gaskets
   - A/C compressor and tighten the bolts to 16 ft. lbs. (22 Nm)
   - Electrical connections to the pressure transducer and compressor
   - Accessory drive belt
   - Right hand splash shield
   - Oil level indicator tube
   - Coolant extension pipe/engine lift bracket bolt and torque to 15 ft. lbs. (20 Nm)
   - O₂S sensor, tighten to 33 ft. lbs. (45 Nm) and attach the electrical connection
   - Negative battery cable
5. Fill the cooling system.
6. Start the vehicle and check for leaks, repair if necessary.

REAR MANIFOLD

1. Before servicing the vehicle, refer to the precautions in the beginning of this section.
2. Drain the cooling system.
3. Remove or disconnect the following:
   - Negative battery cable
   - Air cleaner assembly
   - Purge solenoid from the manifold

- Middle exhaust manifold nut using a boxed end wrench
- Oxygen (O₂S) sensor connection at the rear of the engine
- Rear exhaust manifold pipe
- Right hand stabilizer bar-to-frame bolts
- Left-to-right hand stabilizer bar to lower link nut and discard the nut
- Exhaust manifold pipe heat shield
- Exhaust manifold

To install:

4. Install or connect the following:
   - Exhaust manifold with a new gasket and torque the bolts to 15 ft. lbs. (20 Nm)
   - Manifold heat shield and tighten the retainers to 18 ft. lbs. (25 Nm)
   - Right hand stabilizer bar-to-frame bolts and tighten to 37 ft. lbs. (50 Nm)
   - New left-to-right hand stabilizer bar to lower link nut to 48 ft. lbs. (65 Nm)
   - Rear exhaust manifold pipe
   - O₂S sensor connection at the rear of the engine
   - Middle exhaust manifold nut using a boxed end wrench to 15 ft. lbs. (20 Nm)
   - Purge solenoid to the manifold
   - Air cleaner assembly
   - Negative battery cable
5. Start the vehicle and check for leaks, repair if necessary.

Front Crankshaft Seal

REPLACEMENT

On the 2.2L engines the front crankshaft seal is located in the timing chain front cover. Refer to the timing chain procedure for information about removing the front cover and replacing the seal.

3.0L Engine

1. Before servicing the vehicle, refer to the precautions in the beginning of this section.
2. Remove or disconnect the following:
   - Negative battery cable
   - Timing belt
   - Crankshaft gear
3. Drill a small pilot hole into the steel ring of the seal.
4. Screw in a self taping screw.
5. Use pliers to pull out the oil seal.

To install:

6. Coat the lip of the new oil seal with engine oil.
7. Install the oil seal using a suitable seal installer.
8. Install the crankshaft gear and torque the bolt to 184 ft. lbs. (250 Nm) plus an additional 45 degrees then another 15 degrees.
9. Install the timing belt.

Camshaft and Lifters

REMOVAL & INSTALLATION

2.2L Engine

➡ **Be very careful when working around the camshaft sprockets and timing chain cover during this procedure. If a bolt or washer is accidentally dropped between the front cover and engine assembly, the cover will have to be removed for retrieval.**

1. Before servicing the vehicle, refer to the precautions in the beginning of this section.
2. Relieve the fuel system pressure.
3. Remove or disconnect the following:
   - Negative battery cable
   - Air cleaner assembly
   - Ignition coil
   - Coolant degas hose clips from the fuel rail
   - Ground strap
   - Fuel rail bracket
   - Fuel line
   - Cam cover
   - Purge solenoid from the power steering plate, if removing the intake camshaft
   - Power steering block off the plate, if removing the intake camshaft

➡ **To avoid valve piston contact, the No. 1 cylinder piston must be positioned at 60 degrees Before Top Dead Center (BTDC). The pistons are properly aligned when the diamond shaped hole on the intake camshaft sprocket is located at 12 o'clock.**

4. Remove the upper timing chain guide and front camshaft caps.
5. Install a camshaft sprocket holding tool J43655 through the sprocket holes from the timing chain side. Align the guide pins into the slot on the support head. Torque the pins to 89 inch lbs. (10 Nm).
6. Hold each camshaft in place with a

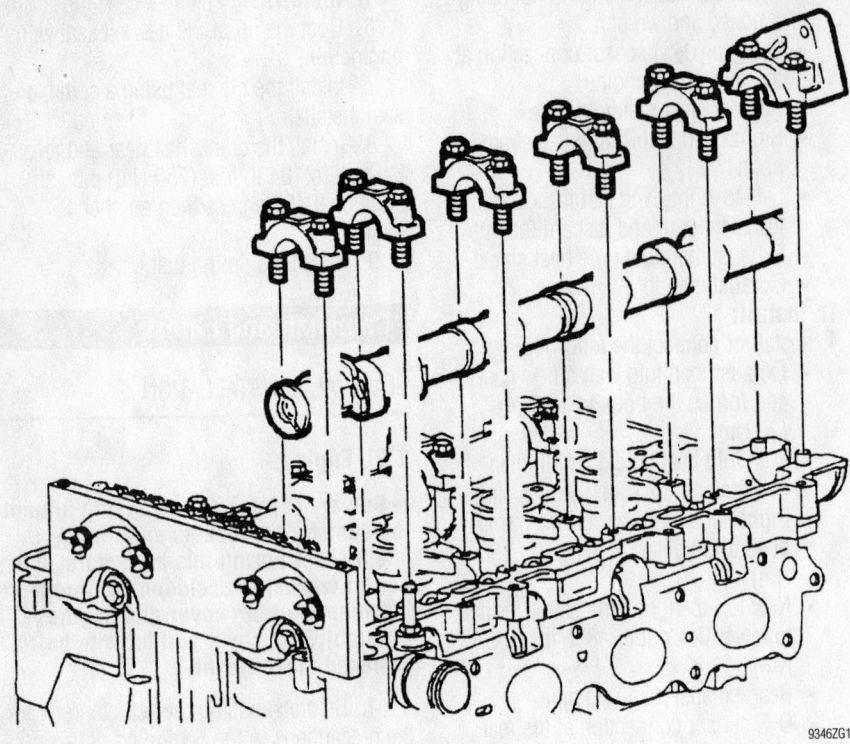

Remove the camshaft and bearing caps–2.2L engine

24mm open end wrench and remove the camshaft timing sprocket retaining bolts and washers. Discard the bolts.

7. Uniformly loosen and remove the remaining camshaft bearing caps.

8. Slide the camshaft sprockets away from the camshafts and remove the camshaft.

To install:

9. Lubricate the camshaft bearing journals with clean engine oil.

10. Install both camshafts and all bearing caps except the front cap on each camshaft.

11. Torque the bearing caps uniformly, except for the front caps and the rear intake cap, to 89 inch lbs. (10 Nm).

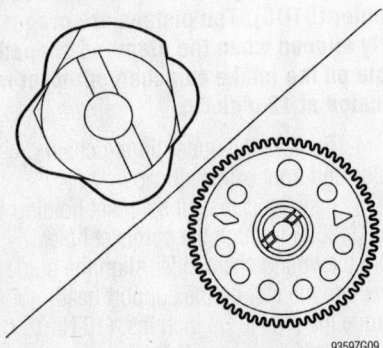

Make certain that the alignment notches are properly positioned with the notches in the camshaft sprockets—2.2L engines

➡**Make certain that the alignment notches are properly positioned with the notches in the camshaft sprockets before final torque is applied. Also, be sure that the timing chain is properly aligned on the fixed guide.**

12. Slide the camshaft sprockets and timing chain on the guide pins toward the camshafts. Rotate the camshafts with a 24mm open end wrench to align the camshaft and sprocket.

13. Install new camshaft sprocket bolts and torque them to 63 ft. lbs. (85 Nm) plus 30 degrees.

14. Remove the camshaft sprocket holding tool.

15. Install or connect the following:
- Front camshaft bearing caps and torque the bolts to 89 inch lbs. (10 Nm)
- Upper timing chain guide and apply Loctite® to the bolts
- Rear intake camshaft bearing cap and torque the bolts 19 ft. lbs. (25 Nm)
- Power steering block off plate and torque the bolts to 19 ft. lbs. (25 Nm)
- Cam cover
- Fuel line and tighten the fitting to 89 inch lbs. (10 Nm)
- Fuel line bracket and tighten to 89 inch lbs. (10 Nm)

- Ground strap
- Coolant degas hose
- Ignition coil
- Air cleaner assembly
- Negative battery cable

16. Start the vehicle and check for leaks, repair if necessary.

3.0L Engine

This engine is equipped with front and rear camshafts.

The front camshaft bearing caps for the cylinder head are marked R1–R8 and the rear cylinder head bearing caps are marked L1–L8.

➡**Be very careful when working around the camshaft sprockets and timing chain cover during this procedure. If a bolt or washer is accidentally dropped between the front cover and engine assembly, the cover will have to be removed for retrieval.**

1. Before servicing the vehicle, refer to the precautions in the beginning of this section.

2. Remove or disconnect the following:
- Negative battery cable
- Air cleaner assembly
- Intake plenum
- Camshaft cover
- Front timing belt cover
- Timing belt

➡**Rotate the crankshaft counterclockwise to 60 degrees Before Top Dead Center (BTDC) to prevent valve to piston contact.**

3. Install a Camshaft Gear Locking Tool J42069-2 into the camshaft gears.

4. Remove or disconnect the following:
- Loosen the camshaft gear bolt, remove the holding tool
- Camshaft gear bolt and discard it
- Camshaft gear
- Loosen the camshaft bearing caps sequentially starting in the center and working outward in a spiral direction
- Camshaft bearing caps
- Camshaft

➡**The bearing caps for the front camshaft bearing caps are marked with an R followed by a number and the rear caps marked with an L followed by a number.**

To install:

5. Lubricate all bearing surfaces with clean engine oil.

6. Install or connect the following:

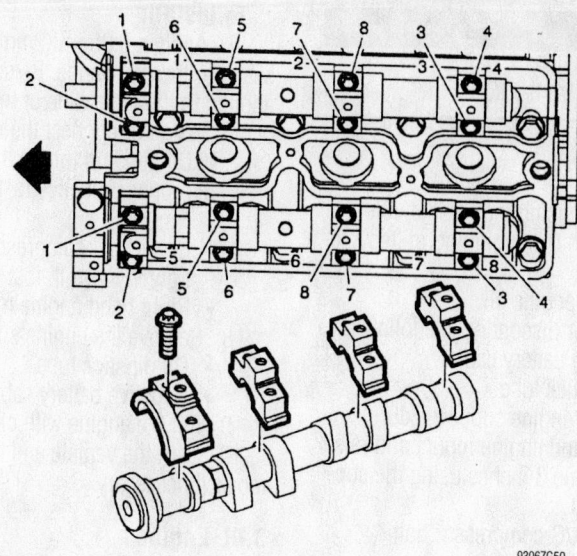

Front camshaft bearing cap removal sequence—3.0L engine

9306ZG50

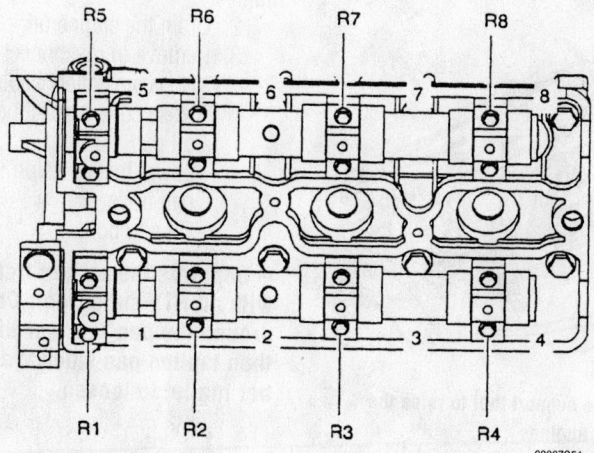

The rear camshaft bearing caps are marked to ensure proper installation—3.0L engine

9306ZG51

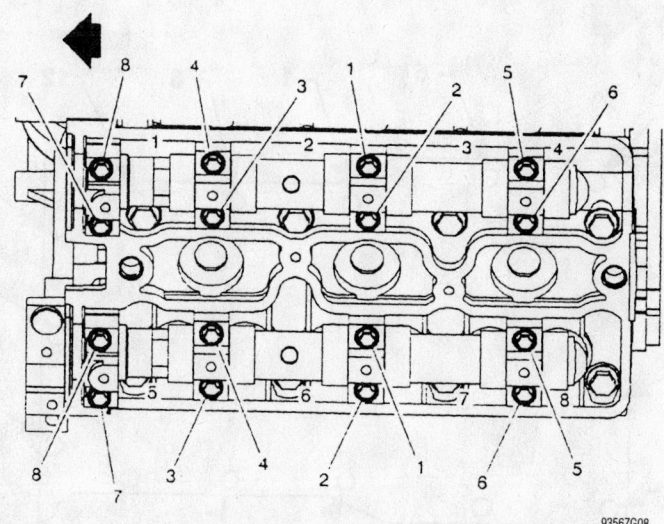

Front and rear camshaft bearing cap installation sequence—3.0L engine

9356ZG08

- Camshaft to the cylinder head and make sure that the pin on the exhaust camshaft is in the 12 o'clock position or that the pin on the intake camshaft is in the 7 o'clock position
- Camshaft bearing caps in their proper position and torque the bolts, starting in the center and working outward, to 71 inch lbs. (8 Nm)
- New camshaft seal lubricated with clean engine oil
- Camshaft gear with a new bolt and torque the bolt to 37 ft. lbs. (50 Nm) plus 60 degrees and an additional 15 degrees using the locking tool to hold the gear in place while tightening the bolt, then remove the locking tool
- Timing belt and adjust as needed
- Timing belt cover
- Camshaft cover
- Intake plenum
- Air cleaner assembly
- Negative battery cable

Valve Lash

ADJUSTMENT

All engines utilize hydraulic lash adjusters; no adjustment is necessary.

Starter Motor

REMOVAL & INSTALLATION

2.2L Engines

1. Before servicing the vehicle, refer to the precautions in the beginning of this section.

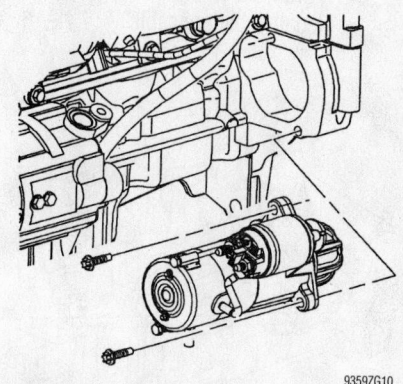

Starter assembly removal—2.2L engines

9359ZG10

For complete Engine Mechanical specifications, see Section 1 of this manual

2. Remove or disconnect the following:
- Negative battery cable

➡ **Spray the starter solenoid electrical connectors with penetrating oil before removal.**

- Starter electrical connections
- Starter bolts
- Starter assembly by pulling it toward the left side of the vehicle

To install:

3. Install or connect the following:
- Starter to the flywheel housing and torque the bolts to 30 ft. lbs. (407 Nm)
- Starter electrical connectors and torque the solenoid ignition wire to 44 inch lbs. (5 Nm) and the positive battery cable to 89 inch lbs. (10 Nm)
- Negative battery cable

3.0L Engine

1. Before servicing the vehicle, refer to the precautions in the beginning of this section.
2. Remove or disconnect the following:
- Negative battery cable
- Right front wheel
- Starter electrical connections
- Loosen the fastener securing the electrical harness bracket to the engine
- Starter bolts
- Starter assembly by pulling it toward the left side of the vehicle

To install:

3. Install or connect the following:
- Starter to the flywheel housing and torque the bolts to 26 ft. lbs. (35 Nm)
- Starter electrical connections and tighten the electrical harness bracket bolt
- Right front wheel
- Negative battery cable

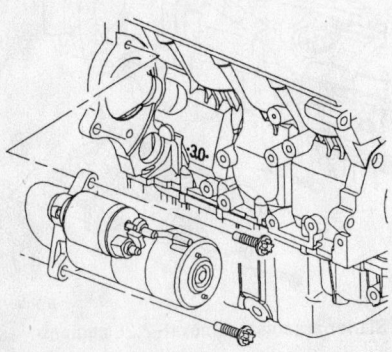

9346ZG19

Starter motor mounting—3.0L engine

Oil Pan

REMOVAL & INSTALLATION

2.2L Engine

1. Before servicing the vehicle, refer to the precautions in the beginning of this section.
2. Drain the engine oil.
3. Remove or disconnect the following:
- Negative battery cable
- Oil dipstick tube
4. Install an engine support fixture.
- Right hand engine mount and raise the engine 3 inches using the support tool
- Lower A/C compressor bolt
- Oil pan bolts
5. Using a flat blade tool, pry the oil pan from the engine block.

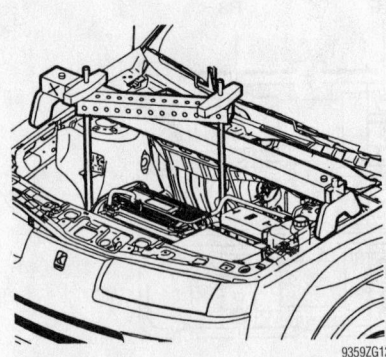

9359ZG12

Use the engine support tool to raise the engine—2.2L engines

To install:

6. Apply a 0.08 in. (2mm) bead of RTV sealer to the pan flange. Be sure the RTV is applied to the inner side of the flange.
7. Install or connect the following:
- Oil pan and torque the bolts in the proper sequence to 11 ft. lbs. (15 Nm)
- Lower A/C compressor bolt and tighten to 15 ft. lbs. (20 Nm)
- Right hand engine mount
8. Remove the engine support fixture.
- Oil dipstick tube
- Negative battery cable
9. Fill the engine with clean oil.
10. Start the vehicle and check for leaks, repair if necessary.

3.0L Engine

1. Before servicing the vehicle, refer to the precautions in the beginning of this section.
2. Drain the engine oil.
3. Remove or disconnect the following:
- Negative battery cable
- Nose cone bracket bolts from the oil pan
- Lower transmission flange-to-oil pan bolts
- Oil pan bolts

➡ **Separate the oil pan from the engine with an RTV cutter tool. Drive the tool around the pan to shear the RTV seam, then tap the pan sideways with a rubber mallet to loosen.**

- Oil pan

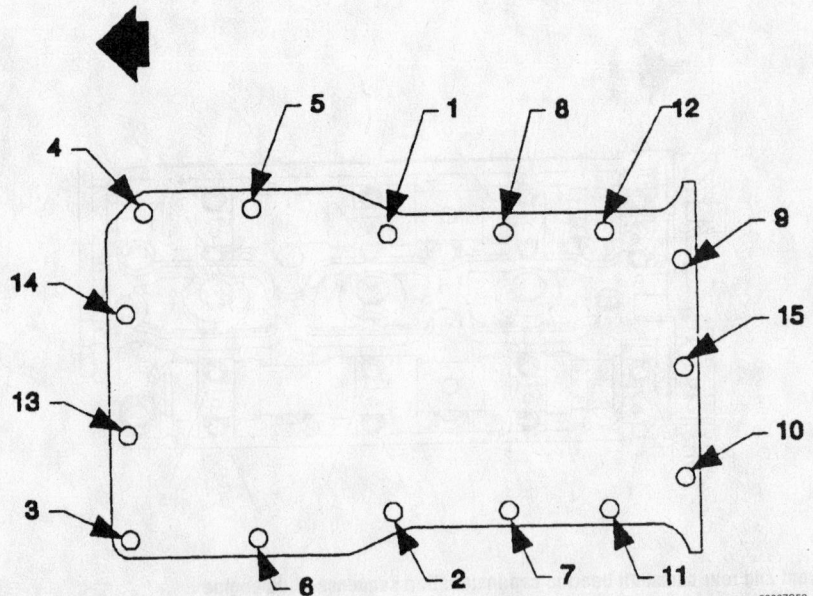

9306ZG52

Oil pan bolts torque sequence—2.2L engine

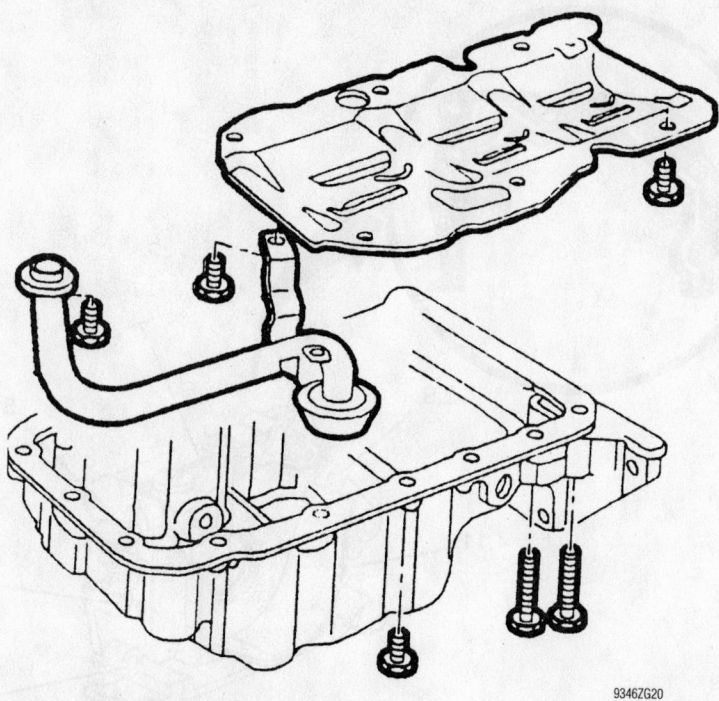

Oil pan and related components—3.0L engine

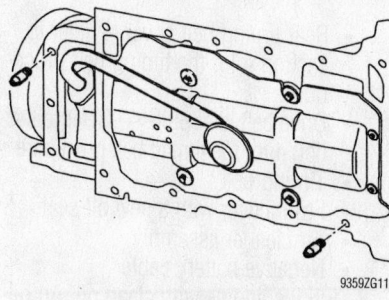

Install the oil pan alignment pins J44715 into the Datum holes—3.0L engines

To install:

➡The alignment of the oil pan is critical in ensuring transaxle flange-to-oil pane sealing. You must apply the proper amount of RTV to ensure a positive seal of the chamfered edge. Once the pan ahs been placed on the bloc, do not allow the pan to move as this will not allow the sealant to properly seal the chamfered edge

4. Apply a 0.08 in. (2mm) bead of RTV sealer to the pan flange. Be sure the RTV is applied to the inner side of the flange.

5. Install oil pan alignment pins J44715 into the Datum holes shown in the accompanying illustration.

6. Apply Loctite® 242 to the oil pan bolts prior to installation.

7. Install or connect the following:
   • Oil pan and finger tighten all bolts, and remove the alignment pins, then torque the oil pan bolts to 11 ft. lbs. (15 Nm) and the transaxle-to-oil pan bolts to 48 ft. lbs. (65 Nm)

8. Connect the negative battery cable.

9. Fill the engine with clean oil.

10. Start the vehicle and check for leaks, repair if necessary.

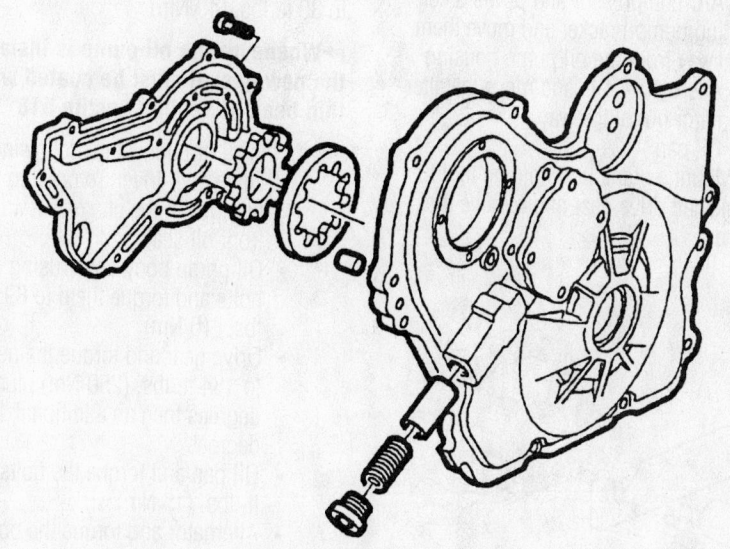

Front cover and oil pump assembly—2.2L engine

Oil Pump

REMOVAL & INSTALLATION

2.2L Engine

1. Before servicing the vehicle, refer to the precautions in the beginning of this section.

2. Drain the engine oil.

3. Remove or disconnect the following:
   • Negative battery cable
   • Air cleaner assembly
   • Right front wheel and splash shield
   • Accessory drive belt
   • Crankshaft damper pulley
   • Belt tensioner

4. Install an engine support fixture.
   • Right front engine mount
   • Front cover bolts and the 13mm bolt under the water pump
   • Front cover
   • Oil pump cover plate
   • Drive rotor and driven rotor
   • Pressure relief valve

To install:

5. Install or connect the following:
   • New relief valve into the cover bore, if removed. Coat the valve with clean engine oil and tap it into the bore. Torque the plug to 30 ft. lbs. (40 Nm).

➡**Whenever the oil pump is installed, the assembly must be packed with petroleum jelly in order to prime the pump.**

For Accessory Drive Belt illustrations, see Section 1 of this manual

- Drive and driven rotors into the pump with the chamfer toward the front oil seal
- Oil pump body cover using new bolts and torque the bolts to 53 inch lbs. (6 Nm)
- Front cover with a new oil seal and torque the perimeter and center bolts to 19 ft. lbs. (25 Nm) and the lower center bolt to 89 inch lbs. (10 Nm)
- Right side engine mount and torque the bolts to 41 ft. lbs. (55 Nm)

6. Remove the engine support fixture.
- Drive belt tensioner and torque the bolts 37 ft. lbs. (50 Nm)
- Crankshaft damper pulley and torque the bolt to 74 ft. lbs. (100 Nm) plus 75 degrees
- Accessory drive belt
- Right front splash shield and wheel
- Air cleaner assembly
- Negative battery cable

7. Fill the engine with clean oil and replace the oil filter.

8. Start the vehicle and check for leaks, repair if necessary.

3.0L Engine

1. Before servicing the vehicle, refer to the precautions in the beginning of this section.

2. Drain the engine oil.

3. Drain the cooling system.

4. Remove or disconnect the following:
- Negative battery cable
- Air cleaner assembly
- Front timing belt cover
- Timing belt
- Rear timing belt cover
- A/C compressor and power steering pump bracket and move them away from the oil pump housing
- Alternator bolts and move the alternator out of the way
- Oil pan

5. Mount a crank hub holding tool to the crankshaft drive gear and remove the drive gear.

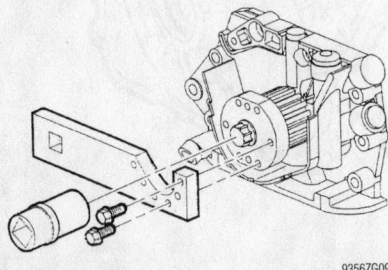

Mount a crank hub holding tool to the crankshaft drive gear—3.0L engine

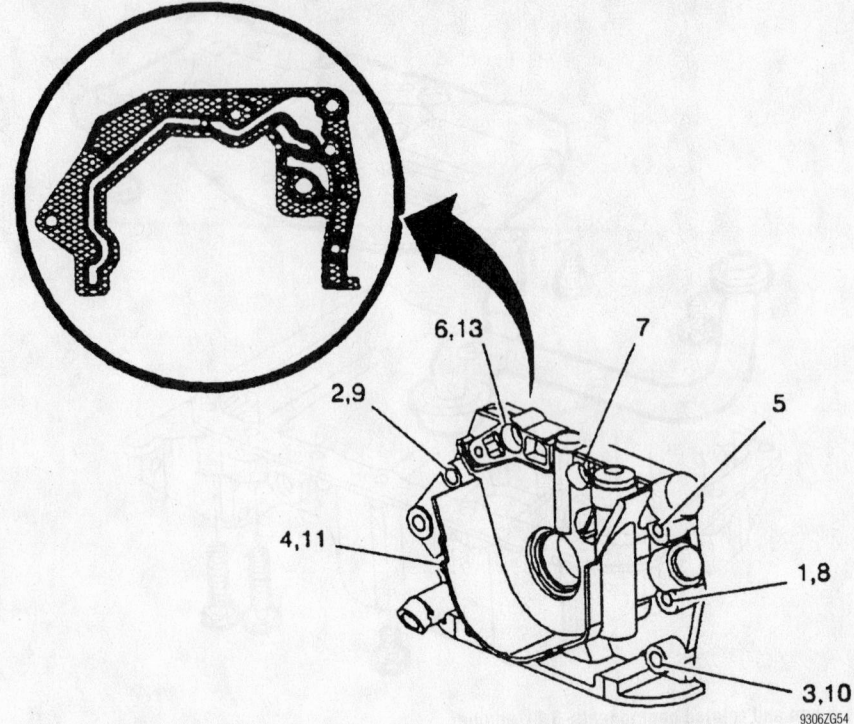

Oil pump bolt tightening sequence–3.0L engine

- Oil pump bolts
- Oil pan housing bolts that thread into the oil pump
- Oil pump
- Front main oil seal and collar
- Pressure relief valve
- Oil pump cover
- Drive rotor and driven rotor

To install:

6. Install the new relief valve into the cover bore (if removed) and torque the plug to 30 ft. lbs. (40 Nm).

➡**Whenever the oil pump is installed, the new gasket must be coated with a thin bead of sealing Loctite 518®.**

7. Install or connect the following:
- Drive and driven rotors into the pump with the chamfer toward the front oil seal
- Oil pump body cover using new bolts and torque them to 89 inch lbs. (10 Nm)
- Drive gear and torque the new bolt to 184 ft. lbs. (250 Nm) plus 45 degrees then an additional 15 degrees
- Oil pan and torque the bolts to 11 ft. lbs. (15 Nm)
- Alternator and torque the bolts to 30 ft. lbs. (40 Nm)
- A/C compressor and power steering pump bracket. Torque the bolts to 30 ft. lbs. (40 Nm).

- Rear timing belt cover—Refer to section 4 for the timing belt procedure.
- Drive belt idler pulley—Refer to section 4 for the timing belt procedure.
- Timing belt
- Front cover with a new oil seal
- Air cleaner assembly
- Negative battery cable

8. Fill the engine with clean oil. An oil filter replacement is also recommended.

9. Fill the cooling system.

10. Start the vehicle and check for leaks, repair if necessary.

Rear Main Seal

REMOVAL & INSTALLATION

2.2L Engine

1. Before servicing the vehicle, refer to the precautions in the beginning of this section.

2. Remove or disconnect the following:
- Negative battery cable
- Transmission
- Clutch/pressure plate assembly, if equipped with a manual transmission
- Flywheel

3. Center punch the steel ring of the oil seal.

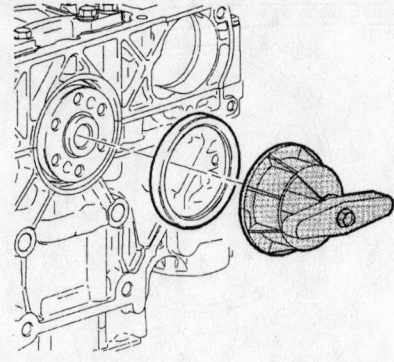

9346ZG23

**Rear oil seal and installation tool
J42067–2.2L engine**

4. Drill a small hole into the steel ring.

5. Install a self-tapping screw and using pliers, pull out the rear main oil seal.

➡ **Be careful not to damage or scratch the seal mounting surfaces.**

To install:

6. Lubricate the new rear main bearing seal with engine oil.

7. Install or connect the following:
- New rear main seal using a Rear Main Bearing Oil Seal Installer Tool J42067 until it is flush with the block
- Flywheel
- Clutch/pressure plate assembly, if equipped with a manual transmission
- Transmission
- Negative battery cable

8. Start the engine and check for leaks, repair if necessary.

3.0L Engine

1. Before servicing the vehicle, refer to the precautions in the beginning of this section.

2. Remove or disconnect the following:
- Negative battery cable
- Transmission
- Clutch/pressure plate assembly, if equipped with a manual transmission
- Flywheel

3. Center punch the steel ring of the oil seal.

4. Drill a small hole into the steel ring.

5. Install a self-tapping screw and using pliers, pull out the rear main oil seal.

➡ **Be careful not to damage or scratch the seal mounting surfaces.**

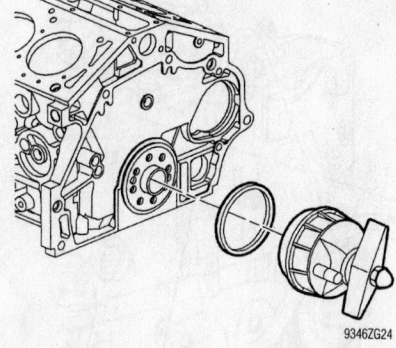

9346ZG24

**Rear main oil seal and installation tool
J42067–3.0L engine**

To install:

6. Lubricate the new rear main oil seal with engine oil.

7. Install or connect the following:
- New rear main seal using a Rear Main Seal Installer Tool J42067 until it is flush with the block
- Flywheel
- Clutch/pressure plate assembly, if equipped with a manual transmission
- Transmission
- Negative battery cable

8. Start the engine and check for leaks, repair if necessary.

Timing Chain, Sprockets, Front Cover and Seal

REMOVAL & INSTALLATION

2.2L Engine

1. Before servicing the vehicle, refer to the precautions in the beginning of this section.

2. Drain the cooling system.

3. Drain the engine oil.

4. Properly relieve the fuel system pressure.

5. Remove or disconnect the following:
- Negative battery cable
- Intake Air Temperature (IAT) sensor connection
- Air cleaner assembly
- Ignition module assembly
- Electronic Control Module (ECM) connections
- Oil dipstick tube bolt
- Throttle body electrical connection
- Electrical connector from the fuel injector harness and attachment at the bottom of the intake manifold

- Electrical connector at the purge solenoid and Manifold Absolute Pressure (MAP) sensor
- Vacumm hose at the brake booster
- Coolant pipe bracket bolts from the cylinder head
- De-gas hose clamp from the cylinder head and fuel rail and position aside
- Ground strap from the rear cam cover
- Fuel rail bracket from the cam cover
- Fuel lines
- Cam cover

6. Position the No. 1 piston 60 degrees Before Top dead center (TDC) using a 24mm wrench to rotate the camshafts in a clockwise motion and make sure the diamond shaped hole on the intake sprocket is at the 12 O'clock position.
- Upper timing guide

➡ **Remove the timing chain tensioner to unload chain tension before removing the timing chain.**

- Fixed timing chain guide access plug
- Upper fixed guide bolt using a magnetic socket

7. Install a three bar engine support fixture.
- Right hand engine mount
- Right hand mount bracket
- Right wheel splash shield
- Install a 1 x 2 x 4 inch block of wood between the oil pan and cradle
- Drive belt
- Drive belt tensioner assembly

8. Install crankshaft pullet holder J38122A..

9359ZG03

Position the No. 1 piston 60 degrees Before Top dead center (TDC) using a 24mm wrench to rotate the camshafts in a clockwise motion and make sure the diamond shaped hole on the intake sprocket is at the 12 O'clock position—2.2L engine

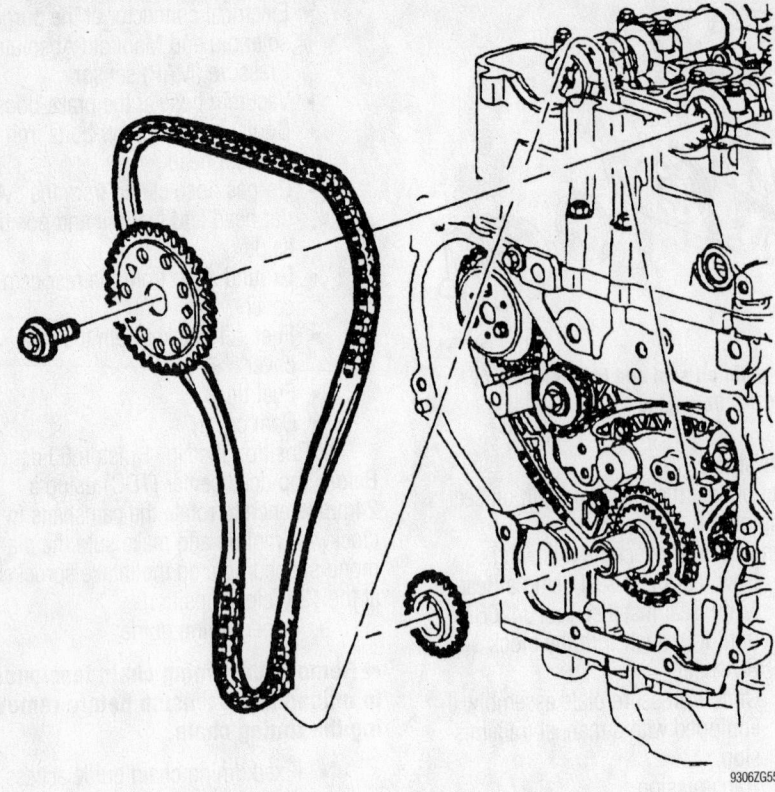

9306ZG55

Remove the timing chain through the top of the cylinder head–2.2L engine

- Crankshaft balancer bolt and pulley
- Front cover bolts
- Lower water pump bolt
- Front cover and gasket
- Lower fixed guide
- Upper radiator hose from the cylinder head
- Exhaust manifold pipe nuts
- Front and rear Oxygen (O2S) sensor electrical connector
- Down pipe-to-intermediate pipe nuts
- Exhaust camshaft sprocket bolts while holding the camshaft with a 24mm wrench and discard the camshaft sprocket bolts
- Exhaust sprocket
- Adjustable guide through the top of the cylinder head
- Intake camshaft sprocket bolts while holding the camshaft with a 24mm wrench and discard the camshaft sprocket bolts
- Intake sprocket
- Timing chain assembly
- Timing drive sprocket from the crankshaft

To install:

➡ Set the crankshaft to 60 degrees Before Top Dead Center (BTDC) to prevent contact between the pistons and valves.

9. Install or connect the following:
10. Position the exhaust camshaft with the offset slot in the 2 o'clock position and the intake camshaft with the offset slot in the 11 o'clock position.

- Timing chain around the intake camshaft sprocket with the copper link aligned with the **INT** diamond timing mark
- Sprocket to the camshaft and align it with the offset slot
- New camshaft sprocket bolt but do not tighten
- Timing chain around the crankshaft sprocket and align the silver link to the timing mark
- Adjustable timing chain guide through the opening on top of the cylinder head and torque the bolt to 89 inch lbs. (10 Nm)
- Timing chain around the exhaust camshaft sprocket with the silver link aligned with the offset slot. Install but do not tighten a new sprocket bolt

✲✲ CAUTION

Make certain that all timing marks and colored links are aligned properly before proceeding to the next step. If the timing chain is not aligned properly, severe engine damage may occur.

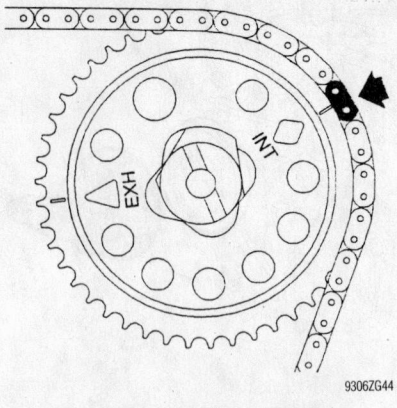

9306ZG44

Align the copper link on the timing chain with the INT diamond timing mark

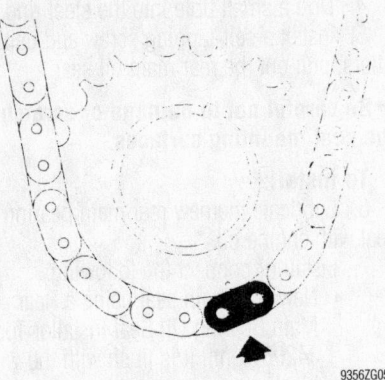

9356ZG05

Route the timing chain around the crankshaft sprocket and align the silver link to the timing mark (5 o'clock position)—2.2L engines

11. Torque the intake and exhaust camshaft bolts to 63 ft. lbs. (85 Nm) plus a 30 degree turn.
12. Install or connect the following:
- Fixed timing guide and torque the bolt to 89 inch lbs. (10 Nm)
- Fixed timing guide bolt access plug after applying Loctite® to the threads and torque it to 30 ft. lbs. (40 Nm)
- Timing chain tensioner and torque the bolts 55 ft. lbs. (75 Nm)
13. Tap the top of the timing chain between the camshaft sprockets to engage the tensioner.
- Upper timing chain guide and torque the bolts to 89 inch lbs. (10 Nm)
- Front cover with a new gasket and torque the bolts to 18 ft. lbs. (25 Nm)
- Water pump bolt and torque it to 18 ft. lbs. (25 Nm)
- Crankshaft damper and torque the bolt to 74 ft. lbs. (100 Nm) plus 75 degrees
- Drive belt tensioner and torque the bolts to 30 ft. lbs. (40 Nm)

- Exhaust manifold pipe to the manifold and tighten the nuts to 22 ft. lbs. (30 Nm).
- Oil dipstick tube bolt to 89 inch lbs. (10 Nm).
- Engine mount bracket and torque the bolts to 66 ft. lbs. (90 Nm)
- Engine mount to body nuts to 81 ft. lbs. (110 Nm)
- Engine mount to the bracket and torque the bolts to 41 ft. lbs. (55 Nm)

14. Remaining components in the reverse order of removal.

15. Fill the engine with clean oil.

16. Fill the cooling system.

17. Prime the fuel system by cycling the ignition **ON** for 5 seconds and **OFF** for 10 seconds a few times without cranking the engine.

18. Start the engine, check for leaks, and repair if necessary.

Piston and Ring

POSITIONING

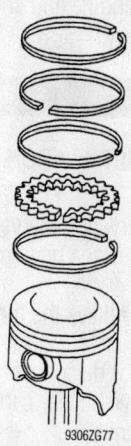

Piston ring positioning—2.2L engine

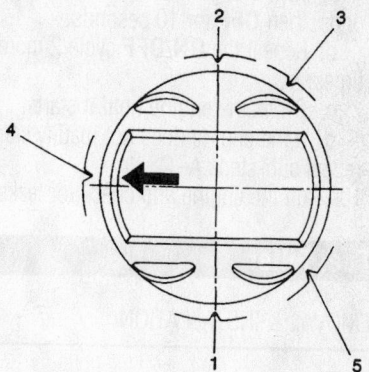

(1) 1st Compression Ring End Gap Location
(2) 2nd Compression Ring End Gap Location
(3) Oil Control Ring Upper Ring End Gap Location
(4) Oil Control Ring Spacer End Gap Location
(5) Oil Control Ring Lower Ring End Gap Location

Piston ring positioning—3.0L engine

FUEL SYSTEM

Fuel System Service Precautions

Safety is the most important factor when performing not only fuel system maintenance but any type of maintenance. Failure to conduct maintenance and repairs in a safe manner may result in serious personal injury or death. Maintenance and testing of the vehicle's fuel system components can be accomplished safely and effectively by adhering to the following rules and guidelines.

- To avoid the possibility of fire and personal injury, always disconnect the negative battery cable unless the repair or test procedure requires that battery voltage be applied
- Always relieve the fuel system pressure prior to disconnecting any fuel system component (injector, fuel rail, pressure regulator, etc.), fitting or fuel line connection. Exercise extreme caution whenever relieving fuel system pressure, to avoid exposing skin, face and eyes to fuel spray. Please be advised that fuel under pressure may penetrate the skin or any part of the body that it contacts
- Always place a shop towel or cloth around the fitting or connection prior to loosening to absorb any excess fuel due to spillage. Ensure that all fuel spillage (should it occur) is quickly removed from engine surfaces. Ensure that all fuel soaked cloths or towels are deposited into a suitable waste container

- Always keep a dry chemical (Class B) fire extinguisher near the work area
- Do not allow fuel spray or fuel vapors to come into contact with a spark or open flame
- Always use a back-up wrench when loosening and tightening fuel line connection fittings. This will prevent unnecessary stress and torsion to fuel line piping
- Always replace worn fuel fitting O-rings with new. Do not substitute fuel hose or equivalent, where fuel pipe is installed

Fuel System Pressure

RELIEVING

1. Before servicing the vehicle, refer to the precautions in the beginning of this section.

2. Unless battery voltage is necessary for testing, disconnect the negative battery cable. This will prevent the fuel pump from running and causing a fuel spill through the disconnected components if the ignition key is accidentally turned **ON**.

3. Remove the air cleaner assembly, for access.

4. Connect gauge bar 53476 to fuel gauge pressure adapter 309725 using a flexible hose from gauge bar set SA9127E. Make sure the needle valve on the pressure adapter is off.

➡**Do not use tools to tighten the adapter to the pressure port. If the adapter will not hand-tighten the seals are defective and needs to be replaced.**

5. Wrap a shop rag around the fuel test port fitting, located at the lower rear of the engine, then remove the cap and connect fuel pressure gauge.

6. Install the bleed hose from the pressure gauge into an approved container and open the valve to bleed the system pressure.

7. After the pressure is bled, remove the gauge from the test port and recap it.

8. Install the air cleaner assembly.

9. After servicing the vehicle, connect the negative battery cable and prime the fuel system as follows:

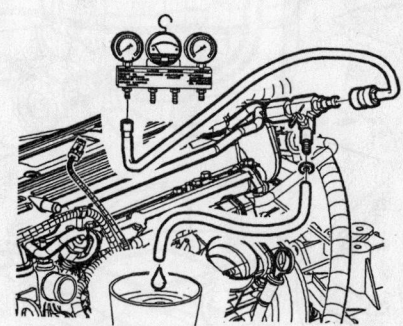

Connect gauge bar 53476 to fuel gauge pressure adapter 309725 using a flexible hose from gauge bar set SA9127E

For Wheel Alignment specifications, see Section 1 of this manual

a. Turn the ignition **ON** for 5 seconds, then **OFF** for 10 seconds.

b. Repeat the **ON/OFF** cycle 2 more times.

c. Crank the engine until it starts.

d. If the engine does not readily start, repeat sub-steps A–C.

10. Run the engine and check for leaks.

Fuel Filter

REMOVAL & INSTALLATION

1. Before servicing the vehicle, refer to the precautions in the beginning of this section.

2. Properly relieve the fuel system pressure.

3. Remove or disconnect the following:
- Negative battery cable
- Fuel filter bracket screw
- Fuel feed and return lines from the filter
- Fuel filter from the bracket

To install:

4. Install or connect the following:
- New fuel filter into the bracket
- New fuel line retainers to the female portion of the quick connect fittings
- Fuel feed and return lines
- Fuel filter bracket attaching screw and torque it to 18 inch lbs. (2 Nm)
- Negative battery cable

5. Prime the fuel system as follows:

a. Turn the ignition **ON** for 5 seconds, then **OFF** for 10 seconds.

b. Repeat the **ON/OFF** cycle 2 more times.

c. Crank the engine until it starts.

d. If it does not start, repeat the 3 above steps.

6. Start the engine and check for leaks, repair if necessary.

Fuel Pump

REMOVAL & INSTALLATION

1. Before servicing the vehicle, refer to the precautions in the beginning of this section.

2. Properly relieve the fuel system pressure.

3. Remove or disconnect the following:
- Negative battery cable
- Fuel tank
- Fuel pump electrical connector
- Fuel lines and hoses from the fuel pump module cover
- Fuel pump module retaining ring with Wrench SA9156E
- Pull the retaining clip toward the float arm and lift up
- Fuel pump straight up from the fuel tank
- Fuel pump tank seal and discard the seal

To install:

4. Install or connect the following:
- Fuel pump with new seal
- Fuel pump cover lockring with Tool SA9156E

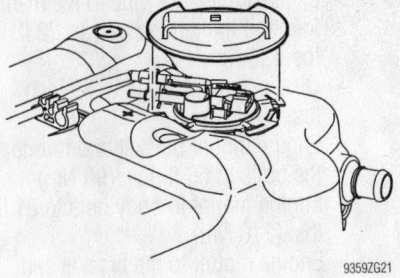

Remove the fuel pump cover lockring

- Fuel lines, hoses and wiring harness
- Fuel tank
- Negative battery cable

5. Start the vehicle and check for leaks, repair if necessary.

Fuel Injector

REMOVAL AND INSTALLATION

2.2L Engines

1. Before servicing the vehicle, refer to the precautions in the beginning of this section.

2. Properly relieve the fuel system pressure.

3. Remove or disconnect the following:
- Negative battery cable
- Air cleaner assembly
- Engine harness and position aside
- Fuel rail bracket from the rear of the can cover
- Fuel feed line
- Fuel injector electrical connectors
- Fuel rail bolts
- Fuel rail with the injectors as an assembly
- Fuel injector retaining clip off the injector
- Fuel injector
- Fuel injector O-rings

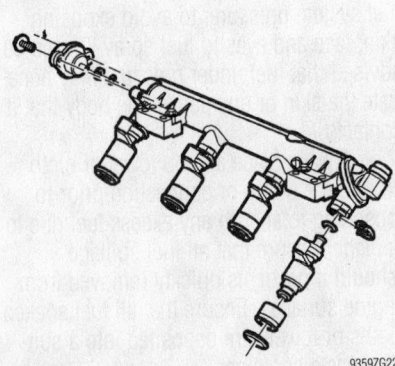

Remove the retainer clip from the fuel injector—2.2L engines

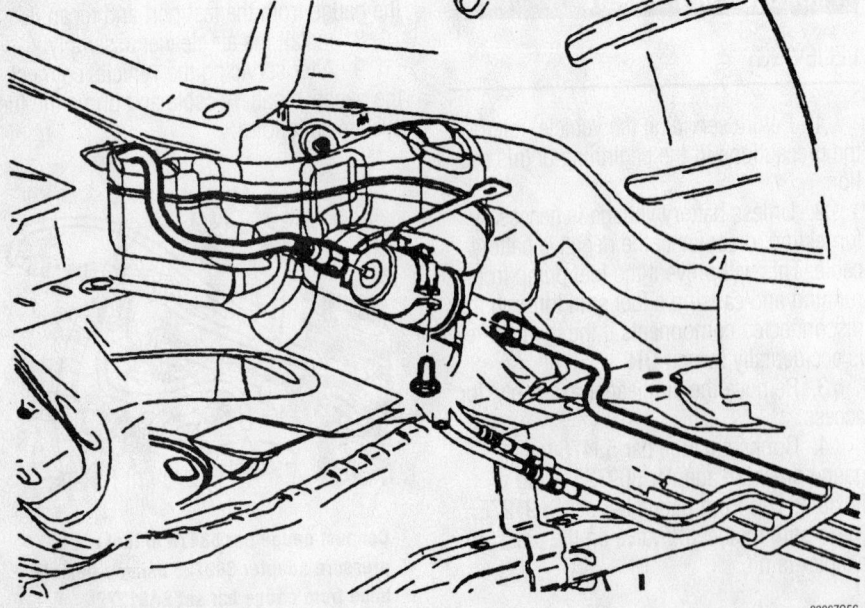

Remove the fuel filter bracket attaching screw 2.2L and 3.0L engines

To install:

4. Lubricate the new fuel injector O-ring with clean engine oil.

5. Install or connect the following:
- New O-ring seals on the fuel injector
- Fuel injector to the fuel rail
- Retaining clip to the fuel injector
- Fuel rail and torque the bolts to 89 inch lbs. (10 Nm)
- Fuel injector electrical connectors
- Fuel feed line and tighten the fitting to 89 inch lbs. (10 Nm)
- Fuel line bracket and tighten the retainer to 89 inch lbs. (10 Nm)
- Air cleaner assembly
- Negative battery cable

6. Start the vehicle and check for leaks, repair if necessary.

3.0L Engines

1. Before servicing the vehicle, refer to the precautions in the beginning of this section.

2. Properly relieve the fuel system pressure.

3. Remove or disconnect the following:
- Negative battery cable
- Air cleaner assembly
- Fuel feed line
- Intake manifold assembly
- Fuel injector electrical connectors
- Fuel rail line bracket bolt
- Vacuum chamber
- Fuel rail bracket bolt at the rear of the cam cover
- Transfer fittings, loosen only using a back-up wrench
- Fuel rail-to-manifold bolts
- Fuel rail with the injectors as an assembly
- Fuel injector retaining clip off the injector
- Fuel injector
- Fuel injector O-rings

To install:

4. Lubricate the new fuel injector O-ring with clean engine oil.

5. Install or connect the following:
- New O-ring seals on the fuel injector
- Fuel injector to the fuel rail
- Retaining clip to the fuel injector aligning the notch in the clip to the tab on the rail
- Fuel rail and torque the bolts to 66 inch lbs. (7.5 Nm)
- Transfer fittings, and tighten to 11 ft. lbs. (15 Nm) using a back-up wrench
- Fuel rail bracket bolt at the rear of the cam cover and tighten to 39 inch lbs. (4.5 Nm)
- Vacuum chamber with a new gasket and tighten the retainers to 18 inch lbs. (2 Nm), then attach the vacuum hose
- Intake manifold top cover and tighten to 35 inch lbs. (4 Nm)
- Fuel injector electrical connectors
- Intake manifold assembly
- Air cleaner assembly
- Negative battery cable

6. Start the vehicle and check for leaks, repair if necessary.

DRIVE TRAIN

Manual Transmission Assembly

REMOVAL & INSTALLATION

1. Before servicing the vehicle, refer to the precautions in the beginning of this section.

2. Drain the transmission fluid.

3. Remove or disconnect the following:
- Both battery cables

4. Install an engine support fixture.

5. Fasten the radiator to the upper radiator support.
- Wheels
- Splash shields
- Ball joints and tie rods from the knuckle
- Lower control arm from the knuckle
- Lower stabilizer bar links
- Steering gear from the steering gear assembly
- Rear transaxle mount-to-cradle bolts
- Rear transaxle mount bracket-to-transaxle bolts
- Front lower mount through bolt from the cradle
- Front air deflector from the body but leave it attached to the cradle

6. Support the cradle with a jack

- Cradle bolts and the cradle
- Shift lever cable from the shift control housing using tool J36346
- Shift lever cable from the bracket
- Pressure line from the clutch actuator by removing the C-clip and pulling it away from the actuator
- Back-up lamp switch connector
- Front transaxle mount from the transaxle
- Right side drive axle from the intermediate drive shaft using removal tool J45341 and slide hammer SA9173G

7. Secure the drive axle from the intermediate drive shaft.

➡ **Remove the retainer ring from the stub shaft before removing the tool and discard the ring.**

- Intermediate drive shaft using removal tool J440177 and axle seal puller SA9133T
- Left side drive axle drive shaft using removal tool J45341 and slide hammer SA9173G and secure the axle aside
- Top transaxle mount bolts

8. Use the engine support fixture to lower the transaxle enough so that the assembly can be removed.

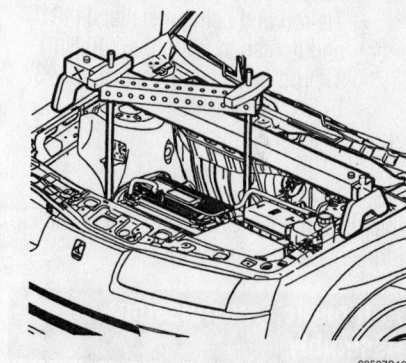

9359ZG12

Install an engine support tool

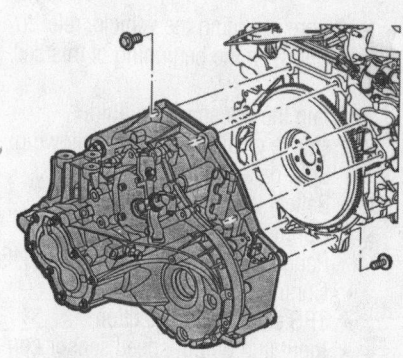

9359ZG23

Remove the engine to manual transmission bolts

9. Attach the transaxle to a suitable transaxle jack.

10. Remove the transaxle bolts on the engine side and the transaxle bolt on the transaxle side.

To install:

11. Installation is the reverse of removal, please note the following torque specifications:

- Transaxle-to-engine and engine-to-transaxle bolts: 55 ft. lbs. (75 Nm)
- Top mount-to-transaxle bolts: 37 ft. lbs. (50 Nm)
- Front transaxle mount bolts : 37 ft. lbs. (50 Nm)
- Rear transaxle mount-to-transaxle bolts: 37 ft. lbs. (50 Nm)
- Cradle-to-body bolts: 140 ft. lbs. (190 Nm)
- Front transaxle mount through bolt: 81 ft. lbs. (110 Nm)
- Rear transaxle mount through bolt: 81 ft. lbs. (110 Nm)
- Sterring gear-to-column bolt: 25 ft. lbs. (34 Nm)
- Lower control arm nut: 89 inch lbs. (10 Nm) plus 150 degrees
- Stabilzer bar link: 48 ft. lbs. (65 Nm)
- Tie rod end using installer J44015 and tighten to 30 ft. lbs. (40 Nm)
- Tie rod-to-steering knuckle nut: 37 ft. lbs. (50 Nm)

12. Fill the transmission to the proper level.

13. Warm the engine and check the transmission fluid. Check and adjust vehicle alignment, as necessary.

Automatic Transmission Assembly

REMOVAL & INSTALLATION

1. Before servicing the vehicle, refer to the precautions in the beginning of this section.

2. Drain the transmission fluid.

3. Remove or disconnect the following:
- Both battery cables
- Battery and battery tray bracket
- Control cable from the Transaxle Range Switch using prytool J36346
- Control cable from the bracket
- TRS electrical connection
- Input and output speed sensor connections
- Positive battery cable from the transaxle stud
- Headlamp fasteners and wire radiator to the core support

4. Install an engine support fixture.
- Transaxle fluid dipstick tube
- Upper left hand transaxle mount
- Wheels
- Splash shields
- Front air deflector
- Transaxle cooler line nut
- Transaxle cooler lines from the transaxle

➡**Mark the position of the Power take Off (PTO) prior to removal.**

- Drive shaft from the PTO marking alignment locations prior to removal
- Lower stabilizer nut from the cradle on both sides
- Lower control arm from the knuckle
- Ball joints and tie rods from the knuckle
- Front pitch restrictor bolts, through bolts and restrictor
- Rear restrictor through bolt from the cradle

5. Position a support table under the cradle.
- Cradle bolts and lower the cradle onto the table
- Left side drive axle drive shaft using removal tool J45341 and slide hammer SA9173G and secure the axle aside

➡**The stub shaft may disengage from the PTO if this occurs, plug the PTO to avoid fluid loss.**

- Right side drive axle from the PTO using removal tool J45341 and slide hammer SA9173G
- PTO-to-engine bracket
- Starter
- Torque converter-to-flexplate bolts through the starter hole
- 3 lower transaxle-to-engine bolts
- 1 engine-to-transaxle bolt located above the PTO

6. Attach the transaxle to a suitable transaxle jack.

7. Separate the engine from the transaxle , lower the assembly and disconnect the PTO hose.

8. Remove the PTO, if necessary.

To install:

9. Installation is the reverse of removal, please note the following torque specifications:

- Transaxle-to-engine bolts: 55 ft. lbs. (75 Nm)
- PTO bracket bolts: 44 ft. lbs. (60 Nm)
- Torque converter-to-flexplate bolts: 44 ft. lbs. (60 Nm)

- Bracket-to-engine bolts: 26 ft. lbs. (35 Nm)
- Cradle bolts: 148 ft. lbs. (200 Nm)
- Rear pitch restrictor through bolt: 81 ft. lbs. (110 Nm)
- Front pitch restrictor bolts: 37 ft. lbs. (50 Nm)
- Front pitch restrictor through bolt: 81 ft. lbs. (110 Nm)
- Tie rod end using installer J44015 and tighten to 30 ft. lbs. (40 Nm)
- Tie rod-to-steering knuckle nut: 37 ft. lbs. (50 Nm)
- Lower control arm nut: 89 inch lbs. (10 Nm) plus 150 degrees
- Rack and pinion bolts: 81 ft. lbs. (110 Nm)
- Lower stabilizer nut: 48 ft. lbs. (65 Nm)
- Drive shaft retainers: 74 ft. lbs. (100 Nm)
- Transaxle oil cooler line assembly: 71 inch lbs. (8 Nm)
- Front axle stub shaft nut: 151 ft. lbs. (205 Nm)
- Upper left hand transaxle mount bolts : 37 ft. lbs. (50 Nm)
- Upper transaxle-to-engine bolts: 55 ft. lbs. (75 Nm)

10. Fill the transmission to the proper level.

11. Warm the engine and check the transmission fluid. Check and adjust vehicle alignment, as necessary.

Clutch

ADJUSTMENT

The hydraulic clutch system is self-adjusting.

REMOVAL & INSTALLATION

1. Before servicing the vehicle, refer to the precautions in the beginning of this section.

2. Remove the transmission from the vehicle.

3. Remove the pressure plate and clutch disc.

4. Inspect the pressure plate, as follows:

a. Check for excessive wear, chatter marks, cracks or overheating (indicated by a blue discoloration). Black random spots on the friction surface of the pressure plate is normal.

b. Check the plate for warpage using a straightedge and a feeler gauge; the maximum allowable warpage is 0.006 in. (0.15mm).

c. Replace the plate, if necessary.

5. Inspect the clutch disc, as follows:

a. Check the disc face for oil or burnt spots.

b. Check the disc for loose damper springs, hub or rivets.

c. Replace the disc, if necessary.

6. Check the flywheel, as follows:

a. Check the ring gear for wear or damage.

b. Check the friction surface for excessive wear, chatter marks, cracks or overheating.

c. Check flywheel thickness; the minimum allowable is 1.102 in. (28mm).

d. Measure flywheel run-out using a dial indicator, positioned for at least 2 flywheel revolutions. Push the crankshaft forward to take up thrust bearing clearance. Maximum flywheel run-out is 0.006 in. (0.15mm).

e. Check the flywheel for warpage using a straight-edge and a feeler gauge; the maximum allowable warpage is 0.006 in. (0.15mm).

f. Replace the flywheel, if necessary.

7. If necessary, remove the flywheel retaining bolts and remove the flywheel from the crankshaft.

To install:

8. Install or connect the following:

- Flywheel (if removed) and torque the bolts in a crisscross pattern to 39 ft. lbs. (53 Nm) plus 25 degrees
- Clutch disc and pressure plate and loosely install the pressure plate bolts
- Clutch alignment tool in the clutch disc, and push in until it bottoms out in the crankshaft

9. Tighten the pressure plate bolts using multiple passes of a crisscross sequence to 11 ft. lbs. (15 Nm) and remove the alignment tool.

10. Lubricate the splines of the input shaft lightly with a high temperature grease.

11. Install the transmission assembly.

12. Connect the negative battery cable.

Hydraulic Clutch System

BLEEDING

Vacuum Bleeding

This procedure outlines how to bleed the hydraulic clutch with the transmission in the vehicle. Only **DOT 3** brake fluid should be added to the system.

1. Before servicing the vehicle, refer to the precautions in the beginning of this section.

2. Remove the reservoir cap and fill the reservoir with new brake fluid.

3. Install a Bleeder Adapter Tool J23738A, to the reservoir and connect a pressure bleeder to the adapter.

4. Charge the pressure bleeder to 15–20 psi (103–138 kPa).

5. Attach a transparent hose over the clutch bleeder screw nipple and submerge the opposite end of the hose in a container of brake fluid.

6. Loosen the bleeder screw on the transmission hydraulic fitting.

7. Bleed the system until no air bubbles are seen in the hose.

8. Tighten the bleeder screw.

9. Check the clutch pedal for a spongy feel. If the pedal feels soft, repeat the bleeding procedure.

10. Remove the bleeder tools and top off the fluid level if necessary.

Manual Bleeding

1. Before servicing the vehicle, refer to the precautions in the beginning of this section.

a. Fill the clutch reservoir with brake fluid. Check the reservoir level frequently and add fluid as needed.

b. Connect one end of a vinyl tube to the bleeder plug on the slave cylinder and submerge the other end into a clear container half-filled with clean brake fluid.

c. Slowly pump the clutch pedal 10–15 times without bring the pedal the full way up.

d. Repeat Steps 2 and 3 until all of the air bubbles are removed from the system.

e. Tighten the bleeder screw to 62 inch lbs. (7 Nm).

f. Refill the master cylinder to the proper level.

g. Check the system for leaks.

Halfshafts

REMOVAL & INSTALLATION

1. Before servicing the vehicle, refer to the precautions in the beginning of this section.

2. Remove the wheel cover or the center cap for access to the halfshaft nut. Have an assistant depress the brake pedal and loosen the front halfshaft nut.

3. Remove or disconnect the following:

- Wheel
- Tie rod end torque prevailing nut and discard it
- Tie rod end from the steering knuckle, using a Tie Rod Separator Tool SA91100C
- Lower control arm to steering knuckle
- Stabilzer bar link nut

➡ **Do not allow the steering knuckle to contact the ball stud seal. Contact may cause the seal to rip and the ball stud will need replacement.**

- Pull the steering knuckle/strut assembly away from the vehicle and pull the halfshaft out of the hub
- Properly support the halfshaft and remove the halfshaft from the transmission or Power Take Off (PTO) unit
- Shaft retaining ring and discard it

To install:

4. Apply Output Shaft Lubricant 7847638, to the splines of the output shaft, if equipped with an automatic transmission.

5. Install or connect the following:

- New stub shaft retaining ring
- Halfshaft to the transmission or PTO after installing a Seal Protector Tool SA91112T

6. Remove the seal protector tool after the splines have passed the oil seal.

7. Install or connect the following:

- Fully seat the halfshaft into the transmission
- Outer end of the halfshaft to the wheel hub with a new washer and nut
- Lower control arm ball stud to the steering knuckle. Torque the fastener to 89 inch lbs. (10 Nm) plus 150 degrees.
- Stabilizer bar link nut to 48 ft. lbs. (65 Nm)
- Tie rod end to the steering knuckle using tool J44015. When seated properly, torque the fastener to 37 ft. lbs. (50 Nm).
- Wheel
- Halfshaft to wheel nut. Torque the nut to 151 ft. lbs. (205 Nm).
- Cotter pin

8. Top off the transmission with the proper fluid.

9. Check and adjust the front end alignment as necessary.

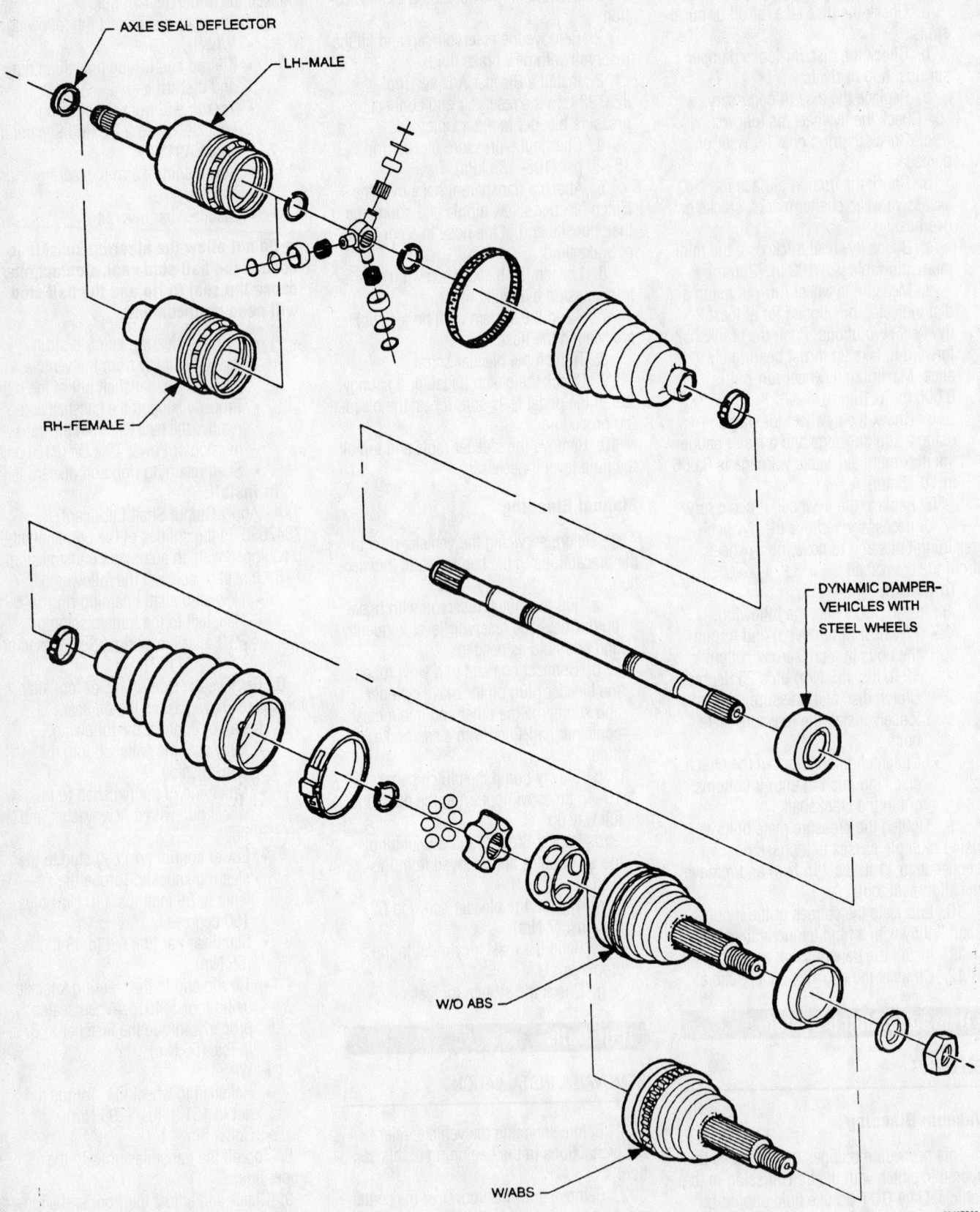

AXLE SEAL DEFLECTOR

LH-MALE

RH-FEMALE

DYNAMIC DAMPER-
VEHICLES WITH
STEEL WHEELS

W/O ABS

W/ABS

9346ZG28

Exploded view of a typical axle shaft assembly

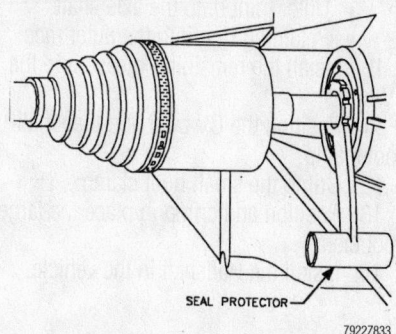

Failure to use a seal protector may allow the halfshaft splines to damage the transaxle seal

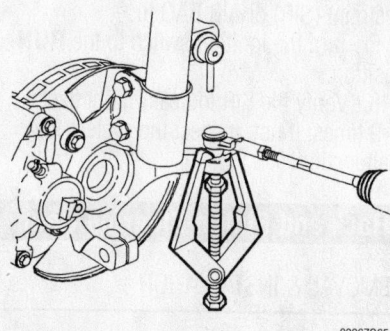

Separate the tie rod end from the steering knuckle with a Tie Rod Separator tool

CV-Joints

OVERHAUL

Constant Velocity (Outer) Joint

1. Before servicing the vehicle, refer to the precautions in the beginning of this section.
2. Remove or disconnect the following:
   - Front wheel
   - Halfshaft
   - Swage ring using a hand grinder
   - Large CV-joint boot clamp
   - CV-joint boot by sliding it away from the tri-pod joint
   - Tri-pod housing from the tri-pod spider
   - Inboard spacer ring slide it rearward on the shaft
   - Outboard retaining ring
   - Tri-pod joint spider assembly
   - Inboard spacer ring and CV-joint boot

To install:

3. Install or connect the following:
   - Swage ring clamp on the CV-joint boot
   - CV-joint boot

4. Position the CV-joint boot seal into the axle shaft's joint seal groove and align the swage ring clamp on the boot.

✳✳ WARNING

Make sure that there are no pinch points on the inboard seal.

5. Crimp the swage ring
6. Install or connect the following:
   - Inboard spacer ring, slide it rearward on the shaft
   - Tri-pod joint spider assembly onto the shaft
   - Outboard retaining ring into the axle shaft groove
   - Tri-pod joint spider assembly, slide it against the outboard retaining ring
   - Inboard spacer ring, seat it in the groove
   - ½ kit grease into the boot
   - ½ kit grease into the tri-pod housing
   - New large seal clamp onto the CV-joint boot
   - Tri-pod housing, slide it over the tri-pod joint spider assembly
   - CV-joint boot/clamp, slide it into place, over the trilobal tri-pod bushing with the seal lip in the groove

➡**Make sure the boot lies flat against the trilobal bushing.**

7. Position the CV-joint boot so it measures 4.9 in. (125mm).
8. Using a Crimp tool, a torque wrench and a breaker bar, crimp the large CV-joint boot clamp to 130 ft. lbs. (176 Nm).
9. Install the halfshaft and the front wheel.

Tri-Pot (Inner) Joint

1. Before servicing the vehicle, refer to the precautions in the beginning of this section.
2. Remove or disconnect the following:
   - Axle shaft from the vehicle
   - Large CV boot retaining clamp
   - Small CV boot retaining clamp
   - CV boot from the joint
   - Axle shaft retaining ring
   - Outer joint from the axle shaft
   - CV boot
3. Disassemble the chrome alloy balls from the CV-joint cage as follows:
   a. Position a brass drift against the CV-joint cage and tap it with a hammer to tilt the cage.
   b. Remove the 1st chrome alloy ball from the cage.

c. Tilt the cage in the opposite direction.
   d. Remove the opposite chrome alloy ball.
   e. Repeat the procedure until all 6 balls are removed.
4. Disassemble the CV-joint cage and inner race as follows:
   a. Pivot the cage and race 90 degrees to the center line of the outer race.
   b. Align the cage windows with outer race lands.
   c. Remove the cage from the outer race.
   d. Rotate the inner race upward and remove it from the cage.

To install:

5. Lubricate the parts with a light coat of grease.
6. Assemble the CV-joint cage and inner race, as follows:
   a. Rotate the inner race 90 degrees to the cage centerline.
   b. Align the cage windows with inner race lands.
   c. Insert the inner race into the cage by rotating the inner race downward.

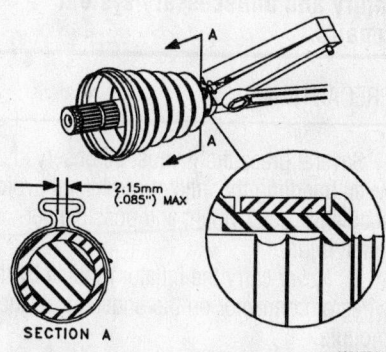

Tighten the small boot clamp to the specification shown

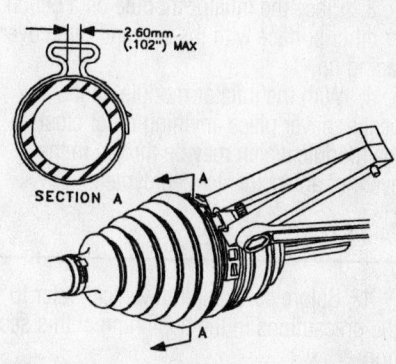

Tighten the large boot clamp to the specification shown

d. Insert the cage/inner race into the outer race.

7. Assemble the chrome alloy balls into the CV-joint cage, as follows:

a. Position a brass drift against the CV-joint cage and tap it with a hammer to tilt the cage.

b. Insert the 1st chrome alloy ball into the cage.

c. Tilt the cage in the opposite direction.

d. Insert the opposite chrome alloy ball.

e. Repeat the procedure until all 6 balls are inserted.

8. Install ½of the grease provided, into the CV-joint.

9. Install or connect the following:
- Small CV boot retaining ring
- CV boot on the halfshaft
- New retaining ring on the halfshaft
- Large ring clamp on the CV boot
- Outer joint onto the axle shaft
- Retaining ring into the outer race

10. Install the remaining grease into the CV boot.

11. Position the CV boot and the small boot clamp.

12. Crimp the small boot clamp.

13. Position and crimp in place the large boot clam

14. Install the Halfshaft in the vehicle.

STEERING AND SUSPENSION

Air Bag

※ CAUTION

All vehicles are equipped with an air bag system. The system must be disabled before performing service on or around system components, steering column, instrument panel components, wiring and sensors. Failure to follow safety and disabling procedures could result in accidental air bag deployment, possible personal injury and unnecessary system repairs.

PRECAUTIONS

Several precautions must be observed when handling the inflator module to avoid accidental deployment and possible personal injury.

1. Never carry the inflator module by the wires or connector on the underside of the module.

2. When carrying a live inflator module, hold securely with both hands, and ensure that the bag and trim cover are pointed away.

3. Place the inflator module on a bench or other surface with the bag and trim cover facing up.

4. With the inflator module on the bench, never place anything on or close to the module which may be thrown in the event of an accidental deployment.

DISARMING

1. Before servicing the vehicle, refer to the precautions in the beginning of this section.

2. Align the steering wheel so the vehicle wheels are pointing in the straight-ahead position.

3. Turn the ignition switch to the **LOCK** position.

4. Remove the SIR or AIR BAG fuse from the fuse block.

5. Disable the passenger side air bag as follows:

a. Locate the SIR connector attached to the HVAC blower motor and disconnect the clip.

➡ Do not remove the upper trim panel as you are able to disable the passenger side air bag from the underside of the instrument panel.

b. Remove the Connector Position Assurance (CPA) device, then disengage the yellow 2-way SIR wiring harness connector.

6. Disable the drivers side air bag as follows:

7. Remove the Connector Position Assurance (CPA) device, then disengage the yellow 2-way SIR wiring harness connector at the base of the steering column.

8. Disable the curtain air bag as follows:

a. Remove the center push pins located in the upper headliner trim panel.

b. Remove the left D-pillar upper trim panel.

c. Using a flat bladed tool, partially remove the right and left coat hook center retainers, then remove the coat hooks.

d. Pull back the headliner gently to access the yellow 2-way connectors.

e. Remove the Connector Position Assurance (CPA) device, then disengage the yellow 2-way SIR wiring harness connector.

ARMING

※ CAUTION

After the repairs, enable the system as follows:

1. Turn the ignition switch to the **LOCK** position.

2. Engage the yellow 2-way connectors for the airbags, then install the CPA device.

3. Install any removed trim pieces.

4. Reinstall the Supplemental Inflatable Restraint (SIR) or AIR BAG fuse.

5. Turn the ignition switch to the **RUN** position.

6. Verify the SIR indicator light flashes 7–9 times, if not, inspect the system for malfunction.

Rack and Pinion Steering Gear

REMOVAL & INSTALLATION

1. Before servicing the vehicle, refer to the precautions in the beginning of this section.

2. Remove or disconnect the following:
- Negative battery cable
- Both front wheels
- Tie rods from the steering knuckles
- Intermediate shaft from the steering gear pinch bolt and discard the bolt
- Shaft from the gear
- Stabilzer bar link nuts, links and swing the bar upwards
- Shift cable clip from the gear housing, if equipped
- Steering gear-to-cradle bolts
- Steering gear by sliding it out the right side of the vehicle

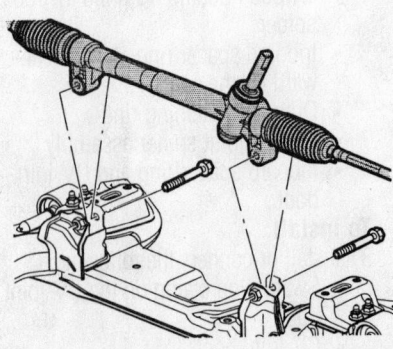

9359ZG32

Steering gear-to-frame assembly bolts

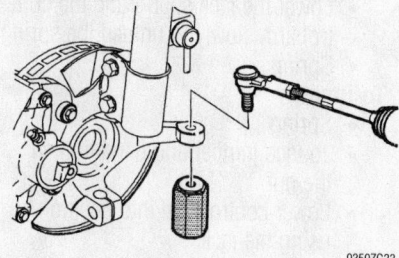

Use installer tool J44015 to attach the tie rods to the steering knuckle

To install:

3. Install or connect the following:
   - Steering gear through the right wheel opening
4. Center the gear mounting bushings into the cradle supports.
   - Steering gear-to-mounting bolts and tighten to 81 ft. lbs. (110 Nm)
   - Shift cable clip to the gear housing, if equipped
   - Intermediate shaft to the steering gear and torque the new pinch bolt to 25 ft. lbs. (34 Nm)
   - Stabilzer bar into position
   - Stabilzer bar links and nuts, then tighten the nuts to 48 ft. lbs. (65 Nm)
   - Tie rods to the steering knuckles using installer tool JJ44015 and tighten to 30 ft. lbs. (40 Nm)
   - New tie rod nut and tighten to 37 ft. lbs. (50 Nm)
   - Front wheels
   - Negative battery cable
5. Check the alignment and adjust if necessary.

Strut

REMOVAL & INSTALLATION

Front

1. Before servicing the vehicle, refer to the precautions in the beginning of this section.
2. Remove or disconnect the following:
   - Strut to body attaching nuts
   - Front wheel
   - Brake hose bracket from the strut assembly
   - Loosen the steering knuckle-to-strut fasteners, but do not remove them
   - Stabilizer bar link to the strut assembly attaching nut and move it toward the rear of the vehicle

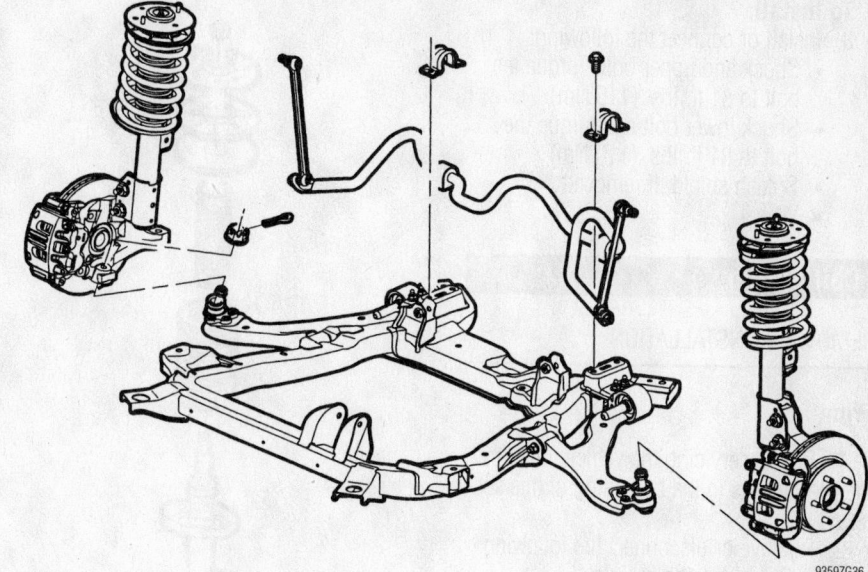

Exploded view of the front suspension

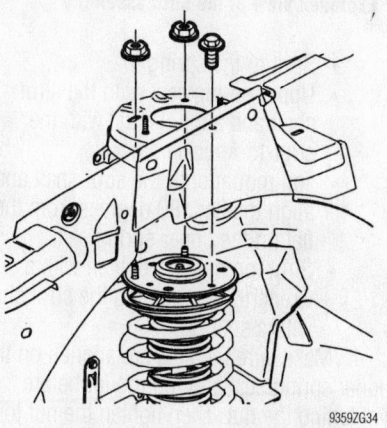

Remove the strut-to-body attaching nut

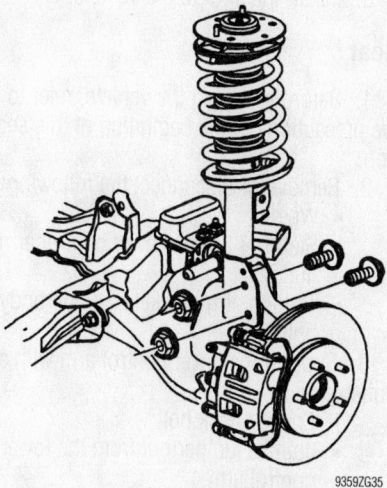

Remove the steering knuckle-to-strut fasteners

- Place a rag over the CV-joint seal to protect it from damage, then remove the 2 steering knuckle-to-strut housing bolts
- Steering knuckle-to-strut fasteners
- Strut assembly from the vehicle

To install:

3. Install or connect the following:
   - Strut to the body and torque the new attaching nuts and bolt to 18 ft. lbs. (25 Nm)
   - Strut to the steering knuckle and torque the new fasteners to 133 ft. lbs. (180 Nm)
   - Stabilizer bar link to the strut and torque the fastener to 48 ft. lbs. (65 Nm)
   - Brake hose bracket to the strut
   - Front wheel
   - Negative battery cable
4. Check and adjust the alignment as necessary.

Shock Absorber

REMOVAL & INSTALLATION

1. Before servicing the vehicle, refer to the precautions in the beginning of this section.
2. Remove or disconnect the following:
   - Wheel
   - Lower shock bolt

➡**If removing the right side shock absorber, remove the splash shield.**

- Upper shock bolt and the shock

To install:

3. Install or connect the following:
- Shock and upper bolt. Torque the bolt to 81 ft. lbs. (110 Nm).
- Shock lower bolt and torque the bolt to 81 ft. lbs. (110 Nm)
- Splash shield, if removed
- Wheel

Coil Spring

REMOVAL & INSTALLATION

Front

1. Before servicing the vehicle, refer to the precautions in the beginning of this section.

2. Remove or disconnect the following:
- Strut from the vehicle
- Place the strut into spring compressor J45400 or a similar compressor.
- Compress the spring enough to completely unload the upper strut mount
- Strut shaft nut while holding the shaft stationary with a TORX® socket
- Release the spring compressor
- Spring from the strut
- Remaining strut assembly components, examine for wear or damage and replaces as necessary

To install:

3. Install or connect the following:
- Strut into spring compressor
- Extend the strut shaft to its full travel

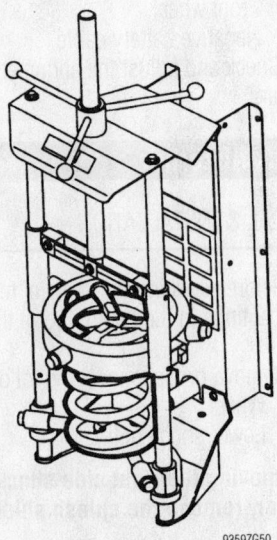

Place the strut sinto spring compressor J45400

9359ZG51

Exploded view of the strut assembly

- Strut to the spring
- Upper spring seat onto the strut shaft and align the flat with the strut-to-knuckle bracket
- Top mount onto the strut shaft and align the flat 180 degrees from the flat on the upper spring seat
- Strut top nut and tighten with a wrench while holding the strut shaft with a socket

4. Make sure the flats are aligned on the upper spring seat and top mount before tightening the nut. Then tighten the nut to 55 ft. lbs. (75 Nm).

5. Release the spring compressor tool.

6. Install the strut to the vehicle.

Rear

1. Before servicing the vehicle, refer to the precautions in the beginning of this section.

2. Remove or disconnect the following:
- Wheel
- Stabilzer link-to-lower control arm nut
- Trailing arm bracket-to-underbody bolts

3. Support the lower control arm with a suitable jack.
- Lower shock bolt
- Jounce jumper nut from the lower control arm
- Lower control arm-to-support frame bolt
- Lower control arm-to-knuckle retainers

- Lower the jack supporting the control arm slowly to unload the spring
- Spring

To install:
- Spring
- Jounce jumper and hand tighten the nut
- Lower control arm into position using the jack
- Lower control arm-to-knuckle fasteners and hand tighten making sure the bolt heads face the rear of the vehicle. Tighten the bolts to 81 ft. lbs. (110 Nm).
- Lower control arm-to-support bolt to 81 ft. lbs. (110 Nm)
- Shock absorber lower bolt and tighten to 81 ft. lbs. (110 Nm)
- Jounce jumper nut to 46 ft. lbs. (63 Nm)
- Stabilzer bar link nut and tighten to 11 ft. lbs. (15 Nm)

4. When installing the training arm, push upward and install the front bolt, then use a drift to align the rear holes and install the bolts. Tighten the bolts to 81 ft. lbs. (110 Nm).
- Wheel

Lower Ball Joint

REMOVAL & INSTALLATION

1. Before servicing the vehicle, refer to the precautions in the beginning of this section.

2. Install or connect the following:
- Lower control arm from the vehicle
- Rivets retaining the ball joint to the control arm using a 5/16 in. (8 mm) drill bit
- Ball joint from the control arm

To install:

3. Install or connect the following:
- Ball joint into the control arm

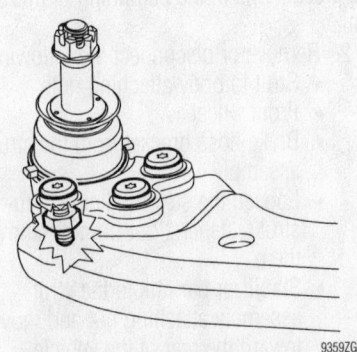

9359ZG38

The new ball stud is bolted into the control arm

- Nuts and bolts (included with new ball joint kit) as shown and torque them to 50 ft. lbs. (68 Nm)
- Control arm to the vehicle

Lower Control Arm

REMOVAL & INSTALLATION

Front

1. Before servicing the vehicle, refer to the precautions in the beginning of this section.
2. Remove or disconnect the following:
   - Wheel
   - Ball stud bolt
   - Separate the ball stud from the steering knuckle by using ball joint removal tool J43828
   - Lower control arm-to-frame retainers
   - Lower control arm

To install:

3. Install or connect the following:
   - Control arm to the frame and torque the rear bolts and nuts to 52 ft. lbs. (70 Nm) and the front arm-to-frame bolt and nut to 89 ft. lbs. (120 Nm)
   - Ball stud to steering knuckle and torque the bolt to 89 inch. lbs. (10 Nm) plus 150 degrees
   - Wheel
   - Check and adjust the alignment, if necessary.

Rear

1. Before servicing the vehicle, refer to the precautions in the beginning of this section.
2. Remove or disconnect the following:
   - Wheel

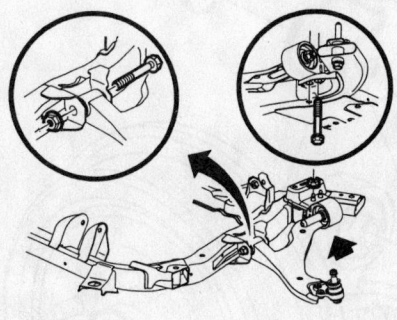

9359ZG39

Remove the lower control arm to frame bolts

- Stabilzer link-to-lower control arm nut
- Trailing arm bracket-to-underbody bolts
3. Support the lower control arm with a suitable jack.
   - Lower shock bolt
   - Jounce jumper nut from the lower control arm
   - Lower control arm-to-support frame bolt
   - Lower control arm-to-knuckle retainers
   - Lower the jack supporting the control arm slowly to unload the spring
   - Lower control arm-to-knuckle retainers
   - Spring
   - Lower control arm-to-support nut and bolt

To install:

   - Lower control arm-to-support and hand tighten the nut
   - Spring
   - Jounce jumper and hand tighten the nut
   - Lower control arm into position using the jack
   - Lower control arm-to-knuckle fasteners and hand tighten making sure the bolt heads face the rear of the vehicle. Tighten the bolts to 81 ft. lbs. (110 Nm).
   - Lower control arm-to-support bolt to 81 ft. lbs. (110 Nm)
   - Shock absorber lower bolt and tighten to 81 ft. lbs. (110 Nm)
   - Jounce jumper nut to 46 ft. lbs. (63 Nm)
   - Stabilzer bar link nut and tighten to 11 ft. lbs. (15 Nm)
4. When installing the training arm, push upward and install the front bolt, then use a drift to align the rear holes and install the bolts. Tighten the bolts to 81 ft. lbs. (110 Nm).
   - Wheel

LOWER CONTROL ARM BUSHING REPLACEMENT

Front Control Arm

FRONT BUSHING

1. Before servicing the vehicle, refer to the precautions in the beginning of this section.
2. Remove or disconnect the following:

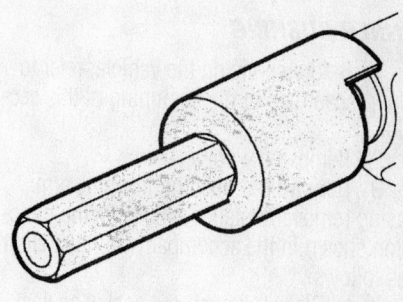

9359ZG40

Press out the front control arm front bushing

   - Wheel
   - Control arm
3. Front bushing by using bearing removal/installer tools J44971 to press out the bushing.

To install:

4. Install or connect the following:
5. Press in the new control arm bushing using bearing removal/installer tools J44971.
   - Control arm

FRONT BUSHING

1. Before servicing the vehicle, refer to the precautions in the beginning of this section.
2. Remove or disconnect the following:
   - Wheel
   - Control arm
   - Rear bushing nut and the bushing

To install:

3. Install or connect the following:
   - Bushing and nut. Tighten the nut to 85 ft. lbs. (115 Nm).
   - Control arm
   - Wheel

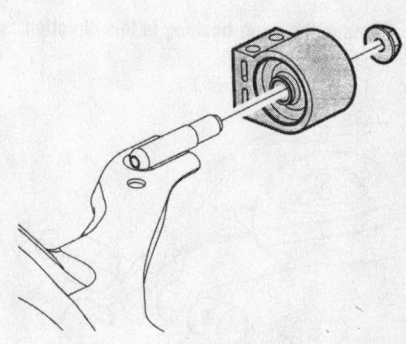

9359ZG41

Remove the rear bushing nut and the bushing

Timing belt service is covered in Section 3 of this manual

Rear Control Arm

INNER BUSHING

1. Before servicing the vehicle, refer to the precautions in the beginning of this section.

2. Remove the control arm.

3. Remove the bushing from the arm using removal/installer J45097 in the direction shown in the accompanying illustration as follows:

    a. Place the push out socket against the bushing from the flanged side of the arm.

    b. Install the through-bolt with the washer and bearing against the push-out socket.

    c. Install a backing socket against the control arm on the opposite side of the flange.

➡ **Apply high pressure lube to the tool threads.**

    d. Install the flat washer and nut, then tighten the nut to remove the bushing.

To install:

4. Install the bushing centering it to align with the center of the control arm approximately 0.079 inch (2mm) beyond the flange.

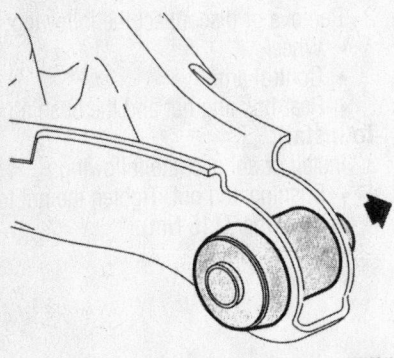

Remove the inner bushing in this direction

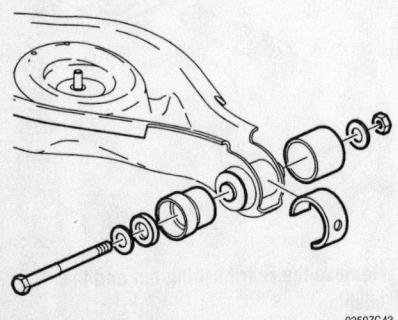

Remove the bushing from the arm using removal/installer J45097

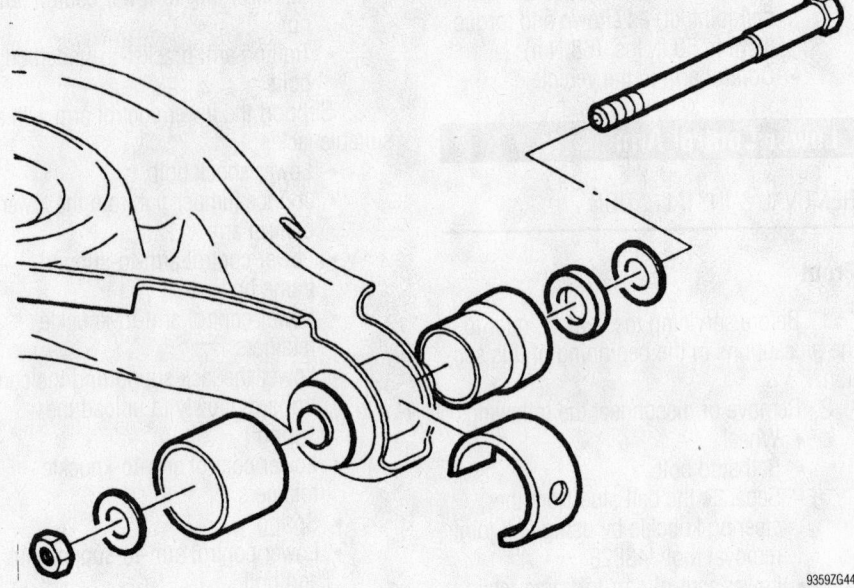

Reverse the installer removal tool and press the bearing into the control arm

5. Reverse the installer removal tool and press the bearing into the control arm

6. Install the control arm.

Upper Control Arm

REMOVAL & INSTALLATION

Rear

1. Before servicing the vehicle, refer to the precautions in the beginning of this section.

2. Remove or disconnect the following:
- Trailing arm bracket-to-underbody bolts

- Anti-lock Brake System (ABS) harness from the upper control arm, if equipped
- Upper control arm-to-knuckle fasteners
- Upper control arm-to-support frame retainers
- Upper control arm

To install:
- Upper control arm

➡ **Make sure the bolt heads face the front of the vehicle.**

- Upper control arm-to-knuckle fasteners and hand tighten
- Upper control arm-to-support frame cam nut

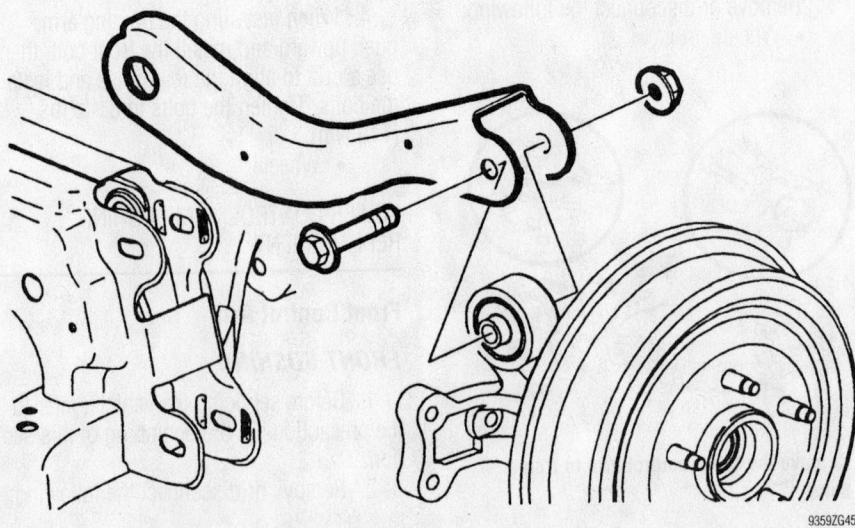

Remove the lower control arm

- Tighten the control arm-to-knuckle bolts to 118 ft. lbs. (160 Nm)
- Tighten the upper control arm-to-support bolt to 81 ft. lbs. (110 Nm)
- Anti-lock Brake System (ABS) harness to the upper control arm, if equipped

3. When installing the training arm, push upward and install the front bolt, then use a drift to align the rear holes and install the bolts. Tighten the bolts to 81 ft. lbs. (110 Nm).

4. Align the vehicle.

Wheel Bearings

ADJUSTMENT

The wheel bearing are sealed at the factory and do not require any adjustment or maintenance.

REMOVAL & INSTALLATION

Front

1. Before servicing the vehicle, refer to the precautions in the beginning of this section.

2. Remove or disconnect the following:
   - Wheel
   - Brake caliper mounting bracket bolts and suspend the assembly from the strut spring with wire
   - Brake rotor
   - ABS sensor electrical connection, if equipped
   - ABS sensor electrical connection from the bracket, if equipped
   - Drive shaft axle nut

3. Support the axle with wire
   - Drive shaft from the knuckle
   - Bearing bolts, the bearing assembly and shield

To install:

➡ **Make sure the drive shaft splines are aligned with the wheel bearing assembly splines. If you do not align the splines you could damage the bearing or drive shaft assemblies.**

4. Install or connect the following:
   - Bearing assembly with the shield.
   - Bearing assembly bolts and tighten to 96 ft. lbs. (130 Nm)
   - Drive shaft axle nut and torque to 92 ft. lbs. (205 Nm)
   - ABS sensor electrical connection to the bracket, if equipped

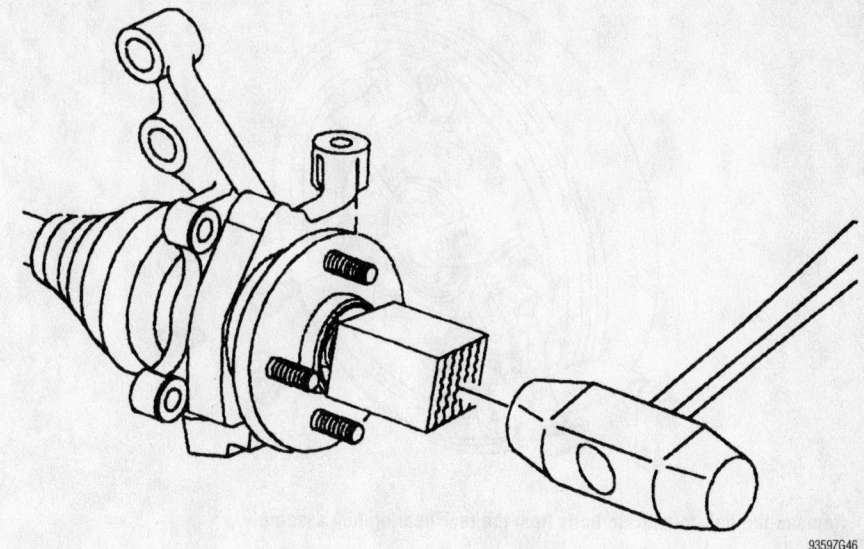

Separate the axle from the wheel hub

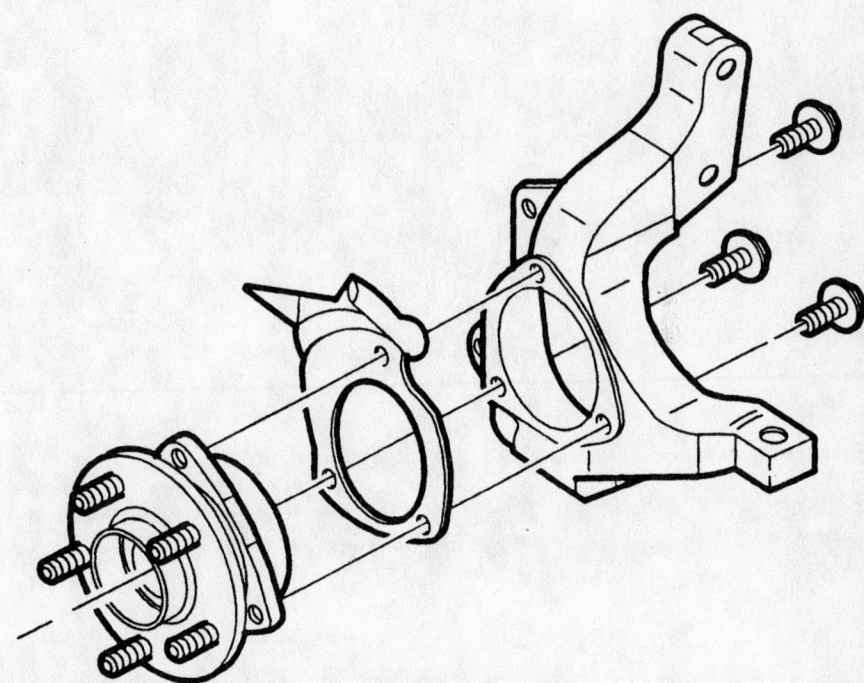

Exploded view of the front bearing/hub assembly

- ABS sensor electrical connection, if equipped
- Brake rotor
- Brake caliper
- Wheel

Rear

1. Before servicing the vehicle, refer to the precautions in the beginning of this section.

2. Remove or disconnect the following:
   - Negative battery cable, if equipped with an Antilock Braking System (ABS)
   - Rear wheel
   - Brake drum
   - Drive axle nut, if equipped with All Wheel Drive (AWD)
   - ABS electrical connector, if equipped

Heater Core replacement is covered in Section 2 of this manual

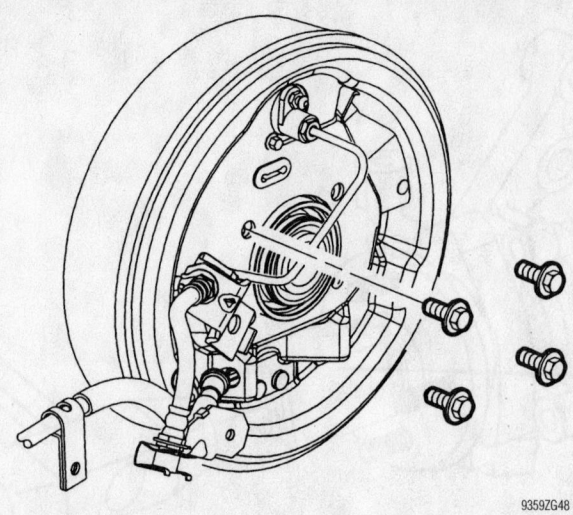

9359ZG48

Remove the hub-to-knuckle bolts from the rear bearing/hub assembly

3. Support the axle shaft with wire.
   - Axle shaft from the hub
   - Hub-to-knuckle bolts
   - Hub

To install:

4. Install or connect the following:
   - Hub to the knuckle and torque the new bolts to 96 ft. lbs. (130 Nm)
   - ABS electrical connector through the hole and seat the rubber grommet, if equipped
   - ABS electrical connector, if equipped
   - Drive axle nut and torque to 81 ft. lbs. (110 Nm), if equipped with All Wheel Drive (AWD)
   - Brake drum
   - Rear wheel

SUBARU

Forester

PRECAUTIONS

Before servicing any vehicle, please be sure to read all of the following precautions, which deal with personal safety, prevention of component damage, and important points to take into consideration when servicing a motor vehicle:

• Never open, service or drain the radiator or cooling system when the engine is hot; serious burns can occur from the steam and hot coolant.

• Observe all applicable safety precautions when working around fuel. Whenever servicing the fuel system, always work in a well-ventilated area. Do not allow fuel spray or vapors to come in contact with a spark, open flame, or excessive heat (a hot drop light, for example). Keep a dry chemical fire extinguisher near the work area. Always keep fuel in a container specifically designed for fuel storage; also, always properly seal fuel containers to avoid the possibility of fire or explosion. Refer to the additional fuel system precautions later in this section.

• Fuel injection systems often remain pressurized, even after the engine has been turned **OFF**. The fuel system pressure must be relieved before disconnecting any fuel lines. Failure to do so may result in fire and/or personal injury.

• Brake fluid often contains polyglycol ethers and polyglycols. Avoid contact with the eyes and wash your hands thoroughly after handling brake fluid. If you do get brake fluid in your eyes, flush your eyes with clean, running water for 15 minutes. If eye irritation persists, or if you have taken brake fluid internally, IMMEDIATELY seek medical assistance.

• The EPA warns that prolonged contact with used engine oil may cause a number of skin disorders, including cancer. You should make every effort to minimize your exposure to used engine oil. Protective gloves should be worn when changing oil. Wash your hands and any other exposed skin areas as soon as possible after exposure to used engine oil. Soap and water, or waterless hand cleaner should be used.

• All new vehicles are now equipped with an air bag system. The system must be disabled before performing service on or around system components, steering column, instrument panel components, wiring and sensors. Failure to follow safety and disabling procedures could result in accidental air bag deployment, possible personal injury, and unnecessary system repairs.

• Always wear safety goggles when working with, or around, the air bag system. When carrying a non-deployed air bag, be sure the bag and trim cover are pointed away from your body. When placing a non-deployed air bag on a work surface, always face the bag and trim cover upward, away from the surface. This will reduce the motion of the module if it is accidentally deployed. Refer to the additional air bag system precautions later in this section.

• Clean, high quality brake fluid from a sealed container is essential to the safe and proper operation of the brake system. You should always buy the correct type of brake fluid for your vehicle. If the brake fluid becomes contaminated, completely flush the system with new fluid. Never reuse any brake fluid. Any brake fluid that is removed from the system should be discarded. Also, do not allow any brake fluid to come in contact with a painted surface; it will damage the paint.

• Never operate the engine without the proper amount and type of engine oil; doing so WILL result in severe engine damage.

• Timing belt maintenance is extremely important. Many models utilize an interference-type, non-freewheeling engine. If the timing belt breaks, the valves in the cylinder head may strike the pistons, causing potentially serious (also time-consuming and expensive) engine damage. Refer to the maintenance interval charts in the front of this manual for the recommended replacement interval for the timing belt, and to the timing belt section for belt replacement and inspection.

• Disconnecting the negative battery cable on some vehicles may interfere with the functions of the on-board computer system(s) and may require the computer to undergo a relearning process once the negative battery cable is reconnected.

• When servicing drum brakes, only disassemble and assemble one side at a time, leaving the remaining side intact for reference.

ENGINE REPAIR

➡**Disconnecting the negative battery cable on some vehicles may interfere with the functions of the on board computer systems and may require the computer to undergo a relearning process, once the negative battery cable is reconnected.**

Distributor

The Forester is equipped with a distributorless ignition system.

Alternator

REMOVAL

1. Before servicing the vehicle, refer to the precautions in the beginning of this section.
2. Remove or disconnect the following:

• Negative battery cable
• Wires
• Belt cover
• Drive belt
• Mounting bolts
• Alternator

INSTALLATION

Install or connect the following:
• Alternator
• Mounting bolts
• Drive belt. Adjust the tension to 0.276–0.354 in. (7–9mm) for new or 0.354–0.433 in. (9–11mm) for used. Torque the slider bolt to 4–6 ft. lbs. (6–10 Nm) and the lockbolt to 16.5 ft. lbs. (19.5 Nm).
• Belt cover
• Wires
• Negative battery cable

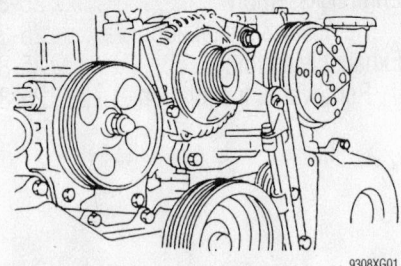

9308XG01

View of alternator mounting

Ignition Timing

ADJUSTMENT

The ignition timing is controlled by the engine control computer and is not adjustable. To check the ignition timing proceed as follows:

1. Before servicing the vehicle, refer to the precautions in the beginning of this section.

2. Warm up the engine, then turn the ignition **OFF**.

3. Connect a timing light to the No. 1 spark plug wire according to the manufactures directions.

4. Start the engine. With the vehicle at idle check the timing.

5. The timing should be 7–23 degrees BTDC at 700 RPM.

6. If the timing is not correct, there could be a problem in the ignition control system.

Engine Assembly

REMOVAL & INSTALLATION

✳✳ CAUTION

The fuel injection system remains under pressure after the engine has been turned OFF. Properly relieve fuel pressure before disconnecting any fuel lines. Failure to do so may result in fire or personal injury.

1. Before servicing the vehicle, refer to the precautions in the beginning of this section.

2. Relieve the fuel system pressure.

3. Drain the engine oil and coolant.

4. Discharge and recover the air conditioning system.

5. Remove or disconnect the following:
- Negative battery cables and battery
- Engine undercover
- Radiator hoses and fan motor harness
- Radiator
- Air conditioning compressor and cap the lines
- Air intake duct
- Air cleaner element and upper cover
- Evaporative Emissions (EVAP) canister and bracket
- Front Oxygen (O₂S) sensor

6. If equipped with California emissions specifications, disconnect the rear O₂S sensor.

7. Remove or disconnect the following:
- Engine ground terminal
- Crankshaft Position (CKP) sensor connector
- Camshaft Position (CMP) sensor connector
- Knock Sensor (KS) connector

- Alternator connector and terminal
- Air conditioning compressor connectors, if equipped
- Accelerator cable
- Cruise control cable, if equipped
- Brake booster hose
- Heater inlet and outlet hoses
- Alternator drive belt
- Wires from the spark plugs on the left side of the engine
- Power steering pump line bracket
- Power steering pump, leaving the lines connected and position it aside
- Exhaust Y-pipe
- Lower starter nuts
- Lower engine-to-transmission nuts
- Front engine mount-to-crossmember nuts
- Starter

8. If equipped with an automatic transmission, perform the following:

a. Remove the torque converter service hole plug.

b. Matchmark the torque converter-to-driveplate.

c. Rotate the engine to remove the torque converter-to-driveplate bolts as they become accessible.

9. Remove or disconnect the following:
- Flywheel cover, if equipped with a manual transmission
- Pitching stopper
- Fuel delivery, return and evaporation hoses

10. Support the engine with a suitable lifting device attached to the engine lifting eyes.

11. Slightly raise the engine.

12. Raise the transmission with a floor jack.

13. If equipped with a manual transmission, pull the engine forward then up and out of the vehicle to clear the transmission mainshaft.

14. If equipped with an automatic transmission, pull the engine forward then up and out of the vehicle.

To install:

15. If equipped with a manual transmission, apply a small amount of grease to the splines of the mainshaft.

16. Position the engine in the engine compartment and align it with the transmission.

17. Install the engine. Torque the upper bolts to 34–40 ft. lbs. (44–54 Nm).

18. Remove the lifting device and floor jack.

19. Install or connect the following:
- Pitching stopper. Torque the bolts to 49 ft. lbs. (67 Nm) on the body side and 40 ft. lbs. (54 Nm) on the bracket side.
- Flywheel cover, if equipped with a manual transmission

20. If equipped with an automatic transmission, perform the following:

a. Align the matchmarks, install the torque converter-to-driveplate bolts while rotating the engine and tighten to 20 ft. lbs. (26 Nm).

b. Install the service hole cover.

21. Install or connect the following:
- EVAP canister and bracket
- Power steering pump. Tighten the

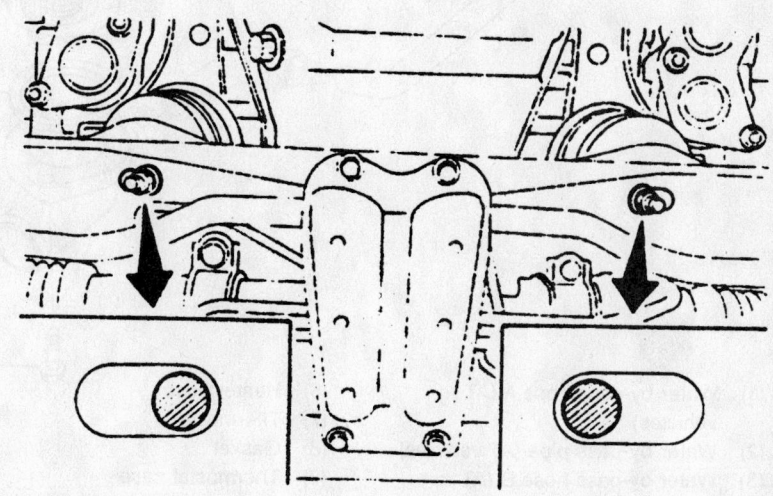

7924XG25

Be sure to tighten the front cushion rubber mounting bolts in the innermost elliptical hole in the front crossmember

retainer bolts to 22–36 ft. lbs. (29–47 Nm).

- Accessory drive belt
- Starter. Torque the bolts to 34–40 ft. lbs. (44–52 Nm).
- Lower engine-to-transmission nuts. Tighten them to 34–40 ft. lbs. (44–52 Nm).
- Lower engine mounting nuts. Tighten them to 61 ft. lbs. (83 Nm) in the inner most elliptical hole in the front crossmember so the clearance is 0.16–0.24 in. (4–6mm).
- Exhaust Y-pipe with new gaskets and nuts
- Brake booster hose
- Heater inlet and outlet hoses
- Accelerator cable
- Cruise control cable, if equipped

- Engine harness connectors
- Engine ground terminal
- CKP sensor connector
- CMP sensor connector
- Knock sensor connector
- Alternator connector and terminal
- Air conditioning compressor connectors, if equipped
- Front O₂S sensor, and if removed, the rear O₂S sensor.
- Air cleaner element and cover
- Air conditioning lines with new O-rings, if equipped. Torque the bolts to 23 ft. lbs. (31 Nm).
- Radiator
- Engine undercover
- Negative battery cable

22. Fill the crankcase to the proper level with clean engine oil.

23. Fill and bleed the cooling system.

24. Charge the air conditioning system using an approved recovery/recycling machine.

25. If equipped, check the automatic transmission fluid level and add Dexron®II if necessary.

26. Start the engine and allow it to reach normal operating temperature. Check for leaks.

Water Pump

REMOVAL & INSTALLATION

1. Before servicing the vehicle, refer to the precautions in the beginning of this section.

2. Drain the coolant into a suitable container.

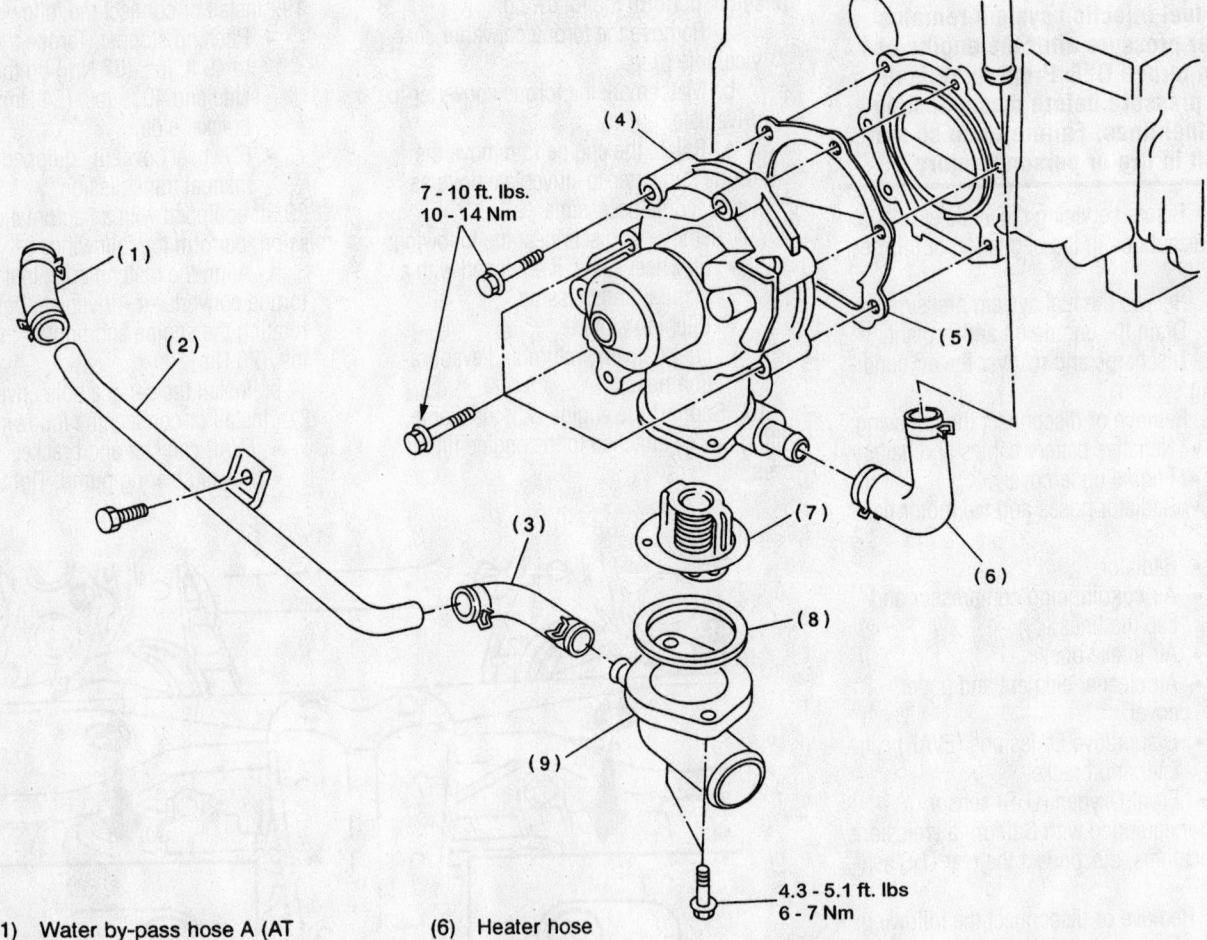

7 - 10 ft. lbs.
10 - 14 Nm

4.3 - 5.1 ft. lbs
6 - 7 Nm

(1) Water by-pass hose A (AT vehicles)
(2) Water by-pass pipe (AT vehicles)
(3) Water by-pass hose B (AT vehicles)
(4) Water pump ASSY
(5) Gasket
(6) Heater hose
(7) Thermostat
(8) Gasket
(9) Thermostat case

Exploded view of the water pump mounting and related components

7924XG01

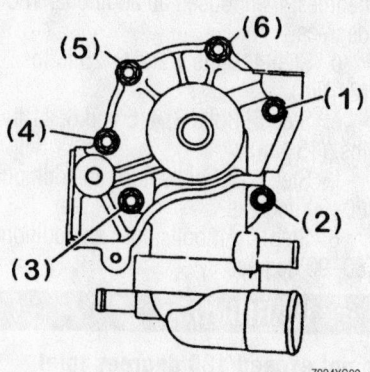

Water pump bolt tightening sequence

3. Remove or disconnect the following:
  • Negative battery cable
  • Engine undercover
  • Radiator outlet hose
  • Radiator fan motor assembly
  • Water pipe bypass pipe retaining bolt
  • Accessory drive belts
  • Timing belt and tensioner
  • Camshaft Position (CMP) sensor
  • Left side camshaft pulleys and left side rear timing belt cover
  • Tensioner bracket
  • Radiator hose and heater hose from the water pump

  • Water pump retainer bolts
  • Water pump
To install:
4. Clean the gasket mating surfaces thoroughly. Always use new gaskets during installation.
5. Install or connect the following:
  • Water pump. Torque the bolts, in sequence, to 84–120 inch lbs. (10–14 Nm). After tightening the bolts once, retighten to the same specification again.
  • Radiator hose and heater hose to the water pump
  • Left side rear timing belt cover, left side camshaft pulleys and tensioner bracket
  • CMP sensor
  • Timing belt and tensioner
  • Accessory drive belts
  • Water pipe bypass pipe retaining bolt
  • Radiator fan motor assembly
  • Radiator outlet hose
  • Engine undercover
  • Negative battery cable
6. Fill the system with coolant.
7. Start the engine and allow it to reach operating temperature.
8. Check for leaks.

Cylinder Head

REMOVAL & INSTALLATION

1. Before servicing the vehicle, refer to the precautions in the beginning of this section.
2. Properly relieve the fuel system pressure.
3. Remove or disconnect the following:

  • Negative battery cable
  • Oxygen (O2S) sensor. If equipped with California emissions, disconnect the rear O2S sensor.
  • Engine undercover
  • Exhaust Y-pipe and lower it just enough to clear the studs in the heads. Do not allow the Y-pipe to hang without support.
  • Accessory drive belts
  • Engine accessories and brackets from the side of the engine the cylinder head is being removed
  • Connector bracket attaching bolt, if necessary
4. On the left cylinder head, remove the CMP sensor.
5. Remove or disconnect the following:

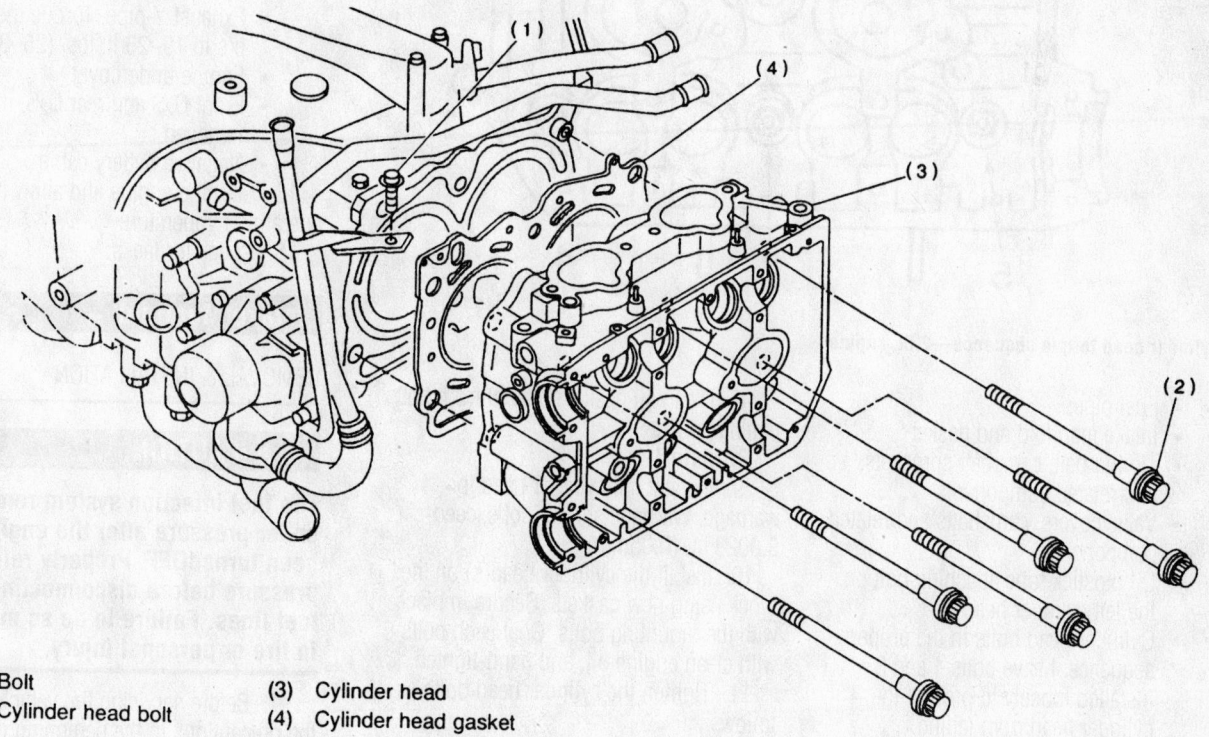

(1) Bolt
(2) Cylinder head bolt
(3) Cylinder head
(4) Cylinder head gasket

Exploded view of the cylinder head mounting

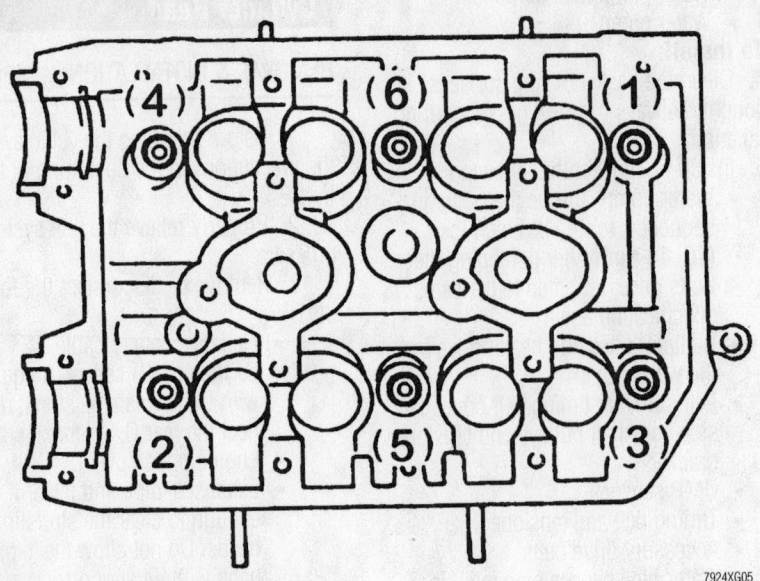

Cylinder head bolt loosening sequence

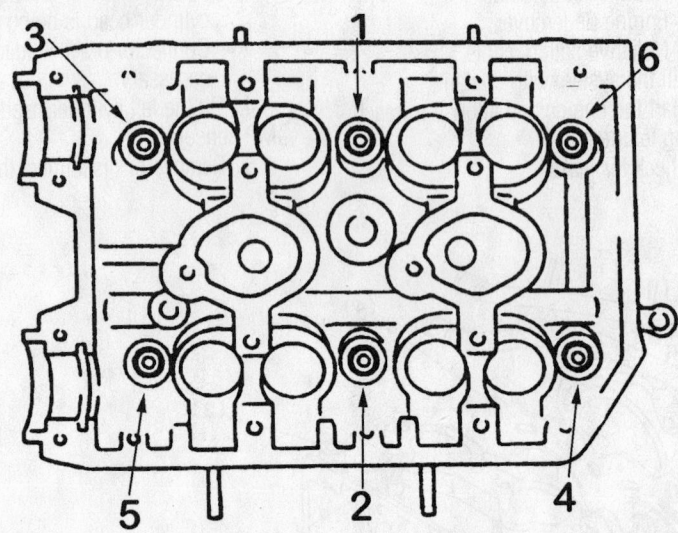

Cylinder head torque sequence—2.5L engine

- Fuel pipes
- Intake manifold and gasket
- Timing belt, camshaft sprockets, and related components
- Valve covers, camshafts and related components
- Oil dipstick tube attaching bolt on the left cylinder head
- Cylinder head bolts in the proper sequence. Leave bolts 1 and 3 installed loosely to prevent the cylinder head from falling.

6. Separate the cylinder head from the block. Use a plastic-faced hammer, if needed.

7. Remove bolts 1 and 3. Remove the cylinder head and gasket.

8. Clean all gasket material from both mating surfaces.

To install:

9. Inspect the cylinder head for warpage. Warpage should not exceed 0.0020 in. (0.05mm).

10. Install the cylinder head(s) on the block using new gaskets. Secure in place with the mounting bolts. Coat each bolts with clean engine oil, and hand-tighten.

11. Tighten the cylinder head bolts as follows:

   a. Step 1: Torque the bolt to 22 ft. lbs. (29 Nm).

   b. Step 2: Torque the bolt to 51 ft. lbs. (69 Nm).

   c. Step 3: Loosen all bolts by 180 degrees, then loosen an additional 180 degrees.

   d. Step 4: Bolts 1 and 2: 25 ft. lbs. (24 Nm).

   e. Step 5: Bolts 3, 4, 5 and 6: 11 ft. lbs. (15 Nm).

   f. Step 6: All bolts: turn an additional 80–90 degrees.

   g. Step 7: All bolts: turn an additional 80–90 degrees.

✱✱ WARNING

Do not exceed 180 degrees total tightening.

12. Install or connect the following:
- Oil dipstick tube attaching bolt on the left cylinder head
- Valve covers, camshafts and related components
- Camshaft sprocket, timing belt, and related components
- Intake manifold. Torque the bolts to 21–25 ft. lbs. (28–34 Nm).
- Fuel delivery pipes

13. On the left cylinder head, install the CMP sensor.

14. Install or connect the following:
- Connector bracket attaching bolt
- Spark plug wires
- Engine accessories and brackets
- Accessory drive belts
- Exhaust Y-pipe. Torque the fasteners to 19–26 ft. lbs. (25–35 Nm).
- Engine undercover
- Front O_2S and rear O_2S, if removed.
- Negative battery cable

15. Start the engine and allow it to reach operating temperature.

16. Check for leaks.

Intake Manifold

REMOVAL & INSTALLATION

✱✱ CAUTION

The fuel injection system remains under pressure after the engine has been turnedOFF. Properly relieve fuel pressure before disconnecting any fuel lines. Failure to do so may result in fire or personal injury.

1. Before servicing the vehicle, refer to the precautions in the beginning of this section.

2. Properly relieve the fuel system pressure.

3. Drain the cooling system.

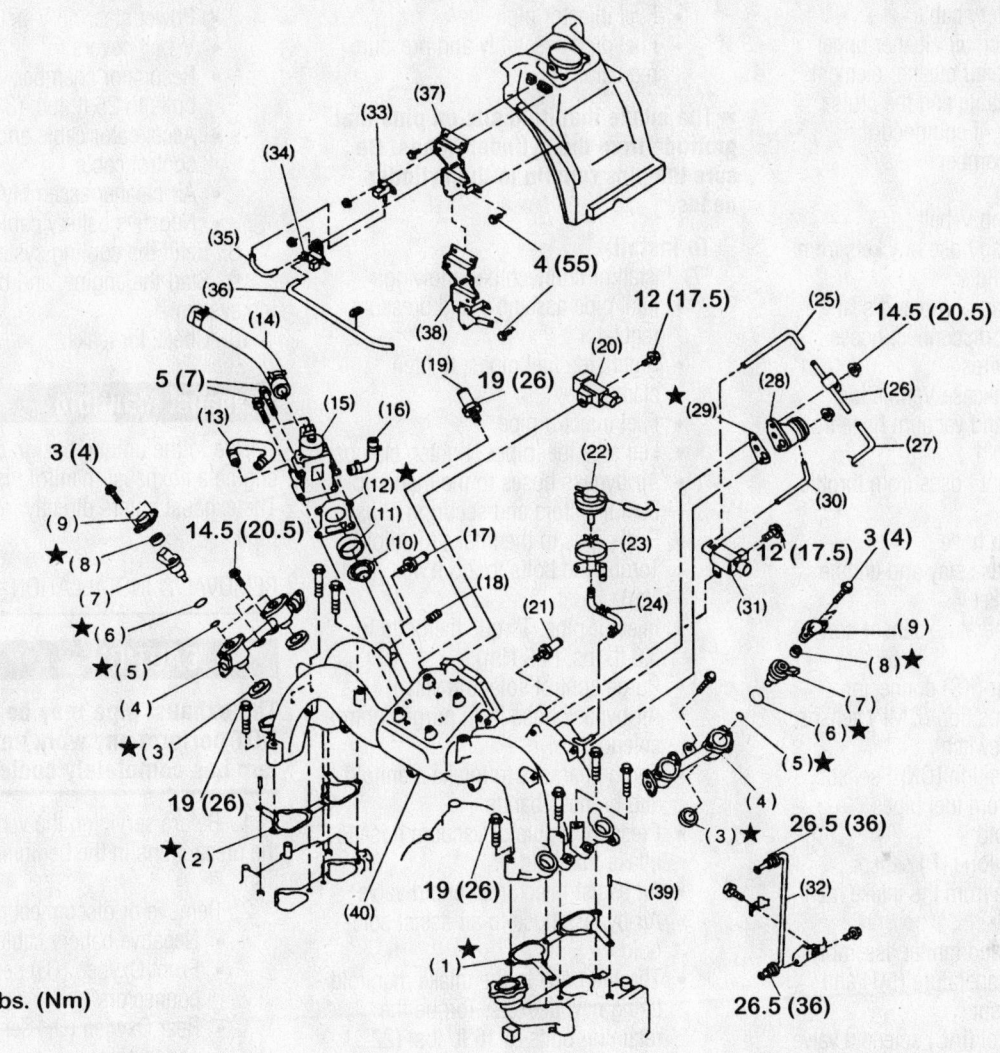

ft. lbs. (Nm)

Exploded view of the intake manifold

(1) Intake manifold gasket LH	(17) Nipple (Equipped cruise control model)	(33) Pressure sensor
(2) Intake manifold gasket RH	(18) Plug	(34) Pressure sources switching sole-noid valve
(3) Fuel injector pipe insulator	(19) PCV valve	(35) Vacuum hose A
(4) Fuel injector pipe	(20) Purge control solenoid valve	(36) Vacuum hose B
(5) O-ring A	(21) Nipple	(37) Bracket (Except Canada spec. vehicles)
(6) O-ring B	(22) BPT	(38) Bracket (For Canada spec. vehicles)
(7) Fuel injector	(23) BPT holder bracket	(39) Collar
(8) Insulator	(24) Back pressure hose	(40) Intake manifold
(9) Fuel injector cap	(25) EGR vacuum hose A	
(10) Plate	(26) EGR vacuum pipe	
(11) Sealing	(27) EGR vacuum hose C	
(12) Gasket	(28) EGR valve	
(13) Engine coolant hose B	(29) Gasket	
(14) Air by-pass hose	(30) EGR vacuum hose B	
(15) Idle air control solenoid valve	(31) EGR solenoid valve	
(16) Engine coolant hose A	(32) EGR pipe	

7924XG06

Timing belt service is covered in Section 3 of this manual

4. Remove or disconnect the following:
- Fuel filler cap
- Negative battery cable
- Air intake duct, air cleaner upper cover and the air cleaner element
- Accelerator cable and the cruise control cable, if equipped
- Resonator chamber
- V-belt covers
- Power steering V-belt
- Power steering hose brackets from intake manifold
- Power steering pump and seat aside, do not disconnect hoses
- Spark plug wires
- Positive Crankcase Ventilation (PCV) hose and vacuum hose from intake manifold
- Engine coolant hoses from throttle body
- Brake booster hose
- Air cleaner case stay and engine harness bracket
- Coolant temperature sensor connector
- Knock Sensor (KS) connector
- Crankshaft Position (CMP) sensor
- Oil pressure switch
- Camshaft Position (CKP) sensor
- Fuel hoses from fuel pipes
- Intake manifold
- Throttle Position (TP) sensor
- Ground cable from the intake manifold
- Ignition coil and igniter assembly
- Intake Air Temperature (IAT) and pressure sensor
- Idle Air Control (IAC) solenoid valve
- Throttle Position (TP) sensor and air control solenoid
- Air bypass from throttle body
- Throttle body from the intake manifold and discard the gasket
- Air bypass hose from air assist solenoid
- Air assist injector solenoid valve
- Pressure regulator vacuum hose from intake manifold
- Fuel injectors
- Purge control solenoid valve
- Air bypass hose from purge control solenoid valve
- Harness bands
- Engine harness from intake manifold
- Purge control solenoid valve
- Injector pipe
- Fuel pipes from the intake manifold
- Fuel injectors and securing clips
- Air bypass hoses from the manifold

5. Loosen the right and left fuel hose to injector clamps.

6. Remove or disconnect the following:
- Left and right side fuel pipes
- Fuel injector pipe
- Fuel pipe assembly and pressure regulator

➡ **The intake manifold sits on pins that protrude from the cylinder heads. Be sure the pins remain in the cylinder heads.**

To install:

7. Install or connect the following:
- Fuel pipe assembly and pressure regulator
- Right side fuel pipes, tighten clamps
- Fuel injector pipe
- Left side fuel pipes, tighten clamps
- Air bypass hoses to the manifold
- Fuel injectors and securing clips
- Fuel pipes to the intake manifold. Torque the bolts to 2.5 ft. lbs. (3.4 Nm).
- Injector pipe. Torque the bolts to 3.6 ft. lbs. (4.9 Nm).
- Purge control solenoid valve
- Air bypass hose from purge control solenoid valve
- Engine harness to intake manifold and harness bands
- Pressure regulator vacuum hose to intake manifold
- Air assist injector solenoid valve
- Air bypass hose to air assist solenoid
- Throttle body to the intake manifold using new gaskets. Torque the retaining bolts to 16 ft. lbs. (22 Nm).
- TP sensor and air control solenoid
- Air bypass hoses to the manifold
- IAC solenoid valve to the intake manifold using a new gasket. Tighten the retaining bolts to 60 inch lbs. (7 Nm).
- IAT and pressure sensor use new O-ring. Torque the bolts to 1.4 ft. lbs. (2.0 Nm).
- Ignition coil and igniter assembly
- Ground cable from the intake manifold
- Air cleaner case stay and engine harness bracket
- Brake booster hose
- Engine coolant hoses to the throttle body.
- PCV hose and vacuum hose to intake manifold
- Spark plug wires
- Power steering pump to intake manifold. Torque the bolts to 15 ft. lbs. (20 Nm).

- Power steering hose brackets to the right side of the intake manifold
- Power steering V-belt
- V-belt covers
- Resonator chamber. Torque the bolts to 25 ft. lbs. (33 Nm).
- Accelerator cable and the cruise control cable
- Air cleaner assembly
- Negative battery cable

8. Refill the cooling system.
9. Start the engine, and bleed the cooling system.
10. Check for leaks.

Exhaust Manifold

Due to the unique design of the Subaru engine an exhaust manifold is not used. The exhaust enters directly into the front Y-pipe.

REMOVAL & INSTALLATION

❋❋ CAUTION

The exhaust pipe may be hot; DO NOT perform any work until the system has completely cooled.

1. Before servicing the vehicle, refer to the precautions in the beginning of this section.
2. Remove or disconnect the following:
- Negative battery cable
- Front Oxygen (O_2) sensor electrical connectors
- Rear Oxygen (O_2) sensor electrical connectors
- Center exhaust pipe from front exhaust pipe
- Nuts that secure the exhaust pipe to the cylinder head
- Front pipe-to-front catalytic converter mounting nuts
3. Discard the gaskets.
To install:
4. Clean all gasket surfaces completely.
5. Install or connect the following:
- Catalytic converter to front exhaust pipe using new gasket. Torque the bolts to 22 ft. lbs. (30 Nm).
- Exhaust pipe to the cylinder head using new gaskets. Torque the mounting nuts to 22 ft. lbs. (30 Nm).
- Exhaust pipe to the center pipe using new gaskets. Torque the mounting nuts to 26 ft. lbs. (35 Nm).
- Rear O_2 sensors electrical connectors

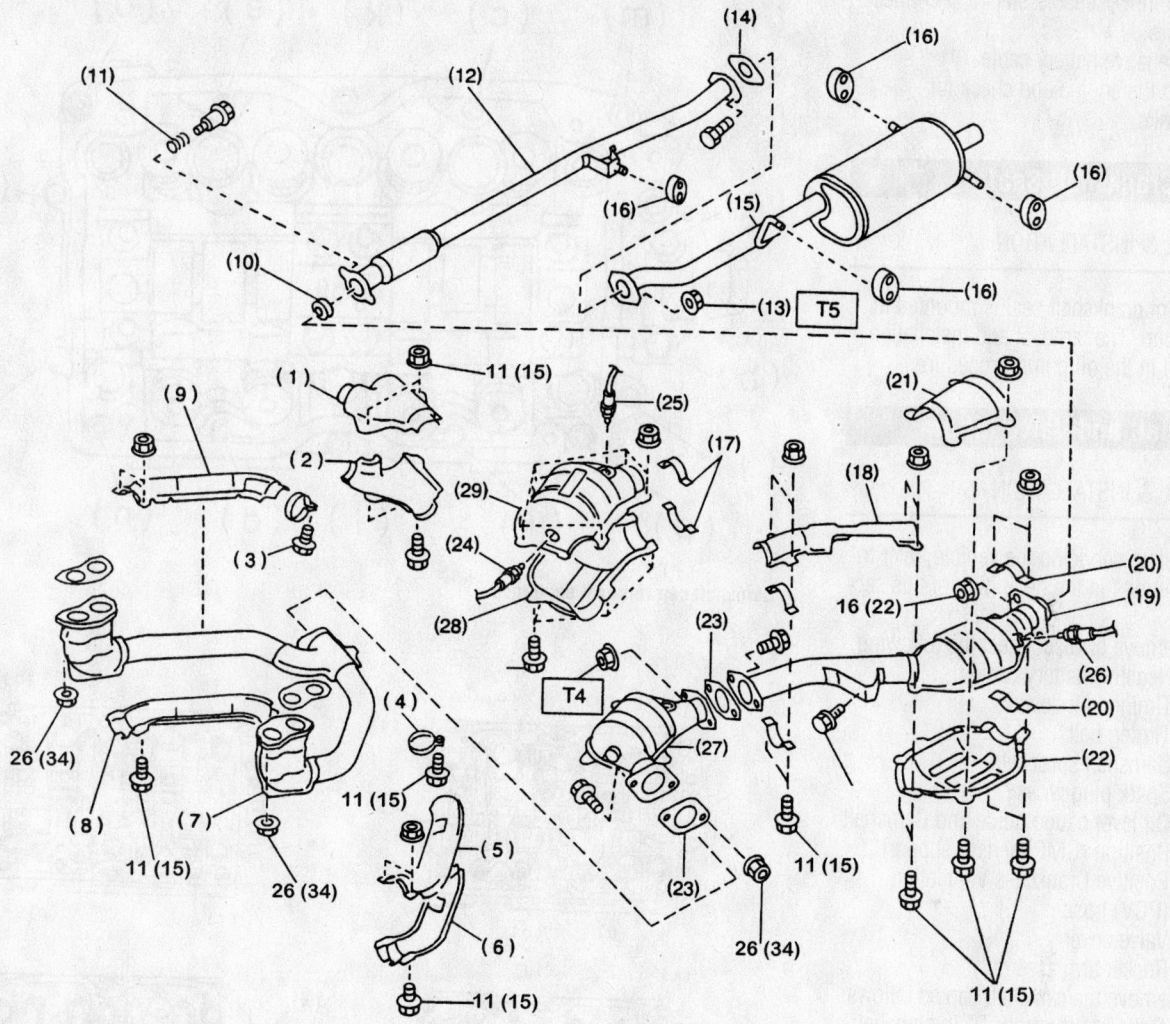

ft. lbs. (Nm)

(1) Upper front exhaust pipe cover CTR
(2) Lower front exhaust pipe cover CTR
(3) Band RH
(4) Band LH
(5) Upper front exhaust pipe cover LH
(6) Lower front exhaust pipe cover LH
(7) Front exhaust pipe
(8) Lower front exhaust pipe cover RH
(9) Upper front exhaust pipe cover RH
(10) Gasket
(11) Spring

(12) Rear exhaust pipe
(13) Self-locking nut
(14) Gasket
(15) Muffler
(16) Cushion rubber
(17) Clamp
(18) Upper center exhaust pipe cover
(19) Center exhaust pipe
(20) Clamp B
(21) Upper rear catalytic converter cover
(22) Lower rear catalytic converter cover
(23) Gasket
(24) Front oxygen sensor
(25) Rear oxygen sensor (California spec. vehicles)

(26) Rear oxygen sensor (Except California spec. vehicles)
(27) Front catalytic converter
(28) Lower front catalytic converter cover
(29) Upper front catalytic converter cover

Exploded view of the exhaust system and related components

7924XG07

Heater Core replacement is covered in Section 2 of this manual

- Front O$_2$sensors electrical connectors
- Negative battery cable

6. Start the engine and check for exhaust leaks.

Front Crankshaft Seal

REMOVAL & INSTALLATION

The front crankshaft seal is mounted in the oil pump. The removal and installation is covered in the oil pump procedure.

Camshaft and Valve Lifters

REMOVAL & INSTALLATION

1. Before servicing the vehicle, refer to the precautions in the beginning of this section.

2. Remove or disconnect the following:
- Negative battery cable
- Timing belt covers
- Timing belt
- Camshaft sprockets
- Spark plug wires
- Oil level gauge guide and Camshaft Position (CMP) sensor support
- Positive Crankcase Ventilation (PCV) hose
- Valve cover
- Rocker arm assembly

3. Remove the camshaft cap as follows:
   a. Bolts "A" through "B" in alphabetical sequence
   b. Loosen bolts "C" through "J" equally all the way in alphabetical sequence
   c. Bolts "K" through "P" in alphabetical sequence using tool 499497000

4. Remove or disconnect the following:
- Camshaft
- Oil seal, if necessary
- Plug from the rear side of the camshaft

To install:

➡ **Lubricate the camshaft bearings prior to camshaft installation.**

5. Install or connect the following:
- Camshaft
- Camshaft cap. Apply liquid gasket on the edge of the cam cap mating surface 0.12 inch (3mm) thick

6. Temporarily tighten bolts "G" through "J" in alphabetical sequence

7. Install the valve rocker assembly. Torque the bolts "A" through "H" to 18 ft. lbs. (25 Nm).

8. Torque bolts "I" through "N" to 13 ft.

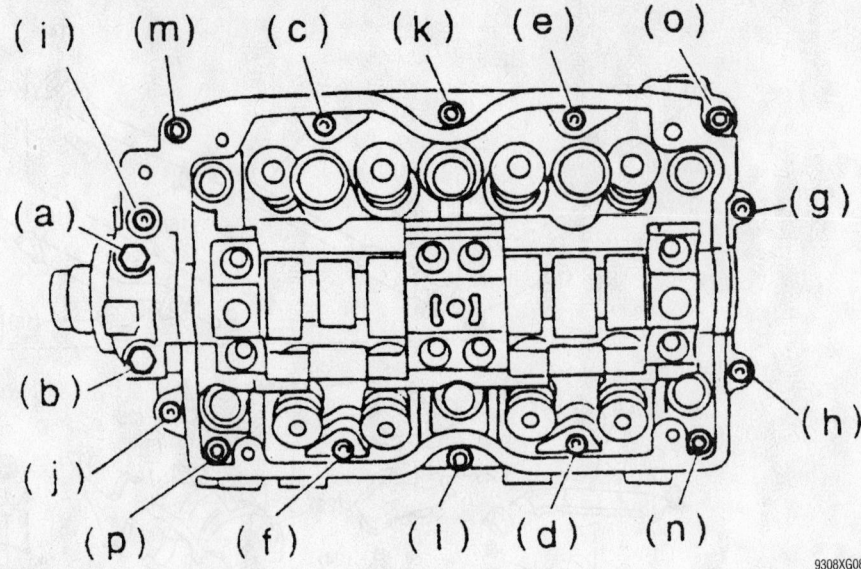

Camshaft cap removal sequence

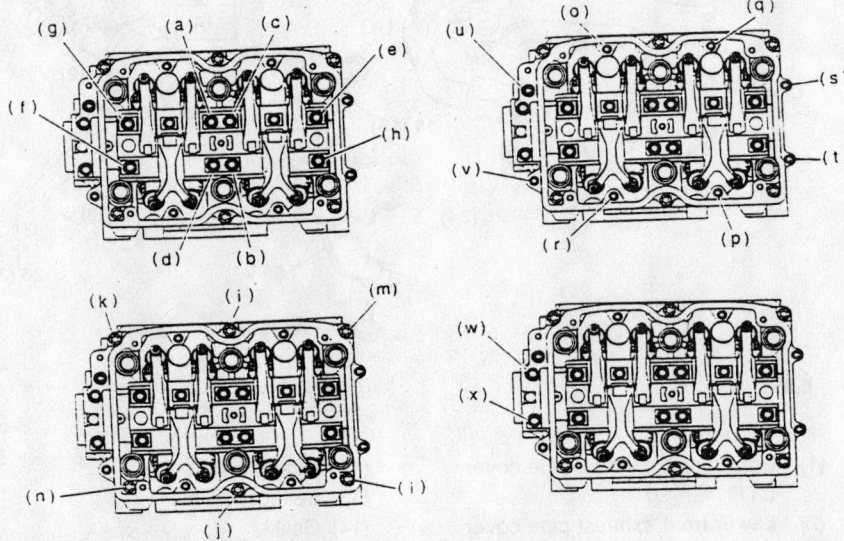

Camshaft cap tightening sequence

lbs. (25 Nm) in alphabetical sequence using tool 499497000

9. Torque bolts "O" through "X" to 7.2 ft. lbs. (10 Nm) in alphabetical sequence

10. Install or connect the following:
- Oil seal to the camshaft using tool 499597000 oil seal guide and 49958700 oil seal installer
- Plug to the rear side of the camshaft using tool 499587700 oil seal installer
- Valve cover
- Oil level gauge guide and CMP sensor support
- PCV house
- Spark plug wires

- Camshaft sprockets. Torque the bolts to 58 ft. lbs. (78 Nm).
- Timing belt
- Timing belt covers. Torque the bolts to 3.6 ft. lbs. (5 Nm).
- Negative battery cable

Valve Lash

ADJUSTMENT

➡ **The valve adjustment should be performed while the engine is cold. A Shim Replace Kit 498187100 will be needed to perform the valve adjustment.**

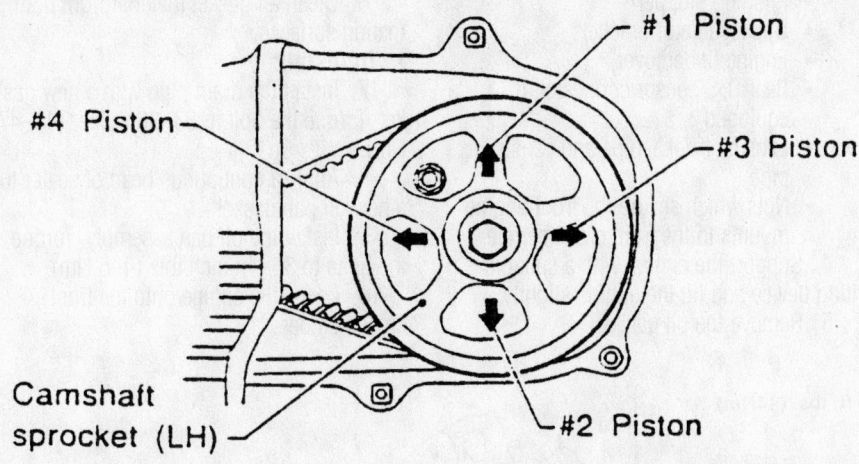

#4 Piston

#1 Piston

#3 Piston

Camshaft sprocket (LH)

#2 Piston

9308XG05

Position the camshaft for adjustment to valves

1. Before servicing the vehicle, refer to the precautions in the beginning of this section.

2. Adjustment should be performed when engine is cold.

3. Remove or disconnect the following:
   - Negative battery cable
   - Engine coolant reservoir tank
   - Timing belt cover on the left hand side

4. When inspecting Nos. 1 and 3 cylinders remove the following:
   - Air intake duct as a unit
   - Resonator chamber
   - Spark plug wires from Nos. 1 and 3 cylinders
   - Blow-by house from valve cover
   - Engine undercover
   - Timing belt cover on the right hand side
   - Valve cover on the right hand side

5. When inspecting Nos. 2 and 4 cylinders remove the following:
   - Battery and battery tray
   - Window washer motor connectors front and rear
   - Rear gate glass washer hose from the washer motor
   - Washer tank mounting bolts and secure out of the way
   - Spark plug wires from Nos. 2 and 4 cylinders
   - Blow by house from valve cover
   - Timing belt cover on the right hand side
   - Valve cover on the left hand side

6. Set No. 1 cylinder to Top Dead Center (TDC).

➡ **When arrow mark on the left hand side comes exactly to the top, No. 1 cylinder piston is brought to TDC of the compression stroke.**

7. Check the valve clearance:
   - Intake valve: 0.0071–0.0087 in. (0.18–0.22mm).
   - Exhaust valve: 0.0090–0.0106 in. (0.23–0.27mm).

8. If any valve needs adjustment, perform the following:

   a. Loosen the valve rocker nut and screw.

   b. Place a thickness gage in at as horizontal a direction as possible with respect to the vale stem and face.

   c. Adjust the screw until proper clearance is obtained.

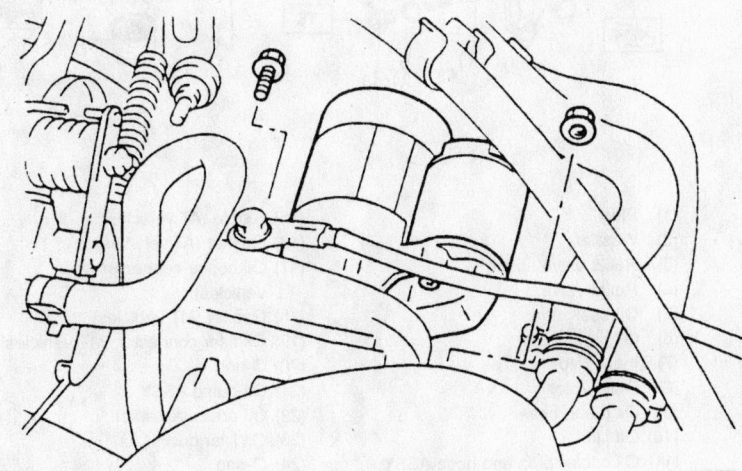

View of the starter mounting

9308XG02

d. Tighten the rocker nut after adjusted.

9. Install or connect the following:
   - Valve covers left and right
   - Timing belt covers
   - Blow-by houses to valve covers
   - Spark plug wires
   - Washer tank
   - Rear gate glass washer hose to the washer motor
   - Washer motor connectors
   - Battery and battery tray
   - Engine undercover
   - Resonator chamber
   - Air intake duct unit
   - Engine coolant reservoir tank

Starter

REMOVAL & INSTALLATION

1. Before servicing the vehicle, refer to the precautions in the beginning of this section.

2. Remove or disconnect the following:
   - Negative battery
   - Air intake duct and assembly
   - Wires
   - Starter

To install:

3. Install or connect the following:
   - Starter. Torque the mounting bolts to 37 ft. lbs. (50 Nm).
   - Wires
   - Air intake duct and assembly
   - Negative battery

Brake service is covered in Section 4 of this manual

Oil Pan

REMOVAL & INSTALLATION

1. Before servicing the vehicle, refer to the precautions in the beginning of this section.

2. Drain the oil from the engine.

3. Remove or disconnect the following:
- Air intake duct
- Oxygen (O₂S) sensor connectors
- Pitching stopper
- Upper radiator brackets
- Engine undercover
- Rear O₂S sensor connector, if equipped
- Exhaust front Y-pipe and center pipe
- Nuts which secure the front engine mounts to the front crossmember

4. Support the engine with a suitable lifting device and lift the engine slightly.

5. Remove the oil pan.

6. Clean all gasket material from both mating surfaces.

To install:

7. Install the drain plug with a new gasket. Torque the bolt to 33–36 ft. lbs. (43–47 Nm).

8. Apply a continuous bead of sealer to a new oil pan gasket.

9. Install the oil pan assembly. Torque the bolts to 36–48 inch lbs. (4–5 Nm).

10. Lower the engine onto the front crossmember.

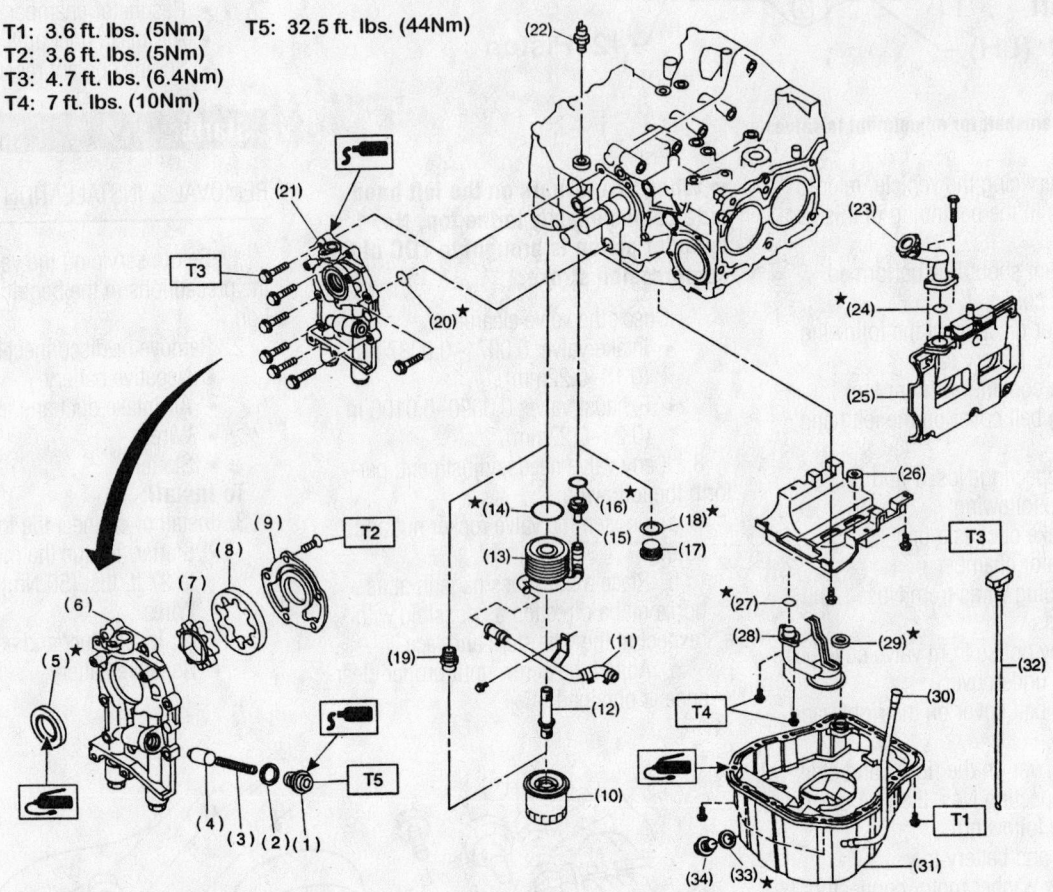

T1: 3.6 ft. lbs. (5Nm)
T2: 3.6 ft. lbs. (5Nm)
T3: 4.7 ft. lbs. (6.4Nm)
T4: 7 ft. lbs. (10Nm)
T5: 32.5 ft. lbs. (44Nm)

(1) Plug
(2) Washer
(3) Relief valve spring
(4) Relief valve
(5) Oil seal
(6) Oil pump case
(7) Inner rotor
(8) Outer rotor
(9) Oil pump cover
(10) Oil filter
(11) Oil cooler pipe and hose ASSY (AT vehicles)
(12) Connector (AT vehicles)
(13) Oil cooler (AT vehicles)
(14) O-ring (AT vehicles)
(15) Nipple (AT vehicles)
(16) Gasket (AT vehicles)
(17) Oil cooler connector (MT vehicles)
(18) Gasket (MT vehicles)
(19) Oil filter connector (MT vehicles)
(20) O-ring
(21) Oil pump ASSY
(22) Oil pressure switch
(23) Oil filler duct
(24) O-ring
(25) Cylinder head cover
(26) Baffle plate
(27) O-ring
(28) Oil strainer
(29) Gasket
(30) Oil level gauge guide
(31) Oil pan
(32) Oil level gauge
(33) Metal gasket
(34) Drain plug

Oil pan and lubrication components

7924XG28

11. Install or connect the following:
- Front engine mount nuts. Torque the nuts to 69 ft. lbs. (51 Nm).
- Exhaust front Y-pipe with new gaskets. Torque the nuts that secure the pipe to the engine to 23 ft. lbs. (30 Nm).
- O_2S sensor connectors
- Pitching stopper. Torque the front bolt to 36 ft. lbs. (49 Nm) and the rear bolt to 42 ft. lbs. (57 Nm).
- Upper radiator brackets
- Air intake duct
- Engine undercover

12. Refill the engine to the proper level with the recommended oil and run the engine.

13. Check for leaks.

Oil Pump

REMOVAL & INSTALLATION

1. Before servicing the vehicle, refer to the precautions in the beginning of this section.
2. Drain the cooling system.
3. Drain the engine oil into a separate container.
4. Remove or disconnect the following:
- Negative battery cable
- Engine undercover
- Water pipe and hose between oil cooler and water pump
- Radiator
- Crankcase Position (CKP) Sensor
- Camshaft Position (CMP) Sensor
- Belt(s) and rear tensioner
- Crankshaft pulley
- Water pump
- Timing belt, and belt guide, if equipped
- Oil pump mounting bolts and carefully pry the pump from the engine block

✳✳ WARNING

Use extreme care not to damage the engine block or the oil pump during removal of the pump.

To install:
5. Apply a continuous bead sealant to the mating surfaces of the oil pump.
6. Install or connect the following:
- New front seal to the oil pump coat the inside of the seal with engine oil
- New O-ring to the oil pump

- Oil pump. Torque the bolts to 56 inch lbs. (6.4 Nm).
- Timing belt, and belt guide, if equipped
- Water pump
- Crankshaft pulley
- Belt(s) and rear tensioner
- CMP Sensor
- CKP Sensor
- Radiator
- Engine coolant pipe
- Engine undercover
- Negative battery cable

7. Refill the cooling system.
8. Refill the engine to the proper level with the recommended oil.

Rear Main Seal

REMOVAL & INSTALLATION

1. Before servicing the vehicle, refer to the precautions in the beginning of this section.
2. Remove or disconnect the following:
- Engine from the vehicle
- Clutch assembly/flywheel, if equipped with manual transmission
- Torque converter flexplate from the crankshaft, if equipped with an automatic transmission

3. Using a seal removal tool, pry the oil seal from the housing.

To install:
4. Utilizing the appropriate seal installer
5. Install or connect the following:
- New oil seal and press it into the housing using the appropriate driver

- Clutch assembly/flywheel, if equipped
- Flywheel/flexplate and tighten the bolts to 53 ft. lbs. (72 Nm), if equipped
- Engine into the vehicle

Piston and Ring

POSITIONING

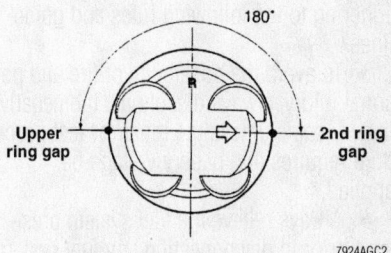

Compression ring end-gap spacing—2.5L engine

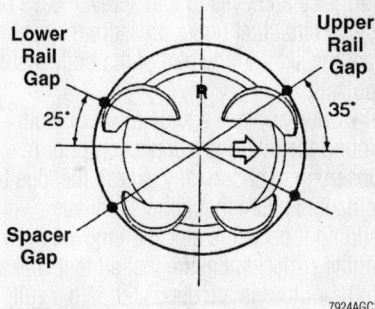

Upper, spacer and lower oil ring end-gap spacing—2.5L engine

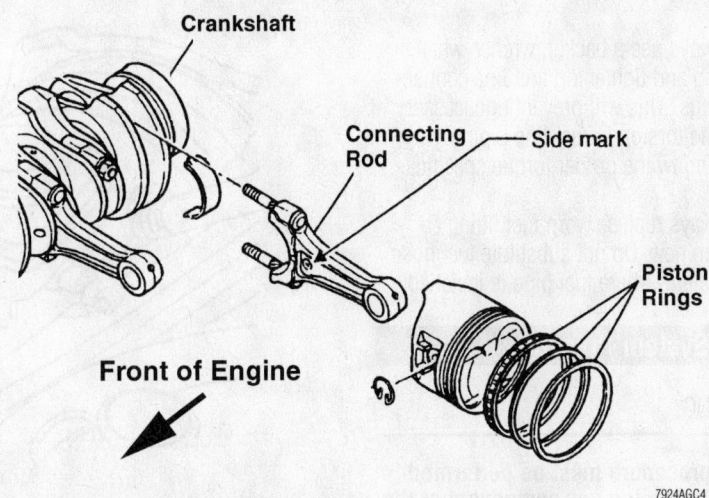

Piston and connecting rod assembly positioning—2.5L engine

For complete Engine Mechanical specifications, see Section 1 of this manual

FUEL SYSTEM

Fuel System Service Precautions

Safety is the most important factor when performing not only fuel system maintenance but any type of maintenance. Failure to conduct maintenance and repairs in a safe manner may result in serious personal injury or death. Maintenance and testing of the vehicle's fuel system components can be accomplished safely and effectively by adhering to the following rules and guidelines.

• To avoid the possibility of fire and personal injury, always disconnect the negative battery cable unless the repair or test procedure requires that battery voltage be applied.

• Always relieve the fuel system pressure prior to disconnecting any fuel system component (injector, fuel rail, pressure regulator, etc.), fitting or fuel line connection. Exercise extreme caution whenever relieving fuel system pressure, to avoid exposing skin, face and eyes to fuel spray. Please be advised that fuel under pressure may penetrate the skin or any part of the body that it contacts.

• Always place a shop towel or cloth around the fitting or connection prior to loosening to absorb any excess fuel due to spillage. Ensure that all fuel spillage (should it occur) is quickly removed from engine surfaces. Ensure that all fuel soaked cloths or towels are deposited into a suitable waste container.

• Always keep a dry chemical (Class B) fire extinguisher near the work area.

• Do not allow fuel spray or fuel vapors to come into contact with a spark or open flame.

• Always use a backup wrench when loosening and tightening fuel line connection fittings. This will prevent unnecessary stress and torsion to fuel line piping. Always follow the proper torque specifications.

• Always replace worn fuel fitting O-rings with new. Do not substitute fuel hose or equivalent, where fuel pipe is installed.

Fuel System Pressure

RELIEVING

➡ **This procedure must be performed prior to servicing any component of the fuel injection system.**

1. Before servicing the vehicle, refer to the precautions in the beginning of this section.
2. Disconnect the fuel pump harness at the fuel pump, under the rear seat access panel.
3. Crank the engine for 5 seconds or more to relieve the fuel pressure. If the engine starts during this time, allow it to run until it stalls.
4. Connect the fuel pump harness after repairs are completed.

Fuel Filter

REMOVAL & INSTALLATION

1. Before servicing the vehicle, refer to the precautions in the beginning of this section.
2. Locate the fuel filter in the engine compartment on the left inside fender.
3. Properly relieve the fuel system pressure.
4. Remove or disconnect the following:
   • Negative battery cable
   • Fuel delivery hoses from the fuel filter
   • Fuel filter from its holder

To install:
5. Install or connect the following:
   • Fuel filter into its mounting bracket
   • Fuel delivery hoses and tighten the hose clamps
   • Negative battery cable

Fuel Pump

REMOVAL & INSTALLATION

1. Before servicing the vehicle, refer to the precautions in the beginning of this section.
2. Remove the rear seat cushion and access panel.
3. Properly relieve the fuel system pressure.
4. Disconnect the negative battery cable.
5. Clean any debris away from the fuel pump mounting to prevent it from entering the tank.
   • Fuel pump electrical connector
   • Fuel delivery and return hoses
   • Fuel pump mounting nuts
   • Fuel pump out of the fuel tank

To install:
6. Replace the sealing gaskets for the fuel pump.
7. Install or connect the following:
   • Fuel pump into the tank. Torque the mounting nuts in sequence to 39 inch lbs. (4.4 Nm).
   • Fuel delivery and return hoses
   • Fuel pump electrical connector
   • Fuel filler cap
   • Negative battery cable
8. Start the vehicle and check for leaks.
9. Install the fuel pump access cover and rear seat cushion.

View of fuel filter mounting

7924XG11

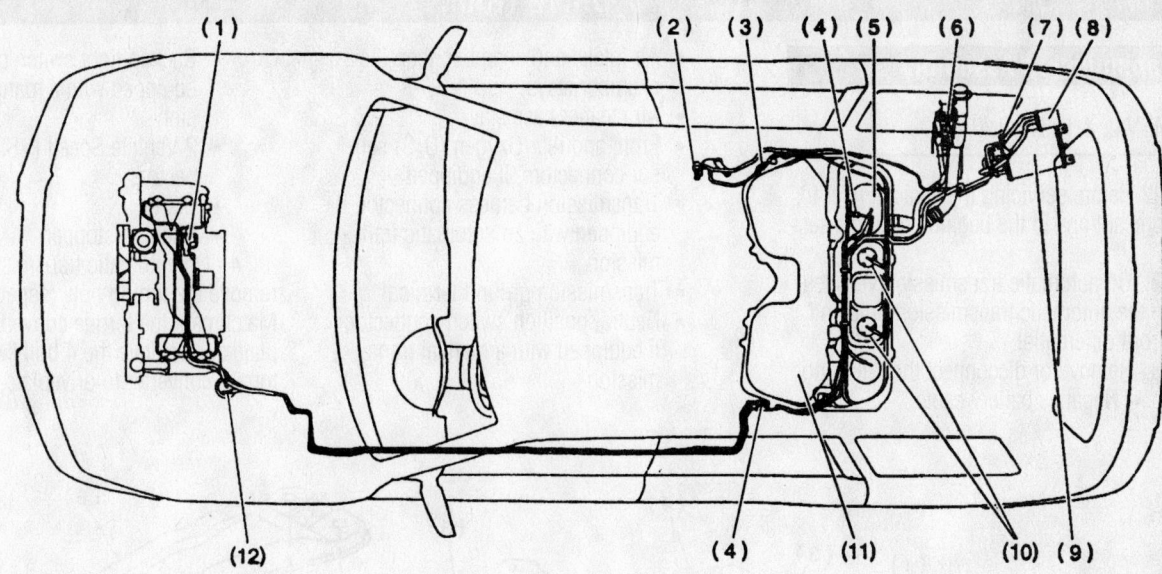

(1) Purge control solenoid valve
(2) Roll over valve
(3) Pressure control solenoid
(4) Quick connector

(5) Fuel pump
(6) Fuel tank pressure sensor
(7) Vent control solenoid valve
(8) Air filter

(9) Canister
(10) Fuel cut valve
(11) Fuel tank
(12) Fuel filter

7924XG12

Fuel system component locations

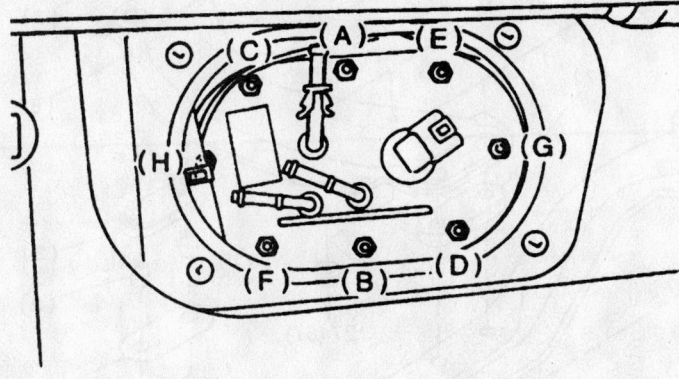

7924XG13

Fuel pump mounting nut tightening sequence

Fuel Injector

REMOVAL & INSTALLATION

1. Before servicing the vehicle, refer to the precautions in the beginning of this section.

2. Properly relieve the fuel system pressure.

3. Remove or disconnect the following:

- Negative battery cable
- Air duct and cleaner assembly
- Resonator chamber
- Spark plug wires No. 1 and 3
- V-belt covers
- Power steering pump belt
- Power steering pipe bracket from manifold
- Power steering pump and seat aside
- Wires from fuel injector

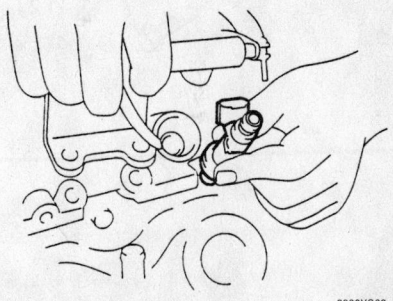

9308XG03

View of fuel injector removal

- Injector pipe from intake manifold
- Fuel injector

To install:

4. Install or connect the following:

- Fuel injector
- Injector pipe to intake manifold
- Wires to fuel injector
- Power steering pump
- Power steering pipe bracket to manifold
- Power steering pump belt
- V-belt covers
- Spark plug wires No. 1 and 3
- Resonator camber
- Air duct and cleaner assembly
- Negative battery cable

For Accessory Drive Belt illustrations, see Section 1 of this manual

DRIVE TRAIN

Transmission Assembly

REMOVAL & INSTALLATION

1. Before servicing the vehicle, refer to the precautions in the beginning of this section.

2. On automatic transmission vehicles, drain the automatic transmission fluid and the front differential.

3. Remove or disconnect the following:
- Negative battery cable
- Air intake and chamber, then the chamber stays
- Air cleaner case stay
- Front and rear Oxygen (O2S) sensor connectors, if equipped
- Transmission harness connector, if equipped with an automatic transmission
- Transmission ground terminal
- Neutral position switch connector, if equipped with a manual transmission
- Backup light switch connector, if equipped with a manual transmission
- 2 Vehicle Speed (VSS) sensor connectors
- Starter
- Pitching stopper

4. On automatic transmission vehicles, remove the timing hole inspection plug. Matchmark the torque converter-to-driveplate and remove the 4 bolts which hold torque converter to driveplate.

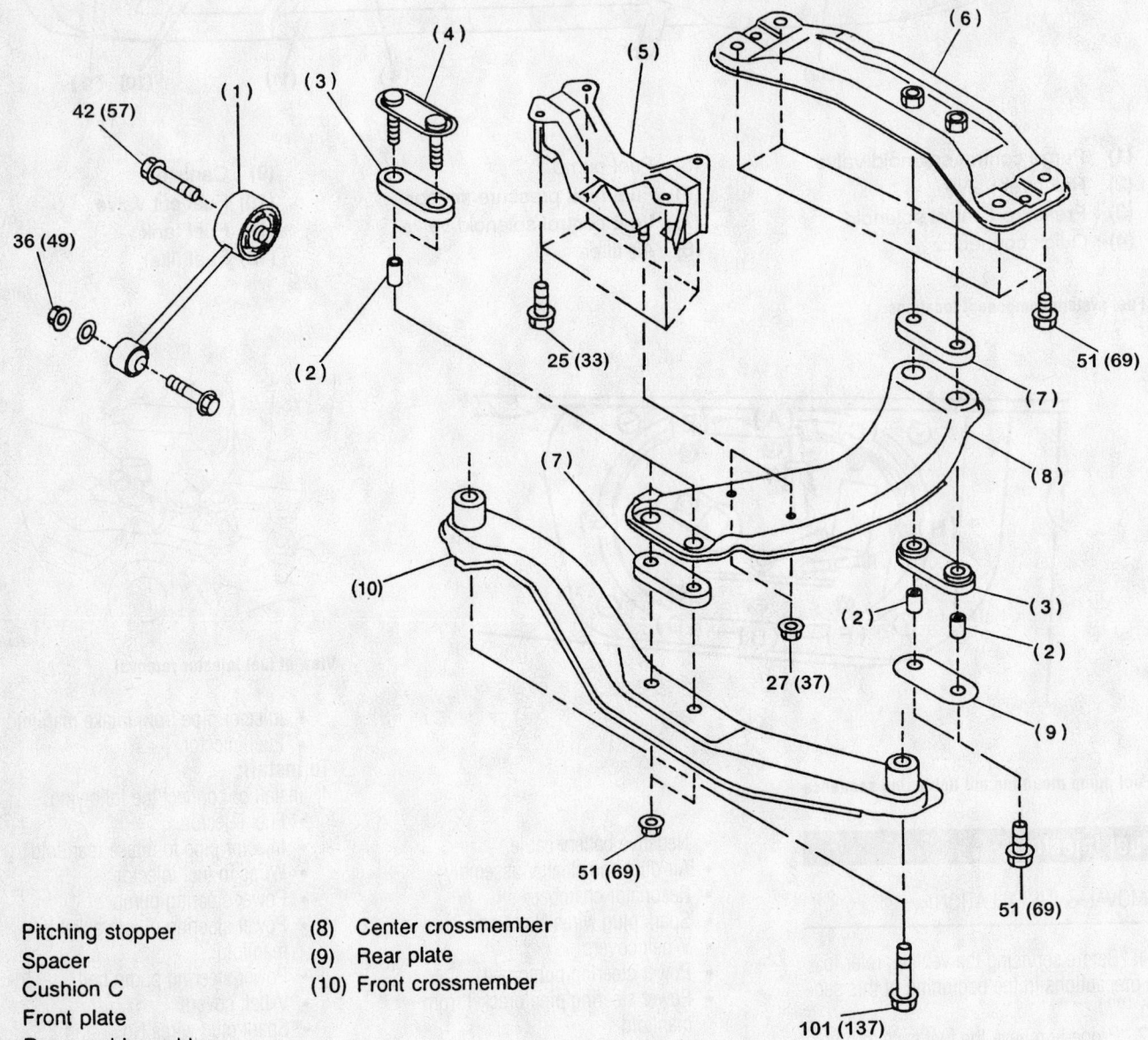

Pitching stopper
Spacer
Cushion C
Front plate
Rear cushion rubber
Rear crossmember
Cushion D
(8) Center crossmember
(9) Rear plate
(10) Front crossmember

7924XG14

Exploded view of the manual transmission mounting

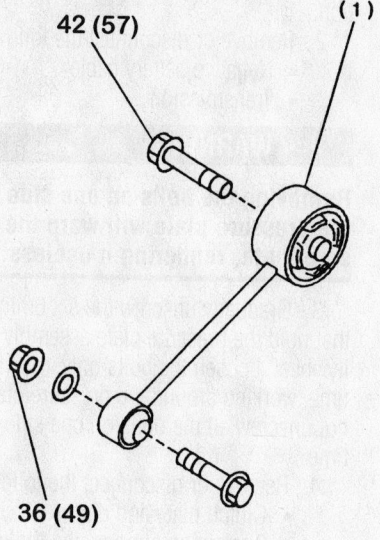
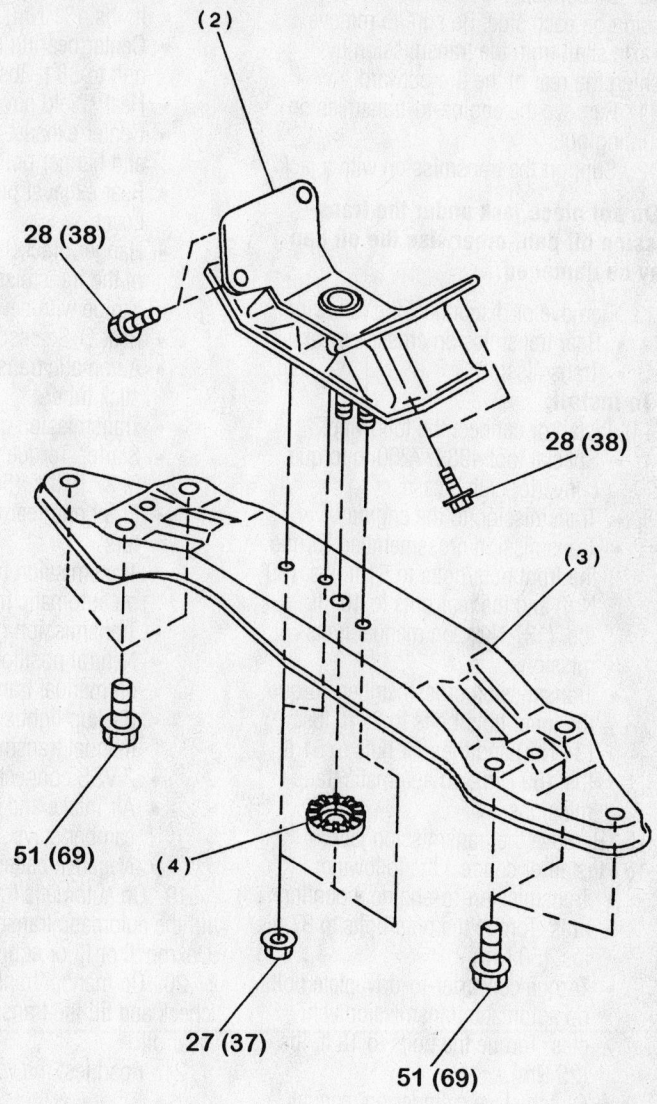

42 (57)
(1)
28 (38)
(2)
28 (38)
36 (49)
(3)
51 (69)
(4)
27 (37)
51 (69)

7924XG15

(1) Pitching stopper
(2) Rear cushion rubber
(3) Crossmember
(4) Stopper

Exploded view of the automatic transmission mounting

5. Remove or disconnect the following:
 • Automatic transmission fluid dip-stick and tube
 • Clutch slave cylinder, on manual transmission vehicles
6. Install engine support assembly ST 41099AA020. (Also available as part no. 927670000).
7. Remove or disconnect the following:
 • Bolt securing the right upper side of the transmission to the engine
 • Engine undercover
 • Front Y-pipe
8. Disconnect connector from rear O2S sensor
9. Remove or disconnect the following:

 • Center exhaust pipe from rear pipe and hanger bolt
 • Rear exhaust pipe and heat shield cover
 • Hanger bracket from the right side of the transmission
 • Transmission cooler lines
 • Rear driveshaft, matchmark for reassembly
 • Center bearing bracket

➡**Plug the opening at the rear of the extension housing to prevent oil from flowing out.**

 • Shifter stay and rod from the transmission, on manual transmission

 • Gear shift cable from the transmission select lever, on automatic transmission
 • Sway bar from the transverse link
 • Parking brake cable bracket from the transverse link and bolt holding the transverse link to the cross-member on each side, lower the transverse link
 • Lower ball joint from knuckle
 • Spring pin and separate the half-shaft from the transmission on each side

➡**Use a small punch to remove the spring pin. Discard old spring pin and always install a new pin.**

For Tire, Wheel and Ball Joint specifications, see Section 1 of this manual

10. Disconnect the halfshaft from transmission on each side. Be sure to remove the axle shaft from the transmission by pushing the rear of the tire outward.

11. Remove the engine-to-transmission mounting nuts

12. Support the transmission with a jack.

→Do not place jack under the transmission oil pan, otherwise the oil pan may be damaged.

13. Remove or disconnect the following:
- Rear transmission crossmember
- Transmission

To install:

14. Install or connect the following:
- Special tool 498277200 to torque converter clutch case
- Transmission to the engine
- Transmission crossmember. Torque the front nuts/bolts to 51 ft. lbs. (69 Nm) and the rear nuts to 101 ft. lbs. (137 Nm), on manual transmissions
- Transmission crossmember. Torque the inner nuts/bolts to 25 ft. lbs. (34 Nm) and the rear nuts to 51 ft. lbs. (69 Nm), on automatic transmissions

15. Remove the transmission jack.

16. Install or connect the following:
- Transmission-to-engine mounting nuts. Torque the nuts/bolts to 37 ft. lbs. (50 Nm).
- Torque converter-to-driveplate bolts on automatic transmission vehicles. Torque the bolts to 18 ft. lbs. (25 Nm).
- Clutch slave cylinder on manual transmission vehicles. Torque the mounting bolts to 27 ft. lbs. (37 Nm).

17. Remove special tool 927670000.

18. Install or connect the following:
- Pitching stopper
- Halfshaft to transmission and spring pin into place

→Always use new spring pin. Be sure to align the axle shaft and shaft from the transmission at chamfered holes and install shaft splines correctly.

- Lower ball joint to knuckle
- Sway bar to the crossmember. Torque the clamp bolts to 18 ft. lbs. (25 Nm).
- Shift control rod, shifter stay to the transmission and the spring, on manual transmission vehicles
- Gear shift cable to the select lever, on automatic transmission vehicles
- Fluid cooler lines

- Driveshaft. Tighten the bolts to 23 ft. lbs. (31 Nm).
- Center bearing bracket. Torque the bolt to 38 ft. lbs. (52 Nm).
- Heat shield cover, if removed
- Center exhaust pipe from rear pipe and hanger bolt
- Rear exhaust pipe and heat shield cover
- Hanger bracket from the right side of the transmission
- Y-pipe with new gaskets and nuts
- Rear O$_2$S sensor connector
- Automatic transmission fluid dipstick tube
- Transmission connector bracket
- Starter. Torque the mounting bolts to 37 ft. lbs. (50 Nm).
- Front and rear O$_2$S sensor connectors
- Transmission harness connector, on automatic transmission vehicles
- Transmission ground terminal
- Neutral position switch connector, on manual transmission vehicles
- Backup light switch connector, on manual transmission vehicles
- 2 VSS connectors
- Air intake and chamber, and the camber stays
- Negative battery cable

19. On automatic transmission vehicles, fill the automatic transmission fluid with Dexron®II or III or equivalent.

20. On manual transmission vehicles, check and fill the transmission with 75W-90 gear oil.

21. Road test the vehicle.

Clutch

ADJUSTMENT

This vehicle is equipped with a hydraulic clutch that is self-adjusting, therefore no adjustment is possible or necessary.

REMOVAL & INSTALLATION

✳✳ CAUTION

The clutch driven disc may contain asbestos, which has been determined to be a cancer-causing agent. Never clean clutch surfaces with compressed air. Avoid inhaling any dust from any clutch surface. When cleaning clutch surfaces, use a commercially available brake cleaning fluid.

1. Before servicing the vehicle, refer to the precautions in the beginning of this section.

2. Remove or disconnect the following:
- Negative battery cable
- Transmission

✳✳ WARNING

Removing the bolts on one side of the pressure plate will warp the pressure plate, rendering it useless.

3. Gradually unscrew the six 6mm bolts that hold the pressure plate assembly on the flywheel. Loosen the bolts only 1 turn at a time, working around the pressure plate. Do not unscrew all the bolts on one side at one time.

4. Remove or disconnect the following:
- Clutch plate and disc
- 2 retaining springs, the throwout bearing and the release fork

→Do not disassemble either the clutch cover or disc. Inspect the parts for wear or damage and replace any parts as necessary. Replace the clutch disc if there is any oil or grease on the facing. Do not wash or attempt to lubricate the throwout bearing, because it is sealed and permanently lubricated. If it requires replacement, the bearing may be removed and a new one installed in the holder by means of a press.

To install:

5. Fit the release fork boot on the front of the transmission housing.

6. Install or connect the following:
- Release fork
- Throwout bearing assembly and secure it with the 2 springs

→Coat the inside diameter of the throwout bearing and the release lever contact points with grease.

- Clutch alignment tool through the clutch cover and disc, then insert the end of the tool into the needle bearing

7. Tighten the pressure plate bolts following the illustrated sequence, 1 turn at a time, until the proper torque is reached. Tighten to 12 ft. lbs. (16 Nm).

✳✳ WARNING

When installing the clutch pressure plate assembly, be sure that the O marks on the flywheel and the clutch pressure plate assembly are at least 120 degrees apart. These marks indi-cate the direction of residual unbal-

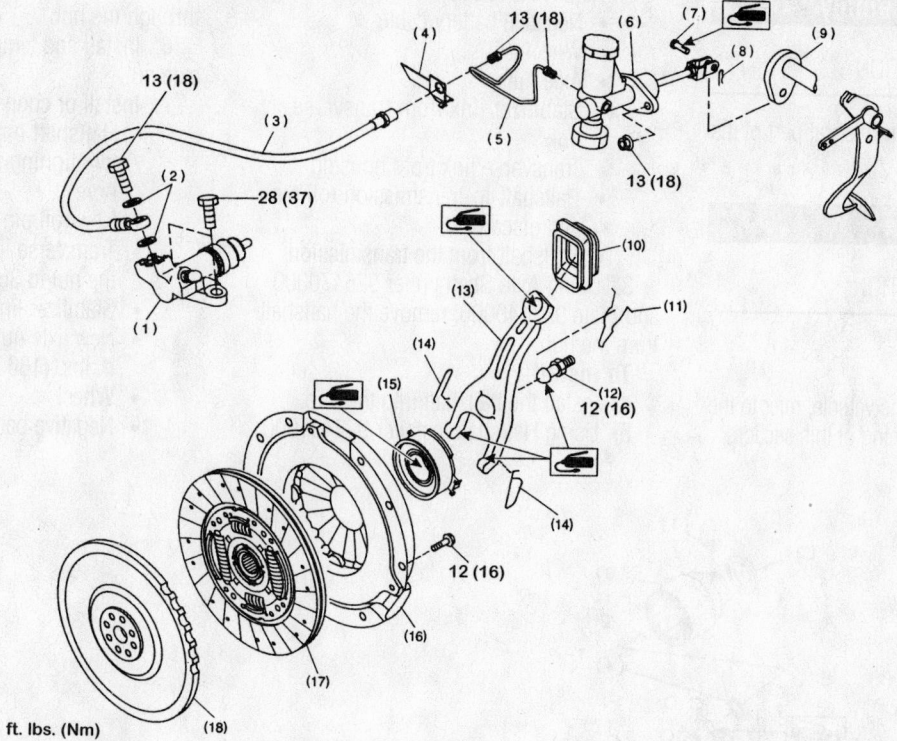

ft. lbs. (Nm)

(1) Operating cylinder	(9) Lever	(17) Clutch disc
(2) Washer	(10) Clutch release lever sealing	(18) Flywheel
(3) Clutch hose	(11) Retainer spring	
(4) Bracket	(12) Pivot	
(5) Pipe	(13) Release lever	
(6) Master cylinder ASSY	(14) Clip	
(7) Clevis pin	(15) Release bearing	
(8) Snap pin	(16) Clutch cover	

7924XG17

Exploded view of clutch system

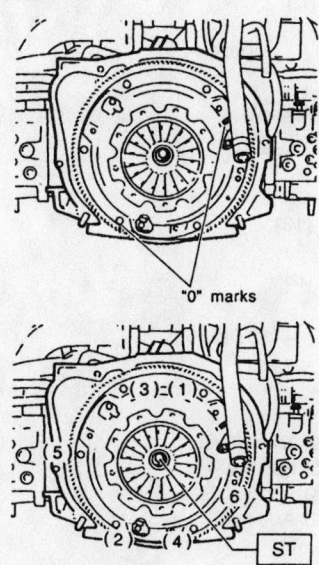

"O" marks

ST

7924XG16

Clutch cover alignment and tightening sequence

ance. Also, be sure that the clutch disc is installed properly, noting the FRONT and REAR markings.

8. Install the transmission.

Hydraulic Clutch System

BLEEDING

→To properly bleed the system, it must be bled at the slave cylinder and at the damper. Each of these has an air bleeder on it.

1. Before servicing the vehicle, refer to the precautions in the beginning of this section.
2. Connect a vinyl tube to the air bleeder on the damper and put the other end in a jar with clean clutch fluid.

→Do not let the fluid level fall too low in the master cylinder. Do not release the pedal with the bleeder open.

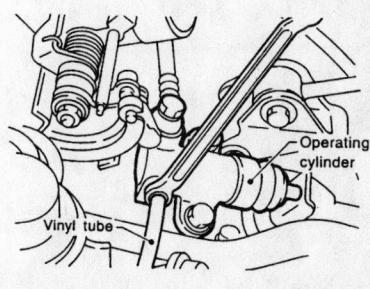

Operating cylinder

Vinyl tube

7924XG18

Bleeding the hydraulic clutch at the slave cylinder

3. With the help of an assistant depressing the clutch pedal, slowly open the bleeder valve. Close the bleeder valve and release the pedal. Repeat this process until no air bubbles appear in the jar.
4. Move the tube to the bleeder on the slave cylinder and repeat the process. Check the operation of the clutch after the bleed procedure is complete.

For Wheel Alignment specifications, see Section 1 of this manual

Transfer Case Assembly

REMOVAL & INSTALLATION

The transfer case is an integral part of the transmission.

Halfshafts

REMOVAL & INSTALLATION

Front

1. Before servicing the vehicle, refer to the precautions in the beginning of this section.

2. Remove or disconnect the following:
- Negative battery cable
- Wheel
- Axle nut
- Stabilizer link from transverse link
- Transverse link from housing
- Halfshaft-to-transmission roll pin and discard it
- Halfshaft from the transmission

3. Using Axle Shaft puller 926470000 and Plate 927140000, remove the halfshaft from the hub.

To install:

4. Install the halfshaft into the hub.
5. Using Halfshaft installer 922431000 and adapter 927390000, pull the halfshaft through the hub.

6. Install and temporarily tighten a new axle nut.

7. Install or connect the following:
- Halfshaft onto the transmission, by aligning the halfshaft roll pin hole
- New roll pin
- Transverse link to housing. Torque the nut to 36 ft. lbs. (49 Nm).
- Stabilizer link
- New axle nut. Torque the nut to 197 ft. lbs. (186 Nm) and stake the nut.
- Wheel
- Negative battery cable

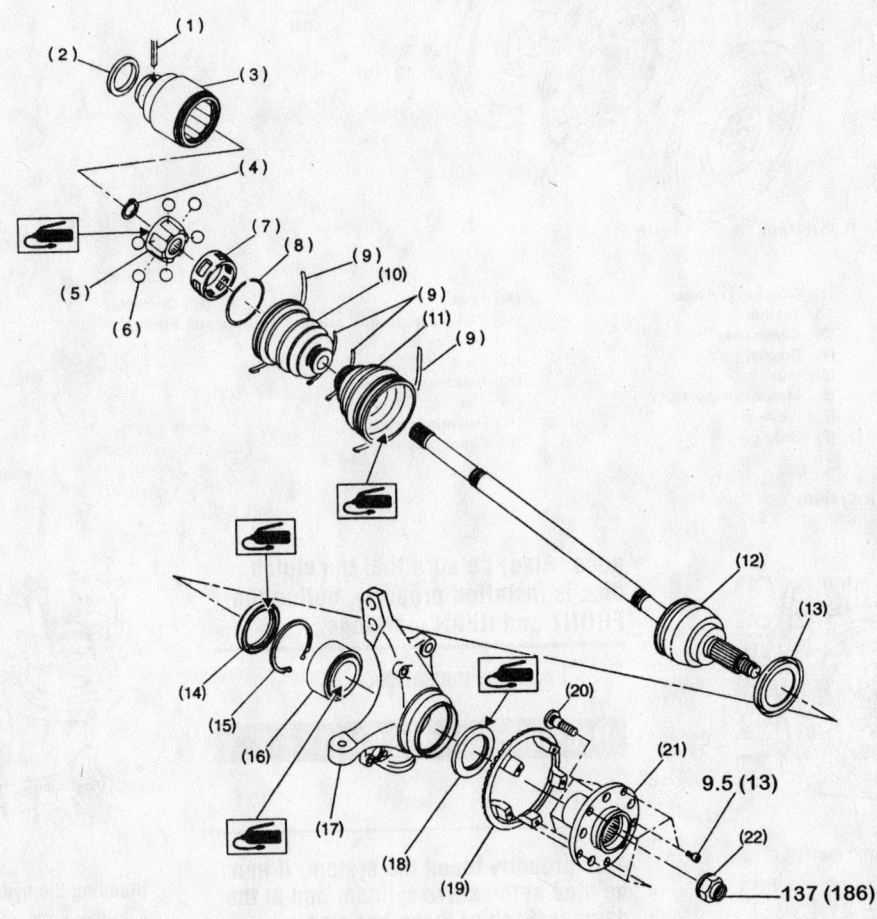

ft. lbs. (Nm)

(1) Spring pin	(10) Boot (DOJ)	(19) Tone wheel
(2) Baffle plate (DOJ)	(11) Boot (BJ)	(20) Hub bolt
(3) Outer race (DOJ)	(12) BJ ASSY	(21) Hub
(4) Snap ring	(13) Baffle plate	(22) Axle nut
(5) Inner race (DOJ)	(14) Oil seal (IN)	
(6) Ball	(15) Snap ring	
(7) Cage	(16) Bearing	
(8) Circlip	(17) Housing	
(9) Boot band	(18) Oil seal (OUT)	

7924XG19

Exploded view of the front halfshaft—manual transmission

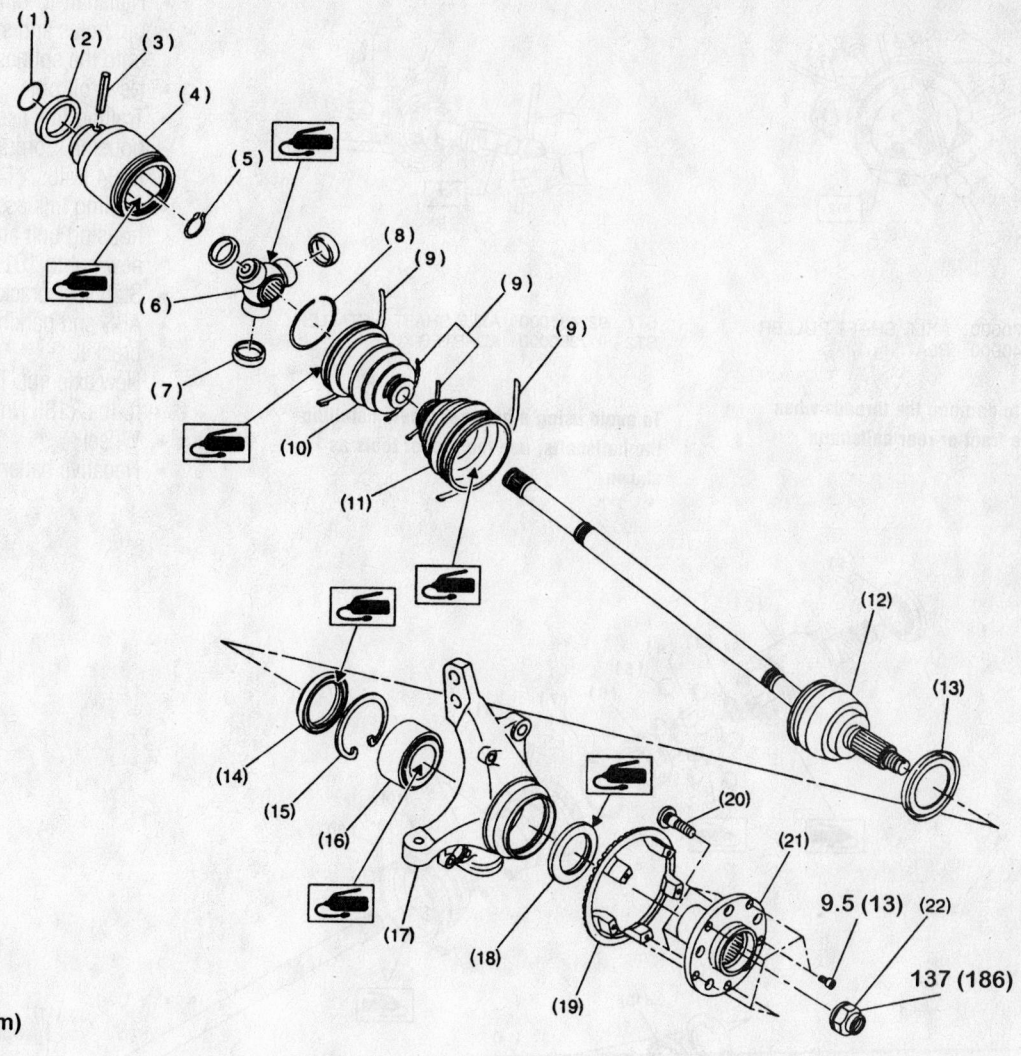

ft. lbs. (Nm)

(1) O-ring
(2) Baffle plate (FTJ)
(3) Spring pin
(4) Outer race (FTJ)
(5) Snap ring
(6) Trunnion
(7) Free ring
(8) Circlip
(9) Boot band

(10) Boot band
(11) Boot (BJ)
(12) BJ ASSY
(13) Baffle plate
(14) Oil seal (IN)
(15) Snap ring
(16) Bearing
(17) Housing
(18) Oil seal (OUT)

(19) Tone wheel
(20) Hub bolt
(21) Hub
(22) Axle nut

7924XG20

Exploded view of the front halfshaft—automatic transmission

Rear

1. Before servicing the vehicle, refer to the precautions in the beginning of this section.
2. Remove or disconnect the following:
   - Negative battery cable
   - Axle nut
   - Anti-lock Brakes (ABS) and parking brake cable bracket
   - Lateral link assembly to rear housing
   - Trailing link assembly from the rear housing bolt and nut
   - Halfshaft-to-differential roll pin
   - Halfshaft from the differential
3. Using Axle Shaft puller 926470000 and Plate 927140000, remove the halfshaft from the hub.

To install:

4. Install the halfshaft into the rear housing.
5. Using Halfshaft Installer 922431000 and adapter 927390000, pull the halfshaft into place.
6. Install and temporarily tighten a new axle nut.
7. Install or connect the following:

For Maintenance Interval recommendations, see Section 1 of this manual

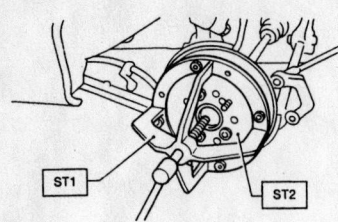

ST1 926470000 AXLE SHAFT PULLER
ST2 927140000 PLATE

7924XG29

Be sure not to damage the threads when removing the front or rear halfshafts

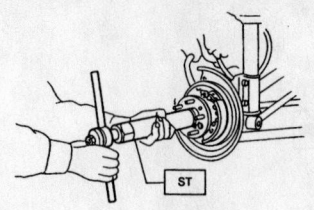

ST1 922431000 AXLE SHAFT INSTALLER
ST2 927390000 ADAPTER

7924XG30

To avoid using a hammer when installing the halfshafts, use the proper tools as shown

- Halfshaft-to-differential align roll pin holes and slide the halfshaft onto the splines
- New roll pin
- Trailing link assembly to the rear housing. Torque bolt and new nut to 84 ft. lbs. (113 Nm).
- Trailing link assembly-to-rear housing bolt and nut. Torque the new nut to 101 ft. lbs. (137 Nm).
- Stabilizer bracket
- ABS and parking brake cable bracket
- New axle nut. Torque the nut to 137 ft. lbs. (186 Nm).
- Wheel
- Negative battery cable

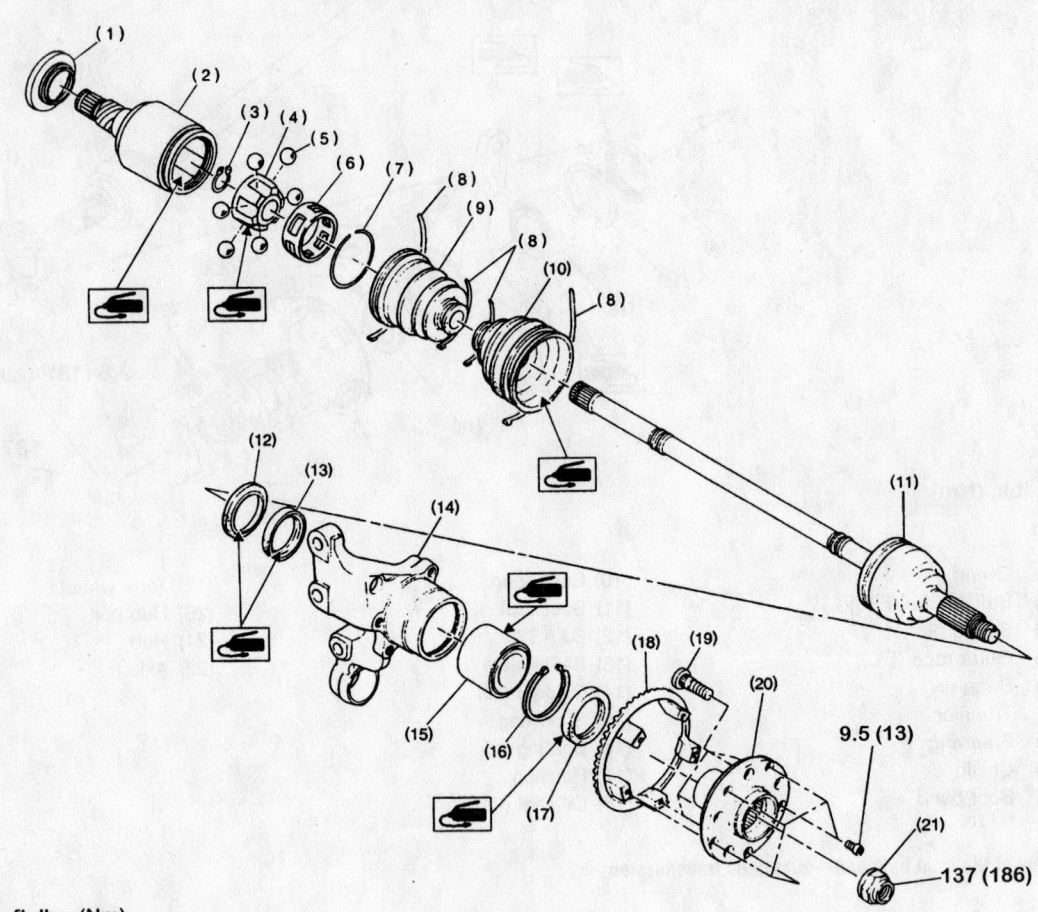

ft. lbs. (Nm)

(1) Baffle plate (DOJ)
(2) Outer race (DOJ)
(3) Snap ring
(4) Inner race
(5) Ball
(6) Cage
(7) Circlip
(8) Boot band
(9) Boot (DOJ)
(10) Boot (BJ)
(11) BJ ASSY
(12) Oil seal (IN. No. 2)
(13) Oil seal (IN. No. 3)
(14) Housing
(15) Bearing
(16) Snap ring
(17) Oil seal (OUT)
(18) Tone wheel
(19) Hub bolt
(20) Hub
(21) Axle nut

Exploded view of the rear halfshaft

7924XG21

CV-Joints

REMOVAL & INSTALLATION

Front

INNER

1. Before servicing the vehicle, refer to the precautions in the beginning of this section.
2. Remove or disconnect the following:
   - Front wheel(s)
   - Halfshaft and place in vise
   - Boot bands and slide boot down
   - Circlip from CV-joint outer race
   - Outer race from shaft, wipe off all grease
3. Matchmark Tri-pot spider assembly for reassembly.
   - Snapring and Tri-pot
   - CV-joint boot

To install:

4. Install or connect the following:
   - CV-joint boot and fill with 1.06–1.41 oz. of grease
   - Tri-pot spider assembly and snapring

- Outer race, fill with 3.53–3.88 oz. of grease
- Circlip to CV-joint outer race
5. Slide boot onto outer race
6. New band on the boot

OUTER

The outer boot is the only thing that is replaceable the outer CV-joint is not disassembled.

Rear

1. Before servicing the vehicle, refer to the precautions in the beginning of this section.
2. Remove or disconnect the following:
   - Front wheel(s)
   - Half shaft and place in vise
   - Boot bands and slide boot down
   - Circlip from CV- joint outer race
   - Outer race from shaft, wipe off all grease
   - 6 balls
   - Snapring and inner race
   - Cage from the shaft
   - CV boot

To install:

3. Install or connect the following:
   - CV-joint boot and fill with 0.76–1.06 oz. of grease
   - Cage to the shaft with the cut out portion facing the shaft end
   - Inner race and snapring
4. Install the cage to the inner race.

➡ **Fit the cage with the protruded part aligned with the track on the inner race then turn by a half pitch.**

5. Install the outer race and snapring; then, fill with 2.82–3.17 oz. of grease
6. Coat the cage pockets with grease.
7. Install or connect the following:
   - 6 balls into cage, align the inner race and cage
   - Outer race and circlip
8. Slide boot on to outer race
9. New band on the boot

OUTER

The outer boot is the only thing that is replaceable the outer CV-joint is not disassembled.

STEERING AND SUSPENSION

Air Bag

✳✳ CAUTION

All vehicles are equipped with an air bag system. The system must be disabled before performing service on or around system components, steering column, instrument panel components, wiring and sensors. Failure to follow safety and disabling procedures could result in accidental air bag deployment, possible personal injury and unnecessary system repairs.

PRECAUTIONS

Several precautions must be observed when handling the inflator module to avoid accidental deployment and possible personal injury.

- Never carry the inflator module by the wires or connector on the underside of the module.
- When carrying a live inflator module, hold securely with both hands, and ensure that the bag and trim cover are pointed away.

- Place the inflator module on a bench or other surface with the bag and trim cover facing up.
- With the inflator module on the bench, never place anything on or close to the module, which may be thrown in the event of an accidental deployment.

DISARMING

1. Before servicing the vehicle, refer to the precautions in the beginning of this section.
2. Disconnect the negative battery cable.
3. Disconnect the positive battery cable.
4. Wait more than 20 seconds to allow the air bag system to deplete its backup power before starting work.
5. To rearm the air bag system, reconnect the positive, then the negative battery cables.

Rack and Pinion Steering Gear

REMOVAL & INSTALLATION

1. Before servicing the vehicle, refer to the precautions in the beginning of this section.

2. Remove or disconnect the following:
   - Negative battery cable
   - Front wheels
   - Engine undercover
   - Y-pipe
   - Tie rod end cotter pin and nut
   - Jack-up plate and front sway bar
   - Fluid lines from the rack and pinion
3. Matchmark the universal joint to the serration in the steering rack for installation reference.
4. Remove or disconnect the following:
   - Universal joint bolts and lift the joint upward disconnecting it from the rack and pinion shaft.
   - Rack and pinion

To install:

5. Install the rack and pinion. Torque the clamp bolts to 43 ft. lbs. (59 Nm).
6. Align the steering rack to the universal joint. Push the long yoke of the joint all the way into the serrated position of the steering shaft, setting the bolt hole in the cut-out. Pull the short yoke all the way out of the serrated portion of the rack and pinion, setting the bolt hole in the cut-out. Insert the bolt through the short yoke. Pull the yoke and ensure the bolt is properly engaged in the cut-out. Fasten the short

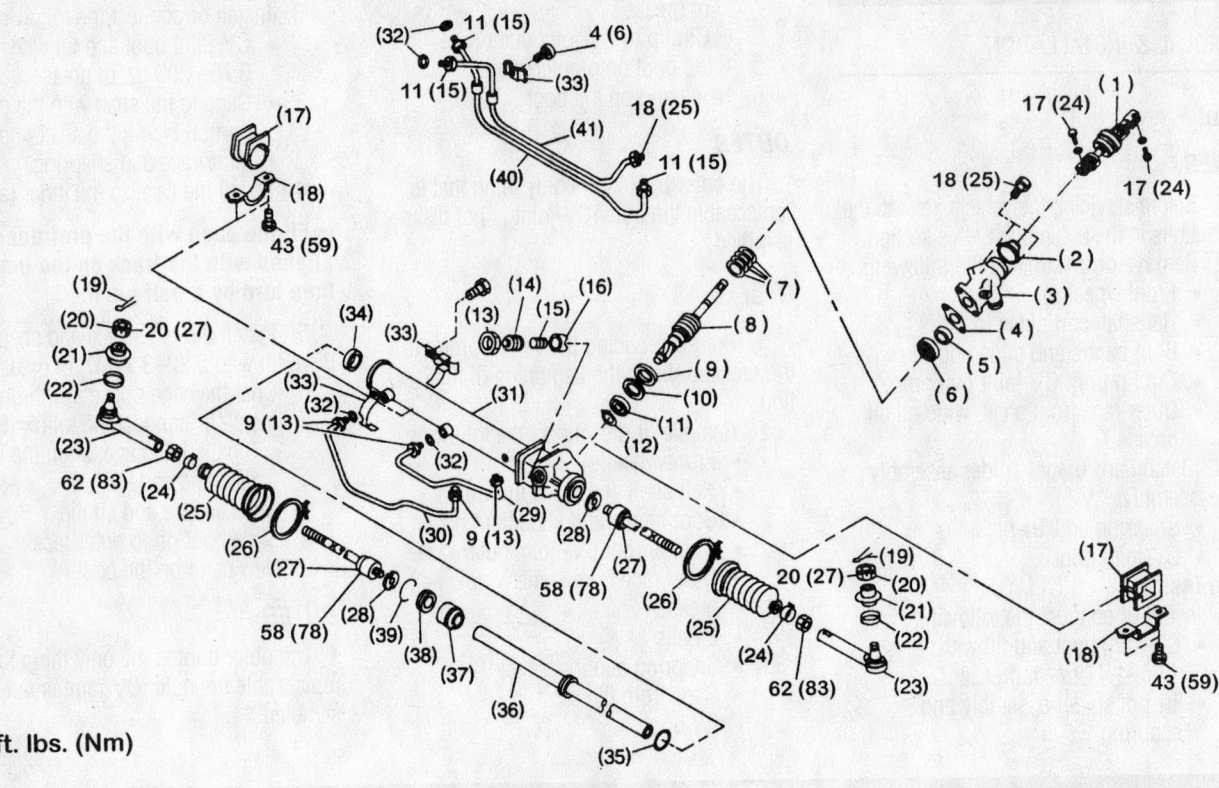

ft. lbs. (Nm)

(1) Universal joint
(2) Dust cover
(3) Valve housing
(4) Gasket
(5) Oil seal
(6) Special bearing
(7) Seal ring
(8) Pinion and valve ASSY
(9) Oil seal
(10) Back-up washer
(11) Ball bearing
(12) Snap ring
(13) Lock nut
(14) Adjusting screw
(15) Spring
(16) Sleeve
(17) Adapter
(18) Clamp

(19) Cotter pin
(20) Castle nut
(21) Dust cover
(22) Clip
(23) Tie-rod end
(24) Clip
(25) Boot
(26) Band
(27) Tie-rod
(28) Lock washer
(29) Pipe B
(30) Pipe A
(31) Housing ASSY
(32) O-ring
(33) Clamp
(34) Oil seal
(35) Piston ring
(36) Rack

(37) Rack bushing
(38) Rack stopper
(39) Circlip
(40) Pipe E
(41) Pipe F

Exploded view of the rack and pinion steering gear

yoke side with the spring washer and bolt, then fasten the yoke side. Tighten the bolts to 17 ft. lbs. (24 Nm).

7. Install or connect the following:
  • Tie rod ends to the steering knuckle
  • Sway bar and jack-up plate
  • Y-pipe with new gaskets and nuts
  • Engine undercover
  • Wheels
8. Fill and bleed the steering system.

Strut

REMOVAL & INSTALLATION

Front

✷✷ CAUTION

Do not remove the large nut on top of the strut assembly unless the coil

spring is properly compressed with a suitable spring compressor.

1. Before servicing the vehicle, refer to the precautions in the beginning of this section.
2. Remove or disconnect the following:
  • Negative battery cable
  • Front wheel assembly
  • Caliper, leaving the line connected and suspend it out of the way
  • Brake line to the strut housing

7924XG22

3. Matchmark the camber adjustment bolt to the strut housing as reference for installation.

4. Remove or disconnect the following:
- Anti-locking Brakes System (ABS) sensor and harness, if equipped
- Strut from the steering knuckle. Notice that the shaft of the top bolt is not round.
- Strut from the body in the engine compartment
- Strut and coil spring assembly

To install:

5. Install the strut and coil assembly. Torque the upper strut retainer nuts to 15 ft. lbs. (20 Nm).

6. Align matchmark on camber adjustment bolt and strut housing.

7. Install or connect the following:
- Lower strut nuts and bolts. Tighten the nuts, while securing the bolts to 112 ft. lbs. (152 Nm).
- ABS sensor and harness. Torque the bolt to 24 ft. lbs. (32 Nm), if equipped.
- Brake line to the strut
- Caliper
- Front wheel
- Negative battery cable

8. Check and adjust the front end alignment.

Rear

> ❊❊ **CAUTION**
>
> **Do not remove the large nut on top of the strut assembly unless the coil spring is properly retained with a spring compressor.**

1. Before servicing the vehicle, refer to the precautions in the beginning of this section.

2. Remove or disconnect the following:
- Strut mount cap located at the right rear interior quarter trim
- Wheel
- Brake hose clip
- Union bolt from the brake caliper, if equipped with disc brakes and move the brake hose out of the way
- Brake hose and pipe from strut and drum, if equipped with drum brakes

3. Support rear with jack.

4. Remove or disconnect the following:

5. Remove or disconnect the following:
- Retainer nuts securing the strut bearing cap to the strut tower, from inside the vehicle

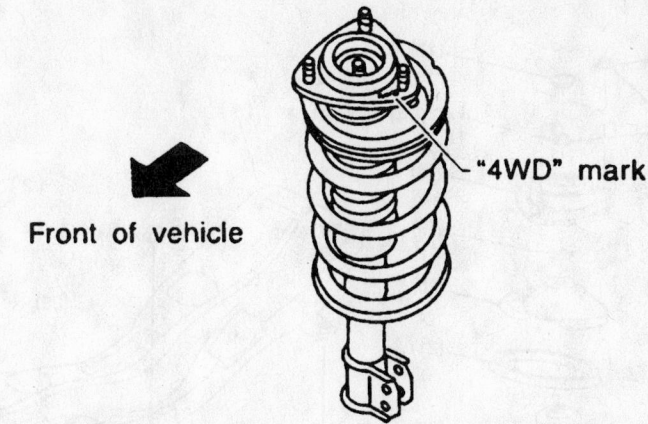

"4WD" mark

Front of vehicle

7924XG31

Position the upper strut bearing as shown

- Lower nuts and bolts securing the strut to the rear wheel housing
- Strut

To install:

6. Install or connect the following:
- Strut on to the vehicle, making sure to position the strut with the "4WD" mark on the strut mount facing the outside of the vehicle as shown in the illustration. Torque the retaining nuts to 15 ft. lbs. (20 Nm).
- Strut and mount cap. Torque the strut mount cap bolts to 14.5 ft. lbs. (20 Nm).
- Strut to the rear wheel knuckle assembly. Torque the retainer nuts/bolts to 145 ft. lbs. (196 Nm).
- Union bolt, if equipped with disc brakes. Torque the bolt and to 13 ft. lbs. (18 Nm).
- Brake hose to brake pipe, if equipped with drum brakes. Torque to 10 ft. lbs. (15 Nm).
- Brake hose clip
- Wheel
- Strut mount cap

7. Bleed the brakes.

Coil Spring

REMOVAL & INSTALLATION

Front and Rear

> ❊❊ **CAUTION**
>
> **Do not remove the large nut on top of the strut assembly unless the coil spring is properly compressed with a spring compressor.**

1. Before servicing the vehicle, refer to the precautions in the beginning of this section.

2. Remove the strut assembly.

3. Place the strut assembly in a vise with a holding tool and install a spring compressor.

4. Compress the spring slightly.

5. Loosen but do not remove the bearing cap locknut.

6. Unload the spring seat using the spring compressor, then remove the locknut.

7. Remove or disconnect the following:
- Strut bearing cap, mounting insulator bracket and upper spring seat
- Coil spring and compressor. If the spring is being replaced, slowly release the spring from the compressor and compress the new coil spring.
- Strut boot and rebound bumper from the strut, inspect and replace if worn
- Strut retainer nut and the strut insert from the assembly

To install:

8. Install or connect the following:
- Strut into the chamber
- Retainer nut. Tighten the nut until snug.
- Rebound bumper and the boot to the strut piston rod
- Coil spring on the strut assembly
- Upper spring seat, mounting insulator and bearing cap
- Locknut. Tighten it to 36–43 ft. lbs. (47–56 Nm).
- Spring compressor from the coil spring
- Strut

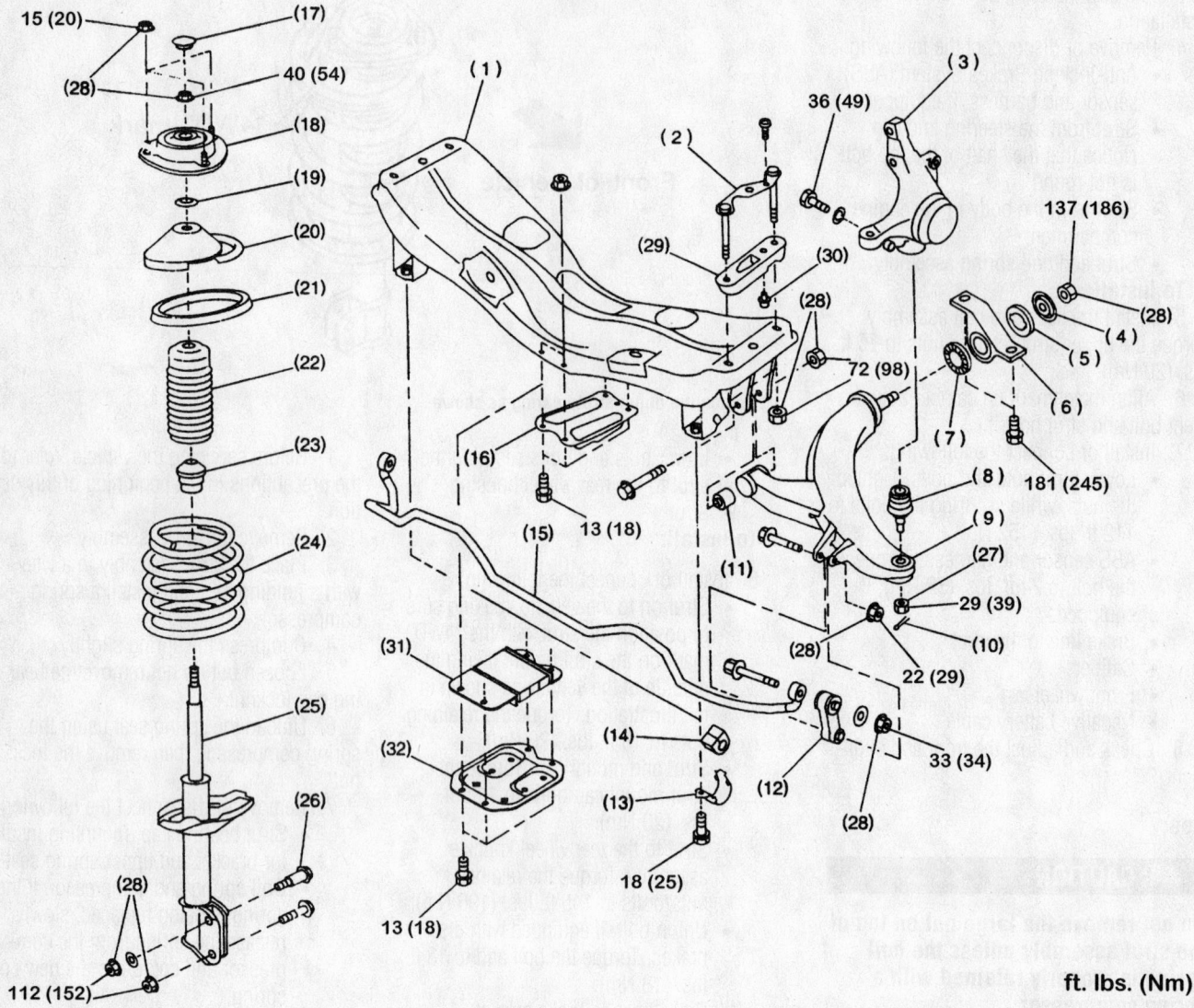

ft. lbs. (Nm)

(1) Front crossmember
(2) Bolt ASSY
(3) Housing
(4) Washer
(5) Stopper rubber (Rear)
(6) Rear bushing
(7) Stopper rubber (Front)
(8) Ball joint
(9) Transverse link
(10) Cotter pin
(11) Front bushing
(12) Stabilizer link
(13) Clamp
(14) Bushing
(15) Stabilizer
(16) Jack-up plate (Except MT model)

(17) Dust seal
(18) Strut mount
(19) Spacer
(20) Upper spring seat
(21) Rubber seat
(22) Dust cover
(23) Helper
(24) Coil spring
(25) Damper strut
(26) Adjusting bolt
(27) Castle nut
(28) Self-locking nut
(29) Adapter front crossmember
(30) Clip
(31) Dynamic damper (MT model)
(32) Jack-up plate (MT model)

Exploded view of the front suspension

7924XG23

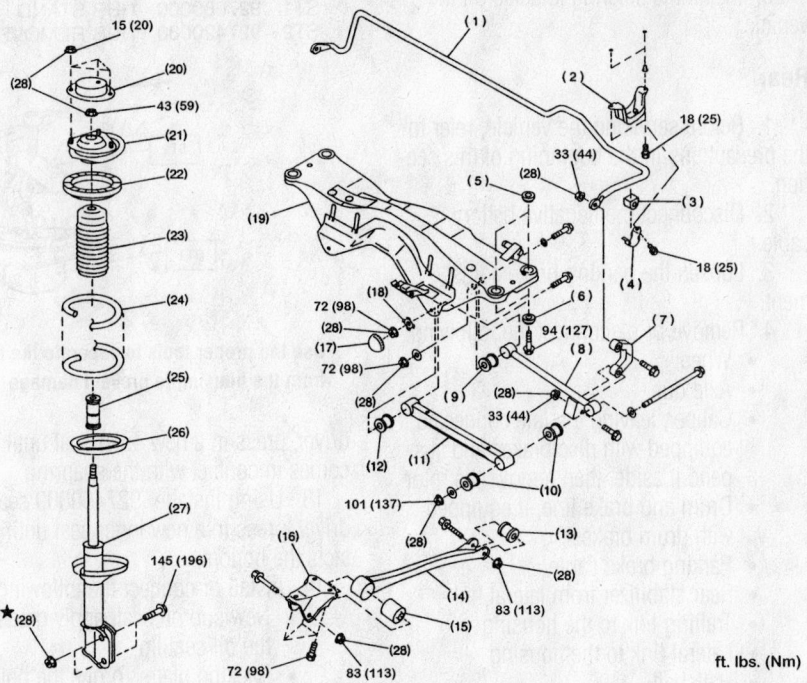

Exploded view of the rear suspension

(1) Stabilizer
(2) Stabilizer bracket
(3) Stabilizer bushing
(4) Clamp
(5) Floating bushing
(6) Stopper
(7) Stabilizer link
(8) Rear lateral link
(9) Bushing (C)
(10) Bushing (A)
(11) Front lateral link
(12) Bushing (B)
(13) Trailing link rear bushing
(14) Trailing link

(15) Trailing link front bushing
(16) Trailing link bracket
(17) Cap (Protection)
(18) Washer
(19) Rear crossmember
(20) Strut mount cap
(21) Strut mount
(22) Rubber seat upper
(23) Dust cover
(24) Coil spring
(25) Helper
(26) Rubber seat lower
(27) Damper strut
(28) Self-locking nut

ft. lbs. (Nm)

7924XG24

Lower Ball Joint

REMOVAL & INSTALLATION

1. Before servicing the vehicle, refer to the precautions in the beginning of this section.
2. Remove or disconnect the following:
   - Negative battery cable
   - Front wheel
   - Ball joint castle nut cotter pin, discard the cotter pin
   - Castle nut
   - Ball joint from the lower control arm assembly
   - Ball joint from the steering knuckle

To install:
3. Install or connect the following:
   - Ball joint to the steering knuckle.

Torque the bolt to 36 ft. lbs. (49 Nm).
   - Ball joint to the lower control arm. Torque the castle nut to 29 ft. lbs. (39 Nm). Then, tighten the castle nut an additional 60 degrees until the slot in the castle nut is aligned with the cotter pin hole in the ball joint.
   - New cotter pin
   - Wheel
   - Negative battery cable

Lower Control Arm

REMOVAL & INSTALLATION

1. Before servicing the vehicle, refer to the precautions in the beginning of this section.

2. Remove or disconnect the following:
   - Wheel assembly
   - Stabilizer link
   - Ball joint from housing
   - Mounting bolts
   - Control arm

To install:
3. Install or connect the following:
   - Control arm to stabilizer. Torque the nut/bolt to 22 ft. lbs. (29 Nm).
   - Control arm to crossmember. Torque the nut/bolt to 72 ft. lbs. (98 Nm).
   - Control arm to rear mount. Torque the bolts to 181 ft. lbs. (245 Nm).
   - Ball joint to housing, Torque the nut to 36 ft. lbs. (49 Nm).
   - Wheel assembly

CONTROL ARM BUSHING REPLACEMENT

Front Bushing

1. Before servicing the vehicle, refer to the precautions in the beginning of this section.
2. Remove or disconnect the following:
   - Wheel
   - Control arm
3. Press the bushing out using Installer/Remover tool 927680000

To install:
4. Press the bushing in using Installer/Remover tool 927680000
5. Install or connect the following:
   - Control arm
   - Wheel

Rear Bushing

1. Before servicing the vehicle, refer to the precautions in the beginning of this section.
2. Remove or disconnect the following:
   - Wheel
   - Control arm and matchmark the bushing for reassembly
   - Nut and bushing

To install:
3. Install or connect the following:
   - Bushing into control arm and align the matchmark. Torque the nut to 137 ft. lbs. (186 Nm).
   - Control arm
   - Wheel

Wheel Bearings

ADJUSTMENT

the wheel bearings are not adjustable.

REMOVAL & INSTALLATION

Front

1. Before servicing the vehicle, refer to the precautions in the beginning of this section.

2. Remove the steering knuckle assembly.

3. Position the steering knuckle in a soft-jawed vise.

4. Press the hub from the steering knuckle. If the inner bearing race remains in the hub, press it out.

5. Remove or disconnect the following:
   - Rotor shield
   - Inner and outer seals
   - Snapring from the steering knuckle

6. Press the inner bearing race to remove the outer bearing.

7. Remove the Anti-lock Brakes (ABS) tone ring, if equipped

8. Press the wheel lugs from the hub.

➡**To prevent deforming the hub, do not hammer the lugs out.**

To install:

9. Press new wheel lugs into the hub.

10. If equipped, clean all foreign material from the hub and tone ring. Install the tone ring.

11. Clean the inside of the steering knuckle.

12. Remove the plastic lock from the inner race and press a new greased bearing into the hub by pressing the outer race.

13. Install the snapring into its groove.

14. Press a new outer oil seal until it contacts the bottom of the housing.

15. Press a new inner oil seal until it contacts the circlip.

16. Apply grease to the oil seal lips.

17. Install the rotor shield and tighten the bolts to 10 ft. lbs. (14 Nm).

18. Attach the hub to the steering knuckle.

19. Press a new bearing into the hub by driving the inner race.

20. Install the steering knuckle on the vehicle.

Rear

1. Before servicing the vehicle, refer to the precautions in the beginning of this section.

2. Disconnect the negative battery cable.

3. Loosen the parking brake adjustment.

4. Remove or disconnect the following:
   - Wheel
   - Axle nut
   - Caliper, leaving the line connected if equipped with disc brakes and suspend it aside, then remove the rotor
   - Drum and brake line, if equipped with drum brakes
   - Parking brake cable
   - Rear stabilizer from lateral link
   - Trailing link to the housing
   - Lateral link to the housing
   - Halfshaft
   - Anti-lock Brakes (ABS), speed sensor from the backing plate, if equipped
   - Strut from the housing
   - Housing assembly

5. Using hub stand 92708000 and Hub Remover 927420000

6. Remove or disconnect the following:
   - Hub from the rear housing
   - Backing plate from the housing.
   - Outer, inner and sub oil seals.
   - Snapring

7. Remove the bearing by pressing the inner race.

To install:

8. Clean the housing thoroughly.

➡**Do not remove the plastic lock from the inner race when installing the bearing.**

9. Install the new bearing into the housing by pressing the outer race.

10. Pack the bearing with grease.

11. Install the snapring.

12. Using installer 927460000 seal

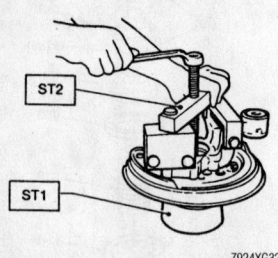

| ST1 | 927080000 | HUB STAND |
| ST2 | 927420000 | HUB REMOVER |

7924XG32

Use the proper tools to separate the hub from the housing to prevent damage

driver, press in a new outer seal until it comes in contact with the snapring.

13. Using installer 927450000 seal driver, press in a new inner seal until it contacts the bottom.

14. Install or connect the following:
   - New sub oil seal, apply grease to the oil seal lip
   - Backing plate. Torque the bolts to 38 ft. lbs. (52 Nm).

15. Using installer 927450000 bearing driver, press in the hub into the housing.

16. Install or connect the following:
   - Housing to the strut. Torque the bolts to 108 ft. lbs. (147 Nm).
   - Halfshaft
   - Lateral link to the housing. Torque the bolt and new nut to 101 ft. lbs. (137 Nm).
   - Trailing link to the housing. Torque the bolt and new nut to 94 ft. lbs. (127 Nm).
   - Stabilizer to rear lateral link
   - Parking brake cable and brake
   - Brake line, if equipped with drum brakes
   - Rotor and caliper, if equipped with disc brakes
   - ABS speed sensor, if equipped
   - New axle nut and tighten it to 137 ft. lbs. (186 Nm). Stake the nut.
   - Wheel
   - Negative battery cable

17. Adjust the parking brake cable.

TOYOTA AND LEXUS

26

Toyota-RAV4 • 4Runner • Highlander • Lexus-RX300

PRECAUTIONS

Before servicing any vehicle, please be sure to read all of the following precautions, which deal with personal safety, prevention of component damage, and important points to take into consideration when servicing a motor vehicle:

• Never open, service or drain the radiator or cooling system when the engine is hot; serious burns can occur from the steam and hot coolant.

• Observe all applicable safety precautions when working around fuel. Whenever servicing the fuel system, always work in a well-ventilated area. Do not allow fuel spray or vapors to come in contact with a spark, open flame, or excessive heat (a hot drop light, for example). Keep a dry chemical fire extinguisher near the work area. Always keep fuel in a container specifically designed for fuel storage; also, always properly seal fuel containers to avoid the possibility of fire or explosion. Refer to the additional fuel system precautions later in this section.

• Fuel injection systems often remain pressurized, even after the engine has been turned **OFF**. The fuel system pressure must be relieved before disconnecting any fuel lines. Failure to do so may result in fire and/or personal injury.

• Brake fluid often contains polyglycol ethers and polyglycols. Avoid contact with the eyes and wash your hands thoroughly after handling brake fluid. If you do get brake fluid in your eyes, flush your eyes with clean, running water for 15 minutes. If eye irritation persists, or if you have taken brake fluid internally, IMMEDIATELY seek medical assistance.

• The EPA warns that prolonged contact with used engine oil may cause a number of skin disorders, including cancer. You should make every effort to minimize your exposure to used engine oil. Protective gloves should be worn when changing oil. Wash your hands and any other exposed skin areas as soon as possible after exposure to used engine oil. Soap and water, or waterless hand cleaner should be used.

• All new vehicles are now equipped with an air bag system. The system must be disabled before performing service on or around system components, steering column, instrument panel components, wiring and sensors. Failure to follow safety and disabling procedures could result in accidental air bag deployment, possible personal injury and unnecessary system repairs.

• Always wear safety goggles when working with, or around, the air bag system. When carrying a non-deployed air bag, be sure the bag and trim cover are pointed away from your body. When placing a non-deployed air bag on a work surface, always face the bag and trim cover upward, away from the surface. This will reduce the motion of the module if it is accidentally deployed. Refer to the additional air bag system precautions later in this section.

• NEVER disconnect the negative battery cable with the ignition **ON** or the engine running. Removing power from the computer control module with the ignition **ON** may destroy the module.

• Clean, high quality brake fluid from a sealed container is essential to the safe and proper operation of the brake system. You should always buy the correct type of brake fluid for your vehicle. If the brake fluid becomes contaminated, completely flush the system with new fluid. Never reuse any brake fluid. Any brake fluid that is removed from the system should be discarded. Also, do not allow any brake fluid to come in contact with a painted surface; it will damage the paint.

• Never operate the engine without the proper amount and type of engine oil; doing so WILL result in severe engine damage.

• Timing belt maintenance is extremely important. Many models utilize an interference-type, non-freewheeling engine. If the timing belt breaks, the valves in the cylinder head may strike the pistons, causing potentially serious (also time-consuming and expensive) engine damage. Refer to the maintenance interval charts in the front of this manual for the recommended replacement interval for the timing belt, and to the timing belt section for belt replacement and inspection.

• Disconnecting the negative battery cable on some vehicles may interfere with the functions of the on-board computer system(s) and may require the computer to undergo a relearning process once the negative battery cable is reconnected.

• When servicing drum brakes, only disassemble and assemble one side at a time, leaving the remaining side intact for reference.

• Only an MVAC-trained, EPA-certified automotive technician should service the air conditioning system or its components.

ENGINE REPAIR

➡Disconnecting the negative battery cable on some vehicles may interfere with the functions of the on board computer system. The computer may undergo a relearning process once the negative battery cable is reconnected.

Distributor

All models are equipped with a distibutorless ignition system.

Alternator

REMOVAL

On some models, the alternator is mounted very low on the engine. On these models, it may be necessary to remove the gravel shield and work from beneath the vehicle in order to gain access to the alternator. Replacing the alternator while the engine is cold is recommended. A hot engine can result in personal injury.

2.0L RAV4

1. Before servicing the vehicle, refer to the precautions in the beginning of this section.
2. Remove or disconnect the following:
   • Electrical wiring from the alternator
   • Loosen the adjusting lockbolt and the pivot bolt.
   • Loosen the adjusting bolt to relieve tension on the drive belt, if equipped with air conditioning
   • Drive belt

➡It may be necessary to remove other belts for access.

   • Pivot bolt and the adjusting lockbolt
   • Alternator

2.4L Highlander

1. Before servicing the vehicle, refer to the precautions in the beginning of this section.
2. Remove or disconnect the following:
   • Electrical wiring from the alternator
   • Drive belt
   • 1 adjusting and 2 mounting bolts
   • Alternator

3.0L Highlander

1. Before servicing the vehicle, refer to the precautions in the beginning of this section.
2. Remove or disconnect the following:
   - Alternator electrical connectors
   - Wiring harness from the clip
   - Pivot bolt
   - Adjuster lockbolt
   - Drive belt
   - Alternator

3.0L RX300

1. Before servicing the vehicle, refer to the precautions in the beginning of this section.
2. Remove or disconnect the following:
   - Alternator electrical connectors
   - Wiring harness from the clip
   - Pivot bolt
   - Plate washer
   - Adjusting lockbolt
   - Drive belt
   - Alternator

2.7L 4Runner

Remove or disconnect the following:
- Negative battery cable
- Alternator wiring
- Alternator lockbolt, pivot bolt, nut and adjusting bolt
- Drive belt

- Wiring harness with the clip
- Alternator

3.4L 4Runner

1. Remove or disconnect the following:
   - Negative battery cable
   - Alternator wiring
   - Alternator locknut, pivot bolt, nut and adjusting bolt
   - Drive belt
   - Alternator

INSTALLATION

2.0L RAV4

1. Install or connect the following:
   - Alternator
   - Adjusting lockbolt and the pivot bolt
   - Drive belt
2. Adjust the drive belt tension to:
   - New belt with A/C: 140–190 lbs.
   - Used belt with A/C: 100–120 lbs.
   - New belt without A/C: 100–150 lbs.
   - Used belt without A/C: 75–115 lbs.
3. Install or connect the following:
   - Tighten the pivot bolt to 38 ft. lbs. (52 Nm).
   - Tighten the adjusting lockbolt to 13 ft. lbs. (18 Nm).
   - Electrical wiring to the alternator

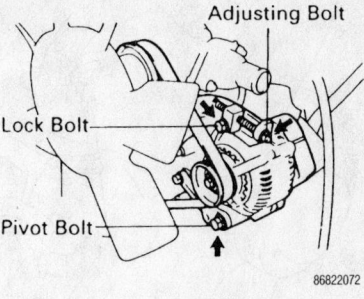

Locations of the adjusting, pivot and lock-bolts—2.7L 4Runner

2.4L Highlander

1. Installation is the reverse of removal. Observe the following torques:
   - M8 bolts: 15 ft. lbs. (21Nm)
   - M10 bolts: 38 ft. lbs. (52Nm)

3.0L Highlander

1. Install or connect the following:
   - Alternator
   - Drive belt
   - Adjusting lockbolt. Tighten the bolt to 13 ft. lbs. (18 Nm).
   - Pivot bolt. Tighten the bolt to 43 ft. lbs. (58 Nm).
   - Wiring harness from the clip
   - Alternator electrical connectors

3.0L RX300

1. Install or connect the following:
   - Alternator
   - Drive belt. Tension the belt to 170–180 lbs. for a new belt or 95–135 lbs. for a used belt.
   - Adjusting lockbolt. Tighten the bolt to 13 ft. lbs. (18 Nm).
   - Plate washer
   - Pivot bolt. Tighten the bolt to 41 ft. lbs. (56 Nm).
   - Wiring harness from the clip
   - Alternator electrical connectors

2.7L 4Runner

Install or connect the following:
- Alternator
- Drive belt; adjust it to the proper tension. Tighten the lockbolt to 21 ft. lbs. (29 Nm) and the pivot bolt to 43 ft. lbs. (59 Nm).
- Wire harness with clip
- Alternator wiring. Tighten the nut to 7 ft. lbs. (10 Nm).

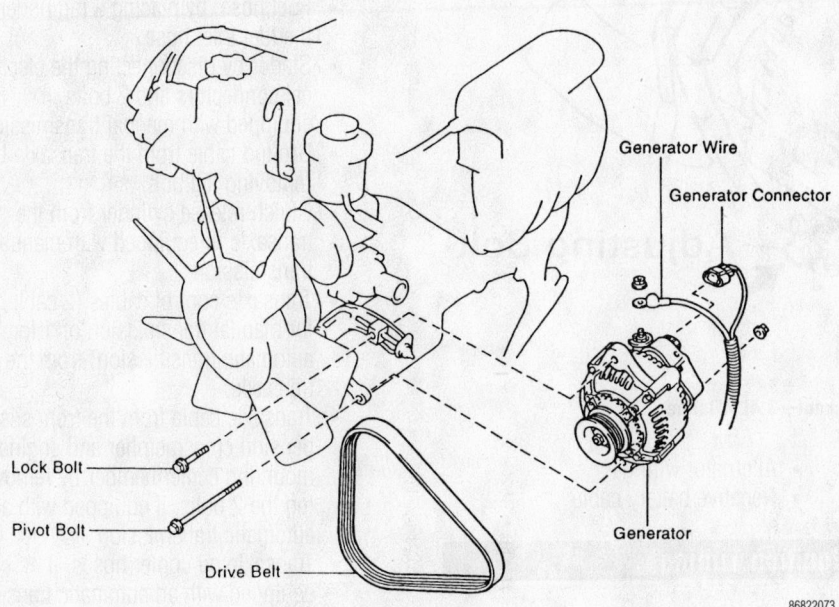

Exploded view of the alternator and drive belt—2.7L 4Runner

For complete service labor times, order Nichols' Chilton Labor Guide

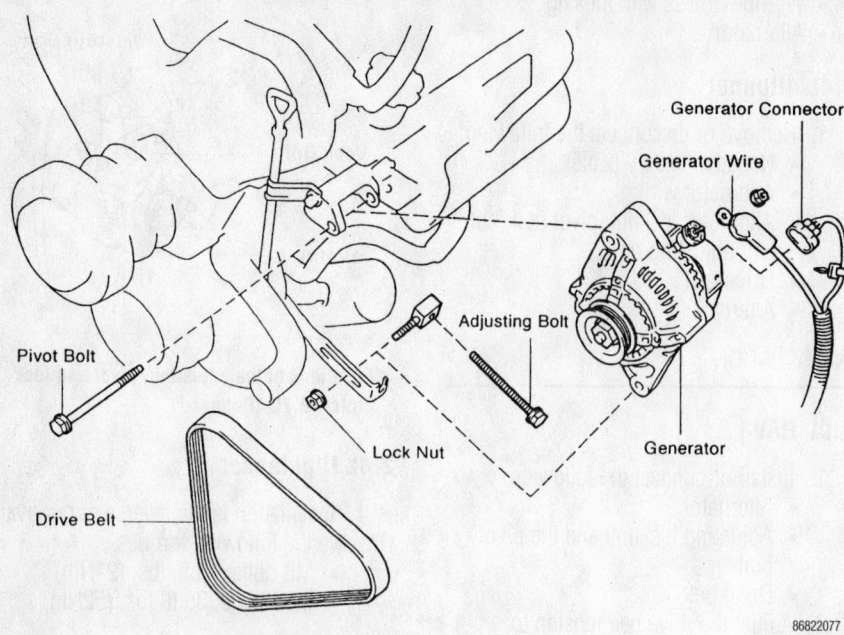

Exploded view of the alternator and drive belt—3.4L 4Runner

86822077

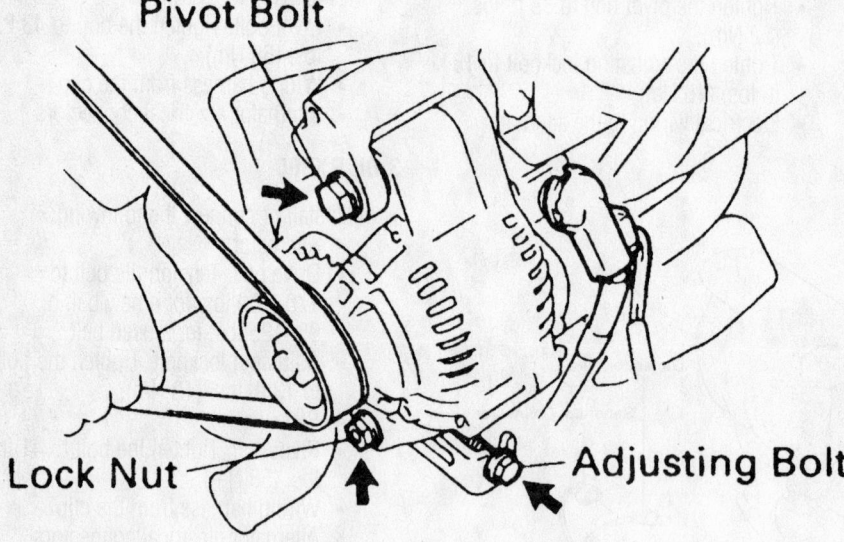

Locations of the adjusting and pivot bolts and the locknut—3.4L 4Runner

86822078

- Rubber boot over the terminal
- Wiring harness connector
- Negative battery cable

3.4L 4Runner

Install or connect the following:
- Alternator
- Drive belt. Tighten the locknut 25 ft. lbs. (33 Nm) and the pivot bolt 38 ft. lbs. (51 Nm).

- Alternator wiring
- Negative battery cable

Ignition Timing

ADJUSTMENT

All engines are equipped with a Distributorless Ignition System (DIS). No timing adjustment is possible.

Engine Assembly

REMOVAL & INSTALLATION

RAV4

1. Before servicing the vehicle, refer to the precautions in the beginning of this section.
2. Relieve the fuel system pressure.
3. Remove or disconnect the following:
- Negative battery cable
- Battery
- Hood
- Engine undercover
4. Drain the engine coolant and oil.
5. Drain the transaxle assembly.
6. Remove or disconnect the following:
- Air cleaner and case
- Accelerator cable from the throttle body, bracket and clamps
7. Disconnect and remove the engine wire from the No. 2 relay block, as follows:
- No. 2 relay block from the body by removing the 2 bolts
- Upper cover to the relay block
- Electrical connectors
- Engine wire, by removing the 2 nuts
- Charcoal canister
- Alternator
- Upper and lower radiator hoses
- Water inlet from the engine by removing the 2 nuts
- Heater hoses
- Fuel hose, by placing a rag under the fuel inlet hose
- Starter by disconnecting the electrical connectors and 2 bolts, if equipped with manual transmission
- Ground cable from the transaxle by removing the bolt
- Clutch release cylinder from the transaxle, if equipped with manual transmission
- Transaxle control cables (2 cable for manual transmission or 1 for automatic transmission) from the transaxle.
- Transaxle cable from the front suspension crossmember and engine mounting centermember by removing the 2 bolts, if equipped with an automatic transmission
- Transaxle oil cooler hoses, if equipped with an automatic transmission or 4WD with manual transmission
8. Detach the following:
- Vapor pressure sensor connector
- Igniter connector

- Ignition coil connector
- Noise filter connector
- Ignition coil wire
- Manifold Absolute Pressure (MAP) sensor connector
- MAP sensor vacuum hose from the gas filter on the intake manifold
- Brake booster hose from the intake manifold
- Differential lock control solenoid connector, if equipped with a 4WD manual transmission
- Ground strap from cowl

9. Detach the engine wire from the passenger compartment, as follows:
   - Right-hand scuff plate
   - Right-hand side trim
   - Right-hand carpet center cover
   - 2 ECM connectors
   - 2 connectors from the bracket connectors
   - No. 4 junction block connector
   - Wire clamp from the bracket
   - Engine wire from the passenger compartment

10. Remove the front exhaust pipe, as follows:
    - 3 nuts and the front exhaust pipe from the exhaust manifold, using a 14mm deep socket wrench; discard the gasket
    - 2 bolts and 2 nuts holding the front exhaust pipe to the catalytic converter
    - Front exhaust pipe and 2 gaskets

11. Remove the compressor from the engine and suspend the compressor securely.

➥**It is not necessary to remove the air conditioning compressor lines in order to remove the engine.**

12. Remove or disconnect the following:
    - Driveshaft, if equipped with 4WD
    - Halfshaft
    - Sway bar

13. Remove the front suspension crossmember assembly, as follows:
    - 2 centermember set nuts holding the centermember to the middle of the crossmember.
    - 2 rack and pinion assembly set bolts/nuts from the crossmember. Securely suspend the steering gear assembly.
    - Catalytic converter with pipe from the ring
    - Support the suspension crossmember with a jack

- 6 bolts from the suspension crossmember
- Suspension crossmember with the lower suspension arms

14. Remove the engine mounting centermember, as follows:
    - 2 bolts holding the centermember to the front engine mounting insulator
    - 2 bolts holding the centermember to the body
    - Centermember

15. Disconnect the power steering pump from the engine, as follows:
    - 2 vacuum hoses from the steering pump
    - Adjusting bolt for the power steering unit. Loosen the pivot bolt to the power steering pump and remove the drive belt. Use Torque Wrench Adapter tool 09249-63010 and a deep socket to loosen the pivot bolt.
    - Power steering pump from the engine by removing the 3 bracket bolts

16. Install a engine hanger to the engine.
17. Attach the engine sling device to the engine hangers.
18. Remove or disconnect the following:
    - Left-hand engine mounting bracket from the mounting insulator by removing the 2 nuts and 2 bolts
    - Ground connector next to the right-hand engine mount
    - Right-hand engine mounting bracket from the mounting insulator by removing the bolt and 2 nuts

19. Lower the engine and transaxle and at the same time, raise the vehicle to gain clearance to the remove the engine.

20. Place the assembly on a stand and separate the engine from the transaxle.

To install:

21. Install or connect the following:
    - Engine and transaxle assembly
    - Left-hand engine mounting bracket to the mounting insulator. Tighten both nuts/bolts to 47 ft. lbs. (64 Nm).
    - Bolt and 2 nuts to hold the right-hand engine mounting bracket to the mounting insulator. Tighten the bolt to 27 ft. lbs. (37 Nm) and both nuts to 38 ft. lbs. (52 Nm).
    - Ground connector next to the right-hand engine mount
    - Engine sling and hanger

22. Install the power steering pump, as follows:

- Pump with the bracket. Tighten the 3 bolts to 32 ft. lbs. (43 Nm).
- Pivot and adjusting bolts. Tighten the pivot bolt to 32 ft. lbs. (43 Nm) and the adjusting bolt to 29 ft. lbs. (39 Nm).
- Drive belt. Adjust the tension.
- Both air hoses to the power steering pump

23. Install or connect the following:
    - Engine mounting centermember to the body; install the 4 bolts but do not tighten the bolts at this time

24. Install or connect the front crossmember, as follows:
    - Suspension crossmember with the lower control arms. Torque both bolts crossmember-to-chassis bolts to 152 ft. lbs. (206 Nm).
    - Rack and pinion. Tighten both nuts/bolts to 83 ft. lbs. (113 Nm).
    - Centermember to the crossmember. Tighten both nuts to 82 ft. lbs. (112 Nm).
    - Tighten the lower control arm rear brackets to 101 ft. lbs. (137 Nm).
    - Tighten both engine mounting centermember-to-front engine mounting insulator bolts to 59 ft. lbs. (80 Nm).
    - Tighten both engine mounting centermember-to-body bolts to 26 ft. lbs. (35 Nm).

25. Install or connect the following:
    - Sway bar
    - Halfshafts
    - Driveshaft, if equipped with 4WD

26. Install the air conditioning compressor. Tighten nut/bolts, as follows:
    - Stud bolt to 34 ft. lbs. (47 Nm)
    - Bolt to 27 ft. lbs. (37 Nm)
    - Nut to 20 ft. lbs. (27 Nm)

27. Install or connect the following:
    - Air conditioning compressor connector
    - Front exhaust pipe with new gaskets. Tighten the 3 nuts to 46 ft. lbs. (62 Nm) and 2 bolts to 35 ft. lbs. (48 Nm).

28. Attach the engine wire to the passenger compartment, as follows:
    - Engine wire through the cowl panel
    - Wire clamp to the bracket
    - Both ECM connectors
    - Both connectors on the bracket
    - No. 4 junction block
    - Right-hand floor carpet center cover
    - Right-hand cowl side trim
    - Right-hand scuff plate

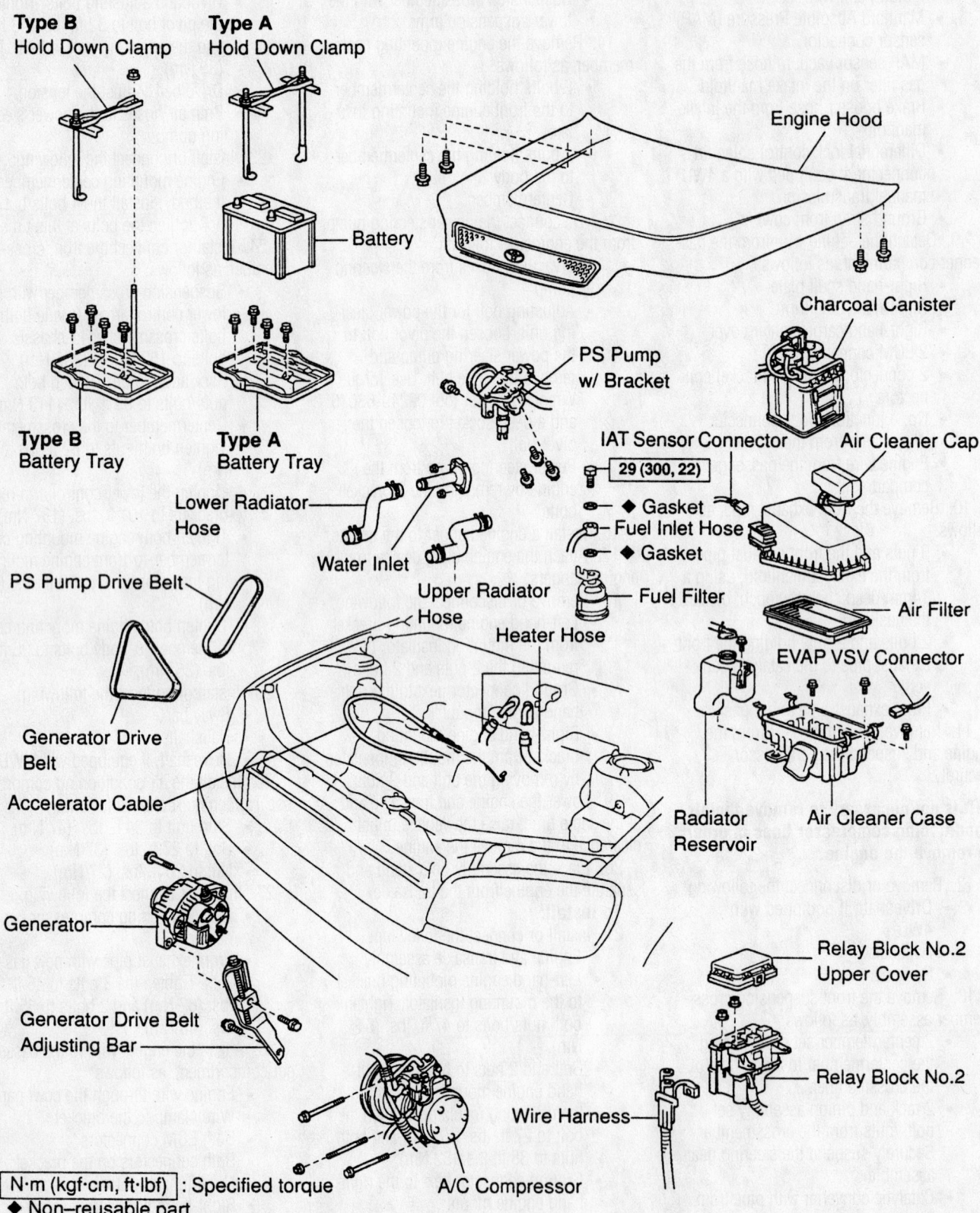

Type B
Hold Down Clamp

Type A
Hold Down Clamp

Engine Hood

Battery

Charcoal Canister

Type B
Battery Tray

Type A
Battery Tray

PS Pump
w/ Bracket

IAT Sensor Connector

Air Cleaner Cap

Lower Radiator
Hose

29 (300, 22)

◆ Gasket

Fuel Inlet Hose

◆ Gasket

Water Inlet

Air Filter

PS Pump Drive Belt

Upper Radiator
Hose

Fuel Filter

EVAP VSV Connector

Heater Hose

Generator Drive
Belt

Accelerator Cable

Radiator
Reservoir

Air Cleaner Case

Generator

Relay Block No.2
Upper Cover

Generator Drive Belt
Adjusting Bar

Relay Block No.2

Wire Harness

N·m (kgf·cm, ft·lbf) : Specified torque
◆ Non–reusable part

A/C Compressor

7924ZG04

Exploded view of the engine accessory removal components—RAV4 model

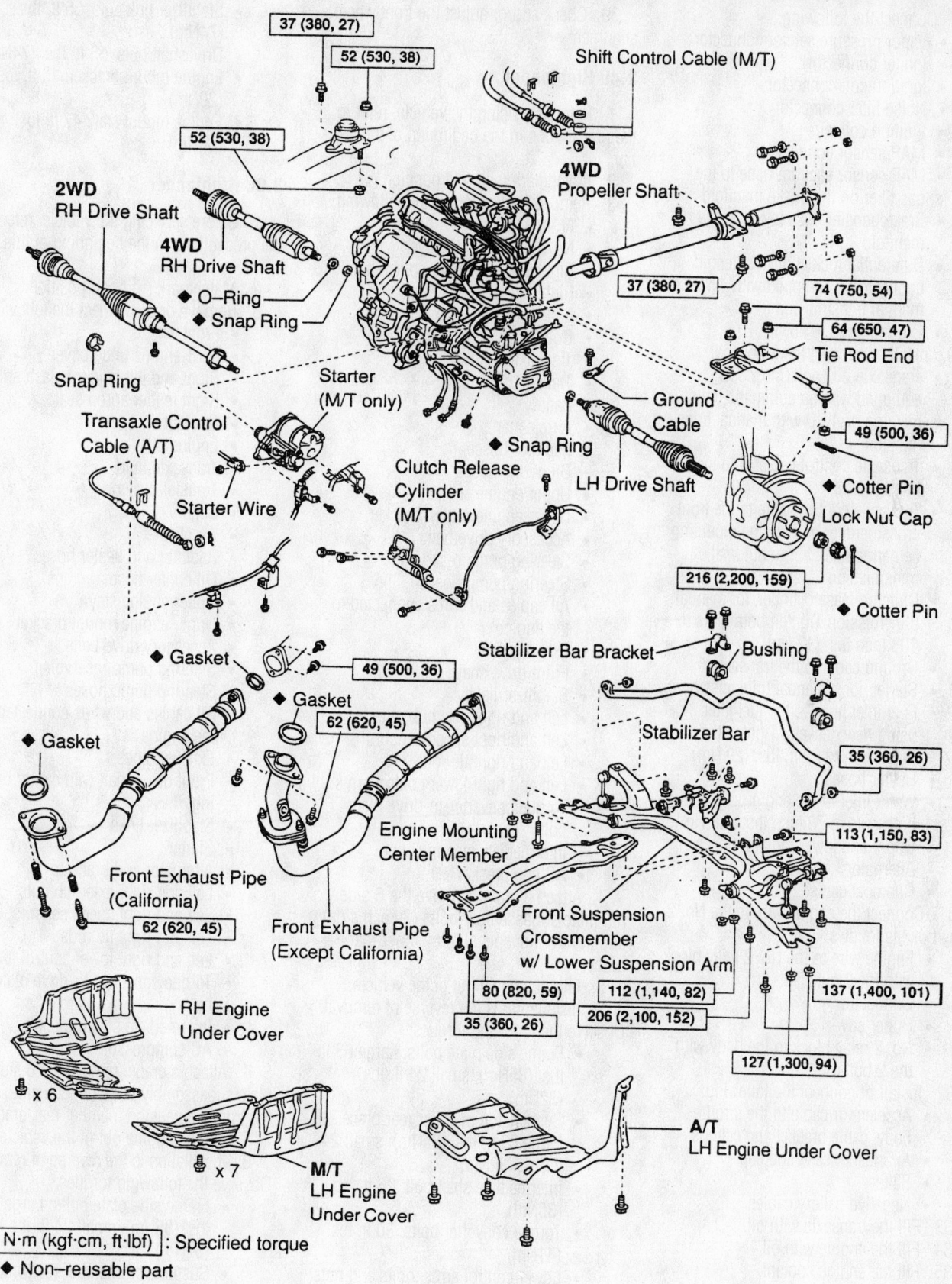

37 (380, 27)

52 (530, 38)

Shift Control Cable (M/T)

52 (530, 38)

2WD
RH Drive Shaft

4WD
RH Drive Shaft

4WD
Propeller Shaft

37 (380, 27)

74 (750, 54)

64 (650, 47)

Tie Rod End

◆ O–Ring

◆ Snap Ring

Snap Ring

49 (500, 36)

Starter
(M/T only)

Ground
Cable

Transaxle Control
Cable (A/T)

◆ Snap Ring

◆ Cotter Pin

Starter Wire

Clutch Release
Cylinder
(M/T only)

LH Drive Shaft

Lock Nut Cap

216 (2,200, 159)

◆ Cotter Pin

◆ Gasket

Stabilizer Bar Bracket

Bushing

49 (500, 36)

◆ Gasket

62 (620, 45)

Stabilizer Bar

35 (360, 26)

◆ Gasket

113 (1,150, 83)

Engine Mounting
Center Member

Front Exhaust Pipe
(California)

Front Exhaust Pipe
(Except California)

Front Suspension
Crossmember
w/ Lower Suspension Arm

62 (620, 45)

80 (820, 59)

112 (1,140, 82)

137 (1,400, 101)

35 (360, 26)

206 (2,100, 152)

RH Engine
Under Cover

127 (1,300, 94)

✖ 6

✖ 7 **M/T**
LH Engine
Under Cover

A/T
LH Engine Under Cover

N·m (kgf·cm, ft·lbf) : Specified torque

◆ Non–reusable part

Exploded view of the engine removal—RAV4 model

7924ZG05

Timing belt service is covered in Section 3 of this manual

29. Connect the following:
- Vapor pressure sensor connector
- Igniter connector
- Ignition coil connector
- Noise filter connector
- Ignition coil wire
- MAP sensor connector
- MAP sensor vacuum hose to the gas filter on the intake manifold
- Brake booster hose to the intake manifold
- Differential lock control solenoid connector, if equipped with 4WD manual transmission
- Ground strap to cowl

30. Install or connect the following:
- Transaxle oil cooler hoses, if equipped with an automatic transmission or 4WD with manual transmission
- Transaxle control cable(s) to the transaxle
- Transaxle control cable to the front crossmember and engine mounting centermember, for an automatic transmission
- Clutch release cylinder, for manual transmission Tighten both bolts to 108 inch lbs. (12 Nm).
- Ground cable to the transaxle
- Starter, for a manual transmission
- Fuel inlet hose to the fuel filter, using new gaskets. Tighten the union bolt to 22 ft. lbs. (29 Nm).
- Heater hoses
- Water inlet to the engine. Tighten both nuts to 78 inch lbs. (8.8 Nm).
- Upper and lower radiator hoses
- Alternator
- Charcoal canister

31. Connect the engine wire to the No. 2 relay box, as follows:
- Engine wire to the No. 2 relay block with the 2 nuts
- Connector
- Upper cover
- No. 2 relay block to the body with the 2 bolts

32. Install or connect the following:
- Accelerator cable to the throttle body, cable bracket and clamps
- Air cleaner case and cap
- Battery
- Negative battery cables

33. Fill the transaxle with oil.
34. Fill the engine with oil.
35. Fill the engine coolant.
36. Start the engine and check for leaks.
37. Install or connect the following:
- Engine undercovers
- Hood
38. Recheck all fluid levels.

39. Check and/or adjust the front wheel alignment.

2.4L Highlander

1. Before servicing the vehicle, refer to the precautions in the beginning of this section.
2. Matchmark the hood position.
3. Remove or disconnect the following:
- Front wheels
- No.1 engine undercover
- Right and left fender splash shields
- Right fender apron seal
- Engine oil
- Coolant
- Transaxle fluid
- Transfer case oil
- Battery
- Air cleaner
- Radiator hoses
- Oil cooler hoses
- Upper engine stay
- Upper engine mount bracket
- Accessory drive belts
- Steering pump reservoir
- Steering pump hoses
- All cables and wires connected to the engine
- Exhaust pipe
- Front drive shaft
- Stabilizer links
- Left and right axle hub nuts
- Left and right speed sensors
- Left and right tie rods
- Left and right lower control arms
- Torque converter-to-drive plate bolts
- Intermediate steering shaft
- AC compressor

4. Attach a crane, remove the 6 side rail plate subassembly bolts (3 each side) and the front suspension member rear brace.
5. Lift the engine out of the vehicle.
6. Installation is the reverse of removal. Observe the following torques:
- Frame side plate bolts: Large 63 ft. lbs. (85Nm); small 24 ft. lbs. (32Nm)
- Suspension member rear brace: Large 63 ft. lbs. (85Nm); small 24 ft. lbs. (32Nm)
- Intermediate shaft bolt: 26 ft. lbs. (35Nm)
- Torque converter bolts: 30 ft. lbs. (41Nm)
- Lower control arms, bolts and nuts: 94 ft. lbs. (127Nm)
- Tie rod nuts: 36 ft. lbs. (49Nm)
- Speed sensors: 71 inch lbs. (8Nm)
- Hub nuts: 217 ft. lbs. (294Nm)

- Stabilizer link nuts: 55 ft. lbs. (74Nm)
- Driveshaft nuts: 55 ft. lbs. (74Nm)
- Engine mount bracket: 15 ft. lbs. (20Nm)
- Engine mount stay: 47 ft. lbs. (64Nm)

3.0L Highlander

1. Before servicing the vehicle, refer to the precautions in the beginning of this section.
2. Matchmark the hood position.
3. Remove or disconnect the following:
- Front wheels
- No.1 engine undercover
- Right and left fender splash shields
- Right fender apron seal
- Coolant
- Engine oil
- Transaxle fluid
- Transfer case oil
- Battery
- Air cleaner
- Radiator and heater hoses
- Oil cooler hoses
- Upper engine stay
- Upper engine mount bracket
- Accessory drive belts
- Steering pump reservoir
- Steering pump hoses
- All cables and wires connected to the engine
- Exhaust pipes
- Front drive shaft with center bearing
- Stabilizer links
- Starter
- Alternator and brackets
- Left and right axle hub nuts
- Left and right speed sensors
- Left and right tie rods
- Left and right lower control arms
- Torque converter-to-drive plate bolts
- Intermediate steering shaft
- AC compressor

4. Attach a crane, remove the 6 side rail plate subassembly bolts (3 each side) and the front suspension member rear brace.
5. Lift the engine out of the vehicle.
6. Installation is the reverse of removal. Observe the following torques:
- Frame side plate bolts: Large 63 ft. lbs. (85Nm); small 24 ft. lbs. (32Nm)
- Suspension member rear brace: Large 63 ft. lbs. (85Nm); small 24 ft. lbs. (32Nm)
- Intermediate shaft bolt: 26 ft. lbs. (35Nm)

- Torque converter bolts: 30 ft. lbs. (41Nm)
- Lower control arms, bolts and nuts: 94 ft. lbs. (127Nm)
- Tie rod nuts: 36 ft. lbs. (49Nm)
- Speed sensors: 71 inch lbs. (8Nm)
- Hub nuts: 217 ft. lbs. (294Nm)
- Stabilizer link nuts: 55 ft. lbs. (74Nm)
- Driveshaft nuts: 55 ft. lbs. (74Nm)
- Engine mount bracket: 15 ft. lbs. (20Nm)
- Engine mount stay: 47 ft. lbs. (64Nm)

RX 300

1. Before servicing the vehicle, refer to the precautions in the beginning of this section.
2. Matchmark the hood position.
3. Remove or disconnect the following:
- Hood
- Wiper and blade assembly
- Top cowl seal and panel
- Window washer hoses from the ventilator louvers
- Left and right ventilator louvers
- Heater air duct
4. Properly relieve the fuel system pressure.
5. Remove or disconnect the following:
- Both battery cables
- Battery and tray
6. Drain the engine coolant.
7. Drain the engine oil.
8. Remove or disconnect the following:
- Intake air cleaner and case assembly
- Cruise control actuator, if equipped
- Upper suspension brace
- Upper and lower radiator hoses
- Radiator
- Automatic transmission oil cooler lines
- Any connectors, hoses and sensors that would interfere with engine removal
- Engine Control Module (ECM) engine wiring harness from inside the glove box; then, pull the harness into the engine compartment
- Compressor

➡**It may be necessary to remove the air conditioning compressor lines in order to remove the engine.**

- Automatic transmission shifter cable from the transaxle
- Header pipes from the exhaust manifolds

- Left and right fender apron seals
- Halfshafts
- Front driveshaft, for 4WD
- Stabilizer links and the steering intermediate shaft
- Power steering pump
- Engine undercover
- Engine hanger to the engine
- Engine sling device to the engine hangers
- Right-hand motor mount and moving control rod
- Front suspension lower braces
9. Lower the engine, transaxle and front suspension member as an assembly from the vehicle.

To install:
10. Raise the engine, transaxle and front suspension member as an assembly into the vehicle.
11. Install the front suspension lower braces, and tighten the fasteners, as follows:
- Bolt A: 134 ft. lbs. (181 Nm)
- Bolt B: 24 ft. lbs. (32 Nm)
- Nut C: 27 ft. lbs. (36 Nm)
12. Install or connect the following:
- Moving control rod. Tighten the bolts to 47 ft. lbs. (64 Nm).
- Right-hand motor mount. Tighten the bolts to 23 ft. lbs. (32 Nm).
- Engine sling device from the engine hangers
- Engine undercover
- Power steering pump hoses
- Stabilizer links and the steering intermediate shaft
- Front driveshaft, for 4WD
- Halfshafts
- Left and right fender apron seals
- Header pipes to the exhaust manifolds
- Automatic transmission shifter cable to the transaxle
- Air conditioning compressor to the engine
13. Push the wiring harness into the glove box.
14. Install or connect the following:
- ECM
- Any connectors, hoses and sensors that were removed
- Automatic transmission oil cooler lines
- Upper and lower radiator hoses and fit the radiator
- Front upper suspension brace. Tighten the nuts to 59 ft. lbs. (80 Nm).
- Cruise control actuator, if removed

- Intake air cleaner and case assembly
15. Fill the engine oil to proper level.
16. Fill the engine with coolant.
17. Install or connect the following:
- Battery tray and battery
- Battery cables
- Heater air duct
- Left and right ventilator louvers
- Window washer hoses from the ventilator louvers
- Top cowl seal and panel
- Wiper and blade assembly
- Hood
- New oil filter
18. Refill the engine with oil.
19. Refill the engine with engine coolant.
20. Install the engine undercovers.
21. Start the engine and check for leaks.

2.7L 4Runner

2WD

1. Before servicing the vehicle, refer to the precautions in the beginning of this section.
2. Properly relieve the fuel system pressure.
3. Remove or disconnect the following:
- Negative battery cable
- Engine undercover
4. Drain the engine coolant.
5. Drain the engine oil and the transmission oil.
6. Remove or disconnect the following:
- Hood
- Radiator
- Drive belt for the alternator and the water pump pulley
- Accelerator cable from the throttle body
- Actuator cover and the cruise control cable from the actuator, if equipped with cruise control
7. Remove or disconnect the air cleaner assembly, as follows:
- Intake Air Temperature (IAT) sensor and the Mass Air Flow (MAF) meter connectors
- 3 wire clamps and the engine wiring harness
- Air cleaner hose clamp, loosen it
- MAF meter, resonator and the air cleaner assembly
8. Install or connect the following:
- Air conditioning compressor, if equipped with air conditioning
- Alternator connector
- Heater hoses

Heater Core replacement is covered in Section 2 of this manual

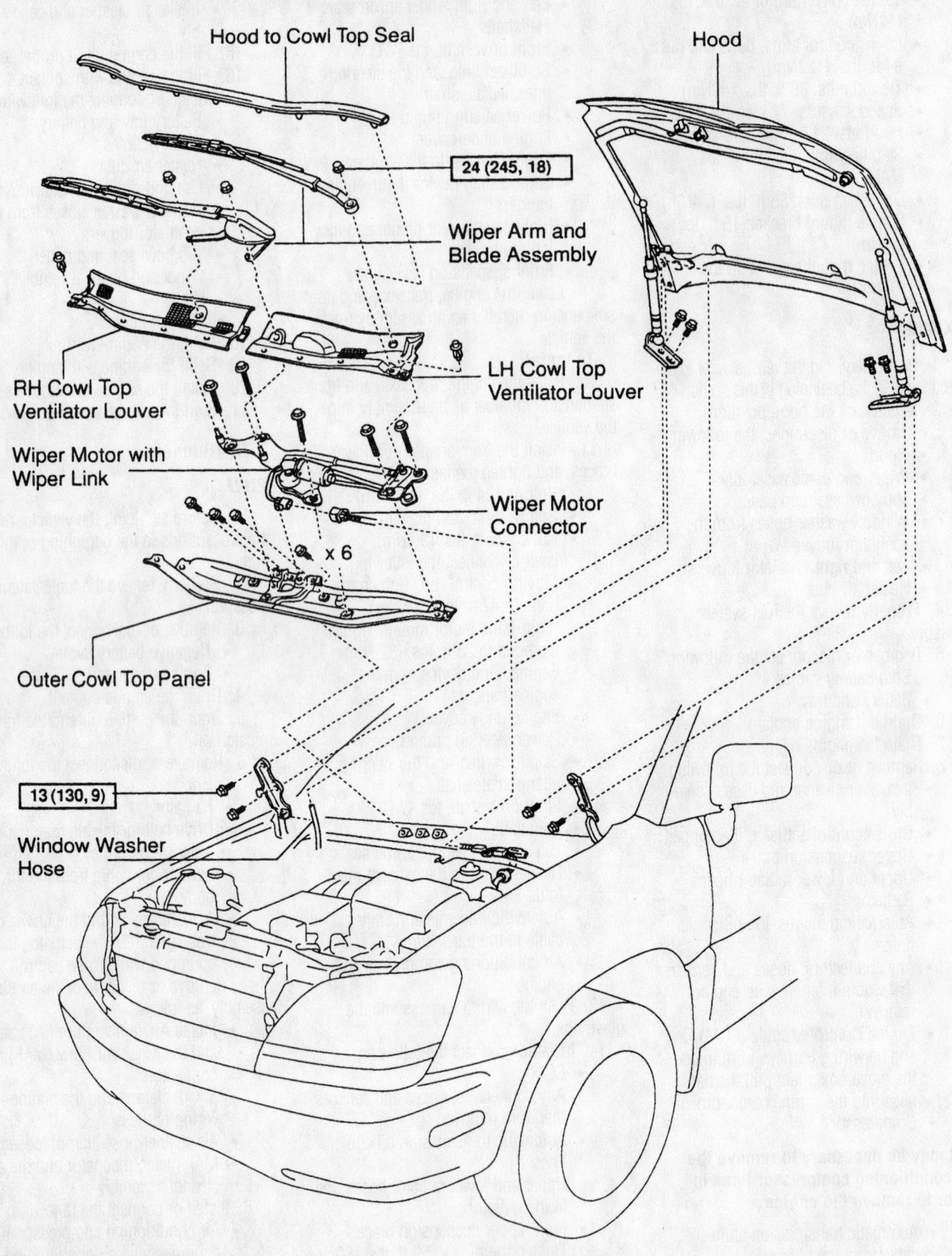

Hood to Cowl Top Seal

Hood

24 (245, 18)

Wiper Arm and
Blade Assembly

RH Cowl Top
Ventilator Louver

LH Cowl Top
Ventilator Louver

Wiper Motor with
Wiper Link

Wiper Motor
Connector

x 6

Outer Cowl Top Panel

13 (130, 9)

Window Washer
Hose

N·m (kgf·cm, ft·lbf) : Specified torque

7924ZG83

Exploded view of the top cowl and related components—RX 300

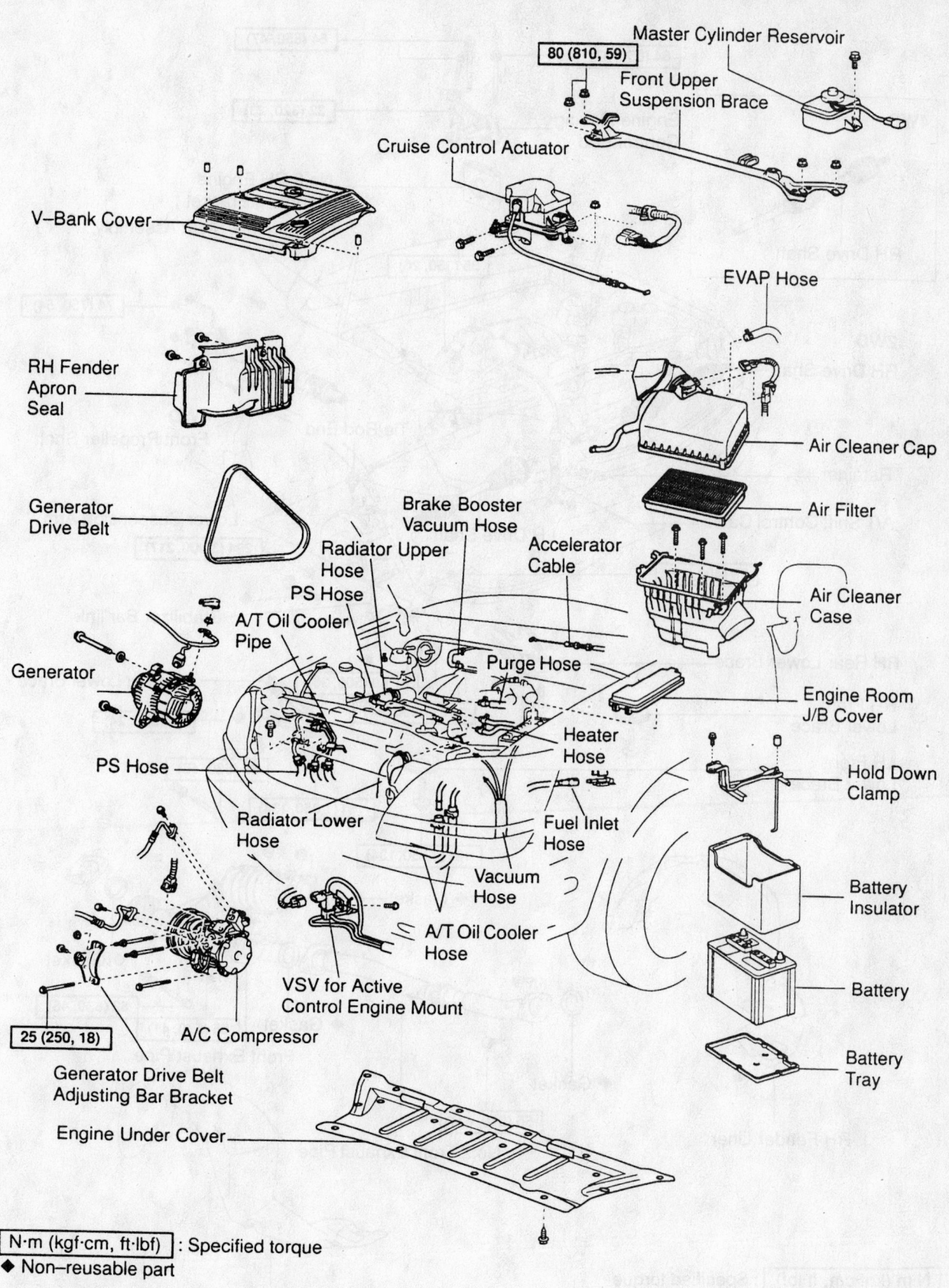

Master Cylinder Reservoir

80 (810, 59)

Front Upper Suspension Brace

Cruise Control Actuator

V–Bank Cover

EVAP Hose

RH Fender Apron Seal

Air Cleaner Cap

Air Filter

Generator Drive Belt

Brake Booster Vacuum Hose

Accelerator Cable

Radiator Upper Hose

PS Hose

Air Cleaner Case

A/T Oil Cooler Pipe

Purge Hose

Generator

Engine Room J/B Cover

PS Hose

Heater Hose

Radiator Lower Hose

Fuel Inlet Hose

Hold Down Clamp

Vacuum Hose

A/T Oil Cooler Hose

Battery Insulator

VSV for Active Control Engine Mount

Battery

25 (250, 18)

A/C Compressor

Generator Drive Belt Adjusting Bar Bracket

Battery Tray

Engine Under Cover

N·m (kgf·cm, ft·lbf) : Specified torque
◆ Non–reusable part

Exploded view of engine pre-removal components—RX 300

79247G84

Brake service is covered in Section 4 of this manual

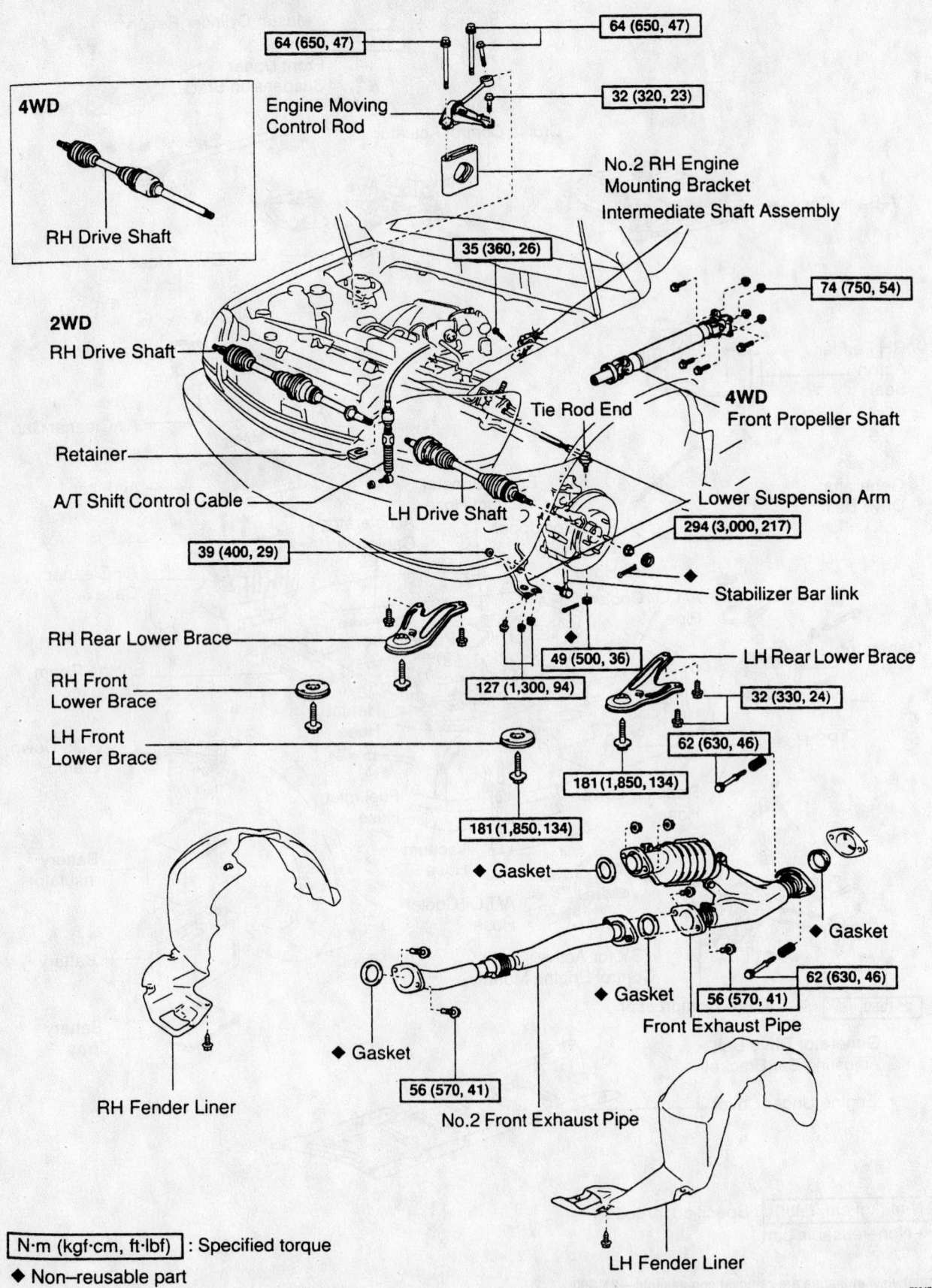

4WD

RH Drive Shaft

64 (650, 47)

64 (650, 47)

Engine Moving
Control Rod

32 (320, 23)

No.2 RH Engine
Mounting Bracket

Intermediate Shaft Assembly

35 (360, 26)

74 (750, 54)

2WD

RH Drive Shaft

Tie Rod End

4WD
Front Propeller Shaft

Retainer

Lower Suspension Arm

A/T Shift Control Cable

LH Drive Shaft

294 (3,000, 217)

39 (400, 29)

Stabilizer Bar link

RH Rear Lower Brace

LH Rear Lower Brace

49 (500, 36)

RH Front
Lower Brace

127 (1,300, 94)

32 (330, 24)

LH Front
Lower Brace

62 (630, 46)

181 (1,850, 134)

181 (1,850, 134)

Gasket

Gasket

62 (630, 46)

Gasket

56 (570, 41)

RH Fender Liner

Front Exhaust Pipe

Gasket

56 (570, 41)

No.2 Front Exhaust Pipe

LH Fender Liner

N·m (kgf·cm, ft·lbf) : Specified torque

◆ Non–reusable part

7924ZG85

Exploded view of engine removal and installation tightening specifications of the related components—RX 300

2WD

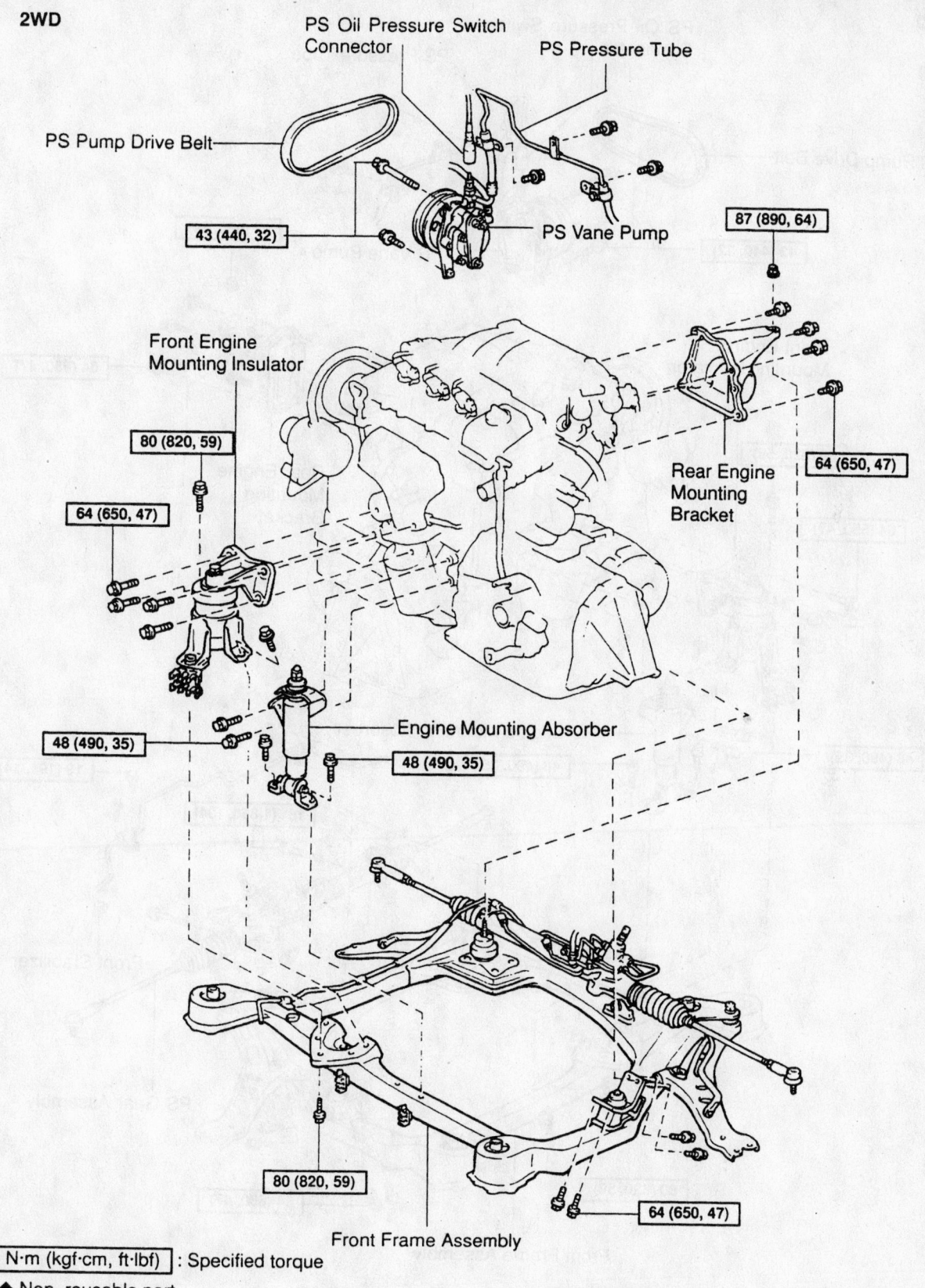

PS Oil Pressure Switch Connector

PS Pressure Tube

PS Pump Drive Belt

87 (890, 64)

43 (440, 32)

PS Vane Pump

Front Engine Mounting Insulator

80 (820, 59)

64 (650, 47)

Rear Engine Mounting Bracket

64 (650, 47)

48 (490, 35)

Engine Mounting Absorber

48 (490, 35)

80 (820, 59)

64 (650, 47)

Front Frame Assembly

N·m (kgf·cm, ft·lbf) : Specified torque

◆ Non−reusable part

7924ZG86

Exploded view of the suspension component removal and installation for engine removal—2WD RX 300

For complete Engine Mechanical specifications, see Section 1 of this manual

4WD

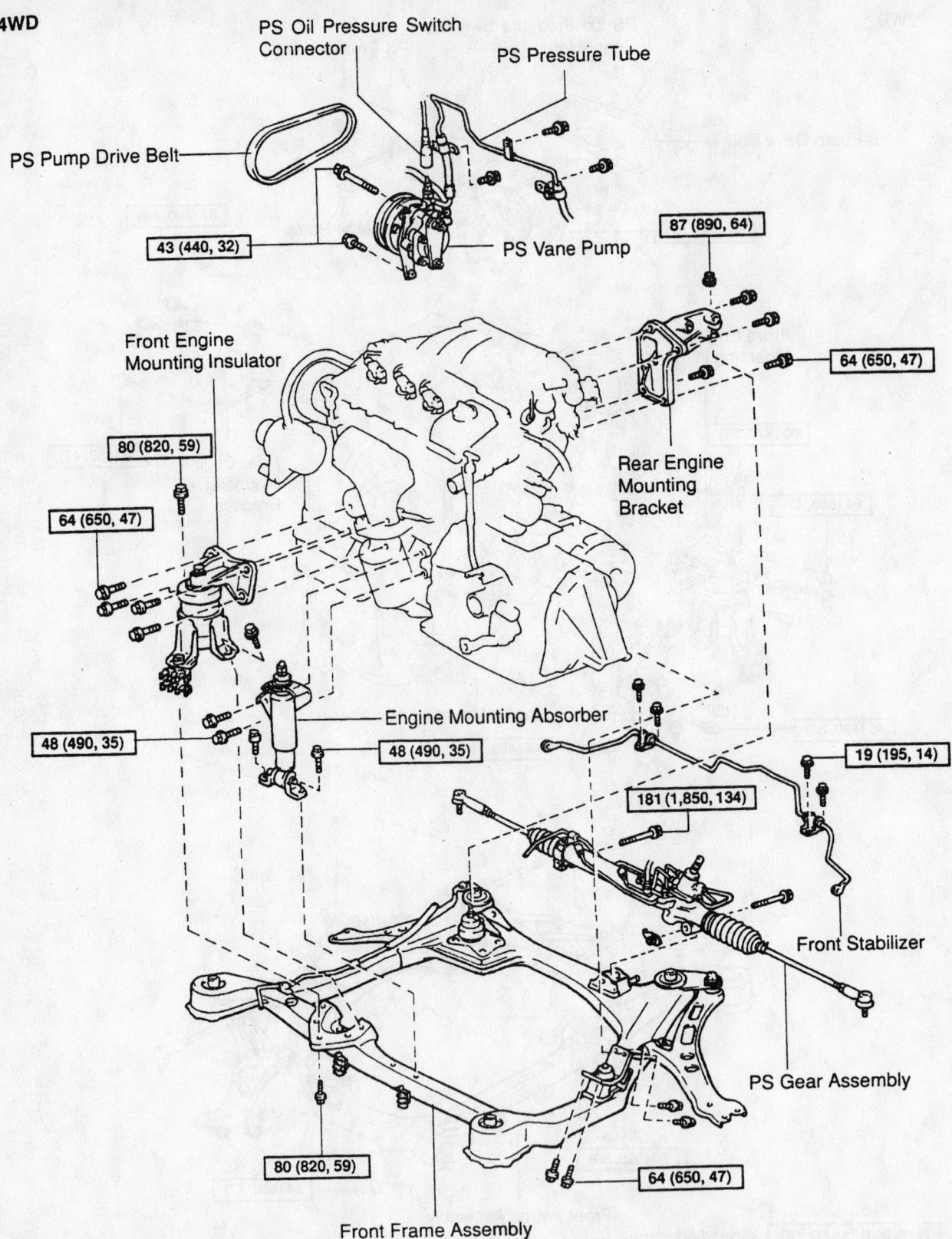

PS Oil Pressure Switch Connector

PS Pressure Tube

PS Pump Drive Belt

PS Vane Pump

43 (440, 32)

87 (890, 64)

Front Engine Mounting Insulator

Rear Engine Mounting Bracket

80 (820, 59)

64 (650, 47)

64 (650, 47)

Engine Mounting Absorber

48 (490, 35)

48 (490, 35)

19 (195, 14)

181 (1,850, 134)

Front Stabilizer

PS Gear Assembly

80 (820, 59)

64 (650, 47)

Front Frame Assembly

N·m (kgf·cm, ft·lbf) : Specified torque

◆ Non–reusable part

7924ZG87

Exploded view of the suspension component removal and installation for engine removal—4WD RX 300

9. Disconnect the following hoses:
- Brake booster vacuum hose
- Evaporative Emissions (EVAP) hose
- 2 air hoses for the power steering idle-up
- Fuel return hose
- Fuel inlet hose

10. Remove the power steering pump from the engine.

11. Disconnect the engine wiring harness, as follows:
- Glove box door
- Finish No. 2 panel, lower it
- 4 ECM connectors
- 2 cassette connectors and the 2 wire clamps from the lower finish panel
- Igniter
- Ground strap from the cowl top panel
- 2 engine wiring harness clamps.
- 2 engine wiring harness retainer-to-cowl panel nuts and the engine wiring harness

12. Install or connect the following:
- Heated Oxygen (HO2S) sensor
- Front exhaust pipe

13. Remove the shift lever assembly for a manual transmission, as follows:

14. Remove or disconnect the following:
- Shift lever knob
- 4 screws and the shift lever boot
- 6 bolts, the shift lever assembly and baffle

15. Remove or disconnect the following:
- Driveshaft
- Speedometer cable
- Clutch release cylinder, if equipped with a manual transmission
- Cross-shaft, if equipped with an automatic transmission
- Starter wire

16. Place a jack under the transmission and remove the engine rear mounting bracket.

17. Install a rear engine hanger in the correct direction.

18. Attach the engine hoist chain to the 2 engine hangers.

19. Remove or disconnect the following:
- Engine front mounting insulators-to-frame bolts/nuts
- Engine/transmission assembly
- Engine

To install:

20. Install or connect the following:
- Transmission to the engine
- Chain hoist to the engine hangers
- Engine/transmission assembly into the engine compartment

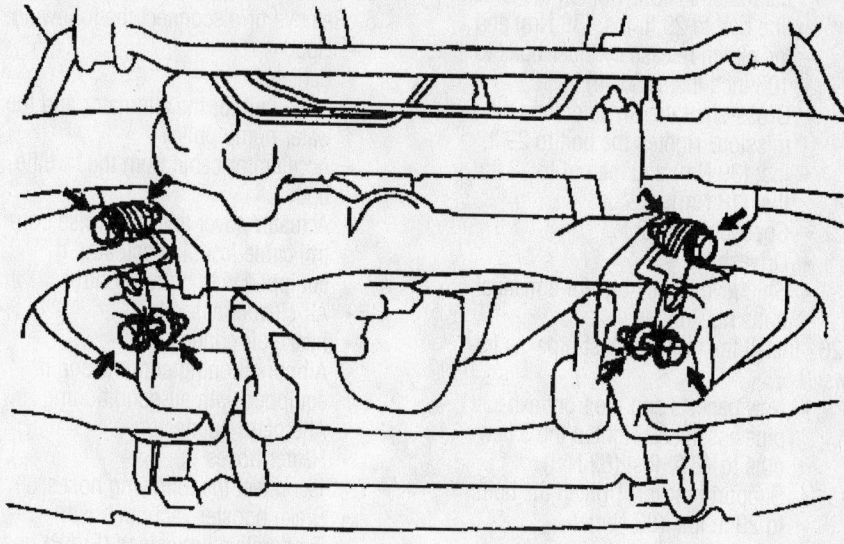

Be sure to support the engine before removing the right and left engine mounts—4Runner 2-wheel drive with 2.7L engine

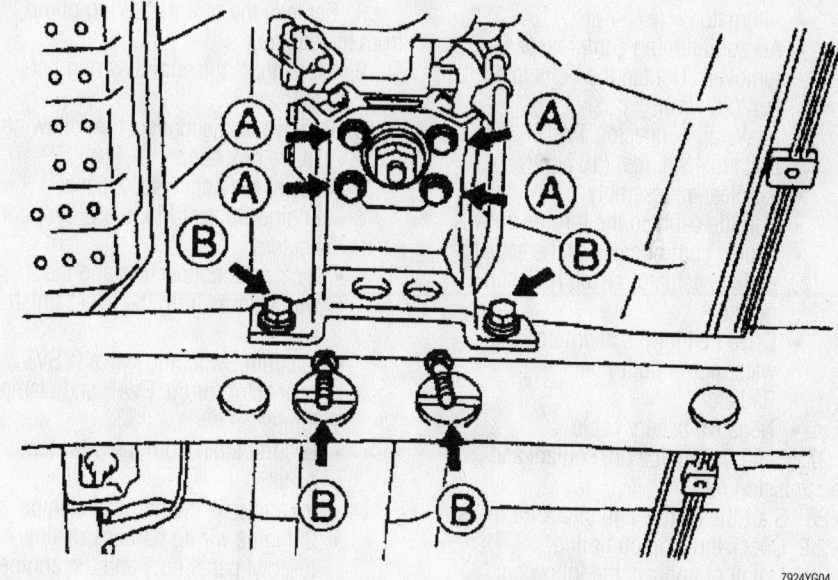

Bolt tightening pattern for the engine rear mounting bracket—4Runner 2-wheel drive with 2.7L engine

➡**Keep the engine level and align the right and left mounting and body mountings.**

- Right and left mounting insulators to the body mountings and temporarily install the bolts/nuts

21. Raise the transmission onto the frame
22. Remove the chain hoist.
23. Remove the bolt and the rear engine hanger.

24. Install the engine rear mounting bracket and tighten to:
- Bolt A: 13 ft. lbs. (19 Nm)
- Bolt B: 19 ft. lbs. (26 Nm)

25. Install or connect the following:
- Tighten the left and right engine mounting insulator bolts and nuts to 28 ft. lbs. (38 Nm).
- Starter wire
- Clutch release cylinder, for a man-

For Accessory Drive Belt illustrations, see Section 1 of this manual

ual transmission. Tighten the clutch line bolt to 29 ft. lbs. (39 Nm) and the clutch release cylinder bolts to 108 inch lbs. (13 Nm).

- Cross-shaft, for an automatic transmission. Tighten the bolt to 29 ft. lbs. (39 Nm) and the nut to 13 ft. lbs. (18 Nm).
- Speedometer cable
- Driveshaft
- Shift lever assembly, for a manual transmission

26. Install the front exhaust pipe, as follows:

- New gaskets and the front exhaust pipe assembly. Tighten the 3 new nuts to 46 ft. lbs. (62 Nm).
- Support bracket. Tighten the bolts to 29 ft. lbs. (39 Nm).
- 3-way catalytic converter with a new gasket to the tail pipe. Tighten to 29 ft. lbs. (39 Nm).
- HO$_2$S sensor
- Engine wiring harness
- Power steering pump
- All hoses previously removed
- Alternator wire
- Air conditioning compressor, if removed. Tighten the bolts to 18 ft. lbs. (25 Nm).
- Intake air connector. Tighten the bolts to 13 ft. lbs. (18 Nm).
- Air cleaner assembly
- Throttle cable to the throttle body
- Cruise control cable to the actuator and the actuator cover, if disconnected
- Drive belt for the alternator and water pump pulley
- Radiator
- Negative battery cable

27. Refill the engine oil, coolant and transmission oil.
28. Start the engine and check for leaks.
29. Check the ignition timing.
30. Install or connect the following:
- Engine undercover
- Hood

31. Road test the vehicle and recheck the fluid levels.

4WD

1. Before servicing the vehicle, refer to the precautions in the beginning of this section.
2. Properly relieve the fuel system pressure.
3. Remove or disconnect the following:
- Negative battery cable
- Transmission
- Engine undercover

4. Drain the engine coolant.

---

5. Drain the engine oil.
6. Remove or disconnect the following:
- Hood
- Radiator
- Drive belt for the alternator and the water pump pulley
- Accelerator cable from the throttle body
- Actuator cover and the cruise control cable from the actuator, if equipped with cruise control
- Air cleaner assembly
- Intake air connector
- Air conditioning compressor, if equipped with air conditioning
- Alternator connector
- Heater hoses

7. Disconnect the following hoses:
- Brake booster vacuum hose
- Evaporative Emissions (EVAP) hose
- 2 air hoses for the power steering idle-up
- With Automatic Disconnecting Differential (ADD) Vacuum hose
- Fuel return hose
- Fuel inlet hose

8. Remove the power steering pump from the engine.
9. Disconnect the engine wiring harness, as follows:
10. Remove or disconnect the following:
- Glove box door
- Lower the finish No. 2 panel
- Engine Control Module (ECM) connectors
- 2 cassette connectors and the 2 wire clamps from the lower finish panel
- Vacuum Switching Valve (VSV) connector for the EVAP and clamp
- Igniter
- Ground strap from the cowl top panel
- 2 engine wiring harness clamps
- 2 engine wiring harness retainer-to-cowl panel nuts and the engine wiring harness

11. Install a rear engine hanger in the correct direction.
12. Attach the engine hoist chain to the 2 engine hangers.
13. Remove or disconnect the following:
- Engine front mounting insulators-to-frame bolts and nuts
- Engine, making sure that the engine is clear of all wiring and hoses
- Engine

To install:

14. Attach a chain hoist to the engine hangers.
15. Install or connect the following:

- Engine into the engine compartment

➡**Keep the engine level and align the right and left mounting and body mountings**

- Right and left mounting insulators to the body mountings and temporarily install the bolts and nuts

16. Remove the chain hoist.
17. Remove the bolt and the rear engine hanger.
18. Install or connect the following:
- Tighten the right and left engine mounting insulator bolts/nuts to 28 ft. lbs. (38 Nm).
- Engine wiring harness
- Alternator wire
- Power steering pump
- All hoses
- Air conditioning compressor. Tighten the bolts to 18 ft. lbs. (25 Nm).
- Intake air connector. Tighten the bolts to 13 ft. lbs. (18 Nm).
- Air cleaner assembly
- Throttle cable to the throttle body
- Cruise control cable to the actuator and the actuator cover, if removed
- Radiator
- Hood

19. Fill with engine with oil and the coolant system with coolant.
20. Install or connect the following:
- Transmission
- Engine undercover
- Negative battery cable

21. Fill the transmission fluid.
22. Check the ignition timing.
23. Test drive the vehicle and check for leaks.
24. Recheck fluid levels.

3.4L 4Runner

2WD

1. Before servicing the vehicle, refer to the precautions in the beginning of this section.
2. Properly relieve the fuel system pressure.
3. Remove or disconnect the following:
- Hood
- Battery
- Engine under covers

4. Drain the engine coolant.
5. Drain the engine oil.
- Radiator
- Fan with the fluid coupling and fan pulleys
- Air cleaner cap
- Mass Air Flow (MAF) meter and the resonator

- Air cleaner case and filter

6. Disconnect the following hoses:
   - Heater hoses
   - Brake booster vacuum hose
   - Evaporative Emissions (EVAP) hose
   - Fuel return hose
   - Fuel inlet hose
7. Detach the starter wire and connectors, as follows:
   - Ground strap, by removing the bolt
   - 3 starter wire clamps and connector
8. Detach the alternator connector and wire.
9. Disconnect the engine wiring harness, as follows:
   - Glove box door
   - Lower the finish No. 2 panel
   - 4 ECM connectors
   - 2 cassette connectors and the 2 wire clamps from the lower finish panel
   - Engine wiring harness clamp
10. Remove or disconnect the following:
    - Igniter connector
    - Ground strap
    - Vacuum Switching Valve (VSV) connector for the Evaporative Emissions (EVAP)
    - Vapor pressure sensor connector and clamp
    - Vapor connector for the vapor pressure sensor and clamp
    - 2 engine wiring harness retainer-to-cowl panel nuts and pull out the engine wiring harness
    - Driveshaft from the transmission
    - Speedometer cable
    - Front exhaust pipe
    - Nut and the control cable
11. Place a jack under the transmission.
12. Remove or disconnect the following:
    - Transmission rear mounting bracket by removing the 8 bolts
    - Bolt and the air conditioning compressor wire clamp, if equipped with air conditioning
13. If necessary, install a No. 2 engine hanger with 2 bolts. Tighten the 2 bolts to 30 ft. lbs. (40 Nm).
14. Attach the engine hoist chain to the 2 engine hangers.
15. Remove or disconnect the following:
    - 4 engine front mounting insulators-to-frame bolts and nuts
    - Engine and transmission

To install:
16. Install or connect the following:
    - Engine
    - Engine mounts to the body mount-

ings. Install the bolts and nuts but do not tighten at this time.
17. Remove the engine chain hoist the No. 2 engine hanger.
18. Install or connect the following:
    - Air conditioning wire with the bolt, if equipped with air conditioning
    - Transmission mounting bracket. Tighten the frame bolts to 43 ft. lbs. (58 Nm) and the mounting insulator bolts to 13 ft. lbs. (18 Nm).
    - Tighten the engine mounting nuts and bolts to 28 ft. lbs. (38 Nm).
    - Control cable
    - Front exhaust pipe
    - Speedometer cable
    - Driveshaft
    - All engine wiring harness, hoses and cables
    - Fan with the fluid coupling and fan pulleys. Tighten the nuts to 48 inch lbs. (5.4 Nm).
    - Air cleaner case and air filter
    - MAF meter, resonator and the air cleaner cap
    - Radiator
19. Fill the engine with oil.
20. Fill the engine and radiator with coolant.
21. Install or connect the following:
    - Engine undercover
    - Battery
    - Hood
22. Start the engine and check for leaks.
23. Make any necessary adjustments and road test the vehicle.

4WD

1. Before servicing the vehicle, refer to the precautions in the beginning of this section.
2. Remove or disconnect the following:
   - Transmission
   - Hood
3. Release the fuel system pressure.
4. Remove or disconnect the following:
   - Battery
   - Engine undercovers
5. Drain the engine coolant.
6. Drain the engine oil.
7. Remove or disconnect the following:
   - Radiator
   - Fan with the fluid coupling and fan pulleys
   - Air cleaner cap
   - Mass Air Flow (MAF) meter and the resonator
8. Disconnect the following hoses:
   - Heater hoses
   - Brake booster vacuum hose

- Evaporative Emissions (EVAP) hose
- Automatic Disconnecting Differential (ADD) vacuum hose
- Fuel return hose
- Fuel inlet hose

9. Detach the starter wire and connectors, as follows:
   - Ground strap, by removing the bolt
   - 3 starter wire clamps and connector
10. Detach the alternator connector and wire.
11. Disconnect the engine wiring harness, as follows:
    - Glove box door
    - Lower the finish No. 2 panel
    - 4 ECM connectors
    - 2 cassette connectors and the 2 wire clamps from the lower finish panel
    - Engine wiring harness clamp
12. Disconnect the following:
    - Igniter connector
    - Ground strap
    - Vacuum Switching Valve (VSV) connector for the EVAP
    - Vapor pressure sensor connector and clamp
    - Vapor connector for the vapor pressure sensor and clamp
13. Remove or disconnect the following:
    - 2 engine wiring harness retainer-to-cowl panel nuts and wiring harness
    - Air conditioning compressor wire clamp, if equipped with air conditioning
14. If necessary, install a No. 2 engine hanger with 2 bolts. Tighten the 2 bolts to 30 ft. lbs. (40 Nm).
15. Attach the engine hoist chain to the 2 engine hangers.
16. Remove or disconnect the following:
    - 4 engine front mounting insulators-to-frame bolts and nuts
    - Engine

To install:
17. Install or connect the following:
    - Engine
    - Engine mounts-to-body mountings. Install the bolts and nuts but do not tighten at this time.
18. Remove the engine chain hoist the No. 2 engine hanger.
19. Install or connect the following:
    - Air conditioning wire with the bolt, if equipped with air conditioning
    - Tighten the engine mounting nuts and bolts to 28 ft. lbs. (38 Nm).
    - Engine wiring harness
    - Engine wiring harness clamp

For Tire, Wheel and Ball Joint specifications, see Section 1 of this manual

- All wires, hoses and cables
- Fan with the fluid coupling and fan pulleys. Tighten the nuts to 48 inch lbs. (5.4 Nm).
- Air cleaner case and air filter
- MAF meter, resonator and the air cleaner cap
- Radiator

20. Fill the engine with oil.
21. Fill the engine and radiator with coolant.
22. Install or connect the following:
- Transmission and refill it with transmission oil
- Engine undercover
- Battery
- Hood

23. Start the engine, make any necessary adjustments and check for leaks.

Water Pump

REMOVAL & INSTALLATION

RAV4

1. Before servicing the vehicle, refer to the precautions in the beginning of this section.
2. Remove or disconnect the following:
- Negative battery cable
- Right-hand engine undercover

3. Drain the engine coolant from the radiator and engine.
4. Remove or disconnect the following:
- Timing belt
- Lower radiator hose from the water inlet
- Timing belt tension spring and the No. 2 idler pulley
- Crankshaft Position (CKP) sensor connector clamp
- Alternator drive belt adjusting bar
- 2 water pump-to-water bypass pipe nuts
- 3 water pump bolts in the sequence
- Water pump cover from the water bypass pipe
- Water pump and water pump cover assembly
- Gasket and 2 O-rings from the water pump and water bypass pipe
- 3 bolts, water pump and gasket, from the water pump cover

To install:
5. Install or connect the following:
- Water pump to the water pump cover, using a new gasket. Tighten the 3 bolts to 78 inch lbs. (9 Nm).
- New O-ring and gasket to the water pump cover

- New O-ring to the water bypass pipe, by applying soapy water to the O-ring
- Water pump cover to the water bypass pipe; do not install the nuts at this time
- Water pump. Tighten the 3 bolts, in sequence, to 78 inch lbs. (9 Nm).
- Water pump cover to the water pump pipe. Tighten the 2 bolts to 82 inch lbs. (9 Nm).
- Alternator drive belt adjusting bar. Tighten the bolt to 20 ft. lbs. (27 Nm).
- CKP connector clamp
- No. 2 idler pulley and timing belt tension spring
- Lower radiator hose
- Timing belt
- Negative battery cable

6. Fill the engine and radiator with engine coolant.
7. Start the engine and check for leaks.
8. Install the right-hand engine undercover.

2.4L Highlander

1. Before servicing the vehicle, refer to the precautions in the beginning of this section.
2. Disconnect the negative battery cable.
3. Drain the engine coolant.
4. Remove or disconnect the following:
- Alternator
- Water pump pulley
- Water pump

5. Installation is the reverse of removal. Torque the pump bolts and nuts to 80 inch lbs. (9Nm) and the pulley bolts to 19 ft. lbs. (26Nm).

3.0L Highlander

1. Before servicing the vehicle, refer to the precautions in the beginning of this section.
2. Disconnect the negative battery cable.
3. Drain the engine coolant.
4. Remove or disconnect the following:
- Right front wheel
- Right fender apron seal
- Accessory drive belts
- Upper engine mount and stay
- Alternator and bracket
- Crankshaft pulley
- Timing belt covers
- Transverse engine mounting bracket
- Timing belt
- Timing belt idler
- Camshaft pulley
- Water pump

5. Installation is the reverse of removal.

Torque the water pump bolts and nuts to 71 inch lbs. (8Nm).

RX 300

1. Before servicing the vehicle, refer to the precautions in the beginning of this section.
2. Disconnect the negative battery cable.
3. Drain the engine coolant.
4. Remove or disconnect the following:
- Wiper and blade assembly
- Top cowl seal and panel
- Window washer hoses, from the ventilator louvers
- Left and right ventilator louvers
- Heater air duct
- Front upper suspension brace
- Timing belt

5. Mark the left and right camshaft pulleys with a touch of paint.
6. Remove or disconnect the following:
- Right and left camshaft pulleys bolts
- Pulleys from the engine

➡ **Be sure not to mix up the pulleys.**

- No. 2 idler pulley by removing the bolt
- 3 clamps and engine wire from the rear timing belt cover
- 6 No. 3 timing belt cover-to-engine bolts
- Water pump nuts/bolts
- Water pump and gasket from the engine

To install:
7. Check that the water pump turns smoothly. Also check the air hole for coolant leakage.
8. Apply liquid sealer to the gasket, water pump and engine block.
9. Install or connect the following:
- Water pump, using a new gasket. Tighten the nuts/bolts to 53 inch lbs. (6 Nm).
- Rear timing belt cover. Tighten the 6 bolts to 74 inch lbs. (9 Nm).
- Engine wire with the 3 clamps to the rear timing belt cover
- No. 2 idler pulley. Tighten the bolt to 32 ft. lbs. (43 Nm).

➡ **After tightening the bolt, be sure the idler pulley moves smoothly.**

- Right-hand camshaft pulley, with the flange side **outward.**

➡ **Be sure to align the knock pin hole on the camshaft pulley with the knock pin on the camshaft.**

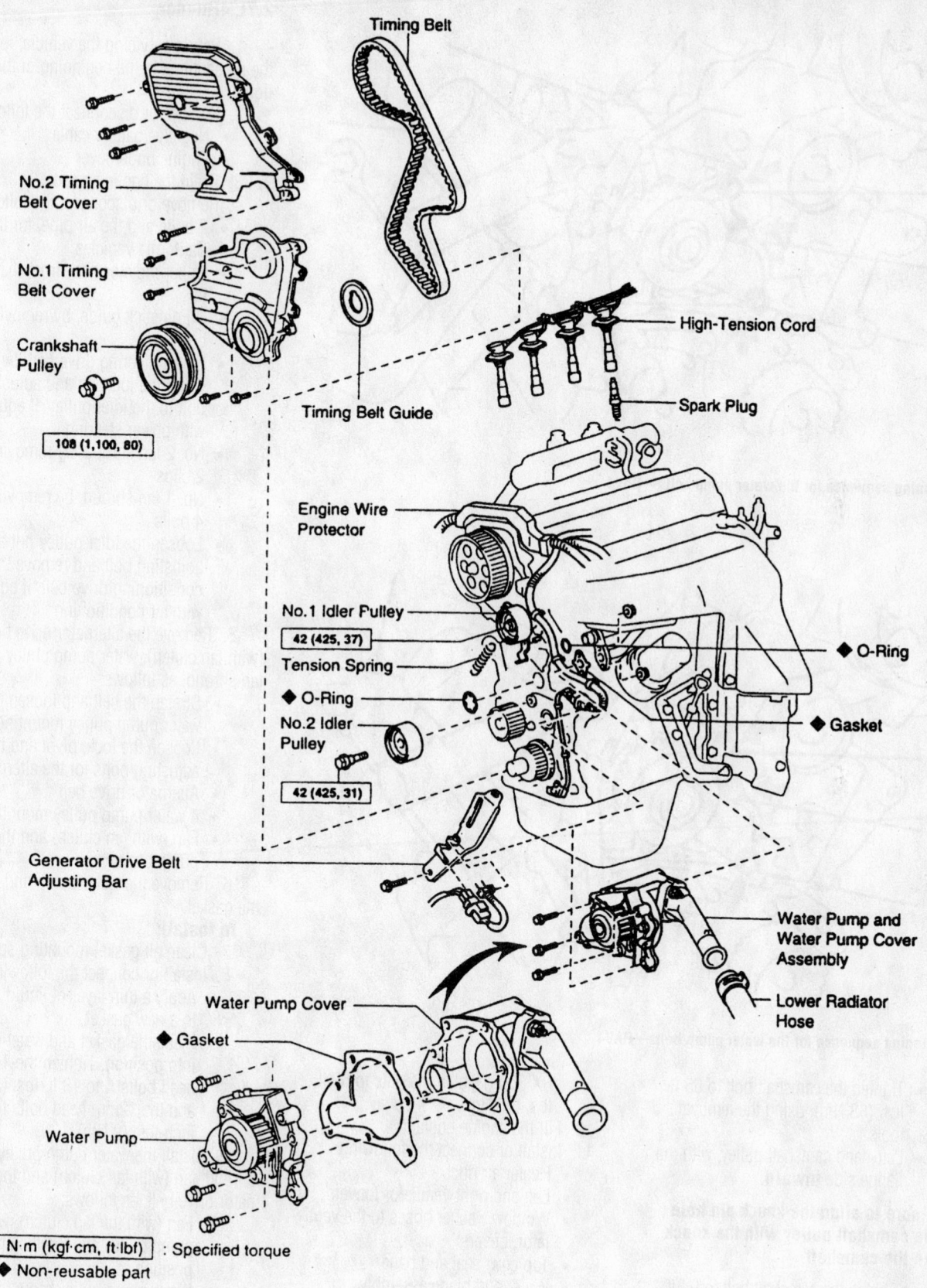

No.2 Timing Belt Cover

No.1 Timing Belt Cover

Crankshaft Pulley

108 (1,100, 80)

Timing Belt

Timing Belt Guide

High-Tension Cord

Spark Plug

Engine Wire Protector

No.1 Idler Pulley

42 (425, 37)

Tension Spring

◆ O-Ring

No.2 Idler Pulley

42 (425, 31)

◆ O-Ring

◆ Gasket

Generator Drive Belt Adjusting Bar

Water Pump and Water Pump Cover Assembly

Lower Radiator Hose

Water Pump Cover

◆ Gasket

Water Pump

N·m (kgf·cm, ft·lbf) : Specified torque
◆ Non-reusable part

7924ZG10

Exploded view of the water pump and related components—RAV4

For Wheel Alignment specifications, see Section 1 of this manual

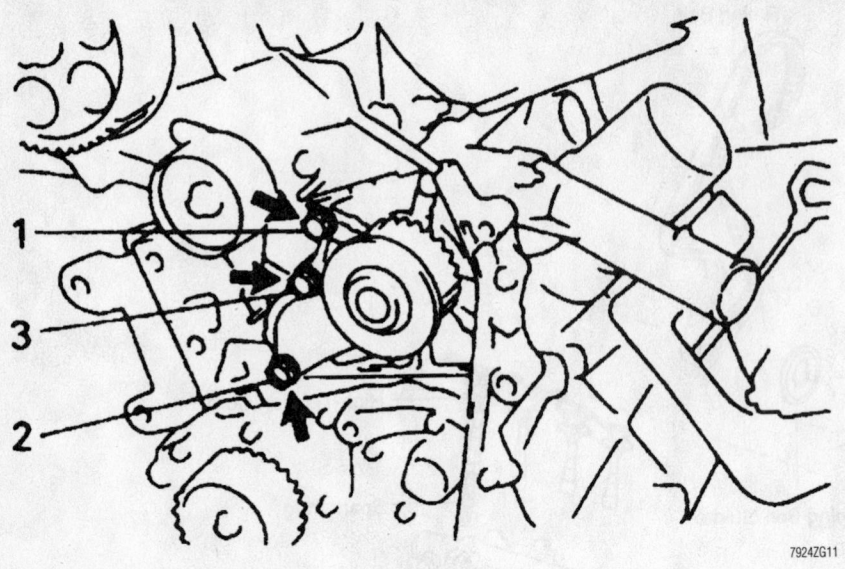

Loosening sequence for the water pump bolts—RAV4

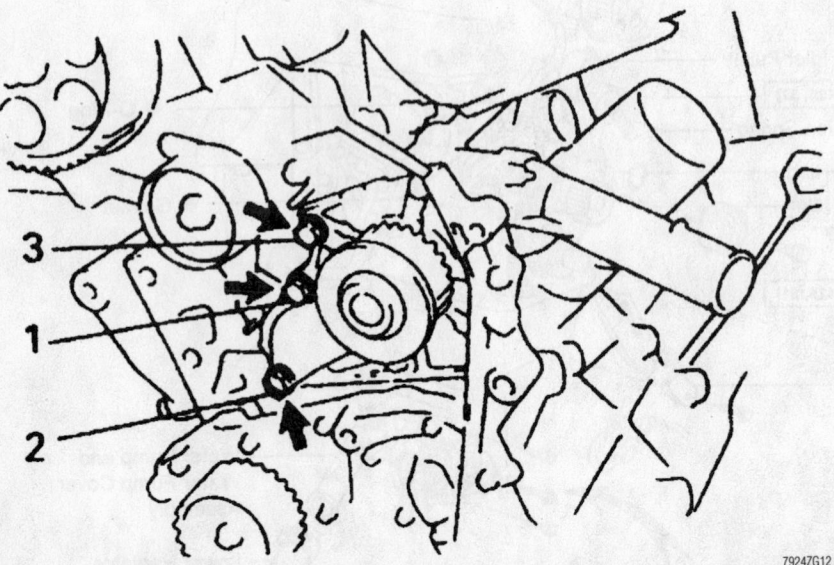

Tightening sequence for the water pump bolts—RAV4

- Tighten the camshaft bolt to 65 ft. lbs. (88 Nm), using the removal tools
- Left-hand camshaft pulley, with the flange side **inward**.

➡**Be sure to align the knock pin hole on the camshaft pulley with the knock pin on the camshaft.**

- Tighten the camshaft bolt to 94 ft. lbs. (125 Nm), using the removal tools
- Timing belt
- Front upper suspension brace, for

RX 300. Tighten the nuts to 59 ft. lbs. (80 Nm).
10. Fill the engine coolant.
11. Install or connect the following:
- Heater air duct
- Left and right ventilator louvers
- Window washer hoses to the ventilator louvers
- Top cowl seal and panel
- Wiper and blade assembly
- Negative battery cable
12. Start the engine.
13. Top off the engine coolant and check for leaks.

2.7L 4Runner

1. Before servicing the vehicle, refer to the precautions in the beginning of this section.
2. Remove or disconnect the following:
- Negative battery cable
- Engine undercover
3. Drain the cooling system.
4. Remove or disconnect the following:
- 2 bolts and the air pipe, for the California vehicles
- Upper radiator hose from the radiator
- Oil dipstick guide, by removing the bolt
- Power steering drive belt, by loosening the lockbolt and adjusting bolt to the idler pulley, if equipped with power steering
- No. 2 fan shroud, by removing the 2 clips
- No. 1 fan shroud, by removing the 4 bolts
- Loosen the idler pulley nut and adjusting bolt and remove the air conditioning drive belt, if equipped with air conditioning
5. Remove the alternator drive belt, fan (with fan clutch), water pump pulley and the fan shroud, as follows:
- Stretch the belt and loosen the water pump pulley mounting nuts
- Loosen the lock, pivot and the adjusting bolts for the alternator
- Alternator drive belt
- 4 water pump pulley mounting nuts
- Fan (with fan clutch) and the water pump pulley
6. Remove the water pump and discard the gasket.

To install:
7. Clean all gasket mounting surfaces.
8. Install or connect the following:
- Apply a thin layer of liquid sealant to a new gasket
- Place the gasket and water pump into position. Tighten the 14mm head bolts **A** to 18 ft. lbs. (25 Nm) and the 12mm head bolts to 78 inch lbs. (9 Nm).
9. Install the water pump pulley, fan shroud, fan (with fan clutch) and the alternator drive belt, as follows:
- Fan (with the fan clutch), water pump pulley and the fan shroud in position
- Water pump pulley mounting nuts but do not tighten the nuts at this time
- Alternator drive belt
- Stretch the alternator belt tight.

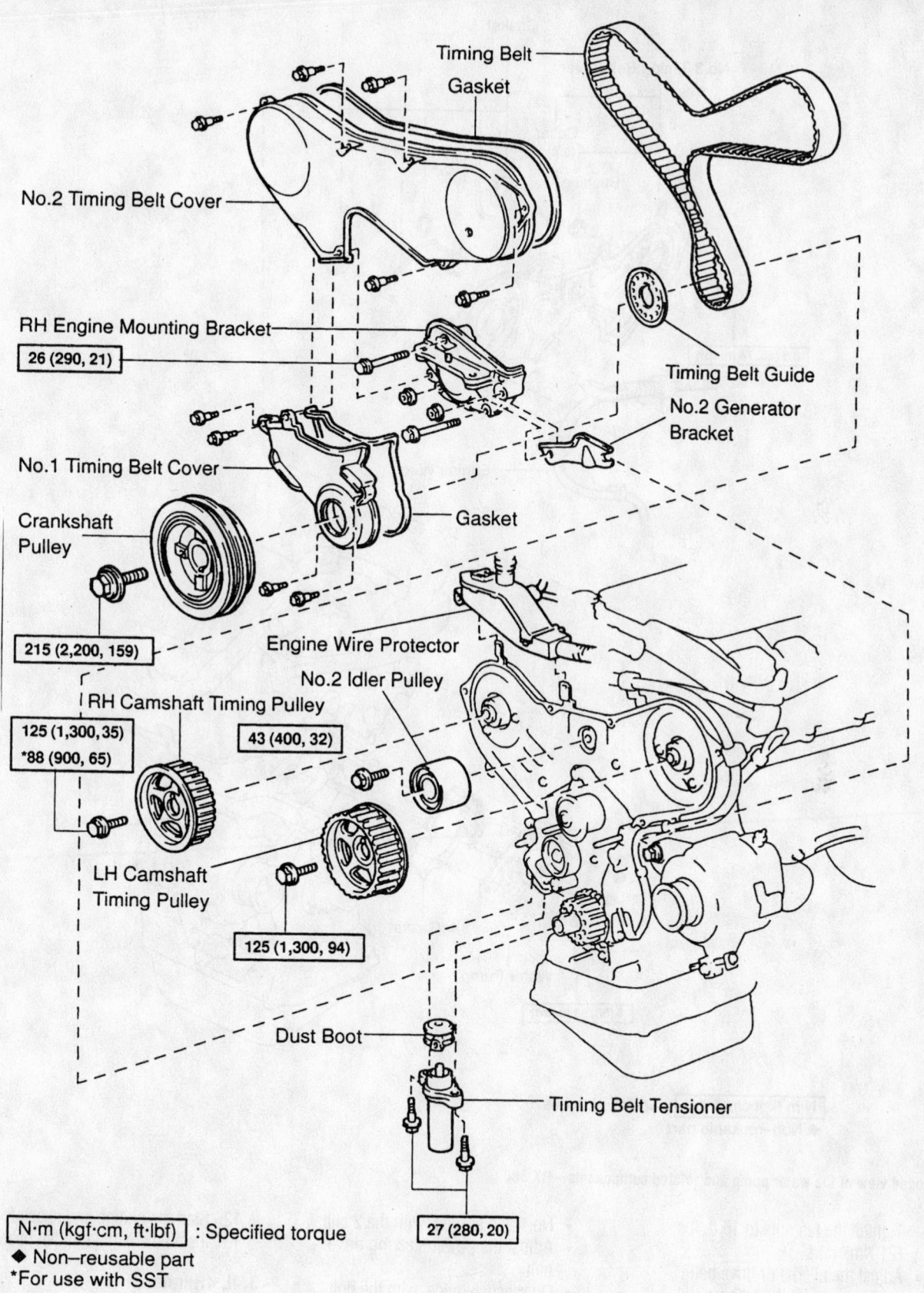

Timing Belt

Gasket

No.2 Timing Belt Cover

RH Engine Mounting Bracket
26 (290, 21)

Timing Belt Guide

No.2 Generator Bracket

No.1 Timing Belt Cover

Crankshaft Pulley

Gasket

215 (2,200, 159)

Engine Wire Protector

No.2 Idler Pulley

RH Camshaft Timing Pulley

125 (1,300, 35)
*88 (900, 65)

43 (400, 32)

LH Camshaft Timing Pulley

125 (1,300, 94)

Dust Boot

Timing Belt Tensioner

27 (280, 20)

N·m (kgf·cm, ft·lbf) : Specified torque
◆ Non-reusable part
*For use with SST

7924ZG15

Exploded view of the components to gain access to the water pump—RX 300

For Maintenance Interval recommendations, see Section 1 of this manual

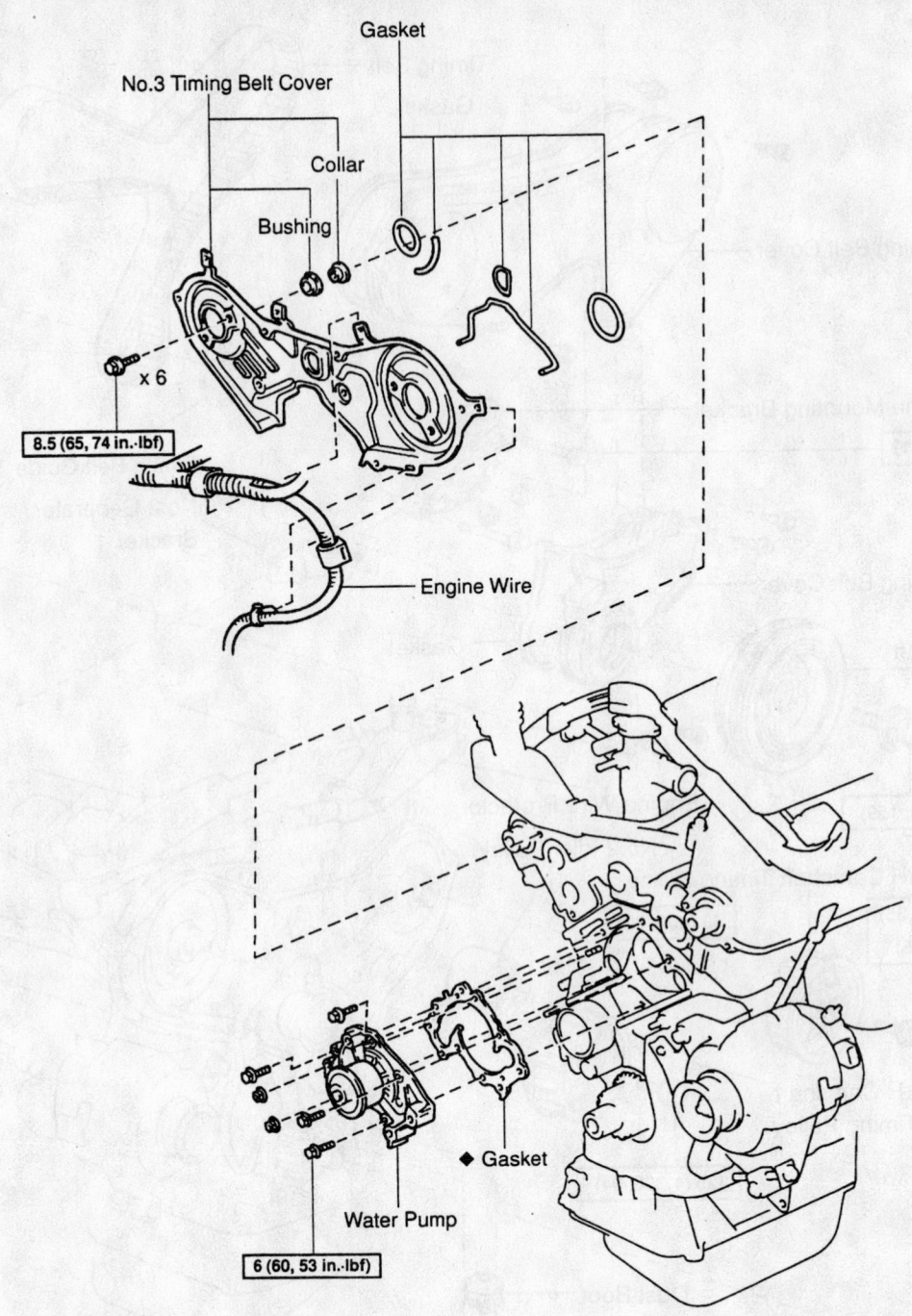

Gasket

No.3 Timing Belt Cover

Collar

Bushing

x 6

8.5 (65, 74 in.-lbf)

Engine Wire

◆ Gasket

Water Pump

6 (60, 53 in.-lbf)

N·m (kgf·cm, ft·lbf) : Specified torque
◆ Non–reusable part

7924ZG16

Exploded view of the water pump and related components—RX 300

Tighten the fan nuts to 16 ft. lbs. (21 Nm).
- Adjust the alternator drive belt
10. Install or connect the following:
  - Adjust the drive belt, if equipped with air conditioning
  - No. 1 fan shroud, by installing the 4 bolts

- No. 2 fan shroud, with the 2 clips
- Adjust the power steering drive belt
- Oil dipstick guide, with the bolt
- Upper radiator hose to the radiator
- Air pipe, If removed
- Negative battery cable
11. Fill and bleed the cooling system.

12. Start the engine and check for leaks.
13. Install the engine undercover.

3.4L 4Runner

1. Before servicing the vehicle, refer to the precautions in the beginning of this section.

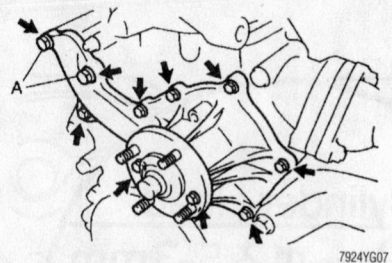

Water pump mounting bolt locations—2.7L 4Runner

2. Disconnect the negative battery cable.
3. Drain the cooling system.
4. Remove or disconnect the following:
   - Timing belt
   - Thermostat
   - No. 2 oil cooler hose from the water pump
   - Water pump
5. Thoroughly clean the mating surfaces.

To install:

6. Apply sealant (PN 08826-00100) to the water pump.

✳✳ WARNING

Parts must be assembled within 5 minutes of application. Otherwise the material must be removed and reapplied.

7. Install or connect the following:
   - Water pump. Tighten the bolts to 14 ft. lbs. (20 Nm).
   - No. 2 oil cooler hose
   - Thermostat
   - Timing belt
   - Negative battery cable
8. Fill the cooling system.
9. Start the engine and check for leaks.

Cylinder Head

REMOVAL & INSTALLATION

Rav4

1. Before servicing the vehicle, refer to the precautions in the beginning of this section.
2. Release the fuel system pressure.
3. Remove or disconnect the following:
   - Negative battery cable
   - Right-hand engine undercover
4. Drain the engine coolant.
5. Remove or disconnect the following:

- Camshafts
- Cylinder head bolts in several passes
- Cylinder head with the intake manifold
- Air hose from the intake manifold
- 2 bolts and the air tube
- Intake manifold and gasket
- Air hose from the cylinder head port
- Air hose
- Fuel delivery pipe and the injectors
- Oil pressure switch

To install:

6. Install or connect the following:
   - Oil pressure switch
   - Fuel injectors and the delivery pipe
   - Air hose to the cylinder head port
   - Intake manifold with new gaskets. Tighten the nut/bolts to 14 ft. lbs. (19 Nm).
   - Air tube with the 2 bolts
   - Air hose to the intake manifold
7. Clean the gasket mating surfaces using care not to damage the aluminum components, replace the gasket; then, lower the cylinder head onto the engine. Be sure the dowel pins are aligned and no hoses or wires are between the head and cylinder block.
8. For 1999–01, tighten the cylinder head bolts as follows:
   a. Apply a light coat of engine oil to the cylinder head bolts.
   b. Tighten the cylinder head bolts, in several passes, in sequence, to 36 ft. lbs. (49 Nm).
   c. Mark the front of the cylinder head bolt with paint.
   d. Retighten the cylinder head bolts by 90 degrees in sequence.

   e. Retighten an additional 90 degrees and be sure that the paint mark is now positioned toward the rear.
9. For 2002, tighten the cylinder head bolts in 3 progressive steps, as follows:
   a. Apply a light coat of engine oil to the cylinder head bolts.
   b. Tighten the cylinder head bolts, in sequence, to 15 ft. lbs. (26 Nm).
   c. Tighten the cylinder head bolts, in sequence, to 30 ft. lbs. (52 Nm).
   d. Mark the front of the cylinder head bolt with paint.
   e. Retighten the cylinder head bolts by 90 degrees in sequence.
10. Install or connect the following:
    - Intake and exhaust camshafts
    - Negative battery cable
11. Refill the engine with coolant, start the engine, warm up and check for leaks.
12. Bleed the cooling system and top off coolant as necessary.
13. Install the right-hand engine undercover.
14. Check ignition timing and road test the vehicle for proper operation.

2.4L Highlander

1. Before servicing the vehicle, refer to the precautions in the beginning of this section.
2. Remove or disconnect the following:
   - Front center suspension brace
   - Timing chain
   - Coolant
   - Transfer case oil
   - Radiator hoses
   - Power steering hoses
   - Heater hoses
   - Fuel rail lines
   - Camshaft timing oil control valve

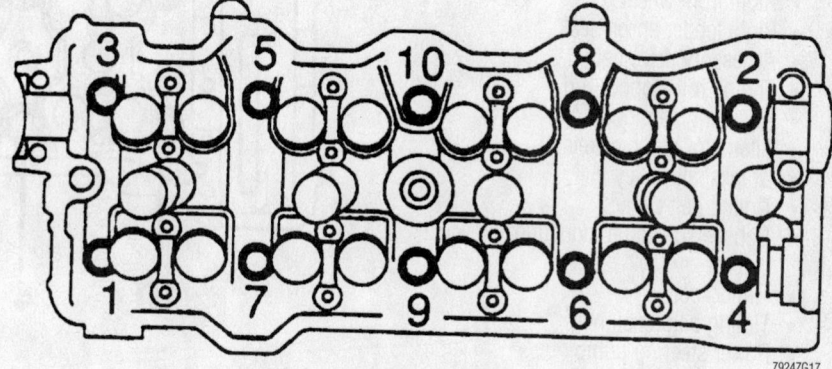

Cylinder head bolts installation sequence—2.0L engine

For Tune-up, Capacities and Firing orders, see Section 1 of this manual

- Front driveshaft
- Rear engine mount insulator (4wd)
- Transverse engine mount bracket (4wd)
- Intake manifold
- All wires and cables connected to the head
- Exhaust manifold
- Camshafts

3. Loosen the 10 head bolts evenly, a little at a time in several passes and lift off the head. Check the head bolt length. Any bolt longer than 6.465 in. (164.2mm) should be replaced.

4. Installation is the reverse of removal. Install the head gasket with the lot number stamp upward. Apply a bead of RTV sealer as shown. The head must be installed within 3 minutes of applying the sealer, and the head bolts must be tightened with 15 minutes. The head bolts must be tightened in sequence, in several passes, to 58 ft. lbs. (79Nm), then, an additional 90 degree turn each.

3.0L Highlander

1. Before servicing the vehicle, refer to the precautions in the beginning of this section.

2. Remove or disconnect the following:
- Coolant
- Engine oil
- Exhaust pipes
- Exhaust manifold
- Camshaft cover
- Upper center front suspension brace
- Air cleaner
- Intake air surge tank
- Fuel rail
- Heater hoses
- Intake manifold
- Radiator hose
- Water outlet
- Right front wheel
- Right fender apron seal
- Accessory drive belts
- Engine roll stopper rod
- Right engine mount
- Alternator and brackets
- Crankshaft pulley
- Timing belt covers
- Transverse engine mounting bracket
- Timing belt
- Timing belt tensioner
- Power steering pump
- Ignition coil pack
- Camshafts
- The hexagonal bolt, using an 8mm hex wrench, then the 8 head bolts
- Cylinder head

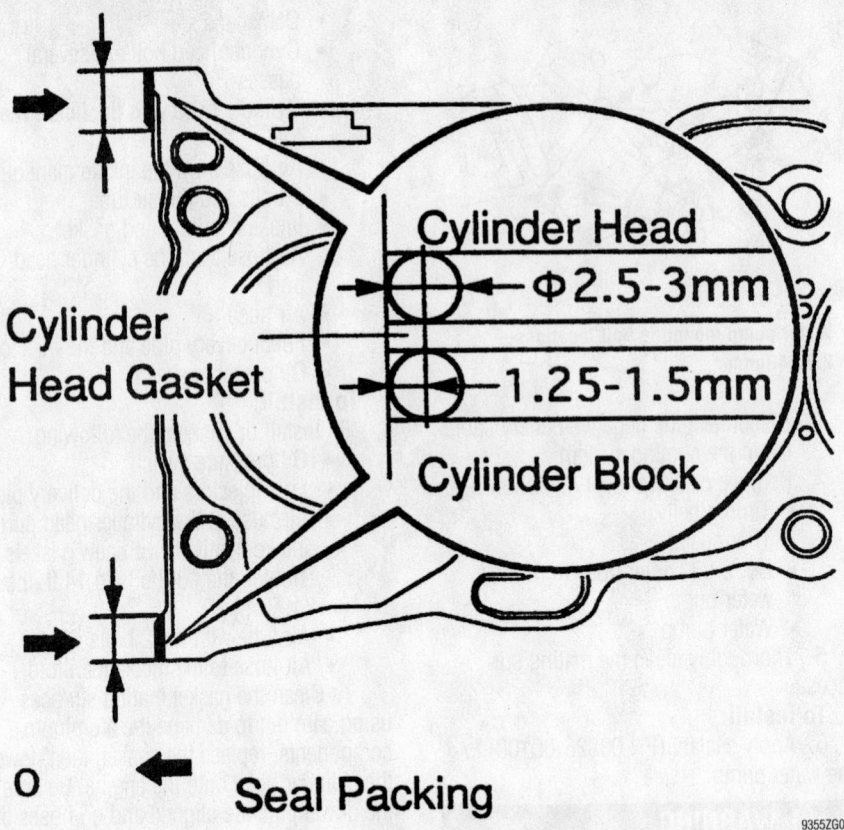

Apply a bead of RTV sealant as shown—2.4L engine cylinder head

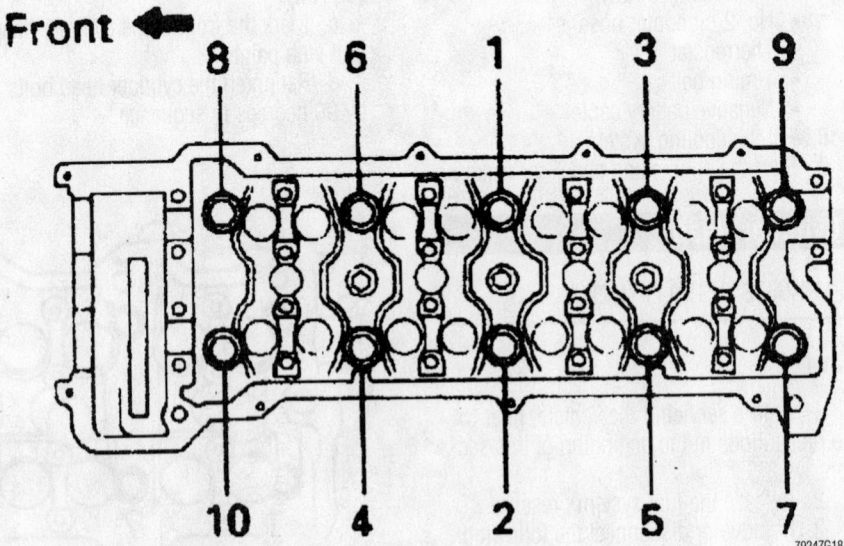

Cylinder head bolt tightening sequence—2.4L engines

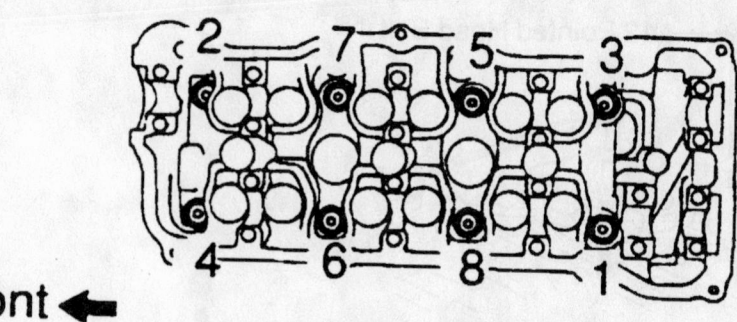

Front ◄

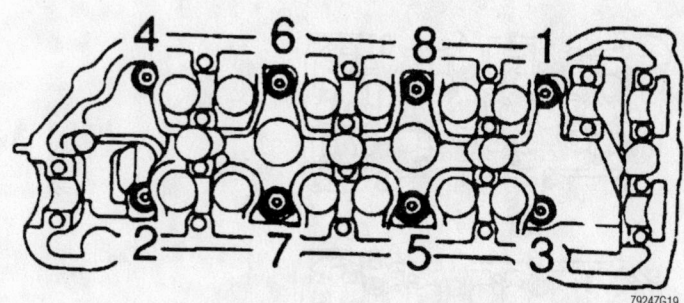

7924ZG19

Cylinder head bolt loosening sequence—3.0L Highlander

3. Installation is the reverse of removal. Measure the head bolts. The minimum diameter of the stretch portion of each bolt should be at least 8.75mm (0.3775 in.). Replace any bolt that does not measure up. The head gasket is installed with the **R** mark upwards. Install the 8 head bolts first, tightened in sequence, in 2 equal steps, to 40 ft. lbs. (54Nm). Then tighten each an additional 90 degrees. Finally, install the hex bolt, torqued to 14 ft. lbs. (19 Nm).

RX 300

1. Before servicing the vehicle, refer to the precautions in the beginning of this section.

2. Remove or disconnect the following:
   • Wiper and blade assembly
   • Top cowl seal and panel
   • Window washer hoses from the ventilator louvers
   • Left and right ventilator louvers
   • Heater air duct

3. Relieve the fuel pressure.

4. Remove or disconnect the following:
   • Turn the ignition key to the **OFF** position
   • Negative battery cable

➡**Wait at least 90 seconds from the time the negative battery was disconnected to start work.**

5. Drain the cooling system.

6. Remove or disconnect the following:
   • Accelerator and throttle cables, if equipped with an automatic transaxle
   • Air cleaner cover, air flow meter and the air duct
   • Front upper suspension brace
   • Cruise control actuator and bracket, if equipped
   • 2 engine ground straps
   • Right engine mounting support
   • Radiator hoses
   • 2 heater hoses
   • Fuel feed and return lines from the fuel rail assembly
   • Pressure hose from the hydraulic motor
   • V-bank cover

7. Disconnect the following vacuum hoses:
   • Fuel pressure control Vacuum Switching Valve (VSV)
   • Fuel pressure regulator
   • Cylinder head rear plate
   • Intake air control valve VSV
   • Exhaust Gas Recirculation (EGR) vacuum modulator
   • EGR valve

8. Disconnect the following wiring and hoses:
   • Intake air control valve

   • Fuel pressure regulator
   • EGR VSV

9. Remove the 2 nuts and the emission control valve set.

10. Disconnect the following hoses;
   • Brake booster vacuum hose
   • PCV hose
   • Intake air control valve vacuum hose

11. Remove or disconnect the following:
   • Data Link Connector (DLC) from the mounting bracket
   • 2 ground straps from the intake chamber
   • Hydraulic motor pressure hose from the intake chamber
   • Right Oxygen (O_2) sensor connector from the power steering pressure tube
   • 2 nuts and the power steering pressure tube from the intake chamber
   • Both power steering air hoses
   • Engine hanger and the intake chamber support
   • EGR pipe and gaskets

12. Disconnect the following wiring:
   • Throttle Position (TP) sensor connector
   • Idle Air Control (IAC) valve connector
   • EGR gas temperature connector
   • Air conditioning idle up connector

13. Disconnect the following vacuum hoses:
   • 2 vacuum hoses from the Thermal Vacuum Valve (TVV)
   • Vacuum hose from the cylinder head rear plate
   • Vacuum hose from the charcoal canister

14. Remove or disconnect the following:
   • Air assist hose and the 2 water bypass hoses
   • Air intake chamber
   • Left engine wiring harness and move it aside
   • Wiring harness from the rear of the engine
   • Right engine wiring harness and move it aside
   • Ignition coils and move them aside
   • Timing belt
   • Camshaft pulleys and the timing belt rear cover
   • Cylinder head rear plate
   • Water inlet pipe
   • Air assist hose and vacuum hose
   • Intake manifold and fuel rail assembly
   • Water outlet

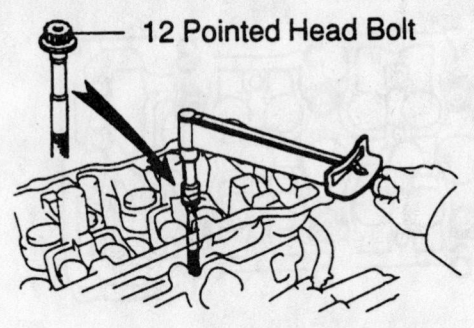

12 Pointed Head Bolt

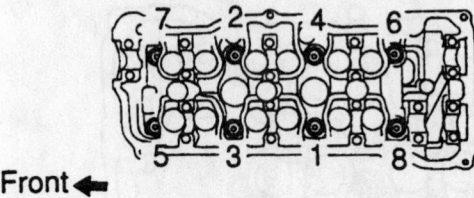

Front ←

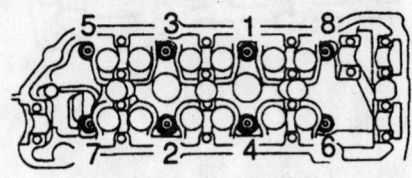

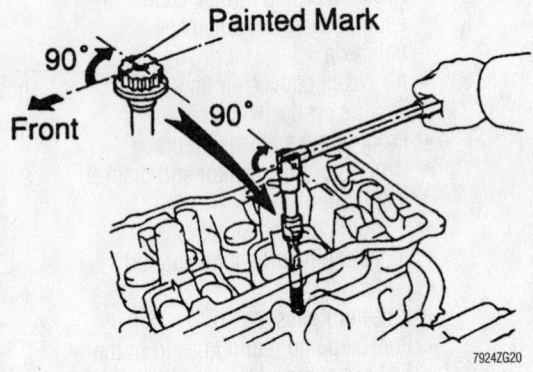

Painted Mark

90°

Front

90°

7924ZG20

Cylinder head bolt tightening sequence—3.0L Highlander

- EGR pipe from the right exhaust manifold
- Front exhaust pipe and exhaust manifolds
- Dipstick assembly and the power steering pump bracket
- Valve covers and the Camshaft Position (CMP) sensor
- Camshafts

15. Be sure the engine is at/or near ambient temperature and remove the 2 (1 on each head) 8mm recessed hex bolts. Loosen and remove the 8 head bolts evenly, in 3 passes, in the reverse order of the installation sequence. Carefully lift the head from the engine; if necessary to pry the head loose, take great care not to damage the mating surfaces. Place the head on wood blocks in a clean work area.

➡ **If the cylinder head bolts are loosened out of sequence, warpage or cracking could result.**

16. Remove the cylinder head gasket. With a gasket scraper, carefully remove all the old gasket material from the cylinder head and engine block surfaces.

To install:

17. Place the new cylinder head gasket onto the cylinder block.

18. Install the cylinder head, in sequence, using several steps, as follows:
- Cylinder head onto the gasket
- Cylinder head bolts lubricated with clean engine oil
- Tighten the bolts in sequence in 3 steps to 40 ft. lbs. (54 Nm).

➡ **If any bolt does not meet the torque, replace it.**

- Mark the forward edge of each bolt with paint, then tighten each bolt, in proper sequence, an additional 90 degrees.
- Check that each painted mark is

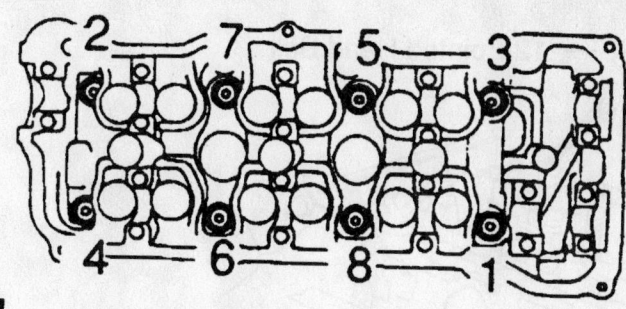

Front ←

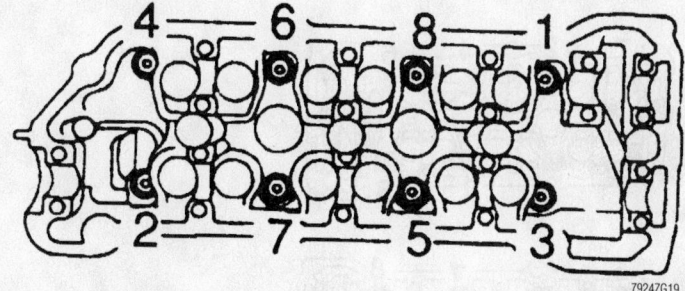

7924ZG19

Cylinder head bolt loosening sequence—RX 300

now at a 90 degrees angle to the front

➡**The paint mark applied to the bolt in the 9 o'clock position and should now be in the 12 o'clock position.**

- Remaining 8mm bolts, lubricated with engine oil. Tighten both bolts to 13 ft. lbs. (18 Nm).
19. Install the camshafts.
20. Check and adjust the valves.
21. Apply sealant to the cylinder heads where the camshaft supports meet the cylinder heads.
22. Install or connect the following:
- Cylinder head covers, using new gaskets
- Dipstick and power steering pump bracket
- Exhaust manifolds. Tighten the nuts to 36 ft. lbs. (49 Nm).
- EGR pipe to the right exhaust manifold
- Water outlet
- Intake manifold and the fuel rail assembly. Tighten the intake manifold nuts/bolts to 11 ft. lbs. (15 Nm).
- Air assist hose and the 2 water bypass hoses
- Water inlet pipe and cylinder head rear plate
- Timing belt rear cover and camshaft pulleys

- Timing belt
- Spark plugs and ignition coils
- Right engine wiring harness
- Wiring harness to the rear of the engine
- Left engine wiring harness
- Air intake chamber
- EGR pipe, using new gaskets
23. Connect the following vacuum hoses:
- The 2 TVV vacuum hoses
- The vacuum hose to the rear cylinder head plate
- Charcoal canister vacuum hose
24. Connect the following electrical wiring:
- TP sensor connector
- IAC valve connector
- EGR gas temperature connector
- Air conditioning idle up connector
25. Install or connect the following:
- Engine hanger and the intake chamber support
- Both power steering air hoses
- Power steering pressure tube to the intake chamber
- O_2 sensor connector to the pressure tube.
- Both ground straps, to the intake chamber
- DLC to the bracket
26. Connect the following hoses:
- Power brake booster vacuum hose
- PCV hose

- IAC valve vacuum hose
27. Install or connect the following:
- Emission control valve set and related vacuum hoses and connectors
- V-bank cover
- Pressure hose to the hydraulic motor
- Fuel lines to the fuel rail assembly
- Heater and radiator hoses
- Right engine mounting support
- both engine ground straps
- Upper front suspension brace, if removed. Tighten the nuts to 59 ft. lbs. (80 Nm).
- Cruise control actuator and bracket
- Air cleaner, air flow meter and air duct assembly
- Accelerator and throttle cables, if equipped with an automatic transaxle
28. Fill the cooling system.
29. Install or connect the following:
- Negative battery cable
- Heater air duct
- Left and right ventilator louvers
- Window washer hoses from the ventilator louvers
- Top cowl seal and panel
- Wiper and blade assembly
30. Start the engine and check for leaks.
31. Bleed the air from the cooling system.
32. Road test the vehicle and check for unusual noise, shock, slippage, correct shift points and smooth operation.
33. Recheck the coolant and engine oil levels.

2.7L 4Runner

1. Before servicing the vehicle, refer to the precautions in the beginning of this section.
2. Release the fuel system pressure.
3. Disconnect the negative battery cable.
4. Drain the engine coolant.
5. Remove or disconnect the following:
- Air cleaner cap
- Mass Air Flow (MAF) meter and resonator
- Accelerator cable from the throttle body, if equipped with a manual transmission
- Accelerator and throttle cables from the throttle body, if equipped with an automatic transmission
- Cruise control cable from the actuator, if equipped with cruise control

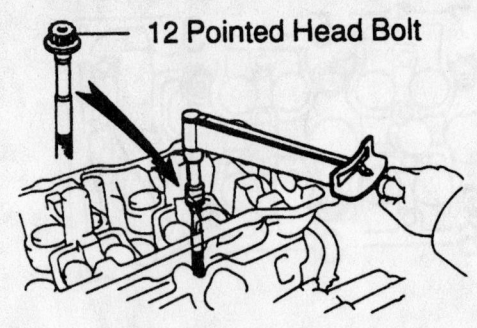

12 Pointed Head Bolt

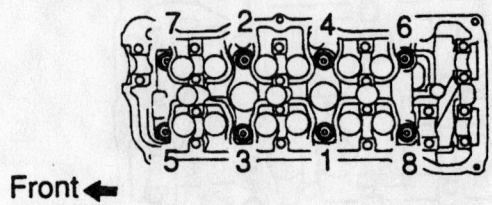

Front ←

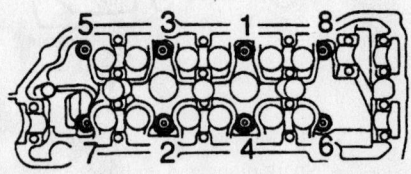

Painted Mark

90°

Front

90°

7924ZG20

Cylinder head bolt tightening sequence—3.0L (1MZ-FE) engine

- Intake air connector
- Air hose for Idle Air Control (IAC)
- Vacuum sensing hose
- Wire clamp for the engine wiring harness
- Oil dipstick guide
- Power steering belt
- Power steering pulley, pump and bracket
- Positive Crankcase Ventilation (PCV) hoses
- Distributor
- Spark plug wires from the spark plugs
- Engine wiring harness

- Air conditioning compressor, if equipped with air conditioning
- Oil pressure sensor
- Engine Coolant Temperature (ECT) sensor connector
- ECT sender gauge connector
- Exhaust Gas Recirculation (EGR) gas temperature sensor connector
- Vacuum Switching Valve (VSV) connector
- 2 vacuum hose from the VSV
- Ground strap from the cowl top panel
- Engine wiring harness from the air intake chamber

- Throttle Position (TP) sensor connector
- IAC valve connector
- Crankshaft Position (CKP) sensor connector
- Knock Sensor (KS) connector
- Data Link Connector 1 (DLC1) from the bracket
- Engine wiring harness clamp
- EGR pipe
- Intake chamber stay
- Air intake chamber assembly

6. Disconnect the following hoses:
   - Evaporative Emissions (EVAP) hose from the throttle body

- Brake booster vacuum hose from the union
- Water bypass hose from the water bypass pipe
- Water bypass hose from the cylinder head rear cover

7. Remove or disconnect the following:
- Injector connectors
- Fuel inlet pipe
- Hoses and the fuel return pipe
- Delivery pipe and injectors
- Intake manifold
- Front exhaust pipe
- Exhaust manifold and gasket
- Water outlet
- Cylinder head rear cover
- Spark plugs
- Front engine hanger
- Engine wiring harness brackets
- Cylinder head cover

8. Set No. 1 cylinder to Top Dead Center (TDC) of the compression stroke. The groove on the crankshaft pulley should align with the **0** mark on the timing chain cover and the timing marks (1 and 2 dots) of the camshaft gears should form a straight line in respect to the cylinder head surface. If not, turn the crankshaft 1 revolution (360 degrees).

9. Remove or disconnect the following:
- Chain tensioner and gasket
- Camshaft timing gear
- Exhaust camshafts

10. Remove the intake camshaft, as follows:

a. Uniformly, loosen and remove the bearing cap bolts in the reverse order of the tightening in several passes, in sequence.

b. Remove the bearing caps and camshaft. Make a note of the bearing cap positions for proper installation.

➡**If the camshaft is not being lifted out straight and level, reinstall the No. 3 bearing cap with the 2 bolts. Then, alternately loosen and remove the 2 bearing cap bolts with the camshaft gear pulled up.**

11. Remove or disconnect the following:
- Valve lifters and shims

➡**Arrange the valve lifters and shims in correct order.**

- Cylinder head, by uniformly loosen and remove the cylinder head bolts in the reverse order of the tightening, in sequence, using several passes

To install:

12. Before installing, thoroughly clean the gasket mating surfaces and check for warpage.

13. Apply sealant (PN 08826-00080) as shown. Place a new head gasket on the block and install the cylinder head.

14. Install the cylinder head as follows:

a. Lightly coat the cylinder head bolts with engine oil.

b. Install the bolts and tighten, in several passes, in the sequence. Tighten all bolts to 29 ft. lbs. (39 Nm).

c. Mark the front of the bolt with paint and retighten bolts 90 degrees in the proper sequence.

d. Retighten an additional 90 degrees. Check that the painted mark is now facing rearward.

15. Install or connect the following:
- Tighten the 2 front mounting bolts to 15 ft. lbs. (21 Nm).
- Valve lifters and shims in their proper locations. Check that the valve lifter rotates smoothly by hand.
- Intake and exhaust camshafts

16. Set No. 1 cylinder to TDC compression stroke. The groove on the crankshaft pulley should align with the **0** mark on the timing chain cover and the timing marks (1 and 2 dots) of the camshaft gears should form a straight line in respect to the cylinder head surface. If not, turn the crankshaft 1 revolution (360 degrees).

17. Install the timing gear, as follows:

a. Place the gear over the straight pin of the intake camshaft.

b. Hold the intake camshaft with a wrench. Install and tighten the bolt to 54 ft. lbs. (74 Nm).

c. Hold the exhaust camshaft and install the bolt and distributor gear. Tighten the bolt to 34 ft. lbs. (46 Nm).

18. Install or connect the following:
- Chain tensioner, using a new gasket (mark toward the front)
- Recheck the valve timing
- Check and adjust the valve clearance
- Spark plugs
- Semi-circular plug

19. Recheck the engine for proper valve timing.

20. Install or connect the following:
- Cylinder head cover, using a new gasket

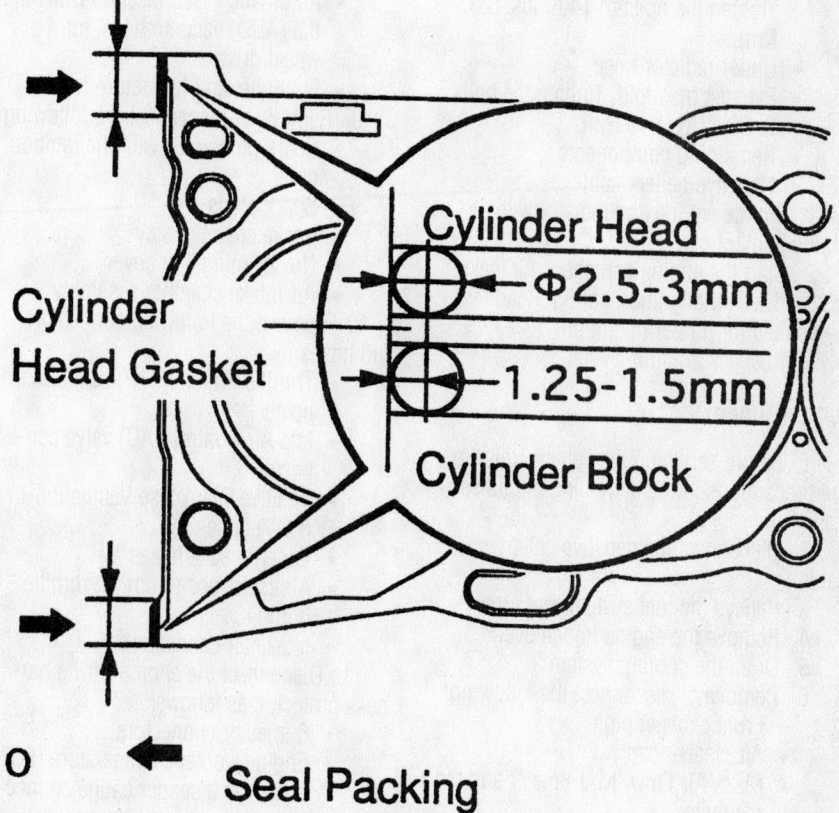

Apply a bead of RTV sealant as shown—2.7L engine cylinder head

93355ZG01

Timing belt service is covered in Section 3 of this manual

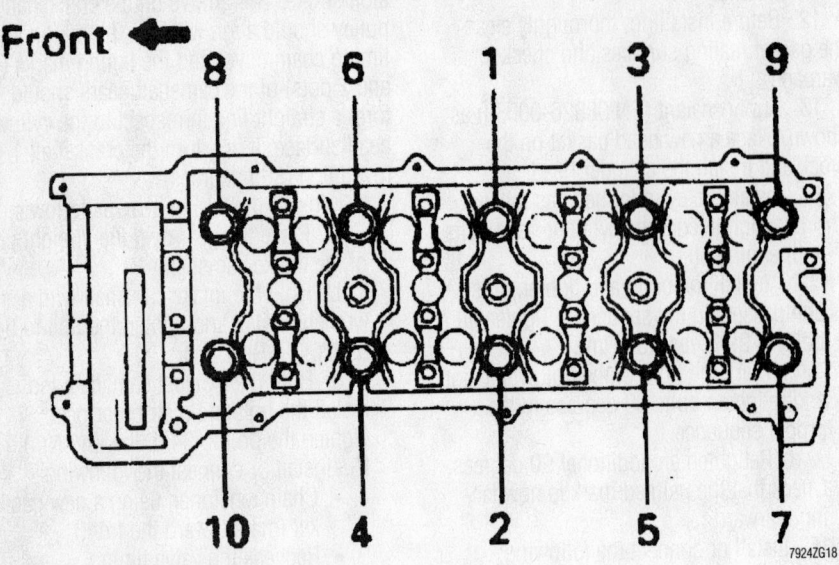

Front ←

8 6 1 3 9

10 4 2 5 7

79242G18

Cylinder head bolt tightening sequence—2.7L engines

- Engine wiring harness brackets
- Front engine hanger. Tighten the bolts to 30 ft. lbs. (42 Nm).
- Cylinder head rear cover. Tighten the bolts to 10 ft. lbs. (13 Nm).
- Water outlet, using a new gasket. Tighten the bolts to 14 ft. lbs. (20 Nm).
- Upper radiator hose
- Exhaust manifold. Tighten the bolts to 36 ft. lbs. (49 Nm).
- Remaining components
- Negative battery cable

21. Fill the engine and radiator with engine coolant.

22. Start the engine and check for leaks.

23. Check the ignition timing. Road test the vehicle for proper operation.

24. Recheck all fluid levels.

3.4L 4Runner

1. Before servicing the vehicle, refer to the precautions in the beginning of this section.

2. Disconnect the negative battery cable.

3. Relieve the fuel system pressure.

4. Remove the engine undercover.

5. Drain the cooling system.

6. Remove or disconnect the following:
- Front exhaust pipe
- Air cleaner cap
- Mass Air Flow (MAF) meter and the resonator

7. Disconnect the following cables:
- Actuator cable with the bracket, if equipped with cruise control
- Accelerator cable

- Throttle cable, if equipped with an automatic transmission

8. Disconnect the following hoses:
- Heater hose
- Brake booster vacuum hose
- Evaporative Emissions (EVAP) hose
- Automatic Disconnecting Differential (ADD) vacuum hose, for 4-wheel drive
- Fuel inlet and fuel return hose

9. Remove or disconnect the following:
- Spark plug wires with the ignition coils
- Spark plugs
- Intake chamber stay
- No. 2 timing belt cover
- Air intake chamber assembly

10. Remove the following connectors and hoses:
- Throttle Position (TP) sensor connector
- Idle Air Control (IAC) valve connector
- Positive Crankcase Ventilation (PCV) hoses
- Water bypass hoses
- Air assist hose from the throttle body
- Intake air connector

11. Disconnect the engine wiring harness protector, as follows:
- 6 injector connectors
- Engine Coolant Temperature (ECT) sensor and sender gauge connectors
- Engine wiring harness protector from the cylinder head

12. Remove or disconnect the following:

- Fuel pressure regulator
- Intake manifold assembly

13. Set the No. 1 cylinder at Top Dead Center (TDC) of the compression stroke, as follows:

a. Turn the crankshaft pulley and align its groove with the timing mark **0** of the No. 1 timing belt cover.

b. Check that the timing marks of the camshaft timing pulleys and the No. 3 timing belt cover are aligned. If not, turn the crankshaft pulley 1 revolution (360 degrees).

14. Remove or disconnect the following:
- Timing belt tensioner, by alternately loosening the 2 bolts
- Timing belt

15. Remove the camshaft timing pulleys, as follows:

a. Using Variable Pin Wrench Set 09960-10010, remove the pulley bolt, the timing pulley and the knock pin.

b. Remove the 2 timing pulleys with the timing belt.

16. Remove or disconnect the following:
- Bolt and the No. 2 idler pulley
- Camshaft Position (CMP) sensor
- No. 3 timing belt cover
- Alternator from the engine
- Alternator bracket
- Power steering pump and move it aside without disconnecting the pump lines
- Exhaust crossover pipe and gaskets, by removing the 6 nuts
- Left-hand exhaust manifold, by removing the heat insulator and 6 nuts
- Right-hand exhaust manifold, by removing the heat insulator and 6 nuts
- 8 bolts, seal washers, cylinder head cover and gasket.

➡**Remove both cylinder head covers**

- Semi-circular plugs
- Right exhaust and intake camshafts
- Left exhaust and intake camshafts
- Valve lifters and shims from the cylinder head; arrange the valve lifters and shims in correct order

17. Remove the cylinder heads, as follows:

- Bolt and ground strap
- Cylinder head (recessed head) bolt on the cylinder head, using an 8mm hexagon wrench; then, repeat the procedure for the other side.
- 8 cylinder head (12-pointed head) bolts, on each cylinder head.

➡**Loosen the bolts in several passes, in the reverse order of the tightening sequence.**

- 16 cylinder head bolts and plate washers
- Cylinder head

To install:

18. Clean all surfaces.
19. Install or connect the following:
- New cylinder head gaskets
- Cylinder heads
20. Apply a light coat of engine oil on the threads and under the heads of the cylinder head bolts.
21. Tighten the cylinder head bolts using several passes, in sequence, as follows:
- Step 1: 25 ft. lbs. (34 Nm)
- Step 2: Mark the front of the cylinder head bolt with paint
- Step 3: Turn 90 degrees
- Step 4: Check that the painted mark is now at a 90 degrees angle to the front
22. Install the recessed head cylinder head bolts, as follows:
- Step 1: Apply a light coat of engine oil on the threads and under the heads of the cylinder head bolts
- Step 2: Tighten the cylinder head bolts, using a 8mm hexagon wrench, to 13 ft. lbs. (18 Nm).
- Bolt and ground strap
23. Install or connect the following:
- Valve lifters and shims

➡**Check that the valve lifter rotates smoothly by hand.**

- Right intake and exhaust camshafts
- Left intake and exhaust camshafts
24. Check and adjust the valve clearance.
25. Install or connect the following:
- Semi-circular plugs
- Cylinder head covers. Uniformly, tighten the bolts, in several passes, to 53 inch lbs. (6 Nm).
- Exhaust manifolds with new gaskets. Tighten the nuts to 30 ft. lbs. (40 Nm).
- Exhaust manifold heat insulators. Tighten the nuts to 71 inch lbs. (8 Nm).
- Exhaust crossover pipe. Tighten the nuts to 33 ft. lbs. (45 Nm).
- Power steering pump
- Alternator bracket. Tighten the fasteners to 14 ft. lbs. (18 Nm).
- Alternator
- No. 3 timing belt cover. Tighten the bolts to 80 inch lbs. (9 Nm).
- CMP sensor. Tighten it to 71 inch lbs. (8 Nm).
- Timing belt
- No. 2 timing belt idler bolt. Tighten the bolt to 30 ft. lbs. (40 Nm).

➡**Check that the pulley bracket moves smoothly.**

- Left camshaft timing pulley
26. Set the No. 1 cylinder to TDC of the compression stroke, as follows:

a. Connect the timing belt to the left camshaft timing pulley.
b. Check that the installation mark on the timing belt is aligned with the end of the No. 1 timing belt cover.
c. Install the right camshaft timing pulley and the timing belt.
d. Set the timing belt tensioner. Alternately, tighten the bolts to 20 ft. lbs. (28 Nm).
27. Using pliers, remove the 1.5mm hexagon wrench from the belt tensioner.
28. Check the valve timing.
29. Install or connect the following:
- New gaskets and the intake manifold assembly. Tighten the bolts and nuts to 13 ft. lbs. (18 Nm).
- Intake manifold stay. Tighten the bolts to 14 ft. lbs. (18 Nm).
- Fuel pressure regulator
30. Connect the engine wiring harness to the intake manifold, as follows:
- Engine wiring harness to the cylinder head
- 3 engine wiring harness clamps
31. Install or connect the following:
- 6 injector connectors
- ECT sender gauge connector
- ECT sensor connector
- Intake air connector
- Air intake chamber assembly. Tighten the bolts and nuts to 13 ft. lbs. (18 Nm).
- Intake chamber stay
- No. 2 timing belt cover. Tighten the bolts to 80 inch lbs. (9 Nm).
- PCV hoses
- Water bypass hoses
- Air assist hose to the throttle body
- IAC valve connector
- TP sensor connector
- CMP sensor connector to the No. 2 timing belt cover
- 3 spark plug wire clamps
32. Connect the following hoses:
- Brake booster vacuum hose
- EVAP hose
- Automatic Disconnecting Differential (ADD) vacuum hose, for 4-wheel drive
- Fuel inlet and fuel return hose
- Heater hose
33. Install or connect the following:
- Oil dipstick and guide, using a new O-ring
- Spark plugs
- Spark plug wires, with the ignition coils
- Alternator drive belt

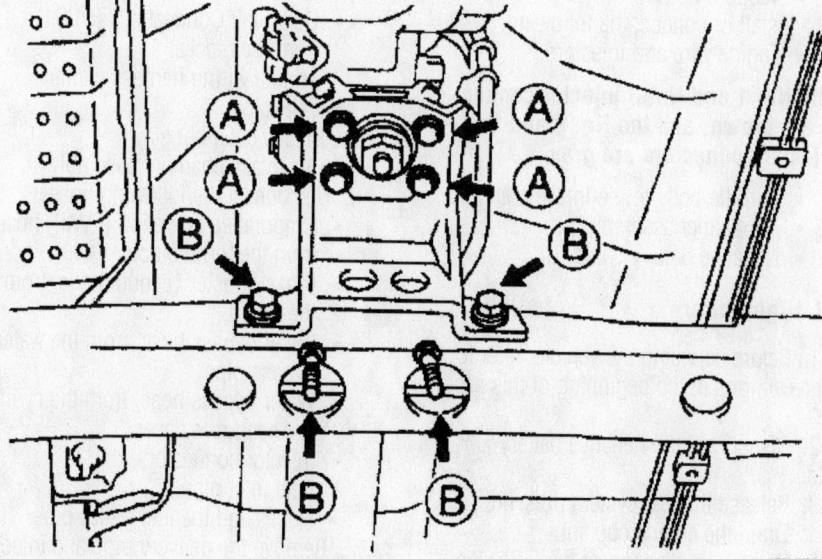

Cylinder head torque sequence—3.4L (5VZ-FE) engine

Heater Core replacement is covered in Section 2 of this manual

34. Connect the following cables:
   - Actuator cable with the bracket, if equipped with cruise control
   - Accelerator cable
   - Throttle cable, if equipped with an automatic transmission
   - MAF meter, resonator and air cleaner cap
   - Front exhaust pipe
   - Negative battery cable
35. Fill the radiator with engine coolant.
36. Start the engine and check for leaks.
37. Check the ignition timing.
38. Install the engine undercover.
39. Road test the vehicle.
40. Recheck all fluid levels.

Intake Manifold

REMOVAL & INSTALLATION

RAV4

1. Before servicing the vehicle, refer to the precautions in the beginning of this section.
2. Properly relieve the fuel system pressure.
3. Remove or disconnect the following:
   - Negative battery cable
   - Air cleaner assembly
   - Throttle body from the intake manifold
4. Disconnect the engine wire from the intake manifold, as follows:
   - 4 injector connectors
   - 2 engine wire clamps from the intake manifold wire brackets
   - Engine wire protector from the right-hand side of the intake manifold by removing the bolt
   - Engine wire from the wire clamp
5. Remove the EGR valve, EGR pipe and modulator, as follows:
   - Both vacuum hoses from the Exhaust Gas Recirculation (EGR) Vacuum Switching Valve (VSV)
   - Vacuum modulator from the clamp on the intake manifold
   - Loosen the cylinder head side of the EGR pipe union nut
   - Both nuts, the EGR valve, pipe assembly and gasket
   - Vacuum modulator
6. Disconnect the following hoses:
   - Fuel filter vacuum sensor hose on the intake manifold
   - Brake booster vacuum hose from the intake manifold
   - Ground strap from the intake manifold

7. Remove or disconnect the following:
   - Intake manifold stay by removing the 2 bolts
   - Control cable from the clamp on the rear side of the intake manifold, if equipped with automatic transmission
   - Air hose from the intake manifold
   - Air tube from the intake manifold, by removing the 2 bolts
   - 6 bolts and 2 nuts from the intake manifold
   - Intake manifold

To install:

8. Install or connect the following:
   - Intake manifold. Tighten the 6 bolts and 2 nuts to 14 ft. lbs. (19 Nm).
   - Air tube with the 2 bolts
   - Air hose to the intake manifold
   - Control cable to the clamp on the rear side of the intake manifold, if equipped with an automatic transmission
   - Intake manifold stay. Tighten both bolts to 31 ft. lbs. (42 Nm).
9. Connect the following hoses:
   - Ground strap to the intake manifold
   - Brake booster vacuum hose to the intake manifold
   - Fuel filter vacuum sensor hose to the intake manifold
10. Install the EGR valve, EGR pipe and the vacuum modulator, as follows:
   - Vacuum modulator
   - EGR valve and pipe. Tighten both nuts to 108 inch lbs. (13 Nm) and the union nut to 43 ft. lbs. (59 Nm).
   - Vacuum hoses
11. Install or connect the following:
   - Engine wire and injectors

➡**The No. 1 and No. 3 injector connectors are brown, and the No. 2 and No. 4 injector connectors are gray.**

   - Throttle body to the intake manifold
   - Air cleaner assembly
   - Negative battery cable

2.4L Highlander

1. Before servicing the vehicle, refer to the precautions in the beginning of this section.
2. Disconnect the negative battery cable.
3. Release the fuel system pressure.
4. Drain the engine coolant.
5. Remove or disconnect the following:
   - Air cleaner cap
   - Mass Air Flow (MAF) meter and the resonator
   - Accelerator cable from the throttle

body, if equipped with a manual transmission
   - Accelerator and throttle cables from the throttle body, if equipped with an automatic transmission
   - Cruise control cable from the actuator, if equipped with cruise control
   - Intake air connector
   - Air hose for Idle Air Control (IAC)
   - Vacuum sensing hose
   - Wire clamp for the engine wiring harness
   - Positive Crankcase Ventilation (PCV) hoses.
   - Engine wiring harness
   - Air conditioning compressor connector, if equipped with air conditioning
   - Oil pressure sensor connector
   - Engine Coolant Temperature (ECT) sensor connector
   - ECT sender gauge connector
   - Exhaust Gas Recirculation (EGR) gas temperature sensor connector
   - Vacuum Switching Valve (VSV) connector, for the EGR
   - 2 vacuum hoses, from the VSV for the EGR
   - Ground strap, from the cowl top panel
   - Engine wiring harness, from the air intake chamber
   - Throttle Position (TP) sensor connector
   - IAC valve connector
   - Crankshaft Position (CKP) sensor connector
   - Knock (KS) sensor connector
   - Data Link Connector 1 (DLC1), from the bracket
   - Engine wiring harness clamp
   - EGR pipe
   - Intake chamber stay
   - Air intake chamber assembly
6. Disconnect the following hoses:
   - Evaporative Emission (EVAP) hose, from the throttle body
   - Brake booster vacuum hose, from the union
   - Water bypass hose, from the water bypass pipe
   - Water bypass hose, from the cylinder head rear cover
   - Injector connectors
   - Fuel inlet pipe
   - Hoses and the fuel return pipe.
7. Remove the delivery pipe and injectors, as follows:
   - Delivery pipe, together with the 4 injectors
   - 4 insulators from the 4 spacers
   - 4 injectors, from the delivery pipe

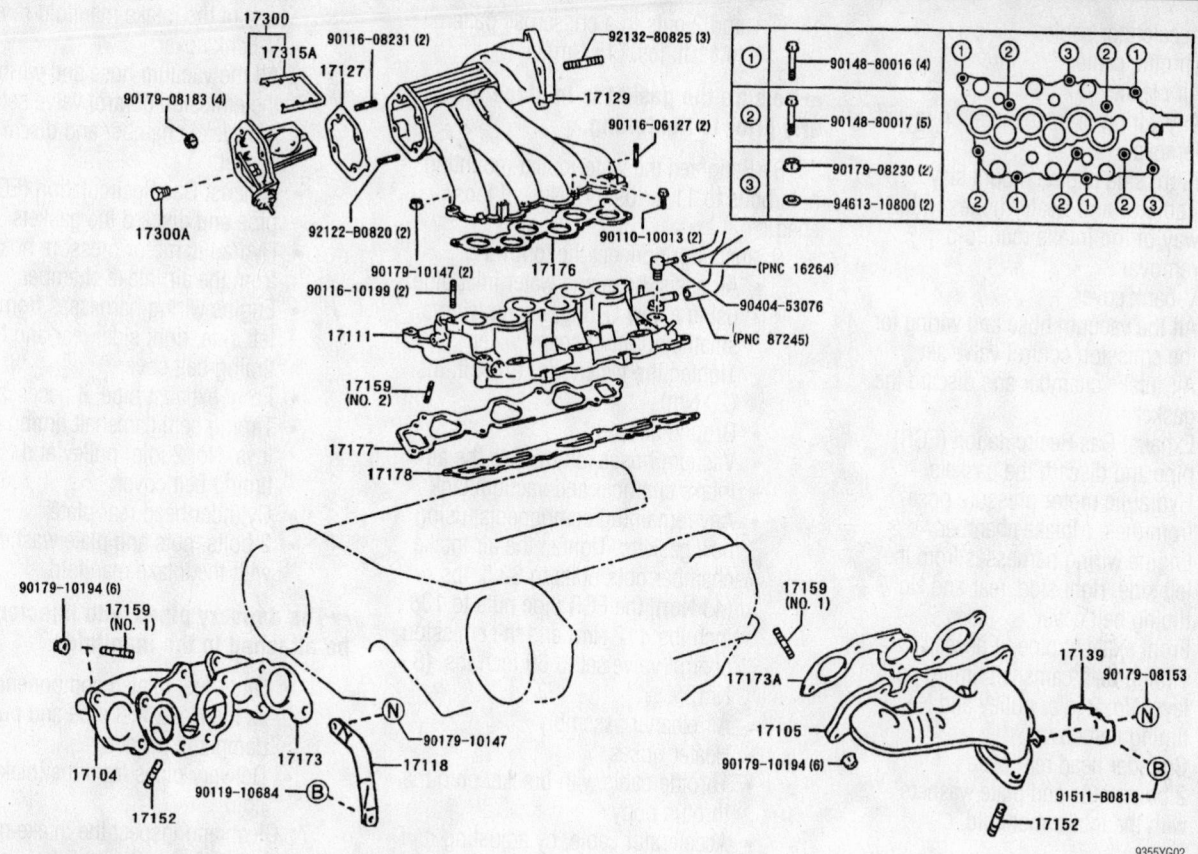

①	⬤ 90148-80016 (4)
②	⬤ 90148-80017 (5)
③	⬤ 90179-08230 (2)
	⬤ 94613-10800 (2)

Intake manifold and related components—2.4L Highlander

9355YG02

- O-ring and grommets, from each injector
- 4 spacers, by carefully prying them out
8. Remove the intake manifold.

To install:

9. Install or connect the following:
- Intake manifold. Tighten the bolts to 22 ft. lbs. (29 Nm).
- Injectors and the delivery pipe
- Fuel return pipe
- Fuel inlet pipe, with a new gasket. Tighten the bolts to 22 ft. lbs. (29 Nm).
- Injector connectors
- Air intake chamber assembly. Tighten the bolts to 15 ft. lbs. (21 Nm).

10. Connect the following hoses:
- Evaporative Emissions (EVAP) hose, to the throttle body
- Brake booster vacuum hose, to the union
- Water bypass hose, to water bypass pipe
- Water bypass hose, to cylinder head rear cover

11. Install or connect the following:

- Air intake chamber stay. Tighten the bolts to 15 ft. lbs. (20 Nm).
- EGR pipe. Tighten bolts to 13 ft. lbs. (18 Nm), nut "A" to 14 ft. lbs. (19 Nm) and nut B to 15 ft. lbs. (20 Nm).
- Air conditioning compressor connector
- Oil pressure sensor connector
- ECT sensor connector
- ECT sender gauge connector
- EGR gas temperature sensor connector
- VSV connector for the EGR
- 2 vacuum hose to the VSV for the EGR
- Ground strap to the cowl top panel
- Engine wiring harness to the air intake chamber
- TP sensor connector
- IAC valve connector
- CKP sensor connector
- KS sensor connector
- DLC1 to the bracket
- Engine wiring harness clamp
- PCV hoses
- Intake air connector. Tighten the bolts to 13 ft. lbs. (18 Nm).

- Cruise control cable to the actuator, if equipped with cruise control
- Accelerator cable to the throttle body, if equipped with a manual transmission
- Accelerator and throttle cables to the throttle body, if equipped with an automatic transmission

12. Fill the engine and radiator with engine coolant.

13. Install or connect the following:
- Air cleaner cap, MAF meter and resonator assembly
- Negative battery cable

14. Start the engine and check for leaks.

15. Road test the vehicle for proper operation.

16. Recheck all fluid levels.

3.0L Highlander

1. Before servicing the vehicle, refer to the precautions in the beginning of this section.

2. Remove or disconnect the following:

3. Properly relieve the fuel system pressure.

4. Drain and recycle the engine coolant.

5. Remove or disconnect the following:

Brake service is covered in Section 4 of this manual

- Accelerator cable
- Throttle cable
- Air cleaner
- Any wiring or hoses interfering with removal
- Right side engine mount stay
- Radiator and heater hoses in the way of the intake manifold removal
- V-bank cover
- All the vacuum hose and wiring for the emission control valve set
- Air intake chamber and discard the gasket
- Exhaust Gas Recirculation (EGR) pipe and discard the gaskets
- Hydraulic motor pressure hose from the air intake chamber
- Engine wiring harnesses from the left side, right side, rear and No. 3 timing belt cover
- Front exhaust pipe, if necessary
- Timing belt, camshaft timing pulleys, No. 2 idler pulley and No. 3 timing belt cover
- Cylinder head rear plate
- 2 bolts, nuts and plate washers with the intake manifold.

➡**The delivery pipes with injectors will be attached to the manifold.**

- Other fuel related components such as the No. 2 fuel pipe and pulsation damper, if needed
- Delivery pipes from the intake manifold

6. Clean and inspect the intake manifold mating surfaces. Scrape all old gasket material off.

To install:

7. Install or connect the following:
- Delivery pipes with injectors to the intake manifold.

➡**Be sure to place 4 spacers in position on the manifold. Temporarily install 4 bolts to retain the delivery pipes to the manifold. Inspect the injectors for smooth rotation.**

- Tighten the delivery pipes bolts to 84 inch lbs. (10 Nm), once the injectors are properly seated
- No. 2 fuel pipe with union bolts and gaskets. Tighten the bolts to 24 ft. lbs. (32 Nm).
- No. 1 fuel pipe with pulsation damper, using 4 new gaskets. Tighten the damper to 35 ft. lbs. (32 Nm) and the bolt to 11 ft. lbs. (15 Nm).
- Fuel pressure regulator, if removed
- Intake manifold. Tighten the 9 bolts

and 2 nuts in a crisscross pattern to 11 ft. lbs. (15 Nm).

➡**Be sure the gasket is in place properly prior to tightening.**

8. Retighten the water outlet mounting nuts/bolts to 11 ft. lbs. (15 Nm), if loosened.

9. Install or connect the following:
- Air assist hose and water inlet pipe, using a new O-ring, by applying a small amount of soapy water. Tighten the fastener(s) to 14 ft. lbs. (20 Nm).
- Ground strap
- Vacuum hoses removed to the air intake chamber and vacuum tank
- Any remaining components, using new gaskets. Tighten the air intake chamber nuts/bolts to 32 ft. lbs. (43 Nm), the EGR pipe nuts to 108 inch lbs. (12 Nm) and the emission control valve set to 69 inch lbs. (8 Nm).
- Air cleaner assembly
- Heater hoses
- Throttle cable with bracket onto the throttle body
- Accelerator cable, by adjusting it, if equipped with an automatic transaxle

10. Refill the cooling system
11. Install or connect the following:
- Negative battery cable
- Heater air duct

12. Start the engine and inspect for leaks.

RX 300

1. Before servicing the vehicle, refer to the precautions in the beginning of this section.

2. Remove or disconnect the following:
- Wiper and blade assembly
- Top cowl seal and panel
- Window washer hoses from the ventilator louvers
- Left and right ventilator louvers
- Heater air duct
- Front upper suspension brace

3. Properly relieve the fuel system pressure.

4. Remove the battery and battery tray.

5. Drain and recycle the engine coolant.

6. Remove or disconnect the following:
- Accelerator cable
- Throttle cable
- Air cleaner cap assembly
- Any wiring or hoses interfering with removal
- Right side engine mount stay
- Radiator and heater hoses in the

way of the intake manifold removal
- V-bank cover
- All the vacuum hose and wiring for the emission control valve set
- Air intake chamber and discard the gasket
- Exhaust Gas Recirculation (EGR) pipe and discard the gaskets
- Hydraulic motor pressure hose from the air intake chamber
- Engine wiring harnesses from the left side, right side, rear and No. 3 timing belt cover
- Front exhaust pipe, if necessary
- Timing belt, camshaft timing pulleys, No. 2 idler pulley and No. 3 timing belt cover
- Cylinder head rear plate
- 2 bolts, nuts and plate washers with the intake manifold.

➡**The delivery pipes with injectors will be attached to the manifold.**

- Other fuel related components such as the No. 2 fuel pipe and pulsation damper, if needed
- Delivery pipes from the intake manifold

7. Clean and inspect the intake manifold mating surfaces. Scrape all old gasket material off.

To install:

8. Install or connect the following:
- Delivery pipes with injectors to the intake manifold.

➡**Be sure to place 4 spacers in position on the manifold. Temporarily install 4 bolts to retain the delivery pipes to the manifold. Inspect the injectors for smooth rotation.**

- Tighten the delivery pipes bolts to 84 inch lbs. (10 Nm), once the injectors are properly seated
- No. 2 fuel pipe with union bolts and gaskets. Tighten the bolts to 24 ft. lbs. (32 Nm).
- No. 1 fuel pipe with pulsation damper, using 4 new gaskets. Tighten the damper to 35 ft. lbs. (32 Nm) and the bolt to 11 ft. lbs. (15 Nm).
- Fuel pressure regulator, if removed
- Intake manifold. Tighten the 9 bolts and 2 nuts in a crisscross pattern to 11 ft. lbs. (15 Nm).

➡**Be sure the gasket is in place properly prior to tightening.**

9. Retighten the water outlet mounting nuts/bolts to 11 ft. lbs. (15 Nm), if loosened.

10. Install or connect the following:
- Air assist hose and water inlet pipe, using a new O-ring, by applying a small amount of soapy water. Tighten the fastener(s) to 14 ft. lbs. (20 Nm).
- Ground strap
- Vacuum hoses removed to the air intake chamber and vacuum tank
- Any remaining components, using new gaskets. Tighten the air intake chamber nuts/bolts to 32 ft. lbs. (43 Nm), the EGR pipe nuts to 108 inch lbs. (12 Nm) and the emission control valve set to 69 inch lbs. (8 Nm).
- Air cleaner assembly
- Heater hoses
- Battery and tray
- Throttle cable with bracket onto the throttle body
- Accelerator cable, by adjusting it, if equipped with an automatic transaxle
- Front upper suspension brace. Tighten the nuts to 59 ft. lbs. (80 Nm).

11. Refill the cooling system
12. Install or connect the following:
- Negative battery cable
- Heater air duct
- Left and right ventilator louvers
- Window washer hoses from the ventilator louvers
- Top cowl seal and panel
- Wiper and blade assembly

13. Start the engine and inspect for leaks.

2.7L 4Runner

1. Before servicing the vehicle, refer to the precautions in the beginning of this section.
2. Disconnect the negative battery cable.
3. Release the fuel system pressure.
4. Drain the engine coolant.
5. Remove or disconnect the following:
- Air cleaner cap
- Mass Air Flow (MAF) meter and the resonator
- Accelerator cable from the throttle body, if equipped with a manual transmission
- Accelerator and throttle cables from the throttle body, if equipped with an automatic transmission
- Cruise control cable from the actuator, if equipped with cruise control
- Intake air connector

- Air hose for Idle Air Control (IAC)
- Vacuum sensing hose
- Wire clamp for the engine wiring harness
- Positive Crankcase Ventilation (PCV) hoses.
- Engine wiring harness
- Air conditioning compressor connector, if equipped with air conditioning
- Oil pressure sensor connector
- Engine Coolant Temperature (ECT) sensor connector
- ECT sender gauge connector
- Exhaust Gas Recirculation (EGR) gas temperature sensor connector
- Vacuum Switching Valve (VSV) connector, for the EGR
- 2 vacuum hoses, from the VSV for the EGR
- Ground strap, from the cowl top panel
- Engine wiring harness, from the air intake chamber
- Throttle Position (TP) sensor connector
- IAC valve connector
- Crankshaft Position (CKP) sensor connector
- Knock (KS) sensor connector
- Data Link Connector 1 (DLC1), from the bracket
- Engine wiring harness clamp
- EGR pipe
- Intake chamber stay
- Air intake chamber assembly

6. Disconnect the following hoses:
- Evaporative Emission (EVAP) hose, from the throttle body
- Brake booster vacuum hose, from the union
- Water bypass hose, from the water bypass pipe
- Water bypass hose, from the cylinder head rear cover
- Injector connectors
- Fuel inlet pipe
- Hoses and the fuel return pipe.

7. Remove the delivery pipe and injectors, as follows:
- Delivery pipe, together with the 4 injectors
- 4 insulators from the 4 spacers
- 4 injectors, from the delivery pipe
- O-ring and grommets, from each injector
- 4 spacers, by carefully prying them out

8. Remove the intake manifold.

To install:
9. Install or connect the following:
- Intake manifold. Tighten the bolts to 22 ft. lbs. (29 Nm).
- Injectors and the delivery pipe
- Fuel return pipe
- Fuel inlet pipe, with a new gasket. Tighten the bolts to 22 ft. lbs. (29 Nm).
- Injector connectors
- Air intake chamber assembly. Tighten the bolts to 15 ft. lbs. (21 Nm).

10. Connect the following hoses:
- Evaporative Emissions (EVAP) hose, to the throttle body
- Brake booster vacuum hose, to the union
- Water bypass hose, to water bypass pipe
- Water bypass hose, to cylinder head rear cover

11. Install or connect the following:
- Air intake chamber stay. Tighten the bolts to 15 ft. lbs. (20 Nm).
- EGR pipe. Tighten bolts to 13 ft. lbs. (18 Nm), nut "A" to 14 ft. lbs. (19 Nm) and nut B to 15 ft. lbs. (20 Nm).
- Air conditioning compressor connector
- Oil pressure sensor connector
- ECT sensor connector
- ECT sender gauge connector
- EGR gas temperature sensor connector
- VSV connector for the EGR
- 2 vacuum hose to the VSV for the EGR
- Ground strap to the cowl top panel
- Engine wiring harness to the air intake chamber
- TP sensor connector
- IAC valve connector
- CKP sensor connector
- KS sensor connector
- DLC1 to the bracket
- Engine wiring harness clamp
- PCV hoses
- Intake air connector. Tighten the bolts to 13 ft. lbs. (18 Nm).
- Cruise control cable to the actuator, if equipped with cruise control
- Accelerator cable to the throttle body, if equipped with a manual transmission
- Accelerator and throttle cables to the throttle body, if equipped with an automatic transmission

For complete Engine Mechanical specifications, see Section 1 of this manual

12. Fill the engine and radiator with engine coolant.

13. Install or connect the following:
- Air cleaner cap, MAF meter and resonator assembly
- Negative battery cable

14. Start the engine and check for leaks.

15. Road test the vehicle for proper operation.

16. Recheck all fluid levels.

3.4L 4Runner

1. Before servicing the vehicle, refer to the precautions in the beginning of this section.

2. Disconnect the negative battery cable.

3. Relieve the fuel system pressure.

4. Remove the engine undercover.

5. Drain the cooling system.

6. Remove or disconnect the following:
- Air cleaner cap
- Mass Air Flow (MAF) meter and the resonator
- Actuator cable with the bracket, if equipped with cruise control
- Accelerator cable
- Throttle cable, if equipped with automatic transmission

7. Disconnect the following hoses:
- Heater hose
- Brake booster vacuum hose
- Evaporative Emissions (EVAP) hose
- Automatic Disconnecting Differential (ADD) vacuum hose, for 4-Wheel drive
- Fuel inlet and fuel return hose

8. Remove or disconnect the following:
- Spark plug wires, with the ignition coils
- Intake chamber stay
- No. 2 timing belt cover
- Air intake chamber assembly
- Throttle Position (TP) sensor connector
- Idle Air Control (IAC) valve connector
- Positive Crankcase Ventilation (PCV) hoses
- Water bypass hoses
- Air assist hose from the throttle body
- Intake air connector
- Engine wiring harness
- Fuel return hose
- Vacuum hose, from the fuel pressure regulator
- Ground strap, from the intake air connector
- Data Link Connector 1 (DLC1), from the bracket
- 6 injector connectors

- Engine Coolant Temperature (ECT) sensor and sender gauge connectors
- Engine wiring harness protector from the cylinder head
- Fuel pressure regulator
- Intake manifold assembly
- Intake manifold stay
- Intake manifold, delivery pipes and the injectors assembly with the gaskets

To install:

9. Install or connect the following:
- New gaskets
- Intake manifold assembly. Tighten the bolts and nuts to 13 ft. lbs. (18 Nm).
- Intake manifold stay. Tighten the bolts to 14 ft. lbs. (18 Nm).
- Fuel pressure regulator
- Engine wiring harness to the cylinder head, by installing the 3 bolts
- 3 engine wiring harness clamps
- 6 injector connectors
- ECT sender gauge connector
- ECT sensor connector
- Intake manifold. Tighten the bolts and nuts to 14 ft. lbs. (18.5 Nm).
- DLC1 to the bracket on the intake manifold
- Ground strap to the intake manifold, by installing the bolt
- Brake booster vacuum hose, to the intake air connector
- 2 fuel return hoses
- Engine wiring harness to the intake manifold
- Air intake chamber assembly to the

engine. Tighten the bolts and nuts to 14 ft. lbs. (18.5 Nm).
- Intake chamber stay. Tighten the bolts to 30 ft. lbs. (40 Nm).
- New O-ring to the oil filler tube
- Oil filler tube end into the tube hole in the oil pan
- Oil filler tube and No. 1 throttle cable clamp
- No. 2 timing belt cover. Tighten the bolts to 80 inch lbs. (9 Nm).
- PCV hoses
- Water bypass hoses
- Air assist hose to the throttle body
- IAC valve connector
- TP sensor connector
- Brake booster vacuum hose
- EVAP hose
- Automatic Disconnecting Differential (ADD) vacuum hose, for 4-Wheel drive
- Fuel inlet and fuel return hose
- 3 spark plug wire clamps to the No. 2 timing belt cover
- CMP connector to the No. 2 timing belt cover
- Spark plug wires with the ignition coils
- Heater hose
- Actuator cable with the bracket, if equipped with cruise control
- Accelerator cable
- Throttle cable, if equipped with automatic transmission
- MAF meter, resonator and the air cleaner cap
- Negative battery cable

10. Fill the radiator with engine coolant.

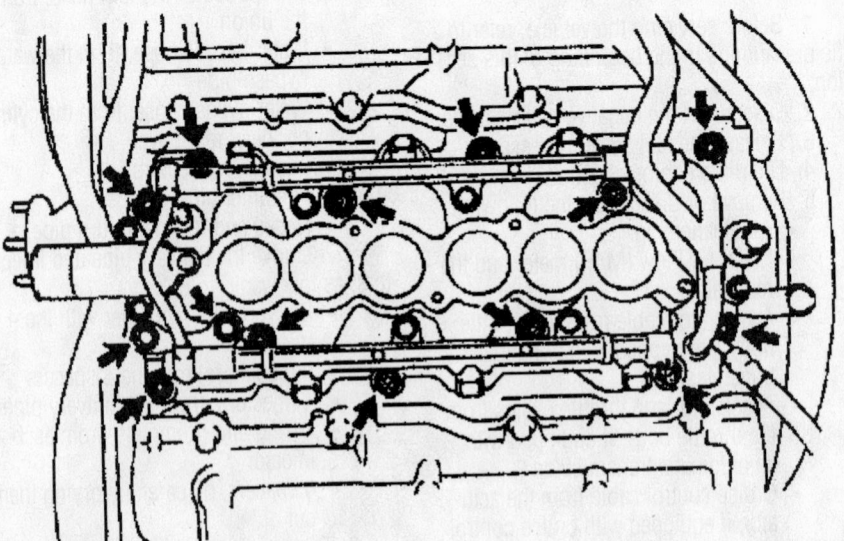

Intake manifold bolts and nuts—4Runner with 3.4L engine

7924YG38

11. Start the engine and check for leaks.
12. Install the engine undercover.
13. Road test the vehicle.
14. Recheck all fluid levels.

Exhaust Manifold

REMOVAL & INSTALLATION

RAV4

1. Before servicing the vehicle, refer to the precautions in the beginning of this section.
2. Remove or disconnect the following:

- Negative battery cable
- Front exhaust pipe from the exhaust manifold, using a 14mm deep socket wrench; discard the gasket
- Main Oxygen (O_2) sensor and the sub Oxygen (O_2) sensor connectors
- 6 bolts and the upper manifold heat insulator
- 2 right-hand exhaust manifold stay-to-cylinder block bolts
- 6 nuts, the exhaust manifold and the Three-Way Catalytic (TWC) converter assembly

- Exhaust manifold and front catalytic converter

To install:
3. Install or connect the following:

- Catalytic converter to the exhaust manifold. Tighten the nuts/bolts to 22 ft. lbs. (29 Nm).
- Exhaust manifold and the front TWC assembly. Tighten the 6 nuts, in several passes to 36 ft. lbs. (49 Nm).
- Right-hand manifold stay. Tighten both bolts to 31 ft. lbs. (42 Nm).
- Manifold upper heat insulator with

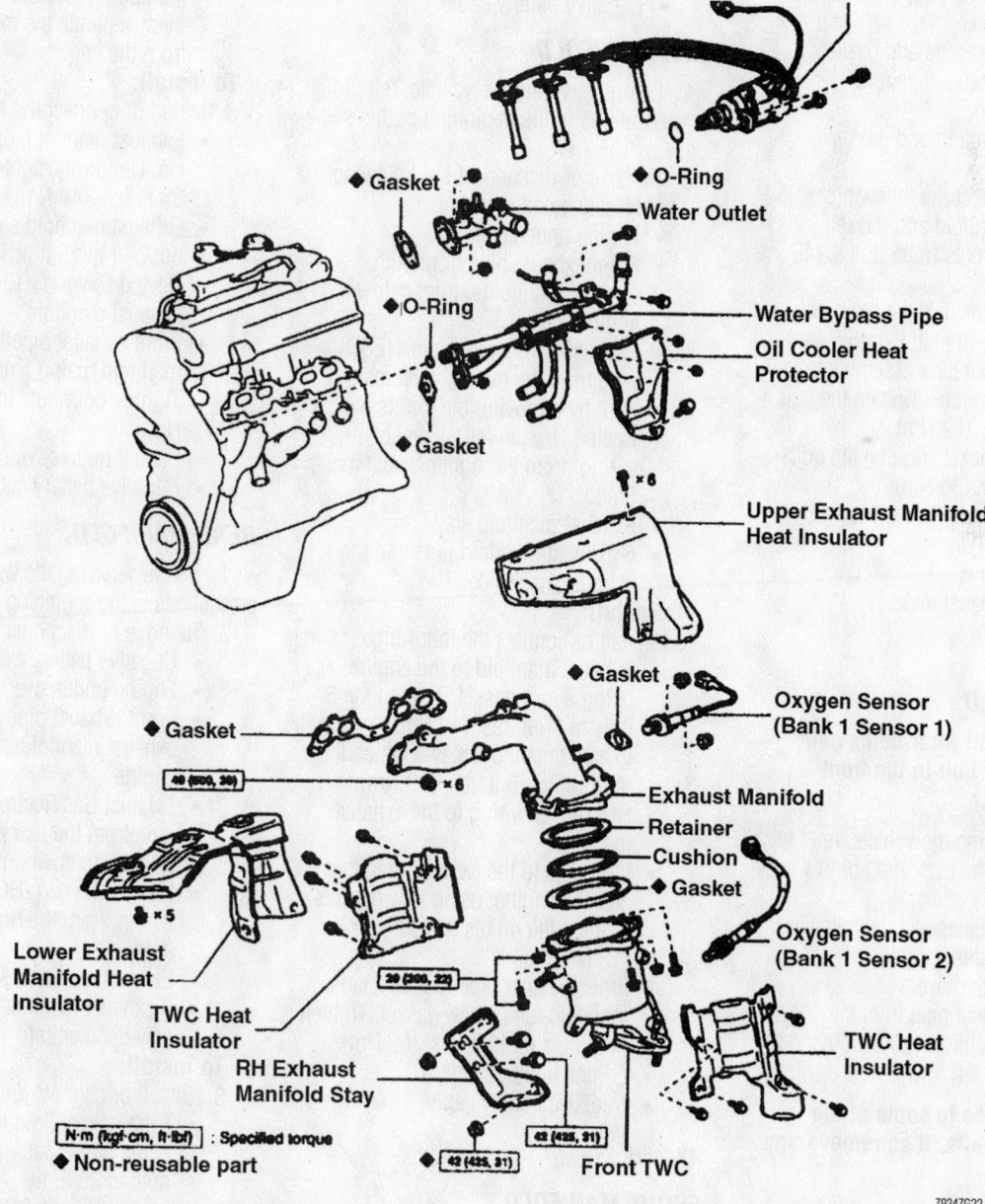

Exploded view of the exhaust manifold and components—rav4

For Accessory Drive Belt illustrations, see Section 1 of this manual

the 6 bolts and attach the main Oxygen (O_2) and the sub Oxygen (O_2) sensor connectors
- Front exhaust pipe to the TWC, using a new gasket. Tighten the 3 nuts to 46 ft. lbs. (62 Nm).
- Negative battery cable

4. Start the engine and be sure that there are no exhaust leaks.

2.4L Highlander

1. Before servicing the vehicle, refer to the precautions in the beginning of this section.
2. Remove or disconnect the following:
   - Clamp from the support bracket
   - Support bracket
   - Front exhaust pipe and gaskets from the exhaust manifold
   - Heat insulator
   - Exhaust manifold and gasket

To install:
3. Install or connect the following:
   - Exhaust manifold and gasket. Tighten the nuts to 36 ft. lbs. (49 Nm).
   - Heat insulator. Tighten the bolts and nuts to 48 inch lbs. (5.5 Nm).
   - Front exhaust pipe assembly to the exhaust manifold. Tighten the nuts to 46 ft. lbs. (62 Nm).
   - Support bracket. Tighten the bolts to 29 ft. lbs. (39 Nm).
   - Clamp. Tighten the bolt to 14 ft. lbs. (19 Nm).
4. Start the engine.
5. Check for exhaust leaks.

3.0L Highlander

FRONT MANIFOLD

➡**Removing the oil filter helps gain access to a lower bolt in the front exhaust manifold.**

1. Before servicing the vehicle, refer to the precautions in the beginning of this section.
2. Remove or disconnect the following:
   - Negative battery cable
   - Engine undercovers
   - Front exhaust pipe from the exhaust manifolds, by removing the nuts

➡**Check for access to some of the manifold lower bolts, if so remove any possible.**

- Heated Oxygen (HO_2) sensor
- Exhaust manifold stay, by removing the bolt and nut
- Remaining exhaust manifold nuts;

then, separate the exhaust manifold from the engine

To install:
3. Install or connect the following:
   - Exhaust manifold, using a new gasket. Uniformly, tighten the bolts to 36 ft. lbs. (49 Nm).
   - Exhaust manifold stay. Tighten the nut/bolt to 15 ft. lbs. (20 Nm).
   - Heated Oxygen (HO_2) sensor to the exhaust manifold
   - Front exhaust pipe to the exhaust manifold, using a new gasket. Tighten both nuts to 46 ft. lbs. (62 Nm).
   - Engine undercovers
   - Negative battery cable

REAR MANIFOLD

1. Before servicing the vehicle, refer to the precautions in the beginning of this section.
2. Remove or disconnect the following:
   - Negative battery cable
   - Engine undercovers
   - Front exhaust pipe from both exhaust manifolds, from below the engine
   - Exhaust Gas Recirculation (EGR) pipe from the rear exhaust manifold, by removing the 4 nuts
   - Heated Oxygen (HO_2) sensor wiring, from the right exhaust manifold
   - Exhaust manifold stay
   - 6 exhaust manifold nuts and the exhaust manifold

To install:
3. Install or connect the following:
   - Exhaust manifold to the engine, using a new gasket. Tighten the 6 nuts to 36 ft. lbs. (49 Nm).
   - Exhaust manifold stay. Tighten the nut/bolt to 15 ft. lbs. (20 Nm).
   - HO_2 sensor wiring to the exhaust manifold
   - EGR pipe to the exhaust manifold and the engine, using new gaskets. Tighten the 4 nuts to 108 inch lbs. (12 Nm).
   - Front exhaust pipe to the exhaust manifold, use a new gasket. Tighten both nuts to 46 ft. lbs. (62 Nm).
   - Engine undercovers
   - Negative battery cable

RX 300

FRONT MANIFOLD

➡**Removing the oil filter helps gain access to a lower bolt in the front exhaust manifold.**

1. Before servicing the vehicle, refer to the precautions in the beginning of this section.
2. Remove or disconnect the following:
   - Negative battery cable
   - Engine undercovers
   - Front exhaust pipe from the exhaust manifolds, by removing the nuts

➡**Check for access to some of the manifold lower bolts, if so remove any possible.**

- Heated Oxygen (HO_2) sensor
- Exhaust manifold stay, by removing the bolt and nut
- Remaining exhaust manifold nuts; then, separate the exhaust manifold from the engine

To install:
3. Install or connect the following:
   - Exhaust manifold, using a new gasket. Uniformly, tighten the bolts to 36 ft. lbs. (49 Nm).
   - Exhaust manifold stay. Tighten the nut/bolt to 15 ft. lbs. (20 Nm).
   - Heated Oxygen (HO_2) sensor to the exhaust manifold
   - Front exhaust pipe to the exhaust manifold, using a new gasket. Tighten both nuts to 46 ft. lbs. (62 Nm).
   - Engine undercovers
   - Negative battery cable

REAR MANIFOLD

1. Before servicing the vehicle, refer to the precautions in the beginning of this section.
2. Remove or disconnect the following:
   - Negative battery cable
   - Engine undercovers
   - Front exhaust pipe from both exhaust manifolds, from below the engine
   - Exhaust Gas Recirculation (EGR) pipe from the rear exhaust manifold, by removing the 4 nuts
   - Heated Oxygen (HO_2) sensor wiring, from the right exhaust manifold
   - Exhaust manifold stay
   - 6 exhaust manifold nuts and the exhaust manifold

To install:
3. Install or connect the following:
   - Exhaust manifold to the engine, using a new gasket. Tighten the 6 nuts to 36 ft. lbs. (49 Nm).
   - Exhaust manifold stay. Tighten the nut/bolt to 15 ft. lbs. (20 Nm).
   - HO_2 sensor wiring to the exhaust manifold

- EGR pipe to the exhaust manifold and the engine, using new gaskets. Tighten the 4 nuts to 108 inch lbs. (12 Nm).
- Front exhaust pipe to the exhaust manifold, use a new gasket. Tighten both nuts to 46 ft. lbs. (62 Nm).
- Engine undercovers
- Negative battery cable

2.7L 4Runner

1. Before servicing the vehicle, refer to the precautions in the beginning of this section.

2. Remove or disconnect the following:

- Clamp from the support bracket
- Support bracket
- Front exhaust pipe and gaskets from the exhaust manifold
- Heat insulator
- Exhaust manifold and gasket

To install:

3. Install or connect the following:

- Exhaust manifold and gasket. Tighten the nuts to 36 ft. lbs. (49 Nm).

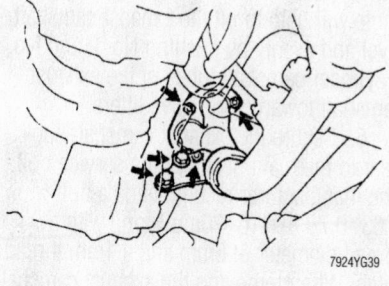

7924YG39

Front exhaust pipe to exhaust manifold nut and bolt locations—4Runner 2.7L

7924YG40

Exhaust manifold nuts—2.7L 4Runner

- Heat insulator. Tighten the bolts and nuts to 48 inch lbs. (5.5 Nm).
- Front exhaust pipe assembly to the exhaust manifold. Tighten the nuts to 46 ft. lbs. (62 Nm).
- Support bracket. Tighten the bolts to 29 ft. lbs. (39 Nm).
- Clamp. Tighten the bolt to 14 ft. lbs. (19 Nm).

4. Start the engine.
5. Check for exhaust leaks.

3.4L 4Runner

1. Before servicing the vehicle, refer to the precautions in the beginning of this section.

2. Remove or disconnect the following:

- Exhaust crossover pipe, from the exhaust manifold by removing the 3 nuts
- Exhaust Gas Recirculation (EGR) pipe, from the exhaust manifold, on the left manifold equipped with an EGR valve
- Exhaust manifold heat insulator, by removing the 3 nuts
- Exhaust manifold

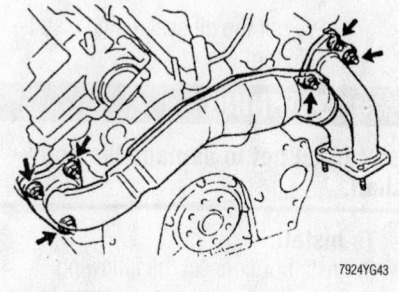

7924YG43

Exhaust crossover pipe mounting nut locations—3.4L 4Runner

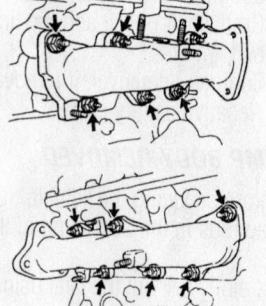

7924YG44

Exhaust manifold nuts—3.4L 4Runner

To install:

3. Install or connect the following:

- Exhaust manifold, using a new gasket. Tighten the nuts to 30 ft. lbs. (40 Nm).
- Exhaust heat insulator. Tighten the nuts to 71 inch lbs. (8 Nm).
- EGR pipe to the exhaust manifold, if equipped with an EGR valve. Tighten the manifold nuts to 14 ft. lbs. (18 Nm) and the clamp nuts to 71 inch lbs. (8 Nm).
- Crossover pipe to the exhaust manifold, using a new gasket. Tighten the nuts to 33 ft. lbs. (45 Nm).

Front Crankshaft Seal

REMOVAL & INSTALLATION

➡**For complete Timing Belt Removal and Installation procedures, see Section 3 of this manual.**

RAV4

1. Before servicing the vehicle, refer to the precautions in the beginning of this section.

➡**The front oil seal can be removed from the engine without removing the oil pump.**

2. Remove or disconnect the following:
- Negative battery cable
- Timing belt covers and the timing belt
- Front crankshaft gear from the crankshaft, using Crankshaft Gear Puller tool 09950-50010

❋❋ WARNING

Be sure not to damage any part of the crankshaft.

- Cut off the oil seal lip
- Oil seal, using a suitable tool. Wrap the edge of the tool with a rag or tape to prevent damaging the crankshaft.

To install:

3. Install or connect the following:
- New seal, by applying a thin layer of liquid sealer to the outside of the seal
- Apply multi purpose grease to the new oil seal lip
- New oil seal, by tapping it in until its surface is flush with the oil pump body edge, using the Oil

---

For Tire, Wheel and Ball Joint specifications, see Section 1 of this manual

Seal Installer tool 09226-00010 and a hammer
- Timing belt and timing belt covers
- All other components
- Negative battery cable

4. Start the engine and check for leaks.

3.0L Highlander and RX 300

1. Before servicing the vehicle, refer to the precautions in the beginning of this section.
2. Remove or disconnect the following:
   - Engine coolant reservoir tank and the alternator belt
   - Right front wheel and the splash shield
   - Power steering pump drive belt, by loosening both bolts
   - Both ground wire connectors
   - Right engine mounting stay
   - Engine moving control rod and the No. 2 right engine mount bracket

➡**To extract the engine bracket and control rod, raise the engine slightly.**

- No. 2 alternator bracket
- Crankshaft pulley bolt, using a prybar and wrench or Crankshaft Pulley Holding tool 09213-54015 and Flange Holding tool 09330-00021
- Crankshaft pulley, using a puller
- No. 1 timing belt cover

3. Remove the No. 2 timing belt cover, as follows:
   - Engine wire protector from the No. 3 (rear) timing belt cover
   - Engine wire protector clamp from the No. 3 timing belt cover
   - 5 bolts from the No. 2 timing belt cover
   - No. 2 cover

To install:
4. Install or connect the following:
   - No. 2 timing belt cover, using a new gasket

➡**Install it evenly to the part of the belt cover shaded black. After installation, press down on it so that the adhesive sticks to the belt cover firmly.**

- No. 2 timing belt cover. Tighten the 5 bolts to 74 inch lbs. (8 Nm).
- Engine wire protector clamp to the No. 3 timing belt cover
- Engine wire protector to the No. 3 timing belt cover with the bolt
- No. 3 timing belt cover, using a new gasket
- Tighten the 4 No. 1 timing belt cover bolts to 74 inch lbs. (8 Nm).

- Crankshaft pulley. Tighten the bolt to 159 ft. lbs. (215 Nm).
- No. 2 alternator bracket. Tighten the nut to 21 ft. lbs. (28 Nm). Do not tighten the pivot bolt at this time.
- No. 2 right engine mounting bracket and the moving control rod
- Right engine mount stay
- Both ground wire connectors
- Drive belts by adjusting them
- Coolant reservoir
- Right front splash shield and wheel
- Negative battery cable

5. Start the vehicle and check for any leaks.
6. Recheck the ignition timing.

3.4L 4Runner

➡**There are 2 methods to replace the oil seal, which are as follows:**

OIL PUMP BODY INSTALLED

1. Before servicing the vehicle, refer to the precautions in the beginning of this section.
2. Remove or disconnect the following:
   - Negative battery cable
   - Timing belt and crankshaft pulley
   - Cut off the oil seal lip, using a knife
   - Pry out the oil seal, using a suitable tool

✳✳ WARNING

Be careful not to damage the crankshaft.

To install:
3. Install or connect the following:
   - Apply multi-purpose grease to the new oil seal lip
   - Tap in the new oil seal until its surface is flush with the oil pump case edge, using Seal Driver tool 09309-37010 and a mallet
   - Crankshaft pulley and the timing belt
   - Engine undercover, if removed
   - Negative battery cable

OIL PUMP BODY REMOVED

1. Before servicing the vehicle, refer to the precautions in the beginning of this section.
2. Carefully pry out the seal using a suitable tool.
3. Apply multi-purpose grease to the new oil seal lip.
4. Using Seal Driver tool 09309-37010, drive the new seal into place.

Camshaft and Valve Lifters

REMOVAL & INSTALLATION

RAV4

1. Before servicing the vehicle, refer to the precautions in the beginning of this section.
2. Remove or disconnect the following:
   - Negative battery cable
   - Cylinder head cover and the upper timing belt cover

3. Rotate the crankshaft to set the engine at Top Dead Center (TDC)/compression for the No. 1 cylinder.

➡**Due to the small thrust clearance on both the intake and exhaust camshafts, the camshafts must be kept level during removal. If the camshafts are removed without being kept level, the camshaft may be caught in the cylinder head causing the head to break or the camshaft to seize.**

4. Remove the camshaft timing sprocket and the timing belt.
5. Set the knock pin of the intake camshaft at 10–45 degrees Before Top Dead Center (BTDC) of camshaft angle. This angle will help to lift the exhaust camshaft level and evenly by pushing No. 2 and No. 4 cylinder camshaft lobes of the exhaust camshaft toward their valve lifters.
6. Secure the exhaust camshaft sub-gear to the main gear using a service bolt. The manufacturer recommends a bolt 0.63–0.79 in. (16–20mm) long with a thread diameter of 6mm and a 1mm thread pitch. When removing the exhaust camshaft, be sure that the torsional spring force of the sub-gear has been eliminated.
7. Remove the No. 1 and No. 2 rear bearing cap bolts and remove the cap. Uniformly loosen and remove bearing cap bolts

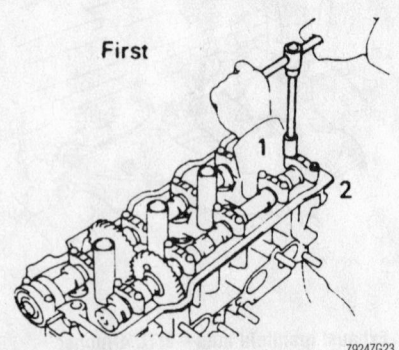

Exhaust camshaft bolt removal: step 1—RAV4 2.0L

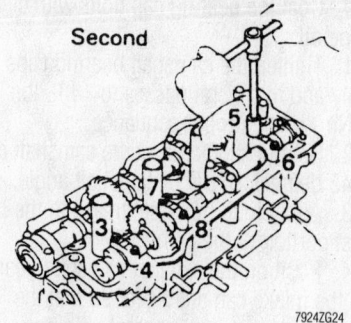

Second

Exhaust camshaft bolt removal: step 2—
RAV4 2.0L

No. 3 to No. 8 in several passes and in the proper sequence. Do not remove bearing cap bolts No. 9 and 10 at this time. Remove the No. 1, 2 and 4 bearing caps.

8. Alternately loosen and remove bearing cap bolts No. 9 and 10. As these bolts are loosened check to see that the camshaft is being lifted out straight and level.

➡ **If the camshaft is not lifting out straight and level retighten No. 9 and 10 bearing cap bolts. Reverse the order of steps 5 through 7 and reset the intake camshaft knock pin to 10–45 degrees BTDC and repeat steps 5 through 7 again. Do not attempt to pry the camshaft from its mounting.**

9. Remove the No. 3 bearing cap and exhaust camshaft from the engine.

10. Set the knock pin of the intake camshaft at 80–115 degrees BTDC of camshaft angle. This angle will help to lift the intake camshaft level and evenly by pushing No. 1 and No. 3 cylinder camshaft lobes of the intake camshaft toward their valve lifters.

11. Remove the No. 1 and No. 2 front bearing cap bolts and remove the front bearing cap and oil seal. If the cap will not come apart easily, leave it in place without the bolts.

12. Uniformly loosen and remove bearing cap bolts No. 3 to No. 8 in several phases and in the proper sequence. Do not remove bearing cap bolts No. 9 and 10 at this time. Remove No. 1, 3 and 4 bearing caps.

13. Alternately loosen and remove bearing cap bolts No. 9 and 10. As these bolts are loosened and after breaking the adhesion on the front bearing cap, check to see that the camshaft is being lifted out straight and level.

Third

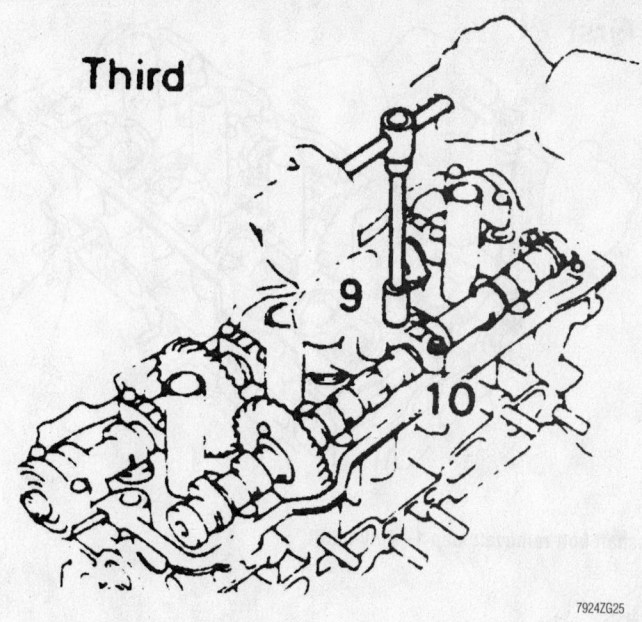

Exhaust camshaft bolt removal: step 3—RAV4 2.0L

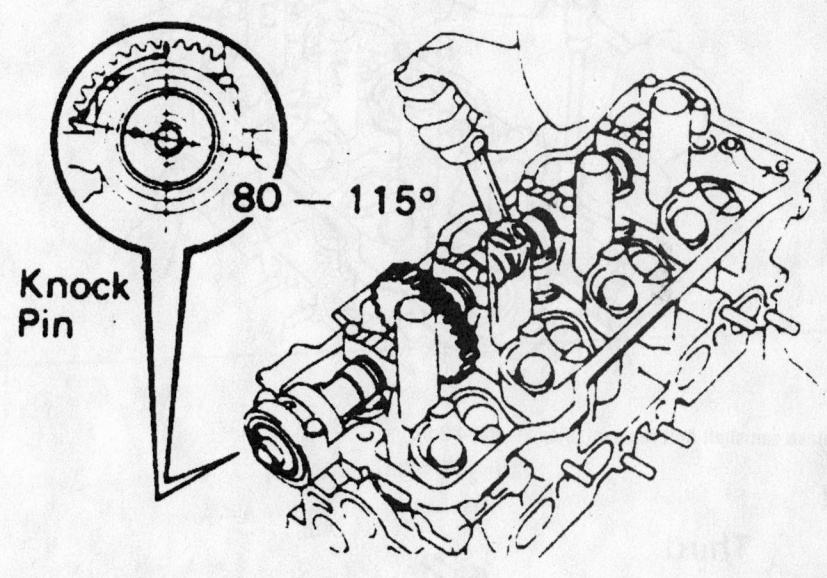

80 — 115°

Knock Pin

Intake camshaft knock pin alignment—RAV4 2.0L

➡ **If the camshaft is not lifting out straight and level retighten No. 9 and 10 bearing cap bolts. Reverse steps 10 through 12, than start over from step 10. Do not attempt to pry the camshaft from its mounting.**

14. Remove the No. 2 bearing cap with the intake camshaft from the engine.

15. Remove the valve adjusting shims from the engine. Be sure to replace the shims to their original location.

To install:

16. Install the valve adjusting shims to the engine.

17. Before installing the intake camshaft, apply multi-purpose grease to the thrust portion of the camshaft.

18. Position the camshaft at 80–115 degrees BTDC of camshaft angle on the cylinder head.

19. Apply sealant to the front bearing cap.

For Wheel Alignment specifications, see Section 1 of this manual

First

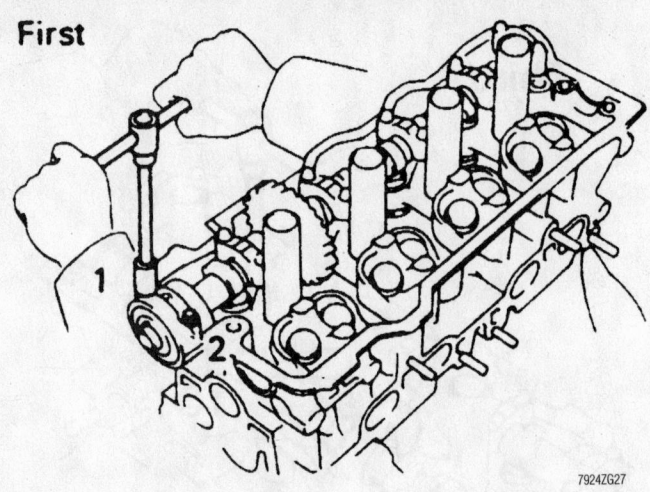

Intake camshaft bolt removal: step 1—RAV4 2.0L

Second

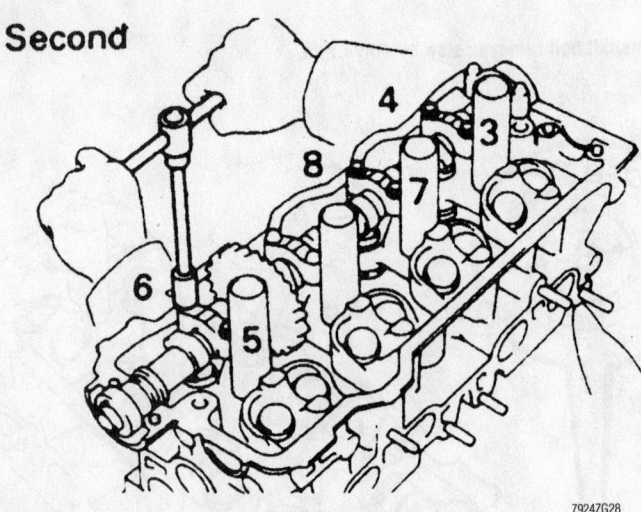

Intake camshaft bolt removal: step 2—RAV4 2.0L

Third

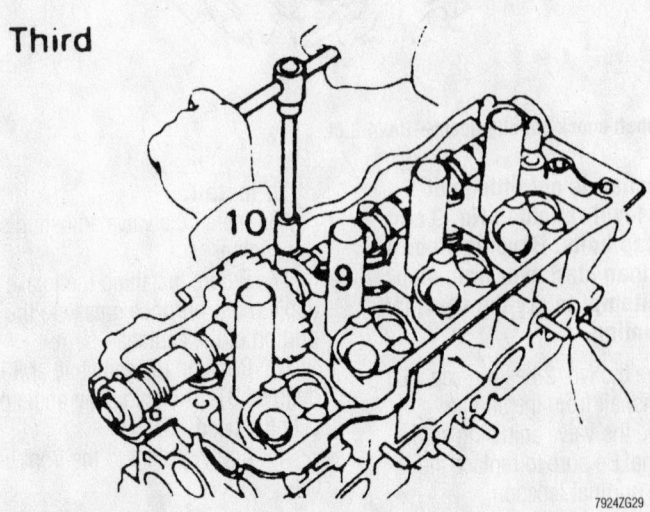

Intake camshaft bolt removal: step 3—RAV4 2.0L

20. Coat the bearing cap bolts with clean engine oil.

21. Tighten the camshaft bearing caps evenly and in several passes to 14 ft. lbs. (19 Nm) in the proper sequence.

22. Set the knock pin of the camshaft at 10–45 degrees BTDC of camshaft angle.

23. Apply multipurpose grease to the thrust portion of the camshaft.

24. Position the exhaust camshaft gear with the intake camshaft gear so that the timing marks are in alignment with one another. Be sure to use the proper alignment marks on the gears. Do not use the assembly reference marks.

25. Turn the intake camshaft clockwise or counterclockwise little by little until the exhaust camshaft sits in the bearing journals evenly without rocking the camshaft on the bearing journals.

26. Coat the bearing cap bolts with clean engine oil.

27. Tighten the camshaft bearing caps evenly and in several passes to 14 ft. lbs. (19 Nm). Remove the service bolt from the assembly.

28. Install the camshaft timing pulleys and the timing belt.

29. Adjust the valve clearance.

30. Install the head cover and the upper timing cover. Reconnect the negative battery cable.

31. Start the engine and check for leaks.

32. Check and adjust the ignition timing.

3.0L Highlander and RX 300

1. Before servicing the vehicle, refer to the precautions in the beginning of this section.

2. Remove or disconnect the following:
- Timing belt and idler pulley
- Camshaft timing pulleys
- Cylinder head covers

➡The thrust clearance on both the intake and exhaust camshafts is very small; the camshafts must be kept level during removal. If the camshafts are removed without being kept level, the camshaft may be caught in the cylinder head, causing the head to break or the camshaft to seize.

3. Remove the exhaust and intake camshafts from the right side cylinder head, as follows:

a. Turn the camshaft with a wrench until the 2 pointed marks drive and driven gears are aligned. (The right camshaft gears have 2 marks apiece; the left side camshaft gears have 1 mark each.)

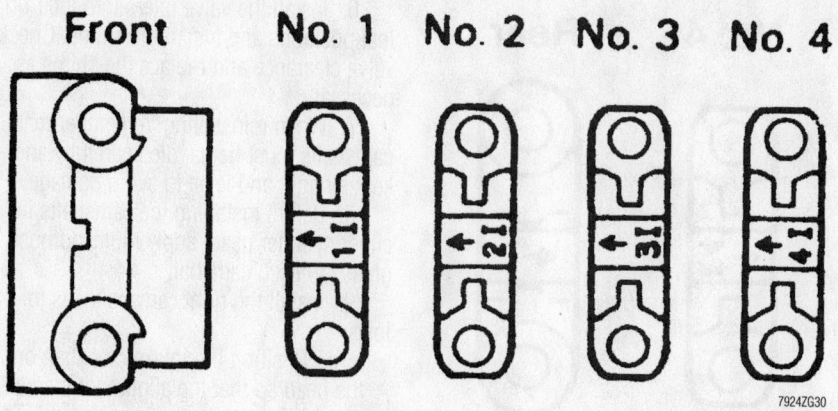

Intake camshaft bearing cap positioning—RAV4 2.0L

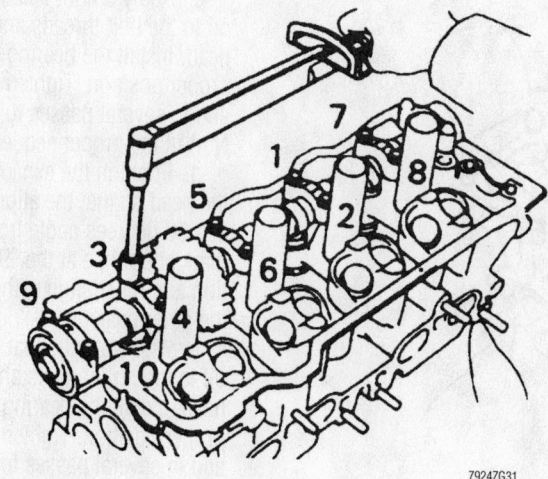

Intake camshaft bolt tightening sequence—RAV4 2.0L

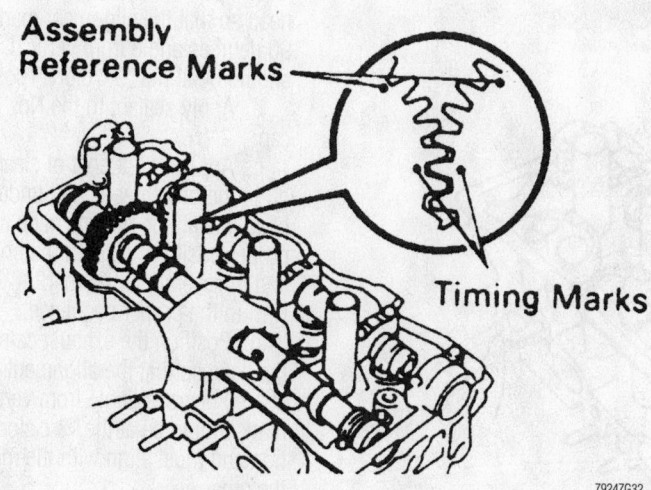

Camshaft timing mark alignment—RAV4 2.0L

b. Secure the exhaust camshaft sub-gear to the main gear using a service bolt. A bolt 0.63–0.79 in. (16–20mm) long with a 6mm thread diameter and a 1mm pitch is recommended. When removing the exhaust camshaft be sure the sub-gear is not loaded; all the force must be eliminated.

c. Uniformly loosen and remove the exhaust camshaft bearing cap bolts in several passes and in the proper sequence. Remove the 8 bearing cap bolts and remove the caps, keeping them in the correct order.

d. Remove the exhaust camshaft from the engine.

e. Uniformly loosen and remove the 10 bearing cap bolts in several passes, in the proper sequence. Remove the bearing caps, keeping them in order, remove the oil seal, then lift out the intake camshaft.

4. Remove the exhaust and intake camshafts from the left side cylinder head, as follows:

a. Turn the camshaft with a wrench until the pointed marks on the drive and driven gears are aligned. (The right camshaft gears have 2 marks apiece; the left side camshaft gears have 1 mark each.)

b. Secure the exhaust camshaft sub-gear to the main gear using a service bolt. A bolt 16–20mm long with a 6mm thread diameter and a 1mm pitch is recommended. When removing the exhaust camshaft be sure the sub-gear is not loaded; all the force must be eliminated.

c. Uniformly loosen and remove the exhaust camshaft bearing cap bolts in several passes and in the proper sequence. Remove the 8 bearing cap bolts and remove the caps. Keep the caps in the correct order.

d. Remove the exhaust camshaft from the engine.

e. Uniformly loosen and remove the 10 bearing cap bolts in several passes, in the reverse order of the installation sequence. Remove the bearing caps, keeping them in order, remove the oil seal, then lift out the intake camshaft.

5. Remove the valve lifter shims and hydraulic lifters. Identify each lifter and shim as it is removed so it can be reinstalled in the same position. If the lifters are to be reused, store them upside down in a sealed container.

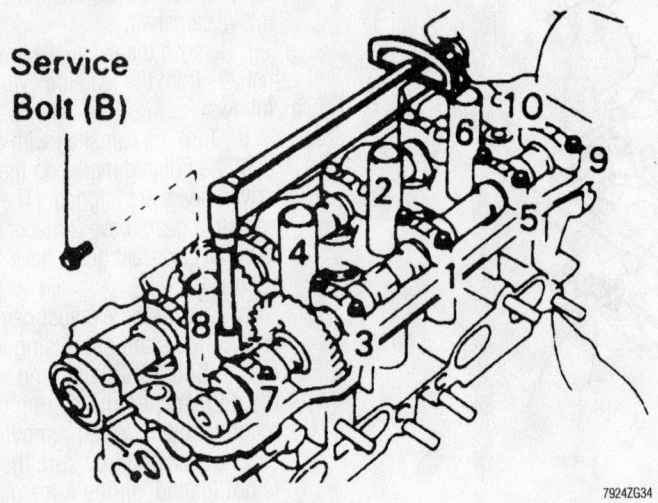

No. 1 No. 2 No. 3 No. 4 Rear

7924ZG33

Exhaust camshaft bearing cap positioning—RAV4 2.0L

Service Bolt (B)

7924ZG34

Exhaust camshaft bolt tightening sequence—RAV4 2.0L

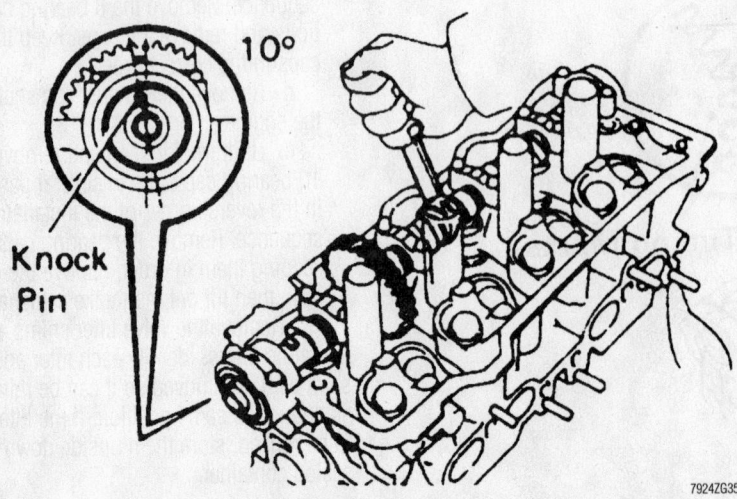

10°

Knock Pin

7924ZG35

Exhaust camshaft knock pin alignment—RAV4 2.0L

To install:

6. Install the valve lifters into their original positions and install the shims. Check valve clearance and replace the shims as necessary.

7. When reinstalling, remember that the camshafts must be handled carefully and kept straight and level to avoid damage.

8. Before installing the camshafts in either cylinder head, apply multi-purpose grease to each camshaft.

9. Install the right camshafts, as follows:

a. Position the intake camshaft on the head so that the alignment marks are at a 90 degrees angle from vertical. The mark should be at the "3 o'clock" position.

b. Apply sealant to the No. 1 bearing cap.

c. Apply a light coat of clean engine oil to the bolt threads and under the bolt head. Install the bearing caps to their proper position. Tighten the bolts evenly and in several passes to 12 ft. lbs. (16 Nm) in the proper sequence.

d. Position the exhaust camshaft on the head so that the alignment marks are at a 90 degrees angle from vertical. The mark should be at the "9 o'clock" position and must align with the marks on the other gear.

e. Apply a light coat of clean engine oil to the bolt threads and under the bolt head. Install the bearing caps to their proper position. Tighten the bolts evenly and in several passes to 12 ft. lbs. (16 Nm) in the proper sequence.

f. Remove the service bolt.

10. Install the left camshafts, as follows:

a. Position the intake camshaft on the head so that the alignment mark is at a 90 degrees angle from vertical. The mark should be at the "9 o'clock" position.

b. Apply sealant to the No. 1 bearing cap.

c. Apply a light coat of clean engine oil to the bolt threads and under the bolt head. Install the bearing caps to their proper position. Tighten the bolts evenly and in several passes to 12 ft. lbs. (16 Nm) in the proper sequence.

d. Position the exhaust camshaft on the head so that the alignment marks are at a 90 degrees angle from vertical. The mark should be at the "3 o'clock" position and must align with the marks on the other gear.

e. Apply a light coat of clean engine oil to the bolt threads and under the bolt head. Install the bearing caps to their proper position. Tighten the bolts evenly

Intake

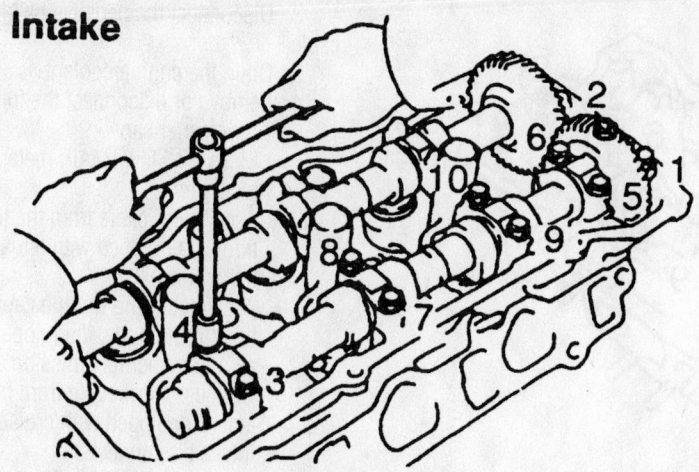

Right intake camshaft bearing cap bolt loosening sequence—3.0L engine

Exhaust

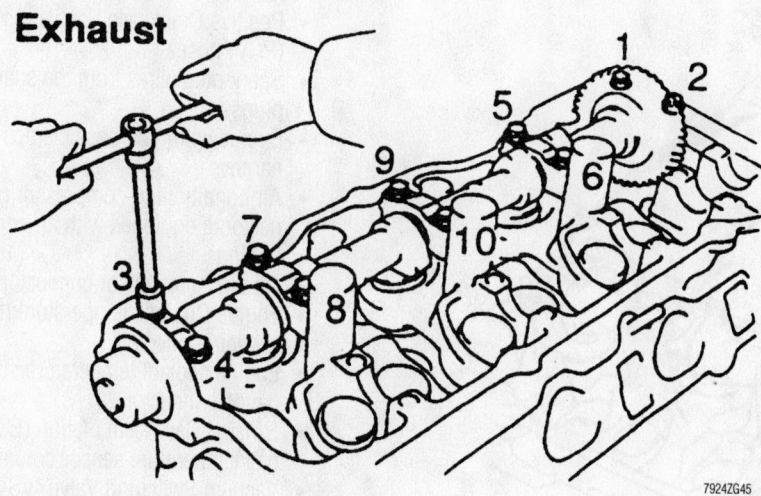

Right side exhaust camshaft bearing cap bolt loosening sequence—3.0L engine

Intake

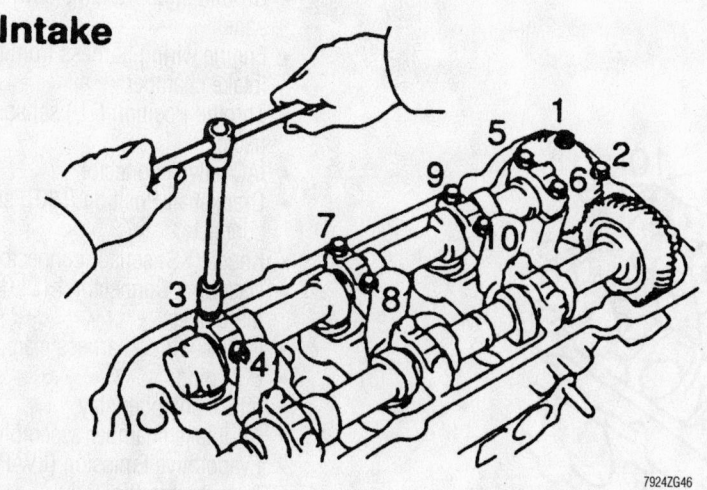

Left intake camshaft bearing cap bolt loosening sequence—3.0L engine

and in several passes to 12 ft. lbs. (16 Nm) in the proper sequence.

   f. Remove the service bolt.

11. Install or connect the following:
- New camshaft oil seals, lubricated with multi-purpose grease
- No. 3 (rear) timing belt cover
- Camshaft timing gears
- Idler pulley, timing belt and covers

12. Check and adjust the valve clearance.

13. Install the cylinder head (valve) covers.

14. Start the engine. Check the ignition timing.

15. Test drive the vehicle.

16. Check all fluid levels.

2.4L Highlander

1. Before servicing the vehicle, refer to the precautions in the beginning of this section.

2. Disconnect the negative battery cable.

3. Drain the engine coolant.

4. Remove or disconnect the following:
- Right front wheel
- Right fender splash shield
- Right fender apron seal
- No.1 engine undercover
- Coil pack
- Cylinder head cover

5. Set the No.1 piston at TDC compression.

6. Remove or disconnect the following:
- Timing chain tensioner No.1

7. Loosen the camshaft timing gear set bolt.

8. Raise the camshaft and remove the set bolt.

9. Remove or disconnect the following:
- Timing gear and chain
- Exhaust camshaft

10. Loosen the intake camshaft cap bolts in several passes, in the sequence shown. Remove the caps. Remove the camshaft.

11. Installation is the reverse of removal. Tighten the cap bolt, in several passes, in the sequences shown, to:
- Front caps: 22 ft. lbs. (30Nm)
- All other caps: 80 inch lbs. (9Nm)

12. See the Timing Chain Removal and Installation procedure.

2.7L 4Runner

1. Before servicing the vehicle, refer to the precautions in the beginning of this section.

For Tune-up, Capacities and Firing orders, see Section 1 of this manual

Exhaust

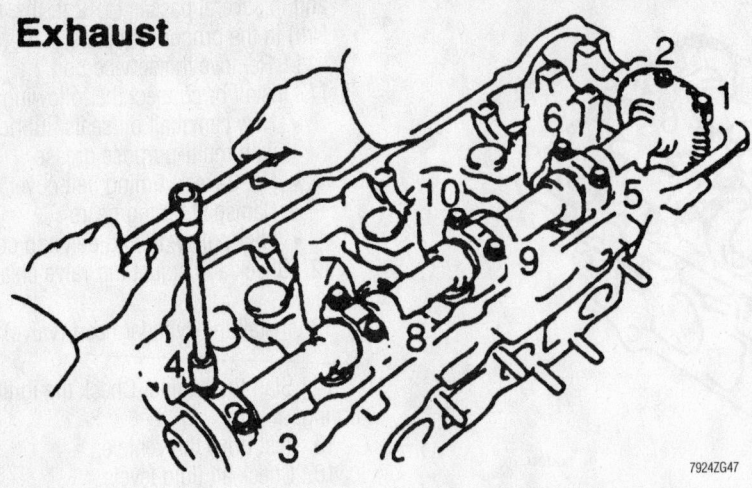

Left side exhaust camshaft bearing cap bolt loosening sequence—3.0L engine

7924ZG47

Exhaust

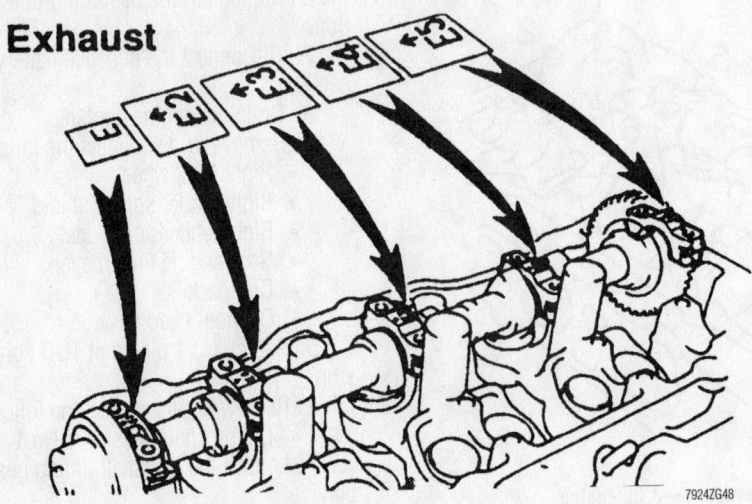

Right exhaust bearing caps must be placed in their proper locations—3.0L engine

7924ZG48

Exhaust

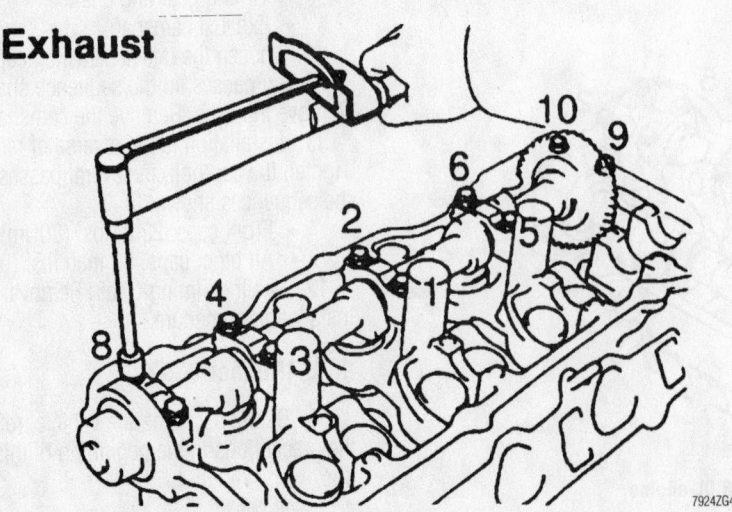

Right exhaust camshaft bearing cap bolt tightening sequence—3.0L engine

7924ZG49

2. Disconnect the negative battery cable.

3. Drain the engine coolant.

4. Remove or disconnect the following:
- Air cleaner cap
- Mass Air Flow (MAF) meter and the resonator
- Accelerator cable from the throttle body, if equipped with a manual transmission
- Accelerator and throttle cables from the throttle body, if equipped with an automatic transmission
- Cruise control cable from the actuator, if equipped with cruise control
- Intake air connector
- Air hose for Idle Air Control (IAC)
- Vacuum sensing hose
- Wire clamp for the engine wiring harness
- Positive Crankcase Ventilation (PCV) hoses
- Spark plug wires from the spark plugs
- Engine wiring harness clamps and harness
- Air conditioning compressor connector, if equipped with air conditioning
- Oil pressure sensor connector
- Engine Coolant Temperature (ECT) sensor connector
- Engine coolant temperature sender gauge connector
- Exhaust Gas Recirculation (EGR) gas temperature sensor connector
- Vacuum Switching Valve (VSV) connector for the EGR
- 2 vacuum hoses from the VSV for the EGR
- Ground strap from the cowl top panel
- Engine wiring harness from the air intake chamber
- Throttle Position (TP) sensor connector
- IAC valve connector
- Crankshaft Position (CKP) sensor connector
- Knock (KS) sensor connector
- Data Link Connector 1 (DLC1) from the bracket
- Engine wiring harness clamp
- EGR pipe
- Intake chamber stay
- Air intake chamber assembly.
- Evaporative Emission (EVAP) hose from the throttle body
- Brake booster vacuum hose from the union
- Water bypass hose from the water bypass pipe

Intake

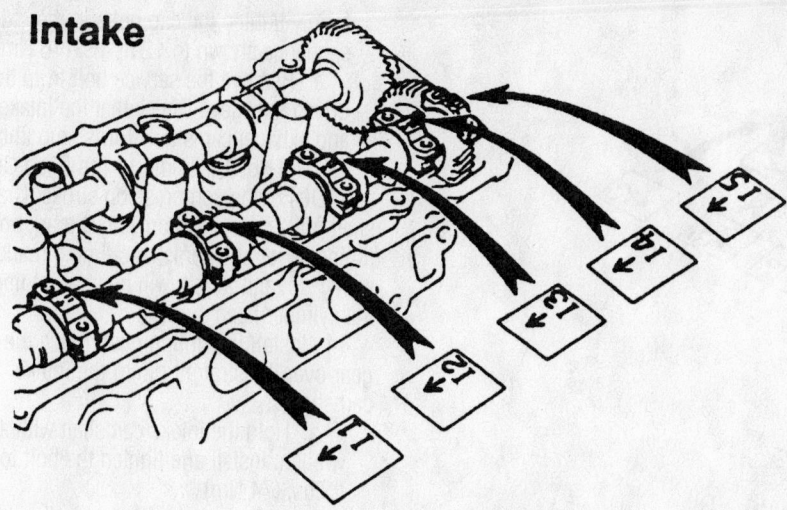

Right intake bearing caps must be placed in their proper locations—3.0L engine

7924ZG50

Intake

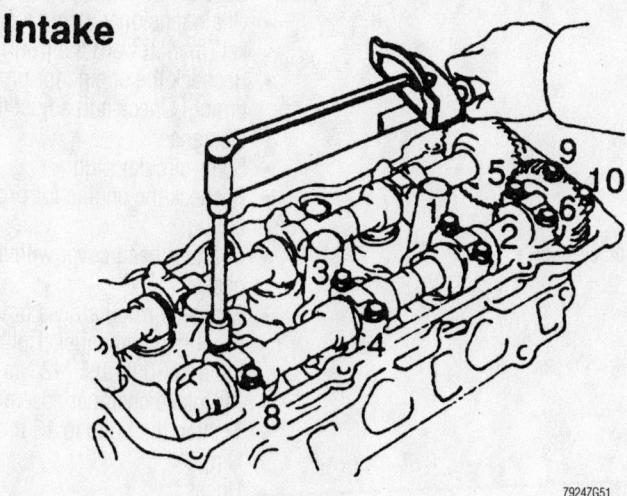

Right intake camshaft bearing cap bolt tightening sequence—3.0L engine

7924ZG51

- Water bypass hose from the cylinder head rear cover
- Front engine hanger
- Engine wiring harness brackets
- Cylinder head cover

5. Set No. 1 cylinder to Top Dead Center (TDC) compression stroke. The groove on the crankshaft pulley should align with the **0** mark on the timing chain cover and the timing marks (1 and 2 dots) of the camshaft gears should form a straight line in respect to the cylinder head surface. If not, turn the crankshaft 1 revolution (360 degrees).

6. Remove the chain tensioner and gasket.

7. Remove the camshaft timing gear as follows:

　a. Remove the 2 semi-circular plugs.

b. Place matchmarks on the camshaft timing gear and No. 1 timing chain.

c. Hold the hexagon head portion of the exhaust camshaft with a wrench and remove the fastener and distributor gear.

d. Hold the hexagon head portion of the intake camshaft with a wrench and remove the bolt.

e. Remove the camshaft timing gear and chain from the intake camshaft and leave on the slipper and damper.

8. Remove exhaust camshafts:

　a. Bring the service bolt hole of the driven sub-gear upward by turning the hexagon head portion of the exhaust camshaft with a wrench.

　b. Secure the exhaust camshaft sub-gear to the driven gear with a service bolt

(6mm diameter, 0.63–0.79 inches in length and 1.0mm in thread pitch).

➡**When removing the camshaft, be sure that the torsional spring force of the sub-gear has been eliminated by the above operation.**

　c. Uniformly loosen and remove the bearing cap bolts in several passes, in the sequence shown.

　d. Remove the bearing caps and camshaft. Make a note of the bearing cap positions for proper installation.

9. Remove or disconnect the following:

- Intake camshaft bearing cap bolts in several passes, in the sequence shown
- Bearing caps and camshaft. Make a note of the bearing cap positions for proper installation.

➡**If the camshaft is not being lifted out straight and level, reinstall the No. 3 bearing cap with the 2 bolts. Then, alternately loosen and remove the 2 bearing cap bolts with the camshaft gear pulled up.**

- Valve lifters and shims from the cylinder head. Arrange the valve lifters and shims in correct order.

To install:

10. Install the valve lifters and shims in their proper locations. Check that the valve lifter rotates smoothly by hand.

11. Install the intake camshaft, as follows:

　a. Apply engine oil to the thrust portion of the intake camshaft.

　b. Position the intake camshaft with the knock pin facing upward.

　c. Install the bearing caps in their proper locations. Apply a light coat of engine oil to the threads and install the cap bolts. Uniformly tighten the cap bolts, in sequence ,to 12 ft. lbs. (16 Nm).

12. Install the exhaust camshaft, as shown:

　a. Apply engine oil to the thrust portion of the intake camshaft.

　b. Engage the exhaust camshaft gear to the intake camshaft gear by matching the timing marks (1 and 2 dots) on each other.

　c. Roll down the exhaust camshaft onto the bearing journals while engaging the gears with each other. Install the bearing caps in their proper locations.

　d. Apply a light coat of engine oil to the threads and install the cap bolts. Uni-

Exhaust

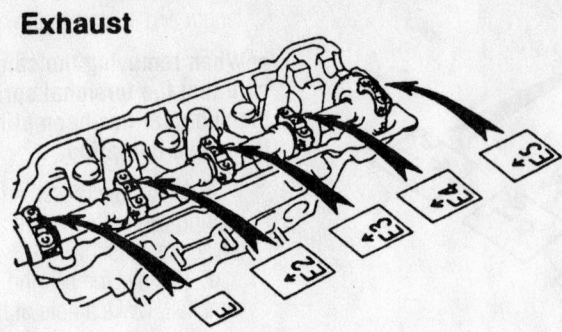

Exhaust

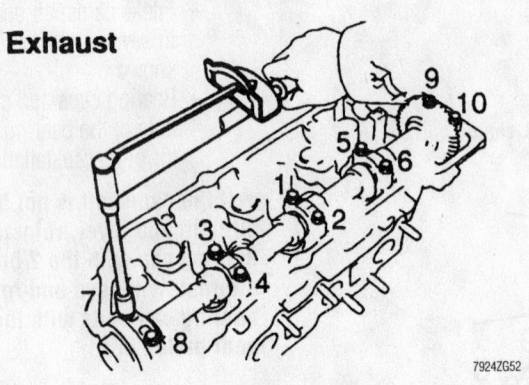

Left exhaust bearing caps locations and bolt tightening sequence—3.0L engine

Intake

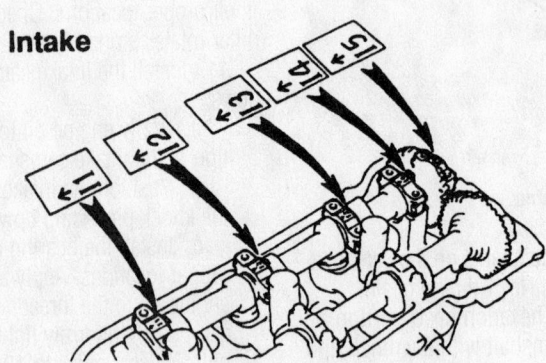

Intake

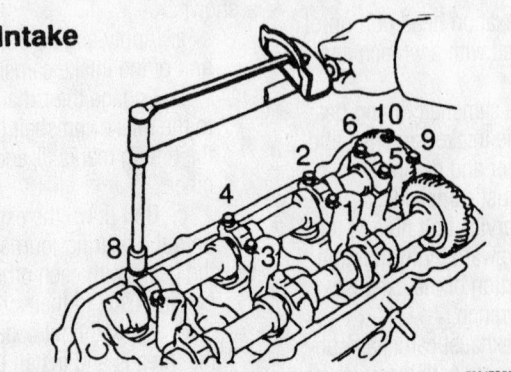

Left intake camshaft bearing cap locations and bolt tightening sequence—3.0L engine

formly tighten the cap bolts in the sequence shown to 12 ft. lbs. (16 Nm).

e. Remove the service bolt from the driven sub-gear. Check that the intake and exhaust camshafts turns smoothly.

13. Set No. 1 cylinder to Top Dead Center (TDC) of the compression stroke: Crankshaft pulley groove align with **0** mark on timing cover and camshafts timing marks with 1 dot and 2 dots will be straight line on the cylinder head surface.

14. Install the timing gear. Place the gear over the straight pin of the intake camshaft.

a. Hold the intake camshaft with a wrench. Install and tighten the bolt to 54 ft. lbs. (74 Nm).

b. Hold the exhaust camshaft and install the bolt and distributor gear. Tighten the bolt to 34 ft. lbs. (46 Nm).

15. Install or connect the following:

- Chain tensioner, using a new gasket (mark toward the front)
- Recheck the engine for proper valve timing. Check and adjust the valve clearance.
- Semi-circular plug
- Recheck the engine for proper valve timing.
- Cylinder head cover with a new gasket
- Engine wiring harness brackets
- Front engine hanger. Tighten the bolts to 30 ft. lbs. (42 Nm).
- Air intake chamber assembly. Tighten the bolts to 15 ft. lbs. (20 Nm).
- Hoses
- Intake chamber stay
- Air intake chamber stay. Tighten the bolts to 15 ft. lbs. (20 Nm).
- EGR pipe. Tighten the bolts to 13 ft. lbs. (18 Nm), nut "A" to 14 ft. lbs. (19 Nm) and nut "B" to 15 ft. lbs. (20 Nm).
- Engine wiring harness
- Spark plug wires to the spark plugs
- PCV hoses
- Intake air connector. Tighten the bolts to 13 ft. lbs. (18 Nm).
- Air hose for the IAC
- Vacuum sensing hose
- Wire clamp for the engine wiring harness
- Cruise control cable to the actuator, if equipped with cruise control
- Accelerator cable to the throttle body, if equipped with a manual transmission
- Accelerator and throttle cables to the throttle body, if equipped with an automatic transmission

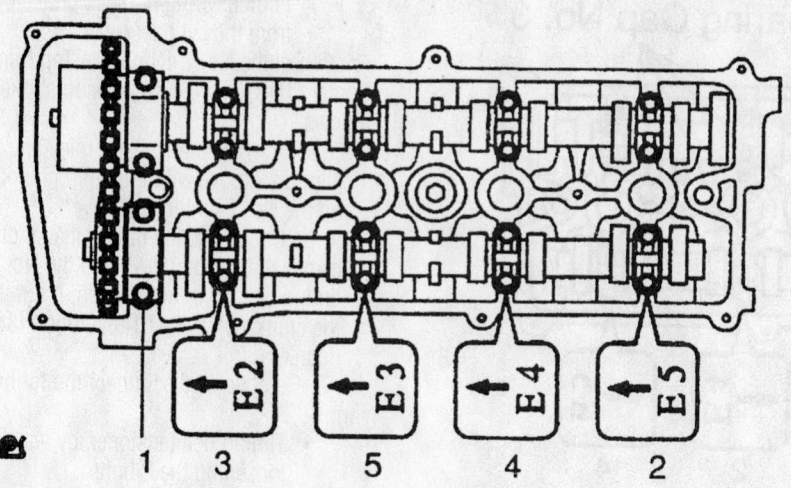

Exhaust camshaft cap bolt loosening sequence—2.4L Highlander

9355YG03

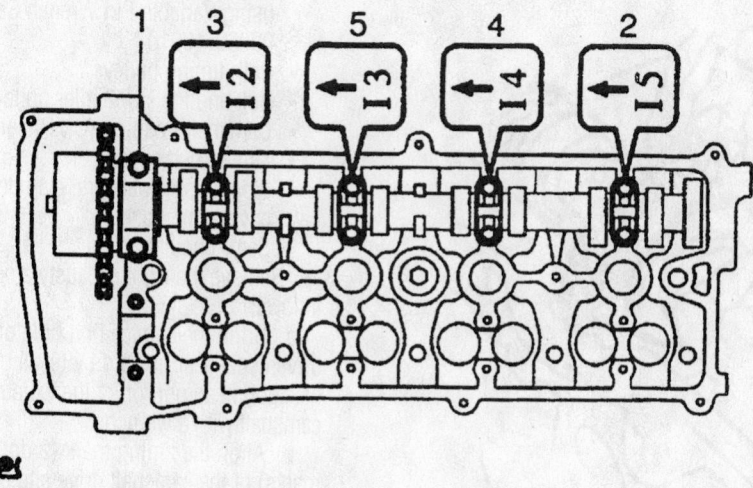

Intake camshaft cap bolt loosening sequence—2.4L Highlander

9355YG04

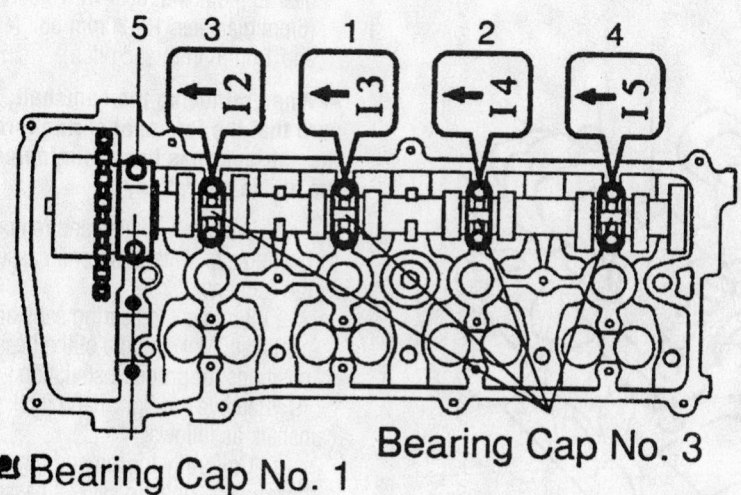

Bearing Cap No. 1

Bearing Cap No. 3

9355YG05

Intake camshaft cap bolt tightening sequence—2.4L Highlander

- Air cleaner cap, MAF meter and the resonator assembly
- Negative battery cable
16. Refill the cooling system.
17. Start the engine and check for leaks.
18. Check the ignition timing. Road test the vehicle for proper operation.
19. Recheck all fluid levels.

3.4L 4Runner

1. Before servicing the vehicle, refer to the precautions in the beginning of this section.
2. Release the fuel pressure.
3. Remove or disconnect the following:
- Negative battery cable
- Engine undercover
4. Drain the cooling system.
5. Remove or disconnect the following:
- Air cleaner cap
- Mass Air Flow (MAF) meter and the resonator
- Actuator cable with the bracket, if equipped with cruise control
- Accelerator cable
- Throttle cable, if equipped with an automatic transmission
- Heater hose
- Brake booster vacuum hose
- Evaporative Emission (EVAP) hose
- Automatic Disconnecting Differential (ADD) vacuum hose, for 4-wheel drive
- Fuel inlet and fuel return hose
- Spark plug wires, with the ignition coils
- Intake chamber stay
- Camshaft Position (CMP) sensor connector from the No. 2 timing belt cover
- 3 spark plug wire clamps from the No. 2 timing belt cover
- 6 bolts and the No. 2 timing belt cover
- Air intake chamber assembly
- Throttle Position (TP) sensor connector
- Idle Air Control (IAC) valve connector
- Positive Crankcase Ventilation (PCV) hoses
- Water bypass hoses
- Air assist hose from the throttle body
- Intake air connector
- 6 injector connectors
- Engine Coolant Temperature (ECT) sensor and sender gauge connectors

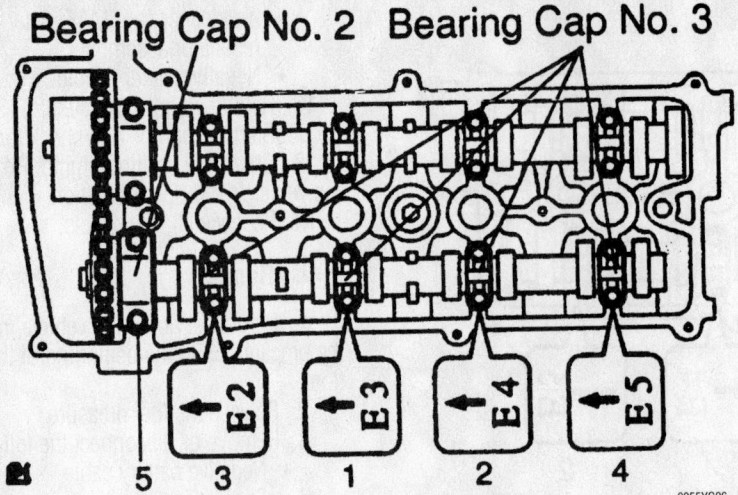

Exhaust camshaft cap bolt tightening sequence—2.4L Highlander

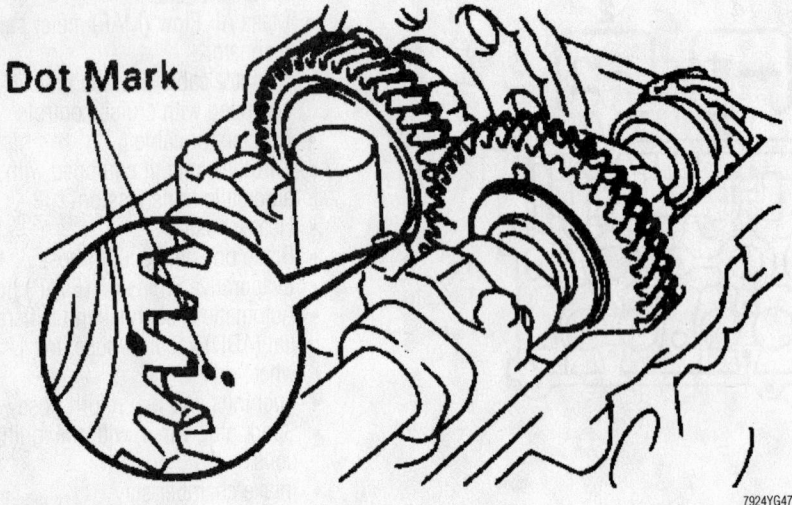

Camshafts TDC/compression timing marks. Marks with 1 and 2 dots will be in straight line on cylinder head surface—2.7L 4Runner

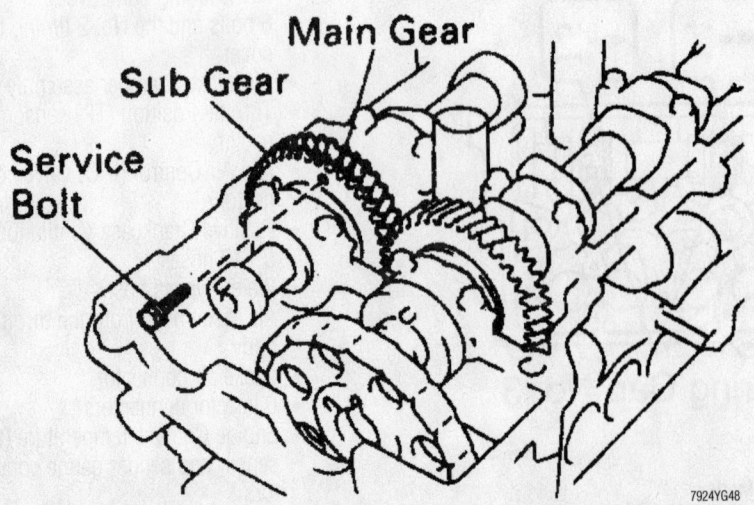

Secure the exhaust camshaft sub-gear to the main gear with a service bolt—2.7L 4Runner

- Engine wiring harness protector, from the cylinder head

6. Set the No. 1 cylinder at Top Dead Center (TDC) of the compression stroke, as follows:

a. Turn the crankshaft pulley and align its groove with the timing mark **0** on the No. 1 timing belt cover.

b. Check that the timing marks of the camshaft timing pulleys and the No. 3 timing belt cover are aligned. If not, turn the crankshaft pulley 1 revolution (360 degrees).

7. Remove or disconnect the following:

- Timing belt tensioner, by alternately loosening the 2 bolts
- Timing belt
- Camshaft timing pulley bolt, the timing pulley and the knock pin, using Variable Pin Wrench Set 09960-10010
- Both timing pulleys
- Bolt and the No. 2 idler pulley
- Camshaft Position (CMP) sensor
- Timing belt cover
- 8 bolts, seal washers, cylinder head covers and gasket
- Semi-circular plugs

8. Remove the right exhaust camshafts, as follows:

a. Bring the service bolt hole of the driven sub-gear upward by turning the hexagon head portion of the exhaust camshaft with a wrench.

b. Align the timing mark (2 dot marks) of the camshaft drive and driven gears by turning the camshaft with a wrench.

c. Secure the exhaust camshaft sub-gear to the driven gear with a service bolt (6mm diameter, 16–20mm bolt length and 1.0mm in thread pitch).

➡ **When removing the camshaft, be sure that the torsional spring force of the sub-gear has been eliminated by the above operation.**

d. Uniformly loosen and remove the bearing cap bolts in several passes, in the sequence.

e. Remove the bearing caps and camshaft. Make a note of the bearing cap positions for proper installation.

9. Remove the right-hand intake camshaft, as follows:

a. Uniformly loosen and remove the bearing cap bolts in several passes, in the sequence.

b. Remove the bearing caps, oil seal and camshaft. Make a note of the bearing cap positions for proper installation.

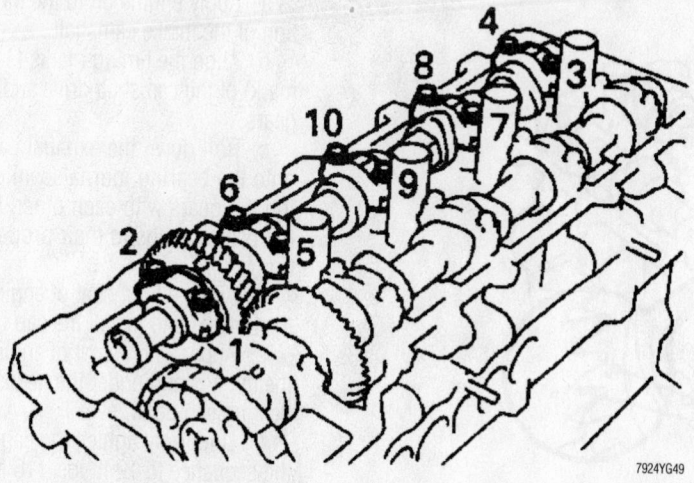

Loosen and remove the exhaust camshaft bearing cap bolts in sequence—2.7L 4Runner

7924YG49

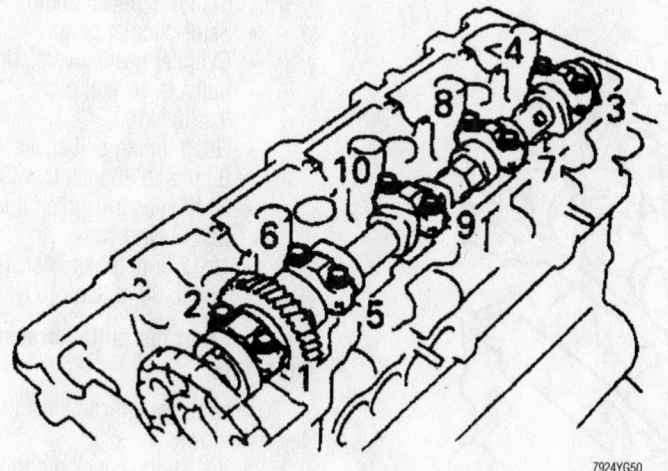

Loosen and remove the intake camshaft bearing cap bolts in sequence—2.7L 4Runner

7924YG50

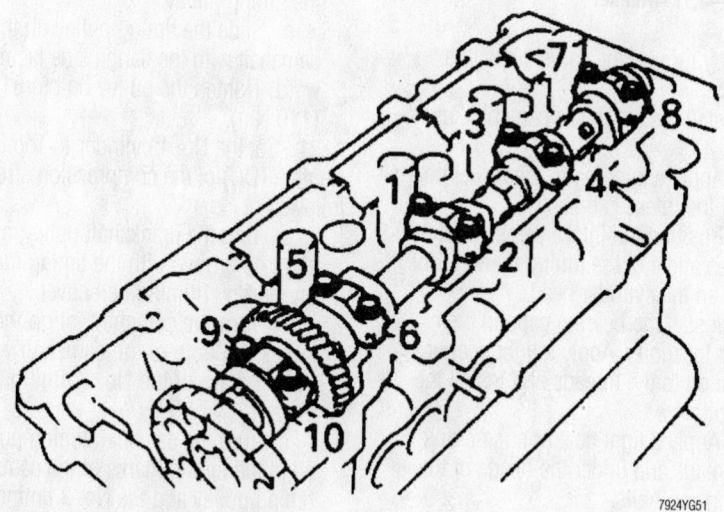

Tighten the intake camshaft bearing cap bolts in sequence—2.7L 4Runner

7924YG51

10. Remove the left exhaust camshafts, as follows:

    a. Align the timing mark (1 dot mark) of the camshaft drive and driven gears by turning the camshaft with a wrench.

    b. Secure the exhaust camshaft sub-gear to the driven gear with a service bolt (6mm diameter, 16–20mm bolt length and 1.0mm in thread pitch).

➡**When removing the camshaft, be sure the torsional spring force of the sub-gear has been eliminated by the above operation.**

    c. Uniformly loosen and remove the bearing cap bolts in several passes, in the sequence.

    d. Remove the bearing caps and camshaft. Make a note of the bearing cap positions for proper installation.

➡**Do not pry on or attempt to force the camshaft with a tool or other object.**

11. Remove the left-hand intake camshaft, as follows:

    a. Uniformly loosen and remove the bearing cap bolts in several passes, in the sequence.

    b. Remove the bearing caps, oil seal and camshaft.

➡**Make a note of the bearing cap positions for proper installation.**

12. Remove the valve lifters and shims from the cylinder head. Arrange the valve lifters and shims in correct order.

To install:

13. Install the valve lifters and shims. Check that the valve lifter rotates smoothly by hand.

14. Install the right intake camshaft, as follows:

    a. Apply engine oil to the thrust portion of the intake camshaft.

    b. Position the intake camshaft at 90 degrees angle of the timing mark (2 dot marks) on the cylinder head.

    c. Install the bearing caps in their proper locations. Apply a light coat of engine oil to the threads and install the cap bolts.

    d. Apply a light coat of engine oil on the threads and under the heads of the bearing cap bolts.

    e. Uniformly tighten the cap bolts in the sequence to 12 ft. lbs. (16 Nm).

15. Install the right exhaust camshaft, as follows:

    a. Apply engine oil to the thrust portion of the intake camshaft.

Timing belt service is covered in Section 3 of this manual

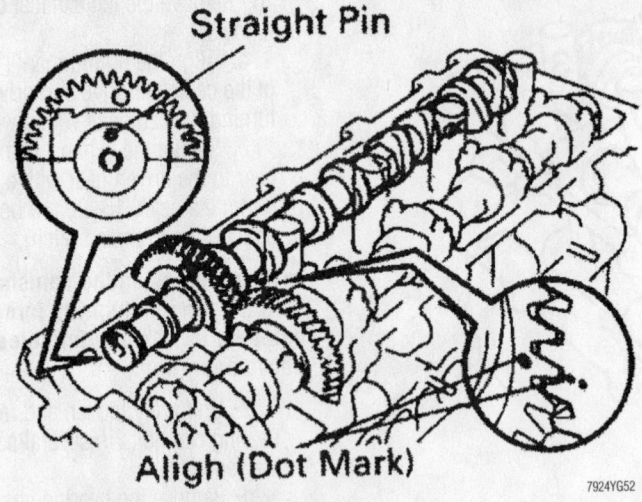

Straight Pin

Aligh (Dot Mark)

7924YG52

Engage both camshaft gears while matching the timing marks—2.7L 4Runner

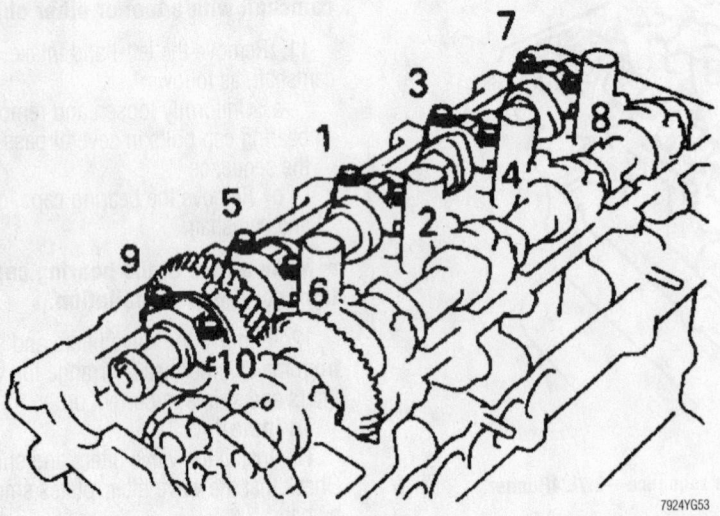

7924YG53

Tighten the exhaust camshaft bearing cap bolts in sequence—2.7L 4Runner

b. Align the timing marks (2 dot marks) of the camshaft drive and driven gears.

c. Roll down the exhaust camshaft onto the bearing journals while engaging the gears with each other. Install the bearing caps in their proper locations.

d. Apply a light coat of engine oil to the threads and install the cap bolts.

e. Apply a light coat of engine oil on the threads and under the heads of the bearing cap bolts.

f. Uniformly tighten the cap bolts in the sequence to 12 ft. lbs. (16 Nm).

g. Remove the service bolt from the driven sub-gear. Check that the intake and exhaust camshafts turn smoothly.

h. Align the timing marks (2 dot marks) of the camshaft drive and driven

gears by turning the camshaft with a wrench.

16. Install the left intake camshaft, as follows:

a. Apply engine oil to the thrust portion of the intake camshaft.

b. Position the intake camshaft at 90 degrees angle of the timing mark (1 dot mark) on the cylinder head.

c. Install the bearing caps in their proper locations. Apply a light coat of engine oil to the threads and install the cap bolts.

d. Apply a light coat of engine oil on the threads and under the heads of the bearing cap bolts.

e. Uniformly tighten the cap bolts in the sequence to 12 ft. lbs. (16 Nm).

17. Install the left exhaust camshaft, as follows:

a. Apply engine oil to the thrust portion of the intake camshaft.

b. Align the timing marks (1 dot mark) of the camshaft drive and driven gears.

c. Roll down the exhaust camshaft onto the bearing journals while engaging the gears with each other. Install the bearing caps in their proper locations.

d. Apply a light coat of engine oil to the threads and install the cap bolts.

e. Apply a light coat of engine oil on the threads and under the heads of the bearing cap bolts.

f. Uniformly tighten the cap bolts in the sequence to 12 ft. lbs. (16 Nm).

g. Remove the service bolt.

18. Check and adjust the valve clearance.

19. Install or connect the following:
- Semi-circular plugs
- Cylinder head covers. Tighten the bolts, in several passes, to 53 inch lbs. (6 Nm).
- No. 3 timing belt cover. Tighten the 6 bolts to 80 inch lbs. (9 Nm).
- CMP sensor. Tighten it to 71 inch lbs. (8 Nm).
- No. 2 timing belt idler. Tighten the bolt to 30 ft. lbs. (40 Nm).

➡ **Check that the pulley bracket moves smoothly.**

20. Install the left camshaft timing pulley, as follows:

a. Install the knock pin to the camshaft.

b. Align the knock pin hose of the camshaft with the knock pin groove of the timing pulley.

c. Slide the timing pulley on the camshaft with the flange side facing outward. Tighten the pulley bolt to 81 ft. lbs. (110 Nm).

21. Set the No. 1 cylinder to Top Dead Center (TDC) of the compression stroke, as follows:

a. Turn the crankshaft pulley, and align its groove with the timing mark **0** on the No. 1 timing belt cover.

b. Turn the camshaft, align the knock pin hole of the camshaft with the timing mark of the No. 3 timing belt cover.

c. Turn the camshaft timing pulley, align the timing marks of the camshaft timing pulley and the No. 3 timing belt cover.

22. Install or connect the following:
- Timing belt to the left camshaft timing pulley. Check that the installa-

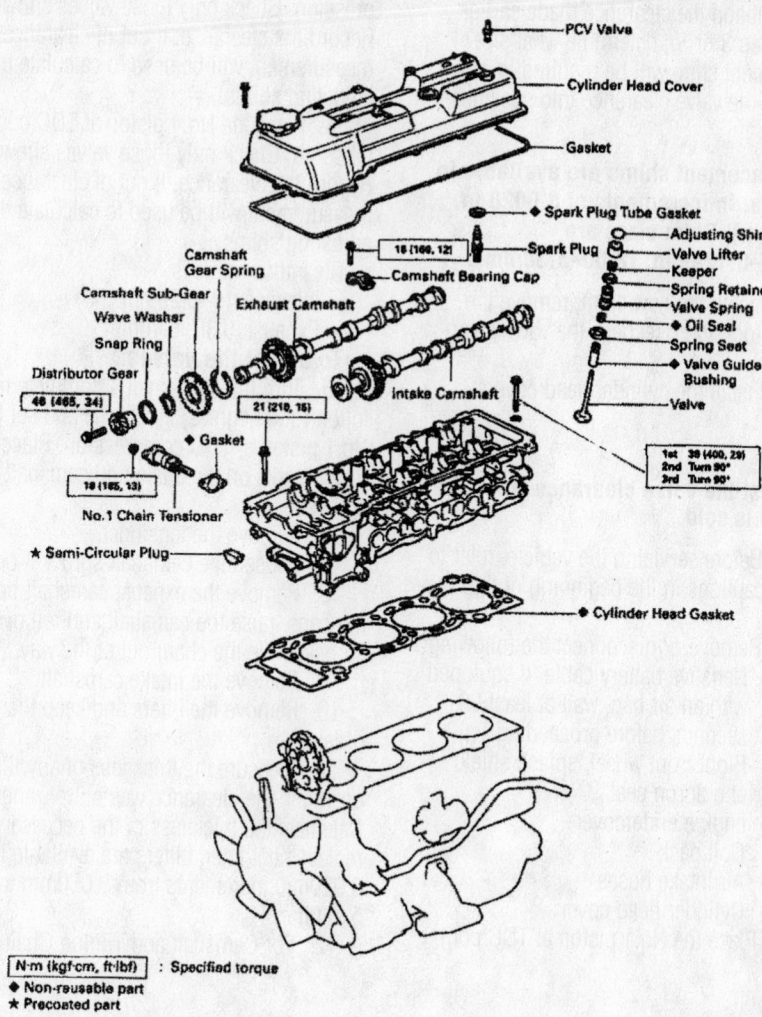

Exploded view of the cylinder head components—2.7L 4Runner

7924YG54

N·m (kgf·cm, ft·lbf) : Specified torque
◆ Non-reusable part
★ Precoated part

Labels in diagram:
- PCV Valve
- Cylinder Head Cover
- Gasket
- ◆ Spark Plug Tube Gasket
- 18 (180, 12)
- Spark Plug
- Camshaft Gear Spring
- Camshaft Bearing Cap
- ◆ Adjusting Shim
- Valve Lifter
- Keeper
- Spring Retainer
- Valve Spring
- ◆ Oil Seal
- Camshaft Sub-Gear
- Exhaust Camshaft
- Wave Washer
- Snap Ring
- Spring Seat
- ◆ Valve Guide Bushing
- Distributor Gear
- 46 (465, 34)
- 21 (210, 15)
- Intake Camshaft
- Valve
- 18 (185, 13)
- ◆ Gasket
- 1st 39 (400, 29)
  2nd Turn 90°
  3rd Turn 90°
- No.1 Chain Tensioner
- ◆ Cylinder Head Gasket
- ★ Semi-Circular Plug

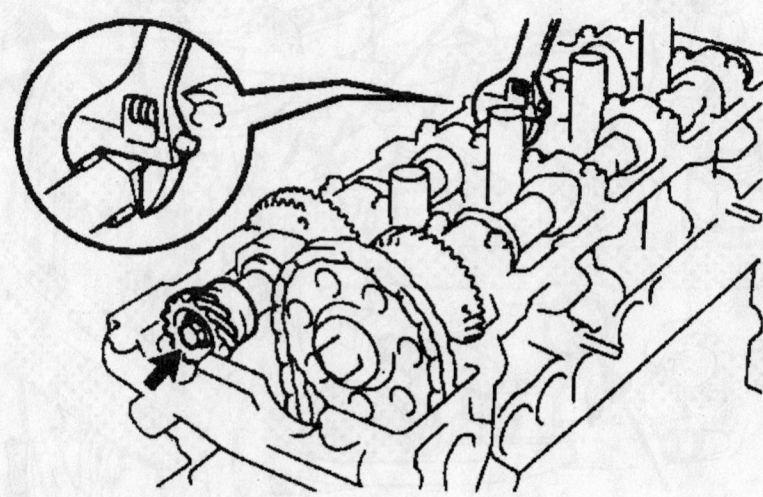

Using a wrench to hold the camshaft—2.7L 4Runner

7924YG55

tion mark on the timing belt is aligned with the end of the No. 1 timing belt cover.
- Right camshaft timing pulley
- Timing belt

23. Set the timing belt tensioner, as follows:

a. Using a press, slowly press in the pushrod using 220–2,205 lbs. (981–9,807 N) of force.

b. Align the holes of the pushrod and housing, pass a 1.5mm hexagon wrench through the holes to keep the setting position of the pushrod.

c. Release the press and install the dust boot to the tensioner.

24. Install the timing belt tensioner and alternately tighten the bolts to 20 ft. lbs. (28 Nm). Using pliers, remove the 1.5mm hexagon wrench from the belt tensioner.

25. Check the valve timing, as follows:

a. Slowly turn the crankshaft pulley 2 revolutions from the TDC-to-TDC. Always turn the crankshaft pulley clockwise.

b. Check that each pulley aligns with the timing marks. If the timing marks do not align, remove the timing belt and reinstall it.

26. Install or connect the following:
- Engine wiring harness to the cylinder head
- 3 engine wiring harness clamps.
- 6 injector connectors
- ECT sender gauge connector
- ECT sensor connector
- Intake air connector
- Air intake chamber assembly. Tighten the 4 bolts and 2 nuts to 13 ft. lbs. (18 Nm).
- Intake chamber stay. Tighten the 2 bolts to 30 ft. lbs. (40 Nm).
- New O-ring to the oil filler tube
- Oil filler tube end into the oil pan tube hole
- Oil filler tube and No. 1 throttle cable clamp
- No. 2 timing belt cover. Tighten the bolts to 80 inch lbs. (9 Nm).
- Remaining components
- Negative battery cable

27. Fill the cooling system.

28. Start the engine and check for leaks.

29. Check the ignition timing.

30. Install the engine undercover.

31. Road test the vehicle.

32. Recheck all fluid levels.

Heater Core replacement is covered in Section 2 of this manual

Valve Lash

ADJUSTMENT

2.0L

1. Before servicing the vehicle, refer to the precautions in the beginning of this section.

2. Remove the cylinder head covers.

3. Use a wrench to turn the crankshaft until the notch in the pulley aligns with timing mark **0** of the No. 1 timing belt cover. This will ensure that the No. 1 piston is at Top Dead Center (TDC) of the compression stroke.

➡**Check that the valve lifters on the No. 1 cylinder are loose and those on the No. 4 cylinder are tight. If not, rotate the crankshaft 1 complete revolution (360 degrees) and then realign the marks.**

4. Using a flat feeler gauge measure the clearance between the camshaft lobe and the valve lifter on the first set of valves shown. This measurement should correspond to specifications.

➡**If the measurement is within specifications, go on to the next step. If not, record the measurement taken for each individual valve.**

5. Rotate the crankshaft 1 complete revolution and realign the timing marks.

6. Measure the clearance of the second set of valves.

➡**If the measurement for this set of valves (and also the previous one) is within specifications, go no further, the procedure is finished. If not, record the measurements and proceed to the next step.**

7. Rotate the crankshaft to position the intake camshaft lobe of the cylinder to be adjusted, facing upward.

➡**Both intake and exhaust valve clearance may be adjusted at the same time, if required.**

8. Using a suitable tool, turn the valve lifter so the notch is easily accessible.

9. Install tool 09248-55010 between both camshaft lobes and turn the handle so the tool presses down both intake and exhaust valve lifters evenly.

10. Using a suitable tool and a magnet, remove the valve shims.

11. Measure the thickness of the old shim with a micrometer. Using this measurement and the clearance made earlier (from Step 3 or 5), determine what size replacement shim will be required in order to bring the valve clearance into specification.

➡**Replacement shims are available in 27 sizes, in increments of 0.0020 in. (0.05mm). Shim sizes are 0.0787–0.1299 in. (2.00–3.30mm).**

12. Install the new shim, remove the special tool; then, recheck the valve clearances.

13. Install the cylinder head covers.

2.4L

➡**Adjust the valve clearance when the engine is cold.**

1. Before servicing the vehicle, refer to the precautions in the beginning of this section.

2. Remove or disconnect the following:
   - Negative battery cable. If equipped with an air bag, wait at least 90 seconds before proceeding.
   - Right front wheel, splash shield and apron seal
   - engine undercover
   - Coil pack
   - Air intake hoses
   - Cylinder head cover

3. Place the No.1 piston at TDC compression. Check only those valves shown. Record the clearance. If out of clearance, the measurement will be used to calculate the adjusting shims.

4. Place the No.4 piston at TDC compression. Check only those valves shown. Record the clearance. If out of clearance, the measurement will be used to calculate the adjusting shims.

Clearance range is:
   - Intake: 0.19–0.29mm
   - Exhaust: 0.30–0.40mm

To adjust the valves:

5. Turn the crankshaft 1 complete revolution (360 degrees) clockwise and set the No.1 piston at TDC compression. Place matchmarks on the chain and camshaft sprocket.

6. Remove the tensioner.

7. Loosen the camshaft sprocket bolt.

8. Remove the exhaust camshaft bearing caps, raise the camshaft and remove the sprocket. Tie the chain out of the way.

9. Remove the intake camshaft.

10. Remove the lifters and keep them in order.

11. Measure the thickness of any lifter on which the clearance was out of range. Calculate the thickness of the necessary replacement lifter. Lifters are available in 0.020mm increments from 5.060mm to 5.740mm.

12. For Camshaft and Timing Chain

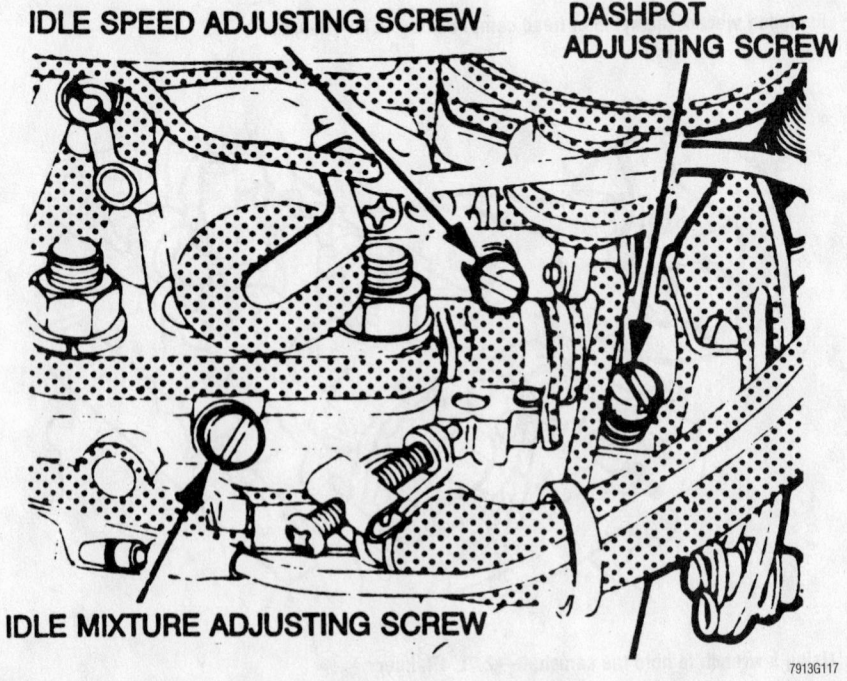

IDLE SPEED ADJUSTING SCREW

DASHPOT ADJUSTING SCREW

IDLE MIXTURE ADJUSTING SCREW

Adjust these valve first—2.0L

7913G117

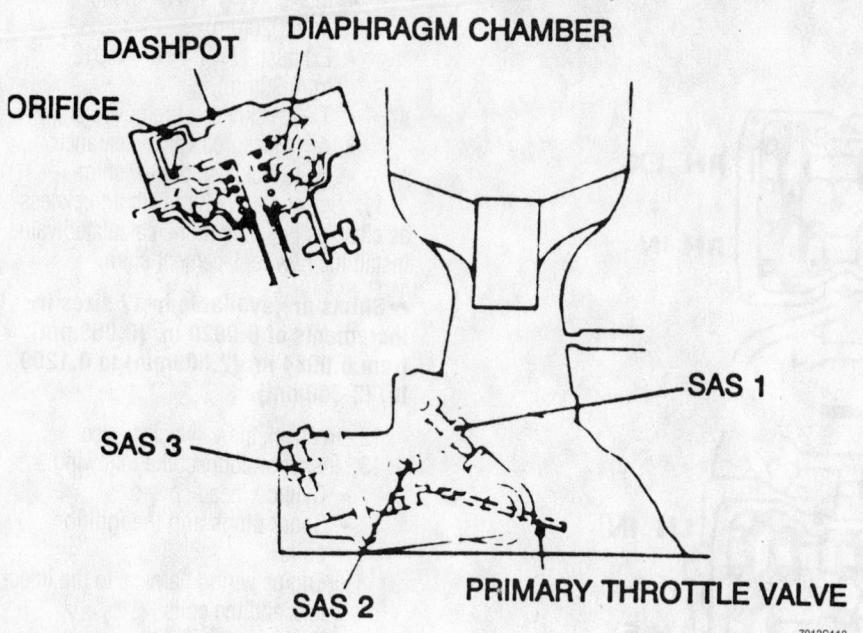

DASHPOT
DIAPHRAGM CHAMBER
ORIFICE
SAS 1
SAS 3
SAS 2
PRIMARY THROTTLE VALVE

7913G118

Adjust these valve second—2.0L

installation, see the respective procedures in this section.

3.0L

➡**Adjust the valve clearance when the engine is cold.**

1. Before servicing the vehicle, refer to the precautions in the beginning of this section.

2. Remove or disconnect the following:
- Negative battery cable. If equipped with an air bag, wait at least 90 seconds before proceeding.
- Accelerator/throttle cable from the throttle linkage
- Air cleaner cover, air flow meter and air duct assembly
- V-bank cover
- Emission control valve set
- Air intake chamber
- Engine harness from the injectors and the ignition coils
- Ignition coils and keep them in order for reassembly
- Spark plugs
- Cylinder head covers

3. Turn the crankshaft pulley and align its groove with the timing mark **0** of the No. 1 timing cover.

4. Check that the valve lifters on the No. 1 intake are loose and the No. 1 exhaust are tight. If not, turn the crankshaft 1 complete revolution (360 degrees).

➡**All measurements should be written down. These recorded measurements will need to be used in conjunction with a mathematical formula to determine the thickness of the replacement shims.**

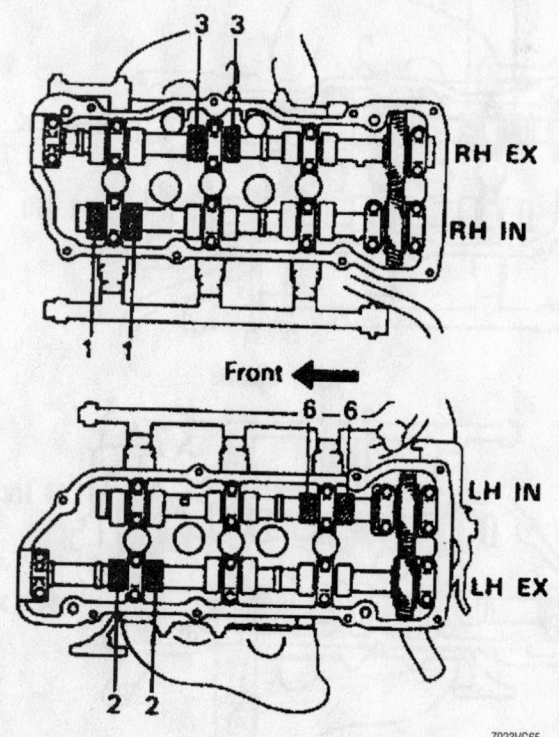

RH EX
RH IN
Front ◄
LH IN
LH EX

7923VG65

Adjust these valves during the 1st step—3.0L

5. Measure the clearance between the valve lifters and the camshaft. Record the measurements on valves No. 1 and 6 intake; No. 2 and 3 exhaust.
  a. The intake valve clearance cold is 0.006–0.010 in. (0.15–0.25mm).
  b. The exhaust valve clearance cold is 0.010–0.014 in. (0.25–0.35mm).

6. Turn the crankshaft ⅔ of a revolution (240 degrees). Record the measurements on valves No. 2 and 3 intake; No. 4 and 5 exhaust.

7. Turn the crankshaft another ⅔ of a revolution. Record the measurements on valves No. 4 and 5 intake; No. 1 and 6 exhaust.

8. Remove the adjusting shim by turning the crankshaft to position the cam lobe of the camshaft in the up position on the valve to be adjusted. Using a small thin flat bladed tool, turn the valve lifter so that the notches are perpendicular to the camshaft. Press down the valve lifter with tool 09248-55010 part A. Place too 09248-55010 part B between the camshaft and the valve lifter; remove part A.

9. Remove the adjusting shim with a magnet and a small screwdriver.

10. Determine the replacement adjusting shim size by either using the charts or the following formulas:

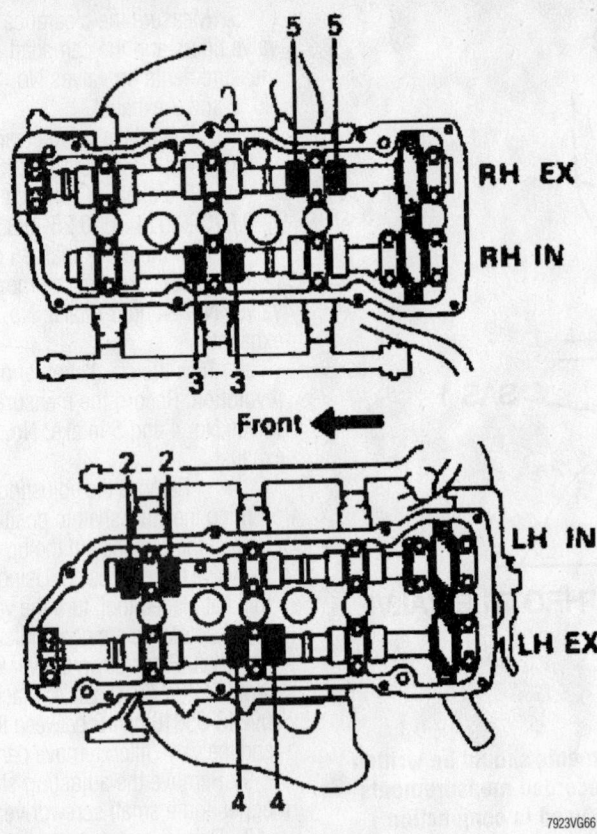

Adjust these valves during the 2nd step—3.0L

7923VG66

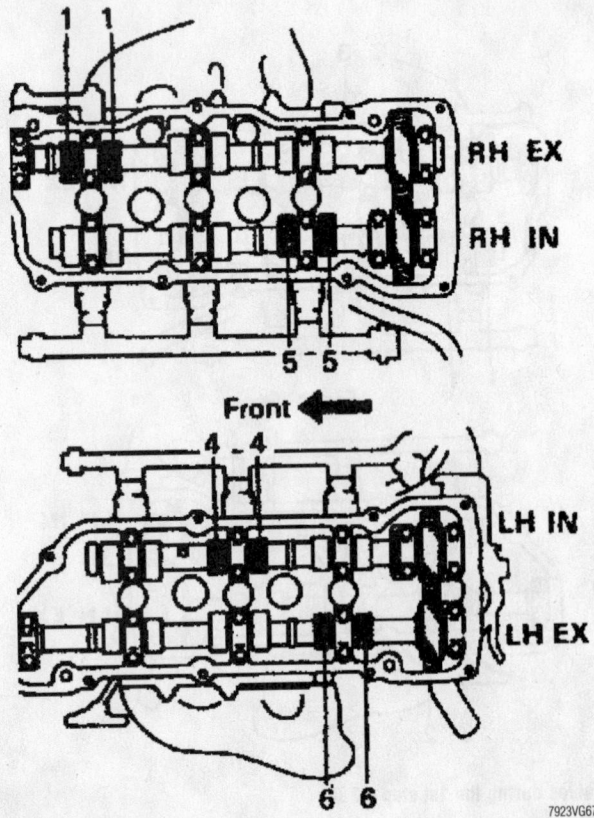

Adjust these valves during the 3rd step 3.0L

7923VG67

- Intake: N = T + (A—0.008 in./0.020mm)
- Exhaust: N = T + (A—0.012 in./0.30mm)
- T = Thickness of removed shim
- A = Measured valve clearance
- N = Thickness of new shim

11. Select a new shim with a thickness as close as possible to the calculated value. Install the new replacement shim.

➡ **Shims are available in 17 sizes in increments of 0.0020 in. (0.050mm), from 0.0984 in. (2.500mm) to 0.1299 in. (3.300mm).**

12. Recheck the valve clearance.
13. Install or connect the following:
- Cylinder head covers
- Spark plugs and the ignition coils
- Engine wiring harness to the injectors and the coils
- Intake chamber
- Emission control valve set
- V-bank cover
- Air flow meter, air duct and air cleaner cover
- Negative battery cable

2.7L Engines

1. Before servicing the vehicle, refer to the precautions in the beginning of this section.
2. Disconnect the negative battery cable.
3. Drain the engine coolant.
4. Remove or disconnect the following:
- Intake air connector
- Positive Crankcase Ventilation (PCV) hoses
- Spark plug wires
- Engine wiring harness clamps and harness
- Air conditioning compressor connector, if equipped with air conditioning
- Oil pressure sensor connector
- Engine Coolant Temperature (ECT) sensor connector
- Distributor connector
- Cylinder head cover

5. Set the No. 1 cylinder to Top Dead Center (TDC) of the compression stroke, as follows:

a. Turn the crankshaft pulley clockwise and align its groove with the **0** mark on the timing chain cover.

b. Check that the timing marks (1 and 2 dots) of the camshaft drive and driven gears are in a straight line on the cylinder head surface. If not, turn the crankshaft 1 revolution (360 degrees) and align the marks.

6. Inspect the valve clearance, as follows:

a. Measure the clearance between the valve lifter and the camshaft. Measure the 1st and 2nd intake and the 1st and 3rd exhaust valves.

b. Turn the crankshaft pulley 1 revolution (360 degrees) and align the marks as above. Measure the 3rd and 4th intake and the 2nd and 4th exhaust valves.

7. Valve clearance cold should be:
- Intake: 0.006–0.010 in. (0.15–0.25mm)
- Exhaust: 0.010–0.014 in. (0.25–0.35mm)

8. Adjust the valve clearance by using adjusting shims, as follows:

a. Turn the camshaft so the cam lobe for the valve to be adjusted faces up.

b. Using SST 09248-55040, press down the valve lifter and place SST 09248-05420, between the camshaft and the valve lifter. Remove SST 09248-55040.

c. Remove the adjusting shim with a small flat prying tool and a magnetic finger.

d. Determine the replacement adjusting shim size according to the following formula or use the adjusting shim charts.

e. Using a micrometer, measure the thickness of the removed shim. Calculate the thickness of a new shim so the valve clearance comes within the specified value.
- T: Thickness of the removed shim
- A: Measured valve clearance
- N: Thickness of the new shim

f. Intake: $N = T + (A - 0.008 \text{ in.} (0.20\text{mm}))$

g. Exhaust: $N = T + (A - 0.012 \text{ in.} (0.30\text{mm}))$

h. Install a new adjusting shim. Place it on the valve lifter. Using the SST 09248-55040, press down the valve lifter and remove SST 09248-05420.

i. Recheck the valve clearance.

9. Install or connect the following:
- Cylinder head cover
- Engine wiring harness and clamps
- Distributor connector
- ECT sensor connector
- Oil pressure sensor connector
- Air conditioning compressor connector, if disconnected
- Spark plug wires
- PCV hoses

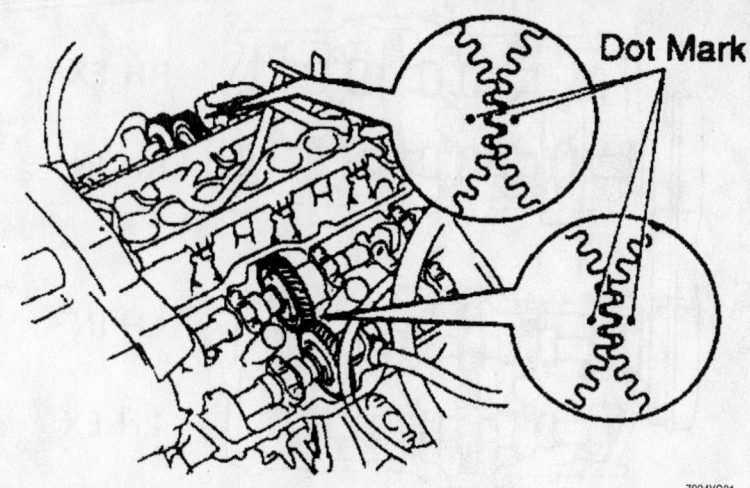

Aligning the timing marks—3.4L engine

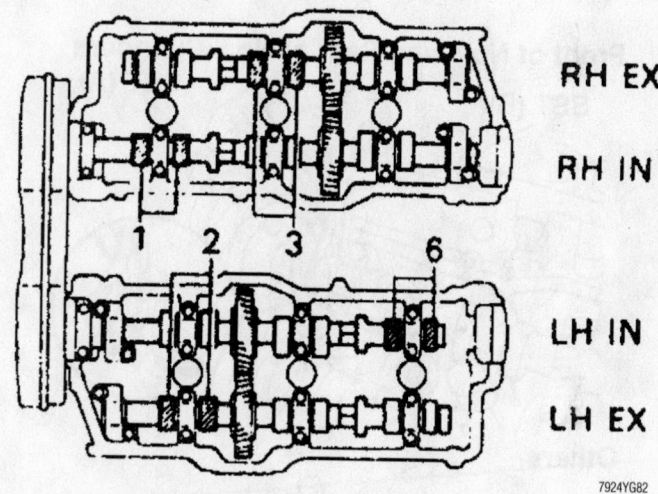

First valve adjustment—3.4L engine

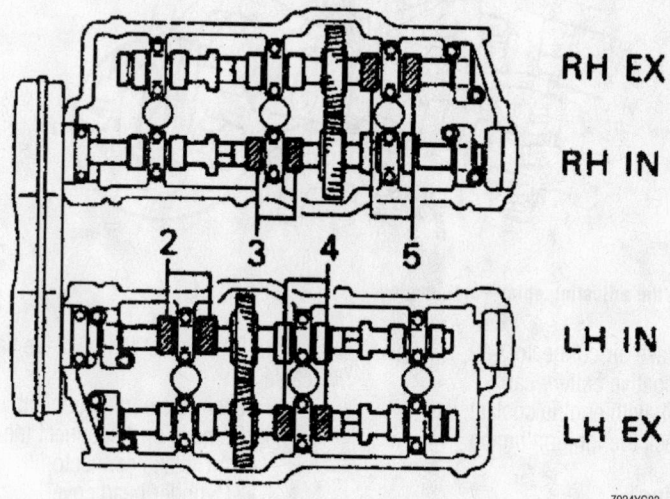

Second valve adjustment—3.4L engine

For complete Engine Mechanical specifications, see Section 1 of this manual

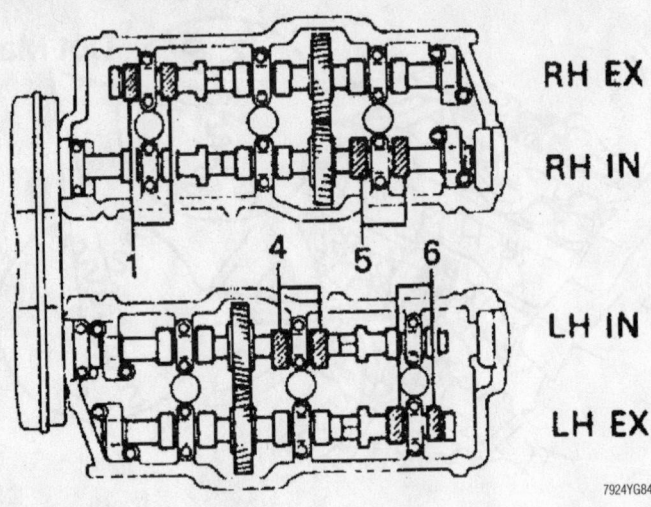

Third valve adjustment—3.4L engine

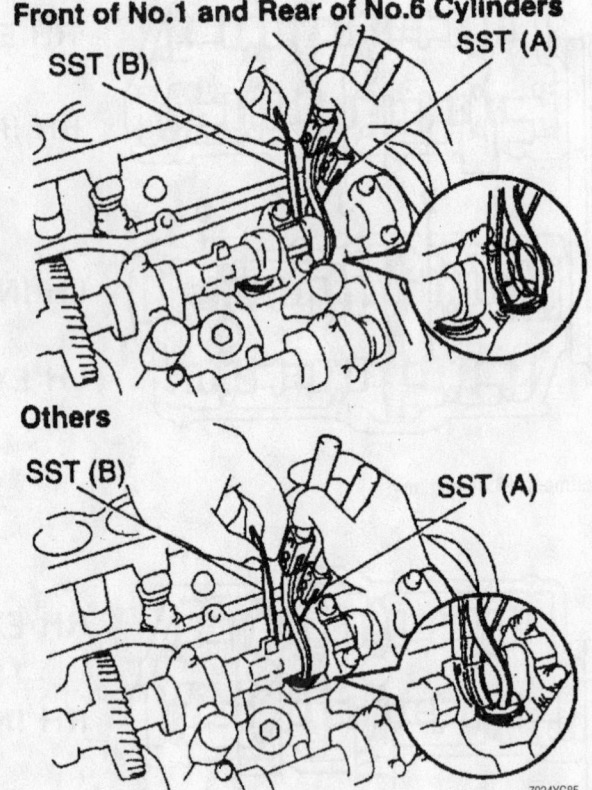

Removing the adjusting shim—3.4L engine

- Intake air connector
- Negative battery cable
10. Refill with engine coolant.
11. Check the ignition timing.

3.4L Engine

1. Before servicing the vehicle, refer to the precautions in the beginning of this section.

2. Disconnect the negative battery cable.

3. Drain the engine coolant.

4. Remove or disconnect the following:
   - Air intake connector
   - Cylinder head cover

5. Set the No. 1 cylinder to Top Dead Center (TDC) of the compression stroke, as follows:
   a. Turn the crankshaft pulley clock-wise and align its groove with the **0** mark on the timing chain cover.

   b. Check that the timing marks (1 and 2 dots) of the camshaft drive and driven gears are in a straight line on the cylinder head surface. If not, turn the crankshaft 1 revolution (360 degrees) and align the marks.

6. Inspect the valve clearance, as follows:

   a. Measure the clearance between the valve lifter and the camshaft. Measure the 1st intake and the 3rd exhaust valves on the right head and the 6th intake and the 2nd exhaust valves on the left head.

   b. Turn the crankshaft ⅔ of a revolution (240 degrees) and adjust the 3rd intake and the 5th exhaust valves on the right head and the 2nd intake and the 4th exhaust valves on the left head.

   c. Turn the crankshaft ⅔ of a revolution (240 degrees) and adjust the 5th intake and the 1st exhaust valves on the right head and the 4th intake and the 6th exhaust valves on the left head.

7. Valve clearance cold should be:
   - Intake: 0.006–0.009 in. (0.13–0.23mm)
   - Exhaust: 0.011–0.014 in. (0.27–0.37mm)

8. Adjust the valve clearance by using adjusting shims, as follows:

   a. Turn the equipment camshaft so that the cam lobe for the valve to be adjusted faces up.

   b. Turn the valve lifter so that the notches are perpendicular to the camshaft.

   c. Using SST 09248-55040, press down the valve lifter and place SST 09248-05420, between the camshaft and the valve lifter. Remove SST 09248-55040.

   d. Remove the adjusting shim with a small flat prying tool and a magnetic finger.

   e. Determine the replacement adjusting shim size according to the following formula or use the adjusting shim charts.

   f. Using a micrometer, measure the thickness of the removed shim. Calculate the thickness of a new shim so that the valve clearance comes within the specified value.
   - T: Thickness of the removed shim
   - A: Measured valve clearance
   - N: Thickness of the new shim

   g. Intake: N = T + (A—0.007 in. (0.18mm))

   h. Exhaust: N = T + (A—0.013 in. (0.32mm))

   i. Install a new adjusting shim. Place

it on the valve lifter. Using the SST 09248-55040, press down the valve lifter and remove SST 09248-05420.

   j. Recheck the valve clearance.

9. Install or connect the following:
- Cylinder head cover
- Intake air connector
- Negative battery cable

10. Refill with engine coolant.

11. Start the engine and check for leaks.

Starter Motor

REMOVAL & INSTALLATION

2.0L RAV4

1. Before servicing the vehicle, refer to the precautions in the beginning of this section.

2. Remove the engine coolant reservoir.

3. Remove the air cleaner cap assembly, as follows:

   a. Disconnect the following:
- Skid control relay connectors
- High tension cord from the air cleaner hose and resonator
- Intake Air Temperature (IAT) sensor connector
- Positive Crankcase Ventilation (PCV) hose from the air cleaner hose
- Air hose from the air cleaner cap assembly

   b. The air cleaner cap assembly from the air cleaner case assembly by removing the 4 clamps.

   c. Loosen the hose clamp and disconnect the air cleaner hose from the throttle body.

   d. Remove the air cleaner cap assembly.

4. Remove or disconnect the following:
- Vacuum Switching Valve (VSV) from the air cleaner case assembly
- 3 bolts and the air cleaner case assembly
- Starter electrical connectors
- Both bolts and the starter

To install:

5. Install or connect the following:
- Starter. Tighten the bolts to 29 ft. lbs. (38 Nm).
- Starter electrical connectors
- Air cleaner case assembly
- VSV to the air cleaner case assembly

6. Install or connect the air cleaner cap assembly, as follows:

   a. Install the air cleaner cap assembly.

   b. Connect the air cleaner hose to the throttle body and tighten the hose clamp.

   c. Install the air cleaner cap assembly to the air cleaner case assembly by installing the 4 clamps.

   d. Connect the following:
- Air hose to the air cleaner cap assembly
- PCV hose to the air cleaner hose
- IAT sensor connector
- High tension cord to the air cleaner hose and resonator
- Skid control relay connectors

7. Install the engine coolant reservoir.

2.4L Highlander

1. Before servicing the vehicle, refer to the precautions in the beginning of this section.

2. Remove or disconnect the following:
- Air cleaner
- starter wiring
- Starter

3. Installation is the reverse of removal. Torque the starter bolts to 29 ft. lbs. (39Nm).

3.0L Highlander

1. Before servicing the vehicle, refer to the precautions in the beginning of this section.

2. Remove or disconnect the following:
- Battery
- Air cleaner
- Starter

3. Installation is the reverse of removal. Torque the starter bolts to 31 ft. lbs. (42Nm).

3.0L RX300

1. Before servicing the vehicle, refer to the precautions in the beginning of this section.

2. Remove or disconnect the following:
- Battery
- Battery tray

3. Remove or disconnect the cruise control actuator, if equipped, as follows:
- Actuator connector and clamp
- 3 bolts and the actuator with the bracket

4. Remove or disconnect the following:
- Automatic transaxle shift control cable
- Engine wiring
- Starter electrical connectors
- Both bolts, shift control cable clamp and the starter

To install:

5. Install or connect the following:
- Starter and the shift control cable clamp. Tighten the bolts to 27 ft. lbs. (37 Nm).
- Starter electrical connectors
- Engine wiring

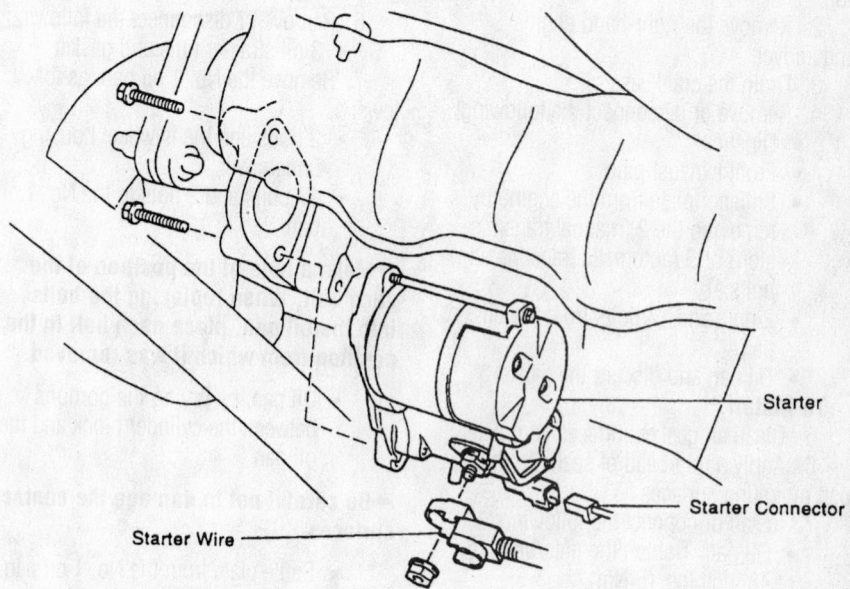

Exploded view of a 2.7L engine starter

86822088

For Accessory Drive Belt illustrations, see Section 1 of this manual

- Automatic transaxle shift control cable
6. Install or connect the following, if equipped with cruise control:
  - 3 bolts and the actuator with the bracket
  - Actuator connector and clamp
7. Install or connect the following:
  - Battery tray
  - Battery

2.7L & 3.4L 4Runner

1. Before servicing the vehicle, refer to the precautions in the beginning of this section.
2. Remove or disconnect the following:
  - Negative battery cable
  - Starter electrical connectors
  - Starter

To install:

3. Install or connect the following:
  - Starter. Tighten the fasteners to 29 ft. lbs. (39 Nm).
  - Electrical connections
  - Negative battery cable

Oil Pan

REMOVAL & INSTALLATION

RAV4

1. Before servicing the vehicle, refer to the precautions in the beginning of this section.
2. Remove the right-hand engine undercover.
3. Drain the crankcase oil.
4. Remove or disconnect the following:
  - Dipstick
  - Front exhaust pipe
  - Stiffener plate from the engine by removing the 2 (manual transmission) or 3 (automatic transmission) bolts.
  - 2 nuts and 17 bolts from the oil pan
  - Oil pan and discard the gasket

To install:

5. Clean all gasket surfaces completely.
6. Apply a thin bead of sealer to the oil pan mounting surfaces.
7. Install or connect the following:
  - Oil pan. Tighten the nuts/bolts to 48 inch lbs. (5 Nm).
  - Stiffener plate. Tighten the bolts to 27 ft. lbs. (37 Nm).
  - Front exhaust pipe
8. Fill the engine with oil to the proper level.

9. Start the engine and check for leaks. Recheck the engine oil level.
10. Install the right engine cover.

2.4L Highlander

1. Before servicing the vehicle, refer to the precautions in the beginning of this section.
2. Remove or disconnect the following:
  - Engine undercover
  - Engine oil
  - Oil pan bolts, nuts and pan
3. Installation is the reverse of removal. Torque the 12 bolts and 2 nuts to 80 inch lbs. (9Nm).

3.0L Highlander and RX 300

1. Before servicing the vehicle, refer to the precautions in the beginning of this section.
2. Remove or disconnect the following:
  - Right front wheel
  - Fender apron seal
  - Engine undercover
3. Drain the engine oil from the engine.
4. Remove or disconnect the following:
  - Front exhaust pipe
  - Front exhaust pipe bracket from the No. 1 oil pan
  - Flywheel housing undercover
  - 10 bolts and 2 nuts to the No. 2 oil pan
5. Insert the blade of the Oil Pan Seal Cutting tool 09032-00100 between the No. 1 and No. 2 oil pans. Clean the surfaces of the oil pans.
6. Remove or disconnect the following:
  - 3 oil strainer nuts and gasket
7. Remove the No. 1 oil pan, as follows:

  - 2 bolts and the flywheel housing undercover
  - 17 bolts and 2 nuts to the No. 1 oil pan

➡**Make a note of the position of the each bolt. When replacing the bolts into the oil pan, place each bolt in the position from which it was removed.**

  - Oil pan, by prying the portions between the cylinder block and the oil pan

➡**Be careful not to damage the contact surfaces.**

  - Baffle plate from the No. 1 oil pan

To install:

8. Clean all mating surfaces of the oil pans.
9. Install the baffle plate to the No. 1 oil pan and tighten to 69 inch lbs. (8 Nm).

10. Install the No. 1 oil pan, as follows:
   a. Using a non residue solvent, clean both sealing surfaces to the oil pan.
   b. Apply liquid sealant to the oil pan and engine block.
   c. Install the oil pan with the 17 bolts and 2 nuts. Uniformly tighten the bolts and nuts in several passes.
   d. Tighten the No. 1 oil pan bolts, as follows:
   - 10mm head bolt: 69 inch lbs. (8 Nm)
   - 12mm head bolt: 14 ft. lbs. (20 Nm)
   - 14mm head bolt: 27 ft. lbs. (37 Nm)
   e. Install the flywheel housing undercover with the 2 bolts. Tighten the bolts to 69 inch lbs. (8 Nm).
11. Install the oil strainer with the 3 nuts. Tighten the nuts to 69 inch lbs. (8 Nm).
12. Install the No. 2 oil pan, as follows:
   a. Using a non residue solvent, clean both sealing surfaces to the oil pan.
   b. Apply liquid sealant to the oil pan and engine block.
   c. Install the No. 2 oil pan with the 10 bolts and 2 nuts. Uniformly tighten the bolts and nuts in several passes. Tighten the bolts to 69 inch lbs. (8 Nm).
13. Install or connect the following:
  - Flywheel housing undercover
  - Front exhaust pipe bracket to the No. 1 oil pan. Tighten the bolts to 15 ft. lbs. (21 Nm).
14. Install the front exhaust pipe, as follows:
  - Temporarily install the 3 new gaskets and the front exhaust pipe with the 2 bolts and 6 nuts
  - Tighten the 4 exhaust manifolds-to-front exhaust pipe nuts to 46 ft. lbs. (62 Nm).
  - Tighten the both front exhaust pipe-to-center exhaust pipe nuts/bolts to 41 ft. lbs. (56 Nm).
  - Bracket. Tighten both bolts to 14 ft. lbs. (19 Nm).
  - Support stay. Tighten both bolts to 22 ft. lbs. (29 Nm).
15. Install or connect the following:
  - Engine undercover
  - Right fender apron seal
  - Right front wheel
16. Fill the engine with oil.
17. Start the engine and check for leaks.

2.7L & 3.4L 4Runner

1. Before servicing the vehicle, refer to the precautions in the beginning of this section.

2. Disconnect the negative battery cable.
3. Drain the engine oil.
4. Remove or disconnect the following:
   - Engine undercover
   - Front differential, if equipped with 4WD
   - Oil pan, separate it from the engine using SST 09032-00100 and a brass bar

To install:
5. Apply seal packing to the oil pan.
6. Install the oil pan to the cylinder block. Tighten the nuts and bolts to:
   - 4Runner: 108 inch lbs. (13 Nm)

※※ WARNING

If parts are not assembled within 5 minutes of applying time, the effectiveness of the seal packing is lost and must be removed and reapplied.

7. Install or connect the following:
   - Front differential, if removed
   - Engine undercover
   - Negative battery cable
8. Fill with engine oil.
9. Start the engine and check for leaks.

Oil Pump

REMOVAL & INSTALLATION

RAV4

1. Before servicing the vehicle, refer to the precautions in the beginning of this section.
2. Remove or disconnect the following:
   - Negative battery cable
   - Hood
   - Right-hand engine undercover
3. Drain the engine oil.
4. Remove or disconnect the following:
   - Front exhaust pipe
   - Rear end stiffener plate
   - Oil dipstick
   - 17 bolts and 2 nuts from the oil pan
5. Insert the blade of the Oil Pan Seal Cutting tool 09032-00100 between the oil pan and the cylinder block; then, cut off the applied sealer and remove the oil pan

➡**Do not use the tool for the oil pump body side and rear oil seal retainer.**

6. Remove the bolts, nuts, oil strainer and gasket.
7. Carefully suspend the engine with a sling device.
8. Remove or disconnect the following:

- Timing belt
- No. 2 idler pulley and crankshaft timing pulley
- Oil pump's pulley, using the Variable Pin Wrench Set 09960-10010
- Crankshaft Position (CKP) sensor
- Oil pump, by discarding the gasket

To install:
9. Install or connect the following:
   - Oil pump, using a new gasket. Tighten the 12 bolts to 82 inch lbs. (9 Nm).

➡**The long bolts are 35mm and all the others are 25mm.**

- CKP sensor
- Oil pump pulley. Tighten the nut to 18 ft. lbs. (24 Nm).
- Crankshaft timing pulley and No. 2 idler pulley
- Timing belt
10. Remove the engine sling.
11. Install the oil strainer with a new gasket. Tighten the nuts/bolts to 48 inch lbs. (5 Nm).
12. Remove any old sealant from the oil pan flange and thoroughly clean both sealing surfaces.
13. Apply a 3–5mm bead of sealant to the oil pan flange.

➡**The pan must be installed within 5 minutes of sealant application or the procedure will have to be repeated.**

14. Install or connect the following:
    - Oil pan. Tighten the 17 bolts and 2 nuts to 48 inch lbs. (5 Nm).
    - Dipstick
    - Rear end stiffener plate. Tighten the bolts to 27 ft. lbs. (37 Nm).
    - Front exhaust pipe
    - Negative battery cable
    - Hood
15. Refill the engine with oil.

※※ WARNING

Be sure to prime the oil pump prior to initial engine start-up or engine damage may occur because of low oil pressure.

16. Start the engine and check for leaks.
17. Recheck the engine oil level.
18. Install the right-hand engine undercover.

2.4L Highlander

1. Before servicing the vehicle, refer to the precautions in the beginning of this section.

2. Remove or disconnect the following:
   - Timing chain
   - Oil pump
3. Installation is the reverse of removal. Torque the pump bolts to 14 ft. lbs. (19Nm).

3.0L Highlander and RX 300

1. Before servicing the vehicle, refer to the precautions in the beginning of this section.
2. Remove or disconnect the following:
   - Oil pan
   - Crankshaft Position (CKP) sensor
   - 9 oil pump bolts

➡**Make a note of the position of the each bolt. When replacing the bolts into the oil pump body, place each bolt in the position from which it was removed.**

- Oil pump body, by prying between the oil pump and main bearing cap
- O-ring from the cylinder block
- Plug, gasket, spring and relief valve from the oil pump body
- 9 screws, pump body cover, drive and driven rotors

To install:
3. Install or connect the following:
   - Driven rotors, drive, pump body cover, using the 9 screws
   - Oil pump relief valve, spring, gasket and the plug to the oil pump body
   - New O-ring on the cylinder block
4. Using a non residue solvent, clean both sealing surfaces to the oil pump.
5. Apply liquid sealant to the oil pump and engine block.
6. Install or connect the following:
   - Oil pump

➡**Be sure to engage the splined teeth of the oil pump drive gear with the large teeth of the crankshaft.**

- 9 oil pump bolts. Tighten the bolts in several passes to 69 inch lbs. (8 Nm), for 10mm or to 14 ft. lbs. (20 Nm), for 12mm.
- CKP sensor. Tighten the bolt to 69 inch lbs. (8 Nm).
- Baffle plate to the No. oil pan. Tighten to 69 inch lbs. (8 Nm).
- No. 1 oil pan, oil strainer and No. 2 oil pan
7. Refill the engine with oil.
8. Start the engine and inspect for leaks.
9. Recheck the engine oil level.

For Tire, Wheel and Ball Joint specifications, see Section 1 of this manual

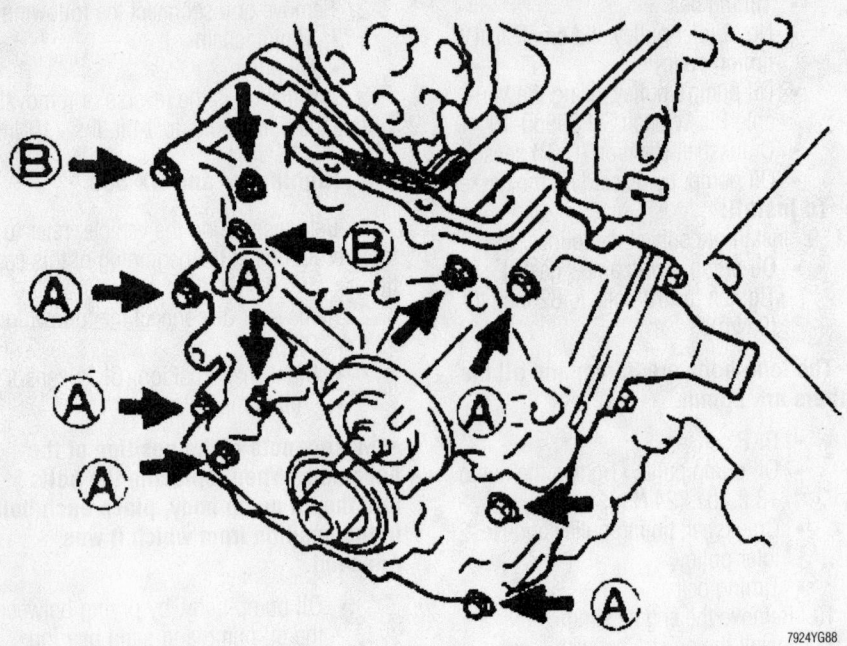

Timing cover bolt pattern—4Runner with 2.7L engine

2.7L 4Runner

1. Before servicing the vehicle, refer to the precautions in the beginning of this section.
2. Remove or disconnect the following:
   - Negative battery cable
   - Cylinder head assembly
   - Water inlet and housing
   - Timing chain cover

- 9 screws and separate the oil pump from the timing chain cover

To install:
3. Install or connect the following:
   - Oil pump assembly to the timing chain cover
   - Timing chain cover
   - Water bypass pipe. Tighten both nuts to 14 ft. lbs. (20 Nm).
   - Cylinder head assembly

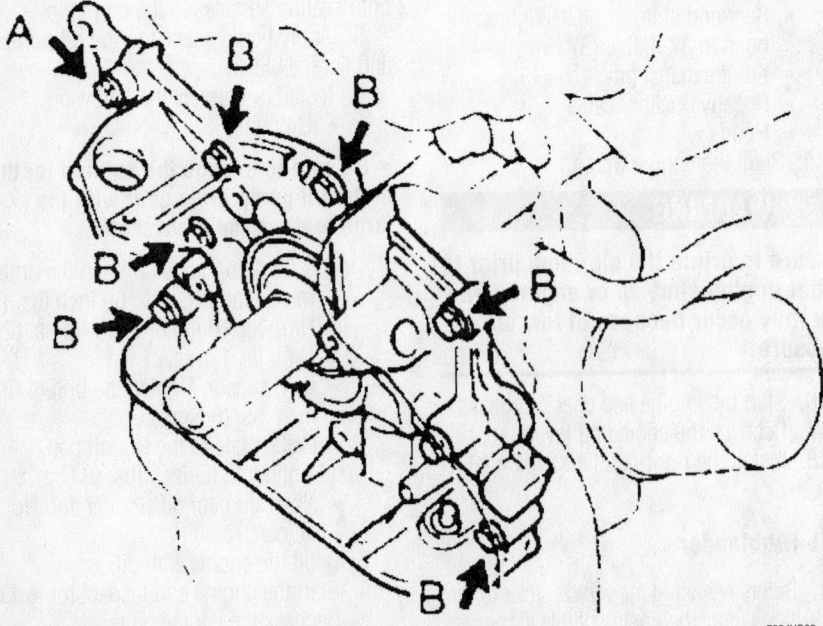

Oil pump bolt identification—4Runner, with 3.4 engine

3.4L 4Runner

1. Before servicing the vehicle, refer to the precautions in the beginning of this section.
2. Remove or disconnect the following:
   - Negative battery cable
   - Engine undercover
   - Crankshaft timing pulley
   - Front differential, if equipped with 4WD
3. Drain the engine oil from the engine.
4. Remove or disconnect the following:
   - Timing belt and crankshaft gear
   - Oil cooler tube and clamp, if equipped with automatic transmission
   - Stiffener plate
   - Flywheel housing undercover and dust cover
   - Rear end cover and dust cover
   - Starter wire clamp
   - Crankshaft Position (CKP) sensor
   - Oil pan

➡**Be careful not to damage the baffle plate flange.**

   - Oil strainer
   - Oil baffle plate
   - Oil pump body by removing the 8 bolts.
   - O-ring from the cylinder block

To install:
5. Install or connect the following:
   - Apply Seal Packing PN 08826-00080 to the oil pump
   - New O-ring into the groove of the cylinder block
   - Oil pump to the crankshaft with the splined teeth of the drive rotor engaged with the large teeth of the crankshaft. Tighten the oil pump bolts "A" 15 ft. lbs. (20 Nm) and bolts "B" 31 ft. lbs. (42 Nm)
   - CKP
   - Oil pan baffle plate
   - Oil strainer with a new gasket. Tighten the bolts to 13 ft. lbs. (18 Nm).
   - Remaining components
   - Negative battery cable
6. Fill with engine oil.
7. Start the engine and check for leaks.

Rear Main Seal

REMOVAL & INSTALLATION

2.0L, 2.4L and 3.0L Engines

If the rear oil seal retainer is not installed to the block, use a tapered ended screw-

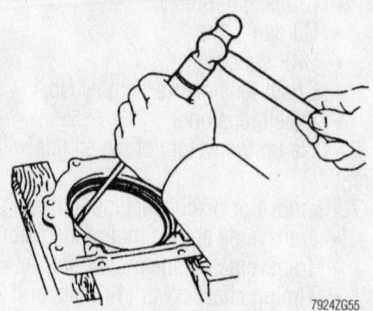

Carefully tap the old seal from the retainer—2.0L, 2.4L and 3.0L Engines

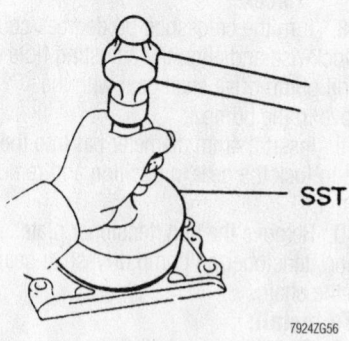

Use the proper sized driver to seat the seal—2.0L, 2.4L and 3.0L Engines

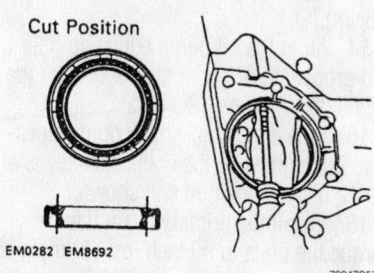

Cut off the oil seal lip, then pry the seal out of the retaining plate—2.0L and 3.0L Engines

driver and hammer to remove the oil seal. Apply multi-purpose grease to the new oil seal lip. Using a seal driver, tap the seal into place. Be careful not to install it slantwise.

1. Before servicing the vehicle, refer to the precautions in the beginning of this section.

If the rear oil seal retainer is installed on the cylinder block, using a knife, cut off the lip of the seal. Using a taped ended prytool, pry the old seal out of the retainer. Inspect

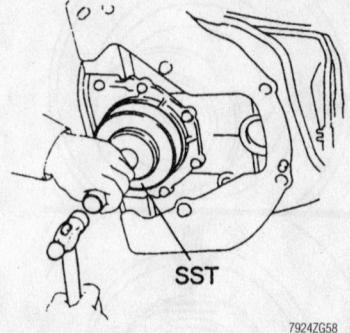

Tap a new seal into place—2.0L and 3.0L Engines

the oil seal lip contacting surface of the crankshaft for cracks or damage. Apply multipurpose grease to the new oil seal, then tap the seal in place with a seal installer. Be careful not to install the seal slantwise.

2.7L & 3.4L Engines

1. Before servicing the vehicle, refer to the precautions in the beginning of this section.
2. Remove the transmission and flywheel from the vehicle.
3. Cut off the rubber lip portion of the seal with a sharp knife.
4. Pry out the oil seal.
To install:
5. Install the rear main seal so that it is flush with the seal retainer housing.
6. Install or connect the following:

• Flywheel/driveplate. Tighten the bolts to 28 ft. lbs. (38 Nm) for engines with a manual transmission, 61 ft. lbs. (83 Nm) for engines with an automatic.
• Transmission

Timing Chain, Sprockets, Front Cover and Seal

REMOVAL & INSTALLATION

2.4L Highlander

1. Before servicing the vehicle, refer to the precautions in the beginning of this section.
2. Disconnect the negative battery cable.
3. Drain the engine coolant.
4. Remove or disconnect the following:
• Hood
• Engine oil
• Right front wheel
• Right fender splash shield
• Right fender apron seal
• No.1 engine undercover
• Engine roll stopper and bracket
• Exhaust pipe
• Upper engine mount, right side
• Accessory drive belts
• Alternator
• Power steering pump
5. Set the No.1 piston at TDC compression.

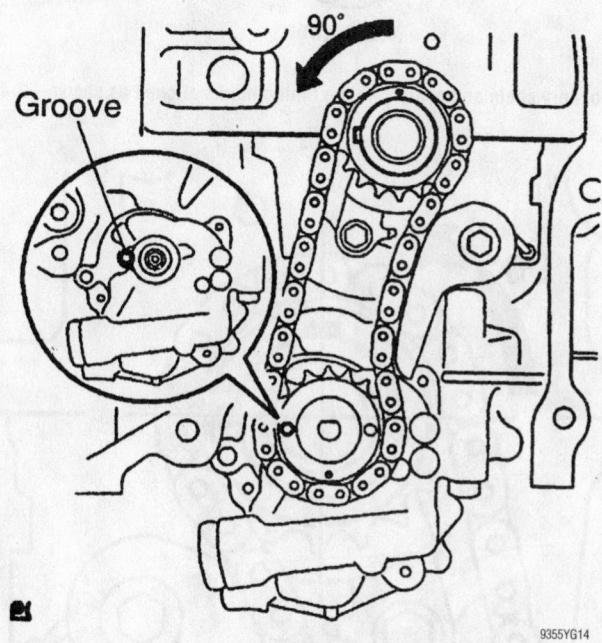

Aligning the adjusting hole and groove—2.4L engine

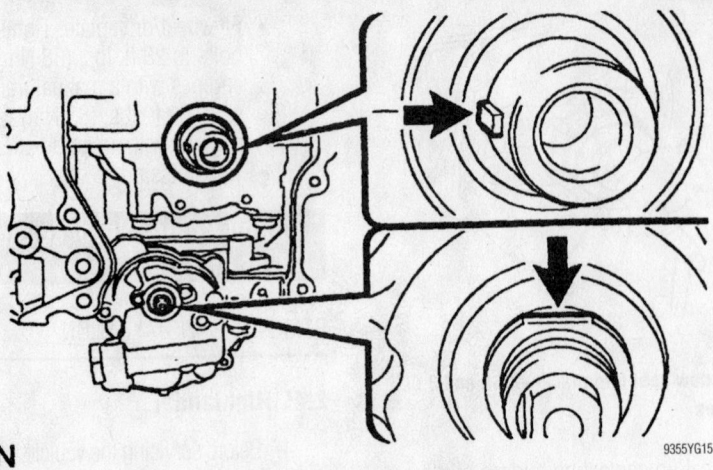

N

Aligning the crankshaft with the key in the left horizontal position—2.4L engine

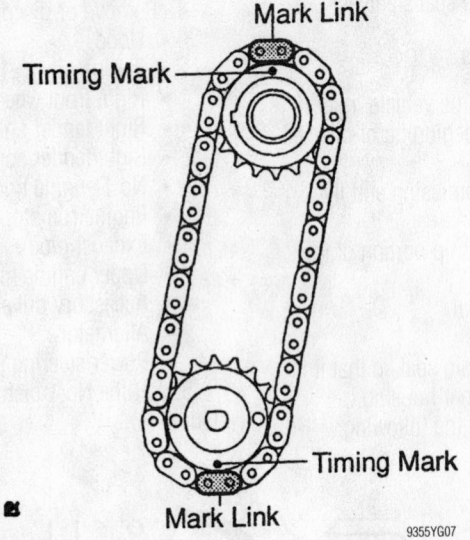

Install the secondary chain and gears with the timing marks aligned as shown—2.4L engine

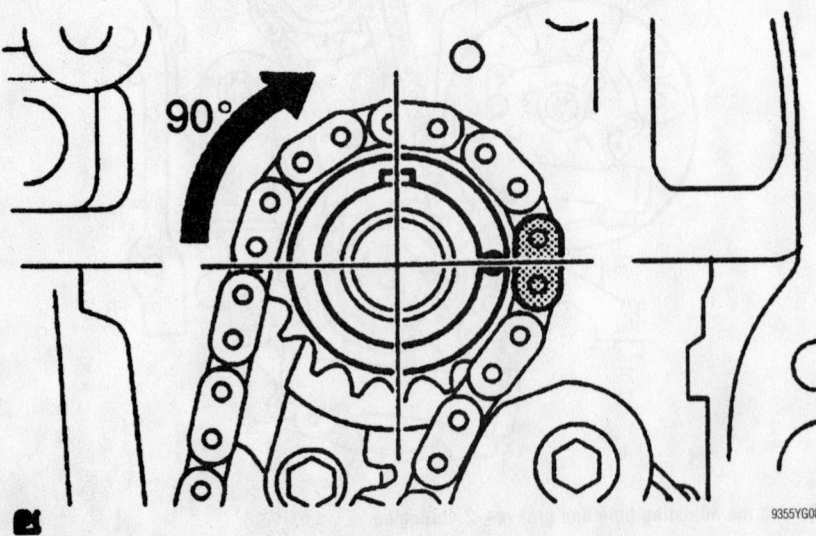

Aligning the crankshaft with the key in the 12 o'clock position—2.4L engine

- Crankshaft pulley
- Oil pan
- CKP sensor
- Chain tensioner assembly No.1
- V-belt tensioner

6. Take up the weight of the engine with a crane.

7. Remove or disconnect the following:
- Transverse engine mount insulator
- Transverse engine mount bracket
- Timing chain cover (14 bolts and 2 nuts
- CKP sensor plate
- Chain tensioner slipper
- Primary chain vibration damper
- Primary chain and crankshaft sprocket

8. Turn the crankshaft 90 degrees counterclockwise and align the adjusting hole of the oil pump drive shaft gear with the groove in the pump.

9. Insert a 4mm diameter bar into the hole to lock the gear in position and remove the nut.

10. Remove the bolt, tensioner plate, spring, tensioner oil pump driveshaft gear and the chain.

To install:

11. Turn the crankshaft so that the key is in the left horizontal position.

12. Install the secondary chain and gears with the timing marks aligned as shown.

13. Install the damper spring and tensioner plate. Torque the nut to 10 ft. lbs. (13Nm).

14. Align the oil pump adjusting hole and groove, lock it with the bar and torque the nut to 22 ft. lbs. (30Nm).

15. Rotate the crankshaft counterclockwise 90 degrees so the crankshaft key is at the 12 o'clock position and shown.

16. Install the primary chain damper. Torque the bolts to 80 inch lbs. (9Nm).

17. Set the No.1 piston at TDC compression with the timing marks aligned as shown.

18. Turn the crankshaft, using the pulley bolt, until the key is at the 12 o'clock position.

19. Install the bottom end of the chain, with sprocket, so that the colored links are aligned as shown.

20. Align the upper end timing marks as shown and install the chain.

21. Install the tensioner slipper. Torque the bolt to 14 ft. lbs. (19Nm).

22. Install the CKP sensor plate with the **F** mark outwards.

➥**When installing the cover, use RTV sealant in the positions shown. The cover must be installed within 3 min-**

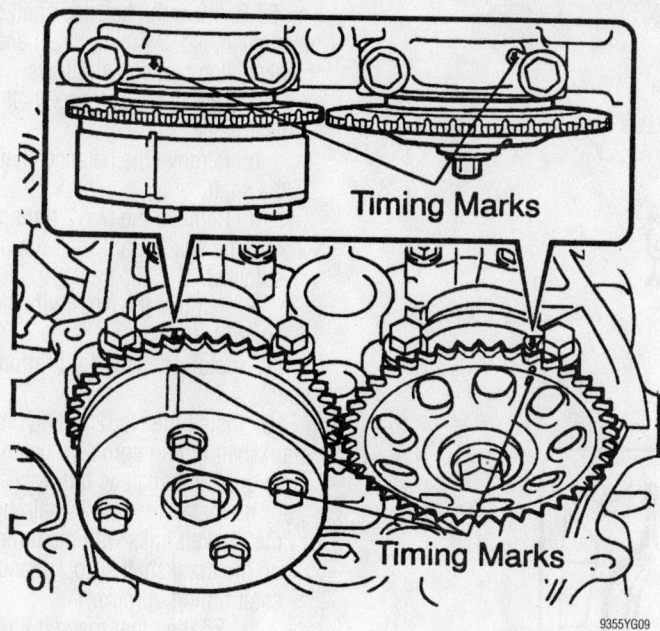

Aligning the timing marks with the No.1 piston at TDC compression—2.4L engine

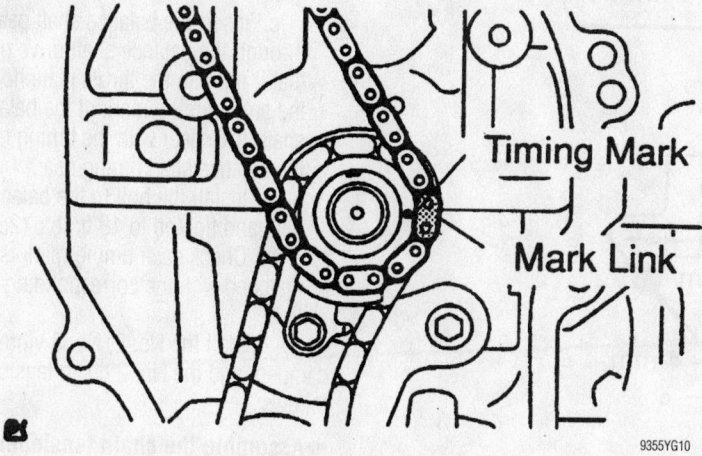

Aligning the timing chain bottom end marks—2.4L engine

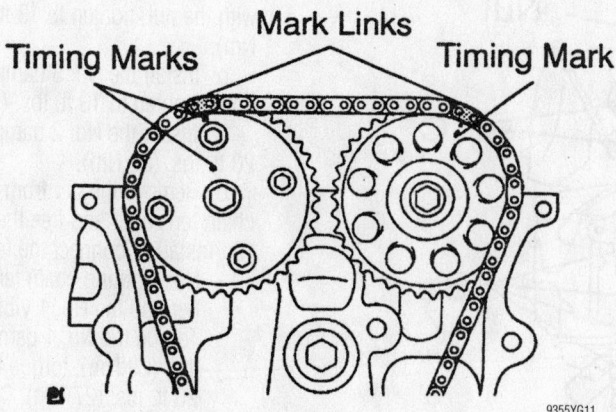

Aligning the timing chain upper end marks—2.4L engine

utes of seal application. Do not start the engine within 2 hours of seal application.

23. Apply the sealant and install the cover. Torque the cover bolts as follows:
- Bolt A: 80 inch lbs. (9Nm)
- Bolts B: 15 ft. lbs. (21 Nm)
- Bolts C: 32 ft. lbs. (43Nm)
- Nuts: 80 inch lbs. (9Nm)

24. The remainder of installation is the reverse of removal.

2.7L 4Runner

1. Before servicing the vehicle, refer to the precautions in the beginning of this section.
2. Disconnect the negative battery cable.
3. Drain the engine coolant.
4. Remove or disconnect the following:
- Engine undercover
- Engine oil
- On 4WD vehicles, the front differential
- Alternator belt, fan with coupling and the water pump pulley
- Cylinder head
- A/C belt, compressor, and the bracket
- Alternator adjusting bar and bracket
- Crankshaft position sensor and O-ring
- On 2WD vehicles, the stiffener plates
- Flywheel housing undercover and dust seal
- Oil pan
- Oil strainer and gasket
- Crankshaft pulley
- Water bypass pipe
- Chain cover assembly. Remove the bolts shown by the arrows.
- No. 1 timing chain and camshaft gear
- Crankshaft timing gear
- No. 1 timing chain tensioner slipper and No. 1 vibration damper
- No. 1 damper

5. Remove the No. 2 and No. 3 vibration dampers and the No. 2 chain tensioner as follows:
a. Install a pin in the No. 2 tensioner and lock the plunger.
b. Remove the bolt and the No. 2 damper.
c. Remove the 2 bolts and the No. 3 damper.
d. Remove the nut and the No. 2 tensioner.

For Maintenance Interval recommendations, see Section 1 of this manual

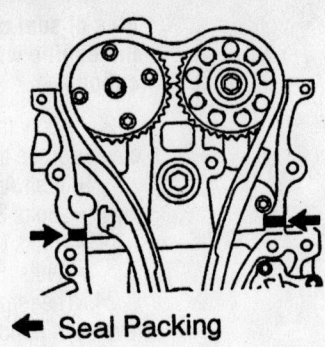

← **Seal Packing**

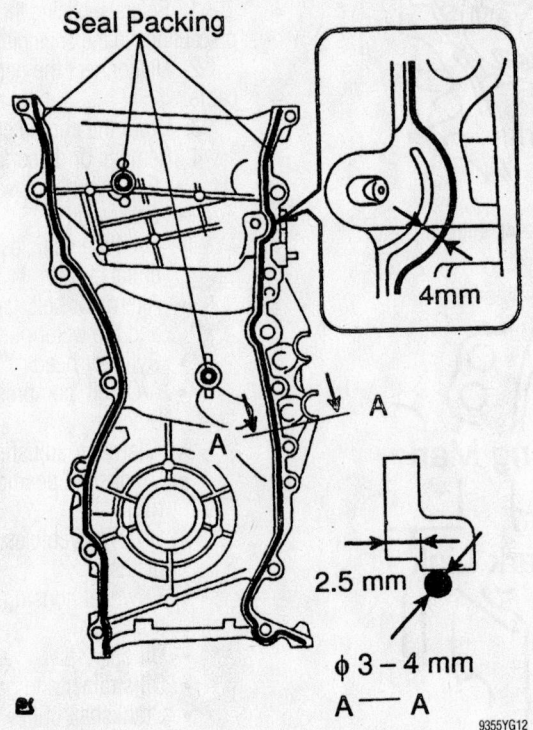

Timing cover sealant application—2.4L engine

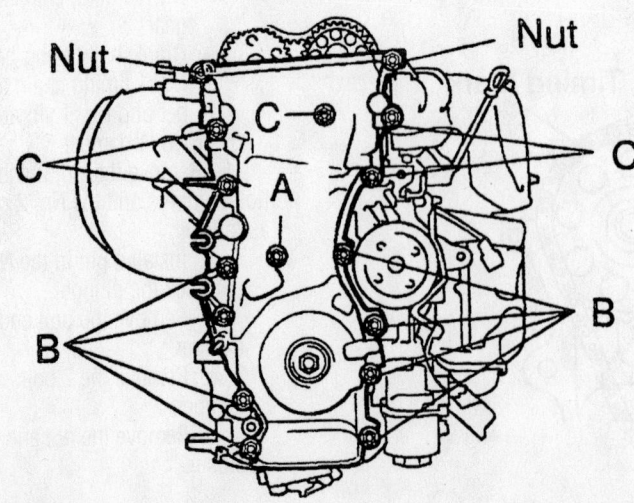

Timing cover bolt positions—2.4L engine

6. Remove the balance shaft driven gear, shaft, No. 2 timing chain and the No. 2 crankshaft sprocket, as follows:

   a. Unbolt the balance shaft driven gear.

   b. Remove the balance shaft gear with the shaft.

   c. Remove the No. 2 timing chain with the No. 2 crankshaft timing sprocket.

   • Remove the No. 4 vibration damper.

To install:

7. Install the No. 4 vibration dampener.

8. Install the No. 2 timing chain, No. 2 crankshaft timing sprocket, balance shaft drive gear and shaft as follows:

   a. Install the No. 2 chain by matching the marked links with the timing marks on the crankshaft sprocket and balance shaft timing sprocket.

   b. Fit the other marked link of the No. 2 chain onto the sprocket behind the large timing mark of the balance shaft gear.

   c. Insert the balance shaft gear shaft through the balance shaft drive gear so that it fits into the thrust plate hole. Align the small timing mark of the balance shaft drive gear with the timing mark of the balance shaft timing gear.

   d. Install the bolt to the balance shaft gear and tighten to 18 ft. lbs. (25 Nm).

   e. Check each timing mark is matched with the corresponding mark link.

9. Install the No. 2, No. 3 vibration dampers and the No. 2 chain tensioner, as follows:

➡ **Assemble the chain tensioner with the pin installed, then remove the pin after assembly.**

   a. Install the No. 2 chain tensioner with the nut, tighten to 13 ft. lbs. (18 Nm).

   b. Install the No. 3 damper with the bolts, tighten to 13 ft. lbs. (18 Nm).

   c. Install the No. 2 damper, tighten to 20 ft. lbs. (27 Nm).

   d. Remove the pin from the No. 2 chain tensioner and free the plunger.

10. Install or connect the following:

   • No. 1 timing chain tensioner slipper and the No. 1 vibration damper. Torque the No. 1 damper to 22 ft. lbs. (29 Nm); torque the slipper to 20 ft. lbs. (27 Nm). Check that the slipper moves smoothly.

   • Crankshaft timing gear.

   • No. 1 timing chain and camshaft timing gear.

11. Align the timing mark between the marked link of the No. 1 timing chain, and install the No. 1 timing chain to the gear.

12. Align the timing mark of the crankshaft timing gear with the mark of the No. 1 timing chain, then install the No. 1 timing chain.

13. Tie the No. 1 chain with a wire or cord, make sure it does not come loose.

14. Install or connect the following:
- Timing chain cover assembly

15. Tighten the following:
- 12mm **A** bolts—14 ft. lbs. (20 Nm)
- 12mm **B** bolts—18 ft. lbs. (25 Nm)
- 14mm bolts—32 ft. lbs. (44 Nm)
- 14mm nut—14 ft. lbs. (20 Nm)

16. Attach the water bypass pipe.

17. Remove the cord or wire from the chain.

18. Install or connect the following:
- Crankshaft pulley, tighten the bolt to 193 ft. lbs. (260 Nm). On A/C vehicles, install the crankshaft pulleys with bolts and tighten to 18 ft. lbs. (25 Nm).
- Oil strainer, tighten to 13 ft. lbs. (18 Nm)
- Oil pan, tighten the mounting bolts to 108 inch lbs. (13 Nm)
- Flywheel housing undercover and dust seal
- Stiffener plates on 2WD vehicles, tighten to 27 ft. lbs. (37 Nm)
- Crankshaft position sensor with a new O-ring
- Alternator, adjusting bar and bracket
- A/C compressor and bracket
- Cylinder head
- Water pump pulley and the fluid coupling with the fan
- On 4WD vehicles, the front differential and driveshaft assemblies

19. Adjust the drive belt tension.

20. Fill with engine coolant and engine oil.

21. Install the engine undercover.

22. Connect the negative battery cable.

SEAL REPLACEMENT

Cover Removed

1. Unbolt the timing chain cover assembly. Be careful to loosen only the correct bolts.

2. Pry out the seal from the cover with a flat-bladed tool.

3. It is a good idea to remove the oil pump from the timing cover and replace the O-ring.

To install:

4. Clean and inspect the timing cover area. Install new gaskets around the dowel areas and pump spline.

5. Apply multi-purpose grease to the new oil seal lip.

6. Tap the seal into place with SST 09223–50010/60010 or equivalent, and a hammer. Do this until the seal surface is flush with the cover edge.

7. Install the cover, tighten the bolts as specified for your engine.

8. If the oil pump was removed, install a new O-ring behind the pump prior to installation.

Cover Installed

1. Unbolt and remove the oil pump.

2. Using a knife, carefully cut off the oil seal lip. With a flat-bladed tool, (preferably with tape around it) pry the seal from the cover.

To install:

3. Apply multi-purpose grease to the new oil seal lip.

4. Tap the seal into place with SST 09223–50010/60011 or equivalent seal driver, and a hammer. Do this until the seal surface is flush with the cover edge.

5. Install the oil pump with a new O-ring.

Piston and Ring

POSITIONING

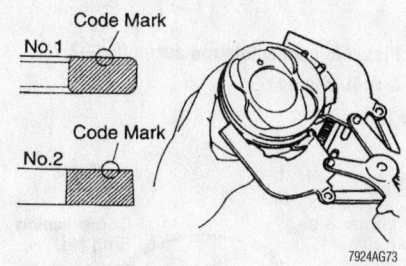

Compression ring identification mark locations–2.0L

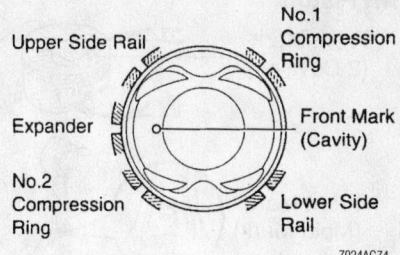

Piston ring end-gap spacing–2.0L

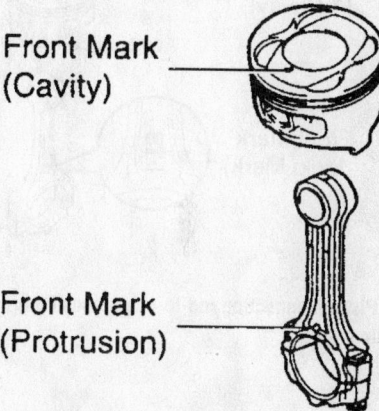

Piston-to-connecting rod assembly–2.0L

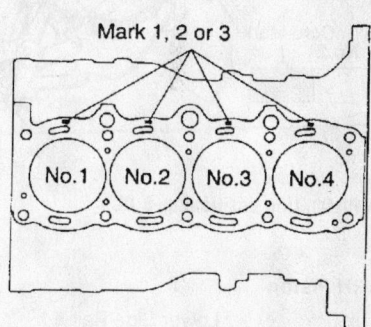

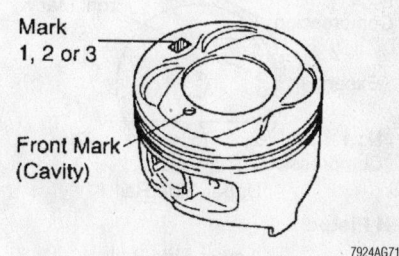

Piston-to-engine installation. Match the number on the piston crown with the number stamped on the block–2.0L

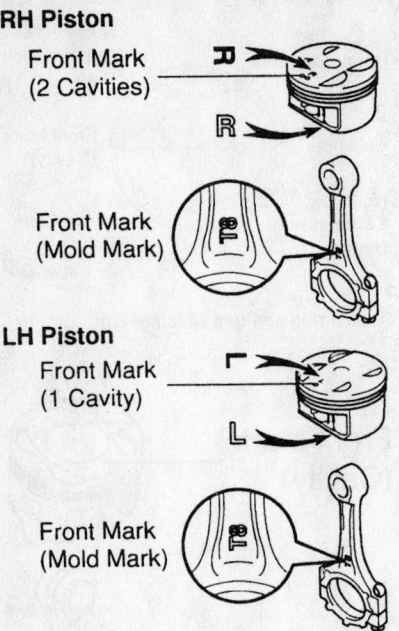

Piston/connecting rod-to-engine positioning–3.0L

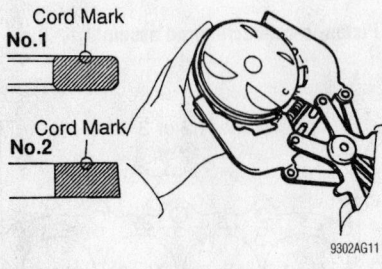

Piston ring positioning–3.0L

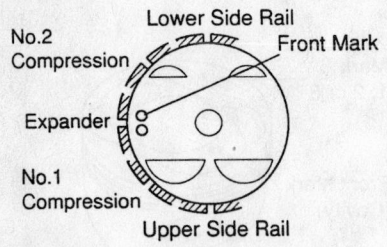

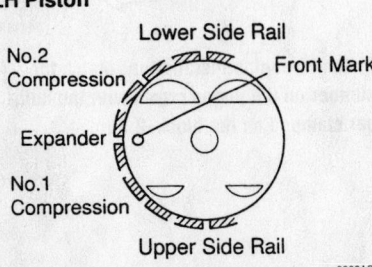

Piston ring identification–3.0L

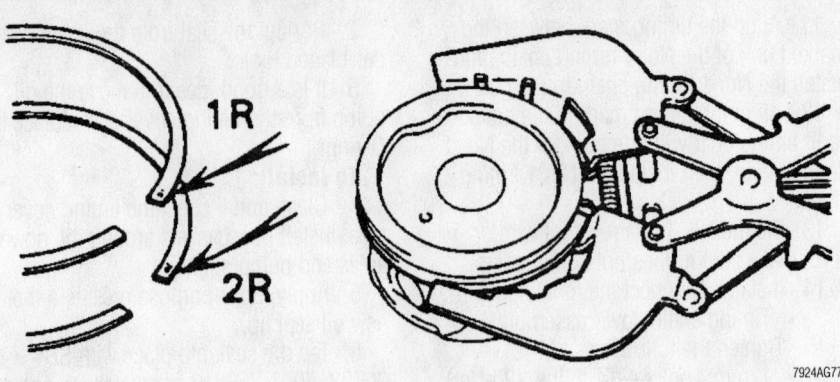

Compression ring identification mark locations—2.7L engine

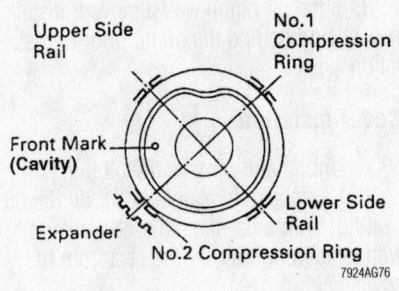

Piston ring end-gap spacing—2.7L

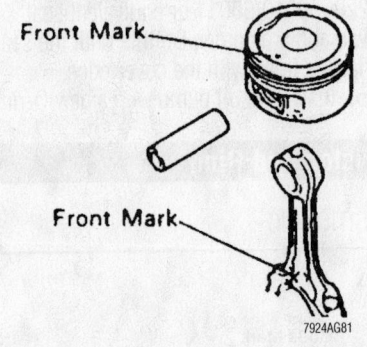

Piston to connecting rod assembly—2.7L & 3.4L engines

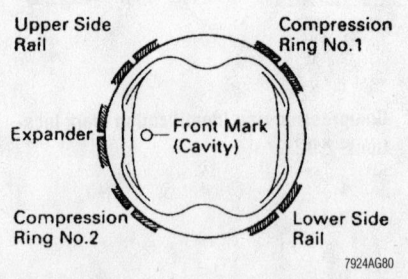

Piston ring end-gap spacing—3.4L engine

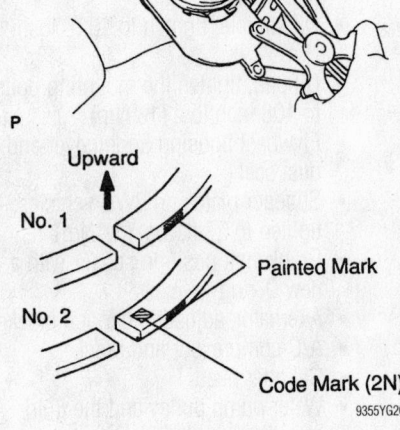

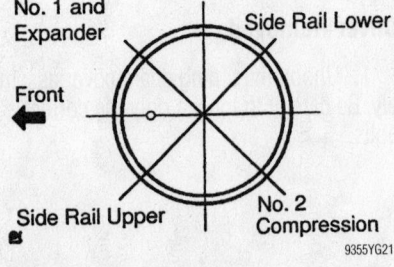

Piston ring identification—2.4L engine

Piston ring installation—2.4L engine

FUEL SYSTEM

Fuel System Service Precautions

Safety is the most important factor when performing not only fuel system maintenance but any type of maintenance. Failure to conduct maintenance and repairs in a safe manner may result in serious personal injury or death. Maintenance and testing of the vehicle's fuel system components can be accomplished safely and effectively by adhering to the following rules and guidelines.

• To avoid the possibility of fire and personal injury, always disconnect the negative battery cable unless the repair or test procedure requires that battery voltage be applied.

• Always relieve the fuel system pressure prior to disconnecting any fuel system component (injector, fuel rail, pressure regulator, etc.), fitting or fuel line connection. Exercise extreme caution whenever relieving fuel system pressure, to avoid exposing skin, face and eyes to fuel spray. Please be advised that fuel under pressure may penetrate the skin or any part of the body that it contacts.

• Always place a shop towel or cloth around the fitting or connection prior to loosening to absorb any excess fuel due to spillage. Ensure that all fuel spillage (should it occur) is quickly removed from engine surfaces. Ensure that all fuel soaked cloths or towels are deposited into a suitable waste container.

• Always keep a dry chemical (Class B) fire extinguisher near the work area.

• Do not allow fuel spray or fuel vapors to come into contact with a spark or open flame.

• Always use a back-up wrench when loosening and tightening fuel line connection fittings. This will prevent unnecessary stress and torsion to fuel line piping.

• Always replace worn fuel fitting O-rings with new. Do not substitute fuel hose or equivalent, where fuel pipe is installed.

Fuel System Pressure

RELIEVING

RAV4 and RX300

1. Before servicing the vehicle, refer to the precautions in the beginning of this section.

2. Disconnect the negative battery terminal.

3. Place a catch-pan under the joint to be disconnected. A large quantity of fuel may be released when the joint is opened.

4. Wear eye or full face protection.

5. Place a shop towel over the area and slowly loosen the joint using a wrench of the correct size. Use a back-up wrench if needed.

6. Allow the fuel left in the line to bleed off slowly before fully disconnecting the joint.

7. Plug the opened lines immediately to prevent fuel spillage or the entry of dirt.

8. Dispose of the released fuel properly.

9. After connecting fuel lines, connect the negative battery cable and start the engine.

10. Check for leaks and repair as needed.

4Runner

1. Before servicing the vehicle, refer to the precautions in the beginning of this section.

2. Disconnect the negative battery terminal.

3. Place a catch-pan under the joint to be disconnected. A large quantity of fuel may be released when the joint is opened.

4. Wear eye or full face protection.

5. Place a shop towel over the area and slowly loosen the joint using a wrench of the correct size. Use a back-up wrench if needed.

6. Allow the fuel left in the line to bleed off slowly before fully disconnecting the joint.

7. Plug the opened lines immediately to prevent fuel spillage or the entry of dirt.

8. Dispose of the released fuel properly.

9. After connecting fuel lines, connect the negative battery cable and start the engine.

10. Check for leaks and repair as needed.

Highlander

Disconnect the fuel pump connector at the tank. Start the engine and let it run out of fuel. Turn the switch off. Disconnect the battery ground. Reconnect the fuel pump connector.

Fuel Filter

REMOVAL & INSTALLATION

RAV4

1. Before servicing the vehicle, refer to the precautions in the beginning of this section.

2. Properly release fuel system pressure.

3. Remove or disconnect the following:
   • Negative battery cable
   • Fuel filter's protective shield

4. Place a pan under the delivery pipe to catch the dripping fuel and slowly loosen the union bolt or flare nut to bleed off the fuel pressure.

5. Drain the remaining fuel.

6. Remove or disconnect the following:
   • Inlet and outlet lines
   • Fuel filter

To install:

7. Coat the flare nut, union nut and bolt threads with engine oil.

8. Hand-tighten the inlet line to the fuel filter.

➡ **When tightening the fuel line bolts to the fuel filter, use a torque wrench. The tightening torque is very important, as under or over tightening may cause fuel leakage. Insure that there is no fuel line interference and that there is sufficient clearance between it and any other parts.**

9. Install or connect the following:
   • Fuel filter. Tighten the inlet bolts to 22 ft. lbs. (30 Nm).
   • Delivery pipe using new gaskets. Tighten the union bolt to 22 ft. lbs. (30 Nm).

10. Run the engine for a few minutes and check for any fuel leaks.

11. Install the protective shield.

RX300

1. Before servicing the vehicle, refer to the precautions in the beginning of this section.

2. Disconnect the negative battery cable.

3. Relieve the fuel system pressure.

➡ **The fuel filter is located in the engine compartment, at the inlet line to the fuel rail.**

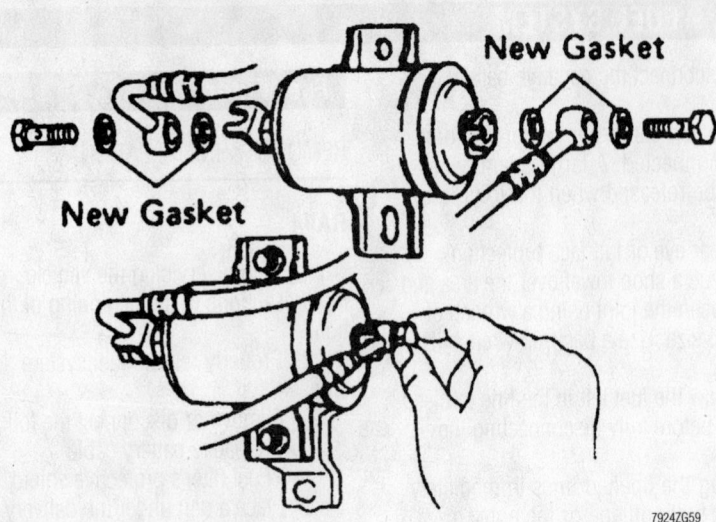

Exploded view of the fuel filter—RX300

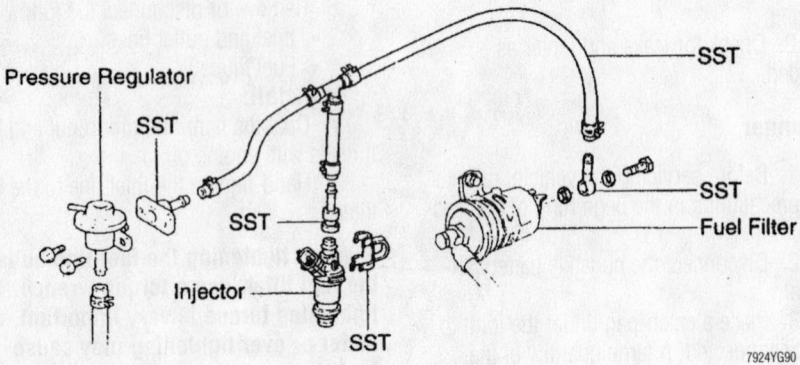

Exploded view of the fuel delivery components—2.7L engines

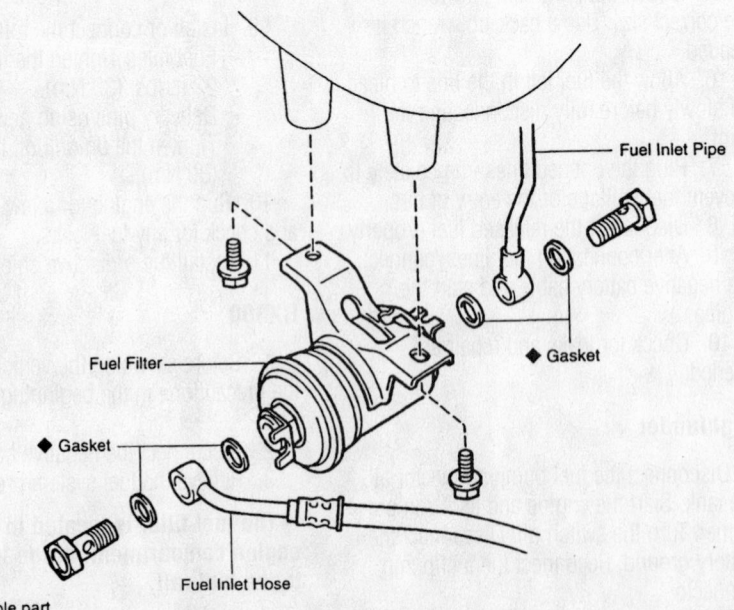

◆ Non-reusable part

Always use new gaskets when replacing the fuel filter—3.4L

4. Remove or disconnect the following:
- Inlet and outlet lines from the filter
- Fuel filter

To install:

5. Install or connect the following:
- Fuel filter, using new O-rings. Tighten the lines to 22 ft. lbs. (29 Nm).
- Negative battery cable
6. Start the engine and check for leaks.

Highlander

The fuel filter is in the tank as part of the pump assembly and is not a normal maintenance item.

4Runner

1. Before servicing the vehicle, refer to the precautions in the beginning of this section.
2. Relieve the fuel system pressure.
3. Remove or disconnect the following:
- Negative battery cable

➡The fuel filter is located in the engine compartment, at the inlet line to the fuel rail.

- Plug the filter inlet and outlet lines
- Fuel filter
- Bracket from the fuel filter

To install:

4. Install or connect the following:
- Fuel filter bracket to the fuel filter
- Fuel filter. Tighten the 2 bolts to 14 ft. lbs. (20 Nm).
- New gaskets. Tighten the union bolts to 22 ft. lbs. (30 Nm).
- Negative battery cable
5. Start the engine and check for leaks.

Fuel Pump

REMOVAL & INSTALLATION

RAV4 and RX 300

1. Before servicing the vehicle, refer to the precautions in the beginning of this section.
2. Relieve the fuel system pressure.
3. Remove or disconnect the following:
- Negative battery cable
- Left-hand rear seat assembly
- Floor service hole by pulling back the carpet; then, remove the 4 screws
- Fuel pump and sender gauge connector

➡Loosen the fuel cap to relieve any fuel pressure within the tank.

- Fuel pipe union bolt and both gaskets
- Fuel pump outlet pipe
- Return vent hose from the fuel pump
- 8 fuel pump bolts and the pump assembly from the tank

To install:

4. Install or connect the following:
- Fuel pump to the fuel tank. Tighten the 8 bolts to 31 inch lbs. (3.5 Nm).
- Return vent hose to the fuel pump
- Outlet pipe to the fuel pump, using new gaskets. Tighten the union bolts to 22 ft. lbs. (29 Nm).
- Fuel pump and sender gauge connector
- Floor hole cover with the 4 screws
- Carpet
- Left rear seat assembly
- Negative battery cable
- Fuel cap
5. Start the vehicle and check for leaks.

4Runner

1. Before servicing the vehicle, refer to the precautions in the beginning of this section.
2. Relieve the fuel pressure.
3. Disconnect the negative battery cable from the battery.
4. Drain the fuel from the fuel tank.
5. Remove or disconnect the following:
- Fuel tank
- Fuel pump connector from the clamp
- Access plate bolts, then pull out the fuel pump assembly from the fuel tank
- Gasket(s) from the pump bracket
- Fuel pump connector
- Bracket from the lower side of the fuel pump
- Fuel pump from the fuel hose
- Rubber cushion, the clip and the fuel filter at the bottom of the fuel pump

To install:

6. Install or connect the following:
- Fuel pump filter to the fuel pump with a new clip
- Fuel pump to the fuel pump bracket
- Fuel hose to the outlet port of the fuel pump
- Fuel pump connector
- Fuel pump assembly with a new gasket(s). Tighten the bolts to 31 inch lbs. (4 Nm).

- Fuel pump connector to the clamp
- Fuel tank
- All electrical and fuel connections
- Negative battery cable
7. Refill the fuel tank and check for leaks.

Highlander

1. Before servicing the vehicle, refer to the precautions in the beginning of this section.
2. Relieve the fuel pressure.
3. Remove or disconnect the following:
- Both rear seats
- Carpet
- Service hole cover
- Connector
- Joint clip and pull out the fuel tube
- Vent tube set plate
- Pump and gauge assembly
4. Installation is the reverse of removal.

Fuel Injector

REMOVAL & INSTALLATION

2.0L

1. Before servicing the vehicle, refer to the precautions in the beginning of this section.
2. Remove or disconnect the following:
- Air cleaner assembly
- Cylinder head cover
- Throttle body from the intake manifold
- Distributor
3. Remove or disconnect the engine wire from the intake manifold, as follows:
- 4 injector connectors
- Both engine wire clamps from the intake manifold wire brackets
- Engine wire protector from the right side of the intake manifold
- Engine wire clamp
4. Remove or disconnect the Exhaust Gas Recirculation (EGR) valve, as follows:
- Vacuum hose from port **E** of the Vacuum Switching Valve (VSV)
- EGR hose from the vacuum modulator
- Loosen the EGR pipe nut from the cylinder head
- Both nuts, EGR valve, pipe assembly and gasket
5. Disconnect the engine compartment R/B No. 2
6. Remove or disconnect the fuel inlet hose and delivery pipe, as follows:

- Union bolt, both gaskets and the fuel inlet hose from the fuel filter outlet
- Air assist hose from the intake manifold port
- Air assist hose
- Loosen both delivery pipe-to-cylinder head bolts
- Delivery pipe from the 4 injectors
- Delivery pipe and fuel inlet hose assembly
7. Remove or disconnect the following:
- 4 injectors and spacers

✳✳ WARNING

Be careful not to drop the injectors and spacers.

- O-rings, insulator and grommet from each injector

To install:

8. Install or connect the following:
- New O-rings, insulator and grommet, lubricated with gasoline, to each injector
- 4 injectors and spacers
9. Install or connect the fuel inlet hose and delivery pipe, as follows:
- Delivery pipe and fuel inlet hose assembly
- Delivery pipe to the 4 injectors. Tighten both delivery pipe-to-cylinder head bolts to 9 ft. lbs. (13 Nm).
- Air assist hose
- Air assist hose to the intake manifold port
- Union bolt, new gaskets and the fuel inlet hose to the fuel filter outlet. Tighten the union bolt to 22 ft. lbs. (29 Nm).
10. Connect the engine compartment R/B No. 2
11. Install or connect the EGR valve, as follows:
- New gasket, pipe assembly and EGR valve. Tighten the nut to 9 ft. lbs. (13 Nm) and the union nut to 43 ft. lbs. (59 Nm).
- EGR hose to the vacuum modulator
- Vacuum hose from port **E** of the VSV
12. Install or connect the engine wire to the intake manifold, as follows:
- Engine wire clamp
- Engine wire protector to the right side of the intake manifold
- Both engine wire clamps to the intake manifold wire brackets
- 4 injector connectors

13. Install or connect the following:
- Distributor
- Throttle body to the intake manifold
- Cylinder head cover
- Air cleaner assembly

2.4L

1. Before servicing the vehicle, refer to the precautions in the beginning of this section.
2. Relieve the fuel system pressure.
3. Remove or disconnect the following:
- Air cleaner and hoses
- Fuel line from the rail
- Injector connectors
- Injector wiring harness
- Fuel rail with injectors
- Injector spacers from the head

4. Installation is the reverse of removal. Coat the new o-rings with clean fuel. Before tightening the fuel rail bolts, make sure that each injector rotates smoothly. Tighten the bolts to 15 ft. lbs. (20Nm).

3.0L Highlander

1. Before servicing the vehicle, refer to the precautions in the beginning of this section.
2. Relieve the fuel system pressure.
3. Remove or disconnect the following:
- Coolant
- V-bank cover
- Battery
- Air cleaner assembly
- Upper front suspension brace
- Intake air surge tank
- Fuel supply line
- Injector connectors
- Fuel rail with injectors

4. Installation is the reverse of removal. Coat the new o-rings with clean fuel. Before tightening the fuel rail bolts, make sure that each injector rotates smoothly. Tighten the bolts to 84 inch lbs. (10Nm).

RX300

1. Before servicing the vehicle, refer to the precautions in the beginning of this section.
2. Remove or disconnect the following:
- Outer front cowl top panel assembly
- Air cleaner cap with hose
- Negative battery cable. Work must be started approximately 90 seconds or longer after the negative battery cable has been disconnected, if equipped with an air bag.
- Coolant
- Accelerator and throttle cables
- V-bank cover

- Emission valve control set
- No. 2 EGR pipe
- Hydraulic motor pressure pipe from the water inlet and air inlet chamber
- Air intake chamber assembly
- Injector wiring
- Air assist pipe from the bracket on the No. 1 fuel pipe
- Air assist hoses from the intake manifold
- Fuel return hose from the No. 1 fuel pipe
- Fuel inlet hose for the fuel filter
- 2 union bolts holding the No. 2 fuel pipe to the delivery pipes
- Fuel return hose from the fuel pressure regulator
- Union bolt for the right hand delivery pipe, 2 gaskets, 2 bolts, left hand delivery pipe together with the 3 injectors and the No. 2 fuel pipe
- Union bolt for the delivery pipe and 2 gaskets from the No. 2 fuel pipe
- The 3 bolts, right hand delivery pipe together with the 3 injectors and the No. 1 fuel pipe
- The 4 spacers from the intake manifold
- The 6 injectors from the delivery pipes
- The two O-rings and two grommets from each injector

To install:

3. Install or connect the following:
- 2 new grommets to each injector
- New O-rings, with a light coat of fuel, to each injector
- Injectors
- The 4 spacers on the intake manifold
- Right hand delivery pipe and the No. 1 fuel pipe together with the 3 injectors in position on the intake manifold
- Bolt holding the right side delivery pipe, temporarily, to the intake manifold
- Left hand delivery pipe and the No. 2 fuel pipe together with the 3 injectors in position on the intake manifold
- Fuel return hose to the fuel pressure regulator

4. Temporarily install the 2 bolts holding the left hand delivery pipe to the intake manifold.
5. Temporarily install the No. 2 fuel pipe to the left side delivery pipe with the union bolt and 2 new gaskets.
6. Check that the injectors rotate smoothly. If they do not, replace the O-rings.

7. Position the injector connector outward. Tighten the 4 bolts holding the delivery pipes to the intake manifold and tighten to 7 ft. lbs. (10 Nm). Tighten the bolt holding the No. 1 fuel pipe to the intake manifold to 14 ft. lbs. (20 Nm). Tighten the 2 union bolts holding the no. 2 fuel pipe to the delivery pipes to 24 ft. lbs. (32 Nm).
8. Install or connect the following:
- Fuel inlet and return hoses. Union bolt: 22 ft. lbs. (30 Nm)
- Fuel return hose to the No. 1 fuel pipe. Pass the fuel return hose under the heater hoses.
- Air assist hoses to the intake manifold
- Air assist pipe to the bracket on the No. 1 fuel pipe
- Fuel injector wiring connectors
- Air intake chamber assembly
- Hydraulic motor pressure pipe to the intake chamber. Bolts: 69 inch lbs. (8 Nm)
- No. 2 EGR pipe with new gaskets, tighten to 9 ft. lbs. (12 Nm)
- Emission control valve set
- V-bank cover
- Air cleaner hose
- Throttle and accelerator cables
- Coolant
- Air cleaner cap with hose
- Outer front cowl top panel assembly
- Negative battery cable

2.7L Engines

1. Before servicing the vehicle, refer to the precautions in the beginning of this section.
2. Relieve the fuel system pressure.
3. Remove or disconnect the following:
- Throttle body
- Fuel injector electrical connectors
- Crankshaft Position (CKP) sensor connector
- Knock Sensor (KS) connector
- Data Link Connector 1 (DLC1) and wire clamp from the brackets
- Vacuum line from the fuel pressure regulator
- Fuel return hose from the pressure regulator
- Union bolt and gaskets
- Fuel inlet pipe from the fuel rail
- Fuel rail with the injectors attached

❊❊ WARNING

The injectors are only retained by their O-rings and will tend to drop out of the fuel rail.

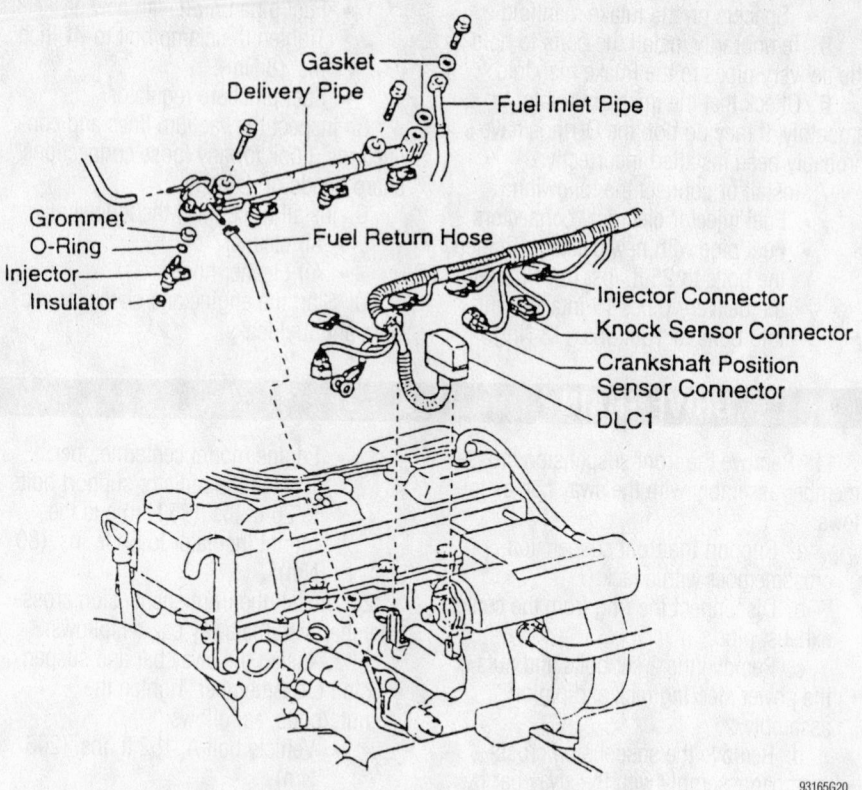

Fuel injector arrangement and related components—2.7L engines

93165G20

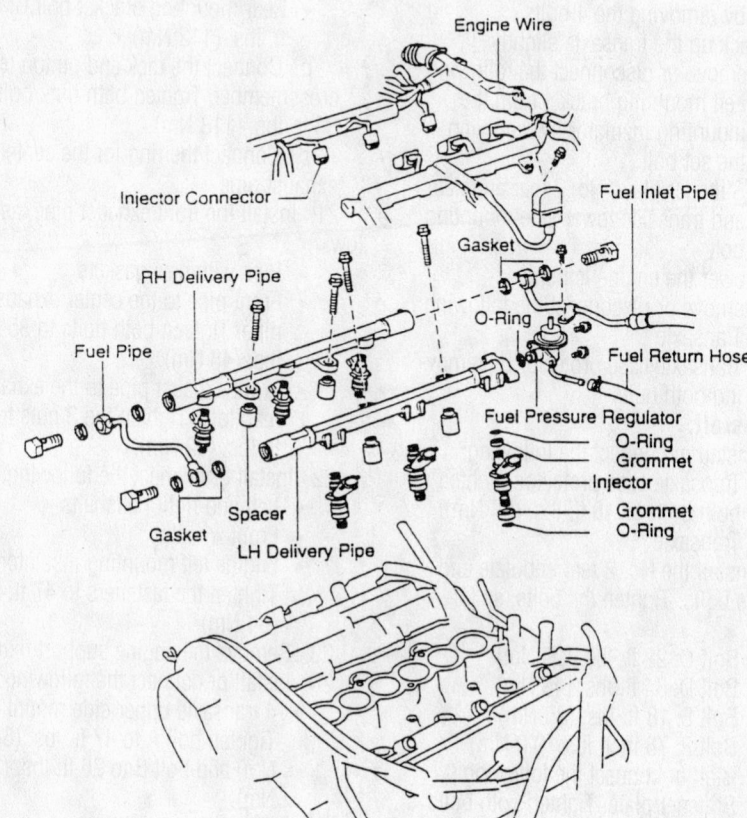

Fuel injector arrangement and related components—3.4L engine

93165G21

- 4 insulators from the four spacers
- Fuel injectors from the fuel rail
- O-ring and grommet, discard them

To install:

4. Install or connect the following:
- New grommets and O-rings on each injector, lubricated with a light coat of gasoline
- Fuel injectors, with the electrical connector facing upwards
- New insulators and spacers on the intake manifold

5. Temporarily install the bolts holding the delivery pipe to the intake manifold.

6. Check that the injectors rotate smoothly.

7. Install or connect the following:
- Tighten the delivery pipe-to-intake manifold bolts to 15 ft. lbs. (21 Nm).
- Injector electrical connectors
- Fuel inlet pipe with new gaskets. Tighten the union bolt to 22 ft. lbs. (29 Nm) and the bolt to 14 ft. lbs. (20 Nm).
- Fuel return pipe to the fuel pressure regulator
- Vacuum line to the pressure regulator
- Throttle body
- Negative battery cable

3.4L Engine

1. Before servicing the vehicle, refer to the precautions in the beginning of this section.

2. Depressurize the fuel system.

3. Remove or disconnect the following:
- Air cleaner hose
- Upper half of the intake manifold
- Fuel pressure regulator
- Fuel inlet pipe
- Fuel injector electrical connections

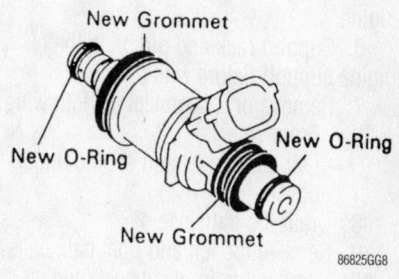

Install new O-rings and grommets on each injector—3.4L engine

86825GG8

Timing belt service is covered in Section 3 of this manual

- Fuel rail with the injectors
- Spacers from the intake manifold
- Injectors from the delivery pipes
- O-rings and grommets, discard them

To install:

4. Install or connect the following:
- New grommets and O-rings on each injector, lubricated with a light coat of gasoline
- Fuel injector with the electrical connector facing outward

- Spacers on the intake manifold

5. Temporarily install the bolts to hold the delivery pipes to the intake manifold.

6. Check that the injectors rotate smoothly. If they do not, the O-rings have probably been installed incorrectly.

7. Install or connect the following:
- Fuel injector electrical connectors
- Fuel pipe with new gaskets. Tighten the bolts to 25 ft. lbs. (34 Nm) and the delivery pipes-to-intake manifold bolts to 10 ft. lbs. (13 Nm).

- Fuel pipe union with new gaskets. Tighten the clamp bolt to 71 inch lbs. (8 Nm).
- Fuel pressure regulator

8. Inspect the vacuum lines and connections. Look for any loose connections, sharp bends or damage.

9. Install or connect the following:
- Air cleaner
- Air cleaner hose

10. Start the engine and check for vacuum and fuel leaks.

DRIVE TRAIN

Manual Transmission Assembly

REMOVAL & INSTALLATION

2WD RAV4

1. Before servicing the vehicle, refer to the precautions in the beginning of this section.

2. Remove or disconnect the following:
- Negative battery cable
- Air cleaner case assembly with hose
- Engine coolant reservoir tank
- Engine wire clamp set nut
- Starter

3. Remove the clutch release cylinder, as follows:
- Clutch line bracket-to-transaxle set bolts
- Release cylinder and line

4. Remove or disconnect the following:
- Ground cable from the transaxle
- Vehicle Speed Sensor (VSS) and backup light switch connector
- Control cable by removing the 4 clips and washers
- 4 upper side transaxle-to-engine bolts
- Left mount insulator

5. Install a engine support to the engine.

6. Support rack and pinion to the engine support fixture with a rope.

7. Remove or disconnect the following:
- Front wheels
- Left and right-hand engine undercovers

8. Drain the transaxle oil.

9. Remove the left and right halfshafts.

10. Remove the front exhaust pipe, as follows:
- 3 exhaust manifold nuts and gasket
- Both exhaust pipe-to-center exhaust pipe bolts
- Exhaust pipe

11. Remove the front suspension crossmember assembly with the sway bar, as follows:

   a. Support the front suspension crossmember with a jack.

   b. Disconnect the ring from the center exhaust pipe.

   c. Remove the 2 set bolts and nuts of the power steering rack and pinion assembly.

   d. Remove the suspension crossmember assembly with the sway bar by removing the 2 nuts and 6 bolts.

12. Remove the engine mounting centermember by removing the 4 bolts.

13. Jack up the transaxle slightly.

14. Remove or disconnect the following:
- Left mounting bracket from the mounting insulator by removing the set bolt
- Stiffener plate, No. 2 rear endplate and transaxle lower side mounting bolt

15. Lower the engine left side

16. Remove or disconnect the following:
- Transaxle
- Transaxle case protector by removing both bolts

To install:

17. Install or connect the following:
- Transaxle case protector. Tighten both bolts to 18 ft. lbs. (25 Nm).
- Transaxle

18. Install the No. 2 rear endplate and transaxle bolts. Tighten the bolts, as follows:
- Bolt C: 22 ft. lbs. (29 Nm)
- Bolt D: 34 ft. lbs. (46 Nm)
- Bolt E: 18 ft. lbs. (25 Nm)
- Bolt F: 78 inch lbs. (9.0 Nm)

19. Install or connect the following:
- Stiffener plate. Tighten both bolts to 27 ft. lbs. (37 Nm).
- Engine left mounting insulator to the left mounting bracket. Tighten the bolt to 47 ft. lbs. (64 Nm).

- Engine mount centermember. Tighten the radiator support bolts to 26 ft. lbs. (35 Nm) and the mount insulator to 59 ft. lbs. (80 Nm).

20. Install the front suspension crossmember with the sway bar, as follows:

   a. Install the sway bar and suspension crossmember. Tighten the nuts/bolts, as follows:
- Vehicle bolt A: 152 ft. lbs. (206 Nm)
- Lower control arm bracket bolt B: 101 ft. lbs. (137 Nm)
- Rear mounting bracket bolt C: 82 ft. lbs. (112 Nm)

   b. Connect the rack and pinion to the crossmember. Tighten both nuts/bolts to 83 ft. lbs. (113 Nm).

   c. Connect the ring for the center exhaust pipe.

21. Install the front exhaust pipe, as follow:
- Pipe with new gaskets
- Front pipe to the center exhaust pipe. Tighten both bolts to 35 ft. lbs. (48 Nm).
- Front exhaust pipe to the exhaust manifold. Tighten the 3 nuts to 46 ft. lbs. (62 Nm).

22. Install or connect the following:
- Left and right halfshafts
- Front wheels
- Engine left mounting insulator. Tighten the fasteners to 47 ft. lbs. (64 Nm).

23. Remove the engine support fixture.

24. Install or connect the following:
- 4 transaxle upper side mount bolts. Tighten bolt A to 47 ft. lbs. (64 Nm) and bolt B to 26 ft. lbs. (35 Nm).
- Ground cable with the clips and washers
- VSS and backup light switch connectors

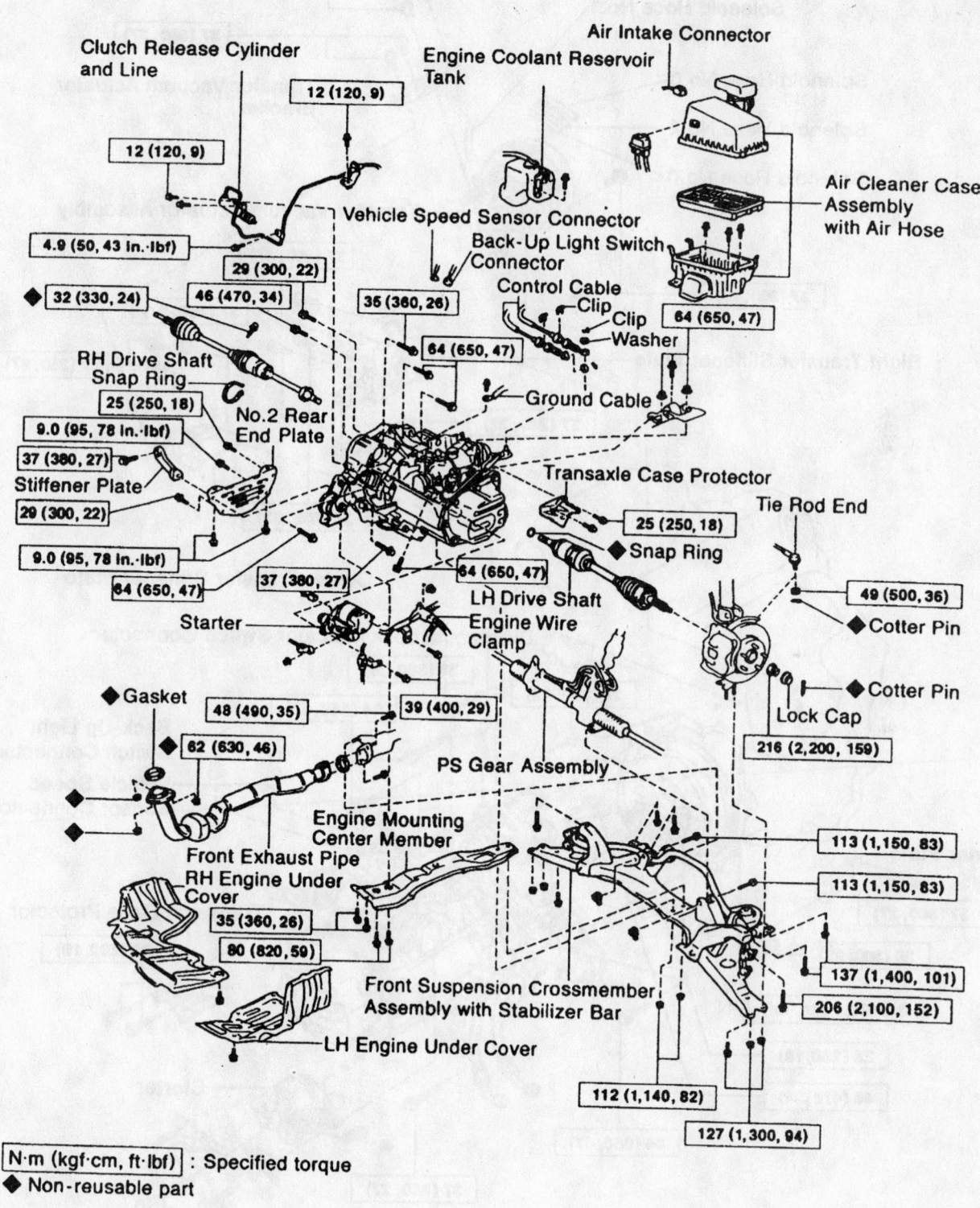

Clutch Release Cylinder and Line

12 (120, 9)

12 (120, 9)

Engine Coolant Reservoir Tank

Air Intake Connector

Air Cleaner Case Assembly with Air Hose

4.9 (50, 43 in.·lbf)

Vehicle Speed Sensor Connector

Back-Up Light Switch Connector

Control Cable

Clip

Clip

Washer

29 (300, 22)

◆ 32 (330, 24)

46 (470, 34)

35 (360, 26)

64 (650, 47)

64 (650, 47)

RH Drive Shaft Snap Ring

Ground Cable

25 (250, 18)

No.2 Rear End Plate

9.0 (95, 78 in.·lbf)

37 (380, 27)

Stiffener Plate

29 (300, 22)

Transaxle Case Protector

25 (250, 18)

Tie Rod End

◆ Snap Ring

Lock Cap

49 (500, 36)

◆ Cotter Pin

◆ Cotter Pin

9.0 (95, 78 in.·lbf)

64 (650, 47)

37 (380, 27)

64 (650, 47)

LH Drive Shaft

Engine Wire Clamp

216 (2,200, 159)

Starter

◆ Gasket

48 (490, 35)

39 (400, 29)

◆ 62 (630, 46)

PS Gear Assembly

Engine Mounting Center Member

Front Exhaust Pipe

RH Engine Under Cover

35 (360, 26)

80 (820, 59)

Front Suspension Crossmember Assembly with Stabilizer Bar

LH Engine Under Cover

113 (1,150, 83)

113 (1,150, 83)

137 (1,400, 101)

206 (2,100, 152)

112 (1,140, 82)

127 (1,300, 94)

N·m (kgf·cm, ft·lbf) : Specified torque
◆ Non-reusable part

7924ZG61

Transaxle exploded view—2WD RAV4

Heater Core replacement is covered in Section 2 of this manual

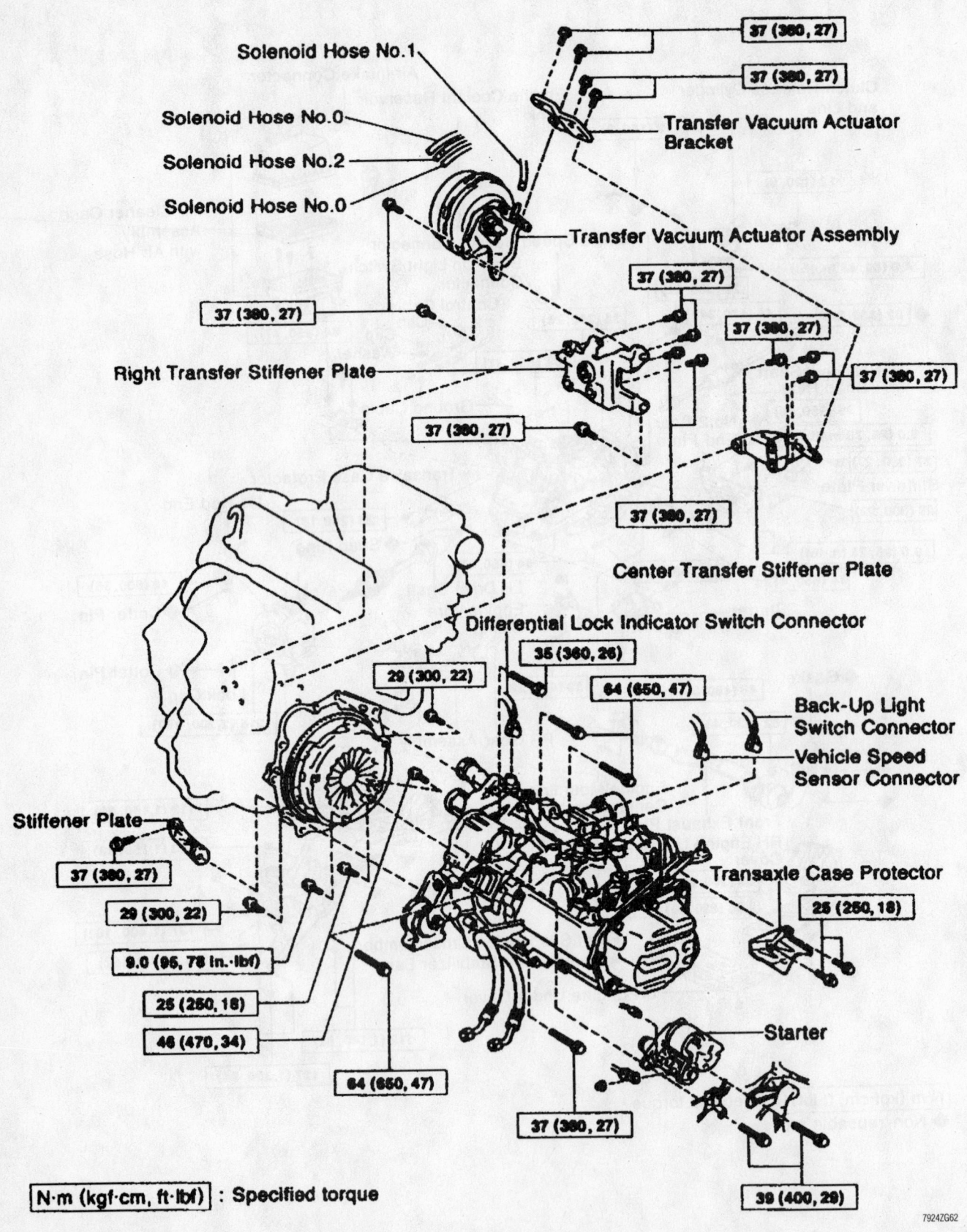

Solenoid Hose No.1

Solenoid Hose No.0

Solenoid Hose No.2

Solenoid Hose No.0

37 (380, 27)

37 (380, 27)

Transfer Vacuum Actuator Bracket

Transfer Vacuum Actuator Assembly

37 (380, 27)

37 (380, 27)

37 (380, 27)

37 (380, 27)

37 (380, 27)

Right Transfer Stiffener Plate

37 (380, 27)

37 (380, 27)

Center Transfer Stiffener Plate

Differential Lock Indicator Switch Connector

35 (360, 26)

64 (650, 47)

29 (300, 22)

Back-Up Light Switch Connector

Vehicle Speed Sensor Connector

Stiffener Plate

37 (380, 27)

29 (300, 22)

Transaxle Case Protector

9.0 (95, 78 in.·lbf)

25 (250, 18)

25 (250, 18)

46 (470, 34)

Starter

64 (650, 47)

37 (380, 27)

39 (400, 29)

N·m (kgf·cm, ft·lbf) : Specified torque

Manual transaxle exploded view—4WD RAV4

7924ZG62

- Ground cable to the transaxle
- Clutch release cylinder and line
- Starter. Tighten both bolts to 29 ft. lbs. (39 Nm).
- Engine wire clamp with the nut
- Engine coolant reservoir tank
- Air cleaner case assembly with the air hose
- Negative battery cable

25. Fill the transaxle with fluid. Check all fluids.

4WD Rav4

1. Before servicing the vehicle, refer to the precautions in the beginning of this section.
2. Remove or disconnect the following:
   - Transaxle/engine assembly
   - Transaxle case protector, by removing the 2 bolts
   - Starter
   - Transfer vacuum actuator bracket, by removing the 4 bolts
3. Remove the transfer vacuum actuator assembly, as follows:
   - 4 solenoid hoses from the transfer vacuum actuator assembly
   - Transfer vacuum actuator assembly, by removing the 2 bolts
4. Remove or disconnect the following:
   - Right transfer stiffener plate, by removing the 5 bolts
   - Center transfer stiffener plate, by removing the 3 bolts
   - Stiffener plate by removing the 2 bolts
   - Transaxle from the engine, by removing the 9 transaxle mount bolts

To install:

5. Connect the transaxle to the engine. Tighten the 9 bolts, as follows:
   - Bolt A: 47 ft. lbs. (64 Nm)
   - Bolt B: 26 ft. lbs. (35 Nm)
   - Bolt C: 22 ft. lbs. (29 Nm)
   - Bolt D: 34 ft. lbs. (46 Nm)
   - Bolt E: 18 ft. lbs. (25 Nm)
   - Bolt F: 78 inch lbs. (9.0 Nm)
6. Install or connect the following:
   - Stiffener plate. Tighten both bolts to 27 ft. lbs. (37 Nm).
   - Center transfer stiffener plate. Tighten the 3 bolts to 27 ft. lbs. (37 Nm).
   - Right transfer stiffener plate. Tighten the 5 bolts to 27 ft. lbs. (37 Nm).
7. Install the transfer vacuum actuator assembly, as follows:

- Transfer vacuum actuator assembly. Tighten both bolts to 27 ft. lbs. (37 Nm).
- 4 solenoid hoses to the transfer vacuum actuator assembly
8. Install or connect the following:
   - Transfer vacuum actuator bracket. Tighten the 4 bolts to 27 ft. lbs. (37 Nm).
   - Starter. Tighten both bolts to 29 ft. lbs. (39 Nm).
   - Transaxle case protector. Tighten both bolts to 18 ft. lbs. (25 Nm).
   - Transaxle/engine assembly

4Runner

W59 AND R150 TRANSMISSIONS

1. Before servicing the vehicle, refer to the precautions in the beginning of this section.
2. Remove or disconnect the following:
   - Negative battery cable
   - Transmission shift lever from the inside of the vehicle
   - Front console box
   - Shift lever boot retainer and the shift lever boot
   - Shift lever cap, by covering it with a cloth, pressing down on it and rotating it counterclockwise
   - Transfer shift lever, using snapring pliers to pull out the shift lever from the transfer case, on 4WD
3. Drain the transmission and the transfer oil.
4. Remove or disconnect the following:
   - No. 1 and No. 2 engine undercover
   - Front and rear driveshafts
   - Vehicle Speed Sensor (VSS), back-up light switch and the 4WD position switch connectors
   - L4 position switch connector, if equipped with Anti-lock Brake System (ABS) and/or differential lock
   - Clutch release cylinder and move it aside without disconnecting the clutch line
   - Exhaust pipe bracket
   - Rear end plate
5. Support the transmission rear side.
6. Remove or disconnect the following:
   - 4 bolts from the engine rear mount
   - 4 bolts, nuts and the crossmember
   - 4 bolts and the engine rear mounting from the transfer case
7. Using a transmission jack, support the transmission.
8. Remove or disconnect the following:

- Starter
- Wiring harness from the transmission
- Transmission-to-engine bolts and lower the transmission with the transfer case, on 4WD
- Transfer case from the transmission, on 4WD

To install:

9. Install or connect the following:
   - Transfer case to the transmission, on 4WD. Tighten the bolts to 17 ft. lbs. (24 Nm).

✵✵ WARNING

Be careful not to damage the oil seal by the input gear spline when installing the transfer.

- Transmission with the transfer case, on 4WD. Tighten the engine-to-transmission bolts to 53 ft. lbs. (72 Nm).
- Starter. Tighten both bolts to 29 ft. lbs. (39 Nm).
- Engine rear mount. Tighten the 4 bolts to 48 ft. lbs. (65 Nm).
- Crossmember. Tighten the 4 bolts to 48 ft. lbs. (65 Nm).
- Engine rear mount. Tighten the 4 bolts to 14 ft. lbs. (19 Nm).
- Rear end plate. Tighten the 4 bolts and nuts to 27 ft. lbs. (37 Nm).
- Front exhaust pipe. Tighten the bracket bolts to 52 ft. lbs. (71 Nm), the exhaust pipe-to-catalytic converter bolts to 35 ft. lbs. (48 Nm) and the support bracket bolts to 14 ft. lbs. (19 Nm).
- Clutch release cylinder. Tighten the 2 bolts to 108 inch lbs. (13 Nm).
- L4 position switch connector, if disconnected
- VSS, back-up light switch and the 4WD position switch connectors
- Front and rear driveshafts
- No. 1 and No. 2 engine under covers
10. Refill the transmission to the correct level.
11. Apply MP grease to the transfer shift lever.
12. Install or connect the following:
    - Transfer shift lever and snapring
13. Install the transmission shift lever, as follows:
    a. Apply MP grease to the transmission shift lever.
    b. Align the groove of the shift lever

cap and the pin par of the case cover. Cover the shift lever cap with a cloth. Pressing down on the shift lever cap, rotate it clockwise to install.

14. Install or connect the following:
- Shift lever boot retainer with the 4 screws
- Front console box with the 4 screws
- Negative battery cable

15. Start the engine and check for leaks.

16. Road test the vehicle for proper operation and recheck all the fluid levels.

Clutch Assembly

ADJUSTMENT

There is no adjustment for the 4Runner clutch.

Free-Play

RAV4

1. Before servicing the vehicle, refer to the precautions in the beginning of this section.

2. Check that the pedal height is correct. Pedal height from the floor panel should be: 6.889–7.283 in. (175–185mm)

3. If necessary to adjust the pedal

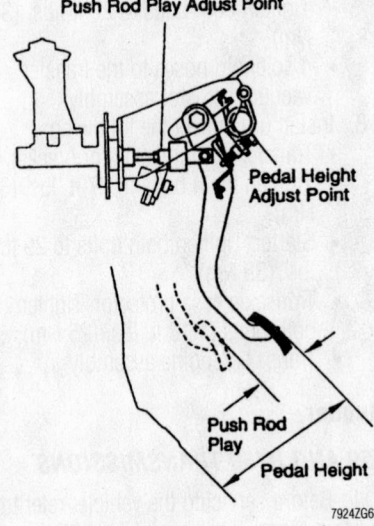

Clutch pedal height measurement location—RAV4

height, loosen the locknut and turn the stopper bolt until the height is correct. Tighten the locknut.

4. Push the pedal inward until the beginning of the clutch resistance is felt. Free-play should be 0.197–0.591 in. (5–15mm).

5. Gently push on the pedal until the resistance begins to increase a little.

Pushrod play at the pedal top should be 0.039–0.197 in. (1–5mm).

6. If necessary, adjust the pedal free-play and the pushrod play, as follows:
   a. Loosen the locknut and turn the push the rod until the free-play and pushrod play are correct.
   b. Tighten the locknut.

REMOVAL & INSTALLATION

RAV4

1. Before servicing the vehicle, refer to the precautions in the beginning of this section.

2. Remove or disconnect the following:
- Negative battery cable
- Transaxle

3. Matchmark the clutch cover to the flywheel.

4. Remove the clutch pressure plate retaining bolts in small amounts and in a crisscross pattern to relieve the clutch disc spring tension.

5. At the clutch cover, loosen each bolt 1 turn until spring tension is released.

6. Remove or disconnect the following:
- Clutch cover set bolts and pull off the clutch cover with the clutch disc

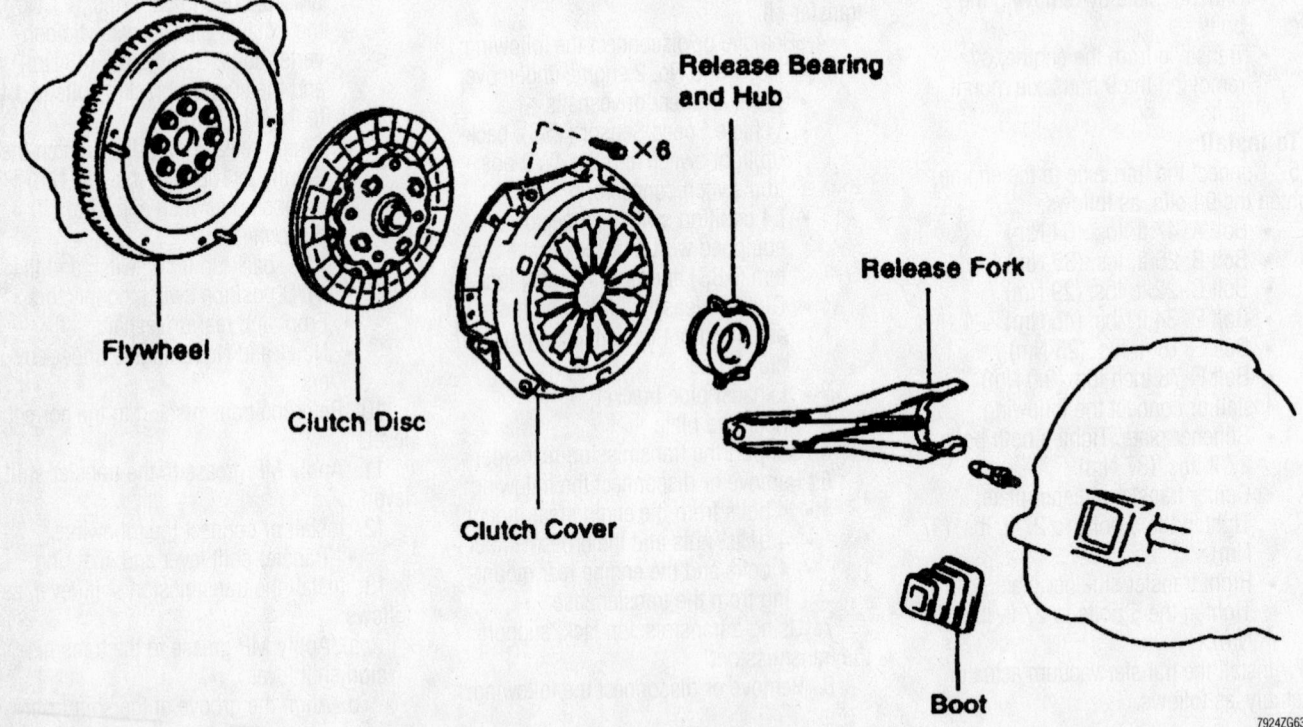

Clutch component assembly—Exploded view

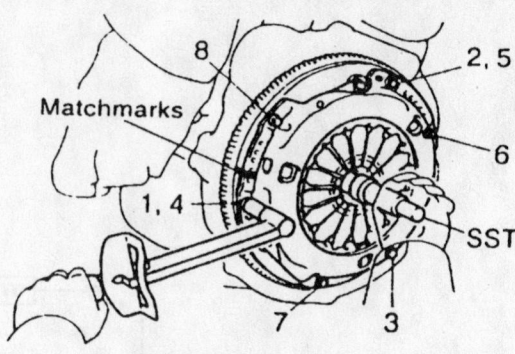

Torque sequence for the clutch cover—RAV4

- Clutch cover-to-flywheel bolts

7. If the clutch release bearing is to be replaced, perform the following:

    a. Remove the bearing retaining clip(s), the bearing and hub.

    b. Remove the release fork and the boot.

    c. The bearing is press fitted to the hub.

    d. Clean all parts and lightly grease the input shaft splines and all of the contact points.

    e. Install the bearing/hub assembly, the fork, the boot and the retaining clip(s) in their original locations.

To install:

8. Inspect the flywheel surface for cracks, heat scoring (blue marks) and warpage. Replace or resurface the flywheel, if any damage is present.

➡ **Before installing any new parts, be sure they are clean. During installation, do not get grease or oil on any of the components, as this will shorten clutch life considerably.**

9. Using a clutch alignment tool, position the clutch disc against the flywheel. The raised center section of the disc faces the transaxle.

10. Install or connect the following:
- Clutch cover onto the flywheel by aligning the matchmarks
- Clutch cover. Tighten the bolts in a crisscross pattern to 14 ft. lbs. (19 Nm).

11. Lubricate the release fork pivot and contact points, release bearing, bearing hub and input shaft spline surfaces with a suitable molybdenum disulfide lithium based or multi-purpose grease.

12. Install or connect the following:
- Boot, release fork, hub and the bearing assemblies

- Transaxle
- Negative battery cable

4Runner

1. Before servicing the vehicle, refer to the precautions in the beginning of this section.

2. Remove or disconnect the following:
- Negative battery cable
- Transmission assembly

3. Matchmark the clutch cover to the flywheel.

4. At the clutch cover, loosen each bolt 1 turn until spring tension is released.

5. Remove or disconnect the following:
- Clutch cover set bolts and the clutch cover with the clutch disc.
- Release bearing retaining clip and withdraw the it
- Release fork and boot assembly

To install:

6. Install or connect the following:
- Clutch disc onto the flywheel, using a clutch disc alignment tool
- Clutch cover, position it onto the flywheel and if reusing the old pressure plate, align the matchmarks.

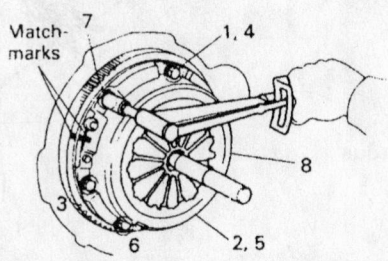

Bolt tightening sequence for the clutch cover—4Runner

- Clutch cover. Tighten the bolts in a crisscross pattern to 14 ft. lbs. (19 Nm).

7. Lubricate the release fork pivot and contact points, the release bearing, bearing hub and input shaft spline surfaces with a suitable molybdenum disulfide lithium based or multi-purpose grease.

8. Install or connect the following:
- Boot, release fork, hub and the bearing assemblies
- Transmission
- Negative battery cable

Hydraulic Clutch System

BLEEDING

1. Before servicing the vehicle, refer to the precautions in the beginning of this section.

2. Fill the clutch reservoir with brake fluid. Check the reservoir level frequently and add fluid as needed.

3. Connect one end of a vinyl tube to the bleeder plug on the slave cylinder and submerge the other end into a clear container half-filled with brake fluid.

4. Slowly pump the clutch pedal several times.

5. Have an assistant hold the clutch pedal down and loosen the bleeder plug until fluid and/or air starts to run out of the bleeder plug. Close the bleeder plug while the pedal is held to the floor.

6. Repeat Steps 2 and 3 until all the air bubbles are removed from the system.

7. Tighten the bleeder plug when all the air is gone.

8. Refill the master cylinder to the proper level as required.

9. Check the system for leaks.

Automatic Transmission/ Transaxle Assembly

REMOVAL & INSTALLATION

4WD RAV4

1. Before servicing the vehicle, refer to the precautions in the beginning of this section.

2. Remove or disconnect the following:
- Negative battery cable
- Engine/transaxle assembly
- Starter
- Stiffener plate, by removing the 3 bolts

For complete Engine Mechanical specifications, see Section 1 of this manual

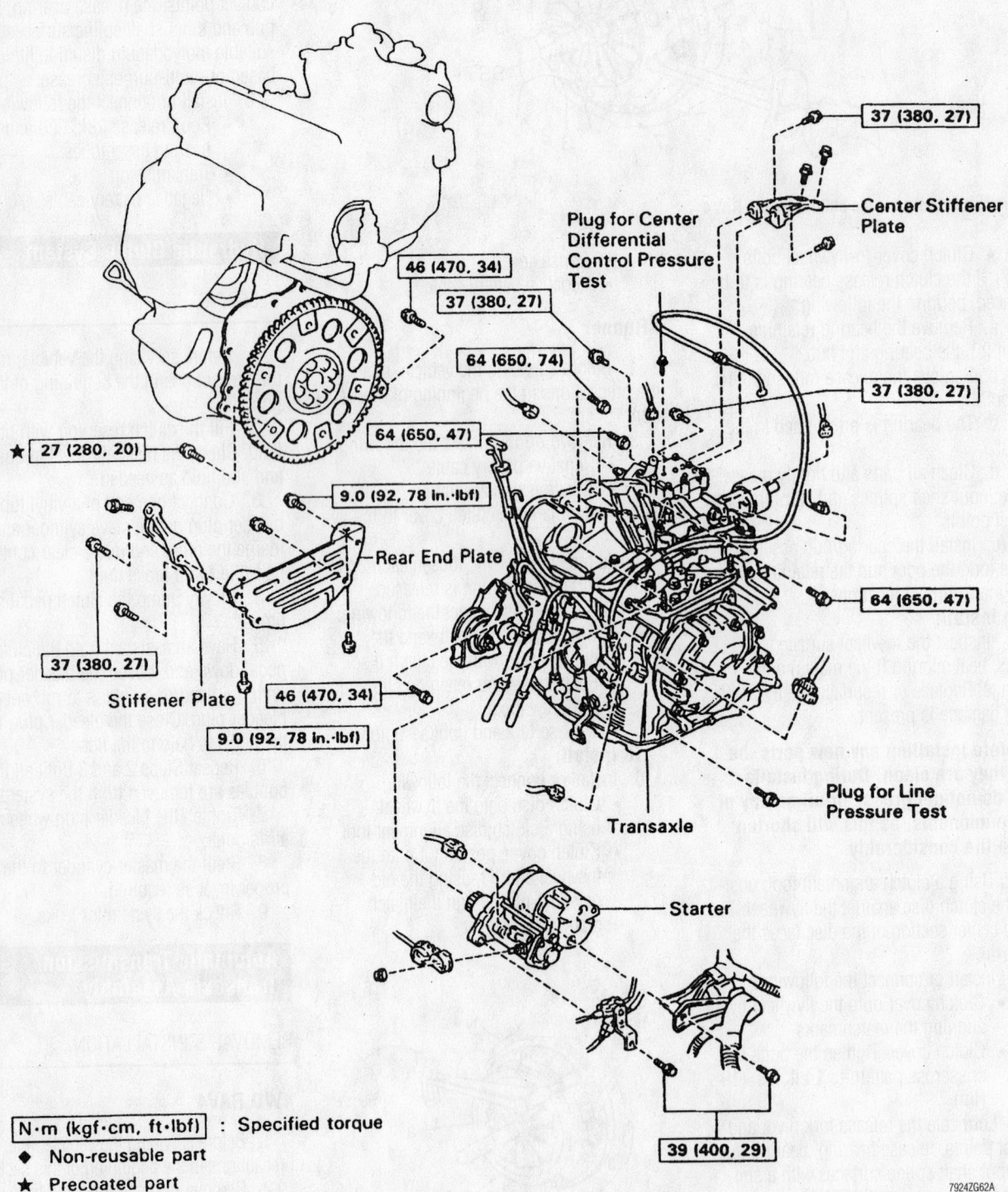

37 (380, 27)

Center Stiffener Plate

Plug for Center Differential Control Pressure Test

46 (470, 34)

37 (380, 27)

64 (650, 74)

37 (380, 27)

64 (650, 47)

★ 27 (280, 20)

64 (650, 47)

9.0 (92, 78 in.·lbf)

Rear End Plate

37 (380, 27)

Stiffener Plate

46 (470, 34)

9.0 (92, 78 in.·lbf)

Plug for Line Pressure Test

Transaxle

Starter

39 (400, 29)

N·m (kgf·cm, ft·lbf) : Specified torque

◆ Non-reusable part

★ Precoated part

79242G62A

Automatic transaxle exploded view—4WD RAV4

- Rear endplate, by removing the 4 bolts
- 6 torque converter clutch mounting bolts
- Connectors and wiring harness, from the transaxle
- Center stiffener plate, by removing the 4 bolts

3. Remove the transaxle with the transfer assembly, as follows:
- 2 bolts
- 5 transaxle mounting bolts
- Transaxle from the engine

To install:

4. Install the transaxle. Tighten the 14mm head bolts to 47 ft. lbs. (64 Nm) and the 12mm head bolts to 34 ft. lbs. (46 Nm).

5. Install or connect the following:
- Tighten both bolts to 27 ft. lbs. (37 Nm).
- Center stiffener plate. Tighten the 4 bolts to 27 ft. lbs. (37 Nm).
- Connectors and the wiring harness to the transaxle
- Torque converter clutch mounting bolts. Tighten each bolt to 20 ft. lbs. (27 Nm).

➡**Coat the threads of the bolts with an approved locking compound.**

- Rear endplate. Tighten the 4 bolts to 80 inch lbs. (9.0 Nm).
- Stiffener plate. Tighten the 3 bolts to 27 ft. lbs. (37 Nm).
- Starter. Tighten both bolts to 29 ft. lbs. (39 Nm).
- Engine/transaxle assembly
- Negative battery cable

2WD RAV4

1. Before servicing the vehicle, refer to the precautions in the beginning of this section.

2. Remove or disconnect the following:
- Negative battery cable
- Throttle cable
- Engine coolant reservoir tank
- Air cleaner assembly
- Ground cable from the transaxle
- Set nut of the engine wire clamp

3. Remove the starter, as follows:
- Connector and nut from the starter
- 2 bolts and the engine wire
- Starter

4. Remove the 3 upper side transaxle mounting bolts.

5. Install an engine support fixture.

6. Remove or disconnect the following:
- 2 bolts and 2 nuts from the left engine mount

- Engine undercovers

7. Drain the fluid from the transaxle.

8. Remove the left and right half-shafts.

9. Remove the front exhaust pipe, as follows:
- 2 front exhaust pipe-to-center exhaust pipe bolts and gasket
- 3 front exhaust pipe-to-exhaust manifold nuts and gasket
- Exhaust manifold

10. Disconnect the shift control cable from the transaxle and frame, as follows:
- Control shaft lever nut
- Clip and the control cable from the transaxle
- 2 shift control cable-to-centermember bolts
- Crossmember

11. Detach the following connectors:
- Shift solenoid valve connector
- Park/Neutral Position (PNP) switch connector
- Vehicle Speed Sensor (VSS) connector

12. Remove or disconnect the following:
- Oil cooler hoses from the transaxle
- Rack and pinion from the crossmember by removing both nuts/bolts

13. Support the rack and pinion.

14. Support the suspension crossmember with a floor jack.

15. Remove or disconnect the following:
- Crossmember-to-centermember fasteners
- Crossmember with the sway bar
- Stiffener plate by removing the 3 bolts
- Rear endplate by removing the 4 bolts
- 6 torque converter bolts
- Both rear side transaxle mounting bolts
- Transaxle

To install:

16. Install or connect the following:
- Transaxle
- Both rear side transaxle mounting bolts. Tighten the top bolt to 18 ft. lbs. (25 Nm) and the lower bolt to 34 ft. lbs. (46 Nm).
- Torque converter bolts. Tighten the bolts to 20 ft. lbs. (27 Nm).

➡**First install the gray bolt; then, install the 5 black bolts.**

- Rear endplate. Tighten the engine bolts to 78 inch lbs. (9.0 Nm) and

---

the transaxle bolts to 14 ft. lbs. (19 Nm).
- Stiffener plate. Tighten the 3 bolts to 27 ft. lbs. (37 Nm).

17. Install the front suspension crossmember and centermember with the sway bar. Tighten the nuts/bolts, as follows:
- Bolt A: 152 ft. lbs. (206 Nm)
- Bolt B: 101 ft. lbs. (137 Nm)
- Bolt C: 26 ft. lbs. (35 Nm)
- Bolt D: 53 ft. lbs. (72 Nm)
- Nut: 54 ft. lbs. (73 Nm)

18. Install or connect the following:
- Rack and pinion to the crossmember. Tighten the nuts to 83 ft. lbs. (113 Nm).
- Oil cooler hoses with both clips

19. Connect the following connectors:
- Shift solenoid valve connector
- PNP switch connector
- VSS connector

20. Install or connect the following:
- Shift control cable to the transaxle. Tighten the nut to 10 ft. lbs. (13 Nm).

21. Install the front exhaust pipe, as follows:
- Front exhaust pipe, using new gaskets
- Front exhaust pipe to the exhaust manifold. Tighten the 3 nuts to 46 ft. lbs. (62 Nm).
- Front exhaust pipe to the center exhaust pipe. Tighten both bolts to 35 ft. lbs. (48 Nm).

22. Install or connect the following:
- Left and right halfshafts
- Engine undercovers
- Left engine mount. Tighten the both nuts/bolts to 47 ft. lbs. (64 Nm).

23. Remove the engine fixture.

24. Install or connect the following:
- Upper side transaxle mount. Tighten the 3 bolts to 47 ft. lbs. (64 Nm).
- Starter. Tighten both bolts to 29 ft. lbs. (39 Nm).
- Engine wire
- Starter wire with the nut
- Engine wire clamp set nut
- Ground cable to the transaxle. Tighten the bolt to 14 ft. lbs. (19 Nm).
- Air cleaner assembly
- Engine coolant reservoir tank
- Throttle cable
- Negative battery cable

25. Check all fluids.

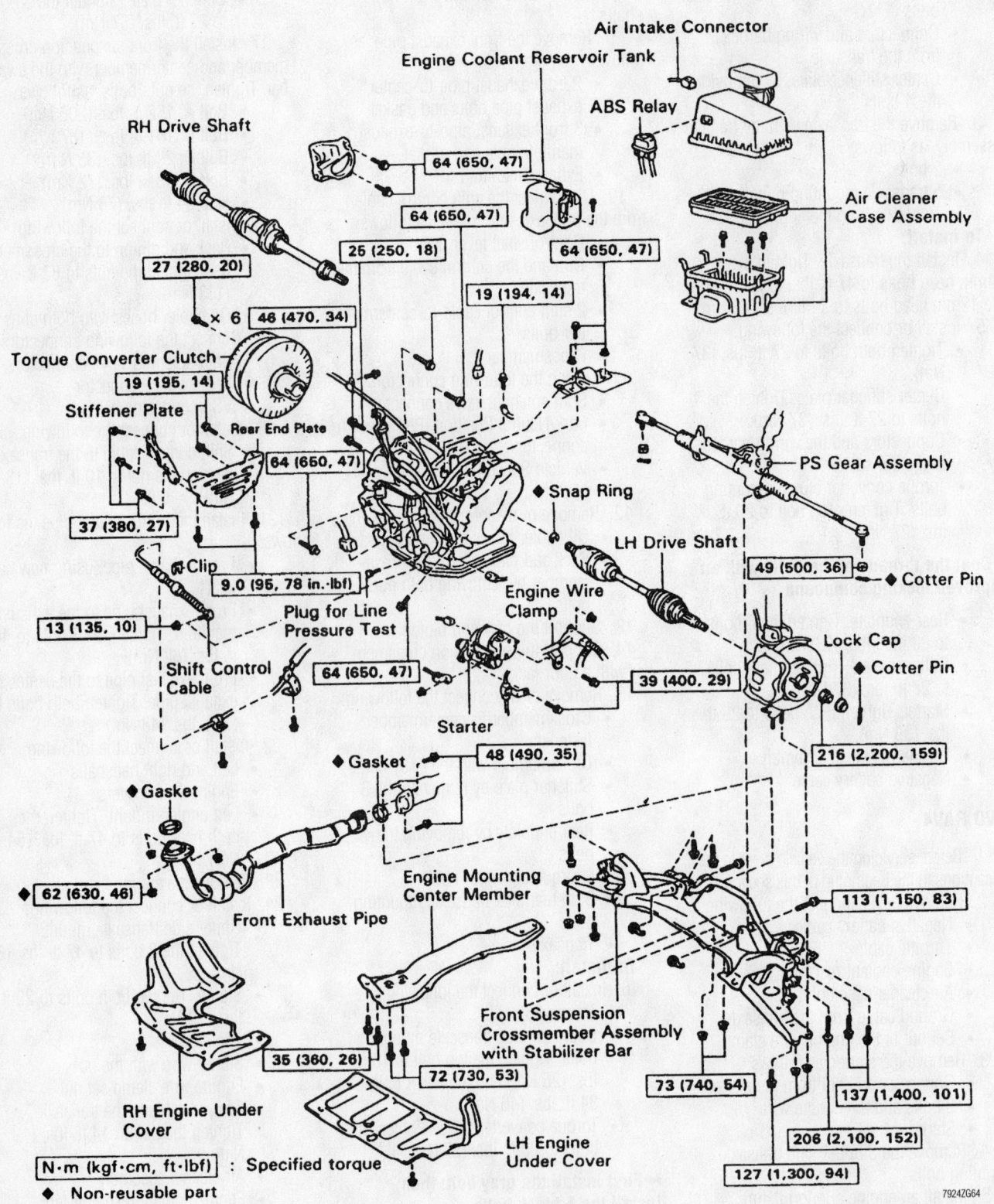

Air Intake Connector

Engine Coolant Reservoir Tank

ABS Relay

Air Cleaner Case Assembly

RH Drive Shaft

64 (650, 47)

64 (650, 47)

25 (250, 18)

64 (650, 47)

27 (280, 20)

19 (194, 14)

× 6   46 (470, 34)

Torque Converter Clutch

19 (195, 14)

Rear End Plate

Stiffener Plate

PS Gear Assembly

◆ Snap Ring

LH Drive Shaft

64 (650, 47)

49 (500, 36)   ◆ Cotter Pin

37 (380, 27)

◆ Clip

9.0 (95, 78 in.·lbf)

Plug for Line Pressure Test

Engine Wire Clamp

Lock Cap

13 (135, 10)

◆ Cotter Pin

Shift Control Cable

64 (650, 47)

39 (400, 29)

Starter

216 (2,200, 159)

48 (490, 35)

◆ Gasket

◆ Gasket

62 (630, 46)

Engine Mounting Center Member

113 (1,150, 83)

Front Exhaust Pipe

Front Suspension Crossmember Assembly with Stabilizer Bar

35 (360, 26)

72 (730, 53)

73 (740, 54)

137 (1,400, 101)

RH Engine Under Cover

206 (2,100, 152)

LH Engine Under Cover

127 (1,300, 94)

N·m (kgf·cm, ft·lbf) : Specified torque

◆ Non-reusable part

Transaxle exploded view—2WD RAV4

7924ZG64

Highlander

The transaxle is removed along with the engine. See Engine Removal and Installation, in this section.

RX 300

1. Before servicing the vehicle, refer to the precautions in the beginning of this section.
2. Remove or disconnect the following:
   - Hood
   - Wiper and blade assembly
   - Top cowl seal and panel
   - Window washer hoses, from the ventilator louvers
   - Left and right ventilator louvers
   - Heater air duct
   - Battery and tray
   - Throttle cable
   - Front upper suspension brace
   - Cruise control actuator with its bracket, if equipped
   - Starter
   - Shift control cable
   - Driveshaft, for 4WD
   - Body-to-engine ground strap
   - Park/Neutral Position (PNP) switch, solenoid and ATF temperature connectors
   - 5 upper transaxle-to-engine mounting bolts
   - Front wheel
   - Engine undercover
   - Halfshafts
   - Front exhaust pipe
   - Stabilizer bar
   - Both steering gear mounting bolts and support it in the vehicle
   - Shift control cable from its bracket
   - Power steering pipe and the oil cooler clamps from the frame
   - Both left-side transaxle mounting nuts
   - Rear-side engine mounting nuts
   - Engine shock absorber mounting bolts
   - 3 front-side engine mounting bolts
3. Attach an engine sling to the engine hangers in order to support the engine weight.
4. Remove or disconnect the following:
   - Front frame mounting bolts and the frame
   - Transaxle oil cooler lines
5. Support the transaxle with a transmission jack.
   - Torque converter access cover
   - 6 torque converter mounting bolts

   - 3 lower transaxle-to-engine mounting bolts
   - Engine from the transaxle

To install:

6. Install or connect the following:
   - Transaxle
   - 3 lower transaxle-to-engine mounting bolts and tighten to the illustrated value.
   - Torque converter-to-flexplate bolts, starting with the black bolt, then the other 5.
7. The rest of installation is the reverse of the removal referring to the illustrations for the tightening specifications.

4Runner

MODEL A340D, A340E AND A340H TRANSMISSIONS

➡ **The transfer case and the transmission should be removed as an assembly.**

1. Before servicing the vehicle, refer to the precautions in the beginning of this section.
2. Remove or disconnect the following:
   - Negative battery cable
   - Air cleaner assembly, if necessary
   - Transmission throttle cable from the throttle body
   - Engine undercover
3. Drain the transmission and transfer case (if applicable) fluid.
4. Remove or disconnect the following:
   - Wiring connectors from the transmission and transfer case, if applicable.
   - Starter
5. Matchmarks on the front and rear driveshaft flanges and the differential pinion flanges. These marks must be aligned during installation.
6. Remove or disconnect the following:
   - Front and rear driveshaft flanges.
   - Center bearing bracket bolts, if equipped with a 2-piece driveshaft
   - Driveshaft
   - Speedometer cable
   - Front exhaust pipe and bracket
   - Transmission oil cooler lines, at the transmission
   - Oil cooler lines bracket and the transmission oil filler tube, as required
7. Support the transmission, using a jack with a wooden block placed between the jack and the transmission pan. Raise the transmission, just enough to take the weight off of the rear mount.

8. Remove or disconnect the following:
   - Rear engine mount with the bracket, the rear crossmember and the transfer case undercover, if applicable
   - Dynamic damper, for Regular Cab only
   - No. 2 cross-shaft bracket
9. Place a wooden block(s) between the engine oil pan and the front frame crossmember.
10. Slowly, lower the transmission until the engine rests on the wooden block(s).
11. Remove or disconnect the following:
    - Torque converter cover to gain access to the converter bolts
    - Torque converter bolts, by rotating the crankshaft to access the bolts through the service holes
    - Stiffener plates from the transmission
    - Shift control rod and the transfer case shift lever
12. For the A340H transmission remove or disconnect, perform the following:
    - Cross-shaft and the No. 2 shifting rod
    - Front stabilizer bar
    - Differential mount bolts, by supporting the front differential with a jack
    - Transmission and transfer case, by slowly lowering the front differential so there is enough clearance, if applicable
    - Differential, if enough clearance can't be obtained
13. Remove or disconnect the following:
    - Stabilizer bar
    - Auxiliary frame crossmember, if equipped
14. For A340D transmissions, obtain a bolt of the same dimensions as the torque converter bolts. Cut the head off of the bolt and hacksaw a slot in the bolt opposite the threaded end. Thread the guide pin into one of the torque converter bolt holes. The guide pin will help keep the converter with the transmission.

➡ **This modified bolt is used as a guide pin. 2 guide pins are needed to properly install the transmission.**

15. Remove or disconnect the following:
    - Transmission bolts, then move the transmission rearward by prying on the dowel pins through the service hole

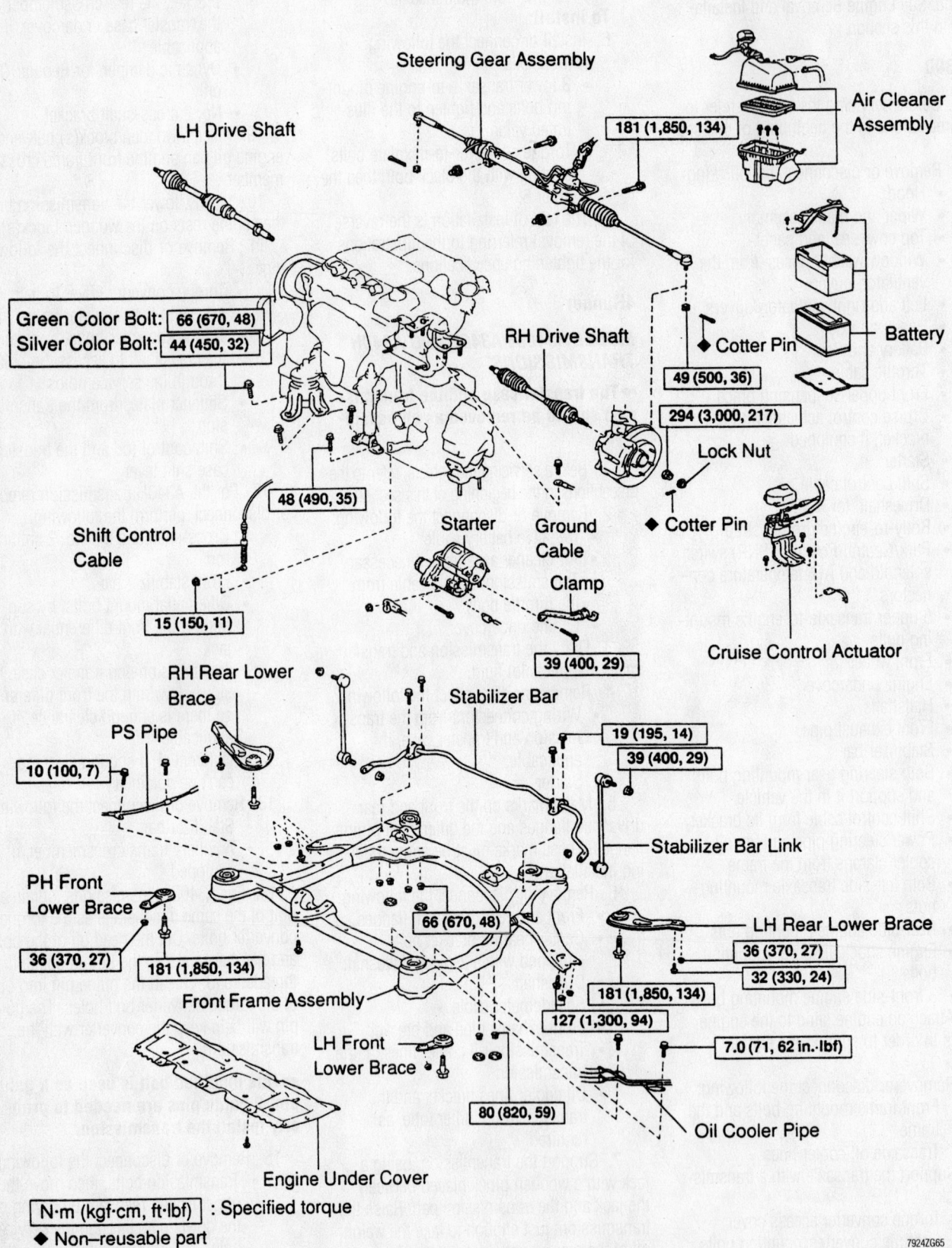

Steering Gear Assembly

Air Cleaner Assembly

LH Drive Shaft

181 (1,850, 134)

Green Color Bolt: 66 (670, 48)
Silver Color Bolt: 44 (450, 32)

RH Drive Shaft

Cotter Pin

Battery

49 (500, 36)

294 (3,000, 217)

Lock Nut

48 (490, 35)

Cotter Pin

Shift Control Cable

Starter   Ground Cable

Clamp

Cruise Control Actuator

15 (150, 11)

39 (400, 29)

RH Rear Lower Brace

Stabilizer Bar

PS Pipe

19 (195, 14)
39 (400, 29)

10 (100, 7)

Stabilizer Bar Link

PH Front Lower Brace

LH Rear Lower Brace

36 (370, 27)

36 (370, 27)

181 (1,850, 134)

66 (670, 48)

32 (330, 24)

Front Frame Assembly

181 (1,850, 134)

LH Front Lower Brace

127 (1,300, 94)

7.0 (71, 62 in.·lbf)

80 (820, 59)

Oil Cooler Pipe

Engine Under Cover

N·m (kgf·cm, ft·lbf) : Specified torque

◆ Non–reusable part

7924ZG65

Exploded view of the transaxle removal and installation components—RX 300 models

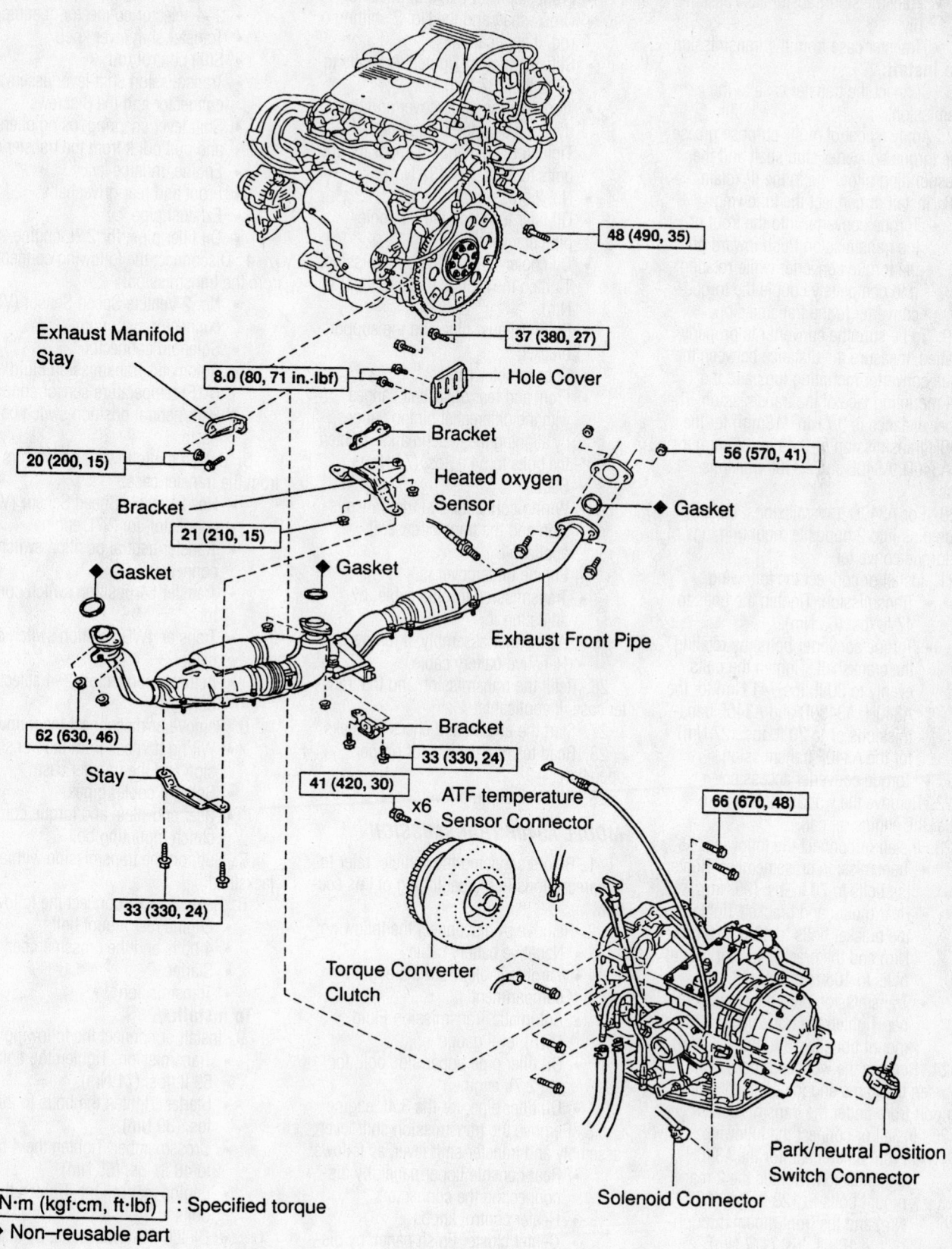

48 (490, 35)

37 (380, 27)

Exhaust Manifold Stay

8.0 (80, 71 in.·lbf)

Hole Cover

Bracket

20 (200, 15)

56 (570, 41)

Bracket

Heated oxygen Sensor

21 (210, 15)

◆ Gasket

◆ Gasket

◆ Gasket

Exhaust Front Pipe

62 (630, 46)

Bracket

33 (330, 24)

Stay

41 (420, 30)

×6

ATF temperature Sensor Connector

66 (670, 48)

33 (330, 24)

Torque Converter Clutch

Park/neutral Position Switch Connector

Solenoid Connector

N·m (kgf·cm, ft·lbf) : Specified torque

◆ Non–reusable part

7924ZG66

Exploded view of the transaxle removal and installation components—RX 300 models, Cont.

For Wheel Alignment specifications, see Section 1 of this manual

- Transmission/transfer case assembly
- Transfer case from the transmission

To install:

16. Connect the transfer case to the transmission.

17. Apply a coat of multi-purpose grease to the torque converter stub shaft and the corresponding pilot hole in the flexplate.

18. Install or connect the following:
- Torque converter into the front of the transmission Push inward on the torque converter while rotating it to completely couple the torque converter to the transmission.

19. To be sure the converter is properly installed, measure the distance between the torque converter mounting lugs and the front mounting face of the transmission. The proper distance is 0.71 in. (18mm) for the A340H transmission or 0.79 in. (20mm) for the A340D, A340E and A340F transmissions.

20. For A340D transmissions, install guide pins into 2 opposite mounting lugs of the torque converter.

21. Install or connect the following:
- Transmission. Tighten the bolts to 47 ft. lbs. (63 Nm).
- Torque converter bolts, by rotating the crankshaft. Tighten the bolts evenly to 30 ft. lbs. (41 Nm) for the A340H, A3430D and A340E transmissions or to 20 ft. lbs. (27 Nm) for the A340F transmission.
- Torque converter access cover

22. Remove the wood block(s) from under the engine oil pan.

23. Install or connect the following:
- Transmission crossmember. Tighten the bolts to 70 ft. lbs. (95 Nm).
- Rear mount and bracket. Tighten the bracket bolts to 43 ft. lbs. (58 Nm) and the bracket-to-rear mount bolts to 108 inch lbs. (13 Nm).
- Transmission onto the crossmember. Tighten the transmission-to-mount bolts to 18 ft. lbs. (25 Nm).

24. Remove the wooden blocks from between the frame and the engine and the support from under the transmission.

25. Install or connect the following:
- Front differential, for the A340H transmission. Tighten the 2 rear mount bolts to 123 ft. lbs. (167 Nm) and the front mount through-bolt to 108 ft. lbs. (147 Nm).

➡ **If the differential oil was drained, refill it at this time.**

- Shift control rod and the transfer case shift lever

- Front stabilizer bar, if applicable
- Cross-shaft and the No. 2 shifting rod, if applicable
- Stiffener plates. Tighten the bolts to 27 ft. lbs. (37 Nm).
- Transfer case undercover and the dynamic damper, if equipped. Tighten the dynamic damper mount bolts to 27 ft. lbs. (37 Nm).
- No. 2 cross-shaft bracket
- Oil filler tube and the oil cooler pipe bracket
- Oil cooler lines to the transmission. Tighten the fittings to 25 ft. lbs. (34 Nm).
- Front exhaust pipe and the support bracket
- Speedometer cable
- Front and rear driveshaft flanges with the differential pinion flanges, by aligning the matchmarks. Tighten the bolts to 54 ft. lbs. (74 Nm).
- Starter
- Wiring connectors to the transmission and the transfer case, if applicable
- Engine undercover
- Transmission throttle cable, by adjusting it
- Air cleaner assembly, if removed
- Negative battery cable

26. Refill the transmission and the transfer case, if applicable.

27. Start the engine and check for leaks.

28. Road test the vehicle for proper operation.

29. Recheck all fluid levels.

MODEL A340F TRANSMISSION

1. Before servicing the vehicle, refer to the precautions in the beginning of this section.

2. Remove or disconnect the following:
- Negative battery cable
- Throttle cable, from the engine compartment
- Automatic Transmission Fluid (ATF) level gauge
- Oil filler pipe upper side bolt, for the 2.7L engine
- Oil filler pipe, for the 3.4L engine

3. Remove the transmission shift lever assembly and transfer shift lever, as follows:
- Rear console upper panel, by disconnecting the connectors
- Heater control knobs
- Center cluster finish panel, by disconnecting the connectors
- Transfer shift lever knob, without the 2–4 selector
- Bolt and the transfer shift lever knob, with the 2–4 selector

- Front console upper panel
- 2–4 selector connector, if equipped
- Transfer shift lever knob
- Shift control rod
- Transmission shift lever assembly connector and the 8 screws
- Shift lever snaping, using pliers and pull out it from the transfer case
- Engine undercover
- Front and rear driveshafts
- Exhaust pipe
- Oil filler pipe, for 2.7L engine

4. Disconnect the following connectors from the transmission:
- No. 2 Vehicle Speed Sensor (VSS) connector
- Solenoid connector
- Automatic Transmission Fluid (ATF) temperature sensor connector
- Park/neutral position switch connector

5. Detach the following connectors from the transfer case:
- No. 1 Vehicle Speed Sensor (VSS) connector, for 2.7L engine
- Transfer neutral position switch connector
- Transfer L4 position switch connector
- Transfer 4WD position switch connector
- Actuator connector (2–4 selector only)

6. Remove or disconnect the following:
- Wiring harness from the transmission and the transfer case
- Both oil cooler pipes
- Rear end-plate and torque converter clutch mounting bolt

7. Support the transmission with a jackstand.

8. Remove or disconnect the following:
- Engine rear mount bolts
- 4 bolts and the crossmember
- Starter
- Transmission

To install:

9. Install or connect the following:
- Transmission. Tighten the bolts to 53 ft. lbs. (71 Nm).
- Starter. Tighten the bolts to 29 ft. lbs. (39 Nm).
- Crossmember. Tighten the 4 bolts to 48 ft. lbs. (65 Nm).
- Engine rear mount. Tighten the 4 bolts to 14 ft. lbs. (19 Nm).
- Clutch converter bolts, by installing the green colored bolt before the other 5. Tighten the bolts to 30 ft. lbs. (41 Nm).
- Rear end-plate. Tighten the bolts to 13 ft. lbs. (18 Nm).

- Both oil cooler pipes. Tighten to 25 ft. lbs. (34 Nm).
- Oil cooler pipe clamps. Tighten the 10mm head bolt to 48 inch lbs. (5 Nm) and the 12mm head bolt to 108 inch lbs. (13 Nm).
- Wiring harness to the transmission and the transfer case
- Remaining components

10. Fill the transmission and transfer case with transmission fluid.
- Throttle cable
- Negative battery cable

Transfer Case Assembly

REMOVAL & INSTALLATION

The transfer case for the RAV4, Highlander and RX300 is part of the transmission/transaxle assembly and is serviced with those units.

4Runner

1. Before servicing the vehicle, refer to the precautions in the beginning of this section.
2. Disconnect the negative battery cable.
3. Drain the transmission and the transfer case.
4. Remove or disconnect the following:

- Transfer case with the transmission
- Breather hose from the transfer upper cover and the transmission control retainer, if equipped with an automatic transmission
- Rear engine mounting
- Dynamic damper

5. Remove the driveshaft upper dust cover and the transfer from the transmissions, as follows:

- Dust cover bolt from the bracket
- Transfer case adapter rear mounting bolts
- Transfer case, by pulling it straight up and away from the transmission.

✳✳ WARNING

Be careful not to damage the adapter rear oil seal with the transfer input gear spline.

To install:

6. Install the transfer case and the driveshaft upper dust cover to the transmission with a new gasket, as follows:

- Shift the 2 shift fork shafts to the high 4 position
- Apply MP grease to the adapter oil seal
- New gasket to the transfer adapter
- Transfer case to the transmission.

✳✳ WARNING

Take care not to damage the oil seal by the input gear spline.

- Transfer case adapter. Tighten the rear bolts to 27 ft. lbs. (37 Nm).
- Dust cover to the bracket. Tighten the bolt to 17 ft. lbs. (23 Nm).

7. Install or connect the following:
- Engine rear mount. Tighten the bolts to 19 ft. lbs. (25 Nm).
- Dynamic damper. Tighten the bolts to 27 ft. lbs. (37 Nm).
- Breather hose, if equipped with an automatic transmission
- Transfer case with the transmission to the engine

8. Fill the transmission and the transfer case with oil.
9. Test drive the vehicle and check the abnormal noise and smooth operation.
10. Recheck the fluid levels.

Halfshaft

REMOVAL & INSTALLATION

RAV4 Front

1. Before servicing the vehicle, refer to the precautions in the beginning of this section.
2. Remove or disconnect the following:
- Negative battery cable
- Engine undercover
3. Drain the transaxle.
4. Remove or disconnect the following:

- Anti-lock Brake System (ABS) sensor by removing the bolt, if equipped
- Cotter pin, lock cap and the locknut holding the halfshaft to the steering knuckle
- Tie rod ends, from the steering knuckle
- Sway bar link, from the lower control arm
- Lower ball joint, from the lower control arm
- Halfshaft from the axle hub, using a plastic hammer

5. If working on a 2WD right-hand half-shaft and the vehicle is equipped with a manual transaxle, perform the following to remove the halfshaft:
- Snapring from the center bearing bracket, using a brass bar and hammer
- Bolt and the center bearing bracket
- Halfshaft with the center halfshaft
- 2 bolts and the center bearing bracket

6. If working on a 2WD right-hand half-shaft and the vehicle is equipped with an automatic transaxle, perform the following to remove the halfshaft:
- 2 bolts of the center bearing bracket and pull out the halfshaft together with the center bearing case and center halfshaft
- 3 bolts and the center bearing bracket

7. If working on a 2WD left-hand, perform the following:
- Halfshaft, using a brass bar and hammer
- Snapring from the transaxle

8. If working of a 4WD right-hand half-shaft, perform the following:
- Halfshaft, using a brass bar and hammer
- Snapring from the transaxle
- O-ring

9. If working on a 4WD left-hand side, perform the following:
- Air cleaner
- Transaxle case protector
- Halfshaft, by prying it out using a hub wrench
- Snapring

To install:

10. If working on a 4WD left-hand side, perform the following:
- Snapring
- Halfshaft to the transaxle
- Transaxle case protector
- Air cleaner

11. If working of a 4WD right-hand half-shaft, perform the following:
- Snapring to the transaxle
- New O-ring
- Halfshaft to the transaxle

12. If working on a 2WD left-hand, perform the following:
- Snapring
- Halfshaft to the transaxle

13. If working on a 2WD right-hand half-shaft and the vehicle is equipped with an automatic transaxle, perform the following to remove the halfshaft:

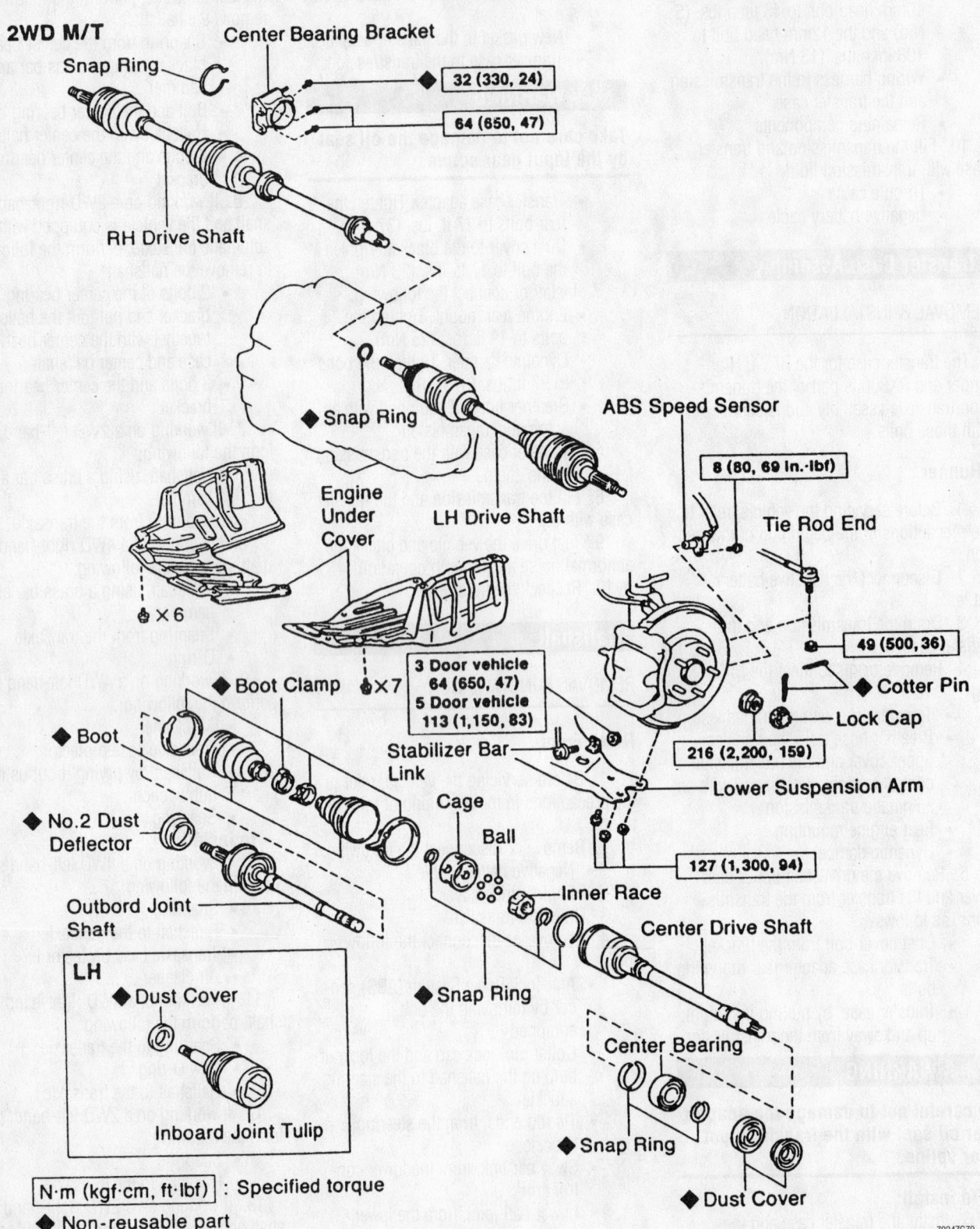

2WD M/T

Snap Ring

Center Bearing Bracket

◆ 32 (330, 24)

◆ 64 (650, 47)

RH Drive Shaft

◆ Snap Ring

Engine Under Cover

◆ ×6

LH Drive Shaft

ABS Speed Sensor

8 (80, 69 in.·lbf)

Tie Rod End

49 (500, 36)

◆ Cotter Pin

Lock Cap

216 (2,200, 159)

Lower Suspension Arm

127 (1,300, 94)

3 Door vehicle
64 (650, 47)
5 Door vehicle
113 (1,150, 83)

Stabilizer Bar Link

◆ Boot Clamp ◆ ×7

◆ Boot

◆ No.2 Dust Deflector

Outbord Joint Shaft

Cage

Ball

Inner Race

◆ Snap Ring

Center Drive Shaft

Center Bearing

◆ Snap Ring

◆ Dust Cover

LH

◆ Dust Cover

Inboard Joint Tulip

N·m (kgf·cm, ft·lbf) : Specified torque

◆ Non-reusable part

7924ZG70

Front halfshaft exploded view (2WD with manual transmission only)—RAV4

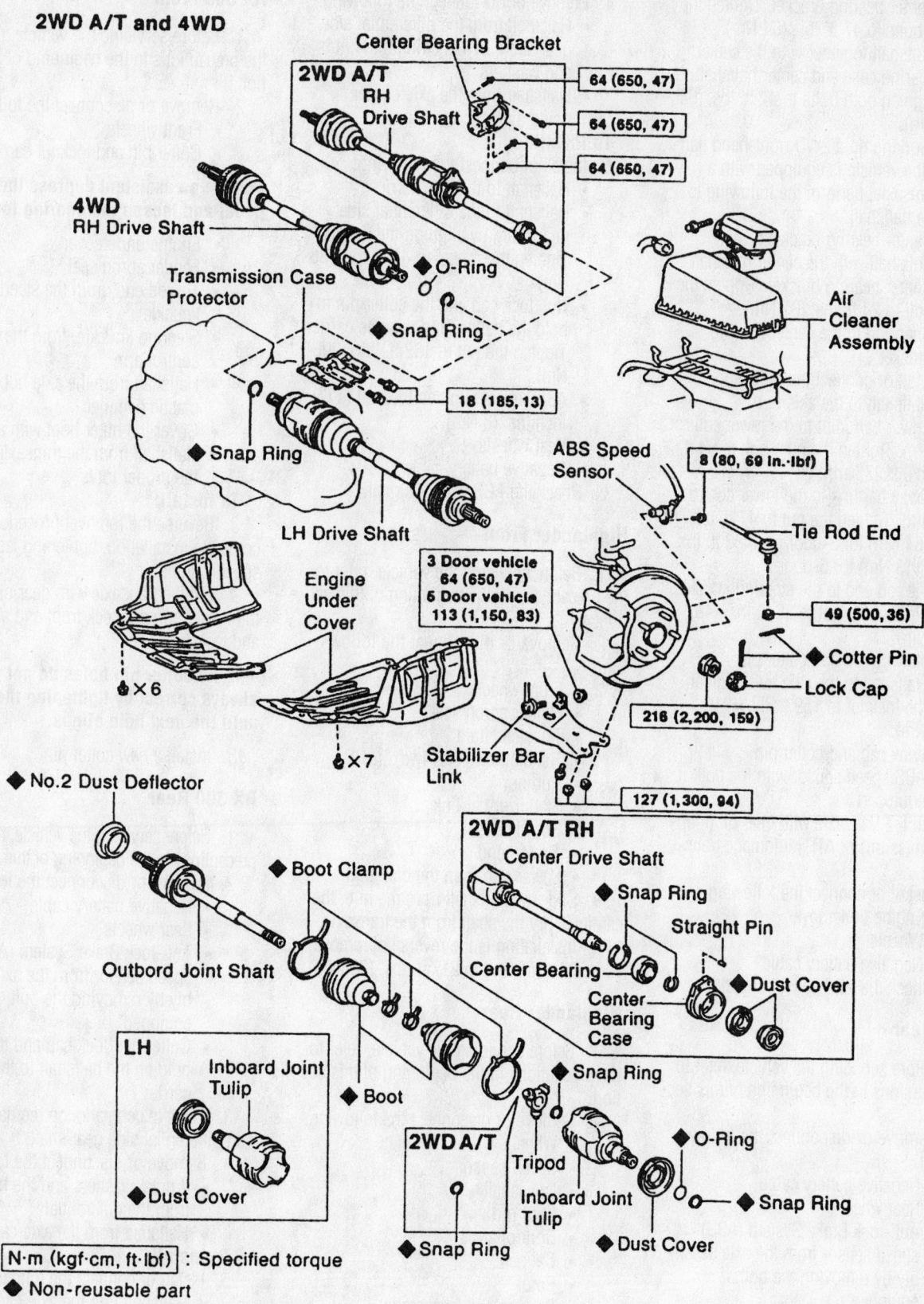

2WD A/T and 4WD

Center Bearing Bracket

2WD A/T RH Drive Shaft

64 (650, 47)
64 (650, 47)
64 (650, 47)

4WD RH Drive Shaft

◆ O-Ring

Transmission Case Protector

◆ Snap Ring

Air Cleaner Assembly

◆ Snap Ring

LH Drive Shaft

ABS Speed Sensor

8 (80, 69 In.·lbf)

18 (185, 13)

Tie Rod End

Engine Under Cover

3 Door vehicle 64 (650, 47)
5 Door vehicle 113 (1,150, 83)

49 (500, 36)

◆ Cotter Pin

Lock Cap

216 (2,200, 159)

×6

×7

Stabilizer Bar Link

127 (1,300, 94)

◆ No.2 Dust Deflector

2WD A/T RH

Center Drive Shaft

◆ Snap Ring

◆ Boot Clamp

Straight Pin

Outbord Joint Shaft

Center Bearing

◆ Dust Cover

Center Bearing Case

◆ Snap Ring

LH

Inboard Joint Tulip

◆ Boot

◆ Dust Cover

2WD A/T

◆ O-Ring

Tripod

◆ Snap Ring

Inboard Joint Tulip

◆ Dust Cover

◆ Snap Ring

N·m (kgf·cm, ft·lbf) : Specified torque

◆ Non-reusable part

7924ZG71

Front halfshaft exploded view (2WD with automatic transmission and 4WD)—RAV4

- Center bearing bracket. Tighten the 3 bolts to 47 ft. lbs. (64 Nm).
- Halfshaft together with the center bearing case and center halfshaft. Tighten both bolts to 47 ft. lbs. (64 Nm).

14. If working on a 2WD right-hand half-shaft and the vehicle is equipped with a manual transaxle, perform the following to remove the halfshaft:

- Center bearing bracket
- Halfshaft with the center halfshaft
- Center bearing bracket. Tighten the bolt to 24 ft. lbs. (32 Nm).
- Snapring to the center bearing bracket

15. Install or connect the following:

- Halfshaft to the axle hub
- Lower ball joint to the lower control arm. Tighten the nuts/bolt to 94 ft. lbs. (127 Nm).
- Sway bar link to the lower control arm. Tighten the nut to 47 ft. lbs. (64 Nm) for 3-door or to 83 ft. lbs. (113 Nm) for 5-door.
- Tie rod end to the steering knuckle. Tighten the nut to 36 ft. lbs. (49 Nm).
- New tie rod end cotter pin
- Halfshaft to the axle hub. Tighten the locknut to 159 ft. lbs. (216 Nm).
- Lock cap and cotter pin
- ABS speed sensor with the bolt, if equipped

16. Fill the transaxle with gear oil (manual transmission) or ATF (automatic transmission).

17. Install or connect the following:

- Engine undercover
- Wheels
- Negative battery cable

18. Check the ABS sensor signal.

RAV4 Rear

1. Before servicing the vehicle, refer to the precautions in the beginning of this section.

2. Remove or disconnect the following:

- Negative battery cable
- Rear wheels
- Anti-lock Brake System (ABS) speed sensor from the axle assembly by removing the bolt, if equipped
- Cotter pin, lock cap and the nut holding the halfshaft to the axle carrier

3. Place matchmarks on the halfshaft and side gear shaft.

4. Remove or disconnect the following:

- Halfshaft from the differential side gear shaft, by removing the 4 nuts and washers
- Halfshaft from the axle carrier, using a plastic hammer

To install:

5. Install or connect the following:

- Halfshaft to the axle carrier
- Halfshaft to the differential side gear shaft, by aligning the marks. Tighten the 4 nuts to 41 ft. lbs. (56 Nm).
- Nut, lock cap and the cotter pin to hold the halfshaft to the axle carrier. Tighten the nut to 152 ft. lbs. (206 Nm).
- ABS sensor. Tighten the bolt to 69 inch lbs. (8 Nm).
- Rear wheels
- Negative battery cable

6. Check the ABS sensor signal.

Highlander Front

1. Before servicing the vehicle, refer to the precautions in the beginning of this section.

2. Remove or disconnect the following:

- Front wheels
- Fender apron seal
- Transaxle fluid
- Transfer case oil (4wd)
- Hub nut
- Stabilizer bar link
- Speed sensor
- Tie rod end
- Lower arm from the ball joint

3. Slide the halfshaft from the hub, then, carefully, pry the shaft from the transaxle.

4. Installation is the reverse of removal. Torque the hub nut to 217 ft. lbs. (294Nm).

Highlander Rear

1. Before servicing the vehicle, refer to the precautions in the beginning of this section.

2. Remove or disconnect the following:

- Wheel
- Speed sensor
- Driveshaft
- Strut rod
- Control arms
- Caliper
- Rotor
- Parking brake assembly
- Strut lower bolts and nuts
- Hub and axle shaft

3. Installation is the reverse of removal. Torque the hub bolts to 59 ft. lbs. (80Nm).

RX 300 Front

1. Before servicing the vehicle, refer to the precautions in the beginning of this section.

2. Remove or disconnect the following:

- Front wheels
- Cotter pin and locknut cap

➡ **Have an assistant depress the brake pedal and loosen the bearing locknut.**

- Engine undercover
- Fender apron seal
- Tie rod end, from the steering knuckle
- Steering knuckle, from the lower control arm
- Halfshaft from the axle hub, using a plastic hammer
- Cover the outer boot with a rag
- Halfshaft from the transaxle, using the proper tools

To install:

3. Reverse the removal procedures to complete installation, tightening fasteners to specifications.

4. Fill the transaxle with gear oil, install the fender apron, check front end alignment and test drive.

➡ **If the cotter pin holes do not align, always correct by tightening the nut until the next hole aligns.**

5. Install a new cotter pin.

RX 300 Rear

1. Before servicing the vehicle, refer to the precautions in the beginning of this section.

2. Remove or disconnect the following:

- Negative battery cable
- Rear wheels
- Anti-lock Brake System (ABS) speed sensor from the axle assembly by removing the bolt, if equipped
- Cotter pin, lock cap and the nut holding the halfshaft to the axle carrier

3. Place matchmarks on the halfshaft and differential side gear shaft.

4. Remove or disconnect the following:

- 4 nuts, washers and the halfshaft from the differential
- Halfshaft from the axle carrier

To install:

5. Install or connect the following:

- Halfshaft into the axle carrier. Tighten the 4 nuts to 51 ft. lbs. (69 Nm).
- Halfshaft. Tighten the locknut to 159 ft. lbs. (216 Nm).

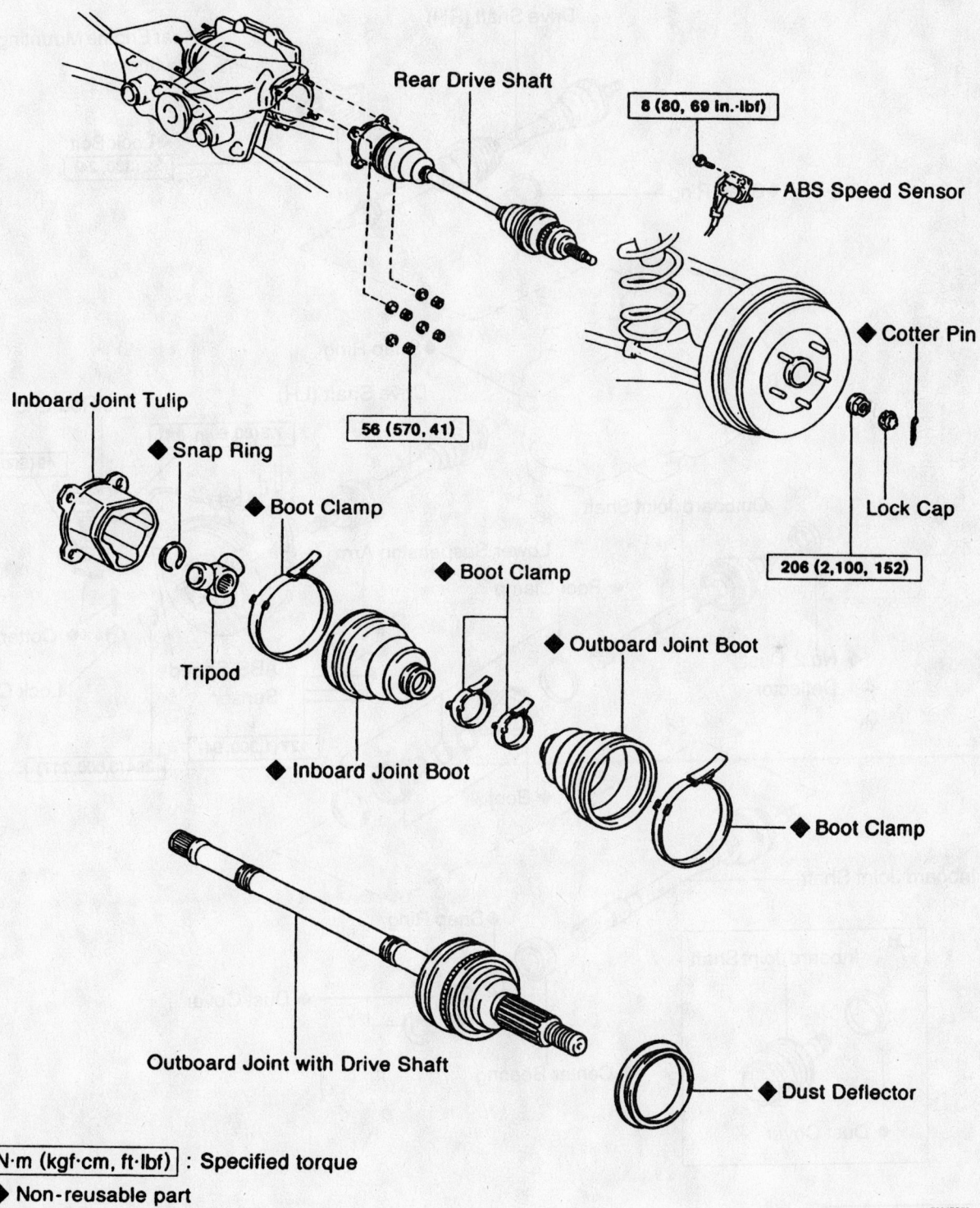

Rear Drive Shaft

8 (80, 69 in.·lbf)

◆ **ABS Speed Sensor**

◆ **Cotter Pin**

Inboard Joint Tulip

◆ **Snap Ring**

◆ **Boot Clamp**

◆ **Boot Clamp**

◆ **Outboard Joint Boot**

Lock Cap

206 (2,100, 152)

56 (570, 41)

Tripod

◆ **Inboard Joint Boot**

◆ **Boot Clamp**

Outboard Joint with Drive Shaft

◆ **Dust Deflector**

N·m (kgf·cm, ft·lbf) : Specified torque

◆ Non-reusable part

7924ZG72

Rear halfshaft removal and installation (4WD only)—RAV4

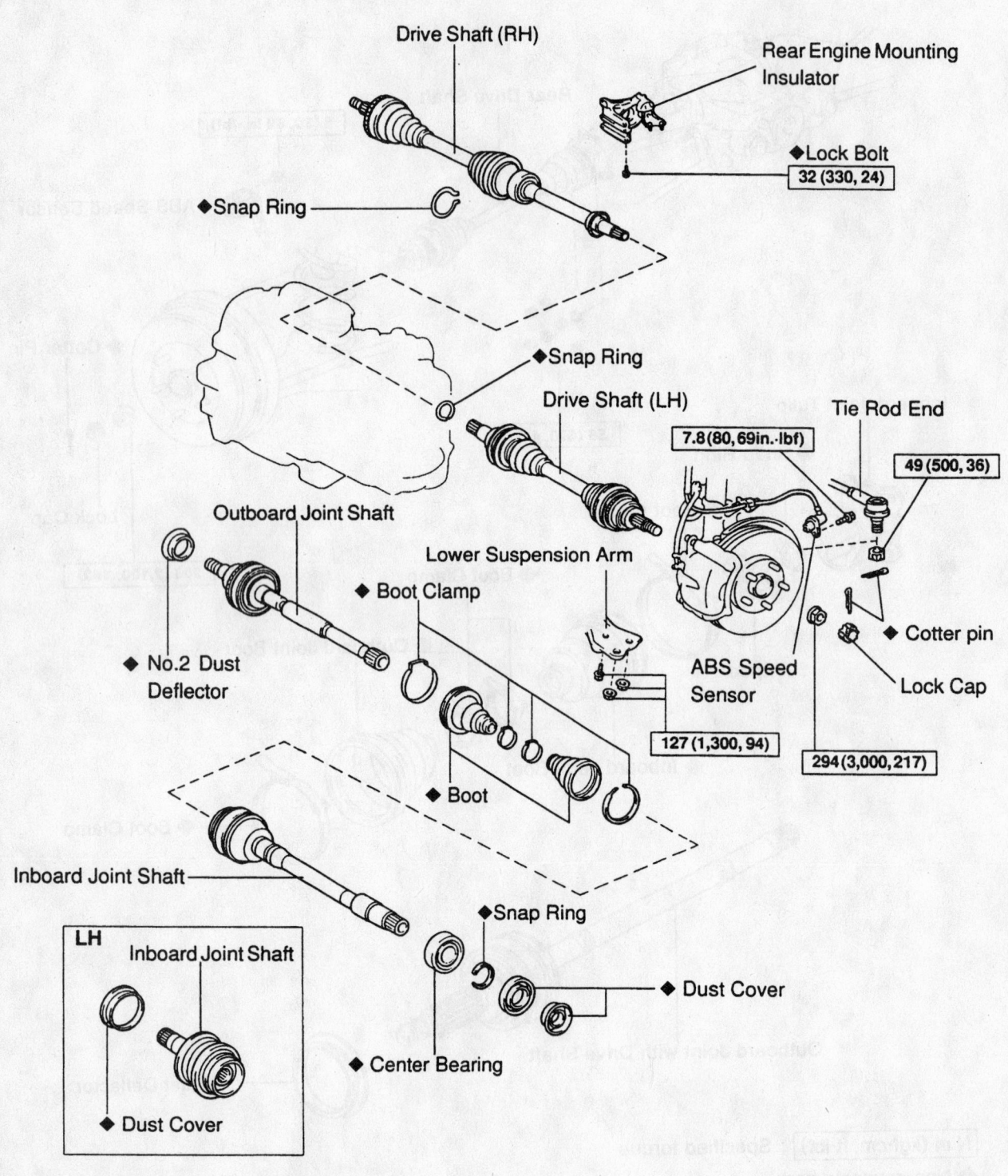

Drive Shaft (RH)

Rear Engine Mounting Insulator

◆ Snap Ring

◆ Lock Bolt
32 (330, 24)

◆ Snap Ring

Drive Shaft (LH)

Tie Rod End

7.8 (80, 69 in.·lbf)

49 (500, 36)

Outboard Joint Shaft

◆ Boot Clamp

Lower Suspension Arm

◆ No.2 Dust Deflector

◆ Boot

ABS Speed Sensor

◆ Cotter pin

Lock Cap

127 (1,300, 94)

294 (3,000, 217)

Inboard Joint Shaft

◆ Snap Ring

LH
Inboard Joint Shaft

◆ Dust Cover

◆ Center Bearing

◆ Dust Cover

N·m (kgf·cm, ft·lbf) : Specified torque

◆ Non-reusable part

7924ZG73

Exploded view of front halfshaft—RX 300

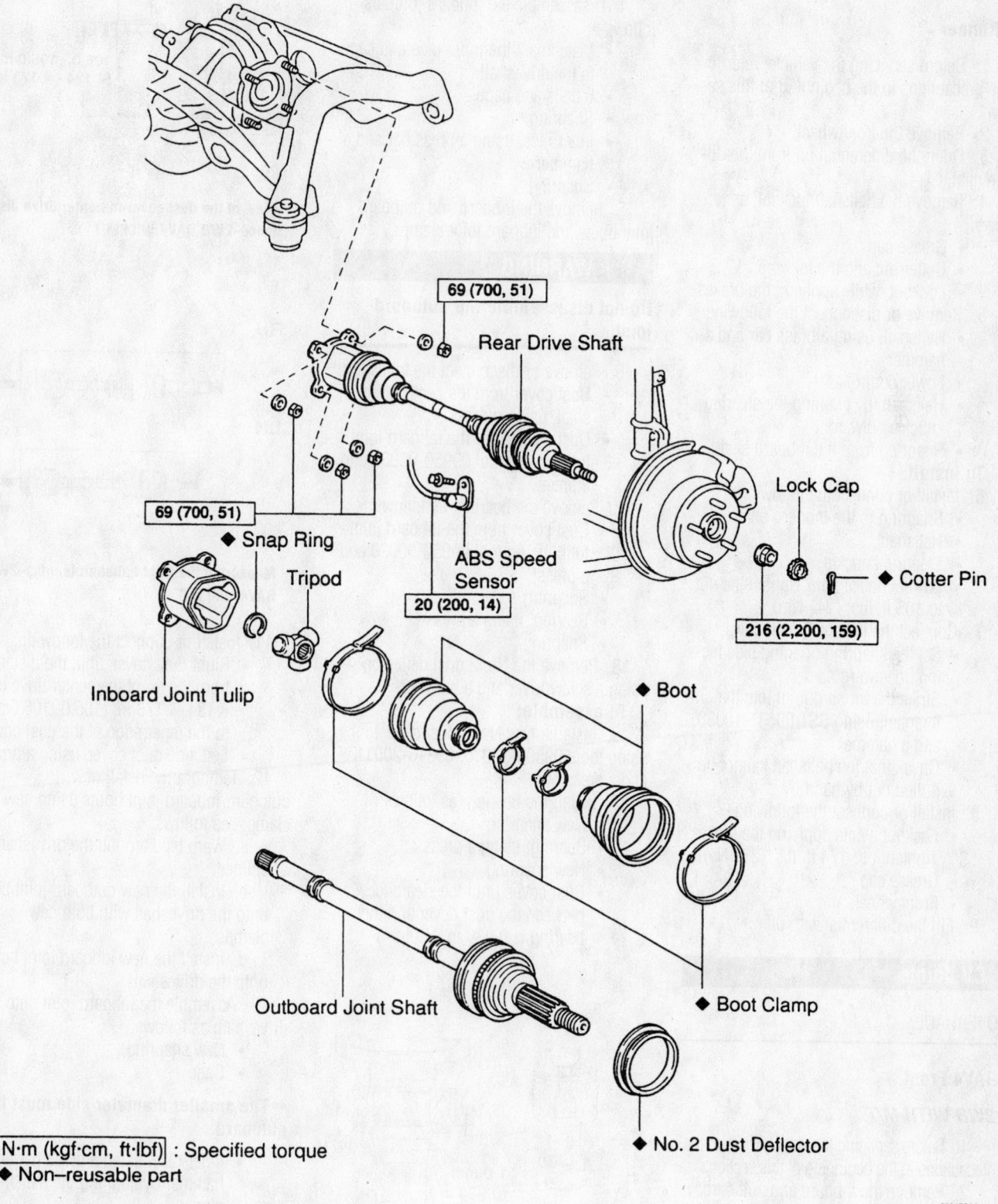

69 (700, 51)

Rear Drive Shaft

69 (700, 51)

♦ Snap Ring

Tripod

ABS Speed
Sensor

20 (200, 14)

Lock Cap

♦ Cotter Pin

216 (2,200, 159)

Inboard Joint Tulip

♦ Boot

Outboard Joint Shaft

♦ Boot Clamp

♦ No. 2 Dust Deflector

N·m (kgf·cm, ft·lbf) : Specified torque
♦ Non–reusable part

7924ZG88

Exploded view of the rear halfshaft—RX 300 model with 4WD

- ABS sensor
- Rear wheels
- Negative battery cable

4Runner

1. Before servicing the vehicle, refer to the precautions in the beginning of this section.
2. Remove the front wheel.
3. Drain the differential oil from the differential.
4. Remove the halfshaft locknut, as follows:
   - Grease cap
   - Cotter pin and the lockcap
   - Locknut, while applying the brakes
5. Remove or disconnect the following:
   - Halfshaft, using a brass bar and a hammer
   - Lower control arm
   - Halfshaft, by pushing the steering knuckle outward
   - Snapring from the inboard shaft

To install:

6. Install or connect the following:
   - Snapring to the inboard shaft
   - Halfshaft
   - Steering knuckle
   - Lower control arm. Tighten the nut to 105 ft. lbs. (142 Nm).
7. Connect the halfshaft, as follows:
   - Set the snapring opening side facing downward
   - Strike the inboard joint into the differential, using SST 09631-10030 and a hammer
   - Check that the halfshaft cannot be pulled out by hand
8. Install or connect the following:
   - Locknut, while applying the brakes. Tighten it to 174 ft. lbs. (235 Nm).
   - Grease cap
   - Front wheel
9. Fill the differential with oil.

CV-Joints

OVERHAUL

RAV4 Front

2WD WITH M/T

1. Before servicing the vehicle, refer to the precautions in the beginning of this section.
2. Remove the inboard and outboard joint boot clamps.
3. Disassemble the inboard joint tulip, as follows:
   - Snapring from the inboard joint tulip (center driveshaft)

- Inboard joint tulip (center driveshaft), by matchmarking it to the shaft
4. Disassemble the inboard joint, as follows:
   - Matchmark the inner race and cage to the driveshaft
   - 6 balls and cage
   - Snapring
   - Inner race, using a brass bar and a hammer
   - Snapring
5. Remove the inboard and outboard joint boots and inboard joint clamps.

❈❈ WARNING

Do not disassemble the outboard joint.

6. Remove or disconnect the following:
   - Dust cover from the center driveshaft, using a press
   - Dust cover from the inboard joint tulip, using tool 09950-00020 and a press
7. Remove the bearing, as follows:
   - Dust cover from the inboard joint tulip, using tool 09950-00020 and a press
   - Snapring
   - Bearing, using a press
   - Snapring
8. Remove the No. 2 dust deflector, using a screwdriver and a hammer.

To assemble:

9. Install a new No. 2 dust deflector, using tools 09309-36010, 09316-20011 and a press.
10. Install the bearing, as follows:
    - New snapring
    - Bearing, using a press
    - New snapring
    - Dust cover, until the clearance between the dust cover and the bearing is 0.039 in. (1.0mm)

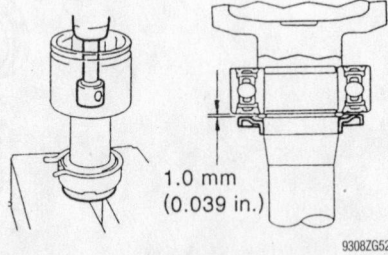

1.0 mm
(0.039 in.)

9308ZG52

View of the bearing-to-dust cover clearance—2WD RAV4 With M/T

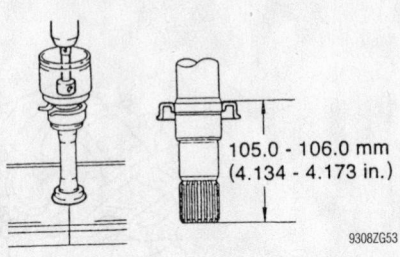

105.0 - 106.0 mm
(4.134 - 4.173 in.)

9308ZG53

View of the dust cover-to-center drive distance—2WD RAV4 With M/T

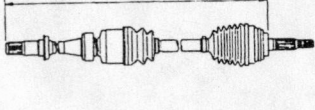

RH

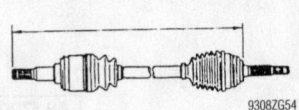

LH

9308ZG54

Measuring the front halfshaft lengths—2WD RAV4 with M/T

11. Install or connect the following:
    - Right dust cover, until the distance from the tip of the center drive is 4.134–4.173 in. (105.0–106.0mm) to the inner edge of the dust cover
    - Left side dust cover, using a press
12. Temporarily install new outboard/inboard joint boots using new clamps, as follows:
    a. Warp tape around the driveshaft splines.
    b. Install the new outboard joint boot onto the driveshaft with both new clamps.
    c. Install the new inboard joint boot onto the driveshaft.
13. Assemble the inboard joint onto the driveshaft, as follows:
    - New snapring
    - Cage

➡ **The smaller diameter side must face outboard.**

   - Inner race, using a brass bar and hammer by aligning the matchmarks

❈❈ WARNING

Be careful not to damage the inner race.

- New snapring

14. Install the outboard joint boot packed with grease from the boot kit.

15. Install the inboard joint tulip, as follows:

- Cage to the inner race by aligning the matchmarks
- 6 cage balls

➡**Lubricate the balls with grease to keep them from falling.**

- Inboard joint tulip, by aligning the matchmarks
- New snapring
- Temporarily, install the inboard joint boot packed with grease from the kit

16. Install the boot clamps to both boots, as follows:

- Both boots to the shaft grooves
- Halfshaft length should be 32.988–33.382 in. (837.9–847.9mm) for the right side or 21.165–21.559 in. (537.6–547.6mm) for the left side
- Both new clamps on the inboard joint boot
- Crimp the new clamps using tool 09521-24010
- Adjust the crimp clearance to 0.047–0.157 in. (1.2–4.0mm)

2WD WITH A/T AND 4WD; HIGHLANDER AND RX300

1. Before servicing the vehicle, refer to the precautions in the beginning of this section.

2. Remove the inboard and outboard joint boot clamps.

3. Disassemble the inboard joint tulip, as follows:

- Matchmark the tri-pot, inboard joint tulip or center driveshaft to the driveshaft

⁂ **WARNING**

Do not use punch marks.

- Inboard joint tulip from the driveshaft

4. Remove the inboard and outboard joint clamps.

5. Remove the tri-pot joint, as follows:

- Snapring
- Matchmark the tri-pot joint to the driveshaft
- Tri-pot joint, using a brass bar and hammer

⁂ **WARNING**

Do not tap the roller.

6. Remove or disconnect the following:

- Inboard and outboard joint boots

➡**Do not disassemble the outboard joint.**

- Dust cover from the center driveshaft, using a press, for 2WD on the right side
- Dust cover from the inboard joint tulip, using tool 09950-00020 and a press, for 2WD on the left side and 4WD

7. Disassemble the center driveshaft, as follows:

- Snapring
- Bearing case, using a press
- Straight pin from the bearing case, using a pin punch and hammer
- Dust cover, using tool 09950-00020 and a press
- Snapring
- Bearing, using a press

8. Remove the No. 2 dust deflector, using a screwdriver and hammer.

To assemble:

9. Install a new No. 2 dust deflector, using a press.

10. Assemble the center driveshaft, as follows:

- Straight pin into the bearing case, using a pin punch and hammer
- New bearing, using tools 09959-60010, 09950-70010 and a press
- New snapring
- Bearing with the bearing case assembly to the center driveshaft,

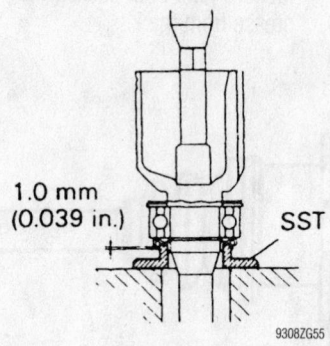

View of the bearing-to-dust cover clearance–RAV4 with 2WD A/T and 4WD

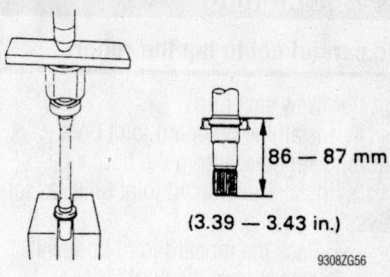

86 — 87 mm
(3.39 — 3.43 in.)
9308ZG56

View of the dust cover-to-center drive distance–RAV4 with 2WD A/T and 4WD

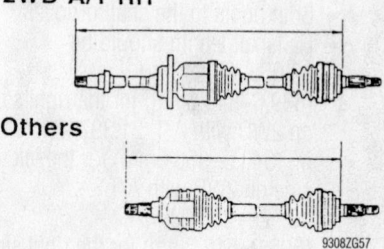

2WD A/T RH

Others

9308ZG57

Measuring the front halfshaft lengths–RAV4 with 2WD A/T and 4WD

using tool 09710-30021 and a press

- New snapring
- New dust cover, until the clearance between the dust cover and the bearing is 0.039 in. (1.0mm)

11. Install or connect the following:

- Right dust cover (2WD), until the distance from the tip of the center drive is 3.39–3.34 in. (86–87mm) to the inner edge of the dust cover
- Left side dust cover (2WD and 4WD), using a press

12. Temporarily install new outboard/inboard joint boots using new clamps, as follows:

a. Warp tape around the driveshaft splines.

b. Install the new outboard joint boot onto the driveshaft.

c. Install the new inboard joint boot onto the driveshaft.

13. Install the tri-pot joint, as follows:

- Tri-pot joint, face the beveled side toward the outboard joint and align the matchmarks
- Tri-pot joint onto the driveshaft, using a press

Timing belt service is covered in Section 3 of this manual

❊❊ WARNING

Be careful not to tap the roller.

- New snapring

14. Install the outboard joint boot packed with grease from the boot kit.

15. Install the inboard joint tulip, as follows:

- Pack the inboard joint boot with grease from the boot kit
- Inboard joint tulip, by aligning the matchmarks
- Temporarily, install the inboard joint boot packed with grease from the kit

16. Install the boot clamps to both boots, as follows:

- Both boots to the shaft grooves
- Halfshaft length should be 33.055–33.449 in. (839.6–849.6mm) for the right side on 2WD with A/T, 21.397–21.791 in. (543.5–553.5mm) for the left side on 2WD with A/T, 19.929–20.323 in. (506.2–516.2mm) for the right side on 4WD or 19.803–20.197 in. (503–511mm) for the left side on 4WD
- Both new boot clamps boot
- Bend the band and lock it using a screwdriver

RAV4 Rear

1. Before servicing the vehicle, refer to the precautions in the beginning of this section.

2. Remove the inboard and outboard joint boot clamps.

3. Disassemble the inboard joint tulip, as follows:

- Matchmark the tri-pot, inboard joint tulip or center driveshaft to the driveshaft

❊❊ WARNING

Do not use punch marks.

- Inboard joint tulip from the driveshaft

4. Remove the tri-pot joint, as follows:
- Snapring
- Matchmark the tri-pot joint to the driveshaft

❊❊ WARNING

Do not use punch marks.

- Tri-pot joint, using a brass bar and hammer

❊❊ WARNING

Do not tap the roller.

5. Remove or disconnect the following:
- Inboard and outboard joint boots

➡ **Do not disassemble the outboard joint.**

- No. 2 dust deflector from the center driveshaft, using a screwdriver and hammer

To assemble:

6. Install a new No. 2 dust deflector, using tools 09309-36010, 09316-20011 and a press.

7. Temporarily install new outboard/inboard joint boots using new clamps, as follows:

    a. Warp tape around the driveshaft splines.

    b. Install the new outboard joint boot onto the driveshaft.

    c. Install the new inboard joint boot onto the driveshaft.

8. Install the tri-pot joint, as follows:

- Tri-pot joint, face the beveled side toward the outboard joint and align the matchmarks
- Tri-pot joint onto the driveshaft, using a brass bar and hammer

❊❊ WARNING

Be careful not to tap the roller.

- New snapring

9. Install the outboard joint boot packed with grease from the boot kit.

10. Install the inboard joint tulip, as follows:

- Pack the inboard joint boot with grease from the boot kit
- Inboard joint tulip, by aligning the matchmarks
- Inboard joint boot packed with grease from the kit

11. Install the boot clamps to both boots, as follows:

- Both boots to the shaft grooves
- Halfshaft length should be 23.392–23.795 in. (594.4–604.4mm) for the right side or 21.590–21.984 in. (548.4–558.4mm) for the left side
- New boot clamps boot
- Bend the band and lock it using a screwdriver

4Runner

OUTBOARD JOINT

The outboard joint is replaced with halfshaft; no overhaul is possible or necessary.

INBOARD (TRI-POT) JOINT

1. Before servicing the vehicle, refer to the precautions in the beginning of this section.

2. Remove the halfshaft from the vehicle.

3. Remove the large clamp from the inboard joint.

4. Remove the small clamp, using side cutters, from the inboard joint.

5. Slide the inboard joint boot toward the outboard joint.

6. Matchmark the inboard joint to the halfshaft.

7. Remove the inboard joint housing from the halfshaft.

8. Remove the snapring from the end of the halfshaft.

9. Matchmark the halfshaft to the tri-pot joint.

10. Remove the tri-pot from the halfshaft, using a brass bar and a hammer.

❊❊ WARNING

Do not tap on the tri-pot joint.

11. Remove the inboard and outboard boots from the halfshaft.

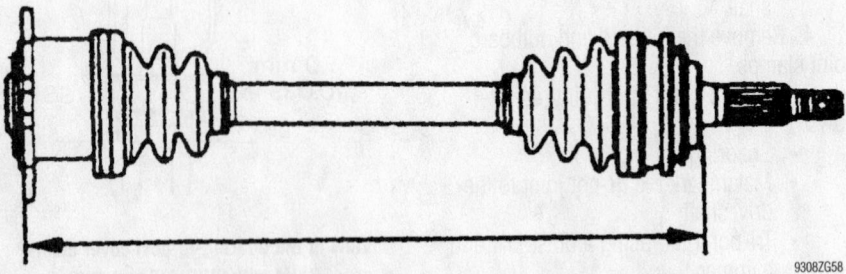

Measuring the rear halfshaft lengths–RAV4

9308ZG58

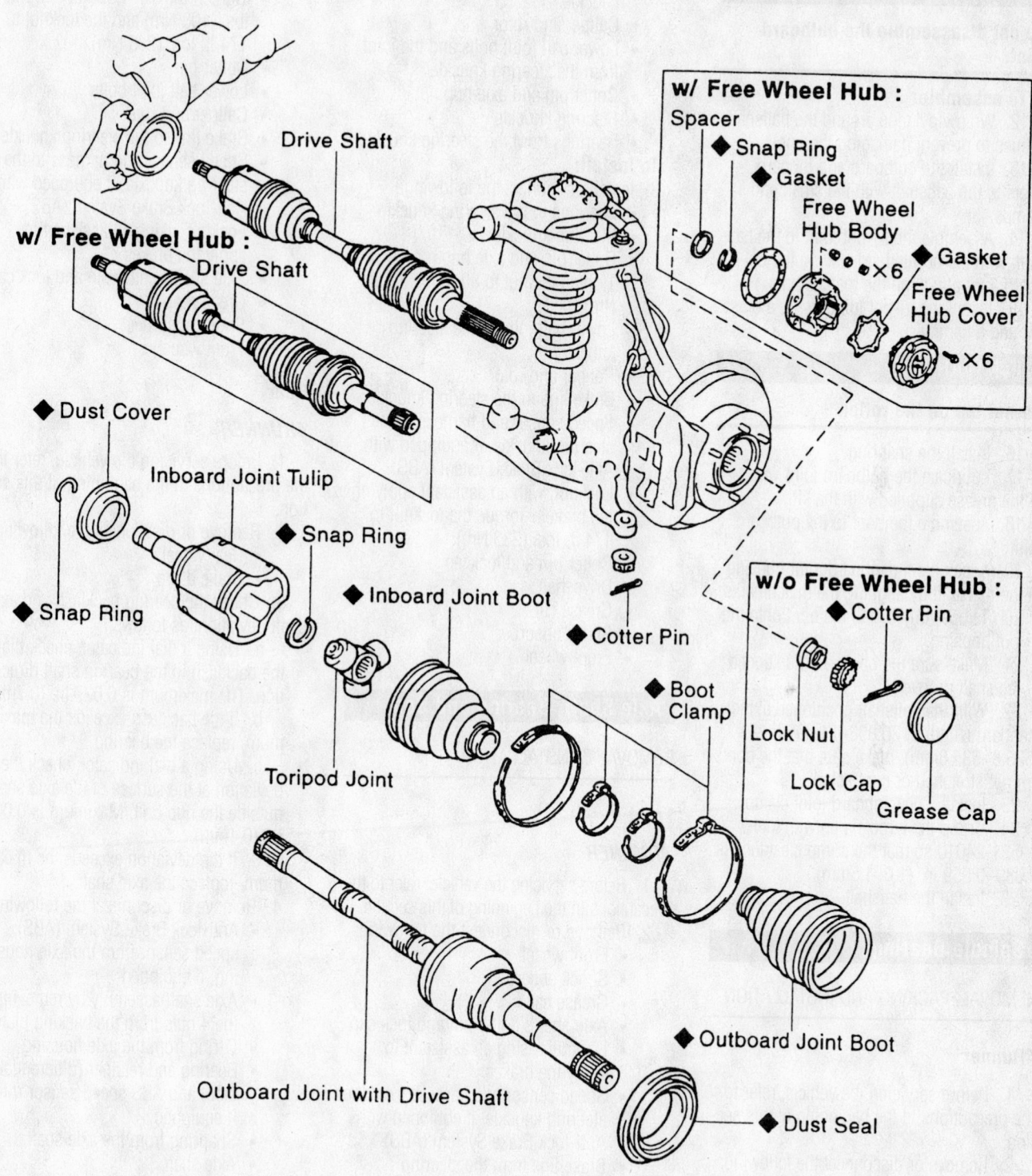

w/ Free Wheel Hub :
- Spacer
- ◆ Snap Ring
- ◆ Gasket
- Free Wheel Hub Body
- ◆ Gasket
- Free Wheel Hub Cover
- ×6
- ×6

Drive Shaft

w/ Free Wheel Hub :
Drive Shaft

◆ Dust Cover

Inboard Joint Tulip

◆ Snap Ring

◆ Snap Ring

◆ Inboard Joint Boot

Toripod Joint

◆ Cotter Pin

◆ Boot Clamp

w/o Free Wheel Hub :
- ◆ Cotter Pin
- Lock Nut
- Lock Cap
- Grease Cap

◆ Outboard Joint Boot

Outboard Joint with Drive Shaft

◆ Dust Seal

◆ Non-reusable part

9308YG07

Exploded view of the halfshaft assembly—4Runner

Heater Core replacement is covered in Section 2 of this manual

✲✲ WARNING

Do not disassemble the outboard joint.

To assemble:

12. Wrap vinyl tape around the halfshaft splines to prevent damaging the boots.

13. Install the outboard and inboard boots to the halfshaft with the small end clamps.

14. Assemble the tri-pot joint to the half-shaft with the beveled side facing the outboard joint and align the matchmarks.

15. Install the tri-pot joint, using a brass bar and a hammer.

✲✲ WARNING

Do not tap on the roller.

16. Install the snapring.

17. Lubricate the outboard joint with ½ of the grease supplied with the kit.

18. Assemble the boot to the outboard joint

19. Assemble the inboard joint housing to the halfshaft by aligning the matchmarks.

20. Temporarily install the boot onto the tri-pot housing.

21. Make sure the boots are positioned in the shaft grooves.

22. With the halfshaft positioned at the standard length of 20.898–21.095 in. (525.8–535.8mm), make sure that the boots are not stretched or contracted.

23. Install a new inboard joint clamp.

24. Crimp the large clamp with tool 09521-24010 so that the crimp clearance is 0.039–0.059 in. (1.0–1.5mm).

25. Install the halfshaft.

Spindle Bearings

REMOVAL, PACKING AND INSTALLATION

4Runner

1. Before servicing the vehicle, refer to the precautions in the beginning of this section.

2. Remove or disconnect the following:
- Front wheel
- Shock absorber
- Grease cap
- Driveshaft
- Cotter pin and lockcap
- Locknut, with an assistant applying the brakes
- Speed sensor and harness from the steering knuckle, if equipped with Anti-lock Brake System (ABS)

- Brake line from the steering knuckle
- Caliper and rotor
- Lower ball joint bolts and the joint from the steering knuckle
- Cotter pin and axle hub nut
- Steering knuckle
- Bearings from the steering knuckle

To install:

3. Install or connect the following:
- Bearings to the steering knuckle
- Steering knuckle
- Cotter pin and axle hub nut. Tighten the nut to 80 ft. lbs. (108 Nm).
- Lower ball joint to the steering knuckle
- Caliper and rotor
- Brake line to the steering knuckle
- Speed sensor and harness to the steering knuckle, if equipped with Anti-lock Brake System (ABS)
- Locknut, with an assistant applying the brakes. Torque the locknut to 174 ft. lbs. (235 Nm).
- Cotter pin and lockcap
- Driveshaft
- Grease cap
- Shock absorber
- Front wheel

Axle Shaft, Bearing and Seal

REMOVAL & INSTALLATION

Front

4RUNNER

1. Before servicing the vehicle, refer to the precautions in the beginning of this section.

2. Remove or disconnect the following:
- Front wheel
- Shock absorber
- Grease cap
- Axle shaft's cotter pin and lock cap
- Locknut, using an assistant to apply the brakes
- Speed sensor and harness from the steering knuckle, if equipped with Anti-lock Brake System (ABS)
- Brake line from the steering knuckle
- Caliper and rotor
- Lower ball joint bolts
- Cotter pin and loosen the axle hub nut
- Steering knuckle
- Axle shaft

To install:

3. Install or connect the following:
- Axle shaft

- Steering knuckle
- Tighten the axle hub nut to 80 ft. lbs. (108 Nm) and the locknut to 174 ft. lbs. (235 Nm).
- Cotter pin
- Lower ball joint bolts
- Caliper and rotor
- Brake line to the steering knuckle
- Speed sensor and harness to the steering knuckle, if equipped with Anti-lock Brake System (ABS)
- Locknut, using an assistant to apply the brakes
- Axle shaft's cotter pin and lock cap
- Grease cap
- Shock absorber
- Front wheel

Rear

4RUNNER

1. Before servicing the vehicle, refer to the precautions in the beginning of this section.

2. Remove or disconnect the following:
- Rear wheel
- Brake drum

3. Check the bearing backlash and axle shaft deviation, as follows:

a. Using a dial indicator, check that the backlash in the bearing shaft direction. The maximum is 0.027 in. (0.7mm).

b. If the backlash exceeds the maximum, replace the bearing.

c. Using a dial indicator, check the deviation at the surface of the axle shaft outside the hub bolt. Maximum is 0.0039 in. (0.1mm).

d. If the deviation exceeds the maximum, replace the axle shaft.

4. Remove or disconnect the following:
- Anti-lock Brake System (ABS) speed sensor from the axle housing, if equipped
- Axle shaft assembly by removing the 4 nuts from the backing plate
- O-ring from the axle housing
- Bearing and retainer (differential side) and ABS speed sensor rotor, if equipped
- Snapring from the axle shaft
- Axle shaft

➡**Inspect the axle shaft and flange run-outs. The axle shaft run-out should be 0.079 in. (2.0mm) and the flange run-out should be 0.004 in. (0.1mm).**

- Outer seal
- Bearing from the axle shaft

To install:

5. Install or connect the following:
- Bearing from the axle shaft

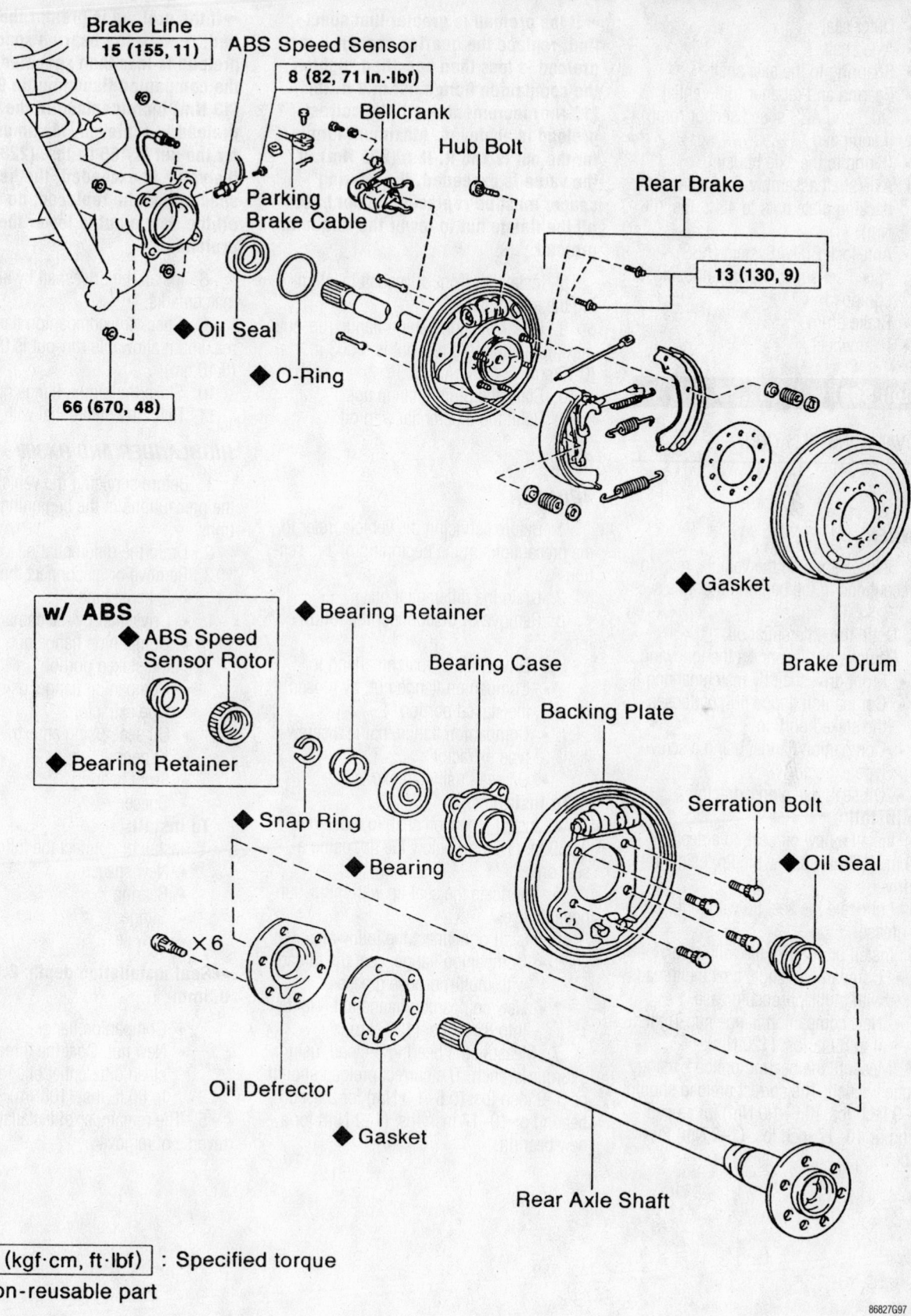

Brake Line
15 (155, 11)
ABS Speed Sensor
8 (82, 71 In.·lbf)
Bellcrank
Hub Bolt
Rear Brake
13 (130, 9)
Parking Brake Cable
◆ Oil Seal
◆ O-Ring
66 (670, 48)
◆ Gasket
Brake Drum

w/ ABS
◆ ABS Speed Sensor Rotor
◆ Bearing Retainer

◆ Bearing Retainer
Bearing Case
Backing Plate
Serration Bolt
◆ Snap Ring
◆ Bearing
◆ Oil Seal
×6
Oil Deflector
◆ Gasket
Rear Axle Shaft

N·m (kgf·cm, ft·lbf) : Specified torque
◆ Non-reusable part

86827G97

Exploded view of the rear axle shaft and components—4Runner

Brake service is covered in Section 4 of this manual

- Outer seal
- Axle shaft
- Snapring to the axle shaft
- Bearing and retainer (differential side) and ABS speed sensor rotor, if equipped
- O-ring to the axle housing
- Axle shaft assembly. Tighten the 4 backing plate nuts to 48 ft. lbs. (66 Nm).
- Anti-lock Brake System (ABS) speed sensor to the axle housing, if equipped
- Brake drum
- Rear wheel

Pinion Seal

REMOVAL & INSTALLATION

Front

4RUNNER

1. Before servicing the vehicle, refer to the precautions in the beginning of this section.
2. Drain the differential oil.
3. Remove or disconnect the following:
- Front driveshaft by matchmarking it
- Companion flange nut, by loosen the staked portion
- Companion flange, using a screw-type extractor
- Oil seal, using an extractor

To install:

4. Install a new oil seal, to a depth of 0.059 in. (1.5mm) below the lip, using a seal driver.
5. Lubricate the seal lip with multi-purpose grease.
6. Install or connect the following:
- Companion flange, coat the threads with multi-purpose grease
- New companion flange nut. Tighten it to 89 ft. lbs. (120 Nm).
7. Measure the bearing preload, using a torque wrench. The correct preload should be 5–9 inch lbs. (0.6–1.0 Nm) for a used bearing or 10–17 inch lbs. (1–2 Nm) for a new bearing.

➡If the preload is greater that specified, replace the bearing spacer. If the preload is less than specified, tighten the companion flange nut in 9 ft. lbs. (13 Nm) increments until the correct preload is achieved. Maximum torque for the nut is 165 ft. lbs. (223 Nm). If the value is exceeded, the bearing spacer must be replaced; do not back off the flange nut to lower the torque or preload.

8. Install the front driveshaft by aligning the matchmarks.
9. Check the companion flange run-out; maximum allowable run-out is 0.003 in. (0.10mm).
10. Stake the pinion flange nut.
11. Refill the differential with oil.

Rear

4RUNNER

1. Before servicing the vehicle, refer to the precautions in the beginning of this section.
2. Drain the differential oil.
3. Remove or disconnect the following:

- Driveshaft by matchmarking it
- Companion flange nut, by loosen the staked portion
- Companion flange, using a screw-type extractor
- Oil seal, using an extractor

To install:

4. Install a new oil seal, to a depth of 0.059 in. (1.5mm) below the lip, using a seal driver.
5. Lubricate the seal lip with multi-purpose grease.
6. Install or connect the following:
- Companion flange, coat the threads with multi-purpose grease
- New companion flange nut. Tighten it to 89 ft. lbs. (120 Nm).
7. Measure the bearing preload, using a torque wrench. The correct preload should be 5–9 inch lbs. (0.6–1.0 Nm) for a used bearing or 10–17 inch lbs. (1–2 Nm) for a new bearing.

➡If the preload is greater that specified, replace the bearing spacer. If the preload is less than specified, tighten the companion flange nut in 9 ft. lbs. (13 Nm) increments until the correct preload is achieved. Maximum torque for the nut is 165 ft. lbs. (223 Nm). If the value is exceeded, the bearing spacer must be replaced; do not back off the flange nut to lower the torque or preload.

8. Install the driveshaft by aligning the matchmarks.
9. Check the companion flange run-out; maximum allowable run-out is 0.003 in. (0.10mm).
10. Stake the pinion flange nut.
11. Refill the differential with oil.

HIGHLANDER AND RX300

1. Before servicing the vehicle, refer to the precautions in the beginning of this section.
2. Drain the differential oil.
3. Remove or disconnect the following:
- Exhaust pipe
- Driveshaft by matchmarking it
- Companion flange nut, by loosen the staked portion
- Companion flange, using a screw-type extractor
- Oil seal, using an extractor
- Slinger
- Front bearing
- Spacer

To install:

4. Install or connect the following
- New spacer
- Bearing
- Slinger
- New seal

➡**Seal installation depth: 2.0mm +/- 0.3mm**

- Companion flange
- New nut. Coat the threads with clean differential oil. Torque the nut to 80 ft. lbs. (108Nm).

5. The remainder of installation is the reverse of removal.

STEERING AND SUSPENSION

Air Bag

❊❊ CAUTION

Some vehicles are equipped with an air bag system. The system must be disarmed before performing service on, or around, system components, the steering column, instrument panel components, wiring and sensors. Failure to follow the safety precautions and the disarming procedure could result in accidental air bag deployment, possible injury and unnecessary system repairs.

PRECAUTIONS

Several precautions must be observed when handling the inflator module to avoid accidental deployment and possible personal injury.

• Never carry the inflator module by the wires or connector on the underside of the module.

• When carrying a live inflator module, hold securely with both hands and ensure that the bag and trim cover are pointed away.

• Place the inflator module on a bench or other surface with the bag and trim cover facing up.

• With the inflator module on the bench, never place anything on or close to the module which may be thrown in the event of an accidental deployment.

DISARMING

To avoid personal injury when working on vehicles equipped with an air bag, the negative battery cable must be disconnected and at least 90 seconds must elapse before working on the system. Failure to do so may result in deployment of the air bag.

Power Rack And Pinion Steering Gear

REMOVAL & INSTALLATION

RAV4

1. Before servicing the vehicle, refer to the precautions in the beginning of this section.

2. Disconnect the negative battery cable.

❊❊ CAUTION

To avoid personal injury when working on air bag equipped vehicles, work must be started after 90 seconds or longer from the time the ignition switch is turned to the LOCK position and the negative battery terminal is disconnected. If the air bag system is disconnected with the ignition switch at the ON or ACC, diagnostic codes will be set. When removing the air bag, take care not to pull the air bag wiring harness. When carrying the wheel pad, carry it with the upper surface facing away. When storing it, keep the upper surface of the pad facing upward.

3. Turn the key to the **LOCK** position and lock the steering wheel in place.

4. Place a drain pan under the steering rack.

5. Remove or disconnect the following:
   • Front wheels
   • Right and left-hand engine undercovers
   • Right and left-hand tie rod ends from the steering knuckle
   • Front exhaust pipe
   • Sway bar with the links

6. Disconnect the No. 2 intermediate shaft from the rack and pinion, as follows:
   a. Loosen the top bolt.
   b. Remove the lower bolt holding the No. 2 intermediate shaft to the rack and pinion.
   c. Shift the No. 2 intermediate shaft and place matchmarks on the control valve shaft and the No. 2 intermediate shaft.
   d. Disconnect the No. 2 shaft from the rack and pinion.

7. Install or connect the following:
   • Pressure feed and return tubes from the rack and pinion, using a line wrench
   • Pressure feed and return tube clamps, by removing the bolt
   • Right and left lower control arms, from the steering knuckle

8. Remove the front suspension crossmember assembly, as follows:
   • Both centermember set nuts, holding the centermember to the middle of the crossmember.

• Both rack and pinion assembly set bolts and nuts from the crossmember.

• Securely suspend the steering gear assembly.

• Support the suspension crossmember with a jack.

• Both bolts from the suspension crossmember

• Suspension crossmember with the lower suspension arms

9. Remove the rack and pinion.

To install:

10. Install or connect the following:
   • Rack and pinion

11. Install the crossmember to the vehicle, as follows:
   • Suspension crossmember with the lower control arms. Tighten both bolts to 152 ft. lbs. (206 Nm).
   • Rack and pinion. Tighten the nuts/bolts to 83 ft. lbs. (113 Nm).
   • Centermember to the crossmember. Tighten both set nuts to 82 ft. lbs. (112 Nm).

12. Install or connect the following:
   • Right and left lower control arms
   • Pressure feed and return tubes clamps
   • Pressure feed and return tubes to the rack and pinion. Tighten the tubes to 26 ft. lbs. (36 Nm), using a torque wrench with a fulcrum length of 11.81 inches (300mm).
   • Steering column No. 2 intermediate shaft to the rack and pinion. Align the marks and tighten the upper and lower pinch bolts to 26 ft. lbs. (35 Nm).
   • Stabilizer bar links. Tighten the nuts to 22 ft. lbs. (29 Nm).
   • Front exhaust pipe with new gaskets. Tighten the bolts to 35 ft. lbs. (48 Nm) and the nuts to 46 ft. lbs. (62 Nm).
   • Right and left-hand tie rod ends to the steering knuckle. Tighten the nuts to 36 ft. lbs. (49 Nm) and install new cotter pins.
   • Right and left-hand engine undercovers
   • Front wheels

13. Fill the power steering unit and bleed the system. Check for leaks.

14. Check and/or adjust the front wheel alignment.

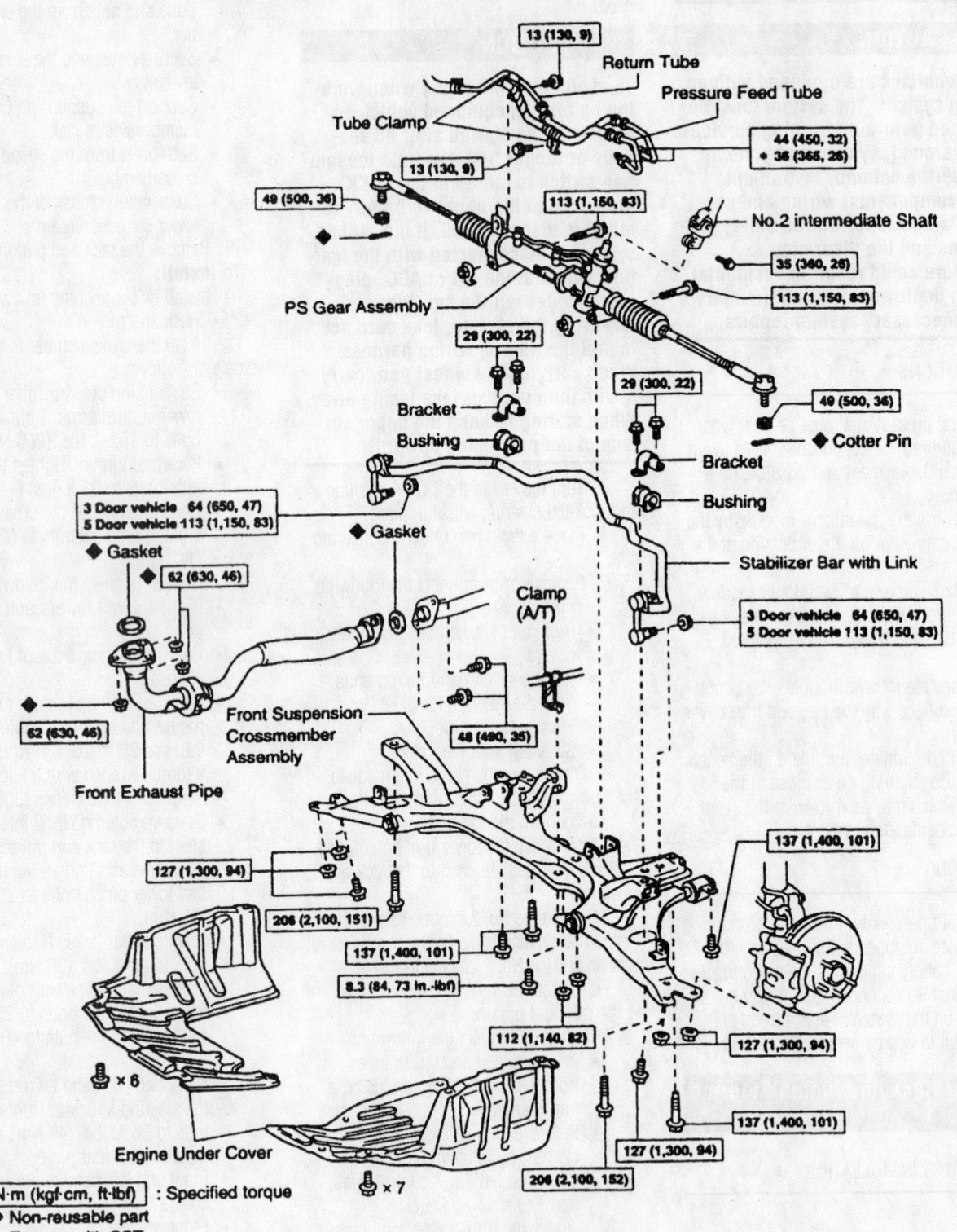

13 (130, 9)

Return Tube

Pressure Feed Tube

Tube Clamp

44 (450, 32)
* 36 (365, 26)

13 (130, 9)

49 (500, 36)

113 (1,150, 83)

No.2 intermediate Shaft

35 (360, 26)

113 (1,150, 83)

PS Gear Assembly

29 (300, 22)

29 (300, 22)

49 (500, 36)

Bracket

Bushing

◆ Cotter Pin

Bracket

Bushing

3 Door vehicle 64 (650, 47)
5 Door vehicle 113 (1,150, 83)

◆ Gasket

◆ Gasket

Stabilizer Bar with Link

◆ Gasket

Clamp (A/T)

62 (630, 46)

3 Door vehicle 64 (650, 47)
5 Door vehicle 113 (1,150, 83)

Front Suspension
Crossmember
Assembly

48 (490, 35)

62 (630, 46)

Front Exhaust Pipe

137 (1,400, 101)

127 (1,300, 94)

206 (2,100, 151)

137 (1,400, 101)

8.3 (84, 73 in.-lbf)

112 (1,140, 82)

127 (1,300, 94)

137 (1,400, 101)

× 6

127 (1,300, 94)

Engine Under Cover

× 7

206 (2,100, 152)

N·m (kgf·cm, ft·lbf) : Specified torque
◆ Non-reusable part
* For use with SST

7924ZG75

Rack and pinion exploded view—RAV4

Highlander

1. Before servicing the vehicle, refer to the precautions in the beginning of this section.
2. Remove or disconnect the following:
   - Negative battery cable

➡ **Wait at least 90 seconds before working on the vehicle to allow the Supplemental Restraint System (SRS) system to disarm.**

   - Steering wheel
   - Front wheels
   - Tie rod ends
   - Intermediate shaft

➡ **Matchmark the shaft and gear.**

   - Stabilizer bar end links
   - Pressure and return lines

   - Steering gear
   - Installation is the reverse of removal. Observe the following torques:
   - Rack mounting bolts: 52 ft. lbs. (70Nm)
   - Stabilizer bar end links: 55 ft. lbs. (74Nm)
   - Intermediate shaft bolt: 26 ft. lbs. (35Nm)
   - Tie rod end nuts: 36 ft. lbs. (49Nm)

RX 300

1. Before servicing the vehicle, refer to the precautions in the beginning of this section.
2. Remove or disconnect the following:
   - Negative battery cable

➡ **Wait at least 90 seconds before working on the vehicle to allow the Supplemental Restraint System (SRS) system to disarm.**

   - Right and left side fender apron seals
   - Right and left tie rod ends
3. Place matchmarks on the intermediate shaft.
4. Remove or disconnect the following:

   - Pinch bolt and the intermediate shaft out from under the vehicle
   - Power steering line clamp
   - Pressure and feed lines
   - Stabilizer bar, unbolt it but do not remove it
   - Heated Oxygen (HO$_2$) sensor

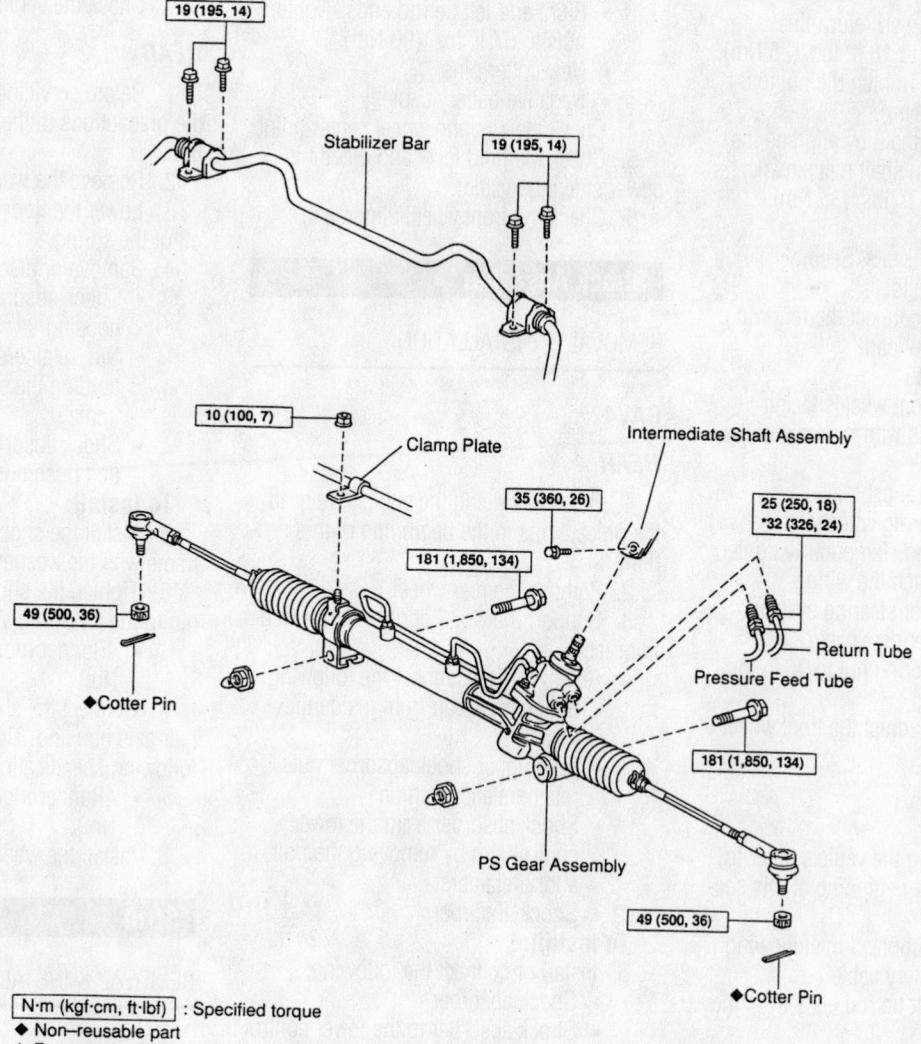

19 (195, 14)   Stabilizer Bar   19 (195, 14)

10 (100, 7)   Clamp Plate   Intermediate Shaft Assembly

35 (360, 26)   25 (250, 18) *32 (326, 24)

181 (1,850, 134)   Return Tube

49 (500, 36)   Pressure Feed Tube

◆Cotter Pin

181 (1,850, 134)

PS Gear Assembly

49 (500, 36)

◆Cotter Pin

N·m (kgf·cm, ft·lbf) : Specified torque
◆ Non–reusable part
* For use with SST

7924ZG76

Exploded view of the power steering gear and related components—RX 300 models

For Accessory Drive Belt illustrations, see Section 1 of this manual

- Both gear assembly set bolts and nuts, by lifting the stabilizer bar
- Gear assembly from the left side of the vehicle

To install:

5. Install or connect the following:
- Gear assembly to the left side of the vehicle

☀☀ WARNING

Be careful not to damage the power steering lines.

- Tighten the gear assembly set bolts and nuts to 134 ft. lbs. (181 Nm), by lifting the stabilizer bar
- HO$_2$sensor
- Stabilizer bar. Tighten the bolt to 14 ft. lbs. (19 Nm) and the nut to 29 ft. lbs. (39 Nm).
- Pressure and feed return lines. Tighten them to 18 ft. lbs. (25 Nm).
- Line clamps. Tighten the nut to 84 inch lbs. (10 Nm).
- Intermediate shaft, by aligning the joint and main shaft matchmarks. Tighten to 26 ft. lbs. (35 Nm).
- Tie rod ends
- Fender apron seals. Securely tighten the bolts.

6. Remove or disconnect the following:
- Steering wheel pad
- Steering wheel

7. Position the front wheels facing straight-ahead. Do this with the front of the vehicle on jackstands.

8. Center the spiral cable.

9. Install the steering wheel at the straight-ahead position. Temporarily tighten the wheel set nut. Attach the wiring.

10. Bleed the power steering system.

11. Check the steering wheel center point. Tighten the steering nut to 26 ft. lbs. (35 Nm).

12. Check and/or adjust the front wheel alignment.

4Runner

1. Before servicing the vehicle, refer to the precautions in the beginning of this section.

2. Remove or disconnect the following:
- Negative battery cable
- Right and left tie rod ends from the knuckle
- Intermediate shaft from the steering rack, by matchmarking it
- Pressure feed and the return tubes, using SST 09631-22020
- Mount bracket and the grommet,

from the power steering rack assembly
- Power steering rack and pinion

To install:

3. Install or connect the following:
- Power steering rack and pinion. Tighten the mounting bolts to 65 ft. lbs. (88 Nm).
- Grommet and mount bracket to the gear assembly. Tighten the bolts to 65 ft. lbs. (88 Nm).
- New O-ring
- Pressure feed and return tubes. Tighten the line fittings to 14 ft. lbs. (19 Nm).
- Intermediate shaft to the steering rack, by aligning the matchmarks

➡ **If installing a new rack assembly, be sure the steering wheel and the rack are centered.**

- Right and left tie rod ends. Tighten nuts to 67 ft. lbs. (90 Nm).
- New cotter pins
- Negative battery cable

4. Check the steering wheel center point.

5. Check the fluid level and bleed the power steering system.

6. Check the front wheel alignment.

Shock Absorber

REMOVAL & INSTALLATION

RAV4

REAR

1. Before servicing the vehicle, refer to the precautions in the beginning of this section.

2. Remove the rear wheel.

3. Support the No. 1 control arm with a floor jack.

4. Remove or disconnect the following:
- Suspension cap from inside the vehicle
- Both upper shock absorber nuts, retainers and cushion
- Shock absorber from the lower control arm by removing the bolt and 2 retainers
- Shock absorber

To install:

5. Install or connect the following:
- Shock absorber
- Shock absorber to the lower control arm retainers. Tighten the bolt to 27 ft. lbs. (37 Nm).
- Shock absorber to the chassis cushion and retainers. Tighten both nuts to 18 ft. lbs. (25 Nm).

- Suspension cap
- Wheel

4Runner

FRONT

1. Before servicing the vehicle, refer to the precautions in the beginning of this section.

2. Remove or disconnect the following:
- Front wheel
- Shock from the lower control arm
- 3 upper nuts
- Shock absorber

To install:

3. Install or connect the following:
- Shock absorber. Tighten the 3 nuts to 47 ft. lbs. (64 Nm).
- Lower shock-to-lower control arm. Tighten the bolt to 101 ft. lbs. (135 Nm).
- Wheels

4. Check the vehicle alignment.

REAR

1. Before servicing the vehicle, refer to the precautions in the beginning of this section.

2. Remove the wheel.

3. Lower the floor jack to take tension off of the spring.

4. Remove or disconnect the following:
- Shock absorber from the rear axle housing
- Nut, retainers and the cushions holding the shock absorber to the frame
- Shock absorber with the washers and bushings

To install:

5. Install the shock absorber to the frame with the washers and bushings.

6. Tighten the shock absorber-to-frame nut to the following values:
- 4Runner models: 14 ft. lbs. (20 Nm)

7. Connect the shock absorber to the rear axle housing. Tighten the bolt to the following specifications:
- 4Runner models: 47 ft. lbs. (64 Nm)

8. Install the wheels.

Strut

REMOVAL & INSTALLATION

RAV4

RAV4 is equipped with front strut and rear shock absorber type suspension arrangement.

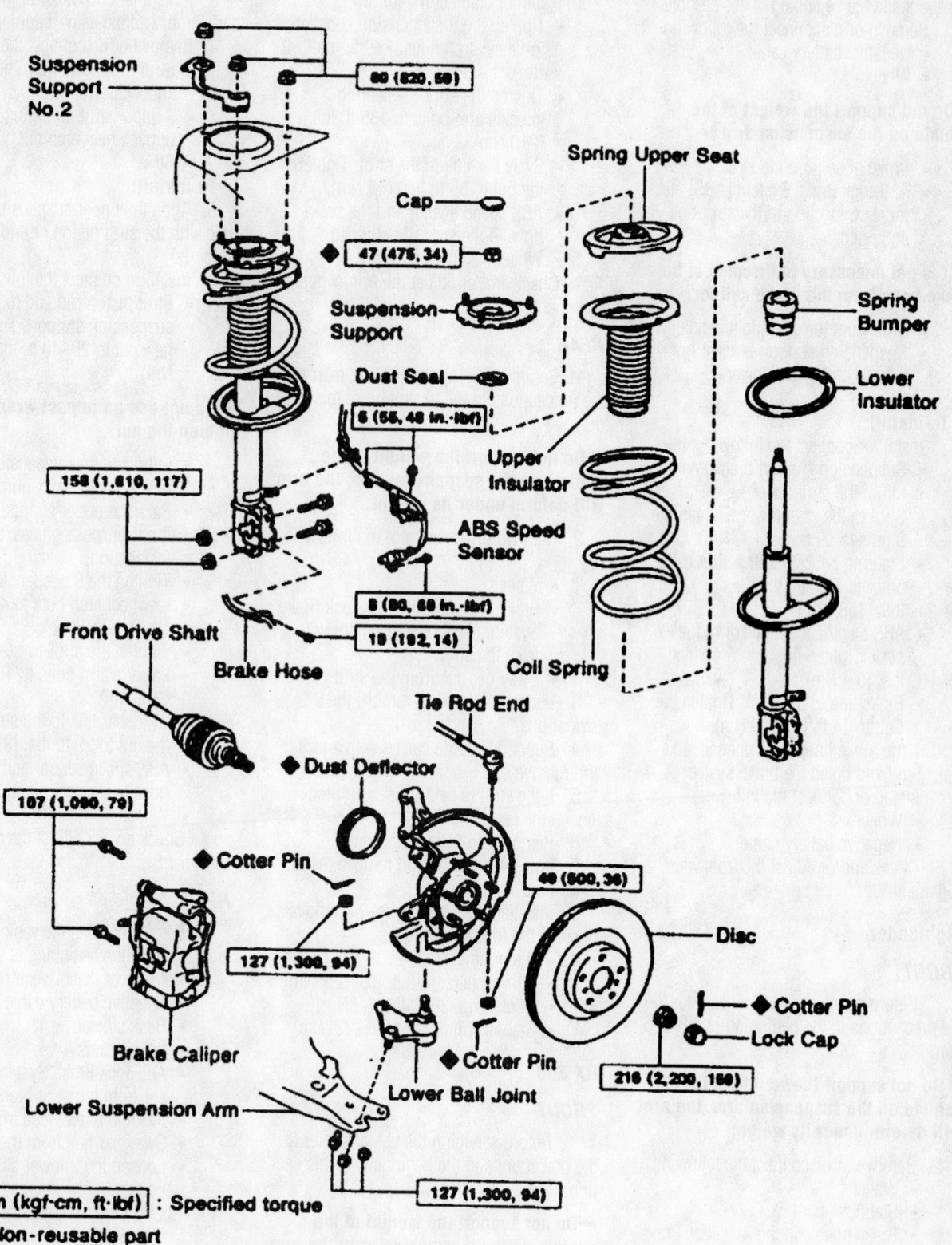

Suspension Support No.2

80 (820, 59)

Spring Upper Seat

Cap

47 (475, 34)

Suspension Support

Spring Bumper

Dust Seal

Lower Insulator

5 (55, 48 in.·lbf)

Upper Insulator

158 (1,610, 117)

ABS Speed Sensor

Front Drive Shaft

5 (50, 68 in.·lbf)

19 (192, 14)

Brake Hose

Coil Spring

Tie Rod End

Dust Deflector

107 (1,090, 79)

◆ Cotter Pin

49 (500, 36)

Disc

127 (1,300, 94)

◆ Cotter Pin

◆ Lock Cap

Brake Caliper

216 (2,200, 159)

◆ Cotter Pin

Lower Ball Joint

Lower Suspension Arm

127 (1,300, 94)

N·m (kgf·cm, ft·lbf) : Specified torque

◆ Non-reusable part

7924ZG78

Strut assembly exploded view—RAV4

For Tire, Wheel and Ball Joint specifications, see Section 1 of this manual

1. Before servicing the vehicle, refer to the precautions in the beginning of this section.
2. Remove or disconnect the following:
   - Negative battery cable
   - Wheel

➡ **Do not support the weight of the vehicle on the suspension arm.**

   - Brake hose from the strut
   - Anti-lock Brake System (ABS) electrical connection to the strut bolt, if equipped

➡ **It is not necessary to disconnect the brake hose from the brake caliper.**

   - Strut from the steering knuckle
   - Suspension support bracket from the top of the strut tower
   - Strut

To install:
3. Install or connect the following:
   - Suspension support bracket to the top of the strut tower
   - Strut to the strut tower. Tighten the 3 nuts to 59 ft. lbs. (80 Nm).
   - Steering knuckle to the strut lower bracket. Tighten the nuts to 117 ft. lbs. (158 Nm).
   - ABS electrical connector to the strut. Tighten the bolt to 48 inch lbs. (5.4 Nm).
   - Brake line to the strut. Tighten the bolt to 14 ft. lbs. (19 Nm).
4. If the brake lines were opened, add brake fluid and bleed the brake system.
5. Install or connect the following:
   - Wheel
   - Negative battery cable
6. Check and/or adjust the front wheel alignment.

Highlander

FRONT

1. Before servicing the vehicle, refer to the precautions in the beginning of this section.

➡ **Do not support the weight of the vehicle on the suspension arm; the arm will deform under its weight.**

2. Remove or disconnect the following:
   - Wheel
   - Stabilizer bar link
   - Brake hose and the Anti-lock Brake System (ABS) speed sensor wire from the strut
   - Strut lower end from the steering knuckle lower arm
   - 3 upper strut mounting plate-to-upper wheel arch nuts
   - Strut

To install:
3. Install or connect the following:
   - Tighten the 3 suspension support-to-wheel arch nuts to 59 ft. lbs. (80 Nm).
   - Tighten the strut-to-steering knuckle arm bolts to 155 ft. lbs. (210 Nm).
   - Sway bar link to the strut. Tighten the nut to 55 ft. lbs. (74 Nm).
   - ABS speed sensor and the brake hose to the strut, if equipped
   - Wheel
4. Check and/or adjust the front wheel alignment.

REAR

1. Before servicing the vehicle, refer to the precautions in the beginning of this section.

➡ **Do not support the weight of the vehicle on the suspension arm; the arm will deform under its weight.**

2. Remove or disconnect the following:
   - Wheel
   - Brake hose and the Anti-lock Brake System (ABS) speed sensor wire from the strut
   - Sway bar link from the strut
3. Loosen, but do not remove the 2 lower bolts.
4. Support the axle carrier with a jack and remove cap.
5. If the strut is being disassembled, loosen the center nut.
6. Remove the 3 mounting nuts.
7. Lower the carrier and remove the 2 lower nuts and bolts.
8. Installation is the reverse of removal. Observe the following torques:
   - 3 mounting nuts: 29 ft. lbs. (39Nm)
   - 2 lower nuts: 188 ft. lbs. (255Nm)
   - Center nut: 36 ft. lbs. (49Nm)
   - Stabilizer link: 29 ft. lbs. (39Nm)

RX 300

FRONT

1. Before servicing the vehicle, refer to the precautions in the beginning of this section.

➡ **Do not support the weight of the vehicle on the suspension arm; the arm will deform under its weight.**

2. Remove or disconnect the following:
   - Wheel
   - Brake hose and the Anti-lock Brake System (ABS) speed sensor wire from the strut

   - Sway bar link from the strut
3. Matchmark on the strut lower bracket and camber adjust cam, if equipped.
4. Remove or disconnect the following:
   - Strut lower end from the steering knuckle lower arm
   - 3 upper strut mounting plate-to-upper wheel arch nuts
   - Strut

To install:
5. Align the upper suspension support hole with the strut piston or end, so they fit properly.
6. Install or connect the following:
   - Strut piston rod end to the upper suspension support. Tighten the new nut to 29–40 ft. lbs. (39–54 Nm).

➡ **Do not use an impact wrench to tighten the nut.**

   - Lubricate the suspension support bearing with multi-purpose grease.
   - Pack the upper support space with multi-purpose grease, also, after installation.
   - Tighten the 3 suspension support-to-wheel arch nuts to 47 ft. lbs. (64 Nm).
   - Tighten the strut-to-steering knuckle arm bolts to 156 ft. lbs. (211 Nm).
   - Sway bar link to the strut. Tighten the nut to 29 ft. lbs. (39 Nm).
   - ABS speed sensor and the brake hose to the strut, if equipped
   - Wheel
7. Check and/or adjust the front wheel alignment.

REAR

1. Before servicing the vehicle, refer to the precautions in the beginning of this section.
2. Remove or disconnect the following:
   - Negative battery cable
   - Deck side cover
   - Rear wheels
   - Anti-lock Brake System (ABS) sensor from the strut bracket
   - Flexible brake hose from the strut
   - Sway bar link from the strut
   - Loosen the 2 lower strut mounting bolts
3. Support the rear axle carrier with a jack.
4. Remove or disconnect the following:
   - 3 upper strut mounting nuts
   - Strut, by lower the rear axle

To install:
5. Install or connect the following:
   - Strut

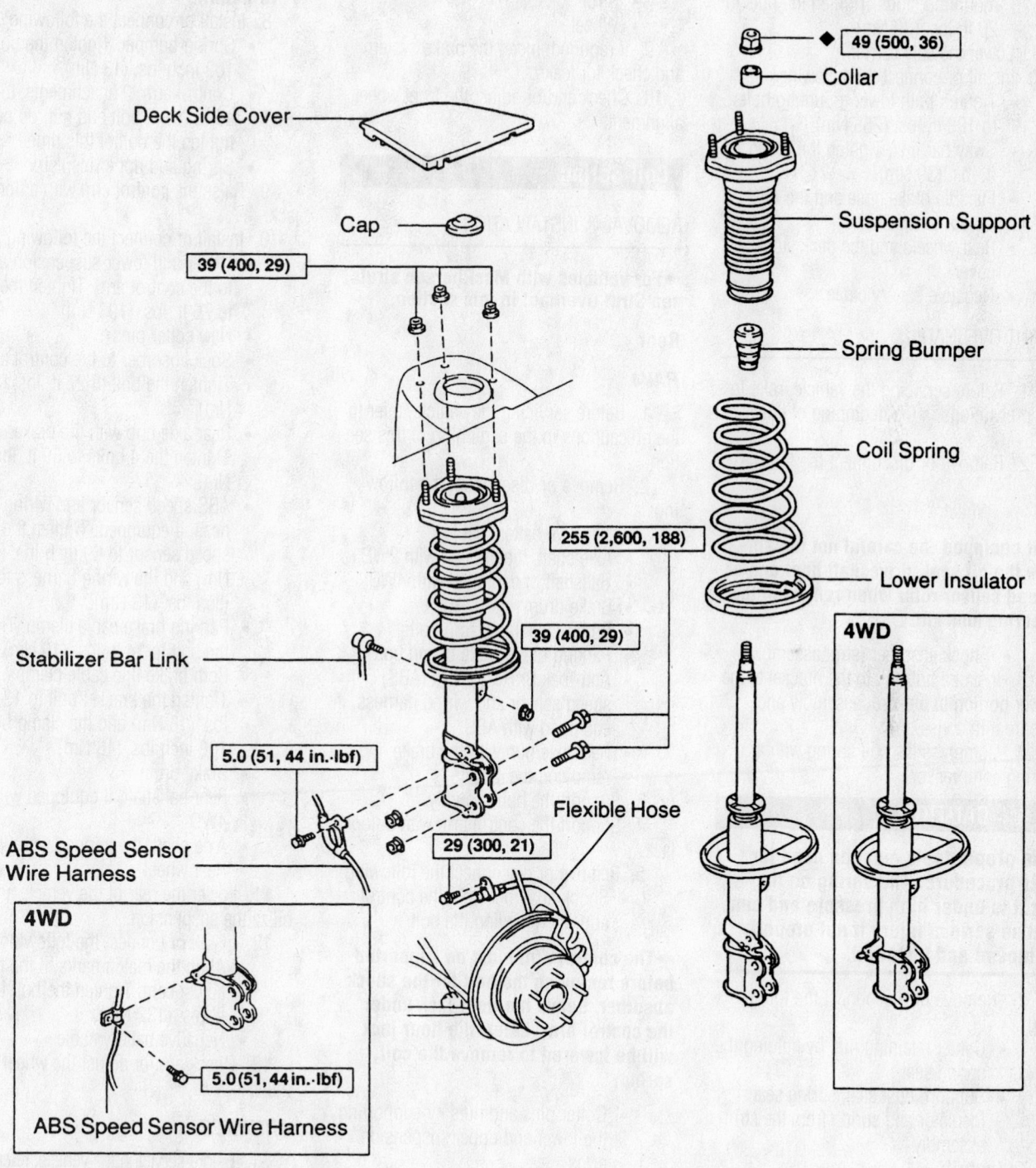

49 (500, 36)

Collar

Suspension Support

Spring Bumper

Coil Spring

Lower Insulator

4WD

Deck Side Cover

Cap

39 (400, 29)

255 (2,600, 188)

39 (400, 29)

Stabilizer Bar Link

5.0 (51, 44 in.·lbf)

ABS Speed Sensor
Wire Harness

Flexible Hose

29 (300, 21)

4WD

5.0 (51, 44 in.·lbf)

ABS Speed Sensor Wire Harness

N·m (kgf·cm, ft·lbf) : Specified torque

◆ Non–reusable part

Exploded view of the rear strut assembly—RX 300

7924ZG89

For Wheel Alignment specifications, see Section 1 of this manual

- Both lower strut mounting bolts, but do not tighten
- Axle carrier by aligning the 3 upper mounting studs. Tighten the nuts to 29 ft. lbs. (39 Nm).
6. Lower the axle carrier.
7. Install or connect the following:
  - Tighten both lower mounting bolts to 188 ft. lbs. (255 Nm).
  - Sway bar link. Tighten the nut to 29 ft. lbs. (39 Nm).
  - Flexible brake hose and the ABS sensor to the strut
  - Rear wheels and the deck side cover
  - Negative battery cable

STRUT OVERHAUL

1. Before servicing the vehicle, refer to the precautions in the beginning of this section.
2. Remove or disconnect the following:
  - Wheel

➡**If equipped, be careful not to damage the oil seal, driveshaft boot and/or speed sensor rotor when removing the steering knuckle.**

  - Shock absorber (strut assembly)
3. Install a nut/bolt to the bracket at the lower portion of the strut assembly and secure it in a vise.
4. Compress the coil spring with a spring compressor.

※※ CAUTION

The proper tools must be used for this procedure. The spring on the strut is under high pressure and can cause serious injury if not properly removed and installed.

5. Remove or disconnect the following:
  - Center retaining nut, by holding the spring seat
  - Support, dust seal, spring seat, insulator and spring from the strut assembly
To install:
6. Install the spring bumper and lower insulator to the strut assembly.
7. Compress the coil spring and fit the lower end of the spring into the spring seat gap.
8. Install or connect the following:
  - Upper insulator, spring seat, dust seal, support and spring seat. Tighten the new retaining nut to 34 ft. lbs. (47 Nm) for the RAV4 and

RX300; 36 ft. lbs. (49Nm) for the Highlander; 18 ft. lbs. (25Nm) for the 4Runner.
  - Strut
  - Wheel
9. If required, bleed the brake system and check for leaks.
10. Check and/or adjust the front wheel alignment.

Coil Spring

REMOVAL & INSTALLATION

➡**For vehicles with MacPherson struts, see Strut Overhaul in this section.**

Rear

RAV4

1. Before servicing the vehicle, refer to the precautions in the beginning of this section.
2. Remove or disconnect the following:
  - Negative battery cable
  - Axle shaft, if equipped with 2WD
  - Halfshaft, if equipped with 4WD
  - Brake drum
  - Both brake line clamp bolts
  - Parking brake cable clamp bolt
  - Anti-lock Brake System (ABS) speed sensor and wiring harness, if equipped with ABS
  - Rear axle hub with the brake, by removing the 4 bolts
3. Support the hub securely.
4. Support the control arm with a floor jack.
5. Remove or disconnect the following:
  - Shock absorber from the control arm, by removing the bolt

➡**The control arm must be supported before removing the bolt for the shock absorber. Leave the floor jack under the control arm. Later, the floor jack will be lowered to remove the coil spring.**

  - Cotter pins and nuts by supporting the lower and upper suspension arms
6. Disconnect the upper and lower control arms from the control arm, using tool 09628-62011
7. Remove the coil spring and control arm, as follows:
  - Matchmark the toe adjust cam and body.
  - Coil spring and upper insulator, by loosening the bolt and lowering the control arm.

  - Bolt, toe-adjust cam, 2 attachments, nut and control arm
  - Bolt and spring bumper
To install:
8. Install or connect the following:
  - Spring bumper. Tighten the bolt to 108 inch lbs. (13 Nm).
  - Control arm, 2 attachments, toe-adjust cam, bolt and nut; do not tighten the bolt at this time
  - Spring and upper insulator
9. Raise the control arm with a floor jack.
10. Install or connect the following:
  - Upper and lower suspension arms to the control arm. Tighten the nuts to 76 ft. lbs. (103 Nm).
  - New cotter pins
  - Sock absorber to the control arm. Tighten the bolt to 27 ft. lbs. (37 Nm).
  - Rear axle hub with the brake. Tighten the 4 bolts to 59 ft. lbs. (80 Nm).
  - ABS speed sensor and wiring harness, if equipped. Tighten the ABS speed sensor to 69 inch lbs. (8 Nm) and the wiring harness to 108 inch lbs. (13 Nm).
  - Parking brake cable clamp. Tighten the bolt to 14 ft. lbs. (19 Nm).
  - Both brake line cable clamps. Tighten the bracket bolt to 13 ft. lbs. (18 Nm) and the clamp bolt to 108 inch lbs. (13 Nm).
  - Brake drum
  - Rear halfshaft, if equipped with 4WD
  - Axle shaft, if equipped with 2WD
  - Rear wheel
11. Lower the rear of the vehicle and stabilize the suspension.
12. Install or connect the following:
  - Align the matchmarks to the toe-adjust cam. Tighten the bolt to 98 ft. lbs. (132 Nm).
  - Negative battery cable
13. Check and/or adjust the wheel alignment

4RUNNER

1. Before servicing the vehicle, refer to the precautions in the beginning of this section.
2. Remove the wheel assemblies.
3. Support the axle housing with a floor jack.
4. Remove or disconnect the following:
  - Brake drum
  - Parking brake cable from the brake shoe
  - Parking brake cable from the axle housing.

5. Place matchmarks on the flanges for the driveshaft and differential.

6. Remove or disconnect the following:

- Driveshaft from the differential
- Brake hose line from the brake hose
- Brake hose-to-brake bracket clip
- Brake hose from the body
- Anti-lock Brake System (ABS) wiring harness bracket, if equipped with ABS
- Shock absorbers from the axle housing
- Lateral control rod nuts/bolts
- Control rod from the suspension
- Coil spring, by lower the rear axle housing

To install:

7. Install or connect the following:

- Coil springs into position and raise the axle housing.

✳✳ CAUTION

Be sure to fit the lower end of the coil spring into the gap of the spring seat on the lower control arm.

- Lateral control rod to the suspension. Tighten the bolts/nuts to 64 ft. lbs. (86 Nm).
- Shock absorbers to the axle housing. Tighten the bolt to 47 ft. lbs. (64 Nm).
- ABS wiring harness bracket, if equipped
- Brake hose to the bracket
- Clip
- Brake line to the brake hose. Tighten the tube.
- Parking brake cable bracket to the axle housing. Tighten to 108 inch lbs. (13 Nm).
- Parking brake cable to the brake shoes
- Driveshaft to the differential, by aligning the matchmarks. Tighten bolts/nuts to 54 ft. lbs. (73 Nm).

8. Fill the differential with the proper amount and type of oil.

9. Install the brake drum.

10. Bleed the brake system.

11. Install the wheel assemblies.

12. Lower the vehicle and bounce the vehicle several times to stabilize the suspension.

13. Tighten the lower control arm to 107 ft. lbs. (145 Nm).

Upper Ball Joint

REMOVAL & INSTALLATION

4Runner

1. Before servicing the vehicle, refer to the precautions in the beginning of this section.

2. Remove or disconnect the following:
- Front wheels
- Strut assembly
- Grease cap

3. If equipped with 4WD, disconnect the halfshaft, as follows:
- Cotter pin and lockcap
- Locknut, while applying the brakes

4. Remove or disconnect the following:
- Anti-lock Brake System (ABS) speed sensor and wiring harness clamp from the steering knuckle, if equipped with ABS
- Brake line bracket from the steering knuckle
- Front brake caliper and the rotor
- Lower ball joint

5. Remove the steering knuckle with the axle hub, as follows:
- Cotter pin and loosen the nut
- Steering knuckle from the upper control arm, using SST 09950-40010
- Steering knuckle
- Upper ball joint
- Wire and the boot
- Snapring
- Upper ball joint, using SST 09950-40010 and a deep socket wrench.

To install:

6. Install the upper ball joint, as follows:
- New ball joint with a new snapring
- New boot secured with a new wire

7. Install or connect the following:
- Steering knuckle with the axle hub to the upper control arm. Tighten the nut to 80 ft. lbs. (108 Nm).
- New cotter pin
- Lower ball joint. Tighten the 4 bolts to 59 ft. lbs. (80 Nm).
- Rotor and caliper. Tighten the caliper bolts to 90 ft. lbs. (123 Nm).
- Brake line bracket to the steering knuckle. Tighten it to 21 ft. lbs. (28 Nm).
- ABS speed sensor and wiring harness clamp to the steering knuckle, if equipped. Tighten the bolts to 72 inch lbs. (8 Nm).

- Halfshaft, if disconnected. Tighten the nut to 174 ft. lbs. (235 Nm).
- Grease cap
- Strut
- Front wheel

8. Check the alignment

Lower Ball Joint

REMOVAL & INSTALLATION

RAV4

1. Before servicing the vehicle, refer to the precautions in the beginning of this section.

2. Remove or disconnect the following:
- Negative battery cable
- Front wheel(s)
- Steering knuckle with the axle hub
- Dust deflector, by prying it from the knuckle
- Cotter pin and nut from the ball joint stud
- Lower ball joint from the steering knuckle, using a 2-jaw puller

To install:

3. Install or connect the following:
- Lower ball joint onto the steering knuckle. Tighten nut to 94 ft. lbs. (127 Nm).
- New cotter pin
- ABS speed sensor, by aligning it the dust deflector hole
- New dust deflector, using a driver
- Steering knuckle and hub
- Front wheel(s)
- Negative battery cable

Highlander

1. Before servicing the vehicle, refer to the precautions in the beginning of this section.

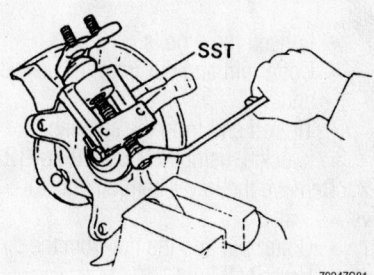

Use a 2-jaw puller to remove the lower ball joint—RAV4

2. Remove or disconnect the following:
- Wheel
- Hub nut
- Caliper and rotor
- Lower control arm from the ball joint
- Tie rod end
- Halfshaft
- Ball joint from the knuckle

3. Installation is the reverse of removal. Observe the following torques:
- Ball stud nut: 90 ft. lbs. (123Nm)
- Lower arm-to-ball joint: 94 ft. lbs. (127Nm)
- Tie rod end: 36 ft. lbs. (49Nm)

RX300

1. Before servicing the vehicle, refer to the precautions in the beginning of this section.

2. Remove or disconnect the following:
- Wheel
- Steering knuckle with the axle hub
- Dust deflector, by prying it from the knuckle
- Cotter pin and nut from the ball joint
- Ball joint from the steering knuckle, by removing the 2 bolts
- Lower ball joint, using a Ball Joint Separator tool 09628-62011

To install:

3. Install or connect the following:
- Lower ball joint. Tighten the nut to 76 ft. lbs. (103 Nm) and both bolts to 94 ft. lbs. (127 Nm).
- New cotter pin
- Wheel

4Runner

1. Before servicing the vehicle, refer to the precautions in the beginning of this section.

2. Remove the wheel.

3. Disconnect the tie rod end, as follows:
- Loosen the 4 bolts
- Cotter pin and nut from the tie rod end
- Tie rod end from the steering knuckle, using SST 09610-20012

4. Remove the lower ball joint, as follows:
- Cotter pin and the nut from the lower ball joint
- Lower ball joint from the lower suspension arm
- Lower ball joint

To install:

5. Install or connect the following:
- Lower ball joint to the lower control

arm. Tighten the 4 bolts to 59 ft. lbs. (80 Nm).
- Nut. Tighten it to 105 ft. lbs. (142 Nm).
- Tie rod end to the steering knuckle. Tighten the nut to 66 ft. lbs. (90 Nm).
- Wheel

Upper Control Arm

REMOVAL & INSTALLATION

4Runner

4WD

1. Before servicing the vehicle, refer to the precautions in the beginning of this section.

2. Remove or disconnect the following:
- Shock and coil spring assembly
- Anti-lock Brake System (ABS) speed sensor wire harness clamp

3. Disconnect the upper ball joint, as follows:
- Cotter pins and loosen the nut
- Upper ball joint from the control arm, using a ball joint separator
- Support the steering knuckle
- Nut

4. Detach the control arm, by removing the nut, bolt, washers and lowering the arm.

To install:

5. Install or connect the following:
- Upper control arm with the washer, bolt and nut. Tighten the nut to 87 ft. lbs. (115 Nm).
- Upper ball joint to the control arm. Tighten the mounting nut to 80 ft. lbs. (105 Nm).
- New cotter pin
- ABS speed sensor wire harness clamp. Tighten it to 71 inch lbs. (8 Nm).
- Shock and coil spring assembly

6. Check and/or adjust the alignment.

CONTROL ARM BUSHING REPLACEMENT

4Runner

1. Before servicing the vehicle, refer to the precautions in the beginning of this section.

2. Remove the upper control arm from the vehicle.

3. Pry up the bushing flange, using a chisel and a hammer.

4. Using tools 09613-26010, 09613-20060 and 09950-00020 and a shop press, remove the bushing.

To install:

5. Using tools 09223-00010, 09506-35010 and a shop press, press the new bushing into the upper control arm.

6. Install the upper control arm to the chassis. Torque the nuts to 87 ft. lbs. (115 Nm).

7. Check and/or adjust the alignment.

Lower Control Arm

REMOVAL & INSTALLATION

4Runner

2WD

1. Before servicing the vehicle, refer to the precautions in the beginning of this section.

2. Remove or disconnect the following:
- Front wheel
- Steering gear assembly
- Stabilizer bar link
- Shock absorber from the lower control arm

3. Support the upper control arm and the steering knuckle securely.

4. Remove or disconnect the following:
- Cotter pin and nut from the lower ball joint
- Lower ball joint from the control arm
- Bolts, nuts, adjusting cams and lower control arm, by placing matchmarks on the front and rear adjusting cams
- 2 spring bumpers, using tool SST 09922-10010

To install:

5. Install or connect the following:
- 2 spring bumpers. Tighten the spring bumpers to 17 ft. lbs. (23 Nm).
- Lower control arm and adjusting cams. Tighten the nuts and bolts to 96 ft. lbs. (130 Nm),
- Lower ball joint to the control arm
- Cotter pin and nut to the lower ball joint. Tighten the nut to 105 ft. lbs. (142 Nm).
- Shock absorber to the lower control arm
- Stabilizer bar link
- Steering gear assembly
- Front wheel

6. Check and/or adjust the alignment.

4WD

1. Before servicing the vehicle, refer to the precautions in the beginning of this section.

2. Remove or disconnect the following:
- Front wheel

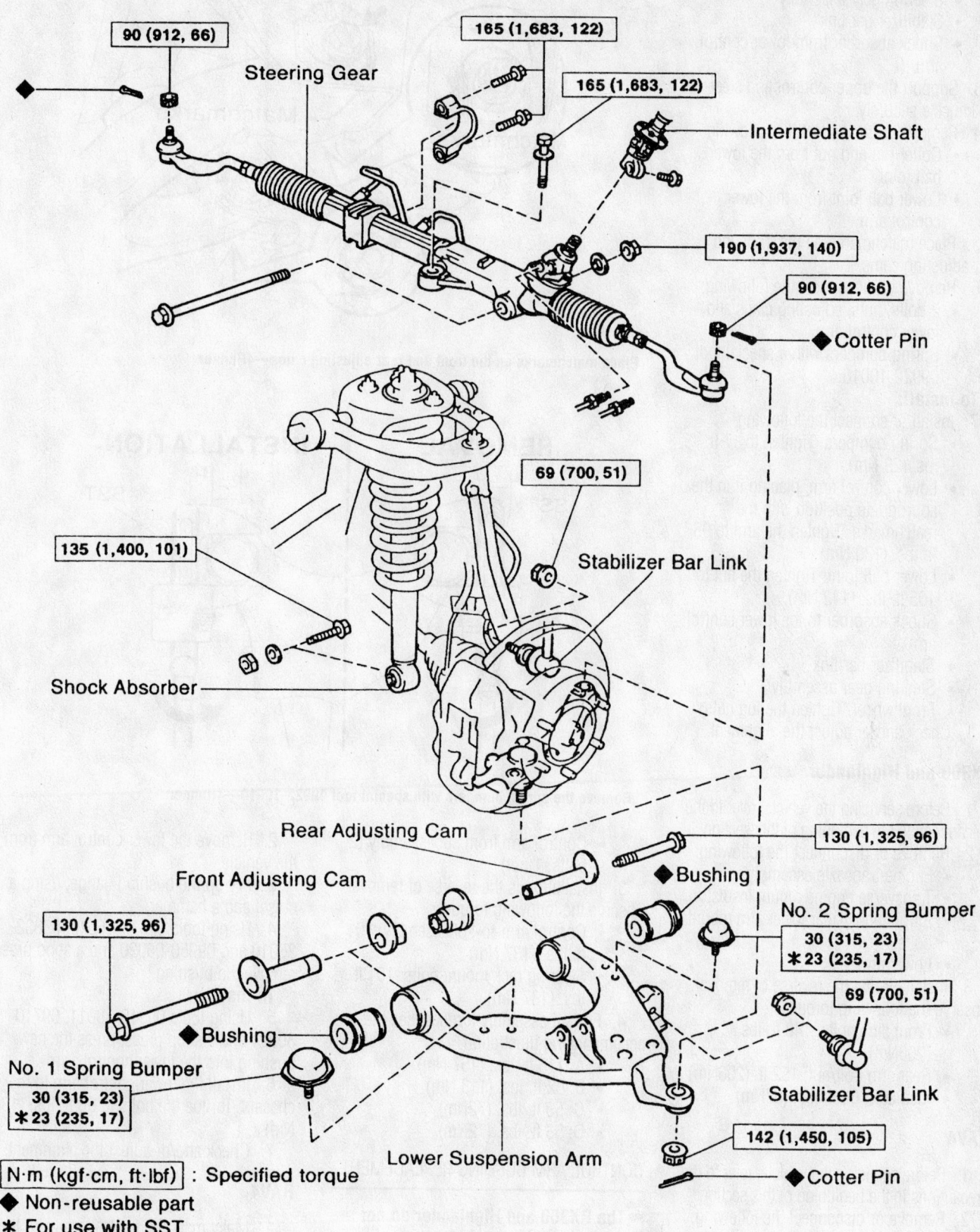

90 (912, 66)

Steering Gear

165 (1,683, 122)

165 (1,683, 122)

Intermediate Shaft

190 (1,937, 140)

90 (912, 66)

Cotter Pin

69 (700, 51)

135 (1,400, 101)

Stabilizer Bar Link

Shock Absorber

Rear Adjusting Cam

130 (1, 325, 96)

Bushing

Front Adjusting Cam

No. 2 Spring Bumper

130 (1, 325, 96)

30 (315, 23)
✱ 23 (235, 17)

69 (700, 51)

Bushing

No. 1 Spring Bumper

Stabilizer Bar Link

30 (315, 23)
✱ 23 (235, 17)

142 (1,450, 105)

Lower Suspension Arm

Cotter Pin

N·m (kgf·cm, ft·lbf) : Specified torque
◆ Non-reusable part
✱ For use with SST

86828G20

Exploded view of the lower control arm and associated components—4Runner

- Steering gear assembly
- Stabilizer bar link
- Shock absorber from lower control arm

3. Support the upper control and steering knuckle securely.

4. Remove or disconnect the following:
- Cotter pin and nut from the lower ball joint
- Lower ball joint from the lower control arm

5. Place matchmarks on the front and rear adjusting cams.

6. Remove or disconnect the following:
- 2 bolts, nuts, adjusting cams and lower control arm
- Spring bumpers with a special tool 09922-10010.

To install:

7. Install or connect the following:
- Spring bumpers. Tighten to 17 ft. lbs. (23 Nm).
- Lower control arm, placing it in the appropriate position with the matchmarks. Tighten the arm to 96 ft. lbs. (130 Nm).
- Lower ball joint. Tighten the nut to 105 ft. lbs. (142 Nm).
- Shock absorber to the lower control arm
- Stabilizer bar link
- Steering gear assembly
- Front wheel. Tighten the lug nuts.

8. Check and/or adjust the alignment.

RX300 and Highlander

1. Before servicing the vehicle, refer to the precautions in the beginning of this section.

2. Remove or disconnect the following:
- Engine/transaxle assembly
- Transverse engine mount insulator
- 2 front and 1 rear lower arm mount bolts
- Lower arm

3. Installation is the reverse of removal. Observe the following torques:
- Front side bolts: 148 ft. lbs. (200Nm)
- Rear arm bolt/nut: 152 ft. (206Nm)
- Insulator: 64 ft. lbs. (87Nm)

RAV4

1. Before servicing the vehicle, refer to the precautions in the beginning of this section.

2. Remove or disconnect the following:
- Wheel
- Sway bar link
- Control arm-to-ball joint bolts
- Steering rack mount bolts
- Suspension member subassembly (5 bolts and 2 nuts)

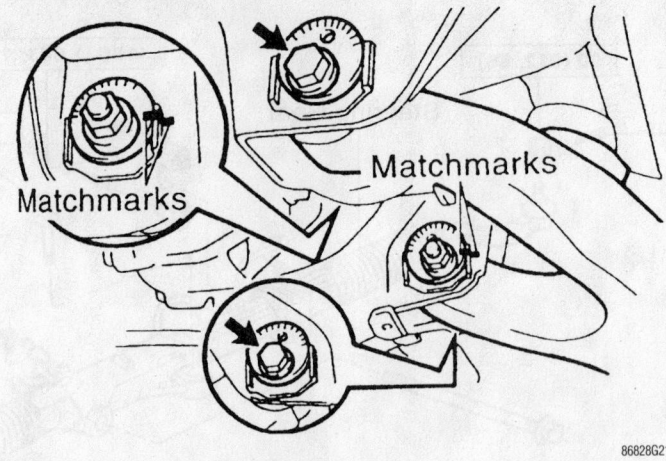

Place matchmarks on the front and rear adjusting cams—4Runner

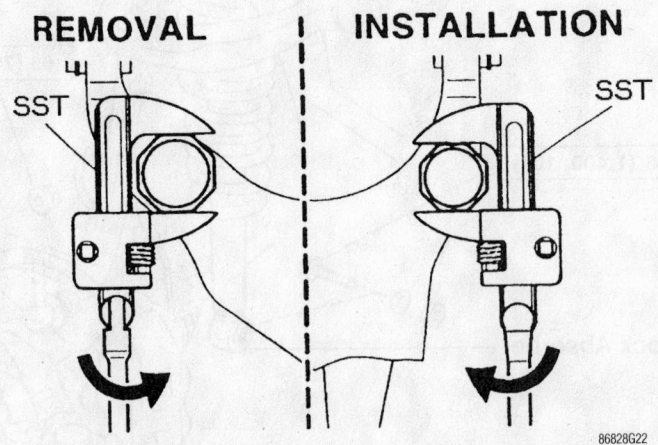

Remove the spring bumpers with special tool 09922-10010—4Runner

- Control arm from subassembly (2 bolts, 1 nut)

3. Installation is the reverse of removal. Observe the following torques:
- Control arm-to-subassembly: 101 ft. lbs. (137 Nm)
- Steering rack mount bolts: 101 ft. lbs. (137 Nm)

4. For subassembly torques, see the accompanying illustration.
- A: 115 ft. lbs. (157 Nm)
- B: 82 ft. lbs. (113 Nm)
- C: 53 ft. lbs. (72Nm)
- D: 53 ft. lbs. (72Nm)

CONTROL ARM BUSHING REPLACEMENT

➡ **The RX300 and Highlander do not have replaceable bushings.**

4Runner

1. Before servicing the vehicle, refer to the precautions in the beginning of this section.

2. Remove the lower control arm from the vehicle.

3. Pry up the bushing flange, using a chisel and a hammer.

4. Using tools 09613-26010, 09632-36010 and 09950-00020 and a shop press, remove the bushing.

To install:

5. Using tools 09316-20011, 09710-30021 and a shop press, press the new bushing into the lower control arm.

6. Install the lower control arm to the chassis. Torque the bolts to 96 ft. lbs. (130 Nm).

7. Check and/or adjust the alignment.

RAV4

1. Matchmark the control arm with the triangle mark on the bushing.

2. Press out the old bushing.

3. Press in the new bushing, aligning the matchmarks.

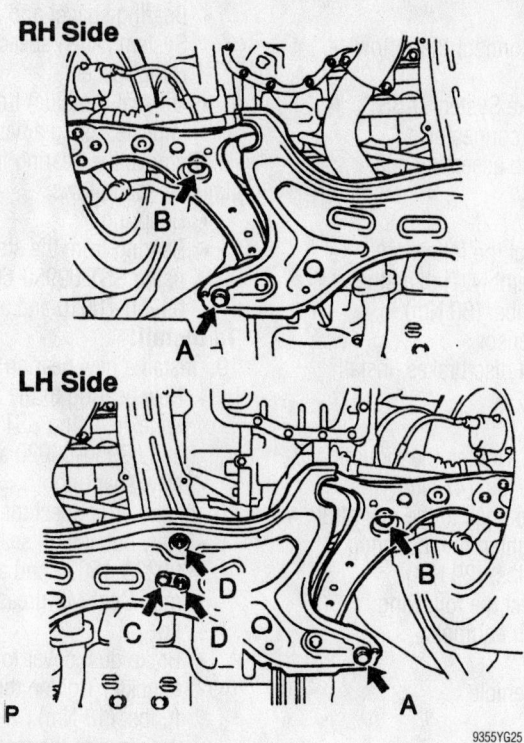

RH Side

LH Side

9355YG25

RAV4 suspension subassembly bolt torques

Wheel Bearing

ADJUSTMENT

RAV4 and RX300

FRONT AND REAR

Check the bearing play in the axial direction and also check the axle hub deviation. The maximum play for both checks should be 0.0020 in. (0.05mm). If greater than the specified maximum, replace the bearing. The wheel bearing is not adjustable.

4Runner

The wheel bearings are sealed unit; no adjustment is possible.

REMOVAL & INSTALLATION

RAV4

FRONT

1. Before servicing the vehicle, refer to the precautions in the beginning of this section.
2. Remove or disconnect the following:
   • Negative battery cable
   • Front wheels
   • Cotter pin and lockcap from the halfshaft end

   • Halfshaft locknut, by applying the front brakes
   • Brake caliper and support it on a wire
3. Matchmark the rotor to the hub.
4. Remove or disconnect the following:
   • Rotor
   • Anti-lock Brake System (ABS) speed sensor from the steering knuckle, if equipped
   • Loosen the strut's lower end nuts
   • Tie rod end from the steering knuckle
   • Lower control arm from the ball joint, by removing the bolt and 2 nuts
   • Halfshaft from the axle hub

➡**Secure the halfshaft aside using a wire. Be careful not to damage the shaft boot or ABS sensor rotor.**

   • Both strut's lower end nuts
   • Steering knuckle
5. Clamp the steering knuckle in a vise with soft jaws to protect the knuckle.
6. Remove or disconnect the following:

   • Dust deflector, by prying it from the hub
   • Ball joint from the steering knuckle
   • Hub from the knuckle, using slide hammer

   • Inner race from the hub, using press and arbor tool
   • 4 bolts and the dust cover
   • Inner oil seal, using Seal Removal tool 09308-00010
   • Outer oil seal, using Seal Removal tool 09308-00010
   • Snapring
7. Install inner race (removed from the hub) on the outside of the bearing
8. Remove the steering knuckle bearing, using a bearing driver

To install:
9. Clean bearing seating surfaces with a clean, dry rag.
10. Install or connect the following:
   • Bearing into the knuckle, using a press and Bearing Installer tool 09608-32010
   • Snapring
   • Dust cover. Tighten the 4 bolts to 74 inch lbs. (8 Nm).
   • New outer oil seal, using a seal driver

➡**Apply multi-purpose grease to the oil seal lip.**

   • Hub into the steering knuckle
   • New inner oil seal, using a seal driver

➡**Apply multi-purpose grease to the oil seal lip.**

   • Lower ball joint to the steering knuckle. Tighten the nut to 94 ft. lbs. (127 Nm).
   • New cotter pin
   • Dust deflector, by aligning it with the ABS speed sensor hole
   • Knuckle to the lower strut and install the bolts
   • Lower ball joint to the lower arm. Tighten the bolts to 94 ft. lbs. (127 Nm).
   • Tie rod end to the steering knuckle. Tighten the nut to 36 ft. lbs. (49 Nm).
   • Halfshaft to the hub and knuckle
   • Tighten the lower strut nuts to 117 ft. lbs. (158 Nm).
   • ABS speed sensor. Tighten the bolt to 69 inch lbs. (8 Nm).
   • Rotor to the hub, by aligning the matchmark
   • Brake caliper. Tighten the mounting bolts to 79 ft. lbs. (107 Nm).
   • Axle locknut, using an assistant to apply the brakes. Tighten the nut to 159 ft. lbs. (216 Nm).

- Lockcap and a new cotter pin
- Wheel
- Negative battery cable

11. Turn the wheel by hand, verify that the wheel turns without noise and without binding.

12. Check the signal from the ABS sensor.

REAR

The wheel bearings are not replaceable. If the bearing has failed, replace the hub/bearing assembly. See the Halfshaft Removal and Installation procedure.

Highlander

FRONT

1. Before servicing the vehicle, refer to the precautions in the beginning of this section.

2. Remove the hub/knuckle assembly. See the Halfshaft Removal and Installation procedure. Mount the assembly in a vise.

3. Remove or disconnect the following:
- Dust deflector
- Snapring
- Hub from the spindle and mount the hub in a vise.
- Dust cover

4. Press out the bearings, inner bearing first.

5. Press in the new bearings.

6. Installation is the reverse of removal.

REAR

The wheel bearings are not replaceable. If the bearing has failed, replace the hub/bearing assembly. See the Halfshaft Removal and Installation procedure.

RX 300

The front wheel bearings and 4wd rear wheel bearings are serviced as a unit with the hub and are not replaceable. See the Halfshaft Removal and Installation procedure.

REAR WITH 2WD

1. Before servicing the vehicle, refer to the precautions in the beginning of this section.

2. Remove or disconnect the following:
- Rear wheel
- Brake drum, if equipped

3. If equipped with disk brakes, remove the following:
- Flexible brake hose from the rear strut assembly
- Brake caliper and support it on using wire

- Brake rotor

4. Remove or disconnect the following:
- Anti-lock Brake System (ABS) speed sensor connector
- 4 rear axle hub assembly nuts
- Hub assembly

To install:

5. Install or connect the following:
- New hub assembly. Tighten the nuts to 59 ft. lbs. (80 Nm).
- ABS speed sensor

6. If equipped with disc brakes, install the following:
- Brake rotor
- Brake caliper. Tighten the mounting bolts to 34 ft. lbs. (47 Nm).
- Flexible brake hose to the rear strut assembly. Tighten the mounting bolt to 21 ft. lbs. (29 Nm).

7. Install or connect the following:
- Brake drum, if equipped
- Wheels

8. Test drive the vehicle.

4Runner

1. Before servicing the vehicle, refer to the precautions in the beginning of this section.

2. Remove or disconnect the following:
- Front wheels
- Shock absorber
- Grease cap

3. On 4WD, remove the halfshaft, as follows:
- Cotter pin and lockcap
- Locknut, while applying the brakes

4. Remove or disconnect the following:
- Anti-lock Brake System (ABS) speed sensor and wiring harness clamp from the steering knuckle, if equipped with ABS
- Brake line bracket from the steering knuckle
- Front brake caliper and rotor
- 4 bolts and the lower ball joint

5. Remove the steering knuckle with the axle hub, as follows:
- Cotter pin and loosen the nut
- Steering knuckle, using SST 09950-40010

6. Clamp the axle hub in a soft jaw vise.

7. Remove or disconnect the following:
- Grease cap, for 2WD
- Inside oil seal, for 4WD
- 4 bolts and shift the brake dust cover towards the hub side
- Axle hub from the steering knuckle, using SST 09710-30021

- Bearing spacer and Anti-lock Brake System (ABS) speed sensor rotor/spacer
- Oil seal (outside) from the steering knuckle, using a flat pry bar

8. Remove the bearing from the steering knuckle, as follows:
- Snapring
- Bearing from the steering knuckle, using SST 09950-60020 and 09950-70010 and a press

To install:

9. Install a new bearing, as follows:
- New bearing to the steering knuckle, using SST 09527-17011 and 09950-60020 and a press
- New snapring

10. Install or connect the following:
- New outside oil seal, using SST 09223-15030 and a plastic hammer. Coat MP grease to the oil seal lip.
- Brake dust cover to the steering knuckle. Tighten the 4 bolts to 13 ft. lbs. (18 Nm).
- Axle hub to the steering knuckle, using a press
- ABS speed sensor rotor/spacer.

⁑ WARNING

Be careful not to scratch the serration of the speed sensor rotor.

- Bearing spacer, using a press
- Grease cap, if removed
- New inside oil seal, if removed, using SST 09527-17011 and a plastic hammer
- Steering knuckle with the axle hub. Tighten the nut to 80 ft. lbs. (108 Nm).
- New cotter pin
- Lower ball joint. Tighten the 4 bolts to 59 ft. lbs. (80 Nm).
- Rotor and the caliper. Tighten the caliper bolts to 90 ft. lbs. (123 Nm).
- Brake line bracket to the steering knuckle. Tighten the fasteners to 21 ft. lbs. (28 Nm).
- ABS speed sensor and wiring harness clamp to the steering knuckle, if removed. Tighten the bolts to 72 inch lbs. (8 Nm).
- Halfshaft, if disconnected. Tighten the nut to 174 ft. lbs. (235 Nm).
- Grease cap
- Shock absorber
- Front wheel
- Negative battery cable

TOYOTA

Matrix

PRECAUTIONS

Before servicing any vehicle, please be sure to read all of the following precautions, which deal with personal safety, prevention of component damage, and important points to take into consideration when servicing a motor vehicle:

• Never open, service or drain the radiator or cooling system when the engine is hot; serious burns can occur from the steam and hot coolant.

• Observe all applicable safety precautions when working around fuel. Whenever servicing the fuel system, always work in a well-ventilated area. Do not allow fuel spray or vapors to come in contact with a spark, open flame, or excessive heat (a hot drop light, for example). Keep a dry chemical fire extinguisher near the work area. Always keep fuel in a container specifically designed for fuel storage; also, always properly seal fuel containers to avoid the possibility of fire or explosion. Refer to the additional fuel system precautions later in this section.

• Fuel injection systems often remain pressurized, even after the engine has been turned **OFF**. The fuel system pressure must be relieved before disconnecting any fuel lines. Failure to do so may result in fire and/or personal injury.

• Brake fluid often contains polyglycol ethers and polyglycols. Avoid contact with the eyes and wash your hands thoroughly after handling brake fluid. If you do get brake fluid in your eyes, flush your eyes with clean, running water for 15 minutes. If

eye irritation persists, or if you have taken brake fluid internally, IMMEDIATELY seek medical assistance.

• The EPA warns that prolonged contact with used engine oil may cause a number of skin disorders, including cancer! You should make every effort to minimize your exposure to used engine oil. Protective gloves should be worn when changing oil. Wash your hands and any other exposed skin areas as soon as possible after exposure to used engine oil. Soap and water, or waterless hand cleaner should be used.

• All new vehicles are now equipped with an air bag system. The system must be disabled before performing service on or around system components, steering column, instrument panel components, wiring and sensors. Failure to follow safety and disabling procedures could result in accidental air bag deployment, possible personal injury and unnecessary system repairs.

• Always wear safety goggles when working with, or around, the air bag system. When carrying a non-deployed air bag, be sure the bag and trim cover are pointed away from your body. When placing a non-deployed air bag on a work surface, always face the bag and trim cover upward, away from the surface. This will reduce the motion of the module if it is accidentally deployed. Refer to the additional air bag system precautions later in this section.

• Clean, high quality brake fluid from a sealed container is essential to the safe and

proper operation of the brake system. You should always buy the correct type of brake fluid for your vehicle. If the brake fluid becomes contaminated, completely flush the system with new fluid. Never reuse any brake fluid. Any brake fluid that is removed from the system should be discarded. Also, do not allow any brake fluid to come in contact with a painted surface; it will damage the paint.

• Never operate the engine without the proper amount and type of engine oil; doing so WILL result in severe engine damage.

• Timing belt maintenance is extremely important! Many models utilize an interference-type, non-freewheeling engine. If the timing belt breaks, the valves in the cylinder head may strike the pistons, causing potentially serious (also time-consuming and expensive) engine damage. Refer to the maintenance interval charts in the front of this manual for the recommended replacement interval for the timing belt, and to the timing belt section for belt replacement and inspection.

• Disconnecting the negative battery cable on some vehicles may interfere with the functions of the on-board computer system(s) and may require the computer to undergo a relearning process once the negative battery cable is reconnected.

• When servicing drum brakes, only disassemble and assemble one side at a time, leaving the remaining side intact for reference.

ENGINE REPAIR

Alternator

REMOVAL & INSTALLATION

1. Before servicing the vehicle, refer to the precautions in the beginning of this section.
2. Remove or disconnect the following:
   • Negative battery cable
   • Drive belt
   • Wire clamp from the clip on the rectifire end frame
   • Rubber clamp and nut
   • Alternator wiring and connector
   • Alternator
To install:
3. Install or connect the following:
   • Alternator. Torque the 12mm bolt to 18 ft. lbs. (25 Nm) and the 14mm bolt to 39 ft. lbs. (54 Nm).

• Alternator connector and wiring
• Rubber clamp and nut
• Wire clamp
• Drive belt
• Negative battery cable

Ignition Timing

ADJUSTMENT

➡**The timing on engines equipped with Distributorless Ignition Systems (DIS) is not adjustable.**

Engine Assembly

REMOVAL & INSTALLATION

1. Before servicing the vehicle, refer to the precautions in the beginning of this section.

2. Relieve the fuel system pressure.
3. Drain the cooling system.
4. Drain the engine oil.
5. Drain the transaxle fluid and transfer fluid, if equipped.
6. Remove or disconnect the following:
   • Negative battery cable. Wait at least 90 seconds before proceeding.
   • Battery
   • Hood
   • Undercovers
   • Radiator inlet and outlet hoses
   • Radiator hose outlet
   • Oil cooler inlet and outlet tubes
   • Upper radiator support and radiator, if equipped with A/C
   • Battery
   • Air cleaner assembly
   • Fuel pipe clamp
   • Fuel tube sub-assembly
   • Accelerator control cable

- Cruise control actuator, if equipped
- Union-to-connector tube hose
- Heater inlet and outlet hoses
- Transmission shift cable(s)
- Clutch release cylinder, on manual transaxle
- Glove compartment door
- Engine relay block cover
- 3 connectors from the relay block
- 2 ground cables
- Engine wire from the Engine Control Module (ECM) and junction block
- Engine wire from the cabin
- Drive belt
- Compressor and magnetic clutch, if equipped with A/C. Unbolt and position aside, DO NOT disconnect the lines.
- Vane pump oil reservoir from the bracket
- Return tube
- Right side front door scuff plate
- Right side cowl side trim plate
- Right side rear door scuff plate, AWD
- Lower right side center pillar garnish, AWD
- Right front seat, AWD
- Column hole cover silencer sheet
- Steering intermediate shaft
- Front floor panel brace, FWD
- Center exhaust pipe, AWD
- Propeller shaft with center bearing shaft, AWD
- Front exhaust pipe
- Front hub nuts
- Tie rod ends from the steering knuckles

7. Separate the front stabilizer links and lower control arm ball joints
- Front halfshafts

8. Remove the engine from the vehicle, as follows:
   a. Set the engine lifter.
   b. Remove the bolts and nuts, then remove the engine mounting insulator.
   c. Remove the through bolt and nut, then detach the engine mounting insulator from the vehicle.
   d. Remove the 6 bolts as shown.
   e. Use a suitable tool to suspend the engine assembly, as shown in the figure.
   f. No. 1 engine hanger: P/N 12281-15040 (1ZZ-FE), 12281-88600 (2ZZ-GE).
   g. No. 2 engine hanger: P/N 12281-22021 (1ZZ-FE), 12281-88600 (2ZZ-GE).
   h. Bolt: P/N 91512-B1016.
   i. Torque the bolts to 28 ft. lbs. (38 Nm).

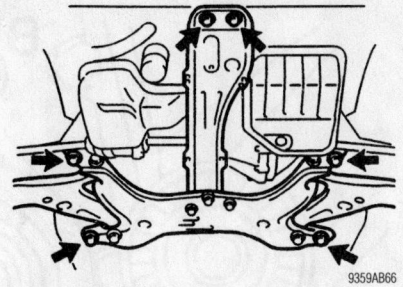

Remove the 6 bolts, as indicated by arrows

✳✳ CAUTION

Do not try to suspend the engine by hooking the chain to any other part.

   j. Attach an engine chain hoist to the hangers.
   k. Using the chain block and sling device, suspend the engine.
   l. Remove the engine and transaxle assembly from the vehicle.

9. Remove or disconnect the following components, as necessary:
- Vane pump
- Steering gear
- Crossmember
- Manifold stay
- Oxygen (O$_2$) sensor
- Exhaust manifold
- Starter
- Transaxle
- Transfer case
- Clutch
- Flywheel
- Alternator
- Ignition coil
- Fuel delivery pipe
- Intake manifold
- Oil level gauge
- Water inlet and bypass pipes
- Thermostat
- Oil pressure switch

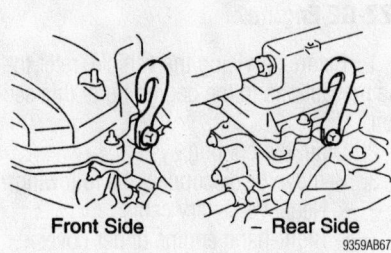

Front Side   **– Rear Side**

Install the engine hangers—1ZZ-FE shown, 2ZZ-GE similar

- Crankshaft Position (CKP) sensor
- Knock Sensor (KS)
- Drive belt tensioner
- Engine mounts and brackets
- Coolant Temperature Sensor (CTS)

To install:

10. Install any removed components to the engine and transaxle assembly.

11. To install the engine:
   a. Place the engine and transaxle on an engine lifter.
   b. Install the engine with the transaxle to the vehicle.
   c. Temporarily install the crossmember and 6 bolts.
   d. Install the left engine mounting insulator. Tighten the bolts to 59 ft. lbs. (80 Nm).
   e. Install the right engine mounting insulator. Tighten the bolts to 38 ft. lbs. (52 Nm).
   f. Insert SST 09670-00010 to the positioning holes of the right handle crossmember and on the right handle of the vehicle. Temporarily tighten bolt A, then bolt B.
   g. Insert SST 09670-00010 to the positioning holes of the left handle

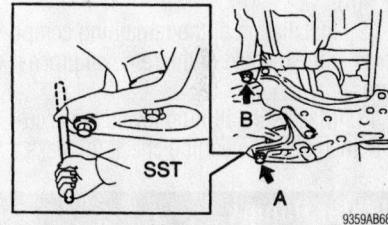

Insert the SST to the positioning holes of the right handle crossmember and on the right handle of the vehicle. Temporarily tighten bolt A, then bolt B

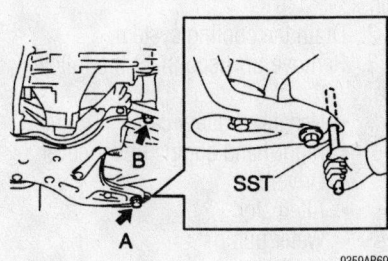

Insert the SST to the positioning holes of the left handle crossmember and on the left handle of the vehicle. Temporarily tighten bolt A, then bolt B

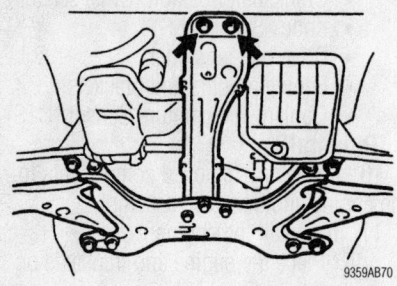

Tighten the 2 crossmember bolts, indicated by arrows

9359AB70

crossmember and on the left handle of the vehicle. Temporarily tighten bolt A, then bolt B.

   h. Insert the SST to the positioning holes on the right-handle crossmember and right handle. Tighten bolt A to 116 ft. lbs. (157 Nm) and bolt B to 83 ft. lbs. (113 Nm).

   i. Insert the SST to the positioning holes on the left-handle crossmember and left handle. Tighten bolt A to 116 ft. lbs. (157 Nm) and bolt B to 83 ft. lbs. (113 Nm).

   j. Tighten the 2 crossmember bolts, shown in the figure, to 29 ft. lbs. (39 Nm).

   12. Installation of the remaining components is the reverse of the removal procedure.

   13. Make sure all fluid levels are accurate, then start the engine check for leaks.

Water Pump

REMOVAL & INSTALLATION

1ZZ-FE Engine

   1. Before servicing the vehicle, refer to the precautions in the beginning of this section.

   2. Drain the cooling system.

   3. Remove or disconnect the following:

   • Negative battery cable
   • Right-hand engine under cover
   • Drive belt
   • Alternator
   • Water pump

To install:

   4. Install or connect the following:

   • Water pump. Torque bolts marked **A** (short) to 80 inch lbs. (9 Nm) and bolts marked **B** (long) to 96 inch lbs. (11 Nm).
   • Alternator

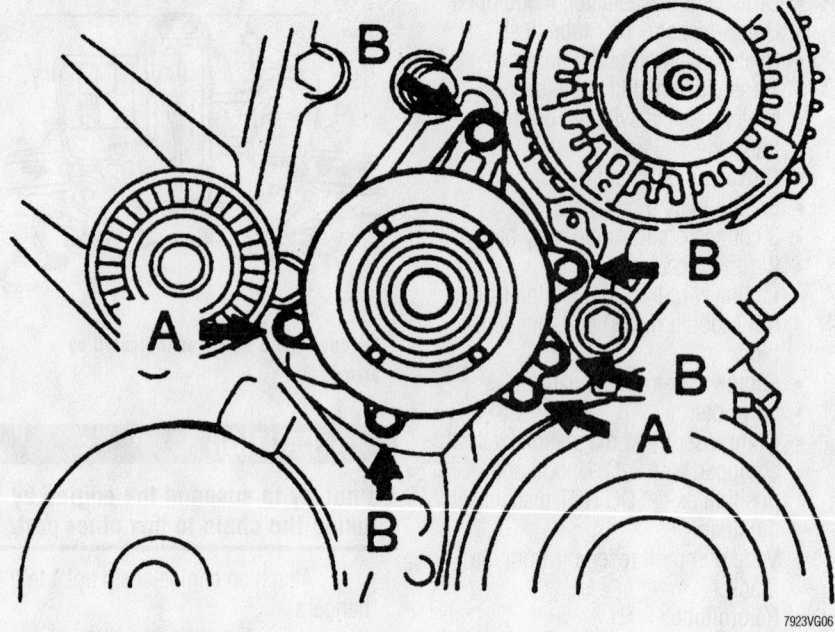

Water pump bolt identification—1.8L (1ZZ-FE) engine

7923VG06

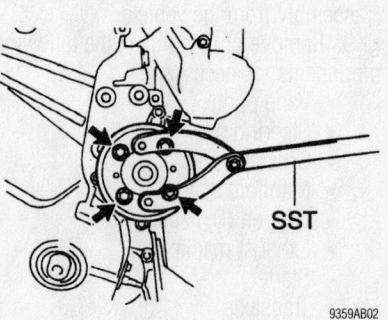

9359AB02

View of the special tool needed to remove and install the water pump pulley

   • Drive belt
   • Right engine under cover
   • Negative battery cable

   5. Fill the cooling system to the proper level.

   6. Start the vehicle, check for leaks and repair if necessary.

2ZZ-GE Engine

   1. Before servicing the vehicle, refer to the precautions in the beginning of this section.

   2. Drain the cooling system.

   3. Remove or disconnect the following:

   • Negative battery cable
   • Right-hand engine under cover
   • Drive belt
   • Alternator
   • Water pump pulley, using SST 09960-10010
   • Water pump and O-ring

To install:

   4. Install or connect the following:

   • Water pump with new O-ring. Torque the bolts to 80 inch lbs. (9 Nm).
   • Water pump pulley, using SST 09960-10010. Torque the bolts to 11 ft. lbs. (15 Nm).
   • Alternator
   • Drive belt
   • Right engine under cover
   • Negative battery cable

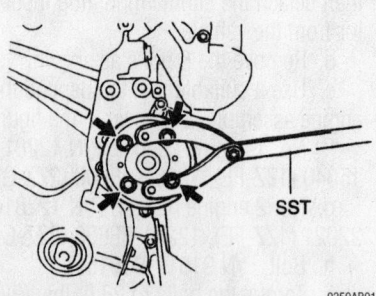

9359AB01

Water pump mounting and bolt locations—2ZZ-GE engine

5. Fill the cooling system to the proper level.

6. Start the vehicle, check for leaks and repair if necessary.

Cylinder Head

REMOVAL & INSTALLATION

1. Before servicing the vehicle, refer to the precautions in the beginning of this section.

2. Drain the cooling system.

3. Remove or disconnect the following:

- Right side engine under cover
- Right front wheel and tire
- Cylinder head cover
- Air cleaner assembly with hose
- Accelerator control cable
- Wire harness clamp and suction hose assembly, 2ZZ-GE engine only
- Water bypass hoses
- Fuel pipe clamp
- Fuel tube sub-assembly
- Union-to-connector tube hose
- Radiator and heater inlet hoses
- Drive belt

4. Separate the vane pipe assembly, but do not disconnect the hose, 1ZZ-FE engine.

- Alternator bracket, 2ZZ-GE
- Alternator

5. Separate the compressor and magnetic clutch, on 2ZZ-GE engines with air conditioning.

- Front exhaust pipe assembly
- Power steering pump reservoir and position it aside, 1ZZ-FE engine

6. Place a jack with a wooden block under the vehicle for support, then remove the 4 bolts and 2 nuts and remove the right side engine mount.

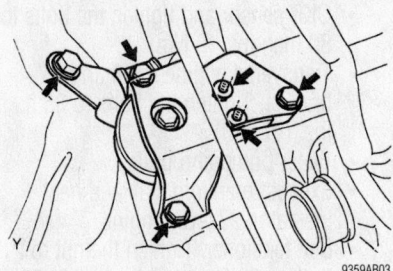

With the engine supported, remove the right side engine mount—1ZZ-FE engine shown, 2ZZ-GE similar

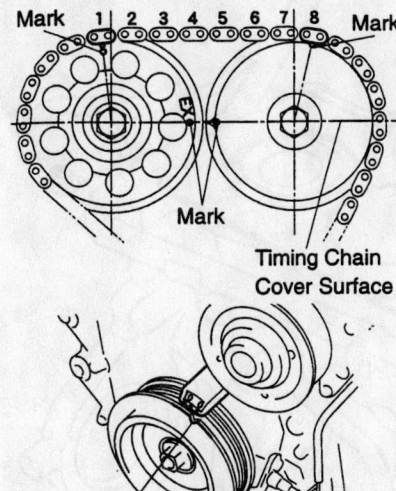

Proper timing mark alignment for TDC

7. Remove the engine wire, on 1ZZ-FE engines as follows:

a. Remove the 5 clamps from the brackets.

b. Detach the connectors.

c. Remove the ignition coil connectors.

d. Bolt and nut holding the engine wire.

8. Remove or disconnect the following:

- Ignition coil assembly
- Positive Crankcase Ventilation (PCV) hoses
- Valve (cylinder head) cover sub-assembly

9. Set the No. 1 cylinder to Top Dead Center (TDC) of the compressor stroke as follows:

a. Turn the crankshaft pulley, and align its groove with the "0" timing mark of the timing chain cover.

b. Make sure the point marks of the camshaft timing sprockets and Variable Valve Timing (VVT) timing sprockets are in a straight line as shown. If not, turn the crankshaft 1 complete revolution (360°) and align the marks.

10. Remove or disconnect the following:

- Crankshaft pulley, using SST 09960-10010
- Belt tensioner
- Exhaust manifold stay and head insulator, 2ZZ-GE engine
- Water pump pulley and pump
- Transverse engine mounting bracket
- Crankshaft Position (CKP) sensor

- No. 1 chain tensioner assembly, making sure not to revolve the crankshaft without the tensioner
- Timing chain or belt cover
- Timing gear cover oil seal
- CKP sensor plate No. 1
- Timing chain tensioner slipper
- Timing chain vibration damper No. 1

➡ **In case you turn the camshafts with the timing chain removed, turn the crankshaft ¼ turn for the valve to avoid contact with the pistons.**

- Timing chain sub-assembly. Remove the chain with the crankshaft gear, using screwdrivers as shown.
- Surge tank stay, 2ZZ-GE engine
- Intake manifold
- Oil level gauge
- Water bypass pipe bolts and pipe, 1ZZ-FE engine
- Camshafts
- Camshaft timing oil control valve, 1ZZ-FE engine
- Manifold stay, 1ZZ-FE engine
- Cylinder head bolts in sequence. To prevent damage to the cylinder head, loosen each bolt about ¼ of a turn during each pass until the bolts are loose.
- Cylinder head

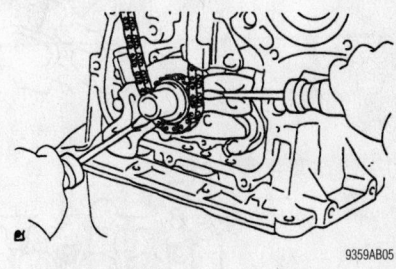

Remove the timing chain with the crankshaft gear

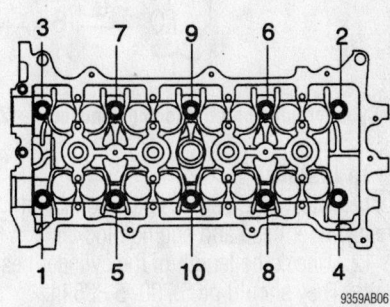

Cylinder head bolt loosening sequence

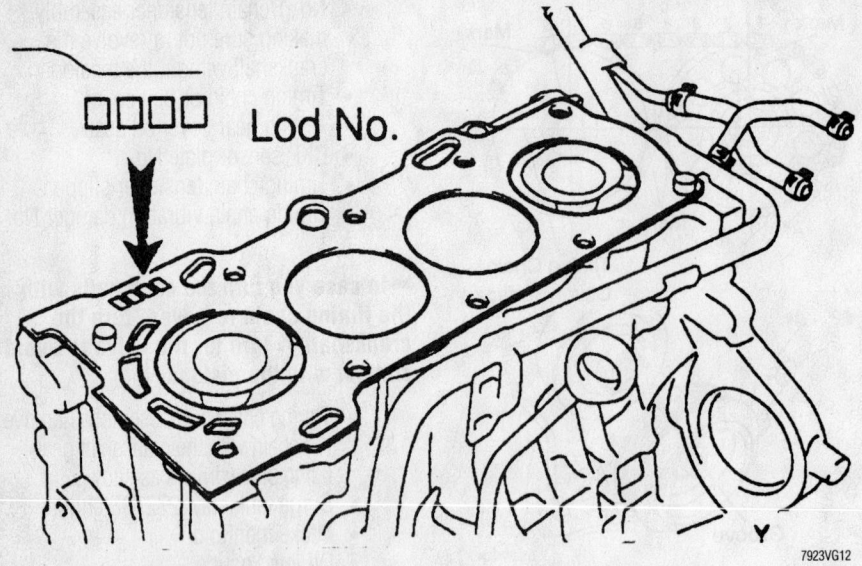

Position the head gasket correctly on the cylinder head—1.8L (1ZZ-FE) engine

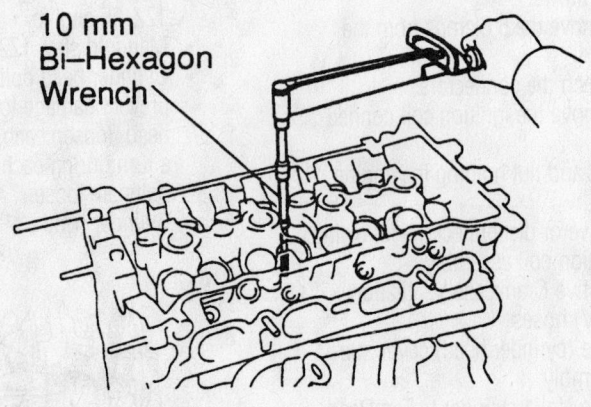

10 mm
Bi–Hexagon
Wrench

Cylinder head bolt tightening sequence—1ZZ-FE and 2ZZ-GE engines

To install:

11. Clean and degrease the surface of the cylinder head and engine block.

12. Check the length of the cylinder head bolts. They should be 5.780–5.835 in. (146.8–148.2mm) long. If they are longer than 5.846 in. (148.5mm), they must be replaced.

13. Install or connect the following:
  • New gasket on the engine block with the Lot No. stamp facing up.
  • Cylinder head
  • Apply a light coat of oil to cylinder head bolt threads and tighten in sequence, in several passes, to 36 ft. lbs. (49 Nm) for 1ZZ-FE engines or to 26 ft. lbs. (35 Nm) for 2ZZ-GE engines.
  • Tighten each head bolt, in sequence, an additional 90 degree turn for 1ZZ-FE engines, or 180 degree turn for 2ZZ-GE engines.
  • Manifold stay, 1ZZ-FE engine. Tighten the bolts to 36 ft. lbs. (49 Nm).
  • Camshaft timing oil control valve, on 1ZZ-FE engines, and tighten to 80 inch lbs. (9 Nm)
  • Camshaft
  • Water by-pass pipe, on 1ZZ-FE engines, and tighten to 80 inch lbs. (9 Nm)
  • Oil level gauge
  • Intake manifold
  • Surge tank stay, 2ZZ-GE engine. Tighten to 18 ft. lbs. (24 Nm).
  • Timing chain
  • Timing chain vibration damper. Tighten the bolts to 80 inch lbs. (9 m).
  • Timing chain tensioner slipper and tighten the bolt to 14 ft. lbs. (19 Nm).
  • Crankshaft position sensor plate, with the "F" mark facing forward.
  • Timing gear cover oil seal
  • Timing cover. For 1ZZ-FE engine, tighten the "A" bolts to 10 ft. lbs. (13 Nm), the "B" bolts to 14 ft. lbs. (19 Nm) and the stud bolt to 84 inch lbs. (9.5 Nm), using a Torx® wrench. For 2ZZ-GE engines, tighten the M8 bolts to 15 ft. lbs. (21 m), the M6 bolts to 8 ft. lbs. (11 Nm) and the stud bolt to 84 inch lbs. (9.5 Nm).

➡ **When installing the tensioner, make sure to set the hook again if the hook releases the plunger.**

  • Timing chain tensioner. Torque the nuts to 80 inch lbs. (9 Nm).
  • CKP sensor and tighten the bolts to 80 inch lbs. (9 Nm)
  • Transverse engine mounting bracket. Tighten the bolts to 35 ft. lbs. (47 Nm).
  • Water pump and pulley
  • Exhaust manifold stay and heat insulator, 2ZZ-GE engine
  • Belt tensioner. Tighten the nut to 21 ft. lbs. (29 Nm) and the bolt to 51 ft. lbs. (69 Nm) on 1ZZ-FE engines or to 74 ft. lbs. (100 Nm) on 2ZZ-GE engines.

14. Install the crankshaft pulley, as follows:

  a. Align the pulley set key with the

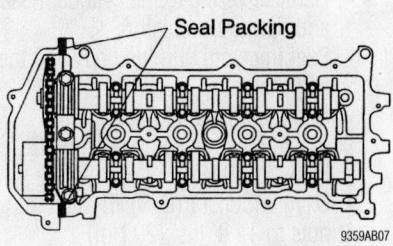

Seal packing installation locations

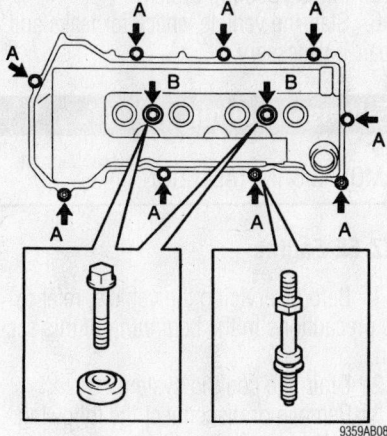

Cylinder head (valve) cover bolt locations—1ZZ-FE engine

key groove of the pulley and slide on the pulley.

b. Use SST 09960-11010 to install the bolt and tighten to 102 ft. lbs. (138 Nm) for 1ZZ-FE engine or to 87 ft. lbs. (118 Nm) on 2ZZ-GE engines.

c. Turn the crankshaft counterclockwise and disconnect the plunger knock pin from the hook.

d. Turn the crankshaft clockwise and check that the slipper is pushed by the plunger. If the plunger does not spring out, press the slipper into the chain tensioner with a screwdriver so that the hook is released from the knock pin and the plunder springs out.

15. Install or connect the following:
- Cylinder head sub-assembly cover. Install seal packing into the locations shown and install within 3 minutes. Tighten the "A" bolts to 8 ft. lbs. (11 Nm) and the "B" bolts to 80 inch lbs. (9 Nm) for 1ZZ-FE engines. For 2ZZ-GE engines, tighten the bolts to 7 ft. lbs. (10 Nm).
- Ignition coil assembly. Torque the bolts to 80 inch lbs. (9 Nm).

- Engine wire and tighten to 80 inch lbs. (9 Nm), 1ZZ-FE
- Right side engine mount. Tighten to 38 ft. lbs. (52 Nm).
- Front exhaust pipe
- Vane pump, 1ZZ-FE
- Compressor and magnetic clutch, 2ZZ-GE
- Alternator bracket, 2ZZ-GE engine
- Alternator
- Suction hose and wire harness clamp, 2ZZ-GE engine
- Air cleaner and hose
- Main cylinder head cover and tighten to 62 inch lbs. (7 Nm)
- Right front wheel and tire. Tighten the lug nuts to 76 ft. lbs. (103 Nm).

16. Fill the cooling system to the proper level.

17. Start the vehicle, check for leaks and repair if necessary.

Intake Manifold

REMOVAL & INSTALLATION

1ZZ-FE Engine

1. Before servicing the vehicle, refer to the precautions in the beginning of this section.
2. Drain the cooling system.
3. Remove or disconnect the following:
- Negative battery cable
- Drive belt and alternator
- Air intake duct
- Accelerator cable

- Exhaust pipe from the manifold.
- Exhaust manifold support bracket
- Spark plug wires, then ignition coils
- Spark plugs
- Positive Crankcase Ventilation (PCV) hoses
- Throttle body assembly
- 2 bolts securing the wiring harness protector
- Wiring connectors and ground wires
- Intake manifold support bracket
- Intake manifold and gasket

To install:

4. Install or connect the following:
- Intake manifold with a new gasket. Torque the bolts to 22 ft. lbs. (30 Nm).
- Harness wiring to the cylinder head and harness protector
- Fuel injectors, throttle body and the PCV hoses
- Spark plugs and ignition coils. Tighten the bolts and nuts to 80 inch lbs. (9 Nm).
- Exhaust manifold and support bracket. Tighten the bolts to 37 ft. lbs. (49 Nm).
- Front exhaust pipe to the manifold. Tighten the bolts to 46 ft. lbs. (62 Nm).
- Oxygen Sensor (O_2S). Tighten the nuts to 14 ft. lbs. (20 Nm).
- Accelerator cable and air intake duct

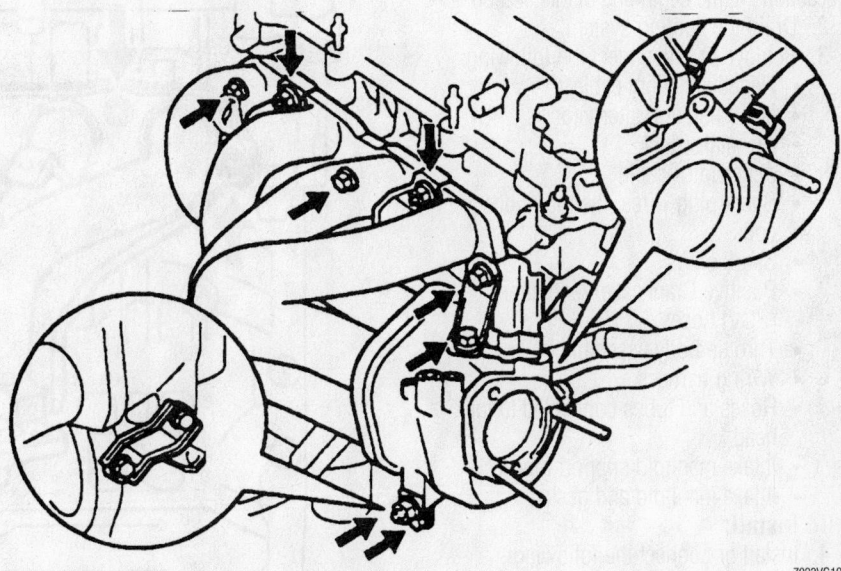

Intake manifold mounting fastener locations—1.8L (1ZZ-FE) engine

Timing belt service is covered in Section 3 of this manual

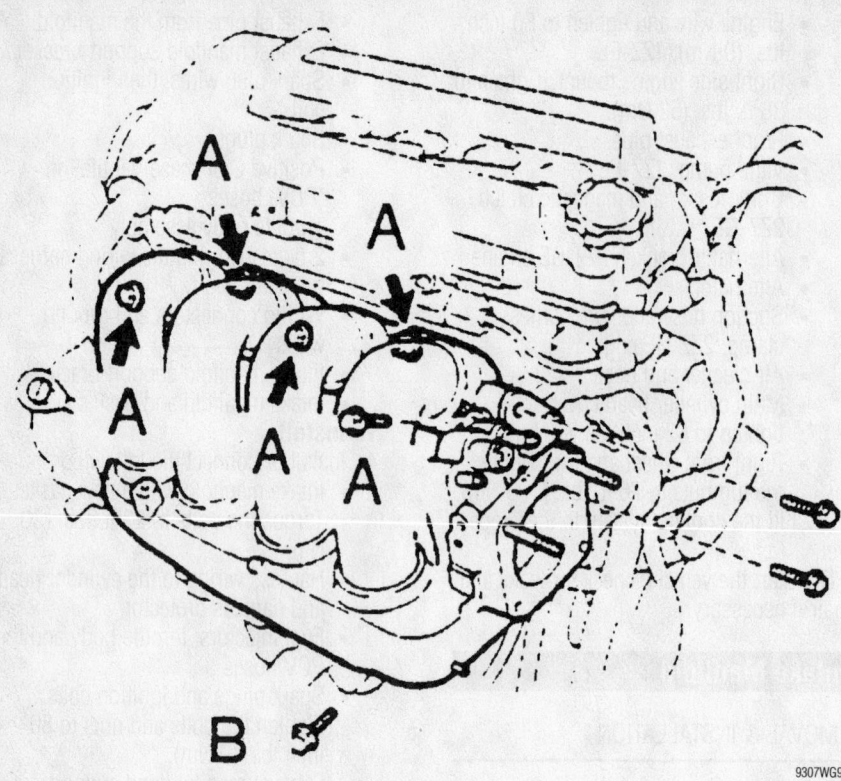

Intake manifold bolt installation—2ZZ-GE engine

9307WG93

- Alternator and drive belt
- Negative battery cable

5. Fill the cooling system.

6. Start the vehicle, check for leaks and repair if necessary.

2ZZ-GE Engine

1. Before servicing the vehicle, refer to the precautions in the beginning of this section.

2. Drain the cooling system.

3. Remove or disconnect the following:
- Negative battery cable
- Drive belt and alternator
- Air intake duct
- Accelerator cable
- Spark plug wires, then ignition coils
- Spark plugs
- Positive Crankcase Ventilation (PCV) hoses
- Throttle body assembly
- Wiring harness
- Hoses and tubes connected to the head
- Intake manifold support bracket
- Intake manifold and gasket

To install:

4. Install or connect the following:
- Intake manifold with a new gasket. Tighten bolts A to 25 ft. lbs. (34 Nm) and bolt B to 34 ft. lbs. (46 Nm)

- Harness wiring to the cylinder head and harness protector
- Fuel injectors, throttle body and the PCV hoses
- Spark plugs and ignition coils. Tighten the bolts and nuts to 80 inch lbs. (9 Nm).
- Oxygen Sensor (O_2S). Tighten the nuts to 14 ft. lbs. (20 Nm).
- Accelerator cable and air intake duct
- Alternator and drive belt
- Negative battery cable

5. Fill the cooling system.

6. Start the vehicle, check for leaks and repair if necessary.

Exhaust Manifold

REMOVAL & INSTALLATION

1ZZ-FE Engine

1. Before servicing the vehicle, refer to the precautions in the beginning of this section.

2. Drain the cooling system.

3. Remove or disconnect the following:
- Negative battery cable

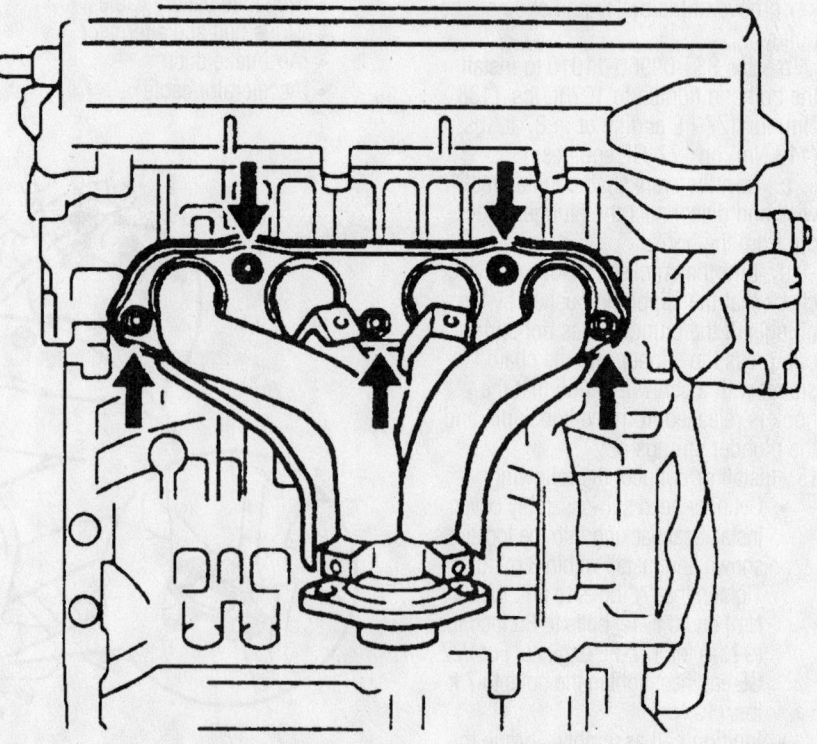

Exhaust manifold mounting nut locations—1.8L (1ZZ-FE) engine

7923VG22

- Drive belt and alternator
- Air intake duct
- Accelerator cable
- Exhaust pipe from the manifold
- Exhaust manifold support bracket
- Heat insulator from the dash panel
- Upper heat insulator
- Exhaust manifold and gasket
- If necessary, the lower heat insulator from the exhaust manifold.

To install:

4. Install or connect the following:
- Lower heat insulator on the exhaust manifold. Tighten the bolts to 108 inch lbs. (12 Nm).
- Exhaust manifold using a new gasket. Tighten the nuts, in several passes, to 27 ft. lbs. (37 Nm).
- Upper heat insulator. Tighten the bolts to 108 inch lbs. (12 Nm).
- Heat insulator on the dash panel
- Exhaust manifold support bracket. Tighten the bolts, in an alternating pattern, to 37 ft. lbs. (49 Nm).
- Front exhaust pipe to the manifold. Tighten the bolts to 46 ft. lbs. (62 Nm).
- Oxygen Sensor (O$_2$S). Tighten the nuts to 14 ft. lbs. (20 Nm).
- Accelerator cable and air intake duct
- Alternator and drive belt
- Negative battery cable

5. Fill the cooling system.
6. Start the vehicle, check for leaks and repair if necessary.

2ZZ-GE Engine

1. Before servicing the vehicle, refer to the precautions in the beginning of this section.
2. Drain the cooling system.
3. Remove or disconnect the following:
- Negative battery cable
- Drive belt and alternator
- Air intake duct
- Accelerator cable
- Exhaust pipe from the manifold
- Exhaust manifold support bracket
- Heat insulator from the dash panel.
- Upper heat insulator
- Exhaust manifold and gasket
- If necessary, the lower heat insulator from the exhaust manifold.

To install:

4. Install or connect the following:
- Lower heat insulator on the exhaust manifold. Tighten the bolts to 15 ft. lbs. (20 Nm).

- Exhaust manifold using a new gasket. Tighten the nuts, in several passes to 37 ft. lbs. (50 Nm).
- Upper heat insulator. Tighten the bolts to 15 ft. lbs. (20 Nm).
- Heat insulator on the dash panel.
- Exhaust manifold support bracket. Tighten the bolts to 37 ft. lbs. (49 Nm).
- Front exhaust pipe to the manifold. Tighten the bolts to 46 ft. lbs. (62 Nm).
- Oxygen Sensor (O$_2$S). Tighten the nuts to 14 ft. lbs. (20 Nm).
- Accelerator cable and air intake duct.
- Alternator and drive belt.
- Negative battery cable

5. Fill the cooling system.
6. Start the vehicle, check for leaks and repair if necessary.

Camshaft(s)

REMOVAL & INSTALLATION

1ZZ-FE Engine

1. Before servicing the vehicle, refer to the precautions in the beginning of this section.
2. Remove or disconnect the following:
- Negative battery cable
- Right side engine under cover
- Cylinder head cover
- Suction hose sub-assembly, 2ZZ-GE engine
- Drive belt
- Power steering pump reservoir and position it aside, 1ZZ-FE engine

3. Place a jack with a wooden block under the vehicle for support, then remove the 4 bolts and 2 nuts and remove the right side engine mount.
4. Remove the engine wire, on 1ZZ-FE engines:
   a. Remove the 5 clamps from the brackets.
   b. Detach the connectors.
   c. Remove the ignition coil connectors.
   d. Bolt and nut holding the engine wire.
5. Remove or disconnect the following:
- Ignition coil assembly
- Positive Crankcase Ventilation (PCV) hoses from the valve cover
- Valve (cylinder head) cover sub-assembly

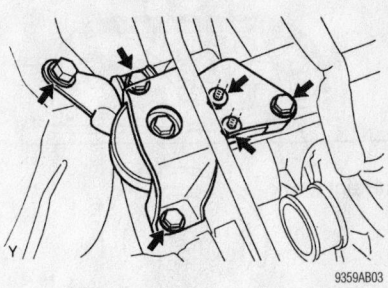

With the engine supported, remove the right side engine mount—1ZZ-FE engine shown, 2ZZ-GE similar

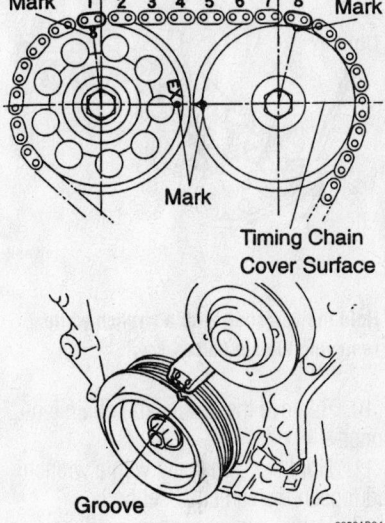

Proper timing mark alignment for TDC

6. Set the No. 1 cylinder to Top Dead Center (TDC) of the compressor stroke as follows:
   a. Turn the crankshaft pulley, and align its groove with the "0" timing mark of the timing chain cover.
   b. Make sure the point marks of the camshaft timing sprockets and VVT timing sprockets are in a straight line as shown. If not, turn the crankshaft 1 complete revolution (360°) and align the marks.
7. Remove the drive belt tensioner.

✴✴ WARNING

Do not turn the crankshaft without the tensioner installed.

8. Make sure the No. 1 cylinder is at TDC of the compression stroke.
9. Matchmark the timing chain and camshaft sprockets

Heater Core replacement is covered in Section 2 of this manual

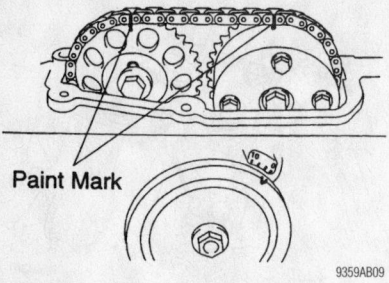

Matchmark the timing chain and cam sprockets

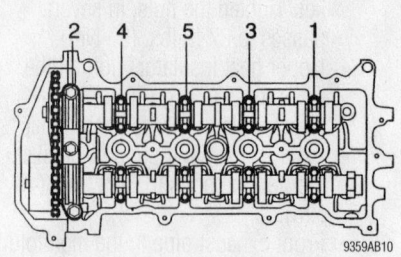

Camshaft bearing cap bolt removal sequence—1ZZ-FE engine

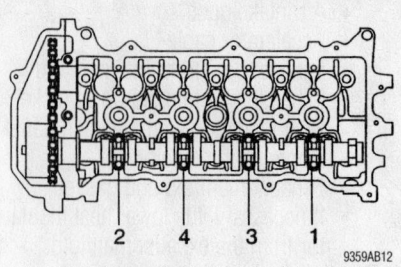

Camshaft bearing cap bolt removal sequence—1ZZ-FE engine

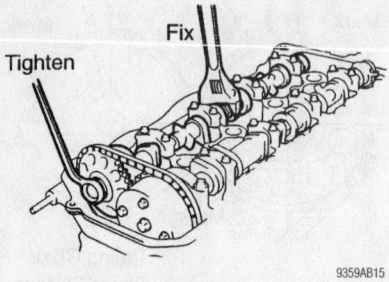

Hold the camshaft with a wrench while removing the set bolt

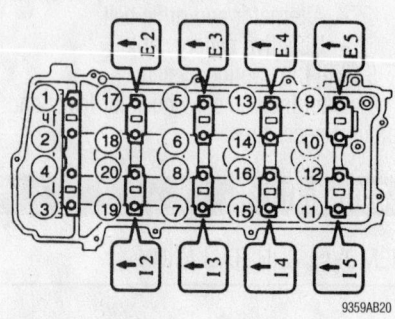

Camshaft bearing cap bolt removal sequence—2ZZ-GE engine

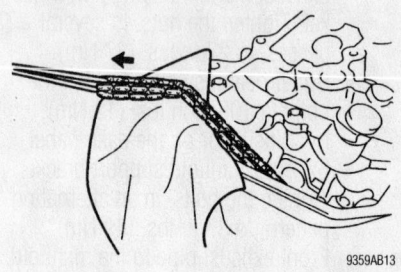

Secure the timing chain with string to prevent it from slipping down into the timing chain cover

10. Remove the 2 nuts and chain tensioner.

11. Hold the camshafts with a wrench and loosen the camshaft set bolt.

12. Using several passes, gradually remove the bearing cap bolts from the No. 2 camshaft, in the proper sequence.

13. Remove the camshaft and timing gear as shown.

14. Using several passes, gradually remove the bearing cap bolts from the other camshaft, in the proper sequence.

15. Remove the camshaft while holding the timing chain.

❈❈ WARNING

Do not let anything drop down into the timing chain cover while the camshafts are removed.

16. Tie the timing chain with a string as shown, to prevent it from dropping down into the timing chain cover.

To install:

17. Position the camshaft on the cylinder head, then install the timing chain on the cam timing gear, with the painted links aligned with the marks on the timing gear.

18. Check the front marks and numbers and torque the camshaft cap bolts, in

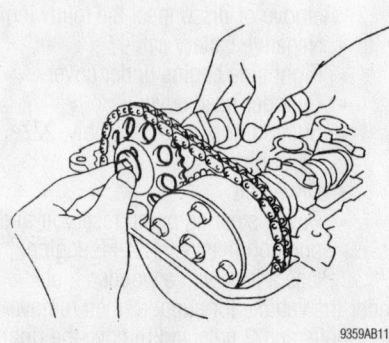

Carefully remove the cam and timing gear

sequence, to 10 ft. lbs. (13 Nm) for 1ZZ-FE engine, or to 14 ft. lbs. (19 Nm) for 2ZZ-GE engines.

19. Put camshaft No. 2 on the cylinder head, with the painted links of the chain aligned with the mark on the timing gear.

20. Tighten the camshaft gear set bolt temporarily.

21. Check the front marks and numbers and torque the camshaft cap bolts, in sequence, to 10 ft. lbs. (13 Nm). Install the No. 1 bearing cap and tighten to 17 ft. lbs. (23 Nm).

22. Hold the camshaft secure with a

wrench and tighten the set bolt to 40 ft. lbs. (54 Nm). Be careful not the damage the lifters.

23. Check to be sure the matchmarks on the timing chain and cam sprockets, and the alignment of the pulley groove with the timing mark on the cover are still aligned.

24. Install the chain tensioner:

a. Make sure the O-ring is clean, then set the hook as shown.

b. Oil the tensioner, then install and tighten to 80 inch lbs. (9 Nm).

➡ **When installing the tensioner, set the hook again if the hook releases the plunger.**

c. Turn the crankshaft counterclockwise, and disconnect the plunger knock pin from the hook.

d. Turn the crankshaft clockwise and check that the slipper is pushed by the plunger. If the plunger does not spring out, press the slipper into the chain tensioner with a screwdriver so that the hook is released from the knock pin and the plunger springs out.

25. Check the valve clearance and make adjustments as needed.

26. Install or connect the following:

- Belt tensioner. Tighten the nut to 21 ft. lbs. (29 Nm) and the bolt to 51 ft. lbs. (69 Nm).

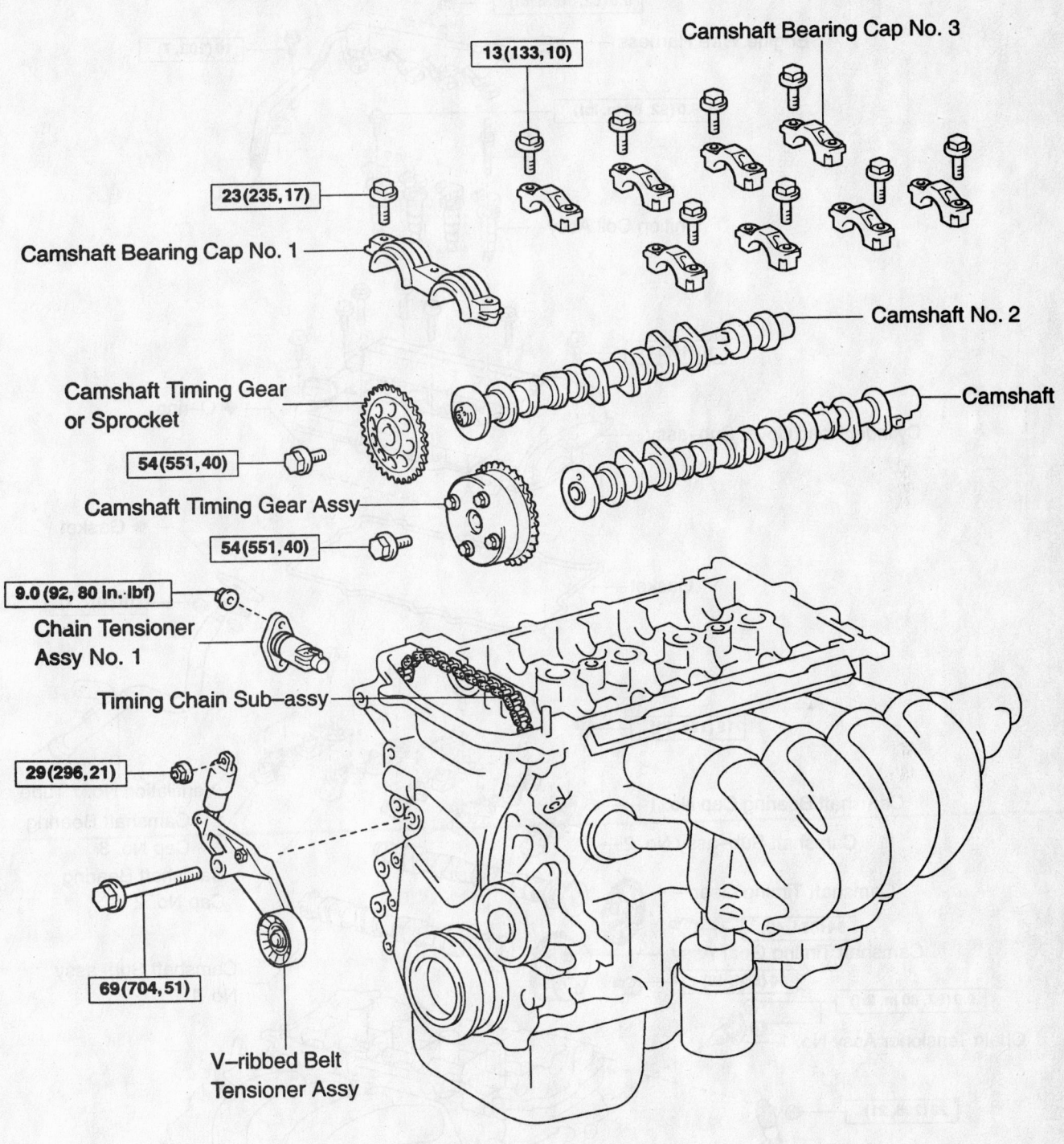

13(133,10)

Camshaft Bearing Cap No. 3

23(235,17)

Camshaft Bearing Cap No. 1

Camshaft No. 2

Camshaft Timing Gear
or Sprocket

Camshaft

54(551,40)

Camshaft Timing Gear Assy

54(551,40)

9.0(92, 80 in.·lbf)

Chain Tensioner
Assy No. 1

Timing Chain Sub–assy

29(296,21)

69(704,51)

V–ribbed Belt
Tensioner Assy

N·m (kgf·cm, ft·lbf) : Specified torque

9359AB21

Exploded view of the camshafts and related components—1ZZ-FE engine

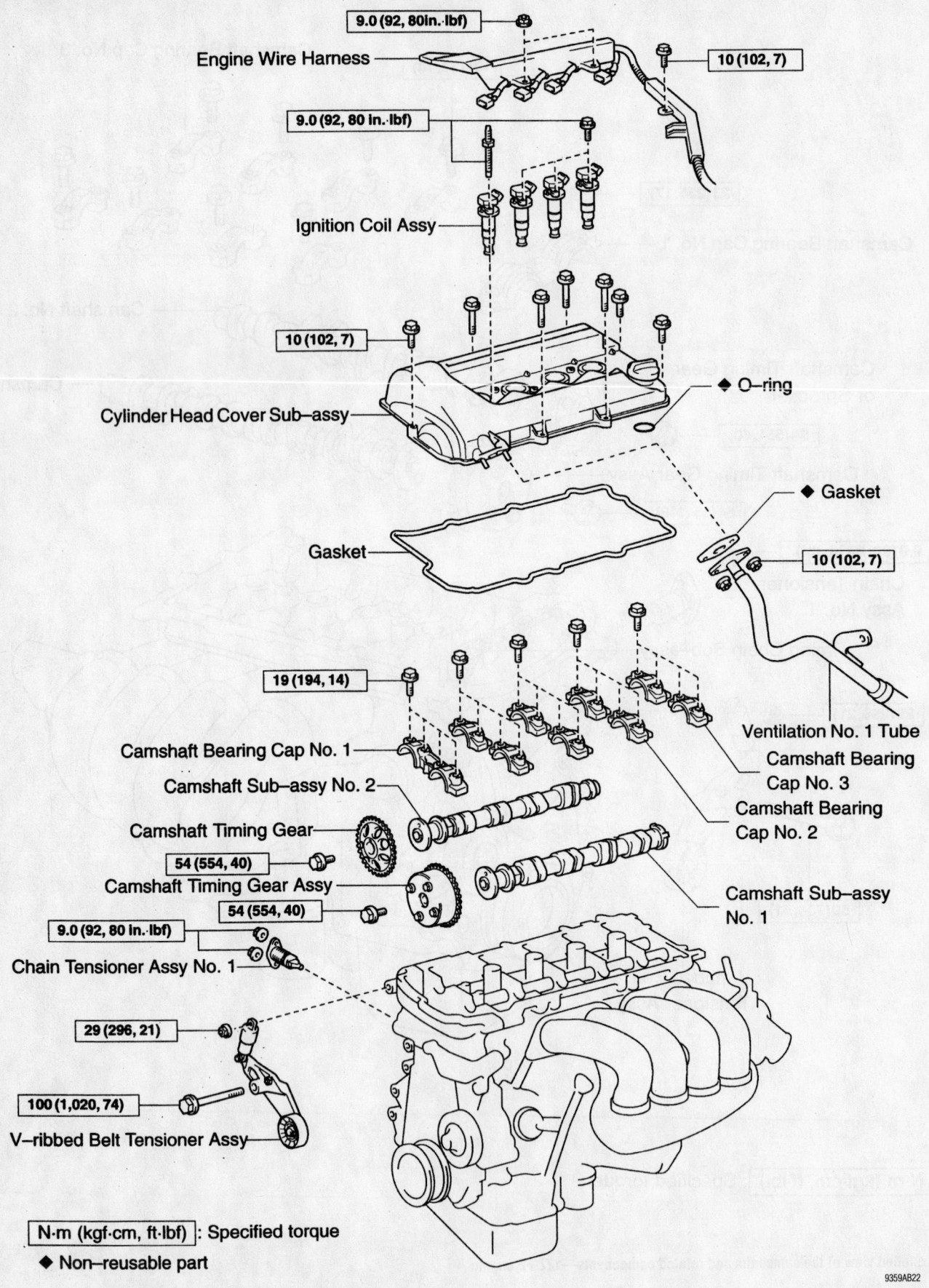

9.0 (92, 80 in.·lbf)

Engine Wire Harness

10 (102, 7)

9.0 (92, 80 in.·lbf)

Ignition Coil Assy

10 (102, 7)

Cylinder Head Cover Sub–assy

◆ O–ring

◆ Gasket

Gasket

10 (102, 7)

Ventilation No. 1 Tube

19 (194, 14)

Camshaft Bearing Cap No. 1

Camshaft Bearing Cap No. 3

Camshaft Sub–assy No. 2

Camshaft Bearing Cap No. 2

Camshaft Timing Gear

54 (554, 40)

Camshaft Timing Gear Assy

Camshaft Sub–assy No. 1

54 (554, 40)

9.0 (92, 80 in.·lbf)

Chain Tensioner Assy No. 1

29 (296, 21)

100 (1,020, 74)

V–ribbed Belt Tensioner Assy

N·m (kgf·cm, ft·lbf) : Specified torque

◆ Non–reusable part

9359AB22

Exploded view of the camshafts and related components—2ZZ-GE engine

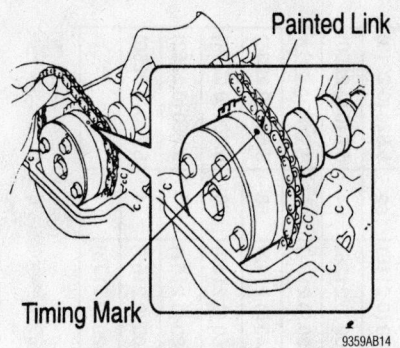

Make sure the alignment marks on the timing chain and camshaft gear match up

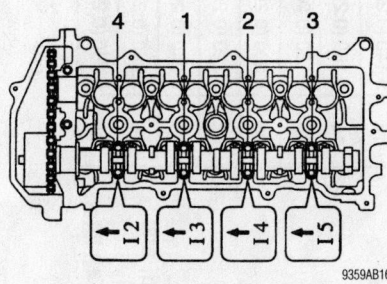

Camshaft cap bolt tightening sequence— 1ZZ-FE engine

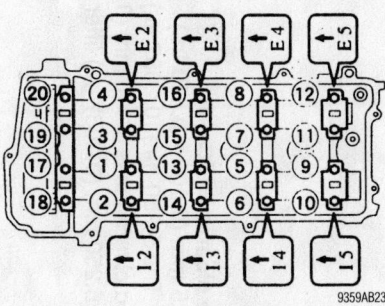

Camshaft cap bolt tightening sequence— 2ZZ-GE engine

- Cylinder head sub-assembly cover. Install seal packing into the locations shown and install within 3 minutes. Tighten the "A" bolts to 8 ft. lbs. (11 Nm) and the "B" bolts to 80 inch lbs. (9 Nm) for 1ZZ-FE engine and to 7 ft. lbs. (10 Nm) for 2ZZ-GE engines.
- Ignition coil assembly. Torque the bolts to 80 inch lbs. (9 Nm).
- Engine wire and tighten to 80 inch lbs. (9 Nm)

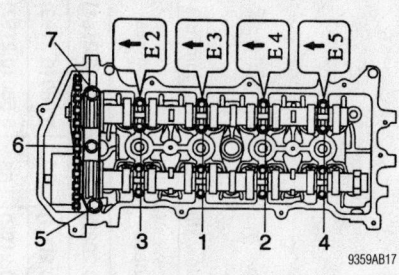

Camshaft cap bolt tightening sequence— 1ZZ-FE

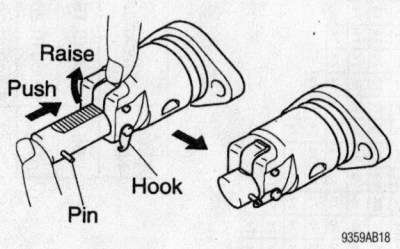

Set the timing chain tensioner hook properly

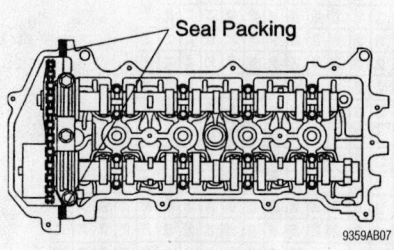

Seal packing installation locations

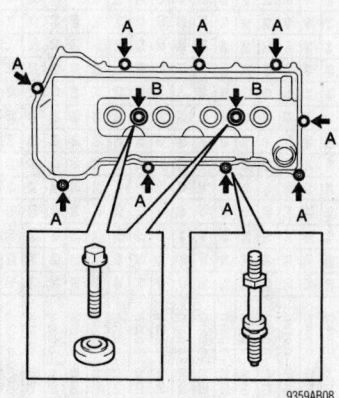

Cylinder head (valve) cover bolt locations—1ZZ-FE engine

- Right side engine mount. Tighten to 38 ft. lbs. (52 Nm).
- Cylinder head (valve) cover
- Negative battery cable

Valve Lash

ADJUSTMENT

1ZZ-FE Engine

➡ **Adjust the valve clearance when the engine is cold.**

1. Before servicing the vehicle, refer to the precautions in the beginning of this section.

2. Remove or disconnect the following:

- Negative battery cable.
- Cylinder head covers
- Engine wire
- Ignition coil
- Positive Crankcase Ventilation (PCV) hoses
- Cylinder head cover sub-assembly

3. Set the No. 1 cylinder to Top Dead Center (TDC) of the compressor stroke as follows:

   a. Turn the crankshaft pulley, and align its groove with the "0" timing mark of the timing chain cover.

   b. Make sure the point marks of the camshaft timing sprockets and VVT timing sprockets are in a straight line as shown. If not, turn the crankshaft 1 complete revolution (360°) and align the marks.

4. Check the valve clearance of the first set of the valves shown:

   a. Use a feeler gauge to measure the clearance between the valve lifter and camshaft. The clearance of the intake valves should be 0.0059–0.0098 in. (0.15–0.25mm). The clearance of the exhaust valves should be 0.0098–0.0138 in. (0.25–0.35mm).

   b. Note the out-of-specification valve clearance measurements. You will need them later to determine the required replacement valve lifter.

   c. Turn the crankshaft 1 revolution (360°) to set the No. 4 cylinder to TDC.

5. Check the valve clearance of the second set of the valves shown:

   a. Use a feeler gauge to measure the clearance between the valve lifter and camshaft. The clearance of the intake valves should be 0.0059–0.0098 in. (0.15–0.25mm). The clearance of the

1ZZ–FE: Valve Lifter Selection Chart (Intake)

New lifter thickness   mm (in.)

Lifter No.	Thickness	Lifter No.	Thickness	Lifter No.	Thickness
06	5.060 (0.1992)	30	5.300 (0.2087)	54	5.540 (0.2181)
08	5.080 (0.2000)	32	5.320 (0.2094)	56	5.560 (0.2189)
10	5.100 (0.2008)	34	5.340 (0.2102)	58	5.580 (0.2197)
12	5.120 (0.2016)	36	5.360 (0.2110)	60	5.600 (0.2205)
14	5.140 (0.2024)	38	5.380 (0.2118)	62	5.620 (0.2213)
16	5.160 (0.2031)	40	5.400 (0.2126)	64	5.640 (0.2220)
18	5.180 (0.2039)	42	5.420 (0.2134)	66	5.660 (0.2228)
20	5.200 (0.2047)	44	5.440 (0.2142)	68	5.680 (0.2236)
22	5.220 (0.2055)	46	5.460 (0.2150)	70	5.700 (0.2244)
24	5.240 (0.2063)	48	5.480 (0.2157)	72	5.720 (0.2252)
26	5.260 (0.2071)	50	5.500 (0.2165)	74	5.740 (0.2260)
28	5.280 (0.2079)	52	5.520 (0.2173)		

The main selection chart cross-references the **installed lifter thickness mm (in.)** (across the top, values from 5.060 (0.1982) to 5.740 (0.2260)) against the **measured clearance mm (in.)** (down the left side) to give the new lifter number.

Measured clearance mm (in.):

- 0.000 – 0.030 (0.0000 – 0.0012)
- 0.031 – 0.050 (0.0012 – 0.0020)
- 0.051 – 0.070 (0.0020 – 0.0028)
- 0.071 – 0.090 (0.0028 – 0.0035)
- 0.091 – 0.110 (0.0036 – 0.0043)
- 0.111 – 0.130 (0.0044 – 0.0051)
- 0.131 – 0.149 (0.0052 – 0.0059)
- 0.150 – 0.250 (0.0059 – 0.0098)
- 0.251 – 0.270 (0.0099 – 0.0106)
- 0.271 – 0.290 (0.0107 – 0.0114)
- 0.291 – 0.310 (0.0115 – 0.0122)
- 0.311 – 0.330 (0.0122 – 0.0130)
- 0.331 – 0.350 (0.0130 – 0.0138)
- 0.351 – 0.370 (0.0138 – 0.0145)
- 0.371 – 0.390 (0.0146 – 0.0154)
- 0.391 – 0.410 (0.0154 – 0.0161)
- 0.411 – 0.430 (0.0162 – 0.0169)
- 0.431 – 0.450 (0.0170 – 0.0177)
- 0.451 – 0.470 (0.0178 – 0.0185)
- 0.471 – 0.490 (0.0185 – 0.0193)
- 0.491 – 0.510 (0.0193 – 0.0201)
- 0.511 – 0.530 (0.0201 – 0.0209)
- 0.531 – 0.550 (0.0209 – 0.0217)
- 0.551 – 0.570 (0.0217 – 0.0224)
- 0.571 – 0.590 (0.0225 – 0.0232)
- 0.591 – 0.610 (0.0233 – 0.0240)
- 0.611 – 0.630 (0.0241 – 0.0248)
- 0.631 – 0.650 (0.0248 – 0.0256)
- 0.651 – 0.670 (0.0256 – 0.0264)
- 0.671 – 0.690 (0.0264 – 0.0272)
- 0.691 – 0.710 (0.0272 – 0.0280)
- 0.711 – 0.730 (0.0280 – 0.0287)
- 0.731 – 0.750 (0.0288 – 0.0295)
- 0.751 – 0.770 (0.0296 – 0.0303)
- 0.771 – 0.790 (0.0304 – 0.0311)
- 0.791 – 0.810 (0.0311 – 0.0319)
- 0.811 – 0.830 (0.0319 – 0.0327)
- 0.831 – 0.850 (0.0327 – 0.0335)
- 0.851 – 0.870 (0.0335 – 0.0343)
- 0.871 – 0.890 (0.0343 – 0.0350)
- 0.891 – 0.910 (0.0351 – 0.0358)
- 0.911 – 0.930 (0.0359 – 0.0366)

Intake valve clearance (Cold):
0.15 – 0.25 mm (0.006 – 0.010 in.)

EXAMPLE: The 5.250 mm (0.2067 in.) lifter is installed, and the measured clearance is 0.400 mm (0.0157 in.).
Replace the 5.250 mm (0.2067 in.) lifter with a new No. 48 lifter.

Adjusting shim chart (intake)—1ZZ-FE engine

9307WG70

1ZZ–FE: Valve Lifter Selection Chart (Exhaust)

9307WG71

New lifter thickness mm (in.)

Lifter No.	Thickness	Lifter No.	Thickness	Lifter No.	Thickness
06	5.060 (0.1992)	30	5.300 (0.2087)	54	5.540 (0.2181)
08	5.080 (0.2000)	32	5.320 (0.2094)	56	5.560 (0.2189)
10	5.100 (0.2008)	34	5.340 (0.2102)	58	5.580 (0.2197)
12	5.120 (0.2016)	36	5.360 (0.2110)	60	5.600 (0.2205)
14	5.140 (0.2024)	38	5.380 (0.2118)	62	5.620 (0.2213)
16	5.160 (0.2031)	40	5.400 (0.2126)	64	5.640 (0.2220)
18	5.180 (0.2039)	42	5.420 (0.2134)	66	5.660 (0.2228)
20	5.200 (0.2047)	44	5.440 (0.2142)	68	5.680 (0.2236)
22	5.220 (0.2055)	46	5.460 (0.2150)	70	5.700 (0.2244)
24	5.240 (0.2063)	48	5.480 (0.2157)	72	5.720 (0.2252)
26	5.260 (0.2071)	50	5.500 (0.2165)	74	5.740 (0.2260)
28	5.280 (0.2079)	52	5.520 (0.2173)		

Exhaust valve clearance (Cold):
0.25 – 0.35 mm (0.010 – 0.014 in.)
EXAMPLE: The 5.340 mm (0.2102 in.) lifter is installed, and the measured clearance is 0.440 mm (0.0173 in.).
Replace the 5.340 mm (0.2102 in.) lifter with a new No. 48 lifter.

Adjusting shim chart (exhaust)—1ZZ-FE engine

For Accessory Drive Belt illustrations, see Section 1 of this manual

Adjusting shim chart (intake) — 2ZZ-GE engine

New Shim thickness mm (in.)

Shim No.	Thickness	Shim No.	Thickness	Shim No.	Thickness
00	2.000 (0.0787)	28	2.280 (0.0898)	56	2.560 (0.1008)
02	2.020 (0.0795)	30	2.300 (0.0906)	58	2.580 (0.1016)
04	2.040 (0.0803)	32	2.320 (0.0913)	60	2.600 (0.1024)
06	2.060 (0.0811)	34	2.340 (0.0921)	62	2.620 (0.1031)
08	2.080 (0.0819)	36	2.360 (0.0929)	64	2.640 (0.1039)
10	2.100 (0.0827)	38	2.380 (0.0937)	66	2.660 (0.1047)
12	2.120 (0.0835)	40	2.400 (0.0945)	68	2.680 (0.1055)
14	2.140 (0.0843)	42	2.420 (0.0953)	70	2.700 (0.1063)
16	2.160 (0.0850)	44	2.440 (0.0961)	72	2.720 (0.1071)
18	2.180 (0.0858)	46	2.460 (0.0969)	74	2.740 (0.1079)
20	2.200 (0.0866)	48	2.480 (0.0976)	76	2.760 (0.1087)
22	2.220 (0.0874)	50	2.500 (0.0984)	78	2.780 (0.1094)
24	2.240 (0.0882)	52	2.520 (0.0992)	80	2.800 (0.1102)
26	2.260 (0.0890)	54	2.540 (0.1000)		

Intake valve clearance (Cold):

0.08 – 0.18 mm (0.0031 – 0.0071 in.)

EXAMPLE: The 2.200 mm (0.0826 in.) shim is installed, and
the measured clearance is 0.400 mm (0.0157 in.).
Replace the 2.600 mm (0.1024 in.) shim with a new No. 60 shim.

Adjusting shim chart — intake (2ZZ-GE engine)

Installed shim thickness mm (in.) across top:
2.000 (0.0787), 2.020 (0.0795), 2.040 (0.0803), 2.060 (0.0811), 2.080 (0.0819), 2.100 (0.0827), 2.120 (0.0835), 2.140 (0.0843), 2.160 (0.0850), 2.180 (0.0858), 2.200 (0.0866), 2.210 (0.0870), 2.220 (0.0874), 2.230 (0.0878), 2.240 (0.0882), 2.250 (0.0886), 2.260 (0.0890), 2.270 (0.0894), 2.280 (0.0898), 2.290 (0.0902), 2.300 (0.0906), 2.310 (0.0909), 2.320 (0.0913), 2.330 (0.0917), 2.340 (0.0921), 2.350 (0.0925), 2.360 (0.0929), 2.370 (0.0933), 2.380 (0.0937), 2.390 (0.0941), 2.400 (0.0945), 2.410 (0.0949), 2.420 (0.0953), 2.430 (0.0957), 2.440 (0.0961), 2.450 (0.0965), 2.460 (0.0969), 2.470 (0.0972), 2.480 (0.0976), 2.490 (0.0980), 2.500 (0.0984), 2.510 (0.0988), 2.520 (0.0992), 2.530 (0.0996), 2.540 (0.1000), 2.550 (0.1004), 2.560 (0.1008), 2.580 (0.1016), 2.600 (0.1024), 2.620 (0.1031), 2.640 (0.1039), 2.660 (0.1047), 2.680 (0.1055), 2.700 (0.1063), 2.720 (0.1071), 2.740 (0.1079), 2.760 (0.1087), 2.780 (0.1094), 2.800 (0.1102)

Measured clearance mm (in.) down side:
0.000 – 0.030 (0.0000 – 0.0012)
0.031 – 0.050 (0.0012 – 0.0020)
0.051 – 0.070 (0.0020 – 0.0028)
0.071 – 0.090 (0.0028 – 0.0035)
0.091 – 0.099 (0.0036 – 0.0039)
0.100 – 0.160 (0.0039 – 0.0063)
0.161 – 0.180 (0.0063 – 0.0071)
0.181 – 0.200 (0.0071 – 0.0079)
0.201 – 0.220 (0.0079 – 0.0087)
0.221 – 0.240 (0.0087 – 0.0094)
0.241 – 0.260 (0.0095 – 0.0102)
0.261 – 0.280 (0.0103 – 0.0110)
0.281 – 0.300 (0.0111 – 0.0118)
0.301 – 0.320 (0.0119 – 0.0126)
0.321 – 0.340 (0.0126 – 0.0134)
0.341 – 0.360 (0.0134 – 0.0142)
0.361 – 0.380 (0.0142 – 0.0150)
0.381 – 0.400 (0.0150 – 0.0157)
0.401 – 0.420 (0.0158 – 0.0165)
0.421 – 0.440 (0.0166 – 0.0173)
0.441 – 0.460 (0.0174 – 0.0181)
0.461 – 0.480 (0.0181 – 0.0189)
0.481 – 0.500 (0.0189 – 0.0197)
0.501 – 0.520 (0.0197 – 0.0205)
0.521 – 0.540 (0.0205 – 0.0213)
0.541 – 0.560 (0.0213 – 0.0220)
0.561 – 0.580 (0.0220 – 0.0228)
0.581 – 0.600 (0.0229 – 0.0236)
0.601 – 0.620 (0.0237 – 0.0244)
0.621 – 0.640 (0.0244 – 0.0252)
0.641 – 0.660 (0.0252 – 0.0260)
0.661 – 0.680 (0.0260 – 0.0268)

9359AB26

9359A827

New Shim thickness mm (in.)

Shim No.	Thickness	Shim No.	Thickness	Shim No.	Thickness
00	2.000 (0.0787)	28	2.280 (0.0898)	56	2.560 (0.1008)
02	2.020 (0.0795)	30	2.300 (0.0906)	58	2.580 (0.1016)
04	2.040 (0.0803)	32	2.320 (0.0913)	60	2.600 (0.1024)
06	2.060 (0.0811)	34	2.340 (0.0921)	62	2.620 (0.1031)
08	2.080 (0.0819)	36	2.360 (0.0929)	64	2.640 (0.1039)
10	2.100 (0.0827)	38	2.380 (0.0937)	66	2.660 (0.1047)
12	2.120 (0.0835)	40	2.400 (0.0945)	68	2.680 (0.1055)
14	2.140 (0.0843)	42	2.420 (0.0953)	70	2.700 (0.1063)
16	2.160 (0.0850)	44	2.440 (0.0961)	72	2.720 (0.1071)
18	2.180 (0.0858)	46	2.460 (0.0969)	74	2.740 (0.1079)
20	2.200 (0.0866)	48	2.480 (0.0976)	76	2.760 (0.1087)
22	2.220 (0.0874)	50	2.500 (0.0984)	78	2.780 (0.1094)
24	2.240 (0.0882)	52	2.520 (0.0992)	80	2.800 (0.1102)
26	2.260 (0.0890)	54	2.540 (0.1000)		

Exhaust valve clearance (Cold):
0.22 – 0.32 mm (0.0087 – 0.0126 in.)

EXAMPLE: The 2.200 mm (0.0862 in.) shim is installed, and the measured clearance is 0.500 mm (0.0197 in.).

Replace the 2.540 mm (0.1000 in.) shim with a new No. 54 shim.

Adjusting shim chart (exhaust)—2ZZ-GE engine

Adjusting shim chart

Installed shim thickness mm (in.) across top:
2.000 (0.0787), 2.020 (0.0795), 2.040 (0.0803), 2.060 (0.0811), 2.080 (0.0819), 2.100 (0.0827), 2.120 (0.0835), 2.140 (0.0843), 2.160 (0.0850), 2.180 (0.0858), 2.200 (0.0866), 2.210 (0.0870), 2.220 (0.0874), 2.230 (0.0878), 2.240 (0.0882), 2.250 (0.0886), 2.260 (0.0890), 2.270 (0.0894), 2.280 (0.0898), 2.290 (0.0902), 2.300 (0.0906), 2.310 (0.0909), 2.320 (0.0913), 2.330 (0.0917), 2.340 (0.0921), 2.340 (0.0921), 2.350 (0.0925), 2.360 (0.0929), 2.370 (0.0933), 2.380 (0.0937), 2.390 (0.0941), 2.400 (0.0945), 2.410 (0.0949), 2.420 (0.0953), 2.430 (0.0957), 2.440 (0.0961), 2.450 (0.0965), 2.460 (0.0969), 2.470 (0.0972), 2.480 (0.0976), 2.490 (0.0980), 2.500 (0.0984), 2.510 (0.0988), 2.520 (0.0992), 2.530 (0.0996), 2.540 (0.1000), 2.550 (0.1004), 2.560 (0.1008), 2.580 (0.1016), 2.600 (0.1024), 2.620 (0.1031), 2.640 (0.1039), 2.660 (0.1047), 2.680 (0.1055), 2.700 (0.1063), 2.720 (0.1071), 2.740 (0.1079), 2.760 (0.1087), 2.780 (0.1094), 2.800 (0.1102)

Measure clearance mm (in.) down side:

Measure clearance mm (in.)
0.000 – 0.030 (0.0000 – 0.0012)
0.031 – 0.050 (0.0012 – 0.0020)
0.051 – 0.070 (0.0020 – 0.0028)
0.071 – 0.090 (0.0028 – 0.0035)
0.091 – 0.110 (0.0036 – 0.0043)
0.111 – 0.130 (0.0044 – 0.0051)
0.131 – 0.150 (0.0052 – 0.0059)
0.151 – 0.170 (0.0059 – 0.0067)
0.171 – 0.190 (0.0067 – 0.0075)
0.191 – 0.210 (0.0075 – 0.0083)
0.211 – 0.230 (0.0083 – 0.0091)
0.231 – 0.239 (0.0091 – 0.0094)
0.301 – 0.320 (0.0119 – 0.0126)
0.321 – 0.340 (0.0126 – 0.0134)
0.341 – 0.360 (0.0134 – 0.0142)
0.361 – 0.380 (0.0142 – 0.0150)
0.381 – 0.400 (0.0150 – 0.0157)
0.401 – 0.420 (0.0158 – 0.0165)
0.421 – 0.440 (0.0166 – 0.0173)
0.441 – 0.460 (0.0174 – 0.0181)
0.461 – 0.480 (0.0181 – 0.0189)
0.481 – 0.500 (0.0189 – 0.0197)
0.501 – 0.520 (0.0197 – 0.0205)
0.521 – 0.540 (0.0205 – 0.0213)
0.541 – 0.560 (0.0213 – 0.0220)
0.561 – 0.580 (0.0221 – 0.0228)
0.581 – 0.600 (0.0229 – 0.0236)
0.601 – 0.620 (0.0237 – 0.0244)
0.621 – 0.640 (0.0244 – 0.0252)
0.641 – 0.660 (0.0252 – 0.0260)
0.661 – 0.680 (0.0260 – 0.0268)
0.681 – 0.700 (0.0268 – 0.0276)
0.701 – 0.720 (0.0276 – 0.0283)
0.721 – 0.740 (0.0284 – 0.0291)
0.741 – 0.760 (0.0292 – 0.0299)
0.761 – 0.780 (0.0300 – 0.0307)
0.781 – 0.800 (0.0307 – 0.0315)
0.801 – 0.820 (0.0315 – 0.0323)

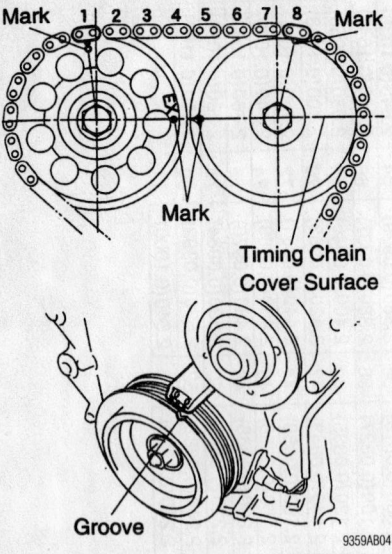

Mark 1 2 3 4 5 6 7 8 Mark

Mark

Timing Chain Cover Surface

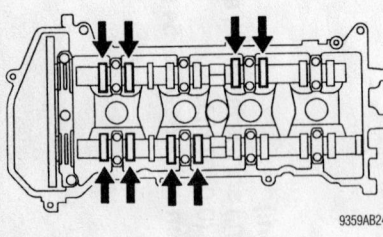

Groove

9359AB04

Proper timing mark alignment for TDC— 1ZZ-FE and 2ZZ-GE engines

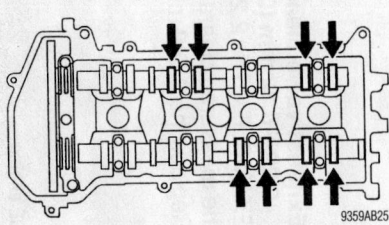

9359AB24

Check the clearance of the 1st set of valves–1ZZ-FE engine

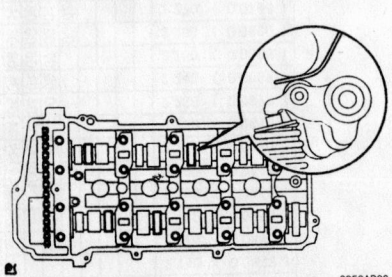

9359AB25

Check the clearance of the 2nd set of valves–1ZZ-FE engine

exhaust valves should be 0.0098–0.0138 in. (0.25–0.35mm).

b. Note the out-of-specification valve clearance measurements. You will need them later to determine the required replacement valve lifter.

6. Remove or disconnect the following:
- Drive belt
- Right side engine mount
- Drive belt tensioner

✳ WARNING

DO NOT turn the crankshaft while the tensioner is removed!

7. Set the No. 1 cylinder to TDC of the compression stroke.
- Camshafts
- Valve lifters.

8. Use a micrometer to measure the thickness of the used lifter. Calculate the thickness of a new lifter. so the valve clearance comes within the specified value:

a. A: Thickness of new lifter.
b. B: Thickness of used lifter.
c. C: Measured valve clearance.
d. Intake valve clearance: $A = B + (C - 0.0079 \text{ in. } (0.20\text{mm})$.
e. Exhaust valve clearance: $A = B + (C - 0.0118 \text{ in. } (0.30\text{mm})$.
f. Select a new lifter with a thickness as close as possible to the calculated values. Lifters come in 35 sizes in increments of 0.0008 in. (0.020mm) from 0.1992–0.2260 in (5.060–5.740mm).

9. Install or connect the following:
- Camshafts
- Drive belt tensioner
- Right hand engine mount
- Cylinder head (valve) cover sub-assembly
- Ignition coil
- Engine wire
- Cylinder head (valve) cover
- Negative battery cable

2ZZ-GE Engine

➡**Adjust the valve clearance when the engine is cold.**

1. Before servicing the vehicle, refer to the precautions in the beginning of this section.

2. Remove or disconnect the following:
- Negative battery cable.
- Right side engine under cover
- Cylinder head cover
- Ignition coil assembly
- Wire harness clamp
- Suction hose sub-assembly
- Cylinder head cover sub-assembly
- Drive belt
- Right side engine mount

3. Set the No. 1 cylinder to Top Dead Center (TDC) of the compressor stroke as follows:

a. Turn the crankshaft pulley, and align its groove with the "0" timing mark of the timing chain cover.

b. Make sure the point marks of the camshaft timing sprockets and VVT timing sprockets are in a straight line as

shown. If not, turn the crankshaft 1 complete revolution (360°) and align the marks.

4. Check the valve clearance of the first set of the valves shown:

a. Use a feeler gauge to measure the clearance between the valve lifter and camshaft. The clearance of the intake valves should be 0.0031–0.0071 in. (0.08–0.18mm). The clearance of the exhaust valves should be 0.0087–0.0126 in. (0.22–0.32mm).

b. Note the out-of-specification valve clearance measurements. You will need them later to determine the required replacement valve lifter.

c. Turn the crankshaft 1 revolution (360°) to set the No. 4 cylinder to TDC.

5. Check the valve clearance of the second set of the valves shown:

a. Use a feeler gauge to measure the clearance between the valve lifter and camshaft. The clearance of the intake valves should be 0.0031–0.0071 in. (0.08–0.18mm). The clearance of the exhaust valves should be 0.0087–0.0126 in. (0.22–0.32mm).

b. Note the out-of-specification valve clearance measurements. You will need them later to determine the required replacement valve lifter.

6. To adjust the intake valve clearance:

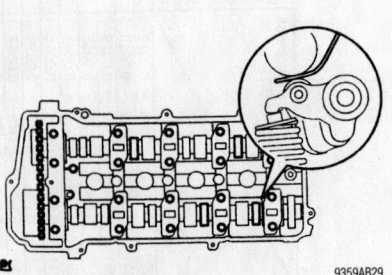

9359AB28

Check the clearance of the 1st set of valves–2ZZ-GE engine

9359AB29

Check the clearance of the 2nd set of valves–2ZZ-GE engine

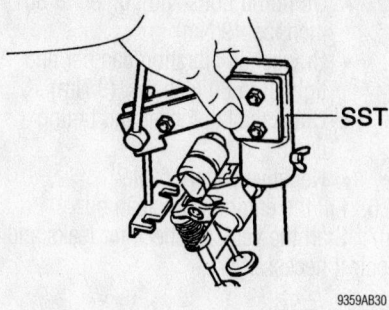

Insert the special tool into the plug tube—2ZZ-GE

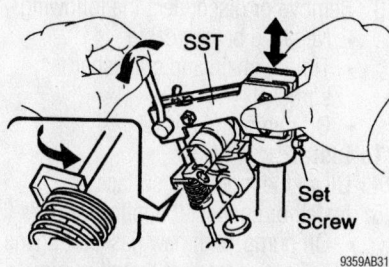

Operate the lever so that the SST's seat surface comes to contact with the valve retainer and lock them with the set screw

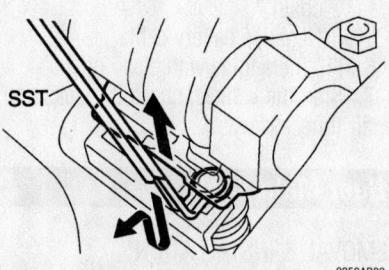

Setting the tool from the right side, makes shim removal easier—2ZZ-GE

a. Set the SST. Turn the crankshaft so the related rocker arm, where the valve clearance is adjusted, is fully pushed down.

➡**Remove he spark plug and take off the compression.**

b. Insert SST 09248-77010 into the plug tube. The tool cannot be inserted unless the set screw is loosened.

c. Operate the lever so that the SST's seat surface comes to contact with the valve retainer and lock them with the set screw. Clearance between the valve retainer and SST's set surface is not allowed. Be careful not to make clearance when inserting the SST, since clearance may unlock the keeper.

d. Lock the set screw on the tube side of the SST.

e. Rotate the crankshaft so that the camshaft is position as shown. During rotation, pay attention to the direction, to prevent the nose of the camshaft from interfering with the SST's shaft. Do not rotate the crankshaft excessively.

f. Lift the rocker arm to make room and remove the adjusting shim using SST 09248-77010.

7. Determine the size of the replaced shim according to the chart or the following formula:

a. Use a dial indicator to measure the thickness of the removed shim.

b. Calculate the thickness of a new shim so that the valve clearance comes within the specified value.

c. A: Thickness of new shim.

d. B: Thickness of used shim.

e. C: Measured valve clearance.

f. Intake: $A = B + (C - 0.005$ in. [0.13mm])

g. Exhaust: $A = B + (C - 0.011$ in. [0.27mm])

h. Select a new shim with a thickness as close as possible to the calculated values. Shims come in 41 sizes in increments of 0.0008 in. (0.020mm) from 0.0787–0.1102 in (2.0–2.8mm).

8. Lift the rocker arm to make room, then install the adjusting shim using the SST. To remove the tool from the shim, push down on the rocker arm.

9. Turn the crankshaft so the related rocker arm, where the valve clearance is adjusted, is fully pushed down.

10. Loosen the 2 set-screws, then remove the SST.

11. Install all components in the reverse of the removal procedure.

Starter

REMOVAL & INSTALLATION

1. Before servicing the vehicle, refer to the precautions in the beginning of this section.

2. Remove or disconnect the following:
   - Negative battery cable
   - Right side engine undercover
   - Starter wiring

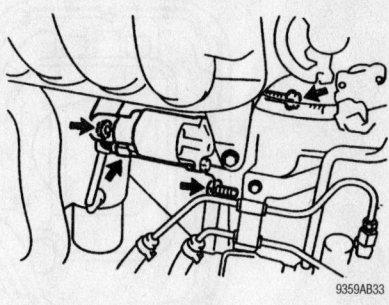

Starter mounting—Matrix

   - Starter

3. Installation is the reverse of removal. Torque the bolts to 27 ft. lbs. (37 Nm) and the nut to 7 ft. lbs. (10 Nm).

Oil Pan

REMOVAL & INSTALLATION

1ZZ-FE Engine

1. Before servicing the vehicle, refer to the precautions in the beginning of this section.

2. Drain the engine oil.

3. Remove or disconnect the following:
   - Negative battery cable
   - Undercovers
   - Front exhaust pipe
   - Oil pan mounting bolts and nuts
   - Oil pan, cutting off the applied sealer.

To install:

4. Remove any old sealant from the oil pan flange and thoroughly clean the sealing surface.

5. Install or connect the following:
   - Oil pan. Tighten the bolts and nuts in several passes to 80 inch lbs. (9 Nm).
   - Front exhaust pipe
   - Negative battery cable
   - Undercovers

6. Fill the engine with clean oil.

7. Start the vehicle, check for leaks and repair if necessary.

2ZZ-GE Engine

1. Before servicing the vehicle, refer to the precautions in the beginning of this section.

2. Drain the engine oil.

3. Remove or disconnect the following:
   - Negative battery cable. On vehicles equipped with an air bag, wait at

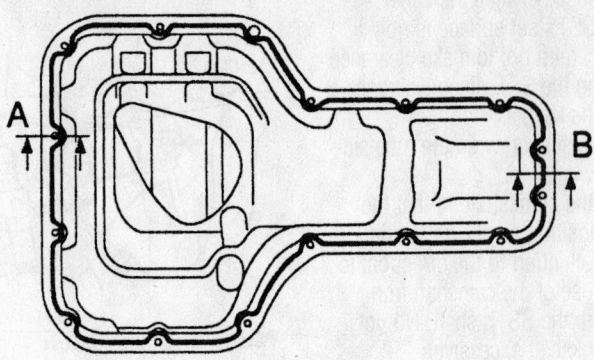

Seal Width
4 – 5 mm

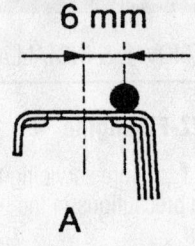
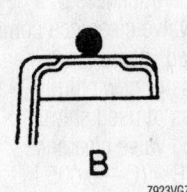

7923VG72

Apply sealant to the oil pan as shown—1.8L (1ZZ-FE) engine

least 90 seconds before proceeding.
- Undercovers
- Front exhaust pipe
- Oil pan mounting bolts and nuts
- Oil pan, cutting off the applied sealer

To install:
4. Remove any old sealant from the oil pan flange and thoroughly clean the sealing surface.
5. Install or connect the following:
- Oil pan. Tighten the bolts and nuts in several passes to 80 inch lbs. (9 Nm).
- Front exhaust pipe
- Negative battery cable
- Undercovers
6. Fill the engine with clean oil.
7. Start the vehicle, check for leaks and repair if necessary.

Oil Pump

REMOVAL & INSTALLATION

1ZZ-FE Engine

1. Before servicing the vehicle, refer to the precautions in the beginning of this section.

2. Drain the engine oil.
3. Remove or disconnect the following:
- Negative battery cable
- Timing chain and crankshaft sprocket
- Timing chain vibration damper
- Oil pump bolts, pump and gasket

To install:
4. Clean the mounting surface.
5. Install or connect the following:
- Oil pump, with new gasket. Engage the spline teeth of the oil pump drive rotor with the larger teeth of the crankshaft, and slide the pump on.

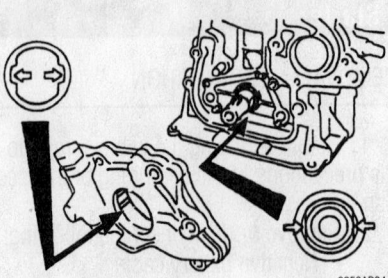

9359AB34

Oil pump mounting—1ZZ-FE and 2ZZ-GE engines

- Oil pump bolts and tighten to 80 inch lbs. (9 Nm)
- Crankshaft vibration damper and tighten to 80 inch lbs. (9 Nm)
- Crankshaft sprocket and timing chain
- Negative battery cable
6. Fill the engine with clean oil.
7. Start the vehicle, check for leaks and repair if necessary.

2ZZ-GE Engine

1. Before servicing the vehicle, refer to the precautions in the beginning of this section.
2. Drain the engine oil.
3. Remove or disconnect the following:
- Negative battery cable
- Timing chain and crankshaft sprocket
- Oil pump and gasket

To install:
4. Clean the mounting surface.
5. Install or connect the following:
- Oil pump, with new gasket. Engage the spline teeth of the oil pump drive rotor with the larger teeth of the crankshaft, and slide the pump on.
- Oil pump bolts and tighten to 80 inch lbs. (9 Nm)
- Crankshaft sprocket and timing chain
- Negative battery cable
6. Fill the engine with clean oil.
7. Start the vehicle, check for leaks and repair if necessary.

Rear Main Seal

REMOVAL & INSTALLATION

1. Remove or disconnect the following:
- Transaxle
- Clutch assembly
- Flywheel or flexplate
2. Use a small sharp knife to cut off the lip of the oil seal. Take great care not to score any metal with the knife.
3. Use a small prytool to pry the old seal from the retaining plate. Be careful not to damage the plate. Protect the tip of the tool with tape and pad the fulcrum point with cloth.
4. Inspect the crankshaft and seal lip contact surfaces for any sign of damage.

To install:
5. Apply a light coat of multi-purpose grease to the lip of a new oil seal. Loosely fit the seal into place by hand, making sure it is not crooked.

6. Use a seal driver of the correct size to install the seal. Tap it into place until the surface of the seal is flush with the edge of the housing.

Timing Chain, Sprockets, Front Cover and seal

REMOVAL & INSTALLATION

1. Before servicing the vehicle, refer to the precautions in the beginning of this section.

2. Drain the cooling system.

3. Remove or disconnect the following:
- Right side engine under cover
- Right front wheel and tire
- Cylinder head cover
- Wire harness clamp and suction hose assembly, 2ZZ-GE engine
- Drive belt

4. Separate the vane pipe assembly, but do not disconnect the hose, 1ZZ-FE engine.
- Alternator bracket, 2ZZ-GE
- Alternator
- Power steering pump reservoir and position it aside, 1ZZ-FE engine

5. Place a jack with a wooden block under the vehicle for support, then remove the 4 bolts and 2 nuts and remove the right side engine mount.

6. Remove the engine wire as follows, on 1ZZ-FE engines:

   a. Remove the 5 clamps from the brackets.

   b. Detach the connectors.

   c. Remove the ignition coil connectors.

   d. Bolt and nut holding the engine wire.

7. Remove the engine wire as follows, on 2ZZ-GE engines:

   a. Detach the ignition coil, oil control

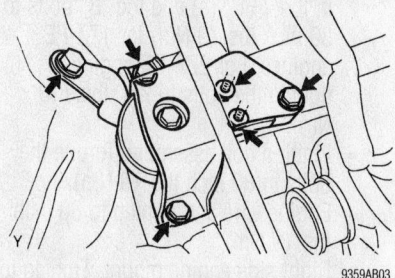

With the engine supported, remove the right side engine mount—1ZZ-FE engine shown, 2ZZ-GE similar

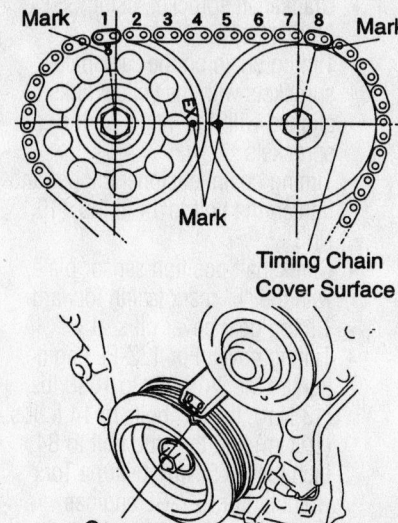

Proper timing mark alignment for TDC

valve and Crankshaft Position Sensor (CKP) sensor electrical connectors.

   b. Bolt and nut for the engine ground, then position the engine wire aside

8. Remove or disconnect the following:
- Ignition coil assembly
- Positive Crankcase Ventilation (PCV) hoses from the cylinder head cover, if necessary
- Cylinder head (valve) cover sub-assembly

9. Set the No. 1 cylinder to Top Dead Center (TDC) of the compressor stroke as follows:

   a. Turn the crankshaft pulley, and align its groove with the "0" timing mark of the timing chain cover.

   b. Make sure the point marks of the camshaft timing sprockets and VVT timing sprockets are in a straight line as shown. If not, turn the crankshaft 1 complete revolution (360°) and align the marks.

- Crankshaft pulley, using SST 09960-10010
- Belt tensioner
- Water pump pulley, if equipped, and pump
- Transverse engine mounting bracket
- Crankshaft Position (CKP) sensor
- No. 1 chain tensioner assembly, making sure not to revolve the crankshaft without the tensioner
- Timing chain cover. The cover is retained with 11 bolts and nuts and a Torx® stud bolt. Pry the cover

between the cylinder head and block to remove it.
- Timing gear cover oil seal
- CKP sensor plate No. 1
- Timing chain tensioner slipper

➡**In case you turn the camshafts with the timing chain removed, turn the crankshaft ¼ turn for the valve to avoid contact with the pistons.**

- Timing chain sub-assembly. Remove the chain with the crankshaft gear, using screwdrivers as shown

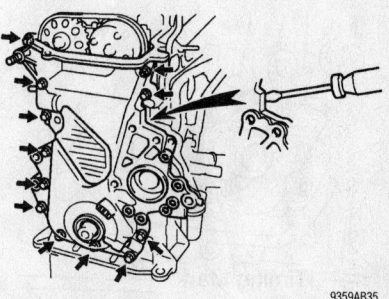

Timing chain cover mounting—1ZZ-FE engine shown, 2ZZ-GE similar

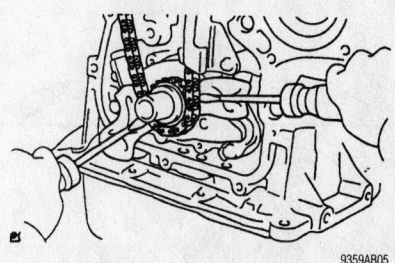

Remove the timing chain with the crankshaft gear

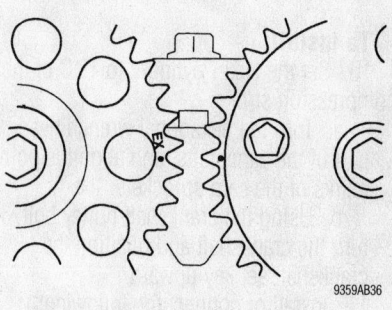

Proper alignment of the camshaft sprockets—1ZZ-FE engine

For Maintenance Interval recommendations, see Section 1 of this manual

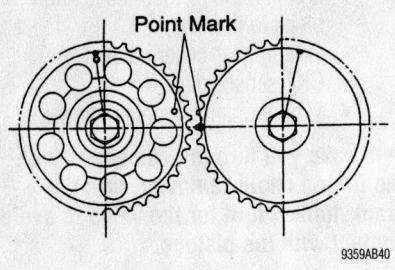

Proper alignment of the camshaft sprockets—2ZZ-GE engine

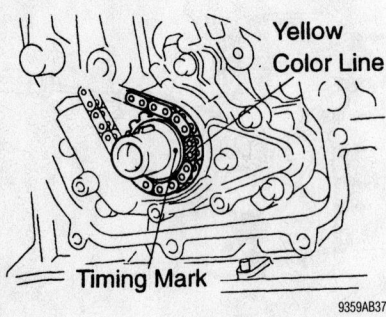

Make sure the yellow link is aligned with the crankshaft sprocket timing mark—1ZZ-FE and 2ZZ-GE engines

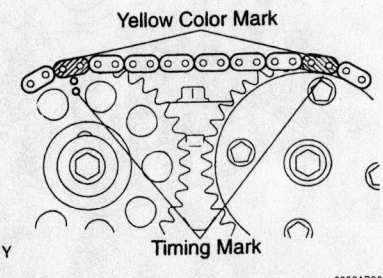

The yellow links of the timing chain must align with the camshaft sprocket timing marks—1ZZ-FE and 2ZZ-GE engines

To install:

10. Set the No. 1 cylinder to TDC of the compression stroke:

a. Turn the hexagonal wrench head part of the camshafts, and align the point marks of the cam sprockets.

b. Using the crankshaft pulley bolt, turn the crankshaft and position the crankshaft set key upward.

11. Install or connect the following:

- Timing chain on the crank sprocket with the yellow link aligned with the mark on the crank sprocket. There are 3 yellow links on the timing chain.

- Crankshaft sprocket, using SST 09223-22010
- Timing chain on the camshaft sprockets with the yellow links aligned with the marks on the cam sprockets
- Timing chain tensioner slipper and tighten the bolt to 14 ft. lbs. (19 Nm)
- Crankshaft position sensor plate, with the "F" mark facing forward
- Timing gear cover oil seal
- Timing cover. For 1ZZ-FE engine, tighten the "A" bolts to 10 ft. lbs. (13 Nm), the "B" bolts to 14 ft. lbs. (19 Nm) and the stud bolt to 84 inch lbs. (9.5 Nm), using a Torx® wrench. For 2ZZ-GE engines, tighten the M8 bolts to 15 ft. lbs. (21 m), the M6 bolts to 8 ft. lbs. (11 Nm) and the stud bolt to 84 inch lbs. (9.5 Nm).

➡ When installing the tensioner, make sure to set the hook again if the hook releases the plunger.

- Timing chain tensioner. Torque the nuts to 80 inch lbs. (9 Nm).
- CKP sensor and tighten the bolts to 80 inch lbs. (9 Nm).
- Transverse engine mounting bracket. Tighten the bolts to 35 ft. lbs. (47 Nm).
- Water pump and pulley
- Drive belt tensioner. Tighten the nut to 21 ft. lbs. (29 Nm) and the bolt to 51 ft. lbs. (69 Nm) on 1ZZ-FE engines or to 74 ft. lbs. (100 Nm) on 2ZZ-GE engines.

12. Install the crankshaft pulley, as follows:

a. Align the pulley set key with the key groove of the pulley and slide on the pulley.

b. Use SST 09960-11010 to install the bolt and tighten to 102 ft. lbs. (138 Nm) for 1ZZ-FE engine or to 87 ft. lbs. (118 Nm) on 2ZZ-GE engines.

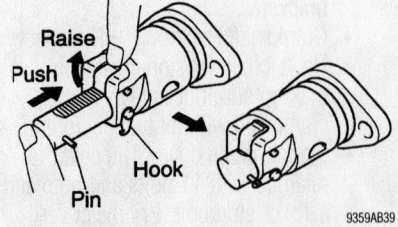

Timing chain tensioner—1ZZ-FE engine

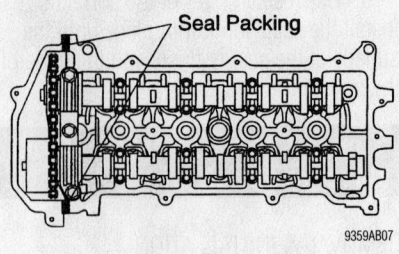

Seal packing installation locations

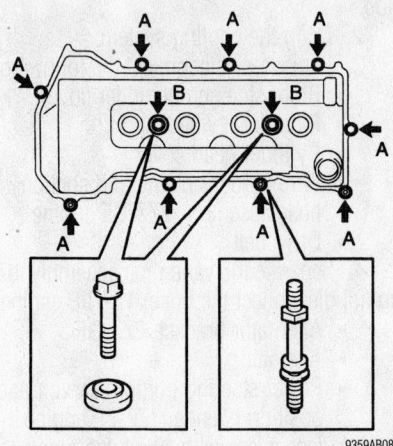

Cylinder head (valve) cover bolt locations—1ZZ-FE engine

c. Turn the crankshaft counterclockwise and disconnect the plunger knock pin from the hook.

d. Turn the crankshaft clockwise and check that the slipper is pushed by the plunger. If the plunger does not spring out, press the slipper into the chain tensioner with a screwdriver so that the hook is released from the knock pin and the plunder springs out.

- Cylinder head sub-assembly cover. Install seal packing into the locations shown and install within 3 minutes. Tighten the "A" bolts to 8 ft. lbs. (11 Nm) and the "B" bolts to 80 inch lbs. (9 Nm) for 1ZZ-FE engines. For 2ZZ-GE engines, tighten the bolts to 7 ft. lbs. (10 Nm).
- Ignition coil assembly. Torque the bolts to 80 inch lbs. (9 Nm).
- Engine wire and tighten to 80 inch lbs. (9 Nm)
- Right side engine mount. Tighten to 38 ft. lbs. (52 Nm).
- Alternator bracket, 2ZZ-GE engine
- Alternator
- Vane pump, 1ZZ-FE
- Main cylinder head cover and tighten to 62 inch lbs. (7 Nm)

- Right front wheel and tire. Tighten the lug nuts to 76 ft. lbs. (103 Nm).
13. Fill the cooling system to the proper level.
14. Start the vehicle, check for leaks and repair if necessary.

Piston and Ring Positioning

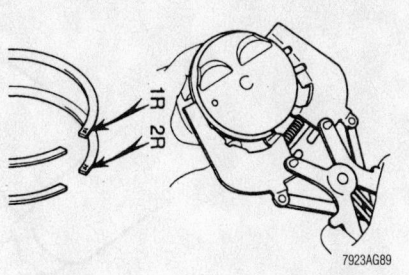

Piston ring identification mark locations— 1ZZ-FE and 2ZZ-GE engines

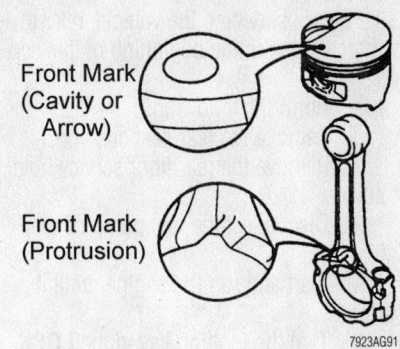

Front Mark (Cavity or Arrow)

Front Mark (Protrusion)

Piston-to-connecting rod assembly —1ZZ-FE and 2ZZ-GE engines

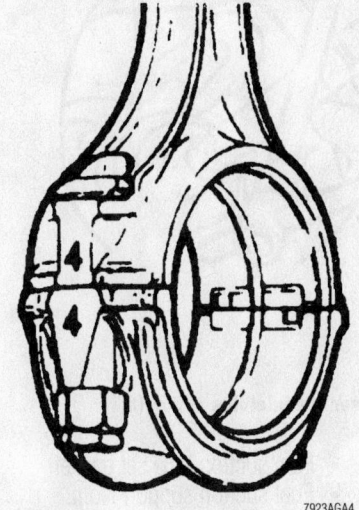

Before removing the caps from the connecting rods, be sure to matchmark them as shown

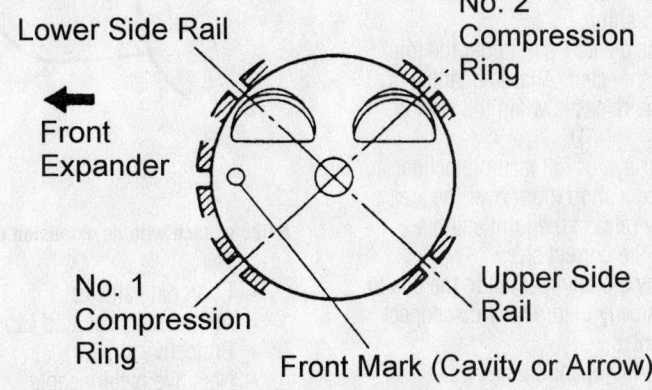

Lower Side Rail

No. 2 Compression Ring

Front Expander

No. 1 Compression Ring

Upper Side Rail

Front Mark (Cavity or Arrow)

Piston ring end-gap spacing —1ZZ-FE and 2ZZ-GE engines

FUEL SYSTEM

Fuel System Service Precautions

Safety is the most important factor when performing not only fuel system maintenance, but any type of maintenance. Failure to conduct maintenance and repairs in a safe manner may result in serious personal injury or death. Work on a vehicle's fuel system components can be accomplished safely and effectively by adhering to the following rules and guidelines.

- To avoid the possibility of fire and personal injury, always disconnect the negative battery cable unless the repair or test procedure requires that battery voltage by applied.
- Always relieve the fuel system pressure prior to disconnecting any fuel system component (injector, fuel rail, pressure regulator, etc.) fitting or fuel line connection. Exercise extreme caution whenever relieving fuel system pressure, to avoid exposing skin, face and eyes to fuel spray. Please be advised that fuel under pressure may penetrate the skin or any part of the body that it contacts.

- Always place a shop towel or rag around the fitting or connection prior to loosening to absorb any excess fuel due to spillage. Ensure that all fuel spillage is quickly remove from engine surfaces. Ensure that all fuel-soaked cloths or towels are deposited into a flame-proof waste container with a lid.
- Always keep a dry chemical (Class B) fire extinguisher near the work area.
- Do not allow fuel spray or fuel vapors to come into contact with a light bulb, spark or open flame.
- Always use a second wrench when loosening or tightening fuel line connections fittings. This will prevent unnecessary stress and torsion to fuel piping. Always follow the proper torque specifications.
- Always replace worn fuel fitting O-rings with new ones. Do not substitute fuel hose where rigid pipe is installed.

Fuel System Pressure

RELIEVING

✳✳ CAUTION

Failure to relieve fuel pressure before repairs or disassembly can cause serious personal injury and/or property damage. Fuel pressure is maintained within the fuel lines, even if the engine is OFF or has not been run in a period of time. This pressure must be safely relieved before any fuel-bearing line or component is loosened or removed. On vehicles equipped with inflatable restraints or air bag systems, wait at least 90 seconds after disconnecting the battery cable before performing any other work. The back-up power will keep the restraint system energized for a period of time after the battery is disconnected.

For Tune-up, Capacities and Firing orders, see Section 1 of this manual

1. Before servicing the vehicle, refer to the precautions in the beginning of this section.

2. Perform the following:

  a. Remove the rear seat cushion.

  b. Remove the rear floor service hole cover.

  c. Disconnect the fuel pump connector.

  d. Start and run the engine, until it stalls.

  e. Turn the ignition key to the **LOCK** position.

  f. Disconnect the negative battery cable.

  g. Connect the fuel pump connector.

  h. Install the service hole cover and rear seat cushion.

  i. Place a catch-pan under the joint to be disconnected. A large quantity of fuel may be released when the joint is opened.

  j. Wear eye or full face protection.

  k. Place a shop towel over the area and slowly release the joint using a wrench of the correct size.

  l. Allow the any fuel left in the line to bleed off slowly before fully disconnecting the joint.

  m. Plug the opened lines.

Fuel Filter

REMOVAL & INSTALLATION

1. Before servicing the vehicle, refer to the precautions in the beginning of this section.
2. Relieve the fuel system pressure.
3. Remove or disconnect the following:
   - Negative battery cable
   - Protective shield for the fuel filter
   - Air cleaner hose and cap, if necessary
   - Charcoal canister, if necessary
   - Slowly loosen the lower flare nut fitting until all the pressure is relieved
   - Banjo fitting and 2 metal gaskets. Discard the gaskets.
   - Fuel line with the flared nut from the filter
   - Filter from the mounting bracket

To install:
4. Install or connect the following:
   - New fuel filter
   - Banjo fitting with a new metal gasket on each side and install the union bolt. Bolt: 22 ft. lbs. (30 Nm).
   - Flare nut to the lower connection. Nut: 22 ft. lbs. (30 Nm).

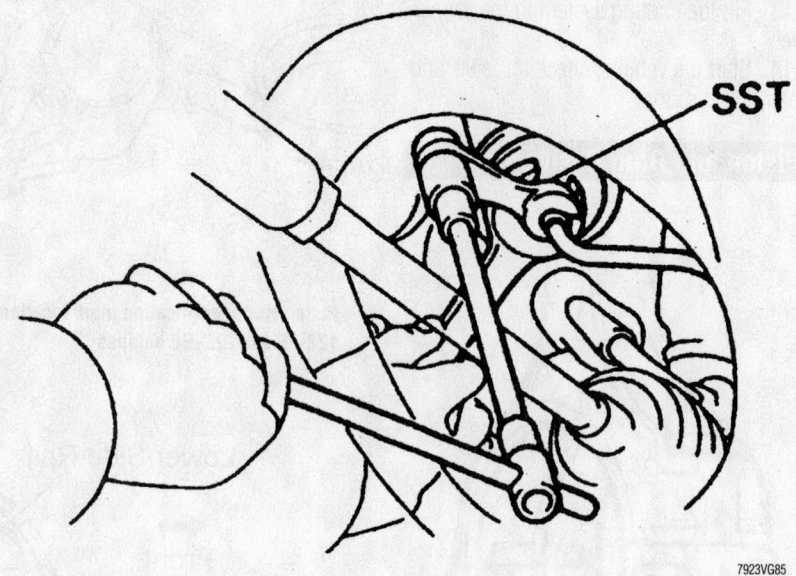

7923VG85

A line wrench with an extension may be needed to loosen the inlet line at the filter

- Charcoal canister
- Air cleaner hose and cap
- Protective shield
- Negative battery cable

Fuel Pump

REMOVAL & INSTALLATION

1. Before servicing the vehicle, refer to the precautions in the beginning of this section.
2. Remove or disconnect the following:
   - Negative battery cable
   - Rear seat cushion and floor service hole cover
   - Fuel pump and vapor pressure sensor connectors
   - Start and run the engine, until it stalls
3. Turn the ignition key to the **LOCK** position.
   - Negative battery cable
4. Connect the fuel pump connector.
   - Fuel tank protector, AWD vehicles
   - Fuel tank main tube sub-assembly
   - Fuel emission tube sub-assembly No. 1, FWD vehicles
   - Fuel tank vent tube set plate. The plate is secured with 8 bolts on FWD vehicles, or 5 bolts on AWD vehicles.
   - Fuel pump assembly, being careful not to damage the filter or bend the arm of the fuel sender gauge

- Fuel suction tube set gasket
- Fuel suction support No. 2
- Fuel pump rubber cushion
- Fuel sender gauge assembly. Unplug the connector, then use a screwdriver to unlock the gauge and slide it to remove.
- Fuel section plate sub-assembly
- Vapor pressure sensor
- Fuel pump harness
- Fuel pump
- Fuel pump filter
- Fuel pressure regulator and O-ring

To install:
5. Install or connect the following:
   - New regulator O-ring and regulator
   - Fuel pump filter
   - Fuel pump
   - Vapor pressure sensor
   - Fuel suction tube set gasket
   - Fuel pump assembly
   - Fuel tank vent tube set plate. Tighten the bolts to 53 inch lbs. (6 Nm).
   - Connect the fuel emission tube sub-assembly
   - Fuel tank main tube sub-assembly
   - Fuel tank protector No. 2, AWD vehicles
   - Negative battery cable. Check for fuel leaks.
   - Floor service hole cover. Use butyl tape to seal the cover.
   - Rear seat cushion

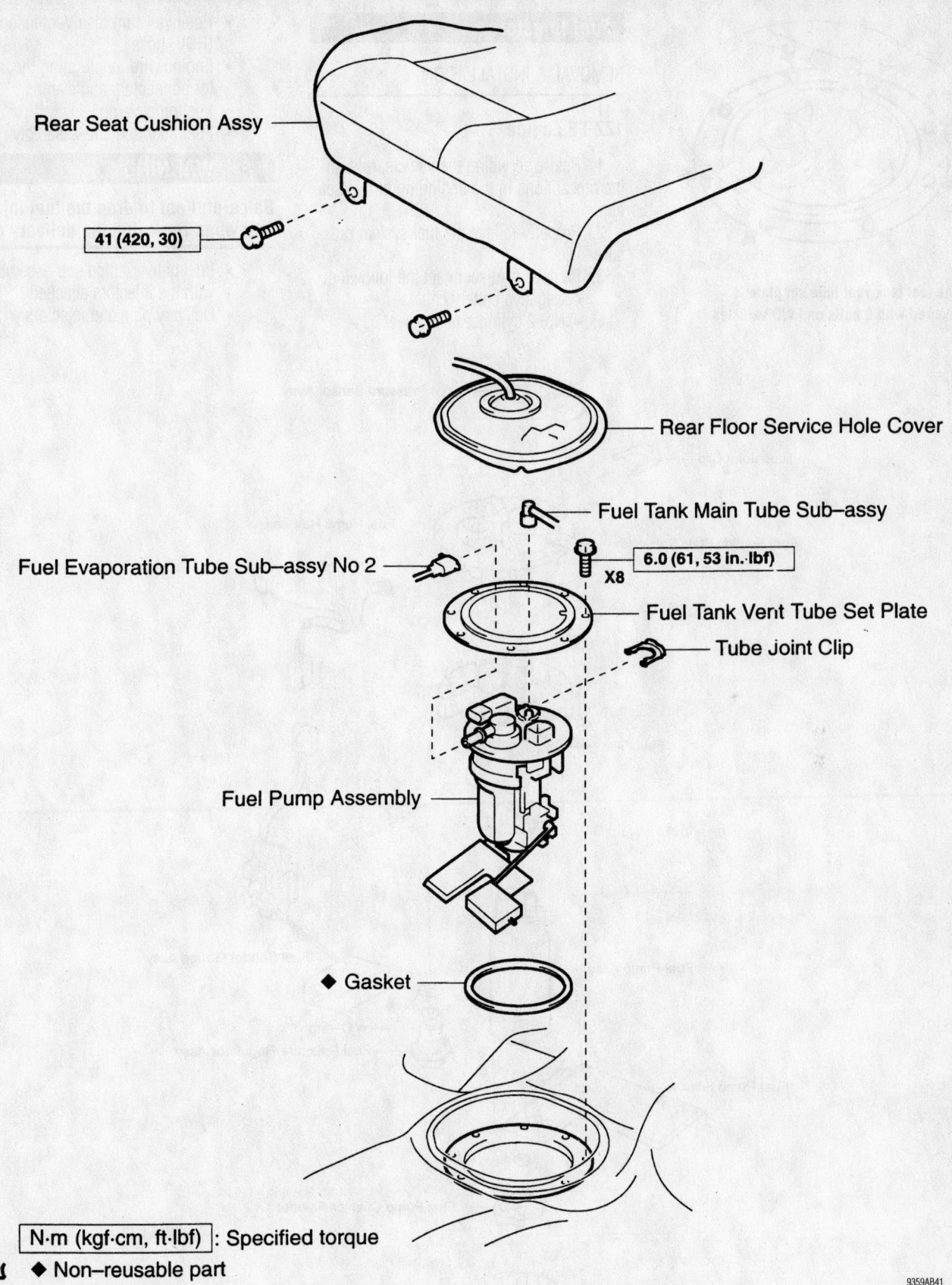

Rear Seat Cushion Assy

41 (420, 30)

Rear Floor Service Hole Cover

Fuel Tank Main Tube Sub–assy

Fuel Evaporation Tube Sub–assy No 2

6.0 (61, 53 in.·lbf)

X8

Fuel Tank Vent Tube Set Plate

Tube Joint Clip

Fuel Pump Assembly

◆ Gasket

N·m (kgf·cm, ft·lbf) : Specified torque

◆ Non–reusable part

9359AB41

Exploded view of the fuel pump mounting—FWD shown, AWD similar

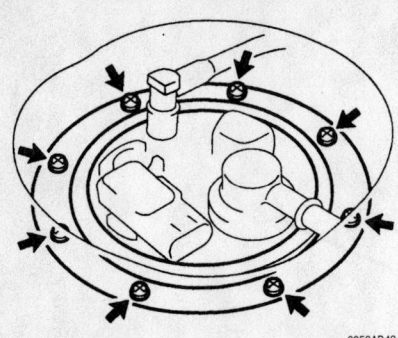

The fuel tank vent tube set plate is secured with 8 bolts on FWD vehicles

Fuel Injectors

REMOVAL & INSTALLATION

1ZZ-FE Engine

1. Before servicing the vehicle, refer to the precautions in the beginning of this section.
2. Properly relieve the fuel system pressure.
3. Remove or disconnect the following:
   - Negative battery cable.
   - No. 2 cylinder head cover
   - Positive Crankcase Ventilation (PCV) hose
   - Engine wire, unplugging the injector connectors and clamps
   - Fuel pipe clamp
   - Fuel line/tube sub-assembly

❋❋ WARNING

Be careful not to drop the fuel injectors when removing the delivery pipe.

   - Fuel delivery pipe sub-assembly with the injectors attached
   - Delivery pipe and injectors

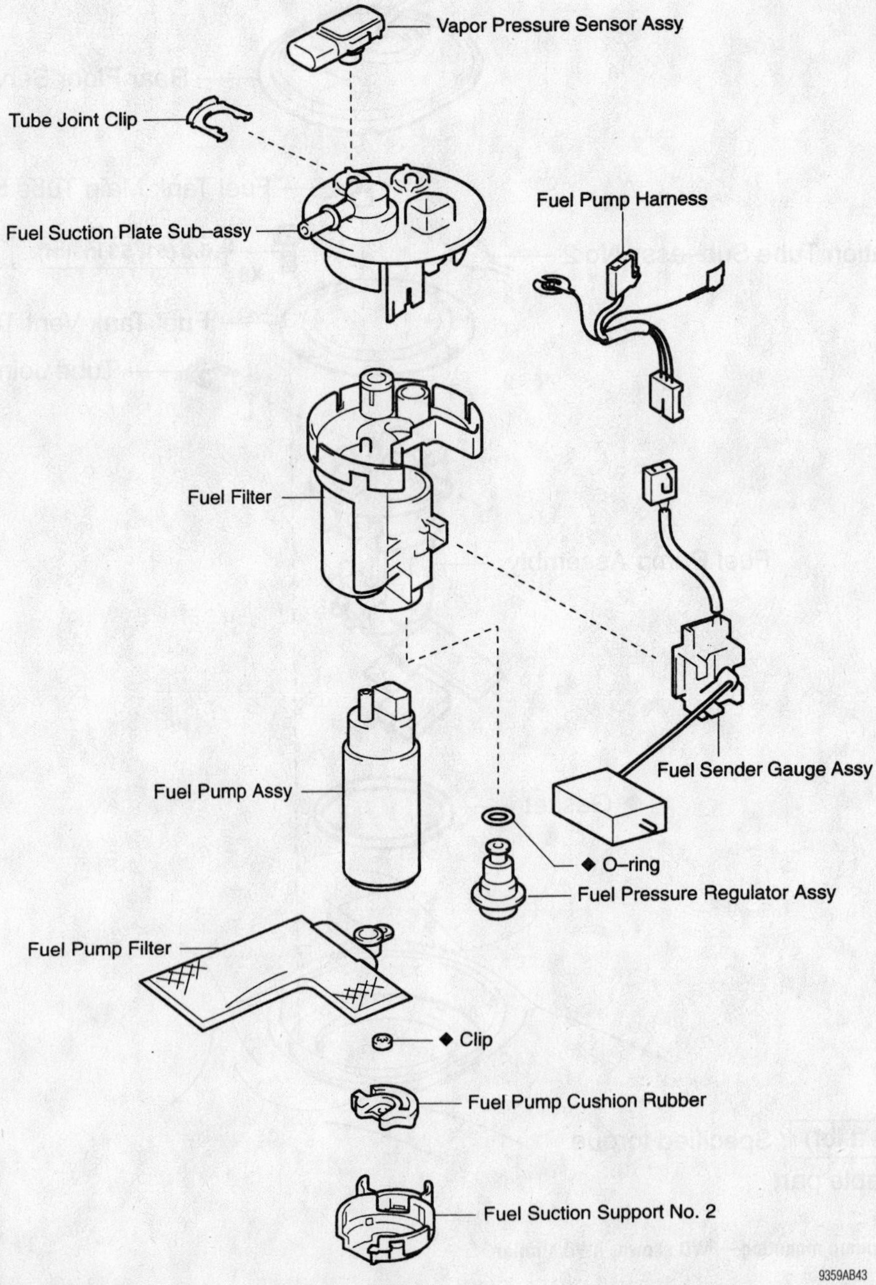

Vapor Pressure Sensor Assy

Tube Joint Clip

Fuel Suction Plate Sub–assy

Fuel Pump Harness

Fuel Filter

Fuel Pump Assy

Fuel Sender Gauge Assy

◆ O–ring

Fuel Pressure Regulator Assy

Fuel Pump Filter

◆ Clip

Fuel Pump Cushion Rubber

Fuel Suction Support No. 2

Fuel pump assembly components—FWD vehicles shown, AWD similar

- Spacers from the head
- Injectors from the delivery pipe
- O-ring and grommet from each injector

To install:

4. Install or connect the following:
- New grommets
- New O-rings coated with light machine oil
- Injectors on the delivery pipe

➡**Coat the contact point on the pipe with light machine oil and twist the** injectors into place. The connector should face outward.

- Spacers

➡**Coat the seats in the head where the injectors contact, with light machine oil.**

- Delivery pipe and injectors

5. Loosely install the hold-down bolts and check that the injectors rotate smoothly. If they don't, the probable cause is incorrect O-ring installation. Torque the delivery pipe hold-down bolts to 14 ft. lbs. (19 Nm) and the fuel pipe bolt to 80 inch lbs. (9 Nm).

- Engine wire, attaching the injector connectors and clamps
- Fuel line/tube sub-assembly
- PCV hose
- No. 2 cylinder head (valve) cover

2ZZ-GE ENGINE

1. Before servicing the vehicle, refer to the precautions in the beginning of this section.

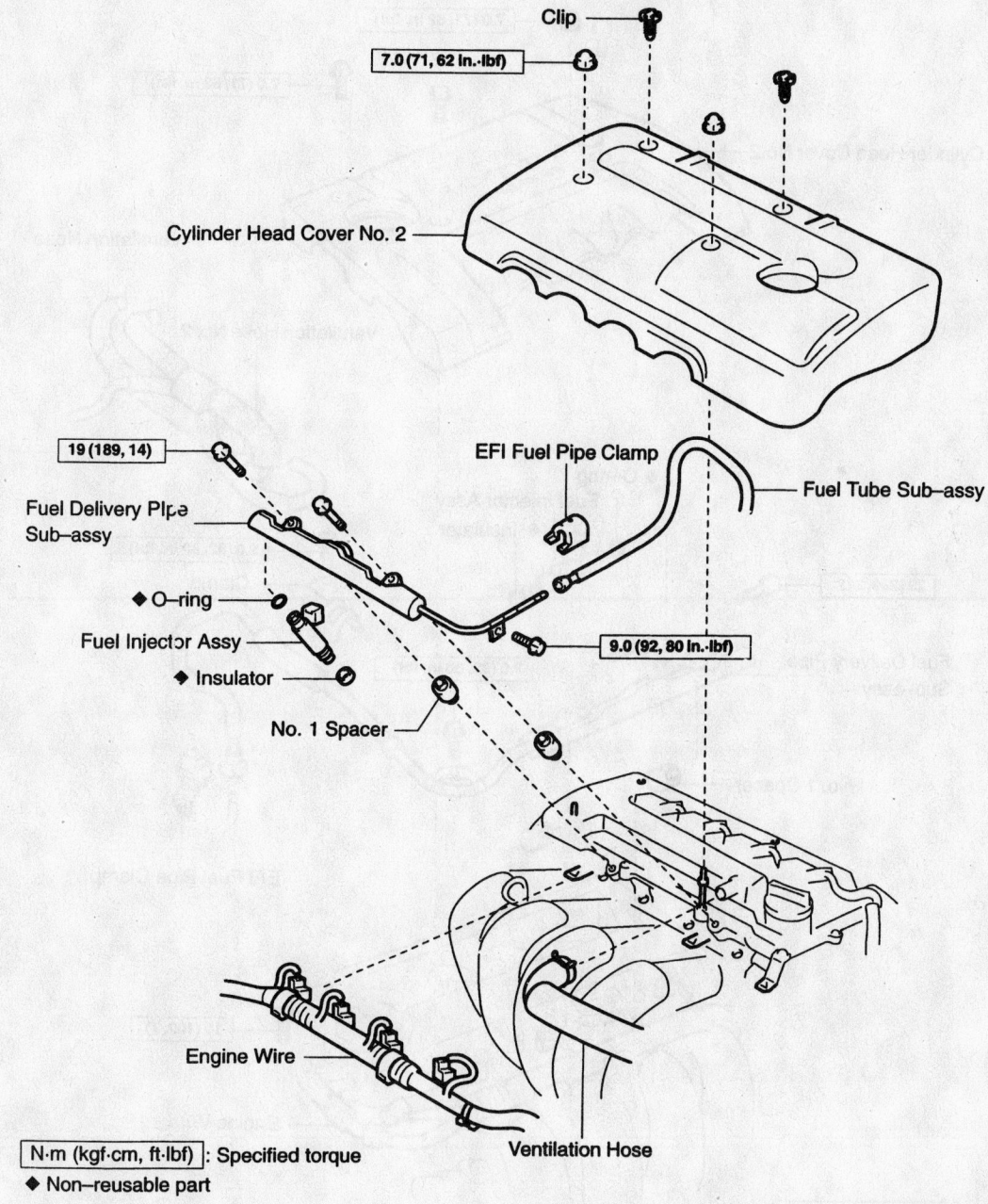

Clip

7.0 (71, 62 in.·lbf)

Cylinder Head Cover No. 2

EFI Fuel Pipe Clamp

Fuel Tube Sub–assy

19 (189, 14)

Fuel Delivery Pipe Sub–assy

◆ O–ring

Fuel Injector Assy

◆ Insulator

No. 1 Spacer

9.0 (92, 80 in.·lbf)

Engine Wire

Ventilation Hose

N·m (kgf·cm, ft·lbf) : Specified torque

◆ Non–reusable part

Fuel injector removal and installation—1ZZ-FE engine

9359AB44

2. Properly relieve the fuel system pressure.

3. Remove or disconnect the following:
   - Negative battery cable.
   - No. 2 cylinder head cover
   - Positive Crankcase Ventilation (PCV) hose
   - Engine wire, by removing the bolt, then unplugging the injector and Camshaft Position (CMP) sensor connectors
   - Fuel pipe clamp

✷✷ WARNING

Be careful not to drop the fuel injectors when removing the delivery pipe.

   - Fuel delivery pipe sub-assembly with the injectors attached
   - Delivery pipe and injectors
   - Spacers from the head
   - Injectors from the delivery pipe
   - O-ring and grommet from each injector

To install:

4. Install or connect the following:
   - New grommets
   - New O-rings coated with light machine oil
   - Injectors on the delivery pipe

➡ **Coat the contact point on the pipe with light machine oil and twist the injectors into place. The connector should face outward.**

   - Spacers

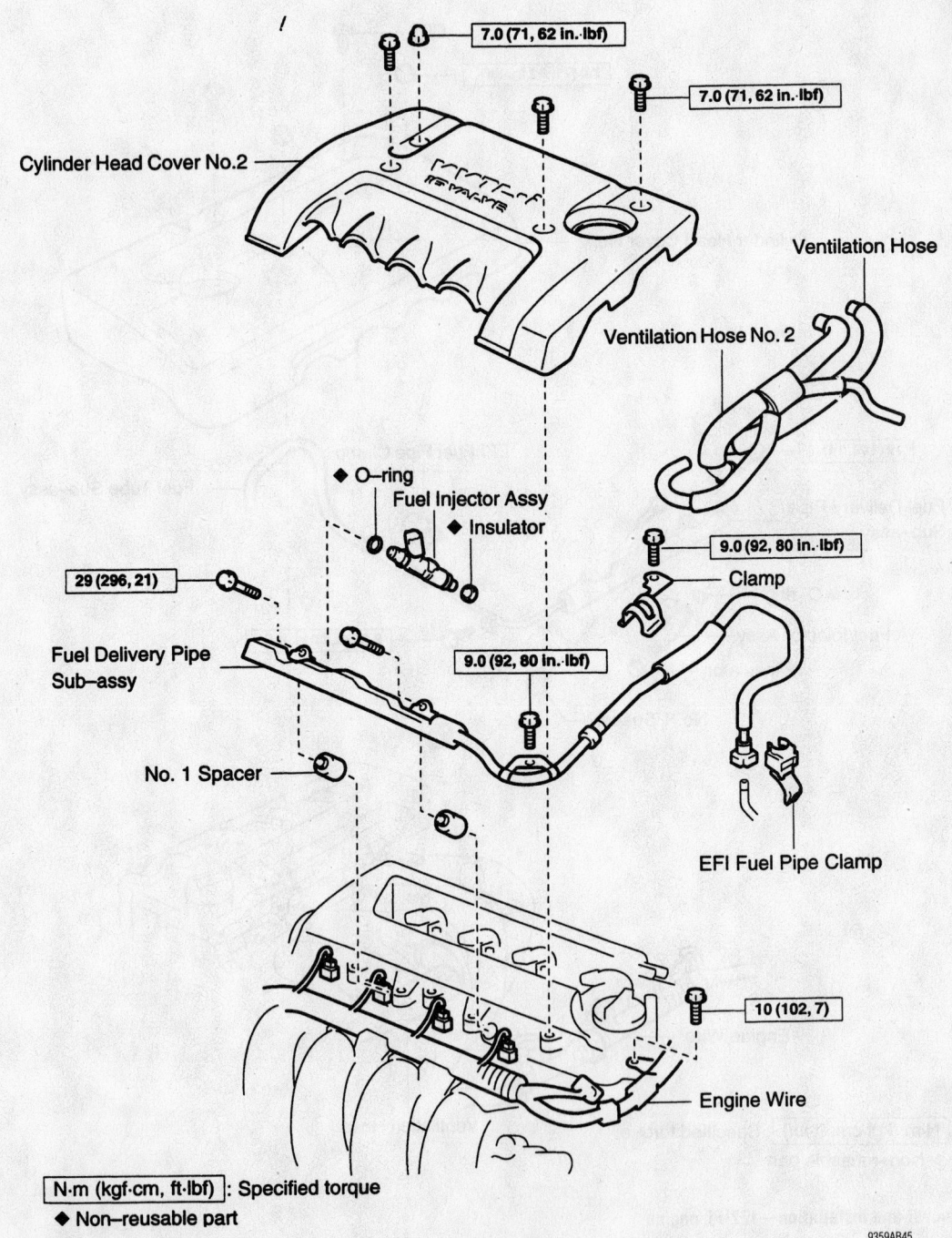

7.0 (71, 62 in.·lbf)

7.0 (71, 62 in.·lbf)

Cylinder Head Cover No.2

Ventilation Hose

Ventilation Hose No. 2

◆ O-ring
Fuel Injector Assy
◆ Insulator

9.0 (92, 80 in.·lbf)

Clamp

29 (296, 21)

Fuel Delivery Pipe Sub-assy

9.0 (92, 80 in.·lbf)

No. 1 Spacer

EFI Fuel Pipe Clamp

10 (102, 7)

Engine Wire

N·m (kgf·cm, ft·lbf) : Specified torque
◆ Non-reusable part

9359AB45

Fuel injector removal and installation—2ZZ-GE engine

➡**Coat the seats in the head where the injectors contact, with light machine oil.**

- Delivery pipe and injectors

5. Loosely install the hold-down bolts

and check that the injectors rotate smoothly. If they don't, the probable cause is incorrect O-ring installation. Torque the delivery pipe hold-down bolts to 14 ft. lbs. (19 Nm) and the fuel pipe bolt to 80 inch lbs. (9 Nm).

- Fuel line/tube sub-assembly

- PCV hose
- Engine wire, by connecting the CMP sensor and injector connectors and installing the bolt. Tighten the bolt to 7 ft. lbs. (10 Nm).
- No. 2 cylinder head (valve) cover

DRIVE TRAIN

Manual Transaxle Assembly

REMOVAL & INSTALLATION

1. Before servicing the vehicle, refer to the precautions in the beginning of this section.

2. Drain the transaxle fluid.

3. Place the front wheels in the straight-ahead position.

4. Remove or disconnect the following:
   - Steering intermediate shaft
   - Front wheel and tires
   - Right and left side undercovers
   - Exhaust pipe
   - Hood
   - Cylinder head (valve) cover
   - Air cleaner assembly
   - Battery clamp, battery, battery tray and battery carrier
   - Cruise control actuator assembly, if equipped

5. Remove the wire harness as follows:
   a. Remove the wire harness clamp, 2 bolts and wire harness brackets.
   b. Remove the 2 bolts and 2 ground cables.

6. Remove or disconnect the following:
   - Back-up lamp switch connector, with ABS
   - Speed sensor connector, without ABS
   - 5 bolts, then separate the release cylinder with the clutch pipes from the transaxle
   - Shift cable clips and washer, then disconnect the cable from the transaxle and bracket
   - Select cable clips and washer, then disconnect the cable from the transaxle and bracket
   - Starter
   - Right and left side tie rod ends
   - Pressure feed tube
   - Front halfshafts

7. Use a suitable tool to suspend the engine assembly, as shown in the illustration:
   a. No. 1 engine hanger: P/N 12281-

22021 (5-speed M/T), 12281-88600 (6-speed M/T).
   b. No. 2 engine hanger: P/N 12281-15040 (5-speed M/T), 12281-88600 (6-speed M/T).
   c. Bolt: P/N 91512-B1016.
   d. Torque the bolts to 28 ft. lbs. (38 Nm).

✳✳ CAUTION

Do not try to suspend the engine by hooking the chain to any other part.

   e. Attach an engine chain hoist to the hangers.

8. Remove or disconnect the following:
   - Front suspension crossmember

9. Support the transaxle with a floor jack.
   - Transverse engine mounting insulator and brackets
   - Manual transaxle assembly
   - Transverse engine mounting brackets from the transaxle, if necessary

To install:
   - Transverse engine mounting brackets to the transaxle, if necessary
   - Manual transaxle, by aligning the input shaft with the clutch disc. Torque the "A" bolts to 47 ft. lbs. (64 Nm), the "B" bolts to 35 ft. lbs. (47 Nm) and the "C" bolts to 17 ft. lbs. (23 Nm).
   - Transverse engine mounting bracket. Tighten to 38 ft. lbs. (52 Nm).

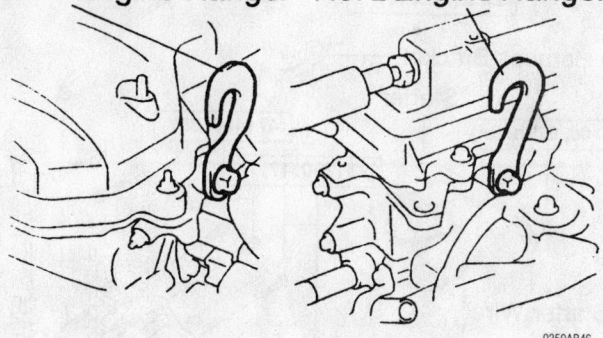

No. 1 Engine Hanger   No. 2 Engine Hanger

9359AB46

Secure the engine using the proper tools—5-speed manual transmission shown

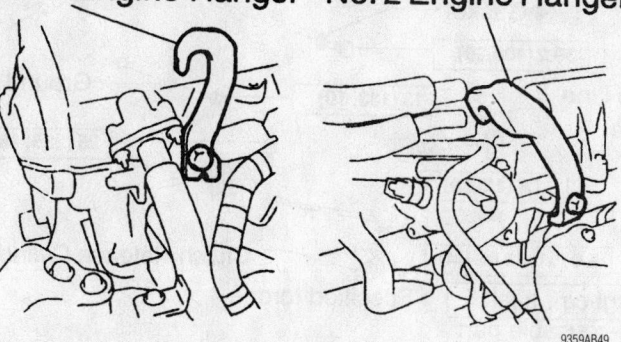

No. 1 Engine Hanger   No. 2 Engine Hanger

9359AB49

Secure the engine using the proper tools—6-speed manual transmission shown

Timing belt service is covered in Section 3 of this manual

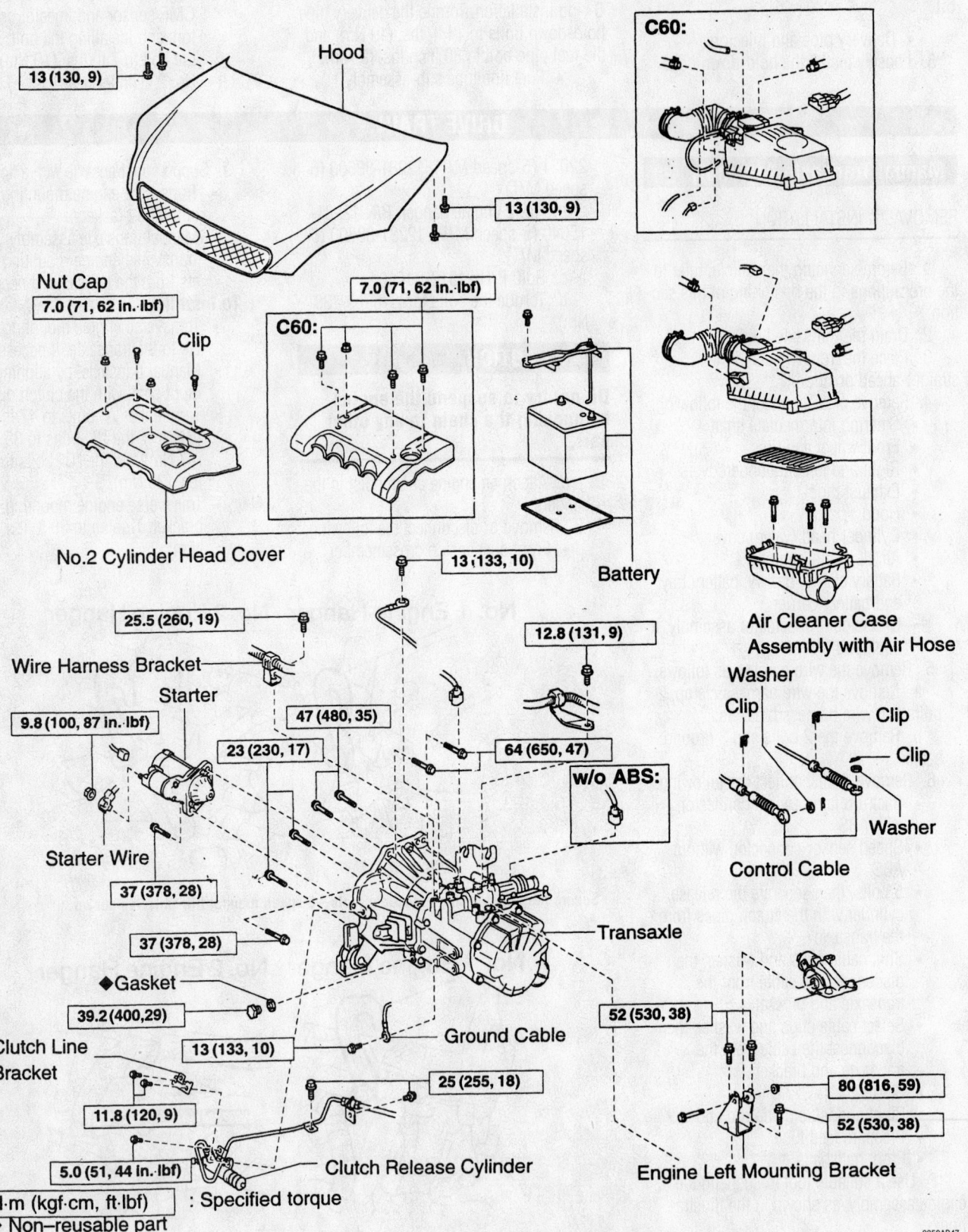

Hood

13 (130, 9)

13 (130, 9)

C60:

Nut Cap
7.0 (71, 62 in. lbf)

Clip

7.0 (71, 62 in. lbf)

C60:

Battery

No.2 Cylinder Head Cover

13 (133, 10)

Air Cleaner Case Assembly with Air Hose

25.5 (260, 19)

Wire Harness Bracket

Starter

12.8 (131, 9)

Washer

Clip

Clip

Clip

9.8 (100, 87 in. lbf)

23 (230, 17)

47 (480, 35)

64 (650, 47)

w/o ABS:

Washer

Starter Wire

Control Cable

Starter Wire

37 (378, 28)

Transaxle

37 (378, 28)

◆Gasket

39.2 (400, 29)

Ground Cable

52 (530, 38)

Clutch Line Bracket

13 (133, 10)

25 (255, 18)

11.8 (120, 9)

80 (816, 59)

5.0 (51, 44 in. lbf)

Clutch Release Cylinder

52 (530, 38)

N·m (kgf·cm, ft·lbf) : Specified torque

Engine Left Mounting Bracket

◆ Non–reusable part

9359AB47

Exploded view of the manual transaxle (1 of 2)

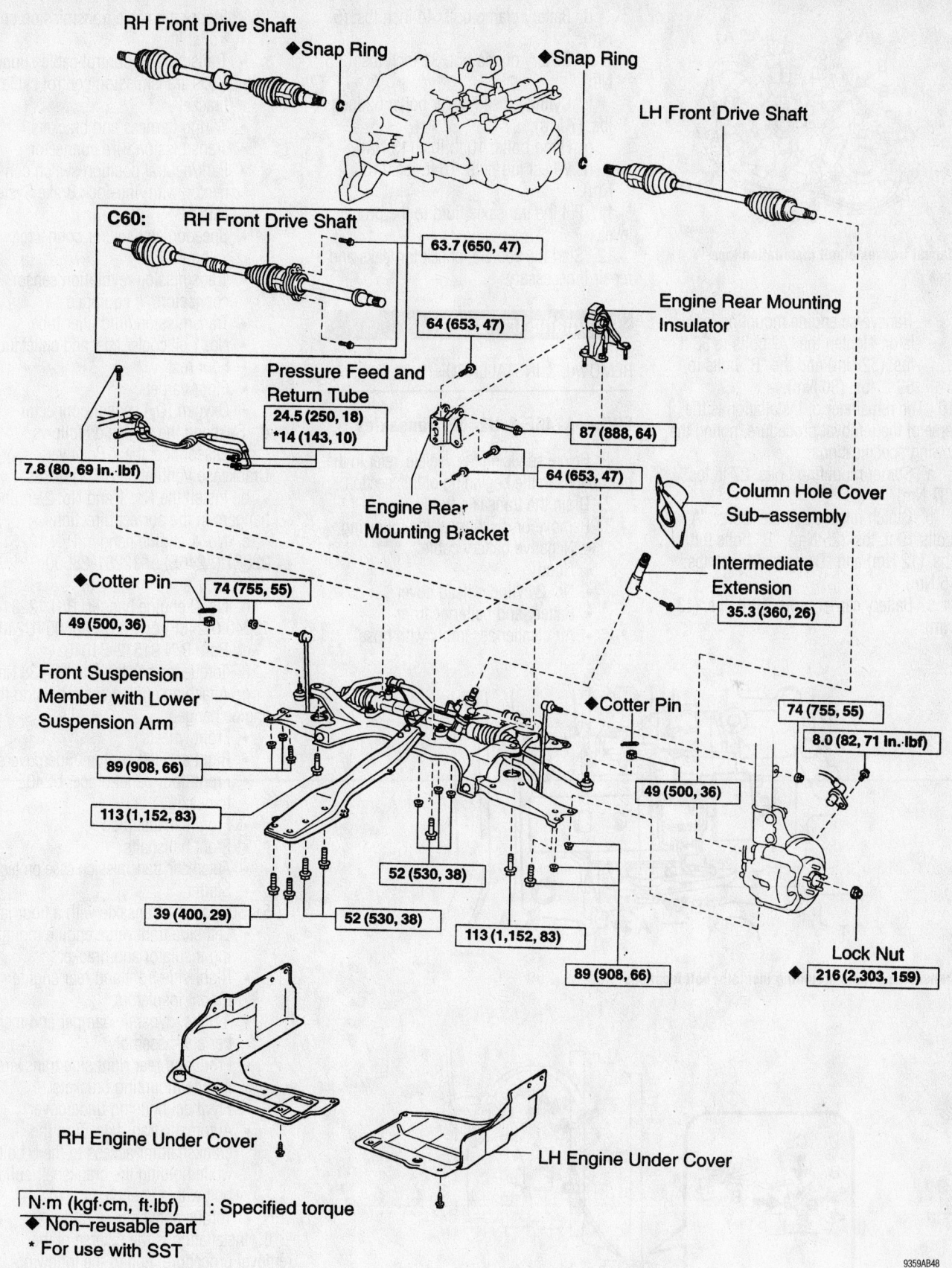

RH Front Drive Shaft

◆Snap Ring

◆Snap Ring

LH Front Drive Shaft

C60: RH Front Drive Shaft

63.7 (650, 47)

64 (653, 47)

Engine Rear Mounting Insulator

Pressure Feed and Return Tube
24.5 (250, 18)
*14 (143, 10)

87 (888, 64)

7.8 (80, 69 in.·lbf)

64 (653, 47)

Engine Rear Mounting Bracket

Column Hole Cover Sub–assembly

◆Cotter Pin

74 (755, 55)

49 (500, 36)

Intermediate Extension

35.3 (360, 26)

Front Suspension Member with Lower Suspension Arm

89 (908, 66)

113 (1,152, 83)

◆Cotter Pin

74 (755, 55)

8.0 (82, 71 in.·lbf)

49 (500, 36)

52 (530, 38)

39 (400, 29)

52 (530, 38)

113 (1,152, 83)

89 (908, 66)

Lock Nut
◆ 216 (2,303, 159)

RH Engine Under Cover

LH Engine Under Cover

N·m (kgf·cm, ft·lbf) : Specified torque
◆ Non–reusable part
* For use with SST

9359AB48

Exploded view of the manual transaxle (2 of 2)

Heater Core replacement is covered in Section 2 of this manual

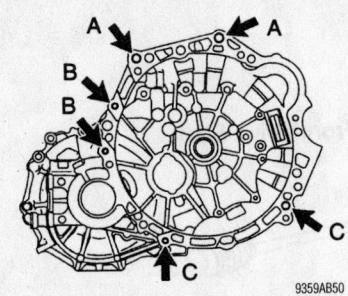

9359AB50

Manual transaxle bolt installation locations

- Transverse engine mounting insulator. Tighten the "A" bolts to 38 ft. lbs. (52 Nm) and the "B" bolts to 59 ft. lbs. (80 Nm).

10. The remainder of installation is the reverse of the removal procedure, noting the following specifications:

   a. Starter mounting bolts: 27 ft. lbs. (37 Nm).

   b. Clutch release cylinder bolts: "A" bolts 19 ft. lbs. (25 Nm), "B" bolts 9 ft. lbs. (12 Nm) and "C" bolts 44 inch lbs. (5 Nm).

   c. Battery carrier bolts: 10 ft. lbs. (13 Nm).

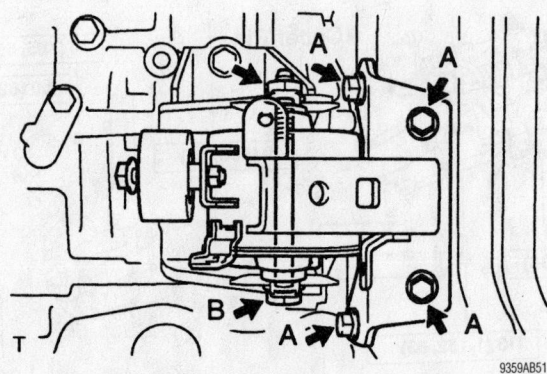

9359AB51

Transverse engine mounting insulator bolt locations

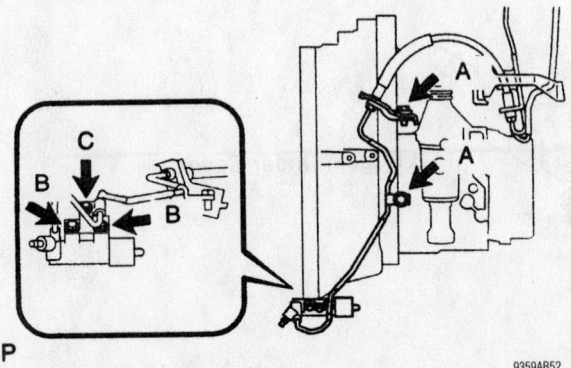

9359AB52

Clutch release cylinder bolt locations

   d. Battery clamp bolt: 44 inch lbs. (5 Nm).

   e. Battery clamp nut: 31 inch lbs. (3.5 Nm).

   f. Cylinder head cover bolts: 62 inch lbs. (7 Nm).

   g. Hood bolts: 10 ft. lbs. (13 Nm).

   h. Wheel lug nuts: 76 ft. lbs. (103 Nm).

11. Fill the transaxle fluid to the proper level.

12. Start the vehicle, check for leaks and repair if necessary.

Automatic Transaxle

REMOVAL & INSTALLATION

FWD—A246E & U240E Transaxles

1. Before servicing the vehicle, refer to the precautions in the beginning of this section.
2. Drain the transaxle fluid.
3. Remove or disconnect the following:
   - Negative battery cable
   - Hood
   - No. 2 cylinder head cover
   - Battery and battery carrier
   - Air cleaner assembly with hose
   - Floor shift cable transmission control shift
   - Transmission control cable support
   - No. 1 transmission control cable bracket
   - Wiring harness and brackets
   - Transmission wire connector
   - Park/neutral position switch connector, with Anti-lock Brake System (ABS)
   - Speedometer sensor connector, without ABS
   - Transmission revolution sensor connectors, if equipped
   - Transmission fluid filler tube
   - No. 1 oil cooler inlet and outlet tubes
   - Foot rest
   - Floor carpet
   - Oxygen (O_2) sensor connector
4. Suspend the engine as follows:
   a. Disconnect the 2 Positive Crankcase Ventilation (PCV) hoses.
   b. Install the No. 1 and No. 2 engine hangers in the correct direction.
   c. No. 1 engine hanger: P/N 12281-22021 (A246E) or 12281-88600 (U240E).
   d. No. 2 engine hanger: P/N 12281-15040 (A246E) or 12281-88600 (U240E).
   e. Bolt: P/N 91512-B1016.
   f. Torque the bolt to 28 ft. lbs. (38 Nm).
   g. Attach an engine chain hoist to the engine hangers.
   - Front wheels
   - Right and left engine undercovers
   - Front floor panel brace, U240E transaxle
   - Front exhaust pipe
   - Front halfshafts
   - Automatic transmission case protector
   - Starter
5. Support the transaxle with a floor jack.
   - Left side transverse engine mounting insulator and bracket
   - Right side front and rear engine mount insulators
   - 4 bolts, dynamic damper and member sub-assembly
   - Front and rear right side transverse engine mounting brackets
   - Flywheel housing undercover
   - Automatic transaxle. Turn the crankshaft for access to the 6 bolts while holding the crankshaft pulley bolt with a wrench.
   - Torque converter clutch
6. Installation is the reverse of the removal procedure, noting the following specifications:
   a. Automatic transaxle: Bolt "A" to 47 ft. lbs. (64 Nm), bolt "B" to 34 ft. lbs. (47 Nm) and bolt "C" to 17 ft. lbs. (23 Nm).

b. Torque converter bolts: 20 ft. lbs. (28 Nm).

c. Front and rear right transverse engine mounting bracket bolts: 47 ft. lbs. (64 Nm).

d. Member sub-assembly center bolts: "A" bolts to 29 ft. lbs. (39 Nm) and "B" bolts to 38 ft. lbs. (52 Nm).

e. Right rear engine mounting insulator-to-engine mounting bracket bolt: 64 ft. lbs. (87 Nm).

f. Right rear engine mount insulator nuts and bolt: 38 ft. lbs. (52 Nm).

g. Left side engine mounting bracket-to-transaxle bolts: 38 ft. lbs. (52 Nm).

h. Left side engine mounting insulator bolts and nut: Bolt "A" to 38 ft. lbs. (52 Nm), Bolt "B" and Nut "B" to 59 ft. lbs. (80 Nm).

i. Front right engine mount insulator-to-mounting bracket bolt and nut: 38 ft. lbs. (52 Nm).

j. Starter bolts: 29 ft. lbs. (39 Nm).

k. Automatic transmission case protector bolts: 14 ft. lbs. (18 Nm).

l. Wheel lug nuts: 76 ft. lbs. (103 Nm).

m. Oil cooler clamp bolts: 49 inch lbs. (5.5 Nm).

n. Oil cooler inlet and outlet tubes: 25 ft. lbs. (34 Nm).

o. Wire harness bracket bolt: 9 ft. lbs. (13 Nm).

p. Transmission control cable bracket bolts: 9 ft. lbs. (12 Nm).

q. Transmission control cable support: 9 ft. lbs. (12 Nm).

r. Battery carrier: 10 ft. lbs. (13 Nm).

s. Air cleaner assembly: 62 inch lbs. (7 Nm).

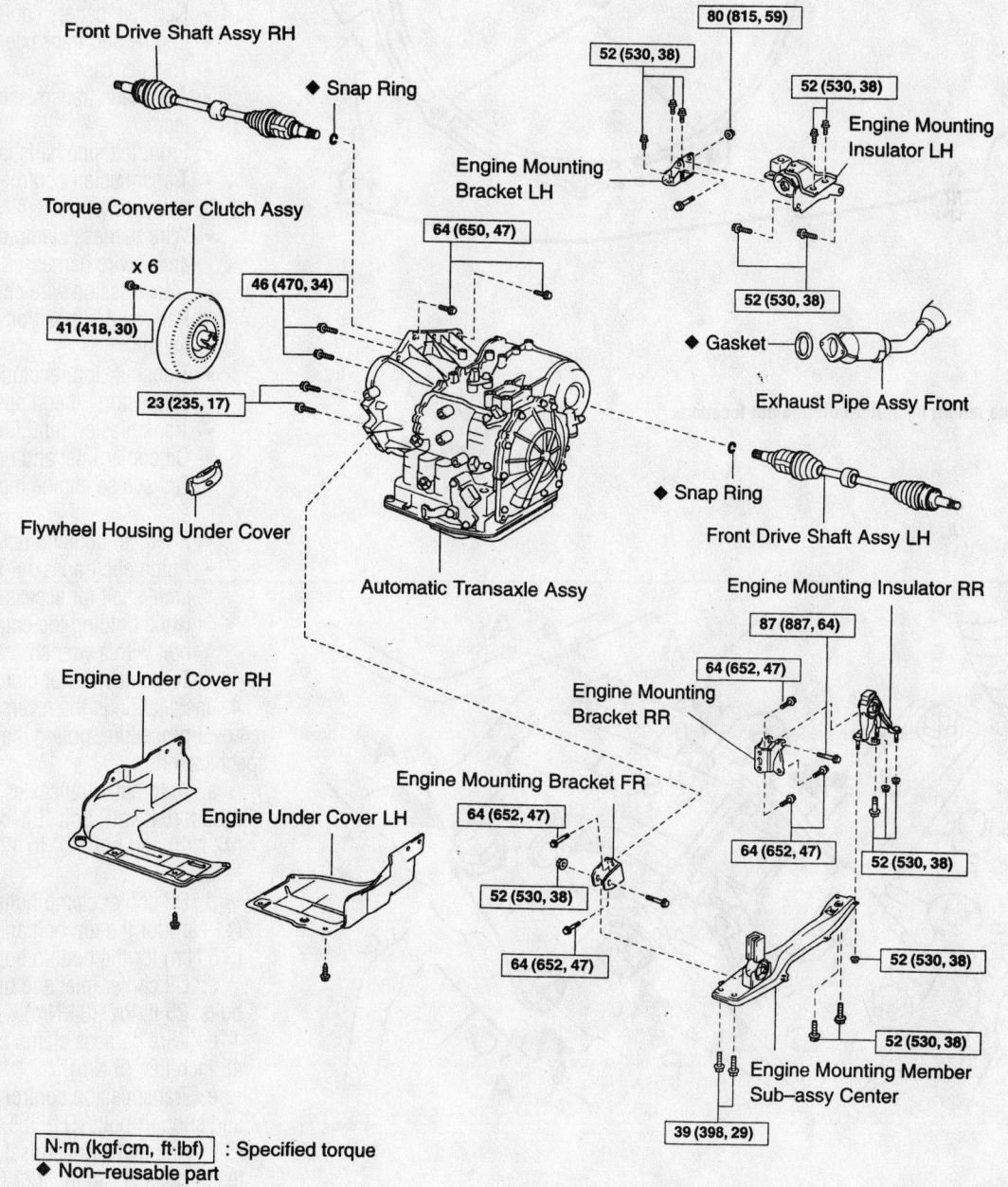

N·m (kgf·cm, ft·lbf) : Specified torque
◆ Non–reusable part

9359AB55

Automatic transaxle and related components—U240E transaxle shown, A246E similar

Brake service is covered in Section 4 of this manual

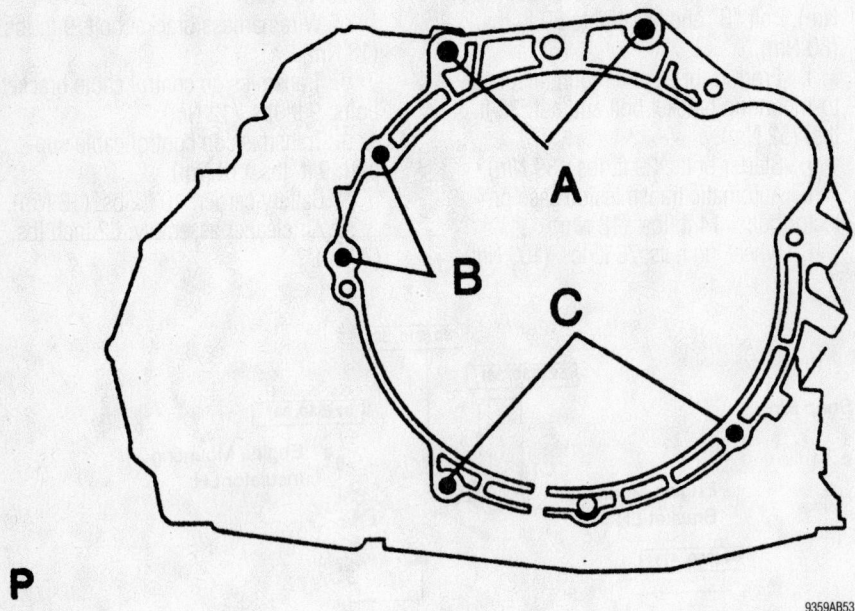

Left side engine mount insulator bolt and nut locations

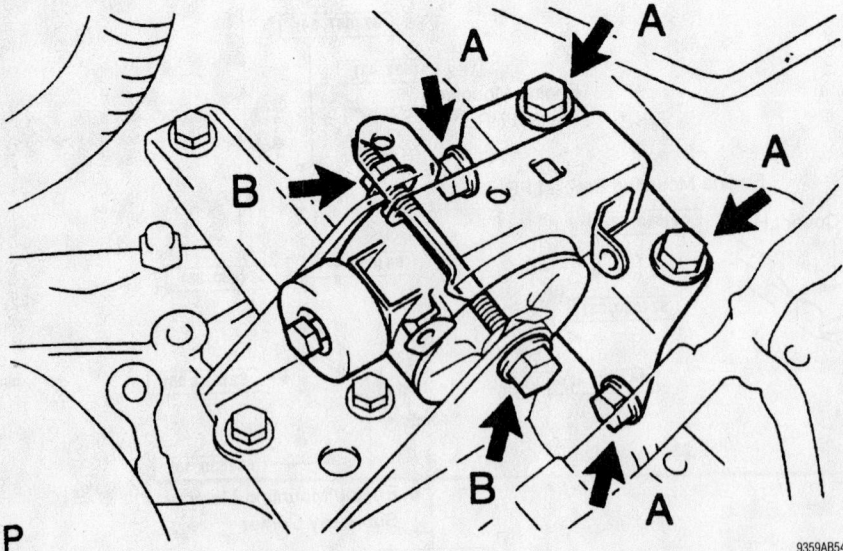

Automatic transaxle bolt locations

t. Cylinder head cover bolts: 62 inch lbs. (7 Nm).

u. Hood bolts: 10 ft. lbs. (13 Nm).

7. Fill the transaxle fluid to the proper level.

8. Start the vehicle, check for leaks and repair if necessary.

AWD—U341F Transaxle

1. Before servicing the vehicle, refer to the precautions in the beginning of this section.

2. Drain the transaxle fluid.

3. Remove or disconnect the following:

- Negative battery cable
- Engine and transaxle assembly
- Transfer case
- Automatic transmission case protector
- Front left side halfshaft
- Transmission control cable support and bracket
- Wire harness clamp bracket, bolts and 2 wire harnesses
- Transmission wire connector
- Park/neutral position switch connector
- Transmission revolution sensor connectors, if equipped
- Transmission fluid filler tube
- Oil cooler inlet and outlet tubes
- Transverse engine mounting brackets
- Flywheel housing undercover
- Automatic transaxle. Turn the crankshaft for access to the 6 bolts while holding the crankshaft pulley bolt with a wrench.
- Torque converter clutch

4. Installation is the reverse of the removal procedure, noting the following specifications:

a. Automatic transaxle: Bolt "A" to 47 ft. lbs. (64 Nm), bolt "B" to 34 ft. lbs. (47 Nm) and bolt "C" to 17 ft. lbs. (23 Nm).

b. Oil cooler clamp bolts: 8 ft. lbs. (11 Nm) for the top bolt and 49 inch lbs. (5.5 Nm) for the bottom bolt

c. Oil cooler inlet and outlet tube bolts: 25 ft. lbs. (34 Nm).

d. Wire harness clamp bracket bolt: 48 inch lbs. (5 Nm).

e. Transmission control cable bracket and support bolts: 9 ft. lbs. (12 Nm).

f. Automatic transmission case protector bolts: 17 ft. lbs. (23 Nm).

5. Fill the transaxle fluid to the proper level.

6. Start the vehicle, check for leaks and repair if necessary.

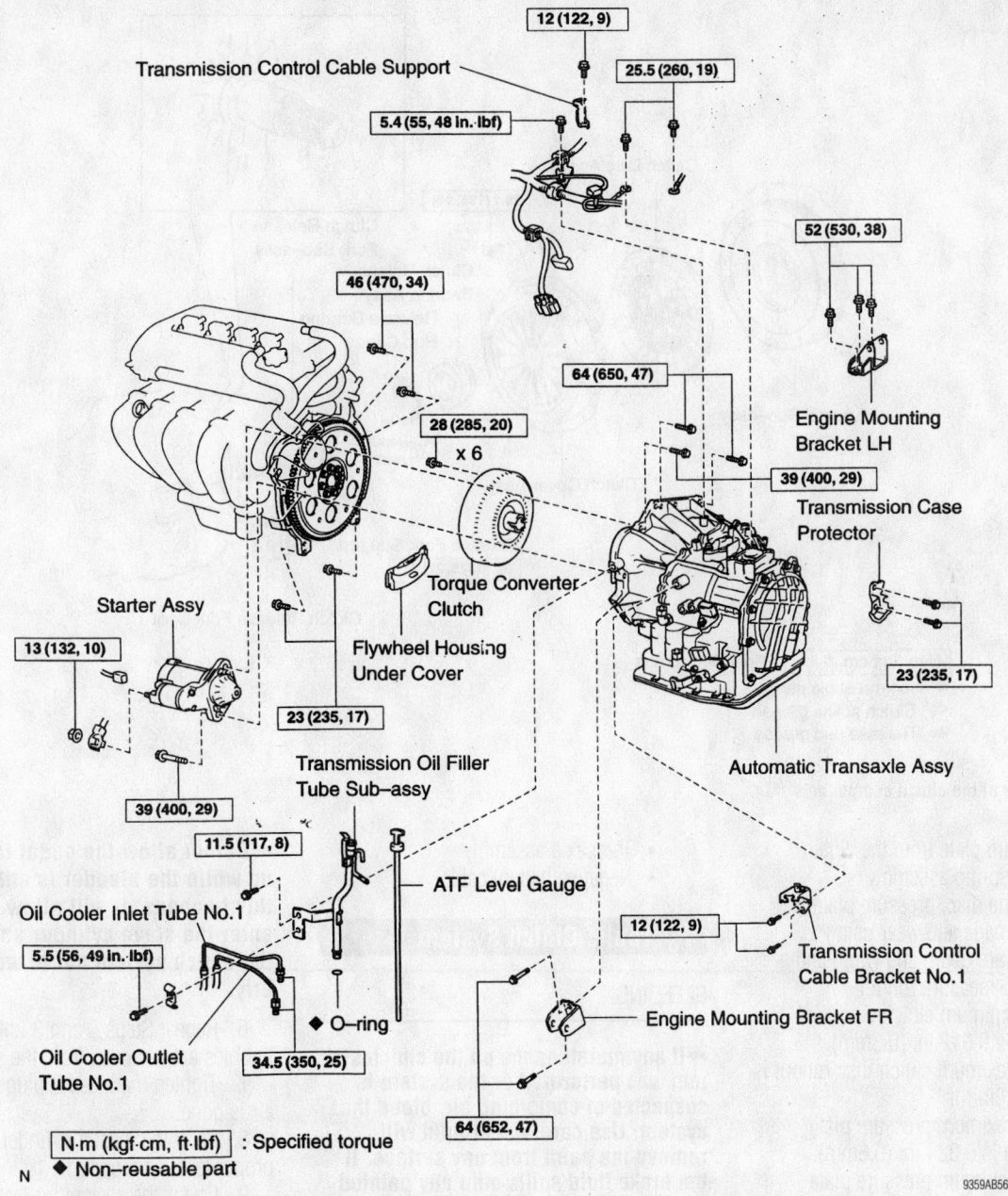

Transmission Control Cable Support

12 (122, 9)

25.5 (260, 19)

5.4 (55, 48 in.·lbf)

52 (530, 38)

46 (470, 34)

64 (650, 47)

28 (285, 20)
x 6

Engine Mounting Bracket LH

39 (400, 29)

Transmission Case Protector

Torque Converter Clutch

Starter Assy

Flywheel Housing Under Cover

13 (132, 10)

23 (235, 17)

23 (235, 17)

Automatic Transaxle Assy

Transmission Oil Filler Tube Sub-assy

39 (400, 29)

11.5 (117, 8)

ATF Level Gauge

Oil Cooler Inlet Tube No.1

12 (122, 9)

Transmission Control Cable Bracket No.1

5.5 (56, 49 in.·lbf)

◆ O-ring

Oil Cooler Outlet Tube No.1

34.5 (350, 25)

Engine Mounting Bracket FR

64 (652, 47)

N·m (kgf·cm, ft·lbf) : Specified torque
N ◆ Non-reusable part

9359AB56

Automatic transaxle and related components—U341F transaxle

Clutch

ADJUSTMENTS

Hydraulic clutch actuating systems used in Toyota vehicles do not require adjustment.

REMOVAL & INSTALLATION

1. Before servicing the vehicle, refer to the precautions in the beginning of this section.

➡ Do not allow grease or oil to get on any part of the disc, pressure plate, or flywheel surfaces.

2. Remove or disconnect the following:
- Negative battery cable. On vehicles equipped with an air bag, wait at least 90 seconds before proceeding
- Transaxle assembly

3. Make matchmarks on the clutch cover (pressure plate) and flywheel so that the pressure plate can be returned to its original position during installation.

4. Remove or disconnect the following:

- Release fork bearing clips
- Release bearing hub, complete with the release bearing
- Release fork and support

❋❋ CAUTION

Slowly unfasten the bolts which attach the pressure plate. Loosen each bolt 1 turn at a time until the spring tension is released. If the bolts are released improperly the clutch assembly could fly apart, causing possible injury.

For complete Engine Mechanical specifications, see Section 1 of this manual

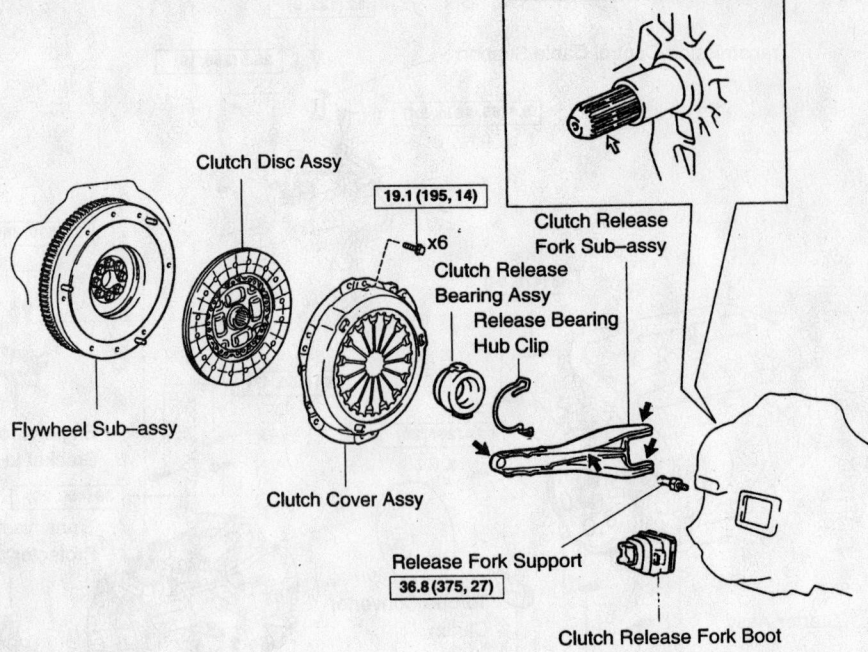

Clutch Disc Assy

19.1 (195, 14)

x6

Clutch Release
Fork Sub–assy

Clutch Release
Bearing Assy

Release Bearing
Hub Clip

Flywheel Sub–assy

Clutch Cover Assy

Release Fork Support
36.8 (375, 27)

Clutch Release Fork Boot

9359AB57

N·m (kgf·cm, ft·lbf) : Specified torque
◆ Non–reusable part
⇐ Clutch spline grease
⇐ Release hub grease

Exploded view of the clutch components

- Pressure plate from the clutch cover/spring assembly

5. Inspect the disc, pressure plate and flywheel for damage and wear using a caliper to measure depth and width and a dial indicator to measure runout.

   a. The minimum clutch disc rivet head depth is 0.012 in. (0.3mm).

   b. The maximum clutch disc runout is 0.031 in. (0.8mm).

   c. The maximum pressure plate spring depth is 0.024 in. (0.6mm).

   d. The maximum pressure plate spring width is 0.197 in. (5.0mm).

   e. The maximum flywheel runout is 0.004 in. (0.1mm).

6. Replace or machine parts as necessary.

To install:

7. When reassembling, apply a thin coating of multipurpose grease to the release bearing hub and release fork contact points. Also, pack the groove inside the clutch hub with multipurpose grease and lubricate the pivot points of the release fork.

8. Install or connect the following:

- Clutch disc and pressure plate. The bolts should be tightened in 2 or 3 steps, gradually and evenly. Final bolt torque is 14 ft. lbs. (19 Nm).
- Release bearing, fork and boot

- Transaxle assembly
- Negative battery cable

Hydraulic Clutch System

BLEEDING

➡ **If any maintenance on the clutch system was performed or the system is suspected of containing air, bleed the system. Use care; brake fluid will remove the paint from any surface. If the brake fluid spills onto any painted surface, wash it off immediately with soap and water.**

1. Before servicing the vehicle, refer to the precautions in the beginning of this section.

2. Fill the clutch reservoir with brake fluid. Check the reservoir level frequently and add fluid as needed.

3. Connect one end of a vinyl tube to the bleeder plug on the slave cylinder and submerge the other end into a clear container half-filled with brake fluid.

4. Slowly pump the clutch pedal several times.

5. Have an assistant hold the clutch pedal down and loosen the bleeder plug until fluid and/or air starts to run out of the bleeder plug. Close the bleeder plug while the pedal is held to the floor.

➡ **Do not allow the pedal to rise back-up while the bleeder is still open. If this happens, it will allow air to re-enter the slave cylinder and cause the clutch system not to work properly.**

6. Repeat Steps 2 and 3 until all the air bubbles are removed from the system.

7. Tighten the bleeder plug when all the air is gone.

8. Refill the master cylinder to the proper level as required.

9. Check the system for leaks.

Transfer Case

REMOVAL & INSTALLATION

1. Before servicing the vehicle, refer to the precautions in the beginning of this section.

2. Drain the transfer case fluid.

3. Remove or disconnect the following:

- Negative battery cable. Due to the air bag system, wait at least 90 seconds before proceeding
- Engine and transaxle assembly
- Separate vane pump
- Steering gear

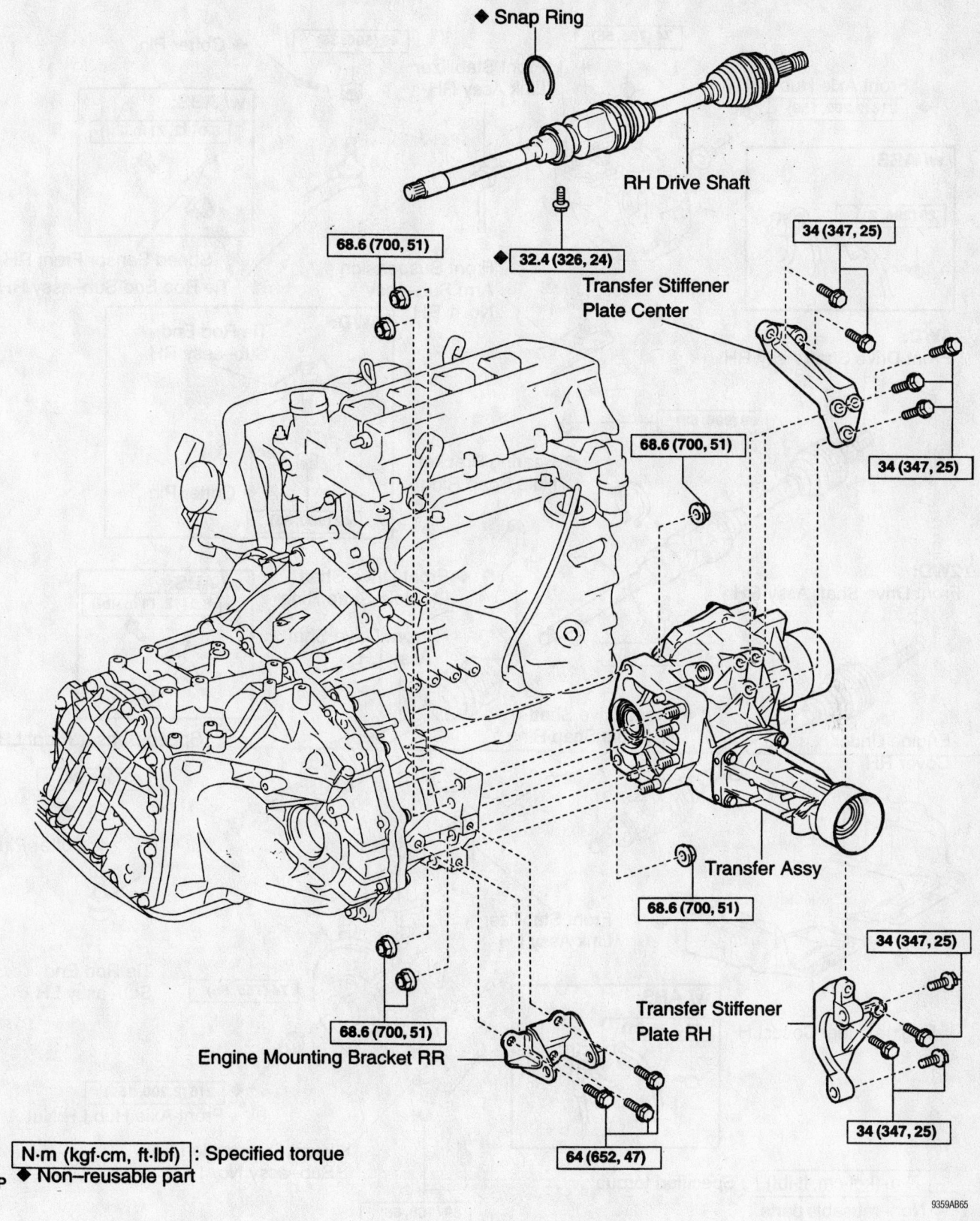

Snap Ring

RH Drive Shaft

68.6 (700, 51)

◆ 32.4 (326, 24)

34 (347, 25)

Transfer Stiffener
Plate Center

68.6 (700, 51)

34 (347, 25)

Transfer Assy

68.6 (700, 51)

34 (347, 25)

Transfer Stiffener
Plate RH

68.6 (700, 51)

Engine Mounting Bracket RR

64 (652, 47)

34 (347, 25)

N·m (kgf·cm, ft·lbf) : Specified torque

P ◆ Non–reusable part

9359AB65

Exploded view of the transfer case mounting

For Accessory Drive Belt illustrations, see Section 1 of this manual

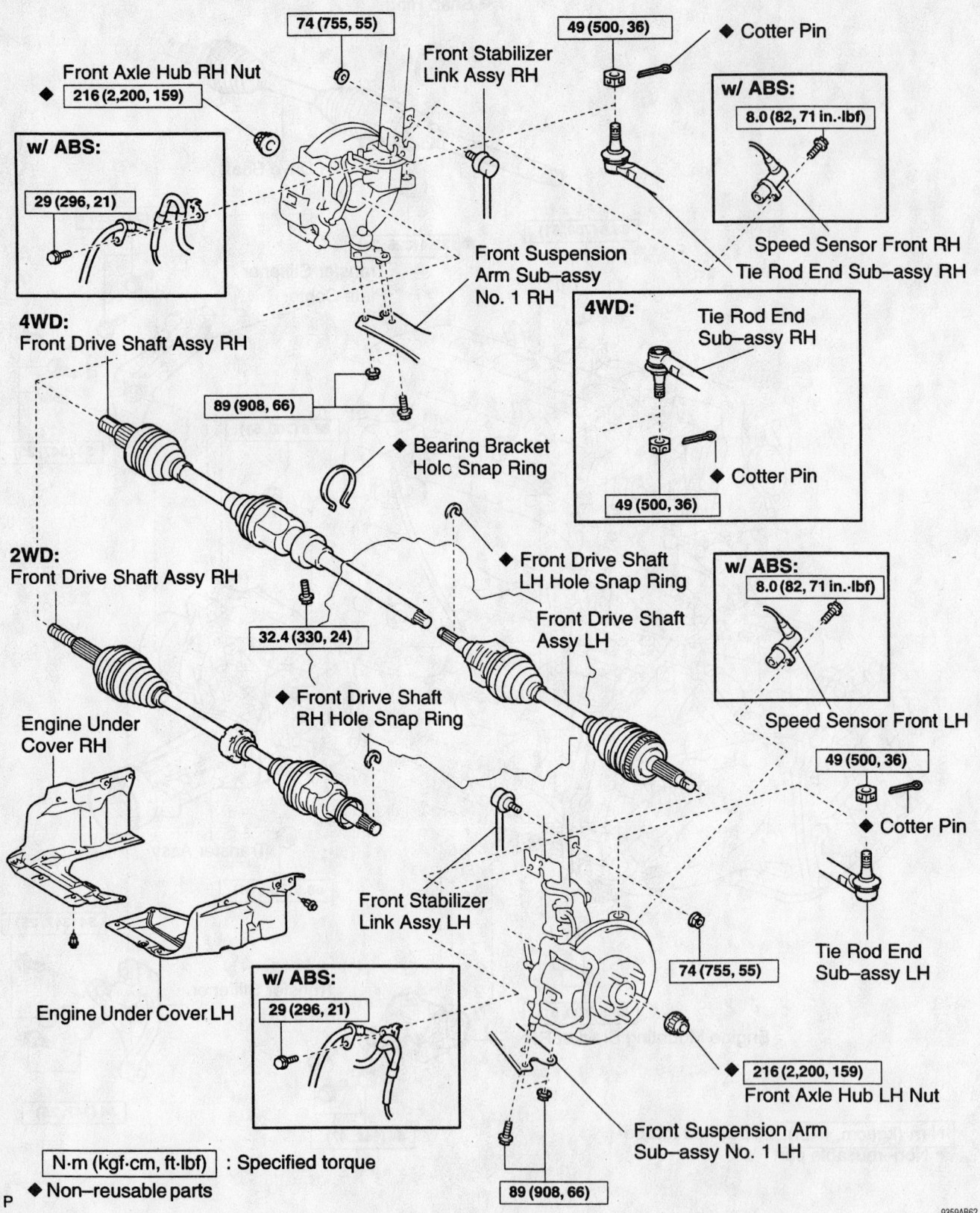

Front Axle Hub RH Nut
◆ 216 (2,200, 159)

74 (755, 55)

Front Stabilizer Link Assy RH

49 (500, 36)

◆ Cotter Pin

w/ ABS:
8.0 (82, 71 in.·lbf)

w/ ABS:
29 (296, 21)

Speed Sensor Front RH

Tie Rod End Sub–assy RH

Front Suspension Arm Sub–assy No. 1 RH

4WD:
Front Drive Shaft Assy RH

4WD:
Tie Rod End Sub–assy RH

89 (908, 66)

◆ Bearing Bracket Hole Snap Ring

◆ Cotter Pin

49 (500, 36)

2WD:
Front Drive Shaft Assy RH

◆ Front Drive Shaft LH Hole Snap Ring

Front Drive Shaft Assy LH

w/ ABS:
8.0 (82, 71 in.·lbf)

32.4 (330, 24)

Engine Under Cover RH

◆ Front Drive Shaft RH Hole Snap Ring

Speed Sensor Front LH

49 (500, 36)

◆ Cotter Pin

Front Stabilizer Link Assy LH

Tie Rod End Sub–assy LH

Engine Under Cover LH

w/ ABS:
29 (296, 21)

74 (755, 55)

◆ 216 (2,200, 159)
Front Axle Hub LH Nut

Front Suspension Arm Sub–assy No. 1 LH

N·m (kgf·cm, ft·lbf) : Specified torque
◆ Non–reusable parts

P

89 (908, 66)

9359AB62

Halfshafts and related components

- Crossmember
- Manifold stay
- Oxygen (O$_2$) sensor
- Exhaust manifold heat shield
- Exhaust manifold
- Starter
- Right side halfshaft
- Transverse engine mounting bracket
- Center and right side transfer stiffener plates

✴✴ WARNING

When removing the transfer case, DO NOT touch the oil seal.

- Transfer case bolts, and transfer assembly, using a mallet to dislodge it from the transaxle

4. Installation is the reverse of the removal procedure, noting the following specifications:

   a. Transfer case stiffener case bolts: 25 ft. lbs. (34 Nm).

   b. Engine mounting bracket bolts: 47 ft. lbs. (64 Nm).

5. Add fluid to the transfer case, and check for leaks.

Halfshaft

REMOVAL & INSTALLATION

➡ **The hub bearing could be damaged if subjected to the full weight of the vehicle, such as if the vehicle is moved without the halfshafts. If it is absolutely necessary to place the full vehicle weight on the hub bearing, first support the bearing with SST No. 09608–16041.**

1. Before servicing the vehicle, refer to the precautions in the beginning of this section.

2. Drain the transaxle fluid.

3. Remove or disconnect the following:

- Negative battery cable. Due to the air bag system, wait at least 90 seconds before proceeding.
- Both front wheels
- Cotter pin, locknut cap, and the hub nut
- Undercovers
- Speed sensors
- Tie rod ball joint from the steering knuckle
- Stabilizer bar link from the lower suspension arm

- Lower ball joint from the lower suspension arm
- Halfshaft from the knuckle

➡ **Be careful not to damage the inner oil seal or the ABS sensor rotor on the halfshaft.**

4. To remove the left side halfshaft, separate the halfshaft from the transaxle.

5. To remove the right side halfshaft perform the following steps:

- Remove the 2 bolts of the center bearing bracket
- Pull the halfshaft out together with the center bearing case and the center halfshaft.
- Remove the center shaft with the right-hand halfshaft from the transaxle through the bearing bracket.

➡ **Do not damage the oil seal lip.**

To install:

6. Install or connect the following:

- Snapring opening side facing downward, on the oiled inboard joint tulip
- Left side halfshaft into the transaxle
- Right side halfshaft, with the bearing case and center shaft, into the transaxle
- Center bearing case (right side).

7. After installing either halfshaft, check that there is 0.08–0.12 in. (2–3mm) of axial play. Check that the halfshaft is making contact with the pinion shaft and that the halfshaft cannot be pulled out.

8. Install or connect the following:

- Halfshaft into the knuckle
- Lower suspension arm to the lower ball joint. Torque the bolt and nuts to 66 ft. lbs. (89 Nm).
- Tie rod end to the steering knuckle. Tighten the nut to 36 ft. lbs. (49 Nm).
- Stabilizer bar link to the lower suspension arm. Torque the nuts to 55 ft. lbs. (74 Nm).
- Front wheels
- Hub nut and washer and tighten to 159 ft. lbs. (216 Nm)
- Negative battery cable
- Locknut cap and a new cotter pin.
- Speed sensors
- Undercover

9. Fill the transaxle fluid to the proper level

10. Start the vehicle, check for leaks and repair if necessary.

CV-Joints

OVERHAUL

1. Before servicing the vehicle, refer to the precautions in the beginning of this section.

2. Remove or disconnect the following:

- Inboard joint boot clips
- Inboard joint tulip from the driveshaft
- Snapring
- Using a brass rod and hammer, the tri-pot joint off the driveshaft without hitting the joint roller
- Inboard joint boot
- Clamp and driveshaft damper
- Clamps and the outboard drive boot. DO NOT disassemble the outboard joint.

To assemble:

3. Install or connect the following:

➡ **Before installing the boot, wrap the spline end of the shaft with masking tape to prevent damage to the boot.**

- Driveshaft damper with a new clamp
- Temporarily, the inboard boot with new clamp to the drive joint

➡ **The inboard boot and clamp are larger than those of the outboard boot.**

- The tri-pot onto the driveshaft with a brass rod and hammer without hitting the joint roller
- The snapring

4. Pack the outboard tulip joint and the outboard boot with about 0.26–0.33 lbs. ounces of grease that was supplied with the boot kit.

5. Install or connect the following:
- Boot onto the outboard joint

6. Pack the inboard tulip joint and boot with ½ lb. of grease that was supplied with the boot kit.

- Inboard tulip joint onto the driveshaft
- Boot onto the driveshaft

7. Before checking the standard length, bend the band and lock it. Make sure that the boot is not stretched or squashed when the driveshaft is at standard length. Standard driveshaft length: LH: 540.2 mm (21.268 in.); RH: 857.4 mm (33.756 in.)

For Tire, Wheel and Ball Joint specifications, see Section 1 of this manual

2WD:

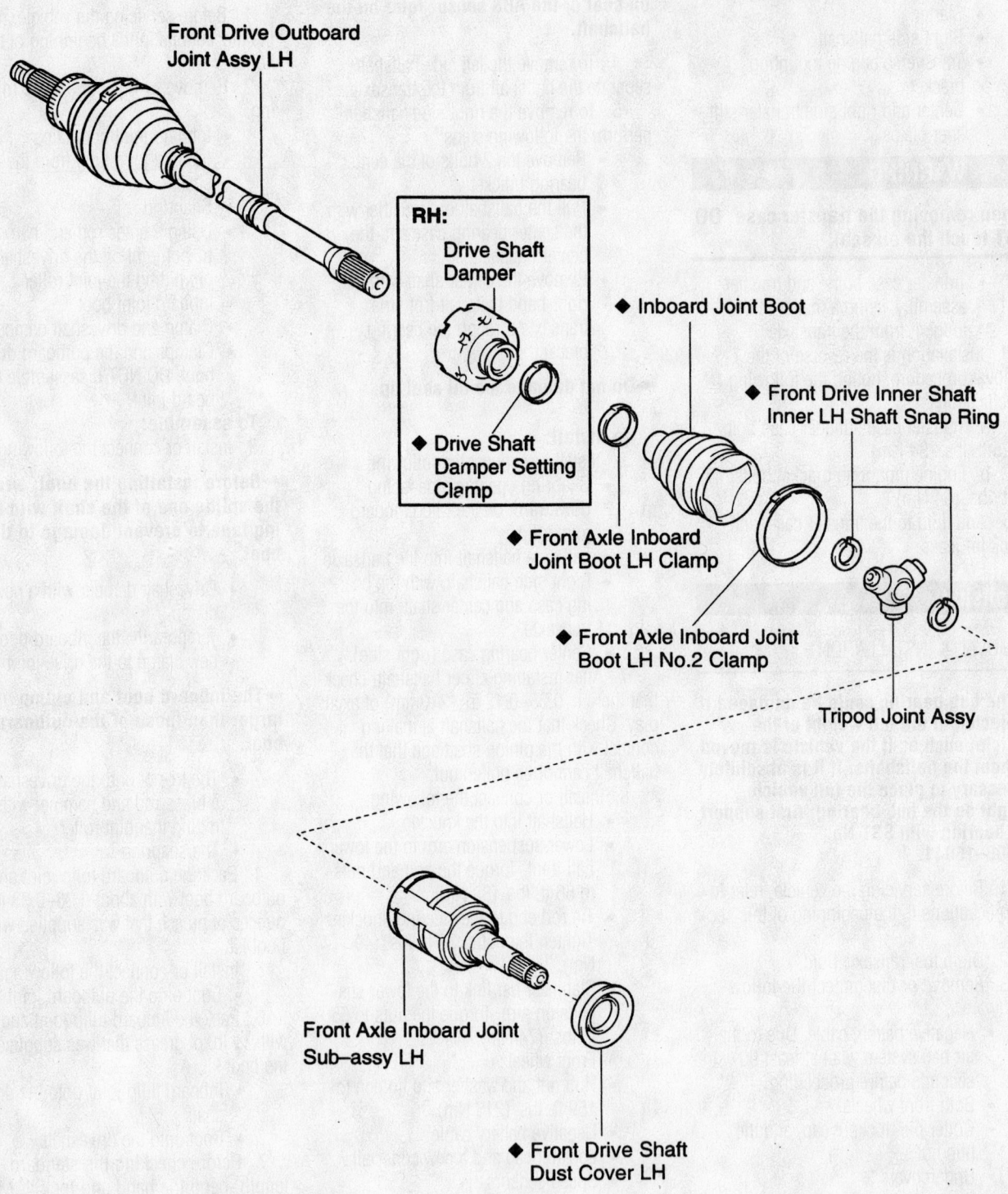

Front Drive Outboard
Joint Assy LH

RH:

Drive Shaft
Damper

◆ Drive Shaft
Damper Setting
Clamp

◆ Inboard Joint Boot

◆ Front Drive Inner Shaft
Inner LH Shaft Snap Ring

◆ Front Axle Inboard
Joint Boot LH Clamp

◆ Front Axle Inboard Joint
Boot LH No.2 Clamp

Tripod Joint Assy

Front Axle Inboard Joint
Sub–assy LH

◆ Front Drive Shaft
Dust Cover LH

◆ **Non–reusable parts**

P

9359AB63

Exploded view of the CV-joint—FWD vehicles

4WD:

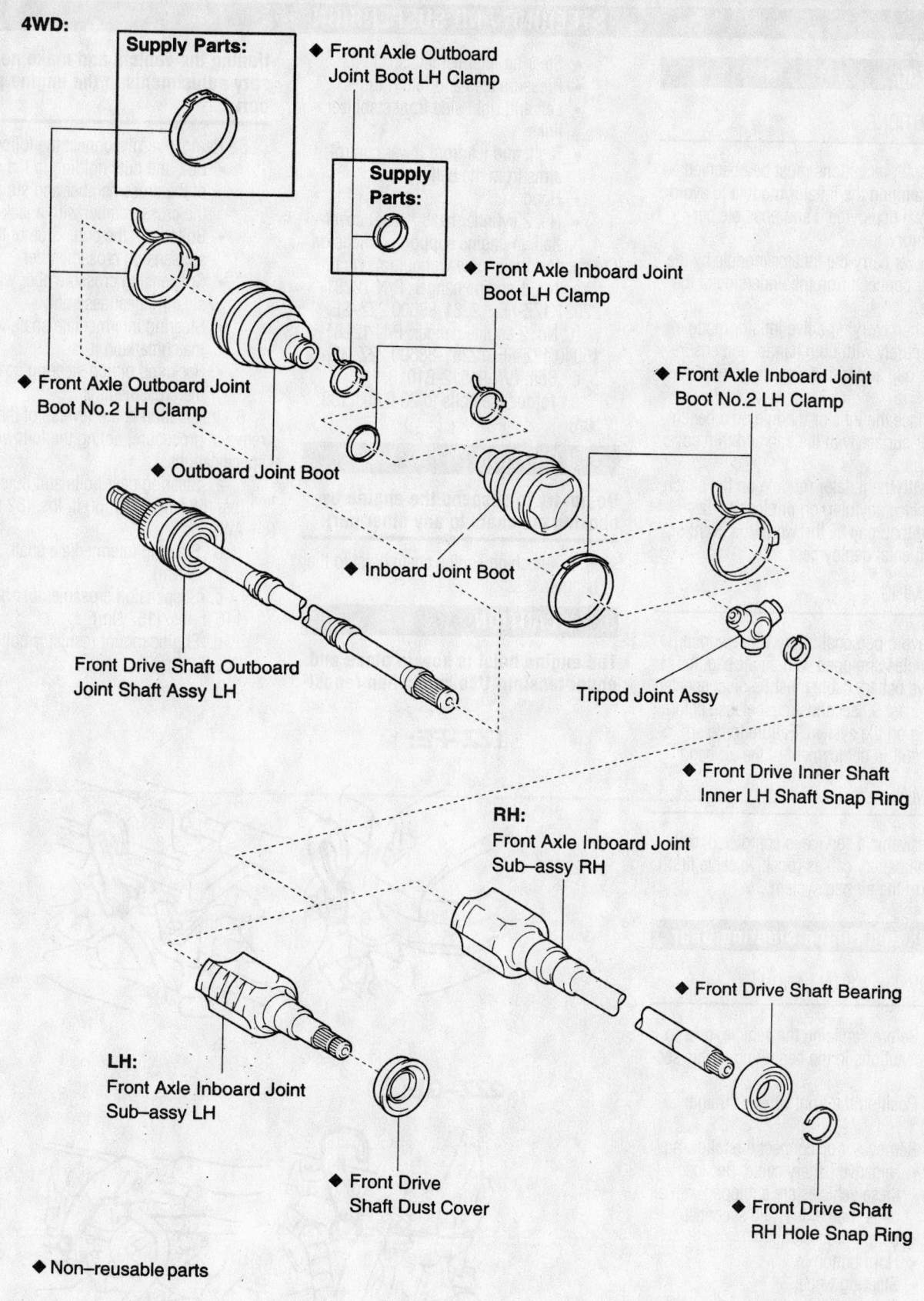

Supply Parts:

◆ Front Axle Outboard
Joint Boot LH Clamp

**Supply
Parts:**

◆ Front Axle Inboard Joint
Boot LH Clamp

◆ Front Axle Outboard Joint
Boot No.2 LH Clamp

◆ Front Axle Inboard Joint
Boot No.2 LH Clamp

◆ Outboard Joint Boot

◆ Inboard Joint Boot

Front Drive Shaft Outboard
Joint Shaft Assy LH

Tripod Joint Assy

◆ Front Drive Inner Shaft
Inner LH Shaft Snap Ring

RH:
Front Axle Inboard Joint
Sub–assy RH

◆ Front Drive Shaft Bearing

LH:
Front Axle Inboard Joint
Sub–assy LH

◆ Front Drive
Shaft Dust Cover

◆ Front Drive Shaft
RH Hole Snap Ring

◆ Non–reusable parts

P

9359AB64

Exploded view of the CV-joint—AWD vehicles

For Wheel Alignment specifications, see Section 1 of this manual

STEERING AND SUSPENSION

Air Bag

PRECAUTIONS

Several precautions must be observed when handling the inflator module to avoid accidental deployment and possible personal injury.

• Never carry the inflator module by the wires or connector on the underside of the module.

• When carrying a live inflator module, hold securely with both hands, and ensure that the bag and trim cover are pointed away.

• Place the inflator module on a bench or other surface with the bag and trim cover facing up.

• With the inflator module on the bench, never place anything on or close to the module that may be thrown in the event of an accidental deployment.

DISARMING

To avoid personal injury when working on vehicles equipped with an air bag, the negative battery cable must be disconnected and at least 90 seconds must elapse before working on the system. Failure to do so may result in deployment of the air bag.

REARMING

After vehicle service is completed, reattach the battery cables (positive cable first!) to rearm the air bag system.

Rack and Pinion Steering Gear

REMOVAL & INSTALLATION

1. Before servicing the vehicle, refer to the precautions in the beginning of this section.

2. Position the front wheels straight ahead.

3. Remove or disconnect the following:

• Negative battery cable. Because these vehicles are equipped with air bags, wait at least 90 seconds before proceeding.
• Horn button
• Steering wheel
• Front wheels
• Left and right engine undercovers
• Left and right tie rod ends
• Column hose cover silencer sheet

• Steering intermediate shaft
• Pressure feed and return tubes
• Left and right side front stabilizer links
• Right and left front lower control arms from the ball joints
• Hood
• No. 2 cylinder head (valve) cover

4. Install an engine support and tension it to support the engine without raising it.

   a. No. 1 engine hanger: P/N 12281-22021 1ZZ-FE, 12281-88600 2ZZ-GE.

   b. No. 2 engine hanger: P/N 12281-15040 1ZZ-FE, 12281-88600 2ZZ-GE.

   c. Bolt: P/N 91512-B1016.

   d. Torque the bolts to 28 ft. lbs. (38 Nm).

> ✳✳ **CAUTION**
>
> **Do not try to suspend the engine by hooking the chain to any other part.**

   e. Attach an engine chain hoist to the hangers.

> ✳✳ **CAUTION**
>
> **The engine hoist is now in place and under tension. Use care when reposi-**

tioning the vehicle and make necessary adjustments to the engine support.

5. Remove or disconnect the following:

• Bolt and nuts holding in the middle of the crossmember and support the crossmember with a jack
• Bolts from the outer side of the suspension crossmember
• Suspension crossmember with the steering gear assembly
• Steering intermediate shaft, after matchmarking it
• Rack and pinion steering gear from the crossmember

6. Installation is the reverse of the removal procedure, noting the following specifications:

   a. Steering gear bolts and nuts: 43 ft. lbs. (58 Nm) FWD, 60 ft. lbs. (82 Nm) AWD.

   b. Steering intermediate shaft: 26 ft. lbs. (35 Nm).

   c. Suspension crossmember bolts: 116 ft. lbs. (157 Nm).

   d. Engine mount insulator bolts: 38 ft. lbs. (52 Nm).

1ZZ-FE:

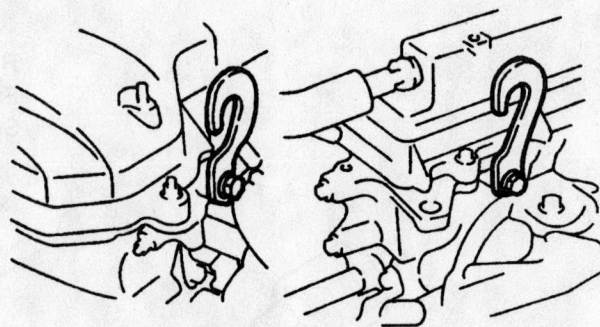

2ZZ-GE:

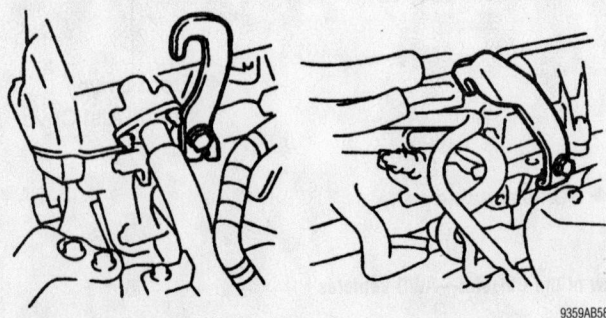

9359AB58

Proper installation of engine hangers

2WD:

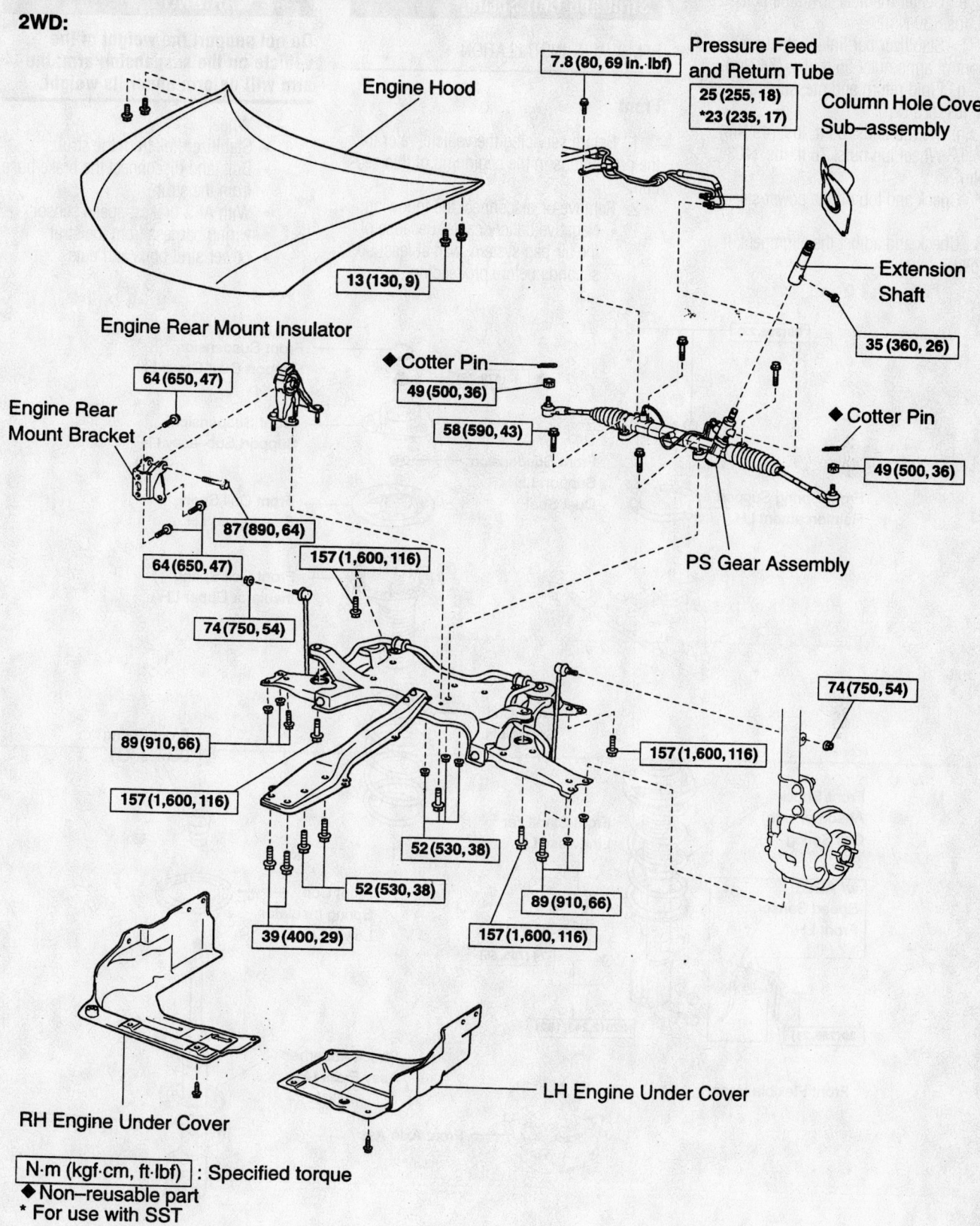

Engine Hood

7.8 (80, 69 in.·lbf)

Pressure Feed and Return Tube

25 (255, 18)
*23 (235, 17)

Column Hole Cover Sub–assembly

13 (130, 9)

Extension Shaft

35 (360, 26)

Engine Rear Mount Insulator

64 (650, 47)

◆ Cotter Pin

49 (500, 36)

Engine Rear Mount Bracket

58 (590, 43)

◆ Cotter Pin

49 (500, 36)

87 (890, 64)

64 (650, 47)

157 (1,600, 116)

PS Gear Assembly

74 (750, 54)

89 (910, 66)

74 (750, 54)

157 (1,600, 116)

157 (1,600, 116)

52 (530, 38)

89 (910, 66)

52 (530, 38)

39 (400, 29)

157 (1,600, 116)

LH Engine Under Cover

RH Engine Under Cover

N·m (kgf·cm, ft·lbf) : Specified torque
◆ Non–reusable part
* For use with SST

9359AB59

Exploded view of a typical power rack and pinion steering gear unit—FWD shown

For Maintenance Interval recommendations, see Section 1 of this manual

e. Center member-to-frame bolts: 29 ft. lbs. (39 Nm).

f. Stabilizer bar link-to-the lower control arms nuts: 55 ft. lbs. (74 Nm).

g. Fluid return and pressure tubes: 17 ft. lbs. (23 Nm).

h. Tie rod ends: 36 ft. lbs. (49 Nm).

i. Wheel lug nuts: 76 ft. lbs. (103 Nm).

7. Check and top off the power steering fluid.

8. Check and adjust the alignment, if needed.

Strut and Coil Spring

REMOVAL & INSTALLATION

Front

1. Before servicing the vehicle, refer to the precautions in the beginning of this section.

2. Remove or disconnect the following:
- Negative battery cable. Because of the air bag system, wait at least 90 seconds before proceeding

✳✳ WARNING

Do not support the weight of the vehicle on the suspension arm; the arm will deform under its weight.

- Wheel
- Stabilizer link from the strut
- Bolt, and disconnect the brake hose from the strut
- With ABS brakes, speed sensor wiring harness from the strut
- Lower strut bolts and nuts

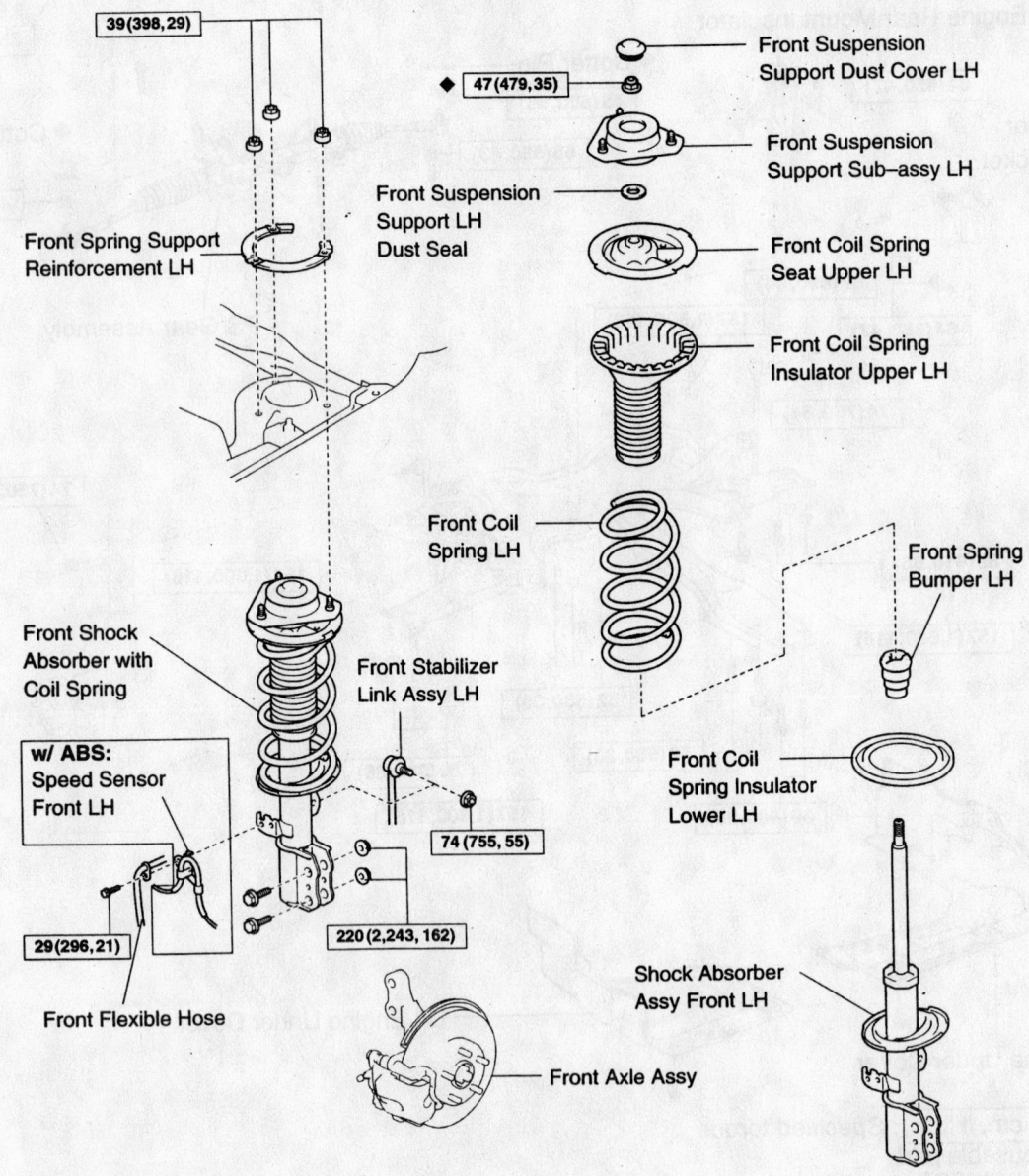

N·m (kgf·cm, ft·lbf): Specified torque

◆ Non–reusable part

Common coil spring and strut component assembly

9359AB60

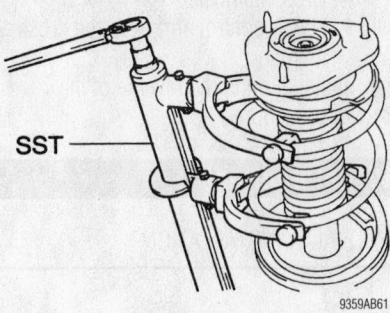

9359AB61

Proper method of supporting the strut in a vise

- Upper strut nuts
- Strut from the steering knuckle
- Strut

3. To disassemble the strut:
- Install a bolt and 2 nuts to the bracket at the lower portion of the strut shell and secure it in a vise
- Compress the coil spring
- Dust cover and hold the spring seat so that it will not turn
- Nut on the top of the strut
- Suspension support, bearing, dust seal, spring seat, spring, insulators and bumper

To install:

4. To assemble the strut:
- Install the spring bumper to piston
5. Using a spring compressor, compress the spring.
- Coil spring to the strut. Fit the lower end of the coil spring into the gap of the lower seat.
- Spring seat with the insulator
- Dust seal on the spring seat
- Suspension support and tighten 35 ft. lbs. (47 Nm). After the nut has been tighten, release the compressor tool tension.
6. Pack multipurpose grease into the suspension support.
- Dust cover.

➡**Do not use an impact wrench to tighten the nut. Also, check that the bearing fits into the recess in the suspension support.**

- Strut
- Nuts holding the strut to the strut tower. Tighten the nuts to 29 ft. lbs. (39 Nm).
- 2 lower strut bolts and nuts. Tighten to 162 ft. lbs. (220 Nm).
- Brake line to the steering knuckle.

Tighten the line bolt to 21 ft. lbs. (29 Nm).
- Secure the wiring harness, if equipped with ABS
- Stabilizer link. Tighten the nut to 55 ft. lbs. (74 Nm).
- Wheel. Tighten the lug nuts to 76 ft. lbs. (103 Nm).
- Negative battery cable
7. Check and adjust the alignment, if needed.

Rear

1. Before servicing the vehicle, refer to the precautions in the beginning of this section.
2. Remove or disconnect the following:
- Negative battery cable. Because of the air bag system, wait at least 90 seconds before proceeding.
- Rear wheel
- Rear deck board, luggage compartment tray and any trim necessary to access the strut towers
- Shock absorber head cover
3. On AWD vehicles, separate the rear stabilizer link.
4. For FWD vehicles:
a. Support the axle beam with a jack.
b. Remove the strut tower nuts and bolt.
c. Remove the lower strut nut, cushion retainer and strut .
5. For AWD vehicles:
a. Support the rear control arm.
b. Remove the bolt and nut from the rear control arm.
c. Remove the strut tower nuts.
d. Remove the 3 rear control arm bolts.
e. Press the rear control arm down to the outside of the vehicle, then remove the strut.
6. To disassemble the strut:
a. Place the strut assembly in a pipe vise or strut vise.

✳✳ WARNING

Do not attempt to clamp the strut assembly in a flat jaw vise as this will result in damage to the strut tube.

b. Compress the spring until the upper suspension support is free of any spring tension. Do not over-compress the spring.
c. Hold the upper support, then remove the nut on the end of the shock piston rod.

d. Remove the support, coil spring, insulator, and bumper.
7. Inspect the strut as follows:
a. Check the shock absorber by moving the piston shaft through its full range of travel. It should move smoothly and evenly throughout its entire travel without any trace of binding or notching.
b. Use a small straightedge to check the piston shaft for any bending or deformation.
c. Inspect the spring for any sign of deterioration or cracking. The waterproof coating on the coils should be intact to prevent rusting.
To install:

➡**Never reuse a self-locking nut. Always replace self-locking nuts and cotter pins as applicable.**

8. Assemble the strut as follows:
a. Loosely assemble all components onto the strut assembly. Be sure the spring end aligns with the hollow in the lower seat.
b. Align the upper suspension support with the piston rod and install the support.
c. Align the suspension support with the strut lower bracket. This assures the spring will be properly seated top and bottom.
d. Compress the spring to expose the strut piston rod threads.
e. Install a new strut piston nut and tighten to 41 ft. lbs. (56 Nm).
f. Remove the spring compressor. Be sure the paint mark on the upper support faces the outside of the strut.
9. Install or connect the following:
- Strut on the vehicle. Tighten the strut-to-strut tower nuts to 59 ft. lbs. (80 Nm).
- Strut to the axle carrier and install the nut and cushion retainer/bolt snug. Do not fully tighten at this time.
- Strut head cover
- Rear control arm (AWD). Tighten the bolts to 48 ft. lbs. (65 Nm).
- Rear stabilizer link (AWD)
- Trunk tray, deckboard and any other trim pieces removed
- Wheel
10. With the vehicle's weight on the suspension, tighten the bolt holding the strut to the axle carrier to 59 ft. lbs. (80 Nm) for FWD vehicles, or 103 ft. lbs. (140 Nm) for AWD vehicles.

- Negative battery cable
11. Check and adjust the rear wheel alignment.

Lower Ball Joint

REMOVAL & INSTALLATION

1. Before servicing the vehicle, refer to the precautions in the beginning of this section.
2. Remove or disconnect the following:
   - Negative battery cable. Wait at least 90 seconds before proceeding.
   - Front wheel
3. Depress the brake pedal and loosen the hub nut
   - ABS speed sensor, if equipped
   - Cotter pin and nut from the tie rod end. Using a tie rod end removal tool, separate the tie rod end from the steering knuckle.
   - Lower control arm ball joint, using a suitable puller
   - Separate the front halfshaft
   - Lower ball joint cotter pin and castle nut
   - Lower ball joint from the steering knuckle using a puller

To install:
4. Install or connect the following:
   - Lower ball joint to the lower arm. Tighten the castle nut to 76 ft. lbs. (103 Nm).
   - New cotter pin
   - Front halfshaft
   - Lower control arm
   - Tie rod end to the knuckle
   - ABS speed sensor
   - Hub nut
   - Wheel
   - Negative battery cable
5. Check and adjust the alignment, if needed.

SST

9359AB71

Removing the ball joint from the knuckle

Upper Control Arm

REMOVAL & INSTALLATION

Rear—AWD Only

1. Before servicing the vehicle, refer to the precautions in the beginning of this section.
2. Remove or disconnect the following:
   - Negative battery cable. Wait at least 90 seconds before proceeding
   - Rear wheel
   - Exhaust pipe
   - Propeller shaft with center bearing shaft
   - Rear stabilizer links
   - Rear hub nuts
   - Rear brake drum
   - Speed sensor
   - Front brake shoe
   - Parking brake shoe strut set
   - Rear brake shoe
   - Parking brake cables
   - Rear brake hoses
   - Separate the rear suspension arms
   - Separate the upper control arm
   - Rear drive axle assembly
   - Rear strut nut and bolt
   - Rear strut
   - Rear suspension arm
   - Rear suspension member
   - Upper control arm assembly. Matchmark the camber adjust cams and rear suspension member prior to removal.
3. Installation is the reverse of the removal procedure.

Lower Control Arm

REMOVAL & INSTALLATION

1. Before servicing the vehicle, refer to the precautions in the beginning of this section.
2. Remove or disconnect the following:
   - Negative battery cable. Wait at least 90 seconds before proceeding..
   - Front wheel
   - Stabilizer link
   - Bolt and nuts and separate the lower control arm from the lower ball joint
   - Bolts and nuts, then separate the steering gear. Loosen the bolt, since the nut cannot be rotated, then suspend the steering gear.
3. Support the engine, using the engine lifting hooks and the procedure under Engine Removal & Installation.

- Crossmember
- Lower control arm from the crossmember
4. Installation is the reverse of the removal procedure.

Wheel Bearings

REMOVAL & INSTALLATION

Front

1. Before servicing the vehicle, refer to the precautions in the beginning of this section.
2. Remove or disconnect the following:
   - Negative battery cable. On vehicles equipped with an air bag, wait at least 90 seconds before proceeding.
   - Wheels
   - Hub nut
   - Front stabilizer link
   - Anti-lock Brake System (ABS) speed sensor
   - Brake caliper
   - Rotor
   - Tie rod end from the steering knuckle
   - Lower control arm ball joint
   - Front halfshaft from the hub, using a mallet to tap it out. Be careful not to damage the boot or speed sensor.
3. Loosen the nuts on the lower side of the strut assembly. Do not remove at this time.
   - Lower ball joint using a puller
   - Tie rod end from the steering knuckle
   - Steering knuckle from the lower control arm
   - Knuckle from the strut assembly
   - Hub

➡**Cover the halfshaft boot with a shop rag to protect it from any damage.**

4. Clamp the steering knuckle in a vise and remove the dust deflector. Remove the nut holding the steering knuckle to the ball joint. Press the ball joint out of the steering knuckle.
5. Remove the inner axle seal.
6. Using a Torx® wrench, remove the bolts securing the dust cover.
7. Using hub puller, remove the hub and backing plate from the steering knuckle.
8. Using a proper sized driver and a press, remove the inner hub race from the axle hub.

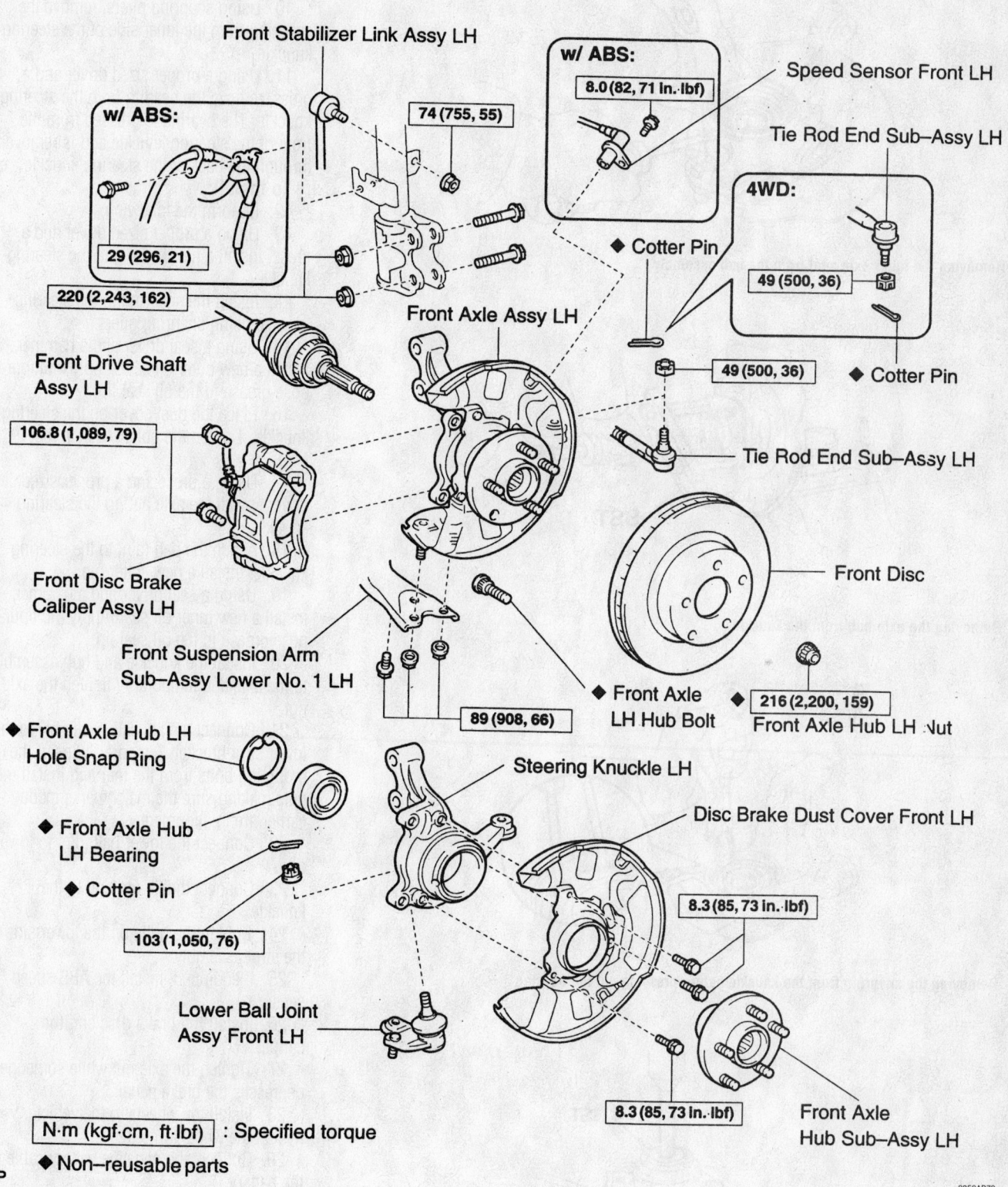

w/ ABS:

Front Stabilizer Link Assy LH

74 (755, 55)

w/ ABS:

8.0 (82, 71 in.·lbf)

Speed Sensor Front LH

Tie Rod End Sub–Assy LH

29 (296, 21)

220 (2,243, 162)

◆ Cotter Pin

4WD:

49 (500, 36)

Front Axle Assy LH

Front Drive Shaft
Assy LH

49 (500, 36)

◆ Cotter Pin

106.8 (1,089, 79)

Tie Rod End Sub–Assy LH

Front Disc Brake
Caliper Assy LH

Front Disc

Front Suspension Arm
Sub–Assy Lower No. 1 LH

89 (908, 66)

◆ Front Axle
LH Hub Bolt

◆ 216 (2,200, 159)
Front Axle Hub LH Nut

◆ Front Axle Hub LH
Hole Snap Ring

Steering Knuckle LH

Disc Brake Dust Cover Front LH

◆ Front Axle Hub
LH Bearing

◆ Cotter Pin

8.3 (85, 73 in.·lbf)

103 (1,050, 76)

Lower Ball Joint
Assy Front LH

8.3 (85, 73 in.·lbf)

Front Axle
Hub Sub–Assy LH

N·m (kgf·cm, ft·lbf) : Specified torque

◆ Non–reusable parts

P

9359AB72

Exploded view of the front hub and bearing, and related components

For complete service labor times, order Nichols' Chilton Labor Guide

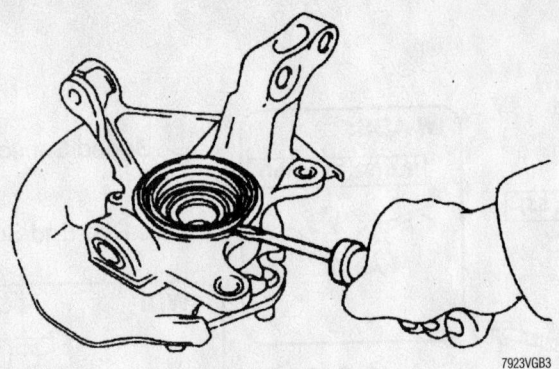

Removing the inner axle seal from the hub assembly

7923VGB3

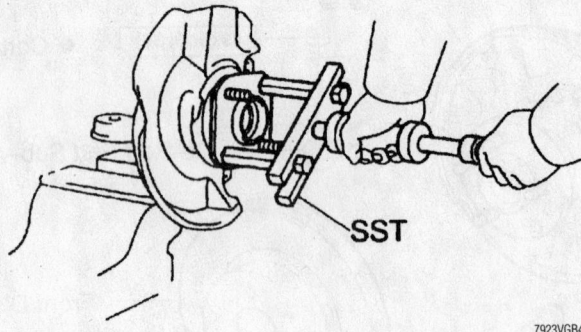

SST

Removing the axle hub from the knuckle

7923VGB4

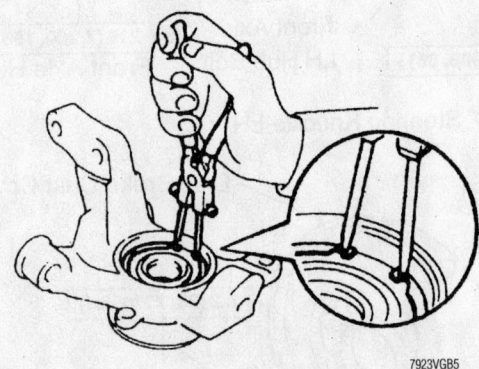

Removing the snapring from the knuckle before pressing out the bearing

7923VGB5

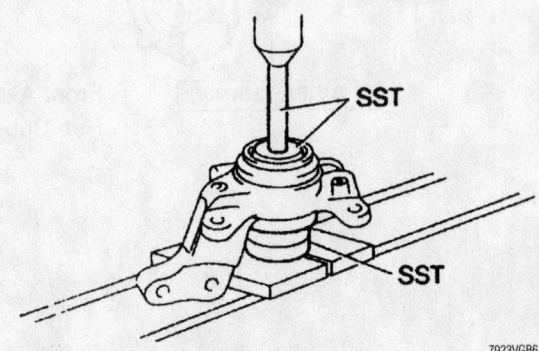

SST

SST

Removing the bearing from the steering knuckle using a press

7923VGB6

9. Using seal removal tool, remove the outer axle seal.

10. Using snapring pliers, remove the snapring from the inner side of the steering knuckle.

11. Using a proper sized driver and a press, remove the bearing from the steering knuckle. The bearing is pressed from the front of the steering knuckle and is removed through the back of the steering knuckle.

To install:

12. Perform the following:

13. Using a proper sized driver and a press, install a new bearing to the steering knuckle.

14. Install the snapring to the steering knuckle using snapring pliers.

15. Using a seal driver and a hammer, install a new outer oil seal. Apply multipurpose grease to the oil seal lip.

16. Place the dust cover on the steering knuckle. Tighten the bolts: 78 inch lbs. (9 Nm).

17. Using a press and a proper sized driver, install the axle hub to the steering knuckle.

18. Attach the ball joint to the steering knuckle. Install a new cotter pin.

19. Using a seal driver and a hammer, install a new inner oil seal. Apply multipurpose grease to the oil seal lip.

20. Install the knuckle and hub assembly to the axle and temporarily tighten the axle nut.

21. Connect the knuckle assembly to the lower strut bracket. Temporarily insert the mounting bolts from the rear and install the nuts making sure the matchmarks made earlier are in alignment.

22. Connect the lower ball joint to lower arm.

23. Connect the tie rod end to the knuckle.

24. Tighten the bolts on the lower side of the strut assembly.

25. If equipped, install the ABS speed sensor.

26. Install the brake disc and the caliper.

27. Tighten the axle nut while someone depresses the brake pedal.

28. Install the wheels to the vehicle. Verify that the wheel turns freely.

29. Connect the negative battery cable to the battery.

30. Check alignment.

Rear

1. Before servicing the vehicle, refer to the precautions in the beginning of this section.

Disc Rear Brake Type:

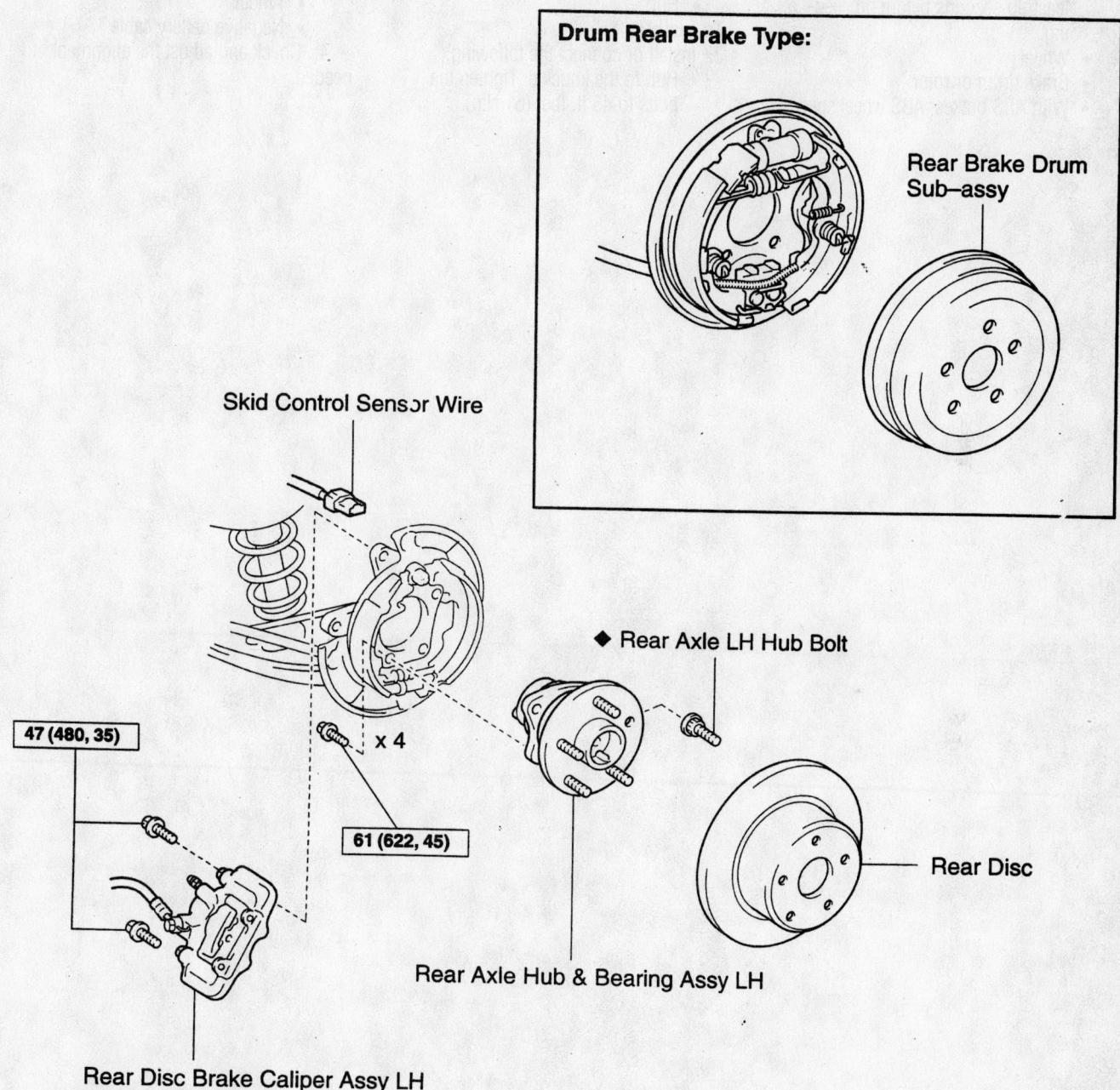

Drum Rear Brake Type:

Rear Brake Drum Sub–assy

Skid Control Sensor Wire

◆ Rear Axle LH Hub Bolt

47 (480, 35)

x 4

61 (622, 45)

Rear Disc

Rear Axle Hub & Bearing Assy LH

Rear Disc Brake Caliper Assy LH

N·m (kgf·cm, ft·lbf) : Specified torque

◆ Non–reusable part

9359AB73

Exploded view of the hub and wheel bearing assembly

2. Remove or disconnect the following:
   - Negative battery cable. On vehicles equipped with an air bag, wait at least 90 seconds before proceeding.
   - Wheel
   - Brake drum or rotor
   - With ABS brakes, ABS wheel speed sensor or skid control sensor, as applicable
   - 4 hub retaining bolts
   - Hub

To install:
3. Install or connect the following:
   - Hub to the knuckle. Tighten the bolts to 45 ft. lbs. (61 Nm).
   - ABS wheel speed or skid control sensor, if equipped
   - Brake drum or rotor
   - Wheel
   - Negative battery cable

4. Check and adjust the alignment, if needed.

TOYOTA AND LEXUS

28

Toyota-Land Cruiser • Sequoia • **Lexus**-LX470

PRECAUTIONS

Before servicing any vehicle, please be sure to read all of the following precautions, which deal with personal safety, prevention of component damage, and important points to take into consideration when servicing a motor vehicle:

• Never open, service or drain the radiator or cooling system when the engine is hot; serious burns can occur from the steam and hot coolant.

• Observe all applicable safety precautions when working around fuel. Whenever servicing the fuel system, always work in a well-ventilated area. Do not allow fuel spray or vapors to come in contact with a spark, open flame, or excessive heat (a hot drop light, for example). Keep a dry chemical fire extinguisher near the work area. Always keep fuel in a container specifically designed for fuel storage; also, always properly seal fuel containers to avoid the possibility of fire or explosion. Refer to the additional fuel system precautions later in this section.

• Fuel injection systems often remain pressurized, even after the engine has been turned **OFF**. The fuel system pressure must be relieved before disconnecting any fuel lines. Failure to do so may result in fire and/or personal injury.

• Brake fluid often contains polyglycol ethers and polyglycols. Avoid contact with the eyes and wash your hands thoroughly after handling brake fluid. If you do get brake fluid in your eyes, flush your eyes with clean, running water for 15 minutes. If eye irritation persists, or if you have taken brake fluid internally, IMMEDIATELY seek medical assistance.

• The EPA warns that prolonged contact with used engine oil may cause a number of skin disorders, including cancer. You should make every effort to minimize your exposure to used engine oil. Protective gloves should be worn when changing oil. Wash your hands and any other exposed skin areas as soon as possible after exposure to used engine oil. Soap and water, or waterless hand cleaner should be used.

• All new vehicles are now equipped with an air bag system. The system must be disabled before performing service on or around system components, steering column, instrument panel components, wiring and sensors. Failure to follow safety and disabling procedures could result in accidental air bag deployment, possible personal injury and unnecessary system repairs.

• Always wear safety goggles when working with, or around, the air bag system. When carrying a non-deployed air bag, be sure the bag and trim cover are pointed away from your body. When placing a non-deployed air bag on a work surface, always face the bag and trim cover upward, away from the surface. This will reduce the motion of the module if it is accidentally deployed. Refer to the additional air bag system precautions later in this section.

• NEVER disconnect the negative battery cable with the ignition **ON** or the engine running. Removing power from the computer control module with the ignition **ON** may destroy the module.

• Clean, high quality brake fluid from a sealed container is essential to the safe and proper operation of the brake system. You should always buy the correct type of brake fluid for your vehicle. If the brake fluid becomes contaminated, completely flush the system with new fluid. Never reuse any brake fluid. Any brake fluid that is removed from the system should be discarded. Also, do not allow any brake fluid to come in contact with a painted surface; it will damage the paint.

• Never operate the engine without the proper amount and type of engine oil; doing so WILL result in severe engine damage.

• Timing belt maintenance is extremely important. Many models utilize an interference-type, non-freewheeling engine. If the timing belt breaks, the valves in the cylinder head may strike the pistons, causing potentially serious (also time-consuming and expensive) engine damage. Refer to the maintenance interval charts in the front of this manual for the recommended replacement interval for the timing belt, and to the timing belt section for belt replacement and inspection.

• Disconnecting the negative battery cable on some vehicles may interfere with the functions of the on-board computer system(s) and may require the computer to undergo a relearning process once the negative battery cable is reconnected.

• When servicing drum brakes, only disassemble and assemble one side at a time, leaving the remaining side intact for reference.

• Only an MVAC-trained, EPA-certified automotive technician should service the air conditioning system or its components.

ENGINE REPAIR

➡ **Disconnecting the negative battery cable on some vehicles may interfere with the functions of the on board computer system. The computer may undergo a relearning process once the negative battery cable is reconnected.**

Distributor

All models are equipped with a distibutorless ignition system.

Alternator

REMOVAL

1. Before servicing the vehicle, refer to the precautions in the beginning of this section.
2. Drain the cooling system.

3. Remove or disconnect the following:
   • Negative battery cable
   • Accessory drive belt
   • Engine under cover
   • Radiator
   • Power steering pump pulley
   • Alternator harness connectors
   • Alternator

INSTALLATION

1. Install or connect the following:
   • Alternator. Tighten the fasteners to 29 ft. lbs. (39 Nm).
   • Alternator harness connectors
   • Power steering pump pulley
   • Radiator
   • Engine under cover
   • Accessory drive belt
   • Negative battery cable

2. Fill the cooling system.
3. Start the engine and check for leaks.

Ignition Timing

ADJUSTMENT

All engines are equipped with a Distributorless Ignition System (DIS). No timing adjustment is possible.

Engine Assembly

REMOVAL & INSTALLATION

Sequoia

1. Before servicing the vehicle, refer to the precautions in the beginning of this section.

2. Relieve the fuel system pressure.
3. Drain the cooling system.
4. Drain the engine oil.
5. Remove or disconnect the following:

- Battery and tray
- Hood
- Engine appearance cover
- Air intake pipe
- Engine under covers
- Coolant recovery tank
- Radiator hoses
- Radiator and fan shroud
- Accessory drive belt
- Cooling fan and pulley
- Powertrain Control Module (PCM) harness connectors and pass the wiring harness through the firewall
- Accelerator cable
- Power steering vacuum hoses
- Alternator harness connectors
- Heater hoses
- Engine control wiring harness and grommet at the firewall
- Ground cable connector
- Fuel lines
- Evaporative Emissions (EVAP) canister hoses
- Wire clamp at right inner fender
- Negative battery cable at the relay box and right inner fender
- Positive battery cable
- Center console
- Transmission shift lever assembly
- Transfer case shift lever and rod
- Exhaust front pipes
- Stabilizer bar
- Front and rear driveshafts
- A/C compressor
- Power steering pump

6. Attach a hoist to the engine lifting eyes.
7. Remove or disconnect the following:

- Transfer case skid plate
- Left and right motor mounts
- Transmission mount crossmember

8. Attach a hoist to the engine lifting eyes and raise the powertrain out of the vehicle.

To install:

9. Lower the powertrain into the vehicle.
10. Install or connect the following:

- Transmission mount crossmember. Tighten the bolts to 37 ft. lbs. (50 Nm) and the nuts to 55 ft. lbs. (74 Nm).
- Transfer case skid plate

- Left and right motor mounts. Tighten the fasteners to 22 ft. lbs. (30 Nm).
- Power steering pump. Tighten the bolts to 13 ft. lbs. (17 Nm).
- A/C compressor. Tighten the bolts to 36 ft. lbs. (49 Nm).
- Front driveshaft. Tighten the fasteners to 59 ft. lbs. (80 Nm).
- Rear driveshaft. Tighten the fasteners to 78 ft. lbs. (106 Nm).
- Stabilizer bar. Tighten the bracket bolts to 13 ft. lbs. (18 Nm) and the link nuts to 18 ft. lbs. (25 Nm).
- Exhaust front pipes
- Transfer case shift lever and rod
- Transmission shift lever assembly
- Center console
- Positive battery cable
- Negative battery cable at the relay box and right inner fender
- Wire clamp at right inner fender
- EVAP canister hoses
- Fuel lines
- Ground cable connector
- Engine control wiring harness and grommet at the firewall
- Heater hoses
- Alternator harness connectors
- Power steering vacuum hoses
- Accelerator cable
- PCM harness connectors
- Cooling fan and pulley
- Accessory drive belt
- Radiator and fan shroud
- Radiator hoses
- Coolant recovery tank
- Engine under covers
- Air intake pipe
- Engine appearance cover
- Hood
- Battery and tray

11. Fill the crankcase to the correct level.
12. Fill the cooling system.
13. Start the engine and check for leaks.

Land Cruiser and LX470

1. Before servicing the vehicle, refer to the precautions in the beginning of this section.
2. Relieve the fuel system pressure.
3. Drain the cooling system.
4. Drain the engine oil.
5. Remove or disconnect the following:

- Battery and tray

- Hood
- Engine appearance cover
- Air intake pipe
- Engine under covers
- Coolant recovery tank
- Radiator hoses
- Radiator and fan shroud
- Accessory drive belt
- Cooling fan and pulley
- Powertrain Control Module (PCM) harness connectors and pass the wiring harness through the firewall
- Accelerator cable
- Power steering vacuum hoses
- Alternator harness connectors
- Heater hoses
- Engine control wiring harness and grommet at the firewall
- Ground cable connector
- Fuel lines
- Evaporative Emissions (EVAP) canister hoses
- Wire clamp at right inner fender
- Negative battery cable at the relay box and right inner fender
- Positive battery cable
- Center console
- Transmission shift lever assembly
- Transfer case shift lever and rod
- Exhaust front pipes
- Stabilizer bar
- Front and rear driveshafts
- A/C compressor
- Power steering pump

6. Attach a hoist to the engine lifting eyes.
7. Remove or disconnect the following:

- Transfer case skid plate
- Left and right motor mounts
- Transmission mount crossmember

8. Attach a hoist to the engine lifting eyes and raise the powertrain out of the vehicle.

To install:

9. Lower the powertrain into the vehicle.
10. Install or connect the following:

- Transmission mount crossmember. Tighten the bolts to 37 ft. lbs. (50 Nm) and the nuts to 55 ft. lbs. (74 Nm).
- Transfer case skid plate
- Left and right motor mounts. Tighten the fasteners to 22 ft. lbs. (30 Nm).
- Power steering pump. Tighten the bolts to 13 ft. lbs. (17 Nm).

- A/C compressor. Tighten the bolts to 36 ft. lbs. (49 Nm).
- Front driveshaft. Tighten the fasteners to 59 ft. lbs. (80 Nm).
- Rear driveshaft. Tighten the fasteners to 78 ft. lbs. (106 Nm).
- Stabilizer bar. Tighten the bracket bolts to 13 ft. lbs. (18 Nm) and the link nuts to 18 ft. lbs. (25 Nm).
- Exhaust front pipes
- Transfer case shift lever and rod
- Transmission shift lever assembly
- Center console
- Positive battery cable
- Negative battery cable at the relay box and right inner fender
- Wire clamp at right inner fender
- EVAP canister hoses
- Fuel lines
- Ground cable connector
- Engine control wiring harness and grommet at the firewall
- Heater hoses
- Alternator harness connectors
- Power steering vacuum hoses
- Accelerator cable
- PCM harness connectors
- Cooling fan and pulley
- Accessory drive belt
- Radiator and fan shroud
- Radiator hoses
- Coolant recovery tank
- Engine under covers
- Air intake pipe
- Engine appearance cover
- Hood
- Battery and tray

11. Fill the crankcase to the correct level.

12. Fill the cooling system.

13. Start the engine and check for leaks.

Water Pump

REMOVAL & INSTALLATION

1. Before servicing the vehicle, refer to the precautions in the beginning of this section.

2. Drain the cooling system.

3. Remove or disconnect the following:
- Negative battery cable
- Timing belt. Refer to the Timing Belt unit repair section.
- No. 2 idler pulley
- Radiator hose
- Bypass hose
- Water inlet housing assembly
- Water pump

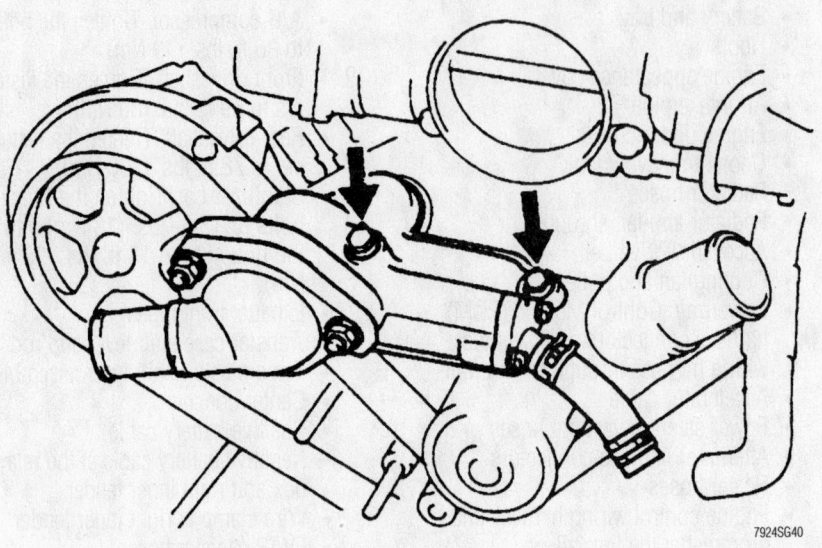

Water inlet housing attaching bolts

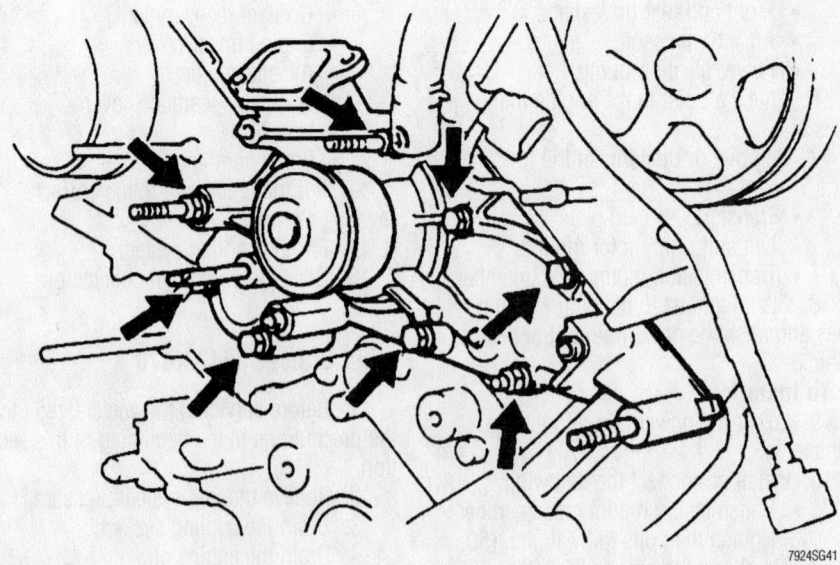

Water pump mounting bolts, stud bolts and nut locations

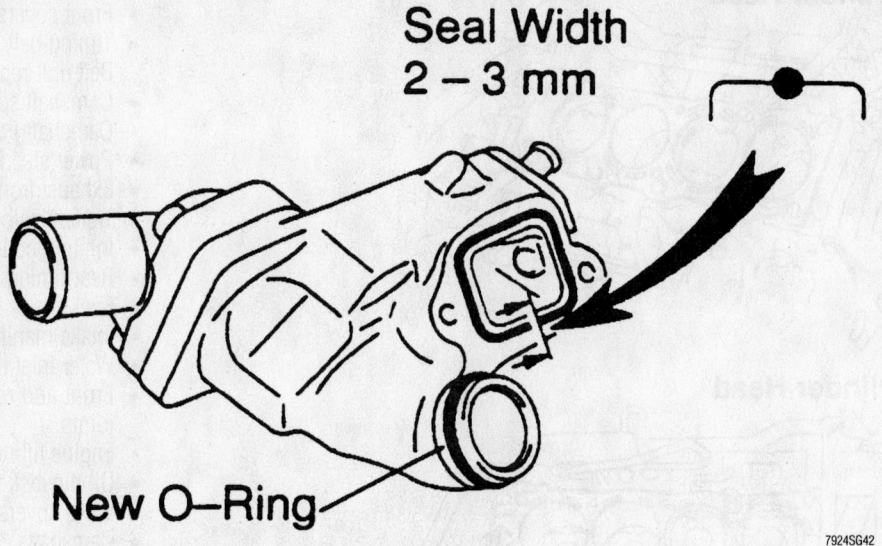

Seal Width
2 – 3 mm

New O–Ring

7924SG42

Water inlet housing sealant application

To install:

4. Install or connect the following:
   - Water pump: Use a new gasket and tighten the bolts to 15 ft. lbs. (21 Nm). Tighten the stud bolt and nut to 13 ft. lbs. (18 Nm).
   - Water inlet housing assembly. Use a new O-ring and apply sealant as shown. Tighten the bolts to 13 ft. lbs. (18 Nm).
   - Bypass hose
   - Radiator hose
   - No. 2 idler pulley
   - Timing belt
   - Negative battery cable
5. Fill the cooling system.
6. Start the engine and check for leaks.

Cylinder Head

REMOVAL & INSTALLATION

1. Before servicing the vehicle, refer to the precautions in the beginning of this section.
2. Drain the cooling system.
3. Relieve the fuel system pressure.
4. Remove or disconnect the following:
   - Battery and tray
   - Engine appearance cover
   - Engine under covers
   - Air intake assembly
   - Accessory drive belt
   - A/C compressor and bracket
   - Cooling fan and bracket
   - Radiator

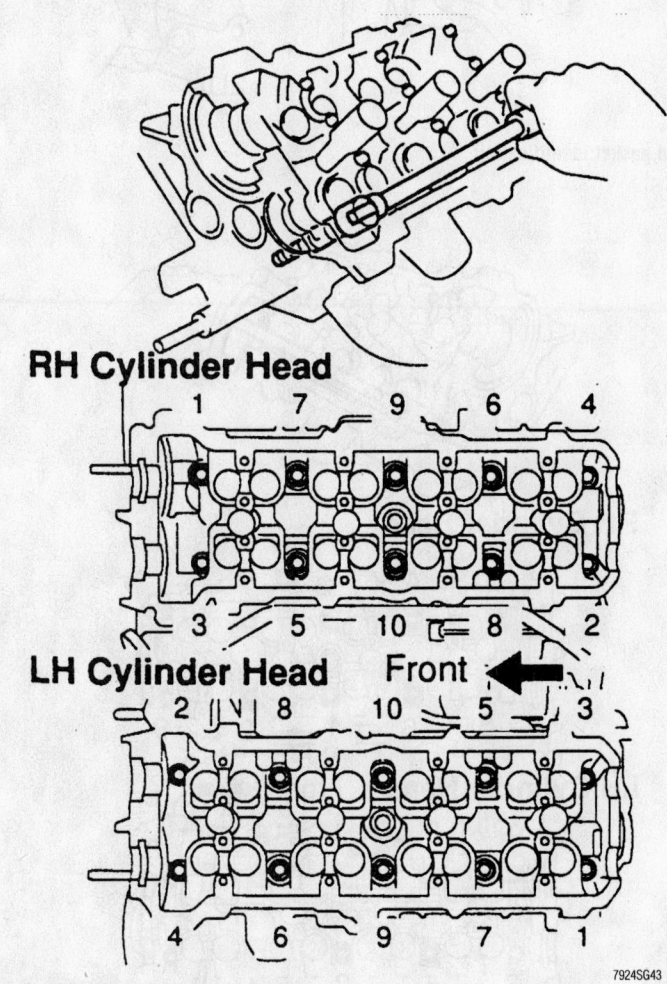

RH Cylinder Head

LH Cylinder Head Front

7924SG43

Cylinder head loosening sequence

RH Cylinder Head

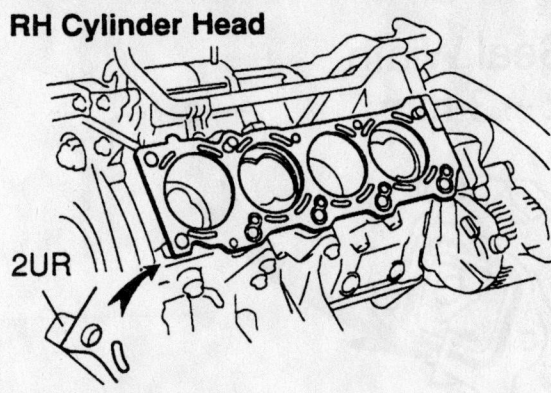

2UR

LH Cylinder Head

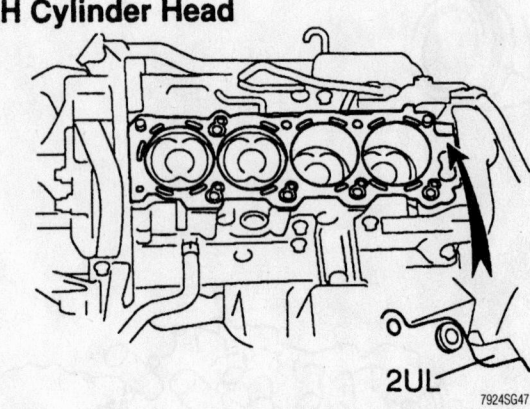

2UL

7924SG47

Cylinder head gasket identification

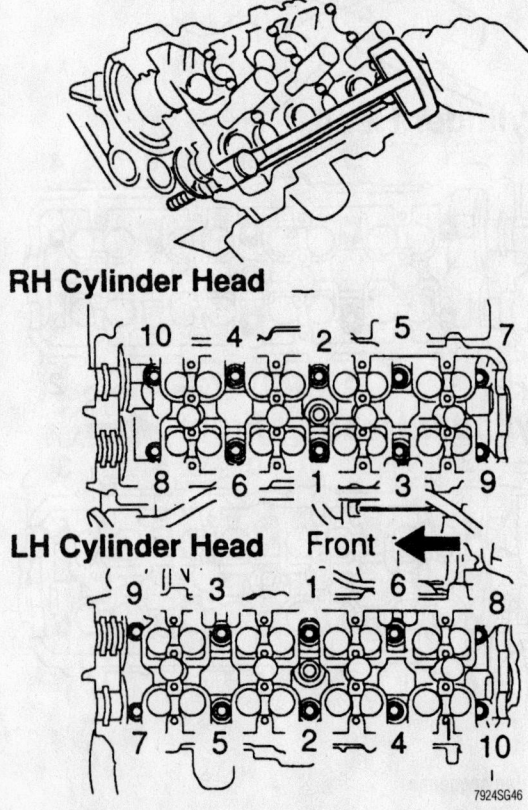

RH Cylinder Head

LH Cylinder Head   Front

7924SG46

Cylinder head torque sequence

- Idler pulley
- Front covers
- Timing belt. Refer to the Timing Belt unit repair section.
- Camshaft sprockets
- Camshaft Position (CMP) sensor
- Power steering pump
- Exhaust front pipes
- Transmission dipstick tube
- Ignition coils
- Rear timing belt covers
- Fuel lines
- Intake manifold
- Water inlet housing assembly
- Front and rear water bypass joints
- Engine lifting eyes
- Oil dipstick tube
- Valve covers
- Camshafts
- Cylinder heads with the exhaust manifolds attached. Loosen the bolts in the sequence shown.

To install:

5. Install the cylinder heads with new gaskets. Tighten the bolts in sequence as follows:
   a. Step 1: 24 ft. lbs. (32 Nm)
   b. Step 2: Plus 180 degrees
6. Install or connect the following:
- Camshafts
- Valve covers
- Oil dipstick tube
- Engine lifting eyes
- Front and rear water bypass joints
- Water inlet housing assembly
- Intake manifold
- Fuel lines
- Rear timing belt covers
- Ignition coils
- Transmission dipstick tube
- Exhaust front pipes
- Power steering pump
- CMP sensor
- Camshaft sprockets
- Timing belt
- Front covers
- Idler pulley
- Radiator
- Cooling fan and bracket
- A/C compressor and bracket
- Accessory drive belt
- Air intake assembly
- Engine under covers
- Engine appearance cover
- Battery and tray
7. Fill the cooling system.
8. Start the engine and check for leaks.

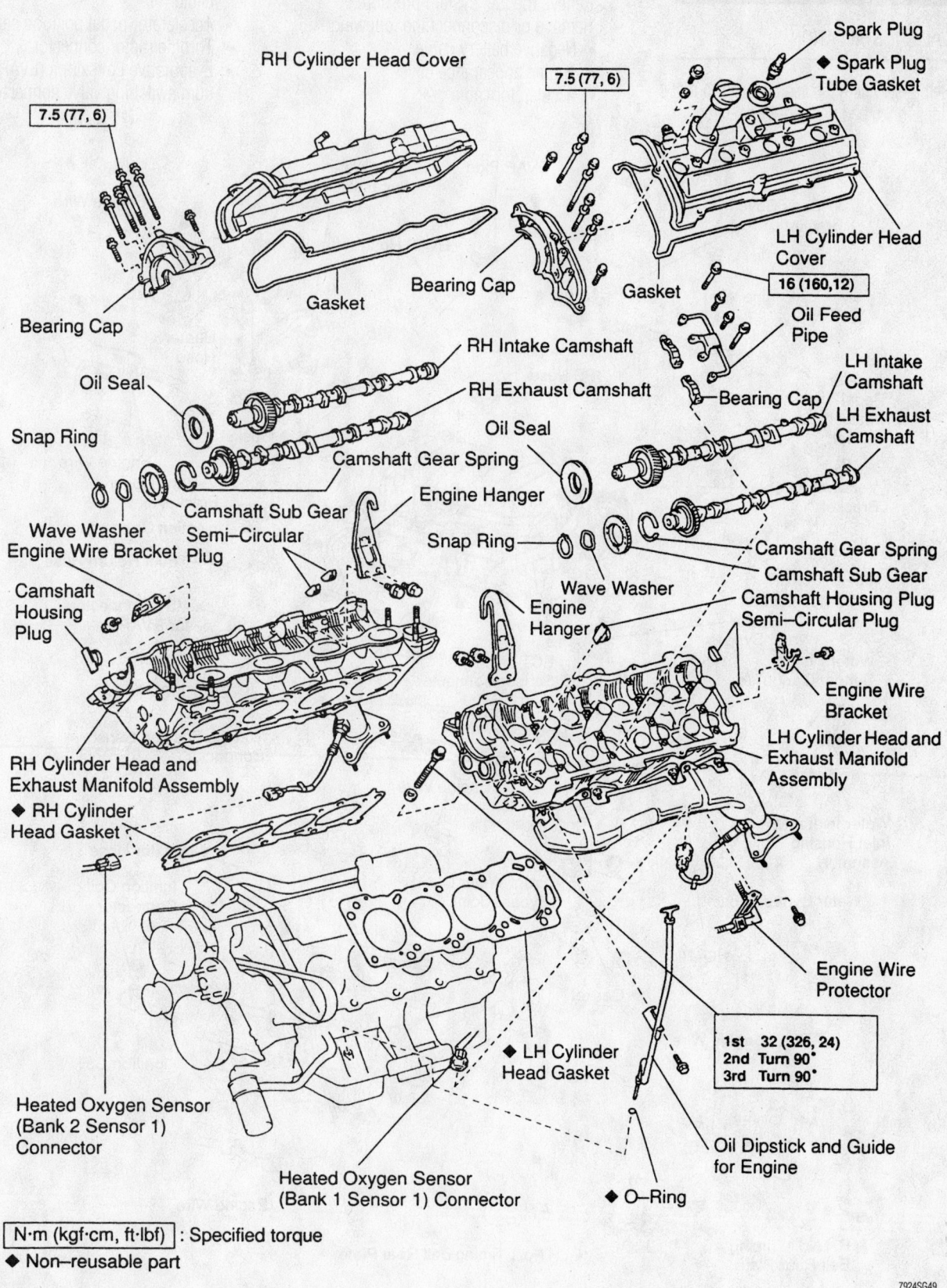

Spark Plug

◆ Spark Plug Tube Gasket

7.5 (77, 6)

RH Cylinder Head Cover

7.5 (77, 6)

7.5 (77, 6)

LH Cylinder Head Cover

16 (160, 12)

Gasket

Bearing Cap

Gasket

Bearing Cap

Oil Feed Pipe

Bearing Cap

RH Intake Camshaft

RH Exhaust Camshaft

LH Intake Camshaft

LH Exhaust Camshaft

Oil Seal

Oil Seal

Snap Ring

Camshaft Gear Spring

Engine Hanger

Snap Ring

Camshaft Gear Spring

Camshaft Sub Gear

Wave Washer
Engine Wire Bracket

Camshaft Sub Gear
Semi–Circular Plug

Wave Washer

Camshaft Housing Plug
Semi–Circular Plug

Camshaft Housing Plug

Engine Hanger

Engine Wire Bracket

RH Cylinder Head and
Exhaust Manifold Assembly

LH Cylinder Head and
Exhaust Manifold Assembly

◆ RH Cylinder
Head Gasket

Engine Wire Protector

1st 32 (326, 24)
2nd Turn 90°
3rd Turn 90°

Heated Oxygen Sensor
(Bank 2 Sensor 1)
Connector

◆ LH Cylinder
Head Gasket

Heated Oxygen Sensor
(Bank 1 Sensor 1) Connector

◆ O–Ring

Oil Dipstick and Guide
for Engine

N·m (kgf·cm, ft·lbf) : Specified torque

◆ Non–reusable part

7924SG49

Exploded view of the cylinder head mounting—4.7L Land Cruiser/LX470

Timing belt service is covered in Section 3 of this manual

Intake Manifold

REMOVAL & INSTALLATION

1. Before servicing the vehicle, refer to the precautions in the beginning of this section.

2. Drain the cooling system.
3. Relieve the fuel system pressure.
4. Remove or disconnect the following:
   - Negative battery cable
   - Engine appearance cover
   - Accelerator cable

   - Throttle Position (TP) sensor connector
   - Accelerator pedal position sensor
   - Throttle motor connector
   - Evaporative Emissions (EVAP) vacuum switching valve connector

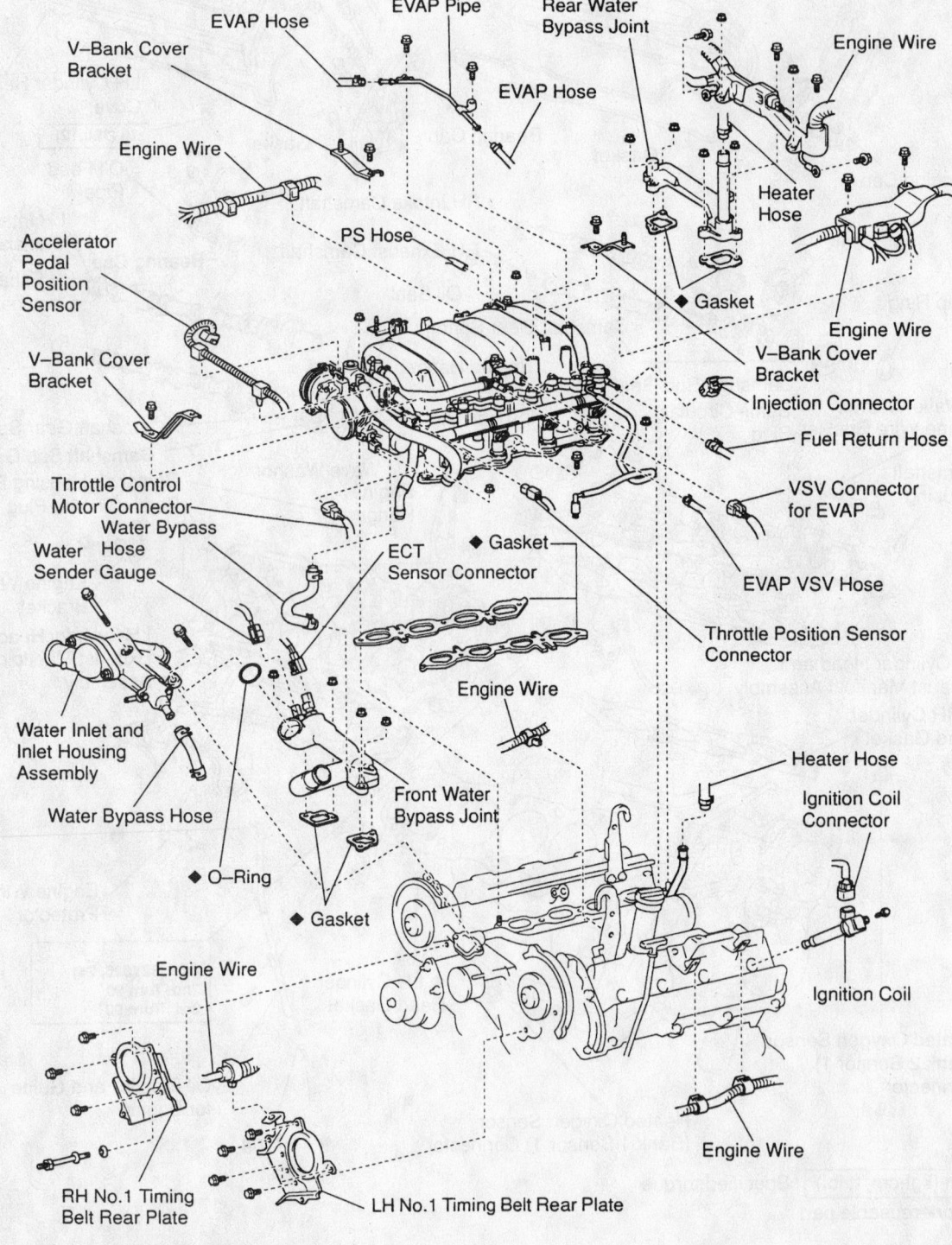

◆ Non-reusable part

Exploded of the intake manifold mounting

7924SG50

- Fuel injector connectors
- Engine Coolant Temperature (ECT) sensor connector
- ETC gauge sender connector
- Heated Oxygen (HO2S) sensor connectors
- Fuel pressure regulator vacuum hose
- Positive Crankcase Ventilation (PCV) valve and hose
- EVAP hoses
- Power steering vacuum hoses
- Water bypass hose
- Engine control wiring harness clamps
- Cylinder head ground cables
- Intake manifold wire harness protector
- EVAP pipe
- Engine appearance cover brackets
- Intake manifold

To install:

5. Install or connect the following:
   - Intake manifold. Tighten the fasteners to 13 ft. lbs. (18 Nm).
   - Engine appearance cover brackets
   - EVAP pipe
   - Intake manifold wire harness protector
   - Cylinder head ground cables
   - Engine control wiring harness clamps
   - Water bypass hose
   - Power steering vacuum hoses
   - EVAP hoses
   - PCV valve and hose
   - Fuel pressure regulator vacuum hose
   - HO2S sensor connectors
   - ETC gauge sender connector
   - ECT sensor connector
   - Fuel injector connectors
   - EVAP vacuum switching valve connector
   - Throttle motor connector
   - Accelerator pedal position sensor
   - TP sensor connector
   - Accelerator cable
   - Engine appearance cover
   - Negative battery cable
6. Fill the cooling system.
7. Start the engine and check for leaks.

Exhaust Manifold

REMOVAL & INSTALLATION

1. Before servicing the vehicle, refer to the precautions in the beginning of this section.

2. Attach a hoist to the engine lifting eyes.
3. Remove or disconnect the following:
   - Negative battery cable
   - Heated Oxygen (HO2S) sensor connectors
   - Exhaust manifold heat shield
   - Exhaust front pipe
   - Motor mount
   - Motor mount bracket
   - Exhaust manifold

To install:

➡**Use new exhaust manifold nuts for assembly.**

4. Install or connect the following:
   - Exhaust manifold. Tighten the nuts to 32 ft. lbs. (44 Nm).
   - Motor mount bracket. Tighten the bolts to 27 ft. lbs. (36 Nm).
   - Motor mount. Tighten the fasteners to 22 ft. lbs. (30 Nm).
   - Exhaust front pipe. Tighten the nuts to 46 ft. lbs. (62 Nm).
   - Exhaust manifold heat shield
   - HO2S sensor connectors
   - Negative battery cable
5. Start the engine and check for leaks.

Front Crankshaft Seal

REMOVAL & INSTALLATION

1. Before servicing the vehicle, refer to the precautions in the beginning of this section.
2. Drain the cooling system.
3. Remove or disconnect the following:
   - Negative battery cable
   - Engine under cover
   - Engine appearance cover
   - Air intake assembly
   - Accessory drive belt
   - Cooling fan and pulley
   - Radiator
   - Drive belt idler pulley
   - Camshaft Position (CMP) sensor connector
   - Upper timing covers
   - Oil cooler pipe
   - Center timing cover
   - A/C compressor
   - Cooling fan bracket
   - Crankshaft pulley
   - Lower timing cover
   - Timing belt. Refer to the Timing Belt unit repair section.
   - Crankshaft timing sprocket
   - Front crankshaft seal

To install:

4. Install the oil seal so that it is flush with the oil pump housing.
5. Install or connect the following:
   - Crankshaft timing sprocket
   - Timing belt
   - Lower timing cover
   - Crankshaft pulley. Tighten the bolt to 181 ft. lbs. (245 Nm).
   - Cooling fan bracket. Tighten the 12mm bolts to 12 ft. lbs. (16 Nm) and the 14mm bolts to 24 ft. lbs. (32 Nm).
   - A/C compressor
   - Center timing cover
   - Oil cooler pipe
   - Upper timing covers
   - CMP sensor connector
   - Drive belt idler pulley. Tighten the bolt to 27 ft. lbs. (37 Nm).
   - Radiator
   - Cooling fan and pulley. Tighten the nuts to 16 ft. lbs. (21 Nm).
   - Accessory drive belt
   - Air intake assembly
   - Engine appearance cover
   - Engine under cover
   - Negative battery cable
6. Fill the cooling system.
7. Start the engine and check for leaks.

Camshaft and Valve Lifters

REMOVAL & INSTALLATION

1. Before servicing the vehicle, refer to the precautions in the beginning of this section.
2. Drain the cooling system.
3. Relieve the fuel system pressure.
4. Remove or disconnect the following:
   - Negative battery cable
   - Engine under covers
   - Engine appearance cover
   - Air intake hose
   - Accessory drive belt
   - Cooling fan
   - Radiator
   - Idler pulley
   - Upper and middle timing belt covers
   - A/C compressor
   - Cooling fan bracket
   - Alternator
   - Accessory drive belt tensioner
5. Set the engine to Top Dead Center (TDC) with the camshaft sprocket timing marks aligned with the rear cover timing marks.

Heater Core replacement is covered in Section 2 of this manual

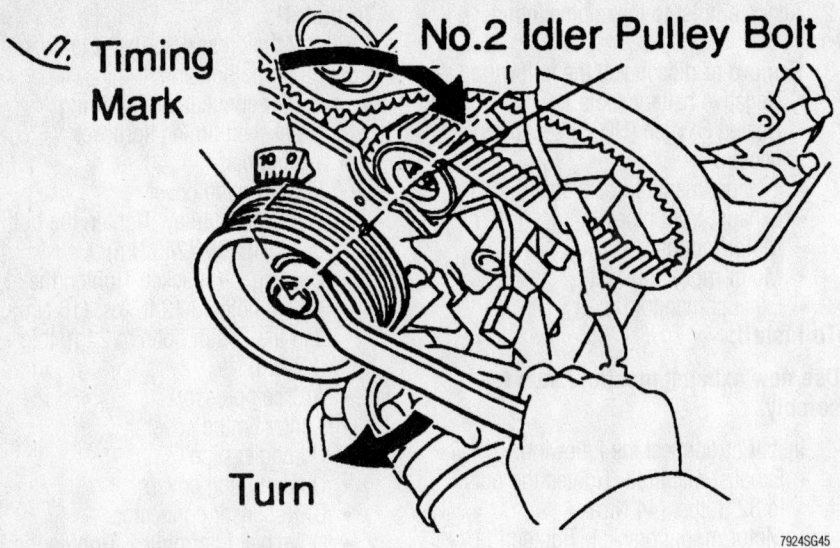

Setting the crankshaft to 50 degrees ATDC

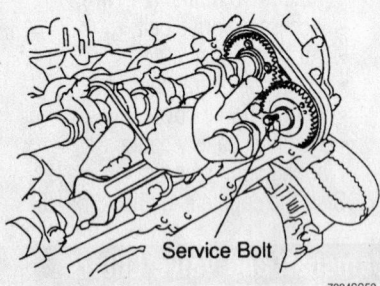

Camshaft service bolt installation

6. Rotate the crankshaft to 50 degrees After TDC as shown. The crankshaft pulley timing mark should align with the center of the No. 2 idler pulley bolt.

7. Remove or disconnect the following:
- Crankshaft pulley
- Lower timing cover
- Timing belt. Refer to the Timing Belt unit repair section.
- Camshaft timing sprockets
- Camshaft Position (CMP) sensor
- Ignition coils
- Valve cover
- Timing belt rear covers

8. Rotate the right bank camshafts as necessary to access the exhaust camshaft sub-gear service bolt hole and install a 6mm x 1.0mm bolt.

➡ **Keep all valvetrain components in order for assembly.**

9. Align the right bank camshaft 1 dot timing marks to a **10** degree angle as shown.

10. Loosen the bearing cap bolts in sequence and in several passes.

11. Remove the right bank camshafts.

12. Rotate the left bank camshafts as necessary to access the exhaust camshaft sub-gear service bolt hole and install a 6mm x 1.0mm bolt.

13. Align the left bank camshaft 2 dot timing marks as shown.

14. Loosen the bearing cap bolts in sequence and in several passes.

15. Remove the left bank camshafts.

16. Remove the valve lifters and shims.

To install:

17. Ensure that the crankshaft is at 50 degrees After TDC.

18. Install or connect the following:
- Valve lifters and shims in their original positions
- Right bank camshafts with the 1 dot timing marks at 10 degrees

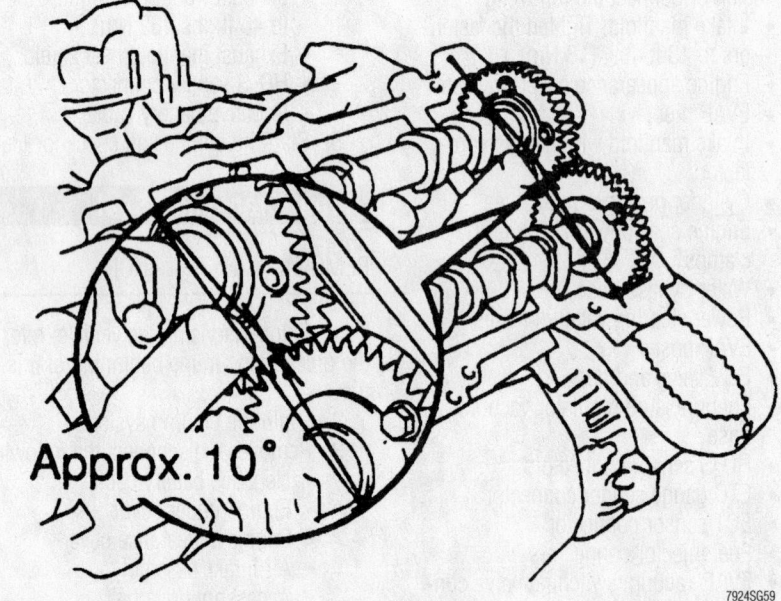

Right bank camshaft timing mark (1 dot marks) alignment

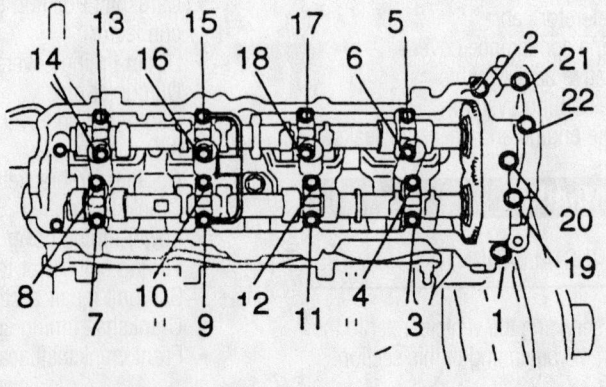

Right bank camshaft bearing cap loosening sequence

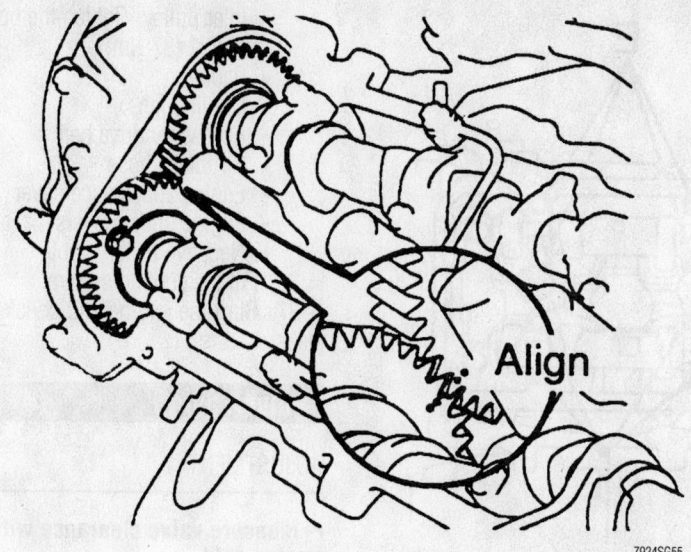

Left bank camshaft timing mark (2 dot marks) alignment

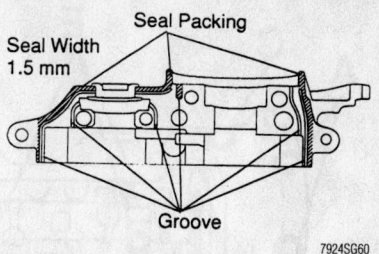

Apply a 1.5mm bead of sealant to the front bearing caps

22. Install oil feed pipes and the bearing cap bolts according to position in the illustrations.

23. Tighten the camshaft bearing bolts in sequence and in several passes to the following specifications:
- Bolt C: 66 inch lbs. (7.5 Nm)
- All others: 12 ft. lbs. (16 Nm)

24. Remove the service bolts from the exhaust camshaft gears.

25. Install or connect the following:
- Timing belt rear covers
- Valve cover
- Ignition coils
- CMP sensor
- Camshaft timing sprockets. Tighten the bolts to 80 ft. lbs. (108 Nm).
- Timing belt
- Lower timing cover
- Crankshaft pulley. Tighten the bolt to 181 ft. lbs. (245 Nm).
- Accessory drive belt tensioner
- Alternator
- Cooling fan bracket
- A/C compressor
- Upper and middle timing belt covers

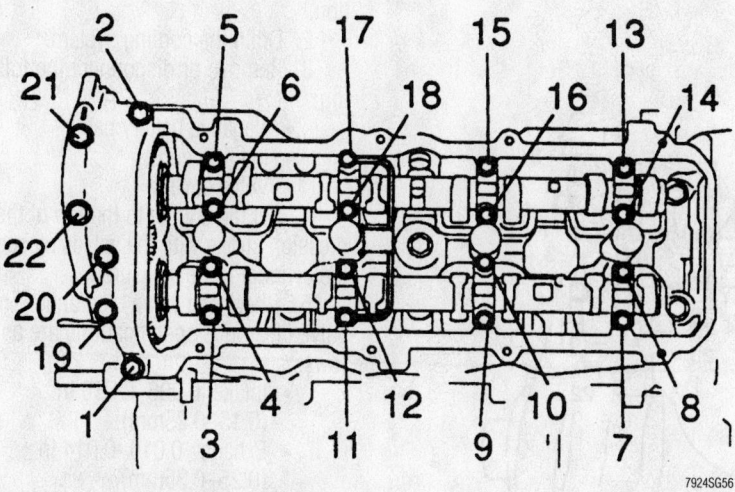

Left bank camshaft bearing cap loosening sequence

- Left bank camshafts with the 2 dot timing marks aligned
- Left and right bank camshaft bearing caps in their original positions. Apply sealant to the front bearing caps as shown.
- Camshaft oil seals

19. The bearing cap bolts vary in length and are identified as follows:
- A: 3.70 inches (94mm)
- B: 2.83 inches (72mm)
- C: 0.98 inches (25mm)
- D: 2.05 inches (52mm)
- E: 1.50 inches (38mm)

20. Bolts in positions **A**, **B** and **C** are installed dry.

21. Lubricate the threads and under the contact flange for bolts in positions **D** and **E**.

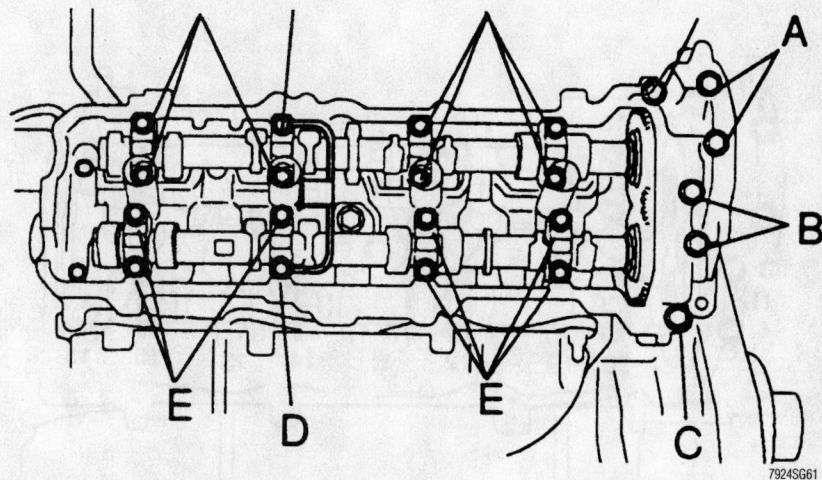

Right bank bearing cap bolt location

Brake service is covered in Section 4 of this manual

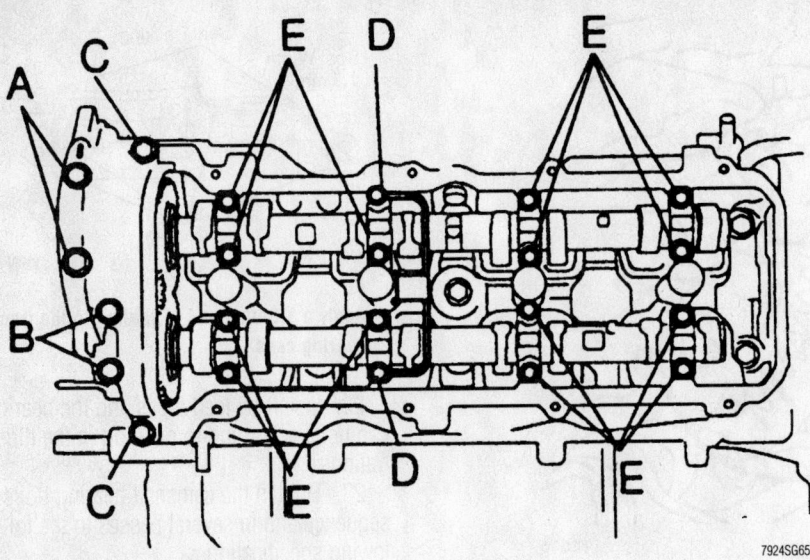

Left camshaft bearing cap bolt locations

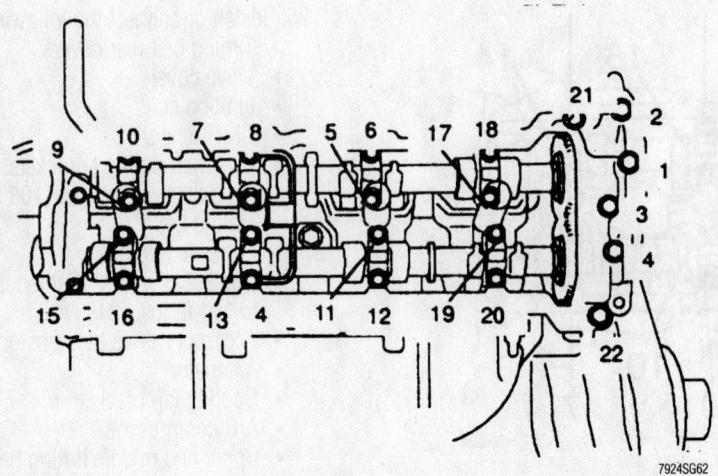

Right bank camshaft bearing cap bolt torque sequence

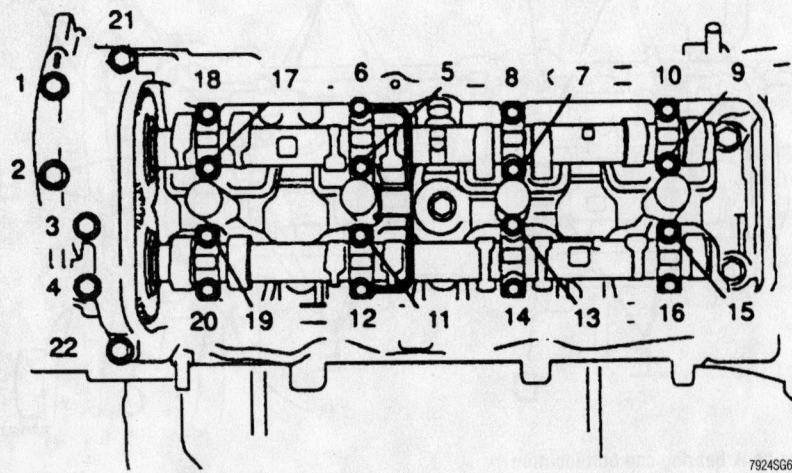

Left bank camshaft bearing cap bolt torque sequence

- Idler pulley. Tighten the bolt to 27 ft. lbs. (37 Nm).
- Radiator
- Cooling fan
- Accessory drive belt
- Air intake hose
- Engine appearance cover
- Engine under covers
- Negative battery cable

26. Fill the cooling system.

27. Start the engine and check for leaks.

Valve Lash

ADJUSTMENT

➡ **Measure valve clearance with the engine cold.**

1. Before servicing the vehicle, refer to the precautions in the beginning of this section.

2. Drain the cooling system.

3. Remove or disconnect the following:

- Negative battery cable
- Ignition coils
- Valve covers

4. Set the engine to the top of the compression stroke with the valves closed for the cylinder to be measured.

5. Check the valve clearance. The valve clearance specifications are as follows:

- Intake: 0.006–0.010 in. (0.15–0.25mm)
- Exhaust: 0.010–0.014 in. (0.25–0.35mm)

6. Record the measurements for each valve.

7. When all valve clearances have been measured, remove the camshafts.

8. Remove the valve shims and measure them. Note this measurement along with the clearance measurement recorded earlier.

9. Using the valve clearance and shim thickness measurements, find replacement shims in the Adjusting Shim Selection charts.

10. Install or connect the following:

- Replacement valve shims
- Camshafts
- Valve covers
- Ignition coils
- Negative battery cable

11. Fill the cooling system.

12. Start the engine and check for leaks.

New shim thickness

mm (in.)

Shim No.	Thickness	Shim No.	Thickness	Shim No.	Thickness
00	2.000 (0.0787)	28	2.280 (0.0898)	56	2.560 (0.1008)
02	2.020 (0.0795)	30	2.300 (0.0906)	58	2.580 (0.1016)
04	2.040 (0.0803)	32	2.320 (0.0913)	60	2.600 (0.1024)
06	2.060 (0.0811)	34	2.340 (0.0921)	62	2.620 (0.1031)
08	2.080 (0.0819)	36	2.360 (0.0929)	64	2.640 (0.1039)
10	2.100 (0.0827)	38	2.380 (0.0937)	66	2.660 (0.1047)
12	2.120 (0.0835)	40	2.400 (0.0945)	68	2.680 (0.1055)
14	2.140 (0.0843)	42	2.420 (0.0953)	70	2.700 (0.1063)
16	2.160 (0.0850)	44	2.440 (0.0961)	72	2.720 (0.1071)
18	2.180 (0.0858)	46	2.460 (0.0969)	74	2.740 (0.1079)
20	2.200 (0.0866)	48	2.480 (0.0976)	76	2.760 (0.1087)
22	2.220 (0.0874)	50	2.500 (0.0984)	78	2.780 (0.1094)
24	2.240 (0.0882)	52	2.520 (0.0992)	80	2.800 (0.1102)
26	2.260 (0.0890)	54	2.540 (0.1000)		

Intake valve clearance (Cold):
0.15 – 0.25 mm (0.006 – 0.010 in.)

EXAMPLE:
The 2.300 mm (0.0906 in.) shim is installed, and the measured clearance is 0.440 mm (0.0173 in.). Replace the 2.300 mm (0.0906 in.) shim with a No. 54 shim.

Intake valve clearance shim selection chart

Installed shim thickness mm (in.):
2.000 (0.0787), 2.020 (0.0795), 2.040 (0.0803), 2.060 (0.0811), 2.080 (0.0819), 2.100 (0.0827), 2.120 (0.0835), 2.140 (0.0843), 2.160 (0.0850), 2.180 (0.0858), 2.200 (0.0866), 2.220 (0.0874), 2.240 (0.0882), 2.260 (0.0890), 2.280 (0.0894), 2.290 (0.0902), 2.300 (0.0906), 2.310 (0.0909), 2.320 (0.0913), 2.330 (0.0917), 2.340 (0.0921), 2.350 (0.0925), 2.360 (0.0929), 2.370 (0.0933), 2.380 (0.0937), 2.390 (0.0941), 2.400 (0.0945), 2.410 (0.0949), 2.420 (0.0953), 2.430 (0.0957), 2.440 (0.0961), 2.450 (0.0965), 2.460 (0.0969), 2.470 (0.0972), 2.480 (0.0976), 2.490 (0.0980), 2.500 (0.0984), 2.510 (0.0988), 2.520 (0.0992), 2.530 (0.0996), 2.540 (0.1000), 2.550 (0.1004), 2.560 (0.1008), 2.570 (0.1012), 2.580 (0.1016), 2.590 (0.1020), 2.600 (0.1024), 2.620 (0.1031), 2.640 (0.1039), 2.660 (0.1047), 2.680 (0.1055), 2.700 (0.1063), 2.720 (0.1071), 2.740 (0.1079), 2.760 (0.1087), 2.780 (0.1094), 2.800 (0.1102)

Measured clearance mm (in.):
0.000 – 0.030 (0.0000 – 0.0012), 0.031 – 0.050 (0.0012 – 0.0020), 0.051 – 0.070 (0.0020 – 0.0028), 0.071 – 0.090 (0.0028 – 0.0035), 0.091 – 0.110 (0.0036 – 0.0043), 0.111 – 0.130 (0.0044 – 0.0051), 0.131 – 0.149 (0.0052 – 0.0059), 0.150 – 0.250 (0.0059 – 0.0098), 0.251 – 0.270 (0.0099 – 0.0106), 0.271 – 0.290 (0.0107 – 0.0114), 0.291 – 0.310 (0.0115 – 0.0122), 0.311 – 0.330 (0.0122 – 0.0130), 0.331 – 0.350 (0.0130 – 0.0138), 0.351 – 0.370 (0.0138 – 0.0146), 0.371 – 0.390 (0.0146 – 0.0154), 0.391 – 0.410 (0.0154 – 0.0161), 0.411 – 0.430 (0.0162 – 0.0169), 0.431 – 0.450 (0.0170 – 0.0177), 0.451 – 0.470 (0.0178 – 0.0185), 0.471 – 0.490 (0.0185 – 0.0193), 0.491 – 0.510 (0.0193 – 0.0201), 0.511 – 0.530 (0.0201 – 0.0209), 0.531 – 0.550 (0.0209 – 0.0217), 0.551 – 0.570 (0.0217 – 0.0224), 0.571 – 0.590 (0.0225 – 0.0232), 0.591 – 0.610 (0.0233 – 0.0240), 0.611 – 0.630 (0.0241 – 0.0248), 0.631 – 0.650 (0.0248 – 0.0256), 0.651 – 0.670 (0.0256 – 0.0264), 0.671 – 0.690 (0.0264 – 0.0272), 0.691 – 0.710 (0.0272 – 0.0280), 0.711 – 0.730 (0.0280 – 0.0287), 0.731 – 0.750 (0.0288 – 0.0295), 0.751 – 0.770 (0.0296 – 0.0303), 0.771 – 0.790 (0.0304 – 0.0311), 0.791 – 0.810 (0.0311 – 0.0319), 0.811 – 0.830 (0.0319 – 0.0327), 0.831 – 0.850 (0.0327 – 0.0335), 0.851 – 0.870 (0.0335 – 0.0343), 0.871 – 0.890 (0.0343 – 0.0350), 0.891 – 0.910 (0.0351 – 0.0358), 0.911 – 0.930 (0.0359 – 0.0366), 0.931 – 0.950 (0.0367 – 0.0374), 0.951 – 0.970 (0.0374 – 0.0382), 0.971 – 0.990 (0.0382 – 0.0390), 0.991 – 1.010 (0.0390 – 0.0398), 1.011 – 1.030 (0.0398 – 0.0406), 1.031 – 1.050 (0.0406 – 0.0413)

For complete Engine Mechanical specifications, see Section 1 of this manual

Exhaust valve clearance (Cold):
0.25 – 0.35 mm (0.010 – 0.014 in.)

EXAMPLE:
The 2.300 mm (0.0906 in.) shim is installed, and the measured clearance is 0.440 mm (0.0173 in.). Replace the 2.300 mm (0.0906 in.) shim with a No. 44 shim.

New shim thickness mm (in.)

Shim No.	Thickness	Shim No.	Thickness	Shim No.	Thickness
00	2.000 (0.0787)	28	2.280 (0.0898)	56	2.560 (0.1008)
02	2.020 (0.0795)	30	2.300 (0.0906)	58	2.580 (0.1016)
04	2.040 (0.0803)	32	2.320 (0.0913)	60	2.600 (0.1024)
06	2.060 (0.0811)	34	2.340 (0.0921)	62	2.620 (0.1031)
08	2.080 (0.0819)	36	2.360 (0.0929)	64	2.640 (0.1039)
10	2.100 (0.0827)	38	2.380 (0.0937)	66	2.660 (0.1047)
12	2.120 (0.0835)	40	2.400 (0.0945)	68	2.680 (0.1055)
14	2.140 (0.0843)	42	2.420 (0.0953)	70	2.700 (0.1063)
16	2.160 (0.0850)	44	2.440 (0.0961)	72	2.720 (0.1071)
18	2.180 (0.0858)	46	2.460 (0.0969)	74	2.740 (0.1079)
20	2.200 (0.0866)	48	2.480 (0.0976)	76	2.760 (0.1087)
22	2.220 (0.0874)	50	2.500 (0.0984)	78	2.780 (0.1094)
24	2.240 (0.0882)	52	2.520 (0.0992)	80	2.800 (0.1102)
26	2.260 (0.0890)	54	2.540 (0.1000)		

Exhaust valve clearance shim selection chart

The main shim selection chart plots Installed shim thickness mm (in.) across the top and Measured clearance mm (in.) down the left side.

Installed shim thickness values (top row), mm (in.):
2.000 (0.0787), 2.020 (0.0795), 2.040 (0.0803), 2.060 (0.0819), 2.080 (0.0811), 2.100 (0.0827), 2.120 (0.0835), 2.140 (0.0843), 2.160 (0.0850), 2.180 (0.0858), 2.200 (0.0866), 2.210 (0.0870), 2.220 (0.0874), 2.230 (0.0878), 2.240 (0.0882), 2.250 (0.0886), 2.260 (0.0890), 2.270 (0.0894), 2.280 (0.0898), 2.290 (0.0902), 2.300 (0.0906), 2.310 (0.0909), 2.320 (0.0913), 2.330 (0.0917), 2.340 (0.0921), 2.350 (0.0925), 2.360 (0.0929), 2.370 (0.0933), 2.380 (0.0937), 2.390 (0.0941), 2.400 (0.0945), 2.410 (0.0949), 2.420 (0.0953), 2.430 (0.0957), 2.440 (0.0961), 2.450 (0.0965), 2.460 (0.0969), 2.470 (0.0972), 2.480 (0.0976), 2.490 (0.0980), 2.500 (0.0984), 2.510 (0.0988), 2.520 (0.0992), 2.530 (0.0996), 2.540 (0.1000), 2.550 (0.1004), 2.560 (0.1008), 2.570 (0.1012), 2.580 (0.1016), 2.590 (0.1020), 2.600 (0.1024), 2.620 (0.1031), 2.640 (0.1039), 2.660 (0.1047), 2.680 (0.1055), 2.700 (0.1063), 2.720 (0.1071), 2.740 (0.1079), 2.760 (0.1087), 2.780 (0.1094), 2.800 (0.1102)

Measured clearance values (left column), mm (in.):
0.000–0.030 (0.0000–0.0012), 0.031–0.050 (0.0012–0.0020), 0.051–0.070 (0.0020–0.0028), 0.071–0.090 (0.0028–0.0035), 0.091–0.110 (0.0036–0.0043), 0.111–0.130 (0.0044–0.0051), 0.131–0.150 (0.0052–0.0059), 0.151–0.170 (0.0059–0.0067), 0.171–0.190 (0.0067–0.0075), 0.191–0.210 (0.0075–0.0083), 0.211–0.230 (0.0083–0.0091), 0.231–0.249 (0.0091–0.0098), 0.250–0.350 (0.0098–0.0138), 0.351–0.370 (0.0138–0.0146), 0.371–0.390 (0.0146–0.0154), 0.391–0.410 (0.0154–0.0161), 0.411–0.430 (0.0162–0.0169), 0.431–0.450 (0.0170–0.0177), 0.451–0.470 (0.0178–0.0185), 0.471–0.490 (0.0185–0.0193), 0.491–0.510 (0.0193–0.0201), 0.511–0.530 (0.0201–0.0209), 0.531–0.550 (0.0209–0.0217), 0.551–0.570 (0.0217–0.0224), 0.571–0.590 (0.0225–0.0232), 0.591–0.610 (0.0233–0.0240), 0.611–0.630 (0.0241–0.0248), 0.631–0.650 (0.0248–0.0256), 0.651–0.670 (0.0256–0.0264), 0.671–0.690 (0.0264–0.0272), 0.691–0.710 (0.0272–0.0280), 0.711–0.730 (0.0280–0.0287), 0.731–0.750 (0.0288–0.0295), 0.751–0.770 (0.0296–0.0303), 0.771–0.790 (0.0304–0.0311), 0.791–0.810 (0.0311–0.0319), 0.811–0.830 (0.0319–0.0327), 0.831–0.850 (0.0327–0.0335), 0.851–0.870 (0.0335–0.0343), 0.871–0.890 (0.0343–0.0350), 0.891–0.910 (0.0351–0.0358), 0.911–0.930 (0.0359–0.0366), 0.931–0.950 (0.0367–0.0374), 0.951–0.970 (0.0374–0.0382), 0.971–0.990 (0.0382–0.0390), 0.991–1.010 (0.0390–0.0398), 1.011–1.030 (0.0398–0.0406), 1.031–1.050 (0.0406–0.0413), 1.051–1.070 (0.0414–0.0421), 1.071–1.090 (0.0422–0.0429), 1.091–1.110 (0.0430–0.0437), 1.111–1.130 (0.0437–0.0445), 1.131–1.150 (0.0445–0.0453)

7924SG72

Starter Motor

REMOVAL & INSTALLATION

1. Before servicing the vehicle, refer to the precautions in the beginning of this section.
2. Drain the cooling system.
3. Relieve the fuel system pressure.
4. Remove or disconnect the following:
   • Negative battery cable
   • Engine appearance cover
   • Air intake tube
   • Intake manifold
   • Starter motor mounting bolts
   • Starter wiring connectors
   • Starter motor

To install:

5. Install or connect the following:
   • Starter motor
   • Starter wiring connectors. Tighten the cable nut to 86 inch lbs. (10 Nm).
   • Starter motor mounting bolts. Tighten the bolts to 29 ft. lbs. (39 Nm).
   • Intake manifold
   • Air intake tube
   • Engine appearance cover
   • Negative battery cable
6. Fill the cooling system.
7. Start the engine and check for leaks.

Oil Pan

REMOVAL & INSTALLATION

1. Before servicing the vehicle, refer to the precautions in the beginning of this section.
2. Remove the engine from the vehicle and mount it on a stand.
3. Remove or disconnect the following:
   • Oil dipstick tube
   • Lower oil pan
   • Oil pan baffle
   • Upper oil pan

To install:

4. The upper oil pan bolts are different lengths and are identified as follows:
   • A: 0.79 inch (20mm) w/10mm head
   • B: 0.98 inch (25mm) w/12mm head
   • C: 2.36 inch (60mm) w/12mm head
   • D: 1.38 inch (35mm) w/10mm head

5. Apply silicone sealant to the upper oil pan as shown.
6. Install the upper oil pan and tighten the fasteners in several passes to the following specifications:
   • 10mm: 66 inch lbs. (7.5 Nm)
   • 12mm: 21 ft. lbs. (28 Nm)
7. Install or connect the following:

• Oil pan baffle. Tighten the fasteners to 66 inch lbs. (7.5 Nm).
• Lower oil pan. Tighten the fasteners in several passes to 66 inch lbs. (7.5 Nm).
• Oil dipstick tube
8. Install the engine.

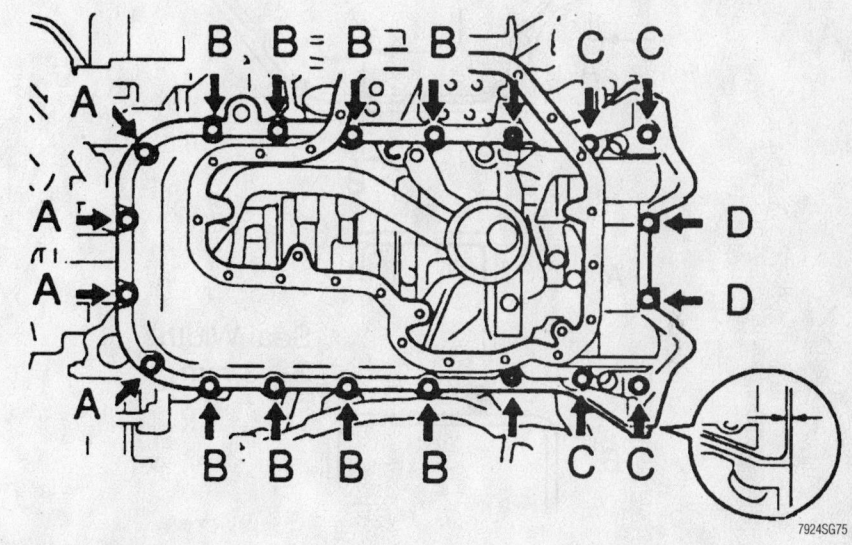

Upper oil pan bolt location

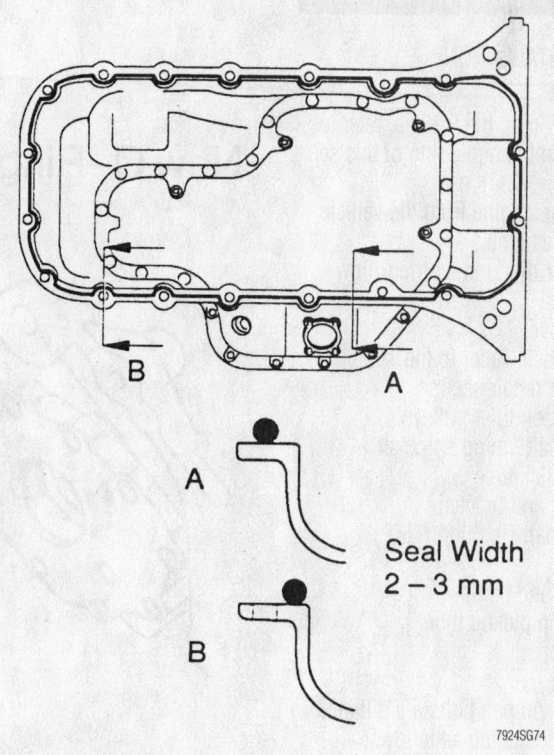

Upper oil pan sealant application

For Accessory Drive Belt illustrations, see Section 1 of this manual

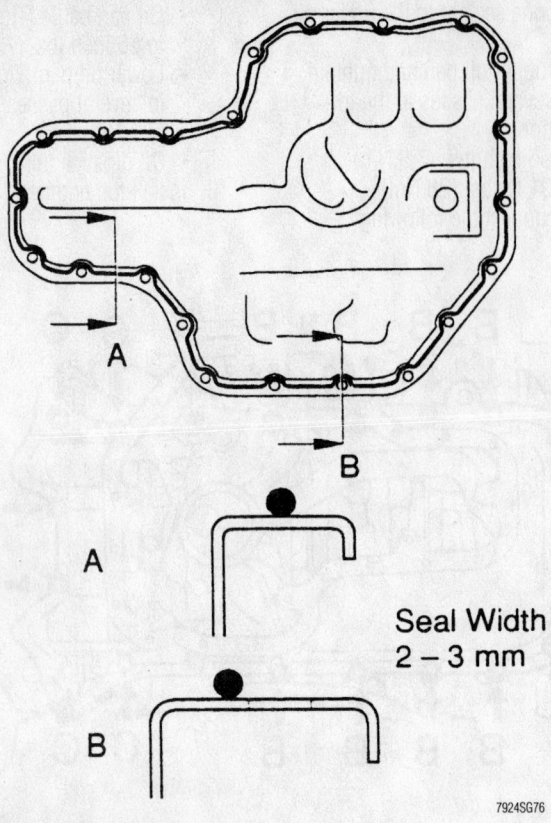

Seal Width
2 – 3 mm

7924SG76

Lower oil pan sealant application

Oil Pump

REMOVAL & INSTALLATION

1. Before servicing the vehicle, refer to the precautions in the beginning of this section.

2. Remove the engine from the vehicle and mount it on a stand.

3. Remove or disconnect the following:

- Front cover
- Timing belt. Refer to the Timing Belt unit repair section.
- Timing belt idler pulleys
- Crankshaft timing sprocket
- Oil dipstick tube
- Oil filter and bracket
- Crankshaft Position (CKP) sensor
- Oil pan and baffle
- Oil pump pickup tube
- Oil pump

To install:

4. The upper oil pan bolts are different lengths and are identified as follows:

- A: 1.38 inch (35mm) w/12mm head
- B: 1.97 inch (50mm) w/12mm head
- C: 4.17 inch (106mm) w/12mm head

- D: 1.57 inch (40mm) w/14mm head
- E: 1.18 inch (30mm) w/6mm hex head

5. Install a new O-ring on the engine block.

6. Apply silicone sealant to the oil pump housing as shown.

7. Install the oil pump. Tighten the bolts in several passes to the following specifications:

- 12mm: 11 ft. lbs. (15.5 Nm)
- 14mm: 22 ft. lbs. (30.5 Nm)
- 6mm Hex: 11 ft. lbs. (15.5 Nm)

8. Install or connect the following:

- Oil pump pickup tube. Tighten the bolts to 66 inch lbs. (7.5 Nm).
- Oil pan and baffle
- CKP sensor
- Oil filter and bracket. Tighten the bolts to 13 ft. lbs. (18 Nm).
- Oil dipstick tube
- Crankshaft timing sprocket
- Timing belt idler pulleys
- Timing belt
- Front cover

9. Install the engine.

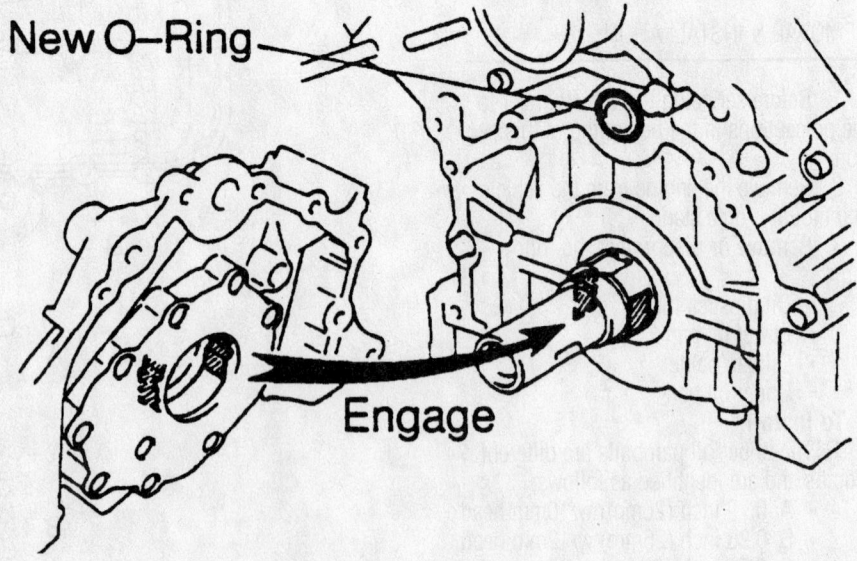

New O-Ring

Engage

9308SG04

Location of the O-ring seal

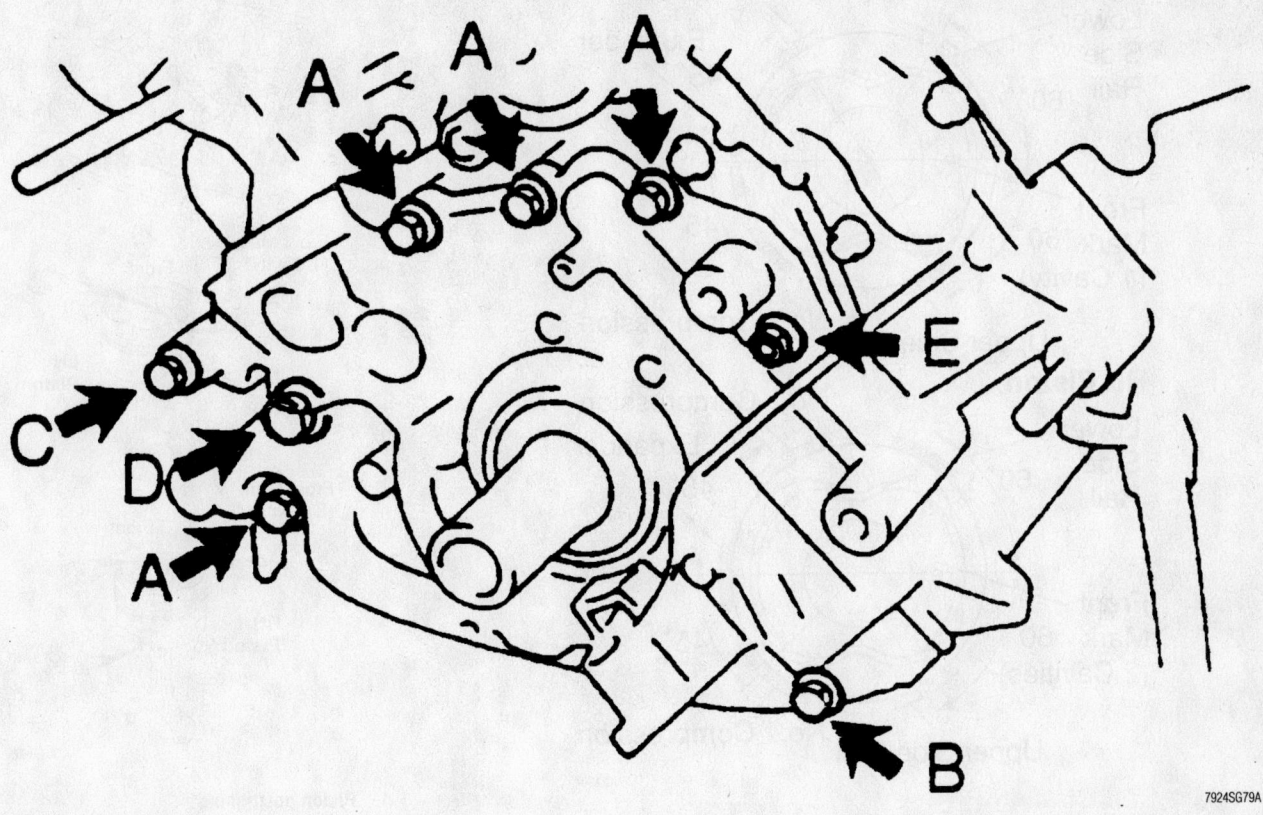

Oil pump bolt location

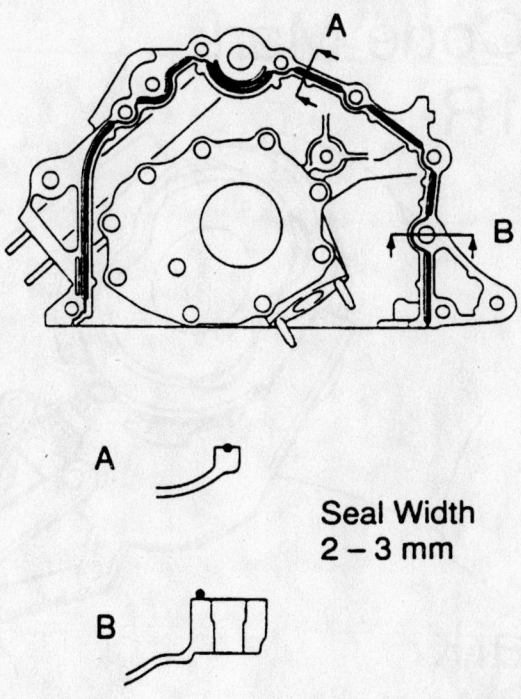

A

B

**Seal Width
2 – 3 mm**

Oil pump housing sealant application

Rear Main Seal

REMOVAL & INSTALLATION

1. Before servicing the vehicle, refer to the precautions in the beginning of this section.
2. Remove the transmission and flywheel from the vehicle.
3. Cut off the rubber lip portion of the seal with a sharp knife.
4. Pry out the oil seal.
To install:
5. Install the rear main seal so that it is flush with the seal retainer housing.
6. Install or connect the following:
  • Flywheel/driveplate. Tighten the bolts to 35 ft. lbs. (48 Nm) plus a 90 degree turn.
  • Transmission

Piston and Ring

POSITIONING

For Tire, Wheel and Ball Joint specifications, see Section 1 of this manual

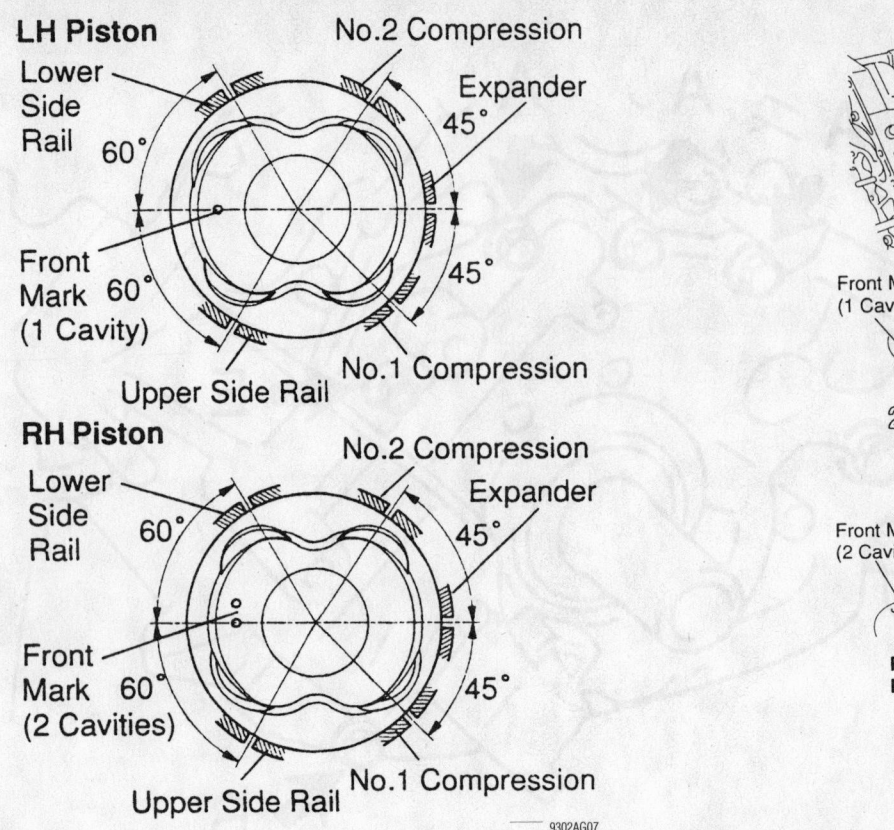

LH Piston

Lower Side Rail — 60°
No.2 Compression
Expander — 45°
45°
Front Mark (1 Cavity) — 60°
No.1 Compression
Upper Side Rail

RH Piston

Lower Side Rail — 60°
No.2 Compression
Expander — 45°
Front Mark (2 Cavities) — 60°
45°
No.1 Compression
Upper Side Rail

9302AG07

Piston ring positioning

Front Mark (1 Cavity)
Front
LH
2L
LH Piston

Front Mark (2 Cavities)
Front
RH
RH Piston
2R

9302AG08

Piston positioning

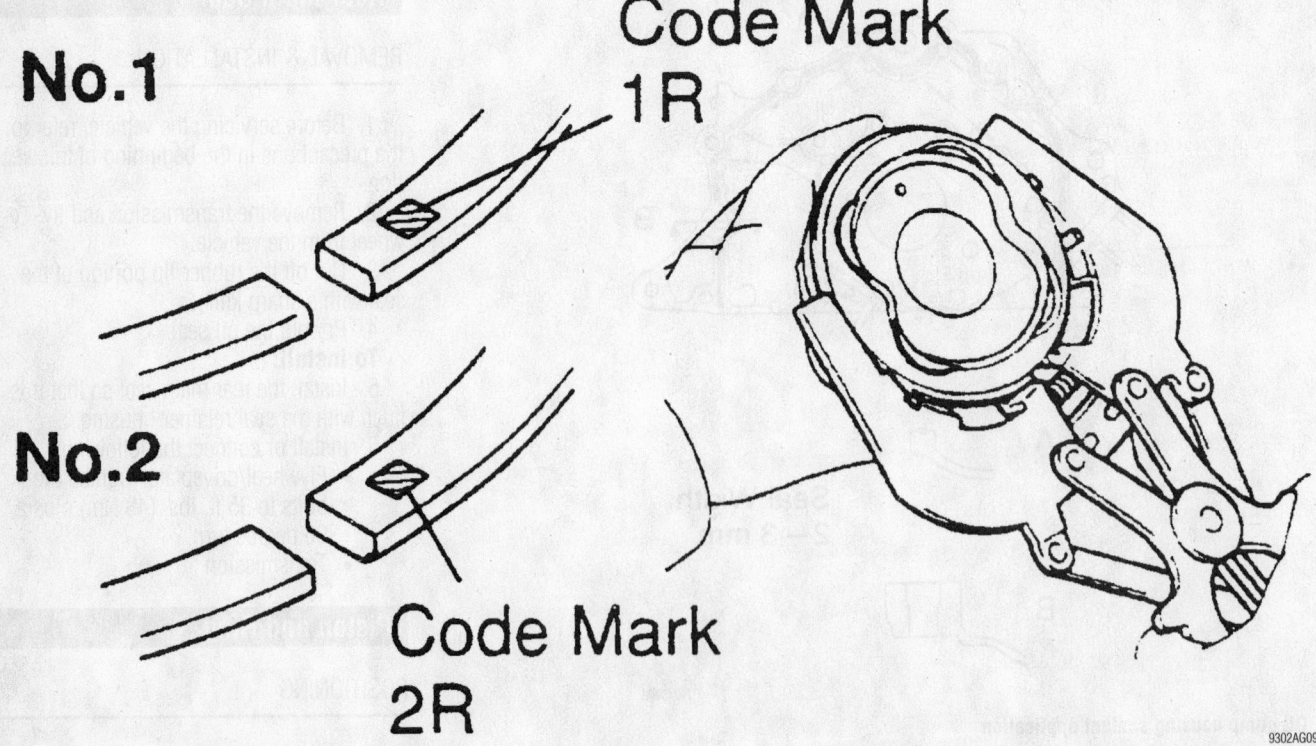

No.1

Code Mark 1R

No.2

Code Mark 2R

9302AG09

Piston ring identification

FUEL SYSTEM

Fuel System Service Precautions

Safety is the most important factor when performing not only fuel system maintenance but any type of maintenance. Failure to conduct maintenance and repairs in a safe manner may result in serious personal injury or death. Maintenance and testing of the vehicle's fuel system components can be accomplished safely and effectively by adhering to the following rules and guidelines.

• To avoid the possibility of fire and personal injury, always disconnect the negative battery cable unless the repair or test procedure requires that battery voltage be applied.

• Always relieve the fuel system pressure prior to disconnecting any fuel system component (injector, fuel rail, pressure regulator, etc.), fitting or fuel line connection. Exercise extreme caution whenever relieving fuel system pressure, to avoid exposing skin, face and eyes to fuel spray. Please be advised that fuel under pressure may penetrate the skin or any part of the body that it contacts.

• Always place a shop towel or cloth around the fitting or connection prior to loosening to absorb any excess fuel due to spillage. Ensure that all fuel spillage (should it occur) is quickly removed from engine surfaces. Ensure that all fuel soaked cloths or towels are deposited into a suitable waste container.

• Always keep a dry chemical (Class B) fire extinguisher near the work area.

• Do not allow fuel spray or fuel vapors to come into contact with a spark or open flame.

• Always use a back-up wrench when loosening and tightening fuel line connection fittings. This will prevent unnecessary stress and torsion to fuel line piping.

• Always replace worn fuel fitting O-rings with new. Do not substitute fuel hose or equivalent, where fuel pipe is installed.

Fuel System Pressure

RELIEVING

1. Before servicing the vehicle, refer to the precautions in the beginning of this section.
2. Disconnect the fuel pump connector near the fuel tank.

3. Start the engine and allow it to run until it stalls. Crank the engine for a few seconds to relieve additional fuel pressure.
4. Disconnect the negative battery cable.
5. When repairs are complete, connect the negative battery cable.

Fuel Filter

REMOVAL & INSTALLATION

1. Before servicing the vehicle, refer to the precautions in the beginning of this section.
2. Relieve the fuel system pressure.
3. Remove or disconnect the following:
   • Negative battery cable
   • Fuel lines
   • Fuel filter

To install:
4. Install the fuel filter.
5. Use new washers and tighten the fuel line bolts to the following specifications:
   • Banjo bolt fittings: 21 ft. lbs. (29 Nm)
   • Flare nut fitting: 28 ft. lbs. (38 Nm)
6. Connect the negative battery cable.
7. Start the engine and check for leaks.

Fuel Pump

REMOVAL & INSTALLATION

Sequoia

1. Before servicing the vehicle, refer to the precautions in the beginning of this section.
2. Relieve the fuel system pressure.
3. Remove or disconnect the following:
   • Negative battery cable
   • Fuel tank
   • Fuel pump harness connector
   • Fuel lines
   • Fuel pump module

To install:
4. Install or connect the following:
   • Fuel pump module. Tighten the bolts to 35 inch lbs. (4 Nm).
   • Fuel lines
   • Fuel pump harness connector
   • Fuel tank
   • Negative battery cable
5. Start the engine and check for leaks.

Land Cruiser & LX470

1. Before servicing the vehicle, refer to the precautions in the beginning of this section.

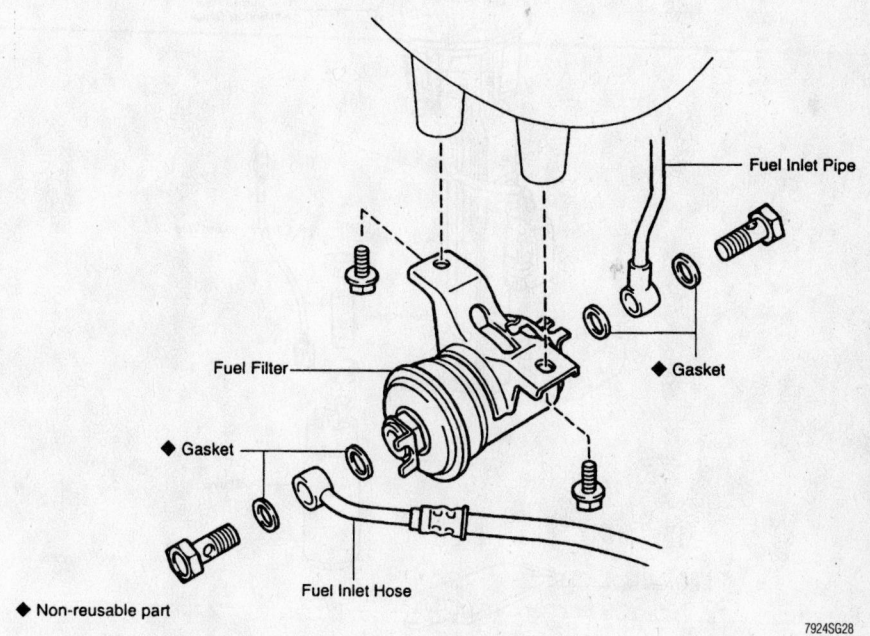

Always use new gaskets when replacing the fuel filter

For Wheel Alignment specifications, see Section 1 of this manual

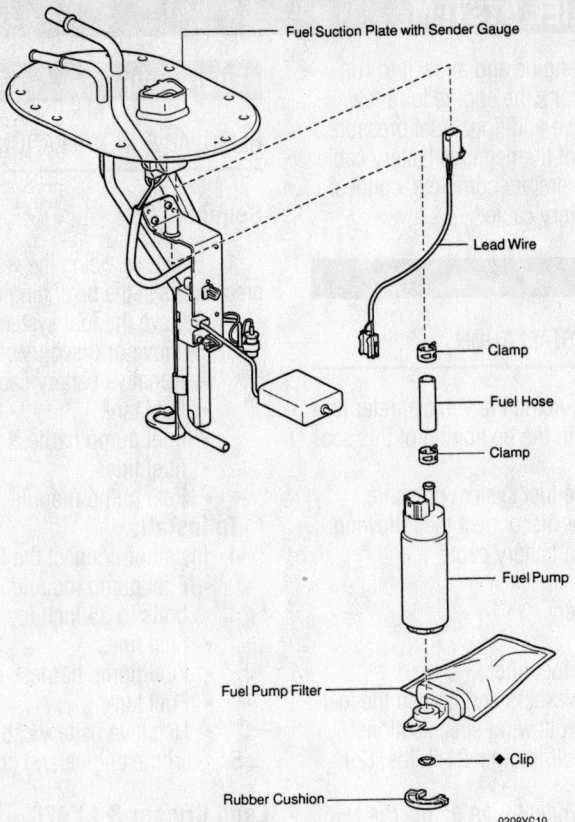

Exploded view of the fuel pump and related components—Sequoia

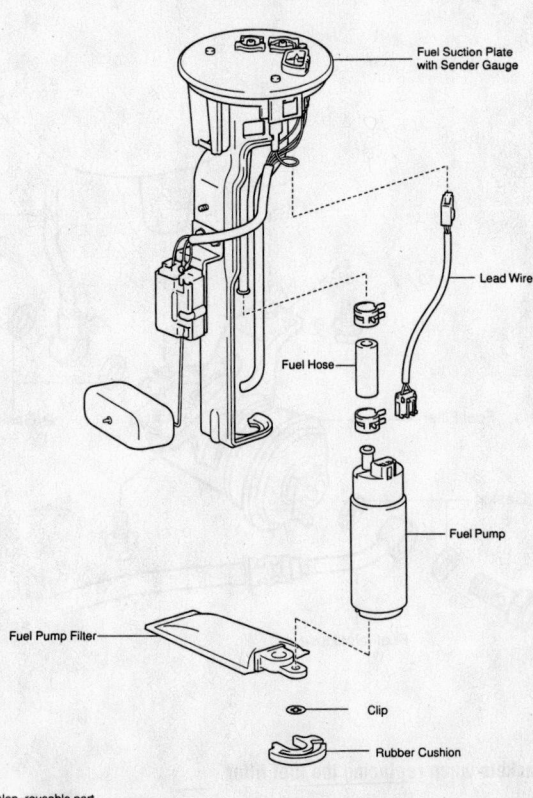

◆ Non–reusable part

7924SG81

Exploded view of the fuel pump and related components— Land Cruiser & LX470

2. Relieve the fuel system pressure.
3. Remove or disconnect the following:
   - Negative battery cable
   - Rear seats
   - Door sill trim plates
   - Carpeting and floor mats
   - Access panel
   - Fuel pump harness connector
   - Fuel lines
   - Fuel pump module

To install:

4. Install or connect the following:
   - Fuel pump module. Tighten the bolts to 35 inch lbs. (4 Nm).
   - Fuel lines
   - Fuel pump harness connector
   - Access panel
   - Carpeting and floor mats
   - Door sill trim plates
   - Rear seats
   - Negative battery cable
5. Start the engine and check for leaks.

Fuel Injector

REMOVAL & INSTALLATION

1. Before servicing the vehicle, refer to the precautions in the beginning of this section.
2. Relieve the fuel system pressure.
3. Remove or disconnect the following:
   - Negative battery cable
   - Engine appearance cover
   - Air intake tube
   - Fuel lines
   - Fuel pulsation damper
   - Fuel pressure regulator vacuum line
   - Accelerator cable and bracket
   - Positive Crankcase Ventilation (PCV) valve and hose
   - Evaporative Emissions (EVAP) vacuum switching valve
   - Engine appearance cover brackets
   - Fuel injector harness connectors
   - Engine harness protector
   - Fuel supply manifold crossover pipe
   - Fuel supply manifolds with injectors attached
   - Fuel injectors

To install:

4. Install the fuel injectors to the supply manifold with new O-ring seals and new grommets.
5. Install new injector insulators to the intake manifold.
6. Install or connect the following:
   - Fuel supply manifolds with injectors attached. Tighten the bolts to 66 inch lbs. (7.5 Nm).

- Fuel supply manifold crossover pipe. Tighten the bolts to 29 ft. lbs. (39 Nm).
- Engine harness protector
- Fuel injector harness connectors
- Engine appearance cover brackets

- EVAP vacuum switching valve
- PCV valve and hose
- Accelerator cable and bracket
- Fuel pressure regulator vacuum line
- Fuel pulsation damper

- Fuel lines
- Air intake tube
- Engine appearance cover
- Negative battery cable

7. Start the engine and check for leaks.

DRIVE TRAIN

Automatic Transmission Assembly

REMOVAL & INSTALLATION

Sequoia

1. Before servicing the vehicle, refer to the precautions in the beginning of this section.
2. Remove or disconnect the following:

- Oil filler pipe
- No. 1 engine undercover
- Exhaust pipes
- Front and rear driveshafts
- Nos. 1&2 vehicle speed sensors
- Solenoid connector
- Shift cable at the transmission
- Overdrive sensor connector
- Oil cooler lines
- ATF temperature sensor
- Park/Neutral switch
- End plate and converter clutch mounting bolts

3. Raise the transmission slightly.
4. Remove or disconnect the following:

- Crossmember
- Rear mount insulator
- Transmission and transfer case as a unit

5. Installation is the reverse of removal. Note the following torques:

- Transmission-to-transfer case bolts: 53 ft. lbs. (71Nm)
- Rear mount insulator-to-transmission: 48 ft. lbs. (65Nm)
- Rear mount insulator-to-crossmember: 13 ft. lbs. (18Nm)
- Crossmember-to-frame: 53 ft. lbs. (71Nm)
- Torque converter clutch: 30 ft. lbs. (41Nm)

➥**Install the green bolt first.**

- Rear end plate: 13 ft. lbs. (18Nm)
- Oil cooler lines: 25 ft. lbs. (34Nm)
- Shift control bracket: 13 ft. lbs. (18Nm)

Land Cruiser & LX470

1. Before servicing the vehicle, refer to the precautions in the beginning of this section.
2. Remove or disconnect the following:

- Battery and tray
- Air intake assembly
- Cooling fan and shroud
- Coolant recovery reservoir
- Transmission dipstick tube
- Center console
- Transmission gear select lever and rod
- Transfer case shift lever and rod
- Engine under covers
- Exhaust front pipes
- Front and rear driveshafts
- Vehicle Speed (VSS) sensor connectors
- Overdrive clutch speed sensor connector
- Solenoid harness connector
- Transmission fluid temperature sensor connector
- Park/Neutral Position (PNP) switch connector
- Center differential lock indicator switch connector
- L4 solenoid valve position switch connector
- Motor actuator connector
- Torque converter
- Transmission oil cooler lines
- Transmission mount crossmember. Support the transmission with a jack.
- Transmission flange bolts
- Transmission

To install:
3. Install or connect the following:

- Transmission. Tighten the flange bolts to 53 ft. lbs. (72 Nm).
- Transmission mount crossmember. Tighten the bolts to 37 ft. lbs. (50 Nm) and the nuts to 54 ft. lbs. (74 Nm).
- Transmission oil cooler lines
- Torque converter. Tighten the bolts to 35 ft. lbs. (48 Nm).

- Motor actuator connector
- L4 solenoid valve position switch connector
- Center differential lock indicator switch connector
- PNP switch connector
- Transmission fluid temperature sensor connector
- Solenoid harness connector
- Overdrive clutch speed sensor connector
- VSS sensor connectors
- Front driveshaft. Tighten the fasteners to 59 ft. lbs. (80 Nm).
- Rear driveshaft. Tighten the fasteners to 78 ft. lbs. (106 Nm).
- Exhaust front pipes
- Engine under covers
- Transfer case shift lever and rod
- Transmission gear select lever and rod
- Center console
- Transmission dipstick tube
- Coolant recovery reservoir
- Cooling fan and shroud
- Air intake assembly
- Battery and tray

4. Check the transmission and transfer case fluid levels and adjust as necessary.

Transfer Case Assembly

REMOVAL & INSTALLATION

Sequoia

1. Before servicing the vehicle, refer to the precautions in the beginning of this section.
2. Drain the transfer case oil.
3. Place the shift lever in the **H** position and the one-touch 2-4 switch **OFF**.
4. Remove or disconnect the following:

- Shift lever
- Skid plate
- Oil from the transfer case
- Exhaust pipes
- Front and rear driveshafts

- Crossmember
- Rear engine mount
- All wiring connectors

5. Support the transfer case, remove the case-to-adapter bolts and lower the transfer case.

6. Installation is the reverse of removal. Note the following torques:

- Transfer case-to-adapter bolts: 17 ft. lbs. (24Nm)
- Engine rear mount-to-adapter: 48 ft. lbs. (65Nm)
- Crossmember: 53 ft. lbs. (72Nm)
- Engine rear mount set bolts: 13 ft. lbs. (18Nm)
- Skid plate: 13 ft. lbs. (18Nm)

Land Cruiser & LX470

1. Before servicing the vehicle, refer to the precautions in the beginning of this section.

2. Drain the transfer case oil.

3. Remove or disconnect the following:

- Transfer case protector
- Front and rear driveshafts
- Transfer case shift lever rod
- Ground cable
- Transmission mount crossmember. Support the transmission with a jack.
- Transfer case vent hose
- Vehicle Speed (VSS) sensor connector
- Center differential lock indicator switch connector
- Motor actuator connectors
- Transfer case adapter bolts
- Transfer case

To install:

4. Install or connect the following:

- Transfer case. Tighten the adapter bolts to 51 ft. lbs. (69 Nm).
- Motor actuator connectors
- Center differential lock indicator switch connector
- VSS sensor connector
- Transfer case vent hose
- Transmission mount crossmember. Tighten the bolts to 37 ft. lbs. (50 Nm) and the nuts to 54 ft. lbs. (74 Nm).
- Ground cable
- Transfer case shift lever rod
- Front driveshaft. Tighten the fasteners to 59 ft. lbs. (80 Nm).
- Rear driveshaft. Tighten the fasteners to 78 ft. lbs. (106 Nm).
- Transfer case protector

5. Fill the transfer case to the correct level.

Halfshaft

REMOVAL & INSTALLATION

Sequoia

1. Before servicing the vehicle, refer to the precautions in the beginning of this section.

2. Remove or disconnect the following:

- Front wheel
- Under cover

3. Drain the differential oil.

4. Remove or disconnect the following:

- Grease cap
- Cotter pin and lock cap
- Halfshaft locknut by applying the brakes

- Lower control arm from the lower ball joint
- Halfshaft from the steering knuckle, using a plastic hammer
- Left strut, for the left halfshaft
- Right halfshaft, using a brass bar and a hammer
- Left halfshaft, using tools 09520-01010 and 09520-24010
- Snapring from the inboard joint shaft

To install:

5. Install or connect the following:

- New snapring, onto the inboard joint shaft with the opening facing downward
- Halfshafts to the differential using a brass bar and a hammer
- Halfshafts to the steering knuckles

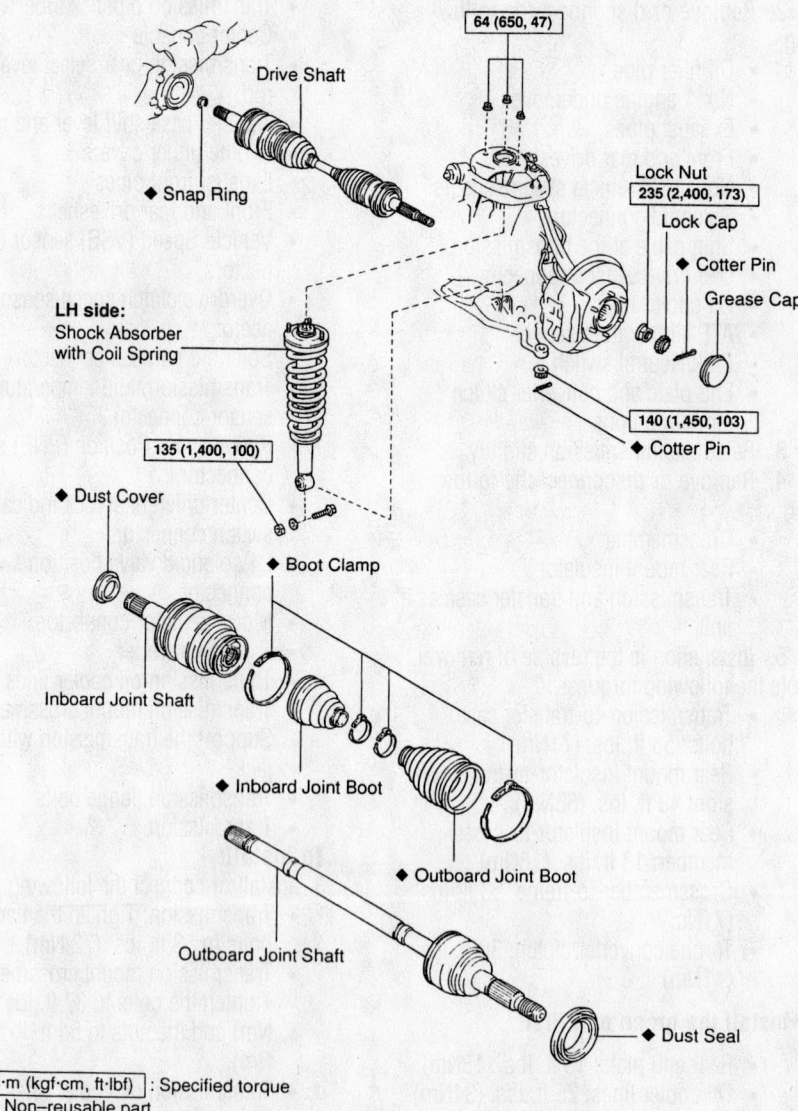

View of the halfshaft and related components—Sequoia

⁂ WARNING

Be careful not to damage the oil seal, boot or dust seal.

- Lower control arm to the lower ball joint using a new cotter pin. Torque the ball joint nut to 103 ft. lbs. (140 Nm).
- Halfshaft locknut by applying the brakes. Torque the nut to 173 ft. lbs. (235 Nm).

LH side:

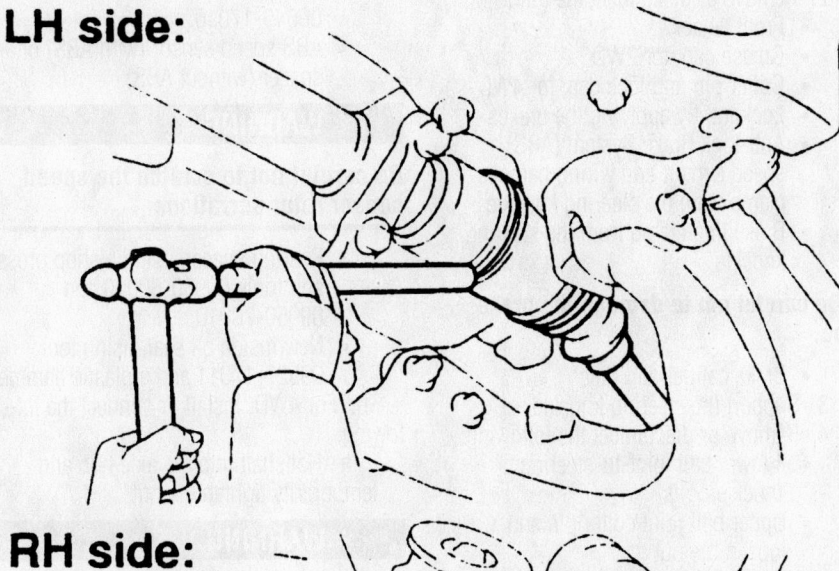

RH side:

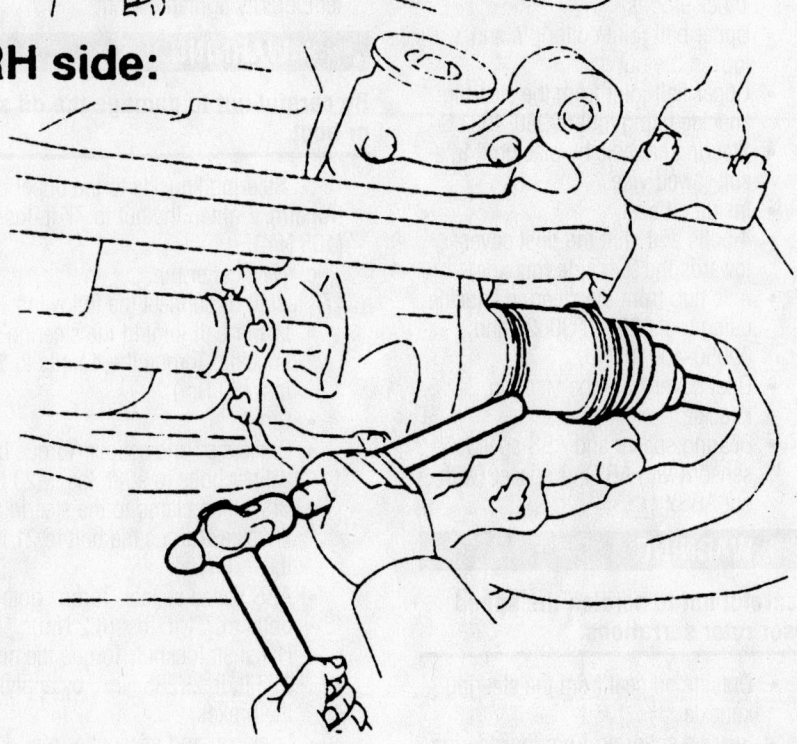

Axle halfshaft removal—Land Cruiser/LX470

- Lock cap and a new cotter pin
- Grease cap
6. Refill the differential with oil.
7. Install or connect the following:
   - Under cover
   - Front wheel

Land Cruiser & LX470

1. Before servicing the vehicle, refer to the precautions in the beginning of this section.

2. Remove or disconnect the following:
   - Front wheel
   - Brake caliper
   - Grease cap
   - Snapring
   - Wheel speed sensor and wire harness
   - Steering knuckle arm
   - Lower ball joint
   - Upper ball joint
   - Steering knuckle
   - Axle halfshaft

To install:

➡ **Use new split pins, snaprings and circlips for assembly.**

3. Install or connect the following:
   - Axle halfshaft
   - Steering knuckle
   - Upper ball joint. Tighten the nut to 81 ft. lbs. (110 Nm).
   - Lower ball joint. Tighten the nut to 117 ft. lbs. (159 Nm).
   - Steering knuckle arm. Tighten the bolts to 108 ft. lbs. (147 Nm).
   - Wheel speed sensor and wire harness
   - Snapring
   - Grease cap
   - Brake caliper
   - Front wheel

CV-Joints

OVERHAUL

Land Cruiser & LX470

OUTER CV-JOINT

The outer CV-joint is serviced with the axle shaft as an assembly. The outer CV-joint boot can be serviced by removing the inner CV-joint.

INNER CV-JOINT

1. Before servicing the vehicle, refer to the precautions in the beginning of this section.

2. Remove or disconnect the following:
   - Halfshaft from the vehicle
   - Grease boot clamps
   - Outer race snapring
   - Outer race
   - Shaft snapring
   - Inner race, cage and balls

To install:

3. Install or connect the following:
   - Inner race, cage and balls
   - Shaft snapring
   - Outer race

9308SG07

- Outer race snapring
4. Fill the outer race and the grease boot with CV-joint grease and tighten the boot clamps.
5. Install the axle halfshaft.

Sequoia

OUTER CV-JOINT

The outer CV-joint is serviced with the axle shaft as an assembly. The outer CV-joint boot can be serviced by removing the inner CV-joint.

INNER CV-JOINT

1. Before servicing the vehicle, refer to the precautions in the beginning of this section.
2. Remove or disconnect the following:
- Halfshaft from the vehicle
- Large boot clamps
- Small boot clamps
3. Matchmark inboard CV-joint to the shaft
4. Remove or disconnect the following:

- Inboard CV-joint from the shaft by expanding the snapring
- Both CV-joint boots
- Outer dust seal, using a shop press and tool 09950-00020
- Outer dust cover, using a shop press and tool 09950-00020

To install:

5. Install or connect the following:
- Outer dust cover, using a screwdriver and a hammer
- Outer dust seal, using a screwdriver and a hammer
6. Wrap the shaft splines with tape to protect the boot from damage.
7. Install or connect the following:
- Both CV-joint boots with clamps, temporarily
- Inboard CV-joint to the shaft by aligning the matchmarks and expanding the snapring
8. Lubricate the outboard joint with 7.23–7.94 oz. (205–225g) grease, provided in the boot kit.
9. Lubricate the inboard joint with 6.70–7.41 oz. (190–210g) grease, provided in the boot kit.
10. Install or connect the following:
- Both joint boots making sure the boots are in the shaft groove
- Standard halfshaft length is 20.531–20.689 in. (521.5–525.5mm) when the shaft is not expanded or contracted
- Large inboard boot clamp
- All other boot clamps using tool

09521-24010. Tighten the crimping tool until the clamp clearance is 0.039–0.059 in. (1.0–1.5mm)
- Halfshaft

Spindle Bearings

REMOVAL, PACKING AND INSTALLATION

Sequoia

1. Before servicing the vehicle, refer to the precautions in the beginning of this section.
2. Remove or disconnect the following:
- Front wheel
- Grease cap, for 2WD
- Cotter pin and lock cap, for 4WD
- Locknut, by applying the brakes
- Anti-lock Brake System (ABS) speed sensor and wiring harness clamp from the steering knuckle
- Brake line clamp from the steering knuckle

➡Be careful not to damage the brake tube.

- Brake caliper and disc
3. Support the steering knuckle
4. Remove or disconnect the following:
- 4 lower ball joint-to-steering knuckle bolts
- Upper ball joint cotter pin and loosen the nut
- Upper ball joint from the steering knuckle using tool 09950-40011
- Steering knuckle by placing it in a soft-jawed vise
- Inside oil seal
- 4 bolts and shift the dust cover towards the hub side (outside)
- Axle hub from the steering knuckle using tools 09710-30021 and 09950-40011
- Dust cover from the steering knuckle
- Bearing spacer and ABS speed sensor (with ABS) or spacer (without ABS)

✳✳ WARNING

Be careful not to scratch the speed sensor rotor serrations

- Outside oil seal from the steering knuckle
- Bearing snapring from the steering knuckle
- Bearing from the steering knuckle, using a shop press and tools 09950-60020 and 09950-70010

To install:

5. Install or connect the following:
- Bearing to the steering knuckle, using a shop press and tools 09527-17011 and 09950-60020
- Bearing snapring to the steering knuckle
- New outside oil seal to the steering knuckle, using tools 09223-15030 and 09527-17011
- Dust cover to the steering knuckle. Torque the 4 bolts to 13 ft. lbs. (18 Nm).
- Axle hub to the steering knuckle using a shop press and tool 09649-17010
- ABS speed sensor (with ABS) or spacer (without ABS)

✳✳ WARNING

Be careful not to scratch the speed sensor rotor serrations

- Bearing spacer, using a shop press and tools 09950-60010 and 09950-70010
- New inside oil seal, using tool 09527-17011 and a plastic hammer
6. For 4WD, install or connect the following:
a. Halfshaft into the axle hub and temporarily tighten the nut

✳✳ WARNING

Be careful not to damage the oil seal or boot.

b. Steering knuckle to the upper control arm. Tighten the nut to 77 ft. lbs. (105 Nm).
c. New cotter pin.
7. Install or connect the following:
- Lower ball joint to the steering knuckle. Torque the 4 bolts to 59 ft. lbs. (80 Nm).
- Strut
- Brake disc and caliper. Torque both caliper bolts to 90 ft. lbs. (123 Nm).
- Brake line clamp to the steering knuckle. Torque the bolt to 21 ft. lbs. (28 Nm).
- ABS speed sensor. Torque both bolts to 7.1 ft. lbs. (8.2 Nm).
- Halfshaft locknut. Torque the nut to 173 ft. lbs. (235 Nm), by applying the brakes.
- Lock cap and new cotter pin
- Grease cap, for 2WD
- Front wheel
8. Depress the brake pedal several times.

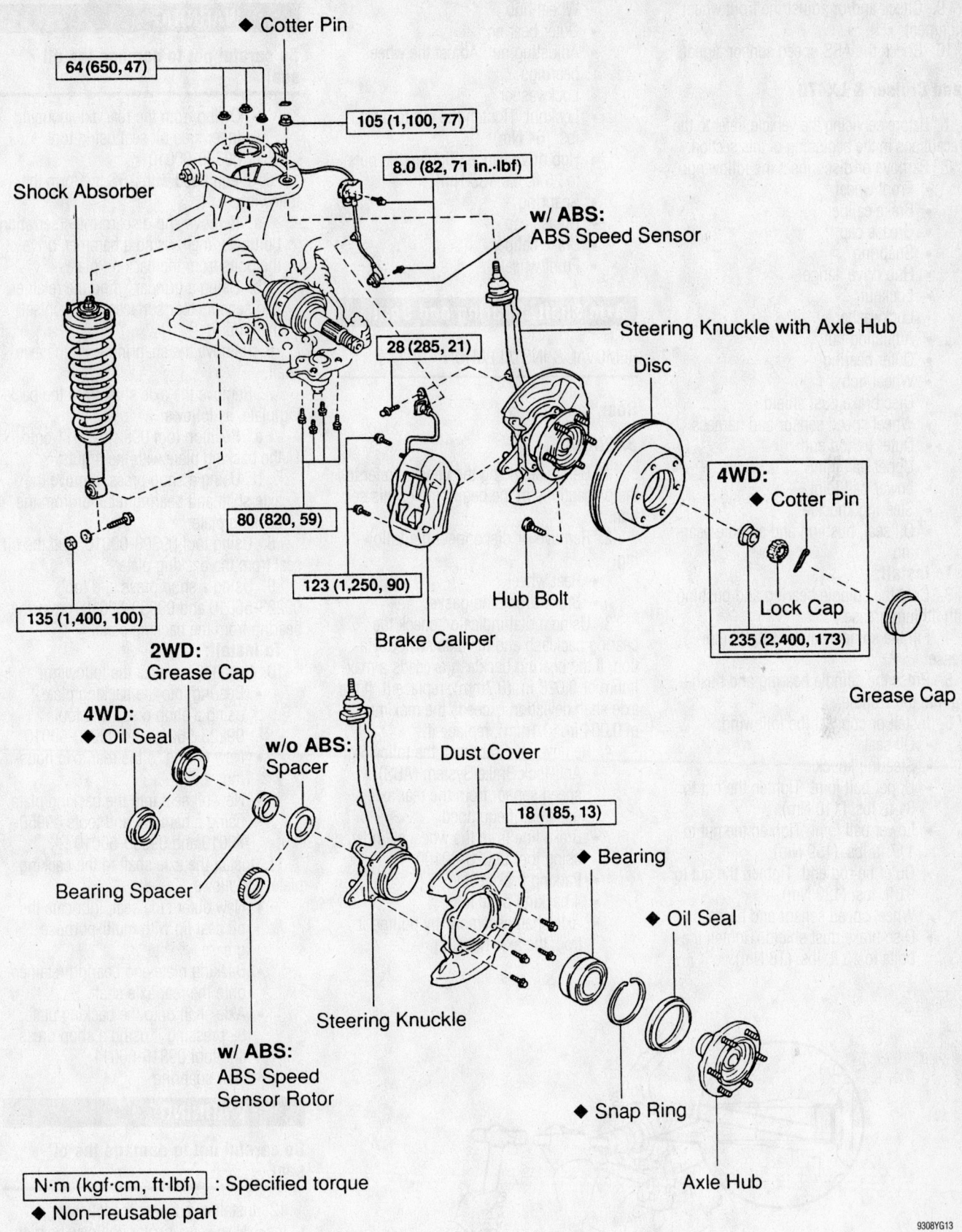

◆ Cotter Pin

64 (650, 47)

105 (1,100, 77)

8.0 (82, 71 in.·lbf)

Shock Absorber

w/ ABS:
ABS Speed Sensor

Steering Knuckle with Axle Hub

Disc

28 (285, 21)

4WD:
◆ Cotter Pin

80 (820, 59)

Lock Cap

123 (1,250, 90)

235 (2,400, 173)

Hub Bolt

135 (1,400, 100)

Brake Caliper

Grease Cap

2WD:
Grease Cap

4WD:
◆ Oil Seal

w/o ABS:
Spacer

Dust Cover

18 (185, 13)

◆ Bearing

Bearing Spacer

◆ Oil Seal

Steering Knuckle

w/ ABS:
ABS Speed
Sensor Rotor

◆ Snap Ring

Axle Hub

N·m (kgf·cm, ft·lbf) : Specified torque

◆ Non−reusable part

9308YG13

Exploded view of the front axle hub and related components—Sequoia

9. Check and/or adjust the front wheel alignment.

10. Check the ABS speed sensor signal.

Land Cruiser & LX470

1. Before servicing the vehicle, refer to the precautions in the beginning of this section.

2. Remove or disconnect the following:
   - Front wheel
   - Brake caliper
   - Grease cap
   - Snapring
   - Hub drive flange
   - Locknut
   - Lockwasher
   - Adjusting nut
   - Outer bearing
   - Wheel hub
   - Disc brake dust shield
   - Wheel speed sensor and harness
   - Outer tie rod end
   - Upper ball joint
   - Lower ball joint
   - Steering knuckle
   - Oil seal, bushing and spindle bearing

To install:

3. Coat the spindle bearing and bushing with lithium grease.

4. Fill the spindle cavity with lithium grease.

5. Press the spindle bearing and bushing into the spindle.

6. Install or connect the following:
   - Oil seal
   - Steering knuckle
   - Upper ball joint. Tighten the nut to 81 ft. lbs. (110 Nm).
   - Lower ball joint. Tighten the nut to 117 ft. lbs. (159 Nm).
   - Outer tie rod end. Tighten the nut to 91 ft. lbs. (122 Nm).
   - Wheel speed sensor and harness
   - Disc brake dust shield. Tighten the bolts to 13 ft. lbs. (18 Nm).

- Wheel hub
- Outer bearing
- Adjusting nut. Adjust the wheel bearings.
- Lockwasher
- Locknut. Tighten the nut to 47 ft. lbs. (64 Nm).
- Hub drive flange. Tighten the nuts to 26 ft. lbs. (35 Nm).
- Snapring
- Grease cap
- Brake caliper
- Front wheel

Axle Shaft, Bearing and Seal

REMOVAL & INSTALLATION

Rear

SEQUOIA

1. Before servicing the vehicle, refer to the precautions in the beginning of this section.

2. Remove or disconnect the following:
   - Rear wheel
   - Brake drum and gasket

3. Using a dial indicator, check the bearing backlash and the axle shaft deviation. If the bearing backlash exceeds a maximum or 0.028 in. (0.7mm), replace it. If the axle shaft deviation exceeds the maximum of 0.004 in. (0.1mm), replace it.

4. Remove or disconnect the following:
   - Anti-lock Brake System (ABS) speed sensor from the rear axle housing, if equipped
   - Brake line from the wheel cylinder, using tool 09023-00100
   - Parking brake cable
   - 4 backing plate nuts
   - Axle shaft assembly, by pulling it from the axle housing

Be careful not to damage the oil seal.

- O-ring from the rear axle housing
- Inner side oil seal using tool 09308-00010

5. If equipped with ABS, perform the following:

   a. Remove and discard the 4 serration bolt nuts; then, using a hammer, drive the bolts from the backing plate.

   b. Using a grinder, grind the retainer and sensor rotor surfaces; then, chisel them out.

6. Remove the snapring from the axle shaft.

7. Remove the axle shaft from the backing plate, as follows:

   a. Position tool 09521-25011 onto the backing plate with the 4 nuts.

   b. Using a shop press, remove the axle shaft and bearing retainer from the backing plate.

8. Using tool 09308-00010, pull the oil seal from the backing plate.

9. Using a shop press and tools 09223-56010 and 09950-60010, press the bearing from the backing plate.

To install:

10. Install or connect the following:
   - Bearing into the backing plate, using a shop press and tools 09223-56010 and 09950-60010
   - New O-ring to the rear axle housing
   - New oil seal into the backing plate, using a hammer and tools 09950-70010 and 09950-60010

11. Install the axle shaft to the backing plate, as follows:
   - New outer side seal, lubricate the oil seal lip with multi-purpose grease
   - Backing plate and bearing retainer onto the rear axle shaft
   - Axle shaft onto the backing plate, by pressing it using a shop press and tool 09316-60011
   - New snapring

Be careful not to damage the oil seal.

12. Install or connect the following:
   - New sensor rotor and new bearing retainer onto the axle shaft, using a shop press and tool 09316-60011 to a standard length of 4.77—4.85

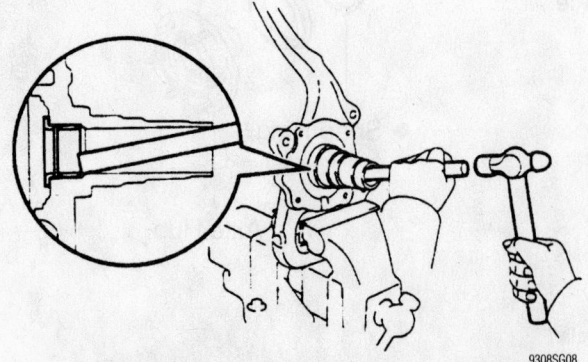

9308SG08

Removing the oil seal, bushing and spindle bearing—Land Cruiser/LX470

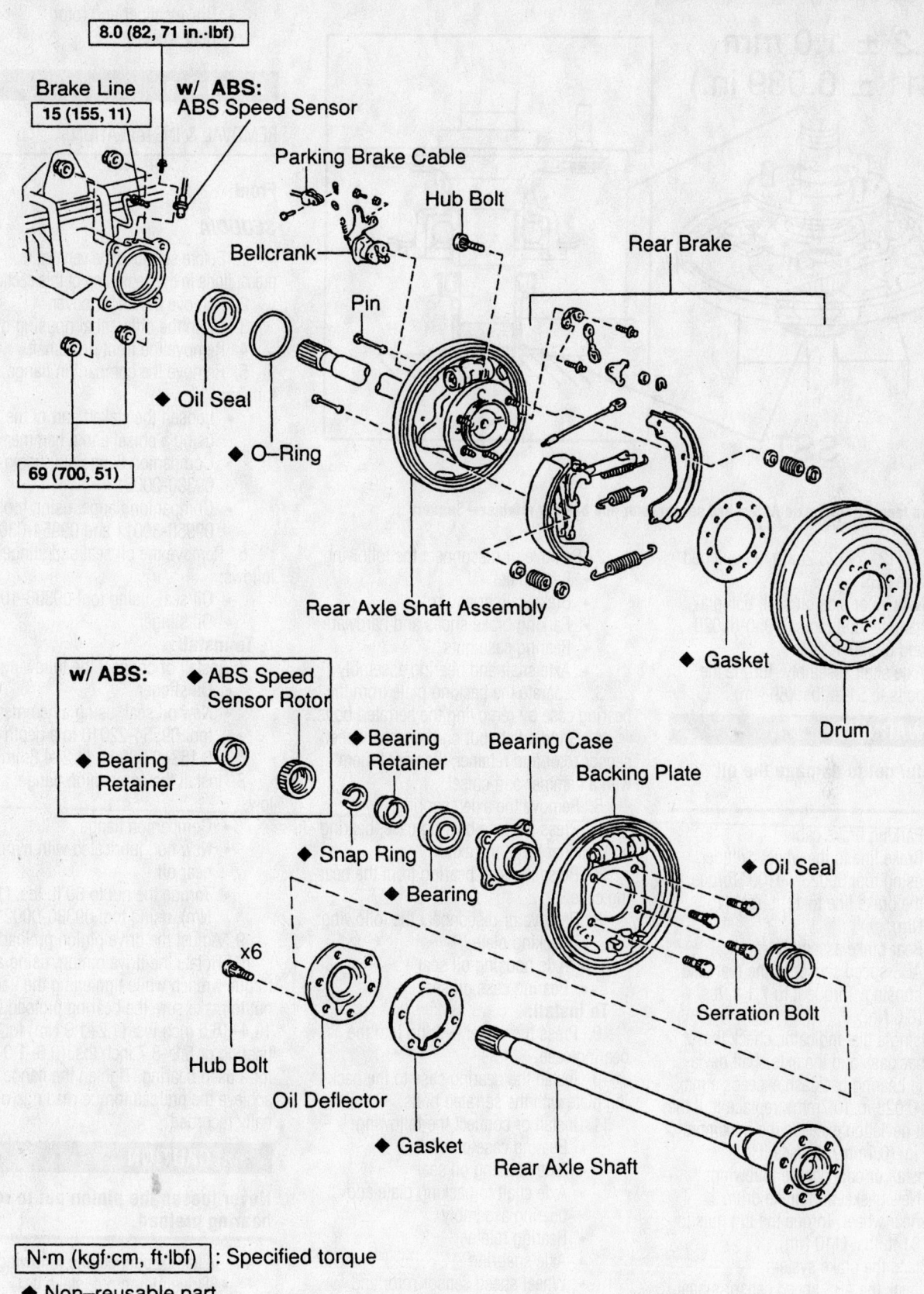

8.0 (82, 71 in.·lbf)

Brake Line
15 (155, 11)

w/ ABS:
ABS Speed Sensor

Parking Brake Cable

Hub Bolt

Rear Brake

Bellcrank

Pin

◆ Oil Seal

◆ O–Ring

69 (700, 51)

Rear Axle Shaft Assembly

◆ Gasket

Drum

w/ ABS: ◆ ABS Speed Sensor Rotor

◆ Bearing Retainer

◆ Bearing Retainer

Bearing Case

Backing Plate

◆ Snap Ring

◆ Bearing

◆ Oil Seal

Serration Bolt

x6

Hub Bolt

Oil Deflector

◆ Gasket

Rear Axle Shaft

N·m (kgf·cm, ft·lbf) : Specified torque

◆ Non–reusable part

Exploded view of the rear axle—Sequoia

9308YG14

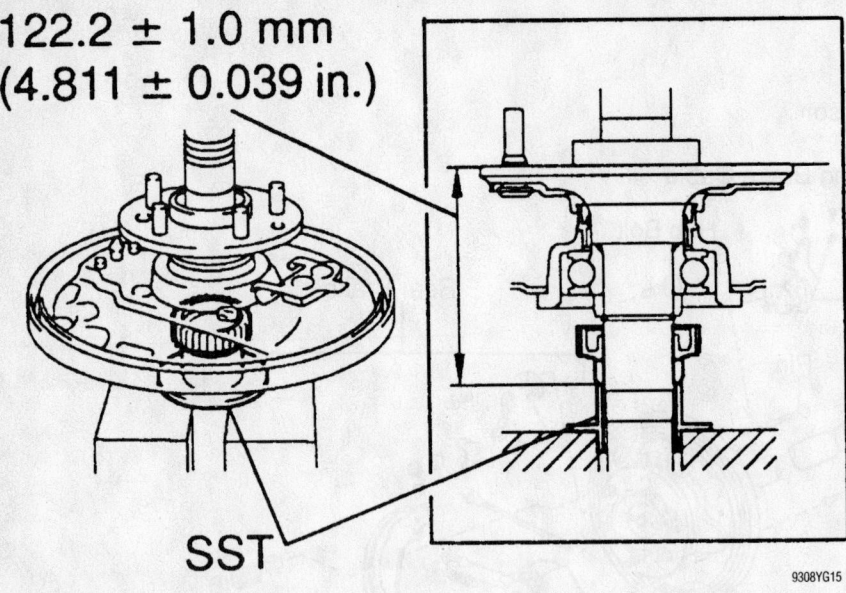

**122.2 ± 1.0 mm
(4.811 ± 0.039 in.)**

SST

9308YG15

Standard length of rear axle ABS speed sensor rotor and bearing retainer—Sequoia

in. (121.2–123.2mm), if equipped with ABS
- New inner side oil seal, using a hammer and tools 09950-60020 and 09950-70010
- Axle shaft assembly. Torque the bolts to 51 ft. lbs. (69 Nm).

❋ WARNING

Be careful not to damage the oil seal.

- Parking brake cable
- Brake line to the wheel cylinder, using tool 09023-00100. Torque the brake line to 11 ft. lbs. (15 Nm).
- Rear brake assembly
- ABS speed sensor to the rear axle housing. Torque it to 7.1 ft. lbs. (8.0 Nm).

13. Using a dial indicator, check the bearing backlash and the axle shaft deviation. If the bearing backlash exceeds a maximum or 0.028 in. (0.7mm), replace it. If the axle shaft deviation exceeds the maximum of 0.004 in. (0.1mm), replace it.

14. Install or connect the following:
- New gasket and brake drum
- Rear wheel. Torque the lug nuts to 81 ft. lbs. (110 Nm).

15. Bleed the brake system.
16. Check the ABS speed sensor signal.

LAND CRUISER & LX470

1. Before servicing the vehicle, refer to the precautions in the beginning of this section.

2. Remove or disconnect the following:
- Rear wheel
- Brake caliper and rotor
- Parking brake shoes and hardware
- Bearing case nuts
- Axle shaft and bearing assembly

3. Separate the backing plate from the bearing case by removing the serrated bolts.

4. Grind a flat spot on the wheel speed sensor rotor and retainer, then split them with a hammer and chisel.

5. Remove the axle snapring.

6. Press the axle bearing case, bearing and retainer off of the axle.

7. Press the axle bearing from the bearing case.

8. Remove or disconnect the following:
- Backing plate
- Axle housing oil seal
- Bearing case oil seal

To install:

9. Press the wheel bearing into the bearing case.

10. Install the bearing case to the backing plate with the serrated bolts.

11. Install or connect the following:
- Bearing case oil seal
- Axle housing oil seal
- Axle shaft to backing plate and bearing assembly
- Bearing retainer
- Axle snapring
- Wheel speed sensor rotor and retainer
- Axle shaft and bearing assembly to the axle housing. Tighten the nuts to 91 ft. lbs. (123 Nm).
- Parking brake shoes and hardware

- Brake caliper and rotor
- Rear wheel

Pinion Seal

REMOVAL & INSTALLATION

Front

SEQUOIA

1. Before servicing the vehicle, refer to the precautions in the beginning of this section.
2. Remove the under cover.
3. Drain the differential housing oil.
4. Remove the front driveshaft.
5. Remove the companion flange, as follows:
- Loosen the staked part of the nut, using a chisel and a hammer
- Companion flange nut, using tool 09330-00021
- Companion flange, using tools 09950-30011 and 09954-03010

6. Remove the oil seal and slinger, as follows:
- Oil seal, using tool 09308-10010
- Oil slinger

To install:

7. Install or connect the following:
- Oil slinger
- New oil seal, using a hammer and tool 09554-22010 to a depth of 0.153–0.189 in. (4.2–4.8mm).

8. Install the companion flange, as follows:
- Companion flange
- New nut, lubricated with hypoid gear oil
- Torque the nut to 80 ft. lbs. (108 Nm), using tool 09330-00021

9. Adjust the drive pinion preload
10. Rotate the drive pinion, using a torque wrench while tightening the flange nut to make sure the bearing preload is 10.4–16.5 inch lbs. (1.2–1.9 Nm) for a new bearing or 5.2–8.7 inch lbs. (0.6–1.0 Nm) for a used bearing. Tighten the flange nut to achieve the preload torque readings originally recorded.

❋ CAUTION

Never loosen the pinion nut to reduce bearing preload.

11. Install or connect the following:
- Drive pinion nut, stake it
- Front driveshaft. Tighten the fasteners to 54 ft. lbs. (74 Nm).
- Under cover

12. Fill the differential with gear lubricant and check for leaks.

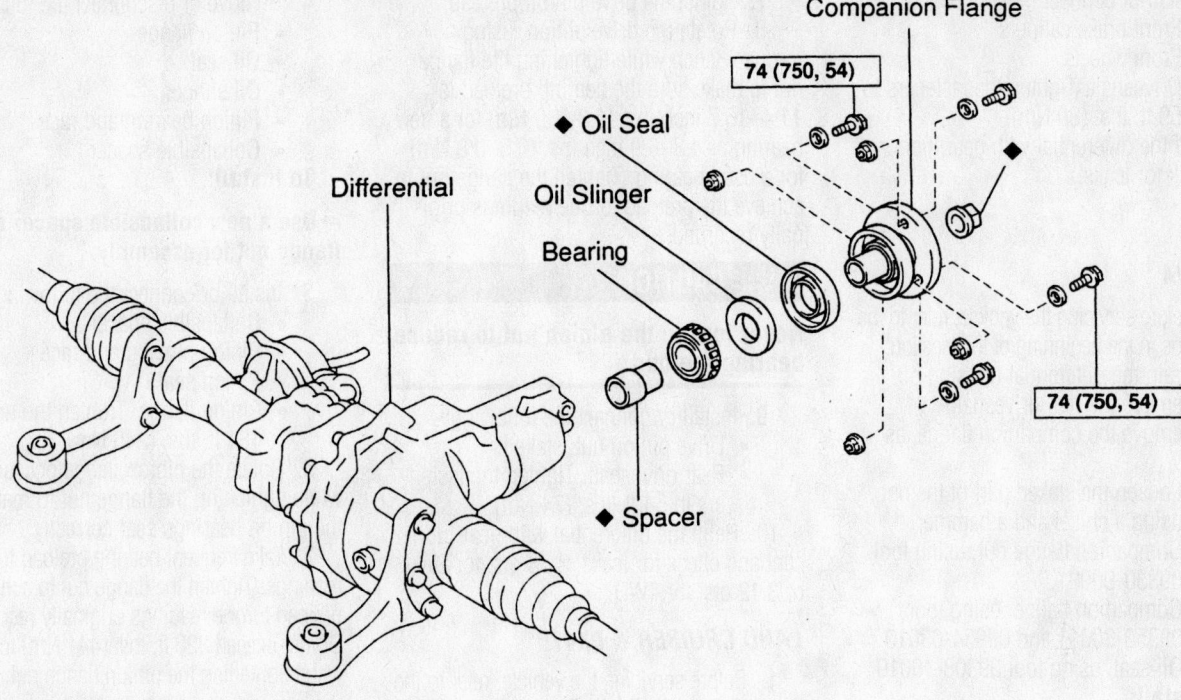

N·m (kgf·cm, ft·lbf) : Specified torque
◆ Non–reusable part

Exploded view of the Sequoia front differential assembly—Rear differential assembly is similar

LAND CRUISER & LX470

1. Before servicing the vehicle, refer to the precautions in the beginning of this section.

2. Remove or disconnect the following:
   • Driveshaft
   • Front wheels
   • Front brake calipers

➡The front brake calipers must be removed so that there is no additional drag when measuring pinion bearing preload.

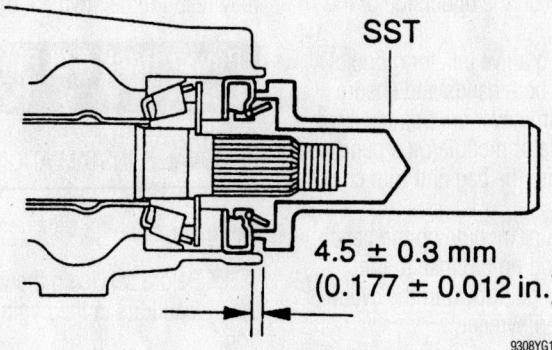

Positioning the Sequoia front pinion seal in the differential housing—Rear differential assembly is similar

3. Use an inch lb. torque wrench and measure the amount of torque required to maintain pinion rotation through several revolutions.

4. Remove or disconnect the following:
   • Pinion flange
   • Oil seal
   • Oil slinger
   • Pinion bearing and race
   • Oil storage ring
   • Collapsible spacer

To install:

➡Use a new collapsible spacer and flange nut for assembly.

5. Install or connect the following:
   • Collapsible spacer
   • Oil storage ring
   • Pinion bearing and race
   • Pinion seal
   • Pinion flange. Tighten the nut to 80 ft. lbs. (108 Nm).

6. Rotate the pinion flange occasionally while tightening the flange nut to make sure the pinion bearings seat correctly.

7. Take frequent bearing preload torque readings. Tighten the flange nut to achieve the preload torque readings originally recorded. Do not exceed 249 ft. lbs. (338 Nm) torque when tightening the pinion flange nut.

❊❊ CAUTION

Never loosen the pinion nut to reduce bearing preload. If it is necessary to reduce bearing preload, install a new collapsible spacer and pinion nut.

Timing belt service is covered in Section 3 of this manual

8. Install or connect the following:
- Front brake calipers
- Front wheels
- Driveshaft. Tighten the fasteners to 59 ft. lbs. (80 Nm).

9. Fill the differential with gear lubricant and check for leaks.

Rear

SEQUOIA

1. Before servicing the vehicle, refer to the precautions in the beginning of this section.

2. Drain the differential housing oil.

3. Remove the rear driveshaft.

4. Remove the companion flange, as follows:
- Loosen the staked part of the nut, using a chisel and a hammer
- Companion flange nut, using tool 09330-00021
- Companion flange, using tools 09950-30011 and 09954-03010
- Oil seal, using tool 09308-10010

To install:

5. Install the new oil seal until it is flush with the housing, using a plastic hammer and tools 09316-12010 and 09649-17010

➡**Use vinyl tape to connect both oil seal installation tools.**

6. Install the companion flange, as follows:
- Companion flange
- New nut, lubricated with hypoid gear oil
- Torque the nut to 109 ft. lbs. (147 Nm), using tool 09330-00021.

7. Adjust the drive pinion preload

8. Rotate the drive pinion, using a torque wrench while tightening the flange nut to make sure the bearing preload is 11.4–16.7 inch lbs. (1.3–1.9 Nm) for a new bearing or 4.3–6.9 inch lbs. (0.5–0.8 Nm) for a used bearing. Tighten the flange nut to achieve the preload torque readings originally recorded.

✳✳ CAUTION

Never loosen the pinion nut to reduce bearing preload.

9. Install or connect the following:
- Drive pinion nut, stake it
- Rear driveshaft. Tighten the fasteners to 54 ft. lbs. (74 Nm).

10. Refill the differential with gear lubricant and check for leaks; 3.33 qts. for 2WD or 3.12 qts. for 4WD.

LAND CRUISER & LX470

1. Before servicing the vehicle, refer to the precautions in the beginning of this section.

2. Remove or disconnect the following:
- Driveshaft
- Rear wheels
- Rear brake calipers

➡**The rear brake calipers must be removed so that there is no additional drag when measuring pinion bearing preload.**

3. Use an inch lb. torque wrench and measure the amount of torque required to maintain pinion rotation through several revolutions.

4. Remove or disconnect the following:
- Pinion flange
- Oil seal
- Oil slinger
- Pinion bearing and race
- Collapsible spacer

To install:

➡**Use a new collapsible spacer and flange nut for assembly.**

5. Install or connect the following:
- Collapsible spacer
- Pinion bearing and race
- Pinion seal
- Pinion flange. Tighten the nut to 181 ft. lbs. (245 Nm).

6. Rotate the pinion flange occasionally while tightening the flange nut to make sure the pinion bearings seat correctly.

7. Take frequent bearing preload torque readings. Tighten the flange nut to achieve the preload torque readings originally recorded. Do not exceed 326 ft. lbs. (441 Nm) torque when tightening the pinion flange nut.

✳✳ CAUTION

Never loosen the pinion nut to reduce bearing preload. If it is necessary to reduce bearing preload, install a new collapsible spacer and pinion nut.

8. Install or connect the following:
- Rear brake calipers
- Rear wheels
- Driveshaft. Tighten the fasteners to 78 ft. lbs. (106 Nm).

9. Fill the differential with gear lubricant and check for leaks.

STEERING AND SUSPENSION

Air Bag

✳✳ CAUTION

Some vehicles are equipped with an air bag system. The system must be disarmed before performing service on, or around, system components, the steering column, instrument panel components, wiring and sensors. Failure to follow the safety precautions and the disarming procedure could result in accidental air bag deployment, possible injury and unnecessary system repairs.

PRECAUTIONS

Several precautions must be observed when handling the inflator module to avoid accidental deployment and possible personal injury.

- Never carry the inflator module by the wires or connector on the underside of the module.
- When carrying a live inflator module, hold securely with both hands and ensure that the bag and trim cover are pointed away.
- Place the inflator module on a bench or other surface with the bag and trim cover facing up.
- With the inflator module on the bench, never place anything on or close to the module which may be thrown in the event of an accidental deployment.

DISARMING

To avoid personal injury when working on vehicles equipped with an air bag, the negative battery cable must be disconnected and at least 90 seconds must elapse before working on the system. Failure to do so may result in deployment of the air bag.

Power Rack And Pinion Steering Gear

REMOVAL & INSTALLATION

Sequoia

1. Before servicing the vehicle, refer to the precautions in the beginning of this section.

2. Position the front wheels in the straight-ahead position.

3. Remove or disconnect the following:
- Engine under cover
- Steering wheel pad

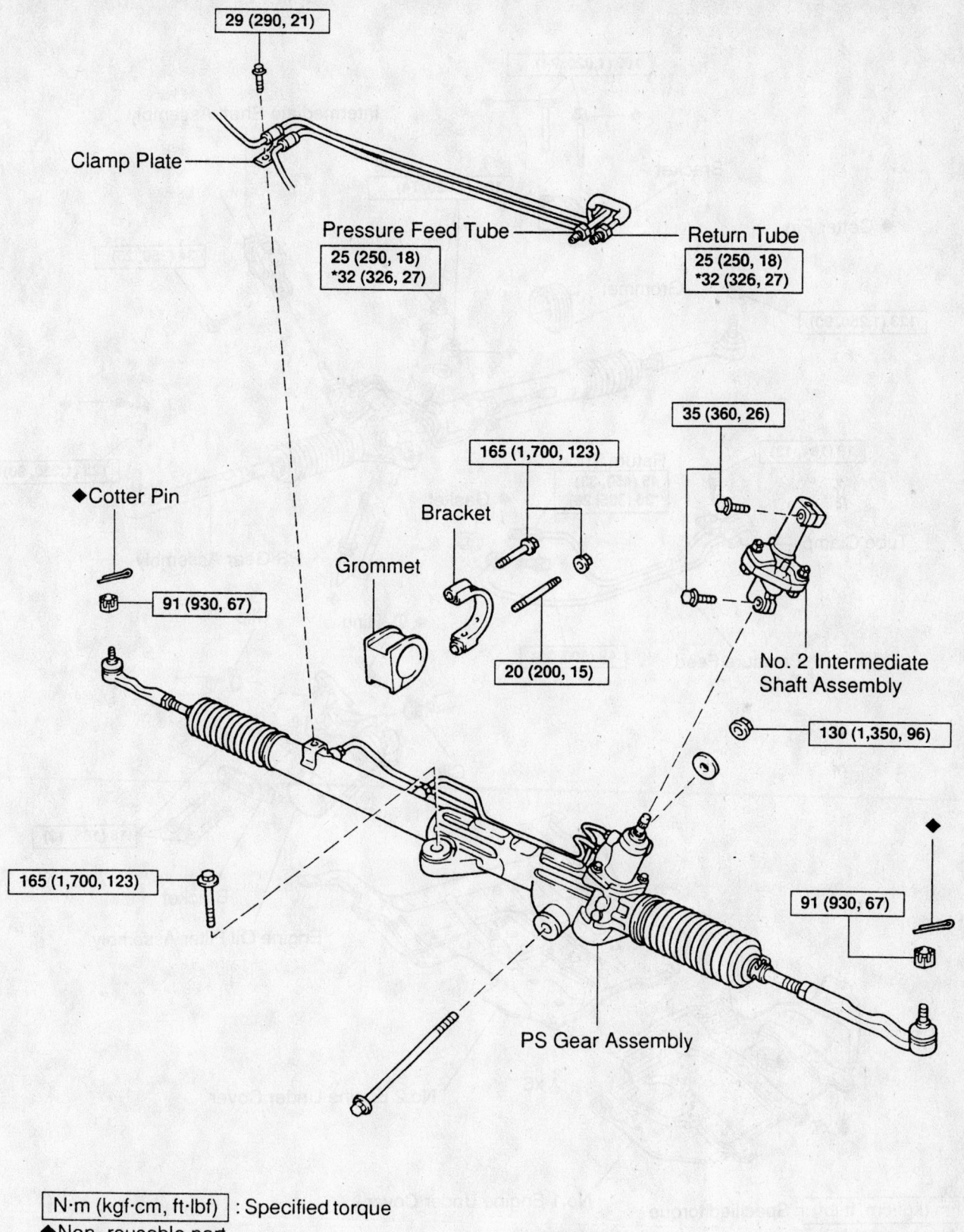

29 (290, 21)

Clamp Plate

Pressure Feed Tube
25 (250, 18)
*32 (326, 27)

Return Tube
25 (250, 18)
*32 (326, 27)

35 (360, 26)

165 (1,700, 123)

◆Cotter Pin

Bracket

Grommet

91 (930, 67)

No. 2 Intermediate
Shaft Assembly

20 (200, 15)

130 (1,350, 96)

165 (1,700, 123)

91 (930, 67)

PS Gear Assembly

N·m (kgf·cm, ft·lbf) : Specified torque
◆Non–reusable part
* For use with SST

9308YG18

Exploded view of the power rack and pinion steering gear mounting—Sequoia

Heater Core replacement is covered in Section 2 of this manual

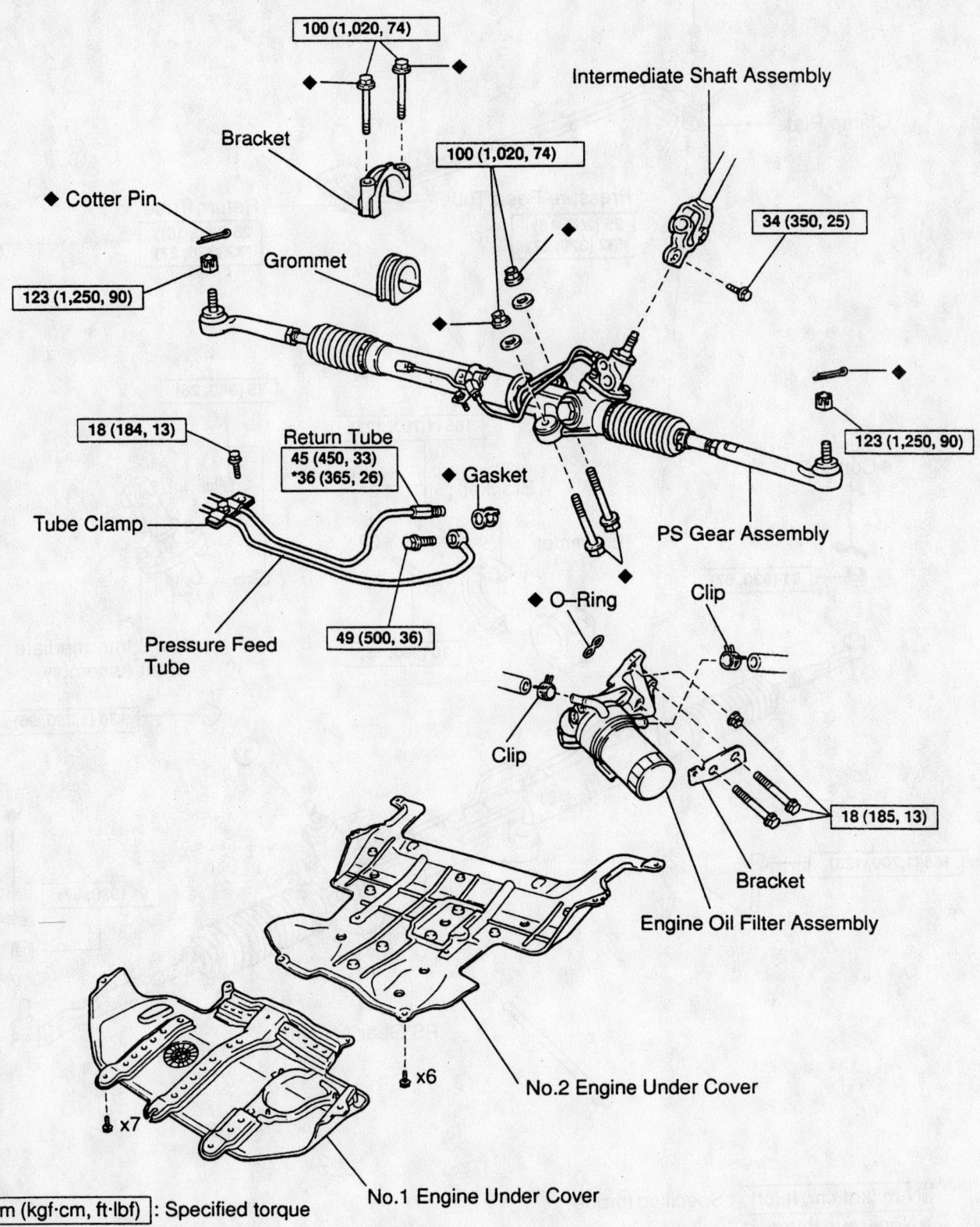

100 (1,020, 74)

Intermediate Shaft Assembly

Bracket

100 (1,020, 74)

34 (350, 25)

◆ Cotter Pin

Grommet

123 (1,250, 90)

123 (1,250, 90)

18 (184, 13)

Return Tube
45 (450, 33)
*36 (365, 26)

◆ Gasket

Tube Clamp

49 (500, 36)

PS Gear Assembly

Pressure Feed Tube

◆ O–Ring

Clip

Clip

18 (185, 13)

Bracket

Engine Oil Filter Assembly

x6

No.2 Engine Under Cover

x7

No.1 Engine Under Cover

N·m (kgf·cm, ft·lbf) : Specified torque
◆ Non–reusable part
* For use with SST

7924SG90

Exploded view of the rack and pinion steering gear mounting—Land Cruiser/LX470

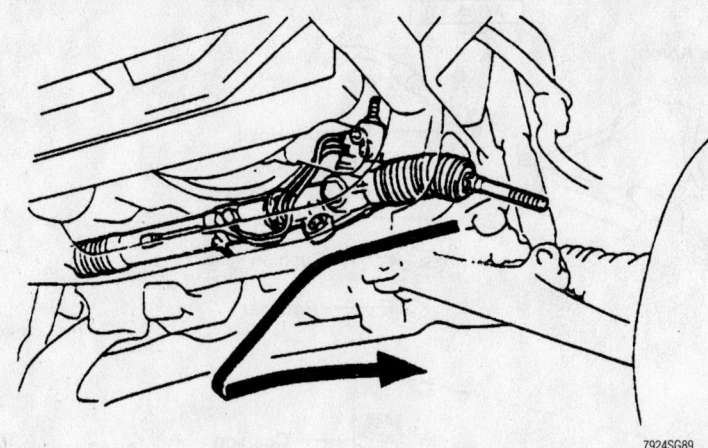

Power rack and pinion steering gear removal—Land Cruiser/LX470

- Steering wheel
- Left and right outer tie-rod ends from the steering knuckles

4. Matchmark the No. 2 intermediate shaft to the steering gear input shaft.

5. Remove or disconnect the following:
- Clamp plate
- Pressure feed and return tubes from the power steering gear, using tool 09631-22020
- Power steering gear assembly

To install:

6. Install or connect the following:
- Power steering gear assembly. Torque the set bolt to 123 ft. lbs. (165 Nm) and the set nut/bolt to 96 ft. lbs. (91 Nm).
- Pressure feed and return tubes to the power steering gear. Torque them to 27 ft. lbs. (32 Nm), using tool 09631-22020.
- Clamp plate. Torque the bolt to 21 ft. lbs. (29 Nm).
- No. 2 intermediate shaft to the steering gear input shaft
- Left and right outer tie-rod ends to the steering knuckles. Torque the nuts to 67 ft. lbs. (91 Nm).
- Steering wheel. Torque the nut to 26 ft. lbs. (35 Nm).
- Steering wheel pad
- Engine under cover

7. Fill and bleed the power steering system.

8. Check and/or adjust the wheel alignment, as necessary.

Land Cruiser & LX470

1. Before servicing the vehicle, refer to the precautions in the beginning of this section.

2. Matchmark the intermediate shaft to the steering gear input shaft.

3. Remove or disconnect the following:
- Negative battery cable
- Engine under covers
- Outer tie rod ends
- Engine oil filter adapter
- Intermediate steering shaft
- Power steering hoses and bracket
- Power steering gear

To install:

4. Install or connect the following:
- Power steering gear. Tighten the fasteners to 74 ft. lbs. (100 Nm).
- Power steering hoses and bracket
- Intermediate steering shaft. Tighten the bolts to 25 ft. lbs. (34 Nm).
- Engine oil filter adapter. Tighten the bolts to 13 ft. lbs. (18 Nm).
- Outer tie rod ends. Tighten the nuts to 90 ft. lbs. (122 Nm).
- Engine under covers
- Negative battery cable

5. Fill the power steering fluid reservoir.

6. Check the wheel alignment and adjust as necessary.

Shock Absorber

REMOVAL & INSTALLATION

Land Cruiser & LX470 Without Active Height Control

FRONT

1. Before servicing the vehicle, refer to the precautions in the beginning of this section.

2. Support the axle with a jackstand.

3. Remove or disconnect the following:
- Front wheel
- Shock absorber

To install:

4. Install or connect the following:
- Shock absorber. Tighten the nut to 51 ft. lbs. (69 Nm) and the bolt to 100 ft. lbs. (135 Nm).
- Front wheel

REAR

1. Before servicing the vehicle, refer to the precautions in the beginning of this section.

2. Support the axle with a jackstand.

3. Remove or disconnect the following:
- Rear wheel
- Shock absorber

To install:

4. Install or connect the following:
- Shock absorber. Tighten the nut to 51 ft. lbs. (69 Nm) and the bolt to 72 ft. lbs. (98 Nm).
- Rear wheel

Land Cruiser & LX470 With Active Height Control

FRONT

⁕⁕ CAUTION

The vehicle ride height may change suddenly when relieving system pressure.

1. Before servicing the vehicle, refer to the precautions in the beginning of this section.

2. Relieve the Active Height Control (AHC) hydraulic pressure as follows:
   a. Connect a hose to the control actuator bleed screw and place the other end in a container.
   b. Open the bleed screw.
   c. When the fluid pressure has dropped and oil stops flowing, close the bleed screw.

3. Remove or disconnect the following:
- Front wheel
- Inner fender liner
- Lower shock absorber mounting bolt
- AHC pressure hose
- Upper shock absorber mounting nut
- Shock absorber

To install:

4. Install or connect the following:
- Shock absorber. Tighten the upper nut to 51 ft. lbs. (68 Nm) and the lower bolt to 101 ft. lbs. (135 Nm).
- AHC pressure hose with new O-ring seals. Tighten the bolts to 13 ft. lbs. (18 Nm).

Brake service is covered in Section 4 of this manual

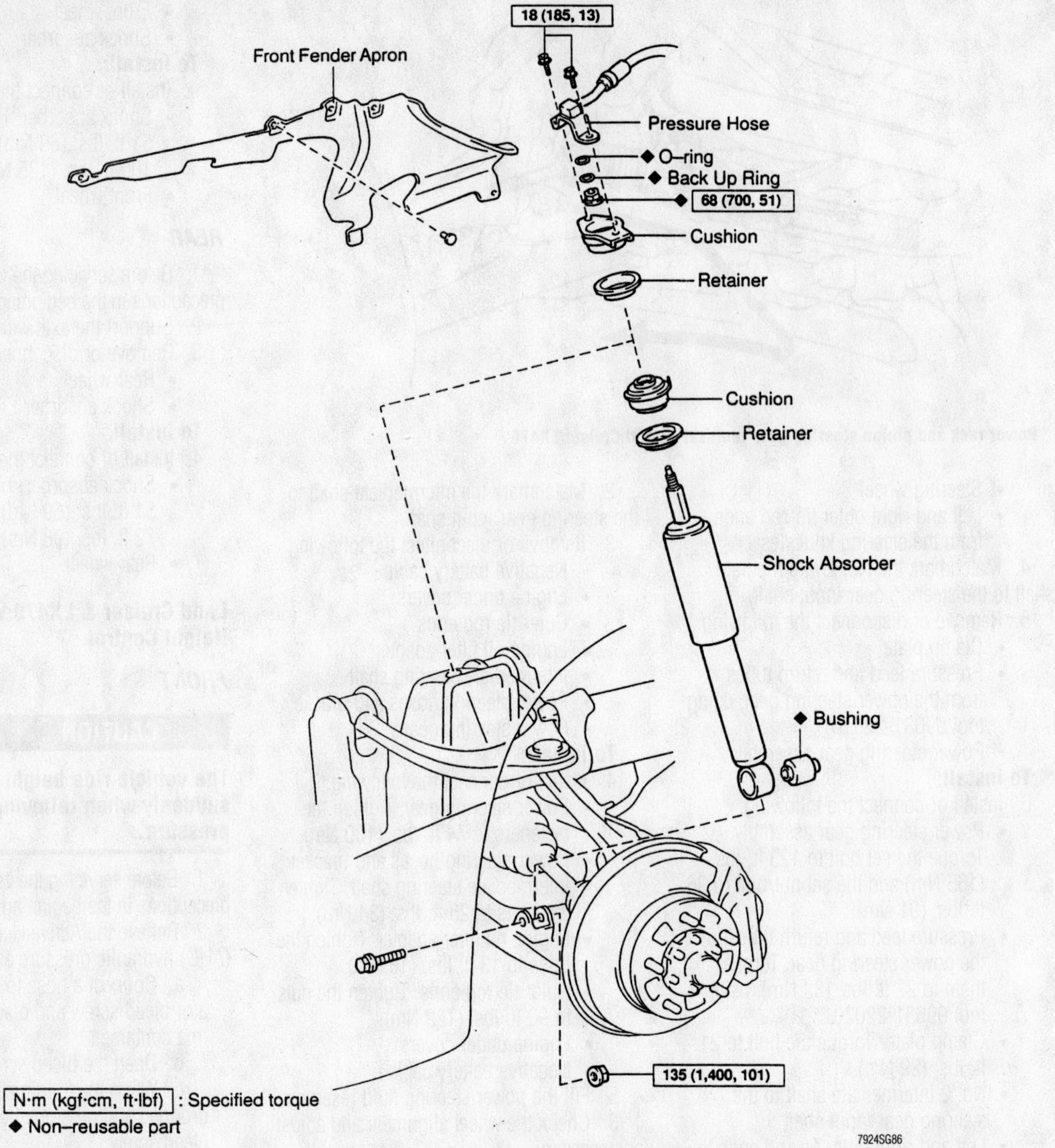
18 (185, 13)
Front Fender Apron
Pressure Hose
◆ O–ring
◆ Back Up Ring
◆ 68 (700, 51)
Cushion
Retainer
Cushion
Retainer
Shock Absorber
◆ Bushing
135 (1,400, 101)
N·m (kgf·cm, ft·lbf) : Specified torque
◆ Non–reusable part
7924SG86

Exploded view of the front shock absorber mounting—Land Cruiser/LX470 models with Active Height Control (AHC)

- Inner fender liner
- Front wheel

➡Do not let the AHC reservoir run empty during this procedure.

5. Bleed the AHC system as follows:
   a. Fill the AHC system reservoir with AHC fluid 08886-01805.
   b. Start the engine and push **N** on the vehicle height select switch.
   c. When the AHC pump stops, turn the engine **OFF**.
   d. Open the bleed screw and allow any air in the system to escape.
   e. Repeat until no air is expelled from the bleed screw.
   f. Fill the AHC reservoir to the correct level.

REAR

⁂ CAUTION

The vehicle ride height may change suddenly when relieving system pressure.

1. Before servicing the vehicle, refer to the precautions in the beginning of this section.
2. Support the rear axle with a jack or stands.
3. Relieve the Active Height Control (AHC) hydraulic pressure as follows:
   a. Connect a hose to the control actuator bleed screw and place the other end in a container.
   b. Open the bleed screw.
   c. When the fluid pressure has dropped and oil stops flowing, close the bleed screw.
4. Remove or disconnect the following:
   - Rear wheel
   - Lower shock absorber mounting bolt

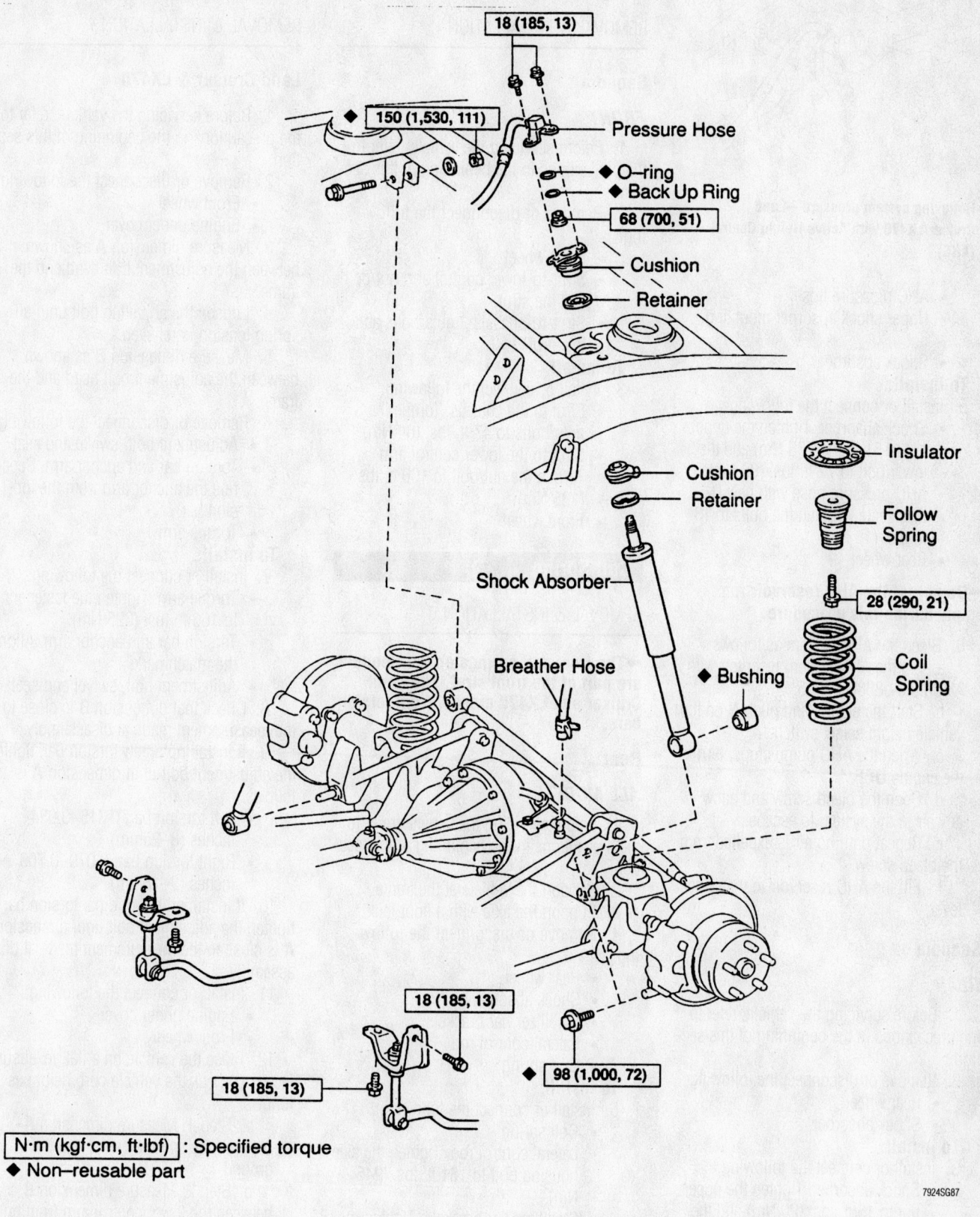

18 (185, 13)

150 (1,530, 111)

Pressure Hose

◆ O–ring

◆ Back Up Ring

68 (700, 51)

Cushion

Retainer

Insulator

Cushion

Retainer

Follow Spring

Shock Absorber

28 (290, 21)

Breather Hose

◆ Bushing

Coil Spring

18 (185, 13)

18 (185, 13)

98 (1,000, 72)

N·m (kgf·cm, ft·lbf) : Specified torque
◆ Non–reusable part

7924SG87

Exploded view of the rear shock absorber mounting—Land Cruiser/LX470 with Active Height Control (AHC)

For complete Engine Mechanical specifications, see Section 1 of this manual

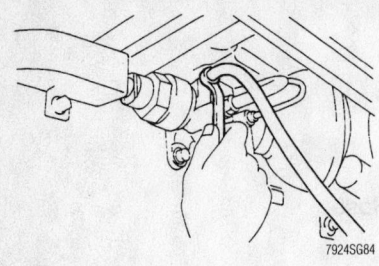

7924SG84

Relieving system pressure—Land Cruiser/LX470 with Active Height Control (AHC)

- AHC pressure hose
- Upper shock absorber mounting nut
- Shock absorber

To install:

5. Install or connect the following:
- Shock absorber. Tighten the upper nut to 51 ft. lbs. (68 Nm) and the lower bolt to 72 ft. lbs. (98 Nm).
- AHC pressure hose with new O-ring seals. Tighten the bolts to 13 ft. lbs. (18 Nm).
- Rear wheel

➡ **Do not let the AHC reservoir run empty during this procedure.**

6. Bleed the AHC system as follows:
   a. Fill the AHC system reservoir with AHC fluid 08886-01805.
   b. Start the engine and push **N** on the vehicle height select switch.
   c. When the AHC pump stops, turn the engine **OFF**.
   d. Open the bleed screw and allow any air in the system to escape.
   e. Repeat until no air is expelled from the bleed screw.
   f. Fill the AHC reservoir to the correct level.

Sequoia

REAR

1. Before servicing the vehicle, refer to the precautions in the beginning of this section.
2. Remove or disconnect the following:
- Rear wheel
- Shock absorber

To install:

3. Install or connect the following:
- Shock absorber. Tighten the upper nut to 15 ft. lbs. (20 Nm) and the lower nut/bolt to 64 ft. lbs. (87 Nm).
- Rear wheel. Torque the lug nuts to 81 ft. lbs. (110 Nm).

Strut

REMOVAL & INSTALLATION

Sequoia

FRONT

1. Before servicing the vehicle, refer to the precautions in the beginning of this section.
2. Remove or disconnect the following:
- Front wheel
- Strut-to-lower control arm nut/bolt and the strut
- Strut-to-chassis 3 nuts/bolts and the strut

To install:

3. Install or connect the following:
- Strut to the chassis. Torque the 3 nuts/bolts to 47 ft. lbs. (64 Nm).
- Strut to the lower control arm. Torque the nut/bolt to 100 ft. lbs. (135 Nm).
- Front wheel

Coil Spring

REMOVAL & INSTALLATION

➡ **The front coil springs on the Sequoia are part of the front strut. The Land Cruiser and LX470 employ front torsion bars.**

Rear

ALL MODELS

1. Before servicing the vehicle, refer to the precautions in the beginning of this section.
2. Support the vehicle at the frame.
3. Support the axle with a floor jack.
4. Remove or disconnect the following:
- Rear wheel
- Shock absorber
- Stabilizer bar brackets
- Lateral control rod
- Coil spring

To install:

5. Install or connect the following:
- Coil spring
- Lateral control rod. Tighten the axle housing bolt to 181 ft. lbs. (245 Nm).
- Stabilizer bar brackets. Tighten the bolts to 13 ft. lbs. (18 Nm)
- Shock absorber
- Rear wheel

Torsion Bars

REMOVAL & INSTALLATION

Land Cruiser & LX470

1. Before servicing the vehicle, refer to the precautions in the beginning of this section.
2. Remove or disconnect the following:
- Front wheel
- Engine under cover
3. Measure dimension **A** as shown between the adjustment bolt head and the frame.
4. Loosen the adjusting bolt until all spring tension is relieved.
5. Measure dimension **B** as shown between the adjustment bolt head and the frame.
6. Remove or disconnect the following:
- Adjustment bolt, swivel and seat
- Torsion bar and anchor arm. Separate the anchor arm from the torsion bar.
- Torque arm

To install:

7. Install or connect the following:
- Torque arm. Tighten the fasteners to 166 ft. lbs. (225 Nm).
- Torsion bar and anchor arm. Align the matchmarks.
- Adjustment bolt, swivel and seat
8. Check that dimension **B** is close to the measurement made at disassembly.
9. If installing a new torsion bar, tighten the adjustment bolt until dimension **A** is as follows:
- Left torsion bar: 0.315–0.984 inches (8–25mm)
- Right torsion bar: 0.079–0.709 inches (2–18mm)
10. If installing the original torsion bar, tighten the adjustment bolt until dimension **A** is close to the measurement made at disassembly.
11. Install or connect the following:
- Engine under cover
- Front wheel
12. Place the vehicle on a flat, level surface and check the vehicle curb height as follows:
   a. Step 1: Measure dimension **A** between the spindle center and the ground.
   b. Step 2: Measure dimension **B** between the lower control arm front bolt center and the ground.
   c. Step 3: Turn the adjusting bolt so that **A** minus **B** is equal to 2.795 inches (71mm).

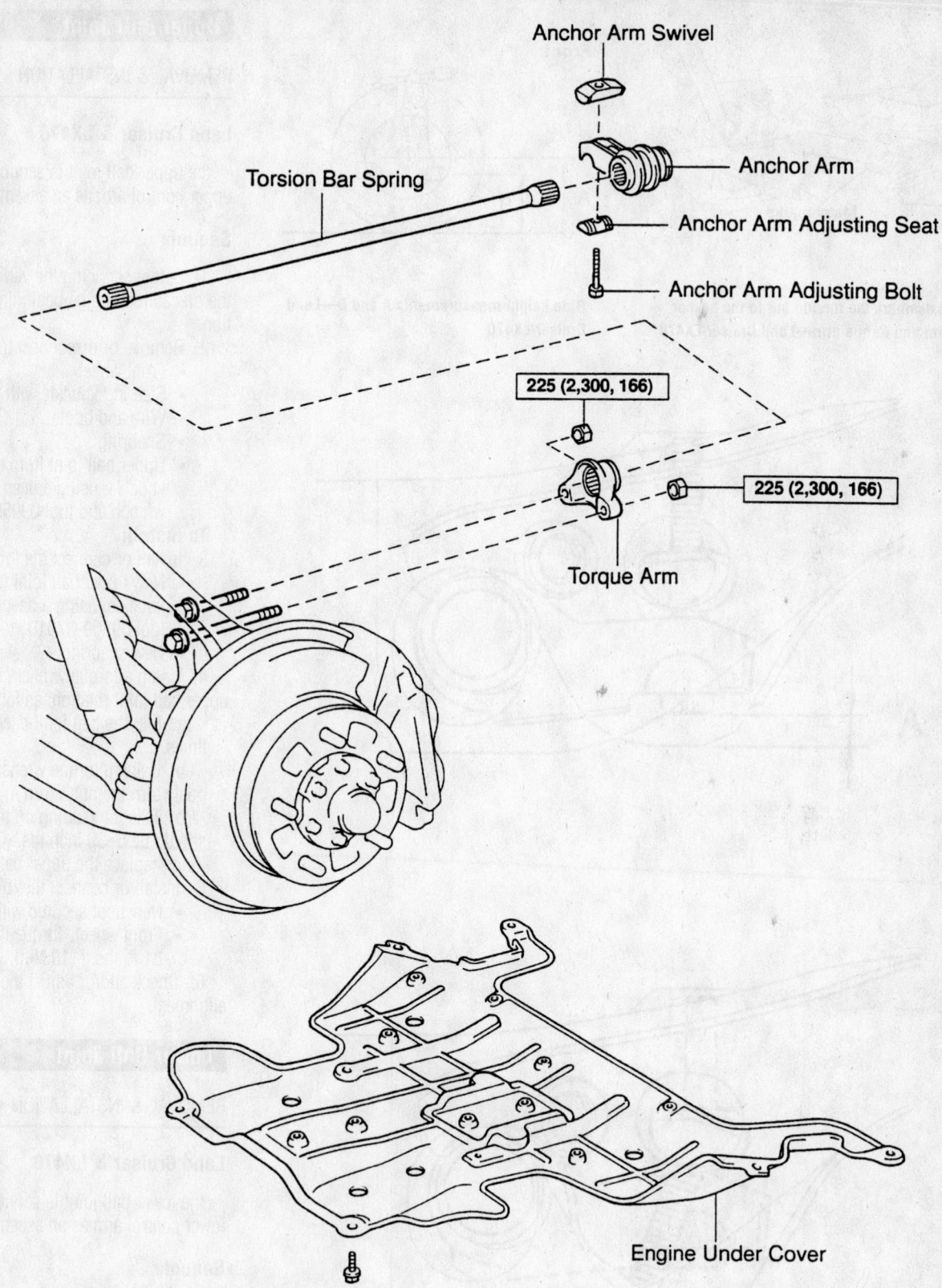

Anchor Arm Swivel

Torsion Bar Spring

Anchor Arm

Anchor Arm Adjusting Seat

Anchor Arm Adjusting Bolt

225 (2,300, 166)

225 (2,300, 166)

Torque Arm

Engine Under Cover

9308SG10

N·m (kgf·cm, ft·lbf) : Specified torque

Torsion bar mounting exploded view—Land Cruiser/LX470

For Accessory Drive Belt illustrations, see Section 1 of this manual

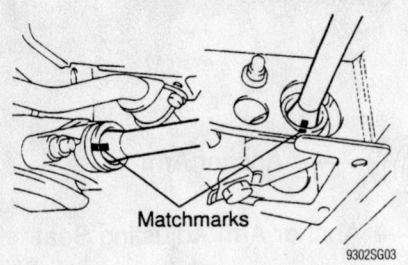

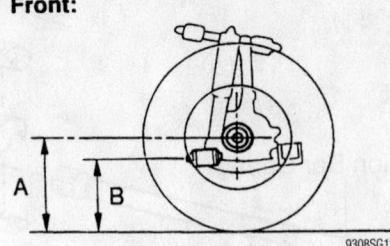

Front:

Matchmark the torsion bar to the anchor arm and torque arm—Land Cruiser/LX470

Ride height measurements A and B—Land Cruiser/LX470

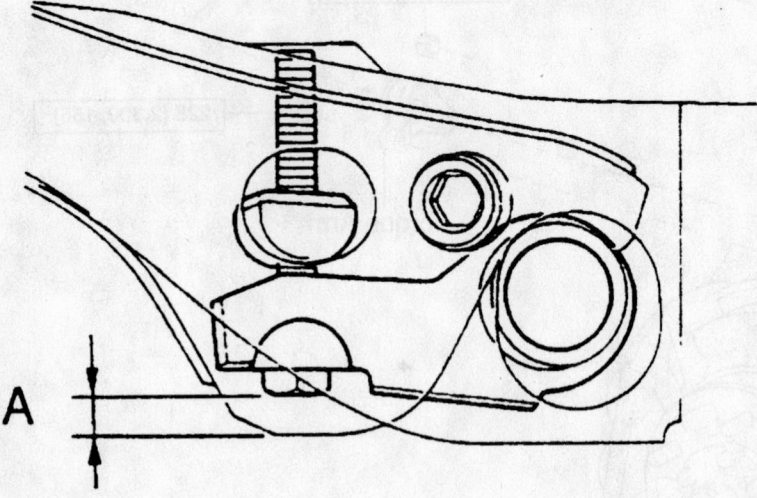

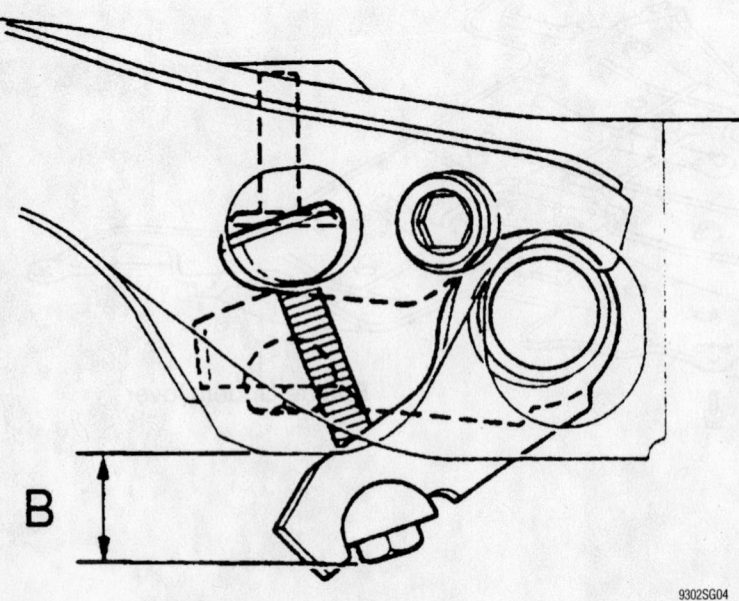

Reference measurements A and B—Land Cruiser/LX470

Upper Ball Joint

REMOVAL & INSTALLATION

Land Cruiser & LX470

the upper ball joint is serviced with the upper control arm as an assembly.

Sequoia

1. Before servicing the vehicle, refer to the precautions in the beginning of this section.
2. Remove or disconnect the following:
   - Front wheel
   - Steering knuckle with the axle hub
   - Wire and boot
   - Snapring
   - Upper ball joint from the steering knuckle, using a deep socket wrench and tool 09050-40011

To install:

3. Install or connect the following:
   - New upper ball joint to the steering knuckle, using a deep socket and tool 09309-37010
   - New snapring
4. Using a torque wrench, inspect the upper ball joint rotation, as follows:
   a. Flip the ball joint back-and-forth 5 times.
   b. Using a torque wrench, continuously turn the nut 1 turn in 2–4 seconds.
   c. Take the reading on the 5th turn; it should be 6–39 inch lbs. (0.7–4.4 Nm). If not, replace the upper ball joint.
5. Install or connect the following:
   - New boot secured with a wire
   - Front wheel. Torque the lug nuts to 81 ft. lbs. (110 Nm).
6. Check and/or adjust the front wheel alignment.

Lower Ball Joint

REMOVAL & INSTALLATION

Land Cruiser & LX470

the lower ball joint is serviced with the lower control arm as an assembly.

Sequoia

1. Before servicing the vehicle, refer to the precautions in the beginning of this section.
2. Remove or disconnect the following:
   - Front wheel
   - 4 lower ball joint set bolts
   - Tie-rod end from the lower ball joint, using tool 09610-20012

- Lower ball joint nut.
- Lower ball joint from the lower control arm, using tool 09628-62011

To install:

3. Install or connect the following:
- New lower ball joint to the lower control. Torque the bolts to 103 ft. lbs. (140 Nm).
- New cotter pin

- Tie-rod end to the lower ball joint. Torque the nut to 67 ft. lbs. (91 Nm).
- Lower ball joint set bolts. Torque the 4 bolts to 59 ft. lbs. (80 Nm).
- Front wheel. Torque the lug nuts to 81 ft. lbs. (110 Nm).

4. Check and/or adjust the front wheel alignment.

Upper Control Arm

REMOVAL & INSTALLATION

Land Cruiser & LX470

1. Before servicing the vehicle, refer to the precautions in the beginning of this section.

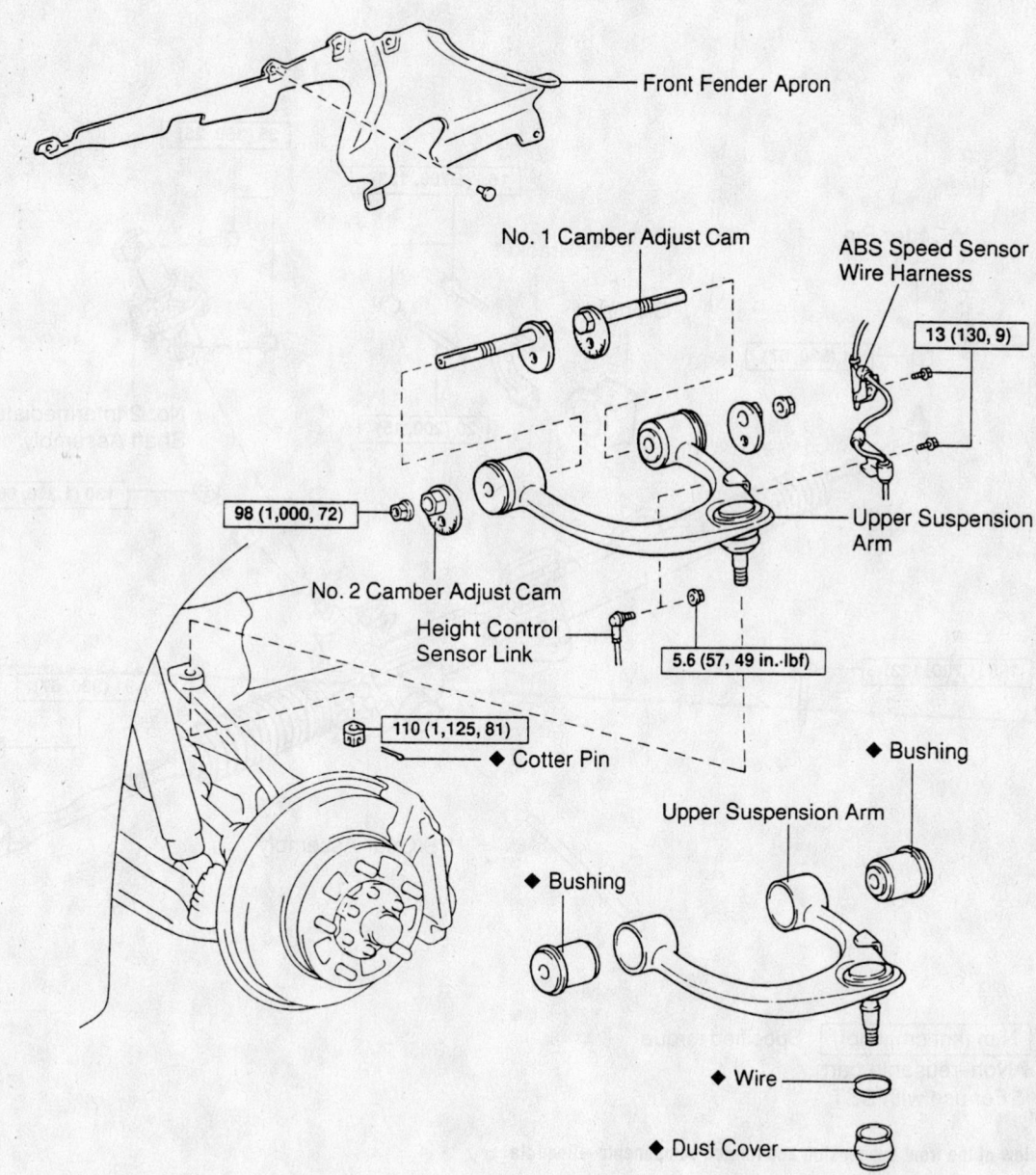

N·m (kgf·cm, ft·lbf) : Specified torque
◆ Non-reusable part

Exploded view of the upper control arm and related components—Land Cruiser/LX470

9302SG01

For Tire, Wheel and Ball Joint specifications, see Section 1 of this manual

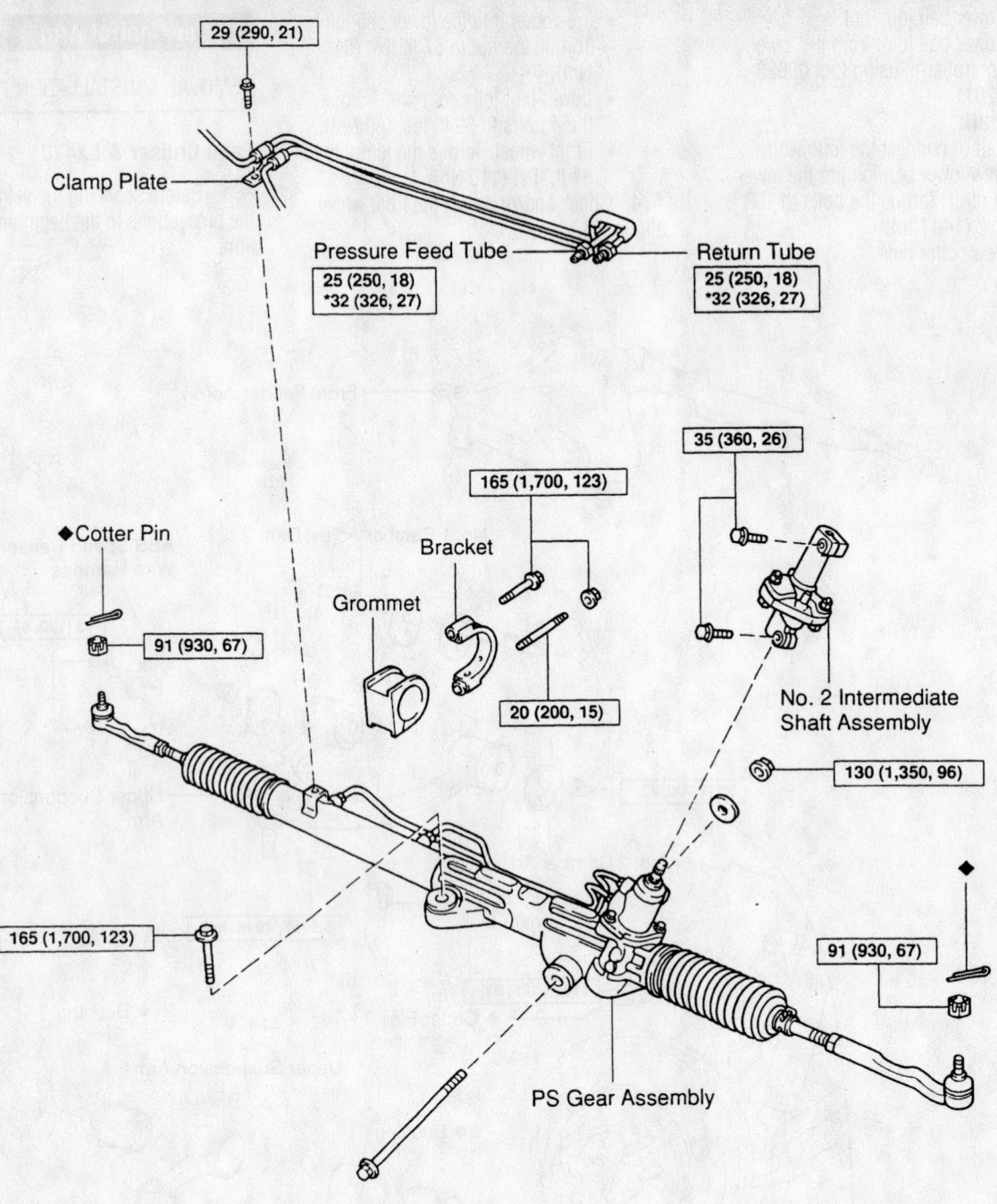

29 (290, 21)

Clamp Plate

Pressure Feed Tube
25 (250, 18)
*32 (326, 27)

Return Tube
25 (250, 18)
*32 (326, 27)

35 (360, 26)

165 (1,700, 123)

◆Cotter Pin

Bracket

Grommet

91 (930, 67)

20 (200, 15)

No. 2 Intermediate
Shaft Assembly

130 (1,350, 96)

165 (1,700, 123)

91 (930, 67)

PS Gear Assembly

N·m (kgf·cm, ft·lbf) : Specified torque
◆Non–reusable part
* For use with SST

9308YG19

Exploded view of the front suspension and related components—Sequoia

2. Remove or disconnect the following:
- Front wheel
- Inner fender liner
- Wheel speed sensor harness
- Upper ball joint
- Adjustment cam bolts
- Upper control arm

To install:

3. Install or connect the following:
- Upper control arm. Tighten the adjustment cam bolts to 72 ft. lbs. (98 Nm).
- Upper ball joint. Tighten the nut to 81 ft. lbs. (110 Nm).

- Wheel speed sensor harness. Tighten the bolts to 10 ft. lbs. (13 Nm).
- Inner fender liner
- Front wheel

4. Check the wheel alignment and adjust as necessary.

Sequoia

1. Before servicing the vehicle, refer to the precautions in the beginning of this section.

2. Remove or disconnect the following:
   - Front wheel
   - Strut
   - Wheel speed sensor harness, if equipped with Anti-lock Brake System (ABS)

3. Upper ball joint, as follows:
   - Cotter pin and loosen the nut
   - Upper ball joint from the upper control arm, using tool 09950-40011
   - Steering knuckle, support it securely
   - Upper ball joint nut

4. Remove or disconnect the following:
   - 4 clips and the fender apron seal
   - Brake/fuel line clamp nut and clamp
   - Both upper control arm-to-chassis nuts/bolts
   - Upper control arm

To install:

5. Install or connect the following:
   - Upper control arm. Torque both upper control arm-to-chassis nuts/bolts to 72 ft. lbs. (98 Nm).
   - Brake/fuel line clamp nut and clamp. Torque the clamp nut to 49 inch lbs. (5.5 Nm).
   - Fender apron seal
   - Upper ball joint. Torque the nut to 77 ft. lbs. (105 Nm).
   - New cotter pin
   - Steering knuckle
   - Wheel speed sensor harness, if equipped with Anti-lock Brake System (ABS). Torque it to 71 inch lbs. (8.0 Nm).
   - Strut
   - Front wheel

6. Check and/or adjust the wheel alignment.

CONTROL ARM BUSHING REPLACEMENT

Land Cruiser & LX470

1. Before servicing the vehicle, refer to the precautions in the beginning of this section.

2. Remove the control arm from the vehicle.

3. Remove the control arm bushings with a hydraulic press.

To install:

4. Lubricate the control arm bushings with liquid soap.

5. Press the bushings into the control arm until the bushing flange contacts the housing edge of the control arm.

6. Install the control arm to the vehicle.

7. Check the wheel alignment and adjust as necessary.

Sequoia

1. Before servicing the vehicle, refer to the precautions in the beginning of this section.

2. Remove the upper control arm from the vehicle.

3. Remove the control arm bushings, as follows:
   - Pry up the bushing flange, using a chisel and a hammer
   - Press the bushing(s) from the upper control arm, using a shop press and tools 09613-26010, 09631-20060 and 09950-00020

To install:

4. Lubricate the new control arm bushings with liquid soap.

5. Press the bushings into the control arm until the bushing flange contacts the housing edge of the control arm, using a shop press, a steel plate and tools 09631-12090 and 09710-30021

6. Install the upper control arm to the vehicle.

7. Check and/or adjust the wheel alignment.

Lower Control Arm

REMOVAL & INSTALLATION

Land Cruiser & LX470

1. Before servicing the vehicle, refer to the precautions in the beginning of this section.

2. Remove or disconnect the following:
   - Front wheel
   - Engine under cover
   - Torsion bar
   - Stabilizer bar link
   - Shock absorber
   - Lower ball joint
   - Lower control arm

To install:

3. Install or connect the following:
   - Lower control arm. Tighten the bolts to 170 ft. lbs. (230 Nm).
   - Lower ball joint. Tighten the nut to 117 ft. lbs. (159 Nm).
   - Shock absorber
   - Stabilizer bar link. Tighten the bolt to 38 ft. lbs. (52 Nm).
   - Torsion bar
   - Engine under cover
   - Front wheel

4. Check the wheel alignment and adjust as necessary.

Sequoia

1. Before servicing the vehicle, refer to the precautions in the beginning of this section.

2. Remove front wheel.

3. Disconnect the tie-rod end, as follows:
   - Cotter pin and nut
   - Tie-rod end from the lower ball joint, using tool 09610-20012

4. Remove or disconnect the following:
   - Power steering gear set bolts and nuts
   - Stabilizer bar link from the lower control arm
   - Strut from the lower control arm

5. Disconnect the lower ball joint, as follows:
   - Cotter pin and nut
   - Lower ball joint from the lower control arm

6. Matchmark both front and rear cam plates and chassis frame.

7. Remove the lower control arm while slightly shifting the power steering gear rearward.

To install:

8. Install or connect the following:
   - Lower control arm while slightly shifting the power steering gear rearward
   - Align both front and rear cam plates and chassis frame matchmarks. Torque both bolts to 96 ft. lbs. (130 Nm).

9. Connect the lower ball joint, as follows:
   - Lower ball joint to the lower control arm. Torque the nut to 103 ft. lbs. (140 Nm).
   - New cotter pin

10. Install or connect the following:
    - Strut to the lower control arm. Torque the nut/bolt to 100 ft. lbs. (135 Nm).
    - Stabilizer bar link to the lower control arm. Torque the nut to 51 ft. lbs. (69 Nm).

For Wheel Alignment specifications, see Section 1 of this manual

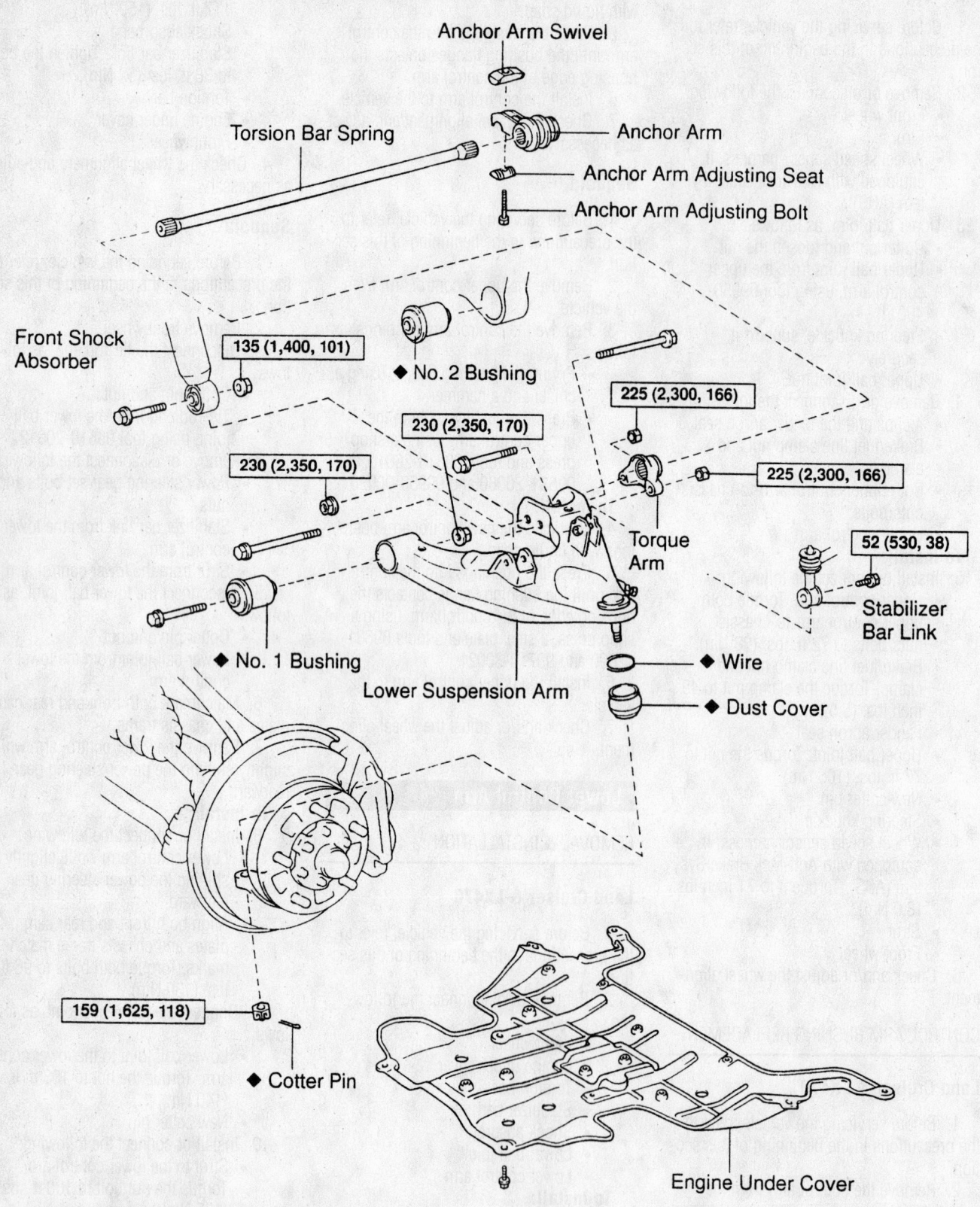

Anchor Arm Swivel

Anchor Arm

Torsion Bar Spring

Anchor Arm Adjusting Seat

Anchor Arm Adjusting Bolt

Front Shock Absorber

135 (1,400, 101)

◆ No. 2 Bushing

225 (2,300, 166)

230 (2,350, 170)

230 (2,350, 170)

225 (2,300, 166)

52 (530, 38)

Torque Arm

Stabilizer Bar Link

◆ No. 1 Bushing

◆ Wire

Lower Suspension Arm

◆ Dust Cover

159 (1,625, 118)

◆ Cotter Pin

Engine Under Cover

N·m (kgf·cm, ft·lbf) : Specified torque

◆ Non–reusable part

9302SG02

Exploded view of the lower control arm and related components—Land Cruiser/LX470

- Power steering gear set bolts and nuts. Torque the set bolt and clamp nut/bolt to 122 ft. lbs. and the set nut/bolt to 96 ft. lbs. (130 Nm)
- Tie-rod end to the lower ball joint. Torque the nut to 67 ft. lbs. (91 Nm).
- New cotter pin
- Front wheel. Torque lug nuts to 81 ft. lbs. (110 Nm).

11. Check and/or adjust the wheel alignment.

CONTROL ARM BUSHING REPLACEMENT

Land Cruiser & LX470

1. Before servicing the vehicle, refer to the precautions in the beginning of this section.
2. Remove the control arm from the vehicle.
3. Remove the control arm bushings with a hydraulic press.

To install:
4. Lubricate the control arm bushings with liquid soap.
5. Press the bushings into the control arm until the bushing flange contacts the housing edge of the control arm.
6. Install the control arm to the vehicle.
7. Check the wheel alignment and adjust as necessary.

Sequoia

1. Before servicing the vehicle, refer to the precautions in the beginning of this section.
2. Remove the lower control arm from the vehicle.
3. Remove the control arm bushings, as follows:
- Pry up the bushing flange, using a chisel and a hammer
- Press the bushing(s) from the upper control arm, using a shop press and tools 09613-26010, 09632-36010 and 09950-00020

To install:
4. Lubricate the new control arm bushings with liquid soap.
5. Press the No. 1 bushing into the control arm until the bushing flange contacts the housing edge of the control arm, using a shop press, a steel plate and tools 09631-12090 and 09502-12010, facing the correct direction.
6. Press the No. 2 bushing into the control arm until the bushing flange contacts

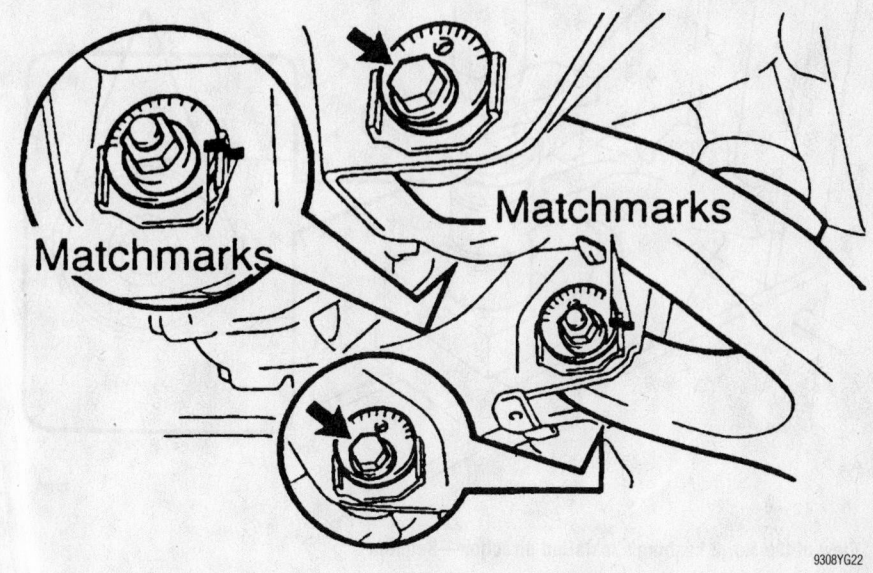

View of the lower control arm's cam plate alignment—Sequoia

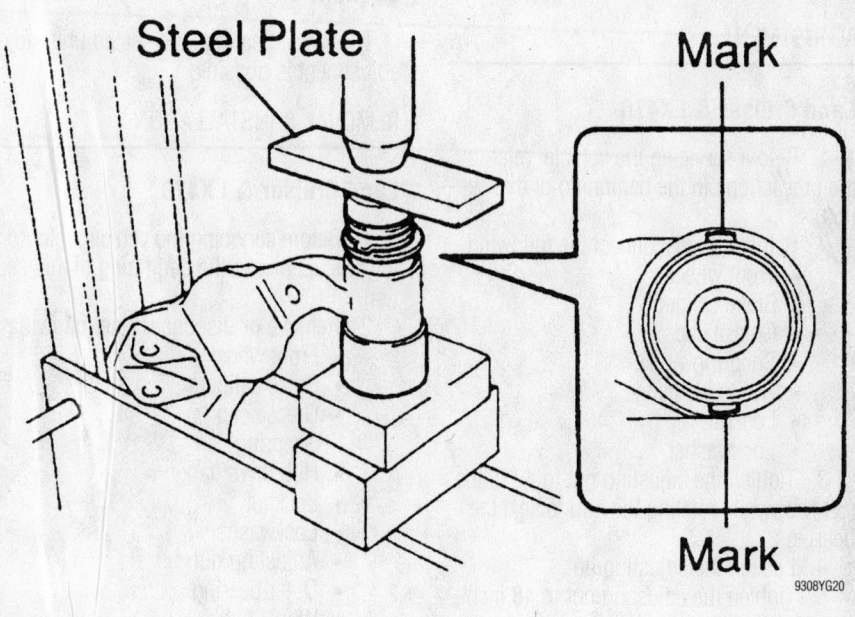

View of the No. 1 bushing's installed direction—Sequoia

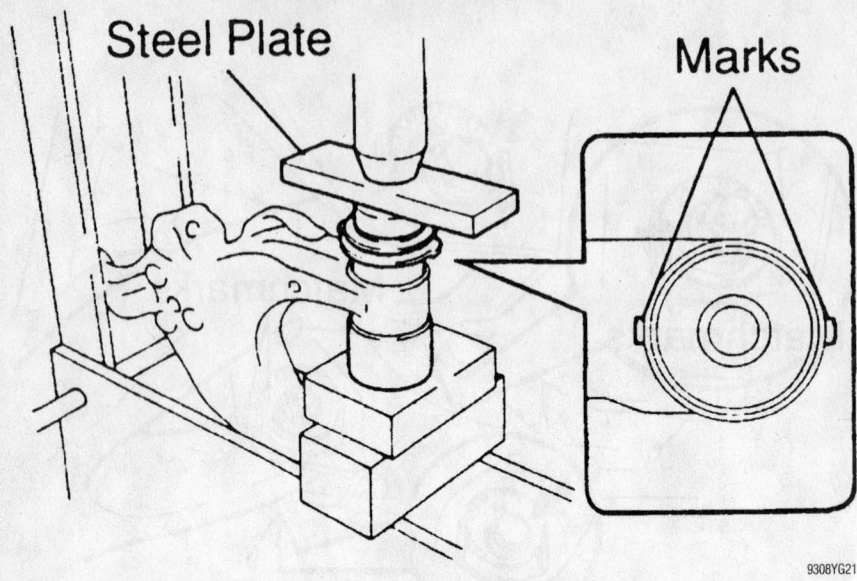

Steel Plate

Marks

9308YG21

View of the No. 2 bushing's installed direction—Sequoia

the housing edge of the control arm, using a shop press, a steel plate and tools 09631-12090 and 09950-60020, facing the correct direction.

7. Install the lower control arm to the vehicle.

8. Check and/or adjust the wheel alignment.

Front Wheel Bearing

ADJUSTMENT

Land Cruiser & LX470

1. Before servicing the vehicle, refer to the precautions in the beginning of this section.

2. Remove or disconnect the following:
- Front wheel
- Brake caliper
- Grease cap
- Snapring
- Hub drive flange
- Locknut
- Lockwasher

3. Tighten the adjusting nut to 43 ft. lbs. (59 Nm) while rotating the hub to seat the bearings.

4. Loosen the adjusting nut.

5. Tighten the adjusting nut to 48 inch lbs. (5.4 Nm) and check that the bearing has no play.

6. Check the bearing preload with a spring tension gauge. The preload should be 6.4–12.6 lbs. (28–56 N).

7. Install or connect the following:

- Lockwasher
- Locknut. Tighten the nut to 47 ft. lbs. (64 Nm).
- Hub drive flange. Tighten the nuts to 26 ft. lbs. (35 Nm).
- Snapring
- Grease cap
- Brake caliper
- Front wheel

Sequoia

The wheel bearings are sealed unit; no adjustment is possible.

REMOVAL & INSTALLATION

Land Cruiser & LX470

1. Before servicing the vehicle, refer to the precautions in the beginning of this section.

2. Remove or disconnect the following:
- Front wheel
- Brake caliper
- Grease cap
- Snapring
- Hub drive flange
- Locknut
- Lockwasher
- Adjusting nut
- Outer bearing
- Wheel hub
- Inner grease seal
- Inner bearing

To install:

3. Install or connect the following:
- Inner bearing
- Inner grease seal

- Wheel hub
- Outer bearing
- Adjusting nut. Adjust the wheel bearings.
- Lockwasher
- Locknut. Tighten the nut to 47 ft. lbs. (64 Nm).
- Hub drive flange. Tighten the nuts to 26 ft. lbs. (35 Nm).
- Snapring
- Grease cap
- Brake caliper
- Front wheel

Sequoia

1. Before servicing the vehicle, refer to the precautions in the beginning of this section.

2. Remove or disconnect the following:
- Front wheel
- Axle hub/steering knuckle assembly and place it in a vise
- Grease cap, for 2WD
- Inner grease seal, for 4WD

3. Remove the axle hub from the steering knuckle

- 4 bolts and shift the dust cover towards the outside (hub side)
- Axle hub from the steering knuckle, using tools 09710-30021 and 09950-40011
- Dust cover from the steering knuckle
- Bearing spacer and Anti-lock Brake System (ABS) speed sensor, if equipped with ABS
- Spacer, if not equipped with ABS

✱✱ WARNING

Be careful not to scratch the speed sensor rotor serrations.

4. Remove the outside oil seal from steering knuckle, using a small prybar

5. Remove the bearing from the steering knuckle, as follows:
- Snapring
- Bearing from the steering knuckle, using tools 09950-60020 and 09950-70010

To install:

6. Install the bearing to the steering knuckle, as follows:
- Bearing to the steering knuckle, using tools 09950-60020 and 09527-17011
- Snapring

7. Install the new outside oil seal to steering knuckle, using a plastic hammer and tools 09223-15030 and 09527-17011

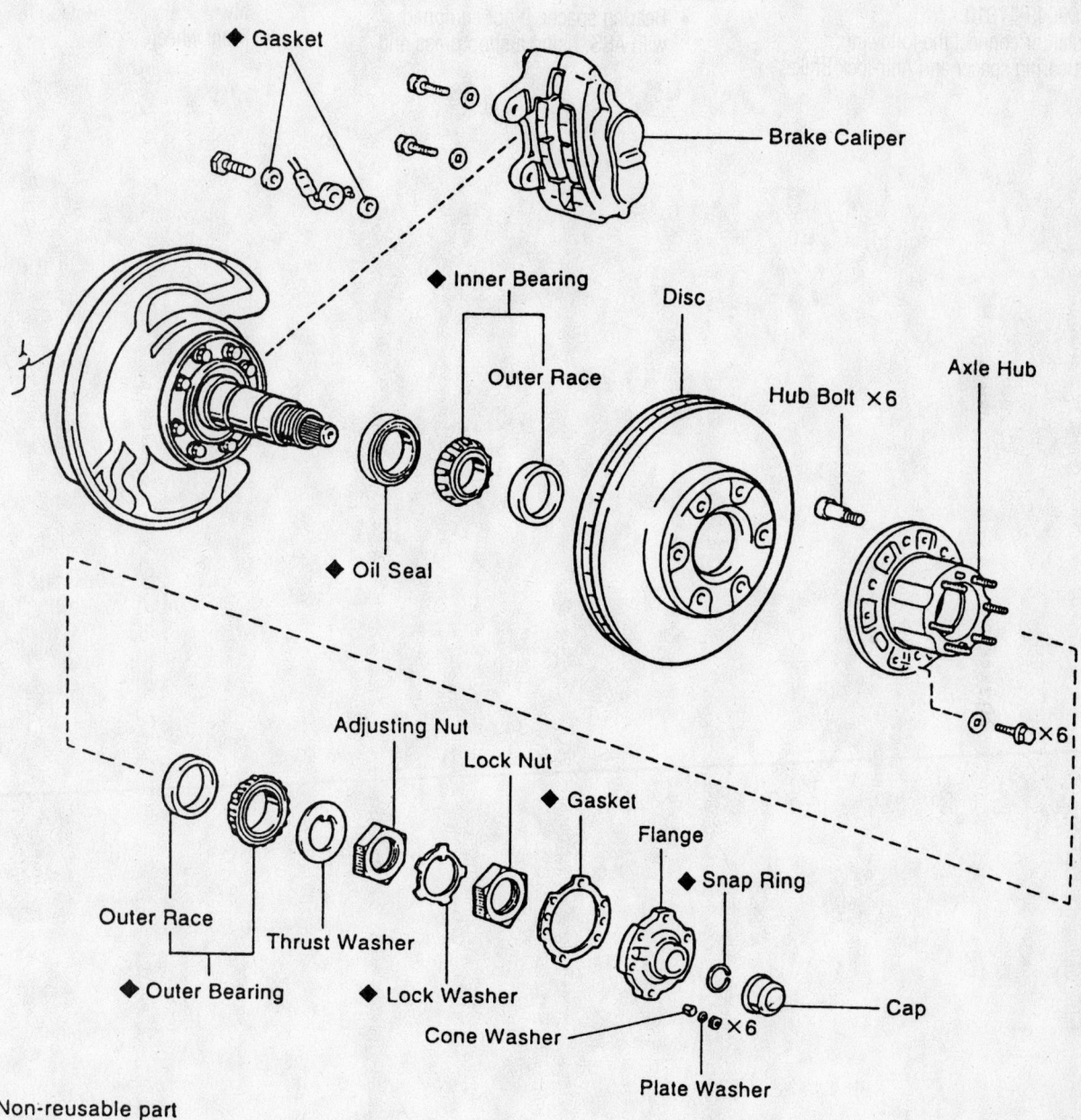

◆ Gasket

Brake Caliper

◆ Inner Bearing

Disc

Outer Race

Hub Bolt ×6

Axle Hub

◆ Oil Seal

◆ ×6

Adjusting Nut

Lock Nut

◆ Gasket

Flange

◆ Snap Ring

Outer Race

Thrust Washer

◆ Snap Ring

◆ Outer Bearing

◆ Lock Washer

Cap

Cone Washer

Plate Washer

×6

◆ Non-reusable part

7924SG31

Exploded view of the front hub and related components—Land Cruiser/LX470

8. Install the axle hub to the steering knuckle, as follows:
   - Dust cover to the steering knuckle. Torque the 4 bolts to 13 ft. lbs. (18 Nm).
   - Axle hub to the steering knuckle, using a shop press and tool 09649-17010
9. Install or connect the following:
   - Bearing spacer and Anti-lock Brake

System (ABS) speed sensor, if equipped with ABS

❋❋ WARNING

Be careful not to scratch the speed sensor rotor serrations.

- Bearing spacer, if not equipped with ABS, using a shop press and

tools 09950-60010 and 09950-70010
- Grease cap, for 2WD
- Inner grease seal, for 4WD, using a plastic hammer and tool 09527-17011
- Axle hub/steering knuckle assembly
- Front wheel

GLOSSARY

ABS: Anti-lock braking system. An electro-mechanical braking system which is designed to minimize or prevent wheel lock-up during braking.

ABSOLUTE PRESSURE: Atmospheric (barometric) pressure plus the pressure gauge reading.

ACCELERATOR PUMP: A small pump located in the carburetor that feeds fuel into the air/fuel mixture during acceleration.

ACCUMULATOR: A device that controls shift quality by cushioning the shock of hydraulic oil pressure being applied to a clutch or band.

ACTUATING MECHANISM: The mechanical output devices of a hydraulic system, for example, clutch pistons and band servos.

ACTUATOR: The output component of a hydraulic or electronic system.

ADVANCE: Setting the ignition timing so that spark occurs earlier before the piston reaches top dead center (TDC).

ADAPTIVE MEMORY (ADAPTIVE STRATEGY): The learning ability of the TCM or PCM to redefine its decision-making process to provide optimum shift quality.

AFTER TOP DEAD CENTER (ATDC): The point after the piston reaches the top of its travel on the compression stroke.

AIR BAG: Device on the inside of the car designed to inflate on impact of crash, protecting the occupants of the car.

AIR CHARGE TEMPERATURE (ACT) SENSOR: The temperature of the airflow into the engine is measured by an ACT sensor, usually located in the lower intake manifold or air cleaner.

AIR CLEANER: An assembly consisting of a housing, filter and any connecting ductwork. The filter element is made up of a porous paper, sometimes with a wire mesh screening, and is designed to prevent airborne particles from entering the engine through the carburetor or throttle body.

AIR INJECTION: One method of reducing harmful exhaust emissions by injecting air into each of the exhaust ports of an engine. The fresh air entering the hot exhaust manifold causes any remaining fuel to be burned before it can exit the tailpipe.

AIR PUMP: An emission control device that supplies fresh air to the exhaust manifold to aid in more completely burning exhaust gases.

AIR/FUEL RATIO: The ratio of air-to-gasoline by weight in the fuel mixture drawn into the engine.

ALDL (assembly line diagnostic link): Electrical connector for scanning ECM/PCM/TCM input and output devices.

ALIGNMENT RACK: A special drive-on vehicle lift apparatus/measuring device used to adjust a vehicle's toe, caster and camber angles.

ALL WHEEL DRIVE: Term used to describe a full time four wheel drive system or any other vehicle drive system that continuously delivers power to all four wheels. This system is found primarily on station wagon vehicles and SUVs not utilized for significant off road use.

ALTERNATING CURRENT (AC): Electric current that flows first in one direction, then in the opposite direction, continually reversing flow.

ALTERNATOR: A device which produces AC (alternating current) which is converted to DC (direct current) to charge the car battery.

AMMETER: An instrument, calibrated in amperes, used to measure the flow of an electrical current in a circuit. Ammeters are always connected in series with the circuit being tested.

AMPERAGE: The total amount of current (amperes) flowing in a circuit.

AMPLIFIER: A device used in an electrical circuit to increase the voltage of an output signal.

AMP/HR. RATING (BATTERY): Measurement of the ability of a battery to deliver a stated amount of current for a stated period of time. The higher the amp/hr. rating, the better the battery.

AMPERE: The rate of flow of electrical current present when one volt of electrical pressure is applied against one ohm of electrical resistance.

ANALOG COMPUTER: Any microprocessor that uses similar (analogous) electrical signals to make its calculations.

ANODIZED: A special coating applied to the surface of aluminum valves for extended service life.

ANTIFREEZE: A substance (ethylene or propylene glycol) added to the coolant to prevent freezing in cold weather.

ANTI-FOAM AGENTS: Minimize fluid foaming from the whipping action encountered in the converter and planetary action.

ANTI-WEAR AGENTS: Zinc agents that control wear on the gears, bushings, and thrust washers.

ANTI-LOCK BRAKING SYSTEM: A supplementary system to the base hydraulic system that prevents sustained lock-up of the wheels during braking as well as automatically controlling wheel slip.

ANTI-ROLL BAR: See stabilizer bar.

ARC: A flow of electricity through the air between two electrodes or contact points that produces a spark.

ARMATURE: A laminated, soft iron core wrapped by a wire that converts electrical energy to mechanical energy as in a motor or relay. When rotated in a magnetic field, it changes mechanical energy into electrical energy as in a generator.

ATDC: After Top Dead Center.

ATF: Automatic transmission fluid.

ATMOSPHERIC PRESSURE: The pressure on the Earth's surface caused by the weight of the air in the atmosphere. At sea level, this pressure is 14.7 psi at 32°F (101 kPa at 0°C).

ATOMIZATION: The breaking down of a liquid into a fine mist that can be suspended in air.

AUXILIARY ADD-ON COOLER: A supplemental transmission fluid cooling device that is installed in series with the heat exchanger (cooler), located inside the radiator, to provide additional support to cool the hot fluid leaving the torque converter.

AUXILIARY PRESSURE: An added fluid pressure that is introduced into a regulator or balanced valve system to control valve movement. The auxiliary pressure itself can be either a fixed or a variable value. (See balanced valve; regulator valve.)

AWD: All wheel drive.

AXIAL FORCE: A side or end thrust force acting in or along the same plane as the power flow.

AXIAL PLAY: Movement parallel to a shaft or bearing bore.

AXLE CAPACITY: The maximum load-carrying capacity of the axle itself, as specified by the manufacturer. This is usually a higher number than the GAWR.

AXLE RATIO: This is a number (3.07:1, 4.56:1, for example) expressing the ratio between driveshaft revolutions and wheel revolutions. A low numerical ratio allows the engine to work easier because it doesn't have to turn as fast. A high numerical ratio means that the engine has to turn more rpm's to move the wheels through the same number of turns.

BACKFIRE: The sudden combustion of gases in the intake or exhaust system that results in a loud explosion.

BACKLASH: The clearance or play between two parts, such as meshed gears.

BACKPRESSURE: Restrictions in the exhaust system that slow the exit of exhaust gases from the combustion chamber.

BAKELITE®: A heat resistant, plastic insulator material commonly used in printed circuit boards and transistorized components.

BALANCED VALVE: A valve that is positioned by opposing auxiliary hydraulic pressures and/or spring force. Examples include mainline regulator, throttle, and governor valves. (See regulator valve.)

BAND: A flexible ring of steel with an inner lining of friction material. When tightened around the outside of a drum, a planetary member is held stationary to the transmission/transaxle case.

BALL BEARING: A bearing made up of hardened inner and outer races between which hardened steel balls roll.

BALL JOINT: A ball and matching socket connecting suspension components (steering knuckle to lower control arms). It permits rotating movement in any direction between the components that are joined.

BARO (BAROMETRIC PRESSURE SENSOR): Measures the change in the intake manifold pressure caused by changes in altitude.

BAROMETRIC MANIFOLD ABSOLUTE PRESSURE (BMAP) SENSOR: Operates similarly to a conventional MAP sensor; reads intake manifold pressure and is also responsible for determining altitude and barometric pressure prior to engine operation.

BAROMETRIC PRESSURE: (See atmospheric pressure.)

BALLAST RESISTOR: A resistor in the primary ignition circuit that lowers voltage after the engine is started to reduce wear on ignition components.

BATTERY: A direct current electrical storage unit, consisting of the basic active materials of lead and sulfuric acid, which converts chemical energy into electrical energy. Used to provide current for the operation of the starter as well as other equipment, such as the radio, lighting, etc.

BEAD: The portion of a tire that holds it on the rim.

BEARING: A friction reducing, supportive device usually located between a stationary part and a moving part.

BEFORE TOP DEAD CENTER (BTDC): The point just before the piston reaches the top of its travel on the compression stroke.

BELTED TIRE: Tire construction similar to bias-ply tires, but using two or more layers of reinforced belts between body plies and the tread.

BEZEL: Piece of metal surrounding radio, headlights, gauges or similar components; sometimes used to hold the glass face of a gauge in the dash.

BIAS-PLY TIRE: Tire construction, using body ply reinforcing cords which run at alternating angles to the center line of the tread.

BI-METAL TEMPERATURE SENSOR: Any sensor or switch made of two dissimilar types of metal that bend when heated or cooled due to the different expansion rates of the alloys. These types of sensors usually function as an on/off switch.

BLOCK: See Engine Block.

BLOW-BY: Combustion gases, composed of water vapor and unburned fuel, that leak past the piston rings into the crankcase during normal engine operation. These gases are removed by the PCV system to prevent the buildup of harmful acids in the crankcase.

BOOK TIME: See Labor Time.

BOOK VALUE: The average value of a car, widely used to determine trade-in and resale value.

BOOST VALVE: Used at the base of the regulator valve to increase mainline pressure.

BORE: Diameter of a cylinder.

BRAKE CALIPER: The housing that fits over the brake disc. The caliper holds the brake pads, which are pressed against the discs by the caliper pistons when the brake pedal is depressed.

BRAKE HORSEPOWER (BHP): The actual horsepower available at the engine flywheel as measured by a dynamometer.

BRAKE FADE: Loss of braking power, usually caused by excessive heat after repeated brake applications.

BRAKE HORSEPOWER: Usable horsepower of an engine measured at the crankshaft.

BRAKE PAD: A brake shoe and lining assembly used with disc brakes.

BRAKE PROPORTIONING VALVE: A valve on the master cylinder which restricts hydraulic brake pressure to the wheels to a specified amount, preventing wheel lock-up.

BREAKAWAY: Often used by Chrysler to identify first-gear operation in D and 2 ranges. In these ranges, first-gear operation depends on a one-way roller clutch that holds on acceleration and releases (breaks away) on deceleration, resulting in a freewheeling coast-down condition.

BRAKE SHOE: The backing for the brake lining. The term is, however, usually applied to the assembly of the brake backing and lining.

BREAKER POINTS: A set of points inside the distributor, operated by a cam, which make and break the ignition circuit.

BRINNELLING: A wear pattern identified by a series of indentations at regular intervals. This condition is caused by a lack of lube, overload situations, and/or vibrations.

BTDC: Before Top Dead Center.

BUMP: Sudden and forceful apply of a clutch or band.

BUSHING: A liner, usually removable, for a bearing; an anti-friction liner used in place of a bearing.

CALIFORNIA ENGINE: An engine certified by the EPA for use in California only; conforms to more stringent emission regulations than Federal engine.

CALIPER: A hydraulically activated device in a disc brake system,

which is mounted straddling the brake rotor (disc). The caliper contains at least one piston and two brake pads. Hydraulic pressure on the piston(s) forces the pads against the rotor.

CAPACITY: The quantity of electricity that can be delivered from a unit, as from a battery in ampere-hours, or output, as from a generator.

CAMBER: One of the factors of wheel alignment. Viewed from the front of the car, it is the inward or outward tilt of the wheel. The top of the tire will lean outward (positive camber) or inward (negative camber).

CAMSHAFT: A shaft in the engine on which are the lobes (cams) which operate the valves. The camshaft is driven by the crankshaft, via a belt, chain or gears, at one half the crankshaft speed.

CAPACITOR: A device which stores an electrical charge.

CARBON MONOXIDE (CO): A colorless, odorless gas given off as a normal byproduct of combustion. It is poisonous and extremely dangerous in confined areas, building up slowly to toxic levels without warning if adequate ventilation is not available.

CARBURETOR: A device, usually mounted on the intake manifold of an engine, which mixes the air and fuel in the proper proportion to allow even combustion.

CASTER: The forward or rearward tilt of an imaginary line drawn through the upper ball joint and the center of the wheel. Viewed from the sides, positive caster (forward tilt) lends directional stability, while negative caster (rearward tilt) produces instability.

CATALYTIC CONVERTER: A device installed in the exhaust system, like a muffler, that converts harmful byproducts of combustion into carbon dioxide and water vapor by means of a heat-producing chemical reaction.

CENTRIFUGAL ADVANCE: A mechanical method of advancing the spark timing by using flyweights in the distributor that react to centrifugal force generated by the distributor shaft rotation.

CENTRIFUGAL FORCE: The outward pull of a revolving object, away from the center of revolution. Centrifugal force increases with the speed of rotation.

CETANE RATING: A measure of the ignition value of diesel fuel. The higher the cetane rating, the better the fuel. Diesel fuel cetane rating is roughly comparable to gasoline octane rating.

CHECK VALVE: Any one-way valve installed to permit the flow of air, fuel or vacuum in one direction only.

CHOKE: The valve/plate that restricts the amount of air entering an engine on the induction stroke, thereby enriching the air/fuel ratio.

CHUGGLE: Bucking or jerking condition that may be engine related and may be most noticeable when converter clutch is engaged; similar to the feel of towing a trailer.

CIRCLIP: A split steel snapring that fits into a groove to hold various parts in place.

CIRCUIT BREAKER: A switch which protects an electrical circuit from overload by opening the circuit when the current flow exceeds a pre-determined level. Some circuit breakers must be reset manually, while most reset automatically.

CIRCUIT: Any unbroken path through which an electrical current can flow. Also used to describe fuel flow in some instances.

CIRCUIT, BYPASS: Another circuit in parallel with the major circuit through which power is diverted.

CIRCUIT, CLOSED: An electrical circuit in which there is no interruption of current flow.

CIRCUIT, GROUND: The non-insulated portion of a complete circuit used as a common potential point. In automotive circuits, the ground is composed of metal parts, such as the engine, body sheet metal, and frame and is usually a negative potential.

CIRCUIT, HOT: That portion of a circuit not at ground potential. The hot circuit is usually insulated and is connected to the positive side of the battery.

CIRCUIT, OPEN: A break or lack of contact in an electrical circuit, either intentional (switch) or unintentional (bad connection or broken wire).

CIRCUIT, PARALLEL: A circuit having two or more paths for current flow with common positive and negative tie points. The same voltage is applied to each load device or parallel branch.

CIRCUIT, SERIES: An electrical system in which separate parts are connected end to end, using one wire, to form a single path for current to flow.

CIRCUIT, SHORT: A circuit that is accidentally completed in an electrical path for which it was not intended.

CLAMPING (ISOLATION) DIODES: Diodes positioned in a circuit to prevent self-induction from damaging electronic components.

CLEARCOAT: A transparent layer which, when sprayed over a vehicle's paint job, adds gloss and depth as well as an additional protective coating to the finish.

CLUTCH: Part of the power train used to connect/disconnect power to the rear wheels.

CLUTCH, FLUID: The same as a fluid coupling. A fluid clutch or coupling performs the same function as a friction clutch by utilizing fluid friction and inertia as opposed to solid friction used by a friction clutch. (See fluid coupling.)

CLUTCH, FRICTION: A coupling device that provides a means of smooth and positive engagement and disengagement of engine torque to the vehicle powertrain. Transmission of power through the clutch is accomplished by bringing one or more rotating drive members into contact with complementing driven members.

COAST: Vehicle deceleration caused by engine braking conditions.

COEFFICIENT OF FRICTION: The amount of surface tension between two contacting surfaces; identified by a scientifically calculated number.

COIL: Part of the ignition system that boosts the relatively low voltage supplied by the car's electrical system to the high voltage required to fire the spark plugs.

COMBINATION MANIFOLD: An assembly which includes both the intake and exhaust manifolds in one casting.

COMBINATION VALVE: A device used in some fuel systems that routes fuel vapors to a charcoal storage canister instead of venting them into the atmosphere. The valve relieves fuel tank pressure and allows fresh air into the tank as the fuel level drops to prevent a vapor lock situation.

COMBUSTION CHAMBER: The part of the engine in the cylinder head where combustion takes place.

COMPOUND GEAR: A gear consisting of two or more simple gears with a common shaft.

COMPOUND PLANETARY: A gearset that has more than the three elements found in a simple gearset and is constructed by combining members of two planetary gearsets to create additional gear ratio possibilities.

COMPRESSION CHECK: A test involving removing each spark plug and inserting a gauge. When the engine is cranked, the gauge will record a pressure reading in the individual cylinder. General operating condition can be determined from a compression check.

COMPRESSION RATIO: The ratio of the volume between the piston and cylinder head when the piston is at the bottom of its stroke (bottom dead center) and when the piston is at the top of its stroke (top dead center).

COMPUTER: An electronic control module that correlates input data according to prearranged engineered instructions; used for the management of an actuator system or systems.

CONDENSER: An electrical device which acts to store an electrical charge, preventing voltage surges.

2. A radiator-like device in the air conditioning system in which refrigerant gas condenses into a liquid, giving off heat.

CONDUCTOR: Any material through which an electrical current can be transmitted easily.

CONNECTING ROD: The connecting link between the crankshaft and piston.

CONSTANT VELOCITY JOINT: Type of universal joint in a halfshaft assembly in which the output shaft turns at a constant angular velocity without variation, provided that the speed of the input shaft is constant.

CONTINUITY: Continuous or complete circuit. Can be checked with an ohmmeter.

CONTROL ARM: The upper or lower suspension components which are mounted on the frame and support the ball joints and steering knuckles.

CONVENTIONAL IGNITION: Ignition system which uses breaker points.

CONVERTER: (See torque converter.)

CONVERTER LOCKUP: The switching from hydrodynamic to direct mechanical drive, usually through the application of a friction element called the converter clutch.

COOLANT: Mixture of water and anti-freeze circulated through the engine to carry off heat produced by the engine.

CORROSION INHIBITOR: An inhibitor in ATF that prevents corrosion of bushings, thrust washers, and oil cooler brazed joints.

COUNTERSHAFT: An intermediate shaft which is rotated by a mainshaft and transmits, in turn, that rotation to a working part.

COUPLING PHASE: Occurs when the torque converter is operating at its greatest hydraulic efficiency. The speed differential between the impeller and the turbine is at its minimum. At this point, the stator freewheels, and there is no torque multiplication.

CRANKCASE: The lower part of an engine in which the crankshaft and related parts operate.

CRANKSHAFT: Engine component (connected to pistons by connecting rods) which converts the reciprocating (up and down) motion of pistons to rotary motion used to turn the driveshaft.

CURB WEIGHT: The weight of a vehicle without passengers or payload, but including all fluids (oil, gas, coolant, etc.) and other equipment specified as standard.

CURRENT: The flow (or rate) of electrons moving through a circuit. Current is measured in amperes (amp).

CURRENT FLOW CONVENTIONAL: Current flows through a circuit from the positive terminal of the source to the negative terminal (plus to minus).

CURRENT FLOW, ELECTRON: Current or electrons flow from the negative terminal of the source, through the circuit, to the positive terminal (minus to plus).

CV-JOINT: Constant velocity joint.

CYCLIC VIBRATIONS: The off-center movement of a rotating object that is affected by its initial balance, speed of rotation, and working angles.

CYLINDER BLOCK: See engine block.

CYLINDER HEAD: The detachable portion of the engine, usually fastened to the top of the cylinder block and containing all or most of the combustion chambers. On overhead valve engines, it contains the valves and their operating parts. On overhead cam engines, it contains the camshaft as well.

CYLINDER: In an engine, the round hole in the engine block in which the piston(s) ride.

DATA LINK CONNECTOR (DLC): Current acronym/term applied to the federally mandated, diagnostic junction connector that is used to monitor ECM/PC/TCM inputs, processing strategies, and outputs including diagnostic trouble codes (DTCs).

DEAD CENTER: The extreme top or bottom of the piston stroke.

DECELERATION BUMP: When referring to a torque converter clutch in the applied position, a sudden release of the accelerator pedal causes a forceful reversal of power through the drivetrain (engine braking), just prior to the apply plate actually being released.

DELAYED (LATE OR EXTENDED): Condition where shift is expected but does not occur for a period of time, for example, where clutch or band engagement does not occur as quickly as expected during part throttle or wide open throttle apply of accelerator or when manually downshifting to a lower range.

DETENT: A spring-loaded plunger, pin, ball, or pawl used as a holding device on a ratchet wheel or shaft. In automatic transmissions, a detent mechanism is used for locking the manual valve in place.

DETENT DOWNSHIFT: (See kickdown.)

DETERGENT: An additive in engine oil to improve its operating characteristics.

DETONATION: An unwanted explosion of the air/fuel mixture in the combustion chamber caused by excess heat and compression, advanced timing, or an overly lean mixture. Also referred to as "ping".

DEXRON®: A brand of automatic transmission fluid.

DIAGNOSTIC TROUBLE CODES (DTCs): A digital display from the control module memory that identifies the input, processor, or output device circuit that is related to the powertrain emission/driveability malfunction detected. Diagnostic trouble codes can be read by the MIL to flash any codes or by using a handheld scanner.

DIAPHRAGM: A thin, flexible wall separating two cavities, such as in a vacuum advance unit.

DIESELING: The engine continues to run after the car is shut off; caused by fuel continuing to be burned in the combustion chamber.

DIFFERENTIAL: A geared assembly which allows the transmission of motion between drive axles, giving one axle the ability to rotate faster than the other, as in cornering.

DIFFERENTIAL AREAS: When opposing faces of a spool valve are acted upon by the same pressure but their areas differ in size, the face with the larger area produces the differential force and valve movement. (See spool valve.)

DIFFERENTIAL FORCE: (See differential areas)

DIGITAL READOUT: A display of numbers or a combination of numbers and letters.

DIGITAL VOLT OHMMETER: An electronic diagnostic tool used to measure voltage, ohms and amps as well as several other functions, with the readings displayed on a digital screen in tenths, hundredths and thousandths.

DIODE: An electrical device that will allow current to flow in one direction only.

DIRECT CURRENT (DC): Electrical current that flows in one direction only.

DIRECT DRIVE: The gear ratio is 1:1, with no change occurring in the torque and speed input/output relationship.

DISC BRAKE: A hydraulic braking assembly consisting of a brake disc, or rotor, mounted on an axle shaft, and a caliper assembly containing, usually two brake pads which are activated by hydraulic pressure. The pads are forced against the sides of the disc, creating friction which slows the vehicle.

DISPERSANTS: Suspend dirt and prevent sludge buildup in a liquid, such as engine oil.

DOUBLE BUMP (DOUBLE FEEL): Two sudden and forceful applies of a clutch or band.

DISPLACEMENT: The total volume of air that is displaced by all pistons as the engine turns through one complete revolution.

DISTRIBUTOR: A mechanically driven device on an engine which is responsible for electrically firing the spark plug at a pre-determined point of the piston stroke.

DOHC: Double overhead camshaft.

DOUBLE OVERHEAD CAMSHAFT: The engine utilizes two camshafts mounted in one cylinder head. One camshaft operates the exhaust valves, while the other operates the intake valves.

DOWEL PIN: A pin, inserted in mating holes in two different parts allowing those parts to maintain a fixed relationship.

DRIVELINE: The drive connection between the transmission and the drive wheels.

DRIVE TRAIN: The components that transmit the flow of power from the engine to the wheels. The components include the clutch, transmission, driveshafts (or axle shafts in front wheel drive), U-joints and differential.

DRUM BRAKE: A braking system which consists of two brake shoes and one or two wheel cylinders, mounted on a fixed backing plate, and a brake drum, mounted on an axle, which revolves around the assembly.

DRY CHARGED BATTERY: Battery to which electrolyte is added when the battery is placed in service.

DVOM: Digital volt ohmmeter

DWELL: The rate, measured in degrees of shaft rotation, at which an electrical circuit cycles on and off.

DYNAMIC: An application in which there is rotating or reciprocating motion between the parts.

EARLY: Condition where shift occurs before vehicle has reached proper speed, which tends to labor engine after upshift.

EBCM: See Electronic Control Unit (ECU).

ECM: See Electronic Control Unit (ECU).

ECU: Electronic control unit.

ELECTRODE: Conductor (positive or negative) of electric current.

ELECTROLYSIS: A surface etching or bonding of current conducting transmission/transaxle components that may occur when grounding straps are missing or in poor condition.

ELECTROLYTE: A solution of water and sulfuric acid used to activate the battery. Electrolyte is extremely corrosive.

ELECTROMAGNET: A coil that produces a magnetic field when current flows through its windings.

ELECTROMAGNETIC INDUCTION: A method to create (generate) current flow through the use of magnetism.

ELECTROMAGNETISM: The effects surrounding the relationship between electricity and magnetism.

ELECTROMOTIVE FORCE (EMF): The force or pressure (voltage) that causes current movement in an electrical circuit.

ELECTRONIC CONTROL UNIT: A digital computer that controls engine (and sometimes transmission, brake or other vehicle system) functions based on data received from various sensors. Examples used by some manufacturers include Electronic Brake Control Module (EBCM), Engine Control Module (ECM), Powertrain Control Module (PCM) or Vehicle Control Module (VCM).

ELECTRONIC IGNITION: A system in which the timing and firing of the spark plugs is controlled by an electronic control unit, usually called a module. These systems have no points or condenser.

ELECTRONIC PRESSURE CONTROL (EPC) SOLENOID: A specially designed solenoid containing a spool valve and spring assembly to control fluid mainline pressure. A variable current flow, controlled by the ECM/PCM, varies the internal force of the solenoid on the spool valve and resulting mainline pressure. (See variable force solenoid.)

ELECTRONICS: Miniaturized electrical circuits utilizing semiconductors, solid-state devices, and printed circuits. Electronic circuits utilize small amounts of power.

ELECTRONIFICATION: The application of electronic circuitry to a mechanical device. Regarding automatic transmissions, electrification is incorporated into converter clutch lockup, shift scheduling, and line pressure control systems.

ELECTROSTATIC DISCHARGE (ESD): An unwanted, high-voltage electrical current released by an individual who has taken on a static charge of electricity. Electronic components can be easily damaged by ESD.

ELEMENT: A device within a hydrodynamic drive unit designed with a set of blades to direct fluid flow.

ENAMEL: Type of paint that dries to a smooth, glossy finish.

END BUMP (END FEEL OR SLIP BUMP): Firmer feel at end of shift when compared with feel at start of shift.

END-PLAY: The clearance/gap between two components that allows for expansion of the parts as they warm up, to prevent binding and to allow space for lubrication.

ENERGY: The ability or capacity to do work.

ENGINE: The primary motor or power apparatus of a vehicle, which converts liquid or gas fuel into mechanical energy.

ENGINE BLOCK: The basic engine casting containing the cylinders, the crankshaft main bearings, as well as machined surfaces for the mounting of other components such as the cylinder head, oil pan, transmission, etc.

ENGINE BRAKING: Use of engine to slow vehicle by manually downshifting during zero-throttle coast down.

ENGINE CONTROL MODULE (ECM): Manages the engine and incorporates output control over the torque converter clutch solenoid. (Note: Current designation for the ECM in late model vehicles is PCM.)

ENGINE COOLANT TEMPERATURE (ECT) SENSOR: Prevents converter clutch engagement with a cold engine; also used for shift timing and shift quality.

EP LUBRICANT: EP (extreme pressure) lubricants are specially formulated for use with gears involving heavy loads (transmissions, differentials, etc.).

ETHYL: A substance added to gasoline to improve its resistance to knock, by slowing down the rate of combustion.

ETHYLENE GLYCOL: The base substance of antifreeze.

EXHAUST MANIFOLD: A set of cast passages or pipes which conduct exhaust gases from the engine.

FAIL-SAFE (BACKUP) CONTROL: A substitute value used by the PCM/TCM to replace a faulty signal from an input sensor. The temporary value allows the vehicle to continue to be operated.

FAST IDLE: The speed of the engine when the choke is on. Fast idle speeds engine warm-up.

FEDERAL ENGINE: An engine certified by the EPA for use in any of the 49 states (except California).

FEEDBACK: A circuit malfunction whereby current can find another path to feed load devices.

FEELER GAUGE: A blade, usually metal, of precisely predetermined thickness, used to measure the clearance between two parts.

FILAMENT: The part of a bulb that glows; the filament creates high resistance to current flow and actually glows from the resulting heat.

FINAL DRIVE: An essential part of the axle drive assembly where final gear reduction takes place in the powertrain. In RWD applications and north-south FWD applications, it must also change the power flow direction to the axle shaft by ninety degrees. (Also see axle ratio).

FIRING ORDER: The order in which combustion occurs in the cylinders of an engine. Also the order in which spark is distributed to the plugs by the distributor.

FIRM: A noticeable quick apply of a clutch or band that is considered normal with medium to heavy throttle shift; should not be confused with harsh or rough.

FLAME FRONT: The term used to describe certain aspects of the fuel explosion in the cylinders. The flame front should move in a controlled pattern across the cylinder, rather than simply exploding immediately.

FLARE (SLIPPING): A quick increase in engine rpm accompanied by momentary loss of torque; generally occurs during shift.

FLAT ENGINE: Engine design in which the pistons are horizontally opposed. Porsche, Subaru and some old VW are common examples of flat engines.

FLAT RATE: A dealership term referring to the amount of money paid to a technician for a repair or diagnostic service based on that particular service versus dealership's labor time (NOT based on the actual time the technician spent on the job).

FLAT SPOT: A point during acceleration when the engine seems to lose power for an instant.

FLOODING: The presence of too much fuel in the intake manifold and combustion chamber which prevents the air/fuel mixture from firing, thereby causing a no-start situation.

FLUID: A fluid can be either liquid or gas. In hydraulics, a liquid is used for transmitting force or motion.

FLUID COUPLING: The simplest form of hydrodynamic drive, the fluid coupling consists of two look-alike members with straight radial varies referred to as the impeller (pump) and the turbine. Input torque is always equal to the output torque.

FLUID DRIVE: Either a fluid coupling or a fluid torque converter. (See hydrodynamic drive units.)

FLUID TORQUE CONVERTER: A hydrodynamic drive that has the ability to act both as a torque multiplier and fluid coupling. (See hydrodynamic drive units; torque converter.)

FLUID VISCOSITY: The resistance of a liquid to flow. A cold fluid (oil) has greater viscosity and flows more slowly than a hot fluid (oil).

FLYWHEEL: A heavy disc of metal attached to the rear of the crankshaft. It smoothes the firing impulses of the engine and keeps the crankshaft turning during periods when no firing takes place. The starter also engages the flywheel to start the engine.

FOOT POUND (ft. lbs., lbs. ft. or sometimes, ft. lb.): The amount of energy or work needed to raise an item weighing one pound, a distance of one foot.

FREEZE PLUG: A plug in the engine block which will be pushed out if the coolant freezes. Sometimes called expansion plugs, they protect the block from cracking should the coolant freeze.

FRICTION: The resistance that occurs between contacting surfaces. This relationship is expressed by a ratio called the coefficient of friction (CL).

FRICTION, COEFFICIENT OF: The amount of surface tension between two contacting surfaces; expressed by a scientifically calculated number.

FRONT END ALIGNMENT: A service to set caster, camber and toe-in to the correct specifications. This will ensure that the car steers and handles properly and that the tires wear properly.

FRICTION MODIFIER: Changes the coefficient of friction of the fluid between the mating steel and composition clutch/band surfaces during the engagement process and allows for a certain amount of intentional slipping for a good "shift-feel".

FRONTAL AREA: The total frontal area of a vehicle exposed to air flow.

FUEL FILTER: A component of the fuel system containing a porous paper element used to prevent any impurities from entering the engine through the fuel system. It usually takes the form of a canister-like housing, mounted in-line with the fuel hose, located anywhere on a vehicle between the fuel tank and engine.

FUEL INJECTION: A system replacing the carburetor that sprays fuel into the cylinder through nozzles. The amount of fuel can be more precisely controlled with fuel injection.

FULL FLOATING AXLE: An axle in which the axle housing extends through the wheel giving bearing support on the outside of the housing. The front axle of a four-wheel drive vehicle is usually a full floating axle, as are the rear axles of many larger (1 ton and over) pick-ups and vans.

FULL-TIME FOUR-WHEEL DRIVE: A four-wheel drive system that continuously delivers power to all four wheels. A differential between the front and rear driveshafts permits variations in axle speeds to control gear wind-up without damage.

FULL THROTTLE DETENT DOWNSHIFT: A quick apply of accelerator pedal to its full travel, forcing a downshift.

FUSE: A protective device in a circuit which prevents circuit overload by breaking the circuit when a specific amperage is present. The device is constructed around a strip or wire of a lower amperage rating than the circuit it is designed to protect. When an amperage higher than that stamped on the fuse is present in the circuit, the strip or wire melts, opening the circuit.

FUSIBLE LINK: A piece of wire in a wiring harness that performs the same job as a fuse. If overloaded, the fusible link will melt and interrupt the circuit.

FWD: Front wheel drive.

GAWR: (Gross axle weight rating) the total maximum weight an axle is designed to carry.

GCW: (Gross combined weight) total combined weight of a tow vehicle and trailer.

GARAGE SHIFT: initial engagement feel of transmission, neutral to reverse or neutral to a forward drive.

GARAGE SHIFT FEEL: A quick check of the engagement quality and responsiveness of reverse and forward gears. This test is done with the vehicle stationary.

GEAR: A toothed mechanical device that acts as a rotating lever to transmit power or turning effort from one shaft to another. (See gear ratio.)

GEAR RATIO: A ratio expressing the number of turns a smaller gear will make to turn a larger gear through one revolution. The ratio is found by dividing the number of teeth on the smaller gear into the number of teeth on the larger gear.

GEARBOX: Transmission

GEAR REDUCTION: Torque is multiplied and speed decreased by the factor of the gear ratio. For example, a 3:1 gear ratio changes an input torque of 180 ft. lbs. and an input speed of 2700 rpm to 540 Ft. lbs. and 900 rpm, respectively. (No account is taken of frictional losses, which are always present.)

GEARTRAIN: A succession of intermeshing gears that form an assembly and provide for one or more torque changes as the power input is transmitted to the power output.

GEL COAT: A thin coat of plastic resin covering fiberglass body panels.

GENERATOR: A device which produces direct current (DC) necessary to charge the battery.

GOVERNOR: A device that senses vehicle speed and generates a hydraulic oil pressure. As vehicle speed increases, governor oil pressure rises.

GROUND CIRCUIT: (See circuit, ground.)

GROUND SIDE SWITCHING: The electrical/electronic circuit control switch is located after the circuit load.

GVWR: (Gross vehicle weight rating) total maximum weight a vehicle is designed to carry including the weight of the vehicle, passengers, equipment, gas, oil, etc.

HALOGEN: A special type of lamp known for its quality of brilliant white light. Originally used for fog lights and driving lights.

HARD CODES: DTCs that are present at the time of testing; also called continuous or current codes.

HARSH(ROUGH): An apply of a clutch or band that is more noticeable than a firm one; considered undesirable at any throttle position.

HEADER TANK: An expansion tank for the radiator coolant. It can be located remotely or built into the radiator.

HEAT RANGE: A term used to describe the ability of a spark plug to carry away heat. Plugs with longer nosed insulators take longer to carry heat off effectively.

HEAT RISER: A flapper in the exhaust manifold that is closed when the engine is cold, causing hot exhaust gases to heat the intake manifold providing better cold engine operation. A thermostatic spring opens the flapper when the engine warms up.

HEAVY THROTTLE: Approximately three-fourths of accelerator pedal travel.

HEMI: A name given an engine using hemispherical combustion chambers.

HERTZ (HZ): The international unit of frequency equal to one cycle per second (10,000 Hertz equals 10,000 cycles per second).

HIGH-IMPEDANCE DVOM (DIGITAL VOLT-OHMMETER): This styled device provides a built-in resistance value and is capable of limiting circuit current flow to safe milliamp levels.

HIGH RESISTANCE: Often refers to a circuit where there is an excessive amount of opposition to normal current flow.

HORSEPOWER: A measurement of the amount of work; one horsepower is the amount of work necessary to lift 33,000 lbs. one foot in one minute. Brake horsepower (bhp) is the horsepower delivered by an engine on a dynamometer. Net horsepower is the power remaining (measured at the flywheel of the engine) that can be used to turn the wheels after power is consumed through friction and running the engine accessories (water pump, alternator, air pump, fan etc.)

HOT CIRCUIT: (See circuit, hot; hot lead.)

HOT LEAD: A wire or conductor in the power side of the circuit. (See circuit, hot.)

HOT SIDE SWITCHING: The electrical/electronic circuit control switch is located before the circuit load.

HUB: The center part of a wheel or gear.

HUNTING (BUSYNESS): Repeating quick series of up-shifts and downshifts that causes noticeable change in engine rpm, for example, as in a 4-3-4 shift pattern.

HYDRAULICS: The use of liquid under pressure to transfer force of motion.

HYDROCARBON (HC): Any chemical compound made up of hydrogen and carbon. A major pollutant formed by the engine as a by-product of combustion.

HYDRODYNAMIC DRIVE UNITS: Devices that transmit power solely by the action of a kinetic fluid flow in a closed recirculating path. An impeller energizes the fluid and discharges the high-speed jet stream into the turbine for power output.

HYDROMETER: An instrument used to measure the specific gravity of a solution.

HYDROPLANING: A phenomenon of driving when water builds up under the tire tread, causing it to lose contact with the road. Slowing down will usually restore normal tire contact with the road.

HYPOID GEARSET: The drive pinion gear may be placed below or above the centerline of the driven gear; often used as a final drive gearset.

IDLE MIXTURE: The mixture of air and fuel (usually about 14:1) being fed to the cylinders. The idle mixture screw(s) are sometimes adjusted as part of a tune-up.

IDLER ARM: Component of the steering linkage which is a geometric duplicate of the steering gear arm. It supports the right side of the center steering link.

IMPELLER: Often called a pump, the impeller is the power input (drive) member of a hydrodynamic drive. As part of the torque converter cover, it acts as a centrifugal pump and puts the fluid in motion.

INCH POUND (inch lbs.; sometimes in. lb. or in. lbs.): One twelfth of a foot pound.

INDUCTANCE: The force that produces voltage when a conductor is passed through a magnetic field.

INDUCTION: A means of transferring electrical energy in the form of a magnetic field. Principle used in the ignition coil to increase voltage.

INITIAL FEEL: A distinct firmer feel at start of shift when compared with feel at finish of shift.

INJECTOR: A device which receives metered fuel under relatively low pressure and is activated to inject the fuel into the engine under relatively high pressure at a predetermined time.

INPUT: In an automatic transmission, the source of power from the engine is absorbed by the torque converter, which provides the power input into the transmission. The turbine drives the input(turbine)shaft.

INPUT SHAFT: The shaft to which torque is applied, usually carrying the driving gear or gears.

INTAKE MANIFOLD: A casting of passages or pipes used to conduct air or a fuel/air mixture to the cylinders.

INTERNAL GEAR: The ring-like outer gear of a planetary gearset with the gear teeth cut on the inside of the ring to provide a mesh with the planet pinions.

ISOLATION (CLAMPING) DIODES: Diodes positioned in a circuit to prevent self-induction from damaging electronic components.

IX ROTARY GEAR PUMP: Contains two rotating members, one shaped with internal gear teeth and the other with external gear teeth. As the gears separate, the fluid fills the gaps between gear teeth, is pulled across a crescent-shaped divider, and then is forced to flow through the outlet as the gears mesh.

IX ROTARY LOBE PUMP: Sometimes referred to as a gerotor type pump. Two rotating members, one shaped with internal lobes and the other with external lobes, separate and then mesh to cause fluid to flow.

JOURNAL: The bearing surface within which a shaft operates.

JUMPER CABLES: Two heavy duty wires with large alligator clips used to provide power from a charged battery to a discharged battery mounted in a vehicle.

JUMPSTART: Utilizing the sufficiently charged battery of one vehicle to start the engine of another vehicle with a discharged battery by the use of jumper cables.

KEY: A small block usually fitted in a notch between a shaft and a hub to prevent slippage of the two parts.

KICKDOWN: Detent downshift system; either linkage, cable, or electrically controlled.

KILO: A prefix used in the metric system to indicate one thousand.

KNOCK: Noise which results from the spontaneous ignition of a portion of the air-fuel mixture in the engine cylinder caused by overly advanced ignition timing or use of incorrectly low octane fuel for that engine.

KNOCK SENSOR: An input device that responds to spark knock, caused by over advanced ignition timing.

LABOR TIME: A specific amount of time required to perform a certain repair or diagnostic service as defined by a vehicle or after-market manufacturer.

LACQUER: A quick-drying automotive paint.

LATE: Shift that occurs when engine is at higher than normal rpm for given amount of throttle.

LIGHT-EMITTING DIODE (LED): A semiconductor diode that emits light as electrical current flows through it; used in some electronic display devices to emit a red or other color light.

LIGHT THROTTLE: Approximately one-fourth of accelerator pedal travel.

LIMITED SLIP: A type of differential which transfers driving force to the wheel with the best traction.

LIMP-IN MODE: Electrical shutdown of the transmission/ transaxle output solenoids, allowing only forward and reverse gears that are hydraulically energized by the manual valve. This permits the vehicle to be driven to a service facility for repair.

LIP SEAL: Molded synthetic rubber seal designed with an outer sealing edge (lip) that points into the fluid containing area to be sealed. This type of seal is used where rotational and axial forces are present.

LITHIUM-BASE GREASE: Chassis and wheel bearing grease using lithium as a base. Not compatible with sodium-base grease.

LOAD DEVICE: A circuit's resistance that converts the electrical energy into light, sound, heat, or mechanical movement.

LOAD RANGE: Indicates the number of plies at which a tire is rated. Load range B equals four-ply rating; C equals six-ply rating; and, D equals an eight-ply rating.

LOAD TORQUE: The amount of output torque needed from the transmission/transaxle to overcome the vehicle load.

LOCKING HUBS: Accessories used on part-time four-wheel drive systems that allow the front wheels to be disengaged from the drive train when four-wheel drive is not being used. When four-wheel drive is desired, the hubs are engaged, locking the wheels to the drive train.

LOCKUP CONVERTER: A torque converter that operates hydraulically and mechanically. When an internal apply plate (lockup plate) clamps to the torque converter cover, hydraulic slippage is eliminated.

LOCK RING: See Circlip or Snapring

MAGNET: Any body with the property of attracting iron or steel.

MAGNETIC FIELD: The area surrounding the poles of a magnet that is affected by its attraction or repulsion forces.

MAIN LINE PRESSURE: Often called control pressure or line pressure, it refers to the pressure of the oil leaving the pump and is controlled by the pressure regulator valve.

MALFUNCTION INDICATOR LAMP (MIL): Previously known as a check engine light, the dash-mounted MIL illuminates and signals the driver that an emission or driveability problem with the powertrain has been detected by the ECM/PCM. When this occurs, at least one diagnostic trouble code (DTC) has been stored into the control module memory.

MANIFOLD ABSOLUTE PRESSURE (MAP) SENSOR: Reads the amount of air pressure (vacuum) in the engine's intake manifold system; its signal is used to analyze engine load conditions.

MANIFOLD VACUUM: Low pressure in an engine intake manifold formed just below the throttle plates. Manifold vacuum is highest at idle and drops under acceleration.

MANIFOLD: A casting of passages or set of pipes which connect the cylinders to an inlet or outlet source.

MANUAL LEVER POSITION SWITCH (MLPS): A mechanical switching unit that is typically mounted externally to the transmission/transaxle to inform the PCM/ECM which gear range the driver has selected.

MANUAL VALVE: Located inside the transmission/transaxle, it is directly connected to the driver's shift lever. The position of the manual valve determines which hydraulic circuits will be charged with oil pressure and the operating mode of the transmission.

MANUAL VALVE LEVER POSITION SENSOR (MVLPS): The input from this device tells the TCM what gear range was selected.

MASS AIR FLOW (MAF) SENSOR: Measures the airflow into the engine.

MASTER CYLINDER: The primary fluid pressurizing device in a hydraulic system. In automotive use, it is found in brake and hydraulic clutch systems and is pedal activated, either directly or, in a power brake system, through the power booster.

MacPherson STRUT: A suspension component combining a shock absorber and spring in one unit.

MEDIUM THROTTLE: Approximately one-half of accelerator pedal travel.

MEGA: A metric prefix indicating one million.

MEMBER: An independent component of a hydrodynamic unit such as an impeller, a stator, or a turbine. It may have one or more elements.

MERCON: A fluid developed by Ford Motor Company in 1988. It contains a friction modifier and closely resembles operating characteristics of Dexron.

METAL SEALING RINGS: Made from cast iron or aluminum, their primary application is with dynamic components involving pressure sealing circuits of rotating members. These rings are designed with either butt or hook lock end joints.

METER (ANALOG): A linear-style meter representing data as lengths; a needle-style instrument interfacing with logical numerical increments. This style of electrical meter uses relatively low impedance internal resistance and cannot be used for testing electronic circuitry.

METER (DIGITAL): Uses numbers as a direct readout to show values. Most meters of this style use high impedance internal resistance and must be used for testing low current electronic circuitry.

MICRO: A metric prefix indicating one-millionth (0.000001).

MILLI: A metric prefix indicating one-thousandth (0.001).

MINIMUM THROTTLE: The least amount of throttle opening required for upshift; normally close to zero throttle.

MISFIRE: Condition occurring when the fuel mixture in a cylinder fails to ignite, causing the engine to run roughly.

MODULE: Electronic control unit, amplifier or igniter of solid state or integrated design which controls the current flow in the ignition primary circuit based on input from the pick-up coil. When the module opens the primary circuit, high secondary voltage is induced in the coil.

MODULATED: In an electronic-hydraulic converter clutch system (or shift valve system), the term modulated refers to the pulsing of a solenoid, at a variable rate. This action controls the buildup of oil pressure in the hydraulic circuit to allow a controlled amount of clutch slippage.

MODULATED CONVERTER CLUTCH CONTROL (MCCC): A pulse width duty cycle valve that controls the converter lockup apply pressure and maximizes smoother transitions between lock and unlock conditions.

MODULATOR PRESSURE (THROTTLE PRESSURE): A hydraulic signal oil pressure relating to the amount of engine load, based on either the amount of throttle plate opening or engine vacuum.

MODULATOR VALVE: A regulator valve that is controlled by engine vacuum, providing a hydraulic pressure that varies in relation to engine torque. The hydraulic torque signal functions to delay the shift pattern and provide a line pressure boost. (See throttle valve.)

MOTOR: An electromagnetic device used to convert electrical energy into mechanical energy.

MULTIPLE-DISC CLUTCH: A grouping of steel and friction lined plates that, when compressed together by hydraulic pressure acting upon a piston, lock or unlock a planetary member.

MULTI-WEIGHT: Type of oil that provides adequate lubrication at both high and low temperatures.

needed to move one amp through a resistance of one ohm.

MUSHY: Same as soft; slow and drawn out clutch apply with very little shift feel.

MUTUAL INDUCTION: The generation of current from one wire circuit to another by movement of the magnetic field surrounding a current-carrying circuit as its ampere flow increases or decreases.

NEEDLE BEARING: A bearing which consists of a number (usually a large number) of long, thin rollers.

NITROGEN OXIDE (NOx): One of the three basic pollutants found in the exhaust emission of an internal combustion engine. The amount of NOx usually varies in an inverse proportion to the amount of HC and CO.

NONPOSITIVE SEALING: A sealing method that allows some minor leakage, which normally assists in lubrication.

O2 SENSOR: Located in the engine's exhaust system, it is an input device to the ECM/PCM for managing the fuel delivery and ignition system. A scanner can be used to observe the fluctuating voltage readings produced by an O2 sensor as the oxygen content of the exhaust is analyzed.

O-RING SEAL: Molded synthetic rubber seal designed with a circular cross-section. This type of seal is used primarily in static applications.

OBD II (ON-BOARD DIAGNOSTICS, SECOND GENERATION): Refers to the federal law mandating tighter control of 1996 and newer vehicle emissions, active monitoring of related devices, and standardization of terminology, data link connectors, and other technician concerns.

OCTANE RATING: A number, indicating the quality of gasoline based on its ability to resist knock. The higher the number, the better the quality. Higher compression engines require higher octane gas.

OEM: Original Equipment Manufactured. OEM equipment is that furnished standard by the manufacturer.

OFFSET: The distance between the vertical center of the wheel and the mounting surface at the lugs. Offset is positive if the center is outside the lug circle; negative offset puts the center line inside the lug circle.

OHM'S LAW: A law of electricity that states the relationship between voltage, current, and resistance. Volts = amperes x ohms

OHM: The unit used to measure the resistance of conductor-to-electrical flow. One ohm is the amount of resistance that limits current flow to one ampere in a circuit with one volt of pressure.

OHMMETER: An instrument used for measuring the resistance, in ohms, in an electrical circuit.

ONE-WAY CLUTCH: A mechanical clutch of roller or sprag design that resists torque or transmits power in one direction only. It is used to either hold or drive a planetary member.

ONE-WAY ROLLER CLUTCH: A mechanical device that transmits or holds torque in one direction only.

OPEN CIRCUIT: A break or lack of contact in an electrical circuit, either intentional (switch) or unintentional (bad connection or broken wire).

ORIFICE: Located in hydraulic oil circuits, it acts as a restriction. It slows down fluid flow to either create back pressure or delay pressure buildup downstream.

OSCILLOSCOPE: A piece of test equipment that shows electric impulses as a pattern on a screen. Engine performance can be analyzed by interpreting these patterns.

OUTPUT SHAFT: The shaft which transmits torque from a device, such as a transmission.

OUTPUT SPEED SENSOR (OSS): Identifies transmission/transaxle output shaft speed for shift timing and may be used to calculate TCC slip; often functions as the VSS (vehicle speed sensor).

OVERDRIVE: (1.) A device attached to or incorporated in a transmission/transaxle that allows the engine to turn less than one full revolution for every complete revolution of the wheels. The net effect is to reduce engine rpm, thereby using less fuel. A typical overdrive gear ratio would be .87:1, instead of the normal 1:1 in high gear. (2.) A gear assembly which produces more shaft revolutions than that transmitted to it.

OVERDRIVE PLANETARY GEARSET: A single planetary gearset designed to provide a direct drive and overdrive ratio. When coupled to a three-speed transmission/transaxle configuration, a four-speed/overdrive unit is present.

OVERHEAD CAMSHAFT (OHC): An engine configuration in which the camshaft is mounted on top of the cylinder head and operates the valve either directly or by means of rocker arms.

OVERHEAD VALVE (OHV): An engine configuration in which all of the valves are located in the cylinder head and the camshaft is located in the cylinder block. The camshaft operates the valves via lifters and pushrods.

OVERRUNCLUTCH: Another name for a one-way mechanical clutch. Applies to both roller and sprag designs.

OVERSTEER: The tendency of some vehicles, when steering into a turn, to over-respond or steer more than required, which could result in excessive slip of the rear wheels. Opposite of under-steer.

OXIDATION STABILIZERS: Absorb and dissipate heat. Automatic transmission fluid has high resistance to varnish and sludge buildup that occurs from excessive heat that is generated primarily in the torque converter. Local temperatures as high as 6000F (3150C) can occur at the clutch plates during engagement, and this heat must be absorbed and dissipated. If the fluid cannot withstand the heat, it burns or oxidizes, resulting in an almost immediate destruction of friction materials, clogged filter screen and hydraulic passages, and sticky valves.

OXIDES OF NITROGEN: See nitrogen oxide (NOx).

OXYGEN SENSOR: Used with a feedback system to sense the presence of oxygen in the exhaust gas and signal the computer which can use the voltage signal to determine engine operating efficiency and adjust the air/fuel ratio.

PARALLEL CIRCUIT: (See circuit, parallel.)

PARTS WASHER: A basin or tub, usually with a built-in pump mechanism and hose used for circulating chemical solvent for the purpose of cleaning greasy, oily and dirty components.

PART-TIME FOUR WHEEL DRIVE: A system that is normally in the two wheel drive mode and only runs in four-wheel drive when the system is manually engaged because more traction is desired. Two or four wheel drive is normally selected by a lever to engage the front axle, but if locking hubs are used, these must also be manually engaged in the Lock position. Otherwise, the front axle will not drive the front wheels.

PASSIVE RESTRAINT: Safety systems such as air bags or automatic seat belts which operate with no action required on the part of the driver or passenger. Mandated by Federal regulations on all vehicles sold in the U.S. after 1990.

PAYLOAD: The weight the vehicle is capable of carrying in addition to its own weight. Payload includes weight of the driver, passengers and cargo, but not coolant, fuel, lubricant, spare tire, etc.

PCM: Powertrain control module.

PCV VALVE: A valve usually located in the rocker cover that vents crankcase vapors back into the engine to be reburned.

PERCOLATION: A condition in which the fuel actually "boils," due to excessive heat. Percolation prevents proper atomization of the fuel causing rough running.

PICK-UP COIL: The coil in which voltage is induced in an electronic ignition.

PING: A metallic rattling sound produced by the engine during acceleration. It is usually due to incorrect ignition timing or a poor grade of gasoline.

PINION: The smaller of two gears. The rear axle pinion drives the ring gear which transmits motion to the axle shafts.

PINION GEAR: The smallest gear in a drive gear assembly.

PISTON: A disc or cup that fits in a cylinder bore and is free to move. In hydraulics, it provides the means of converting hydraulic pressure into a usable force. Examples of piston applications are found in servo, clutch, and accumulator units.

PISTON RING: An open-ended ring which fits into a groove on the outer diameter of the piston. Its chief function is to form a seal between the piston and cylinder wall. Most automotive pistons have three rings: two for compression sealing; one for oil sealing.

PITMAN ARM: A lever which transmits steering force from the steering gear to the steering linkage.

PLANET CARRIER: A basic member of a planetary gear assembly that carries the pinion gears.

PLANET PINIONS: Gears housed in a planet carrier that are in constant mesh with the sun gear and internal gear. Because they have their own independent rotating centers, the pinions are capable of rotating around the sun gear or the inside of the internal gear.

PLANETARY GEAR RATIO: The reduction or overdrive ratio developed by a planetary gearset.

PLANETARY GEARSET: In its simplest form, it is made up of a basic assembly group containing a sun gear, internal gear, and planet carrier. The gears are always in constant mesh and offer a wide range of gear ratio possibilities.

PLANETARY GEARSET (COMPOUND): Two planetary gearsets combined together.

PLANETARY GEARSET (SIMPLE): An assembly of gears in constant mesh consisting of a sun gear, several pinion gears mounted in a carrier, and a ring gear. It provides gear ratio and direction changes, in addition to a direct drive and a neutral.

PLY RATING: A. rating given a tire which indicates strength (but not necessarily actual plies). A two-ply/four-ply rating has only two plies, but the strength of a four-ply tire.

POLARITY: Indication (positive or negative) of the two poles of a battery.

PORT: An opening for fluid intake or exhaust.

POSITIVE SEALING: A sealing method that completely prevents leakage.

POTENTIAL: Electrical force measured in volts; sometimes used interchangeably with voltage.

POWER: The ability to do work per unit of time, as expressed in horsepower; one horsepower equals 33,000 ft. lbs. of work per minute, or 550 ft. lbs. of work per second.

POWER FLOW: The systematic flow or transmission of power through the gears, from the input shaft to the output shaft.

POWER-TO-WEIGHT RATIO: Ratio of horsepower to weight of car.

POWERTRAIN: See Drivetrain.

POWERTRAIN CONTROL MODULE (PCM): Current designation for the engine control module (ECM). In many cases, late model vehicle control units manage the engine as well as the transmission. In other settings, the PCM controls the engine and is interfaced with a TCM to control transmission functions.

Ppm: Parts per million; unit used to measure exhaust emissions.

PREIGNITION: Early ignition of fuel in the cylinder, sometimes due to glowing carbon deposits in the combustion chamber. Preignition can be damaging since combustion takes place prematurely.

PRELOAD: A predetermined load placed on a bearing during assembly or by adjustment.

PRESS FIT: The mating of two parts under pressure, due to the inner diameter of one being smaller than the outer diameter of the other, or vice versa; an interference fit.

PRESSURE: The amount of force exerted upon a surface area.

PRESSURE CONTROL SOLENOID (PCS): An output device that provides a boost oil pressure to the mainline regulator valve to control line pressure. Its operation is determined by the amount of current sent from the PCM.

PRESSURE GAUGE: An instrument used for measuring the fluid pressure in a hydraulic circuit.

PRESSURE REGULATOR VALVE: In automatic transmissions, its purpose is to regulate the pressure of the pump output and supply the basic fluid pressure necessary to operate the transmission. The regulated fluid pressure may be referred to as mainline pressure, line pressure, or control pressure.

PRESSURE SWITCH ASSEMBLY (PSA): Mounted inside the transmission, it is a grouping of oil pressure switches that inputs to the PCM when certain hydraulic passages are charged with oil pressure.

PRESSURE PLATE: A spring-loaded plate (part of the clutch) that transmits power to the driven (friction) plate when the clutch is engaged.

PRIMARY CIRCUIT: The low voltage side of the ignition system which consists of the ignition switch, ballast resistor or resistance wire, bypass, coil, electronic control unit and pick-up coil as well as the connecting wires and harnesses.

PROFILE: Term used for tire measurement (tire series), which is the ratio of tire height to tread width.

PROM (PROGRAMMABLE READ-ONLY MEMORY): The heart of the computer that compares input data and makes the engineered program or strategy decisions about when to trigger the appropriate output based on stored computer instructions.

PULSE GENERATOR: A two-wire pickup sensor used to produce a fluctuating electrical signal. This changing signal is read by the controller to determine the speed of the object and can be used to measure transmission/transaxle input speed, output speed, and vehicle speed.

PSI: Pounds per square inch; a measurement of pressure.

PULSE WIDTH DUTY CYCLE SOLENOID (PULSE WIDTH MODULATED SOLENOID): A computer-controlled solenoid that turns on and off at a variable rate producing a modulated oil pressure; often referred to as a pulse width modulated (PWM) solenoid. Employed in many electronic automatic transmissions and transaxles, these solenoids are used to manage shift control and converter clutch hydraulic circuits.

PUSHROD: A steel rod between the hydraulic valve lifter and the valve rocker arm in overhead valve (OHV) engines.

PUMP: A mechanical device designed to create fluid flow and pressure buildup in a hydraulic system.

QUARTER PANEL: General term used to refer to a rear fender. Quarter panel is the area from the rear door opening to the tail light area and from rear wheel well to the base of the trunk and roof-line.

RACE: The surface on the inner or outer ring of a bearing on which the balls, needles or rollers move.

RACK AND PINION: A type of automotive steering system using a pinion gear attached to the end of the steering shaft. The pinion meshes with a long rack attached to the steering linkage.

RADIAL TIRE: Tire design which uses body cords running at right angles to the center line of the tire. Two or more belts are used to give tread strength. Radials can be identified by their characteristic sidewall bulge.

RADIATOR: Part of the cooling system for a water-cooled engine, mounted in the front of the vehicle and connected to the engine with rubber hoses. Through the radiator, excess combustion heat is dissipated into the atmosphere through forced convection using a water and glycol based mixture that circulates through, and cools, the engine.

RANGE REFERENCE AND CLUTCH/BAND APPLY CHART: A guide that shows the application of clutches and bands for each gear, within the selector range positions. These charts are extremely useful for understanding how the unit operates and for diagnosing malfunctions.

RAVIGNEAUX GEARSET: A compound planetary gearset that features matched dual planetary pinions (sets of two) mounted in a single planet carrier. Two sun gears and one ring mesh with the carrier pinions.

REACTION MEMBER: The stationary planetary member, in a planetary gearset, that is grounded to the transmission/transaxle case through the use of friction and wedging devices known as bands, disc clutches, and one-way clutches.

REACTION PRESSURE: The fluid pressure that moves a spool valve against an opposing force or forces; the area on which the opposing force acts. The opposing force can be a spring or a combination of spring force and auxiliary hydraulic force.

REACTOR, TORQUE CONVERTER: The reaction member of a fluid torque converter, more commonly called a stator. (See stator.)

REAR MAIN OIL SEAL: A synthetic or rope-type seal that prevents oil from leaking out of the engine past the rear main crankshaft bearing.

RECIRCULATING BALL: Type of steering system in which recirculating steel balls occupy the area between the nut and worm wheel, causing a reduction in friction.

RECTIFIER: A device (used primarily in alternators) that permits electrical current to flow in one direction only.

REDUCTION: (See gear reduction.)

REGULATOR VALVE: A valve that changes the pressure of the oil in a hydraulic circuit as the oil passes through the valve by bleeding off (or exhausting) some of the volume of oil supplied to the valve.

REFRIGERANT 12 (R-12) or 134 (R-134): The generic name of the refrigerant used in automotive air conditioning systems.

REGULATOR: A device which maintains the amperage and/or voltage levels of a circuit at predetermined values.

RELAY: A switch which automatically opens and/or closes a circuit.

RELAY VALVE: A valve that directs flow and pressure. Relay valves simply connect or disconnect interrelated passages without restricting the fluid flow or changing the pressure.

RELIEF VALVE: A spring-loaded, pressure-operated valve that limits oil pressure buildup in a hydraulic circuit to a predetermined maximum value.

RELUCTOR: A wheel that rotates inside the distributor and triggers the release of voltage in an electronic ignition.

RESERVOIR: The storage area for fluid in a hydraulic system; often called a sump.

RESIN: A liquid plastic used in body work.

RESIDUAL MAGNETISM: The magnetic strength stored in a material after a magnetizing field has been removed.

RESISTANCE: The opposition to the flow of current through a circuit or electrical device, and is measured in ohms. Resistance is equal to the voltage divided by the amperage.

RESISTOR SPARK PLUG: A spark plug using a resistor to shorten the spark duration. This suppresses radio interference and lengthens plug life.

RESISTOR: A device, usually made of wire, which offers a preset amount of resistance in an electrical circuit.

RESULTANT FORCE: The single effective directional thrust of the fluid force on the turbine produced by the vortex and rotary forces acting in different planes.

RETARD: Set the ignition timing so that spark occurs later (fewer degrees before TDC).

RHEOSTAT: A device for regulating a current by means of a variable resistance.

RING GEAR: The name given to a ring-shaped gear attached to a differential case, or affixed to a flywheel or as part of a planetary gear set.

ROADLOAD: grade.

ROCKER ARM: A lever which rotates around a shaft pushing down (opening) the valve with an end when the other end is pushed up by the pushrod. Spring pressure will later close the valve.

ROCKER PANEL: The body panel below the doors between the wheel opening.

ROLLER BEARING: A bearing made up of hardened inner and outer races between which hardened steel rollers move.

ROLLER CLUTCH: A type of one-way clutch design using rollers and springs mounted within an inner and outer cam race assembly.

ROTARY FLOW: The path of the fluid trapped between the blades of the members as they revolve with the rotation of the torque converter cover (rotational inertia).

ROTOR: (1.) The disc-shaped part of a disc brake assembly, upon which the brake pads bear; also called, brake disc. (2.) The device mounted atop the distributor shaft, which passes current to the distributor cap tower contacts.

ROTARY ENGINE: See Wankel engine.

RPM: Revolutions per minute (usually indicates engine speed).

RTV: A gasket making compound that cures as it is exposed to the atmosphere. It is used between surfaces that are not perfectly machined to one another, leaving a slight gap that the RTV fills and in which it hardens. The letters RTV represent room temperature vulcanizing.

RUN-ON: Condition when the engine continues to run, even when the key is turned off. See dieseling.

SEALED BEAM: A automotive headlight. The lens, reflector and filament from a single unit.

SEATBELT INTERLOCK: A system whereby the car cannot be started unless the seatbelt is buckled.

SECONDARY CIRCUIT: The high voltage side of the ignition system, usually above 20,000 volts. The secondary includes the ignition coil, coil wire, distributor cap and rotor, spark plug wires and spark plugs.

SELF-INDUCTION: The generation of voltage in a current-carrying wire by changing the amount of current flowing within that wire.

SEMI-CONDUCTOR: A material (silicon or germanium) that is neither a good conductor nor an insulator; used in diodes and transistors.

SEMI-FLOATING AXLE: In this design, a wheel is attached to the axle shaft, which takes both drive and cornering loads. Almost all solid axle passenger cars and light trucks use this design.

SENDING UNIT: A mechanical, electrical, hydraulic or electromagnetic device which transmits information to a gauge.

SENSOR: Any device designed to measure engine operating conditions or ambient pressures and temperatures. Usually electronic in nature and designed to send a voltage signal to an on-board computer, some sensors may operate as a simple on/off switch or they may provide a variable voltage signal (like a potentiometer) as conditions or measured parameters change.

SERIES CIRCUIT: (See circuit, series.)

SERPENTINE BELT: An accessory drive belt, with small multiple v-ribs, routed around most or all of the engine-powered accessories such as the alternator and power steering pump. Usually both the front and the back side of the belt comes into contact with various pulleys.

SERVO: In an automatic transmission, it is a piston in a cylinder assembly that converts hydraulic pressure into mechanical force and movement; used for the application of the bands and clutches.

SHIFT BUSYNESS: When referring to a torque converter clutch, it is the frequent apply and release of the clutch plate due to uncommon driving conditions.

SHIFT VALVE: Classified as a relay valve, it triggers the automatic shift in response to a governor and a throttle signal by directing fluid to the appropriate band and clutch apply combination to cause the shift to occur.

SHIM: Spacers of precise, predetermined thickness used between parts to establish a proper working relationship.

SHIMMY: Vibration (sometimes violent) in the front end caused by misaligned front end, out of balance tires or worn suspension components.

SHORT CIRCUIT: An electrical malfunction where current takes the path of least resistance to ground (usually through damaged insulation). Current flow is excessive from low resistance resulting in a blown fuse.

SHUDDER: Repeated jerking or stick-slip sensation, similar to chuggle but more severe and rapid in nature, that may be most noticeable during certain ranges of vehicle speed; also used to define condition after converter clutch engagement.

SIMPSON GEARSET: A compound planetary gear train that integrates two simple planetary gearsets referred to as the front planetary and the rear planetary.

SINGLE OVERHEAD CAMSHAFT: See overhead camshaft.

SKIDPLATE: A metal plate attached to the underside of the body to protect the fuel tank, transfer case or other vulnerable parts from damage.

SLAVE CYLINDER: In automotive use, a device in the hydraulic clutch system which is activated by hydraulic force, disengaging the clutch.

SLIPPING: Noticeable increase in engine rpm without vehicle speed increase; usually occurs during or after initial clutch or band engagement.

SLUDGE: Thick, black deposits in engine formed from dirt, oil, water, etc. It is usually formed in engines when oil changes are neglected.

SNAP RING: A circular retaining clip used inside or outside a shaft or part to secure a shaft, such as a floating wrist pin.

SOFT: Slow, almost unnoticeable clutch apply with very little shift feel.

SOFTCODES: DTCs that have been set into the PCM memory but are not present at the time of testing; often referred to as history or intermittent codes.

SOHC: Single overhead camshaft.

SOLENOID: An electrically operated, magnetic switching device.

SPALLING: A wear pattern identified by metal chips flaking off the hardened surface. This condition is caused by foreign particles, overloading situations, and/or normal wear.

SPARK PLUG: A device screwed into the combustion chamber of a spark ignition engine. The basic construction is a conductive core inside of a ceramic insulator, mounted in an outer conductive base. An electrical charge from the spark plug wire travels along the conductive core and jumps a preset air gap to a grounding point or points at the end of the conductive base. The resultant spark ignites the fuel/air mixture in the combustion chamber.

SPECIFIC GRAVITY (BATTERY): The relative weight of liquid (battery electrolyte) as compared to the weight of an equal volume of water.

SPLINES: Ridges machined or cast onto the outer diameter of a shaft or inner diameter of a bore to enable parts to mate without rotation.

SPLIT TORQUE DRIVE: In a torque converter, it refers to parallel paths of torque transmission, one of which is mechanical and the other hydraulic.

SPONGY PEDAL: A soft or spongy feeling when the brake pedal is depressed. It is usually due to air in the brake lines.

SPOOLVALVE: A precision-machined, cylindrically shaped valve made up of lands and grooves. Depending on its position in the valve bore, various interconnecting hydraulic circuit passages are either opened or closed.

SPRAG CLUTCH: A type of one-way clutch design using cams or contoured-shaped sprags between inner and outer races. (See one-way clutch.)

SPRUNG WEIGHT: The weight of a car supported by the springs.

SQUARE-CUT SEAL: Molded synthetic rubber seal designed with a square- or rectangular-shaped cross-section. This type of seal is used for both dynamic and static applications.

SRS: Supplemental restraint system

STABILIZER (SWAY) BAR: A bar linking both sides of the suspension. It resists sway on turns by taking some of added load from one wheel and putting it on the other.

STAGE: The number of turbine sets separated by a stator. A turbine set may be made up of one or more turbine members. A three-element converter is classified as a single stage.

STALL: In fluid drive transmission/transaxle applications, stall refers to engine rpm with the transmission/transaxle engaged and the vehicle stationary; throttle valve can be in any position between closed and wide open.

STALL SPEED: In fluid drive transmission/transaxle applications, stall speed refers to the maximum engine rpm with the transmission/transaxle engaged and vehicle stationary, when the throttle valve is wide open. (See stall; stall test.)

STALL TEST: A procedure recommended by many manufacturers to help determine the integrity of an engine, the torque converter stator, and certain clutch and band combinations. With the shift lever in each of the forward and reverse positions and with the brakes firmly applied, the accelerator pedal is momentarily pressed to the wide open throttle (WOT) position. The engine rpm reading at full throttle can provide clues for diagnosing the condition of the items listed above.

STALL TORQUE: The maximum design or engineered torque ratio of a fluid torque converter, produced under stall speed conditions. (See stall speed.)

STARTER: A high-torque electric motor used for the purpose of starting the engine, typically through a high ratio geared drive connected to the flywheel ring gear.

STATIC: A sealing application in which the parts being sealed do not move in relation to each other.

STATOR (REACTOR): The reaction member of a fluid torque converter that changes the direction of the fluid as it leaves the turbine to enter the impeller vanes. During the torque multiplication phase, this action assists the impeller's rotary force and results in an increase in torque.

STEERING GEOMETRY: Combination of various angles of suspension components (caster, camber, toe-in); roughly equivalent to front end alignment.

STRAIGHT WEIGHT: Term designating motor oil as suitable for use within a narrow range of temperatures. Outside the narrow temperature range its flow characteristics will not adequately lubricate.

STROKE: The distance the piston travels from bottom dead center to top dead center.

SUBSTITUTION: Replacing one part suspected of a defect with a like part of known quality.

SUMP: The storage vessel or reservoir that provides a ready source of fluid to the pump. In an automatic transmission, the sump is the oil pan. All fluid eventually returns to the sump for recycling into the hydraulic system.

SUN GEAR: In a planetary gearset, it is the center gear that meshes with a cluster of planet pinions.

SUPERCHARGER: An air pump driven mechanically by the engine through belts, chains, shafts or gears from the crankshaft. Two general types of supercharger are the positive displacement and centrifugal type, which pump air in direct relationship to the speed of the engine.

SUPPLEMENTAL RESTRAINT SYSTEM: See air bag.

SURGE: Repeating engine-related feeling of acceleration and deceleration that is less intense than chuggle.

SWITCH: A device used to open, close, or redirect the current in an electrical circuit.

SYNCHROMESH: A manual transmission/transaxle that is equipped with devices (synchronizers) that match the gear speeds so that the transmission/transaxle can be downshifted without clashing gears.

SYNTHETIC OIL: Non-petroleum based oil.

TACHOMETER: A device used to measure the rotary speed of an engine, shaft, gear, etc., usually in rotations per minute.

TDC: Top dead center. The exact top of the piston's stroke.

TEFLON SEALING RINGS: Teflon is a soft, durable, plastic-like material that is resistant to heat and provides excellent sealing. These rings are designed with either scarf-cut joints or as one-piece rings. Teflon sealing rings have replaced many metal ring applications.

TERMINAL: A device attached to the end of a wire or cable to make an electrical connection.

TEST LIGHT, CIRCUIT-POWERED: Uses available circuit voltage to test circuit continuity.

TEST LIGHT, SELF-POWERED: Uses its own battery source to test circuit continuity.

THERMISTOR: A special resistor used to measure fluid temperature; it decreases its resistance with increases in temperature.

THERMOSTAT: A valve, located in the cooling system of an engine, which is closed when cold and opens gradually in response to engine heating, controlling the temperature of the coolant and rate of coolant flow.

THERMOSTATIC ELEMENT: A heat-sensitive, spring-type device that controls a drain port from the upper sump area to the lower sump. When the transaxle fluid reaches operating temperature, the port is closed and the upper sump fills, thus reducing the fluid level in the lower sump.

THROTTLE POSITION (TP) SENSOR: Reads the degree of throttle opening; its signal is used to analyze engine load conditions. The ECM/PCM decides to apply the TCC, or to disengage it for coast or load conditions that need a converter torque boost.

THROTTLE PRESSURE/MODULATOR PRESSURE: A hydraulic signal oil pressure relating to the amount of engine load, based on either the amount of throttle plate opening or engine vacuum.

THROTTLE VALVE: A regulating or balanced valve that is controlled mechanically by throttle linkage or engine vacuum. It sends a hydraulic signal to the shift valve body to control shift timing and shift quality. (See balanced valve; modulator valve.)

THROW-OUT BEARING: As the clutch pedal is depressed, the throwout bearing moves against the spring fingers of the pressure plate, forcing the pressure plate to disengage from the driven disc.

TIE ROD: A rod connecting the steering arms. Tie rods have threaded ends that are used to adjust toe-in.

TIE-UP: Condition where two opposing clutches are attempting to apply at same time, causing engine to labor with noticeable loss of engine rpm.

TIMING BELT: A square-toothed, reinforced rubber belt that is driven by the crankshaft and operates the camshaft.

TIMING CHAIN: A roller chain that is driven by the crankshaft and operates the camshaft.

TIRE ROTATION: Moving the tires from one position to another to make the tires wear evenly.

TOE-IN (OUT): A term comparing the extreme front and rear of the front tires. Closer together at the front is toe-in; farther apart at the front is toe-out.

TOP DEAD CENTER (TDC): The point at which the piston reaches the top of its travel on the compression stroke.

TORQUE: Measurement of turning or twisting force, expressed as foot-pounds or inch-pounds.

TORQUE CONVERTER: A turbine used to transmit power from a driving member to a driven member via hydraulic action, providing changes in drive ratio and torque. In automotive use, it links the driveplate at the rear of the engine to the automatic transmission.

TORQUE CONVERTER CLUTCH: The apply plate (lockup plate) assembly used for mechanical power flow through the converter.

TORQUE PHASE: Sometimes referred to as slip phase or stall phase, torque multiplication occurs when the turbine is turning at a slower speed than the impeller, and the stator is reactionary (stationary). This sequence generates a boost in output torque.

TORQUE RATING (STALL TORQUE): The maximum torque multiplication that occurs during stall conditions, with the engine at wide open throttle (WOT) and zero turbine speed.

TORQUE RATIO: An expression of the gear ratio factor on torque effect. A 3:1 gear ratio or 3:1 torque ratio increases the torque input by the ratio factor of 3. Input torque (100 ft. lbs.) x 3 = output torque (300 ft. lbs.)

TRACTION: The amount of usable tractive effort before the drive wheels slip on the road contact surface.

TORSION BAR SUSPENSION: Long rods of spring steel which take the place of springs. One end of the bar is anchored and the other arm (attached to the suspension) is free to twist. The bars' resistance to twisting causes springing action.

TRACK: Distance between the centers of the tires where they contact the ground.

TRACTION CONTROL: A control system that prevents the spinning of a vehicle's drive wheels when excess power is applied.

TRACTIVE EFFORT: The amount of force available to the drive wheels, to move the vehicle.

TRANSAXLE: A single housing containing the transmission and differential. Transaxles are usually found on front engine/front wheel drive or rear engine/rear wheel drive cars.

TRANSDUCER: A device that changes energy from one form to another. For example, a transducer in a microphone changes sound energy to electrical energy. In automotive air-conditioning controls used in automatic temperature systems, a transducer changes an electrical signal to a vacuum signal, which operates mechanical doors.

TRANSMISSION: A powertrain component designed to modify torque and speed developed by the engine; also provides direct drive, reverse, and neutral.

TRANSMISSION CONTROL MODULE (TCM): Manages transmission functions. These vary according to the manufacturer's product design but may include converter clutch operation, electronic shift scheduling, and mainline pressure.

TRANSMISSION FLUID TEMPERATURE (TFT) SENSOR: Originally called a transmission oil temperature (TOT) sensor, this input device to the ECM/PCM senses the fluid temperature and provides a resistance value. It operates on the thermistor principle.

TRANSMISSION INPUT SPEED (TIS) SENSOR: Measures turbine shaft (input shaft) rpm's and compares to engine rpm's to determine torque converter slip. When compared to the transmission output speed sensor or VSS, gear ratio and clutch engagement timing can be determined.

TRANSMISSION OIL TEMPERATURE (TOT) SENSOR: (See transmission fluid temperature (TFT) sensor.)

TRANSMISSION RANGE SELECTOR (TRS) SWITCH: Tells the module which gear shift position the driver has chosen.

TRANSFER CASE: A gearbox driven from the transmission that delivers power to both front and rear driveshafts in a four-wheel drive system. Transfer cases usually have a high and low range set of gears, used depending on how much pulling power is needed.

TRANSISTOR: A semi-conductor component which can be actuated by a small voltage to perform an electrical switching function.

TREAD WEAR INDICATOR: Bars molded into the tire at right angles to the tread that appear as horizontal bars when 1/16 in. of tread remains.

TREAD WEAR PATTERN: The pattern of wear on tires which can be "read" to diagnose problems in the front suspension.

TUNE-UP: A regular maintenance function, usually associated with the replacement and adjustment of parts and components in the electrical and fuel systems of a vehicle for the purpose of attaining optimum performance.

TURBINE: The output (driven) member of a fluid coupling or fluid torque converter. It is splined to the input (turbine) shaft of the transmission.

TURBOCHARGER: An exhaust driven pump which compresses intake air and forces it into the combustion chambers at higher than atmospheric pressures. The increased air pressure allows more fuel to be burned and results in increased horsepower being produced.

TURBULENCE: The interference of molecules of a fluid (or vapor) with each other in a fluid flow.

TYPE F: Transmission fluid developed and used by Ford Motor Company up to 1982. This fluid type provides a high coefficient of friction.

TYPE 7176: The preferred choice of transmission fluid for Chrysler automatic transmissions and transaxles. Developed in 1986, it closely resembles Dexron and Mercon. Type 7176 is the recommended service fill fluid for all Chrysler products utilizing a lockup torque converter dating back to 1978.

U-JOINT (UNIVERSAL JOINT): A flexible coupling in the drive train that allows the driveshafts or axle shafts to operate at different angles and still transmit rotary power.

UNDERSTEER: The tendency of a car to continue straight ahead while negotiating a turn.

UNIT BODY: Design in which the car body acts as the frame.

UNLEADED FUEL: Fuel which contains no lead (a common gasoline additive). The presence of lead in fuel will destroy the functioning elements of a catalytic converter, making it useless.

UNSPRUNG WEIGHT: The weight of car components not supported by the springs (wheels, tires, brakes, rear axle, control arms, etc.).

UPSHIFT: A shift that results in a decrease in torque ratio and an increase in speed.

VACUUM: A negative pressure; any pressure less than atmospheric pressure.

VACUUM ADVANCE: A device which advances the ignition timing in response to increased engine vacuum.

VACUUM GAUGE: An instrument used for measuring the existing vacuum in a vacuum circuit or chamber. The unit of measure is inches (of mercury in a barometer).

VACUUM MODULATOR: Generates a hydraulic oil pressure in response to the amount of engine vacuum.

VALVES: Devices that can open or close fluid passages in a hydraulic system and are used for directing fluid flow and controlling pressure.

VALVE BODY ASSEMBLY: The main hydraulic control assembly of the transmission/transaxle that contains numerous valves, check balls, and other components to control the distribution of pressurized oil throughout the transmission.

VALVE CLEARANCE: The measured gap between the end of the valve stem and the rocker arm, cam lobe or follower that activates the valve.

VALVE GUIDES: The guide through which the stem of the valve passes. The guide is designed to keep the valve in proper alignment.

VALVE LASH (clearance): The operating clearance in the valve train.

VALVE TRAIN: The system that operates intake and exhaust valves, consisting of camshaft, valves and springs, lifters, pushrods and rocker arms.

VAPOR LOCK: Boiling of the fuel in the fuel lines due to excess heat. This will interfere with the flow of fuel in the lines and can completely stop the flow. Vapor lock normally only occurs in hot weather.

VARIABLE DISPLACEMENT (VARIABLE CAPACITY) VANE PUMP: Slipper-type vanes, mounted in a revolving rotor and contained within the bore of a movable slide, capture and then force fluid to flow. Movement of the slide to various positions changes the size of the vane chambers and the amount of fluid flow. **Note:** GM refers to this pump design as variable displacement, and Ford terms it variable capacity.

VARIABLE FORCE SOLENOID (VFS): Commonly referred to as the electronic pressure control (EPC) solenoid, it replaces the cable/linkage style of TV system control and is integrated with a spool valve and spring assembly to control pressure. A variable computer-controlled current flow varies the internal force of the solenoid on the spool valve and resulting control pressure.

VARIABLE ORIFICE THERMAL VALVE: Temperature-sensitive hydraulic oil control device that adjusts the size of a circuit path opening. By altering the size of the opening, the oil flow rate is adapted for cold to hot oil viscosity changes.

VARNISH: Term applied to the residue formed when gasoline gets old and stale.

VCM: See Electronic Control Unit (ECU).

VEHICLE SPEED SENSOR (VSS): Provides an electrical signal to the computer module, measuring vehicle speed, and affects the torque converter clutch engagement and release.

VESPEL SEALING RINGS: Hard plastic material that produces excellent sealing in dynamic settings. These rings are found in late versions of the 4T60 and in all 4T60-E and 4T80-E transaxles.

VISCOSITY: The ability of a fluid to flow. The lower the viscosity rating, the easier the fluid will flow. 10 weight motor oil will flow much easier than 40 weight motor oil.

VISCOSITY INDEX IMPROVERS: Keeps the viscosity nearly constant with changes in temperature. This is especially important at low temperatures, when the oil needs to be thin to aid in shifting and for cold-weather starting. Yet it must not be so thin that at high temperatures it will cause excessive hydraulic leakage so that pumps are unable to maintain the proper pressures.

VISCOUS CLUTCH: A specially designed torque converter clutch apply plate that, through the use of a silicon fluid, clamps smoothly and absorbs torsional vibrations.

VOLT: Unit used to measure the force or pressure of electricity. It is defined as the pressure

VOLTAGE: The electrical pressure that causes current to flow. Voltage is measured in volts (V).

VOLTAGE, APPLIED: The actual voltage read at a given point in a circuit. It equals the available voltage of the power supply minus the losses in the circuit up to that point.

VOLTAGE DROP: The voltage lost or used in a circuit by normal loads such as a motor or lamp or by abnormal loads such as a poor (high-resistance) lead or terminal connection.

VOLTAGE REGULATOR: A device that controls the current output of the alternator or generator.

VOLTMETER: An instrument used for measuring electrical force in units called volts. Voltmeters are always connected parallel with the circuit being tested.

VORTEX FLOW: The crosswise or circulatory flow of oil between the blades of the members caused by the centrifugal pumping action of the impeller.

WANKEL ENGINE: An engine which uses no pistons. In place of pistons, triangular-shaped rotors revolve in specially shaped housings.

WATER PUMP: A belt driven component of the cooling system that mounts on the engine, circulating the coolant under pressure.

WATT: The unit for measuring electrical power. One watt is the product of one ampere and one volt (watts equals amps times volts). Wattage is the horsepower of electricity (746 watts equal one horsepower).

WHEEL ALIGNMENT: Inclusive term to describe the front end geometry (caster, camber, toe-in/out).

WHEEL CYLINDER: Found in the automotive drum brake assembly, it is a device, actuated by hydraulic pressure, which, through internal pistons, pushes the brake shoes outward against the drums.

WHEEL WEIGHT: Small weights attached to the wheel to balance the wheel and tire assembly. Out-of-balance tires quickly wear out and also give erratic handling when installed on the front.

WHEELBASE: Distance between the center of front wheels and the center of rear wheels.

WIDE OPEN THROTTLE (WOT): Full travel of accelerator pedal.

WORK: The force exerted to move a mass or object. Work involves motion; if a force is exerted and no motion takes place, no work is done. Work per unit of time is called power. Work = force x distance = ft. lbs. 33,000 ft. lbs. in one minute = 1 horsepower

ZERO-THROTTLE COAST DOWN: A full release of accelerator pedal while vehicle is in motion and in drive range.

Commonly Used Abbreviations

2

2WD	Two Wheel Drive

4

4WD	Four Wheel Drive

A

A/C	Air Conditioning
ABDC	After Bottom Dead Center
ABS	Anti-lock Brakes
AC	Alternating Current
ACL	Air cleaner
ACT	Air Charge Temperature
AIR	Secondary Air Injection
ALCL	Assembly Line Communications Link
ALDL	Assembly Line Diagnostic Link
AT	Automatic Transaxle/Transmission
ATDC	After Top Dead Center
ATF	Automatic Transmission Fluid
ATS	Air Temperature Sensor
AWD	All Wheel Drive

B

BAP	Barometric Absolute Pressure
BARO	Barometric Pressure
BBDC	Before Bottom Dead Center
BCM	Body Control Module
BDC	Bottom Dead Center
BPT	Backpressure Transducer
BTDC	Before Top Dead Center
BVSV	Bimetallic Vacuum Switching Valve

C

CAC	Charge Air Cooler
CARB	California Air Resources Board
CAT	Catalytic Converter
CCC	Computer Command Control
CCCC	Computer Controlled Catalytic Converter
CCCI	Computer Controlled Coil Ignition
CCD	Computer Controlled Dwell
CDI	Capacitor Discharge Ignition
CEC	Computerized Engine Control
CFI	Continuous Fuel Injection
CIS	Continuous Injection System
CIS-E	Continuous Injection System - Electronic
CKP	Crankshaft Position
CL	Closed Loop
CMP	Camshaft Position
CPP	Clutch Pedal Position
CTOX	Continuous Trap Oxidizer System
CTP	Closed Throttle Position
CVC	Constant Vacuum Control
CYL	Cylinder

D

DBC	Dual Bed Catalyst
DC	Direct Current
DFI	Direct Fuel Injection
DIS	Distributorless Ignition System
DLC	Data Link Connector
DMM	Digital Multimeter
DOHC	Double Overhead Camshaft
DRB	Diagnostic Readout Box
DTC	Diagnostic Trouble Code
DTM	Diagnostic Test Mode
DVOM	Digital Volt/Ohmmeter

E

EBCM	Electronic Brake Control Module
ECM	Engine Control Module
ECT	Engine Coolant Temperature
ECU	Engine Control Unit or Electronic Control Unit
EDIS	Electronic Distributorless Ignition System
EEC	Electronic Engine Control
EEPROM	Electrically Erasable Programmable Read Only Memory
EFE	Early Fuel Evaporation
EGR	Exhaust Gas Recirculation
EGRT	Exhaust Gas Recirculation Temperature
EGRVC	EGR Valve Control
EPROM	Erasable Programmable Read Only Memory
EVAP	Evaporative Emissions
EVP	EGR Valve Position

F

FBC	Feedback Carburetor
FEEPROM	Flash Electrically Erasable Programmable Read Only Memory
FF	Flexible Fuel
FI	Fuel Injection
FT	Fuel Trim
FWD	Front Wheel Drive

G

GND	Ground

H

HAC	High Altitude Compensation
HEGO	Heated Exhaust Gas Oxygen sensor
HEI	High Energy Ignition
HO2 Sensor	Heated Oxygen Sensor

I

IAC	Idle Air Control
IAT	Intake Air Temperature
ICM	Ignition Control Module
IFI	Indirect Fuel Injection
IFS	Inertia Fuel Shutoff
ISC	Idle Speed Control
IVSV	Idle Vacuum Switching Valve

Commonly Used Abbreviations

K

KOEO	Key On, Engine Off
KOER	Key ON, Engine Running
KS	Knock Sensor

M

MAF	Mass Air Flow
MAP	Manifold Absolute Pressure
MAT	Manifold Air Temperature
MC	Mixture Control
MDP	Manifold Differential Pressure
MFI	Multiport Fuel Injection
MIL	Malfunction Indicator Lamp or Maintenance
MST	Manifold Surface Temperature
MVZ	Manifold Vacuum Zone

N

NVRAM	Nonvolatile Random Access Memory

O

O2 Sensor	Oxygen Sensor
OBD	On-Board Diagnostic
OC	Oxidation Catalyst
OHC	Overhead Camshaft
OL	Open Loop

P

P/S	Power Steering
PAIR	Pulsed Secondary Air Injection
PCM	Powertrain Control Module
PCS	Purge Control Solenoid
PCV	Positive Crankcase Ventilation
PIP	Profile Ignition Pick-up
PNP	Park/Neutral Position
PROM	Programmable Read Only Memory
PSP	Power Steering Pressure
PTO	Power Take-Off
PTOX	Periodic Trap Oxidizer System

R

RABS	Rear Anti-lock Brake System
RAM	Random Access Memory
ROM	Read Only Memory
RPM	Revolutions Per Minute
RWAL	Rear Wheel Anti-lock Brakes
RWD	Rear Wheel Drive

S

SBC	Single Bed Converter
SBEC	Single Board Engine Controller
SC	Supercharger
SCB	Supercharger Bypass
SFI	Sequential Multiport Fuel Injection
SIR	Supplemental Inflatible Restraint
SOHC	Single Overhead Camshaft
SPL	Smoke Puff Limiter
SPOUT	Spark Output
SRI	Service Reminder Indicator
SRS	Supplemental Restraint System
SRT	System Readiness Test
SSI	Solid State Ignition
ST	Scan Tool
STO	Self-Test Output

T

TAC	Thermostatic Air Cleaner
TBI	Throttle Body Fuel Injection
TC	Turbocharger
TCC	Torque Converter Clutch
TCM	Transmission Control Module
TDC	Top Dead Center
TFI	Thick Film Ignition
TP	Throttle Position
TR Sensor	Transaxle/Transmission Range Sensor
TVV	Thermal Vacuum Valve
TWC	Three-way Catalytic Converter

V

VAF	Volume Air Flow, or Vane Air Flow
VAPS	Variable Assist Power Steering
VRV	Vacuum Regulator Valve
VSS	Vehicle Speed Sensor
VSV	Vacuum Switching Valve

W

WOT	Wide Open Throttle
WU-TWC	Warm Up Three-way Catalytic Converter

ENGLISH TO METRIC CONVERSION: TORQUE

To convert foot-pounds (ft. lbs.) to Newton-meters (Nm), multiply the number of ft. lbs. by 1.36
To convert Newton-meters (Nm) to foot-pounds (ft. lbs.), multiply the number of Nm by 0.7376

ft. lbs.	Nm	ft. lbs.	Nm	ft. lbs.	Nm	ft. lbs.	Nm
0.1	0.1	34	46.2	76	103.4	118	160.5
0.2	0.3	35	47.6	77	104.7	119	161.8
0.3	0.4	36	49.0	78	106.1	120	163.2
0.4	0.5	37	50.3	79	107.4	121	164.6
0.5	0.7	38	51.7	80	108.8	122	165.9
0.6	0.8	39	53.0	81	110.2	123	167.3
0.7	1.0	40	54.4	82	111.5	124	168.6
0.8	1.1	41	55.8	83	112.9	125	170.0
0.9	1.2	42	57.1	84	114.2	126	171.4
1	1.4	43	58.5	85	115.6	127	172.7
2	2.7	44	59.8	86	117.0	128	174.1
3	4.1	45	61.2	87	118.3	129	175.4
4	5.4	46	62.6	88	119.7	130	176.8
5	6.8	47	63.9	89	121.0	131	178.2
6	8.2	48	65.3	90	122.4	132	179.5
7	9.5	49	66.6	91	123.8	133	180.9
8	10.9	50	68.0	92	125.1	134	182.2
9	12.2	51	69.4	93	126.5	135	183.6
10	13.6	52	70.7	94	127.8	136	185.0
11	15.0	53	72.1	95	129.2	137	186.3
12	16.3	54	73.4	96	130.6	138	187.7
13	17.7	55	74.8	97	131.9	139	189.0
14	19.0	56	76.2	98	133.3	140	190.4
15	20.4	57	77.5	99	134.6	141	191.8
16	21.8	58	78.9	100	136.0	142	193.1
17	23.1	59	80.2	101	137.4	143	194.5
18	24.5	60	81.6	102	138.7	144	195.8
19	25.8	61	83.0	103	140.1	145	197.2
20	27.2	62	84.3	104	141.4	146	198.6
21	28.6	63	85.7	105	142.8	147	199.9
22	29.9	64	87.0	106	144.2	148	201.3
23	31.3	65	88.4	107	145.5	149	202.6
24	32.6	66	89.8	108	146.9	150	204.0
25	34.0	67	91.1	109	148.2	151	205.4
26	35.4	68	92.5	110	149.6	152	206.7
27	36.7	69	93.8	111	151.0	153	208.1
28	38.1	70	95.2	112	152.3	154	209.4
29	39.4	71	96.6	113	153.7	155	210.8
30	40.8	72	97.9	114	155.0	156	212.2
31	42.2	73	99.3	115	156.4	157	213.5
32	43.5	74	100.6	116	157.8	158	214.9
33	44.9	75	102.0	117	159.1	159	216.2

METRIC TO ENGLISH CONVERSION: TORQUE

To convert foot-pounds (ft. lbs.) to Newton-meters (Nm), multiply the number of ft. lbs. by 1.36
To convert Newton-meters (Nm) to foot-pounds (ft. lbs.), multiply the number of Nm by 0.7376

Nm	ft. lbs.	Nm	ft. lbs.	Nm	ft. lbs.	Nm	ft. lbs.	Nm	ft. lbs.
0.1	0.1	34	25.0	76	55.9	118	86.8	160	117.6
0.2	0.1	35	25.7	77	56.6	119	87.5	161	118.4
0.3	0.2	36	26.5	78	57.4	120	88.2	162	119.1
0.4	0.3	37	27.2	79	58.1	121	89.0	163	119.9
0.5	0.4	38	27.9	80	58.8	122	89.7	164	120.6
0.6	0.4	39	28.7	81	59.6	123	90.4	165	121.3
0.7	0.5	40	29.4	82	60.3	124	91.2	166	122.1
0.8	0.6	41	30.1	83	61.0	125	91.9	167	122.8
0.9	0.7	42	30.9	84	61.8	126	92.6	168	123.5
1	0.7	43	31.6	85	62.5	127	93.4	169	124.3
2	1.5	44	32.4	86	63.2	128	94.1	170	125.0
3	2.2	45	33.1	87	64.0	129	94.9	171	125.7
4	2.9	46	33.8	88	64.7	130	95.6	172	126.5
5	3.7	47	34.6	89	65.4	131	96.3	173	127.2
6	4.4	48	35.3	90	66.2	132	97.1	174	127.9
7	5.1	49	36.0	91	66.9	133	97.8	175	128.7
8	5.9	50	36.8	92	67.6	134	98.5	176	129.4
9	6.6	51	37.5	93	68.4	135	99.3	177	130.1
10	7.4	52	38.2	94	69.1	136	100.0	178	130.9
11	8.1	53	39.0	95	69.9	137	100.7	179	131.6
12	8.8	54	39.7	96	70.6	138	101.5	180	132.4
13	9.6	55	40.4	97	71.3	139	102.2	181	133.1
14	10.3	56	41.2	98	72.1	140	102.9	182	133.8
15	11.0	57	41.9	99	72.8	141	103.7	183	134.6
16	11.8	58	42.6	100	73.5	142	104.4	184	135.3
17	12.5	59	43.4	101	74.3	143	105.1	185	136.0
18	13.2	60	44.1	102	75.0	144	105.9	186	136.8
19	14.0	61	44.9	103	75.7	145	106.6	187	137.5
20	14.7	62	45.6	104	76.5	146	107.4	188	138.2
21	15.4	63	46.3	105	77.2	147	108.1	189	139.0
22	16.2	64	47.1	106	77.9	148	108.8	190	139.7
23	16.9	65	47.8	107	78.7	149	109.6	191	140.4
24	17.6	66	48.5	108	79.4	150	110.3	192	141.2
25	18.4	67	49.3	109	80.1	151	111.0	193	141.9
26	19.1	68	50.0	110	80.9	152	111.8	194	142.6
27	19.9	69	50.7	111	81.6	153	112.5	195	143.4
28	20.6	70	51.5	112	82.4	154	113.2	196	144.1
29	21.3	71	52.2	113	83.1	155	114.0	197	144.9
30	22.1	72	52.9	114	83.8	156	114.7	198	145.6
31	22.8	73	53.7	115	84.6	157	115.4	199	146.3
32	23.5	74	54.4	116	85.3	158	116.2	200	147.1
33	24.3	75	55.1	117	86.0	159	116.9	201	147.8

ENGLISH/METRIC CONVERSION: TEMPERATURE

To convert Fahrenheit (F°) to Celsius (C°), take F° temperature and subtract 32, multiply the result by 5 and divide the result by 9
To convert Celsius (C°) to Fahrenheit (F°), take C° temperature and multiply it by 9, divide the result by 5 and add 32

F°	C°	F°	C°	C°	F°	C°	F°
-40	-40.0	150	65.6	-38	-36.4	46	114.8
-35	-37.2	155	68.3	-36	-32.8	48	118.4
-30	-34.4	160	71.1	-34	-29.2	50	122
-25	-31.7	165	73.9	-32	-25.6	52	125.6
-20	-28.9	170	76.7	-30	-22	54	129.2
-15	-26.1	175	79.4	-28	-18.4	56	132.8
-10	-23.3	180	82.2	-26	-14.8	58	136.4
-5	-20.6	185	85.0	-24	-11.2	60	140
0	-17.8	190	87.8	-22	-7.6	62	143.6
1	-17.2	195	90.6	-20	-4	64	147.2
2	-16.7	200	93.3	-18	-0.4	66	150.8
3	-16.1	205	96.1	-16	3.2	68	154.4
4	-15.6	210	98.9	-14	6.8	70	158
5	-15.0	212	100.0	-12	10.4	72	161.6
10	-12.2	215	101.7	-10	14	74	165.2
15	-9.4	220	104.4	-8	17.6	76	168.8
20	-6.7	225	107.2	-6	21.2	78	172.4
25	-3.9	230	110.0	-4	24.8	80	176
30	-1.1	235	112.8	-2	28.4	82	179.6
35	1.7	240	115.6	0	32	84	183.2
40	4.4	245	118.3	2	35.6	86	186.8
45	7.2	250	121.1	4	39.2	88	190.4
50	10.0	255	123.9	6	42.8	90	194
55	12.8	260	126.7	8	46.4	92	197.6
60	15.6	265	129.4	10	50	94	201.2
65	18.3	270	132.2	12	53.6	96	204.8
70	21.1	275	135.0	14	57.2	98	208.4
75	23.9	280	137.8	16	60.8	100	212
80	26.7	285	140.6	18	64.4	102	215.6
85	29.4	290	143.3	20	68	104	219.2
90	32.2	295	146.1	22	71.6	106	222.8
95	35.0	300	148.9	24	75.2	108	226.4
100	37.8	305	151.7	26	78.8	110	230
105	40.6	310	154.4	28	82.4	112	233.6
110	43.3	315	157.2	30	86	114	237.2
115	46.1	320	160.0	32	89.6	116	240.8
120	48.9	325	162.8	34	93.2	118	244.4
125	51.7	330	165.6	36	96.8	120	248
130	54.4	335	168.3	38	100.4	122	251.6
135	57.2	340	171.1	40	104	124	255.2
140	60.0	345	173.9	42	107.6	126	258.8
145	62.8	350	176.7	44	111.2	128	262.4